DIRECTORY OF BIOMEDICAL AND HEALTH CARE GRANTS

28th Edition

Littleberry Press
West Lafayette, Indiana

We at Littleberry Press work together as a team, with each person contributing different skills and expressing his or her individual interests and opinions to the unity and efficiency of the group in order to achieve common goals. Our aim is to make worthwhile resources available to you so that you and your institution or organization can and will achieve those goals.

Copyright © 2018 by Littleberry Press
Littleberry Press, 1048B, Sagamore Parkway W., #48, West Lafayette, IN 47906
www.littleberrypress.com
Published since 1985, 28th edition 2018

All rights reserved. No part of this publication may be reproduced or transmitted
in any form or by any means, electronic or mechanical, including
photocopying, recording, or by any information storage and
retrieval system, without permission in writing
from Littleberry Press.

Printed and bound in the United States of America

ISBN 978-1-9407503-1-6
1-9407503-1-8

Table of Contents

Introduction iv
How to Use This Directory... v
Government and Organization Acronyms viii
Grant Programs... 1
Subject Index... 651
Program Type Index 831
Geographic Index... 919

Introduction

For more than four decades the GRANTS Database and its print complements, including the annual *Directory of Biomedical and Health Care Grants*, have provided the research community with current, accurate information regarding funding for research-related programs and projects, scholarships, fellowships, conferences, and internships. To meet the increased information needs of professionals, the GRANTS Database, researched and edited by members of the Littleberry Press editorial staff, has grown significantly. This 28th edition of the *Directory of Biomedical and Health Care Grants* offers factual and concise descriptions of nearly 4,000 funding programs. With this new edition both experienced and novice grantseekers will be better able to target only the most applicable programs for their biomedical and healthcare needs.

Directory of Biomedical and Health Care Grants

The *Directory of Biomedical and Healthcare Grants* features listings of programs that offer non-repayable funding for projects ranging from laboratory investigations to those designed to study the needs of society in healthcare delivery. Listings in the main section of the *Directory* contain annotations describing each program's focuses and goals, program requirements explaining eligibility, funding amounts, deadlines, *Catalog of Federal Domestic Assistance* program number (for U.S. government programs only), sponsor name and address, contact information, and the sponsor's Internet Web address. Grantseekers with access to the Internet can use the addresses to locate further information about the organizations and their application procedures. Internet addresses are also provided, when available, within listings. Some of the grant programs listed in the main section have geographic restrictions for applicants.

The following people have devoted a great deal of patience, hard work and thought to keep this project and the database an informative service for the user. It is especially important to mention the GrantSelect team – Louis S. Schafer, Ed.S., and Anita Schafer – for their diligent research, editorial work, development and assignment of index terms, and production – as well as all Littleberry Press staff who contribute to the GRANTS effort.

Indexes. The Subject Index of this *Directory* lists all program titles—with accession numbers—under their applicable subject terms. Other indexes follow, including the Grants by Program Type Index, which lists 44 program categories, such as Basic Research, Fellowships, Travel Grants, etc., with the grants that fall within their scope; and the Geographic Index, which lists programs that have state, regional, or international focus. See How to Use This Directory on pages v-vii for sample index entries.

Using the *Directory* for Grantseeking

By using the *Directory of Biomedical and Health Care Grants*, grantseekers can match the needs of their particular programs with those sponsors offering funding in the researchers' area of interest. The information listed here is meant to eliminate the costs incurred by both grantseekers and grantmakers when inappropriate proposals are submitted for a sponsor's funding program. However, because the GRANTS Database is updated daily basis, with program listings continually added, deleted, and revised, grantseekers using this *Directory* may also search the GRANTS database online through *GrantSelect* (www.grantselect.com).

All new and revised information has been extracted from (1) the sponsor's updated of previously published program statements included in earlier editions of GRANTS publications, (2) questionnaires sent to sponsors whose programs were not listed in previous editions, or (3) other materials published by the sponsor and furnished to Littleberry Press. Updated information for U.S. government programs includes new and revised program information published in the *Federal Register*; the latest edition of the *Catalog of Federal Domestic Assistance*; the *NIH Guide*, published weekly by the National Institutes of Health; and the *NSF E-Bulletin*, a monthly publication of the National Science Foundation. Included in this edition are identifying document numbers from the NIH and NSF publications. Located at the ends of the program descriptions in certain entries, the numbers indicated the ongoing NIH program number (PA) or the request for applications number (RFA). For programs of the National Science Foundation, the *NSF Bulletin* number appears. This information will help users identify the programs when seeking additional information from program staff.

While Littleberry Press has made every effort to verify that all information is both accurate and current within the confines of format and scope, the publisher does not assume and hereby disclaims any liability to any party for loss or damages caused by errors or omissions in this *Directory*, whether such errors or missions result from accident, negligence, or any other cause. Anyone having questions regarding the content, format, or any other aspect of the *Directory of Biomedical and Health Care Grants*, GrantSelect, or the GRANTS Database should contact the Editors of Littleberry Press at editor@littleberrypress.com.

v

How to Use This Directory

The *Directory of Biomedical and Health Care Grants* is designed to allow the user quick and easy access to information regarding funding programs in a researcher's specific area of interest. This *Directory* is composed of a main section, Grant Programs, which lists grant programs in alphabetical order, and three indexes: the Subject Index, the Grants by Program Type Index, and the Geographic Index.

GRANT PROGRAMS

Each listing in this section consists of the following elements: an annotation describing each program's focus and goals, requirements explaining eligibility, funding amounts, application and renewal dates, the *Catalog of Federal Domestic Assistance* program number (for U.S. government programs only), sponsor information, contact information, and Internet address.

GRANT TITLE — **Paul G. Allen Family Foundation Grants**   4298 — **ACCESSION NUMBER**

The single foundation, created through the consolidation of Allen's six previous foundations (The Allen Foundation for the Arts, The Paul G. Allen Charitable Foundation, The Paul G. Allen Foundation for Medical Research, The Paul G. Allen Forest Protection Foundation, The Allen Foundation for Music, and The Paul G. Allen Virtual Education Foundation), will continue to focus on the Allen family's philanthropic interests in the areas of arts and culture, youth engagement, community development and social change, and scientific and technological innovation. The Arts and Culture Program fosters creativity and promotes critical thinking by helping strong arts organizations become sustainable and supporting projects that feature innovative and diverse artistic forms. — **GRANT DESCRIPTION**
The Youth Engagement Program improves the way young people learn by supporting organizations that use innovative teaching strategies and provide opportunities for children to address issues relevant to their lives. The Community Development and Social Change Program promotes individual and community development by supporting initiatives and organizations that provide access to resources and opportunities. The Scientific and Technological Innovation Program advances promising scientific and technology research that has the potential to enhance understanding and stewardship of the world in which we live. Organizations may only receive one grant per year. Organizations must not have any delinquent final reports due to any of the Paul G. Allen Foundations for previous grants. Grantseekers are encouraged to apply through the online application process, where basic organizational and project information will be requested. Guidelines are available online.

REQUIREMENTS — *Requirements* 501(c)3 tax-exempt organizations, status from the Internal Revenue government entities, and IRS-recognized tribes are eligible. Eligible organizations must be located in, or serving populations of, the Pacific Northwest, which includes Alaska, Idaho, Montana, Oregon, and Washington.

RESTRICTIONS — *Restrictions* In general, the foundation will not consider requests for general fund drives, annual appeals, or federated campaigns; special events or sponsorships; direct grants, scholarships, or loans for the benefit of specific individuals; projects of organizations whose policies or practices discriminate on the basis of race, ethnic origin, sex, creed, or sexual orientation; contributions to sectarian or religious organizations whose principle activity is for the benefit of their own members or adherents; loans or debt retirement; projects that will benefit the students of a single school; general operating support for ongoing activities; or projects not aligned with the foundation's specified program areas. 509(a) private foundations are ineligible.

APPLICATION/
DUE DATE — *Date(s) Application Is Due* Mar 31; Sep 30.
Contact Grants Administrator, (206) 342-2030; fax: (206) 342-3030; — **CONTACT**
email: info@pgafamilyfoundation.org

INTERNET ADDRESS — *Internet* http://www.pgafamilyfoundation.org
Sponsor Paul G. Allen Family Foundation — **SPONSOR INFORMATION**
505 Fifth Ave S, Ste 900
Seattle, WA 98104

SUBJECT INDEX

The most effective way to access specific funding programs is through the Subject Index. This index lists the subject terms with applicable grants program titles – and their accession numbers – alphabetically under each term. Terms were assigned to target the specific area of research designated in the description of each program. Cross-references are used to link subjects and assist the user in finding specific grant information.

Following are general guidelines that can make your search of this index more successful. First, check under the specific topic of interest rather than a more general term. For instance, if you are interested in chemical engineering, look under "Chemical Engineering" rather than "Engineering." Items indexed under "Engineering" indicate funding in broad areas of engineering.

Use general headings when you want grants covering broader areas or if you can't find a specific topic. For example, many grants list funding for humanities research, health programs, or science and technology. To find these grants use such headings as "Humanities," "Medical Programs," "Science," or "Technology." For additional grant information on more specific humanities research opportunities, such as in American History or Cultural Anthropology, also check under the topics "United States History" and "Anthropology, Cultural."

Many of the grant programs provide funding for research-related scholarships, faculty fellowships, dissertations, undergraduate education, conferences, or internships. If grant funds are designated for specific disciplines, you will find the items under the specific subject. Scholarships and fellowships are also listed under the terms "Native American Education," "African Americans (Student Support)," "Hispanic Education," "Minority Education," and "Women's Education."

Grants concerning study of a particular country are listed under the name of the country. Grants concerning the history, literature, art and language of a country are listed under the name of the country also, e.g., "Chinese Art" and "Chinese Language/Literature."

SUBJECT TERM ─────── **Education**
    A.L. Mailman Family Foundation Grants, 5
    AAAS Science & Technology Policy Fellowships - Health, Education & Human Services, 22
    AARP Andrus Foundation Grants, 104
    Abbott Laboratories Fund Grants, 125
    ACIE Host University Edmund S. Muskie/Freedom Support Act Graduate Fellowships, 156
    ACT Awards, 189
    Akonadi Foundation Anti-Racism Grants, 351 ─────── PROGRAM TITLE
    Akron Community Foundation Grants, 352
    Albert and Margaret Alkek Foundation Grants, 370
    Albuquerque Community Foundation Grants, 379
    Alcoa Foundation Grants, 380
    Alcon Foundation Grants Program, 382

GEOGRAPHIC INDEX

This index lists programs that have state, regional, or international geographic focus. The Geographic Index is arranged by state, followed by Canadian programs, then by international programs by country, and lists grant program titles and their corresponding accession numbers.

COUNTRY ─────── **United States**
    Alabama ─────── STATE
        3M Fndn Grants, 2
        Alabama Humanities Fndn Grants Program, 368
        Arkema Inc. Fndn Science Teachers Program, 705
        CDC Injury Control Research Centers Grants, 1198
PROGRAM TITLE ─────── DOE Experimental Program to Stimulate Competitive Research (EPSCoR), 1653
        Hill Crest Fndn Grants, 2293
        Linn-Henley Charitable Trust Grants, 2777
        NOAA Community-Based Restoration Program (CRP) Grants, 3843
        Southern Company's Longleaf Pine Reforestation Fund, 4726

PROGRAM TYPE INDEX

This index is broken into 44 categories according to the type of program funded:

- Adult Basic Education
- Adult/Family Literacy Training
- Awards/Prizes
- Basic Research
- Building Construction and/or Renovation
- Capital Campaigns
- Centers: Research/Demonstration/Service
- Citizenship Instruction
- Community Development
- Consulting/Visiting Personnel
- Cultural Outreach
- Curriculum Development/Teacher Training
- Demonstration Grants
- Development (Institutional/Departmental)
- Dissertation/Thesis Research Support
- Educational Programs
- Emergency Programs
- Endowments
- Environmental Programs
- Exchange Programs
- Exhibitions, Collections, Performances, Video/Film Production
- Faculty/Professional Development
- Fellowships
- General Operating Support
- Graduate Assistantships
- Grants to Individuals
- International Exchange Programs
- International Grants
- Job Training/Adult Vocational Programs
- Land Acquisition
- Matching/Challenge Funds
- Materials/Equipment Acquisition (Computers, Books, Tapes, etc.)
- Preservation/Restoration
- Professorships
- Publishing/Editing/Translating
- Religious Programs
- Scholarships
- Seed Grants
- Service Delivery Programs
- Symposia, Conferences, Workshops, Seminars
- Technical Assistance
- Training Programs/Internships
- Travel Grants
- Vocational Education

Government and Organization Acronyms

AAAAI	American Academy of Allergy Asthma and Immunology		ANS	American Numismatic Society
AAAS	American Association for the Advancement of Science		AOA	American Osteopathic Association
			AOCS	American Oil Chemists' Society
AACAP	American Academy of Child and Adolescent Psychiatry		APA	American Psychological Association
			APAP	Association of Performing Arts Presenters
AACN	American Association of Critical Care Nurses		APEAL	Asian Pacific Partners for Empowerment and Leadership
AACR	American Association of Cancer Research			
AAFCS	American Association of Family and Consumer Sciences		APS	Arizona Public Service
			APSA	American Political Science Association
AAF	American Architectural Foundation		ARIT	American Research Institute in Turkey
AAFP	American Academy of Family Physicians Foundation		ARO	Army Research Office
			ASA	American Statistical Association
AAFPRS	American Academy of Facial Plastic and Reconstructive Surgery		ASCSA	American School of Classical Studies at Athens
			ASECS	American Society for Eighteenth-Century Studies
AAP	American Academy of Pediatrics		ASF	American-Scandinavian Foundation
AAR	American Academy in Rome		ASHA	American Speech-Language-Hearing Association
AAS	American Antiquarian Society		ASHRAE	American Society of Heating, Refrigerating, and Air Conditioning Engineers
AASL	American Association of School Libraries			
AAUW	American Association of University Women		ASME	American Society of Mechanical Engineers
ABA	American Bar Association		ASNS	American Society for Nutritional Sciences
ACC	Asian Cultural Council		ASPRS	American Society of Photogrammetry and Remote Sensing
ACF	Administration on Children, Youth and Families			
ACLS	American Council of Learned Societies		ASTA	American String Teachers Association
ACM	Association for Computing Machinery		ATA	Alberta Teachers Association
ACE	American Council on Education		AWHONN	Association of Women's Health, Obstetric, and Neonatal Nurses
ACMP	Amateur Chamber Music Players			
ACS	American Cancer Society		AWU	Associated Western Universities
ADA	American Diabetes Association		AWWA	American Water Works Association
ADHF	American Digestive Health Foundation		BA	British Academy
AF	Arthritis Foundation		BBF	Barbara Bush Foundation
AFAR	American Federation for Aging Research		BCBS	Blue Cross Blue Shield
AFOSR	Air Force Office of Scientific Research		BCBSM	Blue Cross Blue Shield of Michigan
AFUD	American Foundation for Urologic Disease		BCBSNC	Blue Cross Blue Shield of North Carolina
AFUW	Australian Federation of University Women		BWF	Burroughs Wellcome Fund
AGS	American Geriatrics Society		CBIE	Canadian Bureau for International Education
AHA	American Heart Association		CCF	Catholic Community Foundation
AHAF	American Health Assistance Foundation		CCF	Common Council Foundation
AFHMR	Alberta Heritage Foundation for Medical Research		CCFF	Canadian Cystic Fibrosis Foundation
AHRQ	Agency for Healthcare Research and Quality		CCFF	Christopher Columbus Fellowship Foundation
AICR	American Institute for Cancer Research		CDC	Centers for Disease Control and Prevention
AIIS	American Institute for Indian Studies		CDECD	Connecticut Department of Economic and Community Development
AJA	American Jewish Archives			
AJL	American Jewish Libraries		CDI	Children's Discovery Institute
ALA	American Library Association		CEC	Council for Exceptional Children
ALISE	Association for Library and Information Science Education		CEF	Chemical Educational Foundation
			CES	Council for European Studies
			CF	The Commonwealth Fund
AMNH	American Museum of Natural History		CFF	Cystic Fibrosis Foundation
AMS	American Musicological Society		CFFVR	Community Foundation for the Fox Valley Region
ANL	Argonne National Library		CFKF	Classic for Kids Foundation

CFNCR	Community Foundation for the National Capital Region	IUCP	Indiana University Center on Philanthropy
CFPC	College of Family Physicians of Canada	IYI	Indiana Youth Institute
CFUW	Canadian Federation of University Women	JDF	Juvenile Diabetes Foundation International
CHCF	California Health Care Foundation	JMO	John M. Olin Foundation
CHEA	Canadian Home Economics Association	JSPS	Japan Society for the Promotion of Science
CICF	Central Indiana Community Foundation	KFC	Kidney Foundation of Canada
CIES	Council for International Exchange of Scholars	LISC	Local Initiatives Support Corporation
CIUS	Canadian Institute of Ukrainian Studies	LSA	Leukemia Society of America
CLA	Canadian Lung Association	MFRI	Military Family Research Institute
CLF	Canadian Liver Foundation	MHRC	Manitoba Health Research Council
CMS	Centers for Medicare and Medicaid Services	MLA	Medical Library Association
CNCS	Corporation for National and Community Service	MLB	Major League Baseball
CRI	Cancer Research Institute	MMA	Metropolitan Museum of Art
CTCNet	Community Technology Centers Network	MMS	Massachusetts Medical Society
DAAD	Deutscher Akademische Austauschdienst (German Academic Exchange Service)	MSSC	Multiple Sclerosis Society of Canada
		NAA	Newspaper Association of America
DHHS	Department of Health and Human Services	NAACP	National Association for the Advancement of Colored People
DOA	Department of Agriculture	NAGC	National Association for Gifted Children
DOC	Department of Commerce	NAPNAP	National Association of Pediatric Nurse Associates and Practitioners
DOD	Department of Defense		
DOE	Department of Energy	NARSAD	National Alliance for Research on Schizophrenia and Depression
DOI	Department of the Interior		
DOJ	Department of Justice	NASA	National Aeronautics and Space Administration
DOL	Department of Labor	NASE	National Association for the Self-Employed
DOS	Department of State	NASM	National Air and Space Museum
DOT	Department of Transportation	NATO	North Atlantic Treaty Organization
EFA	Epilepsy Foundation of America	NCCAM	National Center for Complementary and Alternative Medicine
EIF	Entertainment Industry Foundation		
EPA	Environmental Protection Agency	NCFL	National Center for Family Literacy
ESF	European Science Foundation	NCI	National Cancer Institute
ETS	Educational Testing Service	NCIC	National Cancer Institute of Canada
FCAR	Formation de Chercheurs et L'Aide a la Recherche	NCRR	National Center for Research Resources
FCD	Foundation for Child Development	NCSS	National Council for the Social Studies
FDA	Food and Drug Administration	NEA	National Education Association
FIC	Fogarty International Center	NEH	National Endowment for the Humanities
GAAC	German American Academic Council	NEI	National Eye Institute
GCA	Garden Club of America	NFID	National Foundation for Infectious Diseases
GEF	Green Education Foundation	NFL	National Football League
GNOF	Greater New Orleans Foundation	NFWF	National Fish and Wildlife Foundation
HAF	Humboldt Area Foundation	NGA	National Gardening Association
HBF	Herb Block Foundation	NHGRI	National Human Genome Research Institute
HHS	Health and Human Services	NHLBI	National Heart, Lung and Blood Institute
HHMI	Howard Hughes Medical Institute	NHSCA	New Hampshire State Council on the Arts
HRSA	Health Resources and Services Administration	NIA	National Institute on Aging
HUD	Department of Housing and Urban Development	NIAF	National Italian American Foundation
ICC	Indiana Campus Compact	NIAAA	National Institute on Alcohol Abuse and Alcoholism
IIE	Institute of International Education		
IRA	International Reading Association	NIAF	National Italian American Foundation
IRC	International Rescue Committee	NIAID	National Institute of Allergy and Infectious Diseases
IREX	International Research and Exchanges Board		

NIAMS	National Institute of Arthritis and Musculoskeletal Skin Diseases	**OSF**	Open Society Foundation
NICHD	National Institute of Child Health and Human Development	**PAS**	Percussive Arts Society
		PCA	Pennsylvania Council on the Arts
NIDA	National Institute on Drug Abuse	**PDF**	Peace Development Fund
NIDCD	National Institute on Deafness and Other Communication Disorders	**PDF**	Parkinson's Disease Foundation
		PhRMA	Pharmaceutical Research and Manufacturers of American Foundation
NIDCR	National Institute of Dental and Craniofacial Research	**PHSC**	The Photographic Historical Society of Canada
NIDDK	National Institute of Diabetes, and Digestive and Kidney Diseases	**PSEG**	Public Service Enterprise Group
		RCF	Richland County Foundation
NIDRR	National Institute on Disability and Rehabilitation Research	**RCPSC**	Royal College of Physicians and Surgeons of Canada
NIEHS	National Institute of Environmental Health Sciences	**RSC**	Royal Society of Canada
		RWJF	Robert Wood Johnson Foundation
NIGMS	National Institute of General Medical Sciences	**SAMHSA**	Substance Abuse and Mental Health Services Administration
NIH	National Institutes of Health		
NIJ	National Institute of Justice	**SLA**	Special Libraries Association
NIMH	National Institute of Mental Health	**SME**	Society of Manufacturing Engineers
NINDS	National Institute of Neurological Disorders and Strokes	**SOCFOC**	Sisters of Charity Foundation of Cleveland
		SORP	Society of Biological Psychiatry
NINR	National Institute of Nursing Research	**SSHRC**	Social Sciences and Humanities Research Council of Canada
NIOSH	National Institute for Occupational Safety and Health		
		SSRC	Social Science Research Council
NIST	National Institute of Standards and Technology	**STTI**	Sigma Theta Tau International
NJSCA	New Jersey State Council on the Arts	**SVP**	Social Venture Partners
NKF	National Kidney Foundation	**SWE**	Society of Women Engineers
NL	Newberry Library	**TAC**	Tennessee Arts Commission
NLM	National Library of Medicine	**TOMF**	Tucson Osteopathic Medical Foundation
NMF	National Medical Fellowships, Inc.	**TRCF**	Three Rivers Community Fund
NMSS	National Multiple Sclerosis Society	**TSYSF**	Teemu Selanne Youth Sports Foundation
NNEDVF	National Network to End Domestic Violence Fund	**UPS**	United Parcel Service
NOAA	National Oceanic and Atmospheric Administration	**USHMM**	United States Holocaust Memorial Museum Research Institute
NRA	National Rifle Association		
NRC	National Research Council	**USIA**	United States Information Agency
NSERC	Natural Sciences and Engineering Research Council of Canada	**USAID**	United States Agency for International Development
		USDA	United States Department of Agriculture
NSF	National Science Foundation	**USFA**	United States Fencing Association
NSTA	National Science Teachers Association	**USGA**	United States Golf Association
NYCH	New York Council for the Humanities	**USIP**	United States Institute of Peace
NYCT	New York Community Trust	**USTA**	United States Tennis Association
NYFA	New York Foundation for the Arts	**UUA**	Unitarian Universalist Association
NYSCA	New York State Council on the Arts	**WAWH**	Western Association of Woman Historians
OAH	Organization of American Historians	**WHO**	Women Helping Others
ODKF	Outrigger Duke Kahanamoku Foundation		
OJJDP	Office of Juvenile Justice and Delinquency Prevention		
ONF	Oncology Nursing Foundation		
ONR	Office of Naval Research		
OREF	Orthopaedic Research and Education Foundation		
ORISE	Oak Ridge Institute for Science and Education		

Grant Programs

1 in 9: Long Island Breast Cancer Action Coalition Grants   1
The Coalition's mission is to promote awareness of the breast cancer epidemic through education, outreach, advocacy, and direct support of research which is being done to find the causes of and cures for breast cancer and other related cancers. The Coalition's major goals are to: raise awareness of the epidemic of breast cancer and keep the disease in the forefront; obtain funding for research; investigate the link between pesticides and breast cancer since Long Island is a farm area; increase detection of breast cancer; and find the causes, a cure and how to prevent breast cancer.
Geographic Focus: New Jersey, New York, Pennsylvania
Contact: Geri Barish, President; (516) 374-3190; fax (516) 569-1894; info@1in9.org
Internet: http://www.1in9.org/
Sponsor: Long Island Breast Cancer Action Coalition
86 East Rockaway Road
Hewlett, NY 11557

1st Source Foundation Grants   2
Established in 1952 in Indiana and administered by the 1st Source Bank, the Foundation supports community foundations, youth clubs and organizations involved with television, education, health, and human services. The Foundation provides support to organizations working in the following areas: social welfare and human services; education; culture and the arts; and community, civic and neighborhood involvement. Giving is primarily centered in Indiana, and the major type of funding given is for general operating support. Since there are no specific applications forms required or deadlines with which to adhere, applicants should send a letter of request detailing the project and the amount of funding needed. Most recent awards have ranged from $500 to $60,000.
Requirements: Any 501(c)3 serving the residents of Indiana communities where 1st Source Banks are located are eligible to apply.
Geographic Focus: Indiana
Amount of Grant: 500 - 60,000 USD
Samples: Bishop Dwenger High School, Fort Wayne, Indiana, $25,000 - general operating support (2014); Five Star Life, Elkhart, Indiana, $30,000 - charitable contribution (2014); Urban Enterprise Association, South Bend, Indiana, $50,000 - general operating support (2014).
Contact: Renée Fleming, Director; (574) 235-2790 or (574) 235-2119
Internet: https://www.1stsource.com/about-us/community-involvement
Sponsor: 1st Source Foundation
100 N Michigan Street, P.O. Box 1602
South Bend, IN 46601-1630

1st Touch Foundation Grants   3
Established by Derek L. Lee in 2006 in an effort to help his daughter, Jada, giving is centered around the testing and researching of Leber's Congenital Amaurosis (LCA), an extremely rare disease which causes severe vision loss and blindness. The Foundation's primary long-range mission is to help researchers find a cure and eradicate this disease. It will also continue to provide grants for higher education, and help children and youth organizations within its local communities. There are no specific guidelines, application forms, or deadlines with which to adhere, so applicants should contact the Foundation office directly.
Geographic Focus: All States
Contact: Mark Martella, (415) 421-0535 or (818) 501-4421; Mmartella@1sttouch.org
Internet: http://1sttouch.org/
Sponsor: 1st Touch Foundation
5098 Foothills B0ulevard, Suite 3, #492
Roseville, CA 95747

2 Depot Square Ipswich Charitable Foundation Grants   4
In December of 2005, Ipswich Co-operative Bank established the 2 Depot Square Ipswich Charitable Foundation with an initial contribution of $200,000. Charitable giving has been a cornerstone of the Bank's business philosophy for many years. The Foundation plays a vital role in supporting economic development and improving the quality of life in the communities that it serves. The Foundation focuses its giving in the following areas: economic and community empowerment--includes programs which focus on the promotion and development of access to safe and affordable housing, and programs which support community revitalization efforts; youth development--initiatives that encourage youth through social, educational, athletic or cultural programs; arts and culture--programs and organizations which provide art and cultural programs that enrich communities; and health and human services--organizations which strive to enhance the health and well-being of children and families in its communities. The Grant Committee meets in April and November. Completed applications are due April 1st and November 1st.
Requirements: The Foundation awards funds to non-profit organizations based in Ipswich and the surrounding communities. Applications for grants will only be accepted from qualified 501(c)3 or 501(c)1 organizations.
Geographic Focus: Massachusetts
Date(s) Application is Due: Apr 1; Nov 1
Contact: Tammy Roeger; (978) 462-3106; fax (978) 462-1980; tammy.roeger@ifs-nbpt.com
Internet: http://www.institutionforsavings.com/site/charitable_2depot_about.html
Sponsor: 2 Depot Square Ipswich Charitable Foundation
2 Depot Square
Ipswich, MA 01938-1914

2COBS Private Charitable Foundation Grants   5
The Foundation, established in Colorado in 2005, supports a variety of causes, which include: health care; cancer services; community services; and local community foundations. Although giving is centered around the Durango, Colorado, region, it is not uncommon for support to be offered to 501(c)3 programs across the country. There are no specific application formats or deadlines with which to adhere, and applicants should send a letter of request outlining the program need and overall budget.
Geographic Focus: All States
Amount of Grant: 250 - 1,000 USD
Contact: Christopher J. O'Brien, Chairperson; (970) 247-7828 or (970) 385-1740
Sponsor: 2COBS Private Charitable Foundation
10 Town Plaza, Suite 100
Durango, CO 81301-6910

3M Company Foundation Community Giving Grants   6
The 3M Company Foundation supports organizations involved with arts and culture, K-12 education, higher education, the environment, and health and human services. Special emphasis is directed toward programs designed to help prepare individuals and families for success. Fields of interest are: arts; arts education; business education; disaster relief, preparedness, and services; economics; elementary and secondary education; employment training; engineering; environmental causes; family services; federated giving programs; health care; higher education; human services; mathematics; minorities; science programs; youth development; and youth, services. Types of support include: building construction and renovation, capital campaigns, curriculum development, employee matching gifts, general operating support, in-kind gifts, program development, and scholarship funds. The foundation utilizes an invitational Request For Proposal (RFP) process for organizations located in Minneapolis and St. Paul, Minnesota, and Austin, Texas. Application forms are not required.
Requirements: Established 501(c)3 organizations in all 3M communities are eligible.
Restrictions: The 3M Company Foundation does not accept unsolicited proposals in St. Paul/Minneapolis, Minnesota, and Austin, Texas. No support for religious organizations, conduit agencies, political groups, fraternal organizations, social groups, or veteran organizations. No funding for hospitals, K-12 schools, military organizations, animal-related organizations, or disease-specific organizations. No grants to individuals, or for endowments, emergency operating support, advocacy and lobbying efforts, fundraising events and associated advertising, travel, publications, start-up needs, non -equipment, debt reduction, conferences, athletic events, or film or video production; no loans or investments.
Geographic Focus: Alabama, Alaska, Arkansas, California, Connecticut, Georgia, Hawaii, Illinois, Indiana, Iowa, Kentucky, Massachusetts, Michigan, Minnesota, Missouri, Nebraska, New York, Ohio, South Carolina, South Dakota, Texas, Utah, Wisconsin
Contact: Cynthia F. Kleven, Secretary; (651) 733-0144 or (651) 736-8146; fax (651) 737-3061; cfkleven@mmm.com
Internet: http://www.3m.com/3M/en_US/gives-us/community/
Sponsor: 3M Company Foundation
3M Center, Building 225-01-S-23
Saint Paul, MN 55144-1000

3M Company Foundation Health and Human Services Grants   7
Parallel to its Foundation, the 3M Company makes charitable contributions to nonprofit health and human services organizations directly. Support is given on a national basis. Most giving is initiated through a Request for Proposal process that allows the company to focus our giving and maximize results. Areas of interest for Health and Human Services include: to increase resiliency in youth through prevention efforts from early childhood to 12th grade; and to build and sustain healthy communities. Types of support include: building and renovation; capital campaigns; donated equipment; donated land; donated products; employee volunteer services; general operating support; in-kind gifts; internship funds; program development; seed money; technical assistance; and use of facilities. The Company also reaches out to bring assistance and help communities prepare for disaster.
Restrictions: No support is offered for: political, fraternal, social, veterans, or military organizations; propaganda or lobbying organizations; religious organizations not of direct benefit to the entire community; animal-related organizations; or disease-specific organizations. No grants are given to individuals, or for electronic media promotion or sponsorships, athletic events, non-3M equipment, endowments, emergency needs, conferences, seminars, workshops, symposia, fund raising or testimonial events, travel, or film or video production; no cause-related marketing.
Geographic Focus: Alabama, Alaska, Arkansas, California, Connecticut, Georgia, Hawaii, Illinois, Indiana, Iowa, Kentucky, Massachusetts, Michigan, Minnesota, Missouri, Nebraska, New Jersey, New York, Ohio, South Carolina, South Dakota, Texas, Utah, Wisconsin
Contact: Cynthia F. Kleven, Secretary; (651) 733-0144 or (651) 736-8146; fax (651) 737-3061; cfkleven@mmm.com

2 | Grant Programs

Internet: http://www.3m.com/3M/en_US/gives-us/
Sponsor: 3M Company Foundation
3M Center, Building 225-01-S-23
Saint Paul, MN 55144-1000

25th Anniversary Foundation Grants 8

Established by Arista Records, which was founded by Clive Jay Davis in the late 1970s, the 25th Anniversary Foundation was named to honor a quarter-century of company number one hits. The Foundation's primary field of interest is supporting AIDS research at nationally recognized organizations. There are no particular application formats or deadlines with which to adhere, and potential applicants should first approach the Foundation in writing, outlining their overall need.
Geographic Focus: All States
Amount of Grant: 10,000 USD
Samples: City of Hope, Los Angeles, California, $10,000--for AIDS research; T.J. Martel Foundation, New York, New York, $10,000--for AIDS research.
Contact: Robert J. Sorrentino; (212) 782-1137 or (212) 782-1000; (212)782-1010
Sponsor: 25th Anniversary Foundation (formerly Arista Records Foundation)
1745 Broadway
New York, NY 10019

49ers Foundation Grants 9

The San Francisco 49ers foundation is designed to enhance the educational, health, social, and cultural needs of the San Francisco Bay Area community. The current funding priority concentrates on supporting organizations and programs that focus on tackling violence and that help establish life goals; encourage positive character building choices; teach life skills; and demonstrate respect, tolerance and appreciation of diversity. Grant proposals are accepted January through April each year. Submit proposals of three or less pages.
Requirements: 49ers Foundation grants will be awarded only to 501(c)3 organizations in the greater San Francisco Bay Area that address violence prevention.
Geographic Focus: California
Amount of Grant: Up to 5,000 USD
Contact: Shauna Standart, Director; (408) 562-4949; fax (408) 727-4937; 49ersfoundation@niners.nfl.com
Internet: http://www.49ers.com/community/foundation.html
Sponsor: 49ers Foundation
4949 Centennial Boulevard
Santa Clara, CA 95054-1229

100 Mile Man Foundation Grants 10

Established in 2004 and formerly known as the Itzler Family Foundation, the 100 Mile Man Foundation currently offers support to organizations across the country. Its primary fields of interest include: health organizations and agencies; social services groups; Jewish agencies and synagogues; and medical research groups. There are no specific guidelines, applications, or deadlines with which to adhere, and interested parties should contact the Foundation directly. Generally, recent grants have ranged from $1,000 to $10,000.
Requirements: Giving is limited to 501(c)3 organizations serving the residents of the greater New York metropolitan area.
Geographic Focus: All States
Amount of Grant: 1,000 - 10,000 USD
Samples: Harold Robinson Foundation, Los Angeles, California, $10,000 - unrestricted operating support (2014); Cancer Foundation, Norwood, New Jersey, $10,000 - unrestricted operating support (2014); Karen Mumford Cancer Foundation, Cleveland Heights, Ohio, $4,000 - unrestricted operating support (2014).
Contact: Jesse Itzler; (212) 997-0500 or (212) 202-3230; info@the100mileman.com
Internet: http://the100mileman.com/
Sponsor: 100 Mile Man Foundation
1350 Avenue of the Americas, 15th Floor
New York, NY 10019-4700

1675 Foundation Grants 11

The foundation is a private family foundation dedicated to improving the quality of life for individuals and families through the support of non-profit organizations working in the areas of health, human services, education, the environment, and history. Priority is given to organizations serving Chester County, southeastern Pennsylvania, the Greater Boston area, and other geographic areas of interest to the Trustees. Grants are made for operating support, special projects, endowment, and capital. Grants range from $2,000 to $50,000 and are made twice a year at the discretion of the Trustees. Requests should be in writing and mailed to the executive director.
Requirements: 501(c)3 nonprofits and public charities under IRS Code 509(a) are eligible.
Restrictions: The foundation does not make grants to individuals, nor are they made for political purposes.
Geographic Focus: Massachusetts, Pennsylvania
Date(s) Application is Due: Mar 1; Oct 1
Amount of Grant: 2,000 - 50,000 USD
Contact: Marge Brennan, Grants Manager; (610) 896-3868; fax (610) 896-3869; mbrennan@1675foundation.org
Internet: http://www.1675foundation.org/guidelines.htm
Sponsor: 1675 Foundation
16 East Lancaster Avenue, Suite 102
Ardmore, PA 19003-2228

1976 Foundation Grants 12

Established in Pennsylvania in 1976, the 1976 Foundation gives primarily to organizations that support services for the blind, pediatric medicine, health care, and research. The average grant ranges up to $10,000, though larger grants will be considered. Though there are no specific guidelines or application forms, the annual deadline is June 30. Applicants should contact the office in writing, explaining the project and amount needed.
Geographic Focus: Pennsylvania
Date(s) Application is Due: Jun 30
Amount of Grant: Up to 10,000 USD
Contact: Nathaniel Peter Hamilton, President; (610) 254-9401; fax (610) 254-9404
Sponsor: 1976 Foundation
200 Eagle Road, Suite 308
Wayne, PA 19087-3115

A-T Children's Project Grants 13

The A-T Children's Project strives to assist respected scientists in developing a clearer understanding of ataxia-telangiectasia. The Project is determined to find a timely cure, or life-improving treatments, for this serious disease. Grant awards are made through a careful and detailed selection process. The members of the Scientific Advisory Board examine each proposal and make their independent recommendations to its Board of Directors which then votes on each proposed project. Proposals from junior investigators, from scientists in related disciplines, and from individuals with innovative new ideas for A-T research are particularly encouraged, as are laboratories and teams working together from industry as well as teaching universities. The Project provides competitive grant awards for basic and translational research grants related to A-T. One- and two-year projects are funded up to a maximum total direct cost of $75,000 per year. Grants of $75,000 per year, however, are rare; grants in the $25,000 to 50,000 per year range are much more common. Budgets for up to $150,000 for a two-year project are acceptable.
Requirements: Proposals from junior investigators, from scientists in related disciplines, and from individuals with innovative new ideas for A-T research are particularly encouraged, as are laboratories and teams working together from industry as well as teaching universities. A Letter of Intent is not required. However, prior to submission of a full-length proposal, applicants seeking Scientific Advisory Board input may submit a Letter of Intent directly to the A-TCP Science Coordinator. This letter (not to exceed two pages) should include a brief abstract describing the proposed research, specific aims and an estimated budget. Applicants must submit an electronic copy of their Proposal in either MSWord or PDF formats to grants@atcp.org.
Restrictions: The sponsor does not pay for administrative overhead and indirect costs. The sponsor does not pay for institutional construction or renovation; purchase of major capital equipment other than directly needed for proposed experiments; office equipment or furniture; travel (except as required to perform the project); tuition fees; journal subscriptions; dues or memberships; and printing or publishing costs.
Geographic Focus: All States
Date(s) Application is Due: Mar 1; Sep 1
Amount of Grant: Up to 150,000 USD
Contact: Cynthia Rothblum-Oviatt, PhD, Science Coordinator; (703) 765-1223 or (954) 481-6611; fax (954) 725-1153; cynthia@atcp.org or grants@atcp.org
Internet: http://www.communityatcp.org/page.aspx?pid=3538
Sponsor: A-T Children's Project
5300 W. Hillsboro Boulevard, Suite 105
Coconut Creek, FL 33073

A-T Children's Project Post Doctoral Fellowships 14

The A-T Children's Project strives to assist respected scientists in developing a clearer understanding of ataxia-telangiectasia. Post doctorals with one year experience or less post degrees are eligible and must be nominated for this award by their principal investigator (PI). Any interested PI who would like to nominate a new post doc for the A-T Post Doctoral Fellowship Award must send a Letter of Intent to the A-T Children's Project. This letter (not to exceed two pages) should include a brief abstract describing the proposed research, specific aims and an estimated budget. Fellowship award applications are subject to the same conditions and guidelines as the Project's regular investigator initiated grants. However, the level of funding will be in the range of $30,000 to $40,000 per year for two years.
Restrictions: The sponsor does not pay for administrative overhead and indirect costs. The sponsor does not pay for institutional construction or renovation; purchase of major capital equipment other than directly needed for proposed experiments; office equipment or furniture; travel (except as required to perform the project); tuition fees; journal subscriptions; dues or memberships; and printing or publishing costs.
Geographic Focus: All States
Date(s) Application is Due: Mar 1; Sep 1
Amount of Grant: 30,000 - 40,000 USD
Contact: Cynthia Rothblum-Oviatt, Science Coordinator; (703) 765-1223 or (954) 481-6611; fax (954) 725-1153; cynthia@atcp.org or grants@atcp.org
Internet: http://www.communityatcp.org/page.aspx?pid=3539
Sponsor: A-T Children's Project
5300 W. Hillsboro Boulevard, Suite 105
Coconut Creek, FL 33073

A-T Medical Research Foundation Grants 15

The Foundation is a national 501(c) tax-exempt organizations offering a comprehensive program of information and support services for patients and their families on ataxia-telangiectasia, a lethal genetic disease that attacks children, causing progressive loss of muscle control, immune system problems, and a strikingly high rate of cancer, especially

leukemia and lymphoma. The Foundation also offers patient and professional information and education materials, sponsors meetings and scientific workshops, funds research, and provides referrals to chapters and support groups. Seed grants and research grants are available.
Geographic Focus: All States
Amount of Grant: 35,000 - 255,000 USD
Contact: Pamela Smith, President; (818) 906-2861; fax (818) 906-2870; becca4435@aol.com or atmrf@aol.com
Sponsor: Ataxia Telangiectasia Medical Research Foundation
16224 Elisa Place
Encino, CA 91436-3320

A. Gary Anderson Family Foundation Grants 16
The foundation supports nonprofits in its areas of interest, including arts and culture, higher education, children and youth, and social services. Requests from California receive priority. There are no application deadlines with which to adhere, but application forms are required. Applicants should begin by initially forwarding a letter outlining the proposed project.
Requirements: Funding is primarily given for higher education, and arts and culture in the State of California.
Restrictions: Grants are not made to individuals.
Geographic Focus: California
Amount of Grant: 1,000 - 25,000 USD
Samples: Hoag Memorial Hospital Presbyterian (Newport Beach, CA)--for its capital campaign for the Hoag Women's Pavilion, $5 million.
Contact: Erin J. Lastinger, President; (714) 685-3990; fax (714) 685-3994
Sponsor: A. Gary Anderson Family Foundation
300 S Harbor Boulevard, Suite 1010
Anaheim, CA 92805-3721

A.O. Smith Foundation Community Grants 17
The A.O. Smith Foundation supports nonprofit organizations in communities where the A.O. Smith Corporation has facilities. The foundation offers funding to: elementary and secondary school projects that focus on the quality of educational programs and curriculum development; civic, cultural, and social welfare of communities; and medical research and improved local health services. Types of support include continuing support, annual campaigns, building and renovation, scholarship funds, and employee matching gifts. Proposals should describe the project's benefits and constituency, budget information including other sources of funding, and how results will be reported. Proposals to be considered for the following year's foundation budget must be received by October 30.
Requirements: Nonprofit organizations in company areas may apply.
Restrictions: Grants do not support political organizations or organizations whose chief purpose is to influence legislation.
Geographic Focus: All States
Date(s) Application is Due: Oct 30
Amount of Grant: 500 - 20,000 USD
Contact: Roger M. Smith, President; (414) 359-4000; fax (414) 359-4064
Internet: http://www.aosmith.com/Foundation/
Sponsor: A.O. Smith Foundation
P.O. Box 245008
Milwaukee, WI 53224-9510

A/H Foundation Grants 18
The A/H Foundation was established in California in 1996, and was named after the initials of its two founders, Albert Friedman and Harvey Friedman. Its primary mission is to support: cancer research, community and economic development; law enforcement and police agencies; community health organizations; public and private universities; human services; and Jewish agencies and synagogues. There are no specific deadlines with which to adhere, and initial approach should be by letter.
Requirements: Giving is primarily in the State of California to 501(c)3 organizations (although there is some giving outside of California).
Geographic Focus: California
Amount of Grant: Up to 25,000 USD
Samples: Concern Foundation, Beverly Hills, California, $24,500; Loyola Marymount University, Los Angeles, California, $12,500; Calicinto Ranch, San Jacinto, California, $400.
Contact: Keith J. Rosen, (818) 920-9888; fax (818)920-9388; info@wantacpa.com
Sponsor: A/H Foundation
15545 Devonshire Street, Suite 210
Mission Hills, CA 91345

AAAAI Allied Health Professionals Recognition Award 19
The recipient for this award must be an AAAAI Allied Health member or AAAAI Member or Fellow who is a non-physician (MD, DO) and who has made a substantial contribution in service or substance to allied health education, research, or to the practice of allergy/immunology. This contribution to allergy/immunology need not be directly related to programs sponsored by the AAAAI, but is preferred.
Requirements: Nomination materials should include a CV, a nomination letter and two letters of reference from AAAAI physician or non-physician members. The letter of nomination should clearly detail the candidate's contributions to the field of allergy and immunology. Nominations will be presented to the Allied Health Committee Co-chairs, Vice Co-chairs and Co-Immediate Past Chairs, who will select the award recipient. This selection must also be approved by the AAAAI Awards, Memorials, and Commemorative Lectureships Committee and the AAAAI Board of Directors.
Restrictions: Current Allied Health Professional Assembly Co-chairs, Vice Co-chairs and Co-Immediate Past Chairs are not eligible for the award.
Geographic Focus: All States
Date(s) Application is Due: Jan 14
Contact: Dr. Stuart L. Abramson, Chair; (414) 272-6071; fax (414) 276-3349
Internet: http://www.aaaai.org/professional-education-and-training/grants---awards/past-aaaaiI-honorary-award-recipients.aspx#ahra
Sponsor: American Academy of Allergy, Asthma, and Immunology
555 East Wells Street, Suite 1100
Milwaukee, WI 53202-3823

AAAAI ARTrust Mini Grants for Allied Health 20
Grants are awarded to AAAAI members for research in allergy and immunology. Awards and deadlines vary according to availability. Applicants should refer to the website for the most current information and application process.
Requirements: Applicants must be members of the AAAAI.
Geographic Focus: All States
Contact: Jerome Schultz; (414) 272-6071; fax (414) 276-6070; jschultz@aaaai.org
Internet: http://www.aaaai.org/professional-education-and-training/grants-awards.aspx
Sponsor: American Academy of Allergy, Asthma, and Immunology
555 East Wells Street, Suite 1100
Milwaukee, WI 53202-3823

AAAAI ARTrust Mini Grants for Allied Health Travel 21
The AAAAI abstract submission and travel grant application provides a new process for growth and learning. The Program includes: presentation of original research; attendance at AAAAI Annual Meeting postgraduate program and symposia; participation in a special program that includes basic research presentations by AAAAI members who are NIH funded researchers in allergy/immunology; information about career opportunities in allergy, asthma and immunology research; and funds for travel (including a stipend for ground transportation), hotel accommodation, and meeting registration. Additional application instructions are available at the AAAAI website.
Requirements: Eligible candidates include first-, second-, and third-year RNs in training.
Restrictions: Individuals with a faculty appointment are not eligible.
Geographic Focus: All States
Date(s) Application is Due: Nov 12
Amount of Grant: Up to 750 USD
Contact: Jerome Schultz; (414) 272-6071; fax (414) 276-3349; jschultz@aaaai.org
Internet: http://www.aaaai.org/professional-education-and-training/grants---awards.aspx
Sponsor: American Academy of Allergy, Asthma, and Immunology
555 East Wells Street, Suite 1100
Milwaukee, WI 53202-3823

AAAAI ARTrust Mini Grants for Clinical Research 22
The ARTrust invites AAAAI members to apply for mini-grants for for projects to advance the knowledge and/or treatment of allergy, asthma, and immunology for: small clinical or research projects; new educational initiatives for professional and public audiences; and other research including equipment needs, supplemental funding, and FIT projects. Funding is available for up to $15,000 for one year.
Requirements: Applicants must be U.S. citizens or permanent residents and AAAAI members in good standing. Those with a membership status of Fellow, Member, or In-training must be ABAI certified or ABAI eligible. Applicants with a membership status of Allied Health may apply. Applicants must submit a one page demographic/CV, budget, and a 1-2 page proposal with the following as section headers: hypothesis (if applicable); rational; methods; and expected outcome. Proposals are submitted online and must not exceed two pages.
Restrictions: Funding is not available for indirect costs, or the salary and benefits for the project's principal investigator. Individuals currently receiving an ARTrust grant are not eligible.
Geographic Focus: All States
Date(s) Application is Due: Nov 12
Amount of Grant: Up to 15,000 USD
Contact: Jerome Schultz; (414) 272-6071; fax (414) 276-6070; jschultz@aaaai.org
Internet: http://www.aaaai.org/Aaaai/media/MediaLibrary/PDF%20Documents/Education%20and%20Training/Awards/2013-ARTrust-Mini-Grants.pdf
Sponsor: American Academy of Allergy, Asthma, and Immunology
555 East Wells Street, Suite 1100
Milwaukee, WI 53202-3823

AAAAI ARTrust Mini Grants in Faculty Development 23
The Faculty Development Grant offers an individual three years of financial support to concentrate his/her research efforts in either basic or clinical allergy and immunology. The objectives of this funding are to recognize and support young researchers and promote the specialty of allergy/immunology. The recipient is expected to participate in a seminar at the AAAAI annual meeting, where awardees present their research.
Requirements: The award is open to researchers of junior faculty status--either instructor, assistant professor, or the equivalent grade. Funding is directed toward MDs, but in unusual circumstances, PhDs will be considered. Applicants must be U.S. or Canadian citizens or permanent residents who are AAAAI members (or have applied for membership by September 1).
Restrictions: Past recipients are ineligible.
Geographic Focus: All States
Date(s) Application is Due: Nov 15

4 | Grant Programs

Amount of Grant: 150,000 USD
Contact: Jerome Schultz; (414) 272-6071; fax (414) 276-6070; jschultz@aaaai.org
Internet: http://www.aaaai.org/global/ARTrust/award-recipients.aspx
Sponsor: American Academy of Allergy, Asthma, and Immunology
555 East Wells Street, Suite 1100
Milwaukee, WI 53202-3823

AAAAI Distinguished Clinician Award 24

The Distinguished Clinician Award is given to a person or persons whose clinically relevant activities have made a significant constructive impact on the field of allergy and clinical immunology. The awardee(s) should have a primary commitment to patient care of clinical investigation focused on diagnosis, treatment and prevention of asthma and allergic disease. The individual(s) should have a record of substantial clinical research as evidenced by papers published in referenced journals (such as JACI); evidence of scholarship by presentations on the programs of national or international meetings, published books, editorship of supplements to referenced journals or equivalent activity. Especially desirable, but not mandatory, the individual should show dedicated service to the field of allergy and clinical immunology, e.g. service to the American Board of Allergy/Immunology, a voluntary foundation or government agency, or public education, outreach, or community activities in the field of allergy\immunology. Alternatively, the recipient(s) could have worked actively in the AAAAI or similar professional society, serving on major committees, participating in workshops, and/or serving as an officer in the AAAAI.
Requirements: Nomination forms are available at the AAAAI website. The nomination letter, along with a copy of the nominee's CV and two letters of recommendation, should be mailed to the Awards Subcommittee at the AAAAI corporate office.
Geographic Focus: All States
Date(s) Application is Due: Jan 14
Contact: Dr. Stuart L. Abramson, Chair; (414) 272-6071; fax (414) 276-6070
Internet: http://www.aaaai.org/professional-education-and-training/grants---awards/past-aaaaI-honorary-award-recipients.aspx#dca
Sponsor: American Academy of Allergy, Asthma, and Immunology
555 East Wells Street, Suite 1100
Milwaukee, WI 53202-3823

AAAAI Distinguished Layperson Award 25

The Distinguished Layperson Award is granted once a year to the citizen who has made the most substantial contribution to the field of allergy. The AAAAI recognizes that many scientists and physicians who are not clinical allergists have made contributions to the understanding and practice of the specialty. Non-medical citizens have worked in various ways to promote the field. Many are members of voluntary health organizations dedicated to public education and fundraising activities to support research. The nomination instructions are available at the AAAAI website.
Requirements: The recipient should be a non-medical person who has made a substantial contribution in service or substance to education, research, or to the practice. The contribution need not be directly related to programs sponsored by the AAAAI. To be eligible, a candidate for the award should still be living at the time the award is made. At least one member of the AAAAI should have first-hand knowledge of the candidate's contribution.
Restrictions: Employees of physicians' or health institutions shall not be eligible for the award unless their contribution has been made outside of their duties of employment.
Geographic Focus: All States
Date(s) Application is Due: Jan 14
Contact: Dr. Stuart L. Abramson, Chair; (414) 272-6071; fax (414) 276-3349
Internet: http://www.aaaai.org/professional-education-and-training/grants---awards/past-aaaaI-honorary-award-recipients.aspx#dla
Sponsor: American Academy of Allergy, Asthma, and Immunology
555 East Wells Street, Suite 1100
Milwaukee, WI 53202-3823

AAAAI Distinguished Scientist Award 26

The Distinguished Scientist Award is presented to recognize scientific contributions to the field of allergy/immunology that have advanced research in the specialty, in addition to leadership contributions to the specialty. Nomination forms are available at the AAAAI website.
Requirements: Nominations should include a letter of nomination, with a copy of the nominee's CV and two letters of recommendation.
Geographic Focus: All States
Date(s) Application is Due: Jan 14
Contact: Dr. Stuart L. Abramson, Chair; (414) 272-6071; fax (414) 276-3349
Internet: http://www.aaaai.org/professional-education-and-training/grants---awards/past-aaaaI-honorary-award-recipients.aspx#dsca
Sponsor: American Academy of Allergy, Asthma, and Immunology
555 East Wells Street, Suite 1100
Milwaukee, WI 53202-3823

AAAAI Distinguished Service Award 27

The Distinguished Service Award recognizes long and distinguished service to the AAAAI rather than primarily for scientific achievement. Past-Presidents are eligible for this award. The number of Distinguished Service Awards should be limited to one per year.
Requirements: Nomination forms are available at the AAAAI website. Nomination letters must include the nominee's CV and two letters of recommendation.
Geographic Focus: All States
Date(s) Application is Due: Jan 14
Contact: Dr. Stuart L. Abramson, Chair; (414) 272-6071; fax (414) 276-3349
Internet: http://www.aaaai.org/professional-education-and-training/grants---awards/past-aaaaI-honorary-award-recipients.aspx#dsa
Sponsor: American Academy of Allergy, Asthma, and Immunology
555 East Wells Street, Suite 1100
Milwaukee, WI 53202-3823

AAAAI Fellows-in-Training Abstract Award 28

Fellows-in-Training are eligible for the Abstract Award. Accepted abstracts are presented at the Annual Meeting. Authors of accepted abstracts will receive a reduced registration rate for the National Conference. Additional information is available at the website, including instructions for online abstract submission.
Requirements: Abstracts must be submitted using the online abstract submission process. Submission by members and non-members are welcome, and participation is open to health professionals in any field. However, some sections require sponsors for papers whose authors do not include a member of the respective section. The text of each abstract is limited to 450 words. Images and tables are permitted. Images will not count toward the word limit, but any words or numbers within an included table will be counted accordingly. All authors will be required to complete conflict of interest disclosures. Authors whose abstracts are accepted for presentation will be expected to attend the meeting and make their presentation.
Restrictions: An abstract may be submitted to one section or council only.
Geographic Focus: All States
Date(s) Application is Due: Sep 12
Contact: Mariana Duran; (414) 272-6071; fax (414) 276-6070; mduran@aaaai.org
Internet: http://www.aaaai.org/professional-education-and-training/fellows-in-training.aspx
Sponsor: American Academy of Allergy, Asthma, and Immunology
555 East Wells Street, Suite 1100
Milwaukee, WI 53202-3823

AAAAI Fellows-in-Training Grants 29

Fellowship status in the AAAAI recognizes the professional achievements in a member's A/I specialty and provides several benefits available only to Fellows. Benefits include: voting, proposing motions, and holding office on the Board of Directors; recognition as a leader in the specialty's premier A/I society; first opportunity to register for seminars, workshops, and courses; complimentary subscriptions to professional journals and magazines; and networking opportunities with other A/I professionals around the world. Applications are accepted year round and reviewed in March, June, and November of each year. Applicants are notified of their status within four weeks after the meeting. The application is available at the corporate website or http://www.aaaai.org/Aaaai/media/MediaLibrary/PDF%20Documents/Member%20Applications/Fellowship-Application.pdf. No application fee is required for current members.
Requirements: All applications must be accompanied by a letter of intent and CV from the applicant. Applicants must also be sponsored by one Fellow of the AAAAI. Applicants should meet the following criteria for consideration: full membership with AAAAI for at least three years; certification by a subspeciality board, the American Board of Allergy and Immunology, or its foreign equivalent; and continuing training and experience in the field of A/I.
Geographic Focus: All States, All Countries
Contact: Mariana Duran, Fellows-in-Training Contact; (414) 272-6071; fax (414) 276-6070; mduran@aaaai.org or membership@aaaai.org
Internet: http://www.aaaai.org/professional-education-and-training/fellows-in-training.aspx
Sponsor: American Academy of Allergy, Asthma, and Immunology
555 East Wells Street, Suite 1100
Milwaukee, WI 53202-3823

AAAAI Fellows-in-Training Travel Grants 30

The AAAAI will provide funding to assist Fellows-in-Training (FITs) and non-faculty PhD Post Doc FITS in Allergy/Immunology (A/I) to attend AAAAI's Annual Meeting. Grant awards vary according to the level of the applicant. First year in-training FITs without an accepted abstract for presentation will receive up to $800. Third or fourth year in-training FITs without an accepted abstract will receive up to $650. In-Training or Post Doc FITs who are first author and present their accepted abstract will receive up to $1,100. Candidates will apply through the online application process at the AAAAI website.
Requirements: First, second, third, or fourth year FITS enrolled in an A/I program approved by the ACGME or by the Royal College of Physicians and Surgeons of Canada may apply. Accepted abstracts are not required (although funding will typically be higher for accepted abstracts). Post-Doctoral Research FITs may apply if their training has taken place in the U.S. or Canada, and involves A/I related research. Abstracts must be submitted and accepted for presentation at the annual meeting. Fellows-in-Training in a medical or surgical specialty other A/I may apply if they have submitted an abstract and it has been accepted for presentation.
Restrictions: Only AAAAI members may apply. A maximum of three FIT awards can be received during the fellowship training years.
Geographic Focus: All States
Date(s) Application is Due: Nov 2
Amount of Grant: 650 - 1,100 USD
Contact: Mariana Duran; (414) 272-6071; fax (414) 276-6070; mduran@aaaai.org
Internet: http://www.aaaai.org/professional-education-and-training/fellows-in-training.aspx
Sponsor: American Academy of Allergy, Asthma, and Immunology
555 East Wells Street, Suite 1100
Milwaukee, WI 53202-3823

AAAAI Mentorship Award 31
The AAAAI Mentorship Award recognizes individuals who, by serving as mentors, make a difference in the careers of trainees and colleagues. Mentoring is broadly defined as someone who makes a difference in another person's career by guidance, education, listening, being accessible, providing a safe haven, serving as an advocate, easing training, helping career development, and serving as a role model. Additional information and the nomination form, with specific instructions about the materials required with the nomination letter, are listed at the AAAAI website.
Requirements: The awardees are expected to have shown dedication to mentoring over a sustained period of time and to have impacted the careers of more than one mentee. Mentors from all areas of the A/I specialty are eligible including internists, pediatricians, allergists, immunologists, clinical immunologists, academic faculty, private practitioners, basic/clinical/ or translational researchers, as well as allergists/immunologists in government or industry worldwide.
Geographic Focus: All States
Date(s) Application is Due: Jan 14
Contact: Jerome Schultz; (414) 272-6071; fax (414) 276-3349; jschultz@aaaai.org
Internet: http://www.aaaai.org/professional-education-and-training/grants---awards/past-aaaaiI-honorary-award-recipients.aspx#ma
Sponsor: American Academy of Allergy, Asthma, and Immunology
555 East Wells Street, Suite 1100
Milwaukee, WI 53202-3823

AAAAI Outstanding Volunteer Clinical Faculty Award 32
This award is given to outstanding volunteer clinical faculty. Volunteer faculty is defined as being primarily in private practice without university remuneration for participation in the allergy/immunology program. A clinical faculty appointment is desirable, as well as demonstration of outstanding contributions to the education of fellows in allergy/immunology for a minimum of five years. The award is limited to one recipient a year. Multiple recipients will be accepted under extenuating circumstances at the discretion of the review committee.
Requirements: Nomination submission should include a nomination letter from the program director and the nominee's curriculum vitae, as well as two letters of recommendation from other faculty members and two letters of recommendation from former fellows. Emphasis should be on quality of contribution, durations of participation, documentation of excellence, innovation in teaching or curriculum planning or program administration, and participation in any clinical research or data collection/assessment contributing to the education of trainees.
Geographic Focus: All States
Date(s) Application is Due: Jan 14
Contact: Dr. Stuart L. Abramson, Chair; (414) 272-6071; fax (414) 276-3349
Internet: http://www.aaaai.org/professional-education-and-training/grants---awards/past-aaaaiI-honorary-award-recipients.aspx#ovcfa
Sponsor: American Academy of Allergy, Asthma, and Immunology
555 East Wells Street, Suite 1100
Milwaukee, WI 53202-3823

AAAAI RSLAAIS Leadership Award 33
The RSLAAIS Leaderships Award recognizes an outstanding leader in a state, regional or local allergy, asthma, immunology society that is part of the Federation of Regional, State & Local Allergy, Asthma and Immunology Societies (RSLAAIS). This award was created to honor active leadership in a member society of the RSLAAIS, long-term community involvement, and clinical teaching at the local level. Additionally, contributions to the mission of the AAAAI as a speaker, volunteer or leader will also be considered as supplementary factors. Award candidates will have a demonstrated involvement in the local society directed toward building, strengthening, and contributing to that society toward the local medical and patient communities, the specialty of allergy/immunology, and the American AAAAI of Allergy, Asthma & Immunology (AAAAI).
Requirements: Award nominations will include a letter of nomination from the local society leadership, two letters of reference from society members, and the award candidate's C.V. to be submitted to the RSLAAIS Board of Governors. The Governors will review the nominations and determine the award candidates. The final roster of candidates will be presented to the Awards, Memorials and Commemorative Lectureships (AMCL) Committee who will select the award candidate. The AMCL Committee will present the candidate nomination to the AAAAI Board of Directors at their June meeting for their final approval.
Geographic Focus: All States
Date(s) Application is Due: Jan 14
Contact: Dr. Stuart L. Abramson, Chair; (414) 272-6071; fax (414) 276-3349
Internet: http://www.aaaai.org/professional-education-and-training/grants---awards/past-aaaaiI-honorary-award-recipients.aspx#rslaais
Sponsor: American Academy of Allergy, Asthma, and Immunology
555 East Wells Street, Suite 1100
Milwaukee, WI 53202-3823

AAAAI Special Recognition Award 34
The AAAAI Special Recognition Award showcases individuals who have made a specific contribution benefiting the allergy/immunology community. This award may be issued annually. Additional information and a list of previous recipients is available at the AAAAI website.
Requirements: Nomination letters are submitted to the AAAAI corporate office. Nominations must also include the nominee's CV and two letters of recommendation.
Geographic Focus: All States
Date(s) Application is Due: Jan 14
Contact: Dr. Stuart L. Abramson, Chair; (414) 272-6071; fax (414) 276-3349
Internet: http://www.aaaai.org/professional-education-and-training/grants---awards/past-aaaaiI-honorary-award-recipients.aspx#sra
Sponsor: American Academy of Allergy, Asthma, and Immunology
555 East Wells Street, Suite 1100
Milwaukee, WI 53202-3823

AAAS/Subaru SB&F Prize for Excellence in Science Books 35
The AAAS/Subaru SB&F Prize for Excellence in Science Books celebrates outstanding science writing and illustration for children and young adults. The prizes, established in 2005, are meant to encourage the writing and publishing of high-quality science book for all age groups. Solely supported by Subaru since their inception, the prizes recognize recently published works that are scientifically sound and foster an understanding and appreciation of science in young readers. Prizes will be awarded in the following categories: Children's Science Picture Book (prize to author and illustrator); Middle Grades Science Book (prize to author); Young Adult Science Book (prize to author); and Hands-On Science Book (prize to author). The annual deadline is September 5. A judging panel made up of scientists, experts in the field of science literature, librarians, and AAAS staff, will assess the entries based on these several criteria.
Geographic Focus: All States
Contact: Stephen Nelson, (202) 326-6617 or (202) 326-6700; snelson@aaas.org
Internet: http://www.sbfonline.com/Subaru/Pages/CurrentWinners.aspx
Sponsor: American Association for the Advancement of Science
1200 New York Avenue NW
Washington, D.C. 20005-3920

AAAS Award for Scientific Freedom and Responsibility 36
The award is presented to honor scientists and engineers whose exemplary actions have served to foster scientific freedom and responsibility and recognizes scientists and engineers who have: acted to protect the public's health, safety, or welfare; or focused public attention on important potential impacts of science and technology on society by their responsible participation in public policy debates; or established important new precedents in carrying out the social responsibilities or in defending the professional freedom of scientists and engineers. Nominations should include nominator's name, address, and phone number; and the name and address of the nominee; a summary of the action or actions that form the basis for the nomination (about 250 words); a longer statement (no more than three pages) providing additional details of the actions for which the candidate is nominated; at least two letters of support with addresses and phone numbers; and the candidate's vita (no more than three pages). Any documentation--books, articles, etc.--that illuminates the significance of the nominee's achievements may also be submitted.
Requirements: The award is open to all regardless of nationality or citizenship. Eligible nominees are scientists and engineers who have acted to protect the public's health, safety, or welfare; or to focus public attention on potentially serious impacts of science and technology on society by their responsible participation in public policy debates; or to establish important new precedents in social responsibility or in defending the professional freedom of scientists and engineers.
Restrictions: All materials become the property of AAAS.
Geographic Focus: All States, All Countries
Date(s) Application is Due: Sep 1
Amount of Grant: 5,000 USD
Contact: Deborah Runkle, (202) 326-6794; fax (202) 289-4950; drunkle@aaas.org
Internet: http://www.aaas.org/aboutaaas/awards/freedom/
Sponsor: American Association for the Advancement of Science
1200 New York Avenue NW
Washington, D.C. 20005-3920

AAAS Early Career Award for Public Engagement with Science 37
The AAAS Early Career Award for Public Engagement with Science, established in 2010, recognizes early-career scientists and engineers who demonstrate excellence in their contribution to public engagement with science activities. A monetary prize of $5,000, a commemorative plaque, complimentary registration to the AAAS Annual Meeting, and reimbursement for reasonable hotel and travel expenses to attend the AAAS Annual Meeting to receive the prize are given to the recipient. For the purposes of this award, public engagement activities are defined as the individual's active participation in efforts to engage with the public on science- and technology-related issues and promote meaningful dialogue between science and society, as highlighted in this video. The award will be given at the AAAS Annual Meeting.
Requirements: Nominee must be an early-career scientist or engineer in academia, government or industry actively conducting research in any scientific discipline (including social sciences and medicine). Groups or institutions will not be considered for this award. AAAS employees are ineligible. One scientist or engineer will be chosen to receive the award on an annual basis. Nominee will have demonstrated excellence in his/her contribution to public engagement with science activities, with a focus on interactive dialogue between the individual and a non-scientific, public audience(s).
Geographic Focus: All States
Date(s) Application is Due: Oct 15
Amount of Grant: 5,000 USD
Contact: Linda Cendes, (202) 326-6656; fax (202) 289-4950; lcendes@aaas.org
Internet: http://www.aaas.org/aboutaaas/awards/public_engagement/
Sponsor: American Association for the Advancement of Science
1200 New York Avenue NW
Washington, D.C. 20005-3920

6 | GRANT PROGRAMS

AAAS Eppendorf and Science Prize for Neurobiology 38
The Eppendorf and Science Prize for Neurobiology acknowledges the increasingly active and important role of neurobiology in advancing our understanding of the functioning of the brain and the nervous system -- a quest that seems destined for dramatic expansion in the coming decades. This international prize, established in 2002, encourages the work of promising young neurobiologists by providing support in the early stages of their careers. It is awarded annually for the most outstanding neurobiological research by a young scientist, as described in a 1,000-word essay based on research performed during the past three years. The winner is awarded money and publication of his or her essay in 'Science'. The essay and those of up to three finalists are also published on 'Science Online'. The award is announced and presented at a ceremony at the annual meeting of the Society for Neuroscience. Eppendorf provides financial support to help enable the grand prize winner to attend the meeting.
Requirements: Entrants must be a neurobiologist with an advanced degree and not older than 35 years. The research described in the entrant's essay must be based on the methods of molecular and cell biology. The entrant must have performed or directed the work described in the essay. The research must have been performed during the previous three years.
Restrictions: Employees of Eppendorf AG, its subsidiaries, Science and AAAS, and their relatives are not eligible for the prize.
Geographic Focus: All States
Date(s) Application is Due: Jun 15
Amount of Grant: 25,000 USD
Contact: Monica M. Bradford, Executive Editor; (202) 326-6550; fax (202) 289-7562; mmadrid@aaas.org or eppendorfscienceprize@aaas.org
Internet: http://www.sciencemag.org/site/feature/data/prizes/eppendorf/
Sponsor: American Association for the Advancement of Science
1200 New York Avenue NW
Washington, D.C. 20005-3920

AAAS GE and Science Prize for Young Life Scientists 39
The GE and Science Prize for Young Life Scientists has been established to provide support to scientists at the beginning of their careers, because Science/AAAS and GE Healthcare believe that such support is critical for continued scientific progress. The prize will recognize outstanding graduate students in molecular biology from all regions of the world. This international prize will be awarded for the outstanding thesis in the general area of molecular biology as described in a 1000-word essay. The winning essay will be published in Science; essays of the regional award winners will appear in the online version of Science. For judging purposes, the essays will be grouped according to the geographic location of the degree-granting institution: North America, Europe, Japan, and all other countries. Initial screening of the submissions will be done by regional judges. Essays will be judged on the quality of the research and the entrant's ability to articulate the contribution of the research to the field of molecular biology. The top five essays from each geographic region will be forwarded to a panel of judges. All regional winners will compete for the grand prize of US$25,000. The regional winners who do not receive the grand prize will be awarded US$5,000. Winners will be announced in Science and the prize will be awarded in a location to be announced. The grand prize essay will be published in 'Science', and essays of the regional winners will be published on the online version of 'Science'.
Requirements: Entrants must have been awarded their Ph.D. between January, 1 and, December 31, of the previous year. Candidates for M.D./Ph.D. degrees are eligible to compete for the prize in either the year the Ph.D. is awarded or the year the final degree is awarded. The research described in the entrant's thesis must be in the field of molecular biology as described above. The prize will recognize only work that was performed while the entrant was a graduate student. The prize will be awarded without regard to sex, race, or nationality.
Restrictions: Employees of GE, Science and AAAS, and their relatives are not eligible for the prize.
Geographic Focus: All States
Date(s) Application is Due: Aug 1
Amount of Grant: 25,000 USD
Contact: Sylvia Kihara, Prize Coordinator; (202) 326-6507; fax (202) 289-7562; skihara@aaas.org or gescienceprize@aaas.org
Internet: http://www.sciencemag.org/site/feature/data/prizes/ge/index.xhtml
Sponsor: American Association for the Advancement of Science
1200 New York Avenue NW
Washington, D.C. 20005-3920

AAAS Martin and Rose Wachtel Cancer Research Award 40
AAAS and Science Translational Medicine invite applications for the inaugural AAAS Martin and Rose Wachtel Cancer Research Award. This annual award, funded by an endowment established through a generous bequest from Martin L. Wachtel, will honor an early-career investigator who has performed outstanding work in the field of cancer research. The award winner will be invited to deliver a public lecture on his or her research and will receive a cash award of $25,000. The award winner's Essay will be published as a Perspective in Science Translational Medicine. The deadline for application is February 15.
Requirements: Entrants must be a researcher in the field of cancer, with an advanced degree (Ph.D., M.D., D.V.M.) received within the last 10 years. The research must have been performed during the previous 10 years. The entrant must have performed or personally directed the work described in the Essay.
Restrictions: Employees of AAAS and their relatives are not eligible for the prize.
Geographic Focus: All States
Amount of Grant: 25,000 USD
Contact: Stephen Nelson, (202) 326-6617 or (202) 326-6700; snelson@aaas.org
Internet: http://www.aaas.org/aboutaaas/awards/wachtel/
Sponsor: American Association for the Advancement of Science
1200 New York Avenue NW
Washington, D.C. 20005-3920

AAAS Science and Technology Policy Fellowships: Global Health and Development 41
The AAAS Science and Technology Policy Fellowships have partnered with the Bill and Melinda Gates Foundation to create the Global Health & Development (GHD) Fellowships. This new fellowship opportunity places Fellows at the Gates Foundation in Seattle, Washington, for one year, with the potential to renew for a second year. The GHD Fellowships provide opportunities for Fellows to learn from and contribute to the international development initiatives of the Foundation, by helping to identify and support strategies and programs in which science and technology can be harnessed to advance development in designated countries and/or regions.
Requirements: This fellowship opportunity is open to current and alumni AAAS Science & Technology Policy Fellows. All applicants must have a doctoral-level degree (PhD, MD, DVM, DSc, PharmD, and other terminal degrees), in any physical, biological, health/medical or social science, any field of engineering, or any relevant interdisciplinary field (individuals with a master's degree in engineering and at least three years of post-degree professional experience also may apply). Note: All requirements for the degree must be completed by the application deadline. A prospective fellow must demonstrate exceptional competence in their specialty appropriate to their career stage, and have the strong endorsement of three references. Show an understanding of the opportunities for science and engineering to support a broad range of non-scientific issues, and display a commitment to apply their scientific or technical expertise to serve society; Exhibit awareness and sensitivity to the political, economic and social issues that influence policy; Be articulate communicators, both verbally and in writing, to decision-makers and non-scientific audiences, and have the ability to work effectively with individuals and groups outside the scientific community; Demonstrate integrity and good judgment, and the flexibility and willingness to address policy issues outside their scientific realm.
Restrictions: Federal employees are ineligible. All applicants must be U.S. citizens (dual citizenship from the United States and another country is acceptable). Some program areas and host agencies seek additional qualifications. Holding a Title 42 position is considered federal employment and therefore renders an applicant ineligible for a AAAS Science and Technology Policy Fellowship. Individuals working in postdoctoral positions at government labs or agencies, for private contractors, and in state government posts are not federal employees and therefore are eligible for the AAAS Fellowships.
Geographic Focus: All States
Date(s) Application is Due: Dec 5
Amount of Grant: 70,000 - 90,000 USD
Contact: Sage Russell, Associate Director; (202) 326-6700; fax (202) 289-4950; srussell@aaas.org or fellowships@aaas.org
Internet: http://fellowships.aaas.org/02_Areas/02_index.shtml#GHD
Sponsor: American Association for the Advancement of Science
1200 New York Avenue NW
Washington, D.C. 20005-3920

AAAS Science Prize for Online Resources in Education 42
The Science Prize for Online Resources in Education (SP.O.RE) has been established to encourage innovation and excellence in education, as well as to encourage the use of high-quality on-line resources by students, teachers, and the public. In 2009, the prize recognized outstanding projects from all regions of the world that brought freely available online resources to bear on science education. The project can be targeted to students or teachers at the precollege or college level, or it can serve the informal education needs of the general public. Nominations are welcome from all sources. Both self- and independent-nominations are welcome. Nominations must be received by March 31.
Requirements: The project must focus on science education. The resources described must be freely available on the Internet. The Internet resources must be in English or include an English translation.
Restrictions: The prize will be awarded without regard to sex, race, nationality, geographical location of the project, or membership in AAAS. Funding partners of AAAS, previous SP.O.RE winners, employees of Science and AAAS and their subsidiaries, and immediate family relatives of employees, are not eligible for the prize.
Geographic Focus: All States
Date(s) Application is Due: Mar 31
Contact: Deborah Runkle, (202) 326-6794 or (202) 326-6700; drunkle@aaas.org
Internet: http://www.sciencemag.org/site/special/spore/
Sponsor: American Association for the Advancement of Science
1200 New York Avenue NW
Washington, D.C. 20005-3920

AABB Dale A. Smith Memorial Award 43
The AABB Dale A. Smith Memorial Award, created in 2002 and funded by Fenwal, Inc., honors Dale A. Smith, a long-time Baxter Healthcare executive who was responsible for establishing the Fenwal Division of Baxter. This award will be presented to an individual or institution to recognize groundbreaking work performed in the application of technology to the practice of transfusion medicine or cellular therapies. The recipient will receive a $7,500 honorarium. The Award is presented annually.
Requirements: Award recipients are not required to be members of AABB. A nomination form must be completed for each nomination submission.
Geographic Focus: All States

Amount of Grant: 7,500 USD
Contact: Awards Coordinator; (301) 907-6977; fax (301) 907-6895; aabb@aabb.org
Internet: http://www.aabb.org/development/awardsscholarships/awards/Pages/descriptions.aspx#smith
Sponsor: American Association of Blood Banks
8101 Glenbrook Road
Bethesda, MD 20814-2749

AABB Hemphill-Jordan Leadership Award   44
The AABB Hemphill-Jordan Leadership Award, renamed in 2005 after Bernice Hemphill, W. Quinn Jordan, and Joel Solomon, honors leaders from the transfusion medicine and cellular therapy community. The award recognizes an individual who made significant contributions in the areas of administration, quality programs, law and/or government affairs. The individual shall have demonstrated leadership qualities and a consistent willingness to lend his/her expertise to his/her peers. It may recognize one particular act or an accumulation of years of contributions. Acceptance of the award requires attendance at the Annual Meeting and the presentation of a lecture. The recipient will receive a $5,000 honorarium. The Award is presented annually and the recipient is chosen one year in advance.
Geographic Focus: All States
Amount of Grant: 5,000 USD
Contact: Awards Coordinator; (301) 907-6977; fax (301) 907-6895; aabb@aabb.org
Internet: http://www.aabb.org/development/awardsscholarships/awards/Pages/descriptions.aspx#hemphill
Sponsor: American Association of Blood Banks
8101 Glenbrook Road
Bethesda, MD 20814-2749

AABB John Elliott Memorial Award   45
The AABB John Elliott Memorial Award, established in 1956, recognizes an individual who has given outstanding service to AABB by demonstrating a willingness to lend his/her expertise to the association through work on committees, the AABB Board of Directors and other areas. The recipient will receive a $1,000 honorarium. The Award is presented in odd years only, and the recipient is selected by a Joint Committee composed of leaders from the Cellular Therapies Section coordinating committee and Transfusion Medicine Section coordinating committee.
Requirements: Award recipients are not required to be members of AABB. A nomination form must be completed for each nomination submission.
Geographic Focus: All States
Amount of Grant: 1,000 USD
Contact: Awards Coordinator; (301) 907-6977; fax (301) 907-6895; aabb@aabb.org
Internet: http://www.aabb.org/development/awardsscholarships/awards/Pages/descriptions.aspx#elliott
Sponsor: American Association of Blood Banks
8101 Glenbrook Road
Bethesda, MD 20814-2749

AABB Karl Landsteiner Memorial Award and Lectureship   46
Initiated in 1954 to honor Karl Landsteiner, MD, whose lifetime research laid the foundation for modern blood transfusion therapy, the AABB Karl Landsteiner Memorial Award and Lectureship recognizes a scientist whose original research resulted in an important contribution to the body of scientific knowledge. The scientist who receives the award shall have an international reputation in transfusion medicine or cellular therapies. The recipient must agree to lecture at the AABB Annual Meeting. The recipient will receive a $7,500 honorarium. The Award and Lectureship will be awarded as distinguished candidates are identified. The recipient is chosen one year in advance by AABB's Board of Directors.
Requirements: Award recipients are not required to be members of AABB. A nomination form must be completed for each nomination submission.
Geographic Focus: All States
Amount of Grant: 7,500 USD
Contact: Awards Coordinator; (301) 907-6977; fax (301) 907-6895; aabb@aabb.org
Internet: http://www.aabb.org/development/awardsscholarships/awards/Pages/descriptions.aspx#landsteiner
Sponsor: American Association of Blood Banks
8101 Glenbrook Road
Bethesda, MD 20814-2749

AABB Sally Frank Memorial Award and Lectureship   47
The Sally Frank Memorial Award and Lectureship was established in 1982 in memory of Ms. Sally Frank and her dedication to red cell serology and education. This award recognizes an individual who is, or has been, a medical technologist involved with these fields and has demonstrated quality research, teaching and/or service abilities in the technical aspects of immunohematology. Recipient will receive a $1,000 honorarium. The Award and Lectureship is presented annually and the recipient is chosen one year in advance by AABB's Transfusion Medicine Section Coordinating Committee with Formal Approval by AABB's Board of Directors.
Requirements: Award recipients are not required to be members of AABB. A nomination form must be completed for each nomination submission.
Geographic Focus: All States
Amount of Grant: 1,000 USD
Contact: Awards Coordinator; (301) 907-6977; fax (301) 907-6895; aabb@aabb.org
Internet: http://www.aabb.org/development/awardsscholarships/awards/Pages/descriptions.aspx#frank
Sponsor: American Association of Blood Banks
8101 Glenbrook Road
Bethesda, MD 20814-2749

AABB Tibor Greenwalt Memorial Award and Lectureship   48
The AABB Tibor Greenwalt Memorial Award and Lectureship honors Tibor Greenwalt, MD, who was the first registrant at the first AABB Annual Meeting and founding editor of Transfusion. The award recognizes an individual who made major scientific or clinical contributions to hematology, transfusion medicine or cellular therapies, and succinctly communicated these advances. Acceptance of this award requires attendance at the AABB Annual Meeting and the presentation of a lecture on a current topic of interest. Recipient will receive a $1,000 honorarium. The Award and Lectureship is presented annually and the recipient is chosen one year in advance by a Joint Committee Composed of Leaders from the Cellular Therapies Section Coordinating Committee and Transfusion Medicine Section Coordinating Committee.
Requirements: Award recipients are not required to be members of AABB. A nomination form must be completed for each nomination submission.
Geographic Focus: All States
Amount of Grant: 1,000 USD
Contact: Awards Coordinator; (301) 907-6977; fax (301) 907-6895; aabb@aabb.org
Internet: http://www.aabb.org/development/awardsscholarships/awards/Pages/descriptions.aspx#greenwalt
Sponsor: American Association of Blood Banks
8101 Glenbrook Road
Bethesda, MD 20814-2749

AACAP-NIDA K12 Career Development Awards   49
The grant provides up to five years of salary support and mentored addiction research training for qualified child and adolescent psychiatrists (CAPs) who intend to establish careers as independent investigators in mental health and addiction research. Letters of Intent are required and due on or before January 5. Full applications are due March 30, with new grants awarded on June 1. Applicants are strongly encouraged to contact program staff regarding the program and candidate qualifications before submitting Letters of Intent.
Requirements: Eligible applicants must be a child and adolescent psychiatrist with a commitment and goal of becoming an independent investigator in addiction-related research focused on children and adolescents. Applicants must also be U.S. citizens, non-citizen nationals, and/or individuals with permanent residence status. Women and minority candidates are especially encouraged to apply.
Geographic Focus: All States
Date(s) Application is Due: Jan 5
Amount of Grant: Up to 140,000 USD
Contact: Stacia Fleisher, (202) 966-7300; fax (202) 966-2891; sfleisher@aacap.org
Internet: http://www.aacap.org/cs/root/research_and_training_awards/aacap_nida12
Sponsor: American Academy of Child and Adolescent Psychiatry
3615 Wisconsin Avenue, NW
Washington, D.C. 20016-3007

AACAP Beatrix A. Hamburg Award for the Best New Research Poster by a Child and Adolescent Psychiatry Resident   50
This award recognizes the author of the best new research poster presented at the AACAP Annual Meeting. The award provides a $1,000 honorarium. The recipient will receive a plaque at the Young Leader Awards Ceremony during the AACAP Annual Meeting. There is no application process for this award. The AACAP Program Committee will review all new research poster submissions by child and adolescent psychiatry residents to determine the recipient of this award. The recipient will be notified through an acceptance letter. Non-recipients will not be notified.
Requirements: The author must be a child and adolescent psychiatry resident at an accredited institution and AACAP member. Applicants must: be a child and adolescent psychiatry resident; attend an accredited institution; be an AACAP member; and attend and present a new research poster at the AACAP Annual Meeting.
Geographic Focus: All States
Date(s) Application is Due: Jun 14
Amount of Grant: 1,000 USD
Contact: Stacie Hall; (202) 966-7300, ext. 113; fax (202) 966-2891; shall@aacap.org
Internet: http://www.aacap.org/cs/root/research_and_training_awards/aacap_beatrix_a_hamburg_award_for_the_best_new_research_poster_by_a_child_and_adolescent_psychiatry_resident_at_the_aacap_annual_meeting
Sponsor: American Academy of Child and Adolescent Psychiatry
3615 Wisconsin Avenue, NW
Washington, D.C. 20016-3007

AACAP Child and Adolescent Psychiatry (CAP) Teaching Scholarships   51
The program is offered through the Harvard Macy Institute for Physician Educators Program, and is designed to fashion master teachers who will inspire medical students and residents to become child and adolescent psychiatrists. To facilitate this objective, the field of child and adolescent psychiatry needs educators trained in state-of-the-art teaching principles, methodologies, and technologies of adult education and learning. The AACAP Teaching Scholars Program addresses this challenge by creating a group of highly trained medical educators. This program is an opportunity for CAP faculty who are committed to teaching careers to train at the prestigious Harvard Medical School, gain research and teaching experience, and to give back to future generations of CAP faculty.

8 | GRANT PROGRAMS

Requirements: Applicants must: have a mid-level academic appointment (Senior Assistant Professor, Associate Professor or early Professor); have a commitment to an active career in medical student/resident teaching; have been accepted by the joint AACAP-Harvard Macy Institute admissions group; attend at two meetings in Boston, MA (12 days in January and 6 days in May); have active participation in a collaborative CAP project spanning January through May; be potential participants as program faculty during a subsequent second year; participate in AACAP Annual Meetings and Mid-Year Institutes to bring modern teaching and educational methods to the AACAP membership; participate in an AACAP College of Teaching Scholars; and possess strong support from the applicant's School of Medicine, Department of Psychiatry and Division of Child and Adolescent Psychiatry (letter of institutional commitment and some release time required).
Geographic Focus: All States
Date(s) Application is Due: Jul 15
Amount of Grant: Up to 8,500 USD
Contact: Gabe Robbins, (202) 966-7300, ext. 117; fax (202) 966-2891; grobbins@aacap.org
Internet: http://www.aacap.org/cs/root/research_and_training_awards/child_and_adolescent_psychiatry_teaching_scholars_program
Sponsor: American Academy of Child and Adolescent Psychiatry
3615 Wisconsin Avenue, NW
Washington, D.C. 20016-3007

AACAP Educational Outreach Program for Child and Adolescent Psychiatry Residents — 52

The Educational Outreach Program (EOP) funding provides the opportunity for up to 50 child and adolescent psychiatry residents to receive a formal overview to the field of child and adolescent psychiatry, establish child and adolescent psychiatrists as mentors and experience the AACAP Annual Meeting. Participants will be exposed to the breadth and depth of the field of child and adolescent psychiatry, including research opportunities, access to mentors, and various networking opportunities. Participation in this program provides up to $1,000 for travel expenses, which includes airfare, hotel, and meals (maximum $75/day).
Requirements: Applicants must: be child and adolescent psychiatry residents at the time of the AACAP Annual Meeting; be currently enrolled in a residency program in the United States; be residents in their first or second year of child fellowship training are eligible (Triple Boarders in their fourth or fifth year of training in their triple board programs are eligible); either be members of the AACAP or have a membership application pending at the time of application; and attend all AACAP Annual Meeting events specified by AACAP.
Geographic Focus: All States
Date(s) Application is Due: Jul 12
Amount of Grant: Up to 1,000 USD
Contact: Ashley Partner, (202) 966-7300, ext. 117; apartners@aacap.org
Internet: http://www.aacap.org/cs/residents/eop-capfellows
Sponsor: American Academy of Child and Adolescent Psychiatry
3615 Wisconsin Avenue, NW
Washington, D.C. 20016-3007

AACAP Educational Outreach Program for General Psychiatry Residents — 53

The Educational Outreach Program (EOP) for General Psychiatry residents provides the opportunity for up to 20 general psychiatry residents to receive a formal overview to the field of child and adolescent psychiatry, establish child and adolescent psychiatrists as mentors and experience the AACAP Annual Meeting. Participants will be exposed to the breadth and depth of the field of child and adolescent psychiatry, including research opportunities, access to mentors, and various networking opportunities. Participation in this program provides up to $1,500 for travel expenses to the AACAP Annual Meeting.
Requirements: Applicants must: be general psychiatry residents at the time of the AACAP Annual Meeting; be currently enrolled in a residency program in the United States; be residents in their first, second or third year of general psychiatry training (Triple Boarders in their first, second or third year of training in their triple board programs are eligible); either be members of the AACAP or have a membership application pending at the time of application; and attend all AACAP Annual Meeting events specified by AACAP.
Geographic Focus: All States
Date(s) Application is Due: Jul 12
Amount of Grant: Up to 1,500 USD
Contact: Ashley Partner, (202) 966-7300, ext. 117 or (202) 587-9663; fax (202) 966-2891; apartner@aacap.org or training@aacap.org
Internet: http://www.aacap.org/cs/residents/eop-generalresidents
Sponsor: American Academy of Child and Adolescent Psychiatry
3615 Wisconsin Avenue, NW
Washington, D.C. 20016-3007

AACAP Educational Outreach Program for Residents in Alcohol Research — 54

The Educational Outreach Program (EOP) for Residents in Alcohol Research provides the opportunity for up to 10 child and adolescent psychiatry and general psychiatry residents to pursue advanced knowledge in adolescent alcohol research. The program also allows residents to expand their knowledge of adolescent alcohol-related issues by attending the AACAP Annual Meeting. EOP recipients will meet with an assigned alcohol research mentor and with a representative from NIAAA at the AACAP Annual Meeting to encourage mentorship connections and career development. Participants will be exposed to the field of child and adolescent psychiatry, including research opportunities, access to mentors, and various networking opportunities. Participation in this program provides up to $1,000 for travel expenses to the AACAP Annual Meeting.
Requirements: Applicants must be child and adolescent psychiatry or general psychiatry residents at the time of the AACAP Annual Meeting. Participants must: be currently enrolled in a child and adolescent psychiatry, general psychiatry, or triple board residency program in the United States; either be members of the AACAP or have a membership application pending at the time of application; and attend all Annual Meeting events specified by AACAP.
Geographic Focus: All States
Date(s) Application is Due: Jul 12
Amount of Grant: Up to 1,000 USD
Contact: Ashley Partner, (202) 966-7300, ext. 117 or (202) 587-9663; fax (202) 966-2891; apartner@aacap.org or training@aacap.org
Internet: http://www.aacap.org/cs/award_opportunities/details/aacap_educational_outreach_program_for_residents_in_alcohol_research
Sponsor: American Academy of Child and Adolescent Psychiatry
3615 Wisconsin Avenue, NW
Washington, D.C. 20016-3007

AACAP Elaine Schlosser Lewis Award for Research on Attention-Deficit Disorder — 55

This annual award acknowledges outstanding leadership and continuous contributions in the field of research by giving $5,000 for the best paper published in the Journal on attention-deficit disorder written by a child and adolescent psychiatrist. The award winner will be recognized at a Distinguished Awards Luncheon and make an Honors Presentation about his or her work during the AACAP Annual Meeting.
Requirements: Nominations must be accompanied by a CV for the individual nominated.
Geographic Focus: All States
Date(s) Application is Due: May 7
Amount of Grant: 5,000 USD
Contact: Alyssa Sommer, Research Coordinator; (202) 966-7300, ext. 157; fax (202) 364-5925; asommer@aacap.org
Internet: http://www.aacap.org/cs/root/research_and_training_awards/aacap_journal_awards
Sponsor: American Academy of Child and Adolescent Psychiatry
3615 Wisconsin Avenue, NW
Washington, D.C. 20016-3007

AACAP George Tarjan Award for Contributions in Developmental Disabilities — 56

The AACAP George Tarjan Award for Contributions in Developmental Disabilities recognizes a child and adolescent psychiatrist and Academy member who has made significant contributions in a lifetime career or single seminal work to the understanding or care of those with mental retardation and developmental disabilities. These contributions must have national and/or international stature and clearly demonstrate lasting effects. The contributions may be in areas of teaching, research, program development, direct clinical service, advocacy or administrative commitment. A cash prize of up to $1,000 will be awarded. The award winner will be recognized at a Distinguished Awards Luncheon and make an Honors Presentation about his or her work during the AACAP Annual Meeting.
Requirements: Nominations must be accompanied by a CV for the individual nominated.
Geographic Focus: All States
Date(s) Application is Due: Apr 30
Amount of Grant: 1,000 USD
Contact: Bryan King; (202) 966-7300; fax (202) 966-2891; bking@aacap.org
Internet: http://www.aacap.org/cs/distinguishedmembers
Sponsor: American Academy of Child and Adolescent Psychiatry
3615 Wisconsin Avenue, NW
Washington, D.C. 20016-3007

AACAP Irving Philips Award for Prevention — 57

The AACAP Irving Philips Award for Prevention recognizes a child and adolescent psychiatrist and Academy member who has made significant contributions in a lifetime career or single seminal work to the prevention of mental illness in children and adolescents. These contributions must have national and/or international stature and clearly demonstrate lasting effects. The contributions may be in the areas of teaching, research, program development, direct clinical service, advocacy or administrative commitment. The award pays $2,500 to the winner and a $2,000 donation to a prevention program or center of the awardee's choice. The award winner will be recognized at a Distinguished Awards Luncheon and make an Honors Presentation about his or her work during the AACAP Annual Meeting.
Requirements: Nominations must be accompanied by a CV for the individual nominated.
Geographic Focus: All States
Date(s) Application is Due: Apr 30
Amount of Grant: 4,500 USD
Contact: James Hudziak; (202) 966-7300; fax (202) 966-2891; jhudziak@aacap.org
Internet: http://www.aacap.org/cs/distinguishedmembers
Sponsor: American Academy of Child and Adolescent Psychiatry
3615 Wisconsin Avenue, NW
Washington, D.C. 20016-3007

AACAP Jeanne Spurlock Lecture and Award on Diversity and Culture — 58

The Jeanne Spurlock Lecture and Award on Diversity and Culture reflects the spirit of the nomination for outstanding contributions to diversity and culture and in some way encourages individuals from diverse cultural backgrounds to become child and adolescent psychiatrists. Nominees should be individuals who have made contributions in the areas of social awareness including: civil rights; spirituality and/or religion; social welfare; public information; scientific research; education and mentoring; and the arts (literature, theatre, music, painting, sculpture or photography). The award includes an honorarium of

$2,500, with the award winner being recognized at a Distinguished Awards Luncheon where he or she will make an honors award presentation about his or her work during the AACAP Annual Meeting.
Geographic Focus: All States
Date(s) Application is Due: Apr 30
Amount of Grant: 2,500 USD
Contact: Adriano Boccanelli, Clinical Practice Manager; (202) 966-7300, ext. 133; fax (202) 966-9518; aboccanelli@aacap.org or clinical@aacap.org
Internet: http://www.aacap.org/cs/distinguishedmembers
Sponsor: American Academy of Child and Adolescent Psychiatry
3615 Wisconsin Avenue, NW
Washington, D.C. 20016-3007

AACAP Jeanne Spurlock Minority Medical Student Clinical Fellowship in Child and Adolescent Psychiatry 59

Up to fourteen (14) fellowships are available each year for minority medical students to explore a career in child and adolescent psychiatry, gain valuable work experience, and meet leaders in the child and adolescent psychiatry field. The fellowship opportunity provides up to $3,500 for 12 weeks of clinical training under a child and adolescent psychiatrist mentor. Research assignments may include responsibility for part of the observation or evaluation, developing specific aspects of the research mechanisms, conducting interviews or tests, use of rating scales, and psychological or cognitive testing of subjects. Contact AACAP for applications.
Requirements: African American, Asian American, Native American, Alaskan Native, Mexican American, Hispanic, and Pacific Islander students in accredited U.S. medical schools are eligible.
Geographic Focus: All States
Date(s) Application is Due: Mar 1
Amount of Grant: 3,500 USD
Contact: Gabe Robbins, (202) 966-7300, ext. 117; grobbins@aacap.org
Internet: http://www.aacap.org/cs/students/opportunities/SpurlockClinical
Sponsor: American Academy of Child and Adolescent Psychiatry
3615 Wisconsin Avenue, NW
Washington, D.C. 20016-3007

AACAP Jeanne Spurlock Research Fellowship in Substance Abuse & Addiction for Minority Medical Students 60

The Jeanne Spurlock Research Fellowship in Substance Abuse and Addiction for Minority Medical Students offers a unique opportunity for minority medical students to explore a research career in substance abuse in relation to child and adolescent psychiatry, gain valuable work experience, and meet leaders in the child and adolescent psychiatry field. The fellowship opportunity provides up to $4,000 for 12 weeks of summer research under a child and adolescent psychiatrist researcher/mentor. The research training plan must provide for significant contact between the student and the mentor and for exposure to state-of-the-art drug abuse and addiction research. The plan should include program planning discussions, instruction in research planning and implementation, regular meetings with the mentor, laboratory director, and the research group, and assigned readings. Research assignments may include responsibility for part of the observation or evaluation, developing specific aspects of the research mechanisms, conducting interviews or tests, use of rating scales, and psychological or cognitive testing of subjects.
Requirements: All fellowship participants must attend the AACAP Annual Meeting. Applications are considered from African-American, Native American, Alaskan Native, Mexican American, Hispanic, Asian, and Pacific Islander students in accredited U.S. medical schools.
Geographic Focus: All States
Date(s) Application is Due: Feb 15
Amount of Grant: Up to 4,000 USD
Contact: Ashley Partner, (202) 966-7300, ext. 117; fax (202) 966-2891; apartner@aacap.org or training@aacap.org
Internet: http://www.aacap.org/cs/students/opportunities/SpurlockResearch
Sponsor: American Academy of Child and Adolescent Psychiatry
3615 Wisconsin Avenue, NW
Washington, D.C. 20016-3007

AACAP Junior Investigator Awards 61

The AACAP Junior Investigator Award offers two awards of up to $30,000 a year for two years for child and adolescent psychiatry junior faculty (assistant professor level or equivalent). The program is intended to facilitate innovative research. The research may be basic or clinical in nature but must be relevant to our understanding, treatment and prevention of child and adolescent mental health disorders. The award also includes the cost of attending the AACAP Annual Meeting for five days.
Requirements: Recipients are required to submit a poster or oral presentation on his or her research for the AACAP's Annual Meeting. Applicants must: be board eligible or certified in child and adolescent psychiatry; have a doctoral level degree and be in a faculty or independent research position; and have an on-site mentor who has had experience in the type of research that is being proposed that will normally include work with children and adolescents. Candidates must either be AACAP members or have a membership application pending (not paid by the award).
Restrictions: Applicants who have served as Principal Investigator on an NIH R01 grant are not eligible.
Geographic Focus: All States
Date(s) Application is Due: Mar 1
Amount of Grant: Up to 30,000 USD
Contact: Alyssa Sommer, Research Coordinator; (202) 966-7300, ext. 157; fax (202) 364-5925; asommer@aacap.org or research@accap.org
Internet: http://www.aacap.org/cs/root/research_and_training_awards/aacap_junior_investigator_awards
Sponsor: American Academy of Child and Adolescent Psychiatry
3615 Wisconsin Avenue, NW
Washington, D.C. 20016-3007

AACAP Klingenstein Third Generation Foundation Award for Research in Depression or Suicide 62

This annual award acknowledges outstanding leadership and continuous contributions in the field of research by giving $5,000 for the best paper on suicide and/or depression published in the Journal during the past year. The award winner will be recognized at a Distinguished Awards Luncheon and make an Honors Presentation about his or her work during the AACAP Annual Meeting.
Requirements: Nominations must be accompanied by a CV for the individual nominated.
Geographic Focus: All States
Date(s) Application is Due: May 7
Amount of Grant: 5,000 USD
Contact: Gabe Robbins, (202) 966-7300, ext. 117; fax (202) 966-2891; grobbins@aacap.org
Internet: http://www.aacap.org/cs/root/research_and_training_awards/aacap_journal_awards
Sponsor: American Academy of Child and Adolescent Psychiatry
3615 Wisconsin Avenue, NW
Washington, D.C. 20016-3007

AACAP Life Members Mentorship Grants for Medical Students 63

The Life Members Mentorship Grants for Medical Students (MGM) provides the opportunity for seven medical students to attend the AACAP Annual Meeting and receive an introduction into the field of child and adolescent psychiatry through the AACAP Mentorship Program. MGM recipients also participate in programs sponsored by the Life Members, which is a group of the oldest and most distinguished members of AACAP, all having been members for at least 30 years. Many of those in this group served as AACAP leadership and also pioneered many of the significant discoveries and developments in the field of child and adolescent psychiatry. Partnered with the Mentorship Program, this program provides participants with networking opportunities, exposure to varying specialties, and interaction with Life Members. Participation in this program provides up to $1,000 for travel expenses to the AACAP Annual Meeting, which includes airfare, hotel, and meals (maximum $75/day).
Requirements: Applicants must be enrolled in a medical school in the United States at the time of the AACAP Annual Meeting. Recipients are required to: share the AACAP recruitment video with fellow medical students in their program within the six months following the Annual Meeting; and write a follow-up report on the experience that they would be encouraged to submit to their program's listserve and/or website and share a copy of this with AACAP. Participants must: be in good standing at their medical school; and attend all AACAP Annual Meeting events specified by AACAP.
Geographic Focus: All States
Date(s) Application is Due: Jul 12
Amount of Grant: Up to 1,000 USD
Contact: Ashley Partner, (202) 966-7300, ext. 117 or (202) 587-9663; fax (202) 966-2891; apartner@aacap.org or training@aacap.org
Internet: http://www.aacap.org/cs/students/opportunities/life_members_mentorship_grants_for_medical_students
Sponsor: American Academy of Child and Adolescent Psychiatry
3615 Wisconsin Avenue, NW
Washington, D.C. 20016-3007

AACAP Mary Crosby Congressional Fellowships 64

The Mary Crosby Congressional Fellowships program is designed to educate policy makers and Congressional staff about child and adolescent psychiatry, and to foster awareness of children's mental health issues. This experience will allow the Fellow to develop a keen understanding of how public policy affects patient care, education, and health insurance issues. The AACAP Mary Crosby Congressional Fellow will be placed with a Congressional office or Committee where they will gain invaluable experience as they assist in the development of legislative and public policy initiatives. The award includes an $85,000 stipend, up to $3,000 in relocation or moving expenses, and health insurance for the length of the fellowship program.
Requirements: Successful applicants must: be an AACAP member; be a PGY-4 psychiatry resident or beyond; have completed training in or be scheduled to complete training in one of the child and adolescent psychiatry programs (CAP training program, Combined General Psychiatry/CAP Program, or Triple Board Program); have a demonstrated interest in public policy issues; exhibit an interest in applying medical and scientific knowledge to the resolution of public policy issues; and be a U.S. citizen or permanent resident.
Geographic Focus: All States
Amount of Grant: 85,000 - 88,000 USD
Contact: Ashley Partner, (202) 966-7300, ext. 117; fax (202) 966-2891; apartner@aacap.org or training@aacap.org
Internet: http://www.aacap.org/cs/root/legislative_action/2008_congressional_fellowship_program
Sponsor: American Academy of Child and Adolescent Psychiatry
3615 Wisconsin Avenue, NW
Washington, D.C. 20016-3007

10 | GRANT PROGRAMS

AACAP Norbert and Charlotte Rieger Award for Scientific Achievement 65
This award is given annually for the most significant article published in the Journal of the American Academy of Child and Adolescent Psychiatry (JAACAP) in the past year (July-May). The article must be written by a psychiatrist specializing in child and/or adolescent psychiatry and cannot be a review.
Requirements: The only submissions accepted are papers from child and adolescent psychiatrists that have been published in the Journal during the past year.
Geographic Focus: All States
Date(s) Application is Due: May 7
Amount of Grant: Up to 4,500 USD
Samples: Gail Bernstein, M.D., University of Minnesota, Minneapolis, MN--School-Based Interventions for Anxious Children.
Contact: Gabe Robbins, (202) 966-7300, ext. 117; fax (202) 966-2891; grobbins@aacap.org
Internet: http://www.aacap.org/cs/root/research_and_training_awards/aacap_journal_awards
Sponsor: American Academy of Child and Adolescent Psychiatry
3615 Wisconsin Avenue, NW
Washington, D.C. 20016-3007

AACAP Pilot Research Award for Attention-Deficit Disorder 66
The AACAP Pilot Research Award for Attention Disorders, supported by The Elaine Schlosser Lewis Fund, offers $15,000 for child and adolescent psychiatry residents and junior faculty who have an interest in beginning a career in child and adolescent mental health research. By providing one award to a child and adolescent psychiatry junior faculty member or resident for pilot research on attention disorders, the AACAP supports a young investigator at a critical stage, encouraging a future career in child and adolescent psychiatry research. The recipient has the opportunity to submit a poster presentation on his or her research for AACAP's Annual Meeting, and present at the Elaine Schlosser Lewis Luncheon.
Requirements: Candidates must be board eligible, certified in child and adolescent psychiatry, or enrolled in a child psychiatry residency or fellowship program. Candidates must also have a faculty appointment in an accredited medical school or be in a fully accredited child and adolescent psychiatry clinical research or training program.
Restrictions: At the time of application, candidates may not have more than two years experience following graduation from residency/fellowship training. Candidates must not have any previous significant, individual research funding in the field of child and adolescent mental health. These include the following: NIMH/NIH Funding (Small Grants, T Award or R-01) or similar foundation or industry research funding.
Geographic Focus: All States
Date(s) Application is Due: Apr 29
Amount of Grant: 15,000 USD
Contact: Ashley Partner, Research Coordinator; (202) 966-7300, ext. 117; fax (202) 364-5925; apartner@aacap.org
Internet: http://www.aacap.org/cs/root/research_and_training_awards/pilot_research_award_for_attention_disorder_for_a_junior_faculty_or_child_psychiatry_resident_supported_by_the_elaine_schlosser_lewis_fund
Sponsor: American Academy of Child and Adolescent Psychiatry
3615 Wisconsin Avenue, NW
Washington, D.C. 20016-3007

AACAP Pilot Research Award for Learning Disabilities, Supported by the Elaine Schlosser Lewis Fund 67
The Award for Learning Disabilities and Psychiatric Disorders is available for a junior faculty or child psychiatry resident, and is supported by the Elaine Schlosser Lewis Fund. The award encourages work with a child and adolescent psychiatric investigator with expertise in his or her particular area of interest. Work must be completed within one year of receipt of the award. Please note that travel to both the Annual Meeting and the Research Update Luncheon is not included in the award and is the responsibility of the recipient.
Requirements: Candidates must be board eligible, certified in child and adolescent psychiatry, or enrolled in a child psychiatry residency or fellowship program. Candidates must also have a faculty appointment in an accredited medical school or be in a fully accredited child and adolescent psychiatry clinical research or training program.
Restrictions: Candidates must not have any previous significant, individual research funding in the field of child and adolescent mental health. These include the following: NIMH/NIH Funding (Small Grants, T Award or R-01) or similar foundation or industry research funding. The recipient must be an AACAP member at the time of application and agree to present his or her research.
Geographic Focus: All States
Date(s) Application is Due: May 7
Amount of Grant: 15,000 USD
Contact: Alyssa Sommer; (202) 966-7300, ext. 157; asommer@aacap.org
Internet: http://www.aacap.org/cs/root/research_and_training_awards/pilot_research_award_for_learning_disabilities_for_a_junior_faculty_or_child_psychiatry_resident_supported_by_the_elaine_schlosser_lewis_fund
Sponsor: American Academy of Child and Adolescent Psychiatry
3615 Wisconsin Avenue, NW
Washington, D.C. 20016-3007

AACAP Pilot Research Awards, Supported by Eli Lilly and Company 68
Eight awards are available each year to junior faculty and child psychiatry fellows for pilot research with child and adolescent psychiatric researchers. Award recipients will be matched with child and adolescent psychiatric investigators working in their particular area of interest to act as consultants or mentors during the course of the award. Work must be completed within one year of receipt of the award.
Requirements: Candidates must be board eligible, certified in child and adolescent psychiatry, or enrolled in a child psychiatry residency or fellowship program. Candidates must have a faculty appointment in an accredited medical school or be in a fully accredited child and adolescent psychiatry clinical research or training program.
Restrictions: At the time of application, candidates may not have more than two years experience following graduation from residency/fellowship training. Candidates must not have any previous significant, individual research funding in the field of child and adolescent mental health. These include the following: NIMH/NIH Funding (Small Grants, T Award or R-01) or similar foundation or industry research funding.
Geographic Focus: All States
Date(s) Application is Due: May 7
Amount of Grant: 15,000 USD
Contact: Alyssa Sommer, Research Coordinator; (202) 966-7300, ext. 157; fax (202) 364-5925; asommer@aacap.org
Internet: http://www.aacap.org/cs/root/research_and_training_awards/aacap_pilot_research_award_for_junior_faculty_and_child_psychiatry_fellows_supported_by_lilly_usa_llc
Sponsor: American Academy of Child and Adolescent Psychiatry
3615 Wisconsin Avenue, NW
Washington, D.C. 20016-3007

AACAP Quest for the Test Bipolar Disorder Pilot Research Award 69
The AACAP Quest for the Test Bipolar Disorder Pilot Research Award, supported by The Ryan Licht Sang Bipolar Foundation, offers $15,000 for child and adolescent psychiatry residents and junior faculty who have an interest in beginning a career in child and adolescent mental health research. Proposed research projects should address questions concerning the pathology or diagnosis of children and adolescents with bipolar disorder. The award also includes the cost of attending the AACAP Annual Meeting for five days.
Requirements: The recipient is required to submit a poster presentation on his or her research for the AACAP's Annual Meeting. Candidates must be board eligible, certified in child and adolescent psychiatry, or enrolled in a child psychiatry residency or fellowship program. Candidates must also have a faculty appointment in an accredited medical school or be in a fully accredited child and adolescent psychiatry clinical research or training program.
Restrictions: At the time of application, candidates may not have more than two years experience following graduation from residency/fellowship training. Candidates must not have any previous significant, individual research funding in the field of child and adolescent mental health. These include the following: NIH Funding (K or R level grants) or similar foundation or industry research funding.
Geographic Focus: All States
Date(s) Application is Due: May 7
Amount of Grant: 15,000 USD
Contact: Alyssa Sommer, Research Coordinator; (202) 966-7300, ext. 157; fax (202) 364-5925; asommer@aacap.org
Internet: http://www.aacap.org/cs/root/research_and_training_awards/aacap_quest_for_the_test_bipolar_disorder_pilot_research_award
Sponsor: American Academy of Child and Adolescent Psychiatry
3615 Wisconsin Avenue, NW
Washington, D.C. 20016-3007

AACAP Rieger Psychodynamic Psychotherapy Award 70
The AACAP Rieger Psychodynamic Psychotherapy Award recognizes the best published or unpublished paper, written by an AACAP member that addresses the use of psychodynamic psychotherapy in clinical practice and fosters development, teaching, and practice of psychodynamic psychotherapy within child and adolescent psychiatry. Papers that express a novel hypothesis, raise questions about existing theory, or integrate new neuroscience and developmental psychotherapy research with psychodynamic principles may be nominated. Unpublished, new papers and papers published within the last three years may be submitted by their authors. Published papers may be nominated by any member of the AACAP. Authors may be senior or junior faculty members or residents. Delivery of the winning paper will be made at the AACAP Annual Meeting Honors Presentation.
Geographic Focus: All States
Date(s) Application is Due: Apr 30
Amount of Grant: 4,500 USD
Contact: Tim Dugan, Co-Chair; (202) 966-7300; fax (202) 966-2891; tdugan@aacap.org or clinical@aacap.org
Internet: http://www.aacap.org/cs/awards/riegerpsychotherapy
Sponsor: American Academy of Child and Adolescent Psychiatry
3615 Wisconsin Avenue, NW
Washington, D.C. 20016-3007

AACAP Rieger Service Program Award for Excellence 71
The AACAP Rieger Service Program Award for Excellence recognizes innovative programs that address prevention, diagnosis, or treatment of mental illnesses in children and adolescents, and serve as model programs to the community. This award of $4,500 is shared among the awardee and his or her service program. The award winner will be recognized at a Distinguished Awards Luncheon and make an Honors Presentation about his or her work during the AACAP Annual Meeting.
Requirements: Nomination letters must be accompanied by a CV and any support materials for the individual or organization nominated.
Geographic Focus: All States
Date(s) Application is Due: Apr 30
Amount of Grant: 4,500 USD
Contact: Kaye McGinty, Co-Chair; (202) 966-7300; fax (202) 966-2891; kmcginty@aacap.org or clinical@aacap.org

Internet: http://www.aacap.org/cs/distinguishedmembers
Sponsor: American Academy of Child and Adolescent Psychiatry
3615 Wisconsin Avenue, NW
Washington, D.C. 20016-3007

AACAP Robert Cancro Academic Leadership Award 72
The AACAP Cancro Academic Leadership Award recognizes a currently serving General Psychiatry Training Director, Medical School Dean, CEO of a Training Institution, Chair of a Department of Pediatrics or Chair of a Department of Psychiatry for his or her contributions to the promotion of child and adolescent psychiatry. Named in honor of Robert Cancro, M.D., Chairman at New York University, this award offers a $2,000 honorarium to the awardee. The award is presented in even-numbered years. Nominations for the award may be made by Child and Adolescent Training Directors or Division Directors and must include a CV for the individual nominated. The recipient of this award will receive a plaque and be recognized at the AACAP Annual Meeting in San Francisco, CA. The award recipient will be honored at the Distinguished Awards Luncheon, Residency Program Directors Luncheon, and will provide an Honors Presentation on his or her work during the Annual Meeting.
Requirements: Nominations must include a CV for the individual nominated
Geographic Focus: All States
Date(s) Application is Due: Apr 30
Amount of Grant: 2,000 USD
Contact: Jeffrey Hunt, Co-Chair; (202) 966-7300; fax (202) 966-2891; jhunt@aacap.org or training@aacap.org
Internet: http://www.aacap.org/cs/distinguishedmembers
Sponsor: American Academy of Child and Adolescent Psychiatry
3615 Wisconsin Avenue, NW
Washington, D.C. 20016-3007

AACAP Sidney Berman Award for the School-Based Study and Intervention for Learning Disorders and Mental Ilness 73
The AACAP Sidney Berman Award for the School-Based Study and Treatment of Learning Disorders and Mental IIlness recognizes an individual or program that has shown outstanding achievement in the school-based study or delivery of intervention for learning disorders and mental illness. A cash prize of $4,500 will be awarded. The award winner will be recognized at a Distinguished Awards Luncheon and make an Honors Presentation about his or her work during the AACAP Annual Meeting.
Requirements: Nomination letters must be accompanied by a CV for the individual nominated or program information.
Geographic Focus: All States
Date(s) Application is Due: Apr 30
Amount of Grant: 4,500 USD
Contact: Sheryl Kataoka, Co-Chair; (202) 966-7300; fax (202) 966-2891; skataoka@aacap.org or clinical@aacap.org
Internet: http://www.aacap.org/cs/distinguishedmembers
Sponsor: American Academy of Child and Adolescent Psychiatry
3615 Wisconsin Avenue, NW
Washington, D.C. 20016-3007

AACAP Simon Wile Leadership in Consultation Award 74
The Simon Wile Leadership in Consultation Award recognizes innovative programs that address prevention, diagnosis, or treatment of mental illnesses in children and adolescents, and serve as model programs to the community. This award of $4,500 is shared among the awardee and his or her service program. Nomination letters must be accompanied by a CV and any support materials for the individual or organization nominated. The recipient will be recognized at a Distinguished Awards Luncheon and make an Honors Presentation about his or her work during the AACAP Annual Meeting.
Requirements: Practicing child and adolescent psychiatrists are eligible for nomination.
Geographic Focus: All States
Date(s) Application is Due: Apr 30
Amount of Grant: 500 USD
Contact: Kaye McGinty, Co-Chair; (202) 966-7300; fax (202) 966-2891; kmcginty@aacap.org or clinical@aacap.org
Internet: http://www.aacap.org/cs/distinguishedmembers
Sponsor: American Academy of Child and Adolescent Psychiatry
3615 Wisconsin Avenue, NW
Washington, D.C. 20016-3007

AACAP Summer Medical Student Fellowships 75
AACAP Summer Medical Student Fellowships offer a chance for medical students to explore a career in child and adolescent psychiatry, gain valuable work experience, and meet leaders in the child and adolescent psychiatry field. The fellowship opportunity provides up to $3,500 for 12 weeks of clinical or research training under a child and adolescent psychiatrist mentor. Fellowship stipends will be distributed in two installments. Upon receipt of the fellowship, $2,500 will be sent to the student's mentor to be disbursed at the onset of the summer fellowship. Upon meeting all program requirements, the last installment will be sent directly to the recipient and will be pro-rated according to the total amount of time spent completing the fellowship. Fellowships lasting the full 12 weeks will receive the maximum amount. Participants are required to attend the AACAP Annual Meeting. The application deadline for this fellowship is February 15.
Requirements: Applicants must be students enrolled in accredited U.S. medical schools.
Geographic Focus: All States
Date(s) Application is Due: Feb 15

Amount of Grant: 3,500 USD
Contact: Ashley Partner, (202) 966-7300, ext. 117; fax (202) 966-2891; apartner@aacap.org or training@aacap.org
Internet: http://www.aacap.org/cs/students/summerfellowship
Sponsor: American Academy of Child and Adolescent Psychiatry
3615 Wisconsin Avenue, NW
Washington, D.C. 20016-3007

AACAP Systems of Care Special Program Scholarships 76
The American Academy of Child and Adolescent Psychiatry (AACAP) is pleased to announce the opportunity for child and adolescent psychiatry residents to apply for the 2012 Systems of Care Special Program Scholarship. The Special Program is a day-long event that explores the psychological, psychosocial and psycho-educational needs of children and adolescents as manifest in their experiences in the schools. Policy and best practice considerations will be reviewed as regards health promotion, screening, assessment and treatment intervention for students and student groups, with the goal of advancing the knowledge, skills and attitudes of child psychiatrists and other mental health providers so that they can better function as service providers and systems consultants for youth and their school systems. The scholarship includes $750 for travel expenses to the AACAP's 59th Annual Meeting. This includes airfare, hotel, and meals, maximum $75 per day.
Geographic Focus: All States
Date(s) Application is Due: Jul 12
Amount of Grant: 750 USD
Contact: Adriano Boccanelli, Clinical Practice Assistant; (202) 966-7300, ext. 133 or (202) 587-9671; fax (202) 966-2891; aboccanelli@aacap.org or clinical@aacap.org
Internet: http://www.aacap.org/cs/root/medical_students_and_residents/residents/2012_systems_of_care_special_program_scholarship
Sponsor: American Academy of Child and Adolescent Psychiatry
3615 Wisconsin Avenue, NW
Washington, D.C. 20016-3007

AACN-Edwards Lifesciences Nurse-Driven Clinical Practice Outcomes Grants 77
Co-sponsored by Edwards Lifesciences and the American Association of Critical-Care Nurses, each grant supports a nurse experienced in research who is conducting a clearly articulated research study that relates to use of protocol-based care driven by nurses. A commitment is required to produce results that will be available and disseminated in a short timeframe. Proposals submitted by an interdisciplinary team or an experienced researcher mentoring a novice will also be considered. Funds may be used for original research or replication of existing research. Especially desirable are proposals that focus on any of the following at any point in the continuum of acute or critical care: implementation of nurse-driven protocols using technology, including minimally invasive devices, with evaluation of patient outcomes and nurses' role in implementing the protocols; or development and clinical evaluation of nurse-driven protocols. Two awards are available each year. Each proposal can be funded up to of $5,000.
Requirements: Principal Investigators must be nurses holding current AACN membership and must remain AACN members throughout the life of the grant funding.
Restrictions: Funds may not be used to pay salaries or institutional overhead or augment funding from formal grants. Principal investigators who have received funding from AACN are ineligible to receive additional funding from AACN during the lifetime of the original award.
Geographic Focus: All States
Date(s) Application is Due: Oct 1
Amount of Grant: Up to 5,000 USD
Contact: Pamela Shellner, Clinical Practice Specialist; (949) 362-2050, ext. 321 or (800) 394-5995, ext. 321; fax (949) 362-2020; research@aacn.org or info@aacn.org
Internet: http://www.aacn.org/WD/Practice/Content/grant-Edwards.pcms?pid=1@menu=practice
Sponsor: American Association of Critical-Care Nurses
101 Columbia
Aliso Viejo, CA 92656-4109

AACN-Philips Medical Systems Clinical Outcomes Grants 78
Co-sponsored by Philips Medical Systems and the American Association of Critical-Care Nurses, each grant supports experienced nurses in conducting clearly articulated research studies. The research must be relevant to clinical nursing practice in acute or critical care. Proposals submitted by an interdisciplinary team or an experienced researcher mentoring a novice will also be considered. Funds may be used for original research or replication of existing research. Qualified proposals should seek to achieve improved outcomes and/or system efficiencies in care of acute or critically ill individuals of any age. Especially desirable are proposals for hospital-based inquiry that focuses on any of the following: intervention strategies addressing technology integration into patient care; use of computerized medical record systems in assessing patient outcomes and managing care; improving specific patient care outcomes (e.g., clinical, safety, financial); use of simulation in clinical education of nurses; and one or more of the 2006 National Patient Safety Goals issued by the Joint Commission on Accreditation of Healthcare Organizations.
Requirements: Principal Investigators must be nurses holding current AACN membership and must remain AACN members throughout the life of the grant funding.
Restrictions: Funds may not be used to pay salaries or institutional overhead or augment funding from formal grants. Principal investigators who have received funding from AACN are ineligible to receive additional funding from AACN during the lifetime of the original award.
Geographic Focus: All States
Date(s) Application is Due: Oct 1

Grant Programs

Amount of Grant: Up to 10,000 USD
Contact: Pamela Shellner, Clinical Practice Specialist; (949) 362-2050, ext. 321 or (800) 394-5995, ext. 321; fax (949) 362-2020; research@aacn.org or info@aacn.org
Internet: http://www.aacn.org/WD/Practice/Content/grant-Philips.pcms?pid=1@menu=practice
Sponsor: American Association of Critical-Care Nurses
101 Columbia
Aliso Viejo, CA 92656-4109

AACN-Sigma Theta Tau Critical Care Grant 79

The grant, cosponsored by AACN and Sigma Theta Tau International, supports critical care nursing research and is awarded annually for a study relevant to critical care nursing practice. The proposed study may be used to meet requirements of an academic degree.
Requirements: Principal investigators must be a member of AACN and/or of Sigma Theta Tau International. The PI must have a minimum Masters Degree.
Restrictions: Funds may not be used to pay salaries or institutional overhead or augment funding from formal grants. Principal investigators who have received funding from AACN are ineligible to receive additional funding from AACN during the lifetime of the original award.
Geographic Focus: All States
Date(s) Application is Due: Oct 1
Amount of Grant: Up to 10,000 USD
Contact: Pamela Shellner, Clinical Practice Specialist; (949) 362-2050, ext. 321 or (800) 394-5995, ext. 321; fax (949) 362-2020; research@aacn.org or info@aacn.org
Internet: http://www.aacn.org/WD/Practice/Content/grant-STT.pcms?pid=1@menu=practice
Sponsor: American Association of Critical-Care Nurses
101 Columbia
Aliso Viejo, CA 92656-4109

AACN Clinical Inquiry Fund Grants 80

The fund provides small awards to qualified individuals carrying out clinical research projects that directly benefit patients and/or families. Interdisciplinary projects are encouraged. Funds may be awarded for new projects, projects in progress, and projects required for an academic degree as long as all other project criteria are met. The funds may be used to cover direct project expenses--i.e., printed materials, small equipment, and supplies including computer software.
Requirements: Principal Investigators must be nurses holding current AACN membership and must remain AACN members throughout the life of the grant funding.
Restrictions: Funds may not be used to pay salaries or institutional overhead or augment funding from formal grants.
Geographic Focus: All States
Date(s) Application is Due: Oct 1
Amount of Grant: 500 USD
Contact: Pamela Shellner, Clinical Practice Specialist; (949) 362-2050, ext. 321 or (800) 394-5995, ext. 321; fax (949) 362-2020; research@aacn.org or info@aacn.org
Internet: http://www.aacn.org/WD/Practice/Docs/Grant_Table09-10.pdf
Sponsor: American Association of Critical-Care Nurses
101 Columbia
Aliso Viejo, CA 92656-4109

AACN Clinical Practice Grants 81

This $6,000 grant supports research focused on one or more AACN Research priorities. Research conducted in fulfillment of an academic degree is acceptable. Funds may be awarded for new projects, projects in progress and projects required for an academic degree as long as all other project criteria are met. Eligible projects may include research utilization studies, CQI projects or outcomes evaluation projects. Interdisciplinary and collaborative projects are encouraged and may involve interdisciplinary teams, multiple nursing units, home health, subacute and transitional care, other institutions or community agencies. Funds may be used to cover direct project expenses, such as printed materials, small equipment or supplies including computer software. Six awards are available each year. One award is funded up to $6,000 each year.
Requirements: Principal Investigators must be nurses holding current AACN membership and must remain AACN members throughout the life of the grant funding.
Restrictions: Funds may not be used to pay salaries or institutional overhead or augment funding from formal grants. Principal investigators who have received funding from AACN are ineligible to receive additional funding from AACN during the lifetime of the original award.
Geographic Focus: All States
Date(s) Application is Due: Oct 1
Amount of Grant: Up to 6,000 USD
Contact: Pamela Shellner, Clinical Practice Specialist; (949) 362-2050, ext. 321 or (800) 394-5995, ext. 321; fax (949) 362-2020; research@aacn.org or info@aacn.org
Internet: http://www.aacn.org/WD/Practice/Content/grant-CP.pcms?pid=1@menu=practice
Sponsor: American Association of Critical-Care Nurses
101 Columbia
Aliso Viejo, CA 92656-4109

AACN Critical Care Grants 82

AACN's research agenda promotes the creation of cultures of inquiry, broad sharing and evidence-based practice. This $15,000 grant supports research focused on one or more of AACN Research priorities. Research priority areas include: effective and appropriate use of technology to achieve optimal patient assessment, management, and/or outcomes; creating a healing, humane environment; processes and systems that foster the optimal contribution of critical care nurses; effective approaches to symptom management; and prevention and management of complication.
Requirements: Principal Investigators must be nurses holding current AACN membership and must remain AACN members throughout the life of the grant funding.
Restrictions: The proposed research may not be used to meet the requirements of an academic degree. Funds may not be used to pay salaries or institutional overhead or augment funding from formal grants. Principal investigators who have received funding from AACN are ineligible to receive additional funding from AACN during the lifetime of the original award.
Geographic Focus: All States
Date(s) Application is Due: Oct 1
Amount of Grant: 15,000 USD
Contact: Pamela Shellner, Clinical Practice Specialist; (949) 362-2050, ext. 321 or (800) 394-5995, ext. 321; fax (949) 362-2020; research@aacn.org or info@aacn.org
Internet: http://www.aacn.org/WD/Practice/Content/grant-CC.pcms?pid=1@menu=practice
Sponsor: American Association of Critical-Care Nurses
101 Columbia
Aliso Viejo, CA 92656-4109

AACN End of Life/Palliative Care Small Projects Grants 83

The end of life/palliative care grant provides an award to a qualified individual(s) carrying out a project focusing on end of life and/or palliative care outcomes in acute and/or critical care settings. Eligible projects may focus on any age group (neonatal to elderly) in the acute or critical care arena. They may include patient education programs, staff development programs, competency-based educational programs, CQI/outcomes evaluation projects, or small clinical research studies. Funds may be awarded for new projects, projects in progress and projects required for an academic degree as long as all other project criteria are met. Collaborative projects are encouraged and may involve interdisciplinary teams, multiple nursing units, home health, subacute and transitional care, other institutions or community agencies. Funds may be used to cover direct project expenses, such as printed materials, small equipment, supplies including computer software. Two awards are available each year. Each proposal can be funded for up to $500.
Requirements: Principal Investigators must be nurses holding current AACN membership and must remain AACN members throughout the life of the grant funding.
Restrictions: Funds may not be used to pay for salaries, speaker honorarium or costs, institutional overhead nor augment funding from formal large grants.
Geographic Focus: All States
Date(s) Application is Due: Oct 1
Amount of Grant: Up to 500 USD
Contact: Pamela Shellner, Clinical Practice Specialist; (949) 362-2050, ext. 321 or (800) 394-5995, ext. 321; fax (949) 362-2020; research@aacn.org or info@aacn.org
Internet: http://www.aacn.org/WD/Practice/Content/grant-EOL.pcms?pid=1@menu=Practice
Sponsor: American Association of Critical-Care Nurses
101 Columbia
Aliso Viejo, CA 92656-4109

AACN Evidence-Based Clinical Practice Grants 84

The Evidence-Based Clinical Practice Grant provides awards to grants that stimulate the use of patient-focused data and/or previously generated research findings to develop, implement and evaluate changes critical and acute care nursing practice. Funds may be awarded for new projects, projects in progress and projects required for an academic degree as long as all other project criteria are met. Eligible projects may include research utilization studies, CQI projects or outcomes evaluation projects. Interdisciplinary and collaborative projects are encouraged and may involve interdisciplinary teams, multiple nursing units, home health, subacute and transitional care, other institutions or community agencies. Funds may be used to cover direct project expenses, such as printed materials, small equipment or supplies including computer software. Six awards are available each year. Each proposal can be funded for up to $1,000.
Requirements: Principal Investigators must be nurses holding current AACN membership and must remain AACN members throughout the life of the grant funding.
Restrictions: Funds may not be used to pay salaries or institutional overhead or augment funding from formal grants. Principal investigators who have received funding from AACN are ineligible to receive additional funding from AACN during the lifetime of the original award.
Geographic Focus: All States
Date(s) Application is Due: Oct 1
Amount of Grant: Up to 1,000 USD
Contact: Pamela Shellner, Clinical Practice Specialist; (949) 362-2050, ext. 321 or (800) 394-5995, ext. 321; fax (949) 362-2020; research@aacn.org or info@aacn.org
Internet: http://www.aacn.org/WD/Practice/Content/grant-EBP.pcms?pid=1@menu=Practice
Sponsor: American Association of Critical-Care Nurses
101 Columbia
Aliso Viejo, CA 92656-4109

AACN Mentorship Grant 85

This award, cosponsored by Mallinckrodt Inc. and AACN, facilitates critical care nursing practice research between a novice and an experienced researcher. The novice researcher is a beginning researcher with limited or no research experience in the area of the proposed investigation. The novice researcher is also a registered nurse with current AACN membership. The grant may be used to fund research for an academic degree. The mentor must show strong evidence of research expertise in the proposed area of research to be pursued by the novice investigator.

Requirements: Principal Investigators must be nurses holding current AACN membership and must remain AACN members throughout the life of the grant funding.
Restrictions: The mentor may not be designated mentor in two consecutive years and may not be conducting research as part of an academic degree.
Geographic Focus: All States
Date(s) Application is Due: Oct 1
Amount of Grant: Up to 10,000 USD
Contact: Pamela Shellner, Clinical Practice Specialist; (949) 362-2050, ext. 321 or (800) 394-5995, ext. 321; fax (949) 362-2020; research@aacn.org or info@aacn.org
Internet: http://www.aacn.org/WD/Practice/Content/grant-Mentor.pcms?pid=1@menu=practice
Sponsor: American Association of Critical-Care Nurses
101 Columbia
Aliso Viejo, CA 92656-4109

AACN Physio-Control Small Projects Grants 86
The program provides awards to qualified individuals carrying out projects focusing on aspects of acute myocardial infarction, resuscitation, or sudden cardiac death. Such projects may include the use of defibrillation, synchronized cardioversion, noninvasive pacing, or interpretive 12-lead electrocardiogram. Eligible projects may include patient education programs, staff development programs, competency-based educational programs, continuous quality improvement projects, outcomes evaluation projects, or small clinical research studies. Funds may be awarded for new projects, projects in progress, and projects required for an academic degree as long as all other project criteria are met. Collaborative projects are encouraged and may involve interdisciplinary teams, multiple nursing units, home health, subacute and transitional care, other institutions, or community agencies. Funds may be used to cover direct project expenses, such as printed materials, small equipment, or supplies including computer software.
Requirements: Principal Investigators must be nurses holding current AACN membership and must remain AACN members throughout the life of the grant funding.
Restrictions: Funds may not be used to pay for salaries, speaker honorarium or costs, institutional overhead nor augment funding from formal large grants.
Geographic Focus: All States
Date(s) Application is Due: Oct 1
Amount of Grant: Up to 1,500 USD
Contact: Pamela Shellner, Clinical Practice Specialist; (949) 362-2050, ext. 321 or (800) 394-5995, ext. 321; fax (949) 362-2020; research@aacn.org or info@aacn.org
Internet: http://www.aacn.org/WD/Practice/Docs/Grant_Table09-10.pdf
Sponsor: American Association of Critical-Care Nurses
101 Columbia
Aliso Viejo, CA 92656-4109

AACR-American Cancer Society Award for Research Excellence in Cancer Epidemiology and Prevention 87
The AACR and the American Cancer Society established this Award in 1992 to honor outstanding research accomplishments in the fields of cancer epidemiology, biomarkers, and prevention. Nominations may be made via letter from any scientist, whether an AACR member or nonmember, who is now or has been affiliated with any institution involved in cancer research, cancer medicine, or cancer-related biomedical science. The winner will receive an honorarium of $5,000 and give a 50-minute lecture during the AACR Annual Meeting.
Requirements: Candidacy is open to all cancer researchers who are affiliated with any institution involved in cancer research, cancer medicine, or cancer-related biomedical science anywhere in the world. Such institutions include those in academia, industry, or government.
Restrictions: Institutions or organizations are not eligible for the Award. Candidates may not nominate themselves.
Geographic Focus: All States
Date(s) Application is Due: Oct 15
Amount of Grant: 5,000 USD
Contact: Monique P. Eversley, Program Associate; (267) 646-0576 or (215) 440-9300, ext. 1400; fax (267) 646-0576; monique.eversley@aacr.org
Internet: http://www.aacr.org/home/scientists/scientific-achievement-awards/acs-award.aspx
Sponsor: American Association for Cancer Research
615 Chestnut Street, 17th Floor
Philadelphia, PA 19106-4404

AACR-Colorectal Cancer Coalition Fellows Grant 88
The Fellowship program supports innovative research by a meritorious young investigator by presenting the Fellow with research funds to pursue an independent line of investigation within the context of his/her current placement. This is a one-year grant of $30,000. Research projects are restricted to translational or clinical cancer research that has an ultimate goal of developing or improving therapeutic interventions for patients with metastatic colorectal cancer. One grant will be awarded this cycle.
Requirements: Candidates must be, at the start of the grant term, in their third, fourth, or fifth year of their fellowship status.
Geographic Focus: All States
Date(s) Application is Due: Dec 18
Amount of Grant: 30,000 USD
Contact: Hanna Hopfinger; (267) 646-0665; hanna.hopfinger@aacr.org
Internet: http://www.aacr.org/home/scientists/research-funding--fellowships/fellows-grants.aspx
Sponsor: American Association for Cancer Research
615 Chestnut Street, 17th Floor
Philadelphia, PA 19106-4404

AACR-FNAB Fellows Grant for Translational Pancreatic Cancer Research 89
The Fellows Grants support innovative research by a meritorious young investigator by presenting the Fellow with research funds to pursue an independent line of investigation within the context of his/her current placement. This is a one-year grant of $35,000. Research projects are restricted to translational pancreatic cancer research that includes the use of human tissue and has implications for therapeutic application and individualized medicine.
Requirements: Candidates must be, at the start of the grant term, in their third, fourth, or fifth year of their fellowship status.
Geographic Focus: All States
Date(s) Application is Due: Dec 18
Amount of Grant: 35,000 USD
Contact: Julia Laurence, Program Assistant; (215) 440-9300, ext. 102 or (267) 646-0655; fax (215) 440-9372; julia.laurence@aacr.org or laurence@aacr.org
Internet: http://www.aacr.org/home/scientists/research-funding--fellowships/fellows-grants.aspx
Sponsor: American Association for Cancer Research
615 Chestnut Street, 17th Floor
Philadelphia, PA 19106-4404

AACR-FNAB Foundation Career Development Award for Translational Cancer Research 90
Career development awards are two-year awards that support research by scientists who are engaged in meritorious cancer research at academic institutions. These awards provide important transitional support for direct research expenses as researchers move from the ranks of early career scientists to faculty status. The Awards are open to junior faculty who completed postdoctoral studies or clinical fellowships between specified dates and who are at an academic or medical institution. Research projects in this area should have direct applicability to translational cancer and be focused on any individualized therapeutic area. Projects should have implications for individualized cancer treatment and make use of human biopsies or samples, such as needle biopsies or circulating cancer cells. One grant will be presented annually. The submission deadline is December 3, with the decision date March 31.
Requirements: Candidates must have acquired a doctoral degree in a related field and may not currently be a candidate for a further doctoral or professional degree. If a candidate has obtained an equivalent degree at a foreign institution, information on the nature of the degree must be provided at the time of application.
Restrictions: Awards are restricted to investigators or institutions within the U.S. Employees of a national government and employees of private industry are ineligible.
Geographic Focus: All States
Date(s) Application is Due: Dec 3
Amount of Grant: 50,000 USD
Samples: Tara A. Young, M.D., Ph.D., University of California, Los Angeles, Los Angeles, CA, $50,000 - Project: High Resolution Cytogenetic Study of Archival Metastatic Choroidal Melanoma (2008-2010).
Contact: Julia Laurence, Program Assistant; (215) 440-9300, ext. 102 or (267) 646-0655; fax (215) 440-9372; julia.laurence@aacr.org or awards@aacr.org
Internet: http://www.aacr.org/home/scientists/research-funding--fellowships/research-funding--fellowships/career-development-award-recipients.aspx
Sponsor: American Association for Cancer Research
615 Chestnut Street, 17th Floor
Philadelphia, PA 19106-4404

AACR-Genentech BioOncology Career Development Award for Cancer Research on the HER Family Pathway 91
Career development awards are two-year awards that support research by scientists who are engaged in meritorious cancer research at academic institutions. These awards provide important transitional support for direct research expenses as researchers move from the ranks of early career scientists to faculty status. The Awards are open to junior faculty who completed postdoctoral studies or clinical fellowships between specified dates and who are at an academic or medical institution. Research projects in this area should have direct applicability to basic, translational, or clinical cancer research focused on the HER Family Pathway. One grant will be presented annually. The submission deadline is December 3, with the decision date March 31.
Requirements: Candidates must be affiliated with an institution in the U.S. and must hold an M.D. or combined M.D./Ph.D. If a candidate has obtained an equivalent degree at a foreign institution, information on the nature of the degree must be provided at the time of application.
Restrictions: Employees of a national government and employees of private industry are ineligible.
Geographic Focus: All States
Date(s) Application is Due: Dec 3
Amount of Grant: 50,000 USD
Contact: Julia Laurence, Program Assistant; (215) 440-9300, ext. 102 or (267) 646-0655; fax (215) 440-9372; julia.laurence@aacr.org or awards@aacr.org
Internet: http://www.aacr.org/
Sponsor: American Association for Cancer Research
615 Chestnut Street, 17th Floor
Philadelphia, PA 19106-4404

AACR-Genentech BioOncology Fellowship for Cancer Research in Angiogenesis 92
AACR Research Fellowships are open to Postdoctoral Fellows and Clinical Research Fellows at an academic facility, teaching hospital, or research institution who will be in the 1st, 2nd, or 3rd year of their postdoctoral training at the start of the fellowship term on July 1. This is a two-year term with a grant of $35,000 per year. One Fellowship will be awarded.

14 | Grant Programs

Requirements: The recipients of the Fellowships must attend the AACR Annual Meeting to accept the award.
Restrictions: Research projects are restricted to cancer research with direct applicability to angiogenesis.
Geographic Focus: All States
Date(s) Application is Due: Dec 3
Amount of Grant: 35,000 USD
Contact: Hanna Hopfinger, Program Assistant; (267) 646-0665 or (215) 440-9300; fax (215) 440-9372; hanna.hopfinger@aacr.org or awards@aacr.org
Internet: http://www.aacr.org/
Sponsor: American Association for Cancer Research
615 Chestnut Street, 17th Floor
Philadelphia, PA 19106-4404

AACR-GlaxoSmithKline Clinical Cancer Research Scholar Awards 93

These awards provide support of $4,000 which may be used by the recipient to support attendance at the Annual Meeting and any other AACR meeting within a two-year period. Up to $2,000 per meeting may be used, and reimbursement checks are provided after the meeting upon submission of proper documentation of travel, housing, registration, and/or subsistence expenses. Award recipients are still required to pay the Annual Meeting registration fee and make their own travel and housing arrangements. Selection is made by the Program Committee based upon the novelty, quality, and significance of the abstract submitted. Award recipients will receive notification separate from abstract acceptance and scheduling information.
Requirements: Graduate students, medical students and residents, clinical fellows or equivalent, and postdoctoral fellows (AACR members and nonmembers) who are presenters of a meritorious abstract in clinical cancer research are eligible.
Restrictions: Employees or subcontractors of private industry are not eligible.
Geographic Focus: All States
Date(s) Application is Due: Nov 28
Amount of Grant: 2,000 - 4,000 USD
Samples: Manisha Bhutani, M.D., UT M.D. Anderson Cancer Center; David W. Cescon, M.D., Princess Margaret Hospital, University of Toronto; Zeshaan A. Rasheed, M.D., Ph.D., Sidney Kimmel Comprehensive Cancer Center, Johns Hopkins University School of Medicine.
Contact: Mona E. Shater, Staff Assistant; (215) 440-9300 or (267) 646-0654; fax (215) 440-9372; mona.shater@aacr.org
Internet: http://www.aacr.org/home/scientists/meetings--workshops/travel-grants/aacr-glaxosmithkline-clinical-cancer-research-scholar-awards.aspx
Sponsor: American Association for Cancer Research
615 Chestnut Street, 17th Floor
Philadelphia, PA 19106-4404

AACR-Minorities in Cancer Research Jane Cooke Wright Lectureship Awards 94

The AACR-Minorities in Cancer Research Jane Cooke Wright Lectureship was first presented in 2006. The Lectureship is intended to give recognition to an outstanding scientist who has made meritorious contributions to the field of cancer research and who has, through leadership or by example, furthered the advancement of minority investigators in cancer research. The recipient will present the AACR-Minorities in Cancer Research Jane Cooke Wright Lectureship during the AACR Annual Meeting, receive an honorarium and commemorative plaque, and receive support for the winner and a guest to attend the Annual Meeting.
Requirements: Candidacy is open to all cancer researchers who are affiliated with any institution involved in cancer research, cancer medicine, or cancer-related biomedical science anywhere in the world. Such institutions include those in academia, industry, or government. Nominations may be made by any scientist, whether an AACR member or nonmember, who is now or has been affiliated with any institution involved in cancer research, cancer medicine, or cancer-related biomedical science. Candidates may not nominate themselves. Nominations must be submitted online at https://proposalcentral.altum.com, no later than 4:00 p.m. United States Eastern Time on October 15th. Paper nominations will not be accepted.
Restrictions: Institutions or organizations are not eligible for the Lectureship.
Geographic Focus: All States
Date(s) Application is Due: Oct 15
Contact: Monique P. Eversley, Staff Associate; (267) 646-0576 or (215) 440-9300, ext. 1400; fax (215) 440-9372; monique.eversley@aacr.org or awards@aacr.org
Internet: http://www.aacr.org/home/scientists/scientific-achievement-awards/micr-wright-lectureship.aspx
Sponsor: American Association for Cancer Research
615 Chestnut Street, 17th Floor
Philadelphia, PA 19106-4404

AACR-National Brain Tumor Foundation Fellows Grant 95

The AACR-National Brain Tumor Society Fellowship, in memory of Bonnie Brooks, is open to Postdoctoral Fellows and Clinical Research Fellows at an academic facility, teaching hospital, or research institution. This fellowship provides a one-year grant of $40,000 to support the salary and benefits of Fellows engaged in Glioblastoma multiforme (GBM) research. A partial amount of funds may be designated for research support. It is anticipated that two fellowships will be awarded. The AACR requires applicants to submit both an online and a paper application. The online application must be submitted by 12:00 noon (United States Eastern Time) on August 10, the application must be submitted using the proposalCENTRAL website at https://proposalcentral.altum.com. The paper application with original signatures and all supporting documents must be postmarked and sent no later then August 12.
Requirements: Candidates must be, at the start of the grant term, in their third, fourth, or fifth year of their fellowship status.
Restrictions: Investigators holding the rank of Instructor, Adjunct Professor, Assistant Professor, Research Assistant Professor, the equivalent or higher are not eligible.
Geographic Focus: All States
Date(s) Application is Due: Aug 10; Aug 12
Amount of Grant: 40,000 USD
Contact: Hanna Hopfinger, Grants Program Assistant; (267) 646-0665; fax (215) 440-9372; hanna.hopfinger@aacr.org
Internet: http://www.aacr.org/home/scientists/research-funding--fellowships/postdoctoral-fellowships.aspx
Sponsor: American Association for Cancer Research
615 Chestnut Street, 17th Floor
Philadelphia, PA 19106-4404

AACR-NCI International Investigator Opportunity Grants 96

Recognizing that there is no substitute for the collegial interaction and scholarly discussion that is a cornerstone of scientific advancement, these grants provide significant financial support for meeting registration, housing, travel, and incidentals to attend the annual American Association for Cancer Research Meeting. Grants are available for ten cancer researchers from countries where such opportunities are limited.
Requirements: Applicants must be nationals or permanent residents of, and must reside and conduct cancer research within, low and middle-income countries as defined by the World Bank. Applicants should also hold an doctorate or equivalent degree and be actively engaged in a program of cancer research with a record of cancer-focused publications in peer-reviewed scientific journals.
Restrictions: Funds may not be used for entertainment expenses (such as for tours or in-room movies), meals for other individuals, souvenirs, and other non-meeting expenses.
Geographic Focus: All States
Date(s) Application is Due: Sep 28
Samples: Hakan Akbulut, M.D., Professor, Medical Oncology, Ankara University, Ankara, Turkey; Soo H. Teo, D.Phil., Chief Executive, Cancer Research Initiatives Foundation, Subang Jaya, Malaysia; Jozsef Tovari, Ph.D., Lab Head, Department of Tumor Progression, National Institute of Oncology, Budapest, Hungary.
Contact: Susan Waskey, Program Assistant; (215) 440-9300 or (866) 423-3965; fax (215) 440-9313; susan.waskey@aacr.org
Internet: http://www.aacr.org/home/scientists/meetings--workshops/travel-grants.aspx
Sponsor: American Association for Cancer Research
615 Chestnut Street, 17th Floor
Philadelphia, PA 19106-4404

AACR-Pancreatic Cancer Action Network Career Development Award for Pancreatic Cancer Research 97

The Pancreatic Cancer Action Network-AACR Career Development Award represents a joint effort to encourage and support early career scientist who are in the first four years of a faculty appointment to conduct pancreatic cancer research and establish successful career paths in this field. The research proposed for funding may be basic, translational, clinical or epidemiological in nature and must have direct applicability and relevance to pancreatic cancer. The Award provides a two year grant of $200,000 ($100,000 per year) for direct research expenses, which may include salary and benefits of the grant recipient, postdoctoral or clinical research fellows, and/or research assistants, research/laboratory supplies and equipment. It is anticipated that three Career Development Awards will be funded.
Requirements: Candidates must have acquired a doctoral degree in a related field and may not currently be a candidate for a further doctoral or professional degree. If a candidate has obtained an equivalent degree at a foreign institution, information on the nature of the degree must be provided at the time of application.
Restrictions: Awards are restricted to investigators or institutions within the U.S. Employees of a national government and employees of private industry are ineligible.
Geographic Focus: All States
Date(s) Application is Due: Oct 28
Amount of Grant: 200,000 USD
Contact: Julia Laurence, Program Assistant; (215) 440-9300, ext. 102 or (267) 646-0655; fax (215) 440-9372; julia.laurence@aacr.org or awards@aacr.org
Internet: http://www.aacr.org/home/scientists/research-funding--fellowships/career-development-awards.aspx
Sponsor: American Association for Cancer Research
615 Chestnut Street, 17th Floor
Philadelphia, PA 19106-4404

AACR-Pancreatic Cancer Action Network Innovative Grants 98

The AACR-PanCAN Pilot Grants represent a joint effort of the AACR and the Pancreatic Cancer Action Network to support innovative research focusing on the early detection or treatment of pancreatic cancer. Research projects may have direct application to pancreatic cancer or may demonstrate relevance to pancreatic cancer in order to encourage investigators with experience in other areas of research to apply their ideas to this challenging field. Proposals will be accepted for innovative research projects that have a direct application or demonstrate relevance to the early detection or treatment of pancreatic cancer; projects may be in any discipline of basic, translational, clinical, or epidemiological cancer research. Proposals that will develop preliminary data necessary to prepare and submit a competitive research grant application to a major federal funding agency will also be accepted. Special emphasis will be placed on research that is not duplicative of other efforts and has the

potential for national application. Researchers who are affiliated with any institution involved in cancer research, cancer medicine, or cancer-related biomedical science within the United States may apply. There are no residency status restrictions. Each grantee will be awarded a total of $200,000, paid over the two-year term.
Requirements: Applicants must have acquired a doctoral degree in a related field. If an applicant has obtained an equivalent degree at a foreign institution, information on the nature of the degree must be provided at the time of application.
Restrictions: Employees or subcontractors of a national government or the for-profit private industry are not eligible to apply. Members of the Scientific Review Committee are not eligible for these grants. Current or past recipients of a PanCAN Pilot Grant are not eligible.
Geographic Focus: All States
Date(s) Application is Due: Oct 5; Nov 30
Amount of Grant: 200,000 USD
Contact: Julia Laurence, Program Assistant; (215) 440-9300, ext. 102 or (267) 646-0655; fax (215) 440-9372; julia.laurence@aacr.org or laurence@aacr.org
Internet: http://www.aacr.org/home/scientists/research-funding--fellowships/pancreatic-cancer-action-network-aacr-innovative-grants.aspx
Sponsor: American Association for Cancer Research
615 Chestnut Street, 17th Floor
Philadelphia, PA 19106-4404

AACR-Prevent Cancer Foundation Award for Excellence in Cancer Prevention Research 99

he Award for Excellence in Cancer Prevention Research will be given to a scientist residing in any country in the world for his or her seminal contributions to the field of cancer prevention. Such investigations must have been conducted in basic, translational, clinical, epidemiological, or behavioral science in cancer prevention research. Further, these studies must have had not only a major impact on the field, but must also have stimulated new directions in this important area. The recipient of the Award will receive a $5,000 honorarium and will present a lecture at the Annual AACR International Conference. Nominations for the Award must be submitted electronically no later than 4:00 p.m. United States Eastern Time on Monday, July 31.
Requirements: All cancer researchers who are affiliated with any institution involved in cancer research, cancer medicine, or cancer-related biomedical science anywhere in the world. Such institutions include those in academia, industry, or government may be nominated for the Award. Candidates must currently maintain an active research program, have a record of recent publications, and be able to present the Award lecture at the Conference.
Restrictions: Institutions or organizations are not eligible for the Award. Paper nominations will not be accepted.
Geographic Focus: All States
Date(s) Application is Due: Jul 31
Amount of Grant: 5,000 USD
Contact: Monique P. Eversley, Staff Associate; (267) 646-0576 or (215) 440-9300, ext. 1400; fax (215) 440-9372; monique.eversley@aacr.org or awards@aacr.org
Internet: http://www.aacr.org/home/scientists/scientific-achievement-awards/prevent-cancer-award.aspx
Sponsor: American Association for Cancer Research
615 Chestnut Street, 17th Floor
Philadelphia, PA 19106-4404

AACR-WICR Scholar Awards 100

AACR-WICR Scholar Awards are awarded to ten members of Women in Cancer Research who are scientists-in-training and presenters of meritorious scientific papers at AACR Annual Meetings.
Requirements: Eligible candidates for these awards must fit the following criteria, candidates must be: members of Women in Cancer Research (WICR); full time scientists-in-training who are graduate students, medical students, residents, clinical fellows or equivalent, or postdoctoral fellows; travelling from within the United States or abroad; first authors on abstracts submitted for consideration for presentation at the AACR Annual Meeting.
Geographic Focus: All States
Date(s) Application is Due: Dec 1
Contact: Lauren Medvetz; (267) 646-0689; fax (215) 440-9372; lauren.medvets@aacr.org
Internet: http://www.aacr.org/home/scientists/meetings--workshops/travel-grants/wicr-scholar-awards.aspx
Sponsor: American Association for Cancer Research
615 Chestnut Street, 17th Floor
Philadelphia, PA 19106-4404

AACR-Women in Cancer Research Charlotte Friend Memorial Lectureship Awards 101

The AACR-Women in Cancer Research Charlotte Friend Memorial Lectureship was established in 1998 in honor of renowned virologist and discoverer of the Friend virus, Dr. Charlotte Friend, for her pioneering research on viruses, cell differentiation, and cancer. The lecture is intended to give recognition to an outstanding female or male scientist who has made meritorious contributions to the field of cancer research and who has, through leadership or by example, furthered the advancement of women in science. The winner of the Lectureship will present a 50-minute lecture during the AACR Annual Meeting, receive an honorarium and commemorative plaque, and receive support for the winner and a guest to attend the Annual Meeting.
Requirements: Candidacy is open to all cancer researchers who are affiliated with any institution involved in cancer research, cancer medicine, or cancer-related biomedical science anywhere in the world. Such institutions include those in academia, industry, or government.

Restrictions: Nominations may be made by any scientist, whether an AACR member or nonmember, who is now or has been affiliated with any institution involved in cancer research, cancer medicine, or cancer-related biomedical science. Candidates may not nominate themselves. Nominations must be submitted online at https://proposalcentral.altum.com, no later than 4:00 p.m. United States Eastern Time on October 15th. Paper nominations will not be accepted. Institutions or organizations are not eligible for the Lectureship.
Geographic Focus: All States
Date(s) Application is Due: Oct 15
Contact: Lauren Medvetz; (267) 646-0689; fax (215) 440-9372; lauren.medvets@aacr.org
Internet: http://www.aacr.org/home/scientists/scientific-achievement-awards/wicr-friend-lectureship.aspx
Sponsor: American Association for Cancer Research
615 Chestnut Street, 17th Floor
Philadelphia, PA 19106-4404

AACR Award for Lifetime Achievement in Cancer Research 102

The AACR Award for Lifetime Achievement in Cancer Research was established and first given in 2004 to honor an individual who has made significant fundamental contributions to cancer research, either through a single scientific discovery or a body of work. These contributions, whether they have been in research, leadership, or mentorship, must have had a lasting impact on the cancer field and must have demonstrated a lifetime commitment to progress against cancer. Nominations may be made by any scientist, whether an AACR member or nonmember, who is now or has been affiliated with any institution involved in cancer research, cancer medicine, or cancer-related biomedical science. Nominations for the Award must be submitted electronically to the AACR no later than 4:00 p.m. United States Eastern Time on November 16.
Requirements: Nominations may be made on behalf of individuals who are living at the time of the nomination. Candidates need not be members of the AACR. Candidates need not be currently engaged in cancer research.
Restrictions: Institutions or organizations are not eligible for the Award. Paper nominations will not be accepted.
Geographic Focus: All States
Date(s) Application is Due: Nov 16
Contact: Monique P. Eversley, Staff Associate; (267) 646-0576 or (215) 440-9300, ext. 1400; fax (215) 440-9372; monique.eversley@aacr.org or awards@aacr.org
Internet: http://www.aacr.org/home/scientists/scientific-achievement-awards/lifetime-achievement-award.aspx
Sponsor: American Association for Cancer Research
615 Chestnut Street, 17th Floor
Philadelphia, PA 19106-4404

AACR Award for Outstanding Achievement in Cancer Research 103

This award is given in recognition of a young investigator on the basis of meritorious achievement in cancer research. The winner receives an honorarium of $5,000, presents a 50-minute lecture, and is given full support for the winner and a guest to attend the AACR Annual Meeting.
Requirements: Candidacy is open to all cancer researchers who are affiliated with any institution involved in cancer research, cancer medicine, or cancer-related biomedical science anywhere in the world. Such institutions include those in academia, industry, or government. In accordance with the wishes of the donor, the recipient must be no more than 40 years of age by the time the award is received.
Restrictions: Institutions or organizations are not eligible for the Award.
Geographic Focus: All States
Date(s) Application is Due: Oct 15
Amount of Grant: 5,000 USD
Contact: Monique P. Eversley, Program Associate; (267) 646-0576 or (215) 440-9300, ext. 1400; fax (267) 646-0576; monique.eversley@aacr.org
Internet: http://www.aacr.org/home/scientists/scientific-achievement-awards/outstanding-achievement-award.aspx
Sponsor: American Association for Cancer Research
615 Chestnut Street, 17th Floor
Philadelphia, PA 19106-4404

AACR Award for Outstanding Achievement in Chemistry in Cancer Research 104

The Award will be given for outstanding, novel, and significant chemistry research, which has led to important contributions to the fields of basic cancer research; translational cancer research; cancer diagnosis; the prevention of cancer; or the treatment of patients with cancer. Such research may include, but is not limited to, drug discovery and design; structural biology; proteomics, metabolomics and biological mass spectrometry; chemical aspects of carcinogenesis; imaging agents and radiotherapeutics; and chemical biology. The winner of the award will give a 50-minute lecture during the AACR Annual Meeting, will receive a commemorative plaque, and a $10,000 honorarium, and receive support for the winner and a guest to attend the Annual Meeting. Nominations for the Award must be submitted electronically no later than 4:00 p.m. United States Eastern Time on October 15.
Requirements: Candidacy is open to all researchers who are affiliated with any institution involved in cancer research, cancer medicine, or cancer-related biomedical science anywhere in the world. Such institutions include those in academia, industry, or government.
Restrictions: Institutions or organizations are not eligible for the Award. Paper nominations will not be accepted.
Geographic Focus: All States
Date(s) Application is Due: Oct 15
Amount of Grant: 10,000 USD
Contact: Lauren Medvetz; (267) 646-0689; fax (215) 440-9372; lauren.medvets@aacr.org

16 | Grant Programs

Internet: http://www.aacr.org/home/scientists/scientific-achievement-awards/chemistry-in-cancer-research-award.aspx
Sponsor: American Association for Cancer Research
615 Chestnut Street, 17th Floor
Philadelphia, PA 19106-4404

AACR Basic Cancer Research Fellowships — 105

The AACR Basic Cancer Research Fellowships consist of two Fellowships, each one-year grants. The AACR Anna D, Barker Fellowship in Basic Cancer Research, funds $40,000, and the AACR-Astellas USA Foundation Fellowships in Basic Cancer Research provides funding in the amount of $30,000. These grants are for salary and benefits to the Fellows. A partial amount of funds may be designated for direct research support. The recipients must attend the AACR Annual Meeting to accept the grants. The AACR requires applicants to submit both an online and a paper application. The online application must be completed by 12:00 noon (United States Eastern Time) on December 10, using the proposalCENTRAL website at http://propasalcentral.altum.com. The paper application with original signatures and all supporting documents must be postmarked and sent no later then December 15.
Requirements: Applications are open to Postdoctoral Fellows and Clinical Research Fellows at an academic facility, teaching hospital, or research institution who will be in the 1st, 2nd, or 3rd year of their postdoctoral training at the start of the fellowship term.
Restrictions: Academic faculty holding the rank of assistant professor or higher, graduate or medical students, government employees, and employees of private industry are not eligible.
Geographic Focus: All States
Date(s) Application is Due: Dec 10; Dec 15
Amount of Grant: 30,000 - 40,000 USD
Contact: Hanna Hopfinger, Grants Program Assistant; (267) 646-0665; fax (215) 440-9372; hanna.hopfinger@aacr.org
Internet: http://www.aacr.org/home/scientists/research-funding--fellowships/postdoctoral-fellowships.aspx
Sponsor: American Association for Cancer Research
615 Chestnut Street, 17th Floor
Philadelphia, PA 19106-4404

AACR Brigid G. Leventhal Scholar in Cancer Research Awards — 106

The purpose of this award program is to enhance the education and training of early career scientists by providing financial support for their participation in AACR Annual Meetings and Special Conferences. The AACR-WICR Brigid G. Leventhal Scholar Award Committee makes selection for these competitive awards after careful consideration of the candidate's application and accompanying materials.
Requirements: Candidates must be: Women in Cancer Research members; full time scientists in training who are graduate students, medical students, residents, clinical fellows or equivalent, or postdoctoral fellows; and first authors on abstracts submitted for consideration for presentation at the AACR Annual Meeting or Special Conference which the applicant wants to attend.
Geographic Focus: All States
Date(s) Application is Due: Sep 10
Contact: Mona E. Shater, Staff Assistant; (215) 440-9300 or (267) 646-0654; fax (215) 440-9372; mona.shater@aacr.org or wicr@aacr.org
Internet: http://www.aacr.org/home/membership-/association-groups/women-in-cancer-research.aspx
Sponsor: American Association for Cancer Research
615 Chestnut Street, 17th Floor
Philadelphia, PA 19106-4404

AACR Career Development Awards for Pediatric Cancer Research — 107

Two Career Development Awards will be presented, the AACR-Aflac Incorporated Career Development Award for Pediatric Cancer Research and the AACR Centennial Career Development Award for Childhood Cancer Research. These Awards are open to junior faculty who are in their first full-year, faculty appointment. The Awards provide two-year grants of $100,000 ($50,000 per year) for direct research expenses and/or salary support. Research projects must have direct applicability to pediatric cancer.
Requirements: Candidates must have acquired a doctoral degree and have completed postdoctoral studies or clinical fellowships on or after July 2 (two years prior to the start of the grant term). Candidates must be in their first full-time, faculty appointment and hold the title of Instructor, Research Assistant Professor, Assistant Professor, or an equivalent full-time faculty appointment at the start of the grant term (July 1).
Restrictions: Non AACR members must submit a satisfactory application for Active Membership with the grant application by December 10th. Employees or subcontractors of a government or a for-profit private industry are not eligible.
Geographic Focus: All States
Date(s) Application is Due: Dec 18
Amount of Grant: 100,000 USD
Contact: Julia Laurence, Program Assistant; (215) 440-9300, ext. 102 or (267) 646-0655; fax (215) 440-9372; julia.laurence@aacr.org or awards@aacr.org
Internet: http://www.aacr.org/home/scientists/research-funding--fellowships/career-development-awards.aspx
Sponsor: American Association for Cancer Research
615 Chestnut Street, 17th Floor
Philadelphia, PA 19106-4404

AACR Centennial Postdoctoral Fellowships in Cancer Research — 108

AACR Centennial Postdoctoral Research Fellowships in Cancer Research will provide $60,000 per year for up to three years to provide salary support and research funding for clinical and postdoctoral fellows in the first, second, or third year of their fellowship status. In addition to financial flexibility these fellowships also provide career flexibility. Fellowship funds may be transferred to move with the grantee if the grantee's fellowship is completed during the grant term - a valuable commodity for a newly-independent investigator. Under this award mechanism, the postdoctoral trainee is considered the Lead Investigator and should write the proposal independently with appropriate direction from the mentor. Proposals may be submitted in any area of cancer research whether laboratory-based, translational, clinical, epidemiological, behavioral, or other. Applications must be completed online using the proposalCENTRAL website, with one paper copy submitted to the AACR office. Application instructions and program guidelines are available online.
Requirements: Applicants must: be in the 1st, 2nd, or 3rd year of their postdoctoral training at the start of the fellowship term; be members of the AACR or must submit an application for membership prior to the fellowship application deadline; have acquired a doctoral degree in a biological sciences-related field; and hold the title of Postdoctoral Fellow or Clinical Research Fellow at an academic facility, teaching hospital, or research institution.
Restrictions: Employees, trainees, or subcontractors of a national government or private industry are not eligible.
Geographic Focus: All States
Date(s) Application is Due: Feb 1
Amount of Grant: 60,000 - 180,000 USD
Contact: Hanna Hopfinger, Program Assistant; (267) 646-0665 or (215) 440-9300; fax (215) 440-9372; hanna.hopfinger@aacr.org or awards@aacr.org
Internet: http://www.aacr.org/home/scientists/research-funding--fellowships/research-funding.aspx
Sponsor: American Association for Cancer Research
615 Chestnut Street, 17th Floor
Philadelphia, PA 19106-4404

AACR Centennial Pre-doctoral Fellowships in Cancer Research — 109

The AACR Centennial Pre-doctoral Research Fellowships will provide full-time graduate students a rare opportunity to invest in their future cancer research careers. The grant terms are flexible, allowing applicants to tailor their grant funds to fit their individual research funding and salary needs. Grantees may also transfer their funds to a postdoctoral fellowship position, should they advance in their careers during the grant term. Under this award mechanism, the pre-doctoral trainee is considered the Lead Investigator and should write the proposal independently with appropriate direction from the mentor. Proposals may be submitted in any area of cancer research whether laboratory-based, translational, clinical, epidemiological, behavioral, or other. A grant of $30,000 per year over a maximum period of three years will be awarded for salary support and research funding.
Requirements: Applicants must: be full-time graduate students who are pursuing a doctoral degree and engaged in a program of cancer research; be members of the AACR or must submit an application for membership prior to the fellowship application deadline; have completed at least one year of research training at the graduate level or have equivalent research experience; and be engaged in basic, clinical, translational or epidemiological cancer research. At the time of award acceptance the awardee must have advanced to degree candidacy.
Restrictions: Summer student positions and undergraduate summer studentships do not qualify. Employees, trainees, or subcontractors of a national government or private industry are not eligible.
Geographic Focus: All States
Date(s) Application is Due: Feb 1
Amount of Grant: 30,000 - 90,000 USD
Contact: Hanna Hopfinger, Program Assistant; (267) 646-0665 or (215) 440-9300; fax (215) 440-9372; hanna.hopfinger@aacr.org or awards@aacr.org
Internet: http://www.aacr.org/home/scientists/research-funding--fellowships/research-funding.aspx
Sponsor: American Association for Cancer Research
615 Chestnut Street, 17th Floor
Philadelphia, PA 19106-4404

AACR Clinical and Translational Research Fellowships — 110

The AACR Clinical and Translational Research Fellowships consist of three different Fellowships: AACR-Astellas USA Foundation Fellowship in Clinical/Translational Cancer Research, which is a one-year grant of $30,000; AACR-AstraZeneca Fellowship for Translational Lung Cancer Research, which is a two-year grant of $90,000; and, AACR-Bristol-Myers Squibb Oncology Fellowship in Clinical Cancer Research, which is a one-year grant of $40,000. These fellowships support the salary and benefits of the Fellow. A partial amount of funds may be designated for direct research support. The AACR requires applicants to submit both an online and a paper application. The online application deadline is December 10th, the application must be submitted using the proposalCENTRAL website at https://proposalcentral.altum.com. The paper application with original signatures and all supporting documents must be postmarked and sent no later then December 15.
Requirements: Applications are open to Postdoctoral Fellows and Clinical Research Fellows at an academic facility, teaching hospital, or research institution who will be in the 1st, 2nd, or 3rd year of their postdoctoral training at the start of the fellowship term.
Restrictions: Academic faculty holding the rank of assistant professor or higher, graduate or medical students, government employees, and employees of private industry are not eligible.

Geographic Focus: All States
Date(s) Application is Due: Dec 10; Dec 15
Amount of Grant: 30,000 - 90,000 USD
Contact: Hanna Hopfinger, Program Assistant; (267) 646-0665 or (215) 440-9300; fax (215) 440-9372; hanna.hopfinger@aacr.org or awards@aacr.org
Internet: http://www.aacr.org/home/scientists/research-funding--fellowships/postdoctoral-fellowships.aspx
Sponsor: American Association for Cancer Research
615 Chestnut Street, 17th Floor
Philadelphia, PA 19106-4404

AACR Fellows Grants 111

The Fellowship program supports innovative research by a meritorious young investigator by presenting the Fellow with research funds to pursue an independent line of investigation within the context of his/her current Fellowship placement. By allowing a Fellow to acquire the equipment and supplies needed to pursue a new direction in his/her research program, the Fellows Grant assists the Fellow in developing preliminary data to support a future project or investigating a new technique that otherwise would not be possible in the absence of this funding. The Selection Committee favors proposals that provided evidence of an applicant's research initiative and creativity, and the support of a scientific mentor. The Committee also weighed the potential of the candidate to make meaningful contributions to the field, and promise of success as a cancer researcher in future years.
Requirements: Candidates must be, at the start of the grant term, in their third, fourth, or fifth year of their fellowship status.
Geographic Focus: All States
Date(s) Application is Due: Dec 18
Amount of Grant: 30,000 - 40,000 USD
Contact: Julia Laurence, Program Assistant; (215) 440-9300, ext. 102 or (267) 646-0655; fax (215) 440-9372; julia.laurence@aacr.org or awards@aacr.org
Internet: http://www.aacr.org/home/scientists/research-funding--fellowships/fellows-grants.aspx
Sponsor: American Association for Cancer Research
615 Chestnut Street, 17th Floor
Philadelphia, PA 19106-4404

AACR G.H.A. Clowes Memorial Award 112

The AACR and Eli Lilly and Company established this Award in 1961 to honor Dr. G.H.A. Clowes, who was a founding member of the AACR and a research director of Eli Lilly. The Clowes Award recognizes an individual with outstanding recent accomplishments in basic cancer research. Nominations for the Award must be submitted electronically to the AACR no later than 4:00 p.m. United States Eastern Time on October 15. Paper nominations will not be accepted.
Requirements: Candidacy is open to all cancer researchers who are affiliated with any institution involved in cancer research, cancer medicine, or cancer-related biomedical science anywhere in the world. Such institutions include those in academia, industry, or government.
Restrictions: Institutions or organizations are not eligible for the Award. Paper nominations will not be accepted.
Geographic Focus: All States
Date(s) Application is Due: Oct 15
Amount of Grant: 10,000 USD
Contact: Monique P. Eversley, Staff Associate; (267) 646-0576 or (215) 440-9300, ext. 1400; fax (215) 440-9372; monique.eversley@aacr.org or awards@aacr.org
Internet: http://www.aacr.org/home/scientists/scientific-achievement-awards/clowes-award.aspx
Sponsor: American Association for Cancer Research
615 Chestnut Street, 17th Floor
Philadelphia, PA 19106-4404

AACR Gertrude Elion Cancer Research Award 113

This prestigious award provides a one-year research grant to an assistant professor at an academic or nonprofit research institute worldwide for salary and benefits, laboratory supplies, and limited domestic travel to support research in cancer etiology, diagnosis, treatment, or prevention (basic, translational, or clinical cancer research). The AACR requires applicants to submit both an online and a paper application. The online application must be submitted by 12:00 noon (United States Eastern Daylight Time) on February 27, using the proposalCENTRAL website at https://proposalcentral.altum.com. The paper application with original signatures and all supporting documents must be postmarked and sent no later then March 3.
Requirements: Candidates must be nominated by an AACR member and submit a detailed application. Candidates must have completed postdoctoral studies or clinical fellowships no later than July 1 of the award year and ordinarily not more than five years earlier. Candidates must be tenure-tracked scientists at the level of assistant professor at an academic institution anywhere in the world.
Restrictions: Tenured faculty in academia, employees of a national government, and employees of private industry are not eligible.
Geographic Focus: All States
Date(s) Application is Due: Feb 27; Mar 3
Amount of Grant: 50,000 USD
Contact: Julia Laurence, Program Assistant; (215) 440-9300, ext. 102 or (267) 646-0655; fax (215) 440-9372; julia.laurence@aacr.org or laurence@aacr.org
Internet: http://www.aacr.org/home/scientists/research-funding--fellowships/elion-award.aspx
Sponsor: American Association for Cancer Research
615 Chestnut Street, 17th Floor
Philadelphia, PA 19106-4404

AACR Henry Shepard Bladder Cancer Research Grants 114

These two-year Grants will provide up to $250,000 in total support for innovative cancer research projects designed to accelerate the discovery, development, and application of new agents to treat bladder cancer and/or for pre-clinical research with direct therapeutic intent. Laboratory-based projects must present plans with clinical collaborators indicating how the work will be translated into the clinic. Similarly, clinical studies must show how the work was derived from basic preclinical work and how results will be channeled back to laboratory-based collaborators. Four grants will be awarded. The submission deadline is July 25, at 12:00 noon, U.S. Eastern Time.
Requirements: Independent investigators who are affiliated with any institution involved in cancer research, cancer medicine, or cancer-related biomedical science anywhere in the world may apply. There are no geographic, national, or residency status restrictions. Applicants must have acquired a doctoral degree in a related field.
Restrictions: Pre- and Postdoctoral fellows are not eligible. Employees or subcontractors of a national government or the for-profit private industry are not eligible to serve as the Principal Investigator for the purpose of these Grants. However, collaborations with such individuals are encouraged. Neither the members of the Scientific Review Committee nor the members of their individual laboratories are eligible for these Grants.
Geographic Focus: All States
Date(s) Application is Due: Jul 25
Amount of Grant: Up to 250,000 USD
Contact: Hanna Hopfinger, Program Assistant; (267) 646-0665 or (215) 440-9300; fax (215) 440-9372; hanna.hopfinger@aacr.org or awards@aacr.org
Internet: http://www.aacr.org/home/scientists/research-funding--training-grants/research-funding/research-funding--fellowships/bladder-cancer-research-grants.aspx
Sponsor: American Association for Cancer Research
615 Chestnut Street, 17th Floor
Philadelphia, PA 19106-4404

AACR Joseph H. Burchenal Memorial Award 115

This Award was established in 1996 to recognize outstanding achievements in clinical cancer research. It is named for the late Dr. Joseph H. Burchenal, Honorary Member and Past President of the AACR, and a major figure in clinical cancer research and chemotherapy. The winner of the Award for Outstanding Achievement in Clinical Cancer Research will give a 50-minute lecture during the AACR Annual Meeting, and receive an honorarium and support for attendance at the meeting.
Requirements: Candidacy is open to all cancer researchers who are affiliated with any institution involved in cancer research, cancer medicine, or cancer-related biomedical science anywhere in the world. Such institutions include those in academia, industry, or government.
Restrictions: Institutions or organizations are not eligible for the Award. Candidates may not nominate themselves.
Geographic Focus: All States
Date(s) Application is Due: Oct 15
Samples: Joseph R. Bertino, M.D., Interim Director & Chief Scientific Officer, The Cancer Institute of New Jersey.
Contact: Monique P. Eversley, Program Associate; (267) 646-0576 or (215) 440-9300, ext. 1400; fax (267) 646-0576; monique.eversley@aacr.org
Internet: http://www.aacr.org/home/scientists/scientific-achievement-awards/burchenal-award.aspx
Sponsor: American Association for Cancer Research
615 Chestnut Street, 17th Floor
Philadelphia, PA 19106-4404

AACR Judah Folkman Career Development Award for Anti-Angiogenesis Research 116

The AACR Judah Folkman Career Development Award for Angiogenesis Research is open to junior faculty who completed postdoctoral studies or clinical fellowships no more then three years prior to the start of the grant term (Dec 1); who are in their first full-time, faculty appointment and hold the title of Instructor, Research Assistant Professor, Assistant Professor, or an equivalent full-time faculty appointment; and who are at an academic, medical, or research Institution. The Career Development Award provides a two-year grant of $100,000 ($50,000 per year) for direct research expenses and/or salary support. One Award will be presented.
Requirements: Applicants must have acquired a medical degree or hold a combined M.D./Ph.D and may not currently be a candidate for a further doctoral or professional degree. The Award is open to AACR members. Non-members must submit a satisfactory application for Active Membership and additional Curriculum Vitae with the grant application.
Restrictions: Employees, trainees, or subcontractors of a national government or private industry are not eligible. Exceptions may apply if an applicant holds a full-time position at a Veterans' Hospital or national laboratory in the U.S. Current recipients of any AACR grant are also not eligible.
Geographic Focus: All States
Date(s) Application is Due: Aug 10
Amount of Grant: 100,000 USD
Contact: Julia Laurence, Program Assistant; (215) 440-9300, ext. 102 or (267) 646-0655; fax (215) 440-9372; julia.laurence@aacr.org or awards@aacr.org
Internet: http://www.aacr.org/home/scientists/research-funding--fellowships/career-development-awards.aspx
Sponsor: American Association for Cancer Research
615 Chestnut Street, 17th Floor
Philadelphia, PA 19106-4404

18 | GRANT PROGRAMS

AACR Kirk A. Landon and Dorothy P. Landon Foundation Prizes — 117

These two major international awards recognize outstanding scientists who have made seminal basic and translational cancer research discoveries at the cutting edge of scientific novelty and significance, which have accelerated progress against cancer and have implications for future discoveries and contributions to cancer research. The Kirk A. Landon-AACR Prize for Basic Cancer Research recognizes significant, fundamental contributions to laboratory research. The Dorothy P. Landon-AACR Prize for Translational Cancer Research recognizes extraordinary achievement in translational cancer research--the interface between basic research and its application to the clinic in the areas of diagnosis, treatment, or the prevention of cancer. Both prizes bring heightened public attention to landmark achievements in the continuing effort to prevent and cure cancer through the presentation of dynamic lectures during the AACR annual meeting; and promote and reward continued, productive cancer research.
Requirements: The prizes are open to all cancer researchers who are affiliated with any institution involved in cancer research, cancer medicine, or cancer-related biomedical science anywhere in the world. Such institutions include those in academia, industry, or government. Candidates must be active researchers and have a record of recent publications.
Restrictions: Institutions or organizations are ineligible.
Geographic Focus: All States
Date(s) Application is Due: Aug 25
Amount of Grant: 100,000 USD
Samples: Arnold J. Levine, Ph.D., Professor, The Cancer Institute of New Jersey--Kirk A. Landon-AACR Prize for Basic Cancer Research; John Mendelsohn, M.D., President, UT M.D. Anderson Cancer Center, Houston, Texas--Dorothy P. Landon-AACR Prize for Translational Cancer Research.
Contact: Monique P. Eversley, Staff Associate; (267) 646-0576 or (215) 440-9300, ext. 1400; fax (215) 440-9372; monique.eversley@aacr.org or awards@aacr.org
Internet: http://www.aacr.org/home/scientists/scientific-achievement-awards/landon-prizes.aspx
Sponsor: American Association for Cancer Research
615 Chestnut Street, 17th Floor
Philadelphia, PA 19106-4404

AACR Margaret Foti Award for Leadership and Extraordinary Achievements in Cancer Research — 118

This Award will recognize a true champion of cancer research, who embodies the sustained commitment of Margaret Foti to the prevention and cure of cancer. The Award will be given to an individual whose leadership and extraordinary achievements in cancer research have made a major impact in the field. Such achievements may include contributions to the acceleration of progress in cancer research; raising national or international awareness of cancer research; or in other ways demonstrating a sustained commitment to cancer research. Nominations for the Award must be submitted electronically no later than 4:00 p.m. U.S. Eastern Time on November 16. The recipient will receive an honorarium, a commemorative plaque, and support for the winner to attend the AACR Annual Meeting.
Requirements: Candidacy is open to all individuals who are affiliated with any organization whose mission supports cancer research or any institution involved in cancer research, cancer medicine, or cancer-related biomedical science anywhere in the world. Such institutions include those in academia, industry, or government.
Restrictions: Institutions or organizations are not eligible for the Award. Paper nominations will not be accepted.
Geographic Focus: All States
Date(s) Application is Due: Nov 16
Contact: Monique P. Eversley, Program Associate; (267) 646-0576 or (215) 440-9300, ext. 1400; fax (215) 440-9372; monique.eversley@aacr.org or awards@aacr.org
Internet: http://www.aacr.org/home/scientists/scientific-achievement-awards/margaret-foti-award.aspx
Sponsor: American Association for Cancer Research
615 Chestnut Street, 17th Floor
Philadelphia, PA 19106-4404

AACR Minority-Serving Institution Faculty Scholar in Cancer Research Awards — 119

The AACR is pleased to announce the availability of Scholar Awards in Cancer Research for full-time faculty members of Minority-Serving Institutions [Historically Black Colleges and Universities (HBCUs), Hispanic Serving Institutions (HSIs), and Tribal Colleges and Universities and other post-secondary institutions as defined by the U.S. Department of Education]. The purposes of this Award program is to increase the scientific knowledge base of faculty members at Minority-Serving Institutions, and to encourage them and their students to pursue careers in cancer research. AACR Minority-Serving Institution Faculty Scholar in Cancer Research Awards are presented by the American Association for Cancer Research to scientists at the level of Assistant Professor or above at a Minority-Serving Institution, who are engaged in meritorious basic, clinical, or translational cancer research.
Requirements: Candidates must have completed doctoral studies or clinical fellowships and hold full-time faculty status at an institution designated as a Minority-Serving Institution. Candidates must have acquired doctoral degrees in fields relevant to cancer research. Candidates must be citizens or permanent residents of the United States or Canada.
Geographic Focus: All States
Date(s) Application is Due: Jul 20
Samples: Joseph Agyin, Ph.D., UT Health Science Center, San Antonio, Texas; Wellington Ayensu, M.D., Jackson State University, Jackson, Mississippi; Rebecca Hartley, Ph.D., University of New Mexico, Albuquerque, New Mexico.
Contact: Mona E. Shater, Staff Assistant; (215) 440-9300 or (267) 646-0654; fax (215) 440-9372; mona.shater@aacr.org or micr@aacr.org
Internet: http://www.aacr.org/home/scientists/meetings--workshops/special-conferences/metabolism-and-cancer/financial-support-for-attendance.aspx
Sponsor: American Association for Cancer Research
615 Chestnut Street, 17th Floor
Philadelphia, PA 19106-4404

AACR Minority Scholar in Cancer Research Awards — 120

The AACR administers this award program that is supported by a grant from the Comprehensive Minority Biomedical Branch of the National Cancer Institute to provide funds for participation of meritorious minority scientists in the AACR Special Conferences Series (U.S. domestic conferences only). These awards are intended to enhance the education and training of minority researchers and to increase the visibility and recognition of minorities involved in cancer research.
Requirements: Candidates must be full-time graduate students, medical students, residents, clinical or postdoctoral fellows, or junior faculty members who are either engaged in cancer research or who have the training and potential to make contributions to this field. Candidates must also be citizens or permanent residents of the United States or Canada.
Restrictions: This program applies only to racial/ethnic minority groups identified by the National Cancer Institute as being traditionally underrepresented in cancer and biomedical research. These groups include African Americans/Blacks, Alaskan Natives, Hispanic Americans, Native Americans, and Native Pacific Islanders. Only citizens of the United States or Canada or scientists who are permanent residents of these countries may receive one of these awards.
Geographic Focus: All States
Date(s) Application is Due: Jul 20
Samples: Irene Alvarez, B.S., University of Arizona, Tucson, Arizona; Cristina M. Contreras, B.S., UT Southwestern Medical Center, Dallas, Texas; Vivian A. Galacia, B.S., University of Southern California, Los Angeles, California.
Contact: Mona E. Shater, Staff Assistant; (215) 440-9300 or (267) 646-0654; fax (215) 440-9372; mona.shater@aacr.org or micr@aacr.org
Internet: http://www.aacr.org/home/scientists/meetings--workshops/special-conferences/metabolism-and-cancer/financial-support-for-attendance.aspx
Sponsor: American Association for Cancer Research
615 Chestnut Street, 17th Floor
Philadelphia, PA 19106-4404

AACR Outstanding Investigator Award for Breast Cancer Research — 121

The AACR Outstanding Investigator Award for Breast Cancer Research, funded by Susan G. Komen for the Cure, will recognize an investigator of no more than 50 years of age whose novel and significant work has had or may have a far-reaching impact on the etiology, detection, diagnosis, treatment, or prevention of breast cancer. Such work may involve any discipline across the continuum of biomedical research, including basic, translational, clinical, and epidemiological studies. The recipient of the Award will receive a $10,000 honorarium and present a 25-minute lecture at the Annual San Antonio Breast Cancer Symposium. Nominations may be made by any scientist, whether an AACR member or nonmember, who is now or has been affiliated with any institution involved in cancer research, cancer medicine, or cancer-related biomedical science. Candidates may not nominate themselves. Nominations must be submitted online at https://proposalcentral.altum.com, no later than 4:00 p.m. United States Eastern Time on May 15th.
Requirements: All cancer researchers who are affiliated with any institution involved in cancer research, cancer medicine, or cancer-related biomedical science anywhere in the world may be nominated. Such institutions include those in academia, industry, or government. Candidates must be no more than 50 years of age at the time the Award is received.
Restrictions: Institutions or organizations are not eligible for the Award. Paper nominations will not be accepted.
Geographic Focus: All States
Date(s) Application is Due: May 15
Amount of Grant: 10,000 USD
Contact: Monique P. Eversley, Program Associate; (267) 646-0576 or (215) 440-9300, ext. 1400; fax (215) 440-9372; monique.eversley@aacr.org or awards@aacr.org
Internet: http://www.aacr.org/home/scientists/scientific-achievement-awards/outstanding-investigator-award.aspx
Sponsor: American Association for Cancer Research
615 Chestnut Street, 17th Floor
Philadelphia, PA 19106-4404

AACR Princess Takamatsu Memorial Lectureship — 122

This Lectureship will recognize an individual scientist whose novel and significant work has had or may have a far-reaching impact on the detection, diagnosis, treatment, or prevention of cancer, and who embodies the dedication of the Princess to multinational collaborations. The Lecturer will receive an unrestricted cash award of $10,000, support to attend the AACR Annual Meeting, and a commemorative item serving as tangible witness to the singular honor of his/her selection. Nominations must be submitted electronically no later than 4:00 p.m. U.S. Eastern Time on November 16.
Requirements: Candidacy is open to all cancer researchers who are affiliated with any institution involved in cancer research, cancer medicine, or cancer-related biomedical science anywhere in the world. Such institutions include those in academia, industry, or government.
Restrictions: Institutions or organizations are not eligible for the Lectureship. Paper nominations will not be accepted.
Geographic Focus: All States
Date(s) Application is Due: Nov 16
Amount of Grant: 10,000 USD

Contact: Monique P. Eversley, Staff Associate; (267) 646-0576 or (215) 440-9300, ext. 1400; fax (215) 440-9372; monique.eversley@aacr.org or awards@aacr.org
Internet: http://www.aacr.org/home/scientists/scientific-achievement-awards/takamatsu-lectureship.aspx
Sponsor: American Association for Cancer Research
615 Chestnut Street, 17th Floor
Philadelphia, PA 19106-4404

AACR Richard and Hinda Rosenthal Memorial Awards 123
AACR and the Rosenthal Family Foundation established this Award in 1977 to recognize research that has made, or promises to soon make, a notable contribution to improved clinical care in the field of cancer. In its desire to honor and provide incentive to investigators relatively early in their careers, the Foundation has stipulated that recipients not be more than 50 years of age at the time the Award is received. The winner of the AACR Richard and Hinda Rosenthal Memorial Award will receive an honorarium of $10,000, give a 50-minute lecture during the AACR Annual Meeting and be given support for the winner and a guest to attend the Annual Meeting.
Requirements: Candidacy is open to cancer researchers, engaged in the practice of medicine, who are affiliated with any institution involved in cancer research, cancer medicine, or cancer-related biomedical science. Such institutions include those in academia, industry, or government.
Restrictions: Candidates may not nominate themselves. Nominations must be submitted online at https://proposalcentral.altum.com, no later than 4:00 p.m. United States Eastern Time on October 15th. Paper nominations will not be accepted. Institutions or organizations are not eligible for the Award.
Geographic Focus: All States
Date(s) Application is Due: Oct 15
Amount of Grant: 10,000 USD
Contact: Monique P. Eversley, Staff Associate; (267) 646-0576 or (215) 440-9300, ext. 1400; fax (215) 440-9372; monique.eversley@aacr.org or awards@aacr.org
Internet: http://www.aacr.org/home/scientists/scientific-achievement-awards/clowes-award.aspx
Sponsor: American Association for Cancer Research
615 Chestnut Street, 17th Floor
Philadelphia, PA 19106-4404

AACR Scholar-in-Training Awards 124
Scholar-in-Training Awards provide financial support for early-career scientists to attend the AACR Annual Meeting. Since its inception in 1986, the program has provided over 3,200 grants and has received support from more than 30 cancer research foundations, corporations, individuals, and other organizations dedicated to the fight against cancer. A stipend (from $400-$2,000 depending on the geographic location of the recipient or the type of award) will be presented to recipients on-site. Recipients may pick up their a check and welcome package at the Associate Member Resource and Career Center.
Requirements: Eligible candidates are graduate students, medical students and residents, clinical fellows or equivalent, and postdoctoral fellows. Eligible candidates must be the presenter of a proffered paper. Eligible candidates may be traveling within the U.S. or from abroad. Some awards are specifically designated for those traveling from Asia, Europe, and countries with emerging economies.
Restrictions: Employees or subcontractors of private industry are not eligible.
Geographic Focus: All States
Date(s) Application is Due: Jul 20
Amount of Grant: 400 - 2,000 USD
Contact: Mona E. Shater, Staff Assistant; (215) 440-9300 or (267) 646-0654; fax (215) 440-9372; mona.shater@aacr.org
Internet: http://www.aacr.org/home/scientists/meetings--workshops/special-conferences/metabolism-and-cancer/financial-support-for-attendance.aspx
Sponsor: American Association for Cancer Research
615 Chestnut Street, 17th Floor
Philadelphia, PA 19106-4404

AACR Scholar-in-Training Awards - Special Conferences 125
The AACR offers Scholar-in-Training Awards to enhance the education and training of early career scientists by providing financial support for their attendance at AACR Special Conferences. These awards are provided to offset a portion of the registration, travel, and subsistence expenses incurred in attending these conferences. A stipend (typically $400-$1,400) will be presented to award recipients at the conference.
Requirements: All graduate and medical students, postdoctoral fellows, and physicians-in-training who submit abstracts may be considered for a Scholar-in-Training Award.
Restrictions: Employees or subcontractors of private industry are not eligible.
Geographic Focus: All States
Amount of Grant: 400 - 1,400 USD
Contact: Mona E. Shater, Staff Assistant; (215) 440-9300 or (267) 646-0654; fax (215) 440-9372; mona.shater@aacr.org or programs@aacr.org
Internet: http://www.aacr.org/home/scientists/meetings--workshops/travel-grants/scholar-in-training-awards---special-conferences.aspx
Sponsor: American Association for Cancer Research
615 Chestnut Street, 17th Floor
Philadelphia, PA 19106-4404

AACR Team Science Award 126
The Award has been established by the AACR and Eli Lilly and Company to acknowledge and catalyze the growing importance of interdisciplinary teams to the understanding of cancer and/or the translation of research discoveries into clinical cancer applications. In addition, through the presentation of this Award, the AACR and Eli Lilly seek to effect change within the traditional cancer research culture by recognizing those institutions that value and foster interdisciplinary team science. These institutions will have demonstrated their support of a team science environment by creating mechanisms to enhance the required infrastructure, such as through pilot funding, technology transfer offices, shared resources, etc., and by presenting awards, honors, appointments, and promotions to those who participate in interdisciplinary teams. Nominations for the 2009 Award will be accepted until November 16.
Requirements: For the purpose of this Award, a team is defined as a group of individuals representing interdisciplinary expertise, each of whom have made substantive and quantifiable contributions to the research being recognized. Team members may be working within the same institution or at several institutions. Candidacy is open to all cancer researchers who are affiliated with any institution involved in cancer research, cancer medicine, or cancer-related biomedical science anywhere in the world. Such institutions include those in academia, industry, or government.
Geographic Focus: All States
Date(s) Application is Due: Nov 16
Amount of Grant: 50,000 USD
Contact: Monique P. Eversley, Program Associate; (267) 646-0576 or (215) 440-9300, ext. 1400; fax (215) 440-9372; monique.eversley@aacr.org or awards@aacr.org
Internet: http://www.aacr.org/home/scientists/scientific-achievement-awards/team-science-award.aspx
Sponsor: American Association of Cancer Research
615 Chestnut Street, 17th Floor
Philadelphia, PA 19106-4404

AACR Thomas J. Bardos Science Education Awards 127
AACR-Thomas J. Bardos Science Education Awards for Undergraduate Students are available to full-time third-year undergraduate students majoring in science. The purpose of these awards is to inspire young science students to enter the field of cancer research. The award consists of a waiver of registration fees for two annual AACR meetings and $1500/year for research and travel expenses. The award is for a two-year period.
Requirements: Candidates must be full-time, third year undergraduate students majoring in science. Applications from students who are not yet committed to cancer research are welcome.
Geographic Focus: All States
Date(s) Application is Due: Dec 7
Amount of Grant: 1,500 USD
Contact: Mona E. Shater, Staff Assistant; (215) 440-9300 or (267) 646-0654; fax (215) 440-9412; mona.shater@aacr.org or scienceeducation@aacr.org
Internet: http://www.aacr.org/home/scientists/meetings--workshops/travel-grants/aacr-thomas-j-bardos-science-education-awards-for-undergraduate-students-.aspx
Sponsor: American Association for Cancer Research
615 Chestnut Street, 17th Floor
Philadelphia, PA 19106-4404

AAFA Investigator Research Grants 128
AAFA is dedicated to finding the causes, new treatments and cures for asthma and allergic diseases. AAFA offers a program that sponsors seed grants for investigators wishing to explore new areas of scientific merit related to asthma and allergic diseases. Support has enabled 80% of the researchers who have received the AAFA research grant to win additional research funding from other sources totaling more than $18 million. The Asthma and Allergy of Foundation of America is the only patient organization in the United States that funds research grants for both asthma and allergies. At this time, the application process for grants is through an NIH referral system.
Geographic Focus: All States
Samples: Stephen C. Dreskin, MD, PhD, University of Colorado, Boulder, Colorado; Anthony A. Horner, MD, University of California, San Diego, CA--Analysis of airway; David Lewis, MD, Stanford University School of Medicine--Mechanisms of viral induced asthma.
Contact: Mary Brasler, (202) 466-7643, ext. 238; mbrasler@aafa.org
Internet: http://www.aafa.org/display.cfm?id=6&sub=64
Sponsor: Asthma and Allergy Foundation of America
8201 Corporate Drive, Suite 100
Landover, MD 20785

AAFCS International Graduate Fellowships 129
The Association encourages graduate study in family and consumer sciences and its subspecialties, such as textiles and clothing, nutrition, institutional management/food service systems administration, consumer studies, Cooperative Extension, and family and consumer sciences communication, education, and administration. International graduate fellowships are awarded to individuals who have exhibited the potential to make contributions to the family and consumer sciences profession.
Requirements: Applicants must be citizens or permanent residents of the United States. Applicant must show clearly defined plans for full-time graduate or undergraduate study during the time for which the fellowship is awarded. Each request for appropriate application form must be accompanied by an application fee of $40.
Geographic Focus: All States
Contact: Fellowships Committee; (703) 706-4600; fax (703) 706-4663; staff@aafcs.org
Internet: http://www.aafcs.org/programs/scholarships.htm
Sponsor: American Association of Family and Consumer Sciences
400 N Columbus Street, Suite 202
Alexandria, VA 22314

AAFCS National Graduate Fellowships — 130
AAFCS annually awards national graduate fellowships to support graduate study in areas such as family and consumer sciences, nutrition, textiles, and home economics. The fellowships are awarded to individuals who have exhibited the potential for making contributions to these professions. It is recommended that applicants first complete at least one year of professional family and consumer science experience by the time of application.
Requirements: Applicants must be citizens or permanent residents of the United States. Applicant must show clearly defined plans for full-time graduate or undergraduate study during the time for which the fellowship is awarded. Each request for appropriate application form must be accompanied by an application fee of $40.
Geographic Focus: All States
Date(s) Application is Due: Jan 7
Amount of Grant: Up to 5,000 USD
Contact: Fellowships Committee; (703) 706-4600; fax (703) 706-4663; staff@aafcs.org
Internet: http://www.aafcs.org/programs/fellowshipseven.html
Sponsor: American Association of Family and Consumer Sciences
400 N Columbus Street, Suite 202
Alexandria, VA 22314

AAFCS National Undergraduate Scholarships — 131
The Association encourages undergraduate study in family and consumer sciences and its subspecialties, such as textiles and clothing, nutrition, institutional management/food service systems administration, consumer studies, Cooperative Extension, and family and consumer sciences communication, education, and administration. Scholarships are awarded to individuals who have exhibited the potential to make contributions to the family and consumer sciences profession.
Requirements: Applicants must be citizens or permanent residents of the United States. Applicant must show clearly defined plans for full-time graduate or undergraduate study during the time for which the fellowship is awarded. Each request for appropriate application form must be accompanied by an application fee of $40.
Geographic Focus: All States
Date(s) Application is Due: Jan 7
Amount of Grant: 5,000 USD
Contact: Scholarships Committee; (703) 706-4600; fax (703) 706-4663; staff@aafcs.org
Internet: http://www.aafcs.org/programs/scholarships.htm
Sponsor: American Association of Family and Consumer Sciences
400 N Columbus Street, Suite 202
Alexandria, VA 22314

AAFP Foundation Health Literacy State Grants — 132
The AAFP Foundation recognizes that nearly half of all American adults (90 million people) have difficulty reading, understanding and acting upon all types of health information. They are often ashamed and devise methods to mask their difficulty or may be reluctant to ask questions for fear of being perceived as ignorant. To help address literacy concerns, the AAFP Foundation has developed the Health Literacy State Grant Awards Program. The purpose of the program is to support collaborative projects that address health literacy concerns at the state and local level. Five awards in the amount $10,000 each will be offered.
Geographic Focus: All States
Date(s) Application is Due: Mar 26
Amount of Grant: 10,000 USD
Contact: Susie Morantz, Senior Program Manager; (913) 906-6000, ext. 4470 or (800) 274-2237; fax (913) 906-6095; smorantz@aafp.org
Internet: http://www.aafpfoundation.org/x821.xml
Sponsor: American Academy of Family Physicians Foundation
11400 Tomahawk Creek Parkway, Suite 440
Leawood, KS 66211-2672

AAFP Foundation Joint Grants — 133
The foundation supports clinical research projects within family practice that ultimately will result in improved patient care. This program includes a research grant stimulation program that is designed to stimulate research in family practice by supporting small and first-time research projects. Longer research proposals must be prepared in accordance with standard guidelines for grant proposals.
Requirements: Eligible to apply are individual family physicians, residents in family practice, and medical students planning to pursue the specialty of family practice; educational and health care institutions or organizations that will use the grant exclusively for programs or projects directly involved with family practice or family medicine; departments of family practice; family practice residency programs; and family practice organizations or associations. Research stimulation proposals may be reviewed at any time during the year depending on availability of funds.
Restrictions: Proposals seeking support for the cost of instituting programs or support for such activities as videotape production, designing curriculum, or implementing a project, etc., will be denied.
Geographic Focus: All States
Amount of Grant: Up to 30,000 USD
Contact: Susie Morantz, Senior Program Manager; (913) 906-6000, ext. 4470 or (800) 274-2237; fax (913) 906-6095; smorantz@aafp.org
Internet: http://www.aafpfoundation.org/x270.xml
Sponsor: American Academy of Family Physicians Foundation
11400 Tomahawk Creek Parkway, Suite 440
Leawood, KS 66211-2672

AAFP Foundation Pfizer Teacher Development Awards — 134
The purpose of the awards is to honor community-based physicians who give up time from their practice to teach family medicine students and/or residents part time. Eligible candidates may serve as preceptors or as volunteer teachers at other sites, including family medicine teaching centers. Guidelines are available online.
Requirements: Applicant must have graduated from an ACGME-approved family practice residency program within the last seven years; have entered family medicine; be a member of the American Academy of Family Physicians; have entered or plan to enter part-time family medicine teaching (not less than four hours per month and no more than 32 hours per month--averaged over a year); and teach voluntarily or receive no more than $18,000 compensation for the educational time devoted to residents and/or students.
Geographic Focus: All States
Amount of Grant: 2,000 USD
Contact: Sondra Goodman, Grants Program Manager; (913) 906-6000, ext. 4457 or (800) 274-2237; fax (913) 906-6095; sgoodman@aafp.org
Internet: http://www.aafpfoundation.org/x269.xml
Sponsor: American Academy of Family Physicians Foundation
11400 Tomahawk Creek Parkway, Suite 440
Leawood, KS 66211-2672

AAFP Foundation Wyeth Immunization Awards — 135
This program offers family medicine residency programs the opportunity to receive recognition for identifying and developing creative solutions to overcoming barriers to childhood immunizations, thus increasing immunization rates, and to share the best practices that have benefited the communities in which they live. Applications are accepted annually for one of the following award track categories: Best Practices Award--overall achievement with systems already in place to overcome immunization barriers and achieve high rates within a certain time frame; Most Improved Award--overcame barriers and other challenges to greatly enhance immunization rates; and Implement New System Grant--residency programs seeking to implement a new system to help increase immunization rates in under-served children. Grant winners will receive: one of 10 monetary awards (four at $5,000 and six at $10,000) determined by rank of application as scored by review panel; an education scholarship for $1,000 to send a resident to the annual AAFP National Conference of Family Medicine Residents and Medical Students; an additional travel scholarship to send a resident to the AAFP Scientific Assembly (top six); and a plaque of recognition.
Requirements: Family medicine residency programs that have increased immunization rates within a defined time frame and can demonstrate the best practices implemented to achieve the higher compliance are eligible.
Geographic Focus: All States
Date(s) Application is Due: Apr 21
Amount of Grant: 1,000 - 10,000 USD
Contact: Sondra Goodman, Grants Program Manager; (913) 906-6000, ext. 4457 or (800) 274-2237; fax (913) 906-6095; sgoodman@aafp.org
Internet: http://www.aafpfoundation.org/x655.xml
Sponsor: American Academy of Family Physicians Foundation
11400 Tomahawk Creek Parkway, Suite 440
Leawood, KS 66211-2672

AAFP Research Stimulation Grants — 136
This Program provides relatively small grants to stimulate research in family practice in the areas of preliminary efforts leading to a larger research project, pilot projects, and data collection for a larger research project. The applicant must include a discussion of how the proposed project is anticipated to lead to a larger project. Grant applications will be accepted twice a year: by March 1 and September 1.
Requirements: Eligible to apply are individual family physicians, family practice/medicine organizations or associations, residents in family practice, departments of family medicine, and educational and health care institutions directly involved in family practice/medicine.
Geographic Focus: All States
Date(s) Application is Due: Mar 1; Sep 1
Amount of Grant: 7,500 USD
Contact: Susie Morantz, Senior Program Manager; (913) 906-6000, ext. 4470 or (800) 274-2237; fax (913) 906-6095; smorantz@aafp.org
Internet: http://www.aafpfoundation.org/x446.xml
Sponsor: American Academy of Family Physicians Foundation
11400 Tomahawk Creek Parkway, Suite 440
Leawood, KS 66211-2672

AAFPRS Ben Shuster Memorial Award — 137
This award is given annually for the best paper based on clinical work or research in the field of facial plastic surgery by a resident or fellow in training. The paper must have been delivered at a national meeting (or its equivalent) within the preceding two years prior to the February 1 date of submission. Studies prepared during the first year after completion of residency training will be considered provided that the research was conducted during the author's residency training program or fellowship. A certificate and an award of $1,000 are presented.
Requirements: The competition is open to U.S. or Canadian residents or fellows in otolaryngology who are members of AAFPRS. Each entrant must be the sole or senior author and an AAFPRS member.
Geographic Focus: All States
Date(s) Application is Due: Feb 1
Amount of Grant: 1,000 USD

Contact: Awards Coordinator; (703) 299-9291; fax (703) 299-8898; info@aafprs.org
Internet: http://www.aafprs.org/physician/awards_grants/awards.html
Sponsor: American Academy of Facial, Plastic, and Reconstructive Surgery
310 S Henry Street
Alexandria, VA 22314

AAFPRS Bernstein Grant 138
The purpose of the grant is to encourage original research projects that will advance facial plastic and reconstructive surgery. The grant may be used as seed money for research projects. Applicants should submit an application, a detailed proposal, the investigator's curriculum vita, and letters from the appropriate institutional review committees certifying institutional conformity to the U.S. government guidelines for human and animal experiments. The recipient should complete his or her work within three (3) years with interim reports due each year. All research grant monies awarded are intended to fund direct costs related to the research as described in the grant proposal.
Requirements: Applicants should be U.S. or Canadian AAFPRS fellow members. The primary criteria are that the research be original and have direct application to facial plastic and reconstructive surgery.
Geographic Focus: All States
Date(s) Application is Due: Jan 15
Amount of Grant: 25,000 USD
Contact: Coordinator; (703) 299-9291; fax (703) 299-8898; info@aafprs.org
Internet: http://www.aafprs.org/physician/awards_grants/bernstein_grant.html
Sponsor: American Academy of Facial, Plastic, and Reconstructive Surgery
310 S Henry Street
Alexandria, VA 22314

AAFPRS Fellowships 139
The Educational and Research Foundation for the American Academy of Facial Plastic and Reconstructive Surgery fellowship program provides postgraduate training in facial plastic surgery. The objectives of the fellowship program are to: provide an outstanding academic opportunity for the acquisition of specialized knowledge and skills in facial plastic surgery; develop trained specialists who will contribute to the ongoing development of facial plastic and reconstructive surgery; foster development of facial plastic and reconstructive surgery educators, especially in residency programs; and encourage the development of new skills and knowledge in facial plastic and reconstructive surgery through basic research and clinical trials. Eligible physicians may apply for 40 positions available each year.
Requirements: Applicants to the Fellowship Program must be physicians who are in or have completed an otolaryngology or plastic surgery residency program accredited by the Accreditation Council for Graduate Medical Education (ACGME) or Royal College of Physicians and Surgeons of Canada (RCPSC) or board-certified in otolaryngology-head and neck surgery or plastic surgery. Applicants must be members of the AAFPRS before submitting an application.
Restrictions: Applicants should not be full-time faculty members holding the rank of assistant professor or higher at the institution where the fellowship will take place.
Geographic Focus: All States
Date(s) Application is Due: Feb 1
Contact: Fellowship Coordinator; (703) 299-9291; fax (703) 299-8898; info@aafprs.org or fsanders@aafprs.org
Internet: http://www.aafprs.org/physician/benefits/ben_fellow.html
Sponsor: American Academy of Facial, Plastic, and Reconstructive Surgery
310 S Henry Street
Alexandria, VA 22314

AAFPRS Investigator Development Grant 140
The purpose of the Investigator Development Grant is to support the work of a young faculty member in facial plastic surgery conducting significant clinical or laboratory research and involved in the training of resident surgeons in research. One $15,000 grant may be awarded each year upon receipt of the signed agreement. Recipients should complete their work within three (3) years with interim reports due each year. An extension is possible upon written request to the Research Committee. The original proposal and all other requirements should be submitted to C.O.R.E. no later than January 15 or the next business day thereafter.
Requirements: An applicant must be an AAFPRS member at any level involved in the training of resident surgeons.
Geographic Focus: All States
Date(s) Application is Due: Jan 15
Amount of Grant: Up to 15,000 USD
Contact: Coordinator; (703) 299-9291; fax (703) 299-8898; info@aafprs.org
Internet: http://www.aafprs.org/physician/awards_grants/investigator_grant.html
Sponsor: American Academy of Facial, Plastic, and Reconstructive Surgery
310 S Henry Street
Alexandria, VA 22314

AAFPRS Ira J. Tresley Research Award 141
This annual award is given to a U.S. or Canadian academy member for the best paper based on any research study in the field of facial plastic surgery. The competition is open to any U.S. or Canadian physician who is an AAFPRS member and who has been board certified for at least three years. The paper must have been presented at a national meeting or its equivalent within the preceding two years prior to the February 1 date of submission. A certificate and an award of $1,000 are presented.
Geographic Focus: All States
Date(s) Application is Due: Feb 1
Amount of Grant: 1,000 USD
Contact: Awards Coordinator; (703) 299-9291; fax (703) 299-8898; info@aafprs.org
Internet: http://www.aafprs.org/physician/awards_grants/awards.html
Sponsor: American Academy of Facial, Plastic, and Reconstructive Surgery
310 S Henry Street
Alexandria, VA 22314

AAFPRS John Orlando Roe Award 142
This award, named for John Orlando Roe, the surgeon who accomplished the first rhinoplasty in 1887, includes a certificate and an award of $1,000 to be presented each year to an AAFPRS Fellow in an AAFPRS Foundation Fellowship Program who submits the best clinical research paper written during the current fellowship year.
Geographic Focus: All States
Date(s) Application is Due: Feb 1
Amount of Grant: 1,000 USD
Contact: Awards Coordinator; (703) 299-9291; fax (703) 299-8898; info@aafprs.org
Internet: http://www.aafprs.org/physician/awards_grants/awards.html
Sponsor: American Academy of Facial, Plastic, and Reconstructive Surgery
310 S Henry Street
Alexandria, VA 22314

AAFPRS Leslie Bernstein Resident Research Grants 143
Two resident research grants, tenable in the United States or Canada, may be awarded each year to stimulate resident research in facial plastic surgery on projects that are well conceived and scientifically valid. Residents are encouraged to enter early in their training so that their applications may be revised and resubmitted if not accepted the first time. This award may be integrated with other funding to complete a project. Interested residents should write for entry guidelines.
Requirements: U.S. and Canadian residents at any level who are AAFPRS members are eligible to apply, even if the research work will be done during their fellowship year. Grant recipients must complete their work in one (1) to two (2) years with interim reports due each year.
Geographic Focus: All States
Date(s) Application is Due: Jan 15; Dec 15
Amount of Grant: 5,000 USD
Contact: Coordinator; (703) 299-9291; fax (703) 299-8898; info@aafprs.org
Internet: http://www.aafprs.org/physician/awards_grants/grants.html
Sponsor: American Academy of Facial, Plastic, and Reconstructive Surgery
310 S Henry Street
Alexandria, VA 22314

AAFPRS Residency Travel Awards 144
Two Residency Travel Awards may be given each year for two outstanding papers in facial plastic and reconstructive surgery primarily authored by a resident or medical student in training. The paper must be submitted by February 1 for consideration, which will be presented at the Annual Fall Meeting. Entries should conform to the guidelines of the Archives of Facial Plastic Surgery, the official journal of the AAFPRS. Papers submitted will be judged anonymously. Only the cover page should contain the applicant's name or institutional identification.
Geographic Focus: All States
Date(s) Application is Due: Feb 1
Contact: Awards Coordinator; (703) 299-9291; fax (703) 299-8898; info@aafprs.org
Internet: http://www.aafprs.org/physician/awards_grants/awards.html
Sponsor: American Academy of Facial, Plastic, and Reconstructive Surgery
310 S Henry Street
Alexandria, VA 22314

AAFPRS Sir Harold Delf Gillies Award 145
This award is named for Sir Harold Delf Gillies, a British otolaryngologist who in September 1917 described the tubed pedicle flap. Dr. Gillies frequently visited the U.S. and lectured widely to surgeons of various specialties and was given the title Father of Plastic Surgery. A certificate and an award of $1,000 will be presented each year to an AAFPRS Fellow in an AAFPRS Foundation Fellowship Program who submits the best basic science research paper written during the current fellowship year.
Geographic Focus: All States
Date(s) Application is Due: Feb 1
Amount of Grant: 1,000 USD
Contact: Awards Coordinator; (703) 299-9291; fax (703) 299-8898; info@aafprs.org
Internet: http://www.aafprs.org/physician/awards_grants/awards.html
Sponsor: American Academy of Facial, Plastic, and Reconstructive Surgery
310 S Henry Street
Alexandria, VA 22314

AAHN Grant for Historical Research 146
The American Association for the History of Nursing is offering a research grant of $3,000 for new researchers. (Indirect costs of 8% are also available). It is expected that the research and new materials produced by the award winner will help ensure the growth of scholarly work focused on the history of nursing. The deadline for submission of applications is April 1. The AAHN Panel will make its decision about the award by May 15 and the recipient will be notified by June 1 of each year. Funding will start July 1 of the grant year and last for one year. A no-cost extension may be granted on request.

22 | Grant Programs

Requirements: Applicants must be members of AAHN and hold the doctorate. They may be faculty members or independent researchers.
Geographic Focus: All States
Date(s) Application is Due: Apr 1
Amount of Grant: 3,000 USD
Contact: David Stumph; (303) 422-2685; fax (303) 422-8894; aahn@aahn.org
Internet: http://www.aahn.org/grants.html
Sponsor: American Association of the History of Nursing
10200 W 44th Avenue, Suite 304
Wheat Ridge, CO 80033

AAHN Pre-Doctoral Research Grant 147
This grant is designed to encourage and support graduate training and historical research at the Masters and Doctoral levels. The grant will be in the amount of $2,000. Proposals will focus on a significant question in the history of nursing. Selection criteria include the scholarly merit of the proposal, consideration of the student's preparation for this study, the advisor's qualifications for guiding the study and the project's potential for contributing to scholarship in the field of nursing history.
Requirements: The student must be: enrolled in an accredited masters program or doctoral program; and a member of AAHN. The research advisor must be doctorally prepared with scholarly activity in the field of nursing history and prior experience in guidance of research training.
Geographic Focus: All States
Date(s) Application is Due: Apr 1
Amount of Grant: 2,000 USD
Contact: David Stumph; (303) 422-2685; fax (303) 422-8894; aahn@aahn.org
Internet: http://www.aahn.org/grants.html
Sponsor: American Association of the History of Nursing
10200 W 44th Avenue, Suite 304
Wheat Ridge, CO 80033

AAHPERD-AAHE Barbara A. Cooley Scholarships 148
The award will be $1,000 per scholarship. Recipients may spend the award in any manner they desire. The awarded scholarship will be recognized at the annual Awards Luncheon at the AAHE/AAHPERD annual meeting. Recipients will also receive a one-year complimentary student membership in AAHE. Additional guidelines and application are available at the AAHPERD website. Application materials must be received by November 15.
Requirements: The award is open to a master's level student who is currently enrolled in a health education program at an accredited college/university in the United States or a U.S. territory. To be eligible, the applicant must have a minimum current overall grade point average of 3.0 on a 4.0 scale.
Restrictions: Prior AAHE scholarship recipients are not eligible.
Geographic Focus: All States
Date(s) Application is Due: Nov 15
Amount of Grant: 1,000 USD
Contact: Linda Moore; (703) 476-3437; fax (703) 476-6638; aahe@aahperd.org
Internet: http://www.aahperd.org/aahe/events/Barbara-Cooley.cfm
Sponsor: American Alliance for Health, Physical Education, Recreation and Dance
1900 Association Drive
Reston, VA 20191-1599

AAHPERD-AAHE Delbert Oberteuffer Scholarships 149
The award will be $1,500, which may be used at the discretion of the recipient. The scholarship recipient will be recognized at the annual Awards Luncheon at the AAHE/AAHPERD convention. Additional guidelines and application are available at the AAHPERD website. Application materials must be received by November 15.
Requirements: The award is open to a doctoral level student who is currently enrolled in a health education program in the United States or a U.S. territory. To be eligible to apply, you must be an AAHE member and a graduate student at an accredited college/university in a doctoral level course of study leading to preparing you to develop, implement and/or evaluate health education programs for children and youth. The applicant must also have a minimum cumulative grade point average of 3.5 in graduate courses.
Geographic Focus: All States
Date(s) Application is Due: Nov 15
Amount of Grant: 1,500 USD
Contact: Linda Moore; (703) 476-3437; fax (703) 476-6638; aahe@aahperd.org
Internet: http://www.aahperd.org/aahe/events/awards/Delbert-Oberteuffer.cfm
Sponsor: American Alliance for Health, Physical Education, Recreation and Dance
1900 Association Drive
Reston, VA 20191-1599

AAHPERD-AAHE Undergraduate Scholarships 150
AAHE Undergraduate Scholarships are available to an outstanding student officially recognized as an undergraduate health education major at any four-year university/college. These are one year scholarships.
Requirements: Open to undergraduate students who are enrolled full time in a health education program at a 4-year college or university. Applicants must a GPA of 3.25 or higher and be active in health education professional activities and organizations at their school and/or in their community. They must submit a list of extracurricular service activities; an official transcript; 3 letters of recommendation; and an essay that includes what they hope to accomplish as a health educator (during training and in the future) and attributes and aspirations brought to the field of health education. Financial need is not considered in the selection process.

Geographic Focus: All States
Contact: Linda Moore; (703) 476-3437; fax (703) 476-6638; aahe@aahperd.org
Internet: http://www.aahperd.org/whatwedo/scholarships/Abernathy_Scholarship.cfm
Sponsor: American Alliance for Health, Physical Education, Recreation and Dance
1900 Association Drive
Reston, VA 20191-1599

AAHPERD Ruth Abernathy Presidential Scholarships 151
The Ruth Abernathy Presidential Scholarship, developed by the past presidents of AAHPERD to honor deserving students, is awarded to three undergraduate students and two graduate students in January of each year. All scholarships are presented at the AAHPERD National Convention & Exposition held in the spring. Undergraduate awards are $1,250 each and graduate awards are $1,750 each. Recipients also receive a complimentary three-year AAHPERD membership.
Requirements: Applicants must be enrolled in a full-time program at the time that the scholarship recipients are selected and must be majoring in a field related to one or more of the disciplines represented by AAHPERD and its associations (i.e., health, physical education, recreation, dance, sport - see the list of interest areas at the website). Applicants will be asked to describe the field of study/degree area and its relationship to the HPERDS-related fields. Application forms containing detailed application criteria and procedures for submission are available at the website.
Restrictions: Individuals in the following medical programs are not eligible: Nursing, Pharmacology, Physical Therapy, Occupational Therapy, Athletic Training, and Acupuncture and Oriental Medicine. Students pursuing teaching licensure, national board certification, or completing prerequisite coursework in preparation for admission to a master's degree program are are also not eligible to apply.
Geographic Focus: All States
Date(s) Application is Due: Oct 15
Amount of Grant: 1,750 - 1,250 USD
Contact: Deb Callis, Director; (703) 476-3400; dcallis@aahperd.org
Internet: http://www.aahperd.org/whatwedo/scholarships/Abernathy_Scholarship.cfm
Sponsor: American Alliance for Health, Physical Education, Recreation and Dance
1900 Association Drive
Reston, VA 20191-1599

AAHPM Hospice and Palliative Medicine Fellowships 152
The program offers training programs to prepare physicians for careers in academic or community-based palliative medicine. Fellowships provide clinical experience and supervision for palliative medicine in hospitals, homes, hospices, chronic care institutions, and outpatient settings, and include consultational and longitudinal care exposure, as well as training in bereavement care. Some programs also provide opportunities for mentored research or teaching. Most programs begin in July, and clinical training generally lasts one year, allowing for board eligibility upon successful completion. For a complete list of U.S. palliative medicine fellowship training programs, members should visit the web site.
Requirements: Trainees must have already completed residency training in a primary specialty, such as internal or family medicine, or have similar clinical training and experience.
Geographic Focus: All States
Contact: Fellowships Director; (847) 375-4712; fax (847) 375-6475; info@aahpm.org
Internet: http://www.aahpm.org/fellowship/index.html
Sponsor: American Academy of Hospice and Palliative Medicine
4700 W Lake Avenue
Glenview, IL 60025-1485

AAMC Abraham Flexner Award 153
The Abraham Flexner Award for Distinguished Service to Medical Education is granted annually to an individual for extraordinary contributions to the medical education community. The award was first presented by the Association in 1958 and is the AAMC's most prestigious honor. The Flexner Award recognizes the highest standards in medical education and honors individuals whose impact on medical education is national in scope. Nominations may be made by any faculty or staff member of a medical school or teaching hospital or by any member of an academic society. The Award Review Committee may also solicit nominations.
Requirements: Any individual who has made a significant contribution to academic medicine is eligible for nomination.
Geographic Focus: All States
Date(s) Application is Due: May 1
Amount of Grant: 10,000 USD
Contact: Sandra D. Gordon, (202) 828-0472; fax (202) 828-1125; sgordon@aamc.org
Internet: http://www.aamc.org/about/awards/flexner.htm
Sponsor: Association of American Medical Colleges
2450 N Street NW
Washington, D.C. 20037-1127

AAMC Alpha Omega Alpha Robert J. Glaser Distinguished Teacher Awards 154
The award recognize the significant contributions to medical education made by gifted teachers. Up to four awards will be granted each year. Each awardee will receive a $10,000 grant. The awardee's nominating institution will receive $2,500 for teaching activities, and, if the nominating institution has an AOA chapter, the chapter will receive a stipend of $1,000 toward its activities. The awards will be presented during the AAMC Annual Meeting in Washington, D.C.
Requirements: The nomination must be made by the Dean of the nominating institution.
Geographic Focus: All States
Date(s) Application is Due: May 1
Amount of Grant: 12,500 USD

Contact: Henry M. Sondheimer, (202) 828-0680; hsondheimer@aamc.org
Internet: http://www.aamc.org/about/awards/aoa.htm
Sponsor: Association of American Medical Colleges
2450 N Street NW
Washington, D.C. 20037-1127

AAMC Award for Distinguished Research 155
The Award for Distinguished Research in the Biomedical Sciences honors outstanding biomedical research related to health and disease. The research recognized should have contributed to the substance of medicine. Nominations may be made by anyone on the faculty or staff of a medical school or teaching hospital or by a member of an academic society.
Requirements: The nominee must be an individual who serves on the faculty of an AAMC member medical school or teaching hospital.
Geographic Focus: All States
Date(s) Application is Due: May 1
Amount of Grant: 5,000 USD
Contact: Sandra D. Gordon, (202) 828-0472; fax (202) 828-1125; sgordon@aamc.org
Internet: http://www.aamc.org/about/awards/research.htm
Sponsor: Association of American Medical Colleges
2450 N Street NW
Washington, D.C. 20037-1127

AAMC Caring for Community Grants 156
The association, with the support of the Pfizer Medical Humanities Initiative, conducts an institutional grant program to encourage the development of student-initiated services and programs to the community. Eligible programs may range from those that promote awareness about sexually transmitted diseases, to vaccination and literacy programs, to any program that fulfills an unmet need within the community. Grant awards will also be offered to eligible service programs that are currently underway. Up to 10 grant awards are made each year. Program requirements and applications are available online.
Requirements: Medical schools are eligible to receive support for community service-oriented projects in which they explore new ways to serve their local communities.
Geographic Focus: All States
Date(s) Application is Due: Mar 12
Amount of Grant: 2,000 - 15,000 USD
Contact: Ally Anderson, (202) 828-0682; fax (202) 828-1125; aanderson@aamc.org
Internet: http://www.aamc.org/about/awards/cfc/start.htm
Sponsor: Association of American Medical Colleges
2450 N Street NW
Washington, D.C. 20037-1127

AAMC David E. Rogers Award 157
The Award honors Dr. David Rogers, a former president of the Foundation and an exemplar of academic medicine's commitment to meeting the health care needs of our country. The award is granted annually to a member of a medical school faculty who has made major contributions to improving the health and health care of the American people. Presentation of the award and a will be made at the association's annual meeting. Nomination guidelines are available online.
Requirements: Nominations may be made by any faculty or staff member of a medical school or teaching hospital or by any member of an academic society. The award is limited to an individual who has spent the majority of his or her career in academic medicine in the United States.
Geographic Focus: All States
Date(s) Application is Due: May 1
Amount of Grant: 10,000 USD
Contact: Sandra D. Gordon, (202) 828-0472; fax (202) 828-1125; sgordon@aamc.org
Internet: http://www.aamc.org/about/awards/rogers.htm
Sponsor: Association of American Medical Colleges
2450 N Street NW
Washington, D.C. 20037-1127

AAMC Herbert W. Nickens Award 158
The annual award is given to an individual who has made outstanding contributions to promoting justice in medical education and health care for people in the United States. The recipient receives a monetary award and presents the Nickens Memorial Lecture at the association's annual meeting. Nomination guidelines are available online.
Requirements: Nominees may come from a wide range of fields including medicine, public health, education, law, nursing, and the social sciences.
Geographic Focus: All States
Date(s) Application is Due: May 1
Amount of Grant: 10,000 USD
Contact: Juan Amador, (202) 862-6149; fax (202) 828-1125; jamador@aamc.org
Internet: http://www.aamc.org/about/awards/nickensaward.htm
Sponsor: Association of American Medical Colleges
2450 N Street NW
Washington, D.C. 20037-1127

AAMC Herbert W. Nickens Faculty Fellowship 159
The award recognizes an outstanding junior faculty member who has demonstrated leadership in the United States in addressing inequities in medical education and health care; demonstrated efforts in addressing educational, societal, and health care needs of minorities; and is committed to a career in academic medicine. Funding for the fellowship begins on the year after the fellowship is awarded, and it can be used over a two year period. The recipient is required to submit final narrative and financial reports. Nomination guidelines are available onnline.
Requirements: A medical school may nominate one faculty member for this award. A candidate must: be a U.S. citizen or permanent resident; hold the rank of full-time assistant professor in a LCME-accredited U.S. medical school department; have held a faculty position for no more than three years; have received only one appointment as assistant professor; and hold a M.D., Ph.D., or have earned another doctoral degree.
Geographic Focus: All States
Date(s) Application is Due: May 1
Amount of Grant: 15,000 USD
Contact: Juan Amador, (202) 862-6149; fax (202) 828-1125; jamador@aamc.org
Internet: http://www.aamc.org/about/awards/nickensfellowships.htm
Sponsor: Association of American Medical Colleges
2450 N Street NW
Washington, D.C. 20037-1127

AAMC Herbert W. Nickens Medical Student Scholarships 160
The association makes scholarship awards that are given to outstanding third-year medical students who have demonstrated leadership in efforts to eliminate inequities in medical education and health care, and who have demonstrated leadership efforts in addressing educational, societal, and health care needs of minorities in the United States. Each recipient receives a scholarship and an award certificate. Institutions that have a pass/fail grading system and do not assign honor grades should fully discuss the applicant's academic accomplishments in the nomination letter. Nomination guidelines are available online.
Requirements: A medical school may nominate one student. A candidate must be: a U.S. citizen or permanent resident; and entering the third year of study in a LCME-accredited U.S. medical school in the fall semester. Students enrolled in combined degree programs (such as M.D./Ph.D.) are eligible when they are entering their third year of medical school.
Geographic Focus: All States
Date(s) Application is Due: May 1
Amount of Grant: 5,000 USD
Contact: Juan Amador, (202) 862-6149 or (202) 828-0400; fax (202) 828-1125; jamador@aamc.org or nickensawards@aamc.org
Internet: http://www.aamc.org/about/awards/nickensscholarships.htm
Sponsor: Association of American Medical Colleges
2450 N Street NW
Washington, D.C. 20037-1127

AAMC Humanism in Medicine Award 161
The award recognizes a medical school faculty physician who exemplifies the qualities of a caring and compassionate mentor in the teaching and advising of medical students. The goal of the award is to emphasize and reinforce among medical school faculty and students the importance of humanistic qualities in physicians, and to enhance interactions between medical school faculty and students. Nominations and award information were sent to medical school's organization of student representatives and student affairs deans. The award will be presented at the association's annual meeting.
Requirements: Each medical school in the United States may nominate one physician faculty member.
Geographic Focus: All States
Date(s) Application is Due: Apr 16
Amount of Grant: Up to 5,000 USD
Samples: Yasmin S. Meah, M.D., Assistant Professor of Medical Education and Medicine, The Mount Sinai Visting Doctors Program, Mount Sinai School of Medicine; Robert J. Paeglow, M.D., Albany Medical College; Melissa A. Warfield, M.D., Eastern Virginia Medical School (EVMS); Sharad Jain, M.D., University of California, San Francisco School of Medicine.
Contact: Denine Hales, (202) 828-0681; fax (202) 828-1125; dhales@aamc.org
Internet: http://www.aamc.org/about/awards/humanism.htm
Sponsor: Association of American Medical Colleges
2450 N Street NW
Washington, D.C. 20037-1127

AANS Neurosurgery Research & Education Foundation/Spine Section Young Clinician Investigator Award 162
Established by the Neurosurgery Research and Education Foundation (NREF) of the American Association of Neurological Surgeons (AANS), the Joint Section on Disorders of the Spine and Peripheral Nerves Award grants support to young faculty who are pursuing careers in the area of spinal cord, vertebral column or peripheral nerve disorders. The purpose of the award is to fund pilot studies that provide preliminary data used to strengthen applications for more permanent funding from other sources. The one-year award totals $40,000.
Requirements: Applicants must be physicians or neurosurgeons who are full-time faculty in North American teaching institutions and in the early years of their careers.
Restrictions: Applicant may not accept another award for the same project during the same time period. No more than one award per year will be awarded to the same institution.
Geographic Focus: All States
Date(s) Application is Due: Oct 31
Contact: Michele S. Gregory, Director of Development; (847) 378-0540 or (847) 378-0500; fax (847) 378-0600; msg@aans.org or info@aans.org
Internet: http://www.aans.org/research/fellowship/nref_y.asp
Sponsor: Research Foundation of the American Association of Neurological Surgeons
5550 Meadowbrook Drive
Rolling Meadows, IL 60008-3852

24 | Grant Programs

AANS Neurosurgery Research and Education Foundation Young Clinician Investigator Award 163

This one-year, renewable award is available to North American neurosurgeons who are full-time faculty in teaching institutions and are in the early years of their careers. The purpose of the program is to fund pilot studies that could provide preliminary data to be used for strengthening applications for more permanent funding from other sources. Applications related to any field of neurosurgery are encouraged. The NREF encourages both patient-oriented clinical research and basic science research projects. The one-year award totals $40,000.
Restrictions: Applicant may not accept another award for the same project during the same time period. No more than one award per year will be awarded to the same institution.
Geographic Focus: All States
Date(s) Application is Due: Oct 31
Amount of Grant: 40,000 USD
Contact: Michele S. Gregory; (847) 378-0540 or (847) 378-0500; msg@aans.org
Internet: http://www.aans.org/research/fellowship/nref_y.asp
Sponsor: Research Foundation of the American Association of Neurological Surgeons
5550 Meadowbrook Drive
Rolling Meadows, IL 60008-3852

AAO-HNSF CORE Research Grants 164

CORE Research Grants are available in three different categories: Research Project Grants--an independent or assisted project using existing skills and resources, and is hypothesis-based, with specific aims and timetable; Research Training Grants--a research project that is designed to help you refine existing research skill set or acquire new skills, and requires mentor and evidence of training process; and Career Development Grants--a formal, mentored period of research combined with coursework, designed to help researcher become independent researcher. Applicants should complete a Letter of Intent by December 15th, using the online application system, and should then submit the completed application before 12 o'clock midnight EST on January 15.
Geographic Focus: All States
Date(s) Application is Due: Jan 15; Dec 15
Contact: Stephanie L. Jones; (703) 519-1586 or (703) 836-4444; sljones@entnet.org
Internet: http://www.entnet.org/EducationAndResearch/coreGrants.cfm
Sponsor: American Academy of Otolaryngology-Head and Neck Surgery Foundation
1650 Diagonal Road
Alexandria, VA 22314-2857

AAO-HNSF Health Services Research Grants 165

The purpose of the Health Services Research Grant is to foster research that will improve the effectiveness and appropriateness of medical practice. Projects supported under this program will develop and disseminate scientific information on the effects of otolaryngology services and procedures on patients' survival, health status, functional capacity, and quality of life. The award is intended to promote increased participation by otolaryngologists in the rapidly expanding area of health services research. Proposed projects may be related to any area of otolaryngology-head and neck surgery, but must have direct or potential clinical significance for patients seen by otolaryngologist-head and neck surgeons. Grants range up to $10,000 maximum, are for one year, and are not renewable. Projects must be completed within two years of the award date; no-cost extensions are available upon written request. Letters of intent must be received by December 15, and completed applications must be submitted by January 15.
Requirements: Any otolaryngologist in the United States or Canada is eligible to apply for the Health Services Research Grant. Applicants may be independent practitioners, residents or fellows in an approved training program, or practitioners affiliated with academic or similar institutions. All applicants must be members in good standing of AAO-HNSF; Associate and Corresponding Members are not eligible to apply as Principal Investigator, but may participate actively in the proposed project.
Geographic Focus: All States
Date(s) Application is Due: Jan 15; Dec 15
Contact: Stephanie L. Jones; (703) 519-1586 or (703) 836-4444; sljones@entnet.org
Internet: http://www.entnet.org/EducationAndResearch/coreGrants.cfm
Sponsor: American Academy of Otolaryngology-Head and Neck Surgery Foundation
1650 Diagonal Road
Alexandria, VA 22314-2857

AAO-HNSF Maureen Hannley Research Training Awards 166

The purpose of the Maureen Hannley Research Training Award program is to foster the acquisition of contemporary basic or clinical research skills among new full-time academic otolaryngologist-head and neck surgeons. The award is intended as a preliminary step in clinical investigator career development and is expected to facilitate the recipient's preparation of a more comprehensive individualized research training plan suitable for submission to one of the National Institutes of Health Clinical Investigator Career Development Award (CIA) programs (K series). Grants range up to $15,000 maximum. The award is one year, and is non-renewable. Projects must be completed within two years of the award date; no-cost extensions are available upon written request. Letters of intent must be received by December 15, and completed applications must be submitted by January 15.
Requirements: Applicants in the United States or Canada and members in good standing of the AAO-HNSF are welcome. Applicants must have demonstrated potential for excellence in research and teaching and a serious commitment to an academic research career in otolaryngology-head and neck surgery. Priority will be given to senior residents, fellows or faculty who have completed residencies or fellowships within two years of the application receipt date. All candidates must be sponsored by the Chairperson of his/her Division or Department and by an official representative of the institution which would administer the Award and in whose name the application is formally submitted.
Geographic Focus: All States
Date(s) Application is Due: Jan 15; Dec 15
Amount of Grant: Up to 15,000 USD
Contact: Stephanie L. Jones; (703) 519-1586 or (703) 836-4444; sljones@entnet.org
Internet: http://www.entnet.org/EducationAndResearch/coreGrants.cfm
Sponsor: American Academy of Otolaryngology-Head and Neck Surgery Foundation
1650 Diagonal Road
Alexandria, VA 22314-2857

AAO-HNSF Percy Memorial Research Award 167

The Percy Memorial Research Award is an annual grant-in-aid of a worthy research project proposed in any area within the scope of otolaryngology-head and neck surgery. The Award was established in memory of A. Edward Percy, Jr., M.D. and his parents. Projects must have direct or potential clinical significance for patients seen by otolaryngologists-head and neck surgeons. They must be designed so as to yield useful information within the period of award, but priority will be given to projects which are also innovative and promise to develop into new long range research programs which will attract funding from other sources. Grants range up to $10,000 maximum, are for one year, and are not renewable. Letters of intent must be received by December 15, and completed applications must be submitted by January 15.
Requirements: Applicants in the United States or Canada and members in good standing of AAO-HNSF are welcome to apply. Applicants should be experienced independent otolaryngologist investigators affiliated with academic or similar institutions eligible to apply for and administer Federal research awards. Fellows with substantial research experience are eligible to apply, as are individuals who have already competed successfully for independent research grant support from a private or Federal funding agency.
Restrictions: Funds may not be requested to pay any portions of the salaries of the principal investigator or of any support personnel with strictly secretarial or clerical responsibilities. Funds may also not be used for the purchase of any item of equipment costing more than $500.
Geographic Focus: All States
Date(s) Application is Due: Jan 15; Dec 15
Amount of Grant: Up to 25,000 USD
Contact: Stephanie L. Jones; (703) 519-1586 or (703) 836-4444; sljones@entnet.org
Internet: http://www.entnet.org/EducationAndResearch/coreGrants.cfm
Sponsor: American Academy of Otolaryngology-Head and Neck Surgery Foundation
1650 Diagonal Road
Alexandria, VA 22314-2857

AAO-HNSF Rande H. Lazar Health Services Research Grant 168

The purpose of the Rande H. Lazar Health Services Research Grant is to support the gathering of socioeconomic data for otolaryngology. The award is intended to promote increased participation by otolaryngologists in the rapidly expanding area of health services research. Proposed projects may be related to any area of otolaryngology-head and neck surgery, but must have direct or potential clinical significance for patients seen by otolaryngologist-head and neck surgeons. Grants range up to $10,000 maximum, are for one year, and are not renewable. Letters of intent must be received by December 15, and completed applications must be submitted by January 15.
Requirements: Any otolaryngologist in the United States or Canada is eligible to apply for the Rande H. Lazar Health Services Research Grant. Applicants may be independent practitioners, residents or fellows in an approved training program, or practitioners affiliated with academic or similar institutions. All applicants must be members in good standing of AAO-HNSF; Associate and Corresponding Members are not eligible to apply as Principal Investigator, but may participate actively in the proposed project.
Geographic Focus: All States
Date(s) Application is Due: Jan 15; Dec 15
Amount of Grant: Up to 10,000 USD
Contact: Stephanie L. Jones; (703) 519-1586 or (703) 836-4444; sljones@entnet.org
Internet: http://www.entnet.org/EducationAndResearch/coreGrants.cfm
Sponsor: American Academy of Otolaryngology-Head and Neck Surgery Foundation
1650 Diagonal Road
Alexandria, VA 22314-2857

AAO-HNSF Resident Research Awards 169

The purpose of this grant is to stimulate original resident research in otolaryngology projects that are well-conceived and scientifically valid, with the potential to advance otolaryngology. Proposed projects may be related to any area of otolaryngology-head and neck surgery, and should be designed in collaboration with a preceptor investigator and approved by the candidate's department chairperson and institution. Grants range up to $10,000 maximum. Projects must be completed within one year of the award date; no-cost extensions are available upon written request. Letters of intent must be received by December 15, and completed applications must be submitted by January 15. Up to eight awards are available annually.
Requirements: The award is open to any AAO-HNS member in good standing who is a resident in an accredited otolaryngology - head and neck surgery training program in the U.S. and Canada.
Geographic Focus: All States
Date(s) Application is Due: Jan 15; Dec 15
Amount of Grant: Up to 10,000 USD
Contact: Stephanie L. Jones; (703) 519-1586 or (703) 836-4444; sljones@entnet.org
Internet: http://www.entnet.org/EducationAndResearch/coreGrants.cfm
Sponsor: American Academy of Otolaryngology-Head and Neck Surgery Foundation
1650 Diagonal Road
Alexandria, VA 22314-2857

AAO-HNS Medical Student Research Prize — 170
The Medical Student Research Prize is given annually for a manuscript submitted to the Editor of Otolaryngology-Head and Neck Surgery for review and consideration for publication in the journal. An award winning manuscript will be automatically accepted as an oral or poster presentation at the Translational and Basic Research Presentations Program (previously called the Research Forum). First prize is $1,000, and the award is available in both clinical and basic science categories through the generosity of GYRU.S. ACMI ENT Division.
Requirements: All manuscripts submitted for this competition must have a medical student as senior author and must describe work that has not been placed under active consideration for publication other than in Otolaryngology-Head and Neck Surgery.
Geographic Focus: All States
Date(s) Application is Due: Mar 9
Amount of Grant: 1,000 USD
Contact: Stephanie L. Jones; (703) 519-1586 or (703) 836-4444; sljones@entnet.org
Internet: http://www.entnet.org/educationandresearch/grantsandfellowships.cfm
Sponsor: American Academy of Otolaryngology-Head and Neck Surgery Foundation
1650 Diagonal Road
Alexandria, VA 22314-2857

AAO-HNS Resident Research Prizes — 171
Resident Research Prizes are given annually for manuscripts submitted to the Editor of Otolaryngology-Head and Neck Surgery for review and consideration for publication in the journal. All award winning manuscripts will be automatically accepted as oral or poster presentations at the Translational and Basic Research Presentations Program (previously called the Research Forum). First prize is $2,000, second prize is $1,500, and third prize is $1,000. All awards are available in both clinical and basic science categories through the generosity of GYRU.S. ACMI ENT Division.
Requirements: All manuscripts submitted for this competition must have a resident as senior author and must describe work that has not been placed under active consideration for publication other than in Otolaryngology-Head and Neck Surgery.
Geographic Focus: All States
Date(s) Application is Due: Mar 9
Contact: Stephanie L. Jones; (703) 519-1586 or (703) 836-4444; sljones@entnet.org
Internet: http://www.entnet.org/educationandresearch/grantsandfellowships.cfm
Sponsor: American Academy of Otolaryngology-Head and Neck Surgery Foundation
1650 Diagonal Road
Alexandria, VA 22314-2857

AAOA Foundation/AAO-HNSF Combined Research Grants — 172
The purpose of this award is to support a collaborative AAOA Foundation research project by fostering basic or clinical research related to otolaryngic allergy, rhinology, or related immunology by otolaryngologists-head and neck surgeons. Grants range up to $10,000 maximum. Projects must be completed within two years of the award date; no-cost extensions are available upon written request. Letters of intent must be received by December 15, and completed applications must be submitted by January 15.
Requirements: Applicants must have demonstrated potential for excellence in research. There are no restrictions on career stage, education, or country of residence. AAOA members are preferred but not required. All candidates must be sponsored by the Chairperson of his/her Division or Department and by an official representative of the institution that would administer the Award and in whose name the application is formally submitted.
Geographic Focus: All States
Date(s) Application is Due: Jan 15; Dec 15
Amount of Grant: Up to 10,000 USD
Contact: Stephanie L. Jones; (703) 519-1586 or (703) 836-4444; sljones@entnet.org
Internet: http://www.entnet.org/EducationAndResearch/coreGrants.cfm
Sponsor: American Academy of Otolaryngic Allergy Foundation
1990 M. Street NW, Suite 680
Washington, D.C. 20036

AAOHN Foundation Experienced Researcher Grants — 173
The foundation funds research focused on improving the health and safety of the nation's workforce. Research grants are awarded to support research in the field of occupational and environmental health by an experienced occupational and environmental health registered nurse. Proposals may be submitted that are in the early stages of development (prior to data collection). Two $10,000 grants are awarded annually.
Requirements: Experienced occupational and environmental health registered nurse researchers are eligible to apply.
Restrictions: Completed research projects will not be accepted for consideration.
Geographic Focus: All States
Date(s) Application is Due: Dec 1
Contact: Ann Cox; (770) 455-7757; fax (770) 455-7271; ann@aaohn.org
Internet: http://www.aaohn.org/foundation/grants/index.cfm
Sponsor: American Association of Occupational Health Nurses Foundation
2920 Brandywine Road, Suite 100
Atlanta, GA 30341

AAOHN Foundation New Investigator Researcher Grants — 174
The foundation funds research, scholarship, and leadership development grants focused on improving the health and safety of the nation's workforce. Grants are awarded to encourage research in the field of occupational and environmental health by a new or novice registered nurse principal investigator. Proposals may be submitted that are in the early stages of development (prior to data collection).
Restrictions: Completed research projects are ineligible.
Geographic Focus: All States
Date(s) Application is Due: Dec 1
Amount of Grant: 3,000 USD
Contact: Ann Cox; (770) 455-7757; fax (770) 455-7271; ann@aaohn.org
Internet: http://www.aaohn.org/foundation/grants/index.cfm
Sponsor: American Association of Occupational Health Nurses Foundation
2920 Brandywine Road, Suite 100
Atlanta, GA 30341

AAOHN Foundation Professional Development Scholarships — 175
Annually, the AAOHN Foundation awards grants to support professional development (academic and continuing education) and leadership development in occupational and environmental health. Leadership development scholarship applications, due August 1, support volunteer leadership development in occupational and environmental health nursing. Scholarship applications for academic study, due December 1, are offered to provide opportunities to further professional education for occupational and environmental health professionals. To support occupational and environmental health professionals in attending and successfully completing continuing education activities that will further their professional development and continued competence, the AAOHN Foundation offers continuing education scholarships, with applications due December 1.
Restrictions: These grants are not intended to supplement tuition in an academic program.
Geographic Focus: All States
Date(s) Application is Due: Dec 1
Amount of Grant: 3,000 USD
Contact: Ann Cox; (770) 455-7757; fax (770) 455-7271; ann@aaohn.org
Internet: http://www.aaohn.org/foundation/scholarships/index.cfm
Sponsor: American Association of Occupational Health Nurses Foundation
2920 Brandywine Road, Suite 100
Atlanta, GA 30341

AAP Anne E. Dyson Child Advocacy Awards — 176
The AAP Resident Section Anne E. Dyson Child Advocacy Award, supported by the Dyson Foundation, celebrates the efforts of pediatricians-in-training as they work in their communities to improve the health of children. This award seeks to showcase projects that are designed and implemented by residents, which aim to improve the lives of children. The Award includes: $300 in funds to advance the winning projects' goals; travel and lodging expenses for up to 2 residents per project to the AAP National Conference and Exhibition (NCE); and presentation of the Advocacy Award plaque during the Resident Section assembly meeting.
Requirements: Any resident-sponsored and/or resident led project that seeks to advocate on behalf of children is eligible for this award.
Geographic Focus: All States
Date(s) Application is Due: Jul 15
Amount of Grant: 300 USD
Contact: Kimberley VandenBrook; (800) 433-9016, ext. 7134; kvandenbrook@core.com
Internet: http://www.aap.org/sections/ypn/r/funding_awards/anne_dyson.html
Sponsor: American Academy of Pediatrics
141 Northwest Point Boulevard
Elk Grove Village, IL 60007-1098

AAP Community Access To Child Health (CATCH) Advocacy Training Grants — 177
The Program supports five pediatric faculty-resident pairs (10 people) to attend the AAP Legislative Conference in Washington, D.C. each year. Each faculty-resident pair is required to implement an educational activity on child advocacy in coordination with their local AAP chapter following the conference. Local chapters will receive up to $1,000 toward completing the educational activity in conjunction with the faculty-resident pairs. The next call for applications will be available each fall.
Geographic Focus: All States
Date(s) Application is Due: Nov 9
Amount of Grant: Up to 1,000 USD
Contact: Kathy Kocvara; (847) 434-7085 or (800) 433-9016, ext. 7632; kkocvara@aap.org
Internet: http://www.aap.org/commpeds/cpti/Opportunities.htm
Sponsor: American Academy of Pediatrics
141 Northwest Point Boulevard
Elk Grove Village, IL 60007-1098

AAP Community Access To Child Health (CATCH) Implementation Grants — 178
This program supports pediatricians in the initial and/or pilot stage of developing and implementing a community-based child health initiative. Grants of up to $12,000 are awarded each year on a competitive basis to pediatricians who want to address the local needs of children in the community. Priority will be given to projects serving communities with the greatest demonstrated health care access needs and health disparities. Strong collaborative community partnerships and future sustainability of the project are encouraged.
Requirements: A pediatrician must lead the project and be significantly involved in proposal development and project activities.
Restrictions: Only applicants from the United States and its territories are eligible to apply.
Geographic Focus: All States
Date(s) Application is Due: Jan 29
Amount of Grant: Up to 12,000 USD
Contact: Kathy Kocvara; (847) 434-7085 or (800) 433-9016, ext. 7632; kkocvara@aap.org
Internet: http://www.aap.org/catch/implementgrants.htm
Sponsor: American Academy of Pediatrics
141 Northwest Point Boulevard
Elk Grove Village, IL 60007-1098

26 | GRANT PROGRAMS

AAP Community Access to Child Health (CATCH) Planning Grants   179
The program awards competitive grants to pediatricians who want to plan community-based initiatives to increase access to children's health care. Proposed initiatives must be broad-based community partnerships. Priority will be given to proposals promoting medical homes for under-served children and those with special health care needs; collaborating with SCHIP or Medicaid, and representing new initiatives within the community. Grant funds must be used for planning, not implementation. Planning activities may include needs assessments and community asset mapping, feasibility studies, community meetings, focus groups, and development of grant proposals. If an applicant is not an AAP member, a letter of support from the AAP chapter president in his/her area must be obtained. Applicants must contact their chapter CATCH facilitators for approval of applications prior to submission. Technical assistance for applicants is available from CATCH staff and facilitators. Annual deadline dates may vary; contact program staff for exact dates.
Requirements: U.S. pediatricians and those in U.S. territories are eligible. Every program must be led by, facilitated by, or have the significant involvement of a pediatrician. During the planning phase, the involvement of other community members should be secured in order to ensure local support for the program. Applicants are encouraged to demonstrate collaboration with state child health insurance programs and/or state Medicaid programs.
Restrictions: Only applicants from the United States and its territories are eligible to apply.
Geographic Focus: All States
Date(s) Application is Due: Jul 30
Amount of Grant: 2,500 - 12,000 USD
Contact: Kathy Kocvara, Program Coordinator, CATCH; (847) 434-7632; fax (847) 228-6432; kkocvara@aap.org
Internet: http://www.aap.org/catch/planninggrants.htm
Sponsor: American Academy of Pediatrics
141 Northwest Point Boulevard
Elk Grove Village, IL 60007-1098

AAP Community Access To Child Health Residency Training Grants   180
The Community Pediatrics Training Initiative (CPTI) partners with the Community Access to Child Health (CATCH) Program to provide this grant opportunity targeting pediatric residency training programs. The mission of the program is to provide support to residency programs to build sustainable opportunities for residents to gain experience working on community-based child health initiatives that increase access to medical homes or specific health services not otherwise available. Grants of up to $15,000 will be awarded on a competitive basis to pediatric residency programs which submit proposals to plan and implement community-based child health initiatives as part of the training curriculum over the course of 16 months. The next Call for Proposals will be available in September.
Requirements: A pediatric faculty member must oversee the project and provide mentorship for residents participating in project activities.
Geographic Focus: All States
Date(s) Application is Due: Nov 18
Amount of Grant: Up to 15,000 USD
Contact: Kathy Kocvara, Program Coordinator; (847) 434-7085 or (800) 433-9016, ext. 7632; kkocvara@aap.org or catch@aap.org
Internet: http://www.aap.org/commpeds/cpti/Opportunities.htm
Sponsor: American Academy of Pediatrics
141 Northwest Point Boulevard
Elk Grove Village, IL 60007-1098

AAP Community Access To Child Health (CATCH) Resident Grants   181
The Community Access to Child Health (CATCH) Resident Funds program supports pediatric residents in the planning of community-based child health initiatives. Grants of up to $3,000 are awarded twice each year on a competitive basis for pediatric residents to address the needs of children in their communities. CATCH Resident Funds grant projects must include planning activities and also may include some implementation activities. Resident grants are available twice a year - May to July, during the CATCH Planning Funds grant cycle, and November to January, during the CATCH Implementation Funds grant cycle.
Requirements: A pediatric resident must lead the project and be significantly involved in proposal development and project activities.
Restrictions: Only pediatric residents from the United States and its territories are eligible to apply for CATCH Resident Funds grants.
Geographic Focus: All States
Date(s) Application is Due: Jan 31; Jul 31
Amount of Grant: Up to 3,000 USD
Contact: Kathy Kocvara, Program Coordinator; (847) 434-7085 or (800) 433-9016, ext. 7632; kkocvara@aap.org or catch@aap.org
Internet: http://www.aap.org/catch/residentgrants.htm
Sponsor: American Academy of Pediatrics
141 Northwest Point Boulevard
Elk Grove Village, IL 60007-1098

AAPD Henry B. Betts Award   182
Each year the award honors an individual whose work and scope of influence has significantly improved the quality of life for people with disabilities in the past, and will be a force for change in the future. It is named for Henry B. Betts, MD, in recognition of his pioneering leadership in the field of physical medicine and rehabilitation and decades of dedicated service to the Rehabilitation Institute of Chicago. The awardee receives an unrestricted cash award and a commemorative crystal piece, which is presented publicly at AAPD's Leadership Gala in Washington, D.C..
Requirements: Successful nominees will have demonstrated a strong vision and understanding of how to improve the quality of life for Americans with disabilities; possess a record of efforts and accomplishments that have affected a wide disability population; and have served as a powerful force for change, enhancing the opportunities for people with disabilities to participate fully in all aspects of society.
Geographic Focus: All States
Date(s) Application is Due: Oct 9
Amount of Grant: 50,000 USD
Contact: Tracey Murray; (800) 840-8844 or (770) 232-9001; murr9001@bellsouth.net
Internet: http://www.aapd-dc.org
Sponsor: American Association of People with Disabilities
1629 K Street NW, Suite 503
Washington, D.C. 20006

AAPD Paul G. Hearne Leadership Award   183
The awards are presented to people with disabilities who are emerging as leaders in their respective fields to help them continue their progress as leaders. They will also have an opportunity to meet and network with national disability leaders at the AAPD Leadership Gala in Washington, D.C.. Winners of the Award must demonstrate all of the following: leadership achievements that show a positive impact on the community of people with disabilities or within their area of disability interest; connections they have made between individuals with disabilities and others in their communities; a positive vision for the disability community and a continuing commitment to their leadership activities; and potential to contribute at a national level. AAPD encourages emerging leaders with disabilities of any age to apply. Previous awardees represent a diverse group of people with disabilities aged 11 to 56.
Requirements: U.S. residents with any type of disability are eligible to apply.
Geographic Focus: All States
Date(s) Application is Due: Oct 1
Amount of Grant: 10,000 USD
Sample: Claudia Gordon, a consultant to the National Council on Disability (Washington, D.C.)--award winner, the first black deaf female lawyer in the United States and a long-time advocate for the rights of deaf people, $10,000; Peter Cody Hunt--award winner, a graduate student in rehabilitation science and technology in Pittsburgh, who has developed outreach programs for Asian-Americans with disabilities, $10,000; Sarah Louise Triano--award winner, director of the Yield the Power to the Youth program at Access Living (Chicago), who works to mobilize and educate disabled people and increase their ability to influence public and social policy, $10,000.
Contact: David Hale; (202) 521-4306; dHale@aapd.com or awards@aapd.com
Internet: http://www.aapd.com/DMD/PaulHearneAward.html
Sponsor: American Association of People with Disabilities
1629 K Street NW, Suite 503
Washington, D.C. 20006

AAP International Travel Grants   184
The AAP has set aside a minimum of twelve (12) five-hundred dollar ($500.00) grants to be awarded to categorical pediatric or combined-training pediatric residents who wish to complete a clinical pediatric elective in the developing world during residency. The selection committee is composed of members from the AAP Section on International Child Health and the Section on Residents. After the selection committee reviews the applications the recipients will be notified mid-May and mid-November.
Requirements: Must be a categorical pediatric or combined-training pediatric resident at a U.S. or Canadian university.
Geographic Focus: All States
Date(s) Application is Due: Mar 15; Sep 15
Amount of Grant: 500 USD
Contact: Kimberley VandenBrook; (800) 433-9016, ext. 7134; kvandenbrook@core.com
Internet: http://www.aap.org/sections/ypn/r/funding_awards/international_travel.html
Sponsor: American Academy of Pediatrics
141 Northwest Point Boulevard
Elk Grove Village, IL 60007-1098

AAP Legislative Conference Scholarships   185
The American Academy of Pediatrics Legislative Conference Scholarships is specifically designed to help allay financial difficulties for members interested in attending the annual Legislative Conference, typically help each March. The Academy grants annually a number of scholarships varying from $1,000 to $2,000 on average. Each applicant will be reviewed carefully and awards will be granted to the most deserving.
Geographic Focus: All States
Contact: Kimberley VandenBrook, Grants Administrator; (800) 433-9016, ext. 7134; kvandenbrook@core.com
Internet: http://www.aap.org/sections/ypn/r/funding_awards/
Sponsor: American Academy of Pediatrics
601 13th Street NW, Suite 400 North
Washington, D.C. 20005

AAP Leonard P. Rome Community Access to Child Health (CATCH) Visiting Professorships   186
The purpose of the Leonard P. Rome Community Access to Child Health (CATCH) Visiting Professorships Program is to promote advocacy for children and advance the field of community pediatrics. Beginning with the 2008 grant cycle, CATCH will be collaborating with the Community Pediatrics Training Initiative (CPTI) to implement this program. Four accredited pediatric residency programs will receive up to $4,500 each

to fund a 2-or 3-day educational program focusing on the field of community pediatrics. District CATCH Facilitators (D.C.Fs) and Chapter CATCH Facilitators (CCFs) are available to provide technical assistance.
Geographic Focus: All States
Amount of Grant: Up to 4,500 USD
Contact: Alanna Bailey Whybrew, (847) 434-7085 or (800) 433-9016, ext. 7397; awhybrew@aap.org or catch@aap.org
Internet: http://www.aap.org/catch/vp.htm
Sponsor: American Academy of Pediatrics
141 Northwest Point Boulevard
Elk Grove Village, IL 60007-1098

AAP Nutrition Award   **187**
One award is given annually for outstanding achievements in research relating to nutrition of infants and children. The award is made to an individual or for one project. There is no age restriction for the award; however, it is hoped that younger persons will be considered. The award is made possible by a grant from the Infant Formula Council. In addition to the honorarium, the award includes round-trip airfare and two days' lodging for the recipient and a guest to attend the annual meeting and receive the award.
Requirements: The competition is open to residents of the United States or Canada whose research has been completed and publicly reported. Separate letters should be written for each individual nominated; a letter should contain a description of the nominee's achievements, stating clearly the basis for the recommendation, including references to the literature that describes her/his work. Nominee's bibliography is to be submitted with the nominating letter together with copies of available reprints. Letters supporting the nomination are to be screened by the nominator and forwarded with the nomination.
Restrictions: An individual may not submit more than five letters supporting the nomination. Current members of the AAP Committee on Nutrition are not eligible for this award.
Geographic Focus: All States
Date(s) Application is Due: Mar 1
Amount of Grant: 3,000 USD
Samples: Virginia A Stallings; Dennis M Bier; William Dietz; William W. Hay, Jr..
Contact: Debra Burrows; (847) 434-4000; fax (847) 434-8000; research@aap.org
Internet: http://www.aap.org/visit/nutrannouncemts.htm
Sponsor: American Academy of Pediatrics
141 Northwest Point Boulevard
Elk Grove Village, IL 60007-1098

AAP Program Delegate Awards   **188**
The travel grants are available to residents who serve as program delegates to the AAP. This grant serves to support the program delegate's travel to the AAP Resident Section meeting at the annual National Conference and Exhibition (NCE). The travel grant is available to both categorical and combined-training pediatric residents. However, there is only one grant available per institution. The AAP Program Delegate will automatically receive the $300 travel grant.
Geographic Focus: All States
Date(s) Application is Due: Sep 14
Amount of Grant: 300 USD
Contact: Kimberley VandenBrook; (800) 433-9016, ext. 7134; kvandenbrook@core.com
Internet: http://www.aap.org/sections/ypn/r/funding_awards/nce.html
Sponsor: American Academy of Pediatrics
141 Northwest Point Boulevard
Elk Grove Village, IL 60007-1098

AAP Resident Initiative Fund Grants   **189**
The Program is offering twenty-five (25) grants of up to $1,000 each. The funding will enable residents to educate fellow residents and/or parents on a specific aspect of one of the Academy's national child health priorities (Special Health Care Needs, Foster Care, Oral Health, Disaster Preparedness, Mental Health, Obesity, and Immunizations). The Academy hopes that this program will: increase the number of new, creative and innovative opportunities for residents to apply their leadership and advocacy skills within the AAP structure; increase residents' knowledge of the Academy's child health priorities; and increase resident collaboration with national AAP and its chapters.
Geographic Focus: All States
Date(s) Application is Due: Mar 31
Amount of Grant: 1,000 USD
Contact: Kimberley VandenBrook; (800) 433-9016, ext. 7134; kvandenbrook@core.com
Internet: http://www.aap.org/sections/ypn/r/funding_awards/res_initiative.html
Sponsor: American Academy of Pediatrics
141 Northwest Point Boulevard
Elk Grove Village, IL 60007-1098

AAP Resident Research Grants   **190**
The Program is designed to give pediatric residents with limited research experience an opportunity to initiate and complete research projects related to their professional interests. Projects may be related to the full spectrum of child health research, such as behavioral sciences, biomedical sciences, epidemiology, health services, perinatal/neonatal health, prevention, public health, quality improvement, quality measurement, or basic laboratory-based science. Research projects can be conducted for a maximum of 2 years and should be completed during the residency program.
Requirements: Applicants must be legal residents of the United States or Canada. Although awards are primarily intended for support of first- and second-year pediatric residents, individuals who have secured a position for a third or fourth year of residency beginning in July may also apply.
Geographic Focus: All States
Date(s) Application is Due: Feb 27
Amount of Grant: Up to 2,000 USD
Contact: Jeannine Hess; (800) 433-9016, ext. 7876; fax (847) 434-8000; jhess@aap.org
Internet: http://www.aap.org/sections/ypn/r/funding_awards/research_grants.html
Sponsor: American Academy of Pediatrics
141 Northwest Point Boulevard
Elk Grove Village, IL 60007-1098

Aaron & Cecile Goldman Family Foundation Grants   **191**
The Aaron and Cecile Goldman Family Foundation, established in 1962 in the District of Columbia, offers its primary support to Jewish and local agencies, including education and human services, and to local children's programs and camps. Fields of interest include: arts and humanities; children and youth services; education; human services; Jewish agencies and synagogues; Jewish federated giving programs; recreational programs; and camps. Types of support include building and renovation, capital campaigns, emergency funding, endowments, equipment purchase, fellowships, matching/challenge grants, support for professions, program development, and scholarships. Applicants should forward a letter of proposal to the Foundation office, whose board members meet in the spring and fall of each year.
Requirements: One copy of the proposal letter is required. 501(c)3 organizations supporting the residents of Washington, D.C., and its surrounding area are eligible to apply.
Restrictions: No grants are offered to individuals.
Geographic Focus: District of Columbia
Amount of Grant: Up to 20,000 USD
Contact: Jennifer Margolius Fisher, (202) 332-6600; fax (202) 332-1800; jennifer@themargoliusfirm.com
Sponsor: Aaron and Cecile Goldman Family Foundation
3000 Connecticut Avenue NW, Suite 100
Washington, D.C. 20008-2509

Aaron & Freda Glickman Foundation Grants   **192**
The Aaron and Freda Glickman Foundation, established in New York, offers grant support primarily in the State of New York. Funding typically comes in the form of general operating support for health organizations and associations, as well as Jewish agencies and synagogues. There are no specific deadlines or application forms with which to adhere, and applicants should contact the Foundation in writing, outlining the need and a description of the program.
Requirements: 501(c)3 organizations and Jewish agencies serving the residents of New York are eligible to apply.
Geographic Focus: New York
Amount of Grant: Up to 2,500 USD
Contact: Edwin J. Glickman, Secretary; (561) 470-1434
Sponsor: Aaron and Freda Glickman Foundation
3080 N 36th Avenue, V-61
Hollywood, FL 33021

Aaron Foundation Grants   **193**
Established in Massachusetts in 1951, the Aaron Foundation awards grants to eligible nonprofit organizations in its areas of interest, including: arts and culture; health care; higher education; Jewish services and temples; health care; social services; and youth. There are no application deadlines or forms with which to adhere. Applicants should submit: results expected from proposed grant; a copy of IRS Determination Letter; copy of most recent annual report/audited financial statement or 990; a listing of board of directors, trustees, officers and other key people and their affiliations; and a copy of current year's organizational budget and/or project budget.
Requirements: Connecticut, Massachusetts, and Rhode Island nonprofit organizations are eligible to apply.
Restrictions: Individuals are ineligible to apply.
Geographic Focus: Connecticut, Massachusetts, Rhode Island
Amount of Grant: 1,000 - 100,000 USD
Samples: Beth Israel Deaconess Medical Center, Boston, Massachusetts, $50,000 - to support a capital campaign; United Way of Massachusetts Bay, Boston, Massachusetts, $90,000 - for the annual campaign.
Contact: Avram J. Goldberg, Trustee; (617) 695-1300 or (617) 695-1946
Sponsor: Aaron Foundation
225 Franklin Street, Suite 1450
Boston, MA 02110

Aaron Foundation Grants   **194**
The Aaron Foundation, established in California in 1987 by the Aaron Mortgage Corporation, supports programs for children and youth, health care facilities, and human services agencies. Giving is primarily centered in the Bakersfield, California, region. Applicants should submit: a copy of current year's organizational budget and/or project budget; a copy of IRS Determination Letter; a detailed description of project and amount of funding requested; a statement of the problem that the project will address; and a brief history of organization and description of its mission.
Requirements: The proposed project must benefit the residents of Bakersfield, California.
Geographic Focus: California
Date(s) Application is Due: Nov 30
Contact: Hal E. Aaron, Trustee; (661) 322-6353; fax (661) 322-6120
Sponsor: Aaron Foundation
651 H Street, Suite 100
Bakersfield, CA 93304-1305

28 | Grant Programs

AAUW American Dissertation Fellowships 195

AAUW advances equality for women and girls through advocacy, education, philanthropy, and research. The American Fellowships program has been in existence since 1888, making it the oldest non-institutional source of graduate funding for women in the United States. The program provides fellowships for women writing their dissertations and those pursuing postdoctoral research. Research publication grants are also available to enable scholars to complete manuscripts for publication. The purpose of the Dissertation Fellowship is to offset a scholar's living expenses while she completes her dissertation. Award amount is $20,000, with a November 17 deadline for online submissions of applications.
Requirements: Applicants must be U.S. citizens or permanent residents by the application deadline. When comparing proposals of equal merit, special consideration will be given to women holding junior academic appointments who are seeking research leave, women who have held the doctorate for at least three years, and women whose educational careers have been interrupted.
Restrictions: Candidates may apply for only one of the awards described. Former recipients of these awards are not eligible to apply for additional American Fellowships or publication grants. Students holding any fellowship for writing a dissertation in the year prior to the AAUW Educational Foundation fellowship year are not eligible.
Geographic Focus: All States
Date(s) Application is Due: Nov 17
Amount of Grant: 20,000 USD
Contact: Gloria Blackwell; (202) 785-7700 or (800) 326-2289
Internet: http://aauw-amdissert.scholarsapply.org/
Sponsor: American Association of University Women
1111 16th Street NW
Washington, D.C. 20036

AAUW American Postdoctoral Research Leave Fellowships 196

AAUW advances equality for women and girls through advocacy, education, philanthropy, and research. Postdoctoral Research Leave Fellowships are designed to assist scholars in obtaining tenure and other promotions by enabling them to spend a year pursuing independent research. The primary purpose of the fellowship is to increase the number of women in tenure-track faculty positions and to promote equality for women in higher education.
Requirements: Applicants must be U.S. citizens or permanent residents by the application deadline. Candidates are evaluated on the basis of scholarly excellence; the quality and originality of project design; and active commitment to helping women and girls through service in their communities, professions, or fields of research. The award amount is $30,000, with an annual deadline date for submission of online applications being November 17.
Restrictions: Candidates may apply for only one of the awards described. Former recipients of these awards are not eligible to apply for additional American Fellowships or publication grants. Tenured professors are not eligible.
Geographic Focus: All States
Date(s) Application is Due: Nov 17
Amount of Grant: 30,000 USD
Contact: Gloria Blackwell, Vice President of Fellowships, Grants and Global Programs; (202) 785-7700 or (800) 326-2289
Internet: http://www.aauw.org/learn/fellowships_grants/american.cfm
Sponsor: American Association of University Women
1111 16th Street NW
Washington, D.C. 20036

Abbot and Dorothy H. Stevens Foundation Grants 197

The Abbot and Dorothy H. Stevens Foundation funds Massachusetts non-profits with a emphasis on the greater Lawrence/Merrimack Valley area. Giving primarily for the arts, education, conservation, and health and human services. Fields of interest include: arts; children/youth, services; crime/violence prevention, domestic violence; education; elderly; environment, natural resources; health care; health organizations, association; historic preservation/historical societies; humanities; human services; immigrants; medical school/education; museums. Types of support include: building/renovation; capital campaigns; continuing support; endowments; equipment; general/operating support; management development/capacity building; matching/challenge support; program-related investments/loans; program development; technical assistance.
Requirements: Massachusetts 501(c)3 tax-exempt organizations serving the greater Lawrence and Merrimack Valley are eligible. There is no deadline date when applying for funding, nor is there a application form required. The Foundation will however accept the Associated Grantmakers Common Proposal Form. Applicants should submit a copy of their IRS Determination Letter and one copy of their proposal when applying for funding. The board meets and reviews proposals monthly except for the months of July and August.
Restrictions: Grants do not support national organizations, state or federal agencies, individuals, annual campaigns, deficit financing, exchange programs, internships, professorships, scholarships, or fellowships.
Geographic Focus: Massachusetts
Amount of Grant: 1,000 - 20,000 USD
Samples: Boston Childrens Museum, Boston, MA, $15,000; Boston Symphony Orchestra, Boston, MA, $10,000; Boston Ballet, Boston, MA, $5,000.
Contact: Josh Miner, Executive Director; (978) 688-7211; fax (978) 686-1620; grantprocess@stevensfoundation.org
Sponsor: Abbot and Dorothy H. Suitevens Foundation
P.O. Box 111
North Andover, MA 01845-7211

Abbott-ASM Lifetime Achievement Award 198

The award, sponsored by Abbott Laboratories, is given to honor a distinguished scientist for a lifetime of outstanding contributions in fundamental biomedical research in any of the microbiological sciences. The award consists of a cash prize, framed certificate, medallion, and national or international travel expenses incidental to receiving the award at the ASM general meeting. Guidelines are available online. All nomination materials must be submitted electronically.
Requirements: Mature scientists, both active and retired, from all areas of microbiology are eligible.
Restrictions: Self-nominations are not be accepted.
Geographic Focus: All States
Date(s) Application is Due: Oct 1
Amount of Grant: Up to 20,000 USD
Samples: Norman R. Pace, $20,000--award recipient; R. John Collier, $20,000--award recipient; Jonathan Beckwith, $20,000--award recipient.
Contact: Awards Committee; (202) 942-9226; fax (202) 942-9353; awards@asmusa.org
Internet: http://www.asm.org/Academy/index.asp?bid=2587
Sponsor: American Society for Microbiology
1752 N Street, N.W.
Washington, D.C. 20036-2904

Abbott Fund Access to Health Care Grants 199

Abbott's Access to Health Care programs seek innovative solutions to improve and expand access to health care services for disadvantaged populations. Specific areas of focus include cardiovascular health, diabetes, nutrition, maternal and child health and neonatal care. Many Abbott Fund programs are engaged in closing gaps in ethnic and minority communities, promoting health and nutrition education for families, training health workers, and improving delivery of health services. Complete and submit the Abbott Fund online grant application, posted when available. It should include your organization's Federal Tax ID. At times when the Abbott Fund is accepting unsolicited grant applications, it will acknowledge receipt of an online application via email. The Fund will notify an applicant of its decision on a funding request within six to eight weeks.
Requirements: Grants are made to tax-exempt organizations supporting access to health care.
Restrictions: The Abbott Fund does not accept unsolicited grant applications for projects outside the United States. Contributions will not be made to individuals; for-profit entities; purely social organizations; political parties or candidates; sectarian religious organizations; advertising; symposia, conferences, and meetings; ticket purchases; memberships; business-related purposes; volunteer efforts of non-Abbott employees; or marketing sponsorships.
Geographic Focus: Arizona, California, Illinois, Kansas, Massachusetts, Michigan, New Jersey, New York, North Carolina, Ohio, Puerto Rico, Texas, Utah, Virginia
Amount of Grant: 20,000 - 100,000 USD
Contact: Cindy Schwab; (847) 937-7075; fax (847) 935-5051; cindy.schwab@abbott.com
Internet: http://www.abbottfund.org/tags/access
Sponsor: Abbott Fund
100 Abbott Park Road, Department 379, Building 6D
Abbott Park, IL 60064-3500

Abbott Fund CFCareForward Scholarships 200

The CFCareForward program expands a long-standing tradition of the Abbott Fund offering support to cystic fibrosis families nutritional, financial and educational resources. Recognizing the financial burdens that exist for many CF families, the Fund developed the CFCareForward Scholarship to honor young adults with CF as they pursue goals of higher education. Since 1993, it has awarded scholarship funds through this program totaling more than $2.3 million. Forty recipients are selected annually to receive $2,500 for use during the school year based on established criteria, including academic record and extracurricular activities, essay and creative presentation.
Geographic Focus: All States
Date(s) Application is Due: Sep 11
Amount of Grant: 2,500 USD
Contact: Cindy Schwab; (847) 937-7075; fax (847) 935-5051; cindy.schwab@abbott.com
Sponsor: Abbott Fund
100 Abbott Park Road, Department 379, Building 6D
Abbott Park, IL 60064-3500

Abbott Fund Community Grants 201

The Abbott Fund Community Grant program is active in communities around the world where Abbott has a significant presence. It pursues local partnerships and creative programs that address unmet needs of a community. Emphasis is placed on improving access to health care and promoting science education. The program also supports major civic, arts and other cultural institution programming, primarily in the Chicago metropolitan area where Abbott is headquartered. Complete and submit the Abbott Fund online grant application, posted when available. It should include your organization's Federal Tax ID. At times when the Abbott Fund is accepting unsolicited grant applications, it will acknowledge receipt of an online application via email. The Fund will notify an applicant of its decision on a funding request within six to eight weeks.
Requirements: Grants are made to tax-exempt organizations supporting company operating areas in Arizona, California, Illinois, Kansas, Massachusetts, Michigan, New Jersey, New York, North Carolina, Ohio, Puerto Rico, Texas, and Virginia, and Utah. It also supports the communities of: Abingdon, England; Brockville, Canada; Campoverde, Italy; Clonmel, Ireland; Cootehill, Ireland; Delkenheim, Germany; Kanata, Canada; Katsuyama, Japan; Ludwigshafen, Germany; Queenborough, England; Rio de Janeiro, Brazil; Sligo, Ireland; and Zwolle, the Netherlands.

Restrictions: Contributions will not be made to individuals; for-profit entities; purely social organizations; political parties or candidates; sectarian religious organizations; advertising; symposia, conferences, and meetings; ticket purchases; memberships; business-related purposes; volunteer efforts of non-Abbott employees; or marketing sponsorships.
Geographic Focus: Arizona, California, Illinois, Kansas, Massachusetts, Michigan, New Jersey, New York, North Carolina, Ohio, Puerto Rico, Texas, Utah, Virginia, Brazil, Canada, Germany, Great Britain, Ireland, Italy, Japan, Netherlands
Amount of Grant: 10,000 - 100,000 USD
Contact: Cindy Schwab; (847) 937-7075; fax (847) 935-5051; cindy.schwab@abbott.com
Internet: http://www.abbottfund.org/tags/community/1
Sponsor: Abbott Fund
100 Abbott Park Road, Department 379, Building 6D
Abbott Park, IL 60064-3500

Abbott Fund Global AIDS Care Grants 202
Abbott has been a significant contributor to the fight against HIV/AIDS for more than two decades. Since 2000, Abbott and the Abbott Fund have invested $225 million in grants and product donations targeted to resource poor countries most impacted by HIV/AIDS. The focus of these efforts includes expanding access to care, testing and treatment; strengthening HIV/AIDS health care systems; preventing mother-to-child transmission; and supporting children and families affected by HIV/AIDS. In addition, Abbott and the Abbott Fund have helped pioneer innovative model programs to combat the disease. For example, we helped build the Baylor International Pediatric AIDS Initiative's first pediatric outpatient clinic in Romania in 2001 that has served an average of 600 patients per year. The clinic has reduced pediatric HIV mortality rates by more than 90 percent. Today Baylor has replicated the Romania model in additional clinics throughout Africa and now serves nearly 60,000 children and young people with HIV – including those at two clinics built by and supported by the Abbott Fund in Malawi and Tanzania.
Requirements: Grants are made to tax-exempt organizations supporting global HIV/AIDS programs.
Restrictions: Contributions will not be made to individuals; for-profit entities; purely social organizations; political parties or candidates; sectarian religious organizations; advertising; symposia, conferences, and meetings; ticket purchases; memberships; business-related purposes; volunteer efforts of non-Abbott employees; or marketing sponsorships.
Geographic Focus: Arizona, California, Illinois, Kansas, Massachusetts, Michigan, New Jersey, New York, North Carolina, Ohio, Puerto Rico, Texas, Utah, Virginia
Amount of Grant: 10,000 - 100,000 USD
Contact: Cindy Schwab; (847) 937-7075; fax (847) 935-5051; cindy.schwab@abbott.com
Internet: http://www.abbottfund.org/project/20/30/Helping-Children-and-Young-People-Living-with-HIV-AIDS
Sponsor: Abbott Fund
100 Abbott Park Road, Department 379, Building 6D
Abbott Park, IL 60064-3500

Abbott Fund Science Education Grants 203
The world urgently needs people who are well-trained in science and technology, and the Abbott Fund is committed to doing its part to address this challenge. Serving as a catalyst by stimulating community investment and engagement, Abbott's investment in science education: engages and inspires students, families and teachers in scientific exploration in out-of-school informal settings; encourages young people to be more proficient in science and attracts more scientists to the field; and builds strong partnerships that are systemic, replicable and sustainable for multiple years and multiple locations. Complete and submit the Abbott Fund online grant application, posted when available. It should include your organization's Federal Tax ID. At times when the Abbott Fund is accepting unsolicited grant applications, it will acknowledge receipt of an online application via email. The Fund will notify an applicant of its decision on a funding request within six to eight weeks.
Requirements: Grants are made to tax-exempt organizations supporting company operating areas in Arizona, California, Illinois, Kansas, Massachusetts, Michigan, New Jersey, New York, North Carolina, Ohio, Puerto Rico, Texas, and Virginia, and Utah. It also supports the communities of: Abingdon, England; Brockville, Canada; Campoverde, Italy; Clonmel, Ireland; Cootehill, Ireland; Delkenheim, Germany; Kanata, Canada; Katsuyama, Japan; Ludwigshafen, Germany; Queenborough, England; Rio de Janeiro, Brazil; Sligo, Ireland; and Zwolle, the Netherlands.
Restrictions: Contributions will not be made to individuals; for-profit entities; purely social organizations; political parties or candidates; sectarian religious organizations; advertising; symposia, conferences, and meetings; ticket purchases; memberships; business-related purposes; volunteer efforts of non-Abbott employees; or marketing sponsorships.
Geographic Focus: Arizona, California, Illinois, Kansas, Massachusetts, Michigan, New Jersey, New York, North Carolina, Ohio, Puerto Rico, Texas, Utah, Virginia, Brazil, Canada, Germany, Great Britain, Ireland, Italy, Japan, Netherlands
Amount of Grant: 20,000 - 100,000 USD
Contact: Cindy Schwab; (847) 937-7075; fax (847) 935-5051; cindy.schwab@abbott.com
Internet: http://www.abbottfund.org/tags/science/1
Sponsor: Abbott Fund
100 Abbott Park Road, Department 379, Building 6D
Abbott Park, IL 60064-3500

Abby's Legendary Pizza Foundation Grants 204
As a company, Abby's Legendary Pizza Foundation contributes to many community-based events and programs. Non-profits such as the Children's Miracle Network, the American Cancer Society, the Alzheimer's Association, and many others benefit from the Foundation's partnership as a business that cares about local people and causes. Typically, the Foundation provides support for: athletics, sports, and amateur leagues; child development; and higher education. The Foundation also supports most local high schools in its hometown communities throughout the states of Oregon and Washington. There are no identified annual deadlines for application submission, and interested parties should contact their local Abby's Legendary Pizza. Most recently, awards have ranged from $25 to $5,000.
Requirements: 501(c)3 non-profits either located in, or serving the residents of, Abby's Legendary Pizza communities should apply.
Geographic Focus: Oregon, Washington
Amount of Grant: 25 - 5,000 USD
Contact: B. Mills Sinclair, President; (541) 689-0019
Internet: http://abbys.com/fundraising/
Sponsor: Abby's Legendary Pizza Foundation
1960 River Road
Eugene, OR 97404-2502

Abdus Salam ICTP Federation Arrangement Scheme Grants 205
Federation Arrangements are agreements signed by the ICTP and a scientific institute (Federated Institute) in developing countries. This enables the Institute to send junior scientists - up to the age of 40 - to the ICTP for a total of 60 to 150 days, during the validity of the contract (i.e. three years), on a cost-sharing basis. While living expenses are always covered by the Center, travel expenses are either covered by the Federated Institute or shared with the Center. The Federation Arrangement Scheme gives the opportunity to junior scientists from developing countries to participate in the Center's scientific programs. Their participation is proposed by the Federated Institute and approved by the ICTP. Presently there are five different standard types of Federation Arrangements depending upon geographical areas. While the junior scientists' living expenses are always covered by the Centre, travel expenses are either covered by the Federated Institute or shared with the Center.
Requirements: Scientific institutes from developing countries (to send junior scientists to ICTP) are eligible to apply.
Geographic Focus: All Countries
Contact: Patrizia Passarella, Program Assistant; (+39) 040 2240 389; fax (+39) 040 2240 388; itlabs@ictp.trieste.it or diploma@ictp.it
Internet: http://assoc.ictp.it/federation-scheme/description.html
Sponsor: Abdus Salam International Centre for Theoretical Physics
Strada Costiera 11
Trieste, I - 34014 Italy

Abel Foundation Grants 206
The Abel Foundation awards grants to eligible Nebraska nonprofit organizations in its areas of interest, which include: environmental programs and natural resource conservation; health care; higher education; Protestant religion and churches; arts and culture; and social services. Types of support include building construction and/or renovation, capital campaigns, general operating support, and program grants. Contact the office for application forms. Application deadlines are March 31, July 15, and October 31 each year. Applications qualifying for review will be considered within one to three months of receipt.
Requirements: Nebraska nonprofit organizations are eligible. Preference is given to requests from Lincoln, Nebraska, and the state's southeastern region.
Restrictions: The Foundation does not accept applications from organizations: that have had requests approved or declined in the past 12 months; or that are currently received payments for a multi-year grant.
Geographic Focus: Nebraska
Date(s) Application is Due: Mar 31; Jul 15; Oct 31
Amount of Grant: 50 - 40,000 USD
Contact: J. Ross McCown, Vice President; (402) 434-1212; fax (402) 434-1799; rossm@nebcoinc.com or nebcoinfo@nebcoinc.com
Internet: http://www.abelfoundation.org/grant.htm
Sponsor: Abel Foundation
1815 Y Street, P.O. Box 80268
Lincoln, NE 68501-0268

Abell-Hanger Foundation Grants 207
The Abell-Hanger Foundation makes grants to nonprofit Texas organizations, other than private foundations, that are involved in such undertakings for the public and society benefit, including: arts, cultural, and humanities; education; health; human services; and religion. Types of support include general operating support, continuing support, annual campaigns, capital campaigns, building construction/renovation, equipment acquisition, endowment funds, program development, seed funds, scholarship funds, research grants, and matching funds. Block scholarship grants are made only to institutions of higher education located in Texas. Recipient colleges and universities are free to administer the grants. Education grants are limited generally to institutions of higher education, including religious institutions (Baptist, Christian, Lutheran, Methodist, and Presbyterian). Applicants must seek funding for the same proposal from various sources because sole sponsorship of programs is rarely undertaken. Grant requests are considered and awarded throughout each year. The trustees prefer to consider only one request per applicant each fiscal year. Unsuccessful proposals may not be resubmitted for at least 12 months. Applicant organizations that have never received funding from the foundation should request a pre-proposal questionnaire; the trustees will review the request to determine whether it warrants a complete proposal.
Requirements: Applicant organizations must be located in Texas and be 501(c)3 tax-exempt. National organizations with significant operations in, or providing material benefits to the citizens of, Texas will be considered based on the degree of operations/benefits within the state.
Restrictions: The Foundation does not fund grants, scholarships, or fellowships for individuals.

Geographic Focus: Texas
Date(s) Application is Due: Feb 28; May 31; Aug 31; Nov 6
Amount of Grant: 10,000 - 500,000 USD
Samples: American Fallen Warrior Memorial Foundation, Midland, Texas, $5,000 - for the Texas Permian Basin Honor Flight (2014); Midland Community Theater, Midland, Texas, $45,000 - for general operating support and an out-reach program (2014); Midland ISD Educational Foundation, Midland, Texas, $500,000 - for recruitment and retention of ISD teachers (2014).
Contact: David L. Smith; (432) 684-6655; fax (432) 684-4474; ahf@abell-hanger.org
Internet: http://www.abell-hanger.org/grant-criteria.html
Sponsor: Abell-Hanger Foundation
P.O. Box 430
Midland, TX 79702-0430

Abell Foundation Criminal Justice and Addictions Grants 208
The Foundation seeks to increase access to substance abuse treatment and supportive services such as housing and job training for the uninsured and drug addicted individuals residing in Baltimore City. The Foundation works to increase the impact and effectiveness of treatment services through cutting edge research and support of innovative service models designed to reach under-served populations. The Foundation supports programs and initiatives that increase public safety and reduce recidivism with a special focus on initiatives that address the barriers facing the returning ex-offender. Areas of interest include: substance abuse treatment, prevention, and research; supportive housing; prisoner reentry; criminal justice system reform; and juvenile justice. A particular emphasis is placed on initiatives that provide transitional housing and the necessary wraparound services to support a successful return to the community.
Requirements: 501(c)3 organizations serving Maryland communities, especially in the Baltimore area, may apply. The foundation prefers grantees that show strong fiscal management, their project's benefit to the community, ability to achieve goals, unique work, and other sources of financial support.
Restrictions: The Foundation does not fund educational programs at higher education institutions, medical facilities, individual scholarships, fellowships, annual operating expenses, sponsorships, deficit financing, endowments, travel or memberships.
Geographic Focus: Maryland
Date(s) Application is Due: Jan 1; Mar 1; May 1; Aug 1; Oct 15
Amount of Grant: 1,000 - 50,000 USD
Samples: Maryland Community Health Initiatives, Baltimore, Maryland, $35,000 - purchase of equipment and facility improvements of the Penn North Recovery Community Center (2014).
Contact: Robert C. Embry, Jr.; (410) 547-1300; fax (410) 539-6579; abell@abell.org
Internet: http://www.abell.org/criminal-justice-addictions
Sponsor: Abell Foundation
111 S Calvert Street, Suite 2300
Baltimore, MD 21202-6174

Abell Foundation Health and Human Services Grants 209
Through grants awarded in this area, the Foundation seeks to address societal issues associated with family disintegration, family planning, child support, teenage parenting, domestic violence, children's health and well-being, child abuse and neglect, hunger, food self-sufficiency and homelessness. The Foundation also supports advocacy programs for better health care and social services for children and youth as well as for a comprehensive system of universal health care. Of particular concern is the support of efforts to combat childhood lead paint poisoning and mental health disorders. Furthermore, the Foundation continues to provide opportunities for low-income families to live in quality housing in good neighborhoods in the region. While the Foundation's primary focus is on the development of permanent housing, it also will consider emergency and transitional housing.
Requirements: 501(c)3 organizations serving Maryland communities, especially in the Baltimore area, may apply. The foundation prefers grantees that show strong fiscal management, their project's benefit to the community, ability to achieve goals, unique work, and other sources of financial support.
Restrictions: The Foundation does not fund educational programs at higher education institutions, medical facilities, individual scholarships, fellowships, annual operating expenses, sponsorships, deficit financing, endowments, travel or memberships.
Geographic Focus: Maryland
Date(s) Application is Due: Jan 1; Mar 1; May 1; Aug 1; Oct 15
Amount of Grant: 1,000 - 50,000 USD
Contact: Robert C. Embry, Jr.; (410) 547-1300; fax (410) 539-6579; abell@abell.org
Internet: http://www.abell.org/health-human-services
Sponsor: Abell Foundation
111 S Calvert Street, Suite 2300
Baltimore, MD 21202-6174

Abernethy Family Foundation Grants 210
The Abernethy Family Foundation was established in 2000 in Florida with a mission of supporting elementary and secondary education programs, curriculum development, agricultural programs, and health care within St. Lucie and Indian River counties, Florida. General operating funding is its primary type of support. There are no deadlines with which to adhere, and applicant organizations should begin by requesting an application form directly from the Foundation.
Restrictions: Giving is restricted to eligible organizations within St. Lucie and Indian River counties, Florida.
Geographic Focus: Florida
Contact: Bruce R. Abernethy, Jr; (772) 489-4901; babernethy@bruceapa.com
Sponsor: Abernethy Family Foundation
500 Virginia Avenue, Suite 202
Fort Pierce, FL 34982-5882

Abington Foundation Grants 211
The Abington Foundation awards grants to Ohio nonprofit organizations serving Cuyahoga County in its areas of interest, including pre-primary and higher education, geriatric healthcare and nursing, the promotion or sustenance of individual and family economic independence, and cultural activities. Priority is given to funding requests for specific programs or projects that represent critical periods in a child's development from birth to age 5, and to early adolescence, ages 10-15. Requests for endowment or general operating support are discouraged. Foundation staff is available for consultation during the proposal preparation process.
Requirements: Ohio nonprofit organizations serving Cuyahoga County are eligible.
Geographic Focus: Ohio
Date(s) Application is Due: May 1; Sep 1; Dec 1
Amount of Grant: 5,000 - 50,000 USD
Samples: Cleveland Play House, Cleveland, Ohio, $50,000 - to build the Allen Theatre Complex, a three theatre center for performing arts and arts education, in partnership with Playhouse Square and Cleveland State University; North Coast Health Ministry, Cleveland, Ohio, $33,000 - to hire a chronic care nurse case manager; St. Martin de Porres High School, Cleveland, Ohio, $23,000 - for summer programming and tutoring for incoming freshmen.
Contact: Janet E. Narten; (216) 621-2901 or (216) 621-2632; fax (216) 621-8198
Internet: http://www.fmscleveland.com/abington
Sponsor: Abington Foundation
1422 Euclid Avenue, Suite 627
Cleveland, OH 44115-1952

Able To Serve Grants 212
Able to Serve Inc. is a public charity, providing services and opportunities for persons with mental and physical disabilities. Able primarily serves Wake and Johnston counties in North Carolina with some services extended to surrounding counties when resources are available. There are currently four programs that cover the services Able fund within the local community; Local Ministry Network, Community Partnerships, Van Transportation Services, and Computer Learning/Donations.
Requirements: Able to Serve provides each service and program at no charge to the individual or family receiving services within the Wake and Johnston counties of North Carolina.
Geographic Focus: North Carolina
Contact: Carlton S. McDaniel, President; (919) 779-5545; carlton@abletoserve.com
Internet: www.abletoserve.com
Sponsor: Able To Serve
P.O. Box 334
Garner, NC 27529-0334

Able Trust Vocational Rehabilitation Grants for Individuals 213
The trust supports individuals and non-profit vocational rehabilitation programs throughout Florida with fund-raising, grant making and public awareness of disability issues. Created by the Florida Legislature in 1990, the Florida Endowment Foundation for Vocational Rehabilitation, parent organization of the Able Trust, is 501(c)3 non-profit public/private partnership with a goal of assisting Floridians with disabilities in achieving employment. The trust provides grant funds to Florida not-for-profit agencies and Floridians with disabilities for a wide array of projects leading to the employment of individuals with disabilities. To be considered for funding a proposal must address the employment individuals with disabilities and priority is given to those projects with direct employment placement outcomes during the grant time period. Grants to individuals are typically around $2,500. Historically grants have been made between $500-$4,000 for job accommodations.
Requirements: Individuals currently residing in the state of Florida that need emergency on-the-job accommodations to accept an employment offer, retain or receive a promotion at their current employment and are not currently open clients with a state agency provider may apply to be considered for assistance. There are no deadlines; guidelines, helpful tips and forms can be found at the sponsor's website.
Restrictions: Grants funds may not be used to purchase: vehicles, property, building improvements, capital campaigns, endowments, fellowships, scholarships, travel grants, tuition where state and federal aid is available, lobbying, medical items, incurred debt, and proposed expenses prior to grant approval.
Geographic Focus: Florida
Amount of Grant: 500 - 4,000 USD
Contact: Guenevere Crum; (850) 224-4493; fax (850) 224-4496; guenevere@abletrust.org
Internet: http://www.abletrust.org/grant/booklet.shtml
Sponsor: Able Trust
3320 Thomasville Road, Suite 200
Tallahassee, FL 32308

Abney Foundation Grants 214
The foundation makes grants for innovative and creative projects, and to programs that are responsive to changing community needs in the areas of health, social service, education, and cultural affairs in South Carolina. The foundation's primary focus is on higher education. All requests must be in writing and in accordance with foundation guidelines, which are available upon request, and from the website. Applicants may submit a Letter of Intent (LOI) briefly describing the project before submitting a proposal in order to find out if their ideas are potentially supported by the foundation.

Requirements: Agencies applying for funds should be serving the citizens of South Carolina.
Restrictions: The foundation does not generally fund requests for operating expenses.
Geographic Focus: South Carolina
Date(s) Application is Due: Nov 15
Amount of Grant: 5,000 - 100,000 USD
Samples: Anderson College (Anderson, SC)--for scholarships, $100,000; Anderson College (Anderson, SC)--to establish a scholarship fund, $100,000.
Contact: David C. King; (864) 964-9201; fax (864) 964-9209; info@abneyfoundation.org
Internet: http://www.abneyfoundation.org/guideline.htm
Sponsor: Abney Foundation
100 Vine Street
Anderson, SC 29621-3265

Aboudane Family Foundation Grants 215
Established in Michigan in 2007, the Aboudane Family Foundation offers awards for operating support to Islamic organizations throughout the United States. Primary fields of interest are stated to be: health; education; and community services. There are no specific application forms or annual deadlines, and applicants should contact the Foundation directly before beginning the writing process. Grants generally range from $1,000 to $15,000.
Geographic Focus: All States
Amount of Grant: 1,000 - 15,000 USD
Samples: Grand Blanc Islamin Center, Grand Blanc, Michigan, $8,802 - for operating support; Al-Fajr Institute, Flint, Michigan, $5,000 - for operating support; Islamic Relief, Alexandria, Virginia, $12,000 - for operating support.
Contact: Zakwan A. Aboudane, (810) 742-8770; fax (810) 742-8772
Sponsor: Aboudane Family Foundation
5032 Parkwood Court
Flushing, MI 48433-1390

Abracadabra Foundation Grants 216
The Abracadabra Foundation was established in Oregon in 1997 to provide funding support for Attention Deficit Disorder (ADD) and Attention Deficit Hyperactivity Disorder (ADHD) research in the Portland, Oregon, region. Its primary fields of interest, therefore, are medical research and education. Types of funding generally include internships and research funds. A specific application form is not required, and interested parties should forward a letter directly to the Foundation office. There are no annual deadlines. Most recently, awards have ranged from $2,500 up to $200,000.
Requirements: 501(c)3 non-profits and research facilities, primarily in the Portland, Oregon, metropolitan area, should apply.
Geographic Focus: Oregon
Contact: Steven J. Sharp, President; (503) 615-9408
Sponsor: Abracadabra Foundation
3605 NW Bliss Road
Vancouver, WA 98685-1512

Abramson Family Foundation Grants 217
The foundation awards grants nationwide to Jewish organizations, educational institutions, and health associations in its areas of interest, including arts, education, higher education; hospitals (general); Israel; Jewish agencies, temples, and charitable giving; and social services delivery. The foundation also sponsors a scholarship program. Contact the office for application forms.
Geographic Focus: All States
Contact: Judith Abramson Felgoise
Sponsor: Abramson Family Foundation
376 Regatta Drive
Jupiter, FL 33477

Abundance Foundation International Grants 218
The Abundance Foundation makes grants to organizations aligned with its mission to improve global health through education, economic empowerment and health systems strengthening. It is focused on programs that unlock the potential of local communities particularly in Africa, Central America and Haiti. Its grantee partners train, support and empower local leaders to create new capabilities that result in lasting improvement in quality of life. In addition to direct grant funding that allows for upscaling of successful existing programs, the Foundation raise awareness about local leaders whose vision and heroism are creating positive change in their communities.
Geographic Focus: All States, Haiti
Contact: Stephen Kahn; (510) 841-4123; fax (510) 841-4093; info@abundancefound.org
Internet: http://www.abundancefound.org/grants/
Sponsor: Abundance Foundation
127 University Avenue
Berkeley, CA 94710-1616

ACAAI Foundation Research Grants 219
The Foundation develops, promotes and funds clinical research and educational programs related to allergy, asthma & immunology. The Foundation is dedicated to: strengthening, supporting and funding allergy/immunology training programs; heightening awareness of the critical role of the allergist in cost-effective treatment of asthma and other allergic and immunologic diseases; and supporting asthma camp programs to teach children and their parents how to live with asthma. Programs include clinical fellowship stipends, fellow-in-training research grants, young faculty support awards, scholar's return awards, and support for asthma camps.

Geographic Focus: All States
Amount of Grant: Up to 250,000 USD
Contact: Program Administrator; (847) 427-1200; fax (847) 427-1294; mail@acaai.org
Internet: http://www.acaai.org/Member/ACAAI_Foundation/ACAAI_Foundation.htm
Sponsor: American College of Allergy, Asthma and Immunology Foundation
85 West Algonquin Road, Suite 550
Arlington Heights, IL 60005

AcademyHealth Alice S. Hersh New Investigator Awards 220
The Alice S. Hersh New Investigator Award recognizes the contribution of new scholars to the field of health services research. Eligible nominees are those whose first research appointments were no earlier than January 1, 2002. Post-doctoral positions will be considered research appointments unless these years provided training necessary for entry into the field. In addition, the eligibility period may be extended for a nominee who has taken a formal leave of absence from their employer. The awards will be presented at the Annual Research Meeting in June. All awardees receive complimentary registration and lodging to attend the meeting and a $2,500 prize from the Alice S. Hersh Memorial Fund.
Requirements: Nominators must be members of AcademyHealth although nominees need not be. Nominees from academic and non-academic institutions are encouraged. Nominations must include one letter of nomination (no more than two pages) that reviews the nominee's contribution to the field and specifies the date of the nominee's first research appointment, and must include curriculum vitae.
Geographic Focus: All States
Date(s) Application is Due: Mar 2
Amount of Grant: 2,500 USD
Contact: Jennifer Muldoon, Director; (202) 292-6770; fax (202) 292-6870; jennifer.muldoon@academyhealth.org
Internet: http://www.academyhealth.org/content.cfm?ItemNumber=954@navItemNumber=2342
Sponsor: AcademyHealth
1150 17th Street NW, Suite 600
Washington, D.C. 20036

AcademyHealth Alice S. Hersh Student Scholarships 221
AcademyHealth is pleased to announce a new scholarship program, The Alice S. Hersh Student Scholarship, which provides two free registrations to the National Health Policy Conference (NHPC) and four free registrations to the Annual Research Meeting (ARM). The scholarship is designed to encourage professional and educational development in health services research and policy among student members. This scholarship commemorates the dedication of Alice S. Hersh, the founding executive director of the Association for Health Services Research (AHSR), to supporting the next generation of health services researchers. Application deadline dates are: November 15, for the National Health Policy Conference Scholarships; and March 15, for the Annual Research Meeting Scholarships.
Requirements: Applicants must be current student members of AcademyHealth and must be enrolled in a masters or doctoral program at the time of application. To apply, students must submit the following: a completed scholarship application form (available on the AcademyHealth website); essay outlining how attendance at the ARM or NHPC will assist in your educational and professional pursuits in health services research or policy; current resume; letter from faculty advisor or project coordinator (optional).
Geographic Focus: All States
Date(s) Application is Due: Mar 15; Nov 15
Contact: Jennifer Muldoon, Director; (202) 292-6770; fax (202) 292-6870; jennifer.muldoon@academyhealth.org
Internet: http://www.academyhealth.org/Training/content.cfm?ItemNumber=2064&navItemNumber=2341
Sponsor: AcademyHealth
1150 17th Street NW, Suite 600
Washington, D.C. 20036

AcademyHealth Awards 222
AcademyHealth's awards programs recognize individuals who have made significant contributions to the fields of health services research and health policy. Nominations are requested for four awards: Distinguished Investigator Award; Article-of-the-Year; Alice S. Hersh New Investigator Award; and the Dissertation Award. These awards will be presented at the Annual Research Meeting in June. To obtain a description of each award and guidelines, see AcademyHealth website at: http://www.academyhealth.org/content.cfm?ItemNumber=954@navItemNumber=2342.
Requirements: Nominators must be members of AcademyHealth although nominees need not be. Self-nominations are acceptable. All nominations must include the name and contact information for the nominator and nominee, as well as any additional information specified under the description of each award. Complete nominations must be received at AcademyHealth by March 2. Nominations may be submitted by email or postal mail. All awardees receive complimentary registration and lodging to attend the Annual Research Meeting. The Alice S. Hersh New Investigator Awardee receives a $2,500 prize from the Alice S. Hersh Memorial Fund.
Geographic Focus: All States
Date(s) Application is Due: Mar 2
Contact: Jennifer Muldoon, Director; (202) 292-6770; fax (202) 292-6870; jennifer.muldoon@academyhealth.org
Internet: http://www.academyhealth.org/content.cfm?ItemNumber=954@navItemNumber=2342
Sponsor: AcademyHealth
1150 17th Street NW, Suite 600
Washington, D.C. 20036

AcademyHealth HSR Impact Awards 223
The program requests nominations of health services research that has made a positive impact on health policy and/or practice. The winning impacts will be disseminated widely as part of AcademyHealth's ongoing efforts to promote the field of health services research and communicate its value for health care decision-making. Nominated research may be published or unpublished, a single study or a body of work, the work of an individual or a team. The time frame for when the research was conducted is open, but the impact should have occurred recently. Nominators and nominees are not required to be AcademyHealth members. Self-nominations are accepted. Award winners are announced at the National Health Policy Conference, February 6 and 7, and the winners receive complimentary registration, travel, and lodging to the conference.
Requirements: Each nomination must include a summary describing the research used, the method of translation, and the impact on policy and/or practice (no more than 500 words); a letter of recommendation from the user of this research discussing the impact the research has had; contact information for the lead researcher; and contact information for the nominator (may be the lead researcher).
Geographic Focus: All States
Date(s) Application is Due: Jul 29
Amount of Grant: 2,000 USD
Contact: Jennifer Muldoon; (202) 292-6770; jennifer.muldoon@academyhealth.org
Internet: http://www.academyhealth.org/awards/hsrimpacts.htm
Sponsor: AcademyHealth
1150 17th Street NW, Suite 600
Washington, D.C. 20036

AcademyHealth Nemours Child Health Services Research Awards 224
The Nemours Child Health Services Research Award recognizes the scientific work of emerging scholars in the field of child health services research, particularly research on quality improvement of pediatric health services. Recipients must be within seven years of entry into the field of child health services research as of December 31, 2007. Year of entry will be judged by the nominee's first publication in child health services research. The winner will receive $1,000 in recognition of his/her contribution to child health services research, and the award will be presented at the Child Health Services Research Meeting, held in in Washington, D.C., in conjunction with AcademyHealth's Annual Research Meeting in June.
Requirements: Nominations should include: Letter of nomination that identifies when the candidate first entered the field (i.e., title and year of first publication in child health services research); Curriculum vitae that includes a detailed list of accomplishments that merit this recognition. Self-nominations will be accepted.
Geographic Focus: All States
Date(s) Application is Due: Apr 15
Amount of Grant: 1,000 USD
Contact: Jennifer Muldoon; (202) 292-6770; jennifer.muldoon@academyhealth.org
Internet: http://www.academyhealth.org/content.cfm?ItemNumber=982&navItemNumber=2342
Sponsor: AcademyHealth
1150 17th Street NW, Suite 600
Washington, D.C. 20036

AcademyHealth PHSR Interest Group Student Scholarships 225
Public Health Systems Research (PHSR) is a field of inquiry examining the organization, financing, performance, and impact of health systems, defined as the constellation of governmental and non-governmental actors that influence population health, including health care providers, insurers, purchasers, public health agencies, community-based organizations, and entities that operate outside the traditional sphere of health care. The Student Scholarships recognize graduate students who demonstrate potential to contribute to the field of PHSR through promising research. Scholarship recipients will receive $1,000 to be put toward registration and travel to attend AcademyHealth's Annual Research Meeting (ARM) and the PHSR IG Annual Meeting. Additionally, students will have an opportunity to present their research during the PHSR IG Meeting's poster session.
Requirements: Masters or doctoral students engaged in research that shows potential to advance the understanding of public health systems are encouraged to apply. The deadline for applications is April 15. Applications may be submitted by email or postal mail. To apply, students must submit the following: completed Student Scholarship Form (available at the AcademyHealth website); an abstract for a poster to be presented at the PHSR Interest Group Meeting (this research maybe in-progress); a letter of recommendation from a faculty member or research project director.
Geographic Focus: All States
Date(s) Application is Due: Apr 15
Amount of Grant: 1,000 USD
Contact: Jennifer Muldoon; (202) 292-6770; jennifer.muldoon@academyhealth.org
Internet: http://www.academyhealth.org/Communities/content.cfm?ItemNumber=2535&navItemNumber=2035
Sponsor: AcademyHealth
1150 17th Street NW, Suite 600
Washington, D.C. 20036

AcademyHealth PHSR Research Article of the Year Awards 226
Public Health Systems Research (PHSR) is a field of inquiry examining the organization, financing, performance, and impact of health systems defined as the constellation of governmental and non-governmental actors that influence population health, including health care providers, insurers, purchasers, public health agencies, community-based organizations and entities that operate outside the traditional sphere of health care. The Article of the Year Award recognizes the best scientific work that the field of PHSR has produced and published in the previous year. The awardee will receive complimentary registration for both the PHSR Interest Group Annual Meeting and AcademyHealth's Annual Research Meeting, travel assistance, and a $1,000 cash award. Nominated articles must have been published in a peer-reviewed journal between January 1 and December 31 of the previous year, using either qualitative or quantitative techniques, articles must present, analyze, and comment on new data or synthesize and analyze data that have already been collected. The selection will be based on the article's contribution to the understanding of public health systems; provision of new insights to the field of PHSR; and potential to advance the field and/or challenge current thinking. Nominators must be members of AcademyHealth, although nominees need not be. Self-nominations are acceptable.
Requirements: Nominations must: be received at AcademyHealth by April 15; include the name and contact information for both the nominator and nominee and a copy of the article(s). Nominations may be submitted by email or postal mail. email to: PHSR@academyhealth.org. Mail to: AcademyHealth, Attention: PHSR Article of the Year Committee, 1150 17th Street, NW, Suite 600, Washington, D.C. 20036.
Geographic Focus: All States
Date(s) Application is Due: Apr 15
Amount of Grant: 1,000 USD
Contact: Jennifer Muldoon; (202) 292-6770; jennifer.muldoon@academyhealth.org
Internet: http://www.academyhealth.org/Communities/content.cfm?ItemNumber=2536
Sponsor: AcademyHealth
1150 17th Street NW, Suite 600
Washington, D.C. 20036

ACCP Anticoagulation Training Program Grants 227
The Anticoagulation Traineeship is a minimum of a four-week intensive training program for pharmacy students, residents, and fellows that includes a structured didactic component; extensive clinical experience and participation in ongoing clinical research. The training period can be extended 1 or 2 weeks if needed to meet academic requirements of the trainee's academic institution. Under the current system, the trainee spends five half-days per week actively involved in clinics providing patient care under the direct supervision of one of the ACNA clinicians. Trainees whose usual residence is outside the metropolitan San Antonio area will receive a grant-in-aid of $1000 to partially offset travel and living expenses incurred in conjunction with the traineeship.
Requirements: The program is targeted to doctoral-level pharmacy students enrolled in their final year of professional study, residents, and fellows. The necessary arrangements will be made with the student's home institution so the traineeship qualifies for academic and clerkship credit. Students actively enrolled in a Pharm.D. degree program need not be members of ACCP to be eligible. Residents and fellows must be current members of ACCP to be eligible.
Geographic Focus: All States
Date(s) Application is Due: Nov 1
Amount of Grant: 1,000 USD
Contact: Cathy Englund; (913) 492-3311; fax (913) 492-0088; cenglund@accp.com
Internet: http://www.accp.com/frontiers/research.php#resfel
Sponsor: American College of Clinical Pharmacy
13000 W 87th Street Parkway
Lenexa, KS 66215-4530

ACCP Heart Failure Training Program Grants 228
Funded through an educational grant from Scios this heart failure training program is conducted at six different sites within the U.S; the University of Illinois Medical Center, the University of Michigan Health System, the University of North Carolina Heart Failure Program, the Ohio State University, the University of Southern California Medical Center, and the University of Utah Medical Center UTAH Affiliated Heart Failure Program. Depending on the specific site, the Traineeship is a two to four week, intensive training program that includes extensive clinical experience in either the ambulatory care and/or inpatient setting(s), a structured didactic component, and exposure to ongoing clinical research. The primary goals of the heart failure traineeship are to provide pharmacy practitioners, fellows, and residents with specific knowledge and skills central to the management of patients with heart failure. For practitioners, the traineeship will provide sufficient knowledge and experience such that they should have the basis to establish a heart failure clinic or disease management program within their own practices.
Requirements: Residents, fellows, or practicing clinical pharmacists must be current members of ACCP at the time of application to be eligible.
Geographic Focus: All States
Date(s) Application is Due: May 1; Sep 1
Amount of Grant: 1,000 USD
Contact: Kathy Lemons, Administrative Assistant, Professional Development; (913) 492-3311; fax (913) 492-0088; klemons@accp.com
Internet: http://www.accp.com/
Sponsor: American College of Clinical Pharmacy
13000 W 87th Street Parkway
Lenexa, KS 66215-4530

ACE Charitable Foundation Grants 229
ACE supports the communities around the world in which its employees live and work through its established ACE Foundations and through company-sponsored volunteer initiatives. They focus their philanthropic support in the areas of education, poverty and health, and the environment. Particular consideration is given to opportunities where ACE employees' time and expertise can be utilized in addition to financial support. ACE encourages the development of local and regional initiatives, which reflect their employees' commitment to the needs of the communities in which they live and work. ACE employees direct a significant portion of the company's charitable giving.

Requirements: The mission of the ACE Charitable Foundation is to assist less fortunate individuals and communities in achieving and sustaining productive and healthy lives in geographic areas where ACE employees live and work. The ACE Charitable Foundation strives to accomplish this by focusing the majority of its funds on clearly defined projects that have measurable objectives and outcomes and solve problems in the areas of education, the environment and poverty and health. Submit a brief letter of introduction to the Executive Director for consideration.
Geographic Focus: Arizona, Arkansas, California, Colorado, Connecticut, Delaware, District of Columbia, Florida, Georgia, Hawaii, Illinois, Indiana, Kansas, Louisiana, Maryland, Massachusetts, Michigan, Minnesota, Nevada, New Jersey, New York, North Carolina, Ohio, Oregon, Pennsylvania, Puerto Rico, South Carolina, Texas, Virginia, Washington
Samples: The following is a partial list of organizations with which the ACE Charitable Foundation is associated: American Red Cross; City of Hope Cancer Center; Fairmount Park Conservancy; Kids' Chance; Need in Deed; Philadelphia Cares; Philadelphia Museum of Art; Philadelphia Academies, Inc; United Way
Contact: Eden Kratchman, Executive Director; acefoundation@acegroup.com
Internet: http://www.acelimited.com/AceLimitedRoot/About+ACE/ACE+Philanthropy+Worldwide/The+ACE+INA+Foundation+One+Column.htm
Sponsor: ACE Charitable Foundation
436 Walnut Street, WA 08G
Philadelphia, PA 19106

ACF Foundation Grants 230
The Foundation supports organizations involved with arts and culture, children and youth services, community and economic development, education, health care, higher education, and human services. Giving is limited to areas of company operations, with emphasis on St. Louis and St. Charles, Missouri. There are no specific deadlines or forms, and applicants should forward the entire proposal to the office.
Requirements: 501(c)3 nonprofit organizations, based in or serving the the communities of St. Louis or St. Charles, Missouri, are eligible to apply.
Restrictions: No grants are given to individuals.
Geographic Focus: Missouri
Contact: Nancy Collins; (636) 940-5101 or (636) 940-5000; ncollins@arleasing.com
Sponsor: ACF Foundation
101 Clark Street, Suite 201
St. Charles, MO 63301-2081

ACF Native American Social and Economic Development Strategies Grants 231
Grant awards made under this Funding Opportunity Announcement are for projects that promote economic and social self sufficiency for American Indians, Alaska Natives, Native Hawaiians, and other Native American Pacific Islanders from American Samoa, Guam, and Commonwealth of the Northern Mariana Islands. ANA is particularly interested in projects designed to grow local economies, strengthen Native American families, and decrease the high rate of social challenges caused by the lack of community-based business, and social and economic infrastructure. ANA has identified two major program areas of interest for this funding opportunity announcement, which include social development and economic development. In the area of social development, of most interest are proposals that improve: human services; community living; early childhood development; youth development; community health; arts and culture; safety and security; nutrition and fitness; and the strengthening of families. In the area of economic development, of most interest are proposals that improve: economic stability; economic competitiveness; agriculture; infrastructure; emergency preparedness; subsistence; and commercial trade. Awards will range from $100,000 to $400,000, and the current application deadline is April 6.
Requirements: Eligible applicants include Federally recognized Indian Tribes; consortia of Indian Tribes; incorporated non-Federally recognized Tribes; incorporated non-profit, multi-purpose, community-based Indian organizations; urban Indian centers; National or regional incorporated non-profit Native American organizations with Native American community-specific objectives; Alaska Native villages, as defined in the Alaska Native Claims Settlement Act and/or non-profit village consortia; incorporated non-profit Alaska Native multi-purpose, community-based organizations; non-profit Alaska Native Regional Corporations/Associations in Alaska with village-specfic projects; non-profit native organizations in Alaska with village-specific projects; public and non-profit private agencies serving Native Hawaiians; public and private non-profit agencies serving native peoples from Guam, American Samoa, or the Commonwealth of the Northern Mariana Islands (the populations served may be located on these islands or in the United States); tribally controlled community colleges, tribally controlled post-secondary vocational institutions, and colleges and universities located in Hawaii, Guam, American Samoa, or the Commonwealth of the Northern Mariana Islands which serve Native Pacific Islanders; and non-profit Alaska Native community entities or tribal governing bodies (Indian Reorganization Act or Traditional Councils) as recognized by the Bureau of Indian Affairs. Faith-based and community organizations that meet eligibility requirements are eligible to receive awards under this funding opportunity announcement. Grantees must provide at least 20 percent of the total approved cost of the project.
Restrictions: Individuals, foreign entities, and sole proprietorship organizations are not eligible to compete for, or receive, awards made under this announcement.
Geographic Focus: All States, Guam, Marshall Islands, Northern Mariana Islands, American Samoa
Date(s) Application is Due: Apr 6
Contact: Carmelia Strickland; (877) 922-9262; anacomments@acf.hhs.gov
Internet: http://www.acf.hhs.gov/grants/open/foa/files/HHS-2014-ACF-ANA-NA-0776_2.htm
Sponsor: Administration for Children and Families
370 L'Enfant Promenade SW, Aerospace Center, 2nd Floor-West
Washington, D.C. 20447

ACGT Investigators Grants 232
The Investigators Award in Clinical Translation of Gene Therapy for Cancer distributes funds over 3-5 years, inclusive of a maximum of 10% indirect costs. Funds may be used at the recipient's discretion for salary, technical assistance, supplies, animals or capital equipment, but may not support staff not directly related to the project, e.g. secretaries or administrative assistants. Purchase of equipment is not allowed in the final year of the grant. ACGT is accepting grant applications to produce and release-testing of the clinical trial agents under cGMP, conduct the necessary pre-clinical pharmacological and toxicological studies in appropriate animal models, and/or conducting the clinical translational trials in patients in support of an Investigative New Drug application to the FDA.
Requirements: The candidate must hold an MD, PhD, or equivalent degree and be a tenure-track or tenured faculty. The investigator must be conducting original research as an independent faculty member. ACGT has no citizenship restrictions, and research supported by the award must be conducted at medical schools and research centers only in the United States. While the unambiguous demonstration of preclinical efficacy in cancer treatment by gene therapy is a pre-requisite, entering into the clinical trial during the funding period is also a requirement.
Geographic Focus: All States
Amount of Grant: Up to 1,000,000 USD
Contact: Grace Pedersen; (203) 358-8000, ext. 495; aneslage@acgtfoundation.org
Internet: http://www.acgtfoundation.org/grants.html
Sponsor: Alliance for Cancer Gene Therapy
96 Cummings Point Road
Stamford, CT 06902

ACGT Young Investigator Grants 233
The alliance funds research aimed at furthering the development of gene therapy approaches to the treatment of cancer. The overall objectives of this grant are to advance gene therapy into the causes, treatment, and prevention of all types of cancer by promoting development of novel and innovative studies by young investigators. The emphasis of this initiative is to promote basic, and preclinical research approaches utilizing cells and genes as medicine. The six main areas of research ACGT will support are: tumor-specific replicating viruses and bacteria, anti-angiogenesis, immune-mediated gene therapy and cancer vaccines, oncogenes/suppressor oncogenes/apoptosis, tumor targeting and vector development, and other cancer gene therapy research. The three-year grant may be used at the recipient's discretion for salary, technical assistance, supplies, animals or capital equipment, but may not support staff not directly related to the project, e.g. secretaries or administrative assistants. Purchase of equipment is not allowed in the third year of the grant. Continued support is contingent upon submission and approval of a noncompetitive renewal application each year. Guideline and applications are available online.
Requirements: Candidates must hold an MD, MPH, PhD, or equivalent degree and be a tenure-track assistant professor within five years of their initial appointment to this rank, at the time of award activation. The investigator must be conducting original research as an independent faculty member. ACGT has no citizenship restrictions, and research supported by the award must be conducted at medical schools and research centers only in the United States.
Geographic Focus: All States
Date(s) Application is Due: Sep 21
Amount of Grant: Up to 500,000 USD
Contact: Grace Pedersen, Foundation Administrator; (203) 358-8000, ext. 495; aneslage@acgtfoundation.org
Internet: http://www.acgtfoundation.org/grants.html
Sponsor: Alliance for Cancer Gene Therapy
96 Cummings Point Road
Stamford, CT 06902

Achelis Foundation Grants 234
Elisabeth Achelis was born in Brooklyn Heights in 1880. She used her inheritance from her father Fritz Achelis, who was President of the American Hard Rubber Company, to establish the Achelis Foundation in 1940 to aid and contribute to charitable, benevolent, educational and religious uses and purposes for the moral, ethical, physical, mental and intellectual well-being and progress of mankind; to aid and contribute to methods for the peaceful settlement of international differences; to aid and contribute to the furtherance of the objects and purposes of any charitable, benevolent, educational or religious institution or agency; and to establish and maintain charitable, benevolent and educational institutions and agencies. The Achelis Foundation shares trustees, staff, office space, and even a website with the Bodman Foundation which has a similar mission and geographic area of concentration (both foundations give in New York City, while the Bodman Foundation also gives in New Jersey). Funding is concentrated in six program areas: arts and culture; education; employment; health; public policy; and youth and families. Most recent awards have ranged from $10,000 to $200,000, though typical awards average $20,000 to $50,000.
Requirements: 501(c)3 organizations based in New York City that fall within the foundation's areas of interest are welcome to submit an inquiry or proposal letter by regular mail (initial inquiries by email or fax are not accepted, nor are CDs, DVDs, computer discs, or video tapes). An initial inquiry to the foundation should include only the following items: a proposal letter that briefly summarizes the history of the project, need, objectives, time period, key staff, project budget, and evaluation plan; the applicant's latest annual report and complete set of audited financial statements; and the applicant's IRS 501(c)3 tax-exemption letter. Applications may be submitted at any time during the year. Each request is reviewed by staff and will usually receive a written response within thirty days. Those requests deemed consistent with the interests and resources of the foundation will be evaluated further and more information will be requested. Foundation staff may request a site visit, conference call, or meeting. All grants are reviewed and approved by the Trustees at one of their three board meetings in May, September, or December.

Restrictions: The foundation generally does not make grants for the following purposes or program areas: nonprofit organizations outside of New York; annual appeals, dinner functions, and fundraising events; endowments and capital campaigns; loans and deficit financing; direct grants to individuals; individual day-care and after-school programs; housing; organizations or projects based outside the U.S; films or video projects; small art, dance, music, and theater groups; individual K-12 schools (except charter schools); national health and mental health organizations; and government agencies or nonprofit organizations significantly funded or reimbursed by government agencies. Limited resources prevent the foundations from funding the same organization on an ongoing annual basis.
Geographic Focus: New York
Amount of Grant: 10,000 - 200,000 USD
Samples: American Ballet Theatre, New York, New York, $25,000 - general operating support (2014); East Harlem School at Exodus House, New York, New York, $25,000 - general operating support (2014; International Center for the Disabled, New York, New York, $25,000 - to support the ReadyNow program (2014).
Contact: John B. Krieger; (212) 644-0322; main@achelis-bodman-fnds.org
Internet: http://www.achelis-bodman-fnds.org/guidelines.html
Sponsor: Achelis Foundation
767 Third Avenue, 4th Floor
New York, NY 10017-2023

Acid Maltase Deficiency Association Helen Walker Research Grant   235
The Helen Walker Research Grant, offered through the Acid Maltase Deficiency Association, is an annual award aimed at promoting research and education about Pompe disease. The Acid Maltase Deficiency Association was established in 1995 to assist in funding research and to promote public awareness of the affliction. Pompe disease is one of a family of 49 rare genetic disorders known as Lysosomal Storage Diseases, or LSDs. Pompe disease is also known as Acid Maltase Deficiency or Glycogen Storage Disease type II, and it affects an estimated 5,000 to 10,000 people in the developed world. The research grant is in the amount of $140,000. The annual deadline for application is September 4.
Geographic Focus: All States
Date(s) Application is Due: Sep 4
Amount of Grant: 140,000 USD
Contact: Tiffany L. House; (210) 494-6144; fax (210) 490-7161; TiffanyLHouse@aol.com
Internet: http://www.amda-pompe.org/index.php/main/news/2015_helen_walker_research_grant_for_pompe_disease_now_accepting_applicatio
Sponsor: Acid Maltase Deficiency Association
P.O. Box 700248
San Antonio, TX 78270

Ackerman Foundation Grants   236
The Ackerman Foundation was established as a charitable trust in 1992 by James F. Ackerman, a local entrepreneur and philanthropist. As an Indianapolis based organization, grants are made predominately to central Indiana organizations as well as a few national medical research institutions. Specifically, the foundation focuses on Indiana cultural institutions and organizations benefiting health and human services, community development, and education. Grant requests will be considered for both operating fund purposes and as capital campaigns. The foundation does not have grant application forms. To be considered for assistance, an organization should write a brief one or two page letter describing its proposal. The Trustees of the foundation meet semi-annually on the business day that falls on or closest to June 15 and December 15.
Requirements: Established under the laws of the State of Indiana, the foundation considers grant proposals from eligible organizations which are tax exempt under the United States Internal Revenue Service Code section 501(c)3.
Restrictions: The foundation does not make grants to individuals.
Geographic Focus: All States
Date(s) Application is Due: May 15; Nov 15
Amount of Grant: 500 - 100,000 USD
Contact: John F. Ackerman; (317) 663-0205; fax (317) 663-0215; jdisbro@cardinalep.com
Internet: http://ackermanfoundation.com/
Sponsor: Ackerman Foundation of Indiana
280 E. 96th Street, Suite 350
Indianapolis, IN 46240-3858

ACL Alzheimer's Disease Initiative Grants   237
The purpose of the ADI-SSS program is to provide grants to public and private entities that are working within existing, dementia-capable, long term services and supports systems and are committed to serving populations with the most need and living with or at risk of developing Alzheimer's disease or a related dementia (ADRD) and their caregivers. Successful applicants will propose services designed to expand their existing dementail-capable service system to address the needs of three out of four identified service gaps identified in the Funding Opportunity Announcement. The grantees benefit from targeted technical assistance provided by the National Alzheimer's and Dementia Resource Center which is funded through a contract awarded under the Alzheimer's Disease Supportive Services program It is expected that ten awards of between $800,000 and $1,000,000 will be given annually. Letters of Intent are due each year by June 8, with the annual deadline for application submission being July 24.
Requirements: Eligible applicants include: for profit organizations other than small businesses; Native American tribal governments (Federally recognized); nonprofits that do not have a 501(c)3 status with the IRS, other than institutions of higher education; city or township governments; independent school districts; county governments; Native American tribal organizations (other than Federally recognized tribal governments); state governments; public and state controlled institutions of higher education; private institutions of higher education; special district governments; nonprofits having a 501(c)3 status with the IRS, other than institutions of higher education; small businesses; and public housing authorities/Indian housing authorities.
Geographic Focus: All States
Date(s) Application is Due: Jul 24
Amount of Grant: 800,000 - 1,000,000 USD
Contact: Erin Long, Project Officer; (202) 357-3448 or (202) 401-4634; erin.long@acl.gov or aclinfo@acl.hhs.gov
Internet: http://www.grants.gov/web/grants/view-opportunity.html?oppId=284132
Sponsor: U.S. Department of Health and Human Services
330 C Street SW
Washington, D.C. 20201

ACL Business Acumen for Disability Organizations Grants   238
The purpose of this initiative is to: develop baseline knowledge about the content and infrastructure needs of community-based disability organizations through surveys and feasibility studies; and utilize a learning collaborative model to provide targeted technical assistance to up to fifteen state coalitions of community-based disability organizations that seek to build their business capacity to contract with health care entities (e.g. hospitals, health systems, accountable care organizations, managed/integrated care plans). It is expected that one award of between $700,000 and $750,000 will be given annually. Letters of Intent are due each year by June 1, with the annual deadline for application submission being July 28.
Requirements: Eligible applicants include: Native American tribal governments (Federally recognized); Native American tribal organizations (other than Federally recognized tribal governments); small businesses; nonprofits that do not have a 501(c)3 status with the IRS, other than institutions of higher education; nonprofits having a 501(c)3 status with the IRS, other than institutions of higher education; county governments; and state governments.
Geographic Focus: All States
Date(s) Application is Due: Jul 28
Amount of Grant: 700,000 - 750,000 USD
Contact: Katherine Cargill-Willis, Project Officer; (202) 795-7322 or (202) 401-4634; katherine.cargill-willis@acl.hhs.gov or aclinfo@acl.hhs.gov
Internet: http://www.acl.gov/Funding_Opportunities/Announcements/Index.aspx
Sponsor: U.S. Department of Health and Human Services
330 C Street SW
Washington, D.C. 20201

ACL Centers for Independent Living Competition Grants   239
Centers for Independent Living provide services to assist individuals with disabilities to achieve their maximum potential within their families and communities. CILs are consumer-controlled, community-based, cross-disability, non-residential, private nonprofit agencies. Their work focuses on five core services: information and referrals; independent living skills training; peer counseling; individual and systems advocacy; and services that facilitate transition from nursing homes and other institutions to the community, assistance to individuals at risk of entering institutions, and transition of youth to post-secondary life. Applicants should be prepared to demonstrate that they can comply with standards and assurance as outlined in section 725 of part C of title VII of the Rehabilitation Act of 1973, as amended by the Workforce Innovation and Opportunity Act (Pub.L.113-128), and service areas are consistent with the design included in the State Plan for establishing a network of centers. It is expected that four awards of between $95,096 and $218,997 will be given annually. The annual deadline for application submission is July 26.
Requirements: Eligible applicants are nonprofits having a 501(c)3 status with the IRS, other than institutions of higher education.
Geographic Focus: All States
Date(s) Application is Due: Jul 26
Amount of Grant: 95,096 - 218,997 USD
Contact: Veronica Hogan, Project Officer; (202) 795-7365 or (202) 401-4634; veronica.hogan@acl.hhs.gov or aclinfo@acl.hhs.gov
Internet: http://www.grants.gov/web/grants/view-opportunity.html?oppId=284227
Sponsor: U.S. Department of Health and Human Services
330 C Street SW
Washington, D.C. 20201

ACL Diversity Community of Practice Grants   240
The purpose of the Diversity Community of Practice Grant funding opportunity is to fund one entity to develop a Community of Practice (CoP), designed to: build capacity across and within states through state consortia; and to create and share policies, practices, and systems that support the critical need in the field of intellectual and developmental disabilities for leaders from culturally and linguistically distinct backgrounds. The CoP will assist states in identifying gaps, preparing a plan on how to address needs, and implementing this plan based on the technical assistance provided by the technical assistance provider. It is expected that the award will be between $300,000 and $350,000. Letters of Intent are due each year by June 27, with the annual deadline for application submission being July 26.
Requirements: Eligible applicants include: public and state controlled institutions of higher education; special district governments; city or township governments; nonprofits that do not have a 501(c)3 status with the IRS, other than institutions of higher education; state governments; county governments; nonprofits having a 501(c)3 status with the IRS, other than institutions of higher education; Native American tribal organizations (other than Federally recognized tribal governments); private institutions of higher education; and Native American tribal governments (Federally recognized).

Geographic Focus: All States
Date(s) Application is Due: Jul 26
Amount of Grant: 300,000 - 350,000 USD
Contact: Larissa Crossen, Project Officer; (202) 795-7333 or (202) 401-4634; larissa.crossen@acl.gov or aclinfo@acl.hhs.gov
Internet: http://www.grants.gov/web/grants/view-opportunity.html?oppId=284209
Sponsor: U.S. Department of Health and Human Services
330 C Street SW
Washington, D.C. 20201

ACL Empowering Seniors to Prevent Health Care Fraud Grants   241
The Senior Medicare Patrol (SMP) Program includes 53 project grants in all states, the District of Columbia, Puerto Rico, and Guam. SMP projects have made great progress in recruiting and training volunteers on Medicare errors, fraud and abuse. These volunteers empower and assist Medicare beneficiaries, their families, and caregivers to prevent, detect, and report health care fraud, errors, and abuse through outreach, counseling, and education. SMP projects actively work to disseminate SMP fraud prevention and identification information through the media, outreach campaigns, community events and also working with beneficiaries who present with complex cases such as compromised Medicare numbers. Through these efforts, beneficiaries contact the projects with inquiries and complaints regarding Medicare, Medicaid and other health care or related consumer issues. With this funding opportunity ACL anticipates awarding one new cooperative agreement in the amount of $282,950 to cover the State of Wisconsin. The annual deadline for application submission is July 20.
Requirements: Eligible applicants include: private institutions of higher education; county governments; nonprofits that do not have a 501(c)3 status with the IRS, other than institutions of higher education; nonprofits having a 501(c)3 status with the IRS, other than institutions of higher education; special district governments; Native American tribal organizations (other than Federally recognized tribal governments); state governments; Native American tribal governments (Federally recognized); city or township governments; and public and state controlled institutions of higher education.
Geographic Focus: Wisconsin
Date(s) Application is Due: Jul 20
Amount of Grant: 282,950 USD
Contact: Stacey Platte; (202) 401-4634; stacey.platte@acl.hhs.gov or aclinfo@acl.hhs.gov
Internet: http://www.acl.gov/Funding_Opportunities/Announcements/Index.aspx
Sponsor: U.S. Department of Health and Human Services
330 C Street SW
Washington, D.C. 20201

**ACL Field Initiated Projects Program: Minority-Serving Institution   242
  Development Grants**
The purpose of the Field Initiated (FI) Projects program is to develop methods, procedures, and rehabilitation technology that maximize the full inclusion and integration into society, employment, independent living, family support, and economic and social self-sufficiency of individuals with disabilities, especially individuals with the most severe disabilities. Another purpose of the FI Projects program is to improve the effectiveness of services authorized under the Rehabilitation Act of 1973, as amended. In carrying out a development project under a FI Projects development grant, a grantee must use knowledge and understanding gained from research to create materials, devices, systems, or methods beneficial to the target population, including design and development of prototypes and processes. It is expected that one award of $200,000 will be given annually. Letters of Intent are due each year by May 6, with the annual deadline for application submission being June 14.
Requirements: Minority entities are the only applicants that are eligible to apply for this grant opportunity.
Geographic Focus: All States
Date(s) Application is Due: Jun 14
Amount of Grant: 200,000 USD
Contact: Marlene Spencer, Program Officer; (202) 245-7532 or (202) 401-4634; fax (202) 245-7323; marlene.spencer@acl.hhs.gov or aclinfo@acl.hhs.gov
Internet: http://www.acl.gov/Funding_Opportunities/Announcements/Index.aspx
Sponsor: U.S. Department of Health and Human Services
330 C Street SW
Washington, D.C. 20201

**ACL Field Initiated Projects Program: Minority-Serving Institution Research   243
  Grants**
The purpose of the Field Initiated (FI) Projects program is to develop methods, procedures, and rehabilitation technology that maximize the full inclusion and integration into society, employment, independent living, family support, and economic and social self-sufficiency of individuals with disabilities, especially individuals with the most severe disabilities. Another purpose of the FI Projects program is to improve the effectiveness of services authorized under the Rehabilitation Act of 1973, as amended. In carrying out a development project under a FI Projects development grant, a grantee must use knowledge and understanding gained from research to create materials, devices, systems, or methods beneficial to the target population, including design and development of prototypes and processes. It is expected that one award of $200,000 will be given annually. Letters of Intent are due each year by May 6, with the annual deadline for application submission being June 14.
Requirements: Minority entities are the only applicants that are eligible to apply for this grant opportunity.
Geographic Focus: All States
Date(s) Application is Due: Jun 14
Amount of Grant: 200,000 USD
Contact: Marlene Spencer, Project Officer; (202) 245-7532 or (202) 401-4634; fax (202) 245-7323; marlene.spencer@acl.hhs.gov or aclinfo@acl.hhs.gov
Internet: http://www.acl.gov/Funding_Opportunities/Announcements/Index.aspx
Sponsor: U.S. Department of Health and Human Services
330 C Street SW
Washington, D.C. 20201

ACL Learning Collaboratives for Advanced Business Acumen Skills Grants   244
The purpose of this Funding Opportunity is to expand the readiness of community-based aging and disability organization for contracting with integrated care entities, and prepare state and community-based aging and disability organizations to be active stakeholders and partners in the development and implementation of integrated care systems. To accomplish this, ACL intends to fund one award to achieve the following tasks: organize and conduct three to five topically-based action learning collaboratives to address next generation issues (such as continuous quality improvement, infrastructure and technology, generating and maintaining volume, data pooling, and more); and to provide targeted technical assistance to networks of community-based aging and disability organizations. The grant's primary purpose is to create knowledge and capture insights through these collaboratives, in order to incorporate the findings into future curriculum for national dissemination. It is expected that one award of between $210,350 and $500,000 will be given annually. Letters of Intent are due each year by June 1, with the annual deadline for application submission being July 28.
Requirements: The competition is open to domestic, public or private non-profit entities including state and local governments, Indian tribal governments and organizations (American Indian/Alaskan Native/Native American), faith-based organizations, and community-based organizations.
Geographic Focus: All States
Date(s) Application is Due: Jul 28
Amount of Grant: 210,350 - 500,000 USD
Contact: Lauren Solkowski, Project Officer; (202) 795-7440 or (202) 401-4634; lauren.solkowski@acl.hhs.gov or aclinfo@acl.hhs.gov
Internet: http://www.acl.gov/Funding_Opportunities/Announcements/Index.aspx
Sponsor: U.S. Department of Health and Human Services
330 C Street SW
Washington, D.C. 20201

**ACL Medicare Improvements for Patients and Providers Act Funding for   245
  Beneficiary Outreach and Assistance for Title VI Native American**
Title VI Native American Programs can fill an important role in providing valuable support to eligible Native American elders for the Low Income Subsidy program (LIS), Medicare Savings Program (MSP), Medicare Part D, Medicare prevention benefits and screenings and in assisting beneficiaries in applying for benefits. ACL/AoA seeks certification from Title VI Native American programs that they will use the funds to coordinate at least one community announcement and at least one community outreach event to inform and assist eligible American Indian, Alaska Native or Native Hawaiian elders about the benefits available to them through Medicare Part D, the Low Income Subsidy, the Medicare Savings Program or Medicare prevention benefits and screenings and counsel those who are eligible. The annual deadline for application submission is August 4.
Requirements: Only current Title VI Native American Program grantees are eligible to apply for this funding opportunity.
Geographic Focus: All States
Date(s) Application is Due: Aug 4
Contact: Cecelia Aldridge, Project Officer; (202) 795-7293 or (202) 401-4634; cecelia.aldridge@acl.hhs.gov or aclinfo@acl.hhs.gov
Internet: http://www.acl.gov/Funding_Opportunities/Announcements/Index.aspx
Sponsor: U.S. Department of Health and Human Services
330 C Street SW
Washington, D.C. 20201

ACL Partnerships in Employment Systems Change Grants   246
The purpose of the Partnerships in Employment Systems Change funding opportunity is to encourage state partnerships and systems change efforts that will ultimately contribute to: the development of policies that support competitive employment in integrated settings as the first and desired outcome for youth and young adults with developmental disabilities including intellectual disabilities; the removal of systemic barriers to competitive employment in integrated settings; the implementation of strategies and best practices that improve employment outcomes for youth with intellectual and developmental disabilities; and enhanced statewide collaborations that can facilitate the transition process from secondary and post-secondary school, or other pre-vocational settings, to complete employment in integrated settings. It is expected that six awards of between $200,000 and $250,000 will be given annually. Letters of Intent are due each year by June 27, with the annual deadline for application submission being July 26.
Requirements: Eligible applicants include: public housing authorities/Indian housing authorities; nonprofits that do not have a 501(c)3 status with the IRS, other than institutions of higher education; special district governments; Native American tribal governments (Federally recognized); nonprofits having a 501(c)3 status with the IRS, other than institutions of higher education; county governments; city or township governments; independent school districts; state governments; private institutions of higher education; Native American tribal organizations (other than Federally recognized tribal governments); and public and state controlled institutions of higher education.
Geographic Focus: All States

Date(s) Application is Due: Jul 26
Amount of Grant: 200,000 - 250,000 USD
Contact: Larissa Crossen, Project Officer; (202) 795-7333 or (202) 401-4634; larissa.crossen@acl.gov or aclinfo@acl.hhs.gov
Internet: http://www.grants.gov/web/grants/view-opportunity.html?oppId=284181
Sponsor: U.S. Department of Health and Human Services
330 C Street SW
Washington, D.C. 20201

ACL Self-Advocacy Resource Center Grants 247

The Self-Advocacy Resource Center, guided by an advisory committee, will provide information to several entities including self advocacy organizations. The advisory committee members will be selected by the grantee and 75% of its members will be people with I/DD. The Center will have several activities including: documenting and disseminating tools; collecting success stories and best practices concerning self-advocacy; research activities, policy development, training and technical assistance; small leadership grants; and creating and maintaining a website. It is expected that one award of $400,000 will be given annually. Letters of Intent are due each year by June 14, with the annual deadline for application submission being July 25.
Requirements: Eligible applicants include: nonprofits that do not have a 501(c)3 status with the IRS, other than institutions of higher education; and nonprofits having a 501(c)3 status with the IRS, other than institutions of higher education.
Geographic Focus: All States
Date(s) Application is Due: Jul 25
Amount of Grant: 400,000 USD
Contact: Katherine Cargill-Willis, Project Officer; (202) 357-3436 or (202) 401-4634; katherine.cargill-willis@acl.hhs.gov or aclinfo@acl.hhs.gov
Internet: http://www.grants.gov/web/grants/view-opportunity.html?oppId=284144
Sponsor: U.S. Department of Health and Human Services
330 C Street SW
Washington, D.C. 20201

ACL Training and Technical Assistance Center for State Intellectual and Developmental Disabilities Delivery Systems Grants 248

The purpose of this funding opportunity is to provide training and technical assistance to state intellectual and developmental disabilities state delivery systems and to assist states with building capacity across and within their states to create policies and practices in order to improve competitive integrated employment outcomes for youth and young adults with intellectual and developmental disabilities over a five-year period. Technical assistance will not be limited to states currently funded under the Partnerships in Employment Systems Change grant. The training and technical assistance provider will also incorporate a mentoring component and collaboration with other federal training and technical assistance projects that will assist state with building capacity and systems around provider transformation, school-to-work transition, employer engagement, HCBS final rule, WIOA and other emerging issues for youth and young adults with intellectual and developmental disabilities. It is expected that one award of between $350,000 and $400,000 will be given annually. Letters of Intent are due each year by June 27, with the annual deadline for application submission being July 26.
Requirements: Eligible applicants include: Native American tribal organizations (other than Federally recognized tribal governments); nonprofits having a 501(c)3 status with the IRS, other than institutions of higher education; city or township governments; county governments; private institutions of higher education; nonprofits that do not have a 501(c)3 status with the IRS, other than institutions of higher education; public and state controlled institutions of higher education; state governments; special district governments; Native American tribal governments (Federally recognized); and independent school districts.
Geographic Focus: All States
Date(s) Application is Due: Jul 26
Amount of Grant: 350,000 - 400,000 USD
Contact: Larissa Crossen, Project Officer; (202) 795-7333 or (202) 401-4634; larissa.crossen@acl.gov or aclinfo@acl.hhs.gov
Internet: http://www.grants.gov/web/grants/view-opportunity.html?oppId=284210
Sponsor: U.S. Department of Health and Human Services
330 C Street SW
Washington, D.C. 20201

ACL University Centers for Excellence in Developmental Network Diversity and Inclusion Training Action Planning Grants 249

Funds from this grant will be used to support the UCEDD's provision of a Diversity Training program that equips all levels of the UCEDD with the skills, knowledge, strategies and methods to guide the work of sustaining diversity and cultural and linguistic competence within programs concerned with developmental disabilities. Funds will be utilized to work towards the long term goals of: increasing diversity of leadership, staff and governing bodies across the DD network; building cultural competence capacity with the leadership, staff, and governing bodies across the DD network; increasing the number of persons from underrepresented racial, ethnic groups, people with disabilities, people from disadvantaged backgrounds who benefit from AIDD supported programs; and improving the recruitment and employment of underrepresented groups including racial and ethnic groups, people with disabilities, and people from diverse or disadvantaged backgrounds in the UCEDD. It is expected that one award of between $200,000 and $300,000 will be given annually. Letters of Intent are due each year by June 10, with the annual deadline for application submission being July 25.
Requirements: Entities eligible to apply for funds under this program announcement are the 67 current AIDD grantees that are designated UCEDDs.

Geographic Focus: All States
Date(s) Application is Due: Jul 25
Amount of Grant: 200,000 - 300,000 USD
Contact: Pamela O'Brien, Project Officer; (202) 357-3487 or (202) 401-4634; pamela.obrien@acl.hhs.gov or aclinfo@acl.hhs.gov
Internet: http://www.grants.gov/web/grants/view-opportunity.html?oppId=284165
Sponsor: U.S. Department of Health and Human Services
330 C Street SW
Washington, D.C. 20201

ACS Cancer Control Career Development Awards for Primary Care Physicians 250

ACS annually awards three three-year career development awards to physicians specializing in primary care (e.g., family practice, internal medicine, pediatrics, and obstetrics and gynecology). These awards are intended to encourage and assist in the development of promising candidates who will pursue academic careers in primary care specialties. It is anticipated that physicians trained under these awards will improve cancer control through involvement in primary care practice, education, and research activities related to cancer control. Awards are made for three years with progressive stipends.
Requirements: Candidates must be U.S. citizens or permanent residents; must hold an MD, DO, or equivalent degree; and must have completed residency requirements of the appropriate primary care specialty board. Candidate must hold academic rank from instructor to assistant professor and must not be more than 10 years out of training.
Restrictions: Applicants may not have academic rank above that of assistant professor and must not be tenured or be the section head (or equivalent) in his/her discipline. Additionally, the applicant may not be training as a fellow at the time of the award.
Geographic Focus: All States
Date(s) Application is Due: Oct 15
Amount of Grant: Up to 100,000 USD
Contact: Extramural Grants Department; (404) 329-7558 or (800) 875-2562; fax (404) 321-4669; grants@cancer.org
Internet: http://www.cancer.org/docroot/RES/content/RES_5_2x_Cancer_Control_Career_Development_Awards_for_Primary_Care_Physicians.asp?sitearea=RES
Sponsor: American Cancer Society
3100 Wyman Park Drive, Suite W400
Baltimore, MD 21211

ACS Clinical Research Professor Grants 251

The American Cancer Society offers a limited number of grants to established investigators in mid-career who have made seminal contributions that have changed the direction of clinical, psychosocial, behavioral, health policy or epidemiologic cancer research. Furthermore, it is expected that these investigators will continue to provide leadership in their research area. Up to two awards are made annually for a five-year term that can be renewed once. The award of up to $80,000 per year (direct costs only) may be used for salary or research project support. Interested individuals should submit their curriculum vitae with a complete bibliography and a letter of intent that briefly describes their seminal contributions to research. This information must be submitted via the Research Professor letter of intent on proposalCENTRAL at: https://proposalcentral.altum.com. The Letter of Intent for the Clinical Research Professor Award must be submitted no later than August 1. Upon receipt, the ACS will provide appropriate guidance either by telephone or email. Clinical Research Professor candidates whose Letter of Intent has been approved must submit their application for the award before the October 15 deadline.
Geographic Focus: All States
Date(s) Application is Due: Aug 1; Oct 15
Amount of Grant: Up to 80,000 USD
Contact: John J. Stevens, Vice President; (404) 329-7550; john.stevens@cancer.org
Internet: http://www.cancer.org/docroot/RES/content/RES_5_2x_Clinical_Research_Professorships.asp?sitearea=RES
Sponsor: American Cancer Society
3100 Wyman Park Drive, Suite W400
Baltimore, MD 21211

ACS Doctoral Degree Scholarships in Cancer Nursing 252

Awarded to graduate students pursuing doctoral study in the field of cancer nursing research, and preparing for careers as nurse scientists. Initial awards are made for up to two years with the possibility of a two-year renewal. There is a stipend of $15,000 per year.
Requirements: The applicant must: be currently enrolled in or applying to a doctoral degree program in nursing or a related field of research; meet requirements for doctoral study and must have been accepted by the institution to which s/he has applied at the time of funding; have a current license to practice as a registered nurse; project a program of study that integrates cancer nursing and provides evidence of faculty support for the program of study. Scholarship recipients must take a minimum of 18 credit hours or 6 courses per year (unless coursework has been completed and the student accepted to candidacy).
Restrictions: Students in programs that award the doctorate of nursing practice (DNP) are not eligible for this program.
Geographic Focus: All States
Date(s) Application is Due: Oct 15
Amount of Grant: 15,000 USD
Contact: Virginia Krawiec, (404) 329-7612; fax (404) 321-4669; ginger.krawiec@cancer.org
Internet: http://www.cancer.org/docroot/RES/content/RES_5_2x_Doctoral_Degree_Scholarships_in_Cancer_Nursing.asp?sitearea=RES
Sponsor: American Cancer Society
3100 Wyman Park Drive, Suite W400
Baltimore, MD 21211

ACS Doctoral Scholarships in Cancer Nursing 253
Doctoral degree scholarships in cancer nursing are awarded to graduate students pursuing doctoral study in the following cancer nursing fields: research, education, administration, or clinical practice. Awards are made for up to four years.
Requirements: Applicants must be currently enrolled in or applying to a doctoral degree program and have current licensure to practice as a registered nurse. American citizens or permanent residents are eligible.
Geographic Focus: All States
Date(s) Application is Due: Oct 15
Amount of Grant: 15,000 USD
Contact: Grants Department; (404) 329-7558; fax (404) 321-4669; grants@cancer.org
Internet: http://www.cancer.org/docroot/RES/content/RES_5_2x_Doctoral_Degree_Scholarships_in_Cancer_Nursing.asp?sitearea=RES
Sponsor: American Cancer Society
3100 Wyman Park Drive, Suite W400
Baltimore, MD 21211

ACS Doctoral Training Grants in Oncology Social Work 254
This grant is awarded to qualifying doctoral students at schools of social work that train individuals to conduct research related to the psychosocial needs of persons with cancer and their families. Initial 2-year grant providing a stipend of $20,000 per year with possibility of a 2-year competitive renewal.
Requirements: The following eligibility requirements must be met: the applicant must have master's degree in social work; at least one year of social work experience in a health care setting (oncology/cancer control experience preferred, but not required); a demonstrated commitment to a career in oncology social work as evidenced by recent experience, education, and/or research; the applicant must be currently enrolled in or applying to a doctoral degree program in social work; if an applicant is not yet enrolled in a Ph.D. program, the applicant must meet requirements for doctoral study and must have been accepted by the institution to which s/he has applied at the time of funding; the applicant must project a full-time program of study that integrates oncology social work and provides evidence of faculty support for the program of study; to qualify for funding for the dissertation phase of the doctoral program, the applicant must conduct a dissertation relevant to Oncology Social Work. Successful applicants must also agree to attend the American Cancer Society activities at the annual Society for Social Work Research meeting. (Recipients at the dissertation phase will be expected to present their research at this meeting.)
Geographic Focus: All States
Date(s) Application is Due: Oct 15
Amount of Grant: 20,000 USD
Contact: Virginia Krawiec; (404) 329-7612; fax (404) 321-4669; ginger.krawiec@cancer.org
Internet: http://www.cancer.org/docroot/RES/content/RES_5_2x_Post-Masters_Training_Grants_in_Clinical_Oncology_Social_Work.asp?sitearea=RES
Sponsor: American Cancer Society
3100 Wyman Park Drive, Suite W400
Baltimore, MD 21211

ACS Graduate Scholarships in Cancer Nursing Practice 255
This award supports graduate students pursuing a master's degree in cancer nursing or doctorate of nursing practice (DNP). Awards may be for two years, with stipend of $10,000 per year. The goal of this program is to strengthen nursing practice by providing assistance for advanced preparation in the following fields of cancer nursing: clinical practice, education, and administration.
Requirements: The applicant must: be currently enrolled in or applying to a master's or DNP degree graduate program with demonstrated integration of cancer content; meet requirements for graduate study and at the time of funding have been accepted by the institution to which s/he has applied. The institution must verify the applicant's acceptance into the master's or DNP program; be pursuing an advanced degree and not solely a post master's certificate. All applicants must have a current license to practice as a registered nurse. Students in bridge programs must have passed the N-CLEX examination by the time the award begins.
Geographic Focus: All States
Date(s) Application is Due: Feb 1
Amount of Grant: 10,000 USD
Contact: Virginia Krawiec; (404) 329-7612; fax (404) 321-4669; ginger.krawiec@cancer.org
Internet: http://www.cancer.org/docroot/RES/content/RES_5_2x_Masters_Degree_Scholarships_in_Cancer_Nursing.asp?sitearea=RES
Sponsor: American Cancer Society
3100 Wyman Park Drive, Suite W400
Baltimore, MD 21211

ACS Institutional Research Grants 256
The IRG program provides seed money for the initiation of promising new projects by independent junior investigators so they may obtain preliminary results that will enable them to compete successfully for national research grants.
Requirements: Only full-time, tenure-track JHU faculty members at the ranks of Instructor or Assistant Professor or equivalent who are within the first six years of their first independent research or faculty appointment may apply. Relevance to the cancer problem is important and must be well documented. Within this research scope, a wide breadth of research approaches will be considered; these include basic laboratory studies and clinical research.
Restrictions: Senior investigators, postdoctoral fellows and junior investigators who have competitive national research grants or who have received prior support form the IRG are not eligible.
Geographic Focus: All States
Date(s) Application is Due: Apr 1
Amount of Grant: 90,000 USD
Contact: Grace Bigelow; (410) 516-5256; fax (410) 516-7775; graceb@jhu.edu
Internet: http://www.cancer.org/docroot/RES/content/RES_5_2x_Institutional_Research_Grants.asp?sitearea=RES
Sponsor: American Cancer Society
3100 Wyman Park Drive, Suite W400
Baltimore, MD 21211

ACSM-GSSI Young Scholar Travel Award 257
ACSM invests in the future of sports medicine and exercise science by giving to our student members. Student members are eligible for grants and awards to offset the rising costs of tuition and travel. The ACSM-GSSI (Gatorade Sports Science Institute) Young Scholar Travel Award offers $1,000 for students in the fields of sports nutrition or exercise physiology to support attendance at its annual meeting in Boston. Two awards of $1,000 each are offered. Applicant should submit a written statement detailing their professional goals and describing how attendance at the Annual Meeting will help advance their studies (300 word limit). Applications must be received by February 1.
Requirements: Eligible students must be: 32 years of age or younger (as of June 4); enrolled in an undergraduate or graduate program in sports nutrition or exercise physiology; ACSM members and registered for the Annual Meeting; actively involved in research and are a co-author on at least one scientific paper; and in good academic standing with a 3.0 GPA or higher.
Geographic Focus: All States
Date(s) Application is Due: Feb 1
Amount of Grant: 1,000 USD
Contact: Jane Senior, Research Administration and Programs Department; (317) 637-9200, ext. 143; fax (317) 634-7817; jsenior@acsm.org or foundation@acsm.org
Internet: http://acsm.org/find-continuing-education/awards-grants/student-awards/2016-gssi-acsm-young-scholar-travel-award
Sponsor: American College of Sports Medicine
401 West Michigan Street
Indianapolis, IN 46202-3233

ACS Master's Training Grants in Clinical Oncology Social Work 258
These grants are awarded annually to qualifying institutions that train clinical oncology social workers to provide cancer patients and their families with psychosocial services. Grants are available to second-year students in a master's program and post-master's social workers within five years of graduation. The master's training must introduce social workers to the special needs of cancer patients and their families. Contact the society for application forms.
Requirements: Applications will be considered from institutions that identify clinical oncology social work training activities, cancer programs with defined psychosocial support services, and relationships with schools accredited by the Council on Social Work Education.
Geographic Focus: All States
Date(s) Application is Due: Oct 15
Amount of Grant: Up to 12,000 USD
Contact: Grants Department; (404) 329-7558; fax (404) 321-4669; grants@cancer.org
Internet: http://www.cancer.org/docroot/RES/content/RES_5_2x_Masters_Training_Grants_in_Clinical_Oncology_Social_Work.asp?sitearea=RES
Sponsor: American Cancer Society
3100 Wyman Park Drive, Suite W400
Baltimore, MD 21211

ACSM Carl V. Gisolfi Memorial Fund Grant 259
Carl V. Gisolfi, Ph.D., FACSM, was one of the most influential leaders of ACSM in its 50+ year history. With Gatorade, ACSM created the Gisolfi Memorial Fund to honor his many contributions to ACSM and the exercise science field. This grant is designated to encourage research in thermoregulation, exercise, and hydration. One award will be dispersed to ACSM doctoral students in the amount of $5,000 each for a one-year period.
Requirements: Applicants for student research grants must have graduate student status during the term of the grant to be considered for funding. Applicants must be current members of ACSM at the time of submitting an application to be funded. Grants are open to all ACSM members, including international members.
Geographic Focus: All States, All Countries
Date(s) Application is Due: Jan 15
Amount of Grant: 5,000 USD
Contact: Jane Senior, (317) 637-9200, ext. 143; fax (317) 634-7817; jsenior@acsm.org
Internet: http://acsm.org/find-continuing-education/awards-grants/research-grants/carl-v.-gisolfi-memorial-fund
Sponsor: American College of Sports Medicine
401 West Michigan Street
Indianapolis, IN 46202-3233

ACSM Charles M. Tipton Student Research Award 260
ACSM invests in the future of sports medicine and exercise science by giving to its student members. Student members are eligible for grants and awards to offset the rising costs of tuition and travel. The Charles M. Tipton Student Research Award is presented annually to the student with the most outstanding research project of the year. Criteria for the awards include: the quality of the research project, which is evaluated via; and the extent of the students participation in the project.
Requirements: At the time of application, the applicant must be presently enrolled/accepted into a graduate or professional program in the areas of basic of clinical exercise

science or sports medicine at an accredited university or have graduated from such a program within one year prior to this award's application date. In the latter case, only work accomplished before graduation should be submitted. The applicant must be the first author of the abstract submitted and accepted as a free communication for presentation at the ACSM Annual Meeting. The abstract must be original research in exercise science and should include the following: statement of the program; methods and results and conclusions. The work reported in the abstract must have been completed for academic credit toward completion of a graduate degree. Finally, the applicant must be a current member of the American College of Sports Medicine or one of its Regional Chapters at the time of application and at the time the award is presented.
Geographic Focus: All States
Date(s) Application is Due: Feb 1
Contact: Jane Senior, Research Administration and Programs Department; (317) 637-9200, ext. 143 or (317) 637-9200, ext. 117; fax (317) 634-7817; jsenior@acsm.org or education@acsm.org
Internet: http://acsm.org/find-continuing-education/awards-grants/student-awards/charles-m.-tipton-student-research-awards
Sponsor: American College of Sports Medicine
401 West Michigan Street
Indianapolis, IN 46202-3233

ACSM Coca-Cola Company Doctoral Student Grant on Behavior Research   261
For research addressing behavioral strategies and techniques to promote the adoption and maintenance of physical activity/exercise. Studies that focus on innovative strategies to improve individual-level adoption and/or maintenance of physical activity will be given first priority. The proposed study can focus on youth, adolescents, adults, or older adults 55 and over. One award of $5,000 is available.
Requirements: Applicants for student research grants must have graduate student status during the term of the grant to be considered for funding. Applicants must be current members of ACSM at the time of submitting an application to be funded. Grants are open to all ACSM members, including international members. However, the NASA initiative is open to U.S. residents only.
Geographic Focus: All States, All Countries
Date(s) Application is Due: Jan 17
Amount of Grant: 5,000 USD
Contact: Jane Senior or Michael F. Dell, (317) 637-9200, ext. 143; fax (317) 634-7817; jsenior@acsm.org or mdell@acsm.org
Internet: http://acsm.org/find-continuing-education/awards-grants/research-grants/2011/08/16/the-coca-cola-company-doctoral-student-grant-on-behavior-research
Sponsor: American College of Sports Medicine
401 West Michigan Street
Indianapolis, IN 46202-3233

ACSM Dr. Raymond A. Weiss Research Endowment Grant   262
Wishing to advance research into the health benefits - physical, mental and emotional - of physical activity and sports, the Dr. Raymond A. Weiss Research Endowment has been established for ACSM students studying in this field. One project will be funded for applied rather than for basic research, with the intent of applying the results to programs involving physical activity and sports. The psychological and emotional benefits of physical activity are of particular importance to the benefactors of this endowment, Drs. Raymond and Rosalee Weiss, and proposals addressing those issues will be given priority. One reward of $1,500 is available.
Requirements: Applicants for student research grants must have graduate student status during the term of the grant to be considered for funding. Applicants must be current members of ACSM at the time of submitting an application to be funded. Grants are open to all ACSM members, including international members. However, the NASA initiative is open to U.S. residents only.
Geographic Focus: All States, All Countries
Date(s) Application is Due: Jan 15
Amount of Grant: 1,500 USD
Contact: Jane Senior, (317) 637-9200, ext. 143; fax (317) 634-7817; jsenior@acsm.org
Internet: http://acsm.org/find-continuing-education/awards-grants/research-grants/dr.-raymond-a.-weiss-research-endowment
Sponsor: American College of Sports Medicine
401 West Michigan Street
Indianapolis, IN 46202-3233

ACS MEN2 Thyroid Cancer Professorship Grants   263
The Professorship award is intended for an outstanding senior investigator who is well known for their contributions to the understanding and/or management of MEN2. Appropriate areas of investigation include, but are not limited to: understanding consequences of RET mutations, molecular events underlying the development of MEN2-related tumors, improved animal models of MEN2, new screening and monitoring tools, new imaging approaches, and new pharmacologic and other strategies to blunt the effects of mutations in RET and other genes associated with medullary thyroid cancer. This person will serve as the leader for a consortium consisting of up to seven (7) investigators with Research Scholar and/or Mentored Research Scholar grants and up to five (5) Postdoctoral Fellows. Responsibilities primarily involve serving as the chairman of an annual research meeting of MEN2 investigators. The amount of the award is $80,000 per year for five years and can be budgeted at the recipient's discretion. Only one award will be made and a letter of intent must be submitted. The awardee must be willing to be an occasional spokesperson for the American Cancer Society and for the Consortium. Interested individuals must electronically submit their curriculum vitae with a complete bibliography and a letter of intent that briefly describes their seminal contributions to thyroid cancer research. This information must be submitted via Access Electronic Grant application process at proposalCENTRAL no later than December 31, by applying for the American Cancer Society â€" MEN2 Thyroid Cancer Professorship. Qualified candidates will be invited no later than January 15 to submit full applications by April 1.
Requirements: Candidates must be American citizens or permanent residents with at least 10 years of experience beyond receipt of their terminal degree and in general are expected to hold the rank of full professor. Department chairs or individuals with equivalent administrative positions are eligible.
Geographic Focus: All States
Date(s) Application is Due: Dec 31
Amount of Grant: 80,000 USD
Contact: Charles Saxe, Program Director; (404) 929-6919; charles.saxe@cancer.org
Internet: http://www.cancer.org/docroot/RES/content/RES_5_2x_RFA_for_The_American_Cancer_Society_MEN2_Thyroid_Cancer_Consortium_Professorship.asp?sitearea=RES
Sponsor: American Cancer Society
3100 Wyman Park Drive, Suite W400
Baltimore, MD 21211

ACS Mentored Research Scholar Grant in Applied and Clinical Research   264
This funding opportunity supports mentored research by full-time faculty, typically within the first four years of their appointment, with the goal of becoming independent investigators in clinical, cancer control and prevention, epidemiologic, psychosocial, behavioral, health services and health policy research. Awards are made up to five years and for up to $135,000 per year (direct costs) plus 8% allowable indirect costs. A maximum of $10,000 per year for the mentor(s) (regardless of the number of mentors) is included in the $135,000. These grants are not renewable.
Geographic Focus: All States
Date(s) Application is Due: Apr 1; Oct 15
Amount of Grant: 135,000 USD
Contact: Grants Department; (404) 329-7558; fax (404) 417-5974; grants@cancer.org
Internet: http://www.cancer.org/docroot/RES/content/RES_5_2x_Mentored_Research_Scholar_Grant_in_Applied_and_Clinical_Research.asp?sitearea=RES
Sponsor: American Cancer Society
3100 Wyman Park Drive, Suite W400
Baltimore, MD 21211

ACSM Foundation Clinical Sports Medicine Endowment Grants   265
ACSM's Clinical Sports Medicine Endowment will fund an annual award of $5,000 for research that is directly related to clinical sports medicine. The purpose of the award is to stimulate clinical research in sports medicine and research in clinical sports medicine. Clinical research in sports medicine must be derived from active care of athletes, such as prevention strategies, treatment protocols, or treatment outcomes. Research in clinical sports medicine must be directly related to issues of active care of athletes, such as pathogenesis, pathophysiology, biomechanics, or environmental issues. The proposals must demonstrate the study's direct relationship to one of these purposes. The active patient care must be provided to physically active recreational, competitive or elite athletes. It is anticipated that the completed research will be presented at a highlighted session at the ACSM Annual Meeting.
Requirements: Grants are open to all ACSM members, including international members. This clinical research award is open to MDs, DOs, PTs, ATCs, and other medical professionals (ACSM members only) involved in the conduct of patient-based clinical research. Applicants for student research grants must have graduate student status during the term of the grant to be considered for funding. Applicants must be current members of ACSM at the time of submitting an application to be funded.
Geographic Focus: All States, All Countries
Date(s) Application is Due: Jan 15
Amount of Grant: 5,000 USD
Contact: Jane Senior, (317) 637-9200, ext. 143; fax (317) 634-7817; jsenior@acsm.org
Internet: http://acsm.org/find-continuing-education/awards-grants/research-grants/clinical-sports-medicine-endowment
Sponsor: American College of Sports Medicine
401 West Michigan Street
Indianapolis, IN 46202-3233

ACSM Foundation Doctoral Student Research Grants   266
ACSM Foundation Research Grants for doctoral students awards up to $5,000 for a one-year period. The awards are to be used for experimental subjects, supplies, and small equipment needs. There are two parts to the application. Both Part A and Part B should be submitted as one complete PDF application.
Requirements: Doctoral students enrolled in full-time programs are eligible to apply. Applicants for student research grants must have graduate student status during the term of the grant to be considered for funding. Applicants must be current members of ACSM at the time of submitting an application to be funded. Grants are open to all ACSM members, including international members. However, the NASA initiative is open to U.S. residents only.
Geographic Focus: All States, All Countries
Date(s) Application is Due: Jan 15
Amount of Grant: 5,000 USD
Contact: Jane Senior, (317) 637-9200, ext. 143; fax (317) 634-7817; jsenior@acsm.org
Internet: http://acsm.org/find-continuing-education/awards-grants/research-grants/acsm-foundation-doctoral-student-research-grant

Sponsor: American College of Sports Medicine
401 West Michigan Street
Indianapolis, IN 46202-3233

ACSM Foundation Research Endowment Grants 267
The research endowment has been made possible by charitable funds and individual donations derived from the annual campaign, specifically targeted for research. The intent of the research endowment is to use a portion of the interest derived from these funds to support basic and applied research in exercise science. The primary goal is to fund mechanistic, hypothesis driven, basic and applied research. A $10,000 grant is available and is applicable to all relevant fields of exercise science.
Requirements: Funding is primarily targeted for new or junior investigators, within seven years of attaining a terminal degree (e.g., Ph.D., Ed.D.). It is the intent of this grant to provide seed money support after which further funding would be sought from other sources. Only one application per person is allowed. Applicants for student research grants must have graduate student status during the term of the grant to be considered for funding. Applicants must be current members of ACSM at the time of submitting an application to be funded. Grants are open to all ACSM members, including international members. However, the NASA initiative is open to U.S. residents only.
Geographic Focus: All States, All Countries
Date(s) Application is Due: Jan 15
Amount of Grant: 10,000 USD
Contact: Jane Senior, (317) 637-9200, ext. 143; fax (317) 634-7817; jsenior@acsm.org
Internet: http://acsm.org/find-continuing-education/awards-grants/research-grants/research-endowment
Sponsor: American College of Sports Medicine
401 West Michigan Street
Indianapolis, IN 46202-3233

ACSM International Student Award 268
The American College of Sports Medicine invests in the future of sports medicine and exercise science by giving to its student members. Student members are eligible for grants and awards to offset the rising costs of tuition and travel. The International Student Award, in the amount of $1,000, honors students from countries outside of the United States and helps fund travel expenses for those traveling to the ACSM Annual Meeting to present their scholarly work. Submission of an abstract for presentation at the ACSM Annual Meeting is due November 1, with the completed application deadline three months after, on February 1.
Requirements: Applicants must be a student at the time of the ACSM Annual Meeting and must not hold a completed doctoral degree of any type (Ph.D., Ed.D., M.D., DPH, etc.) at the time of application. Applicants must: be ACSM members in good standing; have been accepted to present research at the ACSM Annual Meeting; be the first author of the accepted abstract and should plan to personally present the research; not be residents of the United States.
Restrictions: Previous International Student Award winners are not eligible to apply.
Geographic Focus: All States
Date(s) Application is Due: Feb 1
Amount of Grant: 1,000 USD
Contact: Jane Senior; (317) 637-9200, ext. 143 or (317) 637-9200, ext. 117; fax (317) 634-7817; jsenior@acsm.org or education@acsm.org
Internet: http://acsm.org/find-continuing-education/awards-grants/student-awards/international-student-award
Sponsor: American College of Sports Medicine
401 West Michigan Street
Indianapolis, IN 46202-3233

ACSM International Student Awards 269
ACSM international awards provide funding for professionals and students to participate in clinical and research exchange opportunities. The International Student Award honors students from countries outside of the United States and helps fund travel expenses for those traveling to the ACSM Annual Meeting to present their scholarly work. Awards will be determined by the quality of research to be presented at the ACSM Annual Meeting. Award recipients will receive $1,000 USD to assist with travel costs for attending ACSM's Annual Meeting.
Requirements: Applicants may not be residents of the United States. Applicants must be a student at the time of the ACSM Annual Meeting and must not hold a completed doctoral degree of any type (Ph.D., Ed.D., M.D., DPH, etc.) at the time of application. Applicants must be ACSM members in good standing and must have been accepted to present research at the ACSM Annual Meeting. Applicants should be the first author of the accepted abstract and should plan to personally present the research.
Restrictions: Previous International Student Award winners are not eligible to apply.
Geographic Focus: All Countries
Date(s) Application is Due: Feb 1
Amount of Grant: 1,000 USD
Contact: Jane Senior or Michael F. Dell, (317) 637-9200, ext. 143; fax (317) 634-7817; jsenior@acsm.org or mdell@acsm.org
Internet: http://acsm.org/find-continuing-education/awards-grants/international-awards/2011/08/16/international-student-award
Sponsor: American College of Sports Medicine
401 West Michigan Street
Indianapolis, IN 46202-3233

ACSM Michael L. Pollock Student Scholarship 270
The American College of Sports Medicine invests in the future of sports medicine and exercise science by giving to its student members. Student members are eligible for grants and awards to offset the rising costs of tuition and travel. The Michael L. Pollock Memorial Fund was established to help pay travel expenses for graduate students traveling to the ACSM Annual Meeting to present their scholarly work. Each scholarship, in the amount of $200, is given to an applicant whose research focuses on health-and-fitness or clinical exercise physiology. The annual deadline for applications is February 1.
Requirements: Applicants must: have been a graduate student in good standing within one year of submitting an application; and be the first author on an accepted research project (free communication or poster).
Geographic Focus: All States
Date(s) Application is Due: Feb 1
Amount of Grant: 200 USD
Contact: Jane Senior; (317) 637-9200, ext. 143 or (317) 637-9200, ext. 117; fax (317) 634-7817; jsenior@acsm.org or education@acsm.org
Internet: http://acsm.org/find-continuing-education/awards-grants/student-awards/michael-l.-pollock-student-scholarship
Sponsor: American College of Sports Medicine
401 West Michigan Street
Indianapolis, IN 46202-3233

ACSM NASA Space Physiology Research Grants 271
ACSM's Foundation Research Grant Program started in 1989 with one research initiative in basic and applied science, and it awarded $50,000 among five doctoral research students. Through the National Aeronautics and Space Administration (NASA), $10,000 is available for research grants in the area of exercise, weightlessness, and musculoskeletal physiology. Doctoral students enrolled in full-time programs are eligible to apply. Grants can range up to $5,000 and are available for a one-year period. There are two parts to the application. Both Part A and Part B should be submitted as one complete PDF application. The annual deadline is January 15.
Requirements: Doctoral students (U.S. residents only) enrolled in full-time programs are eligible to apply. Applicants for student research grants must have graduate student status during the term of the grant to be considered for funding. Applicants must be current members of ACSM at the time of submitting an application to be funded. Grants are open to all ACSM members, including international members. However, the NASA initiative is open to U.S. residents only.
Geographic Focus: All States
Date(s) Application is Due: Jan 15
Amount of Grant: Up to 5,000 USD
Contact: Jane Senior; (317) 637-9200, ext. 143 or (317) 637-9200, ext. 117; fax (317) 634-7817; jsenior@acsm.org or education@acsm.org
Internet: https://www.acsm.org/find-continuing-education/awards-grants/research-grants/nasa-space-physiology-research-grant
Sponsor: American College of Sports Medicine
401 West Michigan Street
Indianapolis, IN 46202-3233

ACSM Oded Bar-Or International Scholar Awards 272
The Oded Bar-Or International Scholar Award allows professionals to gain technical expertise and/or scientific knowledge through an international exchange program. Award recipients from the United States and Canada are required to travel to an institution outside of the United States or Canada. Similarly, award recipients from countries other than Canada and the United States are required to travel to an institution within the United States or Canada. Membership in ACSM is not a requirement. Awards will be determined by the likelihood that the activities will open avenues for professional growth, the intrinsic quality of the proposed research activity and evidence the project can be completed. Award recipients will receive round-trip air transportation to the host institution, a lodging and living stipend, medical health insurance for two months, complimentary registration and lodging for the ACSM Annual Meeting.
Requirements: The award is open to all ACSM members, including international members. Applicants must: be professionals in a discipline recognized by ACSM (minimum of a bachelor's degree is required; plan to engage in a professional activity that will advance the professional standing and reputation of scientists (Though many proposals will be for research-based activities, requests for other types of support will be considered; choose a host scientist who is a member of ACSM (ACSM will not match applicants with potential host scientists).
Geographic Focus: All States, Canada
Date(s) Application is Due: Feb 1
Contact: Jane Senior; (317) 637-9200, ext. 143 or (317) 637-9200, ext. 117; fax (317) 634-7817; jsenior@acsm.org or education@acsm.org
Internet: http://acsm.org/find-continuing-education/awards-grants/research-grants/2011/08/16/oded-bar-or-international-scholar-award
Sponsor: American College of Sports Medicine
401 West Michigan Street
Indianapolis, IN 46202-3233

ACSM Paffenbarger-Blair Fund for Epidemiological Research on Physical Activity Grants 273
Dr. Ralph S. Paffenbarger, Jr., a co-recipient of the first Olympic Prize in Exercise Science and Sports Medicine, made an extraordinarily generous gift to the American College of Sports Medicine by donating his share of the award to the College's capital campaign with the monies added to the College's epidemiology endowment. The interest

on this contribution makes this grant possible. The intent of this award is to encourage researchers early in their career to become involved with physical activity epidemiology. The applications may focus on observational studies of physical activity and health outcomes, or on randomized controlled trials that are clearly focused on physical activity and important public health issues. Applicants are expected to apply within two (2) years of receiving a postgraduate degree or completion of clinical training. The award will be in the amount of $10,000 for a one year period.

Requirements: Applicants for student research grants must have graduate student status during the term of the grant to be considered for funding. Applicants must be current members of ACSM at the time of submitting an application to be funded. Grants are open to all ACSM members, including international members. However, the NASA initiative is open to U.S. residents only.

Geographic Focus: All States, All Countries
Date(s) Application is Due: Jan 15
Amount of Grant: 10,000 USD
Contact: Jane Senior, (317) 637-9200, ext. 143; fax (317) 634-7817; jsenior@acsm.org
Internet: http://acsm.org/find-continuing-education/awards-grants/research-grants/paffenbarger-blair-fund-for-epidemiological-research-on-physical-activity
Sponsor: American College of Sports Medicine
401 West Michigan Street
Indianapolis, IN 46202-3233

ACSM Steven M. Horvath Travel Award 274

ACSM invests in the future of sports medicine and exercise science by giving to its student members. Student members are eligible for grants and awards to offset the rising costs of tuition and travel. The Steven M. Horvath Travel Award, in the amount of $500, funds travel expenses for underrepresented minority graduate students traveling to the ACSM Annual Meeting to present their scholarly work. The annual deadline for applications is February 1.

Requirements: Award requirements include: must be an underrepresented minority (as defined by NIH) graduate student in good standing with a research focus related to sports medicine and/or exercise sciences; and must be presenting a research project at the ACSM Annual Meeting.

Geographic Focus: All States
Date(s) Application is Due: Feb 1
Contact: Jane Senior, Research Administration and Programs Department; (317) 637-9200, ext. 143; fax (317) 634-7817; jsenior@acsm.org or foundation@acsm.org
Internet: http://acsm.org/find-continuing-education/awards-grants/student-awards/steven-m.-horvath-travel-award
Sponsor: American College of Sports Medicine
401 West Michigan Street
Indianapolis, IN 46202-3233

ACS Physician Training Awards in Preventive Medicine 275

The Physician Training Award in Preventive Medicine (PTAPM) is intended to encourage and assist the development of promising individuals who will pursue careers in preventive medicine. This program is designed to create a cadre of preventive medicine specialists who are expert in cancer prevention and control, and with the potential to become leaders in research, education, and intervention in this area. Through the Physician Training Award in Preventive Medicine, the Society seeks to support physicians in accredited residency programs that will lead to eligibility for certification in preventive medicine; such programs must provide cancer prevention and control research and practice opportunities. Awards are for four years in the total amount of $300,000, based on an average of $50,000 per resident training year. These grants are renewable.

Requirements: Requirements for the institution are available on the ACS website. Eligibility requirements for the residents are: nominated residents must be citizens or non-citizen nationals of the United States or its possessions and territories, or must have been lawfully admitted to the United States for permanent residence at the time of application; must have an MD, DO, or equivalent degree; must have completed the clinical year of a residency program in preventive medicine, or have at least one year of postgraduate clinical training; must be accepted by or applying to the sponsoring residency program. It is preferred that the candidates not have completed more than half of the required academic work at the beginning of the award period. Residents that have completed the MPH degree will be considered on a case-by-case basis, and must propose course work in cancer prevention and control.

Geographic Focus: All States
Date(s) Application is Due: Apr 1
Amount of Grant: Up to 300,000 USD
Contact: Grants Department; (404) 329-7558; fax (404) 417-5974; grants@cancer.org
Internet: http://www.cancer.org/docroot/RES/content/RES_5_2x_Physician_Training_Awards_in_Preventive_Medicine.asp?sitearea=RES
Sponsor: American Cancer Society
3100 Wyman Park Drive, Suite W400
Baltimore, MD 21211

ACS Pilot and Exploratory Projects in Palliative Care of Cancer Patients and Their Families Grants 276

This request for application (RFA) provides funding for investigators performing pilot and exploratory research studies whose purpose is to test interventions, develop research methodologies, and explore novel areas of research in palliative care of cancer patients and their families. A condition of funding is a clearly defined plan as to how the investigator will use the results of the project to develop larger, extramurally funded research projects. This RFA is limited to applications that focus on palliative care research projects for seriously ill cancer patients and their families in three specific areas: exploring the relationship of pain and other distressing symptoms on quality and quantity of life, independence, function, and disability and developing interventions directed at their treatment in patients with advanced and chronic illnesses; studying methods of improving communication between adults living with serious illness, their families and their health care providers; evaluating models and systems of care for patients living with advanced illness and their families. As a condition of accepting the award, each recipient of this grant, will agree to: attend the required meetings of the NPCRC in Fall during the award period; present results of the funded research at this required NPCRC meeting; prepare annual progress reports for each year of funding and a final report at the conclusion of the award period; list the American Cancer Society as funding this study on all publications and presentations.

Requirements: Applications may be submitted by not-for-profit institutions located within the United States, its territories and the Commonwealth of Puerto Rico. Applicants must: be United States citizens, non-citizen nationals or permanent residents of the United States; hold a doctorate degree (MD, PhD, or equivalent) and have a full-time faculty position or equivalent at a college, university, medical school, or other fiscally responsible not-for-profit organization within the United States. Independent investigators at all stages of their career are eligible to apply. Applicants must submit their full application electronically and in paper format per guidelines on cancer.org no later than 5pm (Eastern) on October 15th.

Restrictions: Awards may not exceed $60,000 per year (direct costs) plus 20% indirect costs. Salary support for the Principal Investigator may not exceed 20% of the direct costs. Awards may not exceed a period of one to two years duration.

Geographic Focus: All States
Date(s) Application is Due: Oct 15
Amount of Grant: Up to 60,000 USD
Contact: Ronit Elk; (404) 417-5957; fax (404) 321-4669; ronit.elk@cancer.org
Internet: http://www.cancer.org/docroot/RES/content/RES_5_2x_RFA__Pilot_and_Exploratory_Projects_in_Palliative_Care_of_Cancer_Patients_and_Their_Families.asp?sitearea=RES
Sponsor: American Cancer Society
3100 Wyman Park Drive, Suite W400
Baltimore, MD 21211

ACS Postdoctoral Fellowships 277

Fellowships are designed to enable a new investigator to qualify for an independent career in cancer research. Postdoctoral fellowships may be made for one, two, or three years. Stipend payments are made directly to the individual, or the institution if requested, at the beginning of each month. Travel funds are paid to the individual. Institutional allowances are paid annually at the start of the grant and on the anniversary date thereafter. Awards are made for one to three years with progressive stipends.

Requirements: Applicants must be citizens or permanent residents of the United States (the latter must provide notarized evidence of their legal resident alien status) and shall have been awarded a doctoral degree prior to the activation date of the grant. Application must be endorsed by applicant's mentor and the head of the department in which the training will be received. A plan of training must be formulated and agreed upon by the mentor and the applicant and described in detail in the application.

Restrictions: Awards will not be made to applicants who have completed five or more years of postdoctoral training prior to the start date of the fellowship.

Geographic Focus: All States
Date(s) Application is Due: Apr 1; Oct 15
Contact: Grants Department; (404) 329-7558; fax (404) 321-4669; grants@cancer.org
Internet: http://www.cancer.org/docroot/RES/content/RES_5_2x_Postdoctoral_Fellowships.asp?sitearea=RES
Sponsor: American Cancer Society
3100 Wyman Park Drive, Suite W400
Baltimore, MD 21211

ACS Research Professor Grants 278

The American Cancer Society offers a limited number of grants to investigators in mid-career who have made seminal contributions that have changed the direction of cancer research. Furthermore, it is expected that these investigators will continue to provide leadership in their research area. Up to two awards are made annually for a five-year term that can be renewed once. The award of up to $80,000 per year (direct costs only) may be used for salary or research project support. Interested individuals should submit their curriculum vitae with a complete bibliography and a letter of intent that briefly describes their seminal contributions to research. This information must be submitted no later then February 1, via the Research Professor letter of intent on proposalCENTRAL at: https://proposalcentral.altum.com. Upon receipt, ACS will provide appropriate guidance either by telephone or email.

Geographic Focus: All States
Date(s) Application is Due: Feb 1; Apr 1
Amount of Grant: Up to 80,000 USD
Contact: John J. Stevens, Vice President; (404) 329-7550; john.stevens@cancer.org
Internet: http://www.cancer.org/docroot/RES/content/RES_5_2x_Research_Professorships.asp?sitearea=RES
Sponsor: American Cancer Society
3100 Wyman Park Drive, Suite W400
Baltimore, MD 21211

ACS Research Scholar Grants for Health Services and Health Policy Research 279

The grants support research projects centered on health services and health policy initiated by investigators at any stage in their careers. The initial award includes 20 percent for indirect costs for up to four years, and may be renewed once for up to four years.

Geographic Focus: All States
Date(s) Application is Due: Apr 1; Oct 15
Amount of Grant: Up to 200,000 USD
Contact: Grants Department; (404) 329-7558; fax (404) 321-4669; grants@cancer.org
Internet: http://www.cancer.org/docroot/RES/content/RES_5_2x_Research_Scholar_Grants_For_Health_Services_and_Health_Policy_and_Outcomes_Research.asp?sitearea=RES
Sponsor: American Cancer Society
3100 Wyman Park Drive, Suite W400
Baltimore, MD 21211

ACS Research Scholar Grants in Basic, Preclinical, Clinical and Epidemiology Research 280

This grant supports investigator-initiated research projects in basic, preclinical, clinical and epidemiologic research. Awards are for up to four years and for up to $200,000 per year (direct costs), plus 20% allowable indirect costs.
Geographic Focus: All States
Date(s) Application is Due: Apr 1; Oct 15
Amount of Grant: 200,000 USD
Contact: Grants Department; (404) 329-7558; fax (404) 417-5974; grants@cancer.org
Internet: http://www.cancer.org/docroot/RES/content/RES_5_2x_Research_Scholar_Grants_for_Beginning_Investigators.asp?sitearea=RES
Sponsor: American Cancer Society
3100 Wyman Park Drive, Suite W400
Baltimore, MD 21211

ACS Research Scholar Grants in Psychosocial and Behavioral and Cancer Control Research 281

The grants support research projects focusing on the psychosocial and behavioral aspects of cancer by independent investigators at any stage in their careers. Applications are encouraged in which an individual at an early career stage is coprincipal investigator with an established researcher. The initial award supports four years of research and may be renewed once for four additional years.
Requirements: Individuals must not have held an independent position for more than six years at the time the application is submitted.
Geographic Focus: All States
Date(s) Application is Due: Apr 1; Oct 15
Contact: Grants Department; (404) 329-7558; fax (404) 417-5974; grants@cancer.org
Internet: http://www.cancer.org/docroot/RES/content/RES_5_2x_Research_Scholar_Grants_in_Psychological_and_Behavioral_Research_for_Beginning_and_Senior_Investigators.asp?sitearea=RES
Sponsor: American Cancer Society
3100 Wyman Park Drive, Suite W400
Baltimore, MD 21211

Actors Fund Addiction and Recovery Services Grants 282

The Addiction and Recovery Services helps performing arts and entertainment professionals and their families when there is a problem with drug and alcohol abuse or addiction. The Actors Fund goal is to help people achieve and maintain sobriety and help family members, partners, or friends deal with the effects of addiction. Support during early recovery can help professionals and their family members with addiction problems achieve and maintain abstinence, enter into a recovery process, learn effective coping strategies, and take increasing responsibility for managing their lives without the use of mood-altering substances. Whenever possible, social workers meet in person to evaluate the person's substance use. In addition social workers can organize interventions, make referrals to treatment, help coordinate care with treatment programs, and negotiate on the client's behalf with insurance and managed care companies. Financial assistance may be available to help with the cost of treatment and other financial difficulties for eligible members and their family. The Addiction and Recovery Services program also helps to educate and assist industry employers, supervisors, manager, agents, union representatives, etc. in addressing substance abuse issues in the workplace.
Requirements: Eligibility for the Fund's financial assistance program requires an application, documentation of your professional earnings and an interview. In general, eligibility for financial assistance is based on: a minimum of five years of industry employment with earnings of at least $6,500 for three out of the last five years; and, financial need.
Geographic Focus: All States
Contact: Rosalyn Gilbert; (212) 221-7300, ext. 114; tgilbert@actorsfund.org
Internet: http://new.actorsfund.org/services-and-programs/addiction-and-recovery-services
Sponsor: Actors Fund
729 7th Avenue, 10th Floor
New York, NY 10019

Actors Fund HIV/AIDS Initiative Grants 283

The Actors Fund HIV/AIDS Initiative works with men and women in the entertainment industry to create confidential, holistic plans and support systems that will meet each person's emotional, medical and financial needs over the long term. The Initiative's experienced Social Workers provide case management, advocacy, crisis intervention, individual and group counseling, financial assistance, financial management skills development, and referrals to community resources.
Requirements: Eligibility for the Fund's financial assistance program requires an application, documentation of your professional earnings and an interview. In general, eligibility for financial assistance is based on: a minimum of five years of industry employment with earnings of at least $6,500 for three out of the last five years; and, financial need.
Geographic Focus: All States
Contact: Kent Curtis; (212) 221-7300 ext. 142; kcurtis@actorsfund.org
Internet: http://new.actorsfund.org/services-and-programs/hivaids-initiative
Sponsor: Actors Fund
729 7th Avenue, 10th Floor
New York, NY 10019

Acumen East Africa Fellowship 284

The East Africa Fellowships are designed to develop leadership skills. The fellowship is structured like an executive MBA, so the awardee remains on the job for the fellowship period. The fellowship creates a unique opportunity for East Africans to receive world-class business training, leadership development, and mentoring. Highlights of the fellowship include: support to refine the applicant's own projects for a large social impact; leadership skills development; and continual guidance from experienced mentors and trainers in strategy and problem solving. Awardees are most likely to be successful in their fellowship if they have a deep passion and commitment to the East African region; a proven track record of leadership and management responsibilities; and an unrelenting perseverance, personal integrity and critical thinking skills.
Requirements: Applicants must be at least 18 to apply. There is no educational requirement for the program. The most important requirement is that the applicant have been driving a social change project for at least three years. A "social change project" is defined as any project, initiative, or program that addresses a pressing social issue with an innovative solution that has a positive impact in the lives of the community. The fellowship includes five seminars, including one regional trip. A conference at the end of the fellowship is designed and implemented by the fellowship awardees so they may implement the skills they have learned. The selection process is rigorous and takes place over three months. Online applications are reviewed from March 16 to April 15; phone interviews are conducted from April 17 to May 7; selection conference takes place on May 21 in Nairobi; and finalists are announced on May 31.
Restrictions: Because awardees hold jobs during their fellowship, they do not receive stipends or health insurance.
Geographic Focus: Burundi, Djibouti, Eritrea, Ethiopia, Kenya, Rwanda, Somalia, Sudan, Tanzania, Uganda
Contact: Suraj Sudhakar, East Africa Fellows Manager; +254-20-386-1559 or +254-20-386-1561; ssudhakar@acumenfund.org
Internet: http://www.acumenfund.org/fellows/east-africa-fellows-program.html
Sponsor: Acumen Fund
76 Ninth Avenue, Suite 315
New York, NY 10011

Acumen Global Fellowships 285

The Global Fellowship is a full-time one-year fellowship focused on leadership development and operational experience. Fellows come from all over the world and undergo two months of leadership training in New York and are then placed in one of Acumen Fund's global investments for nine months. Applicants are encouraged to review the website videos of past fellowship recipients. Applicants should also refer to the website for current application information.
Requirements: Applicants should submit a resume/curriculum vitae; personal profile information; two professional references; a letter of recommendation; and four short essay questions. Applicants should review the website's key factors as indicators of a successful fellowship: a proven record of leadership and management responsibilities; experience working in emerging markets; unrelenting perseverance, personal integrity, and critical thinking skills; strong passion and commitment; and 3-7 years of work experience, with a graduate degree preferred.
Restrictions: Fellows are expected to travel frequently and on short notice, sometimes to difficult environments. The program is intensive, time-consuming and requires significant commitment and flexibility. When in the field, access to Internet and phone systems will be available wherever possible, but applicants should expect times when frequent communication with family and friends is not possible or affordable.
Geographic Focus: All States, All Countries
Contact: Blair Miller; (646) 747-3961; bmiller@acumenfund.org
Internet: http://www.acumenfund.org/fellows/fellows-programs.html
Sponsor: Acumen Fund
76 Ninth Avenue, Suite 315
New York, NY 10011

ACVIM Foundation Clinical Investigation Grants 286

The Foundation improves the health and well-being of animals by funding humane studies and communicating information about this work to the veterinary community and the general public. It supports leading veterinary scientists as they take on a range of critical animal health issues within the specialties of small and large animal internal medicine, cardiology, neurology, and oncology. This work, in turn, holds promise for better preventive care, better nutrition, and more sensitive screening and improved treatment for such devastating diseases as cancer, epilepsy, cardiomyopathy, and kidney failure. The Foundation's main goal is to support clinical investigations that lead to new diagnostic, treatment, and prevention techniques. Resident applicants may request up to $12,500 per year, while diplomate applicants may request up to $15,000. One $20,000 grant will be awarded to a collaborative/multi-center proposal, with preference toward private practice/academia partnerships. Funding for multi-year projects up to 3 years will be considered, subject to annual non-competitive renewal based on availability of funds and demonstration of adequate progress on the project. The foundation will consider funding for multi-year projects up to three years.
Geographic Focus: All States

42 | Grant Programs

Date(s) Application is Due: Sep 15
Amount of Grant: Up to 20,000 USD
Samples: Dr. Jill Beech, University of Pennsylvania School of Veterinary Medicine--Evaluation of Endogenous Alpha MSH Concentrations at Different Seasons and in Response to TRH in Normal Horses and Those with Pituitary Hyperplasia; Dr. Stephanie Kottler, University of Missouri-Columbia College of Veterinary Medicine--Prevalence of Staphylococcus aureus and MRSA carriage in three populations.
Contact: Angela E. Frimberger, Executive Director; (303) 231-9933 or (800) 245-9081; info@acvimfoundation.org
Internet: http://www.acvimfoundation.org/grants/info.html
Sponsor: American College of Veterinary Internal Medicine
997 Wadsworth Boulevard, Suite A
Lakewood, CO 80214

ADA Foundation Bud Tarrson Dental School Student Community Leadership Awards 287

The American Dental Association Foundation's Bud Tarrson Dental School Student Community Leadership Awards recognize one exemplary volunteer community service project that is organized and/or conducted by a group of dental students enrolled in a pre-doctoral dental education program accredited by the Commission on Dental Accreditation. One application from each dental school is accepted. One award will be made to one dental school on behalf of the student's or mentor's winning application. Programs focused on community outside the U.S. are eligible to qualify for the Thomas A. Zwemer Award, with am additional $5,000 available. Current program information including deadlines can be found at the Foundation's website.
Requirements: Dental student groups enrolled in a pre-doctoral dental education program accredited by the Commission on Dental Accreditation involved in an outreach program to vulnerable communities within the U.S.
Restrictions: The Foundation does not accept unsolicited proposals outside of its request for proposal programs announced via their website.
Geographic Focus: All States
Date(s) Application is Due: Sep 24
Amount of Grant: 5,000 USD
Contact: Cristina Garcia; (312) 440-2763; fax (312) 440-3526; garciac@ada.org
Internet: http://www.ada.org/applyforassistance.aspx#tarrson
Sponsor: American Dental Association Foundation
211 East Chicago Avenue
Chicago, IL 60611

ADA Foundation Dental Student Scholarships 288

The American Dental Association Foundation Dental Student Scholarships help predoctoral dental students defray a part of their professional education expenses. The goal of this program is to facilitate the education of academically gifted dental students. Students in the second year of study at the time of application, and currently attending a dental school accredited by the Commission on Dental Accreditation of the American Dental Association, are eligible to apply. Approximately 25 awards of up to $2,500 each are available. Candidates of these scholarships are also eligible for the Sullivan/Dewhirst Scholarships, which are each worth an additional $2,500. The application and specific guidelines are available at the ADA Foundation website.
Requirements: Eligible students must be enrolled full-time, taking a minimum of 12 credit hours, and in their second year of study at an educational institution; demonstrate a minimum financial need of $2,500; have maintained the minimum grade-point average of 3.5; and submit two reference forms, one from each of to dental school representatives in support of the application.
Geographic Focus: All States
Date(s) Application is Due: Nov 30
Amount of Grant: Up to 2,500 USD
Contact: Cristina Garcia; (312) 440-2763; fax (312) 440-3526; garciac@ada.org
Internet: http://www.ada.org/applyforassistance.aspx#adaf
Sponsor: American Dental Association Foundation
211 East Chicago Avenue
Chicago, IL 60611

ADA Foundation Dentsply International Research Fellowships 289

The American Dental Association Foundation's Dentsply International Research Fellowships may be applied for by finalists in the Annual Dental Student Research Competition conducted by Dentsply International. Fellowship recipients receive a ten-week research opportunity, a $4,000 stipend, and room and board expenses in order to work in the Foundation's Paffenbarger Research Center located in Gaithersburg, Maryland. Applications are available at the dental school research advisor and/or dean's office.
Geographic Focus: All States
Amount of Grant: 4,000 USD
Contact: Dr. Clifton M. Carey; (301) 975-6806; fax (301) 926-9143; clif.carey@nist.gov
Internet: http://www.ada.org/applyforassistance.aspx#research
Sponsor: American Dental Association Foundation
211 East Chicago Avenue
Chicago, IL 60611

ADA Foundation Disaster Assistance Grants 290

The ADA Foundation's Disaster Assistance Grants provide assistance to members of the dental profession who have lost property in a declared disaster and to support emergency dental care in areas affected by disasters. There are two types of assistance. Grants to dentists provide immediate assistance to dentists affected by disaster. Grants to organizations provide needed dental services to communities affected by disaster. The maximum award is based on the applicant's need and proposal submitted. Applications must be sent to the applicant's state dental society by fax or regular mail.
Requirements: To be eligible for funding, a "disaster" must be declared by a governmental agency, or an event meeting the Foundation's definition must have occurred. The Foundation define disaster as "a sudden occurrence which inflicts widespread catastrophic damage to a large geographic area and/or which generally affects a large number of individuals." Disasters can be both natural and caused by human conduct. Examples include, but are not limited to: civil disorders (excluding acts of war), explosions, fires, tornadoes, earthquakes, floods, tidal waves, forest fires and hurricanes. Eligible dentists are those who are a victim of a disaster. Requests must generally be made within two months of the occurrence of the disaster, unless the time is extended at the discretion of the Board. Dentist applying must show that he or she suffered property damages. Eligible organizations are dental-related 501(c)3 organizations that can provide dental services to victims of a disaster in a community where emergency dental care is needed by victims of a disaster.
Restrictions: Applicants will be offered assistance only once per disaster. The Foundation does not accept unsolicited proposals outside of its request for proposal programs announced via their website.
Geographic Focus: All States
Contact: Cristina Garcia, Grants Program Coordinator; (312) 440-2763 or (312) 440-2547; fax (312) 440-3526; garciac@ada.org or adaf@ada.org
Internet: http://www.ada.org/applyforassistance.aspx
Sponsor: American Dental Association Foundation
211 East Chicago Avenue
Chicago, IL 60611

ADA Foundation George C. Paffenbarger Student Research Awards 291

The ADA Foundation George C. Paffenbarger Student Research Awards are offered in collaboration with the Academy of Operative Dentistry. Awards honors George C. Paffenbarger, an internationally recognized authority on dental materials and a pioneer in the development of specifications and standards for testing materials. The Foundation's Paffenbarger Research Center is named in his honor. Awards are open to dental students at all levels who wish to undertake novel research relevant to contemporary operative dentistry. The budget for the proposed research may not exceed $6,000. The awardee will also receive up to $1,000 to attend and present the research findings at the annual meeting. Applications must be submitted electronically to the Research Committee Chair.
Requirements: The application should take the form of a protocol outlining the background, aims and hypothesis of the proposed research, the methodology to be employed, and the anticipated work on the clinical practice of operative dentistry. Application information, including deadline and submission format, can be found at the Foundation's website.
Geographic Focus: All States
Date(s) Application is Due: Jan 13
Amount of Grant: Up to 7,000 USD
Contact: Cristina Garcia; (312) 440-2763; fax (312) 440-3526; garciac@ada.org
Internet: http://www.ada.org/applyforassistance.aspx#tarrson
Sponsor: American Dental Association Foundation
211 East Chicago Avenue
Chicago, IL 60611

ADA Foundation Minority Dental Student Scholarships 292

The American Dental Association Foundation Dental Student Scholarships help predoctoral dental students defray a part of their professional education expenses. The goal of this program is to facilitate the education of academically gifted dental students. The Underrepresented Minority Dental Student Scholarship Program targets African-American, Hispanic, and Native Americans dental students, all of which have been identified as underrepresented minorities in dentistry. The Sullivan and Dewhirst Dental Scholarships, each worth an additional $2,500, will be chosen from a pool of these scholarship recipients. The application and additional information are available at the Foundation website.
Requirements: Students must be full time students (at least 12 credit hours) in their second year of study at the time of the application, and currently attending or enrolled at a dental school accredited by the Commission on Dental Accreditation of the American Dental Association. They must also demonstrate a minimum financial need of $2,500, and hold a grade point average of no less than 3.5. Along with the application, candidates must submit two reference forms, one from each of two dental school representatives (academic advisor or professor) in support of the application.
Geographic Focus: All States
Date(s) Application is Due: Nov 30
Amount of Grant: Up to 2,500 USD
Contact: Cristina Garcia; (312) 440-2763; fax (312) 440-3526; garciac@ada.org
Internet: http://www.ada.org/applyforassistance.aspx#adaf
Sponsor: American Dental Association Foundation
211 East Chicago Avenue
Chicago, IL 60611

ADA Foundation Relief Grants 293

The American Dental Association Foundation Relief Grants provide financial assistance to dentists and their dependents who, because of accidental injury, a medical condition, disability or advanced age, are unable to meet basic living expenses. In determining the amount of the grant, the Foundation will take into consideration the particular circumstances of each applicant, including financial needs, age and physical conditions, opportunity for assistance from immediate members of the family, financial assets, and all other relevant factors. Grant rules and criteria as well as an application can be found at the Foundation's website. Applications are accepted at any time.

GRANT PROGRAMS | 43

Requirements: Eligible applicants are dentists, their dependents and former dependents of deceased dentists if an accidental injury, advanced age, physically debilitating illness or medically-related condition prevents them from gainful employment and results in an inability to be wholly self-sustaining. Dependents are limited to a dentist's current spouse and blood-related or legally adopted children under the age of eighteen. A deceased dentist's spouse, if married at the time of the death, is also considered a dependent.
Restrictions: If a deceased dentist's spouse remarries, the spouse is no longer eligible.
Geographic Focus: All States
Contact: Cristina Garcia; (312) 440-2763; fax (312) 440-3526; garciac@ada.org
Internet: http://www.ada.org/applyforassistance.aspx#research
Sponsor: American Dental Association Foundation
211 East Chicago Avenue
Chicago, IL 60611

ADA Foundation Samuel Harris Children's Dental Health Grants   294
The American Dental Association Foundation Samuel Harris Children's Dental Health Grants provides funding to programs that are designed to improve and maintain children's oral health through outreach, primary prevention, and education. The program currently focuses on funding parent and caregiver education programs to prevent early childhood caries, also known as baby bottle tooth decay, by circumventing primary oral bacterial infection before it can take hold. See the website for the complete request for proposal, application materials, and deadlines.
Requirements: Proposals are accepted from nonprofit 501(c)3 organizations and government agencies for community-based oral health and prevention education programs for parents and caregivers of infants in the United States and its territories.
Restrictions: The Foundation does not accept unsolicited proposals outside of its request for proposal programs announced via their website.
Geographic Focus: All States
Amount of Grant: Up to 5,000 USD
Contact: Cristina Garcia; (312) 440-2763; fax (312) 440-3526; garciac@ada.org
Internet: http://www.ada.org/applyforassistance.aspx#harris
Sponsor: American Dental Association Foundation
211 East Chicago Avenue
Chicago, IL 60611-2616

ADA Foundation Scientist in Training Fellowship   295
The American Dental Association Foundation's Scientist in Training Fellowships are conducted at the Foundation's Paffenbarger Research Center (PRC) in Gaithersburg, Maryland. The program includes a full-time fellow working in conjunction with the PRC's scientific research staff. Fellowships have been made possible by the Great-West Life and Annuity Insurance Company. Individuals interested in learning more about the program, application process or deadlines are invited to contact Paffenbarger Research Center.
Geographic Focus: All States
Contact: Dr. Clifton M. Carey, Director of Administration; (301) 975-6806; fax (301) 926-9143; clif.carey@nist.gov or prc@ada.org
Internet: http://www.ada.org/applyforassistance.aspx#grants
Sponsor: American Dental Association Foundation
211 East Chicago Avenue
Chicago, IL 60611

ADA Foundation Summer Scholars Fellowships   296
The ADA Foundation's Paffenbarger Research Center annually appoints two young investigators to its summer scholars program. The program brings industrial and dental researchers together in an environment outside the dental school. This program is made possible by Dentsply International. Individuals interested in learning more about the program, including the application process and deadlines, are invited to contact the Program Coordinator at the Paffenbarger Research Center.
Geographic Focus: All States
Contact: Gretchen Duppins, Summer Scholars Program Coordinator; (301) 975-6806; fax (301) 926-9143; gretchen@dupins.nist.gov
Internet: http://www.ada.org/applyforassistance.aspx
Sponsor: American Dental Association Foundation
211 East Chicago Avenue
Chicago, IL 60611

ADA Foundation Thomas J. Zwemer Award   297
The Dr. Thomas J. Zwemer Award is offered by the American Dental Association Foundation (ADAF) to dental schools/dental colleges. The goal of the Zwemer Award is to recognize one exemplary volunteer community service project outside of the U.S. that is organized and/or conducted by a group of dental students enrolled in a pre-doctoral dental education program accredited by the Commission on Dental Accreditation. One award of $5,000 per year will be made to the dental school/college in honor of the student outreach program. Award funds must be expressly used to continue to enhance student service learning and outreach to underserved populations outside of the U.S.
Requirements: One application from each dental school will be accepted. The application must be prepared by a student group, and affirmed by the Dean or Associate Dean. Additional guidelines and requirements are available at the Foundation website. Service learning projects that provide academic credit may be submitted. This Award is meant to recognize superior altruistic student efforts. Thus, applications that describe service learning for academic credit should also describe student enhancements to that service learning that go beyond basic academic requirements and which are student directed and volunteer based.
Geographic Focus: All States

Date(s) Application is Due: Nov 26
Amount of Grant: Up to 2,500 USD
Contact: Cristina Garcia, Grants Program Coordinator; (312) 440-2547, ext. 2763; fax (312) 440-3526; garciac@ada.org or adaf@ada.org
Internet: http://www.ada.org/applyforassistance.aspx#grants
Sponsor: American Dental Association Foundation
211 East Chicago Avenue
Chicago, IL 60611-2616

Adam Richter Charitable Trust Grants   298
The Trust, which is administered by the Bank of America in Dallas, Texas, was established in California in 1994. Offering grants on a national basis, its primary fields of interest include: the arts; animal welfare; Christian agencies and churches; the environment; health organizations; higher education; hospitals; and human services. There are no deadlines, though a formal application is required. Interested applicants should begin by sending a letter of inquiry to the Trust office.
Restrictions: Funding is not available for grants to individuals.
Geographic Focus: California
Amount of Grant: 3,000 - 20,000 USD
Contact: Michael Schlebach, Specialty Asset Manager; (866) 461-7281
Sponsor: Adam Richter Charitable Trust
c/o Bank of America, N.A.
Dallas, TX 75201-3115

Adams and Reese Corporate Giving Grants   299
At Adams and Reese, the Corporation takes pride in giving back to its communities and believes success is directly related to the prosperity and the quality of life within the communities it serves. Its corporate philanthropy program, HUGS (Hope, Understanding, Giving, and Support) was founded in 1988 by Partner, Mark Surprenant. Since its inception, the firm has devoted financial resources and thousands of volunteer hours to offer assistance to those in need. A fundamental commitment to volunteerism is the deep-rooted characteristic of the corporation. Primary activities include grants in support of health, youth development, and human services. Fields of interest are: general charitable giving, operating support, health care organizations, legal services, social services, and youth programs.
Requirements: Regions of grant application eligibility include: Birmingham and Mobile, Alabama; Jacksonville, Sarasota, St. Petersburg, Tallahassee, and Tampa, Florida; Baton Rouge and New Orleans, Louisiana; Jackson, Mississippi; Columbia, South Carolina; Chattanooga, Memphis, and Nashville, Tennessee; Houston, Texas; and Washington, D.C..
Geographic Focus: Alabama, District of Columbia, Florida, Louisiana, Mississippi, South Carolina, Tennessee, Texas
Contact: Mark Surprenant, Liaison Partner; (504) 581-3234; fax (504) 566-0210; Mark. Surprenant@arlaw.com
Internet: http://www.adamsandreese.com/community/
Sponsor: Adams and Reese Corporation
701 Poydras Street, Suite 4500
New Orleans, LA 70139-7755

Adams County Community Foundation of Pennsylvania Grants   300
The Adams County Community Foundation was established in October of 2007 as a Pennsylvania corporation to succeed the Adams County Foundation which for 22 years had operated as a trust-based community foundation. The Foundation is a public charity that the IRS has determined to be a 501(c)3 organization. The purpose of this foundation is to inspire people and communities to build and distribute charitable funds for good, for Adams County, forever. After one year of operation, the foundation had acquired nine funds established by individual people or families, members of the Foundation's Board of Directors, groups of business people, a local charity and by a visionary patriot from the 1700s. The Foundation's assets are managed by a professional investment company hired by the Board of Directors and overseen by the Foundation's Investment & Finance Committee. From these funds, distribution of grants is made to qualified, local charities who demonstrate that they are meeting community needs. With a 16 member Board of Directors, Adams County Community Foundation serves as a good steward of the monies of our donors and demonstrates accountability, transparency, confidentiality, compassion, inclusiveness and excellence in its work. The foundation gives to organizations that assist, promote and improve the moral, mental, social and physical well-being of area residents. Fields of interest include: arts; Christian agencies and churches; community and economic development; education; and health organizations.
Requirements: Applicants must serve residents living in Adams County, Pennsylvania.
Geographic Focus: Pennsylvania
Contact: Barbara Ernico, (717) 337-0060 or (717) 337-3353; info@adamscountycf.org
Internet: http://www.adamscountycf.org/receive.html
Sponsor: Adams County Community Foundation of Pennsylvania
101 W Middle Street, P.O. Box 4565
Gettysburg, PA 17325-2109

Adams Foundation Grants   301
Established in 1955, the Adams Foundations was initially funded through a donation by Rolland L. Adams, who became owner of the Bethlehem Globe newspaper in 1929. The newspaper won a Pulitzer Prize for editorial writing in 1972. Abarta, which operates the Foundation, was founded by members of the Adams, Bitzer, Roehr and Taylor families - names which form the acronym of the company title. Grants are currently awarded in both Pittsburgh, Pennsylvania, and Ithaca, New York, along with the surrounding areas of these two communities. It supports food banks and civic centers, as well as organizations involved with arts and culture, education, mental health, and arthritis

treatments. Funding comes in the form of general operating support. Applicants should submit letters of application detailing the project need and funding requested. Though there are no specified annual deadlines, the board meets each February and August. Most recent grant awards have ranged from $3,500 to $125,500.
Requirements: Any 501(c)3 organization serving the residents of Pittsburgh, Pennsylvania, and Ithaca, New York, are eligible to apply.
Geographic Focus: New York, Pennsylvania
Amount of Grant: 3,500 - 150,000 USD
Contact: Shelley M. Taylor, President; (412) 963-1087 or (412) 963-3163
Sponsor: Adams Foundation
1000 Gamma Drive, 5th Floor
Pittsburgh, PA 15238-2929

Adams Rotary Memorial Fund A Grants 302
Established in Indiana, the Adams Rotary Memorial Fund A offers funding in Howard County, Indiana, with a primary goal to support handicapped children throughout the county. With this in mind, its primary fields of interest include: children and youth services; alleviation of disabilities; and money to support families who have a family member with a disability. Money supports philanthropy and volunteerism, and is given to individual applicants. There are no application deadlines or specific formats, and those in need should begin by contacted the Fund trustee directly.
Geographic Focus: Indiana
Contact: Glenn Grundmann, Trustee; (317) 464-8212 or (574) 282-8839
Sponsor: Adams Rotary Memorial Fund
224 N. Main Street
Kokomo, IN 46901

Adams Rotary Memorial Fund B Scholarships 303
Established in Indiana, the Adams Rotary Memorial Fund B offers scholarships in Howard County, Indiana, with a primary goal of supporting the continued education of graduates of high schools throughout the county. Financial support is offered for those students pursuing a degree in medicine or nursing. Applications are available upon request, and the annual deadline is the 3rd Friday of March. Awards range from $1,000 to $3,000, and can be used for tuition, books, and room and board.
Requirements: Any high school graduate of a Howard County, Indiana, school is eligible.
Geographic Focus: Indiana
Date(s) Application is Due: Mar 3
Amount of Grant: 1,000 - 3,000 USD
Contact: Glenn Grundmann, Trustee; (317) 464-8212 or (574) 282-8839
Sponsor: Adams Rotary Memorial Fund
224 N. Main Street
Kokomo, IN 46901

Adaptec Foundation Grants 304
The corporate contributions program is primarily directed, although not exclusively, to organizations and programs serving the communities where its employees work and live, with special emphasis in the Silicon Valley area. Funding priorities include education and research--programs that lower the drop-out rate of youth in accredited schools and colleges, address career development and preparation, job training and placement, educational scholarship assistance, and investment in higher education (primarily in engineering-related university programs); health and human services--research cures for devastating diseases, emergency shelter and subsistence for the homeless, treatment and assistance for abused spouses and children, drug and alcohol rehabilitation, and supporting organizations that strengthen youths and families; helping people enjoy life more fully through music, dance, art museums, opera, theater, and performing arts groups; and $25-$500 employee matching gifts. The Corporate Contributions Committee meets quarterly to consider requests.
Restrictions: In general, Adaptec does not make contributions to any organization that does not have 501(c)3 or comparable status; individuals, except through the Adaptec Scholarship program; churches, synagogues, and other religious groups. (Requests are considered from religious-sponsored and other groups whose activities do not support any specific religious doctrine or ethnic/cultural group.); political or fraternal organizations; underwrite or assist with the development of films, television, or video/radio productions, whether for commercial, independent, or nonprofit ventures, unless such activity is in support of the computer I/O software/hardware industry and is directly related to the business of the company; tickets for raffles, contests, or other fundraising activities (except requests from major fundraising organizations that meet the defined objectives of the contribution program, such as United Way, the Silicon Valley Charity Ball and the San Jose Symphony); organizations whose primary objective is to assist, benefit, and address animal welfare or environmental issues, such as the preservation of endangered species and plants or conservation efforts to preserve the quality of wildlife, water, soil, or air; reduce debts, fund general operating expenses or capital improvement projects, or retroactively fund activities or programs that are already completed; and school sports leagues and events, and other non-educational related activities such as band, debate teams, and goodwill trips.
Geographic Focus: All States
Amount of Grant: 250 - 2,500 USD
Contact: Contribution Program Manager; (408) 945-8600; fax (408) 262-2533
Internet: http://cms.adaptec.com/ko-KR/company/about/_corporate/adaptec_corpcontributions.htm?nc=/ko-KR/company/about/_corporate/adaptec_corpcontributions.htm
Sponsor: Adaptec Foundation
691 South Milpitas Boulevard, MS-15
Milpitas, CA 95035-5473

Addison H. Gibson Foundation Medical Grants 305
The Foundation's principal philanthropic interest is to provide medical care; specifically, to enable working people to continue their employment without the burden of untreated illness, or the financial hardship resulting from loss of income associated with untreated illness. The funds provide such medical treatment to needy, self-supporting, long-term western Pennsylvania residents who otherwise cannot afford the medical aid which they require. Physicians, social service and hospital administrators may apply on behalf of their patients.
Requirements: Applicants for medical grants must be referred by their attending physicians, or via social service or hospital administrators. Medical grants are applied for in advance of prescribed treatment. Previously incurred medical expenses are not eligible for consideration. Patients with a correctable condition only will be considered.
Restrictions: Recipients who benefit from these grants must be long-term residents of western Pennsylvania. Grant monies are paid directly to the providers, never to the patients.
Geographic Focus: Pennsylvania
Contact: Rebecca Wallace, Director; (412) 261-1611; fax (412) 261-5733
Internet: http://www.gibson-fnd.org/Docs/Medical_Grants.htm
Sponsor: Addison H. Gibson Foundation
One PPG Place, Suite 2230
Pittsburgh, PA 15222-5401

Adelaide Breed Bayrd Foundation Grants 306
The Adelaide Breed Bayrd Foundation supports programs and projects of nonprofit organizations serving the Boston, Massachusetts, area with emphasis on the Malden community. Areas of interest include: adult and continuing education; aging centers and services; arts; children and youth services; community and economic development; education; family services; health care; health organizations; hospitals (general); human services; libraries and library science; residential and custodial care; and hospices. The Foundation funds annual campaigns, building and renovation, capital campaigns, emergency requests, equipment, program development, and scholarship funds. Grants have been awarded in the past to support public libraries, hospitals, and the Girl Scouts. The foundation accepts the Associated Grant Makers (AGM) Common Proposal Form available from the AGM website; however an application form is not required. Applicants should submit the following: a copy of their IRS Determination Letter; copies of their most recent annual report, audited financial statements, and 990; a detailed description of their project and the amount of funding requested; and a copy of their organizational or project budget for the current year. Proposals should be submitted before the second Tuesday in February. Awardees will be notified in April or May.
Requirements: 501(c)3 organizations serving the Boston, Massachusetts area are eligible. Preference will be given to organizations serving the community of Malden, Massachusetts.
Restrictions: The foundation does not support the following categories: requests from individuals; the performing arts (except for certain educational programs); matching or challenge grants; demonstration projects; conferences; publications; research or endowment funds; or loans.
Geographic Focus: Massachusetts
Amount of Grant: 1,000 - 150,000 USD
Contact: C. Henry Kezer, President; (781) 324-1231
Sponsor: Adelaide Breed Bayrd Foundation
350 Main Street, Suite 13
Malden, MA 02148-5023

Adler-Clark Electric Community Commitment Foundation Grants 307
Established in Wisconsin in 2004, the Adler-Clark Electric Community Commitment Foundation offers grants in Clark County, Wisconsin. It supports athletics and amateur leagues, fire prevention and control, education, food banks, food services, health care, human services, public libraries, and recreation programs. Types of support include: equipment purchase; general operations; and program development. A formal application is required, and the annual deadline has been identified as December 1. Funding amounts range from $500 to $3,000.
Geographic Focus: Wisconsin
Date(s) Application is Due: Dec 1
Amount of Grant: 500 - 3,000 USD
Contact: Timothy E. Stewart, Trustee; (715) 267-6188 or (800) 272-6188
Internet: http://www.cecoop.com/home
Sponsor: Adler-Clark Electric Community Commitment Foundation
P.O. Box 190
Greenwood, WI 54437-9419

Administaff Community Affairs Grants 308
Five areas qualify for Administaff corporate contribution consideration: the elderly, education, social service, environment, and health. Administaff will consider four types of requests: volunteers and grants; grants; event sponsorships; and in-kind donations--equipment such as computers, copiers and furniture. All contribution requests must be made in writing. Guidelines and sample cover letter are available online.
Requirements: Charitable organizations must have 501(c)3 status, according to IRS standards; operate in one of Administaff's district markets; and submit a written request. Qualifying organizations also must submit a financial statement to the Better Business Bureau before consideration.
Restrictions: Ineligible entities include: religious organizations (i.e., churches, temples, and other houses of worship, or those whose main purpose is to promote a particular faith or creed); political organizations; individuals; athletic groups; or school clubs.
Geographic Focus: All States
Amount of Grant: 5,000 USD

Contact: Betty Collins, Program Director, Community Involvement; (281) 358-8986; fax (281) 312-3559; community_involvement@administaff.com
Internet: http://www.administaff.com/about_asf/grants.asp
Sponsor: Administaff Corporation
19001 Crescent Springs Drive
Kingwood, TX 77339-3802

Administration on Aging Senior Medicare Patrol Project Grants 309
The goal of SMP Projects is to empower beneficiaries/consumers to prevent health care fraud through outreach and education. Program coverage must target vulnerable, hard-to-reach population beneficiaries, their families and other consumers. The purpose of this competition is to provide the opportunity to fund one (1) Project in each of 26 eligible states including the District of Columbia for a project period of up to three (3) years.
Requirements: Domestic public or private and non-profit entities including state, local and Indian tribal governments (American Indian/Alaskan Native/Native American), faith-based organizations, community-based organizations, hospitals, and institutions of higher education are eligible to apply. Through this competition, AoA plans to fund one project in each of the following 26 states and the District of Columbia: Alaska, Arizona, Arkansas, Colorado, Delaware, the District of Columbia, Florida, Georgia, Idaho, Kansas, Kentucky, Maine, Massachusetts, Michigan, Montana, New Jersey, New Mexico, Ohio, Oklahoma, Oregon, Puerto Rico, Tennessee, Texas, Virginia, Washington, and West Virginia.
Geographic Focus: All States
Date(s) Application is Due: Apr 11
Amount of Grant: Up to 345,000 USD
Contact: Barbara Lewis, Project Officer; (202) 357-3532; Barbara.Lewis@aoa.hhs.gov
Internet: http://www.aoa.dhhs.gov/doingbus/fundopp/fundopp.asp
Sponsor: Administration on Aging
1 Massachusetts Avenue NW
Washington, D.C. 20001

Adobe Community Investment Grants 310
Adobe supports strategic programs and partnerships that help make its communities better, stronger, and more vibrant places to live, work and do business. Adobe's focus areas for giving and grants programs are designed to: increase Adobe's impact in the community through support of more organizations; and strengthen Adobe's role as a corporate partner by creating deeper, stronger, and richer partnerships. The Adobe Community Investment Grant program provides multi-year (with annual review), comprehensive support, including cash, software, volunteers, and facilities use through an Adobe-initiated, RFP application process. Grant amounts are at least $20,000 and can be for up to three years. Organizations selected become Adobe Community Investment Partners and are required to sit out one year after the grant ends before reapplying.
Requirements: Adobe is currently accepting grant proposals in the following communities: San Jose/Silicon Valley, California (southern San Mateo County, Santa Clara County, southern Alameda County); San Francisco, California; Seattle/King County, Washington; and Ottawa, Ontario, Canada.
Geographic Focus: California, Washington, Canada
Contact: Lesley Dierks, Community Relations; (408) 536-6000 or (408) 536-3993; fax (408) 537-6000; ldierks@adobe.com or community_relations@adobe.com
Internet: http://www.adobe.com/aboutadobe/philanthropy/commgivingprgrm.html#Community%20giving%20programs
Sponsor: Adobe Systems
345 Park Avenue
San Jose, CA 95110-2704

Adolph Coors Foundation Grants 311
The Coors Foundation supports organizations that promote the western values of self-reliance, personal responsibility, and integrity. The foundation believes these values foster an environment where entrepreneurial spirits flourish and help Coloradans reach their full potential. High priority is placed on programs that help youth to prosper, that encourage economic opportunities for adults, and that advance public policies that uphold traditional American values. Traditional areas of support include one-on-one mentoring programs, job training, and a variety of self-help initiatives. The foundation also has an interest in bringing integrative medicine into the medical mainstream. In each of its giving areas, the foundation seeks evidenced-based results. Civic and cultural programs attracting the Foundation's attention are typically those that enhance our culture and heritage, that demonstrate our creativity as a people and that are likely to be of economic benefit to and broadly used by the communities they serve. Past grants have supported boys and girls clubs and inner-city health programs. Types of support include building funds, general operating budgets, seed money, and special projects. The foundation has moved to an online screening and application system which is accessible from the foundation website. Application deadlines are March 1, July 1, and November 1.
Requirements: All applicants must be classified as 501(c)3 organizations by the Internal Revenue Service and must operate within the United States.
Restrictions: The foundation does not provide support for the following expenses or entities: organizations primarily supported by tax-derived funds; conduit organizations that pass funds to non-exempt organizations; organizations with two consecutive years of operating loss; K-12 schools or the ancillary programs and projects of those schools; individuals; research projects; production of films or other media-related projects; historic renovation; churches or church projects; museums or museum projects; animals or animal-related projects; preschools, day-care centers, nursing homes, extended-care facilities, or respite care; deficit funding or retirement of debt; special events, meetings, or seminars; purchase of computer equipment; adaptive sports programs; and national health organizations. Organizations applying for start-up funding must have been in operation for at least one full year.
Geographic Focus: All States
Date(s) Application is Due: Mar 1; Jul 1; Nov 1
Contact: Jeanne Bistranin, Program Officer; (303) 388-1636; fax (303) 388-1684
Internet: http://www.coorsfoundation.org/Process/index.html
Sponsor: Adolph Coors Foundation
4100 E Mississippi Avenue, Suite 1850
Denver, CO 80246

Advance Auto Parts Corporate Giving Grants 312
The Advance Auto Parts Corporate Giving program was founded on the belief that good business is more than just selling merchandise. Since the company was founded in 1932, it has been guided by the following principles, known as the Advance Values: inspire and build the self-confidence and success of every Team Member; serve customers better than anyone else and help them succeed; and grow the business and profitability with integrity. While the corporation understands that there are a variety of worthy causes in every community, to maximize its giving and make the greatest possible impact, Advance has chosen to focus its charitable efforts on serving those in need through support of the following four impact areas: health – improving the health and well-being of others; education – helping people reach their potential by providing educational opportunities; at-risk children and families – helping children and families to assure their critical needs are met through comprehensive, community-based programs; and disaster relief – providing timely support to those impacted by disasters such as earthquakes, hurricanes, floods and other natural disasters.
Geographic Focus: All States
Contact: Grants Manager; (540) 362-4911; fax (540) 561-1448
Internet: http://corp.advanceautoparts.com/english/about/public.asp
Sponsor: Advance Auto Parts Corporate Giving
5008 Airport Road
Roanoke, VA 24012

Advanced Micro Devices Community Affairs Grants 313
AMD awards grants to nonprofit organizations and schools that serve the communities in which it operates: Sunnyvale, CA, and Austin, TX, and in Europe and Asia. AMD grants funds to agencies that serve the community's health and human service, education and civic needs. AMD K-12 initiatives target programs that increase student interest and/or proficiency in literacy, math, science, and computer technology. AMD also funds programs aimed at developing and supporting effective classroom instruction. AMD also funds basic needs programs to local health and human services organizations. In addition, AMD matches employee gifts to organizations whose activities fall within one of the approved funding areas. AMD also supports engineering education at universities throughout the United States. Generally, funds are granted for specific projects, not for general operating expenses. Requests must be submitted in writing, using appropriate forms, and must include a description of the applicant organization with a statement of its purposes and objectives, a statement regarding the need the grant will address and the geographic area and population served, the names and qualifications of the person(s) who will administer the grant, and amount requested with an explanation of how the funds will be used. Awards are made on a quarterly basis.
Requirements: IRS 501(c)3 tax-exempt organizations and schools based in Sunnyvale, CA, and Austin, TX, are eligible.
Restrictions: AMD will not fund political activities, national programs, advocacy programs, religious or fraternal organizations, individuals, fund-raising events, conferences, seminars, door prizes, registration fees, advertising, or sports leagues.
Geographic Focus: California, Texas, Albania, Andorra, Armenia, Austria, Azerbaijan, Belarus, Belgium, Bosnia & Herzegovina, Bulgaria, Croatia, Cyprus, Czech Republic, Denmark, Estonia, Finland, France, Georgia, Germany, Greece, Hungary, Iceland, Ireland, Italy, Kosovo, Latvia, Liechtenstein, Lithuania, Luxembourg, Macedonia, Malta, Moldova, Monaco, Montenegro, Norway, Poland, Portugal, Romania, Russia, San Marino, Serbia, Slovakia, Slovenia, Spain, Sweden, Switzerland, The Netherlands, Turkey, Ukraine, United Kingdom, Vatican City
Date(s) Application is Due: May 1
Contact: Community Affairs Manager; (408) 749-5373
Internet: http://www.amd.com/us-en/Corporate/AboutAMD/0,,51_52_7697_7702,00.html
Sponsor: Advanced Micro Devices
P.O. Box 3453
Sunnyvale, CA 94088-3453

Advocate HealthCare Post Graduate Administrative Fellowship 314
The program is open to recent graduates from programs in Healthcare Administration or Business Administration, and is designed to provide its fellows with an educational working environment that enables them to use their analytical, project management and communication skills in team-based setting. Under the supervision of senior leaders, fellows complete projects in operational and/or strategic areas. Project assignments are adapted to the individual's needs and interests. The postgraduate fellowship is a one-year program designed to enables the graduate to develop leadership skills and management skills. Fellows are given the opportunity for exposure across Advocate's eight hospitals, three physician practice organizations, home health care agency and corporate offices (Chicago area).
Requirements: In order to be considered, the following must be submitted no later than November 1, the year previous to entry in to the program: (1) Personal statement of interest from explaining qualifications and career objectives; (2) Statement of community involvement; (3) A current resume; (4) Letters of recommendation from applicant's

program director and at least one other professional recommendation; and, (5) Copy of unofficial undergraduate transcripts and most recent graduate school transcripts.
Geographic Focus: All States
Date(s) Application is Due: Nov 1
Contact: Melissa O'Neill, Director; (630) 572-9393
Internet: http://www.advocatehealth.com/body.cfm?id=888
Sponsor: Advocate HealthCare
2025 Windsor Drive
Oak Brook, IL 60523

AEC Trust Grants 315
The AEC Trust is a private foundation established in 1980 as a philanthropic, grantmaking organization. The trust awards grants to eligible nonprofits in Colorado, Florida, Georgia, and Massachusetts in its areas of interest, including: AIDS, arts and culture, community development, elementary education, the environment, health care, higher education, social services, museums, and women. Types of support include building construction/renovation, capital campaigns, challenge/matching grants, equipment acquisition, general operating support, land acquisition, project support, and publications. Request guidelines in writing. There are two annual deadlines: April 1 and September 1.
Requirements: 501(c)3 nonprofit organizations in the communities of Boulder, Colorado, Gainesville, Florida, Atlanta, Georgia, and Amherst, Massachusetts, are eligible.
Restrictions: Grants do not support individuals, international organizations, political organizations, religious organizations, school districts, special events/benefit dinners, state and local government agencies, United Way agencies, national public charities, endowments, sponsorships, or annual fund campaigns.
Geographic Focus: Colorado, Florida, Georgia, Massachusetts
Date(s) Application is Due: Apr 1; Sep 1
Amount of Grant: 5,000 - 50,000 USD
Samples: Center for Human Development, Amherst, Massachusetts, $50,000 - Family Outreach Amherst Project; Conservation Trust for Florida, Micanopy, Florida, $50,000 - Land Conservation and Landowner Education; Feminist Women's Health Center, Atlanta, Georgia, $5,000 - Lifting Latino Voices Initiative.
Contact: Edith Dee Cofrin; (800) 839-1754; requests@foundationsource.com
Internet: https://online.foundationsource.com/public/home/aec
Sponsor: AEC Trust
501 Silverside Road, Suite 123
Wilmington, DE 19809-1377

AEGON Transamerica Foundation Health and Welfare Grants 316
The AEGON Transamerica Foundation will consider favorably grants to established organizations with reputations for excellence and cost-effectiveness. The Foundation's Health and Welfare Grants initiative is interested in supporting programs committed to improving the condition of the human body through nutrition, housing for the homeless, disease prevention and other support services. Types of support include continuing support, matching funds, operating budgets, employee-related scholarships, and special projects. Contributions are normally made on a year-to-year basis with no assurance of renewal of support. In certain cases, pledges may be considered for periods not exceeding three years.
Requirements: Nonprofit organizations within the Foundation's focus areas and mission, and that are designated for a community where there is a significant employee presence are eligible. Requests can be directed to the attention of the AEGON Transamerica Foundation at one of the following locations: Louisville, Kentucky; Atlanta, Georgia; Baltimore, Maryland; Bedford, Texas; Cedar Rapids, Iowa; Exton, Pennsylvania; Harrison, New York; Little Rock, Arkansas; Los Angeles, California; Plano, Texas; and St. Petersburg, Florida.
Restrictions: Individuals, as well as the following types of organizations or programs are not eligible to receive grants from the Foundation: athletes or athletic organizations; conferences, seminars or trips; courtesy or goodwill advertising; fellowships; fraternal organizations; K-12 school fundraisers or events; political parties, campaigns or candidates; religious or denominational organizations except for specific programs broadly promoted and available to anyone and free from religious orientation; or social organizations.
Geographic Focus: Arkansas, California, Florida, Georgia, Iowa, Kentucky, Maryland, New York, Pennsylvania, Texas
Amount of Grant: 1,000 - 50,000 USD
Contact: David Blankenship; (319) 398-8895 or (319) 355-8511; fax (319) 398-8030; david.blankenship@transamerica.com or shaegontransfound@aegonusa.com
Internet: http://www.transamerica.com/about_us/aegon_transamerica_foundation.asp
Sponsor: AEGON Transamerica Foundation
4333 Edgewood Road, NE
Cedar Rapids, IA 52499-0010

AES-Grass Young Investigator Travel Awards 317
To recognize outstanding young investigators conducting research in basic or clinical neuroscience related to epilepsy, the Grass Foundation and the American Epilepsy Society (AES) have partnered to present the Young Investigator Travel Awards for up to eight candidates to help support travel costs to present their research at the annual AES meeting. The AES will announce the awardees and their talk/poster titles in an award ceremony during the annual program. Following the initial review of application abstracts, as part of the application process, a recommendation may be required from one of the applicant's trainers/mentors. This will be solicited by the AES review committee should the application be deemed competitive for the award. This recommendation should include the applicant's research focus, role in the abstract, promise as a future investigator, and any other pertinent information.
Requirements: Candidates will be investigators early in their careers, including graduate students, postdoctoral fellows, residents, clinical fellows or junior faculty (M.D., M.D./Ph.D., Ph.D., or the equivalent), who are conducting basic or clinical research relating to epilepsy. Candidates must also be no more than five years out of postgraduate training. Applicants should also be first authors on an abstract submitted to the AES. Applicants can self-nominate, or nominations from second parties will be accepted. Applicants submit an abstract through the AES online submission program as first author, then indicate the Junior Investigator Travel Award box. Candidates must be present at the AES Annual Meeting.
Geographic Focus: All States
Date(s) Application is Due: Jun 15
Amount of Grant: 1,000 USD
Contact: Cheryl-Ann Tubby; (860) 586-7505, ext. 542; ctubby@aesnet.org
Internet: http://www.aesnet.org/research/aes-sponsored-grant/grass-travel-award/
Sponsor: American Epilepsy Society
342 North Main Street
West Hartford, CT 06117-2507

AES Epilepsy Research Recognition Awards 318
The Epilepsy Research Recognition Awards are designed to encourage and reward clinical and basic science investigators whose research contributes to understanding and conquering epilepsy. Applications are evaluated on pioneering research; originality of research; quality of publications; research productivity; relationship of the candidate's work to problems in epilepsy; training activities; and productivity over the next decade. There are two awards given for $10,000 each to active scientists and clinicians. The awards are intended as prizes made to individuals for their unrestricted use, not to be used for institutional or departmental support. Awards are announced at the annual AES conference. Detailed instructions for application are available at the AES website.
Requirements: Any individual holding a professional degree and whose research impacts on any aspect of epilepsy is eligible for the award. Nominations from outside the U.S. are welcome.
Geographic Focus: All States, All Countries
Date(s) Application is Due: Aug 6
Amount of Grant: 10,000 USD
Contact: Cheryl-Ann Tubby; (860) 586-7505; fax (860) 586-7550; ctubby@aesnet.org
Internet: http://www.aesnet.org/go/research/research-awards/epilepsy-research-recognition-awards-program
Sponsor: American Epilepsy Society
342 North Main Street
West Hartford, CT 06117-2507

AES J. Kiffin Penry Excellence in Epilepsy Care Award 319
The Penry Award honors Dr. Penry's lifelong focus and concern for epilepsy patients. The Award recognizes those whose work has had a major impact on patient care and improved the quality of life for persons with epilepsy. Awardee receives a $3,000 honorarium plus round trip economy airfare, one night lodging at annual conference, reimbursement of per diem expenses, and free Scientific Program registration. In addition, the award winner is highlighted in the Conference Program Book.
Requirements: Members and non-members of AES may be nominated. The nomination form is available at the AES website.
Geographic Focus: All States
Date(s) Application is Due: Aug 5
Amount of Grant: 3,000 USD
Contact: Cheryl-Ann Tubby; (860) 586-7505; fax (860) 586-7550; ctubby@aesnet.org
Internet: http://www.aesnet.org/go/research/research-awards/distinguished-achievement-awards
Sponsor: American Epilepsy Society
342 North Main Street
West Hartford, CT 06117-2507

AES Research and Training Workshop Awards 320
The American Epilepsy Society, through the Research and Training Committee, provides funding for targeted workshops intended for broad-based, national clinical or scientific audiences and on specific collaborative consensus or review topics in neuroscience that are novel and creative. Often, AES will provide support for travel stipends for junior faculty/trainees and for publication or proceedings. The Research and Training Committee provides evaluation, planning and recommendations on AES sponsored research and training programs. The committee is responsible for identifying all funds currently used to support research or training by AES; reviewing other sources of support for epilepsy training and research (including the NIH funding and private funding); making recommendations to the AES board on an annual basis for allocation of research and training funds; and reviewing the outcomes of research and support training by previously funded investigators or previous trainees. Applications for awards are accepted March 9 and October 26, with funding made within eight weeks of the deadline. Applications received after the deadline will be held until the following round of funding. The application is available at the AES website.
Requirements: Applicants must be AES members.
Restrictions: Funds may not be used for indirect costs.
Geographic Focus: All States
Date(s) Application is Due: Mar 9; Oct 26
Contact: Cheryl-Ann Tubby; (860) 586-7505, ext. 542; ctubby@aesnet.org
Internet: http://www.aesnet.org/go/research/aes-sponsored-grant/research-and-training-workshops
Sponsor: American Epilepsy Society
342 North Main Street
West Hartford, CT 06117-2507

AES Research Infrastructure Awards 321
The AES and the Epilepsy Foundation are partnering so that scientists may obtain support for nationwide or international networks of clinical or basic science researchers focused on understanding the causes, consequences, and treatment of epilepsy. Multicenter research programs from around the world can establish centralized databases, common protocols, shared resources, core laboratories and exchange rapidly developing techniques and technologies. Such cooperative efforts are anticipated to hasten the speed of discovery. These funds are meant to be used to support pilot projects and hold organizational and planning sessions with representatives from each center in the planned network. These planning sessions should also be used to develop the research effort and collect results that would be used to prepare and submit a larger application for support from the Federal Government and establish the multicenter research program over the long term. Principal Investigators should be members of the Society. Awards will be based on a proposed budget, and are anticipated to be up to $50,000.00 per year. A one year no-cost extension may be requested with justification. A second year of funding can be requested via a progress report and competitive renewal application after the first year is completed.
Requirements: Scientists interested in applying should send a letter of intent. Selected candidates will then be asked to submit a complete application. Additional application information is available at the AES website.
Geographic Focus: All States
Date(s) Application is Due: Nov 1
Amount of Grant: Up to 50,000 USD
Contact: Cheryl-Ann Tubby, Assistant Executive Director; (860) 586-7505, ext. 542; fax (860) 586-7550; ctubby@aesnet.org
Internet: http://www.aesnet.org/go/research/aes-sponsored-grant/research-infrastructure-awards
Sponsor: American Epilepsy Society
342 North Main Street
West Hartford, CT 06117-2507

AES Research Initiative Awards 322
The American Epilepsy Society has established a grant opportunity for its membership that will provide seed support to encourage innovative basic or clinical research associated with the epilepsy field. Awards from the Research Initiative Fund will be given to AES members who are established investigators. The objective of these awards is to encourage established investigators to think in a unique way and involve other established investigators who may not now be working in the field of epilepsy. Awards will be made based on the proposed budget, with an anticipated award of $30,000-$50,000 per project. Up to three projects will be funded each year. Letters of intent should submitted in August, with selected proposals submitted in November. Additional information is available at the AES website.
Requirements: Only AES members are eligible to apply as PI (although not all participating collaborators must be AES members). Applicants should send a letter of intent, explaining why they want to do the project, why it is not covered by their current research support, and why the intended collaboration is important (no more than five pages). They must also submit a 250 word abstract; a two page bio-sketch of the principal investigator, and a list of current funding.
Restrictions: Members who are candidates for an Epilepsy Foundation Junior Investigator Award or AES and EF research or training fellowships are not eligible.
Geographic Focus: All States
Date(s) Application is Due: Nov 1
Amount of Grant: 30,000 - 50,000 USD
Contact: Cheryl-Ann Tubby; (860) 586-7505, ext. 542; ctubby@aesnet.org
Internet: http://www.aesnet.org/go/research/aes-sponsored-grant/research-initiative-awards
Sponsor: American Epilepsy Society
342 North Main Street
West Hartford, CT 06117-2507

AES Robert S. Morison Fellowship 323
The American Epilepsy Society, in partnership with the Grass Foundation, offers the Robert S. Morison Fellowship for the training of academic clinician scientists in epilepsy. This two-year post-doctoral Fellowship will be awarded to a promising young investigator possessing an M.D. degree who intends to continue training in basic science in an epilepsy research laboratory. The Fellowship was created to honor the contributions of Dr. Morison, one of the founding Trustees of The Grass Foundation. The Fellowship funds $40,000 per year for two years of salary with an additional $10,000 (maximum) per year to be used for institutional fringe benefits plus $1,000 for the Fellow to travel to the AES Annual Meeting to present his/her results. The award is announced and presented during the AES Annual Meeting.
Requirements: Applicants must apply at the Epilepsy Foundation website for either a Post-doctoral Fellowship or a Research and Training Fellowship for Clinicians. Candidates should note in the application their desire to be considered specifically for the Morison Fellowship. Applicants for this Fellowship can at the same time be considered for a Fellowship in the category under which he/she applied.
Geographic Focus: All States, All Countries
Amount of Grant: Up to 50,000 USD
Contact: Cheryl-Ann Tubby; (860) 586-7505, ext. 542; ctubby@aesnet.org
Internet: http://www.aesnet.org/go/research/aes-sponsored-grant-program/grass-foundation/aes-robert-s-morison-fellowship
Sponsor: American Epilepsy Society
342 North Main Street
West Hartford, CT 06117-2507

AES Service Award 324
The Service Award is presented annually to an AES member. The Award recognizes outstanding service in the field, including non-educational and non-scientific, along with exemplary contributions to the welfare of the AES and its members. The Awardee receives a $1,000 honorarium plus round trip economy airfare, one night lodging for the annual conference, reimbursement of per diem expenses, and free Scientific Program registration. In addition, the award winner is highlighted in the Conference Program Book.
Requirements: Nominees must be AES members. The nomination form is available at the AES website.
Geographic Focus: All States
Date(s) Application is Due: Aug 5
Amount of Grant: 1,000 USD
Contact: Cheryl-Ann Tubby; (860) 586-7505, ext. 542; ctubby@aesnet.org
Internet: http://www.aesnet.org/go/research/research-awards/distinguished-achievement-awards
Sponsor: American Epilepsy Society
342 North Main Street
West Hartford, CT 06117-2507

AES Susan S. Spencer Clinical Research Epilepsy Training Fellowship 325
The Spencer Fellowship is a two-year fellowship to support clinical research training in the field of epilepsy. The Fellowship is supported by the American Academy of Neurology Foundation (AAN), the American Epilepsy Society, and the Epilepsy Foundation. Each Fellowship will consist of $55,000 per year for two years, plus $10,000 per year for tuition to support formal education in clinical research methodology at the applicant's institution or elsewhere. Supplementation of the stipend with other grants or by the fellowship institution is permissible, but Fellows may not accept other Fellowships, similar awards, or have another source of support for more than 50% of their research salary. Only direct costs will be funded by this award. Detailed information about the application and its process is available at the AES website. All materials must be submitted online.
Requirements: For the purpose of this Fellowship, clinical research is defined as "patient-oriented research conducted with human subjects, or translational research specifically designed to develop treatments or enhance diagnosis of neurological disease." These areas of research include epidemiologic or behavioral studies, clinical trials, studies of disease mechanisms, the development of new technologies, and health services and outcomes research." Disease related studies not directly involving humans or human tissue also are encouraged if the primary goal is the development of therapies, diagnostic tests, or other tools to prevent or mitigate neurological diseases. Applicants must be AAN members interested in an academic career in clinical research who have completed a residency or a post-doctoral Fellowship (for a Ph.D.) within the past five years. Those early in their clinical research careers will be given priority.
Geographic Focus: All States
Date(s) Application is Due: Oct 1
Amount of Grant: 65,000 USD
Contact: Terry Heinz, Grants Administrator; (651) 695-2746; theinz@aan.com
Internet: http://www.aesnet.org/go/research/aes-sponsored-grant-program
Sponsor: American Epilepsy Society
342 North Main Street
West Hartford, CT 06117-2507

AES William G. Lennox Award 326
The Lennox Award recognizes members of the AES at a senior level, who have a record of lifetime contributions and accomplishments related to epilepsy. The Award was funded by the Lennox and Lombroso Trust for Research and Training to advance and disseminate knowledge concerning epilepsy in all of its aspects -- biological, clinical, and social -- and to promote better care and treatment for those with epilepsy. The Award is not given every year, but depends on the committee's assessment of the suitability of candidates. The Award is presented at the AES annual conference. The Awardee receives a $10,000 honorarium, one night lodging at the annual conference, round trip economy airfare, and free Scientific Program registration. In addition the award winner is highlighted in the Conference Program Book.
Requirements: Nominees must be AES members. The nomination form is available at the AES website. The nominee's CV and a letter of recommendation is also required.
Geographic Focus: All States
Date(s) Application is Due: Jun 1
Amount of Grant: 10,000 USD
Contact: Cheryl-Ann Tubby; (860) 586-7505, ext. 542; ctubby@aesnet.org
Internet: http://www.aesnet.org/go/research/research-awards/distinguished-achievement-awards
Sponsor: American Epilepsy Society
342 North Main Street
West Hartford, CT 06117-2507

Aetna Foundation Health Grants in Connecticut 327
The Aetna Foundation is the independent charitable and philanthropic arm of Aetna Inc. Founded in 1972, the Foundation helps build strong communities by promoting volunteerism, forming partnerships and funding initiatives that improve the quality of life where our employees and customers live and work. With its Health Grants in Connecticut program, the Foundation supports programs designed to enhance the quality of health care, and address disabilities and chronic diseases. Support is limited to Hartford and Middletown, Connecticut. Grants of up to $35,000 are awarded.
Requirements: 501(c)3 tax-exempt organizations serving the residents of Hartford and Middletown, Connecticut, are eligible.
Restrictions: The Foundation generally does not fund: endowment or capital costs, including construction, renovation, or equipment; direct delivery of reimbursable health

care services; basic biomedical research; grants or scholarships to individuals; work for which results and impact cannot be measured; advertising; golf tournaments; advocacy, political causes or events; sacramental or theological functions of religious organizations; operational expenses; or existing deficits.
Geographic Focus: Connecticut
Amount of Grant: Up to 35,000 USD
Contact: Melenie O. Magnotta, Grants Manager; (860) 273-1012 or (860) 273-0123; fax (860) 273-7764; aetnafoundation@aetna.com
Internet: http://www.aetna.com/about-aetna-insurance/aetna-foundation/aetna-grants/connecticut-grants-health.html
Sponsor: Aetna Foundation
151 Farmington Avenue
Hartford, CT 06156-3180

Aetna Foundation Integrated Health Care Grants 328
The Aetna Foundation is advancing an integrated health care agenda that focuses on care coordination and builds on a foundation of strong primary care. The Aetna Foundation seeks to support projects that promote evidence-based models of care coordination that can lead to high-quality, patient-centered health care services, improve health outcomes and lower costs. The Foundation's goal is to demonstrate the key components, best practices and benefits of care coordination that is centered on strong primary care. Examples of grants supported include projects or studies that: develop standards and metrics of care coordination in ambulatory care settings; evaluate models of care coordination that enhance providers' communication with each other and with their patients and lead to improved patient outcomes and experiences with their care; evaluate methods of care coordination for engaging patients as partners in their care, particularly in the management of chronic conditions; identify best practices to align financial and other incentives for achieving well-coordinated care; and identify and assess models of care coordination that reduce the cost of care while improving patient health outcomes.
Requirements: Grants will be made only to nonprofit organizations with evidence of IRS 501(c)3 designation or de facto tax-exempt status.
Restrictions: The Foundation generally does not fund: endowment or capital costs, including construction, renovation, or equipment; direct delivery of reimbursable health care services; basic biomedical research; grants or scholarships to individuals; work for which results and impact cannot be measured; advertising; golf tournaments; advocacy, political causes or events; sacramental or theological functions of religious organizations; operational expenses; or existing deficits.
Geographic Focus: All States
Amount of Grant: 25,000 - 100,000 USD
Contact: Melenie O. Magnotta, Grants Manager; (860) 273-1012 or (860) 273-0123; fax (860) 273-7764; aetnafoundation@aetna.com
Internet: http://www.aetna-foundation.org/foundation/aetna-foundation-programs/integrated-healthcare/index.html
Sponsor: Aetna Foundation
151 Farmington Avenue
Hartford, CT 06156-3180

Aetna Foundation Minority Scholars Grants 329
Diversifying the next generation of health care professionals and researchers is an important part of the Aetna Foundation's efforts to support high-quality health care for all population groups. Founded in 2010, the AcademyHealth/Aetna Foundation Minority Scholars Program provides professional development, mentoring and networking opportunities for graduate-level students, post-doctoral fellows and other researchers to attract more men and women from underrepresented groups to the field of racial and ethnic disparities research in health outcomes and access to health care.
Restrictions: The Foundation generally does not fund: endowment or capital costs, including construction, renovation, or equipment; direct delivery of reimbursable health care services; basic biomedical research; grants or scholarships to individuals; work for which results and impact cannot be measured; advertising; golf tournaments; advocacy, political causes or events; sacramental or theological functions of religious organizations; operational expenses; or existing deficits.
Geographic Focus: All States
Contact: Melenie O. Magnotta, Grants Manager; (860) 273-1012 or (860) 273-0123; fax (860) 273-7764; aetnafoundation@aetna.com
Internet: http://www.aetna-foundation.org/foundation/aetna-foundation-programs/scholars/index.html
Sponsor: Aetna Foundation
151 Farmington Avenue
Hartford, CT 06156-3180

Aetna Foundation National Medical Fellowship in Healthcare Leadership 330
Diversifying the next generation of health care professionals and researchers is an important part of the Aetna Foundation's efforts to support high-quality health care for all population groups. In 2011, in partnership with National Medical Fellowships (NMF), the Aetna Foundation established the Aetna Foundation/NMF Healthcare Leadership program to provide scholarships to second- and third-year medical students from underrepresented minority groups. Recipients commit to practice medicine in medically underserved communities and are distinguished by their community service and leadership potential.
Restrictions: The Foundation generally does not fund: endowment or capital costs, including construction, renovation, or equipment; direct delivery of reimbursable health care services; basic biomedical research; grants or scholarships to individuals; work for which results and impact cannot be measured; advertising; golf tournaments; advocacy, political causes or events; sacramental or theological functions of religious organizations; operational expenses; or existing deficits.
Geographic Focus: All States
Contact: Melenie O. Magnotta, Grants Manager; (860) 273-1012 or (860) 273-0123; fax (860) 273-7764; aetnafoundation@aetna.com
Internet: http://www.aetna-foundation.org/foundation/aetna-foundation-programs/scholars/index.html
Sponsor: Aetna Foundation
151 Farmington Avenue
Hartford, CT 06156-3180

Aetna Foundation Obesity Grants 331
With its Obesity Grants program, the Foundation wants to understand the contributors to obesity, particularly among minority populations, and what supports and sustains better choices that can stave off overeating and reduce inactivity. Grant-making in this area focuses on initiatives that create a better understanding of the root causes of the obesity epidemic. Examples of grants the Foundation would support include projects and/or studies that identify causes of obesity and potential best practices for addressing obesity, such as: domestic food policies and their impact on individual food choices' the impact of our neighborhoods and the built environment on promoting population health and weight loss; assessments of why communities with high rates of food insecurity also are more likely to experience high rates of obesity; how children use recreation time; and how school lunch and food policies impact our children.
Requirements: 501(c)3 tax-exempt organizations are eligible.
Restrictions: The Foundation generally does not fund: endowment or capital costs, including construction, renovation, or equipment; direct delivery of reimbursable health care services; basic biomedical research; grants or scholarships to individuals; work for which results and impact cannot be measured; advertising; golf tournaments; advocacy, political causes or events; sacramental or theological functions of religious organizations; operational expenses; or existing deficits.
Geographic Focus: All States
Amount of Grant: 50,000 - 300,000 USD
Contact: Melenie O. Magnotta, Grants Manager; (860) 273-1012 or (860) 273-0123; fax (860) 273-7764; aetnafoundation@aetna.com
Internet: http://www.aetna-foundation.org/foundation/aetna-foundation-programs/obesity/index.html
Sponsor: Aetna Foundation
151 Farmington Avenue
Hartford, CT 06156-3180

Aetna Foundation Racial and Ethnic Health Care Equity Grants 332
With its Racial and Ethnic Health Care Equity Grant program, the Aetna Foundation focuses philanthropic giving on understanding connections between where people live and receive health care, and the quality and equity of the care they receive. The Aetna Foundation also is interested in how to improve health and health care among the nation's Medicaid population, particularly in settings with large numbers of minority patients. Examples of grants the Foundation would support include projects and/or studies to: explore how a stronger primary care model and relationships with providers could benefit minority populations and close the persistent health care gap; help providers who treat large minority populations become leaders in delivering high-quality care; determine what can be done to reduce the numbers of low-birth weight babies born to mothers at risk; examine, through observational studies, the correlation between a mother's health, stress level, and social supports; and the likelihood of having a healthy baby who lives through its first year of life; and determine, through interventional studies, whether stress-reduction programs (including yoga and meditation) can improve health outcomes for minority patients with chronic conditions, as well as postnatal outcomes for mothers and pregnant women.
Requirements: 501(c)3 tax-exempt organizations are eligible.
Restrictions: The Foundation generally does not fund: endowment or capital costs, including construction, renovation, or equipment; direct delivery of reimbursable health care services; basic biomedical research; grants or scholarships to individuals; work for which results and impact cannot be measured; advertising; golf tournaments; advocacy, political causes or events; sacramental or theological functions of religious organizations; operational expenses; or existing deficits.
Geographic Focus: All States
Amount of Grant: 50,000 - 300,000 USD
Contact: Melenie O. Magnotta, Grants Manager; (860) 273-1012 or (860) 273-0123; fax (860) 273-7764; aetnafoundation@aetna.com
Internet: http://www.aetna-foundation.org/foundation/aetna-foundation-programs/racial-ethnic-healthcare-equity/index.html
Sponsor: Aetna Foundation
151 Farmington Avenue
Hartford, CT 06156-3180

Aetna Foundation Regional Health Grants 333
The Aetna Foundation's Regional Grants fund community wellness initiatives that serve those who are most at risk for poor health - low-income, underserved or minority populations. A healthy diet and regular exercise can help prevent obesity and many chronic conditions. Grants will target communities where healthy food can be difficult to buy, and where social and environmental factors may limit people's ability to be physically active. Types of projects the Foundation seeks to support include: school-based or after-school nutrition and fitness programs that help children learn healthy habits at an early age; community-based nutrition education programs for children and families; efforts to increase the availability or affordability of fresh fruits and vegetables in communities; and community gardening and urban farming activities for children and families. Funding ranges from $25,000 to $40,000.

Requirements: 501(c)3 tax-exempt organizations in regionally designated communities are eligible, including: Phoenix, Arizona; Los Angeles, San Diego, Fresno, and San Francisco, in California; Connecticut; Miami and Tampa, Florida; Atlanta, Georgia; Chicago, Illinois; Maine; New Jersey; New York, New York; Charlotte, North Carolina; Cleveland and Columbus, Ohio; Philadelphia and Pittsburgh, Pennsylvania; Nashville and Memphis, Tennessee; Dallas, Houston, Austin, and San Antonio, in Texas; Washington, D.C; Baltimore, Maryland; Northern Virginia; and Washington State.
Restrictions: The Foundation generally does not fund: endowment or capital costs, including construction, renovation, or equipment; direct delivery of reimbursable health care services; basic biomedical research; grants or scholarships to individuals; work for which results and impact cannot be measured; advertising; golf tournaments; advocacy, political causes or events; sacramental or theological functions of religious organizations; operational expenses; or existing deficits.
Geographic Focus: Arizona, California, Connecticut, District of Columbia, Florida, Georgia, Illinois, Maine, Maryland, New Jersey, New York, North Carolina, Ohio, Pennsylvania, Tennessee, Texas, Virginia, Washington
Date(s) Application is Due: Sep 15
Amount of Grant: 25,000 - 50,000 USD
Contact: Melenie O. Magnotta, Grants Manager; (860) 273-1012 or (860) 273-0123; fax (860) 273-7764; aetnafoundation@aetna.com
Internet: http://www.aetna-foundation.org/foundation/apply-for-a-grant/regional-grants/index.html
Sponsor: Aetna Foundation
151 Farmington Avenue
Hartford, CT 06156-3180

AFAR CART Fund Grants 334
The goal of the CART Fund is to encourage exploratory and developmental AD research projects within the United States by providing support for the early and conceptual plans of those projects that may not yet be supported by extensive preliminary data but have the potential to substantially advance biomedical research. This proposal should be distinct from those projects designed to increase knowledge in a well established area unless it is intended to extend previous discoveries toward new directions or applications. Applications may encompass a project period of up to two years with a combined budget for direct cost up to $250,000. No indirect costs are allowed. Domestic public and private institutions are eligible, such as universities, colleges, hospitals and laboratories. This is for NEW projects only. Up to two $250,000 awards will be available this year. Dr. John Trojanowski will chair a scientific review group that will triage the Letters-of-Intent and select a maximum of fifteen deemed to have the highest merit. Those selected will be invited to submit a subsequent standard grant application from which the final recommendation will be made by the review group. The final selection will be made by the CART Fund selection committee. Notifications of the finalists (maximum of 15) will be mailed by January 7th. Investigators whose proposals are accepted for further consideration will be sent a full application form which must be completed and returned by February 16th.
Requirements: The CART Fund is inviting interested applicants from within the United States to submit a Letter-of-Intent that includes sufficient detail to communicate the importance of your study as well as information on its feasibility. Submit email questions and Letters-of-Intent to Dr. James B. Puryear, Chairman, CART Grants Program: jimpuryear@comcast.net.
Restrictions: Applications will be deemed ineligible from for-profit organizations as well as those already supported by regular or program grants.
Geographic Focus: All States
Date(s) Application is Due: Feb 16
Amount of Grant: 250,000 USD
Contact: Grants Manager; (212) 703-9977 or (888) 582-2327; fax (212) 997-0330; grants@afar.org or info@afar.org
Internet: http://www.afar.org/CART.html
Sponsor: American Federation for Aging Research
55 West 39th Street, 16th Floor
New York, NY 10018

AFAR Medical Student Training in Aging Research Program 335
The Medical Student Training in Aging Research Program (MSTAR) provides medical students, early in their training, with an enriching experience in aging related research and geriatrics, under the mentorship of top experts in the field. This program introduces students to research and academic experiences that they might not otherwise have during medical school. Students participate in an eight to twelve week structured research, clinical, and didactic program in geriatrics, appropriate to their level of training and interests. Students may train at a National Training Center supported by the National Institute on Aging or, for a limited number of medical schools, at their own institution. Refer to the AFAR website for a complete listing of the participating institutions.
Requirements: Any allopathic or osteopathic medical student in good standing, who will have successfully completed one year of medical school at a U.S. institution by June, may apply. Applicants must be citizens or non-citizen nationals of the United States, or must have been lawfully admitted for permanent residence (i.e., in possession of a currently valid Alien Registration Receipt Card I-551, or some other legal verification of such status.)
Restrictions: Individuals on temporary or student visas and individuals holding PhD, MD, DVM, or equivalent doctoral degrees in the health sciences are not eligible.
Geographic Focus: All States
Date(s) Application is Due: Feb 6
Contact: Grants Manager; (212) 703-9977 or (888) 582-2327; fax (212) 997-0330; grants@afar.org or info@afar.org
Internet: http://www.afar.org/medstu.html
Sponsor: American Federation for Aging Research
55 West 39th Street, 16th Floor
New York, NY 10018

AFAR Paul Beeson Career Development Awards in Aging Research for the Island of Ireland 336
The Paul Beeson Career Development Awards In Aging Research Program offers faculty development awards to outstanding junior physician faculty committed to academic careers in aging-related research, teaching, and practice. The goals of the program are: to encourage and assist the development of future leaders in the field of aging by supporting faculty members early in their careers to gain additional research training as needed and to establish independent programs in aging research; to deepen the commitment of research institutions to academic research in aging and to translating research outcomes to geriatric medicine by involving mentor and recipient in establishing and advancing the recipient's career in aging research; to expand medical research on aging broadly defined as including the biology of aging, maintenance of health and independence in old age, diseases and disabilities of old age and issues in their clinical management, and systems of care for the elderly. To maximize the educational/training opportunities of this fellowship program, Beeson Ireland Scholars are encouraged to spend three to six months abroad with a research team in the U.S. in the second or the third year of the award. The goals of this rotation are varied and can be for purposes of training in techniques, data collection, data analysis, etc. to enhance the Scholars' transition to independence. This is unique to the Beeson Ireland Program. To date, the program has provided awards to 149 very promising junior faculty at institutions in the United States as well as 3 junior faculty on the Island of Ireland. The program was established in 1994 by The Atlantic Philanthropies, The John A. Hartford Foundation, The Commonwealth Fund and The Starr Foundation, and administered by the American Federation for Aging Research (AFAR). In 2004, The Atlantic Philanthropies, The John A. Hartford Foundation, The Starr Foundation and AFAR entered a partnership with The National Institute on Aging (NIA.) Until 2006, the program soley funded Scholars at U.S.-based institutions but beginning in 2007, with support from The Atlantic Philanthropies, Scholars are also funded in Ireland (The Republic of Ireland and Northern Ireland.) The scholar will receive a grant for three years. The salary will be at specialty registrar/clinical lecturer level to protect a minimum of 75% of the scholar's time for research, with the remainder available annually for research support, including an institutional overhead on the research support component.
Requirements: Nominations for the Paul Beeson Career Development Awards In Aging Research Program are to be made by the deans of medical schools (or equivalent) in Ireland (the Republic of Ireland and Northern Ireland.) Institutions may submit as many applications as they wish. To be eligible for nomination, a candidate must: be a U.S. physician or an Irish citizen physician of the Republic of Ireland, a UK citizen of Northern Ireland, an EU citizen living and working in the Island of Ireland or a non-EU citizen resident in the island of Ireland with a valid work permit; commit at least 75% of his/her full-time professional effort to the goals of this award; be a medical graduate who has recently completed or is about to complete a doctoral degree (PhD/MD) and be undertaking or have recently completed higher clinical training; have at least one research publication in a high-impact journal. If the candidate is not the first author, he/she must provide a short description of his/her contribution to the research and the writing of the paper; this should be substantial. For each scholar, a senior faculty member at the scholar's institution must be selected to serve as a mentor to help guide the scholar's research and career planning and provide access to organizations, programs, and colleagues helpful to the scholar's efforts. More than one mentor may be selected. Letters of endorsement including specific information on institutional support for the scholar should be provided by the dean, the relevant department chairperson (or equivalent - may be one person), and the mentor. In addition, three letters of reference should be provided by other faculty members and/or senior professionals with whom the scholar has worked and who are well acquainted with his/her capabilities, potential, accomplishments, and commitment. All candidates must submit applications endorsed by the Dean of School of Medicine (or equivalent).
Geographic Focus: All States
Date(s) Application is Due: Jan 22
Contact: Manager; (212) 703-9977 or (888) 582-2327; beeson@afar.org
Internet: http://afar.org/BeesonIreland.html
Sponsor: American Federation for Aging Research
55 West 39th Street, 16th Floor
New York, NY 10018

AFAR Research Grants 337
The major goal of this program is to assist in the development of the careers of junior investigators committed to pursuing careers in the field of aging research. Of particular interest are research projects concerned with understanding the basic mechanisms of aging; projects investigating age-related diseases, especially if approached from the point of view of how basic aging processes may lead to these outcomes; and projects concerning mechanisms underlying common geriatric functional disorders, as long as these include connections to fundamental problems in the biology of aging. Examples of promising areas of research include: aging and immune function; genetic control of longevity; neurobiology and neuropathology of aging; invertebrate or vertebrate animal models; cardiovascular aging; aging and cellular stress resistance; metabolic and endocrine changes; age-related changes in cell proliferation; caloric restriction and aging; DNA repair and control of gene expression; biology of the menopause; and aging and apoptosis. Application and instruction sheet are available online.
Requirements: The applicant must be an independent investigator with assigned independent space and must be within the first four years of a junior faculty appointment (instructor, assistant professor or equivalent) by July 1st.

Restrictions: The program does not provide support for postdoctoral fellows in the laboratory of a senior investigator; investigators who have already received major independent funding for research on aging, such as an R01 grant or a grant of equal to or greater than $100,000 from another private funding source; senior faculty, i.e., at the rank of associate professor level or higher; and projects that deal strictly with clinical problems, such as the diagnosis and treatment of disease, health outcomes, or the social context of aging. Former AFAR grant recipients are not eligible to reapply. Applicants for the Glenn/AFAR Breakthroughs in Gerontology (BIG) program cannot also submit an application for this research grant.
Geographic Focus: All States
Date(s) Application is Due: Dec 16
Amount of Grant: 75,000 USD
Contact: Grants Manager; (212) 703-9977 or (888) 582-2327; fax (212) 997-0330; grants@afar.org or info@afar.org
Internet: http://afar.org/afar99.html
Sponsor: American Federation for Aging Research
55 West 39th Street, 16th Floor
New York, NY 10018

AFB Rudolph Dillman Memorial Scholarship 338
The American Foundation for the Blind offers several scholarships with the same application for a variety of fields of study. The Rudolph Dillman Memorial Scholarship offers $2,500 to four undergraduate or graduate students who are studying in the field of rehabilitation and/or education of persons who are blind or visually impaired. The application is available at the AFB website. The annual application period starts on March 1 and runs through the deadline date of May 31.
Requirements: In addition to the online application, candidates must also submit relevant transcripts, proof of college enrollment, two letters of recommendation, and proof of legal blindness. All application materials must be send in the same package.
Geographic Focus: All States
Date(s) Application is Due: May 31
Amount of Grant: 2,500 USD
Contact: Tara Annis, Information Specialist; (304) 523-8651 or (800) 232-5463; fax (646) 478-9260; tannis@afb.net or afbinfo@afb.net
Internet: https://www.afb.org/section.aspx?Documentid=2962
Sponsor: American Foundation for the Blind
11 Penn Plaza, Suite 300
New York, NY 10121

Affymetrix Corporate Contributions Grants 339
Affymetrix makes charitable donations to nonprofit organizations through its corporate philanthropy program. The company focuses its giving in three main areas: education--programs that support science and math education, focusing on K-12 students and their teachers; ethics--organizations that help foster an ongoing public dialog about genetic-related ethics; and cancer research and advocacy--nonprofit organizations working in the areas of disease research and advocacy, with a specific emphasis on cancer. Requests are reviewed four times per year. The average grant size is approximately $2500.
Requirements: 501(c)3 nonprofit organizations are eligible.
Restrictions: The corporation does not provide funding for advertising journals or booklets; fundraising events such as telethons, walkathons, and races; specific performances or concerts; sporting events; endowment campaigns; film, video, television, or radio projects; grants to individuals; political causes or candidates; or organizations that practice discrimination or limit membership on the basis of race, creed, gender, age, sexual orientation, or national origin.
Geographic Focus: All States
Contact: Contributions Manager, outreach@affymetrix.com
Internet: http://www.affymetrix.com/corporate/outreach/corporate.affx
Sponsor: Affymetrix
3420 Central Expressway
Santa Clara, CA 95051

AFG Industries Grants 340
The corporation awards general operating grants to nonprofits in its headquarters area in the categories of arts and humanities, civic and public affairs, education at all levels, health care, and social services. There are no application deadlines. Submit a brief letter of inquiry.
Requirements: Tennessee nonprofits are eligible.
Geographic Focus: Tennessee
Amount of Grant: 100,000 - 250,000 USD
Contact: Human Resources; (800) 251-0441 or (423) 229-7200; fax (423) 229-7459
Internet: http://www.afgglass.com
Sponsor: AFG Industries
P.O. Box 929
Kingsport, TN 37662

A Friends' Foundation Trust Grants 341
The Foundation, founded as the Hubbard Foundation in 1959 by philanthropist Frank M. Hubbard, primarily serves central Florida with its support of health research and health care, education at all levels, children and youth activities, the arts, and religious agencies. Grants typically range from $5,000 to $35,000, though some higher amounts are given. There are no specific application formats or deadlines with which to adhere, and applicants should send a letter of request to the Foundation address listed.
Requirements: Applicants must be 501(c)3 organizations serving residents of central Florida.
Geographic Focus: Florida
Amount of Grant: 5,000 - 35,000 USD
Samples: Fisher House Foundation, Rockville, Maryland, $5,000 - general operations; Bok Tower Gardens, Lake Wales, Florida, $22,500 - general operations; Adult Literacy League, Orlando, Florida, $2,000 - general operations.
Contact: L. Evans Hubbard, (407) 876-3122; ehubbard@cfl.rr.com
Sponsor: A Friends' Foundation Trust
9000 Hubbard Place
Orlando, FL 32819

A Fund for Women Grants 342
A Fund For Women, a component fund of The Madison Community Foundation was established in 1993 to improve the lives of girls and women in the local community. The fund provides grants to women and girls in the community that enhance education, employment and self-esteem. All grants are driven by the overall goal of helping women and children learn self-reliance and reach self-sufficiency. Under that umbrella, the Fund focuses on the following four key areas of need: keeping elderly or disabled women in their homes, or in community settings; providing services to victims of domestic abuse; helping women achieve economic self sufficiency and increase their earning potential; and reducing homelessness for women and girls. AFFW is particularly interested in innovative programs and services that will reach women and girls from diverse backgrounds in urban, suburban or rural Dane County. Proposals should indicate how the project or program will help connect women or girls to a support network or community that will help them to overcome emotional, social, intellectual, spiritual, occupational and/or physical barriers to self-sufficiency. Applications are due annually by July 1, at 4:30 p.m.
Requirements: Applicants must be non-profit organizations (exempt from Federal income taxes under section 501(c)3 of the Internal Revenue Code), schools, governmental bodies, or under the supervision of such a group. Projects must focus on women and girls in Dane County.
Geographic Focus: Wisconsin
Date(s) Application is Due: Jul 1
Amount of Grant: 2,000 - 25,000 USD
Contact: Jan Gietzel, Executive Director; (608) 441-0630; fax (608) 232-1772; kwoit@madisoncommunityfoundation.org or affw@madisoncommunityfoundation.org
Internet: http://www.affw.org/grants/apply.php
Sponsor: A Fund for Women
2 Science Court, P.O. Box 5010
Madison, WI 53705-0010

AGHE Graduate Scholarships and Fellowships 343
The association awards graduate scholarships and fellowships for research on aging. Gerontology and aging studies faculty are invited to nominate qualified students. This program is funded by the AARP Andrus Foundation. Faculty nomination and student application forms are available online. Annual deadline dates may vary; contact program staff for exact dates.
Requirements: Gerontology and aging studies faculty are invited to nominate qualified students for these awards.
Geographic Focus: All States
Amount of Grant: 2,000 - 15,000 USD
Contact: M. Angela Baker; (202) 289-9806, ext. 125; abaker@aghe.org
Internet: http://www.aghe.org/677858
Sponsor: Association for Gerontology in Higher Education
1220 L Street, NW, Suite 901
Washington, D.C. 20005-4015

Agnes B. Hunt Trust Grants 344
The trust awards grants to eligible Georgia nonprofit organizations in the general areas of health, welfare, and education for the poor and needy in its local community. Apply online; complete guidelines are provided.
Requirements: 501(c)3 nonprofits in Griffin and Spalding County, GA, are eligible.
Geographic Focus: Georgia
Date(s) Application is Due: Nov 1
Amount of Grant: 500 - 5,000 USD
Contact: Joseph Walker
Internet: http://www.agnesbhunttrust.org
Sponsor: Agnes B. Hunt Trust
P.O. Box 1610
Griffin, GA 30224-1610

Agnes Gund Foundation Grants 345
The foundation awards general operating grants in its areas of interest, including health organizations, higher education, and performing arts (dance, music). There are no application deadlines or forms. Submit a letter of inquiry.
Requirements: Although there are no funding restrictions there is a focus on New York City.
Restrictions: Applicants must be a tax-exempt organization.
Geographic Focus: New York
Amount of Grant: 5,000 - 100,000 USD
Samples: Museum of Modern Art (New York, NY)--for general operating support, $1 million; Cleveland Museum of Art (OH)--for general operating support, $200,000; Virginia Museum of Fine Arts (Richmond, VA)--for general operating support, $100,000; Creative Capital Foundation (New York, NY)--for general operating support, $75,000.
Contact: Program Director; (330) 385-3400
Sponsor: Agnes Gund Foundation
517 Broadway, 3rd Floor
East Liverpool, OH 43920

Grant Programs | 51

Agnes M. Lindsay Trust Grants 346
The Trust makes grants to nonprofit organizations in the states of Maine, Massachusetts, New Hampshire and, Vermont to improve the quality of life. Areas of interest include: health & welfare organizations; recreation, camp scholarships (camperships); education; educational scholarships. The Trust awards grants for capital needs; capital campaigns, building renovations, equipment; computers, furniture, etc. Grant proposals are reviewed on a monthly basis, grants average $1,000 - $15,000. Campership applications must be submitted by March 1st and range from $1,000 - $4,000.
Requirements: 501(c)3 Maine, Massachusetts, New Hampshire, and Vermont tax-exempt organizations are eligible to apply for funding. To begin the application process, submit a letter of inquiry prior to submitting a full proposal for funding. In addition to basic information about your organization, include a brief outline of your statement of need and a project budget. You may complete this as a word document and attach it to an email addressed to: adminatlindsaytrustdotorg. Grant proposals should contain: proposal summary document (available at Trust website); narrative of the organization and description of need; financial statements for last two years or Form 990; budget for capital expenditures; estimates or quotes obtained; IRS Determination letter. Submit your proposal and all attachments electronically in either a PDF format and/or Microsoft Word and EXCEL to: proposalsatlindsaytrustdotorg. Applicants requesting a camp scholarships (camperships) provide the following: proposal summary document (available at Trust website); most recent audited financial statements (one calender year) or Form 990; and IRS Determination Letter. Submit your proposal electronically in a PDF file to: proposalsatlindsaytrustdotorg. The trust will accept Microsoft Word documents and Excel documents as attachments. If you are unable to send your proposal or any portion thereof electronically, you may submit it via U.S. mail. A status report for the grant awarded is required after the grant has been expended.
Restrictions: The Trust does not fund: endowments, public entities, awarded to individuals, municipalities, libraries, museums, sectarian organizations, or capital grants to private schools. The Trust very rarely provides funding for operating /program support. If your organization received a grant, wait one year before reapplying for additional funding.
Geographic Focus: Maine, Massachusetts, New Hampshire, Vermont
Date(s) Application is Due: Mar 1
Amount of Grant: 1,000 - 15,000 USD
Samples: Child and Family Services of New Hampshire, Manchester, NH, $50,000--challenge grant to support the Capital Expansion Project to purchase land and furnishings for the new home on Union Street to serve homeless youth; Massachusetts Board of Higher Education, Boston, MA, $30,000--annual scholarship grant; Big Brothers Big Sisters of Bath Brunswick, Brunswick, ME, $5,000--for a new server and computer equipment/software providing the capacity to interface with the national office.
Contact: Susan Bouchard, Administrative Director; (603) 669-1366 or (866) 669-1366; fax (603) 665-8114; admin@lindsaytrust.org
Internet: http://www.lindsaytrust.org/index.html
Sponsor: Agnes M. Lindsay Trust
660 Chestnut Street
Manchester, NH 03104

A Good Neighbor Foundation Grants 347
Established in Cincinnati, Ohio, in 2002, A Good Neighbor Foundation offers support primarily to individuals and families that are in need of immediate aid and services. Its major fields of interest include: cancer; children's services; human services; and emergency aid. Funding comes in the form of general operating support for residents of Cincinnati, Ohio, and northern Kentucky. There are no specific deadlines or application forms, and applicants should contact the office in writing or via telephone.
Geographic Focus: Kentucky, Ohio
Amount of Grant: Up to 400,000 USD
Samples: Redwood School and Rehabilitation Center, Fort Mitchell, Kentucky, $350,000--for capital campaign; Saint Joseph Infant and Maternity Home, Cincinnati, Ohio, $330,000--for capital campaign.
Contact: Michele Kelley, Secretary-Treasurer; (513) 651-9333
Sponsor: A Good Neighbor Foundation
414 Walnut Street, Suite 1014
Cincinnati, OH 45202-3913

Agway Foundation Grants 348
The primary goal of the foundation is to support organizations dedicated to serving the interests of farmers and rural communities in the Northeast, including health care, children and youth services, rural youth organizations, and agriculture. The foundation also supports organizations that contribute to the quality of life in the Agway headquarters area and it actively promotes employee volunteer involvement in community service.
Requirements: 501(c)3 nonprofit organizations are eligible for grant support.
Restrictions: The foundation funds are not used for individuals; political, religious, or labor organizations; scholarships; matching gift programs or capital campaigns of educational institutions; memberships in professional societies or trade associations; or the general operating funds of health care facilities.
Geographic Focus: Connecticut, Maine, Massachusetts, New Hampshire, Rhode Island, Vermont
Amount of Grant: 1,000 - 10,000 USD
Contact: Stephen Hoefer; (315) 449-6474; info@agway.com or chairman@agway.com
Internet: http://www.colebrook-nh.com/Public_Documents/ColebrookNH_BBoard/I008D1C75
Sponsor: Agway Foundation
P.O. Box 4933
Syracuse, NY 13221

AHAF Alzheimer's Disease Research Grants 349
AHAF funds outstanding scientists and physicians in neurobiology, physiology, pathology, molecular and developmental biology, chemistry, pharmacology, epidemiology, and surgery who are conducting research to better understand and/or treat Alzheimer's disease. Junior or senior investigators will be considered. Grants are awarded for up to two years and are renewable. Grant applications are reviewed by a scientific review committee on a competitive peer-review system.
Requirements: Grants are awarded to universities, medical centers, and independent research institutions. AHAF funds grants for research at nonprofit organizations only. The principal investigator must hold the academic rank of assistant professor (or equivalent) or higher.
Restrictions: Funding is not provided for overhead costs, construction, or building expenses.
Geographic Focus: All States
Date(s) Application is Due: Oct 19
Amount of Grant: Up to 300,000 USD
Contact: Kara Hurst, Grants Coordinator; (800) 437-2423 or (301) 948-3244; fax (301) 258-9454; khurst@ahaf.org
Internet: http://www.ahaf.org/alzdis/research/grants.htm
Sponsor: American Health Assistance Foundation
22512 Gateway Center Drive
Clarksburg, MD 20871

AHAF Macular Degeneration Research Grants 350
The foundation awards research grants to advance study of macular degeneration. Applications are evaluated based on the scientific merit of the proposal, the feasibility of the proposed research, the potential of the research to lead to better understanding and treatment of eye diseases, and the demonstrated ability of the investigator to complete the research. Application materials are available on the Web site.
Geographic Focus: All States
Date(s) Application is Due: Jul 10; Oct 29
Amount of Grant: Up to 50,000 USD
Contact: Kara Hurst; (800) 437-2423 or (301) 948-3244; khurst@ahaf.org
Internet: http://www.ahaf.org/macular/research/grants.htm
Sponsor: American Health Assistance Foundation
22512 Gateway Center Drive
Clarksburg, MD 20871

AHAF National Glaucoma Research Grants 351
AHAF funds outstanding scientists and physicians with expertise in cell and molecular biology, physiology, biochemistry, endocrinology, and pharmacology. Grants are awarded on the basis of scientific merit of the proposal, the relevance of the research, and the potential impact of the proposed study on better understanding and/or treatment of glaucoma. Applications are reviewed by a scientific review committee on a competitive peer-review system. AHAF is interested in receiving focused research grant applications from investigators at all stages of their careers. AHAF is particularly interested in new investigators with little or no previous grant support and established investigators with new ideas or directions for their research.
Requirements: AHAF grants are awarded to universities, medical centers, and independent research institutions. AHAF provides grants for research at nonprofit organizations only.
Restrictions: Grants are not made to individuals. Funding is not provided for overhead costs, construction, or building expenses.
Geographic Focus: All States
Date(s) Application is Due: Oct 27
Amount of Grant: Up to 100,000 USD
Contact: Kara Hurst; (800) 437-2423 or (301) 948-3244; khurst@ahaf.org
Internet: http://www.ahaf.org/glaucoma/research/glresrch.htm
Sponsor: American Health Assistance Foundation
22512 Gateway Center Drive
Clarksburg, MD 20871

AHIMA Dissertation Assistance Grants 352
The Dissertation Assistance Award program supports research undertaken as part of an academic program to qualify for a doctorate in areas relevant to health information management (HIM). Each recipient shall be limited to one funded grant per year. Ordinarily, the scope of the proposal shall be such that it can be completed within 18 months from the date of funding. There are no minimum or maximum grant request expectations, but the range of grants historically has been between $5,000 and $10,000. Submissions that address one or more of the AHIMA research priorities will receive priority consideration for funding.
Requirements: To qualify for an award under this program the student (principal investigator) must be enrolled in an accredited doctoral degree program in an area related to HIM (computer science, business management, education, public health, and so forth) and must be an active, associate, or student member of AHIMA. All requirements for the doctoral degree, other than the dissertation, must be completed by the award date.
Geographic Focus: All States
Date(s) Application is Due: Oct 31
Amount of Grant: 5,000 - 10,000 USD
Contact: Carol Nielsen, Director; (312) 233-1175; carol.nielsen@ahimafoundation.org
Internet: http://www.ahimafoundation.org/Scholarships/dissertation.aspx
Sponsor: Foundation of Research and Education of the American Health Information Management Association
233 N Michigan Avenue, 21st Floor
Chicago, IL 60601

AHIMA Faculty Development Stipends — 353

FORE Faculty Development Stipends are intended to encourage excellence in effective teaching and leadership in areas related to e-HIM. Program goals are to: expedite curriculum innovation and development in e-HIM content areas; research and develop new methods of teaching, evaluation, and assessment in the HIM learning process; support faculty in maintaining the highest standards of knowledge and practice in e-HIM; integrate new technologies effectively into the HIM learning process; and strengthen the capacity of HIM academic programs to address emerging issues in design and implementation of the electronic health record at all levels of health care delivery, in support of an electronic health information infrastructure.
Requirements: Application should be for activity planned during the four (4) month period following the expected notification date.
Geographic Focus: All States
Date(s) Application is Due: Mar 14; Jul 11; Oct 3
Contact: Susan H. Fenton, Director of Research; (312) 233-1532 or (312) 233-1100; fax (210) 479-1043; susan.fenton@ahima.org or fore@ahima.org
Internet: http://www.ahimafoundation.org/Scholarships/Faculty.aspx
Sponsor: Foundation of Research and Education of the American Health Information Management Association
233 N Michigan Avenue, 21st Floor
Chicago, IL 60601

AHIMA Grant-In-Aid Research Grants — 354

The FORE Grant-in-Aid program is directed toward supporting the development of HIM professionals as leaders in defining and validating the unique body of knowledge encompassed by HIM. The results of these studies not only provide information for the HIM professional to apply in meeting current and future challenges, but also support policy initiatives and the redefinition of the roles of HIM practitioners. Ordinarily, the scope of the proposal shall be such that it can be completed within 18 months from the date of funding. There are no minimum or maximum grant request expectations, but the range of grants historically has been between $15,000 and $40,000. Submissions that address one or more of the AHIMA research priorities will receive priority consideration for funding.
Requirements: The primary or secondary investigator must be an active, associate, or student member of AHIMA. Each recipient shall be limited to one funded grant per year.
Geographic Focus: All States
Date(s) Application is Due: Sep 19
Amount of Grant: 15,000 - 40,000 USD
Contact: Susan H. Fenton, Director of Research; (312) 233-1532 or (312) 233-1100; fax (210) 479-1043; susan.fenton@ahima.org or fore@ahima.org
Internet: http://www.ahima.org/fore/research/grantinaid.asp
Sponsor: Foundation of Research and Education of the American Health Information Management Association
233 N Michigan Avenue, 21st Floor
Chicago, IL 60601

Ahmanson Foundation Grants — 355

The Foundation reviews grant requests from 501(c)3 organizations that are based in and serving Los Angeles County in the areas of education (elementary and secondary education, higher education, nursing school education, and adult literacy and basic skills), the arts and humanities (including libraries and cultural programs), disadvantaged, domestic violence, health and medicine, human services including youth, and religion (Christian, Episcopal, interdenominational, Jewish, Lutheran, Methodist, Presbyterian, Roman Catholic, and Salvation Army). Types of grant support include capital campaigns, challenge, building/renovations, equipment, endowment funds, matching funds, program development, scholarship funds, and seed money. There are no application forms; the board meets four times annually to consider requests. Full proposals will be invited. Letters of inquiry are to be sent through the postal service following the guidelines listed.
Requirements: 501(c)3 nonprofit organizations that serve or are based in Los Angeles County, California, may apply.
Restrictions: Grants are not awarded to individuals or for continuing support, annual campaigns, professorships or internships, fellowships, or film production.
Geographic Focus: California
Amount of Grant: 10,000 - 50,000 USD
Samples: Optimist Youth Homes and Family Services (Los Angeles, CA)--to help construct its new facility that will expand and enhance academic and therapeutic programs for at-risk youngsters, $100,000; California Lutheran U (Thousand Oaks, CA)--to equip the Exercise Physiology/ Biomechanics Laboratory in the new Gilbert Sports and Fitness Center, $500,000.
Contact: Leonard Walcott Jr.; (310) 278-0770; info@theahmansonfoundation.org
Internet: http://www.theahmansonfoundation.org/fund.html
Sponsor: Ahmanson Foundation
9215 Wilshire Boulevard
Beverly Hills, CA 90210

Ahn Family Foundation Grants — 356

The Ahn Family Foundation, established by Sangwoo and Laura Ahn in 1997, has as its primary purposes to support aging, religion, human and social services, art, culture, education, and the environment. Giving is primarily limited to Connecticut, Massachusetts, and New York, although the Foundation has also supported projects in Maryland and Washington. There are no particular guidelines or deadlines with which to adhere, and applicants should begin by contacting the Foundation in writing. This letter of inquiry should outline the purpose for the request, the mission of the applicant organization, and the amount requested.
Geographic Focus: Connecticut, Massachusetts, New York
Amount of Grant: 1,000 - 20,000 USD
Samples: Holy Apostles Soup Kitchen, New York, New York, $20,000; Saint Lukes LifeWorks, Stamford, Connecticut, $20,000; All Saints Church, Chevy Chase, Maryland, $5,000.
Contact: Alison D. Ahn, Treasurer; fax (203) 869-4875
Sponsor: Ahn Family Foundation
901 Hillsboro Mile
Hillsboro Beach, FL 33062-2801

AHNS/AAO-HNSF Surgeon Scientist Combined Award — 357

The Grant is open to surgeons beginning a clinician-scientist career track to support research in the pathogenesis, pathophysiology, diagnosis, prevention, or treatment of head and neck neoplastic disease. This grant is only available during odd numbered years. Awards are two year, nonrenewable, with a possible $70,000 maximum total costs ($35,000 per year). One award available per year. Letters of intent must be received by December 15, and completed applications must be submitted by January 15.
Requirements: Applicants must be members or candidate members of the American Academy of Otolaryngology-Head and Neck Surgery and/or the American Head and Neck Society.
Geographic Focus: All States
Date(s) Application is Due: Jan 15; Dec 15
Amount of Grant: Up to 70,000 USD
Contact: Stephanie L. Jones; (703) 519-1586 or (703) 836-4444; sljones@entnet.org
Internet: http://www.headandneckcancer.org/research/grants.php
Sponsor: American Head and Neck Society
11300 W Olympic Boulevard, Suite 600
Los Angeles, CA 90064

AHNS/AAO-HNSF Young Investigator Combined Award — 358

The purpose of this award is to support a collaborative AHNS/AAO-HNSF research project by fostering the development of contemporary basic or clinical research skills focused on neoplastic disease of the head and neck among new full-time academic surgeons. The award is intended as a preliminary step in clinical investigator career development and is expected to facilitate the recipient's preparation of a more comprehensive individualized research plan suitable for submission to the National Institutes of Health or comparable funding agency. This is a two year, non-renewable award, offering $40,000 maximum ($20,000 per year). One award is available annually. Letters of intent must be received by December 15, and completed applications must be submitted by January 15.
Requirements: Applicants must be physicians with demonstrated potential for excellence in research and teaching and serious commitment to an academic research career in head and neck surgery. Applicants must be members or candidate members of the American Academy of Otolaryngology-Head and Neck Surgery and/or the American Head and Neck Society. Priority will be given to fellows or junior faculty who have completed residencies or fellowships within four years of the application receipt date. All candidates must be sponsored by the Chair of his/her Division or Department and by an official representative of the institution which would administer the Award and in whose name the application is formally submitted.
Geographic Focus: All States
Date(s) Application is Due: Jan 15; Dec 15
Amount of Grant: Up to 40,000 USD
Contact: Stephanie L. Jones; (703) 519-1586 or (703) 836-4444; sljones@entnet.org
Internet: http://www.headandneckcancer.org/research/grants.php
Sponsor: American Head and Neck Society
11300 W Olympic Boulevard, Suite 600
Los Angeles, CA 90064

AHNS Alando J. Ballantyne Resident Research Pilot Grant — 359

The purpose of this award is to support basic, translational, or clinical research projects in head and neck oncology. Clinical or translational research studies are strongly encouraged and should be specifically related to the prevention, diagnosis, treatment, outcomes, or pathophysiology of head and neck neoplastic disease. Research supported by this award should be specifically directed toward the pathogenesis, pathophysiology, diagnosis, prevention, or treatment of head and neck neoplastic disease, and may be either basic or clinical/translational in approach. While not specifically required, proposals which aim to introduce new knowledge and methodology from other disciplines to research in head and neck disease, or which demonstrate collaborative effort with members of other related disciplines are encouraged. Grants range up to $10,000 maximum, are for one year, and are not renewable. Letters of intent must be received by December 15, and completed applications must be submitted by January 15.
Requirements: This grant is open to resident in U.S. or Canadian training programs. Previous AHNS or AAO-HNS Foundation research grant recipients are eligible to compete for this grant. However, candidates who have successfully obtained funding from a private or federal funding agency for the same research are ineligible. Candidates who have applied for support of the same research from other funding sources, and who are notified of an award from both another agency and from AHNS must choose only one of the awards.
Geographic Focus: All States
Date(s) Application is Due: Jan 15; Dec 15
Amount of Grant: Up to 10,000 USD
Contact: Stephanie L. Jones; (703) 519-1586 or (703) 836-4444; sljones@entnet.org
Internet: http://www.headandneckcancer.org/research/grants.php
Sponsor: American Head and Neck Society
11300 W Olympic Boulevard, Suite 600
Los Angeles, CA 90064

AHNS Pilot Grant 360
The purpose of this award is to support basic, translational, or clinical research projects in head and neck oncology. Clinical or translational research studies are strongly encouraged and should be specifically related to the prevention, diagnosis, treatment, outcomes, or pathophysiology of head and neck neoplastic disease. Research supported by this award should be specifically directed toward the pathogenesis, pathophysiology, diagnosis, prevention, or treatment of head and neck neoplastic disease, and may be either basic or clinical/translational in approach. While not specifically required, proposals which aim to introduce new knowledge and methodology from other disciplines to research in head and neck disease, or which demonstrate collaborative effort with members of other related disciplines are encouraged. Grants range up to $10,000 maximum, are for one year, and are not renewable. Letters of intent must be received by December 15, and completed applications must be submitted by January 15.
Requirements: Candidates for this award should reside in the U.S. or Canada, be medical students, residents, Ph.D.s or faculty members at the rank of associate professor or below.
Geographic Focus: All States
Date(s) Application is Due: Jan 15; Dec 15
Amount of Grant: Up to 10,000 USD
Contact: Stephanie L. Jones; (703) 519-1586 or (703) 836-4444; sljones@entnet.org
Internet: http://www.headandneckcancer.org/research/grants.php
Sponsor: American Head and Neck Society
11300 W Olympic Boulevard, Suite 600
Los Angeles, CA 90064

AHRF Eugene L. Derlacki, M.D. Research Grants 361
The Eugene L. Derlacki, M.D. Grant is awarded for excellence in the field of hearing research. This grant provides $25,000 per year for two years, for a total award of $50,000. It is dedicated to hearing research and is intended for more significant research projects requiring funding that extends beyond the normal AHRF grant of $25,000. Researchers are invited to apply for this grant using the Research Grant Application Guidelines. According to the rules of the grant, the recipient will receive the first year funding of $25,000. After the first year, a progress report must be submitted to the Research Committee. If the report is approved, the second year funding of $25,000 will then be awarded.
Geographic Focus: All States
Date(s) Application is Due: Aug 3
Amount of Grant: 25,000 USD
Samples: Steven H. Green, Ph.D., University of Iowa--for Role of JNK Signaling in the Death of Spiral Ganglion Neurons After Hair Cell Loss (2005-06); Keiko Hirose, M.D., The Cleveland Clinic Foundation, Ohio--for Cellular Repair of the Murine Cochlea After Acoustic Injury.
Contact: Kristen Madhuizen; (312) 726-9670; fax (312) 726-9695; kristen.madhuizen@american-hearing.org or ahrf@american-hearing.org
Internet: http://www.american-hearing.org/research/derlacki_grant.html
Sponsor: American Hearing Research Foundation
8 South Michigan Avenue, Suite 814
Chicago, IL 60603-4539

AHRF Georgia Birtman Grant 362
The American Hearing Research Foundation (AHRF), together with the Northwestern Memorial Foundation (NMF) give the Georgia Birtman Grant, named in honor of long-time supporter of the AHRF. The one-year $75,000 grant ($25,000 each from the Birtman Fund, the AHRF and Northwestern Memorial Foundation) supports the advancement of research and education in otology and neurotology. The grant is awarded to an exceptional researcher in audiology, otology or neurotology who will work in a lab at Northwestern University. The research topic involves some aspect of the diagnosis, treatment, and rehabilitation of hearing and balance disorders related to the inner ear. The research has the potential to generate clinical care innovations, facilitate translational clinical studies, and develop creative educational programs.
Geographic Focus: All States
Date(s) Application is Due: Aug 3
Amount of Grant: Up to 75,000 USD
Samples: Timothy C. Hain, M.D., Northwestern University, Feinberg School of Medicine, Chicago, Illinois--for Vestibular Evoked Myogenic Potentials; Claus-Peter Richter, M.D., Ph.D., Northwestern University, Feinberg School of Medicine, Chicago, Illinois--for Electrical Stimulation of Spiral Ganglion Cells.
Contact: Kristen Madhuizen; (312) 726-9670; fax (312) 726-9695; kristen.madhuizen@american-hearing.org or ahrf@american-hearing.org
Internet: http://www.american-hearing.org/research/birtman_grant.html
Sponsor: American Hearing Research Foundation
8 South Michigan Avenue, Suite 814
Chicago, IL 60603-4539

AHRF Regular Research Grants 363
The American Hearing Research Foundation funds five to ten $20,000 research grants each year. Research Grants should relate to the hearing or balance functions of the ear. Both basic and clinical studies may be proposed. Priority is given to providing start-up funds for new projects. Applications are reviewed by a Research Committee and awards begin in January. Applications are due no later that noon on August 3 of the previous year.
Geographic Focus: All States
Date(s) Application is Due: Aug 3
Amount of Grant: Up to 20,000 USD
Contact: Kristen Madhuizen; (312) 726-9670; fax (312) 726-9695; kristen.madhuizen@american-hearing.org or ahrf@american-hearing.org
Internet: http://www.american-hearing.org/research/grant_guidelines.html
Sponsor: American Hearing Research Foundation
8 South Michigan Avenue, Suite 814
Chicago, IL 60603-4539

AHRF Wiley H. Harrison Memorial Research Award 364
The purpose of this award is to support clinical research projects in otology or neurotology designed to increase understanding of hearing disorders. Research supported by this award should be specifically directed toward the clinical identification, diagnosis, prevention, or treatment of diseases, disorders, or conditions of the ear. While not specifically required, proposals which aim to introduce new knowledge and methodology from other disciplines to research in otology or neurotology, or which demonstrate collaborative effort with members of other related disciplines are encouraged. This is a one year, non-renewable grant of $25,000 maximum, with only one award available annually. Completed applications must be submitted by August 3.
Requirements: Candidates for this award should be physicians (M.D.) at the resident, fellow, or junior faculty stage, or PhD scientists. Previous AHRF or AAO-HNS Foundation research grant recipients are eligible to compete for this grant.
Geographic Focus: All States
Date(s) Application is Due: Aug 3
Amount of Grant: Up to 25,000 USD
Contact: Kristen Madhuizen; (312) 726-9670; fax (312) 726-9695; kristen.madhuizen@american-hearing.org or ahrf@american-hearing.org
Internet: http://www.american-hearing.org/research/harrisongrant.html
Sponsor: American Hearing Research Foundation
8 South Michigan Avenue, Suite 814
Chicago, IL 60603-4539

AHRQ Independent Scientist Award 365
The award is a special salary-only grant designed to provide protected time for newly independent scientists who currently have non-research obligations such as heavy teaching loads, clinical work, committee assignments, service, and administrative duties that prevent them from having a period of intensive research focus. The award is targeted to persons with doctoral degrees who have completed their research training, have independent peer-reviewed research support, and who need a period of protected research time in order to foster their research career development. New applications are due February 12, June 12, and October 12, while renewal applications are due March 12, July 12, and November 12. Up to $90,000 per year of base salary plus fringe benefits is available. No other research development support funds are provided.
Requirements: The following organizations and institutions are eligible to apply: public/state controlled institutions of higher education; private institutions of higher education; nonprofits with 501(c)3 IRS status; nonprofits without 501(c)3 IRS status; small businesses; for-profit organizations; State governments; U.S. territories or possessions; Indian/Native American tribal governments (Federally recognized and other than Federally recognized); Indian/Native American tribally designated organizations; Hispanic-serving institutions; historically Black colleges and universities (HBCUs); tribally controlled colleges and universities (TCCUs); Alaska Native and Native Hawaiian serving institutions; regional organizations; and faith-based or community based organizations. The candidate must have a doctoral degree and peer-reviewed, independent research support at the time the award is made. The candidate must spend a minimum of 75 percent effort conducting research during the period of the award.
Restrictions: The award is not intended for investigators who already have full time to perform research, or have substantial publication records or considerable research support indicating that they are well established in their fields. Foreign institutions are not eligible to apply.
Geographic Focus: All States
Date(s) Application is Due: Feb 12; Mar 12; Jun 12; Jul 12; Oct 12; Nov 12
Contact: Kay Anderson, (301) 427-1555; fax (301) 427-1562; Kay.Anderson@ahrq.hhs.gov
Internet: http://grants.nih.gov/grants/guide/pa-files/PAR-07-444.html
Sponsor: Agency for Healthcare Research and Quality
540 Gaither Road
Rockville, MD 20850

AHRQ Individual Awards for Postdoctoral Fellows Ruth L. Kirschstein National Research Service Awards (NRSA) 366
The purpose of the postdoctoral fellowship award is to provide support to promising postdoctoral applicants who have the potential to become productive and successful independent research investigators. The proposed postdoctoral training must offer an opportunity to enhance the applicant's understanding of health services research and must be responsive to AHRQ's mission, which is to improve the quality, safety, efficiency, and effectiveness of health care for all Americans. The research sponsored and conducted by AHRQ develops and presents scientific evidence regarding all aspects of health care. It addresses issues of organization, delivery, financing, utilization, patient and provider behavior, outcomes, effectiveness and cost. It evaluates both clinical services and the system in which these services are provided. These scientific results improve the evidence base to enable better decisions about health care, including such areas as disease prevention, appropriate use of medical technologies, improving diagnosis and treatment utilizing comparative effectiveness research, and reducing racial and ethnic disparities. In addition, AHRQ is interested in the application of health information technology (health IT), as well as reducing medical errors and improving patient safety.
Requirements: The following organizations and institutions are eligible to apply: public or non-profit private institutions, such as a university, college, or a faith-based or community-based organization; units of local or State government; eligible agencies

54 | Grant Programs

of the Federal government; Indian/Native American tribal governments (Federally recognized); Indian/Native American tribal governments (other than Federally recognized); and Indian/Native American tribally designated organizations.
Restrictions: Awards for fellowship training may only be made to domestic institutions.
Geographic Focus: All States
Date(s) Application is Due: Apr 8; Aug 8; Dec 8
Contact: Shelley M. Benjamin, (301) 427-1528; Shelley.Benjamin@ahrq.hhs.gov
Internet: http://grants.nih.gov/grants/guide/pa-files/PA-09-229.html
Sponsor: Agency for Healthcare Research and Quality
540 Gaither Road
Rockville, MD 20850

Aid for Starving Children Emergency Assistance Fund Grants 367

Aid for Starving Children, formerly known as the African American Self-Help Foundation, has been helping save the lives of children for over thirty years. The foundation belongs to a coalition of organizations on four continents which coordinate both resources and efforts in order to achieve maximum results and efficiency and to save the world - one child at a time. In the United States, Aid For Starving Children sponsors programs that meet critical needs, develop life and work skills, and strengthen relationships with church and community to help African-American single mothers and their children break free of the cycle of poverty. Additionally the organization provides a limited number of small cash grants to help qualified African-American single mothers meet emergency financial needs. Needs can include: overdue rent or mortgage payments; overdue utility bills (gas or electric); car repair; and critical medical needs. Similar type emergencies may also qualify. Applicants for the cash grants must meet the minimum qualifications and submit a completed application with all required documentation in order to be considered for this program. Complete guidelines and the application for the cash grants are available at the Aid for Starving Children website.
Requirements: Applicants who meet all of the following requirements are eligible to apply for small cash grants: currently-employed, African-American single mothers who live in the United States and have at least one child under the age of seventeen living with them (the employment equirement may be waived under emergency conditions); individuals who have documentation that they have sought and been unable to obtain help from family or other local sources; and individuals who have an emergency need such as an overdue rent or mortgage payment or utility bill, a critical car maintenance or repair, or a critical medical need (similar types of emergencies may also qualify).
Restrictions: Requests for help with household and personal expenses (eg. food, clothing, furniture, etc.), education and business expenses, gifts, and credit card payments will not be considered.
Geographic Focus: All States
Contact: Jeff Baugham, U.S. Director; (937) 275--7310; RevJB@donet.com
Internet: http://www.aidforstarvingchildren.org/emergency_assistance
Sponsor: Aid for Starving Children
P.O. Box 2156
Windsor, CA 95492

Aid for Starving Children International Grants 368

Aid for Starving Children, formerly known as the African American Self-Help Foundation, has been helping save the lives of children for over thirty years. The foundation belongs to a coalition of organizations on four continents which coordinate both resources and efforts in order to achieve maximum results and efficiency and to save the world - one child at a time. In sub-Saharan Africa, the Foundation rescues orphaned and abandoned children who have lost their family members to AIDS and/or other devastation, and provides them with new lives of love and hope; provides poor children with support for a healthier and more productive future through feeding programs, medical care and education; and helps poor parents provide a better life for their own children through sustainable agriculture and clean water projects. Organizations interested in partnering with Aid to Starving Children in these programs should call or email the foundation for more information.
Geographic Focus: All States, Algeria, Angola, Benin, Botswana, Burkina Faso, Burundi, Cameroon, Cape Verde, Central African Republic, Chad, Comoros, Congo, Congo, Democratic Republic of, Cote d' Ivoire (Ivory Coast), Djibouti, Egypt, Equatorial Guinea, Eritrea, Ethiopia, Gabon, Gambia, Ghana, Guinea, Guinea-Bissau, Kenya, Lesotho, Liberia, Libya, Madagascar, Malawi, Mali, Mauritania, Mauritius, Morocco, Mozambique, Namibia, Niger, Nigeria, Rwanda, Sao Tome & Principe, Senegal, Seychelles, Sierra Leone, Somalia, South Africa, Sudan, Swaziland
Contact: Jeff Baugham; (937) 275--7310 or (800) 514-3499; RevJB@donet.com
Internet: http://www.aidforstarvingchildren.org/africa
Sponsor: Aid for Starving Children
P.O. Box 2156
Windsor, CA 95492

AIDS Vaccine Advocacy Coalition (AVAC) Fund Grants 369

The Fund functions as a small-scale emergency fund to assist needy clinical sites that require immediate help with purchases such as additional medical or lab supplies not covered by grants or contracts for vaccine research. All AIDS vaccine clinical trial sites in resource-limited countries or needy communities, regardless of sponsor, can apply for grants up to $2,000. A committee consisting of HIV vaccine investigators from around the world reviews proposals and makes recommendations to a special Fund Committee of the AVAC Board.
Geographic Focus: All States
Amount of Grant: 2,000 USD
Samples: Kenya AIDS Vaccine Initiative, Department of Medical Microbiology, Nairobi University, $2,000--for HIV Rapid Test Kits and Emergency Medicines; Brooke Bond Tea Plantation Hospital, $2,000--for general health screenings.
Contact: Marie Semmelbeck, (212) 367-1188 or (212) 367-1279; fax (646) 365-3452; fund@avac.org
Internet: http://www.avac.org/ht/d/sp/i/962/pid/962
Sponsor: AIDS Vaccine Advocacy Coalition
101 West 23rd Street, #2227
New York, NY 10011

AIG Disaster Relief Fund Grants 370

American International Group (AIG) traces its root to 1919, when American Cornelius Vander Starr established a general insurance agency, American Asiatic Underwriters, in Shanghai, China. Since then, the company has become one of the world's leading insurers. AIG is committed to giving back to the communities it serves, including those affected by disasters, through programs and partnerships that leverage the skills, experience, knowledge, and enthusiasm of AIG employees. The AIG Disaster Relief Fund (DRF) was established by AIG to assist victims of natural and man-made disasters around the world. Grant-seekers should contact AIG for information on how to be considered for DRF grants.
Geographic Focus: All States, All Countries
Contact: David Herzog, Chairman; (212) 770-7000
Internet: http://www.aig.com/citizenship_3171_437858.html
Sponsor: American International Group
180 Maiden Lane
New York, NY 10038

AIHP Sonnedecker Visiting Scholar Grants 371

The Sonnedecker Visiting Scholar Program offers assistance for short-term historical research related to the history of pharmacy, including the history of drugs, at the University of Wisconsin-Madison. The program provides assistance for travel, maintaining temporary residence in Madison, and meeting research expenses associated with utilizing the collection. A brochure is available on request that describes the pharmaco-historical collections, which have been developed in Madison during more than a century by the University of Wisconsin-Madison, the State Historical Society of Wisconsin, and the American Institute of the History of Pharmacy. At least $1,000 becomes available annually to defray part of the expenses of a recipient, for whatever period of residence is appropriate. Grants are made throughout the year on the basis of the merit of previous historical work and on the appropriateness of historical resources on the University of Wisconsin campus to the research proposed.
Requirements: Eligible applicants are: Historians, Pharmacists, and other scholars working in the Pharmacy field.
Geographic Focus: All States
Amount of Grant: 1,000 USD
Contact: Gregory Higby, Director; (608) 262-5378; grants@aihp.org
Internet: http://cms.pharmacy.wisc.edu/aihp/programs/sonnedecker
Sponsor: American Institute of the History of Pharmacy
777 Highland Avenue
Madison, WI 53705-2222

AIHP Thesis Support Grants 372

The Program offers a grant-in-aid totaling $2,000 or more annually to a graduate student to reinforce historical investigations of some aspect of pharmacy, whether ancient or modern, to pay research expenses not normally met by the university granting the degree. Any thesis project devoted to the history of pharmacy, history of drugs, or other humanistic study utilizing a pharmaco-historical approach, is eligible if based in an institution of higher learning of the USA. The maximum grant in this program of the Institute will be $2,500.
Requirements: Any graduate student in good standing at an institution of the United States may apply, regardless of the department through which the Doctor of Philosophy degree will be granted. The graduate student need not be an American citizen; nor does the research topic need to be in the field of American history.
Restrictions: Examples of ineligible expenses would be: living expenses of the applicant at the home university; routine typing to produce research notes or the thesis manuscript; routine illustrations for the manuscript; and publication of research results. Indirect expenses, such as overhead and other institution-related costs, may not be included in a grant application.
Geographic Focus: All States
Date(s) Application is Due: Feb 1
Amount of Grant: Up to 2,500 USD
Contact: Gregory Higby, Director; (608) 262-5378; grants@aihp.org
Internet: http://cms.pharmacy.wisc.edu/aihp/programs/thesisgrant
Sponsor: American Institute of the History of Pharmacy
777 Highland Avenue
Madison, WI 53705-2222

AIHS/Mitacs Health Pilot Partnership Internship Grants 373

The AIHS/Mitacs Health Pilot Partnership is a partnership between AIHS, Mitacs and not-for-profit organizations. This is a pilot program created to address the existing gap and lack of funding available to engage highly qualified graduate students in the public and not-for-profit sector, specifically in health-related research. The health pilot partnership will fund up to three internships for graduate and postgraduate health research and innovation trainees in Alberta. The AIHS/Mitacs Health Pilot Partnership provides

a grant of $15,000 for a four-month internship administered by Mitacs-Accelerate and a partner university. The Partnership will meet the following objectives: connect not-for-profit organizations with peer-reviewed research expertise, ensuring high quality research that meets the organization's needs; apply knowledge by interns to real-world issues to create solutions to challenging problems; and provide research labs with a unique opportunity to train graduate and postgraduate trainees in new and applied skills that can be brought back to the lab. Interns may work in the areas of health economics; remote health monitoring; health policy; e-health; chronic disease management; or knowledge transfer/translation. A minimum $10,000 stipend will be provided to the intern, and the remaining $5,000 will be available for other project-related expenses. A longer eight-month double internship can be developed if there is strong justification to accommodate larger projects. Interns will spend approximately 50% of their time with the partnered not-for-profit organization, with the remaining time spent at the university, advancing the research under the guidance of their faculty supervisor.
Requirements: An applicant must be officially accepted into, or currently enrolled in, graduate or postgraduate studies at the University of Alberta or the University of Calgary; able to participate in a four-month single internship or an eight-month double internship with 50% of the intern's time being spent outside of the lab; and have a defined research project that they will undertake with the partnered not-for-profit organization.
Geographic Focus: All States, Canada
Date(s) Application is Due: May 15
Amount of Grant: 15,000 USD
Contact: Dr. Ryan Perry; (780) 423-5727; ryan.perry@albertainnovates.ca
Internet: http://www.aihealthsolutions.ca/grants/training-and-early-career-development-programs/mitacs/
Sponsor: Alberta Innovates - Health Solutions
10104-103 Avenue, Suite 1500
Edmonton, AB T5J 4A7 Canada

AIHS Alberta/Pfizer Translational Research Grants — 374

The Alberta/Pfizer Translational Research Fund Opportunity is a partnership between Pfizer Canada Inc. (Pfizer), Alberta Innovates - Health Solutions (AIHS), Alberta's Ministry of Enterprise and Advanced Education, and Western Economic Diversification Canada. This partnership will provide opportunities to focus on the development and commercialization of innovations in health. The Alberta/Pfizer Translational Research Fund Opportunity will support innovative translational research projects in areas of unmet health or health system need that have a strong likelihood for technology transfer and commercialization within the next two to five years. Grants will support the development and commercialization of innovations in priority areas identified by Pfizer and the province of Alberta. Those areas include the following: neuroscience/Alzheimer's Disease/ neurodegeneration/ neurological diseases; pain and sensory disorders; immunology and autoimmunity; inflammation and remodeling; diabetes and cardiovascular or other chronic disease issues as they relate to the consequences of diabetes; and orphan and genetic diseases. Preference will be given to research projects focused on the development, use and/or application of innovative technological platforms. These platforms include, but are not limited to: metabolomics and other 'omics' platforms that facilitate personalized medicine strategies; drug discovery; information and communication technologies; nanotechnology; and health-systems-based research approaches in the creation of new knowledge and its potential use. Translational research grants will provide up to $200,000 for a period of up to 18 months. Eligible expenses include salaries of research personnel, purchase and/or leasing of equipment, and research material required for the project. Payments will be based on agreed upon milestones and deliverables, and the submission of periodic reports outlining progress.
Requirements: Organizations will submit a letter of intent and full application. Additional guidelines are available at the AIHS website.
Geographic Focus: All States, Canada
Amount of Grant: Up to 200,000 USD
Contact: Tara McCarthy, Grant Coordinator, Programs; (780) 423-5727; fax (780) 429-3509; tara.mccarthy@albertainnovates.ca
Internet: http://www.aihealthsolutions.ca/grants/industry-partnered-translational-fund/pfizer/
Sponsor: Alberta Innovates - Health Solutions
10104-103 Avenue, Suite 1500
Edmonton, AB T5J 4A7 Canada

AIHS Clinical Fellowships — 375

The AIHS Clinical Fellowships program is designed for highly qualified individuals who hold an MD or DDS degree, and who anticipate undertaking a career in health related or clinical research in Alberta. The awards consist of a stipend and a research allowance. A Clinical Fellowship award is normally tenable for a maximum of three years. However, if the trainee is registered in a graduate program, the term may be extended.
Requirements: Normally, an award will be held within the Province of Alberta; however, candidates who are Canadian citizens or permanent residents with records of outstanding performance in postgraduate training may seek research training elsewhere, if sponsored by an Alberta faculty with an expressed interest in future recruitment of the candidate. Candidates must hold an MD or DDS degree and have received a significant portion of their postgraduate training in Alberta.
Restrictions: Clinical fellows are expected to commit to a minimum of two years of full time research.
Geographic Focus: Canada
Date(s) Application is Due: Mar 1; Oct 1
Amount of Grant: 20,000 - 50,000 USD
Contact: Pamela Valentine; (780) 423-5727, ext. 230; fax (780) 429-3509; pamela.valentine@albertainnovates.ca or grants.health@albertainnovates.ca
Internet: http://www.aihealthsolutions.ca/grants/Clin-fellow.php
Sponsor: Alberta Innovates - Health Solutions
10104-103 Avenue, Suite 1500
Edmonton, AB T5J 4A7 Canada

AIHS Collaborative Research and Innovation Grants - Collaborative Program — 376

The AIHS Collaborative Research and Innovation Opportunities (CRIO) aims to bring together experts in different disciplines, fields, and areas to tackle health research problems in areas of strategic priority that would benefit from an interdisciplinary approach. The objectives of the CRIO are to: catalyze and support collaborative, interdisciplinary, multi-sectoral and/or multi-institutional research with a focus on achieving solutions that address complex health problems or issues; support areas of research that are aligned with the thematic priority areas defined in Alberta's Health Research and Innovation Strategy; engage end users in the process to enhance impact on the health of Albertans and/or the health care system; provide opportunities for interdisciplinary research training and mentorship; and encourage the use of the AIHS collaborative opportunities to connect Alberta to national and international initiatives and leverage additional opportunities. A maximum of up to $500,000 per year for five years may be awarded. The Collaborative Program involves three to five principal collaborators who may already have an informal arrangement or shared funding to support interdisciplinary work, or have specific plans to develop such arrangements. The Collaborative Program may be considered for a one time renewal, based on agreed-upon deliverables set up in the management plan for years 1 through 5.
Requirements: The application process involves registration, a letter of intent, and full application. Each CRIO must involve Collaborative Lead(s) who are established individual(s) with demonstrated leadership. Collaborative Leads can lead more than one CRIO Project, Program and/or Team simultaneously. There are no limits on the number of applications a Collaborative Lead can submit. Collaborative members may be defined as needed to address the research objectives, and may include individuals from outside the Alberta area. Interdisciplinary, multi-institutional activity is encouraged. The focus is on a complex program of health research and innovation. A full list of eligibility requirements is available at the AIHS website.
Geographic Focus: All States, Canada
Date(s) Application is Due: Jan 18
Amount of Grant: Up to 500,000 USD
Contact: Kathy Morrison, Grants Coordinator, Programs; (780) 423-5727; fax (780) 429-3509; kathy.morrison@albertainnovates.ca
Internet: http://www.aihealthsolutions.ca/grants/crio/#project
Sponsor: Alberta Innovates - Health Solutions
10104-103 Avenue, Suite 1500
Edmonton, AB T5J 4A7 Canada

AIHS Collaborative Research and Innovation Grants - Collaborative Project — 377

The AIHS Collaborative Research and Innovation Opportunities (CRIO) aims to bring together experts in different disciplines, fields, and areas to tackle health research problems in areas of strategic priority that would benefit from an interdisciplinary approach. The objectives of the CRIO are to: catalyze and support collaborative, interdisciplinary, multi-sectoral and/or multi-institutional research with a focus on achieving solutions that address complex health problems or issues; support areas of research that are aligned with the thematic priority areas defined in Alberta's Health Research and Innovation Strategy; engage end users in the process to enhance impact on the health of Albertans and/or the health care system; provide opportunities for interdisciplinary research training and mentorship; and encourage the use of the AIHS collaborative opportunities to connect Alberta to national and international initiatives and leverage additional opportunities. The CRIO Collaborative Project allows approximately three collaborators to come together to complete a defined project. There is no expectation that the collaboration will continue beyond the duration of this non-renewable grant. The opportunity also provides a framework for collaborators who choose to evolve their project into a more complex Collaborative Program or Collaborative Team. The application process involves registration, a letter of intent, and full application. Up to $250,000 per year may be funded for three years.
Requirements: The Collaborative Lead must be an individual with an established career and proven leadership skills and experience. Collaborative members will have an extensive record of success, be creative and original in their approach to research and its translation, and have experience working in teams. Members may also be located outside the Alberta area. Co-funding from philanthropic, national or international organizations is encouraged and should be declared as part of the submission. A full list of eligibility requirements is available at the AIHS website.
Geographic Focus: Canada
Date(s) Application is Due: Aug 31
Amount of Grant: Up to 250,000 USD
Contact: Kathy Morrison, Grants Coordinator, Programs; (780) 423-5727; fax (780) 429-3509; kathy.morrison@albertainnovates.ca
Internet: http://www.aihealthsolutions.ca/grants/crio/
Sponsor: Alberta Innovates - Health Solutions
10104-103 Avenue, Suite 1500
Edmonton, AB T5J 4A7 Canada

AIHS Collaborative Research and Innovation Grants - Collaborative Team — 378

The AIHS Collaborative Research and Innovation Opportunities (CRIO) aims to bring together experts in different disciplines, fields, and areas to tackle health research problems in areas of strategic priority that would benefit from an interdisciplinary approach. The objectives of the CRIO are to: catalyze and support collaborative,

interdisciplinary, multi-sectoral and/or multi-institutional research with a focus on achieving solutions that address complex health problems or issues; support areas of research that are aligned with the thematic priority areas defined in Alberta's Health Research and Innovation Strategy; engage end users in the process to enhance impact on the health of Albertans and/or the health care system; provide opportunities for interdisciplinary research training and mentorship; and encourage the use of the AIHS collaborative opportunities to connect Alberta to national and international initiatives and leverage additional opportunities. The Collaborative Team fosters and support a large interdisciplinary collaborative team or network with national or international stature. The complexity of the health issues of interest will require the involvement of many stakeholder groups. Up to $1,000,000 per year for five years may be funded.
Requirements: The application process involves registration, a letter of intent, and a full application. Each CRIO must involve a Collaborative Lead() who are established individuals with demonstrated leadership. Collaborative members may be defined as needed to address the research objectives and may include individuals from outside Alberta's knowledge or end users. Interdisciplinary, multi-institutional activity is encouraged. A full list of eligibility requirements is available at the website.
Geographic Focus: All States, All Countries
Date(s) Application is Due: Jan 18
Amount of Grant: Up to 1,000,000 USD
Contact: Tara McCarthy, Grants Coordinator, Programs; (780) 423-5727; fax (780) 429-3509; tara.mccarthy@albertainnovates.ca
Internet: http://www.aihealthsolutions.ca/grants/crio/#project
Sponsor: Alberta Innovates - Health Solutions
10104-103 Avenue, Suite 1500
Edmonton, AB T5J 4A7 Canada

AIHS Fast-Track Fellowships 379
AIHS will accept, at any time, applications from Alberta-based institutions for full-time Fellowship awards for outstanding candidates currently training outside of Alberta. The fast-track program is intended to assist these institutions in recruiting highly qualified trainees to the province. Up to a maximum of 10 awards per calendar year will be awarded through the fast-track review process in each category. Applications to the program will be reviewed on a first come-first serve basis. Fast-track awards are tenable for up to 12 months.
Requirements: Applications must include a letter from the Dean/designate of the faculty indicating that the proposal for a fast-track award has been reviewed and stating why a fast-track application should be considered for the candidate in question. Please refer to the sections on Full-Time Fellowships or Full-Time Studentships for further details regarding eligibility for training awards.
Geographic Focus: Canada
Amount of Grant: 20,000 - 50,000 USD
Contact: Pamela Valentine, Vice President of Research and Innovation; (780) 423-5727, ext. 230; fax (780) 429-3509; pamela.valentine@albertainnovates.ca or grants.health@albertainnovates.ca
Internet: http://www.aihealthsolutions.ca/grants/fasttrack.php
Sponsor: Alberta Innovates - Health Solutions
10104-103 Avenue, Suite 1500
Edmonton, AB T5J 4A7 Canada

AIHS ForeFront Internships 380
The AIHS Internship Program addresses the need for highly trained staff and management of the Alberta-based health, medical products and biotechnology industries. This program is intended to support technology commercialization training and experience for individuals with an appropriate background in science and/or business. Although the work plan should identify the specific area of focus, the intern should receive a broad exposure to all of the following areas: product development; intellectual property strategy; regulatory requirements; market research and strategy; and business planning and development. Interns are awarded a $5,000 training allowance, which may be used for expenses related to educational workshops/seminars/courses necessary for relevant commercialization training, plus a $45,000 to $50,000 stipend. The application and additional guidelines are available at the AIHS website.
Requirements: The applicant must hold, or be in the final year of a degree in business, management, or science. The applicants should have some relevant experience in the business or private sector. Applicants must be sponsored by an Alberta organization actively engaged in the commercialization of medical or health related technology. The organization must have the experience and resources to provide an appropriate training environment, and designate one person within the organization who will be the primary supervisor for the intern. The organization must agree to provide direct and continuous supervision.
Geographic Focus: All States, Canada
Amount of Grant: 45,000 - 50,000 USD
Contact: Carla Weyland, ForeFront Officer; (780) 423-5727; fax (780) 429-3509; carla.weyland@albertainnovates.ca
Internet: http://www.aihealthsolutions.ca/forefront/internship.php
Sponsor: Alberta Innovates - Health Solutions
10104-103 Avenue, Suite 1500
Edmonton, AB T5J 4A7 Canada

AIHS ForeFront MBA Studentship Award 381
This award will enable students with a background in the medical/health/life sciences to apply their knowledge and pursue a career in management in the medical/health industry by supporting their education and hands-on training in the University of Alberta Master of Business Administration program, specializing in Technology Commercialization. The Tech Com MBA builds and delivers a tool kit that emphasizes the importance of capital, management practices, infrastructure, information, and networks. The goal of the Tech Com MBA is to position graduates for a leadership role in today's leading edge, high technology industries. The main objectives are to: encourage Albertans with a background in science to explore careers that will give them the opportunity to apply their knowledge to commercializing medical discoveries that will lead to improved health; and build capacity of knowledgeable, highly-skilled technology managers to meet the needs of the Alberta medical/health industry. The MBA Studentship Award includes a contribution towards tuition fees and a stipend of $2,000 per month for the duration of the student's full-time MBA studies. The MBA Program may be completed in two years on a full time basis. Students have a maximum of six years to complete the course requirements. Any extensions to the two-year full-time term must be approved by AIHS. Additional guidelines and terms of the award are available at the AIHS website.
Requirements: Students must hold a graduate degree in medical/health/life sciences (graduated within the last two years) and meet the minimum eligibility criteria for admission to the MBA program.
Geographic Focus: All States, Canada
Contact: Carla Weyland, Manager, Training and Development Programs; (780) 423-5727; fax (780) 429-3509; carla.weyland@albertainnovates.ca
Internet: http://www.aihealthsolutions.ca/forefront/mba.php
Sponsor: Alberta Innovates - Health Solutions
10104-103 Avenue, Suite 1500
Edmonton, AB T5J 4A7 Canada

AIHS ForeFront MBT Studentship Awards 382
The award will enable students to pursue a career in the medical, health, and biotechnology industries by supporting their education and hands-on training in the Master of Biomedical Technology (MBT) Program at the University of Calgary. The MBT Program takes a multi-disciplinary approach and exposes students to a broad range of courses and cutting-edge technologies that will ensure the students acquire practical competencies and extensive knowledge of concepts in biomedical and bioinformatics disciplines, including the application of this knowledge through commercialization of medical and health-related innovations. The main objectives are to: encourage individuals with a background in science to explore careers that will give them the opportunity to apply their knowledge to commercializing medical discoveries that will lead to improved health; and build capacity of knowledgeable, highly-skilled technology managers to meet the needs of the Alberta medical/health industry. The MBT Studentship Award includes a contribution towards tuition fees and a stipend for the practicum (up to a maximum of $3,000). The MBT Program may be completed in one year on a full time basis. Any extensions to the one-year full-time term must be approved by AIHS. Application information is available at the AIHS website.
Requirements: Candidates must hold at least an undergraduate degree in an area of relevance to the objectives of Alberta Innovates – Health Solutions (AIHS).
Geographic Focus: All States, Canada
Contact: Kathy Morrison, Grants Coordinator; (780) 423-5727; fax (780) 429-3509; kathy.morrison@albertainnovates.ca
Internet: http://www.aihealthsolutions.ca/forefront/mbt.php
Sponsor: Alberta Innovates - Health Solutions
10104-103 Avenue, Suite 1500
Edmonton, AB T5J 4A7 Canada

AIHS Full-Time Fellowships 383
The AIHS Full-Time Fellowships program is designed to enable highly qualified doctoral graduates to prepare for careers in medical or health research as independent investigators. A fellowship award will provide to the host institution funding for one year's stipend, its associated benefits, and a research allowance. Fellows may engage in teaching activities related to their research discipline a maximum of 20 percent of their time. The maximum term is three years
Requirements: Candidates must have a Ph.D., M.D., D.D.S., D.V.M. or D.Pharm. degree. Normally, support will not be provided beyond 6 years after receipt of the Ph.D. degree, or beyond 8 years after receipt of the M.D., D.D.S., D.V.M. or D.Pharm. degrees.
Geographic Focus: Canada
Date(s) Application is Due: Mar 1; Oct 1
Amount of Grant: 35,000 - 50,000 USD
Contact: Pamela Valentine; (780) 423-5727, ext. 230; fax (780) 429-3509; pamela.valentine@albertainnovates.ca or grants.health@albertainnovates.ca
Internet: http://www.aihealthsolutions.ca/grants/FT-fellow.php
Sponsor: Alberta Innovates - Health Solutions
10104-103 Avenue, Suite 1500
Edmonton, AB T5J 4A7 Canada

AIHS Full-Time Health Research Studentships 384
The Health Research Studentship program enables academically superior students to undertake full-time training in health research. The award supports training in: research on the organization and delivery of health care; technology assessment; community health; health promotion; disease prevention; and related disciplines. The Health Research Studentship consists of a stipend and a research allowance. The term of the award is a maximum of five years (three years at the graduate level).
Requirements: Candidates must have been accepted into, or be currently studying in, a full-time, thesis-based, graduate program at an Alberta-based university in a health-related discipline leading to a Master's or doctoral degree. Normally, support will not be provided beyond six years of enrollment in graduate school. Candidates who have interrupted their training for parenting or other reasons and who have consequently exceeded this time limit, may apply for studentship support. In such cases, however,

the candidate is advised to clearly explain the nature of their particular circumstances at the time of application.
Restrictions: The award is not available to students registered in a course-based program.
Geographic Focus: Canada
Date(s) Application is Due: Mar 1; Oct 1
Amount of Grant: 20,000 USD
Contact: Pamela Valentine; (780) 423-5727, ext. 230; fax (780) 429-3509; pamela.valentine@albertainnovates.ca or grants.health@albertainnovates.ca
Internet: http://www.aihealthsolutions.ca/grants/FT-student.php#HRS
Sponsor: Alberta Innovates - Health Solutions
10104-103 Avenue, Suite 1500
Edmonton, AB T5J 4A7 Canada

AIHS Full-Time M.D./Ph.D. Studentships   385
The AIHS Full-Time M.D./Ph.D. Studentships are intended to provide an opportunity for exceptional candidates, who wish to pursue careers as Clinical Investigators, and study simultaneously for the M.D. and the Ph.D. degrees. Support is complementary to the formal M.D./Ph.D. programs at the University of Alberta and the University of Calgary. Awards are tenable only at an Alberta-based university. The term of the award is a maximum of six years.
Requirements: Applicants are limited to those students that formally enter the MD/PhD program no later than the start of their second year of PhD studies at either the University of Alberta or the University of Calgary. AIHS's offer of support is contingent on the granting of complementary stipend support from the Faculty of not less than 15% of the value of the AIHS award, for the duration of the award. Furthermore, it is expected that the universities will undertake to provide administrative support to the offices of the coordinators of the M.D./Ph.D. programs.
Geographic Focus: Canada
Date(s) Application is Due: Mar 1; Oct 1
Amount of Grant: 20,000 USD
Contact: Pamela Valentine; (780) 423-5727, ext. 230; fax (780) 429-3509; pamela.valentine@albertainnovates.ca or grants.health@albertainnovates.ca
Internet: http://www.aihealthsolutions.ca/grants/FT-student.php#MPS
Sponsor: Alberta Innovates - Health Solutions
10104-103 Avenue, Suite 1500
Edmonton, AB T5J 4A7 Canada

AIHS Full-Time Studentships   386
AIHS Full-Time Studentships enable academically superior students to undertake full-time research training in the basic biomedical sciences or in clinical research. The award consists of a stipend and a research allowance. Approved uses of the research allowance include: the purchase of scientific materials, supplies and expendables; the purchase of minor equipment; computer software programs; costs for the use of libraries, or computers; costs associated with the publication of research results; travel expenses to attend scientific meetings; and purchase of books, periodicals and journals. The term of the award is a maximum of five years (three years at the graduate level).
Requirements: Candidates must have been accepted into, or be currently engaged in, a full-time, thesis-based, graduate program at an Alberta-based university in a health-related discipline leading to a Master's or doctoral degree.
Restrictions: This award is not available to students registered in a course-based program. Normally, support will not be provided beyond 6 years of enrollment in graduate school. Candidates who have interrupted their training for parenting or other reasons and who have consequently exceeded this time limit, may apply for studentship support. In such cases, however, the candidate is advised to clearly explain the nature of their particular circumstances at the time of application.
Geographic Focus: Canada
Date(s) Application is Due: Mar 1; Oct 1
Amount of Grant: 20,000 USD
Contact: Pamela Valentine; (780) 423-5727, ext. 230; fax (780) 429-3509; pamela.valentine@albertainnovates.ca or grants.health@albertainnovates.ca
Internet: http://www.aihealthsolutions.ca/grants/FT-student.php#fts_description
Sponsor: Alberta Innovates - Health Solutions
10104-103 Avenue, Suite 1500
Edmonton, AB T5J 4A7 Canada

AIHS Heritage Youth Researcher Summer Science Program   387
The Heritage Youth Researcher Summer Program (HYRS) is an intensive six-week summer science program for high school student in Alberta. The program is designed to give exemplary students from around the province first-hand biomedical and health research experience at a university campus, and introduce them to the many educational and career options available in this field. Students work on health research projects in a variety of work environments of researchers who are on faculty at the University of Alberta, the University of Calgary, and the University of Lethbridge. Schedules vary slightly at each University, but are generally from July 4 through August 15. Students will be supervised by a faculty member or by a member of their research team. Research projects could be in cardiology, cell biology, genetics, biomedical engineering, population health, epidemiology, or other health research areas. The application and additional guidelines are available at the AIHS website.
Requirements: Applicants require at least an 85% standing in each of math 20-1 or math 20-2, biology 20, and one other grade 11 science class, which is the minimum requirement to work in a university laboratory. Applicants must have two reference forms completed by two of their science teachers, or one math and one science teacher. They must also submit a reference form completed by an adult who is not a relative and is from outside their school and who has known the student for at least one year. Additional guidelines are available at the website.
Restrictions: Students must be willing to commit to a full six weeks. Any time taken off will disqualify the student for the program.
Geographic Focus: All States, Canada
Date(s) Application is Due: Mar 23
Contact: Danica Wolkow, HYRS Program Provincial Coordinator; (780) 423-5727, ext 237; fax (780) 429-3509; hyrs@albertainnovates.ca
Internet: http://www.aihealthsolutions.ca/HYRS/
Sponsor: Alberta Innovates - Health Solutions
10104-103 Avenue, Suite 1500
Edmonton, AB T5J 4A7 Canada

AIHS Interdisciplinary Team Grants   388
The AIHS Interdisciplinary Team Grants Program provides opportunities for high-quality, internationally recognized teams of investigators to complete research initiatives with defined health outcomes. Funds available in this competition are to support collaborative, interdisciplinary and multi-institutional teams that address important research questions, health problems or issues in defined areas of research that are aligned with strategic research priorities of the AIHS and Alberta. The Program will provide up to $1 million per year per team for up to five years. The area of research to be pursued must be relevant to one of a number of areas of special interest identified by AIHS in collaboration with other stakeholders in Alberta. Those areas are: maternal, fetal and child health; mental health, mental illness and addictions; health system sustainability; modern lifestyles and health; health issues in rural and remote environments; health, genes and the environment; health and injury; modern techniques and technologies (including IT) and health; health and behavior: disease prevention; health and infectious diseases; and food and health. The LOI and full application must clearly identify under which of these theme areas the proposed program of research falls and how the proposed research is relevant to the theme. More than one theme can be identified as appropriate. Applicants are asked to make special reference as to how their program of research will be relevant to health issues facing vulnerable populations.
Requirements: Each eligible Interdisciplinary Team Grant application will include: A Team Leader--the Team Leader must be an established researcher with proven leadership skills and experience who will act as research program director and who will assume administrative responsibility for the grant. It is expected that the Team Leader will devote a significant and appropriate portion of their time to these tasks. The Team Leader will have their primary academic appointment at an Alberta-based university; at least two additional independent investigators with established research track records. Teams with a nucleus of experienced investigators are encouraged to include promising new investigators as part of their group; team members who collectively have an extensive record of success, are creative and original in their approach to research and its translation, and who have experience working in research teams. The specific contribution of each team member and end-user partner, where applicable, must be described; representation from more than one research discipline and from more than one "research pillar" (i.e. biomedical; clinical science; health systems and services; and the social, cultural and other factors that affect the health of populations.); and representation from more than one Alberta-based university. Multi-institutional collaboration is strongly encouraged in this Program. In addition, the members of the Team may pursue other research in addition to their commitment to the Interdisciplinary Team Grant. However, each individual investigator must contribute sufficient time to the Team Grant research program to ensure the achievement of its research and translational objectives. Collaborators from outside of Alberta who make a substantial intellectual contribution to the research program may be listed as team members. Specific and justified requests that AIHS funding be used for work performed outside of Alberta will be considered. However, it is expected that the great majority of the funds will remain in Alberta.
Geographic Focus: All States, Canada
Amount of Grant: Up to 1,000,000 USD
Contact: Pamela Valentine, Vice President of Research and Innovation; (780) 423-5727; fax (780) 429-3509; pamela.valentine@albertainnovates.ca
Internet: http://www.aihealthsolutions.ca/grants/team_guidelines.php
Sponsor: Alberta Innovates - Health Solutions
10104-103 Avenue, Suite 1500
Edmonton, AB T5J 4A7 Canada

AIHS Knowledge Exchange - Conference Grants   389
Funding under the Conference Grant program is provided to help bring distinguished researchers to Alberta to present lectures and to participate in panel discussions. The process facilitates the dissemination of knowledge to the province's researchers, health professionals and other stakeholders. Funds are offered four times a year, and are provided primarily to defray the costs of travel and accommodation for invited speakers. Funding for satellite conferences is limited to offsetting travel costs (ground transportation and accommodation) from the site of the major conference to Alberta. The application is available at the AIHS website. Applicants must submit the original application package, plus four copies.
Requirements: Conferences must be held in Alberta. It is expected that effort will be made to provide access to the event to students, fellows and other stakeholders from all Alberta institutions and relevant organizations within Alberta's health research and innovation community. Funding requests should be for multiple guest speakers. The conference must demonstrate the intent to promote collaboration and exchange between individuals, research groups, institutions, and organizations within Alberta and beyond.
Restrictions: Applications with only one out-of-province speaker will not be considered.
Geographic Focus: All States, Canada

Contact: Jackie Gettings, Grants Coordinator; (780) 429-6880; fax (780) 429-3509; jackie.gettings@albertainnovates.ca
Internet: http://www.aihealthsolutions.ca/grants/conference.php#confer
Sponsor: Alberta Innovates - Health Solutions
10104-103 Avenue, Suite 1500
Edmonton, AB T5J 4A7 Canada

AIHS Knowledge Exchange - Visiting Professorships 390
Visiting Professorships are awarded to particularly distinguished scientists to contribute to the education and training of students and fellows, and promote the progress of medical or health research in Alberta. Visiting Professorships may be awarded in order to assist the Alberta-based universities in planning or initiating new research directions, or for the purpose of evaluating AIHS-funded programs. The award includes an honorarium and per diem, plus travel costs. The length of the award is expected to be no longer than seven days.
Requirements: Applications should be submitted directly to AIHS at least three months prior to the proposed visit.
Geographic Focus: All States, Canada
Contact: Donna Angus, Director, Knowledge Transfer; (780) 423-5727, ext 208; fax (780) 429-3509; donna.angus@albertainnovates.ca
Internet: http://www.aihealthsolutions.ca/grants/knowledge.php
Sponsor: Alberta Innovates - Health Solutions
10104-103 Avenue, Suite 1500
Edmonton, AB T5J 4A7 Canada

AIHS Knowledge Exchange - Visiting Scientists 391
The Visiting Scientist program is intended to bring productive scientists with expertise not currently available in provincial institutions to Alberta. This expertise may consist of special knowledge, new approaches to research, and the management of new technology. The program is also intended to allow Alberta-based institutions to sponsor Alberta scientists to visit major medical or health research centers, and acquire knowledge abut new research concepts or technology. The visit must result in important contributions to medical or health related science in Alberta, and an increase in the visitor's research productivity. The sponsoring institution must assure the return of the Alberta scientist to their regular position at the completion of the award. Visiting Scientists will receive travel assistance and a research allowance to cover some of the costs of materials, supplies, and expenses critical for their participation in the program. The length of the visit ranges from one month to 12 months.
Requirements: The sponsoring institution must identify the candidate's special expertise, as well as the expected long-term research benefits of the visit. This applies to both visiting scientists to and visiting scientists from Alberta. Applications demonstrating that arrangements have been made for the visitor to meet with scientists in other Alberta universities or institutions will receive preference.
Geographic Focus: All States, Canada
Contact: Donna Angus, Director, Knowledge Transfer; (780) 423-5727, ext 208; fax (780) 429-3509; donna.angus@albertainnovates.ca
Internet: http://www.aihealthsolutions.ca/grants/knowledge.php
Sponsor: Alberta Innovates - Health Solutions
10104-103 Avenue, Suite 1500
Edmonton, AB T5J 4A7 Canada

AIHS Media Summer Fellowship 392
The AIHS Media Fellowship offers University of Alberta and University of Calgary students with strong biomedical science backgrounds the opportunity to spend 12 weeks during the summer (May through August) working with CBC Radio. Media Fellows will work as reporters, researchers and productions assistants with CBC Radio in either Edmonton or Calgary. Through the Media Fellowship, students will have an opportunity to impact the discussion of biomedical science in popular media and will be contributing to the translation of knowledge from the scientists to the media and the general public. The Media Fellowship will strengthen the relationship between scientists and the media; provide young scientists, at a critical stage in their careers, the opportunity to observe and participate in how events and ideas become news; improve the Fellow's skills in communicating complex technical subjects to the public; increase the Fellow's understanding of editorial decision-making and how news is communicated to the public; enhance the coverage of science and technology by the media; and improve public understanding and appreciation of science and medical research. The Media Fellowship provides a stipend to successful applicants, to be determined at the time of acceptance.
Requirements: An applicant must be an undergraduate- or graduate-level university student at either the University of Alberta or the University of Calgary with a strong background in biomedical science; continuing university studies in the fall in any discipline; demonstrate excellent writing skills; and work well independently as well as part of a team.
Geographic Focus: All States, Canada
Date(s) Application is Due: Mar 23
Contact: Dwayne Brunner, Interim Media Relations Manager; (780) 423-5727, 224; fax (780) 429-3509; dwayne.brunner@albertainnovates.ca
Internet: http://www.aihealthsolutions.ca/grants/training-and-early-career-development-programs/media-fellowship/
Sponsor: Alberta Innovates - Health Solutions
10104-103 Avenue, Suite 1500
Edmonton, AB T5J 4A7 Canada

AIHS Part-Time Fellowships 393
The AIHS Part-Time Fellowships enable highly qualified doctoral graduates to prepare for careers in medical or health research as independent investigators doing part-time research training during the regular academic year while enrolled in a full-time professional degree program in Alberta. Part-time awards will consist of a stipend only and will be pro-rated to the amount of time the candidate is prepared to commit to research. The full-time fellowship rate will be used as the basis for this calculation. The maximum term for the award is three years.
Requirements: Candidates must hold a PhD and must be engaged in a professional degree program (usually the MD program).
Restrictions: Part-time fellowship awards are tenable only at an Alberta-based university.
Geographic Focus: Canada
Date(s) Application is Due: Mar 1; Oct 1
Amount of Grant: 5,000 - 30,000 USD
Contact: Pamela Valentine; (780) 423-5727, ext. 230; fax (780) 429-3509; pamela.valentine@albertainnovates.ca or grants.health@albertainnovates.ca
Internet: http://www.aihealthsolutions.ca/grants/PT-fellow.php
Sponsor: Alberta Innovates - Health Solutions
10104-103 Avenue, Suite 1500
Edmonton, AB T5J 4A7 Canada

AIHS Part-Time Studentships 394
The Alberta Innovates - Health Solutions (AIHS) Part-Time Studentships enable students enrolled in a full-time professional degree program (usually the M.D. program) in Alberta to engage in part-time research training during the regular academic year. The award consists of a stipend only. Part-time students may elect to engage in full-time research during the summer; these students should contact AIHS to request that full-time summer trainee status be implemented.
Requirements: Candidates must normally have been accepted into, or be currently engaged in, a full-time graduate program at an Alberta-based university in a health-related discipline leading to a Master's or doctoral degree. It is expected that the amount of time the candidate will commit to research will not be less than the equivalent of one day per week, taken either as one full day or two half days.
Restrictions: The award is tenable at an Alberta-based university only.
Geographic Focus: Canada
Date(s) Application is Due: Mar 1; Oct 1
Contact: Pamela Valentine; (780) 423-5727, ext. 230; fax (780) 429-3509; pamela.valentine@albertainnovates.ca or grants.health@albertainnovates.ca
Internet: http://www.aihealthsolutions.ca/grants/PT-student.php
Sponsor: Alberta Innovates - Health Solutions
10104-103 Avenue, Suite 1500
Edmonton, AB T5J 4A7 Canada

AIHS Proposals for Special Initiatives 395
The AIHS welcomes proposals for innovative initiatives that accelerate the achievement of its overall objectives. Institutions recruiting internationally-recognized researchers to lead new initiatives should consult AIHS, if current programs are inappropriate. AIHS will also consider proposals for pilot projects in patient or population-based research. These projects must have the potential to develop information, and the investigators must have the expertise to attract grants from other agencies. The review process and financial support for special initiatives may be shared by AIHS and other agencies.
Restrictions: Applications that are normally eligible for regular operating grants from other agencies will not be considered under this program.
Geographic Focus: All States, Canada
Contact: Jacques Magnan, Vice President, Research; (780) 423-5727; fax (780) 429-3509; jmagnan@albertainnovates.ca
Internet: http://www.aihealthsolutions.ca/grants/initiatives.php
Sponsor: Alberta Innovates - Health Solutions
10104-103 Avenue, Suite 1500
Edmonton, AB T5J 4A7 Canada

AIHS Research Prize 396
The AIHS Research Prize is intended to maintain and improve AIHS Personnel Awards as a potent vehicle for the recruitment and retention of highly qualified, internationally competitive investigators at Alberta institutions. The Prize will be made to every eligible investigator supported by AIHS (Population Health Investigator, Clinical Investigator, Scholar, Senior Scholar, Scientist). The Research Prize is not determined by institutional rank and salary scale. It is provided in recognition of the individual investigator's research accomplishments, and of his/her success in obtaining an AIHS Independent Investigator award. The Research Prize is a personal prize to the recipient. No application is required.
Requirements: To be eligible, the investigator must be currently supported by an unconditional AIHS investigator award. Investigators receiving AIHS terminal support or those on leaves of absence are not eligible to receive the Research Prize.
Geographic Focus: All States, Canada
Amount of Grant: 10,000 - 20,000 USD
Contact: Pamela Valentine, Vice President of Research and Innovation; (780) 423-5727; fax (780) 429-3509; pamela.valentine@albertainnovates.ca
Internet: http://www.aihealthsolutions.ca/grants/resprize.php
Sponsor: Alberta Innovates - Health Solutions
10104-103 Avenue, Suite 1500
Edmonton, AB T5J 4A7 Canada

AIHS Summer Studentships 397
The AIHS Summer Studentships offer motivated students with exceptional academic records an opportunity to participate in medical or health research in Alberta during the summer months. The award is meant to encourage students to consider pursuing formal training and a career in health research. The award is tenable for a minimum of two months, and a maximum of four months during the period of May to August. The award consists of a stipend of $1300 per month. Applications are available at the AIHS website.
Requirements: Candidates must meet one of the following criteria to be eligible to apply: registered in an Alberta-based undergraduate degree program in a medical or health-related field; registered in an undergraduate degree program outside of Alberta, and desiring to engage in research during the summer at an Alberta institution; registered in an M.D. program, and who may also hold an undergraduate or graduate degree; exceptional high school students with records of participation in the health care system, and a clear interest in pursuing a health research career; or currently in the last term of their undergraduate degrees and who have applied either to medical school or to a graduate program that would start in the coming fall semester.
Restrictions: A faculty supervisor possessing both a record of productive health-oriented research and sufficient resources to ensure the satisfactory conduct of the research must sponsor the candidate. The supervisor may sponsor no more than two summer students per year, and each student must have a separate project. These awards are not renewable, but students may reapply in subsequent years.
Geographic Focus: All States, Canada
Contact: Pamela Valentine, Vice President of Research and Innovation; (780) 423-5727; fax (780) 429-3509; pamela.valentine@albertainnovates.ca
Internet: http://www.aihealthsolutions.ca/grants/Sum-student.php
Sponsor: Alberta Innovates - Health Solutions
10104-103 Avenue, Suite 1500
Edmonton, AB T5J 4A7 Canada

AIHS Sustainability Grants 398
The Sustainability Funding Opportunity provides bridge funding for Alberta health researchers who have submitted renewal applications that, although highly rated, were not funded in the CIHR Open Operating Grants or NSERC Discovery Grant competitions. The objective of these bridge grants is to ensure the maintenance of excellent research programs without loss of momentum, staff, or trainees until the Principal Investigators have the opportunity to resubmit their research proposals. A webinar discussing the grants is available at the AIHS website.
Requirements: The grant recipient must be a full-time university-based independent investigator. The recipient must also have applied to a Tri-Council agency for a peer-reviewed, individual research grant renewal in the CIHR Open Operating Grant or NSERC Discovery Grant competitions within the prior year, and was deemed excellent but was not funded. The applicant must have at least one more opportunity for reapplication of grant renewal with the external agency. The initial renewal application to the Tri-Council agency's operating grant competition was internally peer-reviewed according to processes defined at the institution prior to submission to the Tri-Council agency. Recipients must provide specific plans for a future application to the Vice President Research of their university.
Restrictions: The Sustainability Funding Program will not supplement existing grants or cover shortfalls in approved grant support. Each institution will work through their Office of the Vice President Research for the review of applications and selection of health researchers to be offered support under the Sustainability Funding Opportunity according to the defined eligibility criteria. Applications will be evaluated by the university receiving a Sustainability Funding allocation, not by AIHS.
Geographic Focus: All States, Canada
Amount of Grant: Up to 50,000 USD
Contact: Pamela Valentine, Vice President of Research and Innovation; (780) 423-5727, ext 230; fax (780) 429-3509; pamela.valentine@albertainnovates.ca
Internet: http://www.aihealthsolutions.ca/grants/sustainability.php
Sponsor: Alberta Innovates - Health Solutions
10104-103 Avenue, Suite 1500
Edmonton, AB T5J 4A7 Canada

Air Products and Chemicals Grants 399
The foundation supports nonprofit organizations in company-operating areas in the fields of precollege and higher education; fitness, health, and welfare; community and economic development, arts and culture; and the environment and safety. Types of support include capital grants, employee matching gifts, general operating support, multiyear-continuing grants, project grants, seed grants, fellowships, employee-related scholarships, and donated equipment. There are no application forms; requests must be in writing. Guidelines are available online.
Requirements: 501(c)3 nonprofit organizations in company-operating areas are eligible.
Restrictions: Grants are not made to/for individuals, sectarian or denominational organizations, political candidates or activities, veterans organizations, organizations receiving United Way support, labor groups, elementary or secondary schools, capital campaigns of national organizations, hospital operating expenses, national health organizations, or goodwill advertising.
Geographic Focus: All States
Contact: Kassie Hilgert, Program Manager; hilgerk@airproducts.com
Internet: http://www.airproducts.com/Responsibility/SocialResponsibility
Sponsor: Air Products and Chemicals Corporation
7201 Hamilton Boulevard
Allentown, PA 18195-1501

Alabama Power Foundation Grants 400
The foundation's mission is to improve the lives and circumstances of Alabama residents and to strengthen the communities in which they live. It supports programs that will improve education (by supporting innovative programs and assisting teachers in their crucial responsibilities), strengthen communities, promote arts and culture or restore and enhance the environment. The project must meet several of the guidelines to be considered. The majority of grants are made to support targeted efforts with specific objectives. The foundation also provides general operating assistance to organizations, capital grants for endowment and building, and scholarship funds. Applications should be sent to the local Alabama Power Company office for review and recommendation to the foundation. The amount of the request determines the review frequency.
Requirements: Applications are accepted from Alabama nonprofit organizations whose programs fall within foundation guidelines.
Restrictions: Grants are not made to individuals, for sectarian religious purposes, or for political activities.
Geographic Focus: Alabama
Date(s) Application is Due: Feb 1; May 1; Aug 1; Nov 1
Amount of Grant: 1,000 - 100,000 USD
Contact: William Johnson, President; (205) 257-2508; fax (205) 257-1860
Internet: http://www.alabamapower.com/foundation/grantsandinitiatives.asp
Sponsor: Alabama Power Foundation
600 N 18th Street, P.O. Box 2641
Birmingham, AL 35291-0011

ALA Donald G. Davis Article Award 401
The Donald G. Davis Article Award is presented by the Library History Round Table of the American Library Association every second year to recognize the best article written in English in the field of United States and Canadian library history including the history of libraries, librarianship, and book culture. Entries are judged on quality of scholarship, clarity of style, depth of research, and ability to place research findings in a broad social, cultural, and political context. One award will be given every second year unless the jury does not find a suitable candidate for that biennial period. Any member of the Library History Round Table may nominate one or more articles by sending a recommendation to the Chair of the Davis Award Committee. Also, the editor of the LHRT Newsletter Library History Bibliography will submit a list of candidates. The winner will be announced in a press release on or about June 1st of the award year. Certificates honoring the author(s) and journal will be presented at a Library History Round Table awards ceremony during the American Library Association Annual Conference in the year of the award.
Requirements: Entries for each biennial award must have been published between January 1 and December 31 of the two years preceding the award year.
Restrictions: Papers that have won the Justin Winsor or Jesse Shera Awards are not eligible for consideration.
Geographic Focus: All States
Date(s) Application is Due: Jan 15
Contact: Norman Rose; (312) 280-4283; fax (312) 280-4392; nrose@ala.org
Internet: http://www.ala.org/awardsgrants/awards/138/apply
Sponsor: American Library Association
50 E Huron Street
Chicago, IL 60611-2795

Alaska Airlines Corporate Giving Medical Emergency and Research Grants 402
Alaska Airlines and Horizon Air Corporate Giving support health and human services, arts and cultural programs, as well as education, environmental, and civic organizations. The program focuses on communities served now or in the near future, and where a significant number of its employees live or work. When considering requests, the corporation favors organizations and efforts that are most likely to enhance a community's cultural and economic vitality and improve the quality of life for its citizens. Objectives of the Corporate Giving Program are to: provide support to charitable and cultural organizations within our communities; and maintain a reasonable balance of contributions supporting a wide range of organizations. In the areas of Medical Emergency and Research, the corporation regularly responds to the need for health and human services relief, especially in the state of Alaska, where transportation infrastructure limits access to medical facilities. Through in-kind contributions, the giving program assists hundreds of individual emergency and medical transportation needs. Its partnerships with Angel Flight West and the Shriners' Hospitals provide well over 1,000 complimentary passenger seats each year for those needing treatment in another city.
Requirements: Giving is on a national and international basis in areas of company operations, with emphasis on Alaska, Oregon, and Washington; some further giving occurs in Canada and Mexico.
Restrictions: Alaska Airlines or Horizon Air do not provide support for: loans and grants to individuals or private business; groups that discriminate; endowments; religious organizations (if for sacramental purposes); pageants; capital projects; multi-year commitment or automatic renewal grants; general operating expenses; publicly or privately funded educational institutions; or organizations whose prime purpose is to influence legislation.
Geographic Focus: Alaska, Oregon, Washington, Canada, Mexico
Contact: Tim R. Thompson, Manager, Public Affairs; (907) 266-7230; fax (907) 266-7229; tim.thompson@alaskaair.com
Internet: https://www.alaskaair.com/content/about-us/social-responsibility/corporate-giving.aspx
Sponsor: Alaska Airlines
4750 Old International Airport Road
Anchorage, AK 99502

60 | GRANT PROGRAMS

Alaska Airlines Foundation Grants 403

A small number of cash grants ranging on average from $5,000 to $15,000 are given to 501(c)3 non-profit organizations classified as public charities in Alaska, Hawaii, and Washington. These grants should focus on educational efforts that address a unique need or value to a community. The Foundation supports health and human services, arts and cultural programs, as well as education, environmental, and civic organizations. Letters of inquiry must be received by April 15 for mid-year consideration and by September 15 for consideration before the end of the calendar year. A response will generally be provided within sixty (60) days.
Requirements: 501(c)3 organizations based in Alaska, Hawaii, and Washington are eligible to apply. Unsolicited applications are not encouraged. Those interested in applying to the Foundation should send a letter of inquiry before preparing a full application. Letters of inquiry must be received by September 30 of the calendar year to be considered for possible funding.
Restrictions: Alaska Airlines or Horizon Air do not provide support for: loans and grants to individuals or private business; groups that discriminate; endowments; religious organizations (if for sacramental purposes); pageants; capital projects; multi-year commitment or automatic renewal grants; general operating expenses; publicly or privately funded educational institutions; or organizations whose prime purpose is to influence legislation.
Geographic Focus: Alaska, Hawaii, Washington
Date(s) Application is Due: Apr 15; Sep 15
Amount of Grant: 5,000 - 15,000 USD
Contact: Tim R. Thompson, Executive Director; (907) 266-7230; fax (907) 266-7229; tim.thompson@alaskaair.com
Internet: http://www.alaskaair.com/as/www2/company/csr/as-foundation.asp
Sponsor: Alaska Airlines Foundation
4750 Old International Airport Road
Anchorage, AK 99502

Albany Medical Center Prize in Medicine and Biomedical Research 404

The prize recognizes a physician or biomedical scientist (or group of physicians or scientists) who has made extraordinary and sustained leadership contributions to improving healthcare and patient care; or who has successfully pursued innovative biomedical research with demonstrated translational benefits applied to improved patient care. Each year's prize winner will have demonstrated significant outcomes that offer medical value of national or international importance. Prize winner activities will include but not be limited to disease and injury management, clinical research, and basic science investigations of diseases and injuries, leading to new discoveries and improved clinical outcomes. Nomination guidelines are available online.
Requirements: Any physician or scientist or group whose work has led to significant advances in the fields of health care and scientific research with demonstrated translational benefits applied to improved patient care may be nominated. Those honored will be practitioners and/or scientists whose accomplishments and outcomes have been demonstrated in the past quarter century, with preference to demonstrated accomplishments in the past decade.
Geographic Focus: All States
Date(s) Application is Due: Jan 8
Amount of Grant: 500,000 USD
Contact: Teri A. Cerveny; (518) 262-8043; fax (518) 262-4769; AMCprize@mail.amc.edu
Internet: http://www.amc.edu/Academic/AlbanyPrize/prize_criteria.html
Sponsor: Albany Medical Center
628 Madison Avenue, Center Building, 1st Floor
Albany, NY 12208

Albert and Margaret Alkek Foundation Grants 405

The foundation awards grants to Texas nonprofit organizations to support charitable, religious, scientific (primarily medical), literary, cultural and educational organizations and programs. Preference will be given to research and education-related projects that will pay lasting dividends in terms of new discoveries and improved quality of life. One application per 12-month period will be considered. There are no application deadlines or forms. Applicants should submit a one- to two-page letter of inquiry that includes a brief description of the organization and the project for which funds are being considered, and the amount of funding need in total as well as the amount being requested. Inquiries should be sent via U.S. postal mail; fax or email applications are not accepted.
Requirements: Texas nonprofit organizations are eligible.
Restrictions: The foundation does not make grants to individuals or loans of any type. The foundation does not make direct scholarships to students. All scholarship programs are administered through educational institutions. The foundation prefers not to fund: organizations that in turn make grants to others; grants intended to influence legislation or to support candidates for political office; fund-raising events such as luncheons, dinners, galas, advertising in programs, or other similar activities; charities operated by service clubs; memorials for individuals; student organizations; or purchase of uniforms, equipment, or trips for school-related organizations or sports teams.
Geographic Focus: Texas
Samples: Baylor College of Medicine (Houston, TX)--for research on cancer, cardiovascular science, diabetes, pharmacogenomics, and other biomedical topics, $31.25 million.
Contact: Grants Administrator; (713) 951-0019; fax (713) 951-0043; info@alkek.org
Internet: http://www.alkek.org/grantguidelines.htm
Sponsor: Albert and Margaret Alkek Foundation
1221 McKinney, Suite 4525
Houston, TX 77010-2023

Albert and Mary Lasker Foundation Awards 406

The major purpose of the awards is to honor the individual(s) who have made significant contributions in basic or clinical research in diseases that are the main cause of death and disability. Four categories of awards are made. The Albert Lasker Basic Medical Research Award honors the scientist or scientists who have made fundamental investigations that open new areas of biomedical science. The Lasker-DeBakey Clinical Medical Research Award honors the scientist(s) whose contributions, directly or indirectly, have led to the improvement of the clinical management or treatment of patients and to the alleviation or elimination or one of the major medical causes of disability or death. The Lasker-Koshland Special Achievement Award in Medical Science honors a scientist whose contributions to research are of unique magnitude and immeasurable influence on the course of science, health, or medicine, and whose professional career has engendered extreme respect within the biomedical community. And the Lasker~Bloomberg Public Service Award honors men and women who have helped make possible the federal legislation and funding that supports research, and who have created public communication, public health, and advocacy programs of major importance. Nominations must be received by the listed application deadline.
Geographic Focus: All States
Date(s) Application is Due: Feb 1
Amount of Grant: 25,000 - 50,000 USD
Contact: David Keegan, Senior Program Director; (212) 286-0222; fax (212) 286-0924; dkeegan@laskerfoundation.org
Internet: http://www.laskerfoundation.org/awards/index.htm
Sponsor: Albert and Mary Lasker Foundation
110 East 42nd Street, Suite 1300
New York, NY 10017

Albert and Mary Lasker Foundation Clinical Research Scholars 407

The Lasker Foundation and the National Institutes of Health have joined together in an innovative partnership to nurture the next generation of clinical researchers. The Lasker Clinical Research Scholars Program is an NIH-funded initiative providing up to 11 years of support for medical doctors to bridge the widening gap between cutting-edge research and improved patient care. Under the program, NIH Institutes will specify clinical research topics that merit support and accept applications each year from early-career investigators in those areas. Applications will be reviewed for scientific merit by a peer review group of scientists, convened by the NIH Center for Scientific Review, and make recommendations to the NIH scientific and clinical directors who will annually select outstanding scholars for placement on the NIH tenure-track. The NIH expects to appoint at least three Lasker Scholars in the first year of the program, and to ultimately reach a steady-state of 20-30 active awardees. The Lasker Foundation will assist with providing mentors for the scholars, fund travel costs for NIH lecturers and the attendance by scholars at the Lasker Awards program. The Foundation will also support annual meetings for presentations and discussions on topics related to the program. Scholars' work will receive regular evaluation at NIH. After five to seven years, each scholar will be reviewed again by a group of distinguished clinical investigators. By mutual agreement between the NIH and the candidate, a decision will be made as to whether the scholar will stay on as a tenured senior investigator at the NIH, or whether the candidate will be eligible for outside research grant support worth $500,000 a year plus overhead for four years.
Geographic Focus: All States
Contact: David Keegan, Senior Program Director; (212) 286-0222; fax (212) 286-0924; dkeegan@laskerfoundation.org
Internet: http://www.laskerfoundation.org/programs/index.htm
Sponsor: Albert and Mary Lasker Foundation
110 East 42nd Street, Suite 1300
New York, NY 10017

Alberto Culver Corporate Contributions Grants 408

The company prefers to make small donations to a large number of organizations rather than substantial contributions to a few institutions. Major areas of support are civic and community programs, health and welfare, education, culture and art, and youth activities. Special consideration is given to proposals that benefit a large number of people; directly or indirectly assist women's groups or other purchasers or potential customers of Alberto-Culver products; are involved in programs that rehabilitate, train, teach skills or employ the underprivileged, handicapped or minorities; have received support from Alberto-Culver in the past; and received matching grants for donations. Grants are awarded for one year and may be renewed.
Requirements: Only 501(c)3 organizations are eligible. Requests should be in writing on organization letterhead and should state amount requested, and whether the donation will be used for operating funds, capital expansion, supplies or special projects. The objectives and programs of the organization should be clearly defined.
Restrictions: Grants are not awarded to support United Way affiliates, religious groups, preschools, K-12 schools, tax-supported colleges, projects that duplicate other efforts, and multiyear commitments.
Geographic Focus: All States
Contact: V. James Marino, Director; (708) 450-3000; fax (708) 450-3435
Internet: http://www.alberto.com
Sponsor: Alberto Culver
2525 Armitage Avenue
Melrose Park, IL 60160

Albert Pick Jr. Fund Grants 409

The fund contributes to organizations in the following categories: culture, education, health and human services, civic and community organizations. The fund will consider assistance to new and creative programs within these areas and to operating support. Support will be given to requests from organizations that conduct their programs in Chicago, IL, only. The fund functions on a calendar year; applications are considered four times a year based on dates of board meetings. The fund prefers that prospective grantees call or write for guidelines prior to sending their proposals.
Requirements: Only Illinois organizations may apply.
Restrictions: Funds will not be provided for reduction or liquidation of debts, religious purposes, endowments, long-term commitments, building programs, individuals, political purposes, or advertising/program books.
Geographic Focus: Illinois
Date(s) Application is Due: Jan 21; Apr 1; Jul 1; Oct 1
Amount of Grant: 3,000 USD
Samples: Children First Fund/CPS Foundation (Gale Academy Links to Literacy Program) $6900; Art Resources in Teaching (Chicago, IL)--for general operating support, $7500; Chicago Academy of Sciences/Peggy Notebaert Nature Museum (Chicago, IL)--for general operating support, $10,000; Chicago Opera Theater (Chicago, IL)--for general operating support, $10,000.
Contact: Cleopatra Alexander; (312) 236-1192; cleopatra@albertpickjrfund.org
Internet: http://www.albertpickjrfund.org
Sponsor: Albert Pick Jr. Fund
30 N Michigan Avenue, Suite 1002
Chicago, IL 60602

Albertson's Charitable Giving Grants 410

Albertson's Inc invests in its operating communities and makes corporate contributions to support nonprofits. Areas of charitable giving include health and hunger relief--food banks, churches, and other community-based relief groups; health and nutrition--medical services, flu shots, health screening for diseases such as diabetes and heart disease; and education and development of youth--academic excellence or nurturing efforts. Submit requests in writing and include information about the organization's goals, accomplishments, evaluation plans, leadership, and finances. Requests are accepted throughout the year.
Requirements: Tax-exempt organizations in Arizona, Arkansas, California, Colorado, Delaware, Florida, Georgia, Idaho, Illinois, Indiana, Iowa, Kansas, Louisiana, Maine, Maryland, Massachusetts, Michigan, Minnesota, Mississippi, Missouri, Montana, Nebraska, Nevada, New Hampshire, New Jersey, New Mexico, North Dakota, Oklahoma, Oregon, Pennsylvania, South Dakota, Tennessee, Texas, Utah, Vermont, Washington, Wisconsin, and Wyoming are eligible. Applicants must pass an online eligibility test. Preference will be given to requests that offer volunteerism opportunities.
Restrictions: Contributions cannot be made to churches or religious organizations for purposes of religious advocacy.
Geographic Focus: Arizona, Arkansas, California, Colorado, Delaware, Florida, Georgia, Idaho, Illinois, Indiana, Iowa, Kansas, Louisiana, Maine, Maryland, Massachusetts, Michigan, Minnesota, Mississippi, Missouri, Montana, Nebraska, Nevada, New Hampshire, New Jersey, New Mexico, North Dakota, Oklahoma, Oregon, Pennsylvania, South Dakota, Tennessee, Texas, Utah, Vermont, Washington, Wisconsin, Wyoming
Contact: Community Relations Manager; (877) 932-7948 or (208) 395-6200; fax (208) 395-4382; albertsonscustomercare@albertsons.com
Internet: http://www.albertsons.com/abs_inthecommunity
Sponsor: Albertson's
250 E Parkcenter Boulevard
Boise, ID 83706

Albert W. Rice Charitable Foundation Grants 411

The Albert W. Rice Charitable Foundation was established in 1959 to support and promote quality educational, human-services, and health-care programming for underserved populations. In the area of education, the foundation supports academic access, enrichment, and remedial programming for children, youth, adults, and senior citizens that focuses on preparing individuals to achieve while in school and beyond. In the area of health care, the foundation supports programming that improves access to primary care for traditionally underserved individuals, health education initiatives and programming that impact at-risk populations, and medical research. In the area of human services the foundation tries to meet evolving needs of communities. Currently the foundation's focus is on (but is not limited to) youth development, violence prevention, employment, life-skills attainment, and food programs. Grant requests for general operating support are strongly encouraged. Program support and occasional capital support will also be considered. Special consideration is given to charitable organizations that serve the people of Worcester, Massachusetts, and its surrounding communities. The majority of grants from the Rice Foundation are one year in duration; on occasion, multi-year support is awarded. Applicants must apply online at the grant website. Applicants are strongly encouraged to do the following before applying: review the downloadable state application procedures for additional helpful information and clarifications; review the downloadable online-application guidelines at the grant website; review the foundation's funding history (link is available from the grant website); review the online application questions in advance; and review the list of required attachments. These will generally include: a list of board members, financial statements (audited, reviewed, or compiled by independent auditor); an organization summary; a list of other funding sources; an IRS Determination letter; and other required documents. All attachments must be uploaded in the online application as PDF, Word, or Excel files. The application deadline for the Albert W. Rice Charitable Foundation is 11:59 p.m. on July 1. Applicants will be notified of grant decisions before September 30.
Requirements: Applicants must have 501(c)3 tax-exempt status.
Restrictions: The foundation does not support requests from individuals, organizations attempting to influence policy through direct lobbying, or any political campaigns.
Geographic Focus: Massachusetts
Date(s) Application is Due: Jul 1
Samples: Big Brothers, Big Sisters of Central Massachusetts Metrowest, Worcester, Massachusetts, $30,000, program development; Youth Opportunities Upheld, Worcester, Massachusetts, $20,000, Dynamic Youth Academy; Legal Assistance Corporation of Central Massachusetts, Worcester, Massachusetts, $20,000.
Contact: Michealle Larkins; (866) 778-6859; michealle.larkins@baml.com
Internet: https://www.bankofamerica.com/philanthropic/fn_search.action
Sponsor: Albert W. Rice Charitable Foundation
225 Franklin Street, 4th Floor, MA1-225-04-02
Boston, MA 02110

Albuquerque Community Foundation Grants 412

The foundation seeks to improve the quality of life in the greater Albuquerque, New Mexico, area by providing support for projects and organizations that serve the community in arts and culture, education, environmental and historic preservation, children and youth, and health and human services. Through its grant program, the foundation supports projects that are innovative, meet the needs of underserved segments of the community, encourage matching funds or additional gifts, promote cooperation among agencies, empower the disadvantaged and disabled, and enhance the effectiveness of local charitable organizations. Types of support include continuing support, general operating support, program development, publication, seed grants, scholarships funds and scholarships to individuals, and technical assistance.
Requirements: IRS 501(c)3 organizations based in Albuquerque, NM, are eligible. Proposals are reviewed on the basis of the following priorities: impact, innovation, leverage, management, and nonduplication.
Restrictions: Grants are generally not made to or for individuals, political or religious purposes, debt retirement, payment of interest or taxes, annual campaigns, endowments, emergency funding, to influence legislation or elections, scholarships, awards, or to private foundations and other grantmaking organizations.
Geographic Focus: New Mexico
Date(s) Application is Due: Apr 16; Aug 15
Amount of Grant: Up to 10,000 USD
Contact: R. Randall Royster, Executive Director; (505) 883-6240; fax (505) 883-3629; rroyster@albuquerquefoundation.org
Internet: http://www.albuquerquefoundation.org/grants/grant-home.htm
Sponsor: Albuquerque Community Foundation
P.O. Box 36960
Albuquerque, NM 87176-6960

Alcatel-Lucent Technologies Foundation Grants 413

The Alcatel-Lucent Foundation is the philanthropic arm of Alcatel-Lucent and it leads the company's charitable activities. With a focus on volunteerism, the Foundation's mission is to support the commitment of Alcatel-Lucent to social responsibility by serving and enhancing the communities where its employees and customers live and work. To accomplish its mission, the Foundation manages grants and employee volunteerism on a global level. It receives its income from the corporation - Alcatel-Lucent - whose name it bears. However, legally the Foundation is an independent, charitable, non-profit and private entity and is governed by its own board of trustees that is separate from the corporate board of directors. Global Foundation grants are dedicated to the main focus areas of the Foundation and are managed by the Foundation.
Requirements: Giving is on an international basis.
Geographic Focus: All States, All Countries
Amount of Grant: 500 - 5,000 USD
Contact: Bishalakhi Ghosh, Executive Director; +91-99-58418547 or +91-22-66798700; fax +91-22-26598542; bishalakhi.ghosh@alcatel-lucent.com
Internet: http://www.alcatel-lucent.com/wps/portal/foundation
Sponsor: Alcatel-Lucent Technologies Foundation
600 Mountain Avenue, Room 6F4
Murray Hill, NJ 07974-2008

Alcoa Foundation Grants 414

General priorities of the foundation include safe and healthy children and families--ensuring that children and their families have the tools, the knowledge and the services to remain healthy and safe at home, in the community and in the workplace; conservation and sustainability--educating young leaders on conservation issues, protecting forests, promoting sound public policy research, and understanding the linkages between business and the environment; skills today for tomorrow--providing individuals with critical skills and services to be economically connected, workplace-ready, and productive in a changing economy; business and community partnerships--strengthening the nonprofit sector and developing meaningful partnerships among nonprofits, the private sector, and local government; and global education in science, engineering, technology and business--broadening student participation in areas central to Alcoa to prepare a diverse cross-section of our communities for a global workplace. Types of support include capital grants, building funds, challenge grants, matching gifts, general support, research grants, scholarships, and seed money. Initial contact should be a letter of inquiry.
Requirements: The foundation awards grants to nonprofit public charities in communities where Alcoa has a presence. Local Alcoans work within their communities to evaluate organizations and make recommendations for funding to Alcoa Foundation. Nonprofit organizations that serve localized communities should find the Alcoa facility nearest to

them and write a one-page letter describing their mission, nature of request, connection to the areas of excellence and offering contact information. If interested, the Alcoa location contact will notify the requesting organization and invite them to submit more information. Areas of operation include western Pennsylvania; Davenport, Iowa; Evansville, Indiana; Massena, New York; New Jersey; Cleveland, Ohio; Knoxville, Tennessee; and Rockdale, Texas.
Restrictions: The foundation does not make gifts to local projects other than those near Alcoa plant or office locations; endowment funds, deficit reduction, or operating reserves; hospital capital campaign programs unless the hospital presents a comprehensive area analysis that justifies, on a regional rather than an individual institutional basis, the need for the capital improvement; individuals, except for the scholarship program for children of Alcoa employees; tickets and other promotional activities; trips, tours, or student exchange programs; or documentaries and videos.
Geographic Focus: All States
Amount of Grant: 1,000 - 50,000 USD
Contact: Meg McDonald; (412) 553-2348; fax (412) 553-4498; alcoa.foundation@alcoa.com
Internet: http://www.alcoa.com/global/en/community/foundation.asp
Sponsor: Alcoa Foundation
201 Isabella Street
Pittsburgh, PA 15212-5858

Alcohol Misuse and Alcoholism Research Grants 415
The foundation's research interests include factors influencing transitions in drinking patterns and behavior, effects of moderate use of alcohol on health and well-being, mechanisms underlying the behavioral and biomedical effects of alcohol, biobehavioral/interdisciplinary research on the etiology of alcohol misuse. Applications may be obtained from the Web site or by contacting the office. Grantees are usually notified within two weeks following the advisory council meetings, which are held in April and November.
Requirements: Applications may be submitted by public or private nonprofit organizations such as universities, colleges, hospitals, research institutes and organizations, and governmental research agencies and laboratories in the United States and Canada.
Restrictions: Non-research activities such as education projects, public awareness efforts and treatment or referral services are not eligible for support. The Foundation also does not support the training of pre- and post-doctoral fellows, undergraduates, graduate students, medical students, interns or residents. It does not fund thesis or dissertation research. The Foundation does not encourage applications on treatment of the complications of advanced alcoholism. However, research involving treatment intended to elucidate the pathogenesis of alcohol-related problems will be considered.
Geographic Focus: All States
Date(s) Application is Due: Feb 1; Sep 1
Amount of Grant: Up to 100,000 USD
Samples: Boston U (MA)--to study alcohol abuse among college students, $38,500; Bowman School of Medicine (Winston-Salem, NC)--for research on mechanisms in the transition and development of chronic alcohol consumption-induced cardiomyopathy, $39,000; Cleveland Clinic Foundation (OH)--for research in sub-cellular homeostasis in the heart: effects of chronic ethanol consumption, $40,000.
Contact: Grants Administrator; (410) 821-7066; fax (410) 821-7065; info@abmrf.org
Internet: http://www.abmrf.org/grants.htm
Sponsor: Alcoholic Beverage Medical Research Foundation (ABMRF)
1122 Kenilworth Drive, Suite 407
Baltimore, MD 21204

Alcon Foundation Grants 416
The foundation supports organizations in the fields of health care, leadership programs, research, education and community responsibility. Programs that advance the education and skill levels of eye care professionals are given special consideration. General operating grants to organizations and institutions improving education and research in the areas of specialization of Alcon Laboratories--ophthalmology and vision care. Grants are also awarded to community activities that benefit company employees. Applications may be submitted at any time.
Requirements: Grants are not made for building programs.
Restrictions: Non 501(c)3 organizations, individuals and scholarship programs, religious, veterans or fraternal organizations, political causes, capital campaigns, matching gifts, trips, tournaments and tours, and endowments are not supported.
Geographic Focus: All States
Amount of Grant: 100 - 50,000 USD
Contact: Mary Dulle, Chair; (817) 293-0450; Mary.Dulle@Alconlabs.com
Internet: http://www.alconlabs.com/corporate-responsibility/alcon-foundation.asp
Sponsor: Alcon Foundation
6201 S Freeway
Fort Worth, TX 76134

Alexander & Baldwin Foundation Mainland Grants 417
The foundation awards grants to eligible U.S. nonprofit organizations in its areas of interest, including health and human services, education, the community, culture and arts, the maritime arena and the environment. Grant preferences are given to organizations and projects that address significant community needs, have the active support of A&B employees, are preventive in nature, and have demonstrated support of the community. Start-up, general operating and special project needs, as well as major and minor capital requests are considered. Although the majority of grants range between $1,000 and $5,000, the Foundation considers request upward of $20,000.
Requirements: The foundation gives funding to community-based projects and organizations that qualify with 501(c)3 status.
Restrictions: Grants are not awarded to support United Way agencies for operating support, individuals, events, travel expenses, or scholarships.
Geographic Focus: All States
Date(s) Application is Due: Jan 1; Feb 1; Mar 1; Apr 1; May 1; Jun 1; Jul 1; Aug 1; Sep 1; Oct 1; Nov 1; Dec 1
Amount of Grant: 1,000 - 5,000 USD
Contact: Paul L. Merwin, (707) 421-8121; fax (707) 421-1835; plmifm@aol.com
Internet: http://www.alexanderbaldwinfoundation.org/appguide.htm
Sponsor: Alexander and Baldwin Foundation
555 12th Street
Oakland, CA 94607

Alexander and Baldwin Foundation Hawaiian and Pacific Island Grants 418
The Alexander and Baldwin Foundation awards grants to eligible Hawaii nonprofit organizations in its areas of interest, including health and human services, education, the community, culture and arts, the maritime arena and the environment. Grant preferences are given to organizations and projects that address significant community needs, have the active support of A&B employees, are preventive in nature, and have demonstrated support of the community. Start-up, general operating and special project needs, as well as major and minor capital requests are considered. Deadlines listed are for organizations in Hawaii and the Pacific Islands. Although the majority of grants range between $1,000 and $5,000, the Foundation considers request upward of $20,000.
Requirements: The foundation gives funding to community-based projects and organizations that qualify with 501(c)3 status.
Restrictions: Grants are not awarded to support United Way agencies for operating support, individuals, events, travel expenses, or scholarships.
Geographic Focus: Hawaii
Date(s) Application is Due: Feb 1; Apr 1; Jun 1; Aug 1; Oct 1; Dec 1
Amount of Grant: 1,000 - 5,000 USD
Contact: Linda M. Howe, (808) 525-6642; fax (808) 525-6677; lhowe@abinc.com
Internet: http://alexanderbaldwin.com/corporate-responsibility/commitment/
Sponsor: Alexander and Baldwin Foundation
P.O. Box 3440
Honolulu, HI 96801-3440

Alexander and Margaret Stewart Trust Grants 419
The trust awards grants in the greater Washington, D.C., area for cancer treatment, especially equipment used in diagnosis and treatment; and caring for children who are physically ill, mentally ill, or disabled. Grants also support research, education, and prevention of common childhood diseases, including negative societal behavioral patterns that impact children. Proposals for projects aiding the economically deprived receive preference. The trust awards start-up funding. Applications may be submitted at any time; requests received by the listed deadline date are reviewed by the end of the year.
Requirements: Nonprofits in the greater District of Columbia area are eligible.
Restrictions: Requests for support of endowments, buildings, or capital campaigns are denied.
Geographic Focus: District of Columbia
Date(s) Application is Due: Sep 15
Amount of Grant: 50,000 - 150,000 USD
Sample: Childrens National Medical Center, Washington, D.C., $420,000--for Center for Cancer and Blood Disorders; D.C. Campaign to Prevent Teen Pregnancy, Washington, D.C., $75,000--for general support.
Contact: William J. Bierbower, (202) 785-9892; wbierbower@stewart-trust.org
Internet: http://www.stewart-trust.org/guidelines.htm
Sponsor: Alexander and Margaret Stewart Trust
888 17th Street NW, Brawner Building, Suite 610
Washington, D.C. 20006-3321

Alexander Eastman Foundation Grants 420
The purposes of the Alexander Eastman Foundation are served through grant support to organizations awarded in response to proposals and special initiative commitments planned collaboratively by the Foundation with local service providers. Grants are awarded to support the capital, special projects and operations needs of qualifying organizations. In considering proposals, priority is given to funding activity which serves the Foundation's priority interests. The Alexander Eastman Foundation supports: Education - provide information and community education to improve the health and well-being of residents of the greater Derry area; address goals for healthy individuals and families through a long-term commitment to prevention, health promotion and education of consumers and providers; foster individual responsibility, independence, self-care and healthy life-style choices; Family Systems - strengthen families as the critical unit for community health and well-being; recognize the changing nature of families and provide resources and assistance to reduce stress on families and improve family function; Access - expand access to quality health care and prevention services for people with financial need. Applications are available at the Foundations website.
Requirements: Nonprofit organizations serving Derry, Londonderry, Windham, Chester, Hampstead, and Sandown, NH, are eligible.
Restrictions: Grants are made neither to individuals nor to qualifying organizations to support the cost of services to particular individuals, except through the Alexander Eastman Scholarship Program.
Geographic Focus: New Hampshire
Date(s) Application is Due: Apr 1; Oct 1
Samples: Community Caregivers of Greater Derry, Derry, NH, $10,000 - to support the volunteer-driven program of quality client care in the greater Derry community; The Upper Room, A Family Resource Center, Derry, NH, $28,000 - to support prevention

and intervention for families at risk of having a child placed in foster care; Child and Family Services of New Hampshire, Manchester, NH, $19,760 - to support services to twelve non-Medicaid-eligible teens in the Healthy Family Home Visiting program, to support tuition assistance for low-income, at risk greater Derry area youth to attend a session of residential summer camp at Camp Spaulding.
Contact: Amy Lockwood; (888) 228-1821, ext. 80; alockwood@alexandereastman.org
Internet: http://www.alexandereastman.org/02grants.html
Sponsor: Alexander Eastman Foundation
26 South Main Street, PMB 250
Concord, NH 03301

Alexander Foundation Cancer, Catastrophic Illness and Injury Grants 421
The program is intended to provide support to lesbian, gay, bisexual and transgendered people who are at risk of losing their ability to provide basic life needs due to treatment or complications related to cancer, a catastrophic illness, or a serious illness. Some examples of basic life needs are medical care, housing, food, clothing, phones, utilities, and so on. Unlike the one grant payment provided through the Emergency Grant Program, this program provides a fixed monthly grant for up to 12 months. The amount of the grant is determined by the recipients' needs but not to exceed $2,400.
Requirements: Applicants must have diagnosis and treatment of any form of cancer, catastrophic illness, or serious injury except those occurring simultaneously with HIV/AIDS. Recipients must be gay, lesbian, bisexual or transgendered, diagnosed with AIDS or HIV positive, demonstrate a perceived financial need and have been recommended by a recognized referral agency. Additionally, recipients must reside in Colorado.
Restrictions: Recipients must demonstrate financial need and cannot have received financial assistance from the Foundation during the previous 6 months. The Alexander Foundation will only accept applications from an approved referring agency or source.
Geographic Focus: Colorado
Amount of Grant: Up to 2,400 USD
Contact: Jack Heruska; (303) 331-7733; egrants@thealexanderfoundation.org
Internet: http://www.thealexanderfoundation.org/cancer.html
Sponsor: Alexander Foundation
P.O. Box 1995
Denver, CO 80201-1995

Alexander Foundation Insurance Continuation Grants 422
The program provides funds on a monthly basis to pay medical insurance premiums for those individuals who have been diagnosed with AIDS or HIV positive. After a recipient has been accepted into the program, they remain in the program as long as they can demonstrate a continued need. New applicants may be added to a waiting list until such time as funds become available.
Requirements: Recipients must be gay, lesbian, bisexual or transgendered, diagnosed with AIDS or HIV positive, demonstrate a perceived financial need and have been recommended by a recognized referral agency. Additionally, recipients must reside in Colorado.
Geographic Focus: Colorado
Contact: Jack Heruska; (303) 331-7733; egrants@thealexanderfoundation.org
Internet: http://www.thealexanderfoundation.org/insurance.html
Sponsor: Alexander Foundation
P.O. Box 1995
Denver, CO 80201-1995

Alexander von Humboldt Foundation Georg Forster Fellowships for Experienced Researchers 423
In providing Georg Forster Research Fellowships for experienced researchers, the Alexander von Humboldt Foundation enables highly-qualified scientists and scholars from abroad, who completed their doctorates less than twelve years ago to spend extended periods of research (6-18 months; may be divided up into a maximum of three blocks) in Germany. Candidates are expected to have their own, clearly defined research profile. This means they should usually be working at least at the level of Assistant Professor or Junior Research Group Leader or be able to document independent research work over a number of years. The fellowship is worth 3,150 EUR per month. This includes a mobility lump sum and a contribution towards health and liability insurance. Additional benefits may include: return travel expenses; language fellowship; family allowances; additional extension; a subsidy towards research costs; a Europe allowance; and an extensive alumni sponsorship. There is no closing date for submitting applications. Applications are processed as part of an ongoing procedure.
Requirements: Scientists and scholars from all disciplines from developing countries, emerging economies and transition states may apply. Applicants must have completed their doctorates less than twelve years ago. Scholars in the humanities or social sciences and physicians must have a good knowledge of German if it is necessary to carry out the project successfully; otherwise a good knowledge of English; scientists and engineers must have a good knowledge of German or English.
Restrictions: Short-term study tours, participation in conferences, or educational visits cannot be funded. Applications for extension of research stays already commenced in Germany cannot be considered.
Geographic Focus: All Countries
Date(s) Application is Due: Jun 15
Contact: Dr. Bernhard Fleischer, Chairperson; (+49) 0228-833-0 or (+49) 4042-818-401; fax (+49) 0228-833-212; info@avh.de or bernhard.fleischer@t-online.de
Internet: http://www.humboldt-foundation.de/web/georg-forster-fellowship-experienced.html
Sponsor: Alexander von Humboldt Foundation
Jean-Paul-Street 12
Bonn, D-53173 Germany

Alexander von Humboldt Foundation Georg Forster Fellowships for Postdoctoral Researchers 424
Fellowships for postdoctoral researchers are the instrument with which the Foundation enables highly-qualified scientists and scholars who are just embarking on their academic careers and who completed their doctorates less than four years ago to spend extended periods of research (6-24 months) in Germany. The research proposal should deal with issues of major relevance to the future development of the candidate's country of origin and, in this context, particularly suited to transferring knowledge and methods to developing and threshold countries. Research projects are carried out in cooperation with academic hosts at research institutions in Germany. Candidates choose their own research projects and their host in Germany and prepare their own research plan. The fellowship is worth 2,650 euro per month. This includes a mobility lump sum and a contribution towards health and liability insurance. There is no closing date for submitting applications. Applications are processed as part of an ongoing procedure.
Requirements: Scientists and scholars from all disciplines from developing and threshold countries (excluding People's Republic of China, India and Turkey) may apply.
Geographic Focus: All States
Date(s) Application is Due: Jun 15
Amount of Grant: 31,800 EUR
Contact: Dr. Bernhard Fleischer, Chairperson; (+49) 0228-833-0 or (+49) 4042-818-401; fax (+49) 0228-833-212; info@avh.de or bernhard.fleischer@t-online.de
Internet: http://www.humboldt-foundation.de/web/georg-forster-research-fellowship-postdoc.html
Sponsor: Alexander von Humboldt Foundation
Jean-Paul-Street 12
Bonn, D-53173 Germany

Alexander von Humboldt Foundation Research Fellowships for Postdoctoral Researchers 425
The Research Fellowships for Postdoctoral Researchers are the instrument with which the Foundation enables highly-qualified scientists and scholars from abroad who are just embarking on their academic careers and who completed their doctorates less than four years ago to spend extended periods of research (6-24 months) in Germany. Scientists and scholars from all disciplines and countries may apply. Projects are carried out in cooperation with academic hosts at research institutions in Germany. Candidates choose their own research projects and their host in Germany and prepare their own research plan. Details of the research project and the time schedule must be agreed upon with the prospective host in advance. Candidates are selected solely on the basis of their academic record. There are no quotas for individual disciplines or countries. The Fellowship is worth 2,600 EUR per month. This includes a mobility lump sum and a contribution towards health and liability insurance. The application and additional guidelines are available at the website. Applications may be submitted at any time.
Requirements: Application requirements include the following: a doctorate or comparable academic degree (Ph.D., C.Sc. or equivalent), completed less than twelve years prior to the date of application; the candidate's own research profile documented by a comprehensive list of academic publications reviewed according to international standards and printed in journals and/or by publishing houses; and a good knowledge of German or English if it is necessary to carry out the project successfully.
Restrictions: Short-term visits for study and training purposes or for attending conferences are not eligible for sponsorship.
Geographic Focus: All States, All Countries
Amount of Grant: 31,200 EUR
Contact: Matthias Kleiner; (+49) 0228-833-0; fax (+49) 0228-833-212; info@avh.de
Internet: http://www.humboldt-foundation.de/web/humboldt-fellowship-postdoc.html
Sponsor: Alexander von Humboldt Foundation
Jean-Paul-Street 12
Bonn, D-53173 Germany

Alexis Gregory Foundation Grants 426
The Foundation, established in 1986, gives primarily for education purposes. Interests include: AIDS and AIDS research; arts; France; health organizations; human services; Italy; Jewish agencies and temples; libraries and library science; museums; performing arts; and the United Kingdom. Types of support include program development and scholarships to individuals. The primary region of giving is New York, though grants are sometimes approved for out-of-state organizations. Though there is no specific deadline date, the board generally reviews applications in January. An application form is not required. Applicants should initially send a letter or full proposal.
Geographic Focus: New York
Amount of Grant: Up to 15,000 USD
Contact: Alexis Gregory, President; (212) 737-5297; fax (212) 737-5340
Sponsor: Alexis Gregory Foundation
1334 York Avenue, 3rd Floor
New York, NY 10021

ALFJ Astraea U.S. and International Movement Fund 427
Astraea's Movement Resource Fund provides grants to enhance the capacity and effectiveness of our grantee partners and ally organizations to engage in movement building work. Grants are generally provided in three areas: Technical Assistance, Travel/Peer-to-Peer Learning and Historic Convenings. Letters of inquiry are accepted year round.
Requirements: One page letters of inquiry should be sent to the program officer. Information should include information about the organization, the purpose of the request, and a budget summary of how the funds would be used.
Restrictions: Although any organization that fits the funding criteria may apply, Astraea prioritizes current Astraea grantee partners.

64 | GRANT PROGRAMS

Geographic Focus: All States, All Countries
Contact: Namita Chad; (212) 529-8021; fax (212) 982-3321; nchad@astraeafoundation.org
Internet: http://www.astraeafoundation.org/grants/grant-applications-and-deadlines
Sponsor: Astraea Lesbian Foundation for Justice
116 E 16th Street, 7th Floor
New York, NY 10003

Alfred and Tillie Shemanski Testamentary Trust Grants 428
Alfred Shemanski was an immigrant from Poland with little formal education. He understood the struggles of being a stranger in a strange land and became a champion for both Jewish and secular education. His successes never blinded him to the plight of those less fortunate. Mr. Shemanski exercised philanthropy in the purest definition of the word — love of mankind. The Alfred & Tillie Shemanski Trust was established in 1974 to: improve the capacity of and cooperation among Jewish congregations in the City of Seattle, Washington; support interfaith tolerance and understanding; provide scholarship assistance, primarily to the University of Washington and Seattle University; and support and promote quality educational, human-services, and health-care programming for economically disadvantaged individuals and families. Grant requests for general operating support, start-up funding, and prizes or awards are encouraged. Grants from the Shemanski Trust are one year in duration. Application materials are available for download at the grant website. Applicants are strongly encouraged to review the state application guidelines for additional helpful information and clarifications before applying. Applicants are also encouraged to review the trust's funding history (link is available from the grant website). The application deadline for the Shemanski Trust is October 15. Applicants will be notified of grant decisions before November 30.
Requirements: Applicants must serve residents of the Seattle and Puget Sound area.
Restrictions: Requests for fundraising events or sponsorship opportunities will not be considered. The trust does not support requests from individuals, organizations attempting to influence policy through direct lobbying, or any political campaigns.
Geographic Focus: Washington
Date(s) Application is Due: Oct 15
Samples: Seattle University, Seattle, Washington, $45,000, for Schemanski Scholarship Fund; College Success Foundation, Issaquah, Washington, $10,000, scholarship funds; Temple De Hirsch Sinai, Seattle, Washington, $35,000, education initiative; Mockingbird Society, Seattle, Washington, $10,000, Mockingbird Network (foster-youth-led program)(2010).
Contact: Nancy Atkinson; (800) 848-7177; nancy.l.atkinson@baml.com
Internet: https://www.bankofamerica.com/philanthropic/fn_search.action
Sponsor: Alfred and Tillie Shemanski Testamentary Trust
800 5th Avenue, WA1-501-33-23
Seattle, WA 98104

Alfred Bersted Foundation Grants 429
The Alfred Bersted Foundation was established in 1972 to support and promote quality educational, human services, and health care programming for underserved populations. The Foundation specifically serves the people of DeKalb, DuPage, Kane, and McHenry counties in Illinois. Application materials are available for download at the grant website. Applicants are strongly encouraged to review the state application guidelines for additional helpful information and clarifications before applying. Applicants are also encouraged to review the foundation's funding history (link is available from the grant website). The foundation has a rolling application deadline. In general, applicants will be notified of grant decisions three to four months after proposal submission. The annual deadlines for application submission are January 15 and June 15, with decisions being made by June 30 and December 31, respectively. Awards typically range from $25,000 to $50,000.
Requirements: Applicant organizations must have 501(c)3 tax-exempt status and a physical presence in one of the following counties: DeKalb, DuPage, Kane, or McHenry.
Restrictions: The Alfred Bersted Foundation does not make grants to degree-conferring institutions of higher education, religious houses of worship, or organizations testing for public safety. In general, grant requests for endowment campaigns will not be considered.
Geographic Focus: Illinois
Date(s) Application is Due: Apr 15; Sep 15
Amount of Grant: 25,000 - 50,000 USD
Samples: Older Adult Service and Information System of Fox Valley, Sandwich, Illinois, $30,000 - capacity-building and technical assistance (2014); Habitat for Humanity of Fox Valley, Aurora, Illinois, $50,000 - general operating support (2014); Wayside Cross Rescue Mission, Aurora, Illinois, $40,000 - capacity-building and technical assistance (2014).
Contact: Debra L. Grand; (312) 828-2055; ilgrantmaking@ustrust.com
Internet: https://www.bankofamerica.com/philanthropic/foundation.go?fnId=108
Sponsor: Alfred Bersted Foundation
231 South LaSalle Street, IL1-231-13-32
Chicago, IL 60604

Alfred E. Chase Charitable Foundation Grants 430
The Alfred E. Chase Charitable Foundation was established in 1956 to support and promote quality educational, human-services, and health-care programming for underserved populations. Special consideration is given to charitable organizations that serve the people of the city of Lynn and the North Shore of Massachusetts. The foundation is a generous supporter of the Associated Grant Makers (AGM) Summer Fund which provides operating support for summer camps serving low-income urban youth from Boston, Cambridge, Chelsea, and Somerville. Excluding the grant to the AGM Summer Fund, the typical grant range is $10,000 to $30,000. In the area of education the foundation supports academic access, enrichment, and remedial programming for children, youth, adults, and senior citizens that focuses on preparing individuals to achieve while in school and beyond. In the area of human services the foundation supports organizations meeting the basic needs of all individuals to include but not limited to youth development, violence prevention, employment, life skills attainment, and food programs. In the area of health care, the foundation supports programming that improves access to primary care for traditionally underserved individuals, as well as supporting health education initiatives and programs that impact at-risk populations. Grant requests for general operating support are strongly encouraged. Program support will also be considered. Small, program-related capital expenses may be included in general operating or program requests. The majority of grants from the Chase Charitable Foundation are one year in duration. On occasion, multi-year support is awarded. Applicants must apply online at the grant website. Applicants are strongly encouraged to do the following before applying: review the downloadable Massachusetts state application procedures for additional helpful information and clarifications; review the downloadable online-application guidelines at the grant website; review the foundation's funding history (link is available from the grant website); review the online application questions in advance; and review the list of required attachments. These will generally include: a list of board members, financial statements (audited, reviewed, or compiled by independent auditor); an organization summary; a list of other funding sources; an IRS Determination letter; and other required documents. All attachments must be uploaded in the online application as PDF, Word, or Excel files. The application deadline is 11:59 p.m. on April 1. Applicants will be notified of grant decisions before June 30.
Requirements: Applicants must have 501(c)3 tax-exempt status.
Restrictions: The foundation does not support requests from individuals, organizations attempting to influence policy through direct lobbying, or any political campaigns.
Geographic Focus: Massachusetts
Date(s) Application is Due: Apr 1
Amount of Grant: 10,000 - 30,000 USD
Samples: Uphams Corner Community Center, Dorchester, Massachusetts, $15,000; Massachusetts Advocates for Children, Boston, Massachusetts, $20,000; Union Social Action Foundation, Cambridge, Massachusetts, $15,000.
Contact: Miki C. Akimoto, Vice President; (866) 778-6859; miki.akimoto@baml.com
Internet: https://www.bankofamerica.com/philanthropic/fn_search.action
Sponsor: Alfred E. Chase Charitable Foundation
225 Franklin Street, 4th Floor, MA1-225-04-02
Boston, MA 02110

Alfred P. Sloan Foundation International Science Engagement Grants 431
The International Science Engagement program is an early stage program currently focused on South Asia that seeks to bring scientists and engineers in conflict regions together to collaborate on subjects and projects of mutual interest. A pilot grant was made to the Lee Kuan Yew School of Public Policy/National University of Singapore for a preliminary meeting on rice science involving scientists from eight countries including: India, Pakistan, Bangladesh, Sri Lanka, Nepal, Bhutan, the Philippines, and China. Other grants have supported the Institute for International Education's Scholar Rescue Fund which works to relocate endangered scientists, engineers and mathematicians to research institutes abroad.
Requirements: Concise, well-organized proposals are preferred. In no case should the body of the proposal exceed 20 double-spaced pages.
Restrictions: The Foundation's activities do not normally extend to religion, the creative or performing arts, elementary or secondary education, medical research or health care, the humanities or to activities outside the United States. Grants are not made for endowments or for buildings or equipment.
Geographic Focus: All States, All Countries
Contact: Doron Weber, Program Director; (212) 649-1652 or (212) 649-1649; fax (212) 757-5117; weber@sloan.org
Internet: http://www.sloan.org/major-program-areas/select-issues/international-science-engagement/
Sponsor: Alfred P. Sloan Foundation
630 Fifth Avenue, Suite 2200
New York, NY 10111-0242

Alfred P. Sloan Foundation Research Fellowships 432
Alfred P. Sloan Foundation Research Fellowships stimulate fundamental research by young scholars of outstanding promise at a time in their careers when their creative abilities are especially high and when government or other support is difficult to obtain. Fellows are free to pursue whatever lines of inquiry that are of the most compelling interest to them. Funds are awarded directly to the fellow's institution and may be used by the fellow for such purposes as equipment, technical assistance, professional travel, trainee support, or activities directly related to the fellow's research. The foundation welcomes nominations of all candidates who meet the traditional high standards of this program and strongly encourages the participation of women and members of underrepresented minority groups. Awards are made in the fields of physics, chemistry, neuroscience, economics, pure mathematics, applied mathematics, computer science, and computational and evolutionary molecular biology. Fellowships are awarded for a two-year period, with possible extension for another two years. Each year 126 fellows are selected from the nominations received. Nominations are due September 15 for awards to begin the following September. Nomination forms are available online.
Requirements: Candidates must hold a PhD or equivalent in chemistry, physics, mathematics, computer science, economics, neuroscience or computational and evolutionary molecular biology, or in a related interdisciplinary field, and must also be a faculty member at a U.S. or Canadian university. Candidates must be no more than six years from the completion of the most recent PhD or equivalent as of the year of their nomination.

Restrictions: Direct applications are not accepted. Candidates must be nominated by department heads or other senior scholars.
Geographic Focus: All States, Canada
Date(s) Application is Due: Sep 15
Amount of Grant: 50,000 USD
Contact: Daniel L. Goroff, Program Director; (212) 649-1676 or (212) 649-1649; fax (212) 757-5117; goroff@sloan.org
Internet: http://www.sloan.org/sloan-research-fellowships/
Sponsor: Alfred P. Sloan Foundation
630 Fifth Avenue, Suite 2200
New York, NY 10111-0242

Alfred P. Sloan Foundation Science of Learning STEM Grants 433
Grantmaking in the Science of Learning STEM program aims to improve the quality of higher education in STEM fields through the support of original, high-quality research on the factors affecting undergraduate and graduate student learning and retention in STEM fields. Grants primarily support consortia of colleges, universities, and other educational institutions with plans to develop and to study the impact and effectiveness of new approaches to STEM pedagogy, especially in "gateway" courses, with an explicit commitment to institutionalize successful initiatives. Successful proposals are expected to be hypothesis-driven, sensitive to the heterogeneity of STEM disciplines, attentive to differences in student motivations to choose STEM majors and persist in STEM careers, and concerned with the dissemination and portability of results to other institutions.
Restrictions: The Foundation does not make grants to: individuals (except through its Books program); for-profit institutions; religion, medical research, or for research in the humanities; projects aimed at pre-college students (except through its Civic Initiatives program); projects in the creative or performing arts (except when those projects are related to educating the public about science, technology, or economics); endowments or fundraising drives, including fundraising dinners; support buildings, laboratories, equipment, or instruments unless it can be demonstrated that such capital expenditures are essential to the success of a Foundation-supported research project, research program or educational initiative; or political campaigns, to support political activities, or to lobby for or against particular pieces of legislation.
Geographic Focus: All States
Contact: Elizabeth Boylan, Program Director; (212) 649-1634 or (212) 649-1649; fax (212) 757-5117; boylan@sloan.org
Internet: http://www.sloan.org/major-program-areas/stem-higher-education/the-science-of-learning-stem/
Sponsor: Alfred P. Sloan Foundation
630 Fifth Avenue, Suite 2200
New York, NY 10111-0242

Alfred W. Bressler Prize in Vision Science 434
The Prize recognizes a professional in the field of vision science whose leadership, research and service have resulted in important advancements in the treatment of eye disease or rehabilitation of persons with vision loss. A panel of distinguished vision science professionals will select the winner who receives a prize of $40,000.
Requirements: The application process is open to established professionals in the field of vision science whose contributions have advanced vision care, the treatment of eye disease, or the rehabilitation of persons with visual disabilities or blindness and whose further work is expected to contribute significantly. Candidates from the United States and countries around the world are eligible for the award.
Geographic Focus: All States
Date(s) Application is Due: Dec 31
Amount of Grant: 40,000 USD
Samples: Roy W. Beck, MD, PHD, University of South Florida, Tampa, FL-- for his work in the field of Epidimiology; Jonathan C. Horton, MD, PhD, University of California, San Francisco-- for his work in the field of Neuro - Ophthalmology.
Contact: Program Administrator; (212) 769-7801; bressler@jgb.org
Internet: http://www.jgb.org/programs_bressler.asp
Sponsor: Jewish Guild for the Blind
15 W 65th Street
New York, NY 10023

Alice C. A. Sibley Fund Grants 435
The purpose of the Alice C.A. Sibley Fund is to provide medical eye care to vulnerable populations of all ages in Worcester including: hospital and surgical fees; medical equipment; and eye glasses and corrective treatment. Approximately $24,000 in grants will be awarded annually to one or more organizations that directly support people's eye health.
Requirements: Applicants must be tax-exempt, nonprofit organizations as recognized by IRS code 501(c)3, and must be located in or providing services to residents of Worcester, Massachusetts.
Geographic Focus: Massachusetts
Date(s) Application is Due: Apr 15
Amount of Grant: Up to 24,000 USD
Contact: Lois Smith; (508) 755-0980, ext. 107; lsmith@greaterworcester.org
Internet: http://www.greaterworcester.org/grants/Sibley.htm
Sponsor: Greater Worcester Community Foundation
370 Main Street, Suite 650
Worcester, MA 01608-1738

Alice Fisher Society Fellowships 436
The Barbara Bates Center for the Study of the History of Nursing offers a fellowship of $2,500 to support two weeks in residence at the Center and ongoing collaboration with Center historians. Selection of Alice Fisher Society scholars will be based on evidence of interest in and aptitude for historical research related to nursing. It is expected that the research and new materials produced by Alice Fisher Society scholars will help ensure the growth of scholarly work focused on the history of nursing. Fisher scholars will participate in Center activities and will present their research at a Center seminar.
Requirements: The scholarships are open to individuals with master's and doctoral level preparation. The application should be sent via email to either Patricia D'Antonio, PhD, RN, FAAN (dantonio@nursing.upenn.edu) or Barbra Mann Wall, PhD, RN (wallbm@nursing.upenn.edu).
Geographic Focus: All States
Date(s) Application is Due: Dec 31
Amount of Grant: 2,500 USD
Contact: Dr. Karen Buhler-Wilkerson, Director; (215) 898-4725; fax (215) 573-2168; karenwil@nursing.upenn.edu
Internet: http://www.nursing.upenn.edu/history/Pages/Alice_Fisher_Fellowship.aspx
Sponsor: University of Pennsylvania
School of Nursing, 418 Curie Boulevard
Philadelphia, PA 19104-6020

Alice Tweed Tuohy Foundation Grants 437
The foundation promotes organizations that promote young people; that provide outstanding opportunities for performance, growth, and creativity; that nurture personal integrity and ambition; and that reward high achievement. The foundation assists organizations offering services to children whose choices might otherwise be unfairly restricted by need; supported are activities both academic and extracurricular that challenge young people while encouraging the growth of responsibility and personal integrity. Organizations dedicated to improving the quality of life by meeting the vital needs of the community are also supported. Types of support include building construction/ renovation, scholarship funds, and matching funds. Financing priority is accorded those organizations with the least in-house capacity to raise capital, assisting these groups to surmount critical monetary obstacles and continue productive service to the community. Proposals may be submitted annually between July 1 and September 15. However, the foundation has announced that a three-year partial moratorium period on awarding grants will commence July 1, 2009.
Requirements: Applications are considered only from Santa Barbara, CA public, tax-exempt organizations. Priority consideration is given to applications from organizations serving young people, education, health and medicine, community affairs, and the arts.
Restrictions: Excluded from consideration are applications for the benefit of specific individuals, organizations outside the Santa Barbara area, organizations in overpopulated nonprofit areas, national campaigns, fund-raising normally carried out by the organization, operating expenses, or budgetary support.
Geographic Focus: California
Date(s) Application is Due: Sep 15
Amount of Grant: 750 - 98,000 USD
Contact: Program Contact; (805) 962-6430; fax (805) 962-7135; atuohyfdn@aol.com
Sponsor: Alice Tweed Tuohy Foundation
205 E Carrillo Street, Room 219, P.O. Box 1328
Santa Barbara, CA 93102-1328

Allan C. and Lelia J. Garden Foundation Grants 438
The mission of the Allan C. and Leila J. Garden Foundation is to support charitable organizations that maintain, care for and educate orphan or underprivileged children. It is also the Foundation's intent to support organizations that provide medical, dental, hospital care, nursing and treatment of crippled or physically handicapped children. Grants from Allan C. and Leila J. Garden Foundation are primarily one year in duration. On occasion, multi-year support is awarded. Applicants must apply online at the grant website. Applicants are strongly encouraged to do the following before applying: review the downloadable state application procedures for additional helpful information and clarifications; review the downloadable online-application guidelines at the grant website; review the foundation's funding history (link is available from the grant website); review the online application questions in advance; and review the list of required attachments. These will generally include: a list of board members, financial statements (audited, reviewed, or compiled by independent auditor); an organization summary; a list of other funding sources; an IRS Determination letter; and other required documents. All attachments must be uploaded in the online application as PDF, Word, or Excel files. The Allan C. & Leila J. Garden Foundation application deadline is 11:59 p.m. on June 1. Applicants will be notified of grant decisions by letter within three to four months after the deadline.
Requirements: Applicants must have 501(c)3 tax-exempt status and serve residents of Ben Hill, Irwin, and Wilcox Counties in Georgia. A breakdown of number/percentage of people served by specific counties is required on the online application.
Restrictions: The foundation does not support requests from individuals, organizations attempting to influence policy through direct lobbying, or any political campaigns.
Geographic Focus: Georgia
Date(s) Application is Due: Jun 1
Contact: Mark S. Drake, Vice President; (404) 264-1377; mark.s.drake@ustrust.com
Internet: https://www.bankofamerica.com/philanthropic/fn_search.action
Sponsor: Allan C. and Lelia J. Garden Foundation
3414 Peachtree Road, N.E., Suite 1475, GA7-813-14-04
Atlanta, GA 30326-1113

66 | GRANT PROGRAMS

Allegan County Community Foundation Grants 439
The foundation awards one-year grants to eligible Michigan charitable organizations to improve the quality of life in Allegan County. Areas of interest include education, health, and social services. Types of support include youth grants, general grants, matching/challenge grants, building construction/renovation, program development, and emergency funds. A copy of the IRS tax-exemption letter must be submitted with application. Any youth group, club, school or class can apply for the T.A.G. awards but must have an adult leader who oversees the project and any funds awarded.
Requirements: 501(c)3 tax-exempt organizations serving Allegan County, Michigan, residents are eligible. All applicants are required to schedule an appointment with foundation staff prior to submitting an application.
Restrictions: Individuals are ineligible. All T.A.G. mini-grants must be handwritten by youth 18 or younger.
Geographic Focus: Michigan
Date(s) Application is Due: Dec 4
Contact: Theresa Bray, Executive Director; (269) 673-8344; fax (269) 673-8745
Internet: http://www.alleganfoundation.org/grants.htm
Sponsor: Allegan County Community Foundation
524 Marshall Street
Allegan, MI 49010

Allegheny Technologies Charitable Trust 440
The corporation awards one-year renewable grants to nonprofit organizations to enhance the quality of life for people in company operating locations. Areas of interest include arts and culture, civic and public affairs, education, health, and social services. Types of support include capital grants, general operating grants, program development grants, and employee matching gifts. Applicants should submit a letter of inquiry that describes the organization, purpose of grants, and funds sought. No particular form or information is required. Potential applicants are requested to provide proof of exempt public charity status. There are no deadlines.
Requirements: 501(c)3 organizations in company-operating areas are eligible.
Restrictions: Contributions are made only to public charities. Individuals and private foundations are excluded.
Geographic Focus: Pennsylvania
Amount of Grant: 1,000 - 35,000 USD
Samples: Pittsburgh Symphony Society (Pittsburgh, PA)--grant recipient, $55,000.
Contact: Jon D. Walton, Jr., Trustee; (412) 394-2800; fax (412) 394-3034
Internet: http://www.alleghenytechnologies.com
Sponsor: Allegheny Technologies
1000 Six PPG Place
Pittsburgh, PA 15222

Allegis Group Foundation Grants 441
The foundation awards grants to support organizations providing services or programs in the areas of education, health, and underprivileged children. Giving is national in scope, with an emphasis on Baltimore. There are no application deadlines or forms. Send a letter of inquiry as the initial approach.
Geographic Focus: All States
Contact: Hillary Murray, (410) 579-3509
Internet: http://www.allegisgroup.com
Sponsor: Allegis Group Foundation
7301 Parkway Drive
Hanover, MD 21076

Allen Foundation Educational Nutrition Grants 442
The foundation supports projects that benefit human nutrition in the areas of education, training, and research. Priorities include training programs for children and young adults to improve their health and development; training programs for educators and demonstrators concerned with good nutritional practices; programs for the education and training of mothers during pregnancy and after the birth of their children, so that good nutritional habits can be formed at an early age; and programs that aid in the dissemination of information regarding healthful nutrition practices. The Allen Foundation will consider requests from the following: hospitals or medical clinics; social, religious, fraternal, or community organizations; private foundations; and K-12 public, parochial or private schools. Preference may be given to proposals that include matching funds from the institution or other partners including in-kind contribution. Third party contribution to matching funds such as computer or software donated from a company may be included. Applications are available by fax or mail. The annual deadline for applications is December 31; proposals are reviewed throughout the year.
Requirements: 501(c)3 tax-exempt organizations nationwide may apply. In certain circumstances, the foundation will consider requests from the following: hospitals or medical clinics; social, religious, fraternal, or community organizations; private foundations; and K-12 public, parochial, or private schools.
Geographic Focus: All States
Date(s) Application is Due: Dec 31
Amount of Grant: 5,000 - 250,000 USD
Contact: Dr. Dale Baum, Secretary; (989) 832-5678 or (979) 695-1132; fax (989) 832-8842; dbaum@allenfoundation.org or Lucille@allenfoundation.org
Internet: http://www.allenfoundation.org/commoninfo/aboutus.asp
Sponsor: Allen Foundation
P.O. Box 1606
Midland, MI 48641-1606

Allen P. and Josephine B. Green Foundation Grants 443
The mission of the foundation is to improve the quality of life in Missouri. Grants support programs to bring health care services to people; counseling and educational services for children and adults with physical, mental, or behavioral problems; innovative developmental and educational programs for children; cultural and preservation projects to safeguard our heritage and to promote broader awareness and understanding of such programs and activities on the part of Missourians; and environmental and conservation projects based in Missouri. Grants will be made for one year only. Priority will be given to new projects and to those that have not been funded for at least one year. The board meets in May and November. Guidelines are available online.
Requirements: Grants are awarded to 501(c)3 nonprofit organizations in Missouri.
Restrictions: Grants are not awarded to individuals; to projects and programs located outside of the United States; to charities that are not publicly supported; for social causes or for social activism; for lobbying, propagandizing, or for political campaigns; or to requests deemed as inappropriate for a variety of other reasons. Grant requests that probably will not be approved include projects and programs located outside of Missouri; commitment of funds for more than one year; funding to the same organization for two years in a row; grants for operating funds and annual budgets; unspecified general funds; other than to specific projects or programs; projects with a perceived small chance of success; programs that would seem unlikely to continue without further funding from the Green Foundation; large capital fund drives; large building fund drives; large endowments; to an organization whose primary mission is to raise and distribute funds; or to federated giving programs.
Geographic Focus: Missouri
Date(s) Application is Due: Mar 15; Jun 15; Sep 15; Dec 15
Contact: Walter Staley, Jr.; (573) 581-5568; fax (573) 581-1714; wstaley@greenfdn.org
Internet: http://www.greenfdn.org
Sponsor: Allen P. and Josephine B. Green Foundation
P.O. Box 523
Mexico, MO 65265

Alliance Healthcare Foundation Grants 444
The Foundation provides a public voice for the critical health care needs of its communities. Areas of interest include: Access to Health, Mental Health, Community Health, and HIV Health Care programs. Grants to local organizations support a wide range of programs and services. Priority is given to program and financing strategies that can produce discernible outcomes. All grant applicants should start by submitting a Letter of Intent (LOI), written and addressed according to the Foundation's guidelines. The LOI is the only form of application the Foundation accepts. Letters are reviewed on a monthly basis. Each is carefully evaluated by program staff in relation to other funding requests. Applicants will be notified about the results of the review within four to six weeks.
Restrictions: The Foundation does not fund projects and programs outside San Diego and Imperial Counties. In addition, the Foundation does not fund: research; lobbying; underwriting of medical expenses; general operating expenses we deem to be excessive; construction or renovation; the purchase of costly equipment; development activities, such as fundraising events, capital campaigns or annual fund drives; projects or proposals from individuals; or organizations that do not have 501(c)3 status.
Geographic Focus: California
Samples: Imperial County Office of Education (El Centro, CA)--to establish a family-resource center on the grounds of Seeley Elementary School, $450,000; Point Loma Nazarene U (San Diego, CA)--for its free clinic that provides health services to uninsured residents of San Diego's City Heights neighborhood, $122,880; Shakti Rising (San Diego, CA)--for strategic planning at this organization that uses holistic health practices and counseling to assist women struggling with abuse, addiction, and unhealthy relationships and behavior patterns, $159,415 over two yars; Whittier Institute for Diabetes (La Jolla, CA)--to establish the Child Obesity and Diabetes Prevention Program at Horton Elementary School, in San Diego, $62,657.
Contact: Karen Romero, (858) 614-4888; fax (858) 874 3656; kromero@alliancehf.org
Internet: http://www.alliancehf.org/grants_prog/what_we_fund/access_healthcare.html
Sponsor: Alliance Healthcare Foundation
9325 Sky Park Court, Suite 350
San Diego, CA 92123

Alliant Energy Corporation Contributions 445
As a complement to its Foundation, the Alliant Energy Corporation also makes charitable contributions to local, county, and regional nonprofit organizations directly. Support is given primarily in areas of company operations in Illinois, Iowa, Minnesota, and Wisconsin; giving also to national organizations. The Corporation's primary areas of interest include: breast cancer; community development; diabetes; economic development; employment; energy resources; the environment; financial services; housing development; human services; job training and retraining; and sustainable development. Types of support include: fundraising; general operations; outreach; program development; scholarship funds; sponsorships; technical assistance; and volunteer development. A formal application is required, although there are no identified annual deadlines.
Geographic Focus: Illinois, Iowa, Minnesota, Wisconsin
Contact: Jo Ann Healy, Contributions Manager; (608) 458-5718; fax (608) 458-0134; joannhealy@alliantenergy.com
Internet: http://www.alliantenergy.com/CommunityInvolvement/CommunityOutreach/RequestSponsorshipsAdvertising/index.htm
Sponsor: Alliant Energy Corporation
4902 N. Biltmore Lane, Suite 1000
Madison, WI 53718-2148

Allstate Corporate Giving Grants 446
Allstate is a company of energized people with great ideas. The Corporate Giving program is committed to supporting the communities where company employees live and work by contributing to programs where its experience, partnership and leadership will have the greatest impact. The company offers financial support to a variety of programs and organizations throughout the country that help create strong and vital communities.
Requirements: The Allstate Corporation makes grants to nonprofit, tax-exempt groups under Section 501(c)3 of the Internal Revenue Code.
Geographic Focus: All States
Contact: Executive Director; (847) 402-5000 or (847) 402-5502; fax (847) 326-7517; allfound@allstate.com
Internet: http://www.allstate.com/social-responsibility/social-impact/corporate-contributions.aspx
Sponsor: Allstate Corporation
2775 Sanders Road, Suite F4
Northbrook, IL 60062

Allstate Corporate Hometown Commitment Grants 447
Allstate takes a special interest in the greater Chicagoland community, the company's hometown for more than 75 years. The corporation is particularly invested in this community because it recognizes that a thriving hometown is critical to Allstate's success. The company recruits local talent, relies on local infrastructure, and depends on the city's vibrancy to ensure that its associates have a rich quality of life. By supporting organizations that build strong Chicagoland communities, the company contributes to the city's position as a center of global culture, education and business.
Requirements: The Allstate Corporation makes grants to Chicago area nonprofit, tax-exempt groups under Section 501(c)3 of the Internal Revenue Code.
Geographic Focus: Illinois, Indiana
Contact: Executive Director; (847) 402-5000 or (847) 402-5502; fax (847) 326-7517; allfound@allstate.com
Internet: http://www.allstate.com/social-responsibility/corporate/corporate-giving.aspx
Sponsor: Allstate Corporation
2775 Sanders Road, Suite F4
Northbrook, IL 60062

Allyn Foundation Grants 448
The mission of the Allyn Foundation is to improve the quality of life in Central New York. To accomplish this, the foundation focuses its grant making in the following areas: health facilities and services: promote the delivery of quality health care in our communities; higher education in the Onondaga and Cayuga counties: expand access and success in education beyond high school, particularly for students of low income or other underrepresented backgrounds; basic human services: work to ensure that all people have basic daily needs of food, clothing, and shelter and enable the means to provide for themselves; family planning medical and educational services: promote education in family planning and improved access to service; quality of life for youth: improve quality of life for children and families through organizations that are focused on at risk youth and families.
Requirements: Nonprofit organizations in Onondaga and Cayuga Counties, NY, may submit grant applications.
Restrictions: Grants are not awarded for religious purposes, endowment funds, loans, or to individuals.
Geographic Focus: New York
Contact: Margaret O'Connell; (315) 685-5059; info@allynfoundation.org
Internet: http://www.allynfoundation.org/apply.html
Sponsor: Allyn Foundation
14 West Genesee Street, P.O. Box 22
Skaneateles, NY 13152

AlohaCare Believes in Me Scholarship 449
In late 2003, AlohaCare established the "AlohaCare Believes in Me Scholarship" for University of Hawaii students. Scholarships are for both UH Community Colleges and University campuses in the state. Students who are pursuing a medical or health care profession and have financial needs are considered for scholarships up to $2,500 each year.
Requirements: Applicants must be a Hawaii resident; must be attending to any campus within the University of Hawaii system; enrolled as a full or part-time classified graduate or undergraduate student; and, able to demonstrate financial need. Preferences will be given to students from a neighbor island (other than Oahu) and/or graduates from a neighbor island high school. Special consideration will also be given to students pursuing studies in health care or health-related fields.
Geographic Focus: Hawaii
Date(s) Application is Due: Mar 2
Amount of Grant: Up to 2,500 USD
Contact: Scholarship Coordinator; (808) 956-6625; scholars@hawaii.edu
Internet: http://www.alohacare.org/Communities/Scholarship.aspx
Sponsor: AlohaCare
1357 Kapiolani Boulevard, Suite 1250
Honolulu, HI 96814

Alpha Natural Resources Corporate Giving 450
The Alpha Natural Resources Corporate Giving program makes contributions to nonprofit organizations involved with helping children and families, improving education, strengthening arts and culture programs, and providing social services to those in need of health care and emergency fuel. Its primary fields of interest include: aging centers and services; the arts; children's services; primary education; secondary education; the environment; family services; food services; health care; higher education; housing and shelter; human services; substance abuse prevention; and youth development. Types of support given include: annual campaigns, product donations; employee volunteer services; scholarships; general operating support; and matching funds.
Requirements: Applicants must be 501(c)3 organizations located in, or serving the residents of, areas in which Alpha Natural Resources operates. These include selected regions of Kentucky, Illinois, Pennsylvania, Virginia, and West Virginia.
Geographic Focus: Illinois, Kentucky, Pennsylvania, Virginia, West Virginia
Amount of Grant: Up to 20,000 USD
Contact: Corporate Giving Administrator; (276) 619-4410
Internet: Alpha Natural Resources Corporate Giving
Sponsor: Alpha Natural Resources
1 Alpha Place, P.O. Box 16429
Bristol, VA 24209

Alpha Omega Foundation Grants 451
The Foundation was established to advance the dental profession through research, grants, and scholarships. Fields of interest include: dental care, dental education, and medical research. Grants are given to individuals, program development, and higher education institutions. The Foundation was responsible for dispersing almost $250,000 this past year for projects in Israel and worldwide. The application must be typed and submitted via email by September 1 each year. Notification will be completed by the following December.
Geographic Focus: All States, All Countries
Date(s) Application is Due: Sep 1
Amount of Grant: 1,000 - 50,000 USD
Contact: Heidi Weber, Association Director; (301) 738-6400; fax (301) 738-6403; hweber@ao.org or foundation@ao.org
Internet: http://www.ao.org/index.php?option=com_content&task=view&id=80&Itemid=147
Sponsor: Alpha Omega Foundation
50 W Edmonston Drive, Suite 303
Rockville, MD 20852-1274

Alpha Research Foundation Grants 452
The primary purpose of the Foundation is to encourage and support biomedical research for the welfare of the general public. There are no specific deadlines or application formats with which to adhere. Applicants should contact the foundation prior to submission, in order to discuss their proposed project. If approved, applicants should then submit an outline of the proposed research along with a proposed budget.
Geographic Focus: Maryland
Amount of Grant: Up to 2,000 USD
Contact: Barry Rueben Fierst, (301) 762-8872; fax (301) 762-8874; bfierst@aol.com
Sponsor: Alpha Research Foundation
7118 Glenbrook Road
Bethesda, MD 20814-1238

Alpine Winter Foundation Grants 453
Established in California in 1963, the Alpine Winter Foundation's primary purpose is to support alpine safety, health programs, and education within the Tahoe-Donner, California, region. Applications should be in the form of a written request, and should include the specific purpose and history of the requesting organization. There are no specific deadlines with which to adhere.
Requirements: Giving is limited to organizations described in Section 170(b)(1)(a) and Section 501(c)3(a) that promote alpine safety, heath, and education within the Tahoe-Donner area of California,
Geographic Focus: California
Amount of Grant: Up to 40,000 USD
Samples: Donner Trail School, Soda Springs, California, $1,500--for winter survival training programs; Sugar Bowl Ski Team Foundation, Norden, California, $37,500--for need-based scholarships and a building fund *2008).
Contact: Mary S. Tilden, President; (415) 221-7762
Sponsor: Alpine Winter Foundation
3863 Jackson Street, P.O. Box 591659
San Francisco, CA 94118-1610

ALSAM Foundation Grants 454
The foundation awards grants in its areas of interest, including agriculture, Christian agencies and churches, higher education, health care and medical research, human services, minorities, and the economically disadvantaged. Types of support include building construction/renovation, general operating costs, and scholarships. The board meets in January and October. Contact the office for application forms.
Requirements: Higher education institutions, nonprofit organizations, religious organizations, and research institutions are eligible.
Restrictions: No grants to individuals.
Geographic Focus: All States
Amount of Grant: 5,000 - 50,000 USD
Contact: Ron Cutshall, Chair; (801) 266-4950
Sponsor: ALSAM Foundation
6190 Moffat Farm Lane
Salt Lake City, UT 84121

68 | Grant Programs

Alternatives Research & Development Foundation Alternatives in Education Grants   455

The mission of the Alternatives Research and Development Foundation is to fund and promote the development, validation and adoption of non-animal methods in biomedical research, product testing and education. ARDF has announced a special grant initiative seeking proposals to replace traditional use of animals for education and training purposes. Up to $10,000 in funding is available to support individual projects. the deadline for Letters of Intent is January 8, while invited full proposals are due March 12.
Requirements: Grants from the Foundation are made only to individuals affiliated with a non-profit, tax-exempt, institution, organization, or foreign equivalent.
Geographic Focus: All States
Date(s) Application is Due: Jan 8; Mar 12
Amount of Grant: Up to 10,000 USD
Contact: Sue A. Leary, President and Executive Director; (215) 887-8076; fax (215) 887-0771; grants@ardf-online.org or info@ardf-online.org
Internet: http://www.ardf-online.org/
Sponsor: Alternatives Research and Development Foundation
801 Old York Road, Suite 316
Jenkintown, PA 19046

Alternatives Research and Development Foundation Grants   456

The foundation awards grants to research centers, educational institutions, and other nonprofit organizations exploring non-animal testing. Grants will be awarded for research projects that use human, rather than nonhuman, vertebrae tissue; do not use intact, vertebrate animals; and can be completed within one year. Multiyear projects are considered on a case-by-case basis. The foundation often provides partial grants to initiate projects, with continuation funding available at a later date. The foundation will not consider proposals if program staff uses animals acquired from a shelter, or if individuals are employed who use such animals in their personal research programs. Winners are announced on July 15. Telephone inquiries are discouraged.
Requirements: Individuals attending or employed by U.S. universities and research institutions may apply. Applications from non-U.S. institutions or investigators may be considered.
Restrictions: The foundation does not make grants for in vitro projects that use nonhuman animal serum, indirect costs, purchase of personal computers, salary supplements, fringe benefits, travel expenses, or publication costs for the principal investigators. Phone calls to the foundation are discouraged. Guidelines and application are available online.
Geographic Focus: All States
Date(s) Application is Due: Apr 30
Amount of Grant: Up to 40,000 USD
Sample: Dr. Anthony Hickey, School of Pharmacy, U of North Carolina at Chapel Hill (NC)--for research, Physiologically Relevant Lung Model of Regional Aerosol Deposition and Drug Transport Utilizing Human Respiratory Tract Cells; Dr. Katherine Ralls, Department of Conservation Biology, Smithsonian Institution (Washington, D.C.)--for research, Dogs, Scats, and DNA: A Noninvasive Approach for Carnivore Field Studies; Dr. Carol Reinisch, Marine Biological Laboratory (Woods Hole, MA)--for research, Regulation of p53 Gene Family Expression in Clam Leukemia Cells; Dr. Bingfang Yan, Department of Biomedical Sciences, U of Rhode Island (Kingston, RI)--for research, Prediction of Human CYP3A Induction by a Receptor-Activator Based Method.
Contact: Sue A. Leary, President and Executive Director; (215) 887-8076; fax (215) 887-0771; grants@ardf-online.org or info@ardf-online.org
Internet: http://www.ardf-online.org
Sponsor: Alternatives Research and Development Foundation
801 Old York Road, Suite 316
Jenkintown, PA 19046

Altman Foundation Health Care Grants   457

The Altman Foundation's mission is to support programs, and institutions that enrich the quality of life in New York City, with a particular focus on initiatives that help individuals, families, and communities benefit from the services and opportunities that will enable them to achieve their full potential.
Requirements: IRS 501(c)3 organizations in New York are eligible.
Restrictions: Grants are not awarded to individuals. As a general rule, the foundation does not consider requests for bricks and mortar funds or the purchase of capital equipment.
Geographic Focus: New York
Amount of Grant: 10,000 - 100,000 USD
Samples: Literacy Assistance Venter, Inc., N.Y., $200,000 - to provide support for the Health Literacy Resource Center; The Boys' Club of New York, N.Y., $50,000 - to renew support for the After School Academy; The New York Community Trust, N.Y., to help support the New York City Workforce Development Funders Group $50,000.
Contact: Karen L. Rosa; (212) 682-0970; krosa@altman.org
Internet: http://www.altmanfoundation.org/guide.html
Sponsor: Altman Foundation
521 Fifth Avenue, 35th Floor
New York, NY 10175

Alton Ochsner Award   458

The Award is presented to one or more clinical or basic science investigators, without regard to age, race, gender, or nationality, for outstanding and exemplary original scientific investigations that relate tobacco consumption and health. This scientific work may be clinical, fundamental, epidemiological or preventive in scope. The prime criterion for award selection is its scientific impact on this major health threat. The $15,000 award is presented at the Annual Convocation of the American College of Chest Physicians.
Requirements: All nominations, whatever the category of scientific inquiry, must be supported by letters and copies of peer-reviewed scientific publications.
Geographic Focus: All States
Date(s) Application is Due: Mar 31
Amount of Grant: 15,000 USD
Contact: Edward D. Frohlich, Alton Ochsner Distinguished Scientist; (504) 842-3000; fax (504) 842-3258
Internet: http://www.ochsner.org/homepage.cfm
Sponsor: Ochsner Clinic Foundation
1514 Jefferson Highway
New Orleans, LA 70121

Alvin & Fanny Blaustein Thalheimer Foundation Baltimore Communal Grants   459

The Alvin and Fanny Blaustein Thalheimer Foundation makes grants that strengthen the lives of individuals, families, and communities in the Baltimore region. Program areas include: economic opportunity--technical and business entrepreneurship training, and asset-building strategies; health and human services--improve quality of service, and advocacy and policy initiatives; arts and culture--strengthening education and outreach programs that link arts institutions with communities and schools; strengthening Jewish communities--renewal and development of communities in Eastern Europe and the Former Soviet Union; and addressing threats of anti-Semitism. Proposals and inquiries are accepted on a rolling basis and should be addressed to the President of the Foundation; submissions will be acknowledged. Generally, grant decisions are made within three to four months of proposal submission. Most recent awards have ranged from $4,000 to $2,000,000.
Requirements: The Foundation makes grants to organizations that are tax-exempt under Section 501c(3) of the IRS code, organized and operated for charitable purposes. Grantseekers should apply online or submit a letter of inquiry or short proposal (no more than three to five pages).
Restrictions: The Foundation does not support or make grants in the following areas: individuals; individual scholarship programs; unsolicited proposals for scientific research; finance deficits; annual giving; membership campaigns; or sponsoring fundraising events.
Geographic Focus: Maryland
Amount of Grant: 4,000 - 2,000,000 USD
Samples: Arts Every Day, Baltimore, Maryland, $7,500 - general operating support (2015); Baltimore Children's Museum, Baltimore, Maryland, $20,000 - support for After-School programs for two years (2015); Maryland Institute College of Art, Baltimore, Maryland, $500,000 - endowment grant over five years for Thalheimer Scholarship (2015).
Contact: Rebecca Sirody; (410) 415-7660; fax (410) 580-9250; info@thalheimerfoundation.org
Internet: http://www.thalheimerfoundation.org/baltimorecommunalgiving.html
Sponsor: Alvin and Fanny Blaustein Thalheimer Foundation
6225 Smith Avenue, Suite B-100
Baltimore, MD 21209

ALVRE Casselberry Award   460

The Casselberry Award has been established to encourage the advancement of the art and science of Laryngology and Rhinology. The award is given for outstanding manuscripts or accomplishments in Laryngology and Rhinology and consists of a $1,000 award and a certificate from the Association. Competition for this award will be limited to those persons whose abstracts are selected in consideration for inclusion in the Annual Scientific Program.
Geographic Focus: All States
Date(s) Application is Due: Nov 30
Amount of Grant: 1,000 USD
Contact: Maxine Cunningham, Administrator; (615) 322-6326; fax (615) 322-9102; Maxine@alahns.org or ala-hns@comcast.net
Sponsor: American Laryngological Association / American Laryngological Voice and Research Education Foundation
1215 21st Avenue South, 7302 MCE South
Nashville, TN 37232-8783

ALVRE Grant   461

The purpose of this award is to support basic, translational, or clinical research projects in laryngology, voice, outcomes, and related subjects. Research supported by this award should be specifically directed toward the pathogenesis, pathophysiology, diagnosis, prevention, or treatment of diseases, disorders, or conditions of the larynx and may be either basic or clinical/translational in approach. While not specifically required, proposals which aim to introduce new knowledge and methodology from other disciplines to research in laryngology or neurolaryngology, or which demonstrate collaborative effort with members of other related disciplines are encouraged. Projects must be designed so as to yield useful information within the period of award, but priority will be given to projects that are also innovative with promise to develop into new long-range or expanded research programs capable of attracting funding from other sources. A single, one-year, non-renewable award of $25,000 maximum is available annually. The foundation will consider requests to cover travel expenses up to $1,000 for the principal investigator to present his/her results at the ALA annual meeting.
Requirements: Candidates for this award should be otolaryngologists in the U.S. or Canada who have completed their training at an ACGME accredited program in otolaryngology--head and neck surgery. The prinicipal investigator should be a physician faculty member of a recognized department, division, or section of otolaryngology-head and neck surgery.
Geographic Focus: All States

Amount of Grant: Up to 25,000 USD
Contact: Maxine Cunningham, Administrator; (615) 322-6326; fax (615) 322-9102; Maxine@alahns.org or ala-hns@comcast.net
Internet: http://www.entnet.org/EducationAndResearch/coreGrants.cfm
Sponsor: American Laryngological Association / American Laryngological Voice and Research Education Foundation
1215 21st Avenue South, 7302 MCE South
Nashville, TN 37232-8783

ALVRE Seymour R. Cohen Award 462
The Seymour R. Cohen Award was established with a bequest by the Cohen family during Dr. Cohen's presidency in 1989. In order to qualify for the Award, the candidate must perform basic science research in the area of pediatric laryngology and/or pediatric neurolaryngology and be a citizen of the United States or Canada. The material must be approved and accepted for presentation at the Annual Meeting. Should multiple candidates collaborate in the research, all authors must comply with the citizenship requirements. Dr. Cohen wishes this Award only to be used for the funding of research in pediatric laryngology and/or pediatric neurolaryngology. The Award is to be presented bi-annually.
Requirements: The scope of the Award is limited to citizens of the United States of America and Canada.
Geographic Focus: All States
Contact: Maxine Cunningham, Administrator; (615) 322-6326; fax (615) 322-9102; Maxine@alahns.org or ala-hns@comcast.net
Internet: http://www.alahns.org/i4a/pages/index.cfm?pageid=3332
Sponsor: American Laryngological Association / American Laryngological Voice and Research Education Foundation
1215 21st Avenue South, 7302 MCE South
Nashville, TN 37232-8783

Alzheimer's Association Conference Grants 463
The Alzheimer's Association has a long history of supporting scientific conferences that advance research on Alzheimer's disease. One of the principal goals of the Association from its inception has been to increase public awareness and to facilitate the exchange of information through the scientific and clinical communities. The support of conferences, workshops and meetings has been a key vehicle in achieving this goal. The objectives for conference support are to: facilitate and speed the exchange of information relevant to Alzheimer's disease research; convene experts to address emerging issues in Alzheimer's disease research; offer opportunities for new investigators and graduate students to participate in scientific meetings; facilitate the creation of networks among investigators in related areas; and increase visibility of the research interests and programs of the Alzheimer's Association. At this time, support requests must be limited to no more than $10,000 per conference. Most awarded conference support requests have been in the range of $2000 to $5000.
Requirements: Requests for conference support may be submitted at any time. It is recommended that requests be submitted at least three months before the conference.
Geographic Focus: All States
Date(s) Application is Due: Jan 4; Mar 6; May 1; Jun 7; Aug 7; Oct 2; Nov 1
Amount of Grant: Up to 10,000 USD
Contact: Angela Worlds, (312) 335-5807; fax (312) 335-4034; angela.worlds@alz.org
Internet: http://www.alz.org/research/alzheimers_grants/types_of_grants.asp
Sponsor: Alzheimer's Association
225 North Michigan Avenue, Suite 1700
Chicago, IL 60601-7633

Alzheimer's Association Development of New Cognitive and Functional 464
Instruments Grants
The Alzheimer's Association is launching a new initiative to stimulate the scholarly investigation and development of cognitive or functional evaluation instruments that can capture the earliest changes in the disease, are sensitive to change over time, and/or could be used in clinical trials. This RFA is designed to enable pilot research or proof-of-principle studies that can provide preliminary data for subsequent inquiry. Allowable costs under this award include: purchase and care of laboratory animals; small pieces of laboratory equipment and laboratory supplies; computer equipment if used strictly for data collection; travel (up to $1,000 per year); and salary for the principal investigator, scientific (including post-doctoral fellows) and technical staff (including laboratory technicians and administrative support staff whose work is directly related to the funded project). The Association anticipates funding up to 2 approaches addressing needs in cognitive and functional instrument development. Each award is limited to $400,000 (direct and indirect costs) for two to three years. Requests in any given year may not exceed $200,000 (direct and indirect costs). Indirect costs are capped at 10 percent (rent for laboratory or office space is expected to be covered by indirect costs paid to the institution). Letters of Intent (LOIs) must be received by December 20, with completed applications due by February 7.
Requirements: Researchers with full-time staff or faculty are encouraged to apply.
Restrictions: Applications from post-doctoral candidates will not be accepted. Costs not allowed under this award include: tuition; computer hardware or software for investigators; rent for laboratory or office spaces; or construction or renovation costs.
Geographic Focus: All States
Date(s) Application is Due: Feb 7
Contact: Nico Stanculescu, Program Administrator; (312) 335-5747 or (312) 335-5862; fax (312) 335-4034; nico.stanculescu@alz.org or grantsapp@alz.org
Internet: http://www.alz.org/research/alzheimers_grants/types_of_grants.asp
Sponsor: Alzheimer's Association
225 North Michigan Avenue, Suite 1700
Chicago, IL 60601-7633

Alzheimer's Association Everyday Technologies for Alzheimer Care Grants 465
Everyday Technologies for Alzheimer Care (ETAC) is a cooperative research funding initiative sponsored by the Alzheimer's Association and Intel Corporation. ETAC seeks proposals on personalized diagnostics, preventive tools and interventions for adults coping with the spectrum of cognitive aging and neurodegenerative disease, particularly Alzheimer's disease. ETAC is designed to support exploratory multidisciplinary research that would not typically be funded by national health and science granting foundations. Minor iterations in testing plans or populations will not be considered for funding. Collaboration between social science/medical/public health and computer science/engineering researchers is valued. Mobile computing, high bandwidth sensing, robotics, imaging, face recognition, natural language processing, statistical modeling and a host of other technology advances allow unprecedented opportunities to study disease progression and therapeutic strategies in the context of everyday life. ETAC supports research that integrates such emerging technology capabilities with leading directions in behavioral science and biomedical research. Grants that merely create Internet-based versions of existing services or paper tools will not be considered. Submissions must be original ideas, not continuations of previously funded ETAC projects. Please see links provided below for examples of studies that have been funded by ETAC. Letters of Intent (LOIs) must be received by December 20, with completed applications due by February 7. The Association anticipates funding two awards under this program.
Requirements: Researchers with full-time staff or faculty are encouraged to apply.
Restrictions: The Alzheimer's Association will not accept new research grant applications from currently funded Alzheimer's disease investigators who are delinquent in submitting interim/final scientific or interim/final financial reports on active grants. ETAC applications from post-doctoral candidates will not be accepted. Each total award is limited to $200,000 (direct and indirect costs) for up to three years. Requests in any given year may not exceed $90,000 (direct and indirect costs). Indirect costs are capped at 10 percent (rent for laboratory/office space is expected to be covered by indirect costs paid to the institution).
Geographic Focus: All States
Date(s) Application is Due: Feb 7
Amount of Grant: Up to 200,000 USD
Contact: Nico Stanculescu, Program Administrator; (312) 335-5747 or (312) 335-5862; fax (312) 335-4034; nico.stanculescu@alz.org or grantsapp@alz.org
Internet: http://www.alz.org/research/alzheimers_grants/types_of_grants.asp
Sponsor: Alzheimer's Association
225 North Michigan Avenue, Suite 1700
Chicago, IL 60601-7633

Alzheimer's Association Investigator-Initiated Research Grants 466
The program is structured to provide one to three years of sustained project support for independent, ongoing research. Proposals are solicited for basic, clinical, and social/behavioral research relevant to degenerative brain diseases such as Alzheimer's disease. Allowable costs for this award include the purchase and care of laboratory animals; small pieces of laboratory equipment and laboratory supplies; and salary for the principal investigator, scientific (including postdoctoral fellows) and technical staff (including laboratory technicians and modest secretarial support). It is required that most of the funds awarded under this program be used for direct research support. Allowable costs under this award include: purchase and care of laboratory animals; small pieces of laboratory equipment and laboratory supplies; computer software if used strictly for data collection; salary for the principal investigator, scientific (including post-doctoral fellows) and technical staff (including laboratory technicians and administrative support related directly to the funded project); and travel to scientific and professional meetings, not to exceed $1,000 per year. Letters of Intent (LOIs) must be received by December 20, with completed applications due by February 7.
Requirements: Public, private, domestic and foreign research laboratories, medical centers and hospitals, and universities are eligible to apply. Investigators from all stages of their research career development are encouraged to apply.
Restrictions: Your budget must not exceed the maximum amount of the award, $240,000 or $100,000 per year. Costs not allowed under this award include: tuition; computer hardware or standard software (e.g., Microsoft Office) for investigators; rent for laboratory or office space; or construction or renovation costs.
Geographic Focus: All States
Date(s) Application is Due: Feb 7
Amount of Grant: 100,000 - 240,000 USD
Contact: Nico Stanculescu, Program Administrator; (312) 335-5747 or (312) 335-5862; fax (312) 335-4034; nico.stanculescu@alz.org or grantsapp@alz.org
Internet: http://www.alz.org/research/alzheimers_grants/types_of_grants.asp
Sponsor: Alzheimer's Association
225 North Michigan Avenue, Suite 1700
Chicago, IL 60601-7633

Alzheimer's Association Mentored New Investigator Research Grants to 467
Promote Diversity
The Mentored New Investigator Research Grant to Promote Diversity (MNIGD) is a three-year award intended to be a research-based and mentoring investment in the process of closing the health disparities gap between diverse and non-diverse investigator populations. The Alzheimer's Association feels strongly that the mentoring and involvement of diverse researchers in independently funded Alzheimer's research is a pressing need. The MNIRGD is intended to enhance the capacity of diverse and non-diverse scientists to conduct basic, clinical and social/behavioral research. Each MNIRGD award is limited to $170,000. A total of $150,000 will be awarded for costs related to the proposed research for up to three years (direct and indirect costs). Allowable

costs under this award include: purchase and care of laboratory animals; small pieces of laboratory equipment and laboratory supplies; computer equipment if used strictly for data collection; travel (up to $1,000 per year); and salary for the principal investigator, scientific (including post-doctoral fellows) and technical staff (including laboratory technicians and administrative support related directly to the funded project) Letters of Intent (LOIs) must be received by December 20, with completed applications due by February 7.
Requirements: Eligibility for this grant competition is restricted to investigators who have less than 10 years of research experience after receipt of their terminal degree.
Restrictions: Each MNIRGD award is limited to $170,000. A total of $150,000 will be awarded for costs related to the proposed research for up to three years (direct and indirect costs). Requests in any given year may not exceed $60,000 (direct and indirect costs). Indirect costs are capped at 10 percent (rent for laboratory and/or office space is expected to be covered by indirect costs paid to the institution). Costs not allowed under this award include: tuition; computer hardware or software for investigators; rent for laboratory/office space; or construction or renovation costs.
Geographic Focus: All States
Date(s) Application is Due: Feb 7
Amount of Grant: Up to 170,000 USD
Contact: Nico Stanculescu, Program Administrator; (312) 335-5747 or (312) 335-5862; fax (312) 335-4034; nico.stanculescu@alz.org or grantsapp@alz.org
Internet: http://www.alz.org/research/alzheimers_grants/types_of_grants.asp
Sponsor: Alzheimer's Association
225 North Michigan Avenue, Suite 1700
Chicago, IL 60601-7633

Alzheimer's Association Neuronal Hyper Excitability and Seizures in Alzheimer's Disease Grants 468

The Alzheimer's Association is launching a new initiative to stimulate the development of new pharmacological strategies to prevent or treat seizures and abnormal neural network activity in Alzheimer's disease (AD). The Association's Request for Applications (RFAs) is aimed at the identification, screening and development of therapeutic strategies to reduce seizures and other types of abnormal neural network activity and at the evaluation of drug safety and efficacy at the preclinical and clinical levels. The RFA is designed to enable preliminary pilot research or proof-of-principle studies that can provide data for further research support by other funding agencies. Allowable costs under this award include: purchase and care of laboratory animals; small pieces of laboratory equipment and laboratory supplies; computer equipment if used strictly for data collection; travel (up to $1,000 per year); and salary for the principal investigator, scientific (including post-doctoral fellows) and technical staff (including laboratory technicians and administrative support staff whose work is directly related to the funded project). The Association anticipates funding up to two (2) Neuronal Hyper Excitability and Seizures in AD awards. Each award is limited to $400,000 (direct and indirect costs) for two to three years. Requests in any given year may not exceed $200,000 (direct and indirect costs). Indirect costs are capped at 10 percent (rent for laboratory/office space is expected to be covered by indirect costs paid to the institution). Letters of Intent (LOIs) must be received by December 20, with completed applications due by February 7.
Requirements: Researchers with full-time staff or faculty are encouraged to apply.
Restrictions: Applications from post-doctoral candidates will not be accepted. Costs not allowed under this award include: tuition; computer hardware or software for investigators; rent for laboratory/office spaces; or construction or renovation costs.
Geographic Focus: All States
Date(s) Application is Due: Feb 7
Amount of Grant: Up to 400,000 USD
Contact: Nico Stanculescu, Program Administrator; (312) 335-5747 or (312) 335-5862; fax (312) 335-4034; nico.stanculescu@alz.org or grantsapp@alz.org
Internet: http://www.alz.org/research/alzheimers_grants/types_of_grants.asp
Sponsor: Alzheimer's Association
225 North Michigan Avenue, Suite 1700
Chicago, IL 60601-7633

Alzheimer's Association New Investigator Research Grants 469

The purpose of this program is to provide new investigators with funding that will allow them to develop preliminary or pilot data, to test procedures, and develop hypotheses which will then underpin the preparation of research grant applications to NIH, NSF, and other funding agencies and groups, including the Alzheimer's Association. All applications submitted to the program must focus on a question or questions in interventions for Alzheimer's disease to be considered responsive to the program announcement. Thirty awards will be made under this program. Letters of Intent (LOIs) must be received by December 20, with completed applications due by February 7.
Requirements: Public, private, domestic and foreign research laboratories, medical centers and hospitals, and universities are eligible to apply. Eligibility is restricted to investigators who have less than 10 years of research experience, including postdoctoral fellowships or residencies, after receipt of the doctoral degree. Applications from graduate and doctoral students for research projects, which will be used for the thesis or dissertation, will be accepted.
Restrictions: The Association anticipates funding 45 awards under this competition. Each total award is limited to $100,000 (direct and indirect costs) for up to two years. Requests in any given year may not exceed $60,000 (direct and indirect costs). Indirect costs are capped at 10 percent (rent for laboratory/office space is expected to be covered by indirect costs paid to the institution).
Geographic Focus: All States
Date(s) Application is Due: Feb 7
Amount of Grant: Up to 100,000 USD
Contact: Nico Stanculescu, Program Administrator; (312) 335-5747 or (312) 335-5862; fax (312) 335-4034; nico.stanculescu@alz.org or grantsapp@alz.org
Internet: http://www.alz.org/research/alzheimers_grants/types_of_grants.asp
Sponsor: Alzheimer's Association
225 North Michigan Avenue, Suite 1700
Chicago, IL 60601-7633

Alzheimer's Association New Investigator Research Grants to Promote Diversity 470

The New Investigator Research Grant to Promote Diversity (NIRGD) in Alzheimer's research is a two-year award to investigators who are currently underrepresented at academic institutions in Alzheimer's or related dementias research. The objective of this award is to increase the number of highly trained investigators from diverse backgrounds whose basic, clinical and social/behavioral research interests are grounded in the advanced methods and experimental approaches needed to solve problems related to Alzheimer's and related dementias in general and in health disparities populations. Each NIRGD award is limited to $100,000 (direct and indirect costs) for up to two years. Requests in any given year may not exceed $60,000 (direct and indirect costs). Letters of Intent (LOIs) must be received by December 20, with completed applications due by February 7.
Requirements: Eligibility to apply for this grant competition is restricted to investigators who have less than 10 years of research experience after receipt of their terminal degree.
Restrictions: Costs not allowed under this award include: tuition; computer hardware or software for investigators; rent for laboratory/office space; or construction or renovation costs. Your budget must not exceed the maximum amount of the award, $100,000 or $60,000 per year.
Geographic Focus: All States
Date(s) Application is Due: Feb 7
Amount of Grant: Up to 100,000 USD
Contact: Nico Stanculescu, Program Administrator; (312) 335-5747 or (312) 335-5862; fax (312) 335-4034; nico.stanculescu@alz.org or grantsapp@alz.org
Internet: http://www.alz.org/research/alzheimers_grants/types_of_grants.asp
Sponsor: Alzheimer's Association
225 North Michigan Avenue, Suite 1700
Chicago, IL 60601-7633

Alzheimer's Association U.S.-U.K. Young Investigator Exchange Fellowships 471

The U.S.-U.K. Young Investigator Exchange Fellowship provides a three-year grant to fund quality scientific research into the causes, diagnosis and treatment of Alzheimer's disease. The fellowship aims to address important research questions as well as to help and encourage promising scientists as they establish their careers within Alzheimer's research internationally. It is also hoped that by supporting meaningful scientific collaboration between scientists in the United Kingdom and the United States, there will be mutual benefit to the research output of both countries. In addition to an exchange of ideas, an important aim is to promote the learning of new experimental techniques and methodologies. The purpose of this fellowship is to provide new investigators with funding that will allow them to develop preliminary or pilot data, to test procedures and to develop hypotheses on an international level. The intent is to support international early-career development that will lay the groundwork for future research grant applications to other international funding agencies and groups. Allowable costs under this award include: purchase and care of laboratory animals; small pieces of laboratory equipment and laboratory supplies; computer equipment if used strictly for data collection; travel (up to $1,000 per year); and salary for the principal investigator, scientific (including post-doctoral fellows) and technical staff (including laboratory technicians and administrative support staff whose work is directly related to the funded project). The Alzheimer's Association and the Alzheimer's Reearch United Kingdom anticipate funding up to four awards (two originating from each country) under this competition. Each total award is limited to $300,000 USD. A total of $260,000 USD will be awarded for costs related to the proposed research for up to three years (direct and indirect costs). Requests in any given year may not exceed $90,000 USD (direct and indirect). Indirect costs are capped at 10 percent (rent for laboratory/office space is expected to be covered by indirect costs paid to the institution). The PI must commit to a 75 percent effort toward the proposed project over the funding period. Letters of Intent (LOIs) must be received by December 20, with completed applications due by February 7.
Requirements: Researchers with full-time staff or faculty are encouraged to apply.
Restrictions: Applications from post-doctoral candidates will not be accepted. Costs not allowed under this award include: tuition; computer hardware or software for investigators; rent for laboratory/office space; or construction or renovation costs.
Geographic Focus: All States, United Kingdom
Date(s) Application is Due: Feb 7
Amount of Grant: Up to 300,000 USD
Contact: Nico Stanculescu, Program Director; (312) 335-5747 or (312) 335-5862; fax (312) 335-4034; nico.stanculescu@alz.org or grantsapp@alz.org
Internet: http://www.alz.org/research/alzheimers_grants/types_of_grants.asp
Sponsor: Alzheimer's Association
225 North Michigan Avenue, Suite 1700
Chicago, IL 60601-7633

Alzheimer's Association Zenith Fellows Awards 472

The Zenith Fellows award was initiated in 1991 to provide a vehicle for research support for donors with a substantial personal commitment to the advancement of Alzheimer's disease research. The objective of the Zenith Fellows Awards competition is to provide major support for investigators who have: contributed significantly to the field of Alzheimer's disease research or made significant contributions to other areas of science and are now beginning to focus more directly on problems related to Alzheimer's disease;

and are likely to make substantial contributions in the future. We anticipate funding up to four awards under this competition. Each award is limited to $450,000 (direct and indirect costs) for three years. Requests in any given year may not exceed $250,000 (direct and indirect costs). Indirect costs are capped at 10 percent (rent for laboratory/office space is expected to be covered by indirect costs paid to the institution).
Requirements: Only established independent investigators are eligible as evidenced by: academic appointment; major, peer-reviewed, external multi-year grant support on which the applicant is the principal investigator (PI); independent laboratory operation; and quality and independence of publication record.
Restrictions: Previous recipients of Zenith Awards, Alzheimer's Disease Center Directors (P50 and P30), Medical and Scientific Advisory Council, and members of the National Board of the Alzheimer's Association are not eligible to apply.
Geographic Focus: All States
Date(s) Application is Due: Feb 7
Amount of Grant: Up to 450,000 USD
Contact: Nico Stanculescu, Program Administrator; (312) 335-5747 or (312) 335-5862; fax (312) 335-4034; nico.stanculescu@alz.org or grantsapp@alz.org
Internet: http://www.alz.org/research/alzheimers_grants/types_of_grants.asp
Sponsor: Alzheimer's Association
225 North Michigan Avenue, Suite 1700
Chicago, IL 60601-7633

AMA-MSS Chapter Involvement Grants 473
The AMA-MSS Chapter Involvement Grant (CIG) Program provides each MSS chapter with up to $1,000 per academic year for chapter activities including recruitment and retention, chapter development, education, and community service events. A maximum of $250 or $500 is available for each event, depending on the type of event. Applications are due at least 30 days prior to the event.
Geographic Focus: All States
Amount of Grant: 250 - 500 USD
Contact: Rebecca Gierhahn, Director; (800) 262-3211, ext. 4753; rebecca.gierhahn@ama-assn.org or mss@ama-assn.org
Internet: http://www.ama-assn.org/ama/pub/about-ama/our-people/member-groups-sections/medical-student-section/opportunities/grants-awards-scholarships.shtml
Sponsor: American Medical Association
515 N State Street
Chicago, IL 60654

AMA-MSS Chapter of the Year (COTY) Award 474
The AMA-MSS Chapter of the Year (COTY) Award recognizes the true strength of the AMA-MSS organization, the local chapters. Applicant chapters are judged in a number of areas, including membership, community service, advocacy, innovation, and collaboration. The winning chapter is awarded a $500 grant to be used toward chapter activities and is recognized at the MSS Annual Meeting. All MSS chapters are encouraged to apply.
Geographic Focus: All States
Date(s) Application is Due: Apr 30
Amount of Grant: 500 USD
Samples: University of Texas Medical School at Houston; University of Arkansas for Medical Sciences College of Medicine.
Contact: Rebecca Gierhahn, Director; (800) 262-3211, ext. 4753; rebecca.gierhahn@ama-assn.org or mss@ama-assn.org
Internet: http://www.ama-assn.org/ama/pub/about-ama/our-people/member-groups-sections/medical-student-section/opportunities/grants-awards-scholarships.shtml
Sponsor: American Medical Association
515 N State Street
Chicago, IL 60654

AMA-MSS Government Relations Advocacy Fellowship 475
The AMA-MSS Government Relations Advocacy Fellowship (GRAF) for medical students is currently in its fifth year. One Fellow is selected each spring to work in Washington, D.C., as a full-time paid member of the AMA's Federal advocacy team for one year. The Fellow is responsible for working with the AMA's federal advocacy team to advance the Association's legislative agenda and policies on behalf of physicians, patients and medical students. A key goal for the fellowship is to educate medical student, resident and young physician AMA members about issues in public health and health policy in order to encourage activism and leadership in local communities. Applicants are not expected to be proficient advocates and managers at the start of the fellowship. It is meant to be a learning experience where proficiencies will be developed or enhanced. Prior health policy, legislative, or political experience is helpful. For additional information, please contact the Department of Medical Student Services at mss@ama-assn.org.
Requirements: Any AMA-MSS member in good standing at a medical school in the United States is eligible to apply.
Geographic Focus: All States
Date(s) Application is Due: Jan 31
Contact: Rebecca Gierhahn, Director; (800) 262-3211, ext. 4753; rebecca.gierhahn@ama-assn.org or mss@ama-assn.org
Internet: http://www.ama-assn.org/ama/pub/about-ama/our-people/member-groups-sections/medical-student-section/opportunities/internships-fellowships.shtml
Sponsor: American Medical Association
515 N State Street
Chicago, IL 60654

AMA-MSS Government Relations Internship Programs 476
The Department of Medical Student Services, in conjunction with the Washington, D.C., office of the American Medical Association, offer assistance to students seeking to increase their involvement and education in national health policy and in the national legislative activities of organized medicine. The Government Relations Internship Program (GRIP) provides stipends to assist selected students who are completing summer health policy internships in the Washington, D.C., area. Medical students report that their credibility as AMA-sponsored interns allows them to pursue issues in depth, and in many cases, become experts on a specific topic. Similarly, internship sponsors welcome the enthusiasm and unique perspective that medical students bring to discussions about issues that require medicine's point of view. All students selected to participate in GRIP will receive a stipend. Additionally, GRIP participants completing internships with non-governmental organizations will benefit from attendance at weekly seminars conducted at the AMA Washington Office. These seminars are designed to increase the continuity of the internship experience, promote camaraderie among medical students working in health policy, and facilitate continuing education on important political issues. Upon completion of the program, all GRIP participants will be required to complete and submit a report on their internship experiences. The GRIP application deadline is February 15. However, late applications (through April 15) will be considered on a rolling basis if space remains. If you have any questions, please contact Keith Voogd, MSS Policy Analyst, in the Department of Medical Student Services at (312) 464-4745.
Requirements: To be eligible for the GRIP program, students must be AMA members who have secured policy internships in the Washington, D.C., area during the summer. Eligible internship sites include: National specialty societies seated in the AMA House of Delegates; Public health advocacy groups; Non-clinical international health policy programs (limited spots); Congressional offices*; Federal agencies*.
Restrictions: Please note that internships with narrowly focused lobbying organizations or with specialty organizations that are not seated in the AMA House of Delegates are strongly discouraged. *Due to new Congressional lobbying disclosure rules, interns in Congressional offices or federal agencies will not be able to participate in the weekly educational seminars at the AMA's Washington, D.C., office.
Geographic Focus: All States
Date(s) Application is Due: Feb 15
Contact: Rebecca Gierhahn; (800) 262-3211, ext. 4753; rebecca.gierhahn@ama-assn.org
Internet: http://www.ama-assn.org/ama/pub/about-ama/our-people/member-groups-sections/medical-student-section/opportunities/internships-fellowships.shtml
Sponsor: American Medical Association
515 N State Street
Chicago, IL 60654

AMA-MSS Research Poster Award 477
The MSS/RFS Joint Research Poster Symposium is held annually at the MSS Interim Meeting (November). Research is presented in eight categories (biochemistry/cell biology, cancer biology, cardiology/vascular biology, clinical/epidemiological/health care, immunology/microbiology, neurobiology/neuroscience, radiology/imaging, surgery). The overall winner receives a free trip to the MSS Annual Meeting in Chicago.
Geographic Focus: All States
Date(s) Application is Due: Sep 1
Contact: Rebecca Gierhahn, Director; (800) 262-3211, ext. 4753; rebecca.gierhahn@ama-assn.org or mss@ama-assn.org
Internet: http://www.ama-assn.org/ama/pub/about-ama/our-people/member-groups-sections/medical-student-section/opportunities/grants-awards-scholarships.shtml
Sponsor: American Medical Association
515 N State Street
Chicago, IL 60654

AMA-RFS and AMA Foundation Medical Student & Resident/Fellow 478
Elective-Medicine & the Media
Discovery Communications, the leading global real-world media company, is offering an exclusive four to six week elective for American Medical Association (AMA) medical student and resident/fellow members. As part of Discovery's ongoing dedication to education, Discovery Channel produces continuing medical education (CME) programs that air on the Discovery Channel weekly. During the elective, one AMA medical student and one AMA resident/fellow will each work closely with all members of the Discovery team to develop a medical education program that will air on the network and be viewed by over 2 million people. This exciting opportunity gives the student and resident/fellow hands-on experience in the translation of rigorous scientific data into an entertaining and informational program. The student and resident/fellow will work primarily at Discovery Communications headquarters in Silver Spring, MD, but will also travel to relevant on-site and studio shoots for the program. A stipend will be provided to cover living expenses based on length of rotation ($3,000 for four-week rotation). Dates and length of rotation are flexible. Applications for the Winter elective are due September 30. Applicants not selected for the Winter elective must resubmit for the Summer elective if they wish to be considered for it. Summer applications are due March 15.
Requirements: All medical student and resident/fellow members of the American Medical Association (AMA) are eligible to apply for an exclusive elective in health communications with Discovery Channel in Silver Spring, MD. One AMA medical student and one AMA resident/fellow will be selected.
Geographic Focus: All States
Date(s) Application is Due: Mar 15; Sep 30
Amount of Grant: 3,000 - 4,500 USD
Contact: Jon Fanning, Director; (800) 262-3211, ext. 4978 or (312) 464-4978; fax (312) 464-5845; Jon.Fanning@ama-assn.org or rfs@ama-assn.org

Internet: http://www.ama-assn.org/ama/pub/about-ama/our-people/member-groups-sections/resident-fellow-section/awards-grants.shtml
Sponsor: American Medical Association Resident and Fellow Services
515 N State Street, 14th Floor
CHicago, IL 60654

AMA-RFS Legislative Awareness Internships   479
The Department of Resident and Fellow Services, in conjunction with the Washington, D.C. office of the American Medical Association, is sponsoring a two-week legislative internship program in the spring. Two residents and/or fellows will be selected. The program will afford residents and fellows the unique opportunity to participate in the political process of organized medicine at the national level. Each resident or fellow will receive a $1,000 stipend to help defray program-related expenses (e.g., airfare, hotel, cabs); all other expenses are the responsibility of the resident or fellow. The selected resident or fellow is also responsible for securing his or her own housing accommodations.
Requirements: Applicants must be members of the AMA.
Restrictions: The selected resident or fellow must obtain permission to participate in the program from their program director prior to the selection process.
Geographic Focus: All States
Amount of Grant: 1,000 USD
Contact: Jon Fanning, Director; (800) 262-3211, ext. 4978 or (312) 464-4978; fax (312) 464-5845; Jon.Fanning@ama-assn.org or rfs@ama-assn.org
Internet: http://www.ama-assn.org/ama/pub/about-ama/our-people/member-groups-sections/resident-fellow-section/awards-grants.shtml
Sponsor: American Medical Association Resident and Fellow Services
515 N State Street, 14th Floor
CHicago, IL 60654

AMA-WPC Joan F. Giambalvo Memorial Scholarships   480
The American Medical Association (AMA) Foundation in association with the AMA Women Physicians Congress (WPC) has established the Joan F. Giambalvo Memorial Scholarship Fund with the goal of advancing the progress of women in the medical profession and strengthening the ability of the AMA to identify and address the needs of women physicians and medical students. Proposals for the Joan F. Giambalvo Memorial Scholarship Fund will be accepted between Nov. 1 and Feb. 15. The AMA-WPC Joan F. Giambalvo Memorial Scholarship seeks innovative research proposals focusing on professional work/practice issues that affect women physicians, including, but not limited to: part-time working strategies; leadership training protocols; gender-based physician practice patterns; physician satisfaction or burnout; retention incentives; practice re-entry issues. Proposals for projects with concurrent/complementary funding and/or plans to use this grant as seed money for larger studies to follow will be received favorably, as will proposals for independent new projects. This award is for a maximum of $10,000. A budget submitted with each applicant's proposal should reflect the anticipated use of funds for the amount requested. Proposals will be evaluated and awardee(s) selected based on a variety of factors, including, but not limited to: the innovation, quality and/or feasibility of the idea; project/research methodology; potential to produce action, change or more comprehensive studies; and career goals of the applicant.
Requirements: Applicants/awardees shall be: female or male; working alone or collaboratively on the specific research project; physicians, medical students, other health professionals, or individuals working or doing graduate work in an applicable profession such as public health, sociology, psychology, etc.
Restrictions: Members of the AMA-WPC Governing Council, AMA Foundation Board of Directors, AMA Board of Trustees and Joan F. Giambalvo Memorial Scholarship Fund selection committee, and AMA and AMA Foundation staff are not eligible for this award.
Geographic Focus: All States
Date(s) Application is Due: Feb 15
Amount of Grant: Up to 10,000 USD
Contact: Alice Reed, Scholarship Contact; (312) 464-5523; alice.reed@ama-assn.org
Internet: http://www.ama-assn.org/ama/pub/about-ama/our-people/member-groups-sections/women-physicians-congress/about-wpc/joan-f-giambalvo-memorial-scholarship.shtml
Sponsor: American Medical Association
515 N State Street
Chicago, IL 60654

Amador Community Foundation Grants   481
The foundation supports organizations that enhance the quality of life for the people of Amador County by: encouraging private giving for the public good by providing a flexible, cost-effective and tax-exempt vehicle for donors with varied charitable interests and abilities to give; building and maintaining a permanent endowment fund in order to provide a continuing source of income for grants; making grants that are innovative, strategic and relevant in the support of nonprofit sectors; and serving as a catalyst to address changing and challenging community issues. There are no set project or program areas for funding.
Restrictions: The Foundation does not award grants for political or religious purposes, to retire long-term indebtedness, to influence legislation or elections, to private foundations and other grant-making organizations.
Geographic Focus: California
Contact: Grants Administrator; (209) 223-2148; fax (209) 223-4569; acf@amadorcommunityfoundation.org
Internet: http://www.amadorcommunityfoundation.org/grant.html
Sponsor: Amador Community Foundation
21-B Main Street, P.O. Box 1154
Jackson, CA 95642

AMA Foundation Arthur N. Wilson, MD, Scholarship   482
The Arthur N. Wilson, MD Scholarship provides a $5,000 tuition assistance scholarship to a currently enrolled medical student who graduated from a high school in Southeast Alaska. This scholarship highlights the importance of supporting future physicians in rural communities. The majority of Southeast Alaska is part of the Tongass National Forest, the United States' largest national forest. Major cities are Juneau, Sitka, and Ketchikan. Due to the rural nature of Southeast Alaska, many communities have no road connections outside of their locale, so aircraft and boats are a major means of transport.
Geographic Focus: Alaska
Date(s) Application is Due: Jun 15
Amount of Grant: 5,000 USD
Samples: 2008 - Jodie Totten, University of Washington School of Medicine; 2007 - Mackenzie Slater - University of Washington School of Medicine.
Contact: Steven W. Churchill, Executive Director; (312) 464-4200; fax (312) 464-4142; Steven.Churchill@ama-assn.org
Internet: http://www.ama-assn.org/ama/pub/about-ama/ama-foundation/our-programs/medical-education/arthur-n-wilson.shtml
Sponsor: American Medical Association Foundation
515 N State Street
Chicago, IL 60654

AMA Foundation Dr. Nathan Davis International Awards in Medicine   483
Named for the founder of the AMA, the Dr. Nathan Davis International Award in Medicine recognizes physicians whose influence reach the international patient population and change the future of their medical care. By treating, educating and counseling patients beyond the U.S. border, the physician's work is having a positive impact on health care in the global arena. A $2,500 grant will be given to the institution or organization with which the recipient works. The recipient will also receive travel expenses and accommodations to the Excellence in Medicine Awards Banquet, and the AMA National Advocacy Conference in Washington D.C.
Requirements: To qualify, nominees must: have improved dramatically medical practice, medical education or medical research outside of the United States; embody the values of the medical profession through leadership, service, excellence, integrity and ethical behavior; during a lifetime of service or in a current initiative, has benefited the health and well-being of a specific patient population.
Geographic Focus: All States
Date(s) Application is Due: Nov 16
Amount of Grant: 2,500 USD
Contact: Steven W. Churchill, Executive Director; (312) 464-2593 or (312) 464-4200; fax (312) 464-4142; Steven.Churchill@ama-assn.org or amafoundation@ama-assn.org
Internet: http://www.ama-assn.org/ama/pub/about-ama/ama-foundation/our-programs/public-health/excellence-medicine-awards.shtml
Sponsor: American Medical Association Foundation
515 N State Street
Chicago, IL 60654

AMA Foundation Fund for Better Health Grants   484
The philosophy of the AMA Foundation Fund for Better Health begins with the idea that local communities and organizations have great knowledge and insight into their community's health care issues. Based on this thought, the AMA Foundation, with support from the AMA Alliance, created the Fund for Better Health. Through this program, the AMA Foundation provides seed grants for grassroots, public health projects in communities throughout America. Over the years, the fund has provided over 200 grants totaling nearly $300,000 to projects that address healthy lifestyles, domestic violence prevention, substance abuse prevention, health literacy, patient safety and care for the uninsured. A maximum of $5,000 will be distributed to each grant recipient. The number of grant recipients will be determined by the AMA Foundation after all applications have been received. Typically, the number of grants awarded does not exceed twenty.
Requirements: Organizations are eligible to apply for grants which further the charitable and educational purposes of the AMA Foundation. Grants made in 2009 support programs addressing the issue of healthy lifestyles in the areas of nutrition and physical fitness, alcohol, substance abuse and smoking prevention (and cessation), and violence prevention. The three types of organizations eligible to apply are organizations with annual operating budgets of $1 million or less; new organizations begun in the last 5 years; or established organizations starting a new service or expanding a current service to an underserved population.
Restrictions: None of the funds awarded are to pay for staff salary or overhead expenses.
Geographic Focus: All States
Date(s) Application is Due: Jul 15
Contact: Dina Lindenberg, Program Officer; (312) 464-4193; fax (312) 464-5973; dina.lindenberg@ama-assn.org or amafoundation@ama-assn.org
Internet: http://www.ama-assn.org/ama/pub/about-ama/ama-foundation/our-programs/public-health/fund-better-health.shtml
Sponsor: American Medical Association Foundation
515 N State Street
Chicago, IL 60654

AMA Foundation Health Literacy Grants   485
The foundation awards grants to programs that promote clearer communication between patients and their physicians and other health care providers. They are given to four groups: medical students; residents and fellows; physicians, hospital staffs and medical societies; and AMA Alliance groups and community organizations.
Geographic Focus: All States

Amount of Grant: 500 - 4,000 USD
Contact: Steven W. Churchill, Executive Director; (312) 464-2593 or (312) 464-4200; fax (312) 464-4142; Steven.Churchill@ama-assn.org or healthliteracy@ama-assn.org
Internet: http://www.ama-assn.org/ama/pub/about-ama/ama-foundation/our-programs/public-health/health-literacy-program.shtml
Sponsor: American Medical Association Foundation
515 N State Street
Chicago, IL 60654

AMA Foundation Healthy Communities/Healthy America Grants 486
Through the Healthy Communities/Healthy America program, the AMA Foundation awards $10,000-$25,000 grants to physician-led free clinics. Grants will be awarded to free clinics that: are requesting funds for specific projects, not for activities such as routine operations, maintenance or facility repairs; have regular and considerable operating hours; have significant physician involvement; and provide medical services. Preference will be given to applicants who demonstrate how grant dollars will be leveraged to provide the greatest amount of care.
Restrictions: Clinics that provide both medical and dental services, but are requesting funds for a project that is dental care-specific will not be considered.
Geographic Focus: All States
Amount of Grant: 10,000 - 25,000 USD
Contact: Steven W. Churchill, Executive Director; (312) 464-2593 or (312) 464-4200; fax (312) 464-4142; Steven.Churchill@ama-assn.org or healthliteracy@ama-assn.org
Internet: http://www.ama-assn.org/ama/pub/about-ama/ama-foundation/our-programs/public-health/healthy-communities-healthy.shtml
Sponsor: American Medical Association Foundation
515 N State Street
Chicago, IL 60654

AMA Foundation Jack B. McConnell Awards for Excellence in Volunteerism 487
The Jack B. McConnell, MD, Award for Excellence in Volunteerism recognizes the work of senior physicians who provides treatment to U.S. patients who lack access to health care. After a full career of practice, these physicians remain dedicated to the future of medicine through the spirit of volunteerism. A $2,500 grant will be given to the institution or organization with which the recipient works. The recipient will also receive travel expenses and accommodations to the Excellence in Medicine Awards Banquet, and the AMA National Advocacy Conference in Washington D.C.
Requirements: To qualify, nominees must: have volunteered a significant portion of their medical services while over the age of 55; demonstrate their commitment to health care access by assisting underserved U.S. patients.
Geographic Focus: All States
Date(s) Application is Due: Nov 16
Amount of Grant: 2,500 USD
Contact: Steven W. Churchill, Executive Director; (312) 464-2593 or (312) 464-4200; fax (312) 464-4142; Steven.Churchill@ama-assn.org or amafoundation@ama-assn.org
Internet: http://www.ama-assn.org/ama/pub/about-ama/ama-foundation/our-programs/public-health/excellence-medicine-awards.shtml
Sponsor: American Medical Association Foundation
515 N State Street
Chicago, IL 60654

AMA Foundation Joan F. Giambalvo Memorial Scholarships 488
The American Medical Association (AMA) Women Physicians Congress (WPC) has established the Joan F. Giambalvo Memorial Scholarship Fund and partners with the AMA Foundation with the goal of advancing the progress of women in the medical profession and strengthening the AMA's ability to identify and address the needs and interests of women physicians and medical students.
Requirements: Applicants shall be: Female or male; Working alone or collaboratively on the specific research project; Physicians, medical students, other health professionals, or individuals working or doing graduate work in an applicable profession such as public health, sociology, psychology, etc. Proposals will be evaluated and awardee(s) selected based on a variety of factors, including, but not limited to: the innovation, quality and/or feasibility of the idea; project/research methodology; potential to produce action, change or more comprehensive studies; and career goals of the applicant. AMA membership is not required in order to be eligible for this award.
Restrictions: Members of the AMA-WPC Governing Council, AMA Foundation Board of Directors, AMA Board of Trustees and Joan F. Giambalvo Memorial Scholarship Fund selection committee, and AMA and AMA Foundation staff are not eligible for this award.
Geographic Focus: All States
Date(s) Application is Due: Feb 15
Amount of Grant: Up to 10,000 USD
Contact: Alice Reed, (312) 464-5523; alice.reed@ama-assn.org
Internet: http://www.ama-assn.org/ama/pub/category/15566.html
Sponsor: American Medical Association Foundation
515 N State Street
Chicago, IL 60654

AMA Foundation Jordan Fieldman, MD, Awards 489
The Jordan Fieldman, MD Award was established in the name of Dr. Jordan Fieldman and is sponsored by the AMA Resident and Fellow Section in association with the AMA Foundation. Dr. Jordan Fieldman was an outstanding physician and deeply concerned with helping his patients and making the world of medicine a better place. Unfortunately, Jordan lost his battle with cancer in June 2004. In establishing this award, the AMA-RFS desires to continue to create physicians like Dr. Fieldman and give them the skills and means to do so. The award money will cover resident travel to both the Annual and Interim meeting during the year of the award and necessary expenses while attending these meetings. This award will give a young doctor an opportunity that he or she may otherwise never be able to experience.
Requirements: The Awardee must be: a first-time delegate or attendee to the AMA-RFS meetings; from a state or district that does not have funding available to support resident or fellow travel to attend the AMA-RFS Meetings; interested and active in patient advocacy efforts; involved in organized medicine. Additional guidelines, as well as, an online application are available at, http://www.ama-assn.org/ama/pub/about-ama/ama-foundation/our-programs/public-health/jordan-fieldman-md.shtml.
Geographic Focus: All States
Contact: Sharyn Grose, Program Administrator; (312) 464-4978 or (800) 621-8335; fax (312) 464-5845; sharyn.grose@ama-assn.org or rfs@ama-assn.org
Internet: http://www.ama-assn.org/ama/pub/about-ama/ama-foundation/our-programs/public-health/jordan-fieldman-md.shtml
Sponsor: American Medical Association Foundation
515 N State Street
Chicago, IL 60654

AMA Foundation Leadership Awards 490
An exceptional medical professional goes beyond the medical practice to positively influence health care. Through organized medicine and community activities, individuals from each stage of the physician lifecycle can, and do, make a difference in the quality of health care and the medical environment. The Leadership Awards are presented to 15 medical students, 10 residents/fellows and 5 early career physicians to recognize their strong, nonclinical leadership skills in advocacy, community service and/or education. Award recipients will be invited to attend leadership development training in Washington, D.C. This training will strengthen leadership skills and result in a greater effort to advance health care in America. Directly following the training, recipients will attend the Excellence in Medicine Awards Banquet and stay for the AMA National Advocacy Conference. Airfare and accommodations are provided, in addition to a nominal reimbursement to help defray additional travel-related costs.
Requirements: All Leadership Awards are self-nominated, with applicants responsible for submitting all appropriate materials and documentation. To qualify, applicants must demonstrate outstanding leadership in the areas of advocacy, community service and/or education and be either: a medical student enrolled in an accredited medical school; a resident physician enrolled in an accredited residency program; a fellow physician enrolled in a fellowship program; an early-career physician under the age of 40 or in his or her first eight years of practice following residency/fellowship.
Geographic Focus: All States
Date(s) Application is Due: Nov 16
Contact: Steven W. Churchill, Executive Director; (312) 464-2593 or (312) 464-4200; fax (312) 464-4142; Steven.Churchill@ama-assn.org or amafoundation@ama-assn.org
Internet: http://www.ama-assn.org/ama/pub/about-ama/ama-foundation/our-programs/public-health/excellence-medicine-awards.shtml
Sponsor: American Medical Association Foundation
515 N State Street
Chicago, IL 60654

AMA Foundation Minority Scholars Awards 491
The AMA Foundation, in collaboration with the AMA Minority Affairs Consortium (MAC), with support from Pfizer Inc, offers the Minority Scholars Award. Approximately twelve Minority Scholars Awards are awarded annually, each in the amount of a $10,000 scholarship.
Requirements: You must be a current first or second-year student and a permanent resident or citizen of the U.S. Eligible students of minority background include African American/Black, American Indian, Native Hawaiian, Alaska Native and Hispanic/Latino. Each medical school is invited to submit up to two nominees. Contact your medical school if you are interested in being nominated for the Minority Scholars Award.
Geographic Focus: All States
Date(s) Application is Due: Apr 15
Amount of Grant: 10,000 USD
Contact: Steven W. Churchill, Executive Director; (312) 464-4200; fax (312) 464-4142; Steven.Churchill@ama-assn.org
Internet: http://www.ama-assn.org/ama/pub/category/20116.html
Sponsor: American Medical Association Foundation
515 N State Street
Chicago, IL 60654

AMA Foundation Physicians of Tomorrow Scholarships 492
These $10,000 scholarships reward current third-year medical students, who are entering their fourth-year of study. There will be 12 Physicians of Tomorrow scholarships funded by the AMA Foundation. Multiple scholarships, funded by the AMA Foundation, the Audio-Digest Foundation and the Johnson F. Hammond, MD Fund will be awarded in 2009.
Requirements: The selection of the recipients will be based on academic achievement and financial need. The recipient of the one Physicians of Tomorrow Scholarship funded by the Audio-Digest Foundation should have an interest in 'the communication of science.' Activities such as mentoring and/or teaching are examples of 'communication of science.' The recipient of the one Physicians of Tomorrow Scholarship funded by the Johnson F. Hammond, MD Fund should have an interest in and commitment to a career in medical journalism. Contact your medical school if you are interested in being nominated for the Physicians of Tomorrow Scholarships. Each medical school may submit one nomination for each of these scholarship opportunities. Thus, each school may submit up to three nominations in total.

Geographic Focus: All States
Date(s) Application is Due: May 29
Amount of Grant: 10,000 USD
Samples: 2008: Puya Alikhani, Texas A&M Health Science Center College of Medicine; Diana Badillo, Stanford University School of Medicine; Andrew Barina, Saint Louis University School of Medicine; Karl Bezak, Vanderbilt University School of Medicine; Rozalina Grubina, Johns Hopkins University School of Medicine; Nadia Hernandez, University of Texas Health Science Center at Houston; Stephanie Hu, Harvard Medical School; Arman Kilic, University of Pittsburgh School of Medicine; Youssra Marjoua, Yale University School of Medicine; Monica Patton, University of Vermont College of Medicine; Janae Phelps, Howard University College of Medicine; Dominic Sanford, University of Missouri Columbia; Javay Ross, Charles Drew University/UCLA Medical Education Program; Matthew Bivens, George Washington University School of Medicine and Health Sciences; Helena Hart, University of California, San Francisco School of Medicine.
Contact: Steven W. Churchill, Executive Director; (312) 464-4200; fax (312) 464-4142; Steven.Churchill@ama-assn.org
Internet: http://www.ama-assn.org/ama/pub/category/20119.html
Sponsor: American Medical Association Foundation
515 N State Street
Chicago, IL 60654

AMA Foundation Pride in the Profession Grants 493
The Pride in the Profession Awards honor physicians whose lives encompass the true spirit of being a medical professional: caring for people. By practicing medicine in areas of challenge or crisis, or by devoting their time to volunteerism or public service, these physicians serve as the voice of patients in the United States who otherwise might not be heard. A $2,500 grant will be given to the institution or organization with which the recipient works. The recipient will also receive travel expenses and accommodations to the Excellence in Medicine Awards banquet, and the AMA National Advocacy Conference in Washington D.C.
Requirements: To qualify, nominees must: promote the art and science of medicine and the betterment of the public health; embody the values of the medical profession through leadership, service, excellence, integrity and ethical behavior; enrich patients, colleagues and the community through dedicated medical practice or service; offer better access to quality health care for an underserved patient population in the U.S.
Geographic Focus: All States
Date(s) Application is Due: Nov 16
Amount of Grant: 2,500 USD
Contact: Steven W. Churchill, Executive Director; (312) 464-2593 or (312) 464-4200; fax (312) 464-4142; Steven.Churchill@ama-assn.org or amafoundation@ama-assn.org
Internet: http://www.ama-assn.org/ama/pub/about-ama/ama-foundation/our-programs/public-health/excellence-medicine-awards.shtml
Sponsor: American Medical Association Foundation
515 N State Street
Chicago, IL 60654

AMA Foundation Seed Grants for Research 494
The AMA Foundation established the Seed Grant Research Program to encourage medical students, physician residents and fellows to enter the research field. The program provides $2,500-$5,000 grants to help them conduct small basic science, applied, or clinical research projects. These funds will round out new project budgets, rather than sustain current initiatives. One-year grants will be awarded in the following research categories: Cardiovascular/pulmonary diseases; HIV/AIDS; Leukemia; Neoplastic diseases; Secondhand smoke. Grants in Cardiovascular/Pulmonary Diseases, HIV/AIDS, Leukemia, and Neoplastic Diseases will be $2,500. Grants in the Secondhand smoke category will be $5,000.
Requirements: Applicants must be a medical student, physician resident or fellow of an accredited U.S. medical school or institution; they must also be either a U.S. citizen or a permanent resident of the US. Projects must be applicant-conceived, rather than ongoing research of their mentor or Principal Investigator.
Restrictions: Seed grant funds cannot be used for salary or stipend, indirect/administrative costs, to hire a consultant or contractor, and solely for travel expenses. Seed grants will not be awarded to any applicant who has previously received an AMA Foundation seed grant in the research category in which they are applying.
Geographic Focus: All States
Date(s) Application is Due: Dec 15
Amount of Grant: 2,500 - 5,000 USD
Contact: Program Supervisor; (312) 464-4200; seedgrants@ama-assn.org
Internet: http://www.ama-assn.org/ama/pub/category/7785.html
Sponsor: American Medical Association Foundation
515 N State Street
Chicago, IL 60654

AMA Foundation Worldscopes Program 495
WorldScopes collects stethoscopes from U.S. physicians and distributes them, with the help of humanitarian organizations, to communities around the world where medical supplies are scarce. A stethoscope donation will help a colleague in diagnosing and treating diseases that deprive so many of health, life, and human dignity.
Geographic Focus: All States
Contact: Ethical Force; (312) 464-4075; WorldScopes@ama-assn.org
Internet: http://www.ama-assn.org/ama/pub/category/12768.html
Sponsor: American Medical Association Foundation
515 N State Street
Chicago, IL 60654

AMA Virtual Mentor Theme Issue Editor Grants 496
Virtual Mentor [www.virtualmentor.org] is the American Medical Association's online ethics journal. Its mission is to promote the ethical and professional development of tomorrow's physicians, and its primary audiences are medical students, residents, other physicians, and medical educators. VM serves as an educational resource for a rapidly growing number of readers. VM invites medical students and residents to apply to serve as theme issue editors. Each theme issue editor undertakes, with help from VM staff editors, responsibility for: (1) selecting a theme for one issue and defining the ethical and professionalism concerns inherent in that theme, (2) generating case narratives that provide opportunities for examining those concerns in clinical and educational contexts, (3) identifying, securing, and corresponding with contributors to the issue, (4) editing copy and reviewing page proofs before the issue goes live on the first working day of the month. These tasks will be accomplished at each editor's home location. Theme issue editors, selected in late November, will be flown to Chicago to meet with Virtual Mentor staff some time in February. Each theme issue editor receives a $1,000-stipend.
Requirements: Experience in editing or journalism is desirable. The abilities to critique an argument or article, work well with authors, and meet deadlines are essential. To apply, send a short letter stating your current medical student or resident program status and your reason for wishing to be a theme issue editor for VM. Attach to the letter, the following: a curriculum vitae; a writing sample of not more than 1,000 words; your response to this exercise in 500 words or fewer. Additional guidelines available at: http://virtualmentor.ama-assn.org/site/issue-edscall.html. Send your letter of application and attachments by email to: Faith.Lagay@ama-assn.org. The deadline for applications is midnight CST on November 13.
Geographic Focus: All States
Date(s) Application is Due: Nov 13
Amount of Grant: 1,000 USD
Contact: Faith Lagay, Managing Editor; (312) 464-5438 or (312) 464-5260; fax (312) 464-4799; Faith.Lagay@ama-assn.org or virtualmentor@ama-assn.org
Internet: http://www.ama-assn.org/ama/pub/about-ama/our-people/member-groups-sections/medical-student-section/opportunities/internships-fellowships.shtml
Sponsor: American Medical Association
515 N State Street
Chicago, IL 60654

AMDA Foundation Medical Director of the Year Award 497
The AMDA Foundation's Medical Director of the Year Award recognizes those individuals whose vision, passion, leadership, knowledge, and commitment succeed in taking patient care in the facilities they serve as medical director to exceptional levels of quality, excellence, and innovation. AMDA asks facility staff and their interdisciplinary leaders to identify and nominate outstanding medical directors who are: a physician in good standing with the community and profession; an AMDA member in good standing; a Certified Medical Director (CMD); an experienced medical director (3 or more years); an experienced attending physician in a nursing facility(ies); a proven team leader; a proven clinical leader; involved in community activities; and an effective educator. Nominations for the annual Medical Director of the Year Award begin in July of the preceding year, and the annual deadline is November 9.
Geographic Focus: All States
Date(s) Application is Due: Nov 9
Contact: Christine Ewing; (410) 992-3134; cewing@amdafoundation.org
Internet: http://www.amdafoundation.org/index.php/our-work/recognize-awards/medical-director-of-the-year
Sponsor: American Medical Directors Association Foundation
11000 Broken Land Parkway, Suite 405
Columbia, MD 21044

AMDA Foundation Quality Improvement and Health Outcome Awards 498
The AMDA Foundation's Quality Improvement and Health Outcomes Award (QIHO) program will provide three awards of $1,000 each to facilities that have implemented programs that improved the quality of life for their residents. These awards are based on programs medical directors and care teams have implemented and demonstrated to improve the quality of life for their post-acute/long-term care residents. Types of programs might include: patient safety initiatives (reducing falls, medication errors); reduction of avoidable ER visits and acute hospitalizations; improved consistency of staffing; improved comprehensive advanced care planning; and improved palliative care programs. Each awardee will present their award-winning program at an educational session during the AMDA Annual Conference. Acceptance of applications begin in July, and the annual deadline is November 9.
Requirements: All nursing home facilities are eligible for the awards. Facilities may be for profit or not-for-profit and/or individual facility, regional or national chain. In order for the program to be eligible, the program must be internally generated and funded by the nursing home facility; have demonstrated measurable outcomes and objectives; and have proven sustainability and ability to be replicated in other facilities.
Geographic Focus: All States
Date(s) Application is Due: Nov 9
Amount of Grant: 1,000 USD
Contact: Christine Ewing; (410) 992-3134; cewing@amdafoundation.org
Internet: http://www.amdafoundation.org/index.php/our-work/recognize-awards/amda-foundation-qi-health-outcome-awards
Sponsor: American Medical Directors Association Foundation
11000 Broken Land Parkway, Suite 405
Columbia, MD 21044

GRANT PROGRAMS | 75

AMDA Foundation Quality Improvement Award 499
The AMDA Foundation Quality Improvement Award is a program designed to encourage the development of innovative projects that will help to make a distinct impact on the quality of long term care. The Awards support initiatives that focus on facility staff education, quality improvement programs, research on interventions and treatment, and health literacy to directly enhance the quality of care provided to patients in long term care settings. One project that supports a Quality Improvement program will be selected for the $5,000 award. The winner will present their project results at a future educational session at AMDA's annual symposium. The AMDA Foundation Quality Improvement application should be completed and submitted online by the annual November 9 deadline.
Requirements: The program is open to all: AMDA members (including associate members); residents or fellows in an accredited training program; and mid-career and junior career faculty members of schools of medicine, osteopathy, nursing, or other health-related academic institutions. If the applicant is not an AMDA member the submission must be sponsored by an AMDA member.
Geographic Focus: All States
Date(s) Application is Due: Nov 9
Amount of Grant: 5,000 USD
Contact: Christine Ewing; (410) 992-3134; cewing@amdafoundation.org
Internet: http://www.amdafoundation.org/index.php/our-work/recognize-awards/amda-foundation-qi-awards
Sponsor: American Medical Directors Association Foundation
11000 Broken Land Parkway, Suite 405
Columbia, MD 21044

AMD Corporate Contributions Grants 500
AMD has established two global focus areas: strengthening community and strengthening education. Priority is given to basic needs (food, shelter, and basic health care), education (math and science, teacher development, and college and career awareness), and university education. The k-12 initiatives target programs that increase student interest and/or proficiency in literacy, match, science, and computer technology. AMD also funds programs aimed at developing and supporting effective classroom instruction.
Requirements: Most contributions are made to accredited schools and 501(c)3 nonprofit agencies operating in Austin, Texas (Travis Couunty) or Sunnyvale, California (Santa Clara County) with which AMD has a strong established relationship.
Restrictions: AMD does not consider unsolicited applications for programs outside of the communities in which they operate or outside their focus areas. Also excluded are individuals, medical research, religious, political, service or fraternal organizations, arts or cultural programs, advocacy groups, athletic teams, recreational programs, or individual scouting troops. Organization must be non-discriminatory.
Geographic Focus: All States
Date(s) Application is Due: May 1
Amount of Grant: Up to 25,000,000 USD
Contact: Community Affairs Manager; (800) 538-8450, ext. 45373; fax (408) 749-5373
Internet: http://www.amd.com/us-en/Corporate/AboutAMD/0,,51_52_7697_7702,00.html
Sponsor: AMD Corporation
P.O. Box 3453, M/S 42, 1 AMD Pl
Sunnyvale, CA 94088

Amelia Sillman Rockwell and Carlos Perry Rockwell Charities Fund Grants 501
The Amelia Sillman Rockwell and Carlos Perry Rockwell Charities Fund was established in 1962 to support and promote quality educational, human-services, and health-care programming for underserved populations. Special consideration is given to charitable organizations that serve children or the elderly. Grant requests for general operating support are strongly encouraged. Program support will also be considered. Small, program-related capital expenses may be included in general operating or program requests. The majority of grants from the Rockwell Charities Fund are one year in duration; on occasion, multi-year support is awarded. Applicants must apply online at the grant website. Applicants are strongly encouraged to do the following before applying: review the downloadable state application procedures for additional helpful information and clarifications; review the downloadable online-application guidelines at the grant website; review the foundation's funding history (link is available from the grant website); review the online application questions in advance; and review the list of required attachments. These will generally include: a list of board members, financial statements (audited, reviewed, or compiled by independent auditor); an organization summary; a list of other funding sources; an IRS Determination letter; and other required documents. All attachments must be uploaded in the online application as PDF, Word, or Excel files. The application deadline for the Rockwell Charities Fund is 11:59 p.m. on February 1. Applicants will be notified of grant decisions before May 31.
Requirements: Applicants must have 501(c)3 tax-exempt status.
Restrictions: The trust does not support requests from individuals, organizations attempting to influence policy through direct lobbying, or any political campaigns.
Geographic Focus: Massachusetts
Date(s) Application is Due: Feb 1
Samples: Elizabeth Stone House, Jamaica Plains, Massachusetts, $10,000, general operating support; First Congregational Church, South Windsor, Connecticut, $1,000, for favored charity of Rockwell; Rogerson Communities, Boston, Massachusetts, $10,000, Adult Day Health Programs.
Contact: Miki C. Akimoto, Vice President; (866) 778-6859; miki.akimoto@baml.com
Internet: https://www.bankofamerica.com/philanthropic/fn_search.action
Sponsor: Amelia Sillman Rockwell and Carlos Perry Rockwell Charities Fund
225 Franklin Street, 4th Floor, MA1-225-04-02
Boston, MA 02110

American-Scandinavian Foundation Visiting Lectureship Grants 502
The ASF invites U.S. colleges and universities to apply for funding to host a visiting lecturer from Norway or Sweden. The $20,000 awards are for appointments of one semester. Lectureships should be in the areas of: public policy, conflict resolution, health care, environmental studies or multiculturalism. Letters of Intent are due by November 6, with full applications postmarked by February 5.
Requirements: The competition is open to all American colleges and universities. The award is appropriate not just for Scandinavian studies departments, but for any department or inter-disciplinary program with an interest in incorporating a Scandinavian focus into its course offerings. The lecturer must be a Norwegian or Swedish citizen, and a scholar or expert in a field appropriate to the host department or program.
Geographic Focus: All States
Date(s) Application is Due: Feb 5; Nov 6
Amount of Grant: 20,000 USD
Contact: Ellen McKey; (212) 879-9779; fax (212) 249-3444; grants@amscan.org
Sponsor: American-Scandinavian Foundation
58 Park Avenue
New York, NY 10016

American Academy of Dermatology Camp Discovery Scholarships 503
Every year, the academy sponsors a week of fishing, boating, swimming, water skiing, arts and crafts for young people with a serious skin condition. Under the expert care of dermatologists and nurses, the camp offers the opportunity of spending a week among other young people who have similar skin conditions. Many of the counselors have serious skin conditions as well, and can provide support and advice to campers. The camp fee is covered by full scholarships, including transportation.
Requirements: Campers may attend at any one of four locations selected each year. Campers must meet the age criteria at time of camp (generally between the ages of 8 and 16). Campers must also get a recommendation from a dermatologist who is an academy member. Registration can be completed online at the website.
Geographic Focus: All States
Contact: Jill Mueller, (847) 240-1737; fax (847) 330-8907; jmueller@aad.org
Internet: http://www.campdiscovery.org/
Sponsor: American Academy of Dermatology
930 E Woodfield Road
Schaumburg, IL 60173

American Academy of Dermatology Shade Structure Grants 504
The program awards grants to support the purchase of shade structures designed to provide shade and ultraviolet radiation protection for outdoor areas. Locations can include any area where children and adults gather and are exposed to the harmful rays of the sun, such as playgrounds, pools, bleachers, and eating or recreation areas.
Requirements: The program is open to non-profit organizations or educational institutions that serve children and teenagers, ages 18 and younger. Applicants will be reviewed based on the following: demonstrated commitment to sun safety within the organization and community; sponsorship of application by an academy member dermatologist; and ability to meet the build timeline outlined in grant criteria. Application and guidelines are available online.
Restrictions: Faxed nor emailed applications will not be accepted.
Geographic Focus: All States
Date(s) Application is Due: Apr 12
Amount of Grant: Up to 8,000 USD
Contact: Jennifer Allyn, Program Director; (847) 240-1730; jallyn@aad.org
Internet: http://www.aad.org/public/sun/grants.html
Sponsor: American Academy of Dermatology
930 E Woodfield Road
Schaumburg, IL 60173

American Academy of Nursing Building Academic Geriatric Nursing Capacity Scholarships 505
The Foundation's overall goal is to increase the nation's capacity to provide effective and affordable care to its rapidly increasing older population. Specifically, the Foundation seeks to enhance the training of physicians, nurses, social workers and other health professionals who care for older adults, and promote innovations in the integration and delivery of services. The goal of the Scholarship program is to increase academic geriatric nursing capacity in the United States. BAGNC focuses on the development of academic leadership in gerontological nursing through strong mentorship in the components of academic geriatric nursing (research, teaching and community service); leadership development, a national network of scholars and academic geriatric nurses; and exposure to a wide range of experts in gerontology and geriatrics. Scholars in collaboration with their identified mentor will design and implement a tailored professional development plan designed to support development of new competencies and enhanced effectiveness as an academic leader.
Requirements: Registered nurses who are U.S. citizens or U.S. permanent residents and who hold a degree(s) in nursing are eligible. Predoctoral applicants must: be registered nurses; hold degree(s) in nursing; be United States citizens or permanent U.S. residents; plan an academic and research career; and demonstrate potential for long-term contributions to geriatric nursing.
Geographic Focus: All States
Date(s) Application is Due: Jan 15
Contact: Patricia Archbold; (202) 777-1172; fax (202) 777-0107; parchbold@aannet.org
Internet: http://www.aannet.org/i4a/pages/Index.cfm?pageID=3295
Sponsor: American Academy of Nursing
888 17th Street NW, Suite 800
Washington, D.C. 20006

American Academy of Nursing Claire M. Fagin Fellowships 506

The Fellowship supports two years of full time advanced research and leadership training for doctorally prepared faculty committed to careers in academic geriatric nursing by providing $120,000 for the 2-year fellowship ($60,000 per annum). Program focuses on the development of academic leadership in gerontological nursing through such activities as: research; focused study; networking among scholars, mentors and colleagues in other fields; demonstration of growth in ability to transform self and organizations by moving outside of traditional modes of success; completion and write-up of a significant research project; and by success in achieving funding from other sources. Selected fellows, in collaboration with their mentor, will design and implement an individual professional development plan that will support them in developing new competencies and enhanced effectiveness as an academic leader and researcher. Award programs must begin between July 1st and September 1st of the award year. The program is committed to advancing well-qualified applicants from under-represented minority groups to improve the nation's ability to provide culturally competent care to its increasingly diverse aging population.
Requirements: Applicants must: be doctorally-prepared registered nurses; hold degree(s) in nursing; be United States citizens or permanent U.S. residents; be doctorally-prepared registered nurses,; have the potential to develop into independent investigators; and demonstrate potential for long-term contributions to geriatric nursing. Applications will be accepted from doctoral students who will complete their doctoral program prior to the award. Faculty members in accredited Schools of Nursing who hold the rank of assistant professor or associate professor may apply for fellowships.
Geographic Focus: All States
Date(s) Application is Due: Jan 13
Amount of Grant: 120,000 USD
Contact: Patricia Archbold; (202) 777-1172; fax (202) 777-0107; parchbold@aannet.org
Internet: http://www.geriatricnursing.org/applications/cmf-fellowship.asp
Sponsor: American Academy of Nursing
888 17th Street NW, Suite 800
Washington, D.C. 20006

American Academy of Nursing Mayday Fund Grants 507

The Foundation's overall goal is to increase the nation's capacity to provide effective and affordable care to its rapidly increasing older population. Specifically, the Foundation seeks to enhance the training of physicians, nurses, social workers and other health professionals who care for older adults, and promote innovations in the integration and delivery of services. This program is aimed at candidates whose research includes the study of pain in the elderly. Award programs must begin between July 1st and September 1st.
Requirements: Predoctoral applicants must: be registered nurses; hold degree(s) in nursing; be United States citizens or permanent U.S. residents; plan an academic and research career; and demonstrate potential for long-term contributions to geriatric nursing.
Geographic Focus: All States
Date(s) Application is Due: Jan 9
Amount of Grant: Up to 5,000 USD
Contact: Patricia Archbold; (202) 777-1172; fax (202) 777-0107; parchbold@aannet.org
Internet: http://www.geriatricnursing.org/applications/applications.asp
Sponsor: American Academy of Nursing
888 17th Street NW, Suite 800
Washington, D.C. 20006

American Academy of Nursing MBA Scholarships 508

The Program seeks nurse applicants pursuing a Masters of Business Administration (including Executive MBA) at a highly ranked school of business, and committed to a career focus on the management/leadership of institutions serving older persons. The Scholarship offers grant-funded scholarship support of $50,000 over the period of one or two years to successful candidates who will commence or continue full-time in an MBA program. The Scholarship program includes additional leadership development experiences.
Requirements: Scholarship applicants must: be registered nurses; hold degree(s) in nursing; and be United States citizens or permanent U.S. residents. Applicants must be accepted to, or be currently enrolled in, a highly ranked business school. Applicants must provide specific examples of ability to understand organizational and management challenges as well as strategic approaches to such challenges and provide evidence of previous commitment to gerontology.
Geographic Focus: All States
Date(s) Application is Due: Jan 15
Amount of Grant: Up to 50,000 USD
Contact: Patricia Archbold; (202) 777-1172; fax (202) 777-0107; parchbold@aannet.org
Internet: http://www.geriatricnursing.org/applications/mba-scholarship.asp
Sponsor: American Academy of Nursing
888 17th Street NW, Suite 800
Washington, D.C. 20006

American Academy of Nursing Media Awards 509

The Award is presented to the individual(s) or organization(s) whose use of media has highlighted the unique contribution of nursing and: increased public awareness of the value nursing plays in promoting health and providing health care to the public and/or the need to address the growing nursing shortage; reported the impact of specific public policies on the health status of individuals, communities or the general population; been responsible for motivating specific actions to improve healthcare for diverse groups of people; disseminated nursing research findings nationally (and possibly internationally) using both interdisciplinary and public media; and depicted specific examples of health-enhancing interactions, culturally sensitive health care, health care addressing health disparity issues of under-served and vulnerable populations, health promoting activities or healing actions. Examples of items that may be submitted for the Academy's Media Awards Program include, but are not limited to: newspaper articles, radio programs, motion pictures, public service announcements, books, feature films, television shows, magazine articles, novels, documentaries, popular literature, creative endeavors, and multimedia. Entries are encouraged that highlight nursing practice and/or nursing issues.
Requirements: A non-refundable entry fee of $50.00 is required.
Geographic Focus: All States
Date(s) Application is Due: Jul 27
Contact: Patricia Archbold; (202) 777-1172; fax (202) 777-0107; parchbold@aannet.org
Internet: http://www.aannet.org
Sponsor: American Academy of Nursing
888 17th Street NW, Suite 800
Washington, D.C. 20006

American Chemical Society Alfred Burger Award in Medicinal Chemistry 510

This American Chemical Society Alfred Burger Award in Medicinal Chemistry, supported by GlaxoSmithKline in 1978, is awarded biennially in even-numbered years to recognize outstanding contributions to research in medicinal chemistry. The award is granted without regard to age or nationality, and the recipient presents an award address at the spring meeting of the ACS Division of Medicinal Chemistry. Applications are accepted in odd-numbered years. The award consists of $3,000 and a certificate. Up to $2,500 for travel expenses to the meeting at which the award will be presented will be reimbursed. The award is presented biennially in even-numbered years, and the recipient will present an award address at the spring meeting of the ACS Division of Medicinal Chemistry.
Requirements: Any individual, except a member of the award committee, may submit one nomination or seconding letter for the award in any given year. The nominating documents consist of a letter of not more than 1000 words containing an evaluation of the nominee's accomplishments and a specific identification of the work to be recognized, a biographical sketch including date of birth, and a list of publications and patents authored by the nominee.
Restrictions: Self-nominations are not accepted.
Geographic Focus: All States
Date(s) Application is Due: Nov 1
Amount of Grant: 3,000 - 5,500 USD
Contact: Administrator; (800) 227-5558 or (202) 872-4408; awards@acs.org
Internet: http://webapps.acs.org/findawards/detail.jsp?ContentId=CTP_004491
Sponsor: American Chemical Society
1155 Sixteenth Street, NW
Washington, D.C. 20036-4801

American Chemical Society ANYL Arthur F. Findeis Award for Achievements by a Young Analytical Scientist 511

The purpose of the American Chemical Society ANYL Arthur F. Findeis Award for Achievements by a Young Analytical Scientist is to recognize and encourage outstanding contributions to the fields of analytical chemistry by a young analytical scientist. Evidence shall be presented for one or more of the following outstanding accomplishments: conceptualization and development of unique instrumentation that has had an enabling impact upon analytical chemistry and has substantively advanced the field; development of novel and important analytical methods or methodologies that have found significant beneficial applications in the chemical sciences; elucidation of fundamental events or processes involved in or important to analytical chemistry; authorship of books, patents, and/or research papers that have had an influential role in the development of analytical chemistry; or other significant contributions to the furtherance of analytical chemical sciences role in the use of chemical instrumentation. The Award consists of a plaque and $2,500.
Requirements: The awardee must have earned his or her highest degree within ten years of January 1 of the year of the award. Both the nationality of the young analytical scientist and the arena (e.g., academic, industrial, national laboratory) in which the contributions of the young analytical scientist have been made are unrestricted.
Geographic Focus: All States
Date(s) Application is Due: Nov 1
Amount of Grant: 2,500 USD
Contact: Miquela Sena, (505) 820-0443; fax (505) 989-1073; office@analyticalsciences.org
Internet: http://www.analyticalsciences.org/awards.php#Findeis
Sponsor: American Chemical Society
2019 Galisteo Street, Building I-1
Santa Fe, NM 87505

American Chemical Society ANYL Award for Distinguished Service in the Advancement of Analytical Chemistry 512

The purpose of the American Chemical Society ANYL Award for Distinguished Service in the Advancement of Analytical Chemistry is to honor an individual who through professional service in activities such as teaching, writing, research, and administration has substantially and uniquely enhanced the field of analytical chemistry. Nominations should present evidence for one or more of the following outstanding accomplishments: enhance the positive perception of analytical chemistry in the public eye; foster the development of analytical chemistry research in academic institutions, government laboratories, or in private industries; develop and implement programs that benefit the analytical community (these can be but are not limited to efforts within the Division of Analytical Chemistry or within the American Chemical Society; advance and promote the careers of analytical chemists in any area of employment; and play a central role in improving the way analytical chemistry is practiced role in the use of chemical instrumentation. The Award consists of a plaque and $2,500.
Geographic Focus: All States

Date(s) Application is Due: Nov 1
Amount of Grant: 2,500 USD
Contact: Miquela Sena, (505) 820-0443; fax (505) 989-1073; office@analyticalsciences.org
Internet: http://www.analyticalsciences.org/awards.php#Service
Sponsor: American Chemical Society
2019 Galisteo Street, Building I-1
Santa Fe, NM 87505

American Chemical Society Award for Creative Advances in Environmental Science and Technology — 513

The purpose of the Award is to encourage creativity in research and technology or methods of analysis to provide a scientific basis for informed environmental control decision-making processes, or to provide practical technologies that will reduce health risk factors. The award consists of $5,000 and a certificate. Up to $2,500 for travel expenses to the meeting at which the award will be presented will be reimbursed.
Requirements: Any individual, except a member of the award committee, may submit one nomination or seconding letter for the award in any given year. The nominating documents consist of a letter of not more than 1000 words containing an evaluation of the nominee's accomplishments and a specific identification of the work to be recognized, a biographical sketch including date of birth, and a list of publications and patents authored by the nominee.
Restrictions: The award will be granted without regard to age or nationality. Self-nominations are not accepted.
Geographic Focus: All States
Date(s) Application is Due: Nov 1
Amount of Grant: 5,000 - 7,500 USD
Contact: Administrator; (202) 872-4575 or (202) 872-4408; awards@acs.org
Internet: http://webapps.acs.org/findawards/detail.jsp?ContentId=CTP_004504
Sponsor: American Chemical Society
1155 Sixteenth Street, NW
Washington, D.C. 20036-4801

American Chemical Society Award in Separations Science and Technology — 514

This award is given annually to recognize outstanding accomplishments in fundamental or applied research directed to separations science and technology. The award shall be granted to an individual without regard to age or nationality. The scope of the award is to be as broad as possible, covering all fields where separations science and technology is practiced, including but not limited to biology, chemistry, engineering, geology, and medicine. The award consists of $5,000 and a certificate. Up to $2,500 for travel expenses to the meeting at which the award will be presented will be reimbursed. The recipient will deliver a lecture at the annual Division of Industrial and Engineering Chemistry Separations Science and Technology Symposium.
Requirements: Any individual, except a member of the award committee, may submit one nomination or seconding letter for the award in any given year. The nominating documents consist of a letter of not more than 1000 words containing an evaluation of the nominee's accomplishments and a specific identification of the work to be recognized, a biographical sketch including date of birth, and a list of publications and patents authored by the nominee.
Restrictions: Self-nominations are not accepted.
Geographic Focus: All States
Date(s) Application is Due: Nov 1
Amount of Grant: 5,000 - 7,500 USD
Contact: Administrator; (800) 227-4575 or (202) 872-4408; awards@acs.org
Internet: http://webapps.acs.org/findawards/detail.jsp?ContentId=CTP_004552
Sponsor: American Chemical Society
1155 Sixteenth Street, NW
Washington, D.C. 20036-4801

American Chemical Society Claude S. Hudson Awards — 515

This award, supported by the National Starch and Chemical Company, is given biennially in odd-numbered years to recognize outstanding contributions to carbohydrate chemistry, whether in education, research, or application. The award is granted without regard to age or nationality. Applications are accepted in even-numbered years. The award consists of $5,000 and a certificate. Up to $1,000 for travel expenses to the meeting at which the award will be presented will be reimbursed.
Requirements: Any individual, except a member of the award committee, may submit one nomination or seconding letter for the award in any given year. Nominating documents consist of a letter of not more than 1000 words containing an evaluation of the nominee's accomplishments and a specific identification of the work to be recognized, a biographical sketch including date of birth, and a list of publications and patents authored by the nominee.
Restrictions: Self-nominations are not accepted.
Geographic Focus: All States
Date(s) Application is Due: Nov 1
Amount of Grant: 6,000 USD
Contact: Felicia Dixon, Awards Administrator; (202) 872-4408 or (202) 872-4408; fax (202) 776-8008; f_dixon@acs.org or awards@acs.org
Internet: http://portal.acs.org/portal/acs/corg/content?_nfpb=true&_pageLabel=PP_ARTICLEMAIN&node_id=1319&content_id=CTP_004501&use_sec=true&sec_url_var=region1
Sponsor: American Chemical Society
1155 Sixteenth Street, NW
Washington, D.C. 20036-4801

American Chemical Society GCI Pharmaceutical Roundtable Research Grants — 516

The Roundtable is a partnership with the ACS GCI (Green Chemistry Institute) and pharmaceutical corporations united by a shared commitment to integrate the principles of green chemistry and engineering into the business of drug discovery and production. The program was created to support research in the key green chemistry research areas. (Ten chemical transformations and two process related operations were identified and subsequently published - download the list from the sponsor's website). The primary purpose of this grant is to publish research to make information publicly available. One grant is planned to be awarded. The total award is limited to $150,000 for a grant period of 12 to 24 months.
Requirements: Proposals will only be accepted from public and private institutions of higher education. The grant is not limited to institutions in the United States. Proposals must be submitted through the appropriate institutional office for external funding. Proposed research must focus on the key targeted areas identified by the Roundtable. Proposals must clearly describe how the proposed research will provide improvements over existing technology that are consistent with the goals of Green Chemistry including improved atom economy, use of less hazardous reagents, more energy efficient processes, elimination of by-products, reduced waste generation, improved environmental, health, and safety and life cycle impacts, catalytic versus stoichiometric methods, and development of inherently safer transformations. Illustrations of the relevance of the proposed technology to synthetic and process related problems within the pharmaceutical industry are encouraged. Complete guidelines can be downloaded at the sponsor's website.
Restrictions: Applicants may have only one research grant with the ACS GCI Pharmaceutical Roundtable at a time. Current research grant holders may not apply for a new grant until their active grant or no-cost time extension officially expires.
Geographic Focus: All States
Date(s) Application is Due: Sep 14
Amount of Grant: Up to 150,000 USD
Contact: Gayle Peterman, Finance and Grants Manager; (202) 872-6092 or (202) 872-4481; fax (202) 872-6319; g_peterman@acs.org or gcipr@acs.org
Internet: http://portal.acs.org/portal/acs/corg/content?_nfpb=true&_pageLabel=PP_TRANSITIONMAIN&node_id=1456&use_sec=false&sec_url_var=region1
Sponsor: American Chemical Society
1155 Sixteenth Street, NW
Washington, D.C. 20036-4801

American College of Rheumatology Fellows-in-Training Travel Scholarships — 517

The American College of Rheumatology Fellows-in-Training (FIT) Travel Scholarship is designed to provide funding to assist Fellows- In-Training at U.S. and Canadian programs in rheumatology attend the Annual Meeting. Preference will be given to those applicants who are currently a first or second year adult or first, second or third year pediatric fellows in a training program approved by the Accreditation Council of Graduate Medical Education or the Royal College of Physicians and Surgeons of Canada. Applicants in a third year of an adult or fourth year of a pediatric fellowship will be given consideration after the registration deadline, space permitting on a first come first serve basis.
Requirements: Applicant must be first or second year adult or first, second or third year pediatric fellows in a training program approved by the Accreditation Council of Graduate Medical Education or the Royal College of Physicians and Surgeons of Canada. Applicants must be in an accredited fellowship position at the time of the Annual Meeting.
Restrictions: Individuals with a faculty appointment will not be considered.
Geographic Focus: All States
Date(s) Application is Due: Jul 24
Contact: Coordinator; (404) 633-3777; fax (404) 633-1870; fittravel@rheumatology.org
Internet: http://www.rheumatology.org/fellows/fitscholarship.asp
Sponsor: American College of Rheumatology
1800 Century Place, Suite 250
Atlanta, GA 30345-4300

American College of Surgeons and The Triological Society Clinical Scientist Development Awards — 518

The Triological Society (TRIO) and the American College of Surgeons (ACS) announce a competitive research career development program that will provide supplemental funding to otolaryngologist-head and neck surgeons who receive NIH Mentored Clinical Scientist Development Awards (K08) or Mentored Patient-Oriented Research Development Awards (K23). The TRIO and ACS are offering these awards as a means to facilitate the research career development of otolaryngologists-head and neck surgeons, with the expectation that awardees will have sufficient pilot data to submit a competitive R01 proposal prior to the conclusion of the K awards. Awards will use the K08 or K23 mechanisms. Planning, direction, and execution of the program will be the responsibility of the candidate and her/his mentor on behalf of the applicant institution. The project period will be for a period of up to five years with a minimum of three years. Awards are not renewable. TRIO and ACS will provide salary and fringe benefits for the NIH K08/K23 recipient. A combined total of $80,000 per year in direct costs will be awarded to the institution of the awardee.
Requirements: Applicants must be MD otolaryngologist-head and neck surgeons and members of the American College of Surgeons. Applicants must meet all requirements as set forth in the K08 and K23 application criteria and must be recipients of a new K08 or K23 award or have an existing K08/K23 award with a minimum of three years remaining in the funding period as of June.
Restrictions: Pre-existing applications and awards are not eligible for consideration.
Geographic Focus: All States
Date(s) Application is Due: Apr 15
Amount of Grant: 80,000 USD
Contact: Kate Early; (312) 202-5281; fax (312) 202-5021; kearly@facs.org

Internet: http://www.facs.org/memberservices/acstriol.html
Sponsor: American College of Surgeons
633 N Saint Clair Street
Chicago, IL 60611-3211

American College of Surgeons Australia and New Zealand Chapter Travelling Fellowships 519

The International Relations Committee of the American College of Surgeons announces the availability of a traveling fellowship, the Australia and New Zealand Chapter of the American College of Surgeons Travelling Fellowship. The purpose of this fellowship is to encourage international exchange of information concerning surgical science, practice, and education and to establish professional and academic collaborations and friendships. The Chapter and the College will provide the sum of $8,000 to the successful applicant, who will also be exempted from registration fees for the Annual Scientific Congress. He/she must meet all travel and living expenses.
Requirements: The scholarship is available to a Fellow of the American College of Surgeons in any of the surgical specialties who meets the following *Requirements:* has a major interest and accomplishment in clinical and basic science related to surgery; holds a current full-time academic appointment in Canada or the United States; is under 45 years of age on the date the application is filed; and is enthusiastic and personable and possesses good communication skills. The Fellow is required to spend a minimum of two weeks in Germany. The Fellow is required to spend a minimum of two or three weeks in Australia and New Zealand.
Geographic Focus: All States
Date(s) Application is Due: Nov 16
Amount of Grant: 8,000 USD
Contact: Kate Early; (312) 202-5281; fax (312) 202-5021; kearly@facs.org
Internet: http://www.facs.org/memberservices/traveling.html
Sponsor: American College of Surgeons
633 N Saint Clair Street
Chicago, IL 60611-3211

American College of Surgeons Co-Sponsored K08/K23 Supplement Awards 520

The American College of Surgeons announces a program that will provide supplemental funding to up to five individuals who receive a Mentored Clinical Scientist Development Award (K08/K23). It is directed at surgeon-scientists working in the early stages of their research careers. The award requires co-sponsorship with an approved surgical society of a three, four or five year period of supervised research experience that may integrate didactic studies with laboratory or clinically based research. This award program will offer a means to facilitate the career development of individuals pursuing careers in surgical research by enhancing salary support over and above that offered by the K08/K23 mechanism. The application deadline is June 12, with funding to begin July 1.
Requirements: Awardees must be members in good standing of both the College and a co-sponsoring surgical society. Participating surgical societies include American Association of Plastic Surgeons (AAPS), American Head and Neck Society (AHNS), American Society of Transplant Surgeons (ASTS), American Vascular Association (AVA), Society of Gynecologic Oncologists (SGO), Society of University Surgeons (SUS), and Thoracic Surgery Foundation for Research and Education (TSFRE).
Geographic Focus: All States
Date(s) Application is Due: Jun 12
Contact: Kate Early; (312) 202-5281; fax (312) 202-5021; kearly@facs.org
Internet: http://www.facs.org/memberservices/acs-nih.html
Sponsor: American College of Surgeons
633 N Saint Clair Street
Chicago, IL 60611-3211

American College of Surgeons Faculty Career Development Award for Neurological Surgeons 521

The American College of Surgeons (ACS) and the Neurosurgery Research and Education Foundation of the American Association of Neurological Surgeons (NREF-AANS) are offering a two-year faculty career development award to neurological surgeons. The award is to support the establishment of a new and independent research program in an area of neurological surgery. The award is $40,000 per year for each of two years, to support the research, and is not renewable thereafter.
Requirements: The award is open to surgeons who are members or candidate members in good standing of both the American College of Surgeons (ACS) and the American Association of Neurological Surgeons (AANS); and have completed specialty training within the preceding five years and have received a full-time faculty appointment at a medical school accredited by the Liaison Committee on Medical Education in the United States or by the Committee for Accreditation of Canadian Medical Schools in Canada.
Restrictions: Applicants may not be current recipients of major research grants.
Geographic Focus: All States
Date(s) Application is Due: Dec 1
Amount of Grant: 40,000 USD
Contact: Michele S. Gregory, Director of Development; (847) 378-0540 or (847) 378-0500; fax (847) 378-0600; msg@aans.org or info@aans.org
Internet: http://www.facs.org/memberservices/acsnrefaans07.pdf
Sponsor: American College of Surgeons
633 N Saint Clair Street
Chicago, IL 60611-3211

American College of Surgeons Faculty Research Fellowships 522

The American College of Surgeons is offering two-year faculty research fellowships to surgeons entering academic careers in surgery or a surgical specialty. The fellowship is to assist a surgeon in the establishment of a new and independent research program. Applicants should have demonstrated their potential to work as independent investigators. The fellowship award is $40,000 per year for each of two years, to support the research.
Requirements: The fellowship is restricted to surgeons who have recently (usually within three years) completed formal surgical education and entered the field of full-time academic surgery and received a faculty appointment in a department of surgery, or one of the surgical specialties, at an approved medical school in the United States or Canada. Preference will be given to applicants who directly enter academic surgery following residency or fellowships. Approval of the application is required from the administration (dean or fiscal officer) and the head of the department under whom the recipient will be studying during the fellowship.
Geographic Focus: All States
Date(s) Application is Due: Nov 2
Amount of Grant: 40,000 USD
Contact: Kate Early; (312) 202-5281; fax (312) 202-5021; kearly@facs.org
Internet: http://www.facs.org/memberservices/acsfaculty.html
Sponsor: American College of Surgeons
633 N Saint Clair Street
Chicago, IL 60611-3211

American College of Surgeons Health Policy Scholarships 523

The American College of Surgeons, in partnership with a number of other organizations, is offering annual scholarships to subsidize attendance and participation in the Executive Leadership Program in Health Policy and Management at Brandeis University. The awards are in the amount of $8,000, to be used toward the cost of tuition, travel, housing, and subsistence during the period of the course. The closing date for receipt of all applications is February 2. All applicants will be notified of the outcome of the selection process by March 31.
Requirements: The award is open to surgeons who are members in good standing of both the American College of Surgeons (ACS) and the specific partnering organization for each award. These include: the American Association of Neurological Surgeons (AANS); the American Academy of Ophthalmology (AAO); the American Academy of Otolaryngology, Head and Neck Surgery Foundation (AAO-HNS); the American Association for the Surgery of Trauma (AAST); the American Pediatric Surgical Association (APSA); the American Surgical Association (ASA); the American Society of Breast Surgeons (ASBrS); the American Society of Colon and Rectal Surgeons (ASCRS); the American Society of Plastic Surgeons (ASPS); the American Urogynecologic Society (AUGS); the Society of Thoracic Surgeons (STS); and the Society for Vascular Surgery (SVS). Applicants must be at least 30 years old, but under 55, on the date that the completed application is filed.
Restrictions: Indirect costs are not paid to the recipient or to the recipient's institution.
Geographic Focus: All States
Date(s) Application is Due: Feb 2
Amount of Grant: 8,000 USD
Contact: Michele S. Gregory, Director of Development; (847) 378-0540 or (847) 378-0500; fax (847) 378-0600; msg@aans.org or info@aans.org
Internet: http://www.facs.org/memberservices/acsaans.html
Sponsor: American College of Surgeons
633 N Saint Clair Street
Chicago, IL 60611-3211

American College of Surgeons Health Policy Scholarships for General Surgeons 524

The American College of Surgeons is offering an annual scholarship to subsidize attendance and participation in the Executive Leadership Program in Health Policy and Management at Brandeis University. The award is in the amount of $8,000, to be used toward the cost of tuition, travel, housing, and subsistence during the period of the course. The closing date for receipt of applications is February 2. All applicants will be notified of the outcome of the selection process by March 31.
Requirements: The award is open to general surgeons who are members in good standing of the American College of Surgeons (ACS). Applicants must be at least 30 years old, but under 55, on the date that the completed application is filed. The scholarship must be used in the year for which it is designated.
Restrictions: Indirect costs are not paid to the recipient or to the recipient's institution.
Geographic Focus: All States
Date(s) Application is Due: Feb 2
Amount of Grant: 8,000 USD
Contact: Administrator; (800) 621-4111 or (312) 202-5000; fax (312) 202-5001
Internet: http://www.facs.org/memberservices/acshealthpolicy.html
Sponsor: American College of Surgeons
633 N Saint Clair Street
Chicago, IL 60611-3211

American College of Surgeons International Guest Scholarships 525

The American College of Surgeons offers International Guest Scholarships to competent young surgeons from countries other than the United States or Canada who have demonstrated strong interests in teaching and research. The scholarships, in the amount of $8,000 each, provide the Scholars with an opportunity to visit clinical, teaching, and research activities in North America and to attend and participate fully in the educational opportunities and activities of the American College of Surgeons Clinical Congress.

Requirements: Applicants must: be graduates of schools of medicine; be at least 35 years old, but under 45, on the date that the completed application is filed; submit their applications from their intended permanent location; and have demonstrated a commitment to teaching and/or research in accordance with the standards of the applicant's country.
Geographic Focus: All States
Date(s) Application is Due: Jul 1
Amount of Grant: 8,000 USD
Contact: Kate Early; (312) 202-5281; fax (312) 202-5021; kearly@facs.org
Internet: http://www.facs.org/memberservices/igs.html
Sponsor: American College of Surgeons
633 N Saint Clair Street
Chicago, IL 60611-3211

American College of Surgeons Nizar N. Oweida, MD, FACS, Scholarships 526
The Board of Governors of the American College of Surgeons announces the availability of a scholarship for young rural surgeons, the Nizar N. Oweida, MD, FACS, Scholarship of the American College of Surgeons. The Scholarship provides an award of $5,000 to subsidize the participation of a young rural-based Fellow or Associate Fellow in attendance at the annual Clinical Congress of the American College of Surgeons. This amount is to be used to help defray travel expenses for the Clinical Congress, postgraduate course fees, hotel costs, and per diem expenses during the Clinical Congress.
Requirements: The Oweida Scholarship is available to a member of the American College of Surgeons in any of the surgical specialties who meets the following *Requirements:* serves a rural community in the U.S. or Canada; is a Fellow or Associate Fellow in good standing; and is under 45 years of age on the date the application is filed.
Geographic Focus: All States
Date(s) Application is Due: Dec 15
Amount of Grant: 5,000 USD
Contact: Kate Early; (312) 202-5281; fax (312) 202-5021; kearly@facs.org
Internet: http://www.facs.org/memberservices/oweida.html
Sponsor: American College of Surgeons
633 N Saint Clair Street
Chicago, IL 60611-3211

American College of Surgeons Resident and Associate Society Leadership Scholarships 527
The Resident and Associate Society of the American College of Surgeons (RAS-ACS) is accepting applications for its leadership scholarship for Resident Members and Associate Fellows of the College. The RAS Scholarship will be awarded competitively to young surgeons who exemplify an important goal the RAS has set for itself--developing future leaders in the field of surgery. The award is for the purpose of allowing the young surgeon awardee to attend one of several courses that will be offered for this scholarship. The courses are intended to engage young surgeons and expose them to opportunities to improve their abilities in areas such as leadership, communications skills, and research, skills that are necessary to succeed as leaders in surgery.
Geographic Focus: All States
Date(s) Application is Due: Jan 30
Contact: Peg Haar, Division of Member Services; (312) 202-5312; phaar@facs.org
Internet: http://www.facs.org/ras-acs/resources/scholarship.html
Sponsor: American College of Surgeons
633 N Saint Clair Street
Chicago, IL 60611-3211

American College of Surgeons Resident Research Scholarships 528
ACS scholarships will be awarded each year to encourage residents to pursue careers in academic surgery. Priority will be given to residents beginning full-time investigative activities, with no clinical responsibilities, for the two-year period of the scholarship. Study outside the United States or Canada is permissible. The scholarship is for the personal use of the recipient and is not to diminish or replace the usual or expected compensation; award is made directly to the scholar and not to the institution. Renewal for the second year is contingent upon acceptable progress and study protocol for the second year.
Requirements: The applicant must have completed two postdoctoral years in an accredited surgical training program in the United States or Canada at the time the scholarship is awarded and shall not complete formal residency training before the end of the scholarship. Approval of the application is required from the administration (dean or fiscal officer) and the head of the department under whom the recipient will be studying.
Restrictions: Only in exceptional circumstances will more than one ACS scholarship be granted in a single year to applicants from the same institution.
Geographic Focus: All States
Date(s) Application is Due: Sep 1
Amount of Grant: 30,000 USD
Contact: Administrator; (800) 621-4111 or (312) 202-5000; fax (312) 202-5001
Internet: http://www.facs.org/memberservices/acsresident.html
Sponsor: American College of Surgeons
633 N Saint Clair Street
Chicago, IL 60611-3211

American College of Surgeons Wound Care Management Award 529
The American College of Surgeons is offering a two-year faculty research award through KCI USA, to a general surgeon engaged in a research project addressing wound care management. The purpose of this fellowship will be to acquire knowledge leading to new clinical applications or projects that will provide the medical community with a better understanding of the use of advanced wound healing therapies. The fellowship award is $85,000 per year, with the possibility of an extension for a second year if satisfactory progress is made. Specifically, translational projects such as new methods to mechanically stimulate wound healing, design of clinically applicable skin substitutes, methods to enhance specific, difficult-to-heal wounds related to general or extremity trauma, various soft-tissue injuries, quality limb salvage, or methods to reduce surgical intervention to heal a variety of wounds, are being sought. This award may be used by the recipient for support of his/her research or academic enrichment in any fashion that the recipient deems maximally supportive of his/her investigations. This may include faculty salary replacement, support for clinical research personnel, supplies directly related to the research activity, consumable equipment, and research related travel costs up to $2,000 per year.
Requirements: The fellowship is open to Fellows or Associate Fellows of the College who have: (1) completed the chief residency year or accredited fellowship training within the preceding 10 years; and (2) received a fulltime faculty appointment in a department of surgery or a surgical specialty at a medical school accredited by the Liaison Committee on Medical Education in the United States or by the Committee for Accreditation of Canadian Medical Schools in Canada. Preference will be given to applicants who directly enter academic surgery following residency or fellowship.
Geographic Focus: All States
Date(s) Application is Due: May 1
Amount of Grant: 85,000 USD
Contact: Administrator; (800) 621-4111 or (312) 202-5000; fax (312) 202-5001
Internet: http://www.facs.org/memberservices/woundcareposter2008.pdf
Sponsor: American College of Surgeons
633 N Saint Clair Street
Chicago, IL 60611-3211

American Electric Power Foundation Grants 530
The American Electric Power Foundation complements a tradition of corporate philanthropy exhibited by American Electric Power and its regional utilities in support of AEP's community relations goal: To support and play an active, positive role in the communities where we live and work. The American Electric Power Foundation was created in 2005 to continue to support that goal. The Foundation generally will consider requests of not less than $15,000 or those for multi-year commitments from organizations in the communities served by AEP's regional utilities or where AEP has major facilities. Applicants should review the list of regional contacts at the web site.
Requirements: 501(c)3 nonprofit organizations in Arkansas, Indiana, Kentucky, Louisiana, Michigan, Ohio, Oklahoma, Tennessee, Texas, Virginia and West Virginia are eligible. Grant-seekers should first approach their local AEP operating company.
Restrictions: Grants are not awarded to religious, fraternal, service, and veteran organizations, except for nonsectarian social service activities available to the broader community; organizations with a purpose that is solely athletic in nature; or to individuals.
Geographic Focus: Arkansas, Indiana, Kentucky, Louisiana, Michigan, Ohio, Oklahoma, Tennessee, Texas, Virginia, West Virginia
Amount of Grant: 15,000 - 100,000 USD
Contact: Administrator; (614) 716-1000; mkwalsh@aep.com or mkwalsh@aep.com
Internet: http://www.aep.com/community/AEPFoundation/
Sponsor: American Electric Power Foundation
1 Riverside Plaza, 19th Floor
Columbus, OH 43215

American Express Foundation Community Service Grants 531
Whether it is feeding the hungry, mentoring students, building homes for the homeless or cleaning up the environment, tens of thousands of American Express employees serve their communities through volunteerism and personal financial contributions, and the Foundation views this activity as an expression of the service ethic that lies at the heart of its business. It encourages good citizenship by supporting organizations that cultivate meaningful opportunities for civic engagement by employees and members of the community. The Foundation also serves its communities by supporting immediate and long-term relief and recovery efforts to help victims of natural disasters. Funding also goes to support preparedness programs that allow relief agencies to be better equipped in responding to emergencies as they occur.
Requirements: Eligible organizations must: certify tax-exempt status under Section 501(c)3 and 509(a)1, 2 or 3 of the U.S. Internal Revenue Code. Organizations outside the U.S. must be able to document not-for-profit status.
Restrictions: The program does not fund: individual needs, including scholarships, sponsorships and other forms of financial aid; fund-raising activities, such as galas, benefits, dinners and sporting events; goodwill advertising, souvenir journals or dinner programs; travel for individuals or groups; sectarian activities of religious organizations; political causes, candidates, organizations or campaigns; or books, magazines or articles in professional journals.
Geographic Focus: All States, All Countries
Samples: Association for Persons with Special Needs, Singapore - volunteers were engaged in the development of the center and as mentors to its clients; Feeding America, Chicago, Illinois - funds the expansion of its Store Donation Program which recovers perishable products that do not meet retailers' marketing standards.
Contact: Timothy McClimon, President; (212) 640-5661; fax (212) 693-1033
Internet: http://about.americanexpress.com/csr/comm_serv.aspx
Sponsor: American Express Foundation
200 Vesey Street, 48th Floor
New York, NY 10285-4804

American Foodservice Charitable Trust Grants — 532

American Foodservice Charitable Trust provides funding in the following areas of interest: Christian agencies & churches; community/economic development; higher education; hospitals (general) and; youth development. The types of support available are: general/operating support; program development; scholarship funds. Giving is primarily available in the areas of: King of Prussia, Pennsylvania; Thomasville, Georgia; and Fort Worth, Texas.
Requirements: Qualifying IRA 501(c)3 nonprofit organizations are eligible to apply. There's no application form nor is there a deadline date to adhere to when submitting a proposal to the Foundation.
Restrictions: No grants to individuals.
Geographic Focus: Georgia, Pennsylvania, Texas
Amount of Grant: 200 - 8,000 USD
Samples: Catholic Social Services, West Chester, PA, $200--social services grant; American Red Cross, Philadelphia, PA, $1,000--disaster relief grant; Police Atheletic League, Philadelphia, PA, $2,000--youth program grant.
Contact: Richard S. Downs, Trustee; (610) 933-9792
Sponsor: American Foodservice Charitable Trust
860 First Avenue, Suite 9A
King of Prussia, PA 19406-1404

American Jewish World Service Grants — 533

AJWS' grantmaking supports community-based organizations in the developing world that are undertaking holistic community development programs. These groups design and implement projects that creatively and effectively address economic development, education, healthcare, and sustainable agriculture. All of the initiatives also have strong components of strengthening civil society and/or promoting women's empowerment. Grantmaking links human rights and sustainable development.
Requirements: Grassroots organizations are eligible.
Restrictions: Grants are not awarded to individuals.
Geographic Focus: All States
Amount of Grant: 3,000 - 30,000 USD
Samples: Cross-Cultural Solutions (New Rochelle, NY)--to provide food to rescue workers and to coordinate volunteer efforts at the site of the World Trade Center, $5000; Independent Press Assoc (New York, NY)--to place advertisements in New York ethnic publications on the availability of disaster resources in the wake of the September 11 attacks, $25,500; Mount Sinai Hospital, Disaster Psychiatry Outreach (New York, NY)--to work with the New York City Department of Mental Health in its outreach efforts to counsel rescue workers and victims of the World Trade Center collapse, $10,000; Jewish Board of Family and Children Services (New York, NY)--to provide mental-health services, including drop-in centers, telephone hotlines, and on-site counseling in the wake of the September 11 attacks, $10,000.
Contact: Administrator; (800) 889-7146 or (212) 792-2100; grants@ajws.org
Internet: http://www.ajws.org/index.cfm?section_id=3
Sponsor: American Jewish World Service
45 W 36th Street, 10th Floor
New York, NY 10018-7904

American Legacy Foundation National Calls for Proposals Grants — 534

The American Legacy Foundation provides grants to support innovations in tobacco control. National calls for proposals, such as the Priority Populations Initiative, are issued annually to address a variety of tobacco prevention and control issues. Comprehensive assistance and training are provided to the foundation's grantees as well as to state and local tobacco programs. Application forms, guidelines, and procedures are available online. There are no application deadlines.
Requirements: Funding is available only to state or local political subdivisions and legally constituted tax-exempt 501(c)3 organizations based in the 46 states, the District of Columbia, and five territories (American Samoa, Guam, Northern Mariana Islands, Puerto Rico, and the Virgin Islands) identified in the MSA with tobacco product manufacturers. An Indian reservation, Indian tribe, or tribal organization located within the 46 settling states or a non-governmental entity that serves such a reservation may also apply for funding.
Restrictions: The foundation will not award grants to applicants that are in current receipt of grant monies or in-kind contributions from any tobacco manufacturer, distributor, or other tobacco-related entity.
Geographic Focus: All States
Contact: Katherine Wilson, Assistant Vice President of Grants; (202) 454-5555; fax (202) 454-5599; grantsinfo@americanlegacy.org
Internet: http://www.americanlegacy.org/64.aspx
Sponsor: American Legacy Foundation
1724 Massachusetts Avenue NW
Washington, D.C. 20036

American Legacy Foundation Small Innovative Grants — 535

The program supports projects that advance creative, promising solutions based on sound principles of tobacco control to remedy the harm caused by tobacco use in America. The program was created to seed new projects or enable an organization to pilot a new idea or approach. The proposed project must demonstrate an element of creativity, ingenuity or innovation and must distinguish itself from the large number of solid programs proposed to the foundation in each grant round.
Requirements: Funding is available only to state or local political subdivisions and legally constituted tax-exempt 501(c)3 organizations based in the 46 states, the District of Columbia, and five territories (American Samoa, Guam, Northern Mariana Islands, Puerto Rico, and the Virgin Islands) identified in the MSA with tobacco product manufacturers. An Indian reservation, Indian tribe, or tribal organization located within the 46 settling states or a nongovernmental entity that serves such a reservation may also apply for funding. Successful applications submitted under the Small Innovative Grants Program must: address one or both of Legacy's goals; demonstrate innovative or new tobacco prevention or cessation efforts; demonstrate a strong likelihood for a sustainable effort after the grant period; demonstrate that the project may be replicated; address the Healthy People 2010 risk reduction objectives related to tobacco use, and; incorporate the CD.C.'s Best Practices for Comprehensive Tobacco Control Programs as appropriate. The foundation will give special consideration to applications addressing these current areas of interest.
Restrictions: The foundation will not award a grant to any applicant that is in current receipt of any grant monies or in-kind contribution from any tobacco manufacturer, distributor, or other tobacco-related entity. In addition, the foundation expects that a grantee will not accept any grant monies or in-kind contribution from any tobacco manufacturer, distributor, or other tobacco-related entity over the duration of the grant. Additionally, will not consider applications for: Projects focusing on youth prevention, cessation, activism, or education (up to 18 years old); Nicotine replacement therapy (NRT) and pharmaceuticals (as the sole or primary focus of the grant); Conference support (as the sole or primary focus of the grant); Media or marketing campaigns (as the sole or primary focus of the grant); Research projects EXCEPT community-based participatory research, which is allowed; Projects focusing on substances other than tobacco; Replication of an existing program; Expansion of an existing program; and Replacement funds; Grants to individuals, for religious activities, to build endowments, to support operating deficits, to retire debt, for capital purchases for building improvements, for construction, for lobbying, or for real estate purchase or development.
Geographic Focus: All States
Date(s) Application is Due: Feb 27; Aug 8
Amount of Grant: 20,000 - 100,000 USD
Contact: Katherine Wilson; (202) 454-5555; grantsinfo@americanlegacy.org
Internet: http://www.americanlegacy.org/2530.aspx
Sponsor: American Legacy Foundation
1724 Massachusetts Avenue NW
Washington, D.C. 20036

American Medical Association Awards — 536

The American Medical Association recognizes excellence through the following AMA Awards: Distinguished Service Award; Citation for Distinguished Service; AMA Medal of Valor; Scientific Achievement Award; Benjamin Rush Award for Citizenship and Community Service; President's Citation for Service to the Public; Joseph B. Goldberger Award in Clinical Nutrition; Dr. William Beaumont Award in Medicine; AMA Foundation Award for Health Education; Isaac Hays, MD and John Bell, MD Award for Leadership in Medical Ethics and Professionalism; Medical Executive Lifetime Achievement Award; and Medical Executive Meritorious Achievement Award. Candidates may be nominated for more than one award. Nominations should be emailed by February 23 each year.
Geographic Focus: All States
Date(s) Application is Due: Feb 23
Contact: Roger Brown, (312) 464-4344; fax (312) 464-4505; roger.brown@ama-assn.org
Internet: http://www.ama-assn.org/ama/pub/about-ama/awards/about-ama-awards-program.shtml
Sponsor: American Medical Association
515 N State Street
Chicago, IL 60654

American Philosophical Society Daland Fellowships in Clinical Investigation — 537

A limited number of fellowships are awarded annually for research in clinical medicine including the fields of internal medicine, neurology, psychiatry, pediatrics, and surgery. Patient-oriented research is emphasized. Essentially 100 percent of the fellow's time will be devoted to research. Teaching or clinical service of a limited amount is permitted. Additional salary may be granted by the institution at which the fellow is located, from another fellowship, or from a similar award during the tenure of the fellowship. The term of the fellowship is one year, with possible renewal for another year. Applications are available on the Web site, or may be obtained by written request, stating when the MD or MD/PhD degree was awarded; include a self-addressed mailing label.
Requirements: These fellowships are designed for qualified persons who have held an MD, or MD/PhD degree for less than eight years. The fellowship is generally intended to be the first post-clinical fellowship; but each case will be decided on its merits. Preference is generally given to candidates who have not more than two years of post-doctoral training and research. Applicants must expect to perform their research at an institution in the United States, under the supervision of a scientific adviser. Candidates are to be nominated by their department chairman, in a letter providing assurance that the nominee will work with the guidance of a scientific adviser of established reputation who has guaranteed adequate space, supplies, etc. for the Fellow. The adviser need not be a member of the department nominating the Fellow, nor need the activities of the Fellow be limited to the nominating department. As a general rule, no more than one fellowship will be awarded to a given institution in the same year of competition.
Restrictions: The society does not provide funds for institutional overhead.
Geographic Focus: All States
Date(s) Application is Due: Sep 1
Amount of Grant: Up to 40,000 USD
Contact: Linda Musumeci; (215) 440-3429; LMusumeci@amphilsoc.org
Internet: http://www.amphilsoc.org/grants/daland.htm
Sponsor: American Philosophical Society
104 South Fifth Street
Philadelphia, PA 19106-3387

American Psychiatric Association/AstraZeneca Young Minds in Psychiatry International Awards 538

The program recognizes and supports promising international young psychiatrists within five years of completing a psychiatric residency. Four unrestricted career development awards of $45,000 (USD) will be available. Awards will be made to two promising physicians from the U.S. with one in Bipolar Disorder research and one on research in Schizophrenia. An additional two awards will be made to promising physicians from countries outside the U.S., with one in Bipolar Disorder research and one on research in Schizophrenia. Three other awards of $30,000 (USD) in either Bipolar Disorder or Schizophrenia research will specifically focus on applicants from developing countries whose economies are classified by the World Bank as low income or lower middle income.
Requirements: The U.S. applicants must be citizens or permanent residents of the United States. Applications are evaluated based on evidence of academic promise; how the proposal will advance the applicant's career; and innovative or original concepts, approaches, or methods for developing the applicant's career.
Geographic Focus: All States
Date(s) Application is Due: Nov 30
Amount of Grant: 45,000 USD
Contact: Ernesto Guerra, (703) 907-7300; fax (703) 907-1085; eguerra@psych.org
Internet: http://psych.org/MainMenu/Research/ResearchTrainingandFunding/ResearchFellowships/APAAstraZenecaYoungMindsinPsychiatryInternationalAward.aspx
Sponsor: American Psychiatric Association
1000 Wilson Boulevard, Suite 1825
Arlington, VA 22209-3901

American Psychiatric Association/Bristol-Myers Squibb Fellowships in Public Psychiatry 539

This important collaborative venture between the American Psychiatric Association and the Bristol-Myers Squibb Company provides for the selection of ten residents in psychiatry, based on recommendations from their Departments and evidence of a track record of commitment to and accomplishment in public psychiatry. The purposes of the APA/Bristol-Myers Squibb Fellowship are twofold: (1) to heighten psychiatric residents' awareness of the many activities of psychiatry in the public sector and of the career opportunities in this work, and (2) to provide experiences which will contribute to the professional development of those residents who will play leadership roles within the public sector in future years.
Requirements: Nomination Procedure and Eligibility Criteria: One outstanding resident who has a substantial interest in any area of public sector psychiatry and a significant potential for a leadership role in such work may be nominated by a psychiatric residency training program that is accredited by the Accreditation Council for Graduate Medical Education or the Royal College of Physicians and Surgeons of Canada; submission of a nomination represents an explicit commitment by the residency program to permit the candidate, if selected, time off to attend APA meetings in their entirety during the fellowship term, as well as the candidate's willingness to do so; Psychiatric residents who will be in the PG-3 year of training during the fellowship term are eligible to participate in the program; residents who will be in their PG-4 year will also be considered if they are in a 5-year psychiatric training program; nominee must be an APA member or in the process of becoming a member. To be considered, the following must be submitted: completed application form (submitted electronically or via mail); two-page essay; letter of nomination from the Residency Training Director or Chair, Department of Psychiatry.
Restrictions: Residents who have received (or are scheduled to receive) other APA fellowships or comparable outside fellowships (e.g., Ginsberg, Laughlin) are not eligible. Those residents who currently serve on APA's Committee of Residents and Fellows are not eligible for the fellowship.
Geographic Focus: All States
Date(s) Application is Due: Mar 31
Contact: Nancy Delanoche; (703) 907-8635; ndelanoche@psych.org
Internet: http://www.psych.org/Departments/EDU/ResidentMIT/fellowshipinpublicpsychiatry.aspx
Sponsor: American Psychiatric Association
1000 Wilson Boulevard, Suite 1825
Arlington, VA 22209-3901

American Psychiatric Association/Janssen Resident Psychiatric Research Scholars 540

The program is intended to identify promising PGY-1, PGY-2, and PGY-3 psychiatric residents with the potential to become leaders in clinical and health services research in all areas of psychiatric research. Emphasis will be placed on special mentoring and career enrichment programs both at the APA Annual Meeting and throughout the year. An individual research mentor will be assigned to oversee the resident's fellowship. The mentors will be chosen among nationally known leaders in clinical and health services research. The mentor will advise and encourage the Scholar during the two-year fellowship. The scholar will receive $2,500 during the second year of the fellowship, to assist in their research career development (e.g., for use in developing a pilot research project, obtaining statistical consultation, or visiting potential research training programs). The program also provides funding for travel to the American Psychiatric Association (APA) Annual Meeting during both years of the fellowship.
Geographic Focus: All States
Date(s) Application is Due: Jan 15
Contact: Ernesto Guerra, (703) 907-7300; fax (703) 907-1085; eguerra@psych.org
Internet: http://psych.org/MainMenu/Research/ResearchTrainingandFunding/ResearchFellowships/APIREJanssenResidentPsychiatricResearchScholars.aspx
Sponsor: American Psychiatric Association
1000 Wilson Boulevard, Suite 1825
Arlington, VA 22209-3901

American Psychiatric Association/Merck Co. Early Academic Career Research Award 541

The Award is intended to help support the research of a junior faculty member with an interest in sleep disorders or schizophrenia. Two separate awards will be made to candidates who have completed a psychiatry residency, at least one year of a psychiatry research fellowship, and are seeking to make a commitment to a research career with the end goal of successfully transitioning to that of an independent investigator. The Award is intended to assist in this key transition period by providing one year of funding of $45,000. Salary support provided by this award will allow junior faculty in departments of psychiatry at U.S. academic institutions to devote more time to research.
Requirements: Candidates must be citizens or permanent residents of the U.S. and should also be APA members. Eligible candidates will be trained psychiatrists with an MD, MD/Ph.D. or DO degree who have completed residency training in general psychiatry or child psychiatry in the U.S. Candidates also should have completed at least one year of a psychiatry research fellowship but are not 3 or more years post fellowship completion, and do not currently hold an academic rank higher than Assistant Professor.
Restrictions: Individuals who have obtained K-awards or are considered an independent Principal Investigator on R-01 or R-21 research grants from NIH are not eligible.
Geographic Focus: All States
Date(s) Application is Due: Oct 14
Contact: Ernesto Guerra, (703) 907-7300; fax (703) 907-1085; eguerra@psych.org
Internet: http://www.psych.org/research/apire/res_careerdev/apamerck.cfm
Sponsor: American Psychiatric Association
1000 Wilson Boulevard, Suite 1825
Arlington, VA 22209-3901

American Psychiatric Association Award for Research in Psychiatry 542

The Award is given in recognition of a single distinguished contribution, a body of work, or a lifetime contribution that has had a major impact on the field and/or altered the practice of psychiatry. The Award is intended to cover the full spectrum of psychiatric research. The Award consists of a $5,000 prize and an honorary plaque to be presented at APA's Annual Meeting in May. The Award also includes an honorary lecture by the awardee.
Requirements: Candidates for the Award must be citizens of the United States or Canada and be nominated by a sponsor. Sponsors must be members of the American Psychiatric Association.
Restrictions: Members of the Award Committee are excluded from submitting nominations.
Geographic Focus: All States
Date(s) Application is Due: Aug 28
Contact: Harold Goldstein, (703) 907-8623; goharold@psych.org
Internet: http://www.psych.org/research/apire/training_fund/psychaward.cfm
Sponsor: American Psychiatric Association
1000 Wilson Boulevard, Suite 1825
Arlington, VA 22209-3901

American Psychiatric Association Minority Fellowships Program 543

The APA offers psychiatric residents one-year fellowships in the APA organization to increase the knowledge of cultural factors influencing psychiatric diagnosis and treatment, provide opportunities for these residents to participate in the deliberations and decision-making processes, and provide role models for the fellows. Each fellow is appointed to a component of the APA's organizational structure and attends the association's annual meeting as an observer and active participant. Fellows are selected on the basis of their commitment to serving underrepresented populations, demonstrated leadership abilities, and interest in the interrelationship between mental health/illness and transcultural factors.
Requirements: Psychiatric residents who are starting their second year of psychiatric training, or third year if they are in a four-year residency program, are eligible to apply.
Geographic Focus: All States
Contact: Marilyn King; (703) 907-8653; fax (703) 907-7849; mking@psych.org
Internet: http://www.psych.org/Resources/OMNA/MFP.aspx
Sponsor: American Psychiatric Association
1000 Wilson Boulevard, Suite 1825
Arlington, VA 22209-3901

American Psychiatric Association Minority Medical Student Fellowship in HIV Psychiatry 544

The APA invites ethnic minority medical students who have an interest in psychiatric issues to apply. The program is intended to identify minority medical students who have primary interests in services related to HIV/AIDS and substance abuse and its relationship to the mental health or psychological well being of ethnic minorities.
Requirements: These programs are open to ethnic minority medical students who are U.S. citizens or permanent residents currently enrolled in a U.S. medical school.
Geographic Focus: All States
Date(s) Application is Due: Mar 31
Contact: Diane Pennessi; (703)907-8668; fax (703)907-8668; dpennessi@psych.org
Internet: http://www.psych.org/Resources/OMNA/MFP/HIV-Information-10.aspx
Sponsor: American Psychiatric Association
1000 Wilson Boulevard, Suite 1825
Arlington, VA 22209-3901

American Psychiatric Association Minority Medical Student Summer Externship in Addiction Psychiatry 545

The APA invites ethnic minority medical students who have an interest in psychiatric issues to apply. This clinical shadowing program identifies minority medical students who may have a specific interest in services related to substance abuse treatment/prevention and provide a setting where the student can work closely with a mentor who specializes in addiction psychiatry for one month.

Requirements: These programs are open to ethnic minority medical students who are U.S. citizens or permanent residents currently enrolled in a U.S. medical school.
Geographic Focus: All States
Date(s) Application is Due: Feb 28
Amount of Grant: 1,500 USD
Contact: Marilyn King; (703) 907-8653; fax (703) 907-7852; mking@psych.org
Internet: http://www.psych.org/Resources/OMNA/MFP/minoritymedicalstudentsummerexternshipinaddictionpsychiatry.aspx
Sponsor: American Psychiatric Association
1000 Wilson Boulevard, Suite 1825
Arlington, VA 22209-3901

American Psychiatric Association Minority Medical Student Summer Mentoring Program 546

The APA invites ethnic minority medical students who have an interest in psychiatric issues to apply. This program is intended to identify ethnic minority medical students who have an interest in psychiatric issues and expose students to a setting where they can work closely with a psychiatrist mentor for one month.
Requirements: These programs are open to ethnic minority medical students who are U.S. citizens or permanent residents currently enrolled in a U.S. medical school.
Geographic Focus: All States
Date(s) Application is Due: Feb 28
Amount of Grant: 1,500 USD
Contact: Marilyn King; (703) 907-8653; fax (703) 907-7852; mking@psych.org
Internet: http://www.psych.org/Resources/OMNA/MFP/minoritymedicalstudentsummermentoringprogram.aspx
Sponsor: American Psychiatric Association
1000 Wilson Boulevard, Suite 1825
Arlington, VA 22209-3901

American Psychiatric Association Program for Minority Research Training in Psychiatry (PMRTP) 547

The program is designed to increase the number of underrepresented minority men and women in the field of psychiatric research. Research training offers the opportunity to engage in scientific investigation across the full array of disciplines, from basic neuroscience, genetics, and pharmacology to the cognitive behavioral, and social sciences, clinical psychiatry, and mental health services research. Research exposure can help students and trainees develop sound skills for clinical assessment and treatment planning. The program provides funding for short and long-term training opportunities at three levels: Medical School, Residency, and Post-residency. National competitions also enable qualified mini-fellows to attend research-oriented meetings of psychiatric organizations. For medical students and residents, the duration can be two months to one year. For post-residency fellows, the duration is generally two years. A third year of fellowship support may be available if appropriate to a trainee's career development. Support from PMRTP falls into three categories: stipends, travel, and tuition and fees. Stipends are based on the trainee's years of relevant experience and the length of the research training experience.
Requirements: Preference in selection is given to underrepresented minorities such as American Indians, Asian-Americans, Blacks/African-Americans, Hispanics, Pacific Islanders, or other ethnic or racial group members found to be underrepresented in biomedical or behavioral research.
Geographic Focus: All States
Date(s) Application is Due: Apr 1; Dec 1
Contact: Ernesto Guerra, (703) 907-7300; fax (703) 907-1085; eguerra&@psych.org
Internet: http://www.psych.org/MainMenu/Research/ResearchTrainingandFunding/ResearchFellowships/ProgramforMinorityResearchTraininginPsychiatry.aspx
Sponsor: American Psychiatric Association
1000 Wilson Boulevard, Suite 1825
Arlington, VA 22209-3901

American Psychiatric Association Research Colloquium for Junior Investigators 548

The purpose of the colloquium is to provide guidance, mentorship and encouragement to young investigators in the early phases of their training. Junior investigators will have an opportunity to obtain feedback about their past, present, and future research interests from mentors who are tops in their field in a small group setting as well as general information about research career development and grantsmanship. Candidates whose research interests are similar to those listed below are also encouraged to apply. An all-day workshop for junior psychiatric investigators will focus on these three areas: childhood disorders including autism, ADHD, conduct, eating, and mood disorders; genomics, epigenetics, and proteomics; and treatment of major psychiatric disorders: from substance abuse to schizophrenia.
Requirements: Psychiatrists who are senior residents, fellows, or junior faculty, and who have an interest and potential in developing research careers in the areas of research listed above. Participants should hold a medical degree or be a member of the APA; or be eligible to become members of the APA.
Restrictions: Those with individual federal research awards are not eligible.
Geographic Focus: All States
Date(s) Application is Due: Nov 16
Contact: Ernesto Guerra, (703) 907-7300; fax (703) 907-1085; eguerra@psych.org
Internet: http://psych.org/MainMenu/Research/ResearchTrainingandFunding/ResearchColloquiumforJuniorInvestigators.aspx
Sponsor: American Psychiatric Association
1000 Wilson Boulevard, Suite 1825
Arlington, VA 22209-3901

American Psychiatric Association Travel Scholarships for Minority Medical Students 549

The APA invites ethnic minority medical students who have an interest in psychiatric issues to apply. The program supports travel and related costs for approximately 10 minority medical students interested in psychiatric to attend either the APA annual meeting in May or the Institute on Psychiatric Services (IPS) meeting in October. This program is a way for medical students to witness organized psychiatry at work and to learn more about the field. Not only will students attend sessions for experts and trainees alike, but they will be assigned to a mentor who will help them maximize their annual meeting or IPS experience and discuss career plans and resident training programs.
Requirements: These programs are open to U.S. ethnic minority medical students who are U.S. citizens or permanent residents currently enrolled in a U.S. medical school. Ethnic minorities are: American-Indian, Alaska Native, Native Hawaiian, Asian-American, African-American, and Hispanic/Latino.
Geographic Focus: All States
Date(s) Application is Due: Jan 29
Contact: Marilyn King; (703) 907-8653; fax (703) 907-7852; mking@psych.org
Internet: http://www.psych.org/Resources/OMNA/MFP/MedicalStudentTravelScholarships.aspx
Sponsor: American Psychiatric Association
1000 Wilson Boulevard, Suite 1825
Arlington, VA 22209-3901

American Psychiatric Foundation Alexander Gralnick, MD, Award for Research in Schizophrenia 550

The award will annually acknowledge research achievements in the treatment of schizophrenia, emphasizing early diagnosis and treatment and psychosocial aspects of the disease process. Additional preference will be given to researchers working in a psychiatric facility. The applicant should submit a statement summarizing the nature of the research relevant to the award, emphasizing its internal consistency and scientific implications; a list of publications (or articles accepted for publication) relevant to the award; a current curriculum vita; and a current bibliography. Four copies and the original application should be submitted. The $4,000 award will be presented at the annual meeting.
Geographic Focus: All States
Date(s) Application is Due: Nov 2
Amount of Grant: 4,000 USD
Contact: Paul T. Burke, Executive Director; (703) 907-8518 or (703) 907-8512; fax (703) 907-7851; pburke@psych.org or apf@psych.org
Internet: http://www.psychfoundation.org/GrantAndAwards/AwardsandFellowships/AlexanderGralnickMDAward.aspx
Sponsor: American Psychiatric Foundation
1000 Wilson Boulevard, Suite 1825
Arlington, VA 22209

American Psychiatric Foundation Awards for Advancing Minority Mental Health 551

The program was established by the foundation, with support of Otsuka America Pharmaceutical, to recognize psychiatrists, other health professionals, mental health programs, and other organizations that have undertaken innovative and supportive efforts to raise awareness of mental illness in minority communities, the need for early recognition, the availability of treatment and how to access it, and the cultural barriers to treatment; increase access to quality mental health services for minorities; and improve the quality of care for minorities, particularly those in the public health system or with severe mental illness. The awards, although given to mental health professionals, will focus on patient issues such as raising awareness of mental illness, the need for treatment, and how to access treatment. Programs or efforts that reduce language and cultural barriers, enhance direct community understanding of mental illness, expand efforts to reach out to their community, and support community reintegration are just a few examples of the types of activities that could be recognized by these awards. Winning programs will be featured in publications such as Psychiatric News, and on American Psychiatric Association and American Psychiatric Foundation Web sites. In addition, award winners and their programs will be recognized at the APA Annual Meetings each May. Guidelines are available online.
Requirements: Psychiatrists, other health professionals, mental health programs, and other organizations are eligible. Psychiatrists and other health professionals must be licensed and/or accredited in the state in which you practice. Mental health programs and other organizations must have been in operation for at least two years before the application deadline.
Geographic Focus: All States
Amount of Grant: 5,000 USD
Samples: Family Service of El Paso, El Paso, Texas, $5,000 - for its unique and innovative mental health services provided to individuals and families who have fled the violence in Mexico; Northside Center for Child Development in New York, New York, $5,000 - for the youth mental health programs it runs; Student Stress and Anger Management Program, Los Angeles, California, $5,000 - for its unique services to address the emotional components that contribute to the mental health of middle and high school students.
Contact: Paul T. Burke, Executive Director; (703) 907-8518 or (703) 907-8512; fax (703) 907-7851; pburke@psych.org or apf@psych.org
Internet: http://www.psychfoundation.org/GrantAndAwards/AwardsandFellowships/AAMMH.aspx
Sponsor: American Psychiatric Foundation
1000 Wilson Boulevard, Suite 1825
Arlington, VA 22209

American Psychiatric Foundation Call for Proposals 552
APF support allows organizations across the country to make a difference through unique educational, informational and outreach initiatives that promote the early recognition and treatment of mental illness and encourage leadership in the field of psychiatry. The Call for Proposals program is intended to fund public education programs, as well as information and outreach initiatives that promote the early recognition and treatment of mental illness. Up to $750,000 in grant funds are available over the course of three years. Average grants are in the $50,000 range. Grant making under this program has been temporarily suspended while a new Call for Proposals is being developed. Contact the office for further updates.
Requirements: Organizations that have been in existence for at least two years and currently maintain a 501(c)3 charitable status and American Psychiatric Association District Branches and subsidiaries. Organizations need not be mental health programs.
Geographic Focus: All States
Amount of Grant: Up to 100,000 USD
Contact: Paul T. Burke, Executive Director; (703) 907-8518 or (703) 907-8512; fax (703) 907-7851; pburke@psych.org or apf@psych.org
Internet: http://www.psychfoundation.org/GrantAndAwards/Grants/CallforProposals.aspx
Sponsor: American Psychiatric Foundation
1000 Wilson Boulevard, Suite 1825
Arlington, VA 22209

American Psychiatric Foundation Disaster Recovery Fund for Psychiatrists 553
The Disaster Recovery Fund for Psychiatrists grant program is intended to assist psychiatrists and psychiatric residents who have been adversely affected by disasters such as hurricanes Katrina, Rita and Wilma. Psychiatrists and psychiatric residents who have been affected by disasters are eligible to apply for grant funding up to $2,500. Grants will be made to allow psychiatric residents to continue their medical training and to help psychiatrists rebuild their practices in order to serve the mental health needs of the people in their communities.
Geographic Focus: All States
Amount of Grant: Up to 2,500 USD
Contact: Paul T. Burke, Executive Director; (703) 907-8518 or (703) 907-8512; fax (703) 907-7851; pburke@psych.org or apf@psych.org
Internet: http://www.psychfoundation.org/GrantAndAwards/Grants.aspx
Sponsor: American Psychiatric Foundation
1000 Wilson Boulevard, Suite 1825
Arlington, VA 22209

American Psychiatric Foundation Helping Hands Grants 554
The program was established to encourage medical students to participate in community mental health service activities, particularly those focused on underserved populations. The program seeks to raise awareness of mental illness and the importance of early recognition and builds an interest among medical students in psychiatry and working in underserved communities. The foundation makes grants up to $5,000 to medical schools for mental health service projects that are created and managed by medical students. Projects can be new initiatives conducted in partnership with community agencies or in conjunction with ongoing medical school outreach activities. Examples of fundable activities include screening in community health centers or homeless shelters; outreach and community education about mental health; and health literacy programs.
Requirements: Medical schools are eligible to apply. All projects must be conducted under the supervision of medical faculty. Medical students who participate in the program must be in their second, third, or fourth year of medical school at the time they are engaged in community service. Funds must be expended within one year. Unused funds must be returned to the foundation.
Restrictions: Grants will not fund services beyond basic mental health screenings and referrals.
Geographic Focus: All States
Date(s) Application is Due: May 26
Amount of Grant: Up to 5,000 USD
Contact: Paul T. Burke, Executive Director; (703) 907-8518 or (703) 907-8512; fax (703) 907-7851; pburke@psych.org or apf@psych.org
Internet: http://www.psychfoundation.org/GrantAndAwards/Grants/HelpingHands.aspx
Sponsor: American Psychiatric Foundation
1000 Wilson Boulevard, Suite 1825
Arlington, VA 22209

American Psychiatric Foundation James H. Scully Jr. Educational Grants 555
The fund supports education and research within the American Psychiatric Association (APA), its subsidiary organizations, or affiliated organizations, including district branches, state associations, or other psychiatric organizations that are in partnership with the APA. The fund can be used to support activities that improve the quality of care for patients with mental illness, advance the prevention of mental illness, raise awareness of mental illness and its treatments, and increase access to mental health services. Requests for funding should be sent in writing to the foundation and include the following: identification and significance of need; description of program/activity to address that need; expected outcomes and evaluation process; budget and time line; and background of key project personnel. Funding decisions are made by the foundations board of directors.
Restrictions: The fund will not support public policy or lobbying activities.
Geographic Focus: All States
Contact: Paul T. Burke, Executive Director; (703) 907-8518 or (703) 907-8512; fax (703) 907-7851; pburke@psych.org or apf@psych.org
Internet: http://www.psychfoundation.org/GrantAndAwards/Grants/JamesHScullyEducationalFund.aspx

Sponsor: American Psychiatric Foundation
1000 Wilson Boulevard, Suite 1825
Arlington, VA 22209

American Psychiatric Foundation Jeanne Spurlock Congressional Fellowship 556
The Jeanne Spurlock Congressional Fellowship provides general psychiatry and child psychiatry residents an opportunity to work in a congressional office or committee on federal health policy, particularly policy related to child and minority issues. The recipient will serve a ten-month fellowship in Washington, D.C.. The fellow will be introduced to the structure and development of federal and congressional health policy procedures, with a focus on mental health issues affecting minorities and under-served populations, including children. Fellows traditionally help develop legislative proposals, track and analyze legislative initiatives, arrange hearings, and brief Members of Congress and their staff, and interact with their constituents. During the fellowship, recipients have opportunities to interact with health policymakers and advocacy/professional groups, including the APA.
Requirements: PGY-II and III general psychiatry residents, child psychiatry residents, or child psychiatrists who will be out of training for less than one year at the time of the fellowship may apply. Applicants must be U.S. citizens or permanent residents. The recipient will be required to submit a written summary of the Fellowship experience at the end of the fellowship and may make recommendations or suggestions for improving the fellowship.
Geographic Focus: All States
Date(s) Application is Due: Mar 14
Amount of Grant: 35,000 - 37,500 USD
Contact: Paul T. Burke, Executive Director; (703) 907-8518 or (703) 907-8512; fax (703) 837-5451; pburke@psych.org or apf@psych.org
Internet: http://www.psychfoundation.org/GrantAndAwards/Fellowships/SpurlockFellowship.aspx
Sponsor: American Psychiatric Foundation
1000 Wilson Boulevard, Suite 1825
Arlington, VA 22209

American Psychiatric Foundation Minority Fellowships 557
The Minority Fellowships Program (MFP) endeavors to eliminate racial and ethnic disparities in mental health and substance abuse care by providing specialized training to psychiatry residents and medical students interested in serving minority communities. The MFP is designed to: provide fellowship recipients with enriching training experiences through participation in the APA September and Annual Meetings; provide recipients with resources to support activities that enhance culturally relevant aspects of their training program; stimulate their interest in pursuing training in areas of psychiatry where minority groups are underrepresented, such as research, child psychiatry, and addiction psychiatry; and develop leadership to improve the quality of mental health care for the following federally recognized ethnic minority groups: American Indians, Native Alaskans, Asian Americans, Native Hawaiians, Native Pacific Islanders, African Americans and Hispanics/Latinos.
Geographic Focus: All States
Contact: Paul T. Burke, Executive Director; (703) 907-8518 or (703) 907-8512; fax (703) 837-5451; pburke@psych.org or apf@psych.org
Internet: http://www.psychfoundation.org/GrantAndAwards/Fellowships/MFP.aspx
Sponsor: American Psychiatric Foundation
1000 Wilson Boulevard, Suite 1825
Arlington, VA 22209

American Psychiatric Foundation Typical or Troubled School Mental Health Education Grants 558
The Grant Program (also known as Típico o Problematico in Spanish) provides funding to implement the Typical or Troubled School Mental Health educational model in communities nationwide. Community organizations, high schools and school districts are eligible to receive funding. The educational program is designed for school personnel (teachers, coaches guidance counselors, etc.) to raise their awareness of mental disorders in teens. The program focuses on promoting the importance of early recognition and treatment, recognizing the early warning signs of mental health problems, and encouraging action and appropriate referral to a mental health professional. Grant support is as follows: for implementation in two to four high schools, $1,000; for implementation in five or more high schools, $2,000.
Geographic Focus: All States
Date(s) Application is Due: Mar 30
Amount of Grant: Up to 2,000 USD
Contact: Paul T. Burke, Executive Director; (703) 907-8518 or (703) 907-8512; fax (703) 907-7851; pburke@psych.org or apf@psych.org
Internet: http://www.psychfoundation.org/GrantAndAwards/Grants/TypicalorTroubled.aspx
Sponsor: American Psychiatric Foundation
1000 Wilson Boulevard, Suite 1825
Arlington, VA 22209

American Schlafhorst Foundation Grants 559
The American Schlafhorst Foundation, which was established in 1987, awards grants to eligible North Carolina nonprofit organizations in its areas of interest, including: arts and culture; botanical gardens; children and youth; community college education; the elderly; elementary and secondary education; employment; health care; higher education; hospital care; housing development; museums; science; and social services delivery. Types of support include: building construction and renovation; equipment acquisition; general operating support; research; scholarships; and seed money funding. There are no specific annual application deadlines, so interested parties should begin by contacting

the Foundation directly with a proposed concept. Most recently, awards have ranged from $10,000 to $25,000.
Requirements: 501(c)3 tax-exempt organizations serving the greater Charlotte, North Carolina, area are eligible.
Restrictions: Individuals are not eligible.
Geographic Focus: North Carolina
Amount of Grant: 10,000 - 25,000 USD
Contact: Dan W. Loftis, Wxecutive Director; (704) 554-0800; info@schlafhorst.com
Sponsor: American Schlafhorst Foundation
8801 South Boulevard, P.O. Box 240828
Charlotte, NC 28224-0828

American Society on Aging Graduate Student Research Award 560
The award is presented annually to a graduate student for research relevant to aging and applicable to practice. The winner will be expected to attend the joint conference of the ASA and the National Council on the Aging to present the winning paper. The award comprises a certificate, complimentary registration and one-night's lodging for the conference, presentation of research findings at a highlighted session at the conference, and complimentary one-year membership in ASA. Submission guidelines and application are available online.
Requirements: Only graduate-level research will be considered. Findings must be from completed research; submission of the conceptual framework alone is not sufficient. Applicants must either be currently enrolled in a graduate degree program or must have completed their studies no more than one year before the time of submission. Applicants must be sponsored by a faculty member. Applicants and faculty sponsors need not be ASA members.
Restrictions: Papers that have been published are not eligible for submission.
Geographic Focus: All States
Date(s) Application is Due: Oct 1
Contact: Nancy Decia; (415) 974-9610; fax (415) 974-0300; awards@asaging.org
Internet: http://www.asaging.org/asav2/awards/description_grad.cfm?submenu1=grad
Sponsor: American Society on Aging
833 Market Street, Suite 511
San Francisco, CA 94103-1824

American Society on Aging Mental Health and Aging Awards 561
The awards are made to organizations that have demonstrated high-quality, innovative programs that enhance the health-related quality of life in older adults. The program recognizes outstanding intervention programs developed in recent years to improve the health and health care of the aging. Model programs can include, but are not limited to, wellness/health promotion programs, community-based care, dementia care, mental health, caregiver supports, medication management, chronic illness care, end-of-life care, and programs that expand access to underserved groups of older adults. Winning programs will be honored at the Conference of The National Council on the Aging (NCOA) and the American Society on Aging (ASA). Up to six programs will receive cash awards, one-night's lodging for one person at the conference, and a one-year complimentary membership to ASA and the Healthcare and Aging Network.
Requirements: Applicants must be ASA members. The program must have at least one year's worth of program results and cannot have been in existence longer than five years (or must represent a substantially new program model in the past five years) by the date of submission.
Restrictions: Organizations that have won this award within the past two years are not eligible to apply.
Geographic Focus: All States
Date(s) Application is Due: Oct 31
Amount of Grant: 2,500 USD
Contact: Tobi Abramson, (516) 596-0073; tabramson@excite.com
Internet: http://www.asaging.org/asav2/awards/description_mhan.cfm?submenu1=mhan
Sponsor: American Society on Aging
833 Market Street, Suite 511
San Francisco, CA 94103-1824

American Society on Aging NOMA Award for Excellence in Multicultural Aging 562
Awards will be given to organizations that have demonstrated high-quality, innovative programs that enhance the lives of a multicultural aging population. Using the framework that supports the broader vision of diversity at the American Society on Aging, the NOMA Award for Excellence in Multicultural Aging seeks to identify and recognize best practices in developing and implementing programs and/or services that meet the needs of a multicultural aging population. Up to three programs will receive cash awards of $1,500, one complimentary conference registration, one night's lodging for one person at the conference and a one-year complimentary membership to ASA and the Network on Multicultural Aging (NOMA).
Requirements: Eligible applicants must: be an organization or have an affiliation with an organization that focuses on providing information and services to an aging population; have a program and/or service targeted to a multicultural aging population that has been tested, has a proven and successful track record, and has been in existence for a minimum of one year; and have a membership with ASA.
Geographic Focus: All States
Date(s) Application is Due: Dec 15
Amount of Grant: 1,500 USD
Contact: Mahi Sadeghi; (415) 974-9602; fax (415) 974-0300; awards@asaging.org
Internet: http://www.asaging.org/asav2/awards/description_noma.cfm?submenu1=noma
Sponsor: American Society on Aging
833 Market Street, Suite 511
San Francisco, CA 94103-1824

American Sociological Association Minority Fellowships 563
Through its Minority Fellowship Program (MFP), the American Sociological Association (ASA) supports the development and training of sociologists of color in mental health and drug abuse research. Funded by a training grant sponsored by the National Institute of Mental Health (NIMH) and co-funded by the National Institute of Drug Abuse (NIDA), MFP seeks to attract talented doctoral students to ensure a diverse and highly trained workforce is available to assume leadership roles in research related to the nation's mental health and drug abuse research agendas.
Requirements: American citizens and permanent visa residents including, but not limited to, persons who are African American, Latino/Hispanic (e.g., Chicano, Cuban, Puerto Rican), Native American, and Asian American (e.g., Chinese, Japanese, Korean), and Pacific Islanders (e.g., Hawaiian, Guamanian, Samoan, Filipino) are eligible. The program is open to students beginning or continuing study in graduate sociology departments. New students must qualify for acceptance at accredited institutions of higher learning and express a commitment to sociological research on mental health. Upon completion of their support, recipients are expected to engage in behavioral research or teaching, or a combination thereof, for a period equal to the period of support beyond 12 months.
Geographic Focus: All States
Date(s) Application is Due: Jan 31
Amount of Grant: 20,000 - 25,000 USD
Contact: Margaret Weigers Vitullo, Director; (202) 383-9005, ext. 323; fax (202) 638-0882; spivack@asanet.org or apap@asanet.org
Internet: http://www.asanet.org/cs/root/leftnav/funding/minority_fellowship_program
Sponsor: American Sociological Association
1430 K Street NW, Suite 600
Washington, D.C. 20005

American Trauma Society, Pennsylvania Division Mini-Grants 564
Mini-Grants are made available annually by the American Trauma Society, Pennsylvania Division, to member organizations for the purpose of trauma prevention educational projects. The mini-grant program was designed to encourage member hospitals to promote trauma prevention activities within their own communities. Areas to be considered include, but are not limited to, the following: bicycle safety, impaired driving, home safety, elderly trauma, pedestrian safety, child safety, farm safety, anti-violence prevention, and initiatives surrounding second trauma. Suggested awards do not exceed $1,500.00 each.
Requirements: To receive a mini-grant from ATSPA you must first be a member organization. All mini-grant proposals are reviewed by the ATSPA grant review committee for compliance with grant requirements, feasibility of project, cost-effectiveness, the involvement of other organizations and anticipated project results. Proposals must comply with all federal and state regulations of current funding grants received by the ATSPA.
Geographic Focus: Pennsylvania
Amount of Grant: Up to 1,500 USD
Contact: Ruth Hockley, Grant and Project Coordinator; (717) 766-1616 or (800) 822-2358; fax (717) 766-6989; atspa@atspa.org
Internet: http://www.atspa.org/programs-and-materials/programs/mini-grants-micro-grants/
Sponsor: American Trauma Society, Pennsylvania Division
2 Flowers Drive
Mechanicsburg, PA 17050-1701

Amerigroup Foundation Grants 565
Helping to create healthy communities is the cornerstone of the Amerigroup Foundation's mission. The objective is to serve as a national resource that fosters an environment where there is a continuum of education, access and care, all of which improve the health and well-being of the financially vulnerable and uninsured Americans. The Foundation primarily provides grants in the form of general support, but program development and sponsorships are also available to qualified non-profit organizations. Most recently, grants have ranged from $250 to $35,000.
Requirements: 501(c)3 non-profits are eligible to apply from the following states: Arizona, California, Colorado, Connecticut, Florida, Georgia, Indiana, Kansas, Kentucky, Louisiana, Maine, Maryland, Massachusetts, Missouri, Nevada, New Hampshire, New Jersey, New Mexico, New York, Ohio, South Carolina, Tennessee, Texas, Virginia, Washington, West Virginia, and Wisconsin.
Restrictions: Funding is unavailable for: projects or organizations that offer a direct benefit to the trustees of the Foundation or to employees or directors of Amerigroup; projects or organizations that might in any way pose a conflict with Amerigroup's mission, goals, programs, products or employees; projects or organizations that do not benefit a broad cross section of the community; individuals; political parties, candidates or lobbying activities; benefits, raffles, souvenir programs, trips, tours or similar events; for-profit entities, including start-up businesses.
Geographic Focus: Arizona, California, Colorado, Connecticut, Florida, Georgia, Indiana, Kansas, Kentucky, Louisiana, Maine, Maryland, Massachusetts, Missouri, Nevada, New Hampshire, New Jersey, New Mexico, New York, Ohio, South Carolina, Tennessee, Texas, Virginia, Washington, West Virginia, Wisconsin
Amount of Grant: 250 - 35,000 USD
Samples: College of William and Mary, Williamsburg, Virginia, $35,000 - general operations; Nature Discovery Center, Bellaire, Texas, $10,000 - general operations; Snohomish County Human Services, Everett, Washington, $30,000 - general operations.
Contact: Grants Manager; (757) 490-6900 or (757) 962-6468; fax (757) 222-2360
Internet: http://www.realsolutions.com/company/pages/Foundation.aspx
Sponsor: Amerigroup Foundation
4425 Corporation Lane
Virginia Beach, VA 23462-3103

Amerisure Insurance Community Service Grants 566
The company awards grants nationwide to nonprofit organizations in the areas of education and health. There are no application forms. Send proposals to the headquarters office or the nearest facility. Proposals should contain a detailed description of the project and the amount requested. Junior achievement is supported especially in K-12 classrooms in the Farmington Hills, MI school system.
Requirements: Nonprofit organizations are eligible.
Geographic Focus: All States
Date(s) Application is Due: Dec 31
Contact: Human Resources; (248) 615-9000
Internet: http://www.amerisure.com/au_1d_community.cfm
Sponsor: Amerisure Insurance Company
26777 Halsted Road
Farmington Hills, MI 48331-3586

amfAR Fellowships 567
An amfAR fellowship is a grant that encourages the postdoctoral (M.D., Ph.D., or equivalent) investigator with limited experience in the field to advance a career in HIV/AIDS research. Fellowships are awarded for two years and may not be renewed for additional funding. amfAR fellows and mentors must be affiliated with the same nonprofit institution. The applicant's interest in a career in HIV/AIDS will be demonstrated by previous relevant work at the postdoctoral fellow or instructor level and will be carefully evaluated. Each fellowship is funded at a total of up to $125,000: a maximum of $110,000 is allowed for personnel (salary and fringe benefits) and other research-related direct costs. It is expected that a fellow will devote the decided majority of his or her time to the approved fellowship project. Personnel costs supported by the fellowship must represent a minimum of 85% effort and be consistent with institution policy for other institution personnel of similar rank and title, regardless of source(s) of support. An additional $3,636 is provided to support attendance at amfAR-approved professional development activities, for a direct cost maximum of $113,636. The period of performance for fellowships awarded under this RFP will be for two years starting January 1.
Requirements: The fellowship applicant must be mentored by an experienced investigator who: is qualified to oversee the proposed research; has successfully supervised postdoctoral fellows; and is at the associate professor level or higher.
Restrictions: Institutional indirect costs may not exceed 10% of direct costs.
Geographic Focus: All States
Date(s) Application is Due: Jul 7
Amount of Grant: Up to 125,000 USD
Contact: Administrator; (212) 806-1696 or (212) 806-1600; grants@amfar.org
Internet: http://www.amfar.org/lab/grants/default.aspx?id=7455&terms=Fellowships
Sponsor: American Foundation for AIDS Research
120 Wall Street, 13th Floor
New York, NY 10005-3908

amfAR Global Initiatives Grants 568
Since awarding its first international grant in 1986, amfAR has supported HIV/AIDS research, prevention, education, and training programs in dozens of countries across the developing world. Today, amfAR's international grant making supports two important initiatives. Formed in 2001, TREAT Asia is a network of clinics, hospitals, and research centers working with civil society to ensure the safe and effective delivery of HIV/AIDS treatments across Asia and the Pacific. And through a newly established MSM Initiative, amfAR awards grants to grassroots groups in support of innovative HIV/AIDS services for men who have sex with men in resource-limited countries.
Geographic Focus: All States
Date(s) Application is Due: Jul 21
Amount of Grant: Up to 15,000 USD
Samples: Matthew G. Law, Ph.D., University of New South Wales, Sydney, Australia, $157,465--for Asia Pacific HIV Observational Database (APHOD); Shivananda Khan, Naz Foundation International, London, United Kingdom, $10,000--for Risks and Responsibilities: Male Sexual Health and HIV.
Contact: Administrator; (212) 806-1696 or (212) 806-1600; grants@amfar.org
Internet: http://www.amfar.org/world/grants/default.aspx?id=192
Sponsor: American Foundation for AIDS Research
120 Wall Street, 13th Floor
New York, NY 10005-3908

amfAR Mathilde Krim Fellowships in Basic Biomedical Research 569
amfAR, The Foundation for AIDS Research, is pleased to announce the availability of support for Mathilde Krim Fellowships in Basic Biomedical Research. Each fellowship is funded at a total of up to $125,000 (phase I): a maximum of $110,000 is allowed for personnel (salary and fringe benefits) and other research-related direct costs. It is expected that a Krim fellow will devote the decided majority of his or her time to the approved fellowship project. Personnel costs supported by the fellowship must represent a minimum of 85% effort and be consistent with institution policy for other institution personnel of similar rank and title, regardless of source(s) of support. An additional $3,636 is provided to support attendance at amfAR-approved professional development activities, for a direct cost maximum of $113,636. Contingent upon subsequent application and peer review, phase II funding for an additional consecutive year may be approved to support basic biomedical HIV/AIDS research costs (up to $50,000 total) during the first year of a tenure-track position at any U.S. or international nonprofit research institution. The period of performance for Mathilde Krim Fellowships awarded under this RFP will be from January 1 to December 31 of the following year.
Restrictions: Institutional indirect costs may not exceed 10% of direct costs.
Geographic Focus: All States
Date(s) Application is Due: Jul 7
Amount of Grant: Up to 125,000 USD
Contact: Administrator; (212) 806-1696 or (212) 806-1600; grants@amfar.org
Internet: http://www.amfar.org/lab/grants/default.aspx?id=7455&terms=Fellowships
Sponsor: American Foundation for AIDS Research
120 Wall Street, 13th Floor
New York, NY 10005-3908

amfAR Public Policy Grants 570
AmfAR's public policy grants support the advancement of evidence-based AIDS-related public policies. Many of these awards facilitate the study and implementation of HIV prevention methods, such as harm reduction programs, that are based on sound science and help save lives. Public policy grants also help fund policy-related educational programs, including conferences.
Geographic Focus: All States
Samples: Ronald Bayer, Ph.D., The Evolving Ethics and Politics for HIV Testing, Columbia University, Joseph L. Mailman School of Public Health, New York, NY, $66,000; Paula Santiago, 7th National Harm Reduction Conference, Harm Reduction Coalition, New York, NY, $10,000.
Contact: Administrator; (212) 806-1696 or (212) 806-1600; grants@amfar.org
Internet: http://www.amfar.org/hill/article.aspx?id=3920
Sponsor: American Foundation for AIDS Research
120 Wall Street, 13th Floor
New York, NY 10005-3908

amfAR Research Grants 571
Grants are awarded to support basic and clinical research projects in biomedical, humanistic, and social sciences research relevant to AIDS. In general, funds are applied to direct costs of salaries and fringe benefits for professional and technical personnel, laboratory supplies and equipment, travel, and the publication of findings. Research grants are awarded for one year without assurance of continued funding. $100,000 in direct costs plus up to 20% for indirect costs is available. Investigators are required to submit a pre-application letter of intent (LOI). Any postdoctoral investigator who is affiliated with a nonprofit institution may submit an LOI.
Requirements: Principal investigators for research grants must be faculty-level researchers affiliated with a nonprofit institution. Research grants are given to nonprofit institutions worldwide to support investigator-led projects approved by the Foundation.
Geographic Focus: All States
Amount of Grant: 100,000 USD
Samples: Columbia U (New York, NY)--to support the work of Susana Valente, for studies on microbicide development and AIDS-related research, $90,000.
Contact: Administrator; (212) 806-1696 or (212) 806-1600; grants@amfar.org
Internet: http://www.amfar.org/rfp/
Sponsor: American Foundation for AIDS Research
120 Wall Street, 13th Floor
New York, NY 10005-3908

Amgen Foundation Grants 572
Amgen seeks to: advance science education, improve quality of care and access for patients, and support resources that create sound communities where Amgen staff members live and work. Requests must be received at least 90 days in advance of the desired contribution date. Guidelines are available online.
Requirements: 501(c)3 tax-exempt organizations located in Amgen communities are eligible. Eligible grantees may include public elementary and secondary schools, as well as public colleges and universities, public libraries and public hospitals.
Restrictions: In general, Amgen does not consider requests for the following: support to individuals, fundraising or sports-related events, corporate sponsorship requests, religious organizations unless the program is secular in nature and benefits a broad range of the community, political organization or lobbying activity, labor unions, fraternal, service or veterans' organizations, private foundations, or organizations that are discriminatory.
Geographic Focus: All States
Samples: California State U (Camarillo, CA)--for scientific equipment, $950,000; Institute for Systems Biology (Seattle, WA)--for the endowment and operations, $3 million; International Federation of the Red Cross and Red Crescent Societies (Geneva)--for relief efforts in South Asia and Africa, $1 million.
Contact: Program Contact; (805) 447-4056 or (805) 447-1000; fax (805) 447-1010
Internet: http://wwwext.amgen.com/citizenship/apply_for_grant.html
Sponsor: Amgen Foundation
1 Amgen Center Drive, MS 38-3-B
Thousand Oaks, CA 91320-1799

AMHPS Dr. James A. Ferguson Emerging Infectious Diseases Fellowships 573
The Dr. James A. Ferguson Emerging Infectious Diseases Fellowship is a paid, eight-week, professional-development program (June and July) for racial- and ethnic-minority graduate students with an interest in public health. The fellowship exposes students to research opportunities at the Centers for Disease Control and Prevention (CD.C.). Students gain research experience in infectious-diseases research areas such as antimicrobial resistance, emerging infectious diseases, animal-related diseases, hospital-related infectious issues, etc. In addition to laboratory research, students are exposed to issues concerning public health via bi-weekly seminars (Brown Bag Sessions) and networking activities (Rap Sessions). Fellows are required to submit a progress report, literature review, abstract and final report throughout the course of the fellowship.

Upon completion of all research, students are required to deliver a formal presentation for CD.C. researchers and Association of Minority Health Professions Schools, Inc. (AMHPS) staff. The fellowship provides a $4,000 stipend which is released in two installments – at the beginning and the end of the fellowship. Additionally the fellowship provides roundtrip travel or mileage reimbursement for students who attend school outside of the Atlanta, Georgia area (housing is provided to non-resident students.) The application form with instructions is available for download from the fellowship website and should be received by the Program Officer no later than the deadline date. Applicants are encouraged to start the application process early to give references enough time to fill out and return the recommendation section. The application must be notarized. The deadline date may vary from year to year. Interested applicants are encouraged to check the website or contact the Program Officer to verify current deadline dates.
Requirements: Applicants must be currently enrolled as full-time graduate level students at an AMHPS member institution or in a Non-AMHPS Master of Public Health Program. Applicants must commit to the length of the fellowship.
Restrictions: This internship program is restricted to members of an under-represented minority group as defined by the CD.C. Office of Minority Health & Health Disparities.
Geographic Focus: All States
Date(s) Application is Due: Feb 25
Amount of Grant: 4,000 USD
Contact: Jamel Slaughter, Program Officer; (678) 904-4219; fax (678) 904-4518; jamel.slaughter@amhps.org
Internet: http://www.minorityhealth.org/p-student-drjames.php
Sponsor: Association Of Minority Health Professions Schools, Inc.
3525 Piedmont Road, 7 Piedmont Center, Suite 300
Atlanta, GA 30305

Amica Companies Foundation Grants 574
The Amica Companies Foundation has a mission to harness the power of enduring relationships to help individuals, families and communities become economically independent and strong. Understanding this, the Foundation provides support to programs that align with its charitable giving mission statement. These programs may fall under the following categories: basic needs and community development; education; health; arts and culture; and community and public affairs. Interested parties should submit a proposal via the Foundation's online application system. The review process can take up to eight weeks. When a decision has been made, applicants will receive an email notification.
Requirements: The Foundation supports qualified, federal tax-exempt 501(c)3 organizations, as defined by the Internal Revenue Service Code. Grants are generally restricted to community-based organizations serving Rhode Island residents.
Restrictions: The Foundation generally does not provide support to: capital campaigns; building restorations; or general operating expenses.
Geographic Focus: Rhode Island
Contact: Meredith Gregory, Charitable Giving Coordinator; (800) 622-6422, ext. 2100; fax (401) 334-4241; AmicaCoFoundations@amica.com
Internet: https://www.amica.com/en/about-us/in-your-community/charitable-grants.html
Sponsor: Amica Companies Foundation
100 Amica Way
Lincoln, RI 02865-1167

Amica Insurance Company Community Grants 575
The Amica Insurance Company Citizenship Grant program financially supplements employee volunteer efforts. This program provides annual grants of up to $1,000 maximum to community groups in which Amica employees are involved. The corporation's Matching Gift Fund program offers employee-matching grants of up to $1,000 maximum to elementary and secondary schools, two-year colleges and four-year colleges, and universities and college level graduate schools. The company makes corporate donations to health organizations, hospitals, libraries, schools, museums, and other deserving institutions. Recent recipients include the American Heart Association, the Miami Project to Cure Paralysis, and The Nature Conservancy.
Geographic Focus: Arizona, California, Colorado, Connecticut, Georgia, Illinois, Maine, Maryland, Massachusetts, Michigan, Minnesota, Nevada, New Hampshire, New Jersey, New York, North Carolina, Ohio, Oregon, Pennsylvania, Rhode Island, South Carolina, Tennessee, Texas, Virginia, Washington, Wisconsin
Amount of Grant: Up to 1,000 USD
Contact: Meredith Gregory; (800) 622-6422, ext. 2100; fax (401) 334-4241
Internet: https://www.amica.com/en/about-us/in-your-community/corporate-citizenship.html
Sponsor: Amica Insurance Company
100 Amica Way
Lincoln, RI 02865-1167

AMI Semiconductors Corporate Grants 576
The company makes grants to nonprofit organizations in support of the performing arts, economic development, business education, health cost containment, and social services for senior citizens. Types of support include conferences and seminars, general operating support, matching grants, multiyear grants, professorships, research, and scholarships. There are no application deadlines. Submit a letter of inquiry that includes a description of the organization and program, amount of funds requested, purpose of the request, recently audited financial statement, and proof of tax-exempt status. Contact office for grant availability.
Restrictions: The company does not support political or lobbying groups.
Geographic Focus: All States
Amount of Grant: 1,000 - 2,500 USD
Contact: Tamera Drake, (208) 234-6890; fax (208) 234-6795; tamera_Drake@amis.com
Internet: http://www.amis.com/about
Sponsor: AMI Semiconductors
2300 Buckskin Road
Pocatello, ID 83201

AMSSM Foundation Research Grants 577
The purpose of the AMSSM Foundation Research Grant Awards is to foster original scientific investigations by members of AMSSM. in this program, the Foundation welcomes research grant proposals that investigate issues within the broad discipline of sports medicine, including clinical practice, injury prevention and rehabilitation, basic science, epidemiology, and education. Grant awards are designed to provide partial support of research projects. Grantee institutions are expected to provide all necessary basic facilities and services that normally would be expected to exist in any institution qualified to undertake such research. Overhead or indirect costs will be supported to a maximum of 10% of direct costs. It is anticipated that the requested support will fall into one of three ranges. 1) $500 to $2,500 total costs annually for mailings, supplies, equipment, and technical support of small or pilot projects; 2) $2,500 to $10,000 total costs annually for a project larger in scope than the previous category and may provide partial support of research assistants or study coordinators; and 3) $10,000 to $25,000 total costs annually for a significant project that includes well documented pilot or previous studies and may provide partial salary support for key investigators. The maximum grant award is $25,000.
Requirements: The primary investigator of the grant must be an AMSSM member at the time of grant submission. Resident and fellow AMSSM members may apply as the principal investigator but must have at least one full AMSSM member listed as a co-investigator at the time of application.
Geographic Focus: All States
Date(s) Application is Due: Oct 1
Amount of Grant: Up to 25,000 USD
Contact: Grants Administrator; (913) 327-1415; fax (913) 327-1491; office@amssm.org
Internet: http://www.newamssm.org/Research%20Grant%20Information.html
Sponsor: American Medical Society for Sports Medicine
11639 Earnshaw
Overland Park, KS 66210

ANA/Grass Foundation Award in Neuroscience 578
The ANA/Grass Foundation Award was established to honor outstanding young investigators doing research in basic or clinical neuroscience. The awardee receives a $1,000 honorarium; inclusion in the annual meeting as either a poster or platform presenter; complimentary registration; a commemorative plaque; and up to $1,500 travel reimbursement.
Requirements: Eligible candidates include: physician-scientist neurology faculty members early in their careers (M.D. or M.D./Ph.D.); those who are five years or less out of postgraduate or post fellowship training; and those conducting research in neuroscience. Nominators may send a letter of recommendation and the nominee's CV and biography. Self nominations and second party nominations are accepted.
Geographic Focus: All States
Date(s) Application is Due: Feb 15
Amount of Grant: 2,500 USD
Contact: Laurie Dixon; (952) 545-6284; fax (952) 545-6073; ana@llmsi.com
Internet: http://www.aneuroa.org/i4a/pages/index.cfm?pageid=3360
Sponsor: American Neurological Association
5841 Cedar Lake Road, Suite 204
Minneapolis, MN 55416

ANA Derek Denny-Brown Neurological Scholar Award 579
Each year the American Neurological Association awards $1,000 to a young member of the Association who it is deemed has achieved a significant stature in neurological research, and whose promise of continuing major contributions to the field of neurology is anticipated. The Award may be made to a newly elected member of the Association, to be presented at the time of induction, or who was been admitted within the previous three years. It is anticipated that the awardee will present a paper at the annual ANA meeting.
Geographic Focus: All States
Amount of Grant: 1,000 USD
Contact: Linda Scher; (952) 545-6284; fax (952) 545-6073; lindascher@llmsi.com
Internet: http://www.aneuroa.org/i4a/pages/index.cfm?pageid=3355
Sponsor: American Neurological Association
5841 Cedar Lake Road, Suite 204
Minneapolis, MN 55416

ANA Distinguished Neurology Teacher Award 580
The ANA has established a Teaching Award for outstanding accomplishments in teaching neurology to residents and medical students. Nominees come from the entire field of clinical neurology or neuroscience. The purpose of the Award is to recognize contributions by gifted and talented teachers in neurology. The recipient receives a $1,000 honorarium, and up to $1,000 in expenses to attend the annual meeting where the Award is presented. Nominees must be faculty members in either a tenure or non-tenure track. The nominee's educational accomplishments should be in neurology, and they should be actively teaching. Nominees who have published in the field of medical education and whose educational materials have been adopted by other institutions are given special consideration. An institutional system for selecting candidates that provides for broad student participation and the peer judgment of faculty colleagues will be favored.

Requirements: The nomination packet should include the nominee's CV and complete bibliography. The packet should also include three letters of support from the following people: a former student or trainee no longer at the nominee's institution; a neurology program director or education chair at another institution; a national or international colleague who has worked with the nominee in development or oversight of neurology; and the completed nomination form, signed by the Neurology Department Chair. The nomination form is available at the ANA website. Completed nomination packages should be sent in one email.
Geographic Focus: All States, Canada
Date(s) Application is Due: Mar 1
Amount of Grant: 1,000 USD
Contact: Linda Scher; (952) 545-6284; fax (952) 545-6073; lindascher@llmsi.com
Internet: http://www.aneuroa.org/i4a/pages/index.cfm?pageid=3356
Sponsor: American Neurological Association
5841 Cedar Lake Road, Suite 204
Minneapolis, MN 55416

ANA F.E. Bennett Memorial Lectureship 581
The F.E. Bennett Memorial Lectureship is given to outstanding researchers and educators in neurology. Lecturers receive a $1,000 honorarium, a plaque, and travel expenses to the meeting. The lectureship is not limited to members of the ANA.
Geographic Focus: All States, All Countries
Amount of Grant: 1,000 USD
Contact: Linda Scher; (952) 545-6284; fax (952) 545-6073; lindascher@llmsi.com
Internet: http://www.aneuroa.org/i4a/pages/index.cfm?pageid=3354
Sponsor: American Neurological Association
5841 Cedar Lake Road, Suite 204
Minneapolis, MN 55416

ANA Junior Academic Neurologist Scholarships 582
The American Neurological Association invites nominations of junior faculty members for the Junior Academic Neurologist Scholarships. The Scholarships support up to 10 junior faculty to attend the ANA's annual meeting. As part of ANA's commitment to underrepresented and underserved populations in medicine, consideration will be given to individuals in those populations. The Scholarships include complimentary registration for the meeting and up to four nights lodging.
Requirements: Junior neurologists who have not received a K award, and are one to four years post residency or fellowship will be eligible to be nominated. Nominees must be U.S. citizens with an M.D., committed to a career in academic neurology. Along with the online nomination form, a letter of recommendation and most current CV are also required.
Restrictions: Only one candidate from any single institution is likely to be selected.
Geographic Focus: All States
Date(s) Application is Due: Jun 29
Contact: Laurie Dixon; (952) 545-6284; fax (952) 545-6073; ana@llmsi.com
Internet: http://www.aneuroa.org/i4a/pages/index.cfm?pageID=3315
Sponsor: American Neurological Association
5841 Cedar Lake Road, Suite 204
Minneapolis, MN 55416

ANA Raymond D. Adams Lectureship 583
The Adams Lectureship was established to honor Dr. Raymond D. Adams, emeritus Bullard Professor of Neurology at Harvard Medical School and emeritus Chief of the Neurology Service at the Massachusetts General Hospital. An ANA member presents the lectureship to the recipient at the annual meeting. The recipient receives a $1,000 honorarium and a plaque.
Geographic Focus: All States, All Countries
Amount of Grant: 1,000 USD
Contact: Linda Scher; (952) 545-6284; fax (952) 545-6073; lindascher@llmsi.com
Internet: http://www.aneuroa.org/i4a/pages/index.cfm?pageID=3389
Sponsor: American Neurological Association
5841 Cedar Lake Road, Suite 204
Minneapolis, MN 55416

ANA Soriano Lectureship 584
The Soriano Lectureship is given to a member of the ANA at the annual meeting. It includes a $1,000 honorarium, plaque, and travel expenses to the attend the meeting.
Geographic Focus: All States, All Countries
Amount of Grant: 1,000 USD
Contact: Linda Scher; (952) 545-6284; fax (952) 545-6073; lindascher@llmsi.com
Internet: http://www.aneuroa.org/i4a/pages/index.cfm?pageid=3359
Sponsor: American Neurological Association
5841 Cedar Lake Road, Suite 204
Minneapolis, MN 55416

ANA Wolfe Neuropathy Research Prize 585
The Wolfe Neuropathy Research Prize was designed to honor an outstanding investigator who has identified a new cause or treatment of axonal peripheral neuropathy. The recipient is awarded $2,000 and a plaque.
Geographic Focus: All States
Contact: Linda Scher; (952) 545-6284; fax (952) 545-6073; lindascher@llmsi.com
Internet: http://www.aneuroa.org/i4a/pages/index.cfm?pageid=3358
Sponsor: American Neurological Association
5841 Cedar Lake Road, Suite 204
Minneapolis, MN 55416

Andersen Corporate Foundation 586
The Foundation contributes to: organizations that enhance self-sufficiency for people living in poverty, senior citizens, and people with disabilities; organizations that promote safe and healthy environments, as well as organizations that seek to improve health through prevention and education programs, primarily for young people, senior citizens, and people in vulnerable situations; organizations that offer intellectual and social opportunities with a focus primarily on young people, senior citizens, and people with disabilities; and support that builds, promotes, and preserves communities. Complete guidelines are available online.
Requirements: All grant recipients must meet these giving guidelines: the organization's programs and services are consistent with Andersen Corporate Foundation's mission and values; the organization's purpose and programs fit within Andersen Corporate Foundation's defined program focus areas; the organization can demonstrate sound fiscal management and effective delivery of services; nonprofit corporation is registered under section 501(c)3 of the IRS Code in the United States or for Canadian Charities with the Canadian Revenue Agency.
Restrictions: The Foundation avoids making grants in organizations that are in competition with each other to provide the same service to the community. The Foundation does not support endowments or make grants to individuals. Funding is not granted to national research organizations.
Geographic Focus: Iowa, Minnesota, Virginia, Wisconsin, Canada
Date(s) Application is Due: Apr 15; Jul 15; Oct 15; Dec 15
Contact: Program Director; (888) 439-9508 or (651) 439-1557; fax (651) 439-9480; andersencorpfdn@srinc.biz
Internet: https://www.srinc.biz/bp/index.html
Sponsor: Andersen Corporate Foundation
342 5th Avenue N
Bayport, MN 55003

Andre Agassi Charitable Foundation Grants 587
The foundation awards grants to at-risk youth programs targeting education and recreation in the Las Vegas area. There are no application forms or deadlines. Submitted requests will be considered for funding in the following calendar year. Detailed guidelines are available online.
Requirements: Nevada 501(c)3 nonprofits serving the Las Vegas metro area are eligible.
Restrictions: Grants do not support organizations or projects outside the Las Vegas community, organizations that discriminate, individuals, advertising, religious or sectarian organizations for religious purposes, or political organizations and programs designed to influence legislation or elect candidates to public office.
Geographic Focus: Nevada
Samples: Las Vegas Philharmonic, Las Vegas, NV, $10,000--art program for underprivileged youths; Boys and Girls Club of Nevada, Las Vegas, NV, $334,403--after school program; Center for Independent Living, Las Vegas., NV, $5,000--residential treatment for adolescents.
Contact: Julie Pippenger; (702) 227-5700; fax (702) 866-2928; info@agassi.net
Internet: http://www.agassifoundation.org
Sponsor: Andre Agassi Charitable Foundation
3960 Howard Hughes Parkway, Suite 750
Las Vegas, NV 89169

Andrew Family Foundation Grants 588
The Andrew Family Foundation is a private, philanthropic organization that will consider proposals from public, non-profit organizations under IRS Section 501(c)3 to support projects and organizations that foster individual growth and enhance communities through education, humanitarian efforts, and the arts. Funding primarily in the Illinois with a special interest in the Cook County region. The types of support available include: annual campaigns; building/renovation; capital campaigns; general/operating support; scholarship funds. There is no deadline date when applying for funding. Qualified grant proposals will be reviewed by the Andrew Family Foundation Grant Making Committee prior to quarterly board meetings. The committee will make a recommendation to the Board of Directors of the Foundation.
Requirements: 501(c)3 tax-exempt organizations are eligible to apply for funding. To begin the application process, take the Eligibility Quiz to confirm that you qualify for a grant from the foundation. Upon successful completion of the Eligibility Quiz, you will be invited to complete an online Letter of Inquiry. The Board will review your Letter of Inquiry and may invite you to submit a Full Application for review. Generally, the Board meets in February, May, August and November of each year.
Restrictions: The foundation does not provide funds: to individuals; taxable corporations; religious programs; political organizations; and other private foundations.
Geographic Focus: Illinois
Contact: Connor Humphrey, Grants Administrator; (708) 460-1288 or (602) 828-8471; fax (602) 385-3267; aff@inlignwealth.com or Connor.Humphrey@GenSpring.com
Internet: https://online.foundationsource.com/andrew/board2.htm
Sponsor: Andrew Family Foundation
14628 John Humphrey Drive
Orland Park, IL 60462

Angel Kiss Foundation Grants 589
Angel Kiss Foundation (AKF) is a nonprofit organization dedicated to helping families with children who are being treated for cancer. The Foundation provides financial assistance to families to help with the astronomical expenses associated with the disease. The Foundation works with families on an individual basis and reviews the needs of each family. It provides financial assistance for a variety of expenses including: medical expenses, insurance deductibles, uncovered procedures, household living expenses, mortgage payments, rent, food, utilities, travel expenses related to treatment and other

family expenses causing hardship during the child's cancer treatment. Families can work with their social worker and medical doctor for a referral to Angel Kiss Foundation, call Angel Kiss Foundation for guidance in the application procedure, or access the Web site for an online application.
Geographic Focus: All States
Contact: Donna Breen; (775) 323-7721 or (888) 589-5477; donna@dlbreen.com
Internet: http://angelkissfoundation.org/About/FAQ.html
Sponsor: Angel Kiss Foundation
150 Ridge Street, Suite 3
Reno, NV 89501

Angels Baseball Foundation Grants 590
Since its inception in 2004, the Angels Baseball Foundation has awarded many grants to worthy organizations throughout the community, ranging from music programs to at-risk youth shelters. Currently, it focuses on initiatives aimed to create and improve education, health care, arts and sciences, and community related youth programs, in addition to providing children the opportunity to experience the game of baseball. Giving is limited to the greater Los Angeles, California, area. Fields of interest include: the arts, athletics/sports activities, children and youth, cancer detection, cancer research, community and economic development, education (at all levels), health care, and community service programs. Applicants should begin by contacting the Foundation directly.
Requirements: Applicants must serve the residents of the greater Los Angeles area.
Geographic Focus: California
Contact: Anne Bafus, Community Relations; (714) 940-2174 or (714) 940-2244
Internet: http://losangeles.angels.mlb.com/ana/community/baseball_foundation.jsp
Sponsor: Angels Baseball Foundation
2000 E Gene Autry Way
Anaheim, CA 92806-6100

Anheuser-Busch Foundation Grants 591
Support is provided almost exclusively to causes that are located in communities in which the company has manufacturing facilities. Contributions are made for education, health, social services, minorities and youth, cultural enrichment, and environmental protection programs. Types of support include capital grants, employee matching gifts, equipment and material acquisition, general operating support, and donated products. Full proposals are accepted throughout the year.
Requirements: 501(c)3 tax-exempt organizations in corporation operation areas can apply, which includes California, Colorado, Florida, Georgia, Hawaii, Kentucky, Massachusetts, Missouri, New Hampshire, New Jersey, New York, Ohio, Oklahoma, Texas, and Virginia.
Restrictions: Grants are not made to individuals; political, social, fraternal, religious, or athletic organizations; or hospitals for operating funds.
Geographic Focus: California, Colorado, Florida, Georgia, Hawaii, Kentucky, Massachusetts, Missouri, New Hampshire, New Jersey, New York, Ohio, Oklahoma, Texas, Virginia
Amount of Grant: 25,000 - 100,000 USD
Contact: Assistant Manager; (314) 577-2453; fax (314) 557-3251
Internet: http://anheuser-busch.com/index.php/our-responsibility/community-our-neighborhoods/
Sponsor: Anheuser-Busch Foundation
One Busch Place
Saint Louis, MO 63118-1852

Anna Fitch Ardenghi Trust Grants 592
The Anna Fitch Ardenghi Trust was established in 1981 to support and promote quality educational, cultural, human services, and health care programming for underserved populations living in New Haven, Connecticut. Special preference is given to charitable organizations that focus on the arts or youth-related programming. The deadline for applications is July 1, and applicants will be notified of grant decisions by letter within two to three months after the proposal deadline. Grants from the Ardenghi Trust are one year in duration. Most recent awards range from $2,000 to $3,000.
Requirements: 501(c)3 organizations serving the residents of New Haven, Connecticut, are eligible to apply.
Restrictions: Grant requests for capital projects will not be considered. Applicants will not be awarded a grant for more than 3 consecutive years.
Geographic Focus: Connecticut
Date(s) Application is Due: Jul 1
Samples: Architecture Resource Center, New Haven, Connecticut, $3,000 - in support of the Design Connection Partnership (2014); Columbus House, New Haven, Connecticut, $3,000 - in support of the emergency shelter (2014); Life Haven, New Haven, Connecticut, $2,500 - general operating support (2014).
Contact: Carmen Britt; (860) 657-7019; carmen.britt@baml.com
Internet: https://www.bankofamerica.com/philanthropic/grantmaking.action
Sponsor: Anna Fitch Ardenghi Trust
200 Glastonbury Boulevard, Suite #200, CT2-545-02-05
Glastonbury, CT 06033-4056

Ann and Robert H. Lurie Family Foundation Grants 593
Founded in 1986, the foundation initially had set up six categories: medical services and research; child-related medical organizations; basic services including food and shelter; education; the arts; and so-called wild things. When Ann Lurie's husband, Robert, died in 1990, she was left with a fortune worth hundreds of millions and six kids -- ages 5 to 15. She has given away well over $100-million since Robert H. Lurie died at age 48, generally spurning requests from nonprofit groups with undistinguished records and concentrating her gifts on some of the world's biggest problems, including hunger, cancer, and inadequate health care. Today, the foundation supports educating and providing resources to children so they are better prepared for the future.
Requirements: There are no specific deadlines or application forms to fill out. An organization requesting funding should have finances in good order, demonstrating that it is making good use of its resources. Note: Ann Lurie has a preference for identifying her own projects, rather than responding to solicitations.
Geographic Focus: Illinois, Michigan
Contact: Ann Lurie, President; (312) 466-3750; fax (312) 466-3700
Sponsor: Ann and Robert H. Lurie Family Foundation
2 N Riverside Plaza, Suite 1500
Chicago, IL 60606-2600

Ann Arbor Area Community Foundation Grants 594
The foundation is interested in funding projects that will improve the quality of life for citizens of the Ann Arbor, Michigan, area. Eligible projects generally fall within the categories of education, culture, social service, community development, environmental awareness, health and wellness, and youth and senior citizens. Types of support include: emergency funds, program development, conferences and seminars, publication, seed grants, scholarship funds, research, and matching funds. Higher priority is given to programs that are preventive rather than remedial, increase individual access to community resources, examine and address the underlying causes of local problems, promote independence and personal achievement, attract volunteer resources and support, strengthen the private nonprofit sector, encourage collaboration with other organizations, and build the capacity of the applying organization. Organizations interested in applying are strongly encouraged to discuss their project with the program director prior to submitting an application.
Requirements: 501(c)3 nonprofits in the Ann Arbor, MI, area, which is the area that falls within the boundaries of the Ann Arbor public schools district, are eligible.
Restrictions: The foundation usually does not make grants for construction projects, annual giving campaigns or capital campaigns, normal operating expenses (except for start-up purposes), religious or sectarian purposes, computer hardware equipment, individuals, advocacy or political purposes, multiyear funding, or regranting.
Geographic Focus: Michigan
Date(s) Application is Due: Feb 11; Oct 7
Contact: Phil D'Anieri; (734) 663-0401; fax (734) 663-3514; pdanieri@aaacf.org
Internet: http://www.aaacf.org/grants.asp
Sponsor: Ann Arbor Area Community Foundation
301 N Main Street, Suite 300
Ann Arbor, MI 48104

Anne J. Caudal Foundation Grants 595
The Anne J. Caudal Foundation was established in 2007 to benefit disabled veterans of any time or of any branch of the United States armed forces and to perpetuate the recognition or memory of their accomplishments or sacrifice in time of war or otherwise. Special consideration is given to organizations that serve disabled veterans in New Jersey. Grant requests for general operating support are strongly encouraged. Program support will also be considered. Small, program-related capital expenses may be included in general operating or program requests. The majority of grants from the Caudal Foundation are one year in duration. On occasion, multi-year support is awarded. Application materials are available for download from the grant website. The application deadline for the Anne J. Caudal Foundation is July 1. Applicants are encouraged to review the state application guidelines for additional helpful information and clarification before applying. Applicants are also encouraged to view the foundation's funding history (link is available at the grant website). Applicants will be notified of grant decisions before August 15.
Requirements: Applicants must have 501(c)3 tax-exempt status.
Restrictions: The foundation does not support requests from individuals, organizations attempting to influence policy through direct lobbying, or any political campaigns.
Geographic Focus: All States
Date(s) Application is Due: Jul 1
Contact: Maryann Clemente; (646) 855-0786; maryann.clemente@baml.com
Internet: https://www.bankofamerica.com/philanthropic/fn_search.action
Sponsor: Anne J. Caudal Foundation
One Bryant Park, NY1-100-28-05
New York, NY 10036

Annenberg Foundation Grants 596
The foundation provides support for projects within its grantmaking interests of education and youth development, arts and culture, civic, community and the environment, and health and human services. It encourages the development of more effective ways to share ideas and knowledge. Letters of inquiry may be submitted at all times during the year and there are no deadlines. Please review the grants database for additional types of grants given.
Requirements: 501(c)3 tax-exempt organizations are eligible.
Restrictions: Full proposals are not accepted unless requested by a Foundation representative. The foundation is not presently considering inquiries for: individuals, individual K-12 schools, for-profit organizations, political activities or attempts to influence specific legislation, individual scholarships, projects focused exclusively on research, or programs outside of its grant-making interests.
Geographic Focus: All States
Contact: Leonard Aube; (310) 209-4560; fax (310) 209-1631; info@annenbergfoundation.org
Internet: http://www.annenbergfoundation.org/grants
Sponsor: Annenberg Foundation
2000 Avenue of the Stars, Suite 1000
Los Angeles, CA 90067

Annie Sinclair Knudsen Memorial Fund/Kaua'i Community Grants 597
The priorities of the Fund are projects and services that benefit the Kaua'i community including: culture and arts; education; environment; and health and human services. Priority will be given to programs that are well-defined and likely to be successful, address a community need, demonstrate an ability to deliver, and have an adequate budget. Application information is available online.
Requirements: Any nonprofit, tax-exempt 501(c)3 organization, neighborhood group or project is eligible.
Geographic Focus: Hawaii
Date(s) Application is Due: Jul 16
Amount of Grant: 2,000 - 15,000 USD
Contact: Deborah Rice, (808) 245-4585; drice@hcf-hawaii.org
Internet: http://www.hawaiicommunityfoundation.org/index.php?id=71&categoryID=22
Sponsor: Hawai'i Community Foundation
827 Fort Street Mall
Honolulu, HI 96813

Ann Jackson Family Foundation Grants 598
The foundation awards grants to eligible nonprofit organizations in its areas of interest, including secondary education, healthcare, child welfare, animal welfare, and the disabled. Types of support include building construction/renovation, capital campaigns, and general operating grants. Most grants are awarded to nonprofits in Santa Barbara, California. There are no application forms or deadlines. Contact the foundation for document attachments required.
Geographic Focus: California
Amount of Grant: 5,000 - 50,000 USD
Samples: All Saints By the Sea Episcopal Church, Santa Barbara, CA, $14,000--general purposes grant; Bowl Foundation of Santa Barbara, Santa Barbara, CA, $20,000--building project grant; Dunn School, Los Olivos, CA, $50,000--faculty housing grant.
Contact: Palmer Jackson, President; (805) 969-2258
Sponsor: Ann Jackson Family Foundation
P.O. Box 5580
Santa Barbara, CA 93150-5580

Ann Peppers Foundation Grants 599
The foundation awards grants to eligible California nonprofit organizations in its areas of interest, including arts and cultural programs, disabled, elderly, health care, private education, and social services. Types of support include capital grants, general operating grants on a temporary basis, matching grants, scholarships, and research grants. Grants are initiated by the foundation manager. There are no application deadlines; the board meets quarterly to consider requests.
Requirements: Southern California 501(c)3 nonprofits, colleges, and universities are eligible. Preference is given to requests from Los Angeles County.
Geographic Focus: California
Amount of Grant: 2,000 - 50,000 USD
Contact: Jack Alexander, Secretary; (626) 449-0793
Sponsor: Ann Peppers Foundation
625 S Fair Oaks Avenue
South Pasadena, CA 91030

Annunziata Sanguinetti Foundation Grants 600
The foundation supports charitable organizations in San Francisco, California that are devoted wholly or partially to the care, treatment, rehabilitation, and education of children with physical or mental defects or illnesses. The distribution committee meets annually, usually in November or early December. Proposals are accepted each year between July 1 and October 31.
Requirements: Nonprofit organizations in the San Francisco, California area are eligible to apply. Initial approach should be through a letter of inquiry. There is no application form required when submitting a proposal. The Proposal should include the following items: copy of IRS Determination Letter; copy of most recent annual report/audited financial statement/990; detailed description of project and amount of funding requested; 1 copy of the proposal.
Restrictions: Restricted to organizations which benefit the children residing in San Francisco County, California.
Geographic Focus: California
Date(s) Application is Due: Oct 31
Amount of Grant: 5,000 - 50,000 USD
Contact: Eugene Ranghiasci, Vice President; (415) 396-3215; fax (415) 834-0604
Sponsor: Annunziata Sanguinetti Foundation
420 Montgomery Street, 5th Floor
San Francisco, CA 94106

ANS/AAO-HNSF Herbert Silverstein Otology & Neurotology Research Award 601
The purpose of this award, which is jointly sponsored by the AAO-HNS Foundation and the American Neurotology Society, is to support a clinical or translational research project focused on diseases, disorders, or conditions of the peripheral or central auditory and/or vestibular system among new full-time academic surgeons. The award is intended as a preliminary step in clinical investigator career development and is expected to facilitate the recipient's preparation of a more comprehensive individualized research plan suitable for submission to the National Institutes of Health or comparable funding agency.
Requirements: Applicants must be physicians in the United States or Canada with demonstrated potential for excellence in research and teaching and serious commitment to an academic research career in otology or neurotology. Priority will be given to fellows or junior faculty, who have completed residencies or fellowships within four years of the application receipt date, although otolaryngology-head and neck surgery residents are eligible. All candidates must be sponsored by the Chair of his/her Division or Department and by an official representative of the institution which would administer the Award and in whose name the application is formally submitted.
Geographic Focus: All States
Amount of Grant: Up to 25,000 USD
Contact: Stephanie L. Jones; (703) 519-1586 or (703) 836-4444; sljones@entnet.org
Internet: http://www.americanneurotologysociety.com/funding.html
Sponsor: American Academy of Otolaryngology-Head and Neck Surgery Foundation
1650 Diagonal Road
Alexandria, VA 22314-2857

Anschutz Family Foundation Grants 602
The foundation makes grants in Colorado's rural and urban communities, to assist the elderly, the young, and the economically disadvantaged. Support is given to programs and projects that strengthen families and enable individuals to become productive and responsible citizens of society. Religious organizations, including Christian, Lutheran, Presbyterian, Roman Catholic, and the Salvation Army, also are eligible for funding. Types of support include special projects, general operating budgets, continuing support, seed money, emergency funds, technical assistance, and publications.
Requirements: Nonprofit organizations in Colorado may apply.
Restrictions: Grants are not awarded to individuals, programs outside of Colorado, graduate and post-graduate research, religious organizations for religious purposes, special events, promotions or conferences, candidates for political office, endowments, debt reduction, multi-year grants, and capital campaigns.
Geographic Focus: Colorado
Date(s) Application is Due: Jan 15; Aug 1
Amount of Grant: 2,500 - 10,000 USD
Samples: The Center for Hearing, Speech, and Language (Denver, CO)--to support Kidscreen, a vision, hearing, speech and language developmental screening service for low-income families, $5,000; The Learning Source for Adults and Families (Denver, CO)--general operating support for neighborhood literacy programs for adults and families, $7,500; Seniors Resource Center (Denver, CO)--for general operating support for programs that maximize seniors' independence and dignity, $5000; Women's Resource Center (Durango, CO)--To support the resource and referral program, $5,000.
Contact: Sue Anschutz-Rodgers, President & Executive Director; (303) 293-2338; fax (303) 299-1235; info@anschutzfamilyfoundation.org
Internet: http://www.anschutzfamilyfoundation.org/info.htm
Sponsor: Anschutz Family Foundation
555 17th Street, Suite 2400
Denver, CO 80202

Ansell, Zaro, Grimm & Aaron Foundation Grants 603
The Ansell, Zaro, Grimm and Aaron Foundation was established in New Jersey in 2007 by a leading law firm of the same name. Funding is offered primarily in the State of New Jersey to Jewish agencies and synagogues, health research, and social service agencies in support of the economically disadvantaged. There are no specific deadlines or application forms with which to adhere, and applicants should initially approach the Foundation with a letter of request detailing the purpose of the grant and verification of tax exempt status. Most recently, the maximum amount given has been $7,500.
Requirements: 501(c)3 organizations supporting residents of New Jersey are eligible.
Geographic Focus: New Jersey
Amount of Grant: Up to 7,500 USD
Contact: Jerold L. Zaro; (732) 922-1000; fax (732) 922-6161; jlz@ansellzaro.com
Sponsor: Ansell, Zaro, Grimm and Aaron Foundation
1500 Lawrence Avenue
Ocean, NJ 07712-4023

ANS Neurotology Fellows Award 604
The Neurotology Fellows Award was established by the American Neurotology Society in 1996 to reward scientific excellence. The award has been generously endowed through royalties received for the textbook Neurotology, Mosby Publishers, D. Brackmann and R. Jackler, editors. One or more awards of $500 each is intended to subsidize travel expenses incurred while giving a scientific presentation at the Annual Spring Meeting of the American Neurotology Society. The material presented need not have been performed during the Fellowship year; it may derive from earlier work performed during the residency or extracurricular research experience. Applicants should submit electronically a cover letter, a supporting letter from the Fellowship director to verify qualifications, and electronic submission of the abstract by October 15.
Requirements: Applicants must be full-time participants in a post-residency Neurotology Fellowship in the United States or Canada. In addition, the Fellow must be both podium presenter and first author of the paper submitted for publication.
Geographic Focus: All States
Date(s) Application is Due: Oct 15
Amount of Grant: 500 USD
Contact: Shirley Gossard, Administrator; (352) 751-0932 or (217) 414-4868; fax (352) 751-0696; segossard@aol.com
Internet: http://www.americanneurotologysociety.com/funding.html
Sponsor: American Neurotology Society
3096 Riverdale Road
The Villages, FL 32162

ANS Nicholas Torok Vestibular Award 605
The Nicholas Torok Vestibular Award, established under the aegis of the American Neurotology Society instituting a yearly lecture, is to be presented at the Annual Scientific Spring Meeting of the Society. The subject should be an innovative observation, experience or technique in the field of vestibular basic science or clinical science. The candidate shall be selected preferably from the membership of the Society or upon specific merit of any M.D. or Ph.D. A yearly award of $1,500 will be given for the selected manuscript.
Requirements: The manuscript must be presented electronically in the publication format recommended by Otology & Neurotology, the official journal of the American Neurotology Society.
Geographic Focus: All States
Date(s) Application is Due: Oct 15
Amount of Grant: 1,500 USD
Contact: Shirley Gossard; (352) 751-0932 or (217) 414-4868; segossard@aol.com
Internet: http://www.americanneurotologysociety.com/funding.html
Sponsor: American Neurotology Society
3096 Riverdale Road
The Villages, FL 32162

ANS Trainee Award 606
The American Neurotology Society offers an award for the best clinical or basic science paper in Neurotology submitted by a Resident or Fellow in training in the field of Otolaryngology-Head and Neck Surgery. A clinical study or basic research study is acceptable. The Society has established a cash award of $1,000 for the winning paper. The author will be expected to attend the Annual ANS Spring Meeting and present the paper personally.
Requirements: Residents from any approved residency program or Fellows in Neurotology in the United States or Canada are eligible to compete. The Resident or Fellow must be the primary author. He or she must provide the main inspiration for the paper and do the literature search, data collection, original and final drafts of the article, the discussion, and the conclusions.
Restrictions: The paper must be an original contribution, not previously published.
Geographic Focus: All States
Date(s) Application is Due: Oct 15
Amount of Grant: 1,000 USD
Contact: Shirley Gossard; (352) 751-0932 or (217) 414-4868; segossard@aol.com
Internet: http://www.americanneurotologysociety.com/funding.html
Sponsor: American Neurotology Society
3096 Riverdale Road
The Villages, FL 32162

Anthem Blue Cross and Blue Shield Grants 607
As a multistate entity, the program supports charitable, civic, and nonprofit groups in company operating areas. Categories of giving include healthy minds (educational), healthy bodies (health oriented), and healthy communities (community projects and human services initiatives). Joint sponsorships with hospitals and providers for programs that benefit the community are encouraged. Requests must be submitted in writing. Contact the office for guidelines.
Requirements: 501(c)3 tax-exempt organizations in corporate communities are eligible.
Restrictions: Grants do not support individuals; political organizations/candidates; religious organizations; fraternal, military, or labor organizations; or providers/hospitals.
Geographic Focus: Colorado, Connecticut, Indiana, Kentucky, Maine, Missouri, Nevada, New Hampshire, Ohio, Virginia, Wisconsin
Contact: Vicki Perkins, (317) 488-6216
Internet: http://www.anthem.com/wps/portal/ahpculdesac?content_path=shared/noapplication/f4/s0/t0/pw_018991.htm&na=aboutanthem&rootLevel=3&label=Charity%20Guidelines
Sponsor: Anthem Blue Cross and Blue Shield
120 Monument Circle
Indianapolis, IN 46204

Anthony R. Abraham Foundation Grants 608
For more than 30 years, the Anthony R. Abraham Foundation mission has been to help non-profit organizations worldwide. The foundations strives to: provide programs and services to help people around the world become self-productive and give back to their communities; ensure that no child is denied medical treatment due to a lack of insurance; provide education that breaks barriers; guarantee that research into the cure of catastrophic diseases continues; help raise the quality of life. It's been a privilege for the Foundation to be able to help ease poverty, raise hospitals, build orphanages and further medical research and the Foundation looks forward to doing even more. Some of the organizations helped include: Domestic organizations -St. Jude Children's Research Hospital; Camillus House; Miami Rescue Mission; Habitat for Humanity of Greater Miami; America's Second Harvest; Miami Children's Hospital; Jackson Memorial Hospital; Florida Heart Research Institute; Big Brothers/Big Sisters; Alonzo Mourning Charities; Honey Shine Mentoring Program; Cancer Link; The Miami Lighthouse for the Blind; Overtown Youth Center. International organizations: Brothers of the Good Shepherd; School for the Blind, Lebanon; Rene Moawad Foundation; Haitian Foundation; Children's International Network; Doctors without Borders; Little Sisters of Nazareth; Maronite Order of the Holy Family.
Requirements: Non-profit 501(c)3 organization requesting funding should fill out a funds-request form and submit it at the Foundation's website. The Foundation will contact you, if your organization qualifies under the Foundation's guidelines and the law.
Restrictions: Grants are not made to individuals.
Geographic Focus: All States
Amount of Grant: 100 - 50,000 USD
Contact: Anthony R. Abraham, Chairman; (305) 665-2222
Internet: http://www.abrahamfoundation.com/about
Sponsor: Anthony R. Abraham Foundation
1320 S Dixie Highway, Suite 241
Coral Gables, FL 33146-2937

Antone & Edene Vidinha Charitable Trust Grants 609
The trust provides partial support to programs and projects of tax-exempt, public charities in Hawaii to improve the quality of life in the state, particularly the island of Kauai. Grants of one year's duration are awarded in categories of interest to the trust, including: churches on Kauai; hospitals; health organizations which benefit the people of Kauai; educational scholarships to colleges and universities in the State of Hawaii for deserving students from the Island of Kauai. Types of support include building/renovation; equipment; general/operating support; program development; scholarship funds. Grants average from $2,000 - $80,000.
Requirements: 501(c)3 nonprofit organizations in Hawaii are eligible to apply. The Trust places a special emphasis on the island of Kauai. Contact Paula Boyce to acquire the cover sheet/application forms and any additional guidelines required to begin the application process. Proposals must be submitted by December 1st.
Restrictions: No grants to/for: individuals; endowments; multi-year pledges.
Geographic Focus: Hawaii
Date(s) Application is Due: Dec 1
Amount of Grant: 2,000 - 80,000 USD
Contact: Paula Boyce; (808) 538-4944; fax (808) 538-4647; pboyce@boh.com
Internet: http://www.hawaiicommunityfoundation.org/index.php?id=290
Sponsor: Antone and Edene Vidinha Charitable Trust
Bank of Hawai'i, Foundation Administration Department 758
Honolulu, HI 96802-3170

AOCS A. Richard Baldwin Award 610
The A. Richard Baldwin Award was given to Baldwin in recognition of his lengthy and distinguished service to the AOCS. It is the Society's highest service award. The award includes a $2,000 honorarium, a plaque, and travel expenses to the Annual Meeting. Self nominations are welcomed.
Requirements: Candidates must have made outstanding and extraordinary contributions to the Society or supported the activities of the Society through officer and/or committee participation over a substantial period of time. Candidates must also be current Society members in good standing.
Geographic Focus: All States, All Countries
Date(s) Application is Due: Oct 15
Amount of Grant: 2,000 USD
Contact: Barb Semeraro; (217) 693-4804; fax (217) 693-4849; barbs@aocs.org
Internet: http://www.aocs.org/Membership/content.cfm?ItemNumber=782
Sponsor: American Oil Chemists' Society
2710 South Boulder
Urbana, IL 61802-6996

AOCS Alton E. Bailey Award 611
The Alton E. Bailey Award recognizes outstanding research and exceptional service in the field of lipids and associated products. This prestigious award commemorates Alton E. Bailey's outstanding contributions to the field of fats and oils as a researcher, an author of several standard books in the field and as a leader in the work of the Society. The recipient receives a $750 honorarium, a plaque, and award lecture at the Annual Meeting. The application is available at the website. Self nominations are welcomed.
Requirements: Candidates must be AOCS members in good standing, with outstanding leadership to the Society. They must also have made significant contributions to the science and technology of lipids.
Geographic Focus: All States, All Countries
Date(s) Application is Due: Oct 15
Amount of Grant: 750 USD
Contact: Barb Semeraro; (217) 693-4804; fax (217) 693-4849; barbs@aocs.org
Internet: http://www.aocs.org/Membership/content.cfm?ItemNumber=781
Sponsor: American Oil Chemists' Society
2710 South Boulder
Urbana, IL 61802-6996

AOCS Analytical Division Student Award 612
The Analytical Division Student Award is given to graduate students in the filed of lipid analytical chemistry. Graduate students at any institution of higher learning who are doing research toward a degree are eligible. The Award involves a certificate, $250 honorarium, a $500 travel allowance to present at the Annual Meeting, and an award lecture or poster presentation. Self-nominations are welcomed.
Geographic Focus: All States, All Countries
Date(s) Application is Due: Oct 15
Amount of Grant: 750 USD
Contact: Barb Semeraro; (217) 693-4804; fax (217) 693-4849; barbs@aocs.org
Internet: http://www.aocs.org/Membership/content.cfm?ItemNumber=774
Sponsor: American Oil Chemists' Society
2710 South Boulder
Urbana, IL 61802-6996

AOCS George Schroepfer Medal — 613

The George Schroepfer Medal was established to honor the memory of George J. Schroepfer, Jr., a leader in the sterol and lipid field for more than 40 years. The award aims to foster Schroeper's ideals of personal integrity, high scientific standards, perseverance, and a strong spirit of survival, tempered by charm and wit. The award recognizes major advances in the steroid or sterol field, and is presented every even numbered year. It includes a bronze medal for the recipient, an award lecture at the Annual Meeting, and an honorarium determined by the endowed fund investment return. Self nominations are welcomed.
Requirements: The candidate's work may represent a single major achievement or a series of significant accomplishments, but should be based upon sound primary experimental data. Preference is given to accomplishments in the following: biochemistry and physiology with biomedical application; and interdisciplinary research in which rigorous chemical and analytical methods were applied to elucidate the physiological roles of steroids in animals, plants, or microorganisms. Fundamental advances that are primarily chemical, pharmacological, or analytical will also be considered
Geographic Focus: All States, All Countries
Date(s) Application is Due: Oct 15
Contact: Barb Semeraro; (217) 693-4804; fax (217) 693-4849; barbs@aocs.org
Internet: http://www.aocs.org/Membership/content.cfm?ItemNumber=934
Sponsor: American Oil Chemists' Society
2710 South Boulder
Urbana, IL 61802-6996

AOCS Health and Nutrition Division Student Award — 614

The Health and Nutrition Division of AOCS promotes and facilitates communication and cooperation among professionals whose interests in lipid biochemistry and physiology relate to all aspects of dietary fats and health. The Division encompasses the technical areas of dietary fats and general health. The Health and Nutrition Division Student Award is given to a graduate student in the health and nutrition field. The Award includes a $500 honorarium, certificate, and award lecture or poster presentation at the Annual Meeting. Self nominations are welcomed.
Requirements: Graduate students from any institution of higher learning who are doing research toward a degree are eligible.
Geographic Focus: All States, All Countries
Date(s) Application is Due: Oct 15
Amount of Grant: 500 USD
Contact: Barb Semeraro; (217) 693-4804; fax (217) 693-4849; barbs@aocs.org
Internet: http://www.aocs.org/Membership/content.cfm?ItemNumber=2048
Sponsor: American Oil Chemists' Society
2710 South Boulder
Urbana, IL 61802-6996

AOCS Health and Nutrition Poster Competition — 615

The Health and Nutrition Division promotes and facilitates communication and cooperation among professionals whose interests in lipid biochemistry and physiology relate to all aspects of dietary fats and health; encompasses the technical areas of dietary fats and general health. The Health and Nutrition Division Poster Competition was created to stimulate additional interest in the submission of quality poster student presentations at the Annual Meeting. Awards are given for best overall poster ($300), 1st place ($300), 2nd place ($200), and 3rd place ($100). All awardees also receive certificates, complimentary division dinner tickets, and award ribbons for their poster display.
Requirements: All poster abstracts submitted and selected for the Health and Nutrition Division Poster session are eligible for the competition.
Geographic Focus: All States, All Countries
Date(s) Application is Due: Oct 15
Contact: Barb Semeraro; (217) 693-4804; fax (217) 693-4849; barbs@aocs.org
Internet: http://www.aocs.org/Membership/content.cfm?ItemNumber=18280
Sponsor: American Oil Chemists' Society
2710 South Boulder
Urbana, IL 61802-6996

AOCS Industrial Oil Products Division Student Award — 616

The AOCS Industrial Oil Products Division supports professionals involved in research, development, engineering, marketing, and testing of industrial products and co-products from fats and oil, including glycerine and its derivatives. The Division's Student Award is given to a graduate student for travel to the Annual Meeting to present a paper. The Award includes a certificate for the recipient, an award lecture or poster presentation at the Annual Meeting, and $500 in travel funds for that meeting.
Requirements: Graduate students presenting within the Industrial Oil Products Division technical program at the Annual Meeting are eligible.
Geographic Focus: All States, All Countries
Date(s) Application is Due: Oct 15
Contact: Barb Semeraro; (217) 693-4804; fax (217) 693-4849; barbs@aocs.org
Internet: http://www.aocs.org/Membership/content.cfm?ItemNumber=810
Sponsor: American Oil Chemists' Society
2710 South Boulder
Urbana, IL 61802-6996

AOCS Lipid Oxidation and Quality Division Poster Competition — 617

The Lipid Oxidation and Quality Division supports professionals in lipid oxidation with a major focus in food applications. The Lipid Oxidation and Quality Division Poster Competition is awarded to a student for travel to present a poster at the Annual Meeting. All poster abstracts submitted by students and selected for the Lipid Oxidation and Quality Poster session are considered for the competition. The Award includes a certificate, Award ribbon for poster display, complimentary division dinner ticket, and $300.
Geographic Focus: All States, All Countries
Date(s) Application is Due: Oct 15
Amount of Grant: 300 USD
Contact: Barb Semeraro; (217) 693-4804; fax (217) 693-4849; barbs@aocs.org
Internet: http://www.aocs.org/Membership/content.cfm?ItemNumber=18281
Sponsor: American Oil Chemists' Society
2710 South Boulder
Urbana, IL 61802-6996

AOCS Processing Division Student Award — 618

The AOCS Processing Division advances the processing knowledge and managerial skills by providing a forum of technical information and networking opportunities, to promote and facilitate communication and cooperation between members, and to mentor young professionals. The Processing Division Student Award is given to a graduate student researching oilseed handling preparation and extraction, refining and processing, oil products and packaging, feed ingredients, by-products utilization, safety and health, and environmental concerns. The Award includes a certificate for the recipient, an award lecture or poster presentation at the Annual Meeting, and a $500 honorarium.
Requirements: Graduate students presenting within the Processing Division's technical program at the Annual Meeting are eligible.
Geographic Focus: All States, All Countries
Date(s) Application is Due: Oct 15
Amount of Grant: 500 USD
Contact: Barb Semeraro; (217) 693-4804; fax (217) 693-4849; barbs@aocs.org
Internet: http://www.aocs.org/Membership/content.cfm?ItemNumber=924
Sponsor: American Oil Chemists' Society
2710 South Boulder
Urbana, IL 61802-6996

AOCS Protein and Co-Products Division Student Poster Competition — 619

The Protein and Co-Products Division Student Poster Competition is awarded to a student presenting a poster at the AOCS Annual Meeting. Awards are given to a first place winner for $300, and a second place winner for $200. Each student also receives a certificate, an award ribbon for their poster display, and a complimentary division dinner ticket.
Requirements: All poster abstracts submitted by students and selected for the Protein and Co-Products Poster session are considered for the competition.
Geographic Focus: All States, All Countries
Date(s) Application is Due: Oct 15
Amount of Grant: 500 USD
Contact: Barb Semeraro; (217) 693-4804; fax (217) 693-4849; barbs@aocs.org
Internet: http://www.aocs.org/Membership/content.cfm?ItemNumber=18314
Sponsor: American Oil Chemists' Society
2710 South Boulder
Urbana, IL 61802-6996

AOCS Ralph Holman Lifetime Achievement Award — 620

The AOCS Health and Nutrition Division promotes and facilitates communication and cooperation among professionals whose interests in lipid biochemistry and physiology relate to all aspects of dietary fats and health. The Ralph Holman Lifetime Achievement Award recognizes outstanding performance and meritorious contributions. The Award is named for Ralph Holman in recognition of his lifetime service to the study of essential fatty acids and their impact on health. The Award includes a framed orchid photograph, $1,000 honorarium, and an award lecture at the Annual Meeting.
Requirements: Candidates must have made significant lifetime and meritorious achievements in areas of interest to the Health and Nutrition Division of AOCS.
Geographic Focus: All States, All Countries
Date(s) Application is Due: Oct 15
Amount of Grant: 1,000 USD
Contact: Barb Semeraro; (217) 693-4804; fax (217) 693-4849; barbs@aocs.org
Internet: http://www.aocs.org/Membership/content.cfm?ItemNumber=807
Sponsor: American Oil Chemists' Society
2710 South Boulder
Urbana, IL 61802-6996

AOCS Supelco/Nicholas Pelick-AOCS Research Award — 621

The Supelco/Nicholas Pelick-AOCS Research Award is given to a qualified scientist for outstanding original research on fats, oils, lipid chemistry, and/or biochemistry, the results of which have been published in technical papers of high quality. The recipient receives a plaque and a $10,000 honorarium. The award winner also gives a lecture at the AOCS Annual Meeting. The application is available at the website. Self nominations are welcomed.
Requirements: Candidates must have published technical papers of high quality Preference is given to individuals who are actively associated with research and have made influential discoveries in their fields.
Geographic Focus: All States, All Countries
Date(s) Application is Due: Oct 15
Amount of Grant: 10,000 USD
Contact: Barb Semeraro; (217) 693-4804; fax (217) 693-4849; barbs@aocs.org
Internet: http://www.aocs.org/Membership/content.cfm?ItemNumber=940
Sponsor: American Oil Chemists' Society
2710 South Boulder
Urbana, IL 61802-6996

AOCS Surfactants and Detergents Division Student Award   622

The AOCS Surfactants and Detergents Division encompasses the sciences and technologies associated with the development and production of surfactants, detergents, and soaps. The Surfactants and Detergents Division Student Award is given to a graduate student to help in their travel to present a paper at the Annual Meeting. The Award includes a certificate for the recipient(s), an award lecture or poster presentation at the Annual Meeting, and $500 in travel funds. Self nominations are welcomed.
Requirements: Graduate students presenting within the Surfactants and Detergents Division technical program at the AOCS Annual Meeting are eligible.
Geographic Focus: All States, All Countries
Date(s) Application is Due: Oct 15
Amount of Grant: 500 USD
Contact: Barb Semeraro; (217) 693-4804; fax (217) 693-4849; barbs@aocs.org
Internet: http://www.aocs.org/Membership/content.cfm?ItemNumber=942
Sponsor: American Oil Chemists' Society
2710 South Boulder
Urbana, IL 61802-6996

AON Foundation Grants   623

AON funds community-based initiatives that focus on youths' development through health, education, the arts and cultural activities to schools, school districts and other nonprofit organizations in communities where the company operates, especially the Chicago area. Supported initiatives include community-based projects that focus on youths' development mentally, physically, and academically, and demonstrate positive intervention by including an outcome-evaluation component. The foundation supports higher and other education, social services, community funds, and hospitals and health associations. The board meets three times each year to consider requests. There is no specific application. A letter is acceptable.
Requirements: Nonprofit organizations in AON Corporation domestic and international operating communities, especially organizations in the Chicago area, may apply.
Restrictions: Awards are restricted or limited to charitable, educational (excluding the operation of a secondary educational institution or vocational school) and scientific organizations that qualify as 501(c)3 or 509(a)(1),(2), or (3).
Geographic Focus: All States
Date(s) Application is Due: Apr 1; Jul 1; Oct 1
Amount of Grant: 1,000 - 10,000 USD
Contact: Carolyn Labutka; (312) 381-3549; fax (312) 701-4533
Internet: http://edreform.com/info/grant.htm
Sponsor: AON Foundation
200 East Randolph Street
Chicago, IL 60601

APA Congressional Fellowships   624

The fellowships afford an opportunity for the recipient psychologists to participate in the policy-making process of the U.S. Congress. Trained scientists and practitioners work with the staff members of Congress or congressional committees and are involved in the process of relating social and behavioral science to public policy. The fellowships are sponsored by APA in cooperation with American Association for the Advancement of Science (AAAS). The fellowship is for a period of one year beginning in September. The Fellowship stipend ranges from $60,000 to $75,000, depending upon years of post-doctoral experience. In addition, APA provides $375 per month for health insurance and funding may also be available for professional development and relocation expenses during the Fellowship year.
Requirements: Applicants must have a doctorate in psychology and be members of APA. Persons who are receiving sabbatical funds are also eligible to apply.
Geographic Focus: All States
Date(s) Application is Due: Jan 7
Amount of Grant: 60,000 - 75,000 USD
Contact: Micah Haskell-Hoehl, Program Administrator; (202) 336-5935; fax (202) 336-5935; mhaskell-hoehl@apa.org
Internet: http://www.apa.org/ppo/fellows/
Sponsor: American Psychological Association
750 First Street NE
Washington, D.C. 20002-4242

APA Culture of Service in the Psychological Sciences Award   625

The Award recognizes departments that demonstrate a commitment to service in the psychological sciences. Departments selected for this award will show a pattern of support for service from faculty at all levels, including a demonstration that service to the discipline is rewarded in faculty tenure and promotion. Successful Departments will also demonstrate that service to the profession is an integral part of training and mentoring. Service to the discipline includes such activities as departmental release time for serving on boards and committees of psychological associations; editing journals; serving on a review panel; or chairing an IRB. Other culture of service activities that a department would encourage include mentoring students and colleagues; advocating for psychological science's best interests with state and federal lawmakers; and promoting the value of psychological science in the public eye. The focus of this award is a department's faculty service to the discipline and not their scholarly achievements. Both Undergraduate and Graduate Departments of Psychology are eligible. Self-nominations are encouraged.
Geographic Focus: All States
Amount of Grant: 5,000 USD
Samples: Davidson College, Departments of Psycholog, Davidson, North Carolina; University of Minnesota, Department of Psychology, Minneapolis, Minnesota.
Contact: Suzanne Wandersman, Director for Governance Affairs; (202) 336-6000; fax (202) 336-5953; swandersman@apa.org
Internet: http://www.apa.org/science/dept_award.html
Sponsor: American Psychological Association
750 First Street NE
Washington, D.C. 20002-4242

APA Dissertation Research Awards   626

Dissertation awards are granted to graduate students in psychology programs to help offset the costs associated with dissertation research. The research may be done in any area of psychology. Applicants who have already defended their dissertations are eligible to apply for these funds, as long as they have not yet received doctoral degrees as of the application deadline. APA provides dissertation research awards to approximately 50 students annually. Dissertations must be approved by the students' committees before application. The application deadline falls in mid-September of each year. Awards are announced in late December.
Requirements: APA student affiliates enrolled in a psychology graduate program are eligible to apply.
Restrictions: Departments may not nominate more than three students per year for dissertation awards.
Geographic Focus: All States
Date(s) Application is Due: Sep 15
Amount of Grant: 1,000 - 5,000 USD
Contact: APA Science Directorate, (800) 374-2721 or (202) 336-5500; fax (202) 336-5953; science@apa.org
Internet: http://www.apa.org/science/dissinfo.html
Sponsor: American Psychological Association
750 First Street NE
Washington, D.C. 20002-4242

APA Distinguished Scientific Award for Early Career Contribution to Psychology   627

The Distinguished Scientific Award for Early Career Contribution to Psychology recognizes excellent young psychologists. Nominations of persons who received doctoral degrees during and since 1998 are being sought in these five areas: animal learning and behavior, comparative; cognition/human learning; developmental psychology; health; and psychopathology. These categories should be interpreted broadly and are not meant to be exclusive; all of psychology is of sufficient merit to be considered for awards. The Award consists of a citation and a cash prize, which are presented at the APA Annual Convention.
Geographic Focus: All States
Samples: Robert D. Gray, PhD (applied research), Department of Applied Psychology, Arizona State University; Patrik O. Vuilleumier, MD (behavioral /cognitive neuroscience), Department of Neurology, University of California, Davis; R. Chris Fraley, PhD (individual differences), Department of Psychology at University of Illinois at Urbana-Champaign; Jorn Diedrichsen, PhD (perception/motor performance), School of Psychology, Adeilad Brigantia University, Wales; Matthew D. Lieberman, PhD (social), University of California, Los Angeles.
Contact: Suzanne Wandersman, Director for Governance Affairs; (202) 336-6000; fax (202) 336-5953; swandersman@apa.org
Internet: http://www.apa.org/science/sciaward.html
Sponsor: American Psychological Association
750 First Street NE
Washington, D.C. 20002-4242

APA Distinguished Scientific Award for the Applications of Psychology   628

The Distinguished Scientific Award for the Applications of Psychology honors psychologists who have made distinguished theoretical or empirical advances in psychology leading to the understanding or amelioration of important practical problems. The Award consists of a citation and a cash prize, which are presented at the APA Annual Convention.
Geographic Focus: All States
Samples: Karl G. Joreskog, PhD, Emeritus Professor, University of Upssala, Sweden; Peter M. Bentler, PhD, UCLA Psychology Department, Los. Angeles, CA; John P. Campbell, PhD, Professor and Chair of Psychology and Industrial Relations, University of Minnesota, MN.
Contact: Suzanne Wandersman; (202) 336-6000; swandersman@apa.org
Internet: http://www.apa.org/science/sciaward.html
Sponsor: American Psychological Association
750 First Street NE
Washington, D.C. 20002-4242

APA Distinguished Scientific Contribution Award   629

The Distinguished Scientific Contribution Award honors psychologists who have made distinguished theoretical or empirical contributions to basic research in psychology. The Award consists of a citation and a cash prize, which are presented at the APA Annual Convention.
Geographic Focus: All States
Samples: Marilynn B. Brewer, PhD, Department of Psychology, Ohio State University, Columbus, OH; Jean M. Mandler, PhD, Research Professor, Department of Cognitive Science, University of California, San Diego; Paul Rozin, PhD, Professor of Psychology, University of Pennsylvania.
Contact: Suzanne Wandersman, Director for Governance Affairs; (202) 336-6000; fax (202) 336-5953; swandersman@apa.org
Internet: http://www.apa.org/science/sciaward.html
Sponsor: American Psychological Association
750 First Street NE
Washington, D.C. 20002-4242

APA Distinguished Service to Psychological Science Award 630

This Award recognizes individuals who have made outstanding contributions to psychological science through their commitment to a culture of service. Award recipients will receive an honorarium of $1,000. Nominees will have demonstrated their service to the discipline by aiding in association governance; serving on boards, committees and various psychological associations; editing journals; reviewing grant proposals; mentoring students and colleagues; advocating for psychological science's best interests with state and federal lawmakers; and promoting the value of psychological science in the public eye. Nominees may be involved in one service area, many of the areas, or all of the service areas noted above. An individual's service to the discipline and not a person's scholarly achievements are the focus of this award.
Geographic Focus: All States
Amount of Grant: 1,000 USD
Samples: Robert L. Balster, Ph.D., Professor of Pharmacology, University of Michigan Medical School; Nora Newcombe, Professor of Psychology, Temple University; Robert A. Bjork, Chair of the Psychology Department, University of California, Los Angeles; J. Bruce Overmier, Ph.D., Department of Psychology, University of Minnesota.
Contact: Suzanne Wandersman, Director for Governance Affairs; (202) 336-6000; fax (202) 336-5953; swandersman@apa.org
Internet: http://www.apa.org/science/serv_award.html
Sponsor: American Psychological Association
750 First Street NE
Washington, D.C. 20002-4242

APA Meritorious Research Service Commendation 631

The Commendation recognizes individuals who have made outstanding contributions to psychological science through their service as employees of the federal government or other organizations. Contributions are defined according to service to the field that directly or indirectly advances opportunities and resources for psychological science. This may include staff at federal or non-federal research funding, regulatory or other agencies. Nominees may be active or retired but ordinarily will have a minimum of 10 years of such service. The individual's personal scholarly achievements (i.e., research, teaching, and writing) are not considered in the selection process independent of their service contributions.
Geographic Focus: All States
Samples: Robert B. Huebner, National Institute on Alcohol Abuse and Alcoholism; Jack D. Maser, University of California, San Diego; Robert S. Ruskin, Consortium Research Fellows Program, Universities of the Washington.
Contact: Suzanne Wandersman, Director for Governance Affairs; (202) 336-6000; fax (202) 336-5953; swandersman@apa.org
Internet: http://www.apa.org/science/meritorious.html
Sponsor: American Psychological Association
750 First Street NE
Washington, D.C. 20002-4242

APA Scientific Conferences Grants 632

The program seeks proposals for research conferences in psychology. The purpose of this program is to promote the exchange of important new contributions and approaches in scientific psychology. Conference formats include festschrifts--conferences organized as tributes to distinguished scholars; stand-alone conferences--usually two days long, involves psychologists as the primary organizers; and add-a-day conference--meetings that occur at the beginning or end of a scientific conference other than APA (e.g., Society for Ingestive Behavior or Psychonomics Society). The conference must also be supported by the host institution with direct funds, in-kind support, or a combination of the two.
Requirements: One of the primary organizers must be a member of APA. Only academic institutions accredited by a regional body may apply. Independent research institutions must provide evidence of affiliation with such an accredited institution. Joint proposals from cooperating institutions are encouraged. Conferences may be held only in the United States, its possessions, or Canada.
Restrictions: APA governance groups, APA divisions, and related entities are not eligible for funding under this program.
Geographic Focus: All States
Date(s) Application is Due: Jun 1; Dec 1
Amount of Grant: 500 - 20,000 USD
Contact: Deborah McCall; (202) 336-6000; fax (202) 336-5953; dmccall@apa.org
Internet: http://www.apa.org/science/psa/sep07ann.html
Sponsor: American Psychological Association
750 First Street NE
Washington, D.C. 20002-4242

APA Travel Awards 633

The purpose of the travel award program is to help psychology graduate students travel to the annual APA convention to present their research. The APA awards up to 100 travel awards annually. Multiple authors are accepted, but the student applicant must be the designated presenter of the research. The application deadline falls in mid-April of each year. Awards will be announced in mid-June. Applications are available on the Web site.
Requirements: Applicants must be student affiliates of the APA; students who are not affiliates may apply for affiliation at the time of application.
Restrictions: Departments may not endorse more than three students per year.
Geographic Focus: All States
Date(s) Application is Due: Apr 12
Amount of Grant: 300 USD
Contact: Science Directorate; (800) 374-2721 or (202) 336-5500; fax (202) 336-5963; science@apa.org
Internet: http://www.apa.org/science/funding.html
Sponsor: American Psychological Association
750 First Street NE
Washington, D.C. 20002-4242

APA Young Investigator Grants 634

The program provides financial support to teaching, research, and health care delivery projects in general pediatrics. The number of awards is dependent on available funds and the size of grant requests of selected projects. Proposals should address the goal of the association: to encourage, promote, and facilitate improved patient care, teaching, and research in general pediatrics.
Requirements: The principal investigator must be a member of the APA or have submitted an application for membership. Preference will be given to new investigators, including those in training.
Restrictions: Multiple year funding requests will not be considered.
Geographic Focus: All States
Date(s) Application is Due: Oct 25
Amount of Grant: Up to 10,000 USD
Contact: Administrator; (703) 556-9222; fax (703) 556-8729; info@academicpeds.org
Sponsor: Ambulatory Pediatric Association
6728 Old McLean Village Drive
McLean, VA 22101

APF/COGDOP Graduate Research Scholarships in Psychology 635

The American Psychological Foundation (APF) and the Council of Graduate Departments of Psychology (COGDOP) jointly offer graduate research scholarships, including the Clarence J. Rosecrans Scholarship ($2000), the Ruth G. and Joseph D. Matarazzo Scholarship ($3000), as well as a number of $1,000 scholarships. The scholarships will be given directly to the individual graduate students enrolled in an interim master's program or doctoral program. If a student is currently enrolled in a terminal master's program, the student must intend to enroll in a PhD program. Several fellowships have been reserved for students who, at the time of application, are within the first two years of graduate study in psychology. The purpose of the scholarship program is to assist graduate students of psychology with research costs. The scholarships are administered by APA. Application and guidelines are available online.
Requirements: Each graduate department of psychology that is a member of COGDOP may submit nominations. The number of candidates that each member department is allowed to nominate depends upon the total number of students enrolled in the graduate program. Departments that have 100 or fewer students enrolled in their graduate programs may nominate one candidate; departments that have 101-200 graduate students enrolled may nominate up to two candidates; and departments that have more than 200 graduate students enrolled may nominate up to three candidates.
Geographic Focus: All States
Date(s) Application is Due: Jun 15
Amount of Grant: 1,000 - 3,000 USD
Contact: Science Directorate; (202) 336-6000; fax (202) 336-5953; science@apa.org
Internet: http://www.apa.org/science/apf-cogdop.html
Sponsor: American Psychological Association
750 First Street NE
Washington, D.C. 20002-4242

APHL Emerging Infectious Diseases Fellowships 636

The Emerging Infectious Diseases (EID) Laboratory Fellowship Program, sponsored by the Association of Public Health Laboratories (APHL) and the Centers for Disease Control and Prevention (C.D.C.), trains and prepares scientists for careers in public health laboratories and supports public health initiatives related to infectious disease research. Two types of fellowships are offered: the EID Advanced Laboratory Training Fellowship is a one-year program designed for bachelor's or master's level scientists, with emphasis on the practical application of technologies, methodologies and practices related to emerging infectious diseases; the EID Laboratory Research Fellowship is a two-year program designed for doctoral level (Ph.D., M.D. or D.V.M.) scientists to conduct high-priority research in infectious diseases. Up to 30 pre-doctoral one-year and 15 post-doctoral two-year fellowships will be awarded. Fellows are placed in local, state or federal (C.D.C.) public health laboratories throughout the US. Fellows are provided with a stipend, medical insurance, travel to the host laboratory and a professional development allowance. The fellowship program offers a wide variety of training and research experiences. Examples of projects include research in molecular genetics, pathogenesis, epidemiology and cell biology. Fellows may be trained in specific laboratory techniques such as real-time PCR, DNA sequencing and mass spectrometry. Fellows may receive specialized training and experience with a specific pathogen, or generalized training in influenza or STD surveillance, newborn screening or chemical and biological warfare agents. Other work may include diagnostic virology, bioanalytical chemistry, foodborne disease research including Pulsed Field Gel Electrophoresis (PFGE), clinical and environmental diagnostics and epidemiologic studies of disease outbreaks. Fellows also have opportunities to participate in research seminars, teleconferences, regional and national meetings and other continuing education opportunities. All fellows participate in an orientation session at C.D.C. in Atlanta to gain a general understanding of the public health laboratory system and how it relates to infectious disease surveillance, prevention, research and control. All applicants are required to submit one original single-sided application to APHL by the application deadline. The application may be downloaded from the EID fellowship website in writeable pdf format. All required supporting documents (listed in the application) should be collected by the applicant and forwarded with the application to APHL. Deadline dates may vary from year to year.

Applicants are encouraged to check the EID fellowship website or contact the Fellowship Manager to verify current the deadline date. APHL member laboratories may apply to host one or more EID Fellows. The link to the Host Laboratory Instructions and Application page is available from the EID Fellowship website.
Requirements: EID Advanced Laboratory Training Fellowship applicants must have completed a bachelor's or master's degree in microbiology, virology or a related discipline and/or completed an accredited medical technologist program prior to the start of the fellowship. EID Laboratory Research Fellowship applicants must have received a Ph.D., M.D., or D.V.M., or have completed all requirements for such a degree prior to the start of the fellowship. Applicants without a degree in a physical science will be considered if they have significant laboratory science and/or public health coursework or experience. All bachelor's and master's level candidates must have completed their degree by the time the fellowship begins (mid-September). Postdoctoral candidates may delay their start date to February 2012. Degree requirements must be completed by by this time. All appointees must attend the August Orientation session in Atlanta.
Restrictions: All applicants must be U.S. citizens.
Geographic Focus: All States
Date(s) Application is Due: Feb 11
Amount of Grant: 32,733 - 89,780 USD
Contact: Heather Roney, Manager, Fellowship Program; (240) 485-2778 or (240) 485-2745; fax (240) 485-2700; fellowships@aphl.org or heather.roney@aphl.org
Internet: http://www.aphl.org/mycareer/fellowships/eid/pages/default.aspx
Sponsor: Association of Public Health Laboratories
8515 Georgia Avenue, Suite 700
Silver Spring, MD 20910

Appalachian Regional Commission Health Care Grants 637
Access to comprehensive, affordable health care is vital to social and economic growth in the Appalachian Region. ARC's health projects focus on community-based efforts to encourage health-promotion and disease-prevention activities. Strategies include: using best practices in public health to develop targeted approaches to wellness and disease prevention; supporting partnerships that educate children and families about basic health risks; using telecommunications and other technology to reduce the high cost of health-care services; and encouraging the development and expansion of health professional education services within the Region. ARC health care grants have helped provide equipment for hospitals and rural clinics, training for health care professionals, and support for community-based health education activities. ARC also works with other organizations to address the high incidence of life-threatening diseases in the Region, as in its ongoing partnership with the Centers for Disease Control and Prevention in diabetes and cancer education, prevention, and treatment programs in the Region's distressed counties.
Requirements: States, and through states, public bodies and private nonprofit organizations are eligible to apply.
Restrictions: Generally, ARC grants are limited to 50% of project costs.
Geographic Focus: Alabama, Georgia, Kentucky, Maryland, Mississippi, New York, North Carolina, Ohio, Pennsylvania, South Carolina, Tennessee, Virginia, West Virginia
Contact: Jill Wilmoth, Budget and Program Specialist; (202) 884-7668 or (202) 884-7700; fax (202) 884-7691; jwilmoth@arc.gov
Internet: http://www.arc.gov/funding/ARCProjectGrants.asp
Sponsor: Appalachian Regional Commission
1666 Connecticut Avenue NW, Suite 700
Washington, D.C. 20009-1068

APSAA Fellowships 638
The mission of the Fellowship Program is to encourage interest and involvement in psychoanalysis among the future leaders, researchers and educators of mental health and academia. Early-career psychiatrists, psychologists, social workers and academics are eligible for to apply for the Fellowship. The Fellowship offers the opportunity to attend the biannual meetings of APSAA, to meet analysts and Fellows from across the country, to have a Mentor, and to present their clinical work or research at the meetings. The application for the Fellowship Program is available online September 15.
Requirements: Psychiatrists, psychologists, social workers, and academics who meet the specific eligibility requirements (available online) are eligible. Applicants may be nominated by their department chairs or program directors if applicable. When not applicable, self-nominations are encouraged. Applicants must be training or working in the United States during the fellowship year. All applicants should have demonstrated leadership ability in their discipline, or have special aptitude in research, teaching, and/or clinical endeavors; and have special interest in psychodynamics, psychoanalysis, or applied psychoanalysis.
Geographic Focus: All States
Contact: James Guimaraes; (212) 752-0450; fax (212) 593-0571; jguimaraes@apsa.org
Internet: http://www.apsa.org/FELLOWSHIPPROGRAM/GENERALINFORMATION/tabid/298/Default.aspx
Sponsor: American Psychoanalytic Association
309 East 49th Street
New York, NY 10007-1601

APSAA Foundation Grants 639
The mission of the APF Committee is to raise funds and sponsor programs promoting a better understanding of psychoanalysis and encouraging effective and innovative dissemination of psychoanalytic ideas and services to the public. The APF Committee's objective is to educate the public, the community of mental health workers, and allied disciplines about the relevance of psychoanalysis as a powerful therapeutic and research instrument whose applications span a wide range of understanding of individual behavior and cultural phenomena. One or several of the following elements should be a part of the proposal in order to be considered: community outreach; national focus--not a program that is generally put on as a regular activity by a society or institute; fund raising element--seeks co-funding, matching or other sources of support for project (grant money may be used as seed money for helping Grantee to get other sources of funding); transportable--broad applicability to other psychoanalytic institutes, societies and foundations (the program can serve as a model for other programs); and creative and original.
Geographic Focus: All States
Amount of Grant: 1,000 - 5,000 USD
Samples: Washington Center for Psychoanalysis--to enable the design and implementation of a system of psychodynamic support for Jubilee JumpStart, a new child care and education facility for low-income families, $5,000; Allen Creek Preschool, Ann Arbor, Michigan--to support the Early Childhood Training Initiative, a new outreach program designed to nurture the growth of early childhood educators and day care workers through professional training, workshops, and consultations, $4,000.
Contact: Dean K. Stein; (212) 752-0450; fax (212) 593-0571; deankstein@apsa.org
Internet: http://www.apsa.org/AmericanPsychoanalyticbrFoundationCommittee/tabid/70/Default.aspx
Sponsor: American Psychoanalytic Association
309 East 49th Street
New York, NY 10007-1601

APSAA Mini-Career Grants 640
The Fund conceives of psychoanalytic research along the broadest lines, including: scholarly and empirical investigative contributions that can advance knowledge of psychoanalytic theory, practice, and links between psychoanalysis and neighboring disciplines such as developmental psychology or neuroscience. Awards of up to $15,000 annually for up to two years are intended to support the beginning career of a psychoanalytic investigator. These mini-career awards are intended, for example, to buy time for a junior faculty member or clinician just starting a practice so that they may consult with other investigators, join an investigative team, or attend year long seminars on research methodology or specific methods relevant to their research. Applicants for these awards must provide a year-long career development plan with letters from appropriate faculty and mentors describing the candidate's career development plan and how the award will facilitate the candidate's goals.
Requirements: Applicants should meet the following criteria: proposal relevant to psychoanalysis; applicant is either a new applicant to the Fund or resubmitting a previously reviewed application; applicant has a masters degree or higher; and human subject protection and consent procedures must be specified for research studies involving ongoing data collection.
Restrictions: Applicants who have received $80,000 total from the Fund are ordinarily not eligible to apply again.
Geographic Focus: All States
Date(s) Application is Due: Apr 1; Nov 1
Amount of Grant: Up to 15,000 USD
Contact: Dean K. Stein, Executive Director; (212) 752-0450; fax (212) 593-0571; deankstein@apsa.org or info@apsa.org
Internet: http://www.apsa.org/RESEARCH/FUNDFORPSYCHOANALYTICRESEARCH/GRANTS/tabid/137/Default.aspx
Sponsor: American Psychoanalytic Association
309 East 49th Street
New York, NY 10007-1601

APSAA Research Grants 641
The Fund conceives of psychoanalytic research along the broadest lines, including: scholarly and empirical investigative contributions that can advance knowledge of psychoanalytic theory, practice, and links between psychoanalysis and neighboring disciplines such as developmental psychology or neuroscience. Grants of one or two years duration at a maximum of $20,000 yearly are intended for a specific project building upon psychoanalytic principles or directly investigating the process and/or outcome of psychoanalytically informed treatments. The Board recognizes that this amount of money is not sufficient usually to cover the costs of a large research study, particularly one that may involve a large number of subjects or a longitudinal design. Hence, the Board encourages applicants to consider feasibility of their request, that is, what they can achieve for this amount of money, and also to think about this type of award as supporting the beginning stages of a research project, the gathering of pilot data, or the refinement of design and methods.
Requirements: Applicants should meet the following criteria: proposal relevant to psychoanalysis; applicant is either a new applicant to the Fund or resubmitting a previously reviewed application; applicant has a masters degree or higher; and human subject protection and consent procedures must be specified for research studies involving ongoing data collection.
Restrictions: Applicants who have received $80,000 total from the Fund are ordinarily not eligible to apply again.
Geographic Focus: All States
Date(s) Application is Due: Apr 1; Nov 1
Amount of Grant: Up to 20,000 USD
Contact: Dean K. Stein, Executive Director; (212) 752-0450; fax (212) 593-0571; deankstein@apsa.org or info@apsa.org
Internet: http://www.apsa.org/RESEARCH/FUNDFORPSYCHOANALYTICRESEARCH/GRANTS/tabid/137/Default.aspx
Sponsor: American Psychoanalytic Association
309 East 49th Street
New York, NY 10007-1601

GRANT PROGRAMS | 95

APSAA Small Beginning Scholar Pre-Investigation Grants 642
The Fund conceives of psychoanalytic research along the broadest lines, including: scholarly and empirical investigative contributions that can advance knowledge of psychoanalytic theory, practice, and links between psychoanalysis and neighboring disciplines such as developmental psychology or neuroscience. These small grants of less than $5,000 for one year are intended to permit a beginning scholar to gather pilot data in preparation for the submission of a full grant to the Fund or to another agency. The Fund accepts revisions of a previously submitted application up to four years after the previous submission date. Up to two revised submissions may be submitted within four years of the original submission date.
Requirements: Applicants should meet the following criteria: proposal relevant to psychoanalysis; applicant is either a new applicant to the Fund or resubmitting a previously reviewed application; applicant has a masters degree or higher; and human subject protection and consent procedures must be specified for research studies involving ongoing data collection.
Restrictions: Applicants who have received $80,000 total from the Fund are ordinarily not eligible to apply again.
Geographic Focus: All States
Date(s) Application is Due: Apr 1; Nov 1
Amount of Grant: Up to 5,000 USD
Contact: Dean K. Stein, Executive Director; (212) 752-0450; fax (212) 593-0571; deankstein@apsa.org or info@apsa.org
Internet: http://www.apsa.org/RESEARCH/FUNDFORPSYCHOANALYTICRESEARCH/GRANTS/tabid/137/Default.aspx
Sponsor: American Psychoanalytic Association
309 East 49th Street
New York, NY 10007-1601

APSAA Small Beginning Scholar Visiting and Consulting Grants 643
The Fund conceives of psychoanalytic research along the broadest lines, including: scholarly and empirical investigative contributions that can advance knowledge of psychoanalytic theory, practice, and links between psychoanalysis and neighboring disciplines such as developmental psychology or neuroscience. Small grants of $3,000 are intended to permit a beginning scholar to spend time visiting and consulting with a more experienced, senior investigator who has agreed to help the junior investigator begin their investigative work. Typically these awards are made to permit the beginning scholar the funds to travel and/or to reimburse the senior investigator for travel and consultation. These grants may also be used to support travel to a research training seminar in the U.S. or abroad. The Fund accepts revisions of a previously submitted application up to four years after the previous submission date. Up to two revised submissions may be submitted within four years of the original submission date.
Requirements: Applicants should meet the following criteria: proposal relevant to psychoanalysis; applicant is either a new applicant to the Fund or resubmitting a previously reviewed application; applicant has a masters degree or higher; and human subject protection and consent procedures must be specified for research studies involving ongoing data collection.
Restrictions: Applicants who have received $80,000 total from the Fund are ordinarily not eligible to apply again.
Geographic Focus: All States
Date(s) Application is Due: Apr 1; Nov 1
Amount of Grant: Up to 3,000 USD
Contact: Dean K. Stein, Executive Director; (212) 752-0450; fax (212) 593-0571; deankstein@apsa.org or info@apsa.org
Internet: http://www.apsa.org/RESEARCH/FUNDFORPSYCHOANALYTICRESEARCH/GRANTS/tabid/137/Default.aspx
Sponsor: American Psychoanalytic Association
309 East 49th Street
New York, NY 10007-1601

APSA Congressional Health and Aging Policy Fellowships 644
Supported by the Atlantic Philanthropies and administered by Columbia University, this national program seeks to provide professions in health and aging with the experience and skills necessary to make a positive contribution to the development and implementation of health policies that affect older Americans. The program will offer two different tracks for individual placement: a residential track that includes a nine-to-12-month placement in Washington, D.C. or at a state agency (as a legislative assistant in Congress, a professional staff member in an executive agency or in a policy organization; and a non-residential track that includes a health policy project and brief placement(s) throughout the year at relevant sites. Core program components focused on career development and professional enrichment are provided for fellows in both tracks.
Requirements: The program is open to physicians, nurses and social workers at all career stages (early, mid, and late) with demonstrated commitment to health and aging issues and a desire to be involved in health policy at the federal, state or local level. Other professionals with clinical backgrounds (e.g. pharmacists, dentists, clinical psychologists) working in the field of health and aging are also eligible to apply.
Geographic Focus: All States
Contact: Jeffrey R. Biggs, Program Director; (202) 483-2512, ext. 521; fax (202) 483-2657; jbiggs@apsanet.org or cfp@apsanet.org
Internet: http://www.apsanet.org/content_50835.cfm
Sponsor: American Political Science Association
1527 New Hampshire Avenue, NW
Washington, D.C. 20036-1206

APSA Robert Wood Johnson Foundation Health Policy Fellowships 645
The purpose of this fellowship is to give outstanding mid-career health professionals in academic and community-based settings an opportunity to learn more about the legislative process through direct participation. Fellows actively contribute to the formulation of national health policies and accelerate their careers as leaders in health policy. Initiated in 1973, the program is funded by The Robert Wood Johnson Foundation (RWJF)and conducted by the Institute of Medicine (IOM) of the National Academy of Sciences. The application process typically opens in September and the application deadline typically occurs in mid-November of each year. Fellows must commit 100 percent of their time to program activities during the first 12 months. Up to six grants of up to $165,000 each will be made.
Geographic Focus: All States
Date(s) Application is Due: Nov 14
Amount of Grant: Up to 165,000 USD
Contact: Jeffrey R. Biggs, Program Director; (202) 334-1506 or (202) 483-2512, ext. 521; fax (202) 483-2657; jbiggs@apsanet.org or info@healthpolicyfellows.org
Internet: http://www.apsanet.org/content_3544.cfm
Sponsor: American Political Science Association
1527 New Hampshire Avenue, NW
Washington, D.C. 20036-1206

APS Foundation Grants 646
Through its Corporate Giving and Foundation, APS continues to be the leading Arizona corporate citizen. It supports nonprofit organizations with a 501(c)3 Internal Revenue Service tax-exempt status through cash and/or in kind services. The Foundation support Arizona communities in five strategic areas: health and human services, community development, education, arts and culture, and environment.
Restrictions: APS does not fund individual requests, charter or private schools, religious, political, fraternal, legislative or lobbying efforts or organizations, travel-related or hotel expenses, private or family foundations, private non-profit organizations, salaries and/or debt reduction.
Geographic Focus: Arizona
Amount of Grant: Up to 100,000 USD
Samples: Phoenix Art Museum, Phoenix, Arizona, $100,000; Phoenix Zoo, Phoenix, Arizona, $40,000; Phoenix Indian Center, Phoenix, Arizona, $25,000.
Contact: Teresa DeValle, Contributions Coordinator; (602) 250-2259; fax (602) 250-3066; Teresa.DeValle@aps.com
Internet: http://www.aps.com/general_info/AboutAPS_14_Archive.html
Sponsor: Arizona Public Service Foundation
P.O. Box 53999, MS 8010
Phoenix, AZ 85072-3999

AptarGroup Foundation Grants 647
The foundation awards grants to Illinois nonprofit organizations in its areas of interest, including in order of priority, health and social services, cultural activities and programs, and higher education.
Requirements: Illinois nonprofit organizations are eligible.
Restrictions: Grants do not support nursing homes, animal welfare groups, national or international relief organizations, primary, secondary, or theological schools, or religious, civic, fraternal, veterans', social, or political organizations. Individuals, testimonial dinners, fundraising events, courtesy advertising, or trips or tours will not be supported.
Geographic Focus: Illinois
Date(s) Application is Due: Mar 15; Oct 15
Contact: Lawrence Lowrimore; (815) 477-0424; fax (815) 477-0481; info@aptargroup.com
Sponsor: AptarGroup Charitable Foundation
475 W Terra Cotta Avenue, Suite E
Crystal Lake, IL 60014-9695

Aragona Family Foundation Grants 648
Giving primarily in Austin, Texas, the Aragona Family Foundation provides general operating support grants in the following areas of interest: animals/wildlife, preservation/protection; cancer; children/youth, services; community/economic development; education; food banks; hospitals (general); philanthropy/voluntarism; protestant agencies & churches. Grants range from $2,000 - $100,000.
Requirements: Qualifying 501(c)3 organizations are eligible to apply for funding. There is no: application form required; deadline date to adhere to.
Geographic Focus: Texas
Amount of Grant: 2,000 - 100,000 USD
Samples: St. Stephens Episcopal School, Austin, TX, $100,000--to further the organizations charitable mission; Harvard University, Cambridge, MA, $533,300--to further the organizations charitable mission.
Contact: Joseph C. Aragona, President; (512) 328-2178
Sponsor: Aragona Family Foundation
3311 Westlake Drive
Austin, TX 78746-1901

Aratani Foundation Grants 649
The foundation awards grants to nonprofits in its areas of interest, including education, health care, museums, recreation, and religion. Preference is given to Japanese-American cultural organizations. Types of support include annual campaigns, building construction/renovation, capital campaigns, conferences and seminars, continuing support, curriculum development, endowments, exchange programs, fellowships, general operating support, program development, scholarship funds, and seed grants.

Requirements: The Foundation gives primarily in the state of California but funding is also available in the states of: New York; District of Columbia; Washington; Florida; Rhode Island; Oregon. Application forms are not required, but application outlines are available. Contact the Foundation directly for additional guidelines. Applications maybe submitted in English and Japanese. There are no application deadline dates.
Restrictions: Grants are not made to individuals.
Geographic Focus: California, District of Columbia, Florida, New York, Oregon, Rhode Island, Washington
Amount of Grant: 1,000 - 150,000 USD
Sample: Asian Pacific American Legal Center of South California, Los Angeles, CA, $50,000; Asian Pacific American Institute for Congressional Studies, Washington, D.C., $10,000; Buddhist Churches of America, San Francisco, CA, $200,000.
Contact: George Aratani, President; (310) 530-9900
Sponsor: Aratani Foundation
23505 Crenshaw Boulevard, North 230
Hollywood, CA 90505

Arcadia Foundation Grants 650
The foundation awards grants to Pennsylvania nonprofit organizations to improve the quality of life. Areas of interest include hospitals and hospital building funds, health agencies and services, nursing, hospices, early childhood, adult and higher education, libraries, child development and welfare agencies, youth organizations, and social service and general welfare agencies, including care of the handicapped, aged, and hungry. Also supported are family services, environment and conservation, wildlife and animal welfare, religious organizations, historical preservation, and music organizations. Types of support include general operating support, continuing support, annual campaigns, capital campaigns, building construction/renovation, equipment acquisition, endowment funds, program development, scholarship funds, and research. Applications are accepted between September 1 and November 1.
Requirements: Eastern Pennsylvania organizations whose addresses have zip codes of 18000-19000 are eligible. Application form not required. The initial approach should be a letter or proposal--not exceeding two pages.
Restrictions: Grants are not awarded to support individuals, deficit financing, land acquisition, fellowships, demonstration projects, publications, or conferences.
Geographic Focus: Pennsylvania
Date(s) Application is Due: Nov 1
Contact: Marilyn Lee Steinbright, President; (610) 275-8460
Sponsor: Arcadia Foundation
105 E Logan Street
Norristown, PA 19401

Archer Daniels Midland Foundation Grants 651
The foundation prefers to fund programs that will directly impact the communities where operating units are located. Nearly a third of the funding is directed toward educational institutions, including elementary, secondary, and higher education. The advocacy category covers world affairs and foreign relations groups that deal with international trade; also supported are projects stressing free enterprise and assistance for women. Support of social services goes to minority group development, cultural activities, and hospital and youth agencies. Health service funding supports hospital and disease association programs. Conservation funding supports programs for protecting the environment and beautification as well as conservation. Religion grants support Christian, Jewish, and Roman Catholic organizations, such as churches, colleges and universities, international ministries/missions, Jewish welfare, and the Salvation Army. Applications are accepted at any time; initial contact by letter of inquiry is encouraged.
Requirements: Tax-exempt organizations are eligible. Current United Way recipients and national organizations are also eligible.
Geographic Focus: Arkansas, California, Georgia, Illinois, Indiana, Iowa, Kansas, Kentucky, Massachusetts, Michigan, Minnesota, Mississippi, Missouri, Montana, Nebraska, New Jersey, New York, North Carolina, North Dakota, Ohio, Oklahoma, Pennsylvania, Puerto Rico, South Carolina, Tennessee, Texas, Washington, Wisconsin
Amount of Grant: 10,000 USD
Contact: Brian Peterson; (217) 424-5413 or (217) 424-5200; corpaffairs@admworld.com
Internet: http://www.adm.com/en-US/responsibility/social_investing/Pages/default.aspx
Sponsor: Archer Daniels Midland Foundation
4666 Faries Parkway, P.O. Box 1470
Decatur, IL 62526-5630

ARCO Foundation Education Grants 652
The foundation will concentrate its aid to education in support of the following: precollege programs to improve the quality of teaching and learning in urban public education; programs aimed at decreasing attrition rates among low-income and minority students; programs to motivate low-income and minority students to succeed in college, especially in mathematics-based careers of engineering, science, and business; support for laboratory renovation and scientific equipment in academic disciplines of interest at major research universities; programs to retain the most talented young faculty in academic careers in selected disciplines; selected liberal arts programs at colleges and universities of interest; state associations of private colleges in the states where the company has interests; academic programs relevant to energy interests at regional universities and colleges; and national education associations and organizations that seek to improve education in public high schools and at higher academic levels. Grants are awarded for operating budgets, seed money, equipment, land acquisition, matching funds, employee matching gifts, employee related scholarships, special projects, and technical assistance. Applications are accepted at any time; annual report should be obtained prior to submitting a formal proposal.

Requirements: The foundation is a regional organization funding nonprofit organizations in states where ARCO has facilities and personnel, including Alaska, Arizona, California, Colorado, Nevada, Texas, and Washington. Requests from those states and those nearby should be addressed to the local community affairs managers.
Restrictions: The foundation discourages applications from the following: programs not focused on promoting self-sufficiency and economic development of minority populations; historic preservation or urban development projects not tied to neighborhood economic revitalization; proposals from religious organizations; or funding requests from federal, state, county, and municipal agencies, including school districts. The foundation does not generally consider support of hospital building or endowment campaigns, medical equipment, medical research programs, single-issue health organizations, or health services not directed at low-income people.
Geographic Focus: Alaska, Arizona, California, Colorado, Nevada, Texas, Washington
Amount of Grant: 1,500 - 360,000 USD
Samples: U of Montana (Butte, MT)--to help retain and graduate minority students pursuing engineering degrees, $16,500; U of California (Davis, CA)--to help retain and graduate minority students pursuing engineering degrees, $47,000.
Contact: Virginia Victorin, Grants Administrator; (213) 486-3342; fax (213) 486-0113
Internet: http://www.ntlf.com/html/grants/5977.htm
Sponsor: ARCO Foundation
151 South Flower Street
Los Angeles, CA 90071

Argyros Foundation Grants 653
The foundation annually awards grants for cultural activities, higher education, religious projects, social services, recreation, and health services. Types of support include general operating support and program development. Proposals must give a detailed outline of the project for which funds are requested.
Requirements: Giving is primarily made to organizations in Orange County, CA.
Geographic Focus: California
Date(s) Application is Due: Jun 1
Amount of Grant: 5,000 - 50,000 USD
Contact: Daniel Russo, Argyros Charitable Trusts; (714) 481-5000
Sponsor: Argyros Foundation
949 S Coast Drive, #600
Costa Mesa, CA 92626

Arie and Ida Crown Memorial Grants 654
The program supports programs that offer opportunities to the disadvantaged, strengthens the bond of families, and improves the quality of people's lives. As a general rule, the Foundation funds organizations that serve the greater Chicago area as well as organizations that serve the broader Jewish community. Most grants are awarded to organizations within the city of Chicago. Organizations are supported in the areas of arts and culture (concentrating on educational and enrichment programs for youth), civic affairs, education, health (stressing access to services, hospice and health promotion), and human service (focusing on programs which offer assistance for children and families).
Requirements: Nonprofits in Chicago and Cook County may apply for grant support.
Restrictions: Grants are not made to support individuals, conference expenses, film projects, government programs (50 percent government funded), or research projects.
Geographic Focus: Illinois
Date(s) Application is Due: Jan 31; Jul 31
Amount of Grant: 1,000 - 200,000 USD
Contact: Susan Crown; (312) 236-6300; fax (312) 984-1499; AICM@crown-Chicago.com
Internet: http://www.crownmemorial.org/
Sponsor: Arie and Ida Crown Memorial
222 N LaSalle Street, Suite 2000
Chicago, IL 60601

Arizona Cardinals Grants 655
The National Football League franchise supports programs designed to improve the quality of life and enhance opportunities for children, women, and minorities in the state of Arizona. Specific areas of interest include arts and culture, civic affairs, education, health, science, and social services. The foundation is interested in expanding its giving and looks for new charities to fund. First-time applicant organizations generally will receive grants of $5000 or less.
Requirements: Applicants must be exempt under 501(c)3 of the Internal Revenue Service code.
Geographic Focus: Arizona
Date(s) Application is Due: Aug 1
Amount of Grant: 2,000 - 5,000 USD
Contact: Pat Tankersley, (602) 379-0101; fax (480) 785-7327
Internet: http://www.azcardinals.com/community/charities.php
Sponsor: Arizona Cardinals
P.O. Box 888
Phoenix, AZ 85001-0888

Arizona Community Foundation Scholarships 656
The Arizona Community Foundation sponsors the Dorrance Family Foundation scholarships to provide academic and financial support to Arizona's first generation college students. The foundation awards up to 25 scholarships annually to incoming freshmen who will attend one of three Arizona public universities. The award is renewable for up to three years for a total of eight semesters of full-time undergraduate study. Scholarships are renewed based on academic standing, participation in program events and activities, and community service projects. Any academic major is eligible.

Requirements: Applicants must be Arizona residents; seniors in good standing at an accredited Arizona high school; first generation applicants to attend college; hold a minimum cumulative 3.0 GPA; an SAT score of 1040 or composite ACT score of 22 (excluding writing score); demonstrate financial need; accepted by one of Arizona's three residential public universities (Arizona State, Northern Arizona University, or University of Arizona); and demonstrated leadership and community service. Visit the website for a current application and deadline information.
Restrictions: Applicants must be Arizona residents.
Geographic Focus: Arizona
Amount of Grant: 10,000 USD
Contact: Grant Administrator; (602) 381-1400 or (800) 222-8221; fax (602) 381-1575; nstaylor@azfoundation.org
Internet: http://www.dorrancescholarship.org/index.php?hs=1
Sponsor: Arizona Community Foundation
2201 E Camelback Road, Suite 202
Phoenix, AZ 85016

Arizona Diamondbacks Charities Grants 657
The major-league baseball franchise awards grants in Arizona to support as wide as possible a variety of charitable causes. Priority will to be given to organizations that fall under the foundation's focus areas of health care for the indigent, homelessness and youth education. Types of support include general support, project grants, donated equipment, employee matching gifts, corporate sponsorships, and speakers. Organizations wishing to apply for a grant can fax a request to the community affairs department.
Requirements: Applicants must be 501(c)3 Arizona-based organizations committed to spending grant proceeds in Arizona.
Geographic Focus: Arizona
Samples: Arizona's Children (Phoenix, AZ)--for a program to provide behavioral health services to children and their families; Golden Gate Community Ctr (Phoenix, AZ)--to help replace evaporative coolers in the center's gymnasium; Make Way for Books (Tucson, AZ)--to promote early literacy in low-income area preschools.
Contact: Program Contact; (602) 462-6500; fax (602) 462-6575
Internet: http://arizona.diamondbacks.mlb.com/ari/community/foundation.jsp
Sponsor: Arizona Diamondbacks
P.O. Box 2095
Phoenix, AZ 85001

Arizona Public Service Corporate Giving Program Grants 658
Grants are awarded to support Arizona nonprofit organizations in the areas of health and human services, community development, arts and culture, education, and environment. The foundation awards support for project grants, capital building funds, research, employee matching gifts, in-kind services, conferences and seminars, and operating support. Applications are accepted on an ongoing basis.
Requirements: Arizona 501(c)3 nonprofits are eligible.
Restrictions: APS Corporate Giving does not fund individual request, charter or private schools, religious, political fraternal, legislative or lobbying efforts to organizations, travel-related or hotel expenses, private or family foundation, private non-profit organizations, salaries and/or debt reduction.
Geographic Focus: Arizona
Contact: Cindy Slick; (602) 250-4707; fax (602) 250-2113; Cindy.Slick@aps.com
Internet: http://www.aps.com/main/community/dev/default.html
Sponsor: Arizona Public Service Corporation
P.O. Box 53999, MS 8010
Phoenix, AZ 85072-3999

Arkansas Community Foundation Arkansas Black Hall of Fame Grants 659
Grants support programs and projects that address problems, challenges, and opportunities in black communities in Arkansas. Categories of interest include education; health and wellness; youth development; and small business and/or economic development. Priority goes to proposals that show multiple sponsoring agencies/organizations; include evidence of local financial support; demonstrate collaborative ventures among community organizations; and have promise for sustainability beyond the period of the grant.
Requirements: Tax-exempt entities such as schools, churches, or government entities are eligible. Preference is given to applications submitted by 501(c)3 nonprofit organizations.
Restrictions: Funds cannot be allocated for adult salary support in carrying out the duties of the project; used to support general operating budgets outside of the specific proposal or project; or for scholarships for formal education at any level.
Geographic Focus: Arkansas
Date(s) Application is Due: Apr 10
Amount of Grant: Up to 5,000 USD
Contact: Cecilia Patterson, Program Director; (501) 372-1116; fax (501) 372-1166; cpatterson@arcf.org or arcf@arcf.org
Internet: http://www.arcf.org/page27973.cfm
Sponsor: Arkansas Community Foundation
1400 West Markham, Suite 206
Little Rock, AR 72201

Arkansas Community Foundation Grants 660
The Arkansas Community Foundation (ARCF) serves philanthropic donors and supports non-profits in Arkansas. ARCF makes grants from charitable funds established by individuals, families, corporations, and non-profit organizations. From among over 1100 funds, its donors make grants in the following areas of interest: animal welfare, arts and humanities, community development, education, environment, health, human services, and religion. While the Foundation's discretionary grant cycles are few, grants are usually modest and, generally, fund capital or building campaigns are not funded.
Requirements: 501(c)3 Arkansas nonprofit organizations are eligible.
Geographic Focus: Arkansas
Amount of Grant: 500 - 2,500 USD
Contact: Cecilia Patterson, Program Director; (501) 372-1116; fax (501) 372-1166; cpatterson@arcf.org or arcf@arcf.org
Internet: http://www.arcf.org/page12714.cfm
Sponsor: Arkansas Community Foundation
1400 West Markham, Suite 206
Little Rock, AR 72201

Arkell Hall Foundation Grants 661
The primary mission of the foundation is the operation and maintenance of a home for elderly women. Funds that may become available above the needs of the home may be distributed annually to tax-exempt organizations providing services in the target community--Western Montgomery County, NY--with preference given to those active in service to senior citizens, education (higher education, medical education, adult basic education and literacy), religion (Christian, Christian Reformed Church, Lutheran, Methodist, Roman Catholic, the Salvation Army, and United Methodist), and health care. Types of support include capital, challenge, endowment, general support, matching, and seed money grants. Grants are made on a single-year basis and are awarded annually in October or November. It is recommended that requests be submitted between July 1 and September 15. Initial review of requests is performed upon receipt. Results are forwarded to the applicant within one month.
Requirements: IRS 501(c)3 organizations directly impacting the Western Montgomery County, NY, community are eligible.
Restrictions: Requests that do not include written proof of 501(c)3 status will not be considered. Projects or organizations with large service areas, such as national or regional, will not qualify for funding, nor will projects in which the target community is not the primary area of focus.
Geographic Focus: New York
Date(s) Application is Due: Oct 1
Amount of Grant: 1,000 - 50,000 USD
Samples: Cornell University, College of Human Ecology (Ithaca, NY)--for general support, $20,000; College of Saint Rose (Albany, NY)--to support scholarships for single mothers, $3000.
Contact: Joseph Santangelo, Vice President; (518) 673-5417; fax (518) 673-5493
Sponsor: Arkell Hall Foundation
68 Front Street, P.O. Box 240
Canajoharie, NY 13317-0240

Arlington Community Foundation Grants 662
The foundation awards grants in Arlington to educators and nonprofit organizations for innovative projects that supplement and enrich the learning environment for preschool to adult students. The focus is curriculum enrichment, the arts (musical, dramatic, visual), pursuit of higher education, vocational education, after school and summer programs, life-long learning, environmental issues, parent involvement, and community involvement. The community enhancement grants support arts and humanities, children & families, community improvement, health, housing/homeless & hunger, legal, social services, and senior enrichment. The purpose of the Prompt Response Fund is to enable nonprofit groups in Arlington respond quickly to unanticipated opportunities or unexpected, urgent community needs. Deadlines may vary. Guidelines and applications are available online.
Requirements: Organizations and individuals with projects designed to meet educational needs of Arlington residents are encouraged to apply.
Restrictions: The foundation does not make grants for endowments, capital campaigns, religious purposes, individual debts, or political lobbying.
Geographic Focus: Virginia
Contact: Wanda L. Pierce; (703) 243-4785; fax (703) 243-4796; info@arlcf.org
Internet: http://www.arlcf.org/grants.html
Sponsor: Arlington Community Foundation
2525 Wilson Boulevard
Arlington, VA 22201

Armstrong McDonald Foundation Health Grants 663
The Armstrong McDonald Foundation was incorporated in the State of Nebraska in 1986. The mission of the Foundation is to continue the philanthropic ideals and goals of James M. McDonald, Sr. through prudent and impartial review of all qualifying grant requests received annually to insure that awards are made to soundly conceived and operated non-profit organizations. For its Health category, the Foundation supports a great variety of projects related to health issues ranging from lab supplies for teaching hospitals to multi-drug resistance research. Other examples of projects falling within this category are the need for medical related equipment, renovating areas of hospitals for specialized treatment, sending kids with cancer to camp, recovery programs for burn victims, outpatient therapy for abused and neglected children, programs related to fighting eating disorders, and costs of providing onsite facilities at hospitals providing specialized care for children. The Foundation provides a formal downloadable application which is to be completed by all applicants approved for submission of a grant request. All required application materials need to be postmarked or received by email on or before September 30.
Requirements: The Armstrong McDonald Foundation will only accept unsolicited grant requests from those IRS approved non-profits listed on the Pre-Approved for Grant Submission List. Please note that this list will be updated annually in December with additions and/or deletions. All other IRS approved non-profits desiring to submit a grant

request to this foundation must meet the following three qualifications: be incorporated in either the State of Arizona or Nebraska; have a physical office located in their state of incorporation; and spend any awarded grant funds within their state of incorporation.
Restrictions: Grants do not support advocacy organizations, individuals, international organizations, political organizations, or state and local government agencies. The foundation does not fund capital campaigns, salaries/stipends, and multi-year projects. Organization must have received a grant from the foundation within the last five years unless it is located within the states of Arizona or Nebraska. No organization east of the Mississippi is eligible for a grant.
Geographic Focus: All States
Date(s) Application is Due: Sep 30
Amount of Grant: 1,000 - 80,000 USD
Contact: Laurie L. Bouchard, President; (520) 878-9627; fax (520) 797-3866; info@ArmstrongMcDonaldFoundation.org
Internet: http://www.armstrongmcdonaldfoundation.org/cat.html
Sponsor: Armstrong McDonald Foundation
P.O. Box 70110
Tucson, AZ 85737-0110

Arnold and Mabel Beckman Foundation Beckman-Argyros Award in Vision Research — 664

The Beckman-Argyros Award in Vision Research is an annual award established in 2013 to honor and celebrate a decades-long friendship of the two remarkable men, Dr. Arnold O. Beckman and Ambassador George L. Argyros, and to continue their commitment, dedication and shared vision to make the world a better place. The Beckman-Argyros Award honors Ambassador Argyros for his 22 years of service as Chairman of the Board of the Arnold and Mabel Beckman Foundation and recognizes the special and unique friendship he shared with Arnold O. Beckman for over forty years. The Beckman-Argyros Award in Vision Research is intended to reward individuals who are making significant transformative breakthroughs in vision research; this may include those whose contributions to science in general, or through the development of an innovative technology or fundamental scientific breakthrough have been applied to, aided and/or improved the vision sciences. The recipient of the Beckman-Argyros Award will receive a total of $500,000, as outlined below, along with a solid gold commemorative medallion: $100,000 prize to award recipient; $400,000 grant to support award recipient's research at a university or other qualifying public 501(c)3 charity specifying its designated use.
Geographic Focus: All States
Amount of Grant: 500,000 USD
Contact: Jacqueline Dorrance, Executive Director; (949) 721-2222; fax (949) 721-2225; beckmanargyros@beckman-foundation.com or administration@beckman-foundation.com
Internet: http://www.beckman-foundation.com/beckman-argyros
Sponsor: Arnold and Mabel Beckman Foundation
100 Academy Drive
Irvine, CA 92617

Arnold and Mabel Beckman Foundation Scholars Grants — 665

The Arnold and Mabel Beckman Foundation Scholars program awards are institutional, university or college awards. Each year, the Foundation selects a number of research, doctoral, masters and baccalaureate universities and colleges to be invited to submit applications for the Scholars program. Each institution may submit one application for consideration for an award. If an institution is interested in the Beckman Scholars Program, but has not been invited to participate, it should review the guidelines for eligibility. The amount given for each award is $26,000 for two summers and one academic year.
Restrictions: The Beckman Scholars program does not make awards directly to individuals.
Geographic Focus: All States
Amount of Grant: 26,000 USD
Contact: Jacqueline Dorrance, Executive Director; (949) 721-2222; fax (949) 721-2225; beckmanscholars@beckman-foundation.com or administration@beckman-foundation.com
Internet: http://www.beckman-foundation.com/beckman-scholars
Sponsor: Arnold and Mabel Beckman Foundation
100 Academy Drive
Irvine, CA 92617

Arnold and Mabel Beckman Foundation Young Investigators Grants — 666

The Arnold and Mabel Beckman Foundation makes Young Investigators grants to nonprofit research institutions to promote research in chemistry and the life sciences, broadly interpreted, and particularly to foster the invention of methods, instruments, and materials that will open up new avenues of research in science. The program is intended to provide research support to the most promising young faculty members in the early stages of academic careers in the chemical and life sciences. The program is open to persons with tenure-track appointments in academic and nonprofit institutions that conduct fundamental research in the chemical and life sciences. The program is intended primarily for U.S. institutions. Only proposals of exceptional merit from foreign institutions will receive consideration. Projects are normally funded for a period of two years. When extraordinary circumstances warrant, support may be provided over a one-year or three-year period. All versions of the required Letter of Intent must be submitted prior to September 3. Potential candidates for application submission will be notified in early December, and awardees will be announced in the following July.
Requirements: To be eligible, an applicant should not have completed more than three full years in his or her tenure-track or other comparable independent research appointment on the anniversary date of initial appointment in the year in which application is to be made. Candidates must be citizens or permanent residents of the United States at the time of application.
Restrictions: Funding will not be considered for general institutional expenses or general fund-raising campaign expenses such as dinners and mass mailings. The foundation does not provide funds for overhead or indirect costs. Persons who have applied for permanent residency but have not received their government documentation by the time of application are not eligible. Individuals may not apply more than three times. Applicants from institutions with a currently funded Beckman Young Investigator are not eligible to apply.
Geographic Focus: All States
Date(s) Application is Due: Sep 3
Amount of Grant: 264,000 USD
Contact: Jacqueline Dorrance, Executive Director; (949) 721-2222; fax (949) 721-2225; younginvestigators@beckman-foundation.com or administration@beckman-foundation.com
Internet: http://www.beckman-foundation.com/byi_guides.html
Sponsor: Arnold and Mabel Beckman Foundation
100 Academy Drive
Irvine, CA 92617

ARS Foundation Grants — 667

The mission of the foundation is to make a positive difference in the lives of others. They support human and social services, the arts, children and the elderly, and health care but will consider other organizations. Giving is focused in New York City and Fairfield County, CT, areas. Applicants should submit a cover letter containing a brief history and description of the organization, the program(s), the needs/goals to be addressed, and a description of the project; an itemized budget, current organizational annual budget, and recent audited financial statement; annual report; other funding sources; and 501(c)3 determination letter. There are no application deadlines or forms.
Requirements: Qualified 501(c)3 charitable organizations in the New York City and Fairfield County, CT, areas are eligible.
Restrictions: The foundation does not give grants to individuals, private foundations, or organizations not qualified as charitable organizations.
Geographic Focus: Connecticut, New York
Contact: Administrator; (212) 986-1533; fax (212) 972-2303; info@arsfoundation.com
Internet: http://www.arsfoundation.com
Sponsor: Adolph and Ruth Schnurmacher Foundation
551 Fifth Avenue, Suite 1210
New York, NY 10176

ARS New Investigator Award — 668

The purpose of this award is to support basic, translational, or clinical research projects in rhinology. Proposed projects may be related to any area of rhinology. Proposed project shall be designed in collaboration with a preceptor investigator and approved by the candidate's department chairperson and institution. One award of up to $25,000 over a two-year period will be given annually. Letters of Interest should be received no later than December 15, with full proposals due on January 15.
Requirements: Any member of the American Rhinologic Society who has not received previous, significant outside funding is eligible. The applicant must have, as a mentor, an established researcher who will provide a letter of support stating the extent of involvement in the project and provide a summary of his/her research experience.
Geographic Focus: All States
Date(s) Application is Due: Jan 15; Dec 15
Amount of Grant: Up to 25,000 USD
Contact: Stephanie L. Jones; (703) 519-1586 or (703) 836-4444; sljones@entnet.org
Internet: http://www.entnet.org/EducationAndResearch/coreGrants.cfm
Sponsor: American Rhinology Society
One Prince Street
Alexandria, VA 22314-3357

ARS Resident Research Grants — 669

The purpose of this award is to support basic, translational, or clinical research projects in rhinology. Proposed projects may be related to any area of rhinology. Proposed project shall be designed in collaboration with a preceptor investigator and approved by the candidate's department chairperson and institution. Two awards of up to $8,000 over a one-year period will be given annually. Letters of Interest should be received no later than December 15, with full proposals due on January 15.
Requirements: Any resident in training in an approved program in the U.S. or Canada is eligible to apply for the American Rhinologic Society Research Grant. Resident applicants must have as a co-investigator a supervising faculty who is a member in good standing in the American Rhinologic Society (ARS).
Geographic Focus: All States
Date(s) Application is Due: Jan 15; Dec 15
Contact: Stephanie L. Jones; (703) 519-1586 or (703) 836-4444; sljones@entnet.org
Internet: http://www.entnet.org/EducationAndResearch/upload/2009-ARS-Resideint-Research-Grant.pdf
Sponsor: American Rhinology Society
One Prince Street
Alexandria, VA 22314-3357

Arthur and Rochelle Belfer Foundation Grants — 670

The foundation awards grants to nonprofit organizations of the Jewish faith, with a focus on New York. Grants are targeted toward programs supporting the elderly and women; education/higher education; institutions such as seminaries, synagogues, and temples; hospitals; Jewish welfare; and medical centers. Types of support include general support grants and fellowships. There are no application deadlines. Applicants should send a brief letter of inquiry describing the program.

Restrictions: Grants are not made to individuals.
Geographic Focus: New York
Amount of Grant: 1,000 - 100,000 USD
Samples: Joan S. Brugge, PhD, chair, Department of Cell Biology, arvard Medical School--for breast cancer studies; Dana-Farber Cancer Institute (Boston, MA)--for general support, $1 million; Anti-Defamation League of B'nai B'rith (New York, NY)--for operating support, $22,000; American Friends of Israel Museum (New York, NY)--for operating support, $5000.
Contact: Robert Belfer, President; (212) 508-6020
Sponsor: Arthur and Rochelle Belfer Foundation
767 Fifth Avenue, 46th Floor
New York, NY 10153-0002

Arthur Ashley Williams Foundation Grants 671
AAW Foundation awards grants to charitable organizations to help improve quality of life for those seeking assistance from these charitable organizations. Grant requests will be considered for program support, seed money, challenge grants, and capital improvements.
Requirements: To apply for a grant, applicants must be non-profit 501(c)3 organizations or public schools.
Restrictions: The Foundation will not fund political or sectarian activities.
Geographic Focus: All States
Amount of Grant: 1,000 - 15,000 USD
Samples: The Barry R. Kirschner Wildlife Foundation, Durham, CA, $11,500; Framingham Historical and Natural Historic Society, Framingham, MA, $15,000.
Contact: Program Contact; (508) 893-0757; clambert@rcn.com
Internet: http://www.aawfoundation.org/requests.php
Sponsor: Arthur Ashley Williams Foundation
P.O. Box 6280
Holliston, MA 01746

Arthur F. and Arnold M. Frankel Foundation Grants 672
Established in New York in 1990, the Arthur F. and Arnold M. Frankel Foundation offers financial support in both New York City and throughout the State of California. The Foundation's primary fields of interest include: the arts; health care and health care access; and human service programs. A specific application form is required, and can be secured by contacting the Foundation office. Most recent grant awards have ranged from $1,000 to as much as $155,000. There have been no annual deadlines identified.
Geographic Focus: California, New York
Amount of Grant: 1,000 - 155,000 USD
Samples: Fractured Atlas, New York, New York, $154,700 - general support for area arts program; Center For Independent Doctors, Sharon, Massachusetts, $74,290 - general operating support; Mehadi Foundation, Burbank, California, $107,000 - general operating support for helping veterans heal.
Contact: Jedd H. Wider, Trustee; (212) 362-2703; fax jwider@morganlewis.com
Sponsor: Arthur F. and Arnold M. Frankel Foundation
101 Park Avenue
New York, NY 10178-6000

Arthur M. Blank Family Foundation Inspiring Spaces Grants 673
The Arthur M. Blank Foundation believes that young people, families, and entire communities need healthy, green, inspiring places to grow and develop. The Foundation seeks partners with a passion for parks and green space. Through these partners, it aims to preserve and enhance a new generation of safe, clean, accessible parks and community green spaces. Parks and community green spaces offer social, ecological, and economic benefits to residents and visitors alike. Through this initiative, the Foundation and the Atlanta Falcons Youth Foundation have partnered with the Georgia's State Park system and the Georgia Association of Physician Assistants (GAPA) to sponsor Rx for Fitness. Through the new Rx for Fitness program, physician assistants can prescribe healthy hikes in the great outdoors, and patients can turn in their prescriptions for free park passes. Interested parties should begin by sending a query letter to the Foundation office. Applications are by invitation only.
Requirements: 501(c)3 organizations serving residents of Atlanta, Georgia, are eligible.
Restrictions: The Foundation does not make grants: directly to individuals; for scholarships; or to support houses of worship or religious activity.
Geographic Focus: Georgia
Contact: Penelope McPhee, President and Director; (404) 367-2100
Internet: http://www.blankfoundation.org/inspiring-spaces
Sponsor: Arthur M. Blank Family Foundation
3223 Howell Mill Road, NW
Atlanta, GA 30327

ArvinMeritor Foundation Health Grants 674
The foundation provides grants primarily in company-operating locations in the areas of education and training, civic and health, youth organizations, and arts and culture. In the area of Health, the foundation supports programs designed to promote health research, treatment, education, and awareness; and reduce the cost of illness. Types of support include general operating budgets, building funds, continuing support, equipment, and projects. The committee foundation meets every six to eight weeks to review grant requests. Contact program staff for current guidelines.
Requirements: Nonprofit 501(c)3 organizations should submit a one- to two-page letter outlining the purpose and needs of the program, its budget, duration, goals, leadership, and amount requested.

Restrictions: Ineligibility applies to individuals; organizations that limit participation or services based on race, gender, religion, color, creed, age, or national origin; projects without ties to a community that is home to an ArvinMeritor facility; organizations that pose any conflict with the goals and mission of ArvinMeritor, its employees, communities, or products; operating expenses for United Way local agencies, except through the foundation's support of annual United Way campaigns; sponsorships of fund raising activities by individuals (i.e., walk-a-thons); requests for loans or debt retirement; religious or sectarian programs for religious purposes; labor, political, or veterans organizations; fraternal, athletic, or social clubs; or seminars, conferences, trips, and tours.
Geographic Focus: All States
Contact: Jerry Rush; (248) 435-7907; fax (248) 245-1031; jerry.rush@arvinmeritor.com
Internet: http://www.arvinmeritor.com/community/community.asp
Sponsor: ArvinMeritor Foundation
2135 W Maple Road
Troy, MI 48084

ArvinMeritor Grants 675
ArvinMeritor???s philanthropic programs focus on education and the communities where its employees live and work. Grants support engineering and technical schools worldwide to increase the number of engineers and scientists in the field of automotive engineering, and to stimulate technological development and innovation in this field. The corporation also partners with local, private, and public agencies in order to improve the quality of life in the areas of education, cultural programs, civic responsibility, and health and human services. There are no application deadlines.
Requirements: Nonprofit organizations in ArvinMeritor communities are eligible.
Restrictions: Grants do not support organizations without nonprofit status; individuals; organizations that limit participation or services based on race, gender, religion, color, creed, age, or national origin; projects without ties to a community that is home to an ArvinMeritor facility; organizations that pose any conflict with the goals and mission of ArvinMeritor, its employees, communities, or products; operating expenses for United Way local agencies, except through support of annual United Way campaigns; sponsorships of fund-raising activities by individuals; requests for loans or debt retirement; religious or sectarian programs for religious purposes; labor, political, or veterans organizations; fraternal, athletic, or social clubs; or seminars, conferences, trips, and tours.
Geographic Focus: All States
Amount of Grant: 1,000 - 100,000 USD
Contact: Jerry Rush; (248) 435-7907; fax (248) 245-1031; jerry.rush@arvinmeritor.com
Internet: http://www.arvinmeritor.com/community/guidelines.asp
Sponsor: ArvinMeritor
2135 W Maple Road
Troy, MI 48084-7186

ASA Metlife Foundation MindAlert Awards 676
The awards were established to recognize innovations in mental fitness programming for older adults. Awards recognize programs, products, or tools that promote cognitive fitness in later life. One award will be made to a program in each of three categories: lifelong learning and third age learning programs where mental fitness is implicit; educational programs focused on enhancing mental fitness for the general population of older adults; and programs specifically designed to enhance mental fitness for cognitively-impaired older adults. Applications are judged for their innovation, their basis in research, demonstration of their effectiveness, potential for replicability and the extent to which the programs are accessible to diverse populations of elders. Guidelines and application are available online.
Requirements: Programs and products or tools that promote cognitive function in later life are eligible.
Geographic Focus: All States
Date(s) Application is Due: Oct 1
Amount of Grant: 1,500 USD
Contact: Nancy Decia; (415) 974-9610; fax (415) 974-0300; awards@asaging.org
Internet: http://www.asaging.org/asav2/mindalertaward/index.cfm
Sponsor: American Society on Aging
833 Market Street, Suite 511
San Francisco, CA 94103-1824

ASCO/UICC International Cancer Technology Transfer Fellowships 677
The scholarships are intended primarily for researchers in the early stages of their career as well as for experienced clinical oncologists. The ICRETT objective is to facilitate rapid international transfer of cancer research and clinical technology, and it allows fellows to exchange knowledge and enhance skills in basic, clinical, behavioral and epidemiological areas of cancer research, and cancer control and prevention. Through the program fellows also acquire appropriate clinical management, diagnostic and therapeutic expertise for effective application and use in home organizations upon return. The UICC has also developed the ICRETT for teaching faculty which funds experts to teach at host institutions throughout the world for a duration of 1 week to 1 month.
Requirements: The candidate must possess the appropriate professional qualifications and experience according to the specifications of each fellowship described on the website and must currently be engaged in cancer research, clinical oncology practice, or cancer society work. To permit effective communication at the host institute, the candidate must have adequate fluency in a common language. The candidate must also be on the staff payroll of a university, research laboratory or institute, hospital, oncology unit, or voluntary cancer society (or be accredited volunteers of such societies) to where they will return at the end of a fellowship.
Restrictions: Candidates attached to commercial entities, going to a profit organization or have associations with the tobacco industry not eligible.

Geographic Focus: All States
Amount of Grant: 3,400 USD
Contact: Nancy R. Daly, Executive Director; (571) 483-1432; dalyn@asco.org
Internet: http://www.ascocancerfoundation.org/TACF/Awards/Award+Opportunities/ASCO+%26+UICC+ICRETT+Fellowship
Sponsor: Conquer Cancer Foundation
2318 Mill Road, Suite 800
Alexandria, VA 22314

ASCO Advanced Clinical Research Award in Colorectal Cancer 678

The Advanced Clinical Research Award (ACRA) in Colorectal Cancer is designed to fund investigators who are committed to clinical cancer research and who wish to conduct original research not currently funded. The grant totals $450,000 and is paid in 3 annual increments of $150,000 made on or about July 1 of each year of the grant term. The grant funds will be directed to the sponsoring institution and should be used towards salary support, supplies, equipment, travel, etc. necessary for the pursuit of the recipient's research project.

Requirements: Apppicants must meet the following criteria: be a physician (MD, DO, or international equivalent) who is in the fourth to ninth year of a full-time, primary faculty appointment in a clinical department at an academic medical institution at the time of grant submission; have completed productive post doctoral/post fellowship research and demonstrated the ability to undertake independent investigator-initiated clinical research; be an active member of ASCO or have submitted a membership application with the grant application; be able to commit to 75% of full-time effort in research (applies to total research, not just the proposed project) during the award period.
Geographic Focus: All States
Date(s) Application is Due: Nov 3
Amount of Grant: 450,000 USD
Contact: Nancy R. Daly, Executive Director; (571) 483-1432; dalyn@asco.org
Internet: http://www.ascocancerfoundation.org/TACF/Grants/Grant+Opportunities/Advanced+Clinical+Research+Award+in+Colorectal+Cancer
Sponsor: Conquer Cancer Foundation
2318 Mill Road, Suite 800
Alexandria, VA 22314

ASCO Advanced Clinical Research Awards in Breast Cancer 679

The ASCO Foundation and Genentech BioOncology invite physicians with full-time faculty appointments to apply for a clinical oncology research award in Breast Cancer. The program funds clinical investigators who are committed to clinical cancer research. Their research must have a direct patient-oriented focus including clinical trials and/or translational research involving human subjects. The Advanced Clinical Research Award is designed to provide funding beyond the Clinical Research Career Development Award (CDA). By continuing to support proven clinical researchers who are post-CDA but at a critical stage in their early career, the foundation hopes to expand the cadre of expert clinical oncology researchers who are developing promising research initiatives.

Requirements: The applicant must meet the following criteria at the time of grant award: be a physician (MD or DO) who is 5-10 years post final sub-specialty training; have a full-time faculty appointment in a clinical department at an academic medical center; have completed productive postdoctoral/post-fellowship research and demonstrated the ability to undertake independent investigator-initiated clinical research; be an active member of the American Society of Clinical Oncology (ASCO); and expect to spend 75% of time during the award period dedicated to research.
Geographic Focus: All States
Date(s) Application is Due: Oct 20
Amount of Grant: 450,000 USD
Contact: Nancy R. Daly, Executive Director; (571) 483-1432; dalyn@asco.org
Internet: http://www.ascocancerfoundation.org/TACF/Grants/Grant+Opportunities/Advanced+Clinical+Research+Award+in+Breast+Cancer
Sponsor: Conquer Cancer Foundation
2318 Mill Road, Suite 800
Alexandria, VA 22314

ASCO Advanced Clinical Research Awards in Sarcoma 680

The Advanced Clinical Research Award (ACRA) in Sarcoma is designed to fund investigators who are committed to clinical cancer research and who wish to conduct original research not currently funded. The award is intended to support proposals with a patient-oriented focus, including a clinical research study and/or translational research involving human subjects. The grant totals $450,000 and is paid in three annual increments of $150,000 made on or about January 1 of each year of the grant term. Preference will be given to proposals involving therapeutic vaccine or immunotherapeutic approaches for sarcoma.

Requirements: Applicants must meet the following criteria: be a physician (MD, DO, or international equivalent) who is in the fourth to ninth year of a full-time, primary faculty appointment in a clinical department at an academic medical institution at the time of grant submission; have completed productive post doctoral/post fellowship research and demonstrated the ability to undertake independent investigator-initiated clinical research; be an active member of ASCO or have submitted a membership application with the grant application; be able to commit to 75% of full-time effort in research (applies to total research, not just the proposed project) during the award period.
Restrictions: Proposals with a predominant focus on in vitro or animal studies (even if clinically relevant) are not allowed.
Geographic Focus: All States
Date(s) Application is Due: Aug 6
Amount of Grant: 450,000 USD
Contact: Nancy R. Daly, Executive Director; (571) 483-1432; dalyn@asco.org
Internet: http://www.ascocancerfoundation.org/TACF/Grants/Grant+Opportunities/Advanced+Clinical+Research+Award+in+Sarcoma
Sponsor: Conquer Cancer Foundation
2318 Mill Road, Suite 800
Alexandria, VA 22314

ASCO Long-Term International Fellowships 681

The ASCO Cancer Foundation Long Term International Fellowship (LIFe) provides young oncologists in developing nations the support and resources needed to advance their training by deepening their relationship with a U.S. or Canadian colleague and his or her institution. Through a one or two year fellowship the recipient will earn valuable training and experience with which they can affect change in cancer care in their home country. Recipients must return to their home institutions after completing their fellowship and are expected to disseminate the knowledge they have gained. The ASCO Cancer Foundation (TACF) will administer fundraising and grant distribution for the fellowships. Funds will be sent from TACF to the host institution in January of the award year. The host institution will then be responsible for financial management of the fellowship. Any funds not spent on the fellowship will be returned to TACF. Recipients must attend the ASCO Annual Meeting that falls during the first year of their award term where they will be honored at The ASCO Cancer Foundation Grants Brunch. Recipients will also be expected to submit their research to ASCO via an abstract or publication submission following their fellowship. Applications are due to ASCO by October 15 for the fellowship to begin anytime in the next calendar year. Applications should be completed and returned jointly by the prospective host and fellow. A decision will be distributed by December 15.

Requirements: Hosts must: be an ASCO member in good standing; be employed at a U.S. or Canadian institution with sufficient infrastructure to provide for a 1 year fellowship; independently identify fellowship applicants; agree to cover any costs in excess of ASCO's contribution; complete the host section of the LIFe Application; complete fellowship evaluations as requested by ASCO. Applicants must: be an ASCO member in good standing; reside in a country identified by the World Bank as income level low, lower middle, and upper middle; have the support of their home institution including a letter of support from their supervisor; be a medical doctor who has finished a fellowship in clinical oncology or the equivalent; have less than 10 years experience in the field of oncology; complete the applicant section of the LIFe application including submission of a CV of up to 3 pages in length; commit to return to country of origin at the completion of their fellowship.
Geographic Focus: All States
Date(s) Application is Due: Oct 15
Contact: Nancy R. Daly, Executive Director; (571) 483-1432; dalyn@asco.org
Internet: http://www.ascocancerfoundation.org/TACF/Awards/Award+Opportunities/Long-term+International+Fellowship+%28LIFe%29
Sponsor: Conquer Cancer Foundation
2318 Mill Road, Suite 800
Alexandria, VA 22314

ASCO Merit Awards 682

The Merit Awards are designed to further promote clinical research by young scientists and to provide fellows with an opportunity to present their research and interact with other clinical cancer investigators at ASCO scientific meetings. A select number of Merit Awards are given annually to recognize outstanding abstracts submitted for consideration for presentation at an ASCO scientific meeting. The awards are given to oncology fellows who are first authors on selected abstracts.

Requirements: The first author of an abstract must meet the following requirements to be considered for an Award: be a physician (MD or DO) in an oncology fellowship training program or a doctoral degree candidate (such as PharmD or PhD) enrolled in an approved oncology specialty training program at the time of abstract submission; work in an oncology laboratory or clinical research setting; agree to present the abstract at the ASCO scientific meeting; check the Merit Award consideration box on the online abstract submitter; provide a letter of support from the Training Program Director indicating eligibility for the award; and submit a two-page curriculum vitae.
Geographic Focus: All States
Contact: Nancy R. Daly, Executive Director; (571) 483-1432; dalyn@asco.org
Internet: http://www.ascocancerfoundation.org/TACF/Awards/Award+Opportunities/Merit+Award
Sponsor: Conquer Cancer Foundation
2318 Mill Road, Suite 800
Alexandria, VA 22314

ASCO Young Investigator Award 683

This grant provides funding to promising investigators to encourage and promote quality research in clinical oncology. The purpose of this award is to fund physicians during the transition from a fellowship program to a faculty appointment. Priority consideration will be given to proposals that include patient-oriented and, ultimately, clinical research. Since many awards are supported with restricted grants from outside organizations, the ASCO is particularly interested in identifying young researchers working in the following sub-specialties and emerging disciplines: breast cancer; cancer survivorship; geriatric oncology; health disparities; kidney cancer; multiple myeloma; ovarian cancer; pancreatic cancer; sarcoma; survivorship; and young adult cancer. However, The ASCO Foundation welcomes application submissions in all oncology subspecialties.

Requirements: The recipient must be a physician (MD or DO) who, at the time of grant award is in the final year of his/her final sub-specialty training program or in the first year post his/her final sub-specialty training. The sponsoring facility must be an academic medical institution. The primary mentor must be in the candidate's proposed research field, must assume responsibility, and provide guidance for the research. The applicant must either be a member of ASCO or submit a membership application with the grant application. The applicant should spend at least 60 to 75% of his or her time in research during the award period.
Geographic Focus: All States
Date(s) Application is Due: Sep 9
Amount of Grant: 50,000 USD
Contact: Nancy R. Daly, Executive Director; (571) 483-1432; dalyn@asco.org
Internet: http://www.ascocancerfoundation.org/TACF/Grants/Grant+Opportunities/Young+Investigator+Award
Sponsor: Conquer Cancer Foundation
2318 Mill Road, Suite 800
Alexandria, VA 22314

ASGE / ConMed Award for Outstanding Manuscript by a Fellow/Resident 684
The American Society for Gastrointestinal Endoscopy and ConMed Endoscopic Technologies sponsor the annual ASGE/ConMed Outstanding Manuscript Award. The prize consists of $10,000 cash award to the Fellow or Resident in training in a gastroenterology or a surgical gastrointestinal endoscopic program whose manuscript is accepted in for publication in Gastrointestinal Endoscopy. The award is presented during Digestive Disease Week, the largest and most prestigious meeting in the world for the GI professional.
Geographic Focus: All States
Amount of Grant: 10,000 USD
Contact: Chair; (630) 573-0600; fax (630) 573-0691; grants@asge.org
Internet: http://www.asge.org/index.aspx?id=4892&terms=ConMed+Award
Sponsor: American Society for Gastrointestinal Endoscopy
1520 Kensington Road, Suite 202
Oak Brook, IL 60523

ASGE Don Wilson Award 685
The Award provides Advanced Fellows or Junior Faculty with the opportunity to train outside of their home country with a premier GI endoscopist or group in order to advance their training. The award assists in underwriting reasonable and customary travel and living expenses for a period of one to three months. The award includes a $7,500 cash stipend prior to the recipient's travel. In addition, a 20% disbursement will be made to the host institution. A total of three awards will be available annually for North American and International ASGE members. Note: One training must take place in the United States.
Requirements: Applicants must be the equivalent of Junior Faculty or Advanced Fellows and be proficient in the English Language. Applicants must also: be a current ASGE Member; be a Junior Faculty or Advanced Fellow (3rd or 4th year or international equivalent); have permission from their own institution to undertake the travel; and have permission from an ASGE member at the host institution to accept the application for training.
Restrictions: Individuals applying for 1-2 year advanced fellowships do not qualify.
Geographic Focus: All States
Date(s) Application is Due: Sep 1
Amount of Grant: 7,500 USD
Contact: Chair; (630) 573-0600; fax (630) 573-0691; membership@asge.org
Internet: http://www.asge.org/nspages/research/applications/donWilson.cfm
Sponsor: American Society for Gastrointestinal Endoscopy
1520 Kensington Road, Suite 202
Oak Brook, IL 60523

ASGE Endoscopic Research Awards 686
These research awards are offered to physicians for projects in basic and clinical endoscopic technology research, outcomes and effectiveness of endoscopy research. The primary objective is to foster research in gastrointestinal endoscopy both within and outside of academic centers. Two categories of grants may be used for up to two years of study: grants of $1-15,000 (category A) and grants of $15,001 - 50,000 (category B). Requests for funding seed projects that will lead to further research as well as larger requests for definitive clinical trials will be considered. Funding requests may also include: personnel expenses for research assistant and/or faculty salary support (percentage of time for study should be specified and appropriately justified); study supplies; and equipment essential for study.
Requirements: Candidate must be: an ASGE member; an MD (or have equivalent degree); and current in a gastroenterology-related and endoscopic practice in academic institutions or private practice in North America.
Restrictions: Funding will not be provided for: salary support for trainees; computer purchases (unless a unique application is proposed); standard equipment and supplies needed for appropriate patient care (for example, sclerotherapy needles); travel to meetings; or indirect costs.
Geographic Focus: All States
Date(s) Application is Due: Sep 12
Amount of Grant: Up to 50,000 USD
Contact: Chair; (630) 573-0600; fax (630) 573-0691; grants@asge.org
Internet: http://www.asge.org/index.aspx?id=4892&terms=ConMed+Award
Sponsor: American Society for Gastrointestinal Endoscopy
1520 Kensington Road, Suite 202
Oak Brook, IL 60523

ASGE Endoscopic Research Career Development Awards 687
These awards provide the salary and/or research support necessary for the investigator to enhance his/her career development. The award must be used to acquire new skills for furthering a career in endoscopic research. Examples of such skills include advanced training in endoscopic procedures, training in outcomes research relevant to endoscopy, training in use of large databases such as CORI, and new endoscopic research techniques, including animal models.
Requirements: Applicants must be ASGE members and hold full-time faculty positions at North American (U.S., Canada, or Mexico) universities or professional institutions at the time of application. The award is intended for faculty who have demonstrated promise and have some record of accomplishment in research. Candidates must devote at least 30 percent of their effort to research related to gastrointestinal endoscopy during the period of the award.
Restrictions: The award is not available for fellows. Faculty with principle investigator current federal funding or other concurrent career development support still active at the time of this award are not eligible. Recipients of the ASGE Career Development Awards may not be granted an ASGE Research & Outcomes & Effectiveness Award during their two-year award period.
Geographic Focus: All States
Date(s) Application is Due: Dec 12
Amount of Grant: 75,000 USD
Contact: Chair; (630) 573-0600; fax (630) 573-0691; grants@asge.org
Internet: http://www.asge.org/index.aspx?id=4892&terms=ConMed+Award
Sponsor: American Society for Gastrointestinal Endoscopy
1520 Kensington Road, Suite 202
Oak Brook, IL 60523

ASGE Given Capsule Endoscopy Research Award 688
These research awards are offered to physicians for projects specific to capsule endoscopy both within and outside of academic centers. The award must be used for projects directly relating to capsule endoscopy. Requests for funding seed projects that will lead to further research as well as requests for definitive clinical trials will be considered. Funding requests may include: personnel (research assistant and/or faculty salary support-- percentage of time for study should be specified and appropriately justified); study supplies; and equipment essential for study.
Requirements: Candidate must be an ASGE member, be an MD (or have equivalent degree), and be current in a gastroenterology-related and endoscopic practice in academic institutions or private practice in North America.
Restrictions: Funding will not be provided for: salary support for trainees; computer purchases (unless a unique application is proposed); standard equipment and supplies needed for appropriate patient care (for example, sclerotherapy needles); travel to meetings; publication costs; or indirect costs.
Geographic Focus: All States
Date(s) Application is Due: Dec 7
Amount of Grant: Up to 25,000 USD
Contact: Chair; (630) 573-0600; fax (630) 573-0691; grants@asge.org
Internet: http://www.asge.org/nspages/research/applications/ceapp.cfm
Sponsor: American Society for Gastrointestinal Endoscopy
1520 Kensington Road, Suite 202
Oak Brook, IL 60523

ASGH Award for Excellence in Human Genetics Education 689
The American Society of Human Genetics (ASGH) established the Award for Excellence in Human Genetics Education to recognize outstanding contributions to human genetics education. Nominees should have long-standing involvement in genetics education, contributions in more than one area, and contributions of substantive influence on individuals and/or organizations. Examples of the types of contributions that might qualify a nominee include: producing a set of writings that has had a major influence on human genetics education, developing a curriculum or innovative teaching program that is widely emulated, writing a book that has been adopted by many universities, developing an educational Web site, or directing a fellowship program that has consistently produced exemplary graduates. Nominees may be individuals or groups.
Requirements: Nominees must have made contributions that are recognized nationally or internationally as being of exceptional quality and great importance to human genetics education. All nominees and winners must be current ASHG members. If a group is nominated, at least one nominee in the group must be a current member. A monetary award and plaque will be presented to the recipient(s) at the Society's Annual Meeting. Nominees for the ASHG Excellence in Human Genetics Education Award may emerge from submissions from the membership, from the Awards Committee, or from the Information and Education Committee.
Geographic Focus: All States
Contact: Joann Boughman, Executive Vice-President; (866) 486-4363 or (301) 634-7300; fax (301) 634-7079; jboughman@ashg.org
Internet: http://www.ashg.org/pages/awards_overview.shtml#education
Sponsor: American Society of Human Genetics
9650 Rockville Pike
Bethesda, MD 20814-3998

ASGH C.W. Cotterman Award 690
Each September, the editorial board of The American Journal of Human Genetics selects two articles published in the journal in the previous year that best represent outstanding scientific contributions to the field of human genetics. Two Cotterman Awards will be given annually and a monetary award of $1,000 and a plaque will be presented to the

recipients for the top two papers published in the Journal during the previous year on which the first author was either a pre- or post- doctoral trainee and an ASHG member.
Geographic Focus: All States
Amount of Grant: 1,000 USD
Contact: Joann Boughman, Executive Vice-President; (866) 486-4363 or (301) 634-7300; fax (301) 634-7079; jboughman@ashg.org
Internet: http://www.ashg.org/pages/awards_overview.shtml#cotterman
Sponsor: American Society of Human Genetics
9650 Rockville Pike
Bethesda, MD 20814-3998

ASGH Charles J. Epstein Trainee Awards for Excellence in Human Genetics Research 691

The American Society of Human Genetics (ASGH) honors excellence in research conducted by predoctoral and postdoctoral trainees (including genetic counseling trainees) through merit-based awards that recognize highly competitive abstracts submitted and presented at the ASHG Annual Meeting. These awards were renamed in 2012 to honor the late Dr. Charles Epstein, who was a past president of ASHG, former editor of AJHG, and winner of both the William Allan Award and the McKusick Leadership Award. The Program Committee and Awards Committee, in consultation with the ASHG Board of Directors, determine the number of awards, the categories for which they are given, and the prize amounts of each award (currently $750 plus complimentary registration for approximately 60 semifinalists, plus an additional $1,000 for winners). Abstract scoring is completed by the Program Committee, and the top-ranking abstracts (semifinalists) are submitted to the Awards Committee, which selects 18 finalists prior to the Annual Meeting. Six winners are chosen after the committee members judge the finalists' presentations. The total value of ASHG's Trainee Awards is approximately $70,000 annually.
Geographic Focus: All States
Amount of Grant: 750 - 1,750 USD
Contact: Joann Boughman, Executive Vice-President; (866) 486-4363 or (301) 634-7300; fax (301) 634-7079; jboughman@ashg.org
Internet: http://www.ashg.org/pages/awards_overview.shtml#student
Sponsor: American Society of Human Genetics
9650 Rockville Pike
Bethesda, MD 20814-3998

ASGH McKusick Leadership Award 692

The prestigious McKusick Leadership Award, established by the American Society of Human Genetics (ASHG) in honor of the late Dr. Victor A. McKusick, is presented on behalf of the Society to an individual whose professional achievements have fostered and enriched the development of various human genetics disciplines. Potential recipients should exemplify the enduring leadership and vision required to ensure that the field of human genetics will flourish and successfully assimilate into the broader context of science, medicine, and health. They also may have made major contributions to awareness or understanding of human genetics by policy makers or by the general public. A plaque and a $2,500 prize will be presented in honor of the awardee at the ASHG Annual Meeting.
Geographic Focus: All States
Amount of Grant: 2,500 USD
Contact: Joann Boughman, Executive Vice-President; (866) 486-4363 or (301) 634-7300; fax (301) 634-7079; jboughman@ashg.org
Internet: http://www.grantselect.com/editor/view_grant/133352
Sponsor: American Society of Human Genetics
9650 Rockville Pike
Bethesda, MD 20814-3998

ASGH William Allan Award 693

The William Allan Award is the top prize given by the American Society of Human Genetics (ASHG); it was established in 1961 in memory of William Allan (1881-1943), who was one of the first American physicians to conduct extensive research in human genetics. The Allan Award is presented annually to recognize substantial and far-reaching scientific contributions to human genetics, carried out over a sustained period of scientific inquiry and productivity. The recipient is presented with an engraved medal and a monetary award of $10,000 at the ASHG Annual Meeting. The Allan Award winner is also invited to present a 30-45 minute address at the ASHG Annual Meeting, and it is customary to publish a manuscript of the presentation in The American Journal of Human Genetics. This award is given on a yearly basis, but can be omitted in any given year at the discretion of the Board. Nominations will be solicited from the ASHG Awards Committee and the general membership.
Geographic Focus: All States
Amount of Grant: 1,000 USD
Contact: Joann Boughman, Executive Vice-President; (866) 486-4363 or (301) 634-7300; fax (301) 634-7079; jboughman@ashg.org
Internet: http://www.ashg.org/pages/awards_overview.shtml#allan
Sponsor: American Society of Human Genetics
9650 Rockville Pike
Bethesda, MD 20814-3998

ASGH William Curt Stern Award 694

The American Society of Human Genetics (ASGH) Curt Stern Award honors the memory of Curt Stern (1902-1981) as an outstanding pioneering human geneticist. This award is presented yearly for outstanding scientific achievements in human genetics that occurred in the last 10 years. The work could be a single major discovery or may be a series of contributions on a similar or related topic. A plaque and a monetary award of $2,500 are presented to the recipient(s) at the Society's Annual Meeting. Nominations for the Stern Award will be solicited from the ASHG Awards Committee and the general membership.
Geographic Focus: All States
Amount of Grant: 2,500 USD
Contact: Joann Boughman, Executive Vice-President; (866) 486-4363 or (301) 634-7300; fax 301; jboughman@ashg.org
Internet: http://www.ashg.org/pages/awards_overview.shtml#stern
Sponsor: American Society of Human Genetics
9650 Rockville Pike
Bethesda, MD 20814-3998

ASHA Advancing Academic-Research Careers (AARC) Award 695

This award is given to new faculty in higher education to support their academic and research endeavors in the field of communication sciences and disorders (CSD). Up to five awards of $5,000 each will be granted to support activities such as: Improving teaching skills; Mentoring graduate students; Participating in research activities; Preparing a grant application; Preparing a manuscript or publication; Presenting at a professional meeting (e.g., the ASHA Convention).
Requirements: Candidates must possess: A research doctoral degree (PhD, ScD, CScD, or equivalent, e.g., EdD), which must be conferred by August 1; A full-time faculty appointment; Primary responsibilities of teaching and conducting research; Less than six years of teaching/research experience or are below the level of associate professor; A tenure or non-tenure track appointment. A list of required documents and a downloadable application form are available at the website.
Restrictions: Postdoctoral fellows are not eligible. Only students attending a university in the U.S. are eligible to apply for the award.
Geographic Focus: All States
Date(s) Application is Due: May 15
Amount of Grant: 5,000 USD
Contact: Margaret Rogers, Chief Staff Officer for Science & Research; (800) 498-2071 or (301) 296-5706; fax (301) 296-8580; mrogers@asha.org
Internet: http://www.asha.org/students/awards.htm#aarc
Sponsor: American Speech-Language-Hearing Association
2200 Research Boulevard, #245
Rockville, MD 20850-3289

ASHA Minority Student Leadership Program Awards 696

The MSLP is a leadership development program established for undergraduate seniors and graduate students who are enrolled in communication sciences and disorders programs. Aditionally, ASHA's Focused Initiative on the PhD Shortage in Higher Education will provide funding for 10 eligible candidates enrolled in a research doctoral program. The award package includes complimentary registration to the ASHA Convention held annually; up to five nights lodging; meal stipend; and, program materials. Forty students are chosen to participate - 5 undergraduate seniors, 20 master's students, 5 entry level clinical doctoral students and 10 PhD students.
Requirements: Undergraduate seniors and graduate students who are or will be enrolled in the fall communication sciences and disorders programs and are not members of ASHA are eligible. Preference will be given to students who are members of racial/ethnic minority groups historically under-represented within ASHA, including American Indian or Alaska Native, Asian, Black or African American, Native Hawaiian or other Pacific Islander, and/or Hispanic/Latino. For Graduate Study - The student must be enrolled in, or accepted for, graduate study in a communication sciences and disorders program in the United States. Master's degree candidates and entry level clinical doctoral candidates must be in a program accredited by the Council of Academic Accreditation (CAA). PhD Track - You must be enrolled in a research doctoral degree program. ASHA members may apply. The required application packet is available for download at the website (http://www.asha.org/students/awards.htm#MSLP).
Restrictions: Program participants are responsible for transportation expenses to and from the convention city and personal incidentals.
Geographic Focus: All States
Date(s) Application is Due: May 18
Contact: Melanie Johnson, Program Manager; (301) 296-8681; mjohnson@asha.org
Internet: http://www.asha.org/students/awards.htm#MSLP
Sponsor: American Speech-Language-Hearing Association
2200 Research Boulevard, #245
Rockville, MD 20850-3289

ASHA Multicultural Activities Projects Grants 697

The program is open to clinical and/or school-based speech, language, and hearing programs; university programs, state associations, allied and related professional organizations, and ASHA's Special Interest Divisions. ASHA has allocated funding for projects on multicultural activities with a maximum of $18,000 awarded for a single grant. It is expected that the average individual award given will be between $7,000 and $10,000. Applications must be postmarked by April 4. Proposals receiving funding will be announced in July.
Requirements: Proposals must have a multicultural focus. Multicultural is defined to include issues dealing with race, ethnicity, language, gender or gender identification, age, sexual orientation, and disability. There is particular interest in, but not limited to, proposals that respond to one or more of ASHA's Strategic Initiatives from ASHA's Strategic Pathway to Excellence. Projects must: be compatible with ASHA's mission and vision; have a clear high-quality plan for meeting its objectives; be completed, including evaluation, within 12 months of initiation of the project; be adaptable by other clinical and/or school-based programs, university programs, state associations, and associated and related professional organizations for their own use; and describe what will happen

to the project once ASHA funding has ended. Criteria for evaluating proposals and the required downloadable forms are available at the website.
Restrictions: An ASHA member must serve as project director. Grants will not be awarded for fund raising, governmental lobbying, or awards ceremonies.
Geographic Focus: All States
Amount of Grant: Up to 18,000 USD
Samples: 2007-2008: Development of Standard and Multifrequency Tympanometry Norms for Native American Children (Birth - 5 years) and Young Adults - Samuel R. Atcherson & Paul Brueggeman, University of South Dakota; Digital Treatment Materials for Intervention in Spanish and English - Steven Long, Marquette University; The Validation of a Language Use Questionnaire across Bilingual Clinical Populations - Swathi Kiran, The University of Texas at Austin
Contact: Karen Beverly-Ducker, (800) 638-8255; OMAGrants@asha.org
Internet: http://www.asha.org/about/leadership-projects/multicultural/funding/rfp.htm
Sponsor: American Speech-Language-Hearing Association
2200 Research Boulevard, #245
Rockville, MD 20850-3289

ASHA Research Mentoring-Pair Travel Award   698
This award is designed to foster the professional development of students, clinicians, and emerging scientists who have expressed an interest in research in communication sciences and disorders. Up to 10 travel awards of $1,000 each will be granted to eligible mentor mentee pairs (mentee= $750; mentor= $250) to help defray the cost of attending the annual research symposium. Special consideration will be given to individuals whose research or research interests are (a) relevant to the symposium topic, (b) interdisciplinary in nature, (c) translational in nature, or (d) related to issues of cultural or linguistic diversity.
Requirements: Mentees must be (a) a current student at the bachelor's, master's, AuD, or PhD level, (b) a postdoctoral fellow, (c) a junior level faculty member (less than 6 years teaching/research experience in a tenure-track position or below the level of associate professor), or (d) a clinician. Mentors must be a seasoned investigator in an academic or health care environment. Download the Mentee and Mentory Application Forms at the website.
Geographic Focus: All States
Date(s) Application is Due: Jun 22
Amount of Grant: 1,000 USD
Contact: Margaret Rogers, Chief Staff Officer for Science & Research; (800) 498-2071 or (301) 296-5706; fax (301) 296-8580; mrogers@asha.org
Internet: http://www.asha.org/students/awards.htm#2008%20Research%20Mentoring-Pair%20Travel%20Award
Sponsor: American Speech-Language-Hearing Association
2200 Research Boulevard, #245
Rockville, MD 20850-3289

ASHA Student Research Travel Award   699
The awards are intended to highlight the research activities of budding scientists and encourage careers in science and research. Approximately 40 student research travel awards of $500 each are available for undergraduate, Master's, and doctoral students who are first authors on a paper (all session types are acceptable: technical, poster, one hour seminar, two hour seminar, or short course). Identification of prospective awardees will be based on the recommendation of Convention co-chairs and topic coordinators. Awardees will receive a $500 travel award; have their exceptional paper highlighted in the convention program; and, be recognized at the Researcher-Academic Town Meeting and other designated events.
Requirements: Each award recipient will be required to: register for the Convention at his or her own expense; present his or her paper at the assigned time (as noted in the Convention program booklet); secure a mentor to shadow throughout the convention; and, attend selected research-related sessions during the convention. In order to apply, students must: submit an abstract summarizing their research through the normal Call for Papers process; check the appropriate box (when prompted) indicating that their paper is 'student-authored'; and, check the appropriate box (when prompted) indicating that they desire to have this abstract considered for the Student Research Travel Award.
Restrictions: Papers must be submitted online by the deadline date and must be designated as a student-authored paper.
Geographic Focus: All States
Date(s) Application is Due: Mar 31
Amount of Grant: 500 USD
Contact: Margaret Rogers, Chief Staff Officer for Science & Research; (800) 498-2071 or (301) 296-5706; fax (301) 296-8580; mrogers@asha.org
Internet: http://www.asha.org/students/awards.htm#research
Sponsor: American Speech-Language-Hearing Association
2200 Research Boulevard, #245
Rockville, MD 20850-3289

ASHA Students Preparing for Academic & Research Careers (SPARC) Award   700
The goal of SPARC is to foster students' interest in the pursuit of PhD education and careers in academia in order to fill faculty/researcher vacancies in communication sciences and disorders (CSD). Students will identify a primary faculty mentor and propose a one-year CSD Career Mentoring Plan. As many as ten students will be awarded up to $1,500 each to be used for teaching and research enhancement activities, such as travel to a research, pedagogy conference, or meeting; travel for a visit to an off-campus site that provides learning opportunities in a research lab and/or a college classroom setting; or course registration to support the mentoring plans outlined in the application.
Requirements: Eligible applicants include junior or senior undergraduates, 1st year master's students, and 1st and 2nd year entry-level clinical doctoral (e.g., AuD) students enrolled part-time or full-time in a CSD program in the United States. At the time of the award, eligible recipients' education status for the upcoming academic year must be undergraduate senior, 1st or 2nd year master's, or 1st, 2nd or 3rd year entry-level clinical doctoral (i.e., AuD) in a full-time or part-time CSD program in the United States. Master's degree and entry-level clinical doctoral students must be in a program accredited by the Council for Academic Accreditation in Audiology and Speech-Language Pathology (CAA). Full requirements (essay question, documentation, etc.) are listed at the website, along with the downloadable application form.
Geographic Focus: All States
Date(s) Application is Due: May 15
Amount of Grant: 5,000 USD
Contact: Arlene A. Pietranton; apietranton@asha.org or academicaffairs@asha.org
Internet: http://www.asha.org/students/awards.htm#sparc
Sponsor: American Speech-Language-Hearing Association
2200 Research Boulevard, #245
Rockville, MD 20850-3289

ASHFoundation Clinical Research Grants   701
These $50,000-$75,000 grants support scientists with a research doctorate within the discipline of communication sciences and disorders to support investigations that will advance knowledge of the efficacy of treatment and assessment practices. Project funding is available for mentored treatment research, independent treatment research, or collaborative treatment research as specified in grant guidelines.
Geographic Focus: All States
Amount of Grant: 50,000 - 75,000 USD
Contact: Emily Diaz, Project Assistant; fax (301) 296-8703; ediaz@asha.org
Internet: http://www.ashfoundation.org/Foundation/grants/research_grants.htm
Sponsor: American Speech-Language-Hearing Foundation
2200 Research Boulevard
Rockville, MD 20850-3289

ASHFoundation Graduate Student International Scholarship   702
One award is available for a full-time international graduate student who has demonstrated outstanding academic achievement. Applicants must be studying communications sciences and disorders at a U.S. institution accredited by ASHA's Council of Academic Accreditation. All scholarship recipients will be announced and recognized at the ASHA convention.
Requirements: Applicants must be full-time students.
Geographic Focus: All States
Amount of Grant: Up to 4,000 USD
Samples: Beverly A. Collisson, PhD Candidate, University of Connecticut, $5,000; Beula M. Magimairaj, PhD Candidate, Ohio University, $5,000; Maria V. Ivanova, PhD candidate, Ohio University, $4,000; Victoria Lee, AuD candidate, Northwestern University, $4,000.
Contact: Emily Diaz, Project Assistant; fax (301) 296-8703; ediaz@asha.org
Internet: http://www.ashfoundation.org/Foundation/grants/GraduateScholarships.htm
Sponsor: American Speech-Language-Hearing Foundation
2200 Research Boulevard
Rockville, MD 20850-3289

ASHFoundation Graduate Student Scholarships   703
The foundation invites full-time graduate students to submit applications in competition for one of seven scholarships. Full-time master's or doctoral students in communication sciences and disorders programs demonstrating outstanding academic achievement are eligible to compete for these $4,000 awards. Supported in part by Psi Iota Xi National Philanthropic Organization and the Marni Reisberg Memorial Fund. All scholarship recipients will be announced and recognized at the annual convention held in November.
Requirements: Applicants must be accepted for graduate study in a communication sciences and disorders program (master's degree candidates must be in an ASHA Educational Standards Board accredited program; this is not mandatory for doctoral degree candidates); be enrolled for full-time study (12 or more credit hours or the full-time standard); submit official university transcripts of academic coursework, credits, and grades; and be recommended by a committee of two or more past or present college faculty and colleagues (at least one supervisor) at the student's current place of employment. Annual deadline dates may vary; contact program staff for exact dates.
Restrictions: Applicants must not have received a prior scholarship from the foundation.
Geographic Focus: All States
Amount of Grant: 4,000 USD
Contact: Emily Diaz, Project Assistant; fax (301) 296-8703; ediaz@asha.org
Internet: http://www.ashfoundation.org/Foundation/grants/GraduateScholarships.htm
Sponsor: American Speech-Language-Hearing Foundation
2200 Research Boulevard
Rockville, MD 20850-3289

ASHFoundation Graduate Student Scholarships for Minority Students   704
Racial/ethnic minority students who are U.S. citizens, who are accepted for graduate study in speech-language pathology or audiology, and who demonstrate outstanding academic achievement are eligible to compete for this scholarship. Applicants should submit a formal paper, such as a term paper, in competition for this award. The paper must be either an integrative literature review or an opinion paper on a current professional issue. All scholarship recipients will be announced and recognized at the ASHA convention.
Requirements: Candidates must be U.S. citizens and members of a racial/ethnic minority, including Native American and Alaska Native, Asian and Pacific Islander, African American, and Hispanic. Applicants must be accepted for master's or doctoral study in

a speech-language pathology or audiology program for the upcoming academic year. The student must be enrolled for full-time (12 or more credit hours) study.
Restrictions: Applicants may not have received a prior scholarship from the foundation.
Geographic Focus: All States
Amount of Grant: 2,000 - 4,000 USD
Samples: Tracy Conner, MA candidate, University of Massachusetts-Amherst, $5,000; Rhona Galera, CScD candidate, University of Pittsburgh, $5,000; Derek E. Daniels, PhD candidate, Bowling Green State University, $4,000.
Contact: Emily Diaz, Project Assistant; fax (301) 296-8703; ediaz@asha.org
Internet: http://www.ashfoundation.org/Foundation/grants/GraduateScholarships.htm
Sponsor: American Speech-Language-Hearing Foundation
2200 Research Boulevard
Rockville, MD 20850-3289

ASHFoundation Graduate Student with a Disability Scholarship 705
Full-time graduate students with a disability who are enrolled in a communication sciences and disorders program and demonstrate outstanding academic achievement are eligible to compete for a $4,000 scholarship. Applicants must demonstrate superior academic achievement and submit transcripts, essay and references with official application. The Award is intended for the blind, hearing impaired, physically handicapped, or learning impaired. Master's, but not doctoral, students must be enrolled in an ASHA Council on Academic Accreditation approved program.
Requirements: Candidates must be U.S. citizens and either blind, hearing impaired, physically handicapped, or learning impaired. Applicants must be accepted for master's study in a speech-language pathology or audiology program for the upcoming academic year. The student must be enrolled for full-time (12 or more credit hours) study.
Geographic Focus: All States
Amount of Grant: 4,000 USD
Contact: Emily Diaz, Project Assistant; fax (301) 296-8703; ediaz@asha.org
Internet: http://www.ashfoundation.org/Foundation/grants/GraduateScholarships.htm
Sponsor: American Speech-Language-Hearing Foundation
2200 Research Boulevard
Rockville, MD 20850-3289

ASHFoundation New Century Scholars Program Doctoral Scholarships 706
The Program is a special funding initiative resulting from the ASHFoundation's Dreams and Possibilities Campaign. The Award is for students accepted or currently enrolled in a research doctoral program in communications sciences and disorders. Full-time study will be given priority; part-time status will be considered. Students should be committed to a teacher-investigator career in communication sciences and disorders in the United States.
Requirements: The Award is available to U.S. citizens only. Applicants must be accepted or currently enrolled in a research doctoral program.
Geographic Focus: All States
Amount of Grant: 10,000 USD
Samples: Sophie Ambrose, PhD Candidate, University of Kansas; Megha Bahl, PhD Candidate, University of Arizona; Angela Yarnell Bonino, PhD candidate, University of North Carolina at Chapel Hill; Deanna Britton, PhD Candidate, University of Washington; Jamie L. Desjardins, PhD Candidate, Syracuse University; Kerry Danahy Ebert, PhD Candidate, University of Minnesota, Twin Cities; Aaron M. Johnson, PhD Candidate, University of Wisconsin, Madison; Karen E. Lee, PhD Candidate, University of Connecticut; Jimin Lee, PhD Candidate, University of Wisconsin, Madison; Megan K. McPherson, PhD Candidate, Purdue University; Kimberly M. Meigh, PhD Candidate, University of Pittsburgh.
Contact: Emily Diaz, Project Assistant; fax (301) 296-8703; ediaz@asha.org
Internet: http://www.ashfoundation.org/Foundation/grants/GraduateScholarships.htm
Sponsor: American Speech-Language-Hearing Foundation
2200 Research Boulevard
Rockville, MD 20850-3289

ASHFoundation New Century Scholars Research Grant 707
The Program is a special funding initiative resulting from the ASHFoundation's Dreams and Possibilities Campaign. The Grant is a one-time award intended to advance the knowledge base in communication sciences and disorders. Applicants must have a PhD or equivalent research doctorate within the discipline. Priority will be given to studies that are innovative, groundbreaking, or that meet research needs not yet addressed.
Restrictions: Students enrolled in a degree program or working on dissertation research are not eligible.
Geographic Focus: All States
Amount of Grant: 10,000 USD
Samples: Jungmee Lee, Assistant Professor, University of Arizona--Proposal: Correlation Between Cochlear Tuning and Otoacoustic Emissions: Exploring Scientific and Clinical Implications, $10,000; Chang Liu, Assistant Professor, University of Texas at Austin--Auditory Processing: Comparing Phonologically Disordered and Typically Developing Children, $10,000; John McCarthy, Assistant Professor, Ohio University--Improving Auditory Scanning Interfaces in AAC Devices, $10,000; Valeriy Shafiro, Assistant Professor, Rush University Medical Center--Effects of Environmental Sound Training on the Perception of Environmental Sounds and Speech in Cochlear Implant Patients, $10,000.
Contact: Emily Diaz, Project Assistant; fax (301) 296-8703; ediaz@asha.org
Internet: http://www.ashfoundation.org/Foundation/grants/research_grants.htm
Sponsor: American Speech-Language-Hearing Foundation
2200 Research Boulevard
Rockville, MD 20850-3289

ASHFoundation Research Grant for New Investigators 708
The grants are designed to help further research activities of new investigators and should have particular clinical relevance to speech-language pathology and audiology. The awards are designed to encourage research activities of new scientists who earned their latest degree within the last five years. Proposals, while not limited in topic, are encouraged in the area of treatment research, particularly efficacy and outcome studies. Include abstract, research plan, bibliography, management plan, and budget with application.
Requirements: The individual must have received the master's or doctoral level degree in the last five years; must not have received prior funding for research, with the exception of internal university funding; and the proposal must be for research to be initiated.
Restrictions: Open to individuals with a Master's degree or doctoral degree who are not currently enrolled in a degree program.
Geographic Focus: All States
Amount of Grant: 5,000 USD
Samples: Nina C. Capone, Associate Professor, Seton Hall University--The Effects of Gesture Cues on Object Word Learning by Children with Language Impairments; Gayle L. DeDe Assistant Professor, University of Arizona--On-Line Sentence Comprehension in Aphasia: Is Reading Different than Listening?; Ciara Leydon, Assistant Professor, Brooklyn College--Construction and Characterization of a Novel Model of Vocal Fold Mucosa; Rita R. Patel, Assistant Professor, University of Kentucky--High Speed Digital Analysis of Vocal Fold Vibration in Children; Yasmeen Faroqi Shah, Assistant Professor, University of Maryland, College Park--Retreival of Action Names in Aphasia: An Investigation of the Embodied Cognition Framework; Yana Yunosova, Assistant Professor, University of Toronto--Visual Feedback Systems in Speech Rehabilitation: Defining Vocal Tract Targets.
Contact: Emily Diaz, Project Assistant; fax (301) 296-8703; ediaz@asha.org
Internet: http://www.ashfoundation.org/Foundation/grants/research_grants.htm
Sponsor: American Speech-Language-Hearing Foundation
2200 Research Boulevard
Rockville, MD 20850-3289

ASHFoundation Research Grant in Speech Science 709
ASLH, in conjunction with the Acoustical Society of America, invites new researchers to submit proposals in competition for one grant every other year. The grant is designed to further research activities of new investigators and promulgate the work of the late Dennis Klatt, a noted researcher and professor in the area of speech communication. Priority will be given to areas reflecting Dr. Klatt's broad interests, such as proposals studying speech perception, synthesis, and acoustics, with an emphasis on an interdisciplinary research approach. The grant can be used to initiate new research or supplement an existing research project. Funds may be requested for a variety of purposes, i.e., equipment, subjects, research assistants, or research-related travel. The grant recipient will be announced at the ASHA convention.
Requirements: Individuals having received a doctoral degree within the last five years and who wish to further research activities in the areas of speech communication are eligible to compete for the grant.
Geographic Focus: All States
Amount of Grant: 5,000 USD
Samples: Mary K. Fagan, Indiana University School of Medicine--Vocalization and Sound Exploration in Hearing, Deaf, and Cochlear-Implanted Infants, $5,000; Rajka Smiljanic, Northwestern University--Effect of Clear Speech on Production and Perception of Croatian Rhythm, $4,000; Lori L. Holt, Carnegie Mellon University--An Investigation of the Perceptual and Learning Influences on Phonetic Category Formation, $5,000.
Contact: Emily Diaz, Project Assistant; fax (301) 296-8703; ediaz@asha.org
Internet: http://www.ashfoundation.org/Foundation/grants/research_grants.htm
Sponsor: American Speech-Language-Hearing Foundation
2200 Research Boulevard
Rockville, MD 20850-3289

ASHFoundation Student Research Grants in Audiology 710
Each year the foundation awards one grant for support of research to be initiated in the area of clinical and/or rehabilitative audiology. One $2,000 award is to a master's or doctoral student for research to be initiated in the area of audiology. Applicants must submit up to a ten-page research plan, one-page abstract, two-page management plan and budget, and letter of support. All study must be done in the United States.
Requirements: The competition open to master's or doctoral degree students enrolled in, or accepted for, graduate study in a communication sciences and disorders program in the United States. Master's degree candidates must be in a program accredited by the Council on Academic Accreditation in Audiology and Speech-Language Pathology (CAA); this is not mandatory for doctoral degree candidates. Applicants must be enrolled for full-time study for the full academic year.
Geographic Focus: All States
Amount of Grant: 2,000 USD
Samples: Tara D. Reed, University of Texas at Dallas, Mentor: James F. Jerger--Effects of Aging on Interaural Asymmetry in a Competing Speech; Faith M. Parker, Montclair State University, Mentor: Janet Koehnke--The Effects of Age and Reverberation on Cortical Audiotry Processing; TaskYu-Hsiang Wu, University of Iowa, Mentor: Ruth Bentler--Impact of Visual Cues on Microphone Mode Preference.
Contact: Emily Diaz, Project Assistant; fax (301) 296-8703; ediaz@asha.org
Internet: http://www.ashfoundation.org/Foundation/grants/research_grants.htm
Sponsor: American Speech-Language-Hearing Foundation
2200 Research Boulevard
Rockville, MD 20850-3289

ASHFoundation Student Research Grants in Early Childhood Language Development   711

One $2,000 grant per year is awarded to a graduate or postgraduate student to support research in early childhood language development. The applicant must submit up to a ten-page research plan, one-page abstract, two-page management plan and budget, and letter of support. All study must be done in the United States. The recipient and the recipient's mentor will be announced and recognized at the ASHA Convention.
Requirements: The competition is open to graduate or postgraduate students in the area of communication sciences and disorders. Applicants must submit a proposal according to foundation guidelines, to be received by the indicated deadline date for applications.
Geographic Focus: All States
Amount of Grant: 2,000 USD
Samples: Kathryn Wright Brady, University of Missouri, Mentor: Judith C. Goodman--Clues to Meaning: Exploring Potential Effects of Paired, Congruent Cues on Toddlers' Word Learning; Dawn Vogler-Elias, University of Buffalo, Mentor: Geralyn Timler--A Shared Storybook Reading Intervention for Preschoolers with Autism; Jonathan L. Preston, Syracuse University, Mentor: Mary Louise Edwards--Preliminary Investigation of a Weighted Measure of Speech Sound Accuracy.
Contact: Emily Diaz, Project Assistant; fax (301) 296-8703; ediaz@asha.org
Internet: http://www.ashfoundation.org/Foundation/grants/research_grants.htm
Sponsor: American Speech-Language-Hearing Foundation
2200 Research Boulevard
Rockville, MD 20850-3289

ASHG Public Health Genetics Fellowship   712

The objectives of this co-sponsored Fellowship are to enhance the Center for Disease Control (C.D.C.)'s current activities and develop new approaches in integrating genetics and genomics into public health policy and programs at the national, state, and local levels. A wide range of challenges and opportunities associated with the integration of genetics and genomics are emerging for health policy makers and program staff within these health agencies. Key issues of planning, workforce training, expansion of surveillance systems, health policy development, and changes in disease prevention programs are important program activities. The Fellow is expected to work independently and in conjunction with epidemiologists, geneticists, public health specialists, and communications specialists across C.D.C.. Activities may include: planning and development of current and future projects in genetics and genomics with C.D.C.; providing technical assistance to C.D.C. programs, and collaboratively with C.D.C. programs, providing assistance to state and local health agencies on the integration of genomics into policy and practice around specific diseases; collaborating with national public health organizations and professional organizations, including the American Society of Human Genetics (ASHG), in the planning and development of national public health strategies associated with genomics and disease prevention; participating in ASHG activities related to the work at C.D.C. to promote timely transfer of information about policy and program development; participating in writing reports about the impact of genomics and public health policy and practice; enhancing awareness about the importance of genomics in public health by providing content for the ASHG website and materials for the Mentor network; assisting with the development and delivery of training and educational activities related to genomics and public health. This career development opportunity will offer a broad experience in the emerging role of genomics in public health practice. The Fellow will learn strategies for and practical skills in integrating genomics into specific health policies and programs associated with disease prevention and health promotion. Benefits include access to and cooperation of genetics and public health professionals to enhance collaboration between the fields. The Fellow will work closely with C.D.C. staff and state and local health agencies to assess and develop the role of genomics in public health activities. The Fellow will gain insight and expertise through the training of, and with, public health professionals as well as development and delivery of workshops and case studies appropriate for a variety of audiences. The length of the fellowship is two years for early career candidates and one year for mid-career candidates. The Fellowship starts in July. Deadlines for application may vary from year to year. Interested applicants are encouraged to check the fellowship website and contact program staff for information on how to apply for the fellowship.
Requirements: Successful applicants must have either an M.D. or Ph.D. degree or a master's degree in genetics, genetic counseling, or a related field and experience or interest in public health genetics. Applicants must qualify as early career (3 years experience) or mid-career professionals.
Restrictions: Interested applicants are encouraged to call program staff with any questions on requirements and restrictions.
Geographic Focus: All States, All Countries
Date(s) Application is Due: Apr 23
Contact: Joann Boughman; (866) 486-4363 or (301) 634-7300; jboughman@ashg.org
Internet: http://www.ashg.org/pages/education_cdc.shtml
Sponsor: American Society of Human Genetics
9650 Rockville Pike
Bethesda, MD 20814-3998

Ashland Corporate Contributions Grants   713

The foundation awards grants to eligible Kentucky, Ohio, and West Virginia nonprofit organizations in its areas of interest, including education, arts, communities/civic, disaster relief, environment, and health and human services. Types of support include seed money grants, project grants, endowments, matching gifts, and employee volunteers. The foundation does not provide an application. Most giving is centered on programs that best address the needs of Ashland's employees, stockholders, customers, and other constituencies, and the communities in which they live. Funding requests are not solicited but will be considered. The primary focus is on education, with an emphasis on mentoring, literacy and/or diversity.
Requirements: 501(c)3 tax-exempt organizations in Boyd and Greenup Counties, KY, Lawrence County, OH, and Cabell and Wayne Counties, WV, are eligible. Charitable groups within church or religious organizations are eligible.
Restrictions: The foundation does not support individuals, capital campaigns for building or equipment, endowments, travel, film or video production, tickets, religious or political activities, or goodwill advertising.
Geographic Focus: Kentucky, Ohio, West Virginia
Contact: Program Contact; (859) 815-3333; community@ashland.com
Internet: http://www.ashland.com/commitments/contributions.asp
Sponsor: Ashland
50 East Rivercenter Boulevard, P.O. Box 391
Covington, KY 41012-0391

ASM-PAHO Infectious Diseases Epidemiology and Surveillance Fellowships   714

Funded by the Pan American Health Organization and ASM, this award offers fellowships to young scientists from Bolivia, Colombia, Costa Rica, Dominican Republic, Ecuador, El Salvador, Guatemala, Honduras, Nicaragua, Panama, Paraguay, or Peru, who have obtained their Masters, PhD, or other equivalent academic degree, and who have at least five years of laboratory experience in the area of antimicrobial resistance, to visit a host scientist in the US. The ASM-PAHO Infectious Diseases Epidemiology and Surveillance Fellowship will award a maximum grant of $4,000. The award is not intended to provide travel to obtain a degree at the host institution.
Requirements: The application must be made jointly between a host institution and the visiting fellow. The Investigator must be: a member of ASM or any other national microbiological society; actively involved in research in the microbiological sciences; have obtained, or be in the process of obtaining, their masters, PhD, or other equivalent academic degree within the last five years; a national of a resource-limited Africa, Asian or Latin American country submitting the application from an institution in their home region. Investigators from developed nations, as defined by BOTH the UN high HDI group and the World Bank high income economy group, are ineligible. Investigators submitting an application from the U.S. where they are spending a training period are ineligible; sufficiently proficient in the use of the English language. The Host Scientist must be: an ASM member; actively involved in research in the microbiological sciences in the U.S; actively involved in teaching in the microbiological sciences in the U.S; interested in sustaining international collaborations Preference will be given to: applicants who can prove three years of membership in ASM or any other national microbiological society, and; applicants who have not previously had the opportunity to travel to a facility in another country. The application should demonstrate to the Review Committee: academic excellence of the applicant - Honors & Awards, CV, and Letters of Recommendation; depth of the applicant's research experience - Statement of prior Research Experience; quality and originality of work proposed during the Fellowship - Potential Collaboration and Proposed Research Plan; relevance of the work proposed to the applicant's locale - Potential Collaboration and Proposed Research Plan; excellence of the host, and the research and training environment
Geographic Focus: All States, Argentina, Bolivia, Brazil, Chile, Colombia, Ecuador, Guyana, Paraguay, Peru
Date(s) Application is Due: Apr 15; Oct 15
Amount of Grant: 4,000 USD
Contact: Supervisor; (202) 942-9368; fax (202) 942-9328; international@asmusa.org
Internet: http://www.asm.org/International/index.asp?bid=2778
Sponsor: American Society for Microbiology
1752 N Street, N.W.
Washington, D.C. 20036-2904

ASM-UNESCO Leadership Grant for International Educators   715

This new program, sponsored jointly by ASM and UNESCO, has been developed to enable a select group of educators from resource-limited countries to attend the ASM Conference for Undergraduate Educators (ASMCUE) and a pre-conference workshop to provide leaders in education with the resources to build innovative teaching modules that engage students and lead to enduring understandings in microbiology. Successful applicants will receive financial support to cover airfare (economy class), room, board and registration to the pre-conference workshop and ASMCUE. Participants will be reimbursed for airfare via wire transfer upon submission of a confirmed invoice. Room, board and registration will be arranged by ASM.
Requirements: Preference will be given to ASM members. Applicants must be a citizen of, and residing in, a resource-limited country in Latin America, Asia, Africa, and Central and Eastern Europe. Resource-limited countries are classified as upper-middle income and below by the World Bank Development Group. (To find out if your country fits this criteria, visit http://tinyurl.com/7he3u.) Participants must have a minimum of eight years teaching in microbiology/biology at a university level and must also have regular and sustained access to the Internet.
Restrictions: Members of the ASM International Board, International Education Committee, International Membership Committee, International Laboratory Capacity Building Committee, and the ASM International Ambassador Network are not eligible for this program.
Geographic Focus: All States
Date(s) Application is Due: Dec 1
Contact: Michelle Godinez, Coordinator, Education Programs and Events; (202) 942-9317; fax (202) 942-9328; mgodinez@asmusa.org
Internet: http://www.asm.org/International/index.asp?bid=61306
Sponsor: American Society for Microbiology
1752 N Street, N.W.
Washington, D.C. 20036-2904

ASM/CDC Fellowships in Infectious Disease and Public Health Microbiology 716

The ASM/C.D.C. Postdoctoral Research Fellowship Program is sponsored by the American Society for Microbiology (ASM) and the Centers of Disease Control and Prevention (C.D.C.). Founded in 1899, the American Society for Microbiology (ASM) promotes the microbiological sciences and their application for the common good. The Society accomplishes its mission by disseminating information, stimulating research, promoting education, and advancing the profession. ASM's headquarters office is located in Washington, D.C. The Centers for Disease Control and Prevention (C.D.C.) is the agency of the United States Public Health Service (PHS) responsible for developing and applying disease control and prevention methods and improving the health of the citizens of the United States. The CD.C. has been a leader in efforts to prevent and control infectious diseases for nearly half a century. CD.C. supports surveillance, research, prevention efforts, and training in the area of infectious diseases. The goal of the two-year ASM/CD.C. Fellowships is to support the development of new approaches, methodologies and knowledge in infectious disease prevention and control in areas within the public health mission of the CD.C.. The program awards up to ten fellowships annually. The fellowships are similar to those at the postdoctoral level in universities, industry, and other government laboratories and allow bright, highly motivated, recent doctoral graduates to perform research-in-residence while headquartered at a CD.C. location. Participating CD.C. organizational units are located in: Atlanta, Georgia; Fort Collins, Colorado; and San Juan, Puerto Rico and are comprised of: the Division of Bacterial Diseases; the Division of Foodborne, Bacterial and Mycotic Diseases; the Division of Healthcare Quality Promotion; the Division of HIV/AIDS Prevention; the Division of Influenza; the Division of Viral Hepatitis; the Division of Parasitic Diseases; the Division of Tuberculosis Elimination; the Divisions of Vector-Borne Infectious Diseases; the Division of Viral and Rickettsial Diseases; the Division of Viral Diseases; and the Division of Viral Hepatitis. The fellowship appointments are for two years assuming satisfactory progress. A stipend (up to $45,243 annually), health benefits (up to $3000 annually), relocation benefits (up to $500), and professional development funds (up to $2000 annually) are provided. Fellows are classified as fellows and are not employees of either CD.C. or ASM but rather have the status of a visiting scientist. Applicants must use the ASM online application system to apply for the fellowship. The link to the online system is available at the program website on the top right corner of the screen. Applicants are encouraged to review the "Full Program Catalog" to obtain complete information about the program, research opportunities, and the application process. The link to the catalog is available at the program website. The ASM/CD.C. Postdoctoral Research Fellowship Program is sponsored by the U.S. Department of Health and Human Services.

Requirements: The program is intended for individuals who either earned their doctorate degree or completed a primary residency within three years of their proposed start date (considerations will be given to individuals with more experience if there are compelling reasons). Non-U.S. nationals whose primary language is not English must demonstrate command of the English language through the Test of English as a Foreign Language (TOEFL). These applicants may also be subject to an oral interview. Non-U.S. nationals who are offered awards must have valid visas throughout their tenure. Fellows will be responsible for managing all their own visa matters. Fellows must devote their full-time effort to the approved research project and must be in residence at the sponsoring CD.C. laboratory for the entire two-year period. Fellows must begin the program between October 1 and December 31 during the year in which they are accepted. All fellows must undergo security clearance to work at CD.C.. Support after the first year of the Program is contingent upon satisfactory progress in the proposed research. Interim and final reports and/or presentations are required. Fellows are expected to publish results of their studies in the scientific literature and to present findings at national meetings. Fellows are subject to the general regulations of the laboratory and are expected to conduct research according to the highest scientific and ethical standards and in compliance with all applicable laws, regulations, and policies regarding protection of human research subjects, humane care and use of laboratory animals, and laboratory safety. Participation in CD.C. seminar programs is also expected. After completion of the two-year program, fellows are required to keep ASM informed of their current position, affiliation, and continued professional development for a minimum of ten years. This arrangement will ensure proper assessment and longitudinal studies regarding the effectiveness of such training.

Restrictions: Applicants may not have a faculty position or be enrolled in a graduate degree program during the fellowship. Monies from other appointments, fellowships, private employment, or consulting or contract work are not permitted during the two-year appointment.
Geographic Focus: All States, All Countries
Date(s) Application is Due: Jan 15
Amount of Grant: Up to 100,986 USD
Contact: Tiffani Fonseca, Coordinator, Fellowship Programs; (202) 942-9295 or (202) 942-9283; fax (202) 942-9329; fellowships@asmusa.org or tfonseca@asmusa.org
Internet: http://www.asm.org/asm/index.php/education/asmcdc-program-in-infectious-disease-and-public-health-microbiology.html
Sponsor: American Society for Microbiology
1752 N Street, N.W.
Washington, D.C. 20036-2904

ASM Abbott Laboratories Award in Clinical and Diagnostic Immunology 717

The ASM Abbott Laboratories Award in Clinical and Diagnostic Immunology honors a distinguished scientist in clinical or diagnostic immunology for outstanding contributions to those fields.
Requirements: The nominee must demonstrate significant contributions to the understanding of the functioning of the host immune system in human disease, clinical approaches to diseases involving the immune system or development or clinical application of immunodiagnostic procedures.
Restrictions: Self-nominations will not be accepted.
Geographic Focus: All States
Date(s) Application is Due: Oct 1
Amount of Grant: 2,000 USD
Contact: Awards Committee; (202) 942-9226; fax (202) 942-9353; awards@asmusa.org
Internet: http://www.asm.org/Academy/index.asp?bid=2301
Sponsor: American Society for Microbiology
1752 N Street, N.W.
Washington, D.C. 20036-2904

ASM BD Award for Research in Clinical Microbiology 718

The ASM BD Award for Research in Clinical Microbiology honors a a distinguished scientist for research accomplishments that form the foundation for important applications in clinical microbiology.
Requirements: The nominee must be a distinguished clinical microbiologist. All nomination components must be emailed to awards@asmusa.org. Nominations must consist of the following: curriculum vitae; letter of nomination; letters of support.
Restrictions: Self-nominations will not be accepted.
Geographic Focus: All States
Date(s) Application is Due: Oct 1
Amount of Grant: 2,000 USD
Contact: Awards Committee; (202) 942-9226; fax (202) 942-9353; awards@asmusa.org
Internet: http://www.asm.org/Academy/index.asp?bid=2304
Sponsor: American Society for Microbiology
1752 N Street, N.W.
Washington, D.C. 20036-2904

ASM bioMerieux Sonnenwirth Award for Leadership in Clinical Microbiology 719

Given in memory of Alexander Sonnenwirth, Ph.D., this award recognizes a distinguished microbiologist for the promotion of innovation in clinical laboratory science, dedication to ASM, and the advancement of clinical microbiology as a profession.
Requirements: Nominations will be considered without updating for three years. email (see contact info.) all nomination components. Nominations must consist of the following: curriculum vitae, including a list of nominee's publications; letter of nomination, describe the nominee's innovation in clinical laboratory science, dedication to ASM, role as a leader in the field of diagnostic microbiology, and the advancement of clinical microbiology as a profession; letters of support, two letters of support must come from persons, other than the nominator, who are familiar with the nominee's qualifications and accomplishments. No more than one of the three letters may be from the nominee's institution or the same institution.
Restrictions: Self-nominations will not be accepted.
Geographic Focus: All States
Date(s) Application is Due: Oct 1
Amount of Grant: 2,000 USD
Contact: Awards Committee; (202) 942-9226; fax (202) 942-9353; awards@asmusa.org
Internet: http://www.asm.org/Academy/index.asp?bid=36368
Sponsor: American Society for Microbiology
1752 N Street, N.W.
Washington, D.C. 20036-2904

ASM Congressional Science Fellowships 720

The program will select a postdoctoral to mid-career microbiologist to spend one year on the staff of an individual congressman, congressional committee, or with some other appropriate organizational unit of Congress. The purpose of the program is to make practical contributions to more effective use of scientific knowledge in government, to educate the scientific communities regarding public policy, and to broaden the perspective of both the scientific and governmental communities regarding the value of such science-government interaction. The ASM Fellow will function as special legislative assistant within the congressional staff. The American Association for the Advancement of Science will arrange a carefully structured orientation program, guide the placement process, and coordinate weekly seminars throughout the year for the ASM Fellow, as well as other Congressional Fellows. The period of appointment is from September 1 and extends for one year. The award will include a $60,000 stipend plus health care benefits. The stipend is supported by the Martin Frobisher Fund.
Requirements: Prospective Fellows must be citizens of the United States, be members of ASM for at least one year and must have completed their Ph.D. by the time the fellowship begins in September. Candidates are expected to show competence in some aspect of microbiology, have a broad background in science and technology, and have interest and some experience in applying scientific knowledge toward the solution of social problems. Candidates are expected to be articulate, literate, adaptable, interested in work on a range of public policy problems, and able to work with a variety of people from diverse professional backgrounds. A Complete Application for the Fellowship must include: a letter from the candidate indicating a desire to apply and listing three references; three letters of references; a statement from the candidate about his or her qualifications and career goals.
Restrictions: Candidates selected for interviews must provide travel to and from Washington at their own expense. The candidate is responsible for soliciting the required references, providing the six guidelines for the reference response and seeing that the references are forwarded before the deadline.
Geographic Focus: All States
Date(s) Application is Due: Feb 20
Amount of Grant: 60,000 USD
Contact: Heather Garvey, Program Assistant; (202) 942-9209; fax (202) 942-9335; hgarvey@usmusa.org

Internet: http://www.asm.org/Policy/index.asp?bid=12335
Sponsor: American Society for Microbiology
1752 N Street, N.W.
Washington, D.C. 20036-2904

ASM Corporate Activities Program Student Travel Grants 721
The Corporate Activities Program Student Travel Grant is a $500 grant given a maximum of 160 students who are ASM members and who are the presenting author of a poster at the General Meeting. Students are selected based on a review of their abstract submission for the meeting. The awards are highly competitive and only about 25-30% of those seeking grants will receive them. If a grant recipient cannot present the accepted poster, the grant money must be returned to ASM.
Requirements: To qualify for consideration the student must: be an ASM Member and the presenting author; be accepted to present a poster at the General Meeting; be an undergraduate/pre-doctoral student, or; be a Ph.D. or M.D. whose degree has been or will be awarded after May of the current year and before May of the meeting year; have indicated a desire to be considered for the grant on the abstract submission, and submit a letter of nomination from a faculty member or department head on official letterhead. The letter must state that the student is expected to be active in the training or degree program at the time of the meeting. If the student will complete their terminal degree between the date of submission and the meeting, the letter must indicate the expected date of completion. The abstract tracking number must be included in the body of the letter. The letter should be submitted to ASM not later than December of each year. Exact date will be published in the Call for Abstracts for the General Meeting.
Geographic Focus: All States
Amount of Grant: 500 USD
Contact: Awards Committee; (202) 942-9226; fax (202) 942-9353; awards@asmusa.org
Internet: http://www.asm.org/Awards/index.asp?bid=14968
Sponsor: American Society for Microbiology
1752 N Street, N.W.
Washington, D.C. 20036-2904

ASMCUE/GM Travel Assistance Grants 722
ASM offers travel awards to attend the 2009 ASMCUE to ASM Members who have been accepted to present a poster at ASM General Meeting and ASMCUE.
Requirements: Eligibility: accepted to present a poster at both ASM General Meeting in any division and ASMCUE; be an active ASM member; teach an undergraduate microbiology course or participates in outreach activities related to microbiology; have an interest in pursuing a teaching career at an undergraduate institution; not be a recipient of substantial federal funding or any other ASM faculty enhancement or travel award.
Geographic Focus: All States
Date(s) Application is Due: Mar 20
Amount of Grant: 500 USD
Contact: Awards Committee; (202) 942-9226; fax (202) 942-9353; awards@asmusa.org
Internet: http://www.asm.org/Education/index.asp
Sponsor: American Society for Microbiology
1752 N Street, N.W.
Washington, D.C. 20036-2904

ASM D.C. White Research and Mentoring Award 723
The ASM D.C. White Research and Mentoring Award recognizing distinguished accomplishments in interdisciplinary research and mentoring in microbiology, this award honors D.C. White, who was known for his interdisciplinary scientific approach and for being a dedicated and inspiring mentor.
Requirements: Consideration will be given to the breadth of the nominee's contributions, as well as their originality and overall impact. There are no age restrictions, but the nominee must have a distinguished record of accomplishments in microbiological research. Nominees in all areas of microbiology will be considered. Nominations must consist of the following: curriculum vitae, including a list of publications; list of those mentored; letter of nomination, describe the nominee's accomplishments of unusual merit in interdisciplinary research in the field of microbiology; letters of support, two letters of support must come from persons, other than the nominator, who are familiar with the nominee's qualifications and accomplishments. One letter should be from someone who can comment specifically on the nominee's mentoring. No more than one of the three letters may be from the nominee's institution or the same institution.
Restrictions: Self nominations will not be accepted.
Geographic Focus: All States
Date(s) Application is Due: Oct 1
Contact: Awards Committee; (202) 942-9226; fax (202) 942-9353; awards@asmusa.org
Internet: http://www.asm.org/Academy/index.asp?bid=50931
Sponsor: American Society for Microbiology
1752 N Street, N.W.
Washington, D.C. 20036-2904

ASM Early-Career Faculty Travel Award 724
SM offers travel awards to attend the ASM Conference for Undergraduate Educators (ASMCUE) to increase the participation of early-career undergraduate faculty, post-doctoral scientists or senior level graduate students interested in teaching careers.
Requirements: Eligibility: accepted to present a poster at both ASM General Meeting in any division and ASMCUE; be an active ASM member; teach an undergraduate microbiology course or participates in outreach activities related to microbiology; have an interest in pursuing a teaching career at an undergraduate institution; not be a recipient of substantial federal funding or any other ASM faculty enhancement or travel award.
Geographic Focus: All States
Date(s) Application is Due: Mar 20
Amount of Grant: 500 USD
Contact: Awards Committee; (202) 942-9226; fax (202) 942-9353; awards@asmusa.org
Internet: http://www.asm.org/Education/index.asp
Sponsor: American Society for Microbiology
1752 N Street, N.W.
Washington, D.C. 20036-2904

ASME H.R. Lissner Award 725
The ASME H.R. Lissner Award was established in honor of H.R. Lissner for his pioneering contributions in biomechanics. The award is bestowed annually for outstanding accomplishments in the area of bioengineering. These achievements may be in the form of (1) significant research contributions in bioengineering; (2) development of new methods of measuring in bioengineering; (3) design of new equipment and instrumentation in bioengineering; (4) educational impact in the training of bioengineers; and/or (5) service to the bioengineering community, in general, and to the Bioengineering Division of ASME, in particular. The award includes a bronze medal, certificate, $1,000, and travel expenses to attend the award presentation. Nomination forms and instructions are available at the website. Nomination packets will be held and considered for three years; a new nomination packet must be submitted after that time.
Requirements: Candidate must be an active member of the bioengineering division.
Geographic Focus: All States
Date(s) Application is Due: Sep 1
Amount of Grant: 1,000 USD
Contact: T.P. Andriacchi, Committee Chair; (650) 559-9170; tandriac@stanford.edu
Internet: http://www.asme.org/about-asme/honors-awards/achievement-awards/h-r--lissner-medal
Sponsor: American Society of Mechanical Engineers
Two Park Avenue
New York, NY 10016-5990

ASM Eli Lilly and Company Research Award 726
ASM's oldest and most prestigious prize, it rewards fundamental research of unusual merit in microbiology or immunology by an individual on the threshold of his or her career.
Requirements: The nominee's work is not judged by comparison with the research of more experienced scientists, and special consideration is given to originality and independence of thought. The nominee must not have reached his or her 45th birthday by April 30 of the year the award is given. The nominee must be working in the United States or Canada and be actively engaged in the line of research for which the award is to be made.
Restrictions: Nominations must consist of the following: curriculum vitae, including a list of publications; verification of date of birth. Photocopy or scanned file of driver's license, passport or birth certificate; letter of nomination, describe the nominee's fundamental research of unusual merit in microbiology or immunology; letters of support, two letters of support must come from persons, other than the nominator, who are familiar with the nominee's qualifications and accomplishments. No more than one of the three letters may be from the nominee's institution or the same institution.
Geographic Focus: All States
Date(s) Application is Due: Oct 1
Amount of Grant: 5,000 USD
Contact: Awards Committee; (202) 942-9226; fax (202) 942-9353; awards@asmusa.org
Internet: http://www.asm.org/academy/index.asp?bid=2281
Sponsor: American Society for Microbiology
1752 N Street, N.W.
Washington, D.C. 20036-2904

ASM Faculty Ehancement Program - Travel Awards 727
ASM offers travel awards to attend the ASMCUE to faculty who teach microbiology at 2- or 4-year institutions with a large percentage of historically excluded and underrepresented students.
Requirements: Eligibility: accepted to present a poster at both ASM General Meeting in any division and ASMCUE; be an active ASM member; teach an undergraduate microbiology course or participates in outreach activities related to microbiology; have an interest in pursuing a teaching career at an undergraduate institution; not be a recipient of substantial federal funding or any other ASM faculty enhancement or travel award.
Geographic Focus: All States
Date(s) Application is Due: Mar 20
Amount of Grant: 500 USD
Contact: Awards Committee; (202) 942-9226; fax (202) 942-9353; awards@asmusa.org
Internet: http://www.asm.org/Education/index.asp
Sponsor: American Society for Microbiology
1752 N Street, N.W.
Washington, D.C. 20036-2904

ASM Founders Distinguished Service Award 728
The ASM Founders Distinguished Service Award honors a member of the ASM for outstanding contributions and commitment to the ASM as a volunteer at the national level.
Requirements: Selection is based on commitment to furthering the goals of the ASM, ability to inspire commitment from others, and significance of contributions to the membership of ASM and its various audiences. The nominee must be an ASM member in good standing who has served in a volunteer capacity for ASM at the national level (e.g., as a member of committees or editorial boards or as a workshop leader) for a minimum of five years, and has not held office as ASM President, Secretary, Treasurer, or Chair

of a CPC Board or the American Academy of Microbiology. Nominations must consist of the following: curriculum vitae; letter of nomination, describe the nominee's ASM service and contributions on which the nomination is based; letters of support, two letters of support must come from persons, other than the nominator, who are familiar with the nominee's service and contributions to the ASM. No more than one of the three letters may be from the nominee's institution or the same institution.
Geographic Focus: All States
Date(s) Application is Due: Oct 1
Samples: Self-nominations will not be accepted.
Contact: Awards Committee; (202) 942-9226; fax (202) 942-9353; awards@asmusa.org
Internet: http://www.asm.org/Academy/index.asp?bid=2324
Sponsor: American Society for Microbiology
1752 N Street, N.W.
Washington, D.C. 20036-2904

ASM Gen-Probe Joseph Public Health Award — 729
The ASM Gen-Probe Joseph Public Health Award honors a distinguished microbiologist who has exhibited exemplary leadership and service in the field of public health. This award has been established in memory of J. Mehsen Joseph, Ph.D., who dedicated his life toward the advancement of both microbiology and public health.
Requirements: The nominee must be a microbiologist identified with public health. The recipient will be recognized for significant achievements in integrating the science of microbiology into the practice of public health and for promoting the importance of linking these two disciplines. Nominations must consist of the following: curriculum vitae, including a list of nominee's publications; letter of nomination, describe the nominee's leadership and service in the field of public health; letters of support, two letters of support must come from persons, other than the nominator, who are familiar with the nominee's qualifications and accomplishments, citing in particular the nominee's leadership and service in the field of public health. No more than one of the three letters may be from the nominee's institution or from the same institution.
Geographic Focus: All States
Date(s) Application is Due: Oct 1
Amount of Grant: 2,000 USD
Contact: Awards Committee; (202) 942-9226; fax (202) 942-9353; awards@asmusa.org
Internet: http://www.asm.org/Academy/index.asp?bid=36040
Sponsor: American Society for Microbiology
1752 N Street, N.W.
Washington, D.C. 20036-2904

ASM General Meeting Minority Travel Grants — 730
The American Society for Microbiology (ASM) will offer travel grants to increase the participation of underrepresented minority (URM) groups in the ASM General Meeting. The ASM will select post-doctoral scholars from URM groups in the microbiological sciences or faculty from Minority Serving Institutions. Each grantee will be offered up to $1500 to defray expenses associated with travel to the ASM General Meeting.
Requirements: Eligibility: ASM Member; U.S. citizen or permanent resident; faculty member from a Minority Serving Institution (MSI), Historically Black Colleges and Universities (HBCU), Hispanic Serving institutions (HSI), Tribal Colleges and Universities (TCU) and other institutions with substantial enrollments of minority students are considered to be Minority Serving Institutions (MSI); post-doctoral Scholars from an URM Group
Geographic Focus: All States
Date(s) Application is Due: Jan 30
Amount of Grant: Up to 1,500 USD
Contact: Awards Committee; (202) 942-9226; fax (202) 942-9353; awards@asmusa.org
Internet: http://www.asm.org/Awards/index.asp?bid=38173
Sponsor: American Society for Microbiology
1752 N Street, N.W.
Washington, D.C. 20036-2904

ASM GlaxoSmithKline International Member of the Year Award — 731
The ASM GlaxoSmithKline International Member of the Year Award honors a distinguished microbiologist who exhibited exemplary leadership in the international microbiological community. It recognizes an international ASM member for education, communication, research, and advancement of the profession to the international microbiology community while demonstrating a commitment to the ASM.
Requirements: The nominee can be any international member of ASM who has made major contributions toward the advancement of the microbiological sciences within the international community through education, research, communication, and leadership. The nominee must not have served on the International Board, the International Education Committee or the International Membership Committee within the past two years. Nominations must consist of the following: curriculum vitae, including a list of publications; letter of nomination, describe the nominee's contributions towards the advancement of the microbiological sciences within the international community through education, research, communication, and leadership. Emphasis should be on the nominee's creativity and originality, as well as the nominee's role in ASM-sponsored activities; letters of support, two letters of support must come from persons, other than the nominator, who are familiar with the nominee's contributions and accomplishments. No more than one of the three letters may be from the nominee's institution or the same institution.
Restrictions: Self-nominations will not be accepted.
Geographic Focus: All States
Date(s) Application is Due: Oct 1
Contact: Awards Committee; (202) 942-9226; fax (202) 942-9353; awards@asmusa.org
Internet: http://www.asm.org/Academy/index.asp?bid=36033
Sponsor: American Society for Microbiology
1752 N Street, N.W.
Washington, D.C. 20036-2904

ASM ICAAC Young Investigator Awards — 732
The ASM ICAAC Young Investigator Awards recognize and reward early career scientists for research excellence and potential in microbiology and infectious disease.
Requirements: Nominees must be no more than three years beyond completion of postdoctoral research training in microbiology or infectious diseases at the time of the nomination deadline. Nominations must consist of the following: curriculum vitae, including a list of publications; letter of nomination, describe the nominee's research excellence and potential in microbiology and infectious diseases; letters of support, two letters of support must come from persons, other than the nominator, who are familiar with the nominee's research excellence and potential. No more than one of the three letters may be from the nominee's institution or the same institution.
Restrictions: Self-nominations will not be accepted.
Geographic Focus: All States
Date(s) Application is Due: Apr 1
Amount of Grant: 2,500 USD
Contact: Awards Committee; (202) 942-9226; fax (202) 942-9353; awards@asmusa.org
Internet: http://www.asm.org/academy/index.asp?bid=2278
Sponsor: American Society for Microbiology
1752 N Street, N.W.
Washington, D.C. 20036-2904

ASM Intel Awards — 733
ASM sponsors a special prize in microbiology at the Intel International Science and Engineering Fair. The fair recognizes excellence in science and technology by high school students. Ten cash prizes are given and all laureates receive a one-year student membership in ASM and access to members-only web resources. The first place laureate receives a cash prize of $2,000, the second place laureate receives a cash prize of $1,250, the thrid place laureate receives a cash prize of $750, and the fourth place laureate receives a cash prize of $500. Six honorable mentions of $250 are also awarded.
Geographic Focus: All States
Amount of Grant: 250 - 2,000 USD
Contact: Awards Committee; (202) 942-9226; fax (202) 942-9353; awards@asmusa.org
Internet: http://www.asm.org/Academy/index.asp?bid=36845
Sponsor: American Society for Microbiology
1752 N Street, N.W.
Washington, D.C. 20036-2904

ASM International Fellowship for Latin America and the Caribbean — 734
Funded by ASM this award offers fellowships to promising young investigators in South and Central America and the Caribbean who are within five years of obtaining, or are in the process of obtaining, their Masters, PhD, or other equivalent academic degree, and who are working in any of the microbiology disciplines, to visit a host scientist in the US. The International Fellowship Program provides funding to meet the costs of a visit to one institution for between 6 weeks to 6 months. The International Fellowship Program provides funding to meet the costs of a visit to one institution for between 6 weeks to 6 months.
Requirements: The application must be made jointly between a host institution and the visiting fellow. The Investigator must be:a member of ASM or any other national microbiological society; actively involved in research in the microbiological sciences; have obtained, or be in the process of obtaining, their masters, PhD, or other equivalent academic degree within the last five years; a national of a resource-limited Africa, Asian or Latin American country submitting the application from an institution in their home region. Investigators from developed nations, as defined by BOTH the UN high HDI group and the World Bank high income economy group, are ineligible. Investigators submitting an application from the U.S. where they are spending a training period are ineligible; sufficiently proficient in the use of the English language. The Host Scientist must be: an ASM member; actively involved in research in the microbiological sciences in the U.S; actively involved in teaching in the microbiological sciences in the U.S; interested in sustaining international collaborations Preference will be given to: applicants who can prove three years of membership in ASM or any other national microbiological society, and; applicants who have not previously had the opportunity to travel to a facility in another country. The application should demonstrate to the Review Committee: academic excellence of the applicant - Honors & Awards, CV, and Letters of Recommendation; depth of the applicant's research experience - Statement of prior Research Experience; quality and originality of work proposed during the Fellowship - Potential Collaboration and Proposed Research Plan; relevance of the work proposed to the applicant's locale - Potential Collaboration and Proposed Research Plan; excellence of the host, and the research and training environment
Geographic Focus: All States, Antigua & Barbuda, Argentina, Bahamas, Barbados, Belize, Bolivia, Brazil, Chile, Colombia, Costa Rica, Cuba, Ecuador, Guyana, Paraguay, Peru
Date(s) Application is Due: Apr 15; Oct 15
Amount of Grant: 4,000 USD
Contact: Supervisor; (202) 942-9368; fax (202) 942-9328; international@asmusa.org
Internet: http://www.asm.org/International/index.asp?bid=2778
Sponsor: American Society for Microbiology
1752 N Street, N.W.
Washington, D.C. 20036-2904

ASM International Fellowships for Asia and Africa 735
Funded by ASM this award offers a maximum of one fellowship to a promising young investigator in Asia, as well as to one young investigator in Africa who is within five years of obtaining, or in the process of obtaining, his/her Masters, PhD, or other equivalent academic degree, and who is working in any of the microbiology disciplines, to visit a host scientist in the US. Both International Fellowships will award a maximum grant of $5,500. The award is not intended to provide travel to obtain a degree at the host institution.
Requirements: The application must be made jointly between a host institution and the visiting fellow. The Investigator must be: a member of ASM or any other national microbiological society; actively involved in research in the microbiological sciences; have obtained, or be in the process of obtaining, their masters, PhD, or other equivalent academic degree within the last five years; a national of a resource-limited Africa, Asian or Latin American country submitting the application from an institution in their home region. Investigators from developed nations, as defined by BOTH the UN high HDI group and the World Bank high income economy group, are ineligible. Investigators submitting an application from the U.S. where they are spending a training period are ineligible; sufficiently proficient in the use of the English language. The Host Scientist must be: an ASM member; actively involved in research in the microbiological sciences in the U.S; actively involved in teaching in the microbiological sciences in the U.S; interested in sustaining international collaborations Preference will be given to: applicants who can prove three years of membership in ASM or any other national microbiological society, and; applicants who have not previously had the opportunity to travel to a facility in another country.
Geographic Focus: All States
Date(s) Application is Due: Apr 15; Oct 15
Amount of Grant: 5,500 USD
Contact: Supervisor; (202) 942-9368; fax (202) 942-9328; international@asmusa.org
Internet: http://www.asm.org/International/index.asp?bid=2778
Sponsor: American Society for Microbiology
1752 N Street, N.W.
Washington, D.C. 20036-2904

ASM International Professorship for Latin America 736
The program provides funding support to an ASM member from the U.S. who is scientifically recognized in his/her area to teach a hands-on, highly interactive short course on a single topic in the microbiological sciences at a Latin American institution of higher learning. Up to five Teaching Professorships offered per program year.
Requirements: The application must be made jointly between a hosting institution and the visiting professor. Eligibility for Host Institution: institution with graduate students enrolled in a masters, doctoral or equivalent program, post-doctoral fellows or residents, and teaching faculty; institution where at least 12 students will be enrolled full-time in the short course; commitment to maximizing use of the course, as demonstrated by the applicability of the course's contents to existing programs at the institution or to the field of research of the Research Professor. demonstrated commitment to international collaborations and partnerships; demonstrated commitment to advancing the microbiological sciences through interactions with professional organizations; institutions in developed nations, as defined by the UN high HDI group, are ineligible. Eligibility for Visiting Teaching Professor: scientifically recognized their area of microbiological expertise; actively engaged in teaching at the post-secondary level; commitment to international collaborations and partnerships; must be an ASM member during the term of the Professorship. Preference will be given to: applicants who will be able to make maximal use of the course, as demonstrated either by the applicability of the course's contents to existing programs at the host institution or to the field of research of the visiting Professor; applications that demonstrate a commitment to building an ongoing institutional relationship. It is strongly recommended that the course include a hands-on component, such as a wet lab or other practical activity.
Geographic Focus: All States
Date(s) Application is Due: Apr 15; Oct 15
Amount of Grant: 4,000 USD
Contact: Supervisor; (202) 942-9368; fax (202) 942-9328; international@asmusa.org
Internet: http://www.asm.org/International/index.asp?bid=57785
Sponsor: American Society for Microbiology
1752 N Street, N.W.
Washington, D.C. 20036-2904

ASM International Professorships for Asia and Africa 737
Each program provides funding support to an ASM member from the U.S. who is scientifically recognized in his/her area to teach a hands-on, highly interactive short course on a single topic in the microbiological sciences at an institution of higher learning in Asia and in Africa. The Teaching Professorship for both programs is limited to one per program year.
Requirements: The application must be made jointly between a hosting institution and the visiting professor. Eligibility for Host Institution: institution with graduate students enrolled in a masters, doctoral or equivalent program, post-doctoral fellows or residents, and teaching faculty; institution where at least 12 students will be enrolled full-time in the short course; commitment to maximizing use of the course, as demonstrated by the applicability of the course's contents to existing programs at the institution or to the field of research of the Research Professor. demonstrated commitment to international collaborations and partnerships; demonstrated commitment to advancing the microbiological sciences through interactions with professional organizations; institutions in developed nations, as defined by the UN high HDI group, are ineligible. Eligibility for Visiting Teaching Professor: scientifically recognized their area of microbiological expertise; actively engaged in teaching at the post-secondary level; commitment to international collaborations and partnerships; must be an ASM member during the term of the Professorship. Preference will be given to: applicants who will be able to make maximal use of the course, as demonstrated either by the applicability of the course's contents to existing programs at the host institution or to the field of research of the visiting Professor; applications that demonstrate a commitment to building an ongoing institutional relationship. It is strongly recommended that the course include a hands-on component, such as a wet lab or other practical activity.
Geographic Focus: All States
Date(s) Application is Due: Apr 15; Oct 15
Amount of Grant: 5,500 USD
Contact: Supervisor; (202) 942-9368; fax (202) 942-9328; international@asmusa.org
Internet: http://www.asm.org/International/index.asp?bid=57785
Sponsor: American Society for Microbiology
1752 N Street, N.W.
Washington, D.C. 20036-2904

ASM Maurice Hilleman/Merck Award 738
The award is presented in memory of Maurice R. Hilleman, whose work in the development of vaccines has saved the lives of many throughout the world.
Requirements: The nominee must have made outstanding achievements in pathogenesis, vaccine discovery, vaccine development, and/or control of vaccine-preventable diseases. Nominations must consist of the following: curriculum vitae, including a full list of publications; letter of nomination, describe the nominee's remarkable contributions to pathogenesis, vaccine discovery, vaccine development, and/or the control of vaccine-preventable diseases, including a list of the ten publications most relevant to the award; letters of support, two letters of support must come from persons, other than the nominator, who are familiar with the nominee's achievements. No more than one of the three letters may be from the nominee's institution or the same institution.
Restrictions: Self-nominations will not be accepted.
Geographic Focus: All States
Date(s) Application is Due: Oct 1
Amount of Grant: 20,000 USD
Contact: Awards Committee; (202) 942-9226; fax (202) 942-9353; awards@asmusa.org
Internet: http://www.asm.org/Academy/index.asp?bid=57948
Sponsor: American Society for Microbiology
1752 N Street, N.W.
Washington, D.C. 20036-2904

ASM Merck Irving S. Sigal Memorial Awards 739
The ASM Merck Irving S. Sigal Memorial Awards recognize and award excellence in basic research in medical microbiology and infectious diseases. The awards are presented in memory of Irving S. Sigal, who was instrumental in the early discovery of therapies to treat HIV/AIDS.
Requirements: The nominee must be no more than five years beyond completion of postdoctoral research training in microbiology or infectious diseases at the time of the nomination deadline. Nominations must consist of the following: curriculum vitae, including a list of publications; letter of nomination, describe the nominee's excellence in basic research in medical microbiology and infectious diseases; letters of support, two letters of support must come from persons, other than the nomination, who are familiar with the nominee's qualifications and accomplishments. No more than one of the three letters may be from the nominee's institution or the same institution.
Geographic Focus: All States
Date(s) Application is Due: Oct 1
Amount of Grant: 2,500 USD
Contact: Awards Committee; (202) 942-9226; fax (202) 942-9353; awards@asmusa.org
Internet: http://www.asm.org/academy/index.asp?bid=15056
Sponsor: American Society for Microbiology
1752 N Street, N.W.
Washington, D.C. 20036-2904

ASM Microbiology Undergraduate Research Fellowship 740
The goal of the Microbiology Undergraduate Research Fellowship (MURF) program is to increase the number of underrepresented undergraduate students who wish to, and have demonstrated the ability to pursue graduate careers (Ph.D. or M.D./Ph.D.) in microbiology. Students will have the opportunity to conduct full time summer research with an ASM member at a host institution, and present research results at the Annual Biomedical Research Conference for Minority Students and the ASM General Meeting. Students will: agree to participate in an undergraduate summer research program at a U.S. based institution; conduct a research project for a minimum of 10 weeks; work with a faculty mentor who is an ASM member; submit a research abstract to the Annual Biomedical Research Conference for Minority Students (ABRCMS); submit a research abstract to ASM for presentation at the ASM General Meeting. Students decide the institution, research area, and level of activity for the summer. Based on interests, independence, and ability, students will be placed at a host U.S. Institution of the student's choice to conduct basic science research. From a list provided on the application, students interested in conducting research at a host institution will select three institutions where they would like to conduct their summer research. Every effort will be made to place fellows at their first choice.
Requirements: Eligible student candidates for the fellowship must be from groups that have been determined by the applicant's institution to be underrepresented in the microbiological sciences. The ASM encourages institutions to identify individuals that have been historically underrepresented, and remain underrepresented today in the microbiological sciences nationally. These groups include African-Americans, Hispanics, Native Americans, Alaskan Natives, and Pacific Islanders. In addition, applicants must

also: be U.S. citizen or permanent U.S. resident; be enrolled as full-time matriculating undergraduate students during the academic year at an accredited U.S. institution; be either freshmen with college level research experience, sophomores, juniors, or seniors who will not graduate before the completion date of the summer program; be members of an underrepresented group in microbiology; have taken introductory courses in biology, chemistry, and preferably microbiology prior to submission of the application; have strong interests in obtaining a Ph.D., or M.D./Ph.D. in the microbiological sciences, and have lab research experience.
Restrictions: Travel funds are contingent upon acceptance of an abstract for the General Meeting.
Geographic Focus: All States
Date(s) Application is Due: Feb 1
Amount of Grant: 5,850 USD
Contact: Fellowship Committee; (202) 942-9283; fax (202) 942-9329; fellowships-careerinformation@asmusa.org
Internet: http://www.asm.org/Education/index.asp?bid=4322
Sponsor: American Society for Microbiology
1752 N Street, N.W.
Washington, D.C. 20036-2904

ASM Millis-Colwell Postgraduate Travel Grant 741
This new grant, negotiated between the American Society for Microbiology (ASM USA) and the Australian Society for Microbiology (ASM Australia), allows for the exchange of one student member from each society to present an abstract at the annual General Meeting of the other society and to spend a week at nearby research laboratory. The grant aims to create a long lasting bond between ASM USA and ASM Australia and is designed to benefit PhD students in both countries by giving them the opportunity to travel overseas to present their work and experience the best of microbiology in the partner country. ASM USA will submit the successful applicant's abstract for an oral or poster presentation to ASM Australia for their Annual Scientific Meeting & Exhibition. The applicant will also have negotiated an agreement to visit the research laboratory of an Australian scientist in the period either immediately before or immediately after the ASM Australia annual meeting. The award will be conditional upon ASM Australia's acceptance of the submitted abstract.
Requirements: The successful applicant will be: a U.S. citizen or permanent resident of the U.S; a student member of the American Society for Microbiology; and enrolled in a PhD program at an American University. The award will cover roundtrip economy airfare between the U.S. and Australia, and any necessary internal flights; per diem living allowance for up to 14 days; and, registration fee for the Australian Society for Microbiology's Annual Scientific Meeting & Exhibition.
Geographic Focus: All States
Date(s) Application is Due: Jan 1
Amount of Grant: Up to 5,000 USD
Contact: Program Coordinator; (202) 942-9282; fax (202) 942-9329
Internet: http://www.asm.org/International/index.asp?bid=53729
Sponsor: American Society for Microbiology
1752 N Street, N.W.
Washington, D.C. 20036-2904

ASM Procter & Gamble Award in Applied and Environmental Microbiology 742
The ASM Procter & Gamble Award in Applied and Environmental Microbiology recognizes distinguished achievement in research and development in applied (non-clinical) and environmental microbiology.
Requirements: The nominee must demonstrate outstanding accomplishment in research or development in an appropriate field and be actively engaged in research or development at the time that the award is presented. Nominations must consist of the following: curriculum vitae, including a list of publications; letter of nomination, describe the nominee's outstanding accomplishments in research or development in the appropriate field, including a list of the ten most relevant publications to the award; letters of support, two letters of support must come from persons, other than the nominator, who are familiar with the nominee's outstanding accomplishments. No more than one of the three letters may be from the nominee's institution or the same institution.
Restrictions: Self-nominations will not be accepted.
Geographic Focus: All States
Date(s) Application is Due: Oct 1
Amount of Grant: 2,000 USD
Contact: Awards Committee; (202) 942-9226; fax (202) 942-9353; awards@asmusa.org
Internet: http://www.asm.org/academy/index.asp?bid=2288
Sponsor: American Society for Microbiology
1752 N Street, N.W.
Washington, D.C. 20036-2904

ASM Promega Biotechnology Research Award 743
The ASM Promega Biotechnology Research Award honors outstanding contributions to the application of biotechnology through fundamental microbiological research and development.
Requirements: A nomination can be a single exceptionally significant achievement or the aggregate of a number of exemplary achievements. Nominations must consist of the following: curriculum vitae, including a list of publications; letter of nomination, describe the nominee's outstanding accomplishments in microbiological research and development, including a list of the ten most relevant publications to the award; letters of support, two letters of support must come from persons, other than the nominator, who are familiar with the nominee's accomplishments. No more than one of the three letters may be from the nominee's institution or the same institution.
Restrictions: Neither self-nominations nor co-nominations will be accepted.
Geographic Focus: All States
Date(s) Application is Due: Oct 1
Amount of Grant: 5,000 USD
Contact: Awards Committee; (202) 942-9226; fax (202) 942-9353; awards@asmusa.org
Internet: http://www.asm.org/academy/index.asp?bid=2291
Sponsor: American Society for Microbiology
1752 N Street, N.W.
Washington, D.C. 20036-2904

ASM Public Communications Award 744
The ASM Public Communications Award, sponsored by ASM, recognizes outstanding achievement in increasing public awareness, knowledge and understanding of microbiology. Microbiology is concerned with issues such as the environment, prevention and treatment of infectious diseases, laboratory and diagnostic medicine, and food and water safety, among others.
Requirements: Eligibility is limited to primary authors of information concerning the microbiological sciences appearing in mass media, including print (newspapers and periodicals only) and broadcast media readily available to the general public. Individual items and series are both eligible. Applicants may nominate themselves. International entries are accepted but must be translated into English prior to submission.
Restrictions: Books, websites, and institutionally sponsored publications are not eligible.
Geographic Focus: All States
Date(s) Application is Due: Jan 31
Amount of Grant: 2,500 USD
Contact: Garth Hogan; (202) 942-9226; fax (202) 942-9353; ghogan@asmusa.org
Internet: http://www.asm.org/Media/index.asp?bid=2676
Sponsor: American Society for Microbiology
1752 N Street, N.W.
Washington, D.C. 20036-2904

ASM Raymond W. Sarber Awards 745
The ASM Raymond W. Sarber Awards recognize students at the undergraduate and predoctoral levels for research excellence and potential. The awards honor Raymond W. Sarber for his contributions to the growth and advancement of ASM.
Requirements: Nominees must be a student at the undergraduate or predoctoral level, attending an accredited institution in the United States, in an academic program involving microbiology. Two awards are presented at the ASM General Meeting, one to an undergraduate and one to a graduate student. They will be judged separately.
Restrictions: Nominations must consist of the following: curriculum vitae; statement, no longer than two pages, from the student outlining academic achievements, research accomplishments, and future career goals; letter of nomination, describe the nominee's research excellence and potential; letters of support, two letters of support: one from a supervisor or mentor, and the other letter may be from an academic advisor or colleague.
Geographic Focus: All States
Date(s) Application is Due: Oct 1
Amount of Grant: 1,500 USD
Samples: Self-nominations will not be accepted.
Contact: Awards Committee; (202) 942-9226; fax (202) 942-9353; awards@asmusa.org
Internet: http://www.asm.org/academy/index.asp?bid=2294
Sponsor: American Society for Microbiology
1752 N Street, N.W.
Washington, D.C. 20036-2904

ASM Richard and Mary Finkelstein Travel Grant 746
The Richard and Mary Finkelstein Travel Grant is a $500 grant given to 6 students who are ASM members and who are the presenting author of a poster at the General Meeting. Selection criteria for this award are the same as the Corporate Activities Program Student Travel Grant but the research presented must be in the area of microbial pathogenesis. Potential nominees must first be selected to receive a Corporate Activities Program Student Travel Grant and then the abstracts are reviewed again to establish the quality of the research in the additional area of microbial pathogenesis.
Requirements: To qualify for consideration the student must: be presenting research in the area of microbial pathogenesisbe; be an ASM Member and the presenting author; be accepted to present a poster at the General Meeting; be an undergraduate/pre-doctoral student, or; be a Ph.D. or M.D. whose degree has been or will be awarded after May of the current year and before May of the meeting year; have indicated a desire to be considered for the grant on the abstract submission, and submit a letter of nomination from a faculty member or department head on official letterhead. The letter must state that the student is expected to be active in the training or degree program at the time of the meeting. If the student will complete their terminal degree between the date of submission and the meeting, the letter must indicate the expected date of completion. The abstract tracking number must be included in the body of the letter. The letter should be submitted to ASM not later than December of each year. Exact date will be published in the Call for Abstracts for the General Meeting.
Geographic Focus: All States
Amount of Grant: 500 USD
Contact: Awards Committee; (202) 942-9226; fax (202) 942-9353; awards@asmusa.org
Internet: http://www.asm.org/Awards/index.asp?bid=14968h
Sponsor: American Society for Microbiology
1752 N Street, N.W.
Washington, D.C. 20036-2904

ASM Robert D. Watkins Graduate Research Fellowship 747

The goal of the fellowship is to increase the number of underrepresented groups completing doctoral degrees in the microbiological sciences. The ASM Robert D. Watkins Graduate Research Fellowship is aimed at highly competitive graduate students who are enrolled in a Ph.D. program and who have completed their graduate course work in the microbiological sciences. The fellowship encourages students to continue and complete their research project in the microbiological sciences. Students will be required to submit an abstract each year to ASM for presentation at the annual ASM General Meeting and attend the ASM Kadner Institute one time during the three-year tenure of the fellowship.
Requirements: Eligible candidates must be from groups that have been determined by the applicant's institution to be underrepresented in the microbiological sciences. The ASM encourages institutions to identify individuals that have been historically underrepresented, and remain underrepresented today in the microbiological sciences nationally. These groups include African-Americans, Hispanics, Native Americans, Alaskan Natives, and Pacific Islanders. In addition, applicants must: be formally admitted to a doctoral program in the microbiological sciences in an accredited U.S. institution; have successfully completed the first year of the graduate program (first year graduate students cannot apply); have successfully completed all graduate coursework requirements for the doctoral degree by the date of activation of the fellowship; be a student member of ASM; be mentored by an ASM member; be a U.S. citizen or a permanent resident; not have funding or have funding that will expire by the start date of the fellowship(fellowship cannot run concurrently with other national fellowships from NIH, NSF, HHMI, etc.); three letters of recommendations must be submitted with your application. One letter must be from your research advisor/mentor; Official transcripts from all colleges and universities attended.
Geographic Focus: All States
Date(s) Application is Due: May 1
Amount of Grant: 63,000 USD
Contact: Education Board; (202) 942-9283; fax (202) 942-9329; fellowships-careerinformation@asmusa.org
Internet: http://www.asm.org/Education/index.asp?bid=6278
Sponsor: American Society for Microbiology
1752 N Street, N.W.
Washington, D.C. 20036-2904

ASM Roche Diagnostics Alice C. Evans Award 748

The ASM Roche Diagnostics Alice C. Evans Award recognizes contributions toward the full participation and advancement of women in microbiology. This award was established by the ASM's Committee on the Status of Women in Microbiology, and is given in memory of Alice C. Evans, the first woman to be elected ASM President in 1928.
Requirements: The nominee can be any member of ASM who has made major contributions toward fostering the inclusion, development, and advancement of women in careers in microbiology. The nominee must demonstrate commitment to women in science through mentorship and advocacy and by setting an example through scientific and professional achievement. Nominations must consist of the following: curriculum vitae, including a list of nominee's publications; letter of Nomination, describe the nominee's major contributions toward fostering the inclusion, development, and advancement of women in careers in microbiology; letters of support, two letters of support must come from persons, other than the nominator, who are familiar with the nominee's qualifications and accomplishments. No more than one of the three letters may be from the nominee's institution or from the same institution.
Restrictions: Self-nominations will not be accepted.
Geographic Focus: All States
Date(s) Application is Due: Oct 1
Contact: Awards Committee; (202) 942-9226; fax (202) 942-9353; awards@asmusa.org
Internet: http://www.asm.org/Academy/index.asp?bid=2332
Sponsor: American Society for Microbiology
1752 N Street, N.W.
Washington, D.C. 20036-2904

ASM sanofi-aventis ICAAC Award 749

ASM's premier award in antimicrobial chemotherapy research; it stimulates research and honors outstanding accomplishment in antimicrobial chemotherapy.
Requirements: The nominee must be actively engaged in research involving development of new agents, investigation of antimicrobial action or resistance to antimicrobial agents, and/or the pharmacology, toxicology or clinical use of those agents. The nominee must not have served on the ICAAC Program Committee within the past two years. Nominations must consist of the following: curriculum vitae, including a list of publications; letter of nomination, describe the nominee's outstanding accomplishment in antimicrobial chemotherapy, including a list of the ten most relevant publications to the award; letters of support, two letters of support should come from two people, other than the nominator, who are familiar with the nominee's accomplishments. No more than one of the three letters may be from the nominee's institution or the same institution.
Restrictions: Self-nominations will not be accepted.
Geographic Focus: All States
Date(s) Application is Due: Apr 1
Amount of Grant: 20,000 USD
Contact: Awards Committee; (202) 942-9226; fax (202) 942-9353; awards@asmusa.org
Internet: http://www.asm.org/academy/index.asp?bid=2268
Sponsor: American Society for Microbiology
1752 N Street, N.W.
Washington, D.C. 20036-2904

ASM Scherago-Rubin Award 750

The ASM Scherago-Rubin Award recognizes an outstanding, bench-level clinical microbiologist. The award was established by the late Sally Jo Rubin, an active member of ASM's Clinical Microbiology Division, in honor of her grandfather, Professor Morris Scherago.
Requirements: The nominee must be a non-doctoral-level clinical microbiologist involved primarily in routine diagnostic work, rather than in research, who has distinguished himself or herself with excellent performance in the clinical laboratory. Nominations must consist of the following: curriculum vitae, including a list of the nominee's publications; letter of nomination, a letter from the supervisor describing how the nominee's routine diagnostic work, rather than research, has distinguished himself or herself in the clinical laboratory; letters of support, two letters of support preferably from individuals from different institutions who are familiar with the nominee's laboratory accomplishments.
Restrictions: Self-nominations will not be accepted.
Geographic Focus: All States
Date(s) Application is Due: Oct 1
Amount of Grant: 1,500 USD
Contact: Awards Committee; (202) 942-9226; fax (202) 942-9353; awards@asmusa.org
Internet: http://www.asm.org/academy/index.asp?bid=2335
Sponsor: American Society for Microbiology
1752 N Street, N.W.
Washington, D.C. 20036-2904

ASM Siemens Healthcare Diagnostics Young Investigator Award 751

The ASM Siemens Healthcare Diagnostics Young Investigator Award honors outstanding laboratory research in clinical microbiology or antimicrobial agents and is intended to further the career development of a young clinical scientist and promote awareness of clinical microbiology as a career.
Requirements: The nominee should be conducting outstanding research in clinical microbiology, automation in clinical laboratories, development of novel antimicrobial agents, mechanisms of action of antimicrobial agents or mechanisms of resistance to antimicrobial agents. The nominee must be no more than five years beyond completion of postdoctoral research training in microbiology, infectious diseases or related disciplines at the time of the nomination deadline. Nominations must consist of the following: curriculum vitae, including a list of publications, abstracts, and manuscripts in preparation; letter of nomination, describe the nominee's outstanding clinical research and potential; letters of support, two letters of support must come from persons, other than the nominator, who are familiar with the nominee's qualifications and accomplishments. No more than one of the three letters may be from the nominee's institution or from the same institution.
Restrictions: Self-nominations will not be accepted.
Geographic Focus: All States
Date(s) Application is Due: Oct 1
Amount of Grant: 2,000 USD
Contact: Awards Committee; (202) 942-9226; fax (202) 942-9353; awards@asmusa.org
Internet: http://www.asm.org/Academy/index.asp?bid=2307
Sponsor: American Society for Microbiology
1752 N Street, N.W.
Washington, D.C. 20036-2904

ASM Student and Post Doctoral Fellow Travel Grants 752

The ICAAC Program Committee will award travel grants worth at least $500, to support students' attendance at the 43rd ICAAC. Travel grants are selected on the basis of the scientific quality of the submitted abstracts. Preference is given to students who have not received a travel grant in the past two years, and the Committee strives to award only one travel grant per laboratory. They are sponsored by the ASM Corporate Partners.
Requirements: Criteria: ASM membership at time of application; undergraduate or pre-doctoral student; Ph.D. or M.D. within two years of graduation at the time of the meeting; or an M.D. in a residency or fellowship program within six years of receipt of the M.D. degree at the time of the meeting; cover letter from the applicant, including their contact information, accompanied by letter of nomination from a faculty member on department letterhead, stating that the applicant is expected to be active in the training or degree program at the time of the meeting. If the applicant will have attained their terminal degree at the time of the meeting, the date of graduation must be included in this letter; submission of an abstract for the ICAAC, which is accepted for presentation; letter of nomination, copy of the abstract, and submission control number.
Geographic Focus: All States
Date(s) Application is Due: May 30
Amount of Grant: 500 USD
Contact: Awards Committee; (202) 942-9226; fax (202) 942-9353; awards@asmusa.org
Internet: http://www.asm.org/Meetings/index.asp?bid=15264
Sponsor: American Society for Microbiology
1752 N Street, N.W.
Washington, D.C. 20036-2904

ASM Student Travel Grants 753

ASM provides funds for ttravel grants of $500 each to predoctoral students.
Requirements: To qualify, applicants must be: a member of ASM (student membership in ASM is only $15); a presenter on an accepted abstract; and must be at the predoctoral level at the time of abstract submission. Applicants submitting an abstract via the ASM web site should check the box on the screen when filling out the electronic submission form where indicated. In addition, selected recipients must return a confirmation form sent with notice of award selection, including proof of predoctoral status, either in the form of a copy of the validated student identification card, or a letter on departmental letterhead from their departmental chair which verifies their student status.

Geographic Focus: All States
Amount of Grant: 500 USD
Contact: Awards Committee; (202) 942-9226; fax (202) 942-9353; awards@asmusa.org
Internet: http://www.asm.org/Meetings/index.asp?bid=15264
Sponsor: American Society for Microbiology
1752 N Street, N.W.
Washington, D.C. 20036-2904

ASM TREK Diagnostic ABMM/ABMLI Professional Recognition Award 754
The ASM TREK Diagnostic ABMM/ABMLI Professional Recognition Award recognizes a Diplomate of the American Board of Medical Microbiology (ABMM) or the American Board of Medical Laboratory Immunology (ABMLI) for outstanding contributions to the professional recognition of certified microbiologists and/or immunologists and the work they do. Primary consideration will be given to ABMM or ABMLI Diplomates who have made significant contributions to the advancement and public recognition of the profession over and above scientific achievements or board-related activities.
Requirements: Nominations must consist of the following: curriculum vitae, including a list of nominee's publications; letter of nomination, describe the nominee's outstanding contributions to the professional recognition of clinical microbiologists and/or immunologists; letters of support, two letters of support must come from persons, other than the nominator, who are familiar with the nominee's qualifications and accomplishments. No more than one of the three letters may be from the nominee's institution or from the same institution.
Restrictions: Self-nominations will not be accepted.
Geographic Focus: All States
Date(s) Application is Due: Oct 1
Contact: Awards Committee; (202) 942-9226; fax (202) 942-9353; awards@asmusa.org
Internet: http://www.asm.org/Academy/index.asp?bid=2321
Sponsor: American Society for Microbiology
1752 N Street, N.W.
Washington, D.C. 20036-2904

ASM Undergraduate Research Fellowship 755
The ASM Undergraduate Research Fellowship (URF) is aimed at highly competitive students who wish to pursue graduate careers (Ph.D. or MD/Ph.D) in microbiology. Students will have the opportunity to conduct full time research at their home institutions with an ASM member and present research results at the ASM General Meeting the following year. Students will: conduct a research project for a minimum of 10 weeks beginning in the summer; work with faculty mentors who are ASM members and who are employed at the students home institutions, and submit a research abstract for presentation at the ASM General Meeting. The Fellowship provides: up to $4000 for student stipend; two-year ASM student membership, and up to $1000 in travel support for students to present the results of the research project at the ASM General Meeting. Travel funds are contingent upon acceptance of an abstract for the General Meeting.
Requirements: Eligible student candidates for the fellowship must: be enrolled as full-time matriculating undergraduate students during the academic year at an accredited U.S. Institution; be involved in a research project; have an ASM member at their home institutions willing to serve as a mentor, and not receive financial support for research (i.e., Council for Undergraduate Research, Minority Access to Research Careers, Sigma Xi) during the fellowship. Applicants who do not meet all eligibility requirements will not be considered.
Restrictions: The program requires a joint application from both the student and a faculty mentor. It is student's responsibility to submit a completed application.
Geographic Focus: All States
Date(s) Application is Due: Feb 1
Amount of Grant: 4,000 - 5,000 USD
Contact: Fellowship Committee; (202) 942-9283; fax (202) 942-9329; fellowships-careerinformation@asmusa.org
Internet: http://www.asm.org/Education/index.asp?bid=4319
Sponsor: American Society for Microbiology
1752 N Street, N.W.
Washington, D.C. 20036-2904

ASM USFCC/J. Roger Porter Award 756
This award recognizes outstanding efforts by a scientist who has demonstrated the importance of microbial biodiversity through sustained curatorial or stewardship activities for a major resource used by the scientific community. It honors the memory of Dr. Porter and his remarkable contributions to science.
Requirements: The nominee will have greatly aided other scientists by demonstrating the fundamentals of culture collections and related resources and the rich biodiversity that such collections preserve. These resources may include collections of cells or microorganisms, databases or tools, or major reference works or services as an essential aspect of either one or more major publications or an extensive work effort.
Restrictions: Nominations must consist of the following: curriculum vitae, including a list of publications; letter of nomination, describe the nominee's outstanding efforts demonstrating the importance of microbial biodiversity through sustained curatorial or stewardship activities for a major resource used by the scientific community; letters of support, two letters of support must come from persons, other than the nominator, who are familiar with the nominee's outstanding efforts. No more than one of the three letters may be from the nominee's institution or the same institution.
Geographic Focus: All States
Date(s) Application is Due: Oct 1
Amount of Grant: 2,000 USD
Samples: Self-nominations will not be accepted.

Contact: Awards Committee; (202) 942-9226; fax (202) 942-9353; awards@asmusa.org
Internet: http://www.asm.org/academy/index.asp?bid=2297
Sponsor: American Society for Microbiology
1752 N Street, N.W.
Washington, D.C. 20036-2904

ASM William A Hinton Research Training Award 757
The ASM William A Hinton Research Training Award honors outstanding contributions toward fostering the research training of underrepresented minorities in microbiology. It is given in memory of William A. Hinton, a physician-research scientist and one of the first African-Americans to join the ASM.
Requirements: The nominee must contribute to the research training of undergraduate students, graduate students, postdoctoral fellows or health professional students and efforts leading to the increased participation of underrepresented minorities in microbiology. Nominations must consist of the following: curriculum vitae, including a list of nominee's publications; letter of nomination, describe the nominee's training of undergraduate students, graduate students, postdoctoral fellows, or health professional students. Detail specific examples of excellence in training and mentoring of underrepresented minorities. Describe the duration, extent, and impact of the nominee's training of underrepresented minority students, graduate students, postdoctoral fellows, or health professional students; letters of support, two letters of support must come from persons, other than the nominator, who are familiar with the nominee's outstanding contributions to the training of underrepresented minorities. No more than one of the three letters may be from the nominee's institution or from the same institution.
Geographic Focus: All States
Date(s) Application is Due: Oct 1
Amount of Grant: 2,000 USD
Contact: Awards Committee; (202) 942-9226; fax (202) 942-9353; awards@asmusa.org
Internet: http://www.asm.org/Academy/index.asp?bid=2317
Sponsor: American Society for Microbiology
1752 N Street, N.W.
Washington, D.C. 20036-2904

A Social Corporation Grants 758
The Fund, established in Wisconsin in 1987, has as its primary purpose to support research in medical physics within the Madison, Wisconsin, region. Specific application guidelines are available upon request. There are no particular deadlines with which to adhere, and applicant organizations should contact the fund administrator directly for further details.
Geographic Focus: Wisconsin
Contact: Charles Lescrenier; (608) 831-1188 or (800) 426-6391; fax (608) 836-9201
Sponsor: A Social Corporation
2500 West Beltline Highway at University Avenue, P.O. Box 620327
Middleton, WI 53562-0327

ASPEN Dudrick Research Scholar Award 759
A.S.P.E.N. members are invited to nominate individuals to be considered by the Dudrick Award Committee for the Stanley Dudrick Research Scholar Award. A.S.P.E.N. presents this annual award to recognize and support a mid-career investigator who has shown significant achievements in nutrition support, is an A.S.P.E.N. member, and demonstrates exceptional research productivity and the potential to continue to make contributions in the field of nutrition therapy. The award includes $5,000, plus recognition during the Rhoads Lecture and Awards Symposium at Clinical Nutrition Week.
Requirements: The applicant should: be an A.S.P.E.N. member; have an advanced degree (MD, PharmD, MS, PhD, or equivalent); currently be in a training program or have completed formal training in nutritional research/therapy within the past 10-15 years, e.g. mid-career; have demonstrated exceptional research productivity during and/or following formal training.
Geographic Focus: All States
Date(s) Application is Due: Oct 15
Amount of Grant: 5,000 USD
Contact: Paula Bowen, Research Program Administrator; (301) 587-6315, ext. 132 or (301) 920-9132; fax (301) 587-2365; paulab@aspen.nutr.org
Internet: http://www.nutritioncare.org/wcontent.aspx?id=878
Sponsor: American Society for Parenteral and Enteral Nutrition
8630 Fenton Street, Suite 412
Silver Spring, MD 20910

ASPEN Harry M. Vars Award 760
The Award is made annually for the best research presentation by an investigator at Nutrition Week. The recipient of the award is selected by the Vars Award Selection Subcommittee based upon a review of manuscripts submitted by qualified candidates. The author of the best paper presents his or her paper at the Premier Paper Session and is given a $1,000 cash prize, a travel grant and a plaque during the Research Awards Ceremony at Nutrition Week. The winning manuscript is published in the Journal of Parenteral and Enteral Nutrition (JPEN). The Award is open to all disciplines.
Geographic Focus: All States
Amount of Grant: 1,000 USD
Contact: Michelle Spangenburg, Program Director for Education; (301) 587-6315, ext. 127; fax (301) 587-2365; michelles@aspen.nutr.org
Internet: http://www.nutritioncare.org/wcontent.aspx?id=878
Sponsor: American Society for Parenteral and Enteral Nutrition
8630 Fenton Street, Suite 412
Silver Spring, MD 20910

ASPEN Rhoads Research Foundation Abbott Nutrition Research Grants 761

The ASPEN Rhoads Research Foundation provides annual grant support to nutrition researchers at all stages of their careers. This grant, funded by a donation from the Abbott Foundation, is intended to assist a nutrition investigator by providing preliminary funding for promising new research in the field of nutrition and metabolic support and related areas of clinical nutrition. Persons applying for the grant must be in a post-training position and commit at least 20% of their time to research. Corporate employees are eligible to receive grants if the research project is not part of their normal duties and they have at least a part-time academic appointment or are associated with a professional institute that conducts nutrition research. In making choices among otherwise high-quality applicants, the reviewers will take into account the geographic distribution of grant winners; their ethnic background and gender, should the applicant choose to supply such information; their professional discipline; and other criteria as appropriate to ensure an equitable system. Funds may be used only for technician salary, equipment, supplies, animals, clinical research costs or other expenses directly related to the conduct of the proposed research. Grants can be renewed for a second year of funding, assuming satisfactory progress.
Requirements: Applicants must submit a letter from their supervisor or department head at the institution confirming his/her commitment to the project. If the project involves human subjects, a letter pledging support in recruiting patients from the primary care provider and the institutional review board overseeing human studies is required. For junior faculty and applicants who do not hold faculty positions, three letters of reference are required.
Restrictions: Individuals who have received other research grants in excess of $25,000 are ineligible to apply, as are those who completed training more than 5 years prior to the start of the grant period. This grant is not intended to support pursuit of an additional degree. Grant funds may not be used for: comparison of commercial products; indirect costs or overhead; costs of patient care; constructing or renovating facilities; furniture or office equipment; secretarial services; honoraria or membership dues; textbooks or periodicals; repair or service contract costs on institutional equipment; entertainment; travel; or salary support for the Principal Investigator.
Geographic Focus: All States
Date(s) Application is Due: Sep 8
Amount of Grant: Up to 25,000 USD
Contact: Michelle Spangenburg, Director of Education and Research; (301) 587-6315, ext. 127 or (301) 920-9132; fax (301) 587-2365; michelles@nutritioncare.org
Internet: http://www.nutritioncare.org/Research/ARRF/A_S_P_E_N__Rhoads_Research_Foundation_Grants/
Sponsor: American Society for Parenteral & Enteral Nutrition Rhoads Research Foundation
8630 Fenton Street, Suite 412
Silver Spring, MD 20910

ASPEN Rhoads Research Foundation Baxter Parenteral Nutrition Research Grant 762

The ASPEN Rhoads Research Foundation provides annual grant support to nutrition researchers at all stages of their careers. The Baxter Parenteral Nutrition Research grant is intended to assist a nutrition investigator by providing preliminary funding for promising new research in the field of parenteral nutrition and metabolic support and related areas of clinical nutrition. Priority consideration is given to applications that involve a multidisciplinary team or that investigate the efficacy of parenteral or enteral nutrition. Applicants still in training positions may apply for the Fleming Grant. Corporate employees are eligible to receive grants if the research project is not part of their normal duties and they have at least a part-time academic appointment or are associated with a professional institute that conducts nutrition research. In making choices among otherwise high-quality applicants, the reviewers will take into account the geographic distribution of grant winners; their ethnic background and gender, should the applicant choose to supply such information; their professional discipline; and other criteria as appropriate to ensure an equitable system. Funds may be used only for technician salary, equipment, supplies, animals, clinical research costs or other expenses directly related to the conduct of the proposed research. The grant can be renewed for a second year of funding, assuming satisfactory progress.
Requirements: Applicants must submit a letter from their supervisor or department head at the institution confirming his/her commitment to the project. If the project involves human subjects, a letter pledging support in recruiting patients from the primary care provider and the institutional review board overseeing human studies is required. For junior faculty and applicants who do not hold faculty positions, three letters of reference are required.
Restrictions: Individuals who have received other research grants in excess of $25,000 are ineligible to apply, as are those who completed training more than five years prior to the start of the grant period. This grant is not intended to support pursuit of an additional degree. Grant funds may not be used for: comparison of commercial products; indirect costs or overhead; costs of patient care; constructing or renovating facilities; furniture or office equipment; secretarial services; honoraria or membership dues; textbooks or periodicals; repair or service contract costs on institutional equipment; entertainment; travel; or salary support for the Principal Investigator.
Geographic Focus: All States
Date(s) Application is Due: Sep 8
Amount of Grant: Up to 50,000 USD
Contact: Michelle Spangenburg, Director of Education and Research; (301) 587-6315, ext. 127 or (301) 920-9132; fax (301) 587-2365; michelles@nutritioncare.org
Internet: http://www.nutritioncare.org/Research/ARRF/A_S_P_E_N__Rhoads_Research_Foundation_Grants/
Sponsor: American Society for Parenteral & Enteral Nutrition Rhoads Research Foundation
8630 Fenton Street, Suite 412
Silver Spring, MD 20910

ASPEN Rhoads Research Foundation C. Richard Fleming Grant 763

The ASPEN Rhoads Research Foundation provides annual grant support to nutrition researchers at all stages of their careers. This grant, funded by the donations of ASPEN members, is intended to assist a nutrition investigator by providing preliminary funding for promising new research in the field of nutrition and metabolic support and related areas of clinical nutrition. Applicants still in training positions may apply. Corporate employees are eligible to receive grants if the research project is not part of their normal duties and they have at least a part-time academic appointment or are associated with a professional institute that conducts nutrition research. In making choices among otherwise high-quality applicants, the reviewers will take into account the geographic distribution of grant winners; their ethnic background and gender, should the applicant choose to supply such information; their professional discipline; and other criteria as appropriate to ensure an equitable system. Funds may be used only for technician salary, equipment, supplies, animals, clinical research costs or other expenses directly related to the conduct of the proposed research.
Requirements: Applicants must submit a letter from their supervisor or department head at the institution confirming his/her commitment to the project. If the project involves human subjects, a letter pledging support in recruiting patients from the primary care provider and the institutional review board overseeing human studies is required. For junior faculty and applicants who do not hold faculty positions, three letters of reference are required.
Restrictions: This grant is not intended to support pursuit of an additional degree. Grant funds may not be used for: comparison of commercial products; indirect costs or overhead; costs of patient care; constructing or renovating facilities; furniture or office equipment; secretarial services; honoraria or membership dues; textbooks or periodicals; repair or service contract costs on institutional equipment; entertainment; travel; or salary support for the Principal Investigator.
Geographic Focus: All States
Date(s) Application is Due: Sep 8
Amount of Grant: Up to 10,000 USD
Contact: Michelle Spangenburg, Director of Education and Research; (301) 587-6315, ext. 127 or (301) 920-9132; fax (301) 587-2365; michelles@nutritioncare.org
Internet: http://www.nutritioncare.org/Research/ARRF/A_S_P_E_N__Rhoads_Research_Foundation_Grants/
Sponsor: American Society for Parenteral & Enteral Nutrition Rhoads Research Foundation
8630 Fenton Street, Suite 412
Silver Spring, MD 20910

ASPEN Rhoads Research Foundation Maurice Shils Grant 764

The ASPEN Rhoads Research Foundation provides annual grant support to nutrition researchers at all stages of their careers. This grant, funded by a donation from Baxter Health Care and Nestle Clinical Nutrition, is intended to assist a nutrition investigator by providing preliminary funding for promising new research in the field of nutrition and metabolic support and related areas of clinical nutrition. Persons applying for the grant must be in a post-training position and commit at least 20% of their time to research. Corporate employees are eligible to receive grants if the research project is not part of their normal duties and they have at least a part-time academic appointment or are associated with a professional institute that conducts nutrition research. In making choices among otherwise high-quality applicants, the reviewers will take into account the geographic distribution of grant winners; their ethnic background and gender, should the applicant choose to supply such information; their professional discipline; and other criteria as appropriate to ensure an equitable system. Funds may be used only for technician salary, equipment, supplies, animals, clinical research costs or other expenses directly related to the conduct of the proposed research. Grants can be renewed for a second year of funding, assuming satisfactory progress.
Requirements: Applicants must submit a letter from their supervisor or department head at the institution confirming his/her commitment to the project. If the project involves human subjects, a letter pledging support in recruiting patients from the primary care provider and the institutional review board overseeing human studies is required. For junior faculty and applicants who do not hold faculty positions, three letters of reference are required.
Restrictions: Individuals who have received other research grants in excess of $25,000 are ineligible to apply, as are those who completed training more than 5 years prior to the start of the grant period. This grant is not intended to support pursuit of an additional degree. Grant funds may not be used for: comparison of commercial products; indirect costs or overhead; costs of patient care; constructing or renovating facilities; furniture or office equipment; secretarial services; honoraria or membership dues; textbooks or periodicals; repair or service contract costs on institutional equipment; entertainment; travel; or salary support for the Principal Investigator.
Geographic Focus: All States
Date(s) Application is Due: Sep 8
Amount of Grant: Up to 25,000 USD
Contact: Michelle Spangenburg, Director of Education and Research; (301) 587-6315, ext. 127 or (301) 920-9132; fax (301) 587-2365; michelles@nutritioncare.org
Internet: http://www.nutritioncare.org/Research/ARRF/A_S_P_E_N__Rhoads_Research_Foundation_Grants/
Sponsor: American Society for Parenteral & Enteral Nutrition Rhoads Research Foundation
8630 Fenton Street, Suite 412
Silver Spring, MD 20910

ASPEN Rhoads Research Foundation Norman Yoshimura Grant 765

The ASPEN Rhoads Research Foundation provides annual grant support to nutrition researchers at all stages of their careers. This grant, supported by B. Braun Medical funding, is intended to assist a nutrition investigator by providing preliminary funding for promising new research in the field of nutrition and metabolic support and related

areas of clinical nutrition. Applicants still in training positions may apply. Corporate employees are eligible to receive grants if the research project is not part of their normal duties and they have at least a part-time academic appointment or are associated with a professional institute that conducts nutrition research. In making choices among otherwise high-quality applicants, the reviewers will take into account the geographic distribution of grant winners; their ethnic background and gender, should the applicant choose to supply such information; their professional discipline; and other criteria as appropriate to ensure an equitable system. Funds may be used only for technician salary, equipment, supplies, animals, clinical research costs or other expenses directly related to the conduct of the proposed research.
Requirements: Applicants must submit a letter from their supervisor or department head at the institution confirming his/her commitment to the project. If the project involves human subjects, a letter pledging support in recruiting patients from the primary care provider and the institutional review board overseeing human studies is required. For junior faculty and applicants who do not hold faculty positions, three letters of reference are required.
Restrictions: This grant is not intended to support pursuit of an additional degree. Grant funds may not be used for: comparison of commercial products; indirect costs or overhead; costs of patient care; constructing or renovating facilities; furniture or office equipment; secretarial services; honoraria or membership dues; textbooks or periodicals; repair or service contract costs on institutional equipment; entertainment; travel; or salary support for the Principal Investigator.
Geographic Focus: All States
Date(s) Application is Due: Sep 8
Amount of Grant: Up to 25,000 USD
Contact: Michelle Spangenburg, Director of Education and Research; (301) 587-6315, ext. 127 or (301) 920-9132; fax (301) 587-2365; michelles@nutritioncare.org
Internet: http://www.nutritioncare.org/Research/ARRF/A_S_P_E_N__Rhoads_Research_Foundation_Grants/
Sponsor: American Society for Parenteral & Enteral Nutrition Rhoads Research Foundation
8630 Fenton Street, Suite 412
Silver Spring, MD 20910

ASPEN Scientific Abstracts Awards for Papers or Posters 766
ASPEN will consider all Scientific Abstract submissions for awards. Applicants will be notified if their abstract is a candidate. The first authors of the three highest ranked abstracts from this category (paper or poster) will receive travel grants to attend the Research Workshop, and to present their work at Clinical Nutrition Week. In addition, cash awards in the amounts of $750, $500 and $250 will be offered to the presenters of the three best papers. Awards are presented during the Rhoads Lecture and Research Awards Ceremony at Clinical Nutrition Week.
Geographic Focus: All States
Amount of Grant: 250 - 750 USD
Contact: Paula Bowen, Research Program Administrator; (301) 587-6315, ext. 132 or (301) 920-9132; fax (301) 587-2365; paulab@aspen.nutr.org
Internet: http://www.nutritioncare.org/Index.aspx?id=724
Sponsor: American Society for Parenteral and Enteral Nutrition
8630 Fenton Street, Suite 412
Silver Spring, MD 20910

ASPEN Scientific Abstracts Promising Investigator Awards 767
ASPEN will consider all Scientific Abstract submissions for awards. Applicants will be notified if their abstract is a candidate. These awards are presented to the top three Scientific Abstract authors, who are also in the early stages of their careers. Winners receive $750 in travel funds to offset their travel expenses to and from Clinical Nutrition Week. Vars Award winners are eligible for these awards if they are in the early stages of their careers.
Geographic Focus: All States
Amount of Grant: 750 USD
Contact: Paula Bowen, Research Program Administrator; (301) 587-6315, ext. 132 or (301) 920-9132; fax (301) 587-2365; paulab@aspen.nutr.org
Internet: http://www.nutritioncare.org/wcontent.aspx?id=878
Sponsor: American Society for Parenteral and Enteral Nutrition
8630 Fenton Street, Suite 412
Silver Spring, MD 20910

ASPET-Astellas Awards in Translational Pharmacology 768
The ASPET-Astellas Awards in Translational Pharmacology are intended to extend fundamental research closer to applications directed towards improving human health. The awards will be given to: recognize those individuals whose research has the potential to lead to the introduction of novel pharmacologic approaches or technologies that may offer significant advances in clinical medicine in the future; and to facilitate that translational process. Three (3) awards of $30,000 each will be made to individuals. The money may be used for supplemental research funding, travel, training, or in any way that furthers the goals of the organization. Nominations should be made electronically.
Requirements: Any ASPET member in good standing may nominate an individual for this award.
Restrictions: Self-nominations will not be accepted.
Geographic Focus: All States
Date(s) Application is Due: Sep 15
Contact: Christine Carrico, Executive Officer; (301) 634-7060; fax (301) 634-7061; ccarrico@aspet.org or info@aspet.org
Internet: http://www.aspet.org/public/awards/Astellas_awards.html
Sponsor: American Society for Pharmacology and Experimental Therapeutics
9650 Rockville Pike
Bethesda, MD 20814-3995

ASPET Benedict R. Lucchesi Award in Cardiac Pharmacology 769
The Benedict R. Lucchesi Award in Cardiac Pharmacology was established to honor Dr. Lucchesi's lifelong scientific contributions to our better understanding and appreciation of pharmacological treatment and prevention of cardiovascular disease and for his mentoring of countless prominent functional (in vivo) cardiovascular pharmacologists. The Benedict R. Lucchesi Award is a biennial award, consisting an honorarium of $1,000, a custom-designed crystal bowl depicting the named Lectureship, and up to $2,000 travel expenses including registration to the annual spring ASPET meeting. A recipient will be selected and invited to deliver a state-of-the-art lecture on recent advances in the field of cardiac and electropharmacology at the spring ASPET meeting (Division's programming session). The presentation of his/her research should be of broad interest and contribute to the growth of the Cardiovascular Pharmacology Division.
Requirements: There are no restrictions on institutional affiliation, nationality, or age of the candidate, but the recipient must be a member of the ASPET. Nominations must be made by a member of the ASPET, and no member may nominate more than one candidate per year.
Geographic Focus: All States
Date(s) Application is Due: Sep 15
Amount of Grant: 1,000 - 2,000 USD
Contact: Christine Carrico, Executive Officer; (301) 634-7060; fax (301) 634-7061; ccarrico@aspet.org or info@aspet.org
Internet: http://www.aspet.org/public/divisions/cardiovascular/lucchesi_award.htm
Sponsor: American Society for Pharmacology and Experimental Therapeutics
9650 Rockville Pike
Bethesda, MD 20814-3995

ASPET Bernard B. Brodie Award in Drug Metabolism 770
This biennial award (even-numbered years), established by CIBA-GEIGY Corporation to honor the fundamental contributions of Bernard B. Brodie, is presented to recognize outstanding original research contributions in drug metabolism and disposition, particularly those having a major impact on future research in the field. Only one nominator is necessary although more are acceptable. The award consists of a $2,000 honorarium, a commemorative medal, hotel and economy airfare to the award ceremony at the annual meeting. Nominations must be received by the deadline prior to the award year.
Requirements: Nominees must be members of ASPET; nominators need not be members of ASPET.
Restrictions: Supporting research accomplishments must not have been used to win any other major award.
Geographic Focus: All States
Date(s) Application is Due: Sep 15
Amount of Grant: 2,000 USD
Contact: Christine Carrico, Executive Officer; (301) 634-7060; fax (301) 634-7061; ccarrico@aspet.org or info@aspet.org
Internet: http://www.aspet.org/public/awards/brodie_award.html
Sponsor: American Society for Pharmacology and Experimental Therapeutics
9650 Rockville Pike
Bethesda, MD 20814-3995

ASPET Division for Drug Metabolism Early Career Achievement Award 771
The ASPET Division for Drug Metabolism Early Career Achievement Award has been established to recognize excellent original research by early career investigators in the area of drug metabolism and disposition. The award is presented biennially in odd-numbered years. The award consists of $1,000, a plaque, and complimentary registration plus travel expenses (to a maximum of $1,000) for the winner to attend the award ceremony at the annual meeting. The awardee will deliver a lecture at the annual meeting describing their relevant research accomplishments. The awardee will be invited to publish a review article on the subject matter of the award lecture in Drug Metabolism and Disposition.
Requirements: Nominees for this award must have a doctoral degree (e.g. Ph.D., M.D., Pharm. D., D.V.M.) and must be within 15 years of having received their final degree, as of December 31 of the year of the award. There are no restrictions on institutional affiliation and a candidate need not be a member of ASPET.
Geographic Focus: All States
Date(s) Application is Due: Sep 15
Amount of Grant: 1,000 USD
Contact: Christine Carrico, Executive Officer; (301) 634-7060; fax (301) 634-7061; ccarrico@aspet.org or info@aspet.org
Internet: http://www.aspet.org/public/divisions/drugmetab/early_achievement_award.htm
Sponsor: American Society for Pharmacology and Experimental Therapeutics
9650 Rockville Pike
Bethesda, MD 20814-3995

ASPET Epilepsy Research Award for Outstanding Contributions to the Pharmacology of Antiepileptic Drugs 772
The International League Against Epilepsy (ILAE) has sponsored this biennial award of $2,000 and a Certification of Citation to be awarded by the American Society for Pharmacology and Experimental Therapeutics, for the purpose of recognizing and stimulating outstanding research leading to better clinical control of epileptic seizures. This research may include the basic screening and testing of new therapeutic agents, studies on mechanisms of action, metabolic disposition, pharmacokinetics, and clinical pharmacology studies. The recipient will be selected by the Epilepsy Award Committee appointed by the President of ASPET, with representation of ILAE.
Requirements: Nominations for the Award may be submitted by members of any recognized scientific association, domestic or foreign.

Geographic Focus: All States
Date(s) Application is Due: Sep 15
Amount of Grant: 2,000 USD
Contact: Christine Carrico, Executive Officer; (301) 634-7060; fax (301) 634-7061; ccarrico@aspet.org or info@aspet.org
Internet: http://www.aspet.org/public/awards/epilepsy_award.html
Sponsor: American Society for Pharmacology and Experimental Therapeutics
9650 Rockville Pike
Bethesda, MD 20814-3995

ASPET Goodman and Gilman Award in Drug Receptor Pharmacology 773
This award is given biennially (even-numbered years) to recognize and stimulate outstanding research in the pharmacology of biological receptors. It is the hope that such research might provide a better understanding of the mechanism of biological processes and potentially provide the basis for the discovery of drugs useful in the treatment of diseases. The award includes a monetary prize and travel support for winner and spouse to attend the award ceremony. Nominations must be received September 15 of the year prior to the year in which the award will be made.
Requirements: Nominations must be made by ASPET members; however, there are no restrictions as to age, sex, nationality, or institutional affiliation. Nominee need not be a member of ASPET.
Geographic Focus: All States
Date(s) Application is Due: Sep 15
Amount of Grant: 2,500 USD
Contact: Christine Carrico, Executive Officer; (301) 634-7060; fax (301) 634-7061; ccarrico@aspet.org or info@aspet.org
Internet: http://www.aspet.org/awards/goodman_and_gilman/
Sponsor: American Society for Pharmacology and Experimental Therapeutics
9650 Rockville Pike
Bethesda, MD 20814-3995

ASPET John J. Abel Award in Pharmacology 774
This award, sponsored by Eli Lilly and Company and administered by ASPET, is given annually to stimulate original and outstanding fundamental research in pharmacology and experimental therapeutics by young investigators. Candidates may not have passed their 39th birthday as of April 30 of the year of the award, or previously received an award from the sponsor for the same technical accomplishment.
Requirements: Candidates need not be members of the society; however, nominations must be made by ASPET members.
Geographic Focus: All States
Date(s) Application is Due: Sep 15
Amount of Grant: 2,500 USD
Contact: Christine Carrico, Executive Officer; (301) 634-7060; fax (301) 634-7061; ccarrico@aspet.org or info@aspet.org
Internet: http://www.aspet.org/public/awards/abel_award.html
Sponsor: American Society for Pharmacology and Experimental Therapeutics
9650 Rockville Pike
Bethesda, MD 20814-3995

ASPET Julius Axelrod Award in Pharmacology 775
The Julius Axelrod Award is presented annually for significant contributions to understanding the biochemical mechanisms underlying the pharmacological actions of drugs and for contributions to mentoring other pharmacologists. The award consists of an honorarium of $5,000, a medal, hotel and economy airfare for the winner and spouse to the annual meeting. The formal presentation of this award and medal will be made at the annual meeting of ASPET. The recipient will be invited by the President of the Society to deliver the Julius Axelrod Lecture and organize the Julius Axelrod Symposium at the annual meeting a year hence. The recipient will also be invited by the Catecholamine Club to give a less formal presentation at its annual dinner meeting that year of the award.
Geographic Focus: All States
Date(s) Application is Due: Sep 15
Amount of Grant: 5,000 USD
Contact: Christine Carrico, Executive Officer; (301) 634-7060; fax (301) 634-7061; ccarrico@aspet.org or info@aspet.org
Internet: http://www.aspet.org/public/awards/axelrod_award.html
Sponsor: American Society for Pharmacology and Experimental Therapeutics
9650 Rockville Pike
Bethesda, MD 20814-3995

ASPET P. B. Dews Lifetime Achievement Award for Research in Behavioral Pharmacology 776
ASPET's Division of Behavioral Pharmacology sponsors the P. B. Dews Award for Research in Behavioral Pharmacology to recognize outstanding lifetime achievements in research, teaching and professional service in the field of Behavioral Pharmacology and to honor Peter Dews for his seminal contributions to the development of behavioral pharmacology as a discipline. The Award consists of $750, a plaque, and partial travel expenses to the award ceremony at the ASPET annual meeting. The recipient will be invited by the Chair of the Division of Behavioral Pharmacology to deliver a special lecture on this occasion. The lecture will be published subsequently in an appropriate ASPET-sponsored publication.
Requirements: Nominations may be made by members of ASPET or of any relevant scientific society.
Restrictions: There are no restrictions on nominees for this award.

Geographic Focus: All States
Date(s) Application is Due: Sep 15
Amount of Grant: 750 USD
Contact: Christine Carrico, Executive Officer; (301) 634-7060; fax (301) 634-7061; ccarrico@aspet.org or info@aspet.org
Internet: http://www.aspet.org/public/awards/dews_award.html
Sponsor: American Society for Pharmacology and Experimental Therapeutics
9650 Rockville Pike
Bethesda, MD 20814-3995

ASPET Paul M. Vanhoutte Award 777
The Paul M. Vanhoutte Award in Vascular Pharmacology was established to honor Dr. Vanhoutte's lifelong scientific contributions to our better understanding and appreciation of the importance of endothelial cells and vascular smooth muscle function in health and disease and for his mentoring of countless prominent endothelial and vascular biologists and pharmacologists. This is a biennial award, consisting an honorarium of $1,000, a custom-designed crystal bowl depicting the named Lectureship, and up to $2,000 travel expenses including registration to the annual spring ASPET meeting.
Requirements: There are no restrictions on institutional affiliation, nationality, or age of the candidate, but the recipient must be a member of the ASPET.
Geographic Focus: All States
Date(s) Application is Due: Sep 15
Amount of Grant: 1,000 - 2,000 USD
Contact: Christine Carrico, Executive Officer; (301) 634-7060; fax (301) 634-7061; ccarrico@aspet.org or info@aspet.org
Internet: http://www.aspet.org/public/divisions/cardiovascular/vanhoutte_award.htm
Sponsor: American Society for Pharmacology and Experimental Therapeutics
9650 Rockville Pike
Bethesda, MD 20814-3995

ASPET Torald Sollmann Award in Pharmacology 778
This award, established by Wyeth-Herst Labs, is given in odd-numbered years to commemorate the pioneer work in the United States of Dr. Torald Sollmann in the fields of pharmacological investigation and education. The award is made to a nominee for significant contributions over many years to the advancement and extension of knowledge in the field of pharmacology. The award consists of an honorarium of $3,500, a medal, hotel and economy airfare for the winner and spouse to the annual meeting. The formal presentation of this biennial award and medal will be made at the annual meeting of ASPET. The recipient will be invited by the President of the Society to deliver a lecture to the membership that may be published in an appropriate ASPET journal.
Requirements: Nominations must be made by ASPET members; however, the nominee need not be a member. There are no restrictions as to age or institutional affiliation.
Geographic Focus: All States
Date(s) Application is Due: Sep 15
Amount of Grant: 3,500 USD
Contact: Christine Carrico, Executive Officer; (301) 634-7060; fax (301) 634-7061; ccarrico@aspet.org or info@aspet.org
Internet: http://www.aspet.org/public/awards/sollmann_award.html
Sponsor: American Society for Pharmacology and Experimental Therapeutics
9650 Rockville Pike
Bethesda, MD 20814-3995

ASPET Travel Awards 779
Graduate Student Travel Awards, Minority Graduate Student Travel Awards, and Summer Undergraduate Student Travel Awards are a fixed sum of $600 plus registration. Young Scientist Travel Awards and Minority Young Scientist Travel Awards are a fixed sum of $600 plus registration. All travel awards are inclusive of registration and are given to partially defray the costs of travel and housing. The check for the award and an award certificate will be presented at the annual ASPET Opening Awards Ceremony. Applications will be available in late September.
Requirements: Applications must be submitted online or as email attachments.
Geographic Focus: All States
Date(s) Application is Due: Dec 1
Amount of Grant: 600 USD
Contact: Christine Carrico, Executive Officer; (301) 634-7060; fax (301) 634-7061; ccarrico@aspet.org or travelawards@aspet.org
Internet: http://www.aspet.org/public/awards/awards_fellowships.html#Travel_Awards
Sponsor: American Society for Pharmacology and Experimental Therapeutics
9650 Rockville Pike
Bethesda, MD 20814-3995

ASPH/CDC Allan Rosenfield Global Health Fellowships 780
In recognition of a career dedicated to improving the health of the world's most vulnerable populations, the Association of Schools of Public Health (ASPH), with support from the Centers for Disease Control and Prevention (CD.C.), established a fellowship in honor of Dr. Allan Rosenfield, former Dean of the Columbia University Mailman School of Public Health. As a global health pacesetter Dr. Rosenfield addressed many issues, ranging from women's health to HIV/AIDS to advocating healthcare for the poor and disadvantaged. His commitment to improving lives in neighborhoods not only in America, but in far away places such as rural sub-Saharan Africa and Asia, is an example ASPH hopes students will learn to follow during the period of their fellowships and throughout their public health careers. Students selected as Allan Rosenfield Global Health Fellows are expected to involve themselves, as Dr. Rosenfield did throughout

his career, in all aspects of global health. The training offered through this program will expand the global health protection workforce and provide fellows an opportunity to gain practical, first-hand experience on the front lines of international public health and to learn from leading global health experts. One- or two-year international fellowships will be placed within a global health program at CD.C. headquarters in Atlanta or in one of the fifty foreign countries where CD.C. is working with the Ministry of Health and other Public Health partners. Additional travel during the fellowship period may be required depending on program activities and priorities. Exact geographical locations will not be determined until the final placement process. At that time, successful applicants will be offered positions and notified of all details for that particular position. Applicants must apply through ASPH's online grant application system (click on the APPLICATION tab at the top of the ASPH fellowship website) and mail required hard copy and supporting documents to ASPH. Required application materials include an essay, a resume or curriculum vitae, two letters of recommendation (from the references provided in the online application), official graduate transcripts, and proof of residency. Applicants are encouraged to review requirements for submitting these documents at the fellowship website as well as to read the FAQ section of the website for more information about the fellowship. Deadlines may vary from year to year; however, requests for application are announced in ASPH's weekly electronic newsletter "The Friday Letter" which applicants may register to receive at no cost. The subscription link is available at http://fellowships. asph.org/ from the bottom right hand corner of the screen. Additionally, applicants may check the same website for updates on available fellowships.
Requirements: Applicants must have received their Masters or Doctorate degree no later than the August prior to the beginning of the fellowship or no earlier than May within the last five years. Graduate degrees must come from an ASPH-member graduate school of public health accredited by the Council on Education for Public Health (CEPH). (A list of accredited schools with full ASPH membership can be accessed via the sponsor website.) Prior to leaving for the fellowship assignment locations, fellows will be required to travel to CD.C. in Atlanta, Georgia for a ten-day orientation in mid-September. Fellowship assignments will begin following orientation.
Restrictions: All applicants must be U.S. citizens or hold a visa permitting permanent residence ("Green Card") in the U.S. to be eligible for the fellowship program. Applicants may apply to any number of ASPH fellowship programs; however applicants may apply to a maximum of only three positions within each program.
Geographic Focus: All States, All Countries
Date(s) Application is Due: Feb 16
Amount of Grant: 43,000 - 52,000 USD
Contact: Binley G. Taylor, Coordinator; (202) 296-1099; fax (202) 296-1252; btaylor@asph.org or trainingprograms@asph.org
Internet: https://fellowships.asph.org/programs/details.cfm?programID=!%260%20%20%0A
Sponsor: Association of Schools of Public Health
1900 M Street NW, Suite 710
Washington, D.C. 20036

ASPH/CDC Public Health Fellowships 781
The ASPH/CD.C. Public Health Fellowships were established in 1995 to address emerging needs of public health and to provide leadership and professional opportunities at the Centers for Disease Control and Prevention (CD.C.) for graduate students of members of the Association of Schools of Public Health (ASPH). The fellowships are sponsored cooperatively by ASPH and CD.C.. Positions are full-time opportunities that last one to two years depending on the needs of CD.C. and the fellow. Fellowships typically commence in August and are located at CD.C. headquarters in Atlanta, Georgia. The types of fellowships will vary according to specific areas of research or training within CD.C.'s Centers/Institutes/Offices (CIOs). Selected fellows will have the opportunity to apply graduate education in the field, develop professional experience and expertise, and contribute to public health research and practice. Fellows will receive a stipend payment for the duration of the fellowship on a biweekly basis based on the education level/degree requirement of the fellowship position. In addition to the fellowship stipend, the fellow will have access to a sufficient "fellowship allowance" that is intended to cover health/dental/vision insurance premiums during the fellowship period. In addition, some positions may offer an additional "travel/training allowance" that may be used to cover project-related travel, tuition, journal subscriptions, association dues, etc. (not all positions will have a travel/training allowance). The fellowship stipend is intended to cover all living expenses including housing and project-related travel. Fellows are responsible for their own move to and from the fellowship site (relocation expenses are not covered). Typically, the ASPH application season opens in November and runs through March. Requests for application are announced in ASPH's weekly electronic newsletter "The Friday Letter" which applicants may register to receive at no cost. The subscription link is available at http://fellowships.asph.org/ from the bottom right hand corner of the screen. Additionally, applicants may check the same website for updates on available fellowships. Interested applicants must apply through ASPH's online grant application system (click on the APPLICATION tab at the top of the ASPH fellowship website) and mail required hard copy and supporting documents to ASPH. Required application materials include an essay, a resume or curriculum vitae, two letters of recommendation (from the references provided in the online application), official graduate transcripts, and proof of residency. Applicants are encouraged to review requirements for submitting these documents at the fellowship website.
Requirements: Students must have received an M.P.H. or Doctorate degree from an ASPH-member, CEPH-accredited graduate school of public health prior to the beginning of the fellowship. A list of accredited schools with full ASPH membership can be accessed via the sponsor website. Early career professionals with M.P.H. or Doctorate degrees (within 5 years of graduation) may also apply for the fellowship program. Selected fellows are required to relocate to Atlanta. Fellows are also required to maintain their own health insurance while in the fellowship program.

Restrictions: The ASPH/CD.C. Public Health Fellowships are available only to U.S. citizens or students holding a visa permitting permanent residence in the U.S. Applicants may apply to any number of ASPH fellowship programs; however applicants may apply to a maximum of only three positions within each program.
Geographic Focus: All States
Contact: Binley G. Taylor, Coordinator; (202) 296-1099; fax (202) 296-1252; btaylor@asph.org or trainingprograms@asph.org
Internet: https://fellowships.asph.org/Programs/details.cfm?programID=%22%26M8%20%0A
Sponsor: Association of Schools of Public Health
1900 M Street NW, Suite 710
Washington, D.C. 20036

ASPH/CDC Public Health Internships 782
The ASPH/CD.C. Public Health Internship Program is a collaborative effort between the Association of Schools of Public Health (ASPH) and the Centers for Disease Control and Prevention (CD.C.). The program enables schools of public health to assign graduate students in public health to the CD.C. for full-time internships in which "graduate candidates and early career professionals with graduate degrees can practice applying skills and knowledge learned in the classroom and field." The internships take place for ten weeks over the summer. Internship positions will be located either at CD.C. headquarters in Atlanta, Georgia, or at CD.C. geographical locations throughout the United States. Each intern will receive a training stipend for the ten-week assignment, paid in four equal installments. In addition to possibly fulfilling an internship requirement, students will benefit from: exposure to state-of-the-art information; acquisition of skills and knowledge useful to their careers; and interaction with technical experts in their chosen fields. Upon completion of the internship, the selected applicant will have made a useful contribution to a project of public health importance that is related to the mission of CD.C.. In addition, the selected applicant will have carried out scientific work that, if applicable, may be acceptable to the sponsoring academic institution for degree credit. Typically, the ASPH application season opens in November and runs through March. Requests for application are announced in ASPH's weekly electronic newsletter "The Friday Letter" which applicants may register to receive at no cost. The subscription link is available at http://internships.asph.org/. Additionally, applicants may check the ASPH Internship website for updates on available internships (http://internships.asph.org/). Prospective interns must apply through ASPH's online grant application system (available at the internship website) and mail required hard copy and supporting documents to ASPH. Required application materials include an essay, a resume or curriculum vitae, two letters of recommendation (from the references provided in the online application), official graduate transcripts, and proof of residency. Applicants are encouraged to review requirements for submitting these documents at the internship website.
Requirements: Applicants must be currently enrolled in a Masters or Doctoral program at an ASPH-member graduate school of public health accredited by the Council on Education for Public Health (CEPH). A list of ASPH members is available from the sponsor website. Applicants must be students at the time of application, although they may have graduated prior to entering the program. Applicants must be U.S. citizens or hold visas permitting permanent residence in the U.S. Selected interns are required to relocate to the internship location. Relocation costs of any kind will not be covered by ASPH or CD.C..
Restrictions: Applicants may apply to as many ASPH internships programs as they like (that are accepting applications). Additionally applicants may apply to up to three positions per program.
Geographic Focus: All States
Contact: Binley G. Taylor, Coordinator; (202) 296-1099; fax (202) 296-1252; trainingprograms@asph.org or btaylor@asph.org
Internet: https://internships.asph.org/programs/details.cfm?programID=!%230%20%20%0A
Sponsor: Association of Schools of Public Health
1900 M Street NW, Suite 710
Washington, D.C. 20036

ASP.O. Daiichi Innovative Technology Grant 783
The Daiichi Innovative Technology Grant will fund exploratory or hypothesis-generating projects that are not well-suited to a formal grant application (eg, based on statistical analysis and sample size specification) such as: development of new surgical or diagnostic instruments; survey or quality of life measures; new use of internet technology or computer software such as CDs or DVDs; educational brochures, materials, software for patients or physicians; and other applications of innovative technology for education or research in pediatric otolaryngology. Full patent and copyright control must be retained by the applicant and the applicant's institution. If patented innovations funded by this award generate more than $5,000, the applicant may be required to return funds to the Society. One award of up to $5,000, to be used over a one-year period, will be given annually. Letters of Intent should be received no later than December 15, with full proposals due on January 15.
Requirements: Researchers (MD, PhD, DMD, DO) in disciplines who will conduct research directly relevant to innovative technology in pediatric otolaryngology are eligible to apply. Applications submitted by otolaryngologists or demonstrating collaborations with otolaryngologists are preferred. Participation of an ASP.O. member is not required, but is preferred.
Geographic Focus: All States
Date(s) Application is Due: Jan 15; Dec 15
Amount of Grant: 5,000 USD
Contact: Stephanie L. Jones; (703) 519-1586 or (703) 836-4444; sljones@entnet.org
Internet: http://www.aspo.us/information.php?info_id=12
Sponsor: American Society for Pediatric Otolaryngology
One Prince Street
Alexandria, VA 22314-3357

ASP.O. Fellowships 784
ASP.O. exists to foster excellence in the care of children with otorhinolaryngologic disorders through education and research and thereby enhance the profession of Pediatric Otolaryngology. Its fellowship program intends to: facilitate the creation and dissemination of knowledge about the care of infants and children with ORL disorders; serve as an advocate for infants and children with ORL disorders, ASP.O. members and others with shared goals; and preserve and promote dedication to excellent and humane care for infants and children with ORL disorders. See the web site for details and all available fellowships.
Geographic Focus: All States
Contact: Stephanie L. Jones; (703) 519-1586 or (703) 836-4444; sljones@entnet.org
Internet: http://www.aspo.us/fellowships.php
Sponsor: American Society for Pediatric Otolaryngology
One Prince Street
Alexandria, VA 22314-3357

ASP.O. Research Grants 785
The American Society of Pediatric Otolaryngology (ASP.O.) awards funds annually to support innovative research in pediatric otolaryngology. ASP.O. will consider applications from both individuals and institutions. Preference is given to proposed projects that are to be completed within one year, although exceptional proposals that have duration in excess of one year will be considered. Two awards of up to $15,000, to be used over a one-year period, will be given annually. Letters of Intent should be received no later than December 15, with full proposals due on January 15.
Requirements: Researchers (MD, PhD, DMD) in disciplines who will conduct research directly relevant to pediatric otolaryngology are eligible to apply.
Restrictions: No portion of any grant may be used for travel expenses or for principal investigator salaries.
Geographic Focus: All States
Date(s) Application is Due: Jan 15; Dec 15
Amount of Grant: 15,000 USD
Contact: Stephanie L. Jones; (703) 535-3747 or (703) 836-4444; sljones@entnet.org
Internet: http://www.aspo.us/information.php?info_id=12
Sponsor: American Society for Pediatric Otolaryngology
One Prince Street
Alexandria, VA 22314-3357

Assisi Foundation of Memphis Capital Project Grants 786
The foundation supports organizations in its areas of interest, including health and human services--promote the health and well-being of the Mid-South community and help the health care system respond more effectively to community needs; education and literacy--projects/programs that build organizational capacity of provider agencies, provide professional development to service providers, promote collaboration among provider agencies, and leverage resources (local, state, and federal); social justice/ethics--projects/programs that strengthen ethical values among Mid-South citizens and promote social justice leading to a better understanding of and a more effective response to economic or social threats to the community; and cultural enrichment and the arts--projects/programs that foster an appreciation of the arts in the Greater Memphis community. Religious organizations seeking funding for religious programs also are eligible. Typical capital projects include: building - new construction, addition to existing facility, or renovation; technology - Information Management System installation/upgrade, computer hardware, audio-visual equipment/systems; and furnishings/equipment. Typically, payment of capital project grants is made when the organization begins the construction/renovation. Specific terms and schedule of payments will be based upon the scope of the project, amount of award, duration of the project, and completion of required reporting at appropriate intervals.
Requirements: The Foundation makes grants only to organizations that are classified as tax-exempt under Section 501(c)3 of the Internal Revenue Code and as public charities under Section 509(a) of that Code. The Foundation uses its resources for charitable endeavors that advance the well-being of people and institutions located in Shelby, Fayette, and Tipton Counties in Tennessee; Crittenden County, Arkansas; and Desoto County, Mississippi.
Restrictions: Grants are not made for individuals, national fundraising drives, projects that address the needs of only one congregation, tickets for benefits, political organizations or candidates for public office, lobbying activities, recurring budget deficits, or tournament fees and/or travel for athletic competitions.
Geographic Focus: Arkansas, Mississippi, Tennessee
Date(s) Application is Due: Feb 15; May 17; Aug 16; Nov 15
Contact: Jan Young, Executive Director; (901) 684-1564; jyoung@assisifoundation.org
Internet: http://www.assisifoundation.org/capitalproject.html
Sponsor: Assisi Foundation of Memphis
515 Erin Drive
Memphis, TN 38117

Assisi Foundation of Memphis General Grants 787
The foundation supports organizations in its areas of interest, including health and human services--promote the health and well-being of the Mid-South community and help the health care system respond more effectively to community needs; education and literacy--projects/programs that build organizational capacity of provider agencies, provide professional development to service providers, promote collaboration among provider agencies, and leverage resources (local, state, and federal); social justice/ethics--projects/programs that strengthen ethical values among Mid-South citizens and promote social justice leading to a better understanding of and a more effective response to economic or social threats to the community; and cultural enrichment and the arts--projects/programs that foster an appreciation of the arts in the Greater Memphis community. Religious organizations seeking funding for religious programs also are eligible. Deadlines are set to coordinate with quarterly meetings of the Board of Directors of the Foundation.
Requirements: The Foundation makes grants only to organizations that are classified as tax-exempt under Section 501(c)3 of the Internal Revenue Code and as public charities under Section 509(a) of that Code. The Foundation uses its resources for charitable endeavors that advance the well-being of people and institutions located in Shelby, Fayette, and Tipton Counties in Tennessee; Crittenden County, Arkansas; and Desoto County, Mississippi.
Restrictions: Grants are not made for individuals, national fundraising drives, projects that address the needs of only one congregation, tickets for benefits, political organizations or candidates for public office, lobbying activities, recurring budget deficits, or tournament fees and/or travel for athletic competitions.
Geographic Focus: Arkansas, Mississippi, Tennessee
Date(s) Application is Due: Feb 15; May 17; Aug 16; Nov 15
Amount of Grant: Up to 20,000 USD
Contact: Jan Young, Executive Director; (901) 684-1564; jyoung@assisifoundation.org
Internet: http://www.assisifoundation.org/generalgrants.html
Sponsor: Assisi Foundation of Memphis
515 Erin Drive
Memphis, TN 38117

Assurant Foundation Grants 788
The Foundation awards approximately 15 grants per year to improve the quality of life in the New York metropolitan community. The Foundation is committed to fulfilling the basic needs of the community, be they educational, health, nutrition, or housing.
Requirements: 501(c)3 nonprofits in the New York City metropolitan area are eligible.
Restrictions: As a general policy the foundation does not make grants to individuals; political parties, candidates or lobbying activities; religious organizations; or fundraising events.
Geographic Focus: New York
Amount of Grant: Up to 5,000 USD
Contact: Kristy Hall, (212) 859-7000; Kristy.Hall@assurant.com
Internet: http://www.assurant.com/inc/assurant/community/new-york.html
Sponsor: Assurant Foundation
1 Chase Manhattan Plaza, 41st Floor
New York, NY 10005

Assurant Health Foundation Grants 789
The foundation awards grants in Wisconsin in its areas of interest, including health care access and promotion, disease prevention, the arts and education, as well as the foundation's national public policy initiatives. Types of support include capital campaigns, consulting services, continuing support, employee matching gifts, employee-related scholarships, in-kind gifts, matching/challenge support, and program development.
Requirements: Wisconsin nonprofit organizations are eligible. Preference is given to requests serving southeastern Wisconsin.
Restrictions: Unsolicited requests for funds not accepted. Requests for funds only accepted after the foundation sends out requests for proposals.
Geographic Focus: Wisconsin
Samples: Gerald Ignace Clinic (WI)--$10,000; Task Force on Family Violence (WI)--$10,000; Latino Arts Organization (WI)--$5000.
Contact: Grants Administrator; (414) 299-1348; megan.hindman@assurant.com
Internet: http://www.newsroom.eassuranthealth.com/corp/news/newsroom/articles/press_06092005.htm
Sponsor: Assurant Health Foundation
501 W Michigan Street
Milwaukee, WI 53203

Astellas Foundation Research Grants 790
The Foundation was established with the objectives of contributing to the improvement of national health and welfare, and to the progress of therapeutic medicines. In fulfilling the objectives, the Foundation aims to conduct pioneering research in new fields of diseases and metabolisms of medicines, by elucidating the mechanism of diseases and their treatment, in particular the relationship between therapeutic medicines and in-situ metabolisms. Major activities include: subsidizing research on metabolic disorders related to the diagnosis of diseases and therapeutic medicine; publication and editing of periodicals for research achievements; supporting and production of educational movies; and other projects to achieve the objectives of the Foundation.
Geographic Focus: All States
Contact: Grants Manager; 81 3 3244 3397; fax 81 3 5201 8512
Sponsor: Astellas Foundation
2-3-11 Nihonbashihoncho, Chuo-ku
Tokyo, 103-8411 Japan

ASU Graduate College Reach for the Stars Fellowship 791
Reach for the Stars Fellowship support new and regularly-admitted first-year students who demonstrate academic excellence and are underrepresented in their field of study. The Fellowship provides a $15,000 award for the first academic year plus tuition. In the second year, the academic unit will provide at least a 50% teacher assistant or research assistant position (at the department standard program rate) or similar funding assuming satisfactory academic progress. The nomination form is available at the Fellowship website.
Requirements: The nominee must be regularly admitted into a Master's degree program at the time the award is disbursed; demonstrate academic excellence and contribute to diversity in the field of study and the profession; and be a U.S. citizen or assigned an IP visa. All Fellowship recipients are required to register and participate in the one credit

Interdisciplinary Research Colloquium; enroll for a minimum of nine graduate-level credit hours during fall and spring semesters during the award period; meet regularly with a faculty mentor to discuss program goals, objectives and progress toward degree completion; meet regularly with a peer mentor to discuss getting acclimated to graduate school and student life; maintain satisfactory academic progress as defined by the academic unit and the Graduate College; not be employed as a TA, RA or in any other position at ASU during the year of this award; and submit a progress report signed by the academic mentor at the end of each semester.
Restrictions: Students must be nominated by their admitting academic unit. Mandatory fees, program fee, or any other fees are not covered as part of the fellowship.
Geographic Focus: All States
Date(s) Application is Due: May 15
Amount of Grant: 15,000 USD
Contact: Jennifer Cason; (480) 965-6113; fax (480) 727-6615; Jennifer.Cason@asu.edu
Internet: http://graduate.asu.edu/financing/fellowships/graduate-college-fellowships/stars
Sponsor: Arizona State University Graduate College
Interdisciplinary Building, B-Wing
Tempe, AZ 85287-1003

ASU Graduate College Science Foundation Arizona Bisgrove Postdoctoral Scholars 792
The Bisgrove Postdoctoral Scholars Program (sponsored by Science Foundation Arizona) is designed to attract the nation's best early career scientists and engineers who exhibit the potential for outstanding competence and creativity in their research areas, strong communication skills, a passion for communicating the importance of their research to society, and a keen interest in educational science outreach to the community. Bisgrove Scholars will receive an annual stipend of approximately $60,000, benefits and additional funding for research expenses. The Bisgrove appointment is renewable on a year-to-year basis for a maximum initial term of two years, depending on the availability of funds. The ASU Graduate College Bisgrove Leadership Academy has been launched to provide training and guidance for Bisgrove Postdoctoral Scholars in areas of collaborative interdisciplinary research, user-inspired research, laboratory management and leadership, together with related resources, that will ensure their success as they develop as scientists in a multidisciplinary setting. The major activity of the Academy will include Strategies for Success workshops and seminars led by experts in a range of fields, with participating speakers from ASU, industry and the community. Bisgrove Scholars will also be invited to participate in the Preparing Future Faculty (PFF) program. Specific details about the application package are available at the Fellowship website.
Requirements: United States citizens or permanent residents of the U.S. should have completed a Ph.D. by the time of appointment, with no prior post-doctoral experience. They should also demonstrate research training and potential to transform ideas into value for society and the interest to work at the convergence of several disciplines. Scholars should be prepared to commit time to mentor undergraduate and graduate students in related disciplines; meet regularly with a faculty member to discuss research goals and objectives; submit a progress report every 90 days, summarizing their goals and evidence of progress; and co-write at least one federal grant application with the main research faculty mentor during the second year of the award, assuming satisfactory progress.
Geographic Focus: All States
Date(s) Application is Due: Jan 14
Amount of Grant: 60,000 USD
Contact: Jennifer Cason; (480) 965-8968; Jennifer.Cason@asu.edu
Internet: http://graduate.asu.edu/bisgrove
Sponsor: Arizona State University Graduate College
Interdisciplinary Building, B-Wing
Tempe, AZ 85287-1003

AT&T Foundation Civic and Community Service Program Grants 793
The foundation supports programs that enhance education by integrating new technologies and increasing learning opportunities, improve economic development through technology and local initiatives, provide vital assistance to key community-based organizations, support cultural institutions that make a community unique, and advance the goals and meet the needs of diverse populations.
Requirements: The foundation makes 501(c)3 grants to tax-exempt, nonprofit organizations in certain states. Requirements are available online.
Geographic Focus: All States
Samples: Leader Dogs for the Blind (Rochester, Mich.) received a $25,000 grant for blind and visually impaired individuals. Communities in Schools of San Antonio - received a grant to enhance the technology infrastructure to reach more young people at risk of dropping out of school.
Contact: Marilyn Reznick, (212) 387-6555; fax (212) 387-5097; reznick@att.com
Internet: http://www.att.com/gen/corporate-citizenship?pid=7736&D.C.MP=att_foundation
Sponsor: AT&T Foundation
32 Avenue of the Americas, 24th Floor
New York, NY 10013

Atherton Family Foundation Grants 794
The Foundation was created by Juliette Atherton, daughter of missionaries to Hawaii and widow of industrialist Joseph B. Atherton. The Foundation is now one of the largest endowed private grantmakers in Hawaii. The Foundation supports organizations working in the areas of arts/culture, human services, education, environment, community development, health, religion/spiritual development and youth services. Education and human services are its two largest giving areas. Awards generally range from $2,000 up to $100,000. Annual deadlines for application submission are January 4, April 1, July 1, and October 3.
Requirements: The Foundation makes grants for programs and projects that benefit the people of Hawai'i. Applicants must have 501(c)3 status or must apply through a fiscal sponsor with 501(c)3 status. If an organization applies through a fiscal sponsor, the fiscal sponsor must agree that the purpose of the grant is charitable, to monitor the grant project, control the expenditure of grant funds, and ensure compliance with the terms and conditions of the grant. The Foundation will award no more than one grant to an organization at a time and no more than one grant in any calendar year. A grantee serving as the fiscal sponsor for another organization may receive a second grant for its own project. All previous grants must be completed with the submission of a final report before an organization is eligible to apply for a new grant. If you are unable to submit your proposal online, contact Pam Funai (see contact information below--neighbor islands can call toll-free 888-731-3863 ext. 537). Completed applications must be submitted by 5:00 pm on the deadline date.
Restrictions: The Foundation does not serve as the sole funder of any organization. The Foundation does not make grants to organizations classified under 509(a)3 of the Internal Revenue Code; to the Hawaii State Department of Education schools or public charter schools; to the University of Hawai'i other than an annual grant to the University of Hawai'i Foundation. The Foundation generally does not fund endowments or operating support, nor does it fund: conferences, festivals, and similar one-time events; activities that have already occurred; lobbying; loans; funds for re-granting and grants to individuals or for the benefit of identified individuals (except scholarships through the Juliette M. Atherton Scholarship and the Community Scholarship Fund of the Hawai'i Community Foundation).
Geographic Focus: Hawaii
Date(s) Application is Due: Jan 4; Apr 1; Jul 1; Oct 3
Amount of Grant: 2,000 - 100,000 USD
Samples: Bishop Museum, Honolulu, Hawaii, $100,000 - Hawaiian Hall restoration project; Hawaii Theatre Center, Honolulu, Hawaii, $75,000 - capital improvements; Honolulu Academy of Arts, Honolulu, Hawaii, $45,000 - support of an Ally for Academic Achievements.
Contact: Judith M. Dawson, President; (808) 566-5524 or (808) 566-5537; fax (808) 521-6286; jdawson@hcf-hawaii.org or foundations@hcf-hawaii.org
Internet: http://www.athertonfamilyfoundation.org/grantseekers/
Sponsor: Atherton Family Foundation
827 Fort Street Mall
Honolulu, HI 96813

Athwin Foundation Grants 795
Established in 1956, the Athwin Foundation serves the Minneapolis, St. Paul, Bloomington, Minnesota, Wisconsin area. Primary areas of interest include: arts and culture, church related projects, education, and social services. Providing support in the form of: capital campaigns, general/operating support and, program development. Contact the Foundation by submitting a Letter of Inquiry, if they are interested in your project, a formal request will be made & then a proposal maybe submitted.
Requirements: IRS 501(c)3 non-profit organizations are eligible to apply. The foundation prefers applicants to use the Minnesota Common Grant Application Form that can be obtained by calling the Minnesota Council on foundations (612) 338-1989 or by visiting their web site.
Restrictions: Grants do not support: individuals, scholarships, fellowships, or loans.
Geographic Focus: Minnesota, Wisconsin
Date(s) Application is Due: Mar 1
Amount of Grant: 2,000 - 100,000 USD
Samples: Foundation of Childrens Hospitals and Clinics of Minnesota, Roseville, MN, $100,000-- for general operating support; Neighborhood Involvement Program, Minneapolis, MN, $15,000--for general operating support; Jungle Theater, Minneapolis, MN, $2,000--for general operating support.
Contact: Bruce Bean, Trustee; (952) 915-6165
Sponsor: Athwin Foundation
5200 Wilson Road, Suite 307
Minneapolis, MN 55424-1344

Atlanta Foundation Grants 796
The purpose of the Atlanta Foundation is to assist charitable and educational institutions located in Fulton County or DeKalb County, Georgia that promote education or scientific research; advance care for the sick, aged, or helpless; improve living conditions; provide recreation for all classes; and such other charitable purposes as will improve the mental, moral, and physical life of the inhabitants of Fulton and DeKalb counties regardless of race, color, or creed. The foundation's board meets in April and October. Requests must be received by March 1 or September 1 to be considered.
Requirements: Nonprofit organizations in Georgia's DeKalb and Fulton Counties may apply for grant support.
Restrictions: Grants are not awarded to individuals or for scholarships, fellowships, or loans.
Geographic Focus: Georgia
Date(s) Application is Due: Mar 1; Sep 1
Amount of Grant: 2,500 - 50,000 USD
Contact: Mike Dinnelly; (888) 234-1999; grantadministration@wellsfargo.com
Internet: https://www.wellsfargo.com/private-foundations/atlanta-foundation
Sponsor: Atlanta Foundation
3414 Peachtree Road, 5th Floor, MC GA8023
Atlanta, GA 30326

Atran Foundation Grants 797
The foundation supports Jewish nonprofits with a focus on New York and Israel in the areas of higher education, Jewish education, religious education, community services, Jewish welfare, temples, international ministries and missions, temples, medical centers, and women's affairs. Types of support include conferences and seminars, endowment

funds, general support, matching funds, multiyear/continuing support, project support, research, scholarships, and seed money grants. The foundation requires that proposals be in writing. Proposals should include the nature of the project, its objectives and significance, time estimate, and budget.
Requirements: 501(c)3 tax-exempt Jewish organizations are eligible.
Restrictions: Grants are not made to individuals.
Geographic Focus: All States
Date(s) Application is Due: Sep 30
Amount of Grant: 250 - 100,000 USD
Samples: Yivo Institute for Jewish Research (New York, NY)--for project support, $90,000; Brandeis University (Waltham, MA)--for operating support, $28,000; Folksbiene Yiddish Theater (New York, NY)--for operating support, $15,000.
Contact: Diane Fischer, President; (212) 505-9677
Sponsor: Atran Foundation
23-25 E 21st Street, 3rd Floor
New York, NY 10010

Auburn Foundation Grants 798
The purpose of the Foundation is to stimulate giving and cooperative leadership among the citizens of Auburn; help improve the lives of all community residents, especially those who are most vulnerable; and enrich the cultural environment and community life. Of special interest are projects that bring together all ages and sections of the town, or that contribute to healthy, active living. Application is available online.
Requirements: Any nonprofit that serves residents of Auburn is invited to apply.
Restrictions: Grants will not be awarded to for-profit businesses or expenses already incurred by the applicant.
Geographic Focus: Massachusetts
Date(s) Application is Due: Apr 15
Amount of Grant: 5,000 USD
Samples: Bancroft School (Worcester, MA)--for general support, $45,000; New England Science Ctr, Rutland House (Worcester, MA)--for repairs, $20,000; Tower Hill Botanic Garden (Boylston, MA)--for general support, $30,000.
Contact: Lois Smith; (508) 755-0980, ext. 107; lsmith@greaterworcester.org
Internet: http://www.greaterworcester.org/grants/Auburn.htm
Sponsor: Auburn Foundation
370 Main Street, Suite 650
Worcester, MA 01608-1738

Audrey and Sydney Irmas Charitable Foundation Grants 799
The foundation awards grants to eligible nonprofit organizations in its areas of interest, including arts and culture, higher education, homeless and urban issues, hospitals, and Jewish welfare. Grants support long-term, nonrenewable pledges and general operating grants. Most grants are made to Los Angeles County, CA, nonprofits.
Requirements: 501(c)3 nonprofit organizations are eligible.
Geographic Focus: California
Amount of Grant: 1,000 - 75,000 USD
Samples: Childrens Hospital of Los Angeles (Los Angeles, CA)--$2810.65; Jewish Family Services (Los Angeles, CA)--$500; Downtown Womens Center (Los Angeles, CA)--$200; Carnegie Museum of Art (Pittsburg, PA)--$25,000.
Contact: Robert Irmas; (818) 382-3313; fax (818) 382-3315; robirm@aol.com
Sponsor: Audrey and Sydney Irmas Charitable Foundation
16830 Ventura Boulevard, Suite 364
Encino, CA 91436-2797

AUPHA Corris Boyd Scholarships 800
The Association of University Programs in Health Administration offers the Corris Boyd Scholarship in honor of a healthcare leader who dedicated his life to diversity and excellence in leadership, especially among people of color and women. The program currently offers two $40,000 scholarships to deserving students of color entering a graduate program in healthcare management. The award will be paid in four installments, paid in August and December of the first year, then August and December of the second year. The annual deadline for applicants is May 3.
Requirements: At the time of award, applicants must meet the following eligibility
Requirements: have applied and been accepted into an AUPHA full-member masters degree program for start in the upcoming Fall semester (not before August 1); be an under-represented minority; have a minimum 3.0 GPA (out of 4.0) in undergraduate coursework; and have U.S. citizenship or permanent resident status.
Geographic Focus: All States
Date(s) Application is Due: May 3
Amount of Grant: 40,000 USD
Contact: Lydia Middleton, President and CEO; (703) 894-0940, ext. 115; fax (703) 894-0941; lmiddleton@aupha.org or aupha@aupha.org
Internet: http://www.aupha.org/i4a/pages/Index.cfm?pageID=3541
Sponsor: Association of University Programs in Health Administration
2000 14th Street North, Suite 780
Arlington, VA 22201

AUPHA Foster G. McGaw Scholarships 801
The Association of University Programs in Health Administration offers the Foster G. McGaw Scholarship, which was established in 1975 by Foster G. McGaw, founder of the American Hospital Supply Corporation. McGaw recognized the importance of health administration education and AUPHA's contribution to the field. The award provides financial support to undergraduate and graduate students in health administration.

Scholarship funds are awarded each year to all AUPHA full graduate programs. The faculty within these programs disperse these monies at their discretion to students most deserving of recognition. The annual deadline for applicants is May 3.
Requirements: At the time of award, applicants must meet the following eligibility
Requirements: have applied and been accepted into an AUPHA full-member undergraduate or graduate degree program for start in the upcoming Fall semester (not before August 1); be an under-represented minority; have a minimum 3.0 GPA (out of 4.0) in undergraduate coursework; and have U.S. citizenship or permanent resident status.
Geographic Focus: All States
Date(s) Application is Due: May 3
Contact: Lydia Middleton, President and CEO; (703) 894-0940, ext. 115; fax (703) 894-0941; lmiddleton@aupha.org
Internet: http://www.aupha.org/i4a/pages/index.cfm?pageid=3354
Sponsor: Association of University Programs in Health Administration
2000 14th Street North, Suite 780
Arlington, VA 22201

Aurora Foundation Grants 802
The foundation provides grants to eligible nonprofit organizations in the following categories: education, social services, health care, and the arts and humanities. In general, grants are made for capital purposes only, not for operating expenses. Evidence of tax-exempt status must be submitted with all applications. Preference will be given to requests that demonstrate the most urgent and immediate need for funding. Grants are ordinarily made for one year only.
Requirements: 501(c)3 and 170(b)1a nonprofit agencies located in the foundation's service area, which includes the City of Aurora, southern Kane County, and Kendall County in Illinois, are eligible.
Restrictions: The foundation rarely provides the entire support of a project. Grants are not generally available for those agencies and institutions that are funded primarily through tax support. Telephone the office prior to submitting a formal proposal.
Geographic Focus: Illinois
Date(s) Application is Due: Aug 1
Contact: Grants Administrator; (630) 896-7800; grant@aurorafdn.org
Internet: http://www.aurorafdn.org/grantmaking.html
Sponsor: Aurora Foundation
111 W Downer Place, Suite 312
Aurora, IL 60506

Austin-Bailey Health and Wellness Foundation Grants 803
The foundation awards grants to Ohio nonprofit organizations to support programs that promote the physical and mental well-being of the citizens of Holmes, Stark, Tuscarawas and Wayne Counties in the state of Ohio. The Foundation emphasizes healthcare affordability concerns of the uninsured adn underinsured, the poor, children, single parents and the aging. It also advocates programs that speak to the mental health needs of the individuals and families. Opportunities to work with other foundations and organizations to promote the principles of health and wellness are welcome.
Requirements: Applicant organizations must provide a current copy of IRS determination indicating 501(c)3 and 509(a) tax-exempt status; serve part or all of the four county area of Holmes, Stark, Tuscarawas, and Wayne; provide a description of the organization and its activities; state the need to be addressed; and provide a brief description of the project or program and how it will address the health and wellness needs of the community.
Restrictions: The foundation generally does not support annual appeals or membership drives; fund-raising events, program advertising, lobbying activities, endowment funds or organizations not classified as tax exempt by the IRS.
Geographic Focus: Ohio
Date(s) Application is Due: Jun 30; Dec 17
Amount of Grant: 1,000 - 15,000 USD
Contact: Administrator; (330) 580-2380; fax (330) 580-2381; abfdn@sbcglobal.net
Internet: http://fdncenter.org/grantmaker/austinbailey/guide.html
Sponsor: Austin-Bailey Health and Wellness Foundation
2719 Fulton Road NW, Suite D
Canton, OH 44718

Austin College Leadership Award 804
The award honors an outstanding individual who through his/her life's work has demonstrated the principles of servant leadership by: taking a courageous stand on a public policy issue that advances a humanitarian or educational purpose; serving the youth of a state, nation, or international community to improve the quality of health, educational, or community services; or creating opportunities for young people that help them enhance their educational experience and move to a new level of service to society. The recipient also will visit Austin College to speak to community leaders and interact with students and faculty. Nomination form and guidelines are available online.
Requirements: Nominees will be leaders from all levels whose leadership position will have placed him/her in a unique and special position among leaders throughout the world.
Geographic Focus: All States
Date(s) Application is Due: Jun 1
Amount of Grant: 100,000 USD
Contact: Award Administrator; (903) 813-2000 or (800) 526-4276; fax (903) 813-3199
Internet: http://www.austincollege.edu/category.asp?3523
Sponsor: Austin College
900 North Grand Avenue
Sherman, TX 75090-4400

120 | Grant Programs

Austin S. Nelson Foundation Grants   805
The primary purpose of the Austin S. Nelson Foundation is to support charitable interests throughout Alberta, Canada. Its major interest areas include: medical research; religious organizations and agencies, community services, and human services. In the area of medicine, it will provide funding for pediatrics, palliative care, cancer research, cerebral palsy, heart disease, diabetes, lung disease, kidney disorders, leukemia, multiple sclerosis, Alzheimer disease, and severe burns. It also supports Christian organizations and Anglican churches. Other areas of giving includes support for organizations that help the blind, physically disabled children, animal welfare, emergency shelters, crisis intervention services, the poor, alcohol and drug abuse, young offenders, crime prevention, mental health, domestic violence victims, abused children, and sexual assault victims. Though there are no specified annual deadlines, grant decisions are made primarily in November. Interested applicants should contact the Foundation by mail.
Geographic Focus: Canada
Contact: Director
Sponsor: Austin S. Nelson Foundation
4825 89th Street
Edmonton, AB T6E5L3 Canada

Australasian Institute of Judicial Administration Seed Funding Grants   806
The Australasian Institute of Judicial Administration (AIJA) is a research and educational institute associated with Monash University. It is funded by the Standing Council on Law and Justice (SCLJ) and also from subscription income from its membership. The principal objectives of the Institute include research into judicial administration and the development and conduct of educational programmes for judicial officers, court administrators and members of the legal profession in relation to court administration and judicial systems. The AIJA is interested in working with academics to develop proposals for research projects. It is offering financial assistance to provide an incentive for academics to prepare ARC grant applications in areas of AIJA interest. Broad areas of research identified as relevant to judicial administration are: the proportional use of judicial resources for optimum practical effectiveness and efficiency; and public perceptions and understanding of the justice system. Included in these broad areas are: the examination of the relationship between the principles of adjudicatory independence and administrative accountability; a comparative analysis of reforms in the justice system in other jurisdictions; performance measurement, including the creation of best practice models for the introduction of bench-marking and yardstick competition models in the courts; and proportionality, namely, the proportion of the value of claims expended on legal costs and the expenditure of court resources to meet the instances and areas of greatest need and access to justice.
Geographic Focus: All States, All Countries
Contact: Gregory Reinhardt, Executive Director; +61 3 9600 1311; fax +61 3 9606 0366; Gregory.Reinhardt@monash.edu
Internet: http://www.aija.org.au/index.php/research/seed-funding
Sponsor: Australasian Institute of Judicial Adminstration
555 Lonsdale Street, Ground Floor
Melbourne, VIC 3000 Australia

Australian Academy of Science Grants   807
The objectives of the Academy are to promote science through a range of activities. It has defined four major program areas as their focus. The Academy focus is on recognition of outstanding contributions to science, education and public awareness, science policy, and international relations. Guidelines are available online.
Geographic Focus: All States, Australia
Contact: Executive Secretary; 61-2-6201-9400; fax 61-2-6201-9494; eb@science.org.au
Internet: http://www.science.org.au
Sponsor: Australian Academy of Science
G.P.O. Box 783
Canberra, ACT 2601 Australia

Autauga Area Community Foundation Grants   808
The Autauga Area Community Foundation (AACF) is a public foundation which links charitable resources with community needs and opportunities. Each year, the Foundation awards grants to nonprofits offering projects and programs in Autauga County that, in the opinion of AACF's Advisory Committee, will improve the quality of life in the community. While many factors are considered, priority is given to proposals that meet the following criteria: programs that address issues affecting Autauga County; seed grants to initiate promising new projects addressing underlying causes of community problems; expanding programs representing innovative and efficient approaches to serving community needs and opportunities; programs that maximize resources and leverage other monies; projects reflecting the cooperative efforts of multiple agencies within the community; and programs that can demonstrate funding plans for the continuation of the project beyond initial funding by the AACF. The maximum grant award is $2500, with the average grant ranging from $500 to $1000. Applications will be accepted online.
Requirements: Nonprofits located in, or serving the residents of, Autauga County are eligible.
Restrictions: Grants are not awarded to: individuals, fundraising events, or capital campaigns.
Geographic Focus: Alabama
Date(s) Application is Due: Mar 8
Amount of Grant: Up to 2,500 USD
Contact: Caroline Montgomery Clark, Vice President, Community Services; (334) 264-6223; fax (334) 263-6225; cacfgrants@bellsouth.net
Internet: http://www.cacfinfo.org/aacf/grants.html
Sponsor: Autauga Area Community Foundation
434 N. McDonough Street
Montgomery, AL 36104

Autodesk Community Relations Grants   809
Autodesk plays an active role in the communities where employees live and work. Grants are awarded to local nonprofit organizations in the program's areas of interest, including community development, arts, the disabled, education, environment, health and human services, science and technology, and civic affairs. Grants are awarded on a national basis. Preference is given to organizations where Autodesk does business. There are no application deadlines. Written proposal requirements are available online.
Requirements: 501(c)3 tax-exempt organizations are eligible.
Restrictions: Grants do not support sporting events or athletic teams, churches or religious organizations, political parties or organizations, general advertising, video, film or television productions, individuals, subscription fees or admission tickets, fraternal, veteran or sectarian groups or direct funds to Autodesk employees participating in fund-raising events.
Geographic Focus: All States
Amount of Grant: 1,000 - 3,000 USD
Contact: Community Relations Manager; (415) 507-6138
Internet: http://usa.autodesk.com/adsk/servlet/index?siteID=123112&id=1064603
Sponsor: Autodesk
111 McInnis Parkway
San Rafael, CA 94903

Autzen Foundation Grants   810
The foundation awards grants to Oregon nonprofit organizations in its areas of interest of arts, children/youth, services, environment, health care, higher education, human services, and performing arts. Types of support include building construction/renovation, continuing support, matching/challenge grants, program development, and seed grants.
Requirements: Nonprofit organizations in the Pacific Northwest are eligible with an emphasis on those in Oregon.
Restrictions: Grants do not support individuals, scholarships, fellowships, or loans.
Geographic Focus: Oregon, Washington
Date(s) Application is Due: Mar 15; Aug 15; Nov 15
Amount of Grant: 2,000 - 27,000 USD
Samples: Clackamas Womens Services, Oregon City, OR, $4,000; Christie School, Marylhurst, OR, $15,000; Salvation Army, Cascade Division, Portland, OR, $10,000.
Contact: Kim Freed, Administrator; (503) 226-6051; autzen@europa.com
Sponsor: Autzen Foundation
P.O. Box 3709
Portland, OR 97208-3709

AVDF Health Care Grants   811
Since 1981, the Foundations have focused grants in health care on "caring attitudes." Caring attitudes often involve qualities such as effective communication, expression of compassion and empathy, as well as awareness of and sensitivity to how these factors impact the well-being of a patient. Trustee's current focus is changing the culture of health care to be more attentive to the emotional, psychological and spiritual needs of patients. This includes integrating caring attitudes into the training of health care providers, especially physicians. Programs with emphasis on integrating caring attitudes throughout the delivery system are also important. Particular areas of interest include projects that promote caring attitudes through: inter-professional collaboration within the health care team; training of health care providers in patient-centered care; and improving the culture of physician education at both the undergraduate and graduate levels. Projects designed to leverage improvements in patient care throughout a delivery system will be most competitive. Grants made in this program area normally range from $100,000-$200,000.
Requirements: Preference will be given to institutions and organizations with the visibility and organizational capacity to develop programs likely to be replicated nationally. Proposals should come from the head of an independent nonprofit organization such as a medical center or hospital or, in the case of large universities with multiple schools and colleges, the vice-president or dean of the school of medicine or other health profession. Public and private institutions are eligible to apply. Contributions of the requesting institution and other sources of project support should be identified. Proposals may be submitted via email to dflippin@avdf.org or hard copy. There are no deadlines for proposals and grant applications may be submitted at any time of the year. However, the process of moving from proposal submission to grant approval takes time. Therefore, the Foundations are not able to meet requests for eligible projects requiring immediate or near-term funding.
Geographic Focus: All States
Amount of Grant: 100,000 - 200,000 USD
Contact: Cheryl Tupper, Senior Program Director for Religion and Health Care; (904) 528-0699; fax (904) 359-0675; ctupper@avdf.org
Internet: http://www.avdf.org/FoundationsPrograms/HealthCare.aspx
Sponsor: Arthur Vining Davis Foundations
225 Water Street, Suite 1510
Jacksonville, FL 32202-5185

Avista Foundation Economic and Cultural Vitality Grants   812
The Avista Foundation focuses its giving on grants that strengthen communities and enhance the quality of lives of the people served by Avista Utilities or the Alaska Light and Power Company. In the area of Economic and Cultural Vitality, the Foundation supports projects that help its communities and citizens to grow and prosper. Key examples include: support for capital projects that provide vital civic and social infrastructure and economic growth; and support for signature community events and programs that provide significant community identity and quality of life. The application process should be completed online.
Requirements: Applicants should be an IRS 501(c)3 organization serving the residents of Avista Utilities. Eligible regions include: eastern Washington; Goldendale and

Stevenson, Washington; northern Idaho; southwestern Oregon; La Grande, Oregon; eastern Montana; and the city and borough of Juneau, Alaska.
Restrictions: The Foundation does not support: individuals; team or extra-curricular school events; trips or tours; religious organization; fraternal organization; memorial campaigns; national health organizations (or their local affiliates); or research/disease advocacy groups.
Geographic Focus: Alaska, California, Idaho, Montana, Oregon, Washington
Contact: Kristine Meyer, Executive Director; (509) 495-8156; kristine.meyer@avistacorp.com or contributions@avistacorp.com
Internet: http://www.avistafoundation.com/home/pages/default.aspx
Sponsor: Avista Foundation
P.O. Box 3727
Spokane, WA 99220-3727

Avon Foundation Breast Care Fund Grants 813
The fund awards grants for the development of breast cancer education and outreach activities, particularly breast health education and early detection services for under-served women. AFBCF supports programs that: recruit women for both first time screening and annual screening; develop partnerships between community-based outreach providers and local medical providers; work with health care providers to ensure proper clinical follow-up of abnormal screening results; and educate older women (65+ years old) about Medicare coverage of annual screening mammograms and assist them in obtaining the service from providers who accept Medicare.
Requirements: Private, non-government, nonprofit organizations in the United States and Puerto Rico are eligible. Grants are awarded to community-based programs and/or health care agencies that provide medically under-served women aged 40 and older with direct access to breast cancer education, annual clinical screening services, and prompt follow-up care. All programs must utilize the three-part approach to breast cancer early detection including regular screening mammography, clinical breast examination, and monthly breast self-examination.
Geographic Focus: All States
Date(s) Application is Due: Aug 28
Amount of Grant: 30,000 - 50,000 USD
Contact: Coordinator; (212) 244-5368; fax (212) 695-3081; admin@avonbreastcare.org
Internet: http://www.avonbreastcare.org/fundinginfo.htm
Sponsor: Avon Foundation
505 Eighth Avenue, Suite 1601
New York, NY 10018-6505

Avon Products Foundation Grants 814
The foundation's two-fold focus is to support education, community and social services, and arts organizations and programs that provide economic opportunities for women and girls; and to support breast cancer and other women's health organizations and programs. The foundation awards grants in cities and regions with a large concentration of representatives and business operations, with the majority of funds going to US-based institutions. National, international, and New York metropolitan area programs are administered through the foundation's headquarters in New York. For regional funding support, refer to the Avon Foundation website.
Requirements: Applying organizations must be tax-exempt; national and municipal organizations are eligible. Request the guidelines brochure prior to submitting a formal proposal.
Restrictions: Grants do not support individuals; memberships; lobbying organizations; political activities and organizations; religious, veteran, or fraternal organizations; fundraising events; and journal advertisements.
Geographic Focus: All States
Amount of Grant: Up to 23,000,000 USD
Contact: Grants Administrator; (866) 505-2866; info@avonfoundation.org
Internet: http://www.avoncompany.com/women/avonfoundation
Sponsor: Avon Products Foundation
1345 Avenue of the Americas
New York, NY 10105

Axe-Houghton Foundation Grants 815
The Axe-Houghton Foundation operates in the New York City metropolitan area exclusively for charitable, educational, and scientific purposes to foster and encourage an appreciation of the English language, with major emphasis on the spoken language. Priority is given to projects for the improvement of speech and its uses in the areas of public affairs, education, theater, poetry, debate, and the oral interpretation of literature. A portion of available funds is devoted to speech remediation and to scientific research pertaining to speech. Types of support include program grants and seed money grants. Applications are accepted only from organizations who have been invited to apply. Interested parties can contact the Foundation for information on how to secure an invitation to apply. Funds are limited and awards rarely exceed $5,000.
Requirements: To be eligible to receive a grant from the Foundation, the proposed grantee must be either: an organization which has been determined by the Internal Revenue Service to be a tax-exempt organization described in section 501(c)3 of the Internal Revenue Code, and not a private foundation within the meaning of Section 509(a) of the Internal Revenue Code; a governmental unit as referred to in Section 170(c)1 of the Internal Revenue Code; or an agency or instrumentality of a governmental unit, including, without limitation, a state college or university within the meaning of Section 511(a)2 of the Internal Revenue Code. Initial inquiries should be by email or phone. Invited applicants should include a brief statement of objective, duration of time required to complete this objective, qualifications of personnel responsible for conducting the project, and the general amounts of funds needed. The proposal should also indicate the degree of internal support that would be committed.

Restrictions: Grants are made only to tax-exempt institutions and not to individuals, private foundations or organizations outside of the United States. Grants are not made for scholarships, fellowships, capital, or general support programs.
Geographic Focus: New York
Date(s) Application is Due: Sep 1
Amount of Grant: 1,000 - 7,500 USD
Contact: John Johnson; (800) 839-1754; fax (800) 421-6579; jjohnson@foundationsource.com
Internet: http://www.foundationcenter.org/grantmaker/axehoughton/index.html
Sponsor: Axe-Houghton Foundation
55 Walls Drive, 3rd Floor
Fairfield, CT 06824

Babcock Charitable Trust Grants 816
The Babcock Charitable Trust was established in Pennsylvania in 1957, by way of a donation from Fred C. and Mary A. Babcock. The Trust's primary purpose has always been to support both education and health care throughout the states of Pennsylvania and Florida, although they occasionally give outside of this primary region. With that in mind, the Trust's specified fields of interest include: children and youth services; education; health care programs; higher education; and religion. Application forms are not required, and there are no specific deadlines. Applicants should provide, in written form, a brief overview or history of their organization, a mission statement, a detailed description of the project proposed, and an amount of funding requested. The amount of funding ranges up to $25,000.
Geographic Focus: Florida, Maryland, Massachusetts, New York, Pennsylvania, Wisconsin
Amount of Grant: Up to 25,000 USD
Contact: Courtney B. Borntraeger, Treasurer; (412) 351-3515
Sponsor: Babcock Charitable Trust
1105 N. Market Street, Suite1300
Wilmington, DE 19801

Bacon Family Foundation Grants 817
The foundation awards grants to Colorado nonprofit organizations in its areas of interest, including arts and culture, community development, disabilities, economic development, education, food distribution, environment, health care, historic preservation, homelessness, housing, literacy, recreation/parks, religion, social services, and youth. Types of support include capital campaigns, challenge/matching grants, equipment acquisition, general operating support, project grants, and seed grants. A letter should outline the project or program in brief but sufficient detail for the Foundation to initially evaluate the proposal. There are no application deadlines. The Bacon Family Foundation meets quarterly to consider applications.
Requirements: Colorado nonprofit organizations are eligible. Preference will be given to requests from western Colorado.
Restrictions: No grants are made to individuals.
Geographic Focus: Colorado
Contact: Linda Simpson, (970) 243-3767; lsimpson@wc-cf.org
Internet: http://www.wc-cf.org/bacon.htm
Sponsor: Bacon Family Foundation
P.O. Box 4570
Grand Junction, CO 81502-4570

Bailey Foundation Grants 818
The foundation supports South Carolina nonprofits, primarily in Laurens County, its areas of interest, include: health clinics and organizations involved with theater, education, human services, business promotion, Christianity and awards college scholarships to students graduating from public high schools in Laurens County, South Carolina. Support is offered in the form of : capital campaigns; employee matching gifts; endowments; matching/challenge support; scholarship funds, including individual scholarships; annual campaigns; building/renovation grants. Initial contact may be by telephone or letter. Deadline for scholarship applications is April 15, contact Foundation to request an application form.
Requirements: Nonprofit organizations in Laurens County, South Carolina are eligible.
Restrictions: No grants to individuals (except for scholarships), or for general operating support.
Geographic Focus: South Carolina
Date(s) Application is Due: Apr 15; Oct 1
Amount of Grant: 2,500 - 50,000 USD
Contact: Robert S. Link, Jr.; (864) 938-2632; fax (864) 938-2669
Sponsor: Bailey Foundation
P.O. Box 494
Clinton, SC 29325-0494

Balfe Family Foundation Grants 819
The Balfe Family Foundation, established in Florida in 1999, has as its primary fields of interest: education, health care and research, human services, and social services. Types of support are typically in the form of general operations, service delivery support, and seed funding. There are particular application forms or deadlines with which to adhere. Applicants should contact the Foundation in writing, outlining the program and funding need.
Geographic Focus: All States
Amount of Grant: 50 - 50,000 USD
Contact: Opal M. Balfe, President; (954) 462-6300; fax (954) 462-4607
Sponsor: Balfe Family Foundation
2731 NE 14 Street Causeway, #829
Pomano Beach, FL 33062-3562

Ball Brothers Foundation General Grants 820

Founded in the name of Edmund B. Ball and his brothers, the Foundation seeks to build and sustain a high quality of life in Indiana by awarding grants to nonprofit organizations in broad subject areas, including elementary, secondary, higher, and adult basic education and literacy skills; cultural activities; community betterment; the environment; and health and human services. Usually, Muncie and Delaware Counties receive a higher priority for funding than requests from across the state. Types of support include general operations, annual campaigns, capital campaigns, building construction/renovation, program development, conferences and seminars, professorships, publication, curriculum development, research, fellowships, matching funds, seed grants, and technical assistance. Preference will be given to catalytic grants that will stimulate others to participate in problem solving or in matching fund programs and to innovative approaches for addressing either traditional or emerging community needs. Applications are reviewed by the board of directors in January, May, and September of each calendar year. Proposals are encouraged to be submitted from February to May. Grant seekers may send a preliminary proposal, complete proposal, or ask for a personal visit to discuss a potential grant request.
Requirements: Indiana 501(c)3 nonprofit institutions and organizations are eligible.
Restrictions: The Foundation will not support: direct assistance to individuals or scholarships; applications coming from outside of Indiana; booster organizations; on-going salary requests of staff personnel to support an organization; services that the community-at-large should normally underwrite (i.e. roads, bus transportation, etc.); capital building projects; research projects (except for philanthropic studies); or unsolicited proposals (all requests must begin with a preliminary proposal).
Geographic Focus: Indiana
Date(s) Application is Due: Apr 1; Sep 1
Amount of Grant: Up to 100,000 USD
Samples: Open Door Health Center, Muncie, Indiana, $90,000 - construction of a environmentally-friendly facility; Indiana Council for Economic Education, West Lafayette, Indiana, $5,000 - economic education and financial literacy; Delaware County Soil and Water, Muncie, Indiana, $150,00 - wetlands preserve.
Contact: Donna Munchel, Executive Assistant; (765) 741-5500; fax (765) 741-5518; donna.munchel@ballfdn.org or info@ballfdn.org
Internet: http://www.ballfdn.org
Sponsor: Ball Brothers Foundation
222 South Mulberry Street
Muncie, IN 47305

Baltimore Washington Center Adult Psychoanalysis Fellowships 821

The Adult Psychoanalysis Fellowship Program is a one-year program designed for individuals who have a significant interest in psychoanalysis as a body of knowledge and as a framework with which to understand and carry out therapeutic efforts. Intended for advanced graduate students and residents as well as recent graduates in Psychiatry, Psychology and Social Work, the program at its core consists of monthly seminars presented by psychoanalysts on clinically-related topics. Fellows are also welcome to attend Forums on Psychoanalysis and other conferences offered by the Institute and Society.
Geographic Focus: All States
Date(s) Application is Due: Jul 15
Contact: Elizabeth Manne, Executive Director; (301) 470-3635 or (410) 792-8060; fax (410) 792-4912; admin@bwanalysis.org
Internet: http://www.bwanalysis.org/Fellowship%20Program-Adult.html
Sponsor: Baltimore Washington Center for Psychoanalysis
14900 Sweitzer Lane, Suite 102
Laurel, MD 20707

Baltimore Washington Center Child Psychotherapy Fellowships 822

The Child Psychotherapy Fellowship program is a one-year program specifically designed for child mental health professionals who are interested in deepening their understanding of working with children from a psychoanalytic or psychodynamic perspective. The program is tuition-free. Detailed discussions of actual clinical work with children and reviews of relevant psychoanalytic literature will be used as a framework to enhance understanding and ability to carry out therapeutic work. Fellowship meetings will be held monthly, September through May, on the second Wednesday of every month, at the homes of Baltimore and Washington faculty members.
Requirements: Professionals, including psychiatrists, psychologists, social workers, and counselors currently working with children in psychotherapy should apply. Recent graduates of training programs should also apply.
Geographic Focus: All States
Date(s) Application is Due: Jul 15
Contact: Elizabeth Manne, Executive Director; (301) 470-3635 or (410) 792-8060; fax (410) 792-4912; admin@bwanalysis.org
Internet: http://www.bwanalysis.org/Fellowship%20Program-Child.html
Sponsor: Baltimore Washington Center for Psychoanalysis
14900 Sweitzer Lane, Suite 102
Laurel, MD 20707

BancorpSouth Foundation Grants 823

The BancorpSouth Foundation supports organizations involved with orchestras, secondary and higher education, legal aid, housing, youth development, and human services. Support is available in the form of general/operating grants in areas of operation, that include Arkansas, Mississippi and Tennessee. Applications for funding are accepted on a rolling basis and reviewed quarterly.
Requirements: Arkansas, Mississippi and Tennessee 501(c)3 non-profit organizations are eligible to apply. There is no application deadline nor is there an application form required when applying for funding. Applicants should include a detailed description of project and amount of funding requested in the proposal.
Restrictions: No grants to individuals.
Geographic Focus: Arkansas, Mississippi, Tennessee
Amount of Grant: 5,000 - 15,000 USD
Samples: Southern Arkansas University, Magnolia, AR, $15,000; Alpha House Home For Boys, Tupelo, MS, $5,000; Affordable Housing, Jackson, TN, $5,000.
Contact: Nash Allen, Grants Manager; (662) 680-2000
Sponsor: BancorpSouth Foundation
P.O. Box 789
Tupelo, MS 38802-0789

Banfi Vintners Foundation Grants 824

The foundation awards general operating grants to nonprofits in its areas of interest, including higher education, civic and public affairs, arts and humanities, wildlife protection, health (hospitals and disease research/prevention), religion, science, social services, and international. Grants are made nationwide, with preference given to requests from Massachusetts and the New York, NY, area.
Geographic Focus: Massachusetts, New York
Amount of Grant: 100 - 465,000 USD
Samples: Colgate U (Hamilton, NY)--for operating support, $75,000; Friends for Long Island's Heritage (Syosset, NY--for operating support, $25,000; Huntington Hospital Assoc (Huntington, NY)--for operating support, $25,000; Cornell U (Ithaca, NY)--for operating support, $465,000.
Contact: Philip Calderone, Executive Director; (516) 626-9200
Sponsor: Banfi Vintners Foundation
1111 Cedar Swamp Road
Glen Head, NY 11545

Bank of America Charitable Foundation Matching Gifts 825

The Bank of America Charitable Foundation Matching Gifts program encourages employees to contribute to qualifying charitable organizations. This program supports employee giving by offering a way to double – up to $5,000 per person each calendar year – employees' cash or securities contributions to their favorite charitable organizations and thus improve their communities. Annually, the Bank of America Charitable Foundation provides more than $25 million in matching gifts on behalf of employee donations.
Requirements: Charitable organizations in the United States must be tax-exempt under section 501(c)3 of the Internal Revenue Code and not be classified as a private foundation. Charitable Organizations located in England or Wales must be registered with the Charity Commission. Charitable organizations outside of the United States, England or Wales must be qualified as eligible for donations from CAFAmerica.
Restrictions: The Bank of America Charitable Foundation does not: match charitable gifts to private, family or donor advised funds, or gifts to political or fraternal organizations; or match charitable gifts that benefit students directly or that result in an employee receiving a benefit, including tuition or sponsorships.
Geographic Focus: All States, District of Columbia, Guam, Marshall Islands, Northern Mariana Islands, Puerto Rico, U.S. Virgin Islands, American Samoa, Canada, United Kingdom
Amount of Grant: Up to 5,000 USD
Contact: Anne M. Finucane, Foundation Chairperson/Chief Marketing Officer; (617) 434-9410 or (800) 218-9946; anne.m.finucane@bankofamerica.com
Internet: http://about.bankofamerica.com/en-us/global-impact/matching-gifts-features-and-eligibility.html#fbid=wq9wE7VdEpC
Sponsor: Bank of America Charitable Foundation
100 North Tryon Street
Charlotte, NC 28255

Bank of America Charitable Foundation Volunteer Grants 826

Bank of America employees volunteer thousands of hours globally in our neighborhoods each year. In fact, more than 3,000 charitable organizations benefit from the Foundation's employees' dedication each year. To honor those who give their time and service to causes important to them, the Bank of America Charitable Foundation awards grants, which are up to $500 per employee for each calendar year and are made in the name of the employee, to eligible charitable organizations. An unrestricted grant is made to any eligible nonprofit organization for which an employee or retiree has committed substantial volunteer hours within a calendar year. For 50 hours of volunteer time within a calendar year, Bank of America Charitable Foundation will give a $250 grant; for 100 hours of volunteer time within a calendar year, the grant is $500. Employee hour registration must be completed by January 31 after the year in which the hours were volunteered. Organizations must verify hours by May 15 after the year in which the hours were volunteered.
Requirements: Charitable organizations in the United States must be tax-exempt under section 501(c)3 of the Internal Revenue Code and not be classified as a private foundation. Charitable Organizations located in England or Wales must be registered with the Charity Commission. Charitable organizations outside of the United States, England or Wales must be qualified as eligible for donations from CAFAmerica. Employees must complete an application and have the recipient organization verify the hours.
Geographic Focus: All States, District of Columbia, Guam, Marshall Islands, Northern Mariana Islands, Puerto Rico, U.S. Virgin Islands, American Samoa, Canada, United Kingdom
Date(s) Application is Due: Jan 31
Amount of Grant: 250 - 500 USD
Contact: Anne M. Finucane, Foundation Chairperson/Chief Marketing Officer; (617) 434-9410 or (800) 218-9946; anne.m.finucane@bankofamerica.com

Internet: http://about.bankofamerica.com/en-us/global-impact/volunteer-grants-features-and-eligibility.html#fbid=2SvmqQPvBb7
Sponsor: Bank of America Charitable Foundation
100 North Tryon Street
Charlotte, NC 28255

Baptist-Trinity Lutheran Legacy Foundation Grants 827
The Foundation provides support in the greater Kansas City area for crisis related medical assistance, neighborhood school health grants, and health education programs and services in the service area of Baptist Medical Center and Trinity Lutheran Hospital. Funding priorities include: emergency medical assistance; health education; and health education in neighborhood schools. Highest consideration will be given for funding programs or services that demonstrate diversity, have measurable outcomes and can document continued existence after grant funding. Health Education Grant requests will be reviewed in January, March, May, July, September and November.
Requirements: All applicants must be non-profit, tax exempt organizations.
Restrictions: No grants will be made to individuals, political parties, candidates or political activities. Grants will not be given for special events, annual campaigns, endowments, or construction projects.
Geographic Focus: Missouri
Contact: Becky Schaid, Executive Director; (816) 276-7515 or (816) 276-7555; fax (816) 926-2261; becky@btllf.org
Internet: http://www.btllf.org/funding.html
Sponsor: Baptist-Trinity Lutheran Legacy Foundation
6601 Rockhill Road
Kansas City, MO 64131

Baptist Community Ministries Grants 828
BCM funds grants primarily in four major areas of interest: health; education; public safety; and governmental oversight. Applications not falling in these four major areas of interest are not typically considered. BCM funds transom grants and strategic grants. Transom grants are unsolicited grant proposals submitted by qualified, nonprofit organizations during our two semi-annual open transom cycles, and encourage the development of new ideas and nurture inventive solutions to community problems. Strategic grants generally target the long-range goals in each area of interest and are longer in duration. Strategic grants are usually generated through Request for Proposals or by invitation.
Requirements: Grants are made to eligible organizations in one or more of the five Louisiana parishes of Orleans, Jefferson, St. Bernard, St. Tammany and Plaquemines. Beneficiaries of grants should reside in one or more of the five parishes. It is preferred that the organization receiving a grant also be located in the same five-parish region and be overseen by unpaid volunteers living in the area.
Restrictions: Grants are not made to private foundations described in Section 509(a) of the Internal Revenue Code, private or publicly held corporations, limited liability corporations, partnerships, or sub-chapter S corporations. Funding requests not supported are: capital projects; ongoing general operating expenses; projects of national or statewide scope; single-disease charities; direct support to individuals or fraternal bodies.
Geographic Focus: Louisiana
Date(s) Application is Due: Mar 15; Sep 15
Amount of Grant: 50,000 USD
Contact: Joanne M. Schmidt, Grants Manager; (504) 593-2345 or (504) 593-2323; fax (504) 593-2301; jschmidt@bcm.org or info@bcm.org
Internet: http://www.bcm.org/grantmaking/
Sponsor: Baptist Community Ministries
400 Poydras Street, Suite 2950
New Orleans, LA 70130-3245

Barberton Community Foundation Grants 829
The foundation awards grants to improve the quality of life for the citizens of Barberton. In addition to the regular grants program quarterly deadlines, the Foundation's Small Grant Program accepts applications monthly for grants up to $1,000. Areas of interest include charitable endeavors, education, public health, and public recreation.
Requirements: The eligibility requirements for small grants and large grants differ. See the Application Guidelines for specific requirements.
Restrictions: Grants are not given to: individuals; endowments housed at institutions other than the Barberton Community Foundation; religious organizations for religious purposes; projects that do not exclusively benefit the citizens of Barberton; fund debt reductions, deficits, or previous obligations; fund annual fund raising drives; fund ongoing operating expenses; fund political projects, sabbatical leaves or scholarly research; or fund venture capital for competitive profit-making activities.
Geographic Focus: Ohio
Date(s) Application is Due: Jan 2; Apr 1; Jul 1; Oct 1
Amount of Grant: Up to 1,000 USD
Samples: Barberton All-Town Little League (OH)--to refurbish two ball fields and construct two fields at the U.L. Light Middle School, $14,075; City of Barberton (OH)--for summer concert programs at Lake Anna Park, $55,000; Metro Regional Transit Authority (Akron, OH)--for a transit center in Barberton, $102,000; Barberton Area United Way (OH)--to help needy residents make utility and rent payments, $60,000.
Contact: Chuck Sandstrom, Executive Director; (330) 745-5995; fax (330) 745-3990; csandstrom@bcfcharity.org
Internet: http://www.bcfcharity.org/bcf/grant/grants.shtml
Sponsor: Barberton Community Foundation
460 W Paige Avenue
Barberton, OH 44203

Barker Foundation Grants 830
The foundation offers funding to New Hampshire non-profit organizations in the areas of: children/youth services; education; health organizations, association; hospitals (general); and human services. Giving primarily for health associations, social services and youth. Grants are awarded in the following types: Annual campaigns building/renovation; capital campaigns; continuing support; equipment; general/operating support and; program development.
Requirements: New Hampshire nonprofit organizations are eligible for funding. The Foundation accepts written requests only. No formal application form is required. Send a 1 page concept paper and request for guidelines to the Foundation with a SASE for response.
Restrictions: Grants are not made to individuals.
Geographic Focus: New Hampshire
Amount of Grant: 2,000 - 13,000 USD
Contact: Allan Barker, Treasurer
Sponsor: Barker Foundation
P.O. Box 328
Nashua, NH 03061-0328

Barker Welfare Foundation Grants 831
The mission of the foundation is to make grants to qualified charitable organizations whose initiatives improve the quality of life, with an emphasis on strengthening youth and families and to reflect the philosophy of Catherine B. Hickox, the Founder. Consideration will be given to applications from institutions and agencies operating in the fields of health, welfare, education and literacy, cultural activities, and civic affairs, primarily serving the metropolitan area of New York, NY, and Michigan City, IN. The foundation board meets twice per year to consider requests.
Requirements: In advance of submitting a request, a brief letter or telephone call is suggested to determine if the organization seeking to apply for a grant falls within the current general policy of the Foundation. Before the Foundation sends out its application form, a brief 2-3 page letter describing the organization, the purpose and amount requested should be sent. A copy of the first page of the most recent 990 filed with the IRS and a current budget for the whole organization and a budget for program/project (if requesting program/project support) including income and expense should be sent with the letter of inquiry. Grants are made to tax-exempt organizations which have received a ruling by the Internal Revenue Service that they are organizations described in Section 501(c)3 and classified in Section 509(a)(1),(2), or (3) of the Internal Revenue Code (publicly supported organizations and their affiliates).
Restrictions: Appeals for the following will be declined: organizations not located in nor directly serving the defined areas; national health; welfare; or education agencies; institutions or funds; scholarships; fellowships; loans; student aid; appeals from individuals; medical and scientific research; private elementary and secondary schools; colleges; universities; professional schools; trade organizations; films; program advertising; conferences; seminars; benefits and fund raising costs; start-up organizations; emergency funds; and deficit financing; lobbying-related or legislative activities; endowment funds; and intermediary organizations.
Geographic Focus: Indiana, New York
Date(s) Application is Due: Feb 1; Aug 1
Amount of Grant: 7,500 - 15,000 USD
Contact: Sarane Ross, President; (516) 759-5592; BarkerSMD@aol.com
Internet: http://www.barkerwelfare.org
Sponsor: Barker Welfare Foundation
P.O. Box 2
Glen Head, NY 11545

Barnes Group Foundation Grants 832
The foundation awards grants in the following areas of interest: Education; Health and Welfare; Hospitals; Culture and Art and Civic and Youth Organizations. Highest priority is given to the support of organizations and projects in communities where the company has its executive office, group headquarters offices, plants and other facilities. There are no deadlines for submitting requests. Initial contact should be by letter.
Requirements: In order to be considered for support by the Foundation, a grant seeker must furnish a copy of its ruling from the Internal Revenue Service determining that it is an organization described in Section 501(c)3 and Section 509(a)(1), (2), (3), or (4) of the Internal Revenue Code.
Restrictions: No grants are made to organizations located outside the United States. The Foundation will consider only one grant request per organization during any one year period. The Foundation will not contribute directly or indirectly to any political activities or legislative lobbying efforts.
Geographic Focus: Connecticut, Maine, Massachusetts, New Hampshire, Rhode Island, Vermont
Amount of Grant: 500 - 3,500 USD
Contact: Secretary; (860) 973-2112; fax (860) 589-7466
Internet: http://www.barnesgroupinc.com/about/foundation.html
Sponsor: Barnes Group Foundation
123 Main Street, P.O. Box 489
Bristol, CT 06011-0489

Barra Foundation Community Fund Grants 833
Community Fund grants provide unrestricted contributions to qualified organizations primarily in the Greater Philadelphia area. These grants are generally in amounts between $1,000 and $10,000 per year. The Foundation's categories of funding for these grants are: human services; arts and culture; health; and education. These grants are made at certain times of the year, depending on the category of grant. The categories and time frames for each are as follows: human services--March 1 with grants made in the Spring of each

year; arts and culture--June 1 with grants made in the Summer of each year; and health and education--September 1 with grants made in the fall of each year.
Requirements: Applications are welcome from 501(c)3 organizations serving the greater Philadelphia region.
Geographic Focus: Pennsylvania
Date(s) Application is Due: Mar 1; Jun 1; Sep 1
Amount of Grant: 2,000 - 15,000 USD
Contact: William Harral, III; (215) 233-5115; william.harral@verizon.net
Internet: http://www.barrafoundation.org/grants/index.html
Sponsor: Barra Foundation
8200 Flourtown Avenue, Suite 12
Wyndmoor, PA 19038-7976

Barra Foundation Project Grants 834
Project grants are one-time grants generally for amounts above $10,000. The Foundation considers grants for innovative projects that aid research in advancing the frontiers of human services, arts and culture, health and education. Grants are not made for ongoing or expanding programs where substantial initial support was previously provided from other sources. Three principal criteria are strictly adhered to in judging the merits of a proposed project. They are: innovation; evaluation; and dissemination. Proposals may be submitted at any time of the year. Initial requests for project grants should be in the form of a preliminary concept paper, generally not to exceed two pages, summarizing: the principal focus and objectives of the proposed project; uniqueness of the concept; the overall methodology; estimated timetable; preliminary budget data; and other sources of support for the project.
Restrictions: The foundation does not provide grants for: ongoing operating budgets; staff salaries; budget deficits; endowments; capital campaigns and projects; international programs; environmental and religious organizations; scholarships and fellowships; or audio/video projects, publications, catalogs, and exhibitions.
Geographic Focus: Pennsylvania
Amount of Grant: 10,000 - 500,000 USD
Contact: William Harral, III; (215) 233-5115; william.harral@verizon.net
Internet: http://www.barrafoundation.org/grants/index.html
Sponsor: Barra Foundation
8200 Flourtown Avenue, Suite 12
Wyndmoor, PA 19038-7976

Barr Fund Grants 835
The fund awards grants to Illinois nonprofits in its areas of interest, including services for children and youth, orchestras, higher education, mental health/crisis services, Jewish temples and organizations, and social services. Grants are awarded for general operating support. There are no application forms or deadlines. Submit a letter of request.
Requirements: Illinois nonprofits are eligible.
Restrictions: Individuals are not eligible.
Geographic Focus: Illinois
Amount of Grant: 100 - 75,000 USD
Contact: Donald Lubin, President; (312) 782-4710; fax (312) 876-8000
Sponsor: Barr Fund
230 West Monroe Street, Suite 330
Chicago, IL 60606-4701

Barth Syndrome Foundation Research Grants 836
Barth syndrome is a serious X-linked recessive condition associated with cardiomyopathy, neutropenia, skeletal muscle weakness, exercise intolerance, growth delay, and diverse biochemical abnormalities (including defects in mitochondrial metabolism and phospholipid biosynthesis). Because many clinical and biochemical abnormalities of Barth syndrome remain poorly understood, the program is seeking proposals for research that will advance knowledge on any aspect of the syndrome. The foundation prefers to award seed grants to experienced investigators for testing of initial hypotheses and collection of preliminary data leading to successful long-term funding by NIH and other major granting institutions. The foundation also encourages investigators new to the field of Barth Syndrome research. Send an electronic version of the full application and all the attachments by the listed application deadline. The Foundation anticipates awarding several one- or two-year grants of up to $40,000 each. Funds will be available as soon as the successful grant applicants have been notified.
Requirements: Principal investigators who are affiliated with nonprofit institutions are eligible.
Geographic Focus: All States
Date(s) Application is Due: Oct 31
Amount of Grant: 10,000 - 40,000 USD
Contact: Matthew J. Toth; (617) 469-6769; fax (617) 849-5695; mtoth@barthsyndrome.org
Internet: http://www.barthsyndrome.org/english/View.asp?x=1635
Sponsor: Barth Syndrome Foundation
675 VFW Parkway, #372
Chestnut Hill, MA 02467

Batchelor Foundation Grants 837
The foundation supports food banks and organizations involved with arts and culture, education, the environment, animals and wildlife, health, human services, and economically disadvantaged people. Special emphasis is directed toward programs designed to engage in medical research and provide care for childhood diseases; and promote study, preservation, and public awareness of the natural environment. Funding is available in the form of: capital campaigns; continuing support; endowments; general/operating support; and program development grants. There are no application forms. Initial approach should be a letter that details the grant proposal.
Requirements: Florida area nonprofits are eligible.
Restrictions: Individuals are not eligible.
Geographic Focus: Florida
Amount of Grant: 1,000 - 1,000,000 USD
Samples: Community Partnership for Homeless, Miami, FL, $1,056,600--for general operating support and endowment; Fairchild Tropical Botanic Garden, Coral Gables, FL, $540,000-- For general operating support; University of Miami, Miami, FL, $525,000-- for general operating support.
Contact: Anne Batchelor-Robjohns, Co-C.E.O; (305) 416-9066 or (305) 534-5004; jbatchelor@bellsouth.net
Sponsor: Batchelor Foundation
111 NE 1st Street, Suite 820
Miami, FL 33132

Baton Rouge Area Foundation Grants 838
The Baton Rouge Area Foundation seeks to enhance the quality of life for all citizens of Baton Rouge, Louisiana. The foundation concentrates on projects and programs in the areas of community development, education, environment, health and medical, and religion. Applicants should call the office to determine if their project or program is consistent with the foundation's goals before sending an application.
Requirements: Only nonprofits that are registered as 501(c)3 organizations working in the service region of East and West Baton Rouge Parish, East and West Feliciana, Ascension, Livingston, Iberville and Pointe Coupee are eligible to apply for grants.
Geographic Focus: Louisiana
Contact: John G. Davies; (225) 387-6126; fax (225) 387-6153; jdavies@braf.org
Internet: http://www.braf.org/index.cfm/page/4/n/8
Sponsor: Baton Rouge Area Foundation
402 North Fourth Street
Baton Rouge, LA 70802

Battle Creek Community Foundation Grants 839
The Battle Creek Community Foundation focuses on the greater Battle community, and is most interested in projects that focus on education, health, human services, arts and culture, public affairs or community development. In general, priority is given to those grant ideas that: increases the capacity of the community to participate in identifying needs and developing and implementing solutions; encourages cooperation; demonstrates a clean and convincing need; develops self-reliance; avoids unnecessary duplication of services; targets gaps in services; Mirrors the diversity of our community; promotes equity among its various segments; and has clean, defined goals and/or measurable outcomes. Grant applications and guidelines are available on the Web site.
Requirements: Grants will be considered only from 501(c)3 nonprofit agencies in Michigan offering services to the residents of Battle Creek.
Restrictions: Individuals are not eligible to receive grants.
Geographic Focus: Michigan
Amount of Grant: 5,000 - 10,000 USD
Samples: Drug Free Workplace (Battle Creek, MI)--program support; Battle Creek Health System (Battle Creek, MI)--program support.
Contact: Kelly Boles Chapman, Vice President of Programs; (269) 962-2181; fax (269) 962-2182; bccf@bccfoundation.org
Internet: http://www.bccfoundation.org/grants/
Sponsor: Battle Creek Community Foundation
1 Riverwalk Centre, 34 W Jackson Street
Battle Creek, MI 49017-3505

Batts Foundation Grants 840
Established in 1988, the Batts Foundation supports organizations involved with arts and culture, K-12 and higher education, disease, and human services. Types of support include: annual campaigns, building and renovation; capital campaigns; continuing support; operating support; endowments; matching grants; program development; and scholarship funding. Grants will be awarded primarily in the western Michigan area, particularly the communities of Huron, Zeeland, and Grand Rapids. There are no application forms or deadlines with which to adhere, and applicants should begin by submitting a one page letter summarizing the project.
Requirements: Michigan nonprofit organizations are eligible.
Restrictions: Individuals are ineligible.
Geographic Focus: Michigan
Amount of Grant: 250 - 25,000 USD
Contact: Robert Batts, Director; (616) 956-3053; jsand@battsgroup.com
Sponsor: Batts Foundation
3855 Sparks Drive SE, Suite 222
Grand Rapids, MI 49546-2427

Baxter International Corporate Giving Grants 841
As a complement to its Foundation, the Baxter Corporation makes charitable contributions to nonprofit organizations directly. Primary fields of interest include: disaster preparedness and services; elementary and secondary education; employment services; the environment; health care and health care rights; health organizations; hemophilia; immunology; kidney diseases; mathematics; patients' rights; science; teacher training and education; and youth services. Types of support include: conferences and seminars; curriculum development; donated products; employee volunteer services; general operating support; in-kind and matching gifts; and sponsorships. Support is given primarily in areas of company operations.
Geographic Focus: All States, All Countries

Amount of Grant: Up to 500,000 USD
Contact: Department Chair; (224) 948-2000
Internet: http://www.sustainability.baxter.com/community-support/
Sponsor: Baxter International Corporation
1 Baxter Parkway
Deerfield, IL 60015-4625

Baxter International Foundation Foster G. McGaw Prize 842
Through its prize programs, the Baxter International Foundation celebrates excellence in community service and research. The Foster G. McGaw Prize honors health delivery organizations (hospitals, health systems, integrated networks, or self-defined community partnerships) that have demonstrated exceptional commitment to community service. The Baxter International Foundation takes no role in the selection of recipients or the application process. The annual deadline for applications is April 6.
Requirements: Any health delivery organization that exhibits leadership, Commitment, partnerships, breadth and depth of initiatives, and community involvement is eligible to apply for the Prize.
Geographic Focus: All States, All Countries
Date(s) Application is Due: Apr 6
Amount of Grant: 100,000 USD
Contact: Prize Coordinator; (312) 422-3932 or (847) 948-4605; fdninfo@baxter.com
Internet: http://www.baxter.com/about_baxter/sustainability/international_foundation/foster_mcgaw_prize.html
Sponsor: Baxter International Foundation
One Baxter Parkway
Deerfield, IL 60015-4633

Baxter International Foundation Grants 843
The Baxter International Foundation's grant program is focused on increasing access to healthcare worldwide. The foundation funds initiatives that improve the access, quality and cost-effectiveness of healthcare. Grants awarded most recently fulfilled local needs to increase access to dental care, mental health, and other healthcare services for children, the uninsured, veterans, and the elderly. Funding often comes in the form of salary support and general operations. Focusing on these priorities, the foundation's primary concern is on communities where Baxter has a corporate presence. In Illinois, grants are restricted to Lake, McHenry and Cook counties. The foundation also funds programs throughout the U.S., Asia, Australia, Canada, Europe, Latin America and Mexico.
Requirements: U.S. nonprofits in Lake, McHenry and Cook counties of Illinois are eligible to apply. Internationally, the following regions are eligible to apply: U.S., Asia, Australia, Canada, Europe, Latin America, and Mexico.
Restrictions: In general, The Baxter International Foundation does not make grants to: capital and endowment campaigns (includes requests for infrastructure of any kind, equipment, vehicles, etc.); disease or condition-specific organizations or programs; educational grants/continuing professional education scholarships; educational institutions, except in instances where a grant would help achieve other goals, such as increasing community-based direct health services or the skills and availability of community health-care providers, in areas where there are Baxter facilities; general operating support or maintenence of effort; hospitals; individuals, including scholarships for individuals; lobbying and political organizations; magazines, professional journals, documentary, film, video, radio or website productions; medical missions; organizations seeking travel support for individuals or groups, medical missions or conferences; organizations soliciting contributions for advertising space, tickets to dinners, benefits, social and fund-raising events, sponsorships and promotional materials; organizations with a limited constituency, such as fraternal, veterans or religious organizations; research.
Geographic Focus: All States, Illinois, All Countries
Date(s) Application is Due: Jan 21; Apr 13; Jul 13; Sep 29
Amount of Grant: Up to 100,000 USD
Samples: Access OC, Laguna Hills, California, $49,400 - to support the hiring of a case manager for a new case management program to work with referred patients in Orange County so they can reduce co-morbidities and obtain needed specialty surgeries at their outpatient surgery center; BraveHearts, Harvard, Illinois, $40,000 - to support the salary of a new Volunteer Coordinator needed for all 3 programs within the organization; Kenosha Community Health Center, Kenosha, Wisconsin, $100,000 - upport the hiring of two case managers to be located at the local non-profit agency.
Contact: Foundation Contact; (847) 948-4605; fdninfo@baxter.com
Internet: http://www.baxter.com/about_baxter/sustainability/international_foundation/grants_program.html
Sponsor: Baxter International Foundation
One Baxter Parkway
Deerfield, IL 60015-4633

Baxter International Foundation William B. Graham Prize for Health Services Research 844
Since 1986, The Baxter International Foundation and the Association of University Programs in Health Administration have awarded the William B. Graham Prize for Health Services Research to recognize researchers who have made major contributions to the health of the public through innovative research in health services. The Prize honors the late William B. Graham, longtime chairman and CEO of Baxter International Inc., and is internationally regarded as the premier recognition for individuals conducting health services research. The Prize recognizes individuals who have had a significant impact on the health of the public in one of three primary focus areas: Health Services Management, Health Policy Development and Healthcare Delivery. The Prize includes an award of $25,000 to the individual and $25,000 to a not-for-profit institution that supports the winner's work.

Geographic Focus: All States, All Countries
Amount of Grant: 50,000 USD
Contact: Prize Coordinator; (847) 948-4605; fdninfo@baxter.com
Internet: http://www.baxter.com/about_baxter/sustainability/international_foundation/index.html
Sponsor: Baxter International Foundation
One Baxter Parkway
Deerfield, IL 60015-4633

Bayer Clinical Scholarship Award 845
This award is intended to facilitate the development of specific clinical expertise in the field of hemophilia for applicants who have completed medical training and have an interest in pursuing a career as a hemophilia treater/researcher. The award will support a mentored physician in training for two years. Clinical duties will encompass diagnosis, evaluation, and the planning of management strategies for patients with hereditary bleeding disorders. In addition to the clinical experience, the applicant may pursue a research project in the field of hemostasis. Clinical scholarships will provide funding for two years. Up to five new awards will be made each year. Guidelines and application are available online.
Requirements: The applicant should have earned his/her medical degree within the previous eight years. The commitment of the mentor to the grantee and the research project (if applicable) is a vital element of the application, as is the quality of the clinical environment at which the applicant will undertake the scholarship.
Geographic Focus: All States
Amount of Grant: 70,000 USD
Contact: Administrator; programadministrator@bayer-hemophilia-awards.com
Internet: http://www.bayer-hemophilia-awards.com/awards.cfm#clinical
Sponsor: Bayer HealthCare Corporation
100 Bayer Road
Pittsburgh, PA 15205-9741

Bayer Foundation Grants 846
The Bayer Foundation supports programs that enhance the quality of life, provide unique and enriching opportunities that connect diverse groups and ensure preparedness for tomorrow's leaders. The Foundation welcomes proposals from 501(c)3 organizations whose programming matches at least one of the following areas: civic and social service programs; education; workforce development; arts and culture; and health and human services. The application submission process is online. Applicants will be asked to provide a brief history of the organization, contact information, an executive summary of the proposed project, the requested amount, a proposed budget, program time frame, a letter from the IRS that verifies the organization's 501(c)3 status, a W-9 form, and a listing of other funding sources.
Requirements: 501(c)3 nonprofit organizations in Bayer operating communities are eligible. Submit proposals to regional offices in: Berkley and northern California; Kansas City, Missouri; northern New Jersey; Pittsburgh, Pennsylvania; Raleigh-Durham, North Carolina; Houston and Baytown, Texas; and Shawnee, Kansas. Additionally, the foundation considers proposals from STEM education programs that have a national focus.
Restrictions: The Foundation will not fund: for-profit organizations or those without Internal Revenue Service code 501(c)3 nonprofit, tax-exempt status; organizations that discriminate on the basis of race, color, creed, gender, sexual orientation or national origin; general operating support for United Way affiliated agencies (only support for special projects will be considered); organizations or programs designed to influence legislation or elect candidates to public office; general endowment funds; Deficit reduction or operating reserves; religious organizations; charitable dinners, events or sponsorships; community or event advertising; individuals; student trips or exchange programs; athletic sponsorships or scholarships; telephone solicitations; or organizations outside of the United States or its territories.
Geographic Focus: California, Kansas, New Jersey, North Carolina, Pennsylvania, Texas
Contact: Sarah Toulouse; (412) 777-5725; bayerusafoundation@bayer.com
Internet: http://www.bayerus.com/about/community/i_foundation.html
Sponsor: Bayer USA Foundation
100 Bayer Road, Building 4
Pittsburgh, PA 15205-9741

Bayer Hemophilia Caregivers Education Award 847
This award recognizes the essential role of caregivers and allied health professionals in the care of patients with hemophilia. It is designed to support their role by promoting continuing education. Awards will be provided for education activities in the field of hemophilia. This may include, but is not limited to, the development of educational seminars/symposia, mentored experiences, or training workshops. The Award is not designed for the development of materials for patients and parents. Applications will be judged by: Educational merit of the proposed initiative; Innovation of study rationale and techniques; Impact on the development of the professional and his/her ability to deliver care; Educational environment. The duration of the award is one year. Up to six Awards of up to $25,000 will be made annually. Part of the award may fund salary support. The candidate must spend at least 25% of his/her time on the project in order to request salary support.
Requirements: Requests for these grants must come from caregivers and allied health professionals involved in the care of patients with hemophilia. These may include nurses and nurse practitioners, physical therapists, pharmacists, psychiatrists, psychologists, genetic counselors, social workers, clinical dieticians, dentists and dental hygienists, among others. The Award is not designed for medically qualified hematologists who treat patients with bleeding disorders.
Restrictions: Unauthorized expenses are: Salaries, travel and/or housing related to sabbaticals; Purchase or rental of office equipment; Fees for tuition; Membership dues, congress/meeting registrations, subscriptions.

Geographic Focus: All States
Amount of Grant: 25,000 USD
Contact: Administrator; programadministrator@bayer-hemophilia-awards.com
Internet: http://www.bayer-hemophilia-awards.com/awards.cfm#caregivers
Sponsor: Bayer HealthCare Corporation
100 Bayer Road
Pittsburgh, PA 15205-9741

BBVA Compass Foundation Charitable Grants 848

BBVA Compass is a leading U.S. banking franchise located in the Sunbelt region. Headquartered in Birmingham, Alabama, it operates more than 720 branches throughout Texas, Alabama, Arizona, California, Florida, Colorado and New Mexico. The BBVA Compass Foundation works with local contributions committees in its major markets and city presidents in its community markets, ensuring a localized, focused and inclusive community giving strategy dedicated to making a meaningful, measurable impact in the communities it serves. The Foundation funds eligible 501(c)3 organizations through six focus areas: Community Development (including Financial Education), Education, Health and Human Services, Arts and Culture, Diversity and Inclusion, and Environment and Natural Resources.
Requirements: To be considered for grant funding, organizations must meet the following criteria: have a nonprofit, tax-exempt classification under section 501(c)3 of the Internal Revenue Code; be located or provide service in BBVA Compass' markets; have broad community support and address specific community needs; demonstrate fiscal and administrative stability; and align with one or more of the Foundation's six focus areas. The applicant organization should: exhibit significant support (through the contribution of time or financial resources) from a BBVA Compass employee(s); target individuals or communities with low- to moderate-income levels; help build inclusive and diverse communities; and foster collaborative efforts that leverage our community investments.
Restrictions: The BBVA Compass Foundation will not make contributions to support certain types of requests. These include, but may not be limited to, the following: sponsorships, events or projects for which BBVA Compass and/or its employees receive tangible benefits or privileges. This includes golf tournaments, tables at events and other fundraising activities that include tickets, meals or other benefits; general operating expenses for organizations that already receive substantial support through United Way or other campaigns in which BBVA Compass participates; programs of a national scope that do not specifically benefit communities in our footprint; organizations based outside of the United States political action committees, political causes or candidates; Veteran and fraternal organizations; alumni organizations; religious organizations that are not engaged in a significant project that is nonsectarian and benefits a broad base of the community; an organization or project that discriminates on the basis of age, disability, ethnicity, gender, gender identity, national origin, race, religion, sexual orientation, veteran's status, or any other status or other classification protected by federal, state or local law; private foundations; individual pre-college schools – private, parochial, charter or home schools; individual schools in public school systems (other than through efforts to benefit system-wide programs and initiatives or as a part of a BBVA Compass-sponsored partnership program).
Geographic Focus: Alabama, Arizona, California, Colorado, Florida, Georgia, New Mexico, Texas
Date(s) Application is Due: Sep 30
Contact: Office of Community Giving; (713) 831-5866; grants@bbvacompass.com or corporateresponsibility@bbvacompass.com
Internet: http://www.bbvacompass.com/compass/responsibility/foundations.jsp#grants
Sponsor: BBVA Compass Foundation
15 South 20th Street
Birmingham, AL 35233

BCBSM Building Healthy Communities Engaging Elementary Schools and Community Partners Grants 849

BCBSM (Blue Cross Blue Shield of Michigan) is offering this grant program to strengthen school and community efforts to reduce the risk and prevalence of childhood obesity through prevention and partnership. With this request for proposals Blue Cross invited Michigan elementary schools to collaborate with students, parents and community partners to address childhood obesity's root causes by promoting awareness of and education about physical activity and nutrition. See BCBSM website for all required forms.
Requirements: Electronic notice of intent is to be submitted no later then April 3. The application must be postmarked by April 24 to be eligible and be submitted with three copies of the proposal.
Geographic Focus: Michigan
Date(s) Application is Due: Apr 3; Apr 24
Amount of Grant: 40,000 - 60,000 USD
Contact: BHC Director; (800) 658-6715; buildhealth@bcbsm.com
Internet: http://www.bcbsm.com/buildhealth/index.shtml
Sponsor: Blue Cross Blue Shield of Michigan Foundation
600 East Lafayette Boulevard
Detroit, MI 48226-2998

BCBSM Claude Pepper Award 850

The Claude Pepper Award is presented annually to two outstanding senior citizen advocates who demonstrate a strong concern for the special needs of the elderly. The BCBSM (Blue Cross Blue Shield of Michigan) named the award after the late U.S. Congressman who campaigned for senior citizens' rights. Blue Cross Blue Shield of Michigan's Senior Advisory Council selects two individuals, a retiree volunteer over age 55 and another person of any age who's employed in the senior advocacy field. Each award recipient receives a $1,000 donation for the nonprofit organization of his or her choice. Candidates for this award will have clearly demonstrated their dedication to Michigan's seniors through work that positively impacts older adults, particularly those who don't receive adequate health services due to mental, physical, financial or geographical limitations. To nominate someone for a Claude Pepper Award, submit the nomination online using the form provided at the BCBSM website.
Requirements: Nomination Criteria: awards are given to individuals, not groups or teams of people. However, you may nominate a person for their individual work on behalf of a group to which he or she belongs; volunteers are defined as people who freely choose to provide a service that contributes to the well-being of an individual or a community; nominees can be paid workers; however, the committee is looking for efforts that go above and beyond job responsibilities; the candidate must have performed the services in the State of Michigan; monetary contributions can only to made to 501(c)3 nonprofit organizations.
Restrictions: Employees of Blue Cross Blue Shield of Michigan, its subsidiaries and members of all selection committees are not eligible for this award, nor are their family members or elected political office holders.
Geographic Focus: Michigan
Amount of Grant: 1,000 USD
Contact: Community Affairs Hotline, (800) 733-2583
Internet: http://www.bcbsm.com/home/commitment/claude_pepper_award.shtml
Sponsor: Blue Cross Blue Shield of Michigan Foundation
600 East Lafayette Boulevard
Detroit, MI 48226-2998

BCBSM Corporate Contributions Grants 851

The BCBSM (Blue Cross Blue Shield of Michigan) corporate giving program supports nonprofit 501(3)(c) organizations throughout Michigan. BCBSM invests in local communities by supporting organizations that share a commitment to Michigan. BCBSM charitable contributions support programs that: address health issues specific to Michigan's youth and seniors; extend a nonprofit organization's reach through effective use of volunteers; emphasize collaboration across regions within Michigan. Contribution requests are accepted year round and are reviewed monthly by BCBSM's director of community affairs.
Requirements: To be considered for a charitable contribution from the BCBSM, an organization must be: nonprofit 501(3)(c) status; Michigan based and focused. Requests for contributions must be submitted in writing, and include: Corporate Contributions Cover Sheet (available on BCBSM website); IRS 501(c)3 tax exemption documentation; organization's annual report. Submit requests for contribution one of three ways: mail; fax and; email.
Restrictions: Contributions are not made to: individuals; political organizations, campaigns or candidates; extracurricular school activities; endowments; alumni associations; scholarship programs; research projects; multi-year pledges; operational expenses; group travel expenses; capital campaigns.
Geographic Focus: Michigan
Contact: Cathy Mozham, BCBSM Community Responsibility Director; (313) 225-0539; fax (313) 225-9693; cmozham@bcbsm.com
Internet: http://www.bcbsm.com/home/commitment/corporate_contributions.shtml
Sponsor: Blue Cross Blue Shield of Michigan Foundation
600 East Lafayette Boulevard
Detroit, MI 48226-2998

BCBSM Foundation Community Health Matching Grants 852

The program focuses on access to care for the uninsured and under insured. The program's purpose is to encourage nonprofit community based organizations to form partnerships with health care organizations, research organizations, or governmental agencies to develop and rigorously evaluate new ways of increasing access to care for the under and uninsured, in Michigan. The program offers up to $50,000 per year for two years, contingent on a 25% match.
Requirements: Nonprofit 501(c)3 organizations based in Michigan are eligible. Nonprofits do not need a firm commitment of matching support prior to the submission of the proposal [letters from potential funding partners are encouraged, indicating their support for the project and interest in possible funding partnership(s)]. Proposals must include a rigorous evaluation of the outcome of the initiative in accomplishing its objectives (preferably an independent or university-based evaluation).
Restrictions: In-kind contributions are not accepted as a match. Grants under this program do not pay for the cost of equipment (e.g., personal computers), hardware or software. The BCBSM Foundation does not provide support to for?profit organizations or individuals associated with organizations not located in Michigan.
Geographic Focus: Michigan
Amount of Grant: Up to 100,000 USD
Contact: Nora Maloy, (313) 225-8706; fax (313) 225-7730; foundation@bcbsm.com
Internet: http://www.bcbsm.com/foundation/grant.shtml
Sponsor: Blue Cross Blue Shield of Michigan Foundation
600 East Lafayette Boulevard
Detroit, MI 48226-2998

BCBSM Foundation Excellence in Research Awards for Students 853

The annual Excellence in Research Award for Students acknowledges doctoral candidates or medical students enrolled in Michigan universities who have published research papers that represent significant contributions to health policy or clinical care. Entries will be judged based on the subject matter's potential to make a contribution to health policy and medical care as well as the quality and originality of the research. There will be three awards: 1st place - $1,000; 2nd place - $750; 3rd place - $500.
Requirements: Doctoral candidates or medical students enrolled in Michigan universities are eligible. The nomination must be made by a faculty member at the student's university. If the paper has multiple authors, the student must be the first author.
Restrictions: Basic research papers are not eligible.

Geographic Focus: Michigan
Date(s) Application is Due: Jan 1
Amount of Grant: 500 - 1,000 USD
Contact: Ira Strumwasser; (313) 225-8706; fax (313) 225-7730; foundation@bcbsm.com
Internet: http://www.bcbsm.com/foundation/grant.shtml
Sponsor: Blue Cross Blue Shield of Michigan Foundation
600 East Lafayette Boulevard
Detroit, MI 48226-2998

BCBSM Foundation Investigator Initiated Research Grants 854
The program provides grants for applied research that focuses on quality, cost and appropriate access to health care in Michigan. Grants average $75,000 for one year. Multi-year grants or grants in excess of $75,000 are awarded for exemplary projects. Projects that focus on the quality and cost of health care and appropriate access are considered priority and include: Organization and delivery of health care services; Evaluation of new methods or approaches to containing health care costs; Evaluation of new methods or approaches to providing access to high quality health care; Assessment and assurance of quality care; Identification and validation of clinical protocols and evidence-based practice guidelines. Applications are accepted at any time.
Requirements: Applications will be accepted from medical- and doctoral-level researchers based in hospitals, university settings, non-profit health care organizations, health systems, medical or nursing schools, schools of public health, or other relevant academic disciplines such as psychology, sociology and urban studies.
Restrictions: This program does not support basic science or biomedical research, including drug studies or studies using animals, costs of equipment (e.g., personal computers), hardware, software or on-going operating or developmental expenses. The BCBSM Foundation does not provide support to for-profit organizations or individuals associated with organizations not located in Michigan. Blue Cross and Blue Shield of Michigan employees, members of their immediate families, and employees and immediate family members of any Blue Cross and Blue Shield of Michigan affiliate or subsidiary are not eligible to receive BCBSM Foundation grants.
Geographic Focus: Michigan
Samples: Wei Du, Ph.D., Children's Hospital of Michigan, Department of Pediatrics, $74,994 - Diagnosing Adverse Drug Reactions in Pediatric ICU; Teresa Wehrwein, Ph.D., R.N., Michigan State University – College of Nursing, $82,290 - Preparing Nurses for Roles in Patient-Centered Medical Homes: Case Management and Quality/Safety Management; Faith Hopp, Ph.D., Hospice of Michigan, $89,646 - Evaluation of @HOMe Support Program.
Contact: Nora Maloy; (313) 225-8706; fax (313) 225-7730; foundation@bcbsm.com
Internet: http://www.bcbsm.com/foundation/grant.shtml
Sponsor: Blue Cross Blue Shield of Michigan Foundation
600 East Lafayette Boulevard
Detroit, MI 48226-2998

BCBSM Foundation Physician Investigator Research Awards 855
The program provides seed money to physicians to explore the merits of a particular research idea for further study. Grants are offered of up to $10,000 for projects that include pilot, feasibility or small research studies in clinical or health services research.
Requirements: Applicants must be a physician interested in research, who is licensed and domiciled in the state of Michigan. Applicants may include physicians working in research environments such as medical schools or university-affiliated hospitals, health care systems, or nonprofit agencies. An application may be submitted at any time.
Geographic Focus: Michigan
Amount of Grant: Up to 10,000 USD
Contact: Nora Maloy; (313) 225-8706; fax (313) 225-7730; foundation@bcbsm.com
Internet: http://www.bcbsm.com/foundation/grant.shtml
Sponsor: Blue Cross Blue Shield of Michigan Foundation
600 East Lafayette Boulevard
Detroit, MI 48226-2998

BCBSM Foundation Primary and Clinical Prevention Grants 856
The Blue Cross Blue Shield of Michigan Foundation requests letters of interest from Michigan physicians and members of the research community interested in preventing the three leading causes of death in Michigan: heart disease, cancer and stroke. The purpose of this grant initiative is to increase the use of evidence-based prevention efforts to improve the health of Michigan residents. The BCBSM Foundation has allocated $500,000 to fund several projects for up to two years. Funds will be available for salary support, program costs, supplies, office operations and other costs related to the proposed project. Computer equipment expenses, including PC hardware and software, will not be supported. In addition to the overall quality of the project, the feasibility and appropriateness of the budget is a factor in the funding decisions. The required application cover page and the terms and conditions may be downloaded for the BCBSM web site.
Requirements: Applicants must be doctoral level researchers (including physicians) based at Michigan universities, academic medical settings, community hospitals, health systems and community based nonprofit organizations. The principal investigator must have appropriate research and clinical credentials.
Geographic Focus: Michigan
Date(s) Application is Due: Sep 1
Contact: Nora Maloy; (313) 225-8706; fax (313) 225-7730; nmaloy@bcbsm.com
Internet: http://www.bcbsm.com/foundation/grant.shtml
Sponsor: Blue Cross Blue Shield of Michigan Foundation
600 East Lafayette Boulevard
Detroit, MI 48226-2998

BCBSM Foundation Proposal Development Awards 857
The program is intended to help Michigan-based nonprofit community organizations develop grant proposals for new healthcare initiatives. This award of up to $3,500 is available to help nonprofits obtain grant writing resources so they may secure the funding they need to bring creative community-based healthcare programs to life. Applications are accepted at any time.
Requirements: Applicants must be Michigan-based nonprofit community organizations that deliver health services. Michigan nonprofits are encouraged to apply for all available funding, including Federal grants. Nonprofits may also use the award funds to develop proposals for private, community or corporate foundations, either through local Michigan or national foundations, or from the state of Michigan. Expenditure of award funds is restricted to proposal development costs, such as technical assistance, freelance proposal writers and related proposal development costs.
Restrictions: Award funds are not intended to support any type of project activity or to supplement operational expenses. Proposals already completed are not eligible for funding.
Geographic Focus: Michigan
Amount of Grant: Up to 3,500 USD
Samples: Warren Conner Development Coalition, Wayne County, $3,500 - Promise: 48213; Eastern Huron Ambulance Service, Huron County, $3,500 - Expanding Access to Quality Emergency Medical Services in Rural Michigan; BRAINS Foundation, Kent County, $3,500 - BRAINS Foundation Neuropsychological Training Program.
Contact: Nora Maloy, Dr.P.H., Senior Program Officer; (313) 225-8706; fax (313) 225-7730; foundation@bcbsm.com
Internet: http://www.bcbsm.com/foundation/grant.shtml
Sponsor: Blue Cross Blue Shield of Michigan Foundation
600 East Lafayette Boulevard
Detroit, MI 48226-2998

BCBSM Foundation Student Award Program 858
The program offers a one year stipend to fund a wide range of health care projects, including applied research, pilot programs, or demonstration and evaluation projects.
Requirements: All doctoral and medical students enrolled in Michigan universities are eligible. For consideration, the proposed project must focus geographically on the state of Michigan and address the BCBSM Foundation's objectives.
Restrictions: Completed or substantially completed dissertations or research projects are not eligible. Students who previously received this award are not eligible. Blue Cross and Blue Shield of Michigan affiliates and subsidiaries are not eligible. Investigation of pharmaceutical efficacy, basic research or research involving non-human subjects is not eligible. Grant monies are not intended to support field placements, practica or internships.
Geographic Focus: Michigan
Date(s) Application is Due: Apr 30
Amount of Grant: 3,000 USD
Contact: Ira Strumwasser; (313) 225-8706; fax (313) 225-7730; foundation@bcbsm.com
Internet: http://www.bcbsm.com/foundation/
Sponsor: Blue Cross Blue Shield of Michigan Foundation
600 East Lafayette Boulevard
Detroit, MI 48226-2998

BCBSM Frank J. McDevitt, DO, Excellence in Research Awards 859
The Frank J. McDevitt Excellence in Research Awards for Health Services, Policy and Clinical Care recognize Michigan-based researchers and physicians who have published research in research journals that contributes to improving health and medical care in Michigan. The BCBSM Foundation will award $10,000 for unrestricted research to physician and doctoral level researchers for research on: health policy; health services; clinical care.
Requirements: Applicants must be from Michigan based researchers with terminal research degrees (e.g. PhD, DrPH) and from Michigan based physicians (MD or DO). In the case of multiple authorship, the nominee must be the first author. All authors will be nominated and will constitute one team nomination. The nomination research must have been published, or accepted for publication in a refereed health or medical care journal, within the previous two years. The financial award will be made to the recipient's 501(c)3 non-profit or educational organization.
Restrictions: Only one article per nominee will be accepted for consideration each year.
Geographic Focus: Michigan
Date(s) Application is Due: Jan 1
Amount of Grant: 10,000 USD
Contact: Nora Maloy; (313) 225-8706; fax (313) 225-7730; nmaloy@bcbsm.com
Internet: http://www.bcbsm.com/foundation/grant.shtml
Sponsor: Blue Cross Blue Shield of Michigan Foundation
600 East Lafayette Boulevard
Detroit, MI 48226-2998

BCBSNC Foundation Grants 860
The foundation's primary objective is to fund clearly defined, innovative grants that further the foundation mission of improving the health and well-being of North Carolinians. The foundation funds programs typically possessing the following characteristics: programs and/or services designed to produce measurable, long-term impact; programs that are sustainable and designed to be ongoing, rather than one-time or sporadic events; programs and/or services that are replicable. The three primary focus areas include health of vulnerable populations, healthy active communities and community impact through nonprofit excellence.
Requirements: Applicants must meet the following criteria to be eligible for funding: organization is located within North Carolina; organization is a 501(c)3 organization or an educational or governmental entity with tax-exempt status, that is not a private foundation or Type III supporting organization; organization must be able to provide

its most recent IRS Form 990. Depending on the type of grant and the size of the organization, an audit may be required as part of the submitted proposal.
Restrictions: The foundation will not provide funding for annual campaigns, political campaigns, religious purposes, individuals, endowments, purchase of advertisements, or for the sole purpose of receiving goods or entitlements from a charitable organization.
Geographic Focus: North Carolina
Samples: Diabetes Management Solutions (Raleigh, NC)--to support a pilot program for a physician diabetes day clinic as a model for improved outcomes in diabetes management, $10,980; Interfaith Assistance Ministry (Henderson, NC)--funding for a computer in every office to expedite the access and exchange of information for the MediFind program, $8430; Housing for New Hope (Durham, NC)--to build capacity of a comprehensive and sustainable health program for residents of five housing programs, $10,000; Graham Children???s Health Services of Toe River (Burnsville, NC)--to support preventive care for Hispanic and other uninsured children at school-based health centers, $20,465.
Contact: Grant Review Committee; (919) 765-7347; foundation@bcbsnc.com
Internet: http://www.bcbsnc.com/foundation/grants.html?previouslyOver=true¤tlyOver=true
Sponsor: Blue Cross Blue Shield of North Carolina Foundation
P.O. Box 2291
Durham, NC 27707-0718

BCRF-AACR Grants for Translational Breast Cancer Research   861
The grants will provide direct support for innovative cancer research projects designed to accelerate the discovery, development, and application of new agents to treat breast cancer and/or for pre-clinical research with direct therapeutic intent. Special emphasis will be placed on research that holds promise for leading to individualized therapeutic options for treatment in the near future.
Requirements: Researchers who are affiliated with any institution involved in cancer research, cancer medicine, or cancer-related biomedical science anywhere in the world may apply. There are no geographic, national, or residency status restrictions. Applicants must have acquired a doctoral degree in a related field. If an applicant has obtained an equivalent degree at a foreign institution, information on the nature of the degree must be provided at the time of application.
Restrictions: Employees or subcontractors of a national government or the for-profit private industry are not eligible to serve as the Principal Investigator for the purpose of these grants. However, collaborations with such individuals are encouraged. Neither the members of the Scientific Review Committee nor the members of their individual laboratories are eligible for these grants.
Geographic Focus: All States
Date(s) Application is Due: May 20; Jul 20
Amount of Grant: 200,000 USD
Contact: Hanna Hopfinger, Program Assistant; (267) 646-0665 or (215) 440-9300; fax (215) 440-9372; hanna.hopfinger@aacr.org or awards@aacr.org
Internet: http://www.aacr.org/home/scientists/research-funding--fellowships/bcrf-aacr-grants.aspx
Sponsor: Breast Cancer Research Foundation & the American Association for Cancer Research
615 Chestnut Street, 17th Floor
Philadelphia, PA 19106-4404

BCRF Research Grants   862
The goal of the grant process is a serious peer review of all proposals. Proposals are invited by the Medical Advisory Board, rather than accepted as unsolicited requests. The MAB generally reviews proposals in the late summer. There is no set format, but the BCRF requests a brief narrative (no more than 4-5 pages), lay language summary, annual budget (no more than $225,000 initially, with a maximum of 20% in indirect costs) and budget narrative. Considerable latitude is given to the investigators in terms of the work proposed. Both the Board of Directors and the Medical Advisory Board concur that some of the most important advances in understanding the disease will most likely occur by enabling brilliant minds to pursue some of their most creative theories.
Restrictions: While most of the grants are unrestricted, the Foundation has donors requesting restricted grants for specific purposes. When this occurs, the Medical Advisory Board will meet to determine whether the area of interest merits further exploration. If so, they will identify various research projects meeting the donor's criteria.
Geographic Focus: All States
Amount of Grant: Up to 225,000 USD
Contact: Administrator; (646) 497-2600; fax (646) 497-0890; bcrf@bcrfcure.org
Internet: http://www.bcrfcure.org/action_grantguidelines.html
Sponsor: Breast Cancer Research Foundation
60 East 56th Street, 8th Floor
New York, NY 10022

Bearemy's Kennel Pals Grants   863
The program provides direct support for animals in domestic pet programs including animal welfare organizations, pet rescue and rehabilitation organizations, and therapeutic and humane education pet programs. Grants will be a one-time contribution and generally range from $1,000 to $10,000, but the average grant tends to be $4,000. Programs that will be funded include: (1) Individual Project grants generally for one-time purchases or to fulfill a short-term need. Examples include equipment purchases or spay/neuter events, etc; (2) Organization Program grants - start-up or operational costs for ongoing programs. Examples include spay/neuter programs, humane education initiatives, and service dog training programs.
Requirements: While the geographic focus of the program is broad (United States and Canada), priority is given to organizations located near Build-A-Bear Workshop stores.

United States applicants must be a tax-exempt organization under Section 501(c)3 of the IRS Code, and not a private foundation, within the meaning of Code Sections 509(a)(1) or 509(a)(2), or a state college or university within the meaning of Code Section 511(a)(2)(B) (a Public Charity). In addition, grant recipients must certify that they are not a supporting organization within the meaning of Code Section 509(a)(3). Canadian applicants must be a registered Canadian charity.
Restrictions: Grant types not funded: (1) Capital Campaigns; (2) Construction or new facility expenses; (3) Fundraising or Event Sponsorships; (4) Political Activities; (5) Religious organizations for religious purposes.
Geographic Focus: All States, Canada
Date(s) Application is Due: Mar 31; Jun 30; Sep 31; Dec 31
Amount of Grant: 1,000 - 10,000 USD
Samples: Animal Advocates of South Carolina (Barnwell, SC); Bunny Castle Rabbit Rescue (Bishop, CA); EquiFriends (Snohomish, WA); Hearing and Service Dogs of Minnesota (Minneapolis, MN); Pet Adoption Fund (Canoga Park, CA).
Contact: Maxine Clark, President; (314) 423-8000, ext. 5366; giving@buildabear.com
Internet: http://www.buildabear.com/aboutus/community/bearhugs.aspx
Sponsor: Build-A-Bear Workshop Bear Hugs Foundation
1954 Innerbelt Business Center Drive
Saint Louis, MO 63114

Beatrice Laing Trust Grants   864
The trust was founded for the relief of poverty and for the advancement of the evangelical faith internationally. Grants are awarded for general charitable purposes at the discretion of the Trustees, with an emphasis on social services and relief, and medical aid and research. Grant recipients typically are charities working in deprived sections of the community in the United Kingdom, to missionary societies and, less frequently, to individuals working in the field of missions in United Kingdom and abroad.
Geographic Focus: All States, Albania, Algeria, Andorra, Angola, Armenia, Austria, Azerbaijan, Belarus, Belgium, Benin, Bosnia & Herzegovina, Botswana, Bulgaria, Burkina Faso, Burundi, Cameroon, Cape Verde, Central African Republic, Chad, Comoros, Congo, Congo, Democratic Republic of, Cote d' Ivoire (Ivory Coast), Croatia, Cyprus, Czech Republic, Denmark, Djibouti, Egypt, Equatorial Guinea, Eritrea, Estonia, Ethiopia, Finland, France, Gabon, Gambia, Georgia, Germany, Ghana, Greece, Guinea, Guinea-Bissau, Hungary, Iceland, Ireland, Italy, Kenya, Kosovo, Latvia, Lesotho, Liberia, Libya, Liechtenstein, Lithuania, Luxembourg, Macedonia, Madagascar, Malawi, Mali, Malta, Mauritania, Mauritius, Moldova, Monaco, Montenegro, Morocco, Mozambique, Namibia, Niger, Nigeria, Norway, Poland, Portugal, Romania, Russia, Rwanda, San Marino, Sao Tome & Principe, Senegal, Serbia, Seychelles, Sierra Leone, Slovakia, Slovenia, Somalia, South Africa, Spain, Sudan, Swaziland, Sweden, Switzerland, The Netherlands, Turkey, Ukraine, United Kingdom, Vatican City
Samples: ActionAid, 10,000 pounds to improve access to education for pastoralist families in Somaliland; OxFam, 10,000 pounds to the You Are Here program; Sense International, 15,000 pounds for work with deaf-blind people in Romania.
Contact: Trust Administrator; 020 8238 8890; fax 020 8238 8897
Sponsor: Beatrice Laing Trust
33 Bunns Lane
London, NW7 2DX England

Beazley Foundation Grants   865
The foundation makes grants to Virginia nonprofit organizations engaged in educational, charitable, and religious activities. The focus of support is within the South Hampton Roads cities and counties. Past awards include: music appreciation; educational outreach and mentoring programs; health and wellness; equipment and operating expenses; environmental education programs; renovations and repairs; literacy and community development. Grant applications can be found online.
Requirements: 501(c)3 nonprofit organizations in Virginia are eligible.
Restrictions: Individuals are ineligible.
Geographic Focus: Virginia
Date(s) Application is Due: Jan 4; Mar 1; Jun 1; Oct 1
Samples: The Tidewater Winds, Norfolk, VA, $7,500--operating expenses; The Union Mission Ministries, Norfolk, VA, $125,000--capital campaign; Virginia College Fund, Richmond, VA, $50,000--Beazley Scholarship Endowments; The Academy of Music, Norfolk, VA, $42,000-- Park View Elementary Strings Program.
Contact: Donna Russell; (757) 393-1605; fax (757) 393-4708; info@beazleyfoundation.org
Sponsor: Beazley Foundation
3720 Brighton Street
Portsmouth, VA 23707-3902

Bechtel Group Foundation Building Positive Community Relationships Grants   866
Bechtel Group Foundation was created in 1954 to respond to the needs of the communities around the world in which Bechtel has offices or major projects. These grants support educational, civic and cultural, and social service programs in the communities that host major Bechtel offices and projects. Bechtel offices are located in the following regions/countries: Australia; Brazil; Canada; Chile; Greater China; Egypt; France; India; Indonesia; Japan; Korea; Libya; Malaysia; Peru; Philippines; Poland; Qatar; Russia; Saudi Arabia; Singapore; Taiwan; Thailand; Turkey; United Arab Emirates; United Kingdom - London, England; United States - Maryland, Arizona; Texas; Virginia; New York; Tennessee; Washington; California; Washington, D.C. These grants are typically under U.S. $5,000.
Geographic Focus: Arizona, California, Maryland, Tennessee, Texas, Virginia, Washington, Australia, Brazil, Canada, China, Japan, Peru, Philippines, Poland, Russia, United Kingdom
Amount of Grant: 1,000 - 5,000 USD

Contact: Susan Grisso, Foundation Manager; (415) 768-5444; becfoun@bechtel.com
Internet: http://www.bechtel.com/foundation.html
Sponsor: Bechtel Group Foundation
50 Beale Street
San Francisco, CA 94105

Beckley Area Foundation Grants 867
The Foundation makes annual discretionary grants to charitable organizations in the areas of education, health and human services, the arts, public recreation and civic beautification. Using several general unrestricted endowments which generate income, BAF makes these grants to address the community's current most pressing needs and promising opportunities.
Requirements: To be eligible for a grant from the Beckley Area Foundation, applicants must be a non-profit organizatiion tax-exempt under section 501(c)3 of the Internal Revenue Code, or they must be a public institution. Programs funded through this annual grant must be located in Raleigh County.
Restrictions: As a Community Foundation, BAF cannot accept requests on behalf of individuals or groups that do not have official recognition as a non-profit charity or public entity. Grants are for one year only. Generally grants are not made for: annual campaigns; endowments; sectarian religious programs; political purposes or lobbying; ongoing operating expenses; expenses outside the grant period.
Geographic Focus: West Virginia
Date(s) Application is Due: Dec 15
Amount of Grant: 10,000 USD
Contact: Grants Administrator; (304) 253-3806; fax (304) 253-7304; funds@beckleyareafoundation.com
Internet: http://beckleyareafoundation.com/id7.html
Sponsor: Beckley Area Foundation
129 Main Street, Suite 203
Beckley, WV 25801

Beckman Coulter Foundation Grants 868
Beckman Coulter has a long-standing commitment to its Community Relations program. Within the program, Beckman Coulter associates strive to improve its local communities through a wide range of engagement activities that includes (but is not limited to): service projects; toy drives; blood drives; and walks, runs and bike rides. The Community Relations program also serves as a vehicle that encourages associates to work toward fulfilling the corporate vision of advancing healthcare for every person through our associate-driven fundraising activities that supports science, science education and healthcare-related research to help fight diseases. Grant applications are reviewed by the Foundation's staff on a regular basis. Each request is evaluated to determine its eligibility and conformity to the Beckman Coulter Foundation's giving guidelines.
Requirements: Organizations seeking funding should submit a letter of request providing the following information: full legal name of the organization; brief description of the organization (including a mission statement and services provided); amount requested; and a description of how the funds will be used. Additional attachments should include: history of previous support by Beckman Coulter; description of any involvement by Beckman Coulter employees; statement as to why you consider Beckman Coulter an appropriate donor; and a list of governing board members.
Restrictions: The Foundation does not support the following: advertising; awards and recognition programs; beauty or talent contests; capital or building campaigns, including new construction or renovations; for-profit organizations; fraternal, labor, or veteran's organizations and activities; fund-raisers; galas, banquets, or dinners; golf tournaments; indirect costs, media productions (e.g. radio, T.V., film, webcast, publications); meetings, conferences, workshops, forums, summits and symposiums; organizations that discriminate on the basis of race, color, sex, sexual orientation, marital status, religion, age, national origin, veteran's status, or disability; private foundations; political organizations, campaigns and activities; religious organizations or groups; or sports affiliated activities.
Geographic Focus: All States
Contact: Marci Raudez; (714) 961-6672 or (714) 961-6338; mfraudez@beckman.com
Internet: http://www.beckmancoulterfoundation.org/grant.asp
Sponsor: Beckman Coulter Foundation
250 South Kraemer Boulevard, E1.SE.03
Brea, CA 92821

Becton Dickinson and Company Grants 869
The company supports national and international activities in keeping with its commitment to advance the quality of medical practice and patient care in four areas: the global healthcare fund; the local initiatives fund; the matching gifts program; and the BD product donation program. Call for guidelines.
Geographic Focus: All States
Contact: Community Relations Manager; (201) 847-6800
Internet: http://www.bd.com/responsibility/contributions
Sponsor: Becton Dickinson and Company
1 Becton Drive
Franklin Lakes, NJ 07417

Beirne Carter Foundation Grants 870
The Foundation places an emphasis on health, education, local history, nature, ecology and youth. Except in rare instances, grants are limited to institutions and organizations located in the Commonwealth of Virginia. Requests that are not of a recurring nature are given preference. Any organization that has been awarded a grant is encouraged to wait at least three years before submitting a new grant request. Any organization seeking grant consideration must complete a Preliminary Proposal form which is online. Its content should be limited to the two page format provided online.
Requirements: The Foundation is organized exclusively for charitable purposes. Grants are made only to institutions and organizations which qualify under IRS regulations as tax exempt and which are not private operating foundations as defined by the IRS.
Restrictions: Grants will not normally be made: to endowment funds (including scholarship funds); to organizations supported primarily by government funds (such as public secondary schools and colleges and local municipalities); to churches and related organizations; for ongoing operating expenses (such as salaries, rent, office supplies, etc.); for existing deficits; or for debt reduction. Grants are not made to individuals.
Geographic Focus: Virginia
Date(s) Application is Due: Feb 1; Aug 3
Contact: Peter C. Toms; (804) 521-0272; fax (804) 521-0274; bcarterfn@aol.com
Internet: http://www.bcarterfdn.org
Sponsor: Beirne Carter Foundation
1802 Bayberry Court, Suite 401
Richmond, VA 23226-3773

Beldon Fund Grants 871
The fund focuses project and general support grants in two programs: human health and the environment, and key states. Human health and the environment--the fund seeks proposals that engage new constituencies in exposing the connection between toxic chemicals and human health and in promoting public policies that prevent or eliminate environmental risks to people's health. The program focuses grant making in three areas: new advocates, human exposure to toxic chemicals, and environmental justice. Key states--the fund believes that states hold the key to bringing about rapid, real change on environmental issues and policy in the United States. By strengthening public support for environmental protection in several of these key states, the fund hopes to transform the nation's approach to environmental protection. The fund is currently accepting proposals from Florida, Michigan, Minnesota, Wisconsin, and North Carolina for this program. Proposals do not need to be tied to any particular issue or set of issues, but targeted issues must be those that will build active public support for the environment. From time to time, the fund will add and remove states from this program. Types of support include special projects, seed money, general operating budgets, and technical assistance. One-year and multiyear grants are awarded. Due date applies to the letter of intent; full proposals are by invitation.By supporting effective, nonprofit advocacy organizations, the Beldon Fund seeks to build a national consensus to achieve and sustain a healthy planet. The Fund plans to invest its entire principal and earnings by 2009 to attain this goal.
Requirements: 501(c)3 tax-exempt organizations are eligible.
Restrictions: Grants do not support international efforts, academic or university efforts, school-based environmental education, land acquisition, wildlife or habitat preservation, film or video production, deficit reduction, endowments, capital campaigns, acquisitions of museums, service delivery, scholarshp, publications, or arts/culture.
Geographic Focus: Florida, Michigan, Minnesota, North Carolina, Wisconsin
Date(s) Application is Due: Feb 28; Jun 13
Amount of Grant: 5,000 - 100,000 USD
Samples: Ctr for Public Interest Research (Boston, MA)--for its Campaign Staff Training Institute Program to ameliorate the shortage of skilled activists, $75,000; Institute for Conservation Leadership (Takoma Park, MD)--for its outreach and hands-on support efforts, along with Environmental Support Ctr, $10,000; State Environmental Leadership Program (Madison, WI)--for general support, $175,000; Washington Environmental Alliance for Voter Education (Seattle, WA)--for general support, $50,000.
Contact: Holeri Faruolo; (800) 591-9595 or (212) 616-5600; info@beldon.org
Internet: http://www.beldon.org
Sponsor: Beldon Fund
99 Madison Avenue, 8th Floor
New York, NY 10016

Belk Foundation Grants 872
The Foundation makes grants to a wide variety of community-based nonprofit organizations and institutions whose missions and actions support the advancement of Christian causes and the up-building of mankind. The Foundation supports local and regional organizations by: assisting secondary schools, colleges and universities and their programs; assisting religious institutions and organizations and their programs; supporting area arts and other cultural organizations and their programs; supporting community-based human services organizations and their programs; and aiding hospitals and health care organizations and their programs.
Requirements: 501(c)3 nonprofits in communities in the 14 states where Belk stores are located. Preference is given to organizations in North Carolina.
Restrictions: Grants are not awarded to: individuals, including students; public, government or quasi-governmental programs, agencies or organizations (excluding certain public secondary schools, colleges and universities); or international programs and/or organizations. Additionally, the Foundation does not provide door prizes, gift certificates, merchandise or other giveaways.
Geographic Focus: Alabama, Arkansas, Florida, Georgia, Kentucky, Louisiana, Maryland, Mississippi, North Carolina, South Carolina, Tennessee, Texas, Virginia, West Virginia
Date(s) Application is Due: Apr 15; Oct 15
Contact: Susan C. Blount, Administrator; (704) 426-8396; susan_blount@belk.com
Internet: http://www.belk.com/AST/Misc/Belk_Stores/About_Us/Belk_Community/Belk_Foundation.jsp
Sponsor: Belk Foundation
2801 W Tyvola Road
Charlotte, NC 28217-4500

Bella Vista Foundation Grants 873

The foundation has two main focus areas: early childhood--healthy emotional development in children during the first years of their lives, and parents involvement to help their children deal with the internal issues of attachment and independence; and environment restoration--restoration of land, streams, wetlands, habitat, etc., and acquisition of land for the purpose of preservation, particularly requests that include restoration efforts. Deadlines are January 30 and June 15 for environmental restoration grants, and January 15 and June 15 for early childhood grants. Guidelines are available online.
Requirements: Early childhood grants support public charities located in California's San Francisco, Marin, San Mateo, and Santa Clara counties. Environmental restoration grants support projects in California or Oregon.
Restrictions: Grants will not be made to or for the arts, sectarian religious purposes, individuals, or benefit events, and will only be made for medical research, health care, publications, or video production under special circumstances and only in the early childhood development focus area. The foundation does not make multiyear grants.
Geographic Focus: California, Oregon
Date(s) Application is Due: Jan 15; Jan 30; Jun 15
Amount of Grant: Up to 100,000,000 USD
Samples: Asian Perinatal Advocates, $40,000 to continue home-based and center-based literacy training and resources to Asian families with children 0-3; Blind Babies Foundation, $25,000 for support of early intervention with blind babies and their families through the Off To A Good Start program; Deschutes River Conservancy, $25,000 for pre-restoration planning in the Squaw Creek watershed, a tributary of the Deschutes River, which soon will be reopened to populations of steelhead and wild Chinook; Ecotrust, $50,000 to work with private landowners to design estuary conservation and restoration projects and develop forest thinning and restoration projects with the Siuslaw National Forest; Friends of Corte Madera Creek Watershed, $4,125 to support the restoration of a section of Laurel Grove Creek that runs along the perimeter of an elementary school in Kentfield, CA.
Contact: Mary Gregory; (415) 561-6540; fax (415) 561-6477; mgregory@pfs-llc.net
Internet: http://www.pfs-llc.net/bellavista/index.html
Sponsor: Bella Vista Foundation
Presidio Building 1016, Suite 300, P.O. Box 29906
San Francisco, CA 94129-0906

Belvedere Community Foundation Grants 874

The mission of the Belvedere Community Foundation is to: preserve and enhance the quality of life in Belvedere, California; form an endowment fund with contributions from all of its citizens; and provide grants to support projects and volunteers working to enhance the quality of life throughout the community. Grants are targeted at supporting: preservation and enhancement to historically important structures, as well as the natural beauty of the community; positive community interaction; educational opportunities, particularly those focused on stewardship of natural resources and awareness of cultural heritage; healthy living, particularly as it pertains to the benefits of an active lifestyle; funds for emergency preparedness and public safety; seed funding for new community based projects or to aid projects launched by other locally based, non profits aligned with the goals of the Foundation; and crisis funding to existing community based programs in times of urgent need. The Foundation has two online grant cycles per year, with deadlines on March 1 and September 1. Grant requests should be submitted by the beginning of the relevant grant cycle and all applicants will receive a response within 60 days.
Requirements: Any 501(c)3 organization supporting the residents of Belvedere, California, are eligible to apply.
Geographic Focus: California
Date(s) Application is Due: Mar 1; Sep 1
Amount of Grant: Up to 10,000 USD
Contact: Juli Tantum; (415) 435-3695; info@belvederecommunityfoundation.com
Internet: http://belvederecommunityfoundation.com/grants.htm
Sponsor: Belvedere Community Foundation
P.O. Box 484
Belvedere, CA 94920

Ben B. Cheney Foundation Grants 875

Ben B. Cheney Foundation supports projects in communities where the Cheney Lumber Company was active. These areas include: Tacoma, Pierce, southwestern Washington, southwestern Oregon with a focus on Medford, portions of Del Norte, Humboldt, Lassen, Shasta, Siskiyou, and Trinity counties in California. The Foundation's goal is to improve the quality of life in those communities by making grants to a wide range of activities including: Charity (programs providing for basic needs such as food, shelter, and clothing); Civic (programs improving the quality of life in a community as a whole such as museums and recreation facilities); Culture (programs encompassing the arts); Education (programs supporting capital projects and scholarships, primarily for fourteen pre-selected colleges and universities in the Pacific Northwest); Elderly (programs serving the social, health, recreational, and other needs of older people); Health (programs related to providing health care); Social Services (programs serving people with physical or mental disabilities or other special needs); Youth (programs helping young people to gain the skills needed to become responsible and productive adults). Ben B. Cheney Foundation prefers to fund projects that: develop new and innovative approaches to community problem; facilitate the improvement of services or programs; invest in equipment or facilities that will have a long-lasting impact on community needs. The Foundation's application process always begins with a two to three page proposal letter. The process is the same for past grantees and new grant seekers. There are no deadlines. The Foundation accepts proposal letters throughout the year. It may take six to nine months from the receipt of a proposal letter to consideration of a grant application by the Board of Directors.
Requirements: Ben B. Cheney Foundation makes grants to private, nonprofit organizations that have received their 501(c)3 status from the IRS and that qualify as public charities. In special circumstances proposals from governmental organizations are allowed.
Restrictions: Ben B. Cheney Foundation generally does not make grants for: general operating budgets, annual campaigns, projects which are primarily or normally financed by tax funds, religious organizations for sectarian purposes, basic research, endowment funds, individuals, produce books, films, videos, conferences, seminars, attendance, individual student, or student groups raising money for school related trips.
Geographic Focus: California, Oregon, Washington
Amount of Grant: Up to 300,000 USD
Samples: Kitsap Mental Health Services, Bremerton,Washington, $50,000 - build Keller House for residential and outpatient programs; Ocean Shores Friends of the Library, Oceans Shores, Washington $4,000 - support the conference for Nonprofit Leaders in SW Washington to enhance fundraising; Hoover Elementary PTO, Medford, Oregon, $1,260 - provide the last monies needed to complete the new Boundless community playground at the school; Catholic Community Services, Tacoma, Washington, $40,000 - support operations of the downtown emergency shelter.
Contact: Bradbury F. Cheney; (253) 572-2442; info@benbcheneyfoundation.org
Internet: http://www.benbcheneyfoundation.org
Sponsor: Ben B. Cheney Foundation
3110 Ruston Way, Suite A
Tacoma, WA 98402-5307

Bender Foundation Grants 876

The foundation supports projects for higher education, health agencies, Jewish welfare funds and organizations, Christian youth organizations, and social welfare. Emphasis is support for programs to assist the elderly and aging, children and parenting, and environmental programs. Grants are awarded for challenge/matching grants, endowments, general operations, and scholarships.
Requirements: Grants are made to organizations and institutions in Maryland and Washington, D.C.. An applicant should initially send a brief letter of intent. Full proposals are by invitation.
Restrictions: Grants are not made to individuals.
Geographic Focus: District of Columbia, Maryland
Date(s) Application is Due: Nov 30
Amount of Grant: 100 - 100,000 USD
Sample: Jewish Community Center of Greater Washington (Rockville, MD)--for fitness center, $87,500; Discovery Creek Childrens Museum (Washington, D.C.)--general support, $50,000; Anti-Defamation League of Bnai Brith(Washington, D.C.)--for Concert Against Hate, $25,000.
Contact: Julie Bender Silver, President; (202) 828-9000; fax (202) 785-9347
Sponsor: Bender Foundation
1120 Connecticut Avenue NW, Suite 1200
Washington, D.C. 20036

Benton Community Foundation - The Cookie Jar Grant 877

The Benton Community Foundation - The Cookie Jar Grant was established by women in the Benton County area to fund programs that empower women. The Cookie Jar is a collaboration of women working together to create a stronger community where women and girls are empowered to reach their full potential. Priority is given to programs that improve the overall health or well-being of women and/or girls in Benton County. Priority is also given to programs that: reach as many people as possible; are run by a collaboration of non-profit organizations; and have multiple sources of funding for their project.
Requirements: Organizations should submit the grant application to the Foundation to include the following information: their organizational information and contacts; a summary of their program with amount requested; program details, such as how the program will benefit women or girls in Benton County; the program's anticipated timeline; the organization's top three goals for the project and how they plan to meet them; budget details; and a list of the organization's board of directors.
Restrictions: Foundation grant programs generally do not fund: ongoing operating expenses; individuals; special events such as parades, festivals and sporting events; debt or deficit reduction.
Geographic Focus: Indiana
Date(s) Application is Due: Sep 15
Amount of Grant: 4,000 USD
Contact: Ashley Bice; (765) 884-8022; fax (765) 884-8023; ashley@bentoncf.org
Internet: http://www.bentoncf.org/cookiejar.html
Sponsor: Benton Community Foundation
P.O. Box 351
Fowler, IN 47944

Benton Community Foundation Grants 878

The Foundation grants address the broad needs of Benton County residents. Funding is allocated in the following categories: civic affairs; cultural affairs; education; health and safety; and social services. Priority is given to programs that: reach as many people as possible; improve the ability of the organization to serve the community over the long term; serve Benton County residents; are run by a collaboration of nonprofit organizations; and have multiple sources of funding for the project.
Requirements: Grant requests are accepted year-round. Applicants are encouraged to contact the Foundation prior to submitting a request to confirm that the project is appropriate for Foundation funding. To make a request of $750 or less, organizations should submit a letter of application which includes a description of the organization, statement of the problem or need being addressed, explanation of the project, estimated

expenses, timeframe for completion, and the amount being requested. Grant decisions are made within 30 days of application. Organizations requesting more than $750 are encouraged to contact the Foundation to discuss the proposed project/program prior to submitting an application. They will then complete the Community Grant Application. Grant decisions are made within 90 days of application.
Restrictions: Benton Community Foundation generally does not fund: political organizations or candidates; endowments; ongoing operating expenses; individuals; special events such as parades, festivals, and sporting events; debt or deficit reduction; or projects funding in a previous year, unless invited to resubmit.
Geographic Focus: Indiana
Date(s) Application is Due: Aug 1
Contact: Ashley Bice, Executive Director; (765) 884-8022; fax (765) 884-8023; ashley@bentoncf.org or info@bentoncf.org
Internet: http://www.bentoncf.org/grants_community.html
Sponsor: Benton Community Foundation
P.O. Box 351
Fowler, IN 47944

Benton County Foundation - Fitzgerald Family Scholarships 879
The Benton Community Foundation Fitzgerald Family Scholarships provide funding for students pursuing post-secondary education. Priority is given to students who intend to study in a health-related profession and/or have overcome a specific life challenge.
Requirements: Students should contact their guidance office for the scholarship application and deadline information.
Geographic Focus: Indiana
Contact: Ashley Bice, Executive Director; (765) 884-8022; fax (765) 884-8023; ashley@bentoncf.org or info@bentoncf.org
Internet: http://www.bentoncf.org/scholarships_fitzgerald.html
Sponsor: Benton Community Foundation
P.O. Box 351
Fowler, IN 47944

Benton County Foundation Grants 880
The Foundation's investment philosophy is built on the precepts of diversification, long-term strategic focus, and prudent risk management. The mission is to establish endowments, manage the funds received, and distribute a portion of the earnings each year to benefit youth and community. The Foundation focuses on funding programs that enhance youth education and/or provide positive character building and life skills experience.
Requirements: Organizations must be located in Benton County, Oregon.
Geographic Focus: Oregon
Date(s) Application is Due: Mar 13
Contact: Dick Thompson; (541) 753-1603 or (541) 231-7604; BCF@peak.org
Internet: http://www.bentoncountyfoundation.org/grants/
Sponsor: Benton County Foundation
P.O. Box 911
Corvallis, OR 97330

Berks County Community Foundation Grants 881
The foundation supports a broad range of community projects in this Pennsylvania county, including the arts and culture, economic development, education, the environment, and health and human services. Types of support include general operating grants, capital campaigns, demonstration grants, seed grants, and program grants. Although applicants do not have to be located in Berks County, they must provide programs and services within the county.
Requirements: Tax exempt or public benefit organizations, individuals, associations and private or public agencies are eligible to apply. The grant must be used for charitable purposes only. Organizations are eligible to apply to more than one grant program in the same year.
Geographic Focus: Pennsylvania
Contact: Richard Mappin; (610) 685-2223; fax (610) 685-2240; info@bccf.org
Internet: http://www.bccf.org/pages/grants.html
Sponsor: Berks County Community Foundation
501 Washington Street, Suite 801, P.O. Box 212
Reading, PA 19603-0212

Bernard and Audre Rapoport Foundation Health Grants 882
Bernard and Audre Rapoport Foundation seeks to improve the quality and delivery of healthcare services to all citizens, especially to women, children, and those who do not have access to conventional medical resources. Community-based outreach initiatives such as immunization programs are of interest to the Foundation. The primary focus of the Foundation is on programs that benefit children and youth in Waco and McLennan County, Texas. Proposals that fall outside of this geographical focus are considered as long as they offer imaginative, and when possible, long-range solutions to the problems of the most needy members of society, and ideally solutions that can be replicated in other communities.
Requirements: Program seeking funding must be a catalyst for change and promote both individual competence and social capacity.
Restrictions: Bernard and Audre Rapoport Foundation only supports organizations that are nonprofit 501(c)3 tax-exempt.
Geographic Focus: All States
Date(s) Application is Due: Jun 15; Aug 15
Amount of Grant: 4,000 - 250,000 USD
Samples: University of Texas Health Science Center, Houston, Texas, $20,000 - stroke research; Potter's Vessel Ministry, Waco, Texas, $10,000 - new health care provider; Planned Parenthood of Central Texas, Waco, Texas, $60,000 - general support.
Contact: Carole Jones; (254) 741-0510; fax (254) 741-0092; carole@rapoportfdn.org
Internet: http://www.rapoport-fdn.org/
Sponsor: Bernard and Audre Rapoport Foundation
5400 Bosque Boulevard, Suite 245
Waco, TX 76710

Bernard F. and Alva B. Gimbel Foundation Reproductive Rights Grants 883
The Bernard F. and Alva B. Gimbel Foundation was incorporated in 1943 as a private, family foundation. The Foundation's donors had wide ranging philanthropic interests and the Foundation's original giving guidelines were defined broadly: grants were to be used for charitable, scientific or educational purposes. The Foundation's areas of interest have evolved over the years and are currently more narrowly defined. The Foundation's funding in the area of reproductive rights the Foundation supports domestic programs that seek to protect women's reproductive rights and increase access to reproductive health services. Funded initiatives include those working to achieve these goals through public education efforts, policy advocacy and legal advocacy. Also of interest to the Foundation are efforts to increase reproductive health training opportunities for medical students and healthcare professionals, as well as efforts to support them as public advocates for comprehensive reproductive healthcare. In most cases, the Foundation seeks to fund general operating costs, direct services programs, and advocacy efforts.
Requirements: Grants are made only to tax-exempt 501(c)3 organizations. The Foundation's support for direct services programs is limited to those operating in New York City.
Restrictions: The Foundation does not make grants to: individuals; direct service programs outside of New York City; individual schools; short-term educational programs and workshops; mentoring programs; after-school and summer programs; youth development programs.
Geographic Focus: New York
Amount of Grant: 30,000 - 100,000 USD
Samples: Center for Reproductive Rights, New York, New York, $100,000 - general operating support (2015); National Institute for Reproductive Health, New York, New York, $50,000 - general operating support (2015); Physicians for Reproductive Health, New York, New York, $50,000 - general operating support (2015).
Contact: Leslie Gimbel, President; (212) 684-9110; fax (212) 684-9114
Internet: http://www.gimbelfoundation.org/FundingRestrictionsandPriorities.htm
Sponsor: Bernard F. and Alva B. Gimbel Foundation
271 Madison Avenue, Suite 605
New York, NY 10016

Berrien Community Foundation Grants 884
The community foundation awards grants to nonprofits in Berrien County, Michigan, that address community needs. The Foundation is very interested in providing start-up funding for programs that address our focus areas of nurturing children, building community spirit/arts and culture and youth leadership and development. Higher priority is given to requests that demonstrate community-based collaborative solutions likely to stay in place after Foundation funding concludes. Low priority is given to requests for bricks and mortar, operational funds on a repetitive basis, annual fund drives, equipment, and ongoing programs where alternative funding is not planned to carry a program/project forward following a Foundation grant. Low priority is also given to advertising and capital campaigns and grants to cover deficits or other previously incurred obligations. Applicants must first call the Foundation's program director to discuss the proposed program/project.
Requirements: Grant applications will only be accepted from nonprofit 501(c)3 and grass roots organizations serving Berrien County residents.
Restrictions: Grants are not made for sectarian religious purposes, national fundraising efforts, political organizations or campaigns. Grants are not made to individuals, and form letters/emails are neither reviewed nor acknowledged.
Geographic Focus: Michigan
Date(s) Application is Due: Sep 1
Amount of Grant: Up to 10,000 USD
Contact: Anne McCausland; (269) 983-3304, ext. 2; AnneMcCausland@BerrienCommunity.org
Internet: http://www.berriencommunity.org
Sponsor: Berrien Community Foundation
2900 South State Street, Suite 2 East
Saint Joseph, MI 49085

Bertha Russ Lytel Foundation Grants 885
The foundation awards grants to eligible California nonprofit organizations in its areas of interest, including civic affairs; cultural programs, libraries and museums; elementary education; higher education, including agricultural and nursing scholarships; health organizations, including hospitals and hospice; and social services for the elderly and disabled. The primary focus of the Foundation is programs for seniors. Types of support include building construction and renovation, continuing support, equipment acquisition, general operating grants, matching grants, scholarship funds, and seed grants. There are no specified annual application deadlines. Most recent awards have ranged from $1,000 to $200,000. Interested parties should contact the Foundation Manager for guidelines and application information.
Requirements: California nonprofits in Humboldt County, California, are eligible.
Geographic Focus: California
Amount of Grant: 1,000 - 200,000 USD
Samples: Cal-Poly University, San Luis Obispo, California, $60,000 scholarship endowments (2014); Ferndale Community Church, Ferndale, California, $126,000 - support of the food for seniors program (2014); Ferndale Museum, Ferndale, California, $27,000 - general operating support (2014).
Contact: Donald Hindley, Manager; (707) 786-9236 or (707) 786-4657
Sponsor: Bertha Russ Lytel Foundation
P.O. Box 893
Ferndale, CA 95536

132 | GRANT PROGRAMS

Bert W. Martin Foundation Grants 886
The foundation awards grants to eligible nonprofit organizations in its areas of interest, including higher education, healthcare, sports, and general charitable giving. There are no application forms or deadlines. Proposal letters are accepted by postal mail only. No email proposals accepted.
Requirements: Nonprofit organizations in Florida and Arizona.
Geographic Focus: Arizona, Florida
Samples: Columbia U (New York, NY)--to enable undergraduate students to attend the Earth Semester program, a series of studies in earth systems and policy designed to foster an understanding of critical global issues, at its Biosphere 2 Center, $800,000.
Contact: Grants Administrator, c/o Northern Trust Co.
Sponsor: Bert W. Martin Foundation
50 S LaSalle Street, B-3
Chicago, IL 60675-0002

Besser Foundation Grants 887
The foundation limits its giving to nonprofits in the Alpena, MI area. Areas of interest include education, social services, civic affairs, arts and culture, religion, and international. Types of support include scholarship funds, matching funds, operating budgets, and continuing support. The board meets quarterly to consider requests.
Requirements: Only nonprofits in Michigan may apply.
Restrictions: Unless specifically requested by a Trustee, the Foundation will not consider grant requests from organizations outside of Alpena; nor for endowment funds, to defray meeting or conference expenses, or to pay for travel of individuals or groups. We will not relieve organizations or the public of their responsibilities, nor make grants to individuals for any purpose.
Geographic Focus: Michigan
Contact: J. Richard Wilson; (989) 354-4722; besserfoundation@verizon.net
Sponsor: Besser Foundation
123 North Second Avenue, Suite 3
Alpena, MI 49707-2801

BHHS Legacy Foundation Grants 888
BHHS Legacy Foundation funds efforts that enhance and improve the quality of life and health of children, families and senior citizens that are primarily low-income and under-served in the greater Phoenix and Tri-State regions. The Foundation places a high priority on collaborative projects and programs with other foundations, government agencies and nonprofit entities seeking to strengthen communities and support systemic change in health care access, health improvement and delivery. It also places a strong emphasis on funding nonprofit community organizations to increase their capacity to serve. These may include projects and programs offered by established nonprofit entities as well as those provided by start-up nonprofit organizations. Types of grants include: program/project; education; capacity building; capital equipment; operating; transition; and : Community Assistance Relief for Emergencies (CARE) funding. The foundation recommends discussing project or program concept and proposal with a foundation representative prior to submitting a written request for funding.
Restrictions: Giving is primarily restricted to the greater Phoenix and Bullhead/Laughlin regions, Arizona.
Geographic Focus: Arizona
Date(s) Application is Due: Apr 3; Aug 7; Dec 4
Amount of Grant: 25,000 - 75,000 USD
Contact: Karen E. Orth, Grants Manager; (602) 778-1200; fax (602) 778-1255; korth@bhhslegacy.org or info@bhhslegacy.org
Internet: http://bhhslegacy.org/access_healthcare.aspx
Sponsor: BHHS Legacy Foundation
2999 N 44th Street, Suite 530
Phoenix, AZ 85018

BibleLands Grants 889
The organization was founded to support Christian missions in the lands of the Bible. The program awards grants to hospitals and schools in the Middle Eastern areas, including Egypt, Lebanon, Israel, and Palestine. BibleLands supports schools and colleges that offer good quality Christian-based education for children whose families could not otherwise afford it.
Requirements: Projects must be Christian led.
Restrictions: Grants are not made to individuals or to projects outside of the Middle East.
Geographic Focus: Egypt, Israel, Lebanon, Palestinian Authority, Palestinian Territory
Contact: Jeremy Moodey, Chief Executive; 01494 897 950; fax 01494 897 951; info@biblelands.org.uk or jeremy.moodey@biblelands.org.uk
Internet: http://www.biblelands.org.uk/news/latest_grants.htm
Sponsor: BibleLands
24 London Road West
Amersham, HP7 0EZ United Kingdom

Bikes Belong Foundation Paul David Clark Bicycling Safety Grants 890
The Bikes Belong Foundation, launched in 2006, is a separate, complementary organization to the Bikes Belong Coalition, an organization sponsored by the U.S. bicycle industry. The foundation's focus is on bicycle-safety projects and children's bicycle programs. The foundation offers three distinct grant programs: the REI Grant Program; Bikes Belong Research Grants; and the Paul David Clark Bicycling Safety Fund. The latter is geared toward advocacy efforts to improve bicycling safety. The fund is named for Paul David Clark, one of more than 600 U.S. cyclists killed in motor vehicle collisions and accidents in 2005. Paul was an avid cyclist who loved the outdoors. An attorney, he provided pro bono service to nonprofit conservation groups, including the Natural Resources Defense Council and the Trust for Public Land. In the wake of the accident, Bikes Belong teamed with Paul's brother, Blair, to create a fund to support projects that increase bicycle safety, particularly in northern California. The fund's mission is two-fold: to encourage motorists to be more aware of bicyclists; and to compel motorists and cyclists to respectfully share the road. Interested applicants are encouraged to contact the foundation for information on funding availability and how to apply.
Geographic Focus: All States
Contact: Zoe Kircos; (303) 449-4893 ext. 5; fax (303) 442-2936; zoe@bikesbelong.org
Sponsor: Bikes Belong Foundation
207 Canyon Boulevard Suite 202
Boulder, CO 80302

Bildner Family Foundation Grants 891
The foundation awards grants to nonprofit organizations in its areas of interest, including arts and performing arts, health care and health organizations, Jewish and social service delivery. Types of support include continuing support, general operating support, and program support. The majority of grants are awarded in New Jersey and New York. There are no application forms or deadlines.
Restrictions: The foundation does not support private foundations or individuals.
Geographic Focus: New Jersey, New York
Contact: Allen Bildner, President
Sponsor: Bildner Family Foundation
293 Eisenhower Parkway, Suite 150
Livingston, NJ 07039

Bill and Melinda Gates Foundation Emergency Response Grants 892
The Foundation supports effective relief agencies and local organizations that respond quickly to people's most pressing needs in challenging conditions. The Foundation is interested in proposals that deliver food and clean water; improve sanitation; provide medical attention and shelter; prevent or minimize outbreaks of disease; and support livelihoods through cash-for-work programs. The Foundation currently supports people affected by the global food crisis; people in Sri Lanka and Pakistan displaced by political unrest and violence; victims of the earthquake in Haiti; communities affected by Typhoon Ketsana in the Philippines and Vietnam; and a consortium of leading humanitarian aid organizations.
Requirements: Relief agencies must have extensive experience and local relationships and be able to deliver help within days, when needs are most crucial. The Foundation also funds organizational capacity-building and explores learning opportunities to reinforce emergency response capabilities.
Restrictions: The majority of funding is made to organizations that are independently identified by Foundation staff. Unsolicited proposals are not accepted. Proposals must be made through 501(c)3 or other tax-exempt organizations. The Foundation is unable to make grants directly to individuals. The Gates Foundation will not fund:projects addressing health problems in developed countries; political campaigns and legislative lobbying efforts; building or capital campaigns; or projects that exclusively serve religious purposes.
Geographic Focus: All Countries
Contact: Emergency Coordinator; (206) 709-3140; info@gatesfoundation.org
Internet: http://www.gatesfoundation.org/topics/Pages/emergency-response.aspx
Sponsor: Bill and Melinda Gates Foundation
P.O. Box 23350
Seattle, WA 98102

Bill and Melinda Gates Foundation Policy and Advocacy Grants 893
The Foundation supports proposals that do one or more of the following: promote awareness of global development issues; advocate for least-advantaged populations; draw international attention and commitment; identify and promote powerful solutions; work towards additional and more effective investments; and are capable of lasting progress against global hunger and poverty.
Requirements: The Foundation seeks proposals that are able to produce measurable results; use preventive approaches; promise significant and long-lasting change; leverage support from other sources; and accelerate or are in accordance with work the Foundation already supports.
Restrictions: The majority of funding is made to organizations that are independently identified by our staff. Unsolicited proposals are not accepted. Proposals must be made through 501(c)3 or other tax-exempt organizations. The Foundation is unable to make grants directly to individuals. The Gates Foundation will not fund: projects addressing health problems in developed countries; political campaigns and legislative lobbying efforts; building or capital campaigns; or projects that exclusively serve religious purposes.
Geographic Focus: All Countries
Contact: Geoff Lamb, President; (206) 709-3140; info@gatesfoundation.org
Internet: http://www.gatesfoundation.org/global-development/Pages/overview.aspx
Sponsor: Bill and Melinda Gates Foundation
P.O. Box 23350
Seattle, WA 98102

Bill and Melinda Gates Foundation Water, Sanitation and Hygiene Grants 894
Poor sanitation causes severe diarrhea, which kills 1.5 million children each year. Smart investments in sanitation can reduce disease, increase family incomes, keep girls in school, help preserve the environment, and enhance human dignity. The Foundation is looking to work with partners in an effort to expand affordable access to sanitation. Detailed information is available at the Foundation website.
Requirements: The Gates Foundation seeks proposals that are able to produce measurable results; use preventive approaches; promise significant and long-lasting change; leverage support from other sources; and accelerate or are in accordance with work the foundation already supports.

Restrictions: The majority of funding is made to organizations that are independently identified by Foundation staff. Unsolicited proposals are not accepted. Proposals must be made through 501(c)3 or other tax-exempt organizations. The Foundation is unable to make grants directly to individuals. The Gates Foundation will not fund projects addressing health problems in developed countries; political campaigns and legislative lobbying efforts; building or capital campaigns; or projects that exclusively serve religious purposes.
Geographic Focus: All States, All Countries
Contact: Kellie Sloan, Interim Director of Water, Sanitation, and Hygiene; (206) 709-3140; info@gatesfoundation.org
Internet: http://www.gatesfoundation.org/watersanitationhygiene/Pages/home.aspx
Sponsor: Bill and Melinda Gates Foundation
P.O. Box 23350
Seattle, WA 98102

Bill Hannon Foundation Grants   895
Established in Nevada in 1999, the Bill Hannon Foundation awards grants to eligible organizations. Funding is provided throughout the State of California, though primarily within the greater Los Angeles area. The Foundation's primary areas of interest include: education; Catholic agencies and churches; health care; and human services. There is no formal application required, and an initial approach should be in letter form to the Foundation office. Though there are no identified deadline dates, the Board meets four times each year, in January, April, July, and October.
Requirements: The Foundation limits its grants to programs primarily in the greater Los Angeles area, where Hannon lived and worked. Only IRS certified, non-profit public charities are eligible for grants. There are no application deadlines or forms. Submit a letter of interest; full proposals are by invitation.
Geographic Focus: California
Samples: Loyola Marymount U (Los Angeles, CA)--to construct a library, $10 million; Santa Clara U (CA)--for construction and technology costs for an automated book-retrieval system for the library; and to help students participate in service projects, $4 million and $1 million, respectively; U of San Francisco, Saint Ignatius Institute (San Francisco, CA)--to support visiting scholars, spiritual retreats for students, and a residence hall where students can live under the supervision of a faculty coordinator and resident minister, $100,000.
Contact: Elaine S. Ewen, Chairperson; (310) 207-0303; elaine@comcast.net
Sponsor: Bill Hannon Foundation
11611 San Vicente Boulevard, Suite 530
Los Angeles, CA 90049-6509

Bindley Family Foundation Grants   896
Established in 1997 in Indiana, the Bindley Family Foundation gives primarily in the Indianapolis metropolitan area. Its primary fields of interest include children, education, and health service organizations. There are no specific application forms or deadlines, and applicants should begin by contacting the foundation to offer an overview of their program, project, and budgetary needs. Funding generally is offered in the form of operating support or scholarship endowments. Grants typically range from $2,000 to $15,000, though a small number are significantly higher.
Geographic Focus: Illinois, Indiana
Amount of Grant: 2,000 - 50,000 USD
Samples: Brebeuf Preparatory School, Indianapolis, Indiana, $48,213 - for operating support and scholarships; School on Wheels, Indianapolis, Indiana, $10,000 - operating support; Northwestern University, Evanston, Illinois, $7,500 - for operating expenses.
Contact: James F. Bindley, Executive Director; (317) 704-4770
Sponsor: Bindley Family Foundation
8909 Purdue Road, Suite 500
Indianapolis, IN 46268-3150

Bingham McHale LLP Pro Bono Services   897
The Bingham Greenebaum Doll firm is committed to the communities it serves, including: Indianapolis, Jasper, Evansville, and Vincennes, Indiana; Louisville, Frankfort, and Lexington, Kentucky; and Cincinnati, Ohio. This commitment includes sharing legal services with those who cannot afford legal assistance on their own. As attorneys, the firm recognizes that it has a special responsibility to provide legal services to the underprivileged within the communities where its attorneys work and live. low income families
Requirements: Residents of communities that the Bingham Greenebaum Doll firm serves are eligible.
Geographic Focus: Indiana
Contact: Partners and Associates; (317) 635-8900; fax (317) 236-9907
Internet: http://www.bgdlegal.com/aboutus/xprGeneralContent2.aspx?xpST=CommunityService
Sponsor: Bingham McHale LLP Pro Bono Program
2700 Market Tower, 10 W. Market Street
Indianapolis, IN 46204-4900

Biogen Foundation Fellowships   898
The Biogen Foundation's mission is to improve the quality of people's lives and contribute to the vitality of the communities in which Biogen operates. It is committed to sparking a passion for science and discovery, supporting innovative STEM (science, technology, engineering and math) initiatives, and strengthening efforts to make science accessible to diverse populations. The purpose of Biogen's fellowship program is to support the career development of fellows striving to become specialized in the areas of neurology or hematology. Three distinct fellowships are offered for distinct regions of the world (Europe and Canada, worldwide, and the United States), all of which are committed to improving the lives of people with Multiple Sclerosis (MS) through clinical research. Each fellowship will award up to 75,000 Euro for a single academic year in support,
Geographic Focus: All States, All Countries
Amount of Grant: Up to 75,000 EUR
Contact: Kathryn Bloom, Senior Director, Public Affairs; (617) 914-1299 or (866) 840-1146; fax (617) 679-2617; grantsoffice@biogen.com
Sponsor: Biogen Foundation
225 Binney Street
Cambridge, MA 02142

Biogen Foundation General Donations   899
Biogen corporate philanthropic activities are closely aligned with its mission to increase disease awareness, improve patient access to care, and help patients with unmet medical needs. Toward this mission, the Biogen Foundation provides general donations by collaborating with various patient, medical, and scientific organizations to support organizations' missions that improve the health of the patient community.
Requirements: Donations may be requested by patient, medical, and scientific 501(c)3 organizations that support Biogen's mission.
Restrictions: Biogen will not offer donations with the intent of, directly or indirectly, implicitly or explicitly, influencing or encouraging the recipient to purchase, prescribe, refer, sell, arrange for the purchase or sale, or recommend any Biogen product.
Geographic Focus: All States
Contact: Kathryn Bloom, Senior Director, Public Affairs; (617) 914-1299 or (866) 840-1146; fax (617) 679-2617; grantsoffice@biogen.com
Internet: http://grantsoffice.biogen.com/general-donations
Sponsor: Biogen Foundation
225 Binney Street
Cambridge, MA 02142

Biogen Foundation Healthcare Professional Education Grants   900
Biogen corporate philanthropic activities are closely aligned with its mission to increase disease awareness, improve patient access to care, and help patients with unmet medical needs. Toward this mission, Biogen may provide a Healthcare Professional Education grant to support bona fide, unbiased, fair balanced educational programs or activities designed to educate the medical, scientific or patient community in therapeutic areas of interest to Biogen. Current areas of interest include neurology and hematology. Grants in this area may be awarded for both accredited and non-accredited educational activities. When evaluating a request Biogen will consider: adherence to all relevant regulations, guidelines, and policies; alignment with Biogen areas of interest; robust and quality needs assessment; measurable educational objectives; educational design that incorporates evidence-based adult learning principles and promotes continuous improvement and performance measurement for the targeted audience during the educational intervention; and innovative and cost-effective approaches in designing quality educational activities.
Requirements: Educational grants may be requested by independent accredited educational providers or organizations tax exempt under 501(c)3 of the Internal Service Revenue Code.
Restrictions: Biogen will not offer an Educational grant with the intent of, directly or indirectly, implicitly or explicitly, influencing or encouraging the recipient to purchase, prescribe, refer, sell, arrange for the purchase or sale, or recommend any Biogen product.
Geographic Focus: All States
Contact: Kathryn Bloom, Senior Director, Public Affairs; (617) 914-1299 or (866) 840-1146; fax (617) 679-2617; grantsoffice@biogen.com
Internet: http://grantsoffice.biogen.com/hcp-programs
Sponsor: Biogen Foundation
225 Binney Street
Cambridge, MA 02142

Biogen Foundation Patient Educational Grants   901
Biogen corporate philanthropic activities are closely aligned with its mission to increase disease awareness, improve patient access to care, and help patients with unmet medical needs. Toward this mission, the Foundation provides Patient Educational Grants in its areas of operation as a meaningful way of offering opportunities to learn about personal disease, gain knowledge about the treatment options, and advocate for personal health and well being. When evaluating a request Biogen will consider: adherence to all relevant regulations, guidelines, and policies; alignment with Biogen areas of interest; robust and quality needs assessment; measurable educational objectives; educational design that incorporates evidence-based adult learning principles and promotes continuous improvement and performance measurement for the targeted audience during the educational intervention; and innovative and cost-effective approaches in designing quality educational activities.
Requirements: Patient Educational grants may be requested by independent accredited educational providers or organizations tax exempt under 501(c)3 of the Internal Service Revenue Code.
Restrictions: Biogen will not offer a Patient Educational grant with the intent of, directly or indirectly, implicitly or explicitly, influencing or encouraging the recipient to purchase, prescribe, refer, sell, arrange for the purchase or sale, or recommend any Biogen product.
Geographic Focus: All States
Contact: Kathryn Bloom, Senior Director, Public Affairs; (617) 914-1299 or (866) 840-1146; fax (617) 679-2617; grantsoffice@biogen.com
Sponsor: Biogen Foundation
225 Binney Street
Cambridge, MA 02142

Biogen Foundation Scientific Research Grants 902

Biogen supports non-accredited medical education activities that enhance and expand medical knowledge and skills for researchers. Biogen corporate philanthropic activities are closely aligned with its mission to increase disease awareness, improve patient access to care, and help patients with unmet medical needs. When evaluating a request Biogen will consider: adherence to all relevant regulations, guidelines, and policies; alignment with Biogen areas of interest; robust and quality needs assessment; measurable educational objectives; educational design that incorporates evidence-based adult learning principles and promotes continuous improvement and performance measurement for the targeted audience during the educational intervention; and innovative and cost-effective approaches in designing quality educational activities.
Requirements: Scientific research grants may be requested by independent accredited educational providers or organizations tax exempt under 501(c)3 of the Internal Service Revenue Code.
Restrictions: Biogen will not offer an Scientific Research grant with the intent of, directly or indirectly, implicitly or explicitly, influencing or encouraging the recipient to purchase, prescribe, refer, sell, arrange for the purchase or sale, or recommend any Biogen product.
Geographic Focus: All States
Contact: Kathryn Bloom, Senior Director, Public Affairs; (617) 914-1299 or (866) 840-1146; fax (617) 679-2617; grantsoffice@biogen.com
Internet: http://grantsoffice.biogen.com/scientific-programs
Sponsor: Biogen Foundation
225 Binney Street
Cambridge, MA 02142

Birmingham Foundation Grants 903

The Birmingham Foundation is a private, independent foundation, dedicated to health-related and human services grant-making in the south neighborhoods of Pittsburgh, Pennsylvania, which are part of the 15203, 15210 and 15211 zip code areas. The Board of Directors meets three times a year to act on grant proposals, in March, June and November.
Requirements: 501(c)3 nonprofits in South Pittsburgh communities including Allentown, Arlington and Arlington Heights, Beltzhoover, Bon Air, Carrick, Duquesne Heights, Knoxville, Mount Oliver, Mount Washington, Saint Clair Village, and South Side are eligible.
Restrictions: Grants are not normally made for operating budgets, deficits, fund-raising campaigns, general research, overhead costs, scholarships, political campaigns, or loans. Grants are not made to individuals, to other private foundations, for sectarian religious activities, or for the efforts to influence legislation unless grant-related.
Geographic Focus: Pennsylvania
Amount of Grant: 2,500 - 50,000 USD
Samples: Strong Women Strong Girls, $15,000 - mentoring program; SIDS of PA, $10,000 - SIDS outreach program; Saltworks Theater, $10,000 - bully/drug prevention program.
Contact: Chris Mason, Administrative Assistant; (412) 481-2777; fax (412) 481-2727; info@birmfoundpgh.org
Internet: http://www.birminghamfoundation.org
Sponsor: Birmingham Foundation
2005 Sarah Street, 2nd Floor
Pittsburgh, PA 15203

BJ's Charitable Foundation Grants 904

Established with the goal of creating a positive, long-lasting impact on the communities BJ's serves, the mission of BJ's Charitable Foundation is to enhance and enrich community programs that primarily benefit children and families. The majority of the foundation's giving is focused on organizations that: (a) Promote the safety, security and well-being of children and families; (b) Support education and health programs; (c) Provide community service opportunities; or, (d) Aid in hunger and disaster relief.
Requirements: Organizations that are tax-exempt under 501(c)3 of the Internal Revenue Code and recognized as a 'public charity' by the IRS may apply. The program must align with the foundation's mission of supporting children and families in the specific areas of safety, security and well-being, education, health, community service, hunger/homelessness and disaster relief. And, the program must positively impact communities where BJ's Clubs are located. All organizations are limited to one (1) application per 12-month period. Additional applications will not be considered.
Restrictions: The foundation does not provide funding for: (1) Organizations that discriminate on the basis of race, color, gender, sexual orientation, age, religion, national or ethnic origin or physical disability. (2) Political organizations, fraternal groups or social clubs that engage in any kind of political activity. (3) Religious organizations, unless they serve the general public in a significant non-denominational way. (4) Programs that have been in place for less than one (1) year. (5) Individuals. (6) Organizations located outside BJ's markets. (7) Capital campaigns. (8) Private foundations. (9) Sponsorships for music, film and art festivals. (10) Business expositions/conferences. (11) Journal or program advertisements.
Geographic Focus: Connecticut, Delaware, Florida, Georgia, Maine, Maryland, Massachusetts, New Hampshire, New Jersey, New York, North Carolina, Ohio, Pennsylvania, Rhode Island, South Carolina, Virginia
Date(s) Application is Due: Mar 6; Jul 10; Dec 6; Dec 30
Contact: Community Relations Manager; (508) 651-7400; fax (508) 651-6623
Internet: http://www.bjs.com/about/community/charity.shtml
Sponsor: BJ's Charitable Foundation
1 Mercer Road, P.O. Box 9614
Natick, MA 01760

Blackford County Community Foundation - WOW Grants 905

The Blackford County Community Foundation - WOW Grants offer funding for projects that encourage, educate, and enlighten women and/or children. Applicants may be individuals, groups, or organizations.
Requirements: Applicants should submit the following items to the Foundation: cover page; project narrative with a detailed description of the project, including timeline and results expected; and a detailed budget. All grant recipients will be required to give a brief report about their project at the spring meeting.
Restrictions: The following are not eligible for funding: operating deficits, post-event or after the fact situations; special fundraising events, including endowment campaigns; political endeavors and propaganda; and profit-making enterprises and/or projects for personal gain.
Geographic Focus: Indiana
Date(s) Application is Due: Feb 1
Amount of Grant: 100 - 2,500 USD
Contact: Patricia Poulson; (765) 348-3411; ppoulson@blackfordcounty.org
Internet: http://www.blackfordcofoundation.org/pages.asp?Page=Women%20of%20Worth&PageIndex=411
Sponsor: Blackford County Community Foundation
121 North High Street
Hartford City, IN 47348

Blackford County Community Foundation Grants 906

The Blackford County Community Foundation and its supporting organizations award numerous grants annually. Primary areas of interest include: community and economic development; education; community services planning and coordination; and human services. The Foundation has quarterly deadlines for submitting applications on January 31, March 31, June 30, and September 30.
Requirements: Applicants must contact the Executive Director to determine if their project is suitable for the funding. If the organization is asked to submit a formal proposal, it must consider the following criteria: purpose and definition of the project or program; background of the request office; officers and staff personnel of requesting organization; financial information and budgets; evaluation results; and how the project will be affected if funding is not received. The Foundation will judge the proposal on its merit, priority, and substantive quality.
Restrictions: The Blackford County Community Foundation generally does not fund: profit-making enterprises; political activities; operating budgets of organizations, except for limited experimental or demonstration periods; sectarian or religious organizations operated primarily for the benefit of their own members; endowment purposes; capital grants to building campaigns will only be made when there is evidence that such support is vital to the success of a program meeting priority needs of the community. Grants are not awarded for endowment purposes.
Geographic Focus: Indiana
Date(s) Application is Due: Jan 31; Mar 31; Jun 30; Sep 30
Contact: Patricia D. Poulson, Executive Director; (765) 348-3411; fax (765) 348-4945; ppoulson@blackfordcounty.org or foundation@blackfordcounty.org
Internet: http://blackfordcofoundation.org/Grants
Sponsor: Blackford County Community Foundation
121 North High Street
Hartford City, IN 47348

Blackford County Community Foundation Hartford City Kiwanis Scholarship in Memory of Mike McDougall 907

The Blackford County Community Foundation offers a number of scholarships to seniors of Blackford High School and to Blackford County residents graduating from other secondary educational programs. The Hartford City Kiwanis Scholarship in Memory of Mike McDougall is in memory of a local dentist. Dr. McDougall was active in Masonic Lodge, the Shrine Club, and other civic organizations. This scholarship supports a student planning to attend Indiana University or pursuing a degree in a health care field. The scholarship will be in the amount of $500 annually.
Requirements: Applicants for the scholarship would usually meet some or all of the following criteria: be a graduating senior from a high school or be a student at the undergraduate or graduate level; be accepted to a four-year university or college, a community college, or a vocational school; show achievement in academics, a minimum GPA, athletic, community service, leadership or character, demonstrate financial need, etc; and/or have demonstrated superior work in a particular field of study.
Geographic Focus: Indiana
Date(s) Application is Due: Jan 15
Contact: Patricia D. Poulson, Executive Director; (765) 348-3411 or (765) 348-7560; fax (765) 348-4945; ppoulson@blackfordcounty.org or foundation@blackfordcounty.org
Internet: http://blackfordcofoundation.org/Scholarships/LocalScholarships/MedicalHealthcare
Sponsor: Blackford County Community Foundation
121 North High Street
Hartford City, IN 47348

Blackford County Community Foundation Noble Memorial Scholarship 908

The Blackford County Community Foundation offers a number of scholarships to seniors of Blackford High School and to Blackford County residents graduating from other secondary educational programs. The Noble Memorial Scholarship Fund was established in memory of Chuck Noble, who was a Blackford High School graduate and an EMS. This scholarship supports a local student in pursuit of a postsecondary education in a medical field. The scholarship will be in the amount of $250 annually.
Requirements: Applicants for the scholarship would usually meet some or all of the following criteria: be a graduating senior from a high school or be a student at the undergraduate or graduate level; be accepted to a four-year university or college, a community college, or a vocational school; show achievement in academics, a minimum GPA, athletic, community service, leadership or character, demonstrate financial need, etc; and/or have demonstrated superior work in a particular field of study.

Geographic Focus: Indiana
Date(s) Application is Due: Jan 15
Amount of Grant: 250 USD
Contact: Patricia D. Poulson, Executive Director; (765) 348-3411 or (765) 348-7560; fax (765) 348-4945; ppoulson@blackfordcounty.org or foundation@blackfordcounty.org
Internet: http://blackfordcofoundation.org/Scholarships/LocalScholarships/MedicalHealthcare
Sponsor: Blackford County Community Foundation
121 North High Street
Hartford City, IN 47348

Black River Falls Area Foundation Grants 909
Grant applications are evaluated with consideration to the policies, funding objectives and the mission statement of the Foundation. General categories of support include: education, health services, cultural activities, social services and civic projects. Grants to be considered are those that meet charitable needs of: a new or innovative nature to fulfill unmet needs; one-time projects; capital improvements; equipment needs; start-up expenses; special programs; and emergency funding. Applications will be accepted April 1 through May 15.
Requirements: Any non-profit organization in Jackson County, Wisconsin, may apply for a grant. A copy of the organization's tax-exempt status is required.
Restrictions: Requests to support endowments, operating expenses, debt repayment, religious purposes, individuals, and lobbying expenses are usually discouraged.
Geographic Focus: Wisconsin
Date(s) Application is Due: May 15
Contact: Beth Overlien, Administrative Assistant; (715) 284-3113
Internet: http://www.brfareafoundation.org/application
Sponsor: Black River Falls Area Foundation
P.O. Box 99
Black River Falls, WI 54615

Blanche and Irving Laurie Foundation Grants 910
The Blanche and Irving Laurie Foundation was established in 1983 by New Brunswick philanthropist Irving Laurie. The foundation makes charitable gifts to institutions and nonprofits in broad areas of interest, including the arts, especially theater and music; education; health care; social services; and needs and concerns of the Jewish community. Capital grants, operating support grants, grants for programs/projects, and scholarships are awarded. Applicants should submit seven copies of a written proposal containing the following items: copies of the most recent annual report, audited financial statement, and 990; a detailed description of the project and amount of funding requested; and a copy of the current year's organization budget and/or project budget. The foundation's board meets quarterly to evaluate proposals. Final notification occurs within three to four months from submission. Typically, awards range from $3,000 to $150,000.
Requirements: Nonprofit organizations in New Jersey are eligible to apply, as well as others from around the United States.
Restrictions: Giving is primarily concentrated in New Jersey. The foundation does not support medical research.
Geographic Focus: All States
Amount of Grant: 3,000 - 150,000 USD
Samples: American Repository Ballet, New Brunswick, New Jersey, $10,000 - support of the Dance Power program (2014); Brandis University, Waltham, Massachusetts, $50,000 - in support of the theater department (2014); La Jolla Playhouse, La Jolla, California, $20,000 - production support (2014).
Contact: Gene R. Korf, Executive Director; (973) 993-1583 or (908) 371-1777
Sponsor: Blanche and Irving Laurie Foundation
P.O. Box 53
Roseland, NJ 07068-5788

Blanche and Julian Robertson Family Foundation Grants 911
The Blanche and Julian Robertson Family Foundation is totally committed to the goal of improving the quality of life in Salisbury and Rowan County. The general direction of the Foundation's interest is: programs which address social problems and nurture positive social relationships; efforts aimed at enriching lives through exposure to the cultural arts; neighborhood revitalization programs, especially when such programs encourage development of transitional housing and enable first-time homeowners to purchase homes; programs that improve opportunities for youth at risk and families in crisis; efforts to improve broad-based educational, recreational, and athletic opportunities; efforts which address health and the environment. The Foundation is also interested in programs and projects that demonstrate the attributes of leverage (where a grant will attract matching gifts or other funding), as well as innovation, thoroughness, passion, and commitment. Contact the office for application and guidelines.
Requirements: North Carolina nonprofits serving Salisbury and Rowan County.
Restrictions: Foundation does not make grants outside Salisbury and Rowan County.
Geographic Focus: North Carolina
Date(s) Application is Due: Mar 30
Samples: Rowan Regional Medical Ctr Foundation (Salisbury, NC)--for its capital campaign for a new emergency department, patient-care facility, and community-education center, $1 million.
Contact: David Setzer; (704) 637-0511; fax (704) 637-0177; bjrfoundation@aol.com
Sponsor: Blanche and Julian Robertson Family Foundation
141 East Council Street, P.O. Box 4242
Salisbury, NC 28145-4242

Blowitz-Ridgeway Foundation Early Childhood Development Research Award 912
The foundation supports nonprofit organizations and programs in the areas of medicine, psychology, residential care, and education, and for research in medicine, psychology, social science, and education. Types of support include capital grants, endowments, program development grants, research grants, and scholarships. The Foundation prefers prospective grantees whose programs or services benefit persons who have not yet reached their majority and/or are for the care of individuals or elderly persons who lack sufficient resources to provide for themselves. Although grants may be made to organizations outside the state of Illinois, preference will generally be given to applicants from Illinois. Application information is available on the Web site.
Requirements: Applicants must be classified as 501(c)3 by the IRS.
Restrictions: Grants will not be made for religious or political purposes, nor generally for the production or writing of audio-visual materials.
Geographic Focus: All States
Contact: Laura Romero; (847) 330-1020; fax (847) 330-1028; laura@blowitzridgeway.org
Internet: http://www.blowitzridgeway.org/information/information.html
Sponsor: Blowitz-Ridgeway Foundation
1701 E Woodfield Road, Suite 201
Schaumburg, IL 60173

Blowitz-Ridgeway Foundation Grants 913
The foundation supports nonprofit agencies that provide medical, psychiatric, and psychological care to economically disadvantaged children and adolescents. Program and capital grants are awarded, primarily in Illinois, in support of medical, psychiatric, psychological, and/or residential care; and research programs in medicine, psychology, social science, and education. The foundation supports operating budgets, and applicants may request commitments that extend beyond one year, but requests for annual funding will not be considered. Applications are accepted throughout the year and are reviewed in the order in which they are received. Guidelines and applications are available online.
Requirements: 501(c)3 nonprofit organizations that offer services to people who lack resources to provide for themselves may apply.
Restrictions: Grants will not be awarded to government agencies or to organizations that subsist mainly on third-party funding and have demonstrated no ability or expended little effort to attract private funding. Grants will not be made for religious or political purposes or for the production or writing of audio-visual materials.
Geographic Focus: Illinois
Amount of Grant: 5,000 - 30,000 USD
Samples: Elizabeth Ann Seton Program, $5,000 in general operating support of programs for pregnant and parenting mothers; The Enterprising Kitchen, $15,000 for programs improving self-sufficiency and employability of low-income women; Faith in Action of McHenry County, $7,000 to recruit volunteers; Gospel Rescue Mission, $10,000 for children's social and recreational activities; Guardian Angel Community Services, $5,000 for the Groundwork Domestic Violence Program.
Contact: Serena Moy; (847) 330-1020; fax (847) 330-1028; serena@blowitzridgeway.org
Internet: http://www.blowitzridgeway.org/information/information1.html
Sponsor: Blowitz-Ridgeway Foundation
1701 E Woodfield Road, Suite 201
Schaumburg, IL 60173

Blue Cross Blue Shield of Minnesota Foundation - Health Equity: Building Health Equity Together Grants 914
The Building Health Equity Together Grant is a program focused on Minnesota local governments working in partnership with a local 501(c)3 organization to achieve health equity in their community. The partnership should address one or more of the factors that influence health in low-income communities, including education, employment, income, family and social support, and community safety. The Foundation will fund up to two grants of up to $75,000 each with the opportunity for a second year of funding at the same level based on performance; progress; and identification of practices, policies and partners that can help advance the successes of first-year work into the future. Organizations may apply through the Foundation's online application process. A detailed list of required attachments is available at the Foundation's website. A list of previously funded projects in Minnesota and other locations is also available at the website.
Requirements: Eligible applicants are local units of government (county, statutory or home rule charter city, township or school district) in partnership with a 501(c)3 community-based organization. Tribal governments in partnership with a 501(c)3 community-based organization are also eligible. Proposed projects must address one or more of the following factors: education; employment; income; family and social support; and community safety. At least two departments and a non-profit partner must commit to working jointly. A Memorandum of Understanding (MOU) is part of the application. The grant-funded work must result in a concrete initiative that has documented support from key stakeholders and can be implemented.
Geographic Focus: Minnesota
Date(s) Application is Due: Sep 28
Amount of Grant: Up to 75,000 USD
Contact: Stacey Millett, Senior Program Officer; (866) 812-1593 or (651) 662-1019; fax (651) 662-4266; Stacey_D_Millett@bluecrossmn.com
Internet: http://bcbsmnfoundation.com/pages-programs-program-Building_Health_Equity_Together?oid=19570
Sponsor: Blue Cross Blue Shield of Minnesota Foundation
1750 Yankee Doodle Road, N159
Eagan, MN 55122

Blue Cross Blue Shield of Minnesota Foundation - Healthy Children: Growing Up Healthy Grants 915

The Growing Up Healthy Grants engage community health, early childhood development, housing and environmental organizations, and other community partners to nurture the healthy growth and development of children birth to five years and their families. Through this focus area, the Blue Cross Blue Shield (BCBS) of Minnesota Foundation has improved the quality of housing, reduced children's exposure to harmful chemicals, increased readiness for kindergarten, and increased children's access to healthy foods and safe places to play. Planning grants up to $25,000 are available. Through the planning process, funded organizations and their community partners develop a shared vision of how to improve and protect the health of children through place-based projects (neighborhood, town, region) that address health and at least two of the three determinants: early childhood education, housing, and the environment. At the end of the planning period grantees that have developed a community vision, supported by a written implementation plan, may apply for implementation funding for a period of up to three years. To receive an implementation grant, projects must show broad-based community support, demonstrate innovative approaches and articulate how these approaches will result in healthier communities and children. Letter of inquiry/application instructions, previously funded projects, and an instructional webinar are available at the Foundation website.

Requirements: The Foundation encourages a wide range of organizations to apply for funding, including community- and faith-based organizations; health, environmental, housing, early childhood and civic groups; mutual assistance associations; state, county and municipal agencies; tribal governments and agencies; professional associations or collaboratives; and policy and research organizations. Applicants must be located in Minnesota or serve Minnesotans. Eligible applicants include units of government as those designated as 501(c)3 nonprofit organizations. Organizations are required contact the Foundation to discuss their project idea. Based on the outcome of the conversation, they may then be asked to submit a letter of inquiry with supporting information.

Restrictions: The Foundation is unable to provide funding for the following: individuals; lobbying, political or fraternal activities; legal services; sports events and athletic groups; religious purposes; clinical quality improvement activities; biomedical research; capital purposes (building, purchase, remodeling or furnishing of facilities); equipment or travel, except as related to requests for program support; endowments, fundraising events or development campaigns; retiring debt or covering deficits; payment of services or benefits reimbursable from other sources; supplanting funds already secured for budgeted staff and/or services; or long-term financial support.

Geographic Focus: Minnesota
Amount of Grant: Up to 25,000 USD
Contact: Jocelyn Ancheta, Program Officer; (866) 812-1593 or (651) 662-2894; fax (651) 662-4266; Jocelyn_L_Ancheta@bluecrossmn.com
Internet: http://bcbsmnfoundation.com/pages-grantmaking-initiative-Healthy_Children?oid=13827
Sponsor: Blue Cross Blue Shield of Minnesota Foundation
1750 Yankee Doodle Road, N159
Eagan, MN 55122

Blue Cross Blue Shield of Minnesota Foundation - Healthy Equity: Health Impact Assessment Demonstration Project Grants 916

The Health Impact Project: Advancing Smarter Policies for Healthier Communities, a collaboration of the Robert Wood Johnson Foundation and The Pew Charitable Trusts, encourages the use of health impact assessments (HIA) to help decision-makers identify the potential health effects of proposed policies, projects, and programs, and make recommendations that enhance their health benefits and minimize their adverse effects and any associated costs. The HI Project will support up to five HIA demonstration projects intended to inform decisions on proposed local, tribal, or state policies, projects or programs. This initiative could also fund HIAs that address federal decisions having impacts limited to a specific state, region, or local community, such as permitting a new mine or building a new highway. The HI Project seeks to produce a balanced portfolio of completed HIAs that build a compelling case to policy-makers regarding the utility and potential applications of HIA. The Foundation's call for papers seeks to demonstrate the range of useful applications across a range of sectors, levels of government, geographic regions, and types of applicant organizations. Applicants may request grants from $25,000 to $75,000 for demonstration projects to be completed within 18 months. The application and call for papers are located at the Foundation website. Applicants are also encouraged to access several informational webinars located at the website.

Requirements: Eligible applicant organizations include state, tribal, or local agencies; tax-exempt educational institutions; or tax-exempt organizations described in Section 501(c)3 of the Internal Revenue Code (including public charities and private foundations). All applicant organizations must be located in the U.S. or its territories. The Foundation encourages proposals from organizations representing a range of fields and sectors, such as transportation, education, economic and social policy, agricultural policy, energy, environmental regulation, and natural resource development. Prior experience conducting HIAs is not required. The HI Project will provide tailored training and technical assistance to all grantees throughout each grant. High priority will be giving to HIAs from geographic regions where few HIAs have been completed to date (see map located in the call for papers at the Foundation website).

Restrictions: Many demonstration project applicants will have no prior experience with the HIA process and methods. The Health Impact Project, through partnerships with experienced HIA practitioners, provides HIA training and ongoing technical assistance. Grantees who have not previously conducted an HIA will be expected to work with a technical assistance provider to organize a two-day training for HIA project staff and relevant stakeholders. Technical assistance may include, for example, help developing collaborative partnerships with other stakeholders, guidance on communications strategies, or guidance on developing an effective plan for implementing HIA recommendations. If the applicant and and partners lack the full range of technical expertise needed to complete the proposed scope of work, the HI Project may provide limited support for subject area expertise, such as epidemiological modeling, engaging stakeholders, or another sub-discipline, such as air quality analysis.

Geographic Focus: All States
Date(s) Application is Due: Sep 28
Amount of Grant: Up to 75,000 USD
Contact: Jocelyn Ancheta, Program Officer; (866) 812-1593 or (651) 662-2894; fax (651) 662-4266; Jocelyn_L_Ancheta@bluecrossmn.com
Internet: http://bcbsmnfoundation.com/pages-programs-program-Health_Impact_Assessments?oid=19532
Sponsor: Blue Cross Blue Shield of Minnesota Foundation
1750 Yankee Doodle Road, N159
Eagan, MN 55122

Blue Cross Blue Shield of Minnesota Foundation - Healthy Equity: Health Impact Assessment Program Grants 917

The Health Impact Project supports health impact assessment (HIA) initiatives to enable organizations with previous HIA experience to conduct HIAs and develop sustainable self-supporting HIA programs at the local, state, or tribal level. The project encourages the use of health impact assessments to help decision-makers identify the potential health effects of proposed policies, projects, and programs, and make recommendations that enhance their health benefits and minimize their adverse effects and any associated costs. Up to three program grants will be awarded. Grants will be up to $250,000 each and must be completed within 24 months. The program grants will support organizations that have completed at least one prior HIA to conduct at least two HIAs, and to implement a plan that establishes the relationships, systems, and funding mechanisms needed to maintain a stable HIA program that endures beyond the conclusion of the grant period. The application, how organizations will be selected, evaluated, and monitored, and several informational webinars are available at the Foundation website.

Requirements: All applicant organizations must be located in the United States or its territories. Eligible applicants include state, tribal, or local agencies; tax-exempt educational institutions; or tax-exempt organizations described in Section 501(c)3 of the Internal Revenue Code (including public charities and private foundations). All applicants must have completed at least one previously successful HIA. Recipients of these grants will be responsible for conducting at least two HIAs, and for establishing the systems, relationships, and funding mechanisms to implement a stable HIA program that endures beyond the completion of grant funding. Applicants will be asked to describe how they intend to establish a sustainable, self-supporting HIA program, what actions they will implement to bring this about, and how they will measure success. Samples of previously funded grants are discussed in the call for papers available at the Foundation website.

Restrictions: Applicants must provide $100,000 in matching funds or in-kind support. Grant funds may not be used to subsidize individuals for the costs of health care, support clinical trials of unapproved drugs or devices, construct or renovate facilities, or as a substitute for funds currently being used to support similar activities. The project limits the amount of indirect costs it will support to no more than 10% of salaries and benefits covered directly by the grant; and limits the amount of fringe benefits it will support to no more than 33% of the total staff salaries line item. In addition, no part of the grant can be used to carry on propaganda or otherwise attempt to influence legislation, or a political campaign.

Geographic Focus: All States
Date(s) Application is Due: Dec 14
Amount of Grant: Up to 250,000 USD
Contact: Jocelyn Ancheta, Grants Administrator; (866) 812-1593 or (651) 662-2894; fax (651) 662-4266; Jocelyn_L_Ancheta@bluecrossmn.com
Internet: http://bcbsmnfoundation.com/pages-programs-program-Health_Impact_Assessments?oid=19532
Sponsor: Blue Cross Blue Shield of Minnesota Foundation
1750 Yankee Doodle Road, N159
Eagan, MN 55122

Blue Cross Blue Shield of Minnesota Foundation - Healthy Equity: Public Libraries for Health Grants 918

The Public Libraries for Health program engages public libraries as partners working collectively to improve health for low-income communities and communities of color. As trusted institutions with strong community ties, libraries can work with other organizations in their service area in creative and effective ways. The Foundation awards grants to four libraries across the state for programs or projects that advance health equity. The Foundation will fund up to four grants of up to $50,000 each. Funds may be used for an existing project or a new opportunity. Project activities may occur anywhere in the public library's local community and do not need to take place on public library premises. Organizations are encouraged to review the eligibility checklist before applying. The application and budget worksheet are available at the Foundation website.

Requirements: Organizations must meet the following requirements to apply. They must be classified at a local unit of government with city, county, or state financial support; serve low-income populations; generate library visits as a major portion of service activity beyond electronic, books-by-mail, inter-library loan and other services that do not involve a library visit; provide services that benefit local residents in their service area; retain trained staff to oversee programs and operations; and maintain a physical space for library activities and services that is accessible to the public at least 20 hour per week.

Geographic Focus: Minnesota
Amount of Grant: 50,000 USD

Contact: Jocelyn Ancheta, Program Officer; (651) 662-2894 or (866) 812-1593; fax (651) 662-4266; Jocelyn_L_Ancheta@bluecrossmn.com
Internet: http://bcbsmnfoundation.com/pages-programs-program-Public_Libraries_for_Health?oid=19529
Sponsor: Blue Cross Blue Shield of Minnesota Foundation
1750 Yankee Doodle Road, N159
Eagan, MN 55122

Blue Grass Community Foundation Harrison Fund Grants 919
Blue Grass Community Foundation is part of a network of foundations that meet the National Standards for operational quality, donor service and accountability in the community foundation sector. The Harrison Community Fund was established in 2002 to encourage local philanthropy and to raise charitable dollars for the good of Harrison County. The Fund makes annual competitive grants of up to $10,000 to Harrison County nonprofits.
Requirements: Nonprofit organizations, schools and exempt government entities serving Harrison County are eligible to apply.
Geographic Focus: Kentucky
Amount of Grant: Up to 10,000 USD
Contact: Barbara Fischer; (859) 225-3343; fax (859) 243-0770; bfischer@bgcf.org
Internet: http://bgcf.org/learn/community-funds/harrison-county-community-fund/
Sponsor: Blue Grass Community Foundation
499 East High Street, Suite 112
Lexington, KY 40507

Blue Grass Community Foundation Hudson-Ellis Fund Grants 920
The Hudson-Ellis Fund was established at Blue Grass Community Foundation for the good of Boyle County through a bequest by Lottie Ellis. Lottie Ellis lived her entire life in Danville, Kentucky. When she died in 1999 at the age of 91, she made a gift to her home town that will live forever. Ms. Ellis, a former bookkeeper, lived a quiet, simple, modest life. She was an avid reader. For the last ten years of her life, she was unable to leave her house in downtown Danville but looked forward to weekly visits from the Boyle County Library's book mobile. Few were aware that she'd inherited $4 million from a long-time friend named T. Yates Hudson, Jr. The Hudson-Ellis Discretionary Fund makes annual competitive grants of up to $10,000 to Boyle County nonprofits.
Requirements: Nonprofit organizations, schools and exempt government entities serving Boyle County are eligible to apply.
Geographic Focus: Kentucky
Amount of Grant: Up to 10,000 USD
Contact: Kassie Branham; (859) 225-3343; fax (859) 243-0770; kbranham@bgcf.org
Internet: http://bgcf.org/learn/community-funds/boyle-county/
Sponsor: Blue Grass Community Foundation
499 East High Street, Suite 112
Lexington, KY 40507

Blue Mountain Community Foundation Grants 921
The Foundation administers charitable funds to benefit people of the Blue Mountain Area. Most of the money for discretionary grants is designated by donors for use by agencies serving Walla Walla County. The Foundation's grant making policies are generally directed toward the fields of social and community services, the arts and humanities, education and health. In reviewing grant applications, careful consideration will be given to: potential impact of the program/project on the community and the number of people who will benefit; local volunteer involvement and support; commitment of the organization's Board of Directors; degree to which the applicant works with or complements the services of other community organizations; organization's fiscal responsibility and management skills; possibility of using the grant as seed money for matching funds from other sources; ability of the organization to obtain additional funding and to provide ongoing funding after the term of the grant.
Requirements: Nonprofit organizations serving the Walla Walla Valley, from Dayton to Milton-Freewater are encouraged to submit proposals.
Restrictions: Grants usually will not be made for the following: programs outside the Blue Mountain Area, operating expenses, annual fund drives, field trips, travel to or in support of conferences. No grants will be made for sectarian religious purposes nor to influence legislation or elections.
Geographic Focus: Oregon, Washington
Date(s) Application is Due: Jul 1
Amount of Grant: 125 - 4,000 USD
Samples: Blue Mountain Action Council, WA, $7,000 - Volunteer Adult Literacy Program; Walla Walla Community College Foundation, WA, $1,500 - Parent Education Program Scholarships; Carnegie Art Center, WA, $3,000 - general support.
Contact: Lawson F. Knight, Executive Director; (509) 529-4371; fax (509) 529-5284; BMCF@bluemountainfoundation.org
Internet: http://www.bluemountainfoundation.org/grant-making-programs.php
Sponsor: Blue Mountain Community Foundation
8 South Second, Suite 168, P.O. Box 603
Walla Walla, WA 99362-0015

Blue River Community Foundation Grants 922
The Blue River Community Foundation is a community-based philanthropic organization that identifies, promotes, supports, and manages programs that will enhance the quality of life in Shelby County, Indiana, for this generation and future generations. To this end, the Foundation has established five areas of interest for the competitive grant making program: community and civic--support for community programs designed to improve life in Shelby County; arts and culture--support for programs and facilities that offer wide-spread opportunities for participation and appreciation; education--support for programs at all levels of education; health--support for the promotion of health and well-being for Shelby County residents; and social services--support of human service organization programs. Organizations interested in submitting a grant request should first submit the grant interest form found on the website. If the grant request meets the Foundation's funding guidelines, the organization will be invited to submit a formal Grant Application Form. Grant applications may be submitted at any time, but will only be reviewed during the next upcoming grant cycle.
Requirements: Any 501(c)3 serving the residents of Shelby County, Indiana, may apply.
Geographic Focus: Indiana
Date(s) Application is Due: Feb 1; Jun 1; Oct 1
Contact: Lynne Ensminger, Program Director; (317) 392-7955; fax (317) 392-4545; lensminger@blueriverfoundation.com or brf@blueriverfoundation.com
Internet: http://blueriverfoundation.com/main.asp?SectionID=6&TM=34321.26
Sponsor: Blue River Community Foundation
54 W Broadway Street, Suite 1, P.O. Box 808
Shelbyville, IN 46176

Blue Shield of California Grants 923
Consideration for funding will be given exclusively to organizations that pursue activities directly related to the foundation's program goals, including domestic violence prevention through service provision, education, and outreach; research and education regarding medical best practices and health technologies; and direct or indirect provision of medical insurance or health care to those populations that are uninsured or underinsured, and related policy development.
Requirements: The foundation funds organizations that are non-profit and tax-exempt under 501(c)3 of the Internal Revenue Service Code (IRC) and defined as a public charity under 509(a)1, 2, or 3 (types I, II, or a functionally integrated type III); accredited schools; units of government/public agencies; tribal governments. The foundation will only support projects that meet the following criteria: the mission of the grantee organization is consistent with the goals and mission of the foundation; the grant is used primarily to serve Californians; the grant seeking organization has a reputation for credibility and integrity; the grant seeking organization is pursuing activities directly related to one of the Foundation's three Program Areas: Health Care and Coverage, Health and Technology and Blue Shield Against Violence.
Restrictions: The foundation does not fund award dinners, athletic events, competitions, special events, or tournaments; conferences or seminars; capital construction; television/film/media production; religious organizations for religious purposes; political causes, candidates, organizations or campaigns; capital projects over $50,000; multi-year projects (generally); grants to individuals (with the exception of the regulated Blue Shield of California Employee Scholarship Program); grants to 509(a) 3, type III supporting organizations that are not functionally integrated.
Geographic Focus: California
Samples: U of California at San Francisco Medical Ctr (CA)--to participate in a program that will use cutting-edge monitoring and control technology designed to prevent hospital-borne infections, $90,000; 304 California nonprofit organizations (CA)--for projects to prevent domestic violence, increase access to health-care services, and assess medical technologies that can contribute to improved health, $6.7 million (approximately) divided.
Contact: Grants Administrator; (415) 229-5785; fax (415) 229-6268
Internet: http://blueshieldcafoundation.org/grant-center/index.cfm
Sponsor: Blue Shield of California
50 Beale Street
San Francisco, CA 94105-1808

Blum-Kovler Foundation Grants 924
The Blum-Kovler Foundation was established in 1985 after Everett Kovler retired from his position as President of James Beam Distilling Company. The foundation awards general operating grants to eligible nonprofit organizations in its areas of interest, including social services, Jewish welfare funds, higher education, health services and medical research, and cultural programs. The foundation also supports youth- and child-welfare agencies and public-interest and civic-affairs groups. Grants are awarded primarily in the Chicago metropolitan area and in the Washington, D.C. area. There are no application forms. Applicants should submit a one to two page written proposal with a copy of their IRS determination letter by mid-November to considered for the current year. Typical grant awarded is between $1,000-$5,000.
Requirements: Illinois and District of Columbia nonprofit organizations are eligible.
Geographic Focus: District of Columbia, Illinois
Amount of Grant: 1,000 - 1,000,000 USD
Contact: Hymen Bregar, Secretary; (312) 664-5050
Peter Kovler, Chairperson and Vice President; (312) 664-5050
Sponsor: Blum-Kovler Foundation
875 N Michigan Avenue, Suite 3400
Chicago, IL 60611-1958

Blumenthal Foundation Grants 925
In 1924 Mr. I.D. Blumenthal was a traveling salesman in need of repair to his car's radiator. A local tinsmith in Charlotte, North Carolina, repaired the radiator with a "magic powder". Impressed with the product, I.D. teamed with the tinsmith and Solder Seal became the first product of the Radiator Specialty Company. The Blumenthal Foundation was founded in 1953 and was endowed with the success of the Radiator Specialty Company. The foundation focuses the majority of its grants on programs and projects that have an impact on Charlotte, and the state of North Carolina. The philanthropic efforts of the Foundation are focused in nine areas of grant making: arts,

science and culture; civic and community; education; environment; foundation affiliates; health; Jewish institutions and philanthropies; religious and interfaith; and social sciences. The foundation believes that basic operational funding for non-profits is just as important, if not more so, than support for special programs or projects; consequently, grants are provided for seed money, annual operating costs, capital campaigns, conferences and seminars, special projects, and endowments. Interested organizations may click the Grant Guidelines link at the website for detailed submission instructions. Applications must be mailed. There are no deadlines, and requests are accepted on an ongoing basis. The Board of Trustees meets quarterly to consider grant applications.
Requirements: 501(c)3 organizations and institutions that serve the city of Charlotte and the State of North Carolina in the foundation's areas of interest are eligible to apply.
Restrictions: Grants are not made to individuals for any purpose.
Geographic Focus: North Carolina
Contact: Philip Blumenthal; (704) 688-2305; fax (704) 688-2301; foundation@gunk.com
Internet: http://www.blumenthalfoundation.org/BFGrantListings.htm
Sponsor: Blumenthal Foundation
P.O. Box 34689
Charlotte, NC 28234-4689

BMW of North America Charitable Contributions 926
BMW of North America funds charitable programs that benefit society in the areas of education, road-traffic safety, and the environment. The corporation supports education at all levels and specifically focuses on the following: intercultural learning for K-12 students and their teachers; automotive technology, mechanics, and career and repair programs in high schools, technical schools, and community colleges; and research in the areas of safety design, ergonomics, and new materials. In the area of road traffic safety, the corporation supports driver-education programs geared at teenagers and new drivers; basic auto-maintenance programs for women; consumer education on general road-safety issues; and programs to promote the safety of children and young people on the road. In the area of the environment, BMW is committed to sustainable development and focuses grant making on the following: conservation/preservation of natural resources, in particular park lands and waterways; research and promotion of alternative fuels; and environmental education for K-12 students. In general, grants are awarded for specific projects rather than for general operating support, although some operating and capital grants are given consideration. Interested organizations may download application instructions and guidelines at the grant website. Organizations wishing to be considered for a grant must submit an application; telephone solicitations will not considered.
Requirements: 501(c)3 charities or 501(c)9 organizations are eligible to apply.
Restrictions: The corporate giving program does not support non-tax-exempt organizations; individuals; religious organizations for religious purposes; political candidates or lobbying organizations; organizations with a limited constituency, such as fraternal, labor, or veterans groups; travel by groups or individuals; national or local chapters of disease-specific organizations; national conferences, sports events, and other one-time, short-term events; sponsorships or advertising; anti-business groups; team sponsorships or athletic scholarships; or organizations outside the United States or its territories.
Geographic Focus: All States
Contact: Grants Coordinator; (201) 307-4000; fax (201) 307 3607
Internet: http://www.bmwgroupna.com/philanthropy.htm
Sponsor: BMW of North America
300 Chestnut Ridge Road
Woodcliff Lake, NJ 07677-7731

Bob and Delores Hope Charitable Foundation Grants 927
The foundation is primarily focusing its efforts on the economically disadvantaged populations of California and Texas. Projects of strong interest will be programs designed to help those living at or below the poverty guidelines. Other fields of interest include: the arts; Catholic agencies and churches; education; higher education; human services; medical research; and treatments for substance abuse. There are no specific application forms or deadlines, and applicants should send a letter of inquiry.
Requirements: Proposals are only accepted upon request. Applicants should submit the following: listing of additional sources and amount of support; timetable for implementation and evaluation of project; results expected from proposed grant; brief history of organization and description of its mission; and a detailed description of project and amount of funding requested.
Restrictions: No grants are given to/for: individuals; capital construction; fund raising; deficit financing; conferences; seminars; media events or workshops (unless they are an integral part of a broader program); or loans.
Geographic Focus: California, Texas
Amount of Grant: Up to 100,000 USD
Contact: Linda Hope, Vice-President; (818) 841-2020
Sponsor: Bob and Delores Hope Charitable Foundation
10346 Moorpark Street
North Hollywood, CA 91602-2407

Bodenwein Public Benevolent Foundation Grants 928
The Bodenwein Public Benevolent Foundation was established in 1938 under the will of Theodore Bodenwein, owner and publisher of The Day newspaper, to support and promote quality educational, cultural, human-services, and health-care programming for underserved populations. The Foundation specifically serves the people of Greater New London County where The Day has a substantial circulation. The majority of grants from the Bodenwein Public Benevolent Foundation are one year in duration; on occasion, multi-year support is awarded. Applicants must apply online at the grant website. Applicants are strongly encouraged to do the following before applying: review the downloadable state application procedures for additional helpful information, requirements, and restrictions; review the downloadable online-application guidelines at the grant website; review the foundation's funding history (link is available from the grant website); review the online application questions in advance; and review the list of required attachments. These will generally include: a list of board members, financial statements (audited, reviewed, or compiled by independent auditor); an organization summary; a list of other funding sources; an IRS Determination letter; and other required documents. All attachments must be uploaded in the online application as PDF, Word, or Excel files. The annual application deadline is 11:59 p.m. on November 15. Applicants will be notified of grant decisions by letter within three to four months after the proposal deadline.
Requirements: Nonprofit organizations serving Greater New London County (East Lyme, Groton, Ledyard, Lyme, Montville, Mystic, New London, North Stonington, Old Lyme, Salem, Stonington, and Waterford, Connecticut) are eligible to apply. A breakdown of number/percentage of people served by specific towns will be required in the online application.
Restrictions: The foundation does not support requests from individuals, organizations attempting to influence policy through direct lobbying, or any political campaigns.
Geographic Focus: Connecticut
Date(s) Application is Due: Nov 15
Contact: Amy R. Lynch; (860) 657-7015; amy.r.lynch@ustrust.com
Internet: https://www.bankofamerica.com/philanthropic/fn_search.action
Sponsor: Bodenwein Public Benevolent Foundation
200 Glastonbury Boulevard, Suite # 200
Glastonbury, CT 06033-4056

Bodman Foundation Grants 929
The Bodman Foundation was established by George M. Bodman and his wife Louise Clarke Bodman in 1945. George was born in Toledo, Ohio, in 1882 and died in 1950. Mrs. Bodman was born in Chicago in 1893 and died in 1955. The Bodmans lived for much of their lives in Red Bank, New Jersey, and in New York City, where George Bodman was a senior partner at the investment banking firm of Cyrus J. Lawrence and Sons. The Bodmans were generous supporters of numerous cultural, civic, and service organizations. During World War I Mr. Bodman headed the Intelligence Service of the War Trade Board. During World War II, he served as executive assistant to the Red Cross Commissioner for Great Britain and was regional director in charge of American Red Cross Club operations in England, Scotland, and Ireland. The Bodman Foundation's Certificate of Incorporation states that its funds are to be used for the aid, support or benefit of religious, educational, charitable, and benevolent objects and purposes for the moral, ethical and physical well-being and progress of mankind." The Bodman Foundation shares trustees, staff, office space, and even a website with the Achelis Foundation which has a similar mission and geographic area of concentration (both foundations give in New York City, while the Bodman Foundation also gives in New Jersey). Funding is concentrated in six program areas: arts and culture, education, employment, health, public policy, and youth and families. Most recently, awards have ranged from $15,000 to $200,000.
Requirements: 501(c)3 organizations based in New York City that fall within the foundation's areas of interest are welcome to submit an inquiry or proposal letter by regular mail (initial inquiries by email or fax are not accepted, nor are CDs, DVDs, computer discs, or video tapes). An initial inquiry to the foundation should include only the following items: a proposal letter that briefly summarizes the history of the project, need, objectives, time period, key staff, project budget, and evaluation plan; the applicant's latest annual report and complete set of audited financial statements; and the applicant's IRS 501(c)3 tax-exemption letter. Applications may be submitted at any time during the year. Each request is reviewed by staff and will usually receive a written response within thirty days. Those requests deemed consistent with the interests and resources of the foundation will be evaluated further and more information will be requested. Foundation staff may request a site visit, conference call, or meeting. All grants are reviewed and approved by the Trustees at one of their three board meetings in May, September, or December.
Restrictions: The foundation generally does not make grants for the following purposes or program areas: nonprofit organizations outside of New York; annual appeals, dinner functions, and fundraising events; endowments and capital campaigns; loans and deficit financing; direct grants to individuals; individual day-care and after-school programs; housing; organizations or projects based outside the U.S; films or video projects; small art, dance, music, and theater groups; individual K-12 schools (except charter schools); national health and mental health organizations; and government agencies or nonprofit organizations significantly funded or reimbursed by government agencies. Limited resources prevent the foundations from funding the same organization on an ongoing annual basis.
Geographic Focus: New Jersey, New York
Amount of Grant: 15,000 - 200,000 USD
Contact: John B. Krieger; (212) 644-0322; main@achelis-bodman-fnds.org
Internet: http://www.achelis-bodman-fnds.org/guidelines.html
Sponsor: Bodman Foundation
767 Third Avenue, 4th Floor
New York, NY 10017-2023

Boeing Company Contributions Grants 930
The Boeing U.S. contributions program welcomes applications in five focus areas: education; health and human services; arts and culture; civic; and the environment. Primary fields of interest include: arts; elementary and secondary education; the environment; family services, prevention of domestic violence; health care; public affairs; public safety; substance abuse programs; and general human services. The largest single block of charitable contributions goes toward supporting programs and projects related to education. Boeing also looks for innovative initiatives that promote the economic well-being of the community and neighborhood revitalization. Boeing invests in programs

that promote participation in arts and cultural activities and experiences, programs that increase public understanding of and engagement in the processes and issues that affect communities and programs that protect and conserve the natural environment. Boeing accepts applications for cash grants, in-kind donations, and services.
Requirements: To apply for support you must be a U.S. based IRS 501(c)3 qualified charitable or educational organization or an accredited K-12 educational institution. U.S. grant guidelines and applications are available online.
Restrictions: Grants do not support: an individual person or families; adoption services; political candidates or organizations; religious activities, in whole or in part, for the purpose of further religious doctrine; memorials and endowments; travel expenses; nonprofit and school sponsored walk-a-thons, athletic events and athletic group sponsorships other than Special Olympics; door prizes or raffles; U.S. hospitals and medical research; school-affiliated orchestras, bands, choirs, trips, athletic teams, drama groups, yearbooks and class parties; general operating expenses for programs within the United States; organizations that do not follow our application procedures; follow-on applications from past grantees that have not met our reporting requirements or satisfactorily completed the terms of past grants; fundraising events, annual funds, galas and other special-event fundraising activities; advertising, t-shirts, giveaways and promotional items; documentary films, books, etc; debt reduction; dissertations and student research projects; loans, scholarships, fellowships and grants to individuals; for-profit businesses; gifts, honoraria, gratuities; capital improvements to rental properties.
Geographic Focus: Alabama, Arizona, California, Colorado, District of Columbia, Florida, Georgia, Hawaii, Illinois, Kansas, Maryland, Missouri, Nevada, New Mexico, Ohio, Oklahoma, Oregon, Pennsylvania, South Carolina, Texas, Utah, Washington, Australia, Canada
Contact: Antoinette Bailey, (312) 544-2000; fax (312) 544 - 2082
Internet: http://www.boeing.com/companyoffices/aboutus/community/charitable.htm
Sponsor: Boeing Company Contributions
100 North Riverside
Chicago, IL 60606-1596

Boettcher Foundation Grants 931
Grant support is given to promote the general well-being of humanity. Grants are awarded for arts and culture, community and social service, education and healthcare. Organizations seeking support from the Foundation should send a preliminary letter, describing the organization that wishes to submit a proposal and the project for which funding is being requested. The letter should be signed by the head of the applicant agency and should include a statement related to the priority of the project within the organization's overall plans. Letters of inquiry should be mailed or emailed.
Requirements: Capital grants are made in the form of challenges, conditional on an applicant agency's ability to raise the balance of the funds needed for a project. Although no absolute guidelines have been established, 50 to 75 percent of the goal should already be in hand before the grant request will be considered.
Restrictions: The Foundation does not accept proposals, or provide grants, for the following giving interests: operations; gymnasiums/athletic fields; housing; purchase of tables or tickets for dinner/events; individuals; large urban hospitals; out-of-state projects; media presentations; small business start-ups; open space/parks; conferences, seminars, workshops; organizations that primarily serve animals; debt reduction; pilot programs; endowments; religious groups or organizations for their religious purposes; scholarships; travel.
Geographic Focus: Colorado
Samples: Children's Hospital (Denver, CO)--for its fund-raising campaign to build a new hospital, $5 million; Colorado College (Colorado Springs, CO)--to renovate Palmer Hall, a classroom building, $400,000; Johnson and Wales U (Denver, CO)--for renovations and restoration at its Park Hill Campus, $100,000.
Contact: Administrator; (800) 323-9640 or (303) 534-1937; grants@boettcherfoundation.org
Internet: http://www.boettcherfoundation.org/grants/index.html
Sponsor: Boettcher Foundation
600 Seventeenth Street, Suite 2210 South
Denver, CO 80202-5422

Bonfils-Stanton Foundation Grants 932
Colorado nonprofit organizations are eligible to apply, and funds must be used within the state for the benefit of Colorado citizens. The focus of the foundation is to advance excellence in the areas of arts and culture, community service, and science and medicine. Types of support include operating grants, capital campaigns, and capacity-building grants. Proposals will be reviewed at quarterly meetings. Guidelines and forms are available online.
Requirements: Colorado 501(c)3 organizations are eligible.
Restrictions: Areas generally not eligible for funding include loans, grants, or scholarships to individuals; events, media productions, seminars, conferences, or travel expenses related to meetings; activities or initiatives that have a religious purpose or objective; endowment funding, fellowships, endowed chairs; funding to retire operating debt; requests from organizations outside the State of Colorado or that are not for the benefit of Colorado citizens.
Geographic Focus: Colorado
Date(s) Application is Due: Jan 31; Apr 30; Jul 31; Oct 31
Samples: ArtReach (CO)--for a Web-based ticket reservation system; Capitol Hill Community Services (CO)--to fund meal sites for the homeless; Kids in Need of Dentistry (CO)--to develop a comprehensive technology plan.
Contact: Susan France, Vice President of Programs; (303) 825-3774; fax (303) 825-0802; susan@bonfils-stanton.org
Internet: http://www.bonfils-stantonfoundation.org
Sponsor: Bonfils-Stanton Foundation
1601 Arapahoe Street, Suite 500
Denver, CO 80202

Booth-Bricker Fund Grants 933
The Foundation makes contributions for the purposes of promoting, developing and fostering religious, charitable, scientific, literary and educational programs. Requests are welcomed for capital needs, special projects and other one-time requirements. Applications should be made by letter. There are no forms or deadlines. Requests should include complete information about the applicant organization, including its history, purpose, finances, current operations, governing board and tax status. A detailed explanation of the proposed use of the funds must be provided.
Requirements: Requests are accepted for the funding of projects within the state of Louisiana. Priority is given to the New Orleans area.
Restrictions: The Foundation generally does not provide sustaining (operations and maintenance) funding. No grants are made to individuals.
Geographic Focus: Louisiana
Amount of Grant: 5,000 - 50,000 USD
Samples: YMCA (New Orleans, LA)--program support, $51,700; Tulane U (New Orleans, LA)--for the Medical Ctr Cerise Chair, $30,000; Saint Dominic School (New Orleans, LA)--program support, $15,000.
Contact: Gray S. Parker, Chairperson; (504) 581-2430
Sponsor: Booth-Bricker Fund
826 Union Street, Suite 300
New Orleans, LA 70112

Borkee-Hagley Foundation Grants 934
The foundation awards grants in a wide range of interests, including social services to children and families, religious organizations, and environmental programs. Delaware nonprofits receive preference. The board meets in December to consider requests.
Requirements: Delaware nonprofits are eligible to apply.
Restrictions: No support for specific churches or synagogues. Grants are not made to individuals.
Geographic Focus: Delaware
Date(s) Application is Due: Nov 1
Amount of Grant: 1,000 - 25,000 USD
Samples: Artistic Productions Inc (Hockessin, DE)--$1000; Delaware Hospice (Wilmington, DE)--$13,000; Better Life Outreach Ministries (Newport, DE)--$5000; Children and Families First (Wilmington, DE)--$13,000.
Contact: Henry H. Silliman Jr., President; (302) 652-8616
Sponsor: Borkee-Hagley Foundation
P.O. Box 4590
Wilmington, DE 19807-4590

Bosque Foundation Grants 935
The foundation gives primarily for higher education, human services, Baptist & Protestant agencies/churches and, medical research in Texas. Types of support include capital campaigns, building/renovation and research grants. There are no application forms or deadlines. Applicants should submit a one-page letter of intent that describes the program and request.
Requirements: Texas nonprofit and for-profit organizations are eligible.
Restrictions: Individuals are not eligible.
Geographic Focus: Texas
Amount of Grant: 5,000 - 20,000 USD
Samples: Midland Community Theater, Midland, TX, $1,000; Peoples Missionary Baptist Church, Detroit, MI, $20,000; Dallas Baptist University, Dallas, TX, $10,000.
Contact: Louis A. Beecherl Jr., Trustee; (214) 956-6732; fax (214) 956-6733
Sponsor: Bosque Foundation
5950 Cedar Springs Boulevard, Suite 210
Dallas, TX 75235-6803

Boston Foundation Grants 936
The Boston Foundation has a particular concern for low income and disenfranchised communities and residents and supports organizations and programs whose work helps advance the Foundation's high priorities in a variety of subject areas: Arts and Culture; Civic Engagement; Community Safety, Economic Development; Education/Out-of-School Time, Health and Human Services; Housing and Community Economic Development; the Nonprofit Sector, Urban Environment and Workforce Development. The Foundation generally makes the following types of grants: Project or program support for community-based efforts that improve the quality of life in the community, test new models, and promote collaborative and innovative ventures; advocacy and public policy research that is linked to specific action; support for planning to enable organizations and residents to assess community needs, respond to new challenges and opportunities, and provide for the inclusion of new populations; organizational support to develop and build the capacity of nonprofit organizations - support that helps organizations keep pace with the changing requirements and demands of their communities and broader environments; small grants awarded on a rolling basis for one-time organizational development needs through the Vision Fund. In addition, on a very limited basis, the Foundation will consider development grants and strategic alliances.
Requirements: Grants are made only to tax-exempt organizations in Massachusetts.
Restrictions: The committee does not consider more than one proposal from the same organization within a 12-month period. Discretionary grants are generally not made to the following applicants: city or state government agencies or departments; individuals; medical research; endowments; equipment; replacement of lost/expired government funding or gap funding to cover the full cost of providing services; scholarships and fellowships; video and film production; construction and renovation projects and capital campaigns; programs with religious content; travel; summer camps and lobbying. Activities that are generally lower priorities for the Foundation are conferences, lectures,

one-time events, programs benefiting only a small number of participants or routine service delivery and/or operating expenses.
Geographic Focus: Massachusetts
Date(s) Application is Due: Jan 5; Jul 1
Contact: Corey Davis; (617) 338-1700; fax (617) 338-1604; info@tbf.org
Internet: http://www.tbf.org
Sponsor: Boston Foundation
75 Arlington Street, 10th Floor
Boston, MA 02116

Boston Globe Foundation Grants 937

The foundation concentrates on three focus areas: strengthen the reading, writing, and critical thinking of young people, while fostering their inherent love of learning; strengthen the roads that link people to culture; and strengthen the civic fabric of the city. The foundation also sponsors the Neighbor to Neighbor Initiative, which funds exceptional Dorchester focused nonprofits.
Requirements: Massachusetts nonprofits in the greater Boston area are eligible.
Restrictions: The Foundation will only review one proposal per year from any organization.
Geographic Focus: Massachusetts
Amount of Grant: 5,000 - 15,000 USD
Contact: Leah P. Bailey; (617) 929-2895; fax (617) 929-2041; foundation@globe.com
Internet: http://bostonglobe.com
Sponsor: Boston Globe Foundation
P.O. Box 55819
Boston, MA 02205-55819

Boston Jewish Community Women's Fund Grants 938

The fund, a project of Combined Jewish Philanthropies, invites letters of intent for projects that benefit women and girls. Grants are made in the areas of health, abuse, hunger, education, and empowerment. In Massachusetts, projects funded are sponsored by organizations both within the Jewish community and from the community at large. In Israel, proposals from programs that serve the Haifa community or have a documented history of funding from other North American organizations are invited to apply. Letters of intent should not be more than two pages long and must contain: your organization's mission; a statement of need for the project, how the project will address it, and how it meets BJCWF objectives; a budget narrative including the amount of funding requested, projected major expenditures, and use of BJCWF funds.
Requirements: Grants are made to 501(c)3 organizations in Massachusetts or comparable organizations in Israel.
Geographic Focus: Massachusetts, Israel
Date(s) Application is Due: Oct 15
Amount of Grant: Up to 25,000 USD
Contact: Susan Ebert, (617) 457-8500; susane@cjp.org
Internet: http://www.cjp.org/section_display.html?ID=572
Sponsor: Boston Jewish Community Women's Fund
126 High Street
Boston, MA 02110

Boston Psychoanalytic Society & Institute Fellowship in Child Psychoanalytic Psychotherapy 939

The Child Fellowship Program offers participants an intensive, one year course of study in psychoanalytic approaches to child and adolescent psychotherapy. The curriculum includes both a theoretical course and a clinical course to be taught weekly. The theory courses will introduce the fundamentals of child analytic theories and will review historical as well as modern theoretical approaches. The clinical courses are organized around different developmental stages of childhood. Those who complete the program and wish to continue on in the Child and Adolescent Advanced Training Program will receive credit for their participation and can apply for advanced standing. A printable application form is available for download. Tuition will be $1677 for the academic year.
Requirements: The program is open to mental health clinicians who have completed or are in an advanced phase of their clinical training (residency, graduate internship) and who are interested in further didactic and clinical education in psychoanalytic work with children. Other professionals who work with children and their families such as pediatricians, teachers, and ancillary therapists are also encouraged to apply.
Geographic Focus: All States
Date(s) Application is Due: May 15
Contact: Elizabeth Jordan, Program in Psychoanalytic Studies; (617) 266-0953; fax (617) 266-3466; office@bostonpsychanalytic.org
Internet: http://www.bostonpsychoanalytic.org/child_fellowship
Sponsor: Boston Psychoanalytic Society and Institute
15 Commonwealth Avenue
Boston, MA 02116

Boston Psychoanalytic Society and Institute Fellowship in Psychoanalytic Psychotherapy 940

The Fellowship is a one-year program designed to enhance participants' grasp of the principles and practice of psychoanalytic psychotherapy. It is also well suited for clinicians considering more extensive training in psychoanalytic psychotherapy or training in psychoanalysis, but who are either uncertain or not yet ready for a multi-year training commitment. The Fellowship offers two seminars weekly, one theoretical and one clinical, led by outstanding teachers from the Boston Psychoanalytic Society and Institute. The theoretical seminar focuses on Fundamental Concepts of Psychoanalytic Psychotherapy for the first term, on Comparative Theories of Psychotherapeutic Technique for the second term and on Selected Topics in Psychoanalytic Psychotherapy for the third term. The clinical seminar consists of discussion of case material presented by instructors and class members, with senior faculty as discussants. The seminars meet on Thursday evenings for 30 weeks starting in September. Each Fellow is matched with a faculty advisor who is available to discuss individual training goals and options. There is also an opportunity for individual supervision at reduced fees with a wide choice of experienced supervisors from the Boston Psychoanalytic Society and Institute. Tuition will be $1,677 for the academic year. The application fee is $25.
Requirements: The Fellowship is designed especially for mental health clinicians who have completed or are in an advanced phase of their requisite clinical training (psychiatric residency, psychology internship, MSW, masters in psychiatric nursing, etc.) and who desire further didactic and clinical education in psychoanalytic psychotherapy.
Geographic Focus: All States
Date(s) Application is Due: May 15
Amount of Grant: 1,677 USD
Contact: Elizabeth Jordan, Program in Psychoanalytic Studies; (617) 266-0953; fax (617) 266-3466; office@bostonpsychoanalytic.org
Internet: http://www.bostonpsychoanalytic.org/fellowship
Sponsor: Boston Psychoanalytic Society and Institute
15 Commonwealth Avenue
Boston, MA 02116

Boyd Gaming Corporation Contributions Program 941

From giving generously to a variety of worthy charitable organizations to continually enhancing the effectiveness of our diversity programs, Boyd Gaming Corp. has a long-standing commitment to responsible gaming, and concern for the planet, Boyd strives to make a positive differences in the communities in which it operates. The type of support offered is in: employee volunteer services; general/operating support; in-kind gifts
Requirements: All charitable requests must originate from a 501(c)3 non-profit organization and be from a state in which the company operates.
Restrictions: Boyd Gaming is not unable to act favorably on any request: for an individual, team or school-sponsored endeavor; for programs that discriminate for any reason, including race, color, creed, religion, age, sex or national origin
Geographic Focus: Hawaii, Illinois, Indiana, Louisiana, Mississippi, Nevada, New Jersey
Contact: Corporate Office; (702) 792-7200
Internet: http://www.boydgaming.com/about-boyd/corporate-responsibility
Sponsor: Boyd Gaming Corporation
3883 Howard Hughes Parkway, 9th Floor
Las Vegas, NV 89169

Boyle Foundation Grants 942

The Boyle Foundation was established in Massachusetts in 1990, and works diligently to support its primary fields of interest, including: animal welfare and wildlife; education; and health care. Most often, funding comes in the form of general operating support. An application form is required, though interested parties should begin by forwarding a letter of interest to the office. That letter should include a brief history of organization and description of its mission, along with a detailed description of the project and the amount of funding requested. Most recently, grant awards have ranged from as little as $750 to a maximum of $9,500. There are no specified annual deadlines for submission.
Requirements: Applicants must either be located in, or support the residents of, the State of Massachusetts.
Geographic Focus: Massachusetts
Amount of Grant: 750 - 9,500 USD
Contact: Brian E. Boyle, (508) 349-7955
Sponsor: Boyle Foundation
P.O. Box 786
Truro, MA 02666-0786

BP Foundation Grants 943

The BP Foundation is a charitable organization that helps communities around the world by supporting: science, technology, engineering and math (STEM) education (except in countries where basic literacy is an issue); enterprise development, jobs training and sustainable community projects; programs that further the understanding of and foster practical means of addressing global environmental issues; and emergency humanitarian relief. Of particular interest are the areas of: basic and emergency aid; community and economic development; disaster relief; elementary and secondary education; the environment; foundations endowments; higher education; human services; international development; job training; natural resources; STEM education; and sustainable development.
Requirements: National and international (in areas where BP operates) nonprofit organizations are entitled to apply. Regions include: United States; Australia; Canada; China; all of Europe; Japan; Philippines; Singapore; and the United Kingdom.
Restrictions: The Foundation does not accept unsolicited proposals, but rather reviews requests submitted by BP businesses around the world.
Geographic Focus: All States, Austria, Belgium, Canada, China, Denmark, Estonia, Finland, France, Germany, Great Britain, Greece, Hong Kong, Ireland, Italy, Japan, Luxembourg, Mexico, Netherlands, Norway, Philippines, Poland, Portugal, Russia, Singapore, Spain, Sweden, Switzerland, The Netherlands, United Kingdom
Contact: Iris Cross, Executive Director; (281) 366-2000
Internet: http://www.bp.com/en_us/bp-us/community/bp-foundation.html
Sponsor: BP Foundation
501 Westlake Park Boulevard, 25th Floor
Houston, TX 77079-2604

Bradley-Turner Foundation Grants   944
Incorporated as the W.C. and Sarah H. Bradley Foundation in Georgia in 1943, the Bradley-Turner Foundation uses contributions from the company's success to support the community and region through many different programs and facilities funded in whole or in part by foundation donations. The foundation has a special interest in endeavors related to family and children services, education, religion (Baptist, Christian, interdenominational, Methodist, Presbyterian, Salvation Army, and United Methodist), health, and culture and the arts. Major focus is placed on the vitality and quality of life in Columbus, Georgia, though compelling programs beyond the city's boundaries will also be considered. The foundation is particularly interested in projects that have a broad base of community support. There are no application forms. Applicants are asked to submit a letter of three to five pages describing the project. Grants are reviewed quarterly, in February, May, August, and November, when the board meets.
Requirements: IRS 501(c)3 tax-exempt organizations in Georgia are eligible. Heavily focuses on the southern region of the United States, but the Foundation is not solely limited to this area. It has also donated in Massachusetts, Illinois, and Colorado.
Restrictions: Grants are not made to individuals or to for-profit businesses or corporations.
Geographic Focus: Georgia
Amount of Grant: 2,500 - 300,000 USD
Samples: Andrew College, Cuthbert, Georgia, $50,000 - general operations; Brookstone School, Columbus, Georgia, $301,000 - general operations; Christ Community Health Services, Columbus, Georgia, $25,000 - general operations.
Contact: Phyllis Wagner, Executive Secretary; (706) 571-6040; fax (706) 571-3408
Internet: http://www.wcbradley.com/divisions.asp
Sponsor: Bradley-Turner Foundation
1017 Front Avenue, P.O. Box 140
Columbus, GA 31902-0140

Bradley C. Higgins Foundation Grants   945
The Bradley C. Higgins Foundation was established in Massachusetts in 1961, with its primary fields of interest identified as: arts and culture; education; health and health care; and human services. In the majority of instances, awards are given for either general operating support or program development. Interested parties should begin by forwarding a letter of interest to the Foundation office, explaining their overall program and general budgetary needs. A copy of the IRS determination letter should also be included. There are no annual deadlines. Approximately a dozen awards are approved annually, ranging from $500 to $20,000.
Requirements: Support is limited to Massachusetts-based 501(c)3 organizations.
Geographic Focus: Massachusetts
Amount of Grant: 500 - 20,000 USD
Contact: Sumner B. Tilton Jr., Chairperson; 508-459-8087 or (508) 459-8000; fax (508) 459-8300; stilton@fletchertilton.com
Sponsor: Bradley C. Higgins Foundation
370 Main Street, 12th Floor
Worcester, MA 01608-1779

Bravewell Leadership Award   946
The purpose of the Bravewell Award is to recognize, empower, and support champions medicine in their efforts to transform the culture of healthcare. Recipients of the $100,000 award are physicians or other doctoral level professionals from North America who are catalysts in advancing the field of integrative medicine; have made significant contributions to the field of medicine and have demonstrated positive influence among their colleagues and those they serve; embody and advance the principles of the Declaration for a New Medicine; have a history of collaboration across disciplines and healing philosophies; have a compelling vision for the future of medicine that inspires and encourages others; and are resilient change agents and role models in their communities. Any individual may submit a nomination (of self or other) by completing the online form along with a cover letter and the nominee's curriculum vita to the office. Candidates will be invited to submit additional information.
Requirements: Candidates for the award are physicians and other doctoral level professionals from North America who are catalysts in advancing the field of integrative medicine; have made significant contributions to the field of medicine and have demonstrated positive influence among their colleagues and those they serve; embody and advance the principles of the Declaration for a New Medicine; have a history of collaboration across disciplines and healing philosophies; have a compelling vision for the future of medicine that inspires and encourages others; and are resilient change agents and role models in their communities.
Geographic Focus: All States
Amount of Grant: 100,000 USD
Contact: Jeneen Hartley Sago, Program Officer; (612) 377-8400; info@bravewell.org
Internet: http://www.bravewell.org/current_projects/2011_Leadership/
Sponsor: Bravewell Collaborative
1818 Oliver Avenue South
Minneapolis, MN 55405

Breast Cancer Fund Grants   947
Traditionally, the fund has awarded three types of grants: innovative research grants support projects investigating cutting-edge scientific approaches to the detection, treatment, and prevention of breast cancer; community model grants (non-research grants) to encourage and support the start-up of truly creative, innovative, and replicable U.S. programs that are developing new methods for addressing the support, education, health care access, and advocacy needs of women affected by breast cancer; and discretionary grants that generally fund smaller education, support, and advocacy initiatives. The fund gives preference to projects that address the needs of an ethnically, economically, and geographically diverse representation of women, including persons who are underserved or have low incomes. Initial application for research and community model grants is by letter of intent. If interested, the fund will invite a full proposal. Applicants for discretionary grants may submit an abbreviated application form available from the fund. Contact program staff for project availability.
Requirements: The fund will consider applications from individuals or organizations, whether public or private, for-profit or nonprofit.
Restrictions: No support is available for the direct cost of medical services, and no funds may be used for any lobbying efforts or political expenditures.
Geographic Focus: All States
Contact: Jeanne Rizzo, Executive Director; (415) 346-8223; fax (415) 543-2975; ed@breastcancerfund.org or info@breastcancerfund.org
Internet: http://www.breastcancerfund.org
Sponsor: Breast Cancer Fund
1388 Sutter Street, Suite 400
San Francisco, CA 94109-5400

Brian G. Dyson Foundation Grants   948
The Foundation, established in Atlanta in 1994 by former Coca-Cola Bottling executive, Brian G. Dyson, offers funding to community foundations, higher education, and federated programs in the Atlanta region and throughout Georgia. Grants typically range up to $20,000, and funding supports general operating costs. There are no specific guidelines, application formats, or deadlines with which to adhere, and initial contact should be made in writing.
Requirements: Applicants must be colleges, public or private schools, or other non-profit organizations located within the state of Georgia.
Geographic Focus: Georgia
Amount of Grant: Up to 20,000 USD
Contact: Brian G. Dyson, Director; (404) 364-2940
Sponsor: Brian G. Dyson Foundation
3060 Peachtree Road NW, Suite 1465
Atlanta, GA 30305-2241

Bridgestone/Firestone Trust Fund Grants   949
The trust supports programs and projects of nonprofit organizations in the areas of education, child welfare, and environment and conservation in communities where the company has operations. While primary consideration is given to organizations and causes related to the three major focus points, the Fund recognizes the importance and value in supporting all types of civic, community and cultural activities. Assistance is regularly given for: community and neighborhood improvements; civil rights and equal opportunity; voter registration and education; job training; performing arts programs; public radio and television; cultural programs; non-academic libraries; and museums. Types of support include: annual campaigns; building/renovation; capital campaigns; continuing support; donated equipment; emergency funds; employee matching gifts; employee-related scholarships; endowments; exchange programs; fellowships; general/operating support; matching/challenge support; program development; research; scholarship funds; and sponsorships. Applications must be submitted in writing and should include a description of the organization (two-page maximum) and its record of accomplishment, objectives of the program, whom the program benefits, and proposed method to evaluate the program's success; amount sought from the trust in relation to the total need; exactly how trust fund money would be used; copy of IRS 501(c)3 confirmation letter; list of board of directors and their professional affiliations; previous year's financial report; current year's operating budget; Form 990; list of other contributors and the amount of their donations; and copy of recent audit if available. Proposals are reviewed upon receipt.
Requirements: IRS 501(c)3 nonprofit tax-exempt organizations in Alabama, Arkansas, Colorado, Connecticut, Florida, Kentucky, Illinois, Indiana, Iowa, Louisiana, Michigan, Minnesota, Mississippi, North Carolina, Ohio, Pennsylvania, South Carolina, Tennessee, Texas, Utah, and Wisconsin are eligible. Schools, governmental agencies or other nonprofit, civic organizations are included. Grant proposals should be sent directly to the management of the local Bridgestone Firestone facility.
Restrictions: It is essential that all organizations receiving grants be equal opportunity employers who will operate their programs in support of equal opportunity objectives. Contributions will not be made to groups that discriminate on the basis of race, color, religion, gender, mental or physical disabilities, sexual orientation, national origin, age, citizenship, veteran/reserve/national guard status, or other protected status; partisan political organizations; or groups limited to members of a single religious organization.
Geographic Focus: Alabama, Arkansas, Colorado, Connecticut, Florida, Illinois, Indiana, Iowa, Kentucky, Louisiana, Michigan, Minnesota, Mississippi, Nevada, North Carolina, Ohio, Pennsylvania, South Carolina, Tennessee, Texas, Utah, Wisconsin
Amount of Grant: 50 - 50,000 USD
Contact: Bernice Csaszar, Administrator; (615) 937-1415 or (615) 937-1000; fax (615) 937-1414; CsaszarBernice@bfusa.com or bfstrustfund@bfusa.com
Internet: http://www.bridgestone-firestone.com/trustfund.asp
Sponsor: Bridgestone/Firestone Trust Fund
535 Marriott Drive, P.O. Box 140990
Nashville, TN 37214-0990

Brighter Tomorrow Foundation Grants   950
Every fall, the Brighter Tomorrow Foundation opens its grant process to applicants representing Montgomery County (Ohio) agencies that serve people with developmental disabilities. Consideration will be given to proposals addressing needs in the following priority areas, specified in rank order: housing - assistance ranging from emergency or permanent shelter to enhancement of day-to-day living conditions by providing safe living

environments, improving personal care options, or fulfilling requests for specialized equipment; lasting equipment - adaptive and other equipment that makes daily life easier and safer such as accessible vans for local transportation, battery powered lifts, kids car seats, exercise and therapy equipment, and special tools; education - items such as, computers, software, cameras, projection and recording equipment, art supplies, etc. which encourage and make possible new learning and working options; and recreation - socialization and physical fitness activities for persons with developmental disabilities, including programs that teach families how to incorporate physical exercise for their family member into their daily routines. Also includes funding for field trips, music and dance exploration and artistic creativity. The annual deadline for application submission is November 7, and notification of awards will occur by February 28. The application period extends over two months, from mid-September until the first Friday in November each year. Grant decisions are made in January and grantees are notified at that time.
Requirements: Grants are awarded to 501(c)3 organizations, local governmental agencies and academic institutions that serve individuals with developmental disabilities.
Restrictions: Brighter Tomorrow Foundation will not provide support for: capital campaigns; general operating expenses unrelated to the grant purpose; retroactive funding for activities that have already taken place; basic research; staff development activities' supplanting of projects or activities that have existing funding from other sources; new staff positions; activities and projects directed to individuals with MR/DD outside of Montgomery County; or individual applicants.
Geographic Focus: All States
Date(s) Application is Due: Nov 7
Contact: Kevin Hayde, Manager; (937) 222-3390; fax (937) 222-0636; khayde@brightertomorrowfoundation.org
Internet: http://brightertomorrowfoundation.org/grantprocess.html
Sponsor: Brighter Tomorrow Foundation
500 Kettering Tower
Dayton, OH 45423

Bright Family Foundation Grants 951
Established in 1986, the foundation primarily serves the Stanislaus County area of California. Areas of interest include: religion; education; medical services; medical school/education; internships; human services; children/youth services; arts & culture. The Board meets once a year in December to review proposals. Deadline date for applications is November 1.
Requirements: Stanislaus County, CA, IRS 501(c)3 tax-exempt organizations within a 20 mile radius of Modesto, CA, are eligible to apply.
Geographic Focus: California
Date(s) Application is Due: Nov 1
Amount of Grant: 5,000 - 50,000 USD
Contact: Calvin Bright, President; (209) 526-8242
Sponsor: Bright Family Foundation
1620 North Carpenter Road, Building B
Modesto, CA 95351-1155

Bright Promises Foundation Grants 952
The Bright Promises Foundation's primary activities are identifying the most pressing unmet needs of disadvantaged children in Illinois; calling for individuals, foundations, agencies, legislators, parents and the media to join the foundation in supporting these needs; soliciting grant applications and making grants that support these needs; attracting volunteers and funds to the foundation; and recognizing important role models with awards. Currently, the Bright Promises Foundation's focus is promoting better health among low-income and other at-risk children between the ages of 8-12. The Bright Promises Foundation initiated a four-year grant program called Healthy Children/Healthy Adults: Promoting Health through Better Nutritional Choices. The program responds to the escalating problem of childhood obesity in the state of Illinois. Now in its third year, the Foundation has so far paid and pledged $481,646 to community-based multi-purpose agencies to promote better health among low-income and other at-risk children between the ages of 8-12. The foundation revisits its focus every four years to ensure relevancy.
Requirements: Grant applications are considered annually from a pool of invited applicants. Proposals are evaluated based on criteria including measurable goals and objectives, and sustainability after Bright Promises Foundation funding ends. Grantees are required to report at least twice each year on their measurable objectives, and project coordinators from the Bright Promises Foundation board of directors conduct site visits with staff during the application process and during the grant year.
Geographic Focus: Illinois
Samples: Centers for New Horizons, "Healthy Children/Healthy Adults Project," Chicago, Illinois - program includes nutrition classes delivered to children, parents and staff as well as menu planning, budgeting, food preparation and cooking, developing a working community garden and a peer education program where students will learn to be leaders in their school and to promote health and nutrition school-wide; Children's Home + Aid, "Community Schools Student Health Fitness Project," Austin, Illinois - the program serves 125 children at Howe Elementary School to provide them the foundation for a lifetime of healthy eating and exercise habits; Erie Neighborhood House, "Super H - Healthy Kids Make Happy Kids Project," Chicago, Illinois - the bi-monthly club for children and their parents includes education about the nutritional value of chosen "super foods" and how to prepare them, field trips, staff training and hands-on activities.
Contact: Iris Krieg, Executive Director; (312) 704-8260; info@brightpromises.org
Internet: http://www.brightpromises.org/OurPrograms/Grants/
Sponsor: Bright Promises Foundation
333 N. Michigan Avenue, Suite 510
Chicago, IL 60601

Brinson Foundation Grants 953
The foundation supports education, public health, and scientific research programs that engage, inform, and inspire committed citizens to confront the challenges that face humanity. Grantmaking priorities are education--awareness and outreach, democracy and citizenship, economically disadvantaged, and libraries and literacy; public health--awareness and outreach, and economically disadvantaged; and scientific research--astrophysics, cosmology, geophysics, medical research (i.e., Alzheimer's disease, cancer, Lou Gehrig's Disease (ALS), and stroke). Types of support include general operating grants and project grants. The foundation does not accept unsolicited grant applications. Grantseekers are asked to review the foundation's mission, vision, beliefs, priorities (accessed from the Who We Are link), and guidelines. If a grantseeker believes the request would match one or more of the foundation's grantmaking priorities, they can make an inquiry by completing the online Grantseeker Information Form. The completed form should be emailed to the office. Further application is by invitation.
Requirements: The foundation will consider inviting grant applications from organizations: whose request matches one or more of the Foundation's grantmaking priorities; located in the United States of America that are exempt from tax under Section 501(c)3 of the Internal Revenue Code and are defined as charitable organizations as described in Section 509(a)1, 2 or 3 or 170(b)1A; located outside the United States of America provided they produce a written legal opinion stating that they are a charitable equivalency to a qualifying U.S. organization and/or a written affidavit containing sufficient information for the Foundation to make a reasonable judgment that the organization is charitable. The Foundation's education and public health grants are generally made to organizations that serve individuals and communities in the greater Chicago area. It also considers leading U.S.-based programs that reach broader populations across the U.S. and internationally or have the potential to have a meaningful impact on best practices at the national or international level. The Foundation's physical science research grants are made to leading organizations across the United States. In this priority area, the location of the program is less critical than the match with the Foundation's grantmaking priorities.
Restrictions: The Foundation will not consider grant inquiries from organizations that: discriminate on the basis of race, gender, religion, ethnicity or sexual orientation. The Foundation will not consider grant inquiries that request funding for: activities that attempt to influence public elections; voter registration; political activity; lobbying efforts; promotion of a specific religious faith; medical research involving human cloning. The Foundation discourages grant inquiries requesting funds for: capital improvements; endowments; fundraising events.
Geographic Focus: All States
Samples: Adler Planetarium and Astronomy Museum, Chicago, Illinois, $80,000 - for cosmology and astrophysics research; Lincoln Park Zoo, Chicago, Illinois, $55,000 - for general support; Institute for Humane Studies, Arlington, Virginia, $20,000 - summer seminars for college students.
Contact: Cheryl A. Heads; (312) 799-4500; fax (312) 799-4310; mail@brinsonfoundation.org
Internet: http://www.brinsonfoundation.org/grant_seekers/grantseekers.shtml
Sponsor: Brinson Foundation
737 N Michigan Avenue, Suite 1850
Chicago, IL 60611

Bristol-Myers Squibb / Meade Johnson Award for Distinguished Achievement in Nutrition Research 954
The Award for Distinguished Achievement in Nutrition Research was first presented in 1978. The Award includes a $50,000 cash prize and a silver commemorative medallion. The winners are selected by an independent peer-review selection committee whose members are grant administrators of current Bristol-Myers Squibb/Mead Johnson Unrestricted Nutrition Research Grants. The annual application deadline is in late April.
Geographic Focus: All States
Date(s) Application is Due: Apr 24
Amount of Grant: 50,000 USD
Contact: John L. Damonti; (212) 546-4000 or (800) 332-2056; fax (212) 546-9574
Internet: http://www.bms.com/aboutbms/grants/data/nutrit.html
Sponsor: Bristol-Myers Squibb Foundation
345 Park Avenue, Suite 4364
New York, NY 10154

Bristol-Myers Squibb Clinical Outcomes and Research Grants 955
Bristol-Myers Squibb's mission is to extend and enhance human life. To help achieve that mission, the Company has established programs to support Investigator Sponsored Trials (ISTs). ISTs must be medically appropriate and scientifically valid. While the Company will consider requests for clinical research trials in all clinical and therapeutic areas, it currently gives priority to proposals in the following therapeutic areas: Cardiovascular/Metabolics, Infectious Diseases, Neuroscience, Oncology, Immunology, and Virology. Bristol-Myers Squibb maintains a strict policy of not exercising any influence or control over the design of any investigator initiated clinical research trial supported by BMS.
Requirements: Individuals in the following settings are eligible for support: private practice, hospitals, community health centers, cooperative groups, physician networks, and academic medical centers and universities.
Geographic Focus: All States
Contact: Amit Duggal, (212) 546-4000; fax (212) 546-9574; amit.duggal@bms.com
Internet: http://www.bms.com/responsibility/building_our_communities/Pages/default.aspx
Sponsor: Bristol-Myers Squibb Company
777 Scudders Mill Road
Plainsboro, NJ 08536

Bristol-Myers Squibb Foundation Fellowships   956
The Program provides support for fellows to gain experience in epidemiological and clinical research as it relates to the care of individuals infected with HIV/AIDS and/or HBV. Goals of the Program include: support of studies that will further strengthen the science and knowledge of HIV/AIDS and HBV; development of a foundation for future prospective and retrospective studies; to provide a forum to share research findings; and support of the development of future clinical researchers. Up to 18 fellows will be selected annually, with grant awards up to $20,000 to support research-related expenses for a one-year research period. Applications must be submitted online.
Requirements: Candidates must be: an active Fellow in good standing in an ACGME-accredited Fellows training program; be a senior Fellow not in the first years of the fellowship; desire to enhance knowledge and skill development in the area of HIV/AIDS, HBV, or Oncology research; and identify a faculty member to serve as the project mentor.
Restrictions: Total grant amount is inclusive of indirect costs and associated IRB fees and is not permitted for use towards travel to conferences or for durable equipment.
Geographic Focus: All States
Date(s) Application is Due: Feb 26
Contact: John L. Damonti; (212) 546-4000 or (800) 332-2056; fax (212) 546-9574
Internet: http://www.bms.com/responsibility/grantsandgiving/Corporate_Giving/Pages/fellowships_scholarships_awards.aspx
Sponsor: Bristol-Myers Squibb Foundation
345 Park Avenue, Suite 4364
New York, NY 10154

Bristol-Myers Squibb Foundation Global HIV/AIDS Initiative Grants   957
The intent of this initiative is to develop new models in awareness, in medical care, in community development and in prevention and treatment in poor and resource-limited areas of the world, where the need for all such efforts is greatest. Health care infrastructures must be developed and enhanced, stigmatization must be overcome, health care worker capacity must be built and preserved and local people must be empowered to generate and sustain local solutions to this global problem.
Requirements: The Bristol-Myers Squibb Foundation considers requests for support only from tax-exempt organizations that satisfy the requirements of section 501(c)3 of the U.S. Internal Revenue Code.
Restrictions: The Foundation does not award funds to: individuals; political, fraternal, social or veterans' organizations; religious or sectarian organizations unless engaged in a significant project benefiting the entire community; organizations receiving support through United Way or other federated campaigns; endowments; courtesy advertising; or conferences/special events/videos.
Geographic Focus: All States
Contact: John L. Damonti; (212) 546-4000 or (800) 332-2056; fax (212) 546-9574
Internet: http://www.bms.com/responsibility/grantsandgiving/Pages/default.aspx
Sponsor: Bristol-Myers Squibb Foundation
345 Park Avenue, Suite 4364
New York, NY 10154

Bristol-Myers Squibb Foundation Health Disparities Grants   958
One mission of the Bristol-Myers Squibb Foundation is to reduce health disparities by strengthening community-based health care worker capacity, integrating medical care and community-based supportive services, and mobilizing communities to fight disease. To this end, this Foundation attempts to address health disparities in four strategic disease areas representing major public health burdens and in four highly affected geographies: hepatitis in Asia, HIV/AIDS in Africa, serious mental illness in the U.S., and cancer in Europe. Additional areas of concern include: metabolic diseases, infectious diseases; rheumatoid arthritis; cardiovascular diseases; substance abuse; women's health issues; and overall health care giving.
Requirements: Nonprofit organizations in communities where Bristol-Myers Squibb maintains a facility should submit their requests for company contributions directly to that location. Contact persons are listed at the company website.
Restrictions: The foundation does not support individuals; conferences, special events, or videos; political, fraternal, social, or veterans organizations; religious or sectarian activities, unless they benefit the entire community; organizations funded through federated campaigns; endowments; or courtesy advertising.
Geographic Focus: Connecticut, Indiana, Massachusetts, New Jersey, New York
Contact: John Damonti; (212) 546-4000 or (800) 332-2056; fax (212) 546-9574
Internet: http://www.bms.com/foundation/reducing_health_disparities/Pages/default.aspx
Sponsor: Bristol-Myers Squibb Foundation
345 Park Avenue, Suite 4364
New York, NY 10154

Bristol-Myers Squibb Foundation Health Education Grants   959
The Program supports the development and replication of novel approaches aimed at individuals in the community to learn more about their health and wellbeing so that they can become more informed decision makers about health care and they can actively participate in disease prevention and management. In many of the grants, the Foundation seeks to be a catalyzing force for changes in government or social policies, supporting programs that can demonstrate the value of new approaches to health care through educational efforts. Nearly all the education programs focus on taking on major diseases, like hepatitis B or cancer and creating model programs for either preventing them or helping treat aspects of them in community settings.
Requirements: The Bristol-Myers Squibb Foundation considers requests for support only from tax-exempt organizations that satisfy the requirements of section 501(c)3 of the U.S. Internal Revenue Code.
Restrictions: The Foundation does not award funds to: individuals; political, fraternal, social or veterans' organizations; religious or sectarian organizations unless engaged in a significant project benefiting the entire community; organizations receiving support through United Way or other federated campaigns; endowments; courtesy advertising; or conferences/special events/videos.
Geographic Focus: All States
Contact: John L. Damonti; (212) 546-4000 or (800) 332-2056; fax (212) 546-9574
Internet: http://www.bms.com/responsibility/grantsandgiving/Pages/default.aspx
Sponsor: Bristol-Myers Squibb Foundation
345 Park Avenue, Suite 4364
New York, NY 10154

Bristol-Myers Squibb Foundation Product Donations Grants   960
The Foundation partners with designated non-governmental organizations, donating medical products to support long term health care programs in developing countries as well as addressing immediate needs to provide emergency health care disaster relief. Through this Program, it addresses health disparities and community infrastructure building alongside other Foundation and company charitable contributions programs. There are three main areas where the company's donations of products are used: the Medical Mission Box program; health care infrastructure support; and disaster relief.
Geographic Focus: All States
Contact: John L. Damonti; (212) 546-4000 or (800) 332-2056; fax (212) 546-9574
Internet: http://www.bms.com/responsibility/grantsandgiving/Pages/default.aspx
Sponsor: Bristol-Myers Squibb Foundation
345 Park Avenue, Suite 4364
New York, NY 10154

Bristol-Myers Squibb Foundation Science Education Grants   961
Bristol-Myers Squibb's efforts to improve science education are aimed at strengthening communities and their capabilities to encourage science and math literacy. Changing how science is taught in schools is the central focus. Underlying the program focus is a clearly defined need to increase interest by students in careers in science and to raise the level of science literacy. Ultimately, the program is designed to catalyze and lead systemic reform of science education.
Requirements: The Bristol-Myers Squibb Foundation considers requests for support only from tax-exempt organizations that satisfy the requirements of section 501(c)3 of the U.S. Internal Revenue Code.
Restrictions: The Foundation does not award funds to: individuals; political, fraternal, social or veterans' organizations; religious or sectarian organizations unless engaged in a significant project benefiting the entire community; organizations receiving support through United Way or other federated campaigns; endowments; courtesy advertising; or conferences/special events/videos.
Geographic Focus: All States
Contact: John L. Damonti; (212) 546-4000 or (800) 332-2056; fax (212) 546-9574
Internet: http://www.bms.com/responsibility/grantsandgiving/Pages/default.aspx
Sponsor: Bristol-Myers Squibb Foundation
345 Park Avenue, Suite 4364
New York, NY 10154

Bristol-Myers Squibb Foundation Women's Health Grants   962
Bristol-Myers Squibb Women's Health grants support projects that enhance women's health with strategies that improve education, prevention, diagnosis, treatment and access to care for women worldwide. Support has been given to projects that test innovative outreach programs, cultivate multi-sectoral partnerships and add new information to the existing body of knowledge to help define and achieve improved health for women around the world. The goal of the program is to generate initiatives that will help enhance women's health through novel interdisciplinary strategies that improve education, prevention, diagnosis, treatment and access to care for women worldwide. Since its inception, significant resources have been invested in programs that educate women about diseases and conditions that particularly threaten them as women.
Requirements: The Bristol-Myers Squibb Foundation considers requests for support only from tax-exempt organizations that satisfy the requirements of section 501(c)3 of the U.S. Internal Revenue Code.
Restrictions: The Foundation does not award funds to: individuals; political, fraternal, social or veterans' organizations; religious or sectarian organizations unless engaged in a significant project benefiting the entire community; organizations receiving support through United Way or other federated campaigns; endowments; courtesy advertising; or conferences/special events/videos.
Geographic Focus: All States
Contact: John L. Damonti; (212) 546-4000 or (800) 332-2056; fax (212) 546-9574
Internet: http://www.bms.com/Documents/foundation/women2002.pdf
Sponsor: Bristol-Myers Squibb Foundation
345 Park Avenue, Suite 4364
New York, NY 10154

Bristol-Myers Squibb Patient Assistance Grants   963
The Patient Assistance program was established to provide temporary assistance to qualifying patients with a financial hardship who generally have no private prescription drug insurance and are not enrolled in a prescription drug coverage plan through Medicaid or any other federal, state or local health program. The program provides free medications to indigent patients who qualify in all 50 states, Puerto Rico and the U.S. Virgin Islands. All patients and their health care providers must complete a program application to be considered for assistance. A printable version of the application is available online. Applications can also be obtained by calling the toll free telephone number.

Requirements: Applicants are required to provide information regarding household income, prescription insurance coverage, and citizenship, along with both patient and health care provider signatures.
Geographic Focus: All States
Contact: John L. Damonti; (212) 546-4000 or (800) 332-2056; fax (212) 546-9574
Internet: http://www.bms.com/responsibility/building_our_communities/people_in_need/Pages/default.aspx
Sponsor: Bristol-Myers Squibb Foundation
345 Park Avenue, Suite 4364
New York, NY 10154

Broad Foundation IBD Research Grants 964
The program seeks to stimulate innovative research that will lead to both the prevention and successful therapy of inflammatory bowel disease (IBD), including Chrohn's disease and ulcerative colitis. The foundation's goal is to fund basic or clinical research projects that are in the early stages of exploration; propose new directions or ideas; are creative, novel, cutting edge, and imaginative; and are not ready for funding by other more traditional granting agencies. All proposals must be based on sound scientific evidence and careful evaluation of current knowledge in IBD research. Requests must be preceded by a brief (one to three pages)letter of interest, which may be submitted at any time. Funding will be granted for one year; continued funding is based on progress reports and the perceived value of the findings.
Requirements: Nonprofit institutions worldwide, such as universities, hospitals, and research institutes, are eligible.
Restrictions: It is anticipated that the foundation will not fund projects after sufficient progress and maturity have made them fundable by other agencies.
Geographic Focus: All States
Amount of Grant: 100,000 USD
Samples: Charles N. Bernstein, M.D., University of Manitoba, Winnipeg, Manitoba, Canada--Project Title: A population-based characterization of potential microbial etiologies of inflammatory bowel disease using geographically defined high and low rate prevalence/incidence areas in Manitoba; Cucchiara, Salvatore, M.D., Ph.D., Sapienza University of Rome, Rome, Italy--Project Title: Molecular characterization of mucosa-associated intestinal microbiota and intestinal innate immune response: searching for additional mechanisms in pediatric Crohn's disease; Huang, Emina, M.D., University of Florida, Gainesville, Florida--Project Title: Colitis derived tumorigenic stem cells and the inflammatory bowel.
Contact: Heather Kubinec; (310) 954-5091; fax (310) 954-5092; info@broadmedical.org
Internet: http://www.broadmedical.org/funding.htm
Sponsor: Broad Foundation
10900 Wilshire Boulevard, Suite 1200
Los Angeles, CA 90024-6532

Brookdale Foundation Leadership in Aging Fellowships 965
The Leadership in Aging Fellowships provide two years of support to junior academics to focus on a project that will help establish them in an area of aging research. Fellowships are open to a broad range of disciplines including, but not limited to, medical, biological and basic sciences, nursing, social sciences, and the arts and humanities. The Fellowship is paid to the candidate's sponsoring institution in support of the candidate's research project. The Fellowship amount of up to $125,000 each year is intended to cover 75% of the fellow's time, base salary and fringe benefits. The award could also be used to include the support of a graduate assistant if necessary as long as the total amount does not exceed $125,000. Additional information is available at the website.
Requirements: Fellowships are open to all professionals in the field of aging. Candidates must meet the following criteria: leadership potential; ongoing commitment to a career in aging; a mentor (or mentors); and a willingness to commit at least 75% of his or her time for career development during each of the two years of the Fellowship.
Geographic Focus: All States
Amount of Grant: Up to 125,000 USD
Contact: Anna Condegni; (212) 308-7355; fax (212) 750-0132; annaatbrookdale@aol.com
Internet: http://www.brookdalefoundation.org/Leadership/Fellows/fellows.html
Sponsor: Brookdale Foundation
950 Third Avenue, 19th Floor
New York, NY 10022-3668

Brookdale Foundation National Group Respite Grants 966
The Group Respite Grants fund community-based, social model, day service programs that provide dementia-specific group activities for participants and respite from caregiving tasks for family caregivers. Program goals include the following: to offer opportunities for persons with Alzheimer's disease or a related dementia to engage in a program of meaningful social and recreational activities in a secure and supportive setting in order to maximize their cognitive and social abilities; and to provide relief and support to family members and other primary caregivers of individuals with Alzheimer's disease or a related dementia. Additional program guidelines and examples o previously funded projects are available at the Foundation website.
Requirements: Applicants must be nonprofits with tax-exempt status under Section 501(c)3 of the Internal Revenue Code or public agencies as defined under Section 509(a).
Geographic Focus: All States
Contact: Valerie Hall; (212) 308-7355; fax (212) 750-0132; vah@brookdalefoundation.org
Internet: http://www.brookdalefoundation.org/Respite/respiteprogram2008.html
Sponsor: Brookdale Foundation
950 Third Avenue, 19th Floor
New York, NY 10022-3668

Brook J. Lenfest Foundation Grants 967
The foundation primarily supports programs and organizations operating in urban areas of southeastern and south central Pennsylvania, southern New Jersey, and northern Delaware. Nonprofit organizations in Philadelphia receive preference. The foundation is dedicated to making people aware of positive life choices and providing support and opportunities for those motivated to pursue them. The Foundation will focus mainly on education, job training, mentoring programs, wellness based healthcare and the arts. The Foundation considers grant requests twice annually. To apply, send a short (two to four page) letter of inquiry that includes: amount requested; description of the program; date by which funding should be received; short and long term outcomes information for existing programs; expected outcomes and how the results of the program will be evaluated for new programs; description of how the program will support the mission and interests of the Foundation; total program budget involved in the request; other support for the program received, requested or expected from foundations, corporation, government or other revenue sources.
Requirements: Nonprofit organizations operating in urban areas of southeastern and south central Pennsylvania, southern New Jersey, and northern Delaware with a particular interest in Philadelphia are eligible.
Geographic Focus: Delaware, New Jersey, Pennsylvania
Samples: Pennsylvania State U (University Park, PA)--to provide scholarships to low-income students from Philadelphia public high schools, $1.4 million over five years.
Contact: Grants Administrator; (610) 828-4510; fax (610) 828-0390; lenfestfoundation@lenfestfoundation.org
Internet: http://www.brookjlenfestfoundation.org
Sponsor: Brook J. Lenfest Foundation
Five Tower Bridge, 300 Barr Harbor Drive, Suite 450
West Conshohocken, PA 19428

Brooklyn Community Foundation Caring Neighbors Grants 968
The Caring Neighbors Fund assists vulnerable Brooklyn families and individuals with immediate need for a social safety net and seeks to provide access to health and mental health services. Its goals are to: offer paths out of poverty by supporting the work of emergency food providers and human services agencies; provide access to care for unaddressed physical or mental health needs in accessible community settings; ensure that homeless individuals and families can access safe temporary shelter and support services. The Foundation typically has two grant cycles annually, Letter of Inquiry are accepted then. Contact the Foundation directly for current grant cycles. Organizations invited to submit a complete proposal should expect a final decision from the Foundation within eight to twelve weeks of receipt.
Requirements: Organizations applying for a grant from the Brooklyn Community Foundation must be classified as tax-exempt under Section 501(c)3 of the Internal Revenue Code and as public charities under Section 509(a) of that Code. Fiscally sponsored organizations may also apply. Your organization need not be based in Brooklyn; however grants from the Brooklyn Community Foundation must directly benefit Brooklyn neighborhoods and/or Brooklyn residents. Applying for funding is a two part process: 1st step--begins with a Letter of Inquiry (LOI). This is the opportunity for an organization to provide the Foundation with an overview of the group and its proposed activities, project or program; 2nd step--applicants who have been selected to proceed to the next stage will receive an email from the Foundation's staff indicating that the organization has been approved to submit a complete proposal which best describes in detail the activities, program or project. If selected, your organization will have 30 days to submit a complete a proposal online.
Restrictions: The Foundation does not: fund individuals; support for-profit organizations; purchase tickets for dinners, golf outings or similar fundraising events; make contributions to candidates for elective office or for partisan political purposes; provide funding for religious purposes.
Geographic Focus: New York
Contact: Diane John, (718) 722-5952 or (718) 722-2300; fax (718) 722-5757; info@BrooklynCommunityFoundation.org
Internet: http://www.brooklyncommunityfoundation.org/grants/caring-neighbors
Sponsor: Brooklyn Community Foundation
45 Main Street, Suite 409
Brooklyn, NY 11201

Brown Advisory Charitable Foundation Grants 969
Established in Maryland in 2009, the Brown Advisory Charitable Foundation offers grant funding primarily throughout the State of Maryland. The Foundation has identified the following fields of interest: the arts; primary and secondary education; higher education; human services; and federated giving programs. Typically, funding takes the form of either general operating support or program development. Most recently, grants have ranged from $100 to $40,000. A formal application is required, and can be secured by contacting the Foundation office. There are no annual application submission deadlines.
Requirements: Applicants must be 501(c)3 organizations located in, or serving the residents of, the State of Maryland.
Geographic Focus: Maryland
Amount of Grant: 100 - 40,000 USD
Samples: CollegeTracks, Bethesda, Maryland, $3,000 - general operating support (2014); Maryland Zoo in Baltimore, Baltimore, Maryland, $12,500 - general operating support (2014); Baltimore School for the Arts, Baltimore, Maryland, $40,000 - general operating support (2014).
Contact: Irene Alisa Stesch, Secretary; (410) 537-5503 or (410) 537-5400
Internet: http://www.brownadvisory.com/community/tabid/79/Default.aspx
Sponsor: Brown Advisory Charitable Foundation
901 Bond Street, Suite 400
Baltimore, MD 21231-3340

Brown County Community Foundation Grants 970
The Brown County Community Foundation strives to enhance the lives of the citizens and organizations of the community. It is a non-profit enterprise that seeks to provide the mechanism through which those who desire to help others in the community may carry out their philanthropy. By supporting charitable organizations in broad areas of community need - education, social services, health care, arts and humanities, and environment - it helps build a stronger, healthier Brown County. In general, grants shall be made for capital purposes only, not for operating expenses. Grant applications are generally available in April. Applications are due late May, with decisions announced in late June or early July. Most recently, awards have ranged from $1,500 to $18,000.
Requirements: Evidence of nonprofit tax status must be submitted with all applications. Proposals must serve the residents of Brown County, Indiana.
Restrictions: Preference is generally given to requests that demonstrate the most urgent and immediate need for funding or satisfy an identifiable community need. Grants are ordinarily made for one year only. The Foundation rarely provides the entire support of a project. Grants are not generally available for those agencies and institutions that are funded primarily through tax support.
Geographic Focus: Indiana
Date(s) Application is Due: May 10
Amount of Grant: 1,500 - 18,000 USD
Samples: Brown County Historical Society, Nashville, Indiana, $5,480 - used for a video system for the public meeting room in the new History Center (2014); Brown County Literacy Coalition, Nashville, Indiana, $3,126 - for the implementation of Ready To Learn, a project aimed at fostering development of cognitive ability in children (2014); Indiana Raptor Center, Nashville, Indiana, $18,000 - for a quality used vehicle for transportation to programs (2014).
Contact: Larry Pejeau; (812) 988-4882; fax (812) 988-0299; larry@bccfin.org
Internet: http://browncountygives.org/bcgives/895-2/
Sponsor: Brown County Community Foundation
91 West Mound Street, Unit 4
Nashville, IN 47448

Browning-Kimball Foundation Grants 971
Matt S. Browning and Barbara Kimball Browning founded the Browning Kimball Foundation in 1978. It is a unique private foundation dedicated to reaching out to charitable organizations that improve the quality of living for people and communities throughout Montana. The Foundation is committed to improving the quality of life through support of youth and families, health (with an emphasis on prevention), arts and humanities, and sustainable agriculture in Montana. A preliminary Letter of Intent is required by the annual deadline of May 1. Next, all organizations will be notified and informed if they have been invited to make a formal application. The final application is due by August 1.
Requirements: It is a requirement that applicants have a 501(c)3 designation by the U.S. Treasury Department. Other stipulations are covered in the Foundation's online grant application.
Geographic Focus: Montana
Date(s) Application is Due: Aug 1
Contact: Kari Koster; (405) 755-5571; fax kkoster@foundationmanagementin
Internet: http://www.browningkimballfoundation.com/apply-grant/
Sponsor: Browning-Kimball Foundation
2932 NW 122nd Street, Suite D
Oklahoma City, OK 73120-1955

Bruce and Adele Greenfield Foundation Grants 972
The Foundation offers funding primarily in the forms of ongoing support and capital campaigns. Focus areas include: the arts; community development; health care; higher education; environmental issues; human/social service programs that support families and seniors; and science. Of particular interest are the performing arts, including dance, music, and theater. Though the geographic focus is Florida and Pennsylvania, support is also occasionally given outside of these two states. Initial approach should be by Letter of Interest (LOI), with no specific application forms or deadlines.
Requirements: Grantees must be 501(c)3 organizations with a sound track record.
Restrictions: Generally the Foundation will not provide grants for individuals.
Geographic Focus: Florida, Pennsylvania
Amount of Grant: 100 - 3,000 USD
Samples: Good Shepherd Home, Allentown, Pennsylvania, $500; Pennsylvania Hortocultural Society, Philadelphia, Pennsylvania, $100; Germantown Jewish Center, Philadelphia, Pennsylvania, $3,000.
Contact: Adele G. Greenfield, (508) 428-1762; fax (508) 428-0607
Sponsor: Bruce and Adele Greenfield Foundation
575 Mistic Drive
Marstons Mills, MA 02648-1405

Bullitt Foundation Grants 973
The foundation functions to protect and restore the environment of the Pacific Northwest, including Washington, Oregon, Idaho, western Montana, coastal rainforests in Alaska, and British Columbia, Canada. Program priorities include aquatic ecosystems; terrestrial ecosystems; conservation and stewardship in agriculture; energy and climate change; growth management and transportation; toxic and radioactive substances; training, organizational development, and unique opportunities (including education and public outreach). Areas of interest include air pollution, climate change, endangered species, energy conservation, environmental education and justice, human health, transportation, and tribal communities. The foundation supports challenge/matching, general operating, project/program, seed money, demonstration, and development grants, as well as requests for conferences/seminars and technical assistance support. Grants will be awarded for one year with possible renewal.
Requirements: Nonprofit organizations in the Pacific Northwest, including Washington, Oregon, Idaho, western Montana, coastal rainforests in Alaska, and British Columbia, Canada are eligible.
Geographic Focus: Alaska, Idaho, Montana, Oregon, Washington, Canada
Date(s) Application is Due: May 1; Nov 1
Contact: Program Officer; (206) 343-0807; fax (206) 343-0822; info@bullitt.org
Internet: http://www.bullitt.org
Sponsor: Bullitt Foundation
1212 Minor Avenue
Seattle, WA 98101-2825

Bupa Foundation Medical Research Grants 974
The Bupa Foundation is an independent medical research charity that funds medical research to prevent, relieve and cure sickness and ill health. The Foundation aims to produce long-term benefits that have the potential to improve the health of individuals and populations worldwide. The funding will be available for the following areas: achieving sustained behaviour changes in relation to smoking, diet, physical activity and/or alcohol consumption; facilitating wellbeing and preventing mental ill health; improving patient decision-making through, for example, shared decision-making interventions; and, improving the design of community health activities by using new technologies to cost-effectively organise and interpret health outcome data.
Requirements: All projects should develop, apply or test research-based approaches in community settings - with clear intended impacts on public health. The areas for funding as detailed above are not intended to be exclusive of each other. Health professionals working for public or private organizations may apply for research grants for UK-based projects. If you are unsure of your eligibility, go to http://www.bupafoundation.co.uk/About/Grant-Eligibility.htm to see if your project will qualify.
Restrictions: Study for higher/further degrees, medical electives, educational courses, seminars and conferences, although valuable activities, are not eligible for grants. Similarly the Bupa Foundation does not fund the writing of reviews under the research grants scheme.
Geographic Focus: All States, All Countries
Date(s) Application is Due: Nov 30
Samples: Dr. Richard Chin, Institute of Child Health, University College London - Outcomes 5-10 years after childhood status epilepticus, £189,379; Dr. Alison Heawood, Community Based Medicine, University of Bristol - Negotiating a "therapeutic alliance" within online cognitive behavioural therapy: an analysis of client-therapist interaction, £69,147; Dr. Duncan Young, Nuffield Department of Anaesthetics, John Radcliff Hospital - The Intensive Care Outcome Network (ICON) Study, £193,056.
Contact: Kate Brown, Grants Officer; 020 7656 2591; kate.brown@bupa.com
Internet: http://www.bupafoundation.co.uk/Research-Grants
Sponsor: Bupa Foundation
Bupa House
London, GLONDON WC1A 2BA United Kingdom

Bupa Foundation Multi-Country Grant 975
The Bupa Foundation is an independent medical research charity that funds medical research to prevent, relieve and cure sickness and ill health. The Foundation aims to to fund implementation of translational and action orientated research-based solutions, which: drive tangible action to prevent or alleviate chronic disease or the adverse elements of ageing, and to promote wellbeing; lead to the sustained uptake of healthy lifestyles; measurably improve public health to a significant degree. The Foundation also aims to fund research which leads to the development of such solutions. A total grant of up to £600,000 will be awarded to fund one multi-country initiative of one to three years duration. This is defined as a project carried out in at least two of the following countries: United Kingdom, Spain, Australia, New Zealand, the United States or India.
Requirements: Multi-country grant applications are open to clinicians, researchers and health care professionals based in the countries listed above. Applicants are encouraged to include partnerships with universities, healthcare organizations, research institutes, charities, community groups and other appropriate bodies on their application. Projects should contribute to the Foundation's aims and the following objective: Achieving sustained behavior changes in relation to smoking, diet, physical activity and/or alcohol consumption.
Restrictions: Study for higher/further degrees, medical electives, educational courses, seminars, conterences, etc. are not eligible to apply.
Geographic Focus: All States, Australia, India, New Zealand, Spain, United Kingdom
Date(s) Application is Due: Mar 2
Contact: Kate Brown, Grants Officer; 020 7656 2591; kate.brown@bupa.com
Internet: http://www.bupafoundation.co.uk/Multi-country-Grants
Sponsor: Bupa Foundation
Bupa House
London, GLONDON WC1A 2BA United Kingdom

Burden Trust Grants 976
The trust supports international nonprofit organizations in the fields of medical research and hospitals; schools and training institutions; and care of and homes for the elderly, children, and other individuals in need. Preference will be given to organizations affiliated with the Anglican Church.
Geographic Focus: All States
Date(s) Application is Due: Mar 31
Contact: Patrick O'Conor, 0117 9628611; p.oconor@netgates.co.uk
Sponsor: Burden Trust
51 Downs Park W
Bristol, BS6 7QL England

Burlington Industries Foundation Grants 977

This is a company-sponsored foundation, giving primarily in areas of company operations in North Carolina, South Carolina, and Virginia. The foundation supports organizations involved with arts and culture, education, health, youth development, community development, and civic affairs.
Requirements: 501(c)3 organizations in Burlington communities (North Carolina, South Carolina, and Virginia), are eligible for grant support. Requests for funding must be accompanied by: proof of 501(c)3 tax-exempt status; description of the organization and its objective; justification for the project; evidence that the organization is well established; information about the organization's reputation, efficiency, management ability, financial status and sources of income.
Restrictions: No support for: sectarian or denominational religious organizations, national organizations, private secondary schools, historic preservation organizations, individuals (except for employees in distress), conferences, seminars, workshops, endowments, outdoor dramas, films, documentaries, medical research, loans.
Geographic Focus: North Carolina, South Carolina, Virginia
Amount of Grant: 1,000 - 50,000 USD
Contact: Delores Sides; (336) 379-2903; delores.sides@itg-global.com
Sponsor: Burlington Industries Foundation
P.O. Box 26540
Greensboro, NC 27415-6540

Burlington Northern Santa Fe Foundation Grants 978

The foundation is focused on the communities where the company operates and areas where its railways pass. The foundation supports education, including scholarships for Native Americans and scholarships for children of employees in conjunction with the National Merit Scholarships program; the arts, including museums, performing arts, and libraries; and civic and public affairs. Support goes to the Nature Conservancy and for local fire departments and law enforcement. Types of support include general operating support, continuing support, annual campaigns, capital campaigns, and program development. Health and human services funding concentrates on the United Way. Awards are made for a single year and for continuing support. The company also matches employee funds given to public and private colleges and universities, cultural organizations, and hospitals in the United States. Requests for applications should describe the purpose for the grant. Requests are reviewed every six weeks.
Requirements: Any 501(c)3 organization located in Schaumburg, IL, and communities where the corporation operates, including 28 states and two Canadian provinces, are eligible to apply.
Geographic Focus: All States
Samples: Texas Christian University, Fort Worth, TX, $600,000 - for Career Services Center and Neeley Schools Next Generation Leadership Program; Hastings Rural Fire District, Glenvil, NE, $20,000 - for replacement of rescue truck and grass firefighting rig; Foss Waterway Seaport, Tacoma, WA, $50,000 - for restoration and development of waterfront cultural center and maritime museum.
Contact: Deanna Dugas, Manager Corporate Contributions; (817) 867-6407; fax (817) 352-7924; Deanna.dugas@bnsf.com
Sponsor: Burlington Northern Santa Fe Foundation
2650 Lou Menk Drive, 2nd Floor, P.O. Box 961057
Fort Worth, TX 76131-2830

Bush Foundation Health & Human Services Grants 979

The foundation responds to a broad range of human services proposals. Proposals are reviewed on a case-by-case basis; applicant organizations take the lead in identifying promising solutions to the challenges faced by people who use their programs. In recent years, most grant dollars given to human services organizations have been for programs serving children, youth, and families. The foundation also considers program proposals that will improve the quality, accessibility, and efficiency of health care services in the region.
Requirements: The foundation is most interested in proposals that: (1) Promote opportunities for individuals and communities to become fully contributing members of society by supporting organizational projects that remove barriers to effective education, economic security and good health; (2) Improve the abilities of immigrant and refugee organizations, groups and individuals to obtain basic needs and rights, promote refugee and immigrant civic engagement and enhance their contribution to economic and cultural life; (3) The foundation will also consider proposals for comprehensive capital campaigns for building purchases, major building renovations and new construction to improve physical facilities. To be eligible for consideration, your organization must be a 501(c)3 nonprofit, tax-exempt organization, located in Minnesota, North Dakota or South Dakota, and able to demonstrate that you can take the lead in identifying promising solutions to challenges faced by people who use your programs. The two-step application process begins with a letter of inquiry. Guidelines for writing the letter of inquiry are on the information sheet which is available by contacting the sponsor or by download at the website.
Restrictions: The foundation does not make grants to: Individuals; Government agencies (except in special cases dictated by foundation priorities); Projects not benefiting the three-state region of Minnesota, North Dakota and South Dakota; Or, projects outside the United States. Download the Grant Restrictions file from the website for more details.
Geographic Focus: Minnesota, North Dakota, South Dakota
Date(s) Application is Due: Mar 1; Jul 1; Nov 1
Contact: Program Officer; (651) 227-0891; grants@bushfoundation.org
Internet: http://www.bushfoundation.org/grants/human_services.asp
Sponsor: Bush Foundation
332 Minnesota Street, Suite East 900
St. Paul, MN 55101-1315

Bush Foundation Medical Fellowships 980

The program was established to enhance community health care in Minnesota, North Dakota, South Dakota through the professional and personal development of selected physician leaders. Each year, the program awards approximately 13 fellowships that enable physicians to take a leave of absence from their practices to pursue professional and personal goals that address the health care needs of their communities. Their programs are self designed and self managed; they may last from three to 12 months. During this time, the fellowship provides a monthly stipend, as well as other financial aid.
Requirements: Applicants to the program must be: Physicians currently practicing in Minnesota, North Dakota or South Dakota; At least 35 years old and at least 10 years out of medical or osteopathic college; Able to state clearly their needs and opportunities for application of new skills and knowledge, both for the communities they serve and their own career development; and, able to explain how their programs will benefit an underserved population or need. All applications and references should be submitted via postal or express service. emailed applications and references must be followed by hard copy.
Restrictions: Previous Bush Medical Fellows may not reapply. If a physician (or his or her spouse/partner) is an elected official or public policymaker, the applicant may receive a grant for a degree-granting program only.
Geographic Focus: Minnesota, North Dakota, South Dakota
Date(s) Application is Due: Dec 1
Contact: Michael R. Wilcox, Program Director; (952) 442-2420 or (952) 758-4144; fax (952) 442-5841; bushmed@bushfoundation.org
Internet: http://www.bushfellows.org/medical
Sponsor: Bush Foundation
332 Minnesota Street, Suite East 900
St. Paul, MN 55101-1315

Business Bank of Nevada Community Grants 981

Business Bank of Nevada contributes financial assistance to nonprofit institutions and organizations that enhance the quality of life and promote public interest where the company conducts its business. The four main areas of giving include education, health and human services, community development, and arts and culture.
Requirements: Nevada 501(c)3 tax-exempt organizations in communities where the Bank has a presence. Community Grants are part of Business Bank of Nevada's grants program and must comply with the Bank's overall charitable funding guidelines. Requests must include the following information: a copy of the IRS letter of nonprofit tax-exempt 501C(3) status; a statement of purpose of the organization; the purpose and amount of the grant; the geographic area served; the project budget; a list of current board members with affiliations; a list of sources and amounts of other funding obtained, pledged, or requested; a copy of the latest audited or board approved financial statement.
Geographic Focus: Nevada
Amount of Grant: 250 - 2,500 USD
Contact: Paul Stowell; (702) 952-4415; pstowell@bbnv.com
Internet: http://www.bbnv.com/charitable_giving.php
Sponsor: Business Bank of Nevada
6085 West Twain Avenue
Las Vegas, NV 89103

BWF Ad Hoc Grants 982

To complement the competitive award programs, the Burroughs Wellcome Fund makes modest grants on an ad hoc basis to nonprofit organizations conducting activities intended to improve the general environment for science. These noncompetitive grants are for activities closely related to the Foundation's focus areas. The Foundation places special priority on working with non-profit organizations, including government agencies, to leverage financial support for targeted areas of research, and on encouraging other foundations to support biomedical research. Proposals should be brief and include a description of the focus of the activity, the expected outcomes, qualifications of the organization or individuals involved; provide certification of the sponsor's Internal Revenue Service tax-exempt status; and give the total budget for the activity, including any financial support obtained or promised. Proposals are given careful preliminary review, and those deemed appropriate are presented for consideration by BWF's board of directors.
Geographic Focus: All States, Canada
Contact: Rolly L. Simpson, Jr., Senior Program Officer; (919) 991-5100; fax (919) 991-5160; info@bwfund.org or rsimpson@bwfund.org
Internet: http://www.bwfund.org/grant-programs/ad-hoc-grants
Sponsor: Burroughs Wellcome Fund
21 TW Alexander Drive, P.O. Box 13901
Research Triangle Park, NC 27709-3901

BWF Career Awards at the Scientific Interface 983

BWF's Career Awards at the Scientific Interface (CASI) provide $500,000 over five years to bridge advanced postdoctoral training and the first three years of faculty service. These awards are intended to foster the early career development of researchers who have transitioned or are transitioning from undergraduate and/or graduate work in the physical/mathematical/computational sciences or engineering into postdoctoral work in the biological sciences, and who are dedicated to pursuing a career in academic research. Scientific advances such as genomics, quantitative structural biology, imaging techniques, and modeling of complex systems have created opportunities for exciting research careers at the interface between the physical/computational sciences and the biological sciences. Tackling key problems in biology will require scientists trained in areas such as chemistry, physics, applied mathematics, computer science, and engineering. These awards are open to U.S. and Canadian citizens or permanent residents as well as to U.S. temporary residents. Candidates are expected to draw from their training in a

scientific field other than biology to propose innovative approaches to answer important questions in the biological sciences. Examples of approaches include, but are not limited to, physical measurement of biological phenomena, computer simulation of complex processes in physiological systems, mathematical modeling of self-organizing behavior, building probabilistic tools for medical diagnosis, developing novel imaging tools or biosensors, developing or applying nanotechnology to manipulate cellular systems, predicting cellular responses to topological clues and mechanical forces, and developing a new conceptual understanding of the complexity of living organisms. Proposals that include experimental validation of theoretical models are particularly encouraged.
Requirements: Citizens and non-citizen permanent residents of the U.S. and Canada are eligible. Candidates must: hold a Ph.D. degree in one of the fields of mathematics, physics, chemistry, computer science, statistics, or engineering (this includes related areas of physical, mathematical, computational, theoretical, and engineering science); demonstrate that their work is truly interdisciplinary; have completed at least 12 months but not more than 48 months of postdoctoral research by the date of the full invited application deadline; be committed to a full-time career in research as an independent investigator at a North American degree-granting institution; have at least one first-author publication; including papers on which first authorship is shared; and be based at a non-profit institution in the U.S. or Canada.
Restrictions: Candidates cannot hold nor have accepted, either in writing or verbally, a faculty appointment as a tenure-track assistant professor at the time of application - both pre-proposal and full application. Candidates must not hold nor have accepted a K99 award from the U.S. National Institutes of Health. Temporary residents of Canada are not eligible.
Geographic Focus: All States, Canada
Date(s) Application is Due: Jan 8
Amount of Grant: Up to 500,000 USD
Contact: Rusty Kelley, Ph.D., Senior Program Officer; (919) 991-5120 or (919) 991-5100; fax (919) 991-5160; rkelley@bwfund.org or info@bwfund.org
Internet: http://www.bwfund.org/grant-programs/interfaces-science/career-awards-scientific-interface
Sponsor: Burroughs Wellcome Fund
21 TW Alexander Drive, P.O. Box 13901
Research Triangle Park, NC 27709-3901

BWF Career Awards for Medical Scientists (CAMS) 984
The Career Awards for Medical Scientists (CAMS) is a highly competitive program that provides $700,000 awards over five years for physician-scientists, who are committed to an academic career, to bridge advanced postdoctoral/fellowship training and the early years of faculty service. Awards are made to degree granting institutions in the U.S. or Canada on behalf of the awardee. The postdoctoral/fellowship portion of the award may last up to two years, and allows the awardee the use of up to $95,000 per year ($65,000 per year salary support). BWF will make up to two additional awards to clinically trained psychiatrists who focus on research at the interface between neuroscience and psychiatry. These proposals must clearly demonstrate evidence of integration of neuroscience and psychiatry in project design.
Requirements: The ideal candidate will be two years away from becoming an independent investigator, have at least two years or more of postdoctoral research experience, and have a significant publication record. Candidates must be nominated by their dean or department chair at the degree-granting institution where they will conduct the postdoctoral/fellowship training under the award. Applications must be approved by an official responsible for sponsored programs (generally from the grants office, office of research, or office of sponsored programs) at the degree-granting institution. Candidates must hold an M.D., D.D.S., or D.V.M. degree. Proposals must be in the area of basic biomedical, disease-oriented, or translational research. Proposals in health services research or involving large-scale clinical trials are ineligible. Candidates must be a fellow, resident, or a postdoctoral researcher and have at least two years of postdoctoral research experience at the time of application.
Restrictions: Proposals in health services research or involving large-scale clinical trials are ineligible.
Geographic Focus: All States, Canada
Date(s) Application is Due: Oct 1
Amount of Grant: 700,000 USD
Contact: Rolly L. Simpson, Jr., Senior Program Officer; (919) 991-5100; fax (919) 991-5160; info@bwfund.org or rsimpson@bwfund.org
Internet: http://www.bwfund.org/grant-programs/biomedical-sciences/career-awards-medical-scientists-2
Sponsor: Burroughs Wellcome Fund
21 TW Alexander Drive, P.O. Box 13901
Research Triangle Park, NC 27709-3901

BWF Collaborative Research Travel Grants 985
The Collaborative Research Travel Grant (CRTG) program provides up to $15,000 in support for researchers from degree-granting institutions to travel either domestically or internationally to a laboratory to acquire a new research technique or to facilitate a collaboration. To capitalize on what appears to be an opportunity to provide relatively unrestricted travel funds to academic scientists and trainees and to provide a stimulus for those working or contemplating working at the interface of science, this program provides travel grants that can be used both internationally and domestically to acquire new research techniques, and to promote collaborations.
Requirements: Postdoctoral fellows or faculty at degree-granting institutions in the U.S. or Canada are eligible to apply. Applicants must hold a Ph.D. in mathematics, physics, chemistry, computer science, statistics, or engineering at the time of application; and interested in investigating research opportunities in the biological sciences. Biologists holding a doctorate degree at the time of application who are interested in working with physical scientists, mathematicians, engineers, chemists, statisticians, or computer scientists to incorporate their ideas and approaches to answering biological questions are eligible to apply. Applicants must be citizens or permanent residents of the U.S. or Canada at the time of application.
Restrictions: Grants cannot be used for travel to domestic or international conferences, to cover salary support or purchase computers or equipment. Laboratory or lecture courses are not eligible for this grant. No indirect costs or overhead charges can be applied to this grant. This grant is not bi-directional or reciprocal for the collaborator. Public health, health policy, or epidemiology proposals will not be accepted. This grant is not open to researchers at NIH institutions. Previous travel grant recipients are not eligible.
Geographic Focus: All States, Canada
Date(s) Application is Due: Feb 3
Amount of Grant: Up to 15,000 USD
Contact: Rolly L. Simpson, Jr., Senior Program Officer; (919) 991-5110 or (919) 991-5100; fax (919) 991-5160; rsimpson@bwfund.org or info@bwfund.org
Internet: http://www.bwfund.org/grant-programs/biomedical-sciences/collaborative-research-travel-grants
Sponsor: Burroughs Wellcome Fund
21 TW Alexander Drive, P.O. Box 13901
Research Triangle Park, NC 27709-3901

BWF Innovation in Regulatory Science Grants 986
BWF's Innovation in Regulatory Science Awards provide up to $500,000 over five years to academic investigators developing new methodologies or innovative approaches in regulatory science that will ultimately inform the regulatory decisions the Food and Drug Administration (FDA) and others make. These awards are open to U.S. and Canadian citizens or permanent residents who have a faculty or adjunct faculty appointment at a North American degree-granting institution. Awards are made to degree-granting institutions in the U.S. or Canada on behalf of the awardee. The application process consists of two phases: a preproposal followed by a full proposal invitation. Preproposal applicants selected by the Advisory Committee deemed to meet the goals of this initiative will be invited to submit full proposals.
Requirements: Candidates must: hold an M.D., Ph.D., D.V.M., D.D.S., or M.D.-Ph.D. degree; be based at a non-profit institution [501(c)3 or equivalent] in the U.S. or Canada; hold a faculty position at an accredited, degree-granting institution in the United States or Canada; be an investigator at the adjunct, assistant, associate, or full professor level; be a citizen or permanent resident of the U.S. or Canada at the time of application.
Restrictions: Ineligible applicants include: persons who have applied for permanent resident status but have not received their government documentation by the time of the preproposal application deadline; temporary residents of the U.S. or Canada; postdoctoral students; and candidates who currently hold a BWF award.
Geographic Focus: All States, Canada
Date(s) Application is Due: Apr 1
Amount of Grant: 500,000 USD
Contact: Rusty Kelley, Senior Program Officer; (919) 991-5120 or (919) 991-5100; fax (919) 991-5160; rkelley@bwfund.org or info@bwfund.org
Internet: http://www.bwfund.org/grant-programs/regulatory-science/innovation-regulatory-science
Sponsor: Burroughs Wellcome Fund
21 TW Alexander Drive, P.O. Box 13901
Research Triangle Park, NC 27709-3901

BWF Institutional Program Unifying Population and Laboratory Based Sciences Grants 987
The Institutional Program Unifying Population and Laboratory Based Sciences program offers five-year institutional training awards of $500,000 per year to bridge the gap between the population and computational sciences and the laboratory-based biological sciences. The award will support the training of researchers between existing concentrations of research strength in population approaches to human health and in basic biological sciences. The goal is to establish training programs by partnering researchers working in schools of medicine and schools (or academic divisions) of public health. The programs supported by these awards will develop young researchers who will be equally at home with the ideas, approaches, and insights generated at the molecular scale and at the population scale. Proposals that cross institutional boundaries are encouraged.
Requirements: Degree-granting institutions in the U.S. or Canada may submit applications. Proposals must be driven by core components within medical and public health schools, but beyond those required components, departments or centers located within non-medical parts of a university, existing inter-institutional collaboratives, research museums, free-standing research institutes, and other non-profit institutions that provide advanced-level training are all acceptable as potential additional partners. Dental, osteopathic, and veterinary medical schools are appropriate applicants.
Restrictions: For-profit companies may not participate in the application, but could be valuable partners in such training programs. Research groups working at national laboratories and within the federal government are allowable as partners.
Geographic Focus: All States, Canada
Date(s) Application is Due: May 19
Contact: Victoria McGovern, Ph.D., Senior Program Officer; (919) 991-5112 or (919) 991-5100; fax (919) 991-5160; vmcgovern@bwfund.org or info@bwfund.org
Internet: http://www.bwfund.org/grant-programs/population-sciences/institutional-program-unifying-population-and-laboratory-based
Sponsor: Burroughs Wellcome Fund
21 TW Alexander Drive, P.O. Box 13901
Research Triangle Park, NC 27709-3901

BWF Investigators in the Pathogenesis of Infectious Disease Awards — 988

The Investigators in the Pathogenesis of Infectious Disease Award provides $500,000 over a period of five years to support accomplished investigators at the assistant professor level to study pathogenesis, with a focus on the interplay between human and microbial biology, shedding light on how human and microbial systems are affected by their encounters. The awards are intended to give recipients the freedom and flexibility to pursue new avenues of inquiry and higher-risk research projects that hold potential for significantly advancing the biochemical, pharmacological, immunological, and molecular biological understanding of how microbes and the human body interact. The goal of the program is to provide opportunities for accomplished investigators still early in their careers to study what happens at the points where human and microbial systems connect. The program supports research that sheds light on the fundamentals that affect the outcomes of this encounter: how colonization, infection, commensalism and other relationships play out at levels ranging from molecular interactions to systemic ones.
Requirements: Candidates must: generally have an M.D., D.V.M., or Ph.D. degree; have an established record of independent research and hold a tenure-track position as an assistant professor or equivalent (at the time of application) at a degree-granting institution; and be nominated by accredited, degree-granting institutions in the United States or Canada.
Restrictions: Applications from non-tenure track investigators at tenure-offering, degree-granting institutions will not be accepted.
Geographic Focus: All States, Canada
Date(s) Application is Due: Nov 3
Amount of Grant: 500,000 USD
Contact: Victoria McGovern, Ph.D., Senior Program Officer; (919) 991-5112 or (919) 991-5100; fax (919) 991-5160; vmcgovern@bwfund.org or info@bwfund.org
Internet: http://www.bwfund.org/grant-programs/infectious-diseases-0
Sponsor: Burroughs Wellcome Fund
21 TW Alexander Drive, P.O. Box 13901
Research Triangle Park, NC 27709-3901

BWF Postdoctoral Enrichment Grants — 989

The Postdoctoral Enrichment Program provides a total of $60,000 over three years to support the career development activities for underrepresented minority postdoctoral fellows in a degree-granting institution (or its affiliated graduate and medical schools, hospitals and research institutions) in the United States or Canada whose training and professional development are guided by mentors committed to helping them advance to stellar careers in biomedical or medical research. BWF is committed to funding the next generation of scientists and researchers, thus it has an interest in advancing the careers of underrepresented minority postdoctoral fellows. Up to ten awards will be granted for enrichment activities annually. This grant is meant to supplement the training of postdocs whose research activities are already supported. Funds will be provided to support the following enrichment activities: activities for the postdoctoral fellow to enhance research productivity, e.g. workshops, courses, travel, collaborations, and training in new techniques; activities for the postdoctoral mentor to increase the mentoring of PDEP fellows in university-based programs, including career guidance of the underrepresented minority postdoctoral fellow, research guidance that increases the productivity of the PDEP fellow; and attendance at one annual meeting of mentors hosted by the Burroughs Wellcome Fund (BWF); and participation in a peer network system of underrepresented minority postdoctoral scholars.
Requirements: Applicants must: be nominated by a qualified mentor (see proposal elements) at the degree-granting institution where the applicant will conduct the postdoctoral/fellowship training and applications must be approved by an official responsible for sponsored programs (generally from the grants office, office of research, or office of sponsored programs) at the degree-granting institution; have secured a postdoctoral position with funding (including support by the mentor's existing research grants) at a degree-granting, research-intensive institution in the United States or Canada and must begin the postdoctoral position on or by the designated award commencement date; be underrepresented minorities (i.e. American Indian or Alaska Native, Black or African American, Hispanic, or Native Hawaiian or other Pacific Islander); be citizens of the United States or Canada; and devote at least 75% of time to research.
Restrictions: The Postdoctoral Enrichment grant cannot be used for research. A person with more than 36 months of postdoctoral experience (in a research laboratory) at the time of application or with more than five years from his/her Ph.D. is not eligible for this award.
Geographic Focus: All States, Canada
Amount of Grant: 60,000 USD
Contact: Carr Thompson, Senior Program Officer; (919) 991-5103 or (919) 991-5100; fax (919) 991-5160; cthompson@bwfund.org or info@bwfund.org
Internet: http://www.bwfund.org/grant-programs/diversity-science/postdoctoral-enrichment-program
Sponsor: Burroughs Wellcome Fund
21 TW Alexander Drive, P.O. Box 13901
Research Triangle Park, NC 27709-3901

BWF Preterm Birth Initiative Grants — 990

The Preterm Birth Initiative was created to increase the understanding of the biological mechanisms underlying parturition and spontaneous preterm birth and will provide up to $600,000 over a four-year period ($150,000 per year). The initiative is designed to stimulate both creative individual scientists and multi-investigator teams to approach the problem of preterm birth using creative basic and translation science methods. Postdoctoral fellows nearing their transition to independent investigator status through senior established investigators are encouraged to apply. Molecular and computational approaches such genetics/genomics, immunology, microbiology, evolutionary biology, mathematics, engineering, and other basic sciences hold enormous potential for new insights independently or in conjunction with more traditional areas of parturition research such as maternal fetal medicine, obstetrics, and pediatrics. The formation of new connections between reproductive scientists and investigators who are involved in other areas will give preterm birth research a fresh and unique look, and stimulate a new work face to tackle this challenge. Proposals can be submitted by individual investigators or research teams designating a contact principal investigator. Proposals that cross institutional boundaries (partnerships between multiple universities or collaborations within larger universities) are welcomed.
Requirements: Proposals must be submitted from degree-granting institutions in the United States or Canada. The principal investigator must be a postdoctoral fellow in the final 1-2 years of postdoctoral training or hold faculty appointment at a degree-granting institution in the U.S. or Canada. The principal investigator must be a citizen or permanent resident of the U.S. or Canada.
Restrictions: Prior preterm birth full research grant recipients are not eligible to reapply.
Geographic Focus: All States, Canada
Date(s) Application is Due: Dec 1
Amount of Grant: Up to 600,000 USD
Contact: Rolly L. Simpson, Jr., Senior Program Officer; (919) 991-5110 or (919) 991-5100; fax (919) 991-5160; rsimpson@bwfund.org or info@bwfund.org
Internet: http://www.bwfund.org/grant-programs/reproductive-sciences/preterm-birth-initiative
Sponsor: Burroughs Wellcome Fund
21 TW Alexander Drive, P.O. Box 13901
Research Triangle Park, NC 27709-3901

Byron W. and Alice L. Lockwood Foundation Grants — 991

The foundation awards grants to Washington nonprofit organizations in its areas of interest, including arts, biomedical research, higher education, hospitals and health care organizations, housing and homelessness, museums, religion and religious welfare programs, and social services. Types of support include capital improvements, continuing support, general operating grants, professorships, projects grants, and research grants. There are no application forms; submit a letter of request.
Requirements: Washington nonprofit organizations are eligible. Grants are awarded primarily in Washington's Seattle and Puget Sound areas.
Geographic Focus: All States
Date(s) Application is Due: Oct 31
Amount of Grant: 900 - 56,000 USD
Samples: Seattle's Union Gospel Mission (Seattle) for capital improvements at the men's shelter, $50,000; Youth Suicide Prevention Plan for Washington State,.
Contact: Lee Kraft, Executive Director; (206) 230-8489
Sponsor: Byron W. and Alice L. Lockwood Foundation
P.O. Box 4
Mercer, WA 98040

Cable Positive's Tony Cox Community Fund Grants — 992

Cable Positive's Tony Cox Community Fund is a national grant program that exists to encourage community-based AIDS Service Organizations (ASOs) and cable outlets to partner in joint community outreach efforts, or to produce and distribute new, locally focused HIV/AIDS-related programs and Public Service Announcements (PSAs). Grants are available up to $7,000 for 501(c)3 organizations, with special consideration given to AIDS Service Organizations (ASOs) and cable systems and producers partnering with ASOs. Eligible local community outreach projects include, but are not limited to: World AIDS Day and National HIV Testing Day events, AIDS Rides/Walks, other joint efforts between AIDS organizations and local cable operators, etc. Funding is also available for production costs of HIV/AIDS-related programs and PSAs.
Requirements: In order to be considered, you must partner with a cable system. To find your local cable partner go to www.cableyellowpages.com. If you do not have an existing relationship with your local cable system, contact their Public Affairs or Community Relations department. Successful applicants usually demonstrate the following tactics: PSA or program production; Community partnering-third party participation; Local government involvement; Media outreach (press kits & promotion); and, Marketing campaign. The following three funding levels are available: 1. $3,000-Alpha Cable System (up to 150,000 television households); 2. $5,000-Beta Cable System (150,001 - 300,000 television households); 3. $7,000- Gamma Cable System more than 300,000 television households).
Geographic Focus: Oklahoma, Oregon, Pennsylvania, Rhode Island, South Carolina, South Dakota, Tennessee, Texas, Utah, Vermont, Virginia, Washington, West Virginia, Wisconsin, Wyoming
Date(s) Application is Due: Sep 12
Amount of Grant: Up to 7,000 USD
Contact: Jennifer Medina, (212) 459-1504; Jennifer@cablepositive.org
Internet: http://www.cablepositive.org/programs-tonycox.html
Sponsor: Tony Cox Community Fund
1775 Broadway, Suite 443
New York, NY 10019

Cabot Corporation Foundation Grants — 993

The goal of Cabot Corporation Foundation is to support community outreach objectives, with priority given to science and technology, education, and community and civic improvement efforts in the communities where the company has major facilities or operations. Types of support include capital grants, challenge grants, employee matching gifts, fellowships, general support, professorships, project support, research, scholarships, and seed money. The board meets in January, April, July, and October to consider requests. Applications must be received at least 30 days before a board meeting.

Requirements: The Foundation supports only nonprofit 501(c)3 tax-exempt organizations in areas of company operation. Modest support is available for international organizations that qualify under U.S. tax regulations.
Restrictions: Contributions are not made to individuals; fraternal, political, athletic or veterans organizations; religious institutions; capital and endowment campaigns; sponsorships of local groups/individuals to participate in regional, national, or international competitions, conferences or events; advertising sponsorships; or Tickets or tables at fundraising events.
Geographic Focus: Georgia, Illinois, Louisiana, Massachusetts, New Mexico, Pennsylvania, Texas, West Virginia, Belgium, Canada, China, Switzerland, United Kingdom
Amount of Grant: 2,000 - 75,000 USD
Contact: Cynthia L. Gullotti, Program Manager; (617) 345-0100; fax (617) 342-6312; Cynthia_Gullotti@cabot-corp.com or cabot.corporation.foundation@cabotcorp.com
Internet: http://www.cabot-corp.com/About-Cabot/Corporate-Giving
Sponsor: Cabot Corporation Foundation
Two Seaport Lane, Suite 1300
Boston, MA 02210-2019

Caddock Foundation Grants 994
The foundation supports nonprofit national and international Evangelical Christian religious organizations including churches and religious institutions, community groups, hospitals, international missions and ministries, religious centers and facilities, and youth organizations. Types of support include conferences/seminars, fellowships, and general operating support. Application must be made in writing and include a description of the organization and its objectives and the purpose of the grant. There are no application deadlines.
Restrictions: Grants are made to Evangelical Christian organizations. Grants are not made to individuals.
Geographic Focus: All States
Amount of Grant: 1,800 - 390,000 USD
Contact: Richard E. Caddock, Jr., Treasurer; (951) 683-5361
Sponsor: Caddock Foundation
1717 Chicago Avenue
Riverside, CA 92507

Caesars Foundation Grants 995
Founded as Harrah's Foundation in Nevada in 2002, giving is in the area of company operations. The foundation supports programs designed to help older individuals live longer, healthier, and more fulfilling lives; promote a safe and clean environment; and improve the quality of life in communities where Caesars operates. Fields of interest include: aging centers and services; Alzheimer's disease; developmentally disabled services; the environment; food distribution programs; food services; health care; patient services; higher education; hospitals; human services; mental health services; public affairs; public safety; nutrition; and youth services. Types of support being offered include: building and renovation; capital campaigns; continuing support; general operating support; program development; research; scholarship funding; and sponsorships. There are no specific deadlines or application forms. Caesars Foundation Trustees meet on a quarterly basis, typically around the second week of each quarter. Check with your nearest Caesars Entertainment property for deadlines. The foundation generally funds programs and projects of $10,000 or more.
Requirements: Eligible organizations are 501(c)3 nonprofits operating programs in the communities where Caesars employees and their families live and work. Applying organizations must also: demonstrate diversity by providing services and volunteer opportunities to all without regard to race, ethnicity, gender, religion, sexual orientation, identity or disability; illustrate strong leadership that will significantly strengthen communities in which Caesars operates; show sound administrative and financial condition; provide opportunities for Caesars staff involvement as volunteers and/or opportunity to serve on Board of Directors; and, provide branding opportunities and openly support Caesars Foundation in a public forum.
Restrictions: The Foundation is not designed to react to last-minute requests or event sponsorships--plan the timing of your proposal accordingly. Caesars Foundation does not accept requests for in-kind contributions.
Geographic Focus: Arizona, California, Illinois, Indiana, Iowa, Louisiana, Mississippi, Missouri, Nevada, New Jersey, North Carolina, Pennsylvania
Amount of Grant: 10,000 USD
Contact: Gwen Migita, Community Affairs; (702) 880-4728 or (702) 407-6358; fax (702) 407-6520; caesarsfoundation@caesars.com
Internet: http://www.caesarsfoundation.com/
Sponsor: Caesars Foundation
1 Caesars Palace Drive
Las Vegas, NV 89109-8969

Cailloux Foundation Grants 996
The foundation awards grants to nonprofit organizations in its areas of interest, including civic and cultural, education and youth, family and community service, and health and rehabilitation. Requests for specific projects or programs, technical assistance, and capital projects are considered; general operations are less common and usually small; challenge grants are awarded occasionally; and endowment grants are rare. The general policy of the foundation is not to make grants to an organization more than once within any 12-month period. Applicants must submit a letter of inquiry; instructions are available online.
Requirements: 501(c)3 tax-exempt organizations are eligible. Only grant proposals originating in Kerr and the surrounding communities (Gillespie, Bandera, Edwards, Real, and Kimble Counties) will be considered.
Restrictions: In general, grants are not made for fund raising events, professional conferences, membership drives, competition expenses, or programs/projects normally funded by governmental entities. The foundation does not fund church or seminary construction, or church related entities or activities other than ecumenically oriented projects/programs that otherwise meet foundation guidelines. Grants or loans to individuals are never made.
Geographic Focus: All States
Contact: Grants Administrator; (830) 895-5222; info@cailouxfoundation.org
Internet: http://www.cailouxfoundation.org/grant_guidelines.htm
Sponsor: Cailloux Foundation
P.O. Box 291276
Kerrville, TX 78029-1276

Caleb C. and Julia W. Dula Educational and Charitable Foundation Grants 997
Grants are given in the areas of the arts and humanities (particularly museums and libraries), child welfare, the aged, community funds and appeals, health care, religion, and historical preservation. Most groups receiving foundation grants have an established reputation in their particular field. There are no set requirements. Applicants should submit a letter that describes the organization, project, and amount requested.
Requirements: Grants are given to support projects of tax-exempt organizations.
Restrictions: Support is not available to individuals.
Geographic Focus: All States
Date(s) Application is Due: Apr 1; Oct 1
Amount of Grant: 5,000 - 50,000 USD
Contact: James F. Mauze, (314) 726-2800; fax (314) 863-3821; jfmauze@msn.com
Sponsor: Caleb C. and Julia W. Dula Educational and Charitable Foundation
112 S Hanley Road
Saint Louis, MO 63105

California Community Foundation Health Care Grants 998
The foundation strives to ensure that low-income adults and children have access to regular, sustainable, affordable and quality health care. Priorities include: efforts that serve federally designated Medically Underserved Areas or Health Professional Shortage Areas, and that deliver comprehensive health services (e.g., medical, dental, mental health, etc.) to low-income and underserved populations; efforts that enroll people in health care coverage and connect them with a source of regular care; efforts that integrate strong prevention and early intervention strategies into their health services; and requests that clearly demonstrate knowledge of the health conditions that affect the specific population to be served.
Requirements: Tax-exempt organizations (not private foundations) located within or serving primarily residents of Los Angeles County are eligible.
Restrictions: Grants will not be considered for annual campaigns or special fund-raising events; building campaigns, with the exception of community development grants; endowments; existing obligations; equipment, unless it is an integral part of an eligible project; incurring a debt liability; individuals, with the exception of arts and culture grants; routine operating expenses; sectarian purposes; or programs that will in turn make grants.
Geographic Focus: California
Date(s) Application is Due: Feb 1; Jun 1; Oct 1
Amount of Grant: 100,000 - 1,000,000 USD
Samples: Northeast Valley Health Corporation, San Fernando and Santa Clarita Valleys, CA, $150,000; Planned Parenthood of Pasadena, Northern San Gabriel Valley, CA $125,000; American Heart Association Inc, Los Angeles County, CA, $80,633.
Contact: Tamu Jones, Program Officer; (213) 413-4130, ext. 250; fax (213) 383-2046; tjones@ccf-la.org
Internet: http://www.calfund.org/receive/health_care.php
Sponsor: California Community Foundation
445 S Figueroa Street, Suite 3400
Los Angeles, CA 90071

California Community Foundation Human Development Grants 999
The human development program aims to expand support services for special populations that enable them to acquire the resources and skills that will lead to self-sufficiency. Special populations include aging adults, developmentally and/or physically disabled children and adults, at-risk youth and foster youth. Priorities include: support services for aging adults, developmentally and/or physically disabled children and adults, at-risk youth and foster youth; and efforts that promote civic dialogue and advocate for critical issues affecting aging adults, developmentally and/or physically disabled children and adults, at-risk youth and foster youth.
Requirements: IRS 501(c)3 tax-exempt organizations (not private foundations) located within or serving primarily residents of Los Angeles County are eligible.
Restrictions: Grants will not be considered for annual campaigns or special fund-raising events; building campaigns, with the exception of community development grants; endowments; existing obligations; equipment, unless it is an integral part of an eligible project; incurring a debt liability; individuals, with the exception of arts and culture grants; routine operating expenses; sectarian purposes; or programs that will in turn make grants.
Geographic Focus: California
Date(s) Application is Due: Feb 1; Jun 1; Oct 1
Amount of Grant: 50,000 - 300,000 USD
Contact: Robert Lewis; (213) 413-4130, ext. 273; rlewis@ccf-la.org
Internet: http://www.calfund.org/receive/human_development.php
Sponsor: California Community Foundation
445 S Figueroa Street, Suite 3400
Los Angeles, CA 90071

California Endowment Innovative Ideas Challenge Grants 1000

California Endowment was founded in 1996 as a result of Blue Cross of California's creation of its for-profit subsidiary, WellPoint Health Networks. The Endowment is a private, California-focused, grant-making foundation that advocates for health and health equity. It does this by raising awareness, by expanding access to affordable, high-quality health care for underserved communities, and by investing in fundamental improvements for the health of all Californians. The Endowment supports the statewide Health Happens Here campaign and is currently engaged in a ten-year, one-billion dollar Building Healthy Communities plan. As a part of this strategy, the Endowment's Innovative Ideas Challenge (IIC) grant-making program solicits ideas that can be classified as disruptive innovations. A disruptive innovation is one that brings to market products and services that are more affordable and, ultimately, higher in quality. It improves a product or service in ways that the market does not expect, typically by being lower priced or being designed for a different set of consumers. Proposed ideas should address either emerging or persistent health-related issues impacting underserved California communities. Interested organizations should initially submit 500-word descriptions of their idea through the Endowment's online system. These are due by 5 p.m. Pacific time on May 1st (deadline dates may vary from year to year). Applicants whose ideas are accepted will be asked to submit a full proposal. Further guidance and clarification in the form of downloadable PDFs and an FAQ are available at the Endowment website.
Requirements: California 501(c)3 nonprofits may apply. Awarded projects will demonstrate the following characteristics: be transformative and disruptively innovative; benefit California's underserved individuals and communities; demonstrate cultural and linguistic competency; build partnerships and encourage collaboration; address persistent and/or emerging health challenges; demonstrate organizational capacity to carry out work; have measurable outcomes; and align with one of the 10 Outcomes and/or 4 Big Results of the Endowment's Building Healthy Communities plan.
Restrictions: Funds may not be used for the following purposes: to carry on propaganda or otherwise attempt to influence any legislation; to influence the outcome of any public election or to carry on any voter registration drive; to make any grant which does not comply with Internal Revenue Code Section 4945(d)3 or 4; for fees for any services resulting in substantial personal benefit including membership, alumni dues, subscriptions, or tickets to events or dinners; for capital for building acquisition or renovation; for operating deficits or retirement of debt; for scholarships, fellowships, or grants to individuals; for government and public agencies; or for direct services or core/general operating support.
Geographic Focus: California
Date(s) Application is Due: May 1
Contact: Grants Administration Team; (213) 928-8646 or (818) 703-3311; fax (213) 928-8801; tcegrantreports@calendow.org or questions@calendow.org
Internet: http://www.calendow.org/grants/
Sponsor: California Endowment
1000 North Alameda Street
Los Angeles, CA 90012

California Wellness Foundation Work and Health Program Grants 1001

The foundation focuses its activities on specific priority areas where it has a significant, long-term commitment. Within each priority area, the foundation allocates the majority of its funds toward initiatives. Initiatives are targeted grantmaking programs with distinct objectives and are generally announced through requests for proposals. The foundation awards general grants and project grants. Under general grants, requests for core operating support for organizations that provide direct services to Californians for disease prevention or health promotion are of primary interest. Priority areas under general grants are diversity in the health professions, environmental health, healthy aging, mental health, teenage pregnancy prevention,, violence prevention, women's health, and work and health. Special projects grants are awarded to areas that fall outside the priority areas. Of particular interest are proposals to help California communities respond to cutbacks in federally funded programs. Activities commonly supported under special projects include strengthening traditional safety-net providers, educating consumers about changes in health care systems, advocating for underserved communities in health policy debates, and informing public decision making through policy analysis.
Requirements: Eligible applicants are California 501(c)3 nonprofit organizations, or organizations with a preapproved fiscal sponsor. An organization should first write a succinct letter of interest (one to two pages) that describes the organization, its leadership, the region and population(s) served, and the activities for which funding is needed, including the amount requested.
Geographic Focus: California
Amount of Grant: 5,000 - 200,000 USD
Samples: California Institute for Rural Studies (Davis, CA)--to articulate and distribute research findings on the health of farmworkers to policy makers, opinion leaders, the news media, and the public, $125,000 over three years; California State U, Stanislaus Foundation (Modesto, CA)--to provide holistic, prevention-based health-education services to Southeast Asian women and children living in Stanislaus County, $180,000 over three years; California State U, Auxiliary Services (CA)--to strengthen community organizations working to meet the health needs of residents of East Los Angeles and California's West San Gabriel Valley, $135,000 over three years; California Institute for Nursing and Healthcare (Berkeley, CA)--to develop and disseminate a plan for diversifying the nursing work force in California, $120,000 over three years.
Contact: Grants Administrator; (818) 702-1900; fax (818) 593-6614
Internet: http://www.tcwf.org/grants_program/index.htm
Sponsor: California Wellness Foundation
6320 Canoga Avenue, Suite 1700
Woodland Hills, CA 91367-7111

Callaway Foundation Grants 1002

The Callaway Foundation awards grants for the benefit of projects and people in LaGrange and Troup County, Georgia. Areas of interest, include: arts and entertainment; elementary, higher, and secondary education; libraries; health and hospitals; community funds; care for the aged; community development; historic preservation; and church support. Types of support include: annual campaigns; building construction and renovation; capital campaigns; continuing support; equipment acquisition; general operating support; land acquisition, and matching support. Preference is given to enduring construction projects and capital equipment. The Foundation Board meets four times per year in January, April, July, and October. Grant requests and applications are due on the last day of the month preceding the meetings.
Requirements: IRS 501(c)3 nonprofit organizations in LaGrange and Troup County, Georgia are eligible to apply. Letters of request should briefly cover all aspects of the project, including complete financial planning and costs involved. Copies of budgets and current financial statements should also be included. An application form is available at the Foundation's website.
Restrictions: Grants are usually not made for loans, debt retirement, endowment or operating expenses. Requests from churches located outside Troup County, Georgia, are not considered.
Geographic Focus: Georgia
Date(s) Application is Due: Mar 31; Jun 30; Sep 30; Dec 31
Amount of Grant: 1,000 - 4,000,000 USD
Samples: Chattahoochee Valley Art Museum, LaGrange, GA, $59,000--for operating support; Georgia Cities Foundation, Atlanta, GA, $367,772--for Downtown Revitalization Fund; LaGrange College, LaGrange, GA, $4,000,000--for library building and renovation.
Contact: H. Speer Burdette III, President; (706) 884-7348; fax (706) 884-0201; hsburdette@callaway-foundation.org
Internet: http://www.callawayfoundation.org/grant_policies.php
Sponsor: Callaway Foundation
209 Broome Street, P.O. Box 790
La Grange, GA 30241

Callaway Golf Company Foundation Grants 1003

The foundation strives to support initiatives in communities where Callaway Golf Company employees live and work. The geographic area of focus is primarily North San Diego County, California. The foundation offers support in the form of: matching funds, special projects, and general operating budgets. Areas of interest include but are not limited to: children and youth; biomedical research, with a special interest in the field of cancer; golf; education; boys & girls clubs of America; drug prevention; American Red Cross; housing; veterans; youth programs; food banks; social services; emergency programs; scholarship program for dependents of Callaway Golf employees; grants for training, competition and equipment needs. The foundation does not require the completion of a formal application document but requests a description of the organization and its history, the project at issue including goals and time lines, the qualifications of the leadership personnel involved in the project, and a detailed project budget. Grants are awarded semiannually. The Callaway Golf Company Foundation does not accept unsolicited requests for grants.
Requirements: IRS 501(c)3 nonprofit organizations in California are eligible.
Restrictions: The foundation will not fund applicants that illegally discriminate on the basis of gender, race, color, religion, national origin, ancestry, age, marital status, medical condition, or physical disability, either in the services they provide or in the hiring of staff; or promote political or particular religious doctrines.
Geographic Focus: California
Amount of Grant: 500 - 10,000 USD
Contact: Paul Thompson; (760) 930-8686; cgcfoundation@callawaygolf.com
Internet: http://www.callawaygolf.com/Global/en-US/Corporate/CallawayGolfFoundation.html
Sponsor: Callaway Golf Company Foundation
2180 Rutherford Road
Carlsbad, CA 92008-7328

Cambridge Community Foundation Grants 1004

Cambridge Community Foundation is dedicated to improving the quality of life for the residents of Cambridge, Massachusetts. The CCF serves Cambridge through our support of nonprofit community organizations, by making direct financial grants, providing technical assistance, and forming partnerships among organizations to coordinate services, address gaps, and highlight emerging issues. CCF primarily supports work in: early childhood services; youth service; senior services; community services; emergency outreach; arts; and the environment. See, the Foundations website http://www.cambridgecf.org/grant.html to download Proposal Summary Sheets, and additional guidelines.
Requirements: To be eligible to apply, the agency must be tax-exempt 501(c)3 under the IRS code), and the program must serve the people of Cambridge, Massachusetts.
Restrictions: Support is not provided to municipal, state, or federal agencies. Grants for individuals, scholarships, research studies, conferences, films, capital fund drives, or loans are not eligible.
Geographic Focus: Massachusetts
Date(s) Application is Due: Apr 1; Oct 1
Amount of Grant: 500 - 60,000 USD
Contact: Robert S. Hurlbut, Jr., Executive Director; (617) 576-9966; fax (617) 876-8187; RHurlbut@CambridgeCF.org or info@cambridgecf.org
Internet: http://www.cambridgecf.org
Sponsor: Cambridge Community Foundation
99 Bishop Richard Allen Drive
Cambridge, MA 02139

Camp-Younts Foundation Grants 1005

The foundation supports social services, higher and secondary education, youth organizations, Protestant religion, and hospitals and other health organizations in Florida, Georgia, North Carolina, and Virginia. Applicants should submit a letter describing the program, a copy of the 501(c)3 tax-determination letter, listing of board members, and an audited budget for the previous year.
Requirements: Nonprofit organizations in Florida, Georgia, North Carolina, and Virginia may request grant support.
Geographic Focus: Florida, Georgia, North Carolina, Virginia
Date(s) Application is Due: Sep 1
Amount of Grant: 1,000 - 55,000 USD
Samples: Virginia Baptist General Board (Richmond, VA)--for operating support, $40,000; Saint Edwards School (Bon Air, VA)--for operating support, $25,000.
Contact: Bobby Worrell, Executive Director; (757) 562-3439
Sponsor: Camp-Younts Foundation
P.O. Box 4655
Atlanta, GA 30302

Campbell Hoffman Foundation Grants 1006

The mission of the Campbell Hoffman Foundation is to promote and fund efforts to increase access to comprehensive health care for underserved and uninsured populations in the Northern Virginia region. Northern Virginia is defined as the counties of Arlington, Fairfax, Loudoun and Prince William and the cities of Alexandria, Falls Church, Fairfax, Manassas and Manassas Park. A letter of Inquiry should be submitted, as the initial approach, when approaching the Foundation for funding. Upon review, if the letter of inquiry meets with Foundations criteria, the applicant will be asked to submit a proposal; guidelines will be provided to them when the proposal is requested.
Requirements: Eligible organizations include nonprofit 501(c)3 organizations, government agencies and faith-based organizations. Eligible organizations must be both located in and serve the target populations of Northern Virginia.
Restrictions: The Foundation will not provide funding for capital campaigns, endowment campaigns, special events and/or conferences (including travel to and participation in same), emergency funding, loans, capital projects (including but not limited to building, construction or renovation), land purchases, lawsuits, films, video or publications. The Foundation may choose to provide general operating support.
Geographic Focus: Virginia
Contact: Lyn S. Hainge, Executive Director; (703) 749-1794; fax (703) 442-0846; lhainge@campbellhoffman.org
Internet: http://www.campbellhoffman.org/applicants/default.aspx
Sponsor: Campbell Hoffman Foundation
1420 Spring Hill Road, Suite 600
McLean, VA 22102

Campbell Soup Foundation Grants 1007

Since 1953, the Campbell Soup Foundation has provided financial support to local champions that inspire positive change in communities throughout the United States where Campbell Soup Company employees live and work. The Foundation places particular emphasis on Camden, New Jersey, birthplace of Campbell's flagship soup business and world headquarters. The Campbell Soup Foundation focuses its giving on four key areas: hunger relief-supporting food bank organizations in the communities of operation; wellness-addressing the health of consumers in the communities where they live; education-leveraging the Campbell brand portfolio to support educational programs; community revitalization-enhancing the quality of life in the communities that Campbell operates in. The Foundation only considers applications that meet the following criteria: the proposal must fit one of the key focus areas; the organization must display strong and effective leadership; the proposed plan must be clear and compelling, with measurable and sustainable commitments expressed in terms of real results; the proposed activity must be sufficiently visible to leverage additional support from other funding sources. There is no formal deadline. Proposals are accepted and reviewed on a rolling basis.
Requirements: The Foundation limits grants to nonprofit organizations which are tax-exempt under Section 501(c)3 of the Internal Revenue Code. Grants are made to institutions that serve: Camden, New Jersey; Davis, California; Sacramento, California; Stockton, California; Bloomfield, Connecticut; Norwalk, Connecticut; Lakeland, Florida; Downers Grove, Illinois; Marshall, Michigan; Maxton, North Carolina; Camden, New Jersey; South Plainfield, New Jersey; Napoleon, Ohio; Wauseon, Ohio; Willard, Ohio; Denver, Pennsylvania; Downingtown, Pennsylvania; Aiken, South Carolina; Paris, Texas; Richmond, Utah; Everett, Washington; Milwaukee, Wisconsin. Organizations do not need to be located in these communities in order to qualify for funding. However, the programs to be funded must serve these communities. Proposals must be submitted electronically via email to community_relations@campbellsoup.com. Proposals should be prepared in a concise, narrative form, without extensive documentation.
Restrictions: Grants are not made to the following: organizations that are based outside the United States and its territories; individuals; organizations that limit their services to members of one religious group or whose services propagate religious faith or creed; political organizations and those having the primary purpose of influencing legislation of/or promoting a particular ideological point of view; units of government; events and sponsorships; sports related events, activities and sponsorships. Organizations may not submit the same or similar proposals more than once in a Foundation fiscal year (July 1 - June 30). Proposals submitted via regular mail will not be reviewed.
Geographic Focus: Connecticut, Florida, Illinois, Michigan, New Jersey, North Carolina, Ohio, Pennsylvania, South Carolina, Texas, Utah, Washington, Wisconsin
Samples: Coriell Institute for Medical Research, $12,500; Scotland County Literacy Council, $15,000; Delaware County Christian School, $500.
Contact: Grant Administrator; (856) 342-6423 or (800) 257-8443; fax (856) 541-8185; community_relations@campbellsoup.com
Internet: http://www.campbellsoupcompany.com/community_center.asp
Sponsor: Campbell Soup Foundation
1 Campbell Place
Camden, NJ 08103-1701

Canada-U.S. Fulbright New Century Scholars Program Grants 1008

The Fulbright New Century Scholars Program allows scholars who have attained a level of national or international recognition for demonstrated professional accomplishments pertaining to higher education and its role in national and global economic development. The deadline for this specific award is October 30 for an award to be taken up from May of the following year and conclude the next April. The New Century Scholar award carries a value of US$30,000.
Requirements: Eligibility requirements include: Canadian or American citizenship at time of application (permanent resident status or landed immigrant status are not sufficient); a Ph.D. or equivalent professional/terminal degree by December 31 of the application year or equivalent professional experience; and English language proficiency.
Restrictions: Applicants are not eligible for a Canada-U.S. Fulbright Award if he or she: currently holds permanent residency status in the intended host country; is currently residing in or enrolled at a university in the intended host country; has resided abroad for five or more consecutive years in the six year period immediately preceding the date of application; has received a Fulbright Award within the last five years; has had recent substantial experience in the intended host country (recent substantial experience is defined as study, teaching, research or employment for a period of more than an academic year (nine months) during the past five years); or is a Canadian applicant and you are a dual citizen of Canada and the United States.
Geographic Focus: All States, Canada
Date(s) Application is Due: Oct 30
Amount of Grant: 30,000 USD
Contact: Jennifer Regan, Senior Program Officer; (613) 688-5517 or (613) 688-5540; fax (613) 237-2029; jregan@fulbright.ca
Internet: http://www.fulbright.ca/en/award.asp
Sponsor: Foundation for Educational Exchange between Canada and the USA
Ottawa, ON K1R 1A4 Canada

Canada-U.S. Fulbright Senior Specialists Program Grants 1009

The Fulbright Senior Specialists Program is designed to provide U.S. and Canadian faculty and professionals with opportunities to collaborate on curriculum and faculty development, institutional planning and a variety of other activities. During the course of their grant, Fulbright Senior Specialists may engage in any of the following activities at their Canadian host institution: conduct needs assessments, surveys, institutional or programmatic research; take part in specialized academic programs and conferences; consult with administrators and instructors of post-secondary institutions on faculty development; present lectures at graduate and undergraduate levels; participate in or lead seminars or workshops at overseas academic institutions; develop and/or assess academic curricula or educational materials; and conduct teacher-training programs at the tertiary level. These are short-term grants of two to six weeks to that are offered on a continuous basis.
Requirements: Eligibility requirements include: Canadian or American citizenship at time of application (permanent resident status or landed immigrant status are not sufficient); a Ph.D. or equivalent professional/terminal degree by December 31 of the application year or equivalent professional experience; and English language proficiency.
Restrictions: Applicants are not eligible for a Canada-U.S. Fulbright Award if he or she: currently holds permanent residency status in the intended host country; is currently residing in or enrolled at a university in the intended host country; has resided abroad for five or more consecutive years in the six year period immediately preceding the date of application; has received a Fulbright Award within the last five years; has had recent substantial experience in the intended host country (recent substantial experience is defined as study, teaching, research or employment for a period of more than an academic year (nine months) during the past five years); or is a Canadian applicant and you are a dual citizen of Canada and the United States.
Geographic Focus: All States, Canada
Contact: Jennifer Regan, Senior Program Officer; (613) 688-5517 or (613) 688-5540; fax (613) 237-2029; jregan@fulbright.ca
Internet: http://www.fulbright.ca/en/seniorspecialists.asp
Sponsor: Institute of International Education, Council for Int'l Exchange of Scholars
350 Albert Street, Suite 2015
Ottawa, ON K1R 1A4 Canada

Canada Graduate Scholarships and NSERC Postgraduate Scholarships 1010

Canada Graduate Scholarships and NSERC Postgraduate Scholarships provide financial support to high-calibre scholars who are engaged in master's or doctoral programs in the natural sciences or engineering. The Canada Graduate Scholarships will be offered to the top-ranked applicants at each level (master's and doctoral) and the next tier of meritorious applicants will be offered an NSERC Postgraduate Scholarship. This support allows these scholars to fully concentrate on their studies and to seek out the best research mentors in their chosen fields. NSERC encourages interested and qualified Aboriginal students to apply.
Requirements: To be considered eligible for support, applicants must: (1) be a Canadian citizen or a permanent resident of Canada; (2) hold, or expect to hold (at the time you take up the award), a degree in science or engineering from a university whose standing is acceptable to NSERC (if you have a degree in a field other than science or engineering, NSERC may accept your application at its discretion); (3) intend to pursue in the following year full-time graduate studies and research at the master's or doctoral level in

an eligible program in one of the areas of the natural sciences and engineering supported by NSERC; and (4) have obtained a first-class average (a grade of 'A-') in each of the last two completed years of study (full-time equivalent). Any exceptions to this requirement must be accompanied by supporting comments that justify the submission.
Restrictions: Scholarships are not available to students entering graduate studies in a qualifying year. Students who, after receiving an NSERC scholarship, change their field of study and research to a field that is not supported by NSERC, cease to be eligible and their awards will be cancelled.
Geographic Focus: All States, Canada
Date(s) Application is Due: Oct 15
Amount of Grant: 17,300 - 35,000 CAD
Contact: PGS Program Officer; (613) 995-5521; fax (613) 996-2589; schol@nserc.ca
Internet: http://www.nserc-crsng.gc.ca/Students-Etudiants/PG-CS/BellandPostgrad-BelletSuperieures_eng.asp
Sponsor: Natural Sciences and Engineering Research Council of Canada
350 Albert Street, 14th Floor
Ottawa, ON K1A 1H5 Canada

Canadian Optometric Education Trust Fund Grants   1011
Since 1980, the COETF has received applications and awarded funding to projects covering a vast and varied scope of research. The fund supports quality vision and eye care services for Canadians of all ages. Projects have been funded for improving optometrical education, professional development, and research. The Awards Committee meets annually (generally in March) to consider applications. The deadline for applications is typically in early February. Interested applicants should secure forms from the fund.
Restrictions: The fund customarily does not provide support for travel to and from sites where the results of a project are intended to be presented, i.e., a symposium or continuing education seminar.
Geographic Focus: All States, Canada
Amount of Grant: 3,000 - 5,000 CAD
Contact: Glenn Campbell, Executive Director; (888) 263-4676 or (613) 235-7924; fax (613) 235-2025; gcampbell@opto.ca or info@opto.ca
Internet: http://www.opto.ca/en/public/03_optometry/03_06_coetf.asp#apply
Sponsor: Canadian Optometric Education Trust Fund
234 Argyle Avenue
Ottawa, ON K2P 1B9 Canada

Canadian Patient Safety Institute (CPSI) Patient Safety Studentships   1012
The Canadian Patient Safety Institute (CPSI) is inviting applications from organizations/supervisors interested in enabling students to participate in patient safety work. Funding is to support the work of students in any healthcare profession, as well as other disciplines (including but not limited to social sciences, human kinetics, psychology and human factors engineering) interested in patient safety. The goal is to create new collaborative learning opportunities aligned with the mandate of CPSI. Students are required to be the major resource in carrying out the work for each patient safety project. Any person from organizations such as universities, health regions, and hospitals, non-governmental or not-for-profit private organizations who will provide supervision for a student project may apply for this award. The student selected can be a student of any health care profession as well as other disciplines (including but not limited to social sciences, human kinetics, psychology and human factors) with an interest in patient safety, who are registered in an undergraduate or graduate program (part-time or full-time), and who will be a student during the period of the award. The value of the award from CPSI is $6,000.
Requirements: Applying organizations/supervisors that are interested in accessing these resources must provide matching funding to augment studentship funding that will enable students to participate in patient safety work. Matched funding must be in actual dollars and cannot be in-kind contributions.
Restrictions: CPSI funding is to be used for student stipends only. CPSI will not make funds available for administrative or overhead purposes.
Geographic Focus: All States, Canada
Date(s) Application is Due: Dec 12
Amount of Grant: 6,000 USD
Contact: Laurel Taylor, Director of Operations; (866) 421-6933 or (780) 409-8090; fax (780) 409-8098; ltaylor@cpsi-icsp.ca
Internet: http://www.patientsafetyinstitute.ca/education/Patientsafetystudentship2009.html
Sponsor: Canadian Patient Safety Institute
10235 - 101 Street
Edmonton, AB T5J 3G1 Canada

Canadian Patient Safety Institute (CPSI) Research Grants   1013
The Canadian Patient Safety Institute (CPSI) announces its annual Research Competition to continue its strategic mandate of increasing the scope and scale of research and evaluation activities in patient safety. The primary goal for CPSI Research Competition is to develop knowledge about patient safety that can be helpful in a variety of settings and circumstances in organizations across Canada. It is the applicant team's responsibility to identify and secure matching funding from other contributing organization(s). Applicant teams may obtain matching funding from universities, foundations, voluntary health charities, provider associations, provincial government departments, regional health authorities, hospitals, research centers or the private sector.
Requirements: A 1:1 funding ratio is the acceptable minimum required for any project over $20,000.
Geographic Focus: All States, Canada
Date(s) Application is Due: Dec 2
Amount of Grant: 120,000 USD
Contact: Laurel Taylor, Director of Operations; (866) 421-6933 or (780) 409-8090; fax (780) 409-8098; ltaylor@cpsi-icsp.ca
Internet: http://www.patientsafetyinstitute.ca/news/2008researchcompetition.html
Sponsor: Canadian Patient Safety Institute
10235 - 101 Street
Edmonton, AB T5J 3G1 Canada

Capital Region Community Foundation Grants   1014
The Capital Region Community Foundation awards grants to eligible Michigan nonprofit organizations in its areas of interest, including: education; environment; health care; human services; humanities; arts; and public benefit. Favorable projects: reach a broad segment of the community; encourage matching gifts or additional funding; foster organizational capacity building and sustainability; assist those citizens whose needs are not being met by existing services; meet emerging needs, are innovative and have a high probability of leading to new solutions to community challenges; and are collaborative, comprehensive, promote cooperation among organizations within the region, and have the potential for ongoing community impact. Types of support include: capital grants; program development; and seed money. Grants are awarded for one year and are nonrenewable. Guidelines and application are available online, and application must be made online. Request for proposals are issued each July 1, and full applications are due by the September 1 deadline.
Requirements: Michigan 501(c)3 tax-exempt organizations in Clinton, Eaton, and Ingham Counties are eligible. Churches in the tri-county area are eligible for programs including food banks, after-school programs, programs assisting the needy, and charitable work benefiting the community, there grant applications are due June 1st.
Restrictions: Grants do not support: sectarian or religious programs; individuals; international organizations; endowment funds; administrative costs of fund raising campaigns; annual meetings; routine operating expenses; or existing obligations, debts, or liabilities.
Geographic Focus: Michigan
Date(s) Application is Due: Sep 1
Amount of Grant: 2,000 - 25,000 USD
Contact: Pauline Pasch, Senior Program Officer; (517) 272-2870; fax (517) 272-2871; ppasch@crcfoundation.org
Internet: http://www.crcfoundation.org/Nonprofits-Students/Grants-at-CRCF
Sponsor: Capital Region Community Foundation
330 Marshall Street, Suite 300
Lansing, MI 48912

Caplow Applied Science (CappSci) Carcinogen Prize   1015
According to the American Cancer Society, about 1 in 4 Americans will die from cancer. The disease is caused by mutations of cellular DNA and the abnormally rapid growth that occurs thereafter. This process is triggered by a combination of genetic predisposition, lifestyle factors, and/or exposure to carcinogens. The market for cancer prevention is enormous and universal. Most carcinogens cannot be detected by human senses alone. Whereas lifestyle factors leading to cancer are typically easy to identify, its is much more difficult for a person to control their exposure to carcinogens. The Foundation is requesting that applicants propose an innovative technology to empower people to detect and/or avoid carcinogens. The solution must be accurate, highly portable, affordable to consumers, and suitable for testing in a public science museum. The recipient will receive $100,000 in funding and a 12-18 month Inventor-in-Residence appointment at the Patricia and Phillip Frost Museum of Science in Miami, Florida. Recipients will build-out their technology during their residency, while engaging the public through an interactive exhibit. The initial online application form deadline is May 19, with those selected for second round applications will be asked to provide additional information. Ultimately, three finalists will receive grants of $1,500 each to present their projects to live audiences in November.
Requirements: Proposals will be accepted from non-profits (charities), for profits (companies), government programs, academic institutions, and individuals aged 18 years or older.
Geographic Focus: All States, All Countries
Date(s) Application is Due: May 19
Amount of Grant: 1,500 - 100,000 USD
Contact: Aleyda K. Mejia, Director of Global Health & Social Ventures; (305) 776-0902 or (786) 558-9738; fax (786) 269-2266; aleyda@cappsci.org
Internet: http://www.cappsci.org/prizes/carcinogen-prize/
Sponsor: Caplow Applied Science
3439 Main Highway, Suite 2
Miami, FL 33133

Caplow Applied Science (CappSci) Children's Prize   1016
The CappSci Children's Prize is an open web-based competition focused on under-five child survival with eligibility extending to everyone, individuals and organizations, across the world. Through CappSci's entrepreneurial and scientific approaches, the program is ensuring that more children under the age of five survive and thrive. The Children's Prize seeks the best and most effective project that proposes to save the greatest number of children's lives. One recipient will be awarded the $250,000 prize to directly execute their proposed project, within a two year period. Proposals will be judged according to how many lives they propose to save, how credible the plan and the proposer are, how directly the funds can be applied, the probability of success and the ease of verification. The online application must be received by June 12.
Requirements: The Children's Prize is available to anyone. Proposals will be accepted from non-profits (charities), for profits (companies), government programs, academic institutions, and individuals aged 18 years or older.
Geographic Focus: All States
Date(s) Application is Due: Jun 12

Amount of Grant: 250,000 USD
Contact: Aleyda K. Mejia; (305) 776-0902 or (786) 558-9738; fax (786) 269-2266; aleyda@cappsci.org or childrensprize@cappsci.org
Internet: http://www.childrensprize.org/
Sponsor: Caplow Applied Science
3439 Main Highway, Suite 2
Miami, FL 33133

Cardinal Health Foundation Grants 1017

The foundation's mission is to support employees' interests and to advance and fund programs that improve access to and delivery of health care services in CardinalHealth communities. Additional areas of interest include arts and culture, education, and youth development. Application forms are not required; submit a project summary of no more than three pages. Proposals are by invitation.
Requirements: 501(c)3 organizations in CardinalHealth communities are eligible.
Restrictions: Capital campaigns, endowments, religious organizations or sectarian programs for religious purposes, veteran, labor, and political organizations or campaigns, fraternal, athletic or social clubs, requests for loans or debt retirements, individual endeavors or needs, organizations that discriminate on the basis of age, disability, religion, ethnic origin, gender, or sexual orientation, or organizations with divisive or litigious public agendas will not be supported.
Geographic Focus: All States
Amount of Grant: 250 - 450,000 USD
Contact: Debra Hadley, Executive Director; (614) 757-7450; cardinalfoundation@cardinal.com or communityrelations@cardinalhealth.com
Internet: http://www.cardinal.com/us/en/community/
Sponsor: Cardinal Health Foundation
7000 Cardinal Place
Dublin, OH 43017

Cargill Citizenship Fund-Corporate Giving Grants 1018

Cargill's purpose is to be the global leader in nourishing people. Cargill measures their performance through engaged employees, satisfied customers, profitable growth and enriched communities. Corporate giving is one important way Cargill works to enrich the 1,000 communities where they conduct business. With 149,000 employees in 63 countries, Cargill people are working everyday to nourish the lives of those around us. The Cargill Citizenship Fund provides strategic grants to organizations serving communities where Cargill has a presence. The Fund provides direct grants for regional, national and global partnerships and provides matching grants for selected local projects supported by our businesses. Cargill seeks to build sustainable communities by focusing our human and financial resources in three areas: Nutrition and Health-support for programs and projects that address long-term solutions to hunger, increase access to health education and/or basic health care in developing and emerging countries, and improve youth nutrition and wellness; Education-support for innovative programs that improve academic achievement, develop logic and thinking skills, promote leadership development, and/or increase access to education for socio-economically disadvantaged children. Cargill also supports mutually beneficial partnerships with selected higher education institutions; Environment-support for projects that protect and improve accessibility to water resources; promote biodiversity conservation in agricultural areas; and educate children about conservation and/or proper sanitation. Application and additional guidelines are available at: http://www.cargill.com/wcm/groups/public/@ccom/documents/document/doc-giving-funding-app.pdf
Requirements: Applicants must have 501(c)3 status or the equivalent; and they must be located in communities where Cargill has a business presence. Only under special circumstances will Cargill consider general operating or capital support. Organizations requesting capital or operating support should contact the Cargill Citizenship Fund staff before applying.
Restrictions: Cargil will not fund: organizations without 501(c)3 status or the equivalent; organizations that do not serve communities where Cargill has a business presence; individuals or groups seeking support for research, planning, personal needs or travel; public service or political campaigns; lobbying, political or fraternal activities; benefit dinners or tickets to the same; fundraising campaigns, walk-a-thons, or promotions to eliminate or control; specific diseases; athletic scholarships; advertising or event sponsorships; religious groups for religious purposes; publications, audio-visual productions or special broadcasts; endowments; medical equipment.
Geographic Focus: All States, Albania, Algeria, Andorra, Angola, Armenia, Austria, Azerbaijan, Belarus, Belgium, Benin, Bosnia & Herzegovina, Botswana, Bulgaria, Burkina Faso, Burundi, Cameroon, Cape Verde, Central African Republic, Chad, Comoros, Congo, Congo, Democratic Republic of, Cote d' Ivoire (Ivory Coast), Croatia, Cyprus, Czech Republic, Denmark, Djibouti, Egypt, Equatorial Guinea, Eritrea, Estonia, Ethiopia, Finland, France, Gabon, Gambia, Georgia, Germany, Ghana, Greece, Guinea, Guinea-Bissau, Hungary, Iceland, Ireland, Italy, Kenya, Kosovo, Latvia, Lesotho, Liberia, Libya, Liechtenstein, Lithuania, Luxembourg, Macedonia, Madagascar, Malawi, Mali, Malta, Mauritania, Mauritius, Moldova, Monaco, Montenegro, Morocco, Mozambique, Namibia, Niger, Nigeria, Norway, Poland, Portugal, Romania, Russia, Rwanda, San Marino, Sao Tome & Principe, Senegal, Serbia, Seychelles, Sierra Leone, Slovakia, Slovenia, Somalia, South Africa, Spain, Sudan, Swaziland, Sweden, Switzerland, The Netherlands, Turkey, Ukraine, United Kingdom, Vatican City
Contact: Stacey Smida; (952) 742-4311; stacey_smida@cargill.com
Internet: http://www.cargill.com/wcm/groups/public/@ccom/documents/document/doc-giving-funding-app.pdf
Sponsor: Cargill Corporation
P.O. Box 5650
Minneapolis, MN 55440-5650

Caring Foundation Grants 1019

Blue Cross and Blue Shield of Alabama, Incorporated sponsors The Caring Foundation. The Foundation was established in 1990 in Alabama and focuses it's support primarily in the region. The foundation supports organizations involved with education, federated giving programs, health, hospitals, safety education, children and youth services. The Foundation offers support of a general/operating nature as well, as program development.
Requirements: There is no formal application form, submit a proposal containing the following: name, address and phone number of organization; brief history of organization and description of its mission; detailed description of project and amount of funding requested.
Restrictions: Grants aren't available to individuals, or for capital campaigns.
Geographic Focus: Alabama
Amount of Grant: 10,000 - 900,000 USD
Contact: James M. Brown, Senior Vice President; (205) 220-2500
Sponsor: Caring Foundation
450 Riverchase Parkway, East
Birmingham, AL 35244-2858

Carla J. Funk Governmental Relations Award 1020

The Carla J. Funk Governmental Relations Award was established in 2008 through a contribution from Kent Smith, Fellow of the Medical Library Association, to recognize a medical librarian who has demonstrated outstanding leadership in the area of governmental relations at the federal, state, or local level, and who has furthered the goal of providing quality information for improved health. The award is named in honor of Carla J. Funk, Executive Director of the Medical Library Association, who has worked diligently and passionately to provide visibility for the association and profession and to further the association's government-relations agenda. The recipient receives a certificate at the annual meeting and a cash award of $500 after the annual meeting. The recipient assumes all costs of attending the meeting and the ceremony at which the presentation is made. Guidelines and a nomination form are available at the website. All materials and letters of support must be received by MLA via email, fax, or mail (in order of preference) by November 1. The recipient will be notified in March before the annual meeting. Founded in 1898, the Medical Library Association (MLA) is a nonprofit, educational organization of more than 1,100 institutions and 3,600 individual members in the health sciences information field committed to educating health-information professionals, supporting health-information research, promoting access to the world's health-sciences information, and working to ensure that the best health information is available to all.
Requirements: The nominee must have been a member of MLA at the time the governmental-relations activities occurred. Nominations are accepted from the membership at large; self nominations are encouraged and accepted. The following types of contributions eligible for nomination are as follows: developing information policies affecting existing or pending legislation that are submitted or presented to local, state, or federal governmental bodies; organizing effective campaigns to increase awareness of the views of the medical-library community and the association on legislative and information policies; testifying before governmental bodies and/or meeting with policymakers to articulate important concerns facing the health-information community that affect the provision of quality health information to healthcare providers and the public; and developing new approaches to strengthen the association's governmental-relations network and its voice in the governmental decision-making process.
Geographic Focus: All States
Date(s) Application is Due: Nov 1
Amount of Grant: 500 USD
Contact: Carla Funk, Executive Director; (312) 419-9094, ext. 14; fax (312) 419-8950; mlapd2@mlahq.org or awards@mlahq.org
Internet: http://www.mlanet.org/awards/honors/index.html
Sponsor: Medical Library Association
65 East Wacker Place, Suite 1900
Chicago, IL 60601-7246

Carl and Eloise Pohlad Family Foundation Grants 1021

The mission of the foundation is to improve the lives of economically disadvantaged children and youth and participate in projects that positively impact the quality of life in the Minneapolis/St.Paul area. The foundation awards grants to Minnesota nonprofits in its areas of interest, including arts and culture, economic development, education, environment, health care, housing, and social services. Types of support include general operating support, continuing support, capital campaigns, building construction/renovation, endowments, emergency funds, scholarship funds, and research.
Requirements: Minnesota nonprofits are eligible.
Restrictions: Individuals are not eligible. Capital request are considered only for physical plant improvements or significant technology investments. Capital requests for housing construction, endowment, program start-up or expansion or to establish operating reserves are not considered.
Geographic Focus: Minnesota
Amount of Grant: Up to 32,000,000 USD
Samples: Abbott Northwestern Hospital (Minneapolis, MN)--to construct a cardiology hospital, $3 million.
Contact: Josette Elstad; (612) 661-3910; fax (612) 661-3715; info@pohladfamilygiving.org
Internet: http://www.pohladfamilyfoundation.org/pff/pff_default.aspx
Sponsor: Carl and Eloise Pohlad Family Foundation
60 South Sixth Street, Suite 3900
Minneapolis, MN 55402

Carl B. and Florence E. King Foundation Grants 1022

The lives of Carl B. and Florence E. King were marked by warmth, compassion, and generosity. As they prospered, they believed in giving back. With gracious benevolence, they dedicated themselves to the betterment of individuals, communities, and society through informed giving. Today, the Carl B. and Florence E. King Foundation honors their memory, continues their tradition, and builds upon their vision. The Foundation is principally interested in the following areas: aging population; arts, culture, and history; children and youth; education; indigent; and to build non-profit capacity. The Foundation awards grants twice each year. Applicants must first submit a letter of inquiry, and then a full grant proposal only upon invitation.
Requirements: The King Foundation distributes grants only to entities that serve residents of Arkansas and Texas. Within Texas, the Foundation is principally interested in the Dallas-Fort Worth area and West Texas. Within Arkansas, the foundation focuses on the southern and eastern portions of the state. Applicants must also have a letter of determination from the Internal Revenue Service acknowledging tax-exempt status as described in Section 501(c)3 of the Internal Revenue Code.
Restrictions: The Foundation does not award grants: to individuals; to organizations or programs that do not serve residents of our geographic focus areas in Texas or Arkansas; to organizations that are not tax exempt; for general operating support, annual fund drives, or funds to offset operating losses (including retiring debt incurred to cover operating losses); to create endowments; toward balls, events, or galas benefiting charitable organizations; to efforts to treat or cure a single disease or condition; to church or seminary construction, or religious programs (other than social service-based initiatives); toward the cost of hosting or attending professional conferences or symposia, or participating in amateur sports competitions or similar activities.
Geographic Focus: Arkansas, Texas
Date(s) Application is Due: Jun 15; Dec 15
Amount of Grant: Up to 100,000 USD
Contact: Michelle D. Monse, President; (214) 750-1884; fax (214) 750-1651; michellemonse@kingfoundation.com
Internet: http://www.kingfoundation.com/Grants/King-Foundation-Grants.aspx
Sponsor: Carl B. and Florence E. King Foundation
2929 Carlisle Street, Suite 222
Dallas, TX 75204

Carl C. Icahn Foundation Grants 1023

The foundation awards grants to New York and New Jersey nonprofits in the areas of education, arts and culture, health care, child welfare, and Jewish temples and organizations. Types of support include general operating support, annual campaigns, building construction/renovation, and matching funds. There are no application deadlines or forms.
Requirements: New York and New Jersey nonprofits are eligible to apply.
Restrictions: No grants are provided to individuals.
Geographic Focus: New Jersey, New York
Amount of Grant: 500 - 1,600,000 USD
Samples: Randall's Island Sports Foundation, New York, NY, $1,666,000--to support program/general purposes; Wildlife Conservation Society, Pelham, NY, $10,000--to support program/general purposes; The Ladies' Village Improvement Society, East Hampton, NY, $500--to support program/general purposes.
Contact: Gail Golden-Icahn, Vice Presicent; (212) 702-4300; fax (212) 750-5815
Sponsor: Carl C. Icahn Foundation
767 5th Avenue, 47th Floor
New York, NY 10153-0023

Carl Gellert and Celia Berta Gellert Foundation Grants 1024

The foundation funds religious, charitable, scientific, literary or educational purposes restricted in the nine counties of the greater San Francisco Bay Area (Alameda, Contra Costa, Marin, Napa, San Francisco, San Mateo, Santa Clara, Solano and Sonoma). No grants are made to individuals. Types of support include general operations, annual and capital campaigns, building construction/ renovation, equipment acquisition, debt reduction, program/project development, medical research, publication, and scholarships.
Requirements: California 501(c)3 tax-exempt nonprofit organizations that are not private foundations are eligible.
Restrictions: Grants are not awarded to individuals.
Geographic Focus: California
Date(s) Application is Due: Aug 15
Amount of Grant: 1,000 - 10,000 USD
Samples: U of San Francisco, School of Business and Management (CA)--to renovate and add a new wing to the McLaren Center, home of the School of Business and Management, $500,000.
Contact: Jack Fitzpatrick, Executive Director; (415) 255-2829
Internet: http://home.earthlink.net/~cgcbg
Sponsor: Carl Gellert and Celia Berta Gellert Foundation
1169 Market Street, Suite 808
San Francisco, CA 94103

Carlisle Foundation Grants 1025

The foundation prefers to support proposals that are new, innovative, and/or demonstrate promise as models that might be replicated in other sites. In many instances, grants act as venture capital or seed money through which the applicant organization can demonstrate a new concept or expand its own capacity to deliver services. Although the foundation reviews a wide range of proposals, several areas receive high priority, including substance abuse; domestic and community violence; homelessness/housing; economic development; and services for children, youth, and families. The foundation prefers to make grants for restricted project support rather than general operating support. Applicant organizations should submit a one-page concept paper accompanied by a copy of the 501(c)3 determination letter. There is no time line or deadline for concept letters.
Requirements: Connecticut, Vermont, Rhode Island, Massachusetts, New Hampshire and Maine C501(c)3 organizations are eligible. Applicant organizations should submit a one-page concept paper accompanied by a copy of the 501(c)3 determination letter.
Restrictions: Requests for capital support are rarely considered; requests for support of endowments are never considered. The foundation does not usually support programs with a primary focus on day care, disabilities, education, health services, legal services, mental retardation, or older adults. The foundation also does not provide direct support for individuals.
Geographic Focus: Connecticut, Maine, Massachusetts, New Hampshire, Rhode Island, Vermont
Samples: South County Habitat for Humanity, Shannon, RI, $25,000 - to provide start up capital for the development of Re-Store, a retail outlet designed to distribute surplus building supplies at low cost; Wayside Youth & Family Network, Framingham, MA, $50,000 - community match to a major Robert Wood Johnson grant to develop a comprehensive center, Tempo Young Adult Resource Center, for disenfranchised and underserved youth and young adults; Communities for Restorative Justice, Concord, MA, $7,500 - to help sustain this Juvenile Justice program that provides a mediated settlement to a range of non-violent crimes and that allows the perpetrator to avoid the criminal justice system;
Contact: Richard Goldblatt, Executive Director; (401) 284-0368; fax (401) 284-0390; rag@carlislefoundation.org
Internet: http://www.carlislefoundation.org/annualgrants.htm
Sponsor: Carlisle Foundation
P.O. Box 5549
Wakefield, RI 02880-5549

Carl M. Freeman Foundation FACES Grants 1026

Founded in 2000 in Delaware, FACES stands for Freeman Assists Communities with Extra Support. The FACES program is designed to find and fund the smaller, overlooked projects in it's neighborhoods. The grants are limited to Montgomery County nonprofit organizations with operating budgets of $750,000 or less and Sussex County nonprofit organizations with operating budgets of $500,000 or less. Funding applications are available in five areas of interest: arts/culture; education/environment; health & human services; housing and; other-this may include anything you feel does not fit in the above categories, for example spaying cats/dogs. Additional guidelines and applications are available at: http://www.freemanfoundation.org/carl/CarlMFreemanFoundation/Grants/GrantGuidelines/ApplyForAGrant/tabid/185/Default.aspx.
Requirements: 501(c)3 tax-exempt organizations in Montgomery & Sussex County are eligible. Nonsectarian religious programs also are eligible.
Restrictions: Grants will not be distributed to: individuals; political associations or candidates; organizations that would disperse the funding to others; organizations that discriminate by race, creed, gender, sexual orientation, age, religion, disability or national origin.
Geographic Focus: Delaware, Maryland, West Virginia
Samples: Charles Town Health Right, Inc., Charles Town, WV--to help purchase needed medical and support supplies in order to meet daily medical needs; Olney Community Band, Silver Spring, MD--to support the 2009 Concert Series and help present affordable concerts to diverse crowds; Sussex Technical Adult Division, Georgetown, DE--to support the Even Start Family Literacy Program, which helps improve the literacy skills of low income and low literate families.
Contact: Trish Schechtman, (302) 436-3555; trish@freemanfoundation.org
Internet: http://www.freemanfoundation.org/carl/CarlMFreemanFoundation/Grants/FACES/tabid/204/Default.aspx
Sponsor: Carl M. Freeman Foundation
36097 Sand Cove Road
Selbyville, DE 19975

Carl M. Freeman Foundation Grants 1027

The Carl M. Freeman Foundation has historically emphasized it's support in the following communities: Montgomery County, Maryland; Sussex County, Delaware; and the Eastern Panhandle of West Virginia . Funding is available for a wide variety of community organizations, having supported everything from arts organizations and hunger centers to educational and health related organizations. To simplify the application process, funding applications are available in five areas of interest: arts/culture; education/environment; health & human services; housing and; other-this may include anything you feel does not fit in the above categories, for example spaying cats/dogs. Additional guidelines and applications are available at: http://www.freemanfoundation.org/carl/CarlMFreemanFoundation/Grants/GrantGuidelines/ApplyForAGrant/tabid/185/Default.aspx
Requirements: 501(c)3 tax-exempt organizations in Maryland, Delaware and West Virginia are eligible. Nonsectarian religious programs also are eligible.
Restrictions: Grants will not be distributed to: individuals; political associations or candidates; organizations that would disperse the funding to others; organizations that discriminate by race, creed, gender, sexual orientation, age, religion, disability or national origin.
Geographic Focus: Delaware, Maryland, West Virginia
Amount of Grant: 5,000 - 30,000 USD
Contact: Trish Schechtman; (302) 436-3555; trish@freemanfoundation.org
Internet: http://www.freemanfoundation.org/carl/CarlMFreemanFoundation/Grants/GrantGuidelines/tabid/181/Default.aspx
Sponsor: Carl M. Freeman Foundation
36097 Sand Cove Road
Selbyville, DE 19975

GRANT PROGRAMS | 155

Carl R. Hendrickson Family Foundation Grants 1028
The Carl R. Hendrickson Family Foundation was established in 1991 to support and promote quality education, human-services, and health-care programming for under-served populations. Carl R. Hendrickson was a Chicago entrepreneur who, along with his father and brothers, built the Hendrickson Trucking Company. Carl and his wife, Agnes, had one child, Virginia, who followed in her father's footsteps by leading the family business and by serving as President of the Hendrickson Foundation. Virginia died in 1995, leaving no heirs. The Hendricksons prided themselves on their entrepreneurial spirit, having been in the forefront of the trucking business by inventing the tandem truck. Reflecting the Hendrickson family's strong Christian faith, special consideration is given to charitable organizations that help individuals meet their basic needs while also addressing their spiritual needs. Preference is given to organizations or programs that approach their mission from an entrepreneurial perspective. The majority of grants from the Family Foundation are one year in duration. Application guidelines as well as a link to the downloadable application are given at the grant website. Applicants are also encouraged to review the Illinois state application guidelines and the foundation's funding history before applying. The deadline for application is July 31. Grant decisions will be made by November 1. Most recent awards have ranged from $5,000 to $25,000.
Requirements: Applicants must have 501(c)3 tax-exempt status. Applications must be mailed.
Restrictions: In general, grant requests for individuals, endowment campaigns or capital projects will not be considered. The foundation does not support requests from individuals, organizations attempting to influence policy through direct lobbying, or any political campaigns.
Geographic Focus: Illinois
Date(s) Application is Due: Jul 31
Amount of Grant: 5,000 - 25,000 USD
Contact: Debra L. Grand; (312) 828-2055; ilgrantmaking@bankofamerica.com
Internet: https://www.bankofamerica.com/philanthropic/foundation.go?fnId=109
Sponsor: Carl R. Hendrickson Family Foundation
231 South LaSalle Street, IL1-231-13-32
Chicago, IL 60604

Carlsbad Charitable Foundation Grants 1029
The mission of the Carlsbad Charitable Foundation is to advance philanthropy in its service region in order to: build community excellence; stimulate innovation; and enhance capacity of nonprofits. Its primary purpose is to: meet emerging needs by encouraging and increasing responsible and effective philanthropy by and for the benefit of all who live, work and play in Carlsbad; build a Carlsbad community endowment; provide funding annually to Carlsbad organizations and causes; and give Carlsbad community members a vehicle for legacy planning and gifts that will benefit Carlsbad now and forever. The field of grant giving focus changes annually, so applicants should contact the office directly for specific details. Most recently, the Foundation was interested in funding Senior Health and Human Services programs managed by non-profit or government agencies.
Requirements: Any 501(c)3 organization serving the residents of Carlsbad, California, are eligible to apply. Projects must specifically target seniors over the age of 55 who live in the City of Carlsbad.
Restrictions: The Carlsbad Charitable Foundation does not make grants for: annual campaigns and fund raising events for non-specific purposes; capital campaigns for buildings or facilities; stipends for attendance at conferences; endowments or chairs; for-profit organizations or enterprises; individuals unaffiliated with a qualified fiscal sponsor; projects that promote religious or political doctrine; research projects (medical or otherwise); scholarships; or existing obligations or debt.
Geographic Focus: California
Date(s) Application is Due: Jan 12
Contact: Trudy Amstrong; (619) 814-1384; trudy@sdfoundation.org
Internet: http://www.endowcarlsbad.org/grants.html
Sponsor: Carlsbad Charitable Foundation
2508 Historic Decatur Road, Suite 200
San Diego, CA 92106

Carls Foundation Grants 1030
The foundation has broadly defined charitable purposes, but the principal purpose and mission of the foundation is as follows: children's welfare (primarily in Michigan), including health care facilities and programs, with special emphasis on the prevention and treatment of hearing impairment; recreational, educational, and welfare programs especially for children who are disadvantaged for economic and/or health reasons; and preservation of natural areas, open space, and historic buildings and areas having special natural beauty or significance in maintaining America's heritage and historic ideals, through assistance to land trusts and land conservancies and directly related environmental educational programs. Types of grants include capital grants, limited budget support, start up/seed money, and multi-year grants. The foundation has no formal application for grant requests. The Trustees meet minimally three times per year. Requests are accepted at all times, but organizations are encouraged to submit requests well in advance of scheduled board meetings, in January, May, and September. A letter of inquiry is not required and phone calls are welcome. Guidelines are available online.
Requirements: 501(c)3 tax-exempt organizations are eligible.
Restrictions: Grants are not awarded to individuals or for endowments, publications, conferences, seminars, film, fellowships, educational loans, travel, research, playground structures or athletic facilities, or underwriting special events.
Geographic Focus: Michigan
Date(s) Application is Due: Mar 1; Jul 1; Nov 1
Amount of Grant: 5,000 - 50,000 USD
Contact: Kathy Stenman, Program Officer; (313) 965-0990; fax (313) 965-0547
Internet: http://www.carlsfdn.org
Sponsor: Carls Foundation
333 W Fort Street, Suite 1940
Detroit, MI 48226

Carl W. and Carrie Mae Joslyn Trust Grants 1031
Grants support activities providing services to resident children, elderly, and the disabled in El Paso County, Colorado. Areas of interest include education, medical care, rehabilitation, children and youth services, and aging centers and services. Types of support include general operating support, annual campaigns, building construction and renovation, equipment acquisition, endowment funds, and program development. Application must be in writing and must specifically describe the use of the funds. Grants are not sustaining and new applications must be submitted semiannually for renewal.
Requirements: Nonprofit organizations located in, or serving the residents of, El Paso County, Colorado, are eligible.
Restrictions: Grants are not made to individuals or for research, scholarships, fellowships, loans, or matching gifts.
Geographic Focus: Colorado
Date(s) Application is Due: Apr 30; Oct 31
Amount of Grant: 500 - 15,000 USD
Samples: Pikes Peak Hospice, Colorado Springs, Colorado, $5,000; Silver Key Senior Services, Colorado Springs, Colorado, $5,000; Saint Marys High School, Colorado Springs, Colorado, $3,000.
Contact: Susan Bradt Laabs; (719) 227-6435 or (719) 227-6439; fax (719) 2276448
Sponsor: Carl W. and Carrie Mae Joslyn Charitable Trust
Trust Department, P.O. Box 1699
Colorado Springs, CO 80942

Carnahan-Jackson Foundation Grants 1032
The foundation's areas of interest include higher and other education, libraries, hospitals, youth, the disabled, drug abuse programs, ecology, housing, community development, dance and other performing arts groups, and churches. Types of support include general operating support, continuing support, capital campaigns, building construction/renovation, equipment acquisition, programs/projects, seed grants, curriculum development, scholarship funds, and matching funds.
Requirements: IRS 501(c)3 organizations serving western New York, particularly Chautauqua County, are eligible.
Geographic Focus: New York
Amount of Grant: 2,500 - 125,000 USD
Contact: Stephen E. Sellstrom, (716) 483-1015
Sponsor: Carnahan-Jackson Foundation
13 East 4th Street, P.O. Box 3326
Jamestown, NY 14701-3326

Carnegie Corporation of New York Grants 1033
The Carnegie Corporation of New York provides research, study, and support for projects to improve government at all levels, to increase public understanding of social policy issues, to equalize opportunities for minorities and women, and to increase participation in political and civic life. Also supported are projects that promote electoral reform; education reform from early childhood through higher education; early childhood development; and urban school reform. The foundation will also fund research on the increasing availability and success of after-school and extended service programs for children and teenagers, particularly those in urban areas, that promote high academic achievement. Dissemination of best practices in teacher education will also be emphasized. There is no formal procedure for submitting a proposal. To apply under any of the corporation's grantmaking programs, applicants should submit a full proposal that describes the project's aims, duration, methods, amount of financial support required, and key personnel. The board meets four times a year, in October, February, April, and June.
Requirements: Only full proposals that have been invited for submission will be considered. After a letter of inquiry has been reviewed, applicants may be invited via email to submit a full proposal.
Restrictions: Grants are not made for construction or maintenance of facilities or endowments. The Corporation does not generally make grants to individuals except through the Carnegie Scholars Program, that supports the work of select scholars and experts conducting research in the foundation's fields of interest.
Geographic Focus: All States
Amount of Grant: Up to 4,000,000 USD
Samples: Massachusetts Institute of Technology, Cambridge, Massachusetts, $1,000,000 - international peace and security (2015); Center for Better Schools, Portsmouth, Rhode Island, $650,000 - strengthening teaching and human capital (2015); Citizen Schools, Boston, Massachusetts, $500,000 - strengthening education (2015).
Contact: Nicole Howe Buggs; (212) 371-3200; externalaffairs@carnegie.org
Internet: https://www.carnegie.org/grants/grantseekers/
Sponsor: Carnegie Corporation of New York
437 Madison Avenue
New York, NY 10022

Carolyn Foundation Grants 1034
Priorities for funding include community and environmental grantmaking. In the community focus area needs are addressed only in the communities of interest to the foundation: Minneapolis, Minnesota and, New Haven, Connecticut. There are two community focus areas for funding: economically disadvantaged children and youth; and community and cultural vitality. The foundation works to empower economically

disadvantaged children and youth by supporting their families and others to inspire, nurture, educate and guide them to achieve long-term stability and well-being. In the environmental focus area, the Carolyn Foundation environmental committee is currently most interested in funding renewable energy programs. The Foundation will consider other environmental proposals if funds allow. All proposals submitted must: address root causes and create systemic and sustainable solutions and change; address global issues with local interventions that address local needs, as well as global needs; develop and implement solutions that can be replicated in other areas; collaborate effectively with others in the community: government, non-government, foundations and private parties. The Carolyn Foundation makes grants twice a year, in June and January. Applications must be submitted by January 15 for June grants, and July 15 for January grants. This is a postmark deadline. Grant applications will be reviewed by the Executive Director and a committee of foundation volunteers. Declinations will be sent at the time a decision is made to no longer consider a proposal, typically before the end of the review cycle. Successful applicants will be notified in June and January.
Requirements: IRS 501(c)3 nonprofit organizations in Minnesota and Connecticut may apply for the environmental grants. The Community grants program is limited to the cities of Minneapolis, Minnesota and New Haven, Connecticut. The foundation encourages use of Carolyn Foundation Application Form adapted from the Minnesota Common Grant Application Form. Applicants choosing not to use the common grant form must address the same information as required by the common grant. All proposals must use the Carolyn Foundation Cover Sheet, available at the Foundation's website. It is request that summary information and description of the project be no longer than six pages, printed on one side on 8 1/2 x 11-inch paper. Supporting documents, such as financial information, list of Officers, Directors, and Executive Staff, IRS determination letter may be in addition to the six pages. Do not send bound proposals, cassettes or VCR tapes.
Restrictions: Grants are not awarded to individuals, political organizations or candidates, veterans organizations, fraternal societies or orders, annual fund drives, umbrella organizations, or to deficits already incurred. The foundation does not generally make grants to religious organizations for religious purposes or to organizations in support of operations carried on in foreign countries.
Geographic Focus: Connecticut, Minnesota
Date(s) Application is Due: Jan 15; Jul 15
Amount of Grant: 5,000 - 50,000 USD
Samples: Clean Water Fund, Minneapolis, MN, $25,000--for development and use of safer alternatives for toxic substances; Indian Child Welfare Law Center, Minneapolis, MN, $15,000--for Indian Children's Stability Program; Hiawatha Leadership Academy, Minneapolis, MN, $20,000--for increasing salaries of the teaching staff.
Contact: Becky Erdahl, Executive Director; (612) 596-3279 or (612) 596-3266; fax (612) 339-1951; berdahl@carolynfoundation.org
Internet: http://www.carolynfoundation.org/guidelines.html
Sponsor: Carolyn Foundation
706 2nd Avenue South, Suite 760
Minneapolis, MN 55402

Carpenter Foundation Grants 1035
The foundation's primary areas of interest include the arts, education, public interest, and human services. The foundation is deeply concerned with the well-being of children and families and their relationship to their neighborhoods and communities. Also of concern is the health of the web of agencies and organizations which serve them. Grants are awarded for general operating support, program development, capital campaigns, equipment acquisition, scholarship funds, seed money, matching funds, and technical support. Deadlines are generally about six weeks before the quarterly board meetings, held in January, March, June, and September.
Requirements: Tax exempt agencies in the Jackson and Josephine Counties of Oregon may submit proposals.
Restrictions: Grants are not made to individuals. The foundation rarely makes grants for historical applications, hospital construction or equipment, group or individual trips, or activities for religious purposes.
Geographic Focus: Oregon
Amount of Grant: 250 - 25,000 USD
Samples: Community Health Center (Medford, OR)--fir a program to provide discounted medications to low-income/uninsured patients, $15,000; Science Works Hands-On Museum, (Ashland, OR)--in support of a life science hands-on exhibition for local schools and students, $6,000; Southern Oregon State College (Ashland, OR)--for faculty development opportunities which improve teaching, $25,000; Oregon Water Trust, (Portland, OR)--for a community outreach and education project to locally distribute the General Elections Voter's Guide, $12,000.
Contact: Polly Williams; (541) 772-5732; fax (541) 773-3970; carpfdn@internetcds.com
Internet: http://www.carpenter-foundation.org
Sponsor: Carpenter Foundation
711 E Main Street, Suite 10
Medford, OR 97504

Carrie E. and Lena V. Glenn Foundation Grants 1036
Established in 1971, the Glenn Foundation provides annual grants in the following areas of interest: arts; children/youth, services; Christian agencies & churches; Elementary/secondary education; Environment; and Human services.
Requirements: Federally tax-exempt institutions and not-for-profit agencies that serve Gaston County, NC. agencies or out-of-county agencies whose projects have an impact on Gaston County citizens are eligible to apply.
Restrictions: Funding is not available for: planning grants; grants to individuals; scholarships; capital campaigns; umbrella campaigns; and multi-year grants.
Geographic Focus: North Carolina
Date(s) Application is Due: Mar 1
Amount of Grant: 3,000 - 25,000 USD
Samples: Community Foundation of Gaston County, Gastonia, NC., $25,000 - for purchase and installment of This Little Light of Mine, a painting by John Biggers, recreated as mosaic time mural; Gastonia Potters House, Lowell, NC., $10,000 - for operating expenses for residential drug rehabiliation program for women; Alliance for Children and Youth, Gastonia, NC., $8,885 - for Strengthening Families program for parents and their children;
Contact: Barbara H. Voorhees, Executive Director; (704) 867-0296; fax (704) 867-4496; glennfnd@bellsouth.net
Sponsor: Carrie E. and Lena V. Glenn Foundation
1552 Union Road, Suite D
Gastonia, NC 28054

Carrie Estelle Doheny Foundation Grants 1037
The foundation primarily funds local, not-for profit organizations endeavoring to advance education, medicine and religion, to improve the health and welfare of the sick, aged, incapacitated, and to aid the needy. Educational funding includes support of inner city Catholic schools, and scholarship funds for Catholic high schools and universities. Adult education programs and religious education are also supported. Medical funding is focused in two areas: research and care of the disadvantaged. Religious funding is directed to support the gospel values as expressed in the Roman Catholic faith. Health and welfare funding is directed to organizations who assist individuals to lead independent satisfying lives. Specific areas of interest include adoption and foster care service groups, programs for the disabled, health education programs, and senior programs. Aiding the needy funding includes inner city youth clubs, summer camps, and food banks. Applications accepted anytime, an application form is required for submission and may be downloaded from the foundation web site. The board meets on the last Friday of each month, except for the month of September. Requests should be submitted approximately 6 weeks in advance to allow sufficient time for processing and for distribution to the Board members prior to the meeting. Allow 2 or 3 months for notification of the Board's decision. This is done in writing following the Board of Directors Meeting each month.
Requirements: The Foundation limits its grants to programs located within the fifty states and certified as 501(c)3 non-profit public charities by the Internal Revenue Service. The vast majority of funding is done in the Greater Los Angeles area.
Restrictions: Grant requests are not considered from individuals or from tax-supported entities. Areas also excluded from consideration include support for individuals, endowment funds, publishing books, television or radio programs, travel funds, advertisement, scholarships, or political purposes in any form.
Geographic Focus: All States
Amount of Grant: 5,000 - 150,000 USD
Contact: Shirley Bernard, Senior Grants Administrator; (213) 488-1122; fax (213) 488-1544; doheny@dohenyfoundation.org
Internet: http://www.dohenyfoundation.org/grant/grant.htm
Sponsor: Carrie Estelle Doheny Foundation
707 Wilshire Boulevard, Suite 4960
Los Angeles, CA 90017

Carrier Corporation Contributions Grants 1038
Carrier donates approximately $2 million around the world to registered nonprofit organizations. In the United States, Carrier funds only qualified 501(c)3 organizations that meet its eligibility criteria and operate in locations where the company has a significant employee base. Carrier believes in helping people in the communities where they live, work and do business. To better serve those communities and to better align its corporate contributions with mission and values, it focuses giving on the following these areas: environment and sustainability; civic & community; education; arts & culture; health & human services. All U.S. non-profits are required to complete an online grant application. Applications are accepted from March 1 through June 1 of each year, and are reviewed for funding to be paid the following year. Applicants will receive notification in the first quarter of the calendar year in which funding will occur.
Requirements: Carrier funds only qualified 501(c)3 organizations that meet eligibility criteria and operate in locations where it has a significant employee base.
Restrictions: Carrier will not fund: individuals; religious organizations; alumni groups, sororities or fraternities; booster clubs; political groups; any organization determined by Carrier to have a conflict of interest; any organization whose practices are inconsistent with the company's Code of Ethics
Geographic Focus: Alabama, Arizona, Connecticut, Georgia, Illinois, Indiana, Michigan, Nevada, New York, North Carolina, South Carolina, Tennessee, Texas
Date(s) Application is Due: Jun 1
Contact: Rajan Goel, Vice President; (860) 674-3420; fax (860) 622-0488
Internet: http://www.corp.carrier.com/vgn-ext-templating/v/index.jsp?vgnextoid=6afa80757d7e7010VgnVCM100000cb890b80RCRD
Sponsor: Carrier Corporation
One Carrier Place
Farmington, CT 06034-4015

Carroll County Community Foundation Grants 1039
The Carroll County Community Foundation funds initiatives that improve the quality of life for citizens of Carroll County, Indiana. The Foundation's grant program emphasizes change-oriented issues, with the following areas of interest: health and medical; social services; education; cultural affairs; civic affairs; and community beautification. Proposals: must strive to anticipate the changing needs of the community and be flexible

in responding to them; must be change-oriented and problem-solving in nature with emphasis on "seed" money or pilot project support rather than for ongoing general operating support; support innovative efforts and projects that offer far-reaching gains and widespread community results; may coordinate with other funders and donors where possible, including using matching or challenge grant techniques; closely relate and coordinate with the programs of other sources for funding such as the government, other foundations, and associations; and achieve certain objectives such as become more efficient, increase fund-raising capabilities, and deliver better products. Grants will be made only to organizations: whose programs benefit the residents of the county, with preference given to those projects with high visibility in the community; which provide for a responsible fiscal agent and adequate accounting procedures, with preference given to those projects that generate revenue and/or have plans that sustain the project.
Requirements: A letter of inquiry to the Foundation is required as a pre-qualification. The letter should contain a brief statement of the applicant's needs for assistance, estimate of total cost of the project, and enough factual information to enable the Foundation to determine whether or not the application falls within the guidelines of its grants program. After the organization has received a response to apply for a grant, the grantee will then fill out the online application and submit ten copies of the application and all attachments to the Foundation for approval. Organizations should refer to the Foundation website for further information.
Restrictions: The Foundation typically does not award grants for: normal operating expenses and/or salaries; individuals; seminars or trips except where there are special circumstances which will benefit the larger community; sectarian religious purposes but can be made to religious organizations for general community programs; endowment purposes of recipient organizations; projects which have been proposed by individuals or organizations responsible to advisory bodies or persons; new projects and/or equipment purchased prior to the grant application being approved.
Geographic Focus: Indiana
Date(s) Application is Due: Sep 6
Samples: Burlington Community Park, replacement and updating park playground equipment, $4,500; Book Readers and Horn Blowers, Delphi Elementary School, $1,000; Delphi Public Library, Delphi, IN, digital media lab report, $1,000.
Contact: Ron Harper; (765) 454-7298 or (800) 964-0508; ron@cfhoward.org
Internet: http://cfcarroll.org/newsite/grantprogram.shtml
Sponsor: Carroll County Community Foundation
215 West Sycamore Street
Kokomo, IN 46901

Carylon Foundation Grants 1040
The foundation awards general support grants to nonprofits of the Christian, interdenominational, Jewish, and Presbyterian faiths. Higher education institutions, health care organizations, international missions/ministries, medical centers, religious organizations, and temples receive support. Application may be made by submitting a brief letter describing the organization and program. There are no application deadlines.
Geographic Focus: All States
Amount of Grant: 50 - 50,000 USD
Samples: Rush-Presbyterian-Saint Lukes Medical Center (Chicago, IL)--for educational programs, $25,000; Holocaust Museum (Chicago, IL)--for operating support, $220,500; Weizmann Institute of Science (Palm Beach, FL)--for operating support, $10,000.
Contact: Marcie Mervis, Trustee; (312) 666-7700
Sponsor: Carylon Foundation
2500 W Arthington
Chicago, IL 60612-4108

Cass County Community Foundation Grants 1041
The Cass County Community Foundation (CCCF) assists donors in building enduring sources of charitable assets to promote education, enhance humanity, and advance community development throughout Cass County. The Foundation has the following areas of interest: education; human services; and community development. All applications are reviewed by a committee comprised of CCCF Board members and other volunteers from the community. Non-profit organizations whose projects directly impact the lives of Cass County residents are eligible to apply.
Requirements: The grant application is available online and is also available at the Foundation office. Grant seekers should include their organizational, financial, and project information, and submit seven copies of the application packet to the Foundation. Applicants are encouraged to participate in the free grant writing workshop available each spring.
Restrictions: Public schools, while non-profit, are not 501(c)3 and therefore are not eligible. The Foundation will not consider grants for: existing obligations; services supported by tax dollars; individuals or travel expenses; repeat funding; on-going operating expenses; advocacy; religious purposes or affiliations; or loans or endowments.
Geographic Focus: Indiana
Date(s) Application is Due: Jul 1
Amount of Grant: 5,000 - 14,000 USD
Samples: Galveston, IN, fire department equipment modernization; $5,000; Royal Center, IN, volunteer fire department emergency generator; $5,000; Salvation Army, security and alarm system, ADA compliant restrooms for new building; $14,250.
Contact: Deanna Crispen, Executive Director; (574) 722-2200; fax (574) 753-7501; dcrispen@casscountycf.org or info@casscountycf.org
Internet: http://casscountycf.org/page/Competitive-Grants-Cycle-id-24
Sponsor: Cass County Community Foundation
417 North Street, Suite 102, P.O. Box 441
Logansport, IN 46947

Catherine Holmes Wilkins Foundation Grants 1042
The foundation awards grants in Washington's Puget Sound region to nonprofit organizations in its areas of interest. Funding priorities include medical research and education (medical and academic centers conducting research and training in areas such as cancer, heart disease, and mental illness); physically handicapped and mentally ill (community nonprofit agencies providing direct social services to people with physical disabilities or mental illness); and services for the needy (community-based programs providing immediate support to the needy, with particular emphasis on services for abused women and children). Preference will be given to project support, rather than ongoing operating expenses. Proposals are accepted throughout the year. Grants are awarded quarterly.
Requirements: 501(c)3 nonprofit organizations that operate within or significantly affect the residents of the greater Seattle region (Tacoma to Everett, Seattle, and the Eastside) are eligible.
Restrictions: Grants are not made for multi-year projects, debt retirement, operational deficits, or to individuals or for scholarships.
Geographic Focus: Washington
Date(s) Application is Due: Mar 8; Oct 8
Amount of Grant: 3,000 - 10,000 USD
Contact: Nancy Atkinson, Vice President; (206) 781-3472; OgleFounds@aol.com
Internet: http://fdncenter.org/grantmaker/wilkins
Sponsor: Catherine Holmes Wilkins Foundation
P.O. Box 24565, WA1-501-33-23
Seattle, WA 98124

Catherine Kennedy Home Foundation Grants 1043
The Catherine Kennedy Home Foundation established in 2001, supports nonprofit organizations involved with Aging, centers/services; Christian agencies & churches; developmentally disabled, centers & services; family services; domestic violence; food distribution, meals on wheels; residential/custodial care, hospices; YM/YWCAs & YM/YWHAs. Giving primarily in Wilmington, North Carolina.
Requirements: Initial approach should be made through a letter to the Foundation requesting an application form. The Foundation will provide you with additional guidelines.
Restrictions: Funding not available outside of North Carolina.
Geographic Focus: North Carolina
Date(s) Application is Due: Feb 28
Amount of Grant: 1,000 - 30,000 USD
Contact: Garry A. Garris, Director; (910) 452-0611
Sponsor: Catherine Kennedy Home
P.O. Box 4782
Wilmington, NC 28406-1782

CCFF Chairman's Distinguished Life Sciences Award 1044
The $10,000 Chairmen's Distinguished Life Sciences Award will be presented to a scientist or researcher who is making or has recently made a significant and positive contribution toward the development of a cutting edge innovation in the field of life sciences. The Chairmen's Award will provide incentive for continuing the research. Individuals will compete and their accomplishments will be judged on how they exemplify excellence in life sciences. Winners must attend the awards ceremony in April at Washington, D.C., in order to receive the award.
Requirements: All Nominees must be living United States citizens. The Nomination must demonstrate a program of ongoing work with a specific outcome suggesting that an important achievement will result. The Nomination's proposed achievement must benefit human health in a unique and effective manner, demonstrating originality from other work in the same or related field(s). The proposed achievement must correlate to a tangible outcome to advance human health.
Restrictions: Nominees must be permitted, by their employer or any other relevant authority, to accept a monetary award bestowed by the Foundation, a Federal government agency. Anyone who cannot attend the award ceremony will automatically be disqualified.
Geographic Focus: All States
Date(s) Application is Due: Feb 26
Amount of Grant: 10,000 USD
Contact: Judith M. Shellenberger; (315) 258-0090; judithmscolumbus@cs.com
Internet: http://www.columbusfdn.org/lifesciences/
Sponsor: Christopher Columbus Fellowship Foundation
110 Genesee Street, Suite 390
Auburn, NY 13021

CCFF Life Sciences Student Awards 1045
Two $1,000 Life Sciences Student Awards will be presented to current secondary school students who are making or have recently made significant and positive contributions related to the study of Biology, Chemistry and other life sciences courses. Nominations are strongly encouraged from educators, faculty members and school administrators personally familiar with the student's academic achievements in biology.
Requirements: All Nominees must be living United States citizens. Nominees must demonstrate ongoing work and/or provide proven evidence that distinguishes them in the study of biology.
Geographic Focus: All States
Date(s) Application is Due: Feb 26
Amount of Grant: 1,000 USD
Contact: Judith M. Shellenberger; (315) 258-0090; judithmscolumbus@cs.com
Internet: http://www.columbusfdn.org/lifesciences/
Sponsor: Christopher Columbus Fellowship Foundation
110 Genesee Street, Suite 390
Auburn, NY 13021

CCHD Community Development Grants 1046

CCHD is committed to supporting groups of low-income individuals as they work to break the cycle of poverty and improve their communities. By helping the poor to participate in the decisions and actions that affect their lives, CCHD empowers them to move beyond poverty. The organization's efforts should directly benefit a relatively large number of people rather than a few individuals. The organization should generate cooperation among and within diverse groups in the interest of a more integrated and mutually understanding society. An applicant organization seeking seed or matching monies will also be considered. (If requesting these monies, applicants should present positive documentation that other public and/or private sources will commit their funds to support the organization's efforts.)

Requirements: Only organizations that are not now receiving an organizing grant are required to submit an Eligibility Quiz. Eligibility Quizzes are accepted on a rolling basis between September 1st and November 1st. The sponsor recommends submitting your Eligibility Quiz well in advance of the November 1st deadline, to help with processing and to give eligible applicants more time with the next step in application. To be eligible for CCHD funds, an organization must satisfy ALL the following criteria and guidelines: the activity for which funding is requested must conform to the moral and social teachings of the Catholic Church; the applicant organization must demonstrate both the intention and capacity to effectively work toward the elimination of the root causes of poverty and to enact institutional change; the organization's efforts must benefit people living in poverty. At least 50 percent of those benefiting from the organization's efforts must be people experiencing poverty; people living in poverty must have the dominant voice in the organization (at least 50 percent of those who plan, implement and make policy, hire and fire staff should be persons who are involuntarily poor; the organization should demonstrate ongoing leadership development because it is considered essential to the strength, depth and sustainability of the organization; the organization should demonstrate a clear vision for development of financial capacity that might include membership dues, grassroots fundraising, foundation and/or corporate support; and the organization must be fully nonpartisan when engaging in political activities. Organizations engaged in partisan political activity are not eligible. See the website for further explanations.

Restrictions: The following general classifications do not meet CCHD criteria and/or guidelines for community organizing grants: organizations with primary focus on direct service (e.g., daycare centers, recreation programs, community centers, scholarships, subsidies, counseling programs, referral services, cultural enrichment programs, direct clinical services, emergency shelters and other services, refugee resettlement programs, etc.); advocacy efforts where only staff, a few individuals or middle to upper-income people are speaking for a particular low income constituency without the direct involvement and leadership of low income individuals; organizations controlled by governmental (federal, state, local), educational, or ecclesiastical bodies; research projects, surveys, planning and feasibility studies, etc; individually owned, for-profit businesses; or organizations that would use CCHD money for re-granting purposes or to fund other organizations.

Geographic Focus: All States

Date(s) Application is Due: Nov 1

Amount of Grant: 25,000 - 50,000 USD

Contact: Ralph McCloud, Director; (202) 541-3367 or (202) 541-3210; fax (202) 541-3329; rmccloud@usccb.org or cchdgrants@usccb.org

Internet: http://www.usccb.org/about/catholic-campaign-for-human-development/grants/community-development-grants-program/index.cfm

Sponsor: United States Conference of Catholic Bishops

3211 Fourth Street, NE

Washington, D.C. 20017-1194

CDC-Hubert Global Health Fellowship 1047

The CD.C.-Hubert fellowship provides an opportunity for third- and fourth-year medical and veterinary students to gain public health experience in an international setting. Hubert fellows spend six to twelve weeks in a developing country working on a priority health problem in conjunction with Center of Disease Control (CD.C.) agency staff. Most schools award course credit for the fellowship. Hubert fellows receive a $4,000 stipend to help pay for travel and living expenses while in-country (any cost above $4,000 is the responsibility of the fellow). Deadline dates may vary from year to year. Interested applicants are encouraged to check the website for current information on how and when to apply for the fellowship. Further information on field assignment choices can be found at the grant website. Click on "More about the Program" on the left side of the screen. Examples of past assignments are: "Assessing Barriers to Use of Rapid HIV Tests" (China); "Assessing Health Information Systems for Border Populations" (Thailand); "China National Center for AIDS/STD Prevention and Control" (China); "Descriptive Analysis of HIV/AIDS" (Nigeria); "Field Development and Testing of a Manual to Measure the Burden of Severe Influenza-Related Lower Respiratory Infection" (Thailand); "Pathogen Discovery Focused on Lyssaviruses and Other Emerging Zoonotic Pathogens" and "Knowledge, Attitude and Practice (KAP) Surveys and Risk Assessment within Population at Higher Risk of Exposure to Bats" (Democratic Republic of the Congo, Guatemala or Peru); "Causes of Community-Acquired Pneumonia" (rural Thailand); "Surveillance System for Acute Respiratory Infection" (Egypt); "Hospital Acquired Infections" (Egypt); "Influenza Sentinel Surveillance" (Kenya); "Population-Based Surveillance for Emerging Infectious Diseases" (Kibera Informal Housing Settlement in Nairobi, Kenya); "Population-Based Surveillance for Emerging Infectious Diseases" (rural Lwak Area of Kisumu, Kenya); "A Study of Absenteeism in the Kenyan Health Workforce" (Kenya); "Validating Cortrimoxazole Coverage among People Living with HIV/AIDS" (Ethiopia); and "Surveillance for Influenza Viruses in Domestic Animals" (Kenya). The CD.C.-Hubert Global Health Fellowship is endowed by the O.C. Hubert Charitable Trust.

Requirements: Applicants must be: medical or veterinary students; in their 2nd or 3rd year when they apply; enrolled in a school accredited by the Liaison Committee on Medical Education, the American Osteopathic Association, or the American Veterinary Medical Association; covered by medical insurance during the fellowship; a U.S. citizen or permanent resident; and able to attend the fellowship orientation (unless out of the country on the field assignment). Orientation travel expenses are paid by the CD.C.-Hubert Global Health Fellowship.

Geographic Focus: All States

Date(s) Application is Due: Jan 17

Amount of Grant: 4,000 USD

Contact: Catherine Folowoshele, M.P.H., Program Coordinator; 404-498-6148; fax 404-498-6085; HubertFellowship@CD.C.Foundation.org

Internet: http://www.cdc.gov/HubertFellowship/

Sponsor: Centers for Disease Control and Prevention

1600 Clifton Road

Atlanta, GA 30333

CDC Collegiate Leaders in Environmental Health Internships 1048

The Collegiate Leaders in Environmental Health (CLEH) is a 10-week summer environmental internship (June through August) for undergraduate students who are majoring in environmental studies or environmental, physical, biological, chemical, and/or social sciences; and who are passionate about the environment, interested in human health, and curious about how the two are linked. A healthy environment should be capable of sustaining a healthy population; however, with 6.7 billion people on the earth, there is a need for more sustainable interactions between humans and the environment. Environmental issues such as overpopulation, air pollution, food shortage, natural disasters, water contamination, and exposure to toxic substances provide challenges to human health. Human influence on the environment is the main focus of environmental studies; however there is a growing need to evaluate the effect that the environment has on human populations. Harmful environments can increase the risk of many health conditions: asthma, heart disease, cancer, neurological disease, infections, endocrine dysfunction, injuries, and more. Healthy environments, on the other hand, can promote good health in many ways — protecting people from toxic exposures, providing safe water and clean air, and encouraging healthy behaviors such as outdoor recreation. The link between the environment and health is aptly referred to as "environmental health." Environmental health is the discipline that focuses on the interrelationships between people and their environment, promotes human health and well-being, and fosters safe and healthy living. This branch of public health is concerned with all aspects of the natural and built environment that may affect human health. CLEH Interns will be placed in environmental health programs at the U.S. Centers for Disease Control and Prevention (CD.C.)'s National Center for Environmental Health and the Agency for Toxic Substances and Disease Registry (NCEH/ATSDR) at CD.C.'s Chamblee Campus. Over the course of the summer, interns will be exposed to a broad overview of environmental public health issues at the federal level, participate in environmental health projects, interact with federal officials and scientists, and visit important environmental health sites in and around Atlanta. Other activities include "brown-bag" lunches with CD.C. staff, as well as attending lectures from prominent environmental health leaders in the Atlanta area. In addition, interns will be able to attend the many seminars offered by CD.C. during the summer. CLEH interns receive a weekly stipend of $500 for living expenses. All interns are responsible for the cost of travel to and from Atlanta and for housing during the internship. Information about student housing options at local universities can be found at the internship website. Links to application materials and guidelines are provided at the website along with specific instructions for submitting application materials via email, fax, or U.S. mail. Annual deadline dates may vary; applicants are encouraged to check the website or contact program staff to verify current deadline dates.

Requirements: Applicants must: be U.S. citizens or permanent residents with a green card; be enrolled full time at a college or university as rising juniors or seniors by the fall semester; have achieved a minimum cumulative GPA of 3.0 on a 4.0 scale; and major or have coursework concentration in environmental studies and/or environmental, physical, biological, chemical, or social sciences (if applicants are not majoring in one of these areas they can make the case for why their major or interests are applicable to this internship in essay question #2 of the application materials.) All participants are considered guest researchers and are subject to CD.C. regulations governing visiting scientists, engineers, and other professionals. As a guest researcher, each participant is responsible for payment of income taxes and is advised to become familiar with the relevant sections of the current tax codes. Health-insurance is required for all participants.

Restrictions: Participation in the program is contingent upon the individual's ability to obtain the proper security clearance. All applicants will be subject to a criminal records check and other background investigations conducted by the U.S. Government. These inquiries are conducted to develop information to assess various factors about the applicant, including reliability, trustworthiness, honesty, integrity, character, conduct and loyalty to the United States. Deferrals of the internship will not be allowed. Seniors graduating in Spring 2011 will not be accepted to this program.

Geographic Focus: All States

Date(s) Application is Due: Feb 2

Amount of Grant: 5,000 USD

Contact: Lt. Cory Moore, M.P.H., Environmental Health Specialist; (770) 488-0593; fax (404) 929-2820; CLEH@cdc.gov or cory.moore@cdc.hhs.gov

Internet: http://www.cdc.gov/nceh/cleh/

Sponsor: Centers for Disease Control and Prevention

1600 Clifton Road

Atlanta, GA 30333

CDC Cooperative Agreement for Continuing Enhanced National Surveillance for Prion Diseases in the United States 1049

The purpose of the funding is to continue an active surveillance program similar to that conducted by the National Prion Disease Pathology Surveillance Center since 1997 to monitor the occurrence of potentially emerging human TSEs in the United States. This program

addresses the Healthy People 2010 focus area(s) of Immunizations and Infectious Diseases. Eligible applicants should have experience in conducting prion disease surveillance and must be capable of fulfilling the ongoing needs for enhancing such surveillance at the national level.
Requirements: Eligible applicants include: public nonprofit organizations; private nonprofit organizations; small, minority, women-owned businesses; universities; colleges; research institutions; hospitals; community-based organizations; faith-based organizations; federally recognized Indian tribal governments; Indian tribes; Indian tribal organizations; state and local governments or their Bona Fide Agents (this includes the District of Columbia, the Commonwealth of Puerto Rico, the Virgin Islands, the Commonwealth of the Northern Mariana Islands, American Samoa, Guam, the Federated States of Micronesia, the Republic of the Marshall Islands, and the Republic of Palau); and political subdivisions of States (in consultation with States).
Restrictions: Recipients may: not use funds for research; not use funds for clinical care; only expend funds for reasonable program purposes, including personnel, travel, supplies, and services, such as contractual. Awardees may not generally use HHS/CD.C./ATSDR funding for the purchase of furniture or equipment (any such proposed spending must be identified in the budget). The direct and primary recipient in a cooperative agreement program must perform a substantial role in carrying out project objectives and not merely serve as a conduit for an award to another party or provider who is ineligible.
Geographic Focus: All States
Date(s) Application is Due: Jun 30
Amount of Grant: 100,000,000 USD
Contact: Mattie Jackson; (770) 488-2696; MIJ3@cdc.gov
Internet: http://www.cdc.gov/od/pgo/funding/FOAs.htm
Sponsor: Centers for Disease Control and Prevention
1600 Clifton Road
Atlanta, GA 30333

CDC Cooperative Agreement for Partnership to Enhance Public Health Informatics 1050

The purpose of the program is to strengthen the breadth and depth of the public health workforce by providing training in public health informatics and encouraging new explorations of innovations at the intersection of public health and health informatics. CD.C. wants to partner with a national organization that is established as a leader in clinical and health informatics and with established tutorials and other training in informatics. This program addresses the Healthy People 2010 focus area of Public Health Infrastructure and the CD.C.'s Healthy Protection Goal of People Prepared for Emerging Health Threats to develop and measure how well public health systems perform in critical times.
Requirements: Assistance will be provided only to national non-profit organizations (a national non-profit, tax-exempt Section 501(c)3 agency. Eligible national organizations must have a web-based training and in person tutorials in clinical and public health informatics. These resources should have been established, fully operational, and deemed successful before application is submitted. Additionally, applicants are required to have an active national and international professional membership in informatics. Limiting assistance to national non-profit organizations is necessary to produce the maximum possible enhancement, given available resources.
Restrictions: The following restrictions apply: recipients may not use funds for research; recipients may not use funds for clinical care; recipients may only expend funds for reasonable program purposes, including personnel, travel, supplies, and services, such as contractual; awardees may not generally use HHS/CD.C./ATSDR funding for the purchase of furniture or equipment; the direct and primary recipient in a cooperative agreement program must perform a substantial role in carrying out project objectives and not merely serve as a conduit for an award to another party or provider who is ineligible; reimbursement of pre-award costs is not allowed; and these funds may not be used to set up or support lobbying by interest/advocacy groups.
Geographic Focus: All States
Date(s) Application is Due: Jul 2
Amount of Grant: 500,000 - 2,500,000 USD
Contact: Kaleema McLean; (770) 488-2742; kmclean@cdc.gov
Internet: http://www.cdc.gov/od/pgo/funding/FOAs.htm
Sponsor: Centers for Disease Control and Prevention
1600 Clifton Road
Atlanta, GA 30333

CDC Cooperative Agreement for the Development, Operation, & Evaluation of an Entertainment Education Program 1051

The purpose of this program is raise awareness and behavioral change concerning public health issues through the use of the television media as means of accessing the television viewing public by providing: accurate public health information and public health issues, as well as accurate depictions of healthy living at all stages of life, to entertainment industry leadership for possible inclusion in television story lines; assistance with public health outreach, including providing additional information concerning public health topics depicted on television story lines, and assistance with public service announcements and informational short videos to be aired in conjunction with drama presentations; and evaluation of public health television story lines and the effect on the viewing public.
Requirements: Eligible applicants include: public nonprofit organizations; private nonprofit organizations; universities; colleges; research institutions; federally recognized Indian tribal governments; Indian tribes; Indian tribal organizations; state and local governments or their Bona Fide Agents (this includes the District of Columbia, the Commonwealth of Puerto Rico, the Virgin Islands, the Commonwealth of the Northern Mariana Islands, American Samoa, Guam, the Federated States of Micronesia, the Republic of the Marshall Islands, and the Republic of Palau); and political subdivisions of States (in consultation with States).

Geographic Focus: All States
Date(s) Application is Due: Jun 7
Contact: Sharon Robertson; (770) 488-2748; SRobertson1@cdc.gov
Internet: http://www.cdc.gov/od/pgo/funding/FOAs.htm
Sponsor: Centers for Disease Control and Prevention
1600 Clifton Road
Atlanta, GA 30333

CDC David J. Sencer Museum Adult Group Tour 1052

The David J. Sencer CD.C. (Centers for Disease Control and Prevention) Museum's adult group tour (college, professional, other) is available for groups ranging from ten to thirty people (groups smaller than ten people are asked to visit the museum for a self-guided tour). Upon arrival to the Museum, the group will be greeted and oriented to CD.C. on the Global Symphony platform. The Global Symphony is a media installation featuring four, three–minute stories that describe in depth CD.C.'s contributions to the elimination of polio, the investigation of Legionnaire's disease, the battle to stem the rise of obesity in the United States, and the study of how humans, animals, and the environment interact in the spread of Ebola. Following the orientation, the group will tour the current temporary exhibit and the History of CD.C. exhibit. The group's tour coordinator may also request a CD.C. speaker. Speakers are CD.C. employees who volunteer their time, so speaker availability is not guaranteed. Topics vary and depend on speaker availability. Following the guided tour, visitors will have the opportunity to explore the exhibit area independently. Tour requests may be submitted online at the program website (given above). A complete list of attendees' names must also be submitted via the given fax number or first listed email address. Please note the tour is not scheduled until the group's tour coordinator receives a confirmation email. Tour coordinators are encouraged to visit the program website for more information about current exhibits. The David J. Sencer CD.C. Museum is a Smithsonian Institution Affiliate, which provides access to unique programming, content and expertise, and provides a network and forum for the CD.C. Museum to showcase itself on a national stage and at a national level.
Requirements: Tour coordinators are expected to review driving directions and parking instructions sent in confirmation email. Each group member must bring a government-issued photo ID. Non-U.S. citizens must bring a passport as ID.
Geographic Focus: All States
Contact: Judy Gantt, Curator; (404) 639-0830 or (404) 639-0831; fax (404) 639-0834; museum@cdc.gov or judy.gantt@cdc.hhs.gov
Internet: http://www.cdc.gov/museum/tours/adult.htm
Sponsor: Centers for Disease Control and Prevention
1600 Clifton Road
Atlanta, GA 30333

CDC David J. Sencer Museum Teacher Professional Development Workshops 1053

The David J. Sencer CD.C. (Centers of Disease Control and Prevention) Museum offers free professional-development workshops designed to support middle- and secondary-school science education and the incorporation of public health concepts into existing classroom curriculum. These workshops are a result of museum partnerships with external educational organizations, and divisions and offices within CD.C.. Currently being offered are: "Teach Epidemiology," a five-day workshop for high-school science, health, math, and social studies teachers on teaching epidemiology in the high-school classroom; and "Med Myst: Teaching About the World of Pathogens Through Web Adventures," a two-day workshop for middle-school science teachers who are looking for ways to teach the topics of infectious diseases, pathogens, and the scientific method in an exciting and relevant context. Professional Learning Unit credits are available for attending these workshops. Teachers are encouraged to visit the program website for detailed information on workshop content, presenter qualifications, and lunch arrangements. Application information will be posted at the website in December. The David J. Sencer CD.C. Museum is a Smithsonian Institution Affiliate, which provides access to unique programming, content and expertise, and provides a network and forum for the CD.C. Museum to showcase itself on a national stage and at a national level.
Requirements: All workshops require an application. Participants are expected to arrive on time each day, actively participate, complete workshop assignments and remain in the classroom for the duration of the workshop. PLU credits will only be awarded to participants who meet these expectations.
Restrictions: Teachers who fail to attend a workshop after accepting a slot forfeit their rights to attend future workshops.
Geographic Focus: All States
Contact: Judy Gantt, Curator; (404) 639-0830 or (404) 639-0831; fax (404) 639-0834; museum@cdc.gov or judy.gantt@cdc.hhs.gov
Internet: http://www.cdc.gov/museum/teachers.htm
Sponsor: Centers for Disease Control and Prevention
1600 Clifton Road
Atlanta, GA 30333

CDC Disease Detective Camp 1054

The Centers for Disease Control and Prevention (CD.C.)'s Disease Detective Camp (DD.C.) is an educational program started by CD.C.'s David J. Sencer Museum in 2005 as a mechanism for developing an academic, public-health-day-camp curriculum for state and county health departments. The DD.C. camp is offered to upcoming high school juniors and seniors and is held at CD.C.'s headquarters in Atlanta, Georgia at no charge. The museum will offer two one-week sessions each summer. The CD.C. Disease Detective Camp curriculum is based on contextual and situated-cognition learning principles. By learning through hands–on activities and seminars, high school juniors and seniors at the conclusion of the camp will be able to: identify five careers within public health; demonstrate an understanding of basic epidemiology terms; calculate basic epidemiologic rates given an outbreak scenario and data; recognize how infectious and chronic diseases are tracked in the United States; and understand the role of public health law in protecting the public's health in the United States. Over the course of five days, campers will take on the role of disease detectives and learn first-hand how the CD.C. safeguards the nation's health. Teams will probe a disease outbreak using epidemiologic and laboratory skills and report their findings to a group of CD.C. scientists. Activities may include short lectures by CD.C. experts, a mock press conference in the CD.C. press room, and a look behind the scenes of CD.C.. The application process for the current DD.C. is posted the preceding December at the DD.C. website. Future plans for DD.C. includes creating a curriculum toolkit for local health departments seeking to offer a similar camp experience. Due to the popularity of this camp, there are more interested students then the program can accommodate. For this reason, interested students must apply (parents are encouraged to help interested students with the application, but not apply for them). Applications will be made available at the DD.C. website in December or parents and students may call the second phone listed above to have an application mailed. Camp participants are selected in two phases. Applicants are first selected based on their answers to the essay questions on the application and on the recommendation form. The second phase consists of slots assigned through a lottery system. Twenty–seven high–school juniors and seniors will be selected for each camp session. The David J. Sencer CD.C. Museum is a Smithsonian Institution Affiliate, which provides access to unique programming, content and expertise, and provides a network and forum for the CD.C. Museum to showcase itself on a national stage and at a national level.

Requirements: The CD.C. Disease Detective Camp is open to motivated students who will be high-school juniors or seniors. Applicants must be 16 years old by the first day of the camp in order to comply with CD.C.'s laboratory safety requirements. There is no cost associated with attending the CD.C. Disease Detective Camp, but campers will need to pay for their own lunches. Non–Atlanta residents may apply for the camp, but are responsible for providing their own accommodations and transportation. Campers in past years have stayed with family friends or relatives in Atlanta. Most attendees are from the Atlanta area, but out-of-state students also attend every year. This is a wonderful opportunity to make friends from other schools!
Geographic Focus: All States
Contact: Judy Gantt, Curator; (404) 639-0830 or (404) 639-0831; fax (404) 639-0834; museum@cdc.gov or judy.gantt@cdc.hhs.gov
Internet: http://www.cdc.gov/museum/camp/index.htm
Sponsor: Centers for Disease Control and Prevention
1600 Clifton Road
Atlanta, GA 30333

CDC Epidemic Intelligence Service Training Grants 1055

The Epidemic Intelligence Service (EIS) is a unique two-year post-graduate training program of service and on-the-job learning for health professionals interested in the practice of applied epidemiology. About 75% of EIS graduates remain in public health at the Centers for Disease Control and Prevention (CD.C.) or in state or local health departments. Many become leaders in public health throughout the world. EIS officers work in state or local health departments or at CD.C. including the Center for Global Health (Atlanta, Georgia); the National Center for Chronic Disease Prevention and Health Promotion (Atlanta, Georgia); the National Center for Emerging and Zoonotic Infectious Diseases (Atlanta, Georgia); the National Center for Environmental Health and Agency for Toxic Substances and Disease Registry (Atlanta, Georgia); the National Center for Health Statistics (Hyattsville, Maryland); the National Center for HIV/AIDS, Viral Hepatitis, STD and TB Prevention (Atlanta, Georgia); the National Center for Immunization and Respiratory Diseases (Atlanta, Georgia); the National Center for Injury Prevention and Control (Atlanta, Georgia); the National Center on Birth Defects and Developmental Disabilities (Atlanta, Georgia); the National Institute for Occupational Safety and Health (Anchorage, Alaska; Atlanta, Georgia; Cincinnati, Ohio; Denver, Colorado; Morgantown, West Virginia; Pittsburgh, Pennsylvania; and Spokane, Washington); the Office of Public Health Preparedness and Response (Atlanta, Georgia); the Office of State, Tribal, Local and Territorial Support (Atlanta, Georgia); and the Office of Surveillance, Epidemiology, and Laboratory Services (Atlanta, Georgia). Although international work may be part of any EIS assignment, no assignment currently is based outside the U.S. The Program is modeled after a traditional medical residency program where much of the education occurs through experiential learning. Thus, the EIS program not only organizes training for trainees (called officers), but a major contribution of the program (and work of the officers) is providing service to the CD.C. and its public health partners. EIS officers not only conduct routine research activities that are directly related to fulfilling CD.C.'s mission; but they serve as one of CD.C.'s primary resources for responding to urgent or emergent public health problems nationally and around the world. Recent EIS Investigations include: survey of adoptive parents and medical providers on the health status of Haitian orphans entering the United States after the Haiti earthquake; investigation of potential hepatitis B virus and hepatitis C virus transmission associated with vessel conduits used for solid organ transplantation; community-wide assessment of health impact and public health emergency response following extended disruption of drinking water service during an extreme winter freeze in Alabama; investigation of cases of novel influenza A (H1N1) in multiple states; increased incidence of Haemophilus influenzae type b in Minnesota; investigation of infections with Staphylococcus aureus among patients undergoing procedures at a pain management clinic in West Virginia; outbreak of carbapenem-resistant Klebsiella pneumoniae at a long term care facility in Illinois; and mass exposures to a rabies positive bat carcass in Montana. Classroom instruction includes topics such as applied epidemiology, biostatistics, public health surveillance, scientific writing, and working with the media, as well as emerging public health issues. Each EIS class begins with a 1-month course, starting in July each year in Atlanta. As part of the on-the-job training, EIS officers are required to complete the following core activities of learning (CALs): conduct or participate in a field investigation of a potentially serious public health problem; design, conduct, and interpret an epidemiologic analysis on public health data; design, implement, or evaluate a public health surveillance system; write and submit a scientific manuscript for a peer-reviewed journal; write and submit a report to the Morbidity and Mortality Weekly Report (MMWR); present a paper or poster at the annual EIS Conference; give an oral presentation at CD.C.'s Epidemiology Grand Rounds (Tuesday Morning Seminar) or at a national or international scientific meeting; and respond appropriately to written or oral public health inquiries. During the two-year training program, EIS officers are employees of the CD.C. and receive a salary and benefits. Salaries range from $55,000 to $75,000 per year, based on qualifications and experience. The application period opens in May of each year. Applicants apply online at the program website. The application deadline is midnight of September 1. Required supporting documents must be postmarked September 1. Further details about these may be found at the program website. Accepted applicants must attend an EIS Conference in Atlanta in April, travel expenses to be paid by the EIS. The two-year training program starts in July.

Requirements: The EIS seeks qualified health professionals with an interest in public health and a commitment to public service. Applicants must meet professional degree and licensing requirements, commit to a two-year full-time program starting in July, and be willing to relocate. Physicians (M.D., D.O., M.B.B.S., etc.) must have at least one year of clinical training. U.S citizens and U.S. permanent residents must have an active, unrestricted, U.S. license to practice their clinical specialty. Doctoral-level scientists (Ph.D., Dr.PH., Sc.D., etc.) should have a background in: epidemiology; biostatistics; biological, environmental, social, behavioral, or nutritional sciences; and/or other relevant health science. Other medical professionals such as scientists, nurses, physician assistants, and doctors of pharmacy must have a master of public health (M.P.H.) or equivalent degree. U.S. citizens and U.S. permanent residents must have an active, unrestricted U.S. license to practice their clinical specialty. Veterinarians (D.V.M., V.M.D., etc.) must have either a M.P.H. (or equivalent degree), or relevant public health experience. U.S. citizens and U.S. permanent residents must have an active, unrestricted, U.S. license to practice their clinical specialty. Non-U.S. citizens meeting the above degree requirements are eligible to apply. Non-U.S. citizens must be legal permanent residents or eligible for J-1 status prior to the program's start date. If selected for EIS, CD.C. will sponsor the J1 visa. Because of the domestic nature of the program, only a limited number of non-U.S. citizens are selected. Applicants are encouraged to contact the EIS office with any questions regarding eligibility.
Restrictions: Citizens of countries that have been determined by the U.S. Secretary of State to be State Sponsors of Terrorism will not be able to get security clearance to work at CD.C. and should not apply to the EIS Program. Countries currently on this list include: Cuba, Iran, Syria, and Sudan.
Geographic Focus: All States, All Countries
Date(s) Application is Due: Sep 1
Amount of Grant: 110,000 - 150,000 USD
Contact: Douglas Hamilton, MD, PhD, Director; (404) 498-6110; fax (404) 498.6535; EIS@cdc.gov or douglas.hamilton@cdc.hhs.gov
Internet: http://www.cdc.gov/eis/index.html
Sponsor: Centers for Disease Control and Prevention
1600 Clifton Road
Atlanta, GA 30333

CDC Epidemiology Elective Rotation 1056

The Epidemiology Elective Program (Epi Elective) is a 6 to 8 week rotation sponsored by the Centers of Disease Control and Prevention (CD.C.) for senior medical and veterinary students in their 3rd and 4th years. Epi Elective gives students an introduction to preventive medicine, public health, and the principles of applied epidemiology while working with CD.C. epidemiologists to solve real-world public health problems. Participants learn through hands-on experience working on a current public health project, and are mentored by experienced CD.C. staff. Most schools award course credit. During the elective, participants may: participate in the surveillance of public health problems; analyze public health data for new disease risk factors; become a coauthor on a publication of major health importance; and/or work in the field investigating an outbreak. Selected students are placed in one of the following subject areas within CD.C. or the Indian Health Service: National Center for Emerging and Zoonotic Infectious Diseases (NCEZID), Atlanta, Georgia; Arctic Investigations Program (part of NCEZID), Anchorage, Alaska; Division of Vector-Borne Diseases (part of NCEZID), Ft. Collins, Colorado; Indian Health Service (IHS), Albuquerque, New Mexico; National Center for Chronic Disease Prevention and Health Promotion (NCCDPHP), Atlanta, Georgia; National Center for Environmental Health / Agency for Toxic Substances and Disease Registry (NCEH/ATSDR), Atlanta, Georgia; National Center for Health Statistics (NCHS), Hyattsville, Maryland; National Center for HIV/AIDS, Viral Hepatitis, STD, and TB Prevention (NCHHSTP), Atlanta, Georgia; National Center for Immunization and Respiratory Diseases (NCIRD), Atlanta, Georgia; National Center for Injury Prevention and Control (NCIPC), Atlanta, Georgia; or the National Center on Birth Defects and Developmental Disabilities (NCBDDD), Atlanta, Georgia. Alternatively students may be placed at the National Institute for Occupational Safety and Health (NIOSH) in: Anchorage, Alaska; Atlanta, Georgia; Cincinnati, Ohio; Denver, Colorado; Morgantown, West Virginia; Pittsburgh, Pennsylvania; or Spokane, Washington. Every effort is made to assign students to their preferred public health subject area. Interested students apply online. Supporting documents are required and must be postmarked by the deadline date. Students should check the Epi Elective web site for further details. Application deadlines may vary from year to year. Students are encouraged to check the Epi Elective website or contact the sponsor to verify current deadline dates.

Requirements: Medical and veterinary students in their 3rd year apply for a 6 to 8 week rotation in the fall or spring semester of their 4th year. The rotations must occur between June and December or between January and May. Students must be U.S. citizens or permanent residents and be enrolled in a school accredited by one of the following: the Liaison Committee on Medical Education; the American Osteopathic Association; or the American Veterinary Medical Association. Applicants selected for the program must pay for round-trip transportation to their assigned CD.C. facility and for living expenses during the elective (travel and living expenses related to field investigations are paid by CD.C..)
Geographic Focus: All States
Date(s) Application is Due: Mar 30; May 30
Contact: Renee Amos, CNI; (404) 498-6152 or (800) 232-4636; fax (404) 498-6085; renee.douglas@cdc.hhs.gov or EpiElective@cdc.gov
Internet: http://www.cdc.gov/epielective/index.html
Sponsor: Centers for Disease Control and Prevention
1600 Clifton Road
Atlanta, GA 30333

CDC Evaluation of the Use of Rapid Testing For Influenza in Outpatient Medical Settings   1057

The purpose of this project is to evaluate how rapid tests for influenza are being implemented and used in clinical practice in outpatient medical settings such as community clinics, solo and group practice physician offices, and hospital emergency rooms across the United States. This evaluation will include a determination of the scope of rapid influenza test use, the types of tests in use and how they are selected, the personnel performing testing, the extent to which good laboratory practices and testing guidelines are being followed, how results are reported and interpreted, how results are used for patient care and antiviral and antibiotic prescribing practices, and the presence of linkages between these outpatient settings and the public health system. Additionally, this project may identify potential opportunities to provide guidance to assist sites in making decisions on the appropriate use of these tests and ways to enhance the connectivity with the public health system. Connectivity with public health is especially important in light of the possibility of an influenza pandemic. This project will identify and evaluate practices used in outpatient settings related to processes such as specimen collection, testing, reporting and referral for influenza.
Requirements: Eligible applicants include: public nonprofit organizations; private nonprofit organizations; small, minority, women-owned businesses; universities; colleges; research institutions; hospitals; community-based organizations; faith-based organizations; federally recognized Indian tribal governments; Indian tribes; Indian tribal organizations; state and local governments or their Bona Fide Agents (this includes the District of Columbia, the Commonwealth of Puerto Rico, the Virgin Islands, the Commonwealth of the Northern Marianna Islands, American Samoa, Guam, the Federated States of Micronesia, the Republic of the Marshall Islands, and the Republic of Palau); and political subdivisions of States (in consultation with States).
Restrictions: Recipients may: not use funds for research; not use funds for clinical care; only expend funds for reasonable program purposes, including personnel, travel, supplies, and services, such as contractual. Awardees may not generally use HHS/CD.C./ATSDR funding for the purchase of furniture or equipment (any such proposed spending must be identified in the budget). The direct and primary recipient in a cooperative agreement program must perform a substantial role in carrying out project objectives and not merely serve as a conduit for an award to another party or provider who is ineligible. Reimbursement of pre-award costs is not allowed. Reimbursement of construction costs is not allowed.
Geographic Focus: All States
Date(s) Application is Due: Aug 6
Amount of Grant: 200,000 USD
Contact: Yolanda Sledge; (770) 488-2787; yiso@cdc.gov
Internet: http://www.cdc.gov/od/pgo/funding/FOAs.htm
Sponsor: Centers for Disease Control and Prevention
1600 Clifton Road
Atlanta, GA 30333

CDC Evidence-Based Laboratory Medicine: Quality/Performance Measure Evaluation   1058

The purpose of the program is to evaluate clinical laboratory practice by identifying evidence-based laboratory medicine quality/performance measures associated with the pre- and post-analytic stages of the total testing process, and to identify and address gaps and opportunities for improvement consistent with national health care priorities to improve public health. The primary objectives are to: provide an evidence base which systematically identifies and defines important gaps in laboratory medicine quality related to individuals or populations and/or patient safety health outcomes with demonstrated impacts (i.e., clinical and/or economic); identify, develop, and define laboratory medicine performance/quality measures that can be broadly implemented to evaluate performance associated with patient outcomes; and identify interventions effective in improving performance.
Requirements: Eligible applicants include: public nonprofit organizations; private nonprofit organizations; small, minority, women-owned businesses; universities; colleges; research institutions; hospitals; community-based organizations; faith-based organizations; federally recognized Indian tribal governments; Indian tribes; Indian tribal organizations; state and local governments or their Bona Fide Agents (this includes the District of Columbia, the Commonwealth of Puerto Rico, the Virgin Islands, the Commonwealth of the Northern Marianna Islands, American Samoa, Guam, the Federated States of Micronesia, the Republic of the Marshall Islands, and the Republic of Palau); and political subdivisions of States (in consultation with States).

Restrictions: Recipients may: not use funds for research; not use funds for clinical care; only expend funds for reasonable program purposes, including personnel, travel, supplies, and services, such as contractual. Awardees may not generally use HHS/CD.C./ATSDR funding for the purchase of furniture or equipment (any such proposed spending must be identified in the budget). The direct and primary recipient in a cooperative agreement program must perform a substantial role in carrying out project objectives and not merely serve as a conduit for an award to another party or provider who is ineligible. Reimbursement of pre-award costs is not allowed.
Geographic Focus: All States
Date(s) Application is Due: Jul 17
Amount of Grant: 100,000 - 500,000 USD
Contact: Yolanda Sledge; (770) 488-2787; yiso@cdc.gov
Internet: http://www.cdc.gov/od/pgo/funding/FOAs.htm
Sponsor: Centers for Disease Control and Prevention
1600 Clifton Road
Atlanta, GA 30333

CDC Experience Applied Epidemiology Fellowships   1059

The goal of the CD.C. Experience Applied Epidemiology Fellowship is to increase the number of physicians with a population health perspective by providing medical students with an understanding of applied epidemiology, the role of epidemiology in medicine and health, and the role of physicians in the public health system. The fellowship is held at the Centers of Disease Control and Prevention (CD.C.) headquarters in Atlanta, Georgia. Fellows start in August and work for ten or twelve months. While at CD.C., fellows will: gain training and work experience in applied epidemiology and public health; perform epidemiologic analyses and research; participate in public health field experiences; report findings through written and oral presentations at scientific meetings; write scientific manuscripts for publication; participate in monthly seminars and journal clubs; and attend the annual Epidemic Intelligence Service conference. CD.C. Program areas available to applicants are chronic diseases, emergency preparedness, environmental health, global health, infectious diseases, injury prevention, and reproductive health. Fellows receive a stipend for living expenses. Applicants must apply online. Supporting documents are required and must be postmarked by the deadline date. Applicants should check the CD.C. Experience website for further details. Annual deadline dates may vary; applicants are encouraged to check the website or contact program staff to verify current deadline dates. The CD.C. Experience Applied Epidemiology Fellowship is made possible by a public/private partnership supported by a grant to the CD.C. Foundation from External Medical Affairs, Pfizer Inc., and is considered a valuable experience to those those seeking career advancement in the areas of clinical medicine, clinical epidemiology, health services research, preventive medicine, and public health.
Requirements: Applicants must: be medical students in their 2nd or 3rd year when they apply; be enrolled in a school accredited by the Liaison Committee on Medical Education or the American Osteopathic Association; be a U.S. citizen or permanent resident; and be willing to relocate to Atlanta, Georgia for the duration of the fellowship. Awardees may start the fellowship while in their 3rd or 4th year.
Geographic Focus: All States
Date(s) Application is Due: Dec 2
Contact: Virginia Watson; (404) 498-6151; VWatson1@cdc.gov
Internet: http://www.cdc.gov/CD.C.ExperienceFellowship/index.html
Sponsor: Centers for Disease Control and Prevention
1600 Clifton Road
Atlanta, GA 30333

CDC Foundation Atlanta International Health Fellowships   1060

The Atlanta International Health Fellowships (AIHF) were created in 1984 by Drs. Bob Chen and Katy Irwin to provide financial assistance to international participants in CD.C.'s varied public health training courses. The two young physicians enlisted help from other CD.C. employees and retirees to raise funds and then formed a partnership with Emory University, which provided free tuition to one individual each year to attend Emory's International Course in Applied Epidemiology. Drs. Chen and Irwin also formed a partnership with Villa International to provide free housing to the AIHF recipient. In 1997, a grant from the Tull Charitable Foundation was added to existing funds, and the Atlanta International Health Fellowship became the first endowed fund at the CD.C. Foundation. Since the first fellowships were awarded in 1991, 25 individuals from over 22 countries have received stipends to cover costs of travel, tuition and lodging as they enrich their public health expertise at CD.C.. Interested applicants are encouraged to contact the CD.C. Foundation or Emory University for information on how to apply for the AIHF fellowship.
Geographic Focus: All States, All Countries
Contact: Verla S. Neslund, J.D., Vice President for Programs; (888) 880-4232 or (404) 653-0790; fax (404) 653-0330; vneslund@cdcfoundation.org
Internet: http://www.cdcfoundation.org/what/program/atlanta-international-health-fellowship-endowment
Sponsor: Centers for Disease Control and Prevention (CD.C.) Foundation
55 Park Place NE, Suite 400
Atlanta, GA 30303

CDC Foundation Emergency Response Fund Grants   1061

The CD.C. Foundation's Emergency Response Fund provides immediate, flexible resources to CD.C. experts addressing public health emergencies in the U.S. - whether natural disasters, emerging disease outbreaks or bioterrorist threats. Following the events of September 11, 2001, and the anthrax attacks, the CD.C. Foundation established the Emergency Response Fund to give CD.C. what it needs most in an emergency: flexibility

and access to immediate resources. Federal dollars, even during emergencies, are tied to restrictions and purchasing procedures that can limit CD.C.'s ability to act quickly. The Foundation's Emergency Response Fund gives CD.C. a backup source of funding to fill critical gaps and meet immediate needs. The Fund was activated for the first time in 2005 to support the public health response to Hurricane Katrina in the Gulf Coast region. Donations to the Fund from Kaiser Permanente, the Robert Wood Johnson Foundation, and other organizations and individuals enabled the CD.C. Foundation to immediately respond to requests for help from CD.C. and their public health partners in the gulf coast region. The Foundation was also able to provide new facilities for two public health agencies on the Mississippi coast, replacing buildings that had been destroyed by the storm.
Geographic Focus: All States
Contact: Verla S. Neslund, J.D., Vice President for Programs; (888) 880-4232 or (404) 653-0790; fax (404) 653-0330; vneslund@cdcfoundation.org
Internet: http://www.cdcfoundation.org/response
Sponsor: Centers for Disease Control and Prevention (CD.C.) Foundation
55 Park Place NE, Suite 400
Atlanta, GA 30303

CDC Foundation Tobacco Network Lab Fellowship 1062
According to a Centers for Disease Control and Prevention (CD.C.) morbidity and mortality report, tobacco use is the world's leading single preventable cause of death. Worldwide tobacco-related deaths currently exceed 5 million a year and are expected to exceed 8 million a year by 2030 if left unchecked. In the United States, tobacco use is the single leading preventable cause of disease, disability, and death. In 2003, the World Health Organization (WHO) created the Tobacco Laboratory Network (TobLabNet) of government, academic, and independent laboratories worldwide to strengthen national and regional capacity for testing and research of the contents and emissions of tobacco products. The Network's stated goals are: to establish global capacity to test tobacco products for regulatory compliance; to research and develop harmonized standards for contents and emissions testing; to share tobacco research and testing standards and results; to inform risk assessment activities related to the use of tobacco products; and to develop harmonized reporting of results so that data can be transformed into meaningful trend information that can be compared across countries and over time. The Tobacco Network Lab Fellowship is a collaboration between CD.C. and WHO to place a scientist in CD.C.'s Division of Laboratory Sciences to help TobLabNet address testing and research of tobacco products at the global level in accordance with the tobacco product regulation provisions outlined by the WHO Framework Convention on Tobacco Control. A link to brief information about the fellowship is at the grant website under Chronic Disease & Birth Defects Programs. Interested applicants are encouraged to contact the CD.C. Foundation or the World Health Organization for more information on how to apply for the fellowship.
Geographic Focus: All States, All Countries
Contact: Verla S. Neslund, J.D., Vice President for Programs; (888) 880-4232 or (404) 653-0790; fax (404) 653-0330; vneslund@cdcfoundation.org
Internet: http://www.cdcfoundation.org/what/programs/list#program-518
Sponsor: Centers for Disease Control and Prevention (CD.C.) Foundation
55 Park Place NE, Suite 400
Atlanta, GA 30303

CDC Increasing Breast and Cervical Cancer Screening Services for Urban American Indian/Alaska Native Women 1063
The purpose of this program is to: increase the number of urban American Indian/Alaska Native (AI/AN) women receiving BCCED screening; decrease time to access diagnosis and treatment; increase participation of Urban Indian Health Organizations (UIHOs) in state BCCEDP & Comprehensive Cancer Control (CCC) coalitions; and to provide a model for other state health department programs to work with UIHOs in a collaborative, culturally appropriate manner.
Requirements: Eligible applicants that can apply for this funding are non-profit organizations with experience and expertise in addressing the health care needs of urban Indian populations and providing culturally competent services to them. These include, but are not limited to organizations such as: Indian health boards; inter-tribal councils; American Indian/Alaska native health tribal organizations; inter-tribal councils; inter-tribal consortia; urban organizations; and other non-profit organizations, if incorporated for the primary purpose of improving AI/AN health and represent such interests for the tribes, or urban Indian communities throughout the United States.
Restrictions: Recipients may: not use funds for research; not use funds for clinical care; only expend funds for reasonable program purposes, including personnel, travel, supplies, and services, such as contractual. Awardees may not generally use HHS/CD.C./ATSDR funding for the purchase of furniture or equipment (any such proposed spending must be identified in the budget). The direct and primary recipient in a cooperative agreement program must perform a substantial role in carrying out project objectives and not merely serve as a conduit for an award to another party or provider who is ineligible. Reimbursement of pre-award costs is not allowed.
Geographic Focus: All States
Date(s) Application is Due: Jul 23
Amount of Grant: 350,000 - 1,050,000 USD
Contact: Stephanie Lankford; (770) 488-2936; fzi8@cdc.gov
Internet: http://www.cdc.gov/od/pgo/funding/FOAs.htm
Sponsor: Centers for Disease Control and Prevention
1600 Clifton Road
Atlanta, GA 30333

CDC Preventive Medicine Residency and Fellowship 1064
The Preventive Medicine Residency and Fellowship (PMR/F) is an umbrella that includes two preventive medicine programs that provide hands-on experiences in public health agencies at the federal, state, and local levels. The programs consist of a residency (PMR) for physicians only and a fellowship (PMF) for physicians and other health professionals. The PMR is a 24-month nationally accredited graduate medical education program for physicians. It is accredited by the Accreditation Council for Graduate Medical Education (ACGME) and meets the residency requirement of the American Board of Preventive Medicine (ABPM) in the specialty of Public Health and General Preventive Medicine. Select candidates can be considered for one year of residency training. The PMF is a 12-month program open to veterinarians, dentists, registered nurses, physician assistants, and international medical graduates (physicians) who have no ACGME- or American Osteopathic Association (AOA)- accredited postgraduate clinical training. The residency (PMR) and the fellowship (PMF) both provide service, training, and supervised on-the-job learning in public health and preventive medicine practice. PMR/F activities include: designing, implementing, and evaluating public health programs; developing or analyzing health policy; managing public health and preventive medicine projects; bridging medical and public health sectors to improve wellness; and developing and applying leadership skills. The PMR/F program goals are to prepare clinicians for leadership roles in public health and general preventive medicine at the federal, state or local levels. Residents and fellows will receive mentoring by experienced public health and preventive medicine practitioners; hands-on experience in public health and preventive medicine practice; classroom training to enhance professional growth; and M.P.H. coursework for select physicians in the PMR program. Residents and fellows receive a salary and benefits, moving expenses, travel to PMR/F-sponsored training, and tuition support for an M.P.H. degree for residents (as needed). (Qualified applicants interested in the American Board of Preventive Medicine "complementary pathway to board exam certification" must have completed an ACGME accredited residency in another specialty and have an M.P.H. or equivalent degree per ACGME.) Interested applicants may apply online between July 1 and September 15. Additional supporting documents are required. Applications are due by midnight eastern daylight time on the due date. Supporting documents must also be received on the application due date. Applicants are encouraged to mail supporting documents early. All mail delivery to the CD.C. goes to a central mail facility before being delivered to the PMR/F office. Applicants are encouraged to click on the "Application Information" link at the PMR/F website for additional instructions, guidelines, and information on how to apply and submit the supporting documents.
Requirements: Applicants must: commit to a 1- or 2-year full-time training period (depending on the program) starting in mid-June; be willing to relocate for duration of training; meet the professional and licensing requirements for hiring by the U.S. government according to the U.S. Office of Personnel Management (link provided at the PMR/F website); and have trained in the Epidemic Intelligence Service (EIS) program or have comparable applied epidemiology experience. PMR applicants must be physicians who have completed at least 12 months of ACGME-accredited postgraduate clinical training involving at least 11 months of direct patient care; have a current, full, and unrestricted medical license from a U.S. licensing jurisdiction; and have a Master of Public Health (M.P.H.) or equivalent accredited degree, per ACGME requirements. See the PMR/F website for more information about these requirements. There are some exceptions to the M.P.H. degree requirement. Physicians may be sponsored during the residency to obtain an M.P.H. degree. The completion of necessary courses to earn the degree will be integrated into the training program. M.P.H. sponsored applicants must be Commissioned Corps officers in the U.S. Public Health Service at the time of application with a commitment to an additional 2-year service obligation in the Commissioned Corps following the residency and a willingness to attend the academic institution for the M.P.H. selected by the residency (be willing to relocate). PMF applicants must be veterinarians, dentists, registered nurses, physician assistants, or international medical graduates (physicians who have no ACGME- or American Osteopathic Association- accredited postgraduate clinical training) who have an M.P.H. or equivalent degree from an accredited institution; have a current, full, and unrestricted license from a U.S. licensing jurisdiction in their qualifying clinical specialty; and demonstrate ability in applied epidemiology, public health practice, and written and oral communication for scientific audiences.
Restrictions: Applicants must be U.S. citizens or permanent residents. Physicians who have graduated from a medical school outside the U.S. and do not have a current, full, and unrestricted license from a U.S. licensing jurisdiction may apply.
Geographic Focus: All States
Date(s) Application is Due: Sep 15
Contact: Coordinator, (404) 498-6140; fax (404) 498-6135; PrevMed@cdc.gov
Internet: http://www.cdc.gov/PrevMed/index.html
Sponsor: Centers for Disease Control and Prevention
1600 Clifton Road
Atlanta, GA 30333

CDC Public Health Associates 1065
The Public Health Associate Program (PHAP) is a Centers for Disease Control and Prevention (CD.C.) development program administered through CD.C.'s Office for State, Tribal, Local and Territorial Support (OSTLTS). The program, which began in 1948, provides opportunities for promising future public health professionals to gain broad experience in the day-to-day operation of public health programs. PHAP is geared toward recent baccalaureate college graduates (BA/BS) who are beginning a career in public health. Public Health Associates are employed by CD.C. on term-limited appointments for two years. As "assignees" to agencies in the governmental public health system, most associates hold host-agency positions that are functionally indistinguishable from their local co-workers. In addition to city, county and state assignments, associates may be assigned to territories, tribes and CD.C. quarantine stations in major international airports

within the United States. Assignments are individually tailored to meet local needs in the delivery of public health services, such as case investigation, disease surveillance, health promotion, community outreach, and public health administration. Furthermore, each appointment consists of two different one-year assignments selected by the host agency in one of the following program areas: STD, TB and/or HIV; Other Communicable Diseases; Chronic Disease; Environmental Health; Public Health Preparedness; Global Migration and Quarantine; Immunization; and Injury Prevention. Associate job assignments in the program areas listed above include activities and responsibilities that will lead to the development of specific competency objectives. Both one-year assignments involve substantive activity in the following domains: patient/client interaction to elicit or provide information; investigation of public health problems and issues; data collection, analysis and reporting; surveillance activities; public health education; provision of public health services; community involvement and partnership; and liaison with other public or private providers and partners. While each associate has a different combination of assignments over their two-year employment with CD.C., all associates participate in emergency response activities in support of federal, state or local host agencies. This may include participation in an exercise to prepare for major public health events or an actual response to an on-going event (e.g., hurricanes, H1N1 flu outbreak). Upon completion of the two-year appointment, associates are qualified to compete for entry-level career positions as CD.C. public health advisors and equivalent positions at state, tribal, local, territorial or non-governmental public health organizations. Host sites are responsible for providing the associates with a variety of training opportunities that will not only orientate them to the local or state health agency but also lead to required program competencies. These trainings include both local employment training (e.g., safety, security, information technology, standards of conduct) in accordance with state and local requirements and also program-specific training to enable and enhance work performance. In addition, CD.C. provides all associates with a program of core training in public health concepts and topics designed to support their experiential learning, reinforce the acquisition of the program competencies, and enhance professional growth and development. The core training curriculum involves: CD.C. employment orientation within thirty days of the program start date, a series of required one- to five-day training events hosted by CD.C. semi-annually, and monthly webinars and other e-learning. CD.C. funds annual travel and training expenses for associates to attend orientation and training required by CD.C., many of which are located at CD.C. headquarters in Atlanta. In addition, host sites are obligated to spend local funds to train associates in program-specific areas. Associates are paid every two weeks at the equivalent of a GS-5, step 1 (adjusted for their location) following the federal pay calendar. Associates who successfully complete their first year assignments will then be promoted to the equivalent of a GS-7, step 1 for their final one-year assignment. All associates are given the opportunity to enroll in a choice of full health benefit options (that are partially compensated by the CD.C.) available to all federal employees. Prospective associates may apply for the PHAP online. The link to the online application system will be available at the PHAP website during mid-February through early March. The March deadline will be communicated when the online application link is posted. Additional supporting documents are required, including academic transcripts and letters of recommendation. Additional guidelines, instructions, and downloadable forms for supporting documents are provided at the PHAP website. Applications received after 11:59 p.m. (Eastern) on the due date will not be reviewed. Supporting documents must be postmarked by the closing date of the online application. Prospective associates are encouraged to visit the PHAP website for additional information on the PHAP experience.
Requirements: Applicants should have a keen interest in frontline service careers in public health and in developing strong public health programmatic and operations skills. Applicants must: be U.S. citizens; hold a bachelor's degree from an accredited academic institution by July 1 of the year of application; be willing to commit to full-time work for two years; be willing to work in a state, tribal, local or territorial health agency setting, or CD.C. quarantine station; and, be willing to relocate at their own expense. Note: Applicants with a master's degree will be considered but do not receive preference in hiring, nor a higher pay scale. Individuals holding a master's degree may be eligible for other CD.C. fellowship programs.
Restrictions: Applicants must be U.S. Citizens.
Geographic Focus: All States
Amount of Grant: 76,284 USD
Contact: Lynn Gibbs Scharf; (404) 498-0030; fax (404) 639-9130; phap@cdc.gov
Internet: http://www.cdc.gov/phap/index.html
Sponsor: Centers for Disease Control and Prevention
1600 Clifton Road
Atlanta, GA 30333

CDC Public Health Associates Hosts 1066

The Center of Disease Control and Prevention (CD.C.)'s Office for State, Tribal, Local and Territorial Support (OSTLTS) seeks proposals for field-based assignments for its Public Health Associate Program (PHAP) on an annual basis. PHAP is a two-year workforce development program that provides opportunities for future public health professionals to gain broad experience in the day-to-day operation of public health programs. Each Public Health Associate is assigned to a designated host health agency for two one-year assignments beginning in July of each year. As a training and service program, PHAP offers many benefits to host agencies: a qualified associate for a two-year assignment (the class of 2010 had 1,231 eligible applicants competing for 50 positions in a rigorous screening process); salary and employment benefits for the associate; a core training program in public health concepts and topics designed to support associates' experiential learning, reinforce the acquisition of the program competencies and enhance professional growth and development; and ongoing technical support for host sites in managing the associate, including periodic site visits. The core training curriculum involves: CD.C. employment orientation within 30 days of the associate's start date; a series of required one- to five-day training events hosted by CD.C. semi-annually; and monthly webinars and other e-learning assignments. Costs related to attendance at PHAP core training will be supported by CD.C.. Host sites in their turn must provide all local or long-distance travel and expenses for trainings related to associate work assignments; all local or long distance travel and related expenses for conferences and meetings related to associate work assignments; and a suitable work site, including desk, computer, chair, and other equipment to successfully complete programmatic tasks as named in the application. Host site supervisors oversee the training and field activities of the associate, ensure the associate is familiar with relevant techniques in a given program specialty, and encourage the overall professional development of the associate. Supervisors receive an initial orientation conducted by the CD.C. PHAP office prior to the associate's arrival and participate in conference calls at various intervals during the course of the associateship. The PHAP Host Site Guidelines and Application are available as downloadable PDF documents at the PHAP Host website (click on the "Application Process" link on the left hand side of the screen to access the web page that has the download links). Applications must be submitted electronically as email attachments with the subject line "PHAP Host Site Application – (your health department name)" to the email address given. Applications are due at 11:59 p.m. on the application deadline date. Deadlines may vary from year to year. Prospective hosts are encouraged to check the PHAP Host website or use the email address given to verify current deadlines and to obtain complete information on the PHAP program and application submission requirements.
Requirements: Host sites must be state, tribal, local, and territorial public health agencies, or CD.C. quarantine stations. Host sites are selected based on the scope, quality, and diversity of the experience offered to the associate reflected in the initial host site application. Selection criteria include, but are not limited to, host-site year-one and year-two activities; ability of the host site to support the associate as an employee (e.g., resources, training needs for the associate, travel support for training, and other needs); and demonstrated capacity to provide associates with employment training (e.g., safety, security, information technology, standards of conduct) and program-specific training. Host health agencies and supervisors are strongly encouraged to provide financial support and opportunities for the associate to participate in public health activities which will expand the associate's scope and depth of public health knowledge and / or expand his / her job-related capabilities. Associates are expected to be integrated into the host site and treated as an entry level employee. If employee programs are offered to regular employees, host sites are expected to provide comparable programs and financial support to the associate. Host site supervisors are required to hold a bachelor's degree or above and have prior supervisory experience. In addition, the primary supervisor must devote four hours per week to spend with the associate through the duration of the assignment to provide direct guidance on weekly assignments.
Geographic Focus: All States
Date(s) Application is Due: Feb 21
Contact: Lynn Gibbs Scharf; (404) 498-0030; fax (404) 639-9130; phap@cdc.gov
Internet: http://www.cdc.gov/phap/host/index.html
Sponsor: Centers for Disease Control and Prevention
1600 Clifton Road
Atlanta, GA 30333

CDC Public Health Conference Support Grants 1067

The purpose of the grants is to provide partial support for specific non-federal conferences in the areas of health promotion and disease prevention information/education programs. Funds may be used for direct cost expenditures, such as salaries, speaker fees, rental of necessary equipment, registration fees, and transportation costs (not to exceed economy-class fare) for non-federal individuals. The program continues through March 8, 2009.
Requirements: Public and private organizations, including colleges and universities, and units of state and local government are eligible to apply. Applicants must provide a portion of the conference cost.
Restrictions: Funds may not be used for indirect costs, equipment purchases, honoraria, entertainment, or personal expenses.
Geographic Focus: All States
Contact: Sharon Robertson; (770) 488-2748; SRobertson1@cdc.gov
Internet: http://www.cdc.gov/od/pgo/funding/FOAs.htm
Sponsor: Centers for Disease Control and Prevention
1600 Clifton Road
Atlanta, GA 30333

CDC Public Health Informatics Fellowships 1068

The Public Health Informatics Fellowship Program (PHIFP) provides two-year paid fellowships in public health informatics. The competency-based and hands-on training will allow recipients to apply the disciplines of information/computer science and technology to solve real-world public health problems. Fellows will have the opportunities: to learn about informatics and public health; to work with teams involved in research and development of public health information systems; to lead an informatics project; and to design, develop, implement, evaluate, and manage public health information systems. Fellows will be temporary employees of the Centers of Disease Control and Prevention (CD.C.) and will receive employment benefits (e.g., medical insurance and vacation time). Salaries will be based on federal pay grades GS 11 or 12, depending on applicant qualifications. Applicants should apply online at the PHIFP website (the link to the online application system is available from the Eligibility and Application Information page.) Submissions are due on the deadline date before midnight. All required supporting documents must be mailed in one package and be postmarked by the second deadline date (deadlines may change from year to year; applicants should check the PHIFP website to verify current submission deadlines for proposals and supporting documents). Further details and downloadable instructions for letters of recommendation are available from the PHIFP website.

Requirements: To apply for PHIFP, applicants must meet both the educational and professional requirements for the fellowship, be willing to commit to a two-year, full-time program, and be willing to relocate to Atlanta, Georgia. A doctoral (Ph.D., M.D.) or masters-level degree is required. Qualifying degrees must be from an accredited academic institution in one of the following areas: public health, medicine, health care, or health-services research; computer science, information science, information systems; statistics; epidemiology; or public-health informatics or related discipline. Documented one-year professional experience for doctoral-level candidates and three-year professional experience for masters-level candidates is required in one of the following fields: public-health informatics, health informatics or related field, information systems, information science, computer science, or information technology. Additionally, documented one-year professional experience for doctoral-level candidates and three-year professional experience for masters-level candidates is required in public health or a related healthcare profession (medicine, nursing, veterinary medicine, dentistry, and allied health professions). Applicants should have documented experience in performing research or evaluation during or after academic training (e.g., publication, thesis, poster presentation, or research/evaluation proposal). Non-U.S. citizens must be legal permanent residents or eligible for J-1 status prior to the program's start date.
Restrictions: Citizens of countries that have been determined by the U.S. Secretary of State to be State Sponsors of Terrorism will not be able to get security clearance to work at CD.C. and should not apply to the program (a current list is available at the U.S. Department of State website).
Geographic Focus: All States, All Countries
Date(s) Application is Due: Nov 14; Nov 28
Amount of Grant: 59,987 - 74,297 USD
Contact: Herman Tolentino, M.D., Director, PHIF Program; (404) 498-6168 or (404) 498-6219; fax (404) 498-6135; phifp@cdc.gov or herman.tolentino@cdc.hhs.gov
Internet: http://www.cdc.gov/PHIFP/
Sponsor: Centers for Disease Control and Prevention
1600 Clifton Road
Atlanta, GA 30333

CDC Public Health Prevention Service Fellowships   1069
The Public Health Prevention Service (PHPS) is a 3-year training and service fellowship that prepares master's level public health professionals for leadership positions in local, state, national, and international public health agencies. The fellowship focuses on public health program management and provides hands-on experience and mentorship in program planning, implementation, and evaluation at Centers for Disease Control and Prevention (CD.C.), and in state and local health organizations. PHPS fellows are placed in public health organizations throughout the U.S. after completing one year of training (two six-month rotations) at CD.C.. Examples of rotation activities include: leadership on policy and legislative activities related to nutrition, physical activity, and obesity in the Division of Nutrition, Physical Activity and Obesity; development of case management guidelines for the Division of HIV/AIDS Prevention; and evaluation of reproductive health activities for the World Health Organization. CD.C. rotations typically take place at CD.C. headquarters in Atlanta, Georgia; however, other possible locations include: Washington, D.C; Morgantown, West Virginia; Hyattsville, Maryland; Cincinnati, Ohio; and Denver, Colorado. Examples of field assignments include: development and implementation of HIV/AIDS action plans for Global AIDS programs; creation of a website for physicians and nurses to increase knowledge, improve reporting, and decrease hospitalizations and deaths from pertussis (whooping cough); and coordination and implementation of a multi-cultural media campaign to highlight CD.C.'s National Breast and Cervical Cancer Screening program. Field assignments typically take place at a public health organization in the U.S. or one of its territories. Fellows also have the opportunity to practice public health in an international setting through programs such as the Global AIDS Program (GAP) and Stop Transmission of Polio (STOP). Fellows are CD.C. employees for the duration of the fellowship. They initially earn a salary equivalent to a GS-9 pay grade and advance to GS-11 with geographic adjustments. They receive benefits such as medical insurance and vacation. Applicants must apply online at the PHPS website. Applications are due before midnight on the application deadline. All required supporting documents must be postmarked by the application deadline date. Deadlines may vary from year to year; applicants should check the PHPS website to verify current submission deadlines and to obtain other important submission details for proposals and supporting documents. Organizations wishing to host and mentor a PHPS fellow are encouraged to follow the "hosting and mentoring a PHPS fellow" link provided at the bottom of the screen at the PHPS website.
Requirements: Applicants should have a strong interest in a leadership and management career in public health and must: meet both the educational and professional requirements; be willing to commit to a 3-year full-time program; and be willing to relocate to Atlanta for year one and then again for an assignment at a state or local health organization. To satisfy educational requirements, applicants must have a master's degree in public health or management-related fields from an accredited college or university and attain academic achievement in areas of: epidemiology/biostatistics; biological, physical, and environmental sciences; behavioral and social sciences; health education and promotion; and management and administration. To satisfy professional requirements, applicants must have at least 2,080 hours of paid public health work experience. This is the equivalent of 1-year, full-time work. Multiple paid work experiences qualify. Work experience can include paid internships, externships, and fellowships. Examples of paid, public health work experience include, but are not limited to: special project coordinator in a health department; graduate student researcher or research assistant in a school of public health, assisting with a public health project; community disease investigator; health education specialist; program coordinator for a community-based public health organization; program consultant for a non-profit public health organization; and health educator in the Peace Corps.

Restrictions: Applicants must be U.S. Citizens.
Geographic Focus: All States
Date(s) Application is Due: Feb 1
Amount of Grant: 170,954 USD
Contact: Cindi Melanson, MPH, CHES, Director; (404) 498-6120 or (404) 498-6389; fax (404) 498-6125; PHPS@cdc.gov or cindi.melanson@cdc.hhs.gov
Internet: http://www.cdc.gov/PHPS/
Sponsor: Centers for Disease Control and Prevention
1600 Clifton Road
Atlanta, GA 30333

CDC Public Health Prevention Service Fellowship Sponsorships   1070
The Public Health Prevention Service (PHPS) program offers three-year training and service fellowships for postgraduate masters-level professionals to prepare them for management positions in public health programs. PHPS fellows have advanced degrees in public health or other related fields including (but not limited to) the following areas: public policy; health sciences; health education; public administration; business administration; and social work. A strong interest in a public health leadership and management career, demonstrated achievement in the areas of: epidemiology/biostatistics; biological, physical, and environmental sciences; behavioral and social sciences; health education and promotion; and management and administration, and a prerequisite 2,080 hours of paid public health experience are also required. PHPS fellows complete one year of training at CD.C. and then spend two years in public health organizations throughout the U.S. and its territories. The purpose of the two-year field assignments is to provide PHPS fellows with high-quality work in the area of program management including (but not limited to) the following: development and implementation of management decisions; analysis and formulation of program and policy recommendations; involvement in budget design, preparation, and monitoring; staff and workforce planning and development; implementation and management of science-based interventions; provision of technical assistance and programmatic consultation on epidemiology-based projects and the role of data in program evaluation; evaluation of public health programs; development and implementation of planning processes; communication of public health information to audiences with diverse backgrounds; and development of partnerships to advance public health programs and priorities. All salaries, benefits, and PHPS-related travel expenses will be the responsibility of the PHPS training and service program. In addition, PHPS will provide PHPS fellows with limited assistance for relocation to the field assignment. The health organization must provide the fellow with training and learning opportunities; office space and equipment required for completing the assignment; periodic assessments of performance, portfolios, and professional development; travel expenses for assignment-related activities and professional development opportunities; ongoing mentoring and supervision; and meaningful assignments in program management. Two-year field assignments must provide opportunities for fellows to meet PHPS performance requirements, which are as follows: to conduct a public health assessment; to develop a plan for a public health program or initiative; to implement a public-health program plan or initiative; to evaluate a public health program or initiative; to analyze a public-health-policy-related issue and prepare a written response; to communicate public health information to a lay audience using a variety of media; to communicate public health information to professional audiences; and to participate in various aspects of the funding process. Mentoring is a critical aspect of all assignments. Assignment supervisors should be involved in the fellow's projects. Field supervisors are required to agree to the following: to commit to a minimum of four hours weekly for mentoring activities and performance feedback; to orient the PHPS fellow to program-specific areas, policies, and protocols; to provide technical guidance for all work assignments; to monitor the assignment, ensuring meaningful work-experiences and activities are provided; facilitate and assess successful completion of assignment-specific skill sets (based on the PHPS performance requirements); assist the PHPS fellow in identifying additional professional development opportunities; facilitate mentoring linkages with appropriate staff; submit periodic PHPS assessment and feedback reports to the PHPS office; and participate in ongoing communication with PHPS supervisors. The CD.C. encourages public health organizations to take advantage of the opportunity to provide a practical learning experience for PHPS fellows while addressing the organization's public health priorities. Organizations interested in sponsoring a PHPS fellow should download the updated "Guide for Health Organizations to Request a Fellow" booklet (Adobe PDF file; 21 pages; 265KB) for instructions, samples of a letter of intent, and a full application. A download link for this document is provided at the PHPS Fellows Sponsorship website. In brief, organizations must first submit a Letter of Intent (LOI) by January 10. Accepted applicants will be invited to submit a full application describing the proposed assignment by March 16. If the full application is approved, the organization will be invited to participate in open recruitment. During this two-week period, health organizations are encouraged to discuss their potential assignments with PHPS fellows. After open recruitment, selected health organizations are invited to formal interviews with PHPS fellows in Atlanta, Georgia. Primary supervisors are expected to participate. The health organization is responsible for all travel-related expenses. Application and LOI deadlines may vary from year to year. Applicants are encouraged to check the PHPS Fellows Sponsorship website to verify current deadline dates.
Requirements: Eligible health agencies or organizations (hereafter referred to as organizations) include the following: state and local health departments; U.S. territorial health departments; Indian Health Service area offices; federally-recognized tribes with established public health department structures (or their equivalent) providing public health services to their tribal members; and other health-related organizations (e.g., community-based organizations, foundations, and universities) that have an active collaboration with a state or local health department.
Geographic Focus: All States

Date(s) Application is Due: Mar 16
Contact: Cindi Melanson, MPH, CHES, Director; (404) 498-6120 or (404) 498-6389; fax (404) 498-6125; PHPSAssignments@cdc.gov or cindi.melanson@cdc.hhs.gov
Internet: http://www.cdc.gov/PHPS/FieldAssignments/index.html
Sponsor: Centers for Disease Control and Prevention
1600 Clifton Road
Atlanta, GA 30333

CDC Steven M. Teutsch Prevention Effectiveness Fellowships   1071
The Prevention Effectiveness Fellowship (PEF) is a two-year, highly-selective research fellowship for recent doctoral graduates with a background in economics, policy analysis, operations research, decision sciences, and other quantitative areas, and was established to address demand in the field of public health for economics-based inquiry, quantitative policy analysis, and integrative-health-services research. The PEF allows applicants to apply their academic training in quantitative methods to the science of health protection, health promotion, and disease prevention. Prevention effectiveness uses economic evaluation, policy analysis, and decision modeling to systematically assess the impact of public health interventions, policies, programs, and practices on health outcomes. Econometric, decision, simulation, and operations modeling and methods are employed to determine the efficacy, effectiveness, efficiency, and cost of public health services, policies, and interventions in order to facilitate better decision-making about the allocation of resources around programmatic priorities. At CD.C., prevention effectiveness methods are used to explain: health-related burden and costs; effectiveness and efficiency of health protection, health promotion, and disease prevention programs; health program prioritization based on optimization modeling; disease transmission and intervention modeling; and health system performance. Fellows have the opportunity to work with leading researchers on issues of critical importance to create the expertise, information, and tools that enable people and communities to protect their health through: health promotion; prevention of disease, injury and disability; and preparedness for new health threats. Fellows are assigned to various CD.C. Centers, Institute, and Offices (CIOs) where they assess the public health impact of policies, programs, and practices by determining their cost, effectiveness, and overall impact. The PEF provides initial classroom training and hands-on experience to support the development of valuable research skills. After an introduction to basic public health sciences and prevention-effectiveness methods, fellows are assigned to a CD.C. program area where they take a lead role in prevention-effectiveness studies. Assignment activities include: designing and conducting studies; working closely with national and international experts in public health; presenting and publishing results; and providing technical assistance and service on specific projects or methods. In addition, professional support and career development are provided through didactic training, workshops, research seminars, and peer discussion groups. Fellows are full-time CD.C. employees during the fellowship and will receive the same benefits (such as medical insurance and vacation) as other full-time CD.C. employees. During the first year, fellows are paid the equivalent of a GS 12, Step 1 salary. In the second year, they are eligible for salary equivalent to a GS 12, Step 2 with geographic adjustments. Applicants must apply online at the PEF website (click on "Apply Now" on the left hand side of the screen). Applications are due on January 21 before midnight. All required supporting documents must be postmarked by January 19. Deadlines may vary from year to year; applicants should check the PEF website to verify current submission deadlines for proposals and supporting documents. Further essential details on application and supporting document requirements are available from the PEF website.
Requirements: Applicants must hold a doctoral degree in economics or applied economics, decision sciences, health services research or related health sciences, industrial engineering or operations research, public policy or policy analysis, or a related quantitatively-oriented field. A medical degree with additional relevant training is acceptable. Dissertations must be defended by June 17 and degree requirements must be met by July 15 prior to the start date of the fellowship. Written and spoken proficiency in English is essential. Relocation to Atlanta, Georgia is required. As stated in the FAQ at the PEF website, up to $3,000 is provided to cover relocation expenses. Non-U.S. citizens must be legal permanent residents or eligible for J-1 status prior to the program's start date.
Restrictions: Citizens of countries that have been determined by the U.S. Secretary of State to be State Sponsors of Terrorism will not be able to get security clearance to work at CD.C. and should not apply to the PE Fellowship Program. A list of these may be found at the U.S. Department of State website.
Geographic Focus: All States, All Countries
Date(s) Application is Due: Jan 11
Amount of Grant: 149,198 USD
Contact: Adam G. Skelton, Ph.D., M.P.H., Fellowship Coordinator; (404) 498-6324 or (404) 498-6786; PEF@cdc.gov or adam.skelton@cdc.hhs.gov
Internet: http://www.cdc.gov/PEF/index.html
Sponsor: Centers for Disease Control and Prevention
1600 Clifton Road
Atlanta, GA 30333

CDC Summer Graduate Environmental Health Internships   1072
The Centers for Disease Control and Prevention (CD.C.)'s Graduate Environmental Health (GEH) Program provides paid ten week summer internships (June through August) for graduate students with majors or interests deemed to be in support of environmental health missions at CD.C.'s National Center for Environmental Health (NCEH) and Agency for Toxic Substances and Disease Registry (ATSDR). The mission of the internship is to offer selected students a broad overview of environmental health at the federal level and to foster an interest in environmental health as a career. Applicants should have a passion for the environment and an eagerness to learn about the environment's link to human health. In many cases interested students may receive academic credit for this internship. Academic credit will be considered on a case by case basis. During the program interns will be assigned to projects that utilize the skills they've acquired through graduate studies and personal experiences. Interns will also be able to take advantage of lecture series and other opportunities offered at the CD.C.. GEH interns receive a weekly stipend of $750 for living expenses. All interns are responsible for the cost of travel to and from Atlanta and for housing during the internship. Information about student housing options at local universities can be found at the internship website. Links to application materials and guidelines are provided at the website along with specific instructions for submitting applications and supporting documents via email, fax, or U.S. mail. Annual deadline dates may vary; applicants are encouraged to check the website or contact program staff to verify current deadline dates.
Requirements: Each GEH intern must: possess U.S. Citizenship or Permanent Resident status with a green card; be currently enrolled full-time in a college/university as a graduate student in a degree granting program; have interests and/or studies that are applicable to the field of environmental health; have a minimum cumulative GPA of 3.0 on a scale of 4.0; and have proof of health insurance. All participants are considered guest researchers and are subject to CD.C. regulations governing visiting scientists, engineers, and other professionals. As a guest researcher, each participant is responsible for payment of income taxes and is advised to become familiar with the relevant sections of the current tax codes.
Restrictions: Deferrals of the internship will not be allowed. Participation in the program is contingent upon the individual's ability to obtain the proper security clearance. All applicants will be subject to a criminal records check and other background investigations conducted by the U.S. Government. These inquiries are conducted to develop information to assess various factors about the applicant, including reliability, trustworthiness, honesty, integrity, character, conduct and loyalty to the United States.
Geographic Focus: All States
Date(s) Application is Due: Feb 28
Amount of Grant: 7,500 USD
Contact: Lt. Cory Moore, M.P.H., Environmental Health Specialist; (770) 488-0593; fax (404) 929-2820; geh@cdc.gov or cory.moore@cdc.hhs.gov
Internet: http://www.cdc.gov/nceh/geh/default.htm
Sponsor: Centers for Disease Control and Prevention
1600 Clifton Road
Atlanta, GA 30333

CDC Summer Program In Environmental Health Internships   1073
The Centers for Disease Control and Prevention (CD.C.)'s Summer Program in Environmental Health (SUPEH) is a 10-week internship (June through August) for students majoring in environmental health. Interns participate in activities with the Environmental Health Services Branch of CD.C.'s National Center for Environmental Health (NCEH). In many cases interested students may receive academic credit for this internship and are encouraged to check with their academic advisors or other appropriate school official. Over the summer, interns will explore many aspects of environmental public health and gain an understanding of environmental health work at the local, state, regional, tribal, and federal levels. Activities include: collaborating with federal employees working in the environmental health field and developing mentoring relationships; shadowing senior officials and scientists; spending time in local public health departments; traveling to important environmental health sites in the city of Atlanta and beyond; attending brown-bag lunches with CD.C. staff to discuss environmental health topics; and making presentations at CD.C. staff meetings. CD.C.'s extensive lecture and seminar program is available to interns as well. SUPEH interns and interns in the Collegiate Leaders for Environmental Health (CLEH) program attend field trips and other activities together. SUPEH interns receive a weekly stipend of $500 for living expenses. All interns are responsible for the cost of travel to and from Atlanta and for housing during the internship. Information about student housing options at local universities can be found at the internship website. Applications and supporting materials are due by 5 p.m. on the due date. Applications are available for download at the internship website and may be submitted via email, fax, or U.S. mail. Applicants are advised to check the website for specific instructions for each method of submission. As actual deadlines may vary from year to year, applicants are advised to check the internship website to verify current deadline and other key dates. SUPEH internships are facilitated through the Oak Ridge Institute for Science and Education.
Requirements: Each SUPEH intern must: major in an environmental health program accredited by the National Environmental Health Science and Protection Accreditation Council (EHAC); have U.S. citizenship or Permanent Resident status with a green card; be enrolled full-time as a rising junior, senior, or graduate student at a college or university; have a minimum cumulative GPA of 3.0 on a 4.0 scale; have current health insurance coverage acceptable in the State of Georgia; and be fluent in the English language. Students must return to school to complete degree requirements at the end of their internship OR the internship must serve as needed academic credit to complete degree requirements. Participants in the Junior Commissioned Officer Student Training and Extern Program (JRCOSTEP) are also eligible for the internship. All summer interns are considered guest researchers. As such, interns are subject to CD.C. regulations governing visiting scientists, engineers, and other professionals. As a guest researcher, each intern is responsible for payment of income taxes and is advised to become familiar with the relevant sections of the Internal Revenue Code and other applicable state and federal laws and regulations.
Restrictions: The internship must be accepted in the summer it is offered; internships cannot be deferred. Participation in the program is contingent on the applicant's ability to obtain the proper security clearances. All applicants are subject to a criminal records check and other background investigations conducted by the U.S. government. These inquiries develop information to assess various factors about the applicant, including reliability, trustworthiness, honesty, integrity, character, conduct, and loyalty to the United States.
Geographic Focus: All States

Date(s) Application is Due: Feb 11
Amount of Grant: 5,000 USD
Contact: Lt. Cory Moore, M.P.H., Environmental Health Specialist; (770) 488-0593; fax (404) 929-2820; ehinternship@cdc.gov or cory.moore@cdc.hhs.gov
Internet: http://www.cdc.gov/nceh/ehs/supeh/
Sponsor: Centers for Disease Control and Prevention
1600 Clifton Road
Atlanta, GA 30333

CDI Interdisciplinary Research Initiatives Grants 1074

The Children's Discovery Institute (CDI) is a world-class center for pediatric research and innovation created to encourage researchers to ask bold questions and take bold risks to uncover answers. By funding the work of creative scientists and clinicians in collaborative, multi-disciplinary research aimed at some of the most devastating childhood diseases and disorders, CDI will accelerate the realization of better treatments, cures, and preventions. Interdisciplinary Research Initiatives (II) provide funds for highly innovative and novel projects in need of initial start-up funding to enable procurement of other independent support. Up to $150,000 per year for up to three years is available for Washington University faculty seeking to embark on novel projects in need of initial start-up funding to enable procurement of other independent support. Projects should strive to bring investigators from multiple disciplines together to identify targets for improved diagnosis, prevention or other treatment of a pediatric health problem relevant to the goals of the CDI.
Requirements: Proposals will be accepted from Washington University faculty and postdoctoral trainees. Preference will be given to faculty members prior to achieving tenure, to teams of investigators from multiple disciplines, and to multi-investigator projects to develop interactive research groups. Preference will also be given to projects with potential for acquisition of new knowledge/translational impact and scientific experience; and to programs synergistic with Washington University and St. Louis Children's Hospital. Letters of intent must be submitted by August 1. For investigators invited to submit a proposal, applications will be due on October 15.
Geographic Focus: Missouri
Date(s) Application is Due: Oct 15
Amount of Grant: Up to 450,000 USD
Contact: Angela Mayer (Corless); (314) 286-2711; CDI@kids.wustl.edu
Internet: http://www.childrensdiscovery.org/Grants/FundingMechanismsDueDates.aspx
Sponsor: Children's Discovery Institute
660 S. Euclid Avenue
St. Louis, MO 63110

CDI Postdoctoral Fellowships 1075

The Children's Discovery Institute (CDI) is a world-class center for pediatric research and innovation created to encourage researchers to ask bold questions and take bold risks to uncover answers. By funding the work of creative scientists and clinicians in collaborative, multi-disciplinary research aimed at some of the most devastating childhood diseases and disorders, CDI will accelerate the realization of better treatments, cures, and preventions. Up to $30,000 per year for up to two years is available for especially promising postdoctoral trainees to complete a research project under the direction of a mentor on an issue relevant to the goals of the CDI.
Requirements: Proposals will be accepted from Washington University faculty and postdoctoral trainees. Preference will be given to faculty members prior to achieving tenure, to teams of investigators from multiple disciplines, and to multi-investigator projects to develop interactive research groups. Preference will also be given to projects with potential for acquisition of new knowledge/translational impact and scientific experience; and to programs synergistic with Washington University and St. Louis Children's Hospital. Letters of intent must be submitted by August 1. For investigators invited to submit a proposal, applications will be due on October 15.
Geographic Focus: Missouri
Date(s) Application is Due: Oct 15
Amount of Grant: Up to 60,000 USD
Contact: Angela Mayer (Corless); (314) 286-2711; cdi@kids.wustl.edu
Internet: http://www.childrensdiscovery.org/Grants/FundingMechanismsDueDates.aspx
Sponsor: Children's Discovery Institute
660 S. Euclid Avenue
St. Louis, MO 63110

Cedar Tree Foundation David H. Smith Conservation Research Fellowship 1076

The David H. Smith Conservation Research Fellowship program identifies and supports early-career scientists who will shape the field of applied conservation biology. David H. Smith Conservation Research Fellowships are available to post-doctoral researchers (of any nationality) affiliated with a United States institution. In 1998, to help address the need for post-doctoral opportunities for conservation biology graduates, the foundation and founding partner The Nature Conservancy established the David H. Smith Conservation Research Fellowship Program, devoted exclusively to applied conservation research problems. By fostering the development of promising conservation scientists, the Smith Fellowship Program helps encourage this rapidly expanding field of scientific inquiry and link it to the practice of conservation. Each Fellow will receive an annual salary of $50,000 plus benefits, with the post-doctoral position expected to run for two consecutive years. In addition to the stipend, each Fellow receives a travel budget of $8,000 and a research fund of $32,000 over the 2-year fellowship period.
Requirements: To be eligible individuals must have completed their doctorate within the past five years or by the time the award is made.
Geographic Focus: All States
Date(s) Application is Due: Sep 20

Amount of Grant: 140,000 USD
Contact: Mike Dombeck, Executive Director; (202) 234-4133, ext. 307; info@cedartreefound.org or loi@cedartreefound.org
Internet: http://www.conbio.org/mini-sites/smith-fellows
Sponsor: Cedar Tree Foundation
100 Franklin Street, Suite 704
Boston, MA 02110-1401

Cemala Foundation Grants 1077

The Cemala Foundation is a private family foundation established in 1986 by Martha A. and Ceasar Cone II to continue the family tradition of commitment to enhancing the quality of life of the community through grants to qualified charitable organizations. Areas of interest are: arts/culture; education; health; homelessness; human services; the state of North Carolina and; public interest. Application and additional guidelines are available at the Foundation's website.
Requirements: Grants are made only to non-profit charitable organizations which are tax exempt under Section 501(c)3 of the Internal Revenue Code or to public governmental units. Generally, grants are limited to projects which benefit the citizens of Guilford County, North Carolina. Occasionally, projects which benefit the state of North Carolina as a whole are considered.
Restrictions: The Foundation does not consider support for annual campaigns, endowments, sectarian religious activities, or requests under $1,000. Grants are not made to individuals. Grants from the Cemala Foundation are usually awarded for one year only. Only one grant application may be submitted in any twelve-month period. Organizations receiving grants are required to complete an evaluation report within twelve months after receipt of the funds.
Geographic Focus: North Carolina
Date(s) Application is Due: Mar 1; Sep 1
Amount of Grant: 3,000 - 200,000 USD
Contact: Susan S. Schwartz; (336) 274-3541; fax (336) 272-8153; cemala@cemala.org
Internet: http://www.cemala.org/grant/guidelines.php
Sponsor: Cemala Foundation
330 South Greene Street, Suite 101
Greensboro, NC 27401

Centerville-Washington Foundation Grants 1078

The Centerville-Washington Foundation, a fund family of the Dayton Foundation, accepts and reviews grant requests for programs that meet the following criteria: to launch new projects that represent a unique and unduplicated opportunity for the community, to support established organizations for special purposes and to generate matching funds from government or nonprofit organizations that benefit the citizens of Centerville and Washington Township. Categories of support include: arts and humanities; civic affairs; conservation and the environment; education; health; and social services. The Foundation supports projects that: encourage more efficient use of community resources and promote coordination, cooperation and sharing among organizations and the elimination of duplicated services; test or demonstrate new approaches and techniques in the solution of important community problems; could not be accomplished with other sources of support; promote volunteer participation and citizen involvement in community affairs; and strengthen non-profit agencies and institutions by reducing operating costs increasing public financial support and/or improving internal management. Awards range from $500 to $2,500. There are two annual deadlines for applications: March 1 and September 1.
Requirements: 501(c)3 organizations serving the residents of Centerville, Ohio, are eligible to apply.
Restrictions: The Foundation does not support grants from the discretionary funds to: provide the principal source of support to an organization or activity: organizations located outside the community; establish or add to endowment funds; individuals, except for scholarship purposes; fund specific scientific, medical or academic research; sectarian activities of religious organization; or for profit organizations.
Geographic Focus: Ohio
Amount of Grant: 500 - 2,500 USD
Samples: Friends of the Castle, Centerville, Ohio, $1,000 - transportation costs for the drop-in center serving diagnosed seriously mentally ill disabled adults (2014); Centerville Police Department, Centerville, Ohio, $1,000 - a new patrol bicycle to replace aging/obsolete equipment for the 20-year old specialized unit (2014); Washington Township Recreation Center, Centerville, Ohio, $2,500 - development of a Nature Sounds playground in Countryside Park (2014).
Contact: Robert E. Daley, Trustee; (937) 433-0811; bobdaley888@gmail.com
Internet: http://www.centervillewashingtonfoundation.org/grants.html
Sponsor: Centerville-Washington Foundation
P.O. Box 41125
Centerville, OH 45441

Central Carolina Community Foundation Community Impact Grants 1079

The Foundation's Community Impact grants fund programmatic initiatives of nonprofit organizations. In order to support positive community change and help ensure a program's success, the Grants have a maximum request amount of $10,000. The grant making process is a bi-annual process. Proposals are accepted, April 1 - September 30, for the Fall cycle, October 1 - March 31 for the Spring cycle.
Requirements: Nonprofits serving South Carolina communities are eligible.
Restrictions: Grants do not support routine operating expenses, fund raising projects, debt reduction, endowment development, medical research, conference travel, or conference underwriting or sponsorship. Letter of intent or proposal may not be submitted by email or fax.
Geographic Focus: South Carolina

Date(s) Application is Due: Mar 31; Sep 30
Amount of Grant: Up to 10,000 USD
Samples: Cooperative Ministry, Columbia, SC, $10,000--for the Helping Hands of Hope program, which provides assistance to individuals and families for rent, utilities, food/household, hygiene, transportation and furniture; Project Life: Positeen, Orangeburg, SC, $10,000--to expand its services into two new sites, Rivelon and Bethune-Bowman Elementary Schools in Orangeburg County.
Contact: Veronica L. Pinkett-Barber, Program Officer; (803) 254-5601 ext. 331; fax (803) 799-6663; veronica@yourfoundation.org
Internet: http://www.yourfoundation.org/nonprofits/foundationgrants/impactgrants.aspx
Sponsor: Central Carolina Community Foundation
2711 Middleburg Drive, Suite 213
Columbia, SC 29204

Central Okanagan Foundation Grants 1080
The foundation seeks to enhance the quality of life in Canada and awards grants to nonprofit organizations in its areas of interest, including arts and culture, children, youth, families, education, health, community services, environmental conservation, and heritage preservation. Types of support include emergency funds, equipment acquisition, program development, scholarship funds, and seed money. Contact the foundation by phone to discuss the proposed project. Guidelines and application form will be sent if the project falls within the foundation's guidelines.
Requirements: Nonprofit British Columbia organizations in the Central Okanagan region as defined by the boundaries of BC School District 23 are eligible.
Restrictions: Grants are not awarded to individuals or religious organizations for religious purposes. Grants do not support capital campaigns, debt retirement, endowments, building construction/renovation, or operating expenses.
Geographic Focus: All States, Canada
Date(s) Application is Due: May 1; Oct 1
Contact: Cheryl Miller, Grants Manager; (250) 861-6160; fax (250) 861-6156; cheryl@centralokanaganfoundation.org
Internet: http://www.centralokanaganfoundation.org/pages/grants/cof-grants.php
Sponsor: Central Okanagan Foundation
217-1889 Springfield Road
Kelowna, BC V1Y 5V5 Canada

Cessna Foundation Grants 1081
Grants are made to organizations primarily in the areas where the company operates. Support is given for projects that take a creative approach to such fundamental issues as education, health and human services, youth enrichment, and arts and culture. The foundation values projects which take a creative approach to such fundamental issues as education, neighborhood improvement, youth development, community problem-solving, assistance to people who are disadvantaged, environmental conservation and cultural enrichment. Grants cover projects, building, equipment, capital campaigns, program development, employee matching gifts, and employee-related scholarships. All proposals must be in writing and should include the following information: needs statement and project objectives; purpose of organization; constituency; board members, community, and volunteer involvement; how results of the project will be measured; plans for continued funding of the project, if applicable; one-page project budget and an organizational chart; and primary source of funds. Proposals are accepted at any time.
Requirements: Giving limited to areas of company operations, with emphasis on Wichita.
Restrictions: Grants are not awarded to individuals; national or regional organizations, unless their programs address specific local community needs; programs or initiatives where the primary purpose is the promotion of religious doctrine or tenets; elementary or secondary schools (except to provide special initiatives or programs not provided by regular school budgets); political action or legislative advocacy groups; operational funds; medical or other research organizations; organizations located in or benefiting nations other than the United States and its territories; or fraternal groups, athletic teams, bands, veterans organizations, volunteer firefighters, or similar groups.
Geographic Focus: All States
Amount of Grant: 300 - 10,000 USD
Contact: Rhonda Fullerton, Secretary-Treasurer; (316) 517-7810; fax (316) 517-7812
Internet: http://www.cessna.com/about/corporate-citizenship.html
Sponsor: Cessna Foundation
P.O. Box 7706
Wichita, KS 67277

Cetana Educational Foundation Scholarships 1082
Cetana selects a limited number of scholars, simply because the foundation has limited resources. Scholars from all religious and ethnic groups in Myanmar are eligible for scholarships. Support for scholars covers the cost of education, room and board, travel to and from the university, and associated expenses for the duration of their study abroad. Cetana also provides mentors who are available to offer the scholars advice and practical support in bridging cultures. The Outreach Committee in Myanmar assists the graduate in finding rewarding employment upon return. Once they have finished their studies, scholars return to Myanmar for a period of at least three years so that they may share their new knowledge and experience. Most scholarships are for graduate studies in neighboring SE Asian countries. Only a few scholarships are available in the U.S. where colleges provide full tuition grants.
Requirements: Applicants must be living in Myanmar. Those living outside the country will not be considered. Scholars must be studying in majors that have application to work opportunities in country. Students must pledge to return home for a minimum of 3 years after completion of studies. If the recipient breaks the pledge, he or she must pay back full scholarship so that another student may take his or her place.
Geographic Focus: All States
Contact: Tin May Thein; (240) 242-3997; tinmaythein@cetana.org
Internet: http://www.cetana.org/Cetana/Scholarship_Program.html
Sponsor: Cetana Educational Foundation
12817 Twinbrook Parkway, #207
Rockville, MD 20851

CFF-NIH Funding Grants 1083
The Cystic Fibrosis Foundation (CFF) has developed the overall research grant program to complement the awarding mechanism of the NIH. Support from CFF, through various mechanisms, is intended to provide for the development of sufficient preliminary data to make CF-related grant applications highly competitive in the NIH review process. However, as a result of funding constraints on the NIH, coupled with the growing interest in CF research, occasions arise in which highly meritorious grant applications are submitted to the NIH but are not funded. In an effort to assure that all meritorious CF-related research is supported, CFF has developed the CFF/NIH-unfunded Award mechanism to provide funding. The objective of this award is to support excellent CF-related research projects that have been submitted to and approved by the National Institutes of Health (NIH), but cannot be supported by available NIH funds. Applications must fall within the upper 40th percentile. CFF support ranges from $75,000 to $125,000 per year for up to two years.
Restrictions: Indirect costs are not allowed.
Geographic Focus: All States
Amount of Grant: 75,000 - 125,000 USD
Contact: Grants Management; (800) 344-4823 or (301) 951-4422; grants@cff.org
Internet: http://www.cff.org/research/ForResearchers/FundingOpportunities/ResearchGrants/#Pilot_and_Feasibility_Awards
Sponsor: Cystic Fibrosis Foundation
6931 Arlington Road, 2nd Floor
Bethesda, MD 20814

CFF Clinical Research Grants 1084
Cystic Fibrosis Foundation Therapeutics (CFFT), the nonprofit drug discovery and development affiliate of the Cystic Fibrosis Foundation, offers competitive awards to support clinical research projects directly related to CF treatment and care. Projects may address diagnostic or therapeutic methods related to CF or the pathophysiology of cystic fibrosis, and applicants must demonstrate access to a sufficient number of patients from Foundation-accredited care centers and to appropriate controls. Up to $100,000 per year (plus 8% indirect costs) for a maximum of three years may be requested for single-center clinical research grants. For multi-center clinical research, the potential award is up to $225,000 per year (plus 8% indirect costs) for a maximum of three years.
Requirements: Applications will be accepted only after submission and approval of a letter of intent. CFFT strictly adheres to the letter of intent and award application deadlines.
Geographic Focus: All States
Date(s) Application is Due: Mar 6; Sep 11
Amount of Grant: Up to 225,000 USD
Contact: Grants Management; (800) 344-4823 or (301) 951-4422; grants@cff.org
Internet: http://www.cff.org/research/ForResearchers/FundingOpportunities/ClinicalResearchAwards/
Sponsor: Cystic Fibrosis Foundation
6931 Arlington Road, 2nd Floor
Bethesda, MD 20814

CFF First- and Second-Year Clinical Fellowships 1085
The Cystic Fibrosis Foundation offers competitive clinical fellowships for up to five years for physicians interested in cystic fibrosis and other chronic pulmonary and gastrointestinal diseases of children, adolescents, and adults. The intent of this award is to encourage specialized training early in a physician's career and to prepare well-qualified candidates for careers in academic medicine. Training must take place in one of the foundation's accredited centers and must provide thorough grounding in diagnostic and therapeutic procedures, comprehensive care, and cystic fibrosis-related research. All first- and second-year programs must commit a significant portion (at least 30 percent over two years) to research training. Fellows funded by other sources for their first year of training may apply to the foundation for subsequent training support. Third-year fellowships are available for additional basic and/or clinical research training. Research fellowships are also available from the foundation to support physicians beyond the fellowship experience; salary will be commensurate with experience.
Requirements: Applicants must be U.S. citizens or have permanent U.S. resident visas, and must have completed pediatric training and be eligible for board certification in pediatrics by the time the fellowship begins. Candidates with prior training in internal medicine may also apply but must have completed at least two years of an approved adult pulmonary or GI fellowship; be jointly sponsored by the departments of medicine and pediatrics; and be willing to commit at least 75 percent of their time to cystic fibrosis and related problems of young adults.
Geographic Focus: All States
Date(s) Application is Due: Oct 2
Amount of Grant: 47,600 - 49,250 USD
Contact: Grants Management; (800) 344-4823 or (301) 951-4422; grants@cff.org
Internet: http://www.cff.org/research/ForResearchers/FundingOpportunities/TrainingGrants/
Sponsor: Cystic Fibrosis Foundation
6931 Arlington Road, 2nd Floor
Bethesda, MD 20814

CFF Harry Shwachman Clinical Investigator Award 1086
The three-year CFF Harry Shwachman Clinical Investigator Award provides the opportunity for clinically trained physicians to develop into independent biomedical research investigators who have active involvement in cystic fibrosis-related areas. It is also intended to facilitate the transition from postdoctoral training to a career in academic medicine. Support is available for up to $76,000 per year plus $15,000 for supplies. Support is based on a full-time, twelve-month appointment. Applications are due the second Wednesday in September.
Requirements: U.S. citizenship or permanent resident status is required.
Restrictions: Indirect costs are not allowed.
Geographic Focus: All States
Date(s) Application is Due: Sep 11
Amount of Grant: Up to 91,000 USD
Contact: Grants Management; (800) 344-4823 or (301) 951-4422; grants@cff.org
Internet: http://www.cff.org/research/ForResearchers/FundingOpportunities/ResearchGrants/#Pilot_and_Feasibility_Awards
Sponsor: Cystic Fibrosis Foundation
6931 Arlington Road, 2nd Floor
Bethesda, MD 20814

CFF Leroy Matthews Physician-Scientist Awards 1087
In honor of Dr. LeRoy Matthews' dedication and commitment to cystic fibrosis (CF) research and care, the Cystic Fibrosis Foundation (CFF) announces the Physician Scientist Award named in his memory. Awards will provide up to six years of support for outstanding newly trained pediatricians and internists (MDs and MD/PhDs) to complete sub-specialty training, develop into independent investigators, and initiate research programs. Institutional and individual grants are available. Support ranges from a $48,000 stipend, plus $10,000 for research and development for year one, up to a $76,000 stipend, plus $15,000 for research and development for year six.
Requirements: U.S. citizenship or permanent resident status is required.
Restrictions: Indirect costs are not allowed.
Geographic Focus: All States
Date(s) Application is Due: Sep 11
Amount of Grant: 48,000 - 86,000 USD
Contact: Grants Management; (800) 344-4823 or (301) 951-4422; grants@cff.org
Internet: http://www.cff.org/research/cystic_fibrosis_foundation_grants/research_grants
Sponsor: Cystic Fibrosis Foundation
6931 Arlington Road, 2nd Floor
Bethesda, MD 20814

CFF Pilot and Feasibility Awards 1088
The Cystic Fibrosis Foundation offers feasibility grants for developing and testing new hypotheses and/or new methods, as well as to support promising new investigators as they establish themselves in research areas relevant to cystic fibrosis. Proposed work must be hypothesis driven and must reflect innovative approaches to critical questions in cystic fibrosis research. Funding priority will be placed on those projects proposing to better understand the mechanisms behind disease pathophysiology and to develop strategies to prevent or treat it. Up to $40,000 per year (plus 8% indirect costs) for two years may be requested.
Requirements: Electronic submission is required.
Restrictions: The award is not meant to support continuation of programs begun under other granting mechanisms.
Geographic Focus: All States
Date(s) Application is Due: Sep 11
Amount of Grant: Up to 40,000 USD
Contact: Grants Management; (800) 344-4823 or (301) 951-4422; grants@cff.org
Internet: http://www.cff.org/research/cystic_fibrosis_foundation_grants/research_grants
Sponsor: Cystic Fibrosis Foundation
6931 Arlington Road, 2nd Floor
Bethesda, MD 20814

CFF Postdoctoral Research Fellowships 1089
Competitive postdoctoral research fellowships in basic or clinical research are made on an annual basis, renewable for a second year. The fellowships are awarded in areas of basic cellular and metabolic research or in other problem areas pertinent to cystic fibrosis and related chronic and recurrent pulmonary and gastrointestinal diseases of childhood. Preference is shown to recent graduates or those just beginning their investigative careers. A third year is offered on a limited basis to highly qualified candidates. Interested individuals are encouraged to contact the foundation for application procedures and/or to discuss the potential relevance of their work to the objectives of the foundation. Stipends are $39,000 (first year), $40,100 (second year), and $42,300 (optional third year). Research expenses of $3,750 per year are available, as well.
Requirements: These awards are offered to MDs, PhDs, and MD/PhDs interested in conducting basic or clinical research related to cystic fibrosis. U.S. citizenship or permanent resident status is required.
Restrictions: Electronic submission only. Indirect costs are not allowed.
Geographic Focus: All States
Date(s) Application is Due: Sep 11
Contact: Grants Management; (800) 344-4823 or (301) 951-4422; grants@cff.org
Internet: http://www.cff.org/research/ForResearchers/FundingOpportunities/TrainingGrants/
Sponsor: Cystic Fibrosis Foundation
6931 Arlington Road, 2nd Floor
Bethesda, MD 20814

CFF Research Grants 1090
Grants are offered by the Cystic Fibrosis Foundation in support of high-quality research projects ranging from basic cellular and metabolic mechanisms to therapy of cystic fibrosis and related chronic and recurrent pulmonary and gastrointestinal diseases of childhood. These grants are broadly oriented to the support of projects for developing and initially testing new hypotheses and/or new methods, or those being applied to problems of cystic fibrosis for the first time. The intent of this award is to enable the investigator to collect sufficient data to compete successfully for long-term support from NIH. Interested individuals are encouraged to contact the foundation for guidelines and/or to discuss the potential relevance of their work to the objectives of this program. Support is available for $90,000 per year (plus 8% indirect costs) for a period of two years, at which time a grant may be competitively renewed for an additional year of funding.
Requirements: Investigators who seek support from the Foundation under these funding mechanisms must submit a Letter of Intent (LOI). Applications may be submitted by established investigators or new investigators starting their independent research careers but may not represent support for continuation of a line of research already established by the applicant.
Geographic Focus: All States
Date(s) Application is Due: Sep 11
Amount of Grant: Up to 90,000 USD
Contact: Grants Management; (800) 344-4823 or (301) 951-4422; grants@cff.org
Internet: http://www.cff.org/research/ForResearchers/FundingOpportunities/ResearchGrants/
Sponsor: Cystic Fibrosis Foundation
6931 Arlington Road, 2nd Floor
Bethesda, MD 20814

CFF Student Traineeships 1091
Student Traineeships are offered by the Cystic Fibrosis Foundation to qualified students in order to introduce them to cystic fibrosis research. Each recipient must work with a faculty sponsor on a research project related to cystic fibrosis. Interested individuals are encouraged to contact the foundation for application procedures and/or to discuss the potential relevance of their work to the objectives of the program. Applications are accepted throughout the year. A maximum of $300 of the award may be used for laboratory expenses with the remainder used as a $1,500 stipend for the trainee.
Requirements: Applicants must be students in or about to enter a doctoral program (MD, PhD, or MD/PhD); senior-level undergraduates planning to pursue graduate training may also apply.
Geographic Focus: All States
Amount of Grant: 1,500 USD
Contact: Grants Management; (800) 344-4823 or (301) 951-4422; grants@cff.org
Internet: http://www.cff.org/research/ForResearchers/FundingOpportunities/TrainingGrants/
Sponsor: Cystic Fibrosis Foundation
6931 Arlington Road, 2nd Floor
Bethesda, MD 20814

CFF Third-, Fourth-, and Fifth-Year Clinical Fellowships 1092
The Cystic Fibrosis Foundation offers competitive clinical fellowships for up to five years for physicians interested in cystic fibrosis and other chronic pulmonary and gastrointestinal diseases of children, adolescents, and adults to encourage specialized training early in a physician's career and to prepare well-qualified candidates for careers in academic medicine. The third-, fourth-, and fifth-year fellowship award offers support for additional intense basic and/or clinical research training related to cystic fibrosis. Up to $68,250 may be awarded: $58,250 for stipend and $10,000 for research costs.
Requirements: Applicants and sponsors must submit proposals of the research studies to be undertaken and other specialized training that will be offered during this year. Preference will be given to applicants whose training was supported by the foundation.
Restrictions: Recipients who do not enter a career of academic medicine will be subject to payback provisions. Indirect costs are not allowed.
Geographic Focus: All States
Date(s) Application is Due: Sep 11
Amount of Grant: 58,250 - 68,250 USD
Contact: Grants Management; (800) 344-4823 or (301) 951-4422; grants@cff.org
Internet: http://www.cff.org/research/ForResearchers/FundingOpportunities/TrainingGrants/
Sponsor: Cystic Fibrosis Foundation
6931 Arlington Road, 2nd Floor
Bethesda, MD 20814

CFFVR Alcoholism and Drug Abuse Grants 1093
The Community Foundation for the Fox Valley Region (CFFVR) was established as a public, nonprofit organization in 1986 to enhance the quality of life for all people of the region. Since it was founded, funds within the Foundation have awarded more than $125 million in grants to nonprofit organizations, primarily in Wisconsin's Fox Valley region. The purpose of the Alcoholism and Drug Abuse Grants program is to: supports new or existing programs and projects that address the prevention and/or treatment of alcohol and other drug abuse in the Fox Valley and surrounding area.
Requirements: Organizations serving residents of Outagamie, Calumet, Waupaca, Shawano and northern Winnebago counties are eligible to apply. The grant application form is available at the CFFVR website.
Geographic Focus: Wisconsin
Amount of Grant: 2,500 USD
Contact: Todd Sutton; (920) 830-1290, ext. 28; fax (920) 830-1293; lfilapek@cffoxvalley.org

Internet: https://www.cffoxvalley.org/Page.aspx?pid=652
Sponsor: Community Foundation for the Fox Valley Region
4455 West Lawrence Street, P.O. Box 563
Appleton, WI 54912-0563

CFFVR Basic Needs Giving Partnership Grants   1094
Supported by the U.S. Oil Open Fund for Basic Needs within the Community Foundation and the J.J. Keller Foundation, the partnership assists established charitable organizations with successful programs that address root causes of poverty. Available forms of support include: capacity building; general operating support; project support; project analysis & advocacy. A single organization may request up to $15,000 per year for three years, and collaborative proposals may request up to $100,000 per year for three years. Multiple years of support will be considered only if there is a compelling case for multi-year funding and the project clearly demonstrates how progression shall occur over time.
Requirements: Eligible applicants are well-established charitable organizations that are exempt from federal income taxes under the Internal Revenue Code and have been in operation for a minimum of three years. Wisconsin organizations must serve residents in Outagamie, Calumet, Waupaca, Shawano, or northern Winnebago counties.
Restrictions: Grants from the Basic Needs Giving Partnership will not support the following: technology projects; capital campaigns or building projects; organizational set-up costs; annual fund drives or endowments; lobbying for specific legislation; activities that occur before funding is awarded; organizations with past-due or incomplete grant reports.
Geographic Focus: Wisconsin
Date(s) Application is Due: Feb 15; Sep 15
Contact: Martha Hemwall; (920) 830-1290, ext. 26; mhemwall@cffoxvalley.org
Internet: https://www.cffoxvalley.org/Page.aspx?pid=400
Sponsor: Community Foundation for the Fox Valley Region
4455 West Lawrence Street, P.O. Box 563
Appleton, WI 54912-0563

CFFVR Capital Credit Union Charitable Giving Grants   1095
The Capital Credit Union Charitable Giving Fund was established to support projects and programs that provide for basic needs (food, shelter, clothing and medical care), making a positive impact on the lives of those in the communities served by Capital Credit Union. Grant applications are accepted annually with a submission deadline of December 31. The grant application is available online at the Community Foundation for the Fox Valley Region website.
Requirements: To be eligible to apply for a grant, organizations must: be a public charity, as determined by the IRS and described in 501(c)3 of the tax code (educational institutions and government programs/entities are not eligible for consideration); provide, or propose to develop, services that are focused directly on improving the availability of, or providing for, basic needs in our community; benefit specific communities served by Capital Credit Union. These include the counties of Calumet, Outagamie, and Winnebago in the state of Wisconsin.
Geographic Focus: Wisconsin
Date(s) Application is Due: Dec 31
Contact: Shelly Leadley; (920) 830-1290, ext. 34; sleadley@cffoxvalley.org
Internet: https://www.cffoxvalley.org/Page.aspx?pid=415
Sponsor: Community Foundation for the Fox Valley Region
4455 West Lawrence Street, P.O. Box 563
Appleton, WI 54912-0563

CFFVR Chilton Area Community Foundation Grants   1096
The Chilton Area Community Foundation works to enhance the quality of life for the people of the greater Chilton area. Grants support specific projects or new programs for which a moderate amount of grant money can make a significant impact on an area of need. Grants are made for a broad range of purposes to a wide variety of charitable organizations in the focus areas of health, arts/culture, education and community development. Grants support projects and programs with clear goals and financial accountability, including: creative new activities or services (new programs, one-time projects, events, exhibits, studies or surveys); enhancement or strengthening of existing activities (projects to enhance, expand or strengthen the range, quantity and/or quality of an organization's programs and services); and small capital investments (items that are directly related to program delivery or service to clients, such as a refrigerator for a food pantry or equipment to comply with Americans with Disabilities Act requirements). Applications are available on the website.
Requirements: Organizations serving residents of Outagamie, Calumet, Waupaca, Shawano and northern Winnebago counties are eligible to apply. The grant application form is available at the CFFVR website.
Geographic Focus: Wisconsin
Date(s) Application is Due: Dec 31
Contact: Terry Friederichs; (920) 849-4042; tjfriederichs@charter.net
Internet: https://www.cffoxvalley.org/Page.aspx?pid=415
Sponsor: Community Foundation for the Fox Valley Region
4455 West Lawrence Street, P.O. Box 563
Appleton, WI 54912-0563

CFFVR Clintonville Area Foundation Grants   1097
Clintonville Area Foundation (CAF) grants are awarded from unrestricted funds to support specific projects or new programs for which a moderate amount of grant money can make an impact on an area of need. Grants are made for a broad range of purposes to a wide variety of charitable organizations in the focus areas of health, education and community development. The Foundation is interesting in supporting: creative new activities or services--new programs, one-time projects, events, exhibits, studies or surveys; enhancement or strengthening of existing activities--projects to enhance, expand or strengthen the range, quantity and/or quality of an organization's programs and services; small capital investments--items that are directly related to program delivery or service to clients, such as a refrigerator for a food pantry or equipment to comply with ADA requirements. Grant applications are available online.
Requirements: Wisconsin 501(c)3 nonprofit organizations that serve the residents of the Clintonville area are eligible. General questions can also be directed to the CAF Grants Committee Chair or the Foundation.
Restrictions: The Clintonville Area Foundation will not generally fund the following: general operating expenses not related to the proposed project; annual fund drives or fundraising events; endowment funds; programs with a sectarian or religious purpose that promote a specific journey of faith; major capital projects such as the acquisition of land or buildings; medical research; travel for individuals or groups such as bands, sports teams or classes; activities that occur before funding is awarded; organizations with past-due or incomplete grant reports.
Geographic Focus: Wisconsin
Date(s) Application is Due: Dec 31
Contact: Jenny Goldschmidt, CAF Grants Committee Chair; (715) 823-7125, ext. 2603; clintonvillefoundation@gmail.com
Internet: https://www.cffoxvalley.org/Page.aspx?pid=415
Sponsor: Community Foundation for the Fox Valley Region
4455 West Lawrence Street, P.O. Box 563
Appleton, WI 54912-0563

CFFVR Frank C. Shattuck Community Grants   1098
The Frank C. Shattuck Community Fund supports new or supplements existing services for youth and the elderly and benefits education, the arts and health care in Winnebago and Outagamie counties of Wisconsin. Funding is available for programs, capital expenses or operating expenses of qualifying charitable organizations. Grants to projects may be either one-time payments or multi-year commitments. Application forms are available online.
Requirements: Wisconsin 501(c)3 organizations that serve residents in Outagamie or northern Winnebago counties are eligible to apply. To begin the application process, submit the following prior to the application deadline: grant application form; list of the organization's governing board members, including their professional or community affiliation; current year (board-approved) operating budget.
Restrictions: The Shattuck Fund will not typically support the following: grants for religious or political purposes; grants to support endowment funds of organizations; travel for individuals or groups such as bands, sports teams or classes; reimbursement for previously incurred expenses.
Geographic Focus: Wisconsin
Date(s) Application is Due: Mar 1; Sep 1
Contact: Shelly Leadley; (920) 830-1290, ext. 34; sleadley@cffoxvalley.org
Internet: https://www.cffoxvalley.org/Page.aspx?pid=339
Sponsor: Community Foundation for the Fox Valley Region
4455 West Lawrence Street, P.O. Box 563
Appleton, WI 54912-0563

CFFVR Jewelers Mutual Charitable Giving Grants   1099
Through philanthropy, Jewelers Mutual Insurance Company aspires to achieve a lasting and positive impact on the Fox River Valley. Jewelers Mutual strives to be a responsible corporate citizen and a valued employer by supporting critical needs of the community. Priority areas for giving are: organizations in which Jewelers Mutual employees are actively involved; organizations that address needs in the following areas: basic needs (food, shelter, clothing), as well as programs that reduce/eliminate the root causes of poverty (such as literacy, affordable housing, job training); positive youth development and education; health and wellness, particularly diabetes, cancer, mental health, and affordable access for the disadvantaged; and the vitality of the Fox River area, including gifts to libraries, police or fire departments, parks, the arts, and preservation of the local environment. Letters of request may be sent to the Community Foundation for the Fox Valley Region.
Requirements: Nonprofit, 501(c)3 organizations that serve residents of the Fox River Valley are eligible.
Geographic Focus: Wisconsin
Contact: Shelly Leadley; (920) 830-1290; SLeadley@cffoxvalley.org
Internet: http://www.jewelersmutual.com/information.aspx?id=4305
Sponsor: Community Foundation for the Fox Valley Region
4455 West Lawrence Street, P.O. Box 563
Appleton, WI 54912-0563

CFFVR Myra M. and Robert L. Vandehey Foundation Grants   1100
The Myra M. and Robert L. Vandehey Foundation's mission is support the charitable interests of the Myra and Robert Vandehey family. Areas of interest include: education; children and youth; health care and; family services. Funding opportunities are limited to nonprofit organizations that serve residents in the Fox Cities or Keshena areas of Wisconsin. Contact the Vice President for CFFVR prior to submitting a request to verify that the need aligns with current priorities. Unsolicited grant requests are not accepted from organizations not previously awarded support. Applications may be submitted at any time, and no application form is required.
Requirements: Wisconsin 501(c)3 nonprofit organizations that serve residents in the Fox Cities or Keshena are eligible for funding.
Restrictions: No grants to individuals.
Geographic Focus: Wisconsin

Samples: Fox Valley Technical College Foundation, Appleton, WI, $20,000--educational grant; Boys & Girls Clubs of the Fox Valley, Appleton, WI, $25,000--youth development grant; Nami Fox Valley, Inc., $10,000--mental health and crisis intervention grant.
Contact: Cathy Mutschler, Vice President Community Engagement; (920) 830-1290, ext. 29; cmutschler@cffoxvalley.org
Internet: https://www.cffoxvalley.org/Page.aspx?pid=415
Sponsor: Community Foundation for the Fox Valley Region
4455 West Lawrence Street, P.O. Box 563
Appleton, WI 54912-0563

CFFVR Project Grants   1101

Project grants, support specific projects or new programs for which a moderate amount of grant money can make a significant impact on an area of need. Grants are made for a broad range of purposes to a wide variety of charitable organizations in the focus areas of arts and culture, community development, education, environment, health and human services. Organizations eligible to apply must serve residents in Outagamie, Calumet, Shawano, Waupaca or northern Winnebago counties. Grants for specific projects typically are for no more than $10,000 for one year.
Requirements: Wisconsin 501(c)3 tax-exempt organizations serving residents of Outagamie, Calumet, Waupaca, Shawano and northern Winnebago counties are eligible to apply. The grant application form is available at the CFFVR website.
Restrictions: Project grants typically will not fund the following: general operating expenses not related to the proposed project; annual fund drives or fund raising events; endowment funds; programs with a sectarian or religious purpose that promote a specific journey of faith; major capital projects such as the acquisition of land or buildings; medical research; travel for individuals or groups such as bands, sports teams or classes; activities that occur before funding is awarded; health and safety equipment; playground equipment; organizations with past-due or incomplete grant reports.
Geographic Focus: Wisconsin
Date(s) Application is Due: Feb 1; Aug 1
Amount of Grant: 10,000 USD
Samples: Child Care Resource and Referral, $2,000--to provide at-risk families with young children increased access to supportive community services; Community Clothes Closet, $10,000--to purchase a used cargo van to provide a means of picking up donated clothing; Northeast Wisconsin Land Trust, $4,000--to acquire supplies and equipment necessary to fully implement a newly created land stewardship program.
Contact: Todd Sutton; (920) 830-1290, ext. 28; fax (920) 830-1293; tsutton@cffoxvalley.org
Internet: https://www.cffoxvalley.org/Page.aspx?pid=343
Sponsor: Community Foundation for the Fox Valley Region
4455 West Lawrence Street, P.O. Box 563
Appleton, WI 54912-0563

CFFVR Robert and Patricia Endries Family Foundation Grants   1102

The Robert & Patricia Endries Family Foundation was established for the benefit of people in need, primarily in the Brillion area but with some consideration to the Fox Valley, Lakeshore and Northeastern areas of Wisconsin. Priority areas of giving include: the vitality of the Brillion area; the disadvantaged, particularly the disabled, homeless, low income, single parents, troubled youth, or the chronically or mentally ill; health and human services, particularly diabetes, cancer, cerebral palsy, Alzheimer's disease, kidney disease, or mental health; religious causes or organizations with a spiritual purpose; sports or arts programming and, or sponsorships. Grants are considered for capital campaigns and, or specific capital improvements for the above priority organizations. Gifts will be directed to specific programs or opportunities, not to general operations (with exception for those organizations the foundation has had a long-established relationship with). Matching or challenge gifts are also encouraged, to motivate additional giving by others. Organizations that support needs outside of the Brillion area or that do not yet have an established relationship with the foundation should contact the foundation prior to submission of a formal request.
Requirements: Wisconsin 501(c)3 charitable organizations are eligible to apply. Contact Foundation directly be begin the application process.
Restrictions: The Foundation will not typically support the following: gifts to political organizations or causes; gifts to organizations that are not pro-life supporters or that lack sensitivity to promoting human life in any form (unborn or born); gifts to organizations affiliated with or in support of cloning or embryonic stem-cell research; grants to organizations that receive significant public/government funding; reimbursement for previously incurred expenses.
Geographic Focus: Wisconsin
Date(s) Application is Due: Jan 1; Apr 1; Oct 1
Contact: Shelly Leadley; (920) 830-1290, ext. 34; sleadley@cffoxvalley.org
Internet: https://www.cffoxvalley.org/Page.aspx?pid=415
Sponsor: Community Foundation for the Fox Valley Region
4455 West Lawrence Street, P.O. Box 563
Appleton, WI 54912-0563

CFFVR Schmidt Family G4 Grants   1103

The Schmidt Family G4 grants provide funding to improve the quality of life of those most in need in the Fox Valley, Wisconsin region, with a focus on at-risk youth and self-sufficiency for women. This goal will be accomplished by seeking to address immediate needs and to affect meaningful change in the following areas: at-risk youth--especially those with a physical or mental illness, those who have experienced abuse or those who have significant financial need; adult self-sufficiency--with a priority on issues that affect the stability and independence of women, as well as literacy, job skills training and transitional living for all. The G4 Committee prefers: not to be the sole funder for most projects it considers, unless the amount requested is small and/or a one-time request; to support specific projects or new programs for which a moderate amount of grant money can make a significant impact on an area of need and sustainability. A broad array of requests will be considered, including capital campaigns, existing programs or recurring events as long as they fall within the other listed giving guidelines. Grant awards will typically not exceed $15,000. To assist with the educational aspect of this fund, a formal application is required (available online). Prior to submitting an application, organizations are strongly encouraged to contact Cathy Mutschlerto, discuss the potential proposal and process. Complete and submit the application prior to the March 1, October 1 deadlines.
Requirements: IRS 501(c)3 nonprofit organizations, as well as government agencies are eligible to apply for funding. Organizations that are not public charities may apply through a fiscal sponsor. Organizations must serve Fox Valley residents, particularly in Outagamie, Calumet or northern Winnebago counties of Wisconsin.
Restrictions: The G4 Fund typically will not support the following: organizations that have received funding from the G4 Committee in the most recent 20 months; multi-year requests; programs with a sectarian or religious purpose that promote a specific journey of faith; travel for individuals or groups such as bands, sports teams or classes; reimbursement for previously incurred expenses; endowment funds; fund-raising events; requests from organizations with past-due or incomplete grant reports; a program or need previously declined unless the organization is invited back by the committee; programs or needs that do not serve Fox Valley residents, particularly Outagamie, Calumet or northern Winnebago counties.
Geographic Focus: Wisconsin
Date(s) Application is Due: Mar 1; Sep 1
Amount of Grant: 15,000 USD
Contact: Cathy Mutschler; (920) 830-1290, ext. 29; cmutschler@cffoxvalley.org
Internet: https://www.cffoxvalley.org/Page.aspx?pid=340
Sponsor: Community Foundation for the Fox Valley Region
4455 West Lawrence Street, P.O. Box 563
Appleton, WI 54912-0563

CFFVR Shawano Area Community Foundation Grants   1104

Shawano Area Community Foundation works to preserve and improve the quality of life in Shawano, Wisconsin and, the surrounding area, including communities having economic, educational, cultural and recreational ties with the area. Grant applications are available online.
Requirements: Wisconsin 501(c)3 non-profits in or serving surrounding area of Shawano are eligible to apply.
Geographic Focus: Wisconsin
Date(s) Application is Due: Oct 1
Amount of Grant: 5,000 USD
Samples: City of Shawano/County Library, Shawano, WI, $3,000--reading machine for visually impaired; Shawano Oral Health Fund, Shawano, WI, $1,500--program support for ssecond grade children in Shawano County; St. John's Trinity Lutheran Church, Shawano, WI, $1,000--food purchase grant.
Contact: Susan Hanson; (715) 253-2580; shawanofoundation@granitewave.com
Internet: https://www.cffoxvalley.org/Page.aspx?pid=412
Sponsor: Community Foundation for the Fox Valley Region
4455 West Lawrence Street, P.O. Box 563
Appleton, WI 54912-0563

CFFVR Waupaca Area Community Foundation Grants   1105

Waupaca Area Community Foundation works to preserve and improve the quality of life for the people of the greater Waupaca area of Wisconsin. The Foundation awards grants to Wisconsin non-profits, primarily in the following areas of interest: human services; arts and culture; education; community development. Grant applications are available online.
Requirements: Wisconsin 501(c)3 non-profits in or serving the residents of the Waupaca area are eligible to apply.
Restrictions: The Foundation typically will not award grants to: support routine operating expenses of established organizations, to support annual fund drives of such organizations, or to eliminate their previously incurred deficits; support new or established endowment funds; fund specific research projects; support travel for individuals or groups (such as bands, sports teams, forensics competitors, or the like); support sectarian or religious purposes or causes; reimburse anyone for previously incurred expenses.
Geographic Focus: Wisconsin
Date(s) Application is Due: Jun 30
Amount of Grant: 5,000 USD
Contact: Jack Rhodes, WACF President; (715) 256-1939
Internet: https://www.cffoxvalley.org/Page.aspx?pid=415
Sponsor: Community Foundation for the Fox Valley Region
4455 West Lawrence Street, P.O. Box 563
Appleton, WI 54912-0563

CFFVR Wisconsin King's Daughters and Sons Grants   1106

The Wisconsin Branch of the International Order of King's Daughters and Sons, in alignment with the priorities of its national parent organization, will provide grants to Fox Valley charitable organizations who provide services related to autism and literacy. Application is available online.
Requirements: Wisconsin 501(c)3 non-profit organizations in the Fox Valley region may apply. Organizations that are not public charities may apply through a fiscal sponsor.
Geographic Focus: Wisconsin
Date(s) Application is Due: Oct 15
Amount of Grant: 3,000 USD
Contact: Kathy Mutschler, Director, Donor Engagement; (920) 830-1290, ext. 27; fax (920) 830-1293; cmutschler@cffoxvalley.org

Internet: https://www.cffoxvalley.org/Page.aspx?pid=875
Sponsor: Community Foundation for the Fox Valley Region
4455 West Lawrence Street, P.O. Box 563
Appleton, WI 54912-0563

CFFVR Women's Fund for the Fox Valley Region Grants 1107
The Women's Fund provides grants for programs that inspire women and girls to flourish personally, economically and professionally. Grants have been distributed to programs supporting the following areas: arts & culture; physical and mental health; economic, self-sufficiency; education; parenting and child care; violence prevention. The Women's Fund believes that no project is too small or too new to be considered. Innovative approaches and projects with limited access to other funding are encouraged. Collaborative efforts are welcome. Grant applicants should address one or more of these funding priorities as they relate to women and girls: promotes economic self-sufficiency; improves safety from violence; provides opportunities to develop life skills; promotes physical and/or mental health; enhances dignity and self-worth; promotes leadership development; provides opportunities for artistic development and/or exposure to the arts; provides gender-specific solutions to problems facing women and girls; creates an environment that encourages social change. To apply submit a letter of interest by the deadline.
Requirements: To be eligible for a grant, the project must be consistent with the Women's Fund mission; benefits women and girls in the Wisconsin, Fox Valley region; organization must be a tax-exempt, not-for-profit organization under the Internal Revenue Code, section 501(c)3.
Restrictions: The Women's Fund will not fund: individuals, endowments, government agencies (however educational institutions may qualify), projects with a religious focus, and political parties, candidates or partisan activities.
Geographic Focus: Wisconsin
Date(s) Application is Due: Mar 15
Contact: Becky Boulanger, Program Director; (920) 830-1290, ext. 17; bboulanger@cffoxvalley.org or grants@womensfundfvr.org
Internet: https://www.cffoxvalley.org/Page.aspx?pid=415
Sponsor: Community Foundation for the Fox Valley Region
4455 West Lawrence Street, P.O. Box 563
Appleton, WI 54912-0563

Chamberlain Foundation Grants 1108
The foundation awards grants to Pueblo, CO, nonprofit organizations in its areas of interest, including arts and culture, education, religion, and science. Types of support include general operating support, equipment acquisition, and program development. Submit a brief proposal that includes the purpose of the request and its relevance to the foundation; amount requested; brief history of the organization and its achievements; names, titles, and qualifications of key personnel; name of agency that conducted the last annual audit; copy of the IRS tax-exemption letter; list of other sources of financial support during the past 12 months; and names and affiliations of board members, trustees, and officers of the organization by either of the two deadlines.
Requirements: 501(c)(30 Pueblo, CO, nonprofit organizations are eligible.
Restrictions: Grants do not support individuals; conferences; political activities; religious organizations whose services are limited to members; veteran, labor, fraternal, athletic, or social clubs; national health agencies concerned with specific diseases or health issues; operating expenses for United Way-supported organizations; publications, advertising campaigns, or travel expenses.
Geographic Focus: Colorado
Amount of Grant: 1,000 - 20,000 USD
Samples: YMCA of Pueblo, Pueblo, CO, $20,000; Bessemer Historical Society, Pueblo, CO, $8,000; Animal Welfare and Protection Society, Pueblo, CO, $20,000.
Contact: David Shaw, Foundation Chair; (719) 543-8596; fax (719) 543-8599
Sponsor: Chamberlain Foundation
501 North Main Street, Suite 222
Pueblo, CO 81003

Champ-A Champion For Kids Grants 1109
Grants for children's health and wellness are awarded twice each year. The program provides direct support for children in the areas of health and wellness such as childhood disease research foundations, child safety organizations and organizations that serve children with special needs. The goal is to provide grants to help many programs that are working hard to make the world a healthier and happier place for kids. Programs funded include: (a) Individual Project grants (generally for one-time purchases or to fulfill a short-term need, such as the purchase of materials or equipment); or, (b) Organization Program grants (start-up or operational costs for ongoing programs. Examples include funds for research, health and wellness educational programs, or financial assistance for children and families in-need.)
Requirements: United States applicants must be a tax-exempt organization under Section 501(c)3 of the IRS Code, and 'not a private foundation,' within the meaning of Code Sections 509(a)(1) or 509(a)(2), or a state college or university within the meaning of Code Section 511(a)(2)(B) (a 'Public Charity'). In addition, grant recipients must certify that they are not a supporting organization within the meaning of Code Section 509(a)(3). Canadian applicants must be a registered Canadian charity.
Restrictions: Programs that will not be funded include: (1) Annual Appeals or Capital Campaigns; (2) Construction or New Facility expenses; (3) Fundraising or Event Sponsorships; (4) Political Activities; (5) Religious organizations for religious purposes.
Geographic Focus: All States, Canada
Date(s) Application is Due: Feb 28; Aug 30; Nov 30
Amount of Grant: 1,000 - 10,000 USD
Samples: Angels with Special Needs (Columbia, SC); Building A Generation (Redlands, CA); Hearts and Noses Hospital Clown Troupe (Needham, MA); Catholic Charities of the Diocese of Santa Rosa (Santa Rosa, CA); Hole in the Wall Gang Fund (New Haven, CT); St. Patrick Center (St. Louis, MO); The Children's Hospital Foundation (Omaha, NE); Youth in Need (St. Charles, MO)
Contact: Maxine Clark, President; (314) 423-8000, ext. 5366; giving@buildabear.com
Internet: http://www.buildabear.com/aboutus/community/bearhugs.aspx
Sponsor: Build-A-Bear Workshop Bear Hugs Foundation
1954 Innerbelt Business Center Drive
Saint Louis, MO 63114

Champlin Foundations Grants 1110
The Champlin Foundations are comprised of three foundations, the first established in 1932 by George S. Champlin, Florence C. Hamilton and Hope C. Neaves, who also created The Second Champlin Foundation in 1947. Both of these trusts are administered by the same Distribution Committee. The Third Champlin Foundation was established by George S. Champlin in 1975 and is administered by a separate Distribution Committee. All three foundations share the same management and PNC Bank/Delaware is the trustee of each. The Champlin Foundations are private foundations as defined in Section 509(a) of the Internal Revenue Code and are exempt from Federal income tax under Section 501(c)3. The Foundations areas of interest include: Youth/Fitness; Hospitals/Healthcare; Open Space/Conservation/Environment; Education; Libraries; Social Services; Historic Preservation; Cultural/Artistic; Animal Humane Societies. The Foundations make direct grants to tax exempt organizations, substantially all in Rhode Island, mostly for capital needs. Capital needs may consist of equipment, construction, renovations, the purchase of real property and reduction of mortgage indebtedness. One important goal is to fund tax exempt organizations within Rhode Island that will have the greatest impact on the broadest possible segment of the population. Another important goal is to provide "hands on" equipment and facilities for those being served by these tax-exempt organizations. Grant applications are accepted March 1st through April 30th. No applications will be accepted via facsimile or email. Applications must be postmarked no later than April 30th, or in the case of April 30th falling on a weekend, then the first business day thereafter.
Requirements: 501(c)3 tax exempt organizations of Rhode Island are eligible to apply.
Geographic Focus: Rhode Island
Date(s) Application is Due: Apr 30
Amount of Grant: 25,000 - 65,000 USD
Samples: RiverzEdge Arts Project, Woonsocket, RI, $3,045--technology equipment; Broad Rock Middle School, Wakefield, RI, $91,145-- equip a Cardio Fitness Center; Home & Hospice Care of RI, Pawtucket, RI, $500,000-- towards constructing a new headquarters building with a 24 bed inpatient hospice unit in Providence; Preservation Society of Newport County, Newport, RI, $100,000-- towards the completion of the restoration of the roof at Chateau-sur-Mer; Blackstone Parks Conservancy, Providence, RI, $10,000-- towards the repair of the historic trolley shelter on Blackstone Boulevard; South Kingstown Public Library, Peace Dale, RI, $19,500--security cameras (12,900) and replacement of entrance doors (6,600).
Contact: Keith Lang, Executive Director; (401) 736-0370; fax (401) 736-7248; champlinfdns@worldnet.att.net
Internet: http://www.fdncenter.org/grantmaker/champlin
Sponsor: Champlin Foundations
300 Centerville Road, Suite 300S
Warwick, RI 02886-0226

Changemakers Innovation Awards 1111
The Changemakers Innovation Awards recognize the best social change strategies that emerge from open competitions hosted online every two months. Each competition cycle identifies and refines solutions to a pressing global problem. Winners will be those entries that best address systemic impact, tipping point, replication, sustainability, and innovation. Awards include a cash prize for each winner chosen by vote on the online community. Entries must be submitted in English. Guidelines and nomination forms are available online. Deadlines vary for each contest.
Requirements: To be eligible to win, all project teams/organizations (with the exception of local governments and universities) must enclose a current income statement and a balance sheet. These financial statements need not be audited. Individuals who are partnering with an organization to implement the work must submit a copy of the partnering organization's income financial statement. Individuals without an organizational partner are exempt from the filing requirement.
Geographic Focus: All States, All Countries
Amount of Grant: Up to 5,000 USD
Contact: Awards Director
Internet: http://www.changemakers.com/innovations
Sponsor: Changemakers
1700 North Moore Street, Suite 2000
Arlington, VA 22209

Chapman Charitable Foundation Grants 1112
At Chapman, the ongoing business mission includes supporting the non-profit community through innovation, service and charity. The corporation accomplishes this mission through its corporate endeavors by striving to provide non-profit agencies with comprehensive coverage at the most reasonable price. Likewise, since its inception in 2000, the Chapman Charitable Foundation has donated over $6.75 million dollars to more than 470 California based Social Service Agencies. Primarily, the foundation supports organizations involved with education, forest conservation, health, human services, and religion. Particular fields of interest include: children and youth services, foster care, Christian agencies and churches, education (all levels), the environment, health care access, health care clinics and centers, hospitals, human services, and religion.

Applicants should begin by contacting the Foundation with a one-page letter of inquiry. The foundation utilizes a Recommendation Committee to select potential grantees. Application forms are not required.
Geographic Focus: California
Amount of Grant: Up to 150,000 USD
Samples: AIDS Healthcare Foundation, Los Angeles, California, $10,000 - general operations; Michael Pourson Ministries, San Rafael, California, $108,000 - support programs; Pachamana Alliance, San Francisco, California, $36,000 - general operations.
Contact: Mari Perez, Grants Coordinator; (626) 405-8031; fax (626) 405-0585; mperez@chapmanins.com or info@chapmanins.com
Internet: http://www.chapmanins.com/about/foundation
Sponsor: Chapman Charitable Foundation
265 North San Gabriel Boulevard
Pasadena, CA 91107-3423

CharityWorks Grants 1113
Each year, CharityWorks seeks a partner whose programs improve the quality of life for children and families in the Washington metropolitan area. CharityWorks contributes manpower and money to qualified nonprofit organizations in the Washington Metropolitan area that make a significant impact on the area's most urgent societal and educational needs. Specific focus areas are education, health, and poverty reduction. The beneficiary will receive 80 percent of the proceeds of CharityWork's fundraising efforts. The remaining 20 percent of the proceeds will be divided among the other two finalists.
Geographic Focus: District of Columbia
Date(s) Application is Due: Apr 18
Contact: Administrator; (703) 286-0758; fax (703) 286-0791; charityworks@aol.com
Internet: http://www.charityworksdc.org
Sponsor: CharityWorks
1616 Anderson Road, Suite 209
McLean, VA 22102

Charles A. Frueauff Foundation Grants 1114
The foundation considers proposals 501(c)3 organizations that support private four-year colleges and universities, social service agencies, and health-related agencies and institutions. Types of support include building construction/renovation, capital campaigns, equipment acquisition, general operating support, matching/challenge grants, annual campaigns, and emergency funds. Consideration will be given to programs that support persons leaving welfare, preparing students for employment in non-profit agencies, tutoring at-risk youth, and revitalizing neighborhoods. Applicants are requested to send via postal service a one-page letter of inquiry and include the following information: very brief agency mission and purpose; agency location; brief purpose of request and amount requested; and email address if you wish to receive notification via email.
Requirements: Applicants must be private nonprofit corporations with 501(c)3 status.
Restrictions: The foundation funds nationwide except: Arizona, Alaska, California, Hawaii, Idaho, Iowa, Michigan, Minnesota, Montana, Nevada, New Mexico, North Dakota, Ohio, Oregon, Utah, Washington, Wisconsin, and Wyoming. K-12 schools are ineligible. Grants are not awarded to individuals, provide emergency funds, fund research, or for loans. Multi-year grants, international projects, state supported colleges or universities, primary and secondary schools, churches, or fund raising drives and special events are not supported.
Geographic Focus: Alabama, Arkansas, Colorado, Connecticut, Delaware, District of Columbia, Florida, Georgia, Illinois, Indiana, Kansas, Kentucky, Louisiana, Maine, Maryland, Massachusetts, Mississippi, Missouri, Nebraska, New Hampshire, New Jersey, New Mexico, New York, North Carolina, Oklahoma, Pennsylvania, Rhode Island, South Carolina, South Dakota, Tennessee, Texas, Vermont, Virginia, West Virginia
Date(s) Application is Due: Mar 15; Sep 15
Samples: Melmark Home (Berwyn, PA)--for classroom and office renovation, $50,000; Stetson U (Deland, FL)--to provide scholarships to students majoring in the natural sciences, $100,000.
Contact: Sue Frueauff, (501) 324-2233; fax (501) 324-2236
Internet: http://www.frueauffoundation.com/application/default.asp
Sponsor: Charles A. Frueauff Foundation
200 South Commerce, Suite 100
Little Rock, AR 72201

Charles Delmar Foundation Grants 1115
Established in 1957, the Foundation supports organizations involved with inter-American studies, higher, secondary, elementary, and other education, underprivileged youth, the disadvantaged, the aged, the homeless and housing issues, general welfare organizations, and fine and performing arts. Giving primarily in the Washington, D.C. area in the U.S., and in Europe and South America. There are no specific deadlines with which to adhere. Contact the Foundation for further application information and guidelines.
Restrictions: No grants to individuals, or for building or endowment funds, or matching gifts; no loans.
Geographic Focus: District of Columbia, Maryland, Virginia, West Virginia, Albania, Andorra, Argentina, Armenia, Austria, Azerbaijan, Belarus, Belgium, Bolivia, Bosnia & Herzegovina, Brazil, Bulgaria, Chile, Colombia, Croatia, Cyprus, Czech Republic, Denmark, Ecuador, Estonia, Finland, France, Georgia, Germany, Greece, Guyana, Hungary, Iceland, Ireland, Italy, Kosovo, Latvia, Liechtenstein, Lithuania, Luxembourg, Macedonia, Malta, Moldova, Monaco, Montenegro, Norway, Paraguay, Peru, Poland, Portugal, Romania, Russia, San Marino, Serbia, Slovakia, Slovenia, Spain, Sweden, Switzerland, The Netherlands, Turkey, Ukraine, United Kingdom, Vatican City
Amount of Grant: 500 - 10,000 USD

Sample: James River Association, Mechanicsville, VA, $4,000 - for general support; Pan American Development Foundation, Washington, D.C., $5,000 - for environmental operations; College of Wooster, Wooster, OH, $9,000 - for general support;
Contact: Mareen D. Hughes, President; (703) 534-9109
Sponsor: Charles Delmar Foundation
5205 Leesburg Pike, Suite 209
Falls Church, VA 22041-3858

Charles Edison Fund Grants 1116
The Charles Edison Fund is an endowed philanthropic institution, incorporated in Delaware in 1948, dedicated to the support of worthwhile endeavors generally within the areas of medical research, science education and historic preservation. Funding is available on a national level but, the majority of institutions and, organizations assisted are based principally in the New York-New Jersey Metropolitan area. This concentration of interest facilitates the efforts of the Trustees to evaluate the work of recipient groups, frequently accomplished by personal visitations. The Fund is a seed money organization, primarily starting projects which would not otherwise get off the ground. The Fund meets three times a year, usually in February or March, June and December at which time requests which have been submitted at least three weeks prior to the meeting will be considered.
Requirements: Public and private schools, universities and colleges, and nonprofits are eligible. The Fund recommends that a grant request be submitted on the requesting organization's letterhead and be signed by an official on behalf of the governing board. The Fund does not require or supply application forms. The request should be detailed, complete and include background information about the organization, a full explanation of the project and its costs, and a financial report, current budget and evidence of tax-exempt status of the requesting organization.
Geographic Focus: New Jersey, New York
Contact: John P. Keegan, President; (973) 648-0500; fax (973) 648-0400; info@charlesedisonfund.org
Internet: http://www.charlesedisonfund.org/thefund.html
Sponsor: Charles Edison Fund
One Riverfront Plaza, 4th Floor
Newark, NJ 07102

Charles F. Bacon Trust Grants 1117
The Charles F. Bacon Trust was established in 1928 to support and promote quality educational, human services, and health care programming for underserved populations. Special consideration is given to charitable organizations that serve the needs of elderly women. Grant requests for general operating support are strongly encouraged. Program support will also be considered. Small, program-related capital expenses may be included in general operating or program requests. The application deadline is April 1, and applicants will be notified of grant decisions before June 30.
Requirements: 501(c)3 organizations serving the residents of Massachusetts are eligible.
Restrictions: The majority of Bacon Trust grants are 1 year in duration. On occasion, multi-year support is awarded.
Geographic Focus: Massachusetts
Date(s) Application is Due: Mar 1
Contact: Michealle Larkins; (866) 778-6859; michealle.larkins@baml.com
Internet: https://www.bankofamerica.com/philanthropic/fn_search.action
Sponsor: Charles F. Bacon Trust
225 Franklin Street, 4th Floor, MA1-225-04-02
Boston, MA 02110

Charles H. Dater Foundation Grants 1118
The foundation makes grants to private, nonprofit organizations and public agencies in Greater Cincinnati for programs that benefit children in the region in the areas of arts/culture, education, healthcare, social services, and other community needs. Grants are usually made for one year, and subsequent grants for an extended or ongoing program are based on an evaluation of annual results. Multiple grants to an organization in the foundation's same fiscal year (September through August) are possible, but rare. The foundation looks favorably on applications that leverage a grant to seek additional funding and resources. The foundation directors/officers meet monthly to review grant applications.
Requirements: Nonprofit organizations in the greater Cincinnati area are eligible. This area is defined as the eight-county metropolitan area made up of the counties of Hamilton, Butler, Warren and Clermont in Ohio; Boone, Kenton and Campbell in Northern Kentucky; and Dearborn in Indiana.
Restrictions: The foundation does not make grants to individuals, for scholarships for individuals, for debt reduction, and, with rare exception, for capital fund projects.
Geographic Focus: Indiana, Kentucky, Ohio
Amount of Grant: 10,000 - 20,000 USD
Samples: SON Ministries (Cincinnati, OH)--to provide school supplies and clothing to underprivileged children, $30,000; Children's Home of Cincinnati (OH)--for a summer day camp designed to enhance the creativity and social skills of disadvantaged children, $10,000; Children's Hospital Medical Ctr (Cincinnati, OH)--for salary support of a chaplain in the psychiatry department's Adolescent Medicine Psychiatric Service units, $28,000; Crayons to Computers (Cincinnati, OH)--for programs that collect useful surplus items from businesses and individuals and then redistribute them to schoolteachers, $15,000.
Contact: Beth Broomall; (513) 241-2658; fax (513) 274-2731; bb@DaterFoundation.org
Internet: http://www.daterfoundation.org/grants.php
Sponsor: Charles H. Dater Foundation
602 Main Street, Suite 302
Cincinnati, OH 45202

Charles H. Farnsworth Trust Grants 1119

The Charles H. Farnsworth Trust was established in 1930 to assist older adults to live in dignity and with independence. In describing the purpose of his legacy, Mr. Farnsworth made clear his interest in supporting housing, particularly in developing affordable housing options, and in providing support services to older adults. Program interests include: development of housing, especially housing with support services; services for elderly persons (i.e. health care, homemaker assistance, and nutritional support to enable the elderly to continue living in the community); and research, planning, and communication to better inform individuals, institutions, and the community at large on ways to improve the quality and quantity of housing and support services for seniors. Capital grants related to construction or renovation of housing for older adults is of particular interest. While general operating grants are provided, requests for support for new or special projects/programs are preferred. Planning grants investigating strategies for the development of supportive housing for the elderly are encouraged. Capital grants generally range between $25,000 and $250,000. General operating grants are usually no more than $10,000. Applicants must apply online at the grant website. Applicants are strongly encouraged to do the following before applying: review the downloadable state application procedures for additional helpful information and clarifications; review the downloadable online-application guidelines at the grant website; review the trust's funding history (link is available from the grant website); review the online application questions in advance; and review the list of required attachments. These will generally include: a list of board members, financial statements (audited, reviewed, or compiled by independent auditor); an organization summary; a list of other funding sources; an IRS Determination letter; and other required documents. All attachments must be uploaded in the online application as PDF, Word, or Excel files. The Farnsworth Trust has biannual deadlines of February 1 and October 15. Grant applicants for the February deadline will be notified of grant decisions by May 31 and applicants for the October deadline will be notified of grant decisions by December 31.
Restrictions: Grant opportunities are restricted to the Greater Boston area.
Geographic Focus: Massachusetts
Date(s) Application is Due: Feb 1; Oct 15
Amount of Grant: 10,000 - 250,000 USD
Samples: Hearth, Boston, Massachusetts, $100,000; Greater Boston Chinese Golden Age Center, Boston, Massachusetts, $289,346; Friends of Saint Josephs Food Pantry, Salem, Massachusetts, $10,000.
Contact: Michealle Larkins; (866) 778-6859; michealle.larkins@baml.com
Internet: https://www.bankofamerica.com/philanthropic/fn_search.action
Sponsor: Charles H. Farnsworth Trust
225 Franklin Street, 4th Floor, MA1-225-04-02
Boston, MA 02110

Charles H. Hall Foundation 1120

The Charles H. Hall Foundation was established in 2007 to support and promote educational, health- and human-services, religious, and arts and cultural programming for underserved populations. Special consideration is given to programs whose purpose is the prevention of cruelty to children or animals. Grants from the Charles H. Hall Foundation are one year in duration. Applicants must apply online at the grant website. Applicants are strongly encouraged to do the following before applying: review the downloadable state application procedures for additional helpful information and clarifications; review the downloadable online-application guidelines at the grant website; review the foundation's funding history (link is available from the grant website); review the online application questions in advance; and review the list of required attachments. These will generally include: a list of board members, financial statements (audited, reviewed, or compiled by independent auditor); an organization summary; a list of other funding sources; an IRS Determination letter; and other required documents. All attachments must be uploaded in the online application as PDF, Word, or Excel files. The Charles H. Hall Foundation application deadline is 11:59 p.m. on December 1. Applicants will be notified of grant decisions by letter within three to four months of the deadline.
Requirements: The foundation specifically serves organizations based in Berkshire, Hampden, Hampshire, or Franklin Counties, Massachusetts. Grants will be considered for specific programs or projects with preference given to organizations that provide direct services.
Restrictions: Applicants will not be awarded a grant for more than 3 consecutive years. The foundation does not support requests from individuals, organizations attempting to influence policy through direct lobbying, or any political campaigns.
Geographic Focus: Massachusetts
Date(s) Application is Due: Dec 1
Amount of Grant: 5,000 - 20,000 USD
Contact: Amy Lynch; (860) 657-7015; amy.r.lynch@baml.com
Internet: https://www.bankofamerica.com/philanthropic/fn_search.action
Sponsor: Charles H. Hall Foundation
200 Glastonbury Boulevard, Suite # 200, CT2-545-02-05
Glastonbury, CT 06033-4056

Charles H. Pearson Foundation Grants 1121

The Charles H. Pearson Foundation Fund was established in 1922 to support and promote quality educational, human-services, and health-care programming for underserved populations. In the area of education, the fund supports academic access, enrichment, and remedial programming for children, youth, adults, and senior citizens that focuses on preparing individuals to achieve while in school and beyond. In the area of health care, the fund supports programming that improves access to primary care for traditionally underserved individuals, health education initiatives and programming that impact at-risk populations, and medical research. In the area of human services the fund tries to meet evolving needs of communities. Currently the fund's focus is on (but is not limited to) youth development, violence prevention, employment, life-skills attainment, and food programs. Grant requests for general operating support are strongly encouraged. Program support will also be considered. Small, program-related capital expenses may be included in general operating or program requests. The majority of grants from the Pearson Fund are one year in duration; on occasion, multi-year support is awarded. Applicants must apply online at the grant website. Applicants are strongly encouraged to do the following before applying: review the downloadable state application procedures for additional helpful information and clarifications; review the downloadable online-application guidelines at the grant website; review the foundation's funding history (link is available from the grant website); review the online application questions in advance; and review the list of required attachments. These will generally include: a list of board members, financial statements (audited, reviewed, or compiled by independent auditor); an organization summary; a list of other funding sources; an IRS Determination letter; and other required documents. All attachments must be uploaded in the online application as PDF, Word, or Excel files. The application deadline for the Charles H. Pearson Foundation Fund is 11:59 p.m. on July 1. Applicants will be notified of grant decisions before September 30.
Requirements: Applicants must have 501(c)3 tax-exempt status.
Restrictions: In general, capital requests are not advised. The fund does not support endowment campaigns, events such as galas or award ceremonies, and costs of fundraising events. The fund does not support requests from individuals, organizations attempting to influence policy through direct lobbying, or any political campaigns.
Geographic Focus: Massachusetts
Date(s) Application is Due: Jul 1
Contact: Michealle Larkins; (866) 778-6859; michealle.larkins@baml.com
Internet: https://www.bankofamerica.com/philanthropic/fn_search.action
Sponsor: Charles H. Pearson Foundation Fund
225 Franklin Street, 4th Floor, MA1-225-04-02
Boston, MA 02110

Charles H. Revson Foundation Grants 1122

The foundation awards grants nationwide in its areas of interest, including urban affairs and public policy, education and higher education, biomedical research policy, and Jewish education and philanthropy. Types of support include capital campaigns, continuing support, fellowships, internship funds, program development, and research. Preference is given to requests serving New York, NY. There are no application forms or deadlines. The board meets in April, June, October, and December.
Restrictions: Grants do not support local or national health appeals or direct service programs, individuals, building construction/renovation, book projects, charity events, travel expenses, or budgetary support.
Geographic Focus: All States
Samples: The Legal Aid Society, New York, NY, $100,000--to support the Economic Security Initiative, which provides essential legal assistance to meet the current economic challenges; Jewish Theological Seminary of America, New York, NY, $500,000--to support the creation of the Center for Pastoral Education; Albert Einstein Collefe of Medicine of Yeshiva University, New York, NY, $170,619--to support a fellow in the Charles H. Revson Senior Fellowships in the Life Sciences Program; Fiscal Policy Institute, Somerville, MA, $200,000--to support research, public education, and advocacy work on budget, economic, and related policy issues that affect low- and moderate-income New Yorkers.
Contact: Maria Marcantonio, Grants Administrator; (212) 935-3340; fax (212) 688-0633; info@revsonfoundation.org
Internet: http://www.revsonfoundation.org/guidelines.html
Sponsor: Charles H. Revson Foundation
55 E 59th Street, 23rd Floor
New York, NY 10022

Charles Lafitte Foundation Grants 1123

The foundation is committed to helping groups and individuals foster lasting improvement on the human condition by providing support to education, children's advocacy, medical research, and the arts. Children's advocacy grants support organizations working to improve the quality of life for children, particularly in relation to child abuse, literacy, foster housing, hunger, and after-school programs. Education grants support innovative programs that work to resolve social service issues, address the needs of students with learning disabilities, provide technology and computer-based education, offer leadership skills education, and support at-risk students. Colleges and universities also receive support for research and conferences. The foundation's medical issues and research grants support healthcare studies, with emphasis on cancer research and treatment, children's health, health education, and promoting healthy living and disease prevention. Art grants support emerging artists and educational art programs.
Requirements: 501(c)3 tax-exempt organizations are eligible. The Foundation does not respond to unsolicited submissions, so submit a letter of inquiry prior to preparing an application.
Restrictions: Political organizations or religious-based programs are not supported.
Geographic Focus: All States
Samples: Girl Scouts of the Jersey Shore, $200,000. Bridge of Books Foundation, $2,500. Hand in Hand, $10,000. Jersey Shore University Medical Center, $700,000.
Contact: Jennifer Vertetis, President; jennifer@charleslafitte.org
Internet: http://charleslafitte.org/grants/overview/
Sponsor: Charles Lafitte Foundation
29520 2nd Ave SW
Federal Way, WA 98023

Charles M. and Mary D. Grant Foundation Grants 1124
The Foundation awards grants primarily in the southeast United States in its areas of interest, including community and economic development, health and human services, environment, and education. The Foundation prefers project support, but considers operating support proposals from organizations with budgets of less than $1 million. Grants are made in September. A minimum of three years must elapse between grant awards. Further information is available at the website.
Requirements: Southeast U.S. 501(c)3 tax-exempt organizations are eligible. Specific grant guidelines are available online with the application, which must be filed online.
Restrictions: No grants are made to individuals or for loans. A minimum of three years must elapse between grant awards.
Geographic Focus: Alabama, Florida, Georgia, Kentucky, Mississippi, North Carolina, South Carolina, Tennessee, Virginia, West Virginia
Date(s) Application is Due: Apr 30
Amount of Grant: 20,000 - 40,000 USD
Samples: Collaborative for the 21st Century Appalachia, Inc., Charleston, West Virginia, first installment of a $60,000 grant to launch the e-Resource and Training Center, $30,000; Georgia Legal Services Program, Inc., Atlanta, Georgia, final installment of a $70,000 grant to support affordable housing work in rural Georgia, $30,000; Louisiana Bucket Brigade, New Orleans, Louisiana, final installment of a $40,000 grant for the Environmental Justice Corps, $20,000.
Contact: Casey Castaneda; (212) 464-2487; casey.b.castaneda@jpmchase.com
Internet: http://fdncenter.org/grantmaker/grant
Sponsor: Charles M. and Mary D. Grant Foundation
J.P. Morgan Private Bank, Philanthropic Services
Dallas, TX 75222-7237

Charles Nelson Robinson Fund Grants 1125
The Charles Nelson Robinson Fund was established in 1970 to support and promote quality educational, human-services, and health-care programming for underserved populations in Hartford, Connecticut. Preference is given to organizations that provide human services programming to underserved adults. Grants from the Robinson Fund are one year in duration. Applicants must apply online at the grant website. Applicants are strongly encouraged to do the following before applying: review the downloadable state application procedures for additional helpful information and clarifications; review the downloadable online-application guidelines at the grant website; review the foundation's funding history (link is available from the grant website); review the online application questions in advance; and review the list of required attachments. These will generally include: a list of board members, financial statements (audited, reviewed, or compiled by independent auditor); an organization summary; a list of other funding sources; an IRS Determination letter; and other required documents. All attachments must be uploaded in the online application as PDF, Word, or Excel files. The Charles Nelson Robinson Fund has an annual deadline of 11:59 p.m. on February 15. Applicants will be notified of grant decisions by letter within two to three months of the proposal deadline.
Requirements: Applicant organizations must have 501(c)3 tax-exempt status and have a principal office located in the city of Hartford, Connecticut.
Restrictions: Grant requests for capital projects will not be considered. Applicants will not be awarded a grant for more than three consecutive years. The fund does not support requests from individuals, organizations attempting to influence policy through direct lobbying, or any political campaigns.
Geographic Focus: Connecticut
Date(s) Application is Due: Feb 15
Samples: Hartford Interval House, Hartford, Connecticut, $5,000, general operating support; Loaves and Fishes Ministries, Hartford, Connecticut, $4,000, general operating support; Bulkeley High School, Hartford, Connecticut, $3,000, VOAG Horse Care Program.
Contact: Carmen Britt; (860) 657-7019; carmen.britt@baml.com
Internet: https://www.bankofamerica.com/philanthropic/fn_search.action
Sponsor: Charles Nelson Robinson Fund
200 Glastonbury Boulevard, Suite # 200, CT2-545-02-05
Glastonbury, CT 06033-4056

Charlotte County (FL) Community Foundation Grants 1126
The foundation awards grants from its unrestricted funds to support innovative solutions for the citizens of Charlotte County through the implementation of projects and programs that address specific identified needs of the community and demonstrate a wide-spread positive impact on its residents. The foundation places priority on funding requests from nonprofit organizations whose project can: Provide an unduplicated value and/or service; Leverage dollars and/or people power; Match foundation funds with funds from other sources; Establish cooperative efforts of two or more organizations where such synergy is likely to demonstrate superior results and/or lower costs; Enhance, expand and/or improve the organization's capabilities; Show visible end results within a specific timeline for completion; Demonstrate that the Foundation grant will play an important role; Provide a meaningful outcomes evaluations process with a project completion report; Reach a large number of Charlotte County residents; and, be completed within one year with a request of less than $7,500.
Requirements: Foundation grants are made to tax exempt organizations that have been in existence for at least two years and whose focus is in one of the following areas of interest: Animal Welfare; Arts and Culture; Community Development; Education; Environment; Health, Human, and Social Services; Historic Preservation; Nonprofit Organization Capacity Building. Additionally, organizations must also show that they have: a Board of Directors composed of individuals of diverse backgrounds, at least half of whom reside in Charlotte County; Strong management and leadership qualities; Demonstrated fiscal responsibility; Submitted its prior year IRS Form 990; Benefited Charlotte County residents. To apply, submit a Letter of Intent (LOI) summarizing your request. (See instructions at the foundation's website.) If the foundation feels that the proposed project has merit, you will be asked to complete a grant proposal package to be submitted as your formal request for funding.
Restrictions: The foundation is generally not interested in requests for: General operating support; Building or capital campaigns; Deficit financing and debt reduction; Endowment funds; Fraternal organizations, societies or orders; Loans or assistance to individuals; Religious organizations for sectarian purposes; Lobbying legislators or influencing elections; Political organizations or campaigns; Fund raising events; Basic scientific research; Start up organization funding; Travel expenses for individuals or groups; or, Re-granting.
Geographic Focus: Florida
Date(s) Application is Due: Apr 5; May 5; Jul 6; Sep 16
Amount of Grant: Up to 7,500 USD
Contact: Gregory C. Bobonich, J.D.Chief Executive Officer; (941) 637-0077; gbobonich@charlottecommunityfoundation.org
Internet: http://www.charlottecommunityfoundation.org/index.php?subcat=20&articleid=68&page_num=1#
Sponsor: Charlotte County Community Foundation
1675 West Marion Avenue
Punta Gorda, FL 33950

Chase Paymentech Corporate Giving Grants 1127
The corporation awards grants to nonprofit organizations that provide needed services in the areas of education, health and human services. Preference for funding is given to organizations whose work impacts the citizens of those areas in which Paymentech offices are located, including Westerville, OH; Dallas, TX; Salem, NH; Silver Springs, MD; Tampa, FL; Tempe, AZ; and Toronto, Canada. The Corporate Giving Committee meets each quarter. Organizations are notified about funding status after each meeting.
Requirements: 501(c)3 non-profit organizations that provide needed services in the areas of education and health and human services are eligible. Applicants submit a written request on the organization's letterhead detailing the organization's mission, the amount requested and a detailed explanation of the purpose for which the funds would be used. Organizations should also submit proof of non-profit status.
Restrictions: Paymentech does not make grants to individuals; religious, fraternal, political, or veterans organizations; or colleges and universities. Nor is funding provided for the following purposes: or deficit spending or debt liquidation; annual capital campaigns of hospitals, colleges, universities, grade schools, or high schools.
Geographic Focus: Arizona, Florida, Maryland, New Hampshire, Ohio, Texas, Canada
Date(s) Application is Due: Mar 30; Jun 29; Sep 28; Dec 31
Amount of Grant: Up to 5,000 USD
Contact: Corporate Giving Director
Internet: http://www.chasepaymentech.com/portal/community/chase_paymentech/public/public_website/about_us/company_information_pages/corporate_giving_program
Sponsor: Chase Paymentech
14221 Dallas Parkway, Building Two
Dallas, TX 75254

Chatlos Foundation Grants 1128
The Chatlos Foundation supports nonprofit organizations in the USA and around the globe. Support is provided to organizations currently exempt by the Internal Revenue Service of the United States. The Foundation's areas of interest are: Bible Colleges/Seminaries, Religious Causes, Medical Concerns, Liberal Arts Colleges and Social Concerns.
Requirements: Applicants must be U.S. tax-exempt, nonprofit organizations that provide services in the following areas: bible colleges, religious causes, medical concerns, liberal arts colleges, and social concerns. Proposals must include cover letter, specific request, tax-exemption letter, and budget. If proposal is to be considered at board level, additional information will be requested.
Restrictions: The foundation will not accept requests from individual church congregations, individuals, organizations in existence for less than two years as indicated by IRS tax-exempt letter of determination, for education below the college level, for medical research projects, or for support of the arts.
Geographic Focus: All States
Amount of Grant: 10,000 - 25,000 USD
Contact: C. J. Leff, Administrator; (407) 862-5077; cj@chatlos.org
Internet: http://www.chatlos.org/AppInfo.htm
Sponsor: Chatlos Foundation
P.O. Box 915048
Longwood, FL 32791-5048

Chazen Foundation Grants 1129
The foundation awards grants nationwide to eligible nonprofit organizations in its areas of interest, including arts; business education; higher education; hospitals; human services; Israel; Jewish agencies, temples, and federated giving programs; museums; and music performance. Types of support include building construction/renovation, capital campaigns, general operating support, grants and scholarships to individuals, professorships, and scholarship funds. Education grants are awarded to students of the Rockland County, NY, area. There are no application deadlines.
Geographic Focus: All States
Contact: Grants Administrator; (212) 750-6600
Internet: http://www.chazenscholar.com/project.php3
Sponsor: Chazen Foundation
767 5th Avenue, 26th Floor
New York, NY 10153-2696

Chemtura Corporation Contributions Grants 1130
Chemtura makes charitable contributions to nonprofit organizations involved with education, health care, human services, economic development. Special emphasis is directed towards programs designed to provide educational and economic opportunities for disadvantaged people. Types of support include: building and renovation; general operating funding; and in-kind gifts. Support is given in areas of company operations in California, Connecticut, Georgia, Illinois, Indiana, and Pennsylvania, and on an international basis in areas of company operations.
Geographic Focus: Connecticut, Georgia, Illinois, Indiana, Pennsylvania
Contact: Grants Director; (215) 446-3911
Internet: http://www.chemtura.com/corporatev2/v/index.jsp?vgnextoid=5fe438f220d6d210VgnVCM1000000753810aRCRD&vgnextchannel=5fe438f220d6d210VgnVCM1000000753810aRCRD&vgnextfmt=default
Sponsor: Chemtura Corporation
1818 Market Street, Suite 3700
Philadelphia, PA 19103-3640

Chest Foundation/LUNGevity Foundation Clinical Research in Lung Cancer Grants 1131
The goal of this joint clinical research program is, ultimately, to save the lives of those people afflicted with lung cancer by funding innovative research designed to treat and cure lung cancer. A clinical research award of $75,000 will be granted to a single award recipient for clinical research in lung cancer over a period of 2 years. The amount of $37,500 will be granted after July 1. Upon receipt of the Progress Report received before July 15, the remaining $37,500 will be granted.
Requirements: To be considered a candidate for The Chest Foundation/LUNGevity Foundation Clinical Research Award in Lung Cancer, an individual must: be an ACCP member who has completed at least 2 years of pulmonary or oncology fellowship and be within 5 years of fellowship or no more than 5 years in practice; hold a degree of MD, DO, MB,BCh, PharmD, PhD or its equivalent; can be a citizen of the United States, or have a J-1 or H-1 visa, and be working in a certified United States institution or organization, or be a Canadian or international member who is working in a certified institution or organization in his/her country; and be a clinician in lung cancer research.
Geographic Focus: All States
Date(s) Application is Due: Apr 30
Amount of Grant: 75,000 USD
Contact: Sue Ciezadlo; (847) 498-8363; fax (847) 498-5460; sciezadlo@chestnet.org
Internet: http://www.chestfoundation.org/foundation/clinical/LUNGevityAward.php
Sponsor: Chest Foundation
3300 Dundee Road
Northbrook, IL 60062-2348

Chest Foundation Eli Lilly and Company Distinguished Scholar in Critical Care Medicine Award 1132
The Eli Lilly and Company Distinguished Scholar in Critical Care Medicine will have a 3-year opportunity to examine issues that relate to critically ill patients. The award is intended to permit the investigation of issues that are not easily supported by traditional funding, such as clinical trials or basic science. Rather, the award would support activities such as the development of public policy, patient education models, or economic analysis of treatment or care delivery in this patient group. A stipend of $50,000 annually over the 3-year term of the program will be provided. A fourth year stipend of $10,000 will also be awarded to the individual who will serve in a mentorship role to the fourth Distinguished Scholar position during his/her inaugural year of service.
Requirements: An applicant must be: an FCCP and hold the degree of MD, DO, MBChB, MBBCh, MBBS, DNSc, PharmD, PhD, or EdD; board-certified in critical care medicine; and a recognized clinician and/or scientist with a specialty in sepsis and/or critical care as evidenced by publications, presentations, and/or peers.
Geographic Focus: All States
Date(s) Application is Due: Apr 30
Amount of Grant: Up to 160,000 USD
Contact: Sue Ciezadlo; (847) 498-8363; fax (847) 498-5460; sciezadlo@chestnet.org
Internet: http://www.chestfoundation.org/foundation/clinical/dsCriticalCare.php
Sponsor: Chest Foundation
3300 Dundee Road
Northbrook, IL 60062-2348

Chest Foundation Geriatric Development Research Awards 1133
The two-year award is intended to provide the impetus required for long-term career development focused on integrating geriatrics into the sub-specialties of internal medicine. The program awards the grant to an academic internist to develop and implement a basic, clinical, or health services research project focused on a geriatric aspect of chest medicine. The funding can support the salary of the award recipient and/or the purchase of supplies, the salaries of technical personnel, and other resources necessary for the completion of the research project. The award also includes a one-time travel grant to attend the meetings of the American Geriatrics Society and the American College of Chest Physicians in the second year of the award.
Requirements: To be eligible for the award, applicants must: have U.S. citizenship or permanent resident status; hold a degree of MD or its equivalent; have completed a sub-specialty internal medicine fellowship leading to certification in his/her sub-specialty by the American Board of Internal Medicine and be within the first three years of his/her faculty appointment; possess a faculty appointment by July 1 (this faculty appointment should be documented in the required letters of recommendation from the applicant's department chair and division director); be a member of the American College of Chest Physicians; commit 75 percent of his/her professional effort to research activities; develop and implement a basic, clinical, or health services research project focused on a geriatric aspect of chest medicine; and generate and implement a career development plan focused on the geriatrics aspects of chest medicine. This plan must include organizing and interacting with a mentorship team made up of a minimum of four members--the applicant's research mentor as the leader of the research team, a sub-specialist in the applicant's field of chest medicine, a geriatrician, and one other member at the applicant's discretion.
Restrictions: The funding cannot be used to acquire administrative or clerical support. Funding is to cover total costs; no indirect cost funds will be provided.
Geographic Focus: All States
Date(s) Application is Due: Apr 30
Amount of Grant: 53,000 USD
Contact: Sue Ciezadlo; (847) 498-8363; fax (847) 498-5460; sciezadlo@chestnet.org
Internet: http://www.chestfoundation.org/foundation/clinical/geriatricAward.php
Sponsor: Chest Foundation
3300 Dundee Road
Northbrook, IL 60062-2348

Chest Foundation Grant in Venous Thromboembolism 1134
The Chest Foundation is issuing this Request for Proposal (RFP) that will fund two projects with up to a total grant amount of $180,000 each for a 1-year period. The goal of the grant process is to confer awards to projects that include evidence-based replicable programs and tools to address the current gaps in the management and prophylaxis in venous thromboembolism (VTE). The deadline is February 20.
Requirements: To be considered a candidate for The Chest Foundation Grant in Venous Thromboembolism, an individual must: be a current ACCP member and hold the degree of MD, DO, MBChB, MBBCh, MBBS, DNSc, PharmD, PhD, or EdD; and be a recognized scientist focused on an area of VTE or clinical decision making as evidenced by academic productivity and/or publications and presentations.
Geographic Focus: All States
Date(s) Application is Due: Feb 20
Amount of Grant: Up to 180,000 USD
Contact: Sue Ciezadlo; (847) 498-8363; fax (847) 498-5460; sciezadlo@chestnet.org
Internet: http://www.chestfoundation.org/foundation/clinical/VT_Grant.php
Sponsor: Chest Foundation
3300 Dundee Road
Northbrook, IL 60062-2348

Chicago Institute for Psychoanalysis Fellowships 1135
This is a program for advanced trainees in psychiatry, social work, and psychology who have significant interest in psychoanalysis as a body of knowledge and as a framework with which to understand and carry out therapeutic work. It is jointly sponsored by the Chicago Psychoanalytic Society and the Institute for Psychoanalysis. A mentor is assigned to each Fellow, who is available throughout the year-long program to discuss cases, papers, research, etc. In addition, Fellows are guests at Society meetings and conferences held locally. Opportunities are provided for the Fellows to discuss presented papers with their mentors. A monthly conference is held to acquaint Fellows with basic psychoanalytic clinical concepts through readings and discussion. Each Fellow receives a copy of the Annual of Psychoanalysis and other publications.
Geographic Focus: All States
Date(s) Application is Due: Aug 15
Contact: Christine Susman, Director; (312) 922-7474, ext. 324; csusman@chicagoanalysis.org or admin@chicagoanalysis.org
Internet: http://www.chicagoanalysis.org/fellow.php
Sponsor: Chicago Institute for Psychoanalysis
122 S Michigan Avenue
Chicago, IL 60603

Children's Brain Tumor Foundation Research Grants 1136
The Children's Brain Tumor Foundation (CBTF), a national not-for-profit organization, was founded in 1988 by a group of dedicated parents, physicians and friends to improve the treatment, quality of life, and long-term outlook for children with brain and spinal cord tumors through research, support, education, and advocacy. The number one priority of CBTF is the awarding of grants for research into the causes of and effective treatments for pediatric brain and spinal cord tumors. The grant cycle begins in the spring of each year. Investigators from the U.S. and Canada should submit a pre-application form to be reviewed by an expert panel. Based on this initial review, a selected group of applicants will then be invited to submit a full proposal.
Requirements: Funding is currently restricted to principal investigators at institutions within the United States.
Restrictions: The foundation does not award grants to individuals or to private foundations, and does not fund debt reduction, capital improvements, or travel expenses. Overhead expenses are not to exceed 10 percent of the total project cost.
Geographic Focus: All States
Amount of Grant: Up to 150,000 USD
Contact: Joseph B. Fay, Executive Director; (212) 448-9494 or (866) 228-4673; fax (212) 448-1022; jfay@cbtf.org or info@cbtf.org
Internet: http://www.cbtf.org/grant_info.html
Sponsor: Children's Brain Tumor Foundation
274 Madison Avenue, Suite 1004
New York, NY 10016

Children's Cardiomyopathy Foundation Research Grants   1137
The foundation awards funds to support research related to all forms of cardiomyopathy affecting children under the age of 18 years. The goal of CCF's research program is to advance medical knowledge on the disease and develop more accurate diagnostic methods, life-improving therapies and ultimately a cure. The grant program is designed to provide seed funding to investigators for the testing of initial hypotheses and collecting of preliminary data to help secure long-term funding by the National Institutes of Health and other major granting institutions.
Requirements: Proposals will be accepted on an annual basis for innovative basic, clinical or translational research relevant to the cause or treatment of cardiomyopathy in children. Principal investigators must hold a MD, PhD or equivalent degree and reside in the United States or Canada. The investigator must have a faculty appointment at an accredited U.S. or Canadian institution and have the proven ability to pursue independent research as evidenced by original research in peer-reviewed journals. Guidelines and application forms can be downloaded at the website.
Geographic Focus: All States, Canada
Date(s) Application is Due: Oct 5
Amount of Grant: 25,000 - 50,000 USD
Samples: 2007: Tain-Yen Hsia, MD, Medical University of South Carolina - ($50,000) Extracellular Mechanisms in Pediatric Cardiomyopathy; Anne I. Dipchand, MD, Hospital for Sick Children, Toronto, Canada - ($42,642) Outcome of pediatric patients with cardiomyopathy: a multi-centre review of pediatric patients listed for transplant in the Pediatric Heart Transplant Study. 2006: Tracie Miller, MD, University of Miami, Miami, FL - ($45,000) Exercise Intervention in a Pediatric Population with Cardiomyopathy
Contact: Lisa Yue, Executive Director; (866) 808-2873; fax (201) 227-7016; lyue@childrenscardiomyopathy.org or info@childrenscardiomyopathy.org
Internet: http://www.childrenscardiomyopathy.org/site/grants.php
Sponsor: Children's Cardiomyopathy Foundation (CCF)
P.O. Box 547
Tenafly, NJ 07670

Children's Leukemia Research Association Research Grants   1138
The Children's Leukemia Research Association supports research efforts into the causes and cure of leukemia and gives patient aid to families in need while meeting the expenses incurred in leukemia treatment. A medical advisory committee consisting of prominent internationally known and respected hematologists reviews grant proposals submitted for consideration of projects that are not otherwise funded. Grants are available for start-up funding for laboratory or clinical investigations in leukemia. Although the association prefers to fund new investigators, applications from established investigators for new initiatives are also solicited. Renewal for a second year is considered if other funding for promising projects has not been obtained. Grants are limited to a maximum of $30,000 each. The annual application deadline is June 30.
Requirements: Any doctor at the Ph.D. or MD level who is involved in research towards finding the causes and cure for leukemia may apply.
Restrictions: Previously funded projects are not usually considered for additional funding.
Geographic Focus: All States
Date(s) Application is Due: Jun 30
Amount of Grant: Up to 30,000 USD
Contact: Aileen T. Sullivan, Administrative Assistant; (516) 222-1944; fax (516) 222-0457; info@childrensleukemia.org
Internet: http://www.childrensleukemia.org/researchgrants.html
Sponsor: Children's Leukemia Research Association
585 Stewart Avenue, Suite 18
Garden City, NY 11530

Children's Trust Fund of Oregon Foundation Grants   1139
The mission of the Children's Trust Fund of Oregon Foundation (CTFO) is to foster healthy child development and to support efforts to protect children in Oregon through strategic investments in local, proven or evidence-based child abuse prevention programs. To achieve this goal, CTFO awards grants to programs which can demonstrate their impact in preventing child abuse in Oregon. CTFO prioritizes grants for programs or program components which nurture and protect children by strengthening families and providing parents and caretakers with the education, skills and resources for healthy child development. The annual deadline for applications is April 1, and awards typically range up to a maximum of $20,000. Potential applicants should begin by contacting the grant office.
Requirements: Oregon public and private nonprofit organizations are eligible.
Geographic Focus: Oregon
Date(s) Application is Due: Apr 1
Amount of Grant: Up to 20,000 USD
Contact: Susan Lindauer; (503) 222-7102; fax (503) 222-6975; susan@ctfo.org
Internet: http://ctfo.org/grants/
Sponsor: Children's Trust Fund of Oregon Foundation
1785 NE Sandy Boulevard, Suite 270
Portland, OR 97232

Children's Tumor Foundation Clinical Research Awards   1140
In order to accelerate neurofibromatosis research, the Children's Tumor Foundation launched the Clinical Research Award (CRA) program in 2007. The program's primary purpose is to support early stage pilot clinical trials of candidate therapeutics or interventions for treatment of manifestations of NF1, NF2 and schwannomatosis. In addition, these awards support innovative clinical trial enabling studies. It is hoped that applications for Clinical Research Awards will be broad-thinking, novel ideas. Examples of projects encouraged may include but are not limited to the following: preclinical and clinical collaborative studies (note that all studies must include a clinical element – preclinical-only studies are not eligible); pilot clinical trials (CTF leverages our relationship with industry or larger funding organizations to provide drug at no cost to successful applicants); and clinical enabling studies such as biomarker and patient stratification that may better inform and help accelerate a clinical trial. It is anticipated that the Foundation will fund up to three Clinical Research Awards annually of up to $150,000 each (including overheads and indirect costs). Letters of Intent are required by July 1, with invited full applications due by August 17.
Geographic Focus: All States
Date(s) Application is Due: Aug 17
Amount of Grant: Up to 150,000 USD
Contact: Patrice Pancza, Research Program Director; (212) 344-7291 or (212) 344-6633; fax (212) 747-0004; ppancza@ctf.org or grants@ctf.org
Internet: http://www.ctf.org/CTF-Awards-Grants-and-Contracts/Clinical-Research-Awards/
Sponsor: Children's Tumor Foundation
120 Wall Street, 16th Floor
New York, NY 10005

Children's Tumor Foundation Drug Discovery Initiative Awards   1141
Founded in 1978 as the National Neurofibromatosis Foundation, the Children's Tumor Foundation is a non-profit organization committed to identifying effective drug therapies for neurofibromatosis type 1 (NF1), neurofibromatosis type 2 (NF2) and schwannomatosis, and to improving the lives of those living with these disorders. The Drug Discovery Initiative Awards program supports early stage testing of candidate drug therapies for the treatment of neurofibromatosis (NF): NF1, NF2 and schwannomatosis. These Awards have yielded over $5M in follow-on funding from the federal government and other sources, as well as multiple industry collaborations and publications. DDI Awards is a catalyst program that has helped to fuel the drug pipeline with promising leads. Funding is as follows: up to $40,000 DDI in vitro Awards to fund cell-based preclinical drug testing studies; and up to $85,000 DDI in vivo awards to fund animal-based preclinical drug testing studies. Applications are welcomed from both the academic and private sectors. Applications are due by August 25.
Requirements: Applicants should have an MD, PhD, or equivalent, and have full access to, or identified collaborators with, all required resources including all in vivo and in vitro models. There are no citizenship requirements.
Geographic Focus: All States, All Countries
Date(s) Application is Due: Aug 25
Amount of Grant: Up to 85,000 USD
Contact: Patrice Pancza, Research Program Director; (212) 344-7291 or (212) 344-6633; fax (212) 747-0004; ppancza@ctf.org or grants@ctf.org
Internet: http://www.ctf.org/CTF-Awards-Grants-and-Contracts/Drug-Discovery-Initiative/
Sponsor: Children's Tumor Foundation
120 Wall Street, 16th Floor
New York, NY 10005

Children's Tumor Foundation Schwannomatosis Awards   1142
The Children's Tumor Foundation is focused on ending neurofibromatosis (NF) through research. To date CTF has committed over $35M to NF research, ranging from preclinical drug testing to clinical research and a national NF Clinic Network. Schwannomatosis is the rarest form of NF, affecting 1 in every 40,000 persons, and causing peripheral nerve tumors and unmanageable pain. In 2007 a candidate schwannomatosis gene, INI-1/Smarc-B1/Snf-5 was identified. Harnessing this discovery, over the past several years CTF has: convened a series of Schwannomatosis Workshops to identify priorities for advancing schwannomatosis research; and funded schwannomatosis research totaling over $1.3 million since 2007. Results from this include the first schwannomatosis mouse models, utilizing these for preclinical drug trials, advancing schwannomatosis genetics, and establishing an international Schwannomatosis Database. CTF is offering Schwannomatosis Awards of up to $75,000 each. Schwannomatosis Awards may be requested for research in any area of relevance to advancing schwannomatosis research in the following priority areas: genetics; cell biology and translational research; and clinical research.
Geographic Focus: All States
Contact: Patrice Pancza, Research Program Director; (212) 344-7291 or (212) 344-6633; fax (212) 747-0004; ppancza@ctf.org or grants@ctf.org
Internet: http://www.ctf.org/Research/Schwannomatosis-Awards.html
Sponsor: Children's Tumor Foundation
120 Wall Street, 16th Floor
New York, NY 10005

Children's Tumor Foundation Young Investigator Awards   1143
Founded in 1978, the Children's Tumor Foundation is a non-profit organization committed to identifying effective drug therapies for neurofibromatosis type 1 (NF1), neurofibromatosis type 2 (NF2) and Schwannomatosis, and to improving the lives of those living with these disorders. The Young Investigator (YIA) Awards program provides two-year awards for young scientists early in their careers, bringing them into the NF field and helping to establish them as independent investigators. Though a number of YIAs have made significant research findings and made notable publications the main function of the YIA program has been as a 'seeding mechanism' for researchers who went on to secure larger grants from NIH and CDMRP NFRP. Fellowship amounts range from $32,000 to $54,000, depending upon: level of training at the time of the application; and the request year (1st or 2nd).
Requirements: Applicants must be one of the following: a postdoctoral fellow (MD, PhD, or equivalent) but no more than seven years past the completion of their first doctoral degree; or a graduate student pursuing an MD, PhD, or equivalent.

Geographic Focus: All States
Amount of Grant: 32,000 - 54,000 USD
Contact: Patrice Pancza, Research Program Director; (212) 344-7291 or (212) 344-6633; fax (212) 747-0004; ppancza@ctf.org or grants@ctf.org
Internet: http://www.ctf.org/CTF-Awards-Grants-and-Contracts/CTF-Young-Investigator-Award/
Sponsor: Children's Tumor Foundation
120 Wall Street, 16th Floor
New York, NY 10005

Children Affected by AIDS Foundation Camp Network Grants   1144
The mission of the CAAF Camp Network is to increase the number of children that attend camp every year by providing grants; to foster communication among agencies providing camp experiences for children and their families; and to support the development of new camps. Members of the network gain access to the competitive funding process from the CAAF's yearly financial allocation to the Camp Network.
Requirements: For consideration for membership, Camp Service Providers need to provide proof of the following: (a) Proof of 501(c)3 Not-for-Profit Status; (b) 990 IRS form if applicable; (c) Experience with children infected with or affected by HIV/AIDS and providing camp services. Camps seeking funding from CAAF need to provide proof of the following: (1) Proof of minimum Insurance Coverage's as recommended by the Camp Network; (2) Indemnity for CAAF (Camp Network) from camp agencies, including any employment related claims (Contractual provision, prior to receiving a CAAF grant award.); (3) Compliance with applicable laws such as: employment, health and safety laws and the Americans with Disabilities Act. This will be a contractual provision of the grant agreement, before receiving a CAAF grant award; (4) Release of Liability from Parent or Guardians and Children, using unique identifiers; and, (5) Evidence that the Camp has policies and procedures in place for: (a) Children's safety and emergencies; (b) Pre-screening of paid and non-paid staff; (c) Other information as needed.
Geographic Focus: All States
Date(s) Application is Due: May 16
Amount of Grant: 1,000 - 15,000 USD
Contact: Rolla Bedford, (310) 258-0850, ext. 14; rolla.bedford@caaf4kids.org
Internet: http://www.caaf4kids.org/
Sponsor: Children Affected by AIDS Foundation
6033 W Century Boulevard, Suite 603
Los Angeles, CA 90045

Children Affected by AIDS Foundation Domestic Grants   1145
The foundation awards grants to nonprofit organizations that provide for the medical, emotional, social, and basic needs of children who are infected with HIV or affected by AIDS. Organizations can request funds for their programs that directly impact the lives of children affected by AIDS by submitting a proposal during the foundation's once-a-year application period. Special needs requests are considered throughout the year, depending on funding availability. Funding categories include Basic Needs (clothing, formula, food, diapers, burial services, transportation to/from appointments, emergency child care, and respite services); Social & Recreational (social activities such as arts and crafts, sports, holiday parties, field expeditions, and other forms of recreation); Psychosocial Support (intended to benefit direct services related only to the provision of psychosocial services to children which may include individual and family therapy/counseling, support groups, and life skills classes); Guidelines and application are available online.
Requirements: 501(c)3 nonprofit organizations that provide direct care, support, and assistance to children, ages birth to 13 years, who are HIV-positive or otherwise affected by HIV/AIDS are eligible. This includes children who are HIV-infected, those who are or may become orphaned as a result of losing parents to HIV disease, and those who have siblings, parents, or caregivers who are infected with HIV.
Restrictions: Individuals, research activities, fraternal groups, political or religious activities, and organizations not in good standing with CAAF are not eligible. Additionally, CAAF will not grant funds to pass-through organizations.
Geographic Focus: All States
Date(s) Application is Due: May 16
Amount of Grant: 1,000 - 25,000 USD
Contact: Rolla Bedford, (310) 258-0850, ext. 14; rolla.bedford@caaf4kids.org
Internet: http://www.caaf4kids.org
Sponsor: Children Affected by AIDS Foundation
6033 W Century Boulevard, Suite 603
Los Angeles, CA 90045

Children Affected by AIDS Foundation Family Assistance Emergency Fund Grants   1146
The foundation is making available its Emergency Fund to help HIV/AIDS Service Organizations in the U.S. to meet the needs of HIV-impacted children and their families. Funds can be accessed by qualified HIV/AIDS non-profits with IRS 501(c)3 status. The Foundation will only provide grants to organizations that do not discriminate on the basis of race, color, religion, gender, gender identity, gender expression, age, sexual orientation, marital status, national origin, disability, or other characteristics protected by law. The goal of the Emergency Fund is to meet the pressing, unanticipated needs in the day-to-day life of children, up to 13 years of age, and their families. Requests may be submitted at any time, and will be reviewed shortly after being received.
Requirements: Funds requested must directly improve the well-being of children in the household. Funds can be accessed by qualified HIV/AIDS non-profits with IRS 501(c)3 status. Eligible requests include: Food; Utility assistance; School uniforms, clothing; Transportation to medical/social service appointments; Medications and medical co-pays; Transitional housing assistance; Burial expenses. Other needs not listed may be considered-please contact the Foundation Program staff prior to sending a request. To download the Emergency Fund's guidelines, go to the sponsor's site, click on 'Initiatives' on the home page, then 'Family Assistance.'
Restrictions: The emergency fund cannot be accessed by or for the following purposes: political or fraternal organizations, or for research or religious purposes.
Geographic Focus: All States
Amount of Grant: 50 - 2,000 USD
Contact: Rolla Bedford, (310) 258-0850, ext. 14; rolla.bedford@caaf4kids.org
Internet: http://www.caaf4kids.org/
Sponsor: Children Affected by AIDS Foundation
6033 W Century Boulevard, Suite 603
Los Angeles, CA 90045

Chiles Foundation Grants   1147
The foundation has a deep concern for and confidence in the future of Oregon and the Pacific Northwest. Although the foundation has made a steady commitment to the improvement of the quality of life for those who live and work in this area, it is not restricted in its grant making to the Pacific Northwest. The foundation has traditionally made grants to certain select institutions of higher education for business schools, scholarships, and athletics; supports basic research in certain select medical institutions; supports religion through divinity schools and religious education; and believes that the arts and cultural activities of a community are important and supports certain select, established institutions. Types of support include building construction/renovation, equipment acquisition, and scholarship funds. Annual deadline dates may vary; contact Foundation by a letter of Inquiry. The Foundation does not accept unsolicited proposals.
Requirements: The preferred initial method of contact is a phone call to the grants administration office to determine whether a prospective proposal is within guidelines; if so, an applicant will be invited to submit a one-page written preliminary proposal. An application form will be sent after approval of the preliminary proposal by the executive committee.
Restrictions: No support for projects involving litigation. Grants are not made to individuals, for deficit financing, mortgage retirement, or projects and conferences already completed.
Geographic Focus: California, Oregon, Germany
Amount of Grant: 1,000 - 270,000 USD
Contact: Earle M. Chiles; (503) 222-2143; fax (503) 228-7079; cf@uswest.net
Sponsor: Chiles Foundation
111 SW Fifth Avenue, Suite 4050
Portland, OR 97204-3643

Chilkat Valley Community Foundation Grants   1148
The Chilkat Valley Community Foundation uses proceeds from its growing community permanent fund to award yearly grants to worthwhile programs in the Chilkat Valley. These grants are intended to support organizations and programs in the community that serve the needs of people in such areas as health, education, human services, arts and culture, youth, environment, and community development. Applications are being accepted for the following three (3) categories: operating support; new program and special projects; and capital campaigns. Funding for projects during this grant cycle will range from $500 to $3,500. Most recent awards have ranged from $500 to $2,000. The annual deadline for application submission is September 30.
Requirements: Applications are accepted from qualified 501(c)3 nonprofit organizations, or equivalent organizations located in the state of Alaska and serving the Chilkat Valley region. Equivalent organizations may include tribes, local or state governments, schools, or Regional Educational Attendance Areas. Operating Support Grants may be awarded to sustainable organizations in amounts not to exceed 10% of the organization's secured cash annual budget. Capital Grants may be awarded as the local match to another funding source. New Program and Special Project Grants may be awarded for programs and projects that are not undertaken on an annual basis. A grant requesting $1,000 or more is a challenge grant at a ratio of 1:1 (grantees must raise $1 to receive $1). Grants of $500 to $999 do not require a match. The recipient's match must be raised within twelve months of the award notification and must be raised from at least five (5) different donors.
Geographic Focus: Alaska
Date(s) Application is Due: Sep 30
Amount of Grant: 500 - 3,000 USD
Contact: Ricardo Lopez, Program Officer; (907) 274-6707 or (907) 766-6868; fax (907) 334-5780; rlopez@alaskacf.org
Internet: http://chilkatvalleycf.org/projects/
Sponsor: Chilkat Valley Community Foundation
P.O. Box 1117
Haines, AK 99827

Christensen Fund Regional Grants   1149
The fund (TCF) focuses its grantmaking on maintaining the biological and cultural diversity of the world by focusing on four geographic regions: the greater South West (Southwest United States and Northwest Mexico); Central Asia and Turkey; the African Rift Valley (Ethiopia); and Northern Australia and Melanesia. Grants within these programs are generally directed to organizations based within those regions or, where appropriate, to internationally based organizations working in support of people and institutions on the ground. In general, grants are one year or less; currently grants up to two years are by invitation only.
Requirements: 501(c)3 nonprofit organizations and non-USA institutions with nonprofit or equivalent status in their country of origin are eligible. Partnerships or associations with USA-based nonprofit organizations are preferred.

Restrictions: The fund does not make grants directly to individuals but rather assists individuals through institutions qualified to receive nonprofit support with which such individuals are affiliated.
Geographic Focus: All States
Amount of Grant: Up to 200,000 USD
Contact: Administrator; (415) 644-1600; fax (415) 644-1601; info@christensenfund.org
Internet: http://www.christensenfund.org/index.html
Sponsor: Christensen Fund
260 Townsend Street
San Francisco, CA 94107

Christine and Katharina Pauly Charitable Trust Grants — 1150

The Christine and Katharina Pauly Charitable Trust was established in 1985 to support and promote quality educational, health, and human-services programming for underserved populations. Special consideration is given to charitable organizations that serve the needs of children or older adults. The majority of grants from the Pauly Trust are one year in duration. Applicants must apply online at the grant website. Applicants are strongly encouraged to do the following before applying: review the downloadable state application procedures for additional helpful information and clarifications; review the downloadable online-application guidelines at the grant website; review the trust's funding history (link is available from the grant website); review the online application questions in advance; and review the list of required attachments. These will generally include: a list of board members, financial statements (audited, reviewed, or compiled by independent auditor); an organization summary; a list of other funding sources; an IRS Determination letter; and other required documents. All attachments must be uploaded in the online application as PDF, Word, or Excel files. The application deadline for the Christine and Katharina Pauly Charitable Trust is 11:59 p.m. on September 1. Applicants will be notified of grant decisions by December 31. The Christine and Katharina Pauly Charitable Trust was created under the wills of Ms. Hazel Katharina Pauly and Ms. Frieda Christine Oleta Pauly.
Requirements: Applicants must have 501(c)3 tax-exempt status.
Restrictions: In general, grant requests for individuals, endowment campaigns, or capital projects will not be considered. The Fund will consider requests for general operating support only if the organization's operating budget is less than $1 million. The trust does not support requests from individuals, organizations attempting to influence policy through direct lobbying, or any political campaigns.
Geographic Focus: Missouri
Date(s) Application is Due: Sep 1
Contact: Scott Berghaus; (816) 292-4342; tony.twyman@ustrust.com
Internet: https://www.bankofamerica.com/philanthropic/foundation.go?fnId=6
Sponsor: Christine and Katharina Pauly Charitable Trust
231 South LaSalle Street, IL1-231-13-32
Chicago, IL 60604

Christine O. Gregoire Youth/Young Adult Award for Outstanding Use of Tobacco Industry Documents — 1151

The award recognizes a person 24 years of age or younger who has made a contribution to the health of the public in the recent past through use of tobacco documents. The award also honors innovation in the use and application of tobacco industry documents to improve the public's health and, where applicable, to further the goals of tobacco prevention and control in order to help build a world where young people reject tobacco and anyone can quit. Those nominated should be individuals who have made a notable impact through innovative use of tobacco industry documents as applied to research, policy, or advocacy.
Requirements: At least one of the following two criteria must be met by the nominated individual: Nominees must have made a remarkable research, policy, or advocacy contribution with the use of tobacco industry documents, and/or; Nominees must have employed innovative, creative approaches to the employment of tobacco industry documents that result in an improvement in the health or public awareness of a community or nation. Nominations may be made by individuals such as colleagues, peers, coworkers, instructors/professors, governments, or CBOs that are qualified to make such a nomination because of their familiarity with the nominee's work and contribution.
Restrictions: Nominees must not have any affiliation with the tobacco industry. Members of the awards committee may not nominate potential recipients, although they may provide written support as part of a nomination package. Employees and directors of the American Legacy Foundation and their relatives are not eligible.
Geographic Focus: All States
Date(s) Application is Due: Feb 6
Amount of Grant: 7,500 USD
Contact: Virginia Lockmuller; (202) 454-5555; awards@americanlegacy.org
Internet: http://www.legacyforhealth.org/awards/
Sponsor: American Legacy Foundation
1724 Massachusetts Avenue NW
Washington, D.C. 20036

Christopher & Dana Reeve Foundation Quality of Life Grants — 1152

The Quality of Life Program, conceived by the late Dana Reeve, recognizes the unique and numerous needs of people with paralysis and the importance of providing services and programs that enable them to participate in all areas of life. Quality of Life Grants support organizations that help people with paralysis, their families, and caregivers in ways that more immediately give them increased independence, day-to-day happiness, and improved access. The foundation places some emphasis on paralysis caused by spinal cord injuries, but also considers paralysis from other causes, e.g., stroke, spina bifida, multiple sclerosis, cerebral palsy, and amyotrophic lateral sclerosis (ALS). The foundation accepts applications for Quality of Life grants twice yearly, in March and September. Exact deadlines may vary from year to year. Applicants must download and complete an application and then submit it through an online application system along with their IRS 501(c)3 Letter of Determination (for U.S.-based nonprofits). Prior grantees must also include the final report on their most recent Quality of Life grant. Applications to the Quality of Life Grants program should use "disability-friendly" or "people-first" language. Guidelines for these and all other requirements are available at the website.
Requirements: Nonprofit organizations that serve individuals with physical disabilities, particularly paralysis, and their families are eligible to apply. Quality of Life grants are most often awarded to nonprofit organizations that have IRS 501(c)3 status, but may also be awarded to community parks, schools, veterans hospitals, tribal entities, etc. Most Quality of Life grants are awarded within the United States of America, although the Reeve Foundation does award a small number of grants to nonprofit organizations based outside the United States. The foundation gives special consideration to organizations that serve returning wounded military and their families or to those that provide targeted services to diverse cultural communities.
Restrictions: No grants will be made to individuals, for benefit tickets, or for courtesy advertising.
Geographic Focus: All States, All Countries
Date(s) Application is Due: Mar 9; Sep 4
Amount of Grant: 5,000 - 25,000 USD
Samples: Able Flight, Inc., Chapel Hill, North Carolina, $4,000 - to support aviation career training under the guidelines and course curriculum approved by the FAA; accessAbility Center for Independent Living, Indianapolis, Indiana, $5,000 - to support construction of a community playground that is accessible to children and adults with disabilities (specifically mobility impairments) who reside in the greater Indianapolis area; Church of the Epiphany, Washington, D.C., $10,470 - to support completion of an ADA-compliant ramp at this historic church which has provided outreach services to the poor and homeless since its opening in the mid-1800s.
Contact: Donna Valente, Director, Quality of Life Grants; (800) 225-0292 or (973) 379-2690 outside the U.S; QoL@ChristopherReeve.org or dvalente@ChristopherReeve.org
Internet: http://www.ChristopherReeve.org/qol
Sponsor: Christopher and Dana Reeve Foundation
636 Morris Turnpike, Suite 3A
Short Hills, NJ 07078

Christy-Houston Foundation Grants — 1153

The foundation awards grants to nonprofits in Rutherford County, TN, for health, education, arts and culture, and community development. Areas of interest include hospitals, nursing care, health care and health associations, nutrition, hospice care, and community development. Types of support include building construction/renovation, equipment acquisition, scholarship funds, and matching funds. Application forms are not required, and there are no application deadlines. The board meets in March, June, September, and December to consider requests.
Requirements: 501(c)3 nonprofits in Rutherford County, TN, are eligible.
Restrictions: Grants are not awarded to: religious organizations for religious purposes, veterans organizations, historical societies, individuals, operating expenses or endowments.
Geographic Focus: Tennessee
Amount of Grant: 50,000 - 200,000 USD
Samples: City of Murfreesboro, Murfreesboro, TN, $1,000,000--Chambers of Commerce Building; City of Murfreesboro, Murfreesboro, TN, $88,000-fitness program; Easter Seals, $5,050--camp scholarships.
Contact: Robert B. Mifflin, Executive Director; (615) 898-1140; fax (615) 895-9524
Sponsor: Christy-Houston Foundation
1296 Dow Street
Murfreesboro, TN 37130-2413

Chula Vista Charitable Foundation Grants — 1154

Community leaders throughout Chula Vista are partnering with the San Diego Foundation to create and sustain a local foundation to serve solely the community of Chula Vista. Endow Chula Vista is a community-specific effort to focus on endowment building for Chula Vista, now and for generations to come. It is part of The San Diego Foundation's region-wide initiative, Endow San Diego, to inform and inspire San Diegans regarding the benefits of endowment.
Requirements: Any 501(c)3 organization serving the residents of Chula Vista, California, are eligible to apply.
Restrictions: The Chula Vista Charitable Foundation does not make grants for: annual campaigns and fund raising events for non-specific purposes; capital campaigns for buildings or facilities; stipends for attendance at conferences; endowments or chairs; for-profit organizations or enterprises; individuals unaffiliated with a qualified fiscal sponsor; projects that promote religious or political doctrine; research projects (medical or otherwise); scholarships; or existing obligations or debt.
Geographic Focus: California
Contact: Kerry Helmer, (619) 814-1384; kerry@sdfoundation.org
Internet: http://www.endowchulavista.org/
Sponsor: Chula Vista Charitable Foundation / San Diego Foundation
2508 Historic Decatur Road, Suite 200
San Diego, CA 92106

CICF City of Noblesville Community Grant — 1155

The City of Noblesville Community Grants were established to support the charitable intentions of Noblesville. Other purposes of the grant include, but are not limited to, basic needs, economic stability, health and wellness, education, vitality of neighborhoods and communities, arts and culture, and environment. Priority is given to programs

and projects most likely to have a positive effect on the city's residents. Proposals are accepted in February and July of each year. Applicants are strongly encouraged to view the Grantseeker's Guide posted on the website.
Requirements: The Foundation gives careful consideration to projects that; most benefit Noblesville residents; promote inclusiveness and diversity; respond to basic human needs; connect individuals and families to the community; complement other organizations to eliminate duplication of services; obtain additional funding and provide ongoing funding after the project. Organizations can apply through the online application.
Restrictions: The Foundation does not fund the following: organizations that are not tax exempt; multi-year grants; grants to individuals; projects aimed at promoting a particular religion or construction projects for religious institutions; operating, program, and construction costs at schools, universities, and private academies unless there is significant opportunity for community use or collaboration; organizations or projects that discriminate based on race, ethnicity, age, gender or sexual orientation; political campaigns or direct lobbying efforts by 501(c)3 organizations; post-event, after-the-fact situations or debt retirement; medical, scientific, or academic research; publications, films, audiovisual and media materials, programs produced for artistic purposes or produced for resale; travel for bands, sports teams, classes, and similar groups; annual appeals, galas, or membership contributions; fundraising events such as golf tournaments, walk-a-thons, and fashion shows.
Geographic Focus: Indiana
Amount of Grant: 1,000 - 10,000 USD
Contact: Liz Tate; (317) 843-2479, ext. 302; fax (317) 848-5463; lizt@cicf.org
Internet: http://www.cicf.org/how-to-apply-grantmaking
Sponsor: Central Indiana Community Foundation
615 North Alabama Street, Suite 119
Indianapolis, IN 46204-1498

CICF Indianapolis Foundation Community Grants 1156
CICF's mission is to inspire, support, and practice philanthropy, leadership, and service in the community. Proposed programs should align with any of the Foundation's Seven Elements of a Thriving Community: basic needs; economic stability; health and wellness; education; vitality and connectivity of neighborhoods and communities; arts and culture; and the environment. The application and grant request detail form are available online. Applications are accepted during the months of February and July.
Requirements: CICF welcomes grant applications from charitable organizations that are tax exempt under section 501(c)3 of the Internal Revenue Code, and from governmental agencies. New projects or organizations with pending 501(c)3 status may submit an application with the assistance of a fiscal sponsor. Grant inquiries and proposals will be prioritized using the following criteria: organizations that serve primarily Marion County residents; organizations with a demonstrable track record; programs serving populations disadvantaged due to income, age, ethnicity, language, education, disability, transportation or other adverse conditions; project/program ideas must be fully developed; and projects that strongly connect to existing community initiatives (e.g. the Blueprint to End Homelessness, Indianapolis Cultural Development Initiative, and Family Strengthening Coalition). Application information is available online.
Geographic Focus: Indiana
Contact: Liz Tate, Vice President for Grants; (317) 634-2423, ext. 175; fax (317) 684-0943; liz@cicf.org or program@cicf.org
Internet: http://www.cicf.org/the-indianapolis-foundation
Sponsor: Central Indiana Community Foundation
615 North Alabama Street, Suite 119
Indianapolis, IN 46204-1498

CICF James Proctor Grant for Aged Men and Women 1157
The CICF - James Proctor Grant for Aged Men and Women funds projects that benefit elderly men and women in the fields of arts and culture, basic needs, and health and wellness. Applications are accepted in the months of February and July, and are available at the Foundation website.
Geographic Focus: Indiana
Amount of Grant: 10,000 - 60,000 USD
Contact: Liz Tate; (317) 634-2423, ext. 175; fax (317) 684-0943; liz@cicf.org
Internet: http://www.cicf.org/examples-of-named-funds
Sponsor: Central Indiana Community Foundation
615 North Alabama Street, Suite 119
Indianapolis, IN 46204-1498

CICF Legacy Fund Grants 1158
The CICF - Legacy Fund Grants fund Hamilton County, Indiana nonprofits in the following areas of interest: arts and culture; basic needs; economic stability; education; environment; health and wellness; and vitality of neighborhoods and communities. Applications are accepted during the months of February and July, and are available at the Foundation's website.
Requirements: In reviewing grant applications, the Foundation gives careful consideration to projects that: impact individuals served; promote inclusiveness and diversity and are responsive to basic human needs; connect individuals and families to the community; commit to carrying out the project; complement other organizations and eliminate duplication of services; manage their fiscal responsibilities; match funds; and obtain additional funding now and in the future.
Restrictions: Limited to not-for-profit organizations and for the benefit of the people of Hamilton County, Indiana.
Geographic Focus: Indiana
Amount of Grant: 5,000 - 20,000 USD
Contact: Liz Tate; (317) 634-2423, ext. 175; fax (317) 848-5463; liz@cicf.org
Internet: http://www.cicf.org.php5-17.dfw1-1.websitetestlink.com/how-to-apply-grantmaking
Sponsor: Central Indiana Community Foundation
615 North Alabama Street, Suite 119
Indianapolis, IN 46204-1498

Cigna Civic Affairs Sponsorships 1159
The Cigna Civic Affairs program coordinates the charitable giving and volunteer activities of Cigna and its people, with the overall goal of demonstrating Cigna's commitment to being a socially responsible and responsive corporate citizen. One of the ways Civic Affairs fulfills this mission is through the Cigna Civic Affairs Sponsorships program. These sponsorships support charitable events and activities that enhance the health of individuals and families and the well-being of communities. Its strategy for achieving healthy outcomes around the globe is driven by: promoting wellness—to help individuals and families take ownership of their own health; expanding opportunities—to make health information and services available to everyone; developing leaders—to leverage the education and hands-on life experience that promote personal and professional growth; and embracing communities—to encourage collaborative and sustainable problem-solving approaches.
Geographic Focus: All States
Amount of Grant: 5,000 - 50,000 USD
Contact: Jill Holliday, Program Contact; (860) 226-2094 or (866) 865-5277; jill.holliday@cigna.com
Internet: https://secure16.easymatch.com/cignagive/applications/agency/?Skip=LandingPage&ProgramID=3
Sponsor: Cigna Corporation
1601 Chestnut Street, TL06B
Philadelphia, PA 19192-1540

CIGNA Foundation Grants 1160
The Cigna Foundation has identified four areas for grant consideration: health and human services, education, community and civic affairs, and culture and the arts. Health and education are of primary concern and receive priority. Under education, priority is placed on public secondary education, higher education for minorities, and adult basic education/literacy. The foundation also considers requests from U.S. cultural, educational, and public policy organizations that have international components. Requests are accepted and reviewed throughout the year. Consideration will be given to requests for general operating support, program development, annual campaigns, conferences and seminars, fellowships, scholarship funds and employee-related scholarships, and matching gifts and funds.
Requirements: Organizations with 501(c)3 tax-exempt status are eligible.
Restrictions: The foundation will not consider applications for grants to individuals, organizations operating to influence legislation or litigation, political organizations, or religious activities. In general, the foundation will not consider applications from organizations receiving substantial support through the United Way or other CIGNA-supported federated funding agencies; hospitals' capital improvements; or research, prevention, and treatment of specific diseases.
Geographic Focus: All States
Amount of Grant: 5,000 - 50,000 USD
Contact: Jill Holliday; (860) 226-2094 or (866) 865-5277; jill.holliday@cigna.com
Internet: http://www.cigna.com/aboutus/cigna-foundation
Sponsor: Cigna Foundation
1601 Chestnut Street, TL06B
Philadelphia, PA 19192-1540

CIT Corporate Giving Grants 1161
The corporate giving program focuses its giving to programs that substantially strengthen the communities where CIT has a significant presence; provide services that benefit employees, their families, and communities; and support employee volunteerism. Funding priorities include organizations, institutions, or programs that work to foster excellence in education, particularly among at-risk and underprivileged youth; support local health and social welfare issues; help stabilize and improve neighborhoods by providing viable economic and educational growth opportunities to the community; and support the arts, and their accessibility, as a means of enhancing personal and social development. Proposals are reviewed on an ongoing basis and must be formally submitted by mail.
Requirements: 501(c)3 tax-exempt organizations and Canadian charities registered with the Canada Customs and Revenue Agency (CCRA) are eligible.
Restrictions: The following will not be considered for funding: political organizations, campaigns, or candidates; conferences, memberships, or sports competitions; organizations that discriminate in any way with national equal opportunity policies; organizations whose chief purpose is to influence legislation; veteran and fraternal organizations; religious organizations, unless engaged in a significant, nonsectarian project that benefits a broad base of the community; or individuals.
Geographic Focus: All States
Contact: Stacy Papas, Community Affairs Manager; stacy.papas@cit.com
Internet: http://www.cit.com/about-cit/corporate-giving/index.htm
Sponsor: CIT Group
11 West 42nd Street
New York, NY 10036

Citizens Bank Mid-Atlantic Charitable Foundation Grants 1162
Charitable grants are made only to qualified 501(c)3 Rhode Island-based organizations serving Rhode Island residents. Citizens look for opportunities where moderate funding can affect significant results in the community. Priority consideration is given to programs that encourage the development of innovative responses to basic human needs; promote fair housing and focus on community issues of neighborhood development and economic

self-sufficiency; support the availability of quality, cost-effective, community-based health care, particularly for low-income families and children who are at risk; promote new ways to provide a quality education to populations that are underserved, including job training; promote availability and accessibility in the area of culture and the arts; and promote citizen participation in the development of new and workable solutions for improving and maintaining a healthy environment. Charitable grants are usually for capital funding (to build or renovate a facility) or implementation of a specific program.
Requirements: 501(c)3 Rhode Island-based nonprofits are eligible.
Restrictions: Grants do not support individuals, single disease/issue research organizations, religious organizations for religious purposes, labor or fraternal or veterans groups, political organizations or projects, operating deficits, underwriting of conferences and seminars, governmental public agencies, endowments, annual operating support, trips and tours, payment on bank loans, advertising, or fund-raising events.
Geographic Focus: Rhode Island
Amount of Grant: 1,000 - 100,000 USD
Contact: Jeanne Cola, Senior Vice President; (401) 456-7200; fax (401) 456-7366
Sponsor: Citizens Bank Mid-Atlantic Charitable Foundation
870 Westminster Street
Providence, RI 02903

CJ Foundation for SIDS Program Services Grants 1163
The CJ Foundation for SIDS offers two types of grants to support services related to the following areas of interest: SIDS (Sudden Infant Death Syndrome), SUID (Sudden Unexpected Infant Death), and Infant Safe Sleep. Program Services Grants, which are the first type, are generally for amounts greater than $5,000 and are offered by invitation only. Program Services Mini-Grants (formerly known as Express Grants) are the second type. These are for $5,000 or less and may be applied for without invitation. The purpose of the foundation's program-services grants is to support activities that promote safe sleep for infants, educate about SIDS risk reduction, and provide grief support for parents and others who have experienced the sudden, unexpected death of an infant. The grants support the following types of activities: conferences and meetings; training and workshops; community awareness events; production, purchase, and distribution of educational, bereavement, and resource materials; support groups; peer support; counseling; support of staff and consultants who implement education initiatives or provide bereavement services; and newsletter production and distribution. Interested applicants should contact the Assistant Executive Director (see contact section) to request application materials. Submission deadline for the mini-grants is September 26 (deadlines may vary from year to year).
Requirements: The applicant must demonstrate that services relating to SIDS, SUID, and/or Infant Safe Sleep are an established part of the organization's services.
Restrictions: Projects exclusively serving perinatal and/or neonatal death will not be considered. Data collection and analysis for Fetal Infant Mortality Review (FIMR) and Child Death Review (CDR) will not be considered. The foundation does not provide grants to organizations that discriminate, in policy or practice, against people based on their age, race, color, creed or gender. In general the CJ Foundation does not support the following types of requests: loans; budget requests greater than 100% of the applicant's operating budget; grants to individuals; dues; operating deficits; book publication; capital improvements/building project; chairs or professorships; endowments, annual fund drives, direct mail solicitation, or fundraising events; purchase of advertising space; purchase of products such as t-shirts, cribs, crib sheets, and sleep slacks; activities to influence legislation or support candidates for political office; and re-granting.
Geographic Focus: All States
Date(s) Application is Due: Sep 26
Amount of Grant: 500 - 5,000 USD
Contact: Wendy Jacobs, Assistant Executive Director, Programs & Grants; (866) 314-7437 or (551) 996-5111; fax (551) 996-5326; wendy@cjsids.org or info@cjsids.org
Internet: http://www.cjsids.org/grants/grants-overview.html
Sponsor: CJ Foundation for SIDS
30 Prospect Avenue
Hackensack, NJ 07601

CJ Foundation for SIDS Research Grants 1164
In support of its mission, the CJ Foundation for SIDS (Sudden Infant Death Syndrome) is strongly committed to providing grant support to best-in-class researchers who work toward the ultimate eradication of SIDS and other sleep-related infant and toddler deaths. Applications currently are by invitation only. Preference is given to past grantees and pilot projects that are gathering data for eventual submission to NIH for funding. Applications are reviewed by the Foundation's Medical Advisory Board. Interested applicants should contact the foundation for further information.
Geographic Focus: All States
Contact: Wendy Jacobs, Assistant Executive Director, Programs & Grants; (866) 314-7437 or (551) 996-5111; fax (551) 996-5326; wendy@cjsids.com or info@cjsids.org
Internet: http://www.cjsids.org/grants/grants-overview.html
Sponsor: CJ Foundation for SIDS
30 Prospect Avenue
Hackensack, NJ 07601

Clara Blackford Smith and W. Aubrey Smith Charitable Foundation Grants 1165
The Clara Blackford Smith & W. Aubrey Smith Charitable Foundation was established in 1978. Mrs. Smith was a well-known benefactor of health care. The foundation was established under her will to support and promote quality education, health-care, and human-services programming for underserved populations. Special consideration is given to charitable organizations that serve the people of Grayson County, Texas. The foundation makes an annual grant to Denison High School in Denison, Texas to provide college scholarship assistance for deserving graduates. The majority of grants from the Smith Charitable Foundation are one year in duration; on occasion, multi-year support is awarded. Applicants must apply online at the grant website. Applicants are strongly encouraged to do the following before applying: review the downloadable state application procedures for additional helpful information and clarifications; review the downloadable online-application guidelines at the grant website; review the foundation's funding history (link is available from the grant website); review the online application questions in advance; and review the list of required attachments. These will generally include: a list of board members, financial statements (audited, reviewed, or compiled by independent auditor); an organization summary; a list of other funding sources; an IRS Determination letter; and other required documents. All attachments must be uploaded in the online application as PDF, Word, or Excel files. The Clara Blackford Smith & W. Aubrey Smith Charitable Foundation has four deadlines annually: March 1, June 1, September 1, and December 1. Applications must be submitted by 11:59 p.m. on the deadline dates. Grant applicants are notified as follows: March deadline applicants will be notified of grant decisions by June 30; June applicants will be notified by September 30; September applicants will be notified by December 31; and December applicants will be notified by March 31 of the following year.
Requirements: Applicants must have 501(c)3 tax-exempt status.
Restrictions: The foundation does not support requests from individuals, organizations attempting to influence policy through direct lobbying, or any political campaigns.
Geographic Focus: Texas
Date(s) Application is Due: Mar 1; Jun 1; Sep 1; Dec 1
Samples: Denison Independent School District, Denison, Texas, $55,000, for Smart Boards for DISD classrooms; Loy Park Improvement Association, Denison, Texas, $36,000, installation of ceiling fans in the Major Arena; Grayson County Shelter, Denison, Texas, $50,000, general operating support.
Contact: David Ross, Senior Vice President; tx.philanthropic@baml.com
Internet: https://www.bankofamerica.com/philanthropic/fn_search.action
Sponsor: Clara Blackford Smith and W. Aubrey Smith Charitable Foundation
901 Main Street, 19th Floor, TX1-492-19-11
Dallas, TX 75202-3714

Clarence E. Heller Charitable Foundation Grants 1166
The charitable foundation supports nonprofit organizations, with priority given to proposals from California, in its areas of interest, including environment and health--to prevent serious risk to human health from toxic substances and other environmental hazards by supporting programs in research, education, and policy development; management of resources--to protect and preserve the earth's limited resources by assisting programs that demonstrate how natural resources can be managed on a sustainable and an ecologically sound basis, and supporting initiatives for sustainable agriculture, and for promoting the long-term viability of communities and regions; music--to encourage the playing, enjoyment, and accessibility of symphonic and chamber music by providing scholarship and program assistance at selected community music organizations and schools, and by helping community-based ensembles of demonstrated quality implement artistic initiatives, diversify and increase audiences, and improve fund-raising capacity; and education--to focus on support for programs that improve the teaching skills of educators and artists in environmental and arts education. Types of support include continuing support, general operating expenses, publications, research, seed money, and special projects. The foundation's board meets typically in March, June, and October. Letters of inquiry are accepted at any time.
Requirements: 501(c)3 tax-exempt organizations are eligible.
Restrictions: Grants are not made to individuals.
Geographic Focus: California
Amount of Grant: 5,000 - 600,000 USD
Contact: Bruce Hirsch; (415) 989-9839; fax (415) 989-1909; info@cehcf.org
Internet: http://cehcf.org/app_info.html
Sponsor: Clarence E. Heller Charitable Foundation
44 Montgomery Street, Suite 1970
San Francisco, CA 94104

Clarence T.C. Ching Foundation Grants 1167
The Clarence T. C. Ching Foundation provides funding opportunities to nonprofits primarily in Honolulu, Hawaii. The Foundation's areas of interest include education and health. Types of support available are: scholarship funds; general program support and; building/renovation. Grant applications are accepted year round with no application deadline date.
Requirements: Nonprofits in Honolulu, Hawaii are eligible for funding. When applying for a grant include the following in with your proposal: copy of IRS Determination Letter; detailed description of project and amount of funding requested; listing of additional sources and amount of support; five (5) copies of the proposal.
Restrictions: No grants to individuals.
Geographic Focus: Hawaii
Amount of Grant: 500 - 70,000 USD
Contact: R. Stevens Gilley, Executive Director; (808) 521-0344
Sponsor: Clarence T.C. Ching Foundation
1001 Bishop Street, Suite 960
Honolulu, HI 96813

Clarian Health Critical / Progressive Care Internships 1168
The Clarian Health Critical / Progressive Care Internship Program is a 12 week individualized program. The Internship is offered year round, complete with salary. The program provides experience in the following specialized services: adult critical care; cardiac; comprehensive critical care; cardiovascular critical care; medical intensive care; neuro/trauma critical care; neuro-surgery intensive care; progressive care; surgical intensive care

Geographic Focus: Indiana
Contact: Julie Ruschhaupt; (317) 278-7082 or (877) 354-2996; jruschhaupt@clarian.org
Internet: http://www.clarian.org/clarianjobs/nursing/internships_externships.htm#top
Sponsor: Clarian Health Partners
950 N Meridian Street, Gateway Plaza
Indianapolis, IN 46204

Clarian Health Multi-Specialty Internships 1169
The Clarian Health Multi-Specialty Internship is a 7 to 8 week individualized program, that is offered year round.
Requirements: Contact Joy Fay for additional information to apply.
Geographic Focus: Indiana
Contact: Joy Fay, RN; (317) 962-3788; jfay@clarian.org
Internet: http://www.clarian.org/clarianjobs/nursing/internships_externships.htm#multi
Sponsor: Clarian Health Partners
950 N Meridian Street, Gateway Plaza
Indianapolis, IN 46204

Clarian Health OR Internships 1170
The Clarian Health OR Internship Program is offered twice a year, for a duration of 6 months. The program provides experience in the following specialized services: general; cardio thoracic; cardiovascular; endoscopies; ENT; GYN; laser; neurosurgery; ophthalmology; orthopedics; pediatrics; plastic / Reconstructive; transplants; trauma; urology
Geographic Focus: Indiana
Contact: Shalunda Tyler; (317) 962-9083; styler1@clarian.org
Internet: http://www.clarian.org/clarianjobs/nursing/internships_externships.htm#multi
Sponsor: Clarian Health Partners
950 N Meridian Street, Gateway Plaza
Indianapolis, IN 46204

Clarian Health Scholarships for LPNs 1171
The Clarian Health Scholarships for LPNs Program provides an award of $750 per semester. The scholarship does require a post graduate commitment of 6 months full-time employment for each semester of funding.
Requirements: Applicant must: be employed as an LPN at Clarian; be in good standing to receive the scholarship. A participant whose job performance does not meet the standards of Clarian's performance review program during any time of the employment period/year that the scholarship is in effect will jeopardize his/her opportunity for future funding; be actively involved in Clinical Unit Operations (i.e., attend staff meetings, keep mandatory education requirements up to date, etc.); must maintain a competitive GPA. To complete the application process, you must: complete the online application; schedule appointment to complete paperwork; submit proof of enrollment in an approved nursing school; submit 2 professional recommendations; one from current manager, one from RN co-worker; submit a one-page narrative on why you should be awarded the scholarship
Geographic Focus: Indiana
Date(s) Application is Due: Jul 1; Nov 1
Amount of Grant: 750 USD
Contact: Clarian Nurse Recruitment Line; (877) 354-2996; dspilker@clarian.org
Internet: http://www.clarian.org/clarianjobs/nursing/lpnscholarship.htm
Sponsor: Clarian Health Partners
950 N Meridian Street, Gateway Plaza
Indianapolis, IN 46204

Clarian Health Student Nurse Extern Scholarships 1172
The Clarian Health Student Nurse Extern Scholarship Program offers scholarships for each semester worked as a student nurse extern (SNE) at one of the three downtown Clarian hospitals: Methodist Hospital, Indiana University Hospital and Riley Hospital for Children. The SNE Scholarship Program offers several options to meet scheduling and financial needs. See website for details.
Requirements: Any nursing student in good standing and actively enrolled in a program that results in a RN degree may apply. A nursing student must meet the requirements of the SNE job description by completing at least one semester of nursing clinicals and remain actively enrolled and a student in good standing in the school of nursing they are attending. To apply for a Student Nurse Extern position at Clarian health, please send resume to dspilker@clarian.org. In a cover letter or the objective section of your resume, indicate your interest in the Student Nurse Extern position. The number of sponsorship awards is limited.
Geographic Focus: Indiana
Amount of Grant: 250 - 1,000 USD
Contact: Clarian Nurse Recruitment Line; (877) 354-2996; dspilker@clarian.org
Internet: http://www.clarian.org/portal/patients/education?clarianContentID=/education/nursing_education/SNE.xml
Sponsor: Clarian Health Partners
950 N Meridian Street, Gateway Plaza
Indianapolis, IN 46204

Clark-Winchcole Foundation Grants 1173
The foundation awards grants to nonprofit organizations in the District of Columbia, with an emphasis on higher education, hospitals and health care, cultural programs, youth, the disabled, and religion. Types of support include general operating support, scholarships, and building construction/renovation. Applicants should submit a letter of inquiry that includes a description of the project, amount requested, audited financial report, budget, and proof of tax-exempt status. Further information will be requested by the foundation if interested.
Requirements: Only nonprofits in the District of Columbia are eligible to apply.
Restrictions: No support for individuals and private foundations.
Geographic Focus: District of Columbia
Amount of Grant: 5,000 - 225,000 USD
Sample: Capital Hospice, Falls Church, VA, $75,000--for Patient Care program; Wesley Theological Seminary, Washington, D.C., $125,000--for urban ministry program; Wolf Trap Foundation for the Performing Arts, Vienna, VA, $225,000--for Institute for Early Learning, Opera at Wolf Trap, Fund for Artistic Initiative.
Contact: Vincent Burke, President; (301) 654-3607
Sponsor: Clark-Winchcole Foundation
3 Bethesda Metro Center, Suite 550
Bethesda, MD 20814-5358

Clark and Ruby Baker Foundation Grants 1174
The Clark and Ruby Baker Foundation was established to address a number of charitable concerns. The Baker family cared deeply about the Methodist Church and founded the Foundation to support Methodist affiliated, higher educational institutions in rural or small towns; charitable organizations that serve infirm, deserving, and aged ministers; economically disadvantaged and deserving children; and orphans and orphanages. The Foundation also provides support to charitable organizations that extend financial aid to the sick and infirm receiving medical treatment in any hospital or clinic in the state of Georgia. Capital support may be considered for the following purposes: for construction of educational facilities at a college or university; for clinics and hospitals; for libraries; and for any building with a charitable use. Grants from the Clark and Ruby Baker Foundation are primarily one year in duration; on occasion, multi-year support is awarded. Applicants must apply online at the grant website. Applicants are strongly encouraged to do the following before applying: review the downloadable state application procedures for additional helpful information and clarifications; review the downloadable online-application guidelines at the grant website; review the foundation's funding history (link is available from the grant website); review the online application questions in advance; and review the list of required attachments. These will generally include: a list of board members, financial statements (audited, reviewed, or compiled by independent auditor); an organization summary; a list of other funding sources; an IRS Determination letter; and other required documents. All attachments must be uploaded in the online application as PDF, Word, or Excel files. The Clark and Ruby Baker Foundation application deadline is 11:59 p.m. on June 1. Applicants will be notified of grant decisions by letter within one to two months after the deadline.
Requirements: Applicants must have 501(c)3 tax-exempt status.
Restrictions: The foundation does not support requests from individuals, organizations attempting to influence policy through direct lobbying, or any political campaigns.
Geographic Focus: Georgia
Date(s) Application is Due: Jun 1
Contact: Mark S. Drake, Vice President; (404) 264-1377; mark.s.drake@ustrust.com
Internet: https://www.bankofamerica.com/philanthropic/fn_search.action
Sponsor: Clark and Ruby Baker Foundation
3414 Peachtree Road, N.E., Suite 1475, GA7-813-14-04
Atlanta, GA 30326-1113

Clark County Community Foundation Grants 1175
Since 1999, the Clark County Community Foundation has awarded over $3.6 million to nonprofits serving Clark County. It would be hard to find a Clark County nonprofit that has not received much-need support through its competitive grantmaking program. Funded projects include a five-year pilot program to provide preventive dental care in Clark County schools; support for a free clinic serving the uninsured; and initiatives to increase the number of students who take advanced placement classes at Clark County High School.
Requirements: Clark County, Kentucky, nonprofit organizations are eligible.
Geographic Focus: Kentucky
Date(s) Application is Due: Mar 15; May 15
Amount of Grant: 5,000 - 10,000 USD
Contact: Kassie Branham; (859) 225-3343; fax (859) 243-0770; kbranham@bgcf.org
Internet: http://bgcf.org/learn/community-funds/clark-county community foundation/
Sponsor: Clark County Community Foundation
499 East High Street, Suite 112
Lexington, KY 40507

Claude Bennett Family Foundation Grants 1176
Established in Alabama in 1993, the Claude Bennett Family Foundation has as its primary interest areas: health care, health organizations, higher education, human services, and Protestant churches and agencies. The range of funding is up to $15,000, and the Foundation has supported agencies all across the U.S. There are no specific applications or deadlines with which to adhere, and applicants should contact the Foundation in writing, stating the purpose of their request and offering a detailed budget.
Geographic Focus: Alabama, Colorado
Amount of Grant: Up to 15,000 USD
Samples: Alabama Symphony, Birmingham, Alabama, $5,000 - operating support; Penguin Project, East Peoria, Illinois, $500 - general operations; Colorado State University, Fort Collins, Colorado, $2,000 - operating support.
Contact: Harold I. Apolinsky, Trustee; (205) 930-5122 or (205) 945-4687; fax (205) 212-3888; hapolinsky@sirote.com
Sponsor: Claude Bennett Family Foundation
2311 Highland Avenue South, P.O. Box 130804
Birmingham, AL 35213-0804

Claude Pepper Foundation Grants 1177
The foundation makes grants primarily to support the work of the Claude Pepper Center and the Pepper Institute on Aging and Public Policy, both located at Florida State University, Tallahassee, FL. The foundation makes limited grants to other organizations to continue the work and vision of Claude and Mildred Pepper. The foundation also supports a visiting scholars program and an oratory competition for Florida students. Grants are usually made for a period of one year and except in rare instances no grant will be made for longer than a three-year period of time. Guidelines are available online.
Requirements: 501(c)3 tax-exempt organizations are eligible.
Geographic Focus: All States
Samples: Florida Council on Aging - To support Special Plenary Sessions at the Annual Aging Network Conferences; National Academy of Social Insurance - To sponsor the Academy's annual conference opening dinner in Washington, D.C., which addressed issues surrounding Social Security reform; Tallahassee Senior Center - To co-sponsor the Claude Pepper Senior Walk/Run, which consists of a competition for walkers over the age of 60, and one for runners, with prizes and refreshments.
Contact: John T. Herndon, Executive Director; 850) 644-9309; fax (850) 644-9301; herndon@claudepepperfoundation.org
Internet: http://www.claudepepperfoundation.org/programs_grants.cfm
Sponsor: Claude Pepper Foundation
636 West Call Street
Tallahassee, FL 32306-1122

Claude Worthington Benedum Foundation Grants 1178
Grants are made in the areas of education, health and human services, community and economic development, environment, and the arts. Grants have been awarded to support education reform, teacher education, higher education, workforce development, rural health, professional developing in healthcare, human services, affordable housing, and economic development. Grants are awarded to organizations in West Virginia and southwestern Pennsylvania. Funds are provided for general operations for projects, sometimes including building and equipment, in West Virginia, and for projects in Pittsburgh that address regional problems and needs, that establish demonstration projects with strong potential for replication in West Virginia, or make outstanding contributions to the area. Additional types of support include matching funds, consulting services, technical assistance, capital campaigns, conferences and seminars, research, and seed grants. Organizations wishing to apply should request a copy of the annual report, which includes application guidelines. Applications may be submitted at any time; the board meets for review in March, June, September, and December.
Requirements: Southwestern Pennsylvania and West Virginia nonprofits may apply.
Restrictions: Support is not given for national health and welfare campaigns, medical research, religious activities, fellowships, scholarships, annual campaigns, or travel.
Geographic Focus: Pennsylvania, West Virginia
Contact: Margaret M. Martin, Grants Administrator; (800) 223-5948 or (412) 246-3636; fax (412) 288-0366; mmartin@benedum.org
Internet: http://www.benedum.org/pages.cfm?id=10
Sponsor: Claude Worthington Benedum Foundation
1400 Benedum-Trees Building, 223 Fourth Avenue
Pittsburgh, PA 15222

Clayton Baker Trust Grants 1179
The Clayton Baker Trust, established in 1960, awards grants to Maryland nonprofit organizations for programs targeting the disadvantaged, with an emphasis on the needs of children. Grants are awarded nationally in the areas of: environmental protection; population control; arms control; and nuclear disarmament. Types of support include: capital and infrastructure; general operating support; seed grants; program development; systems reform; and special projects. There are three annual deadlines for application submission: April 5, August 5, and December 5.
Requirements: Nonprofit organizations in Maryland are eligible to apply. The Association of Baltimore Area Grantmakers Common Grant Application Form is required.
Restrictions: Grants do not support the arts, research, higher educational institutions, individuals, building construction/renovation, or endowment funding.
Geographic Focus: Maryland
Date(s) Application is Due: Apr 5; Aug 5; Dec 5
Amount of Grant: 2,000 - 100,000 USD
Samples: Association of Baltimore Area Grantmakers, Baltimore, Maryland, $5,250 - general operating support (2014); New Leaders, Baltimore, Maryland, $45,000 0 general operating support (2014); Arts Education in Maryland Schools, Baltimore, Maryland, $13,000 - general operating support (2014).
Contact: John Powell, Jr., Executive Director; (410) 837-3555; fax (410) 837-7711
Sponsor: Clayton Baker Trust
2 East Read Street, Suite 100
Baltimore, MD 21202

Clayton Fund Grants 1180
The Clayton Fund Trust was established in Texas in 1952, with a primary mission of offering aid to the needy, especially children, the environment, family planning, education, agriculture, and arts and culture. With that in mind, the Fund's current fields of interest have evolved to include: arts and culture; child welfare; diseases; education (both elementary and secondary); family planning; foundation support; higher education; human services; and natural resources. Awards typically take the form of general operating support, program development, continuing support, building campaigns, and endowment funds. Most recent awards have ranged from $5,000 to $45,000. Three annual deadlines for application submission have been identified: February 1, June 1, and October 1.
Requirements: Giving is limited to 501(c)3 organizations that support the residents of Texas, Maryland, and New York.
Restrictions: Grants are not given directly to individuals.
Geographic Focus: Maryland, New York, Texas
Date(s) Application is Due: Feb 1; Jun 1; Oct 1
Amount of Grant: 5,000 - 45,000 USD
Contact: William Garwood, President; (713) 216-1453
Sponsor: Clayton Fund
707 Travis Street, 11th Floor
Houston, TX 77252-3232

Cleveland-Cliffs Foundation Grants 1181
Contributions are made to nonprofit organizations to enhance the quality of life of Cleveland-Cliffs Inc employees and in recognition of a corporate responsibility toward educational, health, welfare, civic, and cultural matters within the communities where the company operates. The foundation was formed for the purpose of making contributions to groups organized and operated exclusively for religious, charitable, scientific, literary, or educational purposes and for the prevention of cruelty to children or animals. Types of support include general operating support, annual campaigns, capital campaigns, building/renovation, professorships, scholarship funds, research, and employee matching gifts. The foundation's major emphasis is on supporting education through a matching gift program and direct contributions to educational institutions. Requests for support must be in writing.
Requirements: Nonprofits in the mining communities in which Cleveland-Cliffs Inc operates, including Alabama, Michigan, Minnesota, Ohio and, West Virginia are eligible.
Geographic Focus: Alabama, Michigan, Minnesota, Ohio, West Virginia
Amount of Grant: 250 - 50,000 USD
Samples: Bell Memorial Hospital, Ishpeming, MI, $400,000; University of Saint Thomas, Minneapolis, MN, $10,000; Michigan Tech Fund, Houghton, MI, $9,550.
Contact: Dana W. Byrne, Vice President; (216) 694-5700; fax (216) 694-4880; publicrelations@cleveland-cliffs.com
Internet: http://www.cliffsnaturalresources.com/Development/CommunityRelations/Pages/Cleveland-CliffsFoundation.aspx
Sponsor: Cleveland-Cliffs Foundation
1100 Superior Avenue, Suite 1500
Cleveland, OH 44114-2589

Cleveland Browns Foundation Grants 1182
The Cleveland Browns Foundation supports the northeast Ohio community by funding programs that improve the lives of disadvantaged children. The foundation's four major focus areas are education, arts and culture, health, and career development as they relate to children. There are no application deadlines. All sponsorships and financial requests should be submitted in writing at least six (6) weeks in advance of an event. All donation requests should be submitted online.
Requirements: Northern Ohio nonprofit organizations are eligible to apply.
Restrictions: Organizations and causes that will not be considered for funding include fund-raising, sponsorship events, or donation requests; religious organizations for sectarian religious purposes; general or annual operation expenses; capital or building funds; or staff salaries or stipends.
Geographic Focus: Ohio
Contact: Dee Bagwell Haslam, President; (440) 891-5063; fax (440) 891-7529
Internet: http://www.clevelandbrowns.com/community/in-kind-support.html
Sponsor: Cleveland Browns Foundation
76 Lou Groza Boulevard
Berea, OH 44017

CLIF Bar Family Foundation Grants 1183
The foundation's mission is to support nonprofits - grassroots organizations in particular - working to promote environmental restoration and conservation, sustainable food and agriculture, people's health, and a wide range of social concerns. It focuses on organizations whose missions support environmental restoration and conservation, sustainable food and agriculture, people's health and youth.
Requirements: The foundation funds nonprofit organizations working on projects that are well informed, have clearly defined objectives, demonstrate strong community ties, and promote values such as compassion, inclusiveness, patience, and positive development both in the program and in its implementation. Emphasis is on grassroots organizations that have the ability to engage local groups, positively impact their communities, and focus most of their resources on useful and positive actions.
Restrictions: Applications that require funding in less then 12 weeks will not be accepted. The foundation will not fund: Deficit financing; Loans or grants to individuals; Capital construction; Research projects, conferences, seminars, media events or workshops, unless they are an integral part of a broader program; State agencies; Religious groups; Individual Sponsorships; Fundraising events for your organization such as fun runs, bike rides, etc. If you are a local/regional branch of a national organization they will not fund grants to local chapters if they fund the national organization. Otherwise, they only take into consideration local offices in Northern California.
Geographic Focus: All States
Samples: The Alameda County Community Food Bank; American Farmland Trust Breast Cancer Fund; New Leaders Initiative/Earth Island Institute; Center for Food Safety; Center for Rural Affairs; Circle of Life; City Slicker Farms; Community Alliance with Family Farmers; Community Resources for Science; Diabetic Youth Foundation; Disability Rights Education & Defense Fund; Ecological Farming Association; Ecology Center; Focus the Nation: Global Warming Solutions for America; Food Alliance Green Belt Movement; Greentreks Grid Alternatives; Habitat for Humanity; Healthy Child

Healthy World; Heifer International; The Land Institute; Leave No Trace; Leukemia & Lymphoma Society; Luna Kids Dance; Lungevity Foundation; KIPP Bay Area; Marin Agricultural Land Trust; NorCal High School Mountain Bike League; Organic Farming Research Foundation; Pedals For Progress; Project Open Hand; Rainforest Action Network; San Francisco AIDS Foundation; Save Mount Diablo; Save the Waves; Saving Teens in Crisis Collaborative; Trips for Kids; Wild Hope; Worldbike; World Neighbors
Contact: Kit Crawford; (510) 859-2283; fax (510) 588-5490; familyfoundation@clifbar.com
Internet: http://clifbarfamilyfoundation.org/subtemplate.php?s=1
Sponsor: CLIF Bar Family Foundation
1610 5th Street
Berkeley, CA 94710

Clinton County Community Foundation Grants 1184
The Clinton County Community Foundation is a catalyst for stimulating and funding initiatives that improve the quality of life for citizens of Clinton County. The Foundation Grants address needs that generally fall into the following categories: health and medical; social services; education; cultural affairs; civic affairs; and community beautification.
Requirements: All applicants must receive pre-qualification prior to submitting an application by submitting a letter of inquiry to the Program Director. The letter should contain a brief statement of the applicant's needs for assistance, estimate of total cost of project, and enough information to enable the Foundation to determine if the application falls within the guidelines of its grants program. If the grant application is decided, organizations must submit one original application, plus ten copies required for review by the grantmaking committee with the following information included: grant application with the cover page; project budget; board list; evidence of board approval; 501(c)3 letter; year-end audit or financial statement; current month and year-to-date financial statement; and when applicable, three estimates must be included, one from a Clinton County business. The organization may be contacted for additional information, an interview with the grant making committee, or a possible site visit, and the time period which a grant decision will likely be made.
Geographic Focus: Indiana
Date(s) Application is Due: May 6; Sep 7
Contact: Kim Abney; (765) 454-7298 or (800) 964-0508; kim@cfhoward.org
Internet: http://www.cfclinton.org/grant_seekers_cl.html
Sponsor: Clinton County Community Foundation
215 West Sycamore Street
Kokomo, IN 46901

Clowes ACS/AAST/NIGMS Jointly Sponsored Mentored Clinical Scientist Development Award 1185
The American College of Surgeons and American Association for the Surgery of Trauma announce a program that will provide supplemental salary funding of up to $75,000 per year to an individual who has received a Mentored Clinical Scientist Development Award (K08/K23) from the National Institute for General Medical Science (NIGMS). This award is directed at surgeon-scientists working in the early stages of their research careers. The award supports a three-, four-, or five-year period of supervised research experience that may integrate didactic studies with laboratory or clinical research. The award program offers a means to facilitate the career development of individuals pursuing careers in trauma surgery research by enhancing salary support over and above that offered by the K08/K23 mechanism.
Requirements: Awardees must be members in good standing of the College and eligible for membership in AAST.
Restrictions: Pre-existing applications and awards are not eligible for consideration.
Geographic Focus: All States
Date(s) Application is Due: Oct 12
Amount of Grant: Up to 75,000 USD
Contact: Kate Early; (312) 202-5281; fax (312) 202-5021; kearly@facs.org
Internet: http://www.facs.org/memberservices/acs-aast-nigms.html
Sponsor: American College of Surgeons
633 N Saint Clair Street
Chicago, IL 60611-3211

CMS Hispanic Health Services Research Grants 1186
The purpose of the grant is to implement Hispanic American health services research activities to meet the needs of diverse CMS beneficiary populations. The program seeks competitive applications for small applied research projects that relate to identifying and evaluating solutions for eliminating health disparities among Hispanic Americans. Funding is available for grants to implement research related to health care delivery and health financing issues affecting Hispanic American communities, including issues of access to health care, utilization of health care services, health outcomes, quality of services, cost of care, health and racial disparities, socio-economic differences, cultural barriers, managed care systems, and active ties related to health screening, prevention, outreach, and education.
Requirements: Investigators should be associated with a university, college, community-based health organization, or a professional association that has a health services research component. Researchers are encouraged to use CMS data as part of their research projects. Researchers are also expected to become involved in the design, implementation, and operation of research projects that address health care issues such as financing, delivery, access, quality, and barriers affecting the Hispanic American community. Applicants must meet one of the following three requirements in order to qualify for funding under this grant program: 1) A health services/disparities researcher at an university or college offering a Ph.D. or Master's Degree Program in one or more of the following disciplines Allied Health, Gerontology, Health Care Administration, Health Education, Health Management, Nursing, Nutrition, Pharmacology, Public Health, Public Policy, Social Work; or 2) a member of a community-based health organization with a Hispanic health services research component; or 3) a member of a professional association focusing on Hispanic health services and health disparities issues. Applications must be submitted electronically (via grants.gov). CMS will accept a hard copy if the applicant is unable to access grants.gov or is having serious problems in sending the application electronically.
Restrictions: Grant funds may not be used for any of the following: to provide direct services to individuals except as explicitly permitted under the grant solicitation; to match any other Federal funds; or, to provide services, equipment, or supports that are already the legal responsibility of another party under Federal law.
Geographic Focus: All States
Date(s) Application is Due: Jul 3
Amount of Grant: 200,000 - 250,000 USD
Contact: Joi Grymes, (410) 786-7251; Joi.Grymes@cms.hhs.gov
Internet: http://www.cms.hhs.gov/ResearchDemoGrantsOpt/03_Hispanic_Health_Services.asp#TopOfPage
Sponsor: Centers for Medicare and Medicaid Services
7500 Security Boulevard, Room C2-21-15
Baltimore, MD 21244-1850

CMS Historically Black Colleges and Universities Health Services Research Grants 1187
The purpose of the grant program is to support researchers in implementing health services research activities to meet the needs of diverse CMS beneficiary populations. The goals of the grant program are to: 1) encourage HBCU health services researchers to pursue research issues which impact the Medicare, Medicaid, and SCHIP (State Children's Health Insurance Program) programs, 2) assist CMS in implementing its mission focusing on health care quality and improvement for its beneficiaries, 3) assist HBCU researchers by supporting extramural research in health care capacity development activities for the African American communities, 4) increase the pool of HBCU researchers capable of implementing the research, demonstration, and evaluation activities of CMS, and 5) assist in fostering interuniversity communication and collaboration regarding African American health disparity issues. Funding is available for grants to implement research related to health care delivery and health financing issues affecting African American communities, including issues of access to health care, utilization of health care services, health outcomes, quality of services, cost of care, health and racial disparities, socio-economic differences, cultural barriers, managed care systems, and activities related to health screening, prevention, outreach, and education.
Requirements: To be eligible for grants under this program, an organization must be an HBCU (Historically Black College or University) and meet one of the following three *Requirements:* 1) offer a Ph.D. or Master's Degree Program in one or more of the following disciplines - Allied Health, Gerontology, Health Care Administration, Health Education, Health Management, Nursing, Nutrition, Pharmacology, Public Health, Public Policy, Social Work; or 2) have a School of Medicine; or 3) a member of the National HBCU Network for Health Services and Health Disparities. All proposals should describe research to be conducted with relevance to the CMS Medicare, Medicaid, and SCHIP programs and which area of Healthy People 2010 is served by this project. Applications must be submitted electronically (via grants.gov). CMS will accept a hard copy if the applicant is unable to access grants.gov or is having serious problems in sending the application electronically.
Restrictions: Grant funds may not be used for any of the following: to provide direct services to individuals except as explicitly permitted under the grant solicitation; to match any other Federal funds; to provide services, equipment, or supports that are already the legal responsibility of another party under Federal law.
Geographic Focus: All States
Date(s) Application is Due: Jul 2
Amount of Grant: 200,000 - 250,000 USD
Contact: Joi Grymes, (410) 786-7251; Joi.Grymes@cms.hhs.gov
Internet: http://www.cms.hhs.gov/ResearchDemoGrantsOpt/02_Historically_Black_Colleges_and_Universities.asp#TopOfPage
Sponsor: Centers for Medicare and Medicaid Services
7500 Security Boulevard, Room C2-21-15
Baltimore, MD 21244-1850

CMS Research and Demonstration Grants 1188
The general purpose of the Centers for Medicare & Medicaid Services' (CMS) research and demonstration program is to conduct and support projects to develop, test, and implement new health care financing and payment policies and to evaluate the impact of the agency's programs on its beneficiaries, providers, States, and other customers and partners. The scope of the agency's activities embraces all areas of health care: costs, access, quality, service delivery models, and financing and payment approaches.
Requirements: The following themes represent the agency's current priorities in research. Note that all projects must fall into the agency's statutory authorities to operate and improve Medicare, Medicaid, and other CMS programs and activities: (1) Monitoring and Evaluating CMS Programs; (2) Strengthening Medicaid, State Children's Health Insurance Program (SCHIP), and State Programs; (3) Expanding Beneficiaries' Choices and Availability of Managed Care Options; (4) Developing FFS Payment and Service Delivery Systems; (5) Improving Quality of Care and Performance Under CMS Programs; (6) Improving the Health of Our Beneficiary Population; (7) Prescription Drugs; (8) Building Research Capacity. Application packet is available online.
Restrictions: Applicants are expected to contribute towards the project costs. Generally 5 percent of the total project costs is considered acceptable. CMS rarely approves grants or cooperative agreements for research or demonstration projects in which the Federal Government covers 100 percent of the project's costs. The budget may not include costs

for construction or remodeling or for project activities that take place before the applicant has received official notification of our approval of the project.
Geographic Focus: All States
Amount of Grant: 25,000 - 1,000,000 USD
Contact: Grant Officer; (410) 786-5130; Jnorris1@cms.hhs.gov
Internet: http://www.cms.hhs.gov/ResearchDemoGrantsOpt/04_Other_CMS-Grant_Opportunities.asp
Sponsor: Centers for Medicare and Medicaid Services
7500 Security Boulevard, Room C2-21-15
Baltimore, MD 21244-1850

CNA Foundation Grants 1189
The CNA Foundation concentrates its support primarily in programs designed to: meet the education needs of children, assist and support children, youth and adults in developing vocational skills. Support economically disadvantaged children and families. Requests for funding are accepted year-round. However, grants will be made only in accordance with the Foundation's budgetary guidelines. Proposals must be clear and brief. The Foundation will contact the organization if more information is needed.
Requirements: IRS 501(c)3 tax-exempt organizations are eligible.
Restrictions: Grants are not made to/for individuals; political causes, candidates, or organizations; veterans, labor, alumni, military, athletic clubs, or social clubs; sectarian organizations or denominational religious organizations; capital improvement or building projects; endowed chairs or professorships; United Way-affiliated agencies; or national groups whose local chapters have already received support.
Geographic Focus: All States
Amount of Grant: 10,000 - 250,000 USD
Sample: Chicago 2016 Exploratory Committee, Chicago, IL, $250,000; National Chamber Foundation, Washington, D.C., $100,000; Starlight Starbright Childrens Foundation Midwest, Chicago, IL, $50,000.
Contact: Marlene Rotstein, Director; (312) 822-7065; marlene.rotstein@cna.com or cna_foundation@cna.com
Internet: http://www.cna.com/portal/site/cna/menuitem.7204aaf0316757e8715f09f6556631a0/?vgnextoid=b1e940fa11056010VgnVCM1000005566130aRCRD
Sponsor: CNA Foundation
333 South Wabash Avenue
Chicago, IL 60604

CNCS AmeriCorps VISTA Project Grants 1190
AmeriCorps VISTA is a national-service program of the Corporation for National and Community Service (the Corporation) which oversees a variety of programs including other AmeriCorps programs, SeniorCorps programs, and Learn and Serve America programs. Designed specifically to fight poverty, VISTA was authorized in 1964 as Volunteers in Service to America. The program was incorporated into the AmeriCorps network of programs in 1993. VISTA supports efforts to alleviate poverty by encouraging volunteers (members), ages 18 years and older, from all walks of life, to engage in a year of full-time service with a sponsoring organization (sponsor) to create or expand programs designed to bring individuals and communities out of poverty. Under this arrangement, the Corporation places a team of VISTA members with a sponsor; the sponsor funds local operating and logistics costs of the project while the Corporation covers member and certain sponsor costs as follows: a biweekly living allowance for members; a Segal AmeriCorps Education Award or post-service stipend for members; health coverage for members; a moving allowance for members relocating to serve; liability coverage for members under the Federal Employees Compensation Act and the Federal Torts Claims Act; childcare (for income-eligible members); FICA; payroll services (members receive their paychecks directly from AmeriCorps VISTA); training in project management and leadership for VISTA members and project supervisors; and assistance for sponsors to recruit VISTA members. Applications for VISTA resources are handled by Corporation State Offices which are federal offices staffed by federal employees in the states. (A list of the offices along with their contact information is available at the AmeriCorps website.) Applying for VISTA resources is a two-step process. As step one, the organization must submit a VISTA Concept Paper. If the concept paper is accepted, the organization must, as step two, submit a VISTA Project Application. Applicants will receive their project-application materials when their concept paper has been approved. The length of the application process varies, but the average length of time from the initial contact to a final decision is three to five months. Both concept papers and project applications are usually submitted using eGrants, the Corporation's web-based system for applications. Organizations must visit the eGrant website (link available at the AmeriCorps website) to create an account prior to submitting concept papers. Organizations that cannot submit using eGrants may submit a paper copy. The forms are included in the downloadable Concept-Paper Instructions document at the AmeriCorps website. As of this writing, VISTA is giving priority to new projects that focus on the areas of housing, financial literacy, and employment.
Requirements: Public organizations such as nonprofit private organizations, Indian Tribes, state and local government organizations, and institutions of higher education can apply to be VISTA sponsors. Eligible nonprofit private organizations are not limited to those with IRS 501(c)3 status, but rather all organizations with IRS 501(c) status that focus on anti-poverty community development. Project sponsors are encouraged (but not required) to provide a financial match; however they must be able to direct the project, supervise the volunteers, and provide necessary administrative support to complete the goals and objectives of the project. Projects must be developed in accordance with all four of the VISTA Core Principles: Anti-Poverty Focus, Community Empowerment, Sustainable Solutions, and Capacity Building. All VISTA resources must be used to create, expand, or enhance projects that lift people out of poverty. Additionally, the Corporation has identified, in its strategic plan, six focus areas for funding: Economic Opportunity; Education; Healthy Futures; Veterans and Military Families; Disaster Services; and Environmental Stewardship. All new VISTA project development must fall within these six focus areas. As of this writing, the Corporation will direct most VISTA resources to the Economic Opportunity and Education focus areas; however, the Corporation will also address the other focus areas, according to the ability of Corporation State Offices to identify opportunities in those areas that can have a direct impact on breaking the cycle of poverty.
Restrictions: Organizations that focus solely on advocacy and lobbying are not eligible. Key legislation and regulations governing the VISTA program are as follows: the Domestic Volunteer Service Act (as amended by Public Law 113-13, April 2009); the National Service Trust Act (as amended by Public Law 113-13, April 2009); the Edward M. Kennedy Serve America Act (Public Law 113-13, April 2009); and the Code of Federal Regulations, Title 45, Parts 1206, 1210-1211, 1216-1220, 1222, and 1226.
Geographic Focus: All States
Contact: Mary Strasser; (202) 606-6943 or (202) 606-5000
Internet: http://www.nationalservice.gov/programs/americorps/americorps-vista
Sponsor: Corporation for National and Community Service
1201 New York Avenue, NW
Washington, D.C. 20525

CNCS Senior Companion Program Grants 1191
The Senior Companions Program (SCP) was established in 1973 under Title II of the Domestic Volunteer Services Act (DVSA) to provide opportunities for older adult volunteers to assist other older adults and persons with disabilities, who, without support, might not be able to live independently. Eighteen model Senior Companion projects were funded initially. Today that number has grown to 223 projects with more than 15,000 volunteers who serve through nonprofit and public organizations (local sponsors) to help home-bound clients with chores such as light housekeeping, paying bills, buying groceries, and finding transportation to medical appointments. SCP volunteers serve from fifteen to forty hours a week and receive hourly stipends. They must be fifty-five or older and meet established income eligibility guidelines. In addition to the stipend, they receive accident, personal-liability, and excess-automobile insurance coverage; assistance with the cost of transportation; an annual physical examination; recognition; and, as feasible, meals during their assignments. Volunteers receive training in how to assist persons diagnosed with Alzheimer's disease, stroke, diabetes, mental illness, etc., as well as when to alert doctors and family members to potential health problems. Currently SCP is administered through the Corporation for National and Community Service (the Corporation)'s Senior Corps Program. The Corporation accepts SCP grant applications only when new funding is available or when it is necessary to replace an existing sponsor. In addition, eligible agencies or organizations may, under a Memorandum of Agreement with the Corporation, receive technical assistance and materials to aid in establishing and operating a non-federally-funded SCP project using local funds. Notices for nationwide competitions for new SCP grants are posted at www.grants.gov and at the Corporation and Senior Corps websites. (Subscription links for receiving RSS feeds on new funding opportunities are also available at the websites.) Notices to apply to replace a sponsor are advertised locally through Corporation State Offices. (A list of the offices along with their contact information is available at the Corporation website.) Grant applications are submitted through the Corporation's eGrants system. For more information, interested applicants may download the SCP Handbook from the Senior Corps and Corporation websites or contact their Corporation State Office.
Requirements: The Corporation awards grants to public agencies, Indian tribes, and secular or faith-based private non-profit organizations in the United States that have authority to accept and the capacity to administer an SCP project. The SCP requires a non-federal share of 10% of the total project cost. SCP projects are generally expected and required to be on-going. SCP Sponsors may apply for continued funding from the Corporation.
Restrictions: The total of cost reimbursements for Senior Companions, including stipends, insurance, transportation, meals, physical examinations, uniforms if appropriate, and recognition must be equal to at least 80 percent of the Corporation's Federal share of the grant. (Federal and non-Federal resources, including excess non-Corporation resources, can be used to make up this sum.) Key legislative pieces enabling and regulating the SCP have been the Domestic Volunteer Service Act (DVSA) of 1973, the National and Community Service Trust Act (1993), 45 C.F.R. § 1216 (non-displacement of contracts and employed workers), the Edward Kennedy Serve America Act, and 45 C.F.R. § 2551. SCP funding generally requires an Office of Management and Budget (OMB) audit.
Geographic Focus: All States
Amount of Grant: 100,000 - 300,000 USD
Contact: Wanda Carney; (202) 606-6934 or (202) 606-5000
Internet: http://www.nationalservice.gov/build-your-capacity/grants/managing-senior-corps-grants
Sponsor: Corporation for National and Community Service
1201 New York Avenue, NW
Washington, D.C. 20525

CNCS Social Innovation Grants 1192
The CNCS Social Innovation Fund is authorized by the Edward M. Kennedy Serve America Act and is administered by the Corporation for National and Community Service (the Corporation), a federal agency that engages more than five million Americans as volunteers through well-known national-service programs like Senior Corps, AmeriCorps, and Learn and Serve America. The Social Innovation Fund is primarily concerned with advancing social innovation as a key strategy for solving critical social challenges. The program's goal is to identify and help spread those innovative and potentially transformative approaches that have been developed at the local level to solve community problems. An approach is considered transformative if it not only produces strong impact, but also if it: has the potential to affect how the same challenge is addressed

in other communities; addresses more than one critical community challenge concurrently; or produces significant cost savings through gains in efficiency. The operating model of the Social Innovation Fund is distinguished by four key elements: reliance on intermediaries with strong skills and track records of success in selecting, validating, and growing high-impact nonprofit organizations; assuring participation from the non-federal stake-holders by requiring each federal dollar be matched 1:1 with money from non-federal sources not only by the intermediaries but also by their subgrantees; requiring that all intermediaries engage each of their subgrantees in formal evaluations of program performance and impact; and requiring each grantee to commit to knowledge sharing and other initiatives that advance social innovation more generally in the nonprofit sector. The SIF makes grant awards of between $1 million and $10 million per year for up to five years to grantmaking intermediaries, selected through a rigorous, open competition. Intermediaries match their federal grants dollar-for-dollar and with those combined funds they then: host open, evidence-based competitions to select nonprofits implementing innovative program models; invest in expanding the capabilities and impact of the nonprofits they select; and support those nonprofits through rigorous evaluation of their programs.
Requirements: Applicants must be an eligible grantmaking institution or partnership in existence at the time of the application. Providing grants to nonprofit community organizations should be central to the applicant's mission and should be clearly reflected in the organization's promotional materials and annual operating budget. Core operations must include conducting open competitive grant competitions, negotiating specific grant requirements with grant recipients; and overseeing and monitoring performance of grant recipients. By statute Social Innovation Fund intermediaries must operate either as geographically-based or as issue-based grantmakers. A geographically-based intermediary will address one or more priority issues withing a single geographic location. An issue-based intermediary will address a single priority issue in multiple geographic locations. At the time of submission, applicants must demonstrate through a letter or other form of documentation that they have either cash-on-hand or commitments (or a combination thereof) toward meeting 50 percent of their first year matching funds.
Restrictions: Intermediaries must distribute at least 80 percent of awarded federal funds to subgrantees, run an open competition that is available to eligible nonprofit organizations beyond the intermediary's own existing grant portfolio or network, and provide sufficient public notice of the availability of Social Innovation Fund subgrants to eligible nonprofit community organizations. Given that innovation funds currently exist in the Departments of Education and Labor to invest specifically in evidence-based programs in education and job training, the Corporation does not intend to make Social Innovation Fund awards to programs in these areas unless they clearly propose a solution to an unmet need as identified in consultation with both Departments. The funding mechanism for Social Innovation Fund awards is a cooperative agreement that provides for substantial involvement by the Corporation with the intermediaries as they carry out approved activities. The assigned Corporation program officer will confer with the grantee on a regular and frequent basis to develop and/or review service delivery and project status, including work plans, budgets, periodic reports, evaluations, etc. In particular the Corporation anticipates having substantial involvement in developing and approving subgrantee selection plans; developing and approving subgrantee evaluation plans; documenting subgrantee growth plans; and documenting and sharing lessons learned through a Corporation-sponsored learning community. Grants under the Social Innovation program are subject to the Cost Principles and Uniform Administration Requirements under the applicable Office of Management and Budget (OMB) Circulars.
Geographic Focus: All States
Date(s) Application is Due: Mar 27
Amount of Grant: 1,000,000 - 5,000,000 USD
Contact: Vielka Garibaldi; (202) 606-5000 or (202) 606-3223; info@cns.gov
Internet: http://www.nationalservice.gov/about/programs/innovation.asp
Sponsor: Corporation for National and Community Service
1201 New York Avenue NW
Washington, D.C. 20525

CNIB Baker Applied Research Fund Grants 1193
The CNIB Baker Applied Research Fund supports research that is focused on the social, educational, cultural and rehabilitative needs of Canadians living with vision loss, and the application of assistive devices to meet these needs. This program is intended both to advance the career development of new investigators, and to encourage senior researchers to develop programs of applied research. Proposals are invited from both. Grants are for a one-year period, with a limit of $35000.
Requirements: Projects must be completed within 12 months. Funds that are unused after 24 months after the start up of the grant must be returned unless a special request for continuance is approved. Applicants must be residents of Canada and research must be conducted primarily in Canada. The applicant must: have an academic appointment at a Canadian university or occupy a supervisory position at a health care facility, or vision rehab facility. All applications will be evaluated with regard to the following criteria: excellence in the health and social sciences as applied to vision care; feasibility of research results translating into improved interventions, programs, and services for Canadians living with vision loss; enhancement of Canada's outstanding research into the needs of Canadians living with vision loss; and creation of training opportunities for students and service providers.
Restrictions: Equipment costs will not generally be covered, but in certain circumstances equipment may be funded on the basis of matching funds from the host facility. Travel and publications costs should not exceed $2,000.
Geographic Focus: Canada
Date(s) Application is Due: Jan 15
Amount of Grant: Up to 35,000 CAD
Contact: Shampa Bose, Executive Assistant and Research Coordinator; (416) 486-2500, ext. 7622 or (800) 563-2642; fax (416) 480-7059; shampa.bose@cnib.ca

Internet: http://www.cnib.ca/en/research/funding/eabaker-applied/
Sponsor: Canadian National Institute for the Blind
1929 Bayview Avenue
Toronto, M4G 3E8 Canada

CNIB Baker Fellowships 1194
CNIB's Baker Fellowships are awarded annually for post-graduate training in ophthalmic sub-specialties. The program offers fellowships for one year in amounts up to $30,000. Applicants must apply online no later than January 15 to be eligible for consideration. Fellowship funds are provided quarterly commencing July 1 of the following year. Successful applicants are normally advised of their award in April.
Requirements: Recipients must begin or return to an academic posting in Canada within three years of completing their training, or return to an area of need as supported by a letter from the Medical Director of the community. An outline of future plans supported by a letter of acceptance to an academic posting or alternate should accompany the application where possible. This letter plays an important role in the eligibility of the candidate and strong efforts should be made to clarify that future position is available to the candidate. If the fellow does not return within a three-year time frame of completion of study to Canada, CNIB requests reimbursement for the funding provided.
Restrictions: Applicants who are expecting to receive funds or stipends from other sources amounting equal or more than $30,000 may not apply for CNIB Baker fellowship.
Geographic Focus: All States, Canada
Date(s) Application is Due: Jan 15
Amount of Grant: Up to 30,000 CAD
Contact: Shampa Bose, Executive Assistant and Research Coordinator; (416) 486-2500, ext. 7622 or (800) 563-2642; fax (416) 480-7059; shampa.bose@cnib.ca
Internet: http://www.cnib.ca/en/research/funding/eabaker-fellowship/
Sponsor: Canadian National Institute for the Blind
1929 Bayview Avenue
Toronto, M4G 3E8 Canada

CNIB Baker New Researcher Fund Grants 1195
The CNIB Baker New Researcher Fund provides one-year grants to encourage new investigations that may lead to the prevention of vision loss. It is intended to benefit new investigators (within 5 years after an academic faculty appointment) by giving them experience and results which can assist them in further grant applications and pilot investigations. A new investigator is someone who is affiliated with a research facility through an academic posting and who has been in the field for less than five years. Grants are for one year, in amounts up to $35,000 (CAD). The application must be submitted online no later than January 15 of each year, with grant decisions made by April 1.
Requirements: Projects should be completed within 12 months. Funds that are unused after 24 months of the start up of the grant must be returned unless a special request for continuance is approved. Applicants must be residents of Canada and research must be conducted primarily in Canada.
Restrictions: Equipment costs will not generally be covered, but in certain circumstances equipment may be funded on the basis of matching funds from the host facility. Travel and publication costs should not exceed $2,000. Applications must be focused and no longer than 10 pages in total.
Geographic Focus: Canada
Date(s) Application is Due: Jan 15
Amount of Grant: Up to 35,000 CAD
Contact: Shampa Bose, Executive Assistant and Research Coordinator; (416) 486-2500, ext. 7622 or (800) 563-2642; fax (416) 480-7059; shampa.bose@cnib.ca
Internet: http://www.cnib.ca/en/research/funding/eabaker-researcher/
Sponsor: Canadian National Institute for the Blind
1929 Bayview Avenue
Toronto, M4G 3E8 Canada

CNIB Barbara Tuck MacPhee Award 1196
The CNIB Barbara Tuck MacPhee Award supports researchers in the field of macular degeneration. The Tuck MacPhee Award provides a one-year grant of up to $35,000. Projects should be completed within 12 months. The investigator will provide CNIB with a progress report six months following the start up of the grant. A second report including a copy of relevant findings of the research is required at the end of the research and prior to releasing the final quarterly payment. All publications and presentations of this research should acknowledge the support of the CNIB Baker Tuck MacPhee Fund. Applicants must apply online no later than January 15 of each year. The application will be reviewed by a multi-disciplinary review committee, and decisions will be finalized by April 1.
Requirements: Funds that are unused after 24 months must be returned unless a special request for continuance is approved. Equipment costs will not generally be covered, but in certain circumstances equipment may be funded on the basis of matching funds from the host facility. This competition is open to all researchers; however applicants must be residents of Canada and research must be conducted primarily in Canada.
Restrictions: Trainees (residents, fellows, post graduate students) are not eligible.
Geographic Focus: All States, All Countries
Date(s) Application is Due: Jan 15
Amount of Grant: Up to 35,000 CAD
Contact: Shampa Bose, Executive Assistant and Research Coordinator; (416) 486-2500, ext. 7622 or (800) 563-2642; fax (416) 480-7059; shampa.bose@cnib.ca
Internet: http://www.cnib.ca/en/research/funding/Tuck-MacPhee/
Sponsor: Canadian National Institute for the Blind
1929 Bayview Avenue
Toronto, M4G 3E8 Canada

CNIB Canada Glaucoma Clinical Research Council Grants 1197
The CNIB-CGCRC Award represents a partnership of CNIB and the Canada Glaucoma Clinical Research Council designed to invest in clinical glaucoma research. This partnership contributes about $225,000 over a three year period. Funding of the CNIB-CGCRC award is via an unrestricted grant from Alcon Canada Inc. The application deadline for the CNIB-CGCRC Award is June 30. There are two levels of funding available: $5,000 or less for investigators with training or practice for less than 5 years and $5000 to $25,000 for investigators with training or practice over 5 years.
Geographic Focus: All States
Amount of Grant: Up to 25,000 CAD
Contact: Shampa Bose, Executive Assistant and Research Coordinator; (416) 486-2500, ext. 7622 or (800) 563-2642; fax (416) 480-7059; shampa.bose@cnib.ca
Internet: http://www.cnib.ca/en/research/funding/cnib-cgrc/
Sponsor: Canadian National Institute for the Blind
1929 Bayview Avenue
Toronto, M4G 3E8 Canada

CNIB Chanchlani Global Vision Research Award 1198
The CNIB Chanchlani Global Vision Research Award was established in 2011 with the goal of encouraging world researchers in the area of vision science and vision rehabilitation. The award is consistent with the goals of CNIB (Canadian National Institute for the Blind) of conducting and funding world-class research to reduce the impact of sight loss in people's lives. The award will be given to vision scientists anywhere in the world who have made a major original contribution to the fields of vision science or vision rehabilitation. The nominations will be evaluated by the CNIB Research Committee, a group of eminent Canadian vision scientists, clinicians and vision rehabilitation practitioners. The evaluation process will depend heavily on the quality of the information provided in the nomination process. The CNIB Chanchlani prize will be valued at $25,000. The award will be to an individual recipient. There will be no shared prizes. The prize winner may be a resident of any country.
Requirements: Nominations may be made by anyone knowing the potential recipient well. This may include the potential recipient his/herself.
Geographic Focus: All States, All Countries
Date(s) Application is Due: Jul 15
Amount of Grant: 25,000 CAD
Contact: Shampa Bose, Executive Assistant and Research Coordinator; (416) 486-2500, ext. 7622 or (800) 563-2642; fax (416) 480-7059; shampa.bose@cnib.ca
Internet: http://www.cnib.ca/en/research/funding/Pages/Chanchlani-Award.aspx
Sponsor: Canadian National Institute for the Blind
1929 Bayview Avenue
Toronto, M4G 3E8 Canada

CNIB E. (Ben) & Mary Hochhausen Access Technology Research Grants 1199
The award was established by Ben and Mary Hochhausen to further their long-time volunteer and financial contribution to CNIB. This international award encourages research in the field of access technology for people living with vision loss. Grants up to $10,000 are made each year at the discretion of the E. (Ben) and Mary Hochhausen Fund trustees. Applications are accepted from any country in the world. Awards may be applied to: research projects; study at centres of excellence in technology; fellowships; development of prototypes; or development costs for bringing important new products to market. Initial application should be received by the E. (Ben) and Mary Hochhausen Fund trustees by September 30.
Requirements: Trustees will consider: students enrolled in a recognized College, University or equivalent program; or researchers from industry, academia or applied science.
Geographic Focus: All States, All Countries
Date(s) Application is Due: Sep 30
Amount of Grant: Up to 10,000 CAD
Contact: Shampa Bose, Executive Assistant and Research Coordinator; (416) 486-2500, ext. 7622 or (800) 563-2642; fax (416) 480-7059; shampa.bose@cnib.ca
Internet: http://www.cnib.ca/en/research/funding/hochhausen/
Sponsor: Canadian National Institute for the Blind
1929 Bayview Avenue
Toronto, M4G 3E8 Canada

CNIB Ross Purse Doctoral Fellowships 1200
The Ross C. Purse Doctoral Fellowship encourages and supports theoretical and practical research and studies at the postgraduate or doctoral level, in the field of vision loss in Canada. The fellowship is awarded for research in the social sciences, engineering, or other fields of study that are immediately relevant to the field of vision loss. One fellowship is awarded annually to a qualified applicant. Successful applicants will be considered for subsequent funding only in exceptional circumstances. Each fellowship is valued at up to CAD$12,500, to be paid in three equal installments. Initial payment is made at the time of the award, and the second installment is paid after receipt of an interim report outlining the development of your research. Final payment is made conditional upon receipt by the Secretariat of the dissertation from the candidate and his/her thesis supervisor or department head. Completed applications must be postmarked not later than April 2 of each year.
Requirements: Applications will be considered from persons studying at a Canadian university or college, or at a foreign university where a commitment to work in the field of vision loss in Canada for at least two years can be demonstrated. Preference will be given to graduates of a Canadian university or college. Applicants must have achieved a high academic standing and must have demonstrated superior intellectual ability and judgment. Recipients may undertake paid employment with the permission of their supervisor of studies.
Geographic Focus: All States, All Countries, Canada
Date(s) Application is Due: Apr 2
Amount of Grant: Up to 12,500 CAD
Contact: Shampa Bose, Executive Assistant and Research Coordinator; (416) 486-2500, ext. 7622 or (800) 563-2642; fax (416) 480-7059; shampa.bose@cnib.ca
Internet: http://www.cnib.ca/en/research/funding/Ross-Purse/
Sponsor: Canadian National Institute for the Blind
1929 Bayview Avenue
Toronto, M4G 3E8 Canada

CNO Financial Group Community Grants 1201
As a company, the CNO Financial Group provides financial support for a number of causes that contribute to the well-being of its communities. As individuals, company personnel invest time and talents in the same causes. Corporate contributions focus on early childhood programs, children-at-risk, and early-childhood education through post-secondary education. Grants support literacy programs, early intervention, and economics education for high-risk students. Interested applicants should contact the office for further direction on guidelines and how to apply.
Geographic Focus: All States
Date(s) Application is Due: Oct 1
Amount of Grant: 1,500 - 10,000 USD
Contact: Carrie Jost, Grant Contact; (312) 396-7673 or (317) 817-3768; fax (317) 817-2179; Carrie.Jost@cnoinc.com
Internet: http://www.cnoinc.com/about-cno/in-the-community
Sponsor: CNO Financial Group
11825 North Pennsylvania Street
Carmel, IN 46032

Coastal Community Foundation of South Carolina Grants 1202
The foundation awards grants to South Carolina nonprofit organizations in its areas of interest, including arts and culture, education, environment, health, religion, justice and equity, and social services. Types of support include general operating support, emergency funds, program development, publication, seed money, scholarship funds, and technical assistance. Deadlines vary per program; check website for exact dates.
Requirements: 501(c)3 South Carolina nonprofits in the following counties are eligible: Beaufort, Berkeley, Charleston, Colleton, Dorchester, Georgetown, Hampton and Jasper.
Restrictions: Grants do not support individuals (except for designated scholarship funds), endowments, deficit financing, dinners, and rarely building funds.
Geographic Focus: South Carolina
Date(s) Application is Due: Jun 1
Amount of Grant: 500 - 10,000 USD
Samples: To each, a $15,000 annual grant for 3 years: Boys & Girls Clubs for Rural Units; Child Abuse Prevention Association; Citizens Opposed to Domestic Abuse; Colleton County Arts Council; Friends of Caroline Hospice; Hope Haven of the Lowcountry (formerly Hope Cottage); Literacy Volunteers of the Lowcountry; Lowcountry Food Bank; Second Helpings;.
Contact: George C. Stevens, President-CEO; (843) 723-3635; fax (843) 577-3671; gstevens@tcfgives.org or info@ccfgives.org
Internet: http://www.ccfgives.org
Sponsor: Coastal Community Foundation of South Carolina
90 Mary Street
Charleston, SC 29403

Coca-Cola Foundation Grants 1203
The foundation has established education as its philanthropic focus and set aside most of its funds to support educational initiatives that address pressing needs. To help prepare youth for life, the foundation gives in three areas: higher education--pipeline programs that connect various levels of education and help students stay in school, scholarships, and minority advancement; classroom teaching and learning--innovative K-12 projects, teacher development, and small projects that deal with classroom activities; and global education--projects that encourage international studies, global understanding, and student exchange. Grants are awarded to both public and private institutions at all levels of education: universities, colleges, and secondary and elementary schools. International educational institutions and health care organizations also receive consideration. Types of support include annual campaigns, donated equipment, employee matching gifts, operating budgets, special projects, capital campaigns, continuing support, fellowships, internships, endowment funds, matching funds, and scholarship funds. Proposals may be submitted at any time.
Requirements: The following are eligible to apply: IRS 501(c)3 non-profits; a foreign organization that has received a ruling from the IRS that it is a section 501(c)3 tax exempt organization; or a foreign organization that is the equivalent of a U.S. charity.
Restrictions: The foundation does not make grants to individuals, religious endeavors, political or fraternal organizations, or organizations without 501(c)3 status.
Geographic Focus: All States
Amount of Grant: 10,000 - 6,000,000 USD
Samples: Holyfield Foundation, Inc., Fairburn, GA, $75,000--contribution to the Giving Disadvantaged Youth a Fighting Chance scholarship program; United Way of Metropolitan Atlanta, Atlanta, GA, $15,000--support for Samaritan House, Impact Group and Atlanta Union Mission; Atlanta Children's Shelter, Inc., Atlanta, GA, $10,000--contribution to support shelter programs.
Contact: Helen Smith Price, Executive Director; (404) 676-2568; fax (404) 676-8804
Internet: http://www.thecoca-colacompany.com/citizenship/our_communities.html
Sponsor: Coca-Cola Foundation
P.O. Box 1734
Atlanta, GA 30301

Cockrell Foundation Grants 1204
The Cockrell Trust was established in 1957 with funds donated by Ernest and Virginia Cockrell. The Foundation's special emphasis is The University of Texas at Austin. Other fields of interest are: youth services, arts, health care, cultural programs, civic, religious and social services. The Foundation awards grants to support: annual campaigns, building funds, capital campaigns, endowment funds, general purposes, matching funds, and special projects. There are no deadlines for making a request to The Cockrell Foundation. The Foundation meetings are in the late spring and late Fall. An application form is not required. See website for recommended guidelines in making a proposal: http://www.cockrell.com/foundation/grant_guidelines.asp .
Requirements: Texas 501(c)3 nonprofit organizations in Houston are eligible.
Restrictions: The Foundation generally makes grants for only one-year periods. An organization should apply only once during any calendar year. If a grant request is denied, the applicant must wait until the following year before submitting a new request. The Foundation does not participate in feasibility studies and generally does not make grants for the following: individuals; mass appeal solicitations; medical or scientific research projects; organizations outside of Houston, Texas and the United States. Grant requests sent via email or fax will not be accepted.
Geographic Focus: Texas
Contact: M. Nancy Williams; (713) 209-7500; foundation@cockrell.com
Internet: http://www.cockrell.com/foundation/grant_guidelines.asp
Sponsor: Cockrell Foundation
1000 Main Street, Suite 3250
Houston, TX 77002

Coeta and Donald Barker Foundation Grants 1205
The foundation awards grants to California and Oregon nonprofit organizations in its areas of interest, including arts, children and youth, community development, disabled, environmental conservation, family services, federated giving, health care and health organizations, heart and circulatory research, higher education, hospitals, mental health, and secondary school education. Types of support include building construction/renovation, equipment acquisition, general operating support, program development, and scholarship funds.
Requirements: California and Oregon nonprofit organizations are eligible.
Restrictions: Grants do not support sectarian religious purposes, federal and tax-dependent organizations, individuals, or endowment funds.
Geographic Focus: California, Oregon
Date(s) Application is Due: Mar 1; Aug 1
Amount of Grant: 100 - 20,450 USD
Samples: Santa Barbara Zoological Gardens (Santa Barbara, CA), $5000.
Contact: Nancy Harris, Executive Administrator; (760) 324-2656; fax (760) 321-8662
Sponsor: Coeta and Donald Barker Foundation
P.O. Box 936
Rancho Mirage, CA 92270

Coleman Foundation Cancer Care Grants 1206
The Foundation has been an advocate for raising the standards of cancer care in the Midwest region and assuring that direct cancer services are available to cancer patients in the Chicago Metro area. Recently, the Foundation has employed an impact framework which provides a basic outline of the goals for the Foundation's grantmaking. For the Cancer program area, the Foundation has identified strategies that can enable health care providers to gain effective tools and resources in providing supportive oncology (care from diagnosis through end of life). The Cancer Impact Plan (see website) is intended to inform potential grantees as to the particular strategies the Foundation seeks to fund. The Foundation welcomes potential grantees to review the Cancer Impact Plan to assess and determine the best possible strategy to meet the Foundation's intended impact.
Requirements: Grants are made only to 501(c)3 or 509(a)1 nonprofit organizations that are not private foundations. The Foundation's primary geographic focus is the Midwest region, particularly the State of Illinois and the Chicago metropolitan area. Only programs within the United States will be considered, which excludes all international programs. Applicants should submit a letter of inquiry first; LOIs are accepted throughout the calendar year. The Foundation will advise you if a full proposal should be submitted for further review. Proposals are presented by Foundation staff and approved by the Board at quarterly meetings, usually in February, May, August and November.
Restrictions: The program does not fund for-profit businesses, individuals, individual scholarships, advertising books, tickets, equipment purchases (including computer hardware or software), or advertising. General solicitations and annual appeals will not be considered.
Geographic Focus: Illinois, Indiana, Iowa, Michigan, Ohio, Wisconsin
Contact: Rosa Berardi; (312) 902-7120; fax (312) 902-7124; rberardi@colemanfoundation.org
Internet: http://www.colemanfoundation.org/what_we_fund/cancer/
Sponsor: Coleman Foundation
651 West Washington Boulevard, Suite 306
Chicago, IL 60661

Coleman Foundation Developmental Disabilities Grants 1207
The Foundation supports an array of programs that historically has supported housing, life skills and supportive employment programs. The intended impact of the Foundation's funding is for individuals in the Chicago metropolitan area with developmental disabilities to experience a higher quality of life and increased self-determination through success in their work, comfort in their home and satisfaction across the varied stages of their lives. As vocational and residential outcomes are central to achieving this vision, Foundation funding concentrates in these areas. The Foundation welcomes potential grantees to review the Developmental Disabilities Impact Plan (see website) to determine the best possible strategy to meet the Foundation's intended impact.
Requirements: Grants are made only to 501(c)3 or 509(a)1 nonprofit organizations that are not private foundations. The Foundation's primary geographic focus is the Midwest region, particularly the State of Illinois and the Chicago metropolitan area. Only programs within the United States will be considered, which excludes all international programs. Applicants should submit a letter of inquiry first; LOIs are accepted throughout the calendar year. The Foundation will advise you if a full proposal should be submitted for further review. Proposals are presented by Foundation staff and approved by the Board at quarterly meetings, usually in February, May, August and November.
Restrictions: The program does not fund for-profit businesses, individuals, individual scholarships, advertising books, tickets, equipment purchases (including computer hardware or software), or advertising. General solicitations and annual appeals will not be considered.
Geographic Focus: Illinois, Indiana, Iowa, Michigan, Ohio, Wisconsin
Contact: Clark McCain, Senior Program Officer; (312) 902-7120; fax (312) 902-7124; cmccain@colemanfoundation.org
Internet: http://www.colemanfoundation.org/what_we_fund/developmental_disabilities/
Sponsor: Coleman Foundation
651 West Washington Boulevard, Suite 306
Chicago, IL 60661

Collective Brands Foundation Grants 1208
The Collective Brands Foundation invests financially in non-profit organizations that align with the Foundation's focus areas. Priority is given to organizations that provide involvement opportunities for team members and employees of Collective Brands, Inc. The Foundation may also consider sponsorships from charitable organizations in the following areas: Eastern Kansas, including Topeka, Lawrence and the Kansas City metropolitan area; New York City; Lexington, Massachusetts and the Greater Boston area; Denver, Colorado; Redlands, California; and Brookville, Ohio.
Requirements: The Collective Brands Foundation will consider requests for monetary grants from 501(c)3 non-profit organizations that manage programs in at least one of the following areas: women's preventative health; children's physical activity and fitness; improving the lives of children and youth in need; preserving the environment; and supporting industry in the United States. Applications must be submitted online at the Foundation's website.
Restrictions: Grants will not be awarded to religious organizations for projects that are sectarian and do not benefit a broad community base. Also, the Foundation will not award grants to: individuals; political causes, candidates or legislative lobbying efforts; or for capital campaigns, debt reduction, travel or conferences.
Geographic Focus: All States
Date(s) Application is Due: Aug 15
Amount of Grant: Up to 3,000 USD
Contact: Michele Gray; (877) 902-4437; grants@greaterhorizons.org
Internet: http://www.collectivebrands.com/foundation
Sponsor: Collective Brands Foundation
3231 SE 6th Avenue
Topeka, KS 66607

Collins C. Diboll Private Foundation Grants 1209
The Collins C. Diboll Private Foundation awards grants to Louisiana nonprofit organizations in the areas of higher education, human services, and youth programs. The Foundation's primary field of interest include: Catholic churches and agencies; education; higher education; human services; art museums; and protestant agencies and churches. Types of support include: building construction and renovation; capital campaigns; endowment funds; and general operating support. Typical grants range from $500 up to a maximum of $200,000. There are no identified annual deadlines for submission, though a formal application is required. Applicants should also submit a detailed description of the project and the amount of funding requested, along with a copy of an IRS determination letter.
Requirements: Louisiana nonprofit organizations are eligible.
Restrictions: Individuals are not eligible to apply.
Geographic Focus: Louisiana
Amount of Grant: 500 - 200,000 USD
Samples: Tulane University, Center for Infectious Diseases, New Orleans, Louisiana, $100,000 - for research purposes; National World War II Museum, New Orleans, Louisiana, $50,000 - in support of the China-Burma-India Display Gallary; New Orleans Botanical Garden Foundation, New Orleans, Louisiana, $125,000 - renovations for the CT Parker Building and the City Park.
Contact: Donald W. Diboll, Chairperson; (504) 582-8103 or (504) 582-8250
Sponsor: Collins C. Diboll Private Foundation
201 Saint Charles Avenue, 50th Floor
New Orleans, LA 70170-5100

Collins Foundation Grants 1210
The Collins Foundation is an independent, private foundation that was created in 1947 by Truman W. Collins, Sr., and other members of the Collins family. The Foundation exists to improve, enrich, and give greater expression to humanitarian endeavors in the state of Oregon, and to assist in improving the quality of life in the state. As a general-purpose, responsive grant maker, the Foundation serves people in urban and rural communities across Oregon through its grants to nonprofit organizations working for the common good. The Foundation's broad areas of interest include: arts and humanities; children and youth; community welfare; education; the environment; health and science; and religion. Most recent awards have ranged from as little as $3,000 to a maximum of $400,000. There are no identified annual submission deadlines for applications.

Requirements: Grants are made to 501(c)3 nonprofit agencies domiciled in Oregon. The proposed project must directly benefit the citizens of Oregon.

Restrictions: Grants are not made to individuals or to organizations sponsoring requests intended to be used by or for the benefit of an individual. Grants normally are not made to elementary, secondary, or public higher education institutions; or to individual religious congregations. Grants normally are not made for development office personnel, annual fundraising activities, endowments, operational deficits, financial emergencies, or debt retirement. The Foundation will consider only one grant request from the same organization in a twelve month period, unless an additional request is invited by the Foundation.

Geographic Focus: Oregon
Amount of Grant: 3,000 - 400,000 USD
Contact: Cynthia G. Addams, Executive Vice President; (503) 227-7171; fax (503) 295-3794; information@collinsfoundation.org
Internet: http://www.collinsfoundation.org/submission-guidelines
Sponsor: Collins Foundation
1618 South West First Avenue, Suite 505
Portland, OR 97201

Colonel Stanley R. McNeil Foundation Grants 1211

Colonel Stanley R. McNeil and his wife Merna created the Colonel Stanley R. McNeil Foundation in 1993 to support and promote quality educational, human-services, and health-care programming for underserved populations. Special consideration is given to charitable organizations that serve the needs of children. During their lifetimes, the McNeils were actively involved in their local church and community. Colonel McNeil also served on a number of boards, including Lake Bluff Children's Home and Ravenswood Hospital. Although the McNeils had no children, they were strong supporters of children's causes. The foundation is particularly interested in funding programs or organizations that focus on children's causes, start-up initiatives within the human-services or arts and culture arenas, and healthcare. To better support the capacity of nonprofit organizations, multi-year funding requests are considered. Grant requests for naming opportunities that honor the donors are strongly encouraged. Applicants must apply online at the grant website. Applicants are strongly encouraged to do the following before applying: review the downloadable state application procedures for additional helpful information and clarifications; review the downloadable online-application guidelines at the grant website; review the foundation's funding history (link is available from the grant website); review the online application questions in advance; and review the list of required attachments. These will generally include: a list of board members, financial statements (audited, reviewed, or compiled by independent auditor); an organization summary; a list of other funding sources; an IRS Determination letter; and other required documents. All attachments must be uploaded in the online application as PDF, Word, or Excel files. The Colonel Stanley R. McNeil Foundation has biannual deadlines of February 1 and June 1. Applicants for the February deadline will be notified of grant decisions by June 30, and applicants for the June deadline will be notified by November 30. Typical awards have ranged from $5,000 to $125,000.

Requirements: Illinois nonprofits serving the Chicago metropolitan area are eligible.

Restrictions: Because requests for support usually exceed available resources, organizations can only apply to either the Lang Burk Fund or the Colonel Stanley McNeil Foundation in the same calendar year. Grant requests to both foundations during the same calendar year will no longer be accepted. The Foundation will consider requests for general operations only if the organization's operating budget is less than $1 million. In general, grant request for individuals, endowment campaigns or capital projects will not be considered.

Geographic Focus: Illinois
Date(s) Application is Due: Feb 1; Jun 1
Amount of Grant: 5,000 - 125,000 USD
Contact: Srilatha Lakkaraju; (312) 828-8166; ilgrantmaking@ustrust.com
Internet: https://www.bankofamerica.com/philanthropic/foundation.go?fnId=84
Sponsor: Colonel Stanley R. McNeil Foundation
231 South LaSalle Street, IL1-231-13-32
Chicago, IL 60604

Colorado Resource for Emergency Education and Trauma Grants 1212

Colorado Resource for Emergency Education and Trauma (CREATE) assists public and private organizations in improving and expanding the emergency medical and trauma system in Colorado. This grant is intended to provide funding for education and training for emergency medical and trauma services. CREATE grants are open-competitive awards with applicants providing a local cash match. The requirements of the grant program state that applicants must provide a 50% cash match, but financial waivers are available. Agencies that can demonstrate financial inability to match at the 50% level may fill out a Financial Waiver form which the committee will consider with their application. A minimum of 10% cash match is required, and any financial waivers that are not approved will result in the non-approval of the entire application.

Requirements: Applicants must have as their purpose the provision of emergency medical and trauma services in the state of Colorado to be eligible. This includes, but is not limited to: ambulance agencies; fire agencies; state and local governing agencies; training facilities; hospitals and clinics; special districts; and other public and private providers of EMS and trauma services. Students affiliated with an eligible agency may be considered for this funding, as well.

Geographic Focus: Colorado
Contact: Lakesha Jones, Grants Coordinator; (720) 248-2742 or (800) 851-6782; fax (303) 832-7496; lj@coruralhealth.org
Internet: http://www.raconline.org/funding/2463
Sponsor: Colorado Resource for Emergency Education and Trauma
3033 S Parker Road, Suite 606
Aurora, CO 80014

Colorado Trust Grants 1213

The Trust's strategic grantmaking supports the development of a coordinated system of policies, programs and services that, expand health coverage, improve and expand health care. The Trust issues Requests for Proposals (RFP) and welcomes responses from nonprofit organizations and governmental entities across Colorado. When a competitive funding opportunity is available, a detailed RFP with related instructions and specific application deadlines is posted to the Funds website. On occasion, The Trust also asks organizations that are focused on strategies specific to achieving access to health to submit individual, non-competitive proposals.

Requirements: The following types of organizations are eligible to apply for grants: nonprofit organizations that are exempt under Section 501(c)3 of the Internal Revenue Code and are classified as not a private foundation under Section 509(a); independent sponsored projects of a nonprofit 501(c)3 organization acting as a fiscal agent; government and public agencies.

Restrictions: The Trust asks for proposals through a Request for Proposal process, rather than accepting unsolicited proposals. Announcements of Requests for Proposals are posted at the website. Grant seekers also may register with the Trust to receive notification of new funding opportunities. The Colorado Trust does not make grants for the following: political campaigns or voter registration drives; capital funding for the purchase, construction or renovation of any facilities or other physical infrastructure; operating deficits or retirement of debt; indirect allocations (excluding fiscal agent fees); religious purposes.

Geographic Focus: Colorado
Contact: Ed Lucero, Senior Program Officer; (888) 847-9140 or (303) 837-1200; fax (303) 839-9034; ed@coloradotrust.org
Internet: http://www.coloradotrust.org
Sponsor: Colorado Trust
1600 Sherman Street
Denver, CO 80203-1200

Columbus Foundation Allen Eiry Fund Grants 1214

The Fund provides support to organizations that serve older adults in Seneca County. Projects supported by the fund include transportation, recreation, homemaker services, hot meals, court ordered guardianships, information and referral programs, library programs for nursing homes, support for capital projects, and other projects that enhance the lives of needy elderly.

Requirements: The Fund welcomes requests from organizations having recognition under Section 501(c)3 of the Internal Revenue Code that serve older adults in Seneca County.

Restrictions: The Fund makes no grants to individuals. Requests for religious purposes, budget deficits, endowments, conferences, scholarly research, or projects that are normally the responsibility of a public agency are generally not funded.

Geographic Focus: Ohio
Date(s) Application is Due: Nov 5
Contact: Dottie Henderson, Executive Assistant; (614) 251-4000; fax (614) 251-4009; dhenderson@columbusfoundation.org
Internet: http://www.columbusfoundation.org/gogrants/targeted_needs/specialized_grants.aspx
Sponsor: Columbus Foundation
1234 East Broad Street
Columbus, OH 43205-1453

Columbus Foundation Ann Ellis Fund Grants 1215

The Fund is used for eye research, with primary emphasis on sponsoring scientific research on the diagnosis, prevention, and treatment of glaucoma. Once this is accomplished, the fund will assist in the research of other hidden eye problems. For more information, please contact the Community Research and Grant Management Department at the Foundation

Restrictions: Individuals are ineligible. Requests for religious purposes, budget deficits, endowments, conferences, or projects that are normally the responsibility of a public agency are generally not funded.

Geographic Focus: All States
Date(s) Application is Due: Oct 1
Contact: Dottie Henderson, Executive Assistant; (614) 251-4000; fax (614) 251-4009; dhenderson@columbusfoundation.org
Internet: http://www.columbusfoundation.org/gogrants/targeted_needs/specialized_grants.aspx
Sponsor: Columbus Foundation
1234 East Broad Street
Columbus, OH 43205-1453

Columbus Foundation Central Benefits Health Care Foundation Grants 1216

The Foundation was established in 1997 with a focus on preventative health care for indigent children and adults; the Trustees have taken a step further by focusing grantmaking for the preventative health care needs of children, prenatal through age six. Proposals for funding are accepted at anytime. Funding is limited to the central Ohio area, giving primarily in Columbus, Ohio.

Requirements: Central Ohio, tax-exempt organizations are eligible to apply.

Restrictions: Individuals are ineligible.

Geographic Focus: Ohio
Samples: Ohio Health Foundation, Columbus, OH, $350,000--health care grant;
Contact: Tamera Durrence, Assistant Vice President and Director of Supporting Foundations; (614) 251-4000; fax (614) 251-4009; tdurrence@columbusfoundation.org
Internet: http://www.columbusfoundation.org/find/support/cenben_usa.aspx
Sponsor: Columbus Foundation
1234 East Broad Street
Columbus, OH 43205-1453

Columbus Foundation Competitive Grants 1217

The foundation's competitive grants are made in the following fields: advancing philanthropy, arts and humanities, conservation, education, health, social services, and urban affairs. The Governing Committee approves distributions from unrestricted and field of interest funds through competitive grants. Competitive grants are the most common way nonprofit organizations request funding from the Foundation. Submit a Letter of Intent or a Full Proposal by accessing our online grant application system.
Requirements: Central Ohio, tax-exempt public charities under Section 501(c)3 of the Internal Revenue Service Code may submit grant requests.
Restrictions: Individuals are ineligible. Requests for religious purposes, budget deficits, endowments, conferences, scholarly research, or projects that are normally the responsibility of a public agency are generally not funded. Funding is not available for projects when funds are available elsewhere.
Geographic Focus: Ohio
Contact: Emily Savors, Director; (614) 251-4000; fax (614) 251-4009; esavors@columbusfoundation.org
Internet: http://www.columbusfoundation.org/gogrants/index.aspx
Sponsor: Columbus Foundation
1234 East Broad Street
Columbus, OH 43205-1453

Columbus Foundation Estrich Fund Grants 1218

The Fund is designed to provide home health care and/or inpatient services to the terminally ill. Funds may be used to provide medical services (excluding doctor visits, in-hospital costs, or care facilities), medical supplies and equipment, transportation, and physical and occupational therapy. Funds may also be used to make available emotional, social, and spiritual support through the use of professionals and/or trained volunteers.
Requirements: Central Ohio, tax-exempt public charities under Section 501(c)3 of the Internal Revenue Service Code may submit grant requests.
Restrictions: Requests for religious purposes, budget deficits, endowments, conferences, scholarly research, or projects that are normally the responsibility of a public agency are generally not funded.
Geographic Focus: Ohio
Date(s) Application is Due: Apr 20
Samples: Fisrtlink, Columbus, OH, $18,725--to support the Dental Options Program, which provides free or low cost dental care to residents of central Ohio; Central Ohio Diabetes Association, Columbus, OH--to support outreach and education programs for the Hispanic community; Impact Safety Programs, Columbus, OH, $20,000--to provide a violence prevention program for residents of Amethyst; Columbus AIDS Task Force (CATF), Columbus, OH, $22,000--to support outreach offices at Neighborhood House and Just For Today.
Contact: Sandi Smith, (614) 251-4000; fax (614) 251-4009; ssmith@columbusfoundation.org
Internet: http://www.columbusfoundation.org/GD/Templates/Pages/TCF/TCFSecondary.aspx?page=66
Sponsor: Columbus Foundation
1234 East Broad Street
Columbus, OH 43205-1453

Columbus Foundation J. Floyd Dixon Memorial Fund Grants 1219

The purpose of the Fund is to provide educational programs for children, health programs for the elderly, and social services programs in Jackson County. An individual grant seldom exceeds $10,000. Applications must be received by May 28, and decisions will be announced three to four months after the application deadline.
Requirements: The fund welcomes requests from organizations having recognition under Section 501(c)3 of the Internal Revenue Code that services the residents of Jackson County.
Restrictions: The Fund makes no grants to individuals, and generally does not fund governmental agencies. Requests for religious purposes, budget deficits, endowments, conferences, scholarly research, or projects that are normally the responsibility of a public agency are generally not funded.
Geographic Focus: Ohio
Date(s) Application is Due: May 28
Amount of Grant: Up to 10,000 USD
Contact: Dottie Henderson, Executive Assistant; (614) 251-4000; fax (614) 251-4009; dhenderson@columbusfoundation.org
Internet: http://www.columbusfoundation.org/gogrants/targeted_needs/specialized_grants.aspx
Sponsor: Columbus Foundation
1234 East Broad Street
Columbus, OH 43205-1453

Columbus Foundation Mary Eleanor Morris Fund Grants 1220

The Fund supports quality-of-life projects, with a focus on the arts, civic affairs, conservation, education, health, and social services that benefits the residents of Logan County. Application deadlines are the first Friday in January and the first Friday in July.
Requirements: The fund welcomes requests from organizations having recognition under Section 501(c)3 of the Internal Revenue Code that services the residents of Logan County, as well as public and private schools.
Restrictions: The Fund makes no grants to individuals. Requests for religious purposes, budget deficits, endowments, conferences, scholarly research, or projects that are normally the responsibility of a public agency are generally not funded.
Geographic Focus: Ohio
Date(s) Application is Due: Jan 1; Jul 2
Contact: Emily Savors; (614) 251-4000; esavors@columbusfoundation.org
Internet: http://www.columbusfoundation.org/gogrants/targeted_needs/small_grants.aspx
Sponsor: Columbus Foundation
1234 East Broad Street
Columbus, OH 43205-1453

Columbus Foundation Paul G. Duke Grants 1221

The Foundation is primarily interested in supporting projects for children, young adults, and the family. Grants are generally made in the fields of the arts, education, health, and social services. The Foundation will also consider making challenge grants for worthwhile projects to encourage matching gifts or additional funding from other donors. Requests may be for general, capital, or specific project support, including seed money for innovative programs. The Foundation accepts proposals twice per year, and support is generally given for one year.
Requirements: Nonprofit organizations having recognition under Section 501(c)3 of the Internal Revenue Code are eligible for grants.
Restrictions: The Foundation generally does not make grants for transportation, computer hardware or software, and research or treatment for specific diseases. The Foundation does not award grants to individuals. Generally, grants are not made for religious purposes, budget deficits, or projects that are normally the responsibility of a public agency.
Geographic Focus: Ohio
Date(s) Application is Due: Apr 1; Aug 1
Contact: Tami Durrence, (614) 251-4000; fax (614) 251-4009; mail@supportingfoundations.org
Internet: http://www.columbusfoundation.org/paul_g_duke.aspx
Sponsor: Paul G. Duke Foundation
1234 East Broad Street
Columbus, OH 43205-1453

Columbus Foundation Robert E. and Genevieve B. Schaefer Fund Grants 1222

The purpose of the Fund is to benefit the residents of Chillicothe and/or Ross County. The funding supports programs that enhance the community in three broad areas: cultural development, health and human welfare, and economic development.
Requirements: The Fund welcomes applications from organizations recognized under Section 501(c)3 of the Internal Revenue Code that serves the residents of Chillicothe and/or Ross County. All organizations with budgets over $75,000 must provide an audit, and those with budgets less than $75,000 must provide their most recent form 990 and compiled financial statements.
Restrictions: The Fund makes no grants to individuals, churches, governmental agencies, or school districts.
Geographic Focus: Ohio
Date(s) Application is Due: Oct 29
Contact: Dottie Henderson, Executive Assistant; (614) 251-4000; fax (614) 251-4009; dhenderson@columbusfoundation.org
Internet: http://www.columbusfoundation.org/gogrants/targeted_needs/specialized_grants.aspx
Sponsor: Columbus Foundation
1234 East Broad Street
Columbus, OH 43205-1453

Columbus Foundation Traditional Grants 1223

The Columbus Foundation's Traditional Grants program creates quality opportunities and meets community need by focusing on two areas: Disadvantaged Children funds programs and projects that meet the diverse needs of at-risk children (priority will be given to programs and projects that are in the home, build relationships with the family, and create a support network around and for the families of disadvantaged children; and Developmental Disabilities funds programs and projects that address the needs of children and adults with physical or cognitive disabilities that impair functions or behavior and that occurred before a person reaches the age of 22 (blindness/visual impairments and deafness/hearing impairments are not considered within this category). If you are implementing your project in a Columbus City School building, or collaborating with the district on a project, and asked to submit a full application, you must request a letter of endorsement from the Office of Development. The annual application deadline is the first Friday in February.
Requirements: Central Ohio, tax-exempt public charities under Section 501(c)3 of the Internal Revenue Service Code may submit grant requests. If you are implementing your project in a Columbus City School building, or collaborating with the district on a project, and asked to submit a full application, you must request a letter of endorsement from the Office of Development. Letter of Intent deadlines are twice a year on the first Friday in February and September.
Restrictions: Operating support will only be considered when the applicant demonstrates continuous innovation that enhances services.
Geographic Focus: Ohio
Date(s) Application is Due: Feb 5
Contact: Emily Savors, Director; (614) 251-4000; fax (614) 251-4009; esavors@columbusfoundation.org or contactus@columbusfoundation.org
Internet: http://columbusfoundation.org/nonprofit-center/grant-opportunities/columbus-foundation-grants/traditional-grants
Sponsor: Columbus Foundation
1234 East Broad Street
Columbus, OH 43205-1453

Comerica Charitable Foundation Grants 1224

The Comerica Charitable Foundation funding priorities support community needs in it's primary markets within Texas, Michigan, California, Arizona, and Florida. Applications are accepted for program support and capital expense of those non-profit organizations that support the foundation's priorities. Economic self-sufficiency for low and moderate income individuals and families will be supported in the areas of financial literacy, job readiness, job creation and retention, small business training and development, and transitional and supportive housing. Neighborhood revitalization areas to be supported are affordable housing and neighborhood business development. Financial literacy programs (K-12 and adult) and scholarships for students with income needs for studies in business, finance and growth industries will be supported. Also a priority is access to health care to include preventive care for the uninsured and under-insured. Also a priority are programs that support diversity and inclusion. Funding deadlines vary according to region. See the Foundation's website for regional offices.
Requirements: 501(c)3 nonprofits are eligible.
Restrictions: Grants are not awarded for United Way organizations, religious and fraternal groups, political parties, charitable golf events, athletic programs, multiyear pledges, or endowment funds.
Geographic Focus: Arizona, California, Florida, Michigan, Texas
Date(s) Application is Due: Mar 15; Jun 15; Sep 15; Nov 15
Contact: Program Contact; (313) 222-7356
Internet: http://www.comerica.com/vgn-ext-templating/v/index.jsp?vgnextoid=374970d75d994010VgnVCM1000004502a8c0RCRD
Sponsor: Comerica Charitable Foundation
P.O. Box 75000, MC 3390
Detroit, MI 48275-3390

Commonwealth Edison Grants 1225

The company recognizes its social responsibility to the area it serves, comprising Chicago and 25 northern Illinois counties, and makes grants and contributions to nonprofit organizations that address the diverse needs of this region and that can most enhance the economic, cultural, educational, and social health of its communities. Financial support is given for general operations as well as for specific projects or purposes that are in the general interest of the company or support general community needs. Contributions to hospitals are generally limited to capital fund drives. Building endowment funds or requests to relieve operating deficits are not supported. Each of the company's six commercial divisions (addresses and phone numbers available upon request) maintains a small philanthropic budget to address local needs. The corporate budget addresses needs of a wider ranging nature, such as support to United Way programs, education, and employee volunteer programs. Applications are accepted at any time. The executive review committee meets and addresses proposals quarterly, normally in February, May, August, and November.
Requirements: Nonprofits serving Chicago and 25 northern Illinois counties are eligible.
Restrictions: Support is not provided for individuals; fraternal or veteran organizations; sectarian religious organizations; non-tax-exempt organizations; municipal, state, or federal agencies; political organizations or campaigns; most United Way agencies; specific elementary or secondary schools or school systems; organizations requesting purchase of ads or sponsorship programs; or national or international organizations that do not have specific business relating to Commonwealth Edison and/or its customers.
Geographic Focus: Illinois
Amount of Grant: 1,000 - 5,000 USD
Contact: Steve Solomon, (312) 394-4321; fax (312) 394-2231
Sponsor: Commonwealth Edison Company
440 S. Lasalle Street, P.O. Box 805379
Chicago, IL 60680-5379

Commonwealth Fund/Harvard University Fellowship in Minority Health Policy 1226

The Commonwealth Fund/Harvard University Fellowship in Minority Health Policy is a one-year, full-time, academic degree-granting program designed to create physician-leaders, particularly minority physician-leaders, who will pursue careers in health policy, public health practice, and academia. It is designed to incorporate the critical skills taught in schools of public health, government, business, and medicine with leadership forums and seminar series conducted by Harvard senior faculty and nationally recognized leaders in minority health and public policy; supervised practicums and shadowing opportunities; and site visits, conferences, and travel. Each fellowship provides: $50,000 stipend; full tuition; health insurance; books; travel; and related program expenses, including financial assistance for a practicum project.
Requirements: Applicants are required to complete applications to both the Commonwealth Fund/Harvard University Fellowship in Minority Health Policy and the Harvard School of Public Health. Applicants must have finished residency and be U.S. citizens.
Geographic Focus: All States
Date(s) Application is Due: Jan 4
Amount of Grant: 50,000 USD
Contact: Joan Y. Reede, Director; (617) 432-2922; mfdp_cfhuf@hms.harvard.edu
Internet: http://www.commonwealthfund.org/Fellowships/Minority-Health-Policy-Fellowship.aspx
Sponsor: Commonwealth Fund
164 Longwood Avenue, 2nd Floor
Boston, MA 02115-5818

Commonwealth Fund Affordable Health Insurance Grants 1227

The Program on Affordable Health Insurance envisions an efficient and equitable health insurance system that makes available to all Americans comprehensive, continuous, and affordable coverage. In support of that vision, the program seeks to: analyze changes in employer-based, private and public insurance coverage for people under age 65, and determine how those changes may affect the affordability and comprehensiveness of coverage, the number of uninsured, the number of under-insured, and churning in and out of coverage; document the consequences of being uninsured, under-insured, and unstably insured, with regard to access to care, health status, personal financial security, and economic productivity; and develop and evaluate federal and state policies to expand and stabilize health insurance, make it more affordable and aligned with incentives to access appropriate and high quality care, and enhance the efficiency with which it is administered.
Requirements: The Commonwealth Fund requests letters of inquiry to initiate the grant application process, and does not wish to review full proposals at this stage. Applicants are encouraged to submit letters of inquiry using our online form.
Restrictions: The Fund makes grants only to tax-exempt organizations and public agencies and does not support: general planning and ongoing activities or existing deficits; endowment or capital costs, including construction, renovation, or equipment; basic biomedical research; conferences, symposia, major media projects, or documentaries, unless they are an outgrowth of one of the Fund's programs; individuals; scholarships; churches or other religious organizations unless the project for which they seek funding is entirely secular in nature; and work for which achievements cannot be measured.
Geographic Focus: All States
Contact: Sara R. Collins; (212) 606-3838 or (212) 606-3800; src@cmwf.org
Internet: http://www.commonwealthfund.org/Content/Program-Areas/Affordable-Health-Insurance.aspx
Sponsor: Commonwealth Fund
1 E 75th Street
New York, NY 10021-2692

Commonwealth Fund Australian-American Health Policy Fellowships 1228

The Australian Government Department of Health and Aging hopes to enrich health policy thinking as Australian-American Health Policy Fellows study how Australia approaches health policy issues, share lessons learned from the United States, and develop an international perspective and network of contacts to facilitate policy exchange and collaboration that extends beyond the fellowship experience. Australian-American Health Policy Fellowships are open to accomplished, mid-career health policy researchers and practitioners, including, academics, physicians, decision makers in managed care and other private sector health care organizations, federal and state health officials, and journalists. The Fellowship provides up to $55,000 (AUD) for terms of six to ten months, with a minimum stay of six months in Australia required. Focused on issues of common concern to Australian and U.S. policymakers, the fellowships are structured around areas of mutual policy interest, for example: health care quality and safety, the private/public mix of insurance and providers, the fiscal sustainability of health systems, the health care workforce, management and efficiency of health care delivery, and investment in preventive care strategies.
Requirements: All applicants must also meet the following criteria: be a citizen of the United States; be a mid-career health services researcher or practitioner (e.g., a physician, decision maker in a managed care organization or other private health care organization, government official or policy analyst, or journalist); have a demonstrated expertise in health policy issues and track record of informing health policy through research, policy analysis, health services, or clinical leadership; have completed a master's degree or doctorate (or the equivalent thereof) in health services research, health administration, health policy, or a related discipline, such as economics or political science; and if academically based, be at a mid-career level (e.g., research fellow to associate professor).
Restrictions: Fellowships are not awarded to support basic research or study for an academic degree.
Geographic Focus: All States
Date(s) Application is Due: Aug 15
Amount of Grant: Up to 55,000 USD
Contact: Robin Osborn; (212) 606-3809 or (212) 606-3800; ro@cmwf.org
Internet: http://www.commonwealthfund.org/Fellowships/Australian-American-Health-Policy-Fellowships.aspx
Sponsor: Commonwealth Fund
1 E 75th Street
New York, NY 10021-2692

Commonwealth Fund Harkness Fellowships in Health Care Policy and Practice 1229

The Harkness Fellowships in Health Care Policy and Practice provide an opportunity for professionals from the Australia, Germany, the Netherlands, New Zealand, Norway, Switzerland, and the United Kingdom to spend four to 12 months in the United States conducting a research study that is relevant to health care policy and practice in both the United States and the fellow's home country, and that is focused on the issues of greatest concern to the fund. Fellowship awards provide up to $107,000, which covers round trip airfare to the United States, a living allowance, funds for project-related travel, research, and conferences, travel to attend the Commonwealth Fund program of fellowship seminars, health insurance, and U.S. and state taxes. A family supplement - including airfare, living allowance, and health insurance - is also provided to Fellows accompanied by a partner and/or children up to age 18.
Requirements: Applicants must be citizens of Australia, Germany, the Netherlands, New Zealand, Norway, Switzerland, and the United Kingdom in their late 20s to early 40s with broad educational backgrounds, not just those in research or academic careers. In order to apply, applicants must be nominated by their institution and submit a formal application, which is available from The Commonwealth Fund in New York City or its representatives in their home country.
Geographic Focus: All States, United Kingdom
Date(s) Application is Due: Sep 15
Amount of Grant: Up to 107,000 USD

Contact: Robin Osborn; (212) 606-3809 or (212) 606-3800; ro@cmwf.org
Internet: http://www.commonwealthfund.org/Fellowships/Minority-Health-Policy-Fellowship.aspx
Sponsor: Commonwealth Fund
1 E 75th Street
New York, NY 10021-2692

Commonwealth Fund Health Care Quality Improvement and Efficiency Grants   1230
The goal of the Fund's Program on Health Care Quality Improvement and Efficiency is to improve the quality and efficiency of health care in the United States. To that end, the program supports projects that: promote the development and widespread adoption of health care quality and efficiency measures; assess and enhance the capacity of health care organizations to provide better care more efficiently; and promote the development and adoption of payment and incentive models that encourage providers to improve quality and efficiency.
Requirements: The Commonwealth Fund requests letters of inquiry to initiate the grant application process, and does not wish to review full proposals at this stage. Applicants are encouraged to submit letters of inquiry using our online form.
Restrictions: The Fund makes grants only to tax-exempt organizations and public agencies and does not support: general planning and ongoing activities or existing deficits; endowment or capital costs, including construction, renovation, or equipment; basic biomedical research; conferences, symposia, major media projects, or documentaries, unless they are an outgrowth of one of the Fund's programs; individuals; scholarships; churches or other religious organizations unless the project for which they seek funding is entirely secular in nature; and work for which achievements cannot be measured.
Geographic Focus: All States
Contact: Anne-Marie J. Audet, M.D; (212) 606-3800; fax (212) 606-3500; ama@cmwf.org
Internet: http://www.commonwealthfund.org/Content/Program-Areas/Health-Care-Quality-Improvement-and-Efficiency.aspx
Sponsor: Commonwealth Fund
1 E 75th Street
New York, NY 10021-2692

Commonwealth Fund Patient-Centered Coordinated Care Program Grants   1231
As defined by the Institute of Medicine, patient-centered care is health care that establishes a partnership among practitioners, patients, and their families (when appropriate), to ensure that decisions respect patients' needs and preferences, and that patients have the education and support they need to make decisions and participate in their own care. In primary care, such care is best provided in a medical home - a primary care practice or health center that ensures patients have enhanced access to their clinicians (for example, through the availability of evening or weekend appointments), coordinates care, and engages in continuous quality improvement. The goal of The Commonwealth Fund's Patient-Centered Coordinated Care Program, established in 2005, is to improve the quality of primary care by making it more patient- and family-centered. The initiative supports projects that: promote the collection of information on patient-centered care and the delivery of care to facilitate public reporting and quality improvement; stimulate adoption of effective practices, models, and tools to make primary care practices patient- and family-centered; and improve policy to encourage patient- and family-centered care in medical homes.
Requirements: The Commonwealth Fund requests letters of inquiry to initiate the grant application process, and does not wish to review full proposals at this stage. Applicants are encouraged to submit letters of inquiry using our online form.
Restrictions: The Fund makes grants only to tax-exempt organizations and public agencies and does not support: general planning and ongoing activities or existing deficits; endowment or capital costs, including construction, renovation, or equipment; basic biomedical research; conferences, symposia, major media projects, or documentaries, unless they are an outgrowth of one of the Fund's programs; individuals; scholarships; churches or other religious organizations unless the project for which they seek funding is entirely secular in nature; and work for which achievements cannot be measured.
Geographic Focus: All States
Contact: Melinda Abrams, (212) 606-3800; fax (212) 606-3500; mka@cmwf.org
Internet: http://www.commonwealthfund.org/Content/Program-Areas/Patient-Centered-Coordinated-Care.aspx
Sponsor: Commonwealth Fund
1 E 75th Street
New York, NY 10021-2692

Commonwealth Fund Payment System Reform Grants   1232
To achieve a high performing health system, the U.S. must curb spending growth and improve the way health care is provided. Payment system reform is critical to accomplishing these objectives. As a nation, we need to align incentives so that health care providers are rewarded for high-value care rather than a high volume of services. Rewarding value over volume will also encourage the development of a more integrated health care delivery system. The Commonwealth Fund, through its Program on Payment System Reform, supports analysis and the development of policy options to accomplish these goals. Areas of interest include: reforming the existing payment structure to improve the alignment of incentives and to provide a base for more comprehensive payment reform; modeling and analyzing the potential impact of alternative options for payment reform in Medicare and throughout the health system; using payment reform to encourage the development of new models of health care delivery that provide better and more coordinated care; and using comparative effectiveness research to support better decision-making by providers, payers, and patients.
Requirements: The Commonwealth Fund requests letters of inquiry to initiate the grant application process, and does not wish to review full proposals at this stage. Applicants are encouraged to submit letters of inquiry using our online form.
Restrictions: The Fund makes grants only to tax-exempt organizations and public agencies and does not support: general planning and ongoing activities or existing deficits; endowment or capital costs, including construction, renovation, or equipment; basic biomedical research; conferences, symposia, major media projects, or documentaries, unless they are an outgrowth of one of the Fund's programs; individuals; scholarships; churches or other religious organizations unless the project for which they seek funding is entirely secular in nature; and work for which achievements cannot be measured.
Geographic Focus: All States
Contact: Stuart Guterman, (202) 292-6735 or (212) 606-3800; sxg@cmwf.org
Internet: http://www.commonwealthfund.org/Content/Program-Areas/Payment-System-Reform.aspx
Sponsor: Commonwealth Fund
1 E 75th Street
New York, NY 10021-2692

Commonwealth Fund Quality of Care for Frail Elders Grants   1233
The Commonwealth Fund Program on Quality of Care for Frail Elders aims to transform the nation's nursing homes and other long-term care facilities into resident-centered organizations that are good places to live and work, capable of providing the highest-quality care. The projects it supports aim to: identify, evaluate, and spread models of resident-centered care, or care delivered in accordance with the needs and desires of the people who live in nursing homes; equip nursing home operators to lead transformational change; and promote policy options that support resident-centered care.
Requirements: The Commonwealth Fund requests letters of inquiry to initiate the grant application process, and does not wish to review full proposals at this stage. Applicants are encouraged to submit letters of inquiry using our online form.
Restrictions: The Fund makes grants only to tax-exempt organizations and public agencies and does not support: general planning and ongoing activities or existing deficits; endowment or capital costs, including construction, renovation, or equipment; basic biomedical research; conferences, symposia, major media projects, or documentaries, unless they are an outgrowth of one of the Fund's programs; individuals; scholarships; churches or other religious organizations unless the project for which they seek funding is entirely secular in nature; and work for which achievements cannot be measured.
Geographic Focus: All States
Contact: Mary Jane Koren, (212) 606-3800; fax (212) 606-3500; mjk@cmwf.org
Internet: http://www.commonwealthfund.org/Content/Program-Areas/Quality-of-Care-for-Frail-Elders.aspx
Sponsor: Commonwealth Fund
1 E 75th Street
New York, NY 10021-2692

Commonwealth Fund Small Grants   1234
The fund awards small grants for projects in its major program areas, which include: affordable health insurance; international health care policy and practice; improving the quality of health care services; and improving insurance coverage and access to care. Types of support include employee-matching gifts, program development, program evaluation, and research. Preference is given to projects that seek to solve problems, especially those affecting vulnerable groups; that analyze the effects of policies and trends on well-defined health issues; and that develop and test practical solutions. Prospective grantees should submit a letter of inquiry via regular or electronic mail. Program staff will contact applicants if more detailed information is required. Proposals recommended by fund staff are reviewed and voted upon by the board of directors, which meets in July, April, and November. Letters of inquiry should be brief, no more than two pages.
Requirements: The fund makes grants only to tax-exempt organizations and public agencies.
Restrictions: The fund does not support general planning and ongoing activities or existing deficits; endowment or capital costs, including construction, renovation, or equipment; basic biomedical research; conferences, symposia, major media projects or documentaries, unless they are an outgrowth of one of the fund's programs; individuals; scholarships; churches or other religious organizations, unless the project is entirely secular in nature; or work for which achievements cannot be measured.
Geographic Focus: All States
Contact: Andrea Landes, Director of Grants Management; (212) 606-3800; fax (212) 606-3508; acl@cmwf.org
Internet: http://www.commonwealthfund.org/Grants-And-Programs/Applicant-and-Grantee-Resources/Applying-for-a-Grant.aspx
Sponsor: Commonwealth Fund
1 E 75th Street
New York, NY 10021-2692

Commonwealth Fund State High Performance Health Systems Grants   1235
The Commonwealth Fund's State High Performance Health Systems program is designed to help states develop the infrastructure needed to improve health care quality and outcomes - and to draw out and share lessons of national import from the experience of states that are moving toward comprehensive health care reforms. To enable states to help local health care providers better meet the needs of their patient populations, the program will support approaches that offer providers shared access to the clinical support and practice-management services essential to achieving high performance. Such resources will often be developed though state-initiated, public-private partnerships. To inform and support state health care leaders, additional projects will facilitate information-sharing among states working toward health care reform.
Requirements: The Commonwealth Fund requests letters of inquiry to initiate the grant application process, and does not wish to review full proposals at this stage. Applicants are encouraged to submit letters of inquiry using our online form.

Restrictions: The Fund makes grants only to tax-exempt organizations and public agencies and does not support: general planning and ongoing activities or existing deficits; endowment or capital costs, including construction, renovation, or equipment; basic biomedical research; conferences, symposia, major media projects, or documentaries, unless they are an outgrowth of one of the Fund's programs; individuals; scholarships; churches or other religious organizations unless the project for which they seek funding is entirely secular in nature; and work for which achievements cannot be measured.
Geographic Focus: All States
Contact: Edward L. Schor; (212) 606-3800; fax (212) 606-3500; els@cmwf.org
Internet: http://www.commonwealthfund.org/Content/Program-Areas/State-High-Performance-Health-System.aspx
Sponsor: Commonwealth Fund
1 E 75th Street
New York, NY 10021-2692

Communities Foundation of Texas Grants 1236
The unrestricted funds of the foundation support programs and projects intended to improve the quality of life for the citizens of the Dallas, TX, area. The funds support education, health and hospitals, social services, youth, and cultural programs. The foundation encourages projects developed in consultation with other agencies and planning groups and that promote coordination, cooperation, and sharing among organizations. Types of support include seed grants, emergency funds, building funds, equipment acquisition, matching funds, technical assistance, research, capital campaigns, and operating budgets. Requests for operating funds generally are not granted.
Requirements: 501(c)3 organizations in the Dallas, TX, area may apply.
Restrictions: Grants are not made to or for individuals, endowments, sectarian religious purposes, political or lobbying efforts, deficit financing, media projects, or for operational expenses of established organizations.
Geographic Focus: Texas
Samples: Undermain Theatre, $25,000--to help restore 90 original Frank Lloyd Wright chairs that comprise the theater's current seating arrangement; Learning Center of North Texas, $25,000--to upgrade technology infrastructure, provide technology training for staff, and engage IT support services for one year; Contact Crisis Line, $16,200--to upgrade the computer hardware and software used by volunteers on the crisis help lines; Boy Scouts of America National Council (Circle Ten Council) $50,000--to fund a new shower and restroom facility at Camp Wisdom as part of the organization's $20 million centennial capital campaign.
Contact: Brent Christopher; (214) 750-4222; fax (214) 750-4210; bchristopher@cftexas.org
Internet: http://www.cftexas.org
Sponsor: Communities Foundation of Texas
5500 Caruth Haven Lane
Dallas, TX 75225-8146

Community Foundation AIDS Endowment Awards 1237
The foundation administers an annual special grants program called the AIDS Endowment Awards. Grants of $1,000 will be awarded to one or two organizations that demonstrate a commitment to working towards the prevention, education and treatment of AIDS so that they may continue their efforts. Services provided could include education programs for the prevention of HIV/AIDS, medical and social services for those living with HIV/AIDS, housing services for those living with HIV/AIDS and medical research for the treatment and prevention of HIV/AIDS.
Requirements: Selected organizations in the metropolitan Richmond area interested in the field of AIDS education, research or community service and care are encouraged to apply. Organizations interested in applying should submit a brief proposal to foundation which should be no more than two single-spaced pages and include: [1] a description of the organization's mission; and [2] a description of the agency's work in the field of AIDS. Include a list of Board of Governors, a copy of the most recent IRS Form 990, an audited financial statement if available, and an IRS Tax Exempt Letter designating your organization a 501(c)3 non-profit.
Geographic Focus: Virginia
Date(s) Application is Due: Feb 15
Amount of Grant: 1,000 USD
Contact: Susan Hallett; (804) 330-7400; fax (804) 330-5992; shallett@tcfrichmond.org
Internet: http://www.tcfrichmond.org/Page2954.cfm
Sponsor: Community Foundation Serving Richmond and Central Virginia
7501 Boulders View Drive, Suite 110
Richmond, VA 23225

Community Foundation Alliance City of Evansville Endowment Fund Grants 1238
The City of Evansville Endowment Fund originated in 1994, when former Mayor Frank McDonald II proposed that $5 million of revenue from the gaming boat be invested in a way that would serve the City of Evansville, Indiana, forever. The City of Evansville Endowment Fund's earnings allow for grant making year after year. During its most recent grant cycle, the Fund allowed for grants totaling more than $180,000 to 21 nonprofit organizations serving the City of Evansville. The Foundation's grant cycle runs from May through October each year. Funding requests are accepted only during the grant cycle. The CEEF grants committee will consider funding requests of any amount. Requests of at least $1,000 are preferred.
Requirements: CEEF serves to provide funds to organizations that qualify as tax-exempt organizations under sections 501(c)3 and 509(a) of the Internal Revenue Code. The organization must serve within the city limits of Evansville, Indiana. At least sixty percent (60%) of grant funding will be distributed for activities that will support or benefit the 4th and 6th wards of the City of Evansville.
Restrictions: Not more than thirty percent (30%) of any grant request may be for personnel costs, travel costs, office supplies and other program operating costs.
Geographic Focus: Indiana
Date(s) Application is Due: Oct 31
Contact: Melinda Waldroup, Program Director; (812) 429-1191 or (877) 429-1191; fax (812) 429-0840; mwaldroup@alliance9.org
Internet: http://www.alliance9.org/city-of-evansville-endowment-fund
Sponsor: Community Foundation Alliance
123 NW Fourth Street, Suite 322
Evansville, IN 47708-1712

Community Foundation for Greater Atlanta AIDS Fund Grants 1239
The Atlanta AIDS Fund is a collaborative funding partnership between The Community Foundation, United Way of Greater Atlanta and Fashion Cares and the Allen Thornell HIV Care and Service Fund. The Community Foundation was an early supporter of HIV/AIDS initiatives, making its first grant in 1981. At that time the disease was known as GRID - Gay Related Immune Disease. The Atlanta community, the country and the world has made incredible progress since that time, but the challenges of HIV/AIDS are still present, particularly in our Atlanta region. The Atlanta AIDS Fund was created in 1991 to support metro Atlanta's HIV/AIDS advocacy, prevention education and service efforts through funding and leadership. Since that time nearly $11 million in grants have been awarded to organizations in the metro area. Currently, the Foundation is working on five goals: offering training to AIDS service organizations and their case management staff to serve as licensed patient navigators; providing funds to AIDS service organizations and/or community centers to facilitate mergers, consolidations and/or joint programming; roviding funds for advocacy efforts to ensure Georgia's Medicaid expansion occurs and includes a full-range of covered services and care for people living with HIV/AIDS (including those at high-risk for infections); providing funds for advocacy efforts to monitor Georgia's AIDS Drug Assistance Program for access issues and challenges; and engaging an evaluator to assess progress and impact.
Requirements: To apply for a grant organizations must: be located and providing services within the Foundation's 23-county service area; spend funds within the 23-county service area; be classified by the U.S. Internal Revenue Service under Section 501(c)3 of the I.R.S. code as a nonprofit, tax-exempt organization, donations to which are deductible as charitable contributions under Section 170(c)2 and the I.R.S. determination must be current; be registered with the Georgia Secretary of State as a nonprofit; have a minimum two-year operating history after the date of receipt of its 501(c)3 classification; have annual operating expenses greater than $100,000 as reflected in the most recently filed IRS form 990; have at least one full-time paid employee (paid minimum wage or more, working 2,080 hours or more) for the 12 months prior to submitting a Letter of Intent; and have a current written strategic or business plan for the whole organization that covers at least 24 months which includes the organization's entire current fiscal year.
Restrictions: The following organizations are not eligible to apply for funding: private and publicly funded schools (K-12) and institutions of higher learning (this does not include nonprofit charter schools); organizations that raise funds for publicly funded schools (K-12), institutions of higher learning and government agencies; organizations that require participation in religious services and/or education as a condition of receiving services; and organizations that have discriminatory policies and/or practices on the basis of race, color, national origin, age, disability, sex/gender, marital status, familial status, parental status, religion, sexual orientation, genetic information or political beliefs. Foundation funds may not be used to support the following: religious services and/or education; fundraising and marketing events; endowment funds; or capital campaign contributions (for building construction or renovation) or use of funds to cover capital campaign feasibility studies or campaign implementation expenses.
Geographic Focus: Georgia
Contact: Kathy Palumbo, Director of Programs; (404) 588-3187 or (404) 688-5525; fax (404) 688-3060; kpalumbo@cfgreateratlanta.org
Internet: http://cfgreateratlanta.org/Community-Leadership/Current-Initiatives/Atlanta-AIDS-Fund.aspx
Sponsor: Community Foundation for Greater Atlanta
191 Peachtree Street NE, Suite 1000
Atlanta, GA 30303

Community Foundation for Greater Atlanta Clayton County Fund Grants 1240
The Clayton Fund of the Community Foundation for Greater Atlanta was established in 1992 to bring together donors, nonprofits, and community members to make philanthropy happen in Clayton County. The Clayton Fund was the first Local Fund of the Community Foundation. It offers an annual grant making program, as well as philanthropic learning opportunities and community awareness activities – all with the goal of building philanthropy and strengthening nonprofit organizations in Clayton County. Organizations receive grants for general operating support, which gives grantees the ability to use funds where they are most needed. The Letter of Intent (LOI) is the first step in the application process. Clayton Fund Advisory Committee and program staff will then review all LOIs and invite organizations to submit final applications.
Requirements: To apply for a grant organizations must: be located and providing services within the Clayton County service area; spend funds within the Clayton County service area; be classified by the U.S. Internal Revenue Service under Section 501(c)3 of the I.R.S. code as a nonprofit, tax-exempt organization, donations to which are deductible as charitable contributions under Section 170(c)2 and the I.R.S. determination must be current; be registered with the Georgia Secretary of State as a nonprofit; have at least one full-time paid employee (paid minimum wage or more, working 2,080 hours or more) for the 12 months prior to submitting a Letter of Intent; and have a current written strategic or business plan for the whole organization that covers at least 24 months which includes the organization's entire current fiscal year.

Geographic Focus: Georgia
Contact: Josh Phillipson, Program Officer; (404) 526-1103 or (404) 688-5525; fax (404) 688-3060; claytonfund@cfgreateratlanta.org
Internet: http://cfgreateratlanta.org/Community-Initiatives/Counties/Clayton.aspx
Sponsor: Community Foundation for Greater Atlanta
191 Peachtree Street NE, Suite 1000
Atlanta, GA 30303

Community Foundation for Greater Atlanta Common Good Funds Grants 1241
Common Good Funds grants are available to support an organization's ongoing administrative and infrastructure costs and/or to maintain existing, effective programs that are offered within the Foundation's 23-county service area. Characteristics of successful applicants include that they have a clear vision of the organization's role in making a difference in the community, demonstrate effectiveness through the investment of time and personnel in measuring and planning for success, and have written strategic plans that cover two or more years, include goals and methods to measure effectiveness, and are used to form an annual work plan. Grants will range from $10,000 to $75,000 for each 12 month period and may be awarded for 12 or 24 months. Each grantee will be asked to identify organizational goals and to report on progress and challenges in either 12- or 24-month intervals.
Requirements: To apply for a grant from the Common Good Funds organizations must: be located and providing services within the Foundation's 23-county service area; spend funds within the 23-county service area; be classified by the U.S. Internal Revenue Service under Section 501(c)3 of the I.R.S. code as a nonprofit, tax-exempt organization, donations to which are deductible as charitable contributions under Section 170(c)2 and the I.R.S. determination must be current; be registered with the Georgia Secretary of State as a nonprofit; have a minimum two-year operating history after the date of receipt of its 501(c)3 classification; have annual operating expenses greater than $100,000 as reflected in the most recently filed IRS form 990; have at least one full-time paid employee (paid minimum wage or more, working 2,080 hours or more) for the 12 months prior to submitting a Letter of Intent; and have a current written strategic or business plan for the whole organization that covers at least 24 months which includes the organization's entire current fiscal year.
Restrictions: The following organizations are not eligible to apply for funding: private and publicly funded schools (K-12) and institutions of higher learning (this does not include nonprofit charter schools); organizations that raise funds for publicly funded schools (K-12), institutions of higher learning and government agencies; organizations that require participation in religious services and/or education as a condition of receiving services; and organizations that have discriminatory policies and/or practices on the basis of race, color, national origin, age, disability, sex/gender, marital status, familial status, parental status, religion, sexual orientation, genetic information or political beliefs. Foundation funds may not be used to support the following: religious services and/or education; fundraising and marketing events; endowment funds; or capital campaign contributions (for building construction or renovation) or use of funds to cover capital campaign feasibility studies or campaign implementation expenses.
Geographic Focus: Georgia
Amount of Grant: 10,000 - 75,000 USD
Contact: Natasha Battle Edwards, Grants Manager; (404) 588-3211 or (404) 688-5525; fax (404) 688-3060; nedwards@cfgreateratlanta.org
Internet: http://cfgreateratlanta.org/Grants-Support/Apply-for-a-Grant/Common-Good-Funds.aspx
Sponsor: Community Foundation for Greater Atlanta
191 Peachtree Street NE, Suite 1000
Atlanta, GA 30303

Community Foundation for Greater Atlanta Managing For Excellence Award 1242
In 1984, The Community Foundation launched its annual Managing for Excellence Award. Each year, local organizations are invited to compete for the prize. This award is presented to an organization that exhibits outstanding nonprofit management. The redesign includes a more robust award package for winners, year-round exposure for the winners to create connections with potential donors and other individuals of influence and presentation as an industry leader at nonprofit events in the community. Winners will be selected from two budget categories (organizations with budgets between under $2,000,000 and organizations with budgets above $2,000,000) and will receive: $75,000 to each of the two winners (an increased from the previous award amount of $25,000); consulting services by Boston Consulting Group; a press release; inclusion in the Foundation's Extra Wish booklet, sent to all Foundation donors; a donor breakfast featuring the Managing for Excellence Winners, hosted by Alicia Philipp and sponsored by The Boston Consulting Group; special events throughout the year featuring Managing for Excellence winners, including a facilitated conversation for donors and connections with professional advisors; recognition at specific Foundation donor events; and a sponsored table at the Association of Fundraising Professionals National Philanthropy Day.
Requirements: To be eligible to receive the Award organizations must: have received at least one discretionary grant during the past ten years from one of the Foundation's discretionary programs; be located and providing services within the 23-county service area; spend funds within the 23-county service area; be classified by the U.S. Internal Revenue Service under Section 501(c)3 of the I.R.S. code as a nonprofit, tax-exempt organization, donations to which are deductible as charitable contributions under Section 170(c)2 and the I.R.S. determination must be current; be registered with the Georgia Secretary of State as a nonprofit; have a minimum two-year operating history after the date of receipt of its 501(c)3 classification; have annual operating expenses greater than $100,000; have at least one full-time paid employee (paid minimum wage or more, working 2,080 hours or more) for the 12 months prior to submitting a Letter of Intent; and have a current written strategic or business plan for the whole organization that covers at least 24 months which includes the organization's entire current fiscal year.
Geographic Focus: Georgia
Amount of Grant: 75,000 USD
Contact: Kristina Morris, Program Officer; (404) 588-3213 or (404) 688-5525; fax (404) 688-3060; Excellence@cfgreateratlanta.org
Internet: http://cfgreateratlanta.org/Managing-For-Excellence-Award.aspx
Sponsor: Community Foundation for Greater Atlanta
191 Peachtree Street NE, Suite 1000
Atlanta, GA 30303

Community Foundation for Greater Atlanta Metropolitan Atlanta An Extra Wish Grants 1243
The Community Foundation of Greater Atlanta's An Extra Wish program provides monetary contributions for specific expenses, not staff or general operations, that contribute in a clear way to the success of organizations and the populations they serve. The Foundation's experience shows that most Extra Wish grants from donor advised funds fall between $500 and $3,500. In the past, the most successful wishes were requests for items that directly help program participants, rather than organizational operating equipment. The Foundation has found the more creative and innovative requests have a better chance of receiving support from a Community Foundation donor. Eligible requests include: resources for program participants or organizations; capital items; and services. Annual application deadlines are February 20 and July 10.
Requirements: For this program, an organization must be located and providing services within the Foundation's 23-county service area. Other eligibility requirements vary by grant program, but common criteria for nonprofits to meet are: have a multi-year written strategic or business plan for the whole organization that includes measureable goals and methods to assess effectiveness; receive funding from at least three different sources such as individuals, foundations, corporations, faith-based organizations and government; have a minimum two-year operating history after receiving its 501(c)3 classification; have audited financial statements for the past two to three completed fiscal years (for organizations with annual budgets greater than $250,000); have at least one full-time paid employee; have a Board of Directors with representation from the community served and a committee structure with diverse areas of expertise; and be registered with the Georgia Secretary of State as a nonprofit.
Geographic Focus: Georgia
Date(s) Application is Due: Feb 20; Jul 10
Amount of Grant: 500 - 3,500 USD
Contact: Natasha Battle Edwards, Grants Manager; (404) 588-3211 or (404) 688-5525; fax (404) 688-3060; nedwards@cfgreateratlanta.org or ExtraWish@cfgreateratlanta.org
Internet: http://cfgreateratlanta.org/Grants-Support.aspx
Sponsor: Community Foundation for Greater Atlanta
191 Peachtree Street NE, Suite 1000
Atlanta, GA 30303

Community Foundation for Greater Atlanta Morgan County Fund Grants 1244
The Morgan Fund of the Community Foundation for Greater Atlanta was established in 2003 to bring together donors, nonprofits and community members to make philanthropy happen in Morgan County. The Morgan Fund offers an annual grant making program, philanthropic learning opportunities, community awareness activities and an annual volunteer service award. It offers an annual grant making program, as well as philanthropic learning opportunities and community awareness activities – all with the goal of building philanthropy and strengthening nonprofit organizations in Morgan County. Organizations receive grants for general operating support, which gives grantees the ability to use funds where they are most needed. The Letter of Intent (LOI) is the first step in the application process. Morgan Fund Advisory Committee and program staff will then review all LOIs and invite organizations to submit final applications.
Requirements: To apply for a grant organizations must: be located and providing services within the Morgan County service area; spend funds within the Morgan County service area; be classified by the U.S. Internal Revenue Service under Section 501(c)3 of the I.R.S. code as a nonprofit, tax-exempt organization, donations to which are deductible as charitable contributions under Section 170(c)2 and the I.R.S. determination must be current; be registered with the Georgia Secretary of State as a nonprofit; have at least one full-time paid employee (paid minimum wage or more, working 2,080 hours or more) for the 12 months prior to submitting a Letter of Intent; and have a current written strategic or business plan for the whole organization that covers at least 24 months which includes the organization's entire current fiscal year.
Geographic Focus: Georgia
Contact: Josh Phillipson, Program Officer; (404) 526-1103 or (404) 688-5525; fax (404) 688-3060; morganfund@cfgreateratlanta.org
Internet: http://cfgreateratlanta.org/Community-Initiatives/Counties/Morgan.aspx
Sponsor: Community Foundation for Greater Atlanta
191 Peachtree Street NE, Suite 1000
Atlanta, GA 30303

Community Foundation for Greater Atlanta Newton County Fund Grants 1245
Newton County, centrally located between the cities of Atlanta, Augusta and Athens, is uniquely positioned for positive expansion and growth. Newton County's strong industrial base continues to grow the level of employment and the local economy. Arts, culture and recreation also play a large role in attracting newcomers to the area. The Newton Fund of The Community Foundation for Greater Atlanta was established in 2000 to bring together donors, nonprofits and community members to make philanthropy happen in Newton County. The Newton Fund offers an annual grantmaking program,

philanthropic learning opportunities, community awareness activities and an annual volunteer service award, the Pat Patrick "Big Heart" Award. Organizations receive grants for general operating support, which gives grantees the ability to use funds where they are most needed. The Letter of Intent (LOI) is the first step in the application process. Newton Fund Advisory Committee and program staff will then review all LOIs and invite organizations to submit final applications.
Requirements: To apply for a grant organizations must: be located and providing services within the Newton County service area; spend funds within the Newton County service area; be classified by the U.S. Internal Revenue Service under Section 501(c)3 of the I.R.S. code as a nonprofit, tax-exempt organization, donations to which are deductible as charitable contributions under Section 170(c)2 and the I.R.S. determination must be current; be registered with the Georgia Secretary of State as a nonprofit; have at least one full-time paid employee (paid minimum wage or more, working 2,080 hours or more) for the 12 months prior to submitting a Letter of Intent; and have a current written strategic or business plan for the whole organization that covers at least 24 months which includes the organization's entire current fiscal year.
Geographic Focus: Georgia
Contact: Josh Phillipson, Program Officer; (404) 526-1103 or (404) 688-5525; fax (404) 688-3060; newtonfund@cfgreateratlanta.org
Internet: http://cfgreateratlanta.org/Community-Initiatives/Counties/Newton.aspx
Sponsor: Community Foundation for Greater Atlanta
191 Peachtree Street NE, Suite 1000
Atlanta, GA 30303

Community Foundation for Greater Atlanta Strategic Restructuring Fund Grants 1246
The Community Foundation for Greater Atlanta is committed to building strong, collaborative nonprofits within the 23-county region. The Foundation holds a common inquiry with other funders and nonprofits to find the best practices to assess, negotiate, implement and evaluate partnership models between nonprofit organizations. The purpose of the Strategic Restructuring Fund is to provide funds and/or management consulting services to support nonprofits as they assess, negotiate, design and/or implement substantive strategic restructuring efforts that seek to promote more effective operations and high-performing programs based on community needs and assets. Currently, the Foundation offers funding in three different stages of a partnership: Partnership Assessment, in which organizations assess their readiness and suitability as potential partners and examine the different types of partnership models; Readiness and Negotiation, in which organizations will examine the pre-existing relationship, if any, between specific potential partners; and Design and/or Implementation, in which organizations finalize the details of their partnership (from timeline to steering committee to changes in staffing structures) and, if appropriate, receive funds to implement their partnership plan.
Requirements: To apply for a grant organizations must: be located and providing services within the Foundation's 23-county service area; spend funds within the 23-county service area; be classified by the U.S. Internal Revenue Service under Section 501(c)3 of the I.R.S. code as a nonprofit, tax-exempt organization, donations to which are deductible as charitable contributions under Section 170(c)2 and the I.R.S. determination must be current; be registered with the Georgia Secretary of State as a nonprofit; have a minimum two-year operating history after the date of receipt of its 501(c)3 classification; have annual operating expenses greater than $100,000 as reflected in the most recently filed IRS form 990; have at least one full-time paid employee (paid minimum wage or more, working 2,080 hours or more) for the 12 months prior to submitting a Letter of Intent; and have a current written strategic or business plan for the whole organization that covers at least 24 months which includes the organization's entire current fiscal year.
Geographic Focus: Georgia
Contact: Kathy Palumbo, Director of Programs; (404) 588-3187 or (404) 688-5525; fax (404) 688-3060; kpalumbo@cfgreateratlanta.org
Internet: http://cfgreateratlanta.org/Grants-Support/Apply-for-a-Grant/Strategic-Restructuring-Fund.aspx
Sponsor: Community Foundation for Greater Atlanta
191 Peachtree Street NE, Suite 1000
Atlanta, GA 30303

Community Foundation for Greater Buffalo Grants 1247
The Community Foundation for Greater Buffalo (CFGB) is a public charity holding more than 800 different charitable funds, large and small, established by individuals, families, nonprofit agencies and businesses to benefit Western New York. Since 1919, the foundation has served the needs of it's community and the wishes of it's donors through personalized service, financial stewardship, local expertise, and community leadership. The Foundation focus's on four main areas of interest: strengthen the region as a center for arts and culture; natural, historic, and architectural resources; reduce racial and ethnic disparities; increase economic self-sufficiency for low-income individuals and families. Special funding will also be available to support the following: AIDS research and its cure ($5,000); programs that serve the visual, speech or hearing impaired ($5,000). Preference will be given to requests that align with one or more of the four focus areas.
Requirements: Applicants must be 501(c)3 not-for-profit organizations located within the eight counties of Western New York: Allegany, Cattaraugus, Chautauqua, Erie, Genesee, Niagara, Orleans, and Wyoming.
Restrictions: The Foundation will not consider competitive funding for: endowments; religious purposes; schools not registered with the New York State Education Department; attendance at or sponsorship of fundraising events for organizations; annual events or festivals; any partisan political activity. Funds from the foundation cannot be used to support or oppose a candidate for political office. Projects and activities that have occurred. The Foundation will not, except in extraordinary cases, provide payment or reimbursement for expenses incurred prior to the funding decision being communicated to the applicant.

Geographic Focus: New York
Date(s) Application is Due: Mar 1
Amount of Grant: 1,000 - 25,000 USD
Samples: Carnegie Art Center, $4,500--to support the continued restoration of the landmark Carnegie Art Center building; Grassroots Gardens, $25,000--to support new community gardens that provide beautification, food and employment; Youth Character Development, $9,100--to aid refugee students in the transition to formal education and encourage college enrollment.
Contact: Jean McKeown; (716) 852-2857, ext. 204; fax (716) 852-2861; jeanm@cfgb.org
Internet: http://www.cfgb.org/page17000.cfm
Sponsor: Community Foundation for Greater Buffalo
712 Main Street
Buffalo, NY 14202

Community Foundation for Greater New Haven $5,000 and Under Grants 1248
The Community Foundation for Greater New Haven (CFFGNH) is a philanthropic institution that was established in 1928. The foundation's mission is to create positive and sustainable change in Connecticut's Greater New Haven region by increasing the amount of and enhancing the impact of community philanthropy. Funding through the CFFGNH $5,000 and Under Grants process are available to any organization with an operating budget of $2 million or less.
Requirements: IRS 501(c)3 nonprofit organizations are eligible to apply in the greater New Haven area, which includes: Ansonia; Bethany; Branford; Cheshire; Derby; East Haven; Guilford; Hamden; Madison; Milford; New Haven; North Branford; North Haven; Orange; Oxford; Seymour; Shelton; Wallingford; West Haven and; Woodbridge.
Restrictions: No organization shall be eligible to receive a grant under this process more often than once in any period of two calendar years. Grants are not made to support religious activities, lobbying, or travel.
Geographic Focus: Connecticut
Amount of Grant: 5,000 USD
Samples: Christ Episcopal Church, $5,000--to support the Kathleen Samela Memorial Food Bank; Connecticut Alliance to Benefit Law Enforcement, Inc., $2,100--to support training for Hamden Fire and EMS personnel; Connecticut Pre-Engineering Program, $4,000--to provide operating support for science, technology, engineering and math (STEM) enrichment programming to students in High School in the Community, East Rock, Clinton Avenue, Micro-Society and Troupe schools.
Contact: Denise Canning, Grants Manager; (203) 777-2386 or (203) 777-7076; fax (203) 787-6584; dcanning@cfgnh.org
Internet: http://www.cfgnh.org/GrantsScholarships/TypesofGrants/tabid/199/Default.aspx
Sponsor: Community Foundation for Greater New Haven
70 Audubon Street
New Haven, CT 06510

Community Foundation for Greater New Haven Responsive New Grants 1249
The Community Foundation for Greater New Haven is a philanthropic institution that was established in 1928. The foundation's mission is to create positive and sustainable change in Connecticut's Greater New Haven region by increasing the amount of and enhancing the impact of community philanthropy. The Responsive New Grants Program is generally awarded to address an agency's general operating, programmatic, capital or technical assistance needs. This funding source is open to all requests for projects and organizational support.
Requirements: IRS 501(c)3 nonprofit organizations are eligible to apply in the greater New Haven area, which includes: Ansonia; Bethany; Branford; Cheshire; Derby; East Haven; Guilford; Hamden; Madison; Milford; New Haven; North Branford; North Haven; Orange; Oxford; Seymour; Shelton; Wallingford; West Haven and; Woodbridge.
Restrictions: Grants are not made to support religious activities, lobbying, or travel.
Geographic Focus: Connecticut
Date(s) Application is Due: Mar 5
Amount of Grant: 10,000 - 185,000 USD
Samples: Arts Council of Greater New Haven, New Haven, CT, $50,000--to provide general operating support; Beth El Center Inc., Milford, CT, $25,000-- general operating support for transitional shelter programs for homeless individuals and families and a soup kitchen/food pantry that feeds individuals and families from Greater New Haven and Lower Naugatuck Valley; Area Congregations Together Inc., Shelton, CT, $150,000--to establish a development program.
Contact: Denise Canning, Grants Manager; (203) 777-2386 or (203) 777-7076; fax (203) 787-6584; dcanning@cfgnh.org
Internet: http://www.cfgnh.org/GrantsScholarships/TypesofGrants/tabid/199/Default.aspx
Sponsor: Community Foundation for Greater New Haven
70 Audubon Street
New Haven, CT 06510

Community Foundation for Greater New Haven Sponsorship Grants 1250
The Community Foundation for Greater New Haven (CFFGNH) is a philanthropic institution that was established in 1928. The foundation's mission is to create positive and sustainable change in Connecticut's Greater New Haven region by increasing the amount of and enhancing the impact of community philanthropy. The available sponsorships are awarded throughout a calendar year, only for events and may not exceed $2,500 per event. It is recommended that nonprofits submit a sponsorship request at least 60 days in advance of their event. Applications are accepted year-round, application available online at: http://www.cfgnh.org/Grantmaking/TypesofGrants/tabid/199/Default.aspx.
Requirements: IRS 501(c)3 nonprofit organizations are eligible to apply in the greater New Haven area, which includes: Ansonia; Bethany; Branford; Cheshire; Derby; East Haven;

Guilford; Hamden; Madison; Milford; New Haven; North Branford; North Haven; Orange; Oxford; Seymour; Shelton; Wallingford; West Haven and; Woodbridge.
Restrictions: Grants are not made to support religious activities, lobbying, or travel.
Geographic Focus: Connecticut
Amount of Grant: 2,500 USD
Contact: Leigh Higgins, Assistant to CEO; (203) 777-2386 or (203) 777-7092; fax (203) 787-6584; lhiggins@cfgnh.org
Internet: http://www.cfgnh.org/Grantmaking/TypesofGrants/tabid/199/Default.aspx
Sponsor: Community Foundation for Greater New Haven
70 Audubon Street
New Haven, CT 06510

Community Foundation for Greater New Haven Valley Neighborhood Grants   1251
The Valley Community Foundation (VCF), a supporting organization of The Community Foundation for Greater New Haven (CFGNH), promotes investment in it's community leaders. The Valley Neighborhood Grant Program is rooted in CFGNH's Valley Neighborhood Small Grants Program but has the purpose of encouraging, identifying, engaging and supporting neighborhood leadership. It provides funding ($100 to $3,000), technical assistance and training to civic groups to assist them in carrying out their civic agenda, and to support development and implementation of projects that will improve the quality of Valley life.
Requirements: The following groups are eligible to apply: Art Groups; Blockwatches; Business Associations; Faith-based Organizations; Groups that serve the elderly; Individuals in partnership with a non-profit; Neighborhood Groups & Associations; Parent Groups; Tenant Associations; Service Clubs; Youth Groups; Partnerships between groups listed above are welcomed; Non-profits that are all-volunteer or have a staff of 4 or less are eligible to apply. If your group does not have legal nonprofit status, establish a relationship with a 501(c)3 organization to serve as your fiduciary agent. Please Note: If you choose to work with a 501(c)3, they must submit a letter of support.
Restrictions: The Neighborhood Program will not fund requests for: salaries or stipends; recreational trips; gas grills or coolers.
Geographic Focus: Connecticut
Date(s) Application is Due: Feb 1
Amount of Grant: 100 - 3,000 USD
Samples: Ansonia Historical Commission, $2,000--to support the costs associated with the repairing and restoring of the monument located outside of the Ansonia Public Library; Ansonia High School Human Relations Club/Doyle Senior Center, $2,000--to support the cost of two intergenerational programs/dinners involving members of the Ansonia High School Human Relations Club and members of the Doyle Senior Center in Ansonia; Assumption Church Parish Nurse Program, $1,500--to support the costs associated with running Bereavement Group, CPR class and Children's Health & Safety Programs.
Contact: Stephanie Sutherland, Associate Philanthropic Officer; (203) 777-2386 or (203) 777-7077; fax (203) 787-6584; ssutherland@cfgnh.org
Internet: http://www.cfgnh.org/Grantmaking/TypesofGrants/tabid/199/Default.aspx
Sponsor: Community Foundation for Greater New Haven
70 Audubon Street
New Haven, CT 06510

Community Foundation for Greater New Haven Women & Girls Grants   1252
The Community Foundation for Greater New Haven (CFFGNH) is a philanthropic institution that was established in 1928. The foundation's mission is to create positive and sustainable change in Connecticut's Greater New Haven region by increasing the amount of and enhancing the impact of community philanthropy. The Community Fund for Women & Girls was created in 1995 by an anonymous woman to provide ongoing support for services important to women and girls in the Greater New Haven and Lower Naugatuck Valley area. Favorable grant proposals would: create and support opportunities for the economic, educational, physical, emotional, social, artistic, and personal growth of women and girls; meet special needs of women and girls and the diverse populations of women in our region; encourage the advancement and full participation of women and girls in the community and in philanthropy; advance the status of women and girls in the core areas of economic security, health, violence, education and political participation.
Requirements: IRS 501(c)3 nonprofit organizations are eligible to apply in the greater New Haven area, which includes: Ansonia; Bethany; Branford; Cheshire, Derby; East Haven; Guilford; Hamden; Madison; Milford; New Haven; North Branford; North Haven; Orange; Oxford; Seymour; Shelton; Wallingford; West Haven and; Woodbridge.
Geographic Focus: Connecticut
Date(s) Application is Due: Jan 15
Amount of Grant: 1,000 - 10,000 USD
Samples: The New Haven Diaper Bank, $10,000--to support a project manager for the Basic Human Needs Policy collaboration between the Liman Project, Wiggin and Dana attorneys and other interested parties. The goal is to secure a sustaining source of state and federal funding for diapers, serving over 1040 children regionally, and potentially all children living in poverty in the U.S.; All Our Kin, Inc., $3,600--to support the Family Child Care Tool kit licensing project. The project will provide six more licensing tool kits to the existing program, allowing six women to become licensed home care providers for 36 children; Hill Health Corporation--$10,000--to support the Prison to Community Peer Support Program, a 44 week peer support group serving 40 newly released formerly incarcerated women, led by women who have experienced and overcome substance abuse and or mental illness and who have made the transition from prison to the community.
Contact: Denise Canning; (203) 777-2386 or (203) 777-7076; dcanning@cfgnh.org
Internet: http://www.cfgnh.org/Grantmaking/TypesofGrants/tabid/199/Default.aspx
Sponsor: Community Foundation for Greater New Haven
70 Audubon Street
New Haven, CT 06510

Community Foundation for Monterey County Grants   1253
The community foundation supports nonprofits in Monterey, CA, in the areas of social services, education, environment, arts, health, historic preservation, and general charitable giving. Collaboratives are highly encouraged. The foundation awards grants through its General Endowment. Technical Assistance grants and Neighborhood Grants also are awarded. Types of support include seed money, emergency funds, general operating support, building funds, equipment acquisition, land acquisition, matching funds, projects, consulting services, and technical assistance. Information on how to apply for each type of grant can be accessed online.
Requirements: 501(c)3 tax-exempt organizations in Monterey, CA, may submit applications.
Restrictions: The foundation does not support individuals, religious activities, scholarships, fellowships, travel, research, salaries and other operating expenses of schools and public agencies, annual campaigns, or special events; create or add to endowment funds; or pay off debt.
Geographic Focus: California
Date(s) Application is Due: Jan 3; May 2; Aug 1
Amount of Grant: 5,000 - 40,000 USD
Contact: Jackie Wendland; (831) 375-9712, ext. 11; fax (831) 375-4731; jackie@cfmco.org
Internet: http://www.cfmco.org/grantsOverview.php
Sponsor: Community Foundation for Monterey County
2354 Garden Road
Monterey, CA 93940

Community Foundation for Muskegon County Grants   1254
The community foundation was established to serve the needs of the people of Muskegon County and nearby western Michigan. The foundation's strategic plan for grant making focuses on the prevention of problems rather than the cure; encourages programs that are collaborative, comprehensive, and have the potential to be continuous; encourages leveraging and matching grant opportunities from multiple funders; and supports seed money opportunities for innovative projects. Major grant interest areas include the arts, community development and urban revitalization, education, environment, health/human services, and needs of young children (ages 0-3). Types of support include seed money grants, special projects, matching funds, equipment, scholarship funds, loans, research, publications, conferences and seminars, endowment funds, consulting services, continuing support, emergency funds, internships, professorships, and renovation projects. Although the foundation generally makes one-year grant commitments, it will consider making longer term commitments for new efforts that show strong promise for positive impact. Applications are accepted on specific dates each year; it is recommended that the applicant contact the foundation to discuss their interests and to obtain the next grant application deadline.
Requirements: IRS 501(c)3 organizations and institutions in Muskegon County and nearby western Michigan are eligible.
Restrictions: Support will not be provided for routine operating expenses; capital equipment, computer hardware and software, and motor vehicles; conferences, publications, videos, films, television, or radio programs; endowment campaigns; special fund-raising events; religious programs that serve specific denominations; existing obligations or debts; individual schools or districts; or individuals.
Geographic Focus: Michigan
Amount of Grant: 250 - 50,000 USD
Contact: Marcy Joy; (231) 722-4538 or (231) 332-4124; mjoy@cffmc.org
Internet: http://www.cffmc.org/grantapply.php
Sponsor: Community Foundation for Muskegon County
425 West Western Avenue, Suite 200
Muskegon, MI 49440

Community Foundation for Northeast Michigan Mini-Grants   1255
Community Foundation for Northeast Michigan Mini-Grants are made to tax-exempt, northeast Michigan charitable agencies and organizations. The foundation looks for projects that prevent community problems, benefit the greatest number of people, help deliver new services or make existing services more efficient, enhance collaboration among organizations, promote youth development, address emerging community needs, try a new approach to a persistent problem, or encourage people to develop new skills and help themselves. Mini-grants can be in any amount up to $500, and application deadlines are January 15, July 15, or October 15.
Requirements: IRS 501(c)3 nonprofit organizations, schools, churches (for non-sectarian purposes), cities, townships, and other governmental units serving the four-county area of Alcona, Alpena, Montmorency, and Presque Isle in northeast Michigan are eligible to apply.
Restrictions: Grants are not given to individuals, except for awards or scholarships from designated donor funds.
Geographic Focus: Michigan
Date(s) Application is Due: Jan 15; Jul 15; Oct 15
Contact: Julie Wiesen; (989) 354-6881 or (877) 354-6881; wiesenj@cfnem.org
Internet: http://www.cfnem.org/grants/community-impact-grants.html
Sponsor: Community Foundation for Northeast Michigan
100 N. Ripley, Suite F, P.O. Box 495
Alpena, MI 49707-0495

Community Foundation for Northeast Michigan Tobacco Settlement Grants   1256
The Community Foundation for Northeast Michigan Tobacco Settlement Grants focus on programs or projects that address tobacco-related issues, and benefit the residents of the northeast Michigan counties of Alcona, Alpena, Montmorency, and Presque Isle. The maximum amount for these awards is $3,000, and the annual grant application deadline is January 15.
Requirements: IRS 501(c)3 nonprofit organizations, schools, churches (for non-sectarian purposes), cities, townships, and other governmental units serving the four-county area of Alcona, Alpena, Montmorency, and Presque Isle in northeast Michigan are eligible to apply.

Restrictions: Grants are not given to individuals, except for awards or scholarships from designated donor funds. Grants are made for future projects only. No funding will be given for projects completed before final board approval. Board approval is usually within six (6) weeks of the grant deadline. Grants are not made for routine operating needs or budget deficits.
Geographic Focus: Michigan
Date(s) Application is Due: Feb 1
Amount of Grant: Up to 3,000 USD
Contact: Julie Wiesen, Program Director; (989) 354-6881 or (877) 354-6881; fax (989) 356-3319; wiesenj@cfnem.org
Internet: http://www.cfnem.org/grants/tobacco-settlement-grants.html
Sponsor: Community Foundation for Northeast Michigan
100 N. Ripley, Suite F, P.O. Box 495
Alpena, MI 49707-0495

Community Foundation for San Benito County Grants 1257
The Foundation serves donors, advances philanthropy and achieves impact by supporting the work of nonprofit organizations. The Foundation continuously monitors the San Benito County community to understand the nature of need, the forces of change, available resources and the capacity for growth. The Foundation provides funding to support impactful programs within and across the following areas of interest: arts and culture; education and youth; health and social services; agriculture and environment; community enhancement; economic development. Application deadlines are announced on the Foundation's website.
Requirements: Selection criteria and priorities are as follows: projects or services which respond to a demonstrated need within San Benito County; effective use and greatest impact of grant funds; initiatives to solve significant community issues; collaboration and coordination of service delivery; demonstrate a level of cooperation with other organizations, including leveraging financial and in-kind support from other groups and individuals; strengthening organizational capacity; addressing diverse community interests; organizational or program sustainability; organizations with demonstrated financial need; and unduplicated services. Awards are generally up to $40,000. Major projects or initiatives over $40,000 must be preliminarily discussed with Foundation staff.
Restrictions: In general the following are not funded: organizations that discriminate on the basis of age, disability, ethnic origin, gender, race or religion; grants to individuals; fraternal or service organizations, unless in support of specific programs open to or benefiting the entire community; salaries and other operating expenses of schools and public agencies; fundraising events such as annual campaigns, walk-a-thons, tournaments, fashion shows, dinners and auctions; organizations and programs designed to support political activities; organizations located outside San Benito County unless for a specific program benefiting residents within San Benito County; pay off existing obligations or enable funding of reserve accounts; endowment funds; and scholarships, fellowships, travel grants and academic, technical or specialized research.
Geographic Focus: California
Samples: Chamberlain's Mental Health Services, Gilroy, California, $100,000 - behavioral management programs for functionally impaired preschool children; SB Collaborative for Homeless Services, Santa Barbara, California, $5,000 - homeless information data program required to enable future HUD funding; and Seniors Council (Foster Grandparents), San Benito, California, $6,159 - support of seniors mentoring and tutoring program for at-risk youth.
Contact: Grants Manager; (831) 630-1924; fax (831) 630-1934; info@cffsbc.org
Internet: http://www.cffsbc.org/grantoverview.php
Sponsor: Community Foundation for San Benito County
829 San Benio Street, Suite 200
Hollister, CA 95023

Community Foundation for Southeast Michigan Grants 1258
The Community Foundation for Southeast Michigan is always looking for effective program and project ideas that can improve life in southeast Michigan, specifically in the seven-county service area of Wayne, Oakland, Macomb, Monroe, Washtenaw, St. Clair and Livingston. Grants are provided considering various local needs and identifying those projects which promise the strongest long-term impact on the region. In general support is provided for projects and programs in the areas of arts and culture, civic affairs, health, human services, neighborhood and regional economic development, work force development, environment and land use. Interested applicants are encouraged to review the grantmaking guidelines at the website and call the Program Officer to discuss proposals. Grants range from $5,000 to $1 million, with the majority ranging from $35,000 to $100,000. Application may be made at any time. Proposals submitted prior to February 15, May 15, August 15 and November 15 will typically receive a response within three or four months.
Requirements: Eligible organizations: are a 501(c)3 tax-exempt organization, a government entity, a school district or a university; have headquarters (or a local partner) located in the seven-county service area; serve residents in the seven-county service area; have Board and/or the CEO approval to submit a proposal; and have a current certified financial audit.
Restrictions: All final reports due to the Foundation for previous grants must be submitted before applying. Requests for sectarian religious programs, individuals, and funding for deficits or other previously incurred obligations are not eligible.
Geographic Focus: Michigan
Contact: Katie Brisson, Senior Program Officer; (888) 933-6369 or (313) 961-6675; fax (313) 961-2886; kbrisson@cfsem.org
Internet: http://cfsem.org/apply-grant
Sponsor: Community Foundation for SE Michigan
333 W Fort Street, Suite 2010
Detroit, MI 48226-3134

Community Foundation for the Capital Region Grants 1259
The Community Foundation for the Capital Region awards grants to eligible New York nonprofit organizations through the administration of more than 260 charitable funds, established by donors to meet their philanthropic objectives and address current and future needs. The Foundation's primary fields of interest include: art and music therapy; basic and remedial instruction; child welfare; disasters and emergency management; elder abuse; employment; the environment; graduate and professional education; health and health care; HIV/AIDS; home health care; homeless services; medical education; music; nursing care; and senior services. Types of support include: fund raising; operating support; capital grants; seed money; technical assistance; and program development. Grants may be held for a maximum of three years. Applicants should submit a letter of inquiry that describes the organization, the program, and the amount sought. Full applications are by invitation.
Requirements: New York 501(c)3 tax-exempt organizations serving Albany, Rensselaer, Schenectady, or Saratoga Counties are eligible.
Geographic Focus: New York
Amount of Grant: 2,500 - 30,000 USD
Contact: Jackie Mahoney, Program Director; (518) 446-9638; fax (518) 446-9708; Jmahoney@cfcr.org, or info@cfcr.org
Internet: http://www.cfcr.org/grantmaking/grantmaking.htm
Sponsor: Community Foundation for the Capital Region
6 Tower Place, Executive Park Drive
Albany, NY 12203-3725

Community Foundation for the National Capital Region Community Leadership Grants 1260
The Community Foundation for the National Capital Region has a particular interest in supporting groups in the metropolitan Washington, D.C., area working in the following issue areas: violence prevention; education; community building; cross-cultural or cultural partnership building; family literacy; and healthcare and dental services for underprivileged children, youth, and families. Types of support include general operating support, program development, technical assistance, and program evaluation. Approximately 30 to 40 grants per year are awarded following a Request-for-Proposal cycle, which occurs "several times per year." Letters of inquiry (three-page maximum) must meet the listed application deadlines.
Requirements: 501(c)3 nonprofits in the metropolitan Washington region, including the District of Columbia, northern Virginia, and suburban Maryland may submit letters of inquiry. Applicants must represent a neighborhood, citywide, or regional coalition effort, with one nonprofit organization serving as project sponsor.
Geographic Focus: District of Columbia, Maryland, Virginia
Contact: Dawnn Leary, Senior Philanthropic Services Officer; (202) 973-2519 or (202) 955-5890; fax (202) 955-8084; dleary@cfncr.org or info@cfncr.org
Internet: http://thecommunityfoundation.org/what-we-do/grantmaking/
Sponsor: Community Foundation for the National Capital Region
1201 15th Street NW, Suite 420
Washington, D.C. 20005

Community Foundation of Abilene Celebration of Life Grants 1261
The Celebration of Life Grant is awarded to a nonprofit organization whose mission seeks to educate and promote the pro-life philosophy. Applicants must submit, in narrative form, a brief history of their organization and its mission, with a description of all programs and services. They must also identify critical challenges or deficiencies, key areas where funding is needed, and measurable outcomes. Organizations must submit 15 copies of the proposal. Detailed attachments are listed at the website. The grant recipient is announced in late September.
Requirements: Applicant must have an IRS 501(c)3 designation and provide services in Abilene, Texas.
Geographic Focus: Texas
Date(s) Application is Due: Aug 3
Amount of Grant: 5,000 - 15,000 USD
Contact: Courtney Vletas; (325) 676-3883, ext 102; cvletas@cfabilene.org
Internet: http://cfabilene.org/grant-information-applications
Sponsor: Community Foundation of Abilene
500 Chestnut, Suite 1634, P.O. Box 1001
Abilene, TX 79604

Community Foundation of Bartholomew County Heritage Fund Grants 1262
The goal of the Heritage Fund's grant program is to achieve the maximum impact with the available resources. The Fund will consider grant applications that: are change-orientated and problem-solving in nature; strive to anticipate the changing needs of the community and to be flexible in responding to them; address the needs of a significant number of community residents and provide the greatest benefit per dollar granted; encourage support from the community by using matching, challenge and other grant techniques; have a broad funding base, with additional support being sought from the government, foundations, associations and other funders; enable grant recipients to achieve certain objectives such as capacity building, and/or increasing efficiency, effectiveness and fundraising capabilities; request technical assistance or specialized help with projects that respond to community needs; or positively impact the Heritage Fund's Areas of Initiative.
Requirements: Grants are made to not-for-profit organizations whose programs benefit the residents of Bartholomew County, Indiana.
Restrictions: Funding is not available for the following: individuals; events, performances, seminars or trips unless there are special circumstances which will benefit the community; individual school needs; faith based organizations unless the project in question is

not religious in nature, is not restricted based on faith, and involves no faith based proselytizing; or agency endowments.
Geographic Focus: Indiana
Date(s) Application is Due: Mar 1; Jun 1; Sep 1; Dec 1
Amount of Grant: Up to 3,000 USD
Contact: Lynda J. Morgan; (812) 376-7772; lmorgan@heritagefundbc.org
Internet: http://www.heritagefundbc.org/grants/process_guidelines_deadlines.php
Sponsor: Community Foundation of Bartholomew County
538 Franklin Street, P.O. Box 1547
Columbus, IN 47202-1547

Community Foundation of Bartholomew County James A. Henderson Award for Fundraising 1263
The Henderson Award recognizes the invaluable role volunteer fundraisers play in advancing the quality of life within Bartholomew County, Indiana. Not-for-profit organizations in the county are invited to nominate volunteers who have performed outstanding fundraising for their organizations. A committee will review the nominations and select the person to be honored. The committee will consider such criteria as innovation, sustainability, creativity, effectiveness, effort, ability to engage others, outreach to new donors, etc. Generally the amount raised will not be a major factor in the scoring. Efforts will be made to recognize unsung heroes in fundraising. The successful nominee will be recognized at the Heritage Fund's Annual Report to the Community. He/she will receive a small gift and the nominating organization will receive a $2,500 grant in honor of the winner. The nomination form is available at the website.
Geographic Focus: Indiana
Date(s) Application is Due: Mar 23
Amount of Grant: 2,500 USD
Contact: Lynda J. Morgan; (812) 376-7772; lmorgan@heritagefundbc.org
Internet: http://www.heritagefundbc.com/grants/award_james_henderson.php
Sponsor: Community Foundation of Bartholomew County
538 Franklin Street, P.O. Box 1547
Columbus, IN 47202-1547

Community Foundation of Bloomington and Monroe County - Precision Health Network Cycle Grants 1264
The Precision Health Network Grants are designed to develop and enhance community health education and outcomes. The Grants are available to applicants serving Monroe, Lawrence, Brown, Greene, Morgan, Owen, Orange, Martin, and Daviess counties in Indiana. Detailed guidelines and application are available at the website.
Requirements: Projects selected to receive grants must meet at least one of the following criteria: promotion of health and healthy lifestyles; prevention of disease; and self-management of chronic disease. Projects must target schools, community centers, or work sites in the defined geographic areas. Grants will be made only to non-profit, 501(c)3 organizations and programs. Programs developed through individual initiative or by the for-profit sector will need to have a not-for-profit serve as the "fiscal sponsor" for the program. The following entities may apply: community service organizations with health-related missions; public entities; employee groups; and employers.
Geographic Focus: Indiana
Date(s) Application is Due: Apr 20
Contact: Renee Chambers, Program Director; (812) 333-9016; renee@cfbmc.org
Internet: http://www.cfbmc.org/page21250.cfm
Sponsor: Community Foundation of Bloomington and Monroe County
101 W Kirkwood Avenue, Suite 321
Bloomington, IN 47404

Community Foundation of Bloomington and Monroe County Grants 1265
Grants will be made for a wide variety of programs and purposes in support of the arts and cultural activities, social and health concerns, educational activities, beautification projects, and for other community development needs that will benefit the citizens of Bloomington and Monroe County, Indiana. Start-up funds for special projects reflecting incremental growth possibilities will be given particular attention. Additional types of support include building construction and renovation, equipment acquisition, program development, seed grants, and scholarship funds. Projects that provide leverage for generating other funds and community resources or are seed money for new programs will also receive special attention. The majority of grants are made for one time only, but multi-year grants are considered in unusual circumstances. Applications are reviewed in March, June, September, and December.
Restrictions: Grants are made only to nonprofit organizations and programs in the city of Bloomington or in Monroe County, Indiana. Grants will not be made for general operating support purposes; endowment campaigns or for previously incurred debts; or for political purposes.
Geographic Focus: Indiana
Contact: Renee M. Chambers; (812) 333-9016; fax (812) 333-1153; renee@cfbmc.org
Internet: http://www.communityfoundation.ws
Sponsor: Community Foundation of Bloomington and Monroe County
101 W Kirkwood Avenue, Suite 321
Bloomington, IN 47404

Community Foundation of Boone County Grants 1266
The Community Foundation of Boone County Grants provides funding in the following areas: arts/culture; community development; education; elderly; health; human services; youth; environment; and recreation. Applications for grants of $10,000 or less should be submitted on the Short Form located on the website, while applications for more than $10,000 should follow instructions in the Content and Format section of the website.

Awards will be made to nonprofit organizations exempt from federal taxation under section 501(c)3 of the Internal Revenue Code. Grants may be allowed to individuals and to non-501(c)3 organizations if there is documented charitable activity benefiting or serving the residents of Boone County, or if the Community Foundation is acting as fiscal agent for the project or program.
Requirements: Grant proposals for over $10,000 must include the following format: abstract or executive summary; description of organization; statement of problem or need; objectives; proposed solution; materials/equipment; staff; facilities; evaluation; budget and its explanation; and appendices. Organizations should review the Foundation website for specific proposal submission instructions.
Restrictions: Non-allowable expenses include support of pre-award costs (i.e., project costs generated during the preparation of a proposal for the same project); existing general fund operating expenses; regular salaries of pre-award permanent staff (unless overload compensation is justified during the life of the funded project); international travel; first-class air fare; luxury accommodations; hospitality for purposes other than those directly related to meeting program objectives as defined in the proposal; alcohol; and indirect or regular existing administrative costs (e.g., telephone, utilities, general maintenance, etc.) of the applicant organization.
Geographic Focus: Indiana
Date(s) Application is Due: Jan 24; Mar 28; May 30; Aug 1; Oct 3; Nov 21
Contact: Barbara J. Schroeder, Program Director; (317) 873-0210 or (765) 482-0024; fax (317) 873-0219; barb@communityfoundationbc.org
Internet: http://www.communityfoundationbc.org/grants.html
Sponsor: Community Foundation of Boone County
60 East Cedar Street
Zionsville, IN 46077

Community Foundation of Central Illinois Grants 1267
The foundation funds programs in the fields of education, arts, human services, community service, or community development, and that advance one or more of the following objectives: address and help resolve important existing or emerging community issues; support new and creative projects and organizations offering the greatest opportunity for positive and significant change; promote cooperation and collaboration among organizations; identify, enhance, and develop leadership in the community through creative and innovative activities that empower individuals; and improve the quality or scope of charitable works in our community. Application guidelines are available online.
Requirements: Nonprofits within a 50-mile radius of Peoria, IL, are eligible.
Restrictions: The foundation will not fund annual campaigns, individuals, or endowments, or make grants for sectarian religious purposes.
Geographic Focus: Illinois
Date(s) Application is Due: Apr 15; Sep 15
Amount of Grant: 2,500 - 3,000 USD
Contact: Kristan Creek, Program Officer; (309) 674-8730; fax (309) 674-8754; Kristan@communityfoundationci.org
Internet: http://www.communityfoundationci.org/grant_guidelines.asp
Sponsor: Community Foundation of Central Illinois
331 Fulton Street, Suite 310
Peoria, IL 61602

Community Foundation of East Central Illinois Grants 1268
The community foundation awards grants from its unrestricted funds to Champaign County, IL, nonprofit organizations to use for charitable purposes. Representative categories include arts and humanities, environmental concerns, education, health, human services, research, urban affairs, church, and youth programs. Types of support include continuing support, annual campaigns, building/renovation, equipment, publication, consulting services, and scholarships.
Requirements: Champaign County, IL, nonprofit organizations are eligible.
Geographic Focus: Illinois
Date(s) Application is Due: Aug 31
Contact: Joan Dixon; (217) 359-0125; fax (217) 352-6494; joandixon@cfeci.org
Internet: http://www.cfeci.org/non-profit/grant-application
Sponsor: Community Foundation of East Central Illinois
404 West Church Street
Champaign, IL 61820-3411

Community Foundation of Eastern Connecticut General Southeast Grants 1269
Within its geographic area, comprising the 42 towns of eastern Connecticut, the community foundation awards grants to assist charitable, educational, and civic institutions; promote health and general welfare; support environmental programs; provide human care services for the needy; secure the care of children and families; encourage artistic and cultural endeavors; and initiate planning of appropriate projects within these areas. High-priority programs include those that strengthen families; improve access to area resources, especially for underserved populations; encourage residents to participate in the cultural life of the community; demonstrate collaborative efforts and inclusive practices; reinforce best practices or show innovative approaches; and add to the general well being of the community. Types of support include building and renovation, equipment, emergency funds, program development, conferences and seminars, publication, seed grants, scholarship funds, technical support, and scholarships to individuals. Foundation grants have ranged from $1,000 to $25,000 in recent years, with most grants in the $5,000 to $15,000 range. Proposals may be submitted between April 1 and November 15 each year.
Requirements: IRS tax-exempt organizations serving Old Lyme, East Lyme, Lyme, Salem, Montville, Waterford, New London, Groton, Ledyard, Stonington, and North Stonington are eligible to apply for this funding.

Restrictions: The foundation does not consider requests for direct financial assistance to individuals; religious or sectarian programs; political or lobbying purposes; fundraising events; or debt retirement. Applications requesting support of normal operating expenses will not be considered.
Geographic Focus: Connecticut
Date(s) Application is Due: Nov 15
Amount of Grant: 1,000 - 25,000 USD
Contact: Jennifer O'Brien, Program Director; (860) 442-3572 or (877) 442-3572; fax (860) 442-0584; jennob@cfect.org
Internet: http://www.cfect.org/ForGrantseekers/Howtoapplyforgrants/tabid/319/Default.aspx
Sponsor: Community Foundation of Eastern Connecticut
68 Federal Street
New London, CT 06320

Community Foundation of Eastern Connecticut Northeast Women and Girls Grants 1270

The Community Foundation of Eastern Connecticut's Northeast Women and Girls Fund aims to remove the inequities that block women from self-sustainability and improve the quality of life for all women and girls living in Northeast Connecticut. The Foundation aims to achieve this by focusing on three priority areas, including: empowerment and personal development programs that help women and girls overcome personal obstacles and take positive control of their lives - areas of focus include, but are not limited to, mentoring programs, life skills, connecting to the community, and confidence-building; positive health and well-being programs that promote good physical and mental health practices and healthy lifestyles - areas of focus for such programs may include the whole being (mind, body, soul, nutrition, exercise, etc.), the prevention of high risk behaviors and relationship violence, and the encouragement of supportive networks; and economic independence programs that include, but are not limited to, access to education, full-time employment, vocational skills, career development, affordable childcare, housing, and building financial literacy skills.
Requirements: IRS tax-exempt organizations serving Brooklyn, Canterbury, Eastford, Hampton, Killingly, Plainfield, Pomfret, Putnam, Sterling, Thompson, and Woodstock are eligible to apply for this funding. Collaboration between service providers is highly encouraged.
Restrictions: The foundation does not consider requests for direct financial assistance to individuals; religious or sectarian programs; political or lobbying purposes; fundraising events; or debt retirement. Applications requesting support of normal operating expenses will not be considered.
Geographic Focus: Connecticut
Date(s) Application is Due: Sep 15
Contact: Alison Woods; (860) 442-3572 or (877) 442-3572; alison@cfect.org
Internet: http://www.cfect.org/ForGrantseekers/Howtoapplyforgrants/tabid/319/Default.aspx
Sponsor: Community Foundation of Eastern Connecticut
68 Federal Street
New London, CT 06320

Community Foundation of Grant County Grants 1271

The Community Foundation of Grant County Grants address the needs of Grant County in the fields of community development, education, and health and human services. Nonprofit organizations, coalitions, community associations, and other civic groups may apply if they provide services within the county. The project must be located in or directly serve the people in Grant County, and meet all other criteria in the guidelines and application. All proposals must be received by the Foundation on the last Friday in April, July, October, and January, with committee reviews in May, August, November, and February.
Requirements: The Board will only accept written proposals for consideration after an applicant has first consulted with the Foundation's staff to find if the project is suitable for funding. The organization should keep the following guidelines in mind when preparing their written proposal: the project's purpose; whether the Foundation will be the sole funder; how many will be served or affected by the project, if there is a broad base of support for it, or if services are duplicated by another source; if an important need has been shown; whether the objectives are realistic and measurable; is there a viable plan for future support; who the key people are and if they are available for the long-term; and if all financial information is included and makes sense.
Restrictions: To be eligible for a grant, the project must be located in or directly serve the people and natural resources in Grant County, Indiana. Funding is not available for: profit making enterprises; political activity; operating budgets, except for limited experimental or demonstration periods; salaries; sectarian or religious purposes; endowment purposes unless for special promotions and/or matching challenges; multi-year funding requests; or capital improvements to church owned facilities or properties.
Geographic Focus: Indiana
Contact: Sherrie Stahl; (765) 662-0065; fax (765) 662-1438; sstahl@comfdn.org
Internet: http://www.comfdn.org/grants.htm
Sponsor: Community Foundation of Grant County
505 West Third Street
Marion, IN 46952

Community Foundation of Greater Birmingham Grants 1272

The community foundation serving the greater Birmingham, AL, area awards grants in arts and culture, education, environment, health, and welfare. Types of support include capital campaigns, operating support, building/renovation, equipment, program development, publication, seed money, curriculum development, and matching funds. Grant requests are considered twice a year at distribution committee meetings. The foundation requires that grant requests be submitted in writing. There is no application form. Applicants are asked to submit a cover letter and a complete statement of the purpose of the grant, including the need for the project and population and number of people to be served, project budget, other funding sources, length of time for which foundation aid is needed, a method to evaluate the project's success, description of applicant organization including annual budget, copy of IRS determination letter, and names of the board of directors or trustees.
Requirements: IRS 501(c)3 organizations that provide services in the greater Birmingham metropolitan area (Jefferson, Shelby, St. Clair, Blount and Walker counties) are eligible.
Restrictions: Grants are not made to or for individuals, operating expenses of organizations, religious organizations for religious purposes, national fund-raising drives, conference or seminar expenses, tickets for benefits, political organizations or candidates for public office, organizations with IRS 501(h) status, budget deficits, replacement of government funding cuts, or scholarships or endowment funds.
Geographic Focus: Alabama
Date(s) Application is Due: Mar 15; Sep 15
Samples: Exchange Club Family Skills Center, $2,200--program support; Catholic Family Services, $4,600-- to expand outpatient psychiatric care for uninsured women and children in metro Birmingham; Birmingham Civil Rights Institute, $20,000-- toward operating costs as the institute continues to serve the general public, tourists and students, particularly in kindergarten through second grade.
Contact: James McCrary, Senior Program Officer; (205) 327-3812 or (205) 327-3800; fax (205) 328-6576; jmccrary@foundationbirmingham.org or guidelines@foundationbirmingham.org
Internet: http://www.foundationbirmingham.org/page30508.cfm
Sponsor: Community Foundation of Greater Birmingham
2100 First Avenue N, Suite 700
Birmingham, AL 35203-4223

Community Foundation of Greater Flint Grants 1273

The community foundation is interested in funding charitable organizations that are able to demonstrate they have planned their projects with respect to the community's needs in the areas of arts and humanities, community services, education, environment and conservation, ethics, health, human and social services, and philanthropy. The foundation's current objectives for the allocation of discretionary funds give top priority to programs that: advance the health and well-being of children; and/or improve the capacity of public education. Types of support include general operating support, program development, seed money, scholarship funds, and technical assistance. Applications are available on the Web site.
Requirements: IRS 501(c)3 organizations with programs of direct relevance to the residents of Genesee County, Michigan are eligible. Types of support include general operating support.
Restrictions: Grants will not be made to individuals. In general, requests for sectarian religious purposes, budget deficits, routine operating expenses of existing organizations, litigation, endowments, and other capital fund drives or projects are not funded.
Geographic Focus: Michigan
Samples: Greater Flint Health Coalition, $40,000--towards the Flint Healthcare Employment Opportunities (FHEO) program, which helps unemployed or low-income individuals in Genesee County to obtain employment opportunities or to advance in the healthcare industry; Genesee Chamber Foundation,$30,000-- towards the Summer Youth Initiative, a summer employment and leadership development program for Genesee County youth ages 14-18; United Way of Genesee County, $66,000- in support of the BEST Project (Building Excellence, Sustainability and Trust) BEST provides technical assistance and resources designed to build the capacity of area nonprofit organizations.
Contact: Lynn Larkin; (810) 767-8270; fax (810) 767-0496; llarkin@cfgf.org
Internet: http://www.cfgf.org/page32610.cfm
Sponsor: Community Foundation of Greater Flint
500 South Saginaw Street
Flint, MI 48502-1206

Community Foundation of Greater Fort Wayne - Community Endowment and Clarke Endowment Grants 1274

The Foundation encourages projects or programs that are developed in consultation with other agencies and planning groups that increase coordination and cooperation among agencies and reduce unnecessary duplication of services. Preference is given to projects or programs that: address priority community concerns; encourage more effective use of community resources; test or demonstrate new approaches and techniques in the solution of community problems; are intended to strengthen the management capabilities of agencies; promote volunteer participation and citizen involvement in community affairs. Contact program staff for current guidelines and deadlines. The deadlines listed are for concept letters, with final invited proposals due by dates provided by the Foundation.
Requirements: Nonprofit organizations in Allen County, Indiana, are eligible to apply. Applicants should mail hardcopies of the application package including the following: concept letter fact sheet; the original concept letter; a detailed program budget or agency budget; current financial statements; a copy of the organization's 501(c)3 IRS determination letter; and a list of the board of directors and their principal affiliations.
Restrictions: Grants do not support: annual fund drives; operating deficits or after-the-fact support; endowment funds, except for endowment-building matching grants for funds held at the Community Foundation; direct or grassroots lobbying; religious purposes; hospitals, medical research, or academic research; public, private, or parochial educational institutions except in special situations when support is essential to projects/programs that meet critical community needs; governmental agencies, including public school systems, except in special situations when support is essential to projects/programs that meet critical community needs; limited, special interest organizations except when such support significantly benefits the disadvantaged; and funding for sponsorships, special events, commercial advertising, films or videos, television programs, conferences, group uniforms, or group trips.
Geographic Focus: Indiana

Date(s) Application is Due: Jan 9; Apr 9; Jul 9
Samples: Salvation Army of Fort Wayne (Fort Wayne, Indiana)--for remodeling of building, $50,000; United Way of Allen County (Fort Wayne, Indiana)--for diversity and special urban initiatives, $100,000; Local Education Fund (Fort Wayne, IN)--for training materials, $38,314.
Contact: Annette Smith; (260) 426-4083; fax (260) 424-0114; asmith@cfgfw.org
Internet: http://www.cfgfw.org/
Sponsor: Community Foundation of Greater Fort Wayne
555 East Wayne Street
Fort Wayne, IN 46802

Community Foundation of Greater Fort Wayne Lilly Endowment Scholarships   1275
The Community Foundation of Greater Fort Wayne administers the Lilly Endowment Scholarships for the Allen County area. High school seniors in the top 40% of their class are eligible to apply. The awards cover full tuition for a four-year degree, and includes an $800 per year book stipend. Any college major is eligible.
Requirements: Students should contact their Allen County high school guidance office for application and deadline information.
Restrictions: Grant recipients must enroll in an Indiana college or university.
Geographic Focus: Indiana
Contact: Lydia Von Uderitz; (260) 426-4083; fax (260) 424-0114; lvon@cfgfw.org
Internet: http://www.cfgfw.org/scholarships/scholar-college.html
Sponsor: Community Foundation of Greater Fort Wayne
555 East Wayne Street
Fort Wayne, IN 46802

Community Foundation of Greater Fort Wayne Scholarships   1276
The Community Foundation of Greater Fort Wayne offers a wide variety of scholarship funding for Allen County students attending public and private schools. Scholarships are available for every school age ranging from elementary school through college, and are categorized by school, major, financial need, GPA, class rank, businesses, and local organizations. Application deadlines vary, but are generally from January to May.
Requirements: Students can contact their guidance office or the Foundation for an application if it is not posted online. Students should consult the individual listings for requirements, information needed, and deadlines.
Restrictions: Most of the scholarships require that students are residents of Allen County.
Geographic Focus: Indiana
Contact: Lydia Von Uderitz; (260) 426-4083; fax (260) 424-0114; lvon@cfgfw.org
Internet: http://www.cfgfw.org/scholarships/scholar_apply.html
Sponsor: Community Foundation of Greater Fort Wayne
555 East Wayne Street
Fort Wayne, IN 46802

Community Foundation of Greater Greensboro Community Grants   1277
The Community Foundation of Greater Greensboro awards community grants from unrestricted and field-of-interest funds, as allocated by the Board of Directors, to support a wide range of community issues. This category of grants are one-time awards given to help nonprofits meet community needs and opportunities by building their capacity to reach their missions. Grants usually range in size from a few hundred dollars up to $10,000, and tend to average $3,000 to $5,000. The current Community Grants program will focus on capacity building for nonprofits. By definition, grants will support activities based upon what different organizations need to more effectively reach their missions. For example: one organization might request dollars for a training program for staff or the board; another organization's needs might generate a grant request for strategic planning; two or more nonprofits might request funds jointly to support a restructuring effort to strategically align services or administrative functions; and other nonprofits might seek support for evaluating a program and assessing future options. Annual deadlines for applications are March 14 and August 15.
Requirements: Applicant must be a 501(c)3 nonprofit organization located in or serving the greater Greensboro, North Carolina, area.
Restrictions: Request may not exceed $10,000 and cannot be used for expenses already incurred. Typically, multi-year grant applications are not considered in this grant program. Public schools or other public agencies will typically not receive grants through this program, although they may be involved as partners in funded efforts. Grants are not awarded to individuals.
Geographic Focus: North Carolina
Date(s) Application is Due: Mar 14; Aug 15
Amount of Grant: Up to 10,000 USD
Contact: Connie Leeper; (336) 379-9100, ext. 130; fax (336) 378-0725; cleeper@cfgg.org
Internet: http://cfgg.org/receive/community-grants-program
Sponsor: Community Foundation of Greater Greensboro
330 S. Greene Street, Suite 100
Greensboro, NC 27401

Community Foundation of Greater Greensboro Women to Women Grants   1278
Women to Women is Greensboro's first permanent grant making endowment to engage women in impacting the lives of our community's women and their families. Each year, the fund awards high-impact grants to programs, organizations and projects that create positive and measurable impact. High-impact grants are awarded annually to projects in the Foundation focus areas, including education, social services, the environment, health and the arts. Women to Women works with existing organizations to find innovative ways to solve the problems facing women and their families. The Foundation already has achieved its initial $3 million fundraising goal, and now is working toward raising an additional $2 million. At $5 million, Women to Women ultimately will be able to award as much as $200,000 each year in high-dollar grants.
Geographic Focus: North Carolina
Amount of Grant: Up to 200,000 USD
Contact: Connie Leeper; (336) 379-9100, ext. 130; fax (336) 378-0725; cleeper@cfgg.org
Internet: http://cfgg.org/women-to-women
Sponsor: Community Foundation of Greater Greensboro
330 S. Greene Street, Suite 100
Greensboro, NC 27401

Community Foundation of Greater Lafayette - Robert and Dorothy Hughes Scholarships   1279
The Community Foundation of Greater Lafayette - Hughes Scholarships are available to help students earn a baccalaureate degree in nursing. Applicants must live or work in Tippecanoe County; plan to attend and obtain a baccalaureate degree from an accredited nursing school; and have financial need.
Requirements: The application is available online. Students must include a completed financial form; a signed certification; high school and/or college transcripts; a two page essay; and three letters of recommendation, then submit the application packet to the Foundation.
Restrictions: The scholarships are renewable, although students must re-apply to be considered for a renewal.
Geographic Focus: Indiana
Date(s) Application is Due: Jan 13
Amount of Grant: 1,000 USD
Contact: Cheryl Ubelhor; (765) 742-9078; fax (765) 742-2428; cheryl@cfglaf.org
Internet: http://www.cfglaf.org/scholarship/index.htm#Hughes
Sponsor: Community Foundation of Greater Lafayette
300 Main Street, Suite 100
Lafayette, IN 47901

Community Foundation Of Greater Lafayette Grants   1280
The Community Foundation of Greater Lafayette Grants help meet the ever-changing needs of the community. Funding priorities include education, children/youth, health, diversity, physical environment, and arts and culture. Charitable organizations that serve Tippecanoe and the surrounding counties are eligible to apply. Most grants are awarded to nonprofit organizations that are located in and serve Tippecanoe County. Priority is given to projects that reach as many people as possible; improve the ability of organizations to serve the community over the long-term; serve the Greater Lafayette area; and are run by non-profit organizations. Grant seekers are encouraged to contact the Foundation office to be certain their proposal is appropriate for consideration. Proposal deadlines are listed according to dollar amount request, so grant seekers should carefully review the website to judge when to submit their proposal.
Requirements: Organizations should submit the following information along with the online application: contact and background information for the organization; a concise narrative about the project; amount requested from the foundation; and a project budget. Also included is a board of directors list; project estimates/bids; and financial statements. Applicants may be asked for further information or a site visit to clarify the request.
Restrictions: The Foundation does not fund: programs that are sectarian or religious in nature; political organizations or candidates; endowments; ongoing operating expenses; government agencies or public institutions; programs that taxpayers would normally support; individuals; special events (i.e. parades, festivals, sporting activities, fundraisers); programs already completed; multi-year grants; debt or deficit reduction; and projects funded in a previous year (unless invited to resubmit).
Geographic Focus: Indiana
Date(s) Application is Due: Apr 1; Sep 1; Dec 1
Contact: Cheryl Ubelhor, Program Director; (765) 742-9078; fax (765) 742-2428; info@cfglaf.org or cheryl@cfglaf.org
Internet: http://www.glcfonline.org/grantseekers/index.htm
Sponsor: Community Foundation of Greater Lafayette
300 Main Street, Suite 100
Lafayette, IN 47901

Community Foundation of Greater New Britain Grants   1281
Grants from the Community Foundation of Greater New Britain support organizations and programs benefiting the residents of Berlin, New Britain, Plainville and Southington, Connecticut. The foundation supports organizations involved with arts, culture and heritage; community and economic development; early childhood development; education; health and human services.
Requirements: Grants are available to non-profit organizations that have tax-exempt status under Section 501(c)3 of the IRS Code or that are a qualified entity eligible to receive grants from community foundations under the IRS Code. The foundation will also consider funding a grassroots group if it consists of at least five people, has a governing body and a fiscal agent with the appropriate tax status. All grant applications must begin with a one or two page Letter of Intent summarizing the program or project for which funding is being sought. Submitted Letters of Intent will be reviewed by staff. Applicants selected by staff to submit complete, formal applications will be contacted directly to begin the proposal.
Restrictions: The Foundation does not make grants for: sectarian or religious activities; previously incurred expenses; annual or endowment campaigns; sponsorships or fundraisers; political activities; direct support of individuals; camperships or scholarships for academic and/or enrichment programs. In addition, the Foundation generally does not make grants for performances or one-time events, conferences or advertising.
Geographic Focus: Connecticut

Date(s) Application is Due: Feb 15; Jun 15; Oct 15
Amount of Grant: 5,000 - 25,000 USD
Samples: New Britain Museum of American Art, New Britain, Connecticut, $60,000 - fund for support of the Museum's Education and Outreach program; Municipal Economic Development Agency, New Britain, Connecticut, $15,000 - to develop an economic development website for the City of New Britain featuring economic development tools and information that will promote business in New Britain; New Britain Public Library/Jefferson Family Resource Center, New Britain, Connecticut, $12,000 - for the Raise a Reader program, aimed at improving literacy skills among hard-to-reach New Britain families with children under age five.;
Contact: Joeline Wruck; (860) 229-6018, ext. 307; jwruck@cfgnb.org
Internet: http://www.cfgnb.org/GrantsScholarships/tabid/68/Default.aspx
Sponsor: Community Foundation of Greater New Britain
74A Vine Street
New Britain, CT 06052

Community Foundation of Greater Tampa Grants 1282
The Community Foundation of Tampa Bay is a nonprofit, tax-exempt organization which administers funds established by individuals, corporations, private foundations, and nonprofit organizations to support the charitable needs of the Tampa Bay area. Program interests of the Foundation include: arts and culture, community development, education, environment, animals, health and, human services.
Requirements: Florida 501(c)3 organizations based in and serving communities in Hillsborough, Pasco, Pinellas and/or Hernando counties are eligible.
Restrictions: The Foundation is generally not interested in requests for: funding of ongoing operating costs; grants for capital campaigns or expenditures; tickets for any fundraising event, conference, or advertising space in programs or other publications; legislative lobbying or political campaigns; medical research; religious or sectarian purposes; loans or assistance to individuals; a multiple year funding commitment.
Geographic Focus: Florida
Date(s) Application is Due: Mar 1; Sep 1
Amount of Grant: Up to 7,500 USD
Samples: Bay Springs--for the equipment, uniforms, shoes, and transportation of adolescent boys in foster care to participate in city and community sports leagues; Salesian Youth Center/Boys and Girls Club--for the Literacy and Technology Initiative at the Salesian Youth Center to offer reading comprehension and computer skills training to children 5-18.
Contact: Ann Berg; (813) 282-1975; fax (813) 282-3119; aberg@cftampabay.org
Internet: http://www.cftampabay.org/nonprofit_resources/grant_app_guidelines.html
Sponsor: Community Foundation of Greater Tampa
550 North Reo Street, Suite 301
Tampa, FL 33609

Community Foundation of Greenville-Greenville Women Giving Grants 1283
Greenville Women Giving (GWG) is a special initiative of Community Foundation of Greenville. GWG is a philanthropic organization, founded on the idea that women, informed about philanthropy and the needs in their community, collectively can make a real difference. GWG is committed to strengthening the community through the collective resources of it's members by awarding high-impact grants in five areas: arts and culture, education, health, human services, and environment. GWG will award grants ranging from a minimum of $40,000 to $100,000 to community organizations in Greenville County, South Carolina. Consideration will be given to applications requesting that funding be spent over a two year period. Additional guidelines and application are available on the GWG website: http://greenvillewomengiving.org/faq.aspx. The completed application should be mailed or delivered to Greenville Women Giving, c/o Community Foundation of Greenville on or before January 15, at 4 p.m.
Requirements: Applicants must qualify as tax-exempt under Section 501(c)3 of the Internal Revenue Code or be classified as a unit of government, located and providing services in Greenville County, South Carolina. Organizations must not discriminate on the basis of age, race, national origin, ethnicity, gender, physical ability, sexual orientation, political affiliation or religious belief.
Restrictions: Support is not available for the following: endowment campaigns; travel expenses and conferences; individuals; projects or programs that promote religious or political views; organizations for re-granting purpose.
Geographic Focus: South Carolina
Date(s) Application is Due: Jan 15
Amount of Grant: 40,000 - 100,000 USD
Contact: Debbie Cooper, Director of Donor Services; (864) 233-5925 or (864) 331-8414; fax (864) 242-9292; dcooper@cfgg.com or Cfgg@cfgg.com
Internet: http://greenvillewomengiving.org/guidelines.aspx
Sponsor: Greenville Women Giving
27 Cleveland Street, Suite 101
Greenville, SC 29601

Community Foundation of Greenville Community Enrichment Grants 1284
The Community Foundation of Greenville supports qualified nonprofit organizations in the areas of the arts and humanities, education, early childhood education, religion, and environmental programs. The Foundation is committed to the responsible allocation of the Foundation's unrestricted funds. As an organization, the Foundation seeks ways to make meaningful, sustained differences in the quality of life for the citizens of Greenville County. The goal is to seek out community-based projects where a modest grant can make a significant impact. The Foundation encourages proposals that: propose practical solutions to community problems; promote cooperation among agencies without duplicating services; use modest funds to make a significant impact; address prevention as well as remediation; make use of matching funds; strengthen an agency's effectiveness and stability; generate community support, both professional and volunteer; state project objectives in measurable terms. The application deadline is 4:00 pm on Friday, March 5.
Requirements: All 501(c)3 non-profit groups can apply for Community Foundation grants to fund innovative projects that have a direct relevance to the Greenville County, South Carolina area. An agency can receive one community enrichment grant per year.
Restrictions: The Foundation will: not make grants to individuals; not fund capacity building; will give low grant making priority to capital fund projects.
Geographic Focus: South Carolina
Date(s) Application is Due: Mar 5
Contact: Debbie Cooper, Director of Donor Services; (864) 233-5925 or (864) 331-8414; fax (864) 242-9292; dcooper@cfgg.com or Cfgg@cfgg.com
Internet: http://www.cfgreenville.org/page18740.cfm
Sponsor: Community Foundation of Greenville
27 Cleveland Street, Suite 101
Greenville, SC 29601

Community Foundation of Greenville Hollingsworth Funds Program/Project Grants 1285
The Hollingsworth Funds, Inc. has a competitive grant process that is administered by the Community Foundation. Consistent with the purposes established by its benefactor, John D. Hollingsworth, Jr., an astute textile executive and real estate investor. The Hollingsworth Funds will serve a broad range of charitable, educational, religious, literary and cultural purposes. The Hollingsworth Funds has adopted initial guidelines in order to assist its evaluation of charitable organizations for funding. The Hollingsworth Funds will make grants: to health and human service agencies dedicated to improving quality of life for residents of Greenville County and particularly for those which deliver services to the poor, homeless, and illiterate; to interdenominational faith-based programs that benefit a broad cross-section of the community; to a wide variety of arts and education related initiatives including high quality preschool and after school programs; for grants for capital projects that provide a public benefit on a nondiscriminatory basis; to increase the organizational capacity of a non-profit organization; to public/private partnerships. The Hollingsworth Funds will give priority to new and innovative projects that are outside of the scope of an organization's ongoing operating budget. To submit a request for funding, a charitable organization must deliver or mail two (2) complete application packets to the Community Foundation of Greenville by 4:00 pm on September 3. All applicants will be notified of their status by December 31.
Requirements: A 501(c)3 charitable organization can submit one grant application per year for an amount not to exceed $50,000 to fund innovative projects for the benefit of charitable uses within Greenville County, South Carolina, or for the benefit of residents of Greenville County.
Restrictions: The Hollingsworth Funds will not make grants: to sponsor special fundraising events, celebration functions, dinners or annual meetings; for areas that are traditionally the primary responsibility of local, state or federal governments; for core operating expenses of public, private or parochial schools; that primarily benefit the religious activities of a church, synagogue, mosque or other house of worship; to pay off existing debts; scholarship awards or grants to individuals.
Geographic Focus: South Carolina
Date(s) Application is Due: Sep 3
Contact: Debbie Cooper, Director of Donor Services; (864) 233-5925 or (864) 331-8414; fax (864) 242-9292; dcooper@cfgg.com or Cfgg@cfgg.com
Internet: http://www.cfgreenville.org/page18741.cfm
Sponsor: Community Foundation of Greenville
27 Cleveland Street, Suite 101
Greenville, SC 29601

Community Foundation of Howard County Grants 1286
The Community Foundation of Howard County Grants fund initiatives that improve the quality of life for citizens of Howard County. The Foundation addresses needs that generally fall into the following categories: health and medical; social services; education; cultural affairs; civic affairs; and community beautification. The Foundation uses the following evaluation criteria: does the project fit the purpose of the organization; is there an established need for the project; how well the project's purpose has been defined; does it fit with the Foundation's guidelines; and what kind of impact the project will have on the community.
Requirements: All applicants must receive pre-qualification prior to submitting an application. A letter of inquiry addressed to the Program Director should contain a brief statement of the applicant's need for assistance, estimate of total cost of project, and enough information so that the Foundation can determine whether the application falls within the guidelines of its grants program. The organization may then submit the online application with the following information: the grant application and cover page; project budget; copy of IRS determination 501(c)3 letter; current month and year-to-date financial statement; year-end financial statements; itemized list of board members; and evidence of board approval.
Restrictions: The Foundation does not award grants for: normal operating expenses and/or salaries; individuals; seminars or trips; sectarian religious purposes; endowment purposes of recipient organizations; projects which have been proposed by individuals or organizations responsible to advisory bodies or persons; and new projects and/or equipment which were purchased prior to the grant application being approved.
Geographic Focus: Indiana
Date(s) Application is Due: Mar 5; May 7; Sep 3; Nov 5
Contact: Kim Abney; (765) 454-7298 or (800) 964-0508; kim@cfhoward.org
Internet: http://www.cfhoward.org/grants.html
Sponsor: Community Foundation of Howard County
215 West Sycamore Street
Kokomo, IN 46901

Community Foundation of Jackson County Grants   1287
The Community Foundation of Jackson County is a community focused organization dedicated to: building a visionary partnership with donors and local service organizations; trustworthy stewardship of gifts; providing funds to enhance the quality of life across Jackson County; and being a catalyst for change in the community. Areas of funding interest include: education - projects that support and enhance educational programs serving a broad spectrum of Jackson County residents; economic development - projects that explore new ways to improve the lives of Jackson County residents through development of the county's economic strength; human services - programs that support human service organizations programs and services for all ages; and arts and culture - programs designed to establish diversified cultural programs that offer widespread opportunities for participation and appreciation of the arts throughout the county. The grant proposal form is available at the website or at the Foundation office.
Requirements: Nonprofits may apply with programs that serve residents of Jackson County.
Restrictions: Funding is not available for the following: individuals or groups of individuals to attend seminars or take trips except where there are special circumstances with a clear benefit to the larger community; political organizations or campaigns; state or national fundraising efforts; programs and/or equipment that were committed to prior to the grant proposal being submitted; endowment purposes of recipient organizations; and programs specifically for sectarian religious purposes. Programs that include both religious components and social needs will be carefully reviewed.
Geographic Focus: Indiana
Date(s) Application is Due: Jul 31
Contact: Lori Miller, Development Associate; (812) 523-4483; fax (812) 523-1433
Internet: http://www.cfjacksoncounty.org/policies.php
Sponsor: Community Foundation of Jackson County
107 Community Drive
Seymour, IN 47274

Community Foundation of Jackson County Seymour Noon Lions Club Grant   1288
The Seymour Noon Lions Club offers an annual grant opportunity for Jackson County. Grants are available in the areas of speech, eye care, hearing, diabetes and youth. Grant proposal applications are due mid-to-late February each year, and are located at the website.
Requirements: Grant proposals must include four copies of the completed proposal form using only the space provided. Applicants must also include four copies of the budget, a list of current governing board members; and a copy of their 501(c)3 IRS determination letter.
Geographic Focus: Indiana
Date(s) Application is Due: Feb 13
Contact: Grant Contact; (812) 523-4483; fax (812) 523-1433
Internet: http://www.cfjacksoncounty.org/lions.php
Sponsor: Community Foundation of Jackson County
107 Community Drive
Seymour, IN 47274

Community Foundation of Louisville AIDS Project Fund Grants   1289
The Community Foundation of Louisville is unique, in that it responds to the evolving needs and opportunities in the community. There is no set agenda, and no pre-determined recipients. Field of Interest Funds support organizations working within a specific geographic region or toward a specific purpose. The Community Foundation makes grants to the most appropriate and effective organizations working in areas such as arts and culture, education, youth, health, and human services. The AIDS Project Fund, administered by the Community Foundation, was created by the dissolution of a separate nonprofit organization, and it provides grants to support HIV prevention, education, and testing. Eligible organizations may apply to receive a grant award of up to $20,000. Applications are available on July 7 each year and may be submitted online only. The annual deadline for submission is August 18 at 4:00 pm.
Requirements: Eligible organizations include those that are: headquartered in Jefferson County, Kentucky (if based outside of the county, then the organization must demonstrate that a majority of beneficiaries are located in Jefferson County, Kentucky); and classified as a 501(c)3 public charity in good standing (organizations with a pending application for 501(c)3 status may apply with proof of Form 1023 receipt from the IRS).
Geographic Focus: Kentucky
Date(s) Application is Due: Aug 18
Contact: Anne McKune; (502) 855-6948 or (502) 585-4649; annemc@cflouisville.org
Internet: http://www.cflouisville.org/grants-partnerships/field-of-interest-grants/
Sponsor: Community Foundation of Louisville
325 W Main Street, Suite 1110, Waterfront Plaza, West Tower
Louisville, KY 40202

Community Foundation of Louisville Anna Marble Memorial Fund for Princeton Grants   1290
The Community Foundation of Louisville is unique, in that it responds to the evolving needs and opportunities in the community. There is no set agenda, and no pre-determined recipients. The Anna Marble Memorial Fund for Princeton, administered by the Community Foundation, was established by Anna Marble, in support of charitable groups helping the residents of Princeton, Kentucky. Eligible organizations may apply to receive a grant award of up to $20,000. Applications are available on July 7 each year and may be submitted online only. The annual deadline for submission is August 18 at 4:00 pm.
Requirements: Eligible organizations include those that are: headquartered in Jefferson County, Kentucky (if based outside of the county, then the organization must demonstrate that a majority of beneficiaries are located in Jefferson County, Kentucky); and classified as a 501(c)3 public charity in good standing (organizations with a pending application for 501(c)3 status may apply with proof of Form 1023 receipt from the IRS).
Geographic Focus: Kentucky
Date(s) Application is Due: Aug 18
Amount of Grant: 100 - 20,000 USD
Contact: Whitney Gentry; (502) 855-6963; fax (502) 585-4649; whitneyg@cflouisville.org
Internet: http://www.cflouisville.org/grants-partnerships/field-of-interest-grants/
Sponsor: Community Foundation of Louisville
325 W Main Street, Suite 1110, Waterfront Plaza, West Tower
Louisville, KY 40202

Community Foundation of Louisville Bobbye M. Robinson Fund Grants   1291
The Community Foundation of Louisville is unique, in that it responds to the evolving needs and opportunities in the community. There is no set agenda, and no pre-determined recipients. Field of Interest Funds support organizations working within a specific geographic region or toward a specific purpose. The Community Foundation makes grants to the most appropriate and effective organizations working in areas such as arts and culture, education, youth, health, and human services. The Bobbye M. Robinson Fund, administered by the Community Foundation, was established by Nancy Klempner Patton, in memory of her mother, to support programs that serve patients with colon cancer. Eligible organizations may apply to receive a grant award of up to $20,000. Applications are available on July 7 each year and may be submitted online only. The annual deadline for submission is August 18 at 4:00 pm.
Requirements: Eligible organizations include those that are: headquartered in Jefferson County, Kentucky (if based outside of the county, then the organization must demonstrate that a majority of beneficiaries are located in Jefferson County, Kentucky); and classified as a 501(c)3 public charity in good standing (organizations with a pending application for 501(c)3 status may apply with proof of Form 1023 receipt from the IRS).
Geographic Focus: Kentucky
Date(s) Application is Due: Aug 18
Amount of Grant: 100 - 20,000 USD
Contact: Anne McKune; (502) 855-6948 or (502) 585-4649; annemc@cflouisville.org
Internet: http://www.cflouisville.org/grants-partnerships/field-of-interest-grants/
Sponsor: Community Foundation of Louisville
325 W Main Street, Suite 1110, Waterfront Plaza, West Tower
Louisville, KY 40202

Community Foundation of Louisville Delta Dental of Kentucky Fund Grants   1292
The Community Foundation of Louisville is unique, in that it responds to the evolving needs and opportunities in the community. There is no set agenda, and no pre-determined recipients. The Delta Dental of Kentucky Foundation, administered by the Community Foundation, was created in 2004 in order to support projects and nonprofit organizations that promote dental health and dental education. Eligible organizations may apply to receive a grant award of up to $10,000. Applications are available on July 7 each year and may be submitted online only. The annual deadline for submission is August 18 at 4:00 p.m.
Requirements: Eligible organizations include those that are: headquartered in Jefferson County, Kentucky (if based outside of the county, then the organization must demonstrate that a majority of beneficiaries are located in Jefferson County, Kentucky); and classified as a 501(c)3 public charity in good standing (organizations with a pending application for 501(c)3 status may apply with proof of Form 1023 receipt from the IRS).
Geographic Focus: Kentucky
Date(s) Application is Due: Aug 18
Amount of Grant: 100 - 10,000 USD
Contact: Whitney Gentry; (502) 855-6963; fax (502) 585-4649; whitneyg@cflouisville.org
Internet: http://www.cflouisville.org/grants-partnerships/field-of-interest-grants/
Sponsor: Community Foundation of Louisville
325 W Main Street, Suite 1110, Waterfront Plaza, West Tower
Louisville, KY 40202

Community Foundation of Louisville Diller B. and Katherine P. Groff Fund for Pediatric Surgery Grants   1293
The Community Foundation of Louisville is unique, in that it responds to the evolving needs and opportunities in the community. There is no set agenda, and no pre-determined recipients. Field of Interest Funds support organizations working within a specific geographic region or toward a specific purpose. The Community Foundation makes grants to the most appropriate and effective organizations working in areas such as arts and culture, education, youth, health, and human services. The Diller B. and Katherine P. Groff Fund for Pediatric Surgery was established in 1990 in support of needs in the areas of pediatric surgery and surgical care of children, as well as education and research in allied disciplines. Typically, eligible organizations may apply to receive a grant award of up to $20,000. Applications are available on July 7 each year and may be submitted online only. The annual deadline for submission is August 18 at 4:00 pm.
Requirements: Eligible organizations include those that are: headquartered in Jefferson County, Kentucky (if based outside of the county, then the organization must demonstrate that a majority of beneficiaries are located in Jefferson County, Kentucky); and classified as a 501(c)3 public charity in good standing (organizations with a pending application for 501(c)3 status may apply with proof of Form 1023 receipt from the IRS).
Geographic Focus: Kentucky
Date(s) Application is Due: Aug 18
Amount of Grant: 100 - 10,000 USD
Contact: Anne McKune; (502) 855-6948; fax (502) 585-4649; annemc@cflouisville.org
Internet: http://www.cflouisville.org/grants-partnerships/field-of-interest-grants/
Sponsor: Community Foundation of Louisville
325 W Main Street, Suite 1110, Waterfront Plaza, West Tower
Louisville, KY 40202

Community Foundation of Louisville Dr. W. Barnett Owen Memorial Fund for the Children of Louisville and Jefferson County Grants 1294

The Community Foundation of Louisville is unique, in that it responds to the evolving needs and opportunities in the community. There is no set agenda, and no pre-determined recipients. Field of Interest Funds support organizations working within a specific geographic region or toward a specific purpose. The Community Foundation makes grants to the most appropriate and effective organizations working in areas such as arts and culture, education, youth, health, and human services. The Dr. W. Barnett Owen Memorial Fund for the Children of Louisville and Jefferson County, administered by the Community Foundation, was established by Dr. Albert P. Williams to honor the memory of Dr. Owen, a children's health care activist. Owen was instrumental in establishing the Kosair Children's Hospital in Louisville, when polio was the scourge of adolescence. The Fund supports children and youth health and other programs. Eligible organizations may apply to receive a grant award of up to $20,000. Applications are available on July 7 each year and may be submitted online only. The annual deadline for submission is August 18 at 4:00 pm.
Requirements: Eligible organizations include those that are: headquartered in Jefferson County, Kentucky (if based outside of the county, then the organization must demonstrate that a majority of beneficiaries are located in Jefferson County, Kentucky); and classified as a 501(c)3 public charity in good standing (organizations with a pending application for 501(c)3 status may apply with proof of Form 1023 receipt from the IRS).
Geographic Focus: Kentucky
Date(s) Application is Due: Aug 18
Amount of Grant: 100 - 20,000 USD
Contact: Anne McKune; (502) 855-6948 or (502) 585-4649; annemc@cflouisville.org
Internet: http://www.cflouisville.org/grants-partnerships/field-of-interest-grants/
Sponsor: Community Foundation of Louisville
325 W Main Street, Suite 1110, Waterfront Plaza, West Tower
Louisville, KY 40202

Community Foundation of Louisville Health Grants 1295

The Community Foundation's Fund for Louisville is unique, in that it responds to the evolving needs and opportunities in the community. There is no set agenda, and no pre-determined recipients. In the area of Health, the fund supports health education programs and organizations that provide direct patient services. The grant program is supported primarily by three funds, including: the AIDS Project Fund; the Irving B. Klempner Fund; and the Bobbye M. Robinson Fund. Eligible organizations may apply to receive a grant award of up to $20,000. Applications are available on July 7 each year and may be submitted online only. The annual deadline for submission is August 21 at 4:00 pm.
Requirements: Eligible organizations include those that are: headquartered in Jefferson County, Kentucky (if based outside of the county, then the organization must demonstrate that a majority of beneficiaries are located in Jefferson County, Kentucky); and classified as a 501(c)3 public charity in good standing (organizations with a pending application for 501(c)3 status may apply with proof of Form 1023 receipt from the IRS).
Geographic Focus: Kentucky
Date(s) Application is Due: Aug 21
Amount of Grant: 100 - 20,000 USD
Contact: Whitney Gentry; (502) 855-6963; fax (502) 585-4649; whitneyg@cflouisville.org
Internet: http://www.cflouisville.org/resources/funds-foundation
Sponsor: Community Foundation of Louisville
325 W Main Street, Suite 1110, Waterfront Plaza, West Tower
Louisville, KY 40202

Community Foundation of Louisville Irving B. Klempner Fund Grants 1296

The Community Foundation of Louisville is unique, in that it responds to the evolving needs and opportunities in the community. There is no set agenda, and no pre-determined recipients. Field of Interest Funds support organizations working within a specific geographic region or toward a specific purpose. The Community Foundation makes grants to the most appropriate and effective organizations working in areas such as arts and culture, education, youth, health, and human services. The Irving B. Klempner Fund, administered by the Community Foundation, was established by Nancy Klempner Patton, in memory of her father, to support programs that serve patients with prostate cancer. Eligible organizations may apply to receive a grant award of up to $20,000. Applications are available on July 7 each year and may be submitted online only. The annual deadline for submission is August 18 at 4:00 pm.
Requirements: Eligible organizations include those that are: headquartered in Jefferson County, Kentucky (if based outside of the county, then the organization must demonstrate that a majority of beneficiaries are located in Jefferson County, Kentucky); and classified as a 501(c)3 public charity in good standing (organizations with a pending application for 501(c)3 status may apply with proof of Form 1023 receipt from the IRS).
Geographic Focus: Kentucky
Date(s) Application is Due: Aug 18
Amount of Grant: 100 - 20,000 USD
Contact: Anne McKune; (502) 855-6948 or (502) 585-4649; annemc@cflouisville.org
Internet: http://www.cflouisville.org/grants-partnerships/field-of-interest-grants/
Sponsor: Community Foundation of Louisville
325 W Main Street, Suite 1110, Waterfront Plaza, West Tower
Louisville, KY 40202

Community Foundation of Louisville Lee Look Fund for Spinal Injury Grants 1297

The Community Foundation of Louisville is unique, in that it responds to the evolving needs and opportunities in the community. There is no set agenda, and no pre-determined recipients. The Lee Look Fund for Spinal Injury, administered by the Community Foundation, was established by Lee Look in support of public charities that help people with spinal injuries. In 1999, while living in Los Angeles, Lee injured his neck in an accident. Following an examination and x-rays at a medical center, the diagnosis provided by the physicians was a sprain. Lee left with instructions to "take Advil." A few weeks later, in Louisville, he had his fifth and sixth vertebrae fused with a steel plate and a bone graft from his own hip. He started the Lee Look Fund for Spinal Injury to help the thousands of people dealing with the challenges of living with such an injury. The fund supports research that will hopefully find more effective treatments, therapies and even cures for the types of disabilities that result from spinal injuries. Eligible organizations may apply to receive a grant award of up to $20,000. Applications are available on July 7 each year and may be submitted online only. The annual deadline for submission is August 18 at 4:00 pm.
Requirements: Eligible organizations include those that are: headquartered in Jefferson County, Kentucky (if based outside of the county, then the organization must demonstrate that a majority of beneficiaries are located in Jefferson County, Kentucky); and classified as a 501(c)3 public charity in good standing (organizations with a pending application for 501(c)3 status may apply with proof of Form 1023 receipt from the IRS).
Geographic Focus: Kentucky
Date(s) Application is Due: Aug 18
Amount of Grant: 100 - 20,000 USD
Contact: Whitney Gentry; (502) 855-6963; fax (502) 585-4649; whitneyg@cflouisville.org
Internet: http://www.cflouisville.org/grants-partnerships/field-of-interest-grants/
Sponsor: Community Foundation of Louisville
325 W Main Street, Suite 1110, Waterfront Plaza, West Tower
Louisville, KY 40202

Community Foundation of Louisville Morris and Esther Lee Fund Grants 1298

The Community Foundation of Louisville is unique, in that it responds to the evolving needs and opportunities in the community. There is no set agenda, and no pre-determined recipients. Field of Interest Funds support organizations working within a specific geographic region or toward a specific purpose. The Community Foundation makes grants to the most appropriate and effective organizations working in areas such as arts and culture, education, youth, health, and human services. The Morris and Esther Lee Fund, administered by the Community Foundation, was established in honor of Morris A. Lee and his wife, Esther B. Lee. Morris was a marketing and sales entrepreneur, operating several small businesses in the Louisville, Cincinnati, and Detroit areas. The businesses provided marketing services and products to community banks and retailers. He retired in 1982. Mr. Lee was known for his story telling skills, great sense of humor, helpfulness and loyalty to friends, customers, and family. The Fund offers support for community organizations that provide human services of all types. Eligible organizations may apply to receive a grant award of up to $20,000. Applications are available on July 7 each year and may be submitted online only. The annual deadline for submission is August 18 at 4:00 pm.
Requirements: Eligible organizations include those that are: headquartered in Jefferson County, Kentucky (if based outside of the county, then the organization must demonstrate that a majority of beneficiaries are located in Jefferson County, Kentucky); and classified as a 501(c)3 public charity in good standing (organizations with a pending application for 501(c)3 status may apply with proof of Form 1023 receipt from the IRS).
Geographic Focus: Kentucky
Date(s) Application is Due: Aug 18
Amount of Grant: 100 - 20,000 USD
Contact: Anne McKune, Program Officer; (502) 855-6948 or (502) 585-4649; fax (502) 855-6173; annemc@cflouisville.org
Internet: http://www.cflouisville.org/grants-partnerships/field-of-interest-grants/
Sponsor: Community Foundation of Louisville
325 W Main Street, Suite 1110, Waterfront Plaza, West Tower
Louisville, KY 40202

Community Foundation of Mount Vernon and Knox County Grants 1299

The Community Foundation of Mount Vernon/Knox County, awards grants and scholarships from a variety of funds. The Foundations mission is to improve the quality of life in Mount Vernon, and Knox County through charitable giving. To assess and respond to emerging and changing community needs in the fields of: education, youth services, recreation, arts and culture, social services, and civic and community development. All grant applications are considered by the Foundation board at regularly scheduled meetings in the months of February, April, June, August, October and December. Applicants should be aware grants are awarded on a competitive basis, and that deadlines for submission of requests must be strictly observed. See the Foundations website for additional guidelines, deadline dates and two different grant applications (small and major.)
Requirements: Most grants are made to tax-exempt, private agencies classified as 501(c)3 organizations, and to public charities as defined by the U.S. Internal Revenue Service. Public schools, governmental entities and Knox County colleges and universities are also eligible to apply for Foundation grants.
Restrictions: Certain projects and organizations are not eligible for funding, including but not limited to: grants to individuals; endowment campaigns; ongoing operating expenses; religious organizations (for religious purposes); purchase of computers or other equipment (that is not part of a broader project); existing obligations, debts or liabilities; police and fire protection; staff positions for government agencies travel (when it is the proposal's primary focus) organizations that do not operate programs in Knox County, or for the benefit of Knox County residents; political campaigns.
Geographic Focus: Ohio
Samples: Knox County Historical Society, $30,000-- for a building addition project; Hospice of Knox County, $75,000-- for its new administrative and training facility; City of Mount Vernon, $3,000-- for its historic district street sign placement project.
Contact: Sam Barone, Executive Director; (740) 392-3270; fax (740) 399-5296; thefoundation@firstknox.com or sbarone@mvkcfoundation.org

Internet: http://www.mvkcfoundation.org/Grant/index.html
Sponsor: Community Foundation of Mount Vernon and Knox County
1 S Main Street, P.O. Box 1270
Mount Vernon, OH 43050

Community Foundation of Muncie and Delaware County Grants 1300
The Community Foundation of Muncie and Delaware County Grants provide funding in the following areas of interest: arts and culture, human services, economic development, education, and community betterment. The Foundation focus on: new and innovative projects and programs for which there is a demonstrable need or community benefit; capital needs of community institutions and organizations; emerging needs of Muncie and Delaware County; establishment of community priorities; monitoring of community services to avoid duplication and ineffective programs; acting as a catalyst for action and community participation. Applications are encouraged for types of projects that address one or more of the following: yield substantial benefits to the community for the resources invested; promote cooperation among agencies without duplicating services; enhance or improve institutional or organizational self-sufficiency; provide "seed money" for innovative community programs; encourage matching gifts or additional funding from other donors; and reach a broad segment of the community with needed services which are presently not provided. The Board of Directors review applications on a quarterly basis in the months of February, May, August, and November.
Requirements: First-time applicants must contact the Foundation to discuss their proposal prior to submission. Organizations are required to submit 18 copies of the application packet, but must submit different items depending on the type of funding requested. See website for submission details.
Restrictions: Requests of $25,000 or more are preferred during the first cycle of each year. The Foundation does not make grants to individuals or grants for religious purposes, budget deficits, for travel, fundraising events, endowments or projects normally the responsibility of a government agency.
Geographic Focus: Indiana
Amount of Grant: Up to USD
Contact: Suzanne Kadinger, Program Officer; (765) 747-7181; fax (765) 289-7770; skadinger@cfmdin.org or info@cfmdin.org
Internet: http://www.cfmdin.org/main/grant-seekers/
Sponsor: Community Foundation of Muncie and Delaware County
201 East Jackson Street
Muncie, IN 47305

Community Foundation of Muncie and Delaware County Maxon Grants 1301
The Community Foundation of Muncie and Delaware County Maxon Grants give back to the community by supporting worthy causes in Delaware County. Organizations are notified within eight weeks whether they have been funded.
Requirements: In addition to the online application, organizations must submit the following: the grant application cover sheet; the organization's mission statement; a list of the board of directors with their affiliations; a copy of the organization's Federal IRS tax exemption letter; and a letter of endorsement from the Board President, Principal or Chief Executive Officer. They must also submit a brief proposal that provides; project goals and objectives; implementation plan; project budget including expected revenue, in-kind contributions and other grants; staff involved in project; community benefits; and the organization's method of evaluation.
Geographic Focus: Indiana
Date(s) Application is Due: Jul 30
Amount of Grant: 5,000 - 15,000 USD
Contact: Suzanne Kadinger, Program Officer; (765) 747-7181; fax (765) 289-7770; skadinger@cfmdin.org or info@cfmdin.org
Internet: http://www.cfmdin.org/main/grant-seekers/
Sponsor: Community Foundation of Muncie and Delaware County
201 East Jackson Street
Muncie, IN 47305

Community Foundation of Randolph County Grants 1302
The Community Foundation of Randolph County seeks to bring people and resources together to enrich the lives of Randolph County, Indiana residents. The Foundation makes grants to increase the capacity of Randolph County's not-for-profit organizations to respond effectively to the needs of the community. In general, the Foundation prefers funding for: start-up costs for new programs; one-time projects or needs; capital needs beyond an applicant's capabilities and means. The Foundation also funds projects in the areas of arts and culture, civic and community development, education and libraries, environmental and historical preservation, health and human services, youth; and the elderly.
Requirements: The Foundation makes grants to tax-exempt 501(c)3 organizations operating or proposing to operate programs for the benefit of Randolph County residents. Proposals shall include: completed application cover; one original and 10 copies of completed application; one list of your organization's/agency's officers or governing body; one copy of your federal tax exemption 501(c)3 letter; one copy of your last financial statement showing income and expenses (annual report); one copy of the total project budget; one copy of all pertinent supporting information. Applications can be obtained by contacting the Foundation office or downloading the form from the Foundation's website. Application forms should be mailed or delivered to the Foundation's office, since applications cannot be submitted online.
Restrictions: The Foundation will usually not fund: individuals other than scholarships; organizations for religious or sectarian purposes; make-up of operating deficits, post-event or after-the-fact situations; endowment campaigns; for any propaganda, political or otherwise, attempting to influence legislation or intervene in any political affairs or campaigns; services such as fire, police, schools, parks, etc. that are the responsibility of government and tax supported. However, the Foundation occasionally supports special projects of these agencies.
Geographic Focus: Indiana
Date(s) Application is Due: Mar 31; Sep 30
Contact: Ruth Mills; (765) 584-9077; fax (765) 584-7710; rmills@cfrandolphcounty.org
Internet: http://cfrandolphcounty.org/cfrc/jsp/GrantCenter/ApplicationForms/main.jsp
Sponsor: Community Foundation of Randolph County
213 South Main Street
Winchester, IN 47394

Community Foundation of Riverside and San Bernardino County Community Impact Grants 1303
The Community Impact Fund was established by the Community Foundation to meet the needs of each community served by the Foundation, as determined by the grants committee and with final approval by the full Board of Directors. The Foundation responds to requests for support across a wide range of community needs and welcomes proposals through this competitive grant process. Primary interest areas are: health and human services; youth and families; art and culture; and civic and public benefit. Grant awards will range up to a maximum of $10,000. Applications should be received by May 2.
Requirements: Eligible applicants include those that: are nonprofit, public benefit organizations with evidence of tax-exempt status under Section 501(c)3 of the Internal Revenue Code and not classified as a private foundation; have been in operation for at least three years; have an annual operating budget of less than $1,000,000; and are located in, and have primary programs delivered in, Riverside or San Bernardino County. Groups or individuals that are not incorporated will be allowed to utilize a fiscal sponsor.
Restrictions: Current grantees of the Community Impact Fund are not eligible to apply. These types of expenses are not eligible: on-going operating/administrative expenses; paying off deficits or existing obligations; retroactive funding for cost already incurred; endowment, capital or annual fund appeals; capital projects (construction of buildings); sectarian programs or fraternal organizations; direct support of individuals; school/college-based extracurricular activities; research or development activities; partisan activities; event sponsorships; or re-granting purposes.
Geographic Focus: California
Date(s) Application is Due: May 2
Amount of Grant: Up to 10,000 USD
Contact: Celia Cudiamat, Executive Vice President of Grant Programs; (951) 241-7777; fax (951) 684-1911; ccudiamat@thecommunityfoundation.net
Internet: http://www.thecommunityfoundation.net/grants/grants/grant-schedule/15-grants/51-the-community-foundation-s-community-impact-fund
Sponsor: Community Foundation of Riverside and San Bernardino County
3700 Sixth Street, Suite 200
Riverside, CA 92501

Community Foundation of Riverside and San Bernardino County Irene S. Rockwell Fund Grants 1304
The Community Foundation's mission is strengthening inland Southern California through philanthropy. The Foundation does this by raising, stewarding, and distributing community assets by providing grants to nonprofit organizations, and working toward their vision of a vibrant, generous and just region—with unlimited opportunities. Field of interest funds are restricted to a specific program area or geographic area, with restrictions made by the fund donors. Specifically, the Irene S. Rockwell Fund supports programs that benefit the residents of the city of Perris, California. Preference is given to projects that: are perceived as a high need in the community; fill a gap in services; benefit a large number of residents; enhance collaboration and/or make the delivery of services more effective and efficient; have clear objectives and can document successful outcomes; expand successful programs to serve additional residents or new geographic areas; and serve remote areas or areas that have received little funding from The Community Foundation. The average award is $3,000, with grants ranging to a maximum of $7,500. The annual deadline for applications is August 12.
Requirements: Eligible applicants include those that: are nonprofit, public benefit organizations with evidence of tax-exempt status under Section 501(c)3 of the Internal Revenue Code and not classified as a private foundation; have been in operation for at least three years; and are located in, and have primary programs delivered in, the community of Perris. Groups or individuals that are not incorporated will be allowed to utilize a fiscal sponsor.
Restrictions: The following types of expenses are not eligible: on-going operating/administrative expenses; paying off deficits or existing obligations; retroactive funding for cost already incurred; endowment, capital or annual fund appeals; capital projects (construction of buildings); sectarian programs or fraternal organizations; direct support of individuals; school/college-based extracurricular activities; research or development activities; partisan activities; event sponsorships; or re-granting purposes.
Geographic Focus: California
Date(s) Application is Due: Aug 12
Amount of Grant: Up to 7,500 USD
Contact: Celia Cudiamat, Executive Vice President of Grant Programs; (951) 241-7777; fax (951) 684-1911; ccudiamat@thecommunityfoundation.net
Internet: http://www.thecommunityfoundation.net/grants/grants/grant-schedule/15-grants/55-field-of-interest-funds
Sponsor: Community Foundation of Riverside and San Bernardino County
3700 Sixth Street, Suite 200
Riverside, CA 92501

Community Foundation of Riverside and San Bernardino County James Bernard and Mildred Jordan Tucker Grants 1305

The Community Foundation's mission is strengthening inland Southern California through philanthropy. The Foundation does this by raising, stewarding, and distributing community assets by providing grants to nonprofit organizations, and working toward their vision of a vibrant, generous and just region—with unlimited opportunities. Field of interest funds are restricted to a specific program area or geographic area, with restrictions made by the fund donors. Specifically, the James Bernard and Mildred Jordan Tucker Fund benefits wheelchair users (ADA compliance projects are eligible). Preference is given to projects that: are perceived as a high need in the community; fill a gap in services; benefit a large number of residents; enhance collaboration and/or make the delivery of services more effective and efficient; have clear objectives and can document successful outcomes; expand successful programs to serve additional residents or new geographic areas; and serve remote areas or areas that have received little funding from the Community Foundation. Most recent grants ranged from $9,000 to $18,000. The annual deadline for applications is August 12.

Requirements: Eligible applicants include those that: are nonprofit, public benefit organizations with evidence of tax-exempt status under Section 501(c)3 of the Internal Revenue Code and not classified as a private foundation; have been in operation for at least three years; and are located in, and have primary programs delivered in, the communities served by the Foundation. Groups or individuals that are not incorporated will be allowed to utilize a fiscal sponsor.

Restrictions: The following types of expenses are not eligible: on-going operating/administrative expenses; paying off deficits or existing obligations; retroactive funding for cost already incurred; endowment, capital or annual fund appeals; capital projects (construction of buildings); sectarian programs or fraternal organizations; direct support of individuals; school/college-based extracurricular activities; research or development activities; partisan activities; event sponsorships; or re-granting purposes.

Geographic Focus: California
Date(s) Application is Due: Aug 12
Amount of Grant: 9,000 - 18,000 USD
Contact: Celia Cudiamat, Executive Vice President of Grant Programs; (951) 241-7777; fax (951) 684-1911; ccudiamat@thecommunityfoundation.net
Internet: http://www.thecommunityfoundation.net/grants/grants/grant-schedule/15-grants/55-field-of-interest-funds
Sponsor: Community Foundation of Riverside and San Bernardino County
3700 Sixth Street, Suite 200
Riverside, CA 92501

Community Foundation of South Alabama Grants 1306

The Community Foundation of South Alabama is the platform for building community in Baldwin, Clarke, Conecuh, Choctaw, Escambia, Mobile, Monroe and Washington counties. The Foundation provides grant support to South Alabama nonprofit organizations in four major program areas: community and civic affairs, education, arts and recreation, and social services (health and human services). Priority consideration is given to projects that clearly provide innovative responses to community needs, are collaborative in nature when appropriate, and potentially affect broad segments of the community. Types of support include general operating support, capital campaigns, endowment funds, program development, and scholarship funds. The board of directors meets annually to consider grant requests. Application forms may be obtained from the office.

Requirements: Nonprofit organizations that have a 501(c)3, government entities and religious organizations that are located in Mobile, Baldwin, Clarke, Conecuh, Washington, Choctaw, Escambia and Monroe counties are eligible.

Restrictions: Grants generally are not provided for or to individuals, recurring requests for the same purpose for which foundation grant funds have already been awarded, research that is noncommunity-related or that does not have short-range results, films, conferences and workshops, lobbying activities, and tickets to fund-raising events.

Geographic Focus: Alabama
Amount of Grant: 2,500 - 28,000 USD
Contact: Janine Phillips, Director and Program Officer; (334) 438-5591; fax (334) 438-5592; info@communityendowment.com or program@communityendowment.com
Internet: http://communityendowment.com/grants/grants.htm
Sponsor: Community Foundation of South Alabama
212 St. Joseph Street, P.O. Box 990
Mobile, AL 36601-0990

Community Foundation of South Puget Sound Grants 1307

The foundation awards grants to a variety of charitable, cultural, education, health, and welfare organizations. Grants from the unrestricted funds are considered for general operational or program support. Typical grants range from $1,000 to $7,500. Emergency grants may be considered on a case-by-case basis. Application and guidelines for grants and scholarships are available online.

Requirements: The foundation awards grants to Washington tax-exempt organizations primarily for use in Thurston, Mason, and Lewis Counties, except on instructions of the donor at the time of the gift or bequest.

Restrictions: Grants do not support religious organizations for religious purposes; individuals; annual campaigns of organizations (direct mail or special events); political or lobbying activities; organizations that discriminate based on race, creed, or ethnic group; capital campaigns for bricks and mortar or endowment funds; or for multiple year commitments.

Geographic Focus: Washington
Date(s) Application is Due: Apr 3; Oct 2
Amount of Grant: 1,000 - 7,500 USD
Samples: Rochester Organization of Families, $7,000--to help maintain ROOF's Kid's Place academic enrichment program for 50 low -income, at-risk children in the first through fifth grades; Harmony Hill, $4,500--to purchase a computer server to maintain its extensive database of clients and donors; which in turn supports their mission to improve the quality of life for those affected by cancer; Thurston Mason Project Access, $4,000--to make donated medical services available to uninsured, low-income Thurston County residents with acute, urgent conditions.
Contact: Norma Schuiteman, Executive Director; (360) 705-3340; fax (360) 705-2656; legacy@thecommunityfoundation.com
Internet: http://www.thecommunityfoundation.com/grants.php
Sponsor: Community Foundation of South Puget
111 Market Street NE, Suite 375
Olympia, WA 98501

Community Foundation of St. Joseph County Lilly Endowment Community Scholarship 1308

The Lilly Endowment Community Scholarships offer four-year full tuition plus book stipend to any Indiana college or university. Four awards are made annually to residents of St. Joseph County. All college majors are eligible to apply.

Requirements: Applicants must be residents of St. Joseph County, Indiana and graduates from an accredited Indiana high school. They must intend to pursue a full time baccalaureate course of study at an accredited public or private Indiana college or university, and possess at least a 3.8 GPA or SAT score of 1800 out of 2400. All applicants must complete and submit the Free Application for Federal Student Aid (FAFSA) to the U.S. Department of Education (see Office of Financial Aid or visit http://www.fafsa.ed.gov for more information). In order for the application to be considered complete, students must forward the Student Aid Report (SAR) listing the Expected Family Contribution (EFC) to the Community Foundation. Applicants should refer to the St. Joseph County Foundation website for current applications and deadlines.

Restrictions: Students eligible to receive a tuition benefit due to their parent(s)/guardian(s) place of employment will not be considered for this scholarship if the benefit meets or exceeds the cost of tuition. In addition, family members of those serving on the scholarship nomination committee or Community Foundation staff are deemed ineligible (defined as spouses, lineal descendants and their spouses, siblings and their spouses).

Geographic Focus: Indiana
Contact: Emily Addis; (574) 232-0041; fax (574) 233-1906; emily@cfsjc.org
Internet: http://www.cfsjc.org/scholarships/lilly_scholarship.html
Sponsor: Community Foundation of St. Joseph County
205 W Jefferson Boulevard, P.O. Box 837
South Bend, IN 46624

Community Foundation of St. Joseph County Scholarships 1309

The Community Foundation of St. Joseph County has more than 50 scholarship funds which award over $450,000 to local students each year. The scholarships were established for a variety of reasons—to assist low-income students, encourage children of employees to pursue their dreams, train medical professionals, teachers, and engineers, or in memory of a lost loved one. The Community Foundation has distributed more than $6 million in scholarships since its inception. Applicants should view the website for a variety of scholarships categorized by the school they attended in St. Joseph County or the college they plan to attend.

Geographic Focus: Indiana
Contact: Emily Addis; (574) 232-0041; fax (574) 233-1906; emily@cfsjc.org
Internet: http://www.cfsjc.org/scholarships/index.html
Sponsor: Community Foundation of St. Joseph County
205 W Jefferson Boulevard, P.O. Box 837
South Bend, IN 46624

Community Foundation of St. Joseph County Special Project Challenge Grants 1310

The Special Project Challenge Grants assist public and other 501(c)3 agencies in their efforts to serve community needs. For every $1 raised by the chosen agency, the Community Foundation will match $1. The foundation encourages projects in the following areas: community development and urban affairs; health and human services; parks, recreation, and environment; and youth and education.

Requirements: In additional to the online application, all applicants must submit the following materials online: up to a two page proposal narrative; a detailed project budget; current board roster with officers identified; fiscal year income statement; proof of nonprofit status. Application materials must be submitted via email to grants@cfsjc.org in word processing format (narrative or budget) or Microsoft Excel (budget). Hard copy applications are no longer accepted.

Restrictions: Grants are not made to fund: operational phases of established programs; endowment campaigns; religious organizations for religious purposes; individuals directly; development or public relations activities (e.g. literature, videos, etc.); retirement of debts; camperships; annual appeals or membership contributions; travel for bands, sports teams, classes, etc; j) computers (unless presented as a necessary component of larger program or objective); and post-event or after-the-fact situations.

Geographic Focus: Indiana
Date(s) Application is Due: Mar 1; Oct 1
Contact: Angela Butiste; (574) 232-0041; fax (574) 233-1906; angela@cfsjc.org
Internet: http://www.cfsjc.org/grants/sproj/special_project_grants.html
Sponsor: Community Foundation of St. Joseph County
205 W Jefferson Boulevard, P.O. Box 837
South Bend, IN 46624

GRANT PROGRAMS | 205

Community Foundation of Switzerland County Grants 1311
The Community Foundation of Switzerland County is a nonprofit organization created to make Switzerland County a better place to live for present and future generations. The Foundation gives priority to applications that focus on the basic needs of the community (food, housing, shelter, health care, clothing, personal care, and transportation). The Foundation also welcomes applications for other programs and projects that benefit Switzerland County. Organizations may request up to $5,000. Applications are reviewed monthly; there are no deadlines. The application and additional guidelines are available at the website.
Requirements: Any organization with a 501(c)3 or any organization that provides a program with charitable intent or has a fiscal agent is eligible to apply.
Geographic Focus: Indiana
Amount of Grant: Up to 5,000 USD
Contact: Pam Acton; (812) 427-9160; fax (812) 427-4033; pacton@cfsci.org
Internet: http://www.cfsci.org/
Sponsor: Community Foundation of Switzerland County
303 Ferry Street, P.O. Box 46
Vevay, IN 47043

Community Foundation of Tampa Bay Grants 1312
The Community Foundation of Tampa Bay awards creative grants from its community fund with the goal of fostering positive changes in the lives of their citizens. The areas of interest include: arts and culture, community development, education, environment and animals, and health and human services. See the Foundations website for the Grant Application Form and guidelines: http://www.cftampabay.org/nonprofit_resources/grant_app_guidelines.html
Requirements: Non-profit organizations based in and serving communities in Hillsborough, Pasco, Pinellas and Hernando counties, with a 501(c)3 status in good standing from the IRS are eligible to apply.
Restrictions: The Foundation is generally not interested in requests for: funding of ongoing operating costs; grants for capital campaigns or expenditures; tickets for any fundraising event or advertising space in programs or other publications; legislative lobbying or political campaigns; medical research; religious or sectarian purposes; loans or assistance to individuals; a multiple year funding commitment, although in exceptional cases, such funding will be considered not to exceed three years.
Geographic Focus: Florida
Date(s) Application is Due: Mar 1; Sep 1
Amount of Grant: Up to 7,500 USD
Contact: Ann Berg; (813) 282-1975; fax (813) 282-3119; aberg@cftampabay.org
Internet: http://www.cftampabay.org/nonprofit_resources/grant_app_guidelines.html
Sponsor: Community Foundation of Tampa Bay
550 North Reo Street, Suite 301
Tampa, FL 33609

Community Foundation of the Eastern Shore Community Needs Grants 1313
The Community Foundation of the Eastern Shore awards grants from its discretionary fund, known as the Lower Shore Fund for Community Needs. Monetary grants are awarded to a wide range of tax exempt organizations in Somerset, Wicomico and Worcester Counties of Maryland, whose programs benefit: health; human services; arts and culture; community affairs; environmental conservation; and historic preservation. Through this grant program, the Foundation identifies high priority needs and seeks opportunities where a relatively modest amount of grant money can make a significant difference in the community. Often the grants provide start-up or short-term funding for innovative, potentially replicable projects that meet newly identified needs or demonstrate new solutions for previously identified needs.
Requirements: To be eligible for a grant from this program, an organization must meet two fundamental criteria: it must be a governmental unit, a religious organization engaged in a non-sectarian activity, or a non-profit, tax exempt public charity, as defined in Section 501(c)3 of the Internal Revenue Code; and it must provide services to benefit the residents of the lower three counties of Maryland's Eastern Shore: Worcester, Wicomico, and Somerset. Programs are not required to serve all three counties, but regional projects are encouraged.
Restrictions: The Community Needs Grant Program does not fund: endowment funds; ongoing/operating expenses; fundraising campaigns; sectarian religious programs; playground equipment; building campaigns; operational deficits; debt retirement; capital requests; political/lobbying programs; office equipment/staff training; and school-based programs. Grants through this program are made to organizations, not individuals.
Geographic Focus: Maryland
Date(s) Application is Due: Feb 1; Aug 1
Amount of Grant: 500 - 7,500 USD
Contact: Erica N. Joseph; (410) 742-9911; fax (410) 742-6638; joseph@cfes.org
Internet: http://www.cfes.org/grants_community_needs.php
Sponsor: Community Foundation of the Eastern Shore
1324 Belmont Avenue, Suite 401
Salisbury, MD 21804

Community Foundation of the Verdugos Educational Endowment Fund Grants 1314
The Educational Endowment Fund provides financial support to innovative as well as traditional educational programs and projects in public and private schools and community organizations in the Crescenta Valley, California region. The goal is to enrich educational opportunities by supporting building, equipment, instruction, guidance, coaching, practical application, classroom activities, in-service, training and/or practice programs. Grants will support: equipment purchase, replacement & modernization; improvement to facilities including athletic facilities; printed materials; fundraising events or capital campaigns; classroom materials; public/private schools and colleges; child day care/development centers; libraries; hospitals; community enrichment projects; salaries. All qualified organizations are invited to submit grant applications once a year for grants that will be awarded at the Crescenta Valley Chamber of Commerce Installation Luncheon in early January. Average grant sizes are from $500 to $3,000. The grant application is available at the Community Foundation of the Verdugos website. The application must be submitted by October 9th.
Requirements: 501(c)3 IRS nonprofit public charities in the Crescenta Valley region of California (includes La Crescenta, Montrose, Tujunga and La Canada) and, organizations located outside the Crescenta Valley but delivering educational programs in the Crescenta Valley are eligible to apply. Please indicate the number of Crescenta Valley residents who will be served by your program. An organization may apply for no more than two grants during a grant cycle. Note, the president, school principal or leader must indicate awareness of multiple requests and each application must be for different types of support.
Restrictions: Grants will not support: uniforms or clothing; travel expenses; fiscal agents; individuals and individual scholarships; endowment funds; feasibility studies or consulting fees; advertising; research; political lobbying, voter registration or political campaigns; insurance or maintenance contracts; faith-based projects.
Geographic Focus: California
Date(s) Application is Due: Oct 9
Amount of Grant: 500 - 3,000 USD
Samples: Crescenta Valley Arts Council, $500; Verdugo Hills Hospital Foundation, $1,000; YMCA of the Foothills & Rosemont School, $750.
Contact: Edna Karinski, Executive Director; (818) 241-8040; fax (818) 241-8045; info@communityfoundationoftheverdugos.org
Internet: http://www.glendalecommunityfoundation.org/grants_endowment.php
Sponsor: Community Foundation of the Verdugos
330 Arden Avenue, Suite 130
Glendale, CA 91203

Community Foundation of the Verdugos Grants 1315
The Community Foundation of the Verdugos welcomes grant requests from public agencies and nonprofit organizations that serve the Verdugo region including Burbank, Glendale, La Canada Flintridge, La Crescenta, Montrose, and Verdugo City of California. Grant recipients must have appropriate fiscal and program accountability. Average grant amounts range from $2,500 to $10,000. Grants are provided in the following areas of interest: arts and culture; civic; health and human services (examples include projects and services for the disabled/handicapped, general community health, or homeless services); education programs; senior services; student aid (scholarships); youth services; environment and; community. Grant application and, application guidelines are available at the Foundation's website. Grants are made three times a year. Deadline for grant applications are February 1, June 1, and September 1. Scholarship deadline is March 5. Only one proposal from an organization is permitted per grant cycle. Approval or denial of your request will be provided to you in writing.
Requirements: The Foundation funds nonprofit organizations that: have IRS nonprofit status; predominantly serve the Verdugo region of California (Glendale, La Canada Flintridge, La Crescenta, Montrose, Verdugo City) and adjacent areas; capital equipment that helps to increase your organization's long-term sustainability or services to increase impact in the Verdugo region; programs (including related overhead, supplies and administrative expenses) responsive to changing community needs and which increase impact in the Verdugo region. Requests that provide significance and impact for the Verdugo region in the areas of arts and culture, children/youth, education and literacy, health and human services, civic activities, senior services, and environmental/animal related programs.
Restrictions: The Foundation does not make grants or loans to individuals unless they are students receiving scholarships. The Foundation also does not provide funds for religious or political purposes, for budget deficits, or projects that are usually the responsibility of a public agency.
Geographic Focus: California
Date(s) Application is Due: Feb 1; Mar 5; Jun 1; Sep 1
Amount of Grant: 25,000 - 10,000 USD
Contact: Edna Karinski, Executive Director; (818) 241-8040; fax (818) 241-8045; info@communityfoundationoftheverdugos.org
Internet: http://www.glendalecommunityfoundation.org/grants.php
Sponsor: Community Foundation of the Verdugos
330 Arden Avenue, Suite 130
Glendale, CA 91203

Community Foundation of Wabash County Grants 1316
The goal of the Community Foundation of Wabash County Grants is to enrich the quality of life in Wabash County, Indiana, by responding to emerging and changing needs of the community. It also seeks to support existing organizations and institutions through grants in support of the following categories: arts and culture; community and civic development; education; environment; health and human services; and recreation. Types of support include: building or renovation; continuing support; curriculum development; endowments; equipment; general operating support; matching/challenge support; program development; program evaluation; scholarship funds; scholarships to individuals; seed money; and technical assistance.
Requirements: Proposals are accepted from organizations serving Wabash County that are defined as tax exempt under Section 501(c)3 of the IRS code or have comparable status and charitable causes. Grant selection is judged on program focus; program design; benefits; reach; and organizational profile. In addition to the online application, organizations must submit the completed cover sheet, proposal budget, a list of member

of the organization's current staff and governing board; current year-end financial statement; and copy of the tax exempt IRS letter. Eight copies of the application and attached documentation are then submitted to the Foundation office for review.
Restrictions: The following are not eligible for funding: national organizations (except for local chapters serving Wabash County); annual fund campaigns; or programs or products produced for resale. Faith-based organizations may apply for program funding, provided there is not a requirement to participate in religious instruction and/or take part in religious activities.
Geographic Focus: Indiana
Date(s) Application is Due: Mar 15; Jul 15; Nov 15
Amount of Grant: Up to USD
Contact: Cathy McCarty; (260) 982-4824; fax (260) 982-8644; cathy@cfwabash.org
Internet: http://www.cfwabash.org/nonprofits-grant-information/guidelines.html
Sponsor: Community Foundation of Wabash County
218 East Main Street
North Manchester, IN 46962-0098

Community Foundation Partnerships - Lawrence County Grants 1317
The Lawrence County Community Foundation funds charitable programs and projects that serve Lawrence County. Funding priorities include education, health and human services, civic and historical affairs, recreation, and arts and culture. Previously funded projects include nutritional programs, park revitalization, programs for children of domestic violence and literacy program. The application is available at the website.
Requirements: Nonprofits organizations such as educational institutions and governmental entities are invited to apply. Priority is given to the following projects or programs: those that reach as many people as possible; are preventative rather than remedial; increase individual access to community resources; promote independence and personal achievement; examine and address the underlying causes of local problems; attract volunteer resources and support; strengthen the private, nonprofit sector; encourage collaboration with other organizations; building the capacity of the applying organizations; and offer services not already provided in the community.
Restrictions: Grants are awarded for short term projects and are not renewable. In order to maximize funding, the Foundation gives lower priority to construction projects, normal operating expenses, computer hardware, multi-year funding, re-granting, or to organizations with an existing tax-base of support.
Geographic Focus: Indiana
Date(s) Application is Due: Apr 23
Amount of Grant: Up to 25,000 USD
Contact: Hope Flores; (812) 279-2215; fax (812) 279-1984; hope@cfpartner.org
Internet: http://cfpartner.org/lccfgrantshowtoapply.htm
Sponsor: Community Foundation Partnership - Lawrence County
1324 K. Street, Suite 150
Bedford, IN 47421

Community Foundation Partnerships - Martin County Grants 1318
Community Foundation Partnership grants are award for programs that address emerging needs in Martin County. Awards are for short-term projects, usually one year, and are not renewable. Priority funding includes education, health and human services, civic and historical affairs, arts and culture, and recreation. Funding is given to projects or programs that provide the following priorities: reach as many people as possible; are preventative rather than remedial; increase individual access to community resources; promote independence and personal achievement; examine and address the underlying causes of local programs; attract volunteer resources and support; strengthen the private, nonprofit sector; encourage collaboration with other organizations; build the capacity of the applying organizations; and offer services not already provided in the community. The application is available at the website.
Requirements: The Foundation welcomes grant requests from nonprofit organizations recognized as 501(c)3 or that are affiliated with another tax exempt organization; educational institutions; government entities; and those located in or provide service to Martin County residents.
Restrictions: In order to maximize the use of funds, the Foundation gives low priority to construction projects; normal operating expenses; computer hardware; multi-year funding; re-granting; and organizations with an existing tax-base of support. The Foundation does not award grants to individuals; debt retirement; political organizations or campaigns; for-profit entities; capital campaigns; programing that promotes religious instruction or doctrine; or endowments.
Geographic Focus: Indiana
Date(s) Application is Due: Sep 14
Contact: Jason T. Jones; (812) 295-1022; fax (812) 295-1042; mccf@rtccom.net
Internet: http://cfpartner.org/mccfgrantshowtoapply.htm
Sponsor: Community Foundation Partnership - Martin County
P.O. Box 28
Loogootee, IN 47553

Community Foundation Serving Riverside and San Bernardino Counties Impact Grants 1319
The Community Foundation Serving Riverside and San Bernardino Counties awards grants aimed at meeting the needs and enhancing the lives of individuals in California's Riverside and San Bernardino Counties. The Community Impact Fund was established by The Community Foundation to meet the needs of each community served by the Foundation, as determined by our grants committee and with final approval by our full Board of Directors. The grants are awarded through a competitive grant process each year. Funding categories include health and human services--promoting access to healthcare for all residents and helping individuals and families obtain basic services to promote an improved quality of life; youth and families--enhancing opportunities that promote academic achievement and positive youth development and developing family support services that foster learning and growth; arts and culture--encouraging creative expression and providing opportunities for enjoyment of cultural activities and art forms; civic and public benefit--building a sense of community and promoting civic participation. Preference is given to projects that are perceived as a high need in the community being served; fill a gap in service; benefit a large number of residents; enhance collaboration and/or make the delivery of services more effective and efficient; have clear objectives and can document successful outcomes; expand successful programs to serve additional residents or new geographic areas within the two counties; serve remote areas or areas that have received little funding from the foundation. Contact the office for application deadlines and forms.
Requirements: Nonprofit, public benefit organizations with evidence of tax-exempt status under Section 501(c)3 of the Internal Revenue Code and nor classified as a private foundation are eligible to apply.
Restrictions: Grants are generally not made for on-going operating expenses; retroactive funding for cost already incurred; paying off deficits or existing obligations; endowment, capital fund, or annual fund appeals; capital projects, i.e. construction of new buildings; direct support of individuals; sectarian programs or fraternal organizations; event sponsorships; research or development activities; school or college-based extracurricular activities; partisan activities; or re-granting purposes.
Geographic Focus: California
Date(s) Application is Due: Feb 25
Amount of Grant: Up to 10,000 USD
Contact: Celia Cudiamat, Vice President of Grants; (951) 241-7777; fax (909) 684-1911; ccudiamat@thecommunityfoundation.net or grant-info@thecommunityfoundation.net
Internet: http://www.thecommunityfoundation.net/grants/grants/grant-schedule/15-grants/51-the-community-foundation-s-community-impact-fund
Sponsor: Community Foundation Serving Riverside and San Bernardino Counties
3700 Sixth Street, Suite 200
Riverside, CA 92501

Comprehensive Health Education Foundation Grants 1320
The foundation awards grants to support programs that address health inequities. The initial grantmaking effort will focus on Clark, Pierce, and Spokane Counties in Washington State. One-year grants of up to $20,000 each will be awarded to culturally appropriate, community-led collaborations to test their best idea on how to make it easier for people who suffer from health inequities to move more and eat healthier. Health inequities are defined as differences in the incidence, prevalence, mortality, and burden of diseases that exist for specific populations in the United States. Low-income individuals and people of color within the United States generally have higher rates of poor health and injury than those who are in higher-income groups and are Caucasian.
Requirements: 501(c)3 tax-exempt organizations located in Clark, Pierce, and Spokane Counties in Washington State and units of government that are nondiscriminatory in policy and practice regarding disabilities, age, sex, sexual orientation, race, ethnic origin, or creed are eligible.
Restrictions: Support will not be provided for building or land acquisitions; equipment or furniture purchases; endowment funds; emergency funds; grants to individuals; fellowships/scholarships; research; debt retirement; fundraising activities; general fund drives; indirect overhead; or CHEF programs or products.
Geographic Focus: Washington
Amount of Grant: 500 - 20,000 USD
Contact: Kari L. Lewis; (800) 323-2433, ext. 1899; fax (206) 824-3072; KariL@chef.org
Internet: http://www.chef.org/about/grants.php
Sponsor: Comprehensive Health Education Foundation
159 S Jackson Street, Suite 510
Seattle, WA 98104

ConAgra Foods Foundation Community Impact Grants 1321
The Community Impact Grants (CIG) program will award grants between $10,000 and $100,000 to impactful, grassroots organizations that leverage innovation and creativity to address childhood hunger and nutrition needs in communities where ConAgra Foods' employees live and work or states where 20% or more of children are food insecure. Organizations that demonstrate a strong alignment with the Foundation's giving strategies (i.e., direct service, capacity building, and advocacy) and core funding priorities have the greatest chance of receiving a grant. The CIG program is a two-step, competitive process that first requires the submission of an Letter of Inquiry (LOI) and then the subsequent completion of a full application if invited to apply for a grant. For more detailed program information and guidelines see: http://www.nourishkidstoday.org/downloads/pdf/CIGGuidelines.pdf.
Requirements: The preference of ConAgra Foods Foundation is to award Community Impact Grants to organizations located in states with a child food insecurity rate of 20% or more according to Feeding America's Child Hunger Study as well as those communities where ConAgra Foods has a significant employee presence, these states include: AR, AZ, CA, CO, D.C., FL, GA, IA, ID, IL, IN, LA, MA, MI, MN, MO, MS, NC, NE, NM, OH, OR, PA, SC, TN, TX, WA, and WI.
Restrictions: ConAgra Foods Foundation does not fund: professional or amateur sports organizations and teams, or athletic events and programs; political organizations; terrorist organizations or those not compliant with the USA Patriot Act; fundraising events; emergency funding; loans, debt reduction or operating deficits; individuals; endowments; capital campaigns (unless solicited at the founder's discretion); memorial campaigns; elementary and secondary education.

Geographic Focus: Arizona, Arkansas, California, Colorado, District of Columbia, Florida, Georgia, Idaho, Illinois, Indiana, Iowa, Louisiana, Massachusetts, Michigan, Minnesota, Mississippi, Missouri, New Mexico, North Carolina, Ohio, Oregon, Pennsylvania, South Carolina, Tennessee, Texas, Washington, Wisconsin
Date(s) Application is Due: Jan 29
Amount of Grant: 10,000 - 100,000 USD
Contact: Program Contact; foundation@conagrafoods.com
Internet: http://www.nourishkidstoday.org/about-us/application-guidelines.jsp
Sponsor: ConAgra Foods Foundation
One ConAgra Drive, CC-304
Omaha, NE 68102-5001

ConAgra Foods Foundation Nourish Our Community Grants 1322
The Foundation awards Nourish Our Community Grants to non-profit organizations based on recommendations from employees. While any organization that is working to address community needs is eligible for funding, preference will be given to those that seek to provide children and their families with access to food and nutrition education. Organizations must be located in the communities where ConAgra Foods employees live and work in order to be considered for a Nourish Our Community grant. Nourish Our Community grants typically range from $5,000 to $25,000, with an average grant of $10,000. The grant requests are reviewed by a committee representative of a cross-section of employees within the company. Organizations can receive funding for up to three consecutive years and then must postpone applying for support for one grant making cycle. Applications for Nourish Our Community grant requests are reviewed annually. The application process for the current fiscal year is June 1-May 31. Additional guidelines are available at: http://www.nourishkidstoday.org/about-us/application-guidelines.jsp
Requirements: Non-profit organizations based in AR, AZ, CA, CO, D.C., FL, GA, IA, ID, IL, IN, LA, MA, MI, MN, MO, MS, NC, NE, NM, OH, OR, PA, SC, TN, TX, WA, and WI are eligible to apply for this grant.
Restrictions: ConAgra Foods Foundation does not fund: professional or amateur sports organizations and teams, or athletic events and programs; political organizations; terrorist organizations or those not compliant with the USA Patriot Act; fundraising events; emergency funding; loans, debt reduction or operating deficits; individuals; endowments; capital campaigns (unless solicited at the founder's discretion); memorial campaigns; elementary and secondary education.
Geographic Focus: Arizona, Arkansas, California, Colorado, Connecticut, District of Columbia, Florida, Georgia, Idaho, Illinois, Indiana, Iowa, Louisiana, Maine, Massachusetts, Michigan, Minnesota, Mississippi, Missouri, New Hampshire, New Mexico, North Carolina, Ohio, Oregon, Pennsylvania, Rhode Island, South Carolina, Tennessee, Texas, Vermont, Washington, Wisconsin
Date(s) Application is Due: May 31
Amount of Grant: 5,000 - 25,000 USD
Contact: Program Contact; foundation@conagrafoods.com
Internet: http://www.nourishkidstoday.org/about-us/application-guidelines.jsp
Sponsor: ConAgra Foods Foundation
One ConAgra Drive, CC-304
Omaha, NE 68102-5001

Cone Health Foundation Grants 1323
The Cone Health Foundation invests in the development and support of activities, programs, and organizations that measurably improve the health of those in the greater Greensboro, North Carolina area. The Foundation awards grants to eligible not-for-profit organizations, government agencies, public schools and academic and/or research institutions, directing resources to four funding priorities: access to necessary health services with particular emphasis on eliminating the barriers often encountered by people in need; adolescent pregnancy prevention; HIV/AIDS and other sexually transmitted infections; and mental health and substance abuse. Current applications and guidelines are available online.
Requirements: In addition to meeting our funding priorities, your organization must serve people in the greater Greensboro area and fall into one of the following categories: not-for-profit organization; government agency; public school; or academic and/or research institution.
Restrictions: The Foundation does not support: activities that exclusively benefit the members of sectarian or religious organizations; annual fund drives; political campaigns or other partisan political activity; direct financial assistance to meet the immediate needs of individuals; endowments; or retirement of debt.
Geographic Focus: North Carolina
Contact: Antonia Monk Reaves, Senior Program Director; (336) 832-9555; fax (336) 832-9559; antonia.reaves@conehealth.com
Sandra Welch Boren, Senior Program Officer; (336) 832-9555; fax (336) 832-9559; sandra.boren@conehealth.com
Internet: http://www.conehealthfoundation.com/home/for-grantseekers/
Sponsor: Cone Health Foundation
721 Green Valley Road, Suite 102
Greensboro, NC 27408

Connecticut Community Foundation Grants 1324
The Connecticut Community Foundation serves the people of Greater Waterbury and Northwest Connecticut by supporting public and nonprofit organizations providing programs and services including those for the arts, human services, health care, environment, youth development and education. The Foundation gives priority to efforts that prevent problems, encourage community solutions or improve the organizational capability and financial stability of nonprofit agencies.
Requirements: Nonprofit organizations in Beacon Falls, Bethlehem, Bridgewater, Cheshire, Goshen, Litchfield, Middlebury, Morris, Naugatuck, New Milford, Oxford, Prospect, Roxbury, Southbury, Thomaston, Warren, Watertown, Washington, Wolcott, and Woodbury, CT, may submit applications.
Restrictions: Grants are not awarded for religious purposes, political activities, deficit financing, continuing support, fund-raising events, annual campaigns, newly established arts organizations, commissioning of new works of art, general operating support, or endowments.
Geographic Focus: Connecticut
Date(s) Application is Due: Jan 9; Mar 27; Aug 28
Amount of Grant: 8,000 - 10,000 USD
Contact: Carol O'Donnell; (203) 753-1315; fax (203) 756-3054; info@conncf.org
Internet: http://www.conncf.org/grants/grants.htm
Sponsor: Connecticut Community Foundation
43 Field Street
Waterbury, CT 06702

Connecticut Health Foundation Health Initiative Grants 1325
The foundation awards grants to organizations and institutions that directly respond to its current priority areas and result in improving the health status of Connecticut's underserved and unserved populations. Program priorities include children???s mental health--projects related to children???s mental health, including research, community grants, creating resources for clinical effective practices, and parent advocacy groups; oral health--improving oral health care access, quality, and utilization; and racial and ethnic health disparities--improving the diversity of the health care workforce, and increasing cultural competency in the existing workforce. The foundation awards two major types of grants: strategic and responsive. Application and guidelines are available online.
Requirements: Connecticut state and local units of government, community health centers, health advocacy organizations, community-based organizations, community and cultural groups, schools, and faith-based organizations are eligible. Applicants must have IRS 501(c)3 tax-exempt status or be public entities. Unincorporated organizations may apply through 501(c)3 fiscal agents.
Restrictions: Foundation grants do not support awards to individuals; construction of buildings; capital projects, endowments, or chairs associated with universities, and medical schools; conferences (unless part of a greater project or program); projects that do not benefit Connecticut residents; lobbying or influencing the outcomes of a proposed piece of legislation or election; and indirect cost for discretionary grants.
Geographic Focus: Connecticut
Date(s) Application is Due: Mar 15; Jun 15; Sep 15; Dec 15
Amount of Grant: 50,000 - 200,000 USD
Samples: 1000 Friends of Connecticut, $10,000 to improve public health through the Smart Growth Education and Communications Campaign; Asian Family Services, Inc., Hartford $25,000 to integrate health-related policy issues into the community dialogue during the political season in 2006; Asian Family Services, Inc., Hartford, $10,000 to merge their services with Community Renewal Team, an anti-poverty multi-service agency that serves families and people throughout the Connecticut River Valley; Association of Yale Alumni in Public Health, New Haven, $30,000 to conduct an assessment of the policy, practices and procedures for recruiting and retaining historically under-represented faculty members at the Yale University School of Public Health.
Contact: Onell Jesus Calderas, Grants Administrator; (860) 224-2200; fax (860) 224-2230; onell@cthealth.org
Internet: http://www.cthealth.org/matriarch
Sponsor: Connecticut Health Foundation
74A Vine Street
New Britain, CT 06052

Connelly Foundation Grants 1326
To achieve its mission to foster learning and improve the quality of life, Connelly Foundation provides grants toward costs associated with programs, direct services, general operations and capital projects to non-profit organizations and institutions working in the following fields: education; health and human services; arts and culture and civic enterprise. The Foundation supports non-profits with strong leadership, sound ideas, future viability, and attainable and well defined goals. It directs its philanthropy toward 501(c)3 organizations and institutions based in and serving Philadelphia and the counties of Bucks, Chester, Delaware, Montgomery and the City of Camden. The Foundation values the proposal process. Given its preference to review a comprehensive package as a primer for discussion, letters of inquiry or requests for pre-proposal discussions are not deemed necessary. Written proposals from nonprofit organizations are accepted and reviewed by the Connelly Foundation throughout the year, there are no deadlines.
Requirements: 501(c)3 organizations and institutions based in and serving in Philadelphia and its surrounding counties of Bucks, Chester, Delaware, and Montgomery in Pennsylvania and in the City of Camden, New Jersey are eligible to apply. There are determined parameters to the Foundation's financial support. It provides non-profit organizations only one grant within a twelve month period. As a general practice, it does not fund advocacy, annual appeals, charter schools, conferences, environmental projects, feasability or planning studies, general solicitations, historic preservation projects, national organizations, organizations focused on a single disease, public schools or research.
Restrictions: The foundation does not award grants to individuals, or political or national organizations; Nor does it respond to annual appeals or general letters of solicitation.
Geographic Focus: Pennsylvania
Contact: Emily C. Riley, Executive Vice President; (610) 834-3222; fax (610) 834-0866; info@connellyfdn.org
Sponsor: Connelly Foundation
1 Tower Bridge, Suite 1450
West Conshohocken, PA 19428

ConocoPhillips Foundation Grants 1327

The foundation makes charitable grants (primarily in the communities where it has operations) in support of education, medical programs, human services, civic, cultural, youth, and other services. Contributions will be considered for organizations such as: federated organizations; educational institutions, both public and private, primarily at the college level; youth organizations; hospital and medical facilities and programs such as hospital buildings and equipment, improvement campaigns and other medical facilities; cultural organizations; civic services; and human service organizations. Applicants may download the application form from the Web site. There are no application deadlines.

Requirements: 501(c)3 tax-exempt organizations and, where appropriate 170(c) organizations, and international nonprofit organizations are eligible. Proof of the exemption must be submitted with grant applications.

Restrictions: Grants do not support religious organizations for religious purposes; war veterans and fraternal service organizations; endowment funds; national health organizations and programs; grants or loans to individuals; fund-raising events; corporate memberships or contributions to chambers of commerce, taxpayer associations and other bodies whose activities are expected to directly benefit the company; or political organizations, campaigns and candidates.

Geographic Focus: Alaska, California, Illinois, Louisiana, Montana, New Jersey, Oklahoma, Pennsylvania, Texas, Washington
Date(s) Application is Due: Aug 1
Amount of Grant: Up to 46,000,000 USD
Contact: Community Relations Manager
Internet: http://www.conocophillips.com/about/Contribution+Guidelines/index.htm
Sponsor: ConocoPhillips Foundation Grants
600 N Dairy Ashford
Houston, TX 77079

Conquer Cancer Foundation of ASCO Career Development Award 1328

The Career Development Award (CDA) is a research grant that provides funding to clinical investigators who have received their initial faculty appointment to establish an independent clinical cancer research program. The cancer research must have a patient-oriented focus, including a clinical research study and/or translational research involving human subjects. Proposals with a predominant focus on in vitro or animal studies (even if clinically relevant) are not allowed. CDA is a three-year cancer grant totaling $200,000, paid in three annual increments to the awardee's institution.

Requirements: The applicant must: be a physician (MD, DO, or international equivalent) who is within the first to third year of a full-time, primary faculty appointment in a clinical department at an academic medical institution at the time of grant submission; have a valid, active medical license at the time of application; have completed productive postdoctoral research and demonstrated the ability to undertake independent investigator-initiated clinical research; be an ASCO member (Full Member or International Corresponding) or have submitted a membership application with the grant application; be able to commit more than 50% of full-time effort in research (applies to total research, not just the proposed project) during the award period; have a mentor from the sponsoring institution who must provide a letter of support. If the mentor is not an ASCO Member, a supporting letter from an ASCO Member from the sponsoring institution must be included; be up-to-date and in compliance with all requirements (e.g. progress reports, final reports, budget summaries, IRB approvals, etc.) of any past grants received from Conquer Cancer Foundation. Applicants holding Instructor/Lecturer appointments are eligible to apply; however, the institutional letter of support must include information about the institution's commitment to support the applicant for the duration of the grant period. The full application must be submitted by 11:59PM EDT on the deadline date.

Restrictions: Eligible physicians are allowed to hold only one grant from the Conquer Cancer Foundation at a time. Applicants should not have any current career development awards (i.e., K23, K08, or any other type of career development award) and have not been a Principal Investigator on any large project grants (i.e., R01 or international equivalent, or private foundation grants). Past recipients of training fellowships (i.e., Young Investigator Award or an F32 grant) are eligible. Applicants with institutional KL2/K12 grants are eligible to apply but will be asked to relinquish their institutional grant at the start of the CDA grant period.

Geographic Focus: All States
Date(s) Application is Due: Sep 25
Amount of Grant: 200,000 USD
Contact: Eileen Melnick; (571) 483-1700; Grants@conquercancerfoundation.org
Internet: http://www.conquercancerfoundation.org/cancer-professionals/funding-opportunities/career-development-award
Sponsor: Conquer Cancer Foundation
2318 Mill Road, Suite 800
Alexandria, VA 22314

Conquer Cancer Foundation of ASCO Comparative Effectiveness Research Professorship in Breast Cancer 1329

The Translational Research Professorship is designed to provide flexible funding to outstanding translational researchers who have made, and are continuing to make, significant contributions that have changed the direction of cancer research and who provide mentorship to future translational researchers. The award is intended to support qualified individuals who are dedicated to bringing advances in basic sciences into the clinical arena and to mentoring other translational researchers. The award totals $500,000 and is paid in five annual increments of $100,000 made on or about July 1 of each year of the grant term. The award is not designed to cover the total cost of a research project or the investigator's entire compensation, but can be used at the discretion of the awardee to support their overall research activities.

Requirements: Applicants must: have an MD, DO, PhD or equivalent degree; have the rank of full professor (or equivalent); have a full-time faculty appointment at an academic medical center; have made significant contributions that have changed the direction of breast cancer research; be serving as a research mentor to one or more researcher(s) in training, and must be planning to continue to provide leadership in this area throughout the award period; lead a research team in conducting research on comparative effectiveness in breast cancer; be an active member of ASCO or have submitted a membership application with the grant application; commit to spending 75% of time during the award period dedicated to research (applies to total research, not just the proposed project) including leading a team of researchers (in the lab and/or the clinic) and mentoring physician-scientists. The sponsoring institution must be a not-for-profit institution.

Geographic Focus: All States
Date(s) Application is Due: Dec 22
Amount of Grant: 500,000 USD
Contact: Eileen Melnick, Director, Grants and Awards; (571) 483-1700; Grants@conquercancerfoundation.org
Internet: http://www.conquercancerfoundation.org/cancer-professionals/funding-opportunities/research-professorships
Sponsor: Conquer Cancer Foundation
2318 Mill Road, Suite 800
Alexandria, VA 22314

Conquer Cancer Foundation of ASCO Drug Development Research Professorship 1330

The Drug Development Research Professorship (DDRP) is a five-year award designed to provide flexible cancer research funding to outstanding researchers who have made, and are continuing to make, significant contributions that have changed the direction of cancer research and who provide mentorship to future researchers. This award is intended to support qualified individuals who are dedicated to bringing advances into the clinical arena by exploring new and promising therapeutic compounds that will lead to improved treatments, and to mentoring other researchers in the scientific and regulatory aspects of drug development. The total award amount is $500,000 payable on July 1 in annual increments of $100,000 over five years. The award is not designed to cover the total cost of a research project or the investigator's entire compensation, but can be used at the discretion of the awardee to support their overall research activities.

Requirements: Applicants must meet the following criteria at the time of grant submission: be a physician (MD, DO, or international equivalent) working in any country with a full-time faculty appointment in a clinical department at an academic medical center; currently hold the rank of full professor (or equivalent) at an academic medical center; made significant contributions to the development of new therapies for cancer; be serving as a research mentor to researcher(s) in training, and must be planning to continue to provide leadership in the area of drug development throughout the award period; lead a research team in the conduct of drug development research that includes clinical trials; be a full member of ASCO in good standing or submit a membership application with grant application; expect to spend 75% of time during the award period dedicated to research and drug development activities, leading a team of researchers (both in the lab and/or in the clinic) and mentoring junior faculty and trainees on research and drug development. Individuals with an endowed professorship from their own institution are eligible to apply for this award. Individuals with administrative responsibilities (such as a Division or Department Chair) must clearly document their time commitment for performing research activities and mentoring activities. The Drug Development Research Professorship is open to international applicants.

Geographic Focus: All States
Date(s) Application is Due: Dec 1
Amount of Grant: 500,000 USD
Contact: Eileen Melnick; (571) 483-1700; Grants@conquercancerfoundation.org
Internet: http://www.conquercancerfoundation.org/cancer-professionals/funding-opportunities/research-professorships
Sponsor: Conquer Cancer Foundation
2318 Mill Road, Suite 800
Alexandria, VA 22314

Conquer Cancer Foundation of ASCO Improving Cancer Care Grants 1331

The Improving Cancer Care Grant (ICCG), funded by Susan G. Komen for the Cure, provides extramural cancer research funding to address important issues regarding access to healthcare, quality of care, and delivery of care, with general applicability to breast cancer. The goal of this cancer grant program is to encourage multi-disciplinary research that will have a major impact on cancer care, with general applicability in the breast cancer arena. Proposals must be focused on developing solutions to current problems, not just describing disparities in care that currently exist. Proposals must be research-focused, with specific aims and hypothesis, not a request for support of a program. ICCG is a three-year cancer research grant totaling $1.35 million, paid in three annual increments of $450,000.

Requirements: Eligible research teams: will focus on implementing and/or evaluating new solutions to existing problems in quality of, access to, and delivery of care, with general applicability to breast cancer; will be led by a single Principal Investigator, who must be an active ASCO member (or have submitted a membership application) with an MD, DO, PhD or equivalent degree; will have a multidisciplinary team of investigators that may include clinicians, nurses, pharmacists, statisticians, epidemiologists, information technologists, and other research experts; will be allowed to obtain expertise not represented in the core team through consultants and/or sub-contracts. All categories

of active ASCO members are eligible; will have a multidisciplinary team of investigators that may include clinicians, nurses, pharmacists, statisticians, epidemiologists, information technologists, and other research experts; will be allowed to obtain expertise not represented in the core team through consultants and/or sub-contracts.
Geographic Focus: All States
Date(s) Application is Due: Dec 21
Amount of Grant: 1,350,000 USD
Contact: Eileen Melnick; (571) 483-1700; Grants@conquercancerfoundation.org
Internet: http://www.conquercancerfoundation.org/cancer-professionals/funding-opportunities/improving-cancer-care-grant
Sponsor: Conquer Cancer Foundation
2318 Mill Road, Suite 800
Alexandria, VA 22314

Conquer Cancer Foundation of ASCO International Development and Education Award in Palliative Care   1332

The International Development and Education Award in Palliative Care (IDEA-PC) provides medical education in palliative care, assists with career development, and helps establish strong relationships with leading ASCO members in the field of palliative care who serve as scientific Mentors to each recipient. IDEA-PC Recipients are expected to share the knowledge and training they receive through the program with colleagues in their home countries once they return. The IDEA-PC program enables Recipients to attend sessions at the ASCO Annual Meeting that address symptom management, communication, end-of-life care, and other relevant topics. The program also includes a visit to the Mentor's U.S. or Canadian institution to experience palliative care performed in a multidisciplinary environment, and provides an opportunity to cultivate long-term professional relationships that will help improve palliative care in the Recipient's home countries.
Requirements: Applicants must: have a current passport that does not expire before December; be a current resident of a country classified by the World Bank as Low-Income, Lower-Middle-Income, or Upper-Middle-Income, and have limited resources to attend the ASCO Annual Meeting; be a full Member, Member in Training, or International Corresponding member of ASCO or willing to submit an application for ASCO membership as part of the IDEA application process; be less than ten years past teir oncology program training; be fluent in English (both writing and speaking). Although applicants are not required to submit an abstract for the ASCO Annual Meeting, applicants who do submit an abstract will be given the strongest consideration.
Restrictions: Applicants cannot: have completed more than one academic year of "formal training" in a country classified by the World Bank as High-Income; have previously received the IDEA award or its predecessor, the International Travel Grant (ITG); have previously attended more than one ASCO meeting in the past five (5) years.
Geographic Focus: All States, All Countries
Date(s) Application is Due: Jan 8
Amount of Grant: 1,160 USD
Contact: Eileen Melnick; (571) 483-1700; Grants@conquercancerfoundation.org
Internet: http://www.conquercancerfoundation.org/cancer-professionals/funding-opportunities/international-development-and-education-award
Sponsor: Conquer Cancer Foundation
2318 Mill Road, Suite 800
Alexandria, VA 22314

Conquer Cancer Foundation of ASCO International Development and Education Awards   1333

The International Development and Education Award (IDEA) provides support for early-career oncologists in developing countries and facilitates the sharing of knowledge between these oncologists and ASCO members. The program pairs IDEA Recipients with a leading ASCO member "Mentor" and enables Recipients to attend the ASCO Annual Meeting, participate in a post-meeting visit to their Mentor's institution, and develop long-term relationships to improve cancer care in their countries and inform ASCO programs in developing countries. The award is not intended to completely cover all travel or daily expenses during a Recipient's trip but will cover much of the expense.
Requirements: Applicants must: have a current passport that does not expire before December; be a current resident of a country classified by the World Bank as Low-Income, Lower-Middle-Income, or Upper-Middle-Income, and have limited resources to attend the ASCO Annual Meeting; be a full Member, Member in Training, or International Corresponding member of ASCO or willing to submit an application for ASCO membership as part of the IDEA application process; be less than ten years past teir oncology program training; be fluent in English (both writing and speaking). Although applicants are not required to submit an abstract for the ASCO Annual Meeting, applicants who do submit an abstract will be given the strongest consideration.
Restrictions: Applicants cannot: have completed more than one academic year of "formal training" in a country classified by the World Bank as High-Income; have previously received the IDEA award or its predecessor, the International Travel Grant (ITG); have previously attended more than one ASCO meeting in the past five (5) years.
Geographic Focus: All States, All Countries
Date(s) Application is Due: Jan 8
Amount of Grant: 1,160 USD
Contact: Eileen Melnick; (571) 483-1700; Grants@conquercancerfoundation.org
Internet: http://www.conquercancerfoundation.org/cancer-professionals/funding-opportunities/international-development-and-education-award
Sponsor: Conquer Cancer Foundation
2318 Mill Road, Suite 800
Alexandria, VA 22314

Conquer Cancer Foundation of ASCO International Innovation Grant   1334

The International Innovation Grant provides research funding in support of novel and innovative projects that can have a significant impact on cancer control in low- and middle-income countries. The International Innovation Grant is a one-year research grant of up to $20,000 that is awarded and paid directly to a nonprofit organization or governmental agency in a low-income or middle-income country. The grant may be used by the organization only for the approved, budgeted costs of the research project. Each grant will have a Principal Investigator who is an ASCO member, is affiliated with the Grantee Organization, and is a resident of the low-income or middle-income country. International Innovation Grants are hypothesis-driven research grants that fund a specific research project that may result in the discovery of new knowledge about how to advance cancer control in a low- or middle-income setting. It is anticipated that novel approaches and clinical designs proposed for this grant may differ from what would be considered standard practice within high-income settings. Grantee Organizations and Principal Investigators will be expected to share and disseminate the knowledge gained during their research project.
Requirements: The Applicant Organization must: be an organization with a charitable purpose registered as a not-for-profit with the relevant national authority or must be a government agency; be located in a country categorized by the World Bank as Low-Income, Lower-Middle-Income or Upper-Middle Income; have been operating for at least one full year, have an acceptable management structure and processes in place, and be solvent with or without the support of the International Innovation Grant; should have experience in carrying out activities with tangible outcomes. The Principal Investigator must: be a member of ASCO or have submitted a membership application with the grant application; be a citizen or permanent resident of a country defined by the World Bank as low-income or middle-income, and currently residing in that country; be affiliated with the applicant organization.
Geographic Focus: All States, All Countries
Date(s) Application is Due: Oct 9
Amount of Grant: Up to 20,000 USD
Contact: Eileen Melnick; (571) 483-1700; Grants@conquercancerfoundation.org
Internet: http://www.conquercancerfoundation.org/cancer-professionals/funding-opportunities/international-innovation-grant
Sponsor: Conquer Cancer Foundation
2318 Mill Road, Suite 800
Alexandria, VA 22314

Conquer Cancer Foundation of ASCO Medical Student Rotation Grants   1335

This funding opportunity designed to facilitate the recruitment and retention of individuals from populations underrepresented in medicine to cancer careers and increase access to quality care for underserved communities. The Medical Student Rotation program was initially developed as part of the Conquer Cancer Foundation/ASCO/Susan G. Komen for the Cure Quality of Care Initiative. The program provides 8- to 10-week clinical or clinical research oncology rotations for U.S. medical students from populations underrepresented in medicine who are interested in pursuing oncology as a career. Recipients receive a $5,000 stipend for the rotation plus $1,500 for future travel to the ASCO Annual Meeting, An additional $2,000 will be provided to support the student's mentor.
Requirements: Applicants must be enrolled in an MD or DO, U.S. medical school program and be of an underrepresented population (For the purposes of this program, underrepresented population is defined according to the Association of American Medical Colleges definition, currently specified as those racial and ethnic populations that are underrepresented in the field of medicine relative to their numbers in the general population. Races/ethnicities that have been identified include but are not limited to: American Indian/Alaska Native, Black/African American, Hispanic/Latino, and Native Hawaiian/Pacific Islander). Candidates must be U.S. citizens, U.S. nationals, or permanent residents. Candidates also must demonstrate an interest in pursuing oncology as a career and have a record of good academic standing. Medical students may choose a mentor and rotation setting on their own, or they can request the Conquer Cancer Foundation assistance in pairing them with a mentor. Mentors are responsible for ensuring that the student receives consistent guidance and supervision throughout the rotation, either directly or by another member of the oncology practice or staff able to perform this role. The mentor is responsible for defining the goals and objectives of the rotation for the student, orienting the student to the office or hospital setting, and for providing regular feedback to the student.
Geographic Focus: All States
Date(s) Application is Due: Dec 17
Amount of Grant: Up to 8,500 USD
Contact: Eileen Melnick; (571) 483-1700; Grants@conquercancerfoundation.org
Internet: http://www.conquercancerfoundation.org/cancer-professionals/funding-opportunities/medical-student-rotation-underrepresented-populations
Sponsor: Conquer Cancer Foundation
2318 Mill Road, Suite 800
Alexandria, VA 22314

Conquer Cancer Foundation of ASCO Resident Travel Award for Underrepresented Populations   1336

The Conquer Cancer Foundation of ASCO Resident Travel Award for Underrepresented Populations (RTA) is designed to facilitate the recruitment and retention of individuals from underrepresented populations in medicine to cancer careers and increase access to quality care for underserved communities. RTA provides financial support for residents from underrepresented populations to attend ASCO's Annual Meeting. The intention of this award is to attract residents from underrepresented populations to a possible career in one of the oncology specialties. This travel award will allow the recipients to travel to the ASCO Annual Scientific Meeting, where they will have an opportunity to meet oncologists and to understand

the career possibilities in this area. The Resident Travel Award for Underrepresented Populations includes a $1,500 travel advance, complimentary Annual Meeting registration, and access to the Annual Meeting housing block and ASCO's travel agent.
Requirements: Applicants must be enrolled in an ACGME-accredited residency program required for future training in a cancer related subspecialty (i.e. Internal Medicine considering Med/Onc, Surgery considering Surg/Onc) and be of an underrepresented population as defined by the program eligibility criteria. (For the purposes of this program, underrepresented population is defined according to the Association of American Medical Colleges definition, currently specified as those racial and ethnic populations that are underrepresented in the field of medicine relative to their numbers in the general population. Races and ethnicities that have been identified include but are not limited to: American Indiana/Alaska Native, Black/African American, Hispanic/Latino, and Native Hawaiian/Pacific Islander.) Candidates must be U.S. citizens, U.S. nationals or permanent residents. Candidates must have a record of good academic standing.
Geographic Focus: All States
Date(s) Application is Due: Dec 13
Amount of Grant: 1,500 USD
Contact: Eileen Melnick; (571) 483-1700; Grants@conquercancerfoundation.org
Internet: http://www.conquercancerfoundation.org/cancer-professionals/funding-opportunities/resident-travel-award-underrepresented-populations
Sponsor: Conquer Cancer Foundation
2318 Mill Road, Suite 800
Alexandria, VA 22314

Conquer Cancer Foundation of ASCO Translational Research Professorships 1337
The Translational Research Professorship is designed to provide flexible funding to outstanding translational researchers who have made and are continuing to make significant contributions that have changed the direction of cancer research and who provide mentorship to future translational researchers. The award is intended to support qualified individuals who are dedicated to bringing advances in basic sciences into the clinical arena and to mentoring other translational researchers. The award totals $300,000 payable on July 1 in annual increments of $100,000 over three years.
Requirements: The applicant must meet the following criteria at the time of grant award (July 1): be a physician (MD, DO or international equivalent) with a full-time faculty appointment in a clinical department at an academic medical center; currently hold the rank of full professor (or equivalent) at an academic medical center. If applicant does not hold the rank of full professor, but meets all other eligibility criteria, he/she must provide written explanation of why his/her current rank at their institution should be considered equivalent; made significant contributions that have changed the direction of cancer research; be serving as a research mentor to one or more translational researcher(s) in training, and must be planning to continue to provide leadership in this area throughout the award period; lead a research team in the conduct of translational research; be a member of the American Society of Clinical Oncology (ASCO); expect to spend 75% of time during the award period dedicated to translational research activities, leading a team of translational researchers (both in the lab and in the clinic) and mentoring physician-scientists. Individuals with an endowed professorship from their own institution are eligible to apply for this award. Individuals with administrative responsibilities (such as a Division or Department Chair) must clearly document their time commitment for performing research activities and mentoring activities. The sponsoring institution must be a not-for-profit institution.
Geographic Focus: All States
Date(s) Application is Due: Jan 13
Amount of Grant: 300,000 USD
Contact: Eileen Melnick; (571) 483-1700; Grants@conquercancerfoundation.org
Internet: http://www.conquercancerfoundation.org/cancer-professionals/funding-opportunities/research-professorships
Sponsor: Conquer Cancer Foundation
2318 Mill Road, Suite 800
Alexandria, VA 22314

Conservation, Food, and Health Foundation Grants for Developing Countries 1338
The Conservation, Food, and Health Foundation's geographic focus is the developing world. Through grants to support research and through targeted grants to help solve specific problems, the foundation helps build capacity within developing countries in three areas of interest: conservation, food, and health. The foundation concentrates its grant-making on research, technical assistance, and training projects of benefit to the Third World; favors grants for pilot projects and special programs that have a potential for replication; prefers to support projects that employ and/or train personnel from the developing world; and favors research concerning problems of importance to the developing world. The foundation has two 4-month funding cycles each year. Concept papers received by January 1 will be considered for eligibility for the March 1 full proposal deadline; concept papers received by July 1 will be considered for eligibility for the September 1 full proposal deadline. Full proposals are by invitation. Concept applications must be submitted through the foundation's online application system. Detailed guidelines, instructions, and faqs are provided at the website.
Requirements: The following organizations are eligible to apply: 501(c)3 organizations that are not private foundations under section 509(a) of the United States Internal Revenue Code; "501(c)3 equivalent" foreign or domestic government units; and nongovernmental foreign organizations which can provide secure evidence of their nongovernmental status and charitable purpose.
Restrictions: The foundation does not provide support for buildings, land purchases, or vehicles; quantity purchases of durable medical equipment; endowments or fundraising activities; famine or emergency relief; direct medical care or treatment; feeding or food distribution programs; films, videos, or web-site production; scholarships, fellowships, or travel grants; conferences; re-granting through intermediaries; general operating support; or individuals (however, the foundation may support an individual engaged in research on a problem of significance to the developing world where the research is sponsored by an established, nonprofit organization such as an educational institution and conducted in close partnership with a local nongovernmental organization). The foundation does not consider the states of the former Soviet Union or former Eastern Bloc countries as within its geographic focus.
Geographic Focus: All States, All Countries
Date(s) Application is Due: Jan 1; Jul 1
Amount of Grant: Up to 25,000 USD
Samples: Fundacion Natura Bolivia, Santa Cruz, Bolivia, $29,242 - to develop a training program that helps municipal watershed authorities in Bolivia to develop payment schemes that help upstream landowners protect forests valuable to downstream communities; One Acre Fund, Falcon Heights, Minnesota, $25,000 - to refine a training, financing, education, and support program that helps rural Kenyan and Rwandan farmers double their harvests and income; Hohenheim University, Stuttgart, Germany, $23,813 - for a study in collaboration with Veld Products Research & Development in Botswana to assess how strip intercropping wild grass with sorghum-cowpea contributes to the field water balance as well as effects on field-water-use efficiency.
Contact: Prentice Zinn; (617) 391-3091 or skype: prentice.zinn; pzinn@gmafoundations.com
Internet: http://cfhfoundation.grantsmanagement08.com/?page_id=5
Sponsor: Conservation, Food, and Health Foundation
77 Summer Street, 8th Floor
Boston, MA 02110-1006

CONSOL Energy Community Development Grants 1339
Through its grant program, the CONSOL Energy Corporation supports employees and their families in its communities of operation. Its overall mission is to become a vital part of the communities where it is located. The Corporation strives to be involved in activities that enjoy wide community support and that benefit the most people, and specifically focus its efforts in the areas of public safety, youth organizations, community organizations, and arts and culture. It the area of community development, CONSOL Energy creates collaboration and partnerships in communities where its employees live and work. The Corporation is involved in a variety of community organizations throughout the year, supporting the local chambers, fairs, festivals, food banks, and more.
Restrictions: CONSOL Energy does not fund: sectarian or denominational religious organizations; individuals; fraternal organizations; organizations outside of the company's service areas; profit-making entities; or individual disease-related fundraising organizations.
Geographic Focus: Ohio, Pennsylvania, Virginia, West Virginia
Contact: Kate O'Donovan, Director of Public Affairs; (724) 485-3097 or (724) 503-8223; kateo'donovan@consolenergy.com
Internet: http://www.consolenergy.com/about-us/corporate-responsibility/consol-in-the-community.aspx
Sponsor: CONSOL Energy
1000 Consol Energy Drive
Canonsburg, PA 15317-6506

CONSOL Military and Armed Services Grants 1340
Through its grant program, the CONSOL Energy Corporation supports employees and their families in its communities of operation. Its overall mission is to become a vital part of the communities where it is located. The Corporation strives to be involved in activities that enjoy wide community support and that benefit the most people, and specifically focus its efforts in the areas of public safety, youth organizations, community organizations, and arts and culture. It the area of military and armed services, CONSOL Energy is a proud supporter of active military and veterans. Through organizations such as Operation Troop Appreciation and Dark Horse Benefits, CONSOL honors the armed forces for their courage and dedication to the United States. The Corporation also involves employees in volunteer activities and fundraising efforts through organizations such as Operation Troop Appreciation.
Restrictions: CONSOL Energy does not fund: sectarian or denominational religious organizations; individuals; fraternal organizations; organizations outside of the company's service areas; profit-making entities; or individual disease-related fundraising organizations.
Geographic Focus: Ohio, Pennsylvania, Virginia, West Virginia
Contact: Kate O'Donovan; (724) 485-3097; kateo'donovan@consolenergy.com
Internet: http://www.consolenergy.com/about-us/corporate-responsibility/consol-in-the-community.aspx
Sponsor: CONSOL Energy
1000 Consol Energy Drive
Canonsburg, PA 15317-6506

Constantin Foundation Grants 1341
The foundation awards grants to nonprofit organizations in Texas, with an emphasis on higher and other education, including secondary school education, vocational education, and adult continuing education. Grants also support human, family, and social service agencies; hospitals and health organizations, including substance abuse; community development; and arts and culture. Grants are awarded for continuing support, capital campaigns, building construction/renovation, equipment acquisition, challenge/matching funds, and to develop programs. Request guidelines prior to applying.
Requirements: Nonprofits in Dallas County, TX, may submit applications.
Restrictions: No support for tax-supported institutions, theater groups, churches, debt retirement, political organizations or second party requesters. No grants to individuals, or for endowments, research, debt retirement, operations, research, special events, fundraisers, or second party requests; no loans.
Geographic Focus: Texas

Date(s) Application is Due: Sep 15
Amount of Grant: 1,000 - 100,000 USD
Contact: Cathy Doyle, Executive Director; (214) 522-9300 or (214) 522-9305; fax (214) 521-7023; constantinfdn@sbcglobal.net
Sponsor: Constantin Foundation
4809 Cole Avenue, LB 127
Dallas, TX 75205-3578

Constellation Energy Corporate Grants 1342
Constellation Energy provides its philanthropic resources to non-profit organizations that make an impact in these key focus areas: energy Initiatives; environment; education; economic development. Applications are accepted at various times throughout the year. Grant, Sponsorship, Banner Hanging, and In-Kind requests under $10,000 are reviewed on a rolling basis. The Corporate Contributions Committee meets bi-annually, in May and October, to review significant financial grant requests of $10,000 or more. Grant requests should be submitted by April 1 and September 1 respectively. Additional guidelines and the online applications are available at the companies website.
Requirements: The company makes donations to 501(c)3 tax-exempt, nonprofit organizations.
Restrictions: Constellation Energy does not make grants to: individuals; churches or religious causes; individual schools; athletic teams or events; programs located outside Constellation Energy communities.
Geographic Focus: All States
Date(s) Application is Due: Apr 1; Sep 1
Contact: Larry McDonnell, Director; (401) 470-7433; media@constellation.com
Internet: http://www.constellation.com/portal/site/constellation/menuitem.94939662e40191875fb60610025166a0/
Sponsor: Constellation Energy Corporate
100 Constellation Way, Suite 1000C
Baltimore, MD 21202

Consumers Energy Foundation 1343
Since its creation in 1990, the Consumers Energy Foundation has touched countless lives and communities through it's grant programs, corporate giving and employee volunteers. The Foundation accepts grant applications from nonprofit organizations for innovative projects and activities creating measurable impact in five areas: Social Welfare, Michigan Growth and Environmental Enhancement, Education, Community and Civic, and Culture/Arts. The Foundation's areas of support include, operating budgets and capital funds.
Requirements: The Consumers Energy Foundation provides financial support primarily to Michigan organizations classified by the Internal Revenue Service as tax-exempt under section 501(c)3 of the Internal Revenue Code.
Restrictions: The following are ineligible for funding: individuals; individual scholarships; individual sponsorship related to fund-raising; organizations that do not qualify as charitable organizations as defined by the Internal Revenue Service; organizations that practice discrimination on the basis of sex, age, height, weight, marital status, race, religion, sexual orientation, creed, color, national origin, ancestry, disability, handicap, or veteran status; organizations whose operating activities are already supported by the United Way (except when the request is approved by the appropriate community United Way organization); political organizations and political campaigns; religious organizations when the contribution will be used for denominational or sectarian purposes; labor or veterans organizations; fraternal orders; social clubs; sports tournaments; talent or beauty contests; loans for small business; debt reduction campaigns.
Geographic Focus: Michigan
Amount of Grant: 500 - 10,000 USD
Samples: Saginaw Basin Land Conservancy, $6,000--for educational outreach efforts supporting wetlands preservation and land conservation within the Saginaw Bay Watershed; American Red Cross Mid-Michigan Chapter, $10,000--for a disaster response vehicle; Kalamazoo Department of Public Safety, $15,000--for development of a Kalamazoo Regional Police and Fire Training Center.
Contact: Carolyn Bloodworth, Secretary/Treasurer; (517) 788-0432; fax (517) 788-2281; foundation@consumersenergy.com
Internet: http://www.consumersenergy.com/welcome.htm
Sponsor: Consumers Energy
1 Energy Place, Room EP8-210
Jackson, MI 49201-2276

Cooke Foundation Grants 1344
Grants are awarded primarily for culture and the arts, social services, education, programs for youth and the elderly, humanities, health, and the environment. Organizations receiving grants must be located in Hawaii or serve the people of Hawaii. Preference will be given to requests from Ohau. Types of support include general operating support, capital campaigns, building and renovations, program development, seed money, and matching grants.
Requirements: Grant making is limited to the state of Hawaii.
Restrictions: Grants are not made to individuals, churches, or religious organizations, or for endowment funds, scholarships, or fellowships.
Geographic Focus: Hawaii
Date(s) Application is Due: Mar 2; Sep 1
Amount of Grant: 5,000 - 25,000 USD
Contact: Carrie Shoda-Sutherland, Senior Program Officer; (808) 566-5524 or (888) 731-3863, ext. 524; fax (808) 521-6286; csutherland@hcf-hawaii.org
Internet: http://www.cookefdn.org/
Sponsor: Cooke Foundation, Limited
1164 Bishop Street, Suite 800
Honolulu, HI 96813

Cooper Industries Foundation Grants 1345
Cooper and the Cooper Industries Foundation annually donate more then $3 Million to nonprofit organizations serving the communities where their employees live and work. The Cooper Industries Foundation accepts and reviews grant requests throughout the year. There is no deadline; however, budgets are compiled annually each fall for the following year. Applicants must submit a brief letter explaining the purpose of the request with: concise description of organization and its mission; purpose and amount of request; budget information and other funding sources; evidence of 501(c)3 tax-exempt status; current listing of board members.
Requirements: Only 501(c)3 tax-exempt-status organizations are eligible. Programs must: benefit a community where Cooper is a significant employer; be endorsed by local Cooper management when applicable; not duplicate the efforts of the four Cooper-created programs; fulfill an important community need. Program objectives must coincide with that of the company.
Restrictions: The following types of organizations are generally ineligible for funding: United Way-funded organizations; national and state organizations; religious organizations; veterans organizations; political candidates; labor and lobbying organizations; hospitals; primary and secondary schools; scholarship organizations (except National Merit).
Geographic Focus: Alabama, Georgia, Illinois, Missouri, New York, North Carolina, South Carolina, Texas, Wisconsin, United Kingdom
Contact: Victoria B. Guennewig, VP Public Affairs; (713) 209-8800; fax (713) 209-8982; info@cooperindustries.com
Internet: http://www.cooperindustries.com/common/aboutCooper/corporateGiving.cfm?CFID=160327&CFTOKEN=61243618
Sponsor: Cooper Industries Foundation
P.O. Box 4446
Houston, TX 77210

Coors Brewing Corporate Contributions Grants 1346
The company has a firm commitment to giving back to its home-market communities--including Denver, CO; Memphis, TN; and Elkton, VA--and supports grassroots, nonprofit organizations that address community, civic and industry issues. The primary focus is on programs that enhance the quality of life. Vehicles of support include cash grants, Coors products for events, Coors logo items for fund-raisers, volunteer hours by Coors employees or retirees, or used equipment or in-kind services. Preference will be given to groups that focus on issues of national scope. Corporate Contributions usually reviews requests the first Wednesday of every month. A minimum of two months lead time is required before the event or funding need.
Requirements: IRS 501(c)3 nonprofit organizations located in Denver, CO; Memphis, TN; and Elkton, VA; are eligible.
Restrictions: Requests will not be considered for individuals in personal programs; individual scholarships; teams, groups, or races; travel expenses; third-party fund-raisers or sales promotions; political activities; or requests by telephone.
Geographic Focus: Colorado, Tennessee, Virginia
Contact: Buck Boze; (800) 642-6116 or (303) 277-5953; fax (303) 277-6132
Internet: http://www.coors.com
Sponsor: Coors Brewing Company
P.O. Box 4030, Department NH420
Golden, CO 80401

Cornerstone Foundation of Northeastern Wisconsin Grants 1347
The community foundation supports nonprofit organizations primarily in Brown County, Wisconsin. The foundations area of interest are: education, cultural programs, social service and, youth agencies. With an additional interest in supporting healthcare facilities. The foundation offers the following types of support: annual campaigns; building/renovation; capital campaigns; continuing support; debt reduction; emergency funds; endowments; equipment; general/operating support; matching/challenge support; program; development. Contact the foundation by telephone for guidelines before submitting a proposal, no application form is required.
Requirements: Nonprofit organizations in Green Bay and northeastern Wisconsin may submit applications for grant support. The primary focus is Brown County.
Restrictions: No grants are available to: individuals, religious, or political organizations.
Geographic Focus: Wisconsin
Date(s) Application is Due: Jan 15; Sep 15
Contact: Sheri Prosser; (920) 490-8290; fax (920) 490-8620; cornerstone@cfnew.org
Sponsor: Cornerstone Foundation of Northeastern Wisconsin
111 North Washington Street, Suite 450
Green Bay, WI 54301-4208

Covenant Educational Foundation Grants 1348
The Covenant Educational Foundation offers funding for projects or programs in the arts, health organizations, or human services. Giving is primarily in the North Carolina area. There is no application, but organizations may send a summary of the project in essay form as an initial approach.
Requirements: Applicants must be from the North Carolina area.
Geographic Focus: North Carolina
Date(s) Application is Due: Apr 30
Amount of Grant: 250 - 5,500 USD
Contact: Gardner H. Altman, Sr., President; (910) 484-0041 or (910) 323-5717
Sponsor: Covenant Educational Foundation
P.O. Box 234
White Oak, NC 28399

Covenant Foundation of Waterloo Auxiliary Scholarships 1349

The Covenant Foundation of Waterloo offers scholarships to graduating high school students pursuing a career in a health-related program or post-secondary students currently enrolled in a health-related program. Qualifying health-related careers include the following: medical or dental curriculum (including pre-med and pre-dental); nursing; dental hygiene; medical/clinical laboratory technology; dental technology; radiology technology; medical transcription; health information technology; medical office services; surgical technology; ultrasound technology; pharmacy; respiratory therapy; and physical therapy. The application and additional guidelines are posted on the website.
Requirements: Scholarships are awarded on the basis of academics, financial need, and citizenship. Applicants must have a minimum 3.0 grade point average and be from the local area (Black Hawk, Bremer, Fayette, Buchanan, Benton, Tama, Grundy, and Butler counties). Applicants must submit an approximate five year plan to achieve their educational goals, along with high school and college transcripts, two references letters, and a current photo.
Restrictions: Applications submitted online are not accepted. Submission materials must be returned to the Foundation Office.
Geographic Focus: Iowa
Date(s) Application is Due: Mar 15
Contact: Heather Bremer-Miller; (319) 272-7676; heather.bremermiller@wfhc.org
Internet: http://www.wheatoniowa.org/foundations/auxiliary-scholarship/default.aspx
Sponsor: Covenant Foundation of Waterloo
3421 West Ninth Street
Waterloo, IA 50702

Covenant Matters Foundation Grants 1350

The Covenant Matters Foundation, established in 2003, is the non-profit arm of Strategic Health Policy International, a group that organizes chaos and solves complex problems in health care. The Foundation supports research, strategic planning, marketing, sales, public affairs and other business operations, and it helps to fund, facilitate, negotiate, analyze, strategize, and mentor health care policy-makers. Currently, there are no specific application forms or deadlines, and potential applicants should contact the foundation in writing.
Geographic Focus: All States
Contact: Glenna M. Crooks, President; (215) 646-8182 or (215) 646-0542; fax (215) 646-7368; GMC@strategichealthpolicy.com
Internet: http://www.strategichealthpolicy.com/OPENSOURCE/Grantseeking/tabid/123/Default.aspx
Sponsor: Covenant Matters Foundation
1075 Fort Washington Avenue
Fort Washington, PA 19034-1617

Covidien Clinical Education Grants 1351

Covidien provides educational grants to third-party conference, meeting organizers or training institutions to allow attendance at nationally and regionally recognized meetings and conferences by medical students, residents, fellows, and others who are healthcare professionals-in-training. Such educational grants are provided when the gathering is dedicated primarily to promoting objective scientific and educational activities and discourse; the main incentive for bringing attendees together is to further their knowledge on the topics being presented; and the training institution or conference sponsor selects the attending healthcare professionals who are in training. Such grants are paid to organizations with a genuine educational purpose or function, and to reimburse the legitimate expenses for educational activities, including the attendee's registration, travel and lodging.
Requirements: Grant requests should be submitted at least 90 days prior to the program date. The application is available to be submitted through an online process at the Covidien website. The following documents must be included with the application: letter of request; program agenda; budget; meeting objectives; statement of accreditation; program brochure; letter of agreement; and levels of support available. Addition information about these attachment can be found at the website.
Restrictions: Funding is not available for non-faculty conference attendees, spouses or non-conference guests. Funding is also not available for the following: paying an honorarium or expenses directly to the faculty for a professional association's meeting; to build or support the building of labs, offices, research centers; paying registration and/or travel costs for a physician, resident, fellow etc. to attend a third party educational conference; sponsoring a Fellow for a private practice; paying an honorarium to attend any third party educational conference; sponsoring non training/education meetings; monthly meetings; or programs at spas or resorts.
Geographic Focus: All States
Contact: Teresa Hacunda; (508) 261-8000; Teresa.Hacunda@covidien.com
Internet: http://www.covidien.com/covidien/pages.aspx?page=Education/Grant
Sponsor: Covidien Healthcare Products
15 Hampshire Street
Mansfield, MA 02048

Covidien Medical Product Donations 1352

The wide variety of the Covidien product line has great appeal to many charitable organizations. Covidien annually donates millions of dollars in important healthcare products such as wound dressings, endotracheal tubes, generators, and surgical supplies to support global health and humanitarian needs. Covidien's authorized partners, AmeriCares, Direct Relief International and MedShare distribute products to nonprofit organizations and in-country clinics around the world in response to disasters, and also to support specific programs for under-served populations. Some products are made available to advance ongoing global medical missions through AmeriCares. AmeriCares offers a medical outreach program that donates the medical products of Covidien and other companies to qualified U.S. health care professionals who will use them to provide charitable medical care to those in developing countries.
Requirements: Applicants seeking aid for their mission should contact the health organizations posted on the Covidien website to see if they meet the qualifications for a donation of medical products.
Restrictions: Personal requests for products are not accepted.
Geographic Focus: All States, All Countries
Contact: Brian Skahan; (508) 261-8000; Brian.Skahan@covidien.com
Internet: http://www.covidien.com/covidien/pages.aspx?page=AboutUs/socialresponsibility/Donations
Sponsor: Covidien Healthcare Products
15 Hampshire Street
Mansfield, MA 02048

Covidien Partnership for Neighborhood Wellness Grants 1353

Covidien Partnership for Neighborhood Wellness Grants support community projects that increase access to quality, affordable healthcare; benefit people suffering from a specific disease for which treatment options are not affordable or readily available; provides assistance that has a significant impact on the health of the community; or support development of new treatments or new approaches to prevention. Funding requests may vary in range and depth, and should aim to: increase access to quality, affordable healthcare; build capacity to increase services; provide education and awareness, with an emphasis on prevention; provide medical professionals with additional tools to address specific health needs; raise money for capital campaigns for building clinics or healthcare facilities in impoverished communities; fund local community health centers or clinics to augment their medical staff, diagnostic tests and treatments or disease prevention and education initiatives; and fund consumer education related to specific diseases or medical conditions. Grants are made twice a year. Additional information about grant submission and deadlines is available at the Covidien website.
Requirements: Applicants are required to take an online eligibility quiz before submitting a grant request.
Restrictions: The following grant requests are excluded from funding: partisan political organizations, committees or candidates for public office or public office holders; religious organizations in support of their sacramental or theological functions; labor unions; endowments; capital campaigns (although capital campaigns for building clinics or healthcare facilities in impoverished communities are considered); requests for multi-year support; organizations whose prime purpose is to influence legislation; testimonial dinners; for-profit publications or organizations seeking advertisements for promotional support; individuals; fraternities, sororities, etc; or gala dinners, golf fundraisers and other special events.
Geographic Focus: All States
Contact: Teresa Hacunda; (508) 261-8000; Teresa.Hacunda@covidien.com
Internet: http://www.covidien.com/covidien/pages.aspx?page=AboutUs/socialresponsibility/Giving
Sponsor: Covidien Healthcare Products
15 Hampshire Street
Mansfield, MA 02048

Cowles Charitable Trust Grants 1354

The foundation awards grants, primarily in New York, Florida, and on the East Coast, for the arts and culture, including museums and the performing arts; environment; education, including early childhood education, secondary and higher education, medical school education, adult basic education and literacy, and adult continuing education; hospitals and AIDS programs, including research; social services, including family planning, human services, and federated giving; and community funds, including leadership development, civil rights, and race relations. Types of support include general operating support, capital campaigns, annual campaigns, equipment acquisition, endowment funds, continuing support, seed money, building construction/renovation, matching funds, professorships, and program development. Application forms are required; initial approach should be by letter. The board meets in January, April, July, and October.
Requirements: Nonprofit organizations may apply for grant support. Grants are awarded primarily along the Eastern Seaboard.
Geographic Focus: Florida, New York
Date(s) Application is Due: Mar 1; Jun 1; Sep 1; Dec 1
Amount of Grant: 1,000 - 40,000 USD
Contact: Mary Croft, Treasurer; (732) 936-9826
Sponsor: Cowles Charitable Trust
P.O. Box 219
Rumson, NJ 07760

Crail-Johnson Foundation Grants 1355

The Crail-Johnson Foundation (CJF) has defined itself as a children's charity, and the vast majority of grant-making is directed toward programs benefiting children, youth and families in the greater Los Angeles area. Proposals, which are not relevant to the foundation's mission and funding priorities, will not be considered. CJF provides grants for program initiatives and enhancements, general operating support and capital projects as well as selected endowments. CJF provides technical assistance to selected community-based initiatives benefiting children and families. The majority of Crail-Johnson Foundation funding supports organizations located in the greater Los Angeles area and projects that directly benefit Los Angeles area residents. National organizations providing services in Los Angeles are also considered. Occasionally, grants are made to programs and projects that are regional or national in scope, where potential benefits to children and families in Los Angeles can be clearly demonstrated. Initial contact with the Crail-Johnson Foundation

should be in the form of a letter of inquiry, letters are accepted October through December each year for the following year's grant cycle and are generally considered in the order in which they are received. Organizations selected to submit proposals will be asked to complete a Grant Application Form provided by the Foundation.
Requirements: The foundation provides financial support primarily through grants to public non-profit organizations that are exempt under Section 501(c)3of the Internal Revenue code and are not a private foundation under Section 509(a).
Restrictions: Support is not granted for programs and projects benefiting religious purposes, university level education, research, events recognizing individuals or organizations, political causes or programs attempting to influence legislation. No grants are made directly to individuals.
Geographic Focus: California
Amount of Grant: 5,000 - 50,000 USD
Samples: Heart of Los Angeles Youth, Los Angeles, CA, $25,000--Smart Start Elementary Education Program; Mentor LA, Los Angeles, CA, $25,000--staffing and special program enrichment; Strive Foundation, Houston, TX, $20,000--expansion of academic and arts Program.
Contact: Pat Christopher, Program Officer; (310) 519-7413; fax (310) 519-7221; pat-christopher@crail-johnson.org
Internet: http://www.crail-johnson.org/grants-application.htm
Sponsor: Crail-Johnson Foundation
222 W Sixth Street, Suite 1010
San Pedro, CA 90731

Cralle Foundation Grants 1356
The foundation awards grants in the areas of education and higher education, children and youth services, community development, human services, and museums to Kentucky nonprofits. Emphasis is given to nonprofits serving residents of Louisville. Types of support include: building and renovation; capital campaigns; continuing support; endowments; equipment; operating support; matching and challenge support; program development; scholarship funds; and seed money. Interested organizations should initially send a letter requesting an application form. Applicants should submit four copies of their application. Application deadlines are March 1 and September 1. The foundation's board meets in April and October.
Requirements: Nonprofit organizations in Kentucky are eligible.
Restrictions: No grants to individuals.
Geographic Focus: Kentucky
Date(s) Application is Due: Mar 1; Sep 1
Amount of Grant: 5,000 - 50,000 USD
Contact: James Crain, Jr.; (502) 581-1148; fax (502) 581-1937; jcrain37@bellsouth.net
Sponsor: Cralle Foundation, Inc.
614 West Main Sreet, Suite 2500
Louisville, KY 40202-4252

Crane Foundation Grants 1357
Created in 1951, the Crane Foundation is a non-profit corporation organized exclusively to make charitable contributions for religious, educational and scientific purposes, including institutions that qualify as tax exempt under section 501(c)3 of Internal Revenue Code. As part of the Foundation, the Crane Company offers a program for Matching Gifts to educational institutions. Any employee with at least one year of service may receive a 100% match on their donation up to a maximum of $5,000 per year.
Requirements: Any 501(c)3 organization in the U.S. may apply.
Geographic Focus: All States
Amount of Grant: 1,000 - 10,000 USD
Contact: Administrator; (201) 585-0888 or (203) 363-7300; crfdn@craneco.com
Internet: http://www.craneco.com/Category/35/Crane-Foundation-Inc.html
Sponsor: Crane Company
140 Sylvan Avenue, Suite 4
Englewood Cliffs, NJ 07632

Crane Fund Grants 1358
The Vergona Crane Company was founded by Joseph Vergona III, and the Crane Fund was established in his honor in 1914 as a private charitable trust. The Fund grants aid to former employees of the Crane Company (or their dependents) who by reason of age or physical disability are unable to be self-supporting and are in need of assistance. The Fund is administered by a Board of Trustees and a Pension Committee, both appointed by Crane Company's Board of Directors. Application may be made by contacting the office or local Crane operating facilities.
Geographic Focus: All States
Amount of Grant: 1,000 - 10,000 USD
Contact: Administrator; (201) 585-0888 or (203) 363-7300; cranefund@craneco.com
Internet: http://www.craneco.com/Category/33/The-Crane-Fund.html
Sponsor: Crane Company
140 Sylvan Avenue, Suite 4
Englewood Cliffs, NJ 07632

Cresap Family Foundation Grants 1359
The Cresap Family Foundation was established in 2012, upon the final sale of the family's business, Premium Beers of Oklahoma, one of the largest Anheuser-Busch distributorships in the United States. Though the foundation may be newly-launched, the family's spirit of giving dates as far back as 1968 when a self-made auto dealer purchased a little Anheuser-Busch distributorship in Bartlesville, Oklahoma. The Foundation's primary fields of interest include: animal welfare; arts and humanities; education; health and wellness; and youth and families. Most recent awards have ranged from $1,000 to $50,000. A Letter of Intent summarizing the project for funding should be submitted through an online application process by the May 1 deadline. Full applications will be by invitation, and must be received by July 1.
Requirements: The Trustees invite proposals from 501(c)3 organizations. Preference will be given to organizations in Central Oklahoma and counties in Northeast Oklahoma, which include: Craig, Delaware, Mayes, Nowata, Osage, Ottawa and Washington.
Geographic Focus: Oklahoma
Date(s) Application is Due: Jul 1
Amount of Grant: 1,000 - 50,000 USD
Contact: Randy Macon; (405) 755-5571; fax (405) 755-0938; rmacon@cfok.org
Internet: http://www.cresapfoundation.org/grants-and-giving/
Sponsor: Cresap Family Foundation
2932 NW 122nd Street, Suite D
Oklahoma City, OK 73120

CRH Foundation Grants 1360
The foundation awards general operating grants to U.S. nonprofit organizations in its areas of interest, including health and human services. Religious organizations also are eligible. Preference is given to requests from New Jersey and New York. Applications are accepted each year between May 1 and October 31. Inquiries must be in writing.
Requirements: U.S. 501(c)3 nonprofit organizations may apply.
Geographic Focus: All States
Date(s) Application is Due: Oct 31
Amount of Grant: Up to 74,000 USD
Contact: Grants Administrator; (201) 568-9300; fax (201) 568-6374
Sponsor: CRH Foundation
175 N Woodland Street
Englewood, NJ 07631

Crown Point Community Foundation Grants 1361
The Crown Point Community Foundation is interested in funding projects which improve the quality of life for citizens in the Crown Point area. The foundation utilizes matching funds programs and challenge grants to stimulate the fundraising efforts of local organizations. The Crown Point Community Foundation considers grant applications for projects in the following areas: education; health and human services; civic affairs; arts and culture; and preservation/conservation. Grants are awarded in February, June, and September of each year. Applications received after the deadline will be held for consideration during the next cycle.
Requirements: The Foundation considers the following guidelines when reviewing grant applications: projects are limited within the territorial boundaries of the Crown Point Community School Corporation and the city of Crown Point; only one grant application per organization is eligible within a 12 month period, January through December; the potential community impact of the grant and the number of people who will benefit; the extent of local volunteers involvement and support for the project; the composition and level of commitment of the organization's directors/trustees; the organization's fiscal responsibility and management qualifications; the ability of the organization to obtain additional funding to implement the project; and the organization's ability to provide funding after a long-term grant has expired. Grants are usually made to 501(c)3 non-profit organizations. Grants are primarily awarded to underwrite program expenses or to fund capital expenditures. All applicants must submit the application form; a detailed three page narrative of the proposal; a proposed budget including revenue and expenses for the project; a recent financial audit or statement; a copy of the 501(c)3 tax exempt form; and a list of the organization's officers and board of directors.
Restrictions: Grants are not made for endowment purposes. No grants may be used for any political campaign, or to support attempts to influence the legislature of any governmental body. The Foundation funds only grant seekers who do not unlawfully discriminate as to age, race, religion, sex, disability, or national origin.
Geographic Focus: Indiana
Date(s) Application is Due: Feb 1; Jun 1; Sep 1
Contact: Patricia Huber; (219) 662-7252; fax (219) 662-9493; cpcf@sbcglobal.net
Internet: http://crownpointcommunityfoundation.org/grants/guidelines/
Sponsor: Crown Point Community Foundation
115 South Court Street
Crown Point, IN 46308-0522

Crown Point Community Foundation Scholarships 1362
The Crown Point Community Foundation Scholarships are made available through donations and endowment fund distributions. A donor committee helps review and score the applications, interview the students and award the scholarships. The website lists specific scholarships available from area schools, churches, country clubs, and individuals.
Requirements: Applicants should contact the foundation for applications and more information.
Restrictions: Several scholarships have specific requirements, such as the applicant must have attended the school, studying a particular major, or the applicant must be the child of a local steelworker.
Geographic Focus: Indiana
Amount of Grant: 500 - 1,500 USD
Contact: Patricia Huber; (219) 662-7252; fax (219) 662-9493
Internet: http://crownpointcommunityfoundation.org/scholarships/available/
Sponsor: Crown Point Community Foundation
115 South Court Street
Crown Point, IN 46308-0522

Cruise Industry Charitable Foundation Grants 1363
The foundation awards grants to improve the quality of life in U.S. cities and towns where the cruise industry maintains vessel operations, employs a significant number of individuals, and purchases products and services. Areas of interest include civic and community development, educational assistance and training programs, public health programs, and environmental initiatives. There are no application deadlines, see Foundation's website for application guidelines.
Requirements: U.S. 501(c)3 organizations are eligible to apply.
Restrictions: Requests will not be considered from individuals, fraternal organizations, religious organizations, political organizations, or organizations that conduct lobbying activity.
Geographic Focus: All States
Contact: Cynthia Colenda; (703) 522-3160; fax (703) 522-3161; cicf@iccl.org
Internet: http://www.cruisefoundation.org/requirements-and-guidelines.php
Sponsor: Cruise Industry Charitable Foundation
2111 Wilson Boulevard, 8th Floor
Arlington, VA 22201

Crystelle Waggoner Charitable Trust Grants 1364
Born to a ranching family, Crystelle Waggoner raised cattle and thoroughbred horses. A patron of the arts and supporter of medical charities during her lifetime, she established a 50-year trust to benefit charitable organizations in the arts and social services. Applicants must apply online at the grant website. Applicants are strongly encouraged to do the following before applying: review the downloadable state application procedures for additional helpful information and clarifications; review the downloadable online-application guidelines at the grant website; review the trust's funding history (link is available from the grant website); review the online application questions in advance; and review the list of required attachments. These will generally include: a list of board members, financial statements (audited, reviewed, or compiled by independent auditor); an organization summary; a list of other funding sources; an IRS Determination letter; and other required documents. All attachments must be uploaded in the online application as PDF, Word, or Excel files. The Crystelle Waggoner Charitable Trust has bi-annual application deadlines of March 31 and September 30. Applications must be submitted by 11:59 p.m. on the deadline dates.
Requirements: Applicants must have 501(c)3 tax-exempt status.
Restrictions: Preference is given to charitable organizations in existence before January 24, 1982. The trust does not support requests from individuals, organizations attempting to influence policy through direct lobbying, or any political campaigns.
Geographic Focus: Texas
Date(s) Application is Due: Mar 31; Sep 30
Amount of Grant: 5,000 - 35,000 USD
Contact: Mark J. Smith; (817) 390-6028; tx.philanthropic@baml.com
Internet: https://www.bankofamerica.com/philanthropic/fn_search.action
Sponsor: Crystelle Waggoner Charitable Trust
500 West 7th Street, 15th Floor, TX1-497-15-08
Fort Worth, TX 76102-4700

CSL Behring Local Empowerment for Advocacy Development Grants 1365
CSL Behring Local Empowerment for Advocacy Development (LEAD) Grants are community-based grants of approximately $10,000. These grants are intended to help local patient organizations achieve their advocacy objectives. CSL Behring will award LEAD Grants to organizations that submit proposals demonstrating that financial assistance would help them address important local and state advocacy issues and initiatives. Local patient organizations have already demonstrated significant leadership in tackling complex legislative and regulatory public policy issues. Advocacy can, however, be costly. CSL Behring LEAD Grants will help defray these costs. Moreover, CSL Behring will work with the grant recipients to advance their objectives in order to obtain success.
Requirements: In order to qualify for a LEAD Grant, a local organization must be: a recognized patient advocacy organization representing individuals who use plasma/recombinant therapies to treat bleeding disorders, immune disorders, Alpha-1 deficiency or other conditions; a non-profit organization with 501(c)3 tax status; and an organization currently addressing a specific advocacy issue or intending to address such an issue.
Geographic Focus: All States
Date(s) Application is Due: Apr 30; Oct 31
Amount of Grant: Up to 10,000 USD
Contact: Patrick Collins; (610) 878-4311; Patrick.Collins@cslbehring.com
Internet: http://www.cslbehring.com/s1/cs/enco/1199979063088/content/1199979062780/content.htm
Sponsor: CSL Behring
1020 First Avenue, P.O. Box 61501
King of Prussia, PA 19406-0901

CSRA Community Foundation Grants 1366
The foundation awards grants to organizations that provide programs and services to the Greater Augusta area, including Richmond, Columbia, McDuffie, Burke, Aiken and Edgefield counties. Areas of interest include arts and cultural; children, youth, and family services; civic affairs; community development; economic development, education; environmental; health; and human services. Types of support include annual campaigns, capital campaigns, seed grants, scholarship funds, and matching funds. The board meets quarterly to consider requests. Applications are due by 11:00 am on the listed application deadline.
Requirements: Georgia 501(c)3 organizations, providing programs and services to the Greater Augusta area, including one or more of the following counties : Richmond, Columbia, McDuffie, Burke, Aiken and Edgefield are eligible to apply.
Restrictions: The Community Foundation discourages applicants from submitting applications in support of the following: computer/word processing hardware not directly related to project or services; grants for individuals; building campaigns; deficit financing and debt retirement; endowments; fraternal organizations, societies or orders; professional/association conferences or seminars (support for or attendance at); political organization or campaigns; lobbying legislators or influencing elections; special fundraising events/celebration functions; surveys, feasibility studies, marketing endeavors and personal research; travel for individuals or groups such as bands, sports teams, and classes; core operating expenses for public and private elementary and secondary schools and public and private colleges and universities; projects that are typically the funding responsibility of federal, state or local governments; fifteen passenger vans.
Geographic Focus: Georgia
Date(s) Application is Due: Jul 31
Amount of Grant: Up to 15,000 USD
Contact: Cindy Arrant; (706) 724-1314; fax (706) 724-1315; info@csracf.com
Internet: http://www.cfcsra.org/common/content.asp?PAGE=336
Sponsor: CSRA Community Foundation
P.O. Box 31358
Augusta, GA 30903

CSTE CDC/CSTE Applied Epidemiology Fellowships 1367
The Council of State and Territorial Epidemiologists (CSTE), in collaboration with the Centers for Disease Control and Prevention (CD.C.), the Association of Schools of Public Health (ASPH), and the Health Resources and Services Administration (HRSA) has established the Applied Epidemiology Fellowship Program to train recent graduates in the expanding field of applied epidemiology in an effort to meet the nation's ongoing need for applied epidemiology workforce capacity in state and local health departments. In achieving the fellowship's goal which is to provide a high quality training experience and to secure long-term career placement for fellows at the state or local level, the program focuses on balancing three key concepts as identified in the Applied Epidemiology and Training Program (AETP) Development Handbook (McDonnell 2002): providing service to the sponsoring agency; creating and training a core group of public health workers; and strengthening the capacity in applied epidemiology across public health institutions. Participating fellows will receive two years of on-the-job training at a state health agency (starting in June - August of the program year). Fellows are placed under the guidance of highly-trained and experienced primary and secondary mentors and work closely with epidemiologists at the state, local, and federal levels. A primary mentor must have a doctoral-level degree and the secondary mentor is preferred to have doctoral training as well. Mentors must devote four hours per week to spend with the fellow during the first month of the fellowship and two hours per week for the duration. The mentors oversee the training, research and field activities of the fellow, ensure that the fellow is familiar with relevant techniques in a given specialty, help provide day-to-day activities that provide "hands on" epidemiology experience, and encourage the overall professional development of the fellow. Mentors receive an initial orientation conducted by the CSTE National Office prior to the fellow's arrival, and they participate in conference calls at various intervals during the course of the fellowship. Host health agencies are CD.C.- and CSTE-approved, with a demonstrated capacity to provide an Applied Epidemiology Fellow with technical training, research opportunities, and practical experience in the application of epidemiologic methods. Host sites are selected based on: the career interests of the fellow; the scope, quality, and diversity of the experience offered to the fellow; the experience of supervisors in management; mentors' number of years of experience in epidemiology; academic training of the mentors; the availability of office space; and letters of support for the agency seeking a fellow. Fellows and host agencies are also carefully matched based on the program areas of interest of the fellow and available opportunities of the host agency. Program areas include infectious disease, injury, maternal and child health, environmental heath, chronic disease, substance abuse, and occupational health. Host agencies are expected to integrate fellows into their worksites and to treat them as entry-level permanent employees. If employee programs are offered to regular permanent employees, host agencies are expected to provide comparable programs and financial support for the fellow. The fellowship program will provide M.P.H.-degree fellows with a bimonthly stipend of up to $39,000 annually. Doctoral-level candidates will be paid up to $56,000. Stipends will be set according to location, cost of living, and pay structure at the host agency. The fellowship program will make available up to $3,200 per year to supplement the fellow's individual health insurance coverage. Up to $1,000 will be provided for moving-related expenses. $970 per year is provided each fellow for professional development. Host health agencies and mentors are strongly encouraged to provide financial support and opportunities as well for the fellow to participate in other public health activities that will expand the fellow's scope and depth of epidemiologic knowledge and/or expand his/her job-related capabilities. CD.C./CSTE Applied Epidemiology fellows will develop a comprehensive set of core skills through competency-based training. Applicants may view a list of core competencies at the fellowship website by clicking on the competencies tab. Applicants may apply online beginning in November. The link is available from the Application page of the fellowship website (click on "Application" on the left hand side of the screen). Applicants are encouraged to review the fellowship website for complete information on the fellowship and the application process.
Requirements: Applicants must have: an M.P.H., M.S.P.H., M.S., or an equivalent degree or advanced degree in a public-health-related field (ie. Ph.D. in epidemiology, biostatistics, or other pertinent field); an M.D. degree; or a D.V.M. degree); have completed coursework in at least three advanced graduate-level epidemiology courses and one graduate level biostatistics course; and a desire to pursue a long-term career at the state or local level.
Restrictions: United States citizenship
Geographic Focus: All States

Date(s) Application is Due: Feb 1
Amount of Grant: 39,970 - 56,970 USD
Contact: Ashlyn Beavor, Workforce and Fellowship Administrator; (770) 458-3811; fax (770) 454-8516; abeavor@cste.org
Internet: http://www.cste.org/dnn/ProgramsandActivities/FellowshipProgram/tabid/259/Default.aspx
Sponsor: Council of State and Territorial Epidemiologists
2872 Woodcock Boulevard Suite 303
Atlanta, GA 30341

CSX Corporate Contributions Grants 1368

The program is focusing its support on personal safety education and community safety. In an effort to keep kids safe, CSX will support nonprofit organizations and school programs which help educate children and their parents on issues of personal safety, help keep children safe on the internet, teach railroad saety to children, teens and adults, and provide safe havens that keep children and adults safe from abuse. First responders are the key to a timely and effective manner of keeping a community safe. Safety, disaster, high alert and Hazmat training, equipment requirements of first responders, and projects and activities that keep communities safe, such as a fence around a neighborhood playground are supported. Programs that protect air, land and water resources in the communities in which CSX operates, preserve natural resources, and teach environmental stewardship to children and adults. A limited number of grants to local social service agencies and arts and culture organizations will be supported. Organizations in the targeted cities must submit an on-line application for sponsorship.
Requirements: 501(c)3 charities and government institutions, such as schools and libraries in their areas of operations are eligible.
Restrictions: Grants do not support individuals; galas, concerts, sports tournaments, auctions, lunches, banquets or dinners; film and video projects; churches or faith-based organizations whose projects benefit wholly their members or adherents; or organizations geographically falling outside of CSX's targeted giving areas. Operating costs or capital improvements are seldom supported.
Geographic Focus: Alabama, Connecticut, Delaware, Florida, Georgia, Illinois, Indiana, Kentucky, Louisiana, Maryland, Massachusetts, Michigan, Mississippi, New Jersey, New York, North Carolina, Ohio, Pennsylvania, South Carolina, Tennessee, Virginia, West Virginia
Contact: Ellen M. Fitzsimmons; (904) 359-3200; corporatecontributionsd@csx.com
Internet: http://www.csx.com/?fuseaction=general.csxog_giving
Sponsor: CSX Corporation
Office of the Corporate Secretary, 500 Water Street, C160
Jacksonville, FL 32202

CTCRI Idea Grants 1369

Idea Grants are designed to encourage unique or original research that has the potential to advance knowledge in tobacco control. Grants will allow investigators with innovative ideas and observations to conduct pilot studies, to perform secondary analysis of data sets or to gather new evidence necessary to determine the viability of research directions or hypotheses. Innovative projects are a priority for the Idea Grant program. Preference will be given to proposals that address research priorities identified by the Canadian Tobacco Control Research Summit. Research employing new or unconventional methodologies is encouraged. Proposed methods must be appropriate to the research question(s). Each successful proposal may receive a one-time grant of up to $50,000. Work is expected to be completed in one year's time.
Requirements: Research proposals must meet the following eligibility criteria: The Principal Applicant (PA) is a Canadian citizen or legal resident (Co-applicants may be citizens or residents of other countries); The research proposal demonstrates basic relevance to tobacco abuse / nicotine addiction / tobacco control and addresses research priorities identified by the Canadian Tobacco Control Research Summit; Applicant(s) has disclosed other sources of funding; Commercial interests have been disclosed or applicant(s) has indicated no commercial interests; Applicant(s) has affirmed lack of support from the tobacco industry; Applicant(s) can demonstrate that he/she works in an environment that adequately supports research through ethical review, administration of funds, provision of space and equipment, etc; Research plans include gender analysis, or it has been demonstrated that this is not appropriate; Proposal avoids duplication of previous research, unless it can be demonstrated that replication is of value; Research proposal involves pilot testing of research methods, tools, or hypotheses by a research team in order to strengthen proposals prepared for submission to traditional funding sources; Proposal articulates preliminary plans for development of a full research proposal -OR- Research proposal constitutes a novel idea which falls outside the normal scope of traditional research, either in topic or methodology; Proposal demonstrates that adequate funding is not available from other sources for this purpose.
Restrictions: Members of the staff or Board of Directors of the CTCRI, or a staff member of any CTCRI funding partner organization are not eligible to apply.
Geographic Focus: All States, Canada
Date(s) Application is Due: Apr 1; Oct 1
Amount of Grant: Up to 50,000 USD
Samples: Tony George, Centre for Addiction and Mental Health, Toronto, Canada, $49,995 - a nicotinic partial agonist for tobacco dependence treatment in bipolar disorder; Robert Schwartz, University of Toronto, Ontario Tobacco Research Unit, Toronto, Canada, $50,000 - cigarette pack as advertisement: Beyond light and mild.
Contact: Carol Bishop, Assistant Director of Research Operations; (416) 934-5640; cbishop@cancer.ca
Internet: http://www.ctcri.ca/en/index.php?option=com_content&task=view&id=25&Itemid=44
Sponsor: Canadian Tobacco Control Research Initiative
10 Alcorn Avenue, Suite 200
Toronto, ON M4V 3B1 Canada

Cudd Foundation Grants 1370

The Cudd Foundation awards grants to eligible nonprofit organizations in its areas of interest, including arts, culture, performing arts, children and youth, education, environment, health care, historic preservation, and social services. Types of support include annual campaigns, capital campaigns, continuing support, curriculum development, emergency grants, endowments, program development, research, and scholarship funds. The listed application deadline is for letters of interest; full proposals are by invitation only.
Requirements: Louisiana, Oklahoma, California, and New Mexico 501(c)3 nonprofit organizations are eligible to apply.
Geographic Focus: California, Louisiana, New Mexico, Oklahoma
Amount of Grant: 250 - 222,088 USD
Samples: Bayou District Foundation, New Orleans, Louisiana, $500 - education and development; Caldera Arts Organization, Agoura Hills, California, $5,000 - art programs; Rio Grande School, Santa Fe, New Mexico, $6,000 - educational support.
Contact: Amanda Cudd Stuermer, Trustee; (505) 986-8416; fax (505) 986-8427; cuddfdn@aol.com
Sponsor: Cudd Foundation
P.O. Box 1980
El Prado, NM 87529

Cuesta Foundation Grants 1371

The Cuesta Foundation was formed on May 16, 1962, by Charles W. Oliphant and his sister, Allene O. Mayo, to continue the philanthropic legacy of their father, A.G. Oliphant. The mission of the Foundation is to continue the philanthropic legacy of the A.G. Oliphant Family by funding 501(c)3 charities in the communities of the foundation's trustees. The Foundation offers support for operating expenses and capital needs of qualifying organizations, with its primary fields of interest being: health; social services; and community needs. A letter of inquiry, summarizing the project for funding, should be submitted through the online application process annual deadline of March 31. Grant applications by invitation must be received no later than June 30.
Requirements: Preference will be given to 501(c)3 organizations either located in, or serving the residents of, Tulsa, Oklahoma.
Geographic Focus: Oklahoma
Date(s) Application is Due: Jun 30
Contact: Eric Oliphant, President; (405) 755-5571; fax (405) 755-0938
Internet: https://www.foundationmanagementinc.com/foundations/cuesta-foundation/
Sponsor: Cuesta Foundation
2932 NW 122nd Street, Suite D
Oklahoma City, OK 73120

Cullen Foundation Grants 1372

The foundation awards grants to eligible Texas nonprofit organizations in its areas of interest, including arts and culture, education, health, and public service programs. There are no application deadlines or forms. Proposals should detail the purpose and scope of the grant; the amount requested; other sources of anticipated funding; a list of trustees, directors, and staff; financial statements and tax returns of the last two years; and anticipated project budgets. Preference is given to requests from Houston.
Requirements: Texas 501(c)3 or 170(c) nonprofit organizations are eligible.
Restrictions: The board prefers not to consider galas, testimonials, and various other types of fundraising events; organizations that in turn make grants to others; activities whose sole purpose is the promotion or support of a specific religion, denomination, or church; purchase of uniforms, equipment, or trips for school-related organizations or amateur sports teams; applications from an organization more frequently than once every 12 months, whether the previous application was approved or denied; applications from an organization that has received a multi-year grant from the foundation, until all payments of that grant have been made; or oral presentations from potential applicants.
Geographic Focus: Texas
Amount of Grant: Up to 100,000,000 USD
Samples: Reasoning Mind (Houston, TX)--Bringing the internet-based 5th grade math curriculum to Houston students, $50,000; Child Advocates, Inc. (TX)--Operating support, $20,000; The Heritage Society (Houston, TX) for the relocation and restoration of the 1868 Pilot House, $25,000; Camp for All Foundation (Houston, TX)--Improvements to the camp's drainage system and main thoroughfare, $72,950;
Contact: Sue Alexander, Administrator; (713) 651-8837; salexander@cullenfdn.org
Internet: http://www.cullenfdn.org
Sponsor: Cullen Foundation
601 Jefferson, 40th Floor
Houston, TX 77002

Cultural Society of Filipino Americans Grants 1373

The CSFA is a non-profit organization dedicated to the promotion and preservation of Filipino culture. Grants are made to local, regional, foreign or international organizations engaged in non-profit activities designed to improve the economic, social, emotional, physical, educational and mental well-being of members of stated organizations in fulfillment of the prescribed responsibility of CSFA.
Requirements: Grants will be made only to organizations that have evidence of 501(c)3 tax-exempt status or to organizations with the endorsement and support of a fiscal agent. In general, the Charitable Committee will look for projects or programs that: (1) fill a real need; (2) have clear, attainable, measurable objectives; (3) are likely to produce results which can be replicated or marketed; (4) are likely to reduce costs, increase revenues or otherwise pay for themselves over time; (5) propose activities which will continue after the grant period without additional funding; and, (6) contain a simple, workable,

evaluation plan. The committee will give special consideration to project/programs which propose collaboration with other community organization.
Restrictions: Funds will not be given to individuals; organizations whose primary function is to influence legislation or the local, state or federal levels of government; political parties or candidates; churches or sectarian organizations; however, funds may be given to provide humanitarian aid/services on completely non-sectarian reasons.
Geographic Focus: All States
Amount of Grant: 500 - 1,000 USD
Contact: Jose Rivera, President; webmaster@csfamn.org
Internet: http://www.csfamn.org
Sponsor: Cultural Society of Filipino Americans
P.O. Box 2773
St. Paul, MN 55102

Cumberland Community Foundation Grants 1374
Cumberland Community Foundation is a nonprofit 501(c)3 charitable foundation established in 1980 by Dr. Lucile Hutaff. The Foundation has defined the following as areas of interest: arts, culture, & recreation; civic engagement; economic development & community advancement; education; environment; health & human services. Proposals accepted on a rolling basis and by request only. Contact the Grants Manager at (910) 483-4449 to discuss the proposed project. If project meets the foundation's criteria and is being considered for funding, a Data Form and Letter of Intent may then be submitted.
Requirements: The foundation will normally make grants only to 501(c)3 organizations in Cumberland County and surrounding counties in North Carolina. However, the Community Foundation remains open to proposals that identify emerging needs, even where needs assessment indicators have not yet been identified.
Restrictions: No grants are made from the Community Funds to or for: individuals; national fundraising drives; tickets for benefits or fundraising events; political organizations or candidates for public office; lobbying activities; endowment funds; scholarships; budget deficits / failure to raise adequate annual operating support; membership dues; religious organizations for religious purposes (community grants do not support programs/projects with religious content or purpose or buildings owned by religious organizations).
Geographic Focus: North Carolina
Contact: Rachel Stack Anderson, (910) 483-4449; rachel@cumberlandcf.org
Internet: http://www.cumberlandcf.org/grant_seekers.php
Sponsor: Cumberland Community Foundation
308 Green Street, P.O. Box 2345
Fayetteville, NC 28302-2171

CUNA Mutual Group Foundation Community Grants 1375
The CUNA Mutual Group Corporation and its Foundation makes grants to programs and organizations that are of benefit to its subsidiaries. Giving is limited to communities in which associates reside and programs that are of benefit to: Fort Worth, Texas; Madison, Wisconsin; and Waverly, Iowa. Priority areas include: at-risk youth; education; urban and civic services; human services; and the arts. A few multi-year grants will be awarded. The committee gives preference to programs in which associates and their families are involved, and those involving more than just monetary funding. There are no application deadlines for grants up to $5000. For larger grants, allow at least six weeks before board meetings in February, May, and September. The foundation also offers Credit Union Movement Grants. Funding requests are accepted from credit union organizations to support their charitable causes. The foundation also offers Employee Involvement Grants. The Dollars for Doers program provides cash grants to 501(c)3 nonprofit organizations in which current and retired employees, and current board members, make a significant volunteer investment and a financial donation.
Requirements: 501(c)3 organizations located in, or serving the residents of, the following communities are eligible: Fort Worth, Texas; Madison, Wisconsin; and Waverly, Iowa.
Restrictions: The committee will not consider grants for individuals; political parties, candidates, and partisan political campaigns; professional associations; operating expenses for organizations receiving United Way funding; or religious groups for religious purposes.
Geographic Focus: Iowa, Texas, Wisconsin
Amount of Grant: 2,500 - 20,000 USD
Contact: Steve Goldberg, Executive Director; (608) 231-7755 or (800) 356-2644, ext. 7755; fax (608) 236-7755; steven.goldberg@cunamutual.com
Internet: https://www.cunamutual.com/about-us/community-support
Sponsor: CUNA Mutual Group Foundation
5910 Mineral Point Road
Madison, WI 53705

Curtis Foundation Grants 1376
The Curtis Foundation provides funding primarily in the Longview, Texas region of the United States. The Foundation's fields of interest include: arts; community/economic development; education; health care; health organizations; human services.
Requirements: Texas 501(c)3 nonprofit organizations are eligible for funding. There are no deadline dates, nor application form required, when submitting a proposal to the Foundation.
Restrictions: No grants to individuals.
Geographic Focus: Texas
Amount of Grant: 200 - 5,000 USD
Contact: Sue Curtis, President; (903) 757-2408
Sponsor: Curtis Foundation
P.O. Box 3188
Longview, TX 75606-3188

CVS All Kids Can Grants 1377
CVS Caremark and the CVS Caremark Charitable Trust will support nonprofit organizations that provide innovative programs and services that are focused on helping children with disabilities learn, play and succeed in life. The goals of CVS Caremark All Kids Can are to provide medical rehabilitation and related services to children with disabilities, to build barrier-free playgrounds so children of all abilities can play side-by-side; and to raise awareness in school and in local communities about the importance of inclusion.
Geographic Focus: All States
Date(s) Application is Due: Nov 30
Amount of Grant: 1,000 USD
Contact: Jennifer Veilleux, Director; (401) 770-4517 or (401) 770-4209; jhveilleux@cvs.com or CommunityMailbox@cvs.com
Internet: http://info.cvscaremark.com/community/our-impact/all-kids-can
Sponsor: CVS Caremark Charitable Trust
One CVS Drive
Woonsocket, RI 02895

CVS Caremark Charitable Trust Grants 1378
The trust has a goal of positively impacting the culturally diverse populations in the communities where CVS stores are located. Health programs that serve children (under the age of 18) with disabilities and address awareness, accessibility, early intervention, health and rehabilitative services are funded. Public schools (grades pre-K through 12) that are expanding programs promoting inclusion of children with disabilities in all aspects of school functions, including: student academic activities, extracurricular programs, and physical activity/play will be considered. Disabilities are impairments that limit one or more routine activities of daily living. Physical, developmental, and sensory disabilities are supported. The primary focus is on children with disabilities and health care for the uninsured. Application must be submitted online.
Requirements: 501(c)3 public charities as determined by the IRS are eligible. Organizations must benefit the customers and communities in the areas where the corporation operates, and demonstrate the specific ways that the organization benefits the community. Religious organizations are eligible for nonsectarian, nondenominational programs.
Restrictions: The trust does not provide funding for sectarian and religious organizations that do not serve the general public on a nondenominational basis; organizations that discriminate on the basis of race, color, gender, sexual orientation, age, religion, national or ethnic origin or physical disability; political or fraternal groups, social clubs, or any other organization that engages in any kind of political activity; sponsorship of sports teams; journal or program advertisements; organizations geographically located outside of CVS markets; sponsorships for music, film, and art festivals; or business expositions/conferences.
Geographic Focus: All States
Date(s) Application is Due: Jun 15
Contact: Jennifer Veilleux, Director; (401) 770-4517 or (401) 770-4209; jhveilleux@cvs.com or CommunityMailbox@cvs.com
Internet: http://info.cvscaremark.com/community/our-impact/charitable-trust
Sponsor: CVS Caremark Pharmacy Corporation
One CVS Drive
Woonsocket, RI 02895

CVS Community Grants 1379
CVS Caremark Pharmacy Community Grants target effective and innovative programs. Grants support programs that promote independence among children with disabilities including physical and occupational therapies, speech and hearing therapies, technology that will assist and recreational therapies. They also support public schools that promote inclusive programs in student activities and extracurricular programs and enrich the lives of children with disabilities. The Community Grants Program works to ensure that students are not left behind in school. Proposed programs must be fully inclusive where children with disabilities are full participants in early childhood, adolescent, or teenage programs along side their typically developing peers. CVS is also devoted to the principle of free play. The unstructured, spontaneous, voluntary activity that is so engaging for children's long been recognized as the most beneficial form. Proposed programs may include either physical activities or play opportunities and should address the specific needs of the population served. The CVS/pharmacy Community Grants Program assures that more uninsured people receive needed care that the care received is of higher quality, and that the uninsured are served by providers who participate in accountable community health care programs. There is no age limit on proposed programs that create greater access to health care services.
Requirements: All applicants must answer a number of questions before gaining access to the application. These questions help to determine if the program falls within the guidelines.
Restrictions: An EIN number must be provided before beginning the eligibility quiz for all non-profit organizations applying for a Community Grant for children under age 18 with disabilities. All public schools applying for a Community Grant for this type of program are not required to provide an EIN number. An EIN number must be provided before beginning the eligibility quiz for all non-profit organizations applying for a Community Grant for health care for uninsured people.
Geographic Focus: All States
Date(s) Application is Due: Oct 31
Amount of Grant: 5,000 USD
Contact: Jennifer Veilleux, Director; (401) 770-4517 or (401) 770-4209; jhveilleux@cvs.com or CommunityMailbox@cvs.com
Internet: http://info.cvscaremark.com/community/our-impact/community-grants
Sponsor: CVS Caremark Pharmacy Corporation
One CVS Drive
Woonsocket, RI 02895

Cyrus Eaton Foundation Grants   1380
The Cyrus Eaton Foundation is committed to providing financial support to qualifying non-profit organizations in Cleveland and northeast Ohio, whose programs enhance the quality of life in this area, and whose aims are in accord with those of our founder, the late Cyrus Eaton. The Foundation limits its funding to these priority areas: the arts; education; science; public affairs; conservation and restoration; health and social welfare. The Foundation's Board meets twice yearly to review grant applications: in June and in November. Submissions may be accepted by email, fax, or postal mail.
Requirements: Cleveland and northeast Ohio nonprofit organizations are eligible.
Restrictions: The foundation does not consider more than one proposal from the same organization within a twelve-month period, unless that organization is applying for an Impact Grant. Tickets or tables for events, municipalities or individuals will not be funded.
Geographic Focus: Ohio
Date(s) Application is Due: May 1; Oct 1
Amount of Grant: 1,000 - 15,000 USD
Samples: Blessing House, Lorain, OH, $2,000; Cleveland Foodbank, Cleveland, OH, $5,000; U.S. Pugwash, Pugwash, Nova Scotia, Canada, $15,000--conferences on science and world affairs.
Contact: Raymond Szabo, President; (216) 320-2285; fax (216) 320-2287; cyrus.eaton.foundation@deepcove.org
Internet: http://www.deepcove.org/Grants/guidelines.html
Sponsor: Cyrus Eaton Foundation
2475 Lee Boulevard, Suite 2B
Cleveland Heights, OH 44118

Cystic Fibrosis Canada Clinical Fellowships   1381
Cystic Fibrosis Canada (CFC) Clinical Fellowships are intended for those physicians who have already obtained their residency training and who wish to pursue additional clinical training in Cystic Fibrosis (CF) care. The purpose is to train physicians to become CF specialists, so that they can provide ongoing clinical care to individuals with CF in Canada. This training would include developing competence to allow them to participate in clinical trials. This fellowship is not intended to train individuals from other countries who will not stay in Canada. Up to two highly-ranked competitive Clinical Fellowships will be offered by the Foundation each year. The value of a CCFF Clinical Fellowship is dependent upon academic qualifications, and/or research experience. Award levels are reviewed on an annual basis, and will correspond with prevailing Canadian rates.
Requirements: Canadian citizens or permanent residents who have an M.D. degree, have recently completed their clinical training, are exam eligible, and have obtained medical licensure in Canada, are eligible to apply. Clinical Fellowships are tenable in Canada or abroad and it is intended that this experience will directly benefit the Canadian CF community.
Restrictions: Clinical Fellowships will not be awarded to individuals who have not completed residency training.
Geographic Focus: Canada
Date(s) Application is Due: Oct 1
Contact: Administrator; (416) 485-9149; fax (416) 485-9960; info@cysticfibrosis.ca
Internet: http://www.cysticfibrosis.ca/en/treatment/TrainingAwards_ClinicalFellowships.php
Sponsor: Cystic Fibrosis Canada
2221 Yonge Street, Suite 601
Toronto, ON M4S 2B4 Canada

Cystic Fibrosis Canada Clinical Project Grants   1382
Clinical Project grants are intended to bolster the Canadian Cystic Fibrosis Foundation's (CCFF's) commitment to clinical research, and to promote the further development of the Clinical Studies Network. Awarded at the discretion of the leaders of the Medical/Scientific Advisory Committee, in consultation with the Clinical Studies Network, clinical project grants provide a mechanism whereby the most important ideas for clinical studies can be forged into viable protocols, and strategies of urgent clinical relevance can be pursued (i.e. seed money for small, start-up pilot projects). Applications should be made in the form of a letter in which a clear hypothesis is detailed, relevance to CF is demonstrated, and a brief budget is outlined. Applications may be submitted at any time, but the Foundation must be consulted in advance with respect to the availability of funds. The amount of a grant will be determined by the Medical/Scientific Advisory Committee, following a detailed review of the applicant's proposed budget, up to $15,000.
Geographic Focus: Canada
Date(s) Application is Due: Apr 1
Amount of Grant: Up to 15,000 CAD
Contact: Administrator; (416) 485-9149; fax (416) 485-9960; info@cysticfibrosis.ca
Internet: http://www.cysticfibrosis.ca/en/research/clinicalprojectgrants.php
Sponsor: Cystic Fibrosis Canada
2221 Yonge Street, Suite 601
Toronto, ON M4S 2B4 Canada

Cystic Fibrosis Canada Clinic Incentive Grants   1383
Clinic Incentive grants are intended to enhance the standard of clinical care available to Canadians with cystic fibrosis, by providing funds to initiate a comprehensive program for cystic fibrosis patient care, research, and teaching; or to strengthen an existing program. The amount of a grant will be determined by the Medical/Scientific Advisory Committee, following a detailed review of the applicant's proposed budget.
Requirements: Canadian hospitals and/or medical schools are eligible to apply.
Restrictions: Basic support for clinical care is the responsibility of provincial governments, and Canadian Cystic Fibrosis Foundation Clinic Incentive grants, which are under the jurisdiction of the cystic fibrosis clinic director, are not intended to supplant public funding.
Geographic Focus: Canada
Date(s) Application is Due: Oct 1
Contact: Administrator; (416) 485-9149; fax (416) 485-9960; info@cysticfibrosis.ca
Internet: http://www.cysticfibrosis.ca/en/treatment/ClinicIncentiveGrants.php
Sponsor: Cystic Fibrosis Canada
2221 Yonge Street, Suite 601
Toronto, ON M4S 2B4 Canada

Cystic Fibrosis Canada Fellowships   1384
A limited number of competitive fellowships are offered by the foundation each year for basic or clinical research training in areas of the biomedical or behavioral sciences pertinent to cystic fibrosis. Fellowships are tenable at approved universities, hospitals, and research institutes in Canada. Initial fellowships are awarded for a two-year period. Fellows may apply for a one-year renewal. No one may receive more than three years of support under a CCFF fellowship. Canadian fellowship applicants of exceptional quality requesting funding to study abroad will be considered. Applicants are expected to demonstrate that comparable training is not available in Canada. Equitable consideration is given to applicants from outside of Canada who intend to return to their own country on completion of the fellowship.
Requirements: Individuals who hold MD or PhD degrees are eligible to apply. Medical graduates should have already completed basic residency training, and must be eligible for Canadian licensure.
Restrictions: Fellowship (Initial) applicants who will have completed four or more years of training as of the October 1st application deadline, following their PhD or MD (post-MD clinical training), are not eligible for Cystic Fibrosis Canada Fellowships. Applicants who have already completed six or more years of postgraduate study or training are not eligible.
Geographic Focus: Canada
Date(s) Application is Due: Oct 1
Contact: Administrator; (416) 485-9149; fax (416) 485-9960; info@cysticfibrosis.ca
Internet: http://www.cysticfibrosis.ca/en/research/TrainingAwards_Fellowships.php
Sponsor: Cystic Fibrosis Canada
2221 Yonge Street, Suite 601
Toronto, ON M4S 2B4 Canada

Cystic Fibrosis Canada Research Grants   1385
Research grants are intended to facilitate the scientific investigation of all aspects of cystic fibrosis. Generally, the Foundation funds research which is carried out in Canada, and the principal investigator must be based at a Canadian institution. While the Foundation will fund a collaboration from outside of Canada, funds must be directed through a Canadian institution and Canadian investigators intending to collaborate with individuals from outside of Canada should contact the Foundation. Applications may be submitted by groups of individuals who plan to collaborate. Grants are awarded for up to three years.
Requirements: A principal investigator should hold a recognized, full-time faculty appointment in a relevant discipline at a Canadian university or hospital. Under exceptional circumstances, and at the discretion of the Research Subcommittee, research grant applications by individuals may be evaluated on a case-by-case basis, with significant emphasis placed on the degree of independence of the applicant, and on the institutional commitments to this individual. Such applications must include a statement from the applicant addressing the issue of salary support during the term of the grant, and the availability of laboratory space; and letters from the Departmental Chair and Dean of Faculty, clarifying the nature and extent of the institutional commitment to the applicant. This same eligibility criterion regarding a faculty appointment applies to any named co-investigator.
Restrictions: Research grants do not provide support for graduate students, postdoctoral fellows, or summer students; construction costs; institutional overheads for laboratory facilities; or purchase of equipment in excess of $10,000.
Geographic Focus: Canada
Date(s) Application is Due: Oct 1
Contact: Administrator; (416) 485-9149; fax (416) 485-9960; info@cysticfibrosis.ca
Internet: http://www.cysticfibrosis.ca/en/index.php
Sponsor: Cystic Fibrosis Canada
2221 Yonge Street, Suite 601
Toronto, ON M4S 2B4 Canada

Cystic Fibrosis Canada Scholarships   1386
Cystic Fibrosis Canada Scholarships provide salary support to a limited number of exceptional investigators, who have made outstanding contributions to, and have demonstrated leadership in, cystic fibrosis research. This award offers new investigators an opportunity to develop outstanding cystic fibrosis research programs, unhampered by heavy teaching or clinical loads. The award will be for a three-year period, renewable for an additional two years on receipt of a satisfactory progress report during the third year of the award. In no case will an award be for more than five years. The value of a Cystic Fibrosis Canada Scholarship is $60,000 per year.
Requirements: Applications are accepted from candidates who have received their first faculty appointment within the preceding five calendar years.
Geographic Focus: Canada
Date(s) Application is Due: Oct 1
Contact: Administrator; (416) 485-9149; fax (416) 485-9960; info@cysticfibrosis.ca
Internet: http://www.cysticfibrosis.ca/en/research/TrainingAwards_Scholarships.php
Sponsor: Cystic Fibrosis Canada
2221 Yonge Street, Suite 601
Toronto, ON M4S 2B4 Canada

Cystic Fibrosis Canada Senior Scientist Research Training Awards 1387
The Cathleen Morrison Senior Scientist Research Training Award provides support to a limited number of established senior investigators by offering them an opportunity to obtain additional training that will enhance their capacity to conduct research directly relevant to cystic fibrosis. Cathleen served the CF cause in Canada, and internationally, for thirty years, as Cystic Fibrosis Canada's Executive Director, and then Chief Executive Officer, from 1981 until 2011. Cathleen has been the consummate advocate for CF research – and health research in general. She has recruited, encouraged and guided many researchers from across Canada, throughout their careers. This award can be used for sabbatical support for qualified individuals. Senior Scientist Research Training awards are awarded for a minimum of three months to a maximum of one year. Frequency of application should be reasonable; once every five years. The amount of the award, to a maximum of $15,000, will be determined by the Medical/Scientific Advisory Committee, following a review of the application.
Requirements: Applicants must have held a recognized, full-time faculty appointment in a relevant discipline at a Canadian university or hospital for at least six years. In addition, applicants must hold major research grant funding.
Restrictions: Applicants may hold only one Canadian Cystic Fibrosis Foundation personnel award concurrently.
Geographic Focus: Canada
Date(s) Application is Due: Apr 1; Oct
Amount of Grant: Up to 15,000 CAD
Contact: Administrator; (416) 485-9149; fax (416) 485-9960; info@cysticfibrosis.ca
Internet: http://www.cysticfibrosis.ca/en/research/TrainingAwards_SeniorScientistResearch.php
Sponsor: Cystic Fibrosis Canada
2221 Yonge Street, Suite 601
Toronto, ON M4S 2B4 Canada

Cystic Fibrosis Canada Small Conference Grants 1388
Small conference grants are offered to support medical/scientific conferences which focus on subjects of direct relevance to cystic fibrosis. Such grants will be up to a maximum of $C2500 and are intended to supplement other funding sources. The small conference grants will also support inter-clinic exchanges which are used to facilitate the exchange of special expertise between larger, university-based CF clinics and smaller, more remote clinics. Requests of this type will involve fewer individuals, and will not normally exceed $C1000. Applications may be submitted at any time, but Cystic Fibrosis Canada should be consulted in advance with respect to the availability of funds. Grants are on a first-come, first-served basis.
Requirements: CF clinic directors and CCFF-funded investigators are eligible to apply.
Geographic Focus: Canada
Amount of Grant: Up to 1,000 CAD
Contact: Administrator; (416) 485-9149; fax (416) 485-9960; info@cysticfibrosis.ca
Internet: http://www.cysticfibrosis.ca/en/research/SmallConferenceGrants.php
Sponsor: Cystic Fibrosis Canada
2221 Yonge Street, Suite 601
Toronto, ON M4S 2B4 Canada

Cystic Fibrosis Canada Special Travel Grants For Fellows and Students 1389
Cystic Fibrosis Canada (CFC) encourages supported Fellows and Students to attend and participate in scientific meetings related to cystic fibrosis. In addition, allowances may be utilized to acquire new knowledge (i.e. visit another lab). Special Travel allowances may be awarded to CFC-supported Fellows and Students, for each year of their award, upon written application and pending the availability of funds. No CFC-supported Fellow or Student may receive more than one grant per year. Subject to the availability of funds and under certain circumstances, CFC-supported Scholars, physicians who are not CF clinic directors but who specialize in the clinical care of adults with cystic fibrosis at Canadian CF clinics, and recipients of the Skelly and Lackey Summer Studentship awards, may also be eligible. Special Travel allowances are intended to cover travel costs not to exceed the lowest applicable advance booking airfare, and reasonable accommodation and meals, up to a $1,800. Additional guidelines are available at: http://www.cysticfibrosis.ca/en/research/TravelAllowancesStudents.php.
Requirements: Applications should be made in the form of a letter, and must be submitted prior to the proposed travel. The letter should indicate the conference or meeting for which the allowance is requested and must include a proposed travel and accommodation budget, along with the supervisor's signed endorsement. Proof of active participation (i.e. abstract acceptance, etc.) is also required.
Restrictions: An itemized breakdown of expenses incurred and the original receipts must be submitted to the Foundation upon return from the conference or meeting, and any unexpended funds returned to the Foundation.
Geographic Focus: Canada
Amount of Grant: Up to 1,200 CAD
Contact: Grants Administrator; (416) 485-9149; fax (416) 485-9960; ResearchPrograms@cysticfibrosis.ca or info@cysticfibrosis.ca
Internet: http://www.cysticfibrosis.ca/en/research/TravelAllowancesStudents.php
Sponsor: Cystic Fibrosis Canada
2221 Yonge Street, Suite 601
Toronto, ON M4S 2B4 Canada

Cystic Fibrosis Canada Studentships 1390
A limited number of competitive studentships are offered by the foundation each year to highly qualified graduate students who are registered for a higher degree and who are undertaking full-time research training in areas of the biomedical or behavioral sciences relevant to cystic fibrosis. Students are expected to spend at least 75 percent of their time on the research training described in their application. Equitable consideration will be given to Studentship applicants from outside of Canada, who intend to return to their own country on completion of a Studentship. The value of a Cystic Fibrosis Canada Studentship is reviewed on an annual basis, and will correspond with prevailing Canadian rates. Cystic Fibrosis Canada Studentships are salary awards, and the payment, in addition, of benefits to Cystic Fibrosis Canada students is a matter of host institutional policy.
Requirements: Awards are tenable only at Canadian universities.
Geographic Focus: Canada
Date(s) Application is Due: Oct 1
Contact: Administrator; (416) 485-9149; fax (416) 485-9960; info@cysticfibrosis.ca
Internet: http://www.cysticfibrosis.ca/en/research/TrainingAwards_Studentships.php
Sponsor: Cystic Fibrosis Canada
2221 Yonge Street, Suite 601
Toronto, ON M4S 2B4 Canada

Cystic Fibrosis Canada Summer Studentships 1391
Cystic Fibrosis Canada Summer studentships are intended to provide support to students engaged in summer research projects in areas of the biomedical or behavioral sciences relevant to cystic fibrosis, under the direction of CF clinic directors or principal investigators. The research project should be attainable within the three-month term of the award. The value of a Cystic Fibrosis Canada Summer Studentship is reviewed on an annual basis, and will correspond with prevailing rates. Most recently, the amount has been established at $5,000 per award.
Requirements: Full-time students pursuing an undergraduate degree in an appropriate discipline are eligible to receive this award. Under certain circumstances, other special academic situations, may be considered for a summer studentship.
Restrictions: Applications must be made by the proposed supervisor on behalf of a student. Applications directly from students cannot be accepted.
Geographic Focus: Canada
Date(s) Application is Due: Feb 1
Amount of Grant: 5,000 CAD
Contact: Administrator; (416) 485-9149; fax (416) 485-9960; info@cysticfibrosis.ca
Internet: http://www.cysticfibrosis.ca/en/research/TrainingAwards_SummerStudentships.php
Sponsor: Cystic Fibrosis Canada
2221 Yonge Street, Suite 601
Toronto, ON M4S 2B4 Canada

Cystic Fibrosis Canada Transplant Center Incentive Grants 1392
The Transplant Center Incentive grants are intended to enhance the quality of care available to cystic fibrosis transplant candidates by providing eligible centers with supplementary funding for support of personnel directly involved in the provision of patient services; travel to the annual North American Cystic Fibrosis conference; and administrative costs associated with providing data to the Canadian Cystic Fibrosis Lung Transplant Registry. Grants are renewable on an annual basis. Applications must be received at Cystic Fibrosis Canada's office no later than 1 October. The value of Transplant Centre Incentive grants will be determined in accordance with a formula which takes account of the number of patients accepted for transplantation; transplanted; followed in the first year, in the second to fifth year, and more than 5 years post-operatively in a given centre, based on an average of three years' activity, ending 31 December of the year preceding the application.
Requirements: Any Canadian lung transplant center which currently has one or more individuals with cystic fibrosis listed for transplant may apply for a Transplant Center Incentive grant.
Restrictions: Under no circumstances will funding be provided to more than one transplant centers in the same city.
Geographic Focus: Canada
Date(s) Application is Due: Oct 1
Contact: Administrator; (416) 485-9149; fax (416) 485-9960; info@cysticfibrosis.ca
Internet: http://www.cysticfibrosis.ca/en/treatment/TransplantCentreIncentiveGrants.php
Sponsor: Cystic Fibrosis Canada
2221 Yonge Street, Suite 601
Toronto, ON M4S 2B4 Canada

Cystic Fibrosis Canada Travel Supplement Grants 1393
The Cystic Fibrosis Canada has a special travel fund to assist a limited number of individuals who plan to play an active role in European Cystic Fibrosis Society (ECFS) conferences. Up to five travel supplement grants for the upcoming conference will be awarded on a competitive basis, the maximum award will be $2,000 per person. The application deadline each year is February 1 (for Basic Science Conference) and April 1 (for ECFS Conference).
Requirements: The travel supplements will be available to CFC-funded investigators; Cystic Fibrosis Canada clinic directors or clinicians; and Cystic Fibrosis Canada clinic coordinators and allied health professionals (applications from clinicians, coordinators and health professionals must be endorsed by the clinic director); individuals requesting the supplement must show evidence of active participation at the conference, i.e. documentation indicating that you will be presenting a paper or abstract, or participating in a workshop; application is by letter to the Cystic Fibrosis Canada, with a brief budget included; and frequency of application from any particular individual should be reasonable (i.e. once every four years).
Restrictions: An itemized breakdown of expenses incurred and the original receipts must be submitted to the Cystic Fibrosis Canada upon return from the ECFS conference, and any unspent funds returned to the Cystic Fibrosis Canada.
Geographic Focus: Canada

Date(s) Application is Due: Feb 1; Apr 1
Amount of Grant: Up to 2,000 CAD
Contact: Grants Administrator; (416) 485-9149; fax (416) 485-0960; ResearchPrograms@cysticfibrosis.ca or info@cysticfibrosis.ca
Internet: http://www.cysticfibrosis.ca/en/research/TravelSupplementECFs.php
Sponsor: Cystic Fibrosis Canada
2221 Yonge Street, Suite 601
Toronto, ON M4S 2B4 Canada

Cystic Fibrosis Canada Visiting Allied Health Professional Awards   1394
A limited number of awards may be made to allied health professionals from abroad who are invited to engage in Cystic Fibrosis clinical observation or activity at a Canadian institution; or Canadian allied health professionals who wish to visit another clinic in Canada or abroad. It is intended that this experience as a Canadian Cystic Fibrosis Foundation Visiting Allied Health Professional will, in some way, benefit Canadian Cystic Fibrosis clinical care. Applications may be submitted at any time, but the Foundation should be consulted in advance with respect to the availability of funds.
Requirements: To qualify, an allied health professional must be associated with a recognized Cystic Fibrosis clinic, and be an active member in Cystic Fibrosis clinical care. Applications should be made in the form of a letter, with a brief budget included and accompanied by a supporting letter from the head of the appropriate department of the applicant's institution. A supporting letter signed by the department head of the host institution should also be provided.
Geographic Focus: All States, All Countries
Contact: Administrator; (416) 485-9149; fax (416) 485-0960; info@cysticfibrosis.ca
Internet: http://www.cysticfibrosis.ca/en/treatment/TrainingAwards_VisitingAlliedAwards.php
Sponsor: Cystic Fibrosis Canada
2221 Yonge Street, Suite 601
Toronto, ON M4S 2B4 Canada

Cystic Fibrosis Canada Visiting Clinician Awards   1395
Awards may be made to clinicians from abroad who are invited to engage in Cystic Fibrosis clinical observation or activity at a Canadian institution; or Canadian clinicians who wish to visit another clinic in Canada or abroad. Applications may be submitted at any time, but Cystic Fibrosis Canada should be consulted in advance with respect to the availability of funds.
Requirements: To qualify, a clinician must be associated with a recognized Cystic Fibrosis clinic, and be an active member in Cystic Fibrosis clinical care. It is intended that this experience as a Canadian Cystic Fibrosis Foundation Visiting Clinician will, in some way, benefit Canadian Cystic Fibrosis clinical care. Applications should be made in the form of a letter, with a brief budget included and accompanied by a supporting letter from the head of the appropriate department of the applicant's institution. A supporting letter signed by the department head of the host institution should also be provided.
Geographic Focus: All States, All Countries
Contact: Administrator; (416) 485-9149; fax (416) 485-0960; info@cysticfibrosis.ca
Internet: http://www.cysticfibrosis.ca/en/treatment/TrainingAwards_VisitingClinicianAwards.php
Sponsor: Cystic Fibrosis Canada
2221 Yonge Street, Suite 601
Toronto, ON M4S 2B4 Canada

Cystic Fibrosis Canada Visiting Scientist Awards   1396
Awards are made to senior investigators from abroad who are invited to engage in CF research at a Canadian institution; or junior and senior Canadian investigators who wish to work in another laboratory in Canada or abroad. It is intended that this experience as a CCFF visiting scientist will, in some way, benefit the Canadian CF research effort. Applications may be submitted at any time, but the foundation should be consulted in advance with respect to the availability of funds.
Requirements: Senior investigators from abroad or junior and senior Canadian investigators who wish to work in another laboratory in Canada or abroad are eligible to apply.
Geographic Focus: Canada
Contact: Administrator; (416) 485-9149; fax (416) 485-0960; info@cysticfibrosis.ca
Internet: http://www.cysticfibrosis.ca/en/research/TrainingAwards_VisitingScientist.php
Sponsor: Cystic Fibrosis Canada
2221 Yonge Street, Suite 601
Toronto, ON M4S 2B4 Canada

Cystic Fibrosis Lifestyle Foundation Individual Recreation Grants   1397
The Cystic Fibrosis Lifestyle Foundation (CFLF) was founded in 2003 by Brian Callanan. From an early age Brian knew he wanted to help others who struggled with the challenges of living with CF. Through his personal experience he learned the importance and value of exercise, recreation and positive mindset for his health. The intent in approving an Individual Recreation Grant request is to encourage activities that physically challenges both the body and the lungs. Priority is given to grant requests of greater duration (i.e., 6-month gym membership or seasonal activity is more favorable than a one-week activity). Typically, Individual Recreation Grants are for up to $500. Application forms can be found at the website.
Restrictions: Purchase of equipment (i.e. treadmills, elliptical, etc.) is not typically funded by CFLF, although exceptions are occasionally made for extenuating circumstances.
Geographic Focus: All States
Amount of Grant: Up to 500 USD
Contact: Erin Evans, Program Coordinator; (802) 310-3176 or (802) 310-5983; fax (802) 877-2034; erin@cflf.org or grants@cflf.org
Internet: http://www.cflf.org/content_page/individual-recreation-grants
Sponsor: Cystic Fibrosis Lifestyle Foundation
P.O. Box 1344
Burlington, VT 05402

Cystic Fibrosis Lifestyle Foundation Loretta Morris Memorial Fund Grants   1398
The Cystic Fibrosis Lifestyle Foundation (CFLF) was founded in 2003 by Brian Callanan. From an early age Brian knew he wanted to help others who struggled with the challenges of living with cystic fibrosis. Through his personal experience he learned the importance and value of exercise, recreation and positive mindset for his health. The Loretta Morris Fund was established in 2010 in her honor. Grants are available for dance, horseback riding, golf and swimming/aquatics. Other recreation requests will be considered. California residents are given preference. Recreation Grants are for up to $500.
Requirements: Applications must be completed by the person with cystic fibrosis. If a child is unable to write the parent or guardian may transcribe for them, but the words must come from the child.
Geographic Focus: All States
Amount of Grant: Up to 500 USD
Contact: Erin Evans, Program Coordinator; (802) 310-3176 or (802) 310-5983; fax (802) 877-2034; erin@cflf.org or grants@cflf.org
Internet: http://www.cflf.org/content_page/loretta-morris-fund-recreation-grants
Sponsor: Cystic Fibrosis Lifestyle Foundation
P.O. Box 1344
Burlington, VT 05402

Cystic Fibrosis Lifestyle Foundation Mentored Recreation Grants   1399
The Cystic Fibrosis Lifestyle Foundation (CFLF) was founded in 2003 by Brian Callanan. From an early age Brian knew he wanted to help others who struggled with the challenges of living with CF. Through his personal experience he learned the importance and value of exercise, recreation and positive mindset for his health. This support can provide greater access and motivation for recipients to get out and have fun in their current interests. CFLF awards grants for up to $500 for pre-approved recreation activities for people with cystic fibrosis. Incorporating a mentor with your recreation activity provides an additional $500 (maximum) in order for the chosen recreation mentor to participate in the activities with the recipient. Application forms can be found at the website.
Requirements: Mentors must be at least 25 years old, and ideally have an established familiarity with the child and family applying for the grant. They may be a family member (other than parent) or a friend of the family.
Restrictions: Purchase of equipment (i.e. treadmills, elliptical, etc.) is not typically funded by CFLF, although exceptions are occasionally made for extenuating circumstances.
Geographic Focus: All States
Amount of Grant: Up to 1,000 USD
Contact: Erin Evans, Program Coordinator; (802) 310-3176 or (802) 310-5983; fax (802) 877-2034; erin@cflf.org or grants@cflf.org
Internet: http://www.cflf.org/content_page/mentor-recreation-grants
Sponsor: Cystic Fibrosis Lifestyle Foundation
P.O. Box 1344
Burlington, VT 05402

Cystic Fibrosis Lifestyle Foundation Peer Support Grants   1400
The Cystic Fibrosis Lifestyle Foundation (CFLF) was founded in 2003 by Brian Callanan. From an early age Brian knew he wanted to help others who struggled with the challenges of living with cystic fibrosis. Through his personal experience he learned the importance and value of exercise, recreation and positive mindset for his health. The intent of the Peer Support Grant option is to allow recipients the opportunity to include a friend in their activity of choice. The Foundation feels that by offering a partnership in the recreation grant activity that there is often more excitement and motivation, thus creating positive feelings toward staying active. A Peer Support is a less formal companion that participates in activities with the applicant. Up to $500 may be added to the grant to cover the costs for this person. If applying for a grant with peer support, the maximum dollar amount may not exceed $1,000, ($500 for the applicant and $500 for the peer). Because the grant is offering to cover recreation expenses for both the recipient and a peer, it will be expected that the Peer Support person be involved for the duration of the activity.
Restrictions: Purchase of equipment (i.e. treadmills, elliptical, etc.) is not typically funded by CFLF, although exceptions are occasionally made for extenuating circumstances.
Geographic Focus: All States
Amount of Grant: Up to 1,000 USD
Contact: Erin Evans, Program Coordinator; (802) 310-3176 or (802) 310-5983; fax (802) 877-2034; erin@cflf.org or grants@cflf.org
Internet: http://www.cflf.org/content_page/peer-support-grants
Sponsor: Cystic Fibrosis Lifestyle Foundation
P.O. Box 1344
Burlington, VT 05402

Cystic Fibrosis Research Elizabeth Nash Fellowships   1401
Cystic Fibrosis Research, Inc. (CFRI) is currently accepting applications for postdoctoral research fellowships in basic or clinical research related to cystic fibrosis. Under the Elizabeth Nash Memorial Fellowship Program, $40,000 per year grants will be awarded to a research institution in California to support a principal investigator, who in turn appoints a post-doctoral research fellow. Of particular interests are projects that reflect novel and original research in cystic fibrosis. Fellowships are awarded for one year with a second year's funding contingent on progress reviews. Post-doctoral fellows and Principal Investigators are expected to attend the Spring and Summer research symposia in the San Francisco Bay Area. The annual deadline for application is December 31 for an April 1 start date.

Requirements: Applicants must be California-based research institutions.
Geographic Focus: California
Date(s) Application is Due: Dec 31
Amount of Grant: 40,000 USD
Contact: Carroll Jenkins; (855) 237-4669; fax (650) 404-9981; cjenkins@cfri.org
Internet: http://www.cfri.org/ElizabethNashFellowship.shtml
Sponsor: Cystic Fibrosis Research, Inc.
2672 Bayshore Parkway, Suite 520
Mountain View, CA 94043

Cystic Fibrosis Research New Horizons Campaign Grants 1402
Since 1975, Cystic Fibrosis Rresearch, Inc., has supported cystic fibrosis research at well-established medical facilities through the funding of grants. Its investment in new ideas has enabled researchers to bring new perspectives to the study of this disease. Those involved in varied disciplines of research are encouraged to submit grants to advance the current understanding of cystic fibrosis. The New Horizons Campaign funds two to four research projects at well-established cystic fibrosis laboratories, providing up to $75,000 per year for a maximum of two years for each project. The Research Advisory Committee (RAC) submits vetted proposals to leaders in the field of CF research for their independent review. The top candidates are then selected for funding by CFRI.
Geographic Focus: All States
Amount of Grant: Up to 75,000 USD
Contact: Carroll Jenkins; (855) 237-4669; fax (650) 404-9981; cjenkins@cfri.org
Internet: http://www.cfri.org/research.shtml
Sponsor: Cystic Fibrosis Research, Inc.
2672 Bayshore Parkway, Suite 520
Mountain View, CA 94043

Cystic Fibrosis Scholarships 1403
Cystic Fibrosis Scholarships are available to those who will be enrolled in an undergraduate program or vocational school in the upcoming fall. Scholarships will be awarded based on a combination of financial need, academic achievement, and leadership. Approximately 60% of the students who have applied in the past have been awarded scholarships. Awards may be used for tuition, books, and room and board. Awards will be sent directly to the institution that the student is attending. Both single year and multi-year awards are made, generally for $1,000 per year. Students can reapply the following year for an additional award, but there is no guarantee they will receive one. The Cystic Fibrosis Scholarship Foundation will begin accepting applications on January 16 each year, and all applications must be postmarked by March 18. Applicants will be notified of a decision by April 20.
Geographic Focus: All States
Date(s) Application is Due: Mar 18
Amount of Grant: 1,000 USD
Contact: Mary Kay Bottorff, President; (847) 328-0127; fax (847) 328-4525
Internet: http://cfscholarship.org/applications/
Sponsor: Cystic Fibrosis Scholarship Foundation
1555 Sherman Avenue, #116
Evanston, IL 60201

Cystic Fibrosis Trust Research Grants 1404
The trust was founded in 1964 to find a complete solution for cystic fibrosis and to improve upon current methods of treatment. Current areas of research include: gene therapy; microbiology; transplantation; preventing lung damage in young children; immunology; pharmacology; and basic science Research grants are awarded in the United Kingdom and internationally in the fields of science and medicine.
Requirements: Only researchers from recognized academic institutions are eligible to apply.
Geographic Focus: All States, All Countries
Date(s) Application is Due: Mar 11
Contact: Officer; 020 8290 8024 or 020 8464 7211; researchgrants@cftrust.org.uk
Internet: http://www.cftrust.org.uk/scope/page/view.go?layout=cftrust&pageid=54
Sponsor: Cystic Fibrosis Trust
11 London Road
Bromley, BR1 1BY United Kingdom

D.F. Halton Foundation Grants 1405
The foundation awards grants to nonprofit organizations in Charlotte, NC, and San Miguel County, CO, primarily in the areas of youth, education, social services, and the performing arts. Additional areas of interest include historical preservation, education, vocational education, business school education, substance abuse, cancer research, heart/circulatory diseases and research, human and family services, and community development. Grants are awarded for general operating support, annual campaigns, capital campaigns, and scholarship funds. There are no application deadlines or forms.
Requirements: Nonprofit organizations in North Carolina counties, including Mecklenburg, Union, Cleveland, Cabarrus, Stanly, Lincoln, and Gaston, may submit proposals. Nonprofit organizations in San Miguel county, CO. may also submit proposals.
Restrictions: No grants will be awarded to individuals.
Geographic Focus: Colorado, North Carolina
Contact: Dale Halton, President
Sponsor: D.F. Halton Foundation
P.O. Box 834
Ophir, CO 81426

D.V. and Ida J. McEachern Charitable Trust Grants 1406
The Trust was established to give a better start in life to all children, both educationally and physically. Most grants are made to social service agencies addressing basic human needs for children and youth. The Trust is also interested in providing creative and positive opportunities to enhance the lives of the region's children through a variety of artistic and cultural programs. The Advisory Committee prefers proposals from groups that clearly describe how their program makes a positive difference in the lives of children, provides a unique service, or addresses problems that are not being met by other agencies in the area. Grants will be made solely for capital projects and items of immediate and direct service to children. There is no letter of inquiry process. If your project fits the guidelines, submit a proposal letter with a narrative no longer than four pages.
Requirements: Grant organizations should: request capital support only (McEachern grants will generally not exceed 5% of a capital campaign budget); provide services to a broad cross-section of the population up to age 18; be established for at least five years (exceptions may be made for groups with strong community leadership and reputation, offering unique enrichment programs for children); serve Puget Sound region (particularly King, Pierce, and Snohomish Counties); have a secure and diverse funding base, with a majority of budget income derived from non-governmental sources.
Restrictions: Grants will generally not be made to endowment campaigns, individuals, private schools, public schools, day-care programs, political candidates, or religious institutions.
Geographic Focus: Washington
Date(s) Application is Due: Mar 8; Sep 8; Dec 8
Amount of Grant: 15,000 - 25,000 USD
Samples: Burke Museum (Seattle, WA)--for renovation and development of a Natural History Demonstration Lab for students, $25,000; Highline Community Hospital Foundation (Tukwila, WA)--for children's area of a new cancer care center, $15,000; Salish Sea Expeditions (Bainbridge, WA)--equipment upgrades for the inquiry-based marine education science program, $2500; YMCA Tacoma Pierce County (Tacoma, WA)--for sports equipment replacement for youth programs, $5000.
Contact: Therese Ogle; (206) 781-3472; fax (206) 784-5987; oglefounds@aol.com
Internet: http://fdncenter.org/grantmaker/mceachern/guide.html
Sponsor: D.V. and Ida J. McEachern Charitable Trust
P.O. Box 3123
Seattle, WA 98114

D. W. McMillan Foundation Grants 1407
The D.W. McMillan Foundation, established in Alabama in 1956, supports organizations involved with children and youth services, health care, health organizations, homelessness, hospitals, human services, mental health and crisis services, residential and custodial care, hospices, people with disabilities, and the economically disadvantaged population. Giving is primarily centered in the states of Alabama and Florida. There are no specific deadlines, though applicants should submit proposals well before the annual board meeting on December 1. Applicants should begin by contacting the Foundation with a letter of inquiry. Final notification of awards are given by December 31 each year.
Geographic Focus: Alabama, Florida
Date(s) Application is Due: Nov 1
Samples: Department of Human Resources, Escambia County, Alabama, $160,000 - general operations; Appleton Volunteer Fire Department, Brewster, Alabama, $5,000 - general operations; Baptist Hospital, Pensacola, Florida, $30,000 - general operations.
Contact: Ed Leigh McMillan II, Treasurer; (251) 867-4881
Sponsor: D.W. McMillan Foundation
329 Belleville Avenue
Brewton, AL 36426-2039

DAAD Research Stays for University Academics and Scientists 1408
These grants and scholarships aim to provide foreign academics and scientists working in higher education or at research institutes with an opportunity to carry out a research project at a state (public) or state-recognized higher education institution or non-university research institute in Germany. Depending on the applicant's work schedule, the research stay will last between one and three months. Depending on the applicant's academic status, the monthly award will amount to 1,840 Euros for assistant lecturers, assistant professors and young lecturers, and 1,990 Euros for professors. In some rare exceptions, 2,240 Euros may be available. In addition to these payments, the DAAD generally will pay an appropriate flat-rate travel allowance, unless these costs are covered by the home country or by another funding source. Applications for research grants are decided twice a year. Applications must have been submitted by 1 August or 15 January. At the earliest, grants can begin 4 months after the date of application.
Requirements: Applications for DAAD research stays are open to excellently-qualified academics and scientists who should generally hold a doctorate/PhD. All applicants must be working in higher education or at a research institute in their home country. Besides their previous academic achievements (for example, recent publications), the most important selection criterion is a convincing and well-planned research project to be completed during the stay in Germany. The application must provide proof of a workplace being provided at the host institute.
Restrictions: DAAD support for a research stay can only be awarded once in any three-year period. No travel expenses can be paid from the grant.
Geographic Focus: All States
Contact: Administrator; (212) 758-3223; fax (212) 755-5780; kim@daad.org
Internet: http://www.daad.de/deutschland/foerderung/stipendiendatenbank/00462.en.html?detailid=39&fachrichtung=15&land=44&status=4&seite=1&daad=1
Sponsor: German Academic Exchange Service (DAAD)
871 United Nations Plaza
New York, NY 10017

Dade Community Foundation Community AIDS Partnership Grants 1409
The Community AIDS Partnership, is a special funding initiative established in 1990 through the National AIDS Fund and local funders to increase the availability of private funds to address gaps in the local HIV/AIDS service system, particularly in the area of prevention. The Community AIDS Partnership represents one of the major sources of private philanthropic funds for HIV prevention in Miami-Dade County. As the result of a public/private partnership with Miami-Dade County, the Foundation also awards $350,000 in county funds, which represents a significant increase in the local public sector's commitment to funding HIV prevention. Through the Community AIDS Partnership, the Foundation supports quality programs that respond to the local population risk profile and that incorporate strategies with the greatest potential to effectively address the priority populations, communities and service needs related to HIV prevention in Miami-Dade County. Grant Size: $30,000 Single Organization, $60,000 for Collaborations.
Requirements: Eligible applicants include nonprofit tax-exempt organizations, as defined by the Internal Revenue Code, which are serving the residents of Miami-Dade County. Preference will be given to or- ganizations based in Miami-Dade County or if located outside the county, are working in partnership with an organization based in Miami-Dade.
Geographic Focus: Florida
Amount of Grant: 30,000 - 60,000 USD
Samples: Camillus House, $30,000 - to support a comprehensive HIV/AIDS program that targets homeless members of racial/ethnic minority communities who are either HIV positive or are at very high risk for HIV infection due to injection drug use and/or heterosexual contact; MUJER, $30,000 - to continue the Nosotras Viviremos project which seeks to increase awareness about HIV/AIDS in migrant farmworker communities in South Miami-Dade County;
Contact: Gianne Ewing-Chow, Senior Program Officer; (305) 371-2711; fax (305) 371-5342; gianne.ewingchow@dadecommunityfoundation.org
Internet: http://www.dadecommunityfoundation.org/Site/wc/wc145.jsp
Sponsor: Dade Community Foundation
200 S Biscayne Boulevard, Suite 505
Miami, FL 33131-5330

Dade Community Foundation Grants 1410
The funding for this program is made available through the Foundation's unrestricted and field of interest funds. This program is designed to honor both the donors interests and address significant community issues such as: education; health; human services; arts and culture; environment; economic development; at-risk youth; abused and neglected children; living with HIV/AIDS; homelessness; social justice; care of animals; heart disease and more.
Requirements: Eligible applicants include nonprofit tax-exempt organizations, as defined by the Internal Revenue Code, which are serving the residents of Miami-Dade County. Preference will be given to organizations based in Miami-Dade County or if located outside the county, are working in partnership with an organization based in Miami-Dade.
Restrictions: The Foundation does not provide grants to individuals, for memberships, fundraising events or memorials. Grants to government agencies are made on a very restricted basis.
Geographic Focus: Florida
Date(s) Application is Due: Nov 15
Amount of Grant: 7,500 USD
Samples: Alliance for Musical Arts Productions, $7,500 - to provide after school and summer camp services to youth from low-income households living in Opa-locka; Chai Lifeline, $6,500 - to support Smile S'more, a unique program designed to bring fun and love to children who are being treated for cancer and other life-threatening illnesses on pediatric specialty units of hospitals; Earth Learning, $10,000 - to create an EarthFest network that will form an umbrella to unite Earth Day celebrations across South Florida under one banner.
Contact: Charisse Grant, Vice President for Programs; (305) 371-2711; fax (305) 371-5342; charisse.grant@dadecommunityfoundation.org
Internet: http://www.dadecommunityfoundation.org/Site/programs/overview.jsp
Sponsor: Dade Community Foundation
200 S Biscayne Boulevard, Suite 505
Miami, FL 33131-5330

Daimler and Benz Foundation Fellowships 1411
This program is addressed to young postdoctoral scientists in Germany, who are in the early phase of their postgraduate research. The program is, in the sense of the foundation's statutes, open for all scientific disciplines and subject areas. Prerequisites are an own research project and an affiliation with a scientific institution. The amount of the fellowship is 20.000 Euros annually, which is granted for a total of two years. The funds are disbursed by the respective research institution's third-party funds administration, where the fellowship holder is doing research. The funds can be used for equipment, especially for computers, for financing scientific assistants, research travel, participation in conferences, as well as for the organization of own conferences. The fellowship can not, however, be used for living expenses, printing costs, or for expendable items (e.g., chemicals, paper, etc.). The Foundation plans to award approximately ten per year scholarships, the exact number of scholarships may vary depending on the funds available. In accordance with the Foundation's statutes, the program is open to all disciplines and subjects.
Requirements: Postdoctoral scientists and assistant professors in the early phase of their scientific work as well as young researchers in a similar position, such as independent director of junior research groups, are eligible to apply. The applicant must be affiliated with a German university or research institution and carry out their own research project. Applications may be submitted in German and/or English. Each application is reviewed by a committee. The Commission consists of the Board of Management of Daimler and Benz Foundation, the managing director and evaluators. After the deadline initially there will be a pre-selection. The selected applicants will be invited to a seminar in the Daimler and Benz Foundation in which they present their science project in the presence of representatives of the Foundation as well as peer reviewers personally. Participation in this seminar and the presentation of the research project are a prerequisite for the subsequent award of a scholarship. Self-nomination for the Daimler and Benz postdoctoral program is possible and is done without special application forms.
Geographic Focus: Germany
Date(s) Application is Due: Oct 15
Amount of Grant: 20,000 EUR
Contact: Dr. Jörg Klein; 06 203-10 92-0; info@daimler-benz-stiftung.de
Internet: http://www.daimler-benz-stiftung.de/cms/index.php?page=home_en
Sponsor: Daimler and Benz Foundation
Dr.-Carl-Benz-Platz 2
Ladenburg, 68526 Germany

DaimlerChrysler Corporation Fund Grants 1412
The fund contributes to organizations grouped under the general categories of education, health and human services, civic and community, religion, and culture and the arts. Within these categories, grants are made available for public welfare or for charitable, scientific, educational, environmental, safety, building, and affirmative action purposes. Higher education grants largely support science and engineering education and business management. A major interest of the corporation is the establishment of national certification standards for elementary and secondary teachers. Another area of concern is the encouragement of early reading skills, and a pilot project has been funded for research in this area. The fund earmarks funds for its future workforce initiatives, which support business and engineering departments, community-based job-skill training, and entry-level work preparation. Types of support include matching gifts, program grants, scholarships, annual campaigns, building construction/renovation, general support, and employee matching gifts. In considering requests, the fund evaluates each applicant organization on its own merits; considered are the programs in which it is engaged, constituencies served, operation procedures, services offered, quality of management, its accountability, finances, and fund-raising practices. Applications are accepted at any time.
Requirements: Eligible for support are nonprofit, tax-exempt educational, health, civic, and cultural organizations primarily in locations where the greatest number of employees of Chrysler and its US-based subsidiaries live and work (Alabama, Delaware, Illinois, Indiana, Michigan, Missouri, New York, Ohio, and Wisconsin). Some support is targeted for national organizations as well.
Restrictions: Grants are not awarded to support endowments, conferences, trips, direct health care delivery, multiyear pledges, capital campaigns, fund-raising activities related to sponsorships, advertising, or debt retirement.
Geographic Focus: All States
Samples: Charles H. Wright Museum of African American History (Detroit, MI)--$1 million to complete an exhibit area; Second Harvest Food Bank of Central Florida, $25,000 and food via convoy of Dodge Rams, and $50,000 to help residents affected by recent tornadoes.
Contact: Brian Glowiak; (248) 512-2502; fax (248) 512-2503; mek@dcx.com
Internet: http://www.fund.daimlerchrysler.com
Sponsor: DaimlerChrysler Corporation Fund
1000 Chrysler Drive
Auburn Hills, MI 48326-2766

Dairy Queen Corporate Contributions Grants 1413
The Foundation's philosophy toward awarding contribution grants is to assist those organizations which have an impact on the company's home state of Minnesota and its communities and have programs geared toward children. Foundation focus is on: culture and the performing arts, education, health care and hospitals, and social welfare and community services. Application information is available online.
Requirements: U.S. nonprofits are eligible. Priority is given to Minnesota organizations.
Restrictions: Funds are not available for: strictly sectarian or denominational religious organizations; direct or indirect use for political purposes; lobbying activities; benefits and fund-raisers; commemorative, courtesy, institutional or any other form of advertising; organizations that tend to be for the benefit of the individual members, rather than the general public, such as fraternities, sororities, social clubs, labor organizations, parties and banquets; conferences, seminars or meetings; individuals or scholarships to individuals.
Geographic Focus: Minnesota
Date(s) Application is Due: Sep 1
Contact: Janelle Ianfolla; (952) 830-0207; fax (952) 830-0480; janelle.ianfolla@idq.com
Internet: http://www.dairyqueen.com/us-en/community/
Sponsor: Dairy Queen Corporation
7505 Metro Boulevard
Minneapolis, MN 55439-0286

Daisy Marquis Jones Foundation Grants 1414
The mission of the Foundation is to improve the well being of residents in Yates and Monroe counties, New York in particular, the City of Rochester. A central concern is disadvantaged children and families and the neighborhoods in which they live. The Foundation believes it can best serve those in need by granting time-limited support with programs or projects that provide access to health care, attend to the needs of young children, assist senior citizens, or help families develop economic security. The Foundation also looks for programs that: give people the tools they need to help themselves; encourage collaboration among agencies and between individuals and agencies; have measurable outcomes; and make long-term commitments to specific neighborhoods. A letter of inquiry is the first step to application and is available on line. The board reviews letters of inquiry once a month except July and August.

Requirements: Grants are awarded to qualified 501(c)3 nonprofit organizations that are located in Monroe and Yates Counties, NY.
Restrictions: Grants are not awarded to fund the arts, endowments, local chapters of national health-related organizations, private schools, religious projects, research, scholarships, projects by individuals, or organizations, projects not in Yates or Monroe Counties, or non 501(c)3 organizations.
Geographic Focus: New York
Samples: Advertising Council of Rochester, $10,000--to help support the Ad Council Academy, Enhancing Professional Know How, an educational program for non-profits; Community Partners for Youth, Inc., $45,000--to help support the Big Brothers Big Sisters program; George Eastman House, $10,000--to support the Senior Citizens Matinee Series at the Dryden Theatre.
Contact: Roger Gardner; (585) 461-4950; fax (585) 461-9752; mail@dmjf.org
Internet: http://www.dmjf.org/funding.asp
Sponsor: Daisy Marquis Jones Foundation
1600 South Avenue, Suite 250
Rochester, NY 14620

Dale and Edna Walsh Foundation Grants 1415
The foundation joins hands with effective charitable organizations to meet human need and promote the common good worldwide. Grants support ministries, religious activities, health, relief efforts, education, community services, and arts organizations. Guidelines are available online.
Requirements: Tax-exempt, nonprofit charities are eligible.
Restrictions: The foundation does not contribute toward normal church operation; culturally liberal activist causes or organizations that primarily seek to influence legislation or government spending; or political parties or associated political organizations.
Geographic Focus: All States
Contact: Administrator; (847) 230-0056; Info@dewfoundation.org
Internet: http://dewfoundation.org
Sponsor: Dale and Edna Walsh Foundation
6461 Valley Wood Drive
Reno, NV 89523

Dallas Mavericks Foundation Grants 1416
The foundation of this National Basketball Association franchise makes grants in the Dallas, Texas, area for projects that provide education, good health, and leadership skills for kids. The foundation supports the programs and organizations in the Dallas-Fort Worth Metroplex that address the community's most pressing problems involving youth, specifically education, good health, and community service activity. Application guidelines are available online.
Requirements: To apply for an award, complete a grant application form which can be obtained on line. Include the supporting data: copy of IRS 501(c)3 final determination letter; fact sheet; current board roster; project timeline; project line-item detail budget; general operating budget. There is one grant cycle annually.
Restrictions: Grants will not be made to individuals, churches, public/private schools, or national organizations that do not have locally, financially independent chapters. Grants also will not be made for medical research, travel, salaries, operational phases of established programs, political campaigns or fundraising events, including the purchase of tables, tickets or advertisements. The Foundation would prefer not to commit funds for continued support of long-term programs lasting more than one year, endowment campaigns, administrative costs, advertising/fund raising drives, salaries for staff/individuals, intermediary funding agencies and/or research. The Foundation does not fund multi-year-grants.
Geographic Focus: Texas
Date(s) Application is Due: Jun 30
Contact: Dawn Holgate, (214) 658-7170
Internet: http://www.nba.com/mavericks/news/00405573.html
Sponsor: Dallas Mavericks Foundation
2909 Taylor Street
Dallas, TX 75226

Dallas Women's Foundation Grants 1417
The Foundation promotes women's philanthropy and raises money to support community programs that help women and girls realize their full potential. Priority will be given to funding programs that include elements of the following: effect long-term, positive changes to help women succeed in reaching their full potential; expanded choices and opportunities for women and girls; evaluation tools that include measurement for effectiveness with clean definition of program success; and programs specifically designed to take into consideration the gender-specific needs and differences of women and girls. The Foundation also encourages projects developed in consultation and collaboration with other agencies, and which promote coordination, cooperation and sharing among organizations. Effective use of volunteers, as well as Board diversity and involvement, are equally encouraged.
Requirements: To be eligible to receive a grant from the Foundation, applicants must meet all of the following criteria: be in receipt of a current 501(c)3 tax-exempt designation from the Internal Revenue Service (dated within the last 10 years); at least 50% of the population served must be residents of Dallas, Denton, or Collin County, with priority given to organizations serving residents of Dallas County; 75% of the clients benefiting from the grant funding must be women and/or girls.
Geographic Focus: Texas
Samples: Avance (Dallas, TX)--for equipment and operating support of this organization that promotes academic achievement among Hispanics in Dallas, $10,000; U of North Texas, Division of Equity and Diversity (Dallas, TX)--for a seminar on critical issues affecting lesbians, $15,000; Interfaith Housing Coalition (Dallas, TX)--to provide transitional housing, training, and support services to Dallas-area homeless families, $20,000; English Language Ministry (Dallas, TX)--to provide free child care and English-language instruction to non-English speaking adults, particularly women, $17,500.
Contact: Pat Alexander, (214) 965-9977; palexander@dallaswomensfoundation.org
Internet: http://www.dallaswomensfoundation.org/grants/highlights.html
Sponsor: Dallas Women's Foundation
4300 MacArthur Avenue, Suite 255
Dallas, TX 75209

Dammann Fund Grants 1418
The fund is committed to using its resources for the support of qualified charitable organizations operating programs in New York City, Southwestern Connecticut and Charlottesville, VA with a focus in any one or more of two areas: mental health - programs that foster living or independent living for the mentally ill; teen parenthood - programs that enable development of parenting skills in young parents. Grant recipients are eligible to receive funding for a single program for up to three consecutive years.
Requirements: Nonprofits in the greater New York City, Southwestern Connecticut and Charlottesville, VA.
Restrictions: Grants are not made to individuals and they are not made for capital or annual campaigns, endowments, loans, conferences or travel.
Geographic Focus: Connecticut, New York, Virginia
Date(s) Application is Due: Jun 30
Samples: Institute for Community Living (New York, NY)--to support residential and rehabilitation services for people with mental and physical disabilities, $12,000; Boston Medical Ctr Corp (Boston, MA)--for the Domestic Violence Guardian and Litem Program, $30,000; Louise Wise Services (New York, NY)--for the Teen Dad Program, $30,000; Northside Ctr for Child Development (New York, NY)--for therapeutic and educational programs for the Harlem community, $25,000.
Contact: Administrator; (212) 956-4118; fax (212) 262-9321; df@engelanddavis.com
Internet: http://www.thedammannfund.com/
Sponsor: Dammann Fund
521 Fifth Avenue, 31st Floor
New York, NY 10175

Dana Brown Charitable Trust Grants 1419
Dana Brown was a well-known personality in St. Louis due in part to his many appearances on television representing Safari brand coffee and sharing his latest adventures while on safari in Africa. He was an extraordinarily successful entrepreneur and philanthropist. Originally from West Virginia, Mr. Brown adopted St. Louis as his home and had a deep affection for the region and many of the institutions that have made this area great. Established in Missouri in 1994, the Dana Brown Charitable Trust awards grants to eligible Missouri nonprofit organizations. The primary purpose of the trust is to provide for the health, education and welfare of underprivileged and economically disadvantaged children in the St. Louis, Missouri metropolitan area. With that as its focus, the Foundation's current areas of interest include: animal welfare; wildlife and the environment; children and youth; health care; education; and human services. Types of support provided include: annual campaigns; building and renovation; capital campaigns; general operating support; and challenge grants. A formal application is required, with annual deadline dates identified as February 15, May 15, August 15, and November 15.
Requirements: Missouri nonprofit organizations are eligible. Preference is given to requests from the greater Saint Louis metropolitan area.
Restrictions: Grants will not be made directly to individuals. In addition, the Trustees wish to make grants directly to the charitable organizations that will utilize the funds and, therefore, generally will not consider requests of Supporting Organizations. Grants will not be awarded for feasibility studies. Multi-year grants will be considered, but the fulfillment of immediate needs will generally be preferred over long-term possibilities.
Geographic Focus: Missouri
Date(s) Application is Due: Feb 15; May 15; Aug 15; Nov 15
Amount of Grant: 5,000 - 250,000 USD
Samples: Saint Louis Zoo Foundation, Saint Louis, Missouri, $250,000 - general operations (2014); Little Bit Foundation, Saint Louis, Missouri, $5,200 - support of the Saint Louis Journal Giving Guide (2014); Center of Creative Arts, Saint Louis, Missouri, $25,000 - general operations (2014).
Contact: Kimberly Livingston; (314) 418-2643; kimberly.livingston@usbank.com
Internet: http://www.danabrowncharitabletrust.org/rules.html
Sponsor: Dana Brown Charitable Trust
P.O. Box 387
St. Louis, MO 63166

Dana Foundation Science and Health Grants 1420
The Foundation supports research in neuroscience, immunology, and the effects of arts training on cognition. Areas of focus are: Brain and Immuno-imaging; Human Immunology; Neuroimmunology; Clinical Neuroscience Research; and Arts and Cognition. All other Science and Health Grants are made solely by invitation.
Requirements: The Foundation requires institutions, in many cases, to share the cost of a project or raise matching funds.
Restrictions: The foundation makes no grants directly to individuals; does not support annual operating budgets of organizations, deficit reduction, capital campaigns, or individual sabbaticals; and does not schedule meetings with applicants, other than by specific invitation initiated by the foundation.
Geographic Focus: All States
Samples: Dartmouth College (Hanover, NH)--to conduct research on how arts education affects learning and the brain, to be carried out by the Dana Arts and Cognition Consortium,

a collaboration of six universities, $1.85 million over three years; Cold Spring Harbor Laboratory (NY)--to support the work of Hollis Cline, for research on fragile X syndrome, a genetic condition that causes mental impairment, $100,000 over 18 months.
Contact: Administrator; (212) 223-4040; fax (212) 317-8721; danainfo@dana.org
Internet: http://www.dana.org/grants
Sponsor: Charles A. Dana Foundation
745 Fifth Avenue, Suite 900
New York, NY 10151-0002

Danellie Foundation Grants 1421
The foundation's primary areas of interest include services for the financially disadvantaged, including housing and social services. The types of support offered are: building/renovation; capital campaigns; continuing support; general/operating support; program development; scholarship funds and; sponsorships. Contact foundation for application guidelines. New York/New Jersey Area Common Application Form and New York/New Jersey Common Report Form accepted. Application form required.
Requirements: Nonprofit organizations in Southern New Jersey, (including Mercer and portions of Monmouth counties), the Greater Philadelphia, Pennsylvania area and, the Balimore, Maryland region are eligible to apply. The foundation also has an international interest in Guatemala and Haiti.
Restrictions: No support for political organizations or professional sports, libraries or museums. No grants to individuals, or for endowments or radio and television.
Geographic Focus: Maryland, New Jersey
Amount of Grant: 4,000 - 180,000 USD
Samples: University of Medicine and Dentistry, Stratford, NJ, $71,000 - for Latino health clinic; Doctors of the World USA, New York, NY, $65,000 - for medical needs in Third World countries; Aid for Friends, Philadelphia, PA, $20,000 - for needy, disabled, and the elderly homebound.
Contact: Nancy Dinsmore, (856) 810-8320; danelliefoundation@verizon.net
Sponsor: Danellie Foundation
P.O. Box 376
Marlton, NJ 08053

Daniel and Nanna Stern Family Foundation Grants 1422
The Daniel and Nanna Stern Family Foundation was established in 2006, with the expressed interest in providing support primarily to programs in New York City. The Foundations major fields of interest include: the arts; education; hospitals; film and video; television; and performing arts centers. Though a formal application is required, interested applicants should begin by forwarding a letter to the Foundation office, offering a detailed description of the project, the amount of funding requested, contact information, and copies of both the IRS determination letter and most recent audit. No annual deadlines for submission have been identified. Most recent awards have ranged from $50,000 to $150,000.
Requirements: Any 501(c)3 organization located in, or serving the residents of, New York City are eligible to apply.
Geographic Focus: New York
Amount of Grant: 50,000 - 150,000 USD
Contact: Anne Colucci, (212) 610-9006 or (212) 610-9054
Sponsor: Daniel and Nanna Stern Family Foundation
650 Madison Avenue, 26th Floor
New York, NY 10022-1029

Daniel Mendelsohn New Investigator Award 1423
The Daniel P. Mendelsohn award is intended to encourage new investigators whose work is compatible with this mission. New investigators in the field of research or education whose work relates to health and well being are eligible for the award. Candidates, who may or may not be members of the FRI organization, must have demonstrated scholarly research potential compatible with the mission of FRI. They must show future promise, evidenced by energy and enthusiasm in research that addresses prevention, treatment or education activities. Candidates for the biennial award are reviewed by the Awards Review Committee of FRI that presents their recommendation to the Board of Directors at their October Annual Meeting. Deadline for recommending candidates is May 1 of even numbered years. The Award is a check for $5000 payable to the recipient for her or his personal or professional use. The award will be presented in the fall of the awarding year.
Geographic Focus: All States
Date(s) Application is Due: May 1
Amount of Grant: 5,000 USD
Samples: Michael S.Gordon, D.P.A., University of Baltimore, Baltimore, MD, $5,000- -drug abuse treatment for criminal justice populations.
Contact: Janet Klein Brown; (410) 823-5116; fax (410) 823-5131; fri@friendsresearch.org
Internet: http://www.friendsresearch.org/mendelsohn_award.htm
Sponsor: Friends Research Institute
505 Baltimore Avenue, P.O. Box 10676
Baltimore, MD 21285

Daniels Fund Grants-Aging 1424
Bill Daniels was a visionary business leader whose compassion for people, and unwavering ethics and integrity earned him respect throughout his life. He grew up during the Great Depression, served his nation as a decorated fighter pilot, and became a driving force in establishing the cable television industry. Mr. Daniel's passion for helping others inspired him to create the Daniels Fund to extend his legacy of generosity far beyond his lifetime. The Daniels Fund provides grants to nonprofit organizations in Colorado, New Mexico, Utah, and Wyoming that fit within its nine distinct funding areas. Mr. Daniels wanted to help seniors maintain personal dignity, remain independent, and be respected. The Daniels Fund supports organizations that share this vision and recognize the exceptional value of older adults to the community. The Fund accepts grant applications any time during the year; there are no submission deadlines.
Requirements: The Daniels Fund makes grants for specific programs or projects, general operating support, or capital campaigns. The organization applying must be classified by the Internal Revenue Service as a 501(c)3 or equivalent. Eligible nonprofit organizations must provide programs or services in Colorado, New Mexico, Utah, or Wyoming. Organizations with a nationwide impact and large institutions (such as a university or school district) should call before applying. Before starting the online application process, the sponsor strongly encourages all potential applicants to call first. The Daniels Fund is rarely the sole provider of funds for a project, and encourages applicants to develop a variety of individual, government, and private funding sources.
Restrictions: The fund generally will not support medical or scientific research; arts, cultural, and museum programs; environmental stewardship programs; historic preservation projects; candidates for political office; sponsorships, tables, or tickets for special events or fundraising events; endowments; fiscal sponsorships; or debt retirement.
Geographic Focus: Colorado, New Mexico, Utah, Wyoming
Contact: Bill Fowler, Senior Vice President, Grants Program; (303) 393-7220; fax (720) 941-4201; grantsinfo@danielsfund.org or BFowler@DanielsFund.org
Internet: http://danielsfund.org/Grants/Goals-Aging.asp
Sponsor: Daniels Fund
101 Monroe Street
Denver, CO 80206

Daphne Seybolt Culpeper Memorial Foundation Grants 1425
The Foundation, established in 1983, supports organizations involved with education, health care and human services. Giving is limited to Fairfield County, Connecticut and Palm Beach County, Florida. There are no specific deadlines with which to adhere. Contact the Foundation for further application information and guidelines.
Requirements: Nonprofits in Connecticut and Florida are eligible.
Restrictions: No grants to individuals, or for endowments, forums, conferences, seminars, gratuities, honorariums, travel, meals or lodging.
Geographic Focus: Connecticut, Florida
Amount of Grant: 1,000 - 100,000 USD
Contact: Nicholas Nardi, Secretary; (203) 762-3984
Sponsor: Daphne Seybolt Culpeper Memorial Foundation
129 Musket Ridge Road, P.O. Box 206
Norwalk, CT 06852-0206

DAR Alice W. Rooke Scholarship 1426
The Alice W. Rooke Scholarship is awarded to students who have been accepted into or are pursuing an approved course of study to become a medical doctor at an approved, accredited medical school. Renewal is conditional upon maintaining a GPA of 3.25. The application is available at the DAR website.
Requirements: All applicants must be citizens of the U.S. and must attend or plan to attend an accredited college or university in the U.S. DAR chapter sponsorship is not required; however, a chapter or state chairman may work with the applicant to put the information together to send to the DAR Scholarship Committee.
Restrictions: Those studying or accepted into pre-med, osteopathic medicine, veterinary sciences, or physician's assistant are not eligible to apply.
Geographic Focus: All States
Date(s) Application is Due: Feb 15
Amount of Grant: Up to 5,000 USD
Contact: Sharon Cothern Nettles, National Chairman, Scholarship Committee; (202) 628-1776; fax (202) 879-3252; scholarships@dar.org
Internet: http://www.dar.org/natsociety/edout_scholar.cfm#margaret
Sponsor: Daughters of the American Revolution
1776 D Street NW
Washington, D.C. 20006

DAR Dr. Francis Anthony Beneventi Medical Scholarship 1427
The Dr. Francis Anthony Beneventi Medical Scholarship offers funding of up to $5,000 to students who have been accepted into or who are pursing an approved course of study to become a medical doctor at an approved, accredited medical school, college or university. The applicant must have a minimum GPA of 3.25. The Scholarship is not automatically renewable; however, recipients may reapply for consideration each year for up to four consecutive years. The application is available at the DAR website.
Requirements: All applicants must be citizens of the U.S. and must attend or plan to attend an accredited college or university in the U.S. DAR chapter sponsorship is not required; however, a chapter or state chairman may work with the applicant to put the information together to send to the DAR Scholarship Committee.
Restrictions: Those enrolled in pre-med, osteopathic medicine, veterinarian, or physician assistant programs are not eligible to apply.
Geographic Focus: All States
Date(s) Application is Due: Feb 15
Amount of Grant: Up to 5,000 USD
Contact: Sharon Cothern Nettles, National Chairman, Scholarship Committee; (202) 628-1776; fax (202) 879-3252; scholarships@dar.org
Internet: http://www.dar.org/natsociety/edout_scholar.cfm#margaret
Sponsor: Daughters of the American Revolution
1776 D Street NW
Washington, D.C. 20006

DAR Irene and Daisy MacGregor Memorial Scholarship   1428

The Irene and Daisy MacGregor Memorial Scholarship is awarded to students of high scholastic standing and character who have been accepted into or are pursuing an approved course of study to become a medical doctor at an approved, accredited medical school. Renewal is conditional upon maintenance of a GPA of 3.25. This scholarship is also available to students who have been accepted into or who are pursing an approved course of study in the field of psychiatric nursing at the graduate level at accredited medical schools, colleges, or universities. Funding of up to $5,000 annually for four consecutive years is available, with a maximum for $20,000. Information is available at the website about how to obtain an application from the candidate's area DAR office. All applications must be mailed to the address of the National Chairman of the Scholarship Committee.
Requirements: All applicants must be citizens of the U.S. and must attend or plan to attend an accredited college or university in the U.S. Chapter sponsorship is not required; however, a chapter or state chairman may work with the applicant to put the information together to send to the DAR Scholarship Committee.
Restrictions: Pre-med, osteopathic medicine, veterinary, and physician's assistant fields of study are not eligible to apply.
Geographic Focus: All States
Amount of Grant: 5,000 - 20,000 USD
Contact: Sharon Cothern Nettles, National Chairman, Scholarship Committee; (202) 628-1776; fax (202) 879-3252; scholarships@dar.org
Internet: http://www.dar.org/natsociety/edout_scholar.cfm
Sponsor: Daughters of the American Revolution
1776 D Street NW
Washington, D.C. 20006

DAR Madeline Pickett Cogswell Nursing Scholarship   1429

The Madeline Pickett Cogswell Nursing Scholarship is awarded to students who are accepted or enrolled in an accredited school of nursing. Applicants must be members, descendents of members or be eligible for membership in NSDAR. The applicant's DAR Member Number must be on the application. This Scholarship is a one-time award of $1000.
Requirements: All applicants must be citizens of the U.S. and must attend or plan to attend an accredited college or university in the U.S. DAR chapter sponsorship is not required; however, a chapter or state chairman may work with the applicant to put the information together to send to the DAR Scholarship Committee.
Geographic Focus: All States
Date(s) Application is Due: Feb 15
Amount of Grant: 1,000 USD
Contact: Sharon Cothern Nettles, National Chairman, Scholarship Committee; (202) 628-1776; fax (202) 879-3252; scholarships@dar.org
Internet: http://www.dar.org/natsociety/edout_scholar.cfm#margaret
Sponsor: Daughters of the American Revolution
1776 D Street NW
Washington, D.C. 20006

DAR Nursing/Physical Therapy Scholarships   1430

The Daughters of the American Revolution has a number of one-time award nursing and physical therapy scholarships available for $1,000 each. It is only necessary to submit one application for consideration for any award in this group. The application and additional information is available at the DAR website.
Requirements: All applicants must be citizens of the U.S. and must attend or plan to attend an accredited college or university in the U.S. DAR chapter sponsorship is not required; however, a chapter or state chairman may work with the applicant to put the information together to send to the DAR Scholarship Committee.
Geographic Focus: All States
Date(s) Application is Due: Feb 15
Amount of Grant: 1,000 USD
Contact: Sharon Cothern Nettles; (202) 628-1776; scholarships@dar.org
Internet: http://www.dar.org/natsociety/edout_scholar.cfm
Sponsor: Daughters of the American Revolution
1776 D Street NW
Washington, D.C. 20006

David Geffen Foundation Grants   1431

The foundation supports nonprofits in Los Angeles, California, and New York, New York, with some giving in Israel as well. The Foundation offers support in its five principal funding areas: AIDS/HIV, civil liberties, the arts, issues of concern to the Jewish community, and health care. Support is provided for general operations and special projects. There are no application deadlines. Applicants are asked to submit a letter of request.
Requirements: 501(c)3 tax-exempt organizations in Los Angeles, CA, and New York, NY, may submit a proposal letter (without folder or binder) including description of the project, objectives, constituents served, evaluation criteria; background on the organization including key staff, volunteers, and board; copy of IRS letter confirming tax-exempt status; financial information including line-item budget for the project; and copy of non-discrimination policy from the organization's hiring guidelines.
Restrictions: The Foundation does not fund individuals, or organizations based outside of the United States. The foundation generally does not support documentaries or other types of audio-publication of books or magazines.
Geographic Focus: California, New York, Israel
Contact: J. Dallas Dishman; (310) 581-5955; ddishman@geffenco.com
Sponsor: David Geffen Foundation
12011 San Vicente Boulevard, Suite 606
Los Angeles, CA 90049-4926

David M. and Marjorie D. Rosenberg Foundation Grants   1432

The David M. and Marjorie D. Rosenberg Foundation was established in Pennsylvania in 1993. David Rosenberg attended Penn State at University Park and graduated with a bachelor of science degree in health and human development. He also attended the Howard University School of Law and received his juris doctor degree. In 2012, he and his wife, Marjorie Rosenberg, made the Penn State Brandywine Laboratory for Civic and Community Engagement possible by making a generous donation. The lab encourages leadership among students and develops scholarship in the community while promoting citizenship on a local-to-global level. The Rosenbergs also founded the David and Marjorie Rosenberg Trustee Scholarship and the Rosenberg Family Trustee Scholarship for students enrolled at Penn State Brandywine. The Rosenberg Foundation is a civic and social organization assisting a wide variety of non-profits with the primary focus in the areas of character development and children and youth. Giving is primarily in Pennsylvania, with some awards going to California. The Foundation's primary fields of interest include: child welfare; diseases; elementary and secondary education; hugher education; hospital care; human services; Judaism; and community nonprofits. There are no identified annual deadlines, and most recent awards have ranged from $50 to $90,000.
Requirements: 501(c)3 organizations serving residents of Pennsylvania, District of Columbia, and California are eligible to apply.
Geographic Focus: California, District of Columbia, Pennsylvania
Amount of Grant: 250 - 90,000 USD
Samples: American Technion Society, Bala Cynwyd, Pennsylvania, $39,000 - general operating support (2014); Children's Hospital Foundation, Philadelphia, Pennsylvania, $85,600 - general operating support (2014); U.S. Holocaust Museum, Washington, D.C., $26,000 - general operating support (2014).
Contact: Marjorie D. Rosenberg, Trustee; (610) 458-4175
Sponsor: David M. and Marjorie D. Rosenberg Foundation
893 Parkes Run Lane
Villanova, PA 19085-1124

David N. Lane Trust Grants for Aged and Indigent Women   1433

The David N. Lane Trust for Aged & Indigent Women was established in 1964 to support and promote quality human services and health care programming for underserved older women living in Fairfield and New Haven Counties, Connecticut. Grants from the Lane Trust are one year in duration. Applicants must apply online at the grant website. Applicants are strongly encouraged to do the following before applying: review the downloadable state application procedures for additional helpful information and clarifications; review the downloadable online-application guidelines at the grant website; review the trust's funding history (link is available from the grant website); review the online application questions in advance; and review the list of required attachments. These will generally include: a list of board members, financial statements (audited, reviewed, or compiled by independent auditor); an organization summary; a list of other funding sources; an IRS Determination letter; and other required documents. All attachments must be uploaded in the online application as PDF, Word, or Excel files. The deadline for application to the David N. Lane Trust for Aged & Indigent Women is July 1. Applications must be submitted by 11:59 p.m. on the deadline date. Applicants will be notified of grant decisions by letter within two to three months after the proposal deadline.
Requirements: Applicant organizations must have 501(c)3 tax-exempt status and serve senior women in the counties of Fairfield or New Haven, Connecticut. A breakdown of number/percentage of individuals served by specific towns will be required in the online application.
Restrictions: Grant requests for capital projects will not be considered. The trust does not support requests from individuals, organizations attempting to influence policy through direct lobbying, or any political campaigns.
Geographic Focus: Connecticut
Date(s) Application is Due: Jul 1
Samples: TEAM, Derby, Connecticut, $10,000; Wesley Heights, Shelton, Connecticut, $9,000; VNA Health at Home, Watertown, Connecticut $3,000.
Contact: Kate Kerchaert; (860) 657-7016; kate.kerchaert@baml.com
Internet: https://www.bankofamerica.com/philanthropic/fn_search.action
Sponsor: David N. Lane Trust for Aged and Indigent Women
200 Glastonbury Boulevard, Suite # 200, CT2-545-02-05
Glastonbury, CT 06033-4056

Daviess County Community Foundation Health Grants   1434

The Foundation considers proposals for grants on a yearly cycle, which begins each May. At the start of each cycle, a notice is mailed to nonprofit organizations that have applied for grants in the past, have received grants in the past, or have otherwise requested notification of the start of each cycle. Grants in the area of health include activities that: improve and promote health outcomes; provide general and rehabilitative health services; offer mental health services; provide crisis intervention programs; strengthen associations or services associated with specific diseases, disorders, and medical disciplines; and support medical research. Samples of previously funded projects are posted on the website.
Requirements: The Foundation welcomes proposals from nonprofit organizations that are deemed tax-exempt under sections 501(c)3 and 509(a) of the Internal Revenue Code and from governmental agencies serving the County of Daviess, Indiana. Proposals from nonprofit organizations not classified as a 501(c)3 public charity may be considered provided the project is charitable and supports a community need.
Restrictions: Funding is not available for the following: religious organizations for strictly religious purposes; political parties or campaigns; endowment creation or debt reduction; operating costs (not directly related to the proposed project or program); capital campaigns; annual appeals or membership contributions; travel requests for groups or individuals such as bands, sports teams, or classes. Not more than 20% of

any grant request may be for personnel costs, office supplies, or other operating costs. Operating costs for any organization must be directly related to the project or program for which funding is being requested.
Geographic Focus: Indiana
Contact: Jeanne Fields, Director; (812) 254-9354; fax (812) 254-9355; jeanne@daviesscommunityfoundation.org
Internet: http://www.daviescommunityfoundation.org/program-areas
Sponsor: Daviess County Community Foundation
320 East Main Street, P.O. Box 302
Washington, IN 47501

Davis Family Foundation Grants 1435
The foundation provides grants primarily to Maine-based educational, medical, and cultural/arts charitable organizations in support of a wide variety of worthwhile projects. Eligible educational organizations include: colleges, universities, and other educational institutions. Eligible medical organizations include: hospitals, clinics and medical research organizations (grant requests will also be considered from other similar health organizations for programs designed to increase the effectiveness or decrease the cost of medical care). Eligible cultural and arts organizations include: those agencies whose customary and primary activity is to promote music, theater, drama, history, literature, the arts or other similar cultural activities. Further guidelines are available online.
Requirements: Eligible educational organizations include colleges, universities, and other educational institutions (grants are not made to public elementary and secondary schools, nor to schools whose financial support is derived primarily from a church or other religious organization. Trustees will consider grant requests from other educational organizations whose purpose is to promote systemic change in education or to provide innovative programs whose objectives are to improve education). Medical organizations eligible for support include hospitals, clinics, and medical research organizations. Grant requests will also be considered from other similar health organizations for programs designed to increase the effectiveness or decrease the cost of medical care. Eligible cultural/arts organizations include organizations whose customary and primary activity is to promote music, theater, drama, history, literature, the arts, or other similar cultural activities.
Restrictions: The Foundation does not make grants to individuals, religious programs, fellowships, or in the form of loans. The Foundation does not normally provide support for annual giving campaigns or general operating needs. Grants to endowment campaigns have a low priority.
Geographic Focus: Maine
Date(s) Application is Due: Feb 10; May 10; Aug 10; Nov 10
Samples: Arthur L. Mann Memorial Library, West Paris, ME, $15,000 - expansion and renovation project; Maine Archaeological Society, Inc., Augusta, ME, $8,300 - Maine Archaeology Classroom Initiative; South Bristol Historical Society, South Bristol, ME, $20,000 - schoolhouse restoration.
Contact: Anne Vaillancourt; (207) 781-5504; info@davisfoundations.org
Internet: http://www.davisfoundations.org/site/family.asp
Sponsor: Davis Family Foundation
4 Fundy Road
Falmouth, ME 04105

Dayton Power and Light Company Foundation Signature Grants 1436
The Dayton Power and Light Company Foundation introduced its Signature Grant program in 2013 to provide nonprofit organizations the opportunity to request funding for larger projects, including but not limited to capital projects. Signature Grants are aimed at special projects (such as a new or innovative program) and long-time partners in the Dayton community efforts. The Foundation commitment for a Signature Grant is for a one-year period. Requests for multi-year grants will be reviewed annually. There will be at least two Signature Grants given each year, one at the $100,000+ level and one at the $50,000+ level.
Requirements: 501(c)3 organizations in the greater Dayton, Ohio, area are eligible.
Geographic Focus: Ohio
Amount of Grant: 50,000 - 200,000 USD
Samples: Culture Works, Dayton, Ohio, $100,000 - to provide matching funds for Power2Give, an online crowd funding initiative empowering individual donors to give directly to a variety of local arts, science and history projects (2014); Dayton Children's Hospital, Dayton, Ohio, $50,000 - to provide car seats and bike helmets to the community and mobile sanitation stations for the safety of hospital staffers and patients (2014); K-12 Gallery for Young People, Dayton, Ohio, $120,000 - to fund new building renovations and to support youth educational programs.
Contact: Ginny Strausburg; (937) 259-7925; ginny.strausburg@dplinc.com
Internet: http://www.dpandl.com/about-dpl/who-we-are/community-investments/#DPLFoundation
Sponsor: Dayton Power and Light Company Foundation
1065 Woodman Drive, P.O. Box 1247
Dayton, OH 45432

Dayton Power and Light Foundation Grants 1437
The Dayton Power and Light Foundation was established in 1985 reinvest in the communities it serves and contribute to the improvement of the overall quality of life. The Foundation focuses its contributions in the following strategic contribution areas: economic development - creating an engaged, vibrant, welcoming community that is seen as a great place to live and work; arts and culture - heightening the impact of arts and local culture in our communities; health and human services - to improve the quality of life for all; and education: - improving educational access and outcomes. Direct donations are also made to civic, cultural, and health and welfare organizations that do not participate in community funds, such as United Way or community chests, but serve a real need. Requests should be made via the online application format, and should include a description of the history, structure, purpose, and program of the organization and a summary of the support needed and how it will be used. The annual deadline is October 1
Requirements: 501(c)3 organizations in the greater Dayton, Ohio, area are eligible. This includes a twenty-four county service area surrounding the city (see map at the website for specific details).
Restrictions: The foundation prefers not to support: capital campaigns; college fund-raising associations; conduit organizations; endowment or development funds; fraternal, labor, or veterans organizations; hospital operating budgets; individual members of federated campaigns; individuals; national organizations outside the DP&L service territory; religious organizations; sports leagues; or telephone or mass-mail solicitations. Grants are rarely made to tax-supported institutions.
Geographic Focus: Ohio
Date(s) Application is Due: Oct 1
Amount of Grant: 1,000 - 20,000 USD
Contact: Ginny Strausburg; (937) 259-7925; ginny.strausburg@dplinc.com
Internet: http://www.waytogo.com/cc/cc.phtml
Sponsor: Dayton Power and Light Company Foundation
1065 Woodman Drive, P.O. Box 1247
Dayton, OH 45432

Daywood Foundation Grants 1438
The Foundation distributes available funds primarily to community service organizations focusing on the arts, and health and human services. Grants are available in the following types: annual campaigns; building/renovation; capital campaigns; continuing support; debt reduction; emergency funds; equipment; general/operating support; matching/challenge support; seed money.
Requirements: Nonprofit 501(c)3 organizations situated in, and benefiting Charleston, Lewisburg, Barbour, Greenbrier and Kanawha counties Of West Virginia are eligible to apply. There is no standardized grant application form. Send proposal to: William W. Booker, 1500 Chase Tower, Charleston, WV 25301.
Restrictions: Grants do not support endowment funds, research, individuals, or individual scholarships or fellowships.
Geographic Focus: West Virginia
Date(s) Application is Due: Sep 15
Contact: William W. Booker, Treasurer; (304) 343-4841
Sponsor: Daywood Foundation
707 Virginia Street E, Suite 1600
Charleston, WV 25301

Deaconess Community Foundation Grants 1439
The foundation awards grants to eligible Ohio nonprofit organizations in its areas of interest, including health, education, welfare, community, and social service activities. Proposals that are of greatest interest to the Foundation are those that have the strongest fit to the mission statement and that have some or all of the following characteristics: projects that have specific measurable outcomes and a tangible ability to evaluate results and measure success; projects that are supported by other funding sources; and projects that have identified potential for ongoing support beyond the life of the grant. Application information is available online.
Requirements: Only qualified non-profit organizations located in Cuyahoga County which are classified by the Internal Revenue Code as tax-exempt 501(c)3 organizations are eligible for funding consideration.
Restrictions: Grant requests for the following will not be considered: individuals, governmental agencies or any other organization that is not a tax exempt 501(c)3 organization; internal operations and capital campaigns of churches; research projects; or endowments. Grant funds may not be used to carry on propaganda or otherwise attempt to influence legislation, participate in, or intervene in, any political campaign on behalf of or in opposition to any candidate for public office, or to conduct, directly or indirectly, any voter registration drive (within the meaning of Section 4945(d)2 of the Internal Revenue Code).
Geographic Focus: Ohio
Date(s) Application is Due: Jan 15; May 15; Sep 15
Sample: Achievement Centers for Children, Highland Hills, Ohio, $20,000 - program support for Family Support Services for children with disabilities and their families; The Center For Nonprofit Excellence, Cleveland, Ohio, $75,000 - program support for Building Nonprofit Excellence in 2013 which includes Needs Assessments and BVU's Consulting Center for nonprofits with missions that are aligned with D.C.F.
Contact: Deborah Vesy; (216) 741-4077; fax (216) 741-6042; dvesy@deacomfdn.org
Internet: http://www.deacomfdn.org/guidelines.html
Sponsor: Deaconess Community Foundation
7575 Northcliff Avenue, Suite 203
Brooklyn, OH 44144

Dean Foods Community Involvement Grants 1440
The company's community support efforts are focused in three main areas: health/nutrition (including hunger); education/arts; and environmental stewardship/conservation. It supports worthy organizations both at the corporate level and locally through its network of processing facilities nationwide. Preference is given to supporting and participating in a meaningful way with a limited number of organizations that support these three focus areas, rather than spreading limited resources more broadly. Specifically, funding goes to programs that provide direct service to individuals and communities in need. The corporation places special emphasis on supporting organizations that assist children, particularly at-risk children or children with disabilities, or that are dedicated to serving their needs. As a point of contact, the corporate headquarters requests that no phone calls be made regarding its giving programs.

Requirements: 501(3)3 tax-exempt organizations are eligible. Applications must provide background information on the organization as well as how it relates to the mission of Dean Foods. The corporation supports initiatives only in the communities in which they operate and where employees live and work.
Geographic Focus: Alabama, California, Colorado, Connecticut, Florida, Georgia, Hawaii, Idaho, Illinois, Indiana, Kentucky, Louisiana, Maine, Maryland, Massachusetts, Michigan, Minnesota, Montana, Nebraska, Nevada, New Jersey, New Mexico, New York, North Carolina, North Dakota, Ohio, Oklahoma, Pennsylvania, South Carolina, South Dakota, Tennessee, Texas, Utah, Virginia, Wisconsin
Contact: Gregg Engles, Chief Executive Officer; (214) 303-3400; fax (214) 303-3499
Internet: http://www.deanfoods.com/our-company/about-us/corporate-responsibility.aspx
Sponsor: Dean Foods
2515 McKinney Avenue, Suite 1200
Dallas, TX 75201

Dearborn Community Foundation City of Lawrenceburg Community Grants   1441
The City of Lawrenceburg has allocated grant funds specifically for not-for-profit organizations that provide a benefit, direct or indirect, to the Lawrenceburg community and Dearborn County. The Dearborn Community Foundation will administer this program. The program consists of two phases, Phase I: applications requesting $5,000 or less will be accepted and considered as long as funds are available. Phase II: applications requesting $5,001 - $100,000 will be accepted to the grant application deadlines. Applications will be considered as long as funds are available. The application and specific guidelines are available at the website.
Restrictions: Funding is not available for the following: individuals; endowment creation; travel expenses; sustain ongoing programs or projects; salaried/contracted positions; political parties, campaigns, or issues; sectarian religious purposes; debt reduction of recipient organizations and programs, expenses and/or equipment committed to prior to the grant award date.
Geographic Focus: Indiana
Date(s) Application is Due: Mar 6; Jun 5; Sep 11
Amount of Grant: 5,000 - 100,000 USD
Contact: Denise Sedler; (812) 539-4115; fax (812) 539-4119; dsedler@dearborncf.org
Internet: http://www.dearborncf.org/grants/G_Lawrence.aspx
Sponsor: Dearborn Community Foundation
322 Walnut Street
Lawrenceburg, IN 47025

Dearborn Community Foundation County Progress Grants   1442
The Dearborn Community Foundation County Progress Grant supports the interests of the community from non-profit organizations for charitable purposes in the fields of community service, social service, education, health, environment, and the arts. The application is available at the website.
Geographic Focus: Indiana
Date(s) Application is Due: Mar 6; Jun 5; Sep 11
Amount of Grant: 10,000 USD
Contact: Denise Sedler; (812) 539-4115; fax (812) 539-4119; dsedler@dearborncf.org
Internet: http://www.dearborncf.org/grants/G_Progress.aspx
Sponsor: Dearborn Community Foundation
322 Walnut Street
Lawrenceburg, IN 47025

Deborah Munroe Noonan Memorial Fund Grants   1443
The Deborah Munroe Noonan Memorial Fund was established in 1949 by her son, Walter Noonan, to support and promote quality educational, human-services, and health-care programming for underserved populations. Grant requests for general operating support are strongly encouraged. Program support will also be considered. Small, program-related capital expenses may be included in general operating or program requests. To better support the capacity of nonprofit organizations, multi-year funding requests are strongly encouraged. Applicants must apply online at the grant website. Applicants are strongly encouraged to do the following before applying: review the downloadable state application procedures for additional helpful information and clarifications; review the downloadable online-application guidelines at the grant website; review the foundation's funding history (link is available from the grant website); review the online application questions in advance; and review the list of required attachments. These will generally include: a list of board members, financial statements (audited, reviewed, or compiled by independent auditor); an organization summary; a list of other funding sources; an IRS Determination letter; and other required documents. All attachments must be uploaded in the online application as PDF, Word, or Excel files. The application deadline for the Deborah Munroe Noonan Memorial Fund is 11:59 p.m. on July 1. Applicants will be notified of grant decisions before September 30.
Requirements: Applicants must have 501(c)3 tax-exempt status and serve the people of Greater Boston.
Restrictions: The fund does not support requests from individuals, organizations attempting to influence policy through direct lobbying, or any political campaigns.
Geographic Focus: Massachusetts
Date(s) Application is Due: Jul 1
Samples: Community Music Center of Boston, Boston, Massachusetts, $21,000, general operating support of CMCB's 100th anniversary and to support a continuum of music education programs and music therapy for children, teens, adults; Family and Childrens Services of Greater Lynn, Lynn, Massachusetts, $25,000, capacity-building for program services that promote early literacy; Brockton Neighborhood Health Center, Brockton, Massachusetts, $29,000, Certified Medical Interpreter Program.
Contact: Miki C. Akimoto, Vice President; (866) 778-6859; miki.akimoto@baml.com
Internet: https://www.bankofamerica.com/philanthropic/fn_search.action
Sponsor: Deborah Munroe Noonan Memorial Fund
225 Franklin Street, 4th Floor, MA1-225-04-02
Boston, MA 02110

Decatur County Community Foundation Large Project Grants   1444
The Decatur County Community Foundation (D.C.CF) encourages, manages, and distributes charitable contributions to improve the quality of life of Decatur County, Indiana residents, now and in the future. The Foundation places high priority to funding projects which are: new and innovative projects or programs, including start-ups; projects which Foundation funds can be used as match, seed money or challenge grant funding from other donors; projects which will make a significant impact in the community; projects which act as a catalyst for action and community participation. The Large Project Community Grants are reviewed twice a year.
Requirements: Each applicant is required to submit a letter of intent to see if the project complies with general guidelines. No application will be sent without a letter of intent. Form letters will neither be reviewed nor acknowledged. Upon acceptance, the applicant will receive a grant application packet. The grant committee will review the completed packet. A member of that committee may contact the applicant or request a site visit. The committee's recommendations are forwarded to the Foundation's Board of Directors, who will make final funding decisions. The Board may choose to fund the grant as written, fund part of the grant or provide no funding at all. All applicants are notified in writing regarding funding decisions.
Restrictions: The Foundation will not fund: individual and team travel expenses; multi-year or long term funding; the creation of an endowment; programs that fall appropriately under government funding; annual appeals; projects considered part of the school curriculum; attendance to conferences or seminars; annual campaigns; projects where the Foundation is the sole funder; or advertising. The Foundation will also not fund: political activities; make-up operating deficits; post event or after the fact situations; ongoing operating expenses; debt reduction; or religious organizations strictly for religious purposes.
Geographic Focus: Indiana
Date(s) Application is Due: Feb 15; Sep 15
Amount of Grant: Up to 15,000 USD
Contact: Sharon Hollowell; (812) 662-6364; fax (812) 662-8704; sharon@dccfound.org
Internet: http://www.dccfound.org/grants.html
Sponsor: Decatur County Community Foundation
101 E Main Street, Suite 1, P.O. Box 72
Greensburg, IN 47240

Decatur County Community Foundation Small Project Grants   1445
The Decatur Foundation Small Project Grants fund organizations seeking a grant of $1,500 or less. The grant must demonstrate that it meets one or more of the following categories and criteria: youth and family enrichment - promote or provide for positive growth and development of young people or strengthen families; community development/civic engagement - promote the development of an increased quality of life within the community and foster stronger relationships among individuals or groups; cultural life - add to or enhance the variety of artistic and cultural opportunities available to all; education - demonstrate an ability to help residents gain knowledge and the skills necessary to better themselves either economically or socially, or focus on ways to allow citizens to develop skills; and health and recreation - demonstrate the ability to help residents develop healthy lifestyles. Highest priority will be given to innovative programs or projects that: include start-up costs, publicity, or specialized equipment; provide direct services to individuals or groups; enhance or enable participation by individuals or groups. Grant applications are accepted at any time. Applications received by the 10th of the month will be reviewed in the following month by the Board of Directors. The application and additional guidelines are available at the website.
Requirements: Organizations seeking grants should be a 501(c)3 nonprofit entity, an educational institution or a governmental entity. If they are not, the organization must find a qualified agency or entity to act as the fiscal agent.
Restrictions: Funding will not be considered for the following: political activities; make-up operating deficits; post-event or after the fact situations; debt reduction; or religious organizations for strictly religious purposes.
Geographic Focus: Indiana
Amount of Grant: Up to 1,500 USD
Contact: Sharon Hollowell, Executive Director; (812) 662-6364; sharon@dccfound.org
Internet: http://www.dccfound.org/grants.html
Sponsor: Decatur County Community Foundation
101 E Main Street, Suite 1, P.O. Box 72
Greensburg, IN 47240

DeKalb County Community Foundation - Garrett Hospital Aid Foundation Grants   1446
The DeKalb County Community Foundation administers the Garrett Hospital Aid Foundation Grants, which award funding to nonprofit organizations that serve residents of Garrett, Indiana. Grant proposals are accepted in January of each year and grants are awarded at the recommendation of the Garrett Hospital Aid Foundation board of directors.
Requirements: Organization should submit the online application to the DeKalb Foundation for consideration. They should include the organization's name and contact information, a description of their project and why they need it, in addition to a first and second priority they would consider.
Restrictions: The Foundation only funds projects in Garrett, Indiana.

Geographic Focus: Indiana
Date(s) Application is Due: Jan 1
Amount of Grant: 200 - 800 USD
Samples: DeKalb Association for the Developmentally Disabled, job training to prepare individuals for workforce and independent living, $250; Junior Achievement programs serving Garrett School District, $500; St. Martin's Healthcare Clinic, medical and dental services, $800.
Contact: Rosie Shinkel; (260) 925-0311; rshinkel@dekalbfoundation.org
Internet: http://www.dekalbfoundation.org/g_garrett.php
Sponsor: DeKalb County Community Foundation
650 West North Street
Auburn, IN 46706

DeKalb County Community Foundation Grants 1447
The DeKalb County Community Foundation Grants support programs for DeKalb County, Indiana citizens that address today's needs and prepare for tomorrow's challenges. Grant guidelines are intentionally broad in order to meet the community's ever-changing charitable needs. Grants are awarded for charitable programs and projects in the following areas of interest: art and culture; community development; education; environment; health and human services; and youth development. Grants are also available for the general operating expenses of organizations that address local charitable needs. Applicants are encouraged to contact the Foundation before submitting a proposal to be certain it follows the grant guidelines. They are also encouraged to attend a free one hour workshop to help them understand the Foundation's grant process and learn basic proposal writing tips. The Foundation gives priority to grant proposals for programs/projects that: will be completed within one year of receiving a grant; strengthen the grant seeking organization; directly relate to the grant seeker's mission; project a high degree of community impact; benefit many local people; and are proactive rather than reactive.
Requirements: After reviewing the grant guidelines, applicants will fill out the online proposal form. Applicants should include their contact information, financial information, a brief summary of the request, their organization's mission statement, and a detailed explanation of the benefits they'll receive from the grant. They should also include their operating expenses, total budget, and their source of funds. Applicants will then email the proposal form to the Foundation contact person or mail a printed copy to the Foundation address.
Restrictions: Grants are less likely to be awarded for: repeat funding for a program/project that has received a Foundation grant within the last two years; or a funding debt. The Foundation grants to religious organizations for charitable purposes but does not award grants for religious purposes.
Geographic Focus: Indiana
Date(s) Application is Due: Jul 1
Samples: American Red Cross of Northeast Indiana, disaster services for DeKalb County; $2,500: St. Martin's Healthcare, operating expenses for medical clinic; $7,000: DeKalb County Council on Aging, exercise equipment; $2,500; Community Care Food Pantry, food pantry/garage infrastructure construction and new commercial freezer/refrigerator equipment; $3,000.
Contact: Rosie Shinkel; (260) 925-0311; rshinkel@dekalbfoundation.org
Internet: http://www.dekalbfoundation.org/g_grantmaking.php
Sponsor: DeKalb County Community Foundation
650 West North Street
Auburn, IN 46706

Delaware Community Foundation-Youth Philanthropy Board for Kent County 1448
The Youth Philanthropy Board (YPB) for Kent County of the Delaware Community Foundation is composed of 19 students from public, independent and diocesan high schools in Kent County. The Board is studying youth issues in their neighborhoods and schools, learning about community service and grant making, and will award a total of $10,000 to schools and qualified 501(c)3 organizations to carry out youth programs for residents of Kent County. This year, the Youth Philanthropy Board for Kent County will consider programs that provide support for minors who are affected by abusive situations. Additional consideration will be given to programs that encompass reinforcement of academic achievement and/or extracurricular activities.
Requirements: 501(c)3 organizations in Kent County, Delaware are eligible to apply.
Restrictions: Programs completed prior to May 1, are ineligible for funding.
Geographic Focus: Delaware
Date(s) Application is Due: Jan 7
Contact: Ann Frazier; (302) 856-4393; fax (302) 856-4367; frazier@delcf.org
Internet: http://www.delcf.org/Apply_4_1.htm
Sponsor: Youth Philanthropy Board for Kent County c/o Delaware Community Foundation
Southern Delaware Office, 36 The Circle
Georgetown, DE 19947

Delaware Community Foundation Grants 1449
The Foundation supports organizations and programs throughout the State of Delaware that address a wide range of community needs, including, but not limited to, health and human services, the arts, humanities and culture, the environment, housing and community development. The foundation awards capital grant and program grants. Capital grants are for construction, major renovation or repair of buildings and/or the purchase of land; and also may include requests for equipment purchases. Capital grants typically range from $5,000 to $20,000, with a maximum award of $25,000. Equipment grants generally range from $2,000 to $7,000, with a maximum award of $10,000. Grant proposals must be submitted on the appropriate D.C.F application form (capital projects or equipment). One application per organization will be accepted. Application forms may be obtained from the D.C.F office or at the D.C.F Web site.
Requirements: IRS 501(c)3 non-profit organizations in Delaware are eligible to apply.
Restrictions: Ineligible for support: endowment; debt reduction; religious organizations for sectarian purposes (However, projects that serve the entire community, regardless of religious affiliation, are eligible for support); annual fundraising campaigns or general operating expenses; projects completed before June 30; sports clubs or leagues; educational institutions; purchase of vehicles; individuals.
Geographic Focus: Delaware
Date(s) Application is Due: Jan 29
Amount of Grant: 2,000 - 25,000 USD
Contact: Beth Bouchelle, Director of Grants; (302) 504-5239; bbouchelle@delcf.org
Internet: http://www.delcf.org/Apply_4_1.htm
Sponsor: Delaware Community Foundation
P.O. Box 1636
Wilmington, DE 19899

Del E. Webb Foundation Grants 1450
The Foundation applies its resources only for the benefit of the residents of Arizona, California and Nevada for improved and expanded medical services, medical research and education. Within these broad areas of interest, the Foundation draws upon the talent and experience of leaders from many walks of life to select organizations of demonstrated competence which have sound programs that will be able to reach and sustain high levels of performance. In choosing particular projects for support, the Foundation's Board acts on the basis of how the public welfare most effectively may be served. Application information is available online.
Requirements: Grants are confined to the support of nonpartisan, non-profit organizations that are operated in the public interest and have tax-exempt status as charitable organizations granted by the Internal Revenue Service and the states in which the respective organizations are incorporated and/or engage in activities.
Restrictions: Grants are not made to the following: governmental agencies or subdivisions; sectarian or religious organizations whose principle activity is exclusively for the benefit of their own members; organizations soliciting funds in support of projects or programs operated by organizations other than the applicant; expenditures before the recipient incurs and makes such expenditures, and, in the absence of compelling circumstances, the Foundation does not make grants to recipients to liquidate or reduce previously incurred obligations or operating deficits. No grants or loans are awarded or made to individuals for any purpose. Grants are not made for scholarships, student aid or medical assistance. Grants may be made to organizations that provide scholarships, student aid or medical assistance when the organization selects recipients in conformity with accepted standards. The Foundation does not participate in the administration of program(s) that the Foundation funds through the award of the application. The Foundation does not fund or underwrite: galas or gala-like events, testimonial or fund-raising luncheons or dinners, advertising in programs or similar fund-raising activities; organizations that in turn make grants to others; purchase of uniforms, equipment, or trips for school related organizations or amateur sports teams; honoraria for guest speakers or panelists; charities operated by service clubs; or educational seminars.
Geographic Focus: Arizona, California, Nevada
Date(s) Application is Due: Feb 28; May 31; Aug 31; Nov 30
Amount of Grant: 5,000 - 100,000 USD
Samples: Grand Canyon National Park Foundation (Flagstaff, AZ)--to renovate the park's education center, $2.5 million; U of Southern California (Los Angeles, CA)--to establish a laboratory suite in the New Institute for Genetic Medicine, $1 million.
Contact: Program Contact; (928) 684-7223; fax (928) 684-5665
Internet: http://www.dewf.com/DEFAULT.shtml
Sponsor: Del E. Webb Foundation
300 E Willis, Suite C
Prescott, AZ 86301-3110

Della B. Gardner Fund Grants 1451
The Della B. Gardner Fund, maintained by the Dayton Foundation, provides grants for the care and aid aid of worthy aged, sick and needy persons residing within the City of Middletown, Ohio. The Fund's primary fields of interest, therefore, are health and religion, with funding coming in the form of support for general operations. Once per year, the Dayton Foundation will issue a Request for Proposal (RFP) to senior-serving organizations in the Middletown area and will award up to $12,000.
Requirements: Applicants must be 501(c)3 organizations in operation for a minimum of two years and be either located in, or serving the residents of, Middletown, Ohio. Furthermore, such organizations must: have a diversity/inclusion policy; demonstrate systemic collaboration; and address needs that are not met fully by existing organizational or community resources.
Restrictions: The Foundation generally does not award discretionary grants for: general organizational operations and ongoing programs, operational deficits or reduced or lost funding; individuals, scholarship, travel; fundraising drives; special events; political activities; public or private schools; endowment funds; hospitals and universities for internal programs; matching grants (unless local dollars are needed to fulfill a condition for a state or federal grant); neighborhood or local jurisdiction projects; newly organized not-for-profit organizations; or publications, scientific, medical or academic research projects.
Geographic Focus: Ohio
Contact: Beth Geiger, Associate Program Officer; (937) 225-9964 or (937) 222-0410; fax (937) 222-0636; bgeiger@daytonfoundation.org
Internet: http://www.daytonfoundation.org/grntfdns.html
Sponsor: Della B. Gardner Fund
40 North Main Street, Suite 500
Dayton, OH 45423

Dell Scholars Program Scholarships 1452

Dell Scholars demonstrate their desire and ability to overcome barriers and to achieve their goals. Applicants will be evaluated on their individual determination to succeed; future goals and plans to achieve them; ability to communicate the hardships they have overcome or currently face; self motivation in completing challenging coursework; and demonstrated need for financial assistance. Students may apply directly through the Scholar website.
Requirements: To be eligible to apply for the Dell Scholarship, applicants must: participate in a Michael and Susan Dell Foundation approved college readiness program for a minimum of two of the last three years; graduate from an accredited high school this academic year; earn a minimum of a 2.4 GPA; demonstrate need for financial assistance; plan to enter a bachelor's degree program at an accredited higher education institution in the fall directly after their graduation from high school; and be a citizen or permanent resident of the U.S.
Restrictions: Applications faxed, mailed, or emailed to the Michael and Susan Dell Foundation will not be considered. The official Dell Scholars Application can only be found at this web site and must be submitted on-line during the specified application period. email attachments will not be opened.
Geographic Focus: All States
Date(s) Application is Due: Jan 15
Contact: Dell Scholar Contact; 800-294-2039; apply@dellscholars.org
Internet: http://www.dellscholars.org/Criteria.aspx
Sponsor: Michael and Susan Dell Foundation
P.O. Box 163867
Austin, TX 78716-3867

Del Mar Healthcare Fund Grants 1453

The Del Mar Healthcare Fund provides grants to nonprofit organizations that serve or support senior citizens in the six-county areas, including Montgomery, Miami, Greene, Darke, Preble and northern Warren counties. Projects may be broad in nature and may include home care options, adult day services, long-term care, senior centers, home-delivered meals, respite care, social activities, etc. The Dayton Foundation's regular discretionary grantmaking guidelines do not apply to this RFP, and organizations may apply to this fund and the Foundation's discretionary process concurrently.
Requirements: To be eligible for a grant, an organization must: be recognized as a 501(c)3 tax-exempt nonprofit organization, according to the Internal Revenue Code, and be established for at least two years and have a track record of sustainability; benefit the citizens in the Greater Dayton Region, (Montgomery, Miami, Greene, Darke, Preble and Warren (north) counties); have a diversity/inclusion policy; demonstrate systemic collaboration; and address needs that are not met fully by existing organizational or community resources.
Restrictions: The Foundation generally does not award discretionary grants for: general organizational operations and ongoing programs, operational deficits or reduced or lost funding; individuals, scholarship, travel; fundraising drives; special events; political activities; public or private schools; endowment funds; hospitals and universities for internal programs; matching grants (unless local dollars are needed to fulfill a condition for a state or federal grant); neighborhood or local jurisdiction projects; newly organized not-for-profit organizations; or publications, scientific, medical or academic research projects.
Geographic Focus: Ohio
Contact: Beth Geiger, Associate Program Officer; (937) 225-9964 or (937) 222-0410; fax (937) 222-0636; bgeiger@daytonfoundation.org or info@daytonfoundation.org
Internet: http://www.daytonfoundation.org/grntfdns.html
Sponsor: Del Mar Healthcare Fund
40 North Main Street, Suite 500
Dayton, OH 45423

Delmarva Power and Light Company Contributions 1454

The Delmarva Power and Light Company Contributions (formerly known as the Conectiv Corporate Giving program) offers support primarily in areas of company operations in Delaware and Maryland, although giving is also to national organizations. The Company makes charitable contributions to nonprofit organizations involved with education, the environment, health care, housing, public safety, youth development, and the military. Therefore, its primary fields of interest include: disasters and emergency management; education; environment; health; heart and circulatory system diseases; housing development; public utilities; scouting programs; and youth development. Types of support offered include general operations and sponsorships. Although a formal application is required, interested parties should begin by forwarding an email detailing the program in need.
Requirements: Nonprofit organizations in both Delaware and Maryland, within the regions of company operations, are welcome to apply.
Geographic Focus: Delaware, Maryland
Contact: Matt Likovich; (410) 860-6203; matthew.likovich@delmarva.com
Internet: http://www.delmarva.com/community-commitment
Sponsor: Delmarva Power and Light Company
401 Eagle Run Road, P.O. Box 17000
Wilmington, DE 19886-7000

Delta Air Lines Foundation Health and Wellness Grants 1455

Established in 1968 as Delta's company-managed giving system, the Delta Air Lines Foundation contributes more than $1 million annually in endowed funds to deserving organizations and programs. Delta's Health and Wellness Grants support organizations that dedicate their efforts to the health and well-being of communities with a focus on broad-reaching research for cures and education around diseases that affect all walks of life. Once an application is received, applicants should allow up to three months before review.
Requirements: For proposals which meet the foundation's area of focus, priority will be given to: programs meeting compelling needs in communities where Delta has a presence; proposals that exhibit clear, reasonable goals, and measurable outcomes; distinctive projects where the foundation's involvement will leave a legacy; projects that include collaboration or cooperation with other nonprofit organizations; projects that offer opportunities for Delta employee involvement. The foundation Board of Trustees reviews and approves funding in March, June, September, and November. The deadline for receiving completed proposals is the first day of each of these months.
Restrictions: The foundation will generally not consider: individual applicant's request for support of personal needs; religious activities; political organizations or campaigns; specialized single-issue health organizations; annual or automatic renewal grants; general operating expenses; endowment campaigns; capital campaigns; multi-year commitments; fraternal organizations, professional associations, or membership groups; fundraising events such as benefits; charitable dinners, or sporting events.
Geographic Focus: All States
Date(s) Application is Due: Jun 1; Sep 1; Nov 1
Contact: Administrator; (404) 715-5487 or (404) 715-2554; fax (404) 715-3267
Internet: http://www.delta.com/about_delta/community_involvement/delta_foundation/
Sponsor: Delta Air Lines Foundation
P.O. Box 20706, Department 979
Atlanta, GA 30320-6001

Delta Air Lines Foundation Prize for Global Understanding 1456

The Delta Prize for Global Understanding, established at the University of Georgia through an endowment from The Delta Air Lines Foundation, honors individuals or groups who, by their own initiatives, have provided opportunities for greater understanding among cultures and nations. Awarded annually, The Delta Prize calls attention to a variety of contributions to peace and cooperation, such as grassroots projects that diminish hostilities in a particular region of the world, international programs that facilitate communication or commerce among different peoples and leadership in the solution of global problems.
Geographic Focus: All States, All Countries
Contact: Administrator; (404) 715-5487 or (404) 715-2554; fax (404) 715-3267
Internet: http://www.delta.com/about_delta/community_involvement/delta_foundation/
Sponsor: Delta Air Lines Foundation
P.O. Box 20706, Department 979
Atlanta, GA 30320-6001

DeMatteis Family Foundation Grants 1457

The foundation makes grants in the New York metropolitan area to eligible institutions whose mission involves education, health and human services, medical research, social services, and the arts. Types of support include facilities construction, expansion, renovation; acquisition of capital equipment; scientific/medical research; projects and programs that enable the applicant to expand its mission through new or expanded programs to reach a greater segment of the community served; project-oriented capital campaigns. In general, most grants are made to cover projects that can be accomplished within one year. For construction and longer duration projects, grants may be structured to conform to identified milestones. For major projects, the payment of the grant may be over a number of years.
Requirements: Metropolitan New York 501(c)3 tax-exempt agencies, institutions, and organizations are eligible.
Restrictions: In general, the foundation does not support grants for operating deficits; general operating support; endowments; loans, or financing of any kind; annual appeals, dinner functions, and other special fund raising events; or unrestricted funds.
Geographic Focus: New York
Contact: Grants Administrator; (516) 705-4974
Internet: http://fdncenter.org/grantmaker/dematteis/about.html
Sponsor: DeMatteis Family Foundation
P.O. Box 25
Glen Head, NY 11545

Dennis and Phyllis Washington Foundation Grants 1458

The foundation supports a broad spectrum of worthy causes benefiting people of all ages primarily serving the State of Montana and surrounding states where the Washington Companies are located. Priorities for funding are direct service, youth oriented programs and the advancement of educational opportunities through scholarships to units of higher education in Montana. The Foundation also focuses on the needs of economical and socially disadvantaged people, troubled or at-risk youth and individuals with special needs.
Requirements: Eligible applicants must be a charitable, nonprofit entity with tax exempt status under Section 501(c)3. Organizations applying for support must be categorized in one of the four giving areas of education, health and human services, community service and arts and culture. The Foundation places particular emphasis on those organizations and programs that provide a direct service to economically and socially disadvantaged youth and their families, at-risk or troubled youth, and individuals with special needs. Preference is given to applicants who are able to demonstrate that a majority, if not all, of Foundation funds will be used for direct services. The Foundation prefers giving to organizations with no or low administrative costs. Organizations may apply for funding only in the year in which funds will be used. The Foundation prefers that organizations show evidence of substantial financial support from their community, constituency groups or other funding sources prior to applying.
Restrictions: Applications will not be considered for the following organizations or purposes: organizations that, in policy or practice, unfairly discriminate against race, ethnic origin, sex, creed, or religion; to fund loans, debt retirement or operational deficits; to fund on-going, general operations; to individuals, unless under an approved educational scholarship program; to sectarian or religious organizations for religious purposes where

the principal activity is for the benefit of their own members or adherents; to veterans or fraternal organizations, unless their programs are available to members of the community as a whole; for travel expenses or trips; to general endowment funds, private or public foundations and most capital campaigns; for operation expenses of tax-supported groups; for sponsorships including auctions, dinners, tickets, advertising, or annual fundraising events; for political action or legislative advocacy groups or influencing legislation or elections; for operational costs or curriculum development at educational institutions; the purchase of motor vehicles or other forms of transportation; to organizations acting on behalf of, but without the authority of, qualified tax exempt organizations.
Geographic Focus: Montana
Contact: Mike Halligan, Executive Director; (406) 523-1325
Internet: http://www.dpwfoundation.org/home.htm
Sponsor: Dennis and Phyllis Washington Foundation
P.O. Box 16630
Missoula, MT 59808-6630

Denton A. Cooley Foundation Grants 1459
The foundation awards grants in Texas in support of health care, health education, hospitals, and medical research. Types of support include endowments, general operating grants, program grants, and research grants. There are no application forms or deadlines. The board meets quarterly to consider requests.
Restrictions: Grants do not support conferences, loans, individuals, scholarships, fellowships, or publication.
Geographic Focus: Texas
Amount of Grant: 100 - 250,000 USD
Samples: U of Texas Medical Branch at Galveston (TX)--to establish a professorship in the division of cardiothoracic surgery, $250,000.
Contact: Grants Administrator; (713) 799-2700
Sponsor: Denton A. Cooley Foundation
6624 Fannin, Suite 1640
Houston, TX 77030

Denver Broncos Charities Fund Grants 1460
The fund supports qualified non-profit organizations that work to impact the quality of life for youth, health and the hungry and homeless. The emphasis of the fund is on programs designed to assist young people in the areas of education and youth football with a particular emphasis on programs aimed at disadvantaged and at-risk youth. The fund will also consider programs devoted to health and hunger issues. Grants may be stand-alone grants or magnet grants that attract other corporate dollars.
Requirements: Nonprofit organizations in the state of Colorado are eligible.
Geographic Focus: Colorado
Date(s) Application is Due: Jun 1
Samples: Denver Public Schools Foundation, $50,000--Prep League Football Program; Inner City Health Center, $50,000--Denver Rescue Mission; Meeker School District, $50,000--Community Fields Project; Darrent Williams Memorial Teen Center, $50,000--Darrent Williams Memorial Grant.
Contact: Charities Fund Manager; (720) 258-3000
Internet: http://www.denverbroncos.com/page.php?id=1157
Sponsor: Denver Broncos Charities Fund
1701 Bryant Street, Suite 1400
Denver, CO 80204

Denver Foundation Community Grants 1461
The Denver Foundation carries out its mission of improving life in Metro Denver by investing in the vision, passion and expertise of hundreds of nonprofit organizations. The Community grants program awards hundreds of grants annually to nonprofit organizations in the seven-county Metro Denver area, which includes: Adams, Arapahoe, Boulder, Broomfield, Denver, Douglas, and Jefferson counties. Grants are awarded in the four major focus areas of arts and culture, health, human services and civics and education.
Requirements: To qualify for a grant an organization must be a 501(c)3 tax-exempt nonprofit organization, serve residents in seven specific Denver counties, and provide a service that falls under one of the four funding areas.
Restrictions: The Foundation will not consider requests to fund the following: requests from organizations that have received funding from the program for the three previous consecutive calendar years; organizations with fund balance deficits in their most recently completed fiscal year; organizations that discriminate on the basis of race, color, religion, gender, age, national origin; disability, marital status, sexual orientation, or military status; debt retirement; endowments or other reserve funds; membership or affiliation campaigns, dinners, or special events; conferences and symposia and related travel; grants that further political doctrine; grants that further religious doctrine; grants to individuals; scholarships or sponsorships; individual medical procedures; medical, scientific, or academic research; grants to parochial or religious schools; grants to governmental agencies, except public schools; requests from individual public schools that have not coordinated the request with their central school district administration; requests from foundations/organizations that raise money for individual public schools; creation or installation of art objects; development, production, or distribution of books, newspapers, or video productions; grants for re-granting programs; requests for capital campaigns that have not met 75% of their goal; activities, projects, or programs that will have been completed before funding becomes available; and multi-year funding requests.
Geographic Focus: Colorado
Date(s) Application is Due: Feb 2; Jun 1; Oct 1
Contact: Jeff Hirota, Vice President; (303) 300-1790, ext. 129; fax (303) 300-6547; jhirota@denverfoundation.org
Internet: http://www.denverfoundation.org/page17823.cfm
Sponsor: Denver Foundation
55 Madison, 8th Floor
Denver, CO 80206

Dept of Ed College Assistance Migrant Program (CAMP) 1462
The College Assistance Migrant Program (CAMP) assists students who are migratory or seasonal farmworkers (or children of such workers) enrolled in their first year of undergraduate studies at an IHE. The funding supports completion of the first year of studies. Competitive five-year grants for CAMP projects are made to IHEs or to nonprofit private agencies that cooperate with such institutions. The grants funded under CAMP grantee serve approximately 2,400 participants each year.
Requirements: Institutions of higher education or private nonprofit agencies in cooperation with institutions of higher education may apply.
Restrictions: Institutions of higher education or a nonprofit private agency in cooperation with an IHE may apply.
Geographic Focus: All States
Contact: David De Soto, (202) 260-8103; fax (202) 205-0089; david.de.soto@ed.gov
Internet: http://web99.ed.gov/GTEP/Program2.nsf/b39cd123fd4a045b8525644400514f2b/07783e5ad0636ec5852563bc00540495?OpenDocument
Sponsor: U.S. Department of Education
400 Maryland Avenue SW, Room 3E225, FB-6
Washington, D.C. 20202-6135

Dept of Ed Rehabilitation Training Grants 1463
The purpose of this program is to ensure that skilled personnel are available to serve the rehabilitation needs of individuals with disabilities assisted through vocational rehabilitation (VR), supported employment and independent living programs. The program supports training and related activities designed to increase the number of qualified personnel trained in providing rehabilitation services. Grants and contracts are awarded to states and public and nonprofit agencies and organizations, including institutions of higher education, to pay all or part of the cost of conducting training programs.
Requirements: Institutions of Higher Education (IHEs), Nonprofit Organizations, Other Organizations and/or Agencies may apply. Applicants may include state and public or nonprofit agencies and organizations and Indian tribes.
Geographic Focus: All States
Contact: Ruth Brannon; (202) 245-7278; fax (202) 245-7591; ruth.brannon@ed.gov
Internet: http://www.ed.gov/programs/rsatrain/index.html
Sponsor: U.S. Department of Education
400 Maryland Avenue SW, Room 5052, PCP
Washington, D.C. 20202

Dept of Ed Safe and Drug-Free Schools and Communities State Grants 1464
This program provides support to SEAs for a variety of drug and violence prevention activities focused primarily on school-age youths. Activities may include: developing instructional materials; providing counseling services and professional development programs for school personnel; implementing community service projects and conflict resolution, peer mediation, mentoring and character education programs; establishing safe zones of passage for students to and from school; acquiring and installing metal detectors; and hiring security personnel.
Requirements: State Education Agencies may apply. Local Education Agencies or intermediate education agencies or consortia must apply to the State Education Agency.
Geographic Focus: All States
Contact: Paul Kesner, Office of Safe and Drug-Free Schools; (202) 205-8134; fax (202) 260-7767; paul.kesner@ed.gov
Internet: http://www.ed.gov/programs/dvpformula/index.html
Sponsor: U.S. Department of Education
400 Maryland Avenue SW, Room 3E230, FB-6
Washington, D.C. 20202-6450

Dept of Ed Special Education--Personnel Development to Improve Services and Results for Children with Disabilities 1465
The objectives of this program are to help address state-identified needs for highly qualified personnel in special education, related services, early intervention, and regular education, to work with children with disabilities; and ensure that those personnel have the skills and knowledge, derived from practices that have been determined, through research and experience, to be successful, that are needed to serve those children. Awards are made to train personnel in the areas of: leadership; early intervention and early childhood; low-incidence; high-incidence; related services; speech/language, and adapted physical education; and programs in minority institutions.
Requirements: Institutions of Higher Education, Local Education Agencies, Nonprofit Organizations, Other Organizations and/or Agencies and State Education Agencies are eligible to apply. In addition to these categories, public charter schools that are LEAs under state law, other public agencies, private nonprofit organizations, for-profit organizations, outlying areas (American Samoa, Guam, the Northern Mariana Islands, and the U.S. Virgin Islands), freely associated states, and Indian tribes or tribal organizations are eligible to apply.
Geographic Focus: All States
Contact: Patricia Wright, (202) 245-7620; fax (202) 245-7619; patricia.wright@ed.gov
Internet: http://www.ed.gov/programs/osepprep/index.html
Sponsor: U.S. Department of Education
400 Maryland Avenue SW, Room 4113, PCP
Washington, D.C. 20202-2600

Dept of Ed Special Education--Studies and Evaluations 1466
This program is designed to assess progress in implementing IDEA, including the effectiveness of state and local efforts to provide: free appropriate public education to children with disabilities; and early intervention services to infants and toddlers with disabilities. This program supports studies, evaluation, and assessments.
Requirements: Institutions of higher education, local education agencies, nonprofit organizations, other organizations and/or agencies, and state education agencies are eligible. Public agencies, outlying areas (American Samoa, Guam, the Northern Mariana Islands, and the U.S. Virgin Islands), freely associated states, Indian tribes, tribal organizations, and for-profit organizations, if appropriate, may apply.
Geographic Focus: All States
Contact: Jeffrey Payne; (202) 245-7473; fax (202) 245-7617; jeffrey.payne@ed.gov
Internet: http://www.ed.gov/programs/osepsae/index.html
Sponsor: U.S. Department of Education
400 Maryland Avenue SW, Room 4064, PCP
Washington, D.C. 20202-2500

Dept of Ed Special Education--Technical Assistance and Dissemination to Improve Services and Results for Children with Disabilities 1467
The purpose of this program is to promote academic achievement and improve results for children with disabilities by providing technical assistance, model demonstration projects, dissemination of useful information, and implementation activities that are supported by scientifically based research. The program supports technical assistance and dissemination activities. Annual deadline dates may vary; contact the office for exact dates.
Requirements: Institutions of Higher Education, Local Education Agencies, Nonprofit Organizations, Other Organizations and/or Agencies, and State Education Agencies are eligible to apply. Public charter schools that are LEAs under state law, for profit organizations, outlying areas (American Samoa, Guam, the Northern Mariana Islands, and the U.S. Virgin Islands), freely associated states, and Indian tribes or tribal organizations may also apply.
Geographic Focus: All States
Amount of Grant: 200,000 - 2,000,000 USD
Contact: Claudette Carey, Office of Special Education and Rehabilitative Services; (202) 245-7291; fax (202) 245-7617; claudette.carey@ed.gov
Internet: http://web99.ed.gov/GTEP/Program2.nsf/a5b8d6c38fdd4ca08525644400514f2c/02df9c7d71171b1785256571007ce61c?OpenDocument
Sponsor: U.S. Department of Education
400 Maryland Avenue SW, Room 4098, PCP
Washington, D.C. 20202

Dept of Ed Special Education-National Activities-Technology and Media Services for Individuals with Disabilities 1468
The purpose of this program is designed to: improve results for children with disabilities by promoting the development, demonstration, and use of technology; support educational media services activities designed to be of education value in the classroom setting to children with disabilities; and provide support for captioning and video description that is appropriate for use in the classroom setting.
Requirements: State Education Agencies; Local Education Agencies; public charter schools that are LEAs under state law; Institutions of Higher Education; other public agencies; private nonprofit organizations; outlying areas (American Samoa, Guam, the Northern Mariana Islands, The U.S. Virgin Islands); freely associated states; Indian tribes or tribal organizations; and for-profit organizations may apply.
Geographic Focus: All States
Contact: Jeffrey Payne; (202) 245-7473; fax (202) 245-7617; jeffrey.payne@ed.gov
Internet: http://www.ed.gov/programs/oseptms/index.html
Sponsor: U.S. Department of Education
400 Maryland Avenue SW, Room 4064, PCP
Washington, D.C. 20202-2500

DeRoy Testamentary Foundation Grants 1469
Established in 1979, the Foundation gives primarily in the state of Michigan. Giving primarily for youth development and services, education, human services, health care, and the arts. The Foundation offers support in the form of: scholarship funds; annual campaigns; building/renovation; continuing support; general/operating support; program development grants. There are no application deadlines. The board meets monthly.
Requirements: Michigan nonprofit organizations are eligible.
Restrictions: Grants are not made to individuals.
Geographic Focus: Michigan
Amount of Grant: 10,000 - 300,000 USD
Contact: Julie Rodecker Holly; (248) 827-0920; fax (248) 827-0922; deroyfdtn@aol.com
Sponsor: DeRoy Testamentary Foundation
26999 Central Park Boulevard, Suite 160N
Southfield, MI 48076

Detlef Schrempf Foundation Grants 1470
The focus of the foundation is on children, youth and families in the Pacific Northwest. This foundation seeks supporting organizations that can help underwrite the costs of events, enabling the maximum amount of proceeds to go directly to the benefiting charity.
Requirements: Typically, a qualifying charity partner will have a $5-million, or smaller, operating budget.
Geographic Focus: Idaho, Oregon, Washington
Contact: Nicole Morrison; (206) 464-0826; fax (206) 464-8020; info@detlef.com
Internet: http://www.detlef.com
Sponsor: Detlef Schrempf Foundation
1904 Third Avenue, Suite 339
Seattle, WA 98101

Deutsche Banc Alex Brown and Sons Charitable Foundation Grants 1471
The Deutsche Banc Alex Brown and Sons Charitable Foundation awards one-year renewable grants to U.S. nonprofit organizations in its areas of interest, including arts and culture, community affairs, education, medicine, and science. Types of support include annual campaigns, capital grants, endowment funds, general operating grants, program development grants, and scholarship funds. Applicants should submit a letter of interest that briefly outlines the purpose of the grant prior to submission of a more detailed application. There are no specified annual deadlines.
Requirements: 501(c)3 tax-exempt organizations in Maryland and the Washington, D.C., area are eligible to apply.
Restrictions: No support for private schools or churches. Grants are not made to individuals.
Geographic Focus: District of Columbia, Maryland
Amount of Grant: 1,000 - 100,000 USD
Samples: Baltimore Symphony Orchestra (MD)--for discretionary use, $1 million; Johns Hopkins Hospital (Baltimore, MD)--for discretionary use, $1 million; Baltimore Zoo (MD)--for discretionary use, $1 million.
Contact: Secretary, c/o Deutsche Bank Alex Brown; (202) 783-5476 or (202) 626-7000
Sponsor: Deutsche Banc Alex Brown and Sons Charitable Foundation
1440 New York Avenue NW
Washington, DE 20005-2111

DHHS Adolescent Family Life Demonstration Projects 1472
This program has multiple components of prevention, care and research. The prevention component focuses on development and testing of abstinence based programs designed to delay the onset of sexual activity reducing the incidence of adolescent pregnancy, STD transmission and HIV/AIDS. The care component is focused on providing comprehensive health, education and social services to pregnant adolescents, adolescent parents, their infants, male partners and their families. The primary focus of research is to improve understanding of the issues surrounding sexuality, pregnancy and parenting by examining the factors that influence adolescent sexual, contraceptive and fertility behaviors, the nature and effectiveness of care services for pregnant and parenting adolescents and why adoption is a little-used alternative among pregnant adolescents. Deadlines are announced in the Federal Register and on the Office of Population Affairs Web site (http://opa.osophs.dhhs.gov).
Requirements: Public and private non-profit organizations may apply.
Geographic Focus: All States
Samples: Father Flanagan High School (Omaha, NE)--for the Boys Town Adolescent Parenting program; Teenage Alternative School (Lincoln Park, MI)--for a school for pregnant teens and adolescent parents; Mount Vernon Public Schools (Mount Vernon, NY)--for its Family Life Education program, a prevention initiative aimed at fifth- through eighth-grade students.
Contact: Andrea Brandon, Chief Grants Management Officer; (301) 594-6554
Internet: http://aspe.hhs.gov/SelfGovernance/inventory/OPA/111.htm
Sponsor: U.S. Department of Health and Human Services
4350 East-West Highway
Bethesda, MD 20814

DHHS Emerging Leaders Program Internships 1473
The Emerging Leaders Program (ELP) is a competitive, two-year paid, federal internship with the U.S. Department of Health and Human Services (HHS). The Program provides a unique opportunity to develop enhanced leadership skills in one of the largest federal agencies in the nation. Upon successful completion of the ELP, participants will be eligible for non-competitive conversion to a permanent appointment. The ELP offers participants the following: formal competency-based leadership training and professional development; challenging developmental rotational assignments; mentorship; fast-paced and diverse work environment; and promotion to a permanent career track that targets the following mission-critical occupation specialties: scientific, social science, human resources, administrative, information technology, public health, and law enforcement. To begin the application process, applicants apply to one or more career track(s) that match their education and/or experience specialty. Later applicants may apply to the specific positions under their chosen career track(s). At the date of this writing, the website states that the ELP is not currently accepting applications, but encourages those interested to check back for future updates.
Requirements: Applicants must qualify for a minimum of one career track to move on to the next phase of the selection process.
Geographic Focus: All States
Contact: Grants Coordinator; (877) 696-6775; ELP@hhs.gov
Internet: http://hhsu.learning.hhs.gov/elp/howtoapply.asp
Sponsor: U.S. Department of Health and Human Services
200 Independence Avenue SW
Washington, D.C. 20201

DHHS Oral Health Promotion Research Across the Lifespan 1474
The National Institute of Dental and Craniofacial Research has invited proposals for improving the oral health of people of all ages. The research team must include someone with extensive experience in health promotion, behavioral and/or social science research. The health promotion intervention proposed for funding must be based on a previously conducted assessment of the epidemiology, social, behavioral and/or environmental factors related to the disease or condition under study. Research could focus on maternal

and child health, adolescent and young adult health, or health of adults with complex diseases. For example, an applicant might propose to study approaches to involving families, social networks, communities, or neighborhoods in behaviors that promote and improve oral health; improving patient-provider communication related to oral preventive measures; or effective ways to train oral health professional students to communicate with diverse patient populations.
Requirements: Any person with the skills, knowledge, and resources necessary to carry out the proposed research as the project director/principal investigator (PD/PI) is invited to work with his/her organization to develop an application for support. Applications must be submitted electronically through Grants.gov (http://www.grants.gov) using the SF424 research and related forms and the SF424 application guide.
Geographic Focus: All States
Date(s) Application is Due: Mar 5; Jul 5; Nov 5
Contact: Maria Teresa Canto, (301) 594-5497; fax (301) 480-8322; maria.canto@nih.gov
Internet: http://grants.nih.gov/grants/guide/pa-files/PA-07-225.html
Sponsor: National Institute of Dental and Craniofacial Research
45 Center Drive, Natcher Building
Bethesda, MD 20892

DHL Charitable Shipment Support 1475
The corporation provides free shipment of materials supporting charitable programs. Applicants must submit a complete web questionnaire.
Requirements: Application by internet only.
Restrictions: Religious organizations are ineligible for religious sectarian activities, and political organizations are not eligible for political purposes.
Geographic Focus: All States
Contact: Technical Support; (800) 527-7298
Internet: http://www.dhl-usa.com
Sponsor: DHL International, Ltd.
1220 South Pine Island Road, Suite 600
Plantation, FL 33324

Dickson Foundation Grants 1476
The foundation awards grants primarily to local nonprofit organizations in its areas of interest, including education (secondary schools, higher education),and health care. Types of support include building construction/renovation and general operating support. There are no application deadlines or forms. Submit a letter of interest for consideration.
Requirements: Giving is primarily in the Southeast.
Restrictions: Grants are not awarded to individuals, or for building or endowment funds.
Geographic Focus: North Carolina
Contact: Susan Patterson, (704) 372-5404; fax (704) 372-6409
Sponsor: Dickson Foundation
301 S Tryon Street, Suite 1800
Charlotte, NC 28202

DIFFA/Chicago Grants 1477
DIFFA/Chicago (The Design Industries Foundation Fighting AIDS) is a not-for-profit fundraising and grant making foundation that distributes funds to Chicago area HIV/AIDS service agencies that provide direct service, preventative education and outreach to people who are HIV positive, living with AIDS or at risk for infection. Founded by volunteers from the fashion industry, interior design, furnishings and architecture, supporters of DIFFA now come from every field associated with fine design. DIFFA has also been an innovative agent in drawing local and national corporations into the fight against the epidemic, and enjoys tremendous support from the business community. The Foundation issues grants annually. In some cases, a special grant may be awarded during the year to meet specific funder, community or emergency needs. Grants fall under two categories: Chapter Grants and Foundation Grants. In both instances, DIFFA does not accept unsolicited grant proposals. Invitations are sent out in the fall of each year to apply. Grants are then approved by the Foundation's trustees at their January Board meeting.
Requirements: Grass-root community based agencies with a 501(c)3 that provide outstanding services, preventive education and outreach to people in the Chicago area who are HIV positive, living with AIDS or at risk of infection are welcome to submit an application. The DIFFA/Chicago granting process selects agencies providing the best care possible in Chicago, and the ability to put dollars where they have the most impact to help the men, women, and children living with HIV/AIDS. A team consisting of a professional in that field and a DIFFA/Chicago board member review each agency applying for funding. Review of proposals is followed by site visits and interviews with the organizations' staff. DIFFA/Chicago funds in the categories of advocacy and education, health and clinical services, meals and nutrition, housing, and support and counseling. The agencies shown to provide the best care receive an unrestricted grant.
Geographic Focus: Illinois
Contact: Todd Baisch; (312) 577-7147; todd_baisch@gensler.com
Internet: http://www.diffachicago.org/grants.html
Sponsor: Design Industries Foundation Fighting AIDS
222 Merchandise Mart Plaza, Suite 1647A
Chicago, IL 60654

Different Needz Foundation Grants 1478
The Different Needz Foundation was created in 2009 and was inspired by the life of Luke Jordan. Luke was born with developmental disabilities which caused many physical disabilities and medical conditions. Despite his disabilities, Luke was able to bring out the best in those who knew him. The Different Needz Foundation helps individuals with developmental disabilities obtain the necessary equipment and medical services they need to have the best quality of life. Grant applications are made available in January of each year. Grant awards are announced each year in May. The Foundation considers future needs and provides payment for medical services or equipment directly to the provider. The Foundation has approved grants for services and equipment such as; physical therapy, occupational therapy, speech therapy, medical equipment, a toilet chair, a wheelchair, a wheelchair lift, summer camp and a tandem bike which was featured by Team Myles in the 2011 Cleveland Triathlon. Initially, grant recipients were mostly from northeast Ohio. Currently, the Foundation's reach has grown to include recipients in Ohio, Wyoming, Massachusetts, Florida, Maryland, Indiana, and California
Geographic Focus: All States
Date(s) Application is Due: Mar 31
Contact: Michelle Petrillo-Carr; (216) 904-5151; info@differentneedzfoundation.org
Internet: http://www.differentneedzfoundation.org/grants/
Sponsor: Different Needz Foundation
8440 East Washington Street #122
Chagrin Falls, OH 44023

Disable American Veterans Charitable Grants 1479
The trust is dedicated to building better lives for U.S. disabled veterans and their families. Grants support physical and psychological rehabilitation programs, aid efforts to meet the special needs of veterans with specific disabilities such as amputation and blindness, and support programs and shelters for homeless veterans. Priority is given to long-term service projects providing direct assistance to disabled veterans and their families. The trust also awards grants to service programs that assist veterans suffering from substance abuse and post traumatic stress disorder, health promotion programs for aging disabled veterans, and nursing and health programs designed to ensure quality healthcare for veterans.
Requirements: Nonprofit organizations are eligible.
Restrictions: Grants generally do not support individuals; goodwill advertising, souvenir journals, or dinner programs; political causes, candidates, organizations, or campaigns; or endowments or capital campaigns.
Geographic Focus: All States
Date(s) Application is Due: Jan 31; May 1; Aug 1; Nov 1
Contact: Administrator; (877) 426-2838, ext. 2057; fax (859) 442-2088; cst@davmail.org
Internet: http://www.dav.org/cst/index.html
Sponsor: Disabled American Veterans Charitable Service Trust
3725 Alexandria Pike
Cold Spring, KY 41076

DOD HBCU/MI Partnership Training Award 1480
This program's goal challenges the scientific community to design innovative research that will foster new directions for, address neglected issues in, and bring new investigators to the field of breast cancer research. The program focuses its funding on innovative projects, particularly those involving multidisciplinary and/or multi-institutional collaborations and alliances that have the potential to make a significant impact on breast cancer. Proposals that address the needs of minority, low-income, rural, and other underrepresented and/or medically underserved populations are strongly encouraged. The listed application deadline is for preproposals. Preproposals must be submitted through the CDMRP eReceipt system. Guidelines are available online.
Requirements: Applicants must be HBCU/MI doctoral-level faculty members. All individuals, regardless of ethnicity, nationality, or citizenship status, may apply as long as they are employed by, or affiliated with, an eligible HBCU/MI institution. Eligible institutions are those approved as HBCU/MIs by the Department of Education.
Geographic Focus: All States
Contact: Commander, U.S. Army Medical Research and Materiel Command; (301) 619-7079; fax (301) 619-7792; cdmrp.pa@det.amedd.army.mil
Internet: http://cdmrp.army.mil/bcrp/default.htm
Sponsor: U.S. Department of Defense
3701 North Fairfax Drive
Arlington, VA 21702-1714

DOJ Gang Free Schools and Communities Intervention Grants 1481
The objectives of the program include support to prevent and reduce the participation of juveniles in the activities of gangs that commit crimes. Such programs and activities may include individual, peer, family, and group counseling, including provision of life skills training and preparation for living independently, which shall include cooperation with social services, welfare, and health care programs; education and social services designed to address the social and developmental needs of juveniles; crisis intervention and counseling to juveniles, who are particularly at risk of gang involvement, and their families; the organization of the neighborhood and community groups to work closely with parents, schools, law enforcement, and other public and private agencies in the community; and training and assistance to adults who have significant relationships with juveniles who are or may become members of gangs, to assist such adults in providing constructive alternatives to participating in the activities of gangs. Deadlines are published in program announcements.
Requirements: Public or private nonprofit agencies, organizations, or individuals may apply.
Geographic Focus: All States
Contact: Heidi Hsia, Office of Juvenile Justice and Delinquency Prevention; (202) 307-5924; hsiah@ojp.usdoj.gov
Internet: http://www.ojp.usdoj.gov/FinGuide/part4chap2.htm#ojjdp
Sponsor: U.S. Department of Justice
633 Indiana Avenue NW
Washington, D.C. 20531

Dolan Children's Foundation Grants 1482
The Dolan Children's Foundation was founded in New York in 1886 by Charles F. and Helen A. Dolan. Charles F. Dolan was the founder of the cable network HBO, and is the owner of Cablevision Systems Corporation, a cable television provider in New York City that also owns Madison Square Garden, Radio City Music Hall, the New York Knicks, and the New York Rangers. The Foundation awards community service grants with a focus on Long Island, New York. Areas of interest include: human services, disability services, mental health, schools, rehabilitation, hospitals, health facilities, and Catholic agencies and churches. Types of support include: general operating support, building and renovations; capital campaigns; equipment purchase and rental; land acquisition; program development; research; and matching grants. There are no application deadlines, though an application form is required. Recent awards have ranged from $25,000 to $3,000,000. Guidelines are available upon request.
Requirements: Schools and other nonprofits in New York City and Long Island are eligible. Applicants outside New York should call or write prior to submitting proposals.
Geographic Focus: New York
Amount of Grant: 25,000 - 3,000,000 USD
Samples: Chaminade High School, Mineola, New York, $3,000,000 - general operating support (2014); Lesley University, Cambridge, Massachusetts, $250,000 - general operating support (2014); Rising Sun Youth Foundation, Plushing, New York, $50,000 - general operating support (2014).
Contact: Robert Vizza, President; (516) 803-9200
Sponsor: Dolan Children's Foundation
340 Crossways Park Drive
Woodbury, NY 11797-2050

Dole Food Company Charitable Contributions 1483
Dole's charitable program is dedicated to bringing about positive change in the area of nutrition education for children. The goal of the program is to help the next generation of adults prevent many diseases by teaching the value of good nutrition to children through improved and interactive teaching curriculum for schools nationwide. Please contact Corporate Contributions for application process.
Requirements: Dole's program makes grants based on the following guidelines: the program is restricted to funding 501(c)3 charitable organizations only; documentation from the I.R.S. confirming 501(c)3 status must be provided; and the organization should provide nutrition education programs that benefit schools on a nationwide basis as opposed to assisting an individual school.
Restrictions: Grants are restricted for the following: individuals; religious, fraternal, sports or political groups; legislative or lobbying efforts or groups who have a primary focus of changing laws; and sporting events and sponsorships.
Geographic Focus: All States
Contact: Contribution Program Manager; (818) 879-6600; fax (818) 879-6615
Internet: http://dolecsr.com/Principles/CharitableGiving/tabid/410/Default.aspx
Sponsor: Dole Food Company
P.O. Box 5132
Westlake Village, CA 91361

DOL Occupational Safety and Health--Susan Harwood Training Grants 1484
Grants provide funds for programs to train workers and employers to recognize, avoid, and prevent safety and health hazards in their workplaces. The program emphasizes three areas: educating workers and employers in small businesses; training workers and employers about new OSHA standards; and training workers and employers about high risk activities or hazards identified by OSHA through its Strategic Management Plan, or as part of an OSHA special emphasis program.
Requirements: Nonprofit organizations, including community-based and faith-based organizations, that are not an agency of a State or local government are eligible to apply. State or local government supported institutions of higher education are eligible to apply in accordance with 29 CFR part 95. Eligible organizations can apply independently for funding or in partnership with other eligible organizations, but in such a case, a lead organization must be identified. Sub-contracts must be awarded in accordance with 29 CFR 95.40-48, including OMB circulars requiring free and open competition for procurement transactions.
Restrictions: Grants may not be used for: any activity that is inconsistent with the goals and objectives of the Occupational Safety and Health Act of 1970; training individuals not covered by the Occupational Safety and Health Act; training workers or employers from workplaces not covered by the Occupational Safety and Health Act; training on topics that do not cover the recognition, avoidance, and prevention of unsafe or unhealthy working conditions; assisting workers in arbitration cases or other actions against employers, or assisting employers and workers in the prosecution of claims against Federal, State or local governments; duplicating services offered by OSHA, a State under an OSHA approved State Plan, or consultation programs provided by State designated agencies under section 21(d) of the Occupational Safety and Health Act; generating membership in the grantee's organization; administrative costs exceeding 25% of the total grant budget.
Geographic Focus: All States
Contact: Cynthia Bencheck, (847) 297-4810; bencheck.cindy@dol.gov
Internet: http://www.osha.gov/dcsp/ote/sharwood.html
Sponsor: U.S. Department of Labor
2020 S Arlington Heights Road
Arlington Heights, IL 60005-4102

Dominion Foundation Grants 1485
Dominion Foundation grants are made in four focus areas, and they support a variety of programs: food banks, homeless shelters, land and habitat preservation, STEM (science, technology, engineering, math) education, and neighborhood revitalization, to name a few. Special consideration is given to programs with an energy conservation or energy efficiency component. In the area of Human Needs, the Foundation focus is on: providing warmth and cooling; alleviating hunger; ensuring energy-efficient shelter; providing access to medicine and basic health care; supporting communities through the united way; and disaster assistance. Since the Dominion Foundation supports a wide range of charitable programs, most grants are in the $1,000 to $15,000 range. Higher amounts may be awarded when a program is an exceptional fit with corporate business or giving priorities, or when there is significant employee involvement in the effort. Requests are considered quarterly by the Foundation's Community Investment Boards – statewide and regional committees comprised of Dominion employees representing key geographic, business and functional areas.
Requirements: Foundation grants are limited to organizations defined as tax-exempt under Section 501(c)3 of the IRS code. Additional grants occasionally may be made directly from the corporation to sponsor special events that benefit a non-profit organization.
Restrictions: Awards are throughout a 10-state area to include: Connecticut, Maryland, Massachusetts, North Carolina, Ohio, Pennsylvania, Rhode Island, Texas, Virginia and West Virginia. Information pertaining to the specific counties served in each state is on the website.
Geographic Focus: Connecticut, Illinois, Indiana, Maryland, Massachusetts, North Carolina, Ohio, Pennsylvania, Rhode Island, Texas, Virginia, West Virginia
Amount of Grant: 1,000 - 250,000 USD
Contact: James C. Mesloh, Executive Director; (800) 730-7217 or (412) 237-2973; fax (412) 690-7608; Educational_Grants@dom.com
Internet: http://www.dom.com/about/education/grants/index.jsp
Sponsor: Dominion Foundation
501 Martindale Street, Suite 400
Pittsburgh, PA 15222-3199

Donald and Sylvia Robinson Family Foundation Grants 1486
The foundation awards grants to U.S. nonprofit organizations in its areas of interest, including animal and wildlife protection, arms control, arts (general, performing arts, and visual arts), environmental conservation and protection, eye diseases and eye research, family planning and human reproductive health, food distribution, international affairs, Israel, Jewish social services, and social service delivery programs. Types of support include annual campaigns, building construction/renovation, capital campaigns, and general operating support. There are no application deadlines or forms.
Requirements: National nonprofit organizations are eligible to apply.
Restrictions: Individuals are ineligible.
Geographic Focus: Pennsylvania
Amount of Grant: 500 - 5,000 USD
Samples: Phipps Conservatory and Botanical Gardens, Pittsburgh, PA, $5,000; Jewish Residential Services, Pittsburgh, PA, $5,000; Jewish Family and Childrens Service of Pittsburgh, Pittsburgh, PA, $500.
Contact: Donald Robinson, Treasurer; (412) 661-1200; fax (412) 661-4645
Sponsor: Donald and Sylvia Robinson Family Foundation
6507 Wilkins Avenue
Pittsburgh, PA 15217

Donald W. Reynolds Foundation Aging and Quality of Life Grants 1487
The Foundation launched its Aging and Quality of Life Program in 1996. Its goal remains improving the quality of life for America's elderly by preparing physicians to provide better care for frail older people. Most physicians today lack adequate training to meet the needs of the frail elderly patient. Such patients typically suffer from interacting physical, social and psychological conditions –both acute and chronic – that limit their independence and threaten their capacity to function in daily life.
Requirements: Proposals are by invitation only. Periodic requests for proposals are announced on the website. Organizations should contact the Foundation to discuss their project before submitting a proposal.
Geographic Focus: All States
Contact: Rani Snyder, Senior Program Officer; (702) 804-6000; fax (702) 804-6099; rani.snyder@dwrf.org or GeneralQuestions@dwrf.org
Internet: http://www.dwreynolds.org/Programs/National/Aging/Aging.htm
Sponsor: Donald W. Reynolds Foundation
1701 Village Center Circle
Las Vegas, NV 89134-6303

Donna K. Yundt Memorial Scholarship 1488
The Donna K. Yundt Memorial Scholarship was established in memory of a teacher in the Lake Central Schools. This one time scholarship is awarded to a student who attended who attended Peifer Elementary. All college majors are eligible.
Requirements: Applicants must be high school graduating seniors who attended Peifer Elementary. The application is available at the Lake Central High School Guidance office, along with the Dollars for Scholars application. Applicants should check with the guidance office for current application deadline.
Geographic Focus: Indiana
Amount of Grant: 2,500 USD
Contact: Yundt Scholarship Contact; (219) 736-1880; fax (219) 736-1940
Internet: http://www.legacyfoundationlakeco.org/scholarships/index.htm
Sponsor: Legacy Foundation
1000 East 80th Place, 302 South
Merrillville, IN 46410

Dora Roberts Foundation Grants 1489
The foundation awards general operating grants to eligible Texas organizations in its areas of interest, Arts, Social Services, Health and Education. Most grants are awarded in Big Spring, TX. There are no application forms. Trustees meet annually in the Fall to review all requests received prior to the deadline.
Requirements: The Foundation can distribute grants only to qualified charitable organizations in the State of Texas. Persons who represent an organization which they believe might qualify for our support are welcome to submit an application in letter form containing: a brief narrative history of the organization's purpose and work; a specific description of the program or project for which support is asked; a statement of the amount of funds requested; proof of tax exempt status; a list of Trustees or Directors and principal staff; and budgetary information pertaining to the requested grant.
Restrictions: The Foundation does not give or lend money to individuals.
Geographic Focus: Rhode Island
Date(s) Application is Due: Sep 30
Contact: Konnie Darrow, c/o JP Morgan Chase Bank, N.A., (817) 884-4772
Sponsor: Dora Roberts Foundation
P.O. Box 2050
Fort Worth, TX 76113

Doree Taylor Charitable Foundation 1490
The mission of the Doree Taylor Charitable Foundation is to support charitable organizations that: provide relief to people the form of basic needs (including the provision of food, housing, shelter); promote the humane care of animals; provide healthcare services for the underserved; and conduct Public radio or television. Occasional support will also be provided to colleges and universities as well as to environmental charitable organizations in Maine. The foundation will make grants throughout Maine, but has a priority for charitable organizations or projects located in the areas of Boothbay Harbor, Southport, and Brunswick. From time to time, the foundation may support organizations with a national scope. Grant requests for general operating support or program support are strongly encouraged and preferred; however, small capital requests may also be considered. The majority of grants from the Taylor Foundation are one year in duration; on occasion, multi-year support is awarded. Applicants must apply online at the grant website. Applicants are strongly encouraged to do the following before applying: review the downloadable state application procedures for additional helpful information and clarifications; review the downloadable online-application guidelines at the grant website; review the foundation's funding history (link is available from the grant website); review the online application questions in advance; and review the list of required attachments. These will generally include: a list of board members, financial statements (audited, reviewed, or compiled by independent auditor); an organization summary; a list of other funding sources; an IRS Determination letter; and other required documents. All attachments must be uploaded in the online application as PDF, Word, or Excel files. The application deadline for the Doree Taylor Charitable Foundation is 11:59 p.m. on April 1. Applicants will be notified of grant decisions before August 31.
Requirements: 501(c)3 charitable organizations and municipalities which meet the mission criteria are eligible to apply.
Restrictions: The foundation will not contribute to endowments or consider grant requests from individuals, organizations attempting to influence policy through direct lobbying, or political campaigns.
Geographic Focus: All States, Maine
Date(s) Application is Due: Apr 1
Contact: Miki C. Akimoto, Vice President; (866) 778-6859; miki.akimoto@baml.com
Internet: https://www.bankofamerica.com/philanthropic/fn_search.action
Sponsor: Doree Taylor Charitable Foundation
225 Franklin Street, 4th Floor, MA1-225-04-02
Boston, MA 02110

Doris and Victor Day Foundation Grants 1491
The foundation supports local community organizations providing food, shelter, clothing, medical care, and education in Illinois/Iowa Quad Cities region. Preference will be given to preventive programs and projects that foster pride in the local community. Types of support include general operating support, building/renovation, equipment, emergency funds, seed grants, and scholarship funds.
Requirements: Illinois/Iowa Quad Cities region nonprofit organizations may apply.
Restrictions: The Foundation is committed to programs that are non-sectarian, and therefore, will not contribute toward programs and capital projects for religious purposes, except for modest contributions to the churches in which the Days held membership. However, clearly non-sectarian, community serving programs of religious organizations will be considered for funding.
Geographic Focus: Illinois, Iowa
Date(s) Application is Due: May 1
Contact: Program Contact; (309) 788-2300; info@dayfoundation.org
Internet: http://www.dayfoundation.org/guide.htm
Sponsor: Doris and Victor Day Foundation Grants
1800 3rd Avenue, Suite 302
Rock Island, IL 61201-8019

Doris Duke Charitable Foundation Clinical Interfaces Award Program 1492
The program seeks to catalyze activity at the interface of clinical and other research disciplines by: supporting the formation of new collaborations and strengthening existing collaborations of outstanding scientists across disciplines; demonstrating successful models for clinical research at the interface of multiple disciplines; and supporting interdisciplinary and inter-institutional endeavors that go beyond the program project mindset. Full grants, awarded over five years, are made to established teams with key investigators from at least three disciplines. Planning grants are awarded to new teams for the development of full proposals over 18 months. The listed application deadline is for pre-proposals; full proposals are by invitation. Though new grants are not being offered continually, potential applicants can sign up to be notified of future competitions.
Requirements: Teams of at least three key investigators whose primary expertise lie in different disciplines are eligible to apply. Key investigators must have advanced degrees (MD, PhD, MD/PhD, or the equivalent), and one of the key investigators must be a clinical researcher. The team leader must work in a U.S. nonprofit institution, such as an academic medical center. The team may include investigators at other institutions in the United States and overseas.
Restrictions: Planning grants will not be awarded in the current competition.
Geographic Focus: All States
Date(s) Application is Due: Nov 2
Amount of Grant: Up to 100,000,000 USD
Contact: Elaine K. Gallin, (212) 974-7104; fax (212) 974-7590; egallin@ddcf.org
Internet: http://www.ddcf.org/page.asp?pageId=299
Sponsor: Doris Duke Charitable Foundation
650 Fifth Avenue, 19th Floor
New York, NY 10019

Doris Duke Charitable Foundation Clinical Research Fellowships for Medical Students 1493
The fellowship program is designed to encourage medical students to pursue careers in clinical research by giving exceptional students the opportunity to take a year to experience clinical research first hand. The CRF program is available at the following medical schools: Columbia University College of Physicians and Surgeons; Harvard Medical School; Mount Sinai School of Medicine; University of California at San Francisco School of Medicine; University of Iowa Roy J. and Lucille A. Carver College of Medicine; University of North Carolina at Chapel Hill School of Medicine; University of Pennsylvania School of Medicine; University of Texas Southwestern Medical Center at Dallas; Washington University Medical School; and Yale University School of Medicine. An additional 11-12 fellowships will be available to medical students interested in conducting clinical research in Africa. Each participating medical school provides medical students with a one-year fellowship experience in clinical research that includes both didactic and research components; matches students to outstanding clinical research mentors; and offers fellowships to at least five students per year.
Requirements: Medical students matriculated at any U.S. medical school who have completed two or more years of medical school prior to the start of the fellowship and who have completed some clinical experience are eligible to apply to any of the participating schools.
Geographic Focus: All States
Date(s) Application is Due: Jan 18
Amount of Grant: 27,000 USD
Contact: Elaine K. Gallin; (212) 974-7104; fax (212) 974-7590; egallin@ddcf.org
Internet: http://www.ddcf.org/page.asp?pageId=292
Sponsor: Doris Duke Charitable Foundation
650 Fifth Avenue, 19th Floor
New York, NY 10019

Doris Duke Charitable Foundation Clinical Scientist Development Award 1494
Grants are awarded to junior physician-scientists to facilitate their transition to independent clinical research careers. The program is designed to help prepare and support new investigators with an MD or MD/PhD as they begin their careers as independent clinical researchers. The program is aimed at conducting clinical research in any disease area. The listed application deadline is for preproposals, which must be submitted electronically.
Requirements: Applicants must: be a physician-scientist conducting clinical research in any disease area; have received an M.D. or a foreign equivalent from an accredited institution; be working in a U.S. degree-granting institution, but do not have to be a U.S. citizen; have a full-time faculty level position not higher than the Assistant Professor level; and have been appointed to their first full-time faculty level position between January 31, 2002 and January 31 of the grant year. (All full-time post-fellowship Instructor level positions will be considered full-time faculty level appointments).
Restrictions: Funds cannot be used on experiments that utilize animals or primary tissues derived from animals. An award will not be made if, prior to the commencement of the award, the applicant becomes the principal investigator on a federal government, peer-reviewed, research or career development award or any non-government award averaging more than $126,000 per year in direct costs and of a duration of three or more years.
Geographic Focus: All States
Date(s) Application is Due: Jan 23
Amount of Grant: 135,000 USD
Contact: Elaine K. Gallin; (212) 974-7104; fax (212) 974-7590; egallin@ddcf.org
Internet: http://www.ddcf.org/page.asp?pageId=291
Sponsor: Doris Duke Charitable Foundation
650 Fifth Avenue, 19th Floor
New York, NY 10019

Doris Duke Charitable Foundation Clinical Scientist Development Award Bridge Grants 1495
The purpose of the bridge grant is to help Doris Duke Clinical Scientist Development Award (CSDA) recipients successfully transition to independent National Institutes of Health (NIH) R01 funding. The Medical Research Program created these grants to bridge the time between when a CSDA investigator learns of a very strong but unfunded score on an NIH application to when s/he is able to successfully obtain an R01 grant.

The bridge grants provide protected time and support for CSDA recipients to continue their research projects, collect supportive data, keep patients enrolled if applicable, and strengthen their proposals to resubmit to the NIH. The foundation hopes that the bridge grants will help keep promising physician-scientists committed to clinical research careers by providing stable support during their critical transition to independent R01 funding. The foundation will award up to four bridge grants of $135,000 each to be spent only on direct costs over 12 to 18 months. Eligible CSDA recipients are invited to submit applications throughout the year up to the deadline of September 1.
Requirements: This competition is open only to former CSDA grantees or current CSDA grantees in the last year of their award (or in a no-cost extension period).
Geographic Focus: All States
Date(s) Application is Due: Sep 1
Amount of Grant: Up to 135,000 USD
Contact: Elaine Gallin; (212) 974-7104; fax (212) 974-7590; egallin@ddcf.org
Internet: http://www.ddcf.org/page.asp?pageId=291
Sponsor: Doris Duke Charitable Foundation
650 Fifth Avenue, 19th Floor
New York, NY 10019

Doris Duke Charitable Foundation Distinguished Clinical Scientist Award 1496
The purpose of the program is to recognize outstanding physician-scientists who are engaged in applying the latest basic science advances to the prevention, diagnosis, treatment, and cure of disease, and to enable the physician-scientist to support and mentor the next generation of physician-scientists conducting translational clinical research. Awards will be granted to physician-scientists conducting translational clinical research in any disease area; this award cycle is not limited to specific disease areas.
Requirements: Grantees must hold an M.D. degree from an accredited institution in the United States (holders of M.D./Ph.D. degrees are also eligible, as are holders of M.D.-equivalent degrees from non-U.S. institutions); hold a full-time university faculty appointment at the level of Associate Professor or above as of the date of nomination; have been appointed to their first full-time, faculty-level position for no more than 15 years; and have an established translational clinical research program.
Restrictions: Experiments that utilize animals or primary tissues derived from animals are not eligible for support through this award program. An award will not be made if, prior to the commencement of the foundation's award, the applicant becomes the principal investigator of a federal government, peer-reviewed, research or career development award or of any nongovernment award averaging $100,000 or more per year and of three years or more duration.
Geographic Focus: All States
Date(s) Application is Due: Feb 14
Amount of Grant: Up to 15,000,000 USD
Contact: Elaine Gallin; (212) 974-7104; fax (212) 974-7590; egallin@ddcf.org
Internet: http://www.ddcf.org/page.asp?pageId=297
Sponsor: Doris Duke Charitable Foundation
650 Fifth Avenue, 19th Floor
New York, NY 10019

Doris Duke Charitable Foundation Operations Research on AIDS Care and Treatment in Africa (ORACTA) Grants 1497
The Program seeks to improve the care and treatment of AIDS patients in resource-limited settings, inform antiretroviral therapy (ART) policy and practice, and improve outcomes of the roll-out and scale-up of ART in Africa. ORACTA grants provide two-year grants of up to $100,000 per year to teams of investigators conducting health operations research on AIDS care and treatment in Africa. Though new grants are offered only periodically, applicants should sign up for the Medical Research Program's mailing list.
Restrictions: Indirect costs are not supported by this award.
Geographic Focus: All States
Amount of Grant: 100,000 USD
Contact: Elaine Gallin; (212) 974-7104; fax (212) 974-7590; egallin@ddcf.org
Internet: http://www.ddcf.org/page.asp?pageId=486
Sponsor: Doris Duke Charitable Foundation
650 Fifth Avenue, 19th Floor
New York, NY 10019

Dorothea Haus Ross Foundation Grants 1498
The foundation awards grants to eligible nonprofit organizations that work to relieve suffering among children who are sick, handicapped, injured, disfigured, orphaned, or otherwise vulnerable. Types of support include direct services, medical research, equipment and supplies, and small renovation projects. There are no application deadlines. The foundation has a preference for small grassroots projects that it can fully fund or nearly fully fund with the small grants that it makes. The Ross Foundation is less interested in larger projects or capital campaigns that are better left to larger foundations and organizations.
Requirements: U.S. Charities may apply if: they have 501(c)3 status; they are listed in the current edition of the Cumulative List of Charities published by the U.S. Department of the Treasury; they are a Catholic organization listed in the current edition of the Catholic Director; or they are listed in the Free Methodist Yearbook, or other Protestant Denomination Directory that has a group ruling for tax exemption from the IRS. Although grants are made internationally, Foundation by-laws prohibit sending money directly to foreign charities. Applicants from foreign countries (outside the United States) are encouraged to call or email the foundation prior to submitting any applications.
Restrictions: The Foundation does not fund day-to-day operations, individuals, conferences, day care, or public education. Although the Foundation makes international grants, there are restrictions in some countries for the following reasons: war, widespread violence, or breakdown of law and order; or countries where grants are restricted by the U.S. Government due to a boycott or other reason.
Geographic Focus: All States
Contact: Wayne S. Cook, Ph.D., Foundation Executive; (585) 473-6006; fax (585) 473-6007; Rossfoundation@frontiernet.net
Internet: http://www.dhrossfoundation.org/index.php?option=com_content&view=article&id=3&catid=1
Sponsor: Dorothea Haus Ross Foundation
1036 Monroe Avenue
Rochester, NY 14620

Dorothy Hooper Beattie Foundation Grants 1499
The Foundation primarily supports higher education, health care, and religious programs in and around the Greenville, South Carolina, region. Its program interests include: camps; cancer treatment; Catholicism; children and youth programs; higher education; health services; literacy; and the environment. Grants range from $620 to $6,500, and there are no specific deadlines or application forms. Initial approach should be by letter, with telephone contacts discouraged. Mail applications to: Bank of America, 7 North Laurens Street, Greenville, SC 29601.
Geographic Focus: South Carolina
Amount of Grant: 620 - 6,500 USD
Contact: Mary W. Green, Bank of America Contact; (864) 271-5789
Sponsor: Dorothy Hooper Beattie Foundation
101 South Tryon Street
Charlotte, NC 28255

Dorothy Rider Pool Health Care Grants 1500
The Foundation's intent is to serve as a means to improve the quality of life in the Lehigh Valley community, to build on community strengths and add to its vitality, and to increase the capacity of the community to serve the needs of all its citizens. Within this objective the Foundation's funding program is focused on education, health and welfare, culture and art and community development. Interested applicants should submit a letter of intent of five pages or less.
Requirements: Allentown, PA, nonprofit organizations are eligible.
Restrictions: The Foundation is restricted from providing funds to individuals, legislative or lobbying efforts, political or fraternal organizations or organizations outside the United States and its territories. The Foundation as a policy does not provide operating or capital funds to Sectarian institutions, organizations or programs in which funds will be used primarily for the propagation of religion, hospitals or United Way member agencies. Further, the Foundation does not underwrite charitable or testimonial dinners, fund-raising events or related advertising or the subsidization of books, mailings or articles in professional journals.
Geographic Focus: Pennsylvania
Date(s) Application is Due: Apr 1; Aug 15
Amount of Grant: 100,000 USD
Samples: Lehigh Valley Hospital and Health (Allentown, PA)--for its department of obstetrics and gynecology, $225,000 over three years; Lehigh Valley Hospital and Health Network (Allentown, PA)--for the department of emergency medicine's sexual assault response team, $150,000 over three years; Lehigh Valley Hospital and Health Network (Allentown, PA)--for the Physician Leaders in the Lehigh Valley program, $120,000 over three years.
Contact: Ronald Dendas; (610) 770-9346; fax (610) 770-9361; drpool@ptd.net
Internet: http://www.pooltrust.com
Sponsor: Dorothy Rider Pool Health Care Trust
1050 S Cedar Crest Boulevard, Suite 202
Allentown, PA 18103

Dorrance Family Foundation Grants 1501
The Dorrance Family Foundation was founded by Bennett Dorrance, co-owner of the Campbell Soup Company. The Foundation gives primarily in the states of Arizona, California, and Hawaii, offering support for projects that work to resolve societal, educational and environmental problems strategically and make communities a better place. Its two primary areas of interest are education and natural resource conservation. In the area of education, the Foundation awards grants for the funding of: academic needs of low income and/or underserved students; first generation graduates; innovation; literacy; primary, secondary, and post-secondary academics; quality teacher training and recruitment; and science and technology programs. In the area of conservation, the Foundations awards grants for the funding of: forests; innovation; marine and coastal areas; rivers, streams, wetlands, and watersheds; sustainable agriculture, land use, and land management; and wildlife habitats. The Foundation also supports arts and culture, children's medical research, science and other community needs. The Dorrance Family Foundation does not accept unsolicited grant applications. If your organization qualifies for a grant based on the Foundation's grantmaking focus and eligibility requirements, please submit a Letter of Inquiry.
Requirements: No formal application form is required, 501(c)3 non-profits operating in Arizona are eligible to apply for these grants. Applicants should submit a proposal consisting of a detailed description of project and amount of funding requested.
Restrictions: No funding available to individuals.
Geographic Focus: Arizona, California, Hawaii
Contact: Carolyn O'Malley, Executive Director; (480) 367-7000; info@dmbinc.com
Internet: http://www.dorrancefamilyfoundation.org/
Sponsor: Dorrance Family Foundation
7600 East Doubletree Ranch Road, Suite 300
Scottsdale, AZ 85258-2137

Dorr Institute for Arthritis Research and Education 1502
Established in California in 2000, the Dorr Institute for Arthritis Research and Education promotes medical treatment, research and education concerning the causes and treatment of arthritis and joint implant surgery procedures. Its primary fields of interest are arthritis and arthritis research. Recognized as a foundation, it currently funds research for improvement in total hip replacement and total knee replacement. Annually, they support two types of fellows who participate in the research; a research fellow is dedicated to investigations; and a clinical fellow participates in the care of the patients and also participates in research.
Geographic Focus: All States
Contact: Jeri Ward, Director; (323) 442-5762 or (213) 977-2280; jeri@drdorr.com
Internet: http://dorrarthritisinstitute.org/research.html
Sponsor: Dorr Institute for Arthritis Research and Education
1520 San Pablo Street, Suite 2000
Los Angeles, CA 90033

Do Something Awards 1503
The Do Something Awards honor dynamic young people for service in the areas of community building, health, and the environment. Award winners are leaders who identify and realize solutions to problems facing local communities across America. Five winners will receive a minimum of $10,000 in community grants and scholarships. Of those five winners, one will be selected by a national, online vote as a Golden Award winner. That Golden Award winner receives a total of $100,000 in community grants, which will enable him or her to take their work to the next level. Winners are announced and recognized at an annual gala in New York City. Application and guidelines are available online.
Requirements: Applicants must be 25 years old or younger. Only winners who are age 18 and under are eligible for a scholarship of $5,000 and a $5,000 community grant. Winners age 19 to 25 receive their entire award in the form of a community grant.
Geographic Focus: All States
Date(s) Application is Due: Dec 15
Amount of Grant: 10,000 - 100,000 USD
Contact: Naomi Hirabayashi, CMO; (212) 254-2390, ext. 240; nhirabayashi@dosomething.org
Internet: http://www.dosomething.org/programs/awards
Sponsor: Do Something
19 West 21st Street, 8th Floor
New York, NY 10010

DPA Promoting Policy Change Advocacy Grants 1504
The Drug Policy Alliance Advocacy Grants seek to broaden public and political support for drug policy reform and will fund strategic and innovative approaches to increase such support. Proposals should be designed to: educate the public and policymakers about the negative consequences of current local, state or national drug policies; promote better awareness and understanding of alternatives to current drug policies; and broaden awareness and understanding of the extent to which punitive prohibitionist policies are responsible for most drug-related problems around the country. Strategic, geographic or thematic collaborations are strongly encouraged. The Alliance prioritizes organizations focused on one or more of the following: public education campaigns and litigation to raise awareness of the negative consequences of current local, state, and national drug policies; and organizing and mobilizing constituencies that raise awareness about the negative consequences of local, state, and national drug policies. The Alliance also favors public education efforts that speak to: the failures and consequences of drug polices in the U.S. and the potential benefits of alternatives to prohibition; reducing over-reliance on the criminal justice system by raising awareness of the need for alternatives to incarceration and/or health-based approaches to drug use; discrimination in employment, housing, student loans and other benefits against those who use drugs or who have been convicted of drug law violations; the negative consequences of current drug policies on human rights; and efforts that mobilize people around the disproportionate impact of the drug war on communities of color and youth.
Requirements: Tax-exempt 501(c)3 organizations and organizations with 501(c)3 fiscal sponsors are eligible. The program provides both general support and project-specific grants. All grantmaking will be directed to organizations working within the United States, and possibly Canada, with particular emphasis on state-based activity. The Alliance will make grants to organizations that have been invited to apply and who demonstrate a clear ability and commitment to educate the public about the need for broad drug policy reform.
Geographic Focus: All States, Canada
Date(s) Application is Due: Jun 18
Amount of Grant: 15,000 - 25,000 USD
Contact: Asha Bandele, Grants Contact; (212) 613-8020; fax (212) 613-8021; abandele@drugpolicy.org or grants@drugpolicy.org
Internet: http://www.drugpolicy.org/about/jobsfunding/grants/index.cfm
Sponsor: Drug Policy Alliance
131 West 33rd Street, 15th Floor
New York, NY 10001

Dr. and Mrs. Paul Pierce Memorial Foundation Grants 1505
The Dr. and Mrs. Paul Pierce Memorial Foundation was established in 1963 to support and promote quality education, human-services, and health-care programming for underserved populations. Special consideration is given to charitable organizations that serve the people of Grayson County, Texas. Grants from the Pierce Memorial Foundation are one year in duration. Applicants must apply online at the grant website. Applicants are strongly encouraged to do the following before applying: review the downloadable state application procedures for additional helpful information and clarifications; review the downloadable online-application guidelines at the grant website; review the foundation's funding history (link is available from the grant website); review the online application questions in advance; and review the list of required attachments. These will generally include: a list of board members, financial statements (audited, reviewed, or compiled by independent auditor); an organization summary; a list of other funding sources; an IRS Determination letter; and other required documents. All attachments must be uploaded in the online application as PDF, Word, or Excel files. The Dr. and Mrs. Paul Pierce Memorial Foundation has four deadlines annually: March 1, June 1, September 1, and December 1. Applications must be submitted by 11:59 p.m. on the deadline dates. Grant applicants are notified as follows: March deadline applicants will be notified of grant decisions by June 30; June applicants will be notified by September 30; September applicants will be notified by December 31; December applicants will be notified by March 31 of the following year.
Requirements: Applicants must have 501(c) tax-exempt status.
Restrictions: The foundation does not support requests from individuals, organizations attempting to influence policy through direct lobbying, or any political campaigns.
Geographic Focus: Texas
Date(s) Application is Due: Mar 1; Jun 1; Sep 1; Dec 1
Samples: Denison Helping Hands, Denison, Texas, $20,000, purchase of food for families living on restricted incomes; Hyde Park Elementary School, Denison, Texas, $5,000, purchase of new books for school library; Grayson County Womens Crisis Line, Sherman, Texas, $10,000, funding in support of the Shopping Card Program.
Contact: David Ross, Senior Vice President; tx.philanthropic@baml.com
Internet: https://www.bankofamerica.com/philanthropic/fn_search.action
Sponsor: Dr. and Mrs. Paul Pierce Memorial Foundation
901 Main Street, 19th Floor, TX1-492-19-11
Dallas, TX 75202-3714

Dr. John Maniotes Scholarship 1506
The Dr. John Maniotes Scholarship was established to benefit the children or grandchildren of current employees of Lake County Government Center in Crown Point. The Scholarship awards two nonrenewable scholarships of $2,500 each year to fund tuition and books. Any college major is eligible.
Requirements: Applicants are eligible if they meet the following criteria if they are: a U.S. citizen and resident of Indiana; a graduating high school senior; a parent or grandparent is a current employee of Lake County Government Center in Crown Point; accepted as an incoming freshman to an accredited public or private college or university; and seeking a four year baccalaureate degree or two year associate degree at a college or university. Selection criteria is based on academic performance (SAT 1400; ACT 19; and GPA 2.5), character, essay, and extracurricular activities. Applicants must submit the completed application along with an original essay, two recommendation letters, the completed Scholastic Profile Form filled out by the guidance counselor; and current high school transcript.
Geographic Focus: Indiana
Date(s) Application is Due: Apr 1
Amount of Grant: 2,500 USD
Contact: Maniotes Scholarship Contact; (219) 736-1880; fax (219) 736-1940; legacy@legacyfoundationlakeco.org
Internet: http://www.legacyfoundationlakeco.org/scholarships/index.htm
Sponsor: Legacy Foundation
1000 East 80th Place, 302 South
Merrillville, IN 46410

Dr. John T. Macdonald Foundation Grants 1507
The Dr. John T. Macdonald Foundation awards grants to eligible Florida nonprofit organizations in Miami-Dade County for medical and health-related programs to community-based programs, with priority given to those serving children, youth, and economically disadvantaged individuals. The foundation supports medical rehabilitation, disease prevention, and health education. Types of support include capital grants, matching/challenge grants, program development grants, seed money grants, and training grants. Deadline listed is for letters of intent; full proposals are by request. Annual deadline dates may vary; contact program staff for exact dates.
Requirements: Florida nonprofit organizations serving the health care needs of people in Miami-Dade County are eligible. Priority will be given to projects in the Coral Gables community.
Restrictions: Grants do not support national projects, multiyear funding requests, for-profit organizations, political candidates or campaigns, religious projects, individuals, or other grantmaking foundations.
Geographic Focus: Florida
Amount of Grant: Up to 550,000 USD
Contact: Kim Greene; (305) 667-6017; fax (305) 667-9135; kgreene@jtmacdonaldfdn.org
Internet: http://jtmacdonaldfdn.org/grants/grants-scholarships/
Sponsor: Dr. John T. Macdonald Foundation
1550 Madruga Avenue, Suite 215
Coral Gables, FL 33146

Dr. John T. Macdonald Foundation Scholarships 1508
The Dr. John T. Macdonald Foundation awards scholarships for nursing and medical students at Barry University, Florida International University, and the University of Miami. Visit the Foundation web site for specific details on how to apply.
Geographic Focus: Florida
Contact: Kim Greene; (305) 667-6017; fax (305) 667-9135; kgreene@jtmacdonaldfdn.org
Internet: http://jtmacdonaldfdn.org/grants/grants-scholarships/
Sponsor: Dr. John T. Macdonald Foundation
1550 Madruga Avenue, Suite 215
Coral Gables, FL 33146

Dr. Leon Bromberg Charitable Trust Fund Grants 1509
The trust awards grants to U.S. nonprofit organizations in its areas of interest, including arts, education, hospitals and health organizations, and social services. Types of support include building construction/renovation, capital campaigns, conferences and seminars, equipment acquisition, professorships, scholarship funds, and social services. Preference will be given to requests from Texas nonprofits in Bryan, Galveston, Crockett, and Houston. Applications must be received by the 10th of each month for consideration; the board meets monthly.
Requirements: U.S. nonprofit organizations are eligible.
Restrictions: Grants are not awarded to individuals or for operating grants, endowment funds, or debt reduction.
Geographic Focus: All States
Samples: U of Texas Medical Branch (Galveston, TX)--to endow a mentor program in the sciences that pairs graduate students, postdoctoral students, and faculty members with high-school students, $250,000.
Contact: Program Contact; (409) 762-5890
Sponsor: Dr. Leon Bromberg Charitable Trust Fund
2200 Market Street, Suite 710
Galveston, TX 77550-1532

Dr. R.T. White Scholarship 1510
The R.T. White Scholarship was established in memory of a beloved principal at Elliott Elementary School. It is awarded each year to graduating seniors from Munster High School who attended Elliott Elementary. The number of awards varies, but the awards are usually $750 each.
Requirements: Applicants must be graduating seniors from Munster High School who attended Elliott Elementary School. Any college major is eligible. The application is available in the Munster High School Guidance Office.
Geographic Focus: Indiana
Date(s) Application is Due: Apr 1
Amount of Grant: 750 USD
Contact: Scholarship Contact; (219) 736-1880; legacy@legacyfoundationlakeco.org
Internet: http://www.legacyfoundationlakeco.org/scholarships/index.htm
Sponsor: Legacy Foundation
1000 East 80th Place, 302 South
Merrillville, IN 46410

Dr. Scholl Foundation Grants 1511
Applications are considered in the following areas: private education, including elementary, secondary, college and university level; programs for children, developmentally disabled, senior citizens, civic and cultural institutions, social service agencies, hospitals and health care, environmental organizations and religious institutions. The general areas of interest are not intended to limit the interest of the foundation from considering other worthwhile projects. An application form is required and may be obtained from the foundation office. All grant applications must be requested in writing. Telephone, email or fax requests will not be accepted.
Requirements: Grants are awarded on an annual basis to valid IRS 501(c)3 organizations. Non-U.S. applicants without a 501(c)3 must complete a notarized affidavit.
Restrictions: Funding is not available for: organizations that do not have a valid IRS 501(c)3 determination letter; political organizations, political action committees, or individual campaigns whose primary purpose is to influence legislation; foundations that are themselves grantmaking bodies; grants to individuals, endowments or capital campaigns; grants for loans, operating deficit reduction, the liquidation of a debt, or general support; event sponsorships including the purchase of tables, tickets or advertisements; or more than one request from the same organization in the same year.
Geographic Focus: All States
Date(s) Application is Due: Mar 1
Contact: Pamela Scholl, President; (847) 559-7430
Internet: http://www.drschollfoundation.com/procedures.htm
Sponsor: Dr. Scholl Foundation
1033 Skokie Boulevard, Suite 230
Northbrook, IL 60062

Dr. Stanley Pearle Scholarships 1512
OneSight Research Foundation (formerly known as the Pearle Vision Foundation) will award ten (10) scholarships in the amount of $2,000 each through the Dr. Stanley Pearle Scholarship Fund. Invitation to apply is extended to second, third, and fourth year full-time students pursuing graduate studies leading to a Doctor of Optometry degree. Scholarships will be awarded on a competitive basis, based on scholastic performance and potential and evidence of commitment to a career in the optometric profession and to community service. Applicants will be required to describe accomplishments they believe would qualify them for recognition in general, and specifically, accomplishments in the areas of leadership, community service, and extracurricular activities.
Requirements: Students accepted to, or enrolled in, a fully accredited college or university that is an Approved Educational Institution pursuing full-time graduate studies in an optometry program accredited by the Accreditation Council on Optometric Education of the American Optometric Association (ACOE) are eligible to apply. Scholarships are not renewable, but an individual may re-apply for a Scholarship. OneSight values diversity. Members of racial and ethnic minorities are especially encouraged to apply. All scholarships will be awarded on an objective, non-discriminatory basis.
Restrictions: Limited to students in graduating class 2010, 2011 or 2012.
Geographic Focus: All States
Date(s) Application is Due: Apr 15
Amount of Grant: 2,000 USD
Contact: Trina Parasiliti; (972) 277-6191; fax (972) 277-6414; tparasil@onesight.org
Internet: http://www.onesight.org/northamerica/na/
Sponsor: OneSight Research Foundation
2465 Joe Field Road
Dallas, TX 75229

Drs. Bruce and Lee Foundation Grants 1513
The Foundation's goal is to advance the general welfare and the quality of all life in the Florence, S.C. area by providing economic support to qualified programs and non-profit organizations. The Foundation will support a broad range of charitable purposes including, but not limited to: medical; health; human services; education; arts; religion; civic affairs; and the conservation, preservation and promotion of cultural, historical and environmental resources. There are no application deadlines. Contact the office for application materials and guidelines.
Requirements: Florence, SC, nonprofit organizations are eligible.
Restrictions: The Foundation does not purchase tickets for fundraising events. Grants to individuals will not be considered.
Geographic Focus: South Carolina
Contact: L. Bradley Callicott; (843) 664-2870; blfound@bellsouth.net
Sponsor: Drs. Bruce and Lee Foundation
181 East Evans Street, BTC Box 022
Florence, SC 29506

DSO Controlling Pathogen Evolution Workshop Grants 1514
The goal of the workshop is to explore: the potential of computational methods to precisely predict pathogen evolution; and vaccination and/or therapeutic strategies to control direction of pathogen evolution. The workshop will investigate tools for predicting protective epitopes, pathogen evolution and protein structure, and then review strategies to force pathogens into an evolutionary trap that prevents escape through mutation. From there, vaccines and therapies can be developed against the particular pathogen strain. Workshop participants are strongly encouraged to prepare posters describing previous or potential research in the area in order to facilitate discussions and/or formation of well-rounded teams.
Requirements: Attendees should have specific expertise in infectious disease, epitope prediction (computational and/or immunomic), vaccinology, mathematical modeling, protein structure, pathogen evolution, drug design, and/or animal modeling.
Geographic Focus: All States
Date(s) Application is Due: Mar 16
Contact: Dr. Michael V. Callahan, Program Manager; (571) 218-4596; fax (571) 218-4553; michael.callahan@darpa.mil
Internet: http://www.darpa.mil/baa/SN07-21.html
Sponsor: Defense Sciences Office within the Defense Advanced Research Projects Agency
3701 N Fairfax Drive
Arlington, VA 22203-1714

DSO Radiation Biodosimetry (RaBiD) Grants 1515
The DSO is seeking proposals for new technologies for rapid, high-throughput, portable and low-cost biodosimeters to determine radiation dose to individuals after acute radiation exposure. This technology would provide rapid identification of individuals who have been exposed to high-dose radiation in order to accurately assess radiation exposure levels. The Program is a single phase, 15-month program with a goal of revolutionizing radiation exposure detection. The first 12 months consist of scientific research and development during which the proposer will develop a non- or minimally-invasive radiological biodosimeter and demonstrate accurate radiological detection from biological samples into quartiles of doses for humans with a detection time less than 10 minutes. This will include radiological dose/detection curves, as well as multipoint data regarding decay of biological signal as a function of time after radiation exposure. A successful proposal will demonstrate a clear path to these deliverables, including a timeline for technology development, experimentation, and delivery of results.
Requirements: All proposers must provide evidence of a competent team capable of developing a new radiation biodosimetry technology meeting all program objectives and milestones. The Program requires that each proposal include a team with demonstrated (or established) capability. At a minimum, proposers are expected to possess expertise or demonstrate collaboration with professionals in the following areas: radiation biologist/radiation oncologist--with expertise in radiation exposure to animal models; and engineering--with developmental expertise needed to translate biological sampling to radiation readout.
Restrictions: Only unclassified proposals will be accepted in response.
Geographic Focus: All States
Date(s) Application is Due: Jul 9
Contact: Dr. Mildred Donlon, Program Manager; (703) 696-2289; fax (703) 741-3896; mildred.donlon@darpa.mil
Internet: http://www.darpa.mil/dso/solicitations/baa07-29.htm
Sponsor: Defense Sciences Office within the Defense Advanced Research Projects Agency
3701 N Fairfax Drive
Arlington, VA 22203-1714

DTE Energy Foundation Health and Human Services Grants 1516
The DTE Energy Foundation is at the core of DTE Energy's commitment to the communities and customers it is privileged to serve. The DTE Energy Foundation is dedicated to strengthening the health and human services sector of these communities. Priority will be given to supporting organizations that are in the forefront of addressing the critical, acute human needs brought on by the economic downturn. Grant amounts generally range from $500 to $100,000, and the application process is distinct for each of

the following ranges: $500 to $2,000; $2,001 to $10,000; and any amount greater than $10,000. Applications must be submitted electronically by the stated deadlines.
Requirements: Eligible applicants must meet all of the following criteria: be located in or provide services to a community in which DTE Energy does business; and be a nonprofit (i.e. be exempt for federal income tax under section 501(c)3 of the Internal Revenue Code and not a private foundation, as defined in Section 509(a) of the Code).
Restrictions: The Foundation does not provide support to: individuals; political parties, organizations or activities; religious organizations for religious purposes; organizations that are not able to demonstrate commitment to equality and diversity; student group trips; national or international organizations, unless they are providing benefits directly to our service-area residents; projects that may result in undue personal benefit to a member of the DTE Energy Foundation board, or to any DTE Energy employee; conferences unless they are aligned with DTE Energy's business interests; single purpose health organizations; hospitals, for building or equipment needs.
Geographic Focus: Michigan
Date(s) Application is Due: Apr 13; Jul 13; Oct 12; Dec 28
Amount of Grant: 500 - 100,000 USD
Contact: Karla D. Hall; (313) 235-9271 or (313) 235-9416; foundation@dteenergy.com
Sponsor: DTE Energy Foundation
One Energy Plaza, 1046 WBC
Detroit, MI 48226-1279

Dubois County Community Foundation Grants 1517
The central purpose of the community foundation is to serve the needs of Dubois County, Indiana, and the philanthropic aims of donors who wish to better their community. The Foundation's fields of interest are: arts, education, environment, beautification programs, health care, human services, recreation, and youth development. Evaluation is based on the project's feasibility, soundness of its implementation plan, viability of subsequent long-term financing, and fulfillment of community need.
Requirements: Organizations should complete the applicant form, agreement, and certification available at the website. Notification of the Board's decision is made approximately four months after the submission deadline.
Restrictions: Giving is concentrated to Dubois County, Indiana. No support is available for the operational expenses of government units or agencies. No grants for operating expenses of nonprofits, funding after the fact, annual fund raising, sponsorship of events, debt retirement, or loans.
Geographic Focus: Indiana
Date(s) Application is Due: Sep 15
Contact: Brad Ward, Chief Executive Officer; (812) 482-5295; fax (812) 482-7461
Internet: http://www.dccommunityfoundation.org/funds/
Sponsor: Dubois County Community Foundation
600 McCrillus Street, P.O. Box 269
Jasper, IN 47547-0269

Duchossois Family Foundation Grants 1518
The foundation focuses its efforts in the Chicago metropolitan area and gives primary consideration to nonprofit organizations that contribute to the community in the area of health. The foundation supports organizations involved with mental health/crisis services, cancer, HIV/AIDS, cancer research, AIDS research and, human services. A one-page summary-request letter should include a description of the organization and its specific needs and purposes, the amount requested, and a list of members of the board of directors and their business/professional affiliations.
Requirements: 501(c)3 tax-exempt public charities serving the Chicago metropolitan area are eligible.
Geographic Focus: Illinois
Contact: Iris Krieg, Executive Director; (312) 641-5765; iriskrieg1@aol.com
Sponsor: Duchossois Family Foundation (Chicago)
203 N Wabash Avenue, Suite 1800
Chicago, IL 60601

Duke Endowment Health Care Grants 1519
The purpose of the program is to: improve the quality and safety of health care in the areas of patient quality and safety programs/intervention, workforce development, and health information technology; improve access to health care in the areas of care for the medically indigent/uninsured, mental health care, and where racial disparities are evident; and expand prevention and early intervention in the areas of child health and early intervention programs, wellness and healthy lifestyle programs, and disease management.
Requirements: The endowment awards grants in North Carolina and South Carolina to: not-for-profit hospitals; academic health centers; health education centers; and select health organizations in counties without an eligible hospital.
Geographic Focus: North Carolina, South Carolina
Date(s) Application is Due: Jun 15; Dec 15
Contact: Mary L. Piepenbring; (704) 376-0291; mpiepenbring@tde.org
Internet: http://www.dukeendowment.org/program-areas/health-care
Sponsor: Duke Endowment
100 N Tryon Street, Suite 3500
Charlotte, NC 28202-4012

Duke University Adult Cardiothoracic Anesthesia and Critical Care Medicine Fellowships 1520
Nine one year fellowship positions are available each year in the ACGME Accredited Fellowship. Second year research and clinical fellowship opportunities are available; programs are individualized to meet the goals of successful applicants. Generally clinical experience is the focus of the first fellowship year, along with involvement in clinical research. Fellows opting for a two year fellowship continue with one clinical day per week in their second year, in addition to a structured plan involving laboratory and/or clinical research and teaching. Second year clinical specialization in pediatric cardiac anesthesiology or intensive care is also available. The Adult cardiothoracic Anesthesia Fellowship is highly recommended as formal training for specialization in cardiac and/or thoracic anesthesiology. The Fellowship is designed to develop clinical, consultant, and in-depth research expertise. Applications are usually considered starting 18 months prior to the start date
Requirements: Complete application (see Duke University website), in addition to the references required, include a letter from your Chairperson/Program Director and two letters from other faculty members who have worked with you closely.
Geographic Focus: All States
Contact: Mark S. Smith; (919) 681-5046 or (919) 416-3853; fax (919) 681-8993
Internet: http://medschool.duke.edu/modules/som_rt/index.php?id=1
Sponsor: Duke University Medical Center
P.O. Box 3094
Durham, NC 27710

Duke University Ambulatory and Regional Anesthesia Fellowships 1521
The Fellowship in Ambulatory/Regional Anesthesia is designed to provide advanced training in anesthesia in the ambulatory surgery setting. This program will be an extension to the previous training with an emphases on the continued educational component to include research, clinical and teaching responsibilities.
Requirements: The selection process for fellows includes the completion of a Fellowship Application (see website) which is to be substantiated by three letters of recommendation from experts in the field. In addition to the letters of reference the application is to include a current CV and a letter of affirmation stating interest in the program.
Geographic Focus: All States
Date(s) Application is Due: Aug 15
Contact: Stephen M. Klein, Director; (919) 668-2056; klein006@mc.duke.edu
Internet: http://anesthesiology.duke.edu/modules/anes_flwshp/index.php?id=1
Sponsor: Duke University Medical Center
P.O. Box 3094
Durham, NC 27710

Duke University Clinical Cardiac Electrophysiology Fellowships 1522
The Clinical Cardiac Electrophysiology fellowship training program is a comprehensive two-year program including training in catheter ablation of supraventricular tachycardias, ventricular tachycardias, and complex arrhythmias such as atrial fibrillation, ischemic VT, and arrhythmias associated with congenital heart disease. Comprehensive training is also provided in lead extraction and device therapy including permanent pacing, implantable defibrillators, and bi-ventricular devices. Fellows are provided with significant opportunity for research activity through protected block time and are expected to participate in and complete a research project with a faculty mentor. Applicants must have completed at least three years in an ACGME approved Cardiology fellowship program.
Requirements: Applications for training to begin July 1, will be accepted beginning January 1; the application deadline will be September 1. To apply, please include with the completed CCE (EP) Application form (available on the Duke University website), a current curriculum vitae, personal statement (not to exceed 250 words), and the $40 (check or money/order no cash please) processing fee.
Geographic Focus: All States
Date(s) Application is Due: Sep 1
Contact: Dr. Patrick Hranitzky, Program Director; (919) 684-2304; fax (919) 684-4322; Patrick.hranitzky@duke.edu
Internet: http://fellowships.medicine.duke.edu/modules/fellows_cardio/index.php?id=8
Sponsor: Duke University School of Medicine
P.O. Box 3878
Durham, NC 27710

Duke University Hyperbaric Center Fellowships 1523
The clinical Hyperbaric fellow works in the Hyperbaric Center and F.G. Hall Environmental Laboratory, both of which are an integral part of Duke University Medical Center. Fellowships are available for 12 months although they may be extended for additional time. During the period of training the Hyperbaric fellow will be trained in the clinical use of hyperbaric oxygen. This includes the emergency treatment of clostridial myonecrosis, necrotizing fasciitis, arterial gas embolism, decompression sickness, carbon monoxide poisoning and certain other less common conditions such as acute vasculitis with ischemia. The fellow will receive training in the appropriate critical care interventions necessary for the treatment of these patients. This will include hemodynamic monitoring, ventilatory support, fluid resuscitation and the interpretation of arterial blood gases during hyperbaric therapy. The fellow will also receive training in the use of hyperbaric oxygen therapy for non-emergency conditions. The fellow will be responsible for all inpatient and outpatient consults, under the direction of the clinical Hyperbaric faculty. During the training period the fellow will also be required to complete a research project. This will be completed under the guidance of one of the faculty members in the Hyperbaric Center. Part of the didactic program includes regular hyperbaric conferences. The fellow will participate in these and normally be required to present at least one topic during his year of training. The fellow will also have the opportunity to attend the annual two-week course in Hyperbaric and Diving Medicine given by the National Oceanic and Atmospheric Administration in Seattle, Washington. In addition to these activities the applicant may participate in numerous other teaching activities within Duke Medical Center, for example those sponsored by the Departments

of Anesthesiology, Medicine, Surgery and Emergency Medicine. Clinical participation under the aegis of those departments can be arranged in some cases, depending upon the training of the individual concerned.
Requirements: Applicants must be eligible for license as a trainee in the state of North Carolina. The Duke Hyperbaric Center Fellowship Application can be downloaded at: http://hyperbaric.mc.duke.edu/education.htm
Geographic Focus: All States
Contact: Tonya Manning, Assistant; (919) 681-1685 or (919) 684-6726
Internet: http://hyperbaric.mc.duke.edu/education.htm
Sponsor: Duke University Medical Center for Hyperbaric Medicine and Environmental Physiology
P.O. Box 3823
Durham, NC 27710

Duke University Interventional Cardiology Fellowships 1524
The Interventional Cardiology Fellowship Training Program at Duke University Medical Center focuses on training leaders in Interventional Cardiology committed to a career in academic cardiology. The length of the fellowship is 12 months. During the fellowship, all fellows will develop both clinical and technical skills in interventional cardiology. All fellows will gain experience in diagnosing, selecting therapies, performing diagnostic and interventional procedures, and judging the effectiveness of treatments for inpatients and outpatients with chronic ischemic heart disease, acute ischemic syndromes, and valvular heart disease. All fellows are expected to participate in research efforts. During the fellowship, each fellow will work with a mentor on a research project which will help foster their long-term goals.
Requirements: Individuals applying for only a 1 year Interventional Cardiology Fellowship position will be required to provide his/her own funding resources for salary support during their training. This funding must be institutional funding (no personal or private funds). A processing fee of $40 (check or money/order made payable to Duke University- no cash please), letter documenting funding support, current curriculum vitae, and personal statement will be required to accompany the completed Interventional Cardiology Application (see Duke University website for application).
Geographic Focus: All States
Date(s) Application is Due: Feb 1
Contact: Sherolyn Patterson, Fellowship Program Coordinator; (919) 684-2304; fax (919) 684-4322; sherolyn.patterson@duke.edu
Internet: http://fellowships.medicine.duke.edu/modules/fellows_cardio/index.php?id=8
Sponsor: Duke University School of Medicine
P.O. Box 3878
Durham, NC 27710

Duke University Obstetric Anesthesia Fellowships 1525
The Department of Anesthesiology at Duke University Medical Center offers 1 - 2 positions of a one year Fellowship in Obstetric Anesthesia. Emphasis is on clinical research, advanced clinical training and teaching. Duke University is a tertiary center with approximately 3,000 deliveries per year. The division is currently made up of 8 faculty staff with diverse international pedigree.
Geographic Focus: All States
Contact: Yemi Olufolabi, Fellowship Director; (919) 681-6535 or (919) 668-6266; fax (919) 668-6265; olufo001@mc.duke.edu
Internet: http://anesthesiology.duke.edu/modules/anes_flwshp/index.php?id=1
Sponsor: Duke University Medical Center
P.O. Box 3094
Durham, NC 27710

Duke University Pain Management Fellowships 1526
The Fellowship Program is based in the Duke Health Center at Morreene Road (MRC). Fellows will: cover the Pain Clinic and Durham VAMC, and Acute Pain Service at Duke North; evaluate new patients; perform comprehensive or neurologic studies; formulate diagnosis and plan; become familiar with procedures; learn intervention of the lumbar spine and its application e.g. facet rhizotomy, discography, and intradiscal annuloplasty; implantation therapies; dictation of clinic and procedure notes. Fellows will be asked to participate in research, journal clubs, round table discussions, grand rounds and lectures. There are opportunities for the fellow to become involved in research projects already in progress or to develop an original project. The clinical objectives are to build knowledge, skills and professionalism.
Requirements: Applicants must meet eligibility specified by the ACGME. Acceptance into the program is based on demonstrated academic qualification, interpersonal skills, and evidence of ability to function in a multidisciplinary setting. Fellowship applications may be downloaded from http://anesthesia.duhs.duke.edu/. Submit application, CV and three letters of reference.
Geographic Focus: All States
Date(s) Application is Due: Jun 30
Contact: Dianne L. Scott, Fellowship Director; (919) 684-6736
Internet: http://anesthesiology.duke.edu/modules/anes_flwshp/index.php?id=1
Sponsor: Duke University
932 Morreene Road, Room 232
Durham, NC 27705

Duke University Pediatric Anesthesiology Fellowship 1527
The pediatric anesthesia fellowship program provides an opportunity for one anesthesia to develop skills in pediatric anesthesia. The goals of the program are to foster the development of the following for the fellow: a broad clinical experience in pediatric anesthesia to allow proficiency in providing care for neonates, infants, children, and adolescents undergoing a wide variety of surgical, diagnostic, and therapeutic procedures; technical skills in pediatric airway management; technical expertise in pediatric invasive monitoring; a comprehensive understanding of the developmental, pharmacological, anatomic, physiologic, and psychological changes that occur with age and disease; an understanding of pediatric pain and its management; an understanding of critical perioperative care and advanced life support; and competence as a perioperative pediatric anesthesiologist and consultant. The goals of the program will be met by providing fellows with: graded clinical experiences; didactic lectures, seminars, and conferences; and clinical electives and research opportunities.
Requirements: The fellowship application can be downloaded from : http://anesthesia.duhs.duke.edu/education/fellows/fellowapp.pdf. Fellowship selection is based on the process used for application to the core residency program. All applicants are considered without regard to race, color, religion, gender, or national origin and must have completed an ACGME accredited program in Anesthesiology or it's equivalent. Documentation of completion will be required. In addition to the completed application, all candidates are required to submit: completed fellowship application; three letters of recommendation; curriculum vitae; one page personal statement; medical school transcript; letter of completion from the Anesthesiology residency chairman or program director or copy of Anesthesiology training certificate.
Geographic Focus: All States
Contact: B. Craig Weldon M.D., Director; (919) 668-0976 or (919) 668-4208; fax (919) 668-4228; weldo@mc.duke.edu
Internet: http://anesthesiology.duke.edu/modules/anes_flwshp/index.php?id=1
Sponsor: Duke University Medical Center
Department of Anesthesiology, DUMC 3094
Durham, NC 27710

Duke University Postdoctoral Research Fellowships in Aging 1528
The goal of the postdoctoral research training program is to produce highly skilled research scientists who have the potential for leadership in gerontological research. In the Duke Research Training Program (RTP), much of the training for each fellow is provided by that person's faculty mentor(s) in a research apprenticeship program. A fellow carries out his/her own research as a junior colleague in the mentor's research program or laboratory. In addition to working in their mentor's programs, all fellows attend a weekly interdisciplinary didactic seminar. The stipend amount is determined by years of relevant postdoctoral experience. The program also provides health insurance for the fellow (using the University's insurance plan), and partial travel support when presenting research at a professional meeting.
Requirements: Applicants must be citizens of the United States, or have been lawfully admitted to the United States for permanent residence. Applicants must have a completed doctoral degree when joining the program. All course work must be completed, and if applying as a Ph.D., final orals must be passed, and the dissertation signed before an applicant can begin the program.
Geographic Focus: All States
Date(s) Application is Due: May 1
Contact: James A. Blumenthal, (919) 660-7517; fax (919) 684-8569; blume003@mc.duke.edu
Internet: http://www.geri.duke.edu/post_doc/
Sponsor: Duke University School of Medicine
P.O. Box 3878
Durham, NC 27710

Duneland Health Council Incorporated Grants 1529
The Duneland Health Council, Inc. is a private foundation focused on improving the health and general welfare of the greater Michigan City, Indiana community.
Restrictions: Funding limited primarily in the metropolitan Michigan City, IN area. No support for religious organizations. No grants to individuals, or for fund-raising, endowments or advertising.
Geographic Focus: Indiana
Contact: Norm Steider; (219) 874-4193; fax (219) 873-2416; normsteider@yahoo.com
Sponsor: Duneland Health Council Incorporated
P.O. Box 9327
Michigan City, IN 46361-9327

Dunspaugh-Dalton Foundation Grants 1530
Since 1963, the Dunspaugh-Dalton Foundation, Inc. (DDF) has assisted qualifying, exempt 501(c)3 organizations in achieving charitable goals. The Foundation awards grants to eligible nonprofit organizations in the areas of higher, secondary, and elementary education; cpmmunity development; social services; youth services and programs; health associations and hospitals; cultural programs; and civic affairs. Types of support include capital campaigns, continuing support, endowment funds, matching funds, operating support, professorships, and special projects. There are no annual deadlines. The board meets monthly to consider requests.
Requirements: U.S. nonprofit organizations are eligible. The foundation primarily supports programs in California, Florida, and North Carolina.
Restrictions: Individuals are not eligible.
Geographic Focus: California, Florida, North Carolina
Amount of Grant: 5,000 - 50,000 USD
Contact: Sarah Lane Bonner, President; (305) 668-4192; fax (305) 668-4247; ddf@dunspaughdalton.org
Internet: http://www.dunspaughdalton.org/application-process.html
Sponsor: Dunspaugh-Dalton Foundation
1500 San Remo Avenue, Suite 103
Coral Gables, FL 33146

DuPage Community Foundation Grants 1531
The DuPage Community Foundation is a community-based philanthropic organization that identifies, promotes, and supports programs that raise the quality of life in DuPage County for this generation and future generations. To this end, the Foundation seeks to address the broad needs of DuPage County through grants in support of the following program categories: Arts and Culture, Education, Environmental, Health, and Human Services.
Requirements: Grant applications will be accepted from non-profit charitable organizations that: are classified as exempt from federal income taxes under section 501(c)3 of the Internal Revenue Code; and are located in DuPage County. Previous grantees must have complied with all reporting requirements and be in good standing before submitting another grant request.
Restrictions: The Foundation does not generally award grants to: organizations located outside of DuPage County (exceptions, if any, will be made by the director of grants); individuals; governmental agencies, including supporting foundations; programs for religious purposes; endowments; disease-specific organizations; private foundations and private operating foundations; or public, parochial, private or charter schools (pre-K through 12th grade).
Geographic Focus: Illinois
Date(s) Application is Due: Feb 2; Aug 3
Samples: To be divided among 44 organizations in DuPage County (IL)--to support nonprofit groups working in the fields of the arts and culture, civic affairs, education, health, and human services, and for two special projects, $288,000; Willowbrook Wildlife Foundation (Glen Ellyn, IL)--to help establish a permanent endowment at the foundation, $10,000 challenge grant.
Contact: Bonnie Heydorn, (630) 665-5556; fax (630) 665-9571; bheydorn@dcfdn.org
Internet: http://www.dcfdn.org
Sponsor: DuPage Community Foundation
2100 Manchester Road, Building A, Suite 303
Wheaton, IL 60187-4584

DuPont Pioneer Community Giving Grants 1532
The DuPont Pioneer corporation supports a number of community-based projects in the areas of agriculture, education, farm safety, and the environment. Priority consideration is given to projects located in Dupont Pioneer facility communities or rural agricultural regions, as well as to organizations with active Dupont Pioneer management/employee participation and company-related expertise and interest. Types of support include: capital campaigns; general operating support; program development; and seed money. The company accepts proposals from nonprofit organizations nationwide, but favors programs in its operating communities. Dupont Pioneer prefers to make direct contributions to organizations, rather than sponsorships, ticket or table purchases. This allows more funding to go directly toward the non-profit organization; however, we are willing to consider sponsorships when DuPont Pioneer employees are actively involved with the organization. The employee must present the request to Community Investment at least one month in advance. Due to the number of non-profit events held annually, there is a $1,000 maximum contribution level per event per year. The company favors proposals that demonstrate cooperation with other community-based programs, broad-based funding, community need, and positive results. Grant proposals are reviewed on a quarterly basis.
Requirements: Nonprofit organizations are eligible. All requests should be directed to the DuPont Pioneer office within the local area. Otherwise, send to the contact listed. Pioneer does not respond favorably to verbal requests.
Restrictions: Grants are not made to individuals, religious or political organizations that promote a particular doctrine, elected officials, company marketing or advertising, or organizations where there is a conflict of interest with Pioneer Hi-Bred.
Geographic Focus: All States
Amount of Grant: Up to 5,000 USD
Contact: Grants Administrator; (800) 247-6803, ext. 3915; fax (515) 334-4415; community.investment@pioneer.com
Internet: https://www.pioneer.com/home/site/about/business/pioneer-giving/community-giving/
Sponsor: DuPont Pioneer
P.O. Box 1000
Johnston, IA 50131-0184

Dyson Foundation Mid-Hudson Valley General Operating Support Grants 1533
General operating support grants are sometimes referred to as core support or unrestricted grants. Many organizations use general operating support grants to cover day-to-day activities or ongoing expenses such as administrative salaries, utilities, office supplies, technology maintenance, etc. Other organizations use this type of funding to cover project costs, capital, technology purchases, and professional development. The use of these funds is totally at the discretion of the organization's board and/or executive staff, although the Foundation expects all organizational expenditures to be part of a board-approved annual budget. The first step in applying for a general operating support grant is to submit a letter of inquiry.
Requirements: 501(c)3 nonprofit organization or libraries based in the Mid-Hudson Valley (Columbia, Dutchess, Greene, Orange, Putnam, and Ulster counties) are eligible. To apply for a general operating support grant, an organization must: review the principles of best practice relative to nonprofit governance, finance, public disclosure, and programming; have been a recipient of a Dyson Foundation project grant within the past three years of their general operating support request; have a mission and programs that are consistent with core funding interests of the Dyson Foundation; and have demonstrated at least three years of stable executive leadership.
Restrictions: The Foundation places no restrictions on the use of these funds, unlike project or management technical assistance grants that are restricted by the Foundation to a particular project or activity. Colleges and universities, hospitals, faith-based institutions, and organizations with annual budgets in excess of $15 million dollars are not eligible to apply for general operating support grants.
Geographic Focus: New York
Amount of Grant: 1,000 - 1,000,000 USD
Contact: Diana M. Gurieva, Executive Vice President; (845) 790-6312 or (845) 677-0644; fax (845) 677-0650; dgurieva@dyson.org or info@dyson.org
Internet: http://www.dysonfoundation.org/grantmaking/general-operating-support-grants
Sponsor: Dyson Foundation
25 Halcyon Road
Millbrook, NY 12545-9611

Dyson Foundation Mid-Hudson Valley Project Support Grants 1534
In 2009 the Dyson Foundation changed its grantmaking priorities to best support the people and communities in its region most vulnerable to the economic downturn. The Foundation's funding now focuses on organizations and activities that address basic needs such as food, housing, health care, and other human services. In limited circumstances, it will also make grants to faith-based organizations, government entities and libraries. The Foundation will also consider limited funding to arts organizations or projects that provide management support or training to other arts organizations, and to arts organizations or projects that can demonstrate the potential to increase local tourism and employment and/or other local economic development as a result of their efforts. Note that there are separate guidelines for Dutchess County and for the other Mid-Hudson Valley counties.
Requirements: IRS 501(c)3 nonprofits that support Hudson Valley, New York, and are not classified as foundations under section 509(a) of the code are eligible. The region is defined as Columbia, Dutchess, Greene, Orange, Putnam, and Ulster counties. Occasionally the foundation awards grants to fiscal sponsors of non-qualifying organizations.
Restrictions: The Foundation is not currently funding certain areas including: the environment, historic preservation, and capital projects. Grants do not support: individuals for any purpose; dinners, fund raising events, tickets, or benefit advertising; direct mail campaigns; service clubs and similar organizations; debt or deficit reduction; governmental units; or international projects or to organizations outside of the United States.
Geographic Focus: New York
Amount of Grant: 1,000 - 1,000,000 USD
Contact: Diana M. Gurieva, Executive Vice President; (845) 790-6312 or (845) 677-0644; fax (845) 677-0650; dgurieva@dyson.org or info@dyson.org
Internet: http://www.dysonfoundation.org/grantmaking/project-grants
Sponsor: Dyson Foundation
25 Halcyon Road
Millbrook, NY 12545-9611

E.B. Hershberg Award for Important Discoveries in Medicinally Active Substances 1535
This award, supported by Schering-Plough Corporation, is given biennially in odd-numbered years to recognize and encourage outstanding discoveries in the chemistry of medicinally active substances. The discovery for which the award is given should have been made during the past two decades. Nominations are accepted in even-numbered years, and the award consists of $3,000 and a certificate.
Requirements: Any individual, except a member of the award committee, may submit one nomination or seconding letter for the award in any given year. The nominating documents consist of a letter of not more than 1000 words containing an evaluation of the nominee's accomplishments and a specific identification of the work to be recognized, a biographical sketch including date of birth, and a list of publications and patents authored by the nominee.
Restrictions: Self-nominations are not accepted.
Geographic Focus: All States
Date(s) Application is Due: Nov 1
Amount of Grant: 4,000 USD
Contact: Felicia Dixon, Awards Administrator; (800) 227-5558 or (202) 872-4408; fax (202) 776-8008; f_dixon@acs.org or awards@acs.org
Internet: http://portal.acs.org/portal/acs/corg/content?_nfpb=true&_pageLabel=PP_ARTICLEMAIN&node_id=1319&content_id=CTP_004515&use_sec=true&sec_url_var=region1
Sponsor: American Chemical Society
1155 Sixteenth Street, NW
Washington, D.C. 20036-4801

E. Clayton and Edith P. Gengras, Jr. Foundation 1536
The E. Clayton and Edith P. Gengras, Jr. Foundation was established in Connecticut in 1986, with an interest in supporting arts and culture, Christianity, elementary and secondary education, higher education, and community nonprofits and charities. Types of funding given include general operating support and re-granting programs. Though no annual deadlines have been identified, an application form is required. This application should include: program need, a list of board members, proof of 501(c)3 status, and budgetary needs. Most recent awards have ranged from $100 to $250,000, though the majority are less than $5,000.
Restrictions: No grants are awarded directly to individuals.
Geographic Focus: All States
Amount of Grant: 100 - 250,000 USD
Samples: University of Saint Joseph, West Hartford, Connecticut, $250,000 - general operating support (2014); Connecticut Children's Medical Center, Hartford, Connecticut, $1,000 - general operating support (2014); Friends of Fenwick Island Lighthouse, Fenwick Island, Delaware, $5,000 - general operating support (2014); Loblollypop Foundation, Hobe Sound, Florida, $5,000 - general operating support (2014).
Contact: Edith P. Gengras, President and Director; (860) 289-3461
Sponsor: E. Clayton and Edith P. Gengras, Jr. Foundation
300 Connecticut Boulevard
East Hartford, CT 06108-3065

E.J. Grassmann Trust Grants　　1537
The E.J. Grassmann Trust awards grants in central Georgia and Union County, New Jersey, in support of higher and secondary education, hospitals and health organizations, historical associations, environmental conservation, and social welfare. Primary fields of interest include: arts; Catholic agencies and churches; elementary and secondary education; the environment; natural resources; health care; higher education; historic preservation; historical societies; hospitals; and human services. Types of support include capital campaigns, building construction and/or renovation, equipment acquisition, and endowment funds. Groups with low administrative costs, that have outside funding, and that encourage self-help are preferred. Grants are awarded in May and November. Written proposals should not be longer than four pages.
Restrictions: Grants are not awarded to individuals or for operating expenses, current scholarship funds, conferences, or workshops.
Geographic Focus: Georgia, New Jersey
Date(s) Application is Due: Apr 20; Oct 15
Amount of Grant: 5,000 - 20,000 USD
Samples: Action Ministries, Atlanta, Georgia, $5,900 - capital fund support; Benedictine Academy, Elizabeth, New Jersey, $11,250 - capital fund support; Greater Newark Conservancy, Newark, New Jersey, $7,500 - capital fund support.
Contact: William V. Engel, Executive Director; (908) 753-2440
Sponsor: E.J. Grassmann Trust
P.O. Box 4470
Warren, NJ 07059-0470

E.L. Wiegand Foundation Grants　　1538
Established in Nevada in 1982, the Foundation provides grants to develop and strengthen programs and projects in Arizona, California, the District of Columbia, Idaho, Nevada, New York, Oregon, Utah, and Washington. Funding is given to educational institutions in the academic areas of science, business, fine arts, and law; and medicine and health organizations in the areas of heart, eye, and cancer surgery, treatment, and research. The Foundation also considers requests for projects that enrich children, communities, public policy, and the arts. Its primary fields of interest include: arts; biology and life sciences; business education; cancer research; chemistry; elementary education; eye research; heart and circulatory diseases and research; higher education; law school education; medical research and institutes; medical school education; museums; performing arts (music and theater); physics; public affairs; secondary education; and visual arts. Grants for educational equipment, including computers and scientific supplies, are also awarded. The board of trustees meets in February, June, and October to choose recipients, but applications may be submitted at any time. Application guidelines are available upon request.
Requirements: Nonprofit organizations in Arizona, California, District of Columbia, Idaho, Nevada, New York, Oregon, Utah, and Washington State are eligible.
Restrictions: Institutions are to be in existence a minimum of five years. Proposals for endowment, debt reduction, ordinary operations, general fund raising, emergency funding, multi-year funding, productions of documentaries, films or media presentations, direct or indirect loans, or for the benefit of specific individuals are excluded from consideration. The Foundation does not award grants to government agencies or to charitable institutions which derive significant support from public tax-funds or United Way.
Geographic Focus: Arizona, California, District of Columbia, Idaho, Nevada, New York, Oregon, Utah, Washington
Amount of Grant: 10,000 - 200,000 USD
Sample: Nevada Museum of Art, Reno, Nevada, $89,000 - exhibit development and support; American Enterprise Institute, Washington, D.C., $500,829 - capital campaign in support of a communications center; Saint Albert the Great School, Reno, Nevada, $500,000 - expansion project.
Contact: Kristen A. Avansino, Executive Director; (775) 333-0310; fax (775) 333-0314
Sponsor: E.L. Wiegand Foundation
165 W Liberty Street, Suite 200
Reno, NV 89501-2902

E. Rhodes and Leona B. Carpenter Foundation Grants　　1539
The E. Rhodes and Leona B. Carpenter Foundation considers grant requests from: public charities which had direct relationships with Leona or Rhodes Carpenter during their lifetime; requests in support of graduate theological education from U.S. and Canadian organizations which are public charities; public charities in communities where the Carpenter Company has had long-time manufacturing facilities; museums which have a permanent collection for the purchase, restoration and conservation of Asian art; charities involved in providing hospice care; and projects (e.g., programs, conferences, resources, etc.) offering support to lesbian, gay, bisexual and transgender persons of faith, or endeavoring to insure faith communities' understanding, affirmation, and inclusion of such persons. Applications should be postmarked or private carrier dated on or before January 31 to be considered at the spring meeting and on or before July 15 for the autumn meeting. Recent awards have ranged from $1,500 to $500,000, with literally hundreds of grants given annually.
Requirements: IRS 501(c)3 organizations are eligible.
Restrictions: Generally, the Foundation will not consider grant requests to support private secondary education, individuals, local religious congregations, or large public charities and will not transfer funds from its endowment to the endowment of another organization.
Geographic Focus: All States
Date(s) Application is Due: Jan 31; Jul 15
Amount of Grant: 1,500 - 500,000 USD
Samples: Shakertown Revisited, Auburn, Kentucky, $85,000 - general operating support; Asian Art Museum Foundation, San Francisco, California, $50,000 - help with Scenes of Banquets and Ceremonies of the Joseon Period exhibit; CenterStage Foundation, Richmond, Virginia, $100,000 - endowment support.
Contact: Joseph A. O'Connor, Jr., Executive Director; (215) 979-3221 or (215) 979-3222; fax (215) 979-3229; admin@carpenterfoundation.us
Internet: http://www.erlbcarpenterfoundation.org/
Sponsor: E. Rhodes and Leona B. Carpenter Foundation
1735 Marker Street, Suite 3420
Philadelphia, PA 19103

E.W. "Al" Thrasher Awards　　1540
The purpose of the E.W. Al Thrasher Awards is to improve children's health through medical research, with an emphasis on projects that have the potential to translate into clinically meaningful results within a few years. The Fund awards grants three times per year, with no fixed number of awards given in each funding cycle or in each year. There are no deadlines for the awards. Concept papers are accepted on a rolling basis, and are considered by the Fund's Executive Committee approximately once per month. If a full proposal is invited, the applicant will have one year to submit a full proposal. Full proposals recommended by peer reviewers for further consideration are considered by the Fund's Committees in February, June, or October.
Requirements: Principal Investigators must be qualified in terms of education and experience to conduct research. A doctoral-level degree is required. There are no citizenship or residency requirements. The Fund is open to applications from institutions both inside and outside the United States. Ideal applications for the E.W. "Al" Thrasher Award address significant health problems that affect children in large numbers and offer the potential for practical solutions to these problems. Such solutions should be innovative and have the potential for rapid and broad applicability with low financial and/or technical barriers to implementation. Hypothesis-driven research is given priority over exploratory, hypothesis-generating research. The grant amount is based on the actual budgetary needs of the project. The duration of the project can be up to three years. Indirect costs of no more than 7% of direct costs will be paid on E.W. Al Thrasher Awards. Investigators in need of salary support for a specific project may apply for modest support (no more than 20 percent, based on a 40-hour work week). The Fund limits the maximum salary upon which an individual can request salary support. The maximum amount will be based on the NIH guidelines for salary support.
Restrictions: Requests for purchase of major equipment (items costing more than $4,000) are discouraged and rarely funded. The Fund does not award grants for educational programs; general operating expenses; general bridge funding for incomplete projects; construction or renovation of buildings or facilities; loans, student aid, scholarships, tuition; or, support of other funding organizations. Grants generally are not awarded for conferences, workshops, or symposia. Proposals in the areas of research on human fetal tissue or behavioral science research will not be considered. Proposals requesting Thrasher to commit money to a larger pool of funding with no distinct project proposed to Thrasher will not be funded.
Geographic Focus: All States, All Countries
Contact: Megan Duncan; (801) 240-4720; DuncanME@thrasherresearch.org
Internet: http://www.thrasherresearch.org/sites/www_thrasherresearch_org/Default.aspx?page=220
Sponsor: Thrasher Research Fund
68 S. Main Street, Suite 400
Salt Lake City, UT 84101

Earl and Maxine Claussen Trust Grants　　1541
The Earl and Maxine Claussen Trust was established in 1984. Though there are no geographic limitations, it was the Claussen's desire to assist the residents of the Grand Island, Nebraska, metropolitan area, which includes Hall, Merrick and Howard counties. The Trust's primary areas of interest include: religion; public and society benefit; human services; health; the environment; animal care; education; and arts, culture and humanities. The average range of funding is $1,000 to $5,000, with approximately eight awards given each year. The annual deadline for application submission is November 30.
Requirements: Grantees must be qualified as public charities under IRS section 501(c)3. Applications must be submitted through the online grant application form or alternative accessible application designed for assistive technology users.
Geographic Focus: All States
Date(s) Application is Due: Nov 30
Amount of Grant: 1,000 - 5,000 USD
Contact: George Weaver, Special Trustee; (888) 234-1999; fax (877) 746-5889; grantadministration@ wellsfargo.com
Internet: https://www.wellsfargo.com/privatefoundationgrants/claussen
Sponsor: Earl and Maxine Claussen Trust
1740 Broadway
Denver, CO 80274

Eastman Chemical Company Foundation Grants　　1542
The Foundation supports organizations and programs that help to promote its efforts of enhancing the quality of life in Eastman communities. Contributions are concentrated on the following areas: education, health and human services, civic and community, and culture and the arts. Grants may be considered outside of these areas.
Requirements: Contributions are generally restricted to organizations which have been granted 501(c)3 tax exempt status and are usually directed toward the communities in which the company has operating units.
Restrictions: Contributions will not be made to the following: individuals, athletic teams or sports related events, choirs, bands, drill teams, labor, veteran, fraternal, social or political organizations; individual agencies supported by the area United Way organization, except capital fund drives which are normally supported by the community, not the United Way; organizations that discriminate on the basis of race, color, sex or

national origin; national organizations whose programs do not directly serve the region's needs; places of worship and institutions devoted solely to religious instruction; travel related, including student trips or tours.
Geographic Focus: Pennsylvania, South Carolina, Tennessee, Texas
Contact: CeeGee McCord, Contributions Manager; (423) 229-2000; fax (423) 229-6974; CeeGeeMcCord@kingsporthousing.org
Internet: http://www.eastman.com/Company/Sustainability/communities/Philanthropy/Pages/Philanthropy.aspx
Sponsor: Eastman Chemical Company
P.O. Box 431
Kingsport, TN 37662

eBay Foundation Community Grants 1543
The mission of the Foundation is to make investments that improve the economic and social well-being of local communities. The Foundation works to fulfill its mission by collaborating with non-profit organizations and funding innovative programs primarily in microenterprise development. The Foundation also provides support to community organizations in areas where employees are located.
Requirements: 501(c)3 nonprofit organizations in communities where eBay has a major employment base, which includes San Jose, CA, and Salt Lake City, UT, are eligible.
Geographic Focus: California, Utah
Amount of Grant: 1,000 - 15,000 USD
Samples: Friends of Farm Drive (San Jose, CA)--for operating support of this organization, which is dedicated to improving the quality of life for residents of a neighborhood rampant with drug dealing and gang activity; U Research Expedition Program (U of California)--for scholarships to send California elementary and high school teachers on UREP trips, where they learn by working with university scientists and local residents in countries around the world, to increase awareness of current events and to foster understanding of different cultures among both students and teachers.
Contact: Grants Administrator; ebayfdn@cfsv.org
Internet: http://pages.ebay.com/community/aboutebay/foundation/grantapp.html
Sponsor: eBay Foundation
60 South Market Street, Suite 1000
San Jose, CA 95113

Eberly Foundation Grants 1544
The foundation awards grants to eligible nonprofit organizations in the areas of: undergraduate educational opportunities; arts programming and institutions; health and human services projects; primary and secondary supplemental educational programming; and economic development activities and organizations that benefit the residents of Fayette County. The initial approach should be submittal of a letter of request.
Requirements: Nonprofit organizations in Pennsylvania are eligible.
Geographic Focus: Pennsylvania
Samples: Pennsylvania State U (University Park, PA)--to construct a community center on the Fayette campus that will house performance spaces, a cafeteria and dining hall, a fitness center, an auditorium, and other features needed for large gatherings, $6.9 million; Indiana U of Pennsylvania, College of Business (PA)--to support technology initiatives, $599,000 challenge grant; Western Pennsylvania Conservancy (Pittsburgh, PA)--for restoration and protection of Fallingwater, $1 million.
Contact: Carolyn Blaney, President; (724) 438-3789; fax (724) 438-3856
Sponsor: Eberly Foundation
2 West Main Street, Suite 101
Uniontown, PA 15401-3448

EBSCO / MLA Annual Meeting Grants 1545
In order to enable more librarians to attend Medical Library Association (MLA)'s annual meetings, EBSCO Information Services has generously donated money to the MLA Scholarship Endowment to pay up to $1,000 each in travel and conference-related expenses for four librarians. Applicants must complete an application form which includes a 200-word statement answering the question, "What do you expect to gain professionally and/or personally by attending the MLA annual meeting?" The essay and information will be distributed to the EBSCO/MLA Annual Meeting Grant Jury, which will make selections. The application form and guidelines are available for download at the website. Applications must be received by MLA via email, fax, or mail by December 1. email applications will be accepted as PDF or MS Word files only.
Requirements: To be considered eligible, applicants must have between two and five years experience as a health-sciences librarian at the time of the annual meeting and be currently employed as a medical librarian. Preference is given to applicants who are first-time attendees, applicants who are presenting (or have submitted) a paper or a poster for the meeting, applicants who have MLA committee, jury, section, or special interest group (SIG) assignments; and applicants who are members of MLA.
Restrictions: No more than one individual from any given institution will be awarded the grant in a single year. An individual may receive this award only once.
Geographic Focus: All States, All Countries
Date(s) Application is Due: Dec 1
Amount of Grant: 1,000 USD
Contact: Carla Funk, Executive Director; (312) 419-9094, ext. 14; fax (312) 419-8950; mlapd2@mlahq.org or grants@mlahq.org
Internet: http://www.mlanet.org/awards/grants/index.html
Sponsor: Medical Library Association
65 East Wacker Place, Suite 1900
Chicago, IL 60601-7246

Echoing Green Fellowships 1546
The program awards full-time fellowships to emerging entrepreneurs to create innovative domestic or international public service projects that seek to catalyze positive social change. The proposed project may be in any public service area, including but not limited to, the environment, arts, education, health, youth service and development, civil and human rights, and community and economic development. The fellowship provides a two-year stipend, health care benefits, online connectivity, access to Echoing Green's network of social entrepreneurs, training, and technical assistance.
Requirements: Applicants must be at least 18 years old and commit to leading the project for at least two years. Partnerships of up to two individuals also are eligible.
Restrictions: Faith-based, research projects, and lobbying activities are not eligible.
Geographic Focus: All States
Date(s) Application is Due: Dec 2
Amount of Grant: 60,000 - 90,000 USD
Contact: Rich Leimsider, Director of Fellow and Alumni Programs; (212) 689-1165; fax (212) 689-9010; Rich@echoinggreen.org or apply@echoinggreen.org
Internet: http://www.echoinggreen.org/fellowship
Sponsor: Echoing Green Foundation
494 Eighth Avenue, Second Floor
New York, NY 10001

Eckerd Corporation Foundation Grants 1547
The Foundation's philanthropic mission is to monetarily support nonprofit organizations that directly benefit health care, education, women and children through programs and events. Funding proposals are submitted to the board for review on a quarterly basis.
Requirements: Nonprofit organizations in company operating areas are eligible.
Geographic Focus: All States
Contact: Tami Alderman; (727) 395-7091; fax (727) 395-7934; service@eckerd.com
Internet: http://www.eckerd.com/content.asp?content=company/news/presscontact
Sponsor: Eckerd Corporation
8333 Bryan Dairy Road
Largo, FL 33777

Eddie C. and Sylvia Brown Family Foundation Grants 1548
The Eddie C. and Sylvia Brown Family Foundation will focus its grant making in three general areas: the arts; education; and health (HIV/AIDS and cancer). The foundation's grant making supports programs that: provide opportunities for youth involvement in meaningful art experiences - this can include support for arts organizations or youth serving organizations that have developed programs which expose young people to the arts (programs that infuse the arts as a tool for leadership development, academic enrichment, and are built on strong youth development principles are encouraged to apply); provide access to educational opportunities for disadvantaged children and youth - this can include local organizations that work to improve PreK-12 public education and to offer enhancement and enrichment opportunities to Baltimore City students (educational opportunities are defined in the broadest sense to include, but are not limited to afterschool programs, mentoring, alternative education, computer technology, scholarship support, as well as traditional in-school K-12 programs); address the impact of HIV/AIDS on the African American community - programs that provide direct services and intervention including testing and counseling, case management strategies, and access to treatment will be considered; and address the prevention and treatment of cancer - this can include local organizations that provide cancer screening, prevention education, and treatment modalities. Grant awards average $2,500, and the annual application submission deadline is August 31.
Requirements: Organizations (or their fiscal agents) serving the Baltimore area that qualify as public charities under section 501(c)3 of the Internal Revenue Code and do not discriminate on the basis of race, creed, national origin, color, physical handicap, gender or sexual orientation can apply.
Restrictions: The Foundation does not make grants for: start-up programs; capital campaigns; individuals; multiple years; institutions of higher education; and organizations outside the Baltimore region.
Geographic Focus: Maryland
Date(s) Application is Due: Aug 31
Amount of Grant: Up to 5,000 USD
Contact: Maya Smith; (410) 332-4172, ext. 142; fax (410) 837-4701; msmith@bcf.org
Internet: http://www.bcf.org/bReceivebNonprofitsandstudents/Grants/EddieCandCSylviaBrownFamilyFoundation/tabid/681/Default.aspx
Sponsor: Eddie C. and Sylvia Brown Family Foundation
2 East Read Street, 9th Floor
Baltimore, MD 21202

Eden Hall Foundation Grants 1549
Eden Hall Foundation is a private foundation established pursuant to the will of Sebastian Mueller, a Pittsburgh philanthropist and vice president and director of the H. J. Heinz Company. During his lifetime, Mr. Mueller gave substantial support to improve conditions of the poor and disadvantaged, the promotion of sound education, and the support of health facilities and projects. Today, the trustees of Eden Hall Foundation continue his stewardship in the areas of social welfare, health, education and the arts. Eden Hall Foundation seeks to improve the quality of life in Pittsburgh and western Pennsylvania through support of organizations whose missions address the needs and concerns of the area. Proposals and projects of primary interest to the Foundation are: educational programs dedicated to the advancement and dissemination of useful knowledge. Support of schools is generally confined to four-year privately funded and controlled colleges, universities and other educational institutions; social welfare, women's issues, and the

improvement of conditions of the poor and disadvantaged; and the advancement of better health through support of organizations dedicated to health issues. Eden Hall Foundation will entertain grant proposals for capital projects, scholarship funds, research projects, programming and, in limited cases, endowments directed toward these objectives. A specific application form is not required. Therefore, grant requests should be submitted by letter proposal to the Foundation. Acknowledgements will be made promptly, advising applicants whether or not their requests meet Eden Hall Foundation's criteria.
Requirements: Requests for grants should include: a brief statement of the background and purpose of the requesting organization; a copy of that organization's most recent financial statements, as prepared by its certified public accountants; the specific purpose for which the requested grant is to be used, including a detailed budget, timetable for implementation and proposed method of evaluation; IRA certification of tax-exempt and charitable status under sections 501(c)3 and 509(c) of the Internal Revenue Code; information concerning the status of grants solicited from other foundations in the past twelve-month period (indicating the name of each foundation and amount requested); and a listing of the organization's officers and directors, their primary occupations and the responsibilities to the organization.
Restrictions: Grants cannot be made to individuals or private foundations. Not eligible for funding are sectarian or denominational religious organizations, except those providing direct educational or health care services to the public, tax-supported organizations, political parties and fraternal organizations. Requests to cover operating expenses, endowments or accumulated deficits are discouraged.
Geographic Focus: Pennsylvania
Contact: Sylvia Fields, Executive Director; (412) 642-6697; fax (412) 642-6698
Internet: http://www.edenhallfdn.org
Sponsor: Eden Hall Foundation
600 Grant Street, Suite 3232
Pittsburgh, PA 15219

Edina Realty Foundation Grants 1550
The Foundation extends financial support to organizations primarily in Minnesota, which provide housing and related services such as counseling and medical care to homeless children, families and individuals. Application forms may be submitted at anytime, see Foundations website for Grant Application forms: http://www.edinarealty.com/Content/Content.aspx?ContentID=148855
Requirements: Nonprofits 501(C)(03) operating in Minnesota are eligible.
Geographic Focus: Minnesota
Amount of Grant: 500 - 2,000 USD
Samples: Home of the Good Shepherd, MN, $15,000 - operates two transitional housing programs in St. Paul for homeless and at-risk women; Spare Key of Minnesota, MN, $13,000 - provides mortgage assistance to Minnesota homeowners with critically ill or seriously injured children by making a mortgage payment directly to the lender on behalf of the family;
Contact: Susan Cowsert, Director; (952) 928-5900; susancowsert@edinarealty.com
Internet: http://www.edinarealty.com/Content/Content.aspx?ContentID=187015
Sponsor: Edina Realty Foundation
6800 France Avenue South, Suite 670
Edina, MN 55435-2017

Edna G. Kynett Memorial Foundation Grants 1551
The purpose of the Foundation is to improve cardiovascular health. Grants support programs and services that educate primary care practitioners and community members and promote cardiovascular health. Grants are made to medical schools, hospitals and other nonprofit health organizations. Programs that have received support in recent years include innovative curricula, lectures and seminars, a tobacco prevention program, a summer Community Health Internship program and clinical research projects. Innovative projects are preferred.
Requirements: Applicants must be nonprofit organizations in the Delaware Valley, including Philadelphia, Bucks, Delaware, Chester, Montgomery, Berks, Lehigh and Camden counties and New Castle County, Delaware. Applicants must be either public entities or tax-exempt under Section 501(c)3 of the Internal Revenue Code and not private foundations as defined under Section 509(a).
Geographic Focus: Pennsylvania
Amount of Grant: Up to 50,000 USD
Sample: Penn Presbyterian Medical Center, Philadelphia, PA, $5,000 - Cardiology update; Drexel University College of Medicine, Philadelphia, PA, $40,000 - Cardiovascular CME programs; FOCU.S. on Health and Leadership for Women (Trustees of the University of Pennsylvania) Philadelphia, PA, $39,000 - Junior Faculty Investigators and Medical Student Fellowship in cardiovascular health.
Contact: Judith L. Bardes; (610) 828-8145; fax (610) 834-8175; judy1@aol.com
Sponsor: Edna G. Kynett Memorial Foundation
P.O. Box 540
Plymouth Meeting, PA 19462-0540

EDS Foundation Grants 1552
The Foundation wishes to ensure information technology champions cultural and civic change which will enrich the education of current and future generations, while enhancing the communities they serve. The Foundation is interested in supporting comprehensive technology solutions that increase performance and productivity in educational institutions and community organizations globally. Grant seekers must complete the EDS Foundation Application for Funding and have a current EDS employee volunteer partnership.
Requirements: Eligible applicants must be verified as a non-profit charitable organization (or equivalent) according to each country's rules and regulations. For the U.S., the organization must have a 501(c)3 IRS tax-exempt code.
Restrictions: Funding will not be provided for: individuals; operating deficits; foundations that are grant-making institutions; sponsorships; trips or tours; local athletic teams or events; fraternal, social or labor organizations; political or partisan organizations; journal or program advertising; private foundations; or organizations that do not comply with all applicable laws prohibiting discrimination and EDS Equal Opportunity policy.
Geographic Focus: California, Colorado, Georgia, Illinois, Michigan, Texas, Virginia
Contact: Diane Spradlin; (972) 605-8429; diane.spradlin-eds@eds.com
Internet: http://www.eds.com
Sponsor: EDS Foundation
5400 Legacy Drive, H3-6F-47
Plano, TX 75024

Educational Foundation of America Grants 1553
The foundation makes grants only for specific projects in its areas of interest, including, but not limited to, the environment, reproductive freedom, theatre, education, medicine, drug policy reform, democracy, peace & national security issues and human services. Important characteristics considered by EFA are an organization's record of achievement, intended broad impact, sound financial practices, increasing independence, and correspondence with EFA objectives.
Requirements: U.S. nonprofits are eligible for grant support. Applicants are required to send the EFA Inquiry Form via email as the first step.
Restrictions: Grants are not made for endowment or endowed faculty chairs, building/capital programs, religious purposes, grants to individuals, annual fund-raising campaigns, indirect costs, overhead or general support. The Foundation prefers not to fund projects located outside the US.
Geographic Focus: All States
Contact: Diane Allison, Executive Director; (203) 226-6498; fax dallison@efaw.org
Internet: http://www.efaw.org/Inquiry%20Guidelines.htm
Sponsor: Educational Foundation of America
35 Church Lane
Westport, CT 06880

Edward and Helen Bartlett Foundation Grants 1554
Established by a single donor, Edward E, Bartlett, in Oklahoma in 1961, the Foundation primarily offers grant support for: education, particularly public schools; community programs and services; health care; children and youth; and social services. There are no specific application forms or deadlines with which to adhere, and applicants should begin by forwarding a letter of application to the contact listed. In the recent past, grant amounts have ranged between $5,000 to $150,000.
Requirements: Preference id given to non-profit 501(c)3 organizations located in, or serving the residents of, Oklahoma.
Restrictions: No grants are to individuals directly.
Geographic Focus: Oklahoma
Amount of Grant: 5,000 - 150,000 USD
Contact: Bruce A. Currie, (918) 586-5273
Sponsor: Edward and Helen Bartlett Foundation
P.O. Box 3038
Milwaukee, WI 53201-3038

Edward and Romell Ackley Foundation Grants 1555
The Edward and Romell Ackley Foundation was established in Oregon in 2003 in support of Portland-based programs. The Foundation's primary areas of interest include: adoption; child welfare; diseases and conditions; human services; philanthropy; special hospital care; and youth development. Awards generally are given for operating support. Most recent grants have ranged from $2,500 to $25,000.
Requirements: Oregon-based 501(c)3 organizations serving the residents of Portland are eligible to apply.
Geographic Focus: Oregon
Contact: Robert H. Depew, Trustee; (503) 464-3580 or (503) 275-6564
Sponsor: Edward and Romell Ackley Foundation
P.O. Box 3168
Portland, OR 97208-3168

Edward Bangs Kelley and Elza Kelley Foundation Grants 1556
The Foundation's interest is primarily in Barnstable County, Massachusetts. The Foundation has been the leader in improving the health and welfare of the community. Grants are made to a great variety of health, social and human service agencies, as well as to cultural and environmental organizations. Town libraries, theatre and art groups, and musical organizations have been supported. Grants are sometimes utilized as seed money by young organizations.
Requirements: Grant applicants must be tax-exempt organizations which are not private foundations and must be located in Barnstable County. The proposed project/program must have a direct benefit to the inhabitants of Barnstable County. If the tax-exempt organization is located outside of Barnstable, but the proposed project/program will have a direct and substantial benefit to the inhabitants of Barnstable, the grant application may be eligible for funding.
Restrictions: Applicants must be residents of Barnstable County, Massachusetts, or must demonstrate very significant ties to Barnstable County.
Geographic Focus: Massachusetts
Contact: Henry Murphy Jr.; (508) 775-3117; contact@kelleyfoundation.org
Internet: http://www.kelleyfoundation.org
Sponsor: Edward Bangs Kelley and Elza Kelley Foundation
243 South Street, P.O. Drawer M
Hyannis, MA 02601

Edward N. and Della L. Thome Memorial Foundation Grants   1557
The Edward N. and Della L. Thome Memorial Foundation was established in 2002 by Robert P. Thome to honor the memory of his parents, Edward and Della Thome. The mission of the foundation is twofold: to advance the health of older adults through: the support of direct service projects; and medical research on diseases and disorders affecting older adults. The goal of the foundation's direct services program is to support organizations in Maryland, Missouri, and Michigan that provide direct services addressing one or more of the following critical issues facing older adults: health care, housing, family services, neighborhood involvement, workforce opportunities, and aging with dignity at home. Grant requests for general operating support are strongly encouraged. Program support will also be considered. Program-related capital expenses may be included in general operating or program requests. To better support the capacity of nonprofit organizations, multi-year funding requests are encouraged. While medical research grants have no annual deadline, all others must be submitted by June 15. Most recent awards have ranged from $20,000 to $1,500,000.
Requirements: The Thome Foundation direct services program is most interested in expanding services for older adults. Requests should clearly state how many previously-unserved older adults will now be served as a result of funding from the Thome Foundation. Applicants must have 501(c)3 tax-exempt status.
Restrictions: The foundation does not support requests from individuals, organizations attempting to influence policy through direct lobbying, or any political campaigns.
Geographic Focus: Maryland, Michigan, Missouri
Date(s) Application is Due: Jun 15
Amount of Grant: 20,000 - 1,500,000 USD
Contact: Srilatha Lakkaraju; (312) 828-8166; ilgrantmaking@ustrust.com
Internet: https://www.bankofamerica.com/philanthropic/foundation.go?fnId=47
Sponsor: Edward N. and Della L. Thome Memorial Foundation
231 South LaSalle Street, IL1-231-13-32
Chicago, IL 60604

Edwards Memorial Trust Grants   1558
The Edwards Memorial Trust awards grants to eligible Minnesota nonprofit organizations in support of: health care for people without health insurance or who are under-insured; preventive health care for children; and programs for the disabled. Areas of support include health care and hospitals, mental health crisis services, social services, children and youth, and the disabled. Types of support include: building construction and renovation; equipment acquisition; general operating grants; and program development. Types of support include: capital campaigns; sustainability;fundraising; general operating support; program development; and re-granting (endowments). Interested parties should begin by contacting the Foundation with a request to apply. A copy of the tax determination letter and most recent audited financial statements must accompany all applications.
Requirements: Minnesota 501(c)3 tax-exempt organizations in the greater Saint Paul area are eligible.
Geographic Focus: Minnesota
Date(s) Application is Due: May 1; Nov 1
Amount of Grant: 2,000 - 50,000 USD
Samples: Regions Hospital Foundation, St. Paul, Minnesota, $100,000 - general operating support (2014); Children's Health Care Foundation, Roseville, Minnesota, $150,000 - general operating support (2014); 180 Degrees, Minneapolis, Minnesota, $150,000 - general operating support (2014).
Contact: Managing Trustee; (651) 466-8731 or (651) 466-8040
Sponsor: Edwards Memorial Trust
P.O. Box 64713, 101 East Fifth Street
Saint Paul, MN 55164-0713

Edward W. and Stella C. Van Houten Memorial Fund Grants   1559
Stella C. Van Houten resided in Bergen County, New Jersey. This foundation, providing funding for health and human services, education, education of medical professionals, and the care of children, was established in 1978 in memory of her husband and herself. The Van Houten's had a particular fondness for the Valley Hospital of Ridgewood, New Jersey and for the Rollins College in Florida. The Foundation continues to honor their preferences with grants to those two organizations in addition to grants to other organizations. The Foundation's mission is to: supports agencies, institutions and services in Passaic and Bergen Counties, New Jersey, having to do with the care or cure of sick or disabled persons or for the care of orphaned children or aged persons; educates students in the medical profession; support for educational purposes; support for the care of children. A target of 10% of the grants each year is for medical scholarships.
Requirements: Passaic and Bergen Counties, New Jersey non-profits are eligible to apply. The application form & guidelines are available online at the Wachovia website. The applications must be submitted by January 31 for a March meeting & August 1 for an October meeting.
Geographic Focus: New Jersey
Date(s) Application is Due: Jan 31; Aug 1
Samples: Therapeutic Learning Center, Ramsey, NJ, $6,000--speech therapy; William Paterson University Foundation, $20,000--scholarships for undergraduate, minority, and/or graduate nursing students; Christian Health Care Center Foundation, $100,000--to build a Great Room at Heritage Manor East.
Contact: Trustee, c/o Wachovia Bank; grantinquiries2@wachovia.com
Internet: https://www.wachovia.com/foundation/v/index.jsp?vgnextoid=00c78689fb0a a110VgnVCM1000004b0d1872RCRD&vgnextfmt=default
Sponsor: Edward W. and Suitella C. Van Houten Memorial Fund
190 River Road, NJ3132
Summit, NJ 07901

Edwin S. Webster Foundation Grants   1560
The policy of the foundation is to support charitable organizations that are well known to the trustees, with emphasis on special projects and capital programs, or operating income for hospitals, medical research, education, youth agencies, cultural activities, and programs addressing the needs of minorities. Types of support include operating budgets, continuing support, annual campaigns, building funds, equipment, land acquisition, endowment funds, matching funds, scholarship funds, professorships, internships, fellowships, special projects, and research.
Requirements: The Foundation confines its grants primarily to the New England area. Grantees must provide evidence of their tax-exempt status. The AGM common proposal format, available on the Internet at http://agmconnect.org is suitable for submission of proposals but not required. There are no set deadlines, but for consideration at the spring meeting, proposals should arrive by May 1 and by November 1 for consideration at the fall meeting. Proposals received after the trustees meet will be held for consideration at the next meeting.
Restrictions: Grants are not made to organizations outside the United States or to individuals.
Geographic Focus: Massachusetts
Date(s) Application is Due: May 1; Nov 1
Amount of Grant: 15,000 - 50,000 USD
Contact: Michelle Jenney, Administrator; (617) 391-3087; fax (617) 426-7080; mjenney@gmafoundations.com
Sponsor: Edwin S. Webster Foundation
GMA Foundations, 77 Summer Street, 8th Floor
Boston, MA 02110-1006

Edwin W. and Catherine M. Davis Foundation Grants   1561
The foundation awards grants to U.S. nonprofit organizations in its areas of interest, including arts, elderly, environment, higher education, housing, mental health, religion, social services,and youth. Types of support include annual campaigns, endowment funds, fellowships, operating grants, research grants, and scholarship funds. There are no application deadlines. The board meets in May or June; submit a proposal that is three pages or less in length.
Requirements: U.S. nonprofit organizations are eligible.
Geographic Focus: Washington
Amount of Grant: 1,000 - 10,000 USD
Samples: Fulfillment Fund, Los Angeles, CA, $150,000; Childrens Health Council, Palo Alto, CA, $21,000; Trust for Public Land, Seattle, WA, $88,000.
Contact: Gayle Roth, Grants Administrator; (651) 215-4408; fax (651) 228-0776
Sponsor: Edwin W. and Catherine M. Davis Foundation
30 East 7th Street, Suite 2000
Saint Paul, MN 55101-1394

Edyth Bush Charitable Foundation Grants   1562
The Foundation is operated exclusively for charitable, religious, literary, and other exempt purposes. Grants are available for challenge and development purposes, construction and renovation, equipment and expansion of functions, pilot projects and seed start-up of new programs, and study or planning grants. The Foundation has broad interests in human service, education, health care and a limited interest in the arts. Requests for grants or other funds should be submitted in writing. Contact the Program Officer before beginning any proposals or applications.
Requirements: The Foundation welcomes grant requests from otherwise eligible tax-exempt organizations under IRS Sections 501(c)3 and Section 509(a) headquartered within the Orlando MSA of Orange, Seminole, and Osceola Counties, Florida. Other grant requests should have special interest or support from one or more of our Directors.
Restrictions: The Foundation will ordinarily deny grant requests: from chiefly tax-supported institutions, or their support foundations; for individual scholarships or for individual research grants even if through an exempt or otherwise qualified educational organization; for alcoholism or drug abuse programs or facilities; for routine operating expenses; to pay off deficits or pre-existing debt; for foreign organizations or for foreign expenditure; for travel projects or fellowships; for chiefly church, sacramental, denominational or inter-denominational purposes, except outreach projects for elderly, indigents, needy, youth, or homeless regardless of belief, race, color, creed, or sex; for endowment funds or other purely revenue generating funds; advocacy organizations or advocacy component funding; for cultural or arts organizations unless their collections, exhibits, projects or performances are of demonstrated nationally recognized quality; from organizations having receipts of revenues from memberships and/or contributions of less than $25,000 in the previous year, or from any organization whose IRS Sec. 509(a) publicly supported status will need renewal in the next six (6) months.
Geographic Focus: Florida
Amount of Grant: 5,000 - 50,000 USD
Samples: Jewish Community Ctr of Greater Orlando (Maitland, FL)--to renovate its playground for toddlers and to help raise additional funds for a project to renovate the elementary school playground, $75,867 partial challenge grant; Community Communications, Inc., WMFE-TV/FM (Orlando, FL)--for hurricane emergency relief effort, $12,000; Holocaust Memorial Resource and Education Center (Maitland, FL)--to support work in addressing intolerance, $2500; Winter Garden Heritage Foundation (Winter Garden, FL)--for leadership training, $350.
Contact: Deborah Hessler, Program Officer; (888) 647-4322 or (407) 647-4322; fax (407) 647-7716; dhessler@edythbush.org
Internet: http://www.edythbush.org
Sponsor: Edyth Bush Charitable Foundation
P.O. Box 1967
Winter Park, FL 32790-1967

Effie and Wofford Cain Foundation Grants 1563
The Effie and Wofford Cain Foundation gives primarily for higher and secondary education, medical research, and public service organizations. Grants also are awarded to religious organizations (Baptist, Christian, Episcopal, Presbyterian, Salvation Army, and United Methodist), and for aid for the handicapped. Additional fields of interest include elementary and secondary education, early childhood development and education, medical school education, nursing school education, hospitals and general health care and health organizations, religious federated giving programs, government and public administration, African Americans, Latinos, the disabled, the aging, and economically disadvantaged and homeless. Types of support include general operating support, continuing support, annual campaigns, capital campaigns, building/renovations, equipment acquisition, endowment funds, program development, seed money, curriculum development, fellowships, internships, scholarship funds, research, and matching funds. Organizations may reapply for funding every other fiscal year.
Requirements: The foundation only makes grants to 501(c)3 organizations in Texas.
Restrictions: Individuals are ineligible.
Geographic Focus: Texas
Amount of Grant: 1,000 - 150,000 USD
Samples: Vuilding Homes for Heroes, Valley Stream, New York, $150,000 - general operating support (2014); Henderson County Food Pantry, Athens, Texas, $25,000 - general operating support (2014); St. Vincent Citizens Nutrition Program, Los Angeles, California, $62,500 - general operating support (2014).
Contact: Frabklin W. Denius, President and Director; (512) 346-7490; fax (512) 346-7491; info@cainfoundation.org
Sponsor: Effie and Wofford Cain Foundation
4131 Spicewood Springs Road, Suite A-1
Austin, TX 78759-7490

Eisner Foundation Grants 1564
The Eisner Foundation exists to provide access and opportunity for disadvantaged children and the aging of Los Angeles County. The Eisner Foundation has been funding innovative and effective non-profit organizations that improve and enrich the lives of underserved children in Southern California since 1996. In 2008, the Foundation recognized that many of the same attributes that the children unfortunately possess physical and emotional vulnerability, extreme poverty, lack of advocacy on their behalf, minimal access to the arts, and general powerlessness also applied to many members of our rapidly aging population in the community. The Foundation elected to broaden the funding focus to include those at both ends of the spectrum of life, the young and the old. The Foundation's now dedicated to bringing about lasting changes in the lives of disadvantaged and vulnerable people starting and ending their lives in Los Angeles County.
Requirements: California nonprofit organizations serving Los Angeles and Orange Counties are eligible to apply. Applicants can apply at any time, but should know that these proposals will be reviewed and approved at the June or December board meetings. Applying to the Foundation is a two step process. The first step is to submit a Letter of Inquiry (see Eisner Foundation website for guidelines). If Letter of Inquiry meets the Foundation's criteria, a full application will be sent to the applicant for completion and submission, completing the second step.
Restrictions: Proposals for endowments are rarely excepted.
Geographic Focus: California
Amount of Grant: 5,000 - 100,000 USD
Samples: LA Scores, Los Angeles, CA, $100,000--to provide project support for an after-school program that combines creative writing, soccer, and community service; California Institute of the Arts, Los Angeles, CA, $1,250,000--to support the Community Arts Partnership program which provides new media arts education programs to local youth; Friendship Circle, Los Angeles, CA, $25,000--to provide support for an organization that addresses the challenges of families with special-needs children in the South Bay; Eisner Pediatric & Family Medical Center, Los Angeles, CA, $1,000,000--to support the endowment campaign for a pediatric and family medical center serving low-income, uninsured and under-insured families; Aquarium of the Pacific, Los Angeles, CA, $100,000--to provide support for a project that increases science learning access to Title I school communities in the South Bay; St. Barnabas Senior Center, Los Angeles, CA, $75,000--to support a program that provides education on sustained nutrition, adult day care, and other health promotion services for 3,000 seniors in the greater downtown Los Angeles area.
Contact: Trent Stamp; (310) 228-6808; trent.stamp@eisnerfoundation.org
Internet: http://www.eisnerfoundation.org/what_we_do/
Sponsor: Eisner Foundation
233 South Beverly Drive
Beverly Hills, CA 90212

Elaine Feld Stern Charitable Trust Grants 1565
Established in Missouri in 1989, the Elaine Feld Stern Charitable Trust is currently managed by trustees of the Blue Ridge Bank and Trust. The Trust's primary fields of interest include: arts and culture, agriculture, food, health organizations, and human services. A formal application, which can be secured from the Trust office, is required. There are no annual application submission deadlines identified. Most recent grant awards have ranged from $1,000 to $25,000.
Requirements: Giving is limited to 501(c)3 organizations located in, or serving the residents of, the Kansas City, Missouri, area.
Restrictions: No support is provided for tax-supported institutions, individuals, for telethons, or conferences.
Geographic Focus: Missouri
Amount of Grant: 1,000 - 25,000 USD
Samples: Harvesters, Kansas City, Missouri, $12,500 - general operating support (2014); Lee's Summit Cares, Lees Summit, Missouri, $10,000 - general operating support, (2014); Kansas City Symphony, Kansas City, Missouri, $1,000 - general operating support (2014).
Contact: J. Bryan Allee, (816) 358-5000 or (816) 795-9933
Sponsor: Elaine Feld Stern Charitable Trust
4200 Little Blue Parkway
Independence, MO 64057

Elizabeth Glaser International Leadership Awards 1566
The International Leadership Award (ILA) is a three-year grant focused on international investigations in preventing mother-to-child transmission (PMTCT) of HIV and HIV care and treatment. Recent awards were granted for research into improving the effectiveness of PMTCT services; strengthening maternal and infant diagnosis of HIV; and examining the impact of antiretroviral drugs on children. The Award identifies and supports internationally based physicians or scientists who have the training and potential to develop pediatric HIV prevention and/or treatment programs, but lack the resources to do so. Award recipients are required to mentor a minimum of three additional people who would benefit from the experience and could help them achieve their goals. These are individuals who would be hired and trained to work on the project proposed by the applicant. The Award provides up to $105,000 in total costs per year for three years for a total of $315,000. Funding for the second and third year is dependent on progress in year one, and is not guaranteed. Funds support the salary of the applicant (up to 100% effort), salaries for those hired and mentored as part of the project, and direct costs related to the project itself such as travel, supplies, and equipment. Indirect cost is limited to 5% (included in the total). Salary support per individual investigator may be a maximum of $45,000 per year, and must be commensurate with the level of experience of the investigator, the salary structure in the country and the level of commitment to the project. Awardees must agree to spend at least 50% of their time on the project related to the Award. Final determination as to the amount of the Award is made by the Advisory Board depending on the scope of the project. Interested applicants are asked to submit a Letter of Intent. Only those Letters of the highest caliber will be asked to submit a full grant application. Additional guidelines, instructions, and summaries of previously funded projects are posted on the website.
Requirements: Applicants must be from a developing country, and must be performing work in a developing country in order to apply. Applicants must be living in and performing work in either Africa or India. The Foundation is especially interested in applicants who have specific training or experience with HIV/AIDS. Applicants must have an MD or PhD, and demonstrate commitment to continuing work in a developing country on HIV/AIDS when the award is complete.
Restrictions: U.S. citizens or European citizens living in developing countries are not eligible to apply. Senior Investigators such as full professors are usually not eligible, unless they are changing the direction of their career to include HIV/AIDS research.
Geographic Focus: All Countries
Amount of Grant: 105,000 USD
Contact: Chris Hudnall, Senior Program Coordinator, Research; (310) 314-3154; fax (310) 314-1469; chris@pedaids.org
Internet: http://www.pedaids.org/What-We-re-Doing/research/grant-programs.aspx
Sponsor: Elizabeth Glaser Pediatric AIDS Foundation
1140 Connecticut Avenue, NW, Suite 200
Washington, D.C. 20036

Elizabeth Glaser Scientist Award 1567
The Elizabeth Glaser Scientist Award is the Foundation's highest scientific honor, and is an investment in the most promising HIV/AIDS researchers at a critical stage in their careers. These scientists, who represent the best and brightest investigators from the international medical science community, are selected on the basis of their knowledge, innovation, and dedication. By providing research funding over a five-year period, this award enables recipients to focus their long-term efforts on issues specific to pediatric HIV/AIDS. The Award fosters an unprecedented spirit of collaboration among these scientists. Each year, the Elizabeth Glaser Scientists meet with the Foundation's internationally renowned advisory board to stimulate ideas, report on current programs, and plan future research. Since the program's inception, the Foundation has built an invaluable network of scientists whose work in vaccine development, immune response, breast milk transmission, and other critical areas impacts the entire field of HIV/AIDS research. Interested applicants should contact the Foundation to request a letter of intent cover sheet. Full applications are by invitation.
Requirements: Candidates must have an MD, PhD, DDS, or DVM degree and be at the assistant professor level or above.
Restrictions: The Foundation will not fund large therapeutic trials or the pharmocologic development of drugs.
Geographic Focus: All States, All Countries
Contact: Doug Horner; (202) 296-9165; fax (202) 296-9185; research@pedaids.org
Internet: http://www.pedaids.org/What-We-re-Doing/research/grant-programs
Sponsor: Elizabeth Glaser Pediatric AIDS Foundation
1140 Connecticut Avenue, NW, Suite 200
Washington, D.C. 20036

Elizabeth McGraw Foundation Grants 1568
The foundation awards grants nationally, with a focus on the U.S. Northeast, in its areas of interest, including cancer research, elementary and secondary education, art museums, performing arts (opera and theater), and culture. There are no application deadlines or specific guidelines. Send letter containing a detailed description of project and amount of funding requested to: John McGraw, Pres., 157 Eel River Road, Osterville, MA 02655

Geographic Focus: Florida, Massachusetts, New York
Contact: Paul J. Bisset; (212) 454-8607
Sponsor: Elizabeth McGraw Foundation
Deutsche Bank Trust Company of New York, P.O. Box 1297
New York, NY 10008-1297

Elizabeth Morse Genius Charitable Trust Grants 1569
Established in 1992, the Elizabeth Morse Genius Charitable Trust honors the memory of Elizabeth Morse Genius, the daughter of Charles Hosmer Morse, a nineteenth century Chicago financier, industrialist, and land developer. The trust supports and promotes charitable organizations that: encourage the principles of individual self-reliance, self-sacrifice, thrift, industry, and humility; relieve human suffering through scientific research and education regarding disease; provide assistance to youths with troubled childhoods and emotional disorders; attend to the care of the elderly; provide assistance to humankind during times of natural and man-made disasters; foster individual self-worth and dignity, with a broad emphasis on the classical fine arts; develop physical health and spiritual well-being through vigorous athletic activity; and promote world peace and understanding through the improvement of national and international means of travel by air, rail, and sea. The majority of grants from the Genius Charitable Trust are one year in duration; on occasion, multi-year support is awarded. Applicants must first submit a letter of inquiry. Downloadable application guidelines are available at the Sponsor's website by clicking on the "Grant Application Process Button", or prospective applicants may also call the second phone number given to obtain application information. A synopsis of the grant, contact information, and a link to the trust's giving history is available at the grant website. The Elizabeth Morse Genius Charitable Trust has a rolling application deadline. In general, applicants will be notified of grant decisions three to four months after proposal submission. Most recent awards have ranged from $4,500 to $300,000.
Requirements: Nonprofit organizations serving Chicago and Cook County are eligible.
Restrictions: The trust generally does not fund individuals; organizations outside the metropolitan Chicago city area; organizations attempting to influence policy through direct lobbying; capital campaigns; endowment campaigns; or political campaigns.
Geographic Focus: Illinois
Amount of Grant: 4,500 - 300,000 USD
Samples: Ingenuity Chicago, Chicago, Illinois, $300,000 - program development (2014); Chicago Symphony Orchestra, Chicago, Illinois, $100,000 - program development (2014); Devin Shafron Memorial Book Fund, Deerfield, Illinois, $25,000 - program development (2014).
Contact: Lauren MacDonald; (312) 828-6753; ilgrantmaking@ustrust.com
Internet: https://www.bankofamerica.com/philanthropic/foundation.go?fnId=116
Sponsor: Elizabeth Morse Genius Charitable Trust
231 South LaSalle Street, IL1-231-13-32
Chicago, IL 60604

Elizabeth Nash Foundation Scholarships 1570
The Elizabeth Nash Foundation awards scholarships to assist persons with cystic fibrosis (CF) to pursue undergraduate and graduate degrees. Scholarships ranging from $1,000 to $2,500 are awarded annually, and are made directly to the academic institution to assist in covering the cost of tuition and fees. Scholarships are made for one year; however, individuals may re-apply for subsequent years. In selecting applicants, the Elizabeth Nash Foundation Scholarship Committee takes into consideration each applicant's scholastic record, character, demonstrated leadership, service to CF-related causes and the broader community, and need for financial assistance. Applications must be postmarked by April 1, and notification of award status will be mailed no later than June 17.
Requirements: The Elizabeth Nash Foundation Scholarship program is open to individuals with CF who are in-going or current undergraduate or graduate students at an accredited US-based college or university. Given limited resources, the program is currently only open to U.S. citizens. Funds to support Associate Degrees are not currently available. Scholarship recipients must be willing to support the Foundation, at the Board of Directors' request by writing an article for publication by the Foundation, writing up to ten thank you notes to donors and/or five to ten fund raising letters in support of ENF, speaking at one local event, or in some other manner nominated by the recipient.
Geographic Focus: All States
Date(s) Application is Due: Apr 1
Amount of Grant: 1,000 - 2,500 USD
Contact: Ann Nash, (408) 399-6919; scholarships@elizabethnashfoundation.org or info@elizabethnashfoundation.org
Internet: http://www.elizabethnashfoundation.org/scholarshipapply.html
Sponsor: Elizabeth Nash Foundation
P.O. Box 1260
Los Gatos, CA 95031-1260

Elizabeth Nash Foundation Summer Research Awards 1571
The Elizabeth Nash Foundation is in the sixth year of a partnership with Children's Hospital and Research Center Oakland (CHORI), an internationally renowned biomedical research institute that bridges basic science and clinical research in the prevention and treatment of human disease. The Elizabeth Nash Foundation Summer Research Award is available to high school seniors and undergraduates who have a strong interest in pursuing cystic fibrosis (CF) research. The award is designed to identify and train the next generation of talented CF researchers. Students receive a stipend from the Foundation and CHORI pairs student with a mentor scientist to guide them through a structured CF research project. Previous recipients have co-authored publications in Free Radical Biology and Medicine and Cellular Physiology and Biochemistry involving their summer research. The deadline for applications for summer is 4:00 pm on April 15. The awardee will be provided a one-time stipend of $2,500 for the nine-week period. Announcement of all awards will be made on May 3.
Geographic Focus: All States
Date(s) Application is Due: Apr 15
Amount of Grant: 2,500 USD
Contact: Ann Nash, (408) 399-6919; info@elizabethnashfoundation.org
Internet: http://www.elizabethnashfoundation.org/CHORI.html
Sponsor: Elizabeth Nash Foundation
P.O. Box 1260
Los Gatos, CA 95031-1260

Elkhart County Community Foundation Fund for Elkhart County 1572
The Fund for Elkhart County is the county's core grant making program and has the broadest guidelines designed to meet the Foundation's mission to produce a brighter future for all people in Elkhart County in Indiana. Grants are awarded from the fund's earnings on a quarterly basis. More than $550,000 has been awarded from this fund to assist non-profits in Elkhart County.
Requirements: The Fund for Elkhart County grants funding to after-school programs, temporary help for abused women and children, and services for the mentally and physically challenged. Other programs include health and dental care for the underprivileged, services for senior citizens, support of local arts and cultural events, environmental and wildlife services, historical preservations, and capacity building. Applicants should carefully review the website's grant guidelines before applying.
Geographic Focus: Indiana
Date(s) Application is Due: Nov 1
Contact: Fund for Elkhart County Contact; (574) 295-8761
Internet: http://www.elkhartccf.org/File/static/grant_seekers/FundForElkhartCounty.shtml
Sponsor: Elkhart County Community Foundation
101 S Main Street, P.O. Box 2932
Elkhart, IN 46615

Elkhart County Community Foundation Grants 1573
The foundation is looking for innovative programs or projects that address community issues in Indiana's Elkhart County. Community collaboration is encouraged, where organizations work together toward a shared goal with shared responsibility, accountability, and resources. Grants are awarded in the areas of arts and culture, community development, education, youth development, board development and succession planning, and health and human services in addition to continuing support, technical assistance, and matching funds. Most grants are single-year awards, although multi-year grants also will be considered. The staff is available for advice in the preparation of a proposal. Applicants are encouraged to read the Foundation website's "Helpful Tips for Grant Writers." Letters of interest are due to the Foundation by January 1, April 1, June 1, and August 1.
Requirements: Elkhart County non-profit organizations, or those seeking such status, and some governmental agencies, such as the public library or public school system, are eligible.
Restrictions: Support generally will not be given for continuing operating costs of established programs, projects, and agencies. The foundation does not offer support for religious or sectarian purposes. No grants are awarded to individuals (other than scholarships), or for operating budgets or budget deficits, annual funds, conferences, scholarly research, endowments, personal travel, or films.
Geographic Focus: Indiana
Date(s) Application is Due: Mar 1; Jun 1; Sep 1; Nov 1
Contact: Jim Siegmann; (574) 295-8761; fax (574) 389-7497; jim@elkhartccf.org
Internet: http://www.elkhartccf.org/File/static/grant_seekers/WhatWeFund.shtml
Sponsor: Elkhart County Community Foundation
101 S Main Street, P.O. Box 2932
Elkhart, IN 46615

Elkhart County Foundation Lilly Endowment Community Scholarships 1574
The Lilly Endowment Community Scholarship Program strives to enhance the quality of life for residents in Indiana by raising the level of educational achievement. The scholarship covers full time, four year tuition with an $800/per year book stipend. All college majors are eligible.
Requirements: Scholarship candidates must be U.S. citizens and high school graduates from Elkhart County. They are first generation college students (parents do not have a four year degree) with plans to enroll in a full time, four year field of study. Applicants should contact the Elkhart Community Foundation for an application and instructions for submitting the application packet to their high school guidance counselor.
Geographic Focus: Indiana
Date(s) Application is Due: Feb 1
Contact: Peter McCown, President; (574) 295-8761; fax (574) 389-7497
Internet: http://www.elkhartccf.org/static/students/lilly_scholarships.shtml
Sponsor: Elkhart County Community Foundation
101 S Main Street, P.O. Box 2932
Elkhart, IN 46615

Elliot Foundation Inc Grants 1575
The Elliot Foundation Inc. formerly known as Elliot Foundation for Medical Research and Education, Inc. is a independent foundation, operating primarily in the state of Indiana. The foundation offers funding in the form of general/operating support. The fields of interest include: Christian agencies & churches; crime/violence prevention; environment; natural resources; child abuses; education.
Requirements: Contact Foundation for more information.

Geographic Focus: Indiana
Contact: Richard E. Bond, Secretary; (575) 293-1165
Sponsor: Elliot Foundation
2210 East Jackson Boulevard
Elkhart, IN 46516-1165

Ellison Medical Foundation/AFAR Julie Martin Mid-Career Award in Aging Research 1576

The Ellison Medical Foundation and AFAR developed this program for outstanding mid-career scientists who propose novel directions of high importance to biological gerontology. Proposals in areas where NIH awards or other traditional sources are unlikely because the research is high risk, are particularly encouraged if they have the potential for leading to major new advances in our understanding of basic mechanisms of aging. Projects investigating age-related diseases are also supported, but only if approached from the point of view of how basic aging processes may lead to these outcomes. Projects concerning mechanisms underlying common geriatric functional disorders are also encouraged, as long as these include connections to fundamental problems in the biology of aging. Two four-year awards of $500,000 will be given, at the level of $125,000 per year. In addition, up to 10% ($50,000) may be requested for administrative/indirect costs. Recipients of this award are expected to attend the AFAR Grantee Conference. The purpose of the meeting is to promote scientific and personal exchanges among recent AFAR grantees and experts in aging research.
Requirements: The applicant must be an Associate Professor who achieved tenured status after December 1st. Non-tenured Associate Professors at institutions with tenure (even if tenure is only offered at the Full Professor level) are not eligible. Applicants at institutions that do not offer tenure must demonstrate that their appointment is equivalent to that of an Associate Professor who received tenure status after December 1st. The proposed research must be conducted at any type of not-for-profit setting in the United States. The deadline of receipt of applications and all supporting materials is December 16th at 5:00 p.m. EST. Refer to the Julie Martin Mid-Career Award instruction sheet and application at: http://afar.org/grants.html for complete application procedures. Incomplete applications cannot be considered.
Restrictions: Projects that deal strictly with clinical problems such as the diagnosis and treatment of disease, health outcomes, or the social context of aging are not eligible. Applicants who are employees in the NIH Intramural program are not eligible.
Geographic Focus: All States
Date(s) Application is Due: Dec 16
Amount of Grant: 125,000 USD
Contact: Grants Manager; (212) 703-9977 or (888) 582-2327; fax (212) 997-0330; grants@afar.org or info@afar.org
Internet: http://afar.org/Ellison%20Mid-Career.html
Sponsor: American Federation for Aging Research
55 West 39th Street, 16th Floor
New York, NY 10018

Ellison Medical Foundation/AFAR Postdoctoral Fellows in Aging Research 1577

The Ellison Medical Foundation, in partnership with the American Federation for Aging Research (AFAR), created the Ellison/AFAR Postdoctoral Fellows in Aging Research Program to encourage and further the careers of postdoctoral fellows with outstanding promise in the basic biological and biomedical sciences relevant to understanding aging processes and age related diseases and disabilities. The award is intended to provide significant support to permit these postdoctoral fellows to become established in the field of aging. Projects concerned with understanding the basic mechanisms of aging will be considered. Projects investigating age related diseases are also supported, if approached from the point of view of how basic aging processes may lead to these outcomes. Projects concerning mechanisms underlying common geriatric functional disorders are also considered. It is anticipated that up to 15 one year grants will be awarded, ranging from $44,850 for a first year fellow, up to $58,850 for a fellow with more than 7 years of training. Of the award, up to $7,850 may be requested for expenses such as research supplies, equipment, health insurance and travel to scientific meetings. Refer to the Ellison/AFAR instruction sheet and application on the AFAR website for complete application procedures. Recipients of this award are expected to attend the AFAR Grantee Conference. The purpose of the meeting is to promote scientific and personal exchanges among recent AFAR grantees and experts in aging research.
Requirements: The applicant must be a postdoctoral fellow (MD and/or PhD degree) at the start date of the award (July 1). Any former Ellison/AFAR postdoctoral award recipient may apply for this award. All candidates must submit applications endorsed by their institutions. The proposed research must be conducted at any type of not-for-profit setting in the United States.
Restrictions: Employees in the NIH Intramural program are not eligible. Fellows may not hold any concurrent foundation or not-for-profit funding.
Geographic Focus: All States
Date(s) Application is Due: Dec 16
Amount of Grant: 44,000 - 58,000 USD
Contact: Grants Manager; (212) 703-9977 or (888) 582-2327; fax (212) 997-0330; grants@afar.org or info@afar.org
Internet: http://www.afar.org/ellisonpostdoc.html
Sponsor: American Federation for Aging Research
55 West 39th Street, 16th Floor
New York, NY 10018

Elmer L. and Eleanor J. Andersen Foundation Grants 1578

The foundation exists to enhance the quality of the civic, cultural, educational, environmental, and social aspects of life in Minnesota, primarily in the metropolitan area of Saint Paul and Minneapolis. Types of support include general operating support, continuing support, annual campaigns, capital campaigns, building construction/renovation, endowment funds, program development, deficit reduction, publication, seed money, curriculum development, research, technical assistance, and matching funds. The board meets four times annually.
Requirements: Minnesota nonprofit organizations are eligible.
Geographic Focus: Minnesota
Date(s) Application is Due: Feb 1; May 1; Aug 1; Nov 1
Amount of Grant: 500 - 75,000 USD
Contact: Mari Oyanagi Eggum; (651) 642-0127; fax (651) 645-4684; eandefdn@mtn.org
Sponsor: Elmer L. and Eleanor J. Andersen Foundation
2424 Territorial Road
Saint Paul, MN 55114

El Paso Community Foundation Grants 1579

The Foundation is a facilitator linking the generosity of the region's donors to local non-profit organizations to meet the charitable needs of the El Paso area. In this capacity, the Foundation helps to provide funding for programs and initiatives of these organizations for the broader good of the El Paso Community. Major areas of interest are: arts and humanities; education; public benefit; health and disabilities; environment;animals and; human services. Types of support offered include: equipment acquisition; general/operating support; management development/capacity building; matching/challenge support; program development; scholarship funds; seed money; technical assistance. Priority is given to: more effective ways of doing things; projects where a moderate amount of grant money can have an impact; and projects that show collaboration with other organizations. Application deadlines are February 1 and August 1.
Requirements: Grant requests will be considered only from agencies located within or offering services to the citizens of our community, which includes far west Texas, southern New Mexico and northern Chihuahua, Mexico. Applicants must be exempt from income taxes under Section 501(c)3 of the Internal Revenue Service Code. Application form is available on the El Paso Community Foundation website.
Restrictions: Funding is not available for/to: individuals; capital campaigns; fundraising events or projects; religious organizations for religious purposes; annual appeals and membership contributions; organizations that are political or partisan in purpose; travel for individuals or groups; organizations outside the El Paso geographic area; endowment funds; past operating deficits.
Geographic Focus: Texas
Date(s) Application is Due: Feb 1; Aug 1
Amount of Grant: 3,000 - 10,000 USD
Samples: American Cancer Society, 42,000--provide local patients transportation to and from treatments; Angelo Catholic School, $11,000-- teachers' salary; Animal Rescue League of El Paso, Inc., $8,000--animal rescue, rehabilitation and placement.
Contact: Bonita Johnson; (915) 533-4020; fax (915) 532-0716; info@epcf.org
Internet: http://www.epcf.org/grant_guidelines
Sponsor: El Paso Community Foundation
P.O. Box 272
El Paso, TX 79943

El Paso Corporate Foundation Grants 1580

El Paso Corporate Foundation funds initiatives that strengthen the communities where employees live, work, and volunteer. The primary focus areas of the foundation are education, health and human services and the community. The Foundation's secondary focus areas are the environment and arts and culture. Organizations must be invited by the Foundation to submit a letter of inquiry for funding consideration.
Requirements: Giving is limited to nonprofits in Alabama, Colorado and Texas.
Restrictions: The following are ineligible requests for funding: organizations without 501(c)3 status; organizations that already have an active El Paso Corporate Foundation grant; capital campaigns; multi-year commitments; religious organizations for religious purposes; individuals; hospitals and medical research; athletics or youth sports organizations; national or state-wide initiatives; war veterans and fraternal service organizations; endowment funds; pass-through grants; fundraising events, such as dinners, luncheons, and golf tournaments; political organizations, campaigns, and candidates; and technology grants, i.e. computers, software, or related hardware.
Geographic Focus: Alabama, Colorado, Texas
Date(s) Application is Due: Jan 31; May 31; Sep 30
Contact: Community Relations Manager; (713) 420-2878; fax (713) 420-5312; foundation@elpaso.com
Internet: http://www.elpaso.com/community/default.shtm
Sponsor: El Paso Corporate Foundation
P.O. Box 2511
Houston, TX 77252-2511

El Pomar Foundation Anna Keesling Ackerman Fund Grants 1581

The El Pomar Foundation is one of the largest and oldest private foundations in the Rocky Mountain West, and it contributes annually through direct grants and community stewardship programs to support Colorado nonprofit organizations. The Anna Keesling Ackerman Fund seeks to continue the charitable intent of Mr. Jasper D. Ackerman, who supported numerous charitable organizations and causes throughout the Pikes Peak Region. The Fund's primary focus is in health, human services, education, arts and humanities, and civic and community initiatives. Grant applications are typically

reviewed twice annually in the spring and at the end of the year. If funded, applicants must wait three years (36 months) before submitting a new application. If declined, applicants must wait one year (12 months) before submitting a new application.
Requirements: Nonprofits with 501(c)3 tax status serving the Pikes Peak Region of El Paso and Teller counties are eligible to apply for funding.
Restrictions: The Foundation does not accept grant applications for: other foundations or nonprofits that distribute money to recipients of its own selection; endowments; individuals; organizations that practice discrimination of any kind; organizations that do not have fiscal responsibility for the proposed project; organizations that do not have an active 501(c)3 nonprofit IRS determination letter; camps, camp programs, or other seasonal activities; religious organizations for support of religious programs; cover deficits or debt elimination; cover travel, conferences, conventions, group meetings, or seminars; influence legislation or support candidates for political office; produce videos or other media projects; fund research projects or studies; primary or secondary schools (K-12).
Geographic Focus: Colorado
Amount of Grant: 500 - 500,000 USD
Contact: Nicole Cook, Grants Administrator; (719) 633-7733 or (800) 554-7711; fax (719) 577-5702; grants@elpomar.org
Internet: https://www.elpomar.org/grant-making/el-pomars-funds/
Sponsor: El Pomar Foundation
10 Lake Circle
Colorado Springs, CO 80906

El Pomar Foundation Grants — 1582
The El Pomar Foundation's competitive process remains the primary grant making vehicle for organizations throughout Colorado to receive funding. El Pomar is a general purpose foundation, which means the Trustees approve grants across a wide spectrum of focus areas including: arts and culture; civic and community initiatives; education; health; and human services. Under this competitive process the Foundation accepts applications for general operating, programs, and capital support. Grant applications are reviewed three times each year. The board meets May 5, July 19, and October 2.
Requirements: Nonprofits with 501(c)3 tax status serving Colorado are eligible to apply for funding. If funded through the competitive process applicants must wait three years (36 months) before submitting a new application. If declined applicants must wait one year (12 months) before submitting a new application.
Restrictions: The Trustees will not consider any capital grant requests exceeding $100,000, unless initiated by the Foundation. The Foundation does not accept grant applications for: other foundations or nonprofits that distribute money to recipients of its own selection; endowments; individuals; organizations that practice discrimination of any kind; organizations that do not have fiscal responsibility for the proposed project; organizations that do not have an active 501(c)3 nonprofit IRS determination letter; camps, camp programs, or other seasonal activities; religious organizations for support of religious programs; cover deficits or debt elimination; cover travel, conferences, conventions, group meetings, or seminars; influence legislation or support candidates for political office; produce videos or other media projects; fund research projects or studies; primary or secondary schools (K-12).
Geographic Focus: Colorado
Amount of Grant: 500 - 500,000 USD
Samples: Bright Futures, Telluride, Colorado, $55,000 - support of education (2016); Adams State University Foundation, Alamosa, Colorado, $222,000 - general operating support (2016); Community Coalition for Families and Children, Divide, Colorado, $75,000 - human services (2016).
Contact: Nicole Cook, Grants Administrator; (719) 633-7733 or (800) 554-7711; fax (719) 577-5702; grants@elpomar.org
Internet: https://www.elpomar.org/grant-making/general-information/
Sponsor: El Pomar Foundation
10 Lake Circle
Colorado Springs, CO 80906

Elsie H. Wilcox Foundation Grants — 1583
The Foundation provides partial support to programs and projects of tax-exempt, public charities in Hawaii to improve the quality of life in the state, particularly the island of Kauai. Areas of interest to the Foundation, include: education, health organizations, people with disabilities; human services, performing arts, theater, religion, YM/YMCAs & YM/YWHAs. Types of support include building/renovation; equipment; general/operating support. Grants average from $5,000 - $15,000.
Requirements: 501(c)3 nonprofit organizations in Hawaii are eligible to apply. The Foundation places a special emphasis on the island of Kauai. Contact Paula Boyce to acquire the cover sheet/application forms and any additional guidelines required to begin the application process. Proposals must be submitted by October 1st.
Restrictions: No grants to individuals, or for endowments.
Geographic Focus: Hawaii
Date(s) Application is Due: Oct 1
Amount of Grant: 5,000 - 15,000 USD
Samples: Girl Scouts of Hawaii, Honolulu, HI, $5,000--scholarships for Kauai girls in grades K-5; Bishop Museum, Honolulu, HI, $5,000--educational program support; American Cancer Society, Hawaii Pacific Division, Honolulu, HI, $5,000--quality of life services for cancer patients on Kauai.
Contact: Paula Boyce; (808) 538-4944; fax (808) 538-4647; pboyce@boh.com
Internet: http://www.hawaiicommunityfoundation.org/index.php?id=290
Sponsor: Elsie H. Wilcox Foundation
Bank of Hawai'i, Foundation Administration Department 758
Honolulu, HI 96802-3170

Elsie Lee Garthwaite Memorial Foundation Grants — 1584
Established in 1943, the Foundation supports organizations primarily in Philadelphia, Chester, Montgomery and Delaware counties of Pennsylvania. Giving to organizations that: provide for the physical and emotional well-being of children and young people; seek to enable young people, particularly the needy, to reach their fullest potential through education, empowerment, and exposure to the arts; are smaller organizations, with budgets under $1 million per year.
Requirements: Non-profits are eligible in the Philadelphia, Chester, Montgomery and Delaware counties, PA. Contact the Foundation at least 30 days prior to deadlines with a Letter of Intent. An application form can be obtained from the Foundation after reviewing the Letter of Intent, unsolicited applications will not be excepted. Contact the Foundation for further guidelines.
Restrictions: No grants to: individuals; public, private, or parochial schools; or colleges and universities.
Geographic Focus: Pennsylvania
Date(s) Application is Due: Mar 31; Aug 31
Amount of Grant: 3,000 - 16,000 USD
Contact: Thomas Kaneda, Secretary; (610) 527-8101; fax (610) 527-7808
Sponsor: Elsie Lee Garthwaite Memorial Foundation
1234 Lancaster Avenue, P.O. Box 709
Rosemont, PA 19010-0709

Elton John AIDS Foundation Grants — 1585
The Elton John AIDS Foundation (EJAF) was established in the United States in 1992 by Sir Elton John, now headquartered in New York City. In 1993, Sir Elton also established his Foundation as a registered charity in the United Kingdom, headquartered in London. Both organizations pursue the same mission – to reduce the incidence of HIV/AIDS through innovative HIV prevention programs, eliminate stigma and discrimination associated with HIV/AIDS, and support direct HIV-related care for people living with HIV/AIDS. The U.S. organization's grant-making program focuses on the following targeted areas: gay men's mobilization for health; youth mobilization for health; black community mobilization for health; ending injection-related HIV transmission; access to healthcare for ex-offenders; and scaled-up of government HIV programming. EJAF conducts an annual open application grant cycle through a Request for Proposals process launched on or around May 1.
Requirements: Applicants located and conducting work corresponding to EJAF's stated grant-making priorities in the United States, Canada, the Caribbean, and Central and South America are eligible to apply for funding by first submitting an online letter of intent (LOI). After the LOIs have been reviewed, selected applicants will be invited to submit a full online proposal. These proposals are extensively reviewed, and final grant awards are approved by EJAF's Board of Directors.
Geographic Focus: All States, Puerto Rico, U.S. Virgin Islands, Anguilla, Antigua & Barbuda, Argentina, Aruba, Bahamas, Barbados, Belize, Bolivia, Brazil, British Virgin Islands, Canada, Caribbean, Cayman Islands, Chile, Colombia, Costa Rica, Cuba, Dominica, Dominican Republic, Ecuador, El Salvador, Grenada, Guadeloupe, Guatemala, Guyana, Haiti, Honduras, Jamaica, Martinique, Mexico, Montserrat, Nicaragua, Panama, Paraguay, Peru, Saint Kitts And Nevis, St. Lucia, St. Vincent and the Grenadines, Suriname, Trinidad and Tobago, Turks and Caicos Islands, Uruguay, Venezuela
Contact: Matt Blinstrubas, Director of Grants; (212) 219-0670; info@ejaf.org
Internet: http://newyork.ejaf.org/ejaf-grants-overview/
Sponsor: Elton John AIDS Foundation
584 Broadway, Suite 906
New York, NY 10012

EMBO Installation Grants — 1586
The aim of this scheme is to strengthen science in participating EMBC member states. At present applications are accepted from Croatia, the Czech Republic, Estonia, Hungary, Portugal, Poland and Turkey. Additional member states have expressed an interest to join in the near future. Grants will help scientists to relocate, set up their labs and rapidly establish a reputation in the European scientific community. The program is entirely funded by the participating EMBC member states and successful applicants receive an annual support of Euro $50,000 for three to five years. Each participating country sets an upper limit to the number of grants available per year. Application forms should be completed by the individual scientists and the receiving institute.
Requirements: Eligible scientists should: have an excellent publication record; have spent at least two consecutive years prior to the application deadline, outside the country in which they are applying to establish their lab; and be negotiating a full-time position at an institute/university in participating member states by the date of application. No age limit applies.
Geographic Focus: All States, Albania, Andorra, Armenia, Austria, Azerbaijan, Belarus, Belgium, Bosnia & Herzegovina, Bulgaria, Croatia, Cyprus, Czech Republic, Denmark, Estonia, Finland, France, Georgia, Germany, Greece, Hungary, Iceland, Ireland, Italy, Kosovo, Latvia, Liechtenstein, Lithuania, Luxembourg, Macedonia, Malta, Moldova, Monaco, Montenegro, Norway, Poland, Portugal, Romania, Russia, San Marino, Serbia, Slovakia, Slovenia, Spain, Sweden, Switzerland, The Netherlands, Turkey, Ukraine, United Kingdom, Vatican City
Date(s) Application is Due: Apr 15
Amount of Grant: 150,000 - 250,000 EUR
Contact: Hermann Bujard, Executive Director; 49 0 6221 8891, ext. 101; fax 49 0 6221 8891, ext. 200; embo@embo.org
Internet: http://www.embo.org/sdig/index.html
Sponsor: European Molecular Biology Organization
Meyerhofstrasse 1
Heidelberg, D-69117 Germany

EMBO Long-Term Fellowships 1587
Long-Term Fellowships are awarded for prolonged visits (12 to 24 months) and are intended for advanced training through research. As the aim of the EMBO is to promote international research, it follows that mobility is a crucial element in deciding the eligibility of an application. All applications must involve a laboratory of origin or a receiving institute or applicant's nationality from one of the member states. Stipend rates depend on the country being visited, marital status, etc.
Requirements: Applicants must hold a doctorate degree or equivalent before the start of the fellowship but not necessarily when applying. For clarity, it is deemed that this requirement is fulfilled only when the thesis has been stamped as submitted for final examination by the university. As a minimum, applicants must have at least one first author publication in press or published in an international peer reviewed journal at the time of application.
Restrictions: Applications are only considered from candidates with a maximum of 3 years postdoctoral experience.
Geographic Focus: All States, Germany
Date(s) Application is Due: Feb 15; Aug 15
Contact: Liselott Maidment, 49 0 6221 8891, ext. 116; fellowships@embo.org
Internet: http://www.embo.org/fellowships/long_term.html
Sponsor: European Molecular Biology Organization
Meyerhofstrasse 1
Heidelberg, D-69117 Germany

EMBO Short-Term Fellowships 1588
Short-term fellowships are established to advance molecular biology research by helping scientists to visit another laboratory with a view to applying a technique not available in the home laboratory. Such fellowships are not awarded for exchanges between two laboratories within the same country, and are intended for joint research work rather than consultations. The fellowships cover travel plus subsistence of the fellow only and not of any dependents. These fellowships are intended for visits of one week to three months duration (for non-European applicants the short-term fellowships are intended for a fixed duration of 90 days). Stipends vary widely from country to country.
Requirements: Applicants must be either: post-doctoral scientists with less than 10 years of professional experience since finishing their Ph.D. degree (or the equivalent); or pre-doctoral scientists who have not yet completed a Ph.D. degree (or the equivalent).
Restrictions: All applications must involve either a laboratory of origin or a receiving institute from one of the EMBC member states.
Geographic Focus: All States, Germany
Contact: Director; 49 0 6221-8891, ext. 115; fellowships@embo.org
Internet: http://www.embo.org/fellowships/short_term.html
Sponsor: European Molecular Biology Organization
Meyerhofstrasse 1
Heidelberg, D-69117 Germany

Emerson Charitable Trust Grants 1589
Established in 1944 in Missouri as the Emerson Electric Manufacturing Company Charitable Trust, the Foundation supports: arts and culture--fine arts and cultural institutions to enrich the diversity, creativity, and liveliness of the community; education--programs designed to promote educational systems at all levels; health and human services--programs designed to help individuals and families in times of need, including sickness, old age, family crisis, and natural disasters; civic affairs--programs designed to protect citizenry; further economic health of the community; and build and maintain assets such as parks and zoos; and youth--programs designed to give young people the opportunity to recognize their potential, confidence, and skills to achieve their dreams. The foundation awards college scholarships to children and step-children of employees of Emerson Electric. No specific application form is required, and initial approach should be the complete proposal.
Geographic Focus: All States
Amount of Grant: 5,000 - 300,000 USD
Samples: Alvin J. Siteman Ctr, Washington U School of Medicine and Barnes-Jewish Hospital (Saint Louis, MO)--for a new cancer-research facility and for cancer research, $6 million challenge grant.
Contact: Jo Ann Harmon, Senior Vice President; (314) 553-3722; fax (314) 553-1605
Internet: http://www.emerson.com/en-us/about_emerson/company_overview/pages/our_approach_to_corporate_philanthropy.aspx
Sponsor: Emerson Charitable Trust
8000 W Florissant Avenue, P.O. Box 4100
Saint Louis, MO 63136

Emerson Electric Company Contributions Grants 1590
Emerson believes it is important to help support organizations and institutions that play significant roles in enhancing the quality of life in the communities where its facilities are located and where its employees and their families live and work. As a complement to its foundation, Emerson makes charitable contributions to nonprofit organizations directly. Support is given on a national basis in areas of company operations, with emphasis on St. Louis, Missouri. Its primary fields of interest include the arts, education, health care, and public affairs. An application form should be obtained from the nearest company facility.
Geographic Focus: All States
Contact: Robert M. Cox, Senior Vice President; (314) 553-2000; fax (314) 553-1605
Internet: http://www.emerson.com/en-us/about_emerson/company_overview/pages/our_approach_to_corporate_philanthropy.aspx
Sponsor: Emerson Electric Company
8000 W Florissant Avenue, P.O. Box 4100
Saint Louis, MO 63136

Emily Davie and Joseph S. Kornfeld Foundation Grants 1591
The Foundation's selection criteria for all programs are: evidence of strong and visionary leadership; development of new initiatives that are replicable and sustainable; creation of new ways to address existing critical issues; interest in building connections and creative partnerships; desire to improve and enrich lives of individuals (rather than capital improvements); and original, unusual, dynamic and focused proposals. Proposal letters should include a background on the organization, a detailed description of the proposed project and financial information.
Requirements: Eligible applicants must provide a copy of a 501(c)3 determination letter.
Geographic Focus: All States
Date(s) Application is Due: Mar 1; Jul 15; Nov 15
Contact: Bobye List; (718) 624-7969; fax (718) 834-1204; office@kornfeldfdn.org
Internet: http://fdncenter.org/grantmaker/kornfeld
Sponsor: Emily Davie and Joseph S. Kornfeld Foundation
41 Schermerhorn Street, Suite 208
Brooklyn, NY 11021

Emily Hall Tremaine Foundation Learning Disabilities Grants 1592
The Emily Hall Tremaine Foundation has been making grants to support children with learning disabilities (LD) and their families since 1992. Left unaddressed, learning disabilities can affect the emotional and physical health and well being of children, in turn preventing a positive experience in the classroom. During the first decade of the foundation's involvement in the field of learning disabilities, support was directed toward public relations campaigns to dispel the stigma often associated with LD. In 2003, the foundation refocused its efforts on the classroom. The foundation identified two key areas where grants can help prepare classrooms to be designed for all types of learners. The foundation's current priorities guiding its grant making are two-fold: early intervention; and technology and teaching. There are no application forms or deadlines. Unsolicited proposals rarely develop into a grant; submit informative letters of inquiry that highlight the organization's mission, goals, history, strategies, and programmatic scope. Awards have recently ranged from $500 to $25,000.
Requirements: Nonprofits educational organizations dealing with learning disabilities may apply for grant support.
Geographic Focus: All States
Amount of Grant: 500 - 25,000 USD
Contact: Nicole Chevalier, Program Director; (203) 639-5544 or (203) 639-5547; fax (203) 639-5545; chevalier@tremainefoundation.org
Internet: http://www.tremainefoundation.org
Sponsor: Emily Hall Tremaine Foundation
171 Orange Street
New Haven, CT 06510

Emma B. Howe Memorial Foundation Grants 1593
The Emma B. Howe Memorial Foundation makes grants through the Minneapolis Foundation. The focus of all grants is to improve: the health and well-being of children, youth and families; opportunities for educational achievement; access to quality affordable housing; and economic vitality throughout the region. Information on the application process is available online.
Requirements: Eligible organizations include 501(c)3 nonprofits, public institutions; and emerging groups organized for nonprofit purposes.
Restrictions: Funds are not available for: individuals; organizations/activities outside of Minnesota; conference registration fees; memberships; direct religious activities; political organizations or candidates; direct fundraising efforts; telephone solicitations; courtesy advertising; financial deficits.
Geographic Focus: Minnesota
Contact: Grants Manager; (612) 672-3836; grants@mplsfoundation.org
Internet: http://www.mplsfoundation.org/partners/emma.htm
Sponsor: Emma B. Howe Memorial Foundation
80 S Eighth Street
Minneapolis, MN 55402

Emma Barnsley Foundation Grants 1594
Emma Elizabeth Barnsley was born on December 10, 1926, in Crane County, Texas, and grew up in the areas of Crane, Midland, and Odessa, Texas. She moved to New York City as a young woman where she spent most of her adult life. Barnsley had a deep love for animals of all kinds. It is because of this passion she established the Emma Barnsley Foundation for the purposes of the prevention of cruelty to animals and the study, care, protection, and preservation of animals, both domestic and wild, and their environment. The Foundation awards grants exclusively for charitable, scientific, and educational purposes as follows: to aid and assist in the prevention of cruelty to animals, both domestic and wild; to aid and assist in the care, protection and preservation of animals, both domestic and wild, and the environment in which such wild animals may live; to provide scholarships for deserving young men and women to assist them in the study of veterinary medicine at any institution of higher learning in the state of Texas; to assist institutions of higher learning in the state of Texas in the operation and maintenance of courses of study in veterinary medicine and in the study of animals and animal life; and to assist in research projects involving ecological, environmental, and wildlife studies, provided no funds shall be given to any organization utilizing animals for research or experimental purposes if such animals are in any way harmed, mutilated, intentionally inflicted with disease or any substance causing disease, or if such animals are killed in the course of such research. Preference may be given to organizations that operate in the Crane/Midland/Odessa, Texas area or the Manhattan, New York area. Applications are accepted year-round. Applications must be submitted by August 31 to be reviewed at the

annual grant meeting that occurs each September. The typical grant ranges from $10,000 to $20,000, and approximately nine awards are provided annually.
Requirements: To be eligible, organizations must qualify as exempt organizations under Section 501(c)3 of the Internal Revenue Code. Applications must be submitted through the online grant application form or alternative accessible application designed for assistive technology users.
Restrictions: The foundation normally does not fund requests for: general operating support; endowments; individuals; political activities; loans; or fundraising events including dinners, benefits, and athletic events.
Geographic Focus: Colorado, New York
Date(s) Application is Due: Aug 31
Amount of Grant: 10,000 - 20,000 USD
Contact: Peggy Toal, Private Client Services; (720) 947-6725 or (888) 234-1999; fax (877) 746-5889; grantadministration@wellsfargo.com
Internet: https://www.wellsfargo.com/privatefoundationgrants/barnsley
Sponsor: Emma Barnsley Foundation
1740 Broadway
Denver, CO 80274

Ensworth Charitable Foundation Grants　1595
The Ensworth Charitable Foundation was established in 1948 to support and promote educational, cultural, human services, religious, and health care programming for underserved populations. The foundation specifically serves the people of Hartford, Connecticut, and its surrounding communities. Grants from the Ensworth Charitable Foundation are primarily one year in duration. On occasion, multi-year support is awarded. Applicants must apply online at the grant website. Applicants are strongly encouraged to do the following before applying: review the downloadable state application procedures for additional helpful information and clarifications; review the downloadable online-application guidelines at the grant website; review the foundation's funding history (link is available from the grant website); review the online application questions in advance; and review the list of required attachments. These will generally include: a list of board members, financial statements (audited, reviewed, or compiled by independent auditor); an organization summary; a list of other funding sources; an IRS Determination letter; and other required documents. All attachments must be uploaded in the online application as PDF, Word, or Excel files. The Ensworth Charitable Foundation application deadline is January 15 at 11:59 p.m. Applicants will be notified of grant decisions by letter within three to four months after the deadline.
Requirements: Applicants must have 501(c)3 tax-exempt status and serve the people of Hartford, Connecticut, and its surrounding communities.
Restrictions: Applicants will not be awarded a grant for more than 3 consecutive years. The foundation does not support requests from individuals, organizations attempting to influence policy through direct lobbying, or any political campaigns.
Geographic Focus: Connecticut
Date(s) Application is Due: Jan 15
Amount of Grant: 1,000 - 25,000 USD
Contact: Amy R. Lynch; (860) 657-7015; amy.r.lynch@ustrust.com
Internet: https://www.bankofamerica.com/philanthropic/fn_search.action
Sponsor: Ensworth Charitable Foundation
200 Glastonbury Boulevard, Suite #200
Glastonbury, CT 06033-4056

Entergy Corporation Micro Grants　1596
The Foundation will bestow a monetary award for projects that effectively impact arts and culture, community improvement and enrichment, education and literacy, and healthy families. Organizations should be located within Entergy's service territory in Arkansas, Louisiana, Mississippi, Massachusetts, Michigan, New Hampshire, New York, Texas, or Vermont. Micro Grant applications are accepted on an ongoing basis. Applicants should allow at least 6 (six) to 8 (eight) weeks for review and notification of the result of a request.
Requirements: 501(c)3 tax-exempt organizations, schools, hospitals, governmental units, and religious institutions are eligible. Each organization may submit only one application per year.
Restrictions: Entergy will not fund: groups without 501(c)3 or similar non-profit status; administrative expenses (e.g., salaries, office equipment) or recurring expenses that exceed 15% of the requested amount; capital project funding (i.e., building campaigns); political candidates or groups; purchase of uniforms or trips for school-related organizations; amateur sports teams; activities whose sole purpose is promotion or support of a specific religion, denomination, or religious institution; grants to individuals or loans of any type; or any organization owned or operated by an employee of Entergy.
Geographic Focus: Arkansas, Louisiana, Massachusetts, Michigan, Mississippi, New Hampshire, New York, Texas, Vermont
Amount of Grant: Up to 1,000 USD
Contact: Jennifer Quezergue, (504) 576-2674 or (504) 576-6980; jquezer@entergy.com Christine Jordan, (504) 576-7705 or (504) 576-6980; cminor@entergy.com
Internet: http://www.entergy.com/our_community/micro_grant_guidelines.aspx
Sponsor: Entergy Corporation
639 Loyola Avenue
New Orleans, LA 70161-1000

Entergy Corporation Open Grants for Healthy Families　1597
Entergy Corporation's Open Grants program focuses on improving communities as a whole. The Corporation believes that children need a good start to grow into healthy, well-adjusted adults. With that in mind, it gives to programs that have a direct impact on children educationally and emotionally. The Corporation is also interested in family programs, like those that better prepare parents to balance the demands of work and home. In considering requests for grants, priority is placed on programs in specific counties/parishes, including areas of: Arkansas, Louisiana, Massachusetts, Michigan, Mississippi, New Hampshire, New York, Texas, and Vermont. Applicants should contact the contributions coordinator in their region, which are listed on the web site. Open grant applications are accepted on an ongoing basis. Applications should be submitted at least three months prior to the time the funding is needed.
Requirements: Grants from the Entergy Corporation will only be made to the following types of organizations: non-profit organizations that are tax exempt under section 501(c)3 of the Internal Revenue Code; or schools, hospitals, governmental units and religious institutions that hold nonprofit status similar to that of 501(c)3 organizations.
Restrictions: Entergy will not fund: groups without 501(c)3 or similar non-profit status; administrative expenses (e.g., salaries, office equipment) or recurring expenses that exceed 15% of the requested amount; capital project funding (i.e., building campaigns); political candidates or groups; purchase of uniforms or trips for school-related organizations; amateur sports teams; activities whose sole purpose is promotion or support of a specific religion, denomination, or religious institution; grants to individuals or loans of any type; or any organization owned or operated by an employee of Entergy.
Geographic Focus: Arkansas, Louisiana, Massachusetts, Michigan, Mississippi, New Hampshire, New York, Texas, Vermont
Amount of Grant: Up to 1,000 USD
Contact: Jennifer Quezergue, (504) 576-2674 or (504) 576-6980; jquezer@entergy.com Christine Jordan, (504) 576-7705 or (504) 576-6980; cminor@entergy.com
Internet: http://www.entergy.com/our_community/Grant_Guidelines.aspx
Sponsor: Entergy Corporation
639 Loyola Avenue
New Orleans, LA 70161-1000

EPA Children's Health Protection Grants　1598
The objectives of this program are to catalyze community-based and regional projects and other actions that enhance public outreach and communication; assist families in evaluating risks to children and in making informed consumer choices; build partnerships that increase a community's long-term capacity to advance the protection of children's environmental health and safety; leverage private and public investments to enhance environmental quality by enabling community efforts to continue past EPA's ability to provide assistance to communities; and promote protection of children from environmental threats through lessons learned. There are no deadline dates.
Requirements: Eligible applicants include community groups, public nonprofit institutions/organizations, tribal governments, specialized groups, profit organizations, private nonprofit institutions/organizations, and municipal and local governments. Potential applicants are strongly encouraged to discuss proposed projects with or submit preapplications to program staff prior to the completion of a full proposal.
Geographic Focus: All States
Amount of Grant: 5,000 - 250,000 USD
Contact: Office of Children's Health Protection; (202) 564-2188; fax (202) 564-2733; fletcher.bettina@epa.gov
Internet: http://yosemite.epa.gov/ochp/ochpweb.nsf/content/grants.htm
Sponsor: Environmental Protection Agency
1200 Pennsylvania Avenue, NW
Washington, D.C. 20460

ERC Starting Grants　1599
ERC Starting Grants aim to support up-and-coming research leaders who are about to establish a proper research team and to start conducting independent research in Europe. The scheme targets promising researchers who have the proven potential of becoming independent research leaders. It will support the creation of excellent new research teams. Each grant awarded will receive up to € 1.5 million (in some circumstances up to € 2 million) for up to 5 years.
Requirements: Eligible applicants include researchers of any nationality with 2-7 years of experience since completion of PhD (or equivalent degree) and scientific track record showing great promise. Research must be conducted in a public or private research organization (known as a Host Institution/HI) located in one of the EU Member State or Associated Countries. The sole evaluation criterion will be the scientific excellence of researcher and research proposal.
Geographic Focus: Albania, Andorra, Armenia, Australia, Azerbaijan, Belarus, Belgium, Bosnia & Herzegovina, Bulgaria, Croatia, Cyprus, Czech Republic, Denmark, Estonia, Finland, France, Germany, Greece, Hungary, Iceland, Ireland, Italy, Kosovo, Latvia, Liechtenstein, Lithuania, Luxembourg, Macedonia, Malta, Moldova, Monaco, Montenegro, Norway, Poland, Portugal, Romania, Russia, San Marino, Serbia, Slovakia, Slovenia, Spain, Sweden, Switzerland, The Netherlands, Turkey, Ukraine, United Kingdom, Vatican City
Date(s) Application is Due: May 13
Amount of Grant: Up to 1,500,000 EUR
Contact: National Contact Point
Internet: http://erc.europa.eu/starting-grants
Sponsor: European Research Council
ERC Executive Agency, COV2
Brussels, BE-1049 Belgium

Erie Chapman Foundation Grants　1600
Established in 2007 by Erie D. Chapman III, the former CEO of Baptist Hospital System in Nashville, Tennessee. Giving primarily in Tennessee and Florida, the Foundation's mission is to support the arts and radical loving care in health care. The Foundation also provides funding for the arts in general, as well as to Christian agencies and churches.

There are no specific application forms or deadlines with which to adhere, and applicants should begin the process by forwarding a letter in inquiry to Erie D. Chapman III.
Geographic Focus: Florida, Tennessee
Amount of Grant: Up to 20,000 USD
Contact: Erie D. Chapman III, Chief Executive Officer and President; erie.chapman@live.com or erie_chapman@hotmail
Internet: www.eriechapmanfoundation.org
Sponsor: Erie Chapman Foundation
500 Madison Avenue
Nashville, TN 37208

Erie Community Foundation Grants 1601
The Erie Community Foundation is a collection of charitable endowments operating under the administrative umbrella of a single public charity. Grants will be awarded in four program areas: quality of life; human services; health; and education. The Foundation's goal is to help local charities more effectively and efficiently accomplish their missions. Application information is available online. Applicants are strongly encouraged to contact the Foundation's program officer prior to submitting an application.
Requirements: Erie, PA, 501(c)3 tax-exempt organizations are eligible.
Restrictions: The Foundation does not fund: program ads; fund raising events/sponsorships; more than 50 percent of the cost of a vehicle; deficit reduction; sectarian religious activities; government funding cuts; start-up organizations; school playgrounds; fire departments or nursing homes.
Geographic Focus: Pennsylvania
Date(s) Application is Due: Feb 2; Aug 3; Oct 16
Amount of Grant: 2,000 - 15,000 USD
Contact: David Gonzalez, Program Officer; (814) 454-0843; fax (814) 456-4965; mbatchelor@eriecommunityfoundation.org
Internet: http://www.eriecommunityfoundation.org/for-grant-seekers/
Sponsor: Erie Community Foundation
459 West 6th Street
Erie, PA 16507

Essex County Community Foundation Discretionary Fund Grants 1602
The grant awards funds across the broad areas of nonprofit activity, including arts and culture, education, environment, health, social and community services and youth services. The Trustees will award Discretionary Fund grants to assist nonprofits with projects which strengthen the capacity of the organization to perform its work more effectively. ECCF is interested in helping organizations improve their infrastructure so that they can better serve their communities and clients. ECCF will consider projects which address capacity building such as: strengthening Board leadership; managing organizational change and growth; providing for strategic organizational planning; supporting leadership transition; strengthening fiscal management; and improving staff skills or other organizational functions which will improve the organization's long-term capacity.
Requirements: Eligible organizations must be Massachusetts 501(c)3 agencies that serve Essex County citizens with operating budgets of less than $500,000.
Restrictions: Funds are not available for: individuals; costs associated with programs or services provided to citizens outside of Essex County; sectarian or religious purposes; political purposes; debt or deficit reduction; capital campaigns for buildings, land acquisition or endowment; or to support academic research. Funding for equipment is limited to purchases that resolve a specific problem, are part of an overall capacity building project and will strengthen the operation of the organization. Equipment for programmatic purposes will not be granted. In general, staff salaries will not be eligible for funding unless the salary is directly tied to developing the capacity of the organization.
Geographic Focus: Massachusetts
Date(s) Application is Due: Feb 10
Amount of Grant: 1,000 - 5,000 USD
Contact: Julie Bishop, Vice President of Grants and Nonprofit Services; (978) 777-8876, ext. 28; fax (978) 777-9454; j.bishop@eccf.org or info@eccf.org
Internet: http://eccf.org/discretionary-fund-42.html
Sponsor: Essex County Community Foundation
175 Andover Street, Suite 101
Danvers, MA 01923

Essex County Community Foundation Merrimack Valley General Fund Grants 1603
The fund responds to new and important community needs as they arise. The Merrimack Valley General Fund (MVGF) considers requests from agencies serving Eastern Merrimack Valley communities in most fields of interest including: arts and culture, education, social and community services, and youth service. Particular interest is in programs which: benefit children, particularly those fostering youth development; provide health care services to those without health care coverage and address the need for affordable housing. Other areas of interest include: early childhood education; adult literacy; homelessness; elder care; job development; and food, arts and culture.
Requirements: Only non-profit organizations, recognized as tax exempt under section 501(c)3 are eligible for consideration. The MVGF provides grants to the communities of: Lawrence, Methuen, Haverhill, Andover, North Andover, Boxford, Georgetown, Groveland, W. Newbury, Merrimac, Amesbury, Newburyport, Newbury and Salisbury.
Restrictions: Funds are not available for: individuals; state or local government agencies; political purposes; and sectarian or religious purposes. Generally grants are not awarded for: debt or deficit reduction; replacing public funding, or for purposes which are generally a public sector responsibility; supporting academic research; and traveling outside the region.
Geographic Focus: Massachusetts
Date(s) Application is Due: Oct 1
Amount of Grant: 1,000 - 5,000 USD
Contact: Julie Bishop, Vice President of Grants and Nonprofit Services; (978) 777-8876, ext. 28; fax (978) 777-9454; j.bishop@eccf.org or info@eccf.org
Internet: http://eccf.org/merrimack-valley-general-fund-44.html
Sponsor: Essex County Community Foundation
175 Andover Street, Suite 101
Danvers, MA 01923

Essex County Community Foundation Webster Family Fund Grants 1604
The Webster Family Fund provides grants to nonprofit organizations in which a substantial majority of clients reside in the following Greater Lawrence Communities: Lawrence, Methuen, Andover or North Andover. The fund's competitive grants are for capital projects. Preference is given to organizations that provide: youth development; health care; and arts and culture programming or services. The basis of the capital project can be an emergency situation or a capital improvement plan and associated capital campaign. Capital projects include large equipment whose useful life is over five years, as well as asset improvements (ie. bricks and mortar). For other capital grant requests, preference is given to projects that are part of an organization's strategic plan or have resulted from a feasibility study or long-term assessment of needs. For emergency capital projects, applicants should note that the grant review process for the Fund takes three and a half months, making this fund unsuitable for extremely time sensitive emergency requests. Only one-year grants are awarded. Grants typically range from $3,000 to $15,000.
Requirements: Only nonprofit organizations, recognized as tax exempt under section 501(c)3 of the Internal Revenue Code, are eligible for consideration.
Restrictions: Grants are not made to individuals, for political purposes, or for sectarian or religious purposes.
Geographic Focus: Massachusetts
Date(s) Application is Due: Sep 1
Amount of Grant: 3,000 - 15,000 USD
Contact: Julie Bishop, Vice President of Grants and Nonprofit Services; (978) 777-8876, ext. 28; fax (978) 777-9454; j.bishop@eccf.org or info@eccf.org
Internet: http://eccf.org/webster-family-fund-48.html
Sponsor: Essex County Community Foundation
175 Andover Street, Suite 101
Danvers, MA 01923

Essex County Community Foundation Women's Fund Grants 1605
The Women's Fund seeks to identify existing programs or projects which offer the most significant impact for the grant dollar. The focus will be to identify programs or projects that serve women and girls in the following fields of funding interest: leadership and empowerment; economic self-sufficiency and security; and health and well being. Priority will be given to organizations that involve women and/or girls in program development and organizational leadership. The Women's Fund Advisory Board will select one field of funding interest per year. By April 1st of each year, the Women's Fund will make grants with a 1 to 3 year commitment to agencies within that year's Field of Funding Interest. Grants will be awarded up to $15,000. Agencies which receive a multiple year grant under one Field of Funding Interest may apply for other grants in subsequent years under other Field of Funding Interests.
Requirements: Non profit organizations located in Essex County which serve women and girls are eligible. Programs sponsored by religious organizations are eligible provided the enrollment is open to all qualified women and girls and the program is free of mandatory sectarian religious instruction.
Restrictions: Generally, funds will not be awarded: for debt or deficit reduction; to individuals; for political purposes; for sectarian or religious purposes; to state and local government agencies; for endowment or capital campaigns; for research; for feasibility studies; or to agencies with pending 501(c)3 status or using fiscal sponsors.
Geographic Focus: Massachusetts
Date(s) Application is Due: Dec 1
Amount of Grant: 1,000 - 15,000 USD
Contact: Julie Bishop, Vice President of Grants and Nonprofit Services; (978) 777-8876, ext. 28; fax (978) 777-9454; j.bishop@eccf.org or info@eccf.org
Internet: http://eccf.org/womens-fund-45.html
Sponsor: Essex County Community Foundation
175 Andover Street, Suite 101
Danvers, MA 01923

Estee Lauder Grants 1606
The company has long supported numerous institutions and programs that reflect the interests and concerns of their consumers and employees around the world. Over the years, corporate and brand philanthropy programs and community initiatives have focused on advancing the activities and growth of numerous organizations dedicated to health and human services, education, the environment and the arts. Contributions take many forms, both global and local. Submit a letter of request on company letterhead that describes the request, organization, and intended use of funds.
Geographic Focus: All States
Amount of Grant: Up to 85,000,000 USD
Contact: Contribution Program Manager; (212) 572-4200; fax (212) 527-7955
Internet: http://www.elcompanies.com/citizenship/corporate_responsibility/corporate_philanthropy.asp
Sponsor: Estee Lauder
767 Fifth Avenue, Box 100
New York, NY 10153

Ethel S. Abbott Charitable Foundation Grants   1607
The Ethel S. Abbott Charitable Foundation, established in 1972, offers support for art and culture, education, health, federated giving, and human service programs. The Board of Trustees will generally consider grant requests for capital projects and endowment projects more favorably than grant requests for operating funds. However, the Board will occasionally make an operating grant for the start up of a new 501(c)3 or a program which will become self-sustaining after the start up period. Typically, grants range from $200,000 up to over $2 million. The application and its guidelines can be downloaded from the web site.
Requirements: The primary grant areas are: Lincoln, Nebraska and the surrounding 50 mile radius; Omaha, Nebraska and the surrounding 50 mile radius up to the Missouri River; and Western, Nebraska, which is defined as all parts of the state in the Mountain Time Zone.
Geographic Focus: Nebraska
Amount of Grant: 200,000 - 2,000,000 USD
Samples: Lincoln Sports Foundation, Lincoln, Nebraska, $2,750,000 - construction and maintenance of the Ethel S. Abbott Sports Complex; Estes Park Medical Center Foundation, Estes Park, Colorado, $750,000 - construction and Endowment Fund for the Estes Park Medical Center; Nebraska State Historical Society, Bayard, Nebraska, $255,000 - construction and Endowment Fund for the Chimney Rock Visitors Center.
Contact: Del Lienemann, Sr., President; (402) 435-4369; fax (402) 435-4371; info@abbottfoundation.org
Internet: http://www.abbottfoundation.org/request_a_grant/index.html
Sponsor: Ethel S. Abbott Charitable Foundation
P.O. Box 81407
Lincoln, NE 68501-1407

Ethel Sergeant Clark Smith Foundation Grants   1608
The activities of the Ethel Sergeant Clark Smith (ESCS) Memorial Fund focuses on grants to organizations located in Southeastern Pennsylvania, with primary emphasis on those serving community needs in Delaware County. Grants will be made for capital projects, operating expenses and special programs in amounts that are meaningful to the success of the individual endeavors of the organizations. However, operating expense grants are typically awarded for charities without capital requirement and under circumstances where continuing funding is not expected. Grants will be made in areas of medical, educational, cultural, arts, health and welfare, and such other areas as the trustee shall identify and determine from time to time, to be responsive to changes in community needs. Application forms are available online and must be submitted by March 1 or September 1 annually.
Requirements: Southeastern Pennsylvania 501(c)3 non-profit organizations with primary emphasis on those serving community needs in Delaware County, Pennsylvania are eligible to apply. Complete applications should include the following: one original copy of the Proposal which includes the purpose and general activities of the organization should be included as well as a description of the proposed project and its justification, a budget and timetable for the project are also required; one copy of audited financial statements for the last fiscal year (or if not audited, Internal Revenue Service form 990) plus an operating budget for the current period and budgets for future period if appropriate; copy of the Internal Revenue Service tax determination letter which shows the organization is tax-exempt under Section 501(c)3 and that it is not a private foundation under section 509 (a) of the Internal Revenue Code. Any organization that is awarded a grant will be required to sign a Grant Agreement Form prior to the distribution of funds. Approximately one year after a grant has been awarded, a Progress Report should be completed by the organization. This information must be submitted prior to the consideration of any new proposals.
Restrictions: Grants will not be considered for the following: deficit financing; construction or renovations to real estate not owned by the charitable entity; salaries; professional fund raiser fees; multi-year grants over three years; to any organization more than once in a given year; to any organization more than three years in succession; any organization receiving a grant over a three year period or in three successive years will not be eligible for a future grant until two years transpire after the three year period.
Geographic Focus: Pennsylvania
Date(s) Application is Due: Mar 1; Sep 1
Contact: Wachovia Bank, N.A., Trustee; grantinquiries4@wachovia.com
Internet: https://www.wachovia.com/foundation/v/index.jsp?vgnextoid=3b3852199c0a a110VgnVCM1000004b0d1872RCRD&vgnextfmt=default
Sponsor: Ethel Sergeant Clark Smith Foundation
620 Brandywine Parkway, Mail Code PA 5042
West Chester, PA 19380

Eugene B. Casey Foundation Grants   1609
The foundation awards grants to nonprofit organizations in the District of Columbia and Maryland. Nonprofit organizations of the Christian and Roman Catholic faiths also are eligible. Organizations receiving grants include colleges and universities, community service organizations, government agencies, medical centers, parochial schools, and religious groups. Types of support include capital grants and general support grants. There are no application forms or deadlines. Applicants should submit written proposals that include an annual report, the purpose for which the funds are requested, and the amount requested compared with the total sought.
Requirements: Nonprofits in Maryland or the District of Columbia are eligible to apply.
Geographic Focus: District of Columbia, Maryland
Amount of Grant: 10,000 - 100,000 USD
Sample: Redlands Christian Migrant Association, Immokalee, FL, $10,000--for general operating support; Cystic Fibrosis Foundation, Bethesda, MD, $20,000--for general operating support; United States Department of State, Washington, D.C., $50,000--for general operating support.
Contact: Betty Brown Casey, Treasurer; (301) 948-4595
Sponsor: Eugene B. Casey Foundation
800 South Frederick Avenue, Suite 100
Gaithersburg, MD 20877-1701

Eugene G. and Margaret M. Blackford Memorial Fund Grants   1610
The Eugene G. and Margaret M. Blackford Memorial Fund was established in 1981 to support and promote quality educational, human-services, and health-care programming for the blind or visually impaired. The deadline for application is October 1. Application materials are available for download at the grant website. Applicants will be notified of the grant decision by letter within 2 to 3 months after the proposal deadline. Applicants are strongly encouraged to review the downloadable state application guidelines before applying. Applicants are also encouraged to view the fund's funding history (link is available at the grant website).
Requirements: Applicant organizations must have 501(c)3 tax-exempt status and serve blind or visually impaired individuals living in Connecticut.
Restrictions: Applicants will not be awarded a grant for more than 3 consecutive years. The fund does not support requests from individuals, organizations attempting to influence policy through direct lobbying, or any political campaigns.
Geographic Focus: Connecticut
Date(s) Application is Due: Oct 1
Contact: Carmen Britt; (860) 657-7019; carmen.britt@ustrust.com
Internet: https://www.bankofamerica.com/philanthropic/fn_search.action
Sponsor: Eugene G. and Margaret M. Blackford Memorial Fund
200 Glastonbury Boulevard, Suite # 200, CT2-545-02-05
Glastonbury, CT 06033-4056

Eugene M. Lang Foundation Grants   1611
The foundation awards grants in New York and Pennsylvania in its areas of interest, including education (early childhood education and higher education), medical and health programs, arts, health organizations, medical research, minorities, and performing arts. Types of support include annual campaigns, conferences and seminars, continuing support, fellowships, general operating support, internship funds, professorships, program development, scholarship funds, and seed money. The foundation favors social services such as those helping homeless or single mothers. Locally based groups wanting support must involve a Lang family member. There are no application deadlines; initial approach should be by letter.
Requirements: Organizations in New York and Pennsylvania are eligible to apply.
Restrictions: Grants are not made to individuals, or for building funds, equipment and materials, capital or endowment funds, deficit financing, publications, or matching gifts.
Geographic Focus: New York, Pennsylvania
Amount of Grant: 500 - 50,000 USD
Samples: New York-Presbyterian Hospital (New York, NY)--for a six-year health-sciences education program for selected inner-city youths, designed to boost their academic success and facilitate their entry into health careers, $1.25 million.
Contact: Program Contact; (212) 949-4100
Sponsor: Eugene M. Lang Foundation
535 5th Avenue, Suite 906
New York, NY 10017

Eugene Straus Charitable Trust   1612
The Eugene Straus Charitable Trust was established in 1974. Eugene Straus was born to a pioneer Dallas family who settled in Dallas County in the 1840's. Mr. Straus was a successful homebuilder and real estate developer. He created this trust under his will for Dallas County charitable institutions to erect and maintain permanent building improvements. The majority of grants from the Straus Charitable Trust are one year in duration; on occasion, multi-year support is awarded. Applicants must apply online at the grant website. Applicants are strongly encouraged to do the following before applying: review the downloadable state application procedures for additional helpful information and clarifications; review the downloadable online-application guidelines at the grant website; review the trust's funding history (link is available from the grant website); review the online application questions in advance; and review the list of required attachments. These will generally include: a list of board members, financial statements (audited, reviewed, or compiled by independent auditor); an organization summary; a list of other funding sources; an IRS Determination letter; and other required documents. All attachments must be uploaded in the online application as PDF, Word, or Excel files. The application deadline for the Eugene Straus Charitable Trust is 11:59 p.m. on July 31. Applicants will be notified of grant decisions before November 30.
Requirements: Applicants must have 501(c)3 tax-exempt status.
Restrictions: Grants are only awarded to organizations for the purpose of erecting and maintaining permanent building improvements in Dallas County. Requests for general operating or program-related grants will not be considered.
Geographic Focus: Texas
Date(s) Application is Due: Jul 31
Samples: Hope Cottage Pregnancy and Adoption Center, Dallas, Texas, $6,000, building improvements; Dallas Children's Advocacy Center, Dallas, Texas, $6,000, building/capital campaign for a new facility; Saint Philips School and Community Center, Dallas, Texas, $6,000, maintenance of campus facilities.
Contact: David Ross, Senior Vice President; tx.philanthropic@baml.com
Internet: https://www.bankofamerica.com/philanthropic/fn_search.action
Sponsor: Eugene Straus Charitable Trust
901 Main Street, 19th Floor, TX1-492-19-11
Dallas, TX 75202-3714

Eva L. and Joseph M. Bruening Foundation Grants 1613
The Foundation gives priority to grant requests that address one of the four program areas: services for vulnerable older adults; services for physically and mentally impaired individuals; secondary/higher education; and social services for the economically disadvantaged. There is no application or proposal form. Information regarding the application process is available online. Applicants are encouraged to contact the office for further clarification of the Foundation's grantmaking policies.
Requirements: Grants are awarded only to tax-exempt, nonprofit organizations located within Cuyahoga County, Ohio, usually for program, capital, and start-up operating support.
Restrictions: Grants are not awarded for endowment, general operating support, research, symposia or seminars. No grants are awarded to individuals, nor does the Foundation respond to mass mailings or annual campaign solicitations.
Geographic Focus: Ohio
Date(s) Application is Due: Mar 1; Jul 1; Oct 1
Samples: Jewish Family Service Assoc (OH)--to create a centralized, relational database system, $50,000; Youth Opportunities Unlimited (OH)--to help match Workforce Investment Act funds for the Jobs for Ohio's Graduates program in East Cleveland, $25,000; Towards Employment, Inc (OH)--to integrate Cleveland Works' legal assistance and ex-offender employment programs into Towards Employment's program structure, $50,000; Lutheran Chaplaincy Service (OH)--to support the full-time development director position, $25,000.
Contact: Janet Narten, Executive Director; (216) 621-2632; fax (216) 621-8198
Internet: http://www.fmscleveland.com/bruening
Sponsor: Eva L. and Joseph M. Bruening Foundation
1422 Euclid Avenue, Suite 627
Cleveland, OH 44115-1952

Evan and Susan Bayh Foundation Grants 1614
Established in 2002 by Senator Evan Bayh and his wife Susan, the foundation gives to various education and human services programs mostly in Indiana.
Requirements: There are no deadlines and no specific forms to use. It is suggested that you contact the foundation prior to submitting an application.
Geographic Focus: Indiana
Amount of Grant: 2,500 - 5,000 USD
Sample: Indiana University, Matching the Promise Campaign - $50,000; American Cancer Society, Atlanta, GA - $2,500; Deaconess Hospital, Evansville, IN - $5,000; Fisher House Foundation, Rockville, MD - $5,000; Munster Medical Research Foundation, Munster, IN - $5,000; Nature Conservancy, Arlington, VA - $2,500; St. Albans School of Public Service, Washington, D.C. - $5,000; Salvation Army, Indianapolis, IN - $2,500; United Way of Central Indiana, Indianapolis, IN - $2,500.
Contact: G. Frederick Glass, (317) 237-0300; fred.glass@bakerd.com
Sponsor: Evan and Susan Bayh Foundation
300 N Meridian Street, Suite 2700
Indianapolis, IN 46204-1750

Evanston Community Foundation Grants 1615
Grants awarded by the Foundation: encourage and support new initiatives and innovative approaches to addressing community needs; build the capacity of local nonprofit organizations to fulfill their missions more effectively; encourage collaborative ventures that will strengthen the community; provide initial support of projects that will have impact beyond the scope and timeline of the proposed project; build community partnerships and resources; and strengthen the area's nonprofit community. Grants typically provide initial seed money to launch new projects, capstone dollars for a larger project that will grow over a longer period, and support a one-time activity or initial phase of a new program. The Foundation's area of interests include: arts and culture; basic human needs; community development; education; environment; health; women and girls and youth and families. The Foundation's RFP and application instructions are available online.
Requirements: IRS 501(c)3 organizations serving Evanston, Illinois, are eligible.
Geographic Focus: Illinois
Date(s) Application is Due: Feb 26
Contact: Sara Schastok, Executive Director; (847) 492-0990; fax (847) 492-0904; schastok@evcommfdn.org or info@evcommfdn.org
Internet: http://www.evcommfdn.org/grant_making.htm
Sponsor: Evanston Community Foundation
1007 Church Street, Suite 108
Evanston, IL 60201

Everyone Breathe Asthma Education Grants 1616
The program was created by Sunovion Pharmaceuticals Inc., in partnership with the Asthma and Allergy Foundation of America (AAFA), to help improve the quality of asthma care and educate students about asthma. Parents or guardians of children with asthma are invited to nominate their child and child's school to receive an Everyone Breathe Asthma Education savings bond and grant. Ten applications will be selected as grand prize winners: the nominated child with asthma will be awarded a $2,500 savings bond; the school nominated in the winning application will receive a $5,000 grant to improve the quality of asthma care and asthma education in the school. Fifty additional applications will be selected as second-place winners and those nominated schools will receive asthma nebulizer kits. The grants and savings bonds will be awarded based on the level of commitment demonstrated by parents and school administrators to implement and sustain the proposed plan to improve the quality of asthma care and education at the schools. Prior to applying, parents are strongly encouraged to reach out to the administrators at their child's school to make them aware of the grant and engage them in the process of determining how the grant funds would be best utilized to improve the quality of asthma care at the school.
Requirements: Entries must be received from U.S. citizens aged 18 years or older who have a child or are a legal guardian of a child with asthma. All children nominated for the $2,500 savings bond must have asthma and be enrolled in the school (grades kindergarten through 12) that was nominated to receive the $5,000 grant. The Everyone Breathe Asthma Education Grants program is open to schools located in the 50 United States and the District of Columbia. Only public and private schools (excluding home schools) that plan to use the funds to improve the quality of asthma care and educate students about the condition are eligible. Winning schools must utilize the full amount of grant money received within 12 months from the date of the award.
Restrictions: Funding will not be considered for the following: groups that qualify as a government agency; religious schools that restrict entry to those of a certain faith or those who belong to a specific denomination or sect; schools that discriminate because of race, color, creed, gender, sexual orientation, gender identity or expression, ancestry, national origin, disability, veteran status, citizenship or any other legally protected category; capital campaigns or endowment funds; political activities or organizations; or Sunovion employees.
Geographic Focus: All States
Date(s) Application is Due: Jul 29
Amount of Grant: 7,500 USD
Contact: Grants Administrator; 888-394-7377; AsthmaAwards@Sunovion.com
Internet: http://www.everyonebreathe.com
Sponsor: Asthma and Allergy Foundation of America
8201 Corporate Drive, Suite 100
Landover, MD 20785

Evjue Foundation Grants 1617
The Foundation contributes each year to worthy educational, cultural and charitable organizations that contribute to the quality of life in Madison and Dane County. The Foundation's interests include: theater, hunger, troubled youth, and universities. The Foundation looks favorably upon projects responding to overall community needs and priorities that do not duplicate existing services. Application and guidelines are available online.
Requirements: Grants are only made to nonprofit organizations.
Restrictions: Grants are not made to individuals. All grants for scholarships are given to educational institutions, which have full responsibility for selecting the individual recipients. Grants are not usually made to establish or add to endowment funds. Grants are not ordinarily made to fund specific medical or scientific research inquiries, nor to support operating expenses and general administrative expenses of organizations.
Geographic Focus: Wisconsin
Date(s) Application is Due: Mar 5
Contact: Arlene Hornung; (608) 252-6401; ahornung@madison.com
Internet: http://host.madison.com/ct/about/evjue/article_36a7a66e-a46a-11de-ac74-001cc4c002e0.html
Sponsor: Evjue Foundation
1901 Fish Hatchery Road
Madison, WI 53713

Ewing Halsell Foundation Grants 1618
The foundation awards grants to eligible Texas nonprofit organizations in its areas of interest, including education, environment, health care and health organizations, medical research, social services, and youth services. Types of support include annual campaigns, building construction/renovation, equipment acquisition, land acquisition, publication, research, seed grants, and technical assistance. There are no application deadlines or forms.
Requirements: Texas nonprofit organizations are eligible. Preference is given to requests from southwestern Texas, particularly San Antonio.
Restrictions: The foundation's grants do not support individuals or requests for deficit financing, emergency funds, general endowments, matching gifts, scholarships, fellowships, demonstration projects, general purposes, conferences, or loans.
Geographic Focus: Texas
Contact: Grants Administrator; (210) 223-2640
Sponsor: Ewing Halsell Foundation
711 Navarro Street, Suite 537
San Antonio, TX 78205

Expect Miracles Foundation Grants 1619
The MMLC Funding Award is given on an annual basis on behalf of the Miracle Maker Leadership Council to cancer fighting causes and patient care programs where such support is not always readily available. The proposed program should have the ability to materially impact an organization or a specific program of a qualified organization.
Requirements: Funding Award recipients must have 501(c)3 status as defined by the IRS. Applicants must provide service in at least one of the following areas: cancer patient care support programs and initiatives; cancer awareness and educational programs; or cancer prevention/health projects. Priority will be given to initiatives demonstrating sustainable benefits to the communities where the Foundation events are held and where Foundation supporters and the Miracle Maker Leadership Council (MMLC) members live.
Restrictions: Although there are no specific guidelines on the size, annual budget, or annual fundraising of a recipient organization, the program is not designed to assist in funding large institutional/national/society entities. The goal of the MMLC is to select programs and causes where funding is not always readily available. MMLC guidelines prohibit funding for the following: private pursuits; political parties, associations and representatives of advocacy groups; organizations that discriminate by race, creed, gender, sexual orientation, age, religion, or national origin; religious causes; advertising, promotion, or sponsorship; and donation to an individual or team fundraising initiative or program.

Geographic Focus: California, Connecticut, District of Columbia, Illinois, Maine, Maryland, Massachusetts, New Hampshire, New Jersey, New York, Pennsylvania, Rhode Island, Vermont
Date(s) Application is Due: Mar 5
Amount of Grant: Up to 27,000 USD
Contact: Alana Chin, (617) 391-9235 or (617) 827-7463; achin@expectmiraclesfoundation.org
Internet: http://www.expectmiraclesfoundation.org/
Sponsor: Expect Miracles Foundation
6 Quail Run
Hingham, MA 02043

Express Scripts Foundation Grants 1620
The Foundation provides financial and volunteer support for numerous causes. Requests are considered from organizations that advance medical and health-related causes, particularly those intended to support youth and strengthen families. Educational activities that enrich education and help create tomorrow's leaders are of particular interest. Programs designed to provide support that enables at-risk children to complete their educations are considered. The Foundation will consider requests from organizations that fall outside the established criteria if they enhance the communities in which the Foundation operates and make strong appeals for community support. Interested applicants should complete an expression of interest form provided online.
Requirements: All projects or programs requesting funding must be tax-exempt as described in both sections 501(c)3 and 509(a)1, 509(a)3 of the Internal Revenue Code. Organizations and programs must have broad community support and a record of fiscal and administrative stability. Preference is given to innovative projects that develop or expand services, or improve the quality of an organization's program. Typically, support will be awarded only for a term of one year; organizations must reapply for any subsequent year in which they are seeking funding.
Restrictions: The Foundation does not fund: organizations that discriminate on the basis of race, color, sex, or national origin; political causes, candidates, organizations or campaigns; fundraising activities such as benefits, charitable dinners, galas, social clubs; athletic teams or events; individual needs, including scholarships, sponsorships and other forms of financial aid; and endowment funds or capital-fund drives.
Geographic Focus: All States
Contact: Grants Administrator
Internet: http://www.express-scripts.com/ourcompany/aboutus/foundation
Sponsor: Express Scripts Foundation
13900 Riverport Drive, Suite 20S
Maryland Heights, MO 63043

ExxonMobil Foundation Malaria Initiative Grants 1621
As a major employer and investor in Africa, ExxonMobil is committed to working with partners — local institutions, international NGOs and governments — to stop the spread of malaria on the continent. Since 2000, the corporation's funding has reached nearly 83 million people through support exceeding $110 million. Funding supports partners who help improve the delivery and use of prevention tools such as bed nets; train healthcare workers to ensure malaria interventions are reaching those who need them most; provide technical assistance to help countries increase their capacity to control malaria; and facilitate the monitoring and promotion of progress through integrated communications programs.
Requirements: Grants are made to tax-exempt organizations within principal company-operating areas.
Restrictions: Grants are not made to individuals for scholarships, fellowships, research, or travel. Organizations that are primarily religious in nature are ineligible.
Geographic Focus: All States, United Kingdom
Amount of Grant: Up to 100,000,000 USD
Contact: Suzanne McCarron, President; (972) 444-1100 or (972) 444-1007; fax (972) 444-1405; contributions@exxonmobil.com
Internet: http://www.exxonmobil.com/Corporate/community_malaria_initiative.aspx
Sponsor: ExxonMobil Foundation
5959 Las Colinas Boulevard
Irving, TX 75039-2298

Eye-Bank for Sight Restoration and Fight for Sight Summer Student Research Fellowship 1622
Beginning in 2011, The Eye-Bank for Sight Restoration and Fight for Sight co-sponsored a Summer Student Research Fellowship. Annual grants of $2,100 will support full-time research conducted from June to August. The goal of this award is to advance the skills needed to initiate and carry out work in a scientific environment.
Requirements: Applicants must be currently enrolled undergraduates, medical students or graduate students studying in the United States who wish to explore ophthalmology or eye research as a career. Students are expected to complete a short, independent project during the summer months under the guidance of a senior scientist or clinician. Candidates are invited to apply by visiting www.fightforsight.org for an application.
Restrictions: Lab employees are not eligible for this fellowship.
Geographic Focus: All States
Date(s) Application is Due: Feb 1
Amount of Grant: 2,100 - 2,100 USD
Contact: Janice Benson; (212) 679-6060; fax (212) 679-4466; janice@fightforsight.org
Internet: http://www.eyedonation.org/scholarship_program.html#fight_sight
Sponsor: Eye-Bank for Sight Restoration
120 Wall Street
New York, NY 10005-3902

Ezra M. Cutting Trust Grants 1623
The Ezra M. Cutting Trust was established in 1965 under the will of Ezra Cutting, who was born in Marlborough, Massachusetts. The Trust supports charitable organizations serving residents of Marlborough. Areas of special interest include agencies serving youth as well as those with programs fostering economic growth and the general quality of life in the City of Marlborough. Applicants must apply online at the grant website. Applicants are strongly encouraged to do the following before applying: review the downloadable state application procedures for additional helpful information and clarifications; review the downloadable online-application guidelines at the grant website; review the trust's funding history (link is available from the grant website); review the online application questions in advance; and review the list of required attachments. These will generally include: a list of board members, financial statements (audited, reviewed, or compiled by independent auditor); an organization summary; a list of other funding sources; an IRS Determination letter; and other required documents. All attachments must be uploaded in the online application as PDF, Word, or Excel files. The Ezra M. Cutting Trust's deadline is 11:59 p.m. on January 15. Applicants will be notified of decisions by mid-March.
Requirements: Applicants must have 501(c)3 tax-exempt status.
Restrictions: Grants are restricted to organizations located in Marlborough, Massachusetts (and to other organizations which serve Marlborough residents) and are generally made for special projects and programs, with a limited number of capital gifts made each year. The trust does not support requests from individuals, organizations attempting to influence policy through direct lobbying, or any political campaigns.
Geographic Focus: Massachusetts
Date(s) Application is Due: Jan 15
Amount of Grant: 2,500 - 10,000 USD
Samples: Friends of the Marlborough Public Library, Marlborough, Massachusetts, $5,000; Young Entrepreneurs Alliance, Roxbury, Massachusetts, $5,000; Employment Options, Marlborough, Massachusetts, $2,500.
Contact: Michealle Larkins; (866) 778-6859; michealle.larkins@baml.com
Internet: https://www.bankofamerica.com/philanthropic/fn_search.action
Sponsor: Ezra M. Cutting Trust
225 Franklin Street, 4th Floor, MA1-225-04-02
Boston, MA 02110

F.M. Kirby Foundation Grants 1624
Foundation grants are made to a wide range of nonprofit organizations in education, health and medicine, the arts and humanities, civic and public affairs, as well as religious, welfare and youth organizations. Grantees are largely in geographic areas of particular interest to the Kirby family. The Foundation has no required application format and applications will be accepted throughout the year. No solicitations by fax or email are accepted.
Requirements: North Carolina, New Jersey, and Pennsylvania tax-exempt organizations are eligible.
Restrictions: Grants are not made to individuals, public foundations or to underwrite fund-raising activities such as benefits, dinners, theater or sporting events.
Geographic Focus: New Jersey, North Carolina, Pennsylvania
Date(s) Application is Due: Oct 31
Samples: Wake Forest U (Winston-Salem, NC)--to construct a wing at the Wayne Calloway School of Business and Accountancy, $5 million; Children's Hospital of Boston, Division of Neuroscience (Boston, MA)--for research on cerebral palsy, epilepsy, Alzheimer's disease, and other diseases and degenerative disorders, $2 million; Carolina Theatre (Durham, NC)--for general operating support, $40,000.
Contact: S. Dillard Kirby, Executive Director; (973) 538-4800
Internet: http://www.fdncenter.org/grantmaker/kirby
Sponsor: F.M. Kirby Foundation
17 DeHart Street, P.O. Box 151
Morristown, NJ 07963-0151

Fairfield County Community Foundation Grants 1625
As a community foundation, FCCF makes discretionary grants to nonprofits in the broad program areas of economic opportunity (including affordable housing, neighborhood development, and workforce development); children, youth and families; health and human services; the environment; arts and culture; and nonprofit organizational effectiveness. The Foundation is particularly interested in proposals focused on: economic opportunity; education and youth development; advancing school readiness in Fairfield County; organizational effectiveness; and regionalism. Applicants must first submit a letter of inquiry. Application information is available online.
Requirements: Nonprofit organizations in Fairfield County, CT, are eligible.
Restrictions: Funds are not used to provide support for religious or political purposes, deficit financing, annual appeals, fundraising events, open space purchases, for-profit, parochial, charter or private schools, or nonprofit endowments. Grants are not given to individuals.
Geographic Focus: Connecticut
Samples: Cardinal Shehan Center, Bridgeport, CT, $20,000--to support the Made for Talent Program that offers youth arts education throughout the year; ACORN Housing Corporation, Bridgeport, CT, $30,000--to support ACORN's foreclosure counseling program in Bridgeport that prevents mortgage foreclosures as well as assists homeowners impacted by the subprime loan crisis; Danbury Youth Services, Danbury, CT, $20,000--to support the Target After School Program that offers after-school education to children who live in Danbury public housing.
Contact: Karen Brown; (203) 750-3200; fax (203) 750-3232; kbrown@fccfoundation.org
Internet: http://www.fccfoundation.org/cm/grantseekers/what_we_fund.html
Sponsor: Fairfield County Community Foundation
383 Main Avenue
Norwalk, CT 06851-1543

Fairlawn Foundation Grants 1626

The Foundation is dedicated to supporting excellence and innovation in health care delivery, education and research in the Worcester area. The Foundation was established to: provide financial support to organizations, both public and private, that have specific, well-defined proposals to improve and/or expand the practice and delivery of medical and allied health care services in the Worcester area; and to support the education of Worcester area residents in health care related areas.

Requirements: Funding preference is given to institutions involved in higher education, medical or scientific research, hospitals and health care providers, and other organizations capable of making a long-term impact on health care through innovation or research. Applicants must be tax-exempt, nonprofit organizations as recognized by IRS code 501(c)3.

Restrictions: Except in unusual circumstances, the Foundation generally will not approve grants: to any program or individual in excess of $120,000; to programs that require more than five years to complete; for projects supported by other funds in which the Fairlawn Foundation contribution would comprise less than half of total funding; or for pre-existing projects as a minor participant.

Geographic Focus: Massachusetts
Date(s) Application is Due: Mar 15; Sep 15
Amount of Grant: Up to 120,000 USD
Contact: Lois Smith; (508) 755-0980, ext. 107; lsmith@greaterworcester.org
Internet: http://www.greaterworcester.org/grants/Fairlawn.htm
Sponsor: Greater Worcester Community Foundation
370 Main Street, Suite 650
Worcester, MA 01608-1738

Fallon OrNda Community Health Fund Grants 1627

The purpose of the health fund is to advance projects that increase access to health care or health promotion services that improve the health status of vulnerable populations. Of particular interest are projects that result in: the support or creation of primary care outreach services to vulnerable populations; the development of continuing managed care services rather than episodic or uncoordinated care; and the removal of barriers that prevent people from receiving services such as lack of transportation, cultural competency of providers, language differences, or others. Grants may be for operational expenditures such as personnel costs, or for construction, renovation, equipment purchase or other physical improvements. Past grants have ranged from $6,500 to $30,000.

Requirements: Funding will not be provided for long-term underwriting of operational costs for any one program.

Geographic Focus: Massachusetts
Date(s) Application is Due: Feb 15
Amount of Grant: 6,500 - 30,000 USD
Contact: Lois Smith; (508) 755-0980, ext. 107; lsmith@greaterworcester.org
Internet: http://www.greaterworcester.org/grants/Fallon.htm
Sponsor: Greater Worcester Community Foundation
370 Main Street, Suite 650
Worcester, MA 01608-1738

FAMRI Clinical Innovator Awards 1628

The purpose of the Clinical Innovator Awards (CIA) from the Flight Attendant Medical Research Institute (FAMRI) is to stimulate novel medical and clinical scientific research studies on the topic of second hand tobacco smoke. Of high priority to FAMRI are new technologies to detect second hand tobacco-caused illnesses and novel therapies to treat these diseases. FAMRI is interested in clinical and translational research, from basic molecular mechanisms of disease to occupational health. These areas include: studies on the etiology and pathogenesis of tobacco-caused respiratory diseases in humans; therapeutic interventions; clinical trials; and research to establish a protocol for inquiry regarding second hand tobacco smoke exposure by all health care professionals. FAMRI plans to provide awards of up to $100,000 per year for direct costs plus no more than 8.5% for indirect costs, subject to the availability of funds (FAMRI cannot guarantee annual and total levels of support). Awards usually will be made for three years with a possibility of a no-cost extension for one year upon request. Applicants must first submit an online letter of intent by the due date. Upon invitation, they may then submit online proposals. Instructions and links for submitting the letter of intent and proposal are available under "Current and Upcoming Awards" at the FAMRI website. Deadlines may vary from year to year. To be kept up to date of all current and future funding opportunities, interested applicants are encouraged to sign up for the FAMRI ListServ (click on "listserv signup" under the researchers tab at the FAMRI website).

Requirements: Applicants must be certain that their institution can agree to the terms of the FAMRI grant agreement. A copy of the agreement is available for review under "Current and Upcoming Awards" at the FAMRI website. Grantees must devote a minimum of 50% of professional time to research during the period of this award; at least 15% of the grantee's time (or a total of 15% of the combined times of the co-grantees) should be devoted to the research project described in the application. Researchers with an advanced degree (MD, PhD, or MD/PhD, or the equivalent) are eligible to apply. Individuals who apply to FAMRI for funding need not be U.S. citizens. They normally apply through and with an institution or organization that is qualified under United States tax laws to receive grants from FAMRI, a US-based private foundation. The applying institution receives and administers FAMRI's funds and supervises the work of the Principal Investigator. If the applying institution is organized under the laws of a country other than the United States, it must provide FAMRI any and all documentation that FAMRI may require, in form and content acceptable to FAMRI, to ensure that any grant from FAMRI shall be a "qualified distribution" under Internal Revenue Service laws and regulations, including but not limited to affiliation with a qualified US-based institution, a current affidavit from the applicant or an opinion of qualified legal counsel stating the facts and circumstances of the applicant's organization and activities and certifying that any grant from FAMRI shall be a "qualified distribution." Research projects that involve human subjects must follow Public Health Service guidelines (or its equivalent in non-U.S. institutions) and must have the approval of the appropriate Institutional Review Boards (IRB) or foreign equivalent prior to the start of the research. The IRB (or foreign equivalent) approval must be maintained and be in effect throughout the research project. All research must be performed in accordance with all relevant institutional and federal policies and guidelines (or foreign equivalent) relating to human subject protection, radiation and environmental health and safety. All FAMRI grantees are encouraged to collaborate with fellow grantees and may be required to participate periodically in FAMRI's collaborative efforts.

Restrictions: The Clinical Innovator Award must not exceed $100,000 per year plus 8.5 % indirect costs for a total of $108,500 per year. FAMRI will only support animal research that utilizes mice or rats. Research involving other animals will not be considered unless the animals will not be sacrificed as a result of the research. Pending Association for Assessment of Laboratory Animal Care (AALAC) approval must be submitted prior to funding of a grant involving animal subjects. To avoid all conflicts, any principal investigator or other participating applicant who is currently accepting tobacco industry funding, has recently applied for tobacco funding, or has any other association or consulting position with a tobacco company or affiliate of the tobacco industry is not eligible for FAMRI funding. All prior funding and/or association with tobacco companies must be disclosed and will be considered on a case-by-case basis. An award will not be made or will be withdrawn if the applicant becomes the Principal Investigator on a peer-reviewed grant with the same general aims as those proposed in the application to FAMRI. FAMRI reserves the right to announce the award publicly in a manner and at a time of its choosing.

Geographic Focus: All States, All Countries
Date(s) Application is Due: Sep 26
Amount of Grant: Up to 325,500 USD
Samples: Diego A. Preciado, MD, PhD, Children's National Medical Center, Washington, D.C. - Tobacco Smoke Exposure Effects on Sinonasal Mucosa; Noam A. Cohen, MD, PhD, Philadelphia Research & Education Foundation, Philadelphia, Pennsylvania - Induction of Bacterial Biofilms by Tobacco Smoke; Jean Kim, MD, PhD, Johns Hopkins University, Baltimore, Maryland - Nasal Epithelial Growth Dysfunction by Second Hand Smoke in Chronic Sinusitis
Contact: Elizabeth Kress; (305) 379-7007; fax (305) 577-0005; ekress@famri.org
Internet: http://www.famri.org/current_awards/index.php
Sponsor: Flight Attendant Medical Research Institute
201 S Biscayne Boulevard, Suite 1310
Miami, FL 33131

FAMRI Young Clinical Scientist Awards 1629

The Young Clinical Scientist Award (YSCA) is intended to prepare and support new clinical investigators with an MD or PhD as they begin their careers as independent researchers in the field of smoking-related disorders. Awards are offered to two groups of scientists: those individuals who are currently faculty members, or who expect to be before the start date of the award (YCSA faculty), and those individuals who do not expect to become faculty until after the start date of the award (YCSA fellow). Applicants must initially submit an online letter of intent (LOI) by 5 p.m. on the LOI due date, and upon subsequent invitation submit a full proposal by 5 p.m. on the proposal due date. Instructions and links for submitting the letter of intent and proposal are available under "Current and Upcoming Awards" at the FAMRI website. YCSA awards can be extended to a total of no more than five years. Renewal forms will be sent to current grantees in early February of the appropriate year and will be due late March. Grantees who are fellows should notify both the Flight Attendant Medical Research Institute (FAMRI) and the American Institute of Biological Sciences (AIBS) of any change in status. Grantees are encouraged to call or email the FAMRI Administrative Assistant to obtain current AIBS contact information. Deadlines may vary from year to year. To be kept up to date of all current and future funding opportunities, interested applicants are encouraged to sign up for the FAMRI ListServ (click on "listserv signup" under the researchers tab at the FAMRI website).

Requirements: Applicants must be certain that their institution can agree to the terms of the FAMRI grant agreement. A copy of the agreement is available for review under "Current and Upcoming Awards" at the FAMRI website. Each individual other than the institutional official must submit a tobacco disclosure form. A tobacco disclosure form is also required from the applicant's institution. Tobacco disclosure forms may be downloaded from the FAMRI website and after they are signed, they may be scanned and uploaded. Researchers with an advanced degree (M.D., Ph.D., or M.D./Ph.D., or the equivalent) are eligible to apply. A YCSA grantee must devote a minimum of 75% of professional time to research during the period of this award and at least 50% of the grantee's time should be devoted to the research project described in the application. Individuals who apply to FAMRI for funding need not be U.S. citizens. They normally apply through and with an institution or organization that is qualified under United States tax laws to receive grants from FAMRI, a US-based private foundation. The applying institution receives and administers FAMRI's funds and supervises the work of the Principal Investigator. If the applying institution is organized under the laws of a country other than the United States, it must provide FAMRI any and all documentation that FAMRI may require, in form and content acceptable to FAMRI, to ensure that any grant from FAMRI shall be a "qualified distribution" under Internal Revenue Service laws and regulations, including but not limited to affiliation with a qualified US-based institution, a current affidavit from the applicant or an opinion of qualified legal counsel stating the facts and circumstances of the applicant's organization and activities and certifying that any grant from FAMRI shall be a "qualified distribution." Research projects that involve human subjects must follow Public Health Service guidelines (or

its equivalent in non-U.S. institutions) and must have the approval of the appropriate Institutional Review Boards (IRB) or foreign equivalent prior to the start of the research. The IRB (or foreign equivalent) approval must be maintained and be in effect throughout the research project. All research must be performed in accordance with all relevant institutional and federal policies and guidelines (or foreign equivalent) relating to human subject protection, radiation and environmental health and safety. All FAMRI grantees are encouraged to collaborate with fellow grantees and may be required to participate periodically in FAMRI's collaborative efforts.
Restrictions: The program is limited to the development of young researchers in smoking-related disorders. The YCSA/fellow award must not exceed $75,000 per year plus 8.5 % indirect costs for a total of $81,375 per year. The YCSA/faculty award must not exceed $100,000 per year plus 8.5 % indirect costs for a total of $108,500 per year. FAMRI will only support animal research that utilizes mice or rats. Research involving other animals will not be considered unless the animals will not be sacrificed as a result of the research. Pending Association for Assessment of Laboratory Animal Care (AALAC) approval must be submitted prior to funding of a grant involving animal subjects. To avoid all conflicts, any principal investigator or other participating applicant who is currently accepting tobacco industry funding, has recently applied for tobacco funding, or has any other association or consulting position with a tobacco company or affiliate of the tobacco industry is not eligible for FAMRI funding. All prior funding and/or association with tobacco companies must be disclosed and will be considered on a case-by-case basis. An award will not be made or will be withdrawn if the applicant becomes the Principal Investigator on a peer-reviewed grant with the same general aims as those proposed in the application to FAMRI. FAMRI reserves the right to announce the award publicly in a manner and at a time of its choosing.
Geographic Focus: All States, All Countries
Date(s) Application is Due: Sep 26
Amount of Grant: 162,750 - 542,500 USD
Contact: Elizabeth Kress; (305) 379-7007; fax (305) 577-0005; ekress@famri.org
Internet: http://www.famri.org/current_awards/index.php
Sponsor: Flight Attendant Medical Research Institute
201 S Biscayne Boulevard, Suite 1310
Miami, FL 33131

Fannie E. Rippel Foundation Grants 1630
The foundation aids, assists, funds, equips, and provides maintenance for corporations, institutions, associations, organizations, or societies maintained for the relief and care of aged women; provides funds for the building, equipping, and maintenance of hospitals; and provides funds for corporations, institutions, and other organizations existing for treatment of and/or research on heart disease or cancer. The foundation gives emphasis to the equipment and programmatic needs of major teaching medical centers and local rural hospitals, particularly where opportunities exist for leveraging the expertise or capabilities of the medical centers/rural hospitals. Programs should reach underserved rural and urban groups, advocate preventive care, present strategies to change behaviors of the people served, and promote humanistic medicine and mind-body-spirit connections in the healing process. Preference also is given to proposed projects where the benefits can be leveraged through challenge grants.
Requirements: Organizations, associations, institutions, and hospitals in the Northeast are eligible.
Restrictions: Grants are not awarded to individuals.
Geographic Focus: Connecticut, Maine, Massachusetts, New Hampshire, Rhode Island, Vermont
Amount of Grant: 50,000 - 300,000 USD
Contact: Leigh Scherrer, Foundation Associate; (973) 540-0101, ext. 305; fax (973) 540-0404; lscherrer@rippelfoundation.org
Internet: http://rippelfoundation.org/
Sponsor: Fannie E. Rippel Foundation
14 Maple Avenue, Suite 200
Morristown, NJ 07960

FAR Fund Grants 1631
The FAR Fund Project is a New Orleans based program exploring Hurricane Katrina's effects on New Orleans therapists and therapeutic practice. It was designed by and for clinicians. The project's mission is two fold: to offer support and concrete help to local clinicians, and to develop a model for better understanding how shared trauma affects therapists and therapy. Through this project, the FAR Fund wishes to unite and revitalize clinician communities following large-scale disasters wherever they occur, starting in New Orleans.
Requirements: The FAR Fund Project is open to New Orleans psychotherapists of all disciplines and theoretical orientations. Their mission is two-fold: to offer support and concrete help to local mental health clinicians, and to develop a psychodynamic model to better understand how shared trauma affects therapists and therapy.
Geographic Focus: Louisiana
Contact: Shirlee Taylor; (212) 982-8400; fax (212) 982-8477; FARFund@mac.com
Sponsor: Far Fund
928 Broadway, Suite 902
New York, NY 10010

Fargo-Moorhead Area Foundation Grants 1632
The Foundation awards grants each year according to current and emerging community needs. The Board also prioritizes grants that: effect a broad segment of our community; leverage support from other sources; promote collaboration, without duplicating services; strengthen organization, self-sufficiency and long term stability; focus on problem solving; and show realistic planning and management. The Foundation's focus is on arts and culture, education, environment and animals, health, human services and public/society benefit. Applicants are required to use the FMAF grant application forms available online.
Requirements: The foundation welcomes grant requests from 501(c)3 organizations in Cass and Clay Counties in North Dakota and Minnesota or those serving residents of these counties.
Restrictions: Grants are not normally funded for: annual appeals or membership drives; capital debt reduction; capital campaigns; individuals or for-profit organizations; organizations with outstanding final reports from previous FMAF grants; organizations with outstanding due diligence reports from FMAF; political projects; religious groups for religious purposes; travel for groups and retroactive funding for any project expenses incurred before the decision date.
Geographic Focus: Minnesota, North Dakota
Date(s) Application is Due: Apr 20; Jun 1
Amount of Grant: 1,000 - 10,000 USD
Samples: Churches United for the Homeless, $10,000--intern/resident pilot program; Fargo Theatre, $7,000--lighting and sound system upgrades/replacements; Community of Care, $5,000--One Stop Service Center.
Contact: Cher Hersrud, Advancement Officer; (701) 234-0756; fax (701) 234-9724; cher@areafoundation.org
Internet: http://www.areafoundation.org/page5345.cfm
Sponsor: Fargo-Moorhead Area Foundation
502 First Avenue N, Suite 202
Fargo, ND 58102

Fargo-Moorhead Area Foundation Woman's Fund Grants 1633
The Women's Fund, of the Fargo-Moorhead (FM) Area Foundation, is an endowment that will fund projects and programs that will improve the lives of women and girls in the FM area. The Fund's top three funding priorities are: children in need of care--i.e, significant percent of high risk behavior in high school girls, 13,000 children in need of after school care (Cass and Clay); 11,000 children between 0 and 5 needing child care (Cass and Clay); economic & physical well-being of women-- i.e, the poverty rate and homelessness of women and households of lead by women, abuse stats; women in leadership to positively impact civic and business policy-- i.e., exec roles in business, representation on boards, etc. lack of women in elected positions. The Women's Fund generally gives priority to projects that: seek to meet identified new or emerging women's or girl's issues identified by research; leverage additional support from other sources; promote cooperation among organizations without duplicating services; use a preventative approach to solving problems; show evidence of realistic organizational planning and management; will be completed within a year's time.
Requirements: The foundation welcomes grant requests from 501(c)3 tax-exempt organizations in Cass County, North Dakota or Clay County, Minnesota. Once an announcement has been made indicating that grant applications are being accepted, the Women's Fund staff will have grant applications packets available to be picked up at the Women's Fund office, accessed electronically at www.areafoundation.org, or can be emailed upon request.
Restrictions: The Women's Fund does not make grants to individuals & ordinarily doe's not support: ongoing operation expenses; annual appeals in membership drives and capital campaigns; religious groups for religious purposes; capital debt reduction; political projects; travel for groups; organizations with outstanding reports or requests; organizations outside Cass County in North Dakota and Clay County in Minnesota. Organizations are not encouraged to submit more than one request, but for those that choose to do so, they must prioritize. Grants in successive years are also discouraged. A narrative and fiscal report will be required upon completion of the project.
Geographic Focus: Minnesota, North Dakota
Amount of Grant: 300 - 2,500 USD
Contact: Cher Hersrud; (701) 234-0756; fax (701) 234-9724; cher@areafoundation.org
Internet: http://www.areafoundation.org/page24016.cfm
Sponsor: Fargo-Moorhead Area Foundation
502 First Avenue N, Suite 202
Fargo, ND 58102

Farmers Insurance Corporate Giving Grants 1634
The corporate community relations program awards grants in the areas of education, public safety, arts and culture, civic improvement, and health and human services. Education giving focuses on literacy programs, mentoring programs, adopt-a-school programs, employee matching grants, and aid-to-education undergraduate scholarships. Public safety awards support tougher laws against drunk driving, drug- and alcohol-free graduation night parties, neighborhood crime prevention, highway safety, and earthquake relief. Arts and culture funding supports children's programs and public television. Civic improvement focuses on recognizing exemplary youth, voter registration drives, adopt-a-highway programs, and community paint-a-thons. Health and human services giving supports March of Dimes, United Way, aid for families with cancer, aid to migrant farmworkers, and feeding the hungry. There are no application deadlines. Requests for contributions should be in the form of a letter outlining the purpose of the organization or program. The letter also should include the amount requested, its intended use, and a description of how Farmers' support will be recognized. Additional information should include a budget, annual report, proof of tax-exempt status, and a roster of the board of directors.
Requirements: 501(c)3 tax-exempt organizations are eligible.
Restrictions: Farmers does not make charitable contributions to individuals, political candidates, or religious groups or for sports events, advertising or raffle tickets, construction projects, or international programs.
Geographic Focus: All States
Contact: Doris Dunn, Director of Community Relations; (888) 327-6335
Internet: http://www.farmers.com/corporate_giving.html
Sponsor: Farmers Insurance Group of Companies
4680 Wilshire Boulevard
Los Angeles, CA 90010

Faye McBeath Foundation Grants 1635

The Faye McBeath Foundation is a private, independent foundation providing grants to tax-exempt nonprofit 501(c)3 organizations in the metropolitan Milwaukee, Wisconsin area, including Milwaukee, Waukesha, Ozaukee and Washington counties. The major areas of interest are: children; aging and elders; health; health education; civic and governmental affairs. The Foundation's Trustees are primarily interested in promising or established programs, operated by well-managed organizations. Benchmarks that will be applied to projects and organizations in all five interest areas are the following: strength of proposal - Foundation guidelines match; quality and creativity in project design, implementation and evaluation; quality of the applicant nonprofit's leadership and organizational capacity; potential to stretch or leverage McBeath funds; specific interests of Trustees and/or staff. The majority of grants support specific programs. On occasion, operating and capacity-building grants are awarded. Capital grants are limited to projects with community-wide impact that reflect the program interests of the Foundation. Any invited requests for operating or capacity-building support must include an organization's results-oriented annual plan, covering both programs and management. The trustees meet four times each year to consider grant proposals. Written notification of the Trustees' decision regarding a grant proposal will be sent within ten days of the meeting. Organizations awarded grants will receive written notification of grant conditions, payment dates and reporting requirements
Requirements: 501(c)3 nonprofit organizations in the metropolitan Milwaukee area are eligible. To begin the application process, submittal of a letter of intent, 1-2 pages long that includes: a brief description of the applicant organization, the program to be funded, the corresponding match between the Foundation's guidelines and the request, the amount requested and the estimated total project budget. Include the name and address of the organization, and the name, phone number and email address of a contact person. There are no deadlines for letters of intent. Upon receipt of a letter of intent, Foundation staff will formally acknowledge by email or letter. If the request is clearly outside the current program or geographical focus of the Foundation, a decline letter will be sent informing you of this determination. If you are formally invited by McBeath staff to submit a full proposal, you will be asked to use the Wisconsin Common Grant Application Form. The Foundation requires that a volunteer serving on the Board sign the application.
Restrictions: The Foundation does not award grants for annual fund drives, scholarships, support of individuals, sponsorship of fundraising events, or emergency funds. The Foundation does not consider or acknowledge general solicitation letters. Basic health sciences research is not funded by the Foundation. Single disease or condition organizations are typically not funded by the Foundation. An exception may be made for a creative project involving one or more of the following: service to multiple populations, collaboration and/or creation of a replicable model. There are no published deadlines for letters of intent or grant proposals. Proposal deadlines are established based upon discussion with Foundation staff. Organizations serving the disabled will be considered only if the project targets children or older adults. Any discussion of the availability of funding in a subsequent year should be initiated by the grantee and discussed with staff at least 60 days in advance of a project anniversary.
Geographic Focus: Wisconsin
Amount of Grant: 1,000 - 50,000 USD
Samples: Aurora Sinai Medical Center, Milwaukee, WI, 45,000--for improvement of access to Health Care Services; Greater Milwaukee Foundation, Milwaukee, WI, $25,000--for the Camps For Kids Forever Fund; NAMI Waukesha, Waukesha, WI, $10,000--for Housing Support and Advocacy Program.
Contact: Scott Gelzer; (414) 272-2626; fax (414) 272-6235; info@fayemcbeath.org
Internet: http://www.fayemcbeath.org/ProgramPriorities/index.html
Sponsor: Faye McBeath Foundation
101 West Pleasant Street, Suite 210
Milwaukee, WI 53212

Fayette County Foundation Grants 1636

The Fayette County Foundation is the community's resource for charitable giving. The Foundation serves the entire Fayette County by assisting donors and meeting community needs. To be eligible to receive funding from the Fayette County Foundation a letter of intent must be submitted by March 1, July 1 or October 1 and followed by a completed grant application by the appropriate grant deadline. All grant seekers must have prior governing board approval for the project seeking funding. The approval of signed minutes and a signed letter from governing board must be available to the Fayette County Foundation. Faith-based organizations may apply as long as the project does not mandate participation in a religious activity as a condition for receiving services. Samples of previously funded projects are available at the website.
Requirements: Projects must have public access for all Fayette County citizens. Completed grant application (original plus eight copies) must be received by the Foundation by the last Friday in March, the last Friday in July or the last Friday in October at Noon to be considered for funding.
Restrictions: Legal requirements forbid staff, trustees, directors, committee members and their families from profiting financially from any philanthropic grant. All persons actively connected with the Foundation will consistently strive to avoid self-interest in the processing and disposition of grant request. Only one successful application per 12 months will be considered. Repeat funding for the same project may be considered five years after the initial project on a case by case basis.
Geographic Focus: Indiana
Date(s) Application is Due: Mar 1; Jul 1; Oct 1
Contact: Loree Crowe; (765) 827-9966; fax (765) 827-5836; info@fayettefoundation.com
Internet: http://www.fayettefoundation.com/default.asp?Page=Grant+Policy&PageIndex=128
Sponsor: Fayette County Foundation
521 N Central Avenue, Suite A, P.O. Box 844
Connersville, IN 47331

FCD New American Children Grants 1637

The foundation awards grants to support programs for children, particularly the disadvantaged, and promote their well-being through basic and policy-relevant research about the factors that promote optimal development of children and adolescents; policy analysis, advocacy, services, and public education to enhance the discussion and adoption of social policies that support families in their important child-raising responsibilities; and leadership development activities linked to the programmatic focus of the foundation. Grants focus on the integration of research, policy, and advocacy in two areas: the availability of and access to early childhood education programs and health care for children. Most grants support research, but a small number of direct service grants are made for New York City-based projects that advance the foundation's research and policy analysis efforts. There are no application deadlines; submit a brief letter of inquiry. Full proposals are by invitation only, following a strict pre-proposal process.
Requirements: Nonprofit organizations are eligible.
Restrictions: The foundation does not consider requests for scholarships or support for individuals, capital campaigns, building purchase or renovation, or equipment purchase. The foundation does not make grants outside the United States.
Geographic Focus: All States
Amount of Grant: Up to 800,000 USD
Samples: New America Foundation, Washington, D.C., $770,000 - continued support of the PreK-3rd Education Reform Initiative and the Federal Education Budget Project (2014); Rutgers University Foundation, New Brunswick, New Jersey, $30,000 - support of the Burns Family Endowment for Teacher Leadership in Early Childhood Education (2014); University of Texas at Austin, Austin, Texas, $349,561 - continued support of the Dual-Generation Strategy Initiative Project (2014).
Contact: Dorothy Pflager, Grants Manager; (212) 867-5777; fax (212) 867-5844; dorothy@fcd-us.org or info@fcd-us.org
Internet: http://fcd-us.org/grants
Sponsor: Foundation for Child Development
295 Madison Avenue, 40th Floor
New York, NY 10017

FDHN Bridging Grants 1638

Awards of $50,000 each are available to assist investigators were not awarded their first R01 or the first renewal of their R01 to continue their gastroenterology related research. Two awards will be made following each of the three annual NIH review cycles and availability of Summary Statements. The primary objective of the award is to provide interim support to AGA Members who have submitted grants to NIH that were approved on the basis of scientific merit, but received priority scores out of the funding range.
Requirements: Applicants must: be AGA members who are applying for their first R01 or their second R01 either as a competitive renewal or a new R01 (this includes investigators going from a K award to their first R01); hold an MD, PhD, or equivalent degree (e.g., MB, ChB, MBBS, DO); hold full-time faculty positions at North American universities or professional institutes at the time of application; and have submitted a proposal to NIH that has undergone the peer review process and was either approved and not funded or received high commendation but not funded.
Restrictions: Applicants cannot receive more than $75,000 of any additional extramural research support. Up to 25% of total amount the award may be used for salary support of the Principal Investigator. Indirect costs, including travel are not allowed.
Geographic Focus: All States
Date(s) Application is Due: Feb 14; Jun 14; Oct 14
Amount of Grant: 50,000 USD
Contact: Awards Manager; (301) 222-4012; fax (301) 652-3890; awards@fdhn.org
Internet: http://www.fdhn.org/wmspage.cfm?parm1=119
Sponsor: Foundation for Digestive Health and Nutrition
4930 Del Ray Avenue
Bethesda, MD 20814

FDHN Centocor International Research Fellowship in Gastrointestinal Inflammation & Immunology 1639

This award provides $50,000 to enable promising young investigators from outside the United States to spend a year at a U.S. institution engaged full-time in research related to the fundamental processes in gastrointestinal inflammation and immunology. The primary objectives of the award are: to provide the opportunity for young investigators from outside the U.S. to participate in basic research on inflammatory disease processes at prominent institutions in the U.S. (the award provides salary support for fulltime research); to initiate future international collaborative research efforts in inflammatory digestive diseases; and to address an important question in such a way as to provide a meaningful answer.
Requirements: Candidates must hold an MD or equivalent degree (eg, MB. ChB, MBBS). Applicants should currently be on the faculty of an academic institution outside the U.S. Applicants must be AGA members or be eligible for and submit an application for membership.
Restrictions: At the time of application, candidates are not allowed to hold any other similar research grant. These awards may not be renewed. No more than one application will be accepted from any institution in a given year.
Geographic Focus: All States
Date(s) Application is Due: Jan 14
Amount of Grant: 50,000 USD
Contact: Awards Manager; (301) 222-4012; fax (301) 652-3890; awards@fdhn.org
Internet: http://www.fdhn.org/wmspage.cfm?parm1=105
Sponsor: Foundation for Digestive Health and Nutrition
4930 Del Ray Avenue
Bethesda, MD 20814

FDHN Designated Outcomes Award in Geriatric Gastroenterology   1640

Two awards in the amount of $35,000 for one year are available to support investigator-initiated outcomes research in geriatric gastroenterology. In general, outcomes studies examine clinical outcomes, patient satisfaction, quality of life, economic evaluation, quality of care, functional status, appropriateness of care, conformance of recommended/desirable standards of performance, or change in practice patterns. The objective of this award is to promote research by young investigators in the area of outcomes, broadly defined above, as it relates to geriatric gastroenterology. Please review the AGA Future Trends Committee Report: Effects of Aging of the Population on Gastroenterology Practice, Education and Research to learn more about research topics encouraged for study. Funds may be used for salary support of personnel and technicians only, supplies and/or equipment/services. Women and minorities are strongly encouraged to apply.
Requirements: Investigators must possess an MD, PhD or equivalent and must hold faculty positions at accredited North American academic institutions by the time of the start date of the award. MD applicants: No more than five years should elapse following the completion of your clinical training (GI fellowship or equivalent) and the start date of this award. Applicants must be AGA Members.
Restrictions: The award is intended for junior faculty; therefore, established investigators are not eligible. Candidates may not hold awards on a similar topic from other agencies. Indirect costs are not allowed.
Geographic Focus: All States
Date(s) Application is Due: Sep 5
Amount of Grant: 35,000 USD
Contact: Awards Manager; (301) 222-4012; fax (301) 652-3890; awards@fdhn.org
Internet: http://www.fdhn.org/wmspage.cfm?parm1=234
Sponsor: Foundation for Digestive Health and Nutrition
4930 Del Ray Avenue
Bethesda, MD 20814

FDHN Designated Research Award in Geriatric Gastroenterology   1641

This award provides $75,000 per year for three years (total $225,000) for young investigators working toward independent careers in academic research related to geriatric gastroenterology. The overall objective is to enable young investigators to develop independent and productive research careers, with a focus on research related to geriatric gastroenterology, by ensuring that a major proportion of their time is protected for research. Non-recipient applicants for this award will be considered for the Research Scholar Awards. Applicants should review the AGA Future Trends Committee Report: Effects of Aging of the Population on Gastroenterology Practice, Education and Research to learn more about research topics encouraged for study. The award is not intended for fellows, but for young faculty who have demonstrated unusual promise and have some record of accomplishment in research.
Requirements: Candidates must hold an MD, PhD, or equivalent degree (e.g., MB, ChB, MBBS, DO). Applicants must hold full-time faculty positions at North American universities or professional institutes at the time award begins. Applicants must be Members of the AGA. # MD applicants: no more than five years shall have elapsed following the completion of your clinical training (GI fellowship or its equivalent) and the start date of this award. PhD applicants: no more than five years shall have elapsed from the completion of your postdoctoral training and the start date of this award. Candidates must devote at least 70 percent of their efforts to research related to geriatric gastroenterology.
Restrictions: Candidates should be in the beginning years of their careers; therefore, established investigators are not appropriate candidates.
Geographic Focus: All States
Date(s) Application is Due: Sep 15
Amount of Grant: 75,000 USD
Contact: Awards Manager; (301) 222-4012; fax (301) 652-3890; awards@fdhn.org
Internet: http://www.fdhn.org/wmspage.cfm?parm1=224
Sponsor: Foundation for Digestive Health and Nutrition
4930 Del Ray Avenue
Bethesda, MD 20814

FDHN Designated Research Award in Research Related to Pancreatitis   1642

The overall objective of this Award is to enable young investigators to develop independent and productive research careers, with a focus on pancreatic disease, by ensuring that a major proportion of their time is protected for research. The Award provides $75,000 per year for three years (total $225,000) for young investigators working toward independent careers in academic research related to understanding, improving treatments or curing Pancreatitis.
Requirements: Candidates must hold an MD, PhD, or equivalent degree (e.g., MB, ChB, MBBS, DO). Applicants must hold full-time faculty positions at North American universities or professional institutes at the time award begins. Applicants must also be members of the AGA. The award is not intended for fellows, but for young faculty who have demonstrated unusual promise and have some record of accomplishment in research. Candidates should be in the beginning years of their careers; therefore, established investigators are not appropriate candidates. MD applicants: No more than five years shall have elapsed following the completion of your clinical training (GI fellowship or its equivalent) and the start date of this award. PhD applicants: No more than five years shall have elapsed from the completion of your postdoctoral training and the start date of this award.
Geographic Focus: All States
Date(s) Application is Due: Apr 17
Contact: Awards Manager; (301) 222-4012; fax (301) 652-3890; awards@fdhn.org
Internet: http://www.fdhn.org/wmspage.cfm?parm1=100
Sponsor: Foundation for Digestive Health and Nutrition
4930 Del Ray Avenue
Bethesda, MD 20814

FDHN Fellow Abstract Prizes   1643

Two awards of $1000 each will be given to fellows who have submitted abstracts chosen to be presented during Digestive Disease Week. Awards will be presented at a ceremony during DDW. The primary objective of this award is to stimulate interest in GI research careers through competition and recognition
Requirements: Qualified candidates are MD or PhD postdoctoral fellows who are trainee members the AGA. Women and minority investigators are strongly encouraged to apply. Applicants must: be sponsored by an AGA member; and be the first author of an abstract accepted for presentation at DDW and provide evidence of abstract acceptance. Individuals with faculty appointments are not eligible. Applicants may only submit one abstract for consideration.
Restrictions: Fellows who have been awarded must present the abstract. No substitute presenters are allowed. Abstracts must have been selected for presentation at DDW. A letter of recommendation from the sponsor is required.
Geographic Focus: All States
Date(s) Application is Due: Mar 21
Amount of Grant: 1,000 USD
Contact: Awards Manager; (301) 222-4012; fax (301) 652-3890; awards@fdhn.org
Internet: http://www.fdhn.org/wmspage.cfm?parm1=124
Sponsor: Foundation for Digestive Health and Nutrition
4930 Del Ray Avenue
Bethesda, MD 20814

FDHN Fellowship to Faculty Transition Awards   1644

This award provides $40,000 per year for two years for current trainees in gastroenterology related fields so they may gain additional research training in gastrointestinal, liver function or related diseases. The objective is to prepare physicians for independent research careers in digestive diseases. The award provides salary support for additional full-time research training in basic science to acquire modern laboratory skills. The additional two years of research training provided by this award should broaden and expand the scope of investigative tools available to the recipient, generally in basic disciplines such as cell or molecular biology or immunology.
Requirements: Applicants must be MDs or MD/PhDs currently in a gastroenterology-related fellowship at an accredited North American institution, committed to academic careers. They will have completed two years of research training at the start of this award. Applicants must be AGA Trainee Members and be sponsored by an AGA Member.
Restrictions: Individuals who hold a PhD are ineligible. Although the institution may supplement the award, the applicant may not concurrently hold a similar training award or grant from another organization, such as the NIH, ALF, CCFA, or the Glaxo Institute of Digestive Health.
Geographic Focus: All States
Date(s) Application is Due: Sep 5
Amount of Grant: 40,000 USD
Samples: John Chang, MD, Fellow in Gastroenterology University of Pennsylvania, Philadelphia, PA; Lee Peng, MD, PhD, Massachusetts University, Boston, MA; Carl B. Rountree, Jr., MD, Children's Hospital, Los Angeles, CA.
Contact: Awards Manager; (301) 222-4012; fax (301) 652-3890; awards@fdhn.org
Internet: http://www.fdhn.org/wmspage.cfm?parm1=102
Sponsor: Foundation for Digestive Health and Nutrition
4930 Del Ray Avenue
Bethesda, MD 20814

FDHN Funderburg Research Scholar Award in Gastric Biology Related to Cancer   1645

This award of $25,000 per year for two years (total $50,000) is awarded to an established investigator working on novel approaches in gastric cancer, including the fields of gastric mucosal regeneration and regulation of cell growth (not as they relate to peptic ulcer disease inflammation (including Helicobacter pylori) as precancerous lesions; genetics of gastric oncogenes in gastric epithelial malignancies; epidemiology of gastric cancer; etiology of malignancies; or clinical research in the diagnosis or treatment of gastric carcinoma. The primary objective of the award is to support an active, established investigator in the field of gastric biology who enhances the fundamental understanding of gastric cancer pathobiology in order to ultimately develop a cure for the disease.
Requirements: Applicants must hold faculty positions at accredited North American institutions and must have established themselves as independent investigators in the field of gastric biology. Women and minority investigators are strongly encouraged to apply. Applicants must be Members of the AGA.
Geographic Focus: All States
Date(s) Application is Due: Sep 5
Amount of Grant: 25,000 USD
Samples: Xiaolu Yang, PhD, Associate Professor, Department of Cancer Biology, Associate Investigator, Abramson Family Cancer Research Institute, University of Pennsylvania School of Medicine, PA; Steven Itzkowitz, MD, FACP, FACG The Dr. Burrill B. Crohn Professor of Medicine, Director, Gastroenterology Fellowship Program, Associate Director, Division of Gastroenterology, Mount Sinai School of Medicine, New York City, NY; JeanMarie Houghton, PhD, University of Massachusetts Medical School, Worcester, MA.
Contact: Awards Manager; (301) 222-4012; fax (301) 652-3890; awards@fdhn.org
Internet: http://www.fdhn.org/wmspage.cfm?parm1=90
Sponsor: Foundation for Digestive Health and Nutrition
4930 Del Ray Avenue
Bethesda, MD 20814

FDHN Graduate Student Awards 1646
Two awards of $20,000 a year for two years are offered to fund graduate students undertaking research in the biology and epidemiology of diseases of the gastrointestinal tract, liver or pancreas. The award includes $18,000 for stipend and $2,000 to be used towards fringe benefits such as medical insurance and travel to a national meeting. The primary objective of the award is to provide a salary stipend for graduate students performing doctoral research related to the gastrointestinal tract, liver or pancreas.
Requirements: Applicants should have completed at least one year and no more than three years of training towards the doctoral degree and have selected and confirmed the laboratory or department in which they will conduct doctoral research. Research is to be conducted at an accredited academic institution within North America and the research advisor must be a Member of the AGA. Applicants should be U.S. citizens, permanent residents or overseas students who have a current visa to pursue education within North America.
Restrictions: Indirect costs, excluding travel, are not allowed. Tuition costs are expected to be covered by the institution or host department and laboratory.
Geographic Focus: All States
Date(s) Application is Due: Mar 14
Amount of Grant: 20,000 USD
Contact: Awards Manager; (301) 222-4012; fax (301) 652-3890; awards@fdhn.org
Internet: http://www.fdhn.org/wmspage.cfm?parm1=154
Sponsor: Foundation for Digestive Health and Nutrition
4930 Del Ray Avenue
Bethesda, MD 20814

FDHN Jon I. Isenberg International Research Scholar Award 1647
This award provides a total of $50,000 for one year of study, $25,000 from the AGA and $25,000 in matching funds from the applicant's national GI society. The award provides for non-U.S. citizen young investigators to spend one year performing GI-related research at an American institution under the tutelage of an AGA member. Four awards will be given annually. The primary objective of this award is to promote international scholarship, increase AGA's involvement within the international GI community and foster international collaboration in training GI investigators.
Requirements: An eligible candidate must be nominated by his or her national GI Society. The national GI society must match the $25,000 contribution from the AGA. Before submitting an application, a candidate must identify a sponsoring American institution and a research preceptor who agrees in writing to supervise his or her training and research. Qualified candidates must possess a doctoral degree, either an MD, or equivalent degree, and/or a PhD degree. In keeping with the intent to support the career development of young investigators, the candidate must be within five (5) years of completing GI training or, if a PhD, within five (5) years of the receipt of the degree, at the time of initiation of the award. A documented parental leave of absence will not be counted towards the five (5) years of eligibility.
Geographic Focus: All States
Date(s) Application is Due: Mar 1
Amount of Grant: 50,000 USD
Samples: Maria Cristina, Almansa Menchero Hospital Clinico Universitario San Carlos, Madrid, Spain, to the Mayo Clinic School of Medicine, Jacksonville, FL; Koji Nozaki, MD, PhD, Department of Gastrointestinal Surgery, The University of Tokyo, Tokyo, Japan; Revital Kariv, MD, Case Western Reserve University, Beachwood, OH.
Contact: Awards Manager; (301) 222-4012; fax (301) 652-3890; awards@fdhn.org
Internet: http://www.fdhn.org/wmspage.cfm?parm1=101
Sponsor: Foundation for Digestive Health and Nutrition
4930 Del Ray Avenue
Bethesda, MD 20814

FDHN June & Donald O. Castell MD, Esophageal Clinical Research Award 1648
One award of $35,000 is made annually to provide research and/or salary support for junior faculty involved in clinical research in esophageal diseases. This award is funded by the June and Donald O. Castell, MD, Gastroenterology Research and Education Trust. The primary objective of the award is to support investigators who have demonstrated high potential to develop independent, productive research careers.
Requirements: Candidates must hold a MD or PhD or equivalent. Applicants must hold a full-time faculty position at a North American universities or professional institute. Applicants must be members of the AGA. The recipient must be at or below the level of assistant professor, and his/her initial appointment to the faculty position must have been within seven (7) years of the time of application. This award is not intended for fellows, but for junior faculty who have demonstrated unusual promise; have some record of accomplishment in research; and have established independent research programs at the time of the award. Candidates must devote at least 50 percent of their efforts to research related to esophageal function or diseases.
Restrictions: If an award recipient receives notification of another award with overlapping scientific objectives, prior to the start date of an AGA or FDHN award, the applicant must choose between the two awards.
Geographic Focus: All States
Date(s) Application is Due: Jan 14
Amount of Grant: 35,000 USD
Samples: John Pandolfino, MD, Northwestern University, Chicago, IL; Marcelo Vela, MD, Medical University of South Carolina, Charleston, SC; Braden Kuo, MD, Massachusetts General Hospital Boston, MA.
Contact: Awards Manager; (301) 222-4012; fax (301) 652-3890; awards@fdhn.org
Internet: http://www.fdhn.org/wmspage.cfm?parm1=104
Sponsor: Foundation for Digestive Health and Nutrition
4930 Del Ray Avenue
Bethesda, MD 20814

FDHN Moti L. & Kamla Rustgi International Travel Awards 1649
This program awards grants to young basic, translational and clinical investigators to support their travel and related expenses to attend Digestive Disease Week. Two awards of $500 each will be given to selected individuals residing outside North America. The primary objective of the award is to enable young investigators outside of North American (U.S. or Canada) institutions to attend Digestive Disease Week and encourage international trainees to become more involved in digestive tract, pancreatic and liver disease research.
Requirements: Candidates must: be MD or PhD or MD PhD postdoctoral fellows who are international trainee members of the AGA; be sponsored by an International member of AGA; be 35 years of age or younger at the time of the meeting; be fluent in English; and have a sufficient number of scientific papers (impact factor) published and/or poster presentations.
Geographic Focus: All States
Date(s) Application is Due: Mar 21
Amount of Grant: 500 USD
Contact: Awards Manager; (301) 222-4012; fax (301) 652-3890; awards@fdhn.org
Internet: http://www.fdhn.org/wmspage.cfm?parm1=126
Sponsor: Foundation for Digestive Health and Nutrition
4930 Del Ray Avenue
Bethesda, MD 20814

FDHN Non-Career Research Awards 1650
This award provides approximately $10,000 (per symposia) for travel support for young investigators and selected established investigators to participate in symposia on gastrointestinalrelated topics. The primary objective of the award is to foster interactions and enhance the exchange of information between clinical and basic science investigators, and established and junior investigators working in gastrointestinal research. Eligibility women and minority organizers are strongly encouraged to apply. Travel support may be provided for: junior investigators who are within 5 years of completing their clinical training in GI or from completion of their PhD doctoral thesis and are at or below the rank of assistant professor; and up to two established investigators who are invited speakers. Biosketches of these individuals and a letter addressing their contribution to the scientific content and atmosphere of the meeting should be submitted with the application.
Requirements: Within 60 days of the conclusion of the meeting, a one-page summary of the highlights of the meeting and an accounting of the use of funds must be submitted. The names of participants and speakers in the symposium, addresses, dates of birth and academic ranks of the attending scientists and their itemized expenses, must be sent to the Foundation. Any unexpended funds must be returned.
Restrictions: Indirect costs are not allowed.
Geographic Focus: All States
Date(s) Application is Due: Feb 1; May 1; Oct 1
Amount of Grant: 10,000 USD
Contact: Awards Manager; (301) 222-4012; fax (301) 652-3890; awards@fdhn.org
Internet: http://www.fdhn.org/wmspage.cfm?parm1=123
Sponsor: Foundation for Digestive Health and Nutrition
4930 Del Ray Avenue
Bethesda, MD 20814

FDHN Non-Career Research Grants 1651
A research initiative grant of $25,000 for one year is offered to investigators to support pilot research projects in gastroenterology- or hepatology-related areas. The primary objective of the award is to provide non-salary funds for new investigators to help them establish their research careers or to support pilot projects that represent new research directions for established investigators. The intent is to stimulate research in gastroenterology- or hepatology-related areas by permitting investigators to obtain new data that can ultimately provide the basis for subsequent grant applications of more substantial funding and duration. Women and minorities are strongly encouraged to apply.
Requirements: Investigators must possess an MD or PhD degree or equivalent and must hold faculty positions at accredited North American institutions. Candidates may not hold awards on a similar topic from other agencies. Applicants must be AGA Members. Applicants for this award may not simultaneously apply for the AGA/Miles and Shirley Fiterman Foundation Basic Research Award or the AGA June and Donald O. Castell, MD, Esophageal Clinical Research Award.
Restrictions: If an award recipient receives notification of another award with overlapping scientfic objectives, prior to the start date of an AGA or FDHN award, the applicant must choose between the two awards.
Geographic Focus: All States
Date(s) Application is Due: Jan 14
Amount of Grant: 25,000 USD
Contact: Awards Manager; (301) 222-4012; fax (301) 652-3890; awards@fdhn.org
Internet: http://www.fdhn.org/wmspage.cfm?parm1=121
Sponsor: Foundation for Digestive Health and Nutrition
4930 Del Ray Avenue
Bethesda, MD 20814

FDHN Research Scholar Awards 1652
These awards provide salary support for young investigators working in any area of gastrointestinal, liver function, or related diseases. The primary intent of the program is to support physician-investigators who have a high potential to develop independent, productive research careers in gastroenterology and hepatology. Candidates must devote at least 70 percent of their effort to research related to the gastrointestinal tract or liver. There must be a strong commitment from the candidate's division and department to support the candidate by protecting time for research and providing adequate laboratory space and facilities.

Requirements: Applicants must hold full-time faculty positions at North American universities or professional institutes. Nonphysician candidates with a PhD will also be considered. Candidates should be early in their research careers and commonly will have recently completed their fellowship training.
Restrictions: Indirect costs are not allowed. Candidates who have been at the assistant professor level or equivalent for more than five years are not eligible. Nor can applicants hold, or have held, an RO1, R29, K11, K08, or VA research award or any award with similar objectives from nonfederal sources (such as ALF, CCFA, or Glaxo Institute of Digestive Health). However, awards or grants obtained after receipt of this award need not be surrendered.
Geographic Focus: All States
Date(s) Application is Due: Sep 5
Amount of Grant: 75,000 USD
Contact: Awards Manager; (301) 222-4012; fax (301) 652-3890; awards@fdhn.org
Internet: http://www.fdhn.org/wmspage.cfm?parm1=103
Sponsor: Foundation for Digestive Health and Nutrition
4930 Del Ray Avenue
Bethesda, MD 20814

FDHN Student Abstract Prizes 1653
Ten travel awards of $500 each will be given to high school, college, graduate and medical students who have submitted abstracts chosen to be presented during Digestive Disease Week. The three best student abstracts submitted will receive a $1,000 prize. Awards will be presented at a ceremony during DDW. The primary objective of the award is to stimulate interest in gastroenterology research careers through competition and recognition.
Requirements: Any high school, undergraduate medical, premedical, pre-doctoral student or medical resident (up to and including postgraduate year three) who has performed original research related to gastroenterology or hepatology. Applicants must be sponsored by an AGA member.
Restrictions: Post doctoral fellows, technicians, visiting scientists and MD research fellows are not eligible for this award. Applicants may only submit one abstract for consideration and must be the designated presenter or first author of the abstract.
Geographic Focus: All States
Date(s) Application is Due: Mar 21
Amount of Grant: 500 - 1,000 USD
Contact: Awards Manager; (301) 222-4012; fax (301) 652-3890; awards@fdhn.org
Internet: http://www.fdhn.org/wmspage.cfm?parm1=152
Sponsor: Foundation for Digestive Health and Nutrition
4930 Del Ray Avenue
Bethesda, MD 20814

FDHN Student Research Fellowships 1654
This program offers financial support for students to spend a minimum of 10 weeks performing research in digestive diseases or nutrition and is intended to stimulate interest in research careers in these areas. The work may take place at any time during the year. Up to 20 fellowships will be available for full-time research with a preceptor, who must be a faculty member who directs a research project in a gastroenterology-related area at an accredited North American institution. A complete financial statement and scientific progress report are required upon completion of the program. The AGA has recognized the need to attract and encourage minority individuals to enter and pursue gastroenterology research careers. In response to this concern, seven of the student research fellowship awards will be reserved for underrepresented minority students. For the purpose of this award, minorities have been defined as African American, Hispanic, Native American/Alaskan Native, and Pacific Islander.
Requirements: Candidates may be high school, undergraduate, medical, or graduate students (not yet engaged in thesis research) in accredited North American institutions. Women and minority students are strongly encouraged to apply. The preceptor must be a faculty member who directs a research project in a gastroenterology-related area at an accredited North American institution.
Restrictions: Candidates may not hold similar salary support from other agencies: e.g., American Liver Foundation, Crohn's and Colitis Foundation.
Geographic Focus: All States
Date(s) Application is Due: Mar 5
Amount of Grant: 2,000 - 3,000 USD
Contact: Awards Manager; (301) 222-4012; fax (301) 652-3890; awards@fdhn.org
Internet: http://www.fdhn.org/wmspage.cfm?parm1=115
Sponsor: Foundation for Digestive Health and Nutrition
4930 Del Ray Avenue
Bethesda, MD 20814

FDHN TAP Endowed Designated Research Award in Acid-Related Diseases 1655
This award provides $75,000 per year for three years (total $225,000) for young investigators working toward independent careers in acid-related diseases. The overall objective is to enable young investigators to develop independent and productive careers in acid-related research by ensuring that a major proportion of their time is protected for research.
Requirements: Candidates must hold an MD, PhD, or equivalent degree (e.g., MB, ChB, MBBS, DO). Applicants must hold full-time faculty positions at North American universities or professional institutes at the time award begins. Applicants must be Members of the AGA. The award is not intended for fellows, but for young faculty who have demonstrated unusual promise and have some record of accomplishment in research. Candidates should be in the beginning years of their careers; therefore, established investigators are not appropriate candidates. Candidates must devote at least 70 percent of their efforts to research related to geriatric gastroenterolgy.

Restrictions: MD applicants: no more than five years shall have elapsed following the completion of your clinical training (GI fellowship or its equivalent) and the start date of this award. PhD applicants: no more than five years shall have elapsed from the completion of your postdoctoral training and the start date of this award.
Geographic Focus: All States
Date(s) Application is Due: Sep 5
Amount of Grant: 75,000 USD
Contact: Awards Manager; (301) 222-4012; fax (301) 652-3890; awards@fdhn.org
Internet: http://www.fdhn.org/wmspage.cfm?parm1=132
Sponsor: Foundation for Digestive Health and Nutrition
4930 Del Ray Avenue
Bethesda, MD 20814

FDHN Translational Research Awards 1656
One award of $100,000 per year for two years will be made annually to support translational research in gastroenterology and/or hepatology. Translational research will be defined as the process of applying ideas, insights and discovery generated through basic science research to the diagnosis, treatment or prevention of human disease. The primary objective of the award is to enhance interaction between researchers with basic science and clinical backgrounds with the goal of accelerating the pace of discovery that is directly applicable to patient care. The creation of teams including both a Ph.D. and an M.D. researcher is particularly encouraged.
Requirements: This award must be applied for jointly by a team of researchers (typically consisting of two members), including at least one researcher with significant training and experience in a basic science discipline (including areas outside the traditional biomedical sciences, such as physical sciences and engineering) and at least one researcher who is qualified to provide direct clinical care to patients. Junior investigators in either or both categories are particularly encouraged to apply. Applicants must be members of the AGA. Research must be conducted at an accredited North American institution.
Restrictions: Candidates may not hold support for the same project from another agency. If a proposed award recipient receives notification of another award with overlapping scientific objectives prior to the start date of an AGA or FDHN award, the applicant must choose only one of the awards to accept.
Geographic Focus: All States
Date(s) Application is Due: Mar 14
Amount of Grant: 100,000 USD
Contact: Awards Manager; (301) 222-4012; fax (301) 652-3890; awards@fdhn.org
Internet: http://www.fdhn.org/wmspage.cfm?parm1=148
Sponsor: Foundation for Digestive Health and Nutrition
4930 Del Ray Avenue
Bethesda, MD 20814

Federal Express Corporate Contributions Program 1657
The mission of the FedEx Social Responsibility department is to actively support the communities served and to strengthen global reputation through strategic investment of people, resources, and network. The foundation directs its philanthropic efforts to health and welfare programs, education, cultural/arts and civic assistance. Charitable shipping is limited to emergency, disaster or life-threatening situations coordinated through a nonprofit organization, disaster relief agency, or agency of the federal, state, or local government. FedEx is especially interested in supporting nonprofit organizations that request 5 percent or less of a total project budget; contingency grants; or seed monies with the thought that other sources will contribute matching amounts. Organizations must show evidence of competent management, low administrative/fundraising expense ratios, and a nondiscriminatory program benefiting broad segments of the community. FedEx Community Relations responds to all requests in writing. Requests are accepted year-round and generally are reviewed within three weeks of receipt.
Requirements: Charities must be registered 501(c)3 organizations in good financial and public standing.
Geographic Focus: All States
Contact: Grants Administrator, Community Relations; (901) 369-3600
Internet: http://www.fedex.com/us/about/responsibility/community/guidelines.html?link=4
Sponsor: Fedex Corporation
3610 Hacks Cross Road, Building A, 1st Floor
Memphis, TN 38125

Ferree Foundation Grants 1658
The foundation is a private family foundation dedicated to promoting and supporting excellence in the arts, culture and history, education, youth engagement, health, human services, and local economic and community development. Paul W. Ware is the Chairman of Ferree Foundation and continues the Ware family tradition of using philanthropy to enhance the lives of families and institutions primarily throughout Lancaster County and, occasionally, Southeastern Pennsylvania and beyond. Grant requests should be in writing and mailed to the executive director. Guidelines are available online.
Requirements: 501(c)3 nonprofit organizations are eligible. Preference is given to organizations serving Pennsylvania's Chester and Lancaster Counties.
Restrictions: The foundation does not make grants to individuals.
Geographic Focus: Pennsylvania
Date(s) Application is Due: Sep 15
Contact: Deb Arrive; (717) 735-8288; darrive@Ferree-Foundation.org
Internet: http://www.Ferree-Foundation.org/apply.html
Sponsor: Ferree Foundation
229 North Duke Street
Lancaster, PA 17602

Fidelity Foundation Grants 1659

Fidelity Investments Chairman Edward C. Johnson 3d and his father, the founder of the company, established the Fidelity Foundation in 1965 with several operating principles in mind. These principles, still current today, guide the Foundation's decisions and grantmaking. The Foundation grant program was designed to strengthen the long-term effectiveness of nonprofit institutions. The types of projects it funds, and the way in which it funds them, are specifically intended to help nonprofits build the organizational capabilities they need to better fulfill their missions and serve their constituencies. The Fidelity Foundation considers Letters of Inquiry from organizations with current IRS 501(c)3 public charity status only. Grants are made to fund only significant, transformative projects usually budgeted at $50,000 or more. The Foundation's primary philanthropic investments are allocated to the following sectors: arts and culture; community development and social services; health; and education.
Requirements: Grants are generally made only to organizations with operating budgets of $500,000 or more. The Fidelity Foundation considers projects from organizations of regional or national importance throughout the United States.
Restrictions: Grants are not awarded to support sectarian organizations, disease-specific associations, or public school systems. Support does not go to individuals or for scholarships, civic or start-up organizations, corporate memberships, operating support, or participation in benefit events, film, or video projects.
Geographic Focus: All States
Amount of Grant: 50,000 - 1,000,000 USD
Contact: Kathleen Ward; (617) 563-6806; info@FidelityFoundation.org
Internet: http://www.fidelityfoundation.org/index.html
Sponsor: Fidelity Foundation
82 Devonshire Street, S2
Boston, MA 02109

Field Foundation of Illinois Grants 1660

The Foundation awards grants only to institutions and agencies operating in the fields of urban and community affairs, culture, education, community welfare, health, and environment, primarily serving the people of Chicago with extremely limited grant making in the metropolitan area. Preference will be given to funding innovative approaches for addressing program areas.
Requirements: Applicants must reside in Illinois. Grant applications are not provided; however, a formal proposal is required. The Foundation does not accept grant requests via email or fax.
Restrictions: No grants will be made to support: endowments; individuals; medical research or national health agency appeals; propaganda organizations or committees whose efforts are aimed at influencing legislation; printed materials, video or computer equipment; fund-raising events or advertising; appeals for religious purposes; other granting agencies or foundations for ultimate distribution to agencies or programs of its own choosing; custodian afterschool programs or tutoring; most disease specific programs, research or activities; or operating support of established neighborhood health centers or clinics, day care centers for children, or small cultural groups. Requests for computer equipment will not be considered.
Geographic Focus: Illinois
Date(s) Application is Due: Jan 15; May 15; Sep 15
Amount of Grant: 50,000 USD
Samples: Catholic Charities of the Archdiocese of Chicago (IL)--to support expansion of the Emergency Assistance Services program, $15,000; Jobs for Youth/Chicago (IL)--to support the Workforce Advancement Initiative, $10,000; Palliative Care Center and Hospice of the North Shore (IL)--to support the capital campaign for a new community facility, $25,000; Puerto Rican Alliance (IL)--to support the Community Arts Program, $10,000.
Contact: Joann Ross, (312) 831-0910; fax (312) 831-0961; jross@fieldfoundation.org
Internet: http://www.fieldfoundation.org
Sponsor: Field Foundation of Illinois
200 S Wacker Drive, Suite 3860
Chicago, IL 60606

Fifth Third Foundation Grants 1661

The Fifth Third Foundation awards grants to eligible nonprofit organizations in its areas of interest, including community development, education, health and human services, and arts and culture. Proposals are favored that are likely to make a substantial difference in the quality of community life; strengthen families and communities; expand meaningful civic engagement and build social capital; use volunteers; help nonprofit organizations build capacity and become more effective; include financial and other strategic commitments from other funding organizations; and leverage change in the capacity of community-wide systems rather than individual organizations. Interested applicants should initially contact the Foundation via a brief letter detailing the organization, its mission, the project it seeks funding for, and an approximate grant amount to be requested.
Requirements: 501(c)3 nonprofit, tax-exempt organizations operating in Fifth Third's geographic regions of Ohio, Kentucky, Indiana, Michigan, Illinois, Tennessee, West Virginia, and Florida are eligible to apply.
Restrictions: The following types of support are ineligible: capital campaigns for individual churches; publicly supported entities, such as public schools or government agencies; elementary schools; and individuals.
Geographic Focus: Florida, Illinois, Indiana, Kentucky, Michigan, Ohio, Tennessee, West Virginia
Contact: Heidi B. Jark, Managing Director; (513) 534-7001 or (513) 534-4397; fax (513) 534-0960; heidi.jark@53.com
Internet: https://www.53.com/site/about/in-the-community/foundation-office-at-fifth-third-bank.html?
Sponsor: Fifth Third Foundation
38 Fountain Square Plaza, MD 1090CA
Cincinnati, OH 45263

Fight for Sight-Streilein Foundation for Ocular Immunology Research Award 1662

Beginning in 2011, Fight for Sight and the Streilein Foundation for Ocular Immunology co-sponsored a new annual Summer Student Research Fellowship, Fight for Sight-Streilein Foundation for Ocular Immunology Research Award, for the study of ocular immunology. Candidates for the award, an annual $2,100 grant for fulltime research conducted from June to August, are invited to apply by visiting www.fightforsight.org for an application.
Requirements: The award is open to undergraduates, graduate and medical students studying in the United States. Applications will be reviewed by the Fight for Sight Scientific Review Committee and Streilein Foundation for Ocular Immunology and a grantee will be selected based upon the committee's recommendations to the Boards of both organizations.
Geographic Focus: All States
Date(s) Application is Due: Feb 1
Amount of Grant: 2,100 - 2,100 USD
Contact: Janice Benson; (212) 679-6060; fax (212) 679-4466; janice@fightforsight.org
Internet: http://www.streilein-foundation.org/fellowships.html
Sponsor: Streilein Foundation for Ocular Immunology
P.O. Box 6104
Boston, MA 02114

Fight for Sight Grants-in-Aid 1663

Fight for Sight was founded in 1946 by Mildred Weisenfeld, a young woman with retinitis pigmentosa, to encourage and fund research in ophthalmology, vision and related sciences. The goal of Fight for Sight is to encourage and facilitate research in detection, understanding, prevention, treatment and cures of visual disorders especially those diseases leading to impaired sight or blindness. To that end, FFS funds Grants-in-Aid to academic researchers within the first three years of their faculty appointments. FFS primarily supports new investigators, promoting the development of scientific skills and enabling the development of preliminary scientific findings and pilot studies necessary to successfully apply for more substantial federal and private funding such as that provided by the National Eye Institute and other divisions of the NIH. Support may be used to defray costs of personnel (but not the applicant), equipment and consumable supplies needed for the specific research project. One year awards of $20,000 are provided and may start between July 1 and September 1.
Requirements: Applications will only be considered from researchers who have received their first faculty or research appointment in eye/vision within the previous three years. Fringe benefits are not included and institutional overhead charges are not covered. Travel costs are generally not supported.
Geographic Focus: All States
Date(s) Application is Due: Jan 1
Amount of Grant: 20,000 - 20,000 USD
Contact: Janice Benson; (212) 679-6060; fax (212) 679-4466; janice@fightforsight.org
Internet: https://www.fightforsight.org/Grants/Research-Award-Types/
Sponsor: Fight for Sight (USA)
381 Park Avenue S, Suite 809
New York, NY 10016

Fight for Sight Post-Doctoral Awards 1664

Post-Doctoral Awards support individuals with a doctorate (Ph.D., M.D., O.D., Dr.PH, or D.V.M.) who are interested in academic careers in basic or clinical research in ophthalmology, vision or related sciences. This funding is intended to offer those interested in an academic career the opportunity to spend a year engaged in vision and eye research under the supervision of a senior scientist/clinician mentor. Clinical post-doctoral researchers are required to spend sufficient time on the funded research project to carry out the proposed objectives while basic researchers are expected to work full-time. One year grants of $20,000 are awarded for start dates between July 1 and September 1.
Requirements: Applications are considered from individuals who are within three years of their doctoral degrees or clinical residency training and have not received a previous Fight for Sight fellowship award. If at the time of application a doctorate has not yet been obtained, a cover letter must be submitted by the conferring institution advising when such degree is expected to be awarded. Recipients may supplement their awards with institutional or other funds however any anticipated supplemental support must be disclosed at the time of application. Total combined salary support must not exceed the annual stipend level set by the NIH for National Research Service Award recipients.
Restrictions: Fringe benefits are not provided by Fight for Sight.
Geographic Focus: All States
Date(s) Application is Due: Feb 1
Amount of Grant: 20,000 - 20,000 USD
Contact: Janice Benson; (212) 679-6060; fax (212) 679-4466; janice@fightforsight.org
Internet: https://www.fightforsight.org/Grants/Research-Award-Types/
Sponsor: Fight for Sight (USA)
381 Park Avenue S, Suite 809
New York, NY 10016

Fight for Sight Summer Student Fellowships 1665

Fight for Sight was founded in 1946 by Mildred Weisenfeld, a young woman with retinitis pigmentosa, to encourage and fund research in ophthalmology, vision and related sciences. The goal of Fight for Sight is to encourage and facilitate research in detection, understanding, prevention, treatment and cures of visual disorders especially

those diseases leading to impaired sight or blindness. Unrestricted awards of $2,100 are given for two to three months of full-time research, usually during June-August.
Requirements: Fight for Sight's primary mission is to support and encourage promising scientists early in their careers. FFS primarily supports new investigators, promoting the development of scientific skills and enabling the development of preliminary scientific findings and pilot studies necessary to successfully apply for more substantial federal and private funding such as that provided by the National Eye Institute and other divisions of the NIH. Summer Student Fellowships are available to undergraduates, graduate and medical students who are interested in pursuing eye-related clinical or basic research. For most students, this is their first exposure to eye or vision research and the experience has resulted in many students choosing academic ophthalmology or eye research as a full-time career.
Restrictions: Students receiving stipends from other sources are generally not eligible.
Geographic Focus: All States
Date(s) Application is Due: Feb 1
Amount of Grant: 2,100 - 2,100 USD
Contact: Janice Benson; (212) 679-6060; fax (212) 679-4466; janice@fightforsight.org
Internet: https://www.fightforsight.org/Grants/Research-Award-Types/
Sponsor: Fight for Sight (USA)
381 Park Avenue S, Suite 809
New York, NY 10016

Finance Factors Foundation Grants 1666
The Finance Factors Family of Companies is committed to supporting the community by providing financial support to many charitable and community-based organizations. Through the Finance Factors Foundation, they direct their support to four key areas that are aligned with community needs and business objectives: education, health and human services, community and civic affairs, and culture and the arts. In addition to employees volunteering time to schools and other organizations in the community, Finance Factors also strives to support the efforts of its employees in the community by providing financial assistance to those organizations that already benefit from its employees' time, talents, enthusiasm and energy.
Requirements: Send a request for support to the Finance Factors Foundation via mail or email.
Geographic Focus: Hawaii
Contact: June Yip, (808) 548-3393; info@financefactors.com
Internet: https://www.financefactors.com/about-us/community/
Sponsor: Finance Factors Foundation
1164 Bishop Street, Suite 1089
Honolulu, HI 96813

Firelight Foundation Grants 1667
Firelight provides small grants to community-based organizations selected for their vision and resourcefulness. The Foundation is often the first funder to an organization. Its grantmaking model is framed by a seven-year partnership model, which is divided into three phases. Firelight grantee-partners develop programs unique to the needs of their community. By working across multiple focus areas, our partners are able to address the needs of traumatized or vulnerable children and families effected by poverty, HIV and AIDS. Its primary goals include: providing basic necessities, including clothing, bedding, personal hygiene and shelter; supporting food production, feeding programs and household food assistance for vulnerable children and families; providing materials, skills and knowledge to caregivers to help them generate income and strengthen household resiliency; enhancing the caring relationships that meet the emotional, social, and recreational needs of children and help build life and coping skills; building a supportive and protective environment that prevents and responds to violence, abuse, and the exploitation of children; offering building a supportive and protective environment that prevents and responds to violence, abuse, and the exploitation of children; and extending primary health care, preventive care and HIV and AIDS-related preventive and palliative care.
Requirements: The Foundation accepts unsolicited proposals from seven countries in sub-Saharan Africa: Lesotho, Malawi, Rwanda, South Africa, Tanzania, Zambia, and Zimbabwe. In addition, Firelight awards grants to CBOs in Ethiopia, Kenya, and Uganda through solicited proposals.
Restrictions: The foundation does not fund: individuals; organizations or programs designed to influence legislation or elect public officials; programs that limit participation based on race, creed, or nationality; academic or medical research; or fundraising drives or endowments.
Geographic Focus: Ethiopia, Kenya, Lesotho, Malawi, Rwanda, South Africa, Tanzania, Uganda, Zambia, Zimbabwe
Amount of Grant: 500 - 10,000 USD
Contact: Evelyn Brown, Grants Administrator; (831) 429-8750; fax (831) 429-2036; evelyn@firelightfoundation.org
Internet: http://www.firelightfoundation.org/programs/grantmaking/
Sponsor: Firelight Foundation
740 Front Street, Suite 380
Santa Cruz, CA 95060

FirstEnergy Foundation Community Grants 1668
The FirstEnergy Foundation's contributions to local nonprofit organizations help strengthen the social and economic fabric of our communities. Funded solely by FirstEnergy, the Foundation extends the corporate philosophy of providing community support. The Foundation traditionally funds these priorities: help improve the vitality of our communities and support key safety initiatives; promotion of local and regional economic development and revitalization efforts; and support of FirstEnergy employee community leadership and volunteer interests.
Requirements: 501(c)3 organizations within the FirstEnergy Corporation operating companies' service areas - Ohio Edison Company, the Cleveland Electric Illuminating Company, the Toledo Edison Company, Pennsylvania Power Company, Metropolitan Edison Company, Pennsylvania Electric Company, Jersey Central Power and Light Company, Monongahela Power Company, the Potomac Edison Company, West Penn Power Company, FirstEnergy Solutions Corporation, FirstEnergy Generation, and FirstEnergy Nuclear Operations - are eligible to apply.
Restrictions: Funding is not considered for: direct grants to individuals, political or legislative activities; organizations that receive sizable public tax funding; fraternal, religious, labor, athletic, social or veterans organizations - unless the contribution is earmarked for an eligible program or campaign open to all beneficiaries, including those not affiliated with the host organization; national or international organizations; organizations supported by federated campaigns, such as United Way; research; equipment purchases; loans or second party giving, such as endowments, debt retirement, or foundations; or public or private schools.
Geographic Focus: New Jersey, Ohio, Pennsylvania
Contact: Dee Lowery; (330) 761-4246 or (330) 384-5022; fax (330) 761-4302
Internet: http://www.firstenergycorp.com/community
Sponsor: FirstEnergy Foundation
76 South Main Street
Akron, OH 44308-1890

Fischelis Grants for Research in the History of American Pharmacy 1669
The institute supports a wide range of scholarly activities as long as they promise to contribute significantly to historical understanding and are clearly related to the modern practice of pharmacy in the United States. Projects must be connected directly to American pharmacy practice. Preference will be given to 20th-century topics, although studies of earlier periods will be considered. Historical discussions of current practice issues are also eligible. An application received after the deadline date will be evaluated individually if any funds remain unawarded; otherwise it will be considered in the next year's program.
Requirements: Each application is referred for evaluation to a committee of the institute. Principal criteria used in evaluating an application are a candidate's qualifications and record relevant to the intended purpose, relevance to the history of modern American pharmacy, and the adequacy of resources to fulfill the stated purpose.
Geographic Focus: All States
Date(s) Application is Due: Mar 1
Amount of Grant: Up to 5,000 USD
Contact: Gregory Higby, Director; (608) 262-5378; grants@aihp.org
Internet: http://www.pharmacy.wisc.edu/aihp
Sponsor: American Institute of the History of Pharmacy
777 Highland Avenue
Madison, WI 53705-2222

Fisher Foundation Grants 1670
The foundation awards grants to Connecticut nonprofits that benefit education, health and human services, housing, community needs, and arts and culture. The majority of grants are single year awards. Before submitting an application, a letter of inquiry or conversation with staff to discuss the specific purpose for which the funds are being requested is strongly recommended. Application information and guidelines are available online. Do not submit applications by fax or email.
Requirements: Nonprofits located in, and/or serving the residents of the Greater Hartford area, including: Andover, Avon, Bloomfield, Bolton, Canton, East Hartford, East Granby, East Windsor, Ellington, Enfield, Farmington, Glastonbury, Granby, Hartford, Hebron, Manchester, Marlborough, Newington, Rocky Hill, Simsbury, Somers, South Windsor, Suffield, Tolland, Vernon, West Hartford, Wethersfield, Windsor, and Windsor Locks are eligible.
Restrictions: Foundation policy does not allow funding for: organizations which are not tax-exempt under IRS Code section 501(c)3; organizations which have the IRS private foundation designation; individuals; performances, conferences, retreats, one-time events, trips; annual campaigns.
Geographic Focus: Connecticut
Date(s) Application is Due: Jan 15; Apr 15; Sep 15
Contact: Beverly Boyle; (860) 570-0221; fax (860) 570-0225; bboyle@fisherfdn.org
Internet: http://www.fisherfdn.org/application/instructions.htm
Sponsor: Fisher Foundation
36 Brookside Boulevard
West Hartford, CT 06107

Fisher House Foundation Newman's Own Awards 1671
Grants are made to implement innovative programs to improve the quality of life for military families and their communities. This program is administered by Fisher House, a charitable organization that provides a home-like place to stay for United States military families while a family member is receiving treatment at a major military or Veterans Administration medical facility. One award will be given to an organization supporting the Army, Navy, Air Force, Marines, or Coast Guard The organization with the most innovative project submitted will receive up to $15,000. The other four organizations will receive lesser amounts, as determined by the judges. Annual deadline dates may vary; contact program staff for exact dates.
Requirements: Eligible applicants must support an Active Duty, National Guard, or Reserve unit or installation, and be tax-exempt, a private organization, or a volunteer organization. Private, nonprofit organizations as defined in Department of Defense (DoD) Instruction 1000.15 and approved for operation on a DoD installation by the installation commander are eligible.
Geographic Focus: All States
Date(s) Application is Due: Apr 30

Amount of Grant: 1,000 - 15,000 USD
Contact: James D. Weiskopf, Executive Vice President; (888) 294-8560 or (301) 294-8560; fax (301) 294-8562; info@fisherhouse.org
Internet: http://www.fisherhouse.org/programs/newmans.shtml
Sponsor: Fisher House Foundation
1401 Rockville Pike, Suite 600
Rockville, MD 20852-1402

Fishman Family Foundation Grants 1672
The Foundation considers grants for: research, education, and cultural development of and for the community; scholarships related to Jewish services, education, social, and community activities; medical and scientific research; providing resources to meet critical needs in Israel; and educational grants and scholarships. Proposals are reviewed in April and October. Application information is available online.
Requirements: 501(c)3 nonprofits are eligible.
Geographic Focus: All States
Date(s) Application is Due: Mar 31; Sep 30
Contact: Betty Fishman, President; info@fishman.org
Internet: http://www.fishman.org/apply.html
Sponsor: Fishman Family Foundation
730 E Cypress Avenue
Monrovia, CA 91016

FIU Global Civic Engagement Mini Grants 1673
The Center for Leadership & Service at Florida International University strives to support community service initiatives lead by students. If you or your organization has a new or ongoing community service project, the Center would like to support your community engagement mission. The Global Civic Engagement Student Advisory Board is a program at the Center for Leadership and Service funded by Wells Fargo to expand global awareness and community engagement by supporting student-led service projects. Students are encouraged to develop projects that address issues regarding any social issue such as: public police; education; the arts; environment; economic development; community beautification; health; and many more. Selected project proposals may receive a mini-grant up to $1,000. Annual deadlines for applications are October 17 and February 20. Applications received after deadline will be reviewed upon available funding or during the next funding cycle.
Requirements: Applicants must be enrolled as an FIU student during the period of the project. Only volunteer projects are eligible.
Geographic Focus: All States
Date(s) Application is Due: Feb 20; Oct 17
Amount of Grant: Up to 1,000 USD
Contact: Dr. Beverly Dalrymple; 305.348.6995 or 305.348.2149; cls@fiu.edu
Internet: http://leadserve.fiu.edu/index.php/global-civic-engagement-mini-grants/
Sponsor: Florida International University
11200 SW 8th Street
Miami, FL 33199

Flextronics Foundation Disaster Relief Grants 1674
The Flextronics Foundation seeks to aid, enrich, engage, educate and empower the communities where its company, suppliers and customers have a business presence. The Disaster Relief program provides aid to relieve human suffering that may be caused by a natural or civil disaster, or an emergency hardship. These disasters may be floods, fires, storms, earthquakes or similar large-scale adversities. Applications are evaluated on a quarterly basis at the end of March, June, September and November annually.
Requirements: Applicants must qualify as a 501(c)3 nonprofit organization or exclusively public institution or comparable charitable organization. Grants are provided for services and programs that match The Flextronics Foundation's priorities and are within the areas in which Flextronics' personnel live and work and/or where Flextronics' suppliers and customers live and work. Preference is given to those charitable organizations whose services and programs resonate with Flextronics employees (e.g., Flextronics employees donate their time and are actively involved). Grants are generally to be expended within one year, without expectation of further support.
Restrictions: The following are not eligible: Organizations that are not a 501(c)3 nonprofit or exclusively public institutions or comparable charitable organization; Religious (sectarian) and political groups; Stand-alone activities that are not an integrated part of a service or program (advertising, athletic events or league sponsorships, conventions, conferences, meetings or seminars, clubs, contests, field trips, film and/or video projects, fundraising activities such as walk-a-thons, marketing, sponsorships, travel and/or travel expenses and other stand alone or isolated activities); For-profit organizations or ventures; Organizations that discriminate based on race, creed, color, religion, gender, ethnicity, national origin, sexual orientation, age, disability or veteran status.
Geographic Focus: All States
Contact: Lori Kenepp, (408) 576-7528; lori.kenepp@flextronics.com
Internet: http://www.flextronics.com/social_resp/Flextronics_Foundation/Pages/Grants.aspx
Sponsor: Flextronics International
6201 America Center Drive
San Jose, CA 95002

Flinn Foundation Scholarships 1675
The Flinn Foundation scholarships, in partnership with Arizona's three state universities, provide enriched educational offerings that expand a recipient's life and career options. Students receive a financial package for their entire undergraduate study that includes free tuition, room, and board, funding for study abroad, mentorship from faculty, exposure to world leaders, and fellowship in a community of current and alumni Scholars. Total dollar value exceeds $50,000, in addition to the cash value of tuition provided by the universities. Scholars begin as a group, with a three-week seminar in Central Europe. Each Scholar also receives a stipend for at least one international summer seminar, or a semester or year at a foreign university.
Requirements: Recipients must rank in the top 5 percent of their high school graduating class with at least a 3.5 grade point average; score a minimum of 29 on the ACT or 1300 on the SAT (critical reading and math sections only); demonstrate leadership in a variety of extracurricular activities; and hold U.S. citizenship and residency in Arizona for the two years prior to application. All majors may apply.
Restrictions: Applicants must apply to one of three Arizona universities: Arizona State University, Northern Arizona University, or University of Arizona.
Geographic Focus: Arizona
Date(s) Application is Due: Oct 21
Amount of Grant: 50,000 USD
Contact: Flinn Scholarship Contact; (602) 744-6800; fscholars@flinn.org
Internet: http://www.flinnscholars.org/news/977
Sponsor: Flinn Foundation
1802 N Central Avenue
Phoenix, AZ 85004-1506

Florence Hunt Maxwell Foundation Grants 1676
The mission of the Florence Hunt Maxwell Foundation is to support charitable organizations that provide for the underserved and indigent community. Grants from the Florence Hunt Maxwell Foundation are primarily one year in duration; on occasion, multi-year support is awarded. Applicants must apply online at the grant website. Applicants are strongly encouraged to do the following before applying: review the downloadable state application procedures for additional helpful information and clarifications; review the downloadable online-application guidelines at the grant website; review the foundation's funding history (link is available from the grant website); review the online application questions in advance; and review the list of required attachments. These will generally include: a list of board members, financial statements (audited, reviewed, or compiled by independent auditor); an organization summary; a list of other funding sources; an IRS Determination letter; and other required documents. All attachments must be uploaded in the online application as PDF, Word, or Excel files. The Florence Hunt Maxwell Foundation application deadline is 11:59 p.m. on April 1. Applicants will be notified of grant decisions by letter within one to two months after the deadline.
Requirements: Applicants must have 501(c)3 tax-exempt status and serve residents of the Metro Atlanta area.
Restrictions: The foundation does not support requests from individuals, organizations attempting to influence policy through direct lobbying, or any political campaigns.
Geographic Focus: Georgia
Date(s) Application is Due: Mar 1
Contact: Mark S. Drake, Vice President; (404) 264-1377; mark.s.drake@ustrust.com
Internet: https://www.bankofamerica.com/philanthropic/fn_search.action
Sponsor: Florence Hunt Maxwell Foundation
3414 Peachtree Road, N.E., Suite 1475, GA7-813-14-04
Atlanta, GA 30326-1113

Florian O. Bartlett Trust Grants 1677
The Florian O. Bartlett Trust was established in 1937 to support and promote quality educational, human services, and health care programming for underserved populations. The application deadline for the Florian O. Bartlett Trust is April 1. Applicants will be notified of grant decisions before June 30. Grant requests for general operating support are strongly encouraged. Program support will also be considered. Small, program-related capital expenses may be included in general operating or program requests. The majority of grants from the Bartlett Trust are 1 year in duration. On occasion, multi-year support is awarded.
Geographic Focus: Massachusetts
Date(s) Application is Due: Apr 1
Contact: Michealle Larkins; (866) 778-6859; michealle.larkins@baml.com
Internet: https://www.bankofamerica.com/philanthropic/fn_search.action
Sponsor: Florian O. Bartlett Trust
225 Franklin Street, 4th Floor, MA1-225-04-02
Boston, MA 02110

Florida BRAIve Fund of Dade Community Foundation 1678
The Florida BrAIve Fund at Dade Community Foundation will award grants to non-profits to assist eligible military personnel and family members residing in Miami-Dade, Monroe, Broward, Palm Beach and Martin counties. BRAIVE Fund grants will address needs of military personnel and their families in the target population related to pre-deployment, during deployment, and their return from duty. To obtain more information and the application forms, go to http://www.dadecommunityfoundation.org/Site/wc/wc143.jsp.
Requirements: General criteria for evaluating applications to The Florida BrAIve Fund of Dade Community Foundation: applicant is a 501(c)3 organization; applicant organization has a proven record of effective service to its clientele; applicant organization demonstrates the management and financial capacity to efficiently achieve the grant purpose; grants will be made to nonprofit organizations, not individuals; programs or services can only benefit current and former military personnel serving in Iraq or Afghanistan and their families; program services must be provided in the area of Florida to be covered by Dade Community Foundation as indicated in The Florida BrAIve Fund map; successful organizations will demonstrate that the agency is currently serving military personnel and their families in Florida or has the experience to address an unmet need.

Geographic Focus: Florida
Date(s) Application is Due: Jun 8
Contact: Claudianna Williams, Administrative Assistant; (305) 371-2711; fax (305) 371-5342; claudianna.williams@dadecommunityfoundation.org
Internet: http://www.dadecommunityfoundation.org/Site/wc/wc143.jsp
Sponsor: Dade Community Foundation
200 S Biscayne Boulevard, Suite 505
Miami, FL 33131-5330

Florida Division of Cultural Affairs Arts In Education Arts Partnership Grants 1679
Arts In Education project grants are designed to cultivate learning and artistic development for all students and teachers by promoting, encouraging, and supporting arts and culture as an integral part of education and lifelong learning for residents and visitors. This includes but is not limited to: the learning and artistic development of pre-kindergarten through grade 12 students and teachers; or initiatives and proposals that help applicants to work as individuals or partners to carry out community programs and school reform through the arts. The Arts Partnership funding category is designed to support projects that will advance arts education and the development of long-term partnerships through effective collaboration between community arts and cultural organizations, social service agencies, and educational entities. Applicants may request up to $25,000 for arts partnership projects that have completed planning and design work and are ready for implementation or expansion. If proposals include computer, video, and technology equipment, applicants must show how technology equipment, systems, and programs are integrated into their specific arts education partnership. Focus areas for the Arts Partnership project may include the following: school-based arts education; programming that integrates the arts into areas not usually associated with the arts (non-arts curriculum, school-to-work initiatives, the criminal justice system, the healthcare system, community care for the elderly, underserved populations, and adult-continuing education programs); programming that brings together different generations; arts and technology programming in music, visual arts, theatre, dance, media and/or literary arts; and community arts education.
Restrictions: Arts Partnership projects are not intended to fund the same project year after year. However panelists have the discretion to recommend funding for on-going projects.
Geographic Focus: Florida
Contact: Laura Lewis Blischke; (850) 245-6475; llblischke@dos.state.fl.us
Internet: http://www.florida-arts.org/documents/guidelines/2012-2013.scp.guidelines.cfm#aie
Sponsor: Florida Division of Cultural Affairs
500 South Bronough Street, 3rd Floor
Tallahassee, FL 32399-0250

Florida High School/High Tech Project Grants 1680
Florida High School/High Tech (HS/HT) is designed to provide high school students with all types of disabilities the opportunity to explore jobs or postsecondary education leading to technology- related careers. HS/HT links youth to a broad range of academic, career development and experiential resources and experiences that will enable them to meet the demands of the 21st century workforce. HS/HT is a community-based partnership made up of students, parents and caregivers, businesses, educators and rehabilitation professionals. It has been shown to reduce the high school dropout rate and increase the overall self-esteem of participating students. If you are interested in participating, there is a role for you. The program provides step-down funding while the program is gaining independent funding for the future of their program.
Requirements: Any established, Florida nonprofit corporation, agency, organization or association that has been granted exemption from federal income tax under Section 501(c)3 of the IRS Tax Code is eligible to submit a proposal for review.
Geographic Focus: Florida
Contact: Guenevere Crum, Senior Vice President; (850) 224-4493; fax (850) 224-4496; guenevere@abletrust.org or info@abletrust.org
Internet: http://www.abletrust.org/hsht/
Sponsor: Able Trust
3320 Thomasville Road, Suite 200
Tallahassee, FL 32308

Floyd A. and Kathleen C. Cailloux Foundation Grants 1681
The foundation awards grants in the areas of civic and cultural, education and youth, family and community service, and health and rehabilitation. Grants support specific projects or programs, technical assistance, and capital projects. Grants for general operations are less common and usually small. Endowment grants are rare. The foundation only reviews grant proposals from applicants whose letters of inquiry have been approved. Applications are taken online.
Requirements: Grants support 501(c)3 nonprofit organizations. Preference is given to requests from Texas Hill Country.
Restrictions: The Foundation does not fund church or seminary construction, or church related entities or activities other than ecumenically oriented projects/programs that otherwise meet Foundation guidelines. The general policy of the Foundation is not to make grants to an organization more than once within any 12-month period. Grants or loans to individuals are never made.
Geographic Focus: All States
Contact: Grants Administrator; (830) 895-5222; info@caillouxfoundation.org
Internet: http://www.caillouxfoundation.org/grant_guidelines.htm
Sponsor: Floyd A. and Kathleen C. Cailloux Foundation
P.O. Box 291276
Kerrville, TX 78029-1276

Fluor Foundation Grants 1682
The Fluor Corporation achieves its contribution objectives through the Fluor Foundation and corporate giving. The Foundation's areas of interest are: education; human services; cultural outreach; and public/civic affairs. The Foundation considers requests for operating, program, capital or endowment support. Priority is given to funding organizations with employee volunteer participation. Application information is available online.
Requirements: Fluor's giving programs focus on community organizations in those locations around the world where the company has a presence. See website for local contact information: http://www.fluor.com/sustainability/community/fluor_giving/Pages/applying_for_fluor_foundation_grants.aspx
Restrictions: Funding is not available for: film production/publishing activities; individuals; sports organizations/programs; veterans, fraternal, labor or religious organizations; or lobbying/political organizations or campaigns.
Geographic Focus: Alaska, California, Louisiana, New Jersey, New York, North Carolina, Pennsylvania, South Carolina, Tennessee, Texas, Virginia, Washington, Albania, Andorra, Armenia, Australia, Austria, Azerbaijan, Belarus, Belgium, Bosnia & Herzegovina, Bulgaria, Canada, Caribbean, China, Croatia, Cyprus, Czech Republic, Denmark, Estonia, Finland, France, Georgia, Germany, Greece, Hungary, Iceland, Ireland, Italy, Japan, Kosovo, Latvia, Liechtenstein, Lithuania, Luxembourg, Macedonia, Malta, Mexico, Moldova, Monaco, Montenegro, New Zealand, Norway, Peru, Philippines, Poland, Poland, Portugal, Romania, Russia, Russia, San Marino, Serbia, Slovakia, Slovenia, Spain, Sweden, Switzerland, The Netherlands, Turkey, Ukraine, United Kingdom, Vatican City, Venezuela
Contact: Suzanne Esber, Executive Director of Community Relations; (949) 349-7847; fax (949) 349-7694; suzanne.esber@fluor.com
Internet: http://www.fluor.com/sustainability/community/fluor_giving/Pages/applying_for_fluor_foundation_grants.aspx
Sponsor: Fluor Foundation
3 Polaris Way
Aliso Viejo, CA 92698

FMC Foundation Grants 1683
The foundation supports education, community improvement, urban affairs, health and human services, and public issues/economic education. Higher education is supported through scholarships and employee matching gifts. Education support tends to be in business, engineering, chemistry, and some minority education programs. Eligible applicants need to contact the FMC in their geograhic area. Each individual FMC location determines how their contributions will be given.
Requirements: Grants are awarded to 501(c)3 organizations in FMC-plant communities. US-based organizations with an international focus are also eligible.
Restrictions: Grants are not made to individuals.
Geographic Focus: Arizona, California, Delaware, Florida, Illinois, Louisiana, Maine, Maryland, Missouri, New Jersey, New York, North Carolina, Pennsylvania, Tennessee, Texas, West Virginia, Wyoming, Canada
Contact: Program Contact; (215) 299-6000; fax (215) 299-6140
Internet: http://www.fmc.com
Sponsor: FMC Foundation
1735 Market Street
Philadelphia, PA 19103

Foellinger Foundation Grants 1684
The foundation awards grants primarily in Fort Wayne and Allen County, IN, to improve the quality of life in the areas of early child development, youth development, strengthening family services, strengthening organizations, and community concerns. Special emphasis is given to projects/programs that help children and their families, particularly those with the greatest economic need and least opportunity. By doing so, the Foundation hopes to help children and their families move from dependence to independence. In addition to programs and projects, the foundation also awards grants for general operating, capital, seed money, renovation projects, conferences and seminars, consulting services, equipment, challenge/matching, and planning purposes.
Requirements: Indiana organizations are eligible to apply.
Restrictions: The foundation does not fund scholarships, travel assistance, conference fees, religious groups, public or private elementary or secondary schools, sponsorships, special events, advertising, or endowments.
Geographic Focus: Indiana
Date(s) Application is Due: Feb 1; May 1; Nov 1
Amount of Grant: 10,000 - 100,000 USD
Contact: Cheryl Taylor; (219) 422-2900; fax (219) 422-9436; info@foellinger.org
Internet: http://www.foellinger.org
Sponsor: Foellinger Foundation
520 E Berry Street
Fort Wayne, IN 46802

Fondren Foundation Grants 1685
The foundation awards grants to Texas-based programs in the areas of education, youth services, health care, and human services. The board meets quarterly to consider requests.
Requirements: Texas nonprofit organizations, with an emphasis on Houston and the Southwest are eligible to apply.
Restrictions: The Foundation does not allow funding to individuals, or for operating or annual fund drives.
Geographic Focus: Texas
Amount of Grant: 10,000 - 100,000 USD
Contact: Melanie Scioneaux, Assistant Secretary

Internet: http://www.tcc.state.tx.us/cancerfunding/regional/05.html
Sponsor: Fondren Foundation
7 TCT 37, P.O. Box 2558
Houston, TX 77252-8037

Ford Family Foundation Grants - Access to Health and Dental Services   1686
The Health and Dental Services grants increase the health of underserved children through improve access to health and dental services, preventative services, and education. Priority is given to organizations that can demonstrate how grant funding will create clear results and increased capacity; those that show integration of "promising" or "evidence-based" best practice; efforts to increase access for very young children, ages 0-10; and those efforts with evidence of strong regional collaboration and coordination. Lower priority is given to capital funding for large hospitals and medical facilities.
Requirements: Grant requests must meet all of the following requirements before consideration will be given; applicant organizations must have current 501(c)3 public charity status from the IRS, or be a governmental entity, or be an IRS-recognized tribe. It may not be a private foundation as defined in Section 509(a) of the Internal Revenue Code; geographical focus of project must be predominately (60% or more) for the benefit of residents of rural Oregon and Siskiyou County, California. Rural is defined as communities with populations of 30,000 or less and not adjacent to or part of an urban or metropolitan area; must include significant collaboration and community buy-in (as evidenced by in-kind and cash contributions from local and regional sources); must have at least 50% of funding (may include in-kind) for the total project budget committed before applying; organization must not be delinquent in filing final reports for previous grants from the Foundation; organization may not be currently receiving other responsive grant funds from the Foundation. If the organization has received prior funding from the Foundation, they must wait 12 months after the completion of the prior grant before applying again for support.
Restrictions: Funds requested may not exceed one third of the project's total budget. The Foundation will usually not consider funding requests for: projects or programs that are indirectly funded through a fiscal agent; endowments or reserve funds; general fund drives, such as United Way; debt retirement or operating deficits; indirect expenses unrelated to the project or program being funded; sponsorship of fundraising events; or propagandizing or influencing elections or legislation.
Geographic Focus: California, Oregon
Amount of Grant: 50,000 - 100,000 USD
Contact: Grant Contact; (541) 957-5574; fax (541) 957-5720; info@tfff.org
Internet: http://www.tfff.org/Grants/GeneralInformation/tabid/81/Default.aspx
Sponsor: Ford Family Foundation
1600 NW Stewart Parkway
Roseburg, OR 97401

Ford Motor Company Fund Grants   1687
The fund supports not-for-profit organizations in three major areas: innovation and education; community development and American legacy, and auto-related safety education. The fund seeks to build partnerships with organizations that have a well-defined sense of purpose, a demonstrated commitment to maximizing available resources, and a reputation for meeting objectives and delivering quality programs and services. Priority is placed on the support and development of organizations that promote diversity and inclusion. Requests for support are accepted and reviewed throughout the year. The fund now implements an online application system. Details are available online.
Restrictions: Ford does not fund: advocacy-directed programs; animal-rights organizations; beauty or talent contests; day-to-day business operations; debt reduction; donation of vehicles; efforts to influence legislation, or the outcome of any elections or any specific election of candidates to public office or to carry on any voter registration drive; endowments; fraternal organizations; general operating support to hospitals and health care institutions; individual sponsorship related to fundraising activities; individuals; labor groups; loans for small businesses; loans to program-related investments; organizations that do not have 501(c)3 status; organizations that unlawfully discriminate in their provision of goods and services based on race, color, religion, gender, gender identity or expression, ethnicity, sexual orientation, national origin, physical challenge, age, or status as a protected veteran; political contributions; private K-12 schools; profit-making enterprises; religious programs or sectarian programs for religious purposes; species-specific organizations; sports teams.
Geographic Focus: All States
Samples: American Red Cross, Southeastern Michigan Chapter (Detroit, MI)--for disaster-relief efforts in Florida related to Hurricane Charley, $50,000.
Contact: Ford Fund Coordinator; (888) 313-0102; Fordfund@ford.com
Internet: http://www.ford.com/en/goodWorks/fundingAndGrants/fordMotorCompanyFund/default.htm
Sponsor: Ford Motor Company Fund
P.O. Box 1899
Dearborn, MI 48121-1899

Forrest C. Lattner Foundation Grants   1688
The Foundation's primary objectives are in six areas of interest: arts and humanities; education; environment; health and social services; historic preservation; and medical research. The Foundation also wishes to encourage the development of innovative model programs. Application guidelines are available online.
Restrictions: The foundation is unable to fulfill grant requests to individuals, or to support programs that are the primary responsibility of the public sector.
Geographic Focus: Florida, Georgia, Kansas, Rhode Island, Texas
Date(s) Application is Due: Mar 1; Sep 1
Samples: Woodmere Art Museum (Philadelphia, PA)--for programs and exhibitions at the Helen Millard Children's Gallery, $20,000.
Contact: Susan Lattner Lloyd; (561) 278-3781; fax (561) 278-3167; lattner@bellsouth.net
Internet: http://www.lattnerfoundation.org
Sponsor: Forrest C. Lattner Foundation
777 E Atlantic Avenue, Suite 317
Delray Beach, FL 33483

Foster Foundation Grants   1689
The Foundation funds effective, nonprofit organizations that can provide tangible benefits and ongoing support to improve the quality of life for individuals, families and communities within the Pacific Northwest. The Foundation's primary field of interests are: basic human welfare issues, education, medical research, treatment and care and cultural activities. The Foundation places special emphasis on meeting the needs of the underserved and disadvantaged segments of the population, especially children, women and seniors.
Requirements: Eligible applicants must: be a nonprofit located in or serving the populations of the Pacific Northwest; address as their mission or project intent one of the Foundation's priority issues for funding; not have any delinquent final reports due to the Foundation from previous grant cycles.
Restrictions: The Foundation will not consider requests for: direct grants, scholarships or loans for the benefit of specific individuals; projects of organizations whose policies or practices discriminate on the basis of race, ethnic origin, sex, creed or sexual orientation; contributions or program support for sectarian or religious organizations whose activities benefit only their members; or loans.
Geographic Focus: Alaska, Idaho, Montana, Oregon, Washington
Contact: Jill Goodsell; (206) 726-1815; fax (206) 903-0628; info@thefosterfoundation.org
Internet: http://thefosterfoundation.org/Grants_Guide.asp
Sponsor: Foster Foundation
601 Union Street, Suite 3707
Seattle, WA 98101

Foundation Fighting Blindness Marjorie Carr Adams Women's Career Development Awards   1690
The Program's primary purpose is to support female clinical research scientists of superior dedication and talent to pursue vigorous research programs to find the cures for retinal degenerative diseases research (RDD) i.e. inherited orphan retinal degenerative diseases and dry age-related macular degeneration. The goal of this program is to facilitate advances in the laboratory and clinical research, to elucidate the mechanisms for the etiology and pathogenesis of RDDs and to develop innovative strategies to prevent, treat and cure these diseases. This award provides support for mentored study and research for clinically trained biomedical research professionals who have the potential to assume leadership roles as clinician-scientists in RDD research.
Requirements: Female clinical scientists possessing an, M.D., D.O., or recognized equivalent foreign degrees and who are in their first, second or third year of a junior faculty appointment are eligible to apply for the Adams WCDA. Applicants do not have to be U.S. citizens.
Geographic Focus: All States
Date(s) Application is Due: Nov 14
Amount of Grant: 65,000 - 325,000 USD
Contact: Cindy Settar, (410) 568-0150 or (800) 683-5555; csettar@Fightblindness.org
Internet: http://www.blindness.org/index.php?option=com_content&view=section&id=9&Itemid=107#career-development
Sponsor: Foundation Fighting Blindness
7168 Columbia Gateway Drive, Suite 100
Columbia, MD 21046

Foundation for a Healthy Kentucky Grants   1691
The Foundation seeks to address the unmet health care needs of the people of Kentucky through strategic grants designed to improve health status and access to care. Funding is concentrated in the following areas: fitness and nutrition for children and families; youth smoking prevention; and youth substance abuse prevention. The Foundation also hopes to enhance access to health care for low-income and uninsured populations; health care for rural populations; and integrated mental health and medical services.
Requirements: The Foundation accepts proposals for funding only in direct response to a specific Request for Proposals (RFP) or Request for Quotes (RFQ) or grant solicitation issued by the Foundation. Organizations should familiarize themselves with the Foundation's What We Fund page, and the Grant Guidelines to determine whether they are eligible to receive funding. They should sign up for the Foundation's mail/email list to receive grant opportunities and free educational forums. Once the organization receives an RFP announcement, they should review it carefully to see that the goals of the RFP fit with their particular project. If it is acceptable, applicants should submit their proposal according to the guidelines and RFP due dates.
Restrictions: The Foundation will not review unsolicited grant requests except for requests for Matching Grant and Conference Support funds. Grants do not support direct patient care, except as part of demonstration or replication projects; capital campaigns or requests for bricks and mortar (although project related equipment may be included in requests); overhead expenses except in limited amounts for specific projects; organizations that discriminate on the basis of race, gender, age, religion, national origin, sexual orientation, disability, military, or marital status in hiring; multi-year commitments; expenses related to registered legislative agents for the purpose of lobbying; endowment funds; individuals; private, for-profit entities; religious organizations for religious purposes; political causes; or retroactive expenses, deficit reduction, or forgiveness.
Geographic Focus: Kentucky

Samples: Home of the Innocents, Louisville, KY, establish a dental clinic for children in state care, children with special health care needs and other children served by the Home, $250,000; St. Joseph Health System, Lexington, KY, establish primary care clinics in two low-income rural communities, $250,000.
Contact: Susan Zepeda, President/Chief Executive Officer; (502) 326-2583 or (877) 326-2583; fax (502) 326-5748; szepeda@healthyky.org or info@healthy-ky.org
Internet: http://www.healthyky.org
Sponsor: Foundation for a Healthy Kentucky
9300 Shelbyville Road, Suite 1305
Louisville, KY 40222

Foundation for Appalachian Ohio Bachtel Scholarships   1692
Dr. Harry Keig grew up to become a successful physician and surgeon in the Omaha area but never forgot his hometown or the teachers who educated him. He also never forgot the life-shaping influence of his friend, Forrest Bachtel, longtime teacher and coach at Middleport High School in Meigs County. When Dr. Keig passed away in 2003, his bequest created a $500,000 endowed fund at the Foundation for Appalachian Ohio in honor of his friend, Coach Bachtel, to provide college scholarships for Meigs High School students. Each scholarship recipient receives $2,500 to assist with college education.
Requirements: Available to Meigs High School graduates, located in Pomeroy, Ohio.
Geographic Focus: Ohio
Amount of Grant: 2,500 USD
Contact: Phyllis Moody; (740) 753-1111; fax (740) 753-3333; pmoody@ffao.org
Internet: http://www.appalachianohio.org/scholarships/index.php?section=258&page=296
Sponsor: Foundation for Appalachian Ohio
36 Public Square, P.O. Box 456
Nelsonville, OH 45764

Foundation for Appalachian Ohio Susan K. Ipacs Nursing Legacy Scholarships   1693
The Susan K. Ipacs Nursing Legacy Scholarship Fund was established at the Foundation for Appalachian Ohio to honor the work and life of Susan Ipacs – nurse, instructor, mother, wife and friend. Sue never lost her love of nursing and never forgot the circumstances that led to her college education. As a nursing instructor and associate dean, Sue strove to ease the burdens and obstacles facing students, while nurturing the passion to help others in all who surrounded her, both students and fellow colleagues. She felt a personal responsibility to recognize and attend to the underprivileged and non-traditional student. As so many depended on Sue for her compassion and dedication to the profession, it is the hope that students pursuing their nursing degree remember the reason behind the studies, and the importance of compassion for people.
Geographic Focus: Ohio
Contact: Phyllis Moody; (740) 753-1111; fax (740) 753-3333; pmoody@ffao.org
Internet: http://www.appalachianohio.org/scholarships/index.php?section=258&page=459
Sponsor: Foundation for Appalachian Ohio
36 Public Square, P.O. Box 456
Nelsonville, OH 45764

Foundation for Appalachian Ohio Zelma Gray Medical School Scholarship   1694
Before her death in 2007, Zelma Gray established her support for medical professionals and care in Guernsey County through a scholarship fund with WesBanco Trust and Investment Services. Designed to support Guernsey County residents' pursuit of medical school education as well as encourage that talent to remain in, or return to, Guernsey County to serve its citizens upon completion of medical training. Funding will cover the cost of tuition, lab fees and text books while pursuing a Doctor of Medicine (M.D.) or Doctor of Osteopathic Medicine (D.O.) at an accredited medical school within the United States, in hopes that he or she will practice medicine in Guernsey County upon completion of his or her degree and training.
Requirements: Current and recent Guernsey County residents are eligible for the Zelma Gray Medical School Scholarship. Awards will be determined through objective criteria, including MCAT scores and college transcripts.
Geographic Focus: Ohio
Contact: Phyllis Moody; (740) 753-1111; fax (740) 753-3333; pmoody@ffao.org
Internet: http://www.appalachianohio.org/scholarships/index.php?section=258&page=442
Sponsor: Foundation for Appalachian Ohio
36 Public Square, P.O. Box 456
Nelsonville, OH 45764

Foundation for Balance and Harmony Grants   1695
The Foundation for Balance and Harmony was established in the State of Virginia in 2002 with the expressed purpose of working to expand the awareness of Kinesiology and Reiki to the general public as a compliment to conventional healing methods. The Foundation's mission was born after our founder Shawn Adler experienced his first Kinesiology therapy session. It strives to make the world a better place and to support everyone in making their life better, happier, and more fulfilling. The foundation has sponsored programs around the world by offering Kinesiology therapy and other healing sessions at no cost to some of the neediest people. Awards range from $1,000 to $2,000. There are no specific application forms, and no annual deadlines.
Geographic Focus: All States
Amount of Grant: 1,000 - 2,000 USD
Contact: Shawn D. Adler, President; (703) 433-2248
Internet: https://www.ffbh.org/
Sponsor: Foundation for Balance and Harmony
800 Lake Windermere Court
Great Falls, VA 22066

Foundation for Health Enhancement Grants   1696
The purpose of the Foundation for Health Enhancement is to improve care in the United States by the development of the science of delivery of all kinds of health care, including medical, surgical, dental, nursing, health education, and other curative and preventive health services. The majority of grants from the Foundation for Health Enhancement are one year in duration. Applicants must apply online at the grant website. Applicants are strongly encouraged to do the following before applying: review the downloadable state application procedures for additional helpful information and clarifications; review the downloadable online-application guidelines at the grant website; review the foundation's funding history (link is available from the grant website); review the online application questions in advance; and review the list of required attachments. These will generally include: a list of board members, financial statements (audited, reviewed, or compiled by independent auditor); an organization summary; a list of other funding sources; an IRS Determination letter; and other required documents. All attachments must be uploaded in the online application as PDF, Word, or Excel files. The Foundation for Health Enhancement has biannual deadlines of April 1 and September 1. Applications should be submitted by 11:59 p.m. on the deadline dates. In general, applicants will be notified of grant decisions 3 to 4 months after proposal submission.
Requirements: The Foundation places special emphasis on: preventive health services; smaller qualifying applicants in an area geographically proximate to Chicago and/or the Midwest; and service organizations as opposed to research organizations.
Geographic Focus: Illinois
Date(s) Application is Due: Apr 1; Sep 1
Contact: George Thorn; (312) 828-4154; ilgrantmaking@bankofamerica.com
Internet: https://www.bankofamerica.com/philanthropic/fn_search.action
Sponsor: Foundation for Health Enhancement
231 South LaSalle Street, IL1-231-13-32
Chicago, IL 60604

Foundation for Pharmaceutical Sciences Herb and Nina Demuth Grant   1697
The purpose of the Foundation for Pharmaceutical Sciences Herb and Nina Demuth grant program is to help alleviate the shortage of skilled medical and pharmaceutical scientists that could hamper the research, development and manufacture of new pharmaceutical and biotech products and delivery technologies. A maximum award of $25,000 is available. To obtain further information concerning the application process, please contact the Foundation office directly via email. Applications should be postmarked by the May 30 annual deadline. emailed or faxed proposals are accepted.
Requirements: The Foundation is not affiliated with the Parenteral Drug Association (PDA). Eligible applicants are graduate students in universities, colleges, or medical centers that need assistance to complete their Doctoral thesis. This includes: U.S. citizens, resident, or foreign students studying in U.S. institutions.
Geographic Focus: All States
Date(s) Application is Due: May 30
Contact: Roger Dabbah; (301) 762-9258; fax (301) 762-5356; rdabbah@verizon.net
Sponsor: Foundation for Pharmaceutical Sciences
5 Eaglebrook Court
Potomac, MD 20854

Foundation for Seacoast Health Grants   1698
The mission of the foundation is to invest its resources to improve the health and well being of Seacoast residents. The foundation considers a very limited number of new grant initiatives which address one or more of the following prioritized health needs: access to affordable mental health services; access to preventative and restorative dental services; access to affordable child care and after school care; access to affordable primary medical care; and coordination and dissemination of health information related to identified priority needs. The deadline for the submission of the letter of intent is March 1 for Infants, Children and Adolescence and June 1 for Promoting Health and Preventing Disease. Annual deadline dates may vary; contact program staff for exact dates.
Requirements: Nonprofit organizations in cities and towns in the New Hampshire/Maine Seacoast area, including Greenland, New Castle, Newington, North Hampton, Portsmouth, and Rye, NH; and Eliot, Kittery, and York, ME, may submit applications.
Restrictions: Grants do not support ongoing general operating expenses, deficit elimination, political activities, travel, conferences, or lodging.
Geographic Focus: Maine, New Hampshire
Date(s) Application is Due: Mar 1; Jun 1
Contact: Susan Bunting, President; (603) 422-8200; ffsh@communitycampus.org
Internet: http://www.ffsh.org/grants.cfm
Sponsor: Foundation for Seacoast Health
100 Campus Drive, Suite 1
Portsmouth, NH 03801

Foundation for the Mid South Community Development Grants   1699
The Foundation for the Mid South was established to bring together the public and private sectors and focus their resources on increasing social and economic opportunity. The foundation's community development work includes five key focuses that, together, enable communities to grow and prosper: Community Enrichment, Economic Development, Leadership Development, Education, and Health and Wellness. Each focus area addresses an essential community element.
Requirements: Contact a program officer to discuss your project prior to submitting a proposal To be eligible for a grant, the applying organization must possess tax-exempt status under section 501(c)3 of the Internal Revenue Code and a certificate from the Mississippi Secretary of State that designates it as a public charity or exempt for the state of Mississippi. Mississippi organizations must register and receive a certificate that designates it as a public charity or exempt for the state of Mississippi when submitting a

full proposal. The application, Form URS, and instructions are available at the sponsor's website. All other organizations must apply and receive a notice of exemption, Form CE, from the Mississippi Secretary of State's office, also available from the website. If the proposed work aligns with the Foundation's priorities and goals, eligible organizations may submit the Grant Inquiry Form found on the website.
Restrictions: The Foundation does not award grants to individuals or make grants for personal needs or business assistance. Additionally, funds are not awarded for lobbying activities; ongoing general operating expenses or existing deficits; endowments; capital costs including construction, renovation, or equipment; or international programs.
Geographic Focus: Arkansas, Louisiana, Mississippi
Contact: Justin A. Burch; (601) 355-8167; fax (601) 355-6499
Internet: http://www.fndmidsouth.org/priorities/community-development
Sponsor: Foundation for the Mid South
134 East Amite Street
Jackson, MS 39201

Foundation for the Mid South Health and Wellness Grants 1700
The Foundation for the Mid South was established to bring together the public and private sectors and focus their resources on increasing social and economic opportunity. The Health and Wellness program supports activities that promote healthy behaviors leading to the reduction of obesity and diabetes and projects that increase mental health awareness and access to services. The program also seeks to expand access to health services, emphasizing the use of technology to serve rural communities.
Requirements: Contact a program officer to discuss your project prior to submitting a proposal To be eligible for a grant, the applying organization must possess tax-exempt status under section 501(c)3 of the Internal Revenue Code and a certificate from the Mississippi Secretary of State that designates it as a public charity or exempt for the state of Mississippi. Mississippi organizations must register and receive a certificate that designates it as a public charity or exempt for the state of Mississippi when submitting a full proposal. The application, Form URS, and instructions are available at the sponsor's website. All other organizations must apply and receive a notice of exemption, Form CE, from the Mississippi Secretary of State's office, also available from the website. If the proposed work aligns with the Foundation's priorities and goals, eligible organizations may submit the Grant Inquiry Form found on the website.
Restrictions: The Foundation does not award grants to individuals or make grants for personal needs or business assistance. Additionally, funds are not awarded for lobbying activities; ongoing general operating expenses or existing deficits; endowments; capital costs including construction, renovation, or equipment; or international programs.
Geographic Focus: Arkansas, Louisiana, Mississippi
Contact: Dwanda Moore; (601) 355-8167; fax (601) 355-6499
Internet: http://www.fndmidsouth.org/priorities/health-wellness/#theoverview
Sponsor: Foundation for the Mid South
134 East Amite Street
Jackson, MS 39201

Foundation of CVPH April LaValley Fund Grants 1701
In December of 1997, Mrs. Caroline LaValley announced the creation of the April LaValley Endowment Fund. She said that it was April's wish that money which was donated to her family to assist them with expenses related to April's care should be used for other families battling cystic fibrosis. Income generated by the fund is used to help pay for transportation and lodging expenses incurred by immediate family members who travel to be with a family member receiving care for cystic fibrosis. Expenses associates with travel and lodging of family members are allowed.
Requirements: Residents of Clinton, Essex and Franklin counties in New York are eligible for consideration. The person for whom medical expenses are incurred must have a diagnosis of cystic fibrosis.
Restrictions: Funds are not intended to be used for direct medical expenditures.
Geographic Focus: New York
Contact: Roger G. Ahrens; (518) 562-7168 or (518) 561-2000; rahrens@cvph.org
Internet: http://www.cvph.org/Foundation/NamedFunds/AprilLaValleyFund/
Sponsor: Foundation of Champlain Valley Medical Center
75 Beekman Street
Plattsburgh, NY 12901

Foundation of CVPH Chelsea's Rainbow Fund Grants 1702
In December of 1999, the Kinblom family announced the creation of the Chelsea's Rainbow Fund. Their seven year old daughter, Chelsea, fought a long courageous battle against brain cancer. Friends throughout the North Country contributed to a fund to assist with expenses not covered by health insurance. The Kinblom family wishes to help other families that face similar battles against childhood cancer. Income generated by this fund will be used to help pay for transportation, food and housing expenses incurred by immediate family members who travel with a child receiving care for cancer. Expenses associates with travel and lodging of family members are allowed.
Requirements: Residents of Clinton, Essex, and Franklin counties in New York are eligible for consideration. Families that have children 18 years old and younger who are battling cancer are eligible for assistance.
Restrictions: Funds are not intended to be used for direct medical expenditures.
Geographic Focus: New York
Contact: Roger G. Ahrens; (518) 562-7168 or (518) 561-2000; rahrens@cvph.org
Internet: http://www.cvph.org/Foundation/NamedFunds/ChelseasRainbow/
Sponsor: Foundation of Champlain Valley Medical Center
75 Beekman Street
Plattsburgh, NY 12901

Foundation of CVPH Melissa Lahtinen-Penfield Organ Donor Fund Grants 1703
The Fund was created in memory of Melissa Lahtinen-Penfield, whose wish was to assist families with expenses related to organ donation and educate the community of its vital importance. In July 2000, William Santa and members of the Hometown WIRY team, established the annual Melissa Lahtinen-Penfield (MLP) Organ Donor Golf Tournament in her memory. Proceeds of the tournament, together with donations made directly to the fund, are used to promote organ donation awareness and defray those costs associated with organ transplants which are not covered by health insurance. Expenses considered for reimbursement include medical expenses not covered by insurance, travel and lodging expenses, and meals not including alcohol. Financial need will be the primary determinant upon which the application will be weighed. The size of grants is limited to the income generated by the fund.
Requirements: Residents of Clinton County, New York, are eligible to receive consideration for grants of up to $2,500/annually for expenses associated with organ transplants not covered by health insurance.
Restrictions: Funds are not intended to be used for direct medical expenditures.
Geographic Focus: New York
Amount of Grant: Up to 2,500 USD
Contact: Roger G. Ahrens, Executive Director; (518) 562-7168 or (518) 561-2000; fax (518) 561-0881; rahrens@cvph.org
Internet: http://www.cvph.org/Foundation/NamedFunds/OrganDonor/
Sponsor: Foundation of Champlain Valley Medical Center
75 Beekman Street
Plattsburgh, NY 12901

Foundation of CVPH Rabin Fund Grants 1704
In December of 2000, Pearlie and Mark Rabin established a fund to assist patients with fundamentals of care at CVPH Medical Center. With a gift which endowed the fund, they outlined their wish to offer patients who experience financial hardship, services such as essential articles of clothing and other inexpensive incidentals, which might make their stay in the hospital more pleasant. Throughout the year, CVPH staff takes care of patients who are not able to afford these types of items. In the past, staff members pooled their personal resources and purchased the item or the patient managed without it. The Rabin's Fund provides about $3,500 annually, which assists friends and neighbors from throughout the North Country.
Requirements: Residents of Clinton, Essex, and Franklin counties in New York are eligible for consideration.
Restrictions: Funds are not intended to be used for direct medical expenditures.
Geographic Focus: New York
Contact: Roger G. Ahrens, Executive Director; (518) 562-7168 or (518) 561-2000; fax (518) 561-0881; rahrens@cvph.org
Internet: http://www.cvph.org/Foundation/NamedFunds/RabinFund/
Sponsor: Foundation of Champlain Valley Medical Center
75 Beekman Street
Plattsburgh, NY 12901

Foundation of CVPH Roger Senecal Endowment Fund Grants 1705
The Roger Senecal Endowment Fund exists because of the North Country's love for children, particularly when a child is in need. Income generated by the fund is used to help pay for transportation and housing expenses incurred by an ill or injured child's family in order to be near the child when vital care is being given. Applications will be evaluated and approved by a committee appointed by the Foundation of CVPH Medical Center. Financial need is the primary determinant upon which application will be weighed. The size of the grants is limited to the income generated by the endowment fund on a quarterly basis.
Requirements: Residents of Clinton, Essex, and Franklin counties in New York are eligible for consideration. The person for whom medical expenses are incurred must be 19 years of age or younger.
Restrictions: Funds are not intended to be used for direct medical expenditures.
Geographic Focus: New York
Contact: Roger G. Ahrens, Executive Director; (518) 562-7168 or (518) 561-2000; fax (518) 561-0881; rahrens@cvph.org
Internet: http://www.cvph.org/Foundation/NamedFunds/RogerSenecal/
Sponsor: Foundation of Champlain Valley Medical Center
75 Beekman Street
Plattsburgh, NY 12901

Foundation of CVPH Travel Fund Grants 1706
In 2001, the Board of Directors of the Foundation decided to provide funding for adults who need to travel outside the community to receive health care. The fund provides reimbursement for travel, food and lodging expenses. In many ways, it mimics the Roger Senecal Endowment Fund established in the mid 1980s to assist the families of children who need to leave the North Country to receive health care.
Requirements: Residents of Clinton, Essex, and Franklin counties in New York are eligible for consideration. The person for whom medical expenses are incurred must be 19 years of age or older.
Restrictions: Only one adult may apply for travel funds in addition to the patient. The fund will not provide funds for children to travel. No health care expenses, medications or medical equipment will be funded. Maximum income for the family is limited to $45,000.
Geographic Focus: New York
Amount of Grant: Up to 1,000 USD
Contact: Roger G. Ahrens, Executive Director; (518) 562-7168 or (518) 561-2000; fax (518) 561-0881; rahrens@cvph.org
Internet: http://www.cvph.org/Foundation/NamedFunds/TravelFund/

Sponsor: Foundation of Champlain Valley Medical Center
75 Beekman Street
Plattsburgh, NY 12901

Foundation of Orthopedic Trauma Research Grants 1707
The Foundation for Orthopedic Trauma was founded and incorporated in December of 2004 with the mission of influencing and enhancing orthopedic traumatology through education, research, mentorship and research founded on firm clinical grounds. The objective of Research Grant program is to encourage investigators by providing seed and start-up funding for promising research projects in the field of orthopedic trauma surgery through grants of up to $20,000 for a research project extending over a maximum of two years. Up to $30,000 is available for multi-center grants. Both laboratory and clinical projects are suitable, but in either case clinical relevance must be explicitly and clearly described. Completed applications should be submitted via email. The deadline for submission is June 1, with notification of awards to successful applicants mailed by August 30.
Requirements: An orthopedic surgeon, FOT member must serve as either the principal or co-principal investigator. Non-trauma/orthopedic surgeon, M.D.'s, Ph.D.'s or D.V.M.'s may serve as the principal or co-principal investigator, as long as they are affiliated with a trauma/orthopedic department with an FOT member orthopedic surgeon as the co-principal investigator.
Geographic Focus: All States
Date(s) Application is Due: Jun 1
Amount of Grant: Up to 30,000 USD
Contact: Lee-Ann Finno, Administrator; (212) 305-6392; Administrator@FOTnorthamerica.org or lee.finno@residentswap.org
Internet: http://www.fotnorthamerica.org/research_grants.asp
Sponsor: Foundation of Orthopedic Trauma
622 West 168th Street, 11th Floor
New York, NY 10032

Foundation of the American Thoracic Society Research Grants 1708
The American Thoracic Society established the ATS Foundation Research program to foster the development and training of future leaders in science and biomedical research. Through collaborations with non-profit and corporate partners the research program is able to provide research and career development grants and mentored fellowships to encourage young investigators to devote their talents and time to generating novel scientific ideas leading to new life-saving therapies for patients and their families. Since its founding in 2002, the Foundation of the ATS has awarded $10 million in grants to 105 young researchers investigating a wide spectrum of lung diseases, ranging from asthma and COPD to pulmonary fibrosis and alpha-1 antitrypsin. The program has also funded the career development of 24 pulmonary and critical care fellows. The grants amounts vary and range from $5,000 a year for 1 year to $50,000 per year for two years.
Restrictions: Grants are not awarded to individuals.
Geographic Focus: All States
Amount of Grant: 5,000 - 50,000 USD
Contact: Alyssa Chase; (212) 315-8600; fax (212) 315-6498; achase@thoracic.org
Internet: http://foundation.thoracic.org/programs/research-program.php
Sponsor: Foundation of the American Thoracic Society
25 Broadway, 18th Floor
New York, NY 10004

Foundations of East Chicago Health Grants 1709
The Foundations of East Chicago are committed to improving the lives of every resident of its city. Conceived by the citizens of East Chicago to be independent, citizen-run, private foundations, it derives funding from East Chicago's local casino, and uses this money to support local churches, schools, and nonprofit organizations who know the community best and put in the money in action where it can do the most good. Health grants are specifically focused on programs or projects which improve the quality of health education, practices and/or services for East Chicagoans.
Requirements: Applicants must be registered 501(c)3 organizations located in East Chicago, Indiana. If an applicant is not located in East Chicago, it may still qualify if it complies with at least one of the following *Requirements:* the program that an applicant is applying for operates in East Chicago; the funding that an applicant is applying for will go toward assisting East Chicago residents.
Restrictions: All applications must be submitted via Foundations of East Chicago website. The Foundations will not accept any applications in person.
Geographic Focus: Indiana
Amount of Grant: Up to 15,000 USD
Contact: Russell G. Taylor, Executive Director; (219) 392-4225; fax (219) 392-4245; grantinfo@foundationsofeastchicago.org
Internet: http://foundationsofeastchicago.org/apply-now
Sponsor: Foundations of East Chicago
100 W Chicago Avenue
East Chicago, IN 46312

Four County Community Foundation General Grants 1710
The Four County Community Foundation is committed to serving the current and emerging needs of the local communities, which includes the villages and cities of Almont, Armada, Capac, Dryden, Imlay City, Metamora and Romeo, Michigan. General grants are intended to support a variety of charitable purposes, providing grants for non-profits, public schools, and governmental agencies. Major areas of interest include: arts and culture; health care; children and youth programs, athletics; community development and outreach; the environment; higher education; human services; and elementary and secondary education. Support is given for: general operations; capital campaigns; and program development. The four annual deadlines for application submission are January 1, April 1, July 1, and October 1.
Requirements: 501(c)3 organizations, public schools, and governmental agencies in the Michigan cities and villages of Almont, Armada, Capac, Dryden, Imlay City, Metamora and Romeo are eligible to apply.
Geographic Focus: Michigan
Date(s) Application is Due: Jan 1; Apr 1; Jul 1; Oct 1
Amount of Grant: 500 - 50,000 USD
Contact: Micaela Boomer; (810) 798-0909; fax (810) 798-0908; program@4ccf.org
Internet: http://4ccf.org/community/
Sponsor: Four County Community Foundation
231 East St. Clair, P.O. Box 539
Almont, MI 48003

Four County Community Foundation Healthy Senior/Healthy Youth Grants 1711
The Four County Community Foundation is committed to serving the current and emerging needs of the local communities, which includes the villages and cities of Almont, Armada, Capac, Dryden, Imlay City, Metamora and Romeo, Michigan. Healthy Senior/Healthy Youth Fund grants are intended to support projects to promote health or provide treatment for health care for senior citizens and youth. Major priorities for seniors include: health and nutrition; pharmaceuticals (e.g., access, costs, medication interactions, etc.); preventing and managing chronic disease; smoking prevention and cessation; long term care alternatives (e.g., community-based long-term care options, assisted living, day care, respite care, and family caregiver support); mental health and aging (suicide, substance abuse, depression); family counseling; and workforce and aging. Major priorities for youth include: violence and conflict resolution; health and nutrition; behavioral risk factors (e.g., substance abuse, alcohol/binge drinking, smoking prevention and cessation, and depression); community alternatives for recreation; day care, preschool, and after school care; early childhood development; access to dental care; and family counseling. Support is typically given for: general operations; capital campaigns; and program development. The four annual deadlines for application submission are January 1, April 1, July 1, and October 1.
Requirements: 501(c)3 organizations, public schools, and governmental agencies in the Michigan cities and villages of Almont, Armada, Capac, Dryden, Imlay City, Metamora and Romeo are eligible to apply.
Geographic Focus: Michigan
Date(s) Application is Due: Jan 1; Apr 1; Jul 1; Oct 1
Amount of Grant: 500 - 50,000 USD
Contact: Micaela Boomer; (810) 798-0909; fax (810) 798-0908; program@4ccf.org
Internet: http://4ccf.org/community/
Sponsor: Four County Community Foundation
231 East St. Clair, P.O. Box 539
Almont, MI 48003

Fourjay Foundation Grants 1712
The Fourjay Foundation, established in 1988, supports only those organizations whose chief purpose is to improve health and/or promote education, within Philadelphia, Montgomery, and Bucks counties in southeastern Pennsylvania. The Foundation requires no specific application form. It will consider proposals that address a well-defined need, offer a concrete plan of action, and request a specific amount, from organizations whose staff has the ingenuity, commitment, and motivation to carry out the proposal's objectives. Requests may be for operating support, project specific funds, or capital funds.
Requirements: Organizations serving Philadelphia, Montgomery, and Bucks Counties in southeastern Pennsylvania are eligible.
Restrictions: Funding is not available for: charities operating outside Montgomery, Bucks, or Philadelphia counties; individuals; elementary or secondary educational institutions; museums, musical groups, theaters, or cultural organizations; religious organizations in support of their sacramental or theological functions; political groups or related think tanks; athletic organizations or alumni associations; libraries; public radio or television; United Way or the YMCA; civic organizations; organizations that have applied or been funded within the last 12 month period.
Geographic Focus: Pennsylvania
Date(s) Application is Due: Mar 1; Jun 1; Sep 1; Dec 1
Amount of Grant: 1,000 - 10,000 USD
Contact: Ann T. Bucci; (215) 830-1437; fax (215) 830-0157; abucci@fourjay.org
Sponsor: Fourjay Foundation
2300 Computer Avenue, Building G, Suite 1
Willow Grove, PA 19090-1753

Four J Foundation Grants 1713
The Four J Foundation was founded in Idaho in 1997, with the major fields of interest including: arts and culture; education; children and youth; health and health access; hospital care; and in-patient medical care. The Foundation caters primarily to academics, economically disadvantaged; low-income families; and students. Funding typically takes the form of: ensuring financial stability; fund raising projects; program development; and support of re-granting charitable organizations. Applicants should begin the process of forwarding a letter to the Foundation, along with a detailed description of the project, amount of funding requested, and overall program budget. There are no annual deadlines for submission. An average of forty organizations receive grant awards annually, ranging from $1,000 to $50,000. The vast majority of awards are equal to $10,000 or less.
Requirements: Any 501(c)3 that provides services to communities where Four J operates in Idaho is eligible to apply.
Geographic Focus: Idaho

Amount of Grant: 1,000 - 50,000 USD
Samples: Camp Rainbow Gold, Boise, Idaho, $5,000 support of educational programs (2014); St. Luke's Children's Playground, Boise, Idaho, $50,000 - support of children's medical programming (2014); University of Idaho, Moscow, Idaho, $2,500 - athletic scholarship fund support (2014).
Contact: Anne M. Goss, Director; (208) 344-7150 or (208) 344-6778
Sponsor: Four J Foundation
877 W. Main Street, Suite 800
Boise, ID 83702

Fragile X Syndrome Postdoctoral Research Fellowships   1714
FRAXA aims to accelerate research aimed at finding a specific treatment for fragile X syndrome, with a primary goal of bringing practical treatment into current medical practice as quickly as possible. Therefore, preference will be given to research projects that have a clear practical application and the results of which will be shared in a timely fashion. Grants in two categories: investigator-initiated program grants are intended for only the most translationally relevant proposals, with basic science proposals streamlined out without review; postdoctoral fellowship program, though these grants are still awarded with weight given to therapeutic potential. FRAXA will offer a limited number of postdoctoral fellowships at a fixed rate of $45,000 per year. Successful applicants for these grants will likely be working in established labs with secure overall funding, which have funding for supplies and any required animal handling costs. If the applicant proposes to work with a particular fragile X model (KO mouse, drosophila, human neural stem cells, etc.) s/he should demonstrate that this model system has already been established in-house.
Requirements: Individuals are nominated by applicant institutions for the fellowships and should have training and experience at least equal to the PhD or MD level.
Geographic Focus: All States
Date(s) Application is Due: May 1; Dec 1
Amount of Grant: 45,000 USD
Contact: Katie Clapp, President and Co-Founder; (978) 462-1866; fax (978) 463-9985; kclapp@fraxa.org or info@fraxa.org
Internet: http://www.fraxa.org/research/apply/
Sponsor: FRAXA Research Foundation
10 Prince Place, Suite 203
Newburyport, MA 01950

France-Merrick Foundation Health and Human Services Grants   1715
The France-Merrick Foundation's grants in the area of health and human services often assist people that face multiple disadvantages, and funding has evolved to support programs that focus on helping individuals improve their own lives. Funding in this area has grown over time, and has been especially informed by the Foundation's work in East Baltimore. The Foundation is specifically interested in improving health and wellness, with a focus on prevention and population health. Furthermore, the Foundation is focused on increasing access to care for disadvantaged populations and strengthening health care provision in community based settings. Due to the Foundation's concurrent interest in community development, the Foundation also supports anchor health institutions in improving neighborhoods around their hospital with dual health and community development benefits. The Foundation has seen the importance of housing as it relates to many issues of poverty and therefore funds the continuum of services for homeless individuals and families, such as emergency shelter, transitional housing, permanent supportive housing, and low-income housing creation or improvement. The Foundation is also interested in increasing access to health care for the homeless and working to help reduce homelessness, especially within families. Types of support include: building and renovation; capital campaigns; equipment purchase; management development; capacity building; matching grants; and technical assistance. Most recently, awards in this area have ranged from $24,000 to $500,000.
Requirements: Grants are made to Maryland nonprofit 501(c)3 organizations and institutions primarily located in the Baltimore metropolitan area.
Geographic Focus: Maryland
Amount of Grant: 24,000 - 500,000 USD
Samples: University of Maryland Medical System, Baltimore, Maryland, $500,000 - to support construction of the R. Adams Cowley Shock Trauma Center (2014); St. Vincent de Paul Society, West Baltimore, Maryland, $250,000 - to support renovation of Sarah's Hope - Mount Street emergency housing program (2014); Maryland Food Bank, Baltimore, Maryland, $100,000 - to support improvements to the Bauer Community Kitchen (2014).
Contact: Amy M. Gross; (410) 464-2004; fax (410) 464-2001; info@france-merrickfdn.org
Internet: http://www.france-merrickfdn.org/health-human-services
Sponsor: France-Merrick Foundation
2 Hamill Road, Suite 302
Baltimore, MD 21210

Frances and John L. Loeb Family Fund Grants   1716
The fund is committed to improving the quality of American life. To that end, its grant-making program seeks to address significant issues of society. The fund has chosen as its initial priorities the fields of education, health and family planning, including the public policy questions relating to them. In making grants, the fund favors proposals that offer new knowledge of and innovative approaches to problems rather than palliative measures. It also prefers applications that call for challenge grants and provide evaluation of results. There is no formal application form. Interested applicants should submit a preliminary request in letter form.
Requirements: Funding is directed primarily to organizations and institutions operating in the New York City metropolitan area that are tax exempt under the provisions of Section 501(c)3 of the Internal Revenue Code. Exceptions to this geographical limitation will be made for proposals that have the promise of national implication or extensive replication.
Restrictions: The fund will not consider proposals for annual or capital campaigns, building or renovation projects or loan or emergency funds. Neither will it make grants to individuals or for sectarian or religious purposes or political activities such as lobbying or propaganda. It will not fund any organization that discriminates on the basis of race, sex, religion, national origin or sexual preference.
Geographic Focus: All States
Contact: John L. Loeb, President; (212) 588-9052; fax (212) 838-6470; loebff@aol.com
Internet: http://www.geocities.com/CapitolHill/5601/family_fund_brochure.html
Sponsor: Frances and John L. Loeb Family Fund
375 Park Avenue, Suite 801
New York, NY 10152

Frances L. and Edwin L. Cummings Memorial Fund Grants   1717
The foundation supports nonprofits in New York City and northern New Jersey in the areas of education, especially programs that serve public school children from disadvantaged backgrounds; social welfare concerns; and campaigns to build endowments through establishment of challenge grants. Other areas of interest include elementary education, secondary education, vocational education, adult basic education and literacy, and higher education; hospitals, medical care, AIDS, and cancer; children, youth, and human services; and community development. Support is provided for endowment funds, seed money, consulting services, matching funds, technical assistance, professorships, and program development. The board meets in June and December. Application forms are not required.
Requirements: Grants are awarded to nonprofit organizations in the metropolitan New York, NY, area, with emphasis on New York City, southern Westchester County, and northern New Jersey.
Restrictions: Grants are not awarded to individuals or for capital building campaigns, general operating support, moving expenses, conferences, surveys, annual fund-raising campaigns, or research conducted by individuals or private institutions.
Geographic Focus: New Jersey, New York
Amount of Grant: 10,000 USD
Samples: Little Sisters of the Assumption Family Health Service (New York, NY)--to establish an endowment fund, $100,000; Neighborhood Initiatives (Bronx, NY)--for staff support to expand a teen center program, $30,000; Korean Community Service of Metropolitan New York (Woodside, NY)--to support a health educator and outreach worker for an HIV/AIDS prevention and education program, $40,000.
Contact: Elizabeth Costas; (212) 286-1778; fax (212) 682-9458
Sponsor: Frances L. and Edwin L. Cummings Memorial Fund
501 Fifth Avenue, Suite 708
New York, NY 10017-6103

Frances W. Emerson Foundation Grants   1718
The Foundation, in the care of Goldman Tax Service of Bellingham, Massachusetts, offers grants throughout Massachusetts and Vinalhaven, Maine. Its primary fields of interest include community development, economic development, education, and health care. There is no application form required or deadlines to adhere to, and applicants should send a letter or the full proposal to Eaton Vance management in Boston.
Geographic Focus: Maine, Massachusetts
Contact: Thomas Huggins, Vice President, Eaton Vance Management; (617) 482-8260
Sponsor: Frances W. Emerson Foundation
255 State Street
Boston, MA 02109

Francis L. Abreu Charitable Trust Grants   1719
The Francis L. Abreu Charitable Trust was established under the will of May Patterson Abreu in honor of her husband, Francis, who died in 1969. The trust supports Atlanta-area nonprofit organizations in its areas of interest, including arts and cultural programs, secondary education, higher education, health associations, human services, and children and youth services. Types of support include capital campaigns, seed money, program development, and matching funds. Requests are reviewed at April and October trustee meetings. Application forms are available online.
Requirements: Georgia nonprofits serving the greater Atlanta area are eligible to apply.
Restrictions: The foundation does not approve requests for operating costs or individuals.
Geographic Focus: Georgia
Date(s) Application is Due: Mar 31; Sep 30
Contact: Peter Abreu, Chairman; (404) 549-6743; fax (404) 549-6752
Internet: http://www.abreufoundation.org
Sponsor: Francis L. Abreu Charitable Trust
P.O. Box 502407
Atlanta, GA 31150

Francis T. & Louise T. Nichols Foundation Grants   1720
The Foundation gives primarily in the areas of education, health care, and human services. Its major fields of interest include: children and youth services; disaster relief; fire prevention and control; elementary and secondary education programs; health care; hospital support; and human/community service programs. There are no deadlines.
Requirements: Applicants should approach the Foundation initially by letter.
Restrictions: Giving primarily in Hancock County, Maine. There are no grants made to individuals.
Geographic Focus: Maine
Amount of Grant: Up to 80,000 USD
Contact: Calvin E. True, Treasurer; (207) 947-0111
Sponsor: Francis T. and Louise T. Nichols Foundation
P.O. Box 1210
Bangor, ME 04402-1210

Frank B. Hazard General Charity Fund Grants 1721

The Frank B. Hazard General Charity Fund was established in 1924 to support charitable organizations that work to improve the lives of "the poor, or the poor sick." Organizations receiving support from the fund must be managed and/or governed by individuals a majority of whom are of the Protestant religious faith. Grant requests for general operating support or program support are strongly encouraged. The majority of grants from the Hazard General Charity Fund are one year in duration. Applicants must apply online at the grant website. Applicants are strongly encouraged to do the following before applying: review the downloadable state application procedures for additional helpful information and clarifications; review the downloadable online-application guidelines at the grant website; review the trust's funding history (link is available from the grant website); review the online application questions in advance; and review the list of required attachments. These will generally include: a list of board members, financial statements (audited, reviewed, or compiled by independent auditor); an organization summary; a list of other funding sources; an IRS Determination letter; and other required documents. All attachments must be uploaded in the online application as PDF, Word, or Excel files. The application deadline for the Frank B. Hazard General Charity Fund is 11:59 p.m. on December 1. Applicants will be notified of grant decisions before January 31 of the following year.
Requirements: Applicants must have 501(c)3 tax-exempt status.
Restrictions: The Hazard General Charity Fund specifically supports charitable organizations that serve the people of Providence, Rhode Island. The fund does not support requests from individuals, organizations attempting to influence policy through direct lobbying, or any political campaigns.
Geographic Focus: Rhode Island
Date(s) Application is Due: Dec 1
Samples: Family Service, Providence, Rhode Island, $10,000; Community Preparatory School, Providence, Rhode Island, $17,000; Rhode Island Free Clinic, Providence, Rhode Island, $10,000.
Contact: Emma Greene, Director; (617) 434-0329; emma.m.greene@baml.com
Internet: https://www.bankofamerica.com/philanthropic/fn_search.action
Sponsor: Frank B. Hazard General Charity Fund
225 Franklin Street, 4th Floor, MA1-225-04-02
Boston, MA 02110

Frank E. and Seba B. Payne Foundation Grants 1722

The foundation awards grants in the Chicago, IL, area and in Pennsylvania in its areas of interest, including AIDS prevention, children and youth, cultural activities, education, and hospitals. Types of support include building construction/renovation, equipment acquisition, and general operating support.
Requirements: Nonprofit organizations in the greater Chicago, IL, metropolitan area and in Pennsylvania are eligible.
Restrictions: Grants are not made to individuals.
Geographic Focus: Illinois, Pennsylvania
Contact: M. Catherine Ryan, c/o Bank of America, (312) 828-1785
Sponsor: Frank E. Payne and Seba B. Payne Foundation
231 S LaSalle Street
Chicago, IL 60697

Frank G. and Freida K. Brotz Family Foundation Grants 1723

The Foundation supports Wisconsin nonprofits of the Christian, Lutheran, and Roman Catholic faiths, including Salvation Army. Types of support include capital support and general support. Organizations eligible to receive support include colleges and universities, community service groups, hospitals, ministries, parochial schools and religious education organizations, religious welfare organizations, and youth organizations. There are no application forms or deadlines. Applicants should submit a brief letter of inquiry that contains a description of the organization, its purpose, purpose of grant, and proof of tax-exempt status.
Requirements: Grants are awarded to publicly supported, Section 170(c) IRS qualified organizations, primarily in the State of Wisconsin.
Restrictions: No grants are awarded to individuals or organizations that require expenditure responsibility under Treasury Regulations.
Geographic Focus: Wisconsin
Contact: Stuart W. Brotz; (920) 458-2121; fax (920) 458-1923
Sponsor: Frank G. and Freida K. Brotz Family Foundation
3518 Lakeshore Road, P.O. Box 551
Sheboygan, WI 53082-0551

Franklin County Community Foundation Grants 1724

The Franklin County Foundation's mission is simple - build substantial endowment of funds for a community through contributions large and small. These contributions are endowed, permanently invested to produce income, and never spent. The income earned is used to help meet the community's charitable needs - from social work to art and culture. The FCCF Grant Cycle begins late summer when Letters of Intent are due in the office. After these letters are reviewed, the Grants Selection Committee will send applications to the groups or organizations who meet our grant guidelines.
Requirements: To be eligible to receive funding from the Foundation, a letter of intent must be submitted followed by a completed grant application by the appropriate grant deadline. All grant seekers must have prior governing board approval for the project seeking funding. The approval of signed minutes and a signed letter from governing board must be available to the Franklin County Community Foundation.
Restrictions: Funding will not be considered for the following: operating deficits; operation budgets (salaries); annual fund campaigns: religious or sectarian purposes; propaganda, political or otherwise, attempting to influence litigation or intervene in any political affairs or campaigns on behalf of any candidate for public office so as to endanger the charitable nature of the community trust; public school services required by state law; standard instructional or regular operation costs of non-public schools; repeat funding of projects previously supported by the Foundation; individuals; travel purposes; any purpose that is not in conformity with the constraints placed upon the Foundation by the IRS.
Geographic Focus: Indiana
Date(s) Application is Due: Oct 2
Contact: Shelly Lunsford, Executive Director; (765) 647-6810 or (765) 265-1427; fax (765) 647-0238; fcfoundation@yahoo.com
Internet: http://www.franklincountyindiana.com/Grants%20page.htm
Sponsor: Franklin County Community Foundation
527 Main Street
Brookville, IN 47012-1284

Franklin H. Wells and Ruth L. Wells Foundation Grants 1725

The Foundation awards grants to support the arts, community and economic development, education, health care, and human services. Funding is also available for emergency funds, equipment, program development, and seed money. The Foundation board meets in April and October, with funding notification in May and November.
Requirements: Applications are not required. Organizations should initially submit a letter of inquiry, and if accepted, one copy of their proposal.
Restrictions: The Foundation gives primarily to Dauphin, Cumberland, and Perry counties in Pennsylvania. Funding is not available for religious activities, individuals, endowments, debts, and capital campaigns.
Geographic Focus: Pennsylvania
Date(s) Application is Due: Mar 15; Sep 15
Contact: Miles Gibbons Jr.; (866) 398-9023; mgibbons989@earthlink.net
Sponsor: Franklin H. Wells and Ruth L. Wells Foundation
One M and T Plaza, 8th Floor
Buffalo, NY 14203-2309

Frank Loomis Palmer Fund Grants 1726

The Frank Loomis Palmer Fund was established in 1936 to support and promote quality educational, cultural, human-services, and health-care programming for underserved populations. The Palmer Fund specifically serves the people of New London, Connecticut. Grants from the Palmer Fund are primarily one year in duration; on occasion, multi-year support is awarded. Applicants must apply online at the grant website. Applicants are strongly encouraged to do the following before applying: review the downloadable state application procedures for additional helpful information and clarifications; review the downloadable online-application guidelines at the grant website; review the fund's funding history (link is available from the grant website); review the online application questions in advance; and review the list of required attachments. These will generally include: a list of board members, financial statements (audited, reviewed, or compiled by independent auditor); an organization summary; a list of other funding sources; an IRS Determination letter; and other required documents. All attachments must be uploaded in the online application as PDF, Word, or Excel files. The application deadline for the Frank Loomis Palmer Fund is 11:59 p.m. on November 15. Applicants will be notified of grant decisions by letter within three to four months after the proposal deadline.
Requirements: Applicants must have 501(c)3 tax-exempt status and serve the people of New London, Connecticut.
Restrictions: Applicants will not be awarded a grant for more than three consecutive years. The fund does not support requests from individuals, organizations attempting to influence policy through direct lobbying, or any political campaigns.
Geographic Focus: Connecticut
Date(s) Application is Due: Nov 15
Contact: Amy Lynch; (860) 657-7015; amy.r.lynch@ustrust.com
Internet: https://www.bankofamerica.com/philanthropic/fn_search.action
Sponsor: Frank Loomis Palmer Fund
200 Glastonbury Boulevard, Suite # 200, CT2-545-02-05
Glastonbury, CT 06033-4056

Frank Reed and Margaret Jane Peters Memorial Fund I Grants 1727

The Frank Reed & Margaret Jane Peters Memorial Fund I was established in 1935 to support and promote quality educational, human-services, and health-care programming for underserved populations. Special consideration is given to charitable organizations that serve youth and children. The Peters Memorial Fund I is a generous supporter of the Associated Grant Makers (AGM) Summer Fund. The AGM Summer Fund is a collaborative group of donors that provides operating support for summer camps serving low-income urban youth from Boston, Cambridge, Chelsea, and Somerville. Excluding the grant made to the AGM Summer Fund, the typical grant range is $10,000 to $40,000. Grant requests for general operating support are strongly encouraged. Program support will also be considered. Small, program-related capital expenses may be included in general operating or program requests. The majority of grants from the Peters Memorial Fund I are one year in duration; on occasion, multi-year support is awarded. Applicants must apply online at the grant website. Applicants are strongly encouraged to do the following before applying: review the downloadable state application procedures for additional helpful information and clarifications; review the downloadable online-application guidelines at the grant website; review the foundation's funding history (link is available from the grant website); review the online application questions in advance; and review the list of required attachments. These will generally include: a list of board members, financial statements (audited, reviewed, or compiled by independent

auditor); an organization summary; a list of other funding sources; an IRS Determination letter; and other required documents. All attachments must be uploaded in the online application as PDF, Word, or Excel files. The application deadline for the Frank Reed & Margaret Jane Peters Memorial Fund I is 11:59 p.m. on September 1. Applicants will be notified of grant decisions before November 30.
Requirements: Applicants must have 501(c)3 tax-exempt status.
Restrictions: The fund does not support requests from individuals, organizations attempting to influence policy through direct lobbying, or any political campaigns.
Geographic Focus: Massachusetts
Date(s) Application is Due: Sep 1
Amount of Grant: 10,000 - 40,000 USD
Samples: Generations, Inc., Boston, Massachusetts, $20,000, general operating support; Boston Scholars Program, Boston, Massachusetts, $15,000, general operating support; Family Service, Lawrence, Massachusetts, $15,000, Youth Mentoring Department program development.
Contact: Miki C. Akimoto, Vice; (866) 778-6859; miki.akimoto@baml.com
Internet: https://www.bankofamerica.com/philanthropic/fn_search.action
Sponsor: Frank Reed and Margaret Jane Peters Memorial Fund I
225 Franklin Street, 4th Floor, MA1-225-04-02
Boston, MA 02110

Frank Reed and Margaret Jane Peters Memorial Fund II Grants 1728
The Frank Reed & Margaret Jane Peters Memorial Fund II was established in 1935 to support and promote quality educational, human-services, and health-care programming for underserved populations. In the area of education, the fund supports academic access, enrichment, and remedial programming for children, youth, adults, and senior citizens that focuses on preparing individuals to achieve while in school and beyond. In the area of health care, the fund supports programming that improves access to primary care for traditionally underserved individuals, health education initiatives and programming that impact at-risk populations, and medical research. In the area of human services the fund tries to meet evolving needs of communities. Currently the fund's focus is on (but is not limited to) youth development, violence prevention, employment, life-skills attainment, and food programs. Grant requests for general operating support are strongly encouraged. Program support will also be considered. Small, program-related capital expenses may be included in general operating or program requests. The majority of grants from the Peters Memorial Fund II are one year in duration; on occasion, multi-year support is awarded. Applicants must apply online at the grant website. Applicants are strongly encouraged to do the following before applying: review the downloadable state application procedures for additional helpful information and clarifications; review the downloadable online-application guidelines at the grant website; review the foundation's funding history (link is available from the grant website); review the online application questions in advance; and review the list of required attachments. These will generally include: a list of board members, financial statements (audited, reviewed, or compiled by independent auditor); an organization summary; a list of other funding sources; an IRS Determination letter; and other required documents. All attachments must be uploaded in the online application as PDF, Word, or Excel files. The application deadline for the Frank Reed & Margaret Jane Peters Memorial Fund II is 11:59 p.m. on July 1. Applicants will be notified of grant decisions before September 30.
Requirements: Applicants must have 501(c)3 tax-exempt status.
Restrictions: In general, capital requests are not advised. The fund does not support endowment campaigns, events such as galas or award ceremonies, and costs of fundraising events. The fund does not support requests from individuals, organizations attempting to influence policy through direct lobbying, or any political campaigns.
Geographic Focus: Massachusetts
Date(s) Application is Due: Jul 1
Samples: Epiphany School, Dorchester, Massachusetts, $20,000, general operating support; Esther R. Sanger Center for Compassion, Wollaston, Massachusetts, $15,000, for general operating support; Boston Learning Center, Boston, Massachusetts, $20,000, general operating support.
Contact: Michealle Larkins; (866) 778-6859; michealle.larkins@baml.com
Internet: https://www.bankofamerica.com/philanthropic/fn_search.action
Sponsor: Frank Reed and Margaret Jane Peters Memorial Fund II
225 Franklin Street, 4th Floor, MA1-225-04-02
Boston, MA 02110

Frank S. Flowers Foundation Grants 1729
The Frank S. Flowers Foundation primarily serves the Gloucester County, New Jersey area. The Foundation's area of interest include: Education--supporting public high schools of Gloucester County, New Jersey, to provide scholarships for college or graduate study, vocational or technical training; Youth--supporting chapters or councils or branches of Y.M.C.A. and Boys Scouts of America located in Gloucester and/or Salem Counties, grants are also considered for organizations having branches or offices in Gloucester County, New Jersey which treat and educate children with special needs; Health-Related--support to non-profit Gloucester County hospitals; Religious organizations--support to churches in the boroughs of Paulsboro and Wenonah, New Jersey. The Foundation also has a specific interest in The Shriner's Hospital for Crippled Children in Philadelphia, Pennsylvania and the Masonic Home Charity Foundation of New Jersey. Grants range from $1,000 - $8,000. Application deadline date is February 15th, application available online. Requestors will receive a letter acknowledging the receipt of their request.
Requirements: Qualifying tax-exempt 501(c)3 organizations are eligible for grants if they meet the purpose of the foundation. Proposals should be submitted in the following format: completed Common Grant Application Form; an original Proposal Statement*; an audited financial report and a current year operating budget; a copy of your official IRS Letter with your tax determination; a listing of your Board of Directors. *Proposal Statement should answer these questions: what are the objectives and expected outcomes of this program/project/request; what strategies will be used to accomplish your objective; what is the timeline for completion; if this is part of an on-going program, how long has it been in operation; what criteria will you use to determine your success; if the request is not fully funded, what other sources can you engage. A Proposal budget should be included if this is for a specific program within your annual budget. Please describe any collaborative ventures.
Restrictions: Grants are not made for political purpose, nor to organizations which discriminate on the basis of race, ethnic origin, sexual or religious preference, age or gender.
Geographic Focus: New Jersey, Pennsylvania
Date(s) Application is Due: Feb 15
Amount of Grant: 1,000 - 8,000 USD
Samples: Holy Trinity Episcopal Church, Parking Lot Rehabilitation--$8,000; Deptford Township High School, $3,000--scholarships; Ronald McDonald House of Southern New Jersey, $5,000--general operations support.
Contact: Gale Y. Sykes; (908) 598-3576; grantinquiries2@wachovia.com
Internet: https://www.wachovia.com/foundation/v/index.jsp?vgnextoid=68d78689fb0aa110VgnVCM1000004b0d1872RCRD&vgnextfmt=default
Sponsor: Frank S. Flowers Foundation
190 River Road
Summit, NJ 07901

Frank Stanley Beveridge Foundation Grants 1730
The foundation welcomes proposals in the areas of: animal care; arts, culture and humanities; civil rights, social action, advocacy; education; employment/jobs; environmental quality, protection and beautification; food, nutrition and agriculture; health; housing; human services; medical research; mental health; philanthropy; safety; recreation; religion; science; social services; and youth development. The board meets in October and April to consider requests. Multiyear grants are rare. Contact the foundation via the Web site only. No phone or written inquiries will be accepted.
Requirements: Applicants must be 501(c)3 nonprofit organizations or foundations in Massachusetts's Hampden and Hampshire Counties.
Restrictions: The Foundation prefers not to support: awards or prizes; commissioning of new artistic work; conferences/seminars; curriculum development; debt reduction; employee matching gifts; employee-related scholarships; endowment funds; exhibitions; faculty/staff development; fellowship funds; fellowships to individuals; film/video/radio production; foundation administered programs; general operating support; grants to individuals; income development; internship funds; management development; performance/production costs; professorships; program-related investment/loans; publications; scholarships to individuals; student aid; technical assistance.
Geographic Focus: Massachusetts
Date(s) Application is Due: Feb 1; Aug 1
Amount of Grant: 50,000 USD
Contact: Philip Caswell, President; (800) 229-9667; fax (561) 748-0644; administrator@beveridge.org or caswell@beveridge.org
Internet: http://www.beveridge.org/
Sponsor: Frank Stanley Beveridge Foundation
1340 U.S. Highway 1, Suite 102
Jupiter, FL 33469

Frank W. and Carl S. Adams Memorial Fund Grants 1731
The Frank W. and Carl S. Adams Memorial Fund was established in 1925 to support and promote quality educational, human services, and health care programming for underserved populations. Annual gifts are also awarded to the Harvard University Medical School and the Massachusetts Institute of Technology for student scholarships. Grant requests for general operating support are strongly encouraged. Program support will also be considered. Small, program-related capital expenses may be included in general operating or program requests. To better support the capacity of nonprofit organizations, multi-year funding requests are strongly encouraged. The application deadline is December 1. Applicants will be notified of grant decisions before March 31.
Requirements: 501(c)3 organizations serving residents of Massachusetts are eligible.
Geographic Focus: Massachusetts
Date(s) Application is Due: Dec 1
Contact: Miki C. Akimoto, Vice President; (866) 778-6859; miki.akimoto@baml.com
Internet: https://www.bankofamerica.com/philanthropic/fn_search.action
Sponsor: Frank W. and Carl S. Adams Memorial Fund
225 Franklin Street, 4th Floor, MA1-225-04-02
Boston, MA 02110

Fraser-Parker Foundation Grants 1732
The Foundation awards funds in its areas of interest, including Christian religion organizations, education and higher education, and hospitals. There are no application forms and no deadlines.
Geographic Focus: All States
Amount of Grant: 5,000 - 50,000 USD
Samples: Visiting Nurse/Hospice Atlanta, Atlanta, Georgia, $25,000, chaplain residency training program; Colonial Williamsburg Foundation, Williamsburg, Virginia, $5,000, operating support; and University of Rochester, Rochester, New York, $65,000, therapeutic programs.
Contact: John Stephenson, Executive Director; (404) 827-6529
Sponsor: Fraser-Parker Foundation
3050 Peachtree Road NW, Suie 270
Atlanta, GA 30305

FRAXA Research Foundation Program Grants   1733
FRAXA aims to accelerate research aimed at finding a specific treatment for fragile X syndrome, with a primary goal of bringing practical treatment into current medical practice as quickly as possible. Therefore, preference will be given to research projects that have a clear practical application and the results of which will be shared in a timely fashion. FRAXA invites investigator-initiated research applications for innovative pilot studies aimed at developing and characterizing new therapeutic approaches for the treatment and ultimate cure of fragile X syndrome. These are flexible grants, but the expectation is that this program will be much more competitive than in the past, since fewer grants will be awarded. Successful applicants will be pursuing advanced translational, preclinical, and clinical research in fragile X. Grants are given based on innovation and translational relevance: clinical trials are our highest priority, followed by preclinical research (i.e. rescue of animal models) and then translational research. The annual deadlines are February 1 and August 1.
Requirements: There is no limit to structure of the grant (can fund PI, postdoc, grad student, technician, supplies, etc.) or time-frame (though all grants over one year still need yearly renewal). Furthermore, there is no limit on amount, but please remember that applications are ranked based on relative value, so smaller requests have an advantage.
Geographic Focus: All States
Date(s) Application is Due: Feb 1; Aug 1
Contact: Katie Clapp, President and Co-Founder; (978) 452-1866; fax (978) 463-9985; kclapp@fraxa.org or info@fraxa.org
Internet: http://www.fraxa.org/research/apply/
Sponsor: FRAXA Research Foundation
10 Prince Place, Suite 203
Newburyport, MA 01950

Fred & Gretel Biel Charitable Trust Grants   1734
The Fred & Gretel Biel Charitable Trust was established in 2004 to support and promote quality educational, human-services, and health-care programming for underserved populations. Special consideration is given to organizations that provide food and clothing to low-income individuals and families. Consideration is also given to organizations that serve the economically disadvantaged through the provision of housing, legal assistance, or day-care services. Grant requests for general operating and capital support are encouraged. Grants from the Biel Charitable Trust are one year in duration. Application materials are available for download at the grant website. Applicants are strongly encouraged to review the state application guidelines for additional helpful information and clarifications before applying. Applicants are also encouraged to review the trust's funding history (link is available from the grant website). The application deadline for the Biel Charitable Trust is May 1. Applicants will be notified of grant decisions by June 30.
Requirements: The Biel Charitable Trust typically supports organizations serving the people of King and Snohomish Counties in the Puget Sound region of Washington. Occasionally grants will be made outside of the Puget Sound area. Applicant organizations must have 501(c)3 tax-exempt status.
Restrictions: Requests to assist with debt retirement or to correct an operating deficit will not be considered. Applicants who have received a grant for three consecutive years must wait two years before reapplying to the trust. The trust does not support requests from individuals, organizations attempting to influence policy through direct lobbying, or any political campaigns.
Geographic Focus: Washington
Date(s) Application is Due: May 1
Samples: Edmonds School District, Lynnwood, Washington, $2,500; Haller Lake Christian Health Clinic, Seattle, Washington, $5,000; North Helpline, Seattle, Washington, $2,500.
Contact: Nancy Atkinson; (800) 848-7177 or (206) 358-0912; nancy.l.atkinson@baml.com
Internet: https://www.bankofamerica.com/philanthropic/fn_search.action
Sponsor: Fred and Gretel Biel Charitable Trust
800 5th Avenue
Seattle, WA 98104

Fred and Louise Latshaw Scholarship   1735
The Fred and Louise Latshaw Scholarship provides funding to employees of Lake Area United Way Agencies (see application for list of eligible agencies) or employees of a nonprofit agency that has an endowment fund with Legacy Foundation (see application for list of eligible agencies). A scholarship of $1,000 is awarded for tuition, fees, and/or books and is renewable by reapplication each year of undergraduate study.
Requirements: Applicants will be considered for the scholarship if they meet the following criteria: they are U.S. citizens and residents of Lake County, Indiana; they are a high school graduate and have deferred their education and wish to return as a part time student or are currently enrolled; they are currently employed by a Lake Area United Way Agency or by a not-for-profit Agency that has an endowment fund with the Legacy Foundation; they are currently enrolled or a newly admitted student at an accredited public or private college, university, or vocational school. All college majors are eligible.
Geographic Focus: Indiana
Date(s) Application is Due: Jun 1
Amount of Grant: 1,000 USD
Contact: Latshaw Scholarship Contact; (219) 736-1880; fax (219) 736-1940; legacy@legacyfoundationlakeco.org
Internet: http://www.legacyfoundationlakeco.org/scholarships/index.htm
Sponsor: Legacy Foundation
1000 East 80th Place, 302 South
Merrillville, IN 46410

Fred Baldwin Memorial Foundation Grants   1736
The Foundation supports programs and projects that benefit the people of Maui County. Projects in the arts, education, environment, health, and human services are of greatest interest. Funding is available for capital improvement projects. There are two annual deadlines, which fall on the first business day in February and August. Application information is available online.
Requirements: Eligible applicants must have 501(c)3 status, or must apply through a fiscal sponsor with 501(c)3 status.
Restrictions: The Foundation does not fund loans or debt service, endowments, funds for re-granting, scholarships, grants to individuals or units of government, or activities that have already occurred.
Geographic Focus: Hawaii
Date(s) Application is Due: Feb 1; Aug 1
Amount of Grant: 10,000 USD
Contact: Carrie Shoda-Sutherland, Senior Program Officer; (808) 566-5524 or (888) 731-3863, ext. 524; fax (808) 521-6286; csutherland@hcf-hawaii.org
Internet: http://www.fredbaldwinfoundation.org/
Sponsor: Fred Baldwin Memorial Foundation
1164 Bishop Street, Suite 800
Honolulu, HI 96813

Fred C. and Katherine B. Andersen Foundation Grants   1737
Fred C. and Katherine B. Andersen Foundation, formerly known as the Andersen Foundation gives on a national basis for higher education. The foundation provides funds locally in Minnesota and western Wisconsin for all areas of interest, which include: arts; health care; higher education; hospitals; and youth development. Funding is available in the forms of: capital campaigns; general/operating support and; program development.
Requirements: 501(c)3 tax-exempt organizations are eligible. Giving is on a national basis although preference is given to requests from Minnesota.
Restrictions: The foundation does not make grants to institutions that receive federal funding.
Geographic Focus: All States
Date(s) Application is Due: Mar 18; Jul 22; Oct 21
Amount of Grant: 5,000 - 12,000,000 USD
Contact: Mary Gillstrom, Director; (651) 264-5150
Sponsor: Fred C. and Katherine B. Andersen Foundation
P.O. Box 8000
Bayport, MN 55003-0080

Freddie Mac Foundation Grants   1738
The Foundation focus is on children whose families have limited resources and who are vulnerable to poor outcomes. The Foundation emphasizes the integration of services that focus on family strengthening and youth development in order to maximize the benefit to children and their families. Grants will be made for direct service projects, general operating support, capacity building, public awareness, planning and capital projects. The following funding priorities will be considered: the early years; elementary school years; junior high and high school years; children and families in crisis; and public awareness education. Proposals must be submitted online.
Requirements: Eligible organizations must be tax exempt under IRS code 501(c)3 and defined as a public charity. The Foundation's grantmaking program services the following metropolitan Washington, D.C. areas: District of Columbia; Virginia - the counties of Arlington, Fairfax, Loudoun and Prince William and the cities of Alexandria, Falls Church, Manassas Park, and Leesburg; Maryland - the counties of Charles, Frederick, Howard, Montgomery, and Prince George's.
Restrictions: The Foundation will not fund organizations that discriminate in the provision of services or in employment practices based on race, color, religion, ethnicity, sex, age, national origin, disability, sexual orientation, marital status, and any other characteristics protected by applicable law. Unless approved by the board, the Foundation does not fund: individuals; training in/promotion of religious doctrine; incurring a debt liability; endowment campaigns.
Geographic Focus: District of Columbia, Maryland, Virginia
Date(s) Application is Due: Sep 9
Amount of Grant: 5,000 - 50,000 USD
Contact: Ralph F. Boyd, President and CEO; fax (703) 918-8888; freddie_mac_foundation@freddiemac.com
Internet: http://www.freddiemacfoundation.org
Sponsor: Freddie Mac Foundation
8250 Jones Branch Drive, MS A40
McLean, VA 22102-3110

Frederick Gardner Cottrell Foundation Grants   1739
The Research Corporation Technologies (RCT) established the Frederick Gardner Cottrell Foundation in December 1998 to provide financial support for scientific research and educational programs at qualified nonprofit organizations. RCT named the foundation in honor of the university professor and inventor who championed the transfer of academic innovation to public use. The Foundation receives its support from donations made by RCT and is a private, non-operating entity. Since its formation, the Foundation has provided nearly $13 million in support of selected scientific and educational programs throughout the United States. The Foundation does not accept unsolicited grant requests, so interested parties should begin the process by contacting the office and providing project information. The Foundation's primary areas of interest include: eye research; higher education; marine science; and science.
Geographic Focus: All States
Amount of Grant: 100,000 - 300,000 USD

Contact: Gary Munsinger; (520) 748-4400; fax (520) 748-0025; munsinger@rctech.com
Internet: http://www.rctech.com/about-us/foundation/
Sponsor: Frederick Gardner Cottrell Foundation
6440 N. Swan Road, Suite 200
Tucson, AZ 85718

Frederick McDonald Trust Grants   1740

The Frederick McDonald Trust was established in 1950 to support and promote quality educational, human-services, and health-care programming for underserved populations. Grant requests for general operating support, program, project and capital support will be considered. The majority of grants from the McDonald Trust are one year in duration. On occasion, multi-year support is awarded. Applicants must apply online at the grant website. Applicants are strongly encouraged to do the following before applying: review the downloadable state application procedures for additional helpful information and clarifications; review the downloadable online-application guidelines at the grant website; review the trust's funding history (link is available from the grant website); review the online application questions in advance; and review the list of required attachments. These will generally include: a list of board members, financial statements (audited, reviewed, or compiled by independent auditor); an organization summary; a list of other funding sources; an IRS Determination letter; and other required documents. All attachments must be uploaded in the online application as PDF, Word, or Excel files. The application deadline for the Frederick McDonald Trust is 11:59 p.m. on May 1. Applicants will be notified of grant decisions before August 31.
Requirements: Applicants must have 501(c)3 tax-exempt status.
Restrictions: Grants are made only to those organizations located in, or serving the people of Albany City. The trust does not support requests from individuals, organizations attempting to influence policy through direct lobbying, or any political campaigns.
Geographic Focus: New York
Date(s) Application is Due: May 1
Samples: Equinox, Albany, New York, $15,000; Albany Center for Economic Success, Albany, New York, $10,000; Girls, Inc. of the Greater Capital Region, Schenectady, New York, $15,000.
Contact: Christine O'Donnell; (646) 855-1011; christine.l.o'donnell@baml.com
Internet: https://www.bankofamerica.com/philanthropic/fn_search.action
Sponsor: Frederick McDonald Trust
One Bryant Park, NY1-100-28-05
New York, NY 10036

Frederick W. Marzahl Memorial Fund Grants   1741

The Frederick W. Marzahl Memorial Fund was established in 1974 to support and promote quality educational, human-services, and health-care programming for underserved populations in Woodbury, Connecticut. Grants from the Marzahl Memorial Fund are one year in duration. Application materials are available for download at the grant website. Applicants are strongly encouraged to review the state application guidelines for additional helpful information and clarifications before applying. Applicants are also encouraged to review the foundation's funding history (link is available from the grant website). The deadline for application to the Frederick W. Marzahl Memorial Fund is November 1. Applicants will be notified of grant decisions by letter within two to three months after the proposal deadline.
Requirements: Applicants must have 501(c)3 tax-exempt status.
Restrictions: Applicants will not be awarded a grant for more than three consecutive years. The fund does not support requests from individuals, organizations attempting to influence policy through direct lobbying, or any political campaigns.
Geographic Focus: Connecticut
Date(s) Application is Due: Nov 1
Samples: Connecticut Junior Republic Association, Litchfield, Connecticut, $7,000 (010); Flanders Nature Center, Woodbury, Connecticut, $5,400; North Congregational Church, Woodbury, Connecticut, $3,000.
Contact: Carmen Britt; (860) 657-7019; carmen.britt@baml.com
Internet: https://www.bankofamerica.com/philanthropic/fn_search.action
Sponsor: Frederick W. Marzahl Memorial Fund
200 Glastonbury Boulevard, Suite # 200, CT2-545-02-05
Glastonbury, CT 06033-4056

Frederic Stanley Kipping Award in Silicon Chemistry   1742

This award, supported by Dow Corning Corporation, is given biennially in even-numbered years to recognize distinguished achievement in research in silicon chemistry and to stimulate the creativity of others toward further advancement of this field of chemistry. A nominee must have accomplished distinguished achievement in research in silicon chemistry during the preceding 10 years. The measure of this achievement should focus primarily on the nominee's significant publications in the field of silicon chemistry but may include consideration of contributions to the related field of organometallic chemistry. There are no limits on age or nationality. Applications are accepted in odd-numbered years. The award consists of $5,000 and a certificate.
Requirements: Any individual, except a member of the award committee, may submit one nomination or seconding letter for the award in any given year. Nominating documents consist of a letter of not more than 1000 words containing an evaluation of the nominee's accomplishments and a specific identification of the work to be recognized, a biographical sketch including date of birth, and a list of publications and patents authored by the nominee.
Restrictions: Self-nominations are not accepted.
Geographic Focus: All States
Date(s) Application is Due: Nov 1
Amount of Grant: 5,000 USD
Samples: T. Don Tilley--award winner, $5,000; Akira Sekiguchi--award winner, $5,000; James E. Mark--award winner, $5,000.
Contact: Felicia Dixon, Awards Administrator; (800) 227-5558 or (202) 872-4408; fax (202) 776-8008; f_dixon@acs.org or awards@acs.org
Internet: http://portal.acs.org/portal/acs/corg/content?_nfpb=true&_pageLabel=PP_ARTICLEMAIN&node_id=1319&content_id=CTP_004523&use_sec=true&sec_url_var=region1
Sponsor: American Chemical Society
1155 Sixteenth Street, NW
Washington, D.C. 20036-4801

Fred L. Emerson Foundation Grants   1743

The foundation gives grants to nonprofits in Auburn, Cayuga County, and upstate New York to improve the quality of life in the areas of education (primarily private, higher education), hospital and health programs, community agencies, churches, cultural institutions, youth and community service programs and social welfare agencies. An application form is not required. Proposals may be submitted in letter form detailing the project for which support is being sought.
Requirements: New York nonprofit organizations in Auburn, Cayuga County, and the upstate area are eligible.
Restrictions: Grants do not support individuals, operating budgets, or loans.
Geographic Focus: New York
Amount of Grant: 200 - 250,000 USD
Contact: Daniel J. Fessenden, Executive Director; (315) 253-9621; dan@emersonfoundation.com
Sponsor: Fred L. Emerson Foundation
5654 South Street Road, P.O. Box 276
Auburn, NY 13021-9602

Fremont Area Community Foundation Amazing X Grants   1744

The Amazing X Charitable Trust, a supporting organizaton of the Fremont Area Community Foundation, was established in the late 1970s by members of the Gerber family to benefit people with disabilities and to address general charitable needs in Newaygo County, Michigan. Grant requests are accepted for: projects or programs that serve people with disabilities; and projects or programs that address general charitable needs. Preferred programs are innovative, collaborative, and have a significant impact on the residents of Newaygo County. Grants range from $1,000 to $60,000, and applications are due each year by July 15.
Requirements: Michigan 501(c)3 organizations located in or supporting Newaygo County are eligible for funding. When submitting your proposal, include your organizations: mission, history, description of current programs, activities, and accomplishments; purpose of the grant (describe in detail and include supporting evidence); grant proposal budget form/narrative (form available at the Foundation's website). The following list of attachments must also be included: a copy of the current IRS 501(c)3 determination letter; roster of current governing board, including addresses and affiliations; finances: organization's current annual operating budget, including all expenses and revenues, audited financial statement (most recently completed), IRS Form 990 (most recently filed), annual report, if available; resumes and job descriptions of the key project personnel; organizational chart; letters of support (up to five).
Restrictions: In order to make the best use of available funds, the Foundation usually will not award grants for the following: grants to individuals; to pay off existing debts; religious programs that require religious affiliation and/or religious instruction to receive services; to further political campaigns; projects that begin prior to notification of Foundation funding; capital improvements on rental or individual private property; or programs or projects that subsidize or supplant funding for services considered general government obligations.
Geographic Focus: Michigan
Date(s) Application is Due: Jul 15
Amount of Grant: 1,000 - 60,000 USD
Samples: Arbor Circle, Newaygo, Michigan, $6,000 - in support of outpatient substance abuse counseling; Disability Connection of West Michigan, Muskegon, Michigan, $51,500 - support for advocacy, empowerment, accessibility, and education; Second Christian Reformed Church, Fremont, Michigan, $2,200 - support of the Fremont Friendship program
Contact: Vonda Carr; (231) 924-5350; fax (231) 924-5391; vcarr@tfacf.org
Internet: http://www.tfacf.org/grants/amazingx.html
Sponsor: Fremont Area Community Foundation
P.O. Box B
Fremont, MI 49412

Fremont Area Community Foundation Elderly Needs Grants   1745

Overall, the Foundation is focusing its grantmaking resources primarily on expanding opportunities that enhance the well being of residents from the Newaygo County, Michigan area. The purpose of the Elderly Needs Fund is to make grants to support health and enrich aging for the elderly of Newaygo County. The strategies of the Fund include: promotion of the physical health of the elderly; promotion of the mental and emotional well being of the elderly and their caregivers; promotion of the social enrichment and prevention of the social isolation of the elderly; and promotion of the provision of basic human services for the elderly. Application deadlines are February 1 and September 1 each year.
Requirements: Michigan 501(c)3 organizations located in or supporting Newaygo County are eligible for funding. When submitting your proposal, include your organizations: mission, history, description of current programs, activities, and accomplishments; purpose of the grant (describe in detail and include supporting evidence); grant proposal

budget form/narrative (form available at the Foundation's website). The following list of attachments must also be included: a copy of the current IRS 501(c)3 determination letter; roster of current governing board, including addresses and affiliations; finances: organization's current annual operating budget, including all expenses and revenues, audited financial statement (most recently completed), IRS Form 990 (most recently filed), annual report, if available; resumes and job descriptions of the key project personnel; organizational chart; letters of support (up to five).
Restrictions: In order to make the best use of available funds, the Foundation usually will not award grants for the following: grants to individuals; to pay off existing debts; religious programs that require religious affiliation and/or religious instruction to receive services; to further political campaigns; projects that begin prior to notification of Foundation funding; capital improvements on rental or individual private property; or programs or projects that subsidize or supplant funding for services considered general government obligations.
Geographic Focus: Michigan
Date(s) Application is Due: Feb 1; Sep 1
Amount of Grant: 10,000 - 250,000 USD
Samples: Catholic Charities West Michigan, Muskegon, Michigan, $56,000 - support for the Newaygo County Companion Program; Newaygo County Commission on Aging, White Cloud, Michigan, $252,000 - support for medical van services, respite, congregate meals, and home delivery of meals; Senior Sing A-Long, Newaygo, Michigan, $28,650 - support for the Tuned In program.
Contact: Vonda Carr; (231) 924-5350; fax (231) 924-5391; vcarr@tfacf.org
Internet: http://www.tfacf.org/grants/elderlyneeds.html
Sponsor: Fremont Area Community Foundation
P.O. Box B
Fremont, MI 49412

Fremont Area Community Foundation General Grants   1746
The Fremont Area Community Foundation is focusing its grantmaking resources that enhance the well being of children, youth and families in Newaygo County, Michigan. The Foundation's areas of interest include: arts and culture; community development; education; the environment; and human services. Types of support offered include: building and renovation; capital campaigns; conferences and seminars; consulting services; continuing support; curriculum development; emergency funds; employee matching gifts; endowments; equipment; general operating support; management development; capacity building; matching or challenge support; program-related investments and loans; program development; program evaluation; scholarship funds; seed money; and technical assistance.
Requirements: Michigan 501(c)3 organizations located in or supporting Newaygo County are eligible for funding. When submitting your proposal, include your organizations: mission, history, description of current programs, activities, and accomplishments; purpose of the grant (describe in detail and include supporting evidence); grant proposal budget form/narrative (form available at the Foundation's website). The following list of attachments must also be included: a copy of the current IRS 501(c)3 determination letter; roster of current governing board, including addresses and affiliations; finances: organization's current annual operating budget, including all expenses and revenues, audited financial statement (most recently completed), IRS Form 990 (most recently filed), annual report, if available; resumes and job descriptions of the key project personnel; organizational chart; letters of support (up to five).
Restrictions: In order to make the best use of available funds, the Foundation usually will not award grants for the following: grants to individuals; to pay off existing debts; religious programs that require religious affiliation and/or religious instruction to receive services; to further political campaigns; projects that begin prior to notification of Foundation funding; capital improvements on rental or individual private property; or programs or projects that subsidize or supplant funding for services considered general government obligations.
Geographic Focus: Michigan
Date(s) Application is Due: Feb 1; Sep 1
Amount of Grant: Up to 500,000 USD
Samples: Arts Center for Newaygo County, Fremont, Michigan, $324,800 - operating support; City of Grant, Grant, Michigan, $75,000 - completion of Safe Routes to School and support of area amphitheater; Feeding America West Michigan Food Bank, Comstock Park, Michigan, $102,020 - underwriting for Newaygo County fixed pantries and mobile pantries.
Contact: Vonda Carr; (231) 924-5350; fax (231) 924-5391; vcarr@tfacf.org
Internet: http://www.tfaf.org/grants.html
Sponsor: Fremont Area Community Foundation
P.O. Box B
Fremont, MI 49412

Fritz B. Burns Foundation Grants   1747
The foundation supports nonprofit organizations primarily in southern California by awarding grants for education, with an emphasis on buildings, equipment, endowments (except for ordinary operating expenses), student scholarship and loan funds, and faculty fellowships; to/for hospitals, hospital equipment, and medical research; and religious organizations (Christian, Jewish, Latter-day Saints, nondenominational, Presbyterian, Protestant, Roman Catholic, and Salvation Army). Proposals should be concise, containing a brief description of what is planned, with a clear statement of the objective sought; IRS letter certifying tax exemption; financial statements; and a list of officers and directors. No formalized application or proposal format is required. Proposals are considered on a quarterly basis.
Requirements: Nonprofit, tax-exempt organizations in southern California may apply.
Restrictions: Grant requests are not considered from individuals nor from tax-supported entities.
Geographic Focus: California
Date(s) Application is Due: Sep 30
Amount of Grant: 10,000 - 250,000 USD
Contact: Joseph Rawlinson, President; (818) 840-8802
Sponsor: Fritz B. Burns Foundation
4001 West Alameda Avenue, Suite 203
Burbank, CA 91505-4338

Fuji Film Grants   1748
Grants are awarded to proposals that have relevance to Fujifilm's strategic interests which are: environment; education; arts and culture and health and human services. Organizations are limited to one grant per calendar year. Applications are reviewed monthly and applicants should anticipate an award or decline within six to eight weeks after submitting a request. Applicants should submit a corporate giving application online.
Requirements: 501(c)3 tax-exempt organizations are eligible. Organizations must serve or be located within the communities where Fujifilm is a member. Preference is given to those proposals that have relevance to the company's strategic interests.
Restrictions: The program does not fund: individuals or personal use; political organizations, causes, candidates or campaigns; legislative advocacy groups; religious groups; military agencies; fraternal, veterans or social organizations.
Geographic Focus: All States
Amount of Grant: 1,000 - 5,000 USD
Samples: Sea World and Busch Gardens Conservation Fund (Clayton, MO)--for the Environmental Excellence Awards program, $125,000; Westchester Community College (Valhalla, NY) and Rhode Island School of Design (Providence, RI)--for scholarships, $25,000 and $50,000, respectively.
Contact: Contribution Program Manager; (800) 755-3854; contributions@fujifilm.com
Internet: http://www.fujifilm.com/JSP/fuji/epartners/givingGuidelines.jsp?nav=2
Sponsor: Fuji Photo Film USA
200 Summit Lake Drive, Floor 2
Valhalla, NY 10595

Fulbright Alumni Initiatives Awards Program Grants   1749
The objective of the AIA program is to help translate the individual Fulbright experience into long-term institutional impact. To this end, the program provides small institutional grants to Fulbright alumni to develop innovative projects that will foster institutionally supported linkages and sustainable, mutually beneficial relationships between the Fulbright scholar's home and host institutions. Just about any activity resulting in the creation or fostering of a sustainable institutional relationship that will have an impact on both the Fulbright alum's home institution and the Fulbright host institution abroad and which both institutions are prepared to support in both the long and short terms will be considered. The program has been temporarily suspended, and applicants should check back for updates.
Requirements: Applicants must be U.S. citizens. The program is open to eligible alumni whose grants occurred at any time from the 1998/1999 academic year through the present academic year.
Restrictions: This program is limited to application by a Fulbright alum (whether U.S. or Visiting Scholar) in partnership with his/her host institution colleague. However, other faculty members from both applicants' institutions may be part of the proposal and participate in the project.
Geographic Focus: All States
Contact: Stacey Bustillos, Program Officer; (202) 686-6252; sbustillos@cies.iie.org
Internet: http://www.cies.org/aia/
Sponsor: Council for International Exchange of Scholars
3007 Tilden Street NW, Suite 5L
Washington, D.C. 20008-3009

Fulbright Distinguished Chairs Awards   1750
The Fulbright Distinguished Chair Awards comprise approximately forty distinguished lecturing, distinguished research and distinguished lecturing/research awards ranging from three to 12 months. Awards in the Fulbright Distinguished Chairs program are viewed as among the most prestigious appointments in the Fulbright Scholar Program. Candidates should be senior scholars and have a significant publication and teaching record. Applicants should submit hard copies of the Distinguished Chairs Application Form (one page), a letter of interest (about three pages), a curriculum vitae (maximum eight pages) and, if required, a sample syllabus (maximum four pages). Chairs are available in Australia, Austria, Brazil, Canada, Denmark, Finland, France, Germany, Hungary, Ireland, Israel, Italy, Netherlands, Poland, Portugal, Russia, and Sweden. Because an objective of the Fulbright Program is to provide an educational exchange experience for those not previously afforded such an opportunity, preference will usually be given to candidates who have not had substantial recent experience in the country to which they are applying.
Requirements: Applicants must meet all of the following eligibility *Requirements:* U.S. citizenship at the time of application (permanent resident status is not sufficient); possess a Ph.D. or equivalent professional terminal degree at the time of application (for professionals and artists outside academe, recognized professional standing and substantial professional accomplishments); have college or university teaching experience at the level and in the field of the proposed lecturing activity as specified in the award description (for Distinguished Chairs awards, candidates should be senior scholars with a significant publication and teaching record); have foreign language proficiency only if specified in the award description or required for the completion of the proposed project; be of sound physical and mental health; and disclose prior conviction or current indictment for commission of a felony.
Restrictions: Previous Fulbright scholar grantees are eligible to apply only if five years will have elapsed between the ending date of one scholar award and the beginning

date of the new scholar award. This rule does not apply if the previous grant was for less than two months. Employees, spouses or dependent children of the United States Department of State or public and private organizations under contract to the United States Department of State are ineligible to apply for a Fulbright grant until one year after the employee's termination.
Geographic Focus: All States
Date(s) Application is Due: Aug 1
Contact: Jordanna Enrich, Director; (202) 686-6233; jenrich@iie.org
Internet: http://www.cies.org/program/fulbright-distinguished-chair-awards
Sponsor: Council for International Exchange of Scholars
3007 Tilden Street NW, Suite 5L
Washington, D.C. 20008-3009

Fulbright German Studies Seminar Grants 1751
The program allows participation in a group seminar on current German society and culture. The program will begin in Berlin and include visits to other cities in Germany. The focus will be on the formation of policies in current issues at the core of modern society such as climate change, food technology, gene technology, stem cell research and the broad scope of education. The seminar will explore how Germany and its European neighbors view the role of science in their societies, examining the factors that result in differing approaches to these issues and the challenges of harmonizing different standards under a uniform EU-policy umbrella. With an eye to the global marketplace and worldwide competition in science, the seminar will further examine the multiple interests which structure relations between national governments, economic corporations, political and supranational bodies as well as research and development institutions. Scholars from U.S. universities, colleges, and community colleges who hold full-time teaching appointments and meet other academic requirements (Ph.D., Ph.D. candidacy or other equivalent degree or qualifications) are eligible. The seminar lasts two weeks, typically during the month of June.
Requirements: Applicants must be U.S. citizens with permanent residence in the United States at the time of application.
Geographic Focus: All States, Germany
Date(s) Application is Due: Nov 1
Contact: Maria Bettua, Assistant Director; (202) 686-6245; mbettua@cies.iie.org
Internet: http://www.iie.org/cies/us_scholars/us_awards/GSS/index.html
Sponsor: Council for International Exchange of Scholars
3007 Tilden Street NW, Suite 5L
Washington, D.C. 20008-3009

Fulbright International Education Administrators Seminar Program Grants 1752
The IEA seminars are designed to introduce participants to the society, culture and higher education systems of these countries through campus visits, meetings with foreign colleagues and government officials, attendance at cultural events and briefings on education. Participants in the International Education Administrators Program gain a firsthand look into the host country's academic infrastructure and culture. They gain new perspective on the need to internationalize U.S. campuses and insight into how it can be done. The deadline for the German program is February 1, while the deadline for both Japan and Korea is November 1.
Requirements: To be eligible, applicants must: be U.S. citizens; be international education professionals and senior university administrators (e.g., deans, provosts, vice presidents) with significant responsibility for international programs and activities; have an affiliation with an accredited college or university or nonprofit international exchange organization administering postsecondary student or faculty exchange; and have a minimum of three years of work experience in international education. Applicants for the Japan program must be affiliated with a four-year college or university, while Germany and Korea will consider applicants from both two- and four-year institutions.
Restrictions: Employees, spouses or dependent children of the United States Department of State or public and private organizations under contract to the United States Department of State are ineligible to apply for a Fulbright grant until one year after the employee's termination. TEFL administrators are ineligible for these seminar programs, but they are encouraged to view a listing of other Fulbright opportunities.
Geographic Focus: All States
Date(s) Application is Due: Feb 1; Nov 1
Contact: Maria Bettua; (202) 686-6245; mbettua@cies.iie.org
Internet: http://www.cies.org/IEA/
Sponsor: Council for International Exchange of Scholars
3007 Tilden Street NW, Suite 5L
Washington, D.C. 20008-3009

Fulbright New Century Scholars (NCS) Program Grants 1753
The NCS Program brings 30 top academics and professionals from around the world together each year to collaborate on an issue of global importance. Of the thirty, approximately one-third will be U.S. citizens while the remaining two thirds will be visiting scholars from countries with an operational Fulbright Scholar Program. NCS will provide a platform for scholars from the U.S. and around the world to engage in debate and dialogue based on multidisciplinary research and to develop new global models for understanding the social context within which nations and communities shape their responses to the many challenges of the 21st century. This particular aspect of the New Century Scholars program is a unique feature that distinguishes it from the core Fulbright Scholar Program.
Requirements: Applicants must be conducting current research relevant to the program's theme and objectives, be open to exploring and incorporating comparative, interdisciplinary approaches in their investigations, and interested in developing collaborative activities with other NCS Scholars. U.S. applicants must have U.S. citizenship and be residing permanently in the United States. For academic applicants, a Ph.D. or equivalent terminal degree in a relevant field is required. For applicants in the professional fields, the appropriate terminal degree in a relevant field is required. Non-U.S. applicants must be citizens or permanent residents of and residing in the country from which they are applying at the time of application. All applicants must have fluency in English.
Restrictions: Non-U.S. applicants holding permanent residency green cards, whether or not they reside in the U.S., are not eligible.
Geographic Focus: All States
Contact: Jonathan Looper; (202) 686-6235; fax (202) 362-3442; jlooper@cies.iie.org
Internet: http://www.cies.org/NCS/
Sponsor: Council for International Exchange of Scholars
3007 Tilden Street NW, Suite 5L
Washington, D.C. 20008-3009

Fulbright Specialists Program Grants 1754
The Program is designed to provide short-term academic opportunities (two to six weeks) for U.S. faculty and professionals. Shorter grant lengths give specialists greater flexibility to pursue a grant that works best with their current academic or professional commitments. Applications for the Program are accepted on a rolling basis, and peer review of applications is conducted eight times per year. Program goals include: to increase the participation of leading U.S. scholars and professionals in Fulbright academic exchanges; to encourage new activities that go beyond the traditional Fulbright activities of lecturing and research; and to promote increased connections between U.S. and non-U.S. post-secondary academic institutions.
Requirements: Applicants must: be U.S. citizen at the time of application (permanent resident status is not sufficient; if a naturalized citizen, applicant must provide actual date of naturalization); possess a Ph.D. or equivalent professional/terminal degree at the time of application; have a minimum of five years of post-doctoral teaching or professional experience in the field in which you are applying (for professionals and artists outside academe, recognized professional standing and substantial professional accomplishments plus a minimum of five years of professional experience in the field in which you are applying); disclose prior conviction or current indictment for commission of a felony; and be residing in the United States at the time they are approved for a grant and intend to return to their U.S. institution after the grant's completion.
Restrictions: Employees, spouses or dependent children of the United States Department of State or public and private organizations under contract to the United States Department of State exchange programs are ineligible to apply for a Fulbright grant until one year after the employee's termination.
Geographic Focus: All States
Date(s) Application is Due: Jan 28; Feb 24; Mar 25; Apr 21; May 20; Jul 15; Sep 9; Nov 4; Dec 30
Contact: Ryan Hathaway; (202) 686-4026; rhathaway@cies.iie.org
Internet: http://www.cies.org/specialists/#top
Sponsor: Council for International Exchange of Scholars
3007 Tilden Street NW, Suite 5L
Washington, D.C. 20008-3009

Fulbright Traditional Scholar Program in Europe and Eurasia 1755
The traditional Fulbright Scholar Program sends 800 U.S. faculty and professionals abroad each year. Grantees lecture and conduct research in a wide variety of academic and professional fields. Distribution of awards to countries in the region will vary annually according to the caliber of the applicants. Grants are available to: Albania, Andorra, Armenia, Austria, Azerbaijan, Belarus, Belgium, Bosnia and Herzegovina, Bulgaria, Croatia, Cyprus, Czech Republic, Denmark, Estonia, European Union, Finland, France, Georgia, Germany, Greece, Hungary, Iceland, Ireland, Italy, Latvia, Lithuania, Luxembourg, Macedonia, Moldova, Netherlands, Norway, Poland, Portugal, Romania, Russia, Slovakia (Slovak Republic), Slovenia, Spain, Sweden, Switzerland, Turkey, Ukraine, and United Kingdom. Only countries Considered to be in the Eastern Europe or Eurasia regions may be part of a multi-country application (2 to 3 countries). Multi-country proposals are not permitted for Western Europe. Language requirements vary by country, and prior knowledge of the local language may not be required, particularly where language is not commonly taught in the U.S.
Requirements: Applicants must: be U.S. citizens at the time of application (permanent residents are not eligible); be in good health (grantees will be required to submit a satisfactory Medical Certificate of Health from a physician); have sufficient proficiency in the written and spoken language of the host country (where required) to communicate with the people and to carry out the proposed study; and hold a B.A. degree or the equivalent before the start of the grant (applicants who have not earned a B.A. degree or the equivalent, but who have extensive professional study and/or experience in fields in which they wish to pursue a project, may be considered). In the creative and performing arts area, four years of professional study and/or experience meets the basic eligibility requirement. All candidates for the Europe and Eurasia region are required to obtain their own affiliations, generally established with an educational and/or research institution in the host country or countries. Candidates are responsible for securing their own research clearance as required by the host country.
Restrictions: The following persons are ineligible: anyone who has already held a U.S. Department of State-funded Fulbright student grant of any type; anyone who has previously received a Department of Education-funded Doctoral Dissertation Research Abroad (Fulbright Hays grant); employees of the U.S. Department of State, and their immediate families, for a period ending one year following termination of such employment; employees of private and public agencies (excluding educational institutions)

under contract to the U.S. Department of State to perform administrative or screening services on behalf of the U.S. Department of State's exchange program, for a period ending one year following the termination of their services for the U.S. Department of State provided such employees have been directly engaged in performing services related to the exchange programs; applicants holding a doctoral degree at the time of application; applicants seeking enrollment in a medical degree program abroad; or applicants currently residing in the countries of Australia, Belgium/Luxembourg, Canada, Chile, Finland, Hungary, Mexico, Netherlands, New Zealand, Sweden, or Switzerland.
Geographic Focus: All States, Albania, Andorra, Armenia, Austria, Azerbaijan, Belarus, Belgium, Bosnia & Herzegovina, Bulgaria, Croatia, Cyprus, Czech Republic, Denmark, Estonia, Finland, France, Georgia, Germany, Greece, Hungary, Iceland, Ireland, Italy, Kosovo, Latvia, Liechtenstein, Lithuania, Luxembourg, Macedonia, Malta, Moldova, Monaco, Montenegro, Norway, Poland, Portugal, Romania, Russia, San Marino, Serbia, Slovakia, Slovenia, Spain, Sweden, Switzerland, The Netherlands, Turkey, Ukraine, United Kingdom, Vatican City
Date(s) Application is Due: Aug 1
Contact: Rachel Holskin, IIE Program Manager; (212) 984-5326; rholskin@iie.org
Internet: http://us.fulbrightonline.org/program_regions_countries.php?id=3
Sponsor: Institute of International Education
1400 K Street, NW, 7th Floor
Washington, D.C. 20005-2403

Fulbright Traditional Scholar Program in Sub-Saharan Africa   1756
The traditional Fulbright Scholar Program sends 800 U.S. faculty and professionals abroad each year. Grantees lecture and conduct research in a wide variety of academic and professional fields. Selection for countries in sub-Saharan Africa, with the exception of South Africa, will be made based on the quality of the applications, rather than per-country quotas. Distribution of awards to countries in the region will vary annually according to the caliber of the applicants. Grants are available to: Benin, Botswana, Burkina Faso, Cameroon, Chad, Eritrea, Ethiopia, Ghana, Guinea, Ivory Coast, Kenya, Madagascar, Malawi, Mali, Mauritius, Mozambique, Namibia, Niger, Nigeria, Senegal, South Africa, Swaziland, Tanzania, Togo, Uganda, Zambia, and Zimbabwe. All countries within the sub-Saharan Africa Region, with the exception of South Africa, may be part of a multi-country application (2 to 3 countries). For unlisted countries, applications may be considered on a case-by-case basis, but for dissertation research only.
Requirements: Applicants must: be U.S. citizens at the time of application (permanent residents are not eligible); be in good health (grantees will be required to submit a satisfactory Medical Certificate of Health from a physician); have sufficient proficiency in the written and spoken language of the host country to communicate with the people and to carry out the proposed study; and hold a B.A. degree or the equivalent before the start of the grant (applicants who have not earned a B.A. degree or the equivalent, but who have extensive professional study and/or experience in fields in which they wish to pursue a project, may be considered). In the creative and performing arts area, four years of professional study and/or experience meets the basic eligibility requirement. All candidates for Africa are required to obtain their own affiliations, generally established with an educational and/or research institution in the host country or countries.
Restrictions: Grants are not available to: Burundi, Central African Republic, Guinea-Bissau, Liberia, Somalia, or Sudan. The following persons are ineligible: anyone who has already held a U.S. Department of State-funded Fulbright student grant of any type; anyone who has previously received a Department of Education-funded Doctoral Dissertation Research Abroad (Fulbright Hays grant); employees of the U.S. Department of State, and their immediate families, for a period ending one year following termination of such employment; employees of private and public agencies (excluding educational institutions) under contract to the U.S. Department of State to perform administrative or screening services on behalf of the U.S. Department of State's exchange program, for a period ending one year following the termination of their services for the U.S. Department of State provided such employees have been directly engaged in performing services related to the exchange programs; applicants holding a doctoral degree at the time of application; applicants seeking enrollment in a medical degree program abroad; or applicants currently residing in the countries of Australia, Belgium/Luxembourg, Canada, Chile, Finland, Hungary, Mexico, Netherlands, New Zealand, Sweden, or Switzerland.
Geographic Focus: All States, Algeria, Angola, Benin, Botswana, Burkina Faso, Burundi, Cameroon, Cape Verde, Central African Republic, Chad, Comoros, Congo, Congo, Democratic Republic of, Cote d' Ivoire (Ivory Coast), Djibouti, Egypt, Equatorial Guinea, Eritrea, Ethiopia, Gabon, Gambia, Ghana, Guinea, Guinea-Bissau, Kenya, Lesotho, Liberia, Libya, Madagascar, Malawi, Mali, Mauritania, Mauritius, Morocco, Mozambique, Namibia, Niger, Nigeria, Rwanda, Sao Tome & Principe, Senegal, Seychelles, Sierra Leone, Somalia, South Africa, Sudan, Swaziland
Date(s) Application is Due: Aug 1
Contact: Jermaine Jones, Program Manager; (212) 984-5341; jjones@iie.org
Internet: http://us.fulbrightonline.org/program_regions_countries.php?id=1
Sponsor: Institute of International Education
1400 K Street, NW, 7th Floor
Washington, D.C. 20005-2403

Fulbright Traditional Scholar Program in the East Asia/Pacific Region   1757
The traditional Fulbright Scholar Program sends 800 U.S. faculty and professionals abroad each year. Grantees lecture and conduct research in a wide variety of academic and professional fields. Distribution of awards to countries in the region will vary annually according to the caliber of the applicants. Grants are available to: Australia, Cambodia, China, Hong Kong, Indonesia, Japan, Korea, Laos, Macau, Malaysia, Mongolia, New Zealand, Philippines, Singapore, Taiwan, Thailand, and Vietnam. All countries within the East Asia/Pacific region, with the exception of China, may be part of a multi-country application (2 to 3 countries). For unlisted countries, applications may be considered on a case-by-case basis, but for dissertation research only. Language requirements vary by country, and prior knowledge of the local language may not be required, particularly where language is not commonly taught in the U.S.
Requirements: Applicants must: be U.S. citizens at the time of application (permanent residents are not eligible); be in good health (grantees will be required to submit a satisfactory Medical Certificate of Health from a physician); have sufficient proficiency in the written and spoken language of the host country to communicate with the people and to carry out the proposed study; and hold a B.A. degree or the equivalent before the start of the grant (applicants who have not earned a B.A. degree or the equivalent, but who have extensive professional study and/or experience in fields in which they wish to pursue a project, may be considered). In the creative and performing arts area, four years of professional study and/or experience meets the basic eligibility requirement. All candidates for the East Asia/Pacific region are required to obtain their own affiliations, generally established with an educational and/or research institution in the host country or countries. Candidates are responsible for securing their own research clearance as required by the host country.
Restrictions: Grants are not available to: Brunei, the Cook Islands, East Timor, Fiji, Myanmar, the Pacific Island Nations, Papua New Guinea, or Western Samoa. The following persons are ineligible: anyone who has already held a U.S. Department of State-funded Fulbright student grant of any type; anyone who has previously received a Department of Education-funded Doctoral Dissertation Research Abroad (Fulbright Hays grant); employees of the U.S. Department of State, and their immediate families, for a period ending one year following termination of such employment; employees of private and public agencies (excluding educational institutions) under contract to the U.S. Department of State to perform administrative or screening services on behalf of the U.S. Department of State's exchange program, for a period ending one year following the termination of their services for the U.S. Department of State provided such employees have been directly engaged in performing services related to the exchange programs; applicants holding a doctoral degree at the time of application; applicants seeking enrollment in a medical degree program abroad; or applicants currently residing in the countries of Australia, Belgium/Luxembourg, Canada, Chile, Finland, Hungary, Mexico, Netherlands, New Zealand, Sweden, or Switzerland.
Geographic Focus: All States, Abkhazia, Afghanistan, Armenia, Australia, Azerbaijan, Bahrain, Bangladesh, Bhutan, British Indian Ocean Territory, Brunei, Burma (Myanmar), Cambodia, China, Christmas Island, Cocos, Cyprus, Georgia, Hong Kong, India, Indonesia, Iran, Iraq, Israel, Japan, Jordan, Kazakhstan, Kuwait, Kyrgyzstan, Laos, Lebanon, Macau, Malaysia, Maldives, Mongolia, Nagorno-Karabakh, Nepal, North Korea, Northern Cyprus, Oman, Pakistan, Palestinian Authority, Philippines, Qatar, Russia, Saudi Arabia, Singapore, South Korea, South Ossetia, Sri Lanka, Syria, Taiwan, Tajikistan, Thailand, Timor-Lester, Turkey, Turkmenistan, United Arab Emirates, Uzbekistan, Vietnam, Yemen
Date(s) Application is Due: Aug 1
Contact: Jonathan Akeley, Program Manager; (212) 984-5487; jakeley@iie.org
Internet: http://us.fulbrightonline.org/program_regions_countries.php?id=2
Sponsor: Institute of International Education
1400 K Street, NW, 7th Floor
Washington, D.C. 20005-2403

Fulbright Traditional Scholar Program in the Near East & North Africa Region   1758
The traditional Fulbright Scholar Program sends 800 U.S. faculty and professionals abroad each year. Grantees lecture and conduct research in a wide variety of academic and professional fields. Selection for countries in the Near East and North Africa region will be made based on the quality of the applications, rather than per-country quotas. Distribution of awards to countries in the region will vary annually according to the caliber of the applicants. Grants are available to: Bahrain, Egypt, India, Jordan, Kuwait, Morocco, Oman, Qatar, Syria, Tunisia, United Arab Emirates, and Yemen. Multi-country applications (2 to 3 countries) are available, except in Egypt, India, Jordan, and Morocco. For unlisted countries, applications may be considered on a case-by-case basis.
Requirements: Applicants must: be U.S. citizens at the time of application (permanent residents are not eligible); be in good health (grantees will be required to submit a satisfactory Medical Certificate of Health from a physician); have sufficient proficiency in the written and spoken language of the host country to communicate with the people and to carry out the proposed study; and hold a B.A. degree or the equivalent before the start of the grant (applicants who have not earned a B.A. degree or the equivalent, but who have extensive professional study and/or experience in fields in which they wish to pursue a project, may be considered). In the creative and performing arts area, four years of professional study and/or experience meets the basic eligibility requirement. Candidates are responsible for securing research clearance, as required. In countries with Fulbright Commissions, assistance may be provided.
Restrictions: Grants are not available to: Algeria, Iran, Iraq, Lebanon, Saudi Arabia, the West Bank, or Gaza. The following persons are ineligible: anyone who has already held a U.S. Department of State-funded Fulbright student grant of any type; anyone who has previously received a Department of Education-funded Doctoral Dissertation Research Abroad (Fulbright Hays grant); employees of the U.S. Department of State, and their immediate families, for a period ending one year following termination of such employment; employees of private and public agencies (excluding educational institutions) under contract to the U.S. Department of State to perform administrative or screening services on behalf of the U.S. Department of State's exchange program, for a period ending one year following the termination of their services for the U.S. Department of State provided such employees have been directly engaged in performing services related to the exchange programs; applicants holding a doctoral degree at the time of application; applicants seeking enrollment in a medical degree program abroad; or applicants currently

residing in the countries of Australia, Belgium/Luxembourg, Canada, Chile, Finland, Hungary, Mexico, Netherlands, New Zealand, Sweden, or Switzerland.
Geographic Focus: All States, Algeria, Angola, Benin, Botswana, Burkina Faso, Burundi, Cameroon, Cape Verde, Central African Republic, Chad, Comoros, Congo, Congo, Democratic Republic of, Cote d' Ivoire (Ivory Coast), Djibouti, Egypt, Equatorial Guinea, Eritrea, Ethiopia, Gabon, Gambia, Ghana, Guinea, Guinea-Bissau, Israel, Kenya, Lesotho, Liberia, Libya, Madagascar, Malawi, Mali, Mauritania, Mauritius, Morocco, Mozambique, Namibia, Niger, Nigeria, Rwanda, Sao Tome & Principe, Senegal, Seychelles, Sierra Leone, Somalia, South Africa, Sudan, Swaziland, Uzbekistan
Date(s) Application is Due: Aug 1
Contact: Jermaine Jones, Program Manager; (212) 984-5341; jjones@iie.org
Internet: http://us.fulbrightonline.org/program_regions_countries.php?id=4
Sponsor: Institute of International Education
1400 K Street, NW, 7th Floor
Washington, D.C. 20005-2403

Fulbright Traditional Scholar Program in the South and Central Asia Region   1759
The traditional Fulbright Scholar Program sends 800 U.S. faculty and professionals abroad each year. Grantees lecture and conduct research in a wide variety of academic and professional fields. Selection for countries in the South and Central Asia region will be made based on the quality of the applications, rather than per-country quotas. Distribution of awards to countries in the region will vary annually according to the caliber of the applicants. Grants are available to: Bangladesh, India, Kazakhstan, Kyrgyz Republic, Nepal, Sri Lanka, Tajikistan, and Uzbekistan. Multi-country applications (2 to 3 countries) are not available. Language requirements vary by country. Applicants to India who are recommended for final review will need to submit research visa applications in January or February.
Requirements: Applicants must: be U.S. citizens at the time of application (permanent residents are not eligible); be in good health (grantees will be required to submit a satisfactory Medical Certificate of Health from a physician); have sufficient proficiency in the written and spoken language of the host country to communicate with the people and to carry out the proposed study; and hold a B.A. degree or the equivalent before the start of the grant (applicants who have not earned a B.A. degree or the equivalent, but who have extensive professional study and/or experience in fields in which they wish to pursue a project, may be considered). In the creative and performing arts area, four years of professional study and/or experience meets the basic eligibility requirement. Candidates are responsible for securing research clearance, as required. In countries with Fulbright Commissions, assistance may be provided.
Restrictions: Grants are not available to: Afghanistan, Bhutan, Pakistan, Republic of Maldives, or Turkmenistan. The following persons are ineligible: anyone who has already held a U.S. Department of State-funded Fulbright student grant of any type; anyone who has previously received a Department of Education-funded Doctoral Dissertation Research Abroad (Fulbright Hays grant); employees of the U.S. Department of State, and their immediate families, for a period ending one year following termination of such employment; employees of private and public agencies (excluding educational institutions) under contract to the U.S. Department of State to perform administrative or screening services on behalf of the U.S. Department of State's exchange program, for a period ending one year following the termination of their services for the U.S. Department of State provided such employees have been directly engaged in performing services related to the exchange programs; applicants holding a doctoral degree at the time of application; applicants seeking enrollment in a medical degree program abroad; or applicants currently residing in the countries of Australia, Belgium/Luxembourg, Canada, Chile, Finland, Hungary, Mexico, Netherlands, New Zealand, Sweden, or Switzerland.
Geographic Focus: All States, Kazakhstan, Uzbekistan
Date(s) Application is Due: Aug 1
Contact: Jonathan Akeley; (212) 984-5487; jakeley@iie.org
Internet: http://us.fulbrightonline.org/program_regions_countries.html?id=6
Sponsor: Institute of International Education
1400 K Street, NW, 7th Floor
Washington, D.C. 20005-2403

Fulbright Traditional Scholar Program in the Western Hemisphere   1760
The traditional Fulbright Scholar Program sends 800 U.S. faculty and professionals abroad each year. Grantees lecture and conduct research in a wide variety of academic and professional fields. Selection for countries in Central America and the Caribbean will be made based on the quality of the applications, rather than per-country quotas. Distribution of awards to countries in the region will vary annually according to the caliber of the applicants. Grants are available to: Argentina, Barbados and the Eastern Caribbean, Bolivia, Brazil, Canada, Chile, Colombia, Costa Rica, Dominican Republic, Ecuador, El Salvador, Guatemala, Honduras, Jamaica, Mexico, Nicaragua, Panama, Paraguay, Peru, Trinidad and Tobago, Uruguay, and Venezuela. English Teaching Assistantships are available in Argentina, Brazil, Chile, Uruguay and Venezuela. Language proficiency may be preferred or required. Applicants for English Teaching Assistantships can apply to only one country. Multi-country applications (2 to 3 countries) are available in all other disciplines throughout the Western Hemisphere. For unlisted countries, applications may be considered on a case-by-case basis, but for dissertation research only.
Requirements: Applicants must: be U.S. citizens at the time of application (permanent residents are not eligible); be in good health (grantees will be required to submit a satisfactory Medical Certificate of Health from a physician); have sufficient proficiency in the written and spoken language of the host country to communicate with the people and to carry out the proposed study; and hold a B.A. degree or the equivalent before the start of the grant (applicants who have not earned a B.A. degree or the equivalent, but who have extensive professional study and/or experience in fields in which they wish to pursue a project, may be considered). In the creative and performing arts area, four years of professional study and/or experience meets the basic eligibility requirement. Candidates are responsible for securing research clearance, as required. In countries with Fulbright Commissions, assistance may be provided.
Restrictions: Grants are not available to: Bahamas, Belize, Cuba, French Guiana, Guyana, Haiti, Martinique, or Suriname. The following persons are ineligible: anyone who has already held a U.S. Department of State-funded Fulbright student grant of any type; anyone who has previously received a Department of Education-funded Doctoral Dissertation Research Abroad (Fulbright Hays grant); employees of the U.S. Department of State, and their immediate families, for a period ending one year following termination of such employment; employees of private and public agencies (excluding educational institutions) under contract to the U.S. Department of State to perform administrative or screening services on behalf of the U.S. Department of State's exchange program, for a period ending one year following the termination of their services for the U.S. Department of State provided such employees have been directly engaged in performing services related to the exchange programs; applicants holding a doctoral degree at the time of application; applicants seeking enrollment in a medical degree program abroad; or applicants currently residing in the countries of Australia, Belgium/Luxembourg, Canada, Chile, Finland, Hungary, Mexico, Netherlands, New Zealand, Sweden, or Switzerland.
Geographic Focus: All States, Antigua & Barbuda, Argentina, Bahamas, Barbados, Belize, Bolivia, Brazil, Canada, Chile, Colombia, Costa Rica, Cuba, Dominica, Dominican Republic, Ecuador, El Salvador, Grenada, Guatemala, Guyana, Haiti, Honduras, Jamaica, Mexico, Nicaragua, Paraguay, Peru
Date(s) Application is Due: Aug 1
Contact: Jody Dudderar, Program Manager in South America, Mexico and Canada; (212) 984-5565; fax (212) 984-5325; jdudderar@iie.org
Internet: http://us.fulbrightonline.org/program_regions_countries.php?id=5
Sponsor: Institute of International Education
1400 K Street, NW, 7th Floor
Washington, D.C. 20005-2403

Fuller E. Callaway Foundation Grants   1761
The Fuller E. Callaway Foundation was founded by the late Fuller E. Callaway, Sr., as a Relief Association chartered by the Superior Court of Troup County, Georgia on December 1, 1917. On his death in 1928, Callaway left a substantial bequest to the Association under his will. The operation of the Association continued under the management of remaining members of Callaway's family. Today, the Foundation awards grants to nonprofit organizations and individuals in LaGrange and Troup County, Georgia. Grants are awarded in the areas of religion, higher and other education, social services, youth, and health. Types of support include: general operating budgets; annual campaigns; building funds; equipment; matching funds; and student aid. Another primary focus of the Foundation is the operation of the historic Callaway home and garden, Hills and Dales Estate, for the education and enrichment of the interested public. Letters of application from organizations are accepted and have deadlines at the end of December, March, June, and September.
Requirements: Nonprofit organizations and individuals in LaGrange and Troup County, Georgia, are eligible for support.
Geographic Focus: Georgia
Amount of Grant: 25,000 - 200,000 USD
Contact: H. Speer Burdette III, President; (706) 884-7348; fax (706) 884-0201; hsburdette@callaway-foundation.org
Internet: http://www.callawayfoundation.org/history.php
Sponsor: Fuller E. Callaway Foundation
209 Broome Street, P.O. Box 790
LaGrange, GA 30241-0014

Fulton County Community Foundation 4Community Higher Education Scholarship   1762
The primary purpose of the 4Community Higher Education Scholarship is to support the charitable intentions of the Fulton County 4Community Partners in promoting higher education for GED program participants. The Scholarship will be awarded to Fulton County residents who are GED recipients. Scholarship recipients must be accepted into an accredited college, university, trade, or vocational school in Indiana. All majors are eligible to apply. There is no deadline; applicants may apply at any time of the year.
Requirements: Applicants must be U.S. citizens who are Fulton County residents. They must have earned a minimum standard score of 520 on the GED exam. Along with the online application, they must also submit: proof of their score; evidence that they have participated in community and work activities; a letter of recommendation from their GED instructor; and an essay expressing their personal reasons for wanting a post-high school education.
Geographic Focus: Indiana
Contact: Corinne Becknell Lucas, Scholarship Coordinator; (574) 223-2227 or (877) 432-6423; corinne@nicf.org
Internet: http://www.nicf.org/fulton/scholarships.html
Sponsor: Fulton County Community Foundation
715 Main Street, P.O. Box 807
Rochester, IN 46975

Fulton County Community Foundation Grants   1763
Fulton County Community Foundation is part of the Northern Indiana Community Foundation (NICF). Grant making areas of interest are: education; health; human services; arts and culture; environment; and civic and recreation. The Foundation favors activities that: reach a broad segment of the community, especially those citizens whose needs are not being met by existing services; request seed money to meet innovative opportunities in the community; stimulate and encourage additional funding; promote

cooperation and avoid duplication of effort; help make charitable organizations more effective, efficient and self-sustaining; and one time projects or needs. Applications are available online. Applicants are encouraged to contact the Program Coordinator to discuss their project before applying.
Restrictions: The foundation will not consider grants for: religious organizations for the sole purpose of furthering that religion; political activities or those designated to influence legislation; national organizations (unless the monies are to be used solely to benefit citizens of Fulton County); grant that directly benefit the donor or the donor's family; fundraising projects; contributions to endowments
Geographic Focus: Indiana
Date(s) Application is Due: Sep 30
Contact: Corinne Becknell Lucas, Program Coordinator; (574) 223-2227 or (877) 432-6423; fax (574) 224-3709; corinne@nicf.org
Internet: http://www.nicf.org/fulton/grants.html
Sponsor: Fulton County Community Foundation
715 Main Street, P.O. Box 807
Rochester, IN 46975

Fulton County Community Foundation Lilly Endowment Community Scholarships   1764
The Lilly scholarships provide full tuition, plus an $800 book stipend for four years of full time undergraduate study, leading to a baccalaureate degree at any accredited Indiana public or private college or university. Applicants must be U.S. citizens and residents of Fulton County. Applications are available at high school guidance offices. Any college major may apply.
Geographic Focus: Indiana
Contact: Corinne Becknell Lucas, Scholarship Coordinator; (574) 223-2227 or (877) 432-6423; fax (574) 224-3709; corinne@nicf.org
Internet: http://www.nicf.org/fulton/availablescholarships.html
Sponsor: Fulton County Community Foundation
715 Main Street, P.O. Box 807
Rochester, IN 46975

Fulton County Community Foundation Paul and Dorothy Arven Memorial Scholarship   1765
The Paul and Dorothy Arven Memorial Scholarship is offered to students pursuing careers in nursing or agriculture. Students must be accepted into an accredited college or university, U.S. citizens and residents of Fulton County. They must possess good citizenship in school, work, and community; good academic accomplishments; and likely to start and complete their course of study with the scholarship's financial assistance.
Requirements: Along with the online application, students must submit their estimated expenses and resources, academic certification form, signature page, two letters of recommendation, the FAFSA's Student Aid Report, and an answer to the essay question in the application.
Restrictions: Public high school students must return application to the guidance department of their high school. All other student should mail, fax or drop off the application to the attention of Corinne Becknell Lucas at the NICF office.
Geographic Focus: Indiana
Date(s) Application is Due: Mar 14
Contact: Corinne Becknell Lucas, Scholarship Coordinator; (574) 223-2227 or (877) 432-6423; fax (574) 224-3709; corinne@nicf.org
Internet: http://www.nicf.org/fulton/scholarships.html
Sponsor: Fulton County Community Foundation
715 Main Street, P.O. Box 807
Rochester, IN 46975

Fulton County Community Foundation Women's Giving Circle Grants   1766
The Women's Giving Circle Grants help fund non-profit organizations in Fulton County. Organizations receiving their funds include Camp-We-Can, the County Council on Aging, and the County Cancer Fund.
Requirements: Applicants may contact Brian Johnson, the director of development at the Fulton County Community Foundation, for information about whether their project qualifies for application.
Geographic Focus: Indiana
Contact: Brian Johnson, Development Director; (574) 224-3223; fulton@nicf.org
Internet: http://www.nicf.org/fulton/grants.html
Sponsor: Fulton County Community Foundation
715 Main Street, P.O. Box 807
Rochester, IN 46975

G.A. Ackermann Memorial Fund Grants   1767
The G. A. Ackermann Memorial Fund was established under the will of Mrs. Mary A. Ackermann in 1937 to support quality health care and human services programming for underserved populations. According to Mrs. Ackermann's wishes, the Ackermann Memorial Fund supports charitable organizations operated or controlled by the Roman Catholic church and/or members of the Roman Catholic Church, and charitable organizations operated or controlled by a Protestant Church and/or its members. Special consideration is given to hospitals. The Memorial Fund has biannual application deadlines of January 15 and June 1. Applicants for the January 15 deadline will be notified of grant decisions by June 30, and applicants for the June 1 deadline will be notified by December 31. To better support the capacity of nonprofit organizations, multi-year funding requests are considered. Most recent awards have ranged from $10,000 to $222,000.
Requirements: Organizations must be geographically located within the city limits of New York City or Chicago.
Restrictions: The Foundation will consider requests for general operating support only if the organization's operating budget is less than $1 million. In general, grant requests for individuals, endowment campaigns, capital projects, or research will not be considered.
Geographic Focus: Illinois, New York
Date(s) Application is Due: Jan 15; Jun 1
Amount of Grant: 10,000 - 222,000 USD
Samples: Swedish Covenant Hospital, Chicago, Illinois, $222,000 - for Chronic Disease Care Transitions Program (2014); Saint Vincent de Paul Center, Chicago, Illinois, $50,000 - for Child Development Program (2014); Missionary Sisters Servants of the Holy Spirit, Chicago, Illinois, $20,000 - for Domestic Abuse program, English as a Second Language program, stipend for programs coordinator volunteer, and purchase of a copy machine (2014).
Contact: Srilatha Lakkaraju; (312) 828-8166; ilgrantmaking@ustrust.com
Internet: https://www.bankofamerica.com/philanthropic/foundation.go?fnId=14
Sponsor: G.A. Ackermann Memorial Fund
231 South LaSalle Street, IL1-231-13-32
Chicago, IL 60604

G.N. Wilcox Trust Grants   1768
The trust provides partial support to programs and projects of tax-exempt, public charities in Hawaii to improve the quality of life in the state, particularly the island of Kauai. Grants of one year's duration are awarded in categories of interest to the trust, including education, literacy programs and adult basic education, health, Protestant religion, delinquency and crime prevention, social services, youth services, and culture and the performing arts. Types of support include general operating grants, capital grants, equipment acquisition, seed grants, scholarship funds, and challenge/matching grants. Deadlines dates for general grants are: January 1; April 1; July 1; October 1. The deadline date for scholarships is February 15th.
Requirements: Giving is limited to Hawaii, with emphasis on the island of Kauai. To begin application process, contact Paula Boyce for additional guidelines.
Restrictions: Grants are not awarded to support government agencies (or organizations substantially supported by government funds), individuals, or for endowment funds, research, deficit financing, or student aid in scholarships or loans.
Geographic Focus: Hawaii
Date(s) Application is Due: Jan 1; Feb 15; Apr 1; Jul 1; Oct 1
Samples: Save our Seas--to support scholarships for youth to participate in the Second Annual Clean Oceans Conference, $2,954; Church of the Crossroads--to support the capital campaigns for renovations, $10,000; Hawaii Public Radio--to support the membership challenge grant, $3,000.
Contact: Paula Boyce, c/o Bank of Hawaii; (808) 538-4944; fax (808) 538-4647; pboyce@boh.com or emoniz@boh.com
Sponsor: G.N. Wilcox Trust
Bank of Hawai'i, Foundation Administration Department 758
Honolulu, HI 96802-3170

Gamble Foundation Grants   1769
Founded in 1968, The Gamble Foundation is primarily interested in supporting organizations that serve disadvantaged children and youth in San Francisco, Marin and Napa counties. Within the field of youth development, the Foundation focuses on literacy, educational and personal enrichment programs designed to open doors of opportunity for at-risk youth in order to help them succeed in school and become productive, self-sufficient members of society. The Foundation is particularly interested in agricultural/environmental education, financial & computer literacy, vocational training and programs that prevent substance abuse and teen violence. To a lesser degree, the Foundation supports environmental organizations that focus on land preservation and sustainability, animal welfare and management, and pollution control. The foundation is interested in promoting green concepts that increase awareness of science based solutions that help reduce consumption of finite resources. The Foundation prefers to fund specific projects rather than annual appeals. Grants range from $5,000 to $20,000.
Requirements: Northern California 501(c)3 nonprofit organizations, with an emphasis on San Francisco, Marin, and Napa Counties, are eligible to apply. The Board meets in the spring each year and makes grants in late summer. The Foundation will accept proposals for the April meeting from January 25 - February 10. The Foundation encourages submission of proposals and attachments by email. For those submitting by email, the proposal and required attachments should be emailed in PDF format only. Send your proposals to Fiona Barrett at fbarrett@pfs-llc.net. If you do not receive an email within 24 hours confirming that your proposal has been received, please contact Fiona at (415) 561-6540, ext. 221. The Foundation will also accept a proposal submitted by mail as long as it is postmarked on or before February 10th. Proposals must include the following in the order listed: cover letter, on organization letterhead with address and phone number, including a brief summary of the request and a list of attachments (not more than one page); proposal narrative (not to exceed 5 pages); concise description of the organization (not more than two pages) including: relevant history, mission, geography and populations served, overview of programs; description of the project (not more than three pages) including: need, purpose, goals, timeline, project budget, including secured and projected sources of funding; financial statement, including actual revenue and expenses for the organization's most recently completed fiscal year; organizational budget for the present year, detailing proposed expenditures and projected sources of funding (not more than two pages); list of major public and private funders, identifying both secured and planned for funds; list of the Board of Directors, with affiliations; copy of the agency's IRS 501(c)3 tax-exempt determination letter. When submitting proposal, clasp the proposal materials with a single binder clip; do not use staples. Do not send audio-visual materials, binders, folders, or pamphlets unless requested. Receipt

of proposals will be acknowledged with a written response within a reasonable period of time. Should additional information be required, applicants will be contacted.
Restrictions: In general, the Foundation does not support medical research, individuals, endowments, or capital improvements.
Geographic Focus: California
Date(s) Application is Due: Feb 10
Amount of Grant: 5,000 - 20,000 USD
Samples: WildCare: Terwilliger Nature Education and Wildlife Rehabilitation, San Rafael, CA, $20,000--for No Child Left Indoors program, providing outdoor education for Marin and San Francisco youth; Canal Alliance, San Rafael, CA, $25,000--for Youth Education and Development program, providing academic, leadership, and personal development for low-income San Rafael youth; Bay Area Discovery Museum, Sausalito, CA, $25,000--for Connections program.
Contact: Eric Sloan; (415) 561-6540, ext. 205; fax (415) 561-6477; esloan@pfs-llc.net
Internet: http://www.pfs-llc.net/gamble/gamble.html
Sponsor: Gamble Foundation
1660 Bush Street, Suite 300
San Francisco, CA 94109

Gardner Foundation Grants 1770
The Gardner Foundation was established in the State of Nebraska in 1990, in support of programs in both the Wakefield, Nebraska, area and throughout South Carolina. The Foundation's primary fields of interest include: K-12 education; higher education; hospital care; orchestral music, and theater. Funding most often comes in the form of: infrastructure support; capital campaigns; endowment contributions; equipment purchases; fundraising efforts; and seed money. Most recently, awards have ranged from $2,000 to $90,000. An application form is required, and can be secured by contacting the Foundation office. The Board meets quarterly and, though there are no specific deadlines, complete applications should be forwarded ninety days prior to the upcoming meeting.
Requirements: 501(c)3 organizations serving a 75-mile radius of Wakefield, Nebraska, or the residents of South Carolina are eligible to apply.
Geographic Focus: Nebraska, South Carolina
Amount of Grant: 2,000 - 90,000 USD
Contact: Leslie A. Bebee, Vice President; (402) 287-2538
Sponsor: Gardner Foundation of Wakefield
P.O. Box 390, 307 Main Street
Wakefield, NE 68784-6026

Gardner Foundation Grants 1771
The Gardner Foundation was established in Kentucky in 1941, and support nonprofit organizations throughout the State of Florida. Awards are generally limited to environmental programs, public and private schools, churches, health care facilities, and arts and culture programs. Most recently disclosed awards have ranged from $10,000 to $25,000. Specific application forms are not required, and interested parties should submit their request in writing. There are no annual deadlines.
Requirements: 501(c)3 nonprofits either located in, or serving residents of, the State of Florida are eligible to apply.
Geographic Focus: Florida
Amount of Grant: 10,000 - 25,000 USD
Samples: Kampong of the National Tropical Botanical Garden, Miami, Florida, $10,000 - in support of the botanical gardens study program; Baptist Health South Florida, Miami, Florida, $25,000 - general operating funds.
Contact: Peter Coffin Gardner, Trustee; (305) 859-8915; fax (305) 441-2975
Sponsor: Gardner Foundation of Miami
P.O. Box 33-1850
Miami, FL 33233-1850

Gardner Foundation Grants 1772
The Gardner Foundation was established in the State of New York in 1947, with a giving emphasis on Milwaukee, Wisconsin. The Foundation's major mission is to provide grant funding for a wide range of organizations supporting arts and culture, education, and youth services. Its primary fields of interest include: child welfare; community and economic development; elementary and secondary education; employment; homeless services; hospice care; human services; mental health care; museums; performing arts; and reproductive health care. That support typically comes in a variety of forms, which include: annual campaigns; capital and infrastructure; capital campaigns; continuing support; emergency funding; and general operating support. Most recent awards have ranged from $1,000 to $6,000. Application forms and application guidelines are provided. Annual deadlines are at minimum one month prior to Board meetings, which are generally scheduled in April, September, and December. April, September, and December
Requirements: 501(c)3 organizations and schools serving the Milwaukee, Wisconsin, area are eligible to apply.
Geographic Focus: Wisconsin
Date(s) Application is Due: Mar 1; Aug 1; Nov 1
Amount of Grant: 1,000 - 6,000 USD
Samples: 88 Nine Radio, Milwaukee, Wisconsin, $6,000 - general operating funds; Beyond Vision, Milwaukee, Wisconsin, $4,000 - general operating funds; Hunger Task Force, Milwaukee, Wisconsin, $4,000 - general operating funds.
Contact: Theodore Friedlander III, President; (414) 273-0308
Sponsor: Gardner Foundation of Milwaukee
322 E. Michigan Street, Suite 250
Milwaukee, WI 53202-5010

Gardner W. and Joan G. Heidrick, Jr. Foundation Grants 1773
The Gardner W. and Joan G. Heidrick, Jr. Foundation was established in the State of Texas in 1998, with the expressed purpose of providing support for nonprofit organizations in Texas. North Carolina, and Illinois. The Foundation's primary fields of interest include: higher education programs; human services; and recreation and sports. Typically, support is given in the form of general operating funds. Most recent grants have ranged from $20 to $1,000, with an average of ten awards given each year. A formal application is required, and can be secured directly from the Foundation office. There are no specified annual deadlines.
Requirements: 501(c)3 organizations serving residents of Texas, North Carolina, or Illinois are eligible to apply.
Geographic Focus: Illinois, North Carolina, Texas
Amount of Grant: 20 - 1,000 USD
Samples: American Diabetes, Charlotte, North Carolina, $100 - general operating funds; University of Chicago, Chicago, Illinois, $1,000 - general operations; University of Texas, Austin, Texas, $1,000 - general operations.
Contact: Gardner W. Heidrick, President; (704) 366-7880
Sponsor: Gardner W. and Joan G. Heidrick, Jr. Foundation
8919 Park Road, Suite 4019
Charlotte, NC 28210-2242

Garland D. Rhoads Foundation 1774
Garland D. Rhoads established the Foundation in 1971 to promote the extension of religion, education, the alleviation of human suffering and the prevention and control of diseases. The grantor also intended for funds to be utilized for the acquisition, construction, maintenance and beautification of public buildings, grounds and/or works for the encouragement of public and civic betterment. Rhoads also wanted to provide for the relief of the indigent and to aid in scientific research for the betterment of mankind. The Foundation is currently managed by the Bank of America Philanthropic Solutions. The annual deadline for applications is September 1, with award notifications by June 30. The average grant award for a grant cycle has been $5,000.
Requirements: Grants are awarded to qualified 501(c)3 charitable organizations exclusively within the state of Texas.
Geographic Focus: Texas
Date(s) Application is Due: Sep 1
Amount of Grant: 2,000 - 8,000 USD
Contact: Jenae Guillory, Philanthropic Relationship Manager; (214) 209-1965 or (214) 209-1370; tx.philanthropic@ustrust.com
Internet: https://www.bankofamerica.com/philanthropic/grantmaking.go
Sponsor: Garland D. Rhoads Foundation
901 Main Street, 19th Floor
Dallas, TX 75202-3714

Gates Award for Global Health 1775
The Gates Global Health Award recognizes an organization that has made a major and lasting contribution to the field of global health. Nominations will be considered by a jury consisting of health professionals from developing countries as well as the Global Health Council's board of directors. A winner will be selected by the jury, and the award is presented in Washington, D.C., at a special awards ceremony during the Global Health Council's annual international conference.
Requirements: Any organization from any country in the world that has substantively improved the health and the lives of people in need may be nominated for the Gates Award, Organizations may be charitable institutions, private companies, or public entities.
Restrictions: Self-nominations are not accepted.
Geographic Focus: All States, All Countries
Amount of Grant: 1,000,000 USD
Contact: William Foege; (206) 709-3140; info@gatesfoundation.org
Internet: http://www.gatesfoundation.org/gates-award-global-health/Pages/overview.aspx
Sponsor: Bill and Melinda Gates Foundation
P.O. Box 23350
Seattle, WA 98102

Gates Millennium Scholars Program 1776
The Gates Foundation funds 1,000 college scholarships for low-income minority students. Students receiving scholarships may major in any field. Graduate scholarships support postgraduate study up to and including doctoral degrees for work in mathematics, science, public health, computer science, engineering, education, and library science. Nominations and recommendations for students are also accepted. Candidates may apply through the online application system. Additional guidelines and a list of frequently asked questions are available at the website.
Requirements: Applicants must be a citizen of the U.S., with a minimum GPA of 3.3, and demonstrated financial need. They must be a high school senior applying to an accredited college or university for the academic year, or a college student planning to continue undergraduate study. They may also be a college senior or college graduate enrolled or about to enroll in graduate school. Undergraduates may enroll in any field, but graduate students must major in engineering, computer science, public health, mathematics, science, education, or library science.
Geographic Focus: All States
Contact: Gates Millennium Scholars Coordinator; (877) 690-4677
Internet: http://www.gmsp.org
Sponsor: Bill and Melinda Gates Foundation
P.O. Box 23350
Seattle, WA 98102

Gebbie Foundation Grants 1777
The Foundation's mission is to support appropriate charitable and humanitarian programs to improve the quality of life, primarily in Chautauqua County, New York by focusing on: children, youth and education; arts; human services; and community development. The foundation's strategic focus is to rejuvenate downtown Jamestown, New York, through economic development. Types of funding include: annual campaigns; building/renovation; capital campaigns; continuing support; endowments; equipment; general/operating support; matching/challenge support; program-related investments/loans; scholarship funds and; seed money.
Requirements: Organizations requesting funding must be approved as (or sponsored by) a 501(c)3 organization. The Board meets quarterly and proposals are accepted throughout the year. Before preparing a complete proposal, applicants are urged to submit a letter of inquiry addressed to the Executive Director. The letter should contain a descriptive title for the project, the project's intent, objectives, outcome measures, short and long term funding needs, and other relevant factual information. After reviewing the letter of inquiry, a detailed proposal may be requested. It should not be assumed that such a request is an indication of funding. The full proposal should include a detailed narrative, budgets, a board list, an IRS determination letter, other funding sources and financial documentation. In addition, the proposal should describe expected future funding sources and should indicate whether and to what extent additional funding by the Foundation will be necessary for the proposed project to be successful. The Foundation will respond to all inquiries from eligible organizations.
Restrictions: Grants are not made to individuals or sectarian or religious organizations. Because the Foundation makes annual contributions to the United Way of Southern Chautauqua County, applications for assistance from United Way-funded agencies will not be considered unless there is a strong link to the strategic focus.
Geographic Focus: New York
Amount of Grant: 1,000 - 1,000,000 USD
Samples: Jamestown Audobon Society, Jamestown, NY, $20,000--attractors grant; City of Jamestown, Jamestown, NY, $188,599--economic development grant; Rodger Tory Peterson Institute, Jamestown, NY, $37,500--attractors grant/loan relief.
Contact: John C. Merino, Executive Director; (716) 487-1062; fax (716) 484-6401; jmerino@gebbie.org or info@gebbie.org
Internet: http://www.gebbie.org
Sponsor: Gebbie Foundation
111 West Second Street, Suite 1100
Jamestown, NY 14701

GEICO Public Service Awards 1778
This program observes the many accomplishments of federal employees. These employees are making tremendous differences in the quality and efficiency of services and are responsible for the success of many scientific, medical, and technical programs. The foundation recognizes four active and one retired federal employee who have made outstanding achievements in one of the four fields of endeavor: Substance Abuse Prevention and Treatment, Fire Prevention and Safety, Physical Rehabilitation, and Traffic Safety and Accident Prevention. Nomination information is available online.
Requirements: All career civil service employees are eligible, including employees of the Library of Congress, the General Accounting Office, the Office of the Architect of the Capital, the Government Printing Office, the Administrative Office of the U.S. Courts, the Smithsonian Institution, the Botanic Garden, and the Office of Homeland Security.
Geographic Focus: All States
Amount of Grant: 2,500 USD
Contact: Program Contact; (877) 206-0215; federal@geico.com
Internet: http://www.geico.com/insproducts/fedpsa.html
Sponsor: GEICO Philanthropic Foundation
One GEICO Plaza
Washington, D.C. 20076

GenCorp Foundation Grants 1779
The Foundation is dedicated to supporting the communities where our employees live, work and volunteer. While the Foundation's primary focus is education, it also supports human services, civic and arts organizations. Emphasis is on science education. Funding preference is given to projects that involve issues important to employees and requests that are recommended by Foundation coordinators at company facilities. Interested applicants should submit a letter of inquiry.
Requirements: Non-profit organizations are eligible to apply. Priority is given to organizations that are in the GenCorp communities: Huntsville, AL; Camden, AR; Sacramento, CA; Socorro, NM; Jonesborough, TN; Clearfield, UT; Gainesville and Orange, VA; or Redmond, WA.
Geographic Focus: Alabama, Arkansas, California, New Mexico, Tennessee, Utah, Virginia, Washington
Contact: Program Contact; (916) 355-3600; gencorp.foundation@gencorp.com
Internet: http://www.gencorp.com/pages/gcfound.html
Sponsor: GenCorp Foundation
P.O. Box 15619
Sacramento, CA 95852-0619

Genentech Corporate Charitable Contributions Grants 1780
The Foundation's primary focus areas are health science education, patient education/advocacy, and community. The Foundation supports nonprofits through its contributions in two primary ways: project-specific or general support funding to organizations whose mission aligns with the Foundation's focus areas; and sponsorships of selected events that fall into the Foundation's focus areas. Interested applicant's are encouraged to submit applications online.
Requirements: Nonprofit organizations recognized by the IRS as tax exempt, public charities, located in the United States are eligible to apply. Eligible grantees may include public elementary and secondary schools, as well as public colleges and universities and public hospitals.
Restrictions: The Foundation does not provide funding to organizations that discriminate on the basis of age, political affiliation, race, national origin, ethnicity, gender, disability, sexual orientation or religious beliefs. Funding is not provided for: advertising journals or booklets; alumni drives; capital campaigns/building funds; continuing medical education; infrastructural requests (e.g. salaries, equipment); memorial funds; memberships; organizations based outside of the United States; political or sectarian organizations; professional sports events or athletes; religious organizations; scholarships; yearbooks.
Geographic Focus: California
Samples: To be distributed among 19 organizations (CA)--for science programs for students in the San Francisco Bay area, $1 million.
Contact: Program Manager; (650) 467-9494; give@gene.com
Internet: http://www.gene.com/gene/about/community/overview.jsp
Sponsor: Genentech
1 DNA Way
South San Francisco, CA 94080-4990

General Mills Champions for Healthy Kids Grants 1781
The program aims to improve youth nutrition and fitness in the United States through grants to community-based groups. The initiative consists of grants to community-based groups as well as three additional components: sponsorship of the President's Active Lifestyle Awards, developing nutrition and fitness mentoring models, and sharing best practices. Grants are intended to encourage the improvement of eating and physical activity patterns of young people, ages two to 20. Priority is given to programs that demonstrate significant potential impact on youth groups with special needs or for having an impact on large populations of youth. Guidelines are available online.
Requirements: U.S. 501(c)3 and 509(a) tax-exempt organizations are eligible. Proposed projects should reflect the practices and concepts recommended by the Dietary Guidelines for American 2000 and the American Dietetic Association, and should include at least one nutritional behavioral objective and at least one physical activity behavioral objective.
Geographic Focus: All States
Date(s) Application is Due: Jan 15
Amount of Grant: Up to 10,000 USD
Samples: Adams Park Community Ctr (Saint Louis, MO)--for Healthy Futures, a diabetes- prevention curriculum that inspires eight- to 11-year olds, from a low-income African-American community to maintain eating habits and physical activity conducive to a healthy lifestyle, $10,000; Fitness Forward Foundation (Durham, NC)--a program in which students earn points by meeting four daily behaviors: limiting TV and computer time to one hour, being physically active for one hour, not drinking sugar-sweetened beverages and getting eight to 11 hours of sleep, $10,000.
Contact: Beth Labrador, Program Contact; (800) 877-1600, ext. 4821; fax (763) 764-4114; blabrador@eatright.org
Internet: http://www.generalmills.com/corporate/commitment/champions.aspx
Sponsor: General Mills Foundation
1 General Mills Boulevard, P.O. Box 1113
Minneapolis, MN 55440

General Mills Foundation Grants 1782
The Foundation was created to focus its philanthropic resources on community needs. Strategic objectives are to: demonstrably improve the quality of life in communities with GM facilities and employees; initiate innovative solutions and approaches to improve youth nutrition and fitness; and to support GM employees and retirees giving to United Way, education, and arts and culture organizations through gift matching. Funding priorities include: social services; youth nutrition and fitness; education; and arts and culture. Priority is given to organizations meeting the following criteria: their mission is closely related to the Foundation's priorities; programs or activities are based in communities with GM operations and employees; programs or activities involve GM employees and retirees; and services create sustainable community improvement. Applications are accepted at any time.
Requirements: U.S. and Canadian charitable 501(c)3 and 509(a) nonprofits in communities where General Mills operates (California, Georgia, Illinois, Indiana, Iowa, Maryland, Massachusetts, Missouri, Minnesota, Missouri, Montana, New Jersey, New York, Ohio, Oklahoma, Pennsylvania, Tennessee and Wisconsin) are eligible.
Restrictions: The Foundation does not support: organizations without 501(c)3 and 509(a) status; organizations that do not comply with the Foundation's non-discrimination policy; individuals; social, labor, veterans, alumni or fraternal organizations serving a limited constituency; travel by groups; recreational, sporting events or athletic associations; religious organizations for religious purposes; basic research; organizations seeking underwriting for advertising; political causes, candidates or legislative lobbying efforts; conferences, seminars and workshops; campaigns to eliminate or control specific diseases; publications, films or television programs; underwriting for program sponsorship.
Geographic Focus: Arizona, Arkansas, California, Georgia, Illinois, Indiana, Iowa, Maryland, Massachusetts, Michigan, Minnesota, Missouri, Montana, New Jersey, New Mexico, New York, Ohio, Oklahoma, Pennsylvania, Tennessee, Wisconsin
Contact: Christina L. Shea, President; (763) 764-2211; fax (763) 764-4114
Internet: http://www.generalmills.com/corporate/commitment/foundation.aspx
Sponsor: General Mills Foundation
1 General Mills Boulevard, P.O. Box 1113
Minneapolis, MN 55440

General Motors Foundation Grants Support Program 1783
With a strong commitment to diversity in all areas, the targeted areas of focus for the Foundation are: education; health and human services; civic and community; public policy; arts and culture; and environment and energy. Primary consideration is given to requests that meet the following criteria: exhibit a clear purpose and defined need in one of the foundation's areas of focus; recognize innovative approaches in addressing the defined need; demonstrate an efficient organization and detail the organization's ability to follow through on the proposal; and, explain clearly the benefits to the foundation and the plant city communities. Paper applications are no longer accepted. Completion of an online eligibility quiz is the first step in the application process.
Requirements: Nonprofit, tax-exempt organizations and institutions are eligible to apply. Applications must be made online.
Restrictions: The Foundation not not support organizations that discriminate on the basis of race, religion, creed, gender, age, veteran status, physical challenge or national origin. Contributions are generally not provided for: individuals; religious organizations; political partics or candidates, U.S. hospitals and health care institutions (general operating support); capital campaigns; endowment funds; conferences, workshops or seminars not directly related to GM's business interests.
Geographic Focus: All States
Sample: American National Red Cross (Washington, D.C.)--for relief efforts in South Asia and Africa, $1 million; Pierre Chambon, College de France (Paris, France) and Ronald Evans, Salk Institute for Biological Studies (La Jolla, CA)--to honor their contributions to the diagnosis, prevention, and treatment of cancer, $250,000 jointly.
Contact: Grant Coordinator; (313) 556-5000
Internet: http://www.gm.com/company/gmability/community/guidelines/index.html
Sponsor: General Motors Foundation
300 Renaissance Center, P.O. Box 300
Detroit, MI 48265-3000

General Service Foundation Human Rights and Economic Justice Grants 1784
The goal of the General Service Foundation's Human Rights and Economic Justice program is to support efforts that protect, promote and create good jobs with living wages for workers, including low-wage workers in the United States and Mexico. The Foundation seeks to: strengthen worker voices; promote public policies that protect labor rights; and democratize corporate power and promote corporate accountability. In assessing potential grantees, the Foundation will apply the following criteria: supporting organizations that work to address needs identified by the underrepresented and low-income communities that area directly impacted, connecting at the local and national level; and concentrating on organizations that take risks and try new ideas to enhance their skills, and reflect the diversity of their constituency. Samples of previously funded projects are posted on the Foundation website.
Requirements: All applicants, regardless of their prior grant history with the Foundation, start their application process by submitting a letter of inquiry in the spring and fall. Organizations should determine if their project is an appropriate fit for the Foundation, then send a letter of inquiry via the Foundation's online application. Letters of inquiry should include: a political and situational analysis, statement of the issues to be addressed under the proposed project, the history and goals of the organization, and an explanation of why the organization or coalition is the entity that is most likely to achieve success; a brief summary of the project, short and long term goals and anticipated outcomes; the approximate starting date and duration of the proposed activities; the total amount of funding needed, the amount requested from the Foundation, a budget, and information about other sources of support; and a copy of IRS tax exempt 501(c)3 letter. If the project is judged appropriate, the organization will be offered an application to submit to the Foundation.
Restrictions: The Foundation does not fund: organizations based outside the United States or Mexico; projects without significant promise of impact beyond a local or state level; research and publications not directly linked to policy outcomes; direct service delivery; and development or relief projects.
Geographic Focus: All States, Mexico
Date(s) Application is Due: Feb 1; Sep 1
Amount of Grant: 5,000 - 35,000 USD
Contact: Holly Bartling, Program Contact-Mexico grants; (202) 232-1005 or (970) 920-6834, ext 4; fax (970) 920-4578; holly@generalservice.org
Internet: http://www.generalservice.org/International%20Peace.htm
Sponsor: General Service Foundation
557 North Mill Street, Suite 201
Aspen, CO 81611

General Service Foundation Reproductive Justice Grants 1785
The General Service Foundation believes that in order to be effective, the reproductive health and rights movement must address the needs of all women, including women of color, low-income women, and young women. The Reproductive Justice Grants are interested in supporting organizations that work on a broad range of reproductive health and rights issues that includes, but is not limited to, abortion. The Foundation supports organizations and projects engaged in leadership development, organizing, education, policy research and advocacy. In assessing potential grantees, the Foundation places priority on organizations and projects that have a broad approach to reproductive health, try new ideas and strategies, reflect the diversity of its constituency, and address needs identified by people directly impacted by the issues and policies. A list of previously funded projects is available at the Foundation website.
Requirements: All applicants, regardless of their prior grant history with the Foundation, start their application process by submitting a letter of inquiry in the spring and fall. Organizations should determine if their project is an appropriate fit for the Foundation, then send a letter of inquiry via the Foundation's online application. Letters of inquiry should include: a political and situational analysis, statement of the issues to be addressed under the proposed project, the history and goals of the organization, and an explanation of why the organization or coalition is the entity that is most likely to achieve success; a brief summary of the project, short and long term goals and anticipated outcomes; the approximate starting date and duration of the proposed activities; the total amount of funding needed, the amount requested from the Foundation, a budget, and information about other sources of support; and a copy of IRS tax exempt 501(c)3 letter. If the project is judged appropriate, the organization will be offered an application to submit to the Foundation.
Restrictions: The Foundation does not fund direct service delivery, unless it is linked to policy and advocacy; projects without significant promise of impact beyond a local or state level; research and publications not directly linked to policy outcomes; projects outside the United States; and NARAL or Planned Parenthood affiliates.
Geographic Focus: All States
Date(s) Application is Due: Feb 1; Sep 1
Amount of Grant: 3,000 - 35,000 USD
Contact: Holly Bartling; (970) 920-6834; fax (970) 920-4578; holly@generalservice.org
Internet: http://www.generalservice.org/Reproductive%20Health.htm
Sponsor: General Service Foundation
557 North Mill Street, Suite 201
Aspen, CO 81611

Genesis Foundation Grants 1786
The Foundation funds projects and program in the areas of health and formal education, benefiting children 0 to 18 years old, as well as training programs in these two areas. Capital, operating, program and general grants are funded. Approximately eighty percent of grant-making is devoted to Colombian entities, with the remainder devoted to projects in the US, specifically in Southern Florida, Washington, D.C. and the New York Metropolitan area.
Requirements: Preliminary criteria to apply for a grant include: must be located in Colombia, Southern Florida, Washington, D.C. or the New York Metropolitan Area; must be incorporated as a non-profit, charitable institution in the appropriate jurisdiction; must have been in operation for a minimum of two (2) full years; must include at least three (3) people in decision-making process at all times; must benefit a minimum of 50 people per year; in Colombia, must have raised at a minimum the equivalent of U.S. $10,000 in the previous two years; in the U.S. must have raised a minimum of $20,000 in the previous two years; must have an independent financial auditor; must be able to provide audited financial statements for the previous three years; must be able to provide a budget for the organization and the project for which the grant is sought; must not be a political organization; and must not have a religious purpose.
Restrictions: Individuals and political or religious causes are ineligible.
Geographic Focus: District of Columbia, Florida, New York, Colombia
Date(s) Application is Due: Mar 15; Sep 15
Amount of Grant: 10,000 - 100,000 USD
Contact: Administrator; (212) 763-3703; genesis@genesid-foundation.org
Internet: http://www.genesis-foundation.org
Sponsor: Genesis Foundation
140 East 45th Street, 2 Grant Central Tower 18th Floor, Suite A
New York, NY 10017

Genuardi Family Foundation Grants 1787
The Foundation supports direct providers of services in the areas of education, health, human services, and culture. The Foundation gives preference to projects which receive broad-based community support and provides for reasonable costs associated with conducting the proposed project or program. Grant proposals may include the allocation of a reasonable percentage of grant monies towards general support of the recipient organization. On a limited basis, the Foundation will also consider grants in support of capital campaigns for facilities or equipment, and organizational capacity building.
Requirements: The Foundation only considers grant requests from organizations designated as tax exempt under Section 501(c)3 of the Internal Revenue Code. Preference will be given to non-profit organizations based in the Greater Philadelphia Area.
Restrictions: The Foundation's funding will not process grant requests from: individuals; fraternal and/or civic organizations; political candidates or to influence legislation; other foundations; general fund raising and endowment drives; debt reduction; environmental issues or initiatives; annual appeals or letters of solicitation; or public, private or parochial schools that serve the general public.
Geographic Focus: Pennsylvania
Date(s) Application is Due: Aug 1
Amount of Grant: 5,000 - 50,000 USD
Contact: Meredith A. Huffman, Executive Director; (610) 834-2030; info@genuardifamilyfoundation.org
Internet: http://www.genuardifamilyfoundation.org/genuardi_priorities.html
Sponsor: Genuardi Family Foundation
Blue Bell Executive Campus, 470 Norristown Road, Suite 300
Auburn, NY 13021

George A. and Grace L. Long Foundation Grants 1788
The George A. and Grace L. Long Foundation was established in 1960 to support and promote quality educational, cultural, human-services, and health-care programming for underserved populations in Connecticut. Grants from the Long Foundation are one year in duration. Applicants must apply online at the grant website. Applicants are strongly encouraged to do the following before applying: review the downloadable state application procedures for additional helpful information and clarifications; review the downloadable online-application guidelines at the grant website; review the foundation's funding history (link is available from the grant website); review the online application

questions in advance; and review the list of required attachments. These will generally include: a list of board members, financial statements (audited, reviewed, or compiled by independent auditor); an organization summary; a list of other funding sources; an IRS Determination letter; and other required documents. All attachments must be uploaded in the online application as PDF, Word, or Excel files. The George A. and Grace L. Long Foundation has biannual deadlines of March 15 and September 15. Applications must be submitted by 11:59 p.m. on the deadline dates. Applicants will be notified of grant decisions by letter within two to three months after each respective proposal deadline.
Requirements: Applicant organizations must have 501(c)3 tax-exempt status and serve the people of Connecticut.
Restrictions: Applicants will not be awarded a grant for more than three consecutive years. The foundation does not support requests from individuals, organizations attempting to influence policy through direct lobbying, or any political campaigns.
Geographic Focus: Connecticut
Date(s) Application is Due: Mar 15; Sep 15
Contact: Carmen Britt; (860) 657-7019; carmen.britt@baml.com
Internet: https://www.bankofamerica.com/philanthropic/fn_search.action
Sponsor: George A. And Grace L. Long Foundation
200 Glastonbury Boulevard, Suite # 200, CT2-545-02-05
Glastonbury, CT 06033-4056

George and Ruth Bradford Foundation Grants 1789
The foundation awards grants to local nonprofit organizations in the San Francisco Bay Area and Mendocino, California region. Areas of interest include: children and youth; families; education/higher education; environmental/wildlife conservation; housing and; social services. Types of support include general operating support and scholarship funds. T
Requirements: California nonprofit organizations in the San Francisco peninsula and Mendocino, California region are eligible to apply. There is no deadline date to adhere to. The Board meets monthly, letters of inquiry may be submitted throughout the year for review.
Restrictions: No grants to individuals.
Geographic Focus: California
Samples: Frank R. Howard Foundation, Willits, CA, $10,000-hospital building fund; Ukiah Valley Cultural & Recreation Center. Ukiah, CA, $5,000--youth recreation grant; MC Aids Volunteer Network, Ukiah, CA, $1,000--health services grant.
Contact: Myrna Oglesby, Director; (707) 462-0141; fax (707) 462-0160
Sponsor: George and Ruth Bradford Foundation
P.O. Box 720
Ukiah, CA 95482-0720

George and Sarah Buchanan Foundation Grants 1790
Established in 2006, the George and Sarah Buchanan Foundation offers funding throughout the state of Virginia. Its primary fields of interest include: health care, health organizations, and religion. Support typically is given for general operations. There are no specified application formats or annual deadlines, and applicants should proceed by forwarding a proposal to the Foundation office. Recent grants have ranged from $250 to $20,000.
Requirements: Applicants should be 501(c)3 organizations either located in, or serving residents of, Virginia. Preference is given to the general Danville, Virginia, region.
Geographic Focus: Virginia
Amount of Grant: 250 - 20,000 USD
Sample: Averett University, Danville, Virginia, $20,000 - general operating support; Danville Cancer Association, Danville, Virginia, $20,000 - general operating support; Smile Train, Washington, D.C., $250 - general operating support.
Contact: George Buchanan, Jr., (434) 797-3543
Sponsor: George and Sarah Buchanan Foundation
400 Bridge Street
Danville, VA 24541-1404

George A Ohl Jr. Foundation Grants 1791
The purpose of the George A Ohl Jr. Foundation is to improve the well-being of the citizens of the State of New Jersey through science, health, recreation, education and increased good citizenship. Grants are made to organizations engaged in such work whether through research, publications, health, school or college activities. The Foundation's mission is the relief of the poor; the improvement of living conditions; the care of the sick, the young, the aged, the homeless, the incompetent and the helpless. The foundation will target: 35% of its grants to Community Redevelopment; 35% to Health and Human Services organizations; 15% for Arts and culture; 15% for Educational requests. Application deadlines are: January 22 for a March meeting and, June 15 for an August meeting. Application forms are available online. Applicants will receive notice acknowledging receipt of the grant request, and subsequently be notified of the grant declination or approval.
Requirements: New Jersey 501(c)3 nonprofit organizations are eligible to apply. Proposals should be submitted in the following format: completed Common Grant Application Form; an original Proposal Statement; an audited financial report and a current year operating budget; a copy of your official IRS Letter with your tax determination; a listing of your Board of Directors. Proposal Statements (second item in the above Format) should answer these questions: what are the objectives and expected outcomes of this program/project/request; what strategies will be used to accomplish your objective; what is the timeline for completion; if this is part of an on-going program, how long has it been in operation; what criteria will you use to measure success; if the request is not fully funded, what other sources can you engage; an Itemized budget should be included; please describe any collaborative ventures. Prior to the distribution of funds, all approved grantees must sign and return a Grant Agreement Form, stating that the funds will be used for the purpose indicated. Progress reports and Completion reports must also be filed as required for your specific grant. All current grantees must be in good standing with required documentation prior to submitting new proposals to any foundation.
Restrictions: Grants are not made for political purposes, nor to organizations which discriminate on the basis of race, ethnic origin, sexual or religious preference, age or gender
Geographic Focus: New Jersey
Date(s) Application is Due: Jan 22; Jun 15
Amount of Grant: 3,000 - 35,000 USD
Contact: Wachovia Bank, N.A., Trustee; grantinquiries2@wachovia.com
Internet: https://www.wachovia.com/foundation/v/index.jsp?vgnextoid=e0f78689fb0a a110VgnVCM1000004b0d1872RCRD&vgnextfmt=default
Sponsor: George A Ohl Jr. Foundation
190 River Road, NJ3132
Summit, NJ 07901

George E. Hatcher, Jr. and Ann Williams Hatcher Foundation Grants 1792
The George E. Hatcher and Ann Williams Hatcher Foundation was created to support charitable organizations that provide for the relief of diseased people and the relief of human suffering which is due to disease, ill health, physical weakness, physical disability and/or physical injury. In addition, the foundation supports organizations that aid in the promotion and prolongation of life and that support the principle of dying with dignity. Grants from George E. Hatcher and Ann Williams Hatcher Foundation are primarily one year in duration; on occasion, multi-year support is awarded. Applicants must apply online at the grant website. Applicants are strongly encouraged to do the following before applying: review the downloadable state application procedures for additional helpful information and clarifications; review the downloadable online-application guidelines at the grant website; review the foundation's funding history (link is available from the grant website); review the online application questions in advance; and review the list of required attachments. These will generally include: a list of board members, financial statements (audited, reviewed, or compiled by independent auditor); an organization summary; a list of other funding sources; an IRS Determination letter; and other required documents. All attachments must be uploaded in the online application as PDF, Word, or Excel files. The George E. Hatcher and Ann Williams Hatcher Foundation application deadline is 11:59 p.m. on May 31. Applicants will be notified of grant decisions by letter within one to two months after the deadline.
Requirements: Disbursements are authorized to institutions or organizations located in the Middle Georgia area (Bibb County & surrounding communities) which provide health care and/or shelter for individuals (especially children) who cannot otherwise obtain such services due to circumstances beyond their control. A breakdown of number/percentage of people served by specific counties is required on the online application.
Restrictions: The foundation does not support requests from individuals, organizations attempting to influence policy through direct lobbying, or any political campaigns.
Geographic Focus: Georgia
Date(s) Application is Due: May 31
Contact: Quanda Allen, Vice President; (404) 264-1377; quanda.allen@baml.com
Internet: https://www.bankofamerica.com/philanthropic/fn_search.action
Sponsor: George E. Hatcher, Jr. and Ann Williams Hatcher Foundation
3414 Peachtree Road, N.E., Suite 1475, GA7-813-14-04
Atlanta, GA 30326-1113

George F. Baker Trust Grants 1793
Grants are awarded nationwide, with preference given to nonprofits in the eastern United States, primarily for K-12, higher, and secondary education; hospitals; social services; private foundations; and zoos/zoological societies. Types of support include general operating support and matching funds. There are no application forms or deadlines. The board meets in June and November to consider requests.
Requirements: An application form is not required. Along with a letter of inquiry and brief outline of a proposal, applicants should submit the following: signature and title of chief executive officer; a copy of the organization's IRS determination letter; a detailed description of the project and amount of funding requested; and a listing of additional sources and amount of support. As a result of the enormous number of applications received and limited number of grants, only those applicants who receive a grant will be notified
Restrictions: Funding is given primarily in Connecticut, Florida, Massachusetts, and New York. Grants are not made to individuals for scholarships or loans.
Geographic Focus: Connecticut, Florida, Massachusetts, New York
Amount of Grant: 1,000 - 50,000 USD
Contact: Rocio Suarez; (212) 755-1890; fax (212) 319-6316; rocio@bakernye.com
Sponsor: George F. Baker Trust
477 Madison Avenue, Suite 1650
New York, NY 10022

George Family Foundation Grants 1794
The mission of the Foundation is to foster wholeness in mind, body, spirit, and community in order to enhance the development of human potential. The Foundation's areas of interest are: integrative medicine, leadership development, educational opportunities, social justice and spirituality. Non- solicited proposals will not be accepted. Invited applicants are encouraged to submit a letter of inquiry or to discuss their request with staff before submitting an application.
Requirements: 501(c)3 tax-exempt organizations are eligible. Primary giving is restricted to the Twin Cities metropolitan area.
Restrictions: Grants do not support endowment or capital campaigns, programs in which the family has no active interest or involvement; individuals; memberships, special events, or other fundraisers; or programs that address debt retirement or recovery of operating losses.

Geographic Focus: Minnesota
Date(s) Application is Due: May 30; Oct 30
Amount of Grant: 1,000 - 200,000 USD
Samples: WATCH (Minneapolis, MN) for efforts to improve the criminal-justice system's handling of cases involving domestic violence and sexual assault, $15,000; Twin Cities RISE (Minneapolis, MN)--for scholarships for participants in its skills-training and job-acquisition programs, $50,000; Plymouth Congregation Church (Minneapolis, MN)--to add an outreach worker to the church's staff roster, $17,000; Sigma Chi Foundation (Evanston, IL)--for the Horizons Leadership project, $25,000.
Contact: Grants Administrator; (612) 377-8400; fax (612) 377-8407
Internet: http://www.georgefamilyfoundation.org/
Sponsor: George Family Foundation
1818 Oliver Avenue South
Minneapolis, MN 55405

George Foundation Grants 1795

The George Foundation strive to support organizations and programs that assist in developing strong, stable families across Fort Bend County region of Texas. The Foundation's areas of interest include: family stability; scholarships; foundation initiated programs, current programs include, Youth in Philanthropy (YIP), Leadership Excellence for Non-Profits, Integrated Mental Health Care, and Transportation. The Foundation prefers to fund the following types of grants to support Fort Bend organizations in their delivery of services to the community (listed in order of priority): program/project support; foundation initiated; general operating; capital. Proposal deadlines for making grant applications to the Foundation are January 15, April 15, July 15 and October 15 of each year. All proposals for capital support will be grouped together for review. The deadline for capital proposals will be October 15 of each year.
Requirements: Non-profits in Fort Bend County, Texas may submit grant proposals.
Restrictions: The Foundation does not fund: grants to organizations that do not have a current 501(c)3 determination letter; churches or other organized religious bodies; grants to another organization that distributes money to recipients of its own selection, i.e., a regranting organization; regional, national or international programs; grants for research or studies; grants for travel, conferences, conventions, group meetings, or seminars; the purchase of event tickets, tables, ads or sponsorships; support to fairs and festivals; religious or private schools; request for funds to develop films, videos, books or other media projects; direct mail campaigns; loans of any kind; grants to individuals; grants to fraternal organizations; political interests of any kind; and institutions that discriminate on the basis of race, creed, gender, national origin, age, disability or sexual orientation in policy or in practice.
Geographic Focus: Texas
Date(s) Application is Due: Jan 15; Apr 15; Jul 15; Oct 15
Samples: Wharton County Junior College, $110,000--to provide scholarships for Fort Bend County residents; YMCA of the Greater Houston Area, $250,000--to assist in acquiring facility to offer outdoor education/recreation experiences to Fort Bend families and children; OakBend Medical Center, $4,000,000--four year grant to provide indigent healthcare to Fort Bend County residents.
Contact: Dee Koch; (281) 342-6109; fax (281) 341-7635; dkoch@thegeorgefoundation.org
Internet: http://www.thegeorgefoundation.org
Sponsor: George Foundation
310 Morton Street, PMB Suite C
Richmond, TX 77469

George Gund Foundation Grants 1796

The Foundation's long-standing interests include: arts; economic development; community revitalization; education; environment; and human services. The Foundation considers global climate change an urgent issue that cuts across all of the Foundation's programs. The Foundation takes seriously it's own responsibility and wants to hear from grant applicants what they are doing or considering to reduce or to eliminate their organizational impact on climate change. The Foundation also supports special projects grants, which currently include: Retinal Degenerative Diseases research grants, making an annual commitment for research on the causes, nature and prevention of inherited retinal degenerative diseases and; philanthropic services grants, offering support to organizations that strengthen the infrastructure of the nonprofit and philanthropic communities. The George Gund Foundation also supports capital requests but only for projects that are clearly aligned with their program priorities and that meet Green Building Council LEED (Leadership in Energy and Environmental Design) certification. The Foundation's green building policy covers both planning and construction grants. In addition, the Foundation supports opportunities that cross program boundaries and that integrate elements of the Foundation's interests. Although the Foundation's focus is centered in the Greater Cleveland, Ohio region, a portion of their grantmaking will continue to support state and national policy making that bolsters their work. Proposals should be mailed directly to the George Gund Foundation. All proposals are screened and evaluated by the staff before presentation at Trustee Meetings. Receipt of proposals will be acknowledged by mail.
Requirements: 501(c)3 nonprofit organizations are eligible to apply for funding, with a special interest in Greater Cleveland, Ohio region. Proposals are accepted three times: March 15, July 15 and November 15. Proposals are due the next business day if a deadline falls on a weekend. The grant application form is available on the George Gund Foundation website. All proposals must include a climate change statement, the Foundation's website includes resources to assist grantees with this task. Applicants also must include a completed Grant Application Cover Sheet, which is signed by the organization's board chair and executive director. Proposals should also include: organizational background; history; mission; types of programs offered; constituencies served; project description; justification of need; specific goals and objectives; activities planned to meet goals and objectives; project time line; qualifications of key personnel; methods of evaluation; project budget; anticipated expenses, including details about how Foundation funds would be used; anticipated income, including information about other sources approached for funding; organizational budget; previous and current year budget and proposed budget for project year(s) showing both income and expenses; the organization's most recent audited financial statement, do not include IRS 990 forms; supporting documents; list of current trustees; letters of support; readily available printed material about organization such as annual reports and brochures; IRS letter confirming Internal Revenue Code 501(c)3 status and classification as a public charity or information confirming status as a government unit or agency. The Foundation also will accept the Ohio Common Grant Form, available at www.ohiograntmakers.org, if organizations are using it to apply to multiple funders. Faxed or electronic proposals are not accepted.
Restrictions: Do not submit proposals in notebooks, binders or plastic folders and print proposals on both sides of each sheet of paper. The Foundation normally does not consider grants for endowments. Capital requests must meet the Foundation's program goals and also adhere to green building standards of environmental sustainability. Details on these requirements are available from the Foundation. Grants are not made for debt reduction or to fund benefit events. The Foundation does not make grants to individuals, nor does it administer programs it supports. Grants are limited to organizations located in the United States.
Geographic Focus: All States
Date(s) Application is Due: Mar 15; Jul 15; Nov 15
Samples: Progressive Arts Alliance Incorporated, Cleveland, OH, $20,000--database capacity building project; Cleveland Restoration Society, Incorporated, Cleveland, OH, $20,000--preservation programs; Center for Community Solutions, Cleveland, OH, $360,000--two year grant for policy and fiscal analysis and GroundWork early childhood campaign.
Contact: David Abbott; (216) 241-3114; fax (216) 241-6560; info@gundfdn.org
Internet: http://www.gundfdn.org/what.asp
Sponsor: George Gund Foundation
1845 Guildhall Building, 45 Prospect Avenue, West
Cleveland, OH 44115

George H.C. Ensworth Memorial Fund Grants 1797

The George H.C. Ensworth Memorial Fund was established in 1949 to support charitable organizations that focus on health and human services, youth services, enjoyment of the natural environment, education, religion, and the arts. Applicants must apply online at the grant website. Applicants are strongly encouraged to do the following before applying: review the downloadable state application procedures for additional helpful information and clarifications; review the downloadable online-application guidelines at the grant website; review the fund's funding history (link is available from the grant website); review the online application questions in advance; and review the list of required attachments. These will generally include: a list of board members, financial statements (audited, reviewed, or compiled by independent auditor); an organization summary; a list of other funding sources; an IRS Determination letter; and other required documents. All attachments must be uploaded in the online application as PDF, Word, or Excel files. The George H.C. Ensworth Memorial Fund has an annual deadline of May 15 at 11:59 p.m. Applicants will be notified of the grant decisions by letter within two months after the proposal deadline.
Requirements: Applicants must have 501(c)3 status and serve the people of Glastonbury, Connecticut. A breakdown of number/percentage of people served will be required in the online application.
Restrictions: Grant requests for capital projects will not be considered. Applicants will not be awarded a grant for more than 3 consecutive years. The fund does not support requests from individuals, organizations attempting to influence policy through direct lobbying, or any political campaigns.
Geographic Focus: Connecticut
Date(s) Application is Due: May 15
Contact: Kate Kerchaert; (860) 657-7016; kate.kerchaert@ustrust.com
Internet: https://www.bankofamerica.com/philanthropic/fn_search.action
Sponsor: George H.C. Ensworth Memorial Fund
200 Glastonbury Boulevard, Suite # 200, CT2-545-02-05
Glastonbury, CT 06033-4056

George H. Hitchings New Investigator Award in Health Research 1798

Provides awards to provide flexible support for young researchers in their second year of graduate school. Students planning to follow a career path in academic teaching and research that is closely tied to the health of North Carolinians and the strength of North Carolina Science are particularly encouraged to apply.
Requirements: Applications/nominations must be submitted by the institution on behalf of the student(s). Nomination applications are accepted from Duke, NCCU, NCSU, and UNC-Chapel Hill.
Restrictions: Only one nomination/application per graduate program/department within the university will be accepted.
Geographic Focus: North Carolina
Date(s) Application is Due: Mar 5
Amount of Grant: 10,000 USD
Contact: Libby Long; (919) 474-8370, ext. 134; fax (919) 941-9208; libby@trialglecf.org
Internet: http://www.trianglecf.org/DatasetRecord.cfm?recordID=10001417&returnURL=%2Fpage10000237%2Ecfm&returntoname=View%20all%20Scholarships&sidepageid=10000237&thetitle=George%20H.%20Hitchings%20New%20Investigator%20Award%20in%20Health%20Research&Ds_PagepropId=1
Sponsor: Triangle Community Foundation
324 Blackwell Street, Suite 1220
Durham, NC 27701

GRANT PROGRAMS | 283

George Kress Foundation Grants   1799
Incorporated in Wisconsin in 1953, the George Kress Foundation awards grants to eligible Wisconsin nonprofit organizations in its areas of interest, including: arts and culture; boys and girls clubs; children and youth services; Christian agencies and churches; community and economic development; education; family services; health organizations; higher education; historic preservation; historical societies; hospitals; human services; libraries; recreation; United Ways and Federated Giving Programs; and YM/YWCAs and YM/YWHAs. Types of support include: annual campaigns; building and renovation; capital campaigns; continuing support; professorships; program development; and research. Preference is given to nonprofit organizations that benefit the communities of Green Bay and Madison. Most recent awards have ranged from $200 to $50,000. There are no specified annual deadlines for submission.
Requirements: Interested applicants should submit a letter of inquiry describing their proposed project.
Geographic Focus: Wisconsin
Amount of Grant: 200 - 50,000 USD
Samples: St. Vincent Health Systems, Little Rock, Arkansas, $25,000 - overall program support; Weidner Center, Green Bay, Wisconsin, $30,000 - overall program support; University of Wisconsin Foundation, Madison, Wisconsin, $50,000 - overall program support.
Contact: John Kress, President; (920) 327-5670 or (920) 433-3109
Sponsor: George Kress Foundation
1700 North Webster Avenue, P.O. Box 12800
Green Bay, WI 54307-2800

George P. Davenport Trust Fund Grants   1800
Established in Maine in 1927, the Trust awards grants for education, religion, temperance and needy children. The focus of the Trust is on the economically disadvantaged. Types of support include: building and renovation, emergency funding, general operating funds, capacity building, matching/challenge funds, and seed money. It is recommended that applicants contact the Trust office to make an initial inquiry regarding projects before beginning the application process. Interested applicants should use the standard Maine Philanthropy Center common grant application which is available at mainephilanthropy.org. Grant applications are accepted all year, with no deadlines.
Requirements: Only nonprofit organizations serving the residents of Bath, Maine, and its surrounding area are eligible to apply.
Geographic Focus: Maine
Amount of Grant: 125 - 10,000 USD
Samples: Bath Elementary PTA, Bath, Maine, $6,550 - camp scholarships for youth; Brunswick Area Respite Care, Topsham, Maine, $10,000 - adult disabilities of aging support; Good Shepherd Food Bank, Auburn, Maine, $5,000 - food for the needy.
Contact: Barry M. Sturgeon, Trustee; (270) 443-3431; fax (800) 665-5510; davenporttrust@verizon.net
Sponsor: George P. Davenport Trust Fund
65 Front Street
Bath, ME 04530-2508

George S. and Dolores Dore Eccles Foundation Grants   1801
The foundation awards grants to eligible Utah organizations in its areas of interest, including arts, children and youth, economics, higher education, hospitals, medical research, performing arts, visual arts, and social services. Types of support include building construction/renovation, capital campaigns, equipment acquisition, general operating grants, matching/challenge grants, professorships, program development, research grants, and scholarship funds. A request for application is available online.
Requirements: Giving primarily in Utah.
Restrictions: Funding requests will not be considered from the following types of organizations: those that have not received a tax exemption letter establishing 501(c)3 status from the Internal Revenue Service, unless they are a unit of government, in which case such a letter is not required; other private foundations; those of a political nature that attempt to influence legislation and/or candidacy of persons for elected public office; conduit organizations, unified funds, or those that use funds to make grants to support other organizations; those that do not have fiscal responsibility for the proposed project. Funds will also not be considered for: contingencies, deficits, or debt reduction; general endowment funds; direct aid to individuals; conferences, seminars, or medical research; requests which do not fall within the Foundation's specified areas of interest.
Geographic Focus: Utah
Contact: Director; (801) 246-5340; fax (801) 350-3510; gseg@gseccles.org
Internet: http://www.gsecclesfoundation.org
Sponsor: George S. and Dolores Dore Eccles Foundation
79 South Main Street, 12th Floor
Salt Lake City, UT 84111

George W. Brackenridge Foundation Grants   1802
The foundation awards grants in Texas to nonprofit colleges and universities, religious education organizations, accredited K-12 schools, and youth organizations such as YMCA/YWCA. Types of support include endowment funds, project grants, research grants, and scholarships. The foundation's Areas of Interest include: arts; arts education; biomedicine research; children/youth, services; christian agencies & churches; education; elementary/secondary education; higher education; human services and; the performing arts.
Requirements: Texas nonprofit organizations in the San Antonio and the surrounding area are eligible to apply. When applying for funding, submit a letter of request on the organization's letterhead. Upon review of letter of request, you may be invited to submit a proposal. Proposal should include: signature and title of chief executive officer; copy of IRS Determination Letter; copy of most recent annual report/audited financial statement/990; detailed description of project and amount of funding requested; organization's charter and by-laws; four copies of the proposal. There are no application deadlines. The board meets in March, June, September, and December each year.
Restrictions: Grants are not made to or for: individuals; general purposes; continuing support; seed money; emergency funds; land acquisition; renovation projects; building funds; operating budgets; annual campaigns; deficit financing; matching gifts; loans.
Geographic Focus: Texas
Contact: Emily D. Thuss, Treasurer; (210) 224-1011
Sponsor: George W. Brackenridge Foundation
711 Navarro Street, Suite 535
San Antonio, TX 78205-1746

George W. Codrington Charitable Foundation Grants   1803
The George W. Codrington Charitable Foundation gives primarily to nonprofit organizations in Ohio, but may consider other areas. The Foundation funds higher education, hospitals, museums, arts groups and performing arts, and youth programs. Types of support include annual and capital campaigns, continuing support, equipment, general/operating support, program development, and research.
Requirements: Application forms are not required. Applicants should submit three copies of the following: their IRS determination letter; a brief history of the organization and description of its mission; the geographic area to be served; a list of the board of directors, trustees, officers, and other key individuals with their affiliations; and a detailed description of the project and amount of funding requested. Proposals should be submitted one month before the board meets in April, June, September, and December. Organizations are notified of funding promptly after the board meeting.
Restrictions: Funding is not available for individuals, endowment funds, or loans.
Geographic Focus: Ohio
Amount of Grant: 1,000 - 50,000 USD
Contact: Craig Martahus; (216) 566-8674; tommie.robertston@thomasonhine.com
Sponsor: George W. Codrington Charitable Foundation
127 Public Square, 39th Floor
Cleveland, OH 44114-1216

George W. Wells Foundation Grants   1804
The George W. Wells Foundation was established in 1934 to support and promote quality educational, human-services, and health-care programming for underserved populations. In the area of education, the fund supports academic access, enrichment, and remedial programming for children, youth, adults, and senior citizens that focuses on preparing individuals to achieve while in school and beyond. In the area of health care, the fund supports programming that improves access to primary care for traditionally underserved individuals, health education initiatives and programming that impact at-risk populations, and medical research. In the area of human services the fund tries to meet evolving needs of communities. Currently the foundation's focus is on (but is not limited to) youth development, violence prevention, employment, life-skills attainment, and food programs. Special consideration is given to charitable organizations that serve the people of Southbridge, Massachusetts, and its surrounding communities. Grant requests for general operating support are strongly encouraged. Program support will also be considered. Small, program-related capital expenses may be included in general operating or program requests. The majority of grants from the Wells Foundation are one year in duration; on occasion, multi-year support is awarded. Applicants must apply online at the grant website. Applicants are strongly encouraged to do the following before applying: review the downloadable state application procedures for additional helpful information and clarifications; review the downloadable online-application guidelines at the grant website; review the foundation's funding history (link is available from the grant website); review the online application questions in advance; and review the list of required attachments. These will generally include: a list of board members, financial statements (audited, reviewed, or compiled by independent auditor); an organization summary; a list of other funding sources; an IRS Determination letter; and other required documents. All attachments must be uploaded in the online application as PDF, Word, or Excel files. The application deadline for the George W. Wells Foundation is 11:59 p.m. on October 15. Applicants will be notified of grant decisions before December 31.
Requirements: Applicants must have 501(c)3 tax-exempt status.
Restrictions: The foundation does not support requests from individuals, organizations attempting to influence policy through direct lobbying, or any political campaigns.
Geographic Focus: Massachusetts
Date(s) Application is Due: Oct 15
Contact: Miki C. Akimoto, Vice President; (866) 778-6859; miki.akimoto@baml.com
Internet: https://www.bankofamerica.com/philanthropic/fn_search.action
Sponsor: George W. Wells Foundation
225 Franklin Street, 4th Floor, MA1-225-04-02
Boston, MA 02110

Georgiana Goddard Eaton Memorial Fund Grants   1805
Established in 1917, the Trust supports organizations involved with improving the lives of low-income individuals of Boston, Massachusetts. The fields of interest are: education; employment; family services; homelessness; housing/shelter; human services; legal services; urban/community development. Contact the Foundation for further application information and guidelines.
Requirements: Non-profits in the Boston, MA area are eligible. Applications should be addressed to: c/o Grants Mgmt. Assocs., 77 Summer Street, 8th Floor, Boston, MA 02110-1006
Restrictions: No grants to individuals (except former employees of Community Workshops, Inc.), or for endowment funds, or matching gifts; no loans.
Geographic Focus: Massachusetts

Date(s) Application is Due: Mar 1; Sep 1
Contact: Philip Hall, Administrator; (617) 426-7080; fax (617) 426-7087; phall@grantsmanagement.com
Sponsor: Georgiana Goddard Eaton Memorial Fund
45 School Street
Boston, MA 0002108-3206

Georgia Power Foundation Grants 1806

Giving is focused on issues that directly affect customers, employees, business and shareholders. These include: improving the quality of education by partnering with organizations to assist students with personal development, mentoring and career exploration; protecting the environment by promoting programs to improve air and water quality, preserve natural resources and protect endangered species; preventing cancer; and promoting diversity. The Foundation gives strong preference to Georgia-based organizations and programs that seek to improve the quality of life for the state's residents. Applicants may apply online or with a written proposal.
Requirements: The foundation makes grants to tax-exempt organizations that seek to improve the quality of life for Georgia's residents.
Restrictions: The Foundation does not provide grants to individuals, private elementary or secondary schools, and religious organizations, nor political campaigns or causes. The Foundation does not provide multi-year funding commitments.
Geographic Focus: Georgia
Date(s) Application is Due: Feb 15; May 15; Aug 15; Nov 15
Amount of Grant: 10,000 USD
Contact: Grants Administrator; (404) 506-6784; gpfoundation@southernco.com
Internet: http://www.georgiapower.com/community/apply.asp
Sponsor: Georgia Power Foundation
241 Ralph McGill Boulevard NE, Bin 10131
Atlanta, GA 30308-3374

Gerber Foundation Grants 1807

The foundation awards grants to support national programs that have a significant impact on issues facing infants and young children. Areas of interest include pediatric health--promoting health and preventing disease, including projects geared toward research or interventions that will reduce the incidence of serious chronic illnesses (e.g., diabetes, heart disease, obesity, or cancer) or improve cognitive, social, and emotional aspects of development; pediatric nutrition--assuring adequate nutrition for infants and young children through projects of research or interventions; and the effects of environmental hazards. The listed deadlines are for letters of inquiry. Guidelines and application are available online.
Requirements: 501(c)3 nonprofit organizations are eligible. Priority is given to projects that improve infant and young children nutrition, care, and development from the first year before birth to three years of age.
Restrictions: The foundation does not make grants or loans to individuals. Outside the West Michigan area, the foundation does not support capital campaigns, operating support, national child welfare programs, international based programs, or food/baby products giveaway programs.
Geographic Focus: All States
Date(s) Application is Due: Jun 1; Dec 1
Amount of Grant: Up to 1,000,000 USD
Contact: Program Contact; (231) 924-3175; fax (231) 924-7906; cobits@ncresa.org
Internet: http://www.gerberfoundation.org
Sponsor: Gerber Foundation
4747 W 48th Street, Suite 153
Fremont, MI 49412-8119

Gertrude and William C. Wardlaw Fund Grants 1808

Established in 1936 in Georgia, the Gertrude and William C. Wardlaw Fund awards general operating grants to Georgia nonprofit organizations in its areas of interest, including: cultural activities; the arts; community development; education and higher education; and health care and hospitals. Specific application forms are not required, and there are no specified annual deadlines. Grants typically range from $2,500 to $50,000.
Requirements: Georgia nonprofit organizations are eligible to apply.
Geographic Focus: Georgia
Amount of Grant: 2,500 - 50,000 USD
Contact: Gregorie Guthrie, Secretary; (404) 419-3260 or (404) 827-6529
Sponsor: Gertrude and William C. Wardlaw Fund
One Riverside Building
Atlanta, GA 30327

Gertrude B. Elion Mentored Medical Student Research Awards 1809

An annual award granted to support women medical students interested in pursuing health-related research projects.
Requirements: Students must have the support of a faculty mentor and must conduct their research at one of the four medical schools in North Carolina. Applications are accepted from Duke University Medical Center, East Carolina University School of Medicine, UNC-Chapel Hill School of Medicine or Wake Forest University Baptist Medical Center only.
Restrictions: Candidates must be women, enrolled as full-time students, and will have completed at least one year of medical school prior to the start of the award. They must conduct their research at the applying institution. Candidates must be citizens or permanent residents of the United States or Canada at the time of application. Documentation of permanent residency status must be provided with the application. Persons who have applied for permanent residency but have not received their government documentation by the time of application are not eligible.
Geographic Focus: North Carolina
Date(s) Application is Due: Mar 5
Amount of Grant: 10,000 USD
Contact: Libby Long, Program Coordinator; (919) 474-8370, ext. 134; fax (919) 941-9208; libby@trianglecf.org
Internet: http://www.trianglecf.org/page10000237.cfm
Sponsor: Triangle Community Foundation
324 Blackwell Street, Suite 1220
Durham, NC 27701

Gertrude E. Skelly Charitable Foundation Grants 1810

The primary mission of the Foundation is to provide educational opportunities, primarily at colleges and universities, and needed medical care for those who cannot afford them. The advisory committee will meet in October to review applications. There is no formal application. Interested applicants will contact the Foundation to receive application instructions and guidelines.
Requirements: 501(c)3 organizations, including colleges and universities, are eligible.
Restrictions: Funding is not available for endowments or capital projects. Multi-year commitments are generally not favored. Grants will not be made to private foundations, non-charitable organizations or directly to individuals.
Geographic Focus: All States
Date(s) Application is Due: Jul 31
Amount of Grant: Up to 50,000 USD
Contact: Bettee M. Collister; (561) 276-1008; fax (561) 272-2793; skelly@erikjoh.com
Sponsor: Gertrude E. Skelly Charitable Foundation
4600 North Ocean Boulevard, Suite 206
Boynton Beach, FL 33435-7365

GFWC of Massachusetts Catherine E. Philbin Scholarship 1811

The Philbin Scholarship offers funding to graduate or undergraduate study in the field of public health. The application is available at the GFWC website.
Requirements: In addition to the application, candidates must submit a copy of their Massachusetts driver's license, a personal statement addressing their professional goals, and how they relate to public health, and a reference letter from their academic advisor.
Geographic Focus: Massachusetts
Date(s) Application is Due: Feb 1
Amount of Grant: 500 USD
Contact: Judith Wilchynski, Scholarship Chairman; (508) 870-1895
Internet: http://www.gfwcma.org/scholarships.html
Sponsor: General Federation of Women's Clubs of Massachusetts
245 Dutton Road
Sudbury, MA 01776-0679

GFWC of Massachusetts Communication Disorder/Speech Therapy Scholarship 1812

The Communication Disorder/Speech Therapy Scholarship offers funding for graduate students in these fields of study. The application is available at the GFWC website.
Requirements: In addition to the application, candidates must submit a copy of their Massachusetts driver's license, a reference letter from the department chair of their major, and a personal statement addressing their personal and professional goals.
Geographic Focus: Massachusetts
Date(s) Application is Due: Feb 1
Amount of Grant: 800 USD
Contact: Judith Wilchynski, Scholarship Chairman; (508) 870-1895
Internet: http://www.gfwcma.org/scholarships.html
Sponsor: General Federation of Women's Clubs of Massachusetts
245 Dutton Road
Sudbury, MA 01776-0679

GFWC of Massachusetts Memorial Education Scholarship 1813

The GFWC Memorial Education Scholarship offers funding for a female graduate student studying education or physical therapy. The application is available at the GFWC website.
Requirements: Applicants must be residences of Massachusetts for at least five years. In addition to the application, candidates must submit a personal statement addressing their professional goals and financial need; official college transcripts; and a reference letter from their college department chair or recent employer. Applicants will meet with the scholarship committee if selected as a finalist.
Restrictions: Applicants should review the scholarship each year, as fields of study change annually.
Geographic Focus: Massachusetts
Date(s) Application is Due: Feb 1
Amount of Grant: Up to 3,000 USD
Contact: Jane Howard, Chairman of Trustees; (781) 891-1326; jhoward@mountida.edu
Internet: http://www.gfwcma.org/scholarships.html
Sponsor: General Federation of Women's Clubs of Massachusetts
245 Dutton Road
Sudbury, MA 01776-0679

Gheens Foundation Grants 1814

The Foundation makes grants in Metro Louisville, Kentucky and LaFourche and Terrebonne Parishes in Louisiana. The Foundation supports a wide variety of endeavors at all levels, including education, economic development, medical, arts, social/health services, handicapped, and mental health programs. Foundation focus is on education and Christian works. There are no deadlines for applications. Guidelines and application information are available online.

Requirements: The Foundation contributes only to 501(c)3 organizations that are not private foundations.
Restrictions: The Foundation does not give to individuals.
Geographic Focus: Kentucky, Louisiana
Samples: Western Kentucky U (Bowling Green, KY)--to provide scholarships to minority teacher-education students who are committed to teaching in Jefferson County, KY, public schools, $80,000 over four years; Louisville Seminary (Louisville, KY)--to construct the seminary's new academic building, $625,000 (2000); U of Louisville (KY)--to create a center where scientists will work to develop new gene-based technologies for studying a wide variety of age-related diseases, $2.5 million over five years.
Contact: Carl Thomas; (502) 584-4650; fax (502) 584-4652; carl@gheensfoundation.org
Internet: http://gheensfoundation.org/grant_application
Sponsor: Gheens Foundation
705 One Riverfont Plaza, 401 West Main Street
Louisville, KY 40202

Giant Eagle Foundation Grants 1815
The foundation awards grants in Pennsylvania in its areas of interest, including community development; health and human services; Jewish agencies, charitable organizations, and temples; performing arts; and philanthropy. Types of support include project grants and employee-related scholarship. There are no application deadlines. The board meets four times each year to review requests.
Requirements: Pennsylvania nonprofit organizations are eligible. Preference will be given to requests from Pittsburgh.
Geographic Focus: Ohio, Pennsylvania
Amount of Grant: 1,000 - 2,000 USD
Samples: U of Pittsburgh (PA)--for educational and community-outreach programs of the Cancer Institute, the College of Arts and Sciences, the Graduate School of Business, the School of Pharmacy, and the School of Social Work, $1.4 million.
Contact: David Shapira, Coordinator; (412) 963-6200
Sponsor: Giant Eagle Foundation
101 Kappa Drive
Pittsburgh, PA 15238

Giant Food Charitable Grants 1816
The Giant Food Charitable Grants provide funding and in-kind assistance to hundreds of charitable events and causes in schools, churches, synagogues, and civic community groups. Giant's focus is hunger relief programs, education, and wellness initiatives. Grant amounts vary, and there are no application deadlines. Submit requests by mail only.
Requirements: Organizations must send requests in writing on their letterhead, along with proof of their 501(c)3 status, and a description of the project.
Restrictions: Funding is available only in areas where Giant Food operates: District of Columbia, Maryland, Virginia, and Delaware.
Geographic Focus: Delaware, District of Columbia, Maryland, Virginia
Contact: Jamie Miller; (301) 341-8776; jmiller@giantofmaryland.com
Internet: http://www.giantfood.com/about_us/community/index.htm
Sponsor: Giant Food Corporation
8301 Professional Place, Suite 115
Landover, MD 20785

Gibson County Community Foundation Women's Fund 1817
The Gibson County Women's Fund will award a grant once a year to a single project serving women in Gibson County. Grant applications are considered on a yearly cycle which runs from June through September. Grant applications are accepted during this time period. email notices are sent to organizations that have signed up to receive information about the grant cycle. Project areas considered for funding must meet at least one of the following criteria: community development - activities that benefit women in the community or address gender equity issues within all segments of the population; education - activities that promote or strengthen the educational attainment of women, both in and out of the classroom; health - activities that improve the health outcomes for women; human services - activities that support public protection, employment/jobs, food and nutrition, agriculture, housing and shelter, public safety, disaster preparedness, and relief for women; other civic endeavors - activities that will improve the quality of life for women in Gibson County. A $5,000 grant is awarded to a single organization identified at the Women's Fund membership annual meeting. The remaining four finalists will each received $500. The application is available at the website.
Requirements: The Women's Fund welcomes applications from nonprofit organizations that are tax exempt under sections 501(c)3 and 509(a) of the Internal Revenue Code and from governmental agencies serving all women of Gibson County. Applications are also accepted from other nonprofit organizations carrying out charitable projects or activities that address issues facing women in Gibson County. In some cases, organizations without the 501(c)3 designation may be required to obtain a fiscal sponsor. Prospective applicants are invited to attend a meeting with the director to receive an overview of the funding opportunity and to ask specific questions.
Restrictions: Funding will not be considered for: requests for funding to reduce or retire debt of the organization; projects that focus solely on the spiritual needs and growth of a church congregation or members of other religious organizations; political parties or campaigns; operating costs not directly related to the proposed project (the organization's general operating expenses including equipment, staff salary, rent, and utilities); event sponsorships, annual appeals, and membership contributions; travel expenses for groups or individuals such as bands, sports teams, or classes; and scholarships or other grants to individuals.
Geographic Focus: Indiana
Date(s) Application is Due: Sep 1
Amount of Grant: 500 - 5,000 USD
Contact: Tami Muckerheide; (812) 386-8082; tami@gibsoncountyfoundation.org
Internet: http://www.gibsoncountyfoundation.org/gibson-womens-fund-grantmaking
Sponsor: Gibson County Community Foundation
109 North Hart Street, P.O. Box 180
Princeton, IN 47670

Gibson Foundation Grants 1818
The Foundation is committed to making the world a better place for children by creating, developing and supporting programs and other non-profit organizations in their efforts to advance education, music and the arts, the environment and health & welfare causes. The Foundation actively seeks out programs that will be a direct mission-fit and further its goals.
Requirements: Applicants must be 501(c)3 organizations.
Restrictions: The Foundation does not support religious or political affiliations, and does not award individual scholarships.
Geographic Focus: All States
Contact: Nina Miller; (615) 871-4500, ext. 2114; nina.miller@gibson.com
Internet: http://www.gibson.com/en-us/Lifestyle/GibsonFoundation/
Sponsor: Gibson Foundation
309 Plus Park Boulevard
Nashville, TN 37217

Gil and Dody Weaver Foundation Grants 1819
Established in Texas in 1980, the Gil and Dody Weaver Foundation offers support throughout the States of Texas, Oklahoma, and Louisiana, with some emphasis on the Dallas-Fort worth area. The Foundation's primary fields of interest include: cancer; children and youth services; education; health organizations; human services; social services; and recreational camps. Major types of support come in the form of: annual campaigns; general operating/continual support; and scholarship funds. Although no formal application is required, the Foundation does provide specific application guidelines. These guidelines require a history of the organization, detailed information about the proposed project, and budgetary needs. The annual deadline is May 31, with final notifications by September 30. Recent grants have ranged from $1,000 to $20,000, with occasional higher amounts for special circumstances.
Requirements: 501(c)3 organization serving the residents of Texas, Oklahoma, and Louisiana, are welcome to apply.
Restrictions: No grants are given to individuals. No applications are accepted from organizations located in states other than Texas, Oklahoma, or Louisiana.
Geographic Focus: California, Colorado, Louisiana, Mississippi, New Mexico, Oklahoma, Texas
Date(s) Application is Due: May 31
Amount of Grant: 1,000 - 20,000 USD
Samples: Cook Children's Health Foundation, Fort Worth, Texas, $12,500 - construction of an urgent care facility (2014); Camp McFadden, Ponca City, Oklahoma, $5,000 - general operation support (2014); Bookspring, Austin, Texas, $2,000 - support of the Reach Out and Read program (2014).
Contact: William R. Weaver, (214) 999-9497 or (214) 999-9494; fax (214) 999-9496
Sponsor: Gil and Dody Weaver Foundation
1845 Woodall Rodgers Freeway, Suite 1275
Dallas, TX 75201-2299

Gilbert Memorial Fund Grants 1820
The J. Roland Gilbert, Mary R. Gilbert and Elizabeth A. Gilbert Memorial Fund was established under the last will of Mary R. Gilbert in memory of herself, her husband, and their daughter. The fund seeks to support organizations performing research in various areas of medicine. The average grant amount is $5,000, and approximately seven awards are made each year. Applications should be submitted by September 30 to be reviewed at the annual grant meeting that occurs in October.
Requirements: To be eligible, organizations must qualify as exempt organizations under Section 501(c)3 of the Internal Revenue Code. Applications must be submitted through the online grant application form or alternative accessible application designed for assistive technology users. Except as otherwise specified when a grant is awarded, a progress report must be submitted within six months after receiving funds.
Geographic Focus: All States
Date(s) Application is Due: Sep 30
Amount of Grant: Up to 8,000 USD
Contact: Jason Craig; (888) 234-1999; grantadministration@ wellsfargo.com
Internet: https://www.wellsfargo.com/privatefoundationgrants/gilbert2
Sponsor: Gilbert Memorial Fund
1740 Broadway
Denver, CO 80274-0001

Gill Foundation - Gay and Lesbian Fund Grants 1821
In partnership with grantees, the Fund works to create a vibrant, healthy Colorado, where all families can thrive, while being a vehicle for cultural change toward equality for all. Programs of interest are: arts and culture; civic leadership; healthy families; and public broadcasting. Grantmaking is carried out in three ways: events sponsorship; invitation only; and opportunity grants. Requests may not exceed 15% of an organization's total budget, or for event sponsorships, may not exceed 50% of the event budget. Application information is available online.
Requirements: Eligible recipients must: be a 501(c)3 public charity organization; have a board-approved employment nondiscrimination policy in the organization's by-laws, employee manual or other official source stating explicitly that employees will not be discriminated against based on sexual orientation; provide services or programming within the state of Colorado.

Restrictions: Funding is not available for: organizations that are not a 501(c)3 public charity; lesbian, gay, bisexual, and transgender, or HIV/AIDS organizations; housing developments; emergency requests; endowments; individuals; major capital campaigns or projects; golf tournaments; film production or promotion.
Geographic Focus: Colorado
Sample: National Assoc of People With AIDS (Washington, D.C.)--for programs related to HIV/AIDS, $15,000; Human Rights Campaign Foundation (Washington, D.C.)--for programs supporting lesbian, gay, bisexual, and transgender people, $100,000; Gay and Lesbian Fund of Colorado (CO)--for leadership-development activities, $26,500 distributed among seven organizations; AIDS Resource Ctr of Wisconsin (Milwaukee, WI)--for HIV/AIDS prevention and human services programs, $60,000.
Contact: Manager; (800) 964-5643 or (719) 473-4455; info@gayandlesbianfund.org
Internet: http://www.gillfoundation.org/glfc_grants/glfc_grants_show.htm?doc_id=408610&cat_id=1394
Sponsor: Gill Foundation
315 East Costilla Street
Colorado Springs, CO 80903

Ginn Foundation Grants 1822
The foundation's mission is to address educational and community-based health care needs through supporting effective programs and services that bring about long-term solutions for individuals and the community, principally in Cuyahoga County, OH. The Foundation will consider not only grants to academic institutions, but also to organizations that meet non-academic educational needs, such as, programs that address issues of disease avoidance, child and family counseling, after-school training, arts, housing, and employment. Preference in all of these areas will be given to organizations and programs that serve low-income recipients. Faxed applications will not be considered. Application information is available online.
Requirements: Nonprofit organizations in Cuyahoga County, OH, may apply. Consideration will also be given to trustee-sponsored grants to similar types of organizations in the Chicago, Washington, D.C., and Minneapolis-St. Paul metropolitan areas.
Restrictions: The foundation will not fund requests for support of advocacy activities. Nor will it make grants to endowment, capital, or annual fund campaigns. The foundation will not fund special events or attendance at conferences or symposia.
Geographic Focus: District of Columbia, Illinois, Minnesota, Ohio
Date(s) Application is Due: Mar 15; Sep 15
Amount of Grant: 5,000 - 30,000 USD
Contact: Walter Pope Ginn, Trustee; info@ginnfoundation.org
Internet: http://www.ginnfoundation.org/index.html
Sponsor: Ginn Foundation
13938 A Cedar Road, P.O. Box 239
Cleveland Heights, OH 44118

Giving in Action Society Children & Youth with Special Needs Grants 1823
Giving in Action Society is a charitable organization established by Vancouver Foundation in 2006. The Society provides grants directly to individuals and families in B.C. The Children and Youth with Special Needs Fund provides grants to families who have children or youth (newborn to 19 years) with special needs living at home. The Fund offers one-time capital grants to help enhance or improve the individual's health, development or ability to participate in daily activities at home, in school and in the community. Eligible expenses may include such things as home renovations and vehicle modifications. This Fund defines children and youth with special needs as those who have significant impairments in one or more of the following areas: health, cognition, communication, sensory motor, social/emotional/behavioural or self help. The Children and Youth with Special Needs Fund helps address family needs that are not currently met through government-funded programs. If a family receives support from other government-funded programs, they can apply to the Fund for grants but only if the funds will complement â€" versus duplicate â€" the government support. The Fund was established with financial assistance from the B.C. Ministry of Children and Family Development.
Requirements: Only residents of B.C. are eligible to apply. The Children and Youth with Special Needs Fund provides grants to families who have children or youth (newborn to 19 years) with special needs living at home. There is a two-stage application process. The first stage, submit a Letter of Inquiry, consisting of an informal proposal to determine basic suitability and eligibility. The second stage is a full application. If your Letter of Inquiry fits within the funding guidelines you will be sent a grant application form with specific instructions for completion. You have 6 months to submit your application.
Geographic Focus: All States, Canada
Contact: Faye Wightman, President; (866) 523-3157 or (604) 683-3157; fax (604) 683-3134; info@vancouverfoundation.ca
Internet: http://www.givinginaction.ca/ourprograms/specialneeds.htm
Sponsor: Giving in Action Society, Vancouver Foundation
Harbour Centre, 555 W Hastings Street, Suite 1200, Box 12132
Vancouver, BC V6B 4N6 Canada

Giving in Action Society Family Independence Grants 1824
Giving in Action Society is a charitable organization established by Vancouver Foundation in 2006. The Society provides grants directly to individuals and families in B.C. The Family Independence Fund helps families throughout the province who have children or adults with developmental disabilities living at home. Grants from the Family Independence Fund help with the ongoing care of the relative by providing support for projects such as home renovations - including lifts, elevators, ramps, flooring, door widening or vehicle modifications - that enable the individual with the developmental disability to live in the family home and access their community. Preference is given to families that indicate their request for support is part of a long-term plan for their relative. Note that families who receive support from other government-funded programs are still eligible for Family Independence Fund grants. However these additional funds must complement - versus duplicate - the government support.
Requirements: Only residents of B.C. are eligible to apply. The Family Independence Fund supports a child/relative who has a developmental disability, defined as: concurrent with impaired adaptive functioning; low IQ; manifests before the age of 18. There is a two-stage application process. The first stage, submit a Letter of Inquiry, consisting of an informal proposal to determine basic suitability and eligibility. The second stage is a full application. If your Letter of Inquiry fits within the funding guidelines you will be sent a grant application form with specific instructions for completion. You have 6 months to submit your application.
Geographic Focus: All States, Canada
Contact: Faye Wightman, President; (866) 523-3157 or (604) 683-3157; fax (604) 683-3134; info@vancouverfoundation.ca
Internet: http://www.givinginaction.ca/ourprograms/familyindependencefund.htm
Sponsor: Giving in Action Society, Vancouver Foundation
Harbour Centre, 555 W Hastings Street, Suite 1200, Box 12132
Vancouver, BC V6B 4N6 Canada

Giving Sum Annual Grant 1825
Giving Sum is an organization of next generation givers working together to identify challenges in the community and to address them through education, engagement, inspiration and grants. Giving Sum is a charitable giving vehicle pooling resources raised through memberships, sponsorships, and fundraisers to establish annual grants for recipients selected by the organization and its members. Each year Giving Sum will grant 100 percent of its membership contributions to a nonprofit organization in Indianapolis working towards change. Members will be proactive in searching out these organizations in the greater Indianapolis area and having them apply.
Requirements: To be eligible, organizations must be recognized by the Internal Revenue Service as a public charity under sections 501(c)3 and 509(a)(l), (2), or (3) of the Internal Revenue Code, and provide a copy of their determination letter from the Internal Revenue Service. Organizations or the proposed project must serve audiences within the greater Indianapolis area, including the counties of: Boone, Hamilton, Hancock, Hendricks, Johnson, Marion, Morgan and Shelby. Eligible nonprofit organizations may submit only one application to request a grant of $50,000 (Money), volunteer hours (Time) and advocacy efforts (Voice) from Giving Sum to support general operations or special projects in one of Giving Sum's five Focus Areas (Arts, Culture & Humanities, Civic & Community Development, Education, Environment, and Health & Human Services). Each organization must select one of the above focus areas in which to apply. Applicants should select the focus area which best describes the primary operations or mission of their organization. If your organization and/or request is multi-disciplinary, select the category which best describes the activities that will be carried out using the Giving Sum grant. Preliminary review of the applications occurs within each of the five focus areas. Preference will be given to applications that meaningfully and creatively engage Giving Sum members in innovative, high-impact initiatives.
Restrictions: Grant funds may not be used to lobby or otherwise attempt to influence legislation, to influence the outcome of any public election, or to carry on any voter registration drive.
Geographic Focus: Indiana
Amount of Grant: 50,000 USD
Samples: 2009 - Starfish Initiative: $50,000 grant for a leadership camp for up to 300 academically promising, economically disadvantaged high school students, attending more than 30 different high schools in greater Indianapolis. 2008 - Coburn Place: $50,000 toward a playground for the children of Coburn Place.
Contact: Ryan Brady, Program Director; (317) 634-2423; grants@givingsum.org
Internet: http://givingsum.org/grants.asp
Sponsor: Giving Sum
615 N Alabama Street, Suite 119
Indianapolis, IN 46204-1498

Gladys Brooks Foundation Grants 1826
At the beginning of each year, the Foundation Board determines a limited scope of activities for which it will consider grant applications. Current giving will be in the fields of libraries, education, hospitals, and clinics. Applications will be considered where: outside funding (including governmental) is not available; the project will be largely funded by the grant unless the grant request covers a discrete component of a larger project; and the funds will be used for capital projects including equipment or endowments. The Foundation's application form must be used and is available online. Electronic submissions are not acceptable.
Requirements: The foundation makes grants to nonprofit, private, publicly supported, tax-exempt organizations located in Connecticut, Delaware, Florida, Illinois, Indiana, Louisiana, Maine, Maryland, Massachusetts, New Hampshire, New Jersey, New York, Ohio, Pennsylvania, Rhode Island, Tennessee, and Vermont.
Restrictions: Applications for direct salary support will not be accepted. No portion of grants shall be appropriated as an administrative or processing fee, for overseeing the project or for its general overhead.
Geographic Focus: Connecticut, Delaware, Florida, Illinois, Indiana, Louisiana, Maine, Maryland, Massachusetts, New Hampshire, New Jersey, New York, Ohio, Pennsylvania, Rhode Island, Tennessee, Vermont
Date(s) Application is Due: Jun 1
Amount of Grant: 50,000 - 100,000 USD
Contact: Jessica Rutledge, (516) 746-6103

Internet: http://www.gladysbrooksfoundation.org
Sponsor: Gladys Brooks Foundation
1055 Franklin Avenue, Suite 208
Garden City, NY 11530

Glaucoma Foundation Grants   1827
The foundation funds research initiatives to determine the causes of glaucoma, to improve methods of treatment, and to develop cures for the various kinds of glaucoma. Two areas of particular focus are: optic nerve rescue and restoration--research into new approaches designed to protect the optic nerve against glaucomatous damage, to restore vision lost to glaucoma and eventually reverse blindness by restoring or regenerating the function of the optic nerve cells, and to explore the feasibility of achieving transplantation of optic nerve cells; and molecular genetics--research into the genetic causes of the various forms of glaucoma, particularly the identification of the responsible genes, with the long-term goal of finding ways to reverse these genetic defects. Grants are awarded for a one-year period and are renewable. Guidelines and application are available online.
Requirements: Applicants must have a full time faculty position or the equivalent.
Restrictions: The foundation does not provide funds for investigator salaries, travel, overhead, or other indirect costs.
Geographic Focus: All States
Date(s) Application is Due: Mar 1; Sep 1
Amount of Grant: 40,000 - 50,000 USD
Contact: Kira A. Zmuda, Director; (212) 651-2509; kzmuda@glaucomafoundation.org
Internet: http://www.glaucomafoundation.org/Grant_Application.htm
Sponsor: Glaucoma Foundation Grants
80 Maiden Lane, Suite 700
New York, NY 10038

Glaucoma Research Pilot Project Grants   1828
These grants provide funds to encourage innovative and pilot research, as well as to aid ongoing studies that seek to protect and restore the optic nerve, accurately monitor glaucoma's progression, find the genes responsible for the glaucoma, understand the intraocular pressure system and develop better treatments, and determine the risk factors for glaucoma damage.
Requirements: Applicants must have a graduate degree.
Restrictions: Funding is not granted for equipment purchases, overhead commercial applications, or indirect costs.
Geographic Focus: All States
Date(s) Application is Due: Aug 1
Amount of Grant: 40,000 USD
Contact: Jennifer Rulon, (800) 826-6693 or (415) 986-3162; fax (415) 986-3763; research@glaucoma.org
Internet: http://www.glaucoma.org/research/index.php
Sponsor: Glaucoma Research Foundation
251 Post Street, Suite 600
San Francisco, CA 94108

GlaxoSmithKline Corporate Grants   1829
In the United States, GlaxoSmithKline Corporate makes charitable contributions through its U.S. Contributions Committee. Requests for US-based community partnerships with nonprofit organizations should address issues in one of four general areas: education--science education, K-12 literacy, teacher professional development; health and human services--child health or prevention and access to health care for women related to breast or gynecologic cancers, targeting the needs of underserved and diverse populations; arts and culture--local organizations in the Greater Philadelphia and the Research Triangle Park areas, based on local needs, focusing on public school educational outreach; and civic and community--local organizations in the Greater Philadelphia and the Research Triangle Park areas, based on local needs.
Requirements: Organizations based in the United States that have a 501(c)3 IRS designation and that meet GSK's corporate criteria for funding may complete an application. Funding for the arts and civic programs is concentrated exclusively in the Greater Philadelphia and Research Triangle Park areas where employees live and work.
Restrictions: As a matter of policy, grants are not provided for general operating expenses or capital building costs, and they are not made to individuals. Grants are not given to political, religious, fraternal, profit-making, discriminatory, hobby-oriented, or tax-subsidized organizations.
Geographic Focus: Pennsylvania
Amount of Grant: Up to 100,000,000 USD
Contact: Mary Anne Rhyne; (919) 483-2319; community.partnership@gsk.com
Internet: http://us.gsk.com/html/community/community-grants-corporate.html
Sponsor: GlaxoSmithKline Corporation
1 Franklin Plaza, FP2130, P.O. Box 7929
Philadelphia, PA 19101-7929

GlaxoSmithKline Foundation IMPACT Awards   1830
The awards reward excellence in the delivery of nonprofit community health care. Awards are made to small and mid-size community-based health care organizations in the greater Philadelphia area and, the Research Triangle Park, North Carolina area. Selection criteria include: commitment to serving people in need; facilitating access to health care delivery, education, creative partnerships, and policy development; and demonstration of a solid record of achievement, management, and leadership. The application process for the GlaxoSmithKline 1st Annual IMPACT Awards in the Research Triangle Park, North Carolina region closed on May 15. July 15 is the deadline date for the greater Philadelphia region.
Requirements: Nominees in the following Philadelphia counties are eligible: Berks, Bucks, Chester, Delaware, Lancaster, Montgomery, or Philadelphia or in the City of Camden, NJ. Nominees in the following Research Triangle Park, North Carolina counties are: Chatham, Durham, Orange and Wake Counties. Organizations must have an operating or program budget of under $2 million and have been in existence at least five years. Nonprofit healthcare organizations must be nominated by individuals knowledgeable of the complexities and challenges of healthcare delivery in their communities.
Geographic Focus: Pennsylvania
Date(s) Application is Due: May 15; Jul 15
Amount of Grant: 40,000 USD
Contact: Mary Anne Rhyne, Corporate Press; (919) 483-2319; fax (919) 483-8765; community.partnership@gsk.com
Internet: http://www.gsk-us.com/html/community/community-healthcare-awards.html
Sponsor: GlaxoSmithKline Foundation
1 Franklin Plaza, FP2130, P.O. Box 7929
Philadelphia, PA 19101-7929

Glenn/AFAR Breakthroughs in Gerontology Awards   1831
The goal of the program is to provide timely support to a small number of pilot research programs that may be of high risk but offer significant promise of yielding transforming discoveries in the fundamental biology of aging. The hope is that one or more of the funded research projects will lead to major new insights into the molecular factors that coordinate aging in multiple cells and tissues, and the ways in which the aging process is differentially timed in long-lived species. Projects that focus on genetic controls of aging and longevity, on delay of aging by pharmacological agents or dietary means, or that elucidate the mechanisms by which alterations in hormones, anti-oxidant defenses, or repair processes promote longevity are all well within the intended scope of this competition. Projects that focus instead on specific diseases or on assessment of health care strategies will receive much lower priority, unless the research plan makes clear and direct connections to fundamental issues in the biology of aging. Studies of invertebrates, mice, human clinical materials, or cell lines are all potentially eligible for funding. Although preliminary data are always helpful for evaluating the feasibility of the experiments proposed, the emphasis in review will be on creativity and the likelihood that the findings will open new vistas and approaches to aging research that might merit intensive follow-up studies. Applications from individuals not previously engaged in aging research are particularly encouraged, as long as the research proposals show high promise for leading to important new discoveries in biological gerontology. The proposed research must be conducted at any type of nonprofit setting in the United States. Guidelines are available online.
Requirements: Applicants must at the time they submit their proposal be full-time faculty members at the rank of assistant professor or higher. A strong record of independent publication beyond the postdoctoral level is a requirement.
Restrictions: Applicants who are employees in the NIH Intramural program are not eligible.
Geographic Focus: All States
Date(s) Application is Due: Dec 16
Amount of Grant: 200,000 USD
Contact: Grants Manager; (212) 703-9977 or (888) 582-2327; fax (212) 997-0330; grants@afar.org or info@afar.org
Internet: http://afar.org/grants.html
Sponsor: American Federation for Aging Research
55 West 39th Street, 16th Floor
New York, NY 10018

Global Fund for Children Grants   1832
The Global Fund for Children's (GFC) mission is to advance the dignity of children and youth around the world through small grants to innovative community-based organizations working with some of the world's most vulnerable children and youth. The Fund selects its grantee partners based on their potential to grow in effectiveness and to become valuable resources or models for others. Selection criteria include exceptional leadership, sound management, strong community participation, innovative and effective programs, and direct engagement with the most vulnerable children. Fund focus is on four specific issues: learning, enterprise, safety, and healthy minds and bodies. Organizations may submit a letter of inquiry at any time. If a letter of inquiry falls within GFC's priorities, GFC will follow up with the organization to solicit a full proposal.
Requirements: Nonprofit organization worldwide are eligible.
Geographic Focus: All States
Amount of Grant: 5,000 - 20,000 USD
Samples: Anandan, Kolkata, India, $8,000--education grant; Benishyaka Association, Kigali, Rwanda, $15,000--academic scholarships; Children in the Wilderness, Lilongwe, Malawi, $15,000--education grant.
Contact: Andrew Barnes; (202) 331-9003; info@globalfundforchildren.org
Internet: http://www.globalfundforchildren.org/applyforagrant/index.html
Sponsor: Global Fund for Children
1101 Fourteenth Street, North West, Suite 420
Washington, D.C. 20005

Global Fund for Women Grants   1833
The Global Fund for Women supports women's groups that advance the human rights of women and girls. Grants support women's groups based outside the United States by providing small, flexible, and timely grants for operating and project expenses. Grantees address issues that include but are not limited to building peace and ending gender-based violence; advancing health and sexual and reproductive rights; expanding civic and political participation; ensuring economic and environmental justice; increasing access to education; and fostering social change philanthropy. In addition urgent requests

for support to organize or participate in local, regional, or international meetings and conferences will be considered outside of the normal grant cycle. These types of requests must come from organizations, not individuals, and must be received at least eight weeks before the event. Funds for these types of grants are limited. Applications and guidelines are available online. Organizations should refer to the staff website for specific contacts according to their applicant origin.

Requirements: The applicant group must be based in a country outside the United States; demonstrate a strong commitment to women's equality and human rights that is clearly reflected in its activities; be a group of women working together; and be governed, directed, and led by women.

Restrictions: Grants do not support individuals; scholarships; academic research; groups based and working primarily or only in the United States; international organizations proposing projects with local partners; groups without a strong women's rights focus; groups headed and managed by men, or without women in the majority of leadership positions; groups whose sole purpose is to generate income or to provide charity to individuals; or political parties or election campaigns.

Geographic Focus: All Countries
Amount of Grant: 5,000 - 50,000 USD
Contact: Program Assistant; (415) 248-4800; fax (415) 202-4801
Internet: http://www.globalfundforwomen.org/apply-for-a-grant/types-of-grant
Sponsor: Global Fund for Women
222 Sutter Street, Suite 500
San Francisco, CA 94108

GNOF Albert N. & Hattie M. McClure Grants   1834

The Albert N. & Hattie M. McClure Fund was established in 1963 and is a donor-advised fund of the Greater New Orleans Foundation (GNOF). The fund supports the emergency bricks and mortar projects (e.g. repairs, replacements, and additions to facilities) of eligible United Way partner agencies. The intent of the McClure Fund is to fill the funding gap between major capital expenses requiring a capital fund drive and minor capital needs which can be planned for and included in an agency's operating budget. Priority consideration will be given to those requests of a capital nature that stem from a crisis and/or emergency. GNOF will consider requests up to $15,000. The application process has no deadline date. Organizations will be notified of funding decisions within sixty days. Organizations can find a downloadable application form at the website and should contact GNOF for mailing instructions.

Requirements: Applicants must be partner agencies of United Way and should contact GNOF for further eligibility requirements.
Geographic Focus: Louisiana
Amount of Grant: Up to 15,000 USD
Contact: Ellen M. Lee, Sr. Vice President, Programs, Community Revitalization Program Director; (504) 598-4663; fax (504) 598-4676; ellen@gnof.org
Internet: http://www.gnof.org/albert-n-hattie-m-mcclure-fund/
Sponsor: Greater New Orleans Foundation
1055 St. Charles Avenue, Suite 100
New Orleans, LA 70130

GNOF Bayou Communities Grants   1835

In 2012 the Greater New Orleans Foundation (GNOF) established the Bayou Communities Foundation (BCF) that has, as its initial focus, the mission to improve life in the Louisiana parishes of Terrebonne and Lafourche, and that will work to strengthen local nonprofit capacity in compassionate and sustainable coastal communities in Louisiana for generations to come. The impetus to form BCF came from parish residents, who first organized among themselves and then approached GNOF, who had set up similar and successful foundations in St. Bernard, Plaquemines, and Jefferson Parishes after Hurricanes Katrina and Rita and the Gulf Oil Spill. BCF is set up as a GNOF donor-advised fund, an affiliate under GNOF's nonprofit umbrella. The new foundation will receive $500,000 in seed money for the next five years from the Gheens Foundation in Lafourche and has committed to raising $1 million in matching dollars. BCF has also committed to putting at least 90% of its first $100,000 from the Gheens Foundation back into the community through grants. As an affiliate foundation, BCF has access to the expertise of GNOF but makes independent decisions about BCF projects. Organizations interested in obtaining grants through BCF can contact GNOF for more information.

Geographic Focus: Louisiana
Contact: Josephine Everly; (504) 598-4663; fax (504) 598-4676; josephine@gnof.org
Internet: http://www.gnof.org/
Sponsor: Greater New Orleans Foundation
1055 St. Charles Avenue, Suite 100
New Orleans, LA 70130

GNOF Exxon-Mobil Grants   1836

Exxon-Mobil, a grant-making partner of the Greater New Orleans Foundation (GNOF) has established the Exxon-Mobil fund to improve the quality of lives for people in the St. Bernard parish and a portion of Algiers. The fund provides the following types of grants: capital-fund grants for new construction or major renovation; seed-money grants to help start new organizations that respond to an important opportunity in the community; bridge grants to sustain organizations experiencing financial hardships; and grants that support new, creative, or beneficial programs. Amounts up to $4,000 are given in Algiers; amounts up to $10,000 are given in St. Bernard Parish and in extraordinary circumstances may exceed this maximum. Grant requests are reviewed by an advisory committee of business and civic leaders that meets annually. Application materials must be emailed to the address provided on this page and received by the foundation by 5:00 p.m. on August 15. Exact deadlines may vary from year to year.

Prospective applicants should verify the current deadline at the GNOF website where they can also obtain complete guidelines and requirements as well as a downloadable application form. Prospective applicants can also subscribe to GNOF's email newsletter for announcements of future funding opportunities. The subscription link is available from GNOF's Apply-for-a-Grant web page which contains a comprehensive listing of GNOF's current and past funding opportunities.

Requirements: While priority is given to nonprofit organizations based in St. Bernard Parish or Algiers, nonprofit organizations servicing these areas will be given consideration. Applicants must have 501(c)3 status or apply through a fiscal agent who has such status.
Geographic Focus: Louisiana
Date(s) Application is Due: Aug 15
Amount of Grant: Up to 10,000 USD
Samples: Los Islenos Heritage and Cultural Society, St. Bernard, Louisiana—install a sewerage system as part of the restoration of the cultural building; Louisiana Philharmonic Orchestra, New Orleans, Louisiana—put on two young-people's concerts in the St. Bernard parish Cultural Arts Center; St. Bernard Wetlands Foundation, Meraux, Louisiana—coordinate and educate student and adult volunteers to maintain and replenish the tree nursery as well as out-planting mature container-grown trees into selected wetland locations.
Contact: Marco Cocito-Monoc, Sr.; (504) 598-4663; fax (504) 598-4676; marco@gnof.org
Internet: http://www.gnof.org/exxon-mobile-fund/
Sponsor: Greater New Orleans Foundation
1055 St. Charles Avenue, Suite 100
New Orleans, LA 70130

GNOF Freeman Challenge Grants   1837

The Freeman Challenge, a donor-advised fund of the Greater New Orleans Foundation (GNOF), is a memorial tribute to one of New Orleans' leading citizens and philanthropists Richard West Freeman who will be remembered as one of the founders of The United Fund in 1952 (now The United Way). The purpose of The Freeman Challenge is to create long-term financial stability for nonprofit organizations serving the Greater New Orleans thirteen-parish region; the Challenge will match one dollar for every two dollars raised by nonprofits to build their own endowments. Nonprofits can elect to receive one of three matching amounts: $5,000, $10,000, or $15,000. Freeman Challenge grant applications are reviewed by a selection committee of volunteers and GNOF staff that meets annually. The committee looks for nonprofits that represent varying areas of service, e.g. education, the arts, human services, etc. Application materials are due no later 5 p.m. on September 14 and must be emailed to the email address given under the Contact Information section. Exact deadlines may vary from year to year. Prospective applicants should verify the current deadline at the GNOF website where they can also obtain a downloadable brochure about the the program as well as a downloadable application form with guidelines and requirements. Prospective applicants can also subscribe to GNOF's email newsletter for announcements of future funding opportunities. The subscription link is available from GNOF's Apply-for-a-Grant web page which contains a comprehensive listing of GNOF's current and past funding opportunities.

Requirements: To be eligible nonprofit organizations must meet the following *Requirements:* have 501(c)3 status; have been in operation for a minimum of five years; be headquartered in one of Greater New Orleans' thirteen parishes (Assumption, Jefferson, Lafourche, Orleans, Plaquemines, St. Bernard, St. Charles, St. James, St. Johns, St. Tammany, Tangipahoa, Terrebonne, or Washington); presently have no endowment with a market value that exceeds $500,000; show evidence of previous fund-raising success; conduct an annual independent audit; have a volunteer board; and have their board's approval to take part in the Freeman Challenge.
Geographic Focus: Louisiana
Date(s) Application is Due: Jun 25
Amount of Grant: 5,000 - 15,000 USD
Contact: Ellen M. Lee, Sr. Vice President, Programs, Community Revitalization Program Director; (504) 598-4663; fax (504) 598-4676; grants@gnof.org
Internet: http://www.gnof.org/the-freeman-challenge/
Sponsor: Greater New Orleans Foundation
1055 St. Charles Avenue, Suite 100
New Orleans, LA 70130

GNOF IMPACT Grants for Arts and Culture   1838

Through the IMPACT Program, the Greater New Orleans Foundation (GNOF) makes grants to organizations serving the Greater New Orleans region. The ultimate goal of the IMPACT Program is to create a resilient, sustainable, vibrant, and equitable region in which individuals and families flourish and in which the special character of the New Orleans region and its people is preserved, celebrated, and given the means to develop. Specifically GNOF hopes to accomplish the following objectives through its IMPACT grants: provide a much needed source of financial and other support to nonprofit organizations that are struggling in the current financial environment and that are important to the health and vibrancy of the region; develop a better sense of the nonprofit organizations serving the region so GNOF can more effectively match donor desires with effective charitable work; identify and nurture promising new leaders and initiatives, especially in those communities that are in greatest need; and gain knowledge that will help nonprofit leaders and GNOF staff develop better long-term strategies for addressing regional needs and taking best advantage of important opportunities. IMPACT grants are awarded in four categories: Arts and Culture, Youth Development, Education, and Health and Human Services. In the category of Arts and Culture GNOF supports organizations and programs that help preserve and grow the rich cultural heritage of the Greater New Orleans region and ensure that the originators and producers

of creative goods and services can continue to enhance community life. Priority will be given to work that has the following goals: to improve the quality of life for artists and performers in the region; to demonstrate the importance of the arts and make the case for increased public support for the arts; and to form alliances and connections between grassroots-based organizations and the business community to expand income-producing opportunities for artists. Interested organizations must submit a letter of intent along with all attachments via one email by 5 p.m. on July 30. GNOF program staff will review all letters of intent and will contact those organizations that are invited to submit a full application for funding. Awards are announced in November. Deadlines may vary from year to year. Interested organizations should verify the current deadline at the GNOF website where they can also obtain complete guidelines and requirements as well as a downloadable application form and cover sheet. Prospective applicants can also subscribe to GNOF's email newsletter for announcements of future funding opportunities. The subscription link is available from GNOF's Apply-for-a-Grant web page which contains a comprehensive listing of GNOF's current and past funding opportunities.
Requirements: Nonprofit, tax-exempt organizations that serve the Greater New Orleans region are eligible to apply for funding. Organizations that are not tax-exempt but have a fiscal agent relationship with a 501(c)3 organization are also eligible.
Restrictions: Through its IMPACT program, the Greater New Orleans Foundation is unable to fund the following types of requests: requests for individual support, either through scholarships or other forms of financial assistance; special events or conferences; programs that promote religious doctrine; endowments; and scientific or medical research.
Geographic Focus: All States
Date(s) Application is Due: Jul 30
Amount of Grant: Up to 20,000 USD
Samples: Arts Council of New Orleans, New Orleans, Louisiana, $20,000—for general operating support; Sweet Home New Orleans, New Orleans, Louisiana, $20,000—for general operating support; Louisiana Cultural Economy Foundation, New Orleans, Louisiana, $25,000—for general operating support.
Contact: Roy Williams; (504) 598-4663; fax (504) 598-4676; grants@gnof.org
Internet: http://www.gnof.org/programs/impact/
Sponsor: Greater New Orleans Foundation
1055 St. Charles Avenue, Suite 100
New Orleans, LA 70130

GNOF IMPACT Grants for Health and Human Services　　　　1839
Through the IMPACT Program, the Greater New Orleans Foundation (GNOF) makes grants to organizations serving the Greater New Orleans region. The ultimate goal of the IMPACT program is to create a resilient, sustainable, vibrant, and equitable region in which individuals and families flourish and in which the special character of the New Orleans region and its people is preserved, celebrated, and given the means to develop. Specifically GNOF hopes to accomplish the following objectives through its IMPACT grants: provide a much needed source of financial and other support to nonprofit organizations that are struggling in the current financial environment and that are important to the health and vibrancy of the region; develop a better sense of the nonprofit organizations serving the region so GNOF can more effectively match donor desires with effective charitable work; identify and nurture promising new leaders and initiatives, especially in those communities that are in greatest need; and gain knowledge that will help nonprofit leaders and GNOF staff develop better long-term strategies for addressing regional needs and taking best advantage of important opportunities. IMPACT grants are awarded in four categories: Arts and Culture, Youth Development, Education, and Health and Human Services. In the category of Health and Human Services, support is available to organizations that work to improve the health and living conditions of low-income families and their children, the disabled, the elderly, and other under served populations and help move them toward self-sufficiency. Priority will be given to work that increases Medicaid/LaCHIP or Medicare enrollment for indigent consumers of health-care services; advocates to preserve access to health care, to provide consumer protections, and/or to expand Medicaid coverage to increase access to comprehensive, quality primary care, mental health care, and preventive care for all; implements health education and outreach efforts to increase use of health-care services by the most under-served populations, African-American males in particular; uses health education to improve health literacy, influence attitudes, and improve health awareness so that indigent consumers of health care services can make better decisions and take preventive actions that will improve personal, family, and community health; and improve communication, coordination and collaboration between social-services providers to serve comprehensively the needs of low-income families, improving their chances of success in achieving self sufficiency. Interested organizations must submit a letter of intent along with all attachments via one email by 5 p.m. on July 30. GNOF program staff will review all letters of intent and will contact those organizations that are invited to submit a full application for funding. Awards are announced in November. Deadlines may vary from year to year. Interested organizations should verify the current deadline at the GNOF website where they can also obtain complete guidelines and requirements as well as a downloadable application form and cover sheet. Prospective applicants can also subscribe to GNOF's email newsletter for announcements of future funding opportunities. The subscription link is available from GNOF's Apply-for-a-Grant web page which contains a comprehensive listing of GNOF's current and past funding opportunities.
Requirements: Nonprofit, tax-exempt organizations that serve the thirteen parishes of Greater New Orleans are eligible to apply for funding. Organizations that are not tax-exempt but have a fiscal agent relationship with a 501(c)3 organization are also eligible.
Restrictions: Through its IMPACT program, the Greater New Orleans Foundation is unable to fund the following types of requests: requests for individual support, either through scholarships or other forms of financial assistance; special events or conferences; programs that promote religious doctrine; endowments; and scientific or medical research.

Geographic Focus: All States
Date(s) Application is Due: Jul 30
Amount of Grant: Up to 20,000 USD
Samples: "Grow Dat" Youth Farm, New Orleans, Louisiana, $25,238—to support the development of the organization's wellness programming; Latino Farmers Cooperative of Louisiana, New Orleans, Louisiana, $41,425—to provide general operating support; St. Anna's Episcopal Church, HNew Orleans, Louisiana, $30,000—to provide support for the St. Anna's Medical Mission medical outreach program.
Contact: Roy Williams, Program Assistant; (504) 598-4663; fax (504) 598-4676; grants@gnof.org or roy@gnof.org
Internet: http://www.gnof.org/programs/impact/
Sponsor: Greater New Orleans Foundation
1055 St. Charles Avenue, Suite 100
New Orleans, LA 70130

GNOF IMPACT Gulf States Eye Surgery Fund　　　　1840
Through the IMPACT Program, the Greater New Orleans Foundation (GNOF) makes grants to organizations serving the Greater New Orleans region. The ultimate goal of the IMPACT program is to create a resilient, sustainable, vibrant, and equitable region in which individuals and families flourish and in which the special character of the New Orleans region and its people is preserved, celebrated, and given the means to develop. Specifically GNOF hopes to accomplish the following objectives through its IMPACT grants: provide a much needed source of financial and other support to nonprofit organizations that are struggling in the current financial environment and that are important to the health and vibrancy of the region; develop a better sense of the nonprofit organizations serving the region so GNOF can more effectively match donor desires with effective charitable work; identify and nurture promising new leaders and initiatives, especially in those communities that are in greatest need; and gain knowledge that will help nonprofit leaders and GNOF staff develop better long-term strategies for addressing regional needs and taking best advantage of important opportunities. IMPACT grants are awarded in four categories: Arts and Culture, Youth Development, Education, and Health and Human Services. In the category of Health and Human Services, special funding is available for organizations that defray the expenses of poor or indigent patients requiring or receiving eye surgery, care, or treatment. Interested organizations must submit a letter of intent along with all attachments via one email by 5 p.m. on July 30 and should indicate on the IMPACT application cover sheet that they are applying for funding from the Gulf States Eye Surgery Fund. GNOF program staff will review all letters of intent and will contact those organizations that are invited to submit a full application for funding. Awards are announced in November. Deadlines may vary from year to year. Interested organizations should verify the current deadline at the GNOF website where they can also obtain complete guidelines and requirements as well as a downloadable application form and cover sheet. Prospective applicants can also subscribe to GNOF's email newsletter for announcements of future funding opportunities. The subscription link is available from GNOF's Apply-for-a-Grant web page which contains a comprehensive listing of GNOF's current and past funding opportunities.
Requirements: Nonprofit, tax-exempt organizations that serve the thirteen parishes of Greater New Orleans are eligible to apply for funding. Organizations that are not tax-exempt but have a fiscal agent relationship with a 501(c)3 organization are also eligible.
Restrictions: Through its IMPACT program, the Greater New Orleans Foundation is unable to fund the following types of requests: requests for support from individuals, either through scholarships or other forms of financial assistance; special events or conferences; programs that promote religious doctrine; endowments; and scientific or medical research.
Geographic Focus: All States
Date(s) Application is Due: Jul 30
Amount of Grant: Up to 20,000 USD
Contact: Roy Williams; (504) 598-4663; fax (504) 598-4676; grants@gnof.org
Internet: http://www.gnof.org/programs/impact/
Sponsor: Greater New Orleans Foundation
1055 St. Charles Avenue, Suite 100
New Orleans, LA 70130

GNOF IMPACT Harold W. Newman, Jr. Charitable Trust Grants　　　　1841
Through the IMPACT Program, the Greater New Orleans Foundation (GNOF) makes grants to organizations serving the Greater New Orleans region. The ultimate goal of the IMPACT program is to create a resilient, sustainable, vibrant, and equitable region in which individuals and families flourish and in which the special character of the New Orleans region and its people is preserved, celebrated, and given the means to develop. Specifically GNOF hopes to accomplish the following objectives through its IMPACT grants: provide a much needed source of financial and other support to nonprofit organizations that are struggling in the current financial environment and that are important to the health and vibrancy of the region; develop a better sense of the nonprofit organizations serving the region so GNOF can more effectively match donor desires with effective charitable work; identify and nurture promising new leaders and initiatives, especially in those communities that are in greatest need; and gain knowledge that will help nonprofit leaders and GNOF staff develop better long-term strategies for addressing regional needs and taking best advantage of important opportunities. IMPACT grants are awarded in four categories: Arts and Culture, Youth Development, Education, and Health and Human Services. In the category of Health and Human Services, special funding is available for organizations that provide health-care assistance to residents of New Orleans whose U.S. adjusted gross income for the preceding tax year, when added to any tax-exempt income and income from a spouse for that same year, is at least $75,000 but not more than $200,000. The health-care assistance must

be for cancer, heart disease, or Alzheimer's. Interested organizations must submit a letter of intent along with all attachments via one email by 5 p.m. on July 30 and should indicate on the IMPACT application cover sheet that they are applying for funding from the Harold W. Newman, Jr. Charitable Trust. GNOF program staff will review all letters of intent and will contact those organizations that are invited to submit a full application for funding. Awards are announced in November. Deadlines may vary from year to year. Interested organizations should verify the current deadline at the GNOF website where they can also obtain complete guidelines and requirements as well as a downloadable application form and cover sheet. Prospective applicants can also subscribe to GNOF's email newsletter for announcements of future funding opportunities. The subscription link is available from GNOF's Apply-for-a-Grant web page which contains a comprehensive listing of GNOF's current and past funding opportunities.
Requirements: Nonprofit, tax-exempt organizations that serve the thirteen parishes of Greater New Orleans are eligible to apply for funding. Organizations that are not tax-exempt but have a fiscal agent relationship with a 501(c)3 organization are also eligible.
Restrictions: Through its IMPACT program, the Greater New Orleans Foundation is unable to fund the following types of requests: requests for support from individuals, either through scholarships or other forms of financial assistance; special events or conferences; programs that promote religious doctrine; endowments; and scientific or medical research.
Geographic Focus: All States
Date(s) Application is Due: Jul 30
Amount of Grant: Up to 20,000 USD
Contact: Roy Williams; (504) 598-4663; fax (504) 598-4676; grants@gnof.org
Internet: http://www.gnof.org/programs/impact/
Sponsor: Greater New Orleans Foundation
1055 St. Charles Avenue, Suite 100
New Orleans, LA 70130

GNOF IMPACT Kahn-Oppenheim Trust Grants **1842**
Through the IMPACT Program, the Greater New Orleans Foundation (GNOF) makes grants to organizations serving the Greater New Orleans region. The ultimate goal of the IMPACT program is to create a resilient, sustainable, vibrant, and equitable region in which individuals and families flourish and in which the special character of the New Orleans region and its people is preserved, celebrated, and given the means to develop. Specifically GNOF hopes to accomplish the following objectives through its IMPACT grants: provide a much needed source of financial and other support to nonprofit organizations that are struggling in the current financial environment and that are important to the health and vibrancy of the region; develop a better sense of the nonprofit organizations serving the region so GNOF can more effectively match donor desires with effective charitable work; identify and nurture promising new leaders and initiatives, especially in those communities that are in greatest need; and gain knowledge that will help nonprofit leaders and GNOF staff develop better long-term strategies for addressing regional needs and taking best advantage of important opportunities. IMPACT grants are awarded in four categories: Arts and Culture, Youth Development, Education, and Health and Human Services. In the category of Health and Human Services, special funding is available for the development and/or improvement of public-health outreach and education programs to inform people about ways to prevent diseases like asthma, diabetes, heart disease, obesity, HIV/AIDS, and others, insofar as these programs involve physical, nutritional, or dietary regimens. Interested organizations must submit a letter of intent along with all attachments via one email by 5 p.m. on July 30 and should indicate on the IMPACT application cover sheet that they are applying for funding from the Kahn-Oppenheim Trust. GNOF program staff will review all letters of intent and will contact those organizations that are invited to submit a full application for funding. Awards are announced in November. Deadlines may vary from year to year. Interested organizations should verify the current deadline at the GNOF website where they can also obtain complete guidelines and requirements as well as a downloadable application form and cover sheet. Prospective applicants can also subscribe to GNOF's email newsletter for announcements of future funding opportunities. The subscription link is available from GNOF's Apply-for-a-Grant web page which contains a comprehensive listing of GNOF's current and past funding opportunities.
Requirements: Nonprofit, tax-exempt organizations that serve the thirteen parishes of Greater New Orleans are eligible to apply for funding. Organizations that are not tax-exempt but have a fiscal agent relationship with a 501(c)3 organization are also eligible.
Restrictions: Through its IMPACT program, the Greater New Orleans Foundation is unable to fund the following types of requests: requests for support from individuals, either through scholarships or other forms of financial assistance; special events or conferences; programs that promote religious doctrine; endowments; and scientific or medical research.
Geographic Focus: All States
Date(s) Application is Due: Jul 30
Amount of Grant: Up to 20,000 USD
Contact: Roy Williams; (504) 598-4663; fax (504) 598-4676; grants@gnof.org
Internet: http://www.gnof.org/programs/impact/
Sponsor: Greater New Orleans Foundation
1055 St. Charles Avenue, Suite 100
New Orleans, LA 70130

GNOF Maison Hospitaliere Grants **1843**
In 1879 Coralie Correjolles organized 30 women into "La Socíeté Hospitaliere des Dames Louisianaises" to provide food and medicine to the needy of New Orleans, many of whom had lost everything during the Civil War. The group became especially concerned by the plight of elderly ladies, who, due to the loss of their husbands in the war, were destitute and living in squalid conditions. Through its collection of 10 cent monthly dues over 14 years, the Socíeté was able to raise the money for its first building, 822 Barracks Street, to provide residence for 20 women. Over the next 113 years Maison Hospitaliere evolved into a skilled nursing facility for both men and women. Hurricane Katrina scattered both residents and staff across the country, and in November 2006 the board decided to close the facility. When the Maison Hospitaliere sold its French Quarter complex for more than $4 million, the proceeds were incorporated into a Supporting Organization of the Greater New Orleans Foundation so that the Maison's mission could continue by making grants to organizations serving women and their families.
Requirements: Grants will be made available to 501(c)3 organizations that provide living assistance and care to indigent women in the Greater New Orleans area. These grants will support direct services to women in the form of either general operating support or program support. Grants will range up to $20,000. Proposals must be submitted by 5:00 pm of the deadline date.
Restrictions: Maison Hospitaliere will not consider capital projects, event sponsorship, or research requests. Faith-based organizations are welcome to apply for support for programs that do not include religious activities, such as religious worship, instruction, or proselytization.
Geographic Focus: Louisiana
Date(s) Application is Due: Sep 1
Amount of Grant: Up to 20,000 USD
Contact: Roy Williams; (504) 598-4663; fax (504) 598-4676; grants@gnof.org
Internet: http://www.gnof.org/maison-hospitaliere/
Sponsor: Greater New Orleans Foundation
1055 St. Charles Avenue, Suite 100
New Orleans, LA 70130

GNOF New Orleans Works Grants **1844**
New Orleans Works (NOW) is a public-private partnership initiative housed at the Greater New Orleans Foundation (GNOF) and funded by a grant from the National Fund for Workforce Solutions (NFWS), a $31-million, five-year effort to fuel high-impact workforce partnerships and to advance 30,000 workers in 32 regions in the U.S., including the Greater New Orleans area. Led by foundations and regional public workforce systems, NOW pools funding, develops strategy, supports economic-sector-based programs and develops workforce partnerships that meet both the career-advancement needs of workers and the workforce needs of employers. Although the people of the Greater New Orleans area are creative, hardworking, and dedicated, the region's economic performance has persistently fallen below its potential. Too often residents are held back by failures in public education, the legacy of racism, and underinvestment in workforce development. Devastating hurricanes in recent years have battered the economy, inflicting costly damage to businesses and infrastructure and spurring the relocation of major corporations and large employers. Furthermore, the economic turmoil caused by the 2010 oil disaster in the Gulf of Mexico demonstrated the perils of the region's over-reliance on too few industries. To attract jobs that provide sustainable family incomes to the area, NOW takes a two-tiered approach, working both with employers and with education and training providers to create a skilled workforce that is aligned with employers' needs, sector by sector. NOW promotes individual, institutional, and system-wide change in order to achieve the following goals: connecting residents with existing and emerging career opportunities; building long-term relationships between employees, employers, and training providers; expanding the minority middle class; creating an enhanced public workforce system; providing a larger pool of skilled workers; and transforming the many lessons learned in community and environmental resilience into economic opportunities and new industries that can bring the region to the forefront of innovation. GNOF accepts NOW applications through its Request for Proposal (RFP) process. Interested organizations are encouraged to subscribe to GNOF's email newsletter for announcements of future funding opportunities. The subscription link is available from GNOF's Apply-for-a-Grant web page which contains a comprehensive listing of GNOF's current and past funding opportunities. NOW's RFP deadlines and economic-sector focus may vary from offering to offering. Interested organizations are encouraged to visit the GNOF website to obtain more detailed and up todate information on the initiative and its RFP process and to contact the NOW Site Director with any questions.
Requirements: Non-profit, not-for-profit, and for-profit organizations are eligible to apply. Job providers should select a training partner that has the sector knowledge and capacity to execute a training intervention that addresses their workforce-development needs. The funding must be awarded to an entity with the financial management system capability to accept federal funding.
Restrictions: Programs and partnerships funded by the NOW initiative must benefit the Greater New Orleans area.
Geographic Focus: All States
Date(s) Application is Due: Oct 12
Amount of Grant: Up to 250,000 USD
Contact: Bonita Robertson, New Orleans Works Interim Site Director; (504) 598-4663 ext. 40; fax (504) 598-4676; bonita@gnof.org
Danny Murphy, (504) 598-4663; fax (504) 598-4676; danny@gnof.org
Internet: http://www.gnof.org/new-orleans-works/
Sponsor: Greater New Orleans Foundation
1055 St. Charles Avenue, Suite 100
New Orleans, LA 70130

GNOF Norco Community Grants **1845**
Shell Chemicals and Motiva Enterprises, grant-making partners of the Greater New Orleans Foundation (GNOF), have established the Norco Community fund to improve the quality of lives for people in Norco, Louisiana. The fund provides the following types of grants: capital-fund grants for new construction or major renovation; seed-money grants to help start new organizations which respond to an important opportunity in the community; bridge grants to sustain organizations experiencing financial hardships; program grants that support new, creative, or beneficial programs; and grants to

organizations with a positive track record. Areas of interest include arts and humanities, community development, education, environment, human services, health care, community building, and youth development. Grant requests are reviewed by an advisory committee of business and civic leaders that meets annually. Application materials must be postmarked no later than September 14 and mailed to the Norco address given under the Contact Information section. If the deadline falls on a weekend or holiday, then the materials must be postmarked by the weekday immediately following; grant requests received after the deadline will be reviewed in the next year's grant cycle. Exact deadlines may vary from year to year. Prospective applicants should verify the current deadline at the GNOF website where they can also obtain complete guidelines and requirements as well as a downloadable application form. Prospective applicants can also subscribe to GNOF's email newsletter for announcements of future funding opportunities. The subscription link is available from GNOF's Apply-for-a-Grant web page which contains a comprehensive listing of GNOF's current and past funding opportunities.
Requirements: Applicants must have 501(c)3 status or apply through a fiscal agent who has such status.
Restrictions: The fund will only consider support for programs that serve the Norco community and its residents.
Geographic Focus: Louisiana
Date(s) Application is Due: Sep 14
Contact: Program Coordinator
Ellen M. Lee, Sr. Vice President, Programs, Community Revitalization Program Director; (504) 598-4663; fax (504) 598-4676; ellen@gnof.org
Internet: http://www.gnof.org/norco-community-fund/
Sponsor: Greater New Orleans Foundation
1055 St. Charles Avenue, Suite 100
New Orleans, LA 70130

GNOF Organizational Effectiveness Grants and Workshops 1846
In the wake of Hurricane Katrina and the levee failures, many new organizations have sprung up in the greater New Orleans region to address the immediate and pressing needs of recovery at the neighborhood level and up. This surge in activism and engagement holds great promise for the region, but significant issues must be addressed to give this work the greatest impact. Often, dedicated nonprofit leaders and professionals struggle with inexperienced boards, overwhelming fundraising responsibilities, and a lack of resources to develop their own infrastructure and talent. In response, the Greater New Orleans Foundation (GNOF) supports emerging leaders and organizations, empowers organizations in the region to be more competitive in their bids for state and federal funding, and serves as a convener to build connections and relationships within the greater New Orleans nonprofit sector. In partnership with the Marguerite Casey Foundation, GNOF provides training programs for staff and board members of nonprofits in the region to learn new ways of conducting fundraising, working with boards, and managing communications. In partnership with the Kellogg Foundation, GNOF provides technical-assistance grants of up to $4,000 to help nonprofit staff and/or board members increase their capacity to lead, manage, and govern their organizations. Following are examples of ways in which these technical-assistance grants can be used: working with a consultant to assist staff and board in development of fundraising or strategic plans for the organization; hiring a facilitator for a board retreat to grow governance abilities; covering expenses for attendance at a workshop or training session on a specific topic, such as evaluation, strategic communications, financial management, or fundraising; or completing an organizational assessment. All of these forms of technical assistance share the outcome that the grantee organizaqtion's staff and/or board members will be actively involved and will acquire new skills or information that will help the organization to grow and improve. GNOF accepts technical-assistance grant requests on a first-come, first-serve basis until the funding runs out. Priority will be given to requests pertaining to GNOF's current area of focus (advocacy, board governance, evaluation, financial management, fundraising, succession planning, partnering and collaboration, or fundraising). To apply for a GNOF technical-assistance grant, applicants should submit a two-page request along with required supporting documents via email to GNOF's Program Officer for Organizational Effectiveness. A decision will be made on the request approximately four weeks from the date that the request letter is received. Complete guidelines for submission are available at the GNOF website as well as links to capacity-building resources. Prospective applicants are welcome to contact the GNOF Program Officer with any additional questions. To receive further information about GNOF's workshops for nonprofits, interested organizations in the region should check the GNOF website or contact GNOF's Vice-President of Organizational Effectiveness.
Requirements: To be eligible to apply for GNOF's technical-assistance grants, organizations must be current or former (within the last two years) recipients of GNOF discretionary grants and have a primary office located within GNOF's 13-parish service area (Assumption, Jefferson, Lafourche, Orleans, Plaquemines, St. Bernard, St. Charles, St. James, St. John the Baptist, St. Tammany, Tangipahoa, Terrebonne, and Washington). Organizations that have previously received a GNOF technical-assistance grant may apply for additional grants in subsequent years (given funding availability); successful recurring requests will show that the applicant has built upon previous strategies.
Restrictions: Technical-assistance grants may not be used to pay a board member or any party whose direct affiliation with the applicant organization could be construed as or would create a conflict of interest. While prospective applicants may have highly competent professional resources on their boards, GNOF would expect these resources to be provided as an in-kind donation. Technical-assistance grants will not be made for activities that have already occurred or are underway at the time the grant is awarded.
Geographic Focus: Louisiana
Amount of Grant: Up to 4,000 USD
Contact: Kellie Chavez Greene; (504) 598-4663; fax (504) 598-4676; kellie@gnof.org
Internet: http://www.gnof.org/organizational-effectiveness-technical-assistance-grants/
Sponsor: Greater New Orleans Foundation
1055 St. Charles Avenue, Suite 100
New Orleans, LA 70130

GNOF Stand Up For Our Children Grants 1847
In New Orleans almost one in two children under the age of five lives at or below the federal poverty level. The Greater New Orleans Foundation (GNOF) believes that parental involvement is the best change agent for improving conditions for children. To help address the issue, GNOF, in partnership and with assistance from the W.K. Kellogg Foundation, formed an initiative called Stand Up for Our Children. It identifies and invests in nonprofit organizations that train parents to develop leadership skills that enable them to become more effective advocates, essentially helping their voices to be heard. In 2012 a total of $575,366 was awarded to ten organizations for their success working with parents and advocating for families. In addition to receiving grants, all grantees participated in a learning community designed to share knowledge, foster coalitions and alliances, and document lessons learned. The second round of grants from the Stand Up for Our Children Initiative will take place in 2013. These grants are by invitation only. Interested organizations should contact the GNOF Program Officer (see Contact Information Section) and check the GNOF website for updates. Prospective applicants can also subscribe to GNOF's email newsletter for announcements of future funding opportunities. The subscription link is available from GNOF's Apply-for-a-Grant web page which contains a comprehensive listing of GNOF's current and past funding opportunities.
Geographic Focus: Louisiana
Amount of Grant: 40,000 - 130,000 USD
Contact: Flint D. Mitchell, Ph.D.; (504) 598-4663; fax (504) 598-4676; flint@gnof.org
Internet: http://www.gnof.org/stand-up-for-our-children-initiative/
Sponsor: Greater New Orleans Foundation
1055 St. Charles Avenue, Suite 100
New Orleans, LA 70130

Goddess Scholars Grants 1848
The purpose of these grants is to develop young researchers at the entry faculty level (Instructor or Assistant Professor) who will focus a research career on studying the unique aspects of stroke in women. Such research may be basic or clinical in orientation, including areas such as mechanisms of ischemic brain injury, neuroprotective treatments, primary or secondary stroke prevention, epidemiological studies, educational programs for patients or health care providers, rehabilitation, and outcome research. The grant provides funding for two years, with potential non-competitive renewal for a third year should satisfactory progress be demonstrated. These grants are intended to provide bridging funds for individuals who are finishing or have completed residency or postdoctoral training and require additional support before their subsequent establishment as independent investigators.
Requirements: Goddess Scholar grants are targeted at individuals who: hold an M.D., Ph.D., or equivalent degree; are at the fellow, instructor, or assistant professor rank; and are within 5 years of completion of residency or post-doctoral fellowship at the time of grant activation. Individuals who are currently enrolled in a fellowship program may apply if there is assurance of a faculty position at the time of the grant's activation.
Geographic Focus: All States
Amount of Grant: 65,000 USD
Contact: Erin Tower; (212) 713-6789; fax (212) 288-2160; erin@thegoddessfund.org
Internet: http://www.thegoddessfund.org/grants.html
Sponsor: Hazel K. Goddess Fund for Stroke Research in Women
785 Park Avenue
New York, NY 10021-3552

Godfrey Foundation Grants 1849
The Foundation, established in Wisconsin in 1945 and supported by a donation from the Fleming Companies, Inc., offers grants primarily in the Milwaukee and Waukesha, Wisconsin, region. Its major field of interest include: education; health care; human services; community development and services; and youth development programs. The major type of support comes in the form of general operations funding. An application form is required, and interested organizations should contact the Foundation for a copy. There are no particular deadlines with which to adhere, and grant amounts have ranged from $500 to $30,000.
Requirements: Applicants must be 501(c)3 organizations serving residents of Wisconsin.
Geographic Focus: Wisconsin
Amount of Grant: 500 - 30,000 USD
Contact: Terry L. Daniels, Director; (262) 275-0458
Sponsor: Godfrey Foundation
680 Kenosha Street, P.O. Box 810
Walworth, WI 53184-0810

Golden Heart Community Foundation Grants 1850
The Golden Heart Community Foundation's grantmaking priorities are still in development, but it will support projects that strengthen the Fairbanks community. The Foundation will include organizations and programs that serve youth, the elderly, recreation, safety, vulnerable populations, and arts and culture. Preference will be given to applications which have the potential to impact a broad range of area residents. Applications should describe measurable outcomes and other sources of support, collaboration and/or cooperation. Applications should also address the sustainability of the proposed program or project for which funding is desired. Currently, no annual deadlines have been established.

Requirements: The Foundation seeks applications from qualified tax-exempt 501(c)3 organizations (or equivalent organizations) in the greater Fairbanks area. Equivalent organizations may include tribes, local or state governments, schools, or Regional Educational Attendance Areas.
Restrictions: Individuals, for profit, and 501(c)4 and (c)6 organizations, non-Alaska based organizations and state or federal government agencies are not eligible for competitive grants. Applications for religious indoctrination or other religious activities, endowment building, deficit financing, fundraising, lobbying, electioneering and activities of a political nature will not be considered, nor will proposals for ads, sponsorships, or special events.
Geographic Focus: Alaska
Contact: Ricardo Lopez; (907) 249-6707; fax (907) 334-5780; rlopez@alaskacf.org
Internet: http://goldenheartcf.org/grants-community-projects/
Sponsor: Golden Heart Community Foundation
P.O. Box 73183
Fairbanks, AK 99707-3183

Golden State Warriors Foundation Grants 1851
The foundation committed to positively impacting the communities of Oakland and the greater San Francisco Bay Area by providing financial assistance and opportunities to other non-profit civic and community organizations that benefit and enrich the lives of children, youth and those in need. By providing financial support and unique resources, the Foundation endeavors to meet the social, educational and cultural needs of the community. Although the Warriors Foundation will not be accepting grant applications until the beginning of the 2007-08 fiscal year, it remains committed to providing assistance and in-kind support to Bay Area nonprofit organizations through the donation of tickets, autographed memorabilia, promotional items, sponsorships, and player and talent appearances. Application information and forms are available on the Web site.
Requirements: 501(c)3 nonprofit organizations that provide programs and services in the greater San Francisco Bay area are eligible.
Restrictions: Grants do not support individuals; schools; political, labor, religious, or fraternal activities; endowments; fundraising events; government agencies; or organizations with discriminatory or unlawful practices.
Geographic Focus: California
Amount of Grant: Up to 375,000 USD
Contact: Angela Cohan, President; (510) 986-5307; community_web@gs-warriors.com
Internet: http://www.nba.com/warriors/community/Warriors_Foundation-011024.html
Sponsor: Golden State Warriors Foundation
1011 Broadway
Oakland, CA 94607-4019

Goldhirsh Foundation Brain Tumor Research Grants 1852
The Goldhirsh Foundation is interested in providing strategic investment in both pediatric and adult brain tumor research to accelerate progress toward more effective treatment for malignant diffuse glioma tumors. Applications are encouraged from investigators working in the continuum between basic research and clinical application, integrating and translating knowledge in various disciplines into meaningful progress for patients. The Initial Proposal Application must be submitted by Thursday, January 7, 12:00 Noon, U.S. and Canada Eastern Time (GMT-5), to the online application system. Applicants who are invited to submit Full Proposals will be notified by email in early March. Up to three investigators will receive grants of $400,000 USD (inclusive of 10% indirect costs) over two years at $200,000 USD per year. The Foundation also makes awards of $100,000 USD (inclusive of 10% indirect costs) for one-year pilot studies. Please note that The Goldhirsh Foundation is not offering the three-year $600,000 USD award for the 2010 Grant Cycle.
Requirements: The Goldhirsh Foundation is accepting Initial Proposal Applications from applicants who meet the following eligibility *Requirements:* have an M.D. and/or Ph.D. degree(s) or equivalent degree; hold a faculty appointment at a nonprofit academic, medical or research institution in the United States, Canada or Israel. The applicant may collaborate with investigators from other institutions and these institutions may include for-profit companies; preference will be given to originality of ideas, regardless of applicant seniority; eligibility is not limited to those investigators currently working in brain tumor research. Investigators from other fields are encouraged to apply with proposals directly relevant to malignant diffuse glioma tumors; projects must be relevant to tumors within the category of malignant diffuse gliomas, i.e., glioblastomas, diffuse astrocytomas (WHO grade III), oligoastrocytomas (WHO grade III), and oligodendrogliomas (WHO grade III).
Restrictions: Projects focusing on WHO grade I-II gliomas, non-glial neuroepithelial tumors, meningeal tumors, nerve sheath tumors, CNS lymphomas, germ-cell tumors and tumors of the sella region are beyond the scope of the grant program. Projects that characterize genes and gene products of normal cellular development are not eligible, nor are epidemiological studies.
Geographic Focus: All States
Date(s) Application is Due: Jan 7
Amount of Grant: Up to 600,000 USD
Samples: Lara Collier, Ph.D., University of Wisconsin-Madison, $600,000-- Genetic Approaches to Identify and Characterize Potential Drug Targets for High-grade Gliomas; Al Charest, Ph.D., Tufts Medical Center, $100,000--Targeting Cancer Stem Cells in a Pre-Clinical Mouse Model of Glioblastoma Multiforme Using Therapeutic Nanoparticles.
Contact: Sally E. McNagny, Director; (617) 279-2254; fax (617) 423-4619; smcnagny@goldhirshfoundation.org
Internet: http://goldhirshfoundation.org/application_information.htm
Sponsor: Goldhirsh Foundation
95 Berkeley Street, Suite 201
Boston, MA 02116

Goldman Philanthropic Partnerships Program 1853
Since its founding, the Program has solicited, validated and helped donors co-fund innovative research with the potential to quickly find cures for life-altering diseases, and supported these donors using for-profit business tools to insure that the research and philanthropic goals are met through proper project and resource management. The Program provides funding research for cures for a range of catastrophic diseases by exploring groundbreaking research in all disciplines of medicine, including those areas of research outside of conventional medicine such as alternative medicine. The Program usually co-funds a venture with at least one organizational, institutional, or corporate partner. The Partnerships maintains a strong interest in diseases of children and young adults, although it will fund research into any patient population where it believes there is a high likelihood of return on disease prevention, treatment or cure.
Geographic Focus: All States
Contact: Dr. Bruce E. Bloom, President; (312) 601-8856 or (312) 780-3440; fax (312) 780-3459; bruce@goldmanpartnerships.org
Internet: http://www.goldmanpartnerships.org/aboutus.html
Sponsor: Goldman Philanthropic Partnerships
70 West Madison Street, Suite 1500
Chicago, IL 60602

Goodrich Corporation Foundation Grants 1854
The Foundation makes charitable grants in four categories: education; arts and culture; civic and community; and health and human services. Preference shall be given to requests for projects or programs in areas having a significant number of employees, employees serving on boards of charitable organizations or other noticeable corporate presence. The Foundation staff accepts and reviews grant requests quarterly. To request funding, applicants will need to complete the application form which is available online. Telephone and email requests or inquiries are not accepted.
Requirements: 501(c)3 tax-exempt organizations are eligible.
Restrictions: The foundation generally will not support: multiyear grants in excess of five years; individuals, private foundations, endowments, churches or religious programs, fraternal/social/ labor/veterans organizations; groups with unusually high fundraising or administrative expenses; political parties, candidates, or lobbying activities; travel funds for tours, exhibitions, or trips by individuals or special interest groups; organizations that discriminate because of race, color, religion, national origin, or areas covered by applicable federal, state, or local laws; local athletic/sports programs or equipment, courtesy advertising benefits, raffle tickets and other fundraising events; organizations that receive sizable portions of their support through municipal, county, state, or federal dollars; individual United Way agencies that already benefit from Goodrich contributions to the United Way; or international organizations.
Geographic Focus: All States
Date(s) Application is Due: Feb 1; Aug 1
Contact: Foundation Contact, Foundation Contact; (704) 423-7011
Internet: http://www.goodrich.com/CDA/GeneralContent/0,1136,59,00.html
Sponsor: Goodrich Foundation
Four Coliseum Center, 2730 West Tyvola Road
Charlotte, NC 28217-4578

Good Samaritan Inc Grants 1855
Good Samaritan, Inc. is an independent foundation, which was incorporated in 1938, in Delaware. This foundation gives primarily in the northeastern region of the United States: Pennsylvania, New Jersey, Delaware and Maryland. The foundation field of interest include: higher education; cancer research; secondary school/education; human services; hospitals (general); and the economically disadvantaged. Support is offered in the form of: endowments; professorships; program development; seed money: and general/operating support grants.
Requirements: 501(c)3 nonprofit organizations are eligible.
Restrictions: No grants to individuals, or for building funds, capital assets or conferences.
Geographic Focus: All States
Date(s) Application is Due: Apr 1; Oct 1
Amount of Grant: 10,000 - 200,000 USD
Contact: Edmund Carpenter II, President; (302) 654-7558; fax (302) 654-2376
Sponsor: Good Samaritan
600 Center Mill Road
Wilmington, DE 19807-1502

Goodyear Tire Grants 1856
Requests for charitable gifts are required to meet guidelines of strategic giving based on safety by focusing on safety programs plus at least one additional category (civic and community; culture and the arts, education, or health and human services) of giving and/or provide volunteer opportunities for company associates which are appropriately aligned with the company's strategy.
Requirements: 501(c)3 tax-exempt organizations in Goodyear communities are eligible. Organizations in the greater Akron, OH area should send requests to: Faith Stewart, Director, Goodyear Tire and Rubber Company, 1144 East Market Street, D/798, Akron, OH, 44316. Organizations in U.S. communities where Goodyear has plants and major facilities should contact the local plant manager's office for mailing instructions.
Restrictions: Grants cannot be made to: organizations outside of communities in which plants or principal offices are located; individuals or to organizations on behalf of an individual; national, political, labor, fraternal, social, veterans organizations; individual schools within public or private school systems; athletic programs/extracurricular activities, travel or exchange programs; endowments or for debt reduction; religious organizations or endeavors; major hospitals or medical center; second parties except specialized organizations.

Geographic Focus: All States
Contact: Faith Stewart, Director
Internet: http://www.goodyear.com/corporate/about/about_community.html
Sponsor: Goodyear Tire and Rubber Company
1144 East Market Street D/798
Akron, OH 44316

Google Grants Beta 1857

The Google Grants program supports organizations sharing their philosophy of community service to help the world in areas such as science and technology, education, global public health, the environment, youth advocacy, and the arts. Google Grants has awarded AdWords advertising to non-profit groups whose missions range from animal welfare to literacy, from supporting homeless children to promoting HIV education. Recipients use their award of free AdWords advertising on Google.com to raise awareness and increase traffic. Application available online only.
Requirements: In the United States - Organizations must have current 501(c)3 status, as assigned by the Internal Revenue Service to be considered for a Google Grant. Outside the U.S - currently accepting applications from eligible charitable organizations based in Australia, Brazil, Canada, Denmark, France, Germany, India, Ireland, Italy, Japan, the Netherlands, Spain, Sweden, Switzerland and the UK.
Restrictions: Organizations already participating in the Google AdSense program are not eligible for Google Grants consideration. In addition, organizations that are either religious or political in nature are not eligible, including those groups focused primarily on lobbying for political or policy change.
Geographic Focus: All States
Sample: Three award recipients have achieved these results: - Room to Read, which educates children in Vietnam, Nepal, India and Cambodia, attracted a sponsor who clicked on its AdWords ad. He has donated funds to support the education of 25 girls for the next 10 years. - The U.S. Fund for UNICEF's e-commerce site, Shop UNICEF, has experienced a 43 percent increase in sales over the previous year. - CoachArt, supporting children with life-threatening illnesses through art and athletics programs, has seen a 60 to 70 percent increase in volunteers.
Contact: Google Grants Team Contact; googlegrants@google.com
Internet: http://www.google.com/grants/
Sponsor: Google
1600 Amphitheatre Parkway
Mountain View, CA 94043

Grace and Franklin Bernsen Foundation Grants 1858

The Grace and Franklin Bernsen Foundation provide grants primarily within the metropolitan Tulsa, Oklahoma area. Areas of interest supported are: religious; charitable; scientific; literary and; educational purposes. Grant applications for programs and projects that will provide a defined benefit such as capital projects, building programs, specific program needs or ongoing operations from time to time are all considered. Grant applications from elementary or secondary education institutions will be considered if they involve programs for at-risk, handicapped or learning-disabled children; or if they are innovative and apply to all schools in the system.
Requirements: Applicant should submit a narrative summary no more than three pages in length. This letter should be addressed to the Trustees of the Foundation, see website for detailed description of narrative requirements. IRS 501(c)3 non-profit organizations in the metropolitan Tulsa, Oklahoma area are eligible to apply. Application may be made at any time for support of activities consistent with the Foundation's guidelines. The Foundation's Board meets monthly to review grant applications. Applications received on or before the 12th day of the month (unless the 12th falls on a Saturday, Sunday, or a holiday, in which event applications are due on the preceding business day) preceding a next regularly scheduled Board meeting are normally considered at such meeting.
Restrictions: Grant applications for individuals are not considered. The Foundation discourages applications for general support or reduction of debt, or for continuing or additional support for the same programs. The Foundation will only review one grant request per agency during our fiscal year
Geographic Focus: Oklahoma
Amount of Grant: 2,000 - 1,000,000 USD
Contact: Margaret Skyles, Administrator; (918) 584-4711; fax (918) 584-4713; mskyles@bernsen.org or info@bernsen.org
Internet: http://www.bernsen.org/grant.html
Sponsor: Grace and Franklin Bernsen Foundation
15 West Sixth Street, Suite 1308
Tulsa, OK 74119-5407

Grace Bersted Foundation Grants 1859

The Grace Bersted Foundation was established in 1986 to support and promote quality education, human services, and health care programming for under-served populations. Special consideration is given to charitable organizations that serve the needs of children or the disabled. Applicants must apply online at the grant website. Applicants are strongly encouraged to do the following before applying: review the downloadable state application procedures for additional helpful information and clarifications; review the downloadable online-application guidelines at the grant website; review the foundation's funding history (link is available from the grant website); review the online application questions in advance; and review the list of required attachments. These will generally include: a list of board members, financial statements (audited, reviewed, or compiled by independent auditor); an organization summary; a list of other funding sources; an IRS Determination letter; and other required documents. All attachments must be uploaded in the online application as PDF, Word, or Excel files. The annual deadline for application to the Foundation is August 1, and grant decisions will be made by November 1. Most recent awards have ranged from $5,000 to $50,000.
Requirements: Applicant organizations must have 501(c)3 tax-exempt status and an office located in one of the following counties: DuPage, Kane, Lake, or McHenry.
Restrictions: The foundation does not support requests from individuals, organizations attempting to influence policy through direct lobbying, or any political campaigns.
Geographic Focus: Illinois
Date(s) Application is Due: Aug 1
Amount of Grant: 5,000 - 50,000 USD
Samples: Saint Martin de Porres High School, Waukegan, Illinois, $50,000 - general operating support (2014); Easter Seals DuPage and the Fox Valley Region, Villa Park, Illinois, $23,000 - program development (2014); Northern Illinois Food Bank. Geneva, Illinois, $20,000 - program development (2014).
Contact: Debra L. Grand; (312) 828-2055; ilgrantmaking@ustrust.com
Internet: https://www.bankofamerica.com/philanthropic/foundation.go?fnId=58
Sponsor: Grace Bersted Foundation
231 South LaSalle Street, IL1-231-13-32
Chicago, IL 60604

Graco Foundation Grants 1860

The Foundation's goal is to help organizations grow their ability to serve community needs through grants specifically aimed at expanding or enhancing services to clients, with particular focus on capital projects and technology needs. The Foundation addresses the needs of the community in the following areas: self-sufficiency - emphasizing educational programs, human service programs that promote self-sufficiency, and sports/youth development programs; and civic projects - focusing on healthy communities. Projects that improve the community will be given special consideration. Application information is available online.
Requirements: IRS 501(c)3 organizations in company-operating areas may apply.
Restrictions: Grants are not awarded for: organizations or causes that do not directly impact Graco communities; political campaigns; individuals; religious organizations for religious purposes; fund-raising; travel; special events, dinners, courtesy advertising; fraternal organizations; national or local campaigns for disease research; first time general operating grants; or product donations.
Geographic Focus: Minnesota, Ohio, South Dakota
Date(s) Application is Due: Feb 1; May 1; Aug 1; Nov 1
Samples: Minneapolis Neighborhood Employment Network--for operating support, $3000; United Negro College Fund--for scholarships, $5000; City Inc--for renovations of its north Minneapolis facility, $125,000.
Contact: Kristin R. Ridley; (612) 623-6684; kridley@graco.com
Internet: http://www.graco.com/Internet/T_Corp.nsf/SearchView/Foundation2006
Sponsor: Graco Foundation
P.O. Box 1441
Minneapolis, MN 55440-1441

Graham and Carolyn Holloway Family Foundation Grants 1861

In 1995 the Graham and Carolyn Holloway Family Foundation was formed, providing a formal vehicle for a legacy of charitable giving that began on November 26, 1955 when Graham and Carolyn were first married. From a background of humble beginnings, the couple lived and taught the principles of generosity even before they knew what the word philanthropy meant. The Holloway Family Foundation has awarded over $2,000,000 in grants since its inception, seeking out agencies that are providing the most good in support of the most needy. The Foundation maintains a personal connection to its grantees, choosing carefully to respect the integrity of the donors' intent. The mission of the Foundation is to enhance the quality of life for those people in its communities who are least likely to be able to do that on their own; aiding primarily, but not exclusively: the elderly, individuals with developmental and/or physical disabilities, the chronically or terminally ill, and disadvantaged children. Awards typically range from $2,500 to $10,000. The deadline for consideration of a June distribution is March 15 and the deadline for consideration of a December distribution is October 15.
Requirements: Applicants must have a 501(c)3 status and serve residents of: the Stae of Texas; Colleyville, Texas; Nashville, Tennessee; or Salisbury, North Carolina. population. Recipients are limited to one grant per calendar year.
Geographic Focus: North Carolina, Tennessee, Texas
Date(s) Application is Due: Mar 15; Oct 15
Amount of Grant: 2,500 - 10,000 USD
Samples: Aberg Center for Literacy, Dallas, Texas, $2,500 - general operating support (2015); Helping Restore Ability, Arlington, Texas, $10,000 - general operating support (2015); Saddle Up!, Nashville, Tennessee, $5,000 - general operating support (2015).
Contact: Valerie Holloway Skinner, Vice President; (817) 313-9379; valerie@hollowayfamilyfoundation.org
Internet: http://www.hollowayfamilyfoundation.org/about.htm
Sponsor: Graham and Carolyn Holloway Family Foundation
P.O. Box 989
Colleyville, TX 76034-0989

Graham Foundation Grants 1862

The Graham Foundation was established in York, Pennsylvania, in 1986, and was primarily funded by support from both the Graham Architectural Products Corporation and Graham Engineering. Giving is centered throughout the State of Pennsylvania, though grants are sometimes approved for national organizations. The Foundation's primary fields of interest include: arts and culture; education; and human services. Awards are typically given in support of: annual campaigns; capital campaigns; infrastructure; general operations;

program development; sponsorships; and scholarships. There are no annual deadlines specified, and most recent awards have ranged from $200 to $140,000.
Requirements: 501(c)3 organizations serving the residents of York County, Pennsylvania, are eligible to apply.
Geographic Focus: Pennsylvania
Amount of Grant: 200 - 140,000 USD
Samples: Camp Tecumseh, Plymouth Meeting, Pennsylvania, $15,000 - scholarship funds; Cultural Alliances of York County, York, Pennsylvania, $11,000 - annual campaign; Eaglebrook School, Deerfield, Massachusetts, $40,000 - capital campaign.
Contact: William H. Kerlin, Jr., Trustee; (717) 849-4001 or (717) 849-4045
Sponsor: Graham Foundation
1420 Sixth Avenue, P.O. Box 1104
York, PA 17405-1104

Grammy Foundation Grants 1863
The foundation awards grants in two program areas. Research Projects grants are made to organizations and individuals to support efforts that advance the research and/or broad- reaching implementations of original research projects related to the impact of music study on early childhood development, the human development, and the medical and occupational well-being of music professionals. Archiving and Preservation Projects grants are made to organizations and individuals to support efforts that advance the archiving and preservation of the music and recorded sound heritage of the Americas. The foundation funds preservation of original, preexisting media and source material; preservation projects that follow the recommended methodology; projects of historical, artistic, cultural, and or/national significance; and archiving projects including the rescue, organization of, and access to preexisting media and materials. Application and guidelines are available online.
Restrictions: The foundation does not fund Recording Academy chapters, trustees, governors, officers, or staff; organizations that discriminate on the basis of race, sex, religion, national origin, disability, or age; projects promoting advocacy issues; a single organization or individual for more than three consecutive years; organizations or individuals not based in the Americas; purchase of collections; recording projects, demo tapes, or performance events; proposals for commercial purposes (i.e., CD reissue or textbook/ A/V package); purchase of repairs of equipment; purchase or repairs of musical instruments; maintenance or upgrading of computer systems; competitions or any expense associated with competitions; work toward academic degrees; music education or in-residence programs; documentaries; endowments and fundraising; buildings and facilities; marketing, publicity, design costs; or projects where copyright status is unknown.
Geographic Focus: All States
Date(s) Application is Due: Oct 1
Amount of Grant: 10,000 - 40,000 USD
Samples: The Institute for Music & Brain Science, Boston, MA, $20,000 - this project will test whether music decreases behavioral, neurophysiological and endocrinological pain and stress caused by medically-necessary procedures. In addition, the hypothesis that humans innately prefer consonant over dissonant music will be tested; Elliott Leib, San Diego, CA, $5,000 - a plan will be developed to digitally preserve material from the Trade Roots Reggae Collection; Bob Moog Memorial Foundation for Electronic Music, Asheville, N.C., $15,000 - musical and historical content relative to the unique legacy of synthesizer pioneer Dr. Robert Moog will be cleaned, restored, re-housed and transferred to digital format for accessibility and long-term storage.
Contact: Kristin Murphy, (310) 392-3777, ext. 8662; grants@grammy.com
Internet: http://www.grammy.com/GRAMMY_Foundation/Grants/
Sponsor: Grammy Foundation
3030 Olympic Boulevard
Santa Monica, CA 90405

Grand Haven Area Community Foundation Grants 1864
The community Foundation awards grants to meet the changing and emerging needs of residents of the tri-cities area in such fields as arts, education, health, the environment, youth, social services and other human needs. The Foundation encourages programs which: are collaborative, comprehensive and have potential for continuity; encourage leveraging and matching grant opportunities involving multiple funding sources; support seed money requests to assist innovative projects; encourages programs and projects focused on problem prevention rather than cure. Applicants are encouraged to call and talk with a staff member before beginning a proposal. Application information is available online.
Requirements: IRS 501(c)3 organizations working to enhance the quality of life for residents in the Tri-Cities area of Grand Haven, Spring Lake and Ferrysburg in West Michigan.
Restrictions: The Foundation does not generally fund: general operating expenses; private individuals; profit making organizations; elimination of existing financial obligations/debts/liabilities; religious programs that serve, or appear to serve, specific religious denominations; fund raising events.
Geographic Focus: Michigan
Date(s) Application is Due: Jan 7; Mar 13; Jun 26; Oct 10
Samples: City of Grand Haven, $20,000 - restoration of the Boardwalk for the 25th Anniversary of the Connector Park Boardwalk; Grand Haven Area Public Schools, $5,000 - to support pre-school scholarships; Tri-Cities Area Habitat for Humanity, $39,719 - seed money for the Grand Haven ReStore.
Contact: Carol Bedient, Grants and Programs Director; (616) 842-6378; fax (616) 842-9518; cbedient@ghacf.org or bpost@ghacf.org
Internet: http://www.ghacf.org/grants.htm
Sponsor: Grand Haven Area Community Foundation
1 South Harbor Drive
Grand Haven, MI 49417

Grand Rapids Area Community Foundation Grants 1865
The Foundation enables donors to make a difference in the community through grants, scholarships and programs. The deadline for applications is September 15th, and grants are distributed during the month of October. Specific funds include: Alzheimer's Family Support Fund; Fund for the Arts; Fund for the Community; Betty Kauppi Music Fund; Streuferrt Peace and Safety Fund; Warren Youngdahl Memorial Fund; and the Fund for Women. Amount of funding ranges from $500 to $3,000
Restrictions: No funding is available for individuals.
Geographic Focus: Minnesota
Date(s) Application is Due: Sep 15
Amount of Grant: Up to 3,000 USD
Contact: Wendy Roy; (218) 999-9100; fax (218) 999-7430; wroy@gracf.org
Internet: http://www.gracf.org/index.php/grants-and-programs
Sponsor: Grand Rapids Area Community Foundation
350 NW First Avenue, Suite E
Grand Rapids, MN 55744

Grand Rapids Area Community Foundation Nashwauk Endowment Grants 1866
The Fund is interested in supporting innovative and effective ideas that support the greater Nashwauk, Minnesota area. This area includes the areas of: Nashwauk, Nashwauk Township, Buck Lake and Pengilly. The main cycle offers grants of up to $15,000, with a deadline for application November 15th. The Mini-Grant cycle offers grants of up to $5,000, with the deadline for applications being September 15th. Nonprofit organizations that serve or benefit these areas are encouraged to submit a Letter of Inquiry.
Restrictions: Grants can only be made to charitable organizations, and cannot be made to individuals.
Geographic Focus: Minnesota
Date(s) Application is Due: Sep 15
Amount of Grant: 5,000 - 15,000 USD
Contact: Wendy Roy; (218) 999-9100; fax (218) 999-7430; wroy@gracf.org
Internet: http://www.gracf.org/index.php/grants-and-programs
Sponsor: Nashwauk Area Endowment Fund
350 NW First Avenue, Suite E
Grand Rapids, MN 55744

Grand Rapids Area Community Foundation Wyoming Grants 1867
The Wyoming Community Foundation gives priority to projects that address the areas of art & culture, community development, education, environment, health, or social needs and that: represent an innovative, start-up effort or are capital in nature (e.g., construction, renovation, equipment); promote cooperation among agencies without duplicating services; obtain the necessary additional funding to implement and maintain the project; serve the greater city of the Wyoming. Michigan area; strengthen or improve agency self-sufficiency or efficiency; yield substantial community benefits for the resources invested; serve a broad segment of the community; encourage additional and permanent funding or matching gifts from other donors; have non-profit, 501(c)3 status. The Foundation will consider proposals for the following only if they are highly unique, collaborative, and community-oriented, have limited access to other resources, demonstrate substantial impact, or will address the needs of a substantial or underserved portion of the community: hospitals; nursing or retirement facilities; K-12 schools; computers; child care centers; motorized vehicles; commonly accepted community services already supported by tax dollars.
Requirements: Michigan IRS 501(c)3 non-profit organizations in or serving the Wyoming, Michigan area are eligible to apply. General applications will be requested in Fall. For more information on how to request funding for projects in the City of Wyoming Michigan, please contact Lillian VanderVeen, Board Chair, at (616) 534-9625.
Restrictions: The Foundation generally does not support: annual fund-raising drives; films, videos or television projects; endowments or debt reduction; capital projects without site control; medical research; individuals; ongoing operating expenses of established institutions; organizations located outside greater Wyoming; political projects, or those that are primarily cause-related; religious organizations for religious purposes; sabbatical leaves; scholarly research; travel, tours or trips; underwriting of conferences; venture capital for competitive profit-making activities.
Geographic Focus: All States
Contact: Lillian VanderVeen, Board Chair; (616) 534-9625; lillian@lengertravel.com
Internet: http://www.grfoundation.org/wyoming
Sponsor: Grand Rapids Area Community Foundation-Wyoming Community Foundation
185 Oakes Street SW
Grand Rapids, MI 49503

Grand Rapids Area Community Foundation Wyoming Youth Fund Grants 1868
The Wyoming Community Foundation Youth Advisory Committee seeks to improve the community by teaching about the giving of time, talent, and treasure while funding programs that help youth gain advantages and build character. The Wyoming Community Foundation Youth Advisory Committee is offering student groups and local non-profit organizations (including schools, churches, and community groups with 501(c)3 non-profit status), an opportunity to apply for grants. These grants will be made with income derived from the Wyoming Community Foundation Youth Fund. Primary consideration will be given to projects that address drug and/or alcohol use, teen pregnancy, stress, smoking, and sexual abuse. Suggested projects or activities include: real life learning experiences, or experiential learning activities; competitive sports; arts, writing, or music-focused activities; programs that explore jobs, career options or job preparation; programs that teach self-defense strategies or martial arts; outdoor experiences, camps, or challenge courses; classes or groups to help quit smoking or using drugs. Typically,

the Request for Proposals is available late December of the year and the proposals are due mid-February. Grant awards are announced the following May.
Requirements: Student groups and local non-profit organizations (including schools, churches, and community groups with 501(c)3 non-profit status), are eligible to apply for these funding opportunities.
Geographic Focus: Michigan
Date(s) Application is Due: Dec 31
Contact: Cris Kooyer; (616) 454-1751, ext. 118; ckooyer@grfoundation.org
Internet: http://www.grfoundation.org/wyoming-yac.php
Sponsor: Grand Rapids Area Community Foundation-Wyoming Community Foundation
185 Oakes Street SW
Grand Rapids, MI 49503

Grand Rapids Community Foundation Grants 1869
The Foundation impacts a wide range of nonprofit organizations in the communities they serve. Funding is in six broad categories: academic achievement; high quality of life; healthy people; economic prosperity; vibrant neighborhoods; and environmental integrity.
Requirements: IRS 501(c)3 organizations supporting residents of Kent County are eligible to apply, applications are available at the Foundation's website.
Restrictions: The Foundation does not fund one-time, special or annual events, annual operating funds, political or religious causes, or endowments.
Geographic Focus: Michigan
Contact: Marcia Rapp, Vice President, Programs; (616) 454-1751, ext. 104; fax (616) 454-6455; mrapp@grfoundation.org or grfound@grfoundation.org
Internet: http://www.grfoundation.org/grants.php
Sponsor: Grand Rapids Community Foundation
185 Oakes Street SW
Grand Rapids, MI 49503

Grand Rapids Community Foundation Ionia County Grants 1870
The Ionia County Community Foundation gives priority to projects that address the areas of art & culture, community development, education, environment, health, or social needs and that: represent an innovative, start-up effort or are capital in nature (e.g., construction, renovation, equipment); promote cooperation among agencies without duplicating services; obtain the necessary additional funding to implement and maintain the project; are located in Ionia County, Michigan; strengthen or improve agency self-sufficiency or efficiency; yield substantial community benefits for the resources invested; serve a broad segment of the community; encourage additional and permanent funding or matching gifts from other donors; have non-profit, 501(c)3 status. The Foundation will consider proposals for the following only if they are highly unique, collaborative, and community-oriented; have limited access to other resources; demonstrate substantial impact; or will address the needs of a substantial or underserved portion of the community: hospitals; nursing or retirement facilities; K-12 schools; computers; child care centers; motorized vehicles; commonly accepted community services already supported by tax dollars.
Requirements: Non-profit, IRS 501(c)3 status organizations, located in Ionia County, Michigan may apply.
Restrictions: The Foundation generally does not support: annual fundraising drives; film, video or television projects; endowments or debt reduction; capital projects without site control; medical research; individuals; ongoing operating expenses of established institutions; organizations located outside Ionia County; political projects, or those that are primarily cause-related; religious organizations for religious purposes; sabbatical leaves; scholarly research; travel, tours or trips; underwriting of conferences; venture capital for competitive profit-making activities.
Geographic Focus: Michigan
Date(s) Application is Due: Dec 31
Contact: Kate Luckert Schmid; (616) 454-1751, ext. 117; kluckert@grfoundation.org
Internet: http://www.grfoundation.org/ionia-general.php
Sponsor: Grand Rapids Community Foundation
185 Oakes Street SW
Grand Rapids, MI 49503

Grand Rapids Community Foundation Ionia County Youth Fund Grants 1871
The Ionia County Community Foundation's Youth Advisory Committee is committed to addressing and improving problems concerning young people, by funding programs directed toward youth. The Youth Advisory Committee is affiliated with the Ionia County Community Foundation. Current funding priorities are: Drug, alcohol, tobacco use; Teen pregnancy; Depression/suicide; Parent education; Service-learning; Guest speakers that address youth issues; Literacy; Obesity; Teen driver safety. The Youth Advisory Committee is especially interested in youth projects that assist students in bridging the achievement gap. Grant proposals should assist the applying organization to implement a project or program that benefits youth. Proposals developed by youth or with youth involvement in planning are encouraged. The proposed project or program should: have clear goals that are specific, measurable, attainable, realistic, and timely; impact a significant number of Ionia County youth from birth to age 18; avoid duplication with other projects and programs in the community; address an issue and make a significant difference for youth. In addition, the applying organization and its participants will not be discriminated against on the basis of, but not limited to, gender, religion, physical disability, sexual orientation, or ethnicity. Grants will be made with income derived from the Ionia County Community Foundation Youth Fund.
Requirements: Student groups and local IRS 501(c)3 nonprofit organizations are eligible apply for grants.
Geographic Focus: Michigan
Date(s) Application is Due: Dec 31
Contact: Cris Kooyer; (616) 454-1751, ext. 118; ckooyer@grfoundation.org
Internet: http://www.grfoundation.org/ionia-yac.php
Sponsor: Grand Rapids Community Foundation
185 Oakes Street SW
Grand Rapids, MI 49503

Grand Rapids Community Foundation Lowell Area Fund Grants 1872
The Lowell Area Community Fund gives grants to organizations that assist in fulfilling its mission: to assure community cooperation and participation that supports a healthy, dynamic community. The fund places an emphasis on broad educational initiatives, but also supports initiatives in the areas of: Arts & Culture, Community Development, Environment, Health, Human Services, and Recreation. The Lowell area is currently defined as the City of Lowell, Michigan, the Township of Lowell, and the Township of Vergennes. Grant applications are accepted three times a year on the third Friday of April, August, and December. Additional guidelines and grant application are available at the Foundation's website.
Requirements: 501(c)3 tax-exempt organizations are in the Lowell area are eligible.
Restrictions: The fund generally does not support: political projects, or those that are primarily cause-related; individuals; religious organizations for religious purposes; or profit-making activities.
Geographic Focus: Michigan
Samples: Tots on Track for School, $74,000--to help families prepare their preschool children to enter school healthy and ready to learn; Lowell Area Schools Education Foundation, $20,000--to fund grants for innovative projects that support and enhance education in the Lowell Area Schools; Lowell Community Wellness, $10,000--to inspire and educate the greater Lowell community to attain a healthier lifestyle.
Contact: Kate Luckert Schmid; (616) 454-1751, ext. 117; kluckert@grfoundation.org
Internet: http://www.grfoundation.org/lowell.php
Sponsor: Grand Rapids Community Foundation
185 Oakes Street SW
Grand Rapids, MI 49503

Grand Rapids Community Foundation Southeast Ottawa Grants 1873
The Southeast Ottawa Community Foundation gives priority to projects that address the areas of art & culture, community development, education, environment, health, or social needs and that: represent an innovative, start-up effort or are capital in nature (e.g., construction, renovation, equipment); promote cooperation among agencies without duplicating services; obtain the necessary additional funding to implement and maintain the project; are located in the greater Georgetown Township, Hudsonville and Jamestown Township area of Michigan; strengthen or improve agency self-sufficiency or efficiency; yield substantial community benefits for the resources invested; serve a broad segment of the community; encourage additional and permanent funding or matching gifts from other donors; have non-profit, 501(c)3 status. The Foundation will consider proposals for the following only if they are highly unique, collaborative, and community-oriented, have limited access to other resources, demonstrate substantial impact, or will address the needs of a substantial or underserved portion of the community: hospitals; nursing or retirement facilities; K-12 schools; computers; child care centers; motorized vehicles; commonly accepted community services already supported by tax dollars.
Requirements: Michigan IRS 501(c)3 non-profit organizations that are located in the greater Georgetown Township, Hudsonville and Jamestown Township area are eligible to apply. General applications are due August 31st. For more information on how to request funding for projects, contact Larry Bergman at (616) 896-8769.
Restrictions: The Foundation generally does not support: annual fund-raising drives; films, videos or television projects; endowments or debt reduction; capital projects without site control; medical research; individuals; ongoing operating expenses of established institutions; organizations located outside greater Georgetown Township, Hudsonville and Jamestown Township area; political projects, or those that are primarily cause-related; religious organizations for religious purposes; sabbatical leaves; scholarly research; travel, tours or trips; underwriting of conferences; venture capital for competitive profit-making activities.
Geographic Focus: Michigan
Date(s) Application is Due: Aug 31
Contact: Larry Bergman, (616) 896-8769; lbergman@charter.net
Internet: http://www.grfoundation.org/seottawa
Sponsor: Grand Rapids Area Community Foundation-Southeast Ottawa Community Foundation
185 Oakes Street SW
Grand Rapids, MI 49503

Grand Rapids Community Foundation Southeast Ottawa Youth Fund Grants 1874
The Southeast Ottawa Community Foundation's Youth Advisory Committee is seeking to fund projects or programs that positively impact youth within Georgetown, Hudsonville, and Jamestown Townships. Groups or organizations made up of (or serving) youth ages 12 - 20 are invited to apply. The committee is particularly interested in programs or activities that address youth issues, particularly: alcohol and drug use (including smoking); impaired driving; coping with external pressures from and conflicting expectations of family and peers; sexual activity, teen pregnancy, and harassment; time management related to academics, employment, and other activities; depression, self-esteem, and body image. Typically, the Request for Proposals is available late December of the year and the proposals are due mid-February. Grant awards are announced the following May.
Requirements: Any non-profit organization with 501(c)3 status located within Georgetown, Hudsonville, and Jamestown Townships of Michigan, that is in need of funding for programs that benefit youth in this area are eligible to apply.

Restrictions: Grants proposals from religious organizations for a religious purpose will not be considered.
Geographic Focus: Michigan
Date(s) Application is Due: Dec 31
Contact: Cris Kooyer; (616) 454-1751, ext. 118; ckooyer@grfoundation.org
Internet: http://www.grfoundation.org/seottawa-yac.php
Sponsor: Grand Rapids Community Foundation
185 Oakes Street SW
Grand Rapids, MI 49503

Grand Rapids Community Foundation Sparta Grants   1875
The Sparta Community Foundation gives priority to projects that address the areas of art & culture, community development, education, environment, health, or social needs and that: represent an innovative, start-up effort or are capital in nature (e.g., construction, renovation, equipment); promote cooperation among agencies without duplicating services; obtain the necessary additional funding to implement and maintain the project; are located in Sparta and its surrounding areas of Michigan; strengthen or improve agency self-sufficiency or efficiency; yield substantial community benefits for the resources invested; serve a broad segment of the community; encourage additional and permanent funding or matching gifts from other donors; have non-profit, 501(c)3 status. The Foundation will consider proposals for the following only if they are highly unique, collaborative, and community-oriented, have limited access to other resources, demonstrate substantial impact, or will address the needs of a substantial or underserved portion of the community: hospitals; nursing or retirement facilities; K-12 schools; computers; child care centers; motorized vehicles; commonly accepted community services already supported by tax dollars.
Requirements: Michigan IRS 501(c)3 non-profit organizations that are located in Sparta and its surrounding areas are eligible to apply. General applications are due September 4th. For more information on how to request funding for projects, contact Becky Cumings, Grant Committee Chair, at (616) 887-8428.
Restrictions: The Foundation generally does not support: annual fund-raising drives; films, videos or television projects; endowments or debt reduction; capital projects without site control; medical research; individuals; ongoing operating expenses of established institutions; organizations located outside greater Georgetown Township, Hudsonville and Jamestown Township area; political projects, or those that are primarily cause-related; religious organizations for religious purposes; sabbatical leaves; scholarly research; travel, tours or trips; underwriting of conferences; venture capital for competitive profit-making activities.
Geographic Focus: Michigan
Date(s) Application is Due: Sep 4
Samples: Senior Neighbors Inc., Sparta, MI, $1,990--equipment acquisition grant.
Contact: Becky Cumings, Chair; (616) 887-8428; cumingst@hotmail.com
Internet: http://www.grfoundation.org/sparta
Sponsor: Grand Rapids Area Community Foundation-Sparta Community Foundation
185 Oakes Street SW
Grand Rapids, MI 49503

Grand Rapids Community Foundation Sparta Youth Fund Grants   1876
The Sparta Community Foundation's Youth Advisory Committee is seeking to fund projects or programs that positively impact youth in the Sparta community. Groups or organizations made up of (or serving) youth ages 12 - 20 are invited to apply. The committee is particularly interested in programs or activities that address the following issues: alcohol and drug use (including smoking); impaired driving; coping with external pressures from and conflicting expectations of family and peers; sexual activity, teen pregnancy, and harassment; time management related to academics, employment, and other activities; depression, self-esteem, and body image. The projects should be youth initiated and youth driven.
Requirements: Any non-profit organization with 501(c)3 status located in or serving the Sparta area that is in need of funding for programs that benefit youth in the Sparta community.
Restrictions: Grants proposals from religious organizations for a religious purpose will not be considered.
Geographic Focus: Michigan
Date(s) Application is Due: Dec 31
Contact: Cris Kooyer; (616) 454-1751, ext. 118; ckooyer@grfoundation.org
Internet: http://www.grfoundation.org/sparta-yac.php
Sponsor: Grand Rapids Community Foundation
185 Oakes Street SW
Grand Rapids, MI 49503

Grass Foundation Marine Biological Lab Advanced Imaging Fellowships   1877
The Grass Foundation Advanced Imaging Fellowships at Marine Biological Laboratory will support individuals working in any area of neuroscience at any stage of their career, ranging from graduate student to professor. This opportunity is intended to allow individuals to bring their preparations to the MBL for up to four weeks, learning and utilizing the available microscopy resources, gathering data and interacting with other imaging specialists. Grass Fellows in Advanced Imaging are integrated into the traditional Grass Fellows program during the summer, providing a rich environment and access to many activities at the MBL. As such, Imaging Fellows will receive bench space in the Grass Laboratory for sample preparation, and are expected to attend Grass Laboratory lab meetings. In addition, Imaging Fellows are encouraged to audit the many classes available including the neurobiology course, which offers lectures on the theory and practice of advanced microscopy. Support will include research supplies, travel, housing and meals at the Marine Biological Laboratory for the individual Fellow. Applicant instructions with detailed information of all attachments needed, such as proposal and vitae, are available at the Grass Foundation website.
Requirements: Applicants at any stage in their academic career are eligible. Priority is given to applicants with an experimental need to use high-end state of the art imaging systems. Prior research experience at the MBL, including previous Grass Fellowships, is neither required nor disqualifying. IRS regulations require that international Fellows (i.e., not U.S. citizens or resident aliens) hold a J-1 visa for the duration of the Fellowship. The duration of the program is two to four weeks during July. Fellows are expected to be in residence at the MBL the entire time. Up to three Fellows will be accepted.
Restrictions: Holders of H1-B visas are not eligible to apply. Funding is available for the individual Fellow only.
Geographic Focus: All States, All Countries
Date(s) Application is Due: Feb 20
Contact: Ann Woolford; (508) 289-7521; fax (508) 289-7931; awoolford@mbl.edu
Internet: http://www.grassfoundation.org/grass-fellowships-in-advanced-imaging
Sponsor: Grass Foundation
P.O. Box 241458
Los Angeles, CA 90024

Grass Foundation Marine Biological Laboratory Fellowships   1878
Grass Fellowships at the Marine Biological Laboratory (MBL) in Woods Hole, Massachusetts, support investigator-designed, independent research projects by scientists early in their career. Supported approaches include neurophysiology, biophysics, integrative neurobiology, neuroethology, neuroanatomy, neuropharmacology, systems neuroscience, cellular and developmental neurobiology, and computational approaches to neural systems. Grass Fellowships provide research support including laboratory space, animals, equipment, and supplies for one summer at the MBL. Additionally, the investigator, his/her spouse or legal domestic partner, and dependent children are provided housing, a daily meal allowance and round-trip travel to the MBL. Fellows function as an intellectual and social group within the MBL scientific community while sharing space in the Grass Laboratory. In a weekly private seminar series, eminent investigators at the MBL discuss their work with the Fellows. In addition, a yearly Forbes Lecturer will spend a portion of the summer in the Grass Lab interacting with Fellows. From 8 to 10 Fellowships are awarded annually. The Fellowship last from late May to early August. The application and supplemental material are available at the Foundation website.
Requirements: Early investigators (late stage predoctoral trainees and beyond) are eligible to apply. This includes applicants with prior experience at MBL or with the Grass Foundation. MBL course alumni are encouraged to apply. Priority is given to applicants with a demonstrated commitment to pursuing a research career. International Fellows must hold either a J-1 or H-1B visa for the entire duration of the Fellowship.
Geographic Focus: All States, All Countries
Date(s) Application is Due: Dec 5
Contact: Ann Woolford; (508) 289-7521; fax (508) 289-7931; awoolford@mbl.edu
Internet: http://www.grassfoundation.org/grass-fellowship-program
Sponsor: Grass Foundation
P.O. Box 241458
Los Angeles, CA 90024

Great-West Life Grants   1879
The program's objectives are to support organizations that meet the most vital needs in the greater Denver metro community, including access to affordable health care, shelter, nutrition and education, and to support our employees' involvement in their communities. Funding priorities are on preventive and ongoing health services, homeless shelters, food services, and keeping children in school. Support is limited to organizations providing direct services, free or charge or on a sliding scale basis, that demonstrate effectiveness in addressing the communities vital issues. There is no formal application process. Interested applicants should submit a brief concept paper.
Requirements: To be eligible for support an organization: must be a U.S. based, charity and tax-exempt under 501(c)3 of the Internal Revenue Code; must not discriminate on the basis of race, religion, gender, age, national origin, disability, or sexual orientation; must show demonstrated success in meeting the program's objectives and fall within the program's focus areas; must serve residents in areas with a Great-West presence.
Restrictions: Funding is not provided for: capital campaigns; debt financing; endowments; advertising; grant-making foundations; individuals; political organizations; exclusively religious organizations; veterans organizations; fraternal organizations.
Geographic Focus: All States
Contact: Human Resources; (800) 537-2033; greatwestcomments@gwl.com
Internet: http://www.greatwest.com/about/community_supp.htm
Sponsor: Great-West Life and Annuity Insurance Company
8515 East Orchard Road
Greenwood Village, CO 80111

**Greater Cincinnati Foundation Priority and Small Projects/Capacity-Building   1880
Grants**
Grants are awarded on a quarterly basis and support projects addressing community needs and priorities in six areas: arts and culture; community progress; education; environment; health; and human services. The Foundation looks for applicants that are established and fiscally sound with strong governance and a demonstrated organizational capacity/track record. Application information is available online. Nonprofits seeking grants of up to $20,000 can get a decision in eight weeks or less; all other grants above $20,000 will be awarded twice a year.
Requirements: The Foundation awards grants to qualified 501(c)3 nonprofit organizations in eight counties: Butler, Clermont, Hamilton, and Warren, Ohio; Boone, Kenton and Campbell, Kentucky; and Dearborn, Indiana.

Restrictions: Funds are not available for: individuals; units of government or government agencies; religious organizations for religious purposes; ongoing operating expenses of nonprofit organizations; annual fund raising drives, event sponsorship or underwriting; stand-alone books, films or videos unless part of a more comprehensive program activity; stand-alone travel expenses for conferences or educational purposes unless part of a more comprehensive program activity; regular operating costs or capital projects for individual public, private or parochial schools, universities, hospitals, nursing or retirement homes; endowments; scholarships; loans; scholarly or medical research; partisan political advocacy; debt retirement or funding of an activity after it is completed.
Geographic Focus: Indiana, Kentucky, Ohio
Amount of Grant: 10,000 - 100,000 USD
Contact: Kay Pennington, Community Investment Coordinator; (513) 241-2880; fax (513) 852-6886; penningtonk@greatercincinnatifdn.org
Internet: http://www.greatercincinnatifdn.org/page10004265.cfm
Sponsor: Greater Cincinnati Foundation
200 West Fourth Street
Cincinnati, OH 45202

Greater Green Bay Community Foundation Grants   1881
The Foundation provides assistance to worthy charities serving the people of Brown, Kewaunee, and Oconto counties, Wisconsin. The Foundation's areas of interest are: arts and culture; community development; cultural and ethnic diversity; health and human services; historic preservation; education; and the environment. The Foundation also awards special funds to projects that address issues involved with Alzheimer's Disease, Diabetes, and Hospice. Applicants should share their ideas with the Foundation before completing a formal application. Application information is available online.
Requirements: Eligible recipients meet the following criteria: are tax exempt and have 501(c)3 status; primarily serve residents of Brown, Oconto, and Kewaunee counties; and operate, choose a governing board, and provide services without discrimination.
Restrictions: Funding is not available for: annual or capital campaigns; projects that promote any religious belief; individuals; debt retirement; lobbying; activities that occur prior to the awarding of the grant; efforts substantially serving people outside of Brown, Kewaunee, and Oconto counties.
Geographic Focus: Wisconsin
Date(s) Application is Due: Jan 15; Apr 15; Jul 15; Oct 15
Amount of Grant: 1,000 - 10,000 USD
Samples: Children's Museum of Green Bay (Green Bay, WI)--for the Pets are People Too Exhibit, $7,000; Greater Green Bay YMCA (Green Bay, WI)--for Swim Smart & Family Nights, $3,367.
Contact: Lora Warner; (920) 432-0800; fax (920) 432-5577; lora@ggbcf.org
Internet: http://www.ggbcf.org
Sponsor: Greater Green Bay Community Foundation
301 W Walnut Street, Suite 350
Green Bay, WI 54303

Greater Kanawha Valley Foundation Grants   1882
The Greater Kanawha Valley Foundation awards grants to projects and programs that enhance the quality of life in the Greater Kanawha Valley. The Foundation awards grants to non-profit and other charitable organizations under the broad category of community development (defined as improving quality of life, promoting economic development, and reducing poverty) in the categories of education, arts and culture, health, human services, public recreation, and land use. Priority is given to proposals that: encourage "bridging," community bonding, and connectedness; generate matching funds, thus leveraging additional support; exhibit coordination and collaboration among organizations; implement new approaches and innovative techniques to solve community problems; and focus on proactive, preventive measures. The application, specific deadlines for each field of interest, and samples of previously funded projects are available at the Foundation website. Applicants are strongly encouraged to attend the workshop for their field of interest.
Requirements: To be eligible for funding, an applicant must be a 501(c)3 as determined by the IRS, a faith-based organization, or a government entity (i.e., libraries, schools, etc.); provide services within the counties of Boone, Clay, Fayette, Kanawha, Lincoln, and Putnam; and be current on all final reports from the Foundation.
Restrictions: Funding is not available for the following: national or statewide proposals that do not focus on the Foundation's six county service area; general operating budgets established for organizations; annual campaigns or membership drives; travel expenses or school uniform purchases; ongoing support for the same project; staff costs only; consultants, consultant fees, conferences or workshop speakers; individuals, student aid or fellowships; endowments; or religious activities of religious organizations. Each organization may apply for only one grant per year.
Geographic Focus: West Virginia
Date(s) Application is Due: Feb 1; May 1; Aug 1
Samples: Appalachian Children's Chorus, Charleston, West Virginia, support educational resources for the support of a four day national children's choir festival, $8,500; Arts in Action, Charleston, West Virginia, support the arts education program, Urban Stage, designed for underprivileged and at-risk youth, $10,000; Thanks! Plain and Simple, Inc., Charleston, West Virginia, funding to support the creation of the Rosie the Riveter Model Community where one region unites the Women Ordinance Workers to collect their stories of the World War II era, $10,000.
Contact: Sheri Ryder; (304) 346-3620; fax (304) 346-3640; sryder@tgkvf.org
Internet: http://www.tgkvf.org/page.aspx?pid=384
Sponsor: Greater Kanawha Valley Foundation
One Huntington Square
Charleston, WV 25301

Greater Milwaukee Foundation Grants   1883
The Foundation places its highest funding priority on supporting creative efforts to address issues of poverty in the community, particularly grants focused on education, employment, and strengthening children, youth and families. Lower priority is given to projects that do not meet the above criteria and/or do not address issues of persistent poverty. In addition, the Foundation places special emphasis on programs that accomplish the following: improving understanding among people of different backgrounds through support of efforts addressing issues of racial, cultural and economic diversity; and strengthening the voluntary sector by supporting efforts to enhance the management capacities of nonprofit organizations, promote philanthropy, and encourage civic involvement and community service. Additional application information is available online.
Requirements: Grants are made only to 501(c)3 nonprofit organizations and, on occasion, to governmental agencies. Geographically, funding for the discretionary grantmaking program is limited to projects that will significantly improve the lives of people living in Milwaukee, Waukesha, Ozaukee and Washington counties.
Restrictions: Grants for ongoing operational costs or to individuals are not eligible for support from the Foundation's discretionary funds. The Foundation does not provide support for debt reduction, sectarian religious purposes, medical or scientific research, fund drives for sustaining support or organizations that are discriminatory in their practices.
Geographic Focus: Wisconsin
Samples: Milwaukee Chamber Orchestra, Milwaukee, Wisconsin, $25,000 - to support staff and programming expenses as part of its rebuilding efforts; Layton Boulevard West Neighbors, Milwaukee, Wisconsin, $25,000 - to partially fund three staff positions that help support the agency's work as part of the Foundation's Milwaukee Healthy Neighborhoods Initiative.
Contact: Fran Kowalkiewicz, Grants Manager; (414) 272-5805; fax (414) 272-6235; fkowalkiewicz@greatermilwaukeefoundation.org or info@greatermilwaukeefoundation.org
Internet: http://www.greatermilwaukeefoundation.org/grant_seekers/
Sponsor: Greater Milwaukee Foundation
101 West Pleasant Street, Suite 210
Milwaukee, WI 53212

Greater Saint Louis Community Foundation Grants   1884
The Greater Saint Louis Community Foundation (GSLCF) was founded in 1915, one year after the first community foundation was established in Cleveland, Ohio. Currently GSLCF administers over 400 individual charitable funds that total $170 million in assets. These funds annually make over $17 million in grants that shape the greater Saint Louis region, touch communities across the nation, and reach across the globe. Historically, the mission of GSLCF has been two-fold: to serve donors and ensure that their dollars work in line with the goals that are important to them; and to promote charitable giving through community investment in nonprofit organizations capable of addressing community issues in measurable ways. To this end GSLCF maintains an online database YOURGivingLink to help donors find deserving nonprofit organizations that match their giving interests. Nonprofits are encouraged to register their organizations with the database; the link is available at the GSLCF website.
Requirements: While the Foundation predominately supports St. Louis area nonprofits, grants are also made to national and international charities.
Geographic Focus: Illinois, Missouri
Contact: Amy Basore Murphy, Director of Scholarships and Donor Services; (314) 588-8200, ext. 139 or (314) 880-4965; fax (314) 588-8088; amurphy@stlouisgives.org
Internet: http://www.stlouisgives.org/charities/
Sponsor: Greater Saint Louis Community Foundation
319 North Fourth Street, Suite 300
Saint Louis, MO 63102-1906

Greater Sitka Legacy Fund Grants   1885
As an organization, the Greater Sitka Legacy Fund's goal is to support projects of importance to the Sitka community that will help its residents become better stewards of the land. The Fund is continually listening and learning to its residents in order to understand important to them. Grant making priorities are still in development, though the Fund will support projects that strengthen the community. Among others, they will include organizations and programs that serve youth, the elderly, recreation, safety, vulnerable populations, and arts and culture. There are no annual deadlines for applications.
Requirements: The Fund board members seek applications from qualified tax-exempt 501(c)3 organizations that support the organizations and programs in the Sitka area and serve the people's needs in such areas as health, education, community heritage, the arts, vulnerable populations, recreation, safety, and community and economic development.
Restrictions: Individuals, for-profit, and 501(c)4 or 501(c)6 organizations, non-Alaska based organizations and state or federal government agencies are not eligible for competitive grants. Applications for religious indoctrination or other religious activities, endowment building, deficit financing, fundraising, lobbying, electioneering and activities of political nature will not be considered, nor will proposals for ads, sponsorships, or special event and any proposals which discriminate as to race, gender, marital status, sexual orientation, age, disability, creed or ethnicity.
Geographic Focus: Alaska
Contact: Ricardo Lopez, Affiliate Program Officer; (907) 274-6707; fax (907) 334-5780; rlopez@alaskacf.org or greatersitka@alaskacf.org
Internet: http://greatersitkalegacyfund.org/
Sponsor: Greater Sitka Legacy Fund
700 Katlian Street, Suite B
Sitka, AK 99835

Greater Tacoma Community Foundation Grants 1886

The Foundation invests in the community through grants to vital nonprofit agencies in Pierce County. While grants are made across all fields, principal funding areas include: arts and culture; civics; education; the environment; health; and human services. The Foundation also makes capital, equipment, project and operating support grants. Application information is available online.
Requirements: Eligible applicants must meet the following criteria: attest to non-discrimination on the basis of race, sex, age, national origin, religion, physical or mental handicap, veteran status or sexual orientation; provide a service primarily for residents of Pierce County; qualify as tax-exempt under section 501(c)3 of the IRS Code; and apply no more than once a year for any one governing or umbrella organization.
Restrictions: Grants will not be made for: fundraising events or fundraising feasibility projects; individuals; religious organizations for sacramental/theological purposes; production of books, videos, films, or other publications; annual campaign appeals; travel; political or lobbying activities; endowments; debt reduction; events or programs that occur prior to the board of directors decision/notification dates.
Geographic Focus: Washington
Samples: Kids in Distressed Situations (New York, NY)--to provide donated products to low-income children and families in Tacoma and Pierce County, WA, through the Emergency Food Network, $25,000.
Contact: Rose Lincoln Hamilton; (253) 383-5622; fax (253) 272-8099; rlincoln@gtcf.org
Internet: http://www.tacomafoundation.org
Sponsor: Greater Tacoma Community Foundation
950 Pacific Avenue, Suite 1220, P.O. Box 1995
Tacoma, WA 98402

Greater Tacoma Community Foundation Ryan Alan Hade Endowment Fund 1887

The Fund is to provide support and care to children who have been through traumatic experiences of abuse and disfigurement. The Foundation is looking for nonprofit partners who serve abused or disfigured children in the Greater Tacoma area to continue or implement programming that will ease these children's lives. Interested applicants should submit a one page letter of intent. LOI information is available online.
Requirements: Nonprofit organizations in the Greater Tacoma area are eligible to apply.
Geographic Focus: Washington
Contact: Kristen Corning, Coordinator; (253) 383-5622; kcorning@gtcf.org
Internet: http://www.tacomafoundation.org/newsarticle.cfm?articleid=126285&ptsidebaroptid=0&returnto=page19554.cfm&returntoname=Apply%20for%20a%20Grant&siteid=1735&pageid=19538&sidepageid=19554&banner1img=banner%5F1%2EJPG&banner2img=banner%5F2%2EJPG&bannerbg=banner
Sponsor: Greater Tacoma Community Foundation
950 Pacific Avenue, Suite 1220, P.O. Box 1995
Tacoma, WA 98402

Greater Worcester Community Foundation Discretionary Grants 1888

Grants are made to nonprofit organizations to build healthy and vibrant communities. Several large field-of-interest funds are also included in this process, providing support for culture and the arts, academic achievement for disadvantaged youth, progressive education, conservation, homelessness and affordable housing, and health and wellness for women and children. These grants enable us to be responsive to community needs and priorities.
Requirements: The Foundation considers applications only from nonprofit, tax-exempt organizations for activities serving the people of the Central Massachusetts region. Organizations not incorporated as tax-exempt may apply through an established organization that agrees to provide fiscal oversight.
Geographic Focus: Massachusetts
Date(s) Application is Due: Mar 15; Sep 15
Amount of Grant: 3,000 - 25,000 USD
Contact: Pamela B. Kane, (508) 755-0980; pkane@greaterworcester.org
Internet: http://www.greaterworcester.org/grants/disc-grants.htm
Sponsor: Greater Worcester Community Foundation
370 Main Street, Suite 650
Worcester, MA 01608-1738

Greater Worcester Community Foundation Jeppson Memorial Fund for Brookfield Grants 1889

The Fund provides money to civic and community projects that help improve the lives of residents and enrich the cultural environment. The Fund provides support for: cultural or artistic performances; public seminars; festivals or exhibitions; services that help frail or vulnerable citizens, or that contribute to public health and safety; opportunities for educational enrichment; youth involvement in recreation, sports and the arts; and projects that foster community awareness and connections among different groups. Application information is available online.
Requirements: Any nonprofit or civic organization that serves the residents of Brookfield may apply.
Restrictions: Grant funds may not be used for expenses already incurred by the applicant. Fund awards are not intended to replace municipal funds.
Geographic Focus: Massachusetts
Date(s) Application is Due: Jul 15
Contact: Pamela B. Kane, (508) 755-0980; pkane@greaterworcester.org
Internet: http://www.greaterworcester.org/grants/Jeppson.htm
Sponsor: Greater Worcester Community Foundation
370 Main Street, Suite 650
Worcester, MA 01608-1738

Green Bay Packers Foundation Grants 1890

The Green Bay Packers organization has enjoyed tremendous fan support through its long and storied history. To give back to the community, the team created the Green Bay Packers Foundation in December 1986. The Foundation assists in a wide variety of activities and programs that benefit education, civic affairs, health services, human services and youth-related programs.
Requirements: 501(c)3 nonprofits in Wisconsin may apply for grant support.
Restrictions: No substantial part of the activities of the organization shall involve carrying on propaganda, or otherwise intervening in any political campaign on behalf of any candidate for public office.
Geographic Focus: Wisconsin
Samples: Packers Scholarship Program, $15,000--$7,500 to Scholarships, Inc., for distribution to students in four-year colleges and $7,500 to Northeast Wisconsin Technical College for distribution to students in two-year associate degree or apprenticeship trades programs; Cooks for Room at the Inn, Green Bay, WI, $2,000--New Community Shelter; Hometown Huddle Program, Green Bay, WI, $5,000--to provide a new playground for the city's Beaumont Elementary School Park.
Contact: Margaret Meyers; (920) 569-7315; fax (920) 569-7309; meyersm@packers.com
Internet: http://www.packers.com/community/packers_foundation
Sponsor: Green Bay Packers Foundation
P.O. Box 10628
Green Bay, WI 54307-0628

Green Diamond Charitable Contributions 1891

Green Diamond Resource Company is a fifth-generation, family-owned forest-products company that has operations in the states of California and Washington. The mission of Green Diamond's contributions program is to improve the quality of life in communities where the company has a significant number of employees living and working; and to serve as a catalyst for employees to become involved and to provide leadership in their communities. The company has supported a broad range of organizations in its operating communities that address key community needs such as education, health services, economic development, the arts, and more. The company also provides support to environmental and conservation projects, such as environmental education, habitat restoration, and research relevant to the timber industry, as well as other projects that impact the timber business. Contributions are generally made in locations where the company has operations. To the extent possible, contributions will support organizations of interest to, or recommended by, Green Diamond employees. Green Diamond prefers to make capital contributions that will benefit the operating communities for the long term as opposed to contributing operating funds. Generally, support is committed for one year at a time and in amounts less than $5,000. The company supports the United Way in all of its operating communities. The company accepts applications twice a year, generally in May and in late summer/early fall. Interested applicants should call or mail the contact person given for their state to request application materials. More detailed information about application deadlines will be communicated at that time.
Requirements: Organizations that serve Pierce, Mason, Thurston, and Grays Harbor counties in Washington or Del Norte and Humboldt counties in California are eligible to apply. Criteria taken into account in determining the amount of any contributions are as follows: degree of support from company employees; relative size and importance of company operations in the community (balance among Green Diamond communities); needs of organization or program for which funding is requested; amount of previous company contributions to the organization; amount committed by other companies, foundations, and/or governments (projects should demonstrate broad-based community support); and proximity of the requesting organization to Green Diamond operations or administrative offices.
Geographic Focus: California, Washington
Amount of Grant: Up to 5,000 USD
Contact: Jacki Deuschle; (707) 668-4488; fax (707) 668-4402; jdeuschle@greendiamond.com
Patti Case, Public Affairs Manager; (360) 426-3381; pcase@greendiamond.com
Internet: http://www.greendiamond.com/charitable-giving/
Sponsor: Green Diamond Resource Company
1301 Fifth Avenue, Suite 2700
Seattle, WA 98101-2613

Greene County Foundation Grants 1892

The Greene County Foundations mission is to work with charitably minded individuals and organizations to strengthen Greene County, Indiana, now and for generations to come. The Foundation administers several funding opportunities. Eligibility, criteria, and supplemental requirements are different for each grant. Contact the Foundation directly to begin application process.
Geographic Focus: Indiana
Contact: Cam Trampke, Executive Director; (812) 659-3142 or (812) 659-3144; fax (812) 659-3142; ctrampke@greenecountyfoundation.org
Internet: http://www.greenecountyfoundation.org/funds_available.php
Sponsor: Greene County Foundation
4513 West State Road 54
Bloomfield, IN 47424

Greenfield Foundation of Maine Grants 1893

The Foundation, established in 1999, offers funding primarily in the form of ongoing general operating support. Focus areas include: federated giving programs; health care; the environment; wildlife; and animal welfare. The geographic focus is primarily in Maine, with an initial approach by Letter of Interest (LOI) and no specific application forms. There are no annual deadlines, and individual grants average $500.

Requirements: Grantees must be 501(c)3 organizations with a sound track record.
Restrictions: No grants to individuals.
Geographic Focus: Maine
Amount of Grant: 50 - 1,000 USD
Samples: Maine Audubon Society, $50; Animal Refuge League, $1,000.
Contact: Margaret Cary Curran, (207) 781-5360
Sponsor: Greenfield Foundation
2 Lady Cove
Falmouth, ME 04105-1956

Green Foundation Human Services Grants — 1894
The Foundation's mission is to uncover new opportunities, encourage growth and ultimately effect positive change within those institutions that best reflect the core focus areas and the communities it serves. Within the human services area, the Foundation focuses on institutions that provide hope and support to those least able to help themselves as well as the general community; including children, adolescents, the elderly, the homeless, and families who struggle with domestic abuse.
Requirements: 501(c)3 nonprofits (as per the IRS Service Code of 1986) are eligible. Most grant making is limited to institutions that serve the Los Angeles community; however the Foundation will consider requests beyond this geographic boundary for those institutions with the potential to impact communities statewide or nationally.
Restrictions: The Foundation does not provide funds for: those with net assets or fund balances of less than $100,000; multi-year commitments; annual meetings, conferences, and/or seminars; religious programs; capital campaigns; direct mail campaigns; conduit institutions, unified funds, fiscal agents, or institutions using grant funds from donors to support other institutions or individuals; private foundations; or individuals.
Geographic Focus: All States
Amount of Grant: 1,000 - 150,000 USD
Contact: Kylie Wright; (626) 793-6200, ext. 1; fax (626) 793-6201; kylies@ligf.org
Internet: http://www.ligf.org/humanservices.php
Sponsor: Green Foundation
225 South Lake Avenue, Suite 1410
Pasadena, CA 91101

Green River Area Community Foundation Grants — 1895
The Green River Area Community Foundation (GRAF) was organized in 1993 with a mission of advancing philanthropy by serving the charitable interests of donors, enabling increased charitable giving and improving communities by being a permanent philanthropic resource for current and future needs. Presently, the Foundation is dedicated to enriching the quality of life for all citizens in the counties of Daviess, Hancock, McLean, Ohio, Union and Webster, Kentucky. GRACF serves the charitable interests of donors who have established charitable funds as part of a permanent, collective, philanthropic resource for the current and future needs of the region. Two primary funds are used: the Owensboro-Daviess County Community Fund, established in 1995, and the James D. Ryan Memorial Fund, founded the same year. The Foundation is an affiliate of the Community Foundation of Louisville, a tax-exempt public charity that administers charitable funds created by individuals, businesses and organizations for the betterment of their communities.
Requirements: 501(c)3 organizations serving the Kentucky county residents of Daviess, Hancock, McLean, Ohio, Union and Webster are eligible to apply.
Geographic Focus: Kentucky
Contact: Dana Johnson; (502) 855-6957 or (502) 585-4649; danaj@cflouisville.org
Internet: https://www.cflouisville.org/about/green-river-area-community-foundation/
Sponsor: Green River Area Community Foundation
325 W. Main Street, Suite 1110
Louisville, KY 40202

Greenspun Family Foundation Grants — 1896
The Greenspun Family Foundation supports many causes with an emphasis on education, health, children, Jewish issues, and the greater Las Vegas community. Preference is given to requests from the Las Vegas area. There are no application forms or deadlines. Applicants should submit a letter of inquiry. Information about current programs supported is available at the Foundation website.
Geographic Focus: Nevada
Amount of Grant: 50,000 - 2,000,000 USD
Contact: Dr. Brian Cram, Director; (702) 259-2323 or (702) 259-4023; fax (702) 259-4019; brian.cram@lasvegassun.com
Internet: http://www.thegreenspuncorp.com/philanthropy.php
Sponsor: Greenspun Family Foundation
901 North Green Valley Parkway, Suite 210
Henderson, NV 89074

Greenwall Foundation Bioethics Grants — 1897
The Foundation provides funding for physicians, lawyers, philosophers, theologians and other professionals to address micro and macro issues in bioethics, providing guidance for those engaged in decision-making at the bedside as well as those responsible for shaping institutional and public policy. The Foundation is especially interested in the work of junior investigators and pilot projects that may lead to NIH support, and it is prepared to address issues regarded by some as sensitive or potentially controversial.
Restrictions: The Foundation is not normally interested in proposals to support equipment purchase, facility construction or renovation, or general operating expenses, and will not normally consider grants to private foundations, endowment funds, or individual applicants.
Geographic Focus: All States
Date(s) Application is Due: Feb 1; Aug 1
Amount of Grant: 5,000 - 50,000 USD
Contact: Sam Teigen; (212) 679-7266; fax (212) 679-7269; steigen@greenwall.org
Internet: http://www.greenwall.org/guidebio.htm
Sponsor: Greenwall Foundation
420 Lexington Avenue, Suite 2500
New York, NY 10170

Gregory B. Davis Foundation Grants — 1898
The Foundation's mission is to enhance the lives of youth and their families through the development and support of health related and educational programs and initiatives. More specifically, GBDF will support healthy life choices and well-being for rural and medically challenged families, caregivers, and youth affected by HIV/AIDS, Alzheimer's Disease, and other debilitating illnesses and those with limited academic and life skills opportunities. There are no specific deadlines, and applicants should forward an entire proposal.
Geographic Focus: All States
Amount of Grant: 500 - 2,000 USD
Contact: Edna Davis-Brown, CEO; (301) 593-0644; GBDF@gbdf.org
Internet: http://www.gbdf.org/programs.html
Sponsor: Gregory B. Davis Foundation
11404 December Way, Suite 403
Silver Springs, MD 20904-3623

Gregory Family Foundation Grants (Florida) — 1899
The Foundation, established in 1999, offers support to cancer research and prevention, Christian churches and agencies, and hospitals. Funding most often comes in the form of a donation for general operations. The Foundation's geographic restrictions are New York and Florida. A formal application form is required, which should include: a copy of most recent annual report, audited financial statement, or 990; a detailed description of project; and amount of funding requested. There are no specific deadlines.
Requirements: Applicants should be a 501(c)3 located in the state of Florida or New York.
Geographic Focus: Florida, New York
Amount of Grant: Up to 2,000 USD
Contact: Dale M. Gregory, President; (561) 362-9868; gregory1@bellsouth.net
Sponsor: Gregory Family Foundation
101 Plaza Reale S, Apt. 618
Boca Raton, FL 33432-4849

Gregory L. Gibson Charitable Foundation Grants — 1900
The Foundation, established in Terre Haute, Indiana, in 2005, is interested in programs that support children, youth, and families. Its primary areas of interest include: community service programs; education; health care and health care access; hospice care; technology; and youth sports. The specific type of support is for general operations. There are no specific deadlines with which to adhere, and applicants should forward a submission letter.
Requirements: 501(c)3 organizations situated in, or serving the community of, Terre Haute, Indiana, are eligible to apply.
Geographic Focus: Indiana
Amount of Grant: Up to 30,000 USD
Contact: Gregory L. Gibson, President; (812) 466-1233
Sponsor: Gregory L. Gibson Charitable Foundation
3200 E Haythorne Avenue
Terre Haute, IN 47805-002108

Greygates Foundation Grants — 1901
The Greygates Foundation was created in 2001 by J. Ronald Gibbs to provide grants to organizations that serve the needs of children, the elderly, the disabled, or the disadvantaged, and to organizations that promote animal welfare or wildlife preservation. The grant award limit is $3,000, and grants are paid in Canadian dollars. The foundation funds the following types of requests: general operating support, capacity building, program support, equipment, and tuition assistance. Proposals are accepted on a rolling basis. Organizations are asked to take an online eligibility quiz before they apply. Application must be made by email. Guidelines are available for download from the trust's website. Applicants are notified promptly when their proposals are received but should expect to wait three to six months for notification of a decision. Questions may be directed to the foundation's administrator.
Requirements: Though funding may be provided for projects in any country, recipient organizations must be Canadian-registered charities. If an organization does not have such status, the foundation will consider making a grant to a Canadian-registered charity acting as a sponsor for the non-recognized organization. The Greygates Foundation grants are generally awarded to smaller nonprofit organizations, but are not limited to such organizations.
Restrictions: The Foundation does not provide funding to individuals or to organizations operated by or receiving significant support from government sources.
Geographic Focus: All States, All Countries
Amount of Grant: Up to 3,000 CAD
Contact: J. Ronald Gibbs, Administrator; (604) 896-1619; beron@telus.net Janet Ferriaolo, Grants Manager; (415) 332-0166; jferraiolo@adminitrustllc.com
Internet: http://www.adminitrustllc.com/the-greygates-foundation/
Sponsor: Greygates Foundation
c/o Adminitrust LLC
Sausalito, CA 94965

Griffin Family Foundation Grants 1902
The Foundation, established in 1999, is primarily interested in supporting non-profit organizations in the state of New York. Fields of interest include: Catholic agencies and churches; education; health care clinics and centers; and hospitals. Funding most often comes in the form of general operating support. There are no particular forms, and applicants should submit a copy of the current year's organizational budget and/or the specific project budget.
Requirements: Applicants should have 501(c)3 status with the IRS.
Geographic Focus: New York
Amount of Grant: Up to 20,000 USD
Contact: William E. Griffin, President; (914) 961-1300; fax (914) 961-9385
Sponsor: Griffin Family Foundation
51 Pondfield Road
Bronxville, NY 10708-3703

Griffin Foundation Grants 1903
The Foundation, established in 1991, serves non-profit organizations in Fort Collins, Colorado, and Laramie, Wyoming. Its major fields of interest include: the arts; health care; higher education; performing arts; orchestras; and substance abuse service programs. Funding most often comes in the forms of building construction, renovation, and scholarships. Applicants should request an application form, and submit a detailed description of the project and budget needed.
Geographic Focus: Colorado, Wyoming
Contact: David L. Wood, President; (970) 482-3030; fax (970) 484-6648; carol.wood@thegriffinfoundation.org
Internet: http://www.thegriffinfoundation.org/index.shtml
Sponsor: Griffin Foundation
303 W Prospect Road
Fort Collins, CO 80526-2003

Grifols Community Outreach Grants 1904
At Grifols, the primary mission is to enhance and enrich lives, which extends well beyond the patient communities it serves. Through a robust community outreach program, Grifols helps foster strong, vibrant communities in regions where its employees live and work. Each year, Grifols employees spend countless hours volunteering their time to improve the lives of people in need. Grifols encourages employee participation in community outreach activities by supporting a range of local and national causes through donations, corporate sponsorships and company-wide volunteer initiatives. The Grifols Community Relations program focuses on the following service areas: programs that improve the health of neighbors; science education and workforce development; and civic and community programs.
Requirements: 501(c)3 tax-exempt organizations in Grifols communities (San Francisco East Bay Area, including Alameda, Contra Costa, and Solano Counties; Los Angeles; Seattle; and Philadelphia) are eligible.
Restrictions: In general, Grifols does not support organizations that do not have 501(c)3 tax status; religious, fraternal, service, or veterans' organizations; civic or cultural organizations that do not serve the areas in which Grifols is located; alumni drives and teacher organizations; memorials; municipal and for-profit hospitals; labor unions; city, municipal, or federal government departments; organizations or causes that do not support the company's commitment to non-discrimination and diversity; projects of national scope or from national organizations not related to health care; matching gifts; individuals, including scholarships (other than those awarded as part of the company-sponsored college scholarship program); travel support; fund-raising activities related to individual sponsorship; and fund-raising dinners other than those for health care or medical research organizations aligned with company research and product interests.
Geographic Focus: California, Pennsylvania, Washington
Amount of Grant: 5,000 - 75,000 USD
Contact: Rebecca Barnes; (919) 316-6590; rebecca.barnes@grifols.com
Internet: http://www.grifols.com/web/eeuu/community_outreach
Sponsor: Grifols
2410 Lillyvale Avenue
Los Angeles, CA 90032-3514

Grotto Foundation Project Grants 1905
The Grotto Foundation works to improve the education and economic, physical and social well-being of citizens, with a special focus on families and culturally diverse groups. The Foundation funds support agencies and institutions dedicated to improving the quality of parenting and well-being of infants and children from birth to six years of age. The Foundation is further interested in increasing public understanding of American cultural heritage, the cultures of nations and the individual's responsibility to fellow human beings.
Requirements: Detailed applications for Native American grants are available online. Applicants are encouraged to contact the Foundation and discuss their early childhood development to be certain it is appropriate for funding.
Restrictions: Policy precludes grants being awarded for capital fund projects, travel, publication of books or manuscripts, undergraduate research projects, or grants to individuals.
Geographic Focus: Minnesota
Date(s) Application is Due: Jan 15; Mar 15; Jul 15; Nov 15
Amount of Grant: Up to 10,000 USD
Contact: Jennifer Kolde; (651) 209-8010; fax (651) 209-8014; jkolde@grottofoundation.org
Internet: http://www.grottofoundation.org
Sponsor: Grotto Foundation
1315 Red Fox Road, Suite 100
Arden Hills, MN 55112

Grover Hermann Foundation Grants 1906
The Grover Hermann Foundation Grants provide funding in the Chicago, Illinois and Monterrey County, California. Funding is largely in the following areas: community - local development and organizations established for youth, the elderly, or the disadvantaged, and for cultural and other community-related activities. In these categories, only organizations from the Chicago area are considered; education - although pre-college programs are eligible for grants, higher education will receive the greatest consideration. Scholarships, fellowships, challenge grants, and grants for capital expenditures are considered. Tax-supported colleges and universities receive less consideration than private institutions; health - medical facilities, basic medical research, disease-specific organizations, and other health programs; public policy - organizations dedicated to the strengthening and improvement of governmental, economic, and social systems that recognize and promote the values of individual liberty, a strong work ethic, free-market competition, and limited government; religious - established religious organizations for assistance in furthering well-defined secular causes, in the Chicago area only. Proposals are accepted at any time. The Board of Directors meets in March, June, September, and December, with funding decisions made within two weeks of board meetings.
Requirements: An application form is not required. Applicants should submit the following: results expected from proposed grant; qualifications of key personnel; statement of problem the project will address; copy of IRS Determination Letter; brief history of organization and description of its mission; copy of most recent annual report/audited financial statement/990; how project's results will be evaluated or measured; listing of board of directors, trustees, officers and other key people and their affiliations; detailed description of project and amount of funding requested; copy of current year's organizational budget and/or project budget; and a listing of additional sources and amount of support. Organizations should also submit a separate letter stating that no change in 501(c)3 status has occurred since it was issued, or is currently anticipated.
Restrictions: Funding is not available for fraternal, athletic, foreign organizations, or private foundation. Funding is also not available for individuals or operating budgets (except for national health organizations). The Foundation discourages the submission of brochures, marketing materials, newspaper or magazine clippings or booklets containing the same.
Geographic Focus: California, Illinois
Amount of Grant: 3,000 - 17,500 USD
Samples: $10,000 for Cathedral Shelter of Chicago, Illinois, payable over one year; $5,000 to Chicago Academy for the Arts, Chicago, Illinois, payable over one year; $3,000 for Jobs for Youth/Chicago, Illinois, payable over one year.
Contact: Paul K. Rhoads, President; (630) 908-7800
Sponsor: Grover Hermann Foundation
908 Kenmare Drive
Burr Ridge, IL 60527-7091

Gruber Foundation Cosmology Prize 1907
The Gruber Foundation honors and encourages educational excellence, social justice and scientific achievements that better the human condition. Foundation is a private, United States-based philanthropic organization established in 1993 under the 501(c)3 section of U.S. Corporate Law. It is funded entirely by Peter and Patricia Gruber, who serve as its Chairman and President, respectively. The Foundation is headquartered at Yale University. A major focus of the Foundation's philanthropy is its International Prize Program, created to recognize excellence in science and humanities by highlighting five fields that create a better world: Cosmology, Genetics, Neuroscience, Justice, and Women's Rights. The Cosmology Prize, cosponsored by the International Astronomical Union since 2001, honors a leading cosmologist, astronomer, astrophysicist or scientific philosopher for theoretical, analytical, conceptual or observational discoveries leading to fundamental advances in our understanding of the universe. The Prize acknowledges and encourages further exploration in a field that shapes the way we perceive and comprehend our universe. In doing so, The Gruber Foundation seeks to extend the pioneering legacy of, among others, Plato and Aristotle; Ptolemy and Copernicus; Brahe, Kepler, and Galileo; Newton and Halley; Einstein and Hubble. Individuals whose achievements have produced fundamental, advances in our understanding of the structure and evolution of the universe may be nominated. A gold medal and unrestricted $500,000 cash prize are presented annually.
Requirements: Nominations for the Cosmology Prize are invited annually from the fields of astronomy, cosmology, mathematics, and the philosophy of science. Individuals, organizations, and institutions that are active in or have an appreciation for contemporary cosmological research and study may submit nominations.
Geographic Focus: All States, All Countries
Date(s) Application is Due: Dec 15
Amount of Grant: 500,000 USD
Contact: Owen Jay Gingerich, Research Professor of Astronomy and History of Science; (617) 495-7216; nominations@gruberprizes.org or ogingerich@cfa.harvard.edu
Internet: http://gruber.yale.edu/cosmology
Sponsor: Gruber Foundation
157 Church Street, 9th Floor
New Haven, CT 06510

Gruber Foundation Gentics Prize 1908
The Gruber Foundation honors and encourages educational excellence, social justice and scientific achievements that better the human condition. Foundation is a private, United States-based philanthropic organization established in 1993 under the 501(c)3 section of U.S. Corporate Law. It is funded entirely by Peter and Patricia Gruber, who serve as its Chairman and President, respectively. The Foundation is headquartered at Yale University. A major focus of the Foundation's philanthropy is its International Prize Program, created to recognize excellence in science and humanities by highlighting five fields that create a better world: Cosmology, Genetics, Neuroscience, Justice, and

Women's Rights. The Genetics Prize, established in 2001, is presented to a leading scientist, or up to three, in recognition of groundbreaking contributions to any realm of genetics research. A gold medal and unrestricted $500,000 cash award is awarded for fundamental insights in the field of genetics. These may include original discoveries in genetic function, regulation, transmission, and variation, as well as in genomic organization. Nominations for the Prize close on December 15.
Requirements: Nominations may be submitted by individuals, organizations, and institutions that are active in or have an appreciation for contemporary genetic research or problems. Individuals who have made original discoveries in the fields of genetic function, regulation, transmission, or variation or in genomic organization are eligible.
Geographic Focus: All States
Date(s) Application is Due: Dec 15
Amount of Grant: 500,000 USD
Contact: Helen Hobbs, Foundation Contact; (214) 645-8300 or (214) 648-3111; nominations@gruberprizes.org or info@gruberprizes.org
Internet: http://gruber.yale.edu/genetics
Sponsor: Gruber Foundation
157 Church Street, 9th Floor
New Haven, CT 06510

Gruber Foundation Neuroscience Prize 1909
The Gruber Foundation Neuroscience Prize, established in 2004, is supported by the Gruber Foundation and honors scientists for major discoveries that have advanced understanding of the nervous system. Recipients receive a $500,000 prize and a gold laureate pin. For the SfN annual meeting, the recipient serves as keynote speaker at The Peter and Patricia Gruber Lecture and receives complimentary registration, transportation (economy air or ground), and two nights hotel accommodations. The SfN president presents the prize at a lecture at the meeting. Nominations for the Prize close on December 15.
Requirements: Individuals who have conducted highly distinguished research in the field of the brain, spinal cord, or peripheral nervous system are eligible for the award. Individuals, organizations, and institutions that are active in, or have an appreciation for, contemporary neuroscience research can submit nominations.
Geographic Focus: All States, All Countries
Date(s) Application is Due: Dec 15
Amount of Grant: 500,000 USD
Contact: Robert H Wurtz; (301) 496-7170; fax (301) 402-0511; bob@lsr.nei.nih.gov
Internet: http://gruber.yale.edu/neuroscience-prize-nomination-criteria
Sponsor: Gruber Foundation
157 Church Street, 9th Floor
New Haven, CT 06510

Gruber Foundation Rosalind Franklin Young Investigator Award 1910
The Gruber Foundation funds the Rosalind Franklin Young Investigator Award – a career development research award of $75,000 over three years - given to two young woman geneticists from anywhere in the world. The award, which honors the groundbreaking contributions of Dr. Rosalind Franklin, is intended to inspire and support new generations of women in the field of genetics. One award will fund genetics research in human and non-human mammals, and the other will fund genetics research in model organisms. This unrestricted cash award is presented every three years. The nomination deadline is June 5.
Requirements: The successful candidate must be in her first one to three years of an independent faculty-level position in any area of genetics.
Geographic Focus: All States, All Countries
Date(s) Application is Due: Jun 5
Amount of Grant: 75,000 USD
Contact: Foundation Contact; (203) 432-6231 or (340) 775-4430; nominations@gruberprizes.org or info@gruberprizes.org
Internet: http://gruber.yale.edu/rosalind-franklin-young-investigator-award
Sponsor: Gruber Foundation
157 Church Street, 9th Floor
New Haven, CT 06510

Gruber Foundation Weizmann Institute Awards 1911
The Weizmann Institute is a center of basic multidisciplinary scientific research and graduate study, addressing crucial problems in technology, medicine and health, energy, agriculture, and the environment. Supported by personal funds donated by Peter and Patricia Gruber, the Gruber Awards are used to fund the research of new scientists at the Weizmann Institute and the scientific travel of selected postdoctoral fellows and doctoral students through an award of $30,000 per year over a three-year period. A faculty committee advises the President of the Institute as to a worthy Gruber Award recipient. The inaugural award was presented to Elad Schneidman, a senior scientist in the Institute's Department of Neurobiology, in May 2007. One new scientist will receive the award annually. The Award provides vital funding for gifted new scientists at the beginning of their research activities, and helps enrich their work.
Geographic Focus: All States, All Countries
Date(s) Application is Due: Dec 15
Amount of Grant: 90,000 USD
Contact: Foundation Contact; (203) 432-6231 or (340) 775-4430; nominations@gruberprizes.org or info@gruberprizes.org
Internet: http://gruber.yale.edu/weizmann-institute-awards
Sponsor: Gruber Foundation
157 Church Street, 9th Floor
New Haven, CT 06510

Grundy Foundation Grants 1912
The Grundy Foundation's endowment, which sustains the operations of both the museum and library, also provides grantmaking. Grant support is given to projects that benefit the people and institutions of Pennsylvania. Upon availability, the Board of Trustees generally restricts grantmaking to Bucks County public charities, with special consideration to those of Bristol Borough. Grant applications are awarded primarily for capital projects to serve a wide area rather than a single neighborhood. Grantmaking activities include community development, arts and culture, education, environment, health, and human services. The average grant is $2,500.
Requirements: There are no applications or specific deadlines. The Foundation accepts and reviews written requests for funding throughout the year. In general, organizations having other support for core operating expenses and long-term costs of new projects are given priority. Organizations are encouraged to contact the Foundation prior to submission of a grant application. Prospective grantees may use the forms developed by the Delaware Valley Grant makers, if preferred. At a minimum, each proposal must include: One-page summary with contact name, executive director's name, organization name, address, telephone, email, and fax; project summary; amount requested; and total project budget amount; detailed proposal with mission and history of the organization; complete description of the proposed project; project budget, including other sources of support, with indication of whether support is in hand, pledged, requested, or to be requested; expected sources of support for this project in the future; and expected arrangements for future maintenance and repairs; the organization's financial report of IRS 990 filing for the most recent fiscal year (audited reports are preferred); copy of IRS 501(c)3 letter of determination or proof that the organization is a government agency; a list of officers and directors; a copy of report of the organization's activities over the most recent fiscal year. Proposals can be mailed, faxed or sent electronically (faxes and email versions require prior Foundation approval). Videotaped proposals will not be accepted.
Restrictions: The Foundation does not make grants to nonpublic schools, individuals, religious organizations, or for endowments, loans, research, or political activities.
Geographic Focus: Pennsylvania
Contact: Eugene Williams; (215) 788-5460; fax (215) 788-0915; info@grundyfoundation.com
Internet: http://www.grundymuseum.org/
Sponsor: Grundy Foundation
680 Radcliffe Street, P.O. Box 701
Bristol, PA 19007

Guido A. and Elizabeth H. Binda Foundation Grants 1913
The Foundation supports nonprofit organizations in Southwest Michigan. Areas of interest are educational projects, culture and arts, and human services. Capital campaigns and endowments are considered on a very limited basis. Initial contact may be made by letter describing the project for which funds are sought.
Requirements: Nonprofit organizations in Battle Creek and southwestern Michigan may request grant support.
Geographic Focus: Michigan
Date(s) Application is Due: May 1; Dec 1
Amount of Grant: 500 - 50,000 USD
Samples: Battle Creek Community Foundation, Battle Creek MI, $1,150--health and human services grant(2008); CIR Community Inclusive Recreation Battle Creek, MI, $25,000--arts and culture grant(2008); Grandville Public Schools, Grandville, MI, $4,000--education grant.
Contact: Nancy Taber; (269) 968-6171; fax (269) 968-5126; grants@bindafoundation.org
Internet: http://www.bindafoundation.org/grants.html
Sponsor: Guido A. and Elizabeth H. Binda Foundation
15 Capital Avenue NE, Suite 205
Battle Creek, MI 49017

Guitar Center Music Foundation Grants 1914
The Guitar Center Music Foundation makes the gift of music available to people across the country by providing funding and resources for music programs. Grants are awarded to music academies, schools, local music programs and national music programs across America. Awards can be made in the following categories: In-school music classes, in which the students make music; After-school music programs that are not run by the school; and, Music therapy programs, in which the participants make the music. The intent of the program must be music instruction, not music appreciation or entertainment, and the participants/students cannot be professional or career musicians. The grant committee reviews all applications three times yearly, and grant awards range from $500 to $5,000 in value.
Requirements: Applications are accepted online throughout the year from 501(c)3 or governmental organizations. Qualifying applicants are established, ongoing and sustainable music programs in the United States, which provide music instruction for people of any age who would not otherwise have the opportunity to make music. Almost all of the grants awarded are traditional instruments and the equipment necessary to play them. In specific instances, microphones and PA systems have also been awarded. In order to be eligible for a grant, the instruments and/or equipment must stay in the program for the life of the instrument/equipment and must be played/used by the participants.
Restrictions: Requests that are specifically ineligible are: (a) General operating expenses, including rent, utilities, supplies, travel expenses, etc; (b) Items or funds to benefit individuals, including scholarships, stipends or salaries; (c) Seed money; (d) Event sponsorship; (e) Music production equipment and/or expenses; (f) Recording studio equipment and/or expenses, including those used for educational purposes; (g) Instrument repairs; (h) Uniforms; (i) Sheet music.
Geographic Focus: Oklahoma, Oregon, Pennsylvania, Puerto Rico, Rhode Island, South Carolina, South Dakota, Tennessee, Texas, Utah, Vermont, Virginia, Washington, West Virginia, Wisconsin, Wyoming

Amount of Grant: 500 - 5,000 USD
Samples: Music in Schools Today - provides therapeutic percussion, world music and movement classes to the most underserved and at-risk youth, in Bay Area Schools and community centers; Hijos de la Musica Latina - Non-profit music program that empowers talented Latino youth that are primarily financially, economically and emotionally at-risk, who have a certified talent and gift for Latin music; Conejo Valley Youth Orchestras - serves an ethnically diverse group of families, with no students being turned away due to socio-economic status or ability, and continues to grow as the demand for music programs continues. The Creative Arts Program at Dana-Farber Cancer Institute Program - offers adult oncology patients and their families the opportunity to receive free music lessons that are especially helpful during difficult times of their treatment. Guitars in the Classroom - provides affordable music instruction in guitar, singing and song leading to educators in 20 states reaching over 3,000 educators and 200,000 students annually.
Contact: Larry E. Thomas, Chairman; (818) 735-8800, ext. 2112; fax (818) 735-7518; info@guitarcentermusicfoundation.org
Internet: http://www.guitarcentermusicfoundation.org/grants/index.cfm?sec=overview
Sponsor: Guitar Center Music Foundation
5795 Lindero Canyon Road
Westlake Village, CA 91362

Gulf Coast Community Foundation Grants 1915
The Foundation is a public charity dedicated to the progressive development of the Mississippi Gulf Coast. Its primary mission is to increase philanthropy to worthy causes by providing donor services, grants and leadership for problem solving. The Foundation carries out donor wishes by making grants in the areas of health and human services, education, arts and culture, historic preservation, and neighborhood enrichment. Grant application and additional guidelines are available online at: http://www.gulfcoastfoundation.org/guidelines.html.
Requirements: Applicants must be non-profit, tax-exempt organizations that are involved in enhancing the quality of life for citizens of South Mississippi. Organizations that have not been recognized as tax-exempt by the IRS may apply if they have a fiscal agent relationship with a 501(c)3 nonprofit organization.
Restrictions: Grants do not support: capital, operating endowment fund drives; political activities; individuals; or religious activities.
Geographic Focus: Mississippi
Contact: Rose Dellenger; (228) 897-4841; fax (228) 897-4843; rdellenger@mgccf.org
Internet: http://www.gulfcoastfoundation.org/grants.html
Sponsor: Gulf Coast Community Foundation
P.O. Box 984
Gulfport, MS 39503

Gulf Coast Foundation of Community Operating Grants 1916
Through grants and strategic community initiatives, Gulf Coast invests in the work of effective nonprofit organizations that improve quality of life in our region. The Foundation will award operating grants over $10,000, with grant funds to be expended within one year of approval. Currently, the Foundation has four grant cycles for operating grants, which fund the core operating needs of nonprofits and help make them stronger. This includes staff and training, database and accounting systems, marketing and fundraising operations. Organizations that want to diversify income streams and generate new revenue sources may apply for an operating – earned revenue grant. Examples of successful proposals in this category include using existing facilities to generate rental income or marketing and selling a packaged program/service to other organizations. Groups that want to reduce operating costs may apply for an operating – efficiency grant. Examples of successful proposals in this category include consolidating services with other nonprofits, implementing new or improved technologies such as databases or financial systems, or "greening" offices.
Requirements: The Foundation will make grants to qualified organizations classified as 501(c)3 tax-exempt public charities by the Internal Revenue Service in the counties of Brevard, Charlotte, Citrus, Collier, DeSoto, Glades, Hardee, Hendry, Hernando, Highlands, Hillsborough, Indian River, Lake, Lee, Manatee, Okeechobee, Orange, Osceola, Pasco, Pinellas, Polk, Sarasota, Seminole, Sumter, and St. Lucie. All operating – earned revenue grant applications must provide realistic revenue estimates that result in a return on investment that exceeds the grant amount. Operating – efficiency grant applications must provide a cost?benefit analysis showing how proposed funding approaches will produce cost savings.
Geographic Focus: Florida
Samples: Special Operations Warrior Foundation, $500,000 - immediate financial assistance for special operations personnel, who have been severely wounded, to have his/her family travel and be bedside; Charlotte Behavioral Health Care, $200,000 - addressing mental health, substance abuse, and related psychosocial needs of veterans and their families.
Contact: Kirstin Fulkerson; (941) 486-4600; fax (941) 486-4699; info@gulfcoastcf.org
Internet: http://www.gulfcoastcf.org/resources.php
Sponsor: Gulf Coast Foundation of Community
601 Tamiami Trail South
Venice, FL 34285

Gulf Coast Foundation of Community Program Grants 1917
Gulf Coast Foundation of Community Program Grants address regional priorities identified through its Environmental Scan. The environmental scanning process looked broadly at trends and conditions in its region through interviews with community leaders and analysis of available data. The findings help Gulf Coast identify the best ways to lift the regional economy and sustain community through our grantmaking. Successful program grant applicants will clearly target important regional challenges or opportunities, provide measurable data that will be the basis for assessing their impact, and share stories of transformation through a variety of social and traditional media. Organizations must have strong leadership and competent staff to execute their program and will significantly leverage their own program dollars so that they have "skin in the game" to maximize their program's impact.
Requirements: The Foundation will make grants to qualified organizations classified as 501(c)3 tax-exempt public charities by the Internal Revenue Service in the counties of Brevard, Charlotte, Citrus, Collier, DeSoto, Glades, Hardee, Hendry, Hernando, Highlands, Hillsborough, Indian River, Lake, Lee, Manatee, Okeechobee, Orange, Osceola, Pasco, Pinellas, Polk, Sarasota, Seminole, Sumter, and St. Lucie.
Geographic Focus: Florida
Contact: Kirstin Fulkerson; (941) 486-4600; fax (941) 486-4699; info@gulfcoastcf.org
Internet: http://www.gulfcoastcf.org/resources.php
Sponsor: Gulf Coast Foundation of Community
601 Tamiami Trail South
Venice, FL 34285

Guy's and St. Thomas' Charity Grants 1918
The Charity's grants program supports both small (under 20,000 British pounds) and large (between 20,000 and 1 million British pounds) projects under three themes: staff benefits and development; clinical and service innovation; and buildings and the environment. Applications for grants under 20,000 British pounds are considered by the Small Grants Committee, and grants over 20,000 British pounds are assessed by the New Services and Innovations Committee. The small grants deadline dates are: August, 14; September, 25; October 16. Deadline dates for the large grants are: December, 1; March, 6; June, 5; September, 4; November, 20.
Requirements: The Charity only considers applications from anyone working in one of the four NHS Trusts in Lambeth or Southwark, although there are some restrictions for mental health staff.
Geographic Focus: All States, United Kingdom
Date(s) Application is Due: Mar 6; Jun 5; Aug 14; Sep 4; Sep 25; Oct 16; Nov 20; Dec 1
Amount of Grant: 20,000 - 1,000,000 GBP
Contact: Anne Rigby, Grants Manager; 020 7188 1227; anne.rigby@gsttcharity.org.uk
Internet: http://www.gsttcharity.org.uk/grants/index.html
Sponsor: Guy's and St. Thomas' Charity
Freepost Lon 15724
London, SE1 9YA England

Guy I. Bromley Trust Grants 1919
The Guy I. Bromley Trust was established in 1964 to support and promote quality educational, cultural, human-services, and health-care programming. In the area of education the trust supports programming that: promotes effective teaching; improves the academic achievement of, or expands educational opportunities for disadvantaged students; improves governance and management; strengthens nonprofit organizations, school leadership, and teaching; and bolsters strategic initiatives of area colleges and universities. In the area of culture the trust supports programming that: fosters the enjoyment and appreciation of the visual and performing arts; strengthens humanities and arts-related education programs; provides affordable access; enhances artistic elements in communities; and nurtures a new generation of artists. In the area of human services, the trust supports programming that: strengthens agencies that deliver critical human services and maintains the community's safety net; and helps agencies respond to federal, state, and local public policy changes. In the area of health the trust supports programming that: improves the delivery of health care to the indigent, uninsured, and other vulnerable populations; and addresses health and health-care problems that intersect with social factors. Grant requests for general operating support and program support will be considered. Grants from the trust are one year in duration. There are no application deadlines for the Bromley Trust. Proposals are reviewed on an ongoing basis. Downloadable application materials are available at the grant website. Applicants are encouraged to review the downloadable state guidelines at the grant website for further information and clarification before applying. Applicants are also encouraged to view the trust's funding history (link is available at the grant website).
Requirements: Applicants must have 501(c)3 tax-exempt status and serve the residents of Atchison, Kansas and the Greater Kansas City Metropolitan area. Applications must be mailed.
Restrictions: Grant requests for capital support will not be considered. The trust does not support requests from individuals, organizations attempting to influence policy through direct lobbying, or any political campaigns.
Geographic Focus: All States
Contact: Spence Heddens; (816) 292-4301; Spence.heddens@baml.com
Internet: https://www.bankofamerica.com/philanthropic/fn_search.action
Sponsor: Guy I. Bromley Trust
1200 Main Street, 14th Floor, P.O. Box 219119
Kansas City, MO 64121-9119

H & R Foundation Grants 1920
The Foundation prefers to support projects and programs that strive for excellence, improve service for clients and strengthen the organization. A major emphasis is placed on support of activities that serve underserved, low-income persons living in Jackson, Clay, and Platte counties in Missouri, and Wyandotte and Johnson counties in Kansas.
Requirements: Grants are made only to organizations that are tax-exempt from federal income taxation pursuant to Section 501(c)3 and that are not classified as private foundations within the code.
Restrictions: Except in the most unusual circumstances, the Foundation does not make grants to: organizations that knowingly discriminate on the basis of race, color, religion,

gender, national origin, sexual orientation, age, disability, marital status, or status as a veteran; organizations that are not in compliance with laws and regulations that govern them; individuals or businesses; publications; projects for which the Foundation must exercise expenditure responsibility; single-disease causes; travel or conferences; historic preservation; telethons, dinners, advertising, sponsorships, or other special events; animal-related causes; sports-related causes.
Geographic Focus: Kansas, Missouri
Date(s) Application is Due: Feb 26; Apr 30; Jul 30; Oct 15
Contact: David Miles, President; (816) 854-4372 or (816) 854-4361; fax (816) 854-8025; davmiles@hrblock.com
Internet: http://www.blockfoundation.org/nonprofit_organizations/faqs.html
Sponsor: H and R Foundation
One H&R Bloack Way
Kansas City, MO 64105

H.A. and Mary K. Chapman Charitable Trust Grants 1921
H. Allen Chapman was born in Colorado in 1919. In 1976, he established the H.A. and Mary K. Chapman Charitable Trust, a perpetual charitable private foundation that maintains endowments to fund charitable grants to public charities. The trustees and staff that administer the foundation also provide public stewardship through service to charitable organizations and causes. A major charitable focus of H. A. Chapman during his life, and the lives of his philanthropic parents, James A. and Leta Chapman, was education and medical research. Though not limited geographically, most grants and public service are within Oklahoma. Grants to human services and civic and community programs and projects are primarily focused in the area of Tulsa. There are two steps in the process of applying for a grant. The first is a Letter of Inquiry from the applicant. This letter is used to determine if the applicant will be invited to take the second step of submitting a formal Grant Proposal.
Requirements: IRS 501(c)3 non-profits are eligible to apply.
Restrictions: Grant requests for the following purposes are not favored: endowments, except as a limited part of a capital project reserved for maintenance of the facility being constructed; deficit financing and debt retirement; projects or programs for which the Chapman Trusts would be the sole source of financial support; travel, conferences, conventions, group meetings, or seminars; camp programs and other seasonal activities; religious programs of religious organizations; project or program planning; start-up ventures are not excluded, but organizations with a proven strategy and results are preferred; purposes normally funded by taxation or governmental agencies; requests made less than nine months from the declination of a previous request by an applicant, or within nine months of the last payment made on a grant made to an applicant; requests for more than one project.
Geographic Focus: Oklahoma
Amount of Grant: Up to 300,000 USD
Samples: Gilcrease Museum Management, Tulsa, Oklahoma, $250,000 - general operating funds; Tulsa Air and Space Museum, Tulsa, Oklahoma, $35,000 - general operating support; Friends of the Fairgrounds Foundation, Tulsa, Oklahoma, $15,000 - general operating funds.
Contact: Andrea Doyle; (918) 496-7882; fax (918) 496-7887; andie@chapmantrusts.com
Internet: http://www.chapmantrusts.org/grants_programs.html
Sponsor: H.A. and Mary K. Chapman Charitable Trust
6100 South Yale, Suite 1816
Tulsa, OK 74136

H.B. Fuller Foundation Grants 1922
The Foundation makes grants within the corporate headquarters community, and other key H.B. Fuller communities throughout the world. Within Minnesota, the Foundation recently served as a catalyst in drawing attention to and support for early family literacy programs, and now focuses on science, technology, engineering, and math education. In other communities and cultures, the foundation provides critical grants in education, arts/culture, and health/human services. As an international company, the foundation's commitments extend well beyond the Minnesota state borders. Making significant community grants to schools and non-governmental agencies (NGOs) in locations across Latin America, as well as key community investments in six North American cities where the H.B. Fuller Company has manufacturing operations. Most recently, expanding philanthropic giving into Asia. Minnesota STEM/Youth Leadership grants are reviewed twice a year with proposals accepted March 1 through March 31 and August 1 through August 31. All other North American grants are accepted March 1 through October 1. Individual grants typically range from $5,000 to $15,000.
Requirements: 501(c)3 organizations serving the communities where the company has operations are eligible. Available funding is directed from Fuller's headquarters in St. Paul, Minnesota, and the following North American locations where the company operates: Aurora, Illinois; Grand Rapids, Michigan; Greater Atlanta, Georgia; Paducah, Kentucky; and Vancouver, Washington. Organizations incorporated in countries other than the United States must qualify for tax-exempt status according to U.S. tax regulations and comply with national and/or state charity laws.
Restrictions: Funding is not available for: individuals, including scholarships for individuals; fraternal or veterans' organizations except for programs which are of direct benefit to the broader community; religious groups for religious purposes; political/lobbying organizations; travel; basic or applied research; disease specific organizations; courtesy, goodwill or public service advertisements; fundraiser events or sponsorships; general support of education institutions; capital campaigns; or endowments.
Geographic Focus: Georgia, Illinois, Kentucky, Michigan, Minnesota, Washington, Abkhazia, Afghanistan, Armenia, Azerbaijan, Bahrain, Bangladesh, Bhutan, British Indian Ocean Territory, Brunei, Burma (Myanmar), Cambodia, China, Christmas Island, Cocos, Cyprus, Hong Kong, India, Indonesia, Iran, Iraq, Israel, Japan, Jordan, Kazakhstan, Kuwait, Kyrgyzstan, Laos, Lebanon, Macau, Malaysia, Maldives, Mongolia, Nagorno-Karabakh, Nepal, North Korea, Northern Cyprus, Oman, Pakistan, Palestinian Authority, Philippines, Qatar, Russia, Saudi Arabia, Singapore, South Korea, South Ossetia, Sri Lanka, Syria, Taiwan, Tajikistan, Thailand, Timor-Lester, Turkey, Turkmenistan, United Arab Emirates, Uzbekistan, Vietnam, Yemen
Date(s) Application is Due: Mar 31; Aug 31; Oct 1
Amount of Grant: 5,000 - 15,000 USD
Contact: Jim Owens, President; (651) 236-5104 or (651) 236-5900
Internet: http://www.hbfuller.com/About_Us/Community/000110.shtml#P0_0
Sponsor: H.B. Fuller Foundation
1200 Willow Lake Boulevard, P.O. Box 64683
Saint Paul, MN 55164-0683

H.J. Heinz Company Foundation Grants 1923
The Foundation is committed to promoting the health and nutritional needs of children and families. Priority is given to programs in communities where Heinz operates with a special focus given to southwestern Pennsylvania. The Foundation proactively donates funds to develop and strengthen organizations that are dedicated to nutrition and nutritional education, youth services and education, diversity, healthy children and families, and quality of life. Application information is available online.
Requirements: Only organizations that have 501(c)3 tax status under the U.S. Internal Revenue Code are eligible for support domestically. International organizations are encouraged to submit a letter of inquiry prior to preparing a full proposal to determine eligibility. International organizations will need to provide a 501(c)3 determination letter from the United States Internal Revenue Service or sufficient documentation to demonstrate that the non-U.S. grantee is the equivalent of a U.S. public charity. Documentation should be provided in English. The Foundation will also accept proposals in the Common Grant Application Format. All proposals must be submitted in writing. Electronic submissions are not accepted at this time.
Restrictions: The Foundation will not provide grants to individuals nor make multi-year pledges except for major capital or grant campaigns. Generally, the Foundation does not make loans and does not provide grants for individuals, equipment, conferences, travel, general scholarships, religious programs, political campaigns, and unsolicited research projects.
Geographic Focus: All States
Contact: Tammy B. Aupperle, Program Director; (412) 456-5773; fax (412) 442-3227; heinz.foundation@hjheinz.com
Internet: http://www.heinz.com/sustainability.aspx/social/heinz-foundation.aspx
Sponsor: H.J. Heinz Company Foundation
P.O. Box 57
Pittsburgh, PA 15230-0057

H. Leslie Hoffman and Elaine S. Hoffman Foundation Grants 1924
The foundation awards general operating grants to California non-profits, primarily focusing on education. Additional funding is also available in the following areas of interest: arts; social services; hospitals; health organizations; children/youth services, including children's hospitals and, social services. There are no application forms or deadlines.
Requirements: California IRS 501(c)3 non-profits are eligible to apply. The majority of grants are funded in the Los Angeles area with a special emphasis on Pasadena.
Restrictions: Individuals are ineligible.
Geographic Focus: California
Amount of Grant: 1,000 - 200,000 USD
Samples: All Saints Church, Pasadena, CA, $10,000--general funding; Hillsides Home For Children, Pasadena, CA, $5,000--general funding; Descanso Gardens, Lacanada, CA, $1,000--annual benefit.
Contact: J. Kristoffer Popovich, Treasurer; (626) 793-0043
Sponsor: H. Leslie Hoffman and Elaine S. Hoffman Foundation
225 S Lake Avenue, Suite 1150
Pasadena, CA 91101-3005

H. Schaffer Foundation Grants 1925
The foundation awards grants to nonprofit organizations in New York in the fields of higher education, hospitals and health care, human services, international relief, theater, aging centers and services, federated giving programs, and Jewish agencies and temples. Eligible activites include recreation, children and youth, services, and general charitable giving. There are no application forms or deadlines.
Requirements: Nonprofits in New York, with an emphasis on Schenectady, are eligible.
Geographic Focus: New York
Amount of Grant: 1,000 - 250,000 USD
Contact: Sonya Stall, President; (518) 580-0188
Sponsor: H. Schaffer Foundation
2 Claire Pass
Saratoga Springs, NY 12307

Hackett Foundation Grants 1926
The Foundation is limited by its giving to the funding of grants for supplies and equipment. Grants are primarily given, but not limited to, Catholic Missions. The primary focus of the Foundation is to provide assistance to those Catholic Organizations which promote the Health, Welfare, Education and Independence of individuals. Application information is available online.
Restrictions: Grants to individuals, scholarships, endowments and fellowships will not be considered. Grants will not be given to supporting organizations having a 509(a)(3) status. Grants will not be used as reimbursement for previously purchased items. Grants are not

given for salaries, administrative expenses or matching funds. Grants will not be considered for demonstration projects and/or capital campaigns. Grants will not be provided for non-U.S. based organizations or for taxes or shipping and handling charges.
Geographic Focus: New Jersey, New York, Pennsylvania
Amount of Grant: 1,000 - 20,000 USD
Contact: Maggie Hackett, Grants Manager; (908) 238-9444
Internet: http://fdncenter.org/grantmaker/hackett
Sponsor: Hackett Foundation
P.O. Box 693
Pittstown, NJ 08867

HAF Barry F. Phelps Fund Grants 1927
The Barry F. Phelps Fund was established by the Phelps family after their son passed away from leukemia at the age of 9. The fund is intended for young survivors of leukemia residing in the Eel River Valley, and assists families with the tremendous cost of an anticipated bone marrow transplant. Applications are available at the website, and may be submitted at any time.
Requirements: Applications must be made through a qualified sponsor, such as a recognized social service agency, school counselor, or medical provider, who will help administer the funds which are granted.
Geographic Focus: California
Contact: Amy Jester, Program Manager, Health and Nonprofit Resources; (707) 442-2993, ext. 374; fax (707) 442-9072; amyj@hafoundation.org
Internet: http://www.hafoundation.org/haf/grants/affiliated-grants.html
Sponsor: Humboldt Area Foundation
373 Indianola Road
Bayside, CA 95524

HAF David (Davey) H. Somerville Medical Travel Fund Grants 1928
The Somerville family established the Somerville Medical Travel Fund after their son passed away from a rare form of cancer. After numerous treatments outside of the area, the Somervilles recognized the need to establish this fund to assist in easing some of those travel expenses. The David H. Somerville Medical Travel Fund provides small travel grants to aid families that need to travel out of Humboldt county for their children's medical care. Applications are available at the website and reviewed at any time.
Requirements: Applications should include a note from the child's doctor with the date of the appointment, name, phone number and location of the hospital or clinic, and doctor the child will be seeing. Applications should also indicate if the family is connected with the services of the American Cancer Society, California Children's Services, or has applied for assistance through the Union Labor Health Foundation Angel Fund.
Geographic Focus: California
Contact: Amy Jester, Program Manager, Health and Nonprofit Resources; (707) 442-2993, ext. 374; fax (707) 442-9072; amyj@hafoundation.org
Internet: http://www.hafoundation.org/haf/grants/affiliated-grants.html
Sponsor: Humboldt Area Foundation
373 Indianola Road
Bayside, CA 95524

HAF JoAllen K. Twiddy-Wood Memorial Fund Grants 1929
The JoAllen K. Twiddy-Wood Memorial Fund provides dental and vision care for children of low and medium income families in Humboldt, Del Norte, Trinity and Shasta Counties who do not qualify for insurance or who are under-insured. Applications are available at the website and are reviewed at any time.
Requirements: Service providers complete a request form for each child in need of assistance. Checks are payable to the business rendering service. Dental requests for uninsured children or those in pain can be submitted for up to $500. Applicants should submit a pretreatment plan with all dental related applications.
Restrictions: Checks cannot be made payable to the family of the child in need, or the entity sponsoring the application. This fund does not assist with medical or dental travel costs.
Geographic Focus: California
Amount of Grant: Up to 500 USD
Contact: Amy Jester, Program Manager, Health and Nonprofit Resources; (707) 442-2993, ext. 374; fax (707) 442-9072; amyj@hafoundation.org
Internet: http://www.hafoundation.org/haf/grants/affiliated-grants.html
Sponsor: Humboldt Area Foundation
373 Indianola Road
Bayside, CA 95524

HAF Phyllis Nilsen Leal Memorial Fund Gifts 1930
The Phyllis Nilsen Leal Memorial Fund provides items to lift the spirits of children receiving cancer treatment or other illnesses. Items may be a piece of equipment, a tutor, books, or toys. Applications may be submitted at any time and are available at the website.
Requirements: Applications must be made through a qualifying sponsor, such as a recognized service agency, school counselor, or medical provider. This sponsor will help administer whatever funds are granted.
Geographic Focus: California
Contact: Amy Jester, Program Manager, Health and Nonprofit Resources; (707) 442-2993, ext. 374; fax (707) 422-9072; amyj@hafoundation.org
Internet: http://www.hafoundation.org/haf/grants/affiliated-grants.html
Sponsor: Humboldt Area Foundation
373 Indianola Road
Bayside, CA 95524

HAF Riley Frazel Memorial Scholarship 1931
The Riley Frazel Memorial Scholarship is offered for graduating high school senior from Arcata or McKinleyville high schools. Recipients may be planning to attend any form of higher education (college, university, vocational or trade school), full or part time, and are not limited on where they may enroll. Any major is eligible. Financial need is considered, but grade point average is not. Community service and extra curricular activities are taken into consideration. The scholarship is worth a one time award of $1,000. The application is available at the website.
Requirements: Applicants must be graduating seniors of either Arcata or McKinleyville high schools. They must also have attended Fieldbrook Elementary School at any time from kindergarten through eighth grade. Candidates will submit the application, a personal statement, their official transcript, and two letter of recommendation.
Geographic Focus: California
Date(s) Application is Due: Mar 15
Amount of Grant: 1,000 USD
Contact: Cassandra Wagner, Program Coordinator; (707) 442-2993, ext. 323; fax (707) 442-3811; cassandraw@hafoundation.org
Internet: http://www.hafoundation.org/index.php?option=content&task=view&id=61
Sponsor: Humboldt Area Foundation
373 Indianola Road
Bayside, CA 95524

HAF Senior Opportunities Grants 1932
Humboldt Area Foundation's Field of Interest Grant Program was created to connect donors with community projects in their specific areas of interest. The program brings together a collection of funds that focus on specific populations, geographical areas and/or causes. Field of Interest grants in the Senior Opportunity category are available September 1 through November 1 for up to $2,000. Senior Opportunities Grants offer several funding sources for those wishing to support projects such as seniors living in their own homes who need meal programs and transportation, or other programs that enhance their quality of life. Applicants should apply specifically to the fund that best meets their project's needs, and may apply individually to each fund. The application and a list of previously funded projects are available at the website.
Requirements: Applicants for any fund must be nonprofit charitable or public benefit (federal tax exempt) organizations, public schools, government agencies, Indian tribal governments, or have a qualified fiscal sponsor.
Restrictions: Funding is not available for deferred maintenance or annual operating costs of public institutions, churches, and services of special tax districts, government, or cemeteries. Grants will not be made for religious activities or projects that exclusively benefit the members of sectarian or religious organizations. Expenses that have already been incurred are not eligible.
Geographic Focus: California
Date(s) Application is Due: Nov 1
Amount of Grant: Up to 2,000 USD
Contact: Cassandra Wagner, Grants Program Coordinator; (707) 422-2993, ext. 323; fax (707) 422-3811; cassandraw@hafoundation.org
Internet: http://www.hafoundation.org/haf/grants/haf-grants.html
Sponsor: Humboldt Area Foundation
373 Indianola Road
Bayside, CA 95524

HAF Technical Assistance Program (TAP) Grants 1933
The Technical Assistance Program is a small award program which provides nonprofits with one-on-one technical assistance in a variety of forms, such as consultants, training workshops, self-pace manuals, software, or any other form identified by applicants. Awards typically range from $500 to $2,000. Funds must be used to improve the organization, its leadership or operations. Areas of assistance may include the following: strategic planning; board development; community organizing training; increasing organizational diversity; financial management, operational assessments, board development; program assessments; personnel issues; collaboration building; executive coaching; succession planning; exit strategy planning; fundraising planning; and involving those who serve in program planning, development and evaluation. Applications and information are confidential, and will not affect other potential Humboldt Area Foundation grant funding. Grant applications are accepted on the first of each month, with grants paid directly to a consultant to cover the costs of their consulting services. Additional information is available at the website.
Requirements: Funding is available to organizations and community groups in Humboldt, Del Norte, and Trinity counties.
Restrictions: TAP will not pay for grant writing, programmatic development or for duties that are typically handled by staff (e.g. data entry). Organizations less than a year old and government entities are not eligible to apply.
Geographic Focus: California
Date(s) Application is Due: Jan 1; Feb 1; Mar 1; Apr 1; May 1; Jun 1; Jul 1; Aug 1; Sep 1; Oct 1; Nov 1; Dec 1
Amount of Grant: 500 - 2,000 USD
Contact: Amy Jester, Program Manager, Health and Nonprofit Resources; (707) 442-2993, ext. 374; fax (707) 442-9072; amyj@hafoundation.org
Internet: http://www.hafoundation.org/index.php?option=content&task=view&id=74
Sponsor: Humboldt Area Foundation
373 Indianola Road
Bayside, CA 95524

Hagedorn Fund Grants 1934
William Hagedorn directed that the remainder of his estate be dedicated to this fund, in memory of his late wife, Tillie Hagedorn, that would support religious or charitable organizations in the New York City region. Funding interests include: health (including cancer, HIV/AIDS, blindness), gardens, social services, youth, education, senior services and housing and community development. All applications to the Hagedorn Fund must be submitted online, additional information is available at the Hagedorn Fund website.
Requirements: New York nonprofit organizations are eligible to apply.
Restrictions: No grants are made to individuals or private foundations or for matching gifts or loans.
Geographic Focus: New York
Date(s) Application is Due: Sep 1
Amount of Grant: 5,000 - 45,000 USD
Contact: Erin K. Hogan; (212) 464-2476; fax (212) 464-2305; erin.k.hogan@jpmorgan.com
Internet: http://fdncenter.org/grantmaker/hagedorn
Sponsor: Hagedorn Fund
270 Park Avenue, 16th Floor
New York, NY 10017

Hall-Perrine Foundation Grants 1935
The Hall-Perrine Foundation Grants funds charitable projects for nonprofit tax-exempt organizations that benefit the Linn County, Iowa community. The Foundation screens grant applications, and those meeting the established criteria are presented to the Board of Directors. The Board reviews proposals approximately four times a year. After a decision has been reached, the organization is notified in writing of whether they have been approved for funding.
Requirements: Potential grantees should first make a preliminary inquiry to determine the Foundation's interest in their request. This communication should briefly describe the background and purposes of the organization and outline the proposed project and its goals. If the Foundation is interested, a grant application will be given to the applicant to be completed in order that the proposal may be properly evaluated. Information requested includes: a description of the organization, including its legal name, history, purposes, and activities; a list of members of the governing board; a clear and detailed description of the purpose for which the grant is requested and the goals to be achieved; the total cost of the project and amount requested with a detailed budget; a list of current and potential sources of financial support; a copy of the organization's most recent audited financial statement or the last IRS Form 990 (income tax return of organization exempt from income tax); and a copy of the IRS determination letter indicating 501(x)3 tax-exempt status.
Geographic Focus: Iowa
Contact: Kristin Novak; (319) 362-9079; fax (319) 362-7220; kristin@hallperrine.org
Internet: http://www.hallperrine.org/
Sponsor: Hall-Perrine Foundation
115 Third Street SE, Suite 803
Cedar Rapids, IA 52401-1222

Hallmark Corporate Foundation Grants 1936
The mission of the Foundation is to help create communities where: all children have the chance to grow up as healthy, productive and caring persons; vibrant arts and cultural experiences enrich the lives of all citizens; there is a strong infrastructure of basic institutions and services, especially for persons in need; and all citizens feel a responsibility to serve their community. Proposals are accepted and reviewed throughout the year. There are no deadlines. The application process can be completed online using the Hallmark General Grant Application, which is available on the Hallmark website.
Requirements: IRS Non-profit 501(c)3 organizations are eligible to apply in the following areas: Kansas City metropolitan area; Center, Texas; Columbus, Georgia; Enfield, Connecticut; Lawrence, Kansas; Leavenworth, Kansas; Liberty, Missouri; Metamora, Illinois and; Topeka, Kansas.
Restrictions: Funding is not available for: individuals for any purpose, including travel, starting a business or paying back loans; religious organizations unless they can demonstrate that services are provided to the community-at-large and separated from religious purposes; fraternal, international and veterans organizations; sports teams and athletic organizations; individual youth clubs, troops, groups or school classrooms; social clubs; disease-specific organizations whose local chapters primarily raise funds for national research; past operating deficits; endowment or foundation funds; conferences; scholarly or health-related research; scholarship funds.
Geographic Focus: Connecticut, Georgia, Illinois, Kansas, Missouri, Texas
Contact: Manager; (816) 545-6906; contributions@hallmark.com.
Internet: http://corporate.hallmark.com/Community/Community-Involvement-Program-Guidelines
Sponsor: Hallmark Corporate Foundation
Mail Drop 323, P.O. Box 419580
Kansas City, MO 64141-6580

Hamilton Company Syringe Product Grant 1937
Hamilton is awarding up to five $1,000 syringe product grants per month. Applications must be completed and received by the last Friday of each month to be considered for that month's award. Hamilton Company will notify the winners by email and/or telephone no later than the second Tuesday of the following month. Each grant is for product credit to be used towards Hamilton syringes and needles. Applications will be evaluated primarily on the potential to positively impact the student's learning experience. Secondary consideration will be given to the demonstrated knowledge of Hamilton products, the proposed use of those products and any general information the applicant would like to share with the grant review board. Lastly, the composition and effort of the written sections will also be considered. Applications may be submitted electronically at any time using the links on the website.
Requirements: Multiple forms may be submitted by the same applicant, college/university or other facility as long as the need demonstrated or scientific application described is different within each form. Grants will be awarded to teaching facilities that conduct labs or classes where Hamilton syringes or needles will be used as learning or teaching aids. There is no need to submit an application each month for the same need or application defined on the initial application. If a grant is not awarded in the first month after submission your application will be kept on file and will be considered again each month for six months.
Restrictions: The grant is void where prohibited by law.
Geographic Focus: All States
Amount of Grant: 1,000 USD
Contact: Donna Baird, Marketing Content Manager; (775) 858-3000, ext. 449 or (800) 648-5950, ext. 449; fax (775) 856-7259; donna.baird@hamiltoncompany.com
Internet: http://www.hamiltongrants.com/
Sponsor: Hamilton Company
4970 Energy Way
Reno, NV 89502

Hammond Common Council Scholarships 1938
The Hammond Common Council Scholarships are awarded to graduating Hammond high school seniors. Fifteen scholarships of $1,000 each are awarded annually. Any college major is eligible.
Requirements: Applicants must reside in Hammond, Indiana. They must also enroll in a college or university of their choice as a full time student with at least a 2.0 in a two or four year program. Applications are available at Hammond high school guidance offices between February 15 and April 15.
Geographic Focus: Indiana
Amount of Grant: 1,000 USD
Contact: Hammond Council Scholarship Contact; (219) 736-1880; fax (219) 736-1940; legacy@legacyfoundationlakeco.org
Internet: http://www.legacyfoundationlakeco.org/scholarships/index.htm
Sponsor: Legacy Foundation
1000 East 80th Place, 302 South
Merrillville, IN 46410

Hampton Roads Community Foundation Developmental Disabilities Grants 1939
The Foundation's mission is to inspire philanthropy and transform the quality of life in southeastern Virginia. Developmental Disabilities Grants are funded by the Laura Turner Fund and the Jennifer Lynn Gray Fund to support organizations and programs that help people with disabilities live better lives. The annual deadline for online applications is October 1. Applicants from the Eastern Shore of Virginia should speak with a Program Officer before submitting a proposal. Awards typically range from $1,000 to $2,000.
Requirements: Nonprofit organizations serving residents of south Hampton Roads (Chesapeake, Franklin, Norfolk, Portsmouth, Suffolk, Virginia Beach and Isle of Wight County) and the Eastern Shore of Virginia are eligible.
Geographic Focus: Virginia
Date(s) Application is Due: Oct 1
Amount of Grant: 1,000 - 2,000 USD
Contact: Linda M. Rice, Vice President of Grantmaking; (757) 622-7951; fax (757) 622-1751; lrice@hamptonroadscf.org or grants@hamptonroadscf.org
Internet: http://www.hamptonroadscf.org/nonprofits/specialInterestGrants.html
Sponsor: Hampton Roads Community Foundation
101 W. Main Street, Suite 4500
Norfolk, VA 23510

Hampton Roads Community Foundation Faith Community Nursing Grants 1940
The Hampton Roads Community Foundation's mission is to inspire philanthropy and transform the quality of life in southeastern Virginia. The Nightingale Fund was established in 2004 to support established Faith Community Nursing programs in the Hampton Roads area of Virginia. Programs were formerly known as Parish Nursing programs. Eligible expenses include: medical and educational equipment; specific programs; educational materials; and supplies. The annual deadline for applications is October 31. Applicants from the Eastern Shore of Virginia should speak with a Program Officer before submitting a proposal. Awards generally range from $500 to $1,500.
Requirements: Eligible churches, synagogues or other religious institutions located in the Hampton Roads area of Virginia must have operated a Faith Community Nursing program for at least one year and be classified as tax exempt by the Internal Revenue Service.
Restrictions: Support is not available for: salaries or stipends; financial assistance to individuals; and activities normally considered to be the responsibility of the requesting religious institution.
Geographic Focus: Virginia
Date(s) Application is Due: Oct 31
Amount of Grant: 500 - 1,500 USD
Contact: Susan Saunders, (757) 622-7951; fax (757) 622-1751; SusanSaundersRN@cox.net or grants@hamptonroadscf.org
Internet: http://www.hamptonroadscf.org/nonprofits/nightingaleGrants.html
Sponsor: Hampton Roads Community Foundation
101 W. Main Street, Suite 4500
Norfolk, VA 23510

Hampton Roads Community Foundation Health and Human Service Grants   1941
The Foundation's mission is to inspire philanthropy and transform the quality of life in southeastern Virginia. Community Grants for Health and Human Services focus on providing opportunities for disadvantaged people to become self-sufficient. The Foundation believes the region will thrive only when its most vulnerable residents have the opportunities and support needed to succeed. It wants to support innovative programs that: prevent and alleviate homelessness; improve delivery of basic human services to those in need; improve access to medical care, including dental care and mental health; develop job skills and employment opportunities; and develop sound financial education and savings programs to help low income people build financial assets for long-term economic well-being. Proposals for program funding must be able to articulate: the program's fit with the Foundation's stated priorities; the program's desired outcomes; and the organization's plan for measuring program effectiveness in reaching outcomes. The annual deadline for online applications is July 1.
Requirements: Nonprofit organizations serving residents of south Hampton Roads (Chesapeake, Franklin, Norfolk, Portsmouth, Suffolk, Virginia Beach and Isle of Wight County) and the Eastern Shore of Virginia are eligible. Proposals must articulate the program's desired outcomes and the measurement of effectiveness in reaching those outcomes.
Restrictions: Funding is generally not available for: individuals; fundraising events (such as tickets, raffles, auctions or tournaments), annual fundraising appeals or agency celebrations; ongoing operating support; capital projects and facilities and equipment upgrades that can be considered routine maintenance or replacements; houses of worship unless applying for the Nightingale Fund for faith community nursing or the E.K. Sloane Piano Fund; religious activities (organizations and activities that require religious participation by those receiving services); political or fraternal organizations; endowment building; existing obligations, debts/liabilities or costs that the agency has already incurred; scholarly research; scholarships, camper fees, fellowships or travel; passenger vans for transporting youth; national or international organizations or purposes; hospitals and similar health-care facilities unless applying for a Special Interest Grant; projects or services normally considered the responsibility of government; private primary or secondary schools, daycare facilities or academies other than those whose primary purpose is for students with special needs; and capital campaign requests exceeding 5% of campaigns valued at $1 million or more.
Geographic Focus: Virginia
Date(s) Application is Due: Jul 1
Amount of Grant: 15,000 - 200,000 USD
Samples: Portsmouth Volunteers for the Homeless, Portsmouth, Virginia, $18,000 - for rapid re-housing and case management for homeless adults in Portsmouth and for services to help others avoid becoming homeless (2014); Samaritan House, Virginia Beach, Virginia, $156,400 - over two years to expand a call center for the Bringing an End to All City Homelessness (2014); Urban League of Hampton Roads, Hampton Roads, Virginia, $90,000 - over three years to create a regional financial empowerment center to help low- and moderate-income families (2014).
Contact: Linda M. Rice, Vice President of Grantmaking; (757) 622-7951; fax (757) 622-1751; lrice@hamptonroadscf.org or grants@hamptonroadscf.org
Internet: http://www.hamptonroadscf.org/nonprofits/communityGrants-Health.html
Sponsor: Hampton Roads Community Foundation
101 W. Main Street, Suite 4500
Norfolk, VA 23510

Hampton Roads Community Foundation Mental Health Research Grants   1942
The Hampton Roads Community Foundation's mission is to inspire philanthropy and transform the quality of life in southeastern Virginia. Funding for Mental Health Research Grants is provided by the Benjamin R. Brown and Charles G. Brown Funds. The annual deadline for online applications is October 1. Applicants from the Eastern Shore of Virginia should speak with a Program Officer before submitting a proposal. Grant awards typically range from $1,500 to $15,000.
Requirements: Nonprofit organizations serving residents of south Hampton Roads (Chesapeake, Franklin, Norfolk, Portsmouth, Suffolk, Virginia Beach and Isle of Wight County) and the Eastern Shore of Virginia are eligible.
Geographic Focus: Virginia
Date(s) Application is Due: Oct 1
Contact: Linda M. Rice, Vice President of Grantmaking; (757) 622-7951; fax (757) 622-1751; lrice@hamptonroadscf.org or grants@hamptonroadscf.org
Internet: http://www.hamptonroadscf.org/nonprofits/specialInterestGrants.html
Sponsor: Hampton Roads Community Foundation
101 W. Main Street, Suite 4500
Norfolk, VA 23510

Hancock County Community Foundation - Field of Interest Grants   1943
Field of Interest Grants are offered to organizations in Hancock County that present programs in one of the following areas: addictions and/or substance abuse education and treatment; arts and culture; Hanson Family Endowment Fund; Kingery Friends of Domestic Animals; Lifelong Learning; and Prevent Child Abuse. Successful applications will: document the program's relevance to the organization's mission, as well as the need in Hancock County for the program; show that the organization has a past record of being financially stable, viable, and able to sustain past the grant's time frame; operate effective programs that benefit Hancock County and its residents; show strong management by organization's Board of Directors and executive staff. Detailed guidelines and the application are available at the website.
Requirements: Organizations must be a certified nonprofit with a physical location in Hancock County in order to apply for any grants. Organizations may apply for funding up to the amount available, but must be in good standing with HCCF.
Restrictions: Organizations are only eligible to receive one Field of Interest Grant. Religious organizations may apply, but only for general community programs. No grants will be made specifically for religious purposes.
Geographic Focus: Indiana
Date(s) Application is Due: Aug 3
Amount of Grant: Up to 25,000 USD
Contact: Alyse Vail; (317) 462-8870, ext. 226; fax (317) 467-3330; avail@hccf.cc
Internet: http://www.hccf.cc/GrantTypes.aspx
Sponsor: Hancock County Community Foundation
312 Main Street
Greenfield, IN 46140

Hannaford Charitable Foundation Grants   1944
The Foundation awards grants to nonprofit organizations in its areas of interest, including health and welfare, educational institutions, civic and cultural organizations, and other local charitable organizations. Types of support include capital campaigns, scholarship funds, and exchange programs. There are no application forms or deadlines. Preference for funding is given to organizations or programs that involve Hannaford associates, are located in Hannaford's marketing area, and have the potential to provide ongoing services for their customers. Small to medium requests are reviewed monthly. Allow three to four months for a response to larger requests ($50,00 or more), as these are reviewed quarterly. Large grants are usually reserved for capital drives by organizations with strong community-impact potential.
Requirements: To apply, the organization or program must: have an active and responsible board of trustees; exhibit ethical publicity methods and solicitation of funds; provide for an appropriate audit to reveal income and disbursements in reasonable detail; demonstrate long-term financial viability; be tax-exempt as described in both sections 501(c)3 and 509(a)1, 509(a)3, or 509(a)3 of the Internal Revenue Code. In addition to a letter of inquiry sent to the Foundation, organizations should also send ten copies of the following information: name, address and telephone number of your organization; contact person and title; amount requested; population and geographic area served; a two- or three-sentence mission statement of your organization, with a brief description of its history; a two- or three-page description of the specific project or program for which you are seeking funding; a list of current and potential funding sources; and a recent statement of revenues and expenses. They should also send one set of the following information: a copy of the organization's tax exemption letter, indicating both sections 501(c)3 and 509(1) status; most recent Form 990 return; and a letter attesting that the organization's tax-exempt status is current.
Restrictions: The Foundation does not offer support for the following: individuals; tax-supported institutions; institutions that, by virtue of their charters, programs or policies, are open to a relatively small or restricted segment of the public; operations of veterans, fraternal or religious organizations, except those that make their services fully available to the community on a nonsectarian basis; program advertising; operating expenses; scholarship programs outside of the Foundation's own; and organizations or events outside of the Foundation's marketing area.
Geographic Focus: Maine, Massachusetts, New Hampshire, New York, Vermont
Contact: Grants Administrator; (507) 931-1682
Internet: http://www.hannaford.com/content.jsp?pageName=charitableFoundation&leftNavArea=AboutLeftNav
Sponsor: Hannaford Charitable Foundation
P.O. Box 1000
Portland, ME 04104

Harden Foundation Grants   1945
The Harden Foundation considers grants in the following focus areas: children, youth, and families; senior citizens; general health; agricultural education; animal welfare; environment; and arts and culture. The goals of the Foundation are to improve the well-being of youth; strengthen families; develop individual self-reliance and health; prevent inappropriate institutionalization of individuals; improve the quality of life through the cultural activities; encourage more humane treatment of animals; and eliminate duplication and improve coordination of social and community services in Monterey County areas, with emphasis on the Salinas Valley. Organizations may refer to the Foundation website for examples of funded projects in the Monterey area.
Requirements: Organizations should download the Common Grant application from the Foundation's website. Applicants must also include the following attachments: the completed grant application checklist; a completed and signed grant application form, with signature of an authorized representative; a detailed list of Board of Directors and staff roster; organizational chart, if available; current annual budget; a recent financial statement; year-to-date organizational financial statements; a detailed project budget; a list of the ten largest financial gifts received in the most recent fiscal year; a letter from the board chairperson or board member indicating approval for this application; and, if applicable, a completed grant report for a previous grant. The organization must also submit a 3-5 page narrative briefly describing the organization's history, major accomplishments, and their current programs and activities. They should also include a description of their constituency, how they are involved in their work, and how they benefit from the program. The Common Impact Evaluation Plan is not required for Harden Foundation applicants, but applicants must also include the following: a problem statement describing what problems, needs or issues are addressed, how this was determined, and how the project addresses and/or changes the underlying or root cause of the problem. If the funding requested is not for operating support, describe the project, why it is being pursued, and whether it is new or an expansion of an existing program. Include a list of all other grant requests, pending and approved, for this project, showing funding source and amount requested. Please refer to the Common Grant Guidelines for

specific instructions on submitting the application. Applications may be mailed, dropped off, or emailed to the Foundation's office. Faxed applications are not accepted.
Restrictions: The Harden Foundation does not make grants to organizations headquartered outside of Monterey County unless they demonstrate service to the Monterey area. The Foundation does not fund: organizations that do not have a current 501(c)3 status; organizations that support sectarian religious programs; creation or addition to endowments; annual events, conferences, or fundraising events; academic or medical research or scholarships to individuals; foundations or associations established for the benefit of an organization which receives substantial tax support; individuals; political parties, candidates or partisan political organizations or activities.
Geographic Focus: California
Date(s) Application is Due: Mar 1; Sep 1
Amount of Grant: 5,000 - 100,000 USD
Contact: Joseph Grainger; (831) 442-3005; fax (831) 443-1429; joe@hardenfoundation.org
Internet: http://www.hardenfoundation.org/grant-information-introduction.html
Sponsor: Harden Foundation
P.O. Box 779
Salinas, CA 93902

Hardin County Community Foundation Grants 1946
The Hardin County Community Foundation was established to be the vehicle for those interested in fostering the present and future needs of the people of all ages in Hardin County, Ohio, in the fields of arts, health, education, recreation and beautification. Most recently, awards have ranged from $400 to $5,000. Applications are accepted annually between February 1 and February 28.
Requirements: Any 501(c)3 organization serving the residents of Hardin County, Ohio, is eligible to apply.
Geographic Focus: Ohio
Date(s) Application is Due: Feb 28
Amount of Grant: 400 - 5,000 USD
Samples: Ada Food Pantry, Ada, Ohio, $900 - support food assistance for the needy (2014); Alger First United Methodist Church- Home Missions, Alger, Ohio, $4,000 - support of mission work (2014); Dolly Parton Imagination Library of Hardin County, Kenton, Ohio, $2,000 - support of children between the ages of birth through 5 to have access to books (2014).
Contact: Saundra Neely, Trustee; hardinfoundation@gmail.com
Internet: http://www.hardinfoundation.org/
Sponsor: Hardin County Community Foundation
P.O. Box 343
Kenton, OH 43326

Harley Davidson Foundation Grants 1947
The mission of the Foundation is to support communities with funding and employee volunteerism. The Foundation reaches out to build healthy, thriving communities by placing an emphasis on education, community revitalization, arts and culture, health, and the environment. The Foundation targets areas of greatest need among under-served populations to enhance the quality of life in their communities. The Foundation also supports selected national causes, including veterans initiatives. In making granting decisions, the Foundation looks closely for the following detail: the proposal's relevance to the Foundation's areas of interest; the proposal's clarity in stating the expected project or program outcomes and the strategy for achieving them; the extent to which the strategy involves a collaborative approach to solving a problem or issue affecting the targeted groups or organizations. The program you are nominating should focus on one of the Foundation's funding priorities: education (core curriculum, academic enhancers), community revitalization, job enablers, neighborhood, social services, arts & culture, health, environment. The Common Application Form is located on the company website.
Requirements: 501(c)3 or 170(c) organizations serving metro Milwaukee, Wauwatosa, Franklin, Menomonee Falls or Tomahawk, WI; York, PA; Talladega, AL; or Kansas City, MO are eligible.
Restrictions: The Foundation does not make grants to individuals, political causes or candidates, operating or endowment funds; athletic teams, or religious organizations (unless engaged in a major project benefiting the greater community). The Foundation does not fund conferences or capital campaigns.
Geographic Focus: Alabama, Missouri, Pennsylvania, Wisconsin
Date(s) Application is Due: Jan 23; May 22; Oct 24
Contact: Mary Anne Martiny, Secretary; (414) 343-4001; fax (414) 343-4254; ma.martiny@harley-davidson.com
Internet: http://www.harley-davidson.com/CO/FOU/en/foundation.asp?locale=en_US&bmLocale=en_US
Sponsor: Harley Davidson Foundation
P.O. Box 653
Milwaukee, WI 53201

Harmony Project Grants 1948
The Harmony Project is an innovative, creative nonprofit foundation established in Ohio in 2001 by Judith Harmony to promote the training of doctors and scientists, as well as research at Children's Hospital in Cincinnati. Currently, its primary mission is to enhance the education and personal development of girls, including self-efficacy, leadership and mother-daughter relationships, and to enrich their cultural and artistic lives. Funding is given for: curriculum development; program development; evaluation; research; and scholarship endowments. That support is provided within the greater Cincinnati area, as well as in northern Kentucky. An application form is required, though an initial approach should be by letter to the Foundation office. There are no specific annual submission deadlines.
Requirements: The Foundation accepts written applications from 509(a)1, 509(a)2, and 509(a)3 organizations supporting the residents of greater Cincinnati and northern Kentucky.
Geographic Focus: Kentucky, Ohio
Contact: Judith Harmony; (513) 861-8490; fax (513) 281-4326; jharmony@fuse.net
Internet: http://www.harmonyproject.org/
Sponsor: Harmony Project
3950 Rose Hill Avenue
Cincinnati, OH 45229-1448

Harold Alfond Foundation Grants 1949
The foundation awards grants to eligible organizations in its areas of interest, including secondary education, higher education, medical research, health care, and general charities. Grants support: education--public and private colleges and universities, as well as private secondary schools, to fund athletically oriented capital projects and scholarship endowments; medical research--individual research projects are considered when sponsored by a recognized medical research center; health--community support for capital campaigns and endowment funds; and general charities--community organizations that support youth, the arts, persons with disabilities, underprivileged, substance abuse rehabilitation, and annual fund drives of national organizations focusing on the above areas. There are no application deadlines or forms.
Restrictions: Grants are not made to individuals.
Geographic Focus: Florida, Maine
Amount of Grant: Up to 2,000,000 USD
Contact: Gregory Powell, (207) 828-7999
Sponsor: Harold Alfond Foundation
Two Monument Square
Portland, ME 04101-4093

Harold and Arlene Schnitzer CARE Foundation Grants 1950
The Foundation's principal purpose is to assist with Jewish, cultural, youth, education, medical, social service, and community activities. Foundation focus is on proposals for projects that enhance the quality of life in Oregon and Southwest Washington. The Foundation funds grant requests for operating expenses, special projects, matching grants, multiple-year grants and capital campaigns as well as for purchase of specific items. There is no formal application form. Interested applicants may contact the Foundation for guidelines.
Requirements: Nonprofit organizations in Oregon and Washington are eligible. Organizations located in Portland are given first funding priority. Approximately 90% of grant funds stay in the Portland metropolitan area.
Restrictions: The Foundation does not provide funds for individuals, non tax-exempt organizations, other private foundations or political groups.
Geographic Focus: Oregon, Washington
Date(s) Application is Due: Feb 28; May 31; Aug 31; Nov 30
Amount of Grant: 1,000 - 5,000,000 USD
Contact: Barbara Hall, Vice President; (503) 973-0286; fax (503) 450-0810
Sponsor: Harold and Arlene Schnitzer Foundation
P.O. Box 2708
Portland, OR 97208-2708

Harold and Rebecca H. Gross Foundation Grants 1951
The Harold and Rebecca H. Gross Foundation was established in 2006 to support and promote charitable organizations that assist persons with physical disabilities to become better adjusted to their environments. Preference is given to organizations that provide direct services to physically disabled people. Grants from the Gross Foundation are primarily one year in duration. Multi-year grants may be considered on a case-by-case basis. The Harold and Rebecca H. Gross Foundation utilizes a two-phase application process. Interested organizations must submit a one-page concept paper as the first step in the process. Concept papers to the Gross Foundation are due on May 1. Applicants will be notified by letter of the Trustee's decision to invite a full proposal within four to six weeks after the concept paper deadline. Applications to the Gross Foundation are by invitation only and are due on October 1. Notification of the grant decision will be made in writing within two to three months after the application deadline. Detailed concept paper guidelines are available at the grant website. A link to the foundation's funding history is also available at the grant website.
Requirements: Applicants must have 501(c)3 tax-exempt status.
Restrictions: Grant requests for research or capital projects will not be considered. The foundation does not support requests from individuals, organizations attempting to influence policy through direct lobbying, or any political campaigns.
Geographic Focus: All States, Connecticut
Date(s) Application is Due: Oct 1
Amount of Grant: 10,000 - 60,000 USD
Contact: Kate Kerchaert; (860) 657-7016 or (860) 952-7405; kate.kerchaert@baml.com
Internet: https://www.bankofamerica.com/philanthropic/fn_search.action
Sponsor: Harold and Rebecca Gross Foundation
200 Glastonbury Boulevard, Suite # 200, CT2-545-02-05
Glastonbury, CT 06033-4056

Harold Brooks Foundation Grants 1952
Harold Brooks, of Braintree, Massachusetts, was a successful business executive and entrepreneur who manufactured and sold prefabricated structures and underground bomb shelters during the Cold War. Mr. Brooks died in 1963 and the Foundation that bears his name was established in 1984. The Harold Brooks Foundation provides assistance to causes/organizations that help the largest possible number of residents of Massachusetts' South Shore communities, especially those that support the basic human needs of South

Shore residents. The foundation supports nonprofit organizations that have the greatest impact on improving the human condition and/or that provide the neediest South Shore residents with "tools" that will help them restore their lives. The foundation focuses on five key areas: Education; Food, Agriculture, & Nutrition; Health; Housing & Shelter; and Mental Health. Multi-year (two-three years maximum) requests are welcome. Applicants must apply online at the grant website. Applicants are strongly encouraged to do the following before applying: review the downloadable Massachusetts state application procedures for additional helpful information and clarifications; review the downloadable online-application guidelines at the grant website; review the foundation's funding history (link is available from the grant website); review the online application questions in advance; and review the list of required attachments. These will generally include: a list of board members, financial statements (audited, reviewed, or compiled by independent auditor); an organization summary; a list of other funding sources; an IRS Determination letter; and other required documents. All attachments must be uploaded in the online application as PDF, Word, or Excel files. The Harold Brooks Foundation has biannual deadlines of April 1 and October 1. Applications must be submitted by 11:59 p.m. on the deadline date. Applicants for the April deadline will be notified of grant decisions before June 30. Applicants for the October deadline will be notified before December 31.
Requirements: Grants are made to 501(c)3 tax-exempt organizations that serve the following South Shore residents: Abington, Braintree, Bridgewater, Brockton, Carver, Cohasset, Duxbury, Hanover, Hanson, Hingham, Holbrook, Hull, Marshfield, Norwell, Pembroke, Plymouth, Quincy, Randolph, Rockland, Scituate, Weymouth, and Whitman.
Restrictions: Grants are made to support program/project expenses. General operating support and support for endowment campaigns are not provided. Only one request will be accepted from an organization during a twelve-month period. The foundation does not support requests from individuals, organizations attempting to influence policy through direct lobbying, or any political campaigns.
Geographic Focus: Massachusetts
Date(s) Application is Due: Apr 1; Oct 1
Contact: Miki C. Akimoto, Vice President; (866) 778-6859; miki.akimoto@baml.com
Internet: https://www.bankofamerica.com/philanthropic/fn_search.action
Sponsor: Harold Brooks Foundation
225 Franklin Street, 4th Floor, MA1-225-04-02
Boston, MA 02110

Harold R. Bechtel Charitable Remainder Uni-Trust Grants 1953
Established in Davenport, Iowa, in 1987, the Harold R. Bechtel Charitable Remainder Uni-Trust supports programs in Scott County, Iowa, with giving primarily for health associations, education, and for children, youth, and social services. The Foundation's primary interest areas include: children and youth services; community and economic development; education; family services; health organizations and associations; heart and circulatory research; higher education; human services; museums; music; and performing arts. Types of support building construction and renovation, as well as program-related funding. There are no specific deadlines, and applicants should contact the office in writing or via telephone to request application forms.
Requirements: 501(c)3 organizations serving residents of Rock County, Iowa, are eligible.
Restrictions: No grants are given to individuals, or for endowment funds, debt retirement, past operating deficit, general or continuing support, or for scholarly research in an established discipline.
Geographic Focus: Iowa
Amount of Grant: Up to 150,000 USD
Contact: R. Richard Bittner; (563) 328-3333; fax (563) 328-3352; loseband@blwlaw.com
Sponsor: Harold R. Bechtel Charitable Remainder Uni-Trust
201 West 2nd Street, Suite 1000
Davenport, IA 52801-1817

Harold R. Bechtel Testamentary Charitable Trust Grants 1954
Established in Davenport, Iowa, in 1989, the Harold R. Bechtel Testamentary Charitable Trust Grants gives primarily to higher education and youth services. Major fields of interest include: children and youth services; community development; economic development; education; human services; and medical research. Applicants should contact the office in writing or via telephone to request application forms.
Requirements: 501(c)3 organizations serving the residents of Davenport, Iowa, Rock Island, Illinois, and Moline, Illinois, are eligible to apply.
Restrictions: No grants are given to individuals, or for endowment funds, debt retirement, past operating deficit, general or continuing support, or for scholarly research in an established discipline.
Geographic Focus: Iowa
Amount of Grant: Up to 150,000 USD
Contact: R. Richard Bittner; (563) 328-3333; fax (563) 328-3352; loseband@blwlaw.com
Sponsor: Harold R. Bechtel Testamentary Charitable Trust
201 West 2nd Street, Suite 1000
Davenport, IA 52801-1817

Harold Simmons Foundation Grants 1955
The foundation awards grants to Texas nonprofit organizations, with emphasis on social services, religion, health, the arts, and youth. Grants also support community programs and projects, child development, and adult basic education/literacy programs. The foundation also supports international development and relief efforts in Third World countries. Grants are awarded for general operating support, annual campaigns, capital campaigns, building construction/renovation, continuing support, seed money, and program development. Application forms are not required, and there are no deadline dates.
Requirements: Dallas, TX, nonprofits are eligible.
Restrictions: Grants are not awarded to support individuals or for endowments or loans.
Geographic Focus: Texas
Samples: Children's Medical Ctr Dallas (TX)--to establish a hospital within the medical center devoted to the needs of child cardiac patients, $5 million; U of Texas Southwestern Medical Ctr (Dallas)--to enhance the Simmons Comprehensive Cancer Center, including supporting the work of its newly recruited director, $15.4 million.
Contact: Lisa Simmons Epstein, President; (972) 233-2134
Sponsor: Harold Simmons Foundation
5430 LBJ Freeway, Suite 1700
Dallas, TX 75240-2697

Harris and Eliza Kempner Fund Grants 1956
The foundation provides grants primarily in the Galveston, TX, area to qualifying organizations in the broad areas of the arts, historic preservation, community development, education, health, and human services. The foundation gives preference to requests for seed money, operating funds, small capital needs, and special projects partnering with other funding sources. Application information is available online.
Requirements: Grants are made primarily to Texas qualifying organizations in the greater Galveston area.
Restrictions: Funding is not available for: fund-raising benefits; direct mail solicitations; grants to individuals; and grants to non-USA based organizations.
Geographic Focus: Texas
Date(s) Application is Due: Mar 15; Oct 15
Samples: Galveston Art League, Galveston, TX, $5,000--matching grant to purchase gallery space; Avenue L Missionary Baptist Church, Galveston, TX, $20,000--renovation and restoration repairs to Fellowship Hall; Galveston Island Swim Team, Galveston, TX, $$4,600--program expansion to include all Galveston Island youth.
Contact: Harrette N. Howard, Grants Administrator; (409) 762-1603; fax (409) 762-5435; information@kempnerfund.org
Internet: http://www.kempnerfund.org/app/programs.html
Sponsor: Harris and Eliza Kempner Fund
2201 Market Street, Suite 601
Galveston, TX 77550-1529

Harrison County Community Foundation Grants 1957
The Harrison County Community Foundation (HCCF) awards grants to eligible Indiana nonprofit organizations in its areas of interest, including arts and culture; human services; recreation; government; historical preservation; community projects; education; health and safety; and environment. The HCCF staff will provide training to all eligible not-for-profits serving the community on the proper completion of our grant application. All applicants are strongly encouraged to attend formal training sessions as announced, typically in May and November.
Requirements: Nonprofit agencies providing services to residents of Harrison County, Indiana, are eligible to apply. Tax-exempt 501(c)3 organizations and schools, religious organizations, and local governmental units are eligible.
Restrictions: The Foundation does not award grants to purchase real estate that has not been identified and an offer accepted; political activities or those designed to influence legislation; individuals; travel associated with a school-sponsored event; or religious organizations for projects that do not serve the general public. Traditional equipment, routine maintenance, or facility improvements will not be funded. Unused funding cannot be carried over into the following year.
Geographic Focus: Indiana
Date(s) Application is Due: Jan 15; Jul 15
Contact: Anna Curts; (812) 738-6668; fax (812) 738-6864; annac@hccfindiana.org
Internet: http://www.hccfindiana.org/grants/
Sponsor: Harrison County Community Foundation
1523 Foundation Way, P.O. Box 279
Corydon, IN 47112

Harrison County Community Foundation Signature Grants 1958
Grant requests from the Harrison County Community Foundation (HCCF) in the amount of $200,000 and over are considered Signature Grants. These applications will be reviewed by all members of the Grants Committee. Decisions for Signature Grants may be announced anytime during the year. A vast majority of grant applications should be planned ahead and submitted during our Spring or Fall Grant Cycles, however, the HCCF is aware that some state or federal grants requiring local matching funds are announced on short notice. To support Harrison County serving not-for-profits with local matching funds, we will consider certain grant requests anytime. Emergency grants are awarded to respond to true emergency needs of our community or those that prevent an agency or program from carrying out primary functions or services.
Requirements: Nonprofit agencies providing services to residents of Harrison County, Indiana, are eligible to apply. Tax-exempt 501(c)3 organizations and schools, religious organizations, and local governmental units are eligible.
Restrictions: The Foundation does not award grants to purchase real estate that has not been identified and an offer accepted; political activities or those designed to influence legislation; individuals; travel associated with a school-sponsored event; or religious organizations for projects that do not serve the general public. Traditional equipment, routine maintenance, or facility improvements will not be funded. Unused funding cannot be carried over into the following year.
Geographic Focus: Indiana
Amount of Grant: 200,000 - 500,000 USD
Contact: Anna Curts; (812) 738-6668; fax (812) 738-6864; annac@hccfindiana.org

Internet: http://www.hccfindiana.org/grants/
Sponsor: Harrison County Community Foundation
1523 Foundation Way, P.O. Box 279
Corydon, IN 47112

Harry A. and Margaret D. Towsley Foundation Grants 1959
In 1959, Margaret Towsley created the Harry A. and Margaret D. Towsley Foundation with an initial gift of $4 Million in Dow Chemical Company stock. While the Foundation's initial goals were typical of general family foundations, its mission later became focused on programs promoting education, health care, shelter, and nutrition for children. As its assets grew, its areas of concentration expanded into college and university education, medical education, planned parenthood, and interdisciplinary programs with the schools of law and social work. These areas reflected Dr. and Mrs. Towsley's common interest in teaching. The foundation currently awards grants to Michigan organizations in its areas of interest, including environment, medical and preschool education, social services, continuing education, and research in the health sciences. Types of support include annual campaigns, building construction and renovation, capital campaigns, continuing support, employee matching gifts, endowments, general operating support, matching/challenge support, professorships, program development, research, and seed grants. There are no application forms; submit a letter of inquiry between January and the listed application deadline.
Restrictions: Grants are not awarded to individuals or for travel, scholarships, fellowships, conferences, books, publications, films, tapes, audio-visual or communication media, or loans.
Geographic Focus: Michigan
Date(s) Application is Due: Mar 31
Contact: Lynn Towsley White, President; (989) 837-1100; fax (989) 837-3240
Sponsor: Harry A. and Margaret D. Towsley Foundation
140 Ashman Street, P.O. Box 349
Midland, MI 48640

Harry B. and Jane H. Brock Foundation Grants 1960
The Harry B. and Jane H. Brock Foundation honors a former Fort Payne, Alabama, native who changed the structure of the banking industry in Alabama. Today, the Foundation awards grants to eligible Alabama organizations to support community development and higher education. Primary areas of interest include community service, community funds, education, the environment, higher education, volunteerism, women's services, social services, and cancer treatment and research. Types of support include: annual campaigns, capital campaigns, endowments, general operating support, program development, and research. Written proposals should describe the project and organization and include a copy of the IRS tax-determination letter. The annual deadline is November 1.
Requirements: 501(c)3 organizations serving the residents of Alabama are eligible.
Restrictions: No grants to individuals are awarded. Giving primarily in Birmingham and Huntsville, Alabama.
Geographic Focus: Alabama
Date(s) Application is Due: Nov 1
Amount of Grant: 1,000 - 25,000 USD
Samples: Alabama Business Hall of Fame, Tuscaloosa, Alabama, $1,700 - general operating support; Alabama Wildlife Rehab, Pelham, Alabama, $2,500 - general operating support; Alexis De Tocqueville Society, Birmingham, Alabama, $25,000 - general administration.
Contact: Harry B. Brock, Jr.; (205) 939-0236 or (205) 918-0833; fax (205) 939-0806
Sponsor: Harry B. and Jane H. Brock Foundation
2101 Highland Avenue, Suite 250, P.O. Box 11643
Birmingham, AL 35202-1643

Harry Bramhall Gilbert Charitable Trust Grants 1961
The Harry Bramhall Gilbert Charitable Trust supports tax-exempt organizations that contribute to the health, education, and cultural life of the Tidewater, Virginia region. The Trust currently and substantially funds nonprofits based in the cities of Norfolk, Chesapeake, and Virginia Beach, Virginia. To apply, submit a letter to the Trust (no application form is required) containing the following information: copy of IRS Determination Letter; copy of most recent annual report/audited financial statement/990; listing of board of directors, trustees, officers and other key people and their affiliations; detailed description of project and amount of funding requested. Include two copies.
Requirements: Virginia nonprofit organizations based in the cities of Norfolk, Chesapeake, and Virginia Beach are eligible.
Restrictions: No support for religious or political organizations. No grants to individuals.
Geographic Focus: Virginia
Date(s) Application is Due: Sep 30
Amount of Grant: 2,500 - 100,000 USD
Contact: Stuart D. Glasser, Treasurer; (757) 204-4858; sdglasser@cox.net
Internet: http://fdncenter.org/grantmaker/gilbert
Sponsor: Harry Bramhall Gilbert Charitable Trust
316 Scone Castle Loop
Chesapeake, VA 23322

Harry Edison Foundation 1962
Harry Edison Foundation was incorporated in Missouri in 2003 by one of the founders of Edison Brothers Stores, among five sons of a Latvian immigrant who had peddled shoes off a pack mule in southern Georgia. Giving is primarily centered around the Saint Louis, Missouri, region in the areas of education, hospitals, human services, Jewish organizations, and medical research. Major types of support include: annual campaigns, building and renovation costs, capital campaigns, professorships, research, and scholarship endowments. There are no specific application forms or annual deadlines, and applicants should begin by contacting the Foundation directly. Most recently, support has ranged from $100 to $120,000.
Requirements: Applicants should be 501(c)3 organizations that support the residents of St. Louis, Missouri, and its surrounding region.
Geographic Focus: Missouri
Amount of Grant: 100 - 120,000 USD
Contact: Bernard A. Edison, President; (314) 331-6504 or (314) 331-6505
Sponsor: Harry Edison Foundation
220 N. 4th Street, Suite A
St. Louis, MO 63102-1905

Harry Frank Guggenheim Foundation Dissertation Fellowships 1963
The Foundation awards ten or more dissertation fellowships each year to individuals who will complete their dissertations within the award year. These fellowships are designed to help doctoral candidates finish writing their dissertations rather than to support dissertation research. Questions that interest the Foundation concern violence and aggression in relation to social change, intergroup conflict, war, terrorism, crime, and family relationships, among other subjects. Priority will also be given to areas and methodologies not receiving adequate attention and support from other funding sources. Applicants may be citizens of any country and studying at colleges or universities in any country. The application deadline is February 1, with final decisions made by the Board of Directors at its June meeting for funding that fall.
Requirements: Applicants must submit the following to apply: title page; abstract page and survey; advisor's letter and abbreviated CV; applicant's CV and graduate school transcript; project description; protection of subjects; and their other support for funding. Applicants should refer to the guidelines for additional information and specific instructions for their abstract page and survey (based on their individual discipline), and contact the Foundation if they have questions.
Restrictions: Dissertations with no relevance to understanding human violence and aggression will not be supported. Applicants should only apply if they are entering the dissertation stage of graduate school (fieldwork or other research is complete and writing has begun). Recipients of the dissertation fellowship must submit a copy of the dissertation, approved and accepted by their institution, within six months after the end of the award year. Any papers, books, articles, or other publications based on the research should also be sent to the foundation.
Geographic Focus: All States, All Countries
Date(s) Application is Due: Feb 1
Amount of Grant: 20,000 USD
Contact: Program Officer; (646) 428-0971; fax (646) 428-0981; info@hfg.org
Internet: http://www.hfg.org/df/guidelines.htm
Sponsor: Harry Frank Guggenheim Foundation
25 West 53rd Street
New York, NY 10019-5401

Harry Frank Guggenheim Foundation Research Grants 1964
The Foundation invites proposals in any of the natural sciences, social sciences or humanities that promise to increase understanding of the causes, manifestations, and control of human violence and aggression. Highest priority is given to research that can increase understanding and improve urgent problems of violence and aggression in the modern world. Questions that interest the foundation concern violence and aggression in relation to social change, intergroup conflict, war, terrorism, crime, and family relationships, among other subjects. Priority will also be given to areas and methodologies not receiving adequate attention and support from other funding sources. Applicants may be citizens of any country. Most awards fall within the range of $15,000 to $40,000 per year for periods of one or two years. Applications for larger amounts and longer durations must be very strongly justified. Applications must be received by August 1, with a decision given in December.
Requirements: Applicants must mail two copies of the typed application in English to the Foundation. Applications may not be faxed or emailed. Along with two copies of the application, applicants must include the following: title page; abstract and survey; budget and its justification; personnel; research plan; other support; protection of subjects; and referee comments. Applicants should refer to the Foundation website for specific instructions on how to submit each of the attachments, and contact the Foundation if they have any questions.
Restrictions: Research with no relevance to understanding human problems will not be supported, nor will proposals to investigate urgent social problems where the Foundation cannot be assured that useful, sound research can be done. While almost all recipients of the Foundation grants possess a Ph.D., M.D., or equivalent degree, there are no formal degree requirements for the grant. The grant, however, may not be used to support research undertaken as part of the requirements for a graduate degree. Applicants need not be affiliated with an institution of higher learning, although most are college or university professors. The foundation awards research grants to individuals (or a few principal investigators at most) for individual projects, but does not award grants to institutions for institutional programs. Individuals who receive research grants may be subject to taxation.
Geographic Focus: All States, All Countries
Date(s) Application is Due: Aug 1
Amount of Grant: 15,000 - 40,000 USD
Contact: Program Officer; (646) 428-0971; fax (646) 428-0981; info@hfg.org
Internet: http://www.hfg.org/rg/guidelines.htm
Sponsor: Harry Frank Guggenheim Foundation
25 West 53rd Street
New York, NY 10019-5401

Harry Kramer Memorial Fund Grants 1965
The Harry Kramer Memorial Fund, established in 1982, supports Jewish organizations in the United States and Israel, involved with the care of the sick or aged, education, and religious organizations. Giving primarily in the southern Florida area. Areas of interest include: aging, centers/services; arts; disasters, 9/11/01; higher education; human services; international terrorism; Jewish federated giving programs; disabled individuals.
Requirements: There are no specific deadlines with which to adhere. Contact the Foundation for further application information and guidelines.
Restrictions: No grants to individuals, or for operating budgets or continuing support.
Geographic Focus: All States, Israel
Amount of Grant: 1,800 - 15,000 USD
Contact: Leslie J. August, Wachovia Bank Trustee; (305) 789-4645
Sponsor: Harry Kramer Memorial Fund c/o Wachovia Bank, N.A.
100 North Main Street, 13th Floor
Winston-Salem, NC 27150-0001

Harry S. Black and Allon Fuller Fund Grants 1966
The Harry S. Black and Allon Fuller Fund was established in 1930 to support quality health-care and human-services programming for underserved populations. The grantmaking focus is in the areas of health care and physical disabilities. The fund supports access to health care; health education; health/wellness promotion and disease prevention; health policy analysis and advocacy; access programs for physically disabled individuals; disability policy analysis and advocacy; workforce development programs; and programs that improve quality of life for the disabled. Emphasis will be placed on programs serving low-income communities. Grant requests for general operating support or program/project support are strongly encouraged. Applicants must apply online at the grant website. Applicants are strongly encouraged to do the following before applying: review the downloadable state application procedures at the grant website; review the downloadable online-application guidelines at the grant website; review the foundation's funding history (link is available from the grant website); review the online application questions in advance; and review the list of required attachments. These will generally include: a list of board members, financial statements (audited, reviewed, or compiled by independent auditor); an organization summary; a list of other funding sources; an IRS Determination letter; and other required documents. All attachments must be uploaded in the online application as PDF, Word, or Excel files. The Harry S. Black and Allon Fuller Fund has a deadline of June 30. Grant decisions will be made by December 31. Most recent awards have ranged from $5,000 to $15,000.
Requirements: Nonprofit organizations must be geographically located within the city limits of New York City or Chicago to be eligible to apply.
Restrictions: The fund generally does not support the following: projects in the areas of health care specific to medical/academic research; organizations or programs that primarily provide mental health services; programs that primarily provide services to either the mentally or developmentally disabled; endowment campaigns; and capital projects. The fund does not support requests from individuals, organizations attempting to influence policy through direct lobbying, or any political campaigns.
Geographic Focus: Illinois, New York
Date(s) Application is Due: Jun 30
Amount of Grant: 5,000 - 15,000 USD
Contact: George Suttles; (646) 743-0425; george.suttles@ustrust.com
Internet: https://www.bankofamerica.com/philanthropic/foundation.go?fnId=62
Sponsor: Harry S. Black and Allon Fuller Fund
114 West 47th Street, NY8-114-10-02
New York, NY 10036

Harry Sudakoff Foundation Grants 1967
The Foundation, established in 1956, offers funding primarily in the forms of ongoing support and capital campaigns. Focus areas include: federated giving programs; higher education; and human/social services. Of particular interest are the performing arts, including dance, music, and theater. The geographic focus is Sarasota, Florida, with an initial approach by Letter of Interest (LOI). Though there are no specific application forms, the annual deadline for application is August 15, with final notification by November 1st.
Requirements: Grantees must be 501(c)3 organizations with a sound track record.
Restrictions: Generally the Foundation will not provide grants for: endowments, deficit financing, debt reduction, or ordinary operating expenses; conferences, seminars, workshops, travel, surveys, advertising, fund-raising costs or research; annual giving campaigns; individuals; or projects that have already been completed.
Geographic Focus: Florida
Date(s) Application is Due: Aug 15
Amount of Grant: 46,875 USD
Contact: Janet L Dickens; (941) 952-2826; fax (941) 952-2768; janet.dickens@ustrust.com
Sponsor: Harry Sudakoff Foundation
Bank of America, 1605 Main Street
Sarasota, FL 34236-5840

Harry W. Bass, Jr. Foundation Grants 1968
The Dallas-based Harry W. Bass, Jr. Foundation seeks to enrich the lives of the citizens of Texas by providing support to qualified organizations in the areas of education, health, human services, civic & community, science, research, arts and culture. Grant applications for specific programs or projects, capital projects or, less often, general operations are considered. Endowment gifts are rare. The Harry W. Bass, Jr. Foundation also considers program-related investments as part of its grant-making activities. The foundation strives to be responsive to the needs of all eligible organizations and considers requests of any amount. There is no formal application form. Grant requests are accepted at any time throughout the year. Each organization is limited to one application within a twelve-month period. Requests are usually processed within three to four months. Electronic grant applications should be submitted to dcalhoun@hbrf.org (no file attachments, please). For more information, contact the Harry W. Bass, Jr. Foundation.
Requirements: Texas 501(c)3 tax-exempt organizations are eligible.
Restrictions: In general, grants are not made for purposes of: church or seminary construction; annual fundraising events or general sustentation drives; professional conferences and symposia; out-of-state performances or competition expenses; to other private foundations. Unsolicited grant requests are restricted to organizations based in the Greater Dallas area. In the event a grant is approved, the recipient organization will be unable to submit another grant request for two years due to the foundation policy of not providing subsequent year grants.
Geographic Focus: Texas
Samples: Southwestern Medical Foundation, $3,000,000--Heart, Lung & Vascular Center of Excellence at UT Southwestern University Hospital - St. Paul; Child Protective Services Community Partners, $10,000--Rainbow Room; Health & Human Services Grant; Wilkinson Center, $20,000--CLIMB After-School and Summer Program.
Contact: F. David Calhoun; (214) 599-0300; fax (214) 599-0405; dcalhoun@hbrf.org
Internet: http://www.harrybassfoundation.org/about.asp
Sponsor: Harry Bass Foundation
4809 Cole Avenue, Suite 252
Dallas, TX 75205

Hartford Courant Foundation Grants 1969
The Foundation established in 1950, supports organizations involved with education, the arts, community development, health and social services, with an emphasis on programs benefiting children, youth and families. Grants in the area of the arts will be made only for outreach or education programs for children or families. Grants may be awarded in support of program, capital and operating expenses, for seed money and as challenge or matching grants. The Hartford Courant Foundation serves: Andover, Avon, Bloomfield, Bolton, Canton, East Granby, East Hartford, East Windsor, Ellington, Enfield, Farmington, Glastonbury, Granby, Hartford, Hebron, Manchester, Marlborough, Middletown, New Britain, Newington, Rocky Hill, Simsbury, Somers, South Windsor, Suffield, Tolland, Vernon, West Hartford, Wethersfield, Windsor, and Windsor Locks.
Requirements: An organization wishing to apply for a grant from The Hartford Courant Foundation must be: within the Foundations service areas; a non-profit organization; submit its request using a Hartford Courant Foundation application form, available on the Foundations website; include supporting documentation.
Restrictions: The Hartford Courant Foundation's policies do not allow funding of: organizations which are not tax exempt under IRS Code section 501(c)3; organizations which have the IRS Private Foundation designation; individuals; endowment funds; religious institutions, other than for their provision of non-sectarian community services; capital projects related to the arts; performances, conferences, trips, one-time events; annual campaigns.
Geographic Focus: Connecticut
Date(s) Application is Due: Mar 15; Jun 15; Sep 15; Dec 15
Contact: Kate Miller; (860) 241-6472; fax (860) 520-6988; kmiller@hcfdn.org
Internet: http://www.hcfdn.org/application/guidelines.htm
Sponsor: Hartford Courant Foundation
285 Broad Street
Hartford, CT 06115

Hartford Foundation Regular Grants 1970
The Regular Grants program supports a variety of broad-based areas that reflect the diverse needs and interests of our region, such as: Arts and culture; Education; Family and social services; Health; and, Housing and economic development. An organization may be eligible to apply for one regular grant every three years, selecting its highest priority from one of the following grant types: Program/Project - Grants that support new programs, demonstration projects, studies, or surveys that do not commit the Foundation to recurring costs. May also be used to enhance, expand or strengthen existing programs, services and organizational capacity; Continuation - Additional funding provided at the end of a grant to help an organization reach project outcomes that were not met within the original schedule; Capital Grants - Supporting capital improvements, such as building purchase, construction or renovation, or capital equipment purchase; or, General Operating Support - Grants that support an organization's ongoing activities and stability, as outlined in its strategic plan.
Requirements: 501(c)3 nonprofit organizations serving residents in the following Connecticut towns are eligible to apply: Andover, Avon, Bloomfield, Bolton, Canton, East Granby, East Hartford, East Windsor, Ellington, Enfield, Farmington, Glastonbury, Granby, Hartford, Hebron, Manchester, Marlborough, Newington, Rocky Hill, Simsbury, Somers, South Windsor, Suffield, Tolland, Vernon, West Hartford, Wethersfield, Windsor, and Windsor Locks. Additionally, the board and staff must be representative of the racial/ethnic diversity of the region served.
Restrictions: The Foundation does not make grants from its unrestricted funds for: sectarian or religious activities; grants directly to individuals; grants to private foundations; endowments or memorials; direct or grass-roots lobbying efforts; conferences; research; or informational activities on topics that are primarily national or international in perspective. In addition, the Foundation generally does not make grants for: federal, state, or municipal agencies or departments supported by taxation; sponsorship of or support for one-time events; liquidation of obligations incurred at a previous date; or sustaining support for recurring operating expenses.
Geographic Focus: Connecticut
Amount of Grant: Up to 500,000 USD

Contact: Erika Frank; (860) 548-1888; fax (860) 524-8346; arivera@hfpg.org
Internet: http://www.hfpg.org/for-nonprofits/types-of-grants/
Sponsor: Hartford Foundation for Public Giving
10 Columbus Boulevard, 8th Floor
Hartford, CT 06106

Harvest Foundation Grants 1971

Formed by the sale of Memorial Health Systems, the foundation is managing an endowment to invest in programs and initiatives that will address local challenges in the areas of health, education, and welfare in Martinsville/Henry County. The foundation is committed to honoring the legacy of Memorial Hospital by emphasizing prevention, safety and access to health care; by facilitating opportunities for local citizens to help their community reach its potential; and by improving the learning environment for citizenship, academic, and vocational preparedness. Application and guidelines are available online.
Requirements: To be eligible for consideration, an organization must: be located within, or have its program focused within Martinsville and/or Henry County; have a letter from the IRS stating its 501c3 status; and propose a project within one of the Foundation's three interest areas of health, education or welfare.
Restrictions: The foundation does not fund organizations that discriminate based upon race, creed, gender, or sexual orientation; scholarships, fellowships, or grants to individuals; sectarian religious activities, political lobbying, or legislative activities; profit-making businesses; emergency needs or extremely time sensitive requests; or direct replacement of discontinued government support.
Geographic Focus: Virginia
Samples: Dan River Basin Association, $8,000-- to enhance the Uptown Spur Trail; Boy Scouts of America, Blue Ridge Mountains Council, $45,300--to launch the Scoutreach Program aimed at recruiting underserved boys in Martinsville/Henry County to join the Boy Scouts; MARC Workshop, Inc., $241,900--funding paid over 3 years to start-up a mobile employment program for disabled adults, creating competitive, minimum wage jobs for special education students graduating out of the public school system.
Contact: Allyson Rothrock, Executive Director; (276) 632-3329; fax (276) 632-1878; arothrock@theharvestfoundation.org
Internet: http://www.theharvestfoundation.org/page.cfm/topic/strategic-map
Sponsor: Harvest Foundation
1 Ellsworth Street, P.O. Box 5183
Martinsville, VA 24115

Harvey Randall Wickes Foundation Grants 1972

The Foundation awards grants to eligible Michigan nonprofit organizations in its areas of interest, including the arts; children/youth services; education; hospitals; human services; libraries; and recreation. Types of support include annual campaigns, building/renovation, equipment, and seed money. There are no application deadline dates. Applications must be received two weeks prior to board meetings in March, June, September, and December. Award notification is given within two weeks of the board meeting.
Requirements: Michigan 501(c)3 organizations in Saginaw County are eligible.
Restrictions: Funding is not available for government where support is forthcoming from tax dollars, individuals, endowments, travel, conferences, loans, or film or video projects.
Geographic Focus: Michigan
Amount of Grant: 2,500 - 30,000 USD
Samples: $13,000, Saginaw, Michigan Junior Achievement, payable over one year; $20,000, Saginaw, Michigan Holy Cross Children's Services, payable over one year.
Contact: Hugo Braun, Jr.; (989) 799-1850; fax (989) 799-3327; hrwickes@att.net
Sponsor: Harvey Randall Wickes Foundation
4800 Fashion Square Boulevard, Plaza N., Suite 472
Saginaw, MI 48604-2677

Hasbro Children's Fund Grants 1973

The Hasbro Children's Fund focuses on three core principles: programs which provide hope to children who need it most; play for children who otherwise would not be able to experience that joy; and the empowerment of youth through service. The mission of the Fund is to assist children triumphing over critical life obstacles as well as bringing the joy of play into their lives. Through the Fund's initiatives, the mission is achieved by supporting programs which provide terminal and seriously ill children respite and access to play, educational programs for children at risk, and basics for children in need. The Fund annually provides local community grants which support programs that deliver; stability for children in crisis; pediatric physical and mental health services; hunger security; educational programs; quality out of school time programming and programs that empower youth through service. Interested applicants are asked to submit a letter of inquiry through the online system.
Requirements: Any 501(c)3 serving residents in the locations where Hasbro has operating facilities are eligible to apply. This includes: Rhode Island; Springfield, Massachusetts; Renton, Washington; and Los Angeles, California.
Restrictions: Funding is not available for: religious organizations; individuals; cash free grants; research; political organizations; scholarships; travel stipends; loans; endowments; goodwill advertising; sponsorship of recreational activities; fundraisers; auctions; and schools.
Geographic Focus: California, Massachusetts, Rhode Island, Washington
Amount of Grant: 1,000 - 10,000 USD
Contact: Karen Davis; (401) 431-8151; fax (401) 431-8455; hcfinfo@hasbro.com
Internet: http://www.hasbro.com/corporate/en_US/community-relations/childrens-fund.cfm
Sponsor: Hasbro Children's Fund
1027 Newport Avenue, P.O. Box 200
Pawtucket, RI 02862-1059

Hasbro Corporation Gift of Play Hospital and Pediatric Health Giving 1974

A hospital stay can be a scary and stressful situation. The Hasbro Corporation Gift of Play program works closely with staff and child life members of the children's hospitals located where Hasbro has an operating facility, to provide a sense of comfort and normalcy to the young patients while trying to lift their spirits and bring a smile to their face with a toy or game. The following are local hospitals Hasbro supports: Hasbro's Children's Hospital in Providence, Rhode Island; Bradley Hospital in East Providence, Rhode Island; Bay State Children's Hospital in Springfield, Massachusetts; Seattle Children's Hospital in Seattle, Washington; and Children's Hospital Los Angeles in Los Angeles, California. Since 2007, Hasbro and the Garth Brooks Teammates for Kids Foundation have teamed up to open nine Child Life Zones throughout the United States. Teammates for Kids opens Child Life Zones in hospitals to ensure the patients and families have a therapeutic play area where they can learn, play and relax.
Requirements: Children's hospital patients in corporate operating areas are eligible. This includes: Rhode Island; Springfield, Massachusetts; Renton, Washington; and Los Angeles, California.
Geographic Focus: California, Massachusetts, Rhode Island, Washington
Contact: Karen Davis; (401) 431-8151; fax (401) 431-8455
Internet: http://www.hasbro.com/corporate/en_US/community-relations/hospital-support.cfm
Sponsor: Hasbro Corporation Gift of Play
1027 Newport Avenue, P.O. Box 200
Pawtucket, RI 02862-1059

Hawaiian Electric Industries Charitable Foundation Grants 1975

Named one of the most charitable companies in the state, HEI Charitable Foundation is focused on community programs aimed at promoting educational excellence, economic growth and environmental sustainability. To fulfill its mission of good corporate citizenship, the Foundation funds programs in the categories of: community development; education; the environment; and family services. Particular consideration will be given to organizations or programs which: demonstrate cost-effectiveness; has or will have a significant presence in company communities; provide recognition and goodwill for the company and further the well-being of the company's employees and their interests. Annual deadlines for application submission are January 1, April 1, July 1, and October 1.
Requirements: Hawaii 501(c)3 tax-exempt organizations may apply.
Restrictions: Funds are not available for: activities to replace government support; programs outside the areas served by HEI companies; religious activities of a particular denomination; veterans, fraternal or labor organizations, unless the purpose benefits all the people in the community; political funds; program advertising; special events, e.g., golf tournaments, dinners or functions; direct support for specific individuals.
Geographic Focus: Hawaii
Date(s) Application is Due: Jan 1; Apr 1; Jul 1; Oct 1
Contact: A.J. Halagao, Director of Corporate and Community Advancement; (808) 543-5889 or (808) 543-7960; fax (808) 203-1390; heicf@hei.com
Internet: http://www.hei.com/phoenix.zhtml?c=101675&p=charitable-foundation
Sponsor: Hawaiian Electric Industries Charitable Foundation
P.O. Box 730
Honolulu, HI 96808-0730

Hawaii Community Foundation Health Education and Research Grants 1976

The program provides support from three distinct funds: Tobacco Prevention & Control Trust Fund; the Leahi Fund Research Fund; and the Medical Research Program. Funding will be provided for medical education and research in the fields of: cancer; heart disease; lung disease and research. Priority is given to projects that: demonstrate a foreseeable benefit to the people of Hawaii; support new investigators in Hawaii; and support collaborative efforts. Proposal submission information is available online.
Requirements: To be eligible for consideration an organization must be a 501(c)3 organization or a unit of government. The Principal Investigator must be based in Hawaii and conducting the research in Hawaii.
Geographic Focus: Hawaii
Date(s) Application is Due: Feb 27; Jul 17; Aug 14; Sep 10
Amount of Grant: 25,000 - 50,000 USD
Contact: Christel Wuerfel, Philanthropic Services Assistant; (808) 537-6333 or (888) 731-3863; cwuerfel@hcf-hawaii.org
Internet: http://www.hawaiicommunityfoundation.org/index.php?id=71&categoryID=23
Sponsor: Hawai'i Community Foundation
827 Fort Street Mall
Honolulu, HI 96813

Hawaii Community Foundation Reverend Takie Okumura Family Grants 1977

The Fund was established by members of the Okumura family to continue the charitable work of Reverend Okumura focusing on the healthy development of Hawaii's young children and youth. Proposals will be accepted in the following areas of focus: youth (ages 6-20 years old) and young children (ages birth to 5 years old). Priorities will be given to programs which: develop the ability to think critically; understand and appreciate one's own culture and those of others; develop the ability to settle differences peacefully; strengthen the early care and education community system. Proposal submission information is available online.
Requirements: Hawai'i organizations that are tax exempt, including nonprofit organizations, 501(c)3 organizations, religious organizations that are exempt from taxation, or units of government are eligible to apply.
Restrictions: Funding is not available for: major capitol projects; endowments; or on-going or general operating costs.
Geographic Focus: Hawaii

Date(s) Application is Due: Aug 5
Amount of Grant: 5,000 - 15,000 USD
Contact: Christel Wuerfel, Philanthropic Services Assistant; (808) 537-6333 or (888) 731-3863; cwuerfel@hcf-hawaii.org
Internet: http://www.hawaiicommunityfoundation.org/index.php?id=71&categoryID=24
Sponsor: Hawai'i Community Foundation
827 Fort Street Mall
Honolulu, HI 96813

Hawaii Community Foundation West Hawaii Fund Grants 1978
The West Hawaii Fund was established at the Hawaii Community Foundation in 1992 by a group of concerned citizens for the purpose of accepting charitable gifts for the benefit of the people and communities of West Hawaii, from North Kohala to Hawaiian Ocean View Estates. Today, the Fund continues to grow through additional contributions and planned gifts from donors in the community to respond to current and emerging needs and improve the quality of life for the residents of West Hawaii. The strongest proposals will be those that meet the following criteria: proposal articulates how the proposed project will benefit the West Hawaii community; proposal clearly describes how the project will be implemented and how these particular activities will address critical issues in the West Hawaii community; there are clearly stated outcomes and a plan for measuring and reporting results; project demonstrates collaboration among different sectors of the community; project budget is realistic, relates to project narrative and is reasonable in cost.
Requirements: 501(c)3 organizations - Grant awards will range up to $10,000. Community organizations not designated as 501(c)3 organizations -Grant awards will range up to $2,000. To be eligible for grants of more than $2,000, organization must be designated 501(c)3 or have a 501(c)3 fiscal sponsor.
Restrictions: Projects not likely to be funded: projects that do not benefit the residents of West Hawaii; funds for endowments or for the benefit of specific individuals; out-of-state travel expenses; start-up costs of a new organization.
Geographic Focus: Hawaii
Date(s) Application is Due: Oct 1
Amount of Grant: 2,000 - 10,000 USD
Contact: Diane Chadwick; (808) 885-2174; fax (808) 885-1857; dchadwick@hcf-hawaii.org
Internet: http://www.hawaiicommunityfoundation.org/index.php?id=71&categoryID=22
Sponsor: Hawai'i Community Foundation
827 Fort Street Mall
Honolulu, HI 96813

Hawn Foundation Grants 1979
The foundation awards grants to Texas arts, education, health care, hospitals, medical research, and social services organizations. Preference will be given to requests from organizations in Dallas, TX.
Restrictions: The Foundation does not award scholarships or fellowships.
Geographic Focus: Texas
Date(s) Application is Due: Jun 1
Amount of Grant: 1,000 - 200,000 USD
Samples: U of Texas Southwestern Medical Ctr (Dallas, TX)--for medical research and for studies on age-related macular degeneration, $100,000.
Contact: Joe Hawn Jr., (214) 696-6595; fax (214) 696-6596
Sponsor: Hawn Foundation
5949 Sherry Lane, Suite 775
Dallas, TX 75225-8043

HCA Foundation Grants 1980
The Foundation is committed to the care and improvement of human life. Grants are made in the areas of health and well being, childhood and youth development, and the arts. Preference will be given to requests from organizations where an HCA employee volunteers or serves on the board. New applicants are asked to send a one-or two-page letter of inquiry to the Foundation, describing the proposed project, its goals and objectives and the approximate level of funding required. Foundation staff will review each request and will notify the organization as to whether the project coincides with funding priorities.
Requirements: 501(c)3 nonprofit organizations are eligible. Because the foundation focuses its giving in Middle Tennessee, all requests outside of the Nashville area should be submitted to the closest HCA facility location or division office. Organizations must have a full updated GivingMatters.com profile to be considered for funding. For more information please go to www.givingmatters.com.
Restrictions: Funding is not available for: individuals or their projects; private foundations; political activities; advertising or sponsorships of events; social events or similar fundraising activities. The Foundation does not ordinarily support: organizations in their first three years of operation and organizations involved in research, sports, environmental, wildlife, civic and international affairs. The Foundation does not accept proposals from individual churches or schools, but will support broad faith-based initiatives consistent with it's mission and guidelines.
Geographic Focus: Tennessee
Date(s) Application is Due: Mar 12; Jun 11; Sep 10; Dec 12
Amount of Grant: 20,000 - 100,000 USD
Contact: Lois Abrams, Grants Manager; (615) 344-2390; fax (615) 344-5722; lois.abrams@hcahealthcare.com
Internet: http://www.hcacaring.org/CustomPage.asp?guidCustomContentID=BBB7D8F2-B906-4302-A164-07643DB2582E
Sponsor: HCA Foundation
1 Park Plaza, Building 1, 4th Floor East
Nashville, TN 37203

Healthcare Foundation for Orange County Grants 1981
The foundation's mission is to improve the health of the neediest and most underserved residents of Orange County, with particular emphasis on central Orange County. The foundation will address ways of improving the health of residents by advancing access to health information, prevention, and basic health care. Grantmaking is divided among initiatives that support nonprofit hospital working with community-based organizations (Partners for Health); grants directly to community-based organizations (Healthy Orange County); and increased understanding of community needs, and collaborative granting strategies (coalition projects.)
Requirements: California 501(c)3 nonprofit organizations in Orange County that administer health services consistent with the foundation's goals of providing aid to uninsured, poor families are eligible. Eighty-two percent of support goes to or through qualified nonprofit hospitals. Government agencies also are eligible.
Restrictions: Grants generally are not awarded for annual fund drives, building campaigns, major equipment, or biomedical research. Activities that exclusively benefit the members of a religious or fraternal organization are not funded.
Geographic Focus: California
Amount of Grant: Up to 10,000 USD
Contact: William B. Stannard; (714) 245-1650; fax (714) 245-1653; dflander@hfoc.org
Internet: http://www.hfoc.org/giving/
Sponsor: Healthcare Foundation for Orange County
1450 N Tustin Avenue, Suite 103
Santa Ana, CA 92705-8641

Healthcare Foundation of New Jersey Grants 1982
The Foundation's funding priorities are the vulnerable populations of the greater Newark New Jersey community; the emergent health needs of serving at-risk individuals and families in the MetroWest Jewish community; and clinical research/medical education initiatives that significantly and directly impact these populations. The Foundation seeks grant proposals that promise innovation and change or a significant enhancement of services. Proposals are accepted on a rolling basis throughout the calendar year. Proposal submission instructions are available online.
Requirements: Nonprofit organizations in the Greater Newark area are eligible to apply. The Foundation strongly suggests that application documents be submitted electronically.
Restrictions: Grants are made only to private nonprofit organizations that have tax-exempt status under Section 501(c)3 of the Internal Revenue Code and that are not private foundations. The Foundation does not make grants to individuals or government agencies. The Foundation does not typically fund the following: organizations outside of Essex, Morris or Union County, New Jersey; programs not related to health care; direct support of an individual's healthcare needs; fundraising events or endowment campaigns; advertising campaigns; lobbying; or scholarships.
Geographic Focus: New Jersey
Contact: Program Contact; (973) 921-1210; fax (973) 921-1274; info@hfnj.org
Internet: http://www.hfnj.org
Sponsor: Healthcare Foundation of New Jersey
60 East Willow Street, 2nd Floor
Millburn, NJ 07041

Health Foundation of Greater Cincinnati Grants 1983
The foundation awards grants to programs and activities that improve health in Cincinnati and 20 surrounding counties in Ohio, Kentucky, and Indiana. Health is broadly defined to include social, behavioral, environmental, and other dimensions beyond the absence of illness. Grant making is focused on strengthening primary care providers to the poor; school-based child health interventions in K-8 school settings; substance abuse; and severe mental illness. Grants also are awarded in response to requests from organizations seeking funds to support a broad range of health-related needs in the community. The Health Foundation of Greater Cincinnati considers projects for funding in three ways: through Requests for Proposals (RFPs), through invitations to submit proposals, and through grantee-initiated requests. In any case, proposals are considered using the same eligibility requirements and proposal evaluation criteria. For more information about the types of proposals, see the Foundations website or call the Grant Manager.
Requirements: 501(c)3 nonprofits in Adams, Brown, Butler, Clermont, Clington, Hamilton, Highland, and Warren Counties in Ohio; Boone, Bracken, Campbell, Gallatin, Grant, Kenton, and Pendleton Counties in Kentucky; and Dearborn, Franklin, Ohio, Ripley, and Switzerland Counties in Indiana are eligible.
Restrictions: The foundation does not normally fund capital campaigns, annual fundraising campaigns, endowments, event sponsorships, clinical research, scholarships, routine operational costs, or direct financial subsidy of health services to individuals or groups.
Geographic Focus: Indiana, Kentucky, Ohio
Contact: Shelly Stolarczyk-George, Grants Manager; (513) 458-6619; fax (513) 458-6610; sstolarczyk@healthfoundation.org
Internet: http://www.healthfoundation.org/grants
Sponsor: Health Foundation of Greater Cincinnati
3805 Edwards Road, Suite 500
Cincinnati, OH 45209-1948

Health Foundation of Greater Indianapolis Grants 1984
The foundation promotes health care for children, youth, and families in Marion County, IN, and seven contiguous Indiana counties. Programs of interest include school-based health, adolescent health, and HIV/AIDS Grants also are awarded to support the development of health careers and to promote cooperative effort between health professionals. Types of support include general operations, building/renovations, equipment, program/project development, conferences and seminars, seed money, technical assistance, and matching

funds. Prior to submitting a proposal, contact Stephen L. Everett, Vice President of Programs, to determine if your program and proposal matches The Health Foundation of Greater Indianapolis' funding priorities and application guidelines.
Requirements: Grants are awarded to neighborhood-based service centers in Indiana's Marion County and the seven contiguous counties of Boone, Hamilton, Hancock, Hendricks, Johnson, Morgan and Shelby. Applicants must be a 501(c)3 group, organization or agency that provides health-related programs or services.
Restrictions: Grants will not be provided for: individuals; sectarian religious organizations; research projects; purchase of advertising or tickets to events; production and design of educational materials already available; endowments; short or long term loans or payment of financial obligations.
Geographic Focus: Indiana
Amount of Grant: 10,000 - 100,000 USD
Contact: Stephen Everett; (317) 630-1805; fax (317) 630-1806; severett@thfgi.org
Internet: http://www.thfgi.org
Sponsor: Health Foundation of Greater Indianapolis
429 East Vermont Street, Suite 300
Indianapolis, IN 46202

Health Foundation of Southern Florida Responsive Grants 1985
The foundation awards Responsive Grants through two grant cycles per year. With exceptions, the foundation focuses on providing one to three year grants that do not exceed $300,000 annually. The majority of grants are funded in the $50,000 to $150,000 range over one or two years. Funding is provided in four categories: Project Planning; Health Services; Organizational Capacity Building and Health System/Health Policy Development.
Requirements: Though the foundation welcomes proposal applications anytime, the applications are reviewed on a semi-annual basis. Applicant organizations must be tax-exempt nonprofit under section 501(c)3 of the Internal Revenue Code or a local or state governmental agency. The project must serve exclusively the residents of Broward, Miami-Dade and/or Monroe counties. Initially, a preliminary proposal is required (see the sponsor's website for specific details). If approved, the sponsor will then invite a full proposal. Download the sponsor's grant guide from the website.
Restrictions: The foundation does not fund: Biomedical research or other research that will not impact local residents within the immediate future (1-3 years) or that does not have a direct application to implementing a community-driven health intervention; Capital campaigns of over $1 million (versus grants toward specific health-related equipment or the 'build out' of a specific health-focused space); Secondary and tertiary services (versus preventive and primary medical, oral and behavioral health care services); Health promotion and/or health care with a high per capita cost (this figure will vary depending upon the type of intervention, but over $1,000 per person/year cost may be a rule of thumb); Service expansion or new projects without viable sustainability (unable to reach sustainability without the foundation's resources within a four-year period).
Geographic Focus: Florida
Date(s) Application is Due: Mar 13; Apr 24
Amount of Grant: Up to 300,000 USD
Contact: Eliane Morales, (305) 374-7200; fax (305) 374-7003; emorales@hfsf.org
Internet: http://www.hfsf.org/ORIGHTML/responsive.html
Sponsor: Health Foundation of South Florida
2 South Biscayne Boulevard
Miami, FL 33131

Health Management Scholarships and Grants for Minorities 1986
This financial assistance program provides scholarships and grants to ethnic minority students for graduate study in health care administration/management. Annual tuition scholarships and grants are made to outstanding students in health services management and related programs. Grants are provided to cover the costs of books, tutorial assistance, and other nontuition educational aids. Students are eligible to receive a combination of awards during the same time period.
Requirements: To be eligible, applicants must be members of an ethnic minority group; must be U.S. citizens or have permanent resident status (birth certificate, green card, or current passport); they must be college juniors, seniors or graduate students; they must be health care administration/management majors; they must have a strong academic background (2.5 GPA on a 4.0 scale); and they must have strong extracurricular and community service activities.
Geographic Focus: All States
Amount of Grant: 500 - 2,500 USD
Contact: Program Contact; (800) 233-0996 or (312) 422-2680; fax (312) 422-4566
Internet: http://www.diversityconnection.org/diversityconnection_app/career-center/internships/Standard-Internship-page.jsp?fll=S4
Sponsor: Institute for Diversity in Health Management
1 N Franklin, 30th Floor
Chicago, IL 60606

Hearst Foundations Health Grants 1987
The Hearst Foundations fund non-profit organizations working in the fields of culture, education, health, and social services. The Foundations have two offices, one in New York which manages funding for organizations headquartered east of the Mississippi River and one in San Francisco which manages funding for organizations to the west. About 80% of the Foundations' total funding goes to prior grantees; the Foundations receive approximately 1,200 grant requests annually. The Foundations' health funding comprises 30% of their total giving; 80% of the Foundations' health funding goes to organizations having budgets over ten-million dollars. In the area of health, the Foundations assist leading regional hospitals, medical centers, and specialized medical institutions providing access to high-quality healthcare for low-income populations. In response to the shortage of healthcare professionals necessary to meet the country's evolving needs, the Foundations also fund programs designed to enhance skills and increase the number of practitioners and educators across roles in healthcare. Because the Foundations seek to use their funds to create a broad and enduring impact on the nation's health, support for medical research and the development of young investigators is also considered. Preference is given to the following types of programs; professional development programs; programs improving access to high-quality healthcare for low-income populations; programs developing and providing specialized care for the complex needs of elderly populations; programs scaling innovative healthcare delivery systems to provide efficient, coordinated care; and research, particularly related to finding new cures and treatments for prevalent diseases, such as cancer. The Foundations provide program, capital, and, on a limited basis, endowment support. Requests are accepted year round. These must be submitted via the Foundations' online application portal. Each request goes through an evaluation process that generally spans four to six weeks. The Foundations conduct a site visit of semi-finalists and may also consult with experts in a given field. Applicants will receive an email confirmation of receipt of submission and can follow the status of their request through the online system. Instructions for using the system, guidelines (in the form of an FAQ), and the link to the Foundations' online portal are at the Foundations' website.
Requirements: Grants are made only to 501(c)3 organizations.
Restrictions: Organizations must wait one year from the date of their notice of decline before the Foundations will consider another request. Grantees must wait a minimum of three years from their grant award date before the Foundations will consider another request. The Foundations do not fund individuals or the following types of requests: those from organizations operating outside the United States; those from organizations with operating budgets under one million dollars; those from organizations involved in publishing, radio, film, or television; those from local chapters of organizations; those from organizations lacking demonstrable long-term impact on populations served; requests to fund tours, conferences, workshops, or seminars; requests to fund advocacy or public-policy research; requests to fund special events, tickets, tables, or advertising for fundraising events; requests for seed money or to fund start-up projects; and request to fund program-related investments.
Geographic Focus: All States
Amount of Grant: Up to 600,000 USD
Contact: Mason Granger; (212) 649-3750; fax (212) 586-1917; hearst.ny@hearstfdn.org
Internet: http://www.hearstfdn.org/funding-priorities/
Sponsor: Hearst Foundations
300 West 57th Street, 26th Floor
New York, NY 10019-3741

Heckscher Foundation for Children Grants 1988
The Heckscher Foundation for Children was founded in 1921 to promote the welfare of children, primarily in New York City. Funding organizations serving youth in the fields of education, family services, job training, health, arts and recreation. The Foundation's giving takes the form of program support, capacity-building, capital projects and general operating support. The Foundation does not participate in annual appeals, endowments, fundraising events or political efforts. The Heckscher Foundation provides support in the following categories: Education & Academic Support; Arts; Social Services; Health; Recreation and; Workforce Development. The Foundation accepts applications online only. To begin the application process. fill out the application template provided online. Do not submit additional information or materials unless asked to do so. All inquiring organizations will be notified of the outcome of the Foundation's review, and any further necessary materials will be requested at that time.
Requirements: 501(c)3 nonprofit organizations are eligible.
Restrictions: Funding is not available for annual appeals, endowments, fundraising events or political efforts.
Geographic Focus: New York
Amount of Grant: 25,000 - 500,000 USD
Samples: University of Pennsylvania, Philadelphia, PA, $150,000--for LEAD Program; Miami Art Museum of Dade County, Miami, FL, $500,000--for outreach and education programs; Saint Vincents Services, Brooklyn, NY, $50,000--for capacity building.
Contact: Virginia Sloane, President; (212) 744-0190; fax (212) 744-2761
Internet: http://fdncenter.org/grantmaker/heckscher
Sponsor: Heckscher Foundation for Children
123 East 70th Street
New York, NY 10021

Hedco Foundation Grants 1989
Incorporated in 1972 in California, the Hedco Foundation gives predominantly to qualified educational and health institutions, support is also available for social services. Types of support including: building/renovation; equipment; land acquisition and; matching/challenge grants.
Requirements: 501(c)3 nonprofit organizations are eligible.
Restrictions: No grants to: individuals, or for general support, operating budgets, endowment funds, scholarships, fellowships, special projects, research, publications, or conferences; no loans.
Geographic Focus: California
Amount of Grant: 10,000 - 1,000,000 USD
Contact: Mary Goriup, Manager; (925) 743-0257
Sponsor: Hedco Foundation
P.O. Box 339
Danville, CA 94526-0339

Heineman Foundation for Research, Education, Charitable and Scientific Purposes   1990

The purpose of the Heineman Foundation is to provide seed money to start-up projects and new projects within existing organizations for a maximum of three to five years. Preference will be given to organizations that we have not previously funded. The average range of our donations is $20,000.00 to $50,000.00, per annum. An organization must have 501(c)3 status and upload copies of corresponding IRS documents to the online application form in order for the application to be considered. The Foundation's general areas of interest are the following (in no particular order): programs that enable economically challenged women to enter and remain in the workplace; on site day care centers for women in the workplace; job training programs for women; language and leadership skills for women; environmental research that will help prevent, reduce and/or eliminate water degradation; music as education and preserver of culture; research into prevention of and treatment for childhood illnesses; programs that enable youth to think, create and communicate effectively; and programs that support and promote high achievement in music, science, and literature. Applications/proposals must be submitted online no later than September 1st.
Geographic Focus: All States
Date(s) Application is Due: Sep 1
Amount of Grant: 20,000 - 50,000 USD
Contact: Simon Rose, President; (212) 493-8000; info@heinemanfoundation.org
Internet: http://www.heinemanfoundation.org/guidelines
Sponsor: Heineman Foundation
140 Broadway
New York, NY 10005-1108

Helena Rubinstein Foundation Grants   1991

The Foundation supports programs in education, community services, arts/arts in education, and health with emphasis on projects which benefit women and children. Grants are primarily targeted to organizations in New York City. Although general operating grants are made, the Foundation prefers to support specific programs. Grant proposals are accepted throughout the year. There is no formal application form; however, the New York Common Application Form may be used. Organizations seeking funds are asked not to make telephone inquiries, but to submit a brief letter outlining the project.
Requirements: U.S. nonprofit organizations are eligible.
Restrictions: Support is not offered to individuals, or for film or video projects. Grants are rarely made to endowment funds and capital campaigns. The foundation does not make loans and cannot provide emergency funds. Funding of new proposals is limited by ongoing commitments and fiscal constraints.
Geographic Focus: New York
Samples: New York City Outward Bound Ctr (NY)--for programs to help high school students develop self-esteem, self-reliance, social responsibility, and improved academic performance, $10,000; Goodwill Industries of Greater New York (NY)--to support the capital campaign, $10,000; Winston Preparatory School (NY)--salary support for a vocational guidance counselor, $5000.
Contact: Diane Moss, President; (212) 750-7310
Internet: http://www.helenarubinsteinfdn.org/guide.html
Sponsor: Helena Rubinstein Foundation
477 Madison Avenue, 7th Floor
New York, NY 10022-5802

Helen Bader Foundation Grants   1992

Throughout her life, Helen Bader sought to help others. She played many roles - student, mother, businesswoman, and social worker - believing that everyone should have the opportunity to reach their fullest potential. Growing up in the railroad town of Aberdeen, South Dakota, Helen learned the value of hard work and self-reliance. The Great Depression and the sacrifices of World War II also taught her the importance of reaching out to those in need. Helen attended Downer College in Milwaukee, earning a degree in botany. She married Alfred Bader, a chemist from Austria, and together they started a family and created a business, the Aldrich Chemical Company. From the 1950s to the 1970s, their hard work helped build one of Wisconsin's most successful start-up enterprises of the era. The Baders' eventual divorce led Helen to again become self-reliant. She subsequently finished her Master of Social Work at the University of Wisconsin-Milwaukee. While doing her field work with the Legal Aid Society of Milwaukee, Helen met and helped many people in need, including single mothers and adults with mental illness. In the process, she gained a deeper appreciation for their everyday struggles. After graduation, she worked at the Milwaukee Jewish Home, where working with older adults brought home the many issues of aging. At a time when Alzheimer's disease was almost a complete mystery, she helped open the resident' minds and hearts through dance and music. Helen felt that the residents' quality of life depended upon the small details, so she was happy to run errands or escort them to the symphony. She found herself touched by the arts and studied the violin and guitar at the Wisconsin Conservatory of Music. Helen eventually faced cancer. As the illness began to sap her physical strength, she shared a wish with her family: to continue to aid those in need. She died in 1989. After her death, patterns of Helen's quiet style of philanthropy became more apparent. When she had come across an organization that impressed her, she would just pull out her checkbook without a lot of fanfare. In her name, the Helen Bader Foundation (HBF) supports worthy organizations working in key areas affecting the quality of life in Milwaukee, the state of Wisconsin, and Israel. The foundation also seeks to inspire the generosity in others, as every individual can make a difference through gifts of time, talent, and resources. The foundation will consider multiple-year requests with 24 or 36 month terms. Multi-year grants are subject to annual review before funds for subsequent years are released. The application deadline for the online preliminary proposal is January 5. The application deadline for organizations invited to complete a full proposal is February 2. The link to the online application system is available at the grant website. Application deadlines may vary from year to year. Prospective applicants are encouraged to visit the grant website to verify current deadline dates.
Requirements: Grants are awarded for projects consistent with one or more of the Helen Bader Foundation's program areas: Alzheimer's and aging (national in scope, with priority given to Wisconsin); economic development (restricted to the city of Milwaukee); community partnerships for youth (restricted to the city of Milwaukee); community initiatives (restricted to greater Milwaukee); arts (restricted to the city of Milwaukee); and directed grants and initiatives such as aid and support to Israel (for which proposals must be staff-solicited). Grants are given only to U.S. organizations which are tax exempt under Section 501(c)3 of the Internal Revenue Code or to government entities; grants will only be approved for foreign entities which meet specific charitable status requirements.
Restrictions: The Foundation does not provide direct support for individuals, such as individual scholarships.
Geographic Focus: All States, All Countries
Date(s) Application is Due: Jan 5; Feb 25
Amount of Grant: 10,000 - 100,000 USD
Samples: Aging & Disability Resource Center of Portage County, Stevens Point, Wisconsin, $12,000, creation of an Early Memory Loss Program to serve central Wisconsin; Arts at Large, Inc., Milwaukee, Wisconsin, $412,000, inclusive arts programming for low-income children attending six Milwaukee Public Schools in grades K-8 and their affiliated Community Learning Centers; Ben-Gurion University of the Negev, Beer Sheva, Israel, $10,000, Summer School Science Camp for 50 seventh grade Bedouin students from Abu Basma schools.
Contact: Tamara Hogans, Grants Manager; (414) 224-6464; fax (414) 224-1441; tammy@hbf.org or info@hbf.org
Internet: http://www.hbf.org/apply.htm
Sponsor: Helen Bader Foundation
233 North Water Street, Fourth Floor
Milwaukee, WI 53202

Helen Gertrude Sparks Charitable Trust Grants   1993

Helen Sparks was born in Fort Worth, TX but went to college in the East. After graduation she returned to Fort Worth where she worked with the Fort Worth Little Theater. A cultured woman, she was interested in local artists and supported them by commissioning works. She also enjoyed literature and needlepoint. The Helen Gertrude Sparks Charitable Trust made its first distributions in 1971, and was created to benefit charitable organizations focused on: the elderly; those who are disabled in any way; children who are disabled, orphaned or disadvantaged; the arts, including performing and nonperforming arts; and education. This trust makes approximately 2-3 awards each year and grants are typically between $5,000 and $15,000. Applicants must apply online at the grant website. Applicants are strongly encouraged to do the following before applying: review the downloadable state application procedures for additional helpful information and clarifications; review the downloadable online-application guidelines at the grant website; review the trust's funding history (link is available from the grant website); review the online application questions in advance; and review the list of required attachments. These will generally include: a list of board members, financial statements (audited, reviewed, or compiled by independent auditor); an organization summary; a list of other funding sources; an IRS Determination letter; and other required documents. All attachments must be uploaded in the online application as PDF, Word, or Excel files. The Helen Gertrude Sparks Charitable Trust application deadline is 11:59 p.m. on September 30.
Requirements: Applicants must have 501(c)3 tax-exempt status.
Restrictions: A preference will be shown for organizations which do not receive the predominate portion of their funds from government sources. The trust does not support requests from individuals, organizations attempting to influence policy through direct lobbying, or any political campaigns.
Geographic Focus: Texas
Date(s) Application is Due: Sep 30
Amount of Grant: 5,000 - 15,000 USD
Samples: Carter BloodCare, Bedford, Texas, $15,000, high school population health initiative; Meals on Wheels of Tarrant County, Fort Worth, Texas, $10,000; Performing Arts Fort Worth, Fort Worth, Texas, $10,000, Children's Education Program.
Contact: Mark J. Smith; (817) 390-6028; tx.philanthropic@baml.com
Internet: https://www.bankofamerica.com/philanthropic/fn_search.action
Sponsor: Helen Gertrude Sparks Charitable Trust
500 West 7th Street, 15th Floor, TX1-497-15-08
Fort Worth, TX 76102-4700

Helen Irwin Littauer Educational Trust Grants   1994

Mrs. Littauer was born in Fort Worth, Texas and was a descendant of the Cetti family, a prominent family in Fort Worth, Texas. She earned a degree in journalism and worked in New York as an editor. In 1952, she moved to Connecticut and became an active community volunteer and was involved in city government. As a result of her passion for teaching, she also worked with troubled youth. She was honored with numerous awards for her civic endeavors. Mrs. Littauer established the Educational Trust in 1969 and remained involved in grant decisions until her death in 1989. The trust is particularly interested in, but not limited to charitable organizations that focus on: scholarships that enable needy, but worthy boys and girls and young adults to attend school, college, or university, with a particular emphasis on making scholarships available for attending schools of journalism; promotion of art, education, and good citizenship; alleviating human suffering; medical care and treatment for all needy persons, including hospitals and clinics; providing care, education, recreation and/or physical training for needy,

orphaned or disabled children; providing care of needy persons who are sick, aged or disabled; and improvement of living and working conditions of all persons. Applicants must apply online at the grant website. Applicants are strongly encouraged to do the following before applying: review the downloadable state application procedures for additional helpful information and clarifications; review the downloadable online-application guidelines at the grant website; review the trust's funding history (link is available from the grant website); review the online application questions in advance; and review the list of required attachments. These will generally include: a list of board members, financial statements (audited, reviewed, or compiled by independent auditor); an organization summary; a list of other funding sources; an IRS Determination letter; and other required documents. All attachments must be uploaded in the online application as PDF, Word, or Excel files. The Helen Irwin Littauer Educational Trust has bi-annual application deadlines of March 31 and September 30. Applications must be submitted by 11:59 p.m. on the deadline dates.
Requirements: Applicants must have 501(c)3 tax-exempt status.
Restrictions: The trust considers requests primarily from charitable organizations that provide services to Tarrant County. The trust does not support requests from individuals, organizations attempting to influence policy through direct lobbying, or any political campaigns.
Geographic Focus: All States
Date(s) Application is Due: Mar 31; Sep 30
Amount of Grant: 1,000 - 25,000 USD
Contact: Mark J. Smith; (817) 390-6028; tx.philanthropic@baml.com
Internet: https://www.bankofamerica.com/philanthropic/fn_search.action
Sponsor: Helen Irwin Littauer Educational Trust
500 West 7th Street, 15th Floor, TX1-497-15-08
Fort Worth, TX 76012-4700

Helen K. and Arthur E. Johnson Foundation Grants 1995
The foundation makes grants to a wide variety of nonprofit organizations in an attempt to solve community problems and enrich the quality of life in Colorado in the following areas: education, youth, health, community and social services, civic and cultural, and senior citizens. Requests are welcomed throughout the year. However, to be considered at the next board meeting, complete written proposals must be received by the listed application deadline dates. Proposal submission information is available online.
Requirements: IRS 501(c)3 organizations serving Colorado residents may apply.
Restrictions: Funding is not available for: loans or fund endowments; individuals; conferences; scholarships to individuals; multiple year grants; fundraising dinners or special events. The Foundation does not support organizations whose primary purpose is to influence (directly or indirectly) the legislative or judicial process in any manner or for any cause. In addition, the Foundation will not consider grant requests that pass through the nominal grant recipient to another organization.
Geographic Focus: Colorado
Date(s) Application is Due: Jan 1; Apr 1; Jul 1; Oct 1
Contact: John H. Alexander, President; (800) 232-9931 or (303) 861-4127; fax (303) 861-0607; info@johnsonfoundation.org
Internet: http://www.johnsonfoundation.org
Sponsor: Helen K. and Arthur E. Johnson Foundation
1700 Broadway, Suite 1100
Denver, CO 80290-1718

Helen Pumphrey Denit Charitable Trust Grants 1996
The Helen Pumphrey Denit Trust for Charitable and Educational Purposes was established in 1988 to support charitable organizations that promote quality education, culture, human service, health service and arts opportunities. The grants are primarily made to organizations in the Baltimore region. Special consideration is given to the following three organizations: Montgomery General Hospital in Onley, Maryland; The George Washington University in Washington, D.C; and The Wesley Theological Seminary in Washington, D.C. Grants for capital and program support are encouraged. Requests for general operating support will be received, but they will be given lower priority than other grants that have more specific purposes. Grants are primarily one year in duration. On occasion, multi-year grants will be awarded. Applicants must apply online at the grant website. Applicants are strongly encouraged to do the following before applying: review the downloadable state application procedures for additional helpful information and clarifications; review the downloadable online-application guidelines at the grant website; review the trust's funding history (link is available from the grant website); review the online application questions in advance; and review the list of required attachments. These will generally include: a list of board members, financial statements (audited, reviewed, or compiled by independent auditor); an organization summary; a list of other funding sources; an IRS Determination letter; and other required documents. All attachments must be uploaded in the online application as PDF, Word, or Excel files. The Helen Pumphrey Denit Trust has an application deadline of 11:59 p.m. on February 1. The applicants will be notified of grant decisions by June 30.
Requirements: Applicants must have 501(c)3 tax-exempt status and serve residents of Baltimore and surrounding communities.
Restrictions: The trust does not support requests from individuals, organizations attempting to influence policy through direct lobbying, or any political campaigns.
Geographic Focus: Maryland
Date(s) Application is Due: Feb 1
Contact: Sarah Kay, Vice President; (804) 788-2673; sarah.kay@baml.com
Internet: https://www.bankofamerica.com/philanthropic/fn_search.action
Sponsor: Helen Pumphrey Denit Charitable Trust
1111 E. Main Street, VA2-300-12-92
Richmond, VA 23219

Helen S. Boylan Foundation Grants 1997
The Helen S Boylan Foundation is a private family foundation established in 1982 to continue the family tradition of commitment to enhancing the quality of life of the community through grants to qualified charitable organizations. In carrying out its mission, the Foundation considers a wide range of proposals within the following areas: arts, education, health, human services, environment, and public interest. Generally, grants are limited to projects that benefit the citizens of Jasper County, Missouri, Smith County, Texas and, Kansas City, Metropolitan area. Occasionally, projects that benefit the state of Missouri as a whole may be considered as well. The Foundation prefers to support proposals for new initiatives, special projects, expansion of current programs, capital improvements or building renovations. Grants from the Foundation are usually awarded for one year only. For projects in those areas in which the Foundation has a special interest, requests for multi-year funding and general operating support may be considered. The Board of Directors meet four times a year to consider grant requests. Applications must be received by March 31, June 30, September 30 or December 31 to be acted upon at the following meeting.
Requirements: IRS 501(c)3 nonprofit organizations operating in the Carthage, Kansas City Metro and Lindale Texas area are eligible to apply for funding. Application form is available online at the Foundation's website.
Restrictions: No support for political organizations or religious activities. No grants to individuals, or for annual campaigns or endowments.
Geographic Focus: Missouri, Texas
Date(s) Application is Due: Mar 31; Jun 30; Sep 30; Dec 31
Amount of Grant: 500 - 50,000 USD
Contact: James R. Spradling; (417) 358-4033; fax (417) 358-5937; spradlinglaw@hotmail.com
Internet: http://www.boylanfoundation.org/
Sponsor: Helen S. Boylan Foundation
320 Grant Street, P.O. Box 731
Carthage, MO 64836

Helen Steiner Rice Foundation Grants 1998
The Foundation awards grants to worthy charitable programs that assist the needy and the elderly. Essential objectives for grant consideration are: basic necessities and human needs for the poor and elderly; preference for meeting the immediate needs of the poor; and innovative approaches. Organizations in the Greater Cincinnati area should use the contact information for Cincinnati. Organizations in Lorain County should contact: Linda Weaver, Community Foundation Center of Lorain County, (440) 277-0142, Ext. 23. Application forms and instructions are available online.
Requirements: Nonprofit organizations in the greater Cincinnati area and Lorain, OH, may submit grant applications.
Restrictions: Funding is not available for building or endowment programs, direct gifts to individuals, or capital fund drives.
Geographic Focus: Ohio
Date(s) Application is Due: Jul 1
Contact: James D. Huizenga, Director; (513) 241-2880; fax (513) 768-6122; huizengai@greatercincinnatifdn.org or hrice@cincymuseum.org
Internet: http://www.helensteinerrice.com/grants.html
Sponsor: Helen Suiteiner Rice Foundation
200 West Fourth Street
Cincinnati, OH 45202-2602

Hendrick Foundation for Children Grants 1999
The foundation supports programs designed to provide health, medical, social welfare, and educational services to benefit children with illness, disease, injury, pain, disability, incapacity, or other disadvantages; and improve quality of life for children with life-threatening or chronic injuries, illness, and disabilities. It is also contemplated that the Foundation will honor grants to other charitable and educational organizations sponsoring programs which are community oriented and focus on improving the quality of life of children with life-threatening or chronic injuries, illnesses and disabilities. All requests shall be given careful and thorough consideration by the Hendrick Foundation for Children Grant Committee. Grant awards will be determined on the merits of the organization, its objectives and mission. All grant requests should be submitted in writing on the organization's letterhead
Geographic Focus: All States
Amount of Grant: 3,500 - 250,000 USD
Contact: Charles V. Ricks, President; (704) 568-5550
Internet: http://www.thehendrickfoundation.org/grants.html
Sponsor: Hendrick Foundation for Children
P.O. Box 240070
Charlotte, NC 28224-0070

Hendricks County Community Foundation Grants 2000
The Hendricks County Community Foundation Grants provide funding for organizations or charitable projects that serve in the following program areas: arts and culture; community development; education; environment; health and human services; and youth. These grants enable organizations to provide effective programs and respond to needs of people in the Hendricks County community.
Requirements: A letter of intent should be submitted to the organization between December 1 and January 11. The Foundation uses the following criteria when reviewing applications: sustainability; effective operations; proven success; strong leadership; innovation and creativity; accessibility; collaboration; and engagement.
Restrictions: The Foundation will not fund: bands, sports teams, or other groups without a philanthropic project; annual appeals, galas or membership contributions; fundraising

events such as golf tournaments, walk-a-thons, and fashion shows; grants to individuals; projects aimed at promoting a particular religion or construction projects for religious institutions; operating, program and construction costs at schools, universities and private academies unless there is a significant opportunity for community use or collaboration; organizations or projects that discriminate based upon race, ethnicity, age, gender, sexual orientation; political campaigns or direct lobbying efforts by 501(c)3 organizations; post-event, after-the-fact situations or debt retirement; medical, scientific or academic research; publications, films, audiovisual and media materials, programs produced for artistic purposes or produced for resale.
Geographic Focus: Indiana
Amount of Grant: 500 - 17,500 USD
Samples: Farmers and Hunters Feeding the Hungry, general operating expenses, $3,000; Hendricks County Senior Services, transportation for the elderly and disabled, $5,000; Kingsway Community Care Center, general operating expenses, $2,500.
Contact: Susan Rozzi; (317) 718-1200; fax (317) 718-1033; janet@hendrickscountycf.org
Internet: http://www.hendrickscountycf.org/grants/oppfund_grants/index.shtml
Sponsor: Hendricks County Community Foundation
5055 East Main Street, Suite A
Avon, IN 46123

Henrietta Lange Burk Fund Grants 2001
The Henrietta Lange Burk Fund is a private foundation created by Mrs. Henrietta Lange Burk in 1994 to support and promote quality arts, cultural, educational, health-care, and human-services programming for underserved populations. Special consideration is given to charitable organizations that address the health concerns of older adults, through either direct programming or research. Mrs. Burk created the Foundation as a memorial to her parents, Mr. and Mrs. Henry G. Lange, as well as to her husband, William Burk, and herself. Mrs. Burk was particularly interested in the performing and cultural arts; age-related health problems, research and care; as well as human services organizations, especially those that were connected to her Protestant background. The foundation will consider requests for general operating support only if the organization's operating budget is less than $1 million. The majority of grants from the Lange Burk Fund are one year in duration. Applicants must apply online at the grant website. Applicants are strongly encouraged to do the following before applying: review the downloadable state application procedures for additional helpful information and clarifications; review the downloadable online-application guidelines at the grant website; review the foundation's funding history (link is available from the grant website); review the online application questions in advance; and review the list of required attachments. These will generally include: a list of board members, financial statements (audited, reviewed, or compiled by independent auditor); an organization summary; a list of other funding sources; an IRS Determination letter; and other required documents. All attachments must be uploaded in the online application as PDF, Word, or Excel files. The Henrietta Lange Burk Fund has biannual deadlines of June 1 and November 1. Applicants for the June deadline will be notified of grant decisions by September 30. Applicants for the November deadline will be notified by March 31. Most recent awards have ranged from $2,500 to $35,000.
Requirements: Applicants must have 501(c)3 tax-exempt status and serve residents of the Chicago Metropolitan area.
Restrictions: In general, grant requests for individuals, endowment campaigns or capital projects will not be considered. Because requests for support usually exceed available resources, organizations can only apply to either the Lang Burk Fund or the Colonel Stanley NcNeil Foundation in the same calendar year. Grant requests to both foundations during the same calendar year will no longer be accepted. The foundation does not support requests from individuals, organizations attempting to influence policy through direct lobbying, or any political campaigns.
Geographic Focus: Illinois
Date(s) Application is Due: Jun 1; Nov 1
Amount of Grant: 2,500 - 35,000 USD
Samples: Bethel New Life, Chicago, Illinois, $35,000 - for Beth Ann Fall Prevention program (2014); North Lawndale Employment Network, Chicago, Illinois, $20,000 - for Sweet Beginnings program (2014); Mudlark Theater Company, Evanston, Illinois, $5,000 - general operating support (2014).
Contact: Srilatha Lakkaraju; (312) 828-8166; ilgrantmaking@ustrust.com
Internet: https://www.bankofamerica.com/philanthropic/foundation.go?fnId=61
Sponsor: Henrietta Lange Burk Fund
231 South LaSalle Street, IL1-231-13-32
Chicago, IL 60604

Henrietta Tower Wurts Memorial Foundation Grants 2002
The Foundation supports organizations which are engaged in helping or caring for people in need, or alleviating the conditions under which they live, primarily for the elderly, women, family and child welfare services. Giving is limited to Philadelphia, Pennsylvania.
Requirements: Non-profits in Philadelphia, Pennsylvania are eligible. The initial approach should be a letter requesting an application form from the Foundation.
Restrictions: No grants to individuals, or for endowment funds, scholarships, fellowships, or matching gifts; no loans. Organizations or programs serving disadvantaged youth and the elderly in Philadelphia, Pennsylvania must have an annual budget of less than 3 million.
Geographic Focus: Pennsylvania
Date(s) Application is Due: Feb 1; May 1; Sep 1
Amount of Grant: 1,000 - 7,000 USD
Contact: Andrew Swinney, President; (215) 563-6417
Sponsor: Henrietta Tower Wurts Memorial Foundation
1234 Market Street, Suite 1800
Philadelphia, PA 19107-3704

Henry A. and Mary J. MacDonald Foundation 2003
Established in 1998 in Pennsylvania, the Henry A. and Mary J. MacDonald Foundation offers funding in its primary field of interest, health care. A formal application is required, though there are no annual deadlines. Applicants should begin by contacting the Foundation directly. Most recently, grants have averaged about $3,000, though there are no limitations on award ceilings other than availability of funds.
Restrictions: No grants are given to individuals.
Geographic Focus: All States
Amount of Grant: 2,000 - 5,000 USD
Contact: James D. Cullen, President; (814) 870-7705
Sponsor: Henry A. and Mary J. MacDonald Foundation
100 State Street, Suite 700
Erie, PA 16507-1498

Henry and Ruth Blaustein Rosenberg Foundation Health Grants 2004
The Henry and Ruth Blaustein Rosenberg Foundation provides support primarily in the Baltimore, Maryland region. In the area of Health, the primary goal is to promote quality treatment and supportive care at selected health institutions serving the Baltimore community. Typically, awards range from $100,000 to $250,000, and the majority of grants are multi-year in nature. There are nor specified annual deadlines or application forms.
Requirements: 501(c)3 tax-exempt charitable organizations located in or primarily serving the metropolitan Baltimore area are eligible to apply. An initial application should include the following: information about the program(s) for which funding is requested; need, purpose, activities, and evaluation plan of the proposed program(s); program budget (including sources of anticipated income as well as expenditures) and timeline; dollar amount of funding requested; history, mission, and key accomplishments of your organization; information on Board members and key staff; current institutional operating budget (including major sources of revenue as well as expenditures); copy of an IRS tax status determination letter or information about your fiscal agent.
Restrictions: The Foundation does not accept unsolicited proposals for health research. The foundation does not: make grants or scholarships to individuals; accept unsolicited proposals for academic, scientific, or medical research; or support direct mail, annual giving, membership campaigns, fundraising and commemorative events. The foundations rarely make capital grants unless there is a prior relationship with the applicant organization.
Geographic Focus: Maryland
Amount of Grant: 100,000 - 250,000 USD
Contact: Henry A. Rosenberg, Jr.; (410) 347-7201; fax (410) 347-7210; info@blaufund.org
Internet: http://www.blaufund.org/foundations/henryandruth_f.html
Sponsor: Henry and Ruth Blaustein Rosenberg Foundation
One South Street, Suite 2900
Baltimore, MD 21202

Henry County Community Foundation Grants 2005
As a community foundation, the Henry County Community Foundation addresses the broad needs in Henry County, Indiana which include, but are not limited to, the following five categories: health and medical; social services; education; cultural affairs and civic affairs. All requests for grants are reviewed by the Foundation's Grants Committee, which is made up of members of the Board of Directors and several outside advisers. Reviews and recommendations are then presented to the full Board of Directors at its regularly scheduled meetings in April and October. However, the Board reserves the right to consider individual requests at any regularly scheduled meeting, if deemed necessary. The Foundation's grant program emphasizes change-oriented and focused types of grants to achieve certain objectives such as becoming more efficient, increasing fundraising capabilities, delivering better products, etc. The guidelines and application are available at the website.
Requirements: Applications from organizations whose programs benefit the residents of Henry County are accepted only from organizations who attend the spring or fall grant workshops held at the Foundation office. Organizations must be a 501(c)3 tax-exempt or be sponsored by a 501(c) tax-exempt organization. Grants will also be accepted from school and government entities.
Restrictions: The following are not eligible for funding: operating cost; individuals; individuals or groups of individuals to attend seminars or take trips except where there are special circumstances which will benefit the community; sectarian religious purposes. No grants will be made exclusively for endowment purposes of recipient organizations; individuals or organizations that are not charitable organizations; and programs and/or equipment which were committed to prior to the submission of grant application.
Geographic Focus: Indiana
Date(s) Application is Due: Mar 30; Aug 31
Contact: Beverly Matthews, Executive Director; (765) 529-2235; fax (765) 529-2284; beverly@henrycountycf.org or info@henrycountycf.org
Internet: http://www.henrycountycf.org/index.php?submenu=Grants&src=gendocs&ref=Grants&category=Grants
Sponsor: Henry County Community Foundation
700 S Memorial Drive, P.O. Box 6006
New Castle, IN 47362

Henry L. Guenther Foundation Grants 2006
The Henry L. Guenther Foundation, established in 1956, awards grants to California nonprofit organizations in its areas of interest, including: creating opportunities for youth; disease research; education; human services; and community services. Grants support hospitals, medical research, and social services. There are two annual deadlines: May 31 and October 31. Application forms and guidelines are available upon request. The board meets in January and July to consider all proposals. Most recent grants have ranged from $5,000 to $500,000.

Requirements: California nonprofit 501(c)3 organizations are eligible to apply. Grants are awarded primarily in southern California.
Restrictions: Grants do not support government agencies, religious organizations for religious purposes, individuals, or debt-reduction requests.
Geographic Focus: California
Date(s) Application is Due: May 31; Oct 31
Amount of Grant: 5,000 - 500,000 USD
Samples: Loma Linda University, Loma Linda, California, $400,000 - for the purchase of a Siemens cyclotron for a new Research Imaging Center to further develop Proton Therapy capabilities in the treatment of neurological diseases (2014); PIH Health Foundation, Whittier, California, $100,000 - to purchase tomosynthesis digital mammography equipment (2014); Sharp Healthcare Foundation, San Diego, California, $250,000 - to build the seventh floor of the Stephen Birch Healthcare Center (2014).
Contact: Sarah C. Milliken, President; (310) 785-0658
Sponsor: Henry L. Guenther Foundation
2029 Century Park D, Suite 4392
Los Angeles, CA 90067-3029

Herbert A. and Adrian W. Woods Foundation Grants 2007
The Herbert A. and Adrian W. Woods Foundation was established on June 9, 1999 upon the death of Adrian W. Woods. Mrs. Woods had a long history of charitable giving in the St. Louis community and wanted to establish this foundation to continue that legacy of giving. The Foundation supports charitable organizations primarily in the greater St. Louis, Missouri, area. The Trustees will consider the donor's past giving history and any special needs of any of the charities she has given to in the past. The Trustees will also consider requests from charitable organizations that fall into one or more of the following categories: abused, neglected, or troubled children; the poor; the Episcopal Church and affiliates, including outreach programs; arts and culture in the Metropolitan St. Louis area; animal welfare (in Missouri); and victims of illness or disability, including research in this area. The following types of requests may be submitted: special projects; capital campaign requests (capital grants are awarded only as a source of support among a broad community of funders); challenge or matching grants; and general operation funding. The application deadline for this foundation is 11:59 p.m. on September 1. Final decisions about grant rewards will be made by November 30. Most recent awards have ranged from $5,000 to $25,000.
Requirements: Applicants must have 501(c)3 tax-exempt status.
Restrictions: The fund will not consider requests for multi-year grants (pledges) or endowment creation and funding. The fund does not support requests from individuals, organizations attempting to influence policy through direct lobbying, or any political campaigns.
Geographic Focus: Missouri
Date(s) Application is Due: Sep 1
Amount of Grant: 5,000 - 25,000 USD
Samples: Great Circle, Webster Groves, Missouri, $25,000 - in support of the Arts Alive! program (2014); Saint Louis Crisis Nursery, Saint Louis, Missouri, $25,000 - in support of the SOS for Kids program (2014); Howard Park Early Intervention Center, Saint Louis, Missouri, $15,000 - for the Intensive Behavioral Intervention (IBI) program (2014).
Contact: Srilatha Lakkaraju; (312) 828-8166; ilgrantmaking@ustrust.com
Internet: https://www.bankofamerica.com/philanthropic/foundation.go?fnId=7
Sponsor: Herbert A. and Adrian W. Woods Foundation
231 South LaSalle Street, IL1-231-13-32
Chicago, IL 60604

Herbert H. and Grace A. Dow Foundation Grants 2008
The Foundation has charter goals to improve the educational, religious, economic and cultural lives of Michigan's people. Priority is given to organizations that: have clearly stated objectives, strong and purposeful management and are publicly accountable; have needs which are in areas not normally funded by governmental or public financing; are not hesitant to explore, initiate, volunteer, or execute original ideas or concepts; are willing to collaborate with other persons or organizations to give synergy to a common objective or goal; have purposes which tend to advance private enterprise and the preservation of a free, open and self-resourceful society. There is no formal application form, though grant seekers will find detailed submission information online.
Requirements: Only organizations in Michigan are eligible to apply.
Restrictions: The Foundation does not make grants directly to individuals. It cannot legally support: organizations to which contributions are not tax deductible, according to Internal Revenue Service regulations; organizations that practice discrimination by race, sex, creed, age or national origin; political organizations or organizations whose purposes are to influence legislation.
Geographic Focus: Michigan
Contact: Macauley Whiting Jr., President; (989) 631-3699; fax (989) 631-0675; grants@hhdowfdn.org
Internet: http://www.hhdowfdn.org/guidelines.html
Sponsor: Herbert H. and Grace A. Dow Foundation
1018 West Main Street
Midland, MI 48640-4292

Hereditary Disease Foundation John J. Wasmuth Postdoctoral Fellowships 2009
Support is offered for research projects that contribute to identifying and understanding the basic defect of Huntington's disease. Areas of interest include trinucleotide expansions, animal models, gene therapy, neurobiology and development of the basal ganglia, cell survival and death, and intercellular signaling in striatal neurons. Awards may be renewed for a third year. To receive an application, submit a letter of intent. Letter of intent forms may be submitted electronically via the Web site.
Geographic Focus: All States
Date(s) Application is Due: Feb 15; Jun 15; Oct 15
Amount of Grant: 45,000 USD
Contact: Dr. Carl D. Johnson, (212) 928-2121; carljohnson@hdfoundation.org
Internet: http://www.hdfoundation.org/funding/postdoct.php
Sponsor: Hereditary Disease Foundation
3960 Broadway, 6th Floor
New York, NY 10032

Hereditary Disease Foundation Lieberman Award 2010
The Hereditary Disease Foundation announces a special Lieberman Award, to catalyze innovative proposals leading to the treatment and cure of Huntington's disease. A Lieberman Award can be funded for two years for up to $75,000 per year. Areas of interest include trinucleotide expansions, animal models, gene therapy, neurobiology and development of the basal ganglia, cell survival and death, and intercellular signaling in striatal neurons.
Geographic Focus: All States
Amount of Grant: 75,000 - 150,000 USD
Contact: Dr. Carl D. Johnson, (212) 928-2121; carljohnson@hdfoundation.org
Internet: http://www.hdfoundation.org/funding/lieberman.php
Sponsor: Hereditary Disease Foundation
3960 Broadway, 6th Floor
New York, NY 10032

Hereditary Disease Foundation Research Grants 2011
The focus of the Hereditary Disease Foundation is on Huntington's disease. Support will be for research projects that contribute to identifying and understanding the basic defect in Huntington's disease. Grants are usually for one year, with the possibility of renewal for up to three years. Areas of interest include trinucleotide expansions, animal models, gene therapy, neurobiology and development of the basal ganglia, cell survival and death, and intercellular signaling in striatal neurons. In addition to its regular grants, the foundation also offers the Lieberman Award for innovative proposals accelerating the discovery of a treatment and cure of Huntington's disease. Grants that receive funding are considered seed money. If the project shows promise, it is hoped that other institutions will fund it thereafter. To obtain an application, submit a one-page letter of intent. Letter of intent forms may be submitted electronically via the Web site.
Geographic Focus: All States
Date(s) Application is Due: Feb 15; Jun 15; Oct 15
Amount of Grant: Up to 50,000 USD
Samples: Dr. Andrew Dwork, Columbia U--for research entitled Ultrastructural Immunohistochemistry of Cortical Biopsies in HD; Dr. Nansheng Chen, U of British Columbia--for research entitled Regulation of NMDA Receptor Function by Mutant Huntingtin in Neostriatal Neurons; Dr. Chris Huang, Massachusetts General Hospital--for research entitled Anti-Huntingtin Aggregation Agents as Potential Therapeutics for Huntingtons Disease; Dr. Lucius Passani, Massachusetts General Hospital--for research entitled Evaluation of HYPA/FBP-11, HYPB and HYPC as Candidates for Involvement in HD Pathogenesis.
Contact: Dr. Carl D. Johnson, (212) 928-2121; carljohnson@hdfoundation.org
Internet: http://www.hdfoundation.org/funding/grants.php
Sponsor: Hereditary Disease Foundation
3960 Broadway, 6th Floor
New York, NY 10032

Herman Goldman Foundation Grants 2012
The Herman Goldman Foundation strives to enhance the quality of life through innovative grants in four main areas: 1) health -- to achieve effective delivery of physical and mental health care services; 2) social justice -- to develop organizational, social, and legal approaches to those who are aid deprived or handicapped; 3) education -- for new or improved counseling for effective preschool, vocational and paraprofessional training; and 4) the arts -- to increase opportunities for talented youth to receive training and for less affluent individuals to attend quality presentations. Grantmaking is primarily to 501(c)3 organizations in the metropolitan New York area. Types of support include: annual campaigns; building and renovation; capital campaigns; continuing support; endowments; general operating support; program development; internship funds; research; and seed money.
Requirements: Applicants should submit a proposal along with a copy of the organization's IRS determination letter. There are no deadlines. The board meets monthly, with grants considered in April, July, and November. Organizations are notified within two to three months of submission.
Restrictions: The Foundation does not fund religious organizations, individuals, or emergency funds.
Geographic Focus: New York
Contact: Richard Baron, Executive Director; (212) 797-9090; goldfound@aol.com
Sponsor: Herman Goldman Foundation
44 Wall Street, Suite 1212
New York, NY 10005-2401

Hershey Company Grants 2013
The Hershey Company remains committed to supporting the communities in which it operates and to society in general. Cash and product contributions are made to support a variety of worthy causes and non-profit organizations which support Education, Health & Human Services, Civic & Community initiatives, Arts & Culture and the Environment. Particular emphasis is placed upon causes that support kids and kids at risk. The Hershey

Company has plants in the following areas: Hilo, HI; Robinson, IL; Hazleton, PA; Hershey, PA; Lancaster, PA; Memphis, TN; Stuarts Draft, VA; Sao Roque, Brazil; Guadalajara, Mexico; Monterrey, Mexico; Mumbai, India; Shanghai, China. Organizations located in communities in which The Hershey Company has manufacturing operations should direct requests for funding to the management of The Hershey Company facility in their area, see the Hershey website for additional guidelines and contact information.
Requirements: Nonprofit organizations are eligible. Preference is given to nonprofits in corporate-operating locations. All requests must be submitted in writing.
Restrictions: The following are ineligible for funding: organizations without an Internal Revenue Code 501(c)3 non-profit, tax-exempt status; individuals; organizations outside the immediate areas of The Hershey Company's manufacturing facilities, with the exception of national and state-wide organizations whose programs complement Hershey's funding priorities; political campaigns, political or lobbying organizations, or those supporting the candidacy of a particular individual; churches or religious organizations, including seminaries, Bible colleges and theological institutions; fraternal organizations; labor organizations; member agencies of United Way. Exception: requests for capital campaign funding will be considered; affiliate organizations of the Cultural Enrichment Fund in Central Pennsylvania. Exception: requests for capital campaign funding will be considered.
Geographic Focus: Hawaii, Illinois, Pennsylvania, Tennessee, Virginia, Brazil, China, Mexico
Contact: Grants Administrator; (800) 468-1714; fax (717) 534-6550
Internet: http://www.thehersheycompany.com/about/responsibility.asp
Sponsor: Hershey Company
Community Relations, 100 Crystal A Drive, P.O. Box 810
Hershey, PA 17033-0810

HHMI-NIBIB Interfaces Initiative Grants 2014
HHMI and the National Institute of Biomedical Imaging and Bioengineering have formed a partnership to support biomedical research institutions in developing graduate-level research training programs in emerging interdisciplinary fields. The primary goal of this initiative is to train a cadre of PhD scientists who possess the knowledge and skills to conduct interdisciplinary research at the interface between the biomedical sciences and the physical science, computational, engineering, or mathematical disciplines. These fields may include, but are not limited to, chemistry, imaging, science, materials science, nanotechnology, and physics. Another goal is to reduce barriers to interdisciplinary graduate science education. Proposals should reflect the unique educational and scientific capabilities and strengths of the applicant institution(s) as well as address the specific goals of the initiative. The initiative consists of two phases: Phase I supports the establishment of new interdisciplinary training programs; Phase II will sustain the training programs through their critical years. Registration letters of intent to apply are due by January 20, with the proposal submission deadline June 15.
Requirements: All U.S. institutions that grant PhD degrees in appropriate science or engineering disciplines are eligible to apply. Collaborative programs between two or more institutions are acceptable.
Geographic Focus: All States
Date(s) Application is Due: Jan 20; Jun 15
Amount of Grant: Up to 1,000,000 USD
Contact: Maryrose Franko; (800) 448-4882, ext. 8880; interdisc@hhmi.org
Internet: http://www.hhmi.org/grants/institutions/nibib.html
Sponsor: Howard Hughes Medical Institute
4000 Jones Bridge Road
Chevy Chase, MD 20815-6789

HHMI-NIH Cloister Research Scholars Program 2015
The goal of this program, a joint venture of HHMI and NIH, is to expand the pool of medically trained researchers by encouraging medical students to pursue research careers. The program provides support for an intensive nine months to one year of full-time research, with a possible one-year extension, at NIH in Bethesda, MD. Most students enter after their second year of medical school. The Cloister, a residential facility for scholars, is available on the NIH campus. The research project is selected upon arrival at NIH, after a round of laboratory visits. Application forms may be downloaded from the Web site or requested from the program officer.
Requirements: Applicant must be in good standing at a medical or dental school in the U.S. or Puerto Rico and must receive permission from the school to participate in the program. The school does not have to be a major academic medical center or research-oriented school.
Restrictions: Enrollees in an M.D./Ph.D. or D.D.S./Ph.D. program are not eligible.
Geographic Focus: All States
Date(s) Application is Due: Jan 10
Amount of Grant: 17,800 USD
Contact: Office of Grants and Special Programs; (800) 424-9924 or (301) 215-8873; fax (301) 215-8888; research_scholars@hhmi.org
Internet: http://www.hhmi.org/research/cloister/program.html
Sponsor: Howard Hughes Medical Institute
4000 Jones Bridge Road
Chevy Chase, MD 20815-6789

HHMI/EMBO Start-up Grants for Central Europe 2016
The program encourages young scientists to obtain independent faculty positions in select countries that have ongoing HHMI and EMBO science programs. Applications are encouraged from scientists to set up their first independent laboratories in a Central European EMBC member state (Croatia, Czech Republic, Estonia, Hungary, Poland, and Slovenia). Application forms should be completed by the individual scientist and the receiving institute. It is expected that the institute make the applicant an offer that goes beyond that of a position and laboratory space. HHMI will contribute $50,000, while the participating member state must provide the remaining $25,000. Application and guidelines are available online.
Requirements: Eligible scientists should have an excellent publication record; have a maximum of 10 years postdoctoral experience; be negotiating a position at an institute/university in Central Europe by the date of the application; and be located outside Central Europe at the time of application.
Geographic Focus: All States, Austria, Czech Republic, Germany, Hungary, Liechtenstein, Poland, Slovakia, Slovenia, Switzerland
Date(s) Application is Due: Aug 1
Amount of Grant: 75,000 USD
Contact: Grants Administrator; (301) 215-8500
Internet: http://www.embo.org/projects/yip/embo_hhmi_startup_grants.html
Sponsor: Howard Hughes Medical Institute
4000 Jones Bridge Road
Chevy Chase, MD 20815-6789

HHMI Biomedical Research Grants for International Scientists: Infectious Diseases and Parasitology 2017
The Program provides five-year grants to support promising basic biomedical research scientists working in the fields of infectious diseases and parasitology. The Institute seeks to build the strongest foundation of scientific knowledge and capability by creating a global network of scientists who make important contributions outside the U.S. research community. It support these leaders in their home countries to facilitate the application and dissemination of their knowledge and to strengthen research environments and educational opportunities.
Requirements: Applicants must hold a full-time appointment or have a pending appointment at a nonprofit scientific research organization in any country other than the United States and the United Kingdom. Researchers may not be citizens or permanent residents of the United States.
Restrictions: Only scientists working in the research areas specified in the program announcement are eligible to apply for grant support. The Program does not consider unsolicited grant proposals or award grants for clinical trials and research on health education, health-care delivery, or health services.
Geographic Focus: All States
Date(s) Application is Due: Sep 15
Amount of Grant: 50,000 - 100,000 USD
Contact: Jill G. Conley; (301) 215-8873; fax (301) 215-8888; parasite@hhmi.org
Internet: http://www.hhmi.org/grants/individuals/idap.html
Sponsor: Howard Hughes Medical Institute
4000 Jones Bridge Road
Chevy Chase, MD 20815-6789

HHMI Biomedical Research Grants for International Scientists in Canada and Latin America 2018
HHMI's International Research Scholars Program provides five-year grants to support promising basic biomedical research scientists working in Canada and Latin America. Eligible epidemiology research includes that directed toward an understanding of disease distribution in populations or of associations that may suggest causality or preventive strategies.
Requirements: The supported research must be conducted in one of the following eight eligible countries: Argentina, Brazil, Canada, Chile, Mexico, Peru, Uruguay, or Venezuela. Applicants must hold a full-time appointment or have a pending full-time appointment at a nonprofit scientific research organization in any one of the eight eligible countries.
Restrictions: Clinical trials and research on health education, health care delivery, or health services are not eligible fields.
Geographic Focus: All States
Date(s) Application is Due: Sep 14
Amount of Grant: 50,000 - 100,000 USD
Contact: Jill G. Conley; (301) 215-8873; fax (301) 215-8888; canlatam@hhmi.org
Internet: http://www.hhmi.org/grants/individuals/canlatam.html
Sponsor: Howard Hughes Medical Institute
4000 Jones Bridge Road
Chevy Chase, MD 20815-6789

HHMI Biomedical Research Grants for International Scientists in the Baltics, Central and Eastern Europe, Russia, and Ukraine 2019
The Program will award five-year grants to support promising basic biomedical research scientists working in the Baltics, Central and Eastern Europe, Russia, and Ukraine. Eligible epidemiology research includes that directed toward an understanding of disease distribution in populations or of associations that may suggest causality or preventive strategies. Allowable expense categories include salaries and stipends for award recipients, graduate students, postdoctoral fellows, and technicians, equipment and supplies, travel, and publication costs. Up to 10 percent of the grant may be used to pay for indirect costs (including administrative costs) of the scientist's institution.
Requirements: The supported research must be conducted in one of the following eligible countries: Bulgaria, Croatia, Czech Republic, Estonia, Hungary, Latvia, Lithuania, Poland, Romania, Russia, Slovak Republic, Slovenia, or Ukraine.
Restrictions: Researchers may not be citizens or permanent residents of the United States. Grant funds may not be used for laboratory renovations.
Geographic Focus: All States, Albania, Andorra, Armenia, Austria, Azerbaijan, Belarus, Belgium, Bosnia & Herzegovina, Bulgaria, Croatia, Cyprus, Czech Republic, Denmark, Estonia, Finland, France, Georgia, Germany, Greece, Hungary, Hungary, Iceland, Ireland, Italy, Kosovo, Latvia, Liechtenstein, Lithuania, Luxembourg, Macedonia,

Malta, Moldova, Monaco, Montenegro, Norway, Poland, Portugal, Romania, Russia, Russia, San Marino, Serbia, Slovakia, Slovenia, Spain, Sweden, Switzerland, The Netherlands, Turkey, Ukraine, Ukraine, United Kingdom, Vatican City
Date(s) Application is Due: Nov 17
Amount of Grant: 50,000 - 100,000 USD
Contact: Jill G. Conley; (301) 215-8873; fax (301) 215-8888; bceeru@hhmi.org
Internet: http://www.hhmi.org/grants/individuals/bceeru.html
Sponsor: Howard Hughes Medical Institute
4000 Jones Bridge Road
Chevy Chase, MD 20815-6789

HHMI Grants and Fellowships Programs 2020
The institute administers a grants program that focuses on improving science education from preschool through postdoctoral training, enhancing the science literacy of the general public, and supporting the research of biomedical scientists in selected foreign countries. Currently, grants are administered through five programs: an international program that supports biomedical scientists outside the United States and provides funding for selected courses and workshops; a precollege science-education program that offers grants for educational activities to science museums, botanical and zoological gardens, etc; a biomedical-research program that supports medical schools and research organizations; a graduate science-education program that provides fellowships for graduate students, medical students, and physicians, and supports special courses; and a biological-sciences education program that awards grants to selected undergraduate institutions.
Geographic Focus: All States
Samples: United Way of New York City (NY)--for the September 11th Fund, established to support emergency-assistance to organizations and other nonprofit health and human-service groups providing disaster-relief services to victims of the terrorist attacks in New York, Pennsylvania, and Washington, $50,000.
Contact: Jill G. Conley; (301) 215-8873; fax (301) 215-8888; grantswww@hhmi.org
Internet: http://www.hhmi.org/grants
Sponsor: Howard Hughes Medical Institute
4000 Jones Bridge Road
Chevy Chase, MD 20815-6789

HHMI International Research Scholars Program 2021
This five year grant program supports non-U.S. biomedical scientists, scientific meetings for grant-supported scientists, and other international educational activities. Grants are awarded to promising biomedical scientists who have made significant contributions to fundamental research. Application information is available online.
Requirements: To be considered, individuals must hold a full-time academic or research appointment at a university, medical school, or other nonprofit scientific institution. They are not permitted to have major administrative responsibilities.
Geographic Focus: All States
Contact: Jill G. Conley, Program Director; (301) 215-8873; fax (301) 215-8888
Internet: http://www.hhmi.org/grants/for_grantees/
Sponsor: Howard Hughes Medical Institute
4000 Jones Bridge Road
Chevy Chase, MD 20815-6789

HHMI Med into Grad Initiative Grants 2022
HHMI will award grants to institutions to improve the understanding of medicine and pathobiology by scientists conducting biomedical research. Grants will be used to modify existing graduate training or initiate new programs to develop a cadre of Ph.D. researchers who understand pathophysiology and are committed to working at the interface of the basic sciences and clinical medicine. Applications will be accepted from any university in the United States that offers Ph.D.-level training in an appropriate science or engineering discipline. It is expected that as much as $10 million will be awarded in this competition. Awards will range from $400,000 to $1 million, and the grant funds will be allocated over four consecutive years.
Requirements: U.S. institutions that grant PhD degrees in the appropriate science and engineering disciplines are eligible to apply.
Geographic Focus: All States
Date(s) Application is Due: Apr 27
Amount of Grant: 400,000 - 1,000,000 USD
Contact: Maryrose Franko, Senior Program Officer; (800) 448-4882, ext. 8880; fax (301) 215-8888; medintograd@hhmi.org
Internet: http://www.hhmi.org/grants/institutions/medintograd.html
Sponsor: Howard Hughes Medical Institute
4000 Jones Bridge Road
Chevy Chase, MD 20815-6789

HHMI Physician-Scientist Early Career Award 2023
Through this competitive grant initiative, HHMI awards five-year grants to selected alumni of the HHMI-NIH Research Scholars Program and the HHMI Research Training Fellowships for Medical Students Program to support these individuals as they begin careers as independent physician-scientists. The award provides $375,000 over a five-year period for direct research costs.
Requirements: Only alumni of the HHMI-NIH Research Scholars Program and the HHMI Research Training Fellowship for Medical Students Program who have received an M.D., M.D./Ph.D., D.D.S, or equivalent degree are eligible to apply.
Restrictions: The funds may not be used for the salary of the awardee or institutional indirect costs.
Geographic Focus: All States
Amount of Grant: 375,000 USD
Contact: Office of Grants and Special Programs; (800) 448-4882, ext. 8889 or (800) 424-9924; fax (301) 215-8888; earlycareer@hhmi.org
Internet: http://www.hhmi.org/grants/individuals/earlycareer.html
Sponsor: Howard Hughes Medical Institute
4000 Jones Bridge Road
Chevy Chase, MD 20815-6789

HHMI Research Training Fellowships 2024
Medical fellowships are intended to strengthen and expand the nation's pool of medically trained researchers. These fellowships provide funds to help meet fellows' research-related expenses and education costs during the research training period. Fellowships are awarded annually to provide support for one year of full-time training in fundamental biomedical, clinical, translational, or applied research. The fellowship includes a stipend, a research allowance to meet research-related expenses, and a fellow's allowance to be used on behalf of the fellow for health care, tuition and fees, and other research-related expenses. Application information is available online.
Requirements: Fellowships are awarded to students enrolled in medical and dental programs in the U.S.
Geographic Focus: All States
Date(s) Application is Due: Jan 11
Contact: Program Officer; (301) 215-8500 or (800) 424-9924; fax (301) 215-8888; medfellows@hhmi.org
Internet: http://www.hhmi.org/grants/individuals/medfellows.html
Sponsor: Howard Hughes Medical Institute
4000 Jones Bridge Road
Chevy Chase, MD 20815-6789

Highmark Corporate Giving Grants 2025
The Corporation awards grants in the following categories: health; human services; community; education; and arts and culture. The Corporation will consider applications from: programs and services that tie to its mission and have a compelling potential impact to the health and well-being of individuals; national organizations seeking support for local programs; and grassroots and faith-based organizations. Support is expressed in cash grants, sponsorships, and in-kind gifts to non-profit organizations. There are no maximum or minimum awards; however, awards to programs and services are granted based on importance to the Corporation's business, social mission and corporate citizen objectives, as well as the initiative's proposed impact. Interested applicants should submit a brief proposal. For additional information or proposal submission, applicants should contact the Community Affairs department in their county of residence. Contact information by county is available online.
Requirements: The Foundation awards grants within its service area to nonprofit organizations that are defined as tax exempt under section 501(c)3 of the Internal Revenue Code and as public charities under section 509(a) of that code. Eligible counties in Pennsylvania include: Adams, Allegheny, Armstrong, Beaver, Bedford, Berks, Blair, Butler, Cambria, Cameron, Centre, Clarion, Clearfield, Columbia, Crawford, Cumberland, Dauphin, Erie, Elk, Fayette, Forest, Franklin, Fulton, Greene, Huntingdon, Indiana, Jefferson, Juniata, Lancaster, Lawrence, Lebanon, Lehigh, McKean, Mercer, Mifflin, Montour, Northampton, Northumberland, Perry, Potter, Schuylkill, Snyder, Somerset, Union, Venango, Warren, Westmoreland, Washington, and York. All counties in West Virginia and Delaware are eligible to apply.
Restrictions: Proposals are not accepted for: multiple-year grants; capital campaigns; individual causes; seed money or start-up organizations; endowment funds; political causes or campaigns; fraternal or civic groups; religious programming; or organizations that discriminate on the basis of race, religion, sex, disability or national origin.
Geographic Focus: Delaware, Pennsylvania, West Virginia
Contact: Mary Ann Papale, (412) 544-4032; mary.papale@highmark.com or GrantsSWPAApply@highmark.com
Internet: https://www.highmark.com/hmk2/responsibility/giving/index.shtml
Sponsor: Highmark
120 5th Avenue, Suite 2112
Pittsburgh, PA 15222-3099

Highmark Physician eHealth Collaborative Grants 2026
The program was created to encourage the adoption of health information technology used at the point of care to improve patient safety, quality of care, and cost efficiency for people in western and central Pennsylvania. Physicians may apply to receive grants, which must be used to acquire and use electronic technology systems such as a personal computer, a PDA, or electronic tablet or digital pen to generate and transmit electronically prescriptions to pharmacies. The collaborative will pay up to 75 percent of the cost for a physician's office to acquire, install, and implement the electronic technology system, with the physician's practice to pay the remaining balance. Physicians must apply online.
Requirements: Physicians must be licensed to practice medicine in Pennsylvania and must be a licensed prescriber.
Geographic Focus: Pennsylvania
Amount of Grant: Up to 7,000 USD
Contact: Grants Administrator; info@highmarkehealth.org
Internet: http://www.highmarkehealth.org
Sponsor: Pittsburgh Foundation
1 PPG Place, 30th Floor
Pittsburgh, PA 15222-5401

High Meadow Foundation Grants 2027

The foundation grants funds in support of the performing arts, especially theater and music, and other cultural organizations. Organizations supported are usually located in Berkshire County, MA. Types of support include continuing support, annual campaigns, capital campaigns, building construction/renovation, equipment acquisition, emergency funds, program development, employee-related scholarships, and matching funds. Applications are accepted at any time. Forms are not necessary; a letter outlining the proposed project should include budget and administrative information.
Requirements: Western Massachusetts 501(c)3 tax-exempt organizations in Berkshire County are eligible.
Geographic Focus: Massachusetts
Contact: Jane Fitzpatrick; (413) 298-5565 or (413) 243-1474; fax (413) 298-4058
Sponsor: High Meadow Foundation
30 Main Street
Stockbridge, MA 01262

Hilda and Preston Davis Foundation Grants 2028

The Foundation provides funds to charitable organizations whose programs advance the development of all areas of the lives of children and young adults. The Foundation places special emphasis on, and channels most of its financial resources toward, those organizations whose attention is concentrated on eating disorders and education for the underprivileged. Grants may range from $10,000 up to $100,000. In unique circumstances, the Foundation does consider a more significant grant for a program having a major impact in one or more of its areas of interest. The Foundation encourages pilot initiatives that test new program models. Of particular interest to the Foundation are organizations that promote partnerships and collaborative efforts among multiple groups and organizations. The Foundation almost always limits grant durations to three years or less. Applications are accepted throughout the year.
Requirements: ll applicants must be an approved tax-exempt non-profit organization as defined by the Internal Revenue Service. Areas of greatest importance to the Foundation are funded on a national basis. However, a good deal of the Foundation's grant making in other program areas is focused in areas of geographic importance to the donors–the northeastern United States (southern New England & the middle Atlantic states), as well as California. Priority will be given to requests that show specific plans for funding beyond the present.
Restrictions: The Foundation will not allow any funds to be earmarked for indirect costs or institutional overhead in cases where the grant relationship was developed independent of that institution's direct involvement. The Foundation generally will not provide grants to the following: organizations not determined to be tax-exempt under section 501(c)3 of the Internal Revenue Code; individuals; general fundraising drives; endowments; government agencies; or organizations that subsist mainly on third party funding and have demonstrated no ability or expended little effort to attract private funding.
Geographic Focus: California, Connecticut, Delaware, District of Columbia, Maryland, Massachusetts, New Hampshire, New Jersey, New York, Pennsylvania, Rhode Island, Vermont, Virginia, West Virginia
Amount of Grant: 10,000 - 100,000 USD
Contact: Grants Administrator; (203) 629-8552; fax (203) 547-6112; davis@fsllc.net
Internet: http://www.hpdavis.org/application.htm
Sponsor: Hilda and Preston Davis Foundation
640 West Putnam Avenue, 3rd Floor
Greenwich, CT 06830

Hilda and Preston Davis Foundation Postdoctoral Fellowships in Eating Disorders Research 2029

The Hilda and Preston Davis Foundation's Postdoctoral Fellowship program is administered and managed by The Medical Foundation. By attracting postdoctoral fellows to the field, dollars allocated to support fellows are leveraged into lifetime contributions to understanding biological causes of eating disorders. The long term goal of the program is to accelerate medical research discoveries that will lead to effective new therapies. Proposals must be focused on relevant aspects of the biological causes of anorexia nervosa and bulimia nervosa as defined by clinical criteria. Up to five three-year fellowships ranging from $43,000 - $63,000 per year, inclusive of a $3,000 expense allowance will be awarded to fellows who have completed no more than three years of postdoctoral research training as of the funding start date. The awards will support postdoctoral fellows working in nonprofit academic or research institutions in the United States.
Requirements: Proposals must be focused on relevant aspects of the biological causes of anorexia nervosa and bulimia nervosa as defined by clinical criteria. Examples of funding areas include but are not limited to neural pathways of feeding behavior in animal models; molecular genetic analysis of relevant neural circuit assembly and function; testing of new chemical compounds that might be used in animal models as experimental treatments; and brain imaging technologies that identify neurochemical pathways in patients with these disorders. Applicants must: have a Ph.D. or equivalent awarded from an accredited domestic or foreign institution by July 1 and commit 90% time to research (applicants with clinical responsibilities must commit at least 70% time to research) or, M.D. or equivalent awarded from an accredited domestic or foreign institution by July 1 and commit at least 70% time to research; have completed no more than three years of full-time postdoctoral research experience by the time funding begins on July 1; and, conduct the proposed research project at a hospital, university or other nonprofit research institution where the applicant holds a postdoctoral fellowship appointment. There are no institutional limitations on the number of applicants who may submit applications. However, a mentor may only support one fellow's application. United States citizenship is not required. Online submissions must be received by 12:00 Noon, U.S. Eastern time of January 10. Mailed submissions must be received by 5:00 pm, U.S. Eastern time of January 15.
Restrictions: Clinical psychotherapeutic studies, medication trials, and obesity research are currently outside the scope of this program.
Geographic Focus: All States
Date(s) Application is Due: Jan 10; Jan 15
Amount of Grant: 129,000 - 189,000 USD
Contact: Program Officer; (203) 629-8552; fax (203) 547-6112; davis@fsllc.net
Internet: http://www.hpdavis.org/application.html
Sponsor: Hilda and Preston Davis Foundation
640 West Putnam Avenue, 3rd Floor
Greenwich, CT 06830

Hill Crest Foundation Grants 2030

The Hill Crest Foundation awards grants to Alabama nonprofits, primarily health associations and human services and education, with some funding for the arts. Types of support include: building and renovation; capital campaigns; endowments; equipment; matching and challenge support; professorships; program development; publication; research; scholarship funds; seed money; and technical assistance. There are no deadlines or applications. The Board of Directors meets quarterly.
Requirements: Application forms are not required. Before submitting a proposal, organizations should submit a letter, detailing the project and the amount requested, along with descriptive literature about their organization.
Restrictions: The Foundation does not fund individuals.
Geographic Focus: Alabama
Amount of Grant: 20,000 - 150,000 USD
Contact: Charles Terry, Chairperson; (205) 425-5800
Sponsor: Hill Crest Foundation
P.O. Box 530507
Mountain Brook, AL 35253-0507

Hillcrest Foundation Grants 2031

The Hillcrest Foundation was created by Mrs. W.W. Caruth, Sr. (Mrs. Earle Clark Caruth) in 1958 to provide financial support to qualified Texas charitable organizations for the advancement of education, the promotion of health, and the relief of poverty. Mrs. W.W. Caruth, Sr. was from a pioneer family who settled in the Dallas area in 1848. Several Caruth family generations owned and managed farms and ranches, which the family later developed into real estate properties as Dallas became a major metropolitan area. Each succeeding generation has been characterized by a pioneering spirit, vision, courage, hard work and generosity. The Hillcrest Foundation was created to carry the Caruth family's generosity to the people of Texas. Approximately 90% of grant funds are paid to organizations in North Texas, with emphasis on charitable services in the Dallas area. The trustees will give priority consideration to the following types of grant requests: construction/improvements of permanent buildings; programs and special projects for education, health and poverty relief; capital campaigns; and buildings, facilities and equipment. Grant requests for capital and program support are strongly encouraged. Grant amounts range from $10,000 to $300,000 (multi-year payment) with an average grant of $35,000. The Trustees favorably consider proposals which are unique, necessary, and of high priority for the charitable organizations, and which do not duplicate other services which are available; proposals for which funding may not be readily available from other sources; and essential projects which are sufficiently described as worthwhile, important and of a substantive nature. Grants to meet challenges or matching funds have a special appeal. The majority of grants from the Hillcrest Foundation are one year in duration. On occasion, multi-year support is awarded. Applicants must apply online at the grant website. Applicants are strongly encouraged to do the following before applying: review the downloadable state application procedures for additional helpful information and clarifications; review the downloadable online-application guidelines at the grant website; review the foundation's funding history (link is available from the grant website); review the online application questions in advance; and review the list of required attachments. These will generally include: a list of board members, financial statements (audited, reviewed, or compiled by independent auditor); an organization summary; a list of other funding sources; an IRS Determination letter; and other required documents. All attachments must be uploaded in the online application as PDF, Word, or Excel files. The Hillcrest Foundation has three deadlines annually: February 28, July 31, and November 30. Applications must be submitted by 11:59 p.m. on the deadline dates. Grant applicants are notified as follows: February deadline applicants will be notified of grant decisions by June 30; July applicants will be notified by November 30; and November applicants will be notified by March 31 of the following year.
Requirements: 501(c)3 organizations in Texas may apply.
Restrictions: Grant requests for general operating support will not be considered. Organizations that receive a one-year grant from the foundation must skip a year before submitting a subsequent application. Organizations that receive a multi-year grant are not eligible to apply until one year after the close of their grant cycle. Organizations whose last request was declined must wait one year before applying again. The foundation does not support requests from individuals, organizations attempting to influence policy through direct lobbying, or any political campaigns.
Geographic Focus: Texas
Date(s) Application is Due: Feb 28; Jul 31; Nov 30
Amount of Grant: 10,000 - 300,000 USD
Contact: David Ross, Senior Vice President; tx.philanthropic@baml.com
Internet: https://www.bankofamerica.com/philanthropic/fn_search.action
Sponsor: Hillcrest Foundation
901 Main Street, 19th Floor, TX1-492-19-11
Dallas, TX 75202-3714

Hillman Foundation Grants 2032

The foundation provides grants to nonprofit organizations in the city of Pittsburgh and the southwestern Pennsylvania region for programs and projects designed to improve the quality of life in the area. The foundation's areas of interest include community affairs, social services, culture and the arts, education at all levels, and youth services including medical and health. Grants range widely in size and are allocated for large and small capital projects, endowment, new and expanding programs, and on a limited basis, operating support. Applications should include an annual budget of the organization, project information, and evidence of tax-exempt status. Application information is available online.
Requirements: The Foundation considers requests only from organizations classified as tax-exempt under Section 501(c)3 of the U.S. Internal Revenue Code and designated as public charities under Section 509(a).
Restrictions: Grants are not made to individuals, organizations located outside of the United States, for travel expenses for groups, or in support of events, sponsorships and meetings such as conferences, institutes and seminars.
Geographic Focus: Pennsylvania
Samples: Strong Women, Strong Girls, INc., $10,000--towards program for at-risk and low-income girls; Greater Pittsburgh Community Food Bank, $65,000--toward food pantry and outreach program; Carnegie Institute/Museum of Natural History, $50,000--towards purchase of bournonite, amazonite, smoky quarts and rhodochrosite mineral specimens.
Contact: David K. Roger; (412) 338-3466; fax (412) 338-3463; foundation@hillmanfo.com
Internet: http://www.hillmanfdn.org/grantprograms.html
Sponsor: Hillman Foundation
330 Grant Street, Suite 2000
Pittsburgh, PA 15219

Hillsdale County Community Found Healthy Senior/Healthy Youth Grants 2033

The Healthy Senior/Healthy Youth Fund provides grants to smoking cessation and prevention programs for young people (under 18) and seniors (over 65) in Hillsdale County. Organizations interested in applying are strongly encouraged to discuss their project with the Executive Director prior to submitting an application. If the project meets the basic eligibility criteria and is consistent with the Foundation's program interests, a formal application will be sent to the applicant. The Grantmaking Committee meets twice a year to review funding requests.
Requirements: The Foundation welcomes applications from the Hillsdale County area or outside Hillsdale County, Michigan, if a significant number of the people to be served reside within Hillsdale County. Applicants shall be tax exempt according to Section 501(c)3 of the Internal Revenue Code.
Geographic Focus: Michigan
Date(s) Application is Due: May 1; Nov 1
Contact: Sharon Bisher; (517) 439-5101; fax (517) 439-5109; s.bisher@abouthccf.org
Internet: http://www.abouthccf.org/funds.asp
Sponsor: Hillsdale County Community Foundation
2 S Howell Street, P.O. Box 276
Hillsdale, MI 49242

Hillsdale County Community General Adult Foundation Grants 2034

General Adult Foundation grant making is made from the Foundation's named unrestricted endowment funds. These grants focus on improving the quality of life for the citizens of Hillsdale County. Eligible projects generally fall within these categories: education, fine arts, social services, community development, recreation, environmental issues, health and wellness, and improvement in the physical, mental, and moral conditions of Hillsdale County residents. The Foundation aims to support creative approaches to community needs and problems that benefit the widest possible range of people.
Requirements: The Foundation requires applicants to call and discuss their project with the Executive Director prior to submitting the application. Applications from the Hillsdale County area or outside Hillsdale County, Michigan are welcome, if a significant number of the people to be served reside within Hillsdale County. Applicants shall be tax exempt according to Section 501(c)3 of the Internal Revenue Code.
Restrictions: The current objectives of the Foundation do not allow grants for: religious or sectarian purposes; individuals; legislative or political purposes; loans; capital campaigns; routine maintenance, including office equipment; administrative costs for maintaining the present operation of an organization, including, but not limited to, staff salaries, wages, and benefits; basic education materials including state mandated/benchmark core curriculum supplies and resources.
Geographic Focus: Michigan
Date(s) Application is Due: May 1; Nov 1
Contact: Sharon Bisher; (517) 439-5101; fax (517) 439-5109; s.bisher@abouthccf.org
Internet: http://www.abouthccf.org/grants.asp
Sponsor: Hillsdale County Community Foundation
2 S Howell Street, P.O. Box 276
Hillsdale, MI 49242

Hillsdale Fund Grants 2035

The fund awards grants to nonprofit organizations in its areas of interest, including arts and culture, education, health care, social services delivery, and religion. Call for meeting dates and application materials.
Requirements: Tax-exempt organizations are eligible.
Restrictions: Grants do not support indirect costs or overhead; routine, recurring operating expenses; conferences and seminars; travel and study; or individuals.
Geographic Focus: North Carolina
Amount of Grant: 3,000 - 50,000 USD
Contact: Edward Doolan, Grants Administrator; (336) 274-5471
Sponsor: Hillsdale Fund
P.O. Box 20124
Greensboro, NC 27420-0124

Hilton Head Island Foundation Grants 2036

The Foundation exists to enhance the quality of life for individuals living and/or working in Southern Beaufort County (Hilton Head Island, Daufuskie Island, Bluffton, Okatie). Grantmaking is in the broad areas of: arts, education, and community development. The Board meets three times per year. Application information is available online.
Requirements: To be eligible for funding, an applicant must be: a nonprofit agency with a tax-exempt status under section 501(c)3 of the Internal Revenue code, or eligible to be classified as such, but are not classified as a private foundation; serve the people who live and/or work in the Hilton Head Island area community.
Restrictions: Funding is not available for: sectarian or religious activities; political activities or organizations; grants directly to individuals; endowments; annual fundraising campaigns; special events or fundraisers; scholarships for students in grades K-12.
Geographic Focus: South Carolina
Date(s) Application is Due: Apr 1; Aug 1; Dec 1
Contact: Cynthia Smith, Vice President for Grantmaking; (843) 681-9100; fax (843) 681-9101; csmith@cf-lowcountry.org
Internet: http://www.cf-lowcountry.org/receive/grants
Sponsor: Community Foundation of the Lowcountry
4 Northridge Drive, Suite A, P.O. Box 23019
Hilton Head Island, SC 29925

Hilton Hotels Corporate Giving Program Grants 2037

Hilton makes charitable contributions to nonprofit organizations involved with K-12 education, youth development, public policy, homelessness, and civic affairs. Support is given primarily in areas of company operations, with emphasis on California, including Los Angeles and San Francisco, and Tennessee, including Memphis; giving also to national organizations. Contact Ellen Gonda for additional information at: corporate_communications@hilton.com.
Requirements: 501(c)3 tax-exempt organizations are eligible.
Restrictions: No support for sport teams, religious organizations not of direct benefit to the entire community, government-supported organizations (over 20 percent of budget), hospitals, private schools, pre-schools, or day care facilities, film and video production, or promotional materials. No grants to individuals (except for employee-related scholarships), or for fellowships, sports activities, debt reduction, capital campaigns or endowments, film or video projects, or promotional merchandise.
Geographic Focus: California, Tennessee
Contact: Ellen Gonda; (310) 278-4321; corporate_communications@hilton.com
Internet: http://www.hiltonworldwide.com
Sponsor: Hilton Hotels Corporation
9336 Civic Center Drive
Beverly Hills, CA 90210-3604

Hoblitzelle Foundation Grants 2038

The Hoblitzelle Foundation was established by Karl and Esther Hoblitzelle in 1942. Grants made by the Directors are usually focused on specific, non-recurring needs of the educational, social service, medical, cultural, and civic organizations in Texas, particularly in the Dallas area. The Board meets three times per year. It is preferred that the initial approach be through a brief narrative letter describing the project for which funds are asked. Guidelines are available online.
Requirements: Nonprofit organizations in the State of Texas, primarily within the Dallas Metroplex, are eligible to apply.
Restrictions: No grants are made for religious purposes or to individuals. No grants are made outside the State of Texas. Contributions are not made toward operating budgets, debt retirement, research, media productions or publications, scholarships or endowments. The Foundation makes no loans.
Geographic Focus: Texas
Date(s) Application is Due: Apr 15; Aug 15; Dec 15
Contact: Paul W. Harris, President & CEO; (214) 373-0462; pharris@hoblitzelle.org
Internet: http://www.hoblitzelle.org/
Sponsor: Hoblitzelle Foundation
5956 Sherry Lane, Suite 901
Dallas, TX 75225

Hogg Foundation for Mental Health Grants 2039

The Foundation accomplishes its mission through the funding of external and internal projects in strategically selected areas. The Foundation operates major initiatives in three priority areas: integrated health care, cultural competence, and workforce development. Funds available through the Foundation's major initiatives and projects are distributed through a Request for Proposals (RFP) process. The Foundation does not accept unsolicited grant proposals on any topic. To be notified of future RFP opportunities, send an email message including your name, affiliation, and mailing address to: HoggCommunications@austin.utexas.edu.
Geographic Focus: Texas
Contact: Program Contact; (888) 404-4336 or (512) 471-5041; fax (512) 471-9608; Hogg-Grants@austin.utexas.edu
Internet: http://www.hogg.utexas.edu/funding_grantmaking.html
Sponsor: Hogg Foundation for Mental Health
University of Texas at Austin, P.O. Box 7998
Austin, TX 78713-7998

Hoglund Foundation Grants 2040

The foundation supports nonprofits working in the areas of education, health science and services, social services, and children's health and development. Grants are made primarily in Dallas, and Houston, Texas, with a limited number of grants awarded outside of this area. Types of support include project support, capital support, and general operating support. Application guidelines are available on the foundation's Web site.
Requirements: To be eligible for consideration, an organization must provide proof that they have received a determination letter from the Internal Revenue Service indicating that it is a tax exempt organization as described in Section 501(c)3 of the Internal Revenue Code of 1986 and is treated as other than a private foundation as within the meaning of Section 509(a) of the Code. An organization may also qualify under Section 170(c)(1) if the grant requested is to be used exclusively for public purposes as described in the Code.
Restrictions: Grants are not made to individuals.
Geographic Focus: Texas
Date(s) Application is Due: Mar 15; Jul 15; Nov 15
Amount of Grant: 1,000 - 200,000 USD
Samples: University of Virginia Fund, Charlottesville, VA, $33,000-Global Scholarship Fund; Vickery Meadow Learning Center, Dallas, TX, $10,000--Family Literacy program; Victims Outreach, Dallas, TX, $5,000--program services for victims of crime & violence; Reasoning Mind, Inc., Houston, TX, $250,000--program expansion and general operating support.
Contact: Kelly Compton; (214) 987-3605; fax (214) 363-6507; info@hoglundfoundation.org
Internet: http://www.hoglundfoundation.org/grants.html
Sponsor: Hoglund Foundation
5910 North Central Expressway, Suite 255
Dallas, TX 75206

Holland/Zeeland Community Foundation Grants 2041

This Foundation funds programs and projects that focus on the arts and culture, education, environmental issues, health and human services, youth, seniors, community and economic development, affordable housing, and others that enhance the well-being of all citizens in the community. Proposals that help organizations deliver services more effectively, enhance cooperation and collaboration, address emerging needs or try new approaches, positively impact a large number of people for the resources invested, or focus on prevention receive priority consideration. The Foundation generally awards grants one-time only for a specific program or project. To apply, first call to discuss the project and confirm it meets the Foundation guidelines.
Requirements: Qualifying organizations include nonprofits that are tax-exempt under Section 501(c)3 of the Internal Revenue Code, schools, municipalities, and other governmental entities that serve a charitable purpose in the greater Holland and Zeeland area.
Restrictions: Funding is not available for the following: requests in excess of $15,000; annual fund-raising drives; services which are commonly recognized as government or school obligations (however the foundation does consider applications from schools for pilot projects and innovative programs); endowments, loans, taxes or debt reduction; conference speakers, fieldtrips, travel or tours, for individuals or groups; religious programs that advocate specific religious doctrines or do not serve the broader community.
Geographic Focus: Michigan
Date(s) Application is Due: Jan 18; Jun 7; Sep 20
Contact: Janet DeYoung; (616) 396-6590; fax (616) 396-3573; janet@cfhz.org
Internet: http://cfhz.org/files/2010Funding_Guidelines.pdf
Sponsor: Community Foundation of the Holland/Zeeland Area
70 West 8th Street, Suite 100
Holland, MI 49423

HomeBanc Foundation Grants 2042

Giving primarily in Florida and Georgia, the HomeBanc Foundation also, gives to national organizations. The Foundation has three main goals: provide support to cancer related causes through funding education, advocacy, and research; support the American dream of homeownership for those who might not reach it on their own; and provide college funding to students who strive to achieve their goals through education beyond high school. Scholarship forms are available online. Send a Letter of Inquiry to the Foundation, unsolicited applications are not excepted. The Foundation will contact you if your proposal fits their criteria.
Requirements: Nonprofits providing services in Georgia and Florida are eligible.
Geographic Focus: Florida, Georgia
Samples: Susan B. Koman Foundation, 5,000--general purposes; American Cancer Society, 75,1000--advocacy, research, education; Junior Acheivement, $5,000--hispanic outreach scholarships.
Contact: Amanda Albertelli; (404) 497-1000; aalbertelli@homebanc.com
Sponsor: HomeBanc Foundation
2002 Summit Boulevard, Suite 100
Atlanta, GA 30319-1497

Home Building Industry Disaster Relief Fund 2043

The Home Building Industry Disaster Relief Fund (HBIDRF) makes funds available for direct contribution to other recognized charities aiming to meet similar primary needs as the HBIDRF - shelter, health care, education. The board welcomes recommendations of charitable organizations that are actively working in your region to rebuild communities. Guidelines and application form are available at the website.
Requirements: Charities can include corporations, funds, or foundations organized and operated exclusively for charitable, scientific, or educational purposes. These charitable organizations may be involved in making direct contributions to adversely impacted individuals through established process, or with coordinated building projects such as one-day builds, blitzes, etc. Other entities that could be eligible to receive funds might include vocational training programs, temporary shelters or clinics and educational ventures intended to help the local home building industry recover from disastrous events. To apply for funding, contact your local home builders association to determine if any partnering opportunities are available at this time.
Restrictions: The fund does not make assistance available directly to individuals.
Geographic Focus: All States
Contact: Jerry Howard, Executive Officer; (800) 368-5242; fax (202) 266-8400
Internet: http://www.nahb.org/generic.aspx?sectionID=842
Sponsor: Home Building Industry Disaster Relief Fund
1201 15th Street NW
Washington, D.C. 20005-2800

Homer Foundation Grants 2044

The Foundation seeks to fund innovative, creative projects that have a high likelihood of success and will have a long-term, positive impact on the community. Funding is available for arts and culture, community development, youth programs, education, health care, scholarships and the environment. Application information is available online.
Requirements: 501(c)3 nonprofit organizations and other qualified nonprofit entities located in the southwestern portion of the Kenai Peninsula, from Ninilchik south and including the communities across Kachemak Bay, are eligible. An individual who applies must be a resident of the region as described above, who will spend the funds on projects completed within this region. Projects taking place outside of this region, but by a resident, will be considered on the merit of the project and how it could benefit the community.
Geographic Focus: Alaska
Amount of Grant: 4,000 USD
Contact: Joy Steward; (907) 235-0541; fax (907) 235-0542; jsteward@homerfund.org
Internet: http://www.homerfund.org/grantmaking.html
Sponsor: Homer Foundation
P.O. Box 2600
Homer, AK 99603

Honda of America Manufacturing Foundation Grants 2045

The Foundation supports community outreach in Ohio. Each year, the Foundation funds various community projects throughout the state. This involvement helps build the community where it does business and where its associates live. The Foundation supports programs and organizations in the areas of education, arts and culture, civic and community, health and human services, and the environment. It places its priorities on the development of programs, and not supporting annual operating budgets.
Requirements: Priority consideration will be given for projects within the West Central Ohio area, including: Allen, Auglaize, Champaign, Clark, Darke, Delaware, Franklin, Hardin, Logan, Madison, Marion, Mercer, Miami, Shelby and Union counties. In addition, Honda considers projects on a case-by-case basis outside this area if the projects are Ohio-based and reflect its priorities of interest.
Restrictions: Honda of America does not make grants to: individuals; religious or political groups; organizations which are not tax-exempt under the U.S. Internal Revenue Code paragraph 501(c)3; or programs outside the state of Ohio. Generally, the Foundation does not provide contributions in support of: conferences or workshops; seminars or pageants; field trips, extra-curricular school activities or sports teams; fraternal or veterans organizations; national health organizations; courtesy advertisements, lobbying organizations, memberships or legal advocacy; staff salaries; or pilot programs.
Geographic Focus: Ohio
Date(s) Application is Due: Mar 31; Jun 30; Sep 30; Dec 31
Contact: Ms. Ginny Milburn, Coordinator; (937) 645-8792; fax (937) 645-8787; ginny_milburn@ham.honda.com
Internet: http://www.ohio.honda.com/community/giving.cfm
Sponsor: Honda of America Manufacturing Foundation
24000 Honda Parkway
Marysville, OH 43040

Horace A. Kimball and S. Ella Kimball Foundation Grants 2046

Although the Horace A Kimball and S. Ella Kimball Foundation was not legally established until July of 1956, its roots along with those of its sister organization, the Phyllis Kimball Johnstone and H. Earle Kimball foundation, go back to the early 1900s. It was then that Horace A. Kimball of Providence, Rhode Island, a retired woolen manufacturer, acquired controlling interest of the Clicquot Club Beverage Company of Millis, Massachusetts. Today, the Kimball Foundation makes grants almost exclusively to Rhode Island operatives (charities) or those benefitting Rhode Island residents and causes. Although, the Foundation considers gifts to all areas in the state, greater emphasis is placed on South County. Areas of interest for the Foundation are: human services; the environment; and health care. Most recent awards have ranged from $2,500 to $25,000. Interested parties can apply either with a mailed hard copy or by using the online application format.
Requirements: The Foundation will consider any organization which has proper 501(c)3 and 509(a) IRS tax classification status.
Restrictions: No support for religious organizations. No grants to individuals, or for feasibility studies, capital projects or multi-year commitments.
Geographic Focus: Rhode Island
Date(s) Application is Due: Jul 15
Amount of Grant: 1,000 - 50,000 USD
Samples: Chorus of Westerly, Westerly, Rhode Island, $13,000 - roof repairs; Education Exchange, Peace Dale, Rhode Island, $25,000 - testing equipment purchase; Wood Valley Health Services, Hope Valley, Rhode Island, $3,700 - general program support.
Contact: Thomas F. Black III, President; (401) 364-3565 or (401) 348-1234

Internet: http://www.hkimballfoundation.org/index2.htm
Sponsor: Horace A. Kimball and S. Ella Kimball Foundation
130 Woodville Road
Hope Valley, RI 02832

Horace Moses Charitable Foundation Grants 2047
The Horace Moses Charitable Foundation was established in 1923 to support and promote quality educational, human-services, and health-care programming for underserved populations. Special consideration is given to charitable organizations that serve the community of Springfield and its surrounding communities. Grant requests for general operating support are strongly encouraged. Program support will also be considered. Small, program-related capital expenses may be included in general operating or program requests. The majority of grants from the Moses Charitable Foundation are one year in duration; on occasion, multi-year support is awarded. Applicants must apply online at the grant website. Applicants are strongly encouraged to do the following before applying: review the downloadable state application procedures for additional helpful information and clarifications; review the downloadable online-application guidelines at the grant website; review the foundation's funding history (link is available from the grant website); review the online application questions in advance; and review the list of required attachments. These will generally include: a list of board members, financial statements (audited, reviewed, or compiled by independent auditor); an organization summary; a list of other funding sources; an IRS Determination letter; and other required documents. All attachments must be uploaded in the online application as PDF, Word, or Excel files. The application deadline for the Horace Moses Charitable Foundation is 11:59 p.m. on April 1. Applicants will be notified of grant decisions before June 30.
Requirements: Applicants must have 501(c)3 tax-exempt status.
Restrictions: The foundation does not support requests from individuals, organizations attempting to influence policy through direct lobbying, or any political campaigns.
Geographic Focus: Massachusetts
Date(s) Application is Due: Apr 1
Contact: Michealle Larkins; (866) 778-6859; michealle.larkins@baml.com
Internet: https://www.bankofamerica.com/philanthropic/fn_search.action
Sponsor: Horace Moses Charitable Foundation
225 Franklin Street, 4th Floor, MA1-225-04-02
Boston, MA 02110

Horizon Foundation for New Jersey Grants 2048
The Foundation's purpose is to promote health, well-being, and quality of life in New Jersey communities. Foundation goals are to improve health by promoting quality health care programs and access and to enhance arts and cultural opportunities. Application information is available online. Only electronic grant applications will be accepted. Application are accepted between January 1 and September 31.
Requirements: 501(c)3 tax-exempt organizations in New Jersey communities located in and served by Horizon Blue Cross Blue Shield of New Jersey are eligible.
Restrictions: Funding is not available for: capital campaigns; endowments; hospitals or hospital foundations; individuals; political causes; political candidates; political organizations; political campaigns.
Geographic Focus: New Jersey
Amount of Grant: 10,000 - 50,000 USD
Samples: ChoiceOne Pregnancy & Sexual Health Resource Centers, Lawrenceville, NJ, $25,000--to support the Sexually Transmitted Diseases Education and Prevention component of its Straight Talk Program; Family Guidance Center, Hamilton, NJ, $15.000--to support its Behavioral Health Services-Depression Recovery Program; American Repertory Ballet, New Brunswick, HJ, $7,500--to support the Dance Power Program.
Contact: Michele Berry, Grants Coordinator; Foundation_Info@horizonblue.com
Internet: http://www.horizon-bcbsnj.com/foundation/about/funding.html?WT.svl=leftnav
Sponsor: Horizon Foundation for New Jersey
Three Penn Plaza East, PP-15V
Newark, NJ 07105-2200

Horizons Community Issues Grants 2049
Grants support program activities or projects serving lesbian, gay, bisexual, and transgender (LGBT) people of all ages in designated California counties focusing on the following issue areas: arts and culture; advocacy, awareness, and civil rights; children, youth, and families; and community and social services. Application information is available online.
Requirements: To be eligible, an organization must: be a nonprofit, 501(c)3 organization, or provide documentation that the organization is sponsored under a fiscal agent umbrella that has 501(c)3 status. Organizations or programs must request support for one or more of the following counties: Alameda; Contra; Costa; Marin; Napa; San Francisco; San Mateo; Santa Clara; Solano; Sonoma.
Restrictions: The following are not eligible for support: costs incurred prior to the date of the grant award; government agencies; capital support, including construction and renovation; fundraising or event sponsorship; individuals; Non-LGBT organizations with budget over $1 million.
Geographic Focus: California
Amount of Grant: 10,000 - 20,000 USD
Contact: Jewelle Gomez, Program Officer; (415) 398-2333, ext. 116; jgomez@horizonsfoundation.org
Internet: http://www.horizonsfoundation.org/page/organizations/ci
Sponsor: Horizons Foundation
870 Market Street, Suite 728
San Francisco, CA 94102

Hormel Foods Charitable Trust Grants 2050
The Charitable Trust was established to benefit various organizations and projects that emphasize education, hunger and quality-of-life initiatives in and around Hormel Foods plant communities. The Program operates as a compliment to its foundation, making charitable contributions to nonprofit organizations directly. Support is given primarily in areas of company operations. Fields of interest include: disaster and preparedness services; education; general charitable giving; and health care. Types of support include: scholarships; matching gifts; general operations; and sponsorships. The Public Relations Department handles giving. Annual donations to a single organization typically do not exceed $15,000. A contributions committee reviews all written requests. Applicants should submit a detailed description of the project and amount of funding requested. The board meets once each month.
Requirements: Only organizations classified as 501(c)3 are eligible for support. Applicants should be located in or serve the communities of: Aurora, Illinois; Austin, Minnesota; Algona, Iowa; Alma, Kansas; Atlanta, Georgia; Beloit, Wisconsin; Chino, California; Fort Dodge, Iowa; Fremont, Nebraska; Knoxville, Iowa; Lathrop, California; Mendota Heights, Minnesota; New Berlin, Wisconsin; Osceola, Iowa; Rochelle, Illinois; Stockton, California; Turlock, California; and Wichita, Kansas.
Restrictions: Giving is primarily limited to areas of company operations.
Geographic Focus: California, Georgia, Illinois, Iowa, Kansas, Minnesota, Nebraska, Wisconsin
Contact: Julie H. Craven; fax (507) 437-5345; media@hormel.com
Internet: http://www.hormelfoods.com/responsibility/philanthropy/default.aspx
Sponsor: Hormel Foods Charitable Trust
1 Hormel Place
Austin, MN 55912-3680

Hormel Foundation Grants 2051
The Foundation, established in 1941, is a non-profit organization dedicated to supporting higher education, health care and research, social services, and community programs located in or serving the city of Austin, Minnesota. Applications for grants must be presented to the Foundation in writing. They should be as brief as appropriate to present the necessary facts about the applying organization and the project for which the grant is being sought. An application form is required, and must be received by the Foundation office by September 1st of each year.
Requirements: Only 501(c)3 organizations in Minnesota should apply.
Geographic Focus: Minnesota
Date(s) Application is Due: Sep 1
Contact: R.L. Knowlton, Chairperson; (507) 437-9800; fax (507) 434-6731
Internet: http://www.thehormelfoundation.com/requirements.asp
Sponsor: Hormel Foundation
301 North Main Street
Austin, MN 55912-3498

Hospital Libraries Section / MLA Professional Development Grants 2052
The purpose of this Medical Libraries Association (MLA) award which is sponsored by the association's Hospital Libraries Section (HLS) is to provide librarians working in hospital and similar clinical settings with the support needed for educational or research activities. These include developing and acquiring the knowledge and skills delineated in "Competencies for Lifelong Learning and Professional Success: The Educational Policy Statement of the Medical Library Association" and "The Research Imperative: The Research Policy Statement of the Medical Library Association." The grant may also be used to support reimbursement for expenses incurred in conducting scientific research, such as professional assistance in survey-research design, statistical analyses, etc. The application form and guidelines are available for download at the website. Applications and supporting documents must be received by MLA via email, fax, or mail by December 1. email applications will be accepted as PDF or MS Word files only. Founded in 1898, the Medical Library Association (MLA) is a nonprofit, educational organization of more than 1,100 institutions and 3,600 individual members in the health sciences information field committed to educating health-information professionals, supporting health information research, promoting access to the world's health-sciences information, and working to ensure that the best health information is available to all.
Requirements: Applicants must have been employed as a health-sciences librarian within the last year in either a hospital or other clinical-care institution. It is preferred that applicants be a member of the Hospital Libraries Section of the MLA.
Restrictions: Previous recipients of the HLS/MLA Professional Development Grant are not eligible. Applicants must not have received an MLA grant, scholarship, or other award within the last year. An applicant can only receive one award per year. An award will be made to no more than one employee per institution per year. Awards will not be given to support work toward a degree or certificate program.
Geographic Focus: All States, Canada
Date(s) Application is Due: Dec 1
Amount of Grant: Up to 800 USD
Contact: Carla Funk, Executive Director; (312) 419-9094, ext. 14; fax (312) 419-8950; mlapd2@mlahq.org or grants@mlahq.org
Internet: http://www.mlanet.org/awards/grants/index.html
Sponsor: Medical Library Association
65 East Wacker Place, Suite 1900
Chicago, IL 60601-7246

Houston Endowment Grants 2053
The Endowment supports nonprofit organizations and educational institutions that improve life for the people of the greater Houston area. The Foundation funds programs in the arts, community enhancement, education, health, human services, the environment

and neighborhood development. Requests for funding are accepted throughout the year. Further application information is available online.
Requirements: Nonprofit organizations that are recognized as charitable organizations by the Internal Revenue Code are eligible to apply. Funds are provided primarily to Harris County and contiguous counties. Grants seldom are given outside of Texas and never are made outside of the United States.
Restrictions: The Endowment does not select scholarship recipients or make direct scholarship awards to individuals. Inquiries about scholarship assistance should be made to financial aid offices at colleges and universities. Funding is not available for other grant-making organizations or to charities operated by service clubs. Funds are not available for religious activities; the purchase of uniform, equipment or trips for school-related organizations or sports teams; honoraria for speakers and panelists; fund-raising activities and galas; and memorials for individuals. The Foundation does not make loans or grants to individuals.
Geographic Focus: Texas
Amount of Grant: 2,500 - 1,000,000 USD
Contact: Harriet W. Garland, Grant Manager; (713) 238-8100; fax (713) 238-8101; info@houstonendowment.org
Internet: http://www.houstonendowment.org/grants/grants_apply.htm
Sponsor: Houston Endowment
600 Travis, Suite 6400
Houston, TX 77002-3000

Howard and Bush Foundation Grants 2054
The foundation's funding is limited to Rensselaer County, NY. with an emphasis on arts and culture, education, civic and urban affairs, legal services, including a women's bar association, and social service and health programs that benefit the areas resident. Requests for equipment acquisition, programs development, matching funds, building/renovation and seed money will receive consideration. There are no deadline dates, funding guidelines are sent upon request. Contact staff prior to submitting a proposal. The board meets twice a year, in April and October. Notification of proposal acceptance will be within 14 days, following the board meeting.
Requirements: Nonprofit organizations benefiting Rensselaer County, NY, and its residents may request grant support.
Restrictions: Grants are not awarded to colleges, schools, churches, or individuals or for endowment purposes, operating budgets, or deficit financing. Generally no support for government or largely tax-supported agencies, or churches not connected with the founders.
Geographic Focus: New York
Amount of Grant: 750 - 75,000 USD
Samples: Friends of Barker Park, $5,000-- for park improvements; East Side Neighborhood Association, $5,000--for security system/modify heating system; Whitney M. Young Jr. Foundation, $10,000--for Troy oral health expansion project.
Contact: Deborah Byers, Program Contact; (518) 271-1134; dogclover@aol.com
Sponsor: Howard and Bush Foundation
2 Belle Avenue
Troy, NY 12180

Howe Foundation of North Carolina Grants 2055
The Howe Foundation, established in 1966, supports organizations involved with arts and culture, education, human services, Christianity, and mentally disabled individuals. Giving limited to the Belmont, North Carolina area.
Requirements: No application form required, submit your proposal to: P.O. Box 749, Belmont, NC 28012.
Restrictions: No grants to individuals.
Geographic Focus: North Carolina
Amount of Grant: 200 - 144,000 USD
Contact: Henry Howe, Treasure; (704) 825-5372
Sponsor: Howe Foundation
P.O. Box 227
Belmont, NC 28012-0227

HRAMF Charles A. King Trust Postdoctoral Research Fellowships 2056
The Charles A. King Trust was established in 1936 to support and promote the investigation of human disease and the alleviation of human suffering through improved treatment. In keeping with these principles, the King Trust today supports postdoctoral fellows in the basic sciences and clinical/health services research. The program is designed to support postdoctoral scientists in non-profit academic, medical or research institutions in Massachusetts. Two-year grants ranging from $43,700 to $53,175 per year, inclusive of a $2,000 expense allowance, will be awarded. Awards are made to nonprofit academic, medical, or research institutions in the State of Massachusetts on behalf of the Awards Recipient.
Requirements: Applicants must be working in an academic or medical research institution in the state of Massachusetts and have the required minimum/maximum years of experience. Applicants must apply for the fellowships under the guidance of a Mentor who is an established investigator with an active research program. Mentors are expected to be involved in the planning, execution, and supervision of the proposed research. Mentors must confirm that degrees obtained outside the United States are equivalent to the M.D., D.M.D., Ph.D. or other doctoral degree. United States citizenship is not required; visa documentation is not required. The Basic Science area requires that, by September 1 of each funding cycle, applicants must have completed at least three years, but not more than five years, of full-time postdoctoral research experience. The Clinical and/or Health Services Research area requires that, by September 1 of each funding cycle, applicants with clinical responsibilities must have no more then five years of postdoctoral research training since completing residency training. Applicants without clinical responsibilities must have completed at least three years, but not more than five years, of full-time postdoctoral research experience.
Restrictions: Only one applicant per Mentor may apply per application cycle; there are not institutional limitations on the number of applicants who may submit applications. Funding begins on July 1. A later start date is permitted but the Award must be activated on or before October 1.
Geographic Focus: Massachusetts
Date(s) Application is Due: Jan 29
Amount of Grant: 43,700 - 53,175 USD
Contact: Erin Johnstone, Program Officer; (617) 279-2240 ext. 710; fax (617) 423-4619; ejohnstone@hria.org
Internet: http://www.hria.org/tmfservices/tmfservices/tmfgrants/king.html
Sponsor: Health Resources in Action Medical Foundation
95 Berkeley Street
Boston, MA 02116

HRAMF Charles H. Hood Foundation Child Health Research Awards 2057
The Charles H. Hood Foundation was incorporated in 1942 to improve the health and quality of life for children through grant support of New England-based pediatric researchers. The intent of the Child Health Research Awards Program is to support newly independent faculty, provide the opportunity to demonstrate creativity, and assist in the transition to other sources of research funding. Two-year grants of $150,000 ($75,000 per year inclusive of 10% indirect costs) are awarded to researchers who are within five years of their first faculty appointment by the funding start date. Application deadlines occur in the spring and fall of each year.
Requirements: Applicants must be working in nonprofit academic, medical or research institutions within the six New England states. Grants support hypothesis-driven clinical, basic science, public health, health services research, and epidemiology projects focused on child health. Applicants must hold a doctoral degree with a demonstrated level of independence confirmed by the Department or Division Chair. The Applicant's potential for a lifetime career as an investigator in pediatric research is also critical in the review process. Applicants are required to devote at least 20% effort to the proposed Hood research project. Applicants who have pending R01s or other large applications to the NIH and other agencies are encouraged to submit proposals to the Hood Foundation. Online submissions must be received by April 8, 12:00 Noon, U.S. Eastern time. Printed copies must be received by April 14, 5:00 pm, U.S. Eastern time.
Restrictions: Applicants are ineligible if they have combined federal and non-federal funding totaling $450,000 or more in direct costs at any time during the two years of the Award. This figure refers to external funding only and not an Applicant's start-up package, other intramural support or the Hood Award. Applicants are ineligible if they are currently or have previously been designated as Principal Investigator or Co-P.I. on an R01, P01, Pioneer Award, New Innovator Award or equivalents from federal agencies such as the National Science Foundation (NSF) or Department of Defense (DOD). Applicants who have completed the R00 phase of a K99/R00 are also ineligible for a Hood Award.
Geographic Focus: Connecticut, Maine, Massachusetts, New Hampshire, Rhode Island, Vermont
Date(s) Application is Due: Mar 21; Sep 21
Amount of Grant: 75,000 USD
Contact: Gay Lockwood; (617) 695-9439; fax (617) 423-4619; glockwood@hria.org
Internet: http://hria.org/tmfservices/tmfgrants/hood.html
Sponsor: Health Resources in Action Medical Foundation
95 Berkeley Street
Boston, MA 02116

HRAMF Deborah Munroe Noonan Memorial Research Grants 2058
The Deborah Munroe Noonan Memorial Research Fund was established in 1947 by her son, Frank M. Noonan, to support innovative projects aimed at improving the quality of life for children and adolescents with physical and developmental disabilities. The Noonan Research Fund supports innovative clinical and service system research, demonstration projects, and pilot studies in the Boston area designed to improve the quality of life for children and adolescents with disabilities. The fund provides one-year grants of up to $80,000 (inclusive of 10% indirect costs).
Requirements: Applicants must hold a position within a nonprofit institution or organization within the Fund's geographic area of interest. New investigators are encouraged to submit proposals. Project must address the target age range of birth through 23 years old. Research projects must be conducted within the Fund's geographic area of interest. A complete listing of the geographic eligibility area is contained in the downloadable guidelines available at the grant website. All application information must be completed online, and the proposal uploaded as a PDF by 12:00 Noon U.S. Easter time of the published deadline date.
Restrictions: Proposals for basic science research will not be considered nor will applications for capital costs such as buildings, renovations, or major equipment items. The Noonan Fund does not support direct service, primary prevention projects, primary medical conditions such as obesity or device development. Drug trials are rarely supported by the Noonan Research Fund.
Geographic Focus: Massachusetts
Date(s) Application is Due: Mar 22
Amount of Grant: Up to 80,000 USD
Contact: Jeanne Brown; (617) 279-2240, ext. 709; fax (617) 423-4619; jbrown@hria.org
Internet: http://www.hria.org/tmfservices/tmfservices/tmfgrants/noonan.html
Sponsor: Health Resources in Action Medical Foundation
95 Berkeley Street
Boston, MA 02116

HRAMF Harold S. Geneen Charitable Trust Awards for Coronary Heart Disease Research 2059

The Harold S. Geneen Charitable Trust Awards Program for Coronary Heart Disease Research supports research in the area of the prevention and control of coronary and circulatory failure. The Awards Program seeks to contribute to the prevention of coronary heart disease or circulatory failure, and improving care for patients with these medical conditions. n accordance with Mr. Geneen's directives the Program seeks to establish "...a more direct and personalized relationship with grant recipients than is normally possible in dealing with the diffuse and bureaucratic administrations through which large organizations are managed, and to support smaller institutions rather than major universities or medical complexes which have a demonstrated capacity to raise funds from the public generally." The award is equal to $280,000 distributed over two years (inclusive of 10% indirect costs). The annual deadline for application submission is August 31 at 12:00 noon.
Requirements: Invited institutions may submit one application to the Program which meets the eligibility requirements for the grant cycle. Applicants must be full-time faculty at an invited nonprofit academic, medical, or research institution. United States citizenship is not required. Junior faculty are encouraged to apply.
Geographic Focus: All States
Date(s) Application is Due: Aug 31
Amount of Grant: 280,000 USD
Contact: Jeanne Brown; (617) 279-2240, ext. 709; fax (617) 423-4619; jbrown@hria.org
Internet: http://www.hria.org/tmfservices/tmfservices/tmfgrants/geneen.html
Sponsor: Health Resources in Action Medical Foundation
95 Berkeley Street
Boston, MA 02116

HRAMF Jeffress Trust Awards in Interdisciplinary Research 2060

The Thomas F. and Kate Miller Jeffress Memorial Trust was established in 1981 to support basic research in chemical, medical, and other scientific research through grants to educational and research organizations in the Commonwealth of Virginia. The trust was established under the will of Mr. Robert M. Jeffress, a business executive and philanthropist of Richmond, Virginia. Grants are made to assist scientists in educational and research institutions to conduct investigations in the natural sciences, generally considered to include chemistry, biology (with the exception of field studies, classification, or other largely observational studies), and the basic medical sciences, such as biochemistry, microbiology and others. The principal evaluation criteria will be the scientific significance of the proposed work and the competence of the investigator. The trust is particularly interested in supporting fundamental research by scientists early in their careers and new areas of research or more speculative projects by established investigators. In general, grants will cover direct-expense items essential to the successful completion of the proposed project. Grants from the Jeffress Memorial Trust are one year in duration. Funding may be renewed at the discretion of the allocations committee. Application materials are available for download at the grant website. Applicants are strongly encouraged to review the application guidelines for additional helpful information and clarifications before applying.
Requirements: Eligible research areas for Jeffress support include astronomy, biosciences, chemistry, computer sciences, engineering, environmental sciences, material science, mathematics and physics. Full-time faculty at institutions in Virginia that are within seven years of their first faculty appointment are eligible to apply as Principal Investigators. Student participation is a requirement of the proposed research plan. Online submissions must be received by January 14, 12:00 Noon, U.S. Eastern time. Mailed hard copies must be received by The Medical Foundation division by January 8, 5:00 pm, U.S. Eastern time.
Restrictions: Research in social, economic and behavioral sciences, including psychology, are currently outside the scope of Jeffress funding. Faculty from medical schools and schools of Osteopathic Medicine may not apply as Principal Investigators; however, they are encouraged to collaborate as Co-Investigators.
Geographic Focus: Virginia
Date(s) Application is Due: Jan 8
Amount of Grant: 100,000 USD
Contact: Jeanne Brown; (617) 279-2240, ext. 709; fax (617) 423-4619; jbrown@hria.org
Internet: http://www.hria.org/tmfservices/tmfservices/?page=jeffress/
Sponsor: Health Resources in Action Medical Foundation
95 Berkeley Street
Boston, MA 02116

HRAMF Ralph and Marian Falk Medical Research Trust Catalyst Awards 2061

The Dr. Ralph and Marian Falk Medical Research Trust was created by Marian Falk in 1979 to support biomedical research. Marian Falk sought to fund medical research to improve treatments of the past and eventually find cures for diseases for which no definite cure is known. Bank of America serves as Trustee for the Trust. Beginning with the 2015 grant cycle, the Medical Foundation division of Health Resources in Action began to manage the grant making for both the Falk Catalyst and Transformational programs on behalf of Bank of America. The Catalyst Research Award program provides one year of seed funding to support high-risk, high-reward projects that address critical scientific and therapeutic roadblocks within the program's principal areas of focus. The program is designed to enable planning and development of projects, teams, tools, techniques and management infrastructure necessary to successfully compete for two-year awards through the Transformational Research Award program.
Requirements: Applicants must meet the following *Requirements:* hold a full-time faculty appointment; and be an independent investigator with demonstrated institutional support and the specialized space and facilities needed to conduct the proposed research. United States citizenship and visa documentation are not required.
Restrictions: Applicants may not have funding for a similar project. A PI may only submit one application.
Geographic Focus: All States
Contact: Jeanne Brown; (617) 279-2240, ext. 709; fax (617) 423-4619; jbrown@hria.org
Internet: http://www.hria.org/tmfservices/tmfservices/tmfgrants/Falk-Catalyst.html
Sponsor: Health Resources in Action Medical Foundation
95 Berkeley Street
Boston, MA 02116

HRAMF Ralph and Marian Falk Medical Research Trust Transformational Awards 2062

The Dr. Ralph and Marian Falk Medical Research Trust was created by Marian Falk in 1979 to support biomedical research. Falk sought to fund medical research to improve treatments of the past and eventually find cures for diseases for which no definite cure is known. Bank of America serves as Trustee for the Falk Medical Research Trust. The Transformational Awards Program provides $1,000,000 for a two year funding cycle to successful Catalyst Award Recipients to continue their work tackling critical scientific and therapeutic roadblocks and thereby opening avenues for treating and curing disease. Transformational Awards will be granted based on both scientific merit and having successfully attained the proposed milestones and benchmarks of the Catalyst Award, thus demonstrating its successful execution within the proposed budget and projected time frame. The program is available to researchers nationally.
Requirements: Only Catalyst Award Recipients are eligible to apply for a Transformational Award. Applicants must meet the following *Requirements:* hold a full-time faculty appointment; and be an independent investigator with demonstrated institutional support and the specialized space and facilities needed to conduct the proposed research. United States citizenship and visa documentation are not required.
Restrictions: Applicants may not have funding for a similar project. A PI may only submit one application.
Geographic Focus: All States
Amount of Grant: 1,000,000 USD
Contact: Jeanne Brown; (617) 279-2240, ext. 709; fax (617) 423-4619; jbrown@hria.org
Internet: http://www.hria.org/tmfservices/tmfservices/tmfgrants/falk.html
Sponsor: Health Resources in Action Medical Foundation
95 Berkeley Street
Boston, MA 02116

HRAMF Smith Family Awards for Excellence in Biomedical Research 2063

For the past quarter century, the Smith Family Foundation has been supporting groundbreaking medical research through the Smith Family Awards Program for Excellence in Biomedical Research. Its mission is to launch the careers of newly independent biomedical researchers with the ultimate goal of achieving medical breakthroughs. Applications focus on all fields of basic biomedical science and may also be submitted by investigators in physics, chemistry and engineering. Three year awards in the amount of $300,000 ($100,000 per year inclusive of 5% indirect costs) target new faculty who are within two years of their first independent faculty appointment as of July 1 of the application year.
Requirements: Applicants, nominated by their institutions, must be full-time faculty at nonprofit academic, medical or research institutions in Massachusetts, at Brown University or at Yale University. Applications from women scientists are encouraged. Applications must be received by 12:00 Noon, U.S. Eastern time of the published deadline date.
Geographic Focus: Massachusetts
Date(s) Application is Due: Sep 3
Amount of Grant: 300,000 USD
Contact: Gay Lockwood, Senior Program Officer; (617) 279-2240, ext. 702; fax (617) 423-4619; glockwood@hria.org
Internet: http://www.hria.org/tmfservices/tmfservices/tmfgrants/smith.html
Sponsor: Health Resources in Action Medical Foundation
95 Berkeley Street
Boston, MA 02116

HRAMF Taub Foundation Grants for MDS Research 2064

The Henry and Marilyn Taub Foundation works with The Medical Foundation, a division of Health Resources in Action (HRiA), to select the most qualified applicants. HRiA is a nonprofit organization in Boston that advances public health and medical research. The Taub Foundation Grants Program for Myelodysplastic Syndromes (MDS) Research was created to support high-impact, innovative translational research to understand the underlying causes of MDS and to advance its treatment and prevention. The Program specifically focuses on MDS research, exclusive of AML and MPN. Studies focusing on molecular genetics, epigenetics, splicing factors, stem cells, the microenvironment and novel therapeutic targets relevant to MDS are encouraged. Three-year awards of $600,000 ($200,000 per year, inclusive of 10% indirect costs) will be made to independent investigators working in non-profit, non-governmental academic, medical, or research institutions within the United States. Collaborative efforts are encouraged.
Requirements: All applicants must hold a faculty appointment at a nonprofit, governmental academic, medical, or research institution in the United States. Applicants do not need to be nominated by their institutions. United States citizenship is not required; visa documentation is not required. Awards are not restricted to investigators currently working in MDS. Applications from investigators in other fields and collaborative efforts are encouraged. The complete initial online application must be received by The Medical Foundation by August 14.
Restrictions: Only one application may be submitted per applicant.
Geographic Focus: All States
Date(s) Application is Due: Aug 14
Amount of Grant: 600,000 USD

Contact: Erin Johnstone; (617) 279-2240. ext. 710; fax (617) 423-4619; ejohnstone@hria.org
Internet: http://www.hria.org/tmfservices/tmfservices/tmfgrants/taub.html
Sponsor: Health Resources in Action Medical Foundation
95 Berkeley Street
Boston, MA 02116

HRAMF Thome Foundation Awards in Age-Related Macular Degeneration Research 2065

The Edward N. & Della L. Thome Memorial Foundation was created in 2002 to advance the health of older adults through the support of direct service projects and medical research on diseases and disorders affecting older adults. In keeping with the Foundation's mission, the goal of the Awards Program is to support translational research that will lead to improved therapies for individuals suffering from age-related macular degeneration (AMD). As steward of the Thome Memorial Foundation, Bank of America N.A., works with The Medical Foundation division's Scientific Review Committees to select the most qualified candidates. Two-year awards of $500,000 ($250,000 per year) will be made to investigators who hold a faculty appointment at a non-profit academic, medical, or research instituion in the United States.
Requirements: Successful research proposals will extend recent basic research findings regarding the underlying mechanisms of AMD. Examples of funding areas include, but are not limited to, genetically engineered animal models, the discovery and testing of small molecule therapies directed at promising targets including immune- or inflammation-related pathways associated with AMD, local drug delivery systems, and neuroprotective strategies. Basic research and clinical trials are outside the scope of this Program. However, clinical studies that involve the evaluation of new imaging modalities with no potential risks to human subjects are eligible. Eligibility is not limited to those investigators currently working in age-related macular degeneration research; scientists who have conducted research exploring the biologic causes of related disorders and/or similar translational research programs are encouraged to apply. All applicants must hold a faculty appointment at a nonprofit academic, medical, or research institution in the United States. Applicants do not need to be nominated by their institutions. United States citizenship is not required; visa documentation is not required. Applications must be submitted by 12:00 Noon, U.S. Eastern time of the published deadline date.
Restrictions: Only one application may be submitted per applicant.
Geographic Focus: All States
Date(s) Application is Due: Sep 15
Amount of Grant: Up to 500,000 USD
Contact: Erin Johnstone; (617) 279-2240, ext. 710; fax (617) 423-4619; ejohnstone@hria.org
Internet: http://www.hria.org/tmfservices/tmfservices/tmfgrants/thomeamd.html
Sponsor: Health Resources in Action Medical Foundation
95 Berkeley Street
Boston, MA 02116

HRAMF Thome Foundation Awards in Alzheimer's Disease Drug Discovery Research 2066

The Edward N. and Della L. Thome Memorial Foundation was created in 2002 to advance the health of older adults through the support of direct service projects and medical research on diseases and disorders affecting older adults. In keeping with the foundation's mission, the goal of the research program is to support pilot studies towards innovative drug discovery research that will lead to improved therapies for individuals suffering from Alzheimer's disease. Examples of funding areas include the design, synthesis, and development of target compounds or the modification of existing compounds to improve drug effectiveness and safety as well as other approaches within the field of medicinal chemistry. Researchers dedicated to the validation and testing of target compounds, small molecule therapies, nanotechnologies, or similar techniques are also encouraged to apply. Two-year awards of up to $200,000 ($100,000 per year) will be made (all awards inclusive of 10% indirect costs). Invited applications must be submitted through the online system and must be received by 12:00 noon, U.S. Eastern time on September 25.
Requirements: All applicants must hold a faculty appointment at a non-profit academic, medical, non-governmental or research institution in the United States. U.S. citizenship is not required. Applicants do not need to be nominated by their institutions. Preference will be given to originality of ideas, regardless of faculty seniority.
Restrictions: Genetic studies, biomarker research, neuroimaging, clinical studies, basic research and new target discovery are currently outside the scope of this program.
Geographic Focus: All States
Date(s) Application is Due: Sep 25
Amount of Grant: Up to 200,000 USD
Contact: Erin Johnstone; (617) 279-2240, ext. 710; fax (617) 423-4619; ejohnstone@hria.org
Internet: http://www.hria.org/tmfservices/tmfservices/tmfgrants/thomead.html
Sponsor: Health Resources in Action Medical Foundation
95 Berkeley Street
Boston, MA 02116

HRK Foundation Health Grants 2067

The HRK Foundation is defined by quiet leadership and philanthropy. The Board seeks to improve the fabric of society by promoting healthy families and healthy communities. In the area of Health, the Foundation funds programs that strengthen families and promote healthy lives for children, as well as programs that provide access to health and human services. Within the Saint Croix Valley and Ashland and Bayfield Counties in Wisconsin, the board will consider local community-specific health projects. Types of support include: general operating support; seed funding; community development; matching/challenge grants; equipment acquisition; and annual campaigns. Requests are considered two times each year, with deadlines on March 31 and September 15. Most recent awards in this area have ranged from $20,000 to $25,000.
Requirements: The foundation makes grants only to qualified IRS 501(c)3 organizations that specifically benefit people in the area surrounding Bayport and Saint Paul, Minnesota, and Ashland and Bayfield counties in Wisconsin. The Foundation supports all families, both traditional and non-traditional.
Restrictions: The foundation does not make loans or provide grants to individuals.
Geographic Focus: Minnesota, Wisconsin
Date(s) Application is Due: Sep 15
Amount of Grant: 20,000 - 25,000 USD
Contact: Kathleen Fluegel, Foundation Director; (651) 298-0550 or (866) 342-5475; fax (651) 298-0551; info@HRKFoundation.org
Internet: http://www.hrkfoundation.org/purpose.html
Sponsor: HRK Foundation
345 Saint Peter Street, Suite 1200
Saint Paul, MN 55102-1639

HRSA Nurse Education, Practice, Quality and Retention Grants 2068

Through its Nurse Education, Practice, Quality and Retention grants program, the Division of Nursing will solicit three-year cooperative agreements that propose to develop and implement innovative career ladder programs that will increase the enrollment, progression, and graduation of Veterans in Bachelor of Science in Nursing programs. This VBSN program supports HRSA's strategic plan to improve access to quality health care and services; strengthen, the nation's healthcare workforce; build healthy communities; and improve health equity. Awarded VBSN applications will complement the collaborative efforts of the Health Resources and Services Administration (HRSA), the Department of Defense (DoD), and the Department of Veteran¿s Affairs (VA) that seek to: reduce barriers that prevent veterans from transitioning into nursing careers; develop BSN career ladder programs targeted to the unique needs of veterans; explore innovative educational models to award academic credit for prior health career experience/training or other relevant military training; improve employment opportunities for veterans through high demand careers training, and address the growing national demand for BSN prepared Registered Nurses. The intermediate program goals are to facilitate the transition of veterans into the field of professional nursing, while building upon skills, knowledge, and training acquired during their military service in order to increase employment opportunities. Sub-goals include increasing and diversifying the health workforce, and ensuring that healthcare providers are trained to provide high quality care that is culturally and linguistically aligned with the communities they will serve. The VBSN project will: provide program participants with the opportunity to receive academic credit for prior military medical training and experience; provide participants with knowledge, skills, and support(s) needed to successfully matriculate through innovative BSN career ladder training programs; and prepare program participants for the National Council Licensing Examination for Registered Nurses (NCLEX- RN).
Requirements: Eligible applicants include accredited schools of nursing as defined in section 801(2), of the Public Health Service Act, a health care facility as defined in section 801(11) of the Public Health Service Act, or partnership of such a school and facility.
Geographic Focus: All States
Date(s) Application is Due: Feb 18
Contact: Rebecca Spitzgo; (301) 443-5794 or (888) 275-4772; fax (301) 443-8586
Internet: http://www.grants.gov/view-opportunity.html?oppId=249594
Sponsor: Health Resources and Services Administration
5600 Fishers Lane
Rockville, MD 20857

HRSA Resource and Technical Assistance Center for HIV Prevention and Care for Black Men who Have Sex with Men Cooperative Agreement 2069

The purpose of this program is to: inventory existing evidence-based interventions and strategies; and identify and disseminate best practices and effective models of care for this population. Under this funding opportunity announcement, there will be one cooperative agreement to, compile, and distribute technical assistance resources designed to promote replication and implementation of effective models for HIV clinical care and treatment across the HIV care continuum and best practices for comprehensive HIV clinical care for Black Men who have Sex with Men (BMSM). The annual deadline for application is February 21, and the funding award ceiling is $1,500,000.
Requirements: Eligible applicants include: for profit organizations other than small businesses; private institutions of higher education; nonprofits having a 501(c)3 status with the IRS, other than institutions of higher education; and nonprofits that do not have a 501(c)3 status with the IRS, other than institutions of higher education.
Geographic Focus: All States
Date(s) Application is Due: Feb 21
Amount of Grant: Up to 1,500,000 USD
Contact: Laura Cheever, Associate Administrator; (301) 443-1993 or (888) 275-4772; fax (301) 443-8586; lcheever@hrsa.gov
Internet: http://www.grants.gov/view-opportunity.html?oppId=250035
Sponsor: Health Resources and Services Administration
5600 Fishers Lane
Rockville, MD 20857

HRSA Ryan White HIV AIDS Drug Assistance Grants 2070

The Health Resources and Services Administration, HIV/AIDS Bureau is accepting applications for the Ryan White HIV/AIDS Program (RWHAP) Part B and ADAP Training and Technical Assistance. The purpose of this grant program is to provide technical assistance to Ryan White HIV/AIDS Program Part B grantees on developing

and maintaining comprehensive systems of care, integrated planning, monitoring subgrantees, and Affordable Care Act (ACA) implementation and to provide technical assistance to ADAPs on the implementation of cost-containment strategies, financial modeling, wait list management, and ACA implementation. The annual deadline is february 26, and the award ceiling is $427,500.
Requirements: Eligible applicants include: nonprofits having a 501(c)3 status with the IRS, other than institutions of higher education; small businesses; for profit organizations other than small businesses; and nonprofits that do not have a 501(c)3 status with the IRS, other than institutions of higher education.
Geographic Focus: All States
Date(s) Application is Due: Feb 26
Amount of Grant: Up to 427,500 USD
Contact: Laura Cheever; (301) 443-1993 or (888) 275-4772; lcheever@hrsa.gov
Internet: http://www.grants.gov/web/grants/view-opportunity.html?oppId=249577
Sponsor: Health Resources and Services Administration
5600 Fishers Lane
Rockville, MD 20857

Huber Foundation Grants 2071
The Foundation focuses its grants program on three specific aspects of reproductive rights: keeping abortion safe, legal, and preferably increasingly rare; contraceptive choice and availability; and relevant education. The Foundation's major interest lies in funding organizations that will impact these issues on a national level. There is no formal application procedure. A letter describing the project, a budget for the project, and proof of tax-exempt status is required. Though the Foundation board meets four times annually, there are no fixed deadline dates.
Requirements: U.S. nonprofits, including advocacy groups, hospitals, legal defense groups, family planning agencies, educational organizations, universities, and women's groups may submit grant applications.
Restrictions: The foundation does not encourage proposals for projects that are local or regional in scope. It will not consider grants to individuals, foreign organizations, capital campaigns, scholarships, endowment funds, research, international projects, or film productions.
Geographic Focus: All States
Amount of Grant: 10,000 - 100,000 USD
Samples: Planned Parenthood Federation of America (New York, NY)--for general support and Responsible Choices Campaign, $700,000; Feminist Majority Foundation (Arlington, VA)--for the National Clinic Access Project, $100,000; Planned Parenthood of Northern New England (Williston, VT)--for general support, $90,000.
Contact: Lorraine Barnhart, Executive Director; (732) 933-7700
Sponsor: Huber Foundation
P.O. Box 277
Rumson, NJ 07760-0277

Hudson Webber Foundation Grants 2072
The purpose of the Foundation is to improve the vitality and quality of life of the metropolitan Detroit community. The Foundation concentrates its giving primarily within the City of Detroit and has a particular interest in the revitalization of the urban core because this area is a focus for community activity and pride and is of critical importance to the vitality of the entire metropolitan community. The Foundation presently concentrates its efforts and resources in support of projects within five program missions: Detroit Physical Revitalization; Economic Development; The Arts; Safe Community; and the Detroit Medical Center. A brief letter signed by a senior officer of the requesting organization is the preferred form of application. Information for completing the letter of request are available online.
Requirements: The foundation concentrates its giving within Detroit.
Geographic Focus: Michigan
Samples: Habitat for Humanity - Metro Detroit, Inc., Detroit, MI, $150,000--Partnership to Build Capacity; Detroit Renaissance Foundation, Inc., Detroit, MI, $100,000--Business Attraction Program; Detroit Science Center, Inc., Detroit, MI, $10,000--General program support; Michigan Community Service Commission, Detroit, MI, $75,000--Winning Futures Program-Mentor Michigan; Harper Hospital, Detroit, MI, $200,000--Renal Transplant Unit.
Contact: Katy Locker; (313) 963-7777; fax (313) 963-2818; HWF@hudson-webber.org
Internet: http://www.hudson-webber.org/HowToApply_Instructions.html
Sponsor: Hudson Webber Foundation
333 West Fort Street, Suite 1310
Detroit, MI 48226-3134

Huffy Foundation Grants 2073
The Huffy Foundation gives primarily to areas in which the company has operations in California, Ohio, Pennsylvania, and Wisconsin. These areas include: the visual arts; museums; performing arts; theater; arts/cultural programs; all levels of education; hospitals (general); health care and associations; recreation; children and youth services; and federated giving (United Way). Types of support include general and operating support; continuing support; annual campaigns; capital campaigns; building/renovation programs; emergency funds; program development; seed money; consulting services; employee matching gifts; and matching funds. Requests should include a description of the organization, its history and purpose, a description of the people it serves, and a summary of total budget and funding.
Requirements: The Foundation provides a brochure which delineates the application and grant guidelines. There is no application form. The Foundation recommends that the initial approach be in the form of a letter or a proposal. One copy of the letter/proposal should be submitted. There are no deadlines. The Board meets in February, May, August and November.
Restrictions: Grants are not made to individuals, in support of political activities or of religious organizations for religious purposes, or organizations that are not tax exempt. Grants are seldom made for medical research; to endowments; or for operating funds for organizations located outside the corporation communities.
Geographic Focus: California, Ohio, Pennsylvania, Wisconsin
Contact: Pam Booher, Secretary; (937) 866-6251
Sponsor: Huffy Foundation
225 Byers Road
Miamisburg, OH 45342

Hugh J. Andersen Foundation Grants 2074
The mission of the Hugh J. Andersen Foundation is to give back to our community through focused efforts that foster inclusivity, promote equality, and lead to increased human independence, self sufficiency and dignity. To fulfill this mission, the Foundation acts as a grantmaker, innovator, and convener. The Foundation's primary geographic area of focus is the St. Croix Valley: Washington County in Minnesota and Pierce, Polk and St. Croix Counties in Wisconsin. Secondarily, there is an interest in St. Paul, Minnesota. From time to time the Foundation may consider programs in other parts of the Metro Area and Greater Minnesota. Please contact the Program Director prior to submitting a proposal to determine if the program might be of interest to the Foundation. The Board generally considers requests in June, September, December and February. Faxed or emailed applications will not be considered.
Requirements: The Hugh J. Andersen Foundation awards grants only to qualified charitable organizations that are designated as tax exempt under Internal Revenue Service Code 501(c)3 and are not classified as private foundations.
Restrictions: The Hugh J. Andersen Foundation: does not make loans, and does not provide grants or scholarships to individuals; does not provide grants for lobbying activities, fundraising dinners and events, or travel; will generally not consider the following types of organizations and programs for funding: agencies/divisions/councils/programs that have counterparts in St. Paul or the St. Croix Valley; arts organizations exclusively focused on music, dance or visual arts; athletic teams; business/economics education; child care centers; civic action groups; debt or after the fact situations; immigration/refugee issues and programs; independent media productions; political/voter education; private or alternative schools; religious institutions. In addition, the Foundation: will generally not fund the entire project budget, but prefers to be part of an effort supported by a number of sources; considers major endowment and capital requests for funding to be a low priority; does not provide funding through fiscal agents; considers letters of inquiry. Letters of inquiry are reviewed at the Foundation's next board meeting. If the board determines that the request falls within its guidelines and interests, a full proposal will be requested from the applicant for review at the following board meeting.
Geographic Focus: Minnesota, Wisconsin
Date(s) Application is Due: Mar 15; Jun 15; Aug 15; Nov 15
Contact: Brad Kruse, Program Director; (651) 275-4489 or (888) 439-9508; fax (651) 439-9480; hjafdn@srinc.biz
Internet: https://www.srinc.biz/hja/documents/HJAGuidelines09-10.pdf
Sponsor: Hugh J. Andersen Foundation
White Pine Building, 342 Fifth Avenue North
Bayport, MN 55003

Huie-Dellmon Trust Grants 2075
The trust awards grants to Louisiana nonprofits in its areas of interest, including hospitals, higher and secondary education, libraries, and Protestant churches and organizations. Types of support include general operating support, capital campaigns, building construction/renovation, equipment acquisition, program development, scholarship funds, research, and matching funds. There are no application forms or deadlines.
Requirements: Central Louisiana nonprofit organizations are eligible.
Geographic Focus: Louisiana
Amount of Grant: 1,000 - 100,000 USD
Contact: Richard Crowell Jr., Trustee; (318) 748-8141
Sponsor: Huie-Dellmon Trust
P.O. Box 330
Alexandria, LA 71309-0330

Huisking Foundation Grants 2076
In the defense of freedom, Francis Robert Huisking paid the ultimate sacrifice during War II. As co-captain of a B-24 bomber, he and his crew were lost when the bomber went down over Italy in 1944. To honor his sacrifice and to perpetuate and preserve as a living memory the sweet nature, the sterling qualities, the grand character, and the noble attributes he displayed during the short time he was with us, his parents Charles and Catherine Huisking founded the Frank R. Huisking Foundation on October 24, 1946. In addition to Charles and Catherine, the original founding Directors and Officers included five of Francis' seven brothers and sisters; Charles Jr., Evelyn, William, Edward, and Richard. These individuals managed the growth, investments, contributions, and distributions of the Foundation for the next twenty plus years. In 1971, the name of the Foundation was officially changed to the Huisking Foundation. Today, the Foundation's primary fields of interest include: animal welfare; Catholicism; communication media; elementary and secondary education; higher education; historic preservation; hospital care; human services; museums; natural resources; nonprofits; and performing arts. The Foundation is small, and dedicated to the areas outlined above. To assure that funds distributed meet these principles, projects are reviewed and recommended by specific Directors who have knowledge of the project or program and thus supports its goals. This process results in the Board having a greater degree of confidence and assurance that the support being provided clearly meets the principles and intent as set out to honor Frank R. Huisking and his family.

Geographic Focus: All States
Amount of Grant: 200 - 60,000 USD
Contact: Frank Huisking, Treasurer; (203) 426-8618; wwh@huiskingfoundation.org
Internet: http://www.huiskingfoundation.org/
Sponsor: Huisking Foundation
291 Peddlers Road
Guilford, CT 06437-2324

Humana Foundation Grants 2077
The Humana Foundation was established to promote worthwhile organizations that improve the health and welfare of communities in its headquarters region of Louisville, Kentucky. It supports nonprofit institutions primarily in the areas of domestic and international health, education, and civic and cultural development. Religious organizations are eligible for project-specific support (e.g., social services outreach) or funds for an accredited, church-affiliated educational institution. Grants also are awarded in Humana market areas outside of the headquarters region. Detailed pplication information is available online.
Requirements: Applicant organizations must be 501(c)3 tax-exempt located in communities where Humana has a meaningful presence.
Restrictions: Grants are not given to organizations for seed money. The foundation does not contribute to social, labor, political, veterans, or fraternal organizations. Funds cannot be used solely to support an organization's salary expenses or other administrative costs. The foundation does not support lobbying efforts or political action committees.
Geographic Focus: Arizona, Colorado, Florida, Georgia, Illinois, Indiana, Kansas, Kentucky, Louisiana, Michigan, Ohio, Tennessee, Texas, Utah, Wisconsin
Date(s) Application is Due: Jan 31
Contact: Virginia K. Judd, Executive Director; (502) 580-4140; fax (502) 580-1256; HumanaFoundation@humana.com
Internet: https://www.humanafoundation.org/
Sponsor: Humana Foundation
500 West Main Street, Suite 208
Louisville, KY 40202-2946

Human Source Foundation Grants 2078
The Human Source Foundation is organized to support charitable and educational initiatives that improve the human condition. The Foundation provides assistance and financial support for programs involved with education, human services, and youth development. Giving is restricted to the state of Texas, primarily in Denton and Fort Worth counties. There are no specific deadlines or application forms, and applicants should begin by forwarding a letter of request.
Requirements: Tax-exempt 501(c)3 organizations serving Texas, primarily Denton and Fort Worth counties, are eligible to apply.
Geographic Focus: Texas
Contact: Mary G. Palko; (817) 926-2799; fax (817) 926-5202; mary@ftw.com
Sponsor: Human Source Foundation
2409 Winton Terrace West, P.O. Box 100423
Fort Worth, TX 76185-0423

Huntington's Disease Society of America Research Fellowships 2079
Fellowships are designed to assist promising young postdoctoral investigators in the early stages of their careers who are engaged in research at the basic and clinical levels relating to the cause and treatment of Huntington's disease. The amount of the award will depend upon the policy of the sponsoring accredited medical school or university and the training and experience of the applicant. Fellowships are awarded for one year with the possibility of renewal. Application forms and requested information on curriculum vita, publications, research proposal, sponsorship, and other sources of funding are provided by the society.
Requirements: All applicants must have an MD or PhD degree or the equivalent and work must in some way be related to Huntington's disease. There is no citizenship requirement.
Geographic Focus: All States
Amount of Grant: Up to 80,000 USD
Samples: Dr. Tsugn Peng, Emory U (Atlanta, GA)--for research on impaired mitochortinal function, $30,000; Dr. Nansheng Chen, U of British Columbia (Vancouver, BC)--for research on the role of glutamate receptor-HD protein interactions, $30,000.
Contact: Robert Graze; (212) 242-1968 or (800) 345-4372, ext. 227; rgraze@hdsa.org
Internet: http://www.hdsa.org/research/grant-applications.html
Sponsor: Huntington's Disease Society of America
505 8th Avenue, Suite 902
New York, NY 10018

Huntington's Disease Society of America Research Grants 2080
The Society awards research grants for one year for support of basic or clinical research related to Huntington's disease. Grant awards are provided as seed monies for new or innovative research projects in the hope that they will develop sufficiently to attract funding from other sources. Application forms are provided by the society and generally follow the format of NIH grant applications.
Requirements: Applicants may be doctorate or graduate students or equivalent professionals. There is no citizenship requirement.
Geographic Focus: All States
Amount of Grant: Up to 100,000 USD
Contact: Robert Graze; (212) 242-1968 or (800) 345-4372, ext. 227; rgraze@hdsa.org
Internet: http://www.hdsa.org/research/grant-applications.html
Sponsor: Huntington's Disease Society of America
505 8th Avenue, Suite 902
New York, NY 10018

Huntington Beach Police Officers Foundation Grants 2081
The Foundation provides funds to various schools and community youth activities, benefits to spouses or children of police officers killed in the line of duty, and medical assistance to children of police officers in need of help. Giving is limited to Huntington Beach, California. There are no specific grant application forms or deadlines with which to adhere. Applicants should contact the Foundation office directly.
Requirements: Support is given only to groups situated in the Huntington Beach, California, region.
Geographic Focus: California
Contact: Kreg Muller; (714) 842-8851; fax (714) 847-0064; hbpof@hbpoa.org
Internet: http://www.hbpoa.org/foundation/hbpof.htm
Sponsor: Huntington Beach Police Officers Foundation
20422 Beach Boulevard, Suite 450
Huntington Beach, CA 92648-8301

Huntington Clinical Foundation Grants 2082
The Foundation, established in 1986, supports health care programs and access, higher education, and medical research. Grant amounts range from approximately $1,500 to $8,000, with some larger amounts available. There are no specific applications or deadlines with which to adhere. The Board meets quarterly, and applicants should forward a detailed description of the project and amount of funding requested.
Requirements: Giving is limited to the Huntington, West Virginia, region.
Geographic Focus: West Virginia
Amount of Grant: 1,500 - 8,000 USD
Samples: Cabell County Substance Abuse Prevention Partnership, Huntington, West Virginia, $5,000 - funding for Cabell County REACH; Children's Place, Huntington, West Virginia, $3,000 - purchase of cribs; Marshall University Joan C Edwards School of Medicine, Huntington, West Virginia, $20,000 - scholarship endowment.
Contact: Don Ray, President; (304) 697-4780 or (304) 523-6120; fax (304) 523-6051
Sponsor: Huntington Clinical Foundation
P.O. Box 117
Huntington, WV 25706-0117

Huntington County Community Foundation Make a Difference Grants 2083
The Huntington County Community Foundation Make a Difference Grants fund charitable projects that make a positive impact on the residents of Huntington County. Grant areas to be considered include: arts and culture; community development; health and human services; education; and other charitable services. These grants are awarded in spring and fall cycles. Grant decisions on based on the following criteria: the project's beneficial impact on Huntington County, immediate and ongoing; known and anticipated community needs; number of persons that benefit from and/or are affected by the project; and a clear, complete, and comprehensible statement of particulars.
Requirements: Applicants should submit one original and five additional paper clipped sets of the online application packet to include the following: grant proposal form; detailed project and organizational budget; current year-end financial statement; strategic or long-range plan; roster of board members; IRS tax exempt status; and articles of incorporation.
Restrictions: Grants will not be awarded to fund: operational or ongoing recurring (within 60 months) cost of the program; political projects or campaigns; or projects of applicants, or project owners, with taxing authority (e.g. school corporations; units of government).
Geographic Focus: Indiana
Date(s) Application is Due: Apr 15; Oct 15
Contact: Michael Howell; (260) 356-8878; fax (260) 356-0921; michael@huntingtonccf.org
Internet: http://www.huntingtonccf.org/hccf_grant_opportunities.html
Sponsor: Huntington County Community Foundation
356 West Park Drive
Huntington, IN 46750

Huntington National Bank Community Affairs Grants 2084
The corporation supports a variety of community focused initiatives that benefit children, education, economic development, health, social service housing, and the arts in each community of its banking offices. A letter of request (one to two pages) can serve as an initial application for both single- and multi-year contributions. The letter should include a brief statement about the organization, its history, objectives, and goals; purpose of the project; visibility for Huntington as a participant in the project; list of officers and directors; target population(s) to be served by contribution; and annual organization budget, total project budget, amount requested, list of other contributors, and amounts of funds anticipated or committed.
Requirements: Tax-exempt organizations in Ohio, Kentucky, Indiana, Michigan, and West Virginia may apply.
Restrictions: Community donations may not be used for political campaigns involving individuals or issues; religious, ethnic, military, fraternal, or labor groups; individuals; organizations outside of the corporate market area; or organizations without IRS 501(c)3 tax-exempt status.
Geographic Focus: Indiana, Kentucky, Michigan, Ohio, West Virginia
Amount of Grant: 1,000 - 10,000 USD
Contact: Elfi Di Bella, Community Affairs Director; (614) 480-4483; fax (614) 480-4973; elfi.dibella@huntington.com
Internet: http://www.huntington.com/us/HNB3150.htm
Sponsor: Huntington Bancshares
41 South High Street, Huntington Center 3413
Columbus, OH 43215

Hutchinson Community Foundation Grants 2085
The Hutchinson Community Foundation provides funding in the counties of Hutchinson and, Reno, Kansas, inluding the immediate area surrounding area. Areas of interest include, but are not, limited to: community needs in areas such as early childhood, youth, education, arts and culutre, and neighborhood development.
Requirements: Nonprofit organizations in Kansas are eligible to apply. Special purpose units of government can apply for support of innovative projects located in Reno County.
Restrictions: Grant proposals from individuals or non-qualifying organizations will not be considered.
Geographic Focus: Kansas
Contact: Audrey Abbott Patterson; (620) 663-5293; aubrey@hutchcf.org
Internet: http://www.hutchcf.org/default.asp?sm=2&si=1
Sponsor: Hutchinson Community Foundation
One North Main, Suite 501, P.O. Box 298
Hutchinson, KS 67504-0298

Hut Foundation Grants 2086
Since 1998, the Hut Foundation has been funding nonprofits primarily in the states of California and Virginia, in the following areas of interest: arts; children/youth services; community/economic development; education; health care; youth development. Unsolicited requests for funds are not accepted, therefore initial contact should be through a Letter of Inquiry. If the proposal follows the Foundations guidelines, the applicant will be invited to submit an application.
Requirements: Must be a 501(c)3 nonprofit.
Restrictions: No grants to individuals.
Geographic Focus: California, Virginia
Amount of Grant: 5,000 - 10,000 USD
Contact: Marcus Guerrero; (415) 834-2464; hutfoundation@gmail.com
Sponsor: Hut Foundation
19 Sutter Street
San Francisco, CA 94104

Hutton Foundation Grants 2087
The foundation awards grants throughout Santa Barbara County, California to nonprofit organizations in its areas of interest, including education, arts and culture, health and human services, and civic and community development. Preference is given to efforts that promise to bring about major change, build leadership and institutional capacity, and achieve lasting results. Types of support include standard, program related investments, marketing, media, and endowment grants. An application form is available online.
Requirements: California nonprofit organizations headquartered in Santa Barbara county are eligible to apply.
Restrictions: Funding is not available to individuals or organizations that discriminate on the basis of age, gender, race, ethnicity, sexual orientation, disability, national origin, political affiliation or religious belief.
Geographic Focus: California
Contact: Arlene R. Craig; (805) 957-4740; fax (805) 957-4743; info@huttonfoundation.org
Internet: Arlene R. Craig, Vice President;
Sponsor: Hutton Foundation
26 West Anapamu Street, 4th Floor
Santa Barbara, CA 93101

I.A. O'Shaughnessy Foundation Grants 2088
The Foundation is concerned that too many schools lack sufficient resources; that students in high-poverty areas have lower achievement scores, higher drop-out rates, and lower rates of college graduation; that low-income families lack the resources to choose better schools; and that the gap between the rich and the poor is increasing. The Foundation has set its current funding interest to help address these critical matters of public concern. The Foundation is currently interested in making grants to support high quality education that prepares students in disadvantaged communities for educational and life success. Priority is given to organizations that provide support networks; remove impediments to student success; are broadly supported by the community, and have a record of demonstrated success. Additional guidelines and application forms are available at the foundation's website.
Requirements: Eligible applicant must be: nonprofit organizations; mission-consistent; fiscally sound; demonstrate a need; and capable and accountable.
Restrictions: The Foundation will not fund: grants that are not consistent with Foundation values; national or umbrella organizations that raise funds through broad-based solicitations to the general public; organizations that become overly dependent on the Foundation for on-going operational support; capital campaign gifts exceeding 20% of the campaign goal; political campaigns, events, or organizations whose purpose is to promote political candidates; lobbying; individuals.
Geographic Focus: Illinois, Kansas, Minnesota, Texas
Contact: Eileen A. O'Shaughnessy; (952) 698-0959; info@iaoshaughnessyfdn.org
Internet: http://www.iaoshaughnessyfdn.org/guidelines.htm
Sponsor: I.A. O'Shaughnessy Foundation
2001 Killebrew Drive, Suite 120
Bloomington, MN 55425

IAFF Burn Foundation Research Grants 2089
The IAFF Burn Foundation Research Grant Program continues to lead to new knowledge and innovative approaches to the prevention and treatment of the physical and psychological problems that impair the quality of life for a burn patient. Research grant proposals are accepted annually and awards are distributed during the American Burn Association Annual Meeting. Topics of interest include by are not limited to pain management, physical and psychological rehabilitation, and wound healing and scarring, especially in children. Clinical projects are recommended. The foundation's Board of Medical Advisors oversees the selection process, reviewing each of the submissions. While all of candidates are deserving, only seven to ten grants are chosen and submitted to the IAFF Burn Foundation for consideration. Typically, five to seven are funded.
Requirements: The foundation welcomes research grant applications from members of the American Burn Association. The areas of funded research fall into three categories: (1) Quality of Life - This area of research is the most common, focusing on issues such as pain management, nutrition, and physical and psychological rehabilitation. The major focus in this area is the care of burned children. (2) Burn Prevention - Research in this area is especially important in shaping prevention programs. (3) Basic Science - Research grants funded in this category surround nutrition, wound healing and scar formation. No animal studies are funded. The required application and detailed guidelines can be found at the sponsor's website.
Restrictions: No animal studies are funded. The grant does not include indirect costs.
Geographic Focus: All States, Canada
Amount of Grant: 25,000 USD
Contact: Patrick Morrison; (202) 824-8620; fax (202) 637-0839; burnfoundation@iaff.org
Internet: http://burn.iaff.org/grants.shtml
Sponsor: International Association of Fire Fighters
1750 New York Avenue NW
Washington, D.C. 20006

IARC Expertise Transfer Fellowship 2090
The Fellowship is intended to enable an established investigator to spend normally from six to twelve months in an appropriate host institute in a low- to medium-resource country in order to transfer knowledge and expertise in a research area relevant for the host country and related to epidemiology, biostatistics, environmental chemical carcinogenesis, cancer etiology and prevention, infection and cancer, molecular cell biology, molecular genetics, molecular pathology and mechanisms of carcinogenesis. Applications should include a proposed collaborative research project, specifying the link to IARC's on-going activities and a letter of support from the host lab giving details of feasibility and anticipated benefit to the receiving institute. Priority will be given to projects directly linked to IARC's on-going research program, involving at least one contact at IARC. There will be an annual remuneration of up to $70,000, which will take into account the on-going salary of the Fellow. This amount may include limited support for the project. The cost of travel will also be met.
Requirements: Applicants should be established cancer researchers actively engaged in the field with appropriate scientific or medical qualifications and an excellent publications' record. They must also belong to the staff of a university or a research institution.
Geographic Focus: All States
Date(s) Application is Due: Nov 30
Amount of Grant: Up to 70,000 USD
Contact: Dr. Paolo Boffetta, +33-(0)472-73-84-48; fel@iarc.fr or vsa@iarc.fr
Internet: http://www.iarc.fr/en/education-training/expertisetransfer.php
Sponsor: International Agency for Research on Cancer
150 cours Albert-Thomas
Lyon Cedex 08, 69372 France

IARC Postdoctoral Fellowships for Training in Cancer Research 2091
Applications for training fellowships are invited from junior scientists from low- or medium-resource countries wishing to complete their training in those aspects of cancer research related to coordinate and conduct both epidemiological and laboratory research into the causes of cancer. Disciplines covered include: epidemiology, biostatistics, environmental chemical carcinogenesis, cancer etiology and prevention, infection and cancer, molecular cell biology, molecular genetics, molecular pathology and mechanisms of carcinogenesis, with emphasis on interdisciplinary projects. The fellowship is for a period of one year, with the possibility of an extension for a second year subject to satisfactory appraisal.
Requirements: Candidates should have spent less than five years abroad (including doctoral studies) and have finished their doctoral degree within five years of the closing date for application or be in the final phase of completing their doctoral degree (M.D. or Ph.D.). They must provide evidence of their ability to return to their home country and keep working in cancer research. The working languages at IARC are English and French. Candidates must be proficient in English at a level sufficient for scientific communication.
Restrictions: Candidates already working as a postdoctoral fellow at the Agency at the time of application or who have had any contractual relationship with IARC during the 6 months preceding the application deadline or who have already spent more than one year at IARC cannot be considered.
Geographic Focus: All States
Date(s) Application is Due: Nov 30
Amount of Grant: 32,000 EUR
Contact: Dr. Paolo Boffetta, +33-(0)472-73-84-48; fel@iarc.fr or vsa@iarc.fr
Internet: http://www.iarc.fr/en/education-training/postdoc.php
Sponsor: International Agency for Research on Cancer
150 cours Albert-Thomas
Lyon Cedex 08, 69372 France

IARC Visiting Scientist Award for Senior Scientists 2092
The IARC is offering this Award for a qualified and experienced investigator with recent publications in international peer-reviewed scientific journals who wishes to spend from six to twelve months at the IARC working on a collaborative project in a research area related to the Agency's programs: epidemiology, biostatistics, environmental chemical carcinogenesis, cancer etiology and prevention, infection and cancer, molecular cell

biology, molecular genetics, molecular pathology and mechanisms of carcinogenesis. There will be an annual remuneration of up to $80,000, which will take into account the on-going salary of the visiting scientist plus the cost of travel.
Requirements: Applicants must belong to the staff of a university or a research institution and should provide written assurance of a post to return to at the end of the period of award.
Geographic Focus: All States
Date(s) Application is Due: Nov 30
Amount of Grant: Up to 80,000 USD
Contact: Dr. Paolo Boffetta, +33-(0)472-73-84-48; vsa@iarc.fr or vsa@iarc.fr
Internet: http://www.iarc.fr/en/education-training/vsa.php
Sponsor: International Agency for Research on Cancer
150 cours Albert-Thomas
Lyon Cedex 08, 69372 France

IATA Research Grants 2093

In order to enhance the National Athletic Trainer's Association's (NATA) Research and Education Foundation, the Illinois Athletic Trainers' Association has provided limited funding for athletic training related research in the state of Illinois. Grant monies from the IATA are awarded for projects concerning research in the domains relevant to athletic training within the state of Illinois. Grant applications will be reviewed and monies will be awarded during the Spring and Fall IATA BOD meetings.
Requirements: Applicants must be Certified/Licensed, members in good standing not only with the NATA but also the IATA. Applications must be submitted 3 months prior to the beginning of the research study, and the research must be relevant to the profession of athletic training within the state of Illinois. All applications will be analyzed bi-annually and as stated above must be submitted prior to the beginning of the project. Please note that there is limited funding for these awards, and you may be turned down due to lack of funding.
Restrictions: There will be no awards granted for recovery of cost involved with research that has already been initiated.
Geographic Focus: Illinois
Contact: Mike Sullivan; (630) 853-0820; webmaster@illinoisathletictrainers.org
Internet: http://www.illinoisathletictrainers.org/Researchgrts.htm
Sponsor: Illinois Athletic Trainers' Association
701 Thomas Road
Wheaton, IL 60187

IATA Scholarships 2094

Two scholarships of $1,000, three scholarships of $500, and two scholarships of $250 are available from the Illinois Athletic Trainers' Association (IATA). They are available on an annual basis but do not have to be awarded annually, depending on the applicant's qualifications. Scholarships will be awarded by the Vice-President at the IATA State Meeting Awards Luncheon. Certificates will be awarded to the scholarship recipient and their sponsoring Athletic Trainer, or in the case of the offspring scholarship, the parental member of the IATA.
Requirements: All applicants must have a grade point average of 'B,' or its equivalent, or above. All undergraduate and graduate applicants must be a member of the IATA. Applicants for the continuing education scholarship must be a licensed member of the IATA for one year prior to application. The offspring applicant does not need to be a member. Application forms are available from the IATA Vice President. Application requests must be written and will be included in the applicant's application packet. Recipients will be selected based on academic achievement, letters of recommendation and the applicant's autobiographical sketch.
Restrictions: If a committee member is a scholarship applicant, he/she will recuse from the evaluation process. The President-elect or Past-president will gather the committee evaluations.
Geographic Focus: Illinois
Date(s) Application is Due: Aug 1
Amount of Grant: 250 - 1,000 USD
Contact: Mike Overturf; (630) 575-6212; moverturf@athletico.com
Internet: http://www.illinoisathletictrainers.org/Scholarinfo.htm
Sponsor: Illinois Athletic Trainers' Association
701 Thomas Road
Wheaton, IL 60187

IBCAT Screening Mammography Grants 2095

The mission of the Indiana Breast Cancer Awareness Trust, Inc. (IBCAT) is to increase awareness and improve access to breast cancer screening and diagnosis throughout Indiana. IBCAT receives funds through sales of breast cancer awareness specialty license plate. Through these sales, monies are available for grants deemed to best address the unmet screening needs of the people of Indiana. Grants are intended to provide funding for Screening Mammograms programs serving low income, medically under-served women living in Indiana. Funding for this project will be for one calendar year, January 1 to December 31. Applications are accepted during the months of September and October, and announcement of awards will be made no later than December 15.
Requirements: Applicants must be a federally tax-exempt entity such as nonprofit organization, government agency, educational institution or Indian tribe. Grant applications must e postmarked no later than the first Friday of October. Organizations may only apply for one (1) grant annually. Grant requests for new programs may request up to $5,000; established screening programs may request up to $15,000. Services are to be within the State of Indiana, and all patients must be Indiana residents. American Cancer Society (ACS) guidelines for screening mammography should be followed.
Geographic Focus: Indiana
Date(s) Application is Due: Oct 4
Amount of Grant: 5,000 - 15,000 USD
Contact: Jalana Eash; (866) 724-2228; fax (812) 868-8773; ibcat@insightbb.com
Internet: http://www.breastcancerplate.org/programsgrants/grant-programs/
Sponsor: Indiana Breast Awareness Trust, Inc.
P.O. Box 8212
Evansville, IN

IBEW Local Union #697 Memorial Scholarships 2096

The International Brotherhood of Electrical Workers (IBEW) Local Union #697 Memorial Scholarships are awarded in memory of deceased members of IBEW Local #697 and benefits the children of Local #697. The number of awards vary, but the individual amount is $1,000. Any college major is eligible.
Requirements: Applicants must be dependent children, age 25 or under, of IBEW workers. Eligibility is determined with a copy of a current tax return indicating the child/stepchild is a dependent. Applicants must be registered for at least 12 hours in the school of his or her choice. The application is available at the Local Union office. Recipients are chosen at the Union meeting held on the third Monday of July.
Geographic Focus: Indiana
Date(s) Application is Due: Jun 1
Amount of Grant: 1,000 USD
Contact: IBEW Memorial Scholarship Contact; (219) 736-1880; fax (219) 736-1940; legacy@legacyfoundationlakeco.org
Internet: http://www.legacyfoundationlakeco.org/scholarships/index.htm
Sponsor: Legacy Foundation
1000 East 80th Place, 302 South
Merrillville, IN 46410

IBRO-PERC InEurope Short Stay Grants 2097

As a result of the successful launch of the FENS/IBRO Program of European Neuroscience Schools, IBRO's Central and Eastern Europe and Western Europe regional committees decided that the collaboration would be further strengthened by the activation of inter-regional networking and mobility of young scientists. InEurope Short Stay Grants fund visits to European institutions by researchers living and working within Europe, so that they can acquire new methods or specific techniques that are necessary for their work. The aim of the programme is to increase intra-European mobility of young researchers by providing grants for short, goal-directed exchanges within European laboratories. Exchanges are limited to no more than four weeks. Funding can be used to cover travel and local expenses. Applications deadlines are in March, June, September and December, and must be submitted through the IBRO online application process. Applicants who are not selected in one round of evaluations will remain in competition for the next round.
Requirements: Applicants must be Ph.D. students or postdoctoral Fellows, 35 years of age or less.
Restrictions: emailed or hard copy applications are not acceptable.
Geographic Focus: Albania, Andorra, Armenia, Austria, Azerbaijan, Belarus, Belgium, Bosnia & Herzegovina, Bulgaria, Croatia, Cyprus, Czech Republic, Denmark, Estonia, Finland, France, Germany, Greece, Hungary, Iceland, Ireland, Italy, Latvia, Liechtenstein, Lithuania, Luxembourg, Macedonia, Malta, Moldova, Monaco, Montenegro, Netherlands, Norway, Poland, Portugal, Romania, Russia, San Marino, Serbia, Slovakia, Slovenia, Spain, Sweden, Switzerland, Turkey, Ukraine, United Kingdom, Vatican City
Amount of Grant: Up to 3,000 EUR
Contact: Marta Hallak, Senior Director of Grants; 33 (0) 1 46 47 92 92; fax 33 (0) 1 46 47 42 50; mhallak@mail.fcq.unc.edu.ar
Internet: http://ibro.info/professional-development/funding-programmes/regional-grants/europe-grants/
Sponsor: International Brain Research Organization
255 rue Saint-Honore
Paris, 75001
France, Metropolitan

IBRO-PERC Support for European Workshops, Symposia and Meetings 2098

With the aim to foster cooperation among regional young neuroscientists, IBRO's Western Europe Regional Committee (WERC) offers partial funding of courses, workshops, symposia, and meetings on important topics in neuroscience organized in European countries. Preference will be given to activities that include younger scientists and offer training for scientists from countries with limited resources for research or teaching. Applications must be submitted through the IBRO online application system. emailed or hard copy applications are not acceptable. Applications are accepted until February 15 each year, and awards range up to a maximum of 4,000 Euro.
Geographic Focus: Austria, Belgium, Denmark, Finland, France, Germany, Greece, Iceland, Ireland, Italy, Luxembourg, Netherlands, Norway, Portugal, Spain, Sweden, Switzerland, United Kingdom
Date(s) Application is Due: Feb 15
Amount of Grant: Up to 4,000 EUR
Contact: Jochen Pflueger, Chairperson; 33 (0) 1 46 47 92 92; fax 33 (0) 1 46 47 42 50; pflueger@neurobiologie.fu-berlin.de
Internet: http://ibro.info/professional-development/funding-programmes/regional-grants/europe-grants/
Sponsor: International Brain Research Organization
255 rue Saint-Honore
Paris, 75001
France, Metropolitan

IBRO-PERC Support for Site Lectures 2099

With the aim to facilitate contact between young students and high profile neuroscientists, experience their enthusiasm for brain research and gain first-hand knowledge on modern topics in neuroscience, IBRO-PERC supports the organization of special lectures. These are two one- to two-day long conferences in developing countries, in cooperation with research centres or universities, in which a renowned European neuroscientist lectures about his/her topic of expertise. They carry the title IBRO-PERC Neuroscience Lectures. Applications are to be submitted only by email. The annual application deadline is February 15.
Requirements: The application must include in a single document: title of the lecture; dates and venue; applicant's curriculum vitae; and preliminary program and budget.
Geographic Focus: Austria, Belgium, Denmark, Finland, France, Germany, Greece, Iceland, Ireland, Italy, Luxembourg, Netherlands, Norway, Portugal, Spain, Sweden, Switzerland, United Kingdom
Date(s) Application is Due: Feb 15
Contact: Jochen Pflueger, Chairperson; 33 (0) 1 46 47 92 92; fax 33 (0) 1 46 47 42 50; pflueger@neurobiologie.fu-berlin.de
Internet: http://ibro.info/professional-development/funding-programmes/regional-grants/europe-grants/
Sponsor: International Brain Research Organization
255 rue Saint-Honore
Paris, 75001
France, Metropolitan

IBRO/SfN International Travel Grants 2100

The Society for Neuroscience offers travel fellowships to their annual meeting for neuroscientists under the age of 35 who are currently working and living in a resource-restricted country – those defined by the World Bank as low, lower-middle, or upper-middle income. Registration fees are waived for successful candidates attending the meeting. Awardees will be notified prior to the SfN abstract deadline of May 15 of the current year as to whether they have been selected for the SfN travel grant. However, this award is contingent upon the awardee's abstract being accepted by SfN and proof of attendance at the meeting (in addition to the requirements). Candidates must apply through the IBRO online application system.
Requirements: Candidates must be a citizen of and reside in a country that the World Bank has classified low, lower-middle, or upper-middle income. The applicant should be the first author of an abstract to be presented at the SfN annual meeting. A copy of the abstract submitted to SfN for a poster or a platform presentation is required.
Restrictions: Candidates from the U.S. and Canada are not eligible to apply.
Geographic Focus: All Countries
Date(s) Application is Due: Feb 1
Amount of Grant: Up to 2,000 EUR
Contact: Marta Hallak; 33 (0) 1 46 47 92 92; mhallak@mail.fcq.unc.edu.ar
Internet: http://ibro.info/professional-development/funding-programmes/travel-grants/
Sponsor: International Brain Research Organization
255 rue Saint-Honore
Paris, 75001
France, Metropolitan

IBRO Asia Regional APRC Exchange Fellowships 2101

The IBRO-APRC Exchange Fellowship aims to allow high-quality junior scientists at the level of postdoctoral fellow or junior faculty (under the age of 45) from diverse geographic and scientific areas to broaden the scope of their training in neuroscience by working four to six months abroad in established laboratories. Senior Ph.D. students with good publication records and strong justification for the exchange program will also be considered. Both the applicant and the host laboratory must be within the Asia Pacific Region. Up to $8,500 per Fellowship is available, although funding may vary depending on the country in which the candidate studies. The Fellowship is used along with matching funds from the host laboratory to defray the cost of travel and subsidize living costs. Applications must be submit through the IBRO online application system. emailed or hard copy applications are not acceptable.
Requirements: Selection criteria include academic credentials of the applicant and the host, as well as the quality of the research proposal with a structured research plan. Priority is given to applicants from less developed and less well-funded countries. Strong justification must be made by the applicant that he/she will return home after the exchange, bringing with them new knowledge and skills to advance neuroscience in their home countries. A firm commitment from his/her parent institution to this effect will be viewed favorably by the selection committee.
Geographic Focus: Afghanistan, Bahrain, Bangladesh, Bhutan, Brunei, Cambodia, China, East Timor, India, Indonesia, Iran, Iraq, Israel, Japan, Jordan, Kazakhstan, Kuwait, Kyrgyzstan, Laos, Lebanon, Malaysia, Maldives, Mongolia, Myanmar (Burma), Nepal, North Korea, Oman, Pakistan, Philippines, Qatar, Russia, Saudi Arabia, Singapore, South Korea, Sri Lanka, Syria, Taiwan, Tajikistan, Thailand, Turkey, Turkmenistan, United Arab Emirates, Uzbekistan, Vietnam, Yemen
Date(s) Application is Due: Oct 15
Amount of Grant: Up to 8,500 EUR
Contact: Marta Hallak, Senior Director of Grants; 33 (0) 1 46 47 92 92; fax 33 (0) 1 46 47 42 50; mhallak@mail.fcq.unc.edu.ar
Internet: http://ibro.info/professional-development/funding-programmes/regional-grants/asiapacific-grants/
Sponsor: International Brain Research Organization
255 rue Saint-Honore
Paris, 75001
France, Metropolitan

IBRO Asia Regional APRC Lecturer Exchange Program Grants 2102

The IBRO-APRC Lecturer Exchange Program supports trips for those who will attend meetings of APRC member societies from other APRC countries to make plenary, special lectures or presentations in symposia. The invited speakers must be Principal Investigators no older than age 45 at the time of the meeting. This helps annual meetings of member societies become more international and core researchers in member societies gain greater exposure to wider communities. Annual deadlines are in December and February. Grant awards range up to a maximum of 2000 Euro.
Geographic Focus: Afghanistan, Bahrain, Bangladesh, Bhutan, Brunei, Cambodia, China, East Timor, India, Indonesia, Iran, Iraq, Israel, Japan, Jordan, Kazakhstan, Kuwait, Kyrgyzstan, Laos, Lebanon, Malaysia, Maldives, Mongolia, Myanmar (Burma), Nepal, North Korea, Oman, Pakistan, Philippines, Qatar, Russia, Saudi Arabia, Singapore, South Korea, Sri Lanka, Syria, Taiwan, Tajikistan, Thailand, Turkey, Turkmenistan, United Arab Emirates, Uzbekistan, Vietnam, Yemen
Date(s) Application is Due: Feb 1; Dec 1
Amount of Grant: 2,000 EUR
Contact: Marta Hallak, Senior Director of Grants; 33 (0) 1 46 47 92 92; fax 33 (0) 1 46 47 42 50; mhallak@mail.fcq.unc.edu.ar
Internet: http://ibro.info/professional-development/funding-programmes/regional-grants/asiapacific-grants/
Sponsor: International Brain Research Organization
255 rue Saint-Honore
Paris, 75001
France, Metropolitan

IBRO Asia Regional APRC Travel Grants 2103

The IBRO APRC Travel Grants support young researchers and graduate students working in Asian/Pacific countries to attend and make presentations at international APRC member society meetings, held in English, within the region. Applications from countries of low, lower-middle and upper-middle income in the region (according to The World Bank) are encouraged. Awards range from 1000 to 1500 Euro.
Requirements: Graduate students and researchers who have obtained a PhD within the last five years are eligible to apply. For countries in which it is extremely difficult to obtain support for trips to foreign meetings, excellent applications by those aged under 40 are also considered.
Geographic Focus: Afghanistan, Bahrain, Bangladesh, Bhutan, Brunei, Cambodia, China, East Timor, India, Indonesia, Iran, Iraq, Israel, Japan, Jordan, Kazakhstan, Kuwait, Kyrgyzstan, Laos, Lebanon, Malaysia, Maldives, Mongolia, Myanmar (Burma), Nepal, North Korea, Oman, Pakistan, Philippines, Qatar, Russia, Saudi Arabia, Singapore, South Korea, Sri Lanka, Syria, Taiwan, Tajikistan, Thailand, Turkey, Turkmenistan, United Arab Emirates, Uzbekistan, Vietnam, Yemen
Amount of Grant: 1,000 - 1,500 EUR
Contact: Marta Hallak, Senior Director of Grants; 33 (0) 1 46 47 92 92; fax 33 (0) 1 46 47 42 50; mhallak@mail.fcq.unc.edu.ar
Internet: http://ibro.info/professional-development/funding-programmes/regional-grants/asiapacific-grants/
Sponsor: International Brain Research Organization
255 rue Saint-Honore
Paris, 75001
France, Metropolitan

IBRO International Travel Grants 2104

The IBRO International Travel Grants aim to foster neuroscience research especially in less well-funded countries by providing support to high quality neuroscientists from diverse geographic and scientific areas (US/Canada Region excluded) who wish to participate at international neuroscience meetings. Priority is given to those who have not obtained funding from this programme within the past three years. Applications should be submitted by March 4 for travel to conferences occurring from July to December of that year. Applications should be submitted by September 1 for travel occurring from January to June of the following year. Candidates must apply through the IBRO online application system.
Restrictions: Applicants cannot apply for more than one category of funding within travel grants.
Geographic Focus: All Countries
Date(s) Application is Due: Mar 4; Sep 1
Amount of Grant: Up to 1,500 EUR
Contact: Marta Hallak, Senior Director of Grants; 33 (0) 1 46 47 92 92; fax 33 (0) 1 46 47 42 50; mhallak@mail.fcq.unc.edu.ar
Internet: http://ibro.info/professional-development/funding-programmes/travel-grants/
Sponsor: International Brain Research Organization
255 rue Saint-Honore
Paris, 75001
France, Metropolitan

IBRO Latin America Regional Funding for Neuroscience Schools 2105

The Latin America Regional Funding for Neuroscience Schools provides financial support for the training activities of advanced students and young researchers in the field of neuroscience in order to increase the quality of neuroscience education in the region; increase the links between young neuroscientists; create a network of former awardees and senior researchers; and establish a mentorship programme of continuous support of alumni. Candidates must submit applications and all supporting materials through the IBRO online application process. Awards range from 15,000 to 20,000 Euro.
Requirements: The Latin America Regional Committee (LARC) will review applications and grant financial support for Schools of Neuroscience with the following characteristics based on the IBRO School Guidelines (posted on the website): uniform high-quality

education with an active involvement of students and teachers during the entire event; three levels of schools (basic, advanced/specialized, and inter-regional); cover broad and topical issues, but also include subjects of general interest, such as guidance on writing papers or grant proposals, and ethical issues; and a well balanced series of lectures and laboratory-type practical activities. Neuroscientists from any country in the region are eligible as prospective school directors. Each school should have no less than 15 and not more than 100 students and young scientists. The level of the school should match the level of selected students. The application should contain the scientific program of the school, a list of local and invited faculty, and the budget. Additional information is available at the IBRO website.
Restrictions: Hard copy or emailed applications are not accepted.
Geographic Focus: Puerto Rico, Antigua & Barbuda, Argentina, Bahamas, Barbados, Belize, Bolivia, Brazil, Chile, Colombia, Costa Rica, Cuba, Dominica, Dominican Republic, Ecuador, El Salvador, Grenada, Guatemala, Guyana, Haiti, Honduras, Jamaica, Mexico, Nicaragua, Panama, Paraguay, Peru, Saint Kitts And Nevis, St. Lucia, St. Vincent and the Grenadines, Suriname, Trinidad and Tobago, Uruguay, Venezuela
Amount of Grant: 15,000 - 20,000 EUR
Contact: Rebecca Hadid, Director of Professional Development Programs; 33 (0) 1 46 47 92 92; fax 33 (0) 1 46 47 42 50; ibrobecky@gmail.com or ibrocentral@gmail.com
Internet: http://ibro.info/professional-development/funding-programmes/regional-grants/latinamerica-grants/#schools
Sponsor: International Brain Research Organization
255 rue Saint-Honore
Paris, 75001
France, Metropolitan

IBRO Latin America Regional Funding for PROLAB Collaborations   2106
The PROLAB program seeks to foster scientific collaboration and development of human resources among Latin American and Caribbean neuroscience research groups. PROLAB uses the neuroscience scientific network of the region, and provides economic support to joint research projects. Its main goal is to diminish the asymmetries in the development of the neurosciences in the continent, thus allowing established research groups to set up collaboration with emerging groups and open a space for horizontal collaboration among established groups within the region. The scientific collaborations will be supported through grants for research missions (exchange trips by doctoral and post doctoral students from the participating institutions) planned to work on scientific questions specifically proposed in the submitted project. Financial support from PROLAB/IBRO/LARC is exclusively aimed at supporting research missions (three to six months) by doctoral and post-doctoral student members of the research groups for activities within the approved project. Applications must be submitted through the IBRO online application process. There are no hard copy application forms. PROLAB applications are taken until March 31 each year.
Requirements: Projects should aim at scientific investigations and training of human resources in research. They should be structured around a scientific research project in neuroscience focused on relevant scientific topics in the forefront of this area of human knowledge. Applications should have a joint coordination by the leaders of the research groups involved. The participant groups will choose one of the research leaders to be the responsible liaison with LARC. The projects should be planned for a two year period, with possibility of extension of financial support for another two years. The application should describe how the collaboration will be organized, and the chronogram of activities for the first two years of collaboration. At the end of each year, the project's liaison to LARC must present a joint report of the collaborative activities that took place among the research groups during that period and an update of the chronogram for the next year. Approval by LARC IBRO is necessary for continuation of the support.
Geographic Focus: Puerto Rico, Antigua & Barbuda, Argentina, Bahamas, Barbados, Belize, Bolivia, Brazil, Chile, Colombia, Costa Rica, Cuba, Dominica, Dominican Republic, Ecuador, El Salvador, Grenada, Guatemala, Guyana, Haiti, Honduras, Jamaica, Mexico, Nicaragua, Panama, Paraguay, Peru, Saint Kitts And Nevis, St. Lucia, St. Vincent and the Grenadines, Suriname, Trinidad and Tobago, Uruguay, Venezuela
Date(s) Application is Due: Mar 31
Amount of Grant: 4,000 EUR
Contact: Dr. Dora Fix Ventura, Coordinator; 33 (0) 1 46 47 92 92; fax 33 (0) 1 46 47 42 50; dventura@usp.br or ibrocentral@gmail.com
Internet: http://ibro.info/professional-development/funding-programmes/regional-grants/latinamerica-grants/prolab-collaboration-in-latin-america/
Sponsor: International Brain Research Organization
255 rue Saint-Honore
Paris, 75001
France, Metropolitan

IBRO Latin America Regional Funding for Short Courses, Workshops, and Symposia   2107
IBRO's Latin America Regional Committee (LARC) will evaluate funding for basic neuroscience courses, workshops or symposia taking place in the region. Activities should foster cooperation among regional neuroscientists (students and researchers). Failure to incorporate participants from the region seriously forfeits chances for approval. Applications must be submitted through the IBRO online application system. Within 60 days after the activity, the coordinator must submit a report on the event and the use of IBRO funds. A summary of up to 200 words in English should also be submitted suitable for posting to the IBRO website. Grant awards range up to 3,500 Euro.
Restrictions: Applications by email will not be accepted.
Geographic Focus: Puerto Rico, Antigua & Barbuda, Argentina, Bahamas, Barbados, Bolivia, Brazil, Chile, Colombia, Costa Rica, Cuba, Dominica, Dominican Republic, Ecuador, El Salvador, Grenada, Guatemala, Guyana, Haiti, Honduras, Jamaica, Mexico, Nicaragua, Panama, Paraguay, Peru, Saint Kitts And Nevis, St. Lucia, St. Vincent and the Grenadines, Suriname, Trinidad and Tobago, Uruguay, Venezuela
Amount of Grant: Up to 3,500 EUR
Contact: Rebecca Hadid, Director of Professional Development Programs; 33 (0) 1 46 47 92 92; fax 33 (0) 1 46 47 42 50; robynn@ibro.info or ibrocentral@gmail.com
Internet: http://ibro.info/professional-development/funding-programmes/regional-grants/latinamerica-grants/#shortcourse
Sponsor: International Brain Research Organization
255 rue Saint-Honore
Paris, 75001
France, Metropolitan

IBRO Latin America Regional Funding for Short Research Stays   2108
The International Brain Research Organization's Latin America Regional Committee (LARC) evaluates applications for funding for short research/training lab visits within the Latin American region. Applications must be submitted through the IBRO online application process. Grant awards range up to a maximum of 1200 Euro. Applications should be submitted online.
Requirements: Candidates should not be older than 40 years of age. Priority will be given to doctoral students and postdocs who wish to attend a foreign lab within the region and who plan to link their lab stay with attendance at a course or congress. When the short lab stay is linked to attendance at a course or congress, an application for LARC Travel Grants must also be submitted. Applications for activities outside the region or within the applicant's own country will be considered only if adequately justified. Applicants must supply an activity plan, a letter of support from his or her director - specifically for the proposed activities - and a letter of acceptance from the host institution.
Restrictions: emailed applications will not be accepted.
Geographic Focus: Puerto Rico, Antigua & Barbuda, Argentina, Bahamas, Barbados, Belize, Bolivia, Brazil, Chile, Colombia, Costa Rica, Cuba, Dominica, Dominican Republic, Ecuador, El Salvador, Grenada, Guatemala, Guyana, Haiti, Honduras, Jamaica, Mexico, Nicaragua, Panama, Paraguay, Peru, Saint Kitts And Nevis, St. Lucia, St. Vincent and the Grenadines, Suriname, Trinidad and Tobago, Uruguay, Venezuela
Amount of Grant: Up to 1,200 EUR
Contact: Rebecca Hadid, Director of Professional Development Programs; 33 (0) 1 46 47 92 92; fax 33 (0) 1 46 47 42 50; ibrobecky@gmail.com or ibrocentral@gmail.com
Internet: http://ibro.info/professional-development/funding-programmes/regional-grants/latinamerica-grants/#shortstay
Sponsor: International Brain Research Organization
255 rue Saint-Honore
Paris, 75001
France, Metropolitan

IBRO Latin America Regional Travel Grants   2109
The International Brain Research Organization's Latin America Regional Committee (LARC) evaluates funding for attending courses/workshops or for presenting results at meetings within the Latin American region. Candidates should not be older than forty (40) years of age. Priority will be given to doctoral students and postdocs who are also planning a short stay in a lab. When the travel grant is linked to a short stay, an application for LARC Short Research Stay Grants must also be submitted. Applications must be submitted through IBRO's online application process. Awards range up to a maximum of 1,200 Euro.
Restrictions: emailed applications are not accepted. Applications for activities outside the region or within the applicant's own country will be considered only if adequately justified.
Geographic Focus: Puerto Rico, Antigua & Barbuda, Argentina, Bahamas, Barbados, Belize, Bolivia, Chile, Colombia, Costa Rica, Cuba, Dominica, Dominican Republic, Ecuador, El Salvador, Grenada, Guatemala, Guyana, Haiti, Honduras, Jamaica, Mexico, Nicaragua, Panama, Paraguay, Peru, Saint Kitts And Nevis, St. Lucia, St. Vincent and the Grenadines, Suriname, Trinidad and Tobago, Uruguay, Venezuela
Amount of Grant: Up to 1,200 EUR
Contact: Rebecca Hadid, Director of Professional Development Programs; 33 (0) 1 46 47 92 92; fax 33 (0) 1 46 47 42 50; ibrobecky@gmail.com or ibrocentral@gmail.com
Internet: http://ibro.info/professional-development/funding-programmes/regional-grants/latinamerica-grants/#travelgrant
Sponsor: International Brain Research Organization
255 rue Saint-Honore
Paris, 75001
France, Metropolitan

IBRO Regional Grants for International Fellowships to U.S. Laboratory Summer Neuroscience Courses   2110
The U.S.-Canada Regional Committee for IBRO (USCRC) provides fellowship support for students from developing nations who are accepted into the eligible high level neuroscience-related summer courses at the Marine Biological Laboratory in Woods Hole, Massachusetts and at the Cold Spring Harbor Laboratory in Cold Spring Harbor, New York. Substantial financial support is available to enable the participation of students from resource-restricted countries in these programs. The Fellowships are an effort to allow the most worthy alumni of IBRO Neuroscience Schools and other outstanding young scientists to gain outstanding experiences in leading training programs. These experiences will benefit scientists who will continue to study and work in less financially advantaged regions such as Latin America, Africa, Central and Eastern Europe, and Asia/Pacific. Candidates should apply directly to MBL or CSHL through their online application process. Each lab will inform candidates if they have been selected. Information about courses offered at each lab is available at the IBRO website.

Restrictions: Applicants should contact each lab directly about whether their chosen courses are eligible for the Fellowship.
Geographic Focus: All Countries
Date(s) Application is Due: Feb 1
Contact: Rebecca Hadid, Director of Professional Development Programs; 33 (0) 1 46 47 92 92; fax 33 (0) 1 46 47 42 50; ibrobecky@gmail.com or ibrocentral@gmail.com
Dana Mock-Muñoz de Luna, Research Award Coordinator; (508) 289-7173 or (508) 548-3705; dmock@mbl.edu or researchprograms@mbl.edu
Internet: http://ibro.info/professional-development/funding-programmes/regional-grants/uscanada-grants/
Sponsor: International Brain Research Organization
255 rue Saint-Honore
Paris, 75001
France, Metropolitan

IBRO Research Fellowships 2111
The IBRO Funding Program aims to foster neuroscience, especially in less well-funded countries, by providing support to high quality neuroscientists from diverse geographic and scientific areas (U.S./Canada region excluded) who wish to broaden the scope of their training in neuroscience by working abroad. IBRO Research Fellowships, which include the John G. Nicholls and Rita Levi-Montalcini Fellowships, support post-doctoral laboratory training to applicants under the age of 45 for up to one year overseas. The funding for a 12-month Fellowship is 35,000 euros. Priority is given to those who have not obtained funding from this programme within the past three years and who, after completion of the Fellowship, are willing to return to their home countries, bringing with them new knowledge and skills to advance neuroscience in their regions.
Requirements: Applicants must meet the following *Requirements:* submit a letter of acceptance from the proposed supervisor (host scientist) indicating that he/she will accept the candidate and agrees on the proposed project; a letter of reference from the present supervisor; and, if available, evidence of an offer that the applicant will have a position to return to his/her home country after the Fellowship. Only applications completed through IBRO's online system will be considered.
Geographic Focus: All Countries
Date(s) Application is Due: Jun 10
Amount of Grant: 35,000 EUR
Contact: Marta Hallak, Senior Director of Grants; 33 (0) 1 46 47 92 92; fax 33 (0) 1 46 47 42 50; mhallak@mail.fcq.unc.edu.ar
Internet: http://ibro.info/professional-development/funding-programmes/research-fellowships/
Sponsor: International Brain Research Organization
255 rue Saint-Honore
Paris, 75001
France, Metropolitan

IBRO Return Home Fellowships 2112
The IBRO Fellowships Program aims to foster neuroscience research, especially in less well-funded countries, by providing support to high quality neuroscientists from diverse geographic and scientific areas (US/Canada Region excluded) who wish to broaden the scope of their laboratory training in neuroscience by working abroad. Priority is given to those who have not obtained funding from this program within the past three years and who, after completion of the fellowship, are willing to return to their home countries, bringing with them new knowledge and skills to advance neuroscience in their regions. The objective of the Return Home Program of IBRO is to coordinate efforts with other organizations to both improve the opportunities for productive neuroscience research within the less advantaged regions of the world and to provide more aid to those researchers trained overseas who wish return to their home countries. Three profiles of scientists will be the main target of this Fellowship: postdoctoral Fellows who have finished research training in neurosciences (including clinical research) in a center of excellence of a developed country; research students who have been trained in a Centre of Excellence in Brain Research (CEBR); and scientists who are developing a successful basic/clinical research career in a developed country and wish to return to their country of origin or to a less developed country for personal or cultural reasons. Only applications completed through the IBRO application system will be considered. The grant amount is up to 20,000 Euros, and the annual deadline for application is September 1.
Requirements: Eligibility requirements include: postdoctoral fellows who have finished research training in neuroscience (including clinical research) in a center of excellence of a developed country; research students who have been trained in a Center of Excellence in Brain Research (CEBR); and scientists who are developing a successful basic/clinical research career in a developed country and wish to return to their country of origin or to a less developed country for personal or cultural reasons.
Geographic Focus: All Countries
Date(s) Application is Due: Sep 1
Amount of Grant: Up to 20,000 EUR
Contact: Marta Hallak, Senior Director of Grants; 33 (0) 1 46 47 92 92; fax 33 (0) 1 46 47 42 50; mhallak@mail.fcq.unc.edu.ar
Internet: http://ibro.info/professional-development/funding-programmes/return-home-fellowships/
Sponsor: International Brain Research Organization
255 rue Saint-Honore
Paris, 75001
France, Metropolitan

Ida Alice Ryan Charitable Trust Grants 2113
The trust awards single-year grants for capital projects to eligible Georgia nonprofit organizations. Grants support programs of the Catholic Church and a wide variety of community causes including health organizations, legal aid societies, youth groups, educational programs, museums, and social services. Application and guidelines are available online.
Requirements: 501(c)3 tax-exempt organizations in the Atlanta area are eligible.
Geographic Focus: Georgia
Date(s) Application is Due: Feb 1; Aug 1
Contact: NA Trustee, c/o Wachovia Bank; grantinquiries8@wachovia.com
Internet: https://www.wachovia.com/foundation/v/index.jsp?vgnextoid=cf578689fb0aa110VgnVCM1000004b0d1872RCRD&vgnextfmt=default
Sponsor: Ida Alice Ryan Charitable Trust
3280 Peachtree Road NE, Suite 400, MC G0141-041
Atlanta, GA 30305

Idaho Community Foundation Eastern Region Competitive Grants 2114
The mission of the Foundation is to enrich life's quality throughout Idaho. Grants are awarded through the regional grant cycle for a wide range of organizations and for a wide range of projects consistent with that mission. Grants are made to fund activities, services, and projects of established organizations, as well as to provide assistance for new organizations to fill unmet and/or emerging community needs. Grant areas include: arts and culture; education; emergency services; libraries; natural sciences; health; recreation; social services; and public projects. Application information is available online.
Requirements: Nonprofit entities in Idaho are eligible to apply.
Restrictions: Funding is not available for: projects which replace school district responsibilities to students or that fund state or federally mandated programs; projects which are considered operating expenses or salaries normally paid by a school district; computer hardware used solely for pre-K through 12th grade educational purposes; religious organizations for the sole purpose of furthering that religion; political activities or those designed to influence legislation; national organizations; or grants that directly benefit a donor to ICF or a donor's family.
Geographic Focus: Idaho
Date(s) Application is Due: Apr 1
Amount of Grant: 5,000 USD
Contact: Administrator; (208) 342-3535; fax (208) 342-3577; grants@idcomfdn.org
Internet: http://www.idcomfdn.org/pages/grant_regional_guidelines.htm
Sponsor: Idaho Community Foundation
P.O. Box 8143
Boise, ID 83707

Idaho Power Company Corporate Contributions 2115
The company awards grants in company operating territories in its areas of interest, including arts and culture, civic and community, education, and health and human services. Types of support include building construction/renovation, capital campaigns, continuing support, equipment, general operating support, matching gifts, scholarships, and sponsorships. The corporate contribution request form is available online. There are no application deadlines.
Requirements: Nonprofits in southern Idaho and eastern Oregon are eligible.
Restrictions: Grants do not support: individuals; loans or investments; churches or religious organizations for purposes of religious advocacy; tickets for contests, raffles, or other prize-oriented activities; organizations that discriminate for any reason, including race, color, religion, creed, age, sex, or national origin; individual school programs or projects with limited participation; fraternal or labor organizations; unrestricted operating funds; special occasion good-will advertising.
Geographic Focus: Idaho, Oregon
Contact: Contribution Program Manager; (208) 388-2200
Internet: http://www.idahopower.com/aboutus/community/corporateContributions.htm
Sponsor: Idaho Power Company
P.O. Box 70
Boise, ID 83707

Ida S. Barter Trust Grants 2116
The Ida S. Barter Trust was established in 1953 to support and promote quality educational, human services, and health care programming for underserved populations. Grant requests for general operating support are strongly encouraged. Program support will also be considered. Small, program-related capital expenses may be included in general operating or program requests. The application deadline is April 1, and applicants will be notified of grant decisions before June 30. The majority of grants from the Barter Trust are 1 year in duration. On occasion, multi-year support is awarded.
Geographic Focus: Massachusetts
Date(s) Application is Due: Apr 1
Contact: Michealle Larkins; (866) 778-6859; michealle.larkins@baml.com
Internet: https://www.bankofamerica.com/philanthropic/fn_search.action
Sponsor: Ida S. Barter Trust
225 Franklin Street, 4th Floor, MA1-225-04-02
Boston, MA 02110

IDPH Carolyn Adams Ticket for the Cure Community Grants 2117
On July 6, 2005, PA 92-0120 was signed into law, creating the Illinois Carolyn Adams Ticket for the Cure Lottery instant ticket. Net revenue from the sale of this ticket will go to the Illinois Department of Public Health (IDPH), Office of Women's Health, which will award grants to public and private entities in Illinois for the purpose of funding breast cancer research, and supportive services for breast cancer survivors and those

impacted by breast cancer and breast cancer education. The OWH and the Ticket for the Cure Advisory Board recognize that breast cancer is the most commonly diagnosed cancer in women and sometimes affects men, as well. Awareness and education regarding early detection needs to be increased in every community, especially for low income, underserved and uninsured women with special emphasis on reaching those who are geographically or culturally isolated, older and/or members of racial/ethnic minorities. The Community Grant Program is designed to address this need.
Geographic Focus: Illinois
Contact: Coordinator; 217-524-6088; fax 217-557-3326; dph.owhline@illinois.gov
Internet: http://www.idph.state.il.us/about/womenshealth/grants/tfc.htm
Sponsor: Illinois Department of Public Health
535 W. Jefferson Street, First Floor
Springfield, IL 62761-0001

IDPH Emergency Medical Services Assistance Fund Grants 2118
This project provides for distribution of moneys in the EMS Assistance Fund to each of the eleven Regions in the State in accordance with protocols established in each Region's EMS Region Plan. Objectives of this grant will be to purchase any equipment requested and complete any education as requested. Expected outcomes and goals of this grant are to help improve EMS services and increase education to EMS personal. Measurements and outcomes of this grant will be met by showing timely response to EMS calls, proper use of equipment, education of new equipment, educational objectives completed. Funds might not be equally divided among the eleven regions; consequently, award decisions will not be made based on financial parity among regions.
Requirements: Any Illinois licensed/designated EMS participant that provides EMS service within the State of Illinois may apply for funds through their Regional EMS Advisory Committee. Programs, services, and equipment funded by the EMS Assistance Fund must comply with the Emergency Medical Services (EMS) Systems Act and the Regional EMS Plan in which the applicant participates. The grant cycle runs from July 1-June 30 of each year. All funds remaining at the end of the period of time grant funds are available for expenditure (June 30 of the fiscal year the grant was awarded) shall be returned to the State within 45 days. All applications from providers must be submitted to their respective Regional EMS Advisory Committee by the deadline required by each Regional Committee. No applications will be accepted by the Department directly from an applicant.
Restrictions: Due to limited amount of grant funds available, the Department will not consider applications for new vehicles, vehicle re-chassis, building projects or grant requests over $5000.
Geographic Focus: Illinois
Date(s) Application is Due: Jan 29
Amount of Grant: Up to 5,000 USD
Contact: Mark Vassmer, PHEP Grant Coordinator; 217-558-0560
Internet: http://www.idph.state.il.us/fundop.htm
Sponsor: Illinois Department of Public Health
525-535 W. Jefferson Street
Springfield, IL 62761-0001

IDPH Hosptial Capital Investment Grants 2119
The program allows qualifying hospitals to apply for grants to fund projects to improve or renovate a hospital's physical plant, or to improve, replace or acquire equipment or technology. Projects can include activities to satisfy building code, safety standards, or life safety code; maintain, improve, renovate, expand, or construct buildings or structures; maintain, establish, or improve health information technology; or maintain or improve safety, quality of care, or access to care. Two types of awards are available: Safety Net Grants ($4,600,000 to $7,000,000) and Community Hospital Grants (approximately $350,000 to $1,000,000).
Requirements: Prior to submitting a grant application, hospitals need to submit a letter of intent to the Department which must be received at least 10 calendar days prior to the submission of a grant application. The letter will help determine if a hospital qualifies to apply and will notify the Department of an impending application. The letter must contain the following information: name of the applicant; name of the hospital where grant funds will be used; site of the proposed project, including the address of the hospital where grant funds will be used; county where the hospital is located; description of the project; hospital's Medicaid inpatient utilization rate for the rate year beginning October 1, 2008; signature and contact information of an authorized official from the hospital; and, information on whether the project requires a CON or COE from the Health Facilities and Services Review Board. (The CON/COE Assessment of Applicability Internet site can assist in this determination: http://www.hfsrb.illinois.gov/pdf/checklist-revised.doc.) A hospital that applies for this grant shall be licensed by the Illinois Department of Public Health in accordance with the Hospital Licensing Act. The license shall be valid and the hospital shall be in operation when the grant application is submitted, when the grant agreement is executed and when the project is complete. Applications must be received by the Department within 120 calendar days of the date of the notice's publication. Applications received after this deadline will not be accepted.
Geographic Focus: Illinois
Date(s) Application is Due: Nov 12
Amount of Grant: 350,000 - 7,000,000 USD
Contact: Grant Coordinator; 217-782-1624; dph.crh@illinois.gov
Internet: http://www.idph.state.il.us/fundop.htm
Sponsor: Illinois Department of Public Health
535 West Jefferson Street, Ground Floor
Springfield, IL 62761-0001

IDPH Local Health Department Public Health Emergency Response Grants 2120
The Illinois Department of Public Health, Office of Preparedness and Response, Division of Disaster Planning and Readiness, is making available approximately $5 million in unspent CD.C. Public Health Emergency Response (PHER) grant funding to Illinois local health departments. The purpose of the funding is to continue previously approved PHER-funded activities, or conduct new activities that address or retest identified gaps in local health department pandemic flu response. Many local health departments have indicated that they could not use the original PHER funding for CD.C.'s intended purpose. Therefore, grants will be awarded on a first-come, first-served basis to applicants that can best justify activities that will advance pandemic planning and preparedness based on the Grant Application Review Criteria. IDPH anticipates making about 20 to 40 awards based on the number and type of applicants. The maximum award for either single or multi-jurisdictional will be $300,000, or $1.70 per capita, per participating local health department, whichever is less.
Requirements: Applicant must be a single Illinois Certified Local Health Department. Regional partnerships may select either one certified local health department or an agent to receive the funding from IDPH and issue subgrants to other participating certified local health departments. Application must clearly describe reasonable and significant programmatic and financial participation by all partners. Funds may be used for the continuation of previously approved PHER-funded activities, or new activities that address or retest identified gaps in pandemic flu response, or new activities that directly advance pandemic planning and preparedness.
Geographic Focus: Illinois
Date(s) Application is Due: Jan 31
Amount of Grant: Up to 300,000 USD
Contact: Mark Vassmer, PHEP Grant Coordinator; 217-558-0560
Internet: http://www.idph.state.il.us/fundop.htm
Sponsor: Illinois Department of Public Health
525-535 W. Jefferson Street
Springfield, IL 62761-0001

IHI Quality Improvement Fellowships 2121
The Institute for Healthcare Improvement (IHI), an independent not-for-profit organization based in Cambridge, Massachusetts, is an innovator in health and health care improvement worldwide. The IHI Fellowship Program is a year-long immersion program for mid-career health professionals. The program has two overarching goals, to: (1) develop health care leaders with the drive, skills, and experience to spread improvement in the United States and globally; and (2) build capability within health care organizations to reach dramatically higher levels of performance. Fellows spend one year at IHI's office in Cambridge, Massachusetts, and return to their home organization to lead transformative change. Fellowships are one year in duration, July through June.
Requirements: The IHI Quality Improvement Fellowship supports fellows who live and work in the United States. The fellowship is designed for mid-careers professionals who will return to senior leadership positions in their home organizations. IHI will, however, consider a very strong application from a candidate earlier in his/her career. All successful candidates must have full-time employment with an organization who will sponsor them for the fellowship (salary and benefits for the year are paid by the home organization) and who shows commitment to leveraging the fellow's learning appropriately post-fellowship. Applicants are not required to have a M.D; IHI encourages applications from nurse leaders, quality improvement professionals, and allied health professionals.
Geographic Focus: All States
Contact: Joelle Baehrend, (617) 301-4816; fax (617) 301-4848; jbaehrend@ihi.org
Internet: http://www.ihi.org/offerings/Training/Fellowships/Pages/default.aspx
Sponsor: Institute for Healthcare Improvement
20 University Road, 7th Floor
Cambridge, MA 02138

IIE 911 Armed Forces Scholarships 2122
The 911 Armed Forces Scholarships provide scholarships for the dependent children of active duty U.S. military personnel who were killed in the September 11, 2001 terrorist attacks. Scholarships are one-time awards for study leading to a certificate, associate's or bachelor's degree in the United States or its equivalent at an accredited post-secondary institution in any country.
Requirements: Dependent children pursuing a certificate or undergraduate degree programs in a college or university in the U.S. or abroad are eligible to apply.
Restrictions: Awards will not be provided for study leading to degrees higher than a bachelor's degree.
Geographic Focus: All States
Contact: Jonah Kokodyniak, Deputy Vice President, Strategic Development; (212) 883-8200; fax (212) 984-5401; development@iie.org
Internet: http://www.iie.org/Programs/911-Armed-Forces-Scholarship-Fund
Sponsor: Institute of International Education
1400 K Street, NW, 7th Floor
Washington, D.C. 20005-2403

IIE Adell and Hancock Scholarships 2123
The Institute of International Education (IIE) sponsors the Adell and Hancock Scholarship. The purpose of the Adell & Hancock Fund is to provide supplemental support to U.S. and international students who are in need of additional funds to carry out their international educational plans. The amount awarded will depend on individual need as determined by IIE/RMRC scholarship committee to not exceed $2,400 per candidate. Applicants should refer to the IIE website for further information and current application information.
Geographic Focus: All States, All Countries

Date(s) Application is Due: Oct 15
Amount of Grant: Up to 2,400 USD
Contact: Lauren Granstrom, Administrative Assistant; (303) 837-0788, ext. 27 or (303) 837-1409; rockymountainscholarships@iie.org or lgranstrom@iie.org
Internet: http://www.rockymountainiie.org/scholarships
Sponsor: Institute of International Education
1400 K Street, NW, 7th Floor
Washington, D.C. 20005-2403

IIE African Center of Excellence for Women's Leadership Grants 2124
The African Center of Excellence for Women's Leadership (ACE) aims to develop the capabilities of four women-led organizations in East Africa to be centers of excellence in the delivery of leadership development programs, research and knowledge generation on reproductive health, economic empowerment and girls' education. The ACE for Women's Leadership Program responds to an articulated need by women across the continent for a strategic leadership development program and for a stronger and more coordinated effort to help women advocate for issues that most concern them. The four organizations are intended to serve as places where women can come together for leadership training, strategizing, researching and sharing of best practices across the region.
Requirements: The program is based in four countries: Ethiopia, Uganda, Kenya and Rwanda.
Geographic Focus: Ethiopia, Kenya, Rwanda, Uganda
Contact: Melat Tekletsadik, Program Director; +251 (0) 91 260 9805; mhaile@iie.org
Internet: http://www.iie.org/Programs/ACE-for-Womens-Leadership
Sponsor: Institute of International Education
P.O. Box 586
Addis Ababa, Code 1110 Ethiopia

IIE AmCham Charitable Foundation U.S. Studies Scholarship 2125
The Institute of International Education sponsors the AmCham Charitable Foundation U.S. Studies Scholarship. The scholarship funds undergraduate study at a U.S. university for high school students from Hong Kong.
Requirements: Applicants must be permanent residents of Hong Kong; they must have received at least five years of continuous education in a Hong Kong high school where they are currently enrolled; and they must have gained admission to an undergraduate degree program in a U.S. college or university for the coming year. Current applications and deadlines are found on the website.
Restrictions: Only current residents of Hong Kong may apply.
Geographic Focus: Hong Kong
Date(s) Application is Due: Apr 15
Amount of Grant: 16,000 HKD
Contact: Linda Pham, Scholarship Contact; (852) 2530-6917; fax (852) 2810-1289; lpham@amcham.org.hk or amcham@amcham.org.hk
Internet: http://www.amcham.org.hk/index.php/U.S.-Studies-Scholarship.html
Sponsor: Institute of International Education
1400 K Street, NW, 7th Floor
Washington, D.C. 20005-2403

IIE Brazil Science Without Borders Undergraduate Scholarships 2126
The Brazilian government's new Science Without Borders Program will provide scholarships to undergraduate students from Brazil for one year of study at colleges and universities in the United States. This program, administered by IIE, is part of the Brazilian government's larger initiative to grant 100,000 scholarships for the best students from Brazil to study abroad at the world's best universities.
Requirements: Applicants should meet the following criteria for eligibility: they should be citizens of Brazil, and reside in the home country at the time of application; they must have completed four semesters of their bachelor's degree; they must be in the first quarter of their graduation class, have superior academic abilities, well-rounded personalities, and participate in extracurricular activities; their transcripts should show a consistently high level of performance; and applicants should submit their TOEFL score report, and present official and translated transcripts from an institution recognized by the Ministry of Higher Education.
Restrictions: Scholarships will be given primarily to students in the Science, Technology, Engineering and Mathematics (STEM) fields. Students in the program will return to Brazil to complete their degrees. Candidates with U.S. or dual citizenship of the United States and Brazil are not eligible for this program.
Geographic Focus: Brazil
Contact: Edward Monks, Enrichment and Professional Development Director; (212) 984-5335; fax (212) 984-5578; undergraduatprogram@iie.org
Internet: http://www.iie.org/Programs/Brazil-Science-Without-Borders
Sponsor: Institute of International Education
1400 K Street, NW, 7th Floor
Washington, D.C. 20005-2403

IIE Central Europe Summer Research Institute Summer Research Fellowship 2127
The Central Europe Summer Research Institute (CESRI) is a fellowship opportunity for U.S. graduate students in science and engineering. With support from the National Science Foundation (NSF) and the German Academic Exchange Service (DAAD), CESRI will provide eight U.S. graduate students with a high-quality international research experience in Austria, the Czech Republic, Germany, Hungary, Poland or Slovakia. Two CESRI grants are specifically for CESRI Fellowships to Germany. The remaining six CESRI grants are funded by NSF for the entire Central Europe region, with preference for grants to countries other than Germany. The program is intended for scientific research projects. The CESRI program will be eight weeks in length, with the first four days spent as a group in Budapest, Hungary in a specially designed cultural and academic orientation to the region. Participants will spend the remaining 7.5 weeks working in individually-arranged placements in university labs or other appropriate sites where they can participate in creative research activities under the supervision of European mentors. In addition to the scholarship, recipients receive round trip airfare to Central Europe, room and board for 8 weeks, health and accident insurance, and a four day academic and cultural orientation in Budapest.
Requirements: All applicants must be U.S. citizens or permanent residents in the U.S., and be current Master's or Ph.D. students at a U.S. university in one of the following fields: Biology, Chemistry, Computer Science, Engineering, Environmental Science, or Mathematics. Candidates should submit the online application, along with the online self-evaluation form and two completed professional reference forms.
Restrictions: Projects involving field work, policy, ethics, or clinical research on human objects are not eligible for consideration.
Geographic Focus: All States
Date(s) Application is Due: Feb 1
Amount of Grant: 2,000 USD
Contact: Fellowship Contact; (212) 883-8200; fax (212) 984-5452; cesri@iie.org
Internet: http://www.iie.org/Programs/CESRI
Sponsor: Institute of International Education
1400 K Street, NW, 7th Floor
Washington, D.C. 20005-2403

IIE Chevron International REACH Scholarships 2128
The Chevron International REACH (Recognizing Excellence and Achievement) Scholarship Program is sponsored by Chevron for the sons and daughters of their employees and retirees. The program was established to recognize and assist outstanding children who plan to pursue post-secondary education. Renewable scholarships are offered each year for full-time undergraduate study at colleges, universities, and vocational schools. Four-year scholarships, ranging in value from $500 to $2,500, and honorariums, ranging from $1,000 to $1,500, will be offered to children of Chevron employees. The scholarships for children of non-U.S.-payroll employees are managed by the Institute of International Education West Coast Center in San Francisco. Application forms are available through the applicant's local Chevron human resources office.
Requirements: Applicants for the scholarship must be children of non-U.S. payroll active employees of Chevron and its wholly owned subsidiaries; children of non-U.S. payroll retirees with a minimum of 75 points as calculated by Human Resources or 20 years of service prior to retirement; the retired parent must not be an employee of competitive energy company; and the applicant must be currently enrolled in the final year of high school and planning to attend a college, university, or vocational-technical school as a full time student.
Restrictions: Students planning pre-university study are not eligible. Military academies are not eligible. Scholarships are intended to assist students as they pursue undergraduate study and are not intended to cover all educational expenses. Recipients are responsible for the balance of funding, including tuition and fees, books and supplies, living expenses, and transportation.
Geographic Focus: Angola, Argentina, Australia, Azerbaijan, Bangladesh, Brazil, Cambodia, Canada, Chad, China, Colombia, Indonesia, Kazakhstan, Kuwait, Netherlands, New Zealand, Nigeria, Philippines, Russia, Saudi Arabia, Singapore, South Africa, South Korea, Thailand, Trinidad and Tobago, United Kingdom, Venezuela, Vietnam
Amount of Grant: 500 - 2,500 USD
Contact: Naoko Dunnigan; (415) 362-6520, ext. 201; Chevron@iie.org
Internet: http://www.iie.org/Programs/Chevron-International-REACH-Scholarship-Program
Sponsor: Institute of International Education
1400 K Street, NW, 7th Floor
Washington, D.C. 20005-2403

IIE David L. Boren Fellowships 2129
Boren Fellowships provide up to $30,000 funding to U.S. graduate students to add an international and language component to their graduate education through specialization in area study, language study, or increased language proficiency. Boren Fellowships support study and research in areas of the world that are critical to U.S. interests, including Africa, Asia, Central & Eastern Europe, Eurasia, Latin America, and the Middle East. Boren Fellowships are funded by the National Security Education Program (NSEP), which focuses on geographic areas, languages, and fields of study that are critical to U.S. national security. Applicants should identify how their projects, as well as their future academic and career goals, will contribute to U.S. national security.
Requirements: Boren Fellowships promote long term linguistic and cultural immersion. Therefore, applicant preference will be given to those proposing overseas programs of 6 months or longer. However, applicants proposing overseas programs of 3-6 months, especially those in the STEM (science, technology, engineering, and mathematics) fields are encouraged to apply. Boren Fellowships are awarded with preference for countries, languages, and fields of study critical to U.S. national security. Preference is also given to students who will study abroad for longer periods of time, and who are highly motivated by the opportunity to work in the federal government. Applicants should refer to the website for current deadlines and detailed criteria for application.
Restrictions: Applicants must commit to a length of study of at last 12 weeks.
Geographic Focus: All States
Date(s) Application is Due: Jan 31
Amount of Grant: Up to 30,000 USD
Contact: Michael Saffle; (800) 618-6737; fax (202) 326-7672; boren@iie.org
Internet: http://www.borenawards.org/boren_fellowship
Sponsor: Institute of International Education
1400 K Street, NW, 7th Floor
Washington, D.C. 20005-2403

IIE Eurobank EFG Scholarships 2130
The Eurobank EFG Scholarship Program aims to identify and honor the best undergraduate students of Serbian state universities. An independent selection committee of experts evaluate all applications. Semi-finalists are invited to a personal interview conducted in English. The winners are publicly recognized at an award ceremony, with their names published in daily newspapers.
Requirements: The ideal candidate has high academic standing, leadership potential, and an interest in community development. Participation in extracurricular activities, good communication skills, and proficiency in foreign languages are also essential qualities. All applicants must have: a grade point average higher than 9.5; extracurricular activities; a demonstrated interest in community development with leadership potential; good communication skills; and proficiency in foreign languages.
Geographic Focus: Serbia
Amount of Grant: 1,000 EUR
Contact: Eurobank EFG Scholarship Team; (+36-1) 472-2283; fax (+36-1) 472-2255; eurobankefg-scholarship@iie.eu
Internet: http://www.iie.org/en/Programs/Eurobank-EFG-Scholarships
Sponsor: Institute of International Education
1400 K Street, NW, 7th Floor
Washington, D.C. 20005-2403

IIE Freeman Foundation Indonesia Internships 2131
The Freeman Indonesia Nonprofit Internship (FINIP) addresses the limited knowledge among Indonesian students about the nonprofit sector. It also addresses the limited understanding and exposure of U.S. students to Indonesia, despite the increasing importance of the country globally. The Freeman Indonesia Nonprofit Internship Program is an opportunity to strengthen the leadership of Indonesia's future nonprofits and deepen ties between America and Indonesia. The shared experience of collaborating to assist a local Indonesian nonprofit is a learning experience and a vehicle for cross-cultural understanding, building ongoing friendships and shared interests in the nonprofit sector in the U.S. and Indonesia.
Requirements: Applicants must meet all of the following eligibility *Requirements:* they must be a U.S. or Indonesian citizen; currently enrolled as a full time sophomore or junior pursuing their first bachelor's degree at a U.S. university; and be in good academic standing at their university. Potential applicants should refer to the website for specific instructions about the application process.
Geographic Focus: All States, Indonesia
Date(s) Application is Due: Mar 1
Contact: FINIP Contact; (212) 984-5542; fax (212) 984-5325; finip@iie.org
Internet: http://www.iie.org/Programs/FINIP
Sponsor: Institute of International Education
1400 K Street, NW, 7th Floor
Washington, D.C. 20005-2403

IIE Hewlett Foundation/IIE Dissertation Fellowship 2132
The Hewlett/IIE Dissertation Fellowship in Population, Reproductive Health, and Economic Development provides financial and research development support for dissertations on topics that examine how population dynamics, family planning, and reproductive health influence economic development. This can include economic growth, poverty reduction, and equity. Dissertations that address population and development issues pertinent to the African continent are especially encouraged. Fellowship recipients are awarded up to $20,000 per year (depending on tuition, research expenses, and cost of living) for a total of two years to cover expenses incurred while working on their dissertation. These expenses must be clearly specified on the budget component of the fellowship application. In addition to financial support, Fellows will actively engage with a network of researchers supported by the Hewlett Foundation, Population Reference Bureau, the Institute of International Education, and other funders. Network activities during the two-year fellowship can include an annual research conference, workshops on advanced methods in population-economic analysis, and writing workshops.
Requirements: The fellowship is intended for doctoral students enrolled in economics, economic demography, geography, and epidemiology, whose dissertation addresses population and development issues. All applicants are required to submit a hard copy and electronic application, an application cover sheet, vitae, statement of intent, budget plan, dissertation schedule, personal essay, and two recommendation letters. Students should refer to the website for specific information regarding the requirements and current due dates.
Restrictions: Students must have completed their graduate coursework by the start of the fellowship, and must be studying at a university in sub-Saharan Africa, the US, or Canada.
Geographic Focus: All States, Angola, Benin, Botswana, Burkina Faso, Cameroon, Canada, Cape Verde, Central African Republic, Chad, Comoros, Congo, Congo, Democratic Republic of, Cote d' Ivoire (Ivory Coast), Djibouti, Equatorial Guinea, Eritrea, Ethiopia, Gabon, Gambia, Ghana, Guinea, Guinea-Bissau, Kenya, Lesotho, Liberia, Madagascar, Malawi, Mali, Mauritania, Mauritius, Mozambique, Namibia, Niger, Nigeria, Reunion, Rwanda, Sao Tome & Principe, Senegal, Seychelles, Sierra Leone, Somalia, South Africa, Sudan, Swaziland, Tanzania, Togo, Uganda, Western Sahara, Zambia, Zimbabwe
Amount of Grant: Up to 20,000 USD
Contact: Nancy Scally; (212) 984-5342; popecondissfellows@iie.org
Internet: http://www.iie.org/Programs/Hewlett-IIE-Fellowship
Sponsor: Institute of International Education
1400 K Street, NW, 7th Floor
Washington, D.C. 20005-2403

IIE Iraq Scholars and Leaders Scholarships 2133
The scholarship is available to Iraqi citizens residing in Iraq. The program supports Bachelor's, Master's, or Doctoral levels of study at a U.S. university or college in one of the following fields: business, economics, education administration, engineering, geology, geophysics, law, public health, or public policy. Preference will be given to candidates pursuing studies in business, engineering or geosciences. Selected scholars will receive the following benefits: full tuition and placement assistance to attend a U.S. college or university; visa and academic support; transportation to the U.S. at the beginning of the program and return to Iraq at the end of the program; an extensive pre-academic program and medical insurance while in the U.S. The application deadline is November 1, with study beginning in the U.S. the following June.
Requirements: Applicants should carefully review the requirements for the scholarship. Candidates should submit a completed nine page application, TOEFL scores, resume, transcript, statement of purpose, field of study essays, and letters of recommendation.
Restrictions: Scholarship policy does not allow for a change in field of study once it has been submitted.
Geographic Focus: Iraq
Date(s) Application is Due: Nov 1
Contact: IScholarship; (713) 621-6300, ext. 32; fax (713) 621-0876; islp@iie.org
Internet: http://www.iie.org/Programs/ISLP
Sponsor: Institute of International Education
1800 West Loop South, Suite 250
Houston, TX 77027

IIE KAUST Graduate Fellowships 2134
The King Abdullah University of Science and Technology (KAUST) engages students, faculty, and researchers in advancing science and technology through collaborative inquiry focused on issues of regional and global significance. Their studies focus on the continued advancement of the global quality of life in all issues associated with energy, water, and food. They also explore environment and the Red Sea, and computational science and engineering (high-performance computing, computing for a small planet) as an enabling technology for their research activities. A KAUST Fellowship includes full tuition, a monthly stipend, housing and a travel benefit.
Requirements: Eligibly fields of study include: applied mathematics and computational science; bioscience; chemical and biological engineering; chemical science; computer science; earth science and engineering; electrical engineering; environmental science and engineering; marine science and engineering; materials science and engineering; and mechanical engineering.To qualify for the KAUST fellowship, applicants must meet the following requirements before completing the online application: a bachelor's degree in a KAUST-relevant field of study, pursuing a graduate degree in a similar field; minimum GPA of 3.5; a TOEFL score of 79 or 6.0 on the IELTS (International English Language Testing System; see website for specific requirements where TOEFL is not required); must submit GRE scores; a statement of purpose; three letters of recommendation; and an official university transcript.
Geographic Focus: All States, All Countries
Date(s) Application is Due: Jan 15
Contact: KAUST Fellowship Contact; +966 (0) 2 803 3428
Internet: http://www.kaust.edu.sa/admissions/tokaust/fellowship.html
Sponsor: King Abdullah University of Science and Technology
Graduate Affairs, Building 9, Suite 4328
Thuwal, Jeddah, 23955-6900 Saudi Arabia

IIE Klein Family Scholarship 2135
The Klein Family Scholarship is available to high school students from Hungary or another central or eastern European countries to pursue their four-year Bachelor's degree at The University of the South, located in the United States in Sewanee, Tennessee. The purpose of the scholarship is to provide full financial support for students to accomplish their educational goals and develop their potential. The Klein Family Scholarship will provide a full, four-year scholarship to one of these students.
Requirements: In order to qualify for the scholarship, candidates must be citizens of Hungary or of another central or eastern European country; be in their last year of high school and be able to pursue a four-year degree at Sewanee the following year; and have high academic achievements and leadership potential. The Institute of International Education's European Office (IIE) is managing the prescreening process for the Klein Family Scholarship. IIE will preselect 4 to 5 promising candidates, who will then be invited to apply to Sewanee directly. The ultimate scholarship winner will be selected by Sewanee's Admissions Office, as part of the college's regular annual admissions process.
Restrictions: Candidates must have a high level of English language skills adequate for study at a U.S. college.
Geographic Focus: Hungary
Date(s) Application is Due: Dec 15
Contact: Nora Nemeth; (+36) 1 472-2297; fax (+36) 1 472-2294; nnemeth@iie.eu
Internet: http://www.iie.org/Programs/Klein-Family-Scholarship
Sponsor: Institute of International Education
1400 K Street, NW, 7th Floor
Washington, D.C. 20005-2403

IIE Leonora Lindsley Memorial Fellowships 2136
The Lindsley Scholarship was established for Leonora Lindsley, an American member of an international all-female combat unit on the Western Front during World War II, who was killed the day before the war officially ended. The scholarship was created in her honor by French citizens who are descendants from Resistance fighters to fund those pursuing educational studies in the United States. An award is granted to one or two students for graduate study for a twelve month period, and is renewable for a second year.

The winner(s) are chosen every other year. The amount of the award varies from year to year, but ranges from $10,000 to $12,000.
Requirements: The Fulbright Commission in Paris gathers applications and sends them to IIE where they are reviewed for completeness and eligibility. Once this is done, a selection panel is convened and finalists chosen. Any major is eligible.
Restrictions: Only descendants of French Resistance fighters are eligible. Grantees must be at an accredited U.S. university.
Geographic Focus: All States, France
Amount of Grant: Up to 12,000 USD
Contact: Leonara Lindsley Fellowship Contact; (212) 984-5552
Internet: http://www.iie.org/Programs/Leonora-Lindsley-Memorial-Fellowship
Sponsor: Institute of International Education
1400 K Street, NW, 7th Floor
Washington, D.C. 20005-2403

IIE Lingnan Foundation W.T. Chan Fellowship 2137
The W.T. Chan Fellowships commemorate Professor Wing-Tsit Chan, former Dean of Lingnan University and distinguished Professor of Chinese Philosophy and Religion. The Fellowships extend the Foundation's commitment to higher education, increased international understanding, and personal growth. In addition, the fellowships address and explore the Lingnan motto "Education for Service." It is a step in a movement towards strengthening linkages between academia and communities.
Requirements: The fellowships are available to upper-level undergraduates and postgraduates of Sun Yat-sen University in Guangzhou and Lingnan University in Hong Kong. Applicants are encouraged to view the Lingnan Foundation web link for detailed videos and blogs about requirements in the application process. After their selection and orientation, each Fellow is assigned to work at non-profit organization in the United States. With support and guidance from program organizers, the internships are designed to help the Fellows gain practical experience in service work and facilitate inter-cultural cooperation. Fellows live in American homes for the duration of the program. While in the U.S., they also attend seminars on non-profit organization management and community development, cultural events, and weekly reflection meetings. Fellows are assigned to one of two sites in California. In Los Angeles, the program is managed by the Dashew Center for International Students and Scholars at the University of California-Los Angeles (UCLA). In Berkeley, the program is hosted by UC Berkeley Cal Corps Public Service Center.
Geographic Focus: China, Hong Kong
Date(s) Application is Due: Jan 31
Contact: Maria Luk, W.T. Chan Fellowships; +85 2 2103 1502; fax +85 2 2603 5765; wtchan!iiehongkong.org or scholarships@iiehongkong.org
Internet: http://www.iie.org/Programs/Lingnan-WT-Chan-Fellowships
Sponsor: Institute of International Education
1400 K Street, NW, 7th Floor
Washington, D.C. 20005-2403

IIE Lotus Scholarships 2138
The LOTU.S. Scholarships offers four year scholarships to high school Egyptian students. These scholarships are available for students to attend private Egyptian universities in fields of study that are important to Egypt's current and future development. The LOTU.S. Scholarships aim to educate young men and women so that they can become future leaders and passionate about their national and local communities. Recipients will attend one of the five private participating universities: Ahram Canadian University in Sixth of October City; British University in Shorouk City; Future University in New Cairo; Modern Sciences and Arts University in Sixth of October City; or Pharos University in Alexandria.
Requirements: The LOTU.S. Scholarship Program seeks Egyptian applicants in their last year of public or Azhari high school who have outstanding credentials and high financial need. All majors are eligible to apply. Candidates should complete an application form in Arabic according to the requirements and guidelines in the application packet.
Restrictions: Egyptian citizens who are also U.S. citizens or holders of a U.S. Green Card are not eligible.
Geographic Focus: Egypt
Date(s) Application is Due: Jun 5
Contact: Lotus Scholarship Contact; +20 (02) 2524-2172 or +20 (02) 2528-5779; fax +20 (02) 2524-2175; MENA@iie-egypt.org or lotus@iie-egypt.org
Internet: http://www.iie.org/Programs/USAID-Lotus-Scholarship-Program-English/About
Sponsor: Institute of International Education
1400 K Street, NW, 7th Floor
Washington, D.C. 20005-2403

IIE Mattel Global Scholarship 2139
The Mattel Children's Foundation sponsors a scholarship program designed to assist and encourage children of Mattel employees worldwide to pursue higher education (see Mattel locations posted at http://corporate.mattel.com/about-us/locations). Scholarships are offered for full-time study at a college, university or vocational school, in any country of the student's choice. The awards granted are based on the number of applications received and the funds available for scholarship distribution.
Requirements: Dependent children of regular full-time and part-time employees who have a minimum of six months employment with Mattel as of the application deadline date are eligible to apply. Students who are currently enrolled in post-secondary programs are eligible to apply. Official proof of admission to a recognized educational institution is required before scholarship funds become available. All fields of study are acceptable under these scholarships if offered at recognized universities or vocational schools.
Restrictions: Children of employees with a title of vice president or high are not eligible.
Geographic Focus: All States
Date(s) Application is Due: Apr 30
Contact: Scholarship Contact; (415) 362-6520, ext. 201; Mattel@iie.org
Internet: http://www.iie.org/Programs/Mattel-Global-Scholarship-Program
Sponsor: Institute of International Education
1400 K Street, NW, 7th Floor
Washington, D.C. 20005-2403

IIE Nancy Petry Scholarship 2140
The Institute of International Education sponsors the Nancy Petry Scholarship. The Nancy Petry Scholarship for Study Abroad provides financial support for students enrolled in graduate programs at universities in Colorado. Students should demonstrate some proficiency with the language of their chosen country if the program proposed is for a non-English speaking country.
Requirements: Applicants should refer to the website for current deadlines and application.
Restrictions: Students are not restricted to certain countries or fields of study, but the committee encourages applicants in the fields of environmental studies, international business, or economics.
Geographic Focus: Colorado
Date(s) Application is Due: Oct 21
Amount of Grant: 5,000 USD
Contact: Lauren Granstrom, Administrative Assistant; (303) 837-0788, ext. 27 or (303) 837-1409; lgranstrom@iie.org
Internet: http://www.rockymountainiie.org/scholarships
Sponsor: Institute of International Education
1400 K Street, NW, 7th Floor
Washington, D.C. 20005-2403

IIE New Leaders Group Award for Mutual Understanding 2141
The Institute of International Education (IIE) sponsors the IIE New Leaders Group Award for Mutual Understanding to recognize the outstanding work of one current Fulbright grantee who actively promotes mutual understanding between the U.S. and another country. The individual whose work is judged to have the most impact will receive the Award to continue an ongoing project. The $5,000 Award can be for work in any field. The intent of this Award is to encourage and recognize bright young professionals for their innovative ideas, valuable knowledge, and dedication to the cause of mutual understanding. The recipient will be honored and asked to speak about the impact of this Award on their project at an IIE New Leaders Group event in the spring.
Requirements: There are no limits on the disciplines or fields supported. Individuals from any academic field or disciplines are eligible. Applicants should send a project description, resume, two references, a news article or brochure about their project (if available), along with the application available online. Current information and deadlines are found on the website.
Restrictions: The project being recognized must be an ongoing activity that is currently underway; it cannot be a proposal for a project to be done in the future. The Award is limited to current Foreign and U.S. Fulbright Scholar grantees.
Geographic Focus: All States, All Countries
Amount of Grant: 5,000 USD
Contact: Award Contact; (212) 984-5456; fax (212) 984-5566; byi@iie.org
Internet: http://www.iie.org/Who-We-Are/Awards/New-Leaders-Group-Award
Sponsor: Institute of International Education
1400 K Street, NW, 7th Floor
Washington, D.C. 20005-2403

IIE Rockefeller Foundation Bellagio Center Residencies 2142
The Bellagio Center Residency program offers scholars, artists, thought leaders, policymakers and practitioners a serene setting conducive to focused, goal-oriented work, and the opportunity to establish new connections with fellow residents across an array of disciplines and geographies. The Center sponsors three kinds of residencies—for scholars, creative artists and practitioners. The Center offers one-month residency for each category, and is able to accommodate a group working on a particular project. The Center is particularly interested in applicants whose work connects in some way to the Foundation's issues areas (basic survival safeguards, global health, climate and environment, urbanization, and social and economic security).
Requirements: Applicants must submit an online application, a detailed project proposal of no more than 1,250 words, an abbreviated curriculum vitae, three reference letters, and work samples.
Geographic Focus: All States
Date(s) Application is Due: Dec 1
Contact: Bellagio Center Committee
Internet: http://www.rockefellerfoundation.org/bellagio-center
Sponsor: Institute of International Education
1400 K Street, NW, 7th Floor
Washington, D.C. 20005-2403

IIE Western Union Family Scholarships 2143
The Western Union Family Scholarships are intended to help two members of the same family move up the economic development ladder through education. Scholarships may be used for tuition for college/university education, language acquisition classes, technical/skill training, and/or financial literacy. Families must have overcome barriers to pursue their educational goals and demonstrate financial need, with specific plans to utilize the scholarship. Recipients are eligible to receive scholarships from $1,000 to $5,000. For example, one recipient may receive $3,500 for college tuition, while the other may receive $1,500 for an ESL course.

Requirements: Requirements for the scholarship include the following criteria: applicants must be at least 18 years of age and living in the U.S. for seven years or less; both applicants must have been born outside the U.S. and be currently living in the U.S; application must include educational providers for primary and secondary award recipients (must be two family members); and applicants must reside in one of the following U.S. locations - Los Angeles, California; San Francisco, California; Denver, Colorado; Chicago, Illinois; New York, New York; Washington, D.C., or Miami, Florida.
Restrictions: Western Union employees, Agents and dependents are not eligible.
Geographic Focus: California, Colorado, District of Columbia, Florida, Illinois, New York
Amount of Grant: Up to 5,000 USD
Contact: Scholarship Contact; (303) 837-0788; fax (303) 837-1409; wufamily@iie.org
Internet: http://corporate.westernunion.com/scholarship.html
Sponsor: Institute of International Education
1400 K Street, NW, 7th Floor
Washington, D.C. 20005-2403

IIE Whitaker International Fellowships and Scholarships 2144
The goal of the Whitaker funding is to assist in the development of outstanding professional engineers and scientists who will lead and serve the profession with an international outlook. The term of the Fellowship awards will be one academic year (as defined by the academic calendar of the host country). The term of the Scholarship awards can be for as little as one academic semester or as long as a full academic year. Benefits include maintenance allowance, airfare, accident and sickness insurance, partial tuition reimbursement (Fellows only), and access to grantee events, as well as membership in an elite alumni resource network. Applicants should consult the website for specific information about host institutions in Australia, Brazil, France, Germany, Ireland, Italy, The Netherlands, South Africa, Spain, Switzerland, and the United Kingdom.
Requirements: U.S. citizens and non-citizen permanent residents may apply. Fellows must have a BS or MS degree in biomedical engineering or bioengineering, and they will not have a doctorate at the time they start the grant. The Fellows applicant's most recent degree may not have been obtained more than two years prior to the start of the grant. Scholars must have either a degree in biomedical engineering and a doctorate which may not have been obtained more than two years prior to the start of the grant, or must be advanced Ph.D. candidates who have been admitted to candidacy and who will receive the Ph.D. prior to departure.
Geographic Focus: All States
Date(s) Application is Due: Oct 8
Amount of Grant: Up to 35,000 USD
Contact: Sabeen Altaf, Program Manager; (212) 984-5442; whitaker@iie.org
Internet: http://www.iie.org/Programs/Whitaker-International-Fellows-and-Scholars-Program
Sponsor: Institute of International Education
1400 K Street, NW, 7th Floor
Washington, D.C. 20005-2403

Ike and Roz Friedman Foundation Grants 2145
Established in Nebraska in 1989, the Ike and Roz Friedman Foundation has a mission of supporting Jewish agencies and federated giving programs, health care, the arts, education, and human and children's services. Its primary fields of interest include: arts and culture; child welfare; disease control and cures; higher education; human services; Judaism; museums; and community nonprofits. Interested applicants should submit a letter, which includes: name, address and phone number of organization; a statement of problem the project will address; and a detailed description of project and amount of funding requested. Though are no specified annual deadlines. Most recent awards have ranged from $250 to $130,000, though the majority average between $250 and $5,000.
Requirements: Any 501(c)3 supporting the residents of Nebraska is eligible to apply.
Geographic Focus: Nebraska
Amount of Grant: 250 - 130,000 USD
Samples: Omaha Children's Museum, Omaha, Nebraska, $2,000 - general operating support (2014); Nebraska Jewish Historical Society, Omaha, Nebraska, $1,000 - general operating support (2014); Klutznick Symposium, Omaha, Nebraska, $7,500 - general operating support (2014).
Contact: Susan Cohn, President; (402) 697-1111
Sponsor: Ike and Roz Friedman Foundation
22804 Hansen Avenue
Elkhorn, NE 68022

Illinois Children's Healthcare Foundation Grants 2146
The Illinois Children's Healthcare Foundation works to ensure that all children in Illinois have access to affordable and quality health care. The Foundation focuses its giving on three specific areas: improving the oral health of underserved children; addressing the mental health needs of children; and increasing the incidence of developmental screening in young children. Types of support include: building/renovation; conferences and seminars; curriculum development; emergency funds; equipment; program development and evaluation; and research. There are no specific deadlines or applications. The Board meets six times per year.
Requirements: Applicants must first contact the Foundation by phone or submit a letter of inquiry to discuss their proposed project.
Restrictions: Giving is limited to Illinois. The following are not eligible for funding: intermediary funding agencies; partisan, lobbying, political or denominational organizations; organizations not determined to be public charities; grants to individuals, endowments general medical research.
Geographic Focus: Illinois
Contact: Tammy Lemke, President; (630) 571-2555; fax (630) 571-2566; tammylemke@ilchf.org or info@ilchf.org
Internet: http://www.ilchf.org/
Sponsor: Illinois Children's Healthcare Foundation
1200 Jorie Boulevard, Suite 301
Oak Brook, IL 60523-2269

Illinois Tool Works Foundation Grants 2147
The corporate foundation awards grants in areas of company operations, with emphasis on Chicago, Ilinoise. Grants support organizations that focus on: education, the arts, health and human services, social welfare, housing, environmental and youth issues. The Foundation contributes financial support to not-for-profit organizations through two major giving programs: a direct-giving program and a three-for-one matching gift program for employees.
Requirements: Not-for-profit organizations are eligible to apply.
Geographic Focus: All States
Amount of Grant: 5,000 - 1,000,000 USD
Samples: United Way of Metropolitan Chicago, Chicago, IL, $1,110,732; Scholarship America, Saint Peter, MN, $478,350; Scott & White Memorial Hospital / Scott Sherwood & Brindley Foundation, Temple, TX, $60,000.
Contact: Mary Ann Mallahan; (847) 724-7500; mmallahan@itw.com
Internet: http://www.itwinc.com/itw/corporate_citizenship/itw_foundation
Sponsor: Illinois Tool Works Foundation
3600 West Lake Avenue
Glenview, IL 60026

Impact 100 Grants 2148
Impact 100 empowers women to dramatically improve lives by collectively funding significant grants that make a lasting impact in the Greater Cincinnati and Northern Kentucky community. During the first few two months of the year, non-profit organizations in the ten-county Greater Cincinnati/Northern Kentucky area region are encouraged to submit grant applications to Impact 100. The Grant Review Committees of Impact 100 review the applications and conduct site visits to narrow their choices. At the end of the review process, a finalist is chosen by each of the five committees (Education, Health & Wellness, Environment/Preservation & Recreation, Culture and Family). These five finalists present their grant requests at the Annual Awards Celebration. After the finalist presentations, each member casts a vote (or has sent an absentee vote in advance). The votes are immediately tabulated, and the Impact 100 grants are awarded at the end of the evening.
Requirements: Applications that are programmatic, endowment, capital, start-up, or research-oriented are excepted. Applications from non-profit organizations headquartered in any of the following counties are available to apply: Ohio - Adams, Brown, Butler, Clermont, Hamilton, Warren; Kentucky - Boone, Campbell, Kenton; Indiana - Dearborn. Every applying agency must have 501(c)3 status with the IRS.
Restrictions: The following types of applications are not available for funding: operating; partisan; individual churches; indigent care subsidy; travel; loans; or for individuals.
Geographic Focus: Indiana, Kentucky, Ohio
Date(s) Application is Due: Mar 1
Amount of Grant: 100,000 USD
Contact: Luann Scherer; (513) 624-9509; Grants@Impact100.org
Internet: http://impact100.org/Grant/Default.aspx
Sponsor: Impact 100
PMB314 2692 Madison Road NI
Cincinnati, OH 45208-1320

Inasmuch Foundation Grants 2149
The foundation was established for charitable, scientific, and educational projects primarily in Oklahoma; however, select projects in Colorado Springs, CO, also are supported. The foundation consistently provides funding and support to educational, health and human service, cultural, artistic, historical, and environmental concerns. This funding is not available to individuals, but is available to formal organizations seeking capital and support for existing programs that meet the emerging needs of the community. In order to initiate a proposal, submit a one-page letter of inquiry by the listed application deadlines. Application information is available online.
Restrictions: The foundation generally does not fund endowments or scholarships.
Geographic Focus: Colorado, Oklahoma
Date(s) Application is Due: Feb 15; Aug 15
Samples: Colorado College (Colorado Springs, CO)--to construct a facility for arts education and performances, $4 million.
Contact: Program Contact; (405) 604-5292
Internet: http://www.inasmuchfoundation.org/application.html
Sponsor: Inasmuch Foundation
210 Park Avenue, Suite 3150
Oklahoma City, OK 73102

Independence Blue Cross Charitable Medical Care Grants 2150
The Independence Blue Cross (IBC) Charitable Medical Care Grant Program awards financial and programmatic grant support to nonprofit, privately funded health clinics that provide free or nominal-fee care to the uninsured and medically underserved communities in Bucks, Chester, Delaware, Montgomery, and Philadelphia counties. See website for grant application.
Requirements: To be eligible for a grant through this program, an applicant must meet the following criteria: the mission of the requesting organization aligns with IBC's Social Mission and its funding priorities; the requesting clinic is a privately funded entity. (Note: A clinic may receive federal funding as part of its revenues.); clinic grantee provides direct

medical care for free or for a nominal fee to uninsured and/or underinsured residents of southeastern Pennsylvania; all grantees must be located in Bucks, Chester, Delaware, Montgomery, or Philadelphia counties; organizations must be classified as tax-exempt nonprofit under Section 501(c)3 of the IRS code.
Restrictions: Funding is unavailable to: individuals; private foundations; fund-raising events; conferences or seminars; political causes, candidates, organizations, or campaigns; sports teams; endowments; social organizations; religious groups for religious purposes; organizations that do not meet IBC's Social Mission.
Geographic Focus: Pennsylvania
Amount of Grant: 10,000 - 100,000 USD
Contact: Courtney Smith, Social Mission Program Analyst; (215) 241-4862; fax (215) 241-3543; courtney.smith@ibx.com
Internet: http://www.ibx.com/social_mission/medical_grants/index.html
Sponsor: Independence Blue Cross
1901 Market Street, 28th Floor
Philadelphia, PA 19103-1480

Independence Blue Cross Nurse Scholars Program 2151
This program was created to address the severe threat posed by the current nursing shortage to the quality and cost of health care in southeastern Pennsylvania. The program provides financial assistance to aspiring nurse educators and undergraduate nursing students. It also includes summer internship programs for nursing students and funding for faculty positions at area nursing schools. Application information can be located at the Pennsylvania Higher Education Foundation Web site.
Geographic Focus: Pennsylvania
Samples: Nursing programs in southeastern Pennsylvania (PA)--to establish a scholarship program for graduate students pursuing careers as nursing instructors, $1.74 million over three years; Pennsylvania Higher Education Fdn (Harrisburg, PA)--for a scholarship program for undergraduate nursing students, $750,000.
Contact: Program Supervisor; (800) 377-4502
Internet: http://www.ibx.com/social_mission/nurse_scholars/index.html
Sponsor: Independence Blue Cross
1901 Market Street, 28th Floor
Philadelphia, PA 19103-1480

Independence Blue Cross Nursing Internships 2152
The Independence Blue Cross (IBC) Nursing Internship Program is an integral part of IBC's Nurse Scholars Program and supports the IBC Social Mission commitment to address the nursing shortage and overall health care crisis in the southeastern Pennsylvania region. The IBC Nursing Internship Program is open to undergraduate nursing students who have completed at least one clinical rotation and who attend one of the 31 accredited nursing programs (see website for a complete listing of the participating nursing programs) supported by the IBC Nurse Scholars Program. Two unique student nursing internship opportunities are offered through this program. One is for nursing students who would like to gain experience at IBC in a health care/insurance administration setting; the other is for nursing students interested in exploring the field of public health by interning on site at a community health center clinic.
Requirements: Each program is open to undergraduate nursing students who have completed at least one clinical rotation and who attend an accredited nursing program in southeastern Pennsylvania supported by the IBC Nurse Scholars program. Applications are available on the IBC website. If you are selected to participate, you must be available for the entire ten week internship program beginning June 8 through August 14.
Geographic Focus: Pennsylvania
Date(s) Application is Due: Mar 31
Contact: Courtney Smith, Social Mission Program Analyst; (215) 241-4862; fax (215) 241-3543; courtney.smith@ibx.com
Internet: http://www.ibx.com/social_mission/nurse_scholars/nursing_internship/index.html
Sponsor: Independence Blue Cross
1901 Market Street, 28th Floor
Philadelphia, PA 19103-1480

Independence Community Foundation Community Quality of Life Grant 2153
The Foundation makes grants to organizations that engage in community-based efforts that support important institutions or causes. Generally, the grants support: disease awareness and prevention; food pantries; neighborhood greening and graffiti removal; and counseling support for vulnerable populations. Interested applicants should submit an initial letter of inquiry. Application information is available online.
Requirements: Program or project grants are made on a competitive basis to nonprofit organizations located within New York City, Nassau, Suffolk and Westchester counties in New York, and Essex, Bergen, Union, Hudson, Middlesex, Ocean, and Monmouth counties in New Jersey.
Restrictions: The Foundation does not fund: individuals; political contributions; funding for religious purposes; purchase tickets for dinners, golf outings, or similar fundraising events.
Geographic Focus: New Jersey, New York
Date(s) Application is Due: Mar 30; Sep 30
Contact: Program Contact; (718) 722-2300; fax (718) 722-5757; inquiries@icfny.org
Internet: http://www.icfny.org/comm_quality.html
Sponsor: Independence Community Foundation
182 Atlantic Avenue
Brooklyn, NY 11201

Indiana 21st Century Research and Technology Fund Awards 2154
The Indiana 21st Century Research and Technology Fund of the Indiana Economic Development Corporation (IED.C.) is open to proposals from all public and private entities for technology-based commercialization activities encompassing science/technology creation, innovation, and transfer intended to have commercial impacts. The fund intends to increase the numbers, and rates of development, of new and expanding technology-based companies by funding promising opportunities that, in some cases, the financial markets might find too risky. The Fund makes awards in two broad categories: Science and Technology Commercialization and Centers of Excellence. In addition, the Fund provides cost-share on behalf of Federal proposals submitted by Indiana-based entities. Generally awards are made in multiples of $50,000 up to $2,000,000. Support for awards in excess of $2,000,000 will be rare.
Requirements: The IED.C. defines a technology-based company as one that is involved in transferring advanced technology into products, developing technologies with the near-term intention of creating products, or using new or advanced technologies in its design, development, and/or manufacturing of products. The Fund emphasizes the creation of academic-sector - commercial-sector partnerships. In making awards, the Fund expects significant leverage from the partners involved in the projects. Important: before applying, contact Fund staff (email preferred) to discuss your interest in submitting a proposal and to discuss your technology and commercialization goals. While not a review criterion, the fund encourages the inclusion of interns from any academic institution, or participating commercial sector partner, in order to increase project-related involvement of students at all levels.
Restrictions: Only direct costs will be supported. Institutions will not be provided indirect (overhead) cost support. Entities with previous Fund awards that are not current with regard to financial or technical reporting requirements will be disqualified from making new submissions to the Fund. Resubmissions of previously declined proposals will be considered only if substantive changes have been made to the proposal. Fund staff will determine whether to review resubmissions.
Geographic Focus: Indiana
Amount of Grant: 500,000 - 2,000,000 USD
Samples: 2K Corporation: $400,000 - Commercialization of a Neutron Based Explosives Detection Device (Car Bomb Detection). Arxan Technologies, Inc.: $1,944,096 - Protecting Critical IP in the 21st Century: Advancing Anti-Tamper Technologies. BioVitesse Inc.: $1,300,000 - Rapid Detection and Identification of Live Bacteria. CIS LLC: $1,075,000 - Commercialization of Satellite Radio for Cell Phone Applications.
Contact: Carla Phelps, Financial Manager; (317) 233- 4336; cphelps@21fund.org
Internet: http://www.21fund.org/
Sponsor: Indiana Economic Development Corporation
1 North Capitol Avenue, Suite 900
Indianapolis, IN 46204

Indiana AIDS Fund Grants 2155
Two grant programs are available to Indiana organizations that provide HIV/AIDS advocacy, prevention and direct care services. The Direct Emergency Financial Assistance Fund gives grants to agencies and organizations specifically to help HIV/AIDS patients access emergency financial help for housing, transportation, food, and/or medical care. The Indiana AIDS Fund gives prevention grants in November each year to agencies and organizations that provide persons living with HIV/AIDS direct care services, health education and risk-reduction programs and counseling, testing, and referral programs. Proposal information is available online.
Requirements: Any Indiana-based non-profit 501(c)3 group, organization or agency that provides HIV/AIDS-related programs or services to local constituencies is eligible to apply for grants from the Indiana AIDS Fund. The applicant must function without discrimination or segregation because of race, gender, age, religion, national origin, sexual orientation, disability, military or marital status, in hiring, termination, assignment and promotion of staff, selection of board members or provision of service.
Restrictions: The fund will not make the following types of grants: to individuals; for sectarian religious purposes; to purchase advertising or tickets to events; to support research; for the purpose of deficit reduction; to produce currently available educational materials, to fund endowments; for short or long term loans; for multiple-year program funding; for mass media campaigns; for capacity building; or for services not directly related to primary or secondary prevention.
Geographic Focus: Indiana
Contact: Grants Administrator; (317) 630-1805; severett@thfgi.org
Internet: http://www.indianaaidsfund.org/index.cfm?navigationid=1844
Sponsor: Indiana AIDS Fund
429 East Vermont Street, Suite 300
Indianapolis, IN 46202

Indiana Minority Teacher/Special Services Scholarships 2156
The Minority Teacher Scholarship was created by the 1988 Indiana General Assembly to address the critical shortage of Black and Hispanic teachers in Indiana. In 1990 the Indiana General Assembly amended the Minority program to include the field of Special Education, and in 1991 the fields of Occupational and Physical Therapy were added. The maximum annual scholarship is $1,000. However, if a minority student applicant demonstrates financial need he/she may be eligible to receive up to $4,000 annually. Colleges will determine the actual amount when reviewing a scholar's financial aid package.
Requirements: To be eligible you must be: 1. A minority student (defined as Black or Hispanic) seeking a teaching certification; or a student seeking a Special Education teaching certification; or a student seeking an Occupational or Physical Therapy certification. 2. An Indiana resident and an U.S. Citizen. 3. Admitted to an eligible institution as a full time student or already attending as a full time student. 4. Pursuing or intend to pursue a course of study that would enable the student upon graduation to teach in an accredited

elementary or secondary school in Indiana. 5. Complete and file the Free Application for Federal Student Aid (FAFSA). 6. Not be in default on a State or Federally-funded Student loan. 7. Meet all minimum criteria established by the Commission. In addition, students who are already enrolled in college, must have a Grade Point Average (GPA) of at least 2.0 on a scale of 4.0 or the equivalent, or meet the minimum GPA requirements established at the college for its school of education, if it is higher. Scholars must agree to: 1. Pursue a program which leads to either a teaching or Occupational or Physical Therapy Certification. 2. Complete the Teacher or Occupational or Physical Therapy certification program within six (6) years from the time the first scholarship is received. 3. Teach on a full time basis in an accredited Indiana elementary or secondary school; or, 4. Practice in the field of Occupational or Physical Therapy in an accredited school, vocational rehabilitation center, community mental retardation or other developmental disabilities center, for 3 years out of the first 5 years after certification. Colleges are responsible for making the actual awards. Financial need may be considered but it is not a requirement to apply. Preference will be given to minority students and students enrolling in college for the first time. Applications should be submitted to the Office of Financial Aid at the school where the applicant plans to attend.
Restrictions: Scholarships are non-transferable between universities. Those who fail to fulfill the terms of their scholarship agreements must reimburse the State of Indiana.
Geographic Focus: Indiana
Amount of Grant: Up to 4,000 USD
Contact: Ada Sparkman; (317) 232-2350; asparkman@ssaci.in.gov
Internet: http://www.in.gov/ssaci/2342.htm
Sponsor: State Student Assistance Commission of Indiana
150 W Market Street, Suite 500
Indianapolis, IN 46204

Indiana Nursing Scholarships 2157
The Nursing Scholarship Fund was created by the 1990 Indiana General Assembly to encourage and promote qualified individuals to pursue a nursing career in Indiana. The scholarship can only be applied towards tuition and fees. Colleges will determine the actual award amount when developing a scholar's financial aid package. The maximum annual scholarship is $5,000. However, the amount of the scholarship may be affected by the level of other tuition specific grants and scholarships aid received by an applicant.
Requirements: To be eligible students must: 1) Be admitted to an approved institution of higher learning as a full time (12 hours or more) or part time (6-11 hours) nursing student; 2) Be an Indiana resident and an U.S. citizen; 3) Agree in writing to work as a nurse in an Indiana health care setting for at least the first two (2) years following graduation; 4) Demonstrate a financial need for the scholarship; 5) Have a minimum Grade Point Average (GPA) of at least 2.0 on a scale of 4.0 or the equivalent, or meet the minimum GPA requirements established for the college's nursing program if it is higher; 6) Not be in default on a state or federally sponsored student loan; 7) Complete and file the Free Application for Federal Student Aid (FAFSA); 8) Meet all other minimum criteria established by the Commission. In order to renew the scholarship, Scholars must: 1) Maintain the cumulative GPA required by the college for admission to its nursing program, or; 2) Maintain a GPA of at least 2.0 on a scale of 4.0 or the equivalent if there is no minimum at the college; and, 3) Reapply each year. Applications should be submitted to the Office of Financial Aid where the applicant plans to attend. Applications should not be submitted to SSACI. Colleges will make the award decisions. For a list of colleges and universities participating in the program, go to http://www.in.gov/ssaci/2368.htm.
Geographic Focus: Indiana
Date(s) Application is Due: Mar 10
Amount of Grant: Up to 5,000 USD
Contact: Ada Sparkman; (317) 232-2350; asparkman@ssaci.in.gov
Internet: http://www.in.gov/ssaci/2343.htm
Sponsor: State Student Assistance Commission of Indiana
150 W Market Street, Suite 500
Indianapolis, IN 46204

Infinity Foundation Grants 2158
The foundation awards grants in a broad array of areas including charitable, scientific, religious, educational, and holistic healing activities by organizations and public agencies that work to better the lives of people. Grants also support individuals' research and development of educational materials to improve the authenticity of the portrayal of Indic traditions in the educational system. Proposed projects could result in one or more of the following: books, curriculum development, articles, conferences, CD-Roms, digital slide shows, Internet presentations, and audio/video materials. Topics covered may include philosophy, history, religion, science, art, and sociology, as they pertain to the educational curricula on Indic traditions. Proposals should be submitted by email.
Requirements: Grantee may be a scholar, teacher, visionary, or spiritual leader whose work in the designated topics would be enhanced by a foundation grant.
Geographic Focus: All States
Contact: Rajiv Malhotra; (609) 683-0548; fax (609) 683-0478; rm.infinity@gmail.com
Internet: http://www.infinityfoundation.com/callforgrantproposals.htm
Sponsor: Infinity Foundation
66 Witherspoon Street, Suite 400
Princeton, NJ 08542

Institute for Agriculture and Trade Policy Food and Society Fellowships 2159
The IATP Food and Society Fellows Program provides fellowships to professionals in food and agriculture from across North America, enabling them to use mass media channels to inform and shape the public agenda. The goal of the program is to create sustainable food systems that promote good health, vibrant communities, environmental stewardship, worker justice and accessibility for all. Fellows come from many disciplines: chefs, farmers, nutritionists, activists, public health professionals, fishers, policy experts and academics. Together they form an interdisciplinary team that works to: use communication to influence the issues that reach the public agenda, thereby creating policy changes at the personal, organizational and public policy levels that advance sustainable food and farming systems; increase the mass media communications on issues around sustainable food and farming systems that produce healthy, green, fair and affordable foods; raise the profile of the fellows as food system experts among media and policymakers; and build capacity, leadership, and cohesiveness in a group of experts who collaborate and communicate using mass media channels to bring sustainable food system issues to the public agenda.
Requirements: Applicants should be U.S. residents actively involved in a professional career that involves two or more of the following: agriculture, health promotion, youth development, food production, and/or policy analysis.
Geographic Focus: All States
Contact: Mark Muller; (612) 870-3420; fax (612) 870-4846; mmuller@iatp.org
Internet: http://www.foodandsocietyfellows.org/about/page/about-us
Sponsor: Institute for Agriculture and Trade Policy
2105 First Avenue South
Minneapolis, MN 55404

Intergrys Corporation Grants 2160
Integrys supports the giving initiatives of Wisconsin Public Service Foundation in Michigan, Minnesota and Wisconsin. Other significant giving by Integrys occurs through programs operated by Peoples Gas and North Shore Gas in Illinois, including the city of Chicago and its 54 suburban communities in northeastern Illinois. Areas if interest include: arts and culture; education (all levels); human services and health; community and neighborhood development; and the environment. Employee volunteers allow staff to give back to the neighborhoods they cherish. Matching funds energize charitable involvement. Categories of giving include health and welfare, civic and community, higher education, and cultural. Most corporate contributions are in the form of unrestricted grants. Types of support include capital grants, endowments, general operating grants, matching gifts, and program grants. Before eligibility is determined, consideration will be given to the background of the organization, the organization's legal status, how the program will benefit the community, whether the organization receives broad community support, the quality of the organization's leadership, and the organization's financial status.
Requirements: Organizations in Illinois, Michigan, Minnesota, and Wisconsin are eligible.
Restrictions: Contributions will not be made to individuals; organizations that discriminate by race, color, creed, or national origin; political organizations or campaigns; organizations whose prime purpose is to influence legislation; religious organizations for religious purposes; agencies owned and operated by local, state, or federal governments; or for trips or tours, or special-occasion or goodwill advertising.
Geographic Focus: Illinois, Michigan, Minnesota, Wisconsin
Amount of Grant: Up to 5,000 USD
Contact: Contributions Officer; (312) 240-7516 or (800) 699-1269; fax (312) 240-4389
Internet: http://www.integrysgroup.com/corporate/corporate_giving.aspx
Sponsor: Intergrys Corporation
130 E Randolph Drive, 18th Floor
Chicago, IL 60601

International Positive Psychology Association Student Scholarships 2161
The International Positive Psychology Association offers scholarships to help fund student participation in the World Congress on Positive Psychology conference. The number and amount of awards are determined by the number of applications received. The scholarships help fund travel, registration, food, and lodging for the conference.
Requirements: All applicants must meet the following *Requirements:* complete the online application; submit an abstract proposal of less than 250 words; attach a letter of recommendation from a professor who works with positive psychology themes; attach a curriculum vitae of no more than two pages; provide proof of university or college affiliation and a promise via email to travel to the U.S. to present a poster; currently enrolled in an accredited university or college or at a community college with intent to transfer to a university/college for the upcoming year; and pursuing an undergraduate, master's, or PhD in psychology or related field. Applicants should refer to the website to check for a current online application.
Geographic Focus: All States, All Countries
Date(s) Application is Due: Jan 31
Contact: Gene Terry; (856) 423-2862; fax (856) 423-3420; ippamtg@talley.com
Internet: http://community.ippanetwork.org/worldcongress/studentscholarshipinformation/
Sponsor: International Positive Psychology Association
19 Mantua Road
Mt. Royal, NJ 08061

Ireland Family Foundation Grants 2162
Established in 2000, the Ireland Family Foundation givings primarily in North Carolina with some funding available in California. The Foundations supports non-profit organizations involved with autism research, human services and education.
Requirements: The initial approach should be in the form of a letter, followed up with a proposal containing : a copy of IRS Determination Letter; a detailed description of project and amount of funding requested.
Restrictions: No grants to individuals.
Geographic Focus: California, North Carolina
Amount of Grant: 10,000 - 60,000 USD
Samples: ARC of Orange County, Chapel Hill, NC., $35,000;

Contact: Lori Ireland, President; (919) 932-3556
Sponsor: Ireland Family Foundation
1434 Arboretum Drive
Chapel Hill, NC 27517-9161

Irving S. Gilmore Foundation Grants 2163
The Irving S. Gilmore Foundation endeavors to develop and to enrich the Greater Kalamazoo community of Michigan and, its residents by supporting the work of nonprofit organizations. The Foundation's funding priorities are: health and well-being; arts; human services; education; community development; culture and humanities. Organizations that are first time Foundation applicants or have not received Foundation funding since 2007 must contact the Foundation at least four weeks prior to an applicable submission deadline.
Requirements: The Foundation supports Kalamazoo County projects, programs, and purposes carried out by charitable institutions, primarily public charities and governmental entities.
Restrictions: Grants are not made to individuals.
Geographic Focus: Michigan
Date(s) Application is Due: Jan 4; Mar 1; May 3; Jul 1; Sep 1; Nov 1
Samples: Michigan Festival of Sacred Music--Operational Support; Hospice Care of Southwest Michigan--Capital Campaign; Volunteer Services of Greater Kalamazoo--BoardConnect Program.
Contact: Janice C. Elliott; (269) 342-6411; fax (269) 342-6465
Internet: http://www.isgilmorefoundation.org/communityinvolvement.htm
Sponsor: Irving S. Gilmore Foundation
136 East Michigan Avenue, Suite 900
Kalamazoo, MI 49007-3912

Irvin Stern Foundation Grants 2164
The Irvin Stern Foundation awards grants to nonprofits, primarily in the following areas of interest: human services; civic affairs; the arts; and Jewish welfare. Other grant requests should be within the Foundation's additional areas of interest which include: aiding the under-served, the poor and disadvantaged; improving the quality of life in urban communities; and enhancing Jewish community, education and spirituality. Since there are no specified application forms, interested applicants should submit an online letter of inquiry or application letters mailed to the Foundation office. The Foundation's two annual deadlines for application submission are March 1 and September 1. Most recently, awards have ranged from $1,000 to $120,000.
Requirements: The Foundation makes grants to 501(c)3 tax-exempt organizations.
Restrictions: The Foundation does not contribute to endowments, capital campaigns, capital construction projects, and academic or medical research programs of any kind.
Geographic Focus: All States
Date(s) Application is Due: Mar 1; Sep 1
Amount of Grant: 1,000 - 120,000 USD
Contact: Christine Flood; (312) 321-9402; christine@irvinstern.org
Internet: http://irvinstern.org/guidelines/
Sponsor: Irvin Stern Foundation
4 East Ohio Street, Studio 6
Chicago, IL 60611

Isabel Allende Foundation Esperanza Grants 2165
The Isabel Allende Foundation was established in 1996, and is guided by a vision of a world in which women have achieved social and economic justice. The vision includes empowerment of women and girls and protection of women and children. The Foundation feels that the way to achieve empowerment is: reproductive self-determination; health care; and education. Grants typically range from $1,000 to $10,000 and are made to support programs for vulnerable women and children in Chile and California.
Requirements: 501(c)3 organizations (and equivalent international organizations) are eligible to apply. Priority is given to programs in the San Francisco Bay Area and Chile.
Restrictions: The foundation does not fund capital campaigns, individual trips or tours, conferences, or events; and projects that benefit political, religious, and/or military organizations. Individuals are not eligible to receive grants and should not apply.
Geographic Focus: California, Chile
Date(s) Application is Due: Jan 1; Apr 1; Jul 1; Oct 1
Amount of Grant: 1,000 - 5,000 USD
Contact: Lori Barra, Executive Director; (415) 332-1313 or (415) 289-0992; fax (415) 289-1154; lori@isabelallendefoundation.org
Sarah Kessler, Associate; (415) 331-0261; sarah@isabelallendefoundation.org
Internet: http://www.isabelallendefoundation.org/iaf.php?l=en&p=application
Sponsor: Isabel Allende Foundation
116 Caledonia Street
Sausalito, CA 94965

Ittleson Foundation AIDS Grants 2166
In regards to AIDS, the foundation is particularly interested in new model, pilot, and demonstration efforts: addressing the needs of underserved at-risk populations and especially those programs recognizing the overlap between such programs; responding to the challenges facing community-based AIDS service organizations and those organizations addressing systemic change; providing meaningful school-based sex education; making treatment information accessible, available and easily understandable to those in need of it; or, addressing the psycho-social needs of those infected and affected by AIDS, especially adolescents.
Requirements: Tax-exempt organizations may apply.

Grant Programs | 341

Restrictions: The foundation generally does not provide funds for capital building projects, endowments, grants to individuals, scholarships or internships (except as part of a program), direct service programs (especially outside New York City), projects that are local in focus and unlikely to be replicated, continuing or general support, projects and organizations that are international in scope or purpose, or biomedical research.
Geographic Focus: All States
Date(s) Application is Due: Sep 1
Sample: AIDS Alliance for Children Youth and Family, Washington, D.C., $5,000--to transform the National Consumer Leadership Corps Training Program into one that can be replicated by local AIDS organizations around the nation; Cesar E. Chavez Institute, New York, NY, $40,000--one-time grant to develop new family interventions and a new family-related model of care to reduce risk for HIV and mental health problems in lesbian, gay, bisexual and transgender (LGBT) youth.
Contact: Anthony C. Wood, Executive Director; (212) 794-2008; fax (212) 794-0351
Internet: http://www.ittlesonfoundation.org/aids.html
Sponsor: Ittleson Foundation
15 E 67th Street, 5th Floor
New York, NY 10021

Ittleson Foundation Mental Health Grants 2167
For this program, the foundation is interested in innovative, pilot, model and demonstration projects that are: fighting the stigma associated with mental illness and working to change the public's negative perception of people who have mental illness; utilizing new knowledge and current technological advances to improve programs and services for people who have mental illness; bringing the full benefits of this new knowledge and technology to those who presently do not have access to them; or, advancing preventative mental health efforts, especially those targeted to youth and adolescents, with a special focus on strategies that involve parents, teachers, and others in close contact with these populations.
Requirements: Tax-exempt organizations may apply.
Restrictions: The foundation generally does not provide funds for capital building projects, endowments, grants to individuals, scholarships or internships (except as part of a program), direct service programs (especially outside New York City), projects that are local in focus and unlikely to be replicated, continuing or general support, projects and organizations that are international in scope or purpose, or biomedical research.
Geographic Focus: All States
Date(s) Application is Due: Sep 1
Sample: Active Minds, Inc., Washington, D.C., $10,000--to dramatically expand the network of chapters on college campuses; Partnership with Children, New York, NY, $50,000--one-time grant to develop the Center for Capacity Building to enable the dissemination of the successful counseling and prevention care program for inner-city boys and girls, Open Heart-Open Mind to over 100 schools by 2012; Horticultural Society of New York, New York, NY, $30,000--to launch their new Nonprofit Partnership for Horticultural Therapy to formalize and expand their ability to help a wide range of organizations working with the mentally and physically ill, formerly homeless, HIV+, victims of substance abuse, the elderly, at-risk juveniles and those re-entering society from incarceration, use horticultural therapy programs to address the needs of these marginalized populations.
Contact: Anthony C. Wood, Executive Director; (212) 794-2008; fax (212) 794-0351
Internet: http://www.ittlesonfoundation.org/mental.html
Sponsor: Ittleson Foundation
15 E 67th Street, 5th Floor
New York, NY 10021

J.B. Reynolds Foundation Grants 2168
The foundation makes grants in the local Kansas City, MO, area, up to a 150-mile radius of the city. Grants are awarded for building and equipment, community development, medical research, social welfare, and the arts and humanities. Some support is given to colleges and universities. Additional types of support include general operating support, continuing support, annual campaigns, endowment funds, publications, and research. The board meets in April and December of each year to consider letters of requests, and only invited proposals are reviewed. All grants are awarded in December.
Requirements: 501(c)3 organizations in the Kansas City area may submit applications.
Geographic Focus: Missouri
Amount of Grant: 5,000 - 50,000 USD
Contact: Richard L. Finn, Secretary-Treasurer; (816) 753-7000; fax (816) 753-1354
Sponsor: J.B. Reynolds Foundation
P.O. Box 219139
Kansas City, MO 64141-6139

J.C. Penney Company Grants 2169
The corporate contributions program awards grants to 501(c)3 nonprofit organizations in the company's areas of interest, including improvement of K- through 12th-grade education through curriculum-based after school care, with a priority on JCPenney Afterschool; support/promotion of associate (employee) volunteerism, primarily through the James Cash Penney Awards for Community Service; and United Way (most of JCPenney's support for health and welfare issues is contributed through local United Ways). Funding for other types of programs and projects is limited. Funding is concentrated on projects and organizations that serve a broad sector of the community; national projects that have a multiplier effect by benefiting local organizations across the country; nonprofits offering direct services; and projects with proven track records. Proposals are accepted year-round. There is no formal grant application. Interested applicants should submit a brief letter of inquiry. Submission information is available online.

Requirements: JCPenney only considers grants to tax-exempt organizations with 501(c)3 status or organizations that are a political subdivision of the state as described 170(c)1 of the IRS.
Restrictions: JCPenney refrains from or limits support of: individuals (including family reunions); individual student exchange/travel programs; membership organizations, unless the project benefits the entire community; religious organizations, unless the project benefits the entire community; journal or program advertising; fundraising dinners, luncheons or other types of benefits; merchandise donations; door prizes, gift certificates, or other giveaways; conferences and seminars; capital campaigns and multiyear pledges; international projects except in countries with corporate business locations; pilot projects; film and video projects; research projects; scholarships, except at colleges and universities with which the company has a recruiting relationship; higher education institutions, unless the company has a business or recruiting relationship; individual K-12 schools, unless the company has a business partnership with the school (including proms, graduations, PTOs and PTAs); or donations of returned, damaged or excess merchandise. JCPenney does not offer a matching gifts program at this time.
Geographic Focus: All States
Contact: Jeannette M. Siegel, Community Relations and Contributions Manager; (972) 431-1349; fax (972) 431-1355; jsiegel@jcpenney.com
Internet: http://www.jcpenney.net/about/social_resp/community/default.aspx
Sponsor: J.C. Penney Company
P.O. Box 10001
Dallas, TX 75301-8101

J.E. and L.E. Mabee Foundation Grants 2170
The purpose of the Foundation is to aid Christian religious organizations, charitable organizations, institutions of higher learning, hospitals and other organizations of a general charitable nature. The Foundation favors organizations that combine sound character and stability with progressiveness and purpose. There is no formal application procedure. Proposal submission information is available online.
Requirements: Nonprofit, tax exempt, and not tax- supported organizations in Arkansas, Kansas, Missouri, New Mexico, Oklahoma, and Texas are eligible.
Restrictions: The foundation does not generally favor grants for deficit financing and debt retirement (except a construction contract that was executed prior to the time of application); operating or program funds or annual fundraising campaigns (except Junior Achievement and United Way in Tulsa, OK, and Midland, TX); reserve purposes or for projects likely to be long delayed; endowments; governmental owned or operated institutions and/or facilities (such as state universities and municipal parks and libraries); educational institutions below the college level; furnishing or equipment (except major medical equipment); or churches.
Geographic Focus: Arkansas, Kansas, Missouri, New Mexico, Oklahoma, Texas
Contact: Thomas R. Brett, Director; (918) 584-4286
Internet: http://www.mabeefoundation.com/policies.htm
Sponsor: J.E. and L.E. Mabee Foundation
Mid-Continent Tower, 401 South Boston Avenue, Suite 3001
Tulsa, OK 74103-4017

J.H. Robbins Foundation Grants 2171
The J.H. Robbins Foundation was established in 1983 in California, and was reclassified as a private operating foundation in 1994. Currently, the Foundation's primary fields of interest include: health care; human services; and safety/disaster program support. The type of support most often given is for general operations. Though a formal application is required, there are no specific annual deadlines for submission, and interested parties should begin by contacting the Foundation directly. Recent awards have ranged from $500 to as much as $7,500.
Requirements: 501(c)3 nonprofit organizations throughout the San Mateo, California, region are eligible to apply.
Geographic Focus: California
Amount of Grant: 500 - 7,500 USD
Samples: CASA of San Mateo County, San Mateo, California, $4,500 - general operating support (2014); Mills Peninsula Hospital, Burlingame, California, $7,500 - general operating support (2014); Pacifica Shool Volunteers, Pacifica, California, $3,500 - general operating support (2014).
Contact: Aron H. Hoffman, Treasurer; (650) 343-5300
Sponsor: J.H. Robbins Foundation
503 Princeton Road
San Mateo, CA 94402

J.L. Bedsole Foundation Grants 2172
The Foundation considers requests that most closely match its overall mission: to improve the quality of life for the citizens of Southwest Alabama and to strengthen the communities in which they live. The Foundation will consider only those grant applications that meet several of the following guidelines. The project addresses needs in at least one of the following areas: education, arts and culture, health and human services and economic development; potential for permanent, enduring benefits that will provide value to the community and the residents of Southwest Alabama; diverse groups are collaborating on the project to achieve common goals; the organization and the project clearly demonstrate sound fiscal management and accountability; the organization attracts multiple sources of support for the project; the project addresses underserved segments of the population, the economically disadvantaged or citizens of rural communities. The Foundation offers a scholarship program as well, see: http://www.jlbedsolefoundation.org/default.asp?ID=7 for details and guidelines.
Requirements: Nonprofit organizations in Mobile, Baldwin, Clarke, Monroe and Washington County, Alabama are eligible to apply.
Restrictions: The Foundation will not support the following: grants that support political activities or attempts to influence action on specific legislation; multiple year pledges nor make grants beyond the current year; grants to endowment funds of other organizations; grants are not made directly to individuals; grants are not made to organizations, programs or projects outside of the State of Alabama.
Geographic Focus: Alabama
Contact: Christopher L. Lee; (251) 432-3369; chrislee@jlbedsolefoundation.org
Internet: http://www.jlbedsolefoundation.org/default.asp?ID=6
Sponsor: J.L. Bedsole Foundation
P.O. Box 1137
Mobile, AL 36633

J.M. Long Foundation Grants 2173
The Foundation is focused on providing support to organizations involved with health care, education, and conservation in the communities of Northern California and Hawaii. Preference will be given for new, innovative projects, which can be completed with the Foundation contribution. Applications for new grant requests are given by invitation only. In order to receive an invitation, an organization will need to submit to the Foundation a single page Request for Invitation letter using organization letterhead. This Request for Invitation letter should describe the organization's objective and the specific project for which the grant would be used. The applicant should be sure to include an estimated amount for the grant, as well as contact information including email address if available.
Requirements: Hawaii and Northern California nonprofits are eligible to apply.
Restrictions: Please note that grants will not be made: for the support of solely religious, sacramental or theological functions; in support of political bodies or campaigns; to individuals or foreign organizations; for loans; for purposes of memorializing an individual, although donations may be made to memorial funds as a means of achieving other purposes; for covering operating deficits; to organizations not qualifying as either a federal tax-exempt non-profit private organization or a qualifying public organization; to supporting organizations with a 509(a)(3) designation.
Geographic Focus: California, Hawaii
Amount of Grant: 1,000 - 50,000 USD
Samples: California State Parks Foundation, Kentfield, CA - Park Education Legacy Program; Children's Discovery Center, Honolulu, HI - Ready, Set, Grow Program; The Creek, A Middle School Youth Center, Walnut Creek, CA - Start up Funding.
Contact: Brenda Kauten, Grants Administrator; (925) 935-4138; fax (925) 935-2092
Internet: http://www.jmlongfoundation.org/Grants.html
Sponsor: J.M. Long Foundation
P.O. Box 3827
Walnut Creek, CA 94598-2827

J.N. and Macie Edens Foundation Grants 2174
Joseph Napoleon (Pole) Edens and Lilac (Macie) Edens were long time residents of Corsicana, Texas. Joseph was a self made man who from a young age managed the family ranch. For years, he bred Herefords, winning numerous prizes throughout the nation. Edens later became bank president of the Corsicana First National Bank. He started 4-H clubs and Future Farmers of America Chapters where his influence is still present today. The Edens' had a strong sense of community and established the J. N. and Macie Edens Foundation to promote the well-being of mankind. Grants are awarded to organizations in Navarro County, Texas for the promotion and extension of religion, education, the alleviation of human suffering, prevention and control of disease, to aid scientific endeavors that contribute to the betterment of mankind, for beautification of public buildings, grounds, or civic betterment and to provide relief to the poor and the indigent. The annual deadline for applications is March 1, with notification of awards on June 30.
Requirements: 501(c)3 organizations in, and serving the residents of, Navarro County, Texas, are eligible to apply.
Geographic Focus: Texas
Date(s) Application is Due: Mar 1
Contact: Debi Allen, Philanthropic Administrative Team Leader; (214) 209-1965 or (214) 209-1370; tx.philanthropic@ustrust.com
Internet: https://www.bankofamerica.com/philanthropic/foundation.go?fnId=152
Sponsor: J.N. and Macie Edens Foundation
901 Main Street, 19th Floor
Dallas, TX 75202-3714

J. Spencer Barnes Memorial Foundation Grants 2175
The J. Spencer Barnes Memorial Foundation was established in Michigan in 1999, with the intent of funding programs in and around Grand Rapids, Michigan. Its primary fields of interest include diabetes, with target populations being children and youth, as well as the general population afflicted with the disease. The major types of support are income development and program development. There are no specific application forms or annual deadlines, and applicants should begin by forwarding a proposal to the office listed. Amounts range from $500 to $2,000.
Requirements: Nonprofits in California are eligible to apply.
Geographic Focus: Michigan
Samples: Doran Foundation, Grand Rapids, Michigan, $2,000 - program support; Indian Trails Camp, Grand Rapids, Michigan, $1,000 - program support; Saint John's Home, Grand Rapids, Michigan, $1,000 - program support.
Contact: Robert C. Woodhouse, Jr., President; (616) 949-4854
Sponsor: J. Spencer Barnes Memorial Foundation
3073 East Fulton Street
Grand Rapids, MI 49506-1813

J.W. Kieckhefer Foundation Grants 2176
The Kieckhefer Foundation awards grants to nation-wide 501(c)3 organizations in support of medical research, hospices, and health agencies; family planning services; social services; higher education; youth and child welfare agencies; ecology and conservation; community funds; and cultural programs. Types of support include the following: annual campaigns; building renovation; conferences and seminars; continuing support; emergency funds; endowments; equipment; general operating support; land acquisition; matching and challenge support; program development; publication; and research.
Requirements: Applications are not accepted. Organizations should submit a letter of inquiry to the Foundation, with a description of their project and amount requested.
Restrictions: Grants are not awarded to individuals.
Geographic Focus: All States
Contact: John I. Kieckhefer, Trustee; (928) 445-4010
Sponsor: J.W. Kieckhefer Foundation
116 East Gurley Street
Prescott, AZ 86301-3821

J. Walton Bissell Foundation Grants 2177
The J. Walton Bissell Foundation Grants support 501(c)3 organizations in Connecticut, with emphasis on Hartford. The Foundation gives primarily to the arts and social services, including child welfare and programs for the blind. Funding requests should be submitted four months before funding is needed. Types of support include general operating support, program development, and seed money.
Requirements: Applicants should submit a letter of inquiry, along with a copy of their IRS determination letter, and a copy of their annual report, audited financial statement, or 990.
Restrictions: The Foundation does not grant funding to individuals or endowments.
Geographic Focus: Connecticut
Contact: J. Danford Anthony, Jr., President; (860) 586-8201
Sponsor: J. Walton Bissell Foundation
P.O. Box 370067
West Hartford, CT 06137

J. Willard Marriott, Jr. Foundation Grants 2178
Established in 1992 in Maryland in the name of the Executive Chairman of Marriott International, the J. Willard Marriott, Jr. Foundation primarily supports residents of New Hampshire, Maryland, and the District of Columbia, although funding is occasionally provided to those outside of this region. Giving is primarily in the areas of health and education, including scholarship awards only to residents of the State of New Hampshire who received their primary and/or secondary education through home schooling and are enrolled in an accredited college or university. Initial approach should be by letter, with all applications being postmarked no later than August 31. Typical awards range from $200 to $5,000.
Geographic Focus: District of Columbia, Maryland, New Hampshire
Date(s) Application is Due: Aug 31
Amount of Grant: 200 - 5,000 USD
Contact: Steven J. McNeil, (301) 380-1765 or (301) 380-3000
Sponsor: J. Willard Marriott, Jr. Foundation
1 Marriott Drive, P.O. Box 925
Washington, D.C. 20058-0003

Jack H. and William M. Light Charitable Trust Grants 2179
The Jack H. and William M. Light Charitable Trust was established in Texas in 1998, and makes grant distributions to San Antonio area non-profits on a semiannual basis. Its primary field of interest is human services, particularly the welfare of children in the broadest sense, with an effort to support organizations involved in the health, mental health, and education of children. Types of support include: annual campaigns; capital and infrastructure; capital campaigns; curriculum development; emergency funding; endowment contributions; equipment purchase; general operations; program development; and research. Most recently, awards have ranged from $2,500 to $30,000. The annual deadlines for online application submission are April 30 and October 31.
Requirements: Grants are directed to 501(c)3 organizations which fall within the principal charitable purposes of the foundation, and which have a direct impact on the residents of Bexar, Denton and Harris counties in Texas.
Geographic Focus: Texas
Date(s) Application is Due: Apr 30; Oct 31
Amount of Grant: 2,500 - 30,000 USD
Samples: Houston Pi Beta Phi Foundation, Houston, Texas, $20,000 - general operating support; Camp Allen Conference and Retreat Center, Navasota, Texas, $25,000 - general operating support; Texas Children's Hospital, Houston, Texas, $30,000 - support of the Care Survival Portal.
Contact: Brian R. Korb; (210) 283-6700 or (210) 283-6500; bkorb@broadwaybank.com
Internet: http://www.broadwaybank.com/wealthmanagement/FoundationWilliamMLight.html
Sponsor: Jack H. and William M. Light Charitable Trust
P.O. Box 17001
San Antonio, TX 78217-0001

Jackson County Community Foundation Unrestricted Grants 2180
The Foundation awards grants to nonprofit organizations covering a full range of charitable activity such as health and human services, education, economic/community development, arts and culture, and environmental quality and protection. All grants are evaluated on a competitive basis and priority is given to those projects that show potential for providing maximum benefit to community members. Application information is available online.
Requirements: The Foundation makes grants to Jackson County, Michigan nonprofit, tax-exempt organizations.
Restrictions: While the Foundation remains flexible in trying to address the needs of the community, it do not make grants to individuals (except scholarships), programs that advocate specific religions, or services which are standard government or school obligations. Applications from organizations not serving Jackson County are not accepted.
Geographic Focus: Michigan
Date(s) Application is Due: Jan 15; May 15; Sep 15
Amount of Grant: 3,000 - 50,000 USD
Samples: Smiles on Wheels Preventative Dental Health Care Project, $5,000--to purchase dental supplies and equipment; Jackson School of the Arts, $10,000--Operational support; Center for Family Health, $30,000--Jackson High Teen Health Center.
Contact: Diane McDonald, Grant & Scholarship Coordinator; (517) 787-1321; fax (517) 787-4333; jcf@jacksoncf.org
Internet: http://www.jacksoncf.org/grants.html
Sponsor: Jackson County Community Foundation
1 Jackson Square, 100 E Michigan Avenue, Suite 308
Jackson, MI 49201-1406

Jacob and Hilda Blaustein Foundation Health and Mental Health Grants 2181
Inspired by Jewish values of tzedakah (the obligation to give to the community), social justice and human rights, the Jacob and Hilda Blaustein Foundation promotes social justice and human rights through its five program areas: Jewish life; Israeli democracy; health and mental health; educational opportunity; and human rights. In the area of Health and Mental Health, the goal is is to promote quality health and mental health care for underserved individuals. There is a geographic focus on Baltimore, where the Foundation seeks to: address the health and mental health needs of low-income children and seniors through sustainable programs designed to meet service gaps; support local and national advocacy and public policy initiatives to ensure access to quality care for low-income families; and address environmental health issues. The Foundation supports innovative service provision in community-based settings, programs using evidence-based practices, and professional development. There are no annual deadlines, and most recent awards have been multi-year grants ranging from $15,000 to $150,000.
Requirements: 501(c)3 nonprofit organizations serving residents of Baltimore, Maryland, are eligible to apply.
Restrictions: Support is unavailable for the following: individuals; scholarships to individuals; unsolicited proposals for academic, scientific, or medical research; direct mail; annual giving; membership campaigns; fundraising; commemorative events.
Geographic Focus: Maryland
Amount of Grant: 15,000 - 150,000 USD
Samples: Adoptions Together, Calverton, Maryland, $50,000 - support to add third-party billing capacity to support the delivery of mental health services to children and families involved with the foster care system (2015); Historic East Baltimore Community Action Coalition, Baltimore, Maryland - two-year award for Healthy Minds at Work, a mental health program for young adults enrolled in the Youth Opportunity Center (2015); Roberta's House, Baltimore, Maryland, $50,000 - two-year award for grief and bereavement counseling for children and families (2015).
Contact: Lara A. Hall, Senior Program Officer; (410) 347-7204 or (410) 347-7201; fax (410) 347-7210; info@blaufund.org
Internet: http://www.blaufund.org/foundations/jacobandhilda_f.html#3
Sponsor: Jacob and Hilda Blaustein Foundation
One South Street, Suite 2900
Baltimore, MD 21202

Jacob and Valeria Langeloth Foundation Grants 2182
The Foundation's grantmaking program is centered on the concepts of health and well-being. The Foundation's purpose is to promote and support effective and creative programs, practices and policies related to healing from illness, accident, physical, social or emotional trauma and to extend the availability of programs that promote healing to underserved populations. The Foundation is particularly interested in funding programs that address the health of individuals who, because of barriers to difficulty accessing care. Preference is given in the areas of: caregivers and; correctional health care.
Requirements: 501(c)3 organizations that promote physical and emotional healing, especially to underserved populations, such as: community-based organizations, health care providers and research institutions in the state of New York are eligible to apply.
Restrictions: The Foundation will not consider proposals for annual or capital campaigns, for building or renovation projects, for budgetary relief, or preventive medicine. The Foundation does not support projects that focus on children or end-of-life issues. Neither will the Foundation make grants to individuals or for sectarian or religious purposes, or for political activities or lobbying. The Foundation will not fund any organization that discriminates on the basis of age, gender, national origin, race, or sexual preference.
Geographic Focus: New York
Date(s) Application is Due: Feb 1; Aug 1
Sample: AARP Foundation, Washington, D.C., $261,000-- for professional partners supporting family caregivers; Foundation for Quality Care, Albany, NY, $171,000--for geriatric nursing assistants career development program; New York Academy of Medicine, New York, NY, $200,000-- for Prison Health Reentry initiative.
Contact: Scott Moyer; (212) 687-1133; fax (212) 681-2628; smoyer@langeloth.org
Internet: http://www.langeloth.org/apply.php
Sponsor: Jacob and Valeria Langeloth Foundation
521 Fifth Avenue, Suite 1612
New York, NY 10175-1699

344 | GRANT PROGRAMS

Jacob G. Schmidlapp Trust Grants 2183
The trust supports tax-exempt organizations in the greater Cincinnati, OH, area. The trust supports charitable or educational purposes; for relief in sickness, suffering, and distress; for the care of young children, the aged, or the helpless; and for the promotion of education to improve living conditions. Support is offered in the form of: endowments; equipment; land acquisition; program development; technical assistance; and seed money grants. An organization interested in submitting a grant proposal should initially contact the Foundation office via a letter of inquiry, an application form is required for proposals. Proposals are reviewed quarterly, March, June, September and, December.
Requirements: Nonprofits in the greater Cincinnati, Ohio area receive the majority of funding, however giving is also available in the surrounding states of Indiana, Kentucky and Michigan.
Restrictions: No support for religious or political purposes. No grants to individuals; no loans.
Geographic Focus: Indiana, Kentucky, Michigan, Ohio
Date(s) Application is Due: Feb 1; May 1; Aug 1; Nov 1
Amount of Grant: 2,000 - 500,000 USD
Samples: Jewish Community Center, Cincinnati, OH, $500,000-- for capital support; Spectrum Health Foundation, Grand Rapids, MI, $100,000-- for capital support; Children Inc., Covington, KY, $100,000-- for project/program support; Xavier University, Cincinnati, OH, $250,000--for scholarship.
Contact: Heidi B. Jark, Manager; (513) 534-4397
Internet: https://www.53.com:443/wps/portal/personal
Sponsor: Jacob G. Schmidlapp Trust
38 Fountain Square Plaza, MD 1090CA
Cincinnati, OH 45263-0001

Jacobs Family Foundation Village Neighborhoods Grants 2184
The Foundation primarily serves the four neighborhoods immediately surrounding The Village at Market Creek: Chollas View, Emerald Hills, Lincoln Park, and Valencia Park; proposals are considered from throughout the Fourth City Council District of southeastern San Diego. The Neighborhood Grants Program funds programs or strategies that: strengthen and expand health, environmental, education and family resources; foster opportunities for ownership and building assets through economic and business development; create vibrant places and spaces which express and enhance the cultures and environment of the community; and/or strengthen the ability of the community to get things done and include the voice of residents in decision-making, advocacy, and planning. The Village Neighborhoods Fund makes grants to resident groups and neighborhood organizations that serve or support The Village at Market Creek. Funding is also available for a limited number of grants that support projects outside the neighborhoods that demonstrate innovative community and economic development strategies and a commitment to share what works, what doesn't, and why. Applications are excepted year round, however unsolicited applications are not. Call the Foundation to discuss your project. If the project meets the Foundations criteria, you may be asked to send a brief letter of interest and/or complete a full Village Neighborhoods Fund application.
Requirements: 501(c)3 non-profits may apply that: work with the Jacobs Center to develop The Village at Market Creek in southeastern San Diego; provide programs or services that benefit the neighborhoods surrounding The Village at Market Creek; or provides neighborhood strengthening strategies or models that can inform and advance The Village at Market Creek work.
Geographic Focus: California
Contact: Program Contact; (619) 527-6161 or (800) 550-6856; fax (619) 527-6162; communications@jacobscenter.org
Internet: http://www.jacobsfamilyfoundation.org/how.htm
Sponsor: Jacobs Family Foundation
Joe & Vi Jacobs Center, 404 Euclid Avenue
San Diego, CA 92114

James & Abigail Campbell Family Foundation Grants 2185
The James & Abigail Campbell Family Foundation embraces the values and beliefs of James and Abigail Campbell by investing in Hawwaii's people and the communities that nuture them. The Foundation supports projects in the following areas: Youth--programs that address the challenges of young people; Education--support for public schools, early childhood education and environmental stewardship; Hawaiian--support for programs that promote values and the health and welfare of Hawaiians. Priority is given to programs located in or serving communities in the following areas of West Oahu: Ewa/Ewa Beach, Kapolei, Makakilo and the Wai'anae Coast. The following types of requests are eligible for consideration: support for special projects that are not part of an organization's ongoing operations; program support when unforeseen circumstances have affected the financial base of an organization; financial assistance to purchase items such as office equipment and to fund minor repairs and renovations. Grants range from $5,000 - $50,000. Your grant application must be postmarked by February 1 for the April/May meeting, August 1 for the October/November meeting.
Requirements: The Foundation will only consider requests from organizations which qualify as non-profit, tax-exempt public charities under Section 501(c)3 and 170(b) of the Internal Revenue Code. To apply for a grant, summarize the following information in a two - three page proposal letter: the nature and purpose of your organization; the objectives of your program, include the grant amount requested and the proposed use of funds; a brief outline on how you plan to accomplish your objectives; a statement of a community problem, need or opportunity that this project will address; the duration for which Foundation funds are needed; other sources of funding currently being sought and future funding sources; methods used to measure the program's effectiveness. In addition to the proposal letter, submit a copy of the following: Internal Revenue Service notification of tax-exempt status; most recent annual financial statement; list of the current Board of Directors; the project's proposed budget; one (1) copy of your complete grant proposal package.

Restrictions: The Foundation will not consider funding for: individuals, endowments, sectarian or religious programs, loans, political activities or highly technical research projects. Only one request per organization will ordinarily be considered in a calendar year. Funds are usually not committed for more than one year at a time.
Geographic Focus: Hawaii
Date(s) Application is Due: Feb 1; Aug 1
Amount of Grant: 5,000 - 50,000 USD
Samples: Kapolei High School, $20,000--scholarships grant; Kahikolu Ohana Hale O Wai'anae, $50,000--Hawaiian culture grant; KAMP Hawai'i, $5,000--youth grant; International Dyslexia Association, $10,000--educational grant.
Contact: D. Keola, Grant Manager; (808) 674-3167; fax (808) 674-3349; keolal@jamescampbell.com
Internet: http://www.campbellfamilyfoundation.org/grant_procedures.cfm
Sponsor: James and Abigail Campbell Family Foundation
1001 Kamokila Boulevard
Kapolei, HI 96707

James A. and Faith Knight Foundation Grants 2186
Primarily serving Jackson and Washtenaw counties, Michigan, the Foundation is dedicated to improving communities by providing grant support to qualified nonprofit organizations including, but not limited to, those that address the needs of women and girls, animals and the natural world, and internal capacity. In general, the foundation supports organizations that believe in good nonprofit practices including sound financial management, developed governance practices, and clarity of mission. Attention to gender, diversity, and outcomes are important. Giving is primarily for human services, including a neighborhood center, women's organizations, and family services; support also for nonprofit management, the United Way, housing, the arts, education, and environmental conservation. Fields of interest include: adult education and literacy; basic skills; GED; arts; environment and natural resources; family services; housing development; human services; nonprofit management; and women's services. There are two annual deadlines, and initial approach should be by letter.
Requirements: Giving is limited to Michigan, with emphasis on Jackson and Washenaw counties.
Restrictions: No support is offered for religious or political organizations. No grants to individuals, or for conferences or special events, or for annual campaigns.
Geographic Focus: Michigan
Date(s) Application is Due: Jan 28; Sep 16
Samples: Center for the Childbearing Year, Ann Arbor, Michigan, $23,000--to increase capacity of the Doulas Care program, with targeted emphasis to improve maternal/infant health outcomes and develop a health career pathway for women; Dahlem Conservancy, Jackson, Michigan, $18,000--to purchase a van to initiate outreach programs.
Contact: Margaret A. Talburtt, Executive Director; (734) 769-5653; fax (734) 769-8383; peg@KnightFoundationMi.org or info@knightfoundationmi.org
Internet: http://www.knightfoundationmi.org/guidelines.htm
Sponsor: James A. and Faith Knight Foundation
180 Little Lake Drive, Suite 6B
Ann Arbor, MI 48103-6219

James Ford Bell Foundation Grants 2187
The Foundation supports organizations primarily in Minnesota. Emphasis is on cultural programs, support is also available for wildlife preservation and conservation, youth agencies, the environment, education, health and human services. A high priority is given to projects with historical connections to the Bell Family. The Trustees meet in the Spring and Fall. Contact the Foundation prior to sending in a proposal. Unsolicited requests for funds are not accepted.
Requirements: Nonprofit organizations in Minnesota are eligible to apply.
Restrictions: No grants are made directly to individuals, nor for scholarships, fellowships, or political campaigns. No funding is available to units of local government. The Foundation does not respond to requests for memberships, annual appeals, or special events and fundraisers.
Geographic Focus: Minnesota
Samples: University of Minnesota Foundation, Minneapolis, MN, $3,000,000 - for general operating support; Minneapolis College of Art and Design, Minneapolis, MN, $1,020,000 - for general operating support; Minnesota Environmental Partnership, Saint Paul, MN, $58,000 - for general operating support.
Contact: Ellen George; (612) 377-8400; fax (612) 377-8407; ellen@fpadvisors.com
Internet: https://www.fpadvisors.com/jamesfordbell/jamesfordbell.htm
Sponsor: James Ford Bell Foundation
1818 Oliver Avenue South
Minneapolis, MN 55405-2208

James G.K. McClure Educational and Development Fund Grants 2188
The fund devotes most of its resources to an ongoing scholarship program for the students of western North Carolina. When funds are available, the trustees make grants to organizations in the region, primarily those that enhance the lives of the rural citizens of the area. The trustees look favorably on requests from institutions helping to improve the lives of the people within its geographic area. Hospitals, colleges, libraries, and various schools are common recipients. The trustees meet twice each year, usually in May and October.
Requirements: Applicants must be residents of the following counties: Alleghany, Ashe, Avery, Buncombe, Burke, Caldwell, Cherokee, Clay, Graham, Haywood, Henderson, Jackson, Macon, Madison, McDowell, Mitchell, Polk, Rutherford, Swain, Transylvania, Watauga, and Yancey.
Restrictions: No student may apply for a scholarship unless he or she resides in one of the specified counties, and no grant money will be made to organizations outside this district.

Geographic Focus: North Carolina
Amount of Grant: 1,000 - 5,000 USD
Contact: John Curtis Ager, Executive Director; (704) 628-2114; jager@ioa.com
Sponsor: James G.K. McClure Educational and Development Fund
11 Sugar Hollow Lane
Fairview, NC 28730

James Graham Brown Foundation Grants 2189
The James Graham Brown Foundation fosters quality of life of Louisville in particular, and the State of Kentucky in general, by helping to make improvements for families and businesses. The Foundation gives priority to projects that: strengthen the impact of core human services and cultural organizations and agencies; support efforts to improve and sustain quality neighborhoods and a thriving downtown area; and support initiatives that strengthen the relationship between business and education to increase growth in human capital and jobs in higher wage, higher knowledge, and high technical areas. It actively supports and funds projects in the fields of education, economic development, health and social services, culture and humanities (excluding performing arts) with an emphasis on community-wide capital campaigns. The Board meets six times annually and determinations are made at the meetings. Funds are dispersed semi-annually.
Requirements: Formal applications must be submitted by the following deadline dates: March 2, Education requests; May 4 and July 6, Quality of life requests. Only organizations that have obtained a tax-exempt designation under Section 501(c)3 of the IRS code may apply. The Foundation grants are limited to the Louisville metropolitan area and the state of Kentucky. Campaigns outside the Jefferson County and Louisville metropolitan area must show evidence of significant local community support before being considered by the foundation for funding.
Restrictions: The Foundation does not support the following: organizations outside of the state of Kentucky; those related either directly or indirectly to the performing arts; requests from religious organizations for religious purposes (including theological seminaries); requests from individuals; requests from political entities; requests from national organizations, even if for local projects.
Geographic Focus: Kentucky
Date(s) Application is Due: May 6; Jul 1
Contact: Tina Walters, Grants Director; (502) 896-2440 or (866) 896-5423; fax (502) 896-1774; grants@jgbf.org or info@jgbf.org
Internet: http://www.jgbf.org/funding-areas/
Sponsor: James Graham Brown Foundation
4350 Brownsboro Road, Suite 200
Louisville, KY 40207

James H. Cummings Foundation Grants 2190
The Foundation, gives exclusively for charitable purposes in advancing medical science, research, and education in selected cities in the U.S. and Canada, and for charitable work among underprivileged boys and girls, and aged and infirm persons in designated areas. Priority is given to medical proposals. The funding is limited to: Toronto, Ontario; Canada; Hendersonville, North Carolina and; Buffalo, New York. Grants are available, in the following types of support: building/renovation; capital campaigns; equipment; land acquisition; matching/challenge support; research and seed money.
Requirements: The giving program is limited to the vicinity of the cities of Buffalo, NY; Hendersonville, NC; and Toronto, ON, Canada.
Restrictions: No support for national health organizations. Grants are not awarded to individuals, nor for: loans, annual campaigns, program support, endowment funds, operating budgets, emergency funds, deficit financing, scholarships, fellowships, publications, conferences, or continuing support.
Geographic Focus: New York, North Carolina, Canada
Amount of Grant: 2,000 - 250,000 USD
Samples: Child and Family Services Conners Children's Ctr (Buffalo, NY)--to renovate this center that provides residential care for severely disturbed boys between the ages of six and 14, $100,000; YMCA of Henderson and Henderson County (NC)--program support, $62,500; Margaret R. Pardee Hospital (NC)--capital support, $100,000.
Contact: William McFarland, Executive Director; (716) 874-0040; fax (716) 874-0040; cummings.foundation@verizon.net
Sponsor: James H. Cummings Foundation
1807 Elmwood Avenue, Room 112
Buffalo, NY 14207

James Hervey Johnson Charitable Educational Trust Grants 2191
The goal of the trust is to expose religion as against reason, publicize nontheistic views of religion, and publicize James Hervey Johnson's views on health, which are centered primarily in vegetarianism, natural hygiene, and alternative medicine. Interested applicants should contact the Foundation prior to submitting a proposal. The Foundation requires an application form and, will provide guidelines. There are no geographic restrictions, but preference is given to requests from San Diego County, CA.
Requirements: 501(c)3 nonprofits that have been in existence for at least two years are eligible to apply.
Geographic Focus: All States
Date(s) Application is Due: Jun 30
Amount of Grant: 5,000 - 100,000 USD
Contact: Kevin V. Munnelly, Treasurer; (619) 297-9036; promotions@home.com
Sponsor: James Hervey Johnson Charitable Educational Trust
P.O. Box 16160
San Diego, CA 92176-6160

James J. and Angelia M. Harris Foundation Grants 2192
The foundation's primary areas of interested include: higher and other education, health services and hospitals, social services, youth, Christian organizations and, Presbyterian churches. Types of support include: annual campaigns, building construction/renovation, capital campaigns, challenge/matching grants, program development, seed money and, scholarship funds. There are no application deadlines or forms. The board meets in May and November to consider requests. Letters of inquiry are due in April and October.
Requirements: Nonprofit organizations in Clarke County, GA, and Mecklenburg County, NC, are eligible.
Geographic Focus: Georgia, North Carolina
Amount of Grant: 1,000 - 100,000 USD
Samples: Covenant Presbyterian Church, Charlotte, NC, $35,000; Billy Graham Evangelistic Association, Charlotte, NC, $100,000; YMCA of Greater Charlotte, Charlotte, NC, $70,000.
Contact: Sherri Harrell, Grants Administrator; fax (704) 364-6046
Sponsor: James J. and Angelia M. Harris Foundation
P.O. Box 220427
Charlotte, NC 28222-0427

James J. and Joan A. Gardner Family Foundation Grants 2193
Established in Cincinnati, Ohio, in 1994, the James J. and Joan A. Gardner Family Foundation was named in honor of James Joseph Gardner, a civic-minded leader, a philanthropist, and an exemplary and loving caregiver to his wife, Joan, who suffers from Parkinson's disease. The Foundation's major purpose is to provide support for higher education, Christian and Roman Catholic churches and organizations, social services, and health organizations. With that in mind, its primary fields of interest include: Catholicism; Christianity; diseases and conditions (particularly Parkinson's); education; higher education; human services; and right to life causes. Currently, the geographic focus is on Ohio and Florida. An application form is required, along with a brief history of the applicant organization, its mission, and budgetary needs. Most recent awards have ranged from $1,000 to $50,000. There are no identified annual deadlines.
Requirements: Nonprofit 501(c)3 organizations and educational programs in both the State of Florida and the Cincinnati, Ohio, region are welcome to apply.
Geographic Focus: All States
Amount of Grant: 1,000 - 50,000 USD
Samples: Boys Hope, Girls Hope, Cincinnati, Ohio, $50,000 - general operating funds; Admiral Farragut Academy, St. Petersburg, Florida, $10,000 - general operating funds; Alex's Lemonade Stand, Wynnewood, Pennsylvania, $1,000 - general operating funds.
Contact: Regina L. Estenfelder, Financial Advisor; (513) 459-1085; fax (513) 573-0778
Sponsor: James J. and Joan A. Gardner Family Foundation
6847 Cintas Boulevard, Suite 120
Mason, OH 45040-9152

James L. and Mary Jane Bowman Charitable Trust Grants 2194
Established in 1997 in Virginia through a donation by James L. Bowman, the James L. and Mary Jane Bowman Charitable Trust Supports both education and libraries within Virginia. The Trust's primary fields of interest include: Christian agencies and churches, higher education, human services, public libraries, and recreation. There are no specific application forms or deadlines with which to adhere, and applicants should forward a letter of application that includes a needs statement, population to be served, and the project budget.
Requirements: 501(c)3 nonprofits Serving residents of Virginia are eligible.
Restrictions: The foundation does not award grants to individuals in the form of scholarships or other direct support; orpolitical causes or candidates.
Geographic Focus: Virginia
Date(s) Application is Due: Apr 1
Amount of Grant: 1,000 - 15,000 USD
Samples: Lord Fairfax Community College, Middletown, Virginia, $19,151 - general operations support; Blue Ridge Hospice, Winchester, Virginia, $10,000 - to provide hospice care for local chronically ill patients; Museum of the Shenandoah Valley, Winchester, Virginia, $6,500 - to preserve regional art, history, and culture.
Contact: Beverley B. Shoemaker, Trustee; (540) 869-1800; fax (540) 869-4225
Sponsor: James L. and Mary Jane Bowman Charitable Trust
P.O. Box 480
Stephens City, VA 22655-0480

James M. Collins Foundation Grants 2195
Established in Texas in 1964 by the late James M. Collins, a Republican who represented the Third Congressional District of Texas from 1968-1983, the Foundation awards grants to Texas nonprofit organizations in its areas of interest, including: the arts; performing arts centers; economic development; health organizations and association; higher education; human services; museums; the Salvation Army; and secondary school. Types of support include: research; program support; and social services. There are no application deadlines or formal applications, and interested partied should begin by contacting the Foundation through an inquiry letter. Most recent funding awards have ranged from $500 up to $50,000.
Requirements: Texas nonprofit organizations are eligible to apply.
Restrictions: Individuals are not eligible.
Geographic Focus: Texas
Amount of Grant: 500 - 50,000 USD
Contact: Dorothy Dann Collins Torbert, President; (214) 691-2032
Sponsor: James M. Collins Foundation
8115 Preston Road, Suite 680
Dallas, TX 75225

James M. Cox Foundation of Georgia Grants 2196
The Foundation awards grants in Cox Enterprises communities in its areas of interest, including journalism, visual and performing arts, child development, education, healthcare, environment, wildlife, and community and family services. Types of support include building construction/renovation, capital campaigns, and program development. There are no application forms or deadlines. Interested applicants should send a request to the Foundation outlining project needs and goals.
Requirements: Nonprofits in Cox Enterprises communities are eligible.
Geographic Focus: All States
Contact: Leigh Ann Launius, Assistant Secretary; (678) 645-0929; fax (678) 645-1708
Internet: http://media.corporate-ir.net/media_files/IROL/76/76341/reports/AR_2003/community.html
Sponsor: James M. Cox Foundation of Georgia
6205 Peachtree Dunwoody Road
Atlanta, GA 30328

James McKeen Cattell Fund Fellowships 2197
The nonrenewable fellowships advance the science of psychology and its useful applications by enabling faculty members to take a full year sabbatical leave for independent study and research. Annual deadline dates may vary; contact program staff for exact dates.
Requirements: Eligibility is limited to residents of North America. The fellowships are intended for tenured faculty of a psychology department who are eligible for sabbatical leave. Matching institutions' support required.
Geographic Focus: All States
Date(s) Application is Due: Dec 1
Amount of Grant: Up to 32,000 USD
Contact: Dr. Christina Williams, Program Contact; williams@psych.duke.edu
Internet: http://www.cattell.duke.edu
Sponsor: James McKeen Cattell Fund
9 Flowers Drive, Box 90086, Duke University, Dept of Psychological & Brain Sciences
Durham, NC 27708-0086

James R. Dougherty Jr. Foundation Grants 2198
The James R. Dougherty, Jr. Foundation, established in 1950, gives to organizations primarily in Texas. Areas of interest include: family services, domestic violence, human services and women. Support is offered in the form of: annual campaigns, building/renovation, capital campaigns, continuing support, curriculum development, endowments, equipment, general/operating support, income development, management development/capacity building, matching/challenge support, program development, program evaluation, research, scholarship funds, seed money and, technical assistance grants. The board meets twice a year, in the Spring and Fall. No application form is required, contact the Foundation, before submitting a proposal.
Requirements: Nonprofit organizations in Texas are eligible to apply.
Geographic Focus: Texas
Date(s) Application is Due: Mar 1; Sep 1
Contact: Daren Wilder, Grants Administrator; (512) 358-3560
Sponsor: James R. Dougherty Jr. Foundation
P.O. Box 640
Beeville, TX 78104-0640

James R. Thorpe Foundation Grants 2199
The Foundation is interested in supporting organizations, programs or projects which address the needs of the Elderly and Youth. The Thorpe Foundation is most likely to make general operating or program support grants. In the area of Youth, the Foundation supports organizations and programs that: engage youth in the arts; encourage character development; support academic and social development; provide safety and support to youth. In the area of the Elderly, the Foundation fosters the vital aging of seniors through support of services which help them to live independently. The Foundation will give special consideration to services which address the needs of economically disadvantaged seniors, immigrant seniors and the frail elderly. The Foundation supports organizations and programs that: provide transportation services; assist seniors in remaining in their own homes; educate seniors about housing options; provide social and recreational opportunities for seniors. .
Requirements: The Thorpe Foundation supports 501(c)3 non-profit organizations located in and serving the Minneapolis and the western Minneapolis metro suburbs of the Twin Cities.
Restrictions: The Foundation does not: make multi-year grants or fund individuals; make grants in the east metro area, greater Minnesota, or outside the State of Minnesota; make grants through fiscal agents; support endowment drives, conferences, seminars, tours, or fundraising events; support organizations with operating budgets over $2 million; support organizations with no paid staff.
Geographic Focus: Minnesota
Date(s) Application is Due: Mar 1; Sep 1
Contact: Kerrie Blevins; (612) 822-3412; kerrieblevins@jamesrthorpefoundation.org
Internet: http://www.jamesrthorpefoundation.org/2010guidelines.html
Sponsor: James R. Thorpe Foundation
318 West 48th Street
Minneapolis, MN 55419

James S. Copley Foundation Grants 2200
Incorporated in California in 1953, the Foundation serves as the philanthropic arm of the Copley Press, Inc., publishers of the Copley newspapers. The foundation offers support in the form of: scholarship funds; capital campaigns; equipment; employee matching gifts; endowments; equipment; building and; renovation grants. Funding organizations involved with arts and culture, education, animals and wildlife, health, recreation, and human services should apply. Giving is restricted primarily in areas of company operations in California, Illinois and Ohio.
Requirements: 501(c)3 tax-exempt organizations in the circulation areas of company newspapers/operations are eligible to apply. These areas include: California, Illinois and Ohio.
Restrictions: Ineligible funding opportunities include: religious, fraternal, or athletic organizations; government agencies, local chapters of national organizations, public elementary, secondary schools, public broadcasting systems, individuals, research, publications, conferences, general operating support, large campaigns and loans.
Geographic Focus: California, Illinois, Ohio
Date(s) Application is Due: Jan 2
Contact: Kim Koch, Secretary; (858) 454-0411, ext. 7671
Sponsor: James S. Copley Foundation
P.O. Box 1530
La Jolla, CA 92038-1530

James S. McDonnell Foundation Brain Cancer Research Collaborative Activity Awards 2201
The Foundation offers Collaborative Activity Awards to initiate interdisciplinary discussions on problems or issues, to help launch interdisciplinary research networks, or to fund communities of researchers/practitioners dedicated to developing new methods, tools, and applications of basic research to applied problems. All proposals submitted to the foundation must clearly link the experimental models and questions to human disease. Proposals primarily intending to characterize basic mechanisms of growth and development that may plausibly but are not yet known to be contributory to human brain cancer are not encouraged. Proposals testing molecules as possible treatment interventions should consider including tests designed to uncover unintended biological effects of such molecules that would disqualify future clinical usefulness. The Foundation is particularly interested in supporting novel research that will generate new knowledge leading to increased rates of survival and improve functional recovery for individuals with brain cancer.
Restrictions: The Foundation does not fund: undergraduate tuition, stipends, scholarships, fellowships, research or travel expenses, or other educational expenses; graduate or postdoctoral stipends, scholarships, fellowships, research or travel expenses, or other educational expenses (exceptions include requests that qualify as allowable budget items as part of specific Research Award applications); expenses tied to projects whose explicit goal is the publication of a book or other bound volume (although publication of a book or special issue of a journal may be one goal or outcome of work funded by the Foundation); expenses tied to the establishment or day-to-day running of a journal or small press; scientific meetings or workshops other than those JSMF puts together or which are affiliated with Collaborative Activity Awards; ongoing operational support for university-based centers, programs, or institutes; professional society meetings or specific sessions at such meetings; charitable functions, museum exhibitions, or similar causes or events; or charitable donations to individuals and organizations.
Geographic Focus: All States
Contact: Cheryl Washington, Grants Manager; (314) 721-1532; fax (314) 721-7421; washington@jmsf.org or info@jmsf.org
Internet: http://www.jsmf.org/programs/bc/index.htm
Sponsor: James S. McDonnell Foundation
1034 S Brentwood Boulevard, Suite 1850
Saint Louis, MO 63117

James S. McDonnell Foundation Complex Systems Collaborative Activity Awards 2202
The Foundation offers Collaborative Activity Awards to initiate interdisciplinary discussions on problems or issues, to help launch interdisciplinary research networks, or to fund communities of researchers/practitioners dedicated to developing new methods, tools, and applications of basic research to applied problems. The Program supports scholarship and research directed toward the development of theoretical and mathematical tools that can be applied to the study of complex, nonlinear systems. It is anticipated that research funded in this program will address issues in fields such as biology, biodiversity, climate, demography, epidemiology, technological change, economic Complex Systems development, governance, or computation. While the program's emphasis is on the development and application of theoretical models used in these research fields and not on particular fields per se, JSMF is particularly interested in projects attempting to apply complex systems approaches to real world problems. Proposals attempting to apply tools and models to problems where such approaches are not yet considered usual or mainstream (for example, differentiating normal physiology from disease) are encouraged.
Restrictions: The Foundation does not fund: undergraduate tuition, stipends, scholarships, fellowships, research or travel expenses, or other educational expenses; graduate or postdoctoral stipends, scholarships, fellowships, research or travel expenses, or other educational expenses (exceptions include requests that qualify as allowable budget items as part of specific Research Award applications); expenses tied to projects whose explicit goal is the publication of a book or other bound volume (although publication of a book or special issue of a journal may be one goal or outcome of work funded by the Foundation); expenses tied to the establishment or day-to-day running of a journal or small press; scientific meetings or workshops other than those JSMF puts together or which are affiliated with Collaborative Activity Awards; ongoing operational support for university-based centers, programs, or institutes; professional society meetings or specific sessions at such meetings; charitable functions, museum exhibitions, or similar causes or events; or charitable donations to individuals and organizations.
Geographic Focus: All States
Contact: Cheryl Washington, Grants Manager; (314) 721-1532; fax (314) 721-7421; washington@jmsf.org or info@jmsf.org
Internet: http://www.jsmf.org/programs/cs/index.htm

Sponsor: James S. McDonnell Foundation
1034 S Brentwood Boulevard, Suite 1850
Saint Louis, MO 63117

James S. McDonnell Foundation Research Grants   2203
The Awards are designed to support research projects with a high probability of generating new knowledge and insights. Projects submitted for funding consideration should be at an early, even preliminary stage of development that intend to break new ground or to challenge commonly-held assumptions. Projects submitted should be sufficiently novel, cross-disciplinary, or heterodox so that they have a strong likelihood of influencing the development of new ways of thinking about important problems. Awards provide adequate, flexible funding over a sufficient time period to allow investigators to pursue and develop innovative directions to their research programs. Funds can be expended over a minimum of 3 years or a maximum of 6 years. Smaller amounts of money expended over shorter amounts of time may be requested to help investigators pursue pilot projects or test the feasibility of an experimental approach. The applicant can apply the grant funds towards any research-based expense, including travel, equipment, and supplies. Funds can be used to support collaborative projects. A percentage of the funds can also be used to support small workshops organized by the applicant where the goal of the workshop is to gather expertise in support of the research objective.
Requirements: Applications for grants are considered only from 501(c)3 organizations which are not private foundations. Applications from foreign organizations will be considered only if they provide written legal opinion that they qualify as tax-exempt under these sections of the U.S. IRS code or if a qualifying American organization is authorized to receive funds for them under this code. Individuals also are eligible.
Geographic Focus: All States
Amount of Grant: Up to 450,000 USD
Contact: Cheryl Washington; (314) 721-1532; fax (314) 721-7421; washington@jmsf.org
Internet: http://www.jsmf.org/apply/research/index.htm
Sponsor: James S. McDonnell Foundation
1034 S Brentwood Boulevard, Suite 1850
Saint Louis, MO 63117

James S. McDonnell Foundation Scholar Awards   2204
The Awards program derives from and is consistent with the foundation's commitment to supporting high quality research and scholarship leading to the generation of new knowledge and its responsible application. Currently, the Awards are only available in the Brain, Mind & Behavior program area, and provide largely unrestricted funding over a sufficient time period to allow investigators to pursue and develop new directions to their research programs.
Requirements: Applications must be sponsored by a nonprofit institution as defined by Section 501(c)3 of the United States Internal Revenue Tax Code. Eligible Scholar Award nominees must have completed all doctoral, postdoctoral, or fellowship training and hold an independent research position. Nominees in the earlier career stages are encouraged.
Restrictions: Researchers with current grant support from JSMF are not eligible to apply for a Scholar Award until all prior JSMF grant funds have been fully expended and a final grant report has been received and approved by the foundation office.
Geographic Focus: All States
Amount of Grant: Up to 600,000 USD
Contact: Cheryl Washington; (314) 721-1532; fax (314) 721-7421; washington@jmsf.org
Internet: http://www.jsmf.org/apply/scholar/index.htm
Sponsor: James S. McDonnell Foundation
1034 S Brentwood Boulevard, Suite 1850
Saint Louis, MO 63117

James S. McDonnell Foundation Understanding Human Cognition Awards   2205
The long-standing interest of the James S. McDonnell Foundation on human mind/brain is reflected in changes made to the 2008 RFA, including the explicit emphasis on understanding human cognition. The program is intended to help investigators pursue experiments designed to answer well-articulated questions. JSMF Scholar Awards support research studying how neural systems are linked to and support cognitive functions and how cognitive systems are related to an organism's (preferably human) observable behavior. Studies with model organisms should justify why such models were selected and how data obtained from models advances our understanding of human cognition. Proposals proposing to use functional imaging to identify the neural correlates of cognitive or behavioral tasks (for example, mapping the parts of the brain that light up when different groups of subjects play chess, solve physics problems, or choose apples over oranges) are not funded through this program. In general, JSMF and its expert advisors have taken an unfavorable view of projects attempting too wide a leap in a single bound. Functional imaging studies using poorly characterized tasks as proxies for complex behavioral issues involving empathy, moral judgments, or social decision-making are generally not appropriate responses to this call for proposals. In past competitions, proposals structured along such lines were eliminated from funding consideration early in the review process.
Requirements: Aspects of proposals appropriate to the JSMF UHC program would include, but are not limited to: characterizing the cognitive operations involved in performing a task; studying how the brain identifies, extracts and uses relevant information from complicated environments; examining how manipulations and/or perturbations at one spatial or temporal scale are meaningful at finer or coarser levels of organization (e.g. does a synaptic change account for a change in network function and vice versa); re-examining common wisdom assumptions (such as the existence of critical periods in human learning); evaluating the usefulness of methodologies or improving the usefulness of methodologies commonly used in mind/brain research; applying approaches and knowledge from cognitive psychology or cognitive science to important problems in education, training, or rehabilitation; taking a comparative, evolutionary approach to characterizing the uniqueness of the human brain and of human cognition.
Restrictions: The Foundation does not fund: undergraduate tuition, stipends, scholarships, fellowships, research or travel expenses, or other educational expenses; graduate or postdoctoral stipends, scholarships, fellowships, research or travel expenses, or other educational expenses (exceptions include requests that qualify as allowable budget items as part of specific Research Award applications); expenses tied to projects whose explicit goal is the publication of a book or other bound volume (although publication of a book or special issue of a journal may be one goal or outcome of work funded by the Foundation); expenses tied to the establishment or day-to-day running of a journal or small press; scientific meetings or workshops other than those JSMF puts together or which are affiliated with Collaborative Activity Awards; ongoing operational support for university-based centers, programs, or institutes; professional society meetings or specific sessions at such meetings; charitable functions, museum exhibitions, or similar causes or events; or charitable donations to individuals and organizations.
Geographic Focus: All States
Contact: Cheryl Washington; (314) 721-1532; fax (314) 721-7421; washington@jmsf.org
Internet: http://www.jsmf.org/programs/uhc/index.htm
Sponsor: James S. McDonnell Foundation
1034 S Brentwood Boulevard, Suite 1850
Saint Louis, MO 63117

Jane's Trust Grants   2206
The Jane's Trust has particular interest in organizations and projects which primarily benefit underserved populations and disadvantaged communities. Jane's Trust will support grants for general operating purposes, as well, as it's Fields of Interest, which are: arts and culture, education, the environment, health and welfare. The Trust will make grants in the states of Florida, with a preference for southwest and central Florida; Massachusetts, with a preference for greater Boston and eastern Massachusetts; and in the northern New England states of Maine, New Hampshire and Vermont. Preference will be given to organizations located in those states for projects which will primarily provide benefits within those states. The application process begins by submitting of a concept paper. Guidelines are available at the Trusts website.
Requirements: 501(c)3 tax-exempt organizations in Florida, Maine, Massachusetts, New Hampshire, and Vermont are eligible.
Restrictions: Jane's Trust will not support: loans to charitable organizations; attempts to influence legislation; requests from individuals. Please note: Jane's Trust will normally not support public entities, such as municipalities, municipal departments, or public schools directly, but will entertain applications from tax-exempt fiscal agents or partners for collaborative projects with municipalities or schools. This does not apply to public colleges and universities.
Geographic Focus: Florida, Maine, Massachusetts, New Hampshire, Vermont
Date(s) Application is Due: Jan 25; Jul 15
Amount of Grant: 50,000 - 1,000,000 USD
Samples: Good Shepherd Food-Bank, Auburn, ME, $30,000--for operating support, payable over one year for agency matching funds project to provide for local food pantries and agencies across Maine; Boca Grande Health Clinic Foundation, Inc., Boca Grande, FL, $150,000--capital support in the amount of $100,000 for the renovation of the Boca Grande Health Clinic Annex, $50,000 in endowment support, for the Boca Grande Health Clinic; The Conservation Fund, Arlington, VA, $100,000--capital support for protection of the Three Sisters Springs in Crystal River, Florida, a wintering manatee habitat; Kimball Union Academy, Meriden, NH, $150,000-- capital support, payable over two years for construction and renovation of school campus; Vermont Symphony Orchestra, Burlington, VT, $100,000--endowment support for Symphony Kids education programs.
Contact: Susan Fish; (617) 227-7940, ext. 775; fax (617) 227-0781; sfish@hembar.com
Internet: http://www.hembar.com/selectsrv/janes/index.html
Sponsor: Hemenway and Barnes LLP
60 State Street
Boston, MA 02109-1899

Jane Beattie Memorial Scholarship   2207
The purpose of the fund is to provide scholarships to subsidize travel to the U.S. for purposes of scholarly activity by a foreign scholar in the area of judgment and decision research, broadly defined. Attendance at the annual SJDM meetings is one example of an activity that would be appropriate for support, but by no means the only one. In most years, the Fund awards one or two scholarships in amounts of $400 - $700 each.
Requirements: Applicants should be scholars living and working in a country other than the U.S. who will use the award to help pay for travel to the U.S. for scholarly activities associated with research in judgment and decision making. It is anticipated that most awards will be granted to faculty or graduate students at colleges and universities, especially recent and soon-to-be Ph.D.'s, but others will also be considered. The required application form can be downloaded from: http://gsbwww.uchicago.edu/fac/joshua.klayman/more/BeattieForm.rtf
Restrictions: U.S. Scholars are not eligible to apply.
Geographic Focus: All States
Date(s) Application is Due: Jul 26
Amount of Grant: 400 - 700 USD
Contact: Joshua Klayman, (773) 834-9134; joshk@uchicago.edu
Internet: http://www.sjdm.org/content/beattie-award
Sponsor: Society for Judgment and Decision Making
College of Business, Florida State University, P.O. Box 3061110
Tallahassee, FL 32306-1110

Jane Bradley Pettit Foundation Health Grants 2208
The Jane Bradley Pettit Foundation awards grants to initiate and sustain projects in the Greater Milwaukee community, with a focus on programs and projects that serve low-income and disadvantaged individuals, women, children, and the elderly. In the area of Health, the Foundation gives priority to community-based health care and prevention programs which address the physical and mental health needs of families, children, persons at-risk and the elderly. Support and advocacy for victims of abuse and neglect are also a priority. An initial application should be in the form of a letter of intent.
Requirements: Milwaukee-area charitable organizations are eligible.
Geographic Focus: Wisconsin
Date(s) Application is Due: Jan 15; May 15; Sep 15
Amount of Grant: 5,000 - 100,000 USD
Contact: Heidi Jones, Director of Administration; (414) 982-2880 or (414) 982-2874; fax (414) 982-2889; hjones@staffordlaw.com
Internet: http://www.jbpf.org/guidelines/index.html
Sponsor: Jane Bradley Pettit Foundation
1200 N. Mayfair Road, Suite 430
Wauwatosa, WI 53226-3282

Janirve Foundation Grants 2209
The Janirve Foundation, established in 1954, givings primarily to colleges and universities in western North Carolina. However, the Foundation also supports organizations involved with: children and youth services; community/economic development; environment; environment, natural resources; environment, plant conservation; family services; health organizations, association; higher education; hospitals (general); housing/shelter, development; human services.
Requirements: Applicants should contact Asheville, North Carolina, office for application procedures. Proposals from colleges and universities are considered only in the front quarter. Application form required. Applicants should submit the following: copy of IRS Determination Letter; detailed description of project and amount of funding requested; listing of additional sources and amount of support; how project's results will be evaluated or measured; how project will be sustained once grant support is completed; copy of current year's organizational budget and/or project budget; copy of most recent annual report/audited financial statement/990; listing of board of directors, trustees, officers and other key people and their affiliations; signature and title of chief executive officer; five copies of the proposal.
Restrictions: No support for public and private elementary schools, or churches and religious programs. No grants to individuals (except for scholarships), or generally for operating budgets, endowments or for research programs, publication of books or printed material, theatrical productions, videos, radio or television programs; no loans.
Geographic Focus: North Carolina
Date(s) Application is Due: Mar 1; Jun 1; Sep 1; Dec 1
Amount of Grant: 10,000 - 750,000 USD
Contact: E. Charles Dyson; (828) 258-1877; fax (828) 258-1837; janirve@charterinternet.com
Sponsor: Janirve Foundation
1 North Pack Square, Suite 416
Asheville, NC 28801-3409

Janson Foundation Grants 2210
The Janson Foundation was established in 1983 to provide capital grants to deserving charitable organizations located in Skagit County, Washington. Edward W. Janson had a longtime career as a public servant in Skagit County and was committed to the citizens of the county. He established his foundation to assist the ongoing capital needs of deserving charitable organizations for the purchase of equipment, the construction of facilities, or the acquisition of land. In addition, the foundation funds the repair of existing facilities and the replacement of aging equipment. The majority of grants from the Janson Foundation are one year in duration; on occasion, multi-year support is awarded. Application materials are available for download at the grant website. Applicants are strongly encouraged to review the state application guidelines for additional helpful information and clarifications before applying. Applicants are also encouraged to review the foundation's funding history (link is available from the grant website). The application deadline for the Janson Foundation is February 28. Applicants will be notified of grant decisions by April 15.
Requirements: Applicants must have 501(c)3 status. Application must be mailed.
Restrictions: Grant requests for general operating or program support will not be considered. The foundation does not support requests from individuals, organizations attempting to influence policy through direct lobbying, or any political campaigns.
Geographic Focus: Washington
Date(s) Application is Due: Feb 28
Contact: Heidi Gordon, Vice President; (800) 848-7177; heidi.e.gordon@baml.com
Internet: https://www.bankofamerica.com/philanthropic/fn_search.action
Sponsor: Janson Foundation
800 5th Avenue, WA1-501-33-23
Seattle, WA 98104

Janus Foundation Grants 2211
The Foundation strives to help communities reach greater levels of self-sufficiency, and impact the lives of many in each community. The Foundation awards grants to nonprofit organizations in its areas of interest, including: at-risk youth through education; community service and volunteerism; and cultural institutions in the Denver-metro area. Interested applicants should complete the application available online. There is no application deadline. Please note that the Foundation accepts grant applications from nonprofit organizations throughout the U.S. for the first two giving areas. The third giving area only applies to cultural institutions that operate in the Denver, CO metro area.
Requirements: 501(c)3 tax-exempt organizations are eligible.
Geographic Focus: All States
Contact: Traci Papantones, Vice President; (303) 333-7863; fax (303) 394-7797; janusfoundation@janus.com
Internet: http://ww4.janus.com/Janus/Retail/StaticPage?jsp=Janushome/JanusFoundation.jsp
Sponsor: Janus Foundation
151 Detroit Street, 4th Floor
Denver, CO 80206-4805

Jasper Foundation Grants 2212
The purpose of the Foundation is to assist donors in creating assets to meet the ongoing and changing charitable interests of those living in Jasper County, Indiana. Special emphasis is placed on programs that enrich the quality of life of our community in five areas: arts and culture; preservation of historic and cultural resources; education; health; and social concerns. Favored grant requests will generally reflect the following characteristics: projects which propose practical solutions to community needs; projects which promote cooperation among existing agencies without duplicating services; projects that will become self-sustaining without requiring on-going Foundation funds. Applications and guidelines are available at the Foundation office or website. The Foundation has two grant cycles in the spring and fall.
Requirements: Applicants must submit 5 complete copies of the grant application to include the following: the grant application and supporting materials; an outline that addresses the project's goals and objectives, plan of implementation, project budget, staff, method of evaluation, and why the proposal is important to the community; a list of the officers and board of directors; a current financial report; and a copy of the organization's 501(c)3 tax exempt IRS letter.
Restrictions: Grants are not made for budget deficits or sectarian religious activities.
Geographic Focus: Indiana
Date(s) Application is Due: Apr 1; Oct 1
Contact: Linda Reiners; (219) 866-5899; fax (219) 866-0555; jasper@liljasper.com
Internet: http://www.jasperfdn.org/grants.htm
Sponsor: Jasper Foundation
301 N Van Rensselaer Street, P.O. Box 295
Rensselear, IN 47978

Jay and Rose Phillips Family Foundation Grants 2213
The Jay and Rose Phillips Family Foundation has five funding priorities that guide it's grant making: strengthening families; supporting training and education for lifelong success; improving health and wellness; promoting independence and inclusion for people with disabilities and the elderly; and fostering good relations and civic participation. These funding priorities are the combined product of responding to community needs, the legacy of our founders and the current passions and interests of it's Trustees. Across all of these funding priorities, the Foundation place's a high value on supporting efforts that serve the most vulnerable in the community and those with the least access to resources. The Foundation has three program grant rounds each year with deadlines of March 15, July 15 and November 15. All proposals be submitted electronically using an online proposal submission process. The online submission deadline is 11:59 p.m. on the deadline date. The Foundation accepts requests for capital grants once per year through an online Letter of Inquiry (LOI) process. LOI's must be submitted by 11:59 p.m. on January 8.
Requirements: The Foundation awards grants only to organizations which are tax-exempt and publicly supported under Section 501(c)3 of the Internal Revenue Service Code. Grants are awarded primarily in the Twin Cities metropolitan area. Unsolicited proposals are not accepted from organizations located outside of Minnesota.
Restrictions: The foundation does not make grants in support of individuals, for political campaigns, or for lobbying efforts to influence legislation.
Geographic Focus: Minnesota
Date(s) Application is Due: Jan 8; Mar 15; Jul 15; Nov 15
Amount of Grant: 10,000 - 100,000 USD
Contact: Dana Jensen, Grants Manager; (612) 623-1652 or (612) 623-1654; fax (612) 623-1653; djensen@phillipsfnd.org
Internet: http://www.phillipsfnd.org/index.asp?page_seq=15
Sponsor: Jay and Rose Phillips Family Foundation
East Bridge Building, 10 Second Street NE, Suite 200
Minneapolis, MN 55413

Jayne and Leonard Abess Foundation Grants 2214
The Jayne and Leonard Abess Foundation was established in Miami, Florida, in 2004, by the son of City National Bank co-founder, Leonard L. Abess, Sr. The Foundation supports a variety of causes both locally and nationally. Areas of interest include: design arts education, writer's, employment training, human services programs, Jewish agencies and temples, children with special needs, and higher education. The primary type of support is general operating funds. There are no specific applications or deadlines with which to adhere, and applicants should contact the Foundation office directly in writing. This initial contact should include a detailed description of the program, and a budget narrative. Most recently, awards have ranged from $1,000 to $100,000.
Geographic Focus: All States
Amount of Grant: 1,000 - 100,000 USD
Samples: The Gunnery, Washington, Connecticut, $100 - general operating support; American Diabetes Association, Miami, Florida, $1,000 - general operating support; Our Pride Academy, Miami, Florida, $5,000 - general operating support.
Contact: Leonard L. Abess; (212) 632-3000 or (212) 632-3200; informed@ftci.com
Sponsor: Jayne and Leonard Abess Foundation
600 Fifth Avenue
New York, NY 10020-2302

Jean and Louis Dreyfus Foundation Grants 2215
The Jean and Louis Dreyfus Foundation was established in 1979 from the estate of Louis Dreyfus, a music publisher, and that of his wife, Jean. The mission of the Foundation is to enhance the quality of life of New Yorkers, particularly the aging and disadvantaged. The Foundation disburses grants mainly within the five boroughs of New York City, and supports programs in the arts, health and social services (including youth agencies, women, and the elderly), and education (including literacy). Support is given for program development and matching funds. Application forms are not required. The board meets each year in the spring and in the fall. Initial inquiries should consist of a one or two page letter describing the organization and outlining the project in question. The January 15th and July 15th deadlines are for Letters of Intent.
Requirements: New York City nonprofits are eligible.
Restrictions: Grants are never made to individuals.
Geographic Focus: New York
Date(s) Application is Due: Jan 15; Jul 15
Contact: Jessica Keuskamp; (212) 599-1931; fax (212) 599-2956; jldreyfusfdtn@hotmail.com
Internet: http://foundationcenter.org/grantmaker/dreyfus/guide.html
Sponsor: Jean and Louis Dreyfus Foundation
420 Lexington Avenue, Suite 626
New York, NY 10170

Jeffrey Thomas Stroke Shield Foundation Research Grants 2216
The Dr. Jeffrey Thomas Stroke Shield Foundation (JTSSF) welcomes submissions of grant requests for funding covering all aspects of stroke prevention, treatment and awareness. JTSSF is actively seeking to provide seed money grants for projects that are important, achievable, and innovative that will ultimately lead to stroke prevention and a better treatment and quality of life for stroke sufferers. From the Foundation's perspective within clinical stroke treatment and research, it understands the importance of directing targeted research funds to the scientists and investigators who can move the best ideas in stroke research prevention and treatment forward. Letters of Intent (LOIs) are required and must be postmarked by February 28. All LOI submissions will undergo a review and selection process by the JTSSF Scientific Advisory Board. Only those projects deemed to be of suitable scientific merit will be invited to submit a full application. If invited to submit an application, applicants will receive additional instructions and information necessary for submission. Final grant applications are due by May 16.
Requirements: Principal investigators who hold appointments/positions in accredited hospitals and clinics, healthcare institutions, academic medical centers, professional associations, member societies and other not?for?profit entities in the U.S. are welcome to apply.
Geographic Focus: All States
Date(s) Application is Due: May 16
Contact: Nancy Weber; (415) 830-6031; info@strokeshieldfoundation.org
Internet: http://strokeshieldfoundation.org/researchers/grant-program/
Sponsor: Jeffrey Thomas Stroke Shield Foundation
3053 Fillmore Street, #268
San Francisco, CA 94123

Jeffris Wood Foundation Grants 2217
The Jeffris Wood Foundation funds grants to community-based organizations working to provide opportunities for urban youth and economically disadvantaged. The Foundation supports programs that: help urban youth improve their futures, explore their creativity and avoid unwanted pregnancies; help Native American youth connect with their cultural traditions; connect low-income urban youth to nature; and provide services for domestic violence survivors. Current deadlines are posted on the website.
Requirements: Organizations must first submit a one-page letter of inquiry that includes a description of the organization and the project. They should also attach a brief budget describing income and expenses, plus complete contact information. If the request is screened for further consideration, the organization will be asked to submit a full proposal. All materials must be sent by U.S. mail.
Restrictions: Grants are not made to: individuals; scholarships; schools; capital expenses or renovation projects; food or shelter programs (except domestic violence); research or publications; video & web productions; religious programs that are not all inclusive; or athletic events and sponsorships.
Geographic Focus: Washington
Amount of Grant: 1,000 - 3,000 USD
Samples: Broadview Emergency Shelter, Seattle, Washington, serving 400 homeless women with children, $3,000; Neighborcare Health Homeless Youth Clinic, Seattle, Washington, outreach program that meets youth where they live or congregate, $3,000.
Contact: Therese Ogle, Grants Consultant; (206) 781-3472; OgleFounds@aol.com
Internet: http://foundationcenter.org/grantmaker/jeffriswood/guide.html
Sponsor: Jeffris Wood Foundation
6723 Sycamore Avenue NW
Seattle, WA 98117

JELD-WEN Foundation Grants 2218
The Foundation seeks to improve the quality of life in company-operating areas and awards grants in the areas of capital campaigns, education, youth activities, community development, health and medical, and arts and humanities. Types of support include building construction/renovation, equipment, general operating support, land acquisition, matching/challenge support, program development, scholarship funds, and seed money. Prescreening and application information are available online.
Requirements: Nonprofit organizations in the Foundation's area of operations.
Geographic Focus: All States
Amount of Grant: 1,000 - 1,000,000 USD
Samples: Knox Community Hospital, Mt. Vernon, OH, $500,000--grant to help build an outpatient cancer treatment center; Crater Lake Science and Learning Center, Crater Lake, OR, $500,000--to help build the Science and Learning Center at Crater lake National Park; Yakima Valley Museum, Yakima, WA, $250,000--to upgrade and improve the museum facility.
Contact: JELD-WEN Foundation Headquarters; (503) 478-4478; fax (503) 478-4474
Internet: http://www.jeld-wenfoundation.org/
Sponsor: Jeld-Wen Foundation
200 SW Market Street, Suite 550
Portland, OR 97201

Jenkins Foundation: Improving the Health of Greater Richmond Grants 2219
The foundation is committed to expanding access to community-based services through programs and organizations that have the potential to make a significant impact on the quality of health, especially for the youth in its local area.
Requirements: The current areas of focus are: Expanding access to health care services for the uninsured and underserved; Substance abuse prevention services that promote healthy lifestyles and increase availability of services; Violence prevention services that promote safe and healthy environments for children and their families and work toward the elimination of violence in the local communities; In addition, the foundation is committed to the long-term viability of the organizations it supports, and will consider capacity building grants that strengthen an agency's ability to better serve its clients. The foundation will also consider a limited number of proposals outside the above stated focus areas.
Restrictions: Proposals will be accepted from charitable organizations, which serve the residents of the City of Richmond and the counties of Chesterfield, Hanover, Henrico, Goochland, and Powhatan.
Geographic Focus: Virginia
Date(s) Application is Due: May 5; Nov 5
Amount of Grant: Up to 50,000 USD
Contact: Elaine Summerfield, Program Officer; (804) 330-7400; fax (804) 330-5992; esummerfield@tcfrichmond.org
Internet: http://www.tcfrichmond.org/Page2954.cfm#Jenkins
Sponsor: Community Foundation Serving Richmond and Central Virginia
7501 Boulders View Drive, Suite 110
Richmond, VA 23225

Jennings County Community Foundation Grants 2220
The Foundation seeks to serve philanthropic and charitable needs in Jennings County, Indiana, by offering endowment services, grant making, scholarships, donor estate and planned gift services to individuals and qualified organizations serving the community of Jennings County, Indiana. The Foundation's fields of interest for all ages include the following: community service; social service; education; health; environment; and the arts. The Foundation reviews proposals in the spring and fall. Applications will be made available at the spring and fall cycle grant explanation meeting as announced in the local media. At this mandatory meeting the grant guidelines and process will be explained. The grant proposal form is available at the website.
Requirements: Nonprofit 501(c)3 organizations, coalitions, community associations, and other civic groups may apply if providing services in Jennings County. Applicants must also attend an explanation meeting of the guidelines for grants.
Restrictions: The Foundation does not fund political activity; sectarian religious activity; endowment purposed except for limited experimental or demonstration periods; operating budgets except for limited experimental or demonstration periods; and salaries.
Geographic Focus: Indiana
Contact: Barb Shaw, Executive Director; (812) 346-5553; jcffdirector@comcast.net
Internet: http://www.jenningsfoundation.net/index.html
Sponsor: Jennings County Community Foundation
111 North State Street
North Vernon, IN 47265-1510

Jennings County Community Foundation Women's Giving Circle Grant 2221
The Women's Giving Circle was established to make a lasting impact in the lives of women and children in Jennings County. Grants are available to nonprofit organizations that have a need for assistance on a community project. Grants are for organizations that meet the needs of women and children in the fields of community service, social service, education, health, environment and the arts. The application is available at the website.
Geographic Focus: Indiana
Contact: Barb Shaw; (812) 523-4483 or (812) 346-5553; jccf@jenningsfoundation.net
Sandy Vance, Grant Contact; (812) 592-1280
Darlene Bradshaw, Grant Contact; (812) 346-1742
Linda Erler, Grant Contact; (812) 873-7421
Internet: http://www.jenningsfoundation.net/php/wgc.php
Sponsor: Jennings County Community Foundation
111 North State Street
North Vernon, IN 47265-1510

Jerome and Mildred Paddock Foundation Grants 2222
The Foundation was established in 1967 to benefit disadvantaged children and elderly populations. Primarily, the foundation is interested in funding programming needs that will improve the quality of life of those less fortunate. Field of interest include: aging; children and youth services; programs for economically disadvantaged; health care support; and human/social services. Though giving is focused in and around Sarasota, the foundation does extend funding beyond Florida. The deadline is January 10th, with final notification by March 15th.

Restrictions: No grants to individuals, or for endowments, debt reduction, operating expenses, conferences or seminars, workshops, travel, surveys, advertising, research, fund raising or for annual campaigns or capital campaigns.
Geographic Focus: All States
Date(s) Application is Due: Jan 10
Amount of Grant: 15,000 USD
Samples: Florida Studio Theatre, Sarasota, Florida, $10,000--to partially fund the VIP Performing Arts Program for special needs youth; Shodair Childrens Hospital, Helena, Montana, $30,000--to purchase a Coulter automated cell counter; Samaritan Counseling Services, Ann Arbor, Michigan, $10,000--to assist low income, underinsured and uninsured clients.
Contact: Joan Greenwood; (941) 361-5803; joan.greenwood@wachovia.com
Sponsor: Jerome and Mildred Paddock Foundation
Wachovia Bank, NA., P.O. Box 267
Sarasota, FL 34230

Jerome Robbins Foundation Grants 2223
Jerome Robbins established the Foundation in 1958, in honor of his mother, with the intent to support dance, theater, and their associative arts. In the 1980's, following the outbreak of AIDS, he directed Foundation resources almost exclusively to the AIDS crisis and still later, in letters left to the board, he conveyed his wish that the Foundation once again extend its resources to the performing arts - dance and theater especially. In line with the founder's life, financial support is offered - primarily in the New York City metro area - for dance, theater and groups dedicated to serving those with HIV and AIDS with an emphasis on the artistic community. Applications may take three to four months to evaluate and funds for specific projects are not granted retroactively. Fiscal sponsorship is acceptable. Unless otherwise specified, Foundation grants apply to general operating costs. Between one hundred and two hundred awards are given annually, ranging from $1,000 to $125,000 each.
Requirements: All suggestions for proposals must be submitted via email, and must be in the form of two pdf files and contain only the following information: no more than two pages outlining the organization's activities and the nature of the request; the organization's or project's latest budget; current funding sources; the organization's address; the organization's IRS statement of tax-exempt status (applies only to new proposals); and the organization's latest audited financial statement or tax return, submitted as a separate and second pdf file. If interested, the Foundation will invite a full proposal.
Geographic Focus: All States
Amount of Grant: 1,000 - 125,000 USD
Samples: Alvin Ailey Dance Foundation, New York, New York, $5,000 general operating support (2014); Brooklyn Academy of Music, Brooklyn, New York, $121,426 - general operating support (2014); Manhattan Theater Club, New York, New York, $7,500 - general operating support (2014).
Contact: Christopher Pennington, Executive Director; (212) 367-8956; fax (212) 367-8966; pennington@jeromerobbins.org
Internet: http://www.jeromerobbins.org/foundation
Sponsor: Jerome Robbins Foundation
156 W. 56th Street, Suite 900
New York, NY 10019

Jessica Stevens Community Foundation Grants 2224
The Jessica Stevens Community Foundation (JSCF) is taking a different approach in encouraging philanthropy. They are offering up to $12,000 in challenge grants ranging from $500 to $3,000 to nonprofits located in or serving the Northern Susitna Valley, which includes the Trapper Creek, Talkeetna, Sunshine, and Caswell areas. JSCF's primary goal is to support organizations and programs in its community that serve the needs of people in areas such as health, education, human services, arts and culture, and the preservation and enjoyment of the natural environment. The Foundation is also striving to encourage continued growth of grantees with the challenge to match grant awards by seeking new donors or increased contributions. If the challenge is met, JSCF will match the donations dollar for dollar for the amount of the grant. Grant applications are accepted from May 11 to the annual deadline date of July 10.
Requirements: Applications are accepted from qualified 501(c)3 nonprofit organizations, or equivalent organizations located in the state of Alaska. Equivalent organizations may include tribes, local or state governments, schools, or Regional Educational Attendance Areas.
Geographic Focus: Alaska
Date(s) Application is Due: Jul 10
Amount of Grant: 500 - 3,000 USD
Contact: Mariko Sarafin, Senior Program Associate; (907) 249-6609 or (907) 334-6700; fax (907) 334-5780; msarafin@alaskacf.org or info@jessicasfoundation.org
Internet: http://alaskacf.org/blog/grants/jessica-stevens-community-foundation-grant
Sponsor: Jessica Stevens Community Foundation
P.O. Box 436
Talkeetna, AK 99676

Jessie B. Cox Charitable Trust Grants 2225
The Trust is dedicated to improving the environment and the quality of life for people living in the six New England States. To achieve its goals, the Trust pursues initiatives in three key fields of interest: education: early learning, out-of-school time; environment: habitat conservation, concentrating on fresh and marine water protection; health: access to health care. In all of these funding areas, the Trust is mindful of its special status as one of the few funders with an interest in all six of the New England states. Applications for funding are reviewed in light of the important issues of the region. The Grants Committee meets twice per year to award grants, in April and October. The Trust has a two-step system of application, beginning with a concept paper and progressing to a full grant proposal. Concept papers are due on March 15 and September 15. Applicants who have submitted a concept paper and been asked to submit a complete proposal must do so by July 1 (March applicants) or January 5 (September applicants).
Requirements: All grant applicants are asked to demonstrate a plan for measuring their success. This plan should identify benchmarks against which progress towards identified goals can be measured. Grantees must report on their progress within a year of receiving support and on project completion, with reference to the initial benchmarks.
Restrictions: The Cox Trust does not normally support: general operating support and ongoing maintenance; buildings, equipment or land acquisition; endowments, scholarship funds, or fundraising activities; grant requests of under $50,000; projects for which the Trust is the sole or predominant source of support; programs for which the public sector normally assumes funding responsibility; core educational programs of public, private, parochial, or charter schools; requests from individuals; sectarian religious activity.
Geographic Focus: Connecticut, Maine, Massachusetts, New Hampshire, Rhode Island, Vermont
Date(s) Application is Due: Jan 5; Mar 15; Jul 1; Sep 15
Contact: Kirstie David; (617) 391-3081; kdavid@gmafoundations.com
Internet: http://www.jbcoxtrust.org/?page_id=28
Sponsor: Jessie B. Cox Charitable Trust
GMA Foundations, 77 Summer Street, 8th Floor
Boston, MA 02110-1006

Jessie Ball Dupont Fund Grants 2226
The fund is a national foundation having a special, though not exclusive, interest in issues affecting the South. Grants awarded include competitive grants--program grants, institutional development, and capacity building; and feasibility grants--smaller grants that enable institutions to carefully explore and develop new concepts and programs. Areas of interest include six focus areas: strengthening the independent sector; organizing and nurturing philanthropy; building assets of people, families, and communities; building the capacity of eligible organizations; stimulating community problem solving; and helping people hold their communities accountable. The fund supports four initiatives: the religion initiative, the nonprofit initiative, the small liberal arts colleges initiative, and the independent schools initiative. Types of support include general operating support, building construction/renovation, equipment acquisition, program development, seed money grants, publication, consulting services, professorships, curriculum development, matching funds, and technical assistance. Grants are awarded generally for one year. The trustees meet in January, March, May, July, September, and November to consider requests.
Requirements: Applying organizations must have received a contribution from Mrs. DuPont between the five-year period of January 1, 1960, and December 31, 1964.
Amount of Grant: 2,000 - 5,000 USD
Samples: Jacksonville Symphony Assoc (FL)--to implement a strategic marketing program, $50,000; Historical Society of Delaware (Wilmington, DE)--to hire a consultant to lead a community-based strategic-planning process, $52,690; Episcopal Diocese of Southern Ohio (Cincinnati, OH)--to support a partnership with Scioto Christian Ministries to conduct a needs-assessment in Ohio's Scioto County, $28,000; U of Richmond (VA)--to support the creation of an Internet-based resource center that will serve community organizations in the Northern Neck region of Virginia, $219,975.
Contact: Geana Potter, Grants Manager; (904) 353-0890 or (800) 252-3452; fax (904) 353-3870; contactus@dupontfund.org
Internet: http://www.dupontfund.org
Sponsor: Jessie Ball Dupont Fund
One Independent Drive, Suite 1400
Jacksonville, FL 32202-0511

Jewish Fund Grants 2227
The Fund awards grants to sustain, enrich, and address the overall health care needs of both the Jewish community and general community in the metropolitan Detroit area. The Fund is particularly interested in supporting projects that: address health care and social welfare needs of vulnerable/at-risk populations within the Jewish community; respond to priority capital and equipment needs of the Detroit Medical Center/Sinai Hospital; improve the health and well-being of vulnerable/at risk populations in the general community; support inclusion of people with special needs into the general activities of the community; enhance positive relationships between the Jewish community and the Detroit community. Highest priority is given to requests for programs that: address a critical need; impact the lives of residents of the Wayne, Oakland and Macomb counties; have a defined plan for sustaining the program beyond the grant period; include a financial or in-kind contribution from the organization; involve collaboration with others; have an outcomes-based evaluation plan; and can be funded and replicated by others. Samples of previously funded grants are available on the Fund website.
Requirements: The Jewish Fund will make grants to 501(c)3 organizations and other non-profits qualified as tax exempt under the Internal Revenue Code. Applicant organizations must provide a current audited financial statement. Applicants are encouraged to contact the Fund and discuss their proposed project with the executive director before applying for funding.
Restrictions: The Fund will usually not support: grants made directly to individuals; loans; grants to support religious activities or sectarian education; overseas projects; capital projects or equipment purchases (except equipment at the DMC/Sinai); endowments, annual fund drives, and fundraising events; and past operating deficits.
Geographic Focus: Michigan
Contact: Margo Pernick, Executive Director; (248) 203-1487; fax (248) 645-7879
Internet: http://thejewishfund.org/grant-request-guidelines.html
Sponsor: Jewish Fund
6735 Telegraph Road
Bloomfield Hills, MI 48301-2030

Jim Moran Foundation Grants   2228

The Jim Moran Foundation Grants award funding to 501(c)3 organizations in Florida. The Foundation seeks to improve the quality of life for youth and families through the support of innovative programs and opportunities that meet the ever-changing needs of the community. The Foundation's funding focuses include: education; elder care programs; family strengthening programs; meaningful after school programs; and youth transitional living programs. Proposals for programs that improve the quality of life for those who are at-risk and economically disadvantaged (without extenuating medical or developmental disabilities) will receive priority consideration. Grants are primarily awarded to the Florida counties of Broward, Palm Beach, and Duval. Applicants submit an online letter of inquiry, and will be notified within 90 days if they qualify to submit the online application.
Requirements: The Foundation will consider only organizations that have received 501(c)3 tax-exempt status under the IRS code. Additionally, the organization must be appropriately recognized by state statutes, laws and regulations that govern tax exempt organizations.
Restrictions: The Foundation will consider requests for operating dollars, but only if they do not exceed 50% of the grant request. The Foundation will not consider requests for capital campaigns, capacity building, healthcare or medical research, or event sponsorships.
Geographic Focus: Florida
Contact: Melanie Burgess, Executive Director; (954) 429-2122; fax (954) 429-2699; information@jimmoranfoundation.org
Internet: http://www.jimmoranfoundation.org/GrantApplication.aspx
Sponsor: Jim Moran Foundation
100 Jim Moran Boulevard
Deerfield Beach, FL 33442

JM Foundation Grants   2229

The foundation awards grants to eligible nonprofit organizations in its areas of interest, including education and research that fosters market-based policy solutions; developing state and national organizations that promote free enterprise, entrepreneurship, and private initiative; and identifying and educating young leaders. Types of support include internships, matching/challenge grants, program grants, publication, research grants, seed grants, and technical assistance. The foundation's board of directors meets bi-annually, usually in May and October. There are no formal proposal deadlines. Inquiries and proposals are processed on an ongoing basis.
Requirements: Public charities, including 501(c)3, 509(a)1, and 170(b)1(a)(vi) nonprofit organizations, that shares the foundation's priority interests is invited to submit a proposal.
Restrictions: Grants do not support individuals, the arts, government agencies, public schools, and international agencies; or requests for operating expenses, annual fundraising campaigns, capital campaigns, equipment, endowment funds, and loans.
Geographic Focus: All States
Amount of Grant: 5,000 - 100,000 USD
Samples: Acton Institute, Grand Rapids, Michigan, $20,000--in support of the Effective Compassion Initiative; Center of the American Experiment, Minneapolis, Minnesota, $40,000--for Intellectual Takeout; National Taxpayers Union Foundation, Alexandria, Virginia, $25,000--toward the National Taxpayers Conference.
Contact: Carl Helstrom, Executive Director; (212) 687-7735; fax (212) 697-5495
Internet: http://foundationcenter.org/grantmaker/jm/guide_jm.html
Sponsor: JM Foundation
654 Madison Avenue, Suite 1605
New York, NY 10065

Joe W. and Dorothy Dorsett Brown Foundation Grants   2230

The Joe W. and Dorothy Dorsett Brown Foundation awards grants to nonprofit organizations in Louisiana and the Gulf Coast of Mississippi. Areas of interest include medical research; housing for the homeless; support for organizations who care for the sick, hungry or helpless; religious and educational institutions; and organizations and groups concerned with improving the local community. Types of support include: operating budgets; research; and student aid. The foundation also supports service learning, a learn-by-doing approach to the curriculum. Students receive practical, hands-on experience in the subject matter studied by meeting identified community needs through active participation. The listed annual deadline date for application submission is August 31.
Requirements: Louisiana and Mississippi nonprofit organizations, with a focus on South Louisiana, the New Orleans area, and the Mississippi Gulf Coast, are eligible. Service Learning grant applications are available yearly to sixth through 12th grades in the following parishes: Orleans, Jefferson, Plaquemines, Saint Bernard, Saint Charles, Tangipahoa, Saint James, Saint John, Saint Tammany, and Washington.
Geographic Focus: Louisiana, Mississippi
Date(s) Application is Due: Aug 31
Amount of Grant: 5,000 - 25,000 USD
Contact: Beth Buscher, (504) 834-3433, ext. 200 or (504) 834-3441; bethbuscher@thebrownfoundation.org
Internet: http://www.thebrownfoundation.org
Sponsor: Joe W. and Dorothy Dorsett Brown Foundation
320 Hammond Highway, Suite 500
Metairie, LA 70005

John Ben Snow Memorial Trust Grants   2231

The mission of the Foundation is to make grants within specific focus areas to enhance the quality of life in Central and Northern New York State. Historically, the Foundation has made grants in the following program areas: arts and culture, community development, education, environment, historic preservation, and journalism. The Foundation responds to the ever-changing needs of various segments of the population, especially to the needs of young people and people who are disadvantaged either physically or economically. It is the Foundation's general policy to give preference to proposals seeking funds for new or enhanced programs, one-time, short-term grants to sustain a program until funding is stabilized, matching grants used to encourage the participation of other donors, and last dollars towards a capital campaign. There are no minimums or maximum grant amounts; however, most grants range from $5,000 to $15,000.
Requirements: Giving is primarily in Maryland, Nevada, and central New York.
Restrictions: The Foundation will not accept proposals from individuals or for-profit organizations. Additionally, the Foundation does not encourage proposals from religious organizations or proposals for endowments, contingency funding, or debt reduction.
Geographic Focus: All States
Date(s) Application is Due: Apr 1
Amount of Grant: 5,000 - 25,000 USD
Contact: Jonathan L. Snow, (315) 471-5256; fax (315) 471-5256
Internet: http://www.johnbensnow.com/jbsmt
Sponsor: John Ben Snow Memorial Trust
50 Presidential Plaza, Suite 106
Syracuse, NY 13202

John Clarke Trust Grants   2232

Dating from April 20, 1676, this historic trust was created under the will of Dr. John Clarke, a Baptist clergyman and physician and one of the co-founders of the first European settlement on Aquidneck Island in 1638. He was born in 1609 and arrived in Boston in 1637 and was one of some three hundred persons that founded the colony on Aquidneck Island. He was the author of the Royal Charter of 1663 that maintained Rhode Island as a colony. Dr. Clarke had no surviving children. He practiced medicine, was a minister, and held public office, although his chief interest was that of the Christian ministry. He was also very interested in education and is thought to have been involved in establishing a free school for Newport in 1640. Dr. Clarke is buried in a small cemetery on West Broadway in Newport. In his will, written on the date of his death, John Clarke established a perpetual charitable trust. He directed that the income from the trust be used "for the relief of the poor or bringing up of children unto learning from time to time forever." He further instructed the trustees "to have a special regard and care to provide for those that fear the Lord." Applicants must apply online at the grant website. Applicants are strongly encouraged to do the following before applying: review the downloadable state application procedures for additional helpful information and clarifications; review the downloadable online-application guidelines at the grant website; review the trust's funding history (link is available from the grant website); review the online application questions in advance; and review the list of required attachments. These will generally include: a list of board members, financial statements (audited, reviewed, or compiled by independent auditor); an organization summary; a list of other funding sources; an IRS Determination letter; and other required documents. All attachments must be uploaded in the online application as PDF, Word, or Excel files. The semi-annual application deadlines are April 1 and November 1. Applications should be submitted by 11:59 p.m. on the deadline dates. Applicants will be notified of grant decisions by letter within two to three months after the proposal deadline. All applications will be acknowledged. Applicants are encouraged to call the second phone number given if they do not receive an acknowledgment within two to three weeks of proposal submission.
Requirements: The trustees of the John Clarke Trust have established a policy of giving preference to organizations located on Aquidneck Island, Rhode Island, and within the East Bay area. However, applications from any Rhode Island 501(c)3 charitable organizations are acceptable.
Restrictions: Capital grant requests will be considered ONLY for Aquidneck Island. The trust does not support requests from individuals, organizations attempting to influence policy through direct lobbying, or any political campaigns.
Geographic Focus: Rhode Island
Date(s) Application is Due: Apr 1; Nov 1
Samples: Martin Luther King Community Center, Newport, Rhode Island, $15,000; Community Preparatory School, Providence, Rhode Island, $12,600; Newport Hospital Foundation, Newport, Rhode Island, $20,000.
Contact: Emma Greene; (617) 434-0329 or (888) 703-2345; emma.m.greene@baml.com
Internet: https://www.bankofamerica.com/philanthropic/fn_search.action
Sponsor: John Clarke Trust
225 Franklin Street, 4th Floor, MA1-225-04-02
Boston, MA 02110

John D. and Katherine A. Johnston Foundation Grants   2233

The John D. and Katherine A. Johnston Foundation was established in 1928 to support charitable organizations that work to improve the lives of physically disabled children and adults. Special consideration is given to organizations that serve low-income individuals. Preference is given to charitable organizations that serve the people of Newport, Rhode Island. Capital requests that fund handicapped assistive devices (wheelchairs, walkers, etc.) or adaptive equipment (lift installation, ramp installation, etc.) are strongly encouraged. Grant requests for general operating or program support will also be considered. The majority of grants from the Johnston Foundation are one year in duration. The Johnston Foundation shares a mission and grantmaking focus with the Vigneron Memorial Fund. Both foundations have the same proposal deadline date of 11:59 p.m. on April 1. Applicants will be notified of grant decisions before May 31. Applicants must apply online at the grant website. Applicants are strongly encouraged to do the following before applying: review the downloadable state application procedures for additional helpful information and clarifications; review the downloadable online-application guidelines at the grant website; review the foundation's funding history (link is available from the grant website); review the online application questions in advance; and review the list of required attachments. These will generally include: a list

of board members, financial statements (audited, reviewed, or compiled by independent auditor); an organization summary; a list of other funding sources; an IRS Determination letter; and other required documents. All attachments must be uploaded in the online application as PDF, Word, or Excel files.
Requirements: Applicants must have 501(c)3 tax-exempt status.
Restrictions: The foundation does not support requests from individuals, organizations attempting to influence policy through direct lobbying, or any political campaigns.
Geographic Focus: Rhode Island
Date(s) Application is Due: Apr 1
Samples: Rhode Island Hospital Foundation, Providence, Rhode Island, $7,500; CranstonArc, Cranston, Rhode Island, $4,000; Meals on Wheels of Rhode Island, Providence, Rhode Island, $3,500.
Contact: Emma Greene, Director; (617) 434-0329; emma.m.greene@baml.com
Internet: https://www.bankofamerica.com/philanthropic/fn_search.action
Sponsor: John D. and Katherine A. Johnston Foundation
225 Franklin Street, 4th Floor, MA1-225-04-02
Boston, MA 02110

John Deere Foundation Grants 2234
The foundation invests in programs in education, health/human services, community improvement, arts and culture. Types of support include annual campaigns, building construction/renovation, continuing support, fellowships, general operating support, scholarship funds, and seed money grants. Foundation interest also includes support for Third World development through US-based nonprofits with international building funds, research grants, general operating purposes, and continuing support. There are no application deadlines.
Requirements: Nonprofit organizations in communities with major John Deere operating units, and employee presence are eligible. Eligible U.S. locations are: Augusta, GA; Quad City Region, IL; Des Moines, IA, Dubuque, IA, Iowa Quad Cities, IA, Ottumwa, IA, Waterloo, IA, Coffeyville, KS, Lenexa, KS, Thibodaux, LA, Springfield, MO, Cary, NC, Fuquay-Varina, NC, Fargo, ND, Greeneville, TN, Madison, WI, Horicon, WI. Exceptions include: accredited colleges and universities; organizations focused on international development initiatives related to John Deere Solutions for World Hunger initiative. Because John Deere dealerships are owned and operated independently, their communities are not included in this geographic scope. Also eligible to apply are organizations and institutions of national or international scope that reflect the foundation's concerns.
Restrictions: Funds are not available for the following organizations or purposes: individual initiatives, including scholarships; sports teams, racing teams, athletic endeavors or scholarships designated for athletes; faith-based organizations for sectarian purposes; political candidates, campaigns or organizations; private clubs, fraternities or sororities; other foundations for purposes of building endowment; tax-supported entities.
Geographic Focus: Georgia, Illinois, Iowa, Kansas, Louisiana, Missouri, North Carolina, Tennessee, Wisconsin, Belarus, Brazil, Canada, Estonia, Latvia, Lithuania, Moldova, Ukraine
Contact: Amy Nimmer, Director; (309) 765-8000
Internet: http://www.deere.com/en_US/globalcitizenship/socialinvestment/index.html
Sponsor: John Deere Foundation
1 John Deere Place
Moline, IL 61265-8098

John Edward Fowler Memorial Foundation Grants 2235
The foundation makes grants to qualified charitable organizations providing grassroots programs for people in need in the Washington, D.C., metropolitan area. Preference is given to programs that address the issues of homelessness, hunger, at-risk children and youth (pre-school through high school), adult literacy, free medical care (prenatal to seniors), seniors aging in place, job training and placement. Types of support include general operating support, building construction/renovation, equipment acquisition, program development, and matching funds. The foundation is interested in supporting smaller, well-managed nonprofit organizations that have innovative ideas about how to help people help themselves. Contact the office for application forms. There are no application deadlines.
Requirements: Organizations in Washington, D.C., and its close Maryland and Virginia suburbs are eligible.
Restrictions: Grants are not made outside the metropolitan Washington, D.C., area, or to/for national health organizations; government agencies; medical research; public school districts; individuals; or arts (except for intensive arts-in-education programs that directly benefit at-risk children and youth).
Geographic Focus: District of Columbia, Maryland, Virginia
Amount of Grant: 5,000 - 20,000 USD
Samples: Carpenter's Shelter, Alexandria, VA, $20,000 - housing with intensive supportive services for homeless adults and children; Bucknell University, Lewisburg, PA, $10,000 - general support; All Souls' Episcopal Church, Mechanicsville VA, $15,000 - Orphans Program of the Maseno Mission in Kenya.
Contact: Suzanne Martin, Grant Consultant; (301) 654-2700
Internet: http://fdncenter.org/grantmaker/fowler/about.html
Sponsor: John Edward Fowler Memorial Foundation
4340 East-West Highway, Suite 206
Bethesda, MD 20814

John G. Duncan Charitable Trust Grants 2236
John G. Duncan was born in Sacramento, California, in 1866, and lived a substantial portion of his life in Henderson, Kentucky, where he was actively involved in the philanthropic community. He later relocated to Denver, Colorado, where he also supported a number of different non-profit organizations. Duncan died in 1955 at the age of 89. The John G. Duncan Charitable Trust was established to award grants in the following areas of interest: arts; education; health care; human services; religion; building and renovation; capital campaigns; emergency funds; equipment; program development; seed money; and research. Applications should be submitted through the online grant application form or alternative accessible application designed for assistive technology users. The average award ranges from $5,000 to $10,000. Applications may be submitted year-round, but must be submitted by the following deadlines to be reviewed at the grant meeting held after each date: January 31, April 30, July 31, and October 31.
Requirements: Organizations must have 501(c)3 designation, and serve and operate in Colorado. Organizations are eligible to apply once per calendar year. If a grant is awarded, the organization will be ineligible to apply in the following calendar year.
Restrictions: The foundation does not fund requests to support: general operating expenses; endowments; organizations located outside of Colorado; other grantmaking organizations; or organizations recognized by the IRS as non-functionally integrated supporting organizations per Section 509(a)3 of the Internal Revenue Code.
Geographic Focus: Colorado
Date(s) Application is Due: Jan 31; Apr 30; Jul 31; Oct 31
Amount of Grant: 5,000 - 10,000 USD
Contact: Jason Craig; (888) 234-1999; grantadministration@wellsfargo.com
Internet: https://www.wellsfargo.com/privatefoundationgrants/duncan
Sponsor: John G. Duncan Charitable Trust
1740 Broadway, MAC C7300-483
Denver, CO 80274-0001

John G. Martin Foundation Grants 2237
The foundation awards grants to eligible Connecticut nonprofit organizations in the areas of education at all levels. Grants may be considered in other areas. Preferential consideration will be given to proposals whose benefits include assistance to the aging and elderly or assistance to youth and adolescents. The foundation is interested in the long-term stabilization and advancement of effective organizations specializing in these areas of interest. Types of support include building construction/renovation, capital campaigns, and matching/challenge grants.
Requirements: Nonprofits serving the Hartford, CT, area may submit applications.
Geographic Focus: Connecticut
Amount of Grant: 10,000 - 65,000 USD
Samples: Northwest Catholic High School, West Hartford, CT, $20,000; West Hartford Public Library, West Hartford, CT, $30,000; Northwest Catholic High School, West Hartford, CT, $20,000;
Contact: Frank Loehmann; (860) 677-4574; fax (860) 674-1490; frank@resmgtcorp.com
Sponsor: John G. Martin Foundation
2 Batterson Park Road
Farmington, CT 06032-2553

John H. and Wilhelmina D. Harland Charitable Foundation Children and Youth Grants 2238
The John H. and Wilhelmina D. Harland Charitable Foundation offers support for: children and youth programs; community services; and arts, culture, and the environment. In the area of children and youth, support is offered for early childhood education, after school and summer programs for Elementary and Middle School Students, and programs that enhance success in public schools. The focus is local rather than regional or national, and priority is given to institutions in metropolitan Atlanta, Georgia. Grants awards support: building and renovation; capital campaigns; equipment; general operating support; challenge support; and scholarship funds. The Foundation prefers a telephone call as opposed to a letter of inquiry for the initial approach. January 10 is the deadline for the spring grant cycle and August 8 is the deadline for the fall grant cycle.
Requirements: Grant support is available to nonprofit organizations in Georgia, with emphasis on the metropolitan Atlanta area.
Restrictions: Grants are not awarded to individuals.
Geographic Focus: Georgia
Date(s) Application is Due: Jan 10; Aug 8
Amount of Grant: 4,000 - 25,000 USD
Samples: Adaptive Learning Center, Atlanta, Georgia, $20,000 - Inclusion Education Program in partnership with Our House; Moving in the Spirit, Atlanta, Georgia, $12,500 - in support of the Girls Leadership Track; Scottdale Child Development Center and Family Resource Center, Atlanta, Georgia, $20,000 - tuition assistance for early childhood education program.
Contact: Jane Hardesty; (404) 264-9912; info@harlandfoundation.org
Internet: http://harlandfoundation.org/index.php?option=com_content&view=article&id=48&Itemid=55
Sponsor: John H. and Wilhelmina D. Harland Charitable Foundation
Two Piedmont Court, Suite 710
Atlanta, GA 30305-1567

John H. Wellons Foundation Grants 2239
Established in 1950, the John H. Wellons Foundation, is a private foundation. The foundation's support is limited to the Dunn, North Carolina, area. The foundation awards student loans to local area residents, provides housing for the elderly and/or handicapped, and contributes to charitable organizations. There are no specific deadlines with which to adhere. Contact the Foundation for further application information and guidelines.
Geographic Focus: North Carolina
Contact: John H. Wellons, President; (910) 892-0436
Sponsor: John H. Wellons Foundation
P.O. Box 1254
Dunn, NC 28335-1254

John I. Smith Charities Grants 2240

The foundation supports nonprofit organizations, primarily in South Carolina. With its areas of interest, including: the visual and performing arts, higher education, medical education, theological education, literacy education and basic skills, Christian agencies and churches, child welfare, and programs for the disabled. Grants support general operations, capital campaigns, endowment funds, scholarship funds, and emergency funds. Application forms are not required. The board meets on a quarterly basis. Mail applications to: P.O. Box 1687, Greer, SC 29652
Requirements: Nonprofit organizations in South Carolina are eligible to apply.
Geographic Focus: South Carolina
Amount of Grant: 2,500 - 200,000 USD
Contact: Jefferson Smith, President; (864) 879-2455
Sponsor: John I. Smith Charities
P.O. Box 40200, FL9-100-10-19
Jacksonville, FL 32203-0200

John J. Leidy Foundation Grants 2241

The John J. Leidy Foundation was established in 1957. The foundation gives primarily in the metropolitan Baltimore, Maryland, area. Funding is available to people with disabilities, education, health, social services, and Jewish organizations in the form of, scholarship funds, building/renovation, equipment, general/operating support, and program development grants.
Requirements: Maryland non-profits are eligible to apply.
Restrictions: No grants to individuals.
Geographic Focus: Maryland
Contact: Robert L. Pierson, President; (410) 821-3006; Leidyfd@attglobal.net
Sponsor: John J. Leidy Foundation
305 W Chesapeake Avenue, Suite 308
Towson, MD 21204-4440

John Jewett and Helen Chandler Garland Foundation Grants 2242

Established in 1959, the John Jewett & Helen Chandler Garland Foundation gives primarily in California, with emphasis on southern California. The foundations area of interest include: the arts, education, health care, children and social services. There are no application deadlines.
Requirements: California nonprofit organizations are eligible.
Restrictions: No telephone inquiries excepted, submit a letter of inquiry as your initial approach.
Geographic Focus: California
Amount of Grant: 2,500 - 610,000 USD
Samples: Huntington Library, Friends of the, San Marino, CA, $800,000 - for general support and educational programs at Botanical Center; Marlborough School, Los Angeles, CA, $25,000 - for general support and minority scholarships; Boys and Girls Clubs of Pasadena, Pasadena, CA, $110,000 - for general support and to make swimming pool accessible year-round.
Contact: Lisa M. Hausler, Manager
Sponsor: John Jewett and Helen Chandler Garland Foundation
P.O. Box 550
Pasadena, CA 91102-0550

John Lord Knight Foundation Grants 2243

The foundation, established in 1957, provides funding for program in the Lucas County, Ohio, region and throughout Florida. Its areas of interest include: cancer research; Christian agencies and churches; education; elementary education; federated giving programs; hospitals; human services; public libraries; and youth services. There are no specific deadlines, and applicants should submit a detailed description of project and amount of funding requested to the listed contact person in care of KeyBank.
Geographic Focus: Florida, Ohio
Contact: Erwin Diener, Key Bank, Trust Officer; (419) 259-8372
Sponsor: John Lord Knight Foundation
Three Seagate, P.O. Box 10099
Toledo, OH 43699-0099

John M. Lloyd Foundation Grants 2244

The foundation has the following funding objectives: to increase funding from public and private sectors to address the HIV/AIDS pandemic, both globally and domestically; to improve domestic and international policies (universal protection of human rights to issues concerning HIV/AIDS, access to HIV/AIDS healthcare and treatment, and access to accurate information about HIV/AIDS); to amplify global awareness of HIV/AIDS and to facilitate broad-based change in attitudes to reduce stigma and change behavior; and to develop the leadership of organizations that fight HIV/AIDS, as well as to foster collaborations among those organizations and leaders. Projects that have promise of making a significant impact and those that are new and innovative receive preference. The listed application deadlines are for concept letters; full proposals are by invitation. Organizations may submit only one concept letter per year.
Requirements: 501(c)3 tax-exempt organizations are eligible. Grants are made to locally focused projects in California, US-based projects with a national or global scope, and international projects.
Restrictions: In general, the foundation does not make contributions more than once per calendar year to any single organization; more than three consecutive years to any single project; to annual campaigns; to operating budgets of established organizations; to capital expenditures (physical plant, equipment, endowment); to indirect costs; to individuals; to locally focused projects in the United States with the exception of locally focused projects in California; for health care or service provision; or for general support
Geographic Focus: All States
Date(s) Application is Due: Aug 15; Dec 15
Amount of Grant: Up to 20,000 USD
Contact: Melanie Havelin; (310) 622-1050; fax (310) 622-1070; Mhavelin@johnmlloyd.org
Internet: http://www.johnmlloyd.org/grant_programs.html
Sponsor: John M. Lloyd Foundation
11777 San Vicente Boulevard, Suite 745
Los Angeles, CA 90049

John Merck Fund Grants 2245

The foundation fosters innovative advocacy and problem solving in the fields of Developmental Disabilities, Environment, Reproductive Health, Human Rights and Job Opportunities. Its objective is to act as a catalyst, supporting organizations that can effect constructive and measurable change in each of these areas.
Requirements: The foundation actively seeks out projects and programs that may merit support, then requests grant applications on behalf of those it finds most promising. It does not encourage the submission of unsolicited proposals. However, organizations interested in obtaining support for work they do in one of the foundation's areas of interest are welcome to send a brief letter of inquiry. The foundation favors: outstanding individuals working on promising projects in organizations that may have difficulty attracting funds; pilot projects with potential for widespread application; advocacy, including litigation, capable of setting or protecting important precedents; smaller organizations, start-ups included; one-year grant requests (though multi-year grants of up to three years occasionally are made); matching-grant opportunities, particularly to help broaden support for fledgling initiatives.
Restrictions: The foundation does not provide grants for: endowment or capital-fund projects; large organizations with well-established funding sources (except those that need help launching promising new projects for which funding is not readily available); general support (except in the case of small organizations whose entire mission coincides with one of foundation's areas of interest); individuals (except if his or her project is sponsored by a domestic or foreign educational, scientific or charitable organization).
Geographic Focus: All States
Contact: Ruth G. Hennig; (617) 556-4120; fax (617) 556-4130; info@jmfund.org
Internet: http://www.jmfund.org/program.html
Sponsor: John Merck Fund
2 Oliver Street, 8th Floor
Boston, MA 02108

John Merck Scholars Awards 2246

Scholars are chosen from the ranks of the most promising assistant professors currently working, or planning to work, in neurobiological and cognitive sciences relating to the biology of mental disability and developmental disabilities, including developmental studies of cognition, perception, language, reading, learning and motor performance. The foundation will accept one application for a neurobiologist and one for a cognitive scientist from major universities and other research centers. Scholars will receive $75,000 per year for a four-year period, subject to an annual review of research progress. The sponsoring institution is responsible for fiscal management.
Requirements: The foundation will fund the most promising young researchers whose work illuminates neurodevelopmental disorders from the perspectives of (i) synapse formation and synaptic plasticity; (ii) learning and memory, and synaptic plasticity; (iii) perception, cognition and behavior; (iv) neurogenesis and pattern formation; and (v) genetics and early development. Also encouraged are proposals that (i) investigate the possible role of environmental chemicals in the origins of developmental disabilities, or that (ii) aim to distinguish subgroups within accepted diagnostic categories through the use of sophisticated behavioral and neuroimaging tests of perception, cognition, and emotions based on concepts from modern cognitive neuroscience. In all cases, they seek proposals from young scientists conducting research that is of the highest quality and that has the greatest chance of increasing understanding of neurodevelopmental disorders. Applicants must have the following: Academic rank in a university or medical school, or equivalent standing in a research institute or medical center; A record of research in areas relating to the foundation's interest in the underlying causes of developmental disabilities; Not more than four years of experience in an independent faculty position; Evidence of a commitment to a career in neuroscience or cognitive science. Request application forms by mail, email or phone.
Restrictions: Applicants may not: hold tenured positions or their equivalents; apply in more than two rounds of competition; apply for continued postdoctoral support. Holding other fellowships concurrently with the John Merck Scholars Award is discouraged. Prior approval by the foundation for an overlapping fellowship is required and will be given only in unusual circumstances.
Geographic Focus: All States
Date(s) Application is Due: Sep 15
Amount of Grant: 75,000 USD
Contact: Jason Bentsman; (617) 556-4120; fax (617) 556-4130; j bentsman@jmfund.org
Internet: http://www.jmfund.org/jm_scholars_program.html
Sponsor: John Merck Fund
2 Oliver Street, 8th Floor
Boston, MA 02108

John P. McGovern Foundation Grants 2247

The John P. McGovern Foundation, established in 1961, supports the charitable interests of the donor to support the activities of established nonprofit organizations, which are of importance to human welfare with special focus on children and family health education and promotion, treatment and disease prevention. The Foundation gives

primarily in Texas with an emphasis on the Houston area, but also provides funding in the Southwest region as well. The types of support offered are: building/renovation; conferences/seminars; continuing support; curriculum development; emergency funds; endowments; general/operating support; matching/challenge support; professorships; publication; research; and scholarship funds.
Requirements: Non-profits in Texas are eligible to apply.
Geographic Focus: Texas
Date(s) Application is Due: Aug 31
Amount of Grant: 10,000 - 100,000 USD
Samples: University of Texas Health Science Center, Houston, TX. $1,500,000 - for John P. McGovern, MD, Center for Health, Humanities, and Human Spirit/Certificate Program Endowment Fund; Community Family Centers, Houston, TX. $200,000- for John P. McGovern Community Sports and Recreation Building; Salvation Army of Houston, Houston, TX. $150,000 - for general operating fund.
Contact: Kathrine G. McGovern, President; (713) 661-4808; fax (713) 661-3031
Sponsor: John P. McGovern Foundation
2211 Norfolk Street, Suite 900
Houston, TX 77098-4044

John P. Murphy Foundation Grants 2248
The Foundation awards grants primarily in Cuyahoga County, Ohio, and its immediately adjacent counties. Areas of interest include education--college level, and primary and secondary education to improve educational programs benefiting low income and minority students, most often in Cleveland public schools; arts and culture; social services; community--community activities, agencies, and events; health--institutions and organizations such as hospitals, clinic, and nursing homes; and religion. Grants to national or regional institutions with broad public support will be made only for start-up programs where time is a factor or where the proposal addresses a specific local community need. Guidelines are available online. Contact the office for an application form.
Requirements: Ohio organizations properly classified by the IRS as being eligible to receive grants may apply. Each applicant must file its most recent IRS letter of classification with each proposal. National or regional institutions with broad public support may submit proposals that address a specific local community need.
Restrictions: No grants will be made to endowment funds or for scholarships.
Geographic Focus: Ohio
Amount of Grant: 5,000 - 75,000 USD
Samples: MetroHealth Foundation (Cleveland, OH)--for salary support of an additional child specialist and an intern who will expand services provided by the Child Life and Education Program, which helps families understand the stress and anxiety that children can experience during an illness or injury, $90,000 over two years.
Contact: Allan J. Zambie, Executive Vice President; (216) 623-4770 623-4771 or (216) 623-4771; fax (216) 623-4773
Internet: http://fdncenter.org/grantmaker/jpmurphy/interest.html
Sponsor: John P. Murphy Foundation
50 Public Square, Suite 924
Cleveland, OH 44113-2203

John R. Oishei Foundation Grants 2249
The foundation's primary mission is to support medical research and care and education, as well as cultural and social needs existing in the Buffalo Niagara region. The foundation favors creative programs that attempt to advance from the status quo, are strategically sound, and are strongly focused on excellence. Programs that provide opportunities for foundation support to be leveraged into greater support from other sources will be especially favored. The foundation generally does not fund operating expenses, though occasional exceptions may be made for organizations that address basic human needs. Requests for capital funds will not be considered unless they are an integral part of an otherwise eligible proposal. Contact the foundation to discuss eligibility before submitting an application. Grant applications are accepted throughout the year.
Requirements: It is the general policy of the foundation to confine its support to activities located in the Buffalo, NY, metropolitan region.
Restrictions: Grants do not support endowments; capital requests (buildings or equipment); deficit funding or loans; individual scholarships or fellowships (except within specific foundation programs); travel, conferences, seminars, or workshops; fundraising events; or 509(a) private foundations.
Geographic Focus: New York
Amount of Grant: 100,000 - 200,000 USD
Contact: Blythe T. Merrill; (716) 856-9490, ext. 3; btmerrill@oisheifdt.org
Internet: http://www.oisheifdt.org/Home/Fund/WhatWeFundOverview
Sponsor: John R. Oishei Foundation
1 HSBC Center, Suite 3650
Buffalo, NY 14203-2805

John S. Dunn Research Foundation Grants and Chairs 2250
Established in 1985, the foundation funding area is limited to the Texas, providing support to health and medical-related organizations, especially hospitals; support also for cancer, other medical research, and freestanding clinics.
Requirements: Grants are made to nonprofit organizations within Texas having 501(c)3 tax-exempt status.
Restrictions: No grants to individuals, or for multi-year or seed money grants.
Geographic Focus: Texas
Amount of Grant: 10,000 - 100,000 USD
Samples: Gulf Coast Consortia, Houston, TX, $1,180,000 - to support drug discovery program; University of Texas M.D. Anderson Cancer Center, Houston, TX, $650,000 - to fund radiological sciences and metabolic lab; Memorial Hermann Healthcare System, Houston, TX, $250,000 - to support life flight services to patients without insurance.
Contact: Dr. Lloyd Gregory; (713) 626-0368; fax (713) 626-3866
Sponsor: John S. Dunn Research Foundation
3355 West Alabama street, Suite 720
Houston, TX 77098-1718

Johns Manville Fund Grants 2251
The Johns Manville Fund, Inc. gives primarily in areas of company operations in Denver, Colarado, Etowah, Tennessee, and in Canada. The Fund offers support in the areas of: the arts; education; health care; human services; youth services and the American Red Cross. The grants are offered in the form of employee-related scholarships, employee volunteer services, general/operating support, and program development. Contact the Johns Manville Fund, Inc. for a Informational brochure, including application guidelines.
Requirements: Non-profits operating in Denver, Colorado, Etowah, Tennessee, and in Canada are eligible.
Restrictions: No support for religious organizations not of direct benefit to the entire community, hospitals, or non-special needs private educational organizations. No grants for special events.
Geographic Focus: Colorado, Tennessee, Canada
Amount of Grant: 1,000 - 27,000 USD
Contact: Community Relations Manager; (303) 978-3863; fax (303) 978-2108
Internet: http://www.jm.com/corporate/careers/1241.htm
Sponsor: Johns Manville Fund
717 17th Street
Denver, CO 80202-3330

Johnson & Johnson Community Health Care Program Grants 2252
The program provides support for community public health initiatives impacting access and delivery of quality health care services for medically underserved populations, with emphasis on women and children (including infants and adolescents). Additional consideration will be given to programs focusing on the education and prevention of diabetes, obesity, and cardiovascular disease. Grants also will assist organizations in developing a broad-based public and private support network by extending public recognition of their efforts. Second-year continuation grants also are available.
Requirements: Grants are awarded to community/public health care organizations/agencies in Alabama (all areas); Arkansas (all areas); Boston, Massachusetts; Florida (all areas); New Jersey (all areas); Ohio (all areas); San Diego, California; and San Francisco, California.
Restrictions: Proposals from foundations, universities, and political advocacy groups will not be considered.
Geographic Focus: Alabama, Arkansas, California, Florida, Massachusetts, New Jersey, Ohio
Date(s) Application is Due: Oct 17
Amount of Grant: 150,000 USD
Contact: Sierra Veale, (443) 287-5138; fax (410) 510-1974; jandj@jhsph.edu
Internet: http://www.jhsph.edu/johnsonandjohnson/index.html
Sponsor: Johnson and Johnson
615 N Wolfe Street, Room W1100
Baltimore, MD 21205

Johnson & Johnson Corporate Contributions Grants 2253
Johnson & Johnson and its many operating companies support community-based programs that improve health and well-being. The Company works with community-based partners that have the greatest insight into the needs of local populations and the strategies that stand the greatest chances of success. Giving focuses on: saving and improving the lives of women and children; building on the skills of people who serve community health needs, primarily through education; and preventing diseases and reducing stigma and disability in underserved communities where we have a high potential for impact.
Requirements: Grants are awarded to nonprofit and tax-exempt local, national, and international organizations and institutions.
Restrictions: Grants are not awarded to individuals, for deficit funding, capital or endowment funds, demonstration projects, or publications.
Geographic Focus: All States
Contact: Shaun Mickus, (732) 524-2086; smickus@corus.jnj.com
Internet: http://www.jnj.com/connect/caring/corporate-giving/
Sponsor: Johnson and Johnson
1 Johnson & Johnson Plaza
New Brunswick, NJ 08933-0001

Johnson and Johnson / SAH Arts and Healing Grants 2254
Johnson & Johnson and the Society for the Arts in Healthcare have provided program funding to 117 organizations in the U.S. and Canada. In 2008, the grant structure was updated to a three-year grant period with a focus on existing programs. This new structure promotes the evaluation and replication of promising models in order to strengthen and expand the arts in healthcare field.
Requirements: To be eligible for consideration, grant applicants must be: a current member of Society for the Arts in Healthcare; located in the U.S. or Canada (partners may be international); a non-profit organization with federal tax exempt status and organized and operated for charitable purposes or a governmental agency (state or local, including education or institutions). Note: U.S. based non-profit organization applicants must be tax exempt under section 501(c)(3) of IRS Code; Canada based applicants must be a registered Canadian charity; and able to show a demonstrated ability to work effectively in partnership with other organizations. To be eligible for consideration,

proposed programs must: utilize a model that has documented positive impact; and be designed with proposed outcomes, which show initial impact. Program goals should be specific, measurable, attainable, realistic and timely; and Be sustainable based on existing resources and organizational capacity (adequate staffing, funding, space, etc.).
Restrictions: Individuals are ineligible to apply for scholarships or other forms of assistance. Members of the Society for the Arts in Healthcare board of directors, advisory board, or staff are ineligible to apply or partner with any applicant for the purposes of this grant.
Geographic Focus: All States, Canada
Date(s) Application is Due: Jan 13; Oct 15
Contact: Anita Boles; (202) 299-9770; fax (202) 299-9887; anita@thesah.org
Internet: http://www.thesah.org/template/page.cfm?page_id=15
Sponsor: Society for the Arts in Healthcare
2437 15th Street NW, B South
Washington, D.C. 20009

Johnson Controls Foundation Health and Social Services Grants 2255
The Johnson Controls Foundation provides financial gifts to select U.S.-based organizations located in the communities in which the company has a presence. Operating support for organizations in the health and social services category largely occurs in communities where Johnson Controls has a local presence and is often directed through contributions to United Way.
Requirements: In evaluating requests for funds, the Advisory Board has developed policies and guidelines for giving in health and social services. Contributions will be given financial assistance to federated drives, hospitals, youth agencies and other health and human service agencies. Ordinarily, operating support of health and social service agencies is reserved for Johnson Controls communities, and generally directed through contributions to United Way.
Restrictions: In general, no grants will be made to any political campaign or organization; any municipal, state, federal agency, or department, or to any organization established to influence legislation; any private individual for support of personal needs; any sectarian institutions or programs whose services are limited to members of any one religious group or whose funds are used primarily for the propagation of a religion; for testimonial dinners, fund raising events, tickets to benefits, shows, or advertising; to provide monies for travel or tours, seminars and conferences or for publication of books and magazines or media productions; for specific medical or scientific research projects; foreign-based institutions nor to institutions or organizations for use outside of the United States; fraternal orders or veteran groups; private foundations or to endowment funds. The foundation does not donate equipment, products or labor.
Geographic Focus: All States
Date(s) Application is Due: Apr 30
Amount of Grant: 1,000 - 150,000 USD
Contact: Charles A. Harvey, President; (414) 524-1200 or (414) 524-2296
Internet: http://www.johnsoncontrols.com/publish/us/en/about/our_community_focus/johnson_controls_foundation.html
Sponsor: Johnson Controls Foundation
5757 North Green Bay Avenue, P.O. Box 591
Milwaukee, WI 53201

Johnson County Community Foundation Grants 2256
The Community Foundation seeks to meet the challenges of the Johnson County, Indiana community as a whole. Special attention is given to requests that increase the capacity of not-for-profit organizations to serve the community and requests that demonstrate community support and in-kind investment. The Foundation generally supports three kinds of requests: projects of service to the general community and pilot projects; seed money to enable projects to demonstrate their potential or enhance services; and emergency funding for community needs. Examples of previously funded projects are available at the website.
Requirements: Applicants must submit a letter of inquiry (2-3 pages), also include a copy of their 501(c)3 Letter of Determination. Letters of inquiry should include: a brief statement of the organization's purpose and goals; a brief description of the project, the need and the target population it addresses; short- and long-term outcomes anticipated and plans for assessing achievements; grant amount needed; a statement about the total agency budget and the project budget; a statement about other funding sources for the agency and/or project, specifying both committed and projected sources of support. Letters of inquiry will be reviewed to determine if the proposed effort fits within the community foundation's grant program. If so, the applicant will be contacted by the Foundation, requesting additional information or a full proposal. The Foundation supports organizations that are tax exempt under Section 501(c)3 of the Internal Revenue Service Code and are not classified as private foundations under Section 509(a) of the Code. In selected cases, it may consider support for projects sponsored by governmental entities.
Restrictions: The Community Foundation does not: take multi-year commitments; support services commonly regarded as the responsibility of government; provide support for political or partisan purposes or for programs in which religious teachings are an integral part; consider more than one proposal from the same organization within a 12-month period; provide discretionary grant support to an organization's ongoing operating budget, building funds, capital campaigns, or endowments; fund raising events and functions; make grants to individuals except for scholarship or special award funds; fund existing obligations or to replenish resources (deficit funding) for such purposes.
Geographic Focus: Indiana
Contact: Kim Minton; (317) 738-2213; fax (317) 738-9113; kimm@jccf.org
Internet: http://www.jccf.org/index.asp?p=37
Sponsor: Johnson County Community Foundation
398 S Main, P.O. Box 217
Franklin, IN 46131-2311

Johnson Foundation Wingspread Conference Support Program 2257
The Johnson Foundation at Wingspread sponsors grants to partially fund conferences focusing on subjects in the public interest, primarily health issues and the environment. Meeting facilities include Wingspread, the home designed by Frank Lloyd Wright, and formerly owned by Herbert Fisk Johnson of the Johnson and Johnson family. Conferences are intensive, one- to four-day meetings of small groups convened in partnership with nonprofit organizations, public agencies, universities, and other foundations. Strategic interests of the Foundation are education, sustainable development and environment, democracy and community, and family. The Foundation's usual contribution to a conference sponsored by one or more other organizations consists of the provision of the full conference facilities of Wingspread, planning and logistical support by the staff, meals and other amenities for the period of the meeting.
Requirements: To be invited to submit a full proposal, applicants first must submit a brief concept letter, consisting of: a clear statement of purpose; a draft agenda; the identification of key participants; and an estimated budget and schedule. The letter should describe how the conference will enhance collaboration and community, include diverse opinions and perspectives, identify solutions, and result in action.
Geographic Focus: All States
Contact: Coordinator; (262) 639-3211; fax (262) 681-3327; info@johnsonfdn.org
Internet: http://www.johnsonfdn.org/guidelines.html
Sponsor: Johnson Foundation
33 East Four Mile Road
Racine, WI 53402

John Stauffer Charitable Trust Grants 2258
The trust awards grants to California higher education institutions and hospitals. Types of support include building construction/renovation, challenge/matching funds, endowments, equipment acquisition, fellowships, professorships, and scholarship funds. There are no application forms or deadlines.
Requirements: California nonprofits are eligible.
Geographic Focus: California
Amount of Grant: 100,000 - 500,000 USD
Samples: California Institute of Technology, Pasadena, CA, $500,000 - for construction of research lab in new chemistry building; Loma Linda University, Loma Linda, CA, $250,000 - for classroom for emergency personnel; Occidental College, Los Angeles, CA, $159,320 - for endowment for summer research.
Contact: H. Jess Senecal; (626) 793-9400; fax (626) 793-5900; jess@lagerlof.com
Sponsor: John Stauffer Charitable Trust
301 N Lake Avenue, 10th Floor
Pasadena, CA 91101-4108

John V. and George Primich Family Scholarship 2259
The John V. and George Primich Family Scholarship was established to benefit the children or grandchildren of current employees of G.W. Berkheimer Company and South Central Company. The number of scholarships depends on the number of applicants, but they are a minimum of $2,500 each. Any college major is eligible. Contact the Legacy Foundation for application and requirement details.
Geographic Focus: Indiana
Date(s) Application is Due: Jun 1
Amount of Grant: 2,500 USD
Contact: Primich Family Scholarship Contact; (219) 736-1880; fax (219) 736-1940; legacy@legacyfoundationlakeco.org
Internet: http://www.legacyfoundationlakeco.org/scholarships/index.htm
Sponsor: Legacy Foundation
1000 East 80th Place, 302 South
Merrillville, IN 46410

John W. Alden Trust Grants 2260
The trust awards grants to Massachusetts nonprofit organizations in its areas of interest, including education and therapy for children who are blind, disabled, retarded, or mentally or physically ill. Grants will normally be made for specific projects rather than general operating purposes. There are no application deadlines. Grant applications should be submitted on-line at www.cybergrants.com/alden. Detailed instructions on the web page will guide you through the process. All questions should be directed to the grants coordinator.
Requirements: Eastern Massachusetts 501(c)3 tax-exempt organizations are eligible.
Restrictions: Grants are not made to individuals.
Geographic Focus: Massachusetts
Date(s) Application is Due: Jan 15; Apr 15; Jul 15; Oct 15
Amount of Grant: Up to 15,000 USD
Contact: Susan Monahan; (617) 951-1108; fax (617) 542-7437; smonahan@rackemann.com
Internet: http://www.cybergrants.com/alden
Sponsor: John W. Alden Trust
160 Federal Street, 13th Floor
Boston, MA 02110-1700

John W. and Anna H. Hanes Foundation Grants 2261
The trust awards grants, primarily for the arts, to nonprofit organizations in North Carolina. Other interests that are funded are children and youth services; education; environment; health care; historic preservation; and human services. Types of support include annual and capital campaigns; building and renovation; emergency funds; endowments and seed money; equipment and land acquisition; matching and challenge support; and program development. The board meets in January, April, July, and October; application forms are required.

Requirements: 501(c)3 North Carolina nonprofit organizations are eligible.
Restrictions: Giving is limited to North Carolina, with emphasis on Forsyth County. No grants are made to individuals, or for operating expenses.
Geographic Focus: North Carolina
Date(s) Application is Due: Mar 15; Jun 15; Sep 15; Dec 15
Amount of Grant: 1,000 - 200,000 USD
Samples: Winston-Salem State University, Winston-Salem, NC, $83,333; Yadkin Arts Council, Yadkinville, NC, $33,333; Salem Academy and College, Winston-Salem, NC, $100,000.
Contact: Christopher Spaugh, Vice President, Wachovia Bank NA; (336) 732-5991
Sponsor: John W. and Anna H. Hanes Foundation c/o Wachovia Bank N.A.
1525 West WT Harris Boulevard
Charlotte, NC 28288-5709

John W. Anderson Foundation Grants 2262

The John W. Anderson Foundation is an independent foundation established in 1967, in Indiana. The trust was established by John W. Anderson, a manufacturing executive and inventor. Mr. Anderson was president of the Anderson Company. Grants support education; organizations serving youth; higher educational institutions; community funds; scientific or medical research for the purpose of alleviating suffering; care of needy, crippled or orphaned children; care of needy persons who are sick, aged or helpless; improving the health, and quality of life of all persons; human services; and the arts and humanities. The Foundation gives primarily in Lake and Porter counties in northwest Indiana.
Requirements: Nonprofit organizations which are not classified as private foundations under Section 509(a) are eligible to apply. Organizations must serve Lake and Porter counties of Northwest Indiana. Application form not required. Applications for grants should include, but not necessarily be limited to, the following information: purpose of organization and proposed use of grant; An organization may submit a request once in a 12-month period. There are no submission deadlines. Each application is reviewed on a timely basis.
Restrictions: Applications sent by fax will not be considered. No support for elementary and secondary schools, or for business or any for-profit organization, or for supporting organizations classified 509(a)3. No grants to individuals, or for endowment funds, multi-year grants, fund raising events, advertising, seed money, deficit financing; no loans.
Geographic Focus: Indiana
Amount of Grant: 5,000 - 50,000 USD
Samples: Brothers' Keeper, Inc., $75,000. Family and Youth Services Bureau of Porter County, $50,000. South Shore Arts, $15,000. The Nazareth Home, $15,000.
Contact: William N. Vinovich, Vice-Chair; (219) 462-4611
Sponsor: John W. Anderson Foundation
402 Wall Street
Valparaiso, IN 46383-2562

John W. Boynton Fund Grants 2263

The John W. Boynton Fund was established in 1952 by Dora Carter Boynton in memory of her husband. He had a thriving business in tinware in Templeton, MA, and later in his life was a resident of Athol, MA. In her will, Mrs. Boynton asked that "organizations which benefit poor, needy, and deserving persons and particularly those of advanced years and gentility" be considered. She also expressed a desire that special consideration be given to charitable organizations serving the Town of Athol. Grant requests for new or special programs and capital projects are preferred. General operating support also will be considered. In come cases, general operating grants are made to make up for a temporary loss of public or private funding. A typical grant is $5,000. Applicants must apply online at the grant website. Applicants are strongly encouraged to do the following before applying: review the downloadable state application procedures for additional helpful information and clarifications at the grant website; review the downloadable online-application guidelines at the grant website; review the foundation's funding history (link is available from the grant website); review the online application questions in advance; and review the list of required attachments. These will generally include: a list of board members, financial statements (audited, reviewed, or compiled by independent auditor); an organization summary; a list of other funding sources; an IRS Determination letter; and other required documents. All attachments must be uploaded in the online application as PDF, Word, or Excel files. The application deadline is 11:59 p.m. on July 15. Applicants will be notified of decisions by the end of September.
Requirements: Applicants must have 501(c)3 tax-exempt status and serve the residents of Greater Boston.
Restrictions: The fund does not support requests from individuals, organizations attempting to influence policy through direct lobbying, or any political campaigns.
Geographic Focus: Massachusetts
Date(s) Application is Due: Jul 15
Amount of Grant: 2,500 - 25,000 USD
Contact: Michealle Larkins; (866) 778-6859; michealle.larkins@baml.com
Internet: https://www.bankofamerica.com/philanthropic/fn_search.action
Sponsor: John W. Boynton Fund
225 Franklin Street, 4th Floor, MA1-225-04-02
Boston, MA 02110

John W. Speas and Effie E. Speas Memorial Trust Grants 2264

The John W. Speas and Effie E. Speas Memorial Trust was established in 1943 to support and promote quality educational, cultural, human-services, and health-care programming. In the area of arts, culture, and humanities, the trust supports programming that: fosters the enjoyment and appreciation of the visual and performing arts; strengthens humanities and arts-related education programs; provides affordable access; enhances artistic elements in communities; and nurtures a new generation of artists. In the area of education, the trust supports programming that: promotes effective teaching; improves the academic achievement of, or expands educational opportunities for, disadvantaged students; improves governance and management; strengthens nonprofit organizations, school leadership, and teaching; and bolsters strategic initiatives of area colleges and universities. In the area of health, the trust supports programming that improves the delivery of health care to the indigent, uninsured, and other vulnerable populations and addresses health and health-care problems that intersect with social factors. In the area of human services, the trust funds programming that: strengthens agencies that deliver critical human services and maintains the community's safety net and helps agencies respond to federal, state, and local public policy changes. In the area of community improvement, the trust funds capacity-building and infrastructure-development projects including: assessments, planning, and implementation of technology for management and programmatic functions within an organization; technical assistance on wide-ranging topics, including grant writing, strategic planning, financial management services, business development, board and volunteer management, and marketing; and mergers, affiliations, or other restructuring efforts. Grant requests for general operating support and program support will be considered. Grants from the foundation are one year in duration. Application materials are available for download at the grant website. Applicants are strongly encouraged to review the state application guidelines for additional helpful information and clarifications before applying. Applicants are also encouraged to review the foundation's funding history (link is available from the grant website). There are no application deadlines for the Speas Memorial Trust. The annual deadlines are March 31 and August 31.
Requirements: Applicants must have 501(c)3 tax-exempt status and serve the residents of the Greater Kansas City Metropolitan area. Applications must be mailed.
Restrictions: Grant requests for capital support will not be considered. The trust does not support requests from individuals, organizations attempting to influence policy through direct lobbying, or any political campaigns.
Geographic Focus: Missouri
Date(s) Application is Due: Mar 31; Aug 31
Amount of Grant: 5,000 - 200,000 USD
Samples: Community Living Opportunities, Kansas City, Missouri, $100,000 - support of the HomeLink Support Technologies for special needs population (2014); Saint Lukes Hospital Foundation, Kansas City, Missouri, $100,000 - for critical capital improvements at Crittenton Children's Center (2014); Horizon Academy, Roeland Park, Kansas, $50,000 - support of the Rays of Success Campaign (2014).
Contact: Scott Berghaus; (816) 292-4300; scott.berghaus@ustrust.com
Internet: https://www.bankofamerica.com/philanthropic/foundation.go?fnId=133
Sponsor: John W. Speas and Effie E. Speas Memorial Trust
1200 Main Street, 14th Floor, P.O. Box 219119
Kansas City, MO 64121-9119

Joseph Alexander Foundation Grants 2265

Established in 1960 in New York, the Foundation supports primarily in the New York region. Fields of interest include the following: higher education; health organizations; medical research, particularly optic nerve research; social services; and Jewish organizations. Types of support also include the following: annual campaigns; building/renovation; capital campaigns; conferences/seminars; curriculum development; endowments; equipment; exchange programs; general/operating support; program development; research; and scholarship funds. Applicants should submit a letter requesting application guidelines before submitting a proposal. There is no deadline, but the board meets in January, April, July, and October.
Requirements: Nonprofit organizations are eligible to apply.
Restrictions: Grants are not made to individuals.
Geographic Focus: New York
Amount of Grant: 5,000 - 50,000 USD
Contact: Robert Weintraub, President; (212) 355-3688
Sponsor: Joseph Alexander Foundation
110 East 59th Street
New York, NY 10022-1304

Joseph and Luella Abell Charitable Trust Scholarships 2266

Established in 2003 in Indiana, the Joseph and Luella Abell Charitable Trust (in the care of the Jackson County Bank) offers scholarship awards to individuals that are nursing school students or graduate students of the Methodist ministry throughout Indiana. The funding, which ranges from $1,000 to $5,000 annually, can be used to defray costs for tuition expenses, books, housing, and other costs associated with attending college. An application form can be secured by contacting the Jackson County Bank, and the annual deadline for application submission is April 1.
Requirements: Only students enrolled in a nursing program or ministry in Indiana may apply.
Geographic Focus: Indiana
Date(s) Application is Due: Apr 1
Amount of Grant: 1,000 - 5,000 USD
Contact: Brandon Hunsley, Trustee Officer; (812) 522-3607
Sponsor: Joseph and Luella Abell Charitable Trust
P.O. Box 1001
Seymour, IN 47274-1001

Joseph Collins Foundation Scholarships 2267

The Joseph Collins Foundation offers a limited number of scholarships to enable men and women whose own resources are inadequate to attend accredited medical schools of their own choice that offer a neurology, psychiatry, or general practice degree, and who demonstrate an interest in cultural pursuits (art, music, theater, writing, etc.) outside of medicine. Consideration will include scholastic record; interest in cultural pursuits;

intention to specialize in neurology, psychiatry, or general medicine; moral character; age; and geographical proximity (applicants within 200 miles of the medical school preferred). Award amounts are generally up to $10,000 per year and are renewable with the recommendation of the Pritzker School of Medicine and the approval of the Joseph Collins Foundation. These awards are tenable at any accredited medical school east of the Mississippi River for one year, and renewable at the discretion of the foundation. Applications should be obtained from appropriate medical school authorities; applications are sent to the school, not to students. The school nominates one student per school year; however, trustees act favorably or unfavorably in their sole discretion. Student must interact with the financial aid officer at his/her school. Request applications in January.
Requirements: Grants are open to anyone attending an accredited medical school, and who intend to specialize in neurology, psychiatry, or general practice. Applicants must have satisfactorily completed the first year of medical school. Awards are not made to premedical or postgraduate medical students.
Geographic Focus: All States
Date(s) Application is Due: Mar 1
Amount of Grant: Up to 10,000 USD
Contact: Jack H. Nusbaum; (212) 728-8060; fax (212) 728-9060; jnusbaum@willkie.com
Internet: http://www.apa.org/about/awards/jcollins-grad.aspx?tab=1
Sponsor: Joseph Collins Foundation
787 Seventh Avenue, Room 3950
New York, NY 10019-7099

Joseph Drown Foundation Grants 2268
The foundation makes contributions in the areas of education; community, health and social services; and arts and humanities. It supports programs dealing with such issues as the high school drop-out rate, teen pregnancy, lack of sufficient health care, substance abuse, and violence. Types of support include general operating support, program development, seed money, scholarship funds, and matching funds. Most grant making is limited to programs or organizations in California. Requests are considered each year at March, June, September, and December meetings. The foundation makes grants for both operating support and program support but does not make multiyear commitments. No special application form is required. Proposal should include a letter with information about the organization and the project, a copy of 501(c)3 determination letter, a budget for the organization and the project, the most recent audited financial statements, a copy of the most recent IRS Form 990, and a list of the current board of directors. Any additional materials, such as an annual report, may be attached. Questions should be directed to the program director; proposals should be sent to the foundation president.
Requirements: California 501(c)3 organizations may apply.
Restrictions: The foundation does not provide funds to individuals, endowments, capital campaigns, or annual funds. The foundation does not underwrite annual meetings, conferences, or special events, nor does it fund religious programs or purchase tickets to fund-raising events.
Geographic Focus: California
Date(s) Application is Due: Jan 15; Apr 15; Jul 15; Oct 15
Amount of Grant: 10,000 - 100,000 USD
Samples: The Henry Mancini Institute, Los Angeles, CA, $15,000 - for support of the Community Outreach Initiative; The Museum of Tolerance, Los Angeles, CA, $50,000 - for Tolerance Youth Education programs; Friends of the Semel Institute for Neuroscience, Los Angeles, CA, $40,000 - for a postdoctoral fellowship; Los Angeles Leadership Academy, Los Angeles, CA, $30,000 - for a literacy coordinator and development director in initial year of operation.
Contact: Alyssa Eichelberger, Program Administrator; (310) 277-4488, ext. 100; fax (310) 277-4573; alyssa@jdrown.org
Internet: http://www.jdrown.org
Sponsor: Joseph Drown Foundation
1999 Avenue of the Stars, Suite 1930
Los Angeles, CA 90067

Joseph H. and Florence A. Roblee Foundation Grants 2269
The foundation awards grants to enable organizations to promote change by addressing significant social issues in order to improve the quality of life and help fulfill the potential of individuals. The foundation arises out of a Christian framework, and values ecumenical endeavors. The foundation particularly supports programs which work to break down cultural, racial, and ethnic barriers. Organizations and churches are encouraged to collaborate in achieving positive change through advocacy, prevention, and systemic improvements.
Requirements: Giving limited to nonprofit organizations in the greater bi-state St. Louis region, and Miami/Dade, FL. Contact Foundation for additional guidelines.
Restrictions: Support is not given to individuals or for annual campaigns, research, or loans.
Geographic Focus: Florida, Illinois, Missouri
Date(s) Application is Due: Jan 15; Jun 15
Amount of Grant: 1,000 - 30,000 USD
Samples: Planned Parenthood of the Saint Louis Region, Saint Louis, MO, $20,000 - for comprehensive sex education programs for area teens; Christian Activity Center, East Saint Louis, IL, $15,000 - for Bridging the Gap after-school homework room program; Switchboard of Miami, Miami, FL, $30,000 - for Kevin Kline Suicide Awareness Initiative;
Contact: Peggy Thomas c/o Bank of America, N.A., Bank of America; (314) 466-1304; kathydc@robleefoundation.org.
Sponsor: Joseph H. and Florence A. Roblee Foundation
P.O. Box 14737, MO2-100-07-19
Saint Louis, MO 63178-4737

Joseph Henry Edmondson Foundation Grants 2270
The Joseph Henry Edmondson Foundation serves the Pikes Peak Region. While grants have been awarded outside this geographic area under unique circumstances, most funding is confined to supporting those nonprofits in this region. The Foundation's primary interests include: welfare of children, the incapacitated, homeless, families, and the elderly; preservation and improvement of the environment and natural resources; the arts; education; health care; community improvements; and charitable outreach. It will consider grant requests for general operations, specific projects, programs, capital needs, and capacity building. The Board of Directors reviews grant proposals at quarterly meetings held in January, April, July, and October. Deadlines for submission are December 1, March 1, June 1, and September 1. Grants range in size from $500 to $100,000 with an average of $10,000.
Requirements: All organizations applying to the Foundation must be a nonprofit, tax-exempt 501(c)3 organization; or be a project or organization under the fiscal agency of a 501(c)3 organization. The Foundation transitioned to an on-line grant application process.
Restrictions: The Foundation does not support the following: organizations that have not secured their legal 501(c)3 designation from the IRS or are not under the fiscal agency of a nonprofit; individuals; special and/or fundraising events; debt reduction; and organizations with an evangelical mission.
Geographic Focus: Colorado
Date(s) Application is Due: Mar 1; Jun 1; Sep 1; Dec 1
Amount of Grant: 500 - 100,000 USD
Contact: Heather L, Carroll, Executive Director; (719) 471-1241
Internet: http://www.jhedmondson.org/grants-seekers
Sponsor: Joseph Henry Edmondson Foundation
10 Lake Circle
Colorado Springs, CO 80906

Josephine G. Russell Trust Grants 2271
The purpose of the trust is for the care, healing, and nursing of the sick and injured, the relief and aid of the poor, the training and education of the young, and any other manner of social service in the city of Lawrence, Massachusetts.
Requirements: Massachusetts nonprofit organizations serving the greater Lawrence metropolitan area are eligible.
Geographic Focus: Massachusetts
Date(s) Application is Due: Jan 31
Amount of Grant: Up to 40,000 USD
Contact: Clifford Elias, Treasurer; (978) 500-3171
Sponsor: Josephine G. Russell Trust
59 Lucerne Drive
Andover, MA 01810-1719

Josephine Goodyear Foundation Grants 2272
The foundation makes grants in the greater Buffalo, NY, area for programs and projects benefiting indigent women and children, particularly with their physical needs. Grants also support hospitals, child welfare organizations, youth agencies, and community funds. Types of support include capital campaigns, building construction/renovation, equipment acquisition, land acquisition, emergency funds, program development, seed grants, research, employee matching gifts, and matching funds. Applicants are requested to submit a two-page proposal including the purpose of the organization, a description of the project/program or need, budget, other sources of funding, and an IRS determination letter.
Requirements: 501(c)3 organizations serving the greater Buffalo, NY, area are eligible.
Restrictions: Grants are not awarded in support of individuals, continuing support, annual campaigns, deficit financing, endowment funds, scholarships, fellowships, or loans.
Geographic Focus: New York
Date(s) Application is Due: Apr 15; Aug 15
Amount of Grant: 500 - 30,000 USD
Contact: E.W. Dann Stevens; (716) 566-1465; ewdstevens@hiscockbarclay.com
Internet: http://www.cfgb.org/index.php/affiliates-and-initiatives/josephine-goodyear-foundation/97-josephine-goodyear-grants
Sponsor: Josephine Goodyear Foundation
3 Fountain Plaza, Suite 1100, M & T Center
Buffalo, NY 14203

Josephine S. Gumbiner Foundation Grants 2273
The charitable foundation functions for the benefit of women and children in the Long Beach area. The funder supports a wide array of programs, such as the arts; day care; education health care intervention, prevention, and direct services; housing; and recreation. Organizations are not be eligible for funding more than once in any 12-month period. As a general rule, the foundation will not grant funding to any organization for more than three consecutive years. The two-step application process begins with requesting a letter of intent questionnaire, which should be requested through email. Full applications are by invitation. There are no application deadlines. The board meets three or four times each year.
Requirements: Southern California nonprofit organizations are eligible.
Restrictions: Grants do not support political campaigns, lobbying efforts, programs that supplant tradition schooling, pass-through organizations, or groups with endowments greater than $5 million.
Geographic Focus: California
Amount of Grant: 5,000 - 50,000 USD
Contact: Grants Administrator; (562) 437-2882; fax (562) 437-4212; julie@jsgf.org
Internet: http://www.jsgf.org/
Sponsor: Josephine S. Gumbiner Foundation
333 West Broadway, Suite 302
Long Beach, CA 90802

Josephine Schell Russell Charitable Trust Grants   2274
The Josephine Schell Russell Charitable Trust was established in Ohio in 1976, with giving limited to supporting the economically disadvantaged in the greater Cincinnati area. The Trust's primary fields of interest include: the arts; children and youth services; health care and health care access; and human services. Types of support include: building and renovation; capital campaigns; equipment purchase or rental; program development; and seed money. Interested applicants should submit one copy of either the Ohio Common Grant Form or the Greater Cincinnati Common Grant Form to the Trust office. Initial approach can also be via telephone contact. The annual quarterly deadlines are February 1, May 1, August 1, and October 1. Most recent awards have ranged from $5,500 to $100,000.
Requirements: 501(c)3 organizations either located in, or serving residents of, the greater Cincinnati area are eligible to apply.
Restrictions: No support is offered for private foundations, or for political, fraternal, labor or advocacy groups. Furthermore, no grants are awarded to individuals, or for endowment funds, operating budgets, continuing support, annual campaigns, deficit financing, scholarships, or conferences.
Geographic Focus: Kentucky, Ohio
Date(s) Application is Due: Feb 1; May 1; Aug 1; Oct 1
Amount of Grant: 5,500 - 100,000 USD
Samples: Lighthouse Youth Services, Cincinnati, Ohio, $100,000 - support of the Sheakley Center (2014); Humanitarian League, Union, Kentucky, $10,000 - support of a bully prevention program (2014); Cancer Support Community, Cincinnati, Ohio, $60,000 - support for the Heart of Wellness program (2014).
Contact: Mary Alice Koch; (513) 651-8463 or (412) 768-5898; mary.koch@pnc.com
Internet: https://www1.pnc.com/pncfoundation/charitable_trusts.html
Sponsor: Josephine Schell Russell Charitable Trust
One PNC Plaza, 249 Fifth Avenue, 20th Floor
Pittsburgh, PA 15222

Joseph P. Kennedy Jr. Foundation Grants   2275
The foundation concentrates its funding in the area of mental retardation. Grants providing seed funding are offered to encourage new methods of service and support for people with mental retardation and their families.
Requirements: Nonprofit organizations may apply.
Restrictions: The foundation will not support capital costs or costs of equipment for projects, nor will it pay for ongoing support or operations of existing programs. Individuals are not eligible.
Geographic Focus: All States
Sample: George Washington U (Washington, D.C.)--for the Ctr for the Study and Advancement of Disability Policy, $200,000.
Contact: Contact; (202) 393-1250; fax (202) 824-0351; eidelman@jpkf.org *Internet:* http://www.jpkf.org/JPKF_Info/GRANT.HTML
Sponsor: Joseph P. Kennedy Jr. Foundation
1133 19th Street NW, 12th Floor
Washington, D.C. 20036-3604

Joshua Benjamin Cohen Memorial Scholarship   2276
The Joshua Benjamin Cohen Memorial Scholarship was established by Max and Susan Cohen in memory of their son. Up to two renewable scholarships are awarded to graduating seniors from Valparaiso High School who demonstrate a high standard of leadership, academic performance, athletic performance, and financial need. Any college major is eligible.
Requirements: The scholarship is awarded to a student who demonstrates a high standard of leadership, with a GPA of 3.0 or better. They must also submit an original essay and demonstrate financial need. Renewal of the scholarship is dependent on the student maintaining a B average. Students are required to provide an official copy of their transcript to the Legacy Foundation after each term. Failure to do so may jeopardize their scholarship. Applications are available online and in the Guidance Office of Valparaiso High School each year.
Geographic Focus: Indiana
Date(s) Application is Due: Mar 5
Amount of Grant: 2,000 USD
Contact: Cohen Memorial Scholarship Contact; (219) 736-1880; fax (219) 736-1940; legacy@legacyfoundationlakeco.org
Internet: http://www.legacyfoundationlakeco.org/scholarships/index.htm
Sponsor: Legacy Foundation
1000 East 80th Place, 302 South
Merrillville, IN 46410

Josiah Macy Jr. Foundation Grants   2277
The foundation awards grants to improve the education of physicians and other health care professionals. Grant making focuses on improving medical and health professionals' education in the context of the changing health care system, increasing the number of minority physicians, boosting teamwork between health care professionals, and using educational strategies to increase care for under-served populations. Current initiatives cover the fields of pediatric residency, inter-professional curriculum, improved clinical training, increasing diversity among health care professions, Macy Scholars program, women's health in medical curricula, women in medicine, and funding for a New York post-baccalaureate program. Applications may be submitted at any time. A preliminary letter of inquiry is useful in assisting the foundation staff in determining whether submission of a full proposal is appropriate.
Requirements: Grants are made to colleges, universities, and other 501(c)3 nonprofit organizations for activities within established program areas. Grant proposals, addressed to the president, should include the name of the sponsoring agency or institution; description of the project; names and qualifications of the person who will be responsible for the project; expected cost and duration of the project including an itemized budget; documents substantiating the tax-exempt status of the sponsoring institution; and a letter of endorsement from the sponsoring institution.
Restrictions: Grants are not awarded to individuals or for general undesignated support or for construction/renovation projects.
Geographic Focus: All States
Amount of Grant: Up to 5,000,000 USD
Contact: George E. Thibault; (212) 486-2424; jmacyinfo@josiahmacy
Internet: http://www.josiahmacyfoundation.org/index.php?section=grant_guidelines
Sponsor: Josiah Macy Jr. Foundation
44 East 64th Street
New York, NY 10021

Josiah W. and Bessie H. Kline Foundation Grants   2278
The foundation was established for charitable, scientific, literary, and educational purposes. The objectives for its grantmaking activities are to aid blind or incapacitated persons or crippled children. Grants are made to Pennsylvania colleges and universities and to hospitals, to institutions for crippled children, or to any other benevolent or charitable institution. Generally, grants are made to such institutions that are located in south-central Pennsylvania. Grants also are made for scientific and medical research. Types of support include continuing support, annual campaigns, capital campaigns, equipment acquisition, building/renovation, curriculum development, scholarship funds, matching funds, research, and emergency funds. The foundation does not have an application form. Instead, a letter of request must be submitted with the following information: a description of the need and purpose of the applicant organization, a project budget, amount requested from the foundation and the dates of the need, and a copy of the 501(c)3 tax-exemption letter. Requests may be submitted at any time. The board generally meets in March and November to consider requests.
Requirements: 501(c)3 organizations serving south-central Pennsylvania are eligible.
Restrictions: The foundation does not make loans and does not make grants to individuals or to normal operational phases of established programs, endowments, campaigns or national organizations, or religious programs.
Geographic Focus: Pennsylvania
Amount of Grant: 500 - 150,000 USD
Contact: John Obrock; (717) 561-0820 or (717) 561-4373; fax (717) 561-0826
Sponsor: Josiah W. and Bessie H. Kline Foundation
515 S 29th Street
Harrisburg, PA 17104

Judith and Jean Pape Adams Charitable Foundation ALS Grants   2279
Jean Pape Adams was interested in supporting charitable organizations actively engaged in discovering the causes of Amyotrophic Lateral Sclerosis (ALS) and in finding a cure. The Foundation encourages research that is collaborative, innovative and aggressive. These initiatives may fill gaps not currently being investigated and may involve novel approaches. The research must be of high scientific merit in finding a cure for ALS. Proposals may be longer than two pages. A lay summary is also required. Applications are reviewed by an Advisory Committee comprised of scientific researchers, neurologists and others who have had close relationships with friends and family members afflicted with ALS. The Committee makes funding recommendations to the Trustees.
Requirements: Public agencies and organizations that are classified as a charitable organization described in Section 501(c)3 of the Internal Revenue Code and as a public charity under Section 509(a) of the Internal Revenue Code may apply.
Restrictions: Support groups, education and medical equipment are not supported.
Geographic Focus: All States
Date(s) Application is Due: Sep 13
Amount of Grant: 1,000 - 250,000 USD
Contact: Shelley L. Carter, Executive Director; (830) 997-7347; fax (918) 591-2433; scarter@jjpafoundation.com
Internet: http://www.jjpafoundation.com/guidelines.html
Sponsor: Judith and Jean Pape Adams Charitable Foundation
7030 South Yale Avenue, Suite 600
Tulsa, OK 74136

Judith and Jean Pape Adams Charitable Foundation Tulsa Area Grants   2280
The Foundation was established in 2004 as a private foundation and is involved in making distributions to charitable organizations on an annual basis. It encompasses two areas of support: organizations and agencies predominantly in Tulsa County, Oklahoma, and national Amyotrophic Lateral Sclerosis (ALS) research. For the former, primary areas of interest include arts and culture, human services and education. Support is given for operations, programs, capital projects, and maintenance reserve funding. The annual deadline is August 15.
Requirements: Public agencies serving the Tulsa, Oklahoma, region that are classified as a charitable organization described in Section 501(c)3 of the Internal Revenue Code and as a public charity under Section 509(a) of the Internal Revenue Code may apply.
Geographic Focus: Oklahoma
Date(s) Application is Due: Aug 15
Amount of Grant: 500 - 250,000 USD
Contact: Marcia Y. Manhart; (830) 997-7347; mmanhart@jjpafoundation.com
Sue Mayhue, (316) 383-1795
Internet: http://www.jjpafoundation.com/guidelines.html
Sponsor: Judith and Jean Pape Adams Charitable Foundation
7030 South Yale Avenue, Suite 600
Tulsa, OK 74136

Judith Clark-Morrill Foundation Grants 2281
Established in Indiana, the Judith Clark-Morrill Foundation has specified its primary fields of interest as: the arts; community and economic development; education; and youth development. An application form is required, and there are two annual deadlines: June 1 and December 1. Amount of awards range from $1,000 to $30,000.
Restrictions: No grants are given to individuals, or for student groups, scholarships, annual campaigns, general operating support, travel, or advertising. There are no loans or multi-year grants.
Geographic Focus: Indiana
Date(s) Application is Due: Jun 1; Dec 1
Amount of Grant: 1,000 - 30,000 USD
Contact: Judith Morrill, President; (260) 357-4141
Sponsor: Judith Clark-Morrill Foundation
P.O. Box 180
Garrett, IN 46738-1350

Julius N. Frankel Foundation Grants 2282
The Julius N. Frankel Foundation supports Chicago-area nonprofits in the areas of: arts and performing arts; children and youth service; higher education; hospitals; human services; and medical school education. Recipients have included hospitals, universities, cultural organizations, and social service providers, with an emphasis on large, established organizations. There are no specified application formats or deadlines, though the Board meets at least five times annually. Most recent grant awards have ranged from $25,000 to $200,000. The initial approach should be by letter, detailing the program and budgetary needs.
Requirements: Chicago-area nonprofits are eligible to apply.
Restrictions: Individuals are not eligible.
Geographic Focus: Illinois
Amount of Grant: 25,000 - 200,000 USD
Samples: Chicago Opera Theater, Chicago, Illinois, $45,000 - general operations; Chicago Symphony Orchestra, Chicago, Illinois, $150,000 - general support; Lawrence Hall Youth Services, Chicago, Illinois, $50,000 - general operating support.
Contact: Hector Ahumada, Trustee; (312) 461-5154
Sponsor: Julius N. Frankel Foundation
111 W. Monroe Street, Tax Division 10C
Chicago, IL 60603-4096

June Pangburn Memorial Scholarship 2283
This scholarship was established in memory of June Pangburn, who served as the Director of Nursing at St. Mary's Hospital (Hobart) and as a mentor to many nursing students. The current application and deadline is available in the Hobart High School Guidance Office.
Requirements: Applicants must be Hobart High School seniors who wish to pursue careers in Nursing or any field in Allied Health.
Geographic Focus: Indiana
Amount of Grant: 1,000 USD
Contact: Pangburn Memorial Scholarship Contact; (219) 736-1880; fax (219) 736-1940; legacy@legacyfoundationlakeco.org
Internet: http://www.legacyfoundationlakeco.org/scholarships/index.htm
Sponsor: Legacy Foundation
1000 East 80th Place, 302 South
Merrillville, IN 46410

K21 Health Foundation Cancer Care Fund Grants 2284
The Kosciusko County Cancer Care Fund, administered by K21 Health Foundation, provides assistance to financially-eligible residents of Kosciusko County who are suffering from cancer. The purpose of the fund is to relieve some of the financial strain that often accompanies that dreaded diagnosis. The assistance provided includes but is not limited to items such as rent or mortgage payments, utilities, insurance, food, car payments, and prescription medications.
Requirements: Eligibility Requirements for Assistance: documented resident of Kosciusko County; verified cancer diagnosis within the last three months; can demonstrate a financial need; completion of the required application (contact the CCF Director for application).
Geographic Focus: Indiana
Contact: Clare Sessa, CCF Director; (574) 372-3500; csessa@hcfkc.org
Internet: http://www.k21foundation.org/cancer-care-fund/index.cfm
Sponsor: K21 Health Foundation
2170 North Pointe Drive, P.O. Box 1810
Warsaw, IN 46582

K21 Health Foundation Grants 2285
K21 Health Foundation exists for the benefit of Kosciusko County, Indiana citizens to ensure health care services are provided, and to advance prevention and healthy lifestyles. This will be accomplished by identifying health needs in our community, and maintaining an endowment so funding is available, through investments and grants, for those needs. Organizations are strongly encouraged to review K21's mission statement as that is the primary tool used to evaluate each application. When submitting an application, organizations must clearly explain how K21's mission will be advanced by supporting the project, program, or service, and the intended benefit(s) to residents of Kosciusko County.
Requirements: To qualify for a grant from K21, your organization must meet the following *Requirements:* must be a non-profit agency with verified IRS tax-exempt status, or a governmental agency; must be in good standing with the Indiana Secretary of State as indicated in a Business Entity Report; your Bylaws must require term limits for Directors and a rotating board is strongly encouraged.
Restrictions: To qualify for a grant from K21, your request must meet the following *Requirements:* the project, program, or service for which you are seeking funding must benefit residents of Kosciusko County; the project, program, or service for which you are seeking funding should provide direct health services or advance prevention and healthy life solutions.
Geographic Focus: Indiana
Date(s) Application is Due: Feb 1; May 1; Aug 1; Nov 1
Contact: Holly Swoverland; (574) 269-5188, ext. 102; fax (574) 269-5193
Internet: http://www.k21foundation.org/apply-for-grant/index.cfm
Sponsor: K21 Health Foundation
2170 North Pointe Drive, P.O. Box 1810
Warsaw, IN 46582

Kahuku Community Fund 2286
The Kahuku Community Fund was established by the Estate of James Campbell in 2005 to be used for charitable and community purposes within the geographic district of Kahuku, bounded by Turtle Bay and Malaekahana. Preference is given to projects that address educational opportunities; recreational opportunities; economic sufficiency; social conditions; health care; strategic action plan around future development of the Kahuku community; housing opportunities; cultural arts, practices and values of the Ko'olauloa moku. Grant range is between $1,000 and $25,000 with an average of $10,000.
Requirements: Nonprofit, 501(c)3 organizations, schools, units of government, neighborhood groups or projects are eligible to apply. Community organizations without 501(c)3 status are eligible to apply for a grant up to $5,000, provided the activities to be supported are charitable. To be eligible for a grant of more than $5,000, a group must be a tax-exempt 501(c)3 organization or have a 501(c)3 fiscal sponsor. Program or project must benefit the Kahuku community.
Restrictions: Projects not likely to be funded: costs relating to establishing a new 501((c)(3) organization; general operating support, although up to 10% for indirect costs will be considered; funds for an endowment; funds for the benefit of specific individuals; except for emergency assistance through a 501(c)3 organization; full personnel costs (staffing costs are less likely to be funded) of a project or program; major capital improvements, although minor capital improvements required to implement a project may be considered; travel out of state.
Geographic Focus: Hawaii
Date(s) Application is Due: Jan 19
Amount of Grant: 1,000 - 25,000 USD
Contact: Amy Luersen, (808) 566-5550 or (888) 731-3863; aluersen@hawaiicommunityfoundation.org or aluersen@hcf-hawaii.org
Internet: http://www.hawaiicommunityfoundation.org/index.php?id=71&categoryID=22
Sponsor: Hawai'i Community Foundation
827 Fort Street Mall
Honolulu, HI 96813

K and F Baxter Family Foundation Grants 2287
The foundation focuses on the special educational challenges encountered by multiracial children; policy recommendations in support of revised school curricula that will be more supportive of multiracial children without sacrificing either basic skills training or appropriate performance standards; and developing, by means of policy studies and analysis, a realistic template for education reform focused on the student rather than the school. In addition, the foundation will consider funding worthwhile projects in the arts, health care, and other areas related to education. The foundation encourages requests from schools desiring to enhance learning and/or increase number of students served through the use of blended learning. A preliminary email describing your project or program is required before proposals are accepted. The initial inquiry must be made at least 3 weeks before proposals are due on March 1 each year.
Requirements: Biracial children grants are awarded to nonprofit organizations nationwide. Educational grants are now made exclusively to schools, and applicant schools must be fully enrolled and at least 50% of students must qualify for free or reduced lunches or be otherwise identified as low-income.
Restrictions: Education grants are limited to schools in California Unified School Districts, including Berkeley, Los Angeles, Oakland, West Contra Costa Counties. Programs must take place during regular school hours. A site visit is required before an award can be granted.
Geographic Focus: All States
Date(s) Application is Due: Mar 1
Amount of Grant: Up to 150,000 USD
Contact: Stacey Bell; (510) 524-8145; fax (510) 524-4101; staceybell@kfbaxterfoundation.com
Internet: http://www.kfbaxterfoundation.com/home.html
Sponsor: K and F Baxter Family Foundation
1563 Solano Avenue, #404
Berkeley, CA 94707

Kansas Health Foundation Major Initiatives Grants 2288
The Kansas Health Foundation is committed to supporting strategies that will make Kansas a healthier place. The Foundation's focus areas are: promoting the healthy behaviors of Kansans; strengthening the public health system; improving access to health care for Kansas children; growing community philanthropy; providing health data and information to policymakers; and building civic leadership.
Requirements: The Foundation accepts one-page letters of inquiry from organizations that believe they have innovative work that meets the Foundation's mission. The letter of inquiry form is located at the Foundation's website. Organizations may also review samples of previously funded major initiatives on the website.
Geographic Focus: Kansas

Samples: University of Kansas School of Medicine, Wichita, Kansas, supporting the Department of Preventative Medicine and Public Health in its efforts to enhance the public health components of its physicians' education curriculum, $400,000; Worksite Wellness Evaluation, to engage local leaders and organizations in communities throughout Kansas to complete the training necessary to develop worksite wellness plans and to fund a final report of the results, $800,000.
Contact: Nancy Claassen; (316) 262-676; fax (316) 262-2044; nclaassen@khf.org
Internet: http://www.kansashealth.org/grant_type/major_initiatives
Sponsor: Kansas Health Foundation
309 East Douglas
Wichita, KS 67202-3405

Kansas Health Foundation Recognition Grants 2289
The Kansas Health Foundation Recognition Grants expand the Foundation's support to a broad range of health-related organizations throughout the state. The Foundation defines health broadly, and looks at all the aspects that affect health, including the social factors that contribute to a healthy population (a state of complete physical, mental, and social well-being and not merely the absence of disease or infirmity). While the majority of the Foundation's funding is through invited proposals, the Recognition Grants program is designed to fund unsolicited requests. It is targeted for organizations and agencies proposing meaningful and charitable projects that fit within the Foundation's mission of improving the health of all Kansans.
Requirements: Kansas 501(c)3 nonprofit health organizations are eligible. The application is at the Foundation website and must be submitted online.
Restrictions: The Foundation does not support: medical research; capital campaigns; operating deficits or retirement of debt; endowment programs not initiated by the Foundation; political advocacy of any kind; vehicles, such as vans or buses; medical equipment; construction projects or real estate acquisitions; direct mental health services; or direct medical services.
Geographic Focus: Kansas
Date(s) Application is Due: Mar 15; Sep 15
Amount of Grant: Up to 25,000 USD
Contact: Nancy Claassen; (316) 262-676; fax (316) 262-2044; nclaassen@khf.org
Gina Hess, Grant Assistant; (316) 262-7676 or (800) 373-7681; rinfo@khf.org
Internet: http://www.kansashealth.org/grantmaking/recognitiongrants
Sponsor: Kansas Health Foundation
309 East Douglas
Wichita, KS 67202-3405

Kate B. Reynolds Charitable Trust Health Care Grants 2290
The Trust responds to health care and wellness needs and invests in solutions that improve the quality of health for financially needy residents throughout North Carolina. The Health Care Division seeks impact through two program areas: providing treatment and supporting prevention. The trust requires advanced consultation by phone or in writing. Application materials are available online, but applications are not accepted electronically.
Requirements: Nonprofit 501(c)3 organizations in North Carolina are eligible.
Restrictions: Grants are not awarded to individuals.
Geographic Focus: North Carolina
Date(s) Application is Due: Mar 15; Sep 15
Amount of Grant: 20,000 - 200,000 USD
Contact: John H. Frank; (336) 397-5502 or (866) 551-0690; john@kbr.org
Internet: http://www.kbr.org/health-care-division-fund.cfm
Sponsor: Kate B. Reynolds Charitable Trust
128 Reynolda Village
Winston-Salem, NC 27106-5123

Katharine Matthies Foundation Grants 2291
The Katharine Matthies Foundation was established in 1987 to support and promote quality educational, human-services, and health-care programming for underserved populations. Special consideration is given to organizations that work to prevent cruelty to children and animals. The majority of grants from the Matthies Foundation are one year in duration; on occasion, multi-year support is awarded. Applicants must apply online at the grant website. Applicants are strongly encouraged to do the following before applying: review the downloadable state application procedures for additional helpful information and clarifications; review the downloadable online-application guidelines at the grant website; review the foundation's funding history (link is available from the grant website); review the online application questions in advance; and review the list of required attachments. These will generally include: a list of board members, financial statements (audited, reviewed, or compiled by independent auditor); an organization summary; a list of other funding sources; an IRS Determination letter; and other required documents. All attachments must be uploaded in the online application as PDF, Word, or Excel files. The deadline for application to the Katherine Matthies Foundation is 11:59 p.m. on May 1. Applicants will be notified of grant decisions by letter within three to four months after the proposal deadline.
Requirements: Applicant organizations must have 501(c)3 tax-exempt status and serve the people of the following Connecticut towns: Seymour, Ansonia, Derby, Oxford, Shelton, or Beacon Falls. A breakdown of number/percentage of people served by specific towns will be required in the online application. Special consideration will be given to organizations that serve the people of Seymour, Connecticut.
Restrictions: The Matthies Foundation specifically serves people of the Lower Naugatuck Valley. The foundation does not support requests from individuals, organizations attempting to influence policy through direct lobbying, or any political campaigns.
Geographic Focus: Connecticut
Date(s) Application is Due: May 1
Samples: Town of Seymour, Seymour, Connecticut, $22,000; Griffin Health Services Corporation, Derby, Connecticut, $66,000; Connecticut Hurricanes, Derby, Connecticut, $17,540.
Contact: Amy Lynch; (860) 657-7015; amy.r.lynch@baml.com
Internet: https://www.bankofamerica.com/philanthropic/fn_search.action
Sponsor: Katharine Matthies Foundation
200 Glastonbury Boulevard, Suite # 200
Glastonbury, CT 06033-4056

Katherine Baxter Memorial Foundation Grants 2292
Established by primary donor Martin B. Ortlieb in California in 1992, the Katherine Baxter Memorial Foundation grants program offers support in the arts, higher education, and YMCAs and YWCAs. Giving is limited to the State of California, with a range of $150 to $20,000. Funding generally supports overall operating costs, and comes in the form of a donation.
Requirements: Applicants must be established 501(c)3 organizations either serving the residents of or located in the State of California.
Geographic Focus: California
Amount of Grant: 150 - 20,000 USD
Samples: Pomona College, Claremont, California, $20,000 - general operations; Palomar Family YMCA, Escondido, California, $10,800 - general operating costs; American Stroke Association, Framingham, Massachusetts, $4,500 - general operating support.
Contact: Randolph Ortlieb, Trustee; (760) 747-3200
Sponsor: Katherine Baxter Memorial Foundation
970 Canterbury Avenue
Escondido, CA 92025-3836

Katherine John Murphy Foundation Grants 2293
The Katherine John Murphy Foundation was established in 1954 in Atlanta, Georgia by Katherine Murphy Riley. The foundation awards grants to Georgia tax-exempt organizations in its areas of interest, including services for children, education, the environment, and human services. Types of support include annual campaigns, building construction and/or renovation, capital campaigns, continuing support, general operating support, seed grants, and project development. Grants are awarded primarily in Atlanta, Georgia, and select areas of Latin America. Submit a letter of request.
Requirements: Organizations in Atlanta, Georgia, are eligible to apply, as well as select areas of Latin America.
Restrictions: No grants to individuals, or for research, or matching gifts.
Geographic Focus: Georgia, Argentina, Bolivia, Brazil, Chile, Colombia, Costa Rica, Cuba, Dominican Republic, Ecuador, El Salvador, Guatemala, Haiti, Honduras, Mexico, Nicaragua, Panama, Paraguay, Peru, Uruguay, Venezuela
Amount of Grant: 250 - 50,000 USD
Contact: Brenda Rambeau, (404) 589-8090; fdnsvcs.ga@suntrust.com
Internet: http://www.kjmurphyfoundation.org
Sponsor: Katherine John Murphy Foundation
50 Hurt Plaza, Suite 1210
Atlanta, GA 30303

Kathryne Beynon Foundation Grants 2294
Founded in California in 1967, the Kathryne Beynon Foundation provides support primarily for: hospitals (with a special interest in Asthma); youth agencies; child welfare; Roman Catholic church; higher education, Types of support include: general operating support; building construction/renovation; endowment funds; and scholarship funds. The Board meets quarterly to review grant requests. Applicants should contact the office in writing, outlining their proposal. Application is by invitation only, and there is no deadline date when submitting grant proposals. Contact the Foundation directly for additional guidelines before submitting a full proposal.
Requirements: 501(c)3 southern California tax-exempt organizations are eligible. Preference is given to requests from Pasadena. There are no: deadline dates; formal application form required to submit proposal.
Restrictions: No support to individuals.
Geographic Focus: California
Amount of Grant: 500 - 50,000 USD
Contact: Robert D. Bannon, Trustee; (626) 584-8800
Sponsor: Kathryne Beynon Foundation
1111 South Arroyo Parkway, Suite 470
Pasadena, CA 91105-3239

Katrine Menzing Deakins Charitable Trust Grants 2295
Katrine Deakins was executive secretary to Amon G. Carter and helped found the Amon G. Carter Foundation. She was the foundation's executive director and throughout her life was very active in numerous professional, social and charitable organizations. Katrine then established her own trust in 1987, the Katrine Menzing Deakins Charitable Trust to benefit several favored charities in addition to other charitable organization requests selected each year by the trustees. Grants are typically between $1,000 and $25,000. The Katrine Menzing Deakins Charitable Trust has bi-annual application deadlines of March 31 and September 30. Applications must be submitted by 11:29 p.m. on the deadline dates. Applicants must apply online at the grant website. Applicants are strongly encouraged to do the following before applying: review the downloadable state application procedures for additional helpful information and clarifications; review the downloadable online-application guidelines at the grant website; review the trust's funding history (link is available from the grant website); review the online application questions in advance; and review the list of required attachments. These will generally

include: a list of board members, financial statements (audited, reviewed, or compiled by independent auditor); an organization summary; a list of other funding sources; an IRS Determination letter; and other required documents. All attachments must be uploaded in the online application as PDF, Word, or Excel files.
Requirements: Applicants must have 501(c)3 tax-exempt status.
Restrictions: The trust does not support requests from individuals, organizations attempting to influence policy through direct lobbying, or any political campaigns.
Geographic Focus: Texas
Date(s) Application is Due: Mar 31; Sep 30
Amount of Grant: 1,000 - 25,000 USD
Samples: Harris Methodist Health Foundation, Fort Worth, Texas, $50,000; Communities in Schools of Greater Tarrant County, Fort Worth, Texas, $25,000; Fort Worth Symphony Orchestra, Fort Worth, Texas, $25,000.
Contact: Mark J. Smith; (817) 390-6028; tx.philanthropic@baml.com
Internet: https://www.bankofamerica.com/philanthropic/fn_search.action
Sponsor: Katrine Menzing Deakins Charitable Trust
500 West 7th Street, 15th Floor, TX1-497-15-08
Fort Worth, TX 76102-4700

Kavli Foundation Research Grants 2296
The Kavli Foundation supports research in the areas of astrophysics, nanoscience, neuroscience, and theoretical physics three scientific frontiers that promise great discoveries in the 21st century and beyond. This mission is implemented through an international program of research institutes, professorships, symposia and workshops, as well as prizes in the fields of astrophysics, nanoscience, and neuroscience. Dedicated to advancing science for the benefit of humanity and promoting increased public understanding and support for scientists and their work, the Foundation also makes available materials on these sciences at no cost to educational institutions.
Geographic Focus: All States
Contact: Grants Administrator; (805) 983-6000; fax (805) 988-4800
Internet: http://www.kavlifoundation.org/areasofscience/
Sponsor: Kavli Foundation
1801 Solar Drive, Suite 250
Oxnard, CA 93031

Kelvin and Eleanor Smith Foundation Grants 2297
The foundation awards grants to northeast Ohio nonprofits in its areas of interest, including nonsectarian education, the performing and visual arts, health care, and environmental conservation and protection. Types of support include general operating support, continuing support, annual campaigns, capital campaigns, building construction/renovation, and equipment acquisition. Since there are no required application forms, each proposal include a cover letter that outlines the reason for the request and the dollar amount. Organizations who have previously received funding from this Foundation may submit a proposal annually. There are no specified annual deadlines.
Requirements: Nonprofit organizations in the greater Cleveland, OH, area are eligible.
Restrictions: Grants are not made in support of individuals or for endowment funds, scholarships, fellowships, matching gifts, or loans.
Geographic Focus: Ohio
Amount of Grant: 3,000 - 150,000 USD
Contact: Carol W. Zett; (216) 591-9111; fax (216) 591-9557; cwzett@kesmithfoundation.org
Internet: http://www.kesmithfoundation.org/grantguidelines.html
Sponsor: Kelvin and Eleanor Smith Foundation
30195 Chagrin Boulevard, Suite 275
Cleveland, OH 44124

Kenai Peninsula Foundation Grants 2298
The Kenai Peninsula Foundation offers a competitive grant award process for unrestricted grants to qualified nonprofits offering programs and services in the central Kenai Peninsula area. The Foundation, an affiliate of the Alaska Community Foundation, seeks applications from qualified tax-exempt 501(c)3 nonprofits that support organizations and programs in the central Kenai Peninsula and serve the people's needs in such areas as health, education, community heritage, the arts, vulnerable populations, recreation, safety, and community development. The maximum award is $500, and the annual deadline for submissions is May 15. Awarded grant proposals must be completed within one year.
Requirements: Applications are accepted from qualified 501(c)3 nonprofit organizations, or equivalent organizations located in the state of Alaska. Equivalent organizations may include tribes, local or state governments, schools, or Regional Educational Attendance Areas.
Restrictions: Individuals, for-profit, and 501(c)(4) or (c)(6) organizations, non-Alaska based organizations and state or federal government agencies are not eligible for competitive grants. Applications for religious indoctrination or other religious activities, endowment building, deficit financing, fundraising, lobbying, electioneering and activities of political nature will not be considered, nor will proposals for ads, sponsorships, or special event and any proposals which discriminate as to race, gender, marital status, sexual orientation, age, disability, creed or ethnicity.
Geographic Focus: All States
Date(s) Application is Due: May 15
Amount of Grant: Up to 500 USD
Contact: Ricardo Lopez; (907) 274-6707; fax (907) 334-5780; rlopez@alaskacf.org
Internet: http://kenaipeninsulafoundation.org/projects/
Sponsor: Kenai Peninsula Foundation
P.O. Box 1612
Soldotna, AK 99669

Kendrick Foundation Grants 2299
Kendrick Foundation was created to support health-related programs in Morgan County, Indiana. Support includes making grants to health-related organizations and charities, which may include community health care programs, hospice programs, health care education and training, and tax-exempt medical and health programs. Prior to submitting a grant application, a Letter of Intent must be submitted. Letters of Intent are available for download for a limited amount of time during the opening of the grant cycle. Letters of Intent for the spring grant cycle must be physically received in the office of the Community Foundation of Morgan County by March 15, no later than 4 p.m., with invited applications due April 30. Letters of Intent for the fall grant cycle must be physically received in the office of the Community Foundation of Morgan County, Inc. by October 15, no later than 4 p.m., with invited applications due November 30.
Requirements: The Kendrick Foundation will only accept full grant applications from those organizations serving Morgan County, Indiana, which were approved through a Letter of Intent.
Geographic Focus: Indiana
Date(s) Application is Due: Mar 15; Apr 30; Oct 15; Nov 30
Contact: Tom Zoss, Executive Director; (317) 831-1232 or (877) 822-6958; fax (317) 831-2854; tzoss@cfmconline.org
Internet: http://www.cfmconline.org/kendrick/Grants/tabid/129/Default.aspx
Sponsor: Kendrick Foundation
250 N Monroe Street
Mooresville, IN 46158

Kenneth T. and Eileen L. Norris Foundation Grants 2300
The foundation supports Los Angeles County nonprofits in the areas of: medicine--to improve access to health care, increase knowledge through research, and provide facilities for those activities to take place; youth--to provide constructive activities, positive role models, and opportunities for disadvantaged, disabled, and misguided children; community--to support law enforcement agencies, good citizenship, and environmental conservation; culture--to support museums, symphony orchestras, and dance and theater companies; and education and science--to focus on private education, especially secondary and college levels. Types of support include general operating support, continuing support, building construction and/or renovation, equipment acquisition, endowment funds, program development, professorships, scholarship funds, research, and matching funds. Education/science and medicine projects are accepted between May 1 and June 30; youth requests are accepted between February 15 and March 31; cultural (the arts) and community requests are accepted between December 1 and January 31; and medicine proposals are due between May 1 and June 30.
Requirements: Grants are awarded to organizations in southern California.
Geographic Focus: California
Date(s) Application is Due: Jan 31; Mar 31; Jun 30
Amount of Grant: 5,000 - 25,000 USD
Contact: Lisa D. Hansen; (562) 435-8444; fax (562) 436-0584; grants@ktn.org
Internet: http://www.norrisfoundation.org/grant.html
Sponsor: Kenneth T. and Eileen L. Norris Foundation
11 Golden Shore, Suite 450
Long Beach, CA 90802

Kent D. Steadley and Mary L. Steadley Memorial Trust 2301
The Trust makes contributions to nonprofit organizations in and near Carthage, Missouri, exclusively to support community/economic development; elementary/secondary education; hospitals; human services; and public libraries.
Requirements: Nonprofit organizations operating to promote community well-being of the Carthage, Missouri, area are eligible. Organizations should first send a letter, requesting application materials. There are no application deadlines, and response is usually within six months.
Restrictions: Funding is not available for individuals or national fundraising events.
Geographic Focus: Missouri
Amount of Grant: 5,000 - 200,000 USD
Contact: Lareta Garnier, Program Contact, c/o Bank of America; (417) 227-6237
Sponsor: Kent D. and Mary L. Steadley Memorial Trust
P.O. Box 8300
Springfield, MO 65801-8300

Kessler Foundation Signature Employment Grants 2302
Kessler Foundation awards Signature Employment Grants yearly to support non-traditional solutions and/or social ventures that increase employment outcomes for individuals with disabilities. Signature Employment Grants are awarded nationally to fund new pilot initiatives, demonstration projects or social ventures that lead to the generation of new ideas to solve the high unemployment and underemployment of individuals with disabilities. Preference is given for interventions that overcome specific employment barriers related to long-term dependence on public assistance, advance competitive employment in a cost-effective manner, or launch a social enterprise or individual entrepreneurship project. Signature grants are not intended to fund project expansions or bring proven projects to new communities, unless there is a significant scale, scope or replicable component. Innovation lies at the core of all signature employment grants. Organizations may apply for up to two years of funding. Yearly funding ranges from $100,000 - $250,000, with maximum project funding at $500,000.
Requirements: The Signature Employment Grant program begins with online concept submission. The concept is scored and reviewed for originality, creativity, feasibility, collaborative stakeholder team. A selected group of candidates will then be invited to submit a full grant proposal. Nonprofit organizations that are tax-exempt according to

the Internal Revenue Code may apply for funding. This includes U.S. based non-profit organizations, public/private schools and public institutions, such as universities and government. Application is open to eligible organizations in any state or territories. Priority is placed on serving individuals with mobility disabilities, traumatic brain injury, spinal cord injury, multiple sclerosis, stroke, cerebral palsy, spina bifida, epilepsy or other related impairments. The Foundation requires that 65% of the target grant population meet these criteria. Although matching funds are not required, applicants with additional cash funding provided by the applicant or collaborator(s) will be scored higher. A proven track record managing collaborative grant projects is also desirable.
Restrictions: Kessler Foundation will not fund projects that discriminate in hiring staff or providing services on basis of race, gender, religion, marital status, sexual orientation, age, or national origin. The foundation does not fund projects for which the primary diagnosis of disability is related to autism, developmental/intellectual disabilities, mental illness, post-traumatic stress, learning disabilities, chemical dependency, or sensory impairments of vision and/or hearing.
Geographic Focus: All States
Date(s) Application is Due: May 24
Amount of Grant: 100,000 - 500,000 USD
Sample: APSE, Rockville, MD - To partner with OfficeMax to create a job training model that will allow individuals with significant disabilities to receive the pre-training necessary to close the "skill gap" that has prevented many individuals from successful employment in the past, although they posses amazing potential: $323,333. The Center for Head Injury Services, St. Louis, MO - To create Destination Desserts, a purpose driven, social enterprise business that will provide opportunities for training and employment for people with brain injuries: $500,000. National Disability Institute, Washington, D.C. - To facilitate the connection between employer and qualified job seeker with a disability by forming a collaborative employment model in the financial services sector, through the use of applicant training and certification, capacity building, and Vocational Rehabilitation Work Try-Out and On-The-Job training: $484,452.
Contact: Elaine Katz, Vice President of Grants; (973) 324-8367; fax (973) 324-8373; KFgrantprogram@KesslerFoundation.org
Internet: http://kesslerfoundation.org/grantprograms/signatureemploymentgrants.php
Sponsor: Kessler Foundation
300 Executive Drive, Suite 70
West Orange, NJ 07052

Ketchikan Community Foundation Grants 2303
The Foundation's vision for the Ketchikan community includes a diversified local economy that provides family-supported employment, access to affordable housing and healthcare, a vibrant arts community, and multiple recreational and quality educational opportunities. The community it envisions will attract and retain multi-generations of citizens, while honoring and recognizing cultural diversity. The primary goal, therefore, is to support projects of importance to the community that will help its residents become better stewards of the land. Grant making priorities are still in development, though the Foundation will support projects that strengthen the community. Among others, they will include organizations and programs that serve youth, the elderly, recreation, safety, vulnerable populations, and arts and culture. No annual deadlines for applications have been identified.
Requirements: The Foundation seeks applications from qualified tax-exempt 501(c)3 organizations that support the organizations and programs in the Ketchikan region that serve the people's needs in such areas as health, education, community heritage, the arts, vulnerable populations, recreation, safety, and community and economic development.
Restrictions: ndividuals, for-profit, and 501(c)4 or 501(c)6 organizations, non-Alaska based organizations and state or federal government agencies are not eligible for competitive grants. Applications for religious indoctrination or other religious activities, endowment building, deficit financing, fundraising, lobbying, electioneering and activities of political nature will not be considered, nor will proposals for ads, sponsorships, or special event and any proposals which discriminate as to race, gender, marital status, sexual orientation, age, disability, creed or ethnicity.
Geographic Focus: Alaska
Contact: Ricardo Lopez, Affiliate Program Officer; (907) 274-6707; fax (907) 334-5780; rlopez@alaskacf.org or ketchikan@alaskacf.org
Internet: http://ketchikancf.org/grants/
Sponsor: Ketchikan Community Foundation
P.O. Box 5256
Ketchikan, AK 99901

Kettering Family Foundation Grants 2304
The Kettering Family Foundation was founded by Eugene W. Kettering, son of Charles F. Kettering, and his wife Virginia W. Kettering in 1956. Today, the Foundation supports a broad range of charitable activities of interest to the Board of Trustees, which is composed of members of the Kettering Family. Because funding decisions are driven by the interests of the trustees, Request Summaries endorsed by a trustee at the time of submission are considered a priority. Primary areas of interest include: arts, culture and humanities; education; environment; health and medical needs; human services; and public and societal benefit. The Foundation strongly recommended that you contact the office to discuss your proposed program before you start the application process
Requirements: 501(c)3 tax-exempt organizations are eligible to apply.
Restrictions: A Request Summary will not be accepted for any of the following purposes: religious organizations for religious purposes; individual public elementary or secondary schools or public school districts; multi-year grants; grants or loans to individuals; tickets, advertising or sponsorships of fundraising events; efforts to carry on propaganda or otherwise attempt to influence legislation; or activities of 509(a)3 Type III Supporting Organizations.
Geographic Focus: All States
Date(s) Application is Due: Mar 15; Jun 15; Sep 15; Dec 15
Contact: Judith M. Thompson, Executive Director; (303) 756-7664 or (937) 228-1021; fax (888) 719-1185; info@ketteringfamilyphilanthropies.org
Internet: http://www.ketteringfamilyfoundation.org/main.html
Sponsor: Kettering Family Foundation
40 North Main Street, #1480
Dayton, OH 45423

Kettering Fund Grants 2305
Charles F. Kettering (1876-1958) was an American inventor, founder of Delco, and head of the General Motors Research Laboratory for twenty-seven years. He was born in Loudonville, Ohio and lived most of his life in Dayton, Ohio. He was involved in many philanthropic projects during his lifetime, a legacy which the Kettering Family adopted and continues. The mission of the Kettering Fund is to support scientific, medical, social and educational studies and research conducted by nonprofit, charitable organizations that are located and/or provide service in Ohio. The Distribution Committee meets on a semi-annual basis to make funding decisions, generally in May and November. Primary areas of support include: arts, culture and humanities; education; environment; health and medical needs; human services; and public and societal benefit. The Fund committee will consider support for the following purposes: capital needs (such as the purchase of land, new construction, renovations and equipment); seed grants for new organizations or programs; project support; endowments; research; and scholarship funds. Annual deadlines for application submission are July 31 and January 31.
Requirements: 501(c)3 nonprofits that provide services in or are located in Ohio may apply.
Restrictions: A Request Summary will not be accepted for any of the following purposes: religious organizations for religious purposes; individual public elementary or secondary schools or public school districts; multi-year grants; grants or loans to individuals; tickets, advertising or sponsorships of fundraising events; efforts to carry on propaganda or otherwise attempt to influence legislation; or activities of 509(a)3 Type III Supporting Organizations.
Geographic Focus: Ohio
Date(s) Application is Due: Jan 31; Jul 31
Amount of Grant: 5,000 - 200,000 USD
Contact: Judith M. Thompson, Executive Director; (937) 228-1021; fax (888) 719-1185; info@ketteringfamilyphilanthropies.org
Internet: https://www.cfketteringfamilies.com/fund/about-us
Sponsor: Kettering Fund
40 North Main Street, #1480
Dayton, OH 45423

Kevin P. and Sydney B. Knight Family Foundation Grants 2306
The Foundation, established in 1997 by the founders of Knight Transportation of Phoenix, Arizona, is focused on the greater Phoenix area. Giving programs are aimed at: children and youth; Catholic agencies and churches; athletics/sports activities (particularly baseball); education; health care; and Mormon affiliated groups. Primary types of support include: community development, general research, and youth programs. There are no specific deadlines or application forms, and initial approach should be by letter.
Restrictions: Funding goes primarily to 501(c)3 organization located in Phoenix, Arizona.
Geographic Focus: Arizona
Amount of Grant: 1,000 - 2,000 USD
Contact: Kevin P. Knight, President; (602) 269-2000; fax (602) 269-8409
Sponsor: Kevin P. and Sydney B. Knight Family Foundation
5601 W Buckeye Road
Phoenix, AZ 85043-4603

KeyBank Foundation Grants 2307
The Foundation's objective is to improve the quality of life and economic vibrancy of the places where KeyCorp customers, employees, and shareholders live and work. The Foundation supports programs with the following funding priorities: financial education - fostering effective financial management and understanding of financial services and tools; workforce development - providing training and placement for people to access job opportunities; diversity - promoting inclusive environments by employing systematic changes to improve the access of individuals of diverse backgrounds. The Foundation typically reviews and decides upon requests quarterly, but there are no deadlines to submit the proposal. Proposal development guidelines and the application are available at the Foundation website. Proposals may be submitted via postal mail or email, but not by fax.
Requirements: Nonprofit 501(c)3 tax-exempt organizations are eligible. Requests for funding from organizations within northeast Ohio are reviewed by the KeyBank Foundation headquarters offices in Cleveland. Other proposals are evaluated by funding committees in district offices throughout the U.S. Requests sent to the KeyBank Foundation headquarters that are more appropriate for district review are forwarded to the appropriate district office.
Restrictions: The Foundation does not contribute to individuals; lobbying or political organizations areas outside its retail operations; selected organizations with IRS 509(a) status; fraternal groups; athletic teams; organizations outside the U.S; or organizations that discriminate in any way with national equal opportunity policies. The Foundation also does not buy journal advertisements or memberships.
Geographic Focus: Alaska, Colorado, Idaho, Indiana, Kentucky, Maine, Michigan, New York, Ohio, Oregon, Utah, Vermont, Washington
Contact: Lorraine Vega; (216) 689-7397; fax (216) 828-7845; key_foundation@keybank.com
Internet: https://www.key.com/about/community/key-foundation-philanthropy-banking.jsp
Sponsor: KeyBank Foundation
127 Public Square, 7th Floor
Cleveland, OH 44114-1217

Kimball International-Habig Foundation Health and Human Services Grants 2308
The Kimball-Habig Foundation was established by company founder, Arnold F. Habig, in 1951 for the purpose of supporting charitable causes within the communities in which Kimball operates, or from which it draws employees. The Foundation is funded by a percentage of profit earnings by the company. In keeping with the corporate philosophy and guiding principles, the Foundation is committed to helping the communities in which they operate to become even better places to live. Supporting that goal, the foundation focuses its funding and resources on grants to organizations and programs that most directly benefit those U.S. communities in which Kimball has operations or facilities, or from which it draws employees. In the area of health and human services, the Foundation's main focus is to improve the human condition and help to alleviate suffering by assisting local community organizations, charities, faith-based initiatives, and social services, in their efforts by considering the following types of requests: care and protection of children and infants; care and protection of at-risk women and families; care and protection of the elderly and infirmed; provision of healthcare, medical, and counseling services; and provision of basic social and support services. Awards are typically for general operating support. Though there are no specific deadlines, the Board meets quarterly, during the last week in March, June, September, and December, to award grants applied for during the previous 90 days.
Requirements: All requests for funding made to the Kimball Foundation must be made using the online request form. All requests must be in writing (via this online form), and absolutely no verbal or phone call requests will be processed or acknowledged. Major requests (those over $2,000) are reviewed, assessed and approved quarterly. Standard requests (those under $2,000) are reviewed and approved monthly. Standard requests are reviewed by the foundation board on or about the 25th of each month.
Geographic Focus: California, Florida, Idaho, Indiana, Kentucky, China, Mexico, Poland
Contact: Dean Vonderheide, President; (812) 482-8255 or (812) 482-8701; habigfoundation@kimball.com
Internet: http://www.kimball.com/foundation.aspx
Sponsor: Kimball International-Habig Foundation
1600 Royal Street
Jasper, IN 47549-1001

Kinsman Foundation Grants 2309
The Kinsman Foundation awards grants to eligible nonprofit organizations in its areas of interest, including: historic preservation; arts, culture, and humanities; and animals and wildlife; and health care policy. Types of support include annual budgetary support, building construction and renovation, capital campaigns, challenge/matching grants, conferences and seminars, consulting services, continuing support, curriculum development, endowments, equipment acquisition, general operating support, internships, program development, publication, research, seed grants, and technical assistance. The Foundation is in the process of eliminating its formal grant application in favor of collecting information in whatever format is most convenient to the applicant. Applications for the Betty Kinsman Fund are due February 15 of each year. For Historic Preservation and Native Wildlife Rehabilitation and Appreciation grants, inquiries less than $10,000 are processed throughout the year.
Requirements: Oregon and southern Washington nonprofit organizations are eligible.
Geographic Focus: Oregon, Washington
Contact: Sara Bailey, Grants Associate; (503) 654-1668; fax (503) 654-1759; sara@kinsmanfoundation.org or grants@kinsmanfoundation.org
Internet: http://www.kinsmanfoundation.org/guidelines/index.htm
Sponsor: Kinsman Foundation
3727 SE Spaulding Avenue
Milwaukie, OR 97267-3938

Klarman Family Foundation Grants in Eating Disorders Research Grants 2310
The Klarman Family Foundation is interested in providing strategic investment in translational research that will accelerate progress in developing effective treatments for anorexia nervosa, bulimia nervosa and binge eating disorder. The program's short-term goal is to support the most outstanding science and expand the pool of scientists whose research explores the basic biology of feeding, anorexia nervosa, bulimia nervosa, and/or binge eating disorder. The long-term goal is to improve the lives of patients suffering from these conditions. Examples of funding areas include but are not limited to: molecular genetic analysis of relevant neural circuit assembly and function; genetic and epigenetic research; animal models created by genetically altering neural circuits; and testing of new chemical entities that might be used in animal models as exploratory treatments. Investigators conducting research in the neuro-circuitry of fear conditioning or reward behavior may also apply but must justify the relevance of their research projects to the basic biology of eating disorders. Two-year awards of $400,000 USD ($200,000 per year inclusive of 10% indirect costs) and one-year pilot studies of up to $150,000 USD (inclusive of 10% indirect costs) are made to investigators with a faculty appointment at a nonprofit academic, medical or research institution in the United States, Canada or Israel.
Requirements: At the time of application, investigators must hold a faculty appointment at a nonprofit academic, medical or research institution in the United States, Canada or Israel. Investigators who previously submitted applications or have concluded their Klarman Family Foundation Award remain eligible to apply. Applicants do not need to be nominated by their institutions. Investigators working outside the field of eating disorders research are encouraged to apply. Examples of funding areas include but are not limited to molecular genetic analysis of relevant neural circuit assembly and function; genetic and epigenetic research; animal models created by genetically altering neural circuits; testing of new chemical entities that might be used in animal models as exploratory treatments; and clinical trials involving medications or other non-psychotherapeutic interventions. Applications must be received by The Medical Foundation by February 11, 1:00 pm U.S. Eastern time.
Restrictions: Only one application may be submitted per applicant. Clinical psychotherapeutic studies, imaging studies involving humans, and research in the medical complications of these disorders are outside the scope of this program.
Geographic Focus: All States, Canada, Israel
Date(s) Application is Due: Feb 11
Amount of Grant: 400,000 USD
Contact: Program Officer; (617) 236-7909; info@klarmanfoundation.org
Internet: http://klarmanfoundation.org/eating-disorders-research/
Sponsor: Klarman Family Foundation
P.O. Box 171627
Boston, MA 02117

Kluge Center David B. Larson Fellowship in Health and Spirituality 2311
The Library of Congress invites qualified scholars to apply for a postdoctoral fellowship in the field of health and spirituality. It seeks to encourage the pursuit of scholarly excellence in the scientific study of the relation of religiousness and spirituality to physical, mental, and social health. The fellowship provides an opportunity for a period of six to 12 months of concentrated use of the collections of the Library of Congress, through full-time residency in the Library's John W. Kluge Center. If necessary, special arrangements may be made with the National Library of Medicine for access to its materials as well. Applicants must submit a formal application packet, including an application form, a two-page curriculum vita that should indicate major prior scholarship, a one-paragraph project summary, a bibliography of basic sources, a research proposal of no more than 1500 words, and three letters of reference (in English) from people who have read the research proposal.
Requirements: Applicants must by U.S. citizens or permanent residents and must possess a doctoral degree (PhD, MD, ScD, DrPH, DSW, PPsy, DST, ThD, and JD) awarded by the deadline date.
Geographic Focus: All States
Date(s) Application is Due: Apr 17
Amount of Grant: 25,200 - 50,400 USD
Contact: Coordinator; (202) 707-3302; fax (202) 707-3595; scholarly@loc.gov
Internet: http://www.loc.gov/loc/kluge/fellowships/larson.html
Sponsor: Library of Congress
101 Independence Avenue SE
Washington, D.C. 20540-4860

Knight Family Charitable and Educational Foundation Grants 2312
The Foundation operates in Michigan, and funding is reserved for elementary and secondary education, religious agencies, and community health and human service programs. Giving is primarily reserved for Michigan-based non-profit organizations. There are no specific deadlines or application forms to adhere to, and final approval is given approximately two months after submission.
Requirements: Applicants should be 501(c)3 organizations based in Michigan.
Geographic Focus: All States
Amount of Grant: 500 - 10,000 USD
Contact: N.J. Glasser, Secretary-Treasurer; (517) 547-6131
Sponsor: Knight Family Foundation
215 North Talbot Street
Addison, MI 49220-9698

Knox County Community Foundation Grants 2313
The Knox County Community Foundation is a nonprofit, public charity created by and for the people of Knox County, Indiana. The Foundation helps nonprofits fulfill their missions by strengthening their ability to meet community needs through grants that assist charitable programs, address community issues, support community agencies, launch community initiatives, and support leadership development. Grant proposals are accepted once each year according to the grant cycle. Proposal requirements may change from year to year; therefore, grant seekers are advised to contact the foundation or see the foundations website, prior to beginning the grant application process. Grants are normally given as one-time support of a project but may be considered for additional support for expansions or outgrowths of an initial project. At the start of each cycle, a notice is mailed to nonprofit organizations that have applied for grants in the past, have received grants in the past, or have otherwise requested notification of the start of each cycle. Program areas considered for funding are: arts and culture; community development; education; health; human services; other civic endeavors, such as the environment, recreation, and youth development. Proposals will be accepted from January through the March deadline. All organizations that have submitted grant proposals will be notified of the outcome of their application by June 1. Samples of previously funded projects are available at the website.
Requirements: The Foundation welcomes proposals from nonprofit organizations that are deemed tax-exempt under sections 501(c)3 and 509(a) of the Internal Revenue Code and from governmental agencies serving the county. Proposals from nonprofit organizations not classified as a 501(c)3 may be considered provided the project is charitable and supports a community need. Proposals submitted by an entity under the auspices of another agency must include a written statement signed by the agency's board president on behalf of the board of directors agreeing to act as the entity's fiscal sponsor, to receive grant monies if awarded, and to oversee the proposed project.
Restrictions: Project areas not considered for funding are: religious organizations for strictly religious purposes; political parties or campaigns; endowment creation or debt reduction; operating costs; capital campaigns; annual appeals or membership contributions; travel requests for groups or individuals such as bands, sports teams, or classes. A six-month progress report and a final report at project completion are required by organizations whose proposals are approved for funding. Instructions and appropriate forms will be provided at the time the grant is awarded.

Geographic Focus: Indiana
Contact: Jamie Neal; (812) 886-0093; jamie@knoxcountyfoundation.org
Internet: http://www.knoxcountyfoundation.org/disc-grants-program
Sponsor: Knox County Community Foundation
20 N Third Street, Suite 301, P.O. Box 273
Vincennes, IN 47591

Kodiak Community Foundation Grants 2314

As an organization, the Kodiak Community Foundation's goal is to support projects of importance to Kodiak residents that will help them become better stewards of the land. As in a maritime community, these residents have strong ties to the natural elements. Many depend on Kodiak's local fish and wildlife for subsistence and livelihoods. Despite the changing landscape of funding in the state of Alaska, Kodiak is aware of the importance of self-reliance. Its citizens have come together to produce local initiatives for a vision-oriented legacy. This is how the Kodiak Community Foundation began; with a dream of providing the means for improvement in the quality of life within the town. Grant making priorities are still in development, though the Foundation will support programs and projects that serve youth, the elderly, recreation, safety, vulnerable populations, and arts and culture. There are no established annual deadlines for applications.
Requirements: The Foundation seeks applications from qualified tax-exempt 501(c)3 organizations that support the organizations and programs in the Kodiak region and serve the people's needs in such areas as health, education, community heritage, the arts, vulnerable populations, recreation, safety, and community and economic development.
Restrictions: Individuals, for-profit, and 501(c)4 or 501(c)6 organizations, non-Alaska based organizations and state or federal government agencies are not eligible for competitive grants. Applications for religious indoctrination or other religious activities, endowment building, deficit financing, fundraising, lobbying, electioneering and activities of political nature will not be considered, nor will proposals for ads, sponsorships, or special event and any proposals which discriminate as to race, gender, marital status, sexual orientation, age, disability, creed or ethnicity.
Geographic Focus: Alaska
Contact: Ricardo Lopez, Affiliate Program Officer; (907) 274-6707; fax (907) 334-5780; rlopez@alaskacf.org or kodiak@alaskacf.org
Internet: http://kodiakcf.org/projects/
Sponsor: Kodiak Community Foundation
P.O. Box 400
Kodiak, AK 99615

Komen Greater NYC Clinical Research Enrollment Grants 2315

Women, especially racial and ethnic minorities, are not adequately represented in cancer research efforts. Multiple barriers — cultural, linguistic, financial, and systematic — hinder their participation in clinical research, and particularly in clinical trials. Through Clinical Research Enrollment Grants, Komen Greater NYC seeks to fund projects that employ effective strategies to overcome these barriers to enrollment and retention in breast cancer clinical research. The funding period is two years, based on the Komen Greater NYC fiscal year — April 1st through March 31st. . Programs applying for the CRE grant may request up to $200,000 over this two-year period. Organizations that have previously received Komen Greater NYC funding are welcome to apply for additional funding.
Requirements: To apply, the organization must be: a not-for-profit organization (a charitable or educational tax-exempt organization), a government agency, or an Indian tribe; conducting National Cancer Institute (NCI) or Department of Defense (DOD) approved breast cancer clinical research (this includes studies through the Clinical Trials Cooperative Group Program including but not limited to ACOSOG, ECOG, SWOG, NSABP, ACRIN, and RTOG; and located in the Komen Greater NYC service area, which includes the five boroughs of New York City, Long Island, Westchester County, and Rockland County.
Geographic Focus: New York
Date(s) Application is Due: Oct 9
Amount of Grant: Up to 200,000 USD
Contact: Michelle Marquez, Director of Development; (212) 461-6186 or (212) 560-9590; fax (212) 560-9598; mmarquez@komennyc.org
Zenia Dacio-Mesina, Grants Program Coordinator; (212) 560-9590; fax (212) 560-9598; zdmesina@komennyc.org or grants@komennyc.org
Internet: http://www.komennyc.org/site/PageServer?pagename=grants_clinicaltrials
Sponsor: Susan G. Komen for the Cure Foundation
470 Seventh Avenue, 7th Floor
New York, NY 10018

Komen Greater NYC Community Breast Health Grants 2316

The mission of Komen Greater NYC is to eradicate breast cancer as a life-threatening disease by advancing research, education, screening, and treatment. The Greater New York City Affiliate is, therefore, soliciting applications from non-profit organizations that have current or planned breast health programs within the five boroughs of New York City, Long Island, Westchester County, and Rockland County. The purpose of the Community Breast Health Grants Program (formerly known as Screening, Treatment, and Education Program- STEP) is to support community-based programs that provide the medically under-served with access to breast health education, screening coordination, diagnostic, treatment, and support services. The funding period is twelve months, based on the Komen Greater NYC fiscal year — April 1st through March 31st. Programs applying for the CBH grant may request up to $75,000 over this one year period. Organizations that have previously received Komen Greater NYC funding are welcome to apply for additional funding. However, all CBH grant awards are for one year only.
Requirements: Applicant organizations must be: a not-for-profit organization (a charitable or educational tax-exempt organization), a government agency, or an Indian tribe; located in the Komen Greater NYC service area, which includes the five boroughs of New York City, Long Island, Westchester County, and Rockland County; and a current or planned provider of breast health services.
Geographic Focus: New York
Date(s) Application is Due: Oct 9
Amount of Grant: Up to 75,000 USD
Contact: Michelle Marquez, Director of Development; (212) 461-6186 or (212) 560-9590; fax (212) 560-9598; mmarquez@komennyc.org
Zenia Dacio-Mesina, Grants Program Coordinator; (212) 560-9590; fax (212) 560-9598; zdmesina@komennyc.org or grants@komennyc.org
Internet: http://www.komennyc.org/site/PageServer?pagename=grants_breasthealthgrants
Sponsor: Susan G. Komen for the Cure Foundation
470 Seventh Avenue, 7th Floor
New York, NY 10018

Komen Greater NYC Small Grants 2317

Komen Greater New York City is currently offering small grants up to $5,000 to support small pilot or capacity building projects in breast health. Small Grant Applications are accepted three times a year. Award notifications are generally made within six to eight weeks after the deadline. These grants are for organizations interested in beginning a new program or test new ideas that will increase the innovative capacity and effectiveness of breast health programs that serve low income and uninsured patients. Capacity Building is enhancing an organization's ability to provide services by redesigning processes, implementing new practices, or developing collaborations or partnerships.
Requirements: Applicants and institutions must conform to the eligibility criteria to be considered for funding. Applicants must ensure that all past and current Komen-funded grants or awards are up-to-date and in compliance with Komen requirements. Institutions must be located in or providing services to one or more of the following locations: the five boroughs of New York City; Long Island (Suffolk and Nassau Counties); Westchester County; or Rockland County. Projects must be specific to breast health and/or breast cancer. The applicant must be a non-profit organization with federal tax exemption.
Restrictions: Individuals may not receive grants. Small Grant applicants are restricted to a maximum combined award of $5,000 per Komen Greater NYC grant year (April through March), per program for any type of Small Grant.
Geographic Focus: New York
Date(s) Application is Due: Jan 4; May 1; Sep 1
Amount of Grant: 5,000 USD
Contact: Zenia Dacio-Mesina, Grants Program Coordinator; (212) 560-9590; fax (212) 560-9598; zdmesina@komennyc.org or grants@komennyc.org
Internet: http://www.komennyc.org/site/PageServer?pagename=grants_small_grants_program
Sponsor: Susan G. Komen for the Cure Foundation
470 Seventh Avenue, 7th Floor
New York, NY 10018

Kosair Charities Grants 2318

In 1923, a committee of members from Louisville's Kosair Shrine Temple, doctors, lawyers, and community leaders formed Kosair Charities. The mission was to provide the highest quality health care possible for children who had nowhere else to turn. Since conception, Kosair Charities has becamed the largest charity in the history of the region that provides for the medical care of children. All of this is possible thanks to the generous support of individuals and companies. Kosair Charities receives no assistance from United Way, the Crusade for Children, or any governmental entity. Annual grants are awarded in the early summer of each year and applications are due no later than May 1.
Requirements: Grants are only made to those non-profit agencies and organizations whose program falls within the above stated mission. Two grant requests applications are available for download on the Kosair Charities website in amounts of $1,500 or less and $1,500 or more. With each completed application, attach the following: Copy of IRS determination letter; current year's budget; income and expense statement last two years; copy of last completed audit; copy of roster of Board of Directors; descriptive literature on agency and its services; if the proposal is for an existing or continuing program, submit latest self-evaluation of program; if the proposal is for one that Kosair Charities has funded in the past, demonstrate use of Kosaid Charities Supported Agency Logo and public acknowledgements; detailed proposal budget; original grant application plus five copies for a total of six copies of all requested information.
Restrictions: Grants are available throughout Kentucky and the southern Indiana region. Grants are made for one year only and do not represent any future commitment on the part of Kosair Charities. Grant recipients are required to use the Kosair Charities Supported Agency logo on all printed and electronic materials and media used by their agency.
Geographic Focus: Indiana, Kentucky
Date(s) Application is Due: May 1
Contact: Jo Barrett; (502) 637-7696 or (888) 454-3752; jbarrett@kosair.org
Internet: http://www.kosair.org/grants.html
Sponsor: Kosair Charities
982 Eastern Parkway, P.O. Box 37370
Louisville, KY 40233-7370

Kosciusko County Community Foundation Grants 2319

The Kosciusko County Community Foundation serves Kosciusko County, Indiana. Nonprofit organizations serving Kosciusko County are eligible to apply in seven areas of interest: arts and culture, human services, civic projects, recreation, environment, health, and education. Grant applications and guidelines can be obtained at the Foundations office or website. Grant awards are announced nine weeks after each deadline.

Requirements: Grant seekers are strongly encouraged to call the Foundation's program staff to discuss a grant proposal before submitting a formal application. Once the proposal has been discussed, complete and submit a grant application with the required attachments: 6 copies of the original application; 1 copy of the IRS determination letter; 7 copies of the board of directors listing with names and addresses for all; 7 copies of staff listing with names and addresses for all; 7 copies of current internal financial statements; and 7 copies of program/project budget. Do not provide copies of news articles, brochures or other miscellaneous supporting information.
Restrictions: The Foundation will not consider grants for: individuals; political activities or those designated to influence legislation; national organizations (unless the monies are to be used solely to benefit citizens of Kosciusko County); fundraising projects;the direct benefit of the donor or the donor's family; religious organizations for the sole purpose of furthering that religion (this prohibition does not apply to funds created by donors who have specifically designated religious organizations as beneficiaries of the funds); contributions to endowments.
Geographic Focus: Indiana
Date(s) Application is Due: Jan 15; May 15; Sep 15
Contact: Stephanie Overbey, Communication & Program Director; (574) 267-1901; fax (574) 268-9780; stephanie@kcfoundation.org
Internet: http://www.kcfoundation.org/grants.html
Sponsor: Kosciusko County Community Foundation
102 E Market Street
Warsaw, IN 46580

Kosciusko County Foundation REMC Operation Round Up Grants 2320
The Kosciusko County Community Foundation serves the residents of Kosciusko County, Indiana. Kosciusko Rural Electric Membership Corporation (REMC) encourages its members to round up their electric bills to the nearest whole dollar. The extra funds are deposited into the Kosciusko REMC Operation Round Up Grants, which supports a variety of charitable causes in communities served by Kosciusko REMC. Applications are due the 15th of the following months: February, April, June, August, October, and December. Grant notifications take place within six weeks after the deadline.
Requirements: Applicants must submit the online application, along with a list of the organization's board of directors, officers, or trustees, and their phone numbers; a one page cover letter that specifies the amount requested and details about how the funds will be used locally; a copy of the organization's 501(c)3 letter; and a copy of the organization's most current financial statements.
Geographic Focus: Indiana
Date(s) Application is Due: Feb 15; Apr 15; Jun 15; Aug 15; Oct 15; Dec 15
Amount of Grant: 500 - 5,000 USD
Samples: Boys and Girls Club of Kosciusko County; provide meals to children, $2,500: Fort Wayne Philharmonic, Inc; support of three concerts, $1,000: College Mentors for Kids; launch a College Mentors for Kids at Grace College, $5,000.
Contact: Stephanie Overbey, Communication and Program Director; (574) 946-0906; fax (574) 946-0971; Stephanie@kcfoundation.org
Internet: http://www.kcfoundation.org/seekingfunds/remc.php
Sponsor: Kosciusko County Community Foundation
102 East Market Street
Warsaw, IN 46580

Kosciuszko Foundation Dr. Marie E. Zakrzewski Medical Scholarship 2321
The Dr. Marie E. Zakrzewski Medical Scholarship of $3,500 is awarded each year to a young woman of Polish ancestry for first, second, or third year of medical studies at an accredited school of medicine in the United States. Funding is for the academic year September through May. Selection is based on academic excellence, the applicant's academic achievements, interest, motivation, interest in Polish subjects and involvement in the Polish American community. Financial need is taken into consideration. First preference is given to residents of the state of Massachusetts. Qualified residents of New England are considered if no first preference candidates apply. The scholarship is non-renewable.
Requirements: United States citizens of Polish descent and Polish citizens with permanent residency status in the United States who are entering first, second or third year of M.D. studies in the upcoming academic year and who have a minimum GPA of 3.0. This scholarship is open to female residents of the state of Massachusetts.
Geographic Focus: Connecticut, Maine, Maryland, Massachusetts, New Hampshire, Vermont
Date(s) Application is Due: Jan 5
Amount of Grant: 3,500 USD
Contact: Addy Tymczyszyn, Program Officer, Scholarship & Grants for Americans; (212) 734-2130, ext. 210; fax (212) 628-4552; addy@thekf.org
Internet: http://www.thekf.org/scholarships/tuition/mzms/
Sponsor: Kosciuszko Foundation
15 East 65th Street
New York, NY 10065

Kosciuszko Foundation Grants for Polish Citizens 2322
The Kosciuszko Foundation annually awards a number of grants to Poles for study and research at universities and other institutions of higher learning in the United States. The Foundation provides a cost-of-living stipend which includes: transatlantic travel, housing allowance, health and accident insurance coverage, and, when warranted, domestic travel. Grants are awarded to those without doctoral degrees. In determining awards, in addition to overall excellence, consideration is given to number of other factors such as scholarly affiliation or geographic location in order to achieve fair and reasonable diversity among awardees. There are no restrictions as to fields of study or research supported, and each proposal is evaluated individually.

Requirements: Only Polish citizens residing permanently in Poland are eligible to apply. This program is not for Poles residing outside of Poland, nor those residing temporarily outside of Poland, whether or not engaged in research or study. Dual citizens (United States and Poland) as well as individuals applying or holding United States permanent residency status (green card) are not eligible. Applicants must have strong English language proficiency to communicate and to carry out proposed projects. The level of English proficiency will be determined during the personal interviews in Warsaw.
Restrictions: There is no allowance for dependents (spouse or child) support. Awardees are solely responsible for all expenses of accompanying dependents.
Geographic Focus: Poland
Date(s) Application is Due: Oct 15
Amount of Grant: 7,650 - 25,500 USD
Contact: Addy Tymczyszyn, Program Officer, Scholarship & Grants for Americans; (212) 734-2130, ext. 210; fax (212) 628-4552; addy@thekf.org
Program Officer; (48) (22) 621-7067; fax (48) (22) 621-7067
Internet: http://www.thekf.org/scholarships/exchange-us/
Sponsor: Kosciuszko Foundation
15 East 65th Street
New York, NY 10065

Kovler Family Foundation Grants 2323
The Kovler Family Foundation awards grants in the areas of the arts, children/youth services, medical research (particularly diabetes), education, human services, higher education, human services, and Jewish federated giving programs. General operating or research grants are awarded primarily in the Chicago metropolitan area. There are no application forms. Applicants should submit a one to two page written proposal letter with a copy of their IRS determination letter by mid-November. Typical grant awarded is between $1,000-$5,000.
Requirements: Illinois nonprofit organizations are eligible to apply.
Restrictions: The Foundation does not award grants to individuals.
Geographic Focus: Illinois
Samples: Columbia College, Chicago, Illinois, $5,000, payable over one year; School of the Art Institute of Chicago, Chicago, Illinois, $1,500, payable over one year; University of Chicago, Chicago, Illinois, $1,000, payable over one year.
Contact: Jonathan Kovler, President and Treasurer; (312) 664-5050
Sponsor: Kovler Family Foundation
875 North Michigan Avenue
Chicago, IL 60611-1958

Kroger Company Donations 2324
Kroger has a long history of bringing help to the communities they serve. They contribute more than $220 million annually in funds, food, and products to support local communities. They focus on feeding the hungry through more than 80 local Feeding America food bank partners, women's health, American troops and their families, and local schools and grassroots organizations. They are also strong supporters of The Salvation Army, American Red Cross, and organizations that promote the advancement of women and minorities.
Requirements: Organizations may contact their locally owned Kroger store (Kroger, Dillon's, Fred Meyer, Fry's, QFC, Ralph's, Smith's, Baker's, City Market, Food4Less, Foods Co., Gerbes, JayC, King Soopers, Owen's, Pay Less, Kwik Shop, Littman Jewelers, Loaf'n Jug, QuikStop, The Little Clinic, Tom Thumb, Turkey Hill, and Fred Meyer Jewelers) or the Fiscal Administrator at Kroger's corporate office for addition information about requesting a donation.
Geographic Focus: All States
Contact: Fiscal Administraor; (513) 762-4449; fax (513) 762-1295
Internet: http://www.kroger.com/community/Pages/default.aspx
Sponsor: Kroger Foundation
1014 Vine Street
Cincinnati, OH 45202-1100

Kroger Foundation Women's Health Grants 2325
The Kroger Foundation provides financial support to local schools, hunger relief agencies, and nonprofit organizations in communities where the company operates stores or manufacturing facilities. Kroger focuses on its charitable giving in the following areas: hunger relief; education; diversity; grassroots community support; and women's health. Types of support includes grants for general operation, capital gains, and seed money. The Foundation focuses on women's health, especially breast cancer, because of its impact on women and their family members. The Foundation makes grants to organizations that provide services to women and their families, in addition to funding important research, such as the American Cancer Society and the Susan G. Komen Breast Cancer Foundation. The Foundation also supports the American Heart Association and Red Dress Campaign to educate women about the importance of heart healthy living. Samples of additional organizations and previously funded projects throughout the U.S. are listed on the Foundation website.
Requirements: Grant applications are not available. Nonprofit organizations may submit grant proposals at any time through the community relations departments of their local Kroger retail store, or contact the fiscal administrator at the corporate office for more information. Proposals must include an IRS tax-exempt letter, a statement of goals and objectives, and a list of the board of trustees. Support is provided only to programs that address a clearly identified need in the community, with specific goals and objectives. Organizations should reflect a strong base of community support.
Restrictions: Only organizations that serve the geographic areas where Kroger owned companies operate are eligible to apply. Funding is not available for the following: national or international organizations; for profit organizations; conventions or conferences luncheons or dinners; other foundations, except those associated with

educational initiatives; endowment campaigns; ongoing operating funding, especially for agencies receiving United Way support (or other federation type support such as a Fine Arts Fund); medical research organizations; sponsorship of sporting events; religious organizations or institutions, if the project is for sectarian purposes; individuals; program advertisements; or membership dues.
Geographic Focus: All States
Contact: Fiscal Administrator; (513) 762-4449; fax (513) 762-1295
Internet: http://www.thekrogerco.com/docs/default-document-library/click-here.pdf
Sponsor: Kroger Foundation
1014 Vine Street
Cincinnati, OH 45202-1100

Kuntz Foundation Grants 2326
Since its inception in 1946, the Kuntz Foundation has contributed over $6 million to charitable organizations in the fields of education, religion, healthcare, civic and cultural affairs, and human services. The foundation will consider grant proposals with a local focus (Dayton Region) for special projects, typically with durations of one year, and faith-based projects with a Catholic emphasis. All grants are one-time awards, unless otherwise stipulated. The Board will generally consider: project funding (vs. operating funding); local focus (Greater Miami Valley); one year funding; and faith-based (Catholic emphasis). Grant proposals must be received by June 30 of a given year. Notification of funding will be provided by December 31 of the same year.
Requirements: Applying organizations must be recognized as tax-exempt under Section 501(c)3 of the Internal Revenue Code.
Restrictions: Grants for individuals will not be considered.
Geographic Focus: Ohio
Date(s) Application is Due: Jun 30
Contact: Rose Ann Eckart, Assistant to the President; (937) 225-9961 or (937) 222-0410; reckart@daytonfoundation.org or info@daytonfoundation.org
Internet: http://www.daytonfoundation.org/grntfdns.html
Sponsor: Kuntz Foundation
40 North Main Street, Suite 500
Dayton, OH 45423

L. A. Hollinger Respiratory Therapy and Nursing Scholarships 2327
The intent of this scholarship is to promote his strong commitment to education with the emphasis on continually improving the quality of patient care and ensuring the continued preparation of future nurses and respiratory therapists to meet that objective. The scholarship is awarded to those who are currently pursuing a Nursing or Respiratory Therapy undergraduate or graduate degree. All applications must be submitted or postmarked by May 15th.
Requirements: All applicants must have a cumulative GPA of 2.5. All nursing applicants must be a recipient of a Trinity Lutheran Hospital School of Nursing Alumnae Scholarship for the current year. Respiratory Therapy applicants must be enrolled in the Respiratory Therapy program at either Johnson County Community College or the University of Kansas.
Geographic Focus: All States
Date(s) Application is Due: May 15
Contact: Becky Schaid; (816) 276-7515 or (816) 276-7555; becky@btllf.org
Internet: http://www.btllf.org/funding.html
Sponsor: Baptist-Trinity Lutheran Legacy Foundation
6601 Rockhill Road
Kansas City, MO 64131

L. W. Pierce Family Foundation Grants 2328
Established in 1997, The Foundation supports organizations involved with health, social, and educational services in the areas of alcohol and drug abuse, hospice care and children's welfare. Giving is limited to the Vero Beach, FL and Philadelphia, PA areas.
Requirements: Non-profits in the Vero Beach and Philadelphia areas are eligible.
Restrictions: No grants to individuals.
Geographic Focus: Florida, Pennsylvania
Date(s) Application is Due: Mar 1
Amount of Grant: 1,000 - 50,000 USD
Contact: Constance Buckley, President; (610) 862-2105; fax (610) 862-2120
Sponsor: L. W. Pierce Family Foundation
8 Tower Bridge, Suite 1060, 161 Washington Street
Conshohocken, PA 19428-2060

LaGrange County Community Foundation Grants 2329
The mission of the LaGrange County Community Foundation (LCCF) is to inspire and sustain leadership, generosity and service. Its purpose is to help community service organizations sponsor plans to meet critically important needs. The Foundation funds grants for innovative and creative projects and programs that are responsive to changing community needs in the areas of, but not restricted to: health and human services, environment, arts and culture, and recreation.
Requirements: Non-profit organizations, schools and qualifying government agencies serving the citizens of LaGrange County are invited to apply for a grant through the foundation's application process. In addition to the online application, applicants should submit a complete list of their organization's board of directors and their occupations; a copy of their specific line item budget with projected income and expenses; a copy of the organization's most recent operating budget; documentation to prove the organization's non-profit status.
Restrictions: The foundation does not make grants to individuals, except in the form of academic scholarships. Grants are generally given one-time only for specific purposes and will typically not be awarded to provide annual operating expenses or support. A grant will not be awarded to replenish funds previously expended. Grants are made with the understanding that the foundation has no obligation or commitment to provide additional support to the grantee. Grants may not be used for any political campaign, or to influence legislature of any government body other than through making available the results of nonpartisan analysis, study, and research.
Geographic Focus: Indiana
Date(s) Application is Due: Aug 1
Contact: Laura Lemings; (260) 463-4363; fax (260) 463-4856; llemings@lccf.net
Internet: http://www.lccf.net/grants.html
Sponsor: LaGrange County Community Foundation
109 E Central Avenue, Suite 3
LaGrange, IN 46761

LaGrange County Community Foundation Scholarships 2330
The LaGrange County, Indiana Community Foundation has developed a scholarship program that offers significant financial support for a wide range of educational and career interests. The Foundation offers a variety of 30 separately funded new scholarships, with several offering multiple awards.
Requirements: Eligibility, criteria, and supplemental requirements vary for each scholarship. Scholarships are categorized by area schools and also include scholarships for non-traditional and home schooled students. See website for detailed information on each scholarship, as well as applications. Additional information can also be obtained by contacting the foundation office.
Geographic Focus: Indiana
Contact: Laura Lemings; (260) 463-4363; fax (260) 463-4856; llemings@lccf.net
Internet: http://www.lccf.net/scholarships.html
Sponsor: LaGrange County Community Foundation
109 E Central Avenue, Suite 3
LaGrange, IN 46761

LaGrange County Lilly Endowment Community Scholarship 2331
The LaGrange County Lilly Endowment Community Scholarship may be used for any area of study and offers full tuition and books in return for a commitment to four years of full-time baccalaureate study at any accredited Indiana college or university. Contact the Foundation office for specific information about the scholarship, including application and deadline.
Requirements: Applicants must have been accepted at an Indiana four-year college or university; taken the SAT and have those scores recorded on their official high school transcript; participated in community service confirmed by an adult sponsor or supervisor.
Restrictions: The scholarship does not include room and board.
Geographic Focus: Indiana
Contact: Laura Lemings; (260) 463-4363; fax (260) 463-4856; lccf@lccf.net
Internet: http://www.lccf.net/scholarships.html
Sponsor: LaGrange County Community Foundation
109 E Central Avenue, Suite 3
LaGrange, IN 46761

Lake County Athletic Officials Association Scholarships 2332
This scholarship was established to award funding to Lake County Athletic Official Association members and their children who are attending any accredited four year institution. The number of scholarships varies, but each scholarship is set at $500. All college majors are eligible. Contact Legacy Foundation about current applications.
Geographic Focus: Indiana
Date(s) Application is Due: Mar 12
Amount of Grant: 500 USD
Contact: Athletic Officials Association Scholarship Contact; (219) 736-1880; fax (219) 736-1940; legacy@legacyfoundationlakeco.org
Internet: http://www.legacyfoundationlakeco.org/scholarships/index.htm
Sponsor: Legacy Foundation
1000 East 80th Place, 302 South
Merrillville, IN 46410

Lake County Community Fund Grants 2333
The Lake County Community Grant was established to be responsive to community projects throughout Lake County in the areas of arts and culture, civic affairs, community development, education, the environment, health, human services, and youth services. The Foundation will make grants to non-profit organizations implementing projects with the most potential to improve the quality of life of a substantial number of residents of Lake County. Geographic distribution may be considered in awarding grants. Grants typically range from $1,000 to $25,000.
Requirements: Funding priorities include projects that: develop or test new solutions to community problems; address prevention as well as remediation; assist underserved community resources; provide a sustained effect for a substantial number of residents; improve the efficiency of non-profit groups; provide a favorable ratio between the amount of money requested and number of people served; facilitate collaboration among organizations without duplicating services; encourage volunteerism, civic engagement, and development. Applicants must submit the online detailed application, along with the grant narrative, project budget, a list of applicant's board of directors, summary of the organization's current fiscal year operating budget as well as financial audit or review; evidence of Board approval of this application, copy of tax exempt status, and organization's profile on GuideStar (www.guidestar).
Restrictions: The Legacy Foundation does not support: general operating expenses; endowment campaigns, annual campaigns, or fundraising events; travel grants; grants for individual schools or sponsorship of sports teams; previously incurred debt or retroactive

funding for current projects; individuals and independent scholarly research projects; and religious or sectarian programs, political parties, or campaigns.
Geographic Focus: Indiana
Date(s) Application is Due: Mar 1; May 1; Sep 1; Nov 1
Amount of Grant: 1,000 - 25,000 USD
Contact: Barry Tyler, Jr., Community Initiatives Officer; (219) 736-1880; fax (219) 736-1940; legacy@legacyfoundationlakeco.org or btyler@legacyfdn.org
Internet: http://www.legacyfoundationlakeco.org/grantsfundingopps.html
Sponsor: Legacy Foundation
1000 East 80th Place, 302 South
Merrillville, IN 46410

Lake County Lilly Endowment Community Scholarships 2334
The Lilly Endowment Community Scholarships were designed to help raise the level of educational attainment in Indiana and to leverage further the ability of Indiana's community foundations to enhance the quality of life of the state's residents. Lilly Endowment Community Scholars will receive scholarships for full tuition, required fees, and a special allocation of up to $800 per year for required books and required equipment for four years of undergraduate study on a full-time basis, leading to a baccalaureate degree at any Indiana public or private college or university accredited by the North Central Association of Colleges and Schools. All college majors are eligible to apply.
Requirements: Applicants: must be U.S. citizens, high school seniors, and residents of Lake County, Indiana; must apply for a full time four-year course of study at an accredited public or private college or university in Indiana; must be available for an interview that will be scheduled for mid-February. Applicants should consult the website and their high school guidance office for the detailed procedures of the application process.
Restrictions: These scholarships do not cover room and board expenses.
Geographic Focus: Indiana
Date(s) Application is Due: Jan 5
Contact: Scholarships Contact; (219) 736-1880; legacy@legacyfoundationlakeco.org
Internet: http://www.legacyfoundationlakeco.org/scholarshipslilly.html
Sponsor: Legacy Foundation
1000 East 80th Place, 302 South
Merrillville, IN 46410

Lalor Foundation Anna Lalor Burdick Grants 2335
Anna Lalor Burdick was born in Villisca, Iowa in 1869. Upon graduating from Iowa University in 1889, she taught at the Decorah High School until 1891, when she married and moved to Denver. In 1894 she accepted the principal's position at Iowa Falls High School. Later she became superintendent of schools in Des Moines. From 1917 until her retirement in 1939, she served in the U.S. Department of Education as Special Agent for Trade and Industrial Education for Girls and Women. She died in 1944. The Anna Lalor Burdick Grant program seeks to educate young women about human reproduction in order to broaden and enhance their options in life. Preference will be given to programs that focus on young women who have inadequate access to information regarding reproductive health, including the subjects of contraception and pregnancy termination. The program emphasizes support for projects and initiatives that demonstrate realistic plans to achieve greater financial self-sufficiency; new or smaller organizations, including grassroots efforts; collaborative efforts among nonprofit organizations; organizations that can demonstrate a proven ability to reach out to, include, and involve young women with inadequate access to information regarding reproductive health; and new ideas, initiatives, and demonstration projects, which, if proven effective, may be successfully replicated or provide multiple benefits. Normally grants are awarded for one year only. Under special circumstances, renewals are considered. The first step in applying for a grant is to submit an online concept paper. Instructions for preparing an online concept application are found at the Application Instructions tab on the left side of this screen. Concept papers are submitted online only. If your concept paper is approved by the trustees, you will be invited to submit a full proposal. The trustees award a small number of grants in the range of $10,000 to $50,000, with an average grant size between $20,000 and $25,000.
Requirements: 501(c)3 tax-exempt organizations not classified as private foundations under section 509(a) of the IRS code are eligible. Applicants not located in the United States and that have not already been classified by the U.S. Internal Revenue Service should contact the office.
Restrictions: Grants are not normally made to full proposals submitted in advance of a concept paper that has been reviewed and approved by the trustees for further consideration; individuals, or for individual research projects and scholarships; requests for endowment or major capital support; prior grantees that have failed to provide grant reports; or organizations with no track record or no personnel known to the trustees or to the staff at Grants Management Associates.
Geographic Focus: All States
Date(s) Application is Due: May 1; Nov 1
Amount of Grant: 10,000 - 50,000 USD
Samples: Children's Nurition Program of Haiti, Chatanooga, Tennessee, $20,000 - to provide sex education in Leogane, Haiti, for in-school groups and youth groups for those out of school or from different districts; NARAL Pro-Choice Virginia Foundation, Alexandria, Virginia, $30,000 - to support the Sex Education Awareness Now Petersburg project in which a coalition of community groups will work to bring comprehensive sex education to students in the Petersburg, Virginia public schools.
Contact: Susan Haff, (617) 426-7080, ext. 323; shaff@gmafoundationst.com
Internet: http://lalorfound.org/?page_id=9
Sponsor: Lalor Foundation
77 Summer Street, 8th Floor
Boston, MA 02110-1006

Lalor Foundation Postdoctoral Fellowships 2336
The Lalor Foundation postdoctoral fellowship program supports promising new researchers in establishing scientific and teaching careers. The mission of the program is to support these researchers early in their work so that they can become independently funded in the field of mammalian reproductive biology as related to the regulation of fertility. The individual nominated by the applicant institution for the postdoctoral fellowship for conduct of the work may be a citizen of any country. Fellowships will be $42,000 per year for coverage of fellowship stipend, fringes and institutional overhead.
Requirements: U.S. institutions must be exempt from federal income taxes under Section 501(c)3 of the U.S. Internal Revenue Code and must submit a determination letter from the Internal Revenue Service stating that it is not a private foundation. The individual nominated should have training and experience at least equal to the Ph.D. or M.D. level and should not have a faculty appointment (i.e., instructor, lecturer or higher). Potential fellows should not have held the doctoral degree more than two years from receipt of the degree.
Restrictions: Institutional overhead may not exceed 10 percent of the total fellowship award.
Geographic Focus: All States
Date(s) Application is Due: Jan 15
Amount of Grant: 42,000 USD
Contact: Susan Haff, (617) 426-7080, ext. 323; fax (617) 426-5441; shaff@gmafoundationst.com or fellowshipmanager@gmafoundations.com
Internet: http://www.lalorfound.org/?page_id=13
Sponsor: Lalor Foundation
77 Summer Street, 8th Floor
Boston, MA 02110-1006

LAM Foundation Established Investigator Awards 2337
The Established Investigator Awards support research directly related to lymphangioleiomyomatosis (LAM) and is targeted at established investigators who focus on the abnormal proliferation of smooth muscle that occurs in the disease. Awards provide a maximum of $50,000 per year, renewable for up to two additional years. The structure and terms of this award are identical to the LAM Fellowship Award except that with the LAM Established Award, faculty level investigators are eligible to receive funding for technician support and supplies.
Requirements: U.S. citizens and foreign nationals with appropriate immigrant visa status are eligible. At the time of application, an applicant must hold a doctoral degree and faculty appointment at the level of assistant or associate professor and be undertaking a project related to LAM.
Geographic Focus: All States
Date(s) Application is Due: Sep 30
Contact: Francis X. McCormack; (513) 558-4831; frank.mccormack@uc.edu
Internet: http://www.thelamfoundation.org/research/apply-for-lam-funding
Sponsor: LAM Foundation
231 Albert Sabin Way, MSB Room 6053
Cincinnati, OH 45267-0564

LAM Foundation Pilot Project Grants 2338
Pilot project awards are granted for one year to enable investigators to gather sufficient preliminary data for more substantial LAM funding. Funding is intended to support faculty-level investigators, or postdoctoral-level investigators who will be performing LAM research in the laboratory of an established scientist who is an expert in areas that are directly pertinent to LAM. Letters of intent (LOI) are required prior to submission of grant proposals to the LAM Foundation. Application materials and details are available online.
Requirements: Applications are accepted from investigators of all nationalities who have MDs or PhDs, but candidates must possess visas that allow for completion of the proposed project in the original laboratory.
Geographic Focus: All States
Date(s) Application is Due: Sep 30
Amount of Grant: Up to 25,000 USD
Contact: Francis X. McCormack; (513) 558-4831; frank.mccormack@uc.edu
Internet: http://www.thelamfoundation.org/research/apply-for-lam-funding
Sponsor: LAM Foundation
231 Albert Sabin Way, MSB Room 6053
Cincinnati, OH 45267-0564

LAM Foundation Postoctoral Fellowships 2339
The foundation offers postdoctoral fellowships for the study of the cellular and molecular basis of the abnormal smooth muscle proliferation that occurs in the disease, Lymphangioleiomyomatosis (LAM). Fellowships are renewable for up to two additional years. The fellowship is intended to support postdoctoral level investigators who will be performing LAM research in the laboratory of an established scientist who is an expert in areas that are directly pertinent to LAM. Letters of intent (LOI) are required prior to submission of grant proposals to the LAM Foundation. Application materials and details are available online.
Requirements: Applications are accepted from investigators of all nationalities who have MDs or PhDs, but candidates must possess visas that allow for completion of the proposed project in the original laboratory.
Restrictions: The award is generally not available to individuals who have attained faculty rank, or who are seeking transition funds for the establishment of a laboratory in a new setting.
Geographic Focus: All States
Date(s) Application is Due: Sep 30
Contact: Francis X. McCormack; (513) 558-4831; frank.mccormack@uc.edu
Internet: http://www.thelamfoundation.org/research/apply-for-lam-funding
Sponsor: LAM Foundation
231 Albert Sabin Way, MSB Room 6053
Cincinnati, OH 45267-0564

LAM Foundation Research Grants   2340
The program provides seed money to investigators seeking to understand the abnormal smooth muscle proliferation that occurs in lymphangioleiomyomatosis (LAM). The grant supports clinical, laboratory, epidemiological, or any other kind of research that directly relates to LAM. Grants are subject to annual review and may be granted for up to two years. Scientists who which to apply for LAM Foundation funding must submit a Letter of Intent (LOI). Letters of Intent must reach The LAM Foundation by July 30th.
Requirements: U.S. citizens and foreign nationals with appropriate immigrant visa status are eligible. At the time of application, an applicant must hold a doctoral degree and faculty appointment with an academic institution and have completed two years of research training.
Restrictions: Residents, interns, postdoctoral fellows, students enrolled in degree-granting programs, and established investigators are ineligible.
Geographic Focus: All States
Date(s) Application is Due: Sep 30
Amount of Grant: Up to 40,000 USD
Contact: Francis X. McCormack; (513) 558-4831; frank.mccormack@uc.edu
Internet: http://www.thelamfoundation.org/research/apply-for-lam-funding
Sponsor: LAM Foundation
231 Albert Sabin Way, MSB Room 6053
Cincinnati, OH 45267-0564

Land O'Lakes Foundation Mid-Atlantic Grants   2341
The Land O'Lakes Foundation Mid-Atlantic Grants were developed specifically for the company's dairy communities in Maryland, New Jersey, New York, Pennsylvania, and Virginia. The program works to improve quality of life by supporting worthy projects and charitable endeavors initiated by our Mid-Atlantic dairy-member leaders. Community organizations applying for grants may be eligible for donations of $500 to $5,000 for local projects and programs. Funds could be used to support such worthwhile projects as: backing local food pantries or emergency feeding efforts; aiding 4-H or FFA programs; building a new park pavilion for the community; establishing a local wetland preserve; or purchasing books for the community library. Application procedures are available online.
Requirements: Applications are initiated by Land O'Lakes farmer-members. An Area Procurement Specialist (APS) review is also part of this process. To be considered, grant proposals must demonstrate how the donation will be used to help improve community quality of life. Mid-Atlantic grants are generally restricted to organizations that have been granted tax-exempt status under Section 501(c)3 of the Internal Revenue Code. The Foundation awards grants to projects that best address the following areas: hunger relief; youth and education; rural leadership; civic improvements; soil and water preservation; and art and culture.
Restrictions: Grants will not be awarded for the following purposes: scholarship funds, gifts or fund raisers for individuals, or non-public religious use.
Geographic Focus: Maryland, New Jersey, New York, Pennsylvania, Virginia
Amount of Grant: 500 - 5,000 USD
Contact: Martha Atkins-Sakry, Executive Assistant; (651) 481-2470 or (651) 481-2212; MLAtkins-Sakry@landolakes.com
Internet: http://www.landolakesinc.com/company/corporateresponsibility/foundation/midatlanticgrants/default.aspx
Sponsor: Land O'Lakes Foundation
P.O. Box 64101
St. Paul, MN 55164-0150

Landon Foundation-AACR Innovator Award for Cancer Prevention Research   2342
This Award was established and first given in 2008 to recognize the outstanding achievement of an early career Assistant Professor, and to provide support for cancer prevention research of significant scientific merit in any discipline across the continuum of research. The goals of the program are to: encourage younger investigators to pursue cancer prevention research of significant scientific merit; provide the support necessary to sustain and enhance highly meritorious cancer prevention research; foster interactions between and among cancer scientists and disseminate the scientific knowledge about cancer prevention research; and contribute to a global impact against cancer. A two-year grant of $100,000 ($50,000 per year) will be provided to the recipient's institution to support direct research expenses; salary and benefits; and attendance at the AACR Annual Meeting, Cancer Prevention Conference, or other AACR meeting relevant to cancer prevention for the purpose of participating in scholarly exchange.
Requirements: Candidates must be scientists at the level of Assistant Professor who completed postdoctoral studies or clinical fellowships on or after July 1. Candidates cannot be tenured or under consideration for tenured academic positions at the time of application. A candidate whose title is not Assistant Professor but who believes that he or she meets the eligibility requirements for this Award must attach a letter from the institution's dean certifying that the position held is equivalent to Assistant Professor. Both AACR members and nonmembers are eligible to apply. However, nonmembers must submit a satisfactory application for AACR Active membership on or before October 31. Applications must be submitted by 12:00 noon (United States Eastern Daylight Time) on November 20, using the proposalCENTRAL website at http://proposalcentral.altum.com. One paper copy with original signatures and all supporting documents must be mailed to the AACR office, and be postmarked no later then November 25.
Restrictions: Employees or subcontractors of a national government or private industry are not eligible for this grant.
Geographic Focus: All States
Date(s) Application is Due: Nov 20; Nov 25
Amount of Grant: 100,000 USD
Contact: Julia Laurence, Program Assistant; (215) 440-9300, ext. 102 or (267) 646-0655; fax (215) 440-9372; julia.laurence@aacr.org or awards@aacr.org
Internet: http://www.aacr.org/home/scientists/research-funding--fellowships/landon-aacr-innovator-award-for-cancer-prevention-research.aspx
Sponsor: American Association for Cancer Research
615 Chestnut Street, 17th Floor
Philadelphia, PA 19106-4404

Landon Foundation-AACR Innovator Award for International Collaboration   2343
This Award will support highly meritorious research that is being conducted collaboratively by investigators in different countries around the world. The goals of the program are to: promote international cancer research collaboration as an effective means to accelerate progress against cancer; provide the support necessary to sustain and enhance highly meritorious international cancer research collaborations; foster interactions between and among cancer scientists and disseminate the scientific knowledge gained from international collaboration; and contribute to a global impact against cancer. The Award of $100,000 will be provided to the recipients' institutions in the form of a two-year grant to support the collaborative research project, and may be used for such purposes including, but not limited to: direct research expenses attributable to the proposed research, which may include the salary of research assistants or technicians; expenses for travel to the institutions of the collaborators relevant to the research project; expenses for training in new techniques that will benefit the research project, such as training received through a scholar exchange/visiting professor program; and expenses related to the presentation of research data at scientific meetings or through other means that will contribute to the dissemination of the scientific knowledge gained.
Requirements: Two or more independent researchers who working within an established international cancer research collaboration involving institutes in multiple countries may apply. Proposals to initiate a collaborative project will not be accepted. The co-investigators may be affiliated with any institution involved in cancer research, cancer medicine, or cancer-related biomedical science anywhere in the world. There is no limit to the number of co-investigators within each of the collaborating institutes. There are no national or residency status restrictions. Applicants need not be members of the AACR. Applications must be completed online using the proposalCENTRAL website, with one paper copy submitted to the AACR office. Online applications must be completed by 12:00 noon (United States Eastern Daylight Time) on November 20. The paper copy with original signatures and all supporting documents must be postmarked no later than November 25.
Restrictions: Scientific investigators or health professionals who are funded by the tobacco industry for any project, or whose named mentors in the case of mentored grants are funded by the tobacco industry for any project, may not apply and will not be eligible for AACR grants.
Geographic Focus: All States
Date(s) Application is Due: Nov 20; Nov 25
Amount of Grant: 100,000 USD
Contact: Hanna Hopfinger; (267) 646-0665; hanna.hopfinger@aacr.org
Internet: http://www.aacr.org/home/scientists/research-funding--fellowships/landon-aacr-innovator-award-for-international-collaboration.aspx
Sponsor: American Association for Cancer Research
615 Chestnut Street, 17th Floor
Philadelphia, PA 19106-4404

Lands' End Corporate Giving Program   2344
The Corporation awards grants to nonprofits for youth and family services programs in their area of company operations in Wisconsin. Areas of interest include education, community development, environment, and health and human services.
Requirements: Wisconsin nonprofits are eligible. Organizations should submit the following: a timetable for implementation and evaluation of the project; statement of the problem that the project will address; population and geographic area to be served; name, address and phone number of organization; copy of IRS determination letter; copy of most recent annual report/audited financial statement/990; how the project's results will be evaluated or measured; list of company employees involved with the organization; detailed description of project and amount of funding requested; contact person; copy of current year's organizational budget and/or project budget; and listing of additional sources and amount of support. Applicants should also include a description of their past involvement with Lands' End, if any.
Restrictions: The Foundation does not consider grants for organizations without nonprofit status; individuals; political organizations, campaigns, or candidates for public office; lobbying groups; advertising in programs, bulletins, yearbooks, or brochures; testimonial/awards dinners; endowments; loans; religious groups for religious purposes; pageants; purchasing of land; salaries; administrative costs; international programs; research programs; or general operating expenses.
Geographic Focus: Wisconsin
Date(s) Application is Due: Mar 31; Jun 30; Sep 30; Dec 31
Contact: Jessica Winzenried, Corporate Giving Manager; (608) 935-6776 or (608) 935-6728; fax (608) 935-6432; donate@landsend.com
Sponsor: Lands' End
2 Lands' End Lane
Dodgeville, WI 53595

Laura B. Vogler Foundation Grants   2345
The Vogler Foundation supports innovative programs and projects in New York City and Long Island in the areas of education, health care, and social services. The Foundation is particularly interested in organizations that serve and support children, the elderly, and the disadvantaged. Types of support include general operating support, program development, research grants, and seed money grants. Grants provide one-time, nonrenewable support. Most recent grant awards have ranged from $2,500 to $5,000. The annual deadlines for applications are March 1, July 1, and November 1.

Requirements: Nonprofit 501(c)3 organizations in New York City and Long Island, New York, may submit proposals.
Restrictions: Grants are not awarded to support building or endowment funds, annual fund-raising campaigns, or matching gifts. Requests for funds for conferences, seminars, or loans are not accepted.
Geographic Focus: New York
Date(s) Application is Due: Mar 1; Jul 1; Nov 1
Amount of Grant: 2,500 - 5,000 USD
Contact: Lawrence L. D'Amato; (718) 423-3000; voglerfound@gmail.com
Internet: https://sites.google.com/site/voglerfoundation/
Sponsor: Laura B. Vogler Foundation
51 Division Street, P.O. Box 501
Sag Harbor, NY 11963

Laura Moore Cunningham Foundation Grants 2346
The Laura Moore Cunningham Foundation is dedicated to advancing the State of Idaho. Priorities include rural healthcare, educational programs for children, programs in underserved communities, and programs for underserved populations. Each year the Foundation accepts applications from throughout the State, allowing organizations of all types to express their need. The Foundation is interested in organizations that run in a cost-effective manner, serving large numbers of people who are truly in need.
Requirements: Eligible applicants must be 501(c)3 Idaho organizations. The Foundation does not limit giving to a certain type of program or need; however administrative costs are not preferred.
Restrictions: Individuals are ineligible.
Geographic Focus: Idaho
Date(s) Application is Due: May 15
Samples: The Peregrine Fund, Boise, Idaho, $100,000 - education programs; Teton Valley Health Care Foundation, Driggs, Idaho, $30,000 - boiler upgrade; and Jerome Public Library, Jerome, Idaho, $4,000 - new library books.
Contact: Harry L. Bettis; (208) 472-4066; lmcf_idaho@msn.com
Internet: http://lauramoorecunningham.org/Applying_for_Grants.html
Sponsor: Laura Moore Cunningham Foundation
P.O. Box 1157
Boise, ID 83701

Lawrence Foundation Grants 2347
The foundation awards grants to nonprofit organizations that address the following areas of interest: education (U.S. headquartered organizations for programs in the United States), environment (U.S. headquartered organizations operating programs in the United States or elsewhere in the world), health (U.S. headquartered organizations operating programs in the United States with national scope or specific programs in the Los Angeles area), disaster relief (U.S. headquartered organizations responding to disasters in the United States or elsewhere in the world on an occasional basis), and other topics (U.S. headquartered organizations operating programs in the United States or elsewhere in the world). Grant amounts vary. The duration of grants will be typically for a single year.
Requirements: Nonprofit organizations that qualify for 501(c)3 public charity status, public schools and libraries are eligible for contributions or grants.
Restrictions: For-profit businesses are not eligible. The foundation does not typically make grants for the following purposes: computers or software; audio or video equipment; designing and producing videos, kiosks, or promotional material; music programs or musical instruments; gardening programs or equipment; physical education programs or equipment; recreational programs; theater arts or performance arts programs or equipment; hospice or old age home programs; religious organizations; religious or charter schools; individuals for any purpose; international organizations that do not have a qualified domestic 501(c)3 representative; political lobbying activities or other political purposes; private foundations; dinners, balls, or other ticketed events; or purposes outside of the foundation's funding priorities.
Geographic Focus: All States, All Countries
Date(s) Application is Due: Apr 30; Nov 1
Amount of Grant: 500 - 100,000 USD
Contact: Lori Mitchell; (310) 451-1567; fax (310) 451-7580; info@thelawrencefoundation.org
Internet: http://www.thelawrencefoundation.org/grants/
Sponsor: Lawrence Foundation
530 Wilshire Boulevard, Suite 207
Santa Monica, CA 90401

Lawrence S. Huntington Fund Grants 2348
The Foundation, established in 1997 by its namesake, Lawrence J. Huntington, is dedicated to supporting a historical seaport, as well as for environmental conservation, hospitals and human service programs. Mr. Huntington, who serves as the Chairman of St. Luke's Roosevelt Hospital, Chairman of the Woods Hole Research Center, and Vice Chairman of the Board of the South Street Seaport Museum, is primarily interested in preserving the environment and historical landmarks. There are no specific applications or deadlines with which to adhere, and applicants should approach the Foundation in writing.
Requirements: Giving is primarily restricted to the states of New York and Massachusetts, as well as the District of Columbia.
Geographic Focus: New York
Amount of Grant: Up to 50,000 USD
Contact: Lawrence S. Huntington, Director; (212) 717-8633 or (212) 486-2424
Sponsor: Lawrence S. Huntington Fund
46 E 70th Street, 4th Floor
New York, NY 10021-4928

Layne Beachley Aim For The Stars Foundation Grants 2349
The Layne Beachley Aim for the Stars Foundation was created to inspire girls and women across Australia to dream and achieve. The Foundation hopes to prevent girls and women alike from having to go through adversity and to encourage, motivate and provide for all aspiring women. Encompassing academic, sport, cultural and community pursuits, Aim for the Stars offers ambitious and dedicated females an opportunity to receive financial and moral support to help them achieve their goals. Support will give them the opportunity to maintain a determined focus on their goal, to achieve their dreams earlier in life and allow them to further their ambitions and aim for the stars. Grants will be awarded, at the discretion of the Foundation, to deserving applicants who meet the criteria and best demonstrate their aspirations to further their educational and/or personal development. Females who aim to achieve their ultimate goal in their field of choice, 12 years of age and above, are eligible to apply.
Requirements: Applicants must be citizens of Australia, a minimum of 12 years old, and female.
Geographic Focus: All States, Australia
Date(s) Application is Due: Nov 15
Amount of Grant: 1,500 - 6,000 USD
Contact: Program Coordinator; (042) 260-0733; info@aimforthestars.com.au
Internet: http://www.aimforthestars.com.au/WhoCanApply/ApplicationChecklist/tabid/1854/language/en-US/Default.aspx
Sponsor: Layne Beachley Aim For the Stars Foundation
P.O. Box 666
Forestville, NSW 2087 Australia

Leahi Fund 2350
The purpose of the Fund is to support programs of research and education in, and the prevention of, pulmonary disease. Highest priority will be given to: research and education with foreseeable clinical application to pulmonary problems that exist in Hawai'i; scientific research in the prevention or treatment of pulmonary diseases and infirmities. Preference will be given to supporting new investigators and to new areas of investigation for established researchers. Application information is available online.
Requirements: Research grants can only be approved for individuals affiliated with tax-exempt organizations classified by the Internal Revenue Services as 501(c)3 type, or with units of government.
Restrictions: Funding is not available for: salary for the principal investigator; patient care; travel; major equipment purchases; printing or publication costs.
Geographic Focus: Hawaii
Date(s) Application is Due: Jun 21
Amount of Grant: 30,000 USD
Contact: Christel Wuerfel; (808) 537-6333 or (888) 731-3863; cwuerfel@hcf-hawaii.org
Internet: http://www.hawaiicommunityfoundation.org/index.php?id=71&categoryID=23
Sponsor: Hawai'i Community Foundation
827 Fort Street Mall
Honolulu, HI 96813

LEGENDS Scholarship 2351
The LEGENDS scholarship celebrates the individual who overcomes challenges and obstacles in the pursuit of a college education and lifelong learning. The LEGENDS Scholar must demonstrate one or more of the virtues of Exploration, Courage, Creativity and/or Innovation. The recipient will receive a one-time amount of $1,000 toward higher education and his or her name will be displayed at the Lake County Convention and Visitor's Bureau as a LEGENDS Scholar. All college majors are eligible.
Requirements: The applicant must be a current student in good standing, have completed a minimum of 24 credit hours, and intends to be enrolled in an undergraduate or graduate degree program at one of the following institutions for the coming academic year: Calumet College of St. Joseph, Indiana University Northwest, Ivy Tech Community College of Indiana Northwest, Purdue University Calumet, Purdue University North Central, or Valparaiso University. Applicants must submit: an application form; two recommendation letters; and an essay that addresses how exploration, courage, creativity, and innovation have enabled them to overcome obstacles in pursuing a college education, then how they will use these values to improve the quality of life in northwest Indiana.
Geographic Focus: Indiana
Date(s) Application is Due: Mar 15
Amount of Grant: 1,000 USD
Contact: LEGENDS Scholarship Contact; (219) 736-1880; fax (219) 736-1940; legacy@legacyfoundationlakeco.org
Internet: http://www.legacyfoundationlakeco.org/scholarships/index.htm
Sponsor: Legacy Foundation
1000 East 80th Place, 302 South
Merrillville, IN 46410

Legler Benbough Foundation Grants 2352
The mission of the Foundation is to improve the quality of life of the people in the City of San Diego. To accomplish that mission, the Foundation focuses on three target areas: providing economic opportunity; enhancing cultural opportunity; and providing a focus for health, education and welfare funding. Interested applicants may submit an initial letter requesting funds. When the Foundation wishes to pursue the request, it will provide an application form to the applicant. Initial letters preceding applications are due by February 15 and August 15. Applications (for invited applicants) are due by March 15 and September 15.
Requirements: Funding focuses on activities in support of San Diego city arts and cultural institutions, scientific or research organizations, and health, education and welfare programs.

Restrictions: Awards are made only in the Foundation's focus areas. The following are not funded: capital projects unless there is a special situation where capital expenditure is the best way to achieve a stated objective; awards to individuals; awards for special events, fundraising or recognition events; and projects in the area of the homeless, AIDS, alcohol or drug rehabilitation or treatment and seniors.
Geographic Focus: California
Contact: Peter Ellsworth; (619) 235-8099; peter@benboughfoundation.org
Internet: http://www.benboughfoundation.org/criteria.php
Sponsor: Legler Benbough Foundation
2550 5th Avenue, Suite 132
San Diego, CA 92103-6622

Lena Benas Memorial Fund Grants 2353
The Lena Benas Memorial Fund was established in 1986 to provide for the health, human services, and housing needs of underserved people living in Litchfield, Connecticut, and its surrounding communities. The annual deadline for application to the Fund is November 1, and applicants will be notified of grant decisions by letter within 2 months after the proposal deadline. Special consideration is given to organizations that provide housing maintenance and human services programming to needy populations.
Requirements: Preference is given to organizations serving the people of Litchfield, Connecticut. Organizations serving the towns contiguous to Litchfield, including Cornwall, Goshen, Harwinton, Morris, Thomaston, Torrington, Warren, Washington, and Watertown will also be considered.
Restrictions: Grants from the Benas Memorial Fund are 1 year in duration.
Geographic Focus: Connecticut
Date(s) Application is Due: Nov 1
Contact: Kate Kerchaert; (860) 657-7016; kate.kerchaert@ustrust.com
Internet: https://www.bankofamerica.com/philanthropic/fn_search.action
Sponsor: Lena Benas Memorial Fund
200 Glastonbury Boulevard, Suite #200, CT2-545-02-05
Glastonbury, CT 06033-4056

Leo Goodwin Foundation Grants 2354
The Leo Goodwin Foundation offers grants in the areas of arts, culture, humanities; education; health; human services; and public benefit. Types of support include: capital campaigns for museums and performing arts centers; literacy programs and educational foundations; community college scholarships; cancer research institutes; boys and girls clubs; and child care organizations. There are no deadlines, and organizations may apply at any time. The trustees meet once a month to assess requests for funding.
Requirements: Applicants must be 501(c)3 nonprofit organizations in the state of Florida. All requests must be submitted with the following information: cover letter stating purpose of program and amount requested; objectives, demographics - social and economic status, age, gender, etc; how funds will be used; operating budget, current audited statement and tax return; IRS 501(c)3 status letter; non-recovation statement; funding sources with amounts received; names and information of governing board members; outcome measures and results; and strategic partners or alliances in delivery of services.
Restrictions: Individuals are not eligible.
Geographic Focus: Florida
Amount of Grant: 1,000 - 25,000 USD
Contact: Helen Furia; (954) 772-6863; fax (954) 491-2051; hfurialgj@bellsouth.net
Internet: http://leogoodwinfoundation.org/
Sponsor: Leo Goodwin Foundation
800 Corporate Drive, Suite 500
Fort Lauderdale, FL 33334-3621

Leon and Thea Koerner Foundation Grants 2355
The purpose of the Koerner Foundation is to foster higher education, cultural activities, and public welfare in British Columbia and to stimulate and invigorate cultural and educational life by enabling institutions and individuals to undertake activities that would not normally be possible. Areas of interest include arts and culture, libraries, adult and continuing education, community colleges and universities, social services, special needs groups, community services, family services, and social issues. Applications are available at the Foundation website. Applications require a brief statement of the proposed project, its scope, time required for completion, results sought, total cost, amount requested from the foundation, specific purposes to which the grant will be applied, concise statement of resources available for the project, and a resume of its present state.
Requirements: Grants are made to organizations registered as educational or charitable institutions and organizations.
Restrictions: Individuals may not apply for grant support. Organizations may only receive funding three times within any five year period.
Geographic Focus: Canada
Amount of Grant: 5,000 - 25,000 CAD
Contact: Grants Contact; (604) 224-2611; fax (604) 224-1059
Internet: http://www.koernerfoundation.ca/eligibility.html
Sponsor: Leon and Thea Koerner Foundation
3695 West 10th Avenue, P.O. Box 39209
Vancouver, BC V6R 4P1 Canada

Leonard and Helen R. Stulman Charitable Foundation Grants 2356
The Leonard and Helen R. Stulman Charitable Foundation supports work in four areas of interest. Ninety five percent (95%) of funding from The Foundation is reserved for programs in greater Baltimore and the State of Maryland. The foundation will make a limited number of new grants each year to projects that address one or more of the following: research as to causes and treatment of mental illness and services for people with mental illness and their families; initiatives for an older adult and aging population and programs for geriatric medicine; programs engaged in health care treatment and prevention as well as those involved in medical research; and higher educational institutions. To apply to the Foundation for funding, applicants should submit a two-page letter of inquiry, with basic background on your organization, the identified needs your project proposes to address, an overview of the proposed project, and the amount you intend to request.
Requirements: Organizations (or their fiscal agents) that qualify as public charities under section 501(c)3 of the Internal Revenue Code and do not discriminate on the basis of race, creed, national origin, color, physical handicap, gender or sexual orientation are eligible to apply.
Geographic Focus: All States
Amount of Grant: 50,000 USD
Contact: Cathy Brill; (410) 332-4171; fax (410) 837-4701; cathybrill@comcast.net
Internet: http://www.bcf.org/ourgrants/ourgrantsdetail.aspx?grid=4
Sponsor: Leonard and Helen R. Stulman Charitable Foundation
2 East Read Street, 9th Floor
Baltimore, MD 21202

Leonard L. and Bertha U. Abess Foundation Grants 2357
The Leonard L. and Bertha U. Abess Foundation, established in Miami in 1949 by the City National Bank co-founder, Leonard L. Abess, Sr., is interested in supporting a number of local programs. Fields of interest include: education at all levels, Jewish agencies and temples, and federated giving programs. The primary type of funding offered is for general operations support. Applicants should forward a letter describing the program and an overall budget need. There are no particular deadlines with which to adhere, and giving is centered in the Miami, Florida, region. Most recent awards have ranged from $2,500 to $50,000.
Geographic Focus: Florida
Amount of Grant: Up to 50,000 USD
Contact: Leonard L. Abess; (212) 632-3000 or (212) 632-3200; informed@ftci.com
Sponsor: Leonard L. and Bertha U. Abess Foundation
600 5th Avenue
New York, NY 10020-2302

Leo Niessen Jr., Charitable Trust Grants 2358
Leo Niessen lived in Abington Township, Montgomery County, Pennsylvania. In 1993, his Foundation was funded from a testamentary bequest. He was a charitable man, who also made substantial philanthropic gifts to Holy Redeemer Hospital during his lifetime. He had a special affinity for Red Cloud Indian School of Pine Ridge, South Dakota. To this day, the Co-trustees of his Foundation continue to support this school, as well as the Hospital and The Society for the Propagation of the Faith. All grants are made in the memory of Leo Niessen and his family. The Foundation also supports organizations: that provide health services for all ages; which educate the needy and educable at all academic levels, without regard to age; working for and on behalf of youth and the elderly, and which provide assistance to the homeless and economically disadvantaged; which provide spiritual and emotional guidance. Application Deadlines are, January 31 and July 31. Application forms are available online. Applicants will receive notice acknowledging receipt of the grant request, and subsequently be notified of the grant declination or approval.
Requirements: Pennsylvania 501(c)3 nonprofit organizations are eligible to apply. Proposals should be submitted in the following format: completed Common Grant Application Form; an original Proposal Statement; an audited financial report and a current year operating budget; a copy of your official IRS Letter with your tax determination; a listing of your Board of Directors. Proposal Statements (second item in the above Format) should answer these questions: what are the objectives and expected outcomes of this program/project/request; what strategies will be used to accomplish your objective; what is the timeline for completion; if this is part of an on-going program, how long has it been in operation; what criteria will you use to measure success; if the request is not fully funded, what other sources can you engage; an Itemized budget should be included; please describe any collaborative ventures. Prior to the distribution of funds, all approved grantees must sign and return a Grant Agreement Form, stating that the funds will be used for the purpose intended. Progress reports and Completion reports must also be filed as required for your specific grant. All current grantees must be in good standing with required documentation prior to submitting new proposals to any foundation.
Restrictions: Grants are not made for political purposes, nor to organizations which discriminate on the basis of race, ethnic origin, sexual or religious preference, age or gender. The Niessen Foundation normally does not consider grants for endowment.
Geographic Focus: Pennsylvania
Date(s) Application is Due: Jan 31; Jul 31
Amount of Grant: 10,000 - 60,000 USD
Contact: Wachovia Bank, N.A., Trustee; grantinquiries3@wachovia.com
Internet: https://www.wachovia.com/foundation/v/index.jsp?vgnextoid=345852199c0a a110VgnVCM1000004b0d1872RCRD&vgnextfmt=default
Sponsor: Leo Niessen Jr., Charitable Trust
Wachovia Bank, N A. PA 1279, 1234 East Broad Street
Philadelphia, PA 19109-1199

Lester E. Yeager Charitable Trust B Grants 2359
Established in 1989 in Kentucky, the Lester E. Yeager Charitable Trust B is an independent foundation. Funding is limited to areas of southern Indiana that are adjacent to Kentucky with emphasis on Daviess and Henderson counties in Indiana. The Trust gives primarily for health and human services. Contact Trust coordinator for additional information and to request an application form.

Requirements: Application form required. Applicants should submit the following: detailed description of project and amount of funding requested; additional materials/documentation; one page cover letter with application; four copies of proposal.
Restrictions: No grants are available to individuals. Grant recipients are expected to file a report when project is completed.
Geographic Focus: Indiana, Kentucky
Amount of Grant: 800 - 13,000 USD
Contact: Trust Coordinator; fax (270) 686-8254
Sponsor: Lester E. Yeager Charitable Trust B
P.O. Box 964
Owensboro, KY 42302-0964

Lester Ray Fleming Scholarships 2360
The North Carolina Governor's Institute on Alcohol and Substance Abuse, Inc. has established scholarships for North Carolina community college students in memory of Lester Ray Fleming, son of Eugene and Mary Fleming and father of Tristan and Curtis. Mr. Fleming was an employee of the Institute at the time of his death in August 2005. It is the hope that these scholarships will benefit and prepare students to educate and serve others in their communities. Those selected will not only receive a $250.00 scholarship but may have the opportunity to attend other venues sponsored by the Institute free of charge where they will gain valuable insight meeting leading researchers, policymakers, and administrators in the field. Scholarship recipients may also have the opportunity to participate in a project at the discretion of the Governor's Institute.
Requirements: Applications can be downloaded from the website.
Geographic Focus: North Carolina
Date(s) Application is Due: Mar 19
Amount of Grant: 250 USD
Contact: Larry Woodard, (919) 256-7412; fax (919) 990-9518; lwoodard1@mindspring.com
Internet: http://www.governorsinstitute.org/index.php?option=com_content&task=view&id=103&Itemid=66
Sponsor: North Carolina Governor's Institute on Alcohol and Substance Abuse
1730 Varsity Drive, Suite 105
Raleigh, NC 27606

Lewis H. Humphreys Charitable Trust Grants 2361
The Lewis H. Humphreys Charitable Trust was established in 2004 to support and promote quality educational, cultural, human-services, and health-care programming for underserved and disadvantaged populations. In the area of arts, culture, and humanities, the trust supports programming that: fosters the enjoyment and appreciation of the visual and performing arts; strengthens humanities and arts-related education programs; provides affordable access; enhances artistic elements in communities; and nurtures a new generation of artists. In the area of education, the trust supports programming that: promotes effective teaching; improves the academic achievement of, or expands educational opportunities for, disadvantaged students; improves governance and management; strengthens nonprofit organizations, school leadership, and teaching; and bolsters strategic initiatives of area colleges and universities. In the area of health, the trust supports programming that improves the delivery of health care to the indigent, uninsured, and other vulnerable populations and addresses health and health-care problems that intersect with social factors. In the area of human services, the trust funds programming that: strengthens agencies that deliver critical human services and maintains the community's safety net and helps agencies respond to federal, state, and local public policy changes. In the area of community improvement, the trust funds capacity-building and infrastructure-development projects including: assessments, planning, and implementation of technology for management and programmatic functions within an organization; technical assistance on wide-ranging topics, including grant writing, strategic planning, financial management services, business development, board and volunteer management, and marketing; and mergers, affiliations, or other restructuring efforts. Grant requests for general operating support, program support, and capital support will be considered. Grant requests for capital support, such as for buildings, land, and major equipment should meet a compelling community need and offer a broad social benefit. Grants from the trust are one year in duration. Application materials are available for download at the grant website. Applicants are strongly encouraged to review the state application guidelines for additional helpful information and clarifications before applying. Applicants are also encouraged to review the trust's funding history (link is available from the grant website). Grant applications can be submitted between August 1 and September 30. Applicants will be notified of grant decisions by November 30.
Requirements: Applicants must have 501(c)3 tax-exempt status and serve the residents of Kansas. Grant application materials must be mailed.
Restrictions: The trust does not support requests from individuals, organizations attempting to influence policy through direct lobbying, or any political campaigns.
Geographic Focus: Kansas
Date(s) Application is Due: Sep 30
Contact: James Mueth, Vice President; (816) 292-4342; james.mueth@baml.com
Internet: https://www.bankofamerica.com/philanthropic/fn_search.action
Sponsor: Lewis H. Humphreys Charitable Trust
1200 Main Street, 14th Floor, P.O. Box 219119
Kansas City, MO 64121-9119

Liberty Bank Foundation Grants 2362
The foundation's charitable giving is focused primarily on organizations that provide meaningful programs and activities that benefit people within Liberty Bank's market area. Of particular interest are programs and activities that provide assistance and opportunities to improve the quality of life for people of low income, especially families in crisis or at-risk. Top priorities for funding include community and economic development--affordable housing for low/moderate-income individuals and families, community and neighborhood capacity-building, and community services targeted to low/moderate-income individuals; education--programs that address the needs of low/moderate-income individuals; health care and human services--outreach and educational programs on health issues, quality child care, homeless shelters and services, services for victims of domestic violence, and transitional housing assistance; and arts and culture--programs that increase access to arts and culture for people of low income who might not otherwise be able to participate in them. Grants generally support specific programs rather than capital projects, equipment, or general operating expenses. Organizations that have received funding in two consecutive calendar years should refrain from reapplying for one calendar year. Contact the office to discuss the project prior to applying. Guideline and application are available online.
Requirements: 501(c)3 tax-exempt organizations are eligible.
Restrictions: Individuals, fraternal groups, and organizations that are not open to the general public are ineligible. Grants do not support annual funds of colleges, universities or hospitals; trips, tours, or conferences; scientific or medical research; deficit spending or debt liquidation; lobbying or otherwise influencing the outcome of the legislative or electoral process; religious groups, except for nonsectarian programs; or endowments or other foundations.
Geographic Focus: Connecticut
Date(s) Application is Due: Mar 31; Jun 30; Sep 30; Dec 31
Contact: Grants Administrator; (860) 704-2181; smurphy@liberty-bank.com
Internet: http://www.liberty-bank.com/liberty_foundation.asp
Sponsor: Liberty Bank Foundation
P.O. Box 1212
Middletown, CT 06457

Libra Foundation Grants 2363
The Libra Foundation awards grants to Maine nonprofits in its areas of interest, including art, culture, and humanities; education; health; human services; environment; justice; public/society benefit; and religion. The Foundation makes grants to organizations that it expects to develop innovative and sustainable Maine-based business initiatives and programs that provide for the welfare and betterment of children. The aforementioned activities comprise the majority of the Foundation's charitable giving. The application is available online.
Requirements: Organizations must be in Maine and 501(c)3 nonprofits to apply.
Restrictions: Individuals are ineligible. The Foundation does not provide funding to supplement annual campaigns, regular operating needs, multi-year projects, individuals, scholarships, or travel.
Geographic Focus: Maine
Date(s) Application is Due: Feb 15; May 15; Aug 15; Nov 15
Samples: Farnsworth Art Museum, Rockland, Maine, general operating support, $10,000; Portland Museum of Art, Portland, Maine, general operating support, $20,000; MaineHealth, Portland, Maine, start-up of a leading center for childhood health, $100,000; Pineland Farms, New Gloucester, Maine, develop, operate and staff Pineland Farms, Inc., for agricultural promotion, education, and research in Maine, $2,895,000.
Contact: Elizabeth Flaherty, Executive Assistant; (207) 879-6280
Internet: http://librafoundation.org/application-procedures
Sponsor: Libra Foundation
Three Canal Plaza, Suite 500
Portland, ME 04112-8516

Lil and Julie Rosenberg Foundation Grants 2364
The Lil and Julie Rosenberg Foundation was established in Connecticut in 1963. It was founded in memory of Julius Rosenberg, founder of Hartley and Parker in 1941, which grew to become one of the premier wholesalers of fine wines and spirits in the State of Connecticut. Julius started his venture in the alcoholic beverage industry in his late teens at Star Liquor Distributors in New York. Owned by his brother, Abraham, Star Liquor Distributors was the first company to be issued a liquor license in New York State after Prohibition. After several years with Star, Julius and his wife, Lil, moved to Bridgeport, Connecticut, to open what would be called Hartley and Parker, situated on Crescent Avenue. In the mid-1950s, Hartley and Parker moved to a larger warehouse on Front Street and, in 1966, Julius built an even larger warehouse facility located in Stratford. Today, the Foundation has a mission of supporting Jewish organizations and public charities in Connecticut and sometimes nationally. Its primary fields of interest include: human services; Judaism; community nonprofits; education; science; literary works; testing for public safety; fostering national or international amateur sports competition (as long as it doesn't provide athletic facilities or equipment); and the prevention of cruelty to children and animals. There are no annual deadlines for application, and interested parties should submit a detailed description of project and amount of funding requested. Typical awards range from $500 to $2,000.
Requirements: 501(c)3 organizations serving residents of Connecticut are eligible.
Geographic Focus: Connecticut
Amount of Grant: 500 - 2,000 USD
Samples: American Cancer Society, Norwalk, Connecticut, $500 - medical research (2014); Anti Defamation League, New Haven Connecticut, $2,000 - social services support (2014); Boy Scouts of America, Irving, Texas, $500 - general operating support (2014).
Contact: Jerry Rosenberg, President; (203) 375-5671; fax (203) 378-1463
Sponsor: Lil and Julie Rosenberg Foundation
100 Browning Street
Stratford, CT 06615-7130

372 | Grant Programs

Lillian S. Wells Foundation Grants   2365
The Lillian S. Wells Foundation established in 1976, primarily supports the Fort Lauderdale, Florida, and Chicago, Illinois regions with funding interests that include: medical research, with emphasis on brain cancer research, women's health, substance abuse, and at-risk youth. Additional funding for education and the arts. Contact the Foundation for an application form, the Board meets quarterly in January, April, July, and October.
Requirements: Nonprofit organizations in Chicago, Illinois and Fort Lauderdale, Florida area are eligible to apply.
Geographic Focus: Florida, Illinois
Date(s) Application is Due: Mar 15; Jun 15; Sep 15; Dec 15
Amount of Grant: 1,000 - 1,900,000 USD
Contact: Patricia F. Mulvaney; patricia.mulvaney@thewellsfamily.net
Sponsor: Lillian S. Wells Foundation
600 Sagamore Road
Fort Lauderdale, FL 33301-2215

Lillian Sholtis Brunner Summer Fellowships for Historical Research in Nursing   2366
These summer fellowships for historical research in nursing are offered by the Center for the Study of the History of Nursing in the School of Nursing at the University of Pennsylvania. Selection of the scholars will be based on evidence of preparation and/or productivity in historical research related to nursing. It is expected that the research and new materials produced by Brunner scholars will help ensure the growth of scholarly work focused on the history of nursing. The fellowship will support up to eight weeks of residential study and use of the center's collections. The application should be sent via email to either Patricia Antonio, PhD, RN, FAAN (dantonio@nursing.upenn.edu) or Barbra Mann Wall, PhD, RN (wallbm@nursing.upenn.edu).
Geographic Focus: All States
Date(s) Application is Due: Dec 31
Amount of Grant: 2,500 USD
Contact: Dr. Karen Buhler-Wilkerson, Director; (215) 898-4725; fax (215) 573-2168; karenwil@nursing.upenn.edu
Internet: http://www.nursing.upenn.edu/history/Pages/Brunner_Fellowship.aspx
Sponsor: University of Pennsylvania
School of Nursing, 418 Curie Boulevard
Philadelphia, PA 19104-6020

Lincoln Financial Foundation Grants   2367
Lincoln Financial is committed to making charitable contributions in the communities where it maintains a strong business presence. These communities include: Chicago, Illinois; Concord, New Hampshire; Fort Wayne, Indiana; Greensboro, North Carolina; Hartford, Connecticut; Omaha, Nebraska; Philadelphia, Pennsylvania. The Lincoln Foundation focuses its contributions in four priority areas: Arts, Education, Human Services, and Workforce and Economic Development. Arts grants support arts education and economic development through increased access to arts and cultural activities. Education grants support closing the achievement gap with an emphasis on improving student achievement. Human Services grants provide for basic needs of food and shelter and promoting self-sufficiency. Economic and Workforce Development grants support adult education, job skills training and opportunities to enhance the workforce with the overarching intent of economic development for the area.
Requirements: Nonprofit organizations that have a 501(c)3 designation from the IRS are eligible to apply. The Lincoln Foundation supports only those organizations that respect and encourage diversity and adhere to nondiscriminatory practices. Women and ethnic minorities should be adequately represented on an applicant organization's board of directors. Education grant applications are due by March 11. Human Services grant applications are due by June 17. Workforce and Economic Development grant applications are due by September 17. Arts grant applications are due by December 3. Applications are reviewed by the local contribution committees within three to four months after the cycle deadline. Notification will be given soon after decisions are rendered. Qualified organizations are eligible for one grant per calendar year.
Restrictions: Qualified organizations are eligible for one grant per calendar year. In general, the Lincoln Foundation will not award grants to: individuals; religious organizations; public or private elementary or secondary schools or school foundations; hospitals or hospital foundations; fraternal, political, or war veteran organizations; general operating support; capital funding; endowments.
Geographic Focus: Connecticut, Illinois, Indiana, Nebraska, New Hampshire, North Carolina, Pennsylvania, United Kingdom
Date(s) Application is Due: Mar 15; Jun 17; Sep 17; Dec 3
Amount of Grant: 5,000 - 75,000 USD
Contact: Anne T. Rogers, Director; (260) 455-5604; fax (260) 455-4004
Internet: https://www.lfg.com/LincolnPageServer?LFGPage=/lfg/lfgclient/abt/csr/emp/index.html&LFGContentID=/lfg/lfgclient/abt/csr/lff/app
Sponsor: Lincoln Financial Foundation
1300 S Clinton Street
Fort Wayne, IN 46802

Linford and Mildred White Charitable Fund Grants   2368
The Linford and Mildred White Charitable Fund was established in 1956 to support and promote quality educational, human-services, and health-care programming for underserved populations within the city of Waterbury, Connecticut, and its surrounding communities. Grants from the White Charitable Fund are one year in duration. Applicants must apply online at the grant website. Applicants are strongly encouraged to do the following before applying: review the downloadable state application procedures for additional helpful information and clarifications; review the downloadable online application guidelines at the grant website; review the foundation's funding history (link is available from the grant website); review the online application questions in advance; and review the list of required attachments. These will generally include: a list of board members, financial statements (audited, reviewed, or compiled by independent auditor); an organization summary; a list of other funding sources; an IRS Determination letter; and other required documents. All attachments must be uploaded in the online application as PDF, Word, or Excel files. The deadline for application to the Linford and Mildred White Charitable Fund is 11:59 p.m. on July 1. Applicants will receive notification of grant decisions by letter within two to three months after the proposal deadline.
Requirements: Applicant organizations must have 501(c)3 status and serve the people of Waterbury, Connecticut, and its vicinity.
Restrictions: Grant requests for capital projects will not be considered. Applicants will not be awarded a grant for more than 3 consecutive years. The fund does not support requests from individuals, organizations attempting to influence policy through direct lobbying, or any political campaigns.
Geographic Focus: Connecticut
Date(s) Application is Due: Jul 1
Samples: Boys and Girls Clubs of Greater Waterbury, Waterbury, Connecticut, $2,000, purchase of a bounce house for Bounce to Be Fit; Covenant to Care for Children, Bloomfield, Connecticut, $1,500, Waterbury Adopt A Social Worker Program; Palace Theater Group, Waterbury, Connecticut, $2,000, Ticket and Travel Subsidy Program for Title One schools.
Contact: Carmen Britt; (860) 657-7019; carmen.britt@baml.com
Internet: https://www.bankofamerica.com/philanthropic/fn_search.action
Sponsor: Linford and Mildred White Charitable Fund
200 Glastonbury Boulevard, Suite # 200, CT2-545-02-05
Glastonbury, CT 06033-4056

Lisa and Douglas Goldman Fund Grants   2369
Established in 1992 the Lisa and Douglas Goldman Fund is a private foundation committed to providing support for charitable organizations that enhance society. As natives of San Francisco, the Goldmans place a high priority on projects that have a positive impact on San Francisco and the Bay Area. Interests and priorities include: children and youth; civic affairs; civil and human rights; education; environmental affairs; health; Jewish affairs; literacy; organizational development; population; social and human services; and sports and recreation.
Requirements: After reviewing the Fund's interests and priorities interested applicants may submit an initial letter of inquiry. Applicants who receive a favorable response will be invited to submit a formal proposal with supporting materials. Applicants are encouraged to contact the Fund directly with questions regarding the appropriateness of a project. There are no deadlines.
Restrictions: The following are ineligible: grants to individuals; documentaries and films; events/conferences; books and periodicals; research; and deficit budgets. Applications for annual support are not accepted. Organizations may submit only one request per year.
Geographic Focus: All States
Samples: Little Kids Rock, Cedar Grove, New Jersey, $10,000 - Bay Area Expansion Initiative, to increase the number of low-income schools participating in the Little Kids Rock music program; Ignite, San Francisco, California, $13,250 - to support the expansion of Ignite's programs at San Francisco schools and foster the development of young women's political knowledge and ambition; and Basel Action Network, Seattle, Washington, $25,000 - for the e-Stewards Initiative, a program to promote responsible recycling of electronic waste through the certification of e-waste recyclers.
Contact: Nancy D. Kami, Executive Director; (415) 771-1717; fax (415) 771-1797
Internet: http://fdncenter.org/grantmaker/goldman
Sponsor: Lisa and Douglas Goldman Fund
One Daniel Burnham Court, Suite 330C
San Francisco, CA 94109-5460

Lisa Higgins-Hussman Foundation Grants   2370
The Lisa Higgins-Hussman Foundation was established in Maryland in 2006, with the expressed purpose of supporting medical breakthroughs. Its primary interests include: research efforts to improve scientific knowledge of, and to develop therapies and find cures for conditions such as autism, addiction, alcoholism and neurological disorders; therapeutic care, particularly for hospice care and therapeutic intervention for individuals with debilitating health or neurological disorders; and issues relating to women's health. Most recent awards have ranged from $500 to $125,000, with an average of thirty to thirty-five grants given each year. There is no annual deadline specified, and interested parties should contact the Foundation in writing with a brief overview of their proposal and budgetary needs.
Requirements: 501(c)3 organizations throughout the U.S. are eligible to apply.
Geographic Focus: All States
Amount of Grant: 500 - 125,000 USD
Contact: Lisa Marie Higgins, President; (410) 480-8264
Sponsor: Lisa Higgins-Hussman Foundation
10215 Tarpley Court
Ellicott City, MD 21042-1681

Littlefield-AACR Grants in Metastatic Colon Cancer Research   2371
The grants will provide support for innovative cancer research projects designed to accelerate the discovery and development of new agents to treat metastatic colon cancer and/or for pre-clinical research with direct therapeutic intent. Special emphasis will be placed on research that holds promise for leading to individualized therapeutic options for treatment in the near future or for developing promising new cancer therapeutics for metastatic colon cancer, which will translate into clinical applications within a one- to two-year period.

Requirements: The prizes are open to all cancer researchers who are affiliated with any institution involved in cancer research, cancer medicine, or cancer-related biomedical science anywhere in the world. Such institutions include those in academia, industry, or government. Candidates must be active researchers and have a record of recent publications.
Geographic Focus: All States
Date(s) Application is Due: Feb 1
Amount of Grant: 500,000 - 1,000,000 USD
Contact: Julia Laurence, Program Assistant; (215) 440-9300, ext. 102 or (267) 646-0655; fax (215) 440-9372; julia.laurence@aacr.org or laurence@aacr.org
Internet: http://www.aacr.org/home/scientists/research-funding--training-grants/research-funding/littlefield-aacr-grants-for-metastatic-colon-cancer-research.aspx
Sponsor: Littlefield 2000 Trust and the American Association for Cancer Research
615 Chestnut Street, 17th Floor
Philadelphia, PA 19106-4404

Little Life Foundation Grants 2372
The Little-Life Foundation is a 501c3 charitable organization founded in 2002 whose primary purpose is to support medical care and research for increasing the survival rates of prematurely born babies. Saint Barnabas Medical Center, where more than 7,500 babies are born each year, will be the primary beneficiary of the Foundation's efforts.
Geographic Focus: New Jersey
Contact: Richard P. Rizzuto, Director; (973) 277-2329; fax (973) 401-9799
Internet: http://www.littlelifefoundation.org/
Sponsor: Little Life Foundation
P.O. Box 765
New Vernon, NJ 07976-0765

Lloyd A. Fry Foundation Health Grants 2373
The Lloyd A. Fry Foundation supports organizations with the strength and commitment to address persistent problems of urban Chicago resulting from poverty, violence, ignorance, and despair. The Foundation's Health Program is committed to increasing access to high quality primary care and reducing health care disparities for Chicago's low-income residents. To accomplish these goals, the Foundation is interested in supporting: efforts to implement medical-home models of care which provide comprehensive integrated primary care services across multi-disciplinary team members in single or multiple settings; high quality primary care services that are not widely available to low-income populations; community outreach to connect hard-to-reach individuals with high-quality primary care; and policy advocacy focused on improving the quality of health care and increasing access to health care for low-income populations in Chicago. Typically, grants in this area range from $18,000 to $100,000.
Requirements: Grants are made only to tax-exempt organizations and are rarely made to organizations outside the Chicago metropolitan area.
Restrictions: The foundation does not fund government entities, political or religious activities, fundraising events, or medical research.
Geographic Focus: Illinois
Date(s) Application is Due: Mar 1; Sep 1; Dec 1
Amount of Grant: 18,000 - 100,000 USD
Samples: Heartland International Health Center, Chicago, Illinois, $100,000 - for Dental Care Services at Senn and Roosevelt High Schools, Hibbard Elementary, and at two community clinics; Lutheran Social Services of Illinois, Des Plaines, Illinois, $40,000 - children and adolescent counseling services; PCC Community Wellness Center, Oak Park, Illinois, $40,000 - for the Maternal and Child Health Services Program.
Contact: Unmi Song; (312) 580-0310; fax (312) 580-0980; usong@fryfoundation.org
Internet: http://www.fryfoundation.org/grants/health/
Sponsor: Lloyd A. Fry Foundation
120 South LaSalle Street, Suite 1950
Chicago, IL 60603-3419

Lockheed Martin Corporation Foundation Grants 2374
The Lockheed Martin Corporation Foundation funds grants that enhance the communities where Lockheed Martin employees work and live. Lockheed Martin will consider grant requests that best support the Corporation's strategic focus areas and reflect effective leadership, fiscal responsibility, and program success. Those focus areas include: education--K-16 science, technology, engineering and math (STEM) education; customer and constituent relations--causes of importance to customers and constituents, including the U.S. military and other government agencies; community relations--building partnerships between employee volunteers and the civic, cultural, environmental, and health and human services initiatives that strengthen the communities where employees work. Applications are accepted year-round. Evaluations are typically performed quarterly.
Requirements: To be considered for grant funding, organizations must meet all of the following criteria: apply through Lockheed Martin's online CyberGrants system; have a non-profit tax exempt classification under Section 501(c)3 of the Internal Revenue Service Code, or equivalent international non-profit classification, or be a public elementary/secondary school, or be a qualifying US-based institute of higher education; align with one or more of Lockheed Martin's three strategic focus areas: delivering standards-based science, technology, engineering and math (STEM) education to students in K-16; investing in programs that support the long term success of the military and their families; and supporting the vitality of the communities where employees live and work; agree to act in accordance with Lockheed Martin's contribution acknowledgement *Requirements:* organization/grantee will comply with all applicable requirements of the Patriot Act and the Voluntary Anti-Terrorist Guidelines and will not use any portion of the grant funds for the support, direct or indirect, of acts of violence or terrorism or for any organization engaged in or supporting such acts; be located or operate in a community in which Lockheed Martin has employees or business interests; demonstrate fiscal and administrative responsibility and have an active, diverse board, effective leadership, continuity and efficiency of administration; be limited to one grant per year, except in unusual circumstances.
Restrictions: Some grant applications may not be able to be considered until the next year's budget cycle, particularly those received in the second half of the year. Grants are generally not made to: organizations that unlawfully discriminate on the basis of race, ethnicity, religion, national origin, age, military veteran's status, ancestry, sexual orientation, gender identity or expression, marital status, family structure, genetic information, or mental or physical disability; private K-12 schools, unless the contribution is in acknowledgement of employee volunteer service provided to the school; home-based child care/educational services; individuals; professional associations, labor organizations, fraternal organizations or social clubs; social events sponsored by social clubs; athletic groups, clubs and teams, unless the contribution is in acknowledgement of employee volunteer service provided to the school; religious organizations for religious purposes; or advertising in souvenir booklets, yearbooks or journals unrelated to Lockheed Martin's business interests.
Geographic Focus: California, Colorado, Florida, Georgia, Louisiana, Maryland, Minnesota, Mississippi, New Jersey, New Mexico, New York, Ohio, Pennsylvania, South Carolina, Texas, Virginia, Canada
Contact: Emily Simone, Corporate Community Relations; (301) 897-6000 or (301) 897-6866; fax (301) 897-6485; david.e.phillips@lmco.com
Internet: http://www.lockheedmartin.com/us/who-we-are/community.html
Sponsor: Lockheed Martin Corporation Foundation
6801 Rockledge Drive
Bethesda, MD 20817-1836

Long Island Community Foundation Grants 2375
The Foundation awards grants to eligible New York nonprofit organizations in the following program areas: arts; community development; education; environment; health; mental health; hunger; technical assistance; and youth violence prevention. Projects are preferred which accomplish specific tasks, solve problems, address needs of the disadvantaged, help a large number of people, and use community resources. Specific guidelines for each category and the application are available on the Foundation website. Applicants are encouraged to view recent grants on the website in order to judge if their project is appropriate for the Foundation.
Requirements: New York nonprofit organizations are eligible.
Restrictions: Grants are not made for the following: individuals; building or capital campaigns; medical or scientific research; equipment purchases; budget deficits; endowments; event sponsorships; re-granting purposes; or religious or political purposes.
Geographic Focus: New York
Date(s) Application is Due: Aug 24
Samples: Long Island Arts Alliance, Long Island, New York, strengthen the Long Island arts infrastructure through arts education advocacy forums, and a study about the economic significance of the nonprofit arts community, $30,000; Citizens Campaign Fund for the Environment, Long Island, New York, to analyze and grade 16 sewage treatment plants on Long Island, $25,000; Mental Health Association of Nassau County, Long Island, New York, to support a geriatric mental health training program that will increase access to quality mental health for the elderly, $15,000.
Contact: Nancy Arnold; (516) 348-0575; fax (516) 348-0570; narnold@licf.org
Internet: http://www.licf.org/grants
Sponsor: Long Island Community Foundation
1864 Muttontown Road
Syosset, NY 11791

Lotus 88 Foundation for Women and Children Grants 2376
The foundation's mission is to promote the empowerment of women and children through supporting their economic, emotional, and spiritual development. The foundation's focus area is American Indian Country. Grants are awarded in Indian Country for two strategic purposes: revitalizing the council tipis as spiritual, cultural, and service centers; and providing the basic needs through community building. Community building grants are intended to promote and support community building in Indian Country to improve basic living conditions and to encourage a more positive future. Tipi project grants are made to help tribal women living on reservation or off reservation in building community. Grants support cultural and social services, tribal gatherings, educational programs, healing and purification ceremonies, and retreats. Each proposal should identify the specific needs and uses for the tipi and should identify a nonprofit program partner working in the tribal area who will work with the foundation on the project. Contact the Foundation directly for application and additional guideline information.
Requirements: Projects in American Indian communities are eligible.
Geographic Focus: All States
Contact: Patricia Stout; (510) 841-4123; fax (510) 841-4093; benita@lotus88.org
Internet: http://lotus88.net/
Sponsor: Lotus 88 Foundation for Women and Children
127 University Avenue, P.O. Box 10728
Berkeley, CA 94710

Louetta M. Cowden Foundation Grants 2377
The Louetta M. Cowden Foundation was established in 1964 to support and promote quality educational, cultural, human-services, and health-care programming. In the area of culture the foundation supports programming that: fosters the enjoyment and appreciation of the visual and performing arts; strengthens humanities and arts-related education programs; provides affordable access; enhances artistic elements in communities; and nurtures a new generation of artists. In the area of education the foundation supports programming that: promotes effective teaching; improves the

academic achievement of, or expands educational opportunities for disadvantaged students; improves governance and management; strengthens nonprofit organizations, school leadership, and teaching; and bolsters strategic initiatives of area colleges and universities. In the area of health the foundation supports programming that: improves the delivery of health care to the indigent, uninsured, and other vulnerable populations; and addresses health and health-care problems that intersect with social factors. In the area of human services, the foundation supports programming that: strengthens agencies that deliver critical human services and maintains the community's safety net; and helps agencies respond to federal, state, and local public policy changes. There are no application deadlines for the Cowden Foundation. Proposals are reviewed on an ongoing basis. Grant requests for general operating support and program support will be considered. Grants from the Foundation are one year in duration. Application materials are available for download at the grant website. Applicants are strongly encouraged to review the downloadable state application guidelines before applying. The annual deadline for the online application submission is July 31.
Requirements: Applicants must have 501(c)3 tax-exempt status and serve the residents of Kansas City, Missouri.
Restrictions: Grant requests for capital support will not be considered. The foundation does not support requests from individuals, organizations attempting to influence policy through direct lobbying, or any political campaigns.
Geographic Focus: Missouri
Date(s) Application is Due: Jul 31
Amount of Grant: 25,000 - 85,000 USD
Samples: Harvesters-The Community Food Network, Kansas City, Missouri, $75,000 - general operating support (2014); Childrens Mercy Hospital, Kansas City, Missouri, $50,000 - capital campaign (2014); Starlight Theater Association, Kansas City, Missouri, $25,000 - program development (2014).
Contact: Scott Berghaus; (816) 292-4300; scott.berghaus@ustrust.com
Internet: https://www.bankofamerica.com/philanthropic/fn_search.action
Sponsor: Louetta M. Cowden Foundation
1200 Main Street, 14th Floor, P.O. Box 219119
Kansas City, MO 64121-9119

Louie M. and Betty M. Phillips Foundation Grants 2378
The Foundation supports a variety of organizations in the fields of health, human services, civic affairs, education, and the arts. Types of support include annual operating grants for selected organizations contributing significantly to the Nashville area; one-year project and program grants for specific projects or equipment; and capital support (five-years maximum) for major capital projects of organizations with strong records of community service. The application and a list of previously funded projects are available at the Foundation website.
Requirements: Nonprofit organizations are eligible. With rare exceptions, grants are limited to organizations in the greater Nashville area.
Restrictions: The Foundation does not support individuals or their projects, private foundations, political activities, advertising, or sponsorships. In general, the Foundation does not support projects, programs, or organizations that serve a limited audience; disease-specific organizations; biomedical or clinical research; organizations whose principal impact is outside the Nashville area; or tax-supported institutions.
Geographic Focus: Tennessee
Date(s) Application is Due: Jun 1; Nov 1
Amount of Grant: 500 - 35,000 USD
Samples: Walden's Puddle Wildlife Rehabilitation Center, Nashville, Tennessee, operating support, $10,000; Safe Haven Family Shelter, Nashville, Tennessee, operating support, $5,000; Men of Valor, Nashville, Tennessee, capital and operating support, $35,000; Boys and Girls Club of Davidson County, Nashville, Tennessee, capital and operating support, $12,500.
Contact: Louie Buntin; (615) 385-5949; fax (615) 385-2507; louie@phillipsfoundation.org
Internet: http://www.phillipsfoundation.org
Sponsor: Louie M. and Betty M. Phillips Foundation
3334 Powell Avenue, P.O. Box 40788
Nashville, TN 37204

Louis and Elizabeth Nave Flarsheim Charitable Foundation Grants 2379
The Louis & Elizabeth Nave Flarsheim Charitable Foundation was established to support and promote quality educational, cultural, human-services, and health-care programming. In the area of arts, culture, and humanities, the foundation supports programming that: fosters the enjoyment and appreciation of the visual and performing arts; strengthens humanities and arts-related education programs; provides affordable access; enhances artistic elements in communities; and nurtures a new generation of artists. In the area of education, the foundation supports programming that: promotes effective teaching; improves the academic achievement of, or expands educational opportunities for, disadvantaged students; improves governance and management; strengthens nonprofit organizations, school leadership, and teaching; and bolsters strategic initiatives of area colleges and universities. In the area of health, the foundation supports programming that improves the delivery of health care to the indigent, uninsured, and other vulnerable populations and addresses health and health-care problems that intersect with social factors. In the area of human services, the foundation funds programming that: strengthens agencies that deliver critical human services and maintains the community's safety net and helps agencies respond to federal, state, and local public policy changes. In the area of community improvement, the foundation funds capacity-building and infrastructure-development projects including: assessments, planning, and implementation of technology for management and programmatic functions within an organization; technical assistance on wide-ranging topics, including grant writing, strategic planning, financial management

services, business development, board and volunteer management, and marketing; and mergers, affiliations, or other restructuring efforts. Grant requests for general operating support and program support will be considered. Grants from the foundation are one year in duration. Application materials are available for download at the grant website. Applicants are strongly encouraged to review the state application guidelines for additional helpful information and clarifications before applying. Applicants are also encouraged to review the foundation's funding history (link is available from the grant website). There are no application deadlines for the Flarsheim Charitable Foundation. Proposals are reviewed on an ongoing basis.
Requirements: The Flarsheim Foundation supports organizations that serve the residents of Kansas City, Missouri. Applications must be mailed.
Restrictions: Grant requests for capital support will not be considered.
Geographic Focus: Missouri
Samples: Paul Mesner Puppets, Kansas City, Missouri, $10,000; Kansas City Chorale, Kansas City, Missouri, $5,000; Jewish Community Center Charitable Supporting Foundation, Overland Park, Kansas, $2,500.
Contact: Spence Heddens; (816) 292-4301; Spence.heddens@baml.com
Internet: https://www.bankofamerica.com/philanthropic/fn_search.action
Sponsor: Louis & Elizabeth Nave Flarsheim Charitable Foundation
1200 Main Street, 14th Floor, P.O. Box 219119
Kansas City, 64121-9119

Louis H. Aborn Foundation Grants 2380
The Louis H. Aborn Foundation was established in Connecticut, in 1975, by the longtime president of Tams-Witmark Music Library, the famed company that licenses Broadway musical scripts and scores to stock, amateur and professional theaters. Giving is primarily centered around public health projects, education and child welfare. Giving is restricted to Connecticut and New York. There are no particular applications or deadlines with which to adhere, and applicants should begin by contacting the Foundation directly.
Requirements: 501(c)3 organizations serving the residents of Connecticut and New York are eligible to apply.
Geographic Focus: Connecticut, New York
Contact: Hermine F. Aborn, Vice-President; (203) 661-4046
Sponsor: Louis H. Aborn Foundation
46 Wilshire Road
Greenwich, CT 06831-2723

Lowe Foundation Grants 2381
The foundation awards grants to organizations in Texas, primarily for programs that support the critical needs of women and children. The arts and higher education also receive support. Grants are awarded to support capital campaigns, building construction/renovation, and general operations. The board meets in April of each year. The December 1 deadline listed is for pre-proposals, with full proposals by request due on December 31.
Requirements: 501(c)3 organizations in Texas may apply for grant support.
Geographic Focus: Texas
Date(s) Application is Due: Dec 1; Dec 31
Amount of Grant: 5,000 - 225,000 USD
Contact: Clayton Maebius, Trustee; (512) 322-0041; fax (512) 322-0061; info@thelowefoundation.org
Internet: http://www.thelowefoundation.org/guidelines.htm
Sponsor: Lowe Foundation
1005 Congress Avenue, Suite 895
Austin, TX 78701

Lowell Berry Foundation Grants 2382
Mr. Lowell W. Berry established The Lowell Berry Foundation in 1950 with the primary purpose of assisting in strengthening Christian ministry at the local church level. Mr. Berry's secondary purpose was that of assisting social service programs in the areas in California where he lived and operated his business. Guidelines are available at the Foundation's website.
Requirements: 501(c)3 tax-exempt organizations may apply. The Foundation provides funding primarily in Contra Costa and Alameda counties of California.
Restrictions: No grants to individuals, or for building or capital funds, equipment, seed money, or land acquisition.
Geographic Focus: California
Amount of Grant: 5,000 - 200,000 USD
Contact: Katherine Sanders; (925) 284-4427; info@lowellberryfoundation.org
Internet: http://www.lowellberryfoundation.org/process.html
Sponsor: Lowell Berry Foundation
3685 Mount Diablo Boulevard, Suite 269
Lafayette, CA 94549-3776

Lubbock Area Foundation Grants 2383
The Lubbock Area Foundation is a nonprofit community foundation that manages a pool of charitable endowment funds, the income from which is used to benefit the South Plains community through grants to nonprofit organizations, educational programs and scholarships. Funding priorities are: art and culture; social services; civic and community; education; social services; health and human services. Grants may be made for start-up funding, general operating support, program support and demonstration programs. Typical grant awards range from $500 - $2,500 with $5,000 as the maximum from the unrestricted funds. Grant application and additional guidelines are available on the Foundation's website.
Requirements: The Foundation restricts its support to organizations in Lubbock and the surrounding South Plains area which are 501(c)3 or the government equivalent.

Restrictions: The Foundation does not make grants to individuals, for political purposes, to retire indebtedness or for payment of interest or taxes.
Geographic Focus: Texas
Date(s) Application is Due: Jan 1; Mar 1; May 1; Jul 1; Sep 1; Nov 1
Amount of Grant: 500 - 5,000 USD
Samples: Louise Hopkins Underwood Center for the Arts Clay Studio, Lubbock, TX, $8,500--Firehouse Theatre and Marketing Staff Salaries; The Haven Animal Care Shelter, Lubbock, TX--$8,000-- funding for veterinary care for animals; Lubbock Children's Health Clinic, Lubbock, TX, $3,600--Pharmacy Assistance Program.
Contact: Kathleen Stocco, Executive Director; (806) 762-8061; fax (806) 762-8551
Internet: http://www.lubbockareafoundation.org/grant.shtml
Sponsor: Lubbock Area Foundation
1655 Main Street, Suite 202
Lubbock, TX 79401

Lubrizol Foundation Grants 2384
The Lubrizol Foundation makes grants in support of education, health care, human services, civic, cultural, youth and environmental activities of a tax-exempt, charitable nature. Scholarships, fellowships, and awards are generally made in the fields of chemistry and chemical and mechanical engineering at colleges and universities. Types of support include the following: annual campaigns; building/renovation; capital campaigns; continuing support; employee matching gifts; employee volunteer services; equipment; fellowships; general/operating support; scholarship funds. Priority is given to the greater Cleveland, Ohio and Houston, Texas areas. The Lubrizol Foundation typically reviews and decides upon requests quarterly. There are no deadlines by which you need to submit your proposal. Proposals may be submitted via postal mail or email, but not by fax. Applicants will receive written notification of the decision on their proposal.
Requirements: Written applications of established Ohio and Texas nonprofit charitable organizations will be considered on a case by case basis. Grant proposals should include the following: a cover letter that summarizes the purpose of the request, signed by the executive officer of the organization or development office; a narrative of specific information related to the subject of the request; current audited financial statements and a specific project budget, if applicable; documentation of the organization's Federal tax-exempt status, e.g., a copy of the 501(c)3 determination letter. Additional descriptive literature (e.g., an annual report, brochures, etc.) that accurately characterizes the overall activities of the organization is appreciated. Upon review, further information may be requested including an interview and site visit.
Restrictions: Grants are not made for religious or political purposes, to individuals nor, generally, to endowments.
Geographic Focus: Ohio, Texas
Amount of Grant: 2,000 - 250,000 USD
Samples: Big Brothers Big Sisters of Greater Cleveland, Cleveland, OH, $2,500 - for operating support; American Red Cross, Houston, Texas, $25,000 - for Hurricane Ike Disaster Relief Fund; Hospice of the Western Reserve, Cleveland, OH, $250,000 - toward construction of new faculty.
Contact: Karen Lerchbacher, Administrator; (440) 347-1797; fax (440) 347-1858; karen.lerchbacher@lubrizol.com
Internet: http://www.lubrizol.com/CorporateResponsibility/Lubrizol-Foundation.html
Sponsor: Lubrizol Foundation
29400 Lakeland Boulevard, 053A
Wickliffe, OH 44092-2298

Lucile Horton Howe and Mitchell B. Howe Foundation Grants 2385
The foundation supports youth organizations, social services, medical research, hospitals, religion, child welfare, drug abuse, education, and family services. Types of support include continuing support, general operating funds, and research. Requests from qualified organizations must be received prior to the second Tuesday of March for the grant year. There is only one consideration meeting a year.
Requirements: Only nonprofit organizations in the Prox-Pasadena area and the San Gabriel Valley of California are eligible. A brief letter and a copy of the organizations 501(c)3 form are the requirements for application.
Restrictions: No restricted grants will be funded by the foundation.
Geographic Focus: California
Date(s) Application is Due: Feb 28
Amount of Grant: 1,000 - 300,000 USD
Samples: Huntington Medical Research Institutes, Pasadena, CA., $60,000; Childrens Hospital Los Angeles, Los Angeles, CA., $12,000; Upward Bound Group Home, Pioneer, CA., $10,000;
Contact: Mitchell B. Howe, President; (626) 792-2771; lhmbhowefoun@earthlink.net
Sponsor: Lucile Horton Howe and Mitchell B. Howe Foundation
180 South Lake Avenue
Pasadena, CA 91101-4932

Lucile Packard Foundation for Children's Health Grants 2386
The vision of the Foundation is that all children in the communities the Foundation serves are able to reach their maximum health potential. The Foundation invests in programs and projects that have the potential to improve California's systems of care for children with special health care needs and their families. Funding is provided for programs that address improving care for children with special needs in the following areas: system reform; care coordination/enhanced medical home; pediatric education; quality measures focused on children with special health care needs; data collection and/or analysis; and advocacy.
Requirements: Generally the Foundation invites applications for grant proposals. However, organizations that have not been invited may submit a letter of inquiry if the organization meets the organizational criteria and if their proposed program falls within the Foundation's focus area. Applicants should review program and organizational criteria first. Eligible organizations: are classified as tax exempt under 501(c)3 or are a public or educational entity, or collaborations of nonprofit and public agencies with a designated fiscal sponsor, or entities that have a charitable purpose; have demonstrated capacity to implement effective, culturally competent programs, reach intended populations, and achieve clear, reasonable and measureable goals and objectives; and promote and maintain nondiscriminatory policies in programs and the workplace. Eligible programs or projects are those that hold promise of improving the systems of care for Children with Special Health Care Needs (CSHCN). Eligible programs should: be focused primarily on California; have potential for national applicability; have the potential to improve the system of care for children with special needs; demonstrate an understanding of the complex system of care for CSHCN and builds on existing knowledge and systems; have potential to affect a large number of CSHCN; have potential for impact within five years; have reasonable evidence of sustainability; have reasonable evidence of replicability; and present opportunities for collaboration.
Restrictions: Unsolicited full proposals should not be submitted. The following are not funded: disease-specific projects; individuals; scholarships; support for candidates for political office; private foundations; religious organizations for religious purposes; fundraising sponsorships; annual fund appeals; capital campaigns; and basic scientific research.
Geographic Focus: California
Contact: Amanda Frederickson, Annual Giving Program Manager; (650) 736-0676; amanda.frederickson@lpfch.org
Internet: http://www.lpfch.org/programs/cshcn/process.html
Sponsor: Lucile Packard Foundation for Children's Health
400 Hamilton Avenue, Suite 340
Palo Alto, CA 94301

Lucy Downing Nisbet Charitable Fund Grants 2387
The Lucy Downing Nisbet Charitable Fund was established in 2002 to support and promote educational, health-and-human-services, and arts programming for underserved populations. The Foundation specifically serves organizations located in and serving the people of Vermont. Special consideration is given to organizations in the area of healthcare/nursing, domestic violence awareness, heart disease, and endangered species, and organizations located in and serving the people of Morrisville, Vermont. Grants are one year in duration. Applicants must apply online at the grant website. Applicants are strongly encouraged to do the following before applying: review the downloadable state application procedures for additional helpful information and clarifications; review the downloadable online-application guidelines at the grant website; review the foundation's funding history (link is available from the grant website); review the online application questions in advance; and review the list of required attachments. These will generally include: a list of board members, financial statements (audited, reviewed, or compiled by independent auditor); an organization summary; a list of other funding sources; an IRS Determination letter; and other required documents. All attachments must be uploaded in the online application as PDF, Word, or Excel files. The Lucy Downing Nisbet Charitable Fund application deadline is 11:59 p.m. on January 15. Applicants will be notified of grant decisions by letter within two to three months of the deadline.
Requirements: The foundation specifically serves organizations located in and serving the people of Vermont. Applicants must have 501(c)3 tax-exempt status.
Restrictions: The trustees do not make grants for deficit financing, annual giving, endowments or capital projects. The fund does not support requests from individuals, organizations attempting to influence policy through direct lobbying, or any political campaigns.
Geographic Focus: Vermont
Date(s) Application is Due: Jan 15
Amount of Grant: 5,000 - 20,000 USD
Samples: Copley Health Systems, Morrisville, Vermont, $58,400, to provide computerized Physician's Order Entry component of Electronic Medical Record Software and to fund Nursing Education; Vermont Foodbank, South Barre Vermont, $20,000, for USDA Summer Food Service Program in high-poverty and extremely rural area; River Arts of Morrisville, Morrisville, Vermont, $8,000, Arts for Everyone archive.
Contact: Amy Lynch; (860) 657-7015; amy.r.lynch@baml.com
Internet: https://www.bankofamerica.com/philanthropic/fn_search.action
Sponsor: Lucy Downing Nisbet Charitable Fund
200 Glastonbury Boulevard, Suite # 200, CT2-545-02-05
Glastonbury, CT 06033-4056

Lucy Gooding Charitable Foundation Trust Grants 2388
The Foundation, established in 1988, supports organizations involved with children and youth, services, education, human services, residential/custodial care, hospices, economically disadvantaged, people with disabilities, and homelessness. The Foundation gives primarily in the five county area surrounding Jacksonville, Florida.
Requirements: Funding preference is for organizations in the five county area surrounding Jacksonville, FL with projects helping children. Contact the Foundation for application form.
Restrictions: No support for private foundations, religious organizations, or for adults-only services, and individuals.
Geographic Focus: Florida
Date(s) Application is Due: Sep 30
Amount of Grant: 5,000 - 1,000,000 USD
Contact: Bonnie H. Smith; (904) 786-4796; fax (904) 786-4796; bhsmith@bellsouth.net
Sponsor: Lucy Gooding Charitable Foundation Trust
P.O. Box 37349
Jacksonville, FL 32236-7349

Ludwick Family Foundation Grants 2389

Founded in 1990 by Arthur and Sarah Ludwick, the Ludwick Family Foundation is a California-based philanthropic organization established exclusively for charitable, scientific, literary, and educational purposes. Grants are awarded to United States or U.S. based international organizations in adherence to the mission of Ludwick Family Foundation and its founding documents. The foundation tends to fund tangible types of items that will remain with and can be used repeatedly by the organization. Since the Foundation only considers invited applications, interested parties should begin by contacting the office with an concept before submitting anything in writing.
Requirements: Eligible applicants include U.S. or U.S.-based international organizations, 501(c)3 nonprofit public charities, and government agencies (any level). All grants are to be used exclusively for charitable, public benefit purposes.
Restrictions: The foundation does not grant requests for salaries, general operating expenses, scholarships, endowment funds, fundraising events or capital campaigns, feasibility studies, consulting fees, or advertising. The foundation will no longer accept any unsolicited requests for research or from public/private schools (K-12), universities/colleges, child daycare/development centers, hospitals, or libraries. Grant consideration will be given only to those organizations that have been invited to submit formal proposals and that have completed the application process.
Geographic Focus: All States
Amount of Grant: 5,000 - 50,000 USD
Contact: Trista Campbell; (626) 852-0092; fax (626) 852-0776; ludwickfndn@ludwick.org
Internet: http://www.ludwick.org
Sponsor: Ludwick Family Foundation
203 South Glendora Avenue, Suite B, P.O. Box 1796
Glendora, CA 91740

Lumosity Human Cognition Grant 2390

Lumosity invites researchers to submit proposals for studies that use functional neuroimaging techniques to investigate mechanisms underlying cognitive processes implicated in Lumosity's games and assessments. These studies would ideally involve both neuroimaging and behavioral methods with healthy adult populations. Examples of types of projects that would be prioritized include, but are not limited to demonstrations of: task-related neural activity within and across brain regions; changes in neural activity that accompany training-related changes in cognitive performance; and neural specificity of cognitive training effects. All applications are encouraged to focus explicitly on the use of neuroimaging as a tool for studying Lumosity's cognitive training platform. A single proposal will be awarded in an amount not to exceed $150,000 for a maximum period of 12 months beginning early next year. The deadline for submission is September 27.
Requirements: Individuals from academic, medical, or research institutions based in the United States are eligible to apply. Institutions may be public or private, and must not be commercial.
Restrictions: Funds are to be used to cover research expenses only and cannot be used for equipment/supplies, travel, conference or publication costs.
Geographic Focus: All States
Date(s) Application is Due: Sep 27
Amount of Grant: Up to 150,000 USD
Contact: Faraz Farzin, Research Scientist; (570) 498-9018; faraz@lumoslabs.com
Internet: http://hcp.lumosity.com/research/from_the_lab/lumosity-announces-human-cognition-grant-for-fall-2013
Sponsor: Lumosity
153 Kearny Street
San Francisco, CA 94108

Lumpkin Family Foundation Healthy People Grants 2391

Historically, the Lumpkin Family Foundation has supported entities that provide services that help keep people in its geographic area healthy. Requests will be evaluated on how well they accomplish one or more of the following: support the creativity of nonprofit organizations by seeding new projects and encouraging experimentation and innovation; support organizations demonstrating outstanding leadership in their field or community; promote the effectiveness of organizations and the nonprofit sector by supporting planning, learning and the professional development of staff and board leaders; facilitate collaboration across traditional organization or sector boundaries for community benefit; and develop public understanding of issues and promote philanthropic support necessary to address issues of community importance.
Requirements: Grants are awarded to nonprofit organizations that serve the community without discrimination on the basis of race, sex, or religion. Special consideration will be given to organizations and programs in East Central Illinois.
Restrictions: Proposals will not be considered from organizations that are not 501(c)3 tax-exempt; organizations whose primary purpose is to influence legislation; political causes, candidates, organizations or campaigns; individuals; or religious organizations, unless the particular program will benefit a large portion of the community and does not duplicate the work of other agencies in the community.
Geographic Focus: All States
Amount of Grant: 1,000 - 50,000 USD
Contact: Bruce Karmazin, (217) 234-5915 or (217) 235-3361; fax (217) 258-8444; Bruce@lumpkinfoundation.org
Internet: http://www.lumpkinfoundation.org/bWHATbwefund/HealthyPeople.aspx
Sponsor: Lumpkin Family Foundation
121 South 17th Street
Mattoon, IL 61938

Lustgarten Foundation for Pancreatic Cancer Research Grants 2392

The foundation aims to make rapid advances in the battle against pancreatic cancer through a combination of cutting-edge technology and novel ideas. Grants can be submitted in one of three areas: novel technologies for pancreatic cancer genetics; screening for the early detection of pancreatic cancer; and novel therapies in pancreatic cancer. Proposals that exhibit leveraging of institutional resources and the attainment of major new programs are encouraged. This initiative encourages, but does not require, the development of integrative and collaborative teams of investigators from within a single institution or among several institutions. Letters of intent must be received on or before 5:00 p.m. Eastern Time on July 31, with full applications due no later than 5:00 p.m. Eastern Time on August 18. Grants will be awarded for a one-year period for a maximum amount of $100,000, of which no more than 10% can be used for indirect costs.
Requirements: Applications will be accepted from individual investigators as well as collaborating investigators. Applicants must have an MD or a PhD degree to become principal investigators. Eligible organizations include nonprofit organizations; public or private institutions, such as universities, colleges, hospitals, and laboratories; and domestic or foreign institutions/organizations.
Geographic Focus: All States
Amount of Grant: Up to 100,000 USD
Samples: Johns Hopkins University, Allison Klein, Ph.D. - Identification of Pancreatic Cancer Susceptibility Genes; University of Michigan Medical School, Chandan Kumar, Ph.D. - Discovery of Recurrent Gene Fusions in Pancreatic Cancer using High-Throughput Sequencing; Johns Hopkins University School of Medicine, Joshua Mendell, M.D., Ph.D. - Ras-regulated microRNAs in Pancreatic Cancer.
Contact: Kerri Kaplan, Executive Director; (516) 803-2304 or (866) 789-1000; fax (516) 803-2303; kkaplan@cablevision.com or lsasso@cablevision.com
Internet: http://www.lustgarten.org/Page.aspx?pid=666
Sponsor: Lustgarten Foundation for Pancreatic Cancer Research
1111 Stewart Avenue
Bethpage, NY 11714

Lydia deForest Charitable Trust Grants 2393

The Lydia Collins deForest Charitable Trust was established in 2002. The deForest Charitable Trust specifically supports: organizations that provide services to those who are visually limited; churches and organizations affiliated with the Protestant Episcopal Church in the United States and other religious organizations in union with or recognized by the Episcopal Church; and organizations that provide services to those who are homeless, unemployed, or substance-dependent. Special consideration is given to the following 3 organizations: The Lighthouse, Inc., in New York, New York; the Calvary Episcopal Church of Summit, New Jersey; and the Salvation Army in Union, New Jersey. Grant requests for general operating support are strongly encouraged. Program support will also be considered. Small, program-related capital expenses may be included in general operating or program requests. To better support the capacity of nonprofit organizations, multi-year funding requests are encouraged. Applicants must apply online at the grant website. Applicants are strongly encouraged to do the following before applying: review the downloadable state application procedures for additional helpful information and clarifications; review the downloadable online-application guidelines at the grant website; review the trust's funding history (link is available from the grant website); review the online application questions in advance; and review the list of required attachments. These will generally include: a list of board members, financial statements (audited, reviewed, or compiled by independent auditor); an organization summary; a list of other funding sources; an IRS Determination letter; and other required documents. All attachments must be uploaded in the online application as PDF, Word, or Excel files. The application deadline for the deForest Charitable Trust is August 31. Applicants will be notified of grant decisions before February 28 of the following year.
Requirements: Applicants must have 501(c)3 tax-exempt status and serve the people of New Jersey. Occasional support is given to organizations within the Metro New York City area.
Restrictions: The Trust does not support requests from individuals, organizations attempting to influence policy through direct lobbying, or any political campaigns.
Geographic Focus: New Jersey
Date(s) Application is Due: Aug 31
Contact: Anne Bridgette Hennessy, Senior Philanthropic Relationship Manager; (646) 855-2270; anne.hennessy@ustrust.com
Internet: https://www.bankofamerica.com/philanthropic/fn_search.action
Sponsor: Lydia Collins deForest Charitable Trust
114 West 47th Street, NY8-114-10-02
New York, NY 10036

Lymphatic Education and Research Network Additional Support Grants for NIH-funded F32 Postdoctoral Fellows 2394

The Lymphatic Education and Research Network (LE&RN) is a 501(c)3 not for profit organization whose mission is to fight lymphatic disease and lymphedema through education, research and advocacy. LE&RN has retained The Medical Foundation, a division of Health Resources in Action (The Medical Foundation division), to manage the administrative aspects of the LE&RN Awards Program. LE&RN seeks to accelerate the prevention, treatment and cure of the disease while bringing patients and medical professionals together to address the unmet needs surrounding lymphatic disorders. The goal of the Additional Support Awards Program is to help foster career interest in the field of lymphatic research by offering additional funds for the F32 postdoctoral research projects. The additional support may be used for research supplies, equipment, health insurance and travel to scientific meetings.
Requirements: Postdoctoral fellows who are supported by the NIHF32 Postdoctoral Fellowship Program, have at least one year remaining of F32 support as of July 1, and

conduct research that clearly advances the mission of LE&RN are eligible to apply. Award recipients with an active F32 Fellowship will receive $10,000 annually for up to two years. The additional support award may be used for salary support of personnel who are assisting in the research project, as well as research supplies, equipment, health insurance and travel to scientific meetings. A budget proposal is not required for the application. Online submissions must be received by the Medical Foundation by January 10.
Geographic Focus: All States
Date(s) Application is Due: Jan 10
Amount of Grant: 20,000 USD
Contact: Erin Johnstone; (617) 279-2240, ext. 710; ejohnstone@hria.org
Internet: http://www.hria.org/tmfservices/tmfgrants/lern32.html
Sponsor: Lymphatic Education and Research Network
261 Madison Avenue
New York, NY 10016

Lymphatic Education and Research Network Postdoctoral Fellowships　2395
The Lymphatic Education and Research Network (LE&RN) is a 501(c) not-for profit organization whose mission is to fight lymphatic disease and lymphedema through education, research and advocacy. LE&RN seeks to accelerate the prevention, treatment and cure of the disease while bringing patients and medical professionals together to address the unmet needs surrounding lymphatic disorders. The LE&RN Postdoctoral Fellowship program is designed to support postdoctoral scientists in academic, medical or research institutions throughout the world. The awards will support investigators who have recently received their doctorates, a critical point in career development when young scientists choose their lifelong research focus. The goal of the Fellowship Program is to expand and strengthen the pool of outstanding junior investigators in the field of lymphatic research. Two-year fellowships ranging from $87,396 to $98,304 will be awarded to fellows who have completed no more than three years of postdoctoral training by July 1 of the LE&RN funding cycle.
Requirements: Each applicant must be working under the supervision of an established investigator who is the designated Mentor. Examples of funding areas include but are not limited to basic, translational or clinical investigations into normal and aberrant lymphatic biology; mechanisms of normal and pathological lymphangiogenesis; imaging and understanding the mechanisms that regulate lymphatic structure and function; animal models of lymphatic system disorders; and the cross-talk between the lymphatic and the immune systems. Applications will be accepted worldwide from postdoctoral fellows who will meet the following eligibility requirements by July 1: received a Ph.D. or equivalent and/or an M.D. or equivalent from an accredited institution; commit at least 90% time for research or, at least 70% time for research if the applicant also has clinical responsibilities; have completed no more than three years of full-time postdoctoral research experience; be a postdoctoral fellow under the supervision of a faculty member; and, conduct the proposed research project at a nonprofit hospital, university or other research institution (there are not institutional limitations on the number of applicants who may submit). Eligibility is not limited to those investigators with past experience in the field. Completed applications must be received by The Medical Foundation by January 10, 1:00 pm, U.S. Eastern time.
Restrictions: Projects in lymphocyte biology, leukemia, lymphoma and conditions secondary to lymphedema such as cutaneous infections are outside the scope of this program.
Geographic Focus: All States, All Countries
Date(s) Application is Due: Jan 10
Amount of Grant: 87,396 - 98,304 USD
Contact: Erin Johnstone; (617) 279-2240 ext. 710; ejohnstone@hria.org
Internet: http://lymphaticnetwork.org/treating-lymphedema/lern-postdoctoral-fellowship-awards/
Sponsor: Lymphatic Education and Research Network
261 Madison Avenue
New York, NY 10016

Lynde and Harry Bradley Foundation Fellowships　2396
The Bradley Fellowship program is a major part of the Foundation's program interest in strengthening America's Intellectual infrastructure at the higher education level. Begun in 1986, the program is a model for influencing the intellectual framework of national life. It helps fulfill the two basic prerequisites for liberal education -- good teaching and good curriculum -- by enabling colleges and universities to offer students in the advanced stages of graduate work the opportunity to pursue serious studies with excellent teachers. For the program, the Foundation chooses institutions of higher education each year to support graduate and post-graduate students, who are designated as Bradley Fellows. The Fellows are chosen by each institution upon the recommendation of one or more distinguished professors in the humanities, social sciences, and the law.
Geographic Focus: All States
Contact: Daniel P. Schmidt; (414) 291-9915; fax (414) 291-9991
Internet: http://www.bradleyfdn.org/bradley_fellows.asp
Sponsor: Lynde and Harry Bradley Foundation
1241 North Franklin Place
Milwaukee, WI 53202-2901

Lynde and Harry Bradley Foundation Grants　2397
The foundation aims to encourage projects that focus on cultivating a renewed, healthier, and more vigorous sense of citizenship among the American people and among peoples of other nations. Projects likely to be supported will generally treat free people as self-governing, personally responsible citizens, not as victims or clients; aim to restore the intellectual and cultural legitimacy of common sense, the wisdom of experience, everyday morality, and personal character; seek to reinvigorate and re-empower the traditional, local institutions--families, schools, churches, and neighborhoods--that provide training in and room for the exercise of genuine citizenship, that pass on everyday morality to the next generation, and that cultivate personal character; and encourage decentralization of power and accountability away from centralized, bureaucratic, national institutions back to states, localities, and revitalized mediating structures. Eligible projects may address any arena of public life--economics, politics, culture, or civil society; the problem of citizenship at home or abroad; Milwaukee and Wisconsin community and state projects that aim to improve the life of the community through increasing cultural and educational opportunities, grassroots economic development, and social and health services; the resuscitation of citizenship in the economic, political, cultural, or social realms; policy research and writing about approaches encouraging that resuscitation; academic research and writing that explore the intellectual roots of citizenship; and popular writing and media projects that illustrate for a broader public audience the themes of citizenship. The foundation supports programs that research the needs of gifted children and techniques of providing education for students with superior skills and/or intelligence; research programs investigating how learning occurs in gifted children; and demonstration programs of instruction.
Requirements: As an initial step, tax-exempt and nonprofit organizations should prepare a brief letter of inquiry presenting a concise description of their project, its objectives and significance, and the qualifications of the organizations and individuals involved. If the project appears to fall within the foundation's mandate, the applicant will be invited to submit a formal proposal. If invited to submit a formal proposal, the applicant should submit another letter. It should include a more-thorough, yet still concise description of the project, its objectives and significance, and the qualifications of the groups and individuals involved in it. It should also include a project budget, the specific amount being sought from Bradley, and a list of its other sources of support, philanthropic or otherwise.
Restrictions: The foundation favors projects that are normally not financed by public funds and will consider requests from religious organizations that are not denominational in character. Grants without significant importance to the foundation's areas of interest will only under special conditions be considered for endowment or deficit financing proposals. Grants will not be made to individuals, for overhead costs, or for fund-raising counsel.
Geographic Focus: All States
Date(s) Application is Due: Feb 1; May 1; Aug 1; Nov 1
Amount of Grant: 25,000 - 300,000 USD
Samples: Bel Canto Chorus of Milwaukee, Milwaukee, Wisconsin, $25,000 - to support general operations; Penfield Children's Center, Milwaukee, Wisconsin, $25,000 - to support general operations; Rawhide, Inc., New London, Wisconsin, $20,000 - to support the Residential Care Assistance Program.
Contact: Daniel P. Schmidt; (414) 291-9915; fax (414) 291-9991
Internet: http://www.bradleyfdn.org/program_interests.asp
Sponsor: Lynde and Harry Bradley Foundation
1241 North Franklin Place
Milwaukee, WI 53202-2901

Lynde and Harry Bradley Foundation Prizes: Bradley Prizes　2398
The Bradley Prizes formally recognize individuals of extraordinary talent and dedication who have made contributions of excellence in areas consistent with The Lynde and Harry Bradley Foundation's mission. Up to four Prizes of $250,000 each are awarded annually to innovative thinkers and practitioners whose achievements strengthen the legacy of the Bradley brothers and the ideas to which they were committed. Each year, Bradley Prize nominations are solicited from a national panel of more than 100 prominent individuals involved in academia, public-policy research, journalism, civic affairs, and the arts. All nominees are carefully evaluated by a distinguished selection committee that makes recommendations to the Foundation's Board of Directors, which selects them. The Prize winners are then honored at a celebratory awards ceremony.
Geographic Focus: All States
Amount of Grant: 250,000 USD
Contact: Daniel P. Schmidt; (414) 291-9915; fax (414) 291-9991
Internet: http://www.bradleyfdn.org/bradley_prizes.asp
Sponsor: Lynde and Harry Bradley Foundation
1241 North Franklin Place
Milwaukee, WI 53202-2901

Lynn and Rovena Alexander Family Foundation Grants　2399
The Lynn and Rovena Alexander Family Foundation was established in Lubbock, Texas, in 2003, and primarily serves the residents of Lexington, Kentucky, and San Angelo, Texas. The Foundation's major fields of interest include: elementary education, secondary education, higher education, human services, and Christian agencies and churches. Along with the formal application, interested parties should submit: a copy of their IRS Determination Letter; a brief history of organization and description of its mission; a copy of most recent annual report or audited financial statement; a detailed description of project and amount of funding requested; and a copy of the current year's organizational budget and project budget. Most recent funding has ranged from $2,500 to $125,000. There are two annual deadlines for applicants, including February 1 and September 1.
Requirements: Organizations located in, or serving the residents of, Lexington, Kentucky, and San Angelo, Texas, are eligible to apply.
Geographic Focus: Kentucky, Texas
Date(s) Application is Due: Feb 1; Sep 1
Amount of Grant: 2,500 - 125,000 USD
Contact: Kimberly L. King, Director; (325) 374-0050 or (806) 795-0470
Sponsor: Lynn and Rovena Alexander Family Foundation
P.O. Box 1160
Nicholasville, KY 40340-1160

M-A-C AIDS Fund Grants 2400
The fund makes grants to registered charitable, nonprofit organizations that provide basic needs, such as food, clothing, housing, or shelter (short-term or transitional); direct services related to health care; social services, transportation (for medical visits, outpatient visits, and other social services); health-related recreational activities; and programs that bring HIV/AIDS education, awareness, and prevention to public attention. Proposals are reviewed quarterly and must arrive at least one month prior to the quarterly review date. Application guidelines and required format are available online.
Requirements: Grants are awarded to tax exempt, non-profit organizations that are 501(c)3 and directly associated with HIV/AIDS. A minimum of 80% of the total organization's budget must be is directed to program.
Restrictions: Grants will not fund: individuals; lobbying activities; ongoing general operating expenses or existing deficits; endowments, unless they provide a direct service to PWA's; capital costs; conferences, summits, briefings, PSA's; or research. With the exception of North American based charities, the Fund does not accept unsolicited international grant proposals.
Geographic Focus: All States
Amount of Grant: 5,000 - 25,000 USD
Contact: Coordinator; (212) 965-6300; fax (212) 372-6171; macaidsf@maccosmetics.com
Internet: http://www.macaidsfund.org/support/givingguidelines.html
Sponsor: M-A-C AIDS Fund
130 Prince Street, 2nd Floor
New York, NY 10012

M.B. and Edna Zale Foundation Grants 2401
The M.B. & Edna Zale Foundation honors the tradition of its founders through grants that stimulate change. To accomplish this mission, the Foundation acts as a catalyst for collaboration and makes grants in communities where the Directors live or have an interest. Grants are made primarily in the communities of Dallas (Dallas County) and Houston (Harris County), Texas; Boca Raton, Florida; Portland, Oregon; and New York, including Long Island. The Foundation has an interest in four areas of funding: community services; health; education; Jewish heritage.
Requirements: 501(c)3 nonprofits organizations in the communities of Dallas (Dallas County) and Houston (Harris County), Texas; Boca Raton, Florida; Portland, Oregon; and New York, including Long Island are eligible. Contact the Foundation for further application information and guidelines.
Restrictions: The Foundation does not ordinarily provide: major support for the arts; grants to individuals; scholarships and fellowships to individuals (except through colleges and universities).
Geographic Focus: Florida, New York, Oregon, Texas
Amount of Grant: 2,000 - 100,000 USD
Contact: Leonard Krasnow; (214) 855-0627; fax (972) 726-7252; mail@zalefoundation.org
Sponsor: M.B. and Edna Zale Foundation
6360 LBJ Highway, Suite 205
Dallas, TX 75240

M. Bastian Family Foundation Grants 2402
The foundation supports nonprofit organizations in its areas of interest, including music, the arts, higher education, health care and health organizations, religion (Christian and Latter-day Saints), social services, and wildlife conservation. Types of support include general operating support and scholarship funds. Contact the office for application forms. There are no application deadlines.
Requirements: Funding focus is primarily in Utah.
Restrictions: Grants are not made to individuals.
Geographic Focus: All States
Amount of Grant: 2,000 - 200,000 USD
Samples: Brigham Young U (Provo, UT)--for scholarship endowment, $200,000; Pediatric AIDS Foundation (Los Angeles, CA)--for general support, $20,000; American Diabetes Assoc (Alexandra, VA)--for research, $100,000.
Contact: McKay Matthews, Program Contact; (801) 225-2455
Sponsor: M. Bastian Family Foundation
51 W Center Street, Suite 305
Orem, UT 84057

M.D. Anderson Foundation Grants 2403
The Anderson Foundation funds projects for the improvement of working class conditions among workers, and for the establishment, support and maintenance of hospitals, homes and institutions for the care of the young, sick, the aged, and the helpless. The Foundation also gives for the improvement of general living conditions and for the promotion of health, science, education, and the advancement of knowledge. Funding is given for aging center and services; education; employment; government/public administration; health care; human services; medical specialties public policy and research; and youth services. Types of support include building and/or renovation, equipment, matching/challenge support, research, and seed money.
Requirements: Organizations should submit a letter of inquiry with the following information: a copy of their IRS determination letter; a detailed description of their project and amount of funding requested; a copy of the current year's organizational budget and/or project budget; and a listing of additional sources and amount of support. Applicants should submit five copies of the proposal. There are no deadlines. The Board meets once a month, with organizations contacted within four weeks.
Restrictions: Funding is given primarily in Texas, with emphasis on the Houston area. Grants are not available to individuals, operating funds, or endowments.
Geographic Focus: Texas
Amount of Grant: 25,000 - 200,000 USD
Samples: Methodist Hospital Foundation, San Antonio, Texas, molecular imaging initiative at the hospital research institute, $200,000 payable over one year; YMCA capital campaign, Houston, Texas, $100,000 payable over one year; Scott and White Memorial Hospital, Temple, Texas, toward support of indigent care program through the development and implementation of the Family Medicine Community Clinic, $25,00 payable over one year.
Contact: Karen Jenkins, Grant Contact; (713) 216-1095
Sponsor: M.D. Anderson Foundation
P.O. Box 2558
Houston, TX 77252-8037

M.E. Raker Foundation Grants 2404
The foundation was established in 1984, serving the Allen County, Indiana area. Areas of interest include: children/youth, services; education; environment; natural resources; health care; historic preservation/historical societies; human services people with disabilities. Support is offered in the following areas: building/renovation; general/operating support; matching/challenge support; Program development. No support for the arts or to individuals.
Requirements: Indiana nonprofit organizations are eligible. Emphasis is given to requests from Fort Wayne. Send a letter requesting an application to the foundation office. Application form required for grant proposals.
Geographic Focus: Indiana
Amount of Grant: 5,000 - 40,000 USD
Contact: Jennifer Pickard, Grants Coordinator; (260) 436-2182
Sponsor: M.E. Raker Foundation
6207 Constitution Drive
Fort Wayne, IN 46804-1517

M.J. Murdock Charitable Trust General Grants 2405
The M.J. Murdock Charitable Trust General Grants provides support for education, arts and culture, and health and human services. The Trust considers educational projects in formal and informal settings, emphasizing enhancement or expansion, as well as new educational approaches. Of special interest to the Trust are the following: performance and visual arts projects which enrich the cultural environment of the region and educational outreach efforts; and programs that emphasize preventative efforts which address physical, spiritual, social, and psychological needs, with a focus on youth. Grants are awarded for capital projects, program initiation, expansion, or for increased organizational capacity. Organizations are encouraged to view the grants awarded link for examples of previously funded grants.
Requirements: Before proceeding, interested parties should review the General Grant Application Guidelines to see if their organization is eligible or their project is appropriate for application to the Trust. After determining eligibility and appropriateness, organizations must submit a letter of inquiry before proceeding with the application. Upon approval, the organization will use the Trust's General Grant Application form and procedures.
Restrictions: The Trust only funds programs in the Pacific Northwest region: Alaska, Idaho, Montana, Oregon, and Washington. The following requests for funding are not considered: specific individuals and/or their personal benefit; individuals unauthorized to act on behalf of a qualified tax-exempt organization; funds that will ultimately be passed through to other organizations; propagandizing or for influencing legislation and elections; institutions that in policy or practice unfairly discriminate against race, ethnic, origin, sex, creed, or religion; sectarian or religious organizations whose principal activity is for the primary benefit of their own members; or for long-term loans, debt retirement, or operational deficits. The following funding requests are rarely considered: normal ongoing operations or the continuation of existing projects; endowments or revolving funds that act as such; continuation of programs previously financed from other external sources; urgent needs, emergency, or gap funding; organizations organized or operating outside any state or territory of the United States.
Geographic Focus: Alaska, Idaho, Montana, Oregon, Washington
Samples: Blachet House of Hospitality, Portland, Oregon, $450,000, to serve Portland homeless; Billings District Council of the Society of St. Vincent de Paul, Billings, Montana, to expand client services, $100,000; Whidbey Institute, Clinton, Washington, for program expansion to enhance civic engagement, $150,000.
Contact: Marybeth Stewart Goon; (360) 694-8415; fax (360) 694-1819
Internet: http://www.murdock-trust.org/grants/general-grants.php
Sponsor: M.J. Murdock Charitable Trust
703 Broadway, Suite 710
Vancouver, WA 98660

Mabel A. Horne Trust Grants 2406
The Mabel A. Horne Trust was established in 1957 to support and promote quality educational, human-services, and health-care programming for underserved populations. Grant requests for general operating support are strongly encouraged. Program support will also be considered. Small, program-related capital expenses may be included in general operating or program requests. The majority of grants from the Horne Trust are one year in duration; on occasion, multi-year support is awarded. Applicants must apply online at the grant website. Applicants are strongly encouraged to do the following before applying: review the downloadable state application procedures for additional helpful information and clarifications; review the downloadable online-application guidelines at the grant website; review the trust's funding history (link is available from the grant website); review the online application questions in advance; and review the list of required attachments. These will generally include: a list of board members, financial statements (audited, reviewed, or compiled by independent auditor); an organization

summary; a list of other funding sources; an IRS Determination letter; and other required documents. All attachments must be uploaded in the online application as PDF, Word, or Excel files. The application deadline for the Mabel A. Horne Trust is 11:59 p.m. on February 1. Applicants will be notified of grant decisions before May 31.
Requirements: Applicants must have 501(c)3 tax-exempt status.
Restrictions: The trust does not support requests from individuals, organizations attempting to influence policy through direct lobbying, or any political campaigns.
Geographic Focus: Massachusetts
Date(s) Application is Due: Dec 1
Contact: Relationship Manager; (866) 778-6859; ma.grantmaking@ustrust.com
Internet: https://www.bankofamerica.com/philanthropic/fn_search.action
Sponsor: Mabel A. Horne Trust
225 Franklin Street, 4th Floor, MA1-225-04-02
Boston, MA 02110

Mabel F. Hoffman Charitable Trust Grants 2407
The Mabel F. Hoffman Charitable Trust was established in 1969 to support and promote quality educational, human-services, and health-care programming for underserved populations. Special consideration is given to charitable organizations that serve the people of Hartford, Connecticut. The trust makes approximately twelve, modest size grants per year. Grants from the Hoffman Charitable Trust are one year in duration. Applicants must apply online at the grant website. Applicants are strongly encouraged to do the following before applying: review the downloadable state application procedures for additional helpful information and clarifications; review the downloadable online-application guidelines at the grant website; review the trust's funding history (link is available from the grant website); review the online application questions in advance; and review the list of required attachments. These will generally include: a list of board members, financial statements (audited, reviewed, or compiled by independent auditor); an organization summary; a list of other funding sources; an IRS Determination letter; and other required documents. All attachments must be uploaded in the online application as PDF, Word, or Excel files. The deadline for application to the Mabel F. Hoffman Charitable Trust is 11:59 p.m. on June 15. Applicants will be notified of grant decisions by letter within two to three months after the proposal deadline.
Requirements: First time applicants should contact the Program Officer before applying (see Contact Info below). Applicants must have 501(c)3 tax-exempt status.
Restrictions: Grant requests for capital projects will not be considered. Applicants will not be awarded a grant for more than three consecutive years. The trust does not support requests from individuals, organizations attempting to influence policy through direct lobbying, or any political campaigns.
Geographic Focus: Connecticut
Date(s) Application is Due: Jun 15
Contact: Kate Kerchaert; (860) 244-4871; kate.kerchaert@ustrust.com
Internet: https://www.bankofamerica.com/philanthropic/fn_search.action
Sponsor: Mabel F. Hoffman Charitable Trust
99 Founders Plaza, CT2-547-05-19, 5th Floor
East Hartford, CT 06108

Mabel H. Flory Charitable Trust Grants 2408
The Mabel H. Flory Charitable Trust was established in Virginia in 1979, and it currently funds organizations in Washington, D.C. metro area and Baltimore, Maryland. Support is available for research designed to develop new cures and methods of treatment for: children with hearing and sight problems; arthritis in both children and adults; and aphasia related problems. Most recent grant awards have ranged from $4,000 to $10,000.
Requirements: Your initial approach should be in the form of a letter, prior to the deadline date.
Restrictions: No grants are given to individuals.
Geographic Focus: District of Columbia, Maryland
Date(s) Application is Due: Aug 31
Amount of Grant: 4,000 - 10,000 USD
Sample: Georgetown University, Washington, D.C., $10,000 - research purposes; Arthritis Foundation, Washington, D.C., $10,000 - research purposes; Foundation Fighting Blindness, Columbia, Maryland, $10,000 - research purposes.
Contact: BB&T Trustee; (703) 531-2053 or (888) 575-4586
Sponsor: Mabel H. Flory Charitable Trust
P.O. Box 2907
Wilson, NC 27894-2907

Mabel Y. Hughes Charitable Trust Grants 2409
The Mabel Y. Hughes Charitable Trust was established under the last will and testament of Mabel Y. Hughes, a resident of Denver, Colorado, who died on April 9, 1969. Her legacy of giving continues through grants to charitable organizations in the state of Colorado. The Trust awards funding to nonprofit organizations in its areas of interest, including: children and youth services; education; family services; health care; higher education; human services; children's and art museums; performing arts centers; performing arts, opera; and reproductive health and family planning. Types of support include: annual campaigns; continuing support; emergency funds; endowments; equipment; general/operating support; program development; research; and seed money. Approximately fifty awards are given each year, with the average grant size ranging from $5,000 to $25,000. Applications must be submitted by March 1, July 1 or November 1 to be reviewed at the grant meeting that occurs after each deadline.
Requirements: Organizations should submit a letter of inquiry to the Trust, and if their project is appropriate for funding, they will be asked to submit a proposal.
Restrictions: The Trust does not support funding for individuals, deficit financing, scholarships, fellowships, or loans.
Geographic Focus: Colorado
Date(s) Application is Due: Mar 1; Jul 1; Nov 1
Amount of Grant: 5,000 - 30,000 USD
Contact: Peggy Toal, Private Client Services; (720) 947-6725 or (888) 234-1999; fax (877) 746-5889; grantadministration@wellsfargo.com
Sponsor: Mabel Y. Hughes Charitable Trust
1740 Broadway
Denver, CO 80274

Macquarie Bank Foundation Grants 2410
The Foundation is one of Australia's oldest and largest corporate foundations contributing more than $150 million to more than 1500 community organizations world-wide since 1985. The Foundation focuses its resources in five core areas - the arts, education, environment, health, and welfare. The Foundation is also committed to projects specifically aimed at supporting indigenous communities. The Foundation's funding criteria is flexible and open. It welcomes applications from a diverse range of community organizations that are working in innovative ways to provide long-term benefits. Funding levels are flexible and are dictated by the needs of the organization and funding availability. Each application is assessed on its individual merit, with priority given to programs which support a broad section of the community at a regional, state or national level; have the involvement or potential for involvement of Macquarie Bank staff through volunteering, fundraising, pro bono work and board and/or management committee involvement; are located in cities/countries where Macquarie Bank staff are located; and deliver long-term benefits and build community sustainability. Prospective applicants are encouraged to check the Foundation website or contact Foundation Staff for more information on how to apply.
Geographic Focus: All States, All Countries
Amount of Grant: 100 - 500,000 USD
Samples: Po Leung Kuk, Hong Kong, $HK190,000 - a charitable organization supporting disadvantaged children; Oxfam Trailwalker, Melbourne, Australia, $A325,000 (2009-2010); Princess Margaret Hospital Foundation, Toronto, Ontario, Canada, $C350,000 - Ride to Conquer Cancer (2009-2010).
Contact: Heather Matwejev, Macquarie Bank Foundation, Asia; +61 2 8232 6951; fax +61 2 8232 0019; heather.matwejev@macquarie.com or foundation@macquarie.com
Internet: http://www.macquarie.com/mgl/com/foundation/about/application-guidelines
Sponsor: Macquarie Bank Foundation
G.P.O. Box 4294
Sydney, NSW 1164 Australia

Maddie's Fund Medical Equipment Grants 2411
Maddie's Fund offers grants to adoption guarantee shelters for the purchase of new medical equipment. Any adoption guarantee shelter is eligible to apply as long as it is located in the U.S. and employs at least one full-time veterinarian who spends at least 50% of his/her time caring for the animals in their shelter. An adoption guarantee shelter saves all the healthy and treatable animals under their care, with euthanasia reserved only for unhealthy and untreatable animals. An adoption guarantee organization can be an animal shelter, rescue group, foster care organization, or sanctuary. An animal organization does not have to say it is an adoption guarantee organization to qualify for this funding opportunity, but it does have to: (1) save all healthy and treatable animals under its care and make public its commitment to doing so; (2) clearly articulate to its community that it is saving all healthy and treatable animals under its care; (3) use the definitions of healthy and treatable as described in the Asilomar Accords; and (4) agree to publish, at least annually, in the organization's primary publications and on its website the organization's shelter statistics using the Maddie's Fund Animal Statistics Table. Samples of previously funded organizations are located on the website.
Requirements: Proposals should include the description and cost of the medical equipment to be purchased; the number of animals to be cared for as a result of the equipment; a general overview of the shelter's current medical program(s); proof of the organization's tax exempt status; evidence that the shelter meets the full-time veterinarian requirement; and annual shelter statistics from the most recent calendar year using the Animal Statistics Table. Grant size is determined by the impact the grant will have on saving the healthy and treatable shelter dogs and cats in the community. Proposals may be mailed or emailed. There are no deadlines and organizations may apply any time.
Restrictions: Applicants may only submit one medical equipment grant request every three years.
Geographic Focus: All States
Contact: Joey Bloomfield; (510) 337-8988; fax (510) 337-8989; grants@maddiefund.org
Internet: http://www.maddiesfund.org/Grant_Giving/Medical_Equipment_Grants.html
Sponsor: Maddie's Fund
2223 Santa Clara Avenue, Suite B
Alameda, CA 94501-4416

Madison County Community Foundation - City of Anderson Quality of Life Grant 2412
The Madison County Community Foundation - Quality of Life Grant is used as economic development quality of life assistance to non-profits. Generated through the City of Anderson's food and beverage tax, the grant is dispersed to those non-profits who have no voting representation on the committee; are not solely supporting non-secular activities with these funds; carry proof of 501(c)3 status; and are requesting funding for projects with outcomes rather than operational expenses. The grants must be used to increase quality of life for all rather than limited access.
Requirements: Application forms and guidelines are available from the Foundation office or may be downloaded from the Foundation website at www.madisonccf.org. The application will also be available at the City of Anderson website at www.cityofanderson.com.

Restrictions: Funding is limited to the city of Anderson, Indiana. Funding is not available for: annual fund campaigns; individuals; capital debt reduction; sectarian religious purposes; gifts to endowments; political campaigns; medical, scientific, or health research; student loans, scholarship/fellowship programs, or travel grants; programs and/or equipment which were committed prior to the grant application being submitted; organizations without responsible fiscal agents and adequate accounting procedures; schools and government agencies; normal operational expenses of the organization. Funding is limited to grants for one year.
Geographic Focus: Indiana
Date(s) Application is Due: Sep 15
Amount of Grant: 1,000 - 10,000 USD
Contact: Tammy Bowman; (765) 644-0002; fax (765) 662-1438; tbowman@madisoncf.org
Internet: http://www.madisoncf.org/index.php?submenu=grantNO&src=gendocs&ref=GrantProcess&category=Non_Profits
Sponsor: Madison County Community Foundation
33 West 10th Street, Suite 600
Anderson, IN 46015-1056

Madison County Community Foundation General Grants 2413
The Madison County Community Foundation General Grants are made to support projects and programs of non-profit agencies located in or serving residents of Madison County. Grants are typically made in the spring and fall and range from $500 to $10,000 with rare exceptions. Proposals are reviewed by a Grants Committee and those chosen are then approved by the Foundation Board. The priorities of the foundation are arts and culture, education, economic development, civic affairs, and health and human services.
Requirements: Organizations should complete the online application and include the following detailed information: their project and cost; organization information; a project narrative that includes the objective, financial need, justification; constituency; evaluation plan and community impact; additional and/or future funding; professional references and collaborative value. They should also include a list of project expenses and project income. Eight copies of the complete application package should be sent to the Program Director, along with one copy each of the board of directors, IRS tax exempt letter, and financial statement.
Geographic Focus: Indiana
Amount of Grant: 500 - 10,000 USD
Contact: Tammy Bowman; (765) 644-0002; fax (765) 662-1438; tbowman@madisoncf.org
Internet: http://www.madisoncf.org/index.php?submenu=Grants&src=gendocs&ref=Grants&category=Non_Profits
Sponsor: Madison County Community Foundation
33 West 10th Street, Suite 600
Anderson, IN 46015-1056

Maggie Welby Foundation Grants 2414
Inspired by the life and memory of Maggie Welby, the Maggie Welby Foundation seeks to aid students (grades K-12) and families in financial need, in order to help fulfill the dreams and hopes that Maggie realized in every person she touched during her life. The Foundation offers grants for children and families that have a financial need for a particular purpose. Grants may extend to children and families in need of help with bills, athletic opportunities, medical needs, or an opportunity that a child would not otherwise have. All grants are awarded to the family, but are paid directly to the specific purpose for which the grant was applied. Organizations that directly benefit children can also receive grants on an annual need. Foundation awards grants twice per year in the months of July and December. The annual deadlines are June 30 and November 30.
Requirements: Applicants must utilize the on-line grant application process.
Restrictions: The Maggie Welby Foundation does not accept applications for Ipads.
Geographic Focus: All States
Date(s) Application is Due: Nov 30
Contact: Jamie Welby; (314) 330-6947; jamie.welby@maggiewelby.org
Internet: http://maggiewelby.org/Grants.html
Sponsor: Maggie Welby Foundation
7 Needle Court
Dardenne Prairie, MO 63368

Maine Community Foundation Charity Grants 2415
The Maine Community Foundation Charity Grants offer funding for the following priorities: start-up money for an organization or project; projects that involve the disabled or economically disadvantaged, as long as the project is not supported by a national campaign or public money; libraries; symphonies; hospice care; projects that are related to the Friendship, Maine area; discrete projects as opposed to general operating support; and small requests from social service organizations for buildings or purchase of necessary equipment. Application is made online from the Foundation's website.
Requirements: Applicants must be 501(c)3 organizations eligible to accept tax-deductible donations as outlined in Section 170(c) of the Internal Revenue Code.
Restrictions: Religious groups are eligible but funding will not be provided for religious purposes. Funding is not provided for the following: political campaigns, or to support attempts to influence legislation of any governmental body other than through making available the results of non-partisan analysis, study and research; ongoing operating support; endowments or capital campaigns; camperships; or capital equipment over $250.
Geographic Focus: Maine
Date(s) Application is Due: Sep 15
Amount of Grant: Up to 5,000 USD
Samples: Big Brothers Big Sisters of Bath/Brunswick, Brunswick, Maine, $1,500, copier and supplies; Community Health and Counseling Services, Bangor, Maine, $4,750, complete a business plan and feasibility study for the development of a hospice house in the greater Bangor area; and Island Community Center, Stonington, Maine $2,500, start-up costs associated with opening The Community Cafe in Stonington.
Contact: Cathy Melio; (877) 700-6800; fax (207) 667-0447; cmelio@mainecf.org
Internet: http://www.mainecf.org/mainecharityfound.aspx
Sponsor: Maine Community Foundation
245 East Main Street
Ellsworth, ME 04605

Maine Community Foundation Hospice Grants 2416
The purpose of the Foundation's Hospice Grants is to support hospice services in southern Maine, in particular by funding programs that support volunteer and bereavement services in Cumberland and York counties. Preference is given to innovative projects that focus on increasing the utilization of end of life services and/or increasing the quality of end-of-life services. Application is made online from the Foundation's website.
Requirements: Applicants must be 501(c)3 organizations eligible to accept tax-deductible donations as outlined in Section 170(c) of the Internal Revenue Code. High priority is given to proposals that support collaborative efforts among hospice programs and other community groups that enhance training, provide recognition, and or increase the recruitment of diverse volunteers for hospice; raise community awareness of hospice services; and increase hospice volunteer coordinators' capacity to interact with volunteers.
Restrictions: Religious groups are eligible but funding will not be provided for religious purposes. Funding is not provided for the following: political campaigns, or to support attempts to influence legislation of any governmental body other than through making available the results of non-partisan analysis, study and research; ongoing operating support; endowments or capital campaigns; camperships; or capital equipment over $250.
Geographic Focus: Maine
Date(s) Application is Due: Feb 15
Amount of Grant: Up to 5,000 USD
Samples: Camp Sunshine at Sebago Lake, Casco, Maine, $5,000, to support five Cumberland and/or York County families participate in the organization's bereavement program; Home Health Visiting Nurses of Southern Maine, Saco, Maine, to support the organization's palliative care program; and Maine Humanities Council, Portland, Maine, to bring the Philoctetes Project to three Maine sites.
Contact: Pam Cleghorn; (877) 700-6800; pcleghorn@mainecf.org
Internet: http://www.mainecf.org/hospicefund.aspx
Sponsor: Maine Community Foundation
245 East Main Street
Ellsworth, ME 04605

Maine Community Foundation Penobscot Valley Health Association Grants 2417
The Penobscot Valley Health Association supports projects that propose innovative ways to strengthen the health and welfare of the greater Bangor community. Priority is given to proposals that involve the targeted population in the design, implementation and evaluation of the project, draw upon the strengths of the community, and foster collaboration between community groups. Applications are available at the Foundation's website, along with a list of previously funded projects.
Requirements: Applicants must be 501(c)3 organizations eligible to accept tax-deductible donations as outlined in Section 170(c) of the Internal Revenue Code.
Geographic Focus: Maine
Date(s) Application is Due: Feb 15
Amount of Grant: Up to 10,000 USD
Samples: 32nd Degree Masonic Learning Centers for Children, Bangor, Maine, $5,000, for tutoring services for children with dyslexia in the greater Bangor area; Evaluation Practice Group, Newburgh, Maine, $9,560, to support the Hooves of Hope equine therapy program for children with disabilities; and Hammond Street Senior Center, Bangor, Maine, $18,828, to support programs for healthy aging in the Bangor region.
Contact: Amy Pollien, Grants Administrator; (877) 700-6800, ext 1109; fax (207) 667-0447; grants@mainecf.org or apollien@mainecf.org
Internet: http://www.mainecf.org/PVHAFund.aspx
Sponsor: Maine Community Foundation
245 East Main Street
Ellsworth, ME 04605

Majors MLA Chapter Project of the Year 2418
The Majors/MLA Chapter Project of the Year Award, sponsored by J.A. Majors, was established by the Medical Library Association (MLA)'s Board of Directors in 1995. The award was proposed by the Platform for Change Implementation Task Force as one of the ways to encourage health-sciences librarians to creatively respond to the challenges of an evolving profession. The recipient chapter will receive a certificate at the annual meeting and a cash award of $500 afterwards. The cash award is earmarked for enhancing the programming of the recipient chapter's annual or regular membership meeting. The recipient chapter or chapter representative assumes all costs of attending the meeting and the ceremony at which the presentation is made. Guidelines and an application form are available at the website. Applications and any supporting documentation must be received by MLA via email, fax, or mail (in order of preference) by November 1. The chair of the recipient chapter will be notified in March before the annual meeting. Founded in 1898, the Medical Library Association is a nonprofit, educational organization of more than 1,100 institutions and 3,600 individual members in the health sciences information field committed to educating health-information professionals, supporting health-information research, promoting access to the world's health-sciences information, and working to ensure that the best health information is available to all.

Requirements: Application is made for excellence in special projects or innovative operation programming demonstrating advocacy, leadership, service, technology, or innovations that contribute to the advancement of the health-sciences librarian. Applicants must satisfy the following criteria: be a chapter of the MLA as defined by the MLA bylaws; demonstrate chapter support for the application, as evidenced by an endorsement from the chapter chair on behalf of the executive committee or officiating body of the chapter; reference the project's goals, objectives, and outcomes or evaluations to provide evidence of the achievements and impact of the project on the profession and the members of the affected chapter; have completed the project within the three years preceding the date of the application or operate an ongoing program that accomplishes the goals set out for the project.
Restrictions: Projects or programs may be resubmitted as long as they continue to meet time frames and other eligibility criteria.
Geographic Focus: All States, All Countries
Date(s) Application is Due: Nov 1
Amount of Grant: 500 USD
Contact: Carla Funk, Executive Director; (312) 419-9094, ext. 14; fax (312) 419-8950; mlapd2@mlahq.org or awards@mlahq.org
Internet: http://www.mlanet.org/awards/honors/index.html
Sponsor: Medical Library Association
65 East Wacker Place, Suite 1900
Chicago, IL 60601-7246

Mann T. Lowry Foundation Grants 2419
Established in Virginia in 1996, the Mann T. Lowry Foundation has specified in primary fields of interest to include: education; health organizations; human services; and social services. Geographic restrictions for giving are primarily in the State of Virginia, and grants come in the form of general operating support. Amounts have varied most recently from $500 to $21,000. Application forms are not required, and there are no identified annual deadlines. Applicants should provide a two- to three-page letter of request, outlining their program and attaching any pertinent brochures or materials.
Geographic Focus: Virginia
Amount of Grant: 500 - 21,000 USD
Samples: Alzheimer's Association, Allen, Virginia, $500 - for general operating support; Joan Grossman Fegely Foundation, Richmond, Virginia, $1,000 - for general operating support; Massey Cancer Center, Richmond, Virginia, $21,000 - for general operating support.
Contact: George R. Hinnant, Director; (804) 643-3512
Sponsor: Mann T. Lowry Foundation
1630 Huguenot Road
Midlothian, VA 23113-2427

Manuel D. and Rhoda Mayerson Foundation Grants 2420
The Manuel and Rhoda Mayerson Foundation Grants supports organizations in its areas of interest, including children and youth, people with disabilities, health and well-being, arts and culture, housing and homelessness, and Jewish life and culture. Types of support include building construction/renovation, conferences and seminars, matching/challenge grants, seed grants, and technical assistance. The board meets quarterly.
Requirements: Grants are only awarded to nonprofit, tax-exempt organizations as defined by Section 501(c)3 of the Internal Revenue Code, to organizations that are public charities under Section 509(a)(1),(2), or (3) of the Internal Revenue Code, and to those that comply with the requirement of Section 4945(d)(3) or (4) of the Code. In making funding decisions, the Foundation is responsive to sound strategic planning, organizational stability, leadership, creativity, entrepreneurial visioning, leveraging of resources, collaboration and empowerment of people. Additionally, there is a focus on providing funding to worthy efforts that otherwise struggle to find vital support. The application is located on the Foundation website.
Restrictions: The Foundation funds projects in Israel, in addition to Cincinnati, Ohio; Berkeley, California; and Boca Raton, Florida. The Foundation does not make grants in support of any non-charitable purpose or to organizations that promote racial, ethnic or religious disharmony, hatred or violence.
Geographic Focus: California, Florida, Ohio, Israel
Amount of Grant: 1,000 - 30,000 USD
Contact: Manuel Mayerson, Foundation President; (513) 621-7500; fax (513) 621-2864; info@mayersonfoundation.org
Internet: http://mayersonfoundation.org/GrantMaking/HowtoApply/tabid/609/Default.aspx
Sponsor: Manuel D. and Rhoda Mayerson Foundation
312 Walnut Street, Suite 3600
Cincinnati, OH 45202

MAP International Medical Fellowships 2421
The MAP International Medical Fellowship encourages lifelong involvement in global health issues by providing selected medical students firsthand exposure in a Christian context to the health, social and cultural characteristics of a developing world community. MAP International, through a grant from the founders of Reader's Digest, DeWitt and Lila Wallace, sponsors fourth year medical students by funding travel arrangements for medical students to their chosen destination. The field experience must be designed to provide the student exposure to community health in a Christian context. Students select a mission agency or hospital that has an outreach among the poor in a rural or urban setting. Students must spend at least eight weeks in the field (six weeks for residents and interns). The fellowship provides 100 percent of the approved round trip airfare to one destination. In most instances, students pay room and board as well as any in-country travel expenses. To obtain an application, check with the dean of your school or the Student Affairs Office. Applications are sent to all medical schools annually. You may also download the application in PDF or Microsoft Word format on the Map International website.
Requirements: Fourth-year medical students, residents and interns are eligible. Applications should be submitted during the academic year prior to travel.
Geographic Focus: All States
Date(s) Application is Due: Mar 1
Contact: Margaret Stevenson, Medical Fellowship Coordinator; (404) 880-0540; fax (912) 265-6170; mstevenson@map.org
Internet: http://www.map.org/site/PageServer?pagename=what_Medical_Fellowship
Sponsor: Map International
4700 Glynco Parkway
Brunswick, GA 31525-6800

Marathon Petroleum Corporation Grants 2422
The Marathon Petroleum Corporation offers grant support within its home-base of Ohio, as well as throughout the states of Illinois, Indiana, Kentucky, Louisiana, Michigan, Texas, and West Virginia. Occasionally, it also gives to national organizations, Its primary purposes are aligned with its core values of health and safety, diversity and inclusion, environmental stewardship and honesty and integrity. With that in mind, special emphasis is also directed toward programs that empower the socially or economically disadvantaged, and provide opportunities for students to reach their full potential. Fields of interest include: the arts, children and youth services, community and economic development, education, the environment, health care, human services, and public affairs. Types of support include: annual campaigns, cause-related marketing, employee matching gifts, general operating support, in-kind donations, and scholarships to individuals.
Geographic Focus: Illinois, Indiana, Kentucky, Louisiana, Michigan, Ohio, Texas, West Virginia
Contact: Bill Conlisk, 419-422-2121; whconlisk@marathonpetroleum.com
Internet: http://www.marathonpetroleum.com/Corporate_Citizenship/
Sponsor: Marathon Petroleum Corporation
539 South Main Street
Findlay, OH 45840-3229

March of Dimes Agnes Higgins Award 2423
Established in 1980, the March of Dimes Agnes Higgins Award honors the late Agnes Higgins of the Montreal Diet Dispensary for her innovation and years of service to the cause of improved maternal nutrition. A pioneer in devising methods of nutritional assessment and counseling, Mrs. Higgins greatly advanced the understanding of eating healthy as a crucial factor in healthy pregnancy and prevention of low birthweight. The Agnes Higgins Award is presented in recognition of distinguished achievement in research, education or clinical services in the field of maternal-fetal nutrition. Winners will receive a $5,000 honorarium and will be honored at a presentation and reception at the American Public Health Association Annual Meeting. Applications must be postmarked by March 30.
Requirements: Candidates must have: been widely involved in maternal-fetal nutrition through teaching, research, or clinical practice for at least five years; shown a demonstrable effect in raising the quality of maternal-fetal nutritional care through scholarly pursuits, research, education, or practice; and demonstrated the ability to apply maternal-fetal nutritional standards of practice or facilitate their implementation by others.
Geographic Focus: All States
Date(s) Application is Due: Mar 30
Amount of Grant: 5,000 USD
Contact: Mary Lavan; (914) 997-4609; fax (914) 997-4560; mlavan@marchofdimes.com
Internet: http://www.marchofdimes.org/professionals/agnes-higgins-award-maternal-fetal-nutrition.aspx
Sponsor: March of Dimes
1275 Mamaroneck Avenue
White Plains, NY 10605

March of Dimes Graduate Nursing Scholarships 2424
Intended to recognize and promote excellence in nursing care of mothers and babies, the Graduate Nursing Scholarship program offers several $5,000 scholarships annually to registered nurses enrolled in graduate programs of maternal-child nursing. The March of Dimes Nurse Advisory Council, a group of distinguished perinatal nurses, chooses the recipients. The annual deadline for applications is February 1, with recipients announced in May
Requirements: Applicant must: be a registered nurse; currently be enrolled in graduate education with a focus on maternal-child nursing at the master's or doctorate level; and have at least one academic term to complete after August of the year in which the scholarship is awarded. In addition, an applicant must be a member of at least one of the following professional organizations: the Association of Women's Health; Obstetric and Neonatal Nurses; the American College of Nurse-Midwives; or the National Association of Neonatal Nurses.
Restrictions: Applicant cannot be: a previous recipient of the March of Dimes graduate nursing scholarship; or an employee of the March of Dimes, a member of its Board of Trustees or a family member of either.
Geographic Focus: All States
Date(s) Application is Due: Feb 1
Amount of Grant: 5,000 USD
Contact: Mary Lavan, Grants Administrator; (914) 997-4609; fax (914) 997-4560; mlavan@marchofdimes.com
Internet: https://www.marchofdimes.org/nursing/index.bm2?cid=00000003&spid=ne_s3_1&tpid=ne_s3_1_3
Sponsor: March of Dimes
1275 Mamaroneck Avenue
White Plains, NY 10605

March of Dimes Newborn Screening Awards 2425
The March of Dimes has created four awards to honor state health officials who put policies in place to address this important issue. The Newborn Screening Awards recognize state health officials who establish newborn screening delivery times of 72 hours, 48 hours or 24 hours. The highest award - known as the Robert Guthrie Newborn Screening Award - is presented to the state health official who achieves 95 percent of all newborn screens delivered from the hospital to the lab in 24 hours. An essential component of these awards is transparency - states must monitor their newborn screening transit times and publicly share the results.
Geographic Focus: All States
Contact: Mary Lavan; (914) 997-4609; fax (914) 997-4560; mlavan@marchofdimes.com
Internet: http://www.marchofdimes.org/professionals/recognizing-state-excellence-in-newborn-screening.aspx
Sponsor: March of Dimes
1275 Mamaroneck Avenue
White Plains, NY 10605

March of Dimes Program Grants 2426
Every year the March of Dimes provides millions of dollars in grants and scholarships. The program supports community programs and education for professionals, with a primary aim to prevent birth defects, premature birth, and infant mortality. The March of Dimes Program Grants fund a number of maternal-child health community programs each year in collaboration with MDs local chapters. External organizations can apply for funding to support programs working to improve the health of mothers and babies. For more information, applicants should contact their nearest local March of Dimes chapter.
Requirements: 501(c)3 organizations throughout the country are eligible to apply.
Geographic Focus: All States
Amount of Grant: Up to 25,000 USD
Contact: Mary Lavan, Grants Administrator; (914) 997-4609; fax (914) 997-4560; researchgrants@marchofdimes.com
Internet: http://www.marchofdimes.org/professionals/scholarships-and-grants.aspx
Sponsor: March of Dimes
1275 Mamaroneck Avenue
White Plains, NY 10605

Marcia and Otto Koehler Foundation Grants 2427
Grants are made to support arts and culture, medical, education, and social organizations in San Antonio, Texas. Proposals must demonstrate leadership in effecting positive change; encourage collaborative effort; serve large and diverse sectors of the population; and demonstrate vision, effectiveness, and good fiscal management. Types of support include general operating support, building construction/renovation, and research. Grants are made for one time only. Applicants must send a letter of request for application by March 1 to receive an application.
Requirements: Grants are awarded only to organizations in Bexar County, Texas.
Restrictions: Grants will not be made to individuals or to support other foundations or endowments; salaries; operating deficits; political organizations; or churches, synagogues, or parishes.
Geographic Focus: Texas
Date(s) Application is Due: Jun 1
Amount of Grant: 2,500 - 50,000 USD
Contact: Thomas K. Killion, Senior Vice President, Bank of America; (210) 270-5422; fax (210) 270-5520; thomas.k.killion@ustrust.com
Sponsor: Marcia and Otto Koehler Foundation
P.O. Box 121
San Antonio, TX 78291-0121

Mardag Foundation Grants 2428
The Mardag Foundation is committed to making grants to qualified nonprofit organizations in Minnesota that help enhance and improve the quality of life, inspire learning, revitalize communities, and promote access to the arts. The Foundation focuses their grantmaking in these priority areas: improving the lives of at-risk families, children, youth, and young adults; supporting seniors to live independently; building the capacity of arts and humanities organizations to benefit their communities; and supporting community development throughout the St. Paul area. Grants normally support; capital projects, program expansion and special projects of a time-limited nature; start-up costs for promising new programs that demonstrate sound management and clear goals relevant to community needs; support for established agencies that have temporary or transitional needs; funds to match contributions received from other sources or to provide a challenge to raise new contributions. Applicants are encouraged to submit a brief summary of their project prior to preparation of a full proposal to see if the project fits the guidelines and interests of the foundation. The Foundation's grantmaking meetings are in April, August, and November. Generally, full proposals must be received three months prior to a meeting date.
Requirements: Nonprofit 501(c)3 organizations are eligible to apply. Organizations must be in the East Metro area of Dakota, Ramsey, or Washington counties.
Restrictions: The Foundation does not fund: programs exclusively serving Minneapolis and the surrounding West Metro area; scholarships and grants to individuals; ongoing annual operating expenses; sectarian religious programs; medical research; federated campaigns; conservation or environmental programs; events and conferences; programs serving the physically, developmentally or mentally disabled; capital campaigns of private secondary schools; and capital and endowment campaigns of private colleges and universities. The Foundation will review, on their own merits, grant applications received from private secondary schools and private colleges and universities for purposes not excluded in the information above.
Geographic Focus: Minnesota
Date(s) Application is Due: May 1; Aug 1; Dec 31
Amount of Grant: 5,000 - 50,000 USD
Contact: Lisa Hansen, Grants Administration Manager; (651) 224-5463 or (800) 875-6167; lisa.hansen@mnpartners.org
Internet: http://www.mardag.org/apply_for_a_grant/
Sponsor: Mardag Foundation
55 Fifth Street East, Suite 600
St. Paul, MN 55101

Margaret T. Morris Foundation Grants 2429
The Margaret T. Morris Foundation awards grants, primarily in Arizona, in its areas of interest, including: animal welfare; arts; children and youth, services; education; environment; higher education; homeless service; human services; marine science; medical research and education; mental health and crisis services; museums; performing arts; reproductive health and family planning; and hospices. The Foundation's types of support include: building renovation; capital campaigns; debt reduction; endowments; general operating support; land acquisition; matching and challenge support; and program development. The Board of Directors meets in August, December, and as needed.
Requirements: Applications are not accepted. Applicants should submit a letter of inquiry with their request for funding and a description of the project.
Geographic Focus: Arizona
Contact: Thomas Polk, Trustee; (928) 445-4010
Sponsor: Margaret T. Morris Foundation
P.O. Box 592
Prescott, AZ 86302-0592

Margaret Wiegand Trust Grants 2430
Established in Wisconsin, the Margaret Wiegand Trust provides grant funding for individuals who are legally blind and need assistant, care, maintenance, and educational needs. The primary fields of interest, therefore, is human services aimed people with visual disabilities. Grants are given either directly to individuals or via a scholarship funding program.
Requirements: Applicants should be residents of Waukesha County, Wisconsin, and referred from the Waukesha Rehabilitation Office or other community service organizations.
Geographic Focus: Wisconsin
Contact: Anne McCullough, (214) 965-2908 or (866) 888-5157
Sponsor: Margaret Wiegand Trust
10 S. Dearborn, IL1-0117
Chicago, IL 60603

Marie C. and Joseph C. Wilson Foundation Rochester Small Grants 2431
The Marie and Joseph Wilson Foundation strives to improve the quality of life through initiating and supporting projects that measurably demonstrate a means of creating a sense of belonging within the family and community. The Foundation considers 501(c)3 organization requests ranging from $1,000 to $25,000. Grant applications are accepted on an ongoing basis. Foundation board members review applications as they are received. The review committee meets once a month except for July and August. Because the Foundation receives a large number of applications, responses may take up to four months. Prior to the receiving funding, grant recipients are required to sign a grant agreement contract. Written progress reports are required at six months and one year following the date of the grant. Samples of the Foundation's previously funded grants are available online.
Requirements: The Foundation review committee looks for one or more of the following conditions in a proposal: the proposal is a well-planned approach to delivering services; Foundation support would be catalytic to the project's success; the proposal is efficient in its use of funds and expenses are reduced by sharing resources with other agencies or groups; and a collaborative network exists that multiplies the impact of the grant. Applicants may contact the Foundation for a current application form.
Restrictions: Grants are limited to 501(c)3 organizations serving the Rochester, New York area. Grants will not be made to individuals, partisan political organizations, or to support lobbying efforts. Requests for capital projects also will not be considered.
Geographic Focus: New York
Amount of Grant: 1,000 - 25,000 USD
Samples: Association for the Blind and Visually Impaired, Rochester, New York, for a full-time children's programming and recreational coordinator, $25,000; Charles Settlement House, Rochester, New York, Teen Clubs, a neighborhood-based program for teens that reduces violent behavior and teen pregnancy while encouraging community service, $10,000; Horizons at Harley, Rochester, New York, a summer enrichment program that offers academic, cultural, wellness, and recreational activities for children from inner-city Rochester, $24,000.
Contact: Megan Bell, Executive Director; (585) 461-4696; fax (585) 473-5206
Internet: http://www.mcjcwilsonfoundation.org/funding.cfm
Sponsor: Marie C. and Joseph C. Wilson Foundation
160 Allens Creek Road
Rochester, NY 14618-3309

Marie H. Bechtel Charitable Remainder Uni-Trust Grants 2432
Established in Scott County, Iowa, in 1987, the Marie H. Bechtel Charitable Remainder Uni-Trust (formerly known as the Marie H. Bechtel Charitable Trust) supports youth services and education. Support is also available for the advancement of health care, maintenance of community cultural activities, and enhancement of the community by restoring its vitality and creating meaningful employment. Contact the Foundation for application forms.
Requirements: 501(c)3 organizations serving residents of Scott County, Iowa, are eligible.

Restrictions: No grants are given for endowment funds, past operating deficit or debt retirement, general or continuing operating support, or basic scholarly research.
Geographic Focus: Iowa
Amount of Grant: Up to 350,000 USD
Contact: R. Richard Bittner; (563) 328-3333; fax (563) 328-3352; loseband@blwlaw.com
Sponsor: Marie H. Bechtel Charitable Remainder Uni-Trust
201 West 2nd Street, Suite 1000
Davenport, IA 52801-1817

Marin Community Foundation Improving Community Health Grants 2433
The Improving Community Health Grants is interested in funding two complimentary priorities: strengthening the delivery of health services with direct health services or community clinics; and addressing the social determinants of health so that everyone has the opportunity to live a long, healthy life regardless of income, education, or racial/ethnic background. Community, neighborhood, or geographically-focused research projects that assess the role of poverty, housing, education, transportation, healthy food, physical activity, and/or other community factors are strongly encouraged to apply. Proposed research projects should contribute to the development and/or implementation of public and/or school policies aimed at addressing social, economic, and/or political factors that contribute to the community's health. The proposed assessments should include leadership, input, and direction from the community. In addition to thorough community/neighborhood assessments, the Foundation is interested in supporting the development, advocacy, and implementation of policies and/or practices that impact the health of a low-income community or community of color. Projects that propose building infrastructure to improve the social conditions, economic opportunities, and/or physical environments in which people live, work, learn, and play will also be considered (e.g., infrastructure and technology for farmers to accept food stamps at the farmers market).
Requirements: To be eligible for funding, organizations must have a nonprofit tax-exempt status or a fiscal sponsor with a nonprofit tax-exempt status. Applicants must serve Marin County. The program director can be emailed directly through the Foundation website.
Geographic Focus: California
Contact: Wendy Todd, Program Director; (415) 464-2541
Internet: http://www.marincf.org/grants-and-loans/grants/community-grants/improving-community-health
Sponsor: Marin Community Foundation
5 Hamilton Landing, Suite 200
Novato, CA 94949

Marion Gardner Jackson Charitable Trust Grants 2434
The Marion Gardner Jackson Charitable Trust was established by Marion Gardner Jackson, the granddaughter of local industrialist, Robert W. Gardner, founder of the Gardner-Denver Company. The Trust funds: the arts and humanities; education; health; religion; and human service organizations. Giving is centered around the Quincy, Illinois, area and its surrounding communities in Adams County. The trust supports capital, program, and operating grants. Grants generally range up to a maximum of $25,000. The application is available at the trust website through Bank of America. The annual deadline for submission is August 31. Most recent awards have ranged from $5,000 to $90,000.
Restrictions: For program support grants, the yearly request may not be more than 50% of the program's budget. Organizations can submit one application per year and will not receive more than one award from the trust in any given year. Organizations receiving a multi-year award from the trust that continues into the next grant year are not eligible to apply for an additional grant until the end of the grant cycle. The trust does not support requests from individuals, organizations attempting to influence policy through direct lobbying, or any political campaigns.
Geographic Focus: Illinois
Date(s) Application is Due: Aug 31
Amount of Grant: 5,000 - 90,000 USD
Samples: Quincy Public Schools Foundation, Quincy, Illinois, $75,000 - program development (2014); Adams County Ambulance and EMS, Mendon, Illinois, $60,000 - capacity-building and technical assistance (2014), Cheerful Home, Quincy, Illinois, $50,000 - program development (2014).
Contact: Debra L. Grand; (312) 828-2055; ilgrantmaking@ustrust.com
Internet: https://www.bankofamerica.com/philanthropic/foundation.go?fnId=109
Sponsor: Marion Gardner Jackson Charitable Trust
231 South LaSalle Street, IL1-231-13-32
Chicago, IL 60604

Marion I. and Henry J. Knott Foundation Discretionary Grants 2435
Founded in 1977, the Marion I. and Henry J. Knott Foundation is a Catholic family foundation committed to honoring its founders' legacy of generosity to strengthen the community within the Archdiocese of Baltimore. Henry J. Knott, the eldest of six boys, grew up in a lively household in the Baltimore area. His father was a hard-working carpenter. Marion Isabel Burk, who was orphaned at the age of eleven, grew up cooking and looking after the children in a small boarding house and received little formal education as a result. Henry and Marion met on a blind date arranged by a good friend in 1926, while Henry was taking classes at Loyola College, and were married in 1928. They went on to build a large family (thirteen children, one lost to cancer) and a thriving construction business. Henry was the first developer in Baltimore to employ the practice of prefabricating wall panels in a factory and then sending them out to construction sites. Projects moved at a blistering pace and eventually led Henry to become the President of Arundel Corporation. Henry and Marion who knew firsthand the challenge of raising a large family always practiced philanthropy. The foundation makes awards in five Program categories: Arts and Humanities; Catholic Activities; Education (Catholic schools, nonsectarian private schools specifically catering to special needs, and private colleges and universities); Health Care; and Human Services. Within the five program categories, the foundation funds within five project categories, including capital expenses, development, new and/or ongoing programs, operating expenses, and technology. In addition to its standard granting program, the Knott Foundation provides a limited number of Discretionary Grants (20-30) throughout the year. These grants, ranging between $500 to $2,500, are designed to increase the Foundation's grant-making options as well as its responsiveness to community needs. Grants are awarded based on the proposed project, the availability of funds, and other current requests for funding. To apply for a discretionary grant, applicants should submit a brief (one page) Letter of Inquiry (LOI) on their organization's letterhead. The LOI should describe the applicant's project or program, detail the applicant's needs, and provide a timeframe for use of the award if granted. In addition to the LOI, the applicant should also submit a 501(c)3 status letter, a project budget if applicable, and a list of the board of directors. Discretionary requests are accepted and awarded on a rolling basis throughout the year. Although not guaranteed, approved funds are usually disbursed within one to two weeks of the discretionary grant's approval date. Interested applicants should visit the website for further details and guidelines.
Requirements: Discretionary grant requests must be in alignment with the foundation's areas of geographic and programmatic giving. Funding is limited to 501(c)3 organizations serving Baltimore City and the following counties in Maryland: Allegheny, Anne Arundel, Baltimore, Carroll, Frederick, Garrett, Harford, Howard, and Washington. Applicants may apply through a fiscal sponsor. The fiscal sponsor must be a 501(c)3 nonprofit organization that has a formal relationship and Memorandum of Understanding (MOU) with the applicant. Selected applicants will need to submit a copy of their most recent IRS 990 and/or audited financials.
Restrictions: The following will not be funded: organizations that have not been in operation for at least one year, scholarships, public education/public sector agencies, pro-choice or reproductive health programs, individuals, annual giving, political activities, one-time only events/seminars/workshops, legal services, environmental activities, medical research, day care centers, endowment funds for arts/humanities, national/local chapters for specific diseases, agencies that redistribute grant funds to other nonprofits, reimbursables or any prior expenses, or government agencies that form 501(c)3 nonprofits to fund public sector projects.
Geographic Focus: Maryland
Amount of Grant: 500 - 2,500 USD
Samples: Museum of Ceramic Art, Baltimore, Maryland, $2,500 - to the after-school Middle School Ceramics Art Program though out Baltimore City; Filbert Street Garden, Baltimore, Maryland, $980 - to suppor their Winter Greens: Student-Supported Agriculture program; Maryland Association of Nonprofits (MANO), Baltimore, Maryland, $5,000 - to support the Leaders Circle's forum on building better relationships between nonprofit executive directors and their board leadership.
Contact: Kathleen McCarthy, Grants Manager; (410) 235-7068; fax (410) 889-2577; knott@knottfoundation.org or info@knottfoundation.org
Internet: http://www.knottfoundation.org/what_we_do/grant_application_process/discretionary_grant_application_process
Sponsor: Marion I. and Henry J. Knott Foundation
3904 Hickory Avenue
Baltimore, MD 21211-1834

Marion I. and Henry J. Knott Foundation Standard Grants 2436
Founded in 1977, the Marion I. and Henry J. Knott Foundation is a Catholic family foundation committed to honoring its founders' legacy of generosity to strengthen the community within the Archdiocese of Baltimore. Henry J. Knott, the eldest of six boys, grew up in a lively household in the Baltimore area. His father was a hard-working carpenter. Marion Isabel Burk, who was orphaned at the age of eleven, grew up cooking and looking after the children in a small boarding house and received little formal education as a result. Henry and Marion met on a blind date arranged by a good friend in 1926, while Henry was taking classes at Loyola College, and were married in 1928. They went on to build a large family (thirteen children, one lost to cancer) and a thriving construction business. Henry was the first developer in Baltimore to employ the practice of prefabricating wall panels in a factory and then sending them out to construction sites. Projects moved at a blistering pace and eventually led Henry to become the President of Arundel Corporation. Henry and Marion who knew firsthand the challenge of raising a large family always practiced philanthropy. The foundation makes both standard and discretionary awards in five Program categories: Arts and Humanities; Catholic Activities; Education (Catholic Schools, Nonsectarian private schools specifically catering to special needs, and private colleges and universities); Health Care; and Human Services. Within the five program categories, the foundation funds within five project categories, including capital expenses, development, new and/or ongoing programs, operating expenses, and technology. The Knott Foundation uses a two-step online application process for its standard-grants program. Step one requires the submission of an online Letter of Inquiry (LOI) along with a Financial Analysis Form. Applicants whose LOIs are approved will move on to step two which requires online submission of a full proposal. Applicants are given the opportunity to submit a draft of their proposal for comments and feedback prior to their final submission. The review process for the foundation's standard grants program takes approximately four months from the date of the LOI submission until a final funding decision is made. The Knott Foundation accepts standard-grant applications three times per year - February, June and October. LOIs and proposals must be received by 5 p.m. on the applicable deadline date. Complete details, guidelines, and links to the online submission system are available at the grant website.
Requirements: Funding is limited to 501(c)3 organizations serving Baltimore City and the following counties in Maryland: Allegheny, Anne Arundel, Baltimore, Carroll, Frederick, Garrett, Harford, Howard, and Washington. Applicants may apply through

a fiscal sponsor. The fiscal sponsor must be a 501(c)3 nonprofit organization that has a formal relationship and Memorandum of Understanding (MOU) with the applicant.
Restrictions: Organizations that are denied funding at the LOI stage of the grant process are eligible to apply again during the next grant cycle; organizations that are denied funding after submitting a full grant proposal must wait one year before reapplying; organizations that receive a grant award must wait two years before reapplying. The following will not be funded: organizations that have not been in operation for at least one year, scholarships, public education/public sector agencies, pro-choice or reproductive health programs, individuals, annual giving, political activities, one-time only events/seminars/workshops, legal services, environmental activities, medical research, day care centers, endowment funds for arts/humanities, national/local chapters for specific diseases, agencies that redistribute grant funds to other nonprofits, reimbursables or any prior expenses, or government agencies that form 501(c)3 nonprofits to fund public sector projects.
Geographic Focus: Maryland
Date(s) Application is Due: Mar 7; Jul 9; Nov 12
Amount of Grant: 35,000 - 45,000 USD
Contact: Kathleen McCarthy, Grants Manager; (410) 235-7068; fax (410) 889-2577; knott@knottfoundation.org or info@knottfoundation.org
Internet: http://www.knottfoundation.org/what_we_do/grant_application_process
Sponsor: Marion I. and Henry J. Knott Foundation
3904 Hickory Avenue
Baltimore, MD 21211-1834

Marjorie C. Adams Charitable Trust Grants 2437
The Marjorie C. Adams Charitable Trust, established in New York in 1987, is administered by the JP Morgan Chase Bank of Newark and supports the care and treatment of deaf and blind people, research into the causes and the treatment of deafness and blindness, and K-12 education. Types of support for these programs include: building and renovation projects, capital campaigns, and endowments. There are no specific application forms required or deadlines with which to adhere, and applicants should begin by contacted the Foundation office directly.
Requirements: 501(c)3 organizations, primarily serving residents of New York and Maryland, are eligible to apply.
Geographic Focus: Maryland, New York
Amount of Grant: Up to 150,000 USD
Contact: Frank Lemma, Vice President; (212) 464-2439 or (212) 648-1477
Sponsor: Marjorie C. Adams Charitable Trust
270 Park Avenue, 16th Floor
New York, NY 10017

Marjorie Moore Charitable Foundation Grants 2438
The Marjorie Moore Charitable Foundation was established in 1957 to support and promote quality educational, cultural, human-services, environmental, and health-care programming for underserved populations. Grants from the Moore Foundation made in support of operations or programming are one year in duration. Multi-year grants for long-term capital projects will be considered on a case-by-case basis. Applicants must apply online at the grant website. Applicants are strongly encouraged to do the following before applying: review the downloadable state application procedures for additional helpful information and clarifications; review the downloadable online-application guidelines at the grant website; review the foundation's funding history (link is available from the grant website); review the online application questions in advance; and review the list of required attachments. These will generally include: a list of board members, financial statements (audited, reviewed, or compiled by independent auditor); an organization summary; a list of other funding sources; an IRS Determination letter; and other required documents. All attachments must be uploaded in the online application as PDF, Word, or Excel files. The Marjorie Moore Charitable Foundation has biannual deadlines of June 1 and December 1. Applications must be submitted by 11:59 p.m. on the deadline dates. Applicants will be notified of grant decisions by letter within two to three months after each respective proposal deadline.
Requirements: Applicant organizations must have 501(c)3 tax-exempt status and serve the people of Kensington or Berlin, Connecticut. A breakdown of number/percentage of people served by specific towns is required on the online application. Preference is given to organizations that provide human services or health care programming.
Restrictions: The foundation does not support requests from individuals, organizations attempting to influence policy through direct lobbying, or any political campaigns.
Geographic Focus: Connecticut
Date(s) Application is Due: Jun 1; Dec 1
Contact: Kate Kerchaert; (860) 657-7016; kate.kerchaert@baml.com
Internet: https://www.bankofamerica.com/philanthropic/fn_search.action
Sponsor: Marjorie Moore Charitable Foundation
200 Glastonbury Boulevard, Suite # 200, CT2-545-02-05
Glastonbury, CT 06033-4056

Marshall County Community Foundation Grants 2439
The Marshall County Community Foundation (MCCF) was established as a 501(c)3 not-for-profit organization with the defined purpose of serving the citizens of Marshall County, Indiana. The Foundation uses the following criteria when reviewing proposals: is there an established need and will the project achieve the desired result; is it appropriate for Marshall County to fund or is it too large; does it fit the County's areas of interest and geography; is it new or innovative; and does it foster collaboration with multiple impacts. Fund decisions are made within 90 days of each submission deadline. The application is available at the Foundation website.
Requirements: Only charitable organizations with a verifiable 501(c)3 status or equivalent will be considered. If 501(c)3 status is not available, organizations must find another organization to host the project or program.
Restrictions: Funding is not available for individuals; sectarian or religious purposes; long term funding; or for events that have already taken place.
Geographic Focus: Indiana
Date(s) Application is Due: Feb 1; Aug 1
Amount of Grant: Up to 10,000 USD
Contact: Linda Yoder, Executive Director; (574) 935-5159; fax (574) 936-8040
Internet: http://www.marshallcountycf.org/grants.htm
Sponsor: Marshall County Community Foundation
2701 North Michigan Street, P.O. Box 716
Plymouth, IN 46563

Mary Black Foundation Active Living Grants 2440
The Mary Black Foundation makes grants to nonprofit organizations in Spartanburg County, South Carolina, region. The Foundation has three applications for active living grants: Programs and Services assist people in becoming more physically active, either for recreation or for transportation; Policies and Places have a direct impact on whether people have the opportunity to be active; and Planning and Capacity Building for organizations that have as part of their core mission to increase active living. Each area of Active Living has different goals and grant submission procedures. The Foundation accepts applications quarterly: March 1, June 1, September 1, and December 1.
Requirements: Nonprofit organizations in South Carolina's Spartanburg County are eligible. Before submitting an application for a grant in Active Living, potential applicants must meet with the Foundation's program staff.
Restrictions: The Foundation does not accept applications from individuals or general fundraising solicitations.
Geographic Focus: South Carolina
Date(s) Application is Due: Mar 1; Jun 1; Sep 1; Dec 1
Amount of Grant: 2,000 - 200,000 USD
Samples: City of Woodruff, Woodruff, South Carolina, $150,000 - to support the Woodruff Greenway Trail; Partners for Active Living, Spartanburg, South Carolina, $75,900 - for the last year of a three-year grant to support a community initiative to increase usage of the 1.9-mile Mary Black Foundation Rail Trail; Spartanburg County School District One, Spartanburg, South Carolina, $198,000 - to support the Inman Trail.
Contact: Amy Page, Grant Consultant; (864) 573-9500; fax (864) 573-5805; apage@maryblackfoundation.org
Internet: http://www.maryblackfoundation.org/active-living/targeted-results
Sponsor: Mary Black Foundation
349 East Main Street, Suite 100
Spartanburg, SC 29302

Mary Black Foundation Community Health Grants 2441
The Mary Black Foundation makes grants to nonprofit organizations in Spartanburg County, South Carolina, region. The Community Health Fund (CHF) is an annual grantmaking opportunity that supports efforts to promote health and wellness. The CHF is for projects outside of the Foundation's Active Living or Early Childhood Development priority areas and represents 10-20% of the Foundation's total grantmaking. The Foundation accepts applications once each year, with an annual deadline of September 4.
Requirements: Nonprofit organizations in South Carolina's Spartanburg County are eligible. Before submitting an application for a grant in Community Health, potential applicants must meet with the Foundation's program staff.
Restrictions: The Foundation does not accept applications from individuals or general fundraising solicitations.
Geographic Focus: South Carolina
Date(s) Application is Due: Sep 4
Amount of Grant: 1,000 - 5,000 USD
Samples: Greenville Area Community Foundation, Greenville, South Carolina, $2,500 - to support the Healing Journeys Conference; Adult Learning Center, Spartanburg, South Carolina, $5,000 - for general operating support, in response to a community challenge; Ellen Hines Smith Girls' Home, Spartanburg, South Carolina, $3,000 - for the last year of a three-year grant to support implementation of the Teaching-Family Model Program.
Contact: Amy Page, Grant Consultant; (864) 573-9500; fax (864) 573-5805; apage@maryblackfoundation.org
Internet: http://www.maryblackfoundation.org/grantmaking/community-health-fund
Sponsor: Mary Black Foundation
349 East Main Street, Suite 100
Spartanburg, SC 29302

Mary Black Foundation Early Childhood Development Grants 2442
The Mary Black Foundation makes grants to nonprofit organizations in Spartanburg County, South Carolina, region. The goals of its investment in early childhood development are: more children in Spartanburg County will enter school ready to learn; and fewer adolescents in Spartanburg County will experience an unintended pregnancy. The Foundation has three applications for early childhood development grants: Programs and Services provide direct assistance, social support, resources, and information to children and teens and their families or to those who work with them; Policies and Places refer to the environmental conditions that affect early childhood development and adolescent pregnancy; and Planning and Capacity Building for organizations that have as part of their core missions the improvement of early childhood development or the reduction of adolescent pregnancy. The Foundation accepts applications quarterly: March 1, June 1, September 1, and December 1.

Requirements: Nonprofit organizations in South Carolina's Spartanburg County are eligible. Before submitting an application for a grant in Early Childhood Development, potential applicants must meet with the Foundation's program staff.
Restrictions: The Foundation does not accept applications from individuals or general fundraising solicitations.
Geographic Focus: South Carolina
Date(s) Application is Due: Mar 1; Jun 1; Sep 1; Dec 1
Amount of Grant: 2,500 - 300,000 USD
Samples: Woodruff Primary School, Spartanburg, South Carolina, $2,500 - to support the NAEYC programs for three- and four-year olds; Spartanburg County School District Seven, Spartanburg, South Carolina, $300,000 - for the second year of a three-year grant to support a high quality child development center program for children ages birth to five; Middle Tyger Community Center, Spartanburg, South Carolina, $91,000 - to support the Adolescent Family Life Program.
Contact: Amy Page, Grant Consultant; (864) 573-9500; fax (864) 573-5805; apage@maryblackfoundation.org
Internet: http://www.maryblackfoundation.org/early-childhood-development/targeted-results
Sponsor: Mary Black Foundation
349 East Main Street, Suite 100
Spartanburg, SC 29302

Mary K. Chapman Foundation Grants 2443

Mary K. Chapman was born in Oklahoma in 1920. She graduated from the University of Tulsa and worked as a nurse before her marriage to Allen Chapman in 1960. After the death of her husband in 1979, Mary Chapman maintained her own personal charitable giving program. Before her death in 2002, she established The Mary K. Chapman Foundation, a charitable trust founded to perpetuate her own charitable giving program. This foundation was fully funded with a bequest from her estate in 2005. Mary K. Chapman was very interested in supporting education, but as a former nurse and a very compassionate person, much of her charity was directed to health, medical research, and educating and caring for the less fortunate and disadvantaged. There are two steps in the process of applying for a grant. The first is a Letter of Inquiry from the applicant. This letter is used to determine if the applicant will be invited to take the second step of submitting a formal Grant Proposal.
Requirements: IRS 501(c)3 non-profits are eligible to apply.
Restrictions: Grant requests for the following purposes are not favored: endowments, except as a limited part of a capital project reserved for maintenance of the facility being constructed; deficit financing and debt retirement; projects or programs for which the Chapman Trusts would be the sole source of financial support; travel, conferences, conventions, group meetings, or seminars; camp programs and other seasonal activities; religious programs of religious organizations; project or program planning; start-up ventures are not excluded, but organizations with a proven strategy and results are preferred; purposes normally funded by taxation or governmental agencies; requests made less than nine months from the declination of a previous request by an applicant, or within nine months of the last payment made on a grant made to an applicant; requests for more than one project.
Geographic Focus: Oklahoma
Amount of Grant: Up to 300,000 USD
Contact: Andie Doyle; (918) 496-7882; fax (918) 496-7887; andie@chapmantrusts.com
Internet: http://www.chapmantrusts.org/grants_programs.php
Sponsor: Mary K. Chapman Foundation
6100 South Yale, Suite 1816
Tulsa, OK 74136

Mary Kay Foundation Cancer Research Grants 2444

In May 2010, the Foundation awarded $1.3 million in grants to select doctors and medical scientists focusing on curing cancers that affect women. These 13 recipients from across the United States received a $100,000 grant to conduct cutting-edge research. Since 1996, the Foundation has given more than $14 million to support this effort. Grants are awarded each year to researchers at medical schools recommended by the The Mary Kay Foundation Research Review Committee, which is composed of prominent doctors who volunteer their time to help the Foundation select the best recipients across the United States. After reviewing these recommendations, the Board of Directors at the Foundation selects the grant recipients. The Mary Kay Foundation accepts grant applications from November to mid-February each year.
Geographic Focus: All States
Amount of Grant: 100,000 USD
Samples: Emily Bernstein, Ph.D., Mount Sinai School of Medicine, $100,000 - Identifying epigenetic drivers of breast cancer; Jeanette Gowen Cook, Ph.D., University of North Carolina at Chapel Hill, $100,000 - Predicting genome instability in breast cancer; Zeng-Quan Yang, Ph.D., Wayne State University, $100,000 - Targeting histone demethylases as a therapeutic strategy in basal breast cancer.
Contact: Lana Rowe, (972) 687-4822 or (877) 652-2737; Lana.Rowe@mkcorp.com or MKCares@marykayfoundation.org
Internet: http://www.marykayfoundation.org/Pages/CancerGrantProgram.aspx
Sponsor: Mary Kay Foundation
P.O. Box 799044
Dallas, TX 75379-9044

Mary L. Peyton Foundation Grants 2445

The private operating foundation awards grants to provide health, welfare, and vocational education benefits to legal residents of El Paso, Texas, who are unable to attain assistance elsewhere. The program supports three areas: Living Assistance (granted for food, clothing, rent/mortgage, any kind of house or car repairs, utilities, car payments/insurance, gasoline, daycare, bus passes, birth certificates, drivers license, or any other living expense); Medical Assistance (granted for dental work and equipment or supplies for physically or mentally handicapped individuals. If an applicant does not have health insurance or money to pay for medical bills or medications, funds are substituted for living assistance, thus enabling the individual to pay for those medical needs); or, Educational Assistance (granted to individuals residing in El Paso County to assist them with an education. This can cover anything that the student needs to go or stay in school, including tuition (full or partial) to a career institution that provides vocational training or to a college/university. Other grants are issued for books, supplies, GED tests, licenses for a job or trade such as a food handler, plumber, electrician, etc., and any form of living assistance such as daycare, rent, gasoline or food).
Requirements: Individuals in El Paso County, Texas, are eligible. Every source from which the applicant could receive needed funds must be exhausted. A comprehensive application with validation of monthly income and expenses is required. The applicant must provide the most recent filed income tax return, proof of the tax stimulus, and, if applicable, documentation of a medical condition, hospital stay, or the current status of an application for Social Security disability.
Restrictions: Support is not given to groups or organizations. Grants are not awarded for building or endowment funds, fellowships, matching gifts, or loans.
Geographic Focus: Texas
Contact: Gloria Perry, Executive Administrator; (915) 533-9698; gp@marylpeyton.org
Internet: http://www.marylpeyton.org/our-program/
Sponsor: Mary L. Peyton Foundation
310 North Mesa Street, # 318
El Paso, TX 79901

Mary Owen Borden Foundation Grants 2446

The Mary Owen Borden Foundation Grants support programs that address the needs of economically disadvantaged youth and their families. This includes needs such as health, family planning, education, counseling, childcare, substance abuse, and delinquency. Other areas of interest for the foundation include affordable housing, conservation, environment, and the arts. Grants average $10,000, and the maximum grant is $15,000. In unique circumstances, the Foundation considers a more significant grant for a program having a major impact in their areas of interests.
Requirements: New Jersey nonprofits in Monmouth and Mercer Counties are eligible. Most of the Foundation's grant go to nonprofit entities in Trenton, Asbury Park, and Long Branch.
Geographic Focus: New Jersey
Date(s) Application is Due: Mar 15; Sep 15
Amount of Grant: Up to 15,000 USD
Contact: Quinn McKean; (732) 741-4645; fax (732) 741-2542; qmckean@aol.com
Internet: http://fdncenter.org/grantmaker/borden/guide.html
Sponsor: Mary Owen Borden Foundation
4 Blackpoint Horseshoe
Rumson, NJ 07760

Mary S. and David C. Corbin Foundation Grants 2447

The Corbin Foundation gives primary consideration to charitable organizations and/or local chapters of national charities located in Akron and Summit County, Ohio, although extremely worthy causes outside of this preferred area may be considered. Areas of interest include arts and culture; civic and community; education; health care; housing; social services; medical research; and youth. The Foundation meets in May and November to consider requests. Grant requests must be received no later than March 1 for consideration in May and September 1 for consideration in November. The application and guidelines are available at the Foundation website.
Requirements: The Foundation has a general application cover sheet which applicants must complete. Organizations should also send a brief letter on their letterhead, submitting one original and one copy of all application materials.
Restrictions: Only written applications will be considered. Telephone and personal interviews are discouraged unless requested by the Foundation. The Foundation does not fund individuals; annual fundraising campaigns; ongoing requests for general operating support (although some repeat grants are made); operating deficits; or organizations which in turn make grants to others.
Geographic Focus: Ohio
Date(s) Application is Due: Mar 1; Sep 1
Contact: Erika J. May; (330) 762-6427; fax (330) 762-6428; corbin@nls.net
Internet: http://foundationcenter.org/grantmaker/corbin/guide.html
Sponsor: Mary S. and David C. Corbin Foundation
Akron Central Plaza
Akron, OH 44308-1830

Mary Wilmer Covey Charitable Trust Grants 2448

The Mary Wilmer Covey Charitable Trust was created to support charitable organizations that promote education including instruction and training that help to build human capabilities. It also supports organizations that focus on relieving human suffering due to disease, ill health, physical weakness, disability, or injury; and supports organizations that work to prolong life and improve the quality of life, especially for children. Grants from the Mary Wilmer Covey Charitable Trust are primarily one year in duration; on occasion, multi-year support is awarded. Applicants must apply online at the grant website. The annual application deadline is September 1 at 11:59 p.m. Applicants will be notified of grant decisions by letter within one to two months after the deadline. Applicants are strongly encouraged to do the following before applying: review the downloadable state-specific application procedures at the grant website; review the downloadable online-application guidelines at the grant website; review the trust's funding history (link is available from the grant website); review the online application

questions in advance; and review the list of required attachments. These will generally include: a list of board members, financial statements (audited, reviewed, or compiled by independent auditor); an organization summary; a list of other funding sources; an IRS Determination letter; and other required documents. All attachments must be uploaded in the online application as PDF, Word, or Excel files.
Requirements: Applicants must have 501(c)3 tax-exempt status. Preference is given to charitable organizations located in Macon, Georgia; Richmond, Virginia; and Chatham Hall, Virginia.
Restrictions: The trust does not support requests from individuals, organizations attempting to influence policy through direct lobbying, or any political campaigns.
Geographic Focus: Georgia, Virginia
Date(s) Application is Due: Sep 1
Contact: Mark S. Drake, Vice President; (404) 264-1377; mark.s.drake@ustrust.com
Internet: https://www.bankofamerica.com/philanthropic/fn_search.action
Sponsor: Mary Wilmer Covey Charitable Trust
3414 Peachtree Road, N.E., Suite 1475, GA7-813-14-04
Atlanta, GA 30326-1113

Massage Therapy Foundation Community Service Grants   2449
The Foundation Community Service grants are awarded to organizations that seek to provide massage therapy to communities or groups who currently have little or no access to such services. This program is designed to promote working partnerships between the massage therapy profession and community-based organizations. It benefits the recipient, the massage therapist, and the sponsoring organization by building stronger relationships between these parties. The Community Service deadline is April 1, annually. The normal award for 12 months is $500-$5000 and must be used in the specific time period for which it has been awarded.
Requirements: These grants are available for organizations or affiliates of organizations that have been in existence for at least one year in the respective state or province; are tax-exempt under schedule 501(c)3 in the U.S., non-profit charitable organization in other countries; currently provide some therapeutic or other service programs to the community; and have designated a qualified staff member to oversee the program.
Geographic Focus: All States
Date(s) Application is Due: Apr 1
Amount of Grant: 500 - 5,000 USD
Contact: Alison Pittas, Program Manager of Research and Grants; (847) 869-5019, ext. 167 or (847) 905-1667; fax (847) 864-1178; apittas@massagetherapyfoundation.org
Internet: http://www.massagetherapyfoundation.org/grants_community.html
Sponsor: Massage Therapy Foundation
500 Davis Street, Suite 900
Evanston, IL 60201

Massage Therapy Foundation Practitioner Case Report Contest   2450
The Foundation has chosen to encourage the writing of case reports to provide an opportunity for massage therapists and bodyworkers to develop research skills and enhance their ability to provide knowledge-based massage to the public. This contest is intended to enhance professional development skills of the practitioner: writing case reports help develop communication skills, critical thinking skills, and could contribute to future research and clinical practice. Cash and publication recognition will be awarded to practitioners submitting the top reports.
Requirements: The Foundation requires that HIPAA guidelines are followed.
Geographic Focus: All States
Date(s) Application is Due: Oct 8
Samples: Erika Larson--for Massage Therapy Effects in a Long-Time Prosthetic User with Fibular Hemimelia; Glenda Keller--for The effects of massage therapy in treatment of chronic plantar fasciitis; Robin B. Anderson--for Reduction and Stabilization in Parkinson-related Peripheral Edema with Therapeutic Massage.
Contact: Alison Pittas, Program Manager of Research and Grants; (847) 869-5019, ext. 167 or (847) 905-1667; fax (847) 864-1178; apittas@massagetherapyfoundation.org
Internet: http://www.massagetherapyfoundation.org/practitionercontest.html
Sponsor: Massage Therapy Foundation
500 Davis Street, Suite 900
Evanston, IL 60201

Massage Therapy Foundation Research Grants   2451
One of the primary purposes of the Foundation is to fund solid research studies investigating the many beneficial applications of massage therapy. Foundation research grants are awarded to individuals or teams conducting studies that promise to advance our understanding of specific therapeutic applications of massage, public perceptions of and attitudes toward massage therapy, and the role of massage therapy in health care delivery. The Research Grant deadline is normally March 1 annually (or the following Monday if this falls on a weekend). The normal award for 12 months is $1,000 to $30,000, and must be used in the specific time period for which it has been awarded. This grant supports high quality, independent research that contributes to the basic science of massage therapy application, including applied research investigating massage therapy as a health/mental health treatment and/or prevention modality.
Requirements: Research Grants are available to investigators who: have experience in the relevant field of research; are presently associated with, or have secured the cooperation of a university, independent research organization, health center, or other institution qualified and willing to function as a Sponsoring Organization for the purpose of this project; or new investigators without prior research experience must document support from an experience investigator willing to act as a collaborator.
Restrictions: Research Grants will not be awarded to spouses, domestic partners, children, descendant, spouses of descendants or any other individual related to any officers or trustees of the Massage Therapy Foundation, or to members of the Research Proposal Review Committee.
Geographic Focus: All States
Date(s) Application is Due: Mar 1
Amount of Grant: Up to 30,000 USD
Contact: Alison Pittas, Program Manager of Research and Grants; (847) 869-5019, ext. 167 or (847) 905-1667; fax (847) 864-1178; apittas@massagetherapyfoundation.org
Internet: http://www.massagetherapyfoundation.org/grants_research.html
Sponsor: Massage Therapy Foundation
500 Davis Street, Suite 900
Evanston, IL 60201

Massage Therapy Research Fund (MTRF) Grants   2452
The purpose of the MTRF is to fund high quality research that investigates the efficacy, and clinical effectiveness of massage therapy and that contributes to our understanding of how massage therapy achieves its effects. A secondary goal is to increase research knowledge and capacity among massage therapists through their engagement in research practices at all levels. The MTRF invites applications in areas indicated above. For both competition cycles, a total funding amount of $50,000 is offered. Grant applications up to $15,000 will be considered.
Requirements: All properly submitted applications will be considered. Studies may be qualitative, quantitative, or mixed method. Initially, funding preference will be given to smaller, pilot, or creative study designs. In this way, the MTRF aims to assist the scientific and massage communities in building a foundation for the eventual construction of more complicated research projects. The MTRF encourages applications that also seek funding from other sources. The principal investigator must be affiliated with a Canadian charitable institution (university or hospital) that is willing to administer the finances of the grant. Funds are awarded to the charitable institution, not to an individual. The research team may include members affiliated with a non-Canadian institution. One member of the research team should be a massage therapist. Students at the masters level and above are encouraged to apply.
Geographic Focus: All States, Canada
Date(s) Application is Due: Mar 1; Sep 30
Amount of Grant: 15,000 USD
Samples: 2008 Awards - University of Saskatchewan ($8,929) Donelda Gowan-Moody, RMT and Anne Leis, PhD: Massage Therapists' Research Utilization and Perceptions Towards Research; Carleton University ($15,000) Diane Sliz, MSc (Cand) and Dr. Shawn Hayley: Psychological and Neuro-immune Effects of a Long-term Massage Therapy Treatment in Individuals Suffering from Major Depressive Disorder; Wilfred Laurier University ($14,981) Peter Tiidus, PhD, Kimberley Dawson, PhD and Lance Dawson, RMT: Effectiveness of regular proactive massage therapy for novice recreational runners.
Contact: Keren Brown, Executive Director; (416) 778-4443 or (866) 778-4443; fax (416) 778-5438; info@holistichealthresearch.ca
Internet: http://www.holistichealthresearch.ca/massagefund.gk
Sponsor: Holistic Health Research Foundation of Canada
80 Carlton Street
Toronto, ON M5B 1L6 Canada

Matilda R. Wilson Fund Grants   2453
Matilda Rausch Dodge Wilson died on September 19, 1967, leaving most of her wealth to the Matilda R. Wilson Fund, a charitable trust she had established in Detroit in 1944. Today, the Matilda R. Wilson Fund awards grants, primarily in southeast Michigan, in support of the arts, youth agencies, higher education, hospitals, and social services. Types of support include building construction/renovation, endowments, equipment acquisition, general operating support, matching/challenge grants, program development, research, and scholarship funds. There are no application forms; initial approach should be a letter of request. The board considers requests at board meetings in January, April, and September.
Requirements: Michigan tax-exempt organizations are eligible.
Restrictions: Grants or loans are not made to individuals.
Geographic Focus: Michigan
Amount of Grant: 10,000 - 100,000 USD
Contact: David P. Larsen; (313) 259-7777; fax (313) 393-7579; roosterveen@bodmanllp.com
Sponsor: Matilda R. Wilson Fund
1901 Saint Antoine Street, 6th Floor
Detroit, MI 48226-2310

Mattel Children's Foundation Grants   2454
Charitable organizations that demonstrate that they directly serve children in need may be eligible for grants of $5,000 up to $20,000. Such organizations must have a mission that focuses on the direct service of children ages zero to 12 years. Applicants or programs must show creative and/or innovative methods to address locally defined need directly impacting children, and must be aligned with Mattel's philanthropic priorities: Learnin--increasing access to education for underserved children and in particular, innovative strategies to promote and address literacy; health--supporting the physical health and well-being of children, with particular emphasis on promoting healthy, active lifestyles; and Girl Empowerment--promoting self-esteem in young girls, up to age 12. Organizations may submit under a fiscal agent as long as the program has been in existence for at least two years.
Requirements: Applicants must be a 501(c)3 Public Charity organizations. If an organization is a project under a fiscal sponsor, that sponsor must have a valid tax exemption status. Organizations must serve children in communities within the U.S. Organizations must not discriminate against a person or a group on the basis of age, political affiliation, race, national origin, ethnicity, gender, disability, sexual orientation or religious beliefs. Organizations must have an annual operating budget of less than $1,000,000.

Restrictions: Organizations may not be affiliated in any way with national organizations, regardless of whether funds are received from the national entity. Examples of organizations that are ineligible to apply include Boys & Girls Clubs, United Ways, YMCA's, and disease-affiliated organizations such as the American Heart Association.
Geographic Focus: All States
Amount of Grant: 5,000 - 20,000 USD
Contact: Administrator; (310) 252-2000; fax (310) 252-2180; foundation@mattel.com
Internet: http://corporate.mattel.com/about-us/philanthropy/programs.aspx
Sponsor: Mattel Children's Foundation
333 Continental Boulevard, M1-1418
El Segundo, CA 90245-5012

Mattel International Grants 2455
The corporation and its foundation award grants to charitable organizations that directly serve children in need. Funding priorities include international organizations or programs that creatively address a defined need directly impacting children in need, and international organizations or programs that align with Mattel's philanthropic priorities, including health--supporting the physical and mental health and well-being of children, and increasing access to health care services for children in need; education--increasing access to education, promoting literacy to children in need, and resources that promote after-school educational achievement; and girls empowerment--promoting self-esteem of girls and increasing access to education, health, and community resources for girls. Types of support include program-specific grants--funding for new programs or expansion of existing programs; and core operating support--support of organizations to sustain their programs. Applications must be submitted online. Applications will not be accepted by fax or mail. Submit questions via email.
Requirements: Organizations must serve children in non-U.S. communities; and must not discriminate against a person or a group on the basis of age, political affiliation, race, national origin, ethnicity, gender, disability, sexual orientation, or religious belief. Pilot projects and new organizations may be considered as long as all eligibility criteria are met; however, preference will be given to organizations that have at least two years of experience. Preference is given to organizations that have an annual operating budget of less than $1 million and are not affiliated with a national organization.
Restrictions: The program does not not fund capital funding for physical property purchase, renovation, or developments; individuals; political parties, candidates, or partisan political organizations; labor organizations, fraternal organizations, athletic clubs, or social clubs; sectarian or denominational religious organizations, except for programs that are available to anyone, broadly promoted, and free from religious orientation; schools and school districts; fundraising events (e.g., dinners, tournaments); advertising or marketing sponsorships; or research overhead and/or indirect costs (including fiscal sponsor fees) that exceed 15 percent of the direct project costs.
Geographic Focus: All States
Amount of Grant: 5,000 - 25,000 USD
Contact: Administrator; (310) 252-2000; fax (310) 252-2180; foundation@mattel.com
Internet: http://corporate.mattel.com/about-us/philanthropy/internationalgrants.aspx
Sponsor: Mattel Children's Foundation
333 Continental Boulevard, M1-1418
El Segundo, CA 90245-5012

Maurice J. Masserini Charitable Trust Grants 2456
The trust awards one-year grants to eligible San Diego County, California nonprofit organizations in its areas of interest, including children and youth, aging, music, higher education, and marine sciences. Types of support include building construction/renovation, equipment acquisition, program development, research grants, matching grants, development grants, internships, and scholarships.
Requirements: San Diego County, California, 501(c)3 tax-exempt organizations are eligible. Interested organizations should contact the trust with a letter of inquiry prior to submitting a formal proposal.
Geographic Focus: California
Amount of Grant: Up to 25,000 USD
Contact: Robert Roszkos, (213) 253-3235
Sponsor: Maurice J. Masserini Charitable Trust
c/o Wells Fargo Bank N.A.
Philadelphia, PA 19106-2112

Max and Victoria Dreyfus Foundation Grants 2457
The Foundation awards grants for: museums; cultural, performing, and visual arts programs, schools; hospitals, educational and skills training projects and programs for youth, seniors, and the disabled. Areas of interest include arts, education, health care, hospitals, social services, and medical research. Submit an inquiry letter (three-page maximum) that describes the nonprofit organization, its annual operating budget, the project to be funded, and the project budget; include proof of tax-exempt status.
Requirements: U.S. nonprofit organizations are eligible.
Restrictions: Foreign charitable organizations and individuals are ineligible.
Geographic Focus: All States
Date(s) Application is Due: Mar 10; Jul 10; Nov 10
Amount of Grant: 5,000 - 20,000 USD
Contact: John W. Hager, (202) 337-3300; fax (202) 337-4925
Sponsor: Max and Victoria Dreyfus Foundation
2233 Wisconsin Avenue NW, Suite 414
Washington, D.C. 20007-4122

Maxon Charitable Foundation Grants 2458
The Maxon Charitable Foundation is a company-sponsored foundation that was established in 1987 in Indiana. The foundation supports hospitals and organizations involved with higher education, natural resources, recreation, human services, community development, and Christianity and awards college scholarships to individuals in the Muncie, Indiana area.
Requirements: Contact the Foundation's office prior to the initial proposal, the office will provide you with the proper Application form.
Geographic Focus: Indiana
Amount of Grant: 500 - 5,000 USD
Contact: Jeffrey R. Lang, Secretary; (765) 284-3304
Sponsor: Maxon Charitable Foundation
201 E 18th Street
Muncie, IN 47302-4124

May and Stanley Smith Charitable Trust Grants 2459
Created in 1989, the May and Stanley Smith Charitable Trust supports organizations serving people in the United States, Canada, the United Kingdom, Australia, the Bahamas, and Hong Kong – places that May and Stanley Smith lived in or spent time in during their lifetimes. The trust supports organizations that offer opportunities to children and youth, elders, the disabled and critically ill, and disadvantaged adults and families which enrich the quality of life, promote self-sufficiency, and assist individuals in achieving their highest potential. The trust will fund requests for general-operating, capacity-building, and program support. All grant seekers (including previously-funded organizations) should follow the step-by-step application process laid out at the website to determine eligibility and fit with the trust's funding goals. Eligible organizations whose projects fall within the trust's areas of interest must submit an online Letter of Inquiry (LOI) from the grant website. The trust's staff will review these and invite selected applicants to submit a full proposal. LOIs may be submitted at any time during the year. Processing a grant application from receipt of the LOI to funding notification generally takes between four and six months.
Requirements: The May and Stanley Smith Charitable Trust has a two-stage application process: an online letter of inquiry (LOI) submission followed by an invited proposal submission. Processing a grant application from receipt of the LOI to funding notification generally takes between four and six months. Please note that proposals are accepted by invitation only. Applications are accepted from organizations meeting the Trust's program area priorities and serving individuals living in British Columbia, Canada and the Western United States: Alaska, Arizona, California, Colorado, Hawaii, Idaho, Montana, Nevada, New Mexico, Oregon, Texas, Utah, Washington, and Wyoming. The Trust makes grants to nonprofit organizations that are tax exempt under Section 501(c)3 of the IRS Code and not classified as a private foundation under Section 509(a) of the Code, and to non-U.S. organizations that can demonstrate that they would meet the requirements for such status. Organizations can also submit applications through a sponsoring organization if the sponsor has 501(c)3 status, is not a private foundation under 509(a), and provides written authorization confirming its willingness to act as the fiscal sponsor. The Trust will only accept proposals sent by regular or express mail services that do not require a signature upon delivery.
Restrictions: The trust rarely supports 100% of a project budget, or more than 25 percent of an organization budget, and takes into account award sizes from other foundations. The trust prefers to fund organizations receiving less than 30% of total revenue from government sources. The trust does not fund the following types of organizations or requests: organizations which are not, or would not qualify as, a 501(c)3 public charity; hospitals or hospital foundations; medical clinics or services; scientific or medical research; building funds or capital projects; schools and universities (except those receiving less than 25% of their operating funds from families and those serving a 100% disabled population); endowment funds; individuals; organizations or programs operated by governments; film or media projects; start-up programs or organizations; proselytizing or religious activities that promote specific religious doctrine or that are exclusive and discriminatory; public policy, research, or advocacy; public awareness, education, or information campaigns/programs; debt reduction; conferences or benefit events; projects which carry on propaganda or otherwise attempt to influence legislation; projects which participate or intervene in any political campaign on behalf of or in opposition to any candidate for public office; projects which conduct, directly or indirectly, any voter registration drive; and organizations that pass through funding to an organization or project that would not be eligible for direct funding as described above.
Geographic Focus: All States, Australia, Bahamas, Canada, Hong Kong, United Kingdom
Contact: Dan Gaff, Grants Manager; (415) 332-0166; grantsmanager@adminitrustllc.com or dgaff@adminitrustllc.com
Internet: http://www.adminitrustllc.com/may-and-stanley-smith-charitable-trust/
Sponsor: May and Stanley Smith Charitable Trust
c/o Adminitrust LLC
Corte Madera, CA 94925

Mayo Clinic Administrative Fellowship 2460
The Mayo Clinic Administrative Fellowship Program is a postgraduate fellowship that fosters the development of outstanding individuals committed to careers in health care administration. Since 1983, the Administrative Fellowship Program has provided participants with practical, skill-building experience through a broad variety of administrative rotations. The Administrative Fellowship Program prepares fellows for health care administration leadership opportunities within the Mayo Clinic system through direct participation in activities designed to build and strengthen essential administrative skills. Mayo Clinic Administrative Fellows will build the following skills: acquire and refine managerial competencies, including analytical, persuasive and human relations skills; develop an awareness and understanding of Mayo Clinic organizational

structure and the interrelation between departments and personnel; receive individualized mentoring from a wide array of administrators and physicians; and establish a performance record through project work and departmental staff responsibilities. Those interested in the fellowship should review the website for available positions. Deadlines vary, but a general timeline for the fellowship recruitment and selection process is as follows: early August - job posted; mid-September - application deadline; October - phone interviews; early November - candidate on-site interviews; mid-November fellow offers extended; December - acceptance letters mailed; and July - fellowship begins. The application and additional information, including a detailed description of a variety of rotations, informational webinars, and a list of frequently asked questions, are available at the website.
Requirements: Diverse candidates from strong master's-level accredited programs are sought for the Administrative Fellowship Program. Qualified candidates are master's-prepared individuals who have completed the degree requirements or are degree candidates in health care administration, business administration, public health, health services administration or related fields of study. Applicants should have experience within the health care industry, such as an internship, externship, fellowship, or applicable employment. Applicants should also possess broad knowledge of current and historical perspectives of the health care industry that can include business management and administration, clinical practice management, clinical research and education, hospital administration and management, finance, human resources and organizational development, information systems, and managed care. Ideal candidates should possess the following traits and behaviors: the need to be challenged at a high level; energetic, hard-working, ambitious and a self-starter; flexibility and willingness to assume a variety of role assignments and deal effectively with ambiguity; objective and insightful; service, team and learning orientation; and understanding and commitment to cultural competency and diversity in the workplace and in the practice of health care.
Restrictions: Mayo Clinic is not able to provide visa sponsorship.
Geographic Focus: All States
Date(s) Application is Due: Sep 15
Contact: Stacy M. Fuhrman; (507) 266-6599; fuhrman.stacy@mayo.edu
Ajani Dunn, Program Director; (904) 953-2000; dunn.ajani@mayo.edu
Internet: http://www.mayoclinic.org/afp
Sponsor: Mayo Clinic
200 First Street SW
Rochester, MN 55905

Mayo Clinic Administrative Fellowship - Eau Claire 2461
The Administrative Fellowship Program is a postgraduate fellowship that fosters the development of outstanding individuals committed to careers in health care administration. The Mayo Clinic Health System in Eau Claire has provided participants with practical experience through a broad variety of administrative interactions, committee involvement, and projects leadership opportunities. The Fellowship Program in Healthcare Administration at Mayo Clinic Health System in Eau Claire is designed to foster the development of master's-prepared individuals committed to a career in healthcare administration. Under the guidance of the chief administrative officer, this one-year management training experience is tailored to meet the needs and interests of the individual candidate. This program allows the fellow to experience the major aspects of managing an integrated healthcare organization. The Fellowship Program offers exposure to all levels of management, allowing fellows to gain an applied understanding of the operations of an integrated healthcare system. Fellows are provided with opportunities to attend various executive-level meetings, and encouraged to participate in committees and other team projects. Examples of previous fellowship experiences include a physician need analysis, a community benefit assessment and a supply chain management project. The Fellow will receive a competitive salary and benefits. The application and additional information, including a detailed list of the objectives and expectations of the Fellowship, are posted at the website.
Requirements: Diverse candidates from strong master's-level accredited programs are sought for the Administrative Fellowship Program. Qualified candidates are master's-prepared individuals who have completed the degree requirements or are degree candidates in health care administration, business administration, public health, health services administration or related fields of study. Applicants should have experience within the health care industry, such as an internship, externship, fellowship, or applicable employment. Applicants should also possess broad knowledge of current and historical perspectives of the health care industry that can include business management and administration, clinical practice management, clinical research and education, hospital administration and management, finance, human resources and organizational development, information systems, and managed care. Ideal candidates should possess the following traits and behaviors: the need to be challenged at a high level; energetic, hard-working, ambitious and a self-starter; flexibility and willingness to assume a variety of role assignments and deal effectively with ambiguity; objective and insightful; service, team and learning orientation; and understanding and commitment to cultural competency and diversity in the workplace and in the practice of health care.
Restrictions: Lodging is the responsibility of the individual. There is no commitment of employment on behalf of the individual or organization beyond the one-year program.
Geographic Focus: All States
Date(s) Application is Due: Oct 1
Contact: Nicole Egan; (715) 838-5081; egan.nicole@mayo.edu
Internet: http://mayoclinichealthsystem.org/locations/eau-claire/careers/for-students/administrative-fellowship
Sponsor: Mayo Clinic
200 First Street SW
Rochester, MN 55905

Mayo Clinic Business Consulting Fellowship 2462
The Business Consulting Fellowship Program at Mayo Clinic is a one-year learning experience designed to develop outstanding individuals committed to a career in health care management. Through this program, recipients will gain practical experience and actively participate in projects within Mayo Clinic. Fellows will assist experienced analysts on large and small projects designed to improve the efficiency, quality, and effectiveness of Mayo Clinic. Sample project topics include the following: business planning; quality improvement; information technology systems implementation; forecasting and access; simulation; patient flow; facilities design; electronic systems usability; and staffing analysis. The Clinic offers competitive compensation and comprehensive benefits for Fellow participants. The program typically begins in June and December. Candidates may apply at the website.
Requirements: Applicants must have recently completed a master's degree and have at least one year of relevant internship or work experience, or have a bachelor's degree and at least four years of work or internship experience. Previous health care experience is preferred. Degrees should be in business or health care administration, industrial engineering, operations research, or a closely related field. Successful candidates will have skills and experience in several of these areas: consulting; workflow and operations analysis; staffing analysis; process design, improvement or re-engineering; simulation; electronic system implementation; and project management. Strong candidates should also have the following skills: objectivity; analytical, innovative and results oriented; comfortable with ambiguity; curious and willing to ask questions; highly skilled communicators; and able to work both independently and as a part of team.
Geographic Focus: All States
Contact: Chad Musolf, Human Resources; (507) 284-4851; musolf.chad@mayo.edu
Internet: http://www.mayoclinic.org/train-sp-rst/
Sponsor: Mayo Clinic
200 First Street SW
Rochester, MN 55905

MBL Albert and Ellen Grass Summer Fellowships 2463
Marine Biological Laboratory (MBL) summer researchers come from all over the world. Visiting researchers find an infrastructure and an informal, interactive scientific community that allows them to launch into research as soon as they arrive. Research awards are available to investigators wishing to do visiting research at the MBL. The Albert and Ellen Grass Fellowship promotes innovative research collaborations, ideally between investigators from different institutions. Preference will be given to junior investigators, who are a required member of any team seeking support from this Fellowship. The MBL Research Awards provide costs for research and housing, and also enable awardees to benefit from the intellectual and interactive environment of the scientific community at the MBL. Funding requires a minimum stay of six weeks at the MBL. Awards can be used to support independent research, collaborative research, or in affiliation with the discovery courses. Proposals for research award support will be considered in, but are not limited to, the following fields of investigation: cell biology; development biology; ecology; evolution; microbiology; neurobiology; physiology; and tissue engineering. Candidates may apply at the MBL website.
Requirements: Research proposed in the resident program or courses requires pre-approval of the host lab and program and the course directors, respectively, in addition to the CASO office.
Geographic Focus: All States, All Countries
Date(s) Application is Due: Dec 15
Contact: Julie Early, Research Award Coordinator; (508) 289-7173; researchawards@mbl.edu or jearly@mbl.edu
Internet: http://hermes.mbl.edu/research/summer/awards_general.html
Sponsor: Marine Biological Laboratory
7 MBL Street
Woods Hole, MA 02543

MBL Ann E. Kammer Memorial Summer Fellowship 2464
Marine Biological Laboratory (MBL) summer researchers come from all over the world. Visiting researchers find an infrastructure and an informal, interactive scientific community that allows them to launch into research as soon as they arrive. Research awards are available to these visiting researchers. In memory of renowned zoologist, Ann E. Kammer, this Fellowship is available to women investigators working in the neurosciences. The MBL Research Awards provide costs for research and housing, and also enable awardees to benefit from the intellectual and interactive environment of the scientific community at the MBL. Funding requires a minimum stay of six weeks at the MBL. Awards can be used to support independent research, collaborative research, or in affiliation with the discovery courses. Proposals for Research award support will be considered in, but are not limited to, the following fields of investigation: cell biology; development biology; ecology; evolution; microbiology; neurobiology; and physiology. Candidates may fill out applications at the MBL online process.
Requirements: Research proposed in the resident program or courses requires pre-approval of the host lab and program and the course directors, respectively, in addition to the CASO office.
Geographic Focus: All States, All Countries
Date(s) Application is Due: Dec 15
Contact: Julie Early, Research Award Coordinator; (508) 289-7173; researchawards@mbl.edu or jearly@mbl.edu
Internet: http://hermes.mbl.edu/research/summer/awards_general.html
Sponsor: Marine Biological Laboratory
7 MBL Street
Woods Hole, MA 02543

GRANT PROGRAMS | 389

MBL Associates Summer Fellowships 2465
Marine Biological Laboratory (MBL) summer researchers come from all over the world. Visiting researchers find an infrastructure and an informal, interactive scientific community that allows them to launch into research as soon as they arrive. Research awards are available to these visiting researchers. The MBL Associates were formed to provide a connection to the Laboratory for friends, both scientists and non-scientists who have an interest in the research, educational and cultural activities of the MBL. The Associates provide fellowship funds to help support the costs of summer research at the MBL. The MBL Research Awards provide costs for research and housing, and also enable awardees to benefit from the intellectual and interactive environment of the scientific community at the MBL. Funding requires a minimum stay of six weeks at the MBL. Awards can be used to support independent research, collaborative research, or in affiliation with the discovery courses. Proposals for Research award support will be considered in, but are not limited to, the following fields of investigation: cell biology; development biology; ecology; evolution; microbiology; neurobiology; physiology; and tissue engineering. Candidates may fill out applications at the MBL online process.
Requirements: Research proposed in the resident program or courses requires pre-approval of the host lab and program and the course directors, respectively, in addition to the CASO office.
Geographic Focus: All States, All Countries
Date(s) Application is Due: Dec 15
Contact: Julie Early; (508) 289-7173; researchawards@mbl.edu or jearly@mbl.edu
Internet: http://hermes.mbl.edu/research/summer/awards_general.html
Sponsor: Marine Biological Laboratory
7 MBL Street
Woods Hole, MA 02543

MBL Baxter Postdoctoral Summer Fellowship 2466
Marine Biological Laboratory (MBL) summer researchers come from all over the world. Visiting researchers find an infrastructure and an informal, interactive scientific community that allows them to launch into research as soon as they arrive. Research awards are available to these visiting researchers. Baxter International, Inc. has established a Fellowship for postdoctoral investigators in the biological and biomedical sciences. This Fellowship supports a scientist undertaking independent research at the MBL. The MBL Research Awards provide costs for research and housing, and also enable awardees to benefit from the intellectual and interactive environment of the scientific community at the MBL. Funding requires a minimum stay of six weeks at the MBL. Awards can be used to support independent research, collaborative research, or in affiliation with the discovery courses. Candidates may fill out applications at the MBL online process.
Requirements: Research proposed in the resident program or courses requires pre-approval of the host lab and program and the course directors, respectively, in addition to the CASO office.
Geographic Focus: All States, All Countries
Date(s) Application is Due: Dec 15
Contact: Julie Early; (508) 289-7173; researchawards@mbl.edu or jearly@mbl.edu
Internet: http://hermes.mbl.edu/research/summer/awards_general.html
Sponsor: Marine Biological Laboratory
7 MBL Street
Woods Hole, MA 02543

MBL Burr and Susie Steinbach Summer Fellowship 2467
Marine Biological Laboratory (MBL) summer researchers come from all over the world. Visiting researchers find an infrastructure and an informal, interactive scientific community that allows them to launch into research as soon as they arrive. Research awards are available to these visiting researchers. The Steinbach Fellowship honors the memory of H. "Burr" and Eleanor "Susie" Steinbach. Dr. Steinbach was a scientist and director of the MBL who made important contributions toward the understanding of bioelectric generators, ionic equilibria across cell membranes, and the enzymology of cell organelles. This Fellowship helps cover laboratory fees, housing expenses, and other appropriate personal expenses. The MBL Research Awards provide costs for research and housing, and also enable awardees to benefit from the intellectual and interactive environment of the scientific community at the MBL. Funding requires a minimum stay of six weeks at the MBL. Awards can be used to support independent research, collaborative research, or in affiliation with the discovery courses. Proposals for Research award support will be considered in, but are not limited to, the following fields of investigation: cell biology; development biology; ecology; evolution; microbiology; neurobiology; physiology; and tissue engineering. Candidates may fill out applications at the MBL online process.
Requirements: Research proposed in the resident program or courses requires pre-approval of the host lab and program and the course directors, respectively, in addition to the CASO office.
Geographic Focus: All States, All Countries
Date(s) Application is Due: Dec 15
Contact: Julie Early; (508) 289-7173; researchawards@mbl.edu or jearly@mbl.edu
Internet: http://hermes.mbl.edu/research/summer/awards_general.html
Sponsor: Marine Biological Laboratory
7 MBL Street
Woods Hole, MA 02543

MBL E.E. Just Summer Fellowship for Minority Scientists 2468
Marine Biological Laboratory (MBL) summer researchers come from all over the world. Visiting researchers find an infrastructure and an informal, interactive scientific community that allows them to launch into research as soon as they arrive. Research awards are available to these visiting researchers. The Just Fellowship honors Dr. Just's contributions to developmental biology and experimental embryology. Dr. Just was the first African American to be professionally recognized for his scientific work, much of it accomplished during the early 1900s. This Fellowship will underwrite an outstanding minority scientist's participation in research at the MBL. The MBL Research Awards provide costs for research and housing, and also enable awardees to benefit from the intellectual and interactive environment of the scientific community at the MBL. Funding requires a minimum stay of six weeks at the MBL. Awards can be used to support independent research, collaborative research, or in affiliation with the discovery courses. Proposals for Research award support will be considered in, but are not limited to, the following fields of investigation: cell biology; development biology; ecology; evolution; microbiology; neurobiology; physiology; and tissue engineering. Candidates may fill out applications at the MBL online process.
Requirements: Research proposed in the resident program or courses requires pre-approval of the host lab and program and the course directors, respectively, in addition to the CASO office.
Geographic Focus: All States, All Countries
Date(s) Application is Due: Dec 15
Contact: Julie Early; (508) 289-7173; researchawards@mbl.edu or jearly@mbl.edu
Internet: http://hermes.mbl.edu/research/summer/awards_general.html
Sponsor: Marine Biological Laboratory
7 MBL Street
Woods Hole, MA 02543

MBL Erik B. Fries Summer Fellowships 2469
Marine Biological Laboratory (MBL) summer researchers come from all over the world. Visiting researchers find an infrastructure and an informal, interactive scientific community that allows them to launch into research as soon as they arrive. Research awards are available to these visiting researchers. The Erik B. Fries Fellowships defray laboratory or library expenses for independent investigators at the MBL. Fries, a long time Woods Hole resident, was a biology professor at City College of New York, and MBL's botany course in the early 1900s. Preference is given to investigator members from Fries' former department at City College. The MBL Research Awards provide costs for research and housing, and also enable awardees to benefit from the intellectual and interactive environment of the scientific community at the MBL. Funding requires a minimum stay of six weeks at the MBL. Awards can be used to support independent research, collaborative research, or in affiliation with the discovery courses. Proposals for Research award support will be considered in, but are not limited to, the following fields of investigation: cell biology; development biology; ecology; evolution; microbiology; neurobiology; physiology; and tissue engineering. Candidates may fill out applications at the MBL online process.
Requirements: Research proposed in the resident program or courses requires pre-approval of the host lab and program and the course directors, respectively, in addition to the CASO office.
Geographic Focus: All States, All Countries
Date(s) Application is Due: Dec 15
Contact: Julie Early; (508) 289-7173; researchawards@mbl.edu or jearly@mbl.edu
Internet: http://hermes.mbl.edu/research/summer/awards_general.html
Sponsor: Marine Biological Laboratory
7 MBL Street
Woods Hole, MA 02543

MBL Eugene and Millicent Bell Summer Fellowships 2470
Marine Biological Laboratory (MBL) summer researchers come from all over the world. Visiting researchers find an infrastructure and an informal, interactive scientific community that allows them to launch into research as soon as they arrive. Research awards are available to these visiting researchers. The Eugene and Millicent Bell Fellowships in Tissue Engineering provide funding to visiting graduate students, postdoctoral fellows, or faculty pursuing research in a variety of biomedical questions that advance knowledge of biological composition and function at levels ranging from molecules to cells, to tissue, to organs and whole organisms. This work is grounded in a solid base of chemistry, physics, math and biology and aims to expand the possibilities for tissue and organ reconstitution and replacement. The MBL Research Awards provide costs for research and housing, and also enable awardees to benefit from the intellectual and interactive environment of the scientific community at the MBL. Funding requires a minimum stay of six weeks at the MBL. Awards can be used to support independent research, collaborative research, or in affiliation with the discovery courses. Candidates may fill out applications at the MBL online process.
Requirements: Research proposed in the resident program or courses requires pre-approval of the host lab and program and the course directors, respectively, in addition to the CASO office.
Geographic Focus: All States, All Countries
Date(s) Application is Due: Dec 15
Contact: Julie Early; (508) 289-7173; researchawards@mbl.edu or jearly@mbl.edu
Internet: http://hermes.mbl.edu/research/summer/awards_general.html
Sponsor: Marine Biological Laboratory
7 MBL Street
Woods Hole, MA 02543

MBL Evelyn and Melvin Spiegal Summer Fellowship 2471
Marine Biological Laboratory (MBL) summer researchers come from all over the world. Visiting researchers find an infrastructure and an informal, interactive scientific community that allows them to launch into research as soon as they arrive. Research awards are available to these visiting researchers. The Spiegal Fellowship support summer research at the MBL in any relevant biomedical discipline. The MBL Research Awards provide costs for research and housing, and also enable awardees to benefit from the

Grant Programs

intellectual and interactive environment of the scientific community at the MBL. Funding requires a minimum stay of six weeks at the MBL. Awards can be used to support independent research, collaborative research, or in affiliation with the discovery courses. Proposals for Research award support will be considered in, but are not limited to, the following fields of investigation: cell biology; development biology; ecology; evolution; microbiology; neurobiology; physiology; and tissue engineering. Candidates may fill out applications at the MBL online process.
Requirements: Research proposed in the resident program or courses requires pre-approval of the host lab and program and the course directors, respectively, in addition to the CASO office.
Geographic Focus: All States, All Countries
Date(s) Application is Due: Dec 15
Contact: Julie Early; (508) 289-7173; researchawards@mbl.edu or jearly@mbl.edu
Internet: http://hermes.mbl.edu/research/summer/awards_general.html
Sponsor: Marine Biological Laboratory
7 MBL Street
Woods Hole, MA 02543

MBL Frank R. Lillie Summer Fellowship — 2472

Marine Biological Laboratory (MBL) summer researchers come from all over the world. Visiting researchers find an infrastructure and an informal, interactive scientific community that allows them to launch into research as soon as they arrive. Research awards are available to these visiting researchers. The Lillie Fellowship was established in memory of Frank R. Lillie, a distinguished developmental biologist and the laboratory's director for many years. The Fellowship provides a research laboratory and funds for supplies. The MBL Research Awards provide costs for research and housing, and also enable awardees to benefit from the intellectual and interactive environment of the scientific community at the MBL. Funding requires a minimum stay of six weeks at the MBL. Awards can be used to support independent research, collaborative research, or in affiliation with the discovery courses. Proposals for Research award support will be considered in, but are not limited to, the following fields of investigation: cell biology; development biology; ecology; evolution; microbiology; neurobiology; physiology; and tissue engineering. Candidates may fill out applications at the MBL online process.
Requirements: Research proposed in the resident program or courses requires pre-approval of the host lab and program and the course directors, respectively, in addition to the CASO office.
Geographic Focus: All States, All Countries
Date(s) Application is Due: Dec 15
Contact: Julie Early; (508) 289-7173; researchawards@mbl.edu or jearly@mbl.edu
Internet: http://hermes.mbl.edu/research/summer/awards_general.html
Sponsor: Marine Biological Laboratory
7 MBL Street
Woods Hole, MA 02543

MBL Frederik B. and Betsy G. Bang Summer Fellowships — 2473

Marine Biological Laboratory (MBL) summer researchers come from all over the world. Visiting researchers find an infrastructure and an informal, interactive scientific community that allows them to launch into research as soon as they arrive. Research awards are available to these visiting researchers. The Bang Fellowships are offered for the study of the immune capability of marine animals and for the use of marine models for research in molecular biology or biomedicine. Funds primarily support laboratory space and research expenses. Funding requires a minimum stay of six weeks at the MBL. Candidates may fill out applications at the MBL online process.
Requirements: Research proposed in the resident program or courses requires pre-approval of the host lab and program and the course directors, respectively, in addition to the CASO office.
Geographic Focus: All States, All Countries
Date(s) Application is Due: Dec 15
Contact: Julie Early; (508) 289-7173; researchawards@mbl.edu or jearly@mbl.edu
Internet: http://hermes.mbl.edu/research/summer/awards_general.html
Sponsor: Marine Biological Laboratory
7 MBL Street
Woods Hole, MA 02543

MBL Fred Karush Library Readership — 2474

Marine Biological Laboratory (MBL) summer researchers come from all over the world. Visiting researchers find an infrastructure and an informal, interactive scientific community that allows them to launch into research as soon as they arrive. A variety of research awards are available to these visiting researchers. The Karush Library Readership provides the fees for desk rental, and full privileges in the MBL/WHOI Library on a monthly basis during the summer. The MBL Research Awards provide costs for research and housing, and also enable awardees to benefit from the intellectual and interactive environment of the scientific community at the MBL. Funding requires a minimum stay of six weeks at the MBL. Awards can be used to support independent research, collaborative research, or in affiliation with the discovery courses. Proposals for Research award support will be considered in, but are not limited to, the following fields of investigation: cell biology; development biology; ecology; evolution; microbiology; neurobiology; physiology; and tissue engineering. Candidates may fill out applications at the MBL online process.
Requirements: Research proposed in the resident program or courses requires pre-approval of the host lab and program and the course directors, respectively, in addition to the CASO office.
Geographic Focus: All States, All Countries
Date(s) Application is Due: Feb 15
Contact: Julie Early; (508) 289-7173; researchawards@mbl.edu or jearly@mbl.edu
Internet: http://hermes.mbl.edu/research/summer/awards_general.html
Sponsor: Marine Biological Laboratory
7 MBL Street
Woods Hole, MA 02543

MBL Gruss Lipper Family Foundation Summer Fellowship — 2475

Marine Biological Laboratory (MBL) summer researchers come from all over the world. Visiting researchers find an infrastructure and an informal, interactive scientific community that allows them to launch into research as soon as they arrive. Research awards are available to these visiting researchers. The Gruss Lipper Fellowship underwrites expenses and travel to bring Israeli investigators and advanced students to the MBL to work independently, collaborate with MBL scientists, or to participate in the MBL summer research and short courses. The MBL Research Awards provide costs for research and housing, and also enable awardees to benefit from the intellectual and interactive environment of the scientific community at the MBL. Funding requires a minimum stay of six weeks at the MBL. Awards can be used to support independent research, collaborative research, or in affiliation with the discovery courses. Proposals for Research award support will be considered in, but are not limited to, the following fields of investigation: cell biology; development biology; ecology; evolution; microbiology; neurobiology; physiology; and tissue engineering. Candidates may fill out applications at the MBL online process.
Requirements: Research proposed in the resident program or courses requires pre-approval of the host lab and program and the course directors, respectively, in addition to the CASO office.
Geographic Focus: Israel
Date(s) Application is Due: Dec 15
Contact: Julie Early; (508) 289-7173; researchawards@mbl.edu or jearly@mbl.edu
Internet: http://hermes.mbl.edu/research/summer/awards_general.html
Sponsor: Marine Biological Laboratory
7 MBL Street
Woods Hole, MA 02543

MBL H. Keffer Hartline and Edward F. MacNichol, Jr. Fellowships — 2476

Marine Biological Laboratory (MBL) summer researchers come from all over the world. Visiting researchers find an infrastructure and an informal, interactive scientific community that allows them to launch into research as soon as they arrive. Research awards are available to these visiting researchers. The Hartline and MacNichol Fellowships were established in memory of Dr. Hartline and his impact on sensory science. Dr. Hartline came to the MBL as a young scientist, initiated his pioneering studies of the eye, and continued this research at the MBL for many years. These Fellowships support scientists carrying out research in sensory neurobiology. The MBL Research Awards provide costs for research and housing, and also enable awardees to benefit from the intellectual and interactive environment of the scientific community at the MBL. Funding requires a minimum stay of six weeks at the MBL. Awards can be used to support independent research, collaborative research, or in affiliation with the discovery courses. Candidates may fill out applications at the MBL online process.
Requirements: Research proposed in the resident program or courses requires pre-approval of the host lab and program and the course directors, respectively, in addition to the CASO office.
Geographic Focus: All States, All Countries
Date(s) Application is Due: Dec 15
Contact: Julie Early; (508) 289-7173; researchawards@mbl.edu or jearly@mbl.edu
Internet: http://hermes.mbl.edu/research/summer/awards_general.html
Sponsor: Marine Biological Laboratory
7 MBL Street
Woods Hole, MA 02543

MBL Herbert W. Rand Summer Fellowship — 2477

Marine Biological Laboratory (MBL) summer researchers come from all over the world. Visiting researchers find an infrastructure and an informal, interactive scientific community that allows them to launch into research as soon as they arrive. A variety of research awards are available to these visiting researchers. The Rand Fellowship includes a summer stipend, and funds or assists with such expenses as travel, supplies, and laboratory fees. The MBL Research Awards provide costs for research and housing, and also enable awardees to benefit from the intellectual and interactive environment of the scientific community at the MBL. Funding requires a minimum stay of six weeks at the MBL. Awards can be used to support independent research, collaborative research, or in affiliation with the discovery courses. Proposals for Research award support will be considered in, but are not limited to, the following fields of investigation: cell biology; development biology; ecology; evolution; microbiology; neurobiology; physiology; and tissue engineering. Candidates may fill out applications at the MBL online process.
Requirements: Research proposed in the resident program or courses requires pre-approval of the host lab and program and the course directors, respectively, in addition to the CASO office.
Geographic Focus: All States, All Countries
Date(s) Application is Due: Dec 15
Contact: Julie Early; (508) 289-7173; researchawards@mbl.edu or jearly@mbl.edu
Internet: http://hermes.mbl.edu/research/summer/awards_general.html
Sponsor: Marine Biological Laboratory
7 MBL Street
Woods Hole, MA 02543

GRANT PROGRAMS | 391

MBL James E. and Faith Miller Memorial Summer Fellowship 2478
Marine Biological Laboratory (MBL) summer researchers come from all over the world. Visiting researchers find an infrastructure and an informal, interactive scientific community that allows them to launch into research as soon as they arrive. Research awards are available to these visiting researchers. The Miller Fellowship provides funds for summer research at the MBL in any relevant field of medical research. The MBL Research Awards provide costs for research and housing, and also enable awardees to benefit from the intellectual and interactive environment of the scientific community at the MBL. Funding requires a minimum stay of six weeks at the MBL. Awards can be used to support independent research, collaborative research, or in affiliation with the discovery courses. Proposals for Research award support will be considered in, but are not limited to, the following fields of investigation: cell biology; development biology; ecology; evolution; microbiology; neurobiology; physiology; and tissue engineering. Candidates may fill out applications at the MBL online process.
Requirements: Research proposed in the resident program or courses requires pre-approval of the host lab and program and the course directors, respectively, in addition to the CASO office.
Geographic Focus: All States, All Countries
Date(s) Application is Due: Dec 15
Contact: Julie Early; (508) 289-7173; researchawards@mbl.edu or jearly@mbl.edu
Internet: http://hermes.mbl.edu/research/summer/awards_general.html
Sponsor: Marine Biological Laboratory
7 MBL Street
Woods Hole, MA 02543

MBL John M. Arnold Award 2479
Marine Biological Laboratory (MBL) summer researchers come from all over the world. Visiting researchers find an infrastructure and an informal, interactive scientific community that allows them to launch into research as soon as they arrive. A variety of research awards are available to these visiting researchers. The John M. Arnold Awards funds a retired or senior researcher in any field of study compatible with the mission of the MBL. The MBL Research Awards provide costs for research and housing, and also enable awardees to benefit from the intellectual and interactive environment of the scientific community at the MBL. Funding requires a minimum stay of six weeks at the MBL. Awards can be used to support independent research, collaborative research, or in affiliation with the discovery courses. Proposals for Research award support will be considered in, but are not limited to, the following fields of investigation: cell biology; development biology; ecology; evolution; microbiology; neurobiology; physiology; and tissue engineering. Candidates may fill out applications at the MBL online process.
Requirements: Research proposed in the resident program or courses requires pre-approval of the host lab and program and the course directors, respectively, in addition to the CASO office.
Geographic Focus: All States, All Countries
Date(s) Application is Due: Dec 15
Contact: Julie Early; (508) 289-7173; researchawards@mbl.edu or jearly@mbl.edu
Internet: http://hermes.mbl.edu/research/summer/awards_general.html
Sponsor: Marine Biological Laboratory
7 MBL Street
Woods Hole, MA 02543

MBL Laura and Arthur Colwin Summer Fellowships 2480
Marine Biological Laboratory (MBL) summer researchers come from all over the world. Visiting researchers find an infrastructure and an informal, interactive scientific community that allows them to launch into research as soon as they arrive. Research awards are available to these visiting researchers. The Colwin Summer Fellowships provide support for independent investigators conducting research in the fields of cell and developmental biology. The MBL Research Awards provide costs for research and housing, and also enable awardees to benefit from the intellectual and interactive environment of the scientific community at the MBL. Funding requires a minimum stay of six weeks at the MBL. Candidates may fill out applications at the MBL online process.
Requirements: Research proposed in the resident program or courses requires pre-approval of the host lab and program and the course directors, respectively, in addition to the CASO office.
Geographic Focus: All States, All Countries
Date(s) Application is Due: Dec 15
Contact: Julie Early; (508) 289-7173; researchawards@mbl.edu or jearly@mbl.edu
Internet: http://hermes.mbl.edu/research/summer/awards_general.html
Sponsor: Marine Biological Laboratory
7 MBL Street
Woods Hole, MA 02543

MBL Lucy B. Lemann Memorial Summer Fellowship 2481
Marine Biological Laboratory (MBL) summer researchers come from all over the world. Visiting researchers find an infrastructure and an informal, interactive scientific community that allows them to launch into research as soon as they arrive. Research awards are available to these visiting researchers. Established in memory of Lucy B. Lemann, a Woods Hole resident, this Fellowship helps support the costs of summer research at the MBL. The MBL Research Awards provide costs for research and housing, and also enable awardees to benefit from the intellectual and interactive environment of the scientific community at the MBL. Funding requires a minimum stay of six weeks at the MBL. Awards can be used to support independent research, collaborative research, or in affiliation with the discovery courses. Proposals for Research award support will be considered in, but are not limited to, the following fields of investigation: cell biology; development biology; ecology; evolution; microbiology; neurobiology; physiology; and tissue engineering. Candidates may fill out applications at the MBL online process.
Requirements: Research proposed in the resident program or courses requires pre-approval of the host lab and program and the course directors, respectively, in addition to the CASO office.
Geographic Focus: All States, All Countries
Date(s) Application is Due: Dec 15
Contact: Julie Early; (508) 289-7173; researchawards@mbl.edu or jearly@mbl.edu
Internet: http://hermes.mbl.edu/research/summer/awards_general.html
Sponsor: Marine Biological Laboratory
7 MBL Street
Woods Hole, MA 02543

MBL M.G.F. Fuortes Summer Fellowships 2482
Marine Biological Laboratory (MBL) summer researchers come from all over the world. Visiting researchers find an infrastructure and an informal, interactive scientific community that allows them to launch into research as soon as they arrive. Research awards are available to these visiting researchers. Established in memory of M. G. F. Fuortes, a renowned neurophysiologist, the Fuortes Summer Fellowships are available to young investigators working in the neurosciences. The MBL Research Awards provide costs for research and housing, and also enable awardees to benefit from the intellectual and interactive environment of the scientific community at the MBL. Funding requires a minimum stay of six weeks at the MBL. Awards can be used to support independent research, collaborative research, or in affiliation with the discovery courses. Proposals for Research award support will be considered in, but are not limited to, the following fields of investigation: cell biology; development biology; ecology; evolution; microbiology; neurobiology; physiology; and tissue engineering. Candidates may fill out applications at the MBL online process.
Requirements: Research proposed in the resident program or courses requires pre-approval of the host lab and program and the course directors, respectively, in addition to the CASO office.
Geographic Focus: All States, All Countries
Date(s) Application is Due: Dec 15
Contact: Julie Early; (508) 289-7173; researchawards@mbl.edu or jearly@mbl.edu
Internet: http://hermes.mbl.edu/research/summer/awards_general.html
Sponsor: Marine Biological Laboratory
7 MBL Street
Woods Hole, MA 02543

MBL Nikon Summer Fellowship 2483
Marine Biological Laboratory (MBL) summer researchers come from all over the world. Visiting researchers find an infrastructure and an informal, interactive scientific community that allows them to launch into research as soon as they arrive. Research awards are available to these visiting researchers. A summer Fellowship at the MBL is available from Nikon, Inc. to a young investigator doing research in an area of biology in which she or he can make extensive use of advanced microscopy or micro-manipulation systems. This Fellowship includes a summer laboratory, housing, and a budget for incidental expenses, equipment rental, and supplies. Requests will also be considered for collaborative research projects at the MBL during the off-season period. The MBL Research Awards provide costs for research and housing, and also enable awardees to benefit from the intellectual and interactive environment of the scientific community at the MBL. Funding requires a minimum stay of six weeks at the MBL. Awards can be used to support independent research, collaborative research, or in affiliation with the discovery courses. Proposals for Research award support will be considered in, but are not limited to, the following fields of investigation: cell biology; development biology; ecology; evolution; microbiology; neurobiology; physiology; and tissue engineering. Candidates may fill out applications at the MBL online process.
Requirements: Research proposed in the resident program or courses requires pre-approval of the host lab and program and the course directors, respectively, in addition to the CASO office.
Geographic Focus: All States, All Countries
Date(s) Application is Due: Dec 15
Contact: Julie Early; (508) 289-7173; researchawards@mbl.edu or jearly@mbl.edu
Internet: http://hermes.mbl.edu/research/summer/awards_general.html
Sponsor: Marine Biological Laboratory
7 MBL Street
Woods Hole, MA 02543

MBL Plum Foundation John E. Dowling Fellowships 2484
Marine Biological Laboratory (MBL) summer researchers come from all over the world. Visiting researchers find an infrastructure and an informal, interactive scientific community that allows them to launch into research as soon as they arrive. Research awards are available to these visiting researchers. The Dowling Fellowships help underwrite the costs of room and board, supplies, and laboratory space rental for outstanding applicants pursuing research on the neurobiology of vision and retinal research. The MBL Research Awards provide costs for research and housing, and also enable awardees to benefit from the intellectual and interactive environment of the scientific community at the MBL. Funding requires a minimum stay of six weeks at the MBL. Awards can be used to support independent research, collaborative research, or in affiliation with the discovery courses. Proposals for Research award support will be considered in, but are not limited to, the following fields of investigation: cell biology; development biology; ecology; evolution; microbiology; neurobiology; physiology; and tissue engineering. Candidates may fill out applications at the MBL online process.

Requirements: Research proposed in the resident program or courses requires pre-approval of the host lab and program and the course directors, respectively, in addition to the CASO office.
Geographic Focus: All States, All Countries
Date(s) Application is Due: Dec 15
Contact: Julie Early; (508) 289-7173; researchawards@mbl.edu or jearly@mbl.edu
Internet: http://hermes.mbl.edu/research/summer/awards_general.html
Sponsor: Marine Biological Laboratory
7 MBL Street
Woods Hole, MA 02543

MBL Robert Day Allen Summer Fellowship   2485
Marine Biological Laboratory (MBL) summer researchers come from all over the world. Visiting researchers find an infrastructure and an informal, interactive scientific community that allows them to launch into research as soon as they arrive. Research awards are available to these visiting researchers. The Robert Day Allen Fellowship honors the memory of Professor Allen, a distinguished investigator of cell motility, and one of the developers of video-enhanced optical microscopy with differential interference contrast. The Fellowship funds a young investigator committed to research on cell motility and cytoarchitecture. The MBL Research Awards provide costs for research and housing, and also enable awardees to benefit from the intellectual and interactive environment of the scientific community at the MBL. Funding requires a minimum stay of six weeks at the MBL. Candidates may fill out applications at the MBL online process.
Requirements: Research proposed in the resident program or courses requires pre-approval of the host lab and program and the course directors, respectively, in addition to the CASO office.
Geographic Focus: All States, All Countries
Date(s) Application is Due: Dec 15
Contact: Julie Early; (508) 289-7173; researchawards@mbl.edu or jearly@mbl.edu
Internet: http://hermes.mbl.edu/research/summer/awards_general.html
Sponsor: Marine Biological Laboratory
7 MBL Street
Woods Hole, MA 02543

MBL Scholarships and Awards   2486
Financial aid for MBL students comes from named scholarship funds at the MBL, the Howard Hughes Medical Institute, and the Grass Foundation. Funding is also available from other course sponsors including the U.S. Public Health Service, National Institute on Drug Abuse, the National Institute of Mental Health, the National Institute of Neurological Disorders and Stroke, the National Science Foundation, and the Department of Energy. A variety of scholarships and awards are available for individual courses (partial or full support) and all levels of researchers (U.S. and international) who work in collaboration with other investigators or independently. Funding is also available for minority investigators and researchers from specific universities. Eligibility varies depending on the scholarship. Students who apply for financial aid to attend MBL courses are automatically considered for all sources of financial assistance. To request aid, candidates must complete the financial aid section of their course application form. Applications are available at the MBL website.
Restrictions: Scholarship funding is only available for courses at the MBL. Research funding is available at another location of the MBL website.
Geographic Focus: All States, All Countries
Contact: Carol Hamel, Admissions Coordinator; (508) 289-7401; chamel@mbl.edu
Internet: http://hermes.mbl.edu/education/admissions/scholarships/index.html
Sponsor: Marine Biological Laboratory
7 MBL Street
Woods Hole, MA 02543

MBL Stephen W. Kuffler Summer Fellowships   2487
Marine Biological Laboratory (MBL) summer researchers come from all over the world. Visiting researchers find an infrastructure and an informal, interactive scientific community that allows them to launch into research as soon as they arrive. Research awards are available to investigators wishing to do visiting research at the MBL. The Stephen W. Kuffler Fellowships are intended to encourage the career development of promising young investigators by helping to support them in the intense intellectual atmosphere of the MBL for the summer. It covers part of the costs of laboratory rental, housing, and other personal expenses. The MBL Research Awards provide costs for research and housing, and also enable awardees to benefit from the intellectual and interactive environment of the scientific community at the MBL. Funding requires a minimum stay of six weeks at the MBL. Awards can be used to support independent research, collaborative research, or in affiliation with the discovery courses. Proposals for Research award support will be considered in, but are not limited to, the following fields of investigation: cell biology; development biology; ecology; evolution; microbiology; neurobiology; physiology; and tissue engineering. The application is available at the MBL website.
Requirements: Research proposed in the resident program or courses requires pre-approval of the host lab and program and the course directors, respectively, in addition to the CASO office.
Geographic Focus: All States, All Countries
Date(s) Application is Due: Dec 15
Contact: Julie Early; (508) 289-7173; researchawards@mbl.edu or jearly@mbl.edu
Internet: http://hermes.mbl.edu/research/summer/awards_general.html
Sponsor: Marine Biological Laboratory
7 MBL Street
Woods Hole, MA 02543

MBL William Townsend Porter Summer Fellowships for Minority Investigators   2488
Marine Biological Laboratory (MBL) summer researchers come from all over the world. Visiting researchers find an infrastructure and an informal, interactive scientific community that allows them to launch into research as soon as they arrive. Research awards are available to these visiting researchers. The Porter Foundation offers fellowship support for young scientists (undergraduates, senior graduate students, and postdoctoral trainees) who are from an under-represented minority group, and are U.S. citizens or permanent residents to do summer research with senior investigators at the MBL. The Fellowship contributes toward the expenses at the MBL during the summer. The MBL Research Awards provide costs for research and housing, and also enable awardees to benefit from the intellectual and interactive environment of the scientific community at the MBL. Funding requires a minimum stay of six weeks at the MBL. Awards can be used to support independent research, collaborative research, or in affiliation with the discovery courses. Proposals for Research award support will be considered in, but are not limited to, the following fields of investigation: cell biology; development biology; ecology; evolution; microbiology; neurobiology; physiology; and tissue engineering. Candidates may fill out applications at the MBL online process.
Requirements: Research proposed in the resident program or courses requires pre-approval of the host lab and program and the course directors, respectively, in addition to the CASO office.
Geographic Focus: All States, All Countries
Date(s) Application is Due: Dec 15
Contact: Julie Early; (508) 289-7173; researchawards@mbl.edu or jearly@mbl.edu
Internet: http://hermes.mbl.edu/research/summer/awards_general.html
Sponsor: Marine Biological Laboratory
7 MBL Street
Woods Hole, MA 02543

McCallum Family Foundation Grants   2489
The McCallum Foundation awards grants to Massachusetts nonprofits in its areas of interest, including health and human services, higher education, and to a United Methodist church. Organizations should call the Foundation or submit a letter of inquiry to see if their project is appropriate for the Foundation.
Requirements: Massachusetts nonprofit organizations are eligible.
Geographic Focus: Massachusetts
Amount of Grant: 1,000 - 10,000 USD
Contact: Donna McCallum, Trustee; (978) 649-9132
Sponsor: McCallum Foundation
134 Middle Street
Lowell, MA 01852

McCarthy Family Foundation Grants   2490
The McCarthy Family Foundation is a small private foundation established in 1988. Organized as a California public benefit nonprofit corporation, it operates from San Diego, California. The foundation's current program interest areas have been funded throughout its almost two decades history. These funding areas are: K-12 science education; HIV/AIDS research, education and direct services; assistance to homeless persons; and child abuse prevention and services for victims and families. The board will occasionally fund a special project outside these categorical areas and/or for regional, national or international charitable purposes. A small portion of the grantmaking budget is provided for director matching grants to encourage and amplify personal philanthropy by the foundation's board of directors. Proposals will be considered in the above general areas or for special project areas established by the Board. Proposals will only be accepted for programs within San Diego County. Multiple year proposals may be considered but it is not the foundation's intention to provide annual support. It is not normally the foundation's desire to be the sole source of funds for a project. Most grants are expected to be small, $5,000-$15,000, reflecting the foundation's limited budget although there are no fixed minimum or maximum amounts. Grant proposals received by March 15 will be considered for decision/funding in June. Grant proposals received by September 15 will be considered for decision/funding in December.
Requirements: Proposals will only be considered for programs within San Diego County, California. All applicants should first submit a letter of inquiry, which can be submitted directly from the Foundation's web site. The letter (1-2 pages) must contain a brief statement describing the applicant and the need for funds with enough information for the foundation to determine whether or not the application falls within its program areas. Proposals submitted without an initial letter of inquiry will not be reviewed by the foundation. Applicants should briefly and clearly provide a statement of their needs and the specific request to the foundation, taking into account other possible sources of funding. Letters of inquiry will be acknowledged upon their receipt, but because the foundation operates without a professional staff, a more detailed response may be delayed. Applicants who receive a favorable response to their initial inquiry will be invited to submit a grant application. The foundation accepts the Common Grant Application of San Diego Grantmakers, which can be downloaded from the Foundation's web site. Five copies of the application (proposal) should be submitted. Only one copy of required attachments need be submitted. The application should be signed by the organization's board chair or the executive director (or equivalent individuals).
Restrictions: The foundation does not consider grants for individuals, scholarship funds, or sectarian religious activities. Normally the foundation does not consider requests for general fundraising drives. It does not make grants intended directly or indirectly to support political candidates or to influence legislation.
Geographic Focus: California
Date(s) Application is Due: Mar 15; Sep 15
Amount of Grant: 500 - 250,000 USD

Samples: United Way of San Diego County, $10,000--general operating support; UCLA AIDS Institute, $50,000--stem cell research for HIV prevention; Elementary Institute of Science, $10,000---science and technology youth education.
Contact: Rachel McCarthy Bender; (858) 485-0129; mail@mccarthyfamilyfdn.org
Internet: http://www.mccarthyfamilyfdn.org/guide.html
Sponsor: McCarthy Family Foundation
P.O. Box 27389
San Diego, CA 92198-1389

McCombs Foundation Grants 2491
The McCombs Foundation, established in 1981, makes grants to Texas nonprofit charitable, philanthropic, educational, and benevolent organizations. Areas of interest include arts, athletics, education and higher education, historic preservation, medical research, philanthropy, recreation and sports, youth services, and voluntarism. There are no application deadlines or forms. Applicants should submit a letter of application stating a brief history of the organization, any available printed support materials, and a budget detail. Final notification takes approximately two weeks.
Requirements: Texas nonprofits are eligible. Application form not required.
Restrictions: No grants are given to individuals.
Geographic Focus: Texas
Amount of Grant: 1,000 - 6,000,000 USD
Samples: University of Texas, Houston, Texas, $5,000,000 - Anderson Cancer Center; Saint Mary's Hall, San Antonio, Texas, $300,000 - general operations.
Contact: Gary Woods, Treasurer; (210) 821-6523
Sponsor: McCombs Foundation
755 East Mulberry, Suite 600
San Antonio, TX 78212

McConnell Foundation Grants 2492
The McConnell Foundation awards grants to California nonprofit organizations in its areas of interest, including arts and culture, community development, recreation, social services, children, youth and education, sustainable/livable communities, and the environment. Grants primarily fund the purchase of equipment or building related projects for small and large projects in each county. Requests of up to $50,000 are accepted for projects of benefit to the giving area. Grants are made at three levels, according to the applicant's location. Each county may apply for $1,000 to $10,000. Grants are made to Modoc and Trinity county for $10,000 to $30,000, and grants are made to Shasta, Siskiyou, and Tehama counties for $10,000 to $50,000. Application forms and materials are available at the website for the Shasta Community Foundation who administers the grants.
Requirements: California nonprofit organizations serving Shasta, Siskiyou, Trinity, Tehama, and Modoc counties are eligible.
Restrictions: Grants do not support individuals, religious organizations, research institutions, endowment funds, annual fund drives, budget deficits, or building purchase/construction.
Geographic Focus: California
Date(s) Application is Due: Sep 5
Amount of Grant: Up to 50,000 USD
Contact: Kerry Caranci, Senior Program and Operations Officer; (530) 244-1219 or (530) 926-5486; kerry@shastarcf.org or info@shastarcf.org
Internet: http://www.mcconnellfoundation.org
Sponsor: McConnell Foundation
800 Shasta View Drive
Redding, CA 96003

McCune Foundation Human Services Grants 2493
The McCune Foundation's grants are assigned to one of four program areas, including: education, human services, humanities, and civic. In the area of human services, the Foundation focuses on health, social services, and community improvement. The interdisciplinary approach to grantmaking in this program area works with community-based and regional institutions to address pressing community needs by supporting the critical work of existing programs, as well as new initiatives and programs that seek to find and ameliorate the root causes of community distress and disinvestment. The human services area supports capital projects, research and development, organizational capacity building, and programming with the following strategic priorities: transfer education and research assets into economic opportunities for the region; leverage public and private dollars for broad-based support of community assets; promote self-sufficiency of residents; increase social and economic stability of communities and the region; and test and support effective prevention programs.
Requirements: The foundation supports 501(c)3 organizations in southwestern Pennsylvania and throughout the country, with emphasis on the Pittsburgh area. This area includes the following counties: Allegheny, Beaver, Butler, Armstrong, Westmoreland, and Washington. To apply, an organization should send a brief (2 to 3 page) initial inquiry, preferably using the Foundation's website. The letter should contain: project overview - describe what the proposed efforts are intended to achieve for the region as well as for the organization; what activities/actions are planned to meet the stated goals; project timeline; resources required - total cost of the project; anticipated income, including private and public funders; amount of funding requested; IRS 501(c)3 determination letter - attach a copy (either scanned via email or hard copy via regular mail); and a copy of the organization's latest audit (either scanned via email or hard copy via regular mail).
Restrictions: Grants are not awarded to individuals or for general operating purposes or loans. Unsolicited proposals from outside the funding area are not accepted.
Geographic Focus: All States
Amount of Grant: 1,000 - 1,000,000 USD
Samples: Hosanna House, Wilkinsburg, Pennsylvania, $500,000 - operations and programs; Mount Lebanon Unites Presbyterian Church, Pittsburgh, Pennsylvania, $7,900 - moving forward action plan; Dallas Children's Advocacy Center, Dallas, Texas, $50,000 - forensic interview wing capital campaign.
Contact: Henry S. Beukema; (412) 644-8779; fax (412) 644-8059; info@mccune.org
Internet: http://www.mccune.org/foundation:Website,mccune,grants
Sponsor: McCune Foundation
750 Sixth PPG Place
Pittsburgh, PA 15222

McGraw-Hill Companies Community Grants 2494
McGraw-Hill focuses its charitable efforts on its goal of financial capability for all. The Corporation partners with nonprofit organizations to help individuals gain the necessary knowledge to make smart savings, credit and spending decisions. McGraw-Hill also extends its support to organizations that share the Corporation's dedication to the arts and culture, education and health and human services. The company gives priority consideration to organizations and projects that: promote and support excellence in education and learning, with a primary emphasis on financial literacy; further financial literacy in the communities and markets where the many diverse businesses of The McGraw-Hill Companies operate; utilize unique applications of innovative and developing technologies; extend their reach globally; can be evaluated and can serve as models elsewhere; are staffed and administered by people with demonstrated competence and experience in their fields; have been determined tax-exempt 501(c)3 organizations and qualify as public charities under IRS rules, or are the local country-specific equivalent of a charitable nonprofit organization.
Requirements: Proposals must include the following: a program narrative (no more than three pages); a detailed organization overview; a concise description of the grant's purpose; and all required attachments. If the request is considered eligible, a meeting may be arranged with the Corporate Responsibility and Sustainability staff. On-site visits may also be made. Proposals are accepted at any time, with funding decisions made quarterly.
Restrictions: McGraw-Hill does not support courtesy advertising, pledges for a walk-a-thon, or similar activities. Funding is also not available for the following: institutions and agencies clearly outside the company's primary geographic concerns and interests; libraries and schools of higher education and K-12; political activities or organizations established to influence legislation; sectarian or religious organizations; member-based organizations, i.e., fraternities, labor, veterans, athletic, social clubs; individuals; endowment funds; or loans. Grants are not renewed automatically. Requests for renewed support must be submitted each year.
Geographic Focus: All States
Contact: Susan Wallman, Manager, Corporate Contributions; (212) 512-6480; fax (212) 512-3611; susan_wallman@mcgraw-hill.com
Internet: http://www.mcgraw-hill.com/site/cr/community/giving#section_2
Sponsor: McGraw-Hill Companies
1221 Avenue of the Americas, 20th Floor
New York, NY 10020-1095

McInerny Foundation Grants 2495
The foundation awards grants to Hawaii nonprofit organizations in its areas of interest, including arts and culture, education, environment, health care, youth services, and social services. Types of support include general operating support, continuing support, building construction/renovation, equipment acquisition, program development, seed money, scholarship funds, and matching funds. The July 1 deadline is for capital funds; the November 15 deadline date is for scholarship funds. All other requests are accepted at any time.
Requirements: Hawaii nonprofit organizations are eligible.
Restrictions: Grants are not awarded in support of religious institutions, individuals, endowment funds, deficit financing, or research.
Geographic Focus: Hawaii
Date(s) Application is Due: Jul 1; Nov 15
Amount of Grant: 3,000 - 250,000 USD
Samples: National Tropical Botanical Garden, Kalaheo, HI, $75,000--capital campaign for construction of Botanical Garden Research Center; Boys and Girls Club of Hawaii, Honolulu, HI, $50,000--for RALLY Project at Central Middle School; Honolulu Symphony Society, Honolulu, HI, $40,000-for youth music education program season.
Contact: Paula Boyce; (808) 538-4944; fax (808) 538-4006; pboyce@boh.com
Sponsor: McInerny Foundation
P.O. Box 3170, Department 758
Honolulu, HI 96802-3170

McKesson Foundation Grants 2496
The foundation seeks to enhance the health and quality of life in communities where McKesson HBOC operates and its employees live. Emphasis is focused on youth, especially health services for underserved populations, educational enrichment, the environment, recreation, and youth development activities. The foundation also funds emergency services for children and families, and a variety of social, educational and cultural programs. For culture and the arts, support is given primarily to organizations that reach out to youth and other populations that do not have easy access to such programs. Grants are made for specific projects and programs, and general operating support will be considered in some circumstances. Grants are awarded for one to three years. The Foundation does not accept uninvited applications. Initial approach should be to contact the Foundation directly.
Requirements: 501(c)3 tax-exempt organizations are eligible.
Restrictions: Grants are not made to endowment campaigns, individuals or individual scholarships, religious organizations for religious purposes, political causes or campaigns, advertising in charitable publications, research studies or health organizations concentrating on one disease.

Geographic Focus: All States
Amount of Grant: 5,000 - 25,000 USD
Samples: LA Theater Work (CA)--for a project that creates hands-on performing arts and literary workshops aimed at finding a positive outlet for the aggression of 3000 at-risk and incarcerated youths, $7500; Massachusetts Coalition for School-Based Health Ctrs (MA)--to hire a coordinator and to continue providing school-based health care, $125,000 over three years.
Contact: Marcia Argyris; (415) 983-8673 or (415) 983-8300; fax (415) 983-7590; marcia.argyris@mckesson.com or community.relations@mckesson.com
Internet: http://www.mckesson.com/en_us/McKesson.com/About%2BUs/Corporate%2BCitizenship/McKesson%2BFoundation.html
Sponsor: McKesson Foundation
1 Post Street, 32nd Floor
San Francisco, CA 94104-5203

McLean Contributionship Grants 2497
Originally established in 1951 as The Bulletin Contributionship for charitable, educational and scientific purposes, the Contributionship became The McLean Contributionship on May 1, 1980 when the association of the McLean family with the Bulletin ended. Independent Publications, also owned by the McLean family, continues as the main financial supporter of the Contributionship. The Contributionship favors projects that stimulate a better understanding of the natural environment and encourage the preservation of its important features; encourage more compassionate and cost-effective care for the ill and aging in an atmosphere of dignity and self-respect; or promote education, medical, scientific, or on occasion, cultural developments that enhance quality of life. In addition, the Trustees from time to time support projects which motivate promising young people to assess and develop their talents despite social and economic obstacles or encourage those in newspaper and related fields to become more effective and responsible in helping people better understand how events in their communities and around the world affect them. The Trustees meet several times a year. The Contribution accepts and processes applications for grants throughout the year. Applications must be received at least six weeks before a meeting date. Interested organizations may view the annual-meeting schedule on the Application Procedure page at the grant website. The Contributionship accepts the common-grant-application form of Delaware Valley Grantmakers Association (link is provided on the Application Procedure web page). Application may also be made by letter (guidelines are included on the Application Procedure web page).
Requirements: Applicants must show evidence of tax-exempt status.
Restrictions: Geographic area is limited to the following locations: Greater Philadelphia area; Nashua, New Hampshire; Dubois Pennsylvania; and Central Florida. The Contributionship does not fund the costs or expenses of existing staff allocated to a project it is asked to support.
Geographic Focus: Florida, New Hampshire, Pennsylvania
Amount of Grant: 2,000 - 100,000 USD
Samples: A Little Taste of Everything, Philadelphia, Pennsylvania, $1,400 - towards maintaining and enhancing their living roof; Visiting Nurse Association of Greater Philadelphia, Philadelphia, Pennsylvania, $50,000 - toward the installation of the Horizon Enterprise Content Management System; National Constitution Center, Philadelphia, Pennsylvania - towards the upgrade of eight poorly-functioning exhibit cases housing artifacts uncovered during the Center's groundbreaking.
Contact: Sandra McLean, Executive Director; (610) 527-6330; fax (610) 527-9733
Internet: http://fdncenter.org/grantmaker/mclean
Sponsor: McLean Contributionship
945 Haverford Road, Suite A
Bryn Mawr, PA 19010-3814

McLean Foundation Grants 2498
The sole purpose of the McLean Foundation is to enhance the quality of life for the people of Humboldt county through grantmaking to qualified organizations. The Foundation directs a significant portion of its efforts toward projects that support children and youth, the elderly, social welfare, health and medical needs, as well as supporting the capacity of the local nonprofit sector. Applicants are encouraged to contact the Foundation for further information.
Requirements: Nonprofit organizations in Humboldt county are eligible to apply. Community organizations that meet the needs of the children and youth, the elderly, social welfare, and health and medical needs are also eligible to apply.
Geographic Focus: California
Contact: Leigh Pierre-Oetker; (707) 725-1722; leigh@mcleanfoundation.org
Internet: http://www.northerncalifornianonprofits.org/content/view/102/92/
Sponsor: Mel and Grace McLean Foundation
1336 Main Street
Fortuna, CA 95540

McMillen Foundation Grants 2499
The McMillen Foundation was incorporated in 1947 and serves the Fort Wayne and Allen Counties of Indiana. Fields of interest include: children/youth services; community/economic development; education; health care; health organizations, association; recreation.
Requirements: The foundation requires no application form and proposals may be submitted at anytime. Applicants should submit the following: copy of IRS Determination Letter; brief history of organization and description of its mission; copy of most recent annual report/audited financial statement/990; listing of board of directors, trustees, officers and other key people and their affiliations; detailed description of project and amount of funding requested; copy of current year's organizational budget and/or project budget; listing of additional sources and amount of support; one copy of the proposal.
Restrictions: Giving limited to Fort Wayne and Allen County, Indiana. No support for churches or religious groups. No grants to individuals.
Geographic Focus: Indiana
Amount of Grant: 4,000 - 400,000 USD
Samples: Girl Scouts of the U.S.A., Fort Wayne, IN, $300,000; Junior Achievement of Northern Indiana, Fort Wayne, IN, $50,000; Fort Wayne Park Foundation, Fort Wayne, IN, $100,000.
Contact: Dorothy J. Robinson, Secretary; (260) 484-8631; fax (260) 484-2141
Sponsor: McMillen Foundation
6610 Mutual Drive
Fort Wayne, IN 46825-4236

MDA Development Grant 2500
The association will consider an application for a research grant from a candidate who may be a member of a research team in the laboratory of a senior investigator under whose guidance the applicant will be given flexibility to conduct a neuromuscular disease research project. Awards are for one to three years. Applications must be requested by the listed application deadlines. Guidelines are available online. Development grants are a maximum of $60,000 per year.
Requirements: To be eligible for a Development Grant, an applicant must: hold a Doctor of Medicine (M.D.), Doctor of Philosophy (Ph.D.), Doctor of Science (D.Sc.) or equivalent degree (i.e. D.O.); be a member of a research team at an appropriate institution; be qualified to conduct a program of original research under the supervision of a Principal Investigator; have an acceptable research plan for a specific disease in MDA's program; have access to institutional resources necessary to conduct the proposed research project; and have eighteen (18) months of post-doctoral research laboratory training at the time of application, but no more than 5 years (60 months). A pre-proposal form must be submitted through proposalCENTRAL to formally request an application for an MDA research grant.
Geographic Focus: All States
Date(s) Application is Due: Jan 15; Jul 15
Amount of Grant: Up to 60,000 USD
Contact: Grants Program Manager; (520) 529-2000; fax (520) 529-5454; grants@mdausa.org
Internet: http://www.mdausa.org/research/guidelines.html#developmentGrant
Sponsor: Muscular Dystrophy Association
3300 E. Sunrise Drive
Tucson, AZ 85718

MDA Neuromuscular Disease Research Grants 2501
MDA supports research aimed at developing treatments for the muscular dystrophies and related diseases of the neuromuscular system. These are the muscular dystrophies (among which are Duchenne and Becker); motor neuron diseases (including ALS and SMA); the peripheral nerve disorders (CMT and Friedreich's ataxia); inflammatory myopathies; disorders of the neuromuscular junction; metabolic diseases of muscle as well as other myopathies. Proposals from applicants outside the United States will be considered for projects of highest priority to MDA and when, in addition to the applicant's having met the requirements noted above the applicant's country of residence may not have adequate sources of financial support for biomedical research. Funding levels for primary Research Grants are unlimited. Awards are for either one, two or three years for all grant types.
Requirements: To be eligible to apply for an MDA research grant, an applicant must: hold a Doctor of Medicine (M.D.), Doctor of Philosophy (Ph.D.), Doctor of Science (D.Sc.) or equivalent degree (i.e. D.O.); be a professional or faculty member (Professor, Associate Professor or Assistant Professor) at an appropriate educational, medical or research institution; be qualified to conduct and mentor a program of original research within their own laboratory; assume both administrative and financial responsibility for the grant; and have access to institutional resources necessary to conduct the proposed research project. A pre-proposal form must be submitted through proposalCENTRAL to formally request an application for an MDA research grant.
Geographic Focus: All States
Date(s) Application is Due: Jan 15; Jul 15
Contact: Program Manager; (520) 529-2000; fax (520) 529-5454; grants@mdausa.org
Internet: http://static.mda.org/research/guidelines.html
Sponsor: Muscular Dystrophy Association
3300 E. Sunrise Drive
Tucson, AZ 85718

Mead Johnson Nutritionals Charitable Giving Grants 2502
Mead Johnson Nutritionals considers requests for charitable giving from a broad range of charitable organizations. Requests for charitable giving must be made in connection with a general fund raising effort by the charitable entity, rather than a request directed only to Mead Johnson Nutritionals. In addition, the request for a contribution must come from the organization itself and not from an individual physician, pharmacist, or health care professional who is not an employee of the charity. Applicants should allow at least three months for the processing of a funding request. Grant applications submitted less than three months prior to an event may be rejected due to insufficient time to process the application. Organizations outside the U.S. that wish to request a charitable gift should refer to Mead Johnson contact information within their country or call the charitable giving department at (812) 429-7800.
Requirements: Charitable giving is permitted if the recipient is a tax-exempt organization under Section 501(c)3 of the Internal Revenue Code.
Restrictions: As with grants, charitable giving may not be tied, in any way, to past, present, or future prescribing, purchasing, or recommending (including formulary recommendations) of any product.

Geographic Focus: All States
Contact: Administrator; (812) 429-7831; meadjohnson.grants1@bms.com
Internet: http://www.mjn.com/app/iwp/MJN/Content2.do?dm=mj&id=/MJN_Home2/mjnBtnSocialResponsibility2/mjnBtnCharitableGivingApplication&iwpst=B2C&ls=0&csred=1&r=3435833953
Sponsor: Mead Johnson Nutritionals
2400 West Lloyd Expressway
Evansville, IN 47721-0001

Mead Johnson Nutritionals Evansville-Area Organizations Grants 2503
Evansville area non-profit organizations of all varieties may submit charitable giving applications throughout the year. However, if a request exceeds $1,000 it is highly recommended that the applicant submit an application between June 1 and August 31 in the year prior to when the funds are needed. Requests greater than $1,000 are generally reviewed as a group, one time a year. If approved, funding will be awarded in the following year. Requests greater than $1,000 received after August 31 might not be considered and the applicant might be asked to reapply the following year.
Requirements: Evansville area 501(c)3 designated organizations are eligible to apply.
Geographic Focus: Indiana
Date(s) Application is Due: Aug 31
Contact: Administrator; (812) 429-7831; meadjohnson.grants1@bms.com
Internet: http://www.mjn.com/app/iwp/MJN/Content2.do?dm=mj&id=/MJN_Home2/mjnBtnSocialResponsibility2/mjnBtnCharitableGivingApplication&iwpst=B2C&ls=0&csred=1&r=3435833953#app
Sponsor: Mead Johnson Nutritionals
2400 West Lloyd Expressway
Evansville, IN 47721-0001

Mead Johnson Nutritionals Medical Education Grants 2504
Mead Johnson Nutritionals has established a Grant program to support medical education and philanthropic initiatives. The program will consider supporting Independent Medical Education (IME) programs and activities designed to enhance the professional skills and knowledge of health care professionals involved in pediatrics, when the program is independently developed and conducted by a qualified medical education provider. Supported programs must be objective, balanced, and scientifically rigorous. Health care provider educational programming can be given for: medical education programs; scientific conferences; creation of health care publications; and other educational and scientific activities. In addition to the U.S. Medical Education Grants, Mead Johnson also funds International Medical Education Grants. Organizations outside the U.S. that wish to request an Independent Medical Education (IME) grant should refer to Mead Johnson contact information within their country or contact Mead Johnson at meadjohnson.grants1@bms.com.
Requirements: Eligible recipients for medical education grants include: academic medical centers; accredited medical education companies; community health centers; hospitals; managed care organizations; medical/professional societies; medical universities; patient advocacy groups; pharmacies; and similar entities.
Restrictions: Grants may not be tied, in any way, to past, present, or future prescribing, purchasing, or recommending (including formulary recommendations) of any product.
Geographic Focus: All States
Contact: Administrator; (812) 429-7831; meadjohnson.grants1@bms.com
Internet: http://www.mjn.com/app/iwp/MJN/Content2.do?dm=mj&id=/MJN_Home2/mjnBtnSocialResponsibility2/mjnBtnMedicalEducationApplications&iwpst=B2C&ls=0&csred=1&r=3435836001
Sponsor: Mead Johnson Nutritionals
2400 West Lloyd Expressway
Evansville, IN 47721-0001

Mead Johnson Nutritionals Scholarships 2505
Mead Johnson Nutritionals provides scholarship grants to medical students, residents, fellows, and other healthcare professionals in training to attend classes and major educational or scientific meetings of a national medical or professional association. The selection of the individuals to receive such support and the grant application must be made by the institution at which they are being trained or the organization hosting the event. Allow at least three months for the processing of your funding application. Independent Medical Education (IME) grant applications submitted less than three months prior to an event may be rejected, due to insufficient time to process the application.
Requirements: Eligible recipients for such grants: academic medical centers; hospitals; medical/professional societies; and medical universities.
Geographic Focus: All States
Contact: Administrator; (812) 429-7831; meadjohnson.grants1@bms.com
Internet: http://www.mjn.com/app/iwp/MJN/Content2.do?dm=mj&id=/MJN_Home2/mjnBtnSocialResponsibility2/mjnBtnMedicalEducationApplications/MedScholarships&iwpst=B2C&ls=0&csred=1&r=3435836794
Sponsor: Mead Johnson Nutritionals
2400 West Lloyd Expressway
Evansville, IN 47721-0001

MeadWestvaco Foundation Sustainable Communities Grants 2506
The Foundation began as the Mead Corporation Foundation in 1957 and the Westvaco Foundation in 1953, anchored in shared values for community and environmental enrichment. Since merging in 2003, it has offered more than $36 million in support to targeted programs. Adding to this, company employee volunteers have donated over 572,000 hours to more than 3,000 qualified organizations. Across its diverse efforts, the Foundation focuses on three key areas for strategic grants and volunteer initiatives. In the area of Sustainable Communities, it partners with organizations that help people and families rely on themselves, and offers support to their neighbors. The Foundation's support of central business districts gives communities a strong center around which to grow. And its work with youth programs nurtures tomorrow's community leaders. The Foundation's primary focus is to address important community needs and improve the quality of life in communities where MeadWestvaco operates.
Requirements: Nonprofit 501(c)3 organizations in or serving the following U.S. areas are eligible to apply: Cottonton and Lanett, Alabama; Bentonville, Arkansas; Chino, Corona, and Tecate, California; District of Columbia; Miami and St. Petersburg, Florida; Atlanta, Roswell, Smyrna, and Waynesboro, Georgia; Bartlett, Chicago, Itasca, Lake in the Hills, Schaumburg, and West Chicago, Illinois; Winfield, Kansas; Wickliffe, Kentucky; DeRidder, Louisiana; Minneapolis, Minnesota; Grandview, Missouri; Reno, Nevada; North Brunswick, Rumson, and Tinton Falls, New Jersey; New York, New York; Mebane, North Carolina; Cincinnati and Powell, Ohio; North Charleston and Summerville, South Carolina; Coppell, Evadale, and Silsbee, Texas; Appomattox, Covington, Low Moor, Raphine, and Richmond, Virginia; and Elkins and Rupert, West Virginia. Most areas worldwide are also eligible.
Geographic Focus: Alabama, Arkansas, California, District of Columbia, Florida, Georgia, Illinois, Kansas, Louisiana, Minnesota, Nevada, New Jersey, New York, North Carolina, Ohio, South Carolina, Texas, Virginia, West Virginia, All Countries
Amount of Grant: 250 - 1,500,000 USD
Samples: Chicago Horticultural Society, Chicago, Illinois, $2,500 - environmental support; Downtown Dayton Partnership, Dayton, Ohio, $$25,000 - civic betterment; Richmond Ballet, Richmond, Virginia, $15,000 - arts and culture.
Contact: Christine W. Hale, Manager, Contributions Programs; (804) 444-2531; fax (804) 444-1971; foundation@mwv.com
Internet: http://www.meadwestvaco.com/corporate.nsf/mwvfoundation/applicationsGuidelines
Sponsor: MeadWestvaco Foundation
501 South 5th Street
Richmond, VA 23219-0501

Mead Witter Foundation Grants 2507
Incorporated in Wisconsin in 1951, the Foundation awards grants to nonprofits in company operating locations, primarily in central and northern Wisconsin. Local community programs are funded, focusing on education, including scholarships in higher education and direct contributions to colleges and universities. Grants also are made to support the arts, health care, human services, youth organizations, environmental programs, and Christian and Roman Catholic organizations for nonreligious purposes. Types of support include general operating support, continuing support, annual campaigns, capital campaigns, building construction and renovation, equipment acquisition, endowment funds, professorships, seed grants, scholarship funds, and employee-matching gifts. The board meets twice annually to consider requests. Full proposal is by invitation only.
Requirements: Wisconsin 501(c)3 organizations may apply.
Restrictions: The foundation does not support religious, athletic, or fraternal groups, except when these groups provide needed special services to the community at large; direct grants or scholarships to individuals; community foundations; or flow-through organizations that redispense funds to other charitable causes.
Geographic Focus: Wisconsin
Amount of Grant: 25,000 - 18,000,000 USD
Samples: Carroll University, Waukesha, Wisconsin, $1,030,000; Medical College of Wisconsin, Milwaukee, $512,000; Beloit College, Beloit, $300,000.
Contact: Cynthia Henke, President; (715) 424-3004; fax (715) 424-1314
Sponsor: Mead Witter Foundation
P.O. Box 39
Wisconsin Rapids, WI 54495-0039

Medical Informatics Section/MLA Career Development Grant 2508
In 1997, the Medical Informatics Section of the Medical Library Association (MLA) established its Career Development Grant. The section awards one individual up to $1,500 to support a career-development activity that will contribute to the advancement of the field of medical informatics. Application guidelines and form are available at the website. The application and supporting materials must be received by MLA via email, fax, or mail by December 1. emailed applications and documents will be accepted as PDF or MS Word files only. Founded in 1898, the Medical Library Association (MLA) is a nonprofit, educational organization of more than 1,100 institutions and 3,600 individual members in the health sciences information field committed to educating health-information professionals, supporting health-information research, promoting access to the world's health-sciences information, and working to ensure that the best health information is available to all.
Requirements: The applicant may be either a recent graduate or a librarian with significant experience in health sciences. Applicants must satisfy the following eligibility criteria: have a Masters of Library Science or equivalent degree; have the potential to make significant contributions to the field of medical informatics; and preferably be a member of the Medical Informatics Section of the MLA.
Geographic Focus: All States, All Countries
Date(s) Application is Due: Dec 1
Amount of Grant: Up to 1,500 USD
Contact: Carla Funk, Executive Director; (312) 419-9094, ext. 14; fax (312) 419-8950; mlapd2@mlahq.org or grants@mlahq.org
Internet: http://www.mlanet.org/awards/grants/index.html
Sponsor: Medical Library Association
65 East Wacker Place, Suite 1900
Chicago, IL 60601-7246

MedImmune Charitable Grants 2509
The program supports educational and charitable activities that ultimately improve patient healthcare and further understanding of disease states and treatment options within the community. While the sponsor will consider initiatives in many areas, priority is given to those in the following areas of interest: (a) health and science education for young people; (b) influenza; (c) immunology; (d) pediatric infectious disease; (e) pediatric respiratory disease.
Requirements: Eligible programs include: (a) Community outreach designed to increase public awareness (e.g, fundraising walk or run, nonprofit community hospital or hospital foundation, etc.); (b) Fellowships; (c) Independent educational/scientific programs, accredited and non-accredited; (d) Medical or Scientific education for Healthcare Professionals; (e) Scholarships; or, (f) Academic Chair Endowments. Applications must be completed online at https://web.medimmune.com/corporategrants/, and must be submitted at least 6 weeks prior to the program date or event.
Restrictions: Applications will not be considered from: (1) individuals, physicians, or physician practices; (2) retroactive support; (3) exhibit or display booths at conventions/conferences; (4) textbooks; (5) operating expenses; (6) fraternal, social, leisure, labor or political organizations; (7) corporate foundations; (8) public or private schools or scholarship funds, except where a grant is specific to a science education initiative.
Geographic Focus: All States
Contact: Grants Office; (866) 396-6235; corporatefunding@medimmune.com
Internet: http://www.medimmune.com/culture_grants.aspx
Sponsor: MedImmune
One MedImmune Way
Gaithersburg, MD 20878

Medtronic Foundation CommunityLink Health Grants 2510
At Medtronic, the Foundation makes it a priority to enhance the vitality of the communities where its employees live and work. The CommunityLink program helps the Foundation accomplish this by supporting health, education and community programs throughout the world. The CommunityLink Health Grants support programs that improve the health and welfare of people, with a focus on Medtronic's areas of expertise. The Foundation gives priority to programs that help people develop and maintain health lifestyles, with particular interest in programs that reduces differences in health care and that support healthy lifestyles, such as nutrition and fitness. The application and guidelines are located on the Foundation website. Guidelines and funding varies depending on location.
Requirements: Only U.S. 501(c)3 nonprofit organizations and equivalent international organizations may apply.
Restrictions: The Foundation does not fund the following: Continuing Medical Education (CME) grants; 501(c)3 type 509(a)3 supporting organizations; capital or capital projects; fiscal agents; fundraising events/activities, social events or goodwill advertising; general operating support; general support of educational institutions; greater Twin Cities United Way supported programs; individuals, including scholarships for individuals; lobbying, political or fraternal activities; long-term counseling or personal development; program endowments; purchases of automatic external defibrillators (AEDs); religious groups for religious purposes; private foundations; or research.
Geographic Focus: Arizona, California, Colorado, Florida, Indiana, Massachusetts, Minnesota, Puerto Rico, Tennessee, Texas, Washington, Canada, Ireland, Japan, Netherlands, Switzerland
Date(s) Application is Due: Nov 16
Contact: Deb Anderson; (763) 514-4000 or (800) 633-876; deb.anderson@medtronic.com
Internet: http://www.medtronic.com/foundation/programs_cl.html
Sponsor: Medtronic Foundation
304 Landmark Center, 75 West Fifth Street
St. Paul, MN 55102

Medtronic Foundation Community Link Human Services Grants 2511
At Medtronic, the Foundation makes it a priority to enhance the vitality of the communities where its employees live and work. The CommunityLink program helps the Foundation accomplish this by supporting health, education and community programs throughout the world. The CommunityLink Human Services Grants support human services in the U.S. that help individuals become more self-sufficient. Programs reach out to a wide range of people, such as: economic literacy programs to help low-income women gain economic stability; peer counseling for teens in crisis situations; homeless shelters; and subsidized childcare programs for low-income parents. The application and current guidelines are available at the website.
Requirements: Only U.S. 501(c)3 nonprofits and equivalent international organizations may apply. Applicants must be located in U.S. areas where Medtronic has employees.
Restrictions: The Foundation does not fund the following: Continuing Medical Education (CME) grants; 501(c)3 type 509(a)3 supporting organizations; capital or capital projects; fiscal agents; fundraising events/activities, social events or goodwill advertising; general operating support; general support of educational institutions; greater Twin Cities United Way supported programs; individuals, including scholarships for individuals; lobbying, political or fraternal activities; long-term counseling or personal development; program endowments; purchases of automatic external defibrillators (AEDs); religious groups for religious purposes; private foundations; or research.
Geographic Focus: Arizona, California, Colorado, Florida, Indiana, Massachusetts, Minnesota, Tennessee, Texas, Washington
Date(s) Application is Due: Nov 16
Contact: Deb Anderson; (763) 514-4000 or (800) 633-876; deb.anderson@medtronic.com
Internet: http://www.medtronic.com/foundation/programs_cl.html
Sponsor: Medtronic Foundation
710 Medtronic Parkway
Minneapolis, MN 55432-5604

Medtronic Foundation Fellowships 2512
The U.S. based Medtronic Fellows Program works to prepare the next generation of scientific innovators. Through the Foundation's partnerships with preselected universities such as Johns Hopkins, Duke University, Stanford University, and Massachusetts Institute of Technology, it supports college and university scholars and fellows in science, engineering and health-related fields. Applicants should review the website for specific university participants and contact the individual universities for more information.
Geographic Focus: All States, United Kingdom
Contact: Kris Fortman, U.S. Programs Manager; (763) 514-4000 or (800) 633-8766; fax (763) 505-2648; kris.fortman@medtronic.com
Internet: http://www.medtronic.com/foundation/programs_msp.html
Sponsor: Medtronic Foundation
710 Medtronic Parkway
Minneapolis, MN 55432-5604

Medtronic Foundation HeartRescue Grants 2513
The HeartRescue Grants fund organizations and government agencies that provide community-based education, training, and awareness programs related to sudden cardiac arrest (SCA) and the use of automatic external defibrillators. Working with select partners, efforts are focused on developing an integrated community response to SCA, coordinating education, training and the application of high-tech treatments among the general public, first responders, emergency medical services, and hospitals. Funds may be used for all aspects of a systems-based approach to SCA, including human resources, data collection and measurement tools, training costs at all levels of the system, and technology. If your organization is interested in participating in the program, please submit a letter of inquiry.
Requirements: Nonprofit organizations and government agencies that provide education, training, and awareness programs related to sudden cardiac death, early defibrillation, and early intervention are eligible. Organizations should send a 1-2 page letter of inquiry with detailed information about their organization, their project, their staff, and plans for implementing their program.
Restrictions: The Foundation does not fund the following: Continuing Medical Education (CME) grants; 501(c)3 type 509(a)3 supporting organizations; capital or capital projects; fiscal agents; fundraising events/activities, social events or goodwill advertising; general operating support; general support of educational institutions; greater Twin Cities United Way supported programs; individuals, including scholarships for individuals; lobbying, political or fraternal activities; long-term counseling or personal development; program endowments; purchases of automatic external defibrillators (AEDs); religious groups for religious purposes; private foundations; or research.
Geographic Focus: All States
Amount of Grant: 300,000 - 500,000 USD
Contact: Deb Anderson, Grants Administrator; (763) 514-4000 or (800) 328-2518; fax (763) 505-2648; deb.anderson@medtronic.com
Internet: http://www.medtronic.com/foundation/programs_hr.html
Sponsor: Medtronic Foundation
710 Medtronic Parkway
Minneapolis, MN 55432-5604

Medtronic Foundation Patient Link Grants 2514
Through Patient Link grants, Medtronic partners with national and international patient organizations that educate, support, and advocate on behalf of patients and their families to improve the lives of people with chronic diseases. Grants are offered in the U.S., Canada, and Europe. In partnership with its Patient Link organizations, the Foundation: empowers patients to become active partners in their health care; engages patients from cultural communities who have not had access to health information and health care; and provides support to patients through what can be difficult emotional times. Each location requires different guidelines and gives a different range of funding.
Requirements: Health grants only fund 501(3)c U.S. or nonprofit international organizations. Public, private, government institutions, and educational institutions are the only exceptions to this rule. U.S. based organizations must gather and complete the application and corresponding documents. Organizations outside the U.S. complete a letter of inquiry to the corresponding contact near their location.
Restrictions: Grants are not available for the following: Continuing Medical Education (CME) grants; 501(c)3 type 509(a)3 supporting organizations; capital or capital projects; fiscal agents; fundraising events/activities, social events or goodwill advertising; general operating support; general support of educational institutions; greater Twin Cities United Way supported programs; individuals, including scholarships for individuals; lobbying, political or fraternal activities; long-term counseling or personal development; program endowments; purchases of automatic external defibrillators; religious groups for religious purposes; private foundations; or research.
Geographic Focus: All States, Belgium, Canada, Denmark, Finland, France, Germany, Great Britain, Iceland, Ireland, Italy, Japan, Lithuania, Luxembourg, Netherlands, Norway, Poland, Portugal, Spain, Sweden, Switzerland, The Netherlands, United Kingdom, Vatican City
Date(s) Application is Due: May 14; Sep 10; Nov 12
Amount of Grant: 10,000 - 175,000 USD
Contact: Deb Anderson, Grants Administrator; (763) 514-4000 or (800) 633-8766; fax (763) 505-2648; deb.anderson@medtronic.com
Internet: http://www.medtronic.com/foundation/programs_pl.html
Sponsor: Medtronic Foundation
710 Medtronic Parkway
Minneapolis, MN 55432-5604

Medtronic Foundation Strengthening Health Systems Grants 2515
Non-communicable diseases (NCDs), including cardiovascular disease, diabetes, cancer, and chronic respiratory disease, are the leading cause of death and disability in the world. But while prevention initiatives and lifesaving treatments have been developed for NCDs, they are not widely available in many low- and middle-income countries. The Strengthening Health Systems Grants expand access to quality healthcare by strengthening health systems and integrating NCDs into primary care in developing countries. The Medtronic Foundation will work with a limited number of chronic disease centers and experts on a global level, and in our priority countries, to address the prevention and management of NCDs. Our emphasis will be on integrating NCDs into primary healthcare, and on improving global understanding of best practices in diabetes and cardiovascular disease. Grants are awarded for specific projects and programs. Being preselected as a potential grantee does not guarantee that a grant will be awarded, as projects are evaluated on their individual merits. Organizations should also consult the website for a detailed list of proposal criteria. Applications are accepted throughout the year, but review of an application usually takes four to five months.
Requirements: Organizations should be located in the countries served by this grant. After reviewing the general guidelines, applicants should send a two- or three-page letter of inquiry. After reviewing the letter, a Foundation representative may recommend that a full proposal is submitted.
Restrictions: Grants for endowment, equipment, or for capital projects are not considered. The Foundation does not support individuals, religious groups for religious purposes, fundraising events or activities, social events or goodwill advertising, reimbursable medical treatment, scientific research, lobbying, or political or fraternal activities.
Geographic Focus: Austria, Belarus, Brazil, Bulgaria, China, Croatia, Czech Republic, Germany, Hungary, India, Liechtenstein, Moldova, Poland, Romania, Russia, Serbia, Slovakia, Slovenia, South Africa, Switzerland, Ukraine
Amount of Grant: 50,000 - 250,000 USD
Contact: Deb Anderson, Grants Administrator; (763) 514-4000 or (800) 633-8766; fax (763) 505-2648; deb.anderson@medtronic.com
Internet: http://www.medtronic.com/foundation/programs_shs_guidelines.html
Sponsor: Medtronic Foundation
710 Medtronic Parkway
Minneapolis, MN 55432-5604

Memorial Foundation Grants 2516
The Memorial Foundation awards grants to support nonprofit organizations that provide services only to people who live in the area served by Nashville Memorial Hospital. The Foundation also strives to respond to immediate, critical needs that arise in the community. With assistance from the Foundation, organizations including The Salvation Army, The American Red Cross, Second Harvest Food Bank, Nashville Tree Foundation, and YWCA have received funds. Special emphasis is given to organizations that focus on health; youth and children; senior citizens; education; human and social services; community services; and substance abuse (alcohol, drugs, and tobacco). Types of support include capital projects; general operating support; and start-up projects for new initiatives that address important, unmet community needs and that demonstrate potential for ongoing operational support from other sources.
Requirements: An applicant organization must be exempt from federal taxation under Section 501(c)3 of the Internal Revenue Code and not be a private foundation as described in Section 509(a) in order to be eligible.
Restrictions: The Memorial Foundation does not fund grants for the following: individuals; newsletters, magazines; churches and religious organizations for projects that primarily benefit their own members (exception: church-based programs with broad community support and separate financial statements); disease-specific organizations seeking support for national research projects and programs; sponsor special events, productions, telethons, performances, or similar fundraising and advertising activities (exceptions may be given for approved educational videos); legislative lobbying or other political purposes; retire accumulated debt; bricks and mortar capital projects for colleges, universities, and private or public school education; computer labs and related technologies to schools; or multi-year grants for operating funds.
Geographic Focus: Tennessee
Contact: Joyce Douglas, Grants Contact; (615) 822-9499; fax (615) 822-7797
Internet: http://www.memfoundation.org
Sponsor: Memorial Foundation
100 Bluegrass Commons Boulevard, Suite 320
Hendersonville, TN 37075

Mercedes-Benz USA Corporate Contributions Grants 2517
The company has committed to conducting programs with the concentration of funds geared toward educational causes and organizations, in particular those that help to empower the next generation and underserved groups; diversity programs; and women's initiatives. Specific programs in the categories of health and human services, and civic and community also are supported. There are no application forms or annual deadlines. Applicants should begin by submitting a proposal idea to the nearest company facility.
Requirements: IRS 501(c)3, 4, 6, or 9 organizations whose primary influences and business operations are in the United States and who enhance the quality of life in Mercedes-Benz communities are eligible. These company communities include: Tuscaloosa, Alabama; Carson, California; Irvine, California; Rancho Cucamonga, California; Jacksonville, Florida; Carol Stream, Illinois; Itasca, Illinois; Rosemont, Illinois; Belcamp, Maryland; Baltimore, Maryland; Montvale, New Jersey; Parsippany, New Jersey; Robinsville, New Jersey; and Fort Worth, Texas.
Geographic Focus: Alabama, California, Florida, Illinois, Maryland, New Jersey, Texas
Amount of Grant: 2,500 - 5,000 USD
Contact: Robert Moran, Director of Communications; (201) 573-2245 or (201) 573-0600; fax (201) 573-4787; robert.moran@mbusa.com
Internet: http://www.mbusa.com/mercedes/about_us/mbcommunity
Sponsor: Mercedes-Benz USA
One Mercedes Drive
Montvale, NJ 07645

Mericos Foundation Grants 2518
The Mericos Foundation primarily awards grants to organizations based in California with emphasis on Santa Barbara. Fields of interest include aging; animals/wildlife; environment; arts and arts education; child development; children and youth; elementary and secondary education; higher education; hospitals; medical care; medical research; libraries; and museums (art and natural history). Types of support include building/renovation, equipment, fellowships, general/operating support, matching/challenge support, and program development. Grants are usually initiated by the foundation. Interested organizations should contact the Vice-President to discuss their project.
Requirements: Nonprofit organizations in California are eligible to apply.
Restrictions: Grants are not made to individuals.
Geographic Focus: California
Amount of Grant: 15,000 - 200,000 USD
Samples: Music Academy of the West, Santa Barbara, California, $125,000; Friends of Independent Schools and Better Education, Tacoma, Washington, $25,000; Los Angeles Children's Chorus, Pasadena, California, $25,000.
Contact: Linda Blinkenberg, Vice President; (626) 441-5188; fax (626) 441-3672
Sponsor: Mericos Foundation
625 South Fair Oaks Avenue, Suite 360
South Pasadena, CA 91030-2630

Meriden Foundation Grants 2519
The foundation awards grants to eligible Connecticut nonprofit organizations in its areas of interest, including arts, children and youth, civic affairs, Christian organizations and churches, health organizations and hospitals, higher education, public libraries, and social services. Types of support include annual campaigns, general operating support, scholarships, and social services delivery. There are no application deadlines. A formal application is required, and the initial approach should be a letter, on organizational letterhead, describing the project and requesting an application.
Requirements: Connecticut nonprofits in the Meriden-Wallingford area are eligible.
Geographic Focus: Connecticut
Amount of Grant: 175 - 10,000 USD
Samples: Best Friends Animal Sanctuary, Kanab, Utah, $3,500 - general operations; Church of the Holy Angels, Meriden, Connecticut, $3,296 - general operations; Franciscan Home Care and Hospice, Meriden, Connecticut, $5,000 - to purchase laptop computers.
Contact: Jeffrey F. Otis, Director; (203) 782-4531; fax (203) 782-4530
Sponsor: Meriden Foundation
123 Bank Street
Waterbury, CT 06702-2205

Merkel Foundation Grants 2520
The Merkel Foundation, established by Daniel A. and Betty Merkel in Sheboygan, Wisconsin, is interested in supporting a variety of Wisconsin organizations. The Foundation's primary interests are research, religion, education, social services, and relief programs. There are no specific application formats or deadlines with which to adhere. Applicants should begin the process by contacting the Foundation directly. Generally, support is given throughout Wisconsin and, ocassionally, across the country. Amounts range from $100 to $10,000.
Geographic Focus: Wisconsin
Amount of Grant: Up to 10,000 USD
Samples: Common Hope, St. Paul, Minnesota, $3000 - support for international relief programs; Wisconsin Foundation for Independent Colleges, Milwaukee, Wisconsin, $1,000 - support for religious programs.
Contact: Betty Merkel; (920) 457-5051; fax (920) 457-1485; info@americanortho.com
Sponsor: Merkel Foundation
3712 Bismarck Circle
Sheboygan, WI 53083

Merrick Foundation Grants 2521
The Merrick Foundation was established in 1948 by Ward S. Merrick, Sr., as a memorial to his father F.W. Merrick and at the request of F.W.'s wife, Elizabeth B. Merrick. The mission of the Foundation is to enhance the quality of life and the improvement of health for Oklahoma residents and their communities, with a primary emphasis on south central Oklahoma. With this goal in mind, the Foundation trustees are committed to furthering the philanthropic vision of Ward S. Merrick, Sr., by awarding grants to charitable organizations that foster independence and achievement, and that stimulate educational, economic, and cultural growth. Primary fields of interest include: the arts; higher education; human services; medical research; and youth services. The majority of awards are given for general operations. Most recent grants have ranged from $500 to $25,000. The grants committee will review all request letters received through the online grant application process and will email instructions on how to complete an application to qualifying charities. Grant applications will be due October 1 each year.
Requirements: The trustees invite proposals from 501(c)3 organizations located in Oklahoma. A letter of request is due by the August 15 deadline, summarizing the project for funding must be submitted through the online grant application process.

Geographic Focus: Oklahoma
Date(s) Application is Due: Oct 1
Amount of Grant: 500 - 25,000 USD
Samples: Arbuckle Life Solutions, Ardmore, Oklahoma, $5,000 - general operation support; CASA of Southern Oklahoma, Oklahoma City, Oklahoma, $10,000 - general operating support; Delta Upsilon Fraternity, Oklahoma City, Oklahoma, $5,000 - general operating support.
Contact: Randy Macon, Executive Director; 405-755-5571; fax (405) 755-0938; fwmerrick@foundationmanagementinc.com
Internet: https://www.foundationmanagementinc.com/foundations/merrick-foundation/
Sponsor: Merrick Foundation
2932 NW 122nd Street, Suite D
Oklahoma City, OK 73120-1955

Merrick Foundation Grants 2522
The goal of Merrick Foundation is to improve the quality of life in the Merrick County area of Nebraska by supporting needs that are not being met in the areas of civic, cultural, health, education and social service. Grants are awarded on the condition that grantees attend Merrick Foundation's Annual Meeting on the 4th Monday of November and provide a poster board display demonstrating the goal, progress or completion of their grant project. Proposals from organizations demonstrating broad community support for their proposed programs are given priority consideration. Most recent awards have ranged from $5,800 to $45,000. Grant requests over $10,000 may be required to attend a regular meeting for explanations and/or clarifications. Grants over $15,000 shall be reviewed in January and June. Deadline dates for submitting applications are due the 10th of each month or the following Monday if the 10th falls on a weekend. Grants are approved every month except December.
Requirements: Only organizations in Merrick County or organizations serving Merrick County residents are eligible to apply.
Restrictions: The Foundation does not make grants to individuals, for religious purposes, or organizations that operate for profit. he Foundation as a general policy gives less consideration to applications from tax-supported institutions, veterans and labor organizations, social clubs and fraternal organizations. A grant cannot be used for political purposes.
Geographic Focus: Nebraska
Date(s) Application is Due: Jan 1; Feb 1; Mar 1; Apr 1; May 1; Jun 1; Jul 1; Aug 1; Sep 1; Oct 1; Nov 1
Amount of Grant: 5,800 - 45,000 USD
Samples: St. Paul's Lutheran Church, Grand Island, Nebraska, $12,000 - community help programs; Merrick County Child Development Center, Central City, Nebraska, $15,000 - general operating costs; Platte Peer Group, Chapman, Nebraska, $45,000 - park development project support.
Contact: Chuck Griffith; (308) 946-3707; merrickfoundation@gmail.com
Internet: http://www.merrick-foundation.org/grants.htm
Sponsor: Merrick Foundation
1532 17th Avenue, P.O. Box 206
Central City, NE 68826

Mervin Bovaird Foundation Grants 2523
The foundation awards grants, with a focus on Tulsa, OK, to nonprofit organizations such as churches, homeless shelters, medical centers, nursing homes, parochial schools, religious welfare organizations, the Salvation Army, and youth organizations. Areas of interest include arts, community development, education, environment, health care and health organizations, and social services. Types of support include general operating support, matching, project, and research grants.
Requirements: Nonprofits in and serving the population of Tulsa, Oklahoma, are eligible. Applicants should submit a brief letter of inquiry, including program and organization descriptions.
Geographic Focus: Oklahoma
Date(s) Application is Due: Nov 15
Amount of Grant: Up to 250,000 USD
Contact: R. Casey Cooper, (918) 592-3300; casey.cooper@cmw-law.com
Sponsor: Mervin Bovaird Foundation
401 South Boston Avenue, Suite 3300
Tulsa, OK 74103

Mesothelioma Applied Research Foundation Grants 2524
The program seeks to stimulate translational research for the treatment of pleural or peritoneal mesothelioma. Grants support projects that apply cutting-edge general cancer research concepts specifically to mesothelioma, and translate directly to improved mesothelioma treatment. Eligible projects may relate to either benchwork research or clinical research, must not be presently funded or pending review, and may be conducted through any not-for-profit academic, medical, or research institution, in the United States or abroad. Grants are awarded for 12 months, renewable for 12 more months.
Requirements: The investigator must be affiliated with a nonprofit academic, medical, or research institution.
Geographic Focus: All States
Date(s) Application is Due: Aug 15
Amount of Grant: 50,000 USD
Contact: Danele Mangone; (805) 563-8400, ext. 7273; dmangone@curemeso.org
Internet: http://www.curemeso.org/site/c.kkLUJ7MPKtH/b.4097447/k.E6A3/Mesothelioma_Research_Grants__Call_for_Applications.htm
Sponsor: Mesothelioma Applied Research Foundation
3944 State Street, Suite 340, P.O. Box 91840
Santa Barbara, CA 93105

Meta and George Rosenberg Foundation Grants 2525
The Meta and George Rosenberg Foundation was established in California in 1991 in honor of Meta and George Rosenberg. Meta, who would become an Emmy-winning executive producer of the durable television series "The Rockford Files," married talent agent George "Rosey" Rosenberg in 1947, and the two of them soon began representing writers and actors. They also sold innovative series to television networks, including "Julia," starring Diahann Carroll as an African American nurse and single mother; "Hogan's Heroes," a comedy set in a World War II German prison camp; and "Ben Casey," among the first series showcasing medicine as drama. Throughout her career, Meta pursued an interest in photography, collecting the works of internationally known photographers such as Henri Cartier-Bresson, Ansel Adams, and Irving Penn, and aiming her own Leica camera at street scenes from Los Angeles to Paris. Currently, the Foundation's primary fields of interest include: arts and culture; children's diseases; elementary and secondary education; AIDS; and human services. Funding is concentrated in southern California, with some grants being awarded in the State of Iowa. The annual deadline for application submission is August 31. Most recent awards have ranged from $10,000 to $140,000.
Geographic Focus: California, Iowa
Date(s) Application is Due: Aug 31
Amount of Grant: 10,000 - 140,000 USD
Samples: Project Grad Los Angeles, North Hollywood, California, $16,800 - general operating support (2014); University of Iowa Foundation, Iowa City, Iowa, $140,000 - general fund contribution (2014).
Contact: Susan J. Ollweiler, (323) 634-2400
Sponsor: Meta and George Rosenberg Foundation
5900 Wilshire Boulevard, Suite 2300
Los Angeles, CA 90036-5050

MetroWest Health Foundation Capital Grants for Health-Related Facilities 2526
The MetroWest Health Foundation offers a limited number of capital grants on a highly competitive basis to community-based health organizations within its twenty-five communities. To be competitive for such a grant, the applicant organization must demonstrate a record of outstanding service to the community, strong program and financial management, significant private support, and a demonstrated need for capital funds that will directly impact on the organization's ability to deliver health services. New construction, additions to an existing structure, and extensive renovation of an existing structure will qualify for consideration. In addition, furnishings, equipment and site work associated with any of the above qualified projects are eligible for funding.
Requirements: The Foundation supports programs that directly benefit the health of those who live and work in one of the 25 communities served by the Foundation: Ashland, Bellingham, Dover, Framingham, Franklin, Holliston, Hopedale, Hopkinton, Hudson, Marlborough, Medfield, Medway, Mendon, Milford, Millis, Natick, Needham, Norfolk, Northborough Sherborn, Southborough, Sudbury, Wayland, Wellesley and Westborough. Such support is limited to organizations that qualify as tax-exempt under Section 501(c)3 of the IRS Code, or organizations that are recognized as instrumentalities of state or local government. Applicants are expected to have completed appropriate organizational, financial and facilities planning prior to applying for a capital grant from the Foundation. The Foundation is not positioned to be the sole or major funding of any individual capital project. Therefore, applicants must demonstrate that other funding for the project has been raised or is available through traditional capital financing mechanisms. Organizations seeking capital grants should send a brief 2-3 page letter outlining the proposed project. Requests for full proposals will sent to only those projects the Foundation wishes to consider for funding. Each capital grant is limited to three years in duration.
Restrictions: No single organization shall receive more than one capital grant from the Foundation at a time. Grants will not be made to: projects that supplant or substitute for government funding from local, state, or federal sources; for endowments or debt retirement; program support or fundraising appeals; projects that have been unsuccessfully submitted to the Foundation before.
Geographic Focus: Massachusetts
Contact: Cathy Glover; (508) 879-7625; fax (508) 879-7628; info@mwhealth.org
Internet: http://www.mwhealth.org/GrantsampScholarships/WhatWeFund/tabid/229/Default.aspx
Sponsor: MetroWest Health Foundation
161 Worcester Road, Suite 202
Framingham, MA 01701

MetroWest Health Foundation Grants--Healthy Aging 2527
Since 1999, the MetroWest Health Foundation has provided over funds to non-profit and government organizations to improve health services within its 25-town service area. The Foundation describes its grantmaking efforts as both reactive and proactive. Its reactive grantmaking includes two annual requests for proposals (spring and fall) where grant applications are solicited from area organizations. These requests for proposals target specific needs or areas of interest, such as access to health care, disease prevention and health promotion. Its proactive grantmaking targets community needs identified by the Foundation. Here the Foundation develops more comprehensive strategies for addressing community health needs. To date, the Foundation's proactive grantmaking has targeted such issues as child obesity, racial and ethnic health disparities, and adolescent substance abuse.
Requirements: The Foundation supports programs that directly benefit the health of those who live and work in one of the 25 communities served by the Foundation: Ashland, Bellingham, Dover, Framingham, Franklin, Holliston, Hopedale, Hopkinton, Hudson, Marlborough, Medfield, Medway, Mendon, Milford, Millis, Natick, Needham, Norfolk, Northborough Sherborn, Southborough, Sudbury, Wayland, Wellesley and Westborough. Such support is limited to organizations that qualify as tax-exempt under Section 501(c)3 of the IRS Code, or organizations that are recognized as instrumentalities of state or local government. The Foundation requires applicants to submit concept papers

prior to a full proposal. Concept papers help the Foundation assess whether or not the proposed project is aligned with its funding priorities. Only a limited number of proposals will be funded with a maximum grant amount of $50,000; funding is limited to one year in duration, and funds cannot be used to supplant ongoing government operations or support. Healthy Aging proposals must address one of the following objectives: (1) Decrease the number of MetroWest older adults who are hospitalized each year from injuries due to falls. Funds can be used to expand the use of evidence-based fall prevention programming within the region. (2) Expand older adults participation in local programs and initiatives that actively engage them in their community, including opportunities for socialization; group activities; volunteering; employment and social and civic engagement. Funds can be sued to develop new or expanded programming that reach older adults, especially when they are at greatest risk for becoming isolated (death of a spouse, loss of mobility, etc.). (3) Improve services and supports for caregivers of older adults in the MetroWest region. Funds can be used to support caregiver training, respite services, information sharing or enhanced care management support.
Restrictions: The Foundation does not provide grants to individuals, nor does it provide funds for endowments, fundraising drives and events, retirement of debt, operating deficits, projects that directly influence legislation, political activities or candidates for public office, or programs that are customarily operated by hospitals in Massachusetts. The Foundation does not award grants to organizations that discriminate in the provision of services on the basis of race, color, religion, gender, age, ethnicity, marital status, disability, sexual orientation or veteran status.
Geographic Focus: Massachusetts
Date(s) Application is Due: Apr 12
Amount of Grant: Up to 50,000 USD
Contact: Cathy Glover, Grants Managment Director; (508) 879-7625; fax (508) 879-7628; cglover@mwhealth.org or cglover@mchcf.org
Internet: http://www.mwhealth.org/GrantsampScholarships/Overview/tabid/180/Default.aspx
Sponsor: MetroWest Health Foundation
161 Worcester Road, Suite 202
Framingham, MA 01701

MetroWest Health Foundation Grants to Reduce the Incidence of High Risk Behaviors Among Adolescents 2528
Since 1999, the MetroWest Health Foundation has provided over funds to non-profit and government organizations to improve health services within its 25-town service area. The Foundation describes its grantmaking efforts as both reactive and proactive. Its reactive grantmaking includes two annual requests for proposals (spring and fall) where grant applications are solicited from area organizations. These requests for proposals target specific needs or areas of interest, such as access to health care, disease prevention and health promotion. Its proactive grantmaking targets community needs identified by the Foundation. Here the Foundation develops more comprehensive strategies for addressing community health needs. To date, the Foundation's proactive grantmaking has targeted such issues as child obesity, racial and ethnic health disparities, and adolescent substance abuse.
Requirements: The Foundation supports programs that directly benefit the health of those who live and work in one of the 25 communities served by the Foundation: Ashland, Bellingham, Dover, Framingham, Franklin, Holliston, Hopedale, Hopkinton, Hudson, Marlborough, Medfield, Medway, Mendon, Milford, Millis, Natick, Needham, Norfolk, Northborough Sherborn, Southborough, Sudbury, Wayland, Wellesley and Westborough. Such support is limited to organizations that qualify as tax-exempt under Section 501(c)3 of the IRS Code, or organizations that are recognized as instrumentalities of state or local government. The Foundation requires applicants to submit concept papers prior to a full proposal. Concept papers help the Foundation assess whether or not the proposed project is aligned with its funding priorities. Only a limited number of proposals will be funded with a maximum grant amount of $25,000; applications may be for one, two or three years in duration, and funds cannot be used to supplant ongoing government operations or support. Applications involving schools must submit a letter signed by the Superintendent indicating support for the request. The Foundation will provide grants to municipalities and nonprofit organizations to implement programs that address youth risk behaviors as reported by the MetroWest Adolescent Health Survey. For this grant round, the focus will only be on: (a) efforts to reduce the rate of marijuana use among adolescents; (b) efforts to reduce the incidence of teenage pregnancy. Preference will be given to interventions that are evidence-based or, if no programs meet this criteria, are research-based or recognized as promising practices. In addition, communities with similar risk behavior data and demographics may consider applying for regional approaches in order to maximize impact, although funding will still be subject to the individual grant maximum.
Restrictions: The Foundation does not provide grants to individuals, nor does it provide funds for endowments, fundraising drives and events, retirement of debt, operating deficits, projects that directly influence legislation, political activities or candidates for public office, or programs that are customarily operated by hospitals in Massachusetts. The Foundation does not award grants to organizations that discriminate in the provision of services on the basis of race, color, religion, gender, age, ethnicity, marital status, disability, sexual orientation or veteran status.
Geographic Focus: Massachusetts
Date(s) Application is Due: Apr 12
Amount of Grant: Up to 25,000 USD
Contact: Cathy Glover, Grants Managment Director; (508) 879-7625; fax (508) 879-7628; cglover@mwhealth.org or cglover@mchcf.org
Internet: http://www.mwhealth.org/GrantsampScholarships/Overview/tabid/180/Default.aspx
Sponsor: MetroWest Health Foundation
161 Worcester Road, Suite 202
Framingham, MA 01701

MetroWest Health Foundation Grants to Schools to Conduct Mental Health Capacity Assessments 2529
Since 1999, the MetroWest Health Foundation has provided over funds to non-profit and government organizations to improve health services within its 25-town service area. The Foundation describes its grantmaking efforts as both reactive and proactive. Its reactive grantmaking includes two annual requests for proposals (spring and fall) where grant applications are solicited from area organizations. These requests for proposals target specific needs or areas of interest, such as access to health care, disease prevention and health promotion. Its proactive grantmaking targets community needs identified by the Foundation. Here the Foundation develops more comprehensive strategies for addressing community health needs. To date, the Foundation's proactive grantmaking has targeted such issues as child obesity, racial and ethnic health disparities, and adolescent substance abuse. In an effort to boost MetroWest public schools' ability to address mental health issues within the context of their primary function as educational organizations, the Foundation will provide grants and technical support to school districts to complete mental health capacity assessments. These assessments will measure policies, systems and activities schools have in place related to mental health prevention, early recognition and referral and intervention efforts.
Requirements: The Foundation supports programs that directly benefit the health of those who live and work in one of the 25 communities served by the Foundation: Ashland, Bellingham, Dover, Framingham, Franklin, Holliston, Hopedale, Hopkinton, Hudson, Marlborough, Medfield, Medway, Mendon, Milford, Millis, Natick, Needham, Norfolk, Northborough Sherborn, Southborough, Sudbury, Wayland, Wellesley and Westborough. Such support is limited to organizations that qualify as tax-exempt under Section 501(c)3 of the IRS Code, or organizations that are recognized as instrumentalities of state or local government. The Foundation requires applicants to submit concept papers prior to a full proposal. Concept papers help the Foundation assess whether or not the proposed project is aligned with its funding priorities. Only a limited number of proposals will be funded with a maximum grant amount of $10,000; funding is limited to one year in duration, although subsequent implementation funding may be available. The Foundation has identified the School Mental Health Capacity Instrument as an appropriate and researched measurement tool. Schools will receive training by Children's Hospital researchers in completing the assessment, feedback on their school's capacity relative to the norm, and recommendations of key areas for improvement. Funding may be used to provide stipends for key school staff to implement the assessment, substitute teacher time where necessary, travel to regional meetings, and other direct costs. Applicants should familiarize themselves with the School Mental Health Capacity Instrument in order to determine their interest and readiness to engage in such a process.
Restrictions: Funds cannot be used to supplant ongoing government operations or support.
Geographic Focus: Massachusetts
Date(s) Application is Due: Apr 12
Amount of Grant: Up to 10,000 USD
Contact: Cathy Glover; (508) 879-7625; fax (508) 879-7628; cglover@mwhealth.org
Internet: http://www.mwhealth.org/GrantsampScholarships/Overview/tabid/180/Default.aspx
Sponsor: MetroWest Health Foundation
161 Worcester Road, Suite 202
Framingham, MA 01701

Metzger-Price Fund Grants 2530
The Metzger-Price Fund, established in New York in 1970, is an independent foundation trust which offers support to the handicapped, health services, child welfare, social service agencies, recreation, and the elderly. Its primary fields of interest include: services and centers for the aged, child and youth services, community and economic development, education, family support services, health care services and health care access, and human services. The Fund's target groups include the elderly, disabled, economically disadvantaged, homeless, women, and other minorities. Funding is directed toward continuing support, general operations, and program development. There are no specific application forms or deadlines, though the board meets to discuss proposals four times each year (January, April, July, and October). An applicant's initial approach should be in the form of an application letter submitted two months prior to each board meeting.
Requirements: 501(c)3 organizations serving the residents of New York, New York, are encouraged to apply.
Restrictions: No grants to individuals, or for capital campaigns or building funds; no multiple grants in single calendar year to same organization.
Geographic Focus: New York
Contact: Isaac A. Saufer; (212) 867-9501 or (212) 867-9500; fax (212) 599-1759
Sponsor: Metzger-Price Fund, Inc.
230 Park Avenue, Suite 2300
New York, NY 10169-0005

Meyer and Stephanie Eglin Foundation Grants 2531
The Eglin Foundation supports programs in the Philadelphia area. Fields of interest include: cancer research; higher education; hospitals; Jewish agencies, synagogues, and federated giving programs; museums; music and the performing arts; and United Way programs. Applications are not required, and there are no deadlines. Organizations should submit a letter of intent.
Requirements: Along with the letter of intent, organizations should submit an annual report/audited financial statement/990; a 501(c)3 tax-exempt letter; descriptive literature about their organization; and a detailed description of the project, with amount requested.
Geographic Focus: Pennsylvania
Contact: Stephanie Eglin, President; (215) 496-9381
Sponsor: Meyer and Stephanie Eglin Foundation
Eglin Square Garage
Philadelphia, PA 19102

Meyer Foundation Benevon Grants 2532

Eugene Meyer was an investment banker, public servant under seven U.S. presidents, and owner and publisher of the Washington Post. His wife Agnes Ernst Meyer was an accomplished journalist, author, lecturer, and citizen activist. Eugene and Agnes created the Meyer Foundation in 1944. For more than sixty-five years the Meyer Foundation has identified, listened to, and invested in visionary leaders and effective community-based nonprofit organizations that work to create lasting improvements in the lives of low-income people in the Washington, D.C. metropolitan region. The foundation offers program, operating, and capital support to eligible organizations in four priority program areas: education, healthy communities, economic security, and a strong nonprofit sector. Additionally Meyer has developed a partnership with Benevon, a firm that provides training and coaching to help nonprofits implement its proprietary model for raising money from individual donors. Meyer will make a limited number of grants each year to current Meyer grantees who are planning to implement the Benevon fundraising model. These grants offset the cost of attending either of Benevon's two-day training programs—$15,000 for Benevon 101 and $22,000 for its follow-up Sustainable Funding Program—plus travel and lodging for a seven-member team. Moreover Meyer will consider up to three years of additional grant support to organizations who demonstrate success with the model so that they may continue to participate in Benevon's Sustainable Funding Program, which provides ongoing training and coaching for five years. The deadlines for submitting Benevon applications occur in January and June; exact dates may vary from year to year. Interested organizations should visit the Meyer website to obtain detailed guidelines, downloadable application forms, and current deadline dates. Applications, along with any required attachments, must be submitted electronically via Meyer's online submission system.

Requirements: 501(c)3 organizations who are current Meyer grantees are eligible to apply. Meyer grantees are usually located within and primarily serve the Washington, D.C. region. The foundation defines the Washington, D.C. region to include the following counties and cities: Washington, D.C; Montgomery and Prince George's counties in Maryland; Arlington, Fairfax, and Prince William counties in Virginia; and the cities of Alexandria, Falls Church, and Manassas Park, Virginia.
Restrictions: Only current Meyer grantees are eligible for Benevon grants. Eligibility extends for two years from the date of an organization's last Meyer grant awarded through the foundation's regular grant-making program.
Geographic Focus: District of Columbia, Maryland, Virginia
Date(s) Application is Due: Jan 10; Jun 6
Amount of Grant: 15,000 - 22,000 USD
Contact: Maegan Scott; (202) 534-1860; mscott@meyerfdn.org
Internet: http://www.meyerfoundation.org/our-programs/Benevon
Sponsor: Meyer Foundation
1250 Connecticut Avenue, Northwest, Suite 800
Washington, D.C. 20036

Meyer Foundation Healthy Communities Grants 2533

Eugene Meyer was an investment banker, public servant under seven U.S. presidents, and owner and publisher of the Washington Post. His wife Agnes Ernst Meyer was an accomplished journalist, author, lecturer, and citizen activist. Eugene and Agnes created the Meyer Foundation in 1944. For more than sixty-five years the Meyer Foundation has identified, listened to, and invested in visionary leaders and effective community-based nonprofit organizations that work to create lasting improvements in the lives of low-income people in the Washington, D.C. metropolitan region. The foundation offers program, operating, and capital support in four priority program areas: education, healthy communities, economic security, and a strong nonprofit sector. In the area of health, the foundation funds programming that facilitates the following outcomes for low-income people in the Washington, D.C. metropolitan area: access to high-quality primary care that integrates mental and behavioral health care and eliminates health disparities; access to affordable places to live, healthful food to eat, and services that promote health and personal safety; public policies at the state and local level that are aimed at strengthening the safety net, reducing poverty, and improving lives. The foundation funds clinics, social-service organizations, community-organizing groups, and multi-issue research and advocacy groups and gives priority to issues such as homelessness, child abuse, domestic violence, and rape. In the case of service organizations, the foundations gives priority to those who track participant outcomes with quantitative and qualitative measures. Letters of Intent (LOIs) may be submitted through the foundation's online application system twice a year, in January and June; exact deadline dates may vary from year to year. Applicants will receive an email confirming receipt of their LOI within two to three weeks of submission and will be notified two months after the LOI deadline whether or not they will be invited to submit a full proposal for the board meetings in April and October. Prospective applicants should visit the foundation's website for detailed funding guidelines and current deadline dates before submitting an LOI.

Requirements: Eligible applicants must be 501(c)3 organizations that are located within and primarily serve the Washington, D.C. region defined by the foundation to include the following geographic areas: Washington, D.C; Montgomery and Prince George's counties in Maryland; Arlington, Fairfax, and Prince William counties in Virginia; and the cities of Alexandria, Falls Church, and Manassas Park, Virginia. The foundation looks for organizations that demonstrate visionary and talented leadership, effectiveness, sustainability, and long-term impact.
Restrictions: The foundation does not fund medical or scientific research, organizations or programs focused on a single disease or medical condition, capital for construction or development of housing, start-up housing developers, operating support for housing developers, AIDS-related programs (the foundation supports these exclusively through the Washington AIDS Partnership), government agencies, for-profit businesses, individuals (including scholarships or other forms of financial assistance), special events or conferences, or endowments.
Geographic Focus: All States
Amount of Grant: 1,500 - 50,000 USD
Samples: Calvary Women's Services, Washington, D.C., $25,000—general operating support; Anacostia Watershed Society, Bladensburg, Maryland, $30,000—to support general operations over two years; Arlington Free Clinic, Arlington, Virginia, $35,000—to support general operations.
Contact: Julie Rogers; (202) 483-8294; fax (202) 328-6850; jrogers@meyerfdn.org
Internet: http://www.meyerfoundation.org/our-programs/grantmaking/healthy-communities
Sponsor: Meyer Foundation
1250 Connecticut Avenue, Northwest, Suite 800
Washington, D.C. 20036

Meyer Foundation Management Assistance Grants 2534

Eugene Meyer was an investment banker, public servant under seven U.S. presidents, and owner and publisher of the Washington Post. His wife Agnes Ernst Meyer was an accomplished journalist, author, lecturer, and citizen activist. Eugene and Agnes created the Meyer Foundation in 1944. For more than sixty-five years the Meyer Foundation has identified, listened to, and invested in visionary leaders and effective community-based nonprofit organizations that work to create lasting improvements in the lives of low-income people in the Washington, D.C. metropolitan region. The foundation offers program, operating, and capital support to eligible organizations in four priority program areas: education, healthy communities, economic security, and a strong nonprofit sector. Additionally the foundation has a Management Assistance Program (MAP) available to current grantees only. MAP provides grants of up to $25,000 to help Meyer grantees strengthen their management and leadership so they can serve the community more effectively. Organizations generally use MAP grants to hire consultants to help board and staff accomplish work that requires time, energy, expertise, and innovative thinking beyond everyday operations. Examples of such work include strengthening executive and board leadership, conducting organizational planning and assessment, and improving financial management and sustainability. MAP grants have proven especially beneficial to groups experiencing significant organization transitions such as shifts in funding sources, the departure of a founder, or rapid growth. MAP application deadlines coincide with those of the foundation's regular grant-making cycle; however, for time-sensitive or out-of-cycle requests (e.g. executive transition, mergers, and financial planning), organizations should use the MAP email address to get in touch with the foundation. The foundation's regular grant-making cycles start in January and in June when Letters of Intent (LOIs) are accepted through the foundation's online application system; exact deadline dates may vary from year to year. The foundation reviews MAP LOIs within one month of receiving the request. If the proposed project meets the criteria for funding and if sufficient funds remain in the budget, foundation staff will schedule a site visit or meeting to discuss the project with the applicant's executive director and key board members and staff. On the basis of the site visit, applicants will be invited to submit a full proposal. If the foundation approves the grant, the program officer will notify the executive director, usually within three months after the LOI was submitted. Prospective applicants should visit the foundation's website for detailed funding guidelines and current deadline dates before submitting an LOI.

Requirements: 501(c)3 organizations who are current Meyer grantees are eligible to apply. Meyer grantees are usually located within and primarily serve the Washington, D.C. region. The foundation defines the Washington, D.C. region to include the following counties and cities: Washington, D.C; Montgomery and Prince George's counties in Maryland; Arlington, Fairfax, and Prince William counties in Virginia; and the cities of Alexandria, Falls Church, and Manassas Park, Virginia. Grantees are responsible for paying a percentage of the total cost of their MAP project based on their annual budget; matches are as follows: 5% for an annual budget less than $250,000; 10% for an annual budget of $250,000 to $500,000; 15% for an annual budget of $500,000 to $1 million; 20% for an annual budget of $1 million to $2 million; 25% for an annual budget of $2 million to $3 million; 30% for an annual budget of $3 million to $4 million; 40% for an annual budget of $4 million to $5 million; and 50% for an annual budget of over $5 million.
Restrictions: Only current Meyer grantees are eligible for management assistance. Eligibility extends for two years from the date of an organization's last Meyer grant awarded through the foundation's regular grant-making program.
Geographic Focus: District of Columbia, Maryland, Virginia
Amount of Grant: 5,000 - 25,000 USD
Contact: Jane Robinson Ward, Grants Manager; (202) 483-8294; fax (202) 328-6850; map@meyerfdn.org or jward@meyerfdn.org
Internet: http://www.meyerfoundation.org/our-programs/management-assistance/
Sponsor: Meyer Foundation
1250 Connecticut Avenue, Northwest, Suite 800
Washington, D.C. 20036

Meyer Memorial Trust Emergency Grants 2535

Emergency Grants are intended for sudden, unanticipated and unavoidable challenges that, if not addressed immediately, could threaten an organization's stability and/or ability to achieve its mission. Examples of emergencies would include: natural disaster; theft or damage to equipment required to operate core programs; or an accident or unexpected occurrence that causes facilities to be inaccessible or programs unable to be operated until the situation is resolved. Emergency proposals can be considered at any program meeting. The application and frequently asked questions are available at the Trust's website.

Requirements: Grants are awarded to 501(c)3 nonprofit organizations in Oregon and Clark County, Washington.
Restrictions: Processing grant requests may take up to 45 days. MMT's Emergency Grants are not intended to address an organization's failure to comply with legal requirements or problems that can be attributed to organizational neglect; failure to plan for likely

contingencies, such as the breakdown of aging equipment; or to replace a gradual loss of organizational funding. In addition, the Emergency Grant program cannot be used solely to expedite the standard processing time for a Responsive or Grassroots Grants application.
Geographic Focus: Oregon, Washington
Amount of Grant: 7,500 - 100,000 USD
Samples: Port Orford Public Library Foundation, Curry, Oregon, $25,000 - to provide core support to offset unexpected emergency expenses; Boys and Girls Clubs of Central Oregon, Deschutes, Oregon, $30,000 - to provide support for operations during temporary club relocation; United Community Action Network, Douglas, Oregon, $15,000 - to replace the central phone system that provides information and referral for basic needs service in Douglas County.
Contact: Maddelyn High, Grants Administrator; (503) 228-5512; maddelyn@mmt.org
Internet: http://www.mmt.org/program/emergency-grants
Sponsor: Meyer Memorial Trust
425 NW 10th Avenue, Suite 400
Portland, OR 97209

Meyer Memorial Trust Grassroots Grants 2536
The Grassroots Grants program is designed to give smaller organizations (often without development departments) an opportunity to compete for grants from MMT. Focus areas include: health and human services; arts and culture; environmental conservation; education; and public affairs. Applications may be submitted at any time but proposals are collected for consideration on the 15th of March, July and October. Grants of $1,000 to $40,000 are made three to four months later: in June, October and February. Grant periods may be one to two years in length. A list of previous funded projects is available at the Trust's website.
Requirements: Grants are awarded to 501(c)3 nonprofit organizations in Oregon and Clark County, Washington. Organizations must apply through the Trust's online application process.
Geographic Focus: Oregon, Washington
Date(s) Application is Due: Mar 15; Jul 15; Oct 15
Amount of Grant: 1,000 - 25,000 USD
Samples: Friends of Latimer Quilt and Textile Center, Tillamook, Oregon, $8,000 - for capital improvements to the exhibition center; Hand 2 Mouth, Portland, Oregon, $20,000 - for infrastructure to support the theater company's artistic quality and expand its reach to smaller rural communities; Redmond Council for Senior Citizens, Redmond, Oregon, $12,000 - for capital repairs to the center.
Contact: Maddelyn High, Grants Administrator; (503) 228-5512; maddelyn@mmt.org
Internet: http://www.mmt.org/program/grassroots-grants
Sponsor: Meyer Memorial Trust
425 NW 10th Avenue, Suite 400
Portland, OR 97209

Meyer Memorial Trust Responsive Grants 2537
Responsive Grants are awarded in the areas of human services; health; affordable housing; community development; conservation and environment; public affairs; arts and culture; and education. Funding ranges from $40,000 to $300,000, with grants periods from one to three years in length. Responsive grants help support many kinds of projects, including core operating support, building and renovating facilities, and strengthening organizations. There are two stages of consideration before Responsive Grants are awarded. Initial Inquiries are accepted at any time through MMT's online grants application. Applicants that pass initial approval are invited to submit full proposals. The full two-step proposal investigation usually takes five to seven months. Final decisions on Responsive Grants are made by trustees monthly, except in January, April and August. Additional information about the application process, along with the online application, is available at the website.
Requirements: Support is available to 501(c)3 nonprofit organizations in Oregon and Clark County, Washington.
Restrictions: Funding is not available for sectarian or religious organizations for religious purposes, or for animal welfare organizations, or projects that primarily benefit students of a single K-12 school (unless the school is an independent alternative school primarily serving low-income and/or special needs populations). Funding is also not available for individuals or for endowment funds, annual campaigns, general fund drives, special events, sponsorships, direct replacement funding for activities previously supported by federal, state, or local public sources, deficit financing, acquisition of land for conservation purposes (except through Program Related Investments), or hospital capital construction projects (except through Program Related Investments).
Geographic Focus: Oregon, Washington
Amount of Grant: 40,000 - 300,000 USD
Contact: Maddelyn High, Grants Administrator; (503) 228-5512; maddelyn@mmt.org
Internet: http://www.mmt.org/program/responsive-grants
Sponsor: Meyer Memorial Trust
425 NW 10th Avenue, Suite 400
Portland, OR 97209

Meyer Memorial Trust Special Grants 2538
From time to time, Meyer Memorial Trust issues Requests for Proposals (RFPs) in targeted, short-term programs that address immediate pressing needs in the nonprofit community. Nonprofits may subscribe to their email announcements to be notified when RFPs are issued. A list of previously funded projects is posted on the Trust's website.
Requirements: Grants are awarded to 501(c)3 nonprofit organizations in Oregon and Clark County, Washington.
Geographic Focus: Oregon, Washington
Amount of Grant: Up to 50,000 USD
Samples: Central Oregon Battering and Rape Alliance Victim Advocacy, Bend County, Oregon, $50,000; Metropolitan Affordable Housing, Eugene County, Oregon, $50,000; South Lane Mental Health Services, Cottage Grove County, Oregon, $50,000.
Contact: Maddelyn High, Grants Administrator; (503) 228-5512; maddelyn@mmt.org
Internet: http://www.mmt.org/program/rfp
Sponsor: Meyer Memorial Trust
425 NW 10th Avenue, Suite 400
Portland, OR 97209

MGFA Post-Doctoral Research Fellowships 2539
The Myasthenia Gravis Foundation of America, Inc. announces a competitive twelve-month post-doctoral research fellowship for clinical or basic research pertinent to myasthenia gravis or related neuromuscular disorders. Research may be concerned with neuromuscular transmission, immunology, molecular or cell biology of the neuromuscular synapse, or the etiology/pathology or diagnosis of myasthenia gravis. See the Foundations website for application and, additional guidelines.
Requirements: The applicant must be: a permanent resident of the United States or Canada who has been accepted to work in the laboratory of an established investigator at an institution in the United States, Canada or abroad deemed appropriate by the Medical/Scientific Advisory Board of the Myasthenia Gravis Foundation of America, Inc; or a foreign national who has been accepted to work in the laboratory of an established investigator at an institution in the United States or Canada deemed appropriate by the Medical/Scientific Advisory Board of the Myasthenia Gravis Foundation of America, Inc.
Geographic Focus: All States
Date(s) Application is Due: Oct 1
Amount of Grant: Up to 50,000 USD
Contact: Tor Holtan, Chief Executive; (800) 541-5454 or (212) 297-2156; fax (212) 370-9047; Tor.Holtan@myasthenia.org or mgfa@myasthenia.org
Internet: http://www.myasthenia.org/hp_fellowships.cfm#nursing
Sponsor: Myasthenia Gravis Foundation of America
355 Lexington Avenue, 15th Floor
New York, NY 10017

MGFA Student Fellowships 2540
These fellowships are awarded annually to current medical students or graduate students interested in the scientific basis of myasthenia gravis or related neuromuscular conditions, serving both to further scientific inquiries into the nature of these disorders and to encourage more research. The stipend is up to $5,000. See the Foundation's website for application form and guidelines.
Requirements: The fellowships are awarded to current medical students or graduate students. Briefly describe, in abstract form, the question that you propose to study, its association to myasthenia gravis or related neuromuscular conditions, and how you will approach the project. Applicants and sponsoring institutions must comply with policies governing the protection of human subjects, the humane care of laboratory animals and the inclusion of minorities in study populations.
Geographic Focus: All States
Date(s) Application is Due: Mar 31
Amount of Grant: 5,000 USD
Contact: Tor Holtan, Chief Executive; (800) 541-5454 or (212) 297-2156; fax (212) 370-9047; Tor.Holtan@myasthenia.org or mgfa@myasthenia.org
Internet: http://www.myasthenia.org/hp_fellowships.cfm#nursing
Sponsor: Myasthenia Gravis Foundation of America
355 Lexington Avenue, 15th Floor
New York, NY 10017

MGM Resorts Foundation Community Grants 2541
The MGM Resorts Foundation invites proposals from nonprofit agencies providing direct services to people living in its communities. All of the funds allocated through the Foundation come from employee contributions and their desire to make a difference in the communities where they live and work. Foundation grant allocations are 100% employee-driven. The Foundation empowers MGM Resorts employees to choose to make direct contributions to the agency of their choice, or to contribute to the Community Grant Funds, which provides grants to nonprofits through an annual Request for Proposal (RFP) process. MGM Resorts Foundation grants are for a one-year period and do not automatically renew. Continued or expanded projects and programs (your organization is currently providing these services) can request a per year maximum grant of $65,000 in Southern Nevada; $10,000 in Northern Nevada, Mississippi and the Detroit, Michigan area. New projects and programs (your organization is not currently providing these services) can request a per year maximum grant of $35,000 in Southern Nevada; $5,000 in Northern Nevada, Mississippi and the Detroit, Michigan area.
Requirements: To receive a grant from the Foundation, your agency must meet the following *Requirements:* Operate as an IRS 501(c)3 organization and have been doing so for a minimum of 36 months; provide service within the regions MGM Resorts employees live, work and care for their families (Nevada, Mississippi, and the greater Detroit, Michigan area); your organization's administrative costs must be 25% or under; provide a human service; and, meet the MGM Resorts diversity policy: open to all people, without regard to race, color, creed, sex, sexual orientation, religion, disability, or national origin. Agencies must request funding for projects/programs that provide services in the following focus areas: Strengthening Neighborhoods (self-sufficiency, revitalization of communities); Strengthening Children (early childhood development, success in school, prevention / intervention); and, Strength in Difficult Times (recovery and counseling services). Proposals must be received by the Foundation by 5:00 pm of the deadline date.

Restrictions: The program does not support the following types of organizations or activities: projects/programs that are exclusively for medical research; public schools or privately funded / tuition-based schools; governmental entities; religious organizations that do not have 501(c)3 status; pass-through agencies (organizations whose staff does not provide direct client services but who allocate funding to subsequent organizations to provide projects/programs and services); sponsorship of special events and/or fundraising activities; capital campaigns or endowment funds; political issues, such as, election campaigns, issue endorsements, bill drafts or legislation reform; organizations that require clients to embrace specific beliefs or traditions; projects/programs that are exclusively recreational or athletic sponsorships; membership-based organizations without a sliding fee scale and scholarship system already in place.
Geographic Focus: Michigan, Mississippi, Nevada
Date(s) Application is Due: May 2
Amount of Grant: 5,000 - 65,000 USD
Contact: Shelley Gitomer; (702) 692-9643; foundation@mgmresorts.com
Internet: http://www.mgmmirage.com/csr/community/foundation.aspx
Sponsor: MGM Resorts Foundation
3260 Industrial Road
Las Vegas, NV 89109

MGN Family Foundation Grants 2542
The MGN Family Foundation makes grants to qualified 501(c)3 organizations specializing in the following areas: education; health care and medical research; children in need; armed service personnel. Areas of particular interest include: colleges, universities and private schools, examples would be: to fund a chair to provide lecturers in literature, philosophy or the arts, and provide scholarships; hospitals and clinics that specialize in excellent patient care and continuing medical research such as Memorial Sloan-Kettering Cancer Center and the Mayo and Cleveland Clinics are other examples, also under consideration would be hospice organizations that provide palliative care to the dying; organizations that support children in need whether due to emotional, physical abuse, neglect or disadvantaged circumstances. Support can be for basic necessities such as food, shelter, education, medical as well as, spiritual and emotional counseling; in light of recent events, this Foundation would like to offer assistance to our servicemen/ servicewomen and their families through those charities which give support to their unique needs. The examples provided above reflect the true mission and goals of the MGN Foundation. There should be no exclusions due to race or creed, provided all applicants have a strong moral base and core values in the areas of education, health care, welfare of children and service personnel and their families. The Foundation Board meets semi-annually in May and November. Applications are due by April 1st and October 1st. Application form is available online.
Requirements: Qualified 501(c)3 organizations specializing in the following areas: education; health care and medical research; children in need; armed service personnel. To apply, submit seven (7) sets of the following items: Grant application form completed, dated, and signed by the Chief Executive Officer or Chairman of the Board of the organization; list of Board of Directors; Financial Statement (audited if available), for the most recent complete fiscal year; copy of IRS 501(c)3 Determination Letter. If you wish, you may submit your proposal in a narrative format of not more than two pages in addition to the completed application form. Optional materials may be submitted but are not required (such as brochures discussing or depicting the activities of the organization).
Restrictions: Do not staple materials or place them in a bound notebook.
Geographic Focus: All States
Date(s) Application is Due: Apr 1; Oct 1
Amount of Grant: 1,000 - 10,000 USD
Samples: Agnes Scott College, $3,000--scholarship; Vermont National Guard Charitable Foundation, Inc., $3,000--wounded veteran program; Center for Molecular Medicine and Immunology/Garden State, $5,000--foster research for new technologies in the diagnosis, detection and treatment of cancer.
Contact: Pamela Nothstein; (843) 937-4614; grantinquiries7@wachovia.com
Internet: https://www.wachovia.com/foundation/v/index.jsp?vgnextoid=61078689fb0aa110VgnVCM1000004b0d1872RCRD&vgnextfmt=default
Sponsor: MGN Family Foundation
16 Broad Street (SC1000)
Charleston, SC 29401

Miami County Community Foundation - Operation Round Up Grants 2543
Miami County Community Foundation - Operation Round Up is a national program that allows participating electric cooperatives to partner with their members to provide charitable giving funds in their local community. Each month the total amount of the participating member's bill is rounded up to the nearest dollar and those extra pennies go into the fund designated for distribution to those in need in the area. Areas of consideration include: cultural, education, recreation, human services, health and medical, community development, and environmental awareness. Grant proposals are reviewed by the following criteria: is there an established need; what is the project solving; is it appropriate for the county; are their adequate resources; its benefit and timing to the community; what are the expected results; and will the organization work with other organizations to achieve their goals.
Requirements: Along with the application, organizations must include a one page budget for the amount requested, with justification; proof of 501(s)3 status; their most recent audited financial statement or annual report; and current organizational budget. A total of nine complete sets of applications and documentation must be submitted. Applications are available at the Foundation website and should be mailed to Miami-Cass REMC, Operation Round Up, 3086 West 100 North, P.O. Box 168, Peru, IN 46970.
Restrictions: Operation Round Up funds will not be used for paying utility bills, or funding political or private interests.
Geographic Focus: Indiana
Date(s) Application is Due: Mar 31; Jun 30; Sep 30; Dec 31
Contact: Mary Alexander, Director of Development; (765) 475-2859 or (877) 432-6423; fax (765) 472-7378; miami@nicf.org
Internet: http://www.nicf.org/miami/grantapplications.html
Sponsor: Miami County Community Foundation
13 East Main Street
Peru, IN 46970

Michael Reese Health Trust Core Grants 2544
The primary focus of the Michael Reese Health Trust is to improve the health status and well-being of vulnerable populations in the Chicago metropolitan area. The Health Trust is committed to supporting community-based health-related services and education that are effective, accessible, affordable, and culturally competent. It is especially interested in efforts to address the barriers that prevent vulnerable groups from accessing quality health care, and in programs that deliver comprehensive, coordinated services. Each year, the Health Trust awards a small number of Core grants. Core grants are larger, multi-year grants designed to strengthen both program quality and organizational capacity. Organizations approved by Health Trust staff may request up to $100,000 a year for each of three years for a total of up to $300,000.
Requirements: Nonprofit organizations operating in the Chicago metropolitan area are eligible to apply, but preference is given to organizations within the City of Chicago. The applicant must have a 501(c)3 and non-private foundation determination letter from the Internal Revenue Service and be designated as a public charity under section 509(a)1 or 509(a)2, of the Internal Revenue Code. Generally, the Health Trust does not provide grants to 509(a)3 "supporting organizations." Organizations must be non-discriminatory in the hiring of staff and in providing services on the basis of race, religion, gender, sexual orientation, age, national origin or disability. Qualified applicants must have prior approval from Health Trust staff to submit a Core grant request. Contact the Program Officer by phone or by email to discuss how your organization would use a Core grant. Once staff approval has been obtained, the sponsor will send an invitation to submit a Letter of Inquiry for a Core grant through its online application process. The Health Trust awards grants twice a year. The submission deadlines for Letters of Inquiry are June 15 (for grants to run January 1 through December 31) and December 15 (for grants to run July 1 through June 30). If the due date falls on a weekend, they will accept submissions until 5:00pm the following business day. Core grants should focus on the following: quality of services; planning for and supporting staff, volunteers and activities fundamental to the organization's health-related mission; mission-related infrastructure needs; and/or sustainability of the agency and its health services. Use of evidence-based practices or using the Core grant to systematically learn about and implement evidence-based practices is encouraged. Participation in this program requires a willingness to share findings and lessons learned in order to assist other Health Trust grantees and others in the field.
Restrictions: Grants do not support: lobbying, propaganda, or other attempts to influence legislation; sectarian purposes (programs that promote or require a religious doctrine); capital needs, such as buildings, renovations, vehicles, and major equipment; durable medical equipment; fundraising events, including sponsorship, tickets, and advertising; or, debt reduction; individual and scholarship support. In general, the Health Trust does not provide endowment support.
Geographic Focus: Illinois
Date(s) Application is Due: Jun 15; Dec 15
Amount of Grant: Up to 300,000 USD
Contact: Jennifer M. Rosenkranz, (312) 726-1008; jrosenkranz@healthtrust.net
Internet: http://www.healthtrust.net/content/how-apply/new-applicants/application-procedures
Sponsor: Michael Reese Health Trust
150 North Wacker Drive, Suite 2320
Chicago, IL 60606

Michael Reese Health Trust Responsive Grants 2545
The primary focus of the Michael Reese Health Trust is to improve the health status and well-being of vulnerable populations in the Chicago metropolitan area. The Health Trust is committed to supporting community-based health-related services and education that are effective, accessible, affordable, and culturally competent. It is especially interested in efforts to address the barriers that prevent vulnerable groups from accessing quality health care, and in programs that deliver comprehensive, coordinated services. Responsive grants generally range from $25,000-$60,000. The Health Trust will entertain requests for program support and general operating and for both one-year and multi-year projects. However, multi-year grants are generally considered for organizations that have received significant prior Health Trust support. Requests may be for continuation or expansion of a current program, or a new program.
Requirements: Nonprofit organizations operating in the Chicago metropolitan area are eligible to apply, but preference is given to organizations within the City of Chicago. The applicant must have a 501(c)3 and non-private foundation determination letter from the Internal Revenue Service and be designated as a public charity under section 509(a)1 or 509(a)2, of the Internal Revenue Code. Generally, the Health Trust does not provide grants to 509(a)3 "supporting organizations." Organizations must be non-discriminatory in the hiring of staff and in providing services on the basis of race, religion, gender, sexual orientation, age, national origin or disability. The Health Trust awards grants twice a year. The submission deadlines for Letters of Inquiry are June 15 (for grants to run January 1 through December 31) and December 15 (for grants to run July 1 through June 30). If the due date falls on a weekend, they will accept submissions until 5:00pm the following business day.
Restrictions: Grants do not support: lobbying, propaganda, or other attempts to influence legislation; sectarian purposes (programs that promote or require a religious doctrine);

capital needs, such as buildings, renovations, vehicles, and major equipment; durable medical equipment; fundraising events, including sponsorship, tickets, and advertising; or, debt reduction; individual and scholarship support. In general, the Health Trust does not provide endowment support.
Geographic Focus: Illinois
Date(s) Application is Due: Jun 15; Dec 15
Amount of Grant: 25,000 - 60,000 USD
Contact: Jennifer M. Rosenkranz, Senior Program Officer for Responsive Grants; (312) 726-1008; fax (312) 726-2797; jrosenkranz@healthtrust.net
Internet: http://www.healthtrust.net/content/how-apply/new-applicants/grantmaking-guidelines
Sponsor: Michael Reese Health Trust
150 North Wacker Drive, Suite 2320
Chicago, IL 60606

Michigan Psychoanalytic Institute and Society SATA Grants　　2546
The purpose of this grant is to partly sponsor one or more trainees in the mental health field in order for them to attend a national meeting of the American Psychoanalytic Association. The winter meeting of the American is held in New York City and the spring meeting is held in different cities. To apply for a partial scholarship, applicants should send a two-paragraph statement telling us about their background, interest in psychoanalytic thinking, and why he/she would like to attend a meeting of the American Psychoanalytic Association.
Geographic Focus: All States
Contact: Kathleen Kunkel; (248) 851-3380, ext. 2 or (248) 865-1164; fax (248) 851-1806
Internet: http://www.mpi-mps.org/joomla/index.php?option=com_content&task=view&id=32&Itemid=109
Sponsor: Michigan Psychoanalytic Institute and Society
32841 Middlebelt Road, Suite 411
Farmington Hills, MI 48334

Michigan Psychoanalytic Institute and Society Scholarships　　2547
The Continuing Education Division of the Michigan Psychoanalytic Institute offers two Programs, one in Ann Arbor and one in Farmington Hills. The Program is a one-year, clinically-based program which meets once weekly. The program is designed to introduce and further develop an understanding of psychoanalytic principles as applied to psychotherapy with adults, children, and adolescents. Participants meet as a class on a weekly basis with instructors from the faculty of the Michigan Psychoanalytic Institute. Each instructor teaches a 5-6 week segment. Instructors present clinical material to illustrate a psychoanalytic approach to clinical work. Discussion of the case material is utilized to examine psychoanalytic concepts and clinical technique as applied to psychoanalytic psychotherapy. Half of the clinical material presented during the year is from work with adult patients. The other half of the year, child analysts present and discuss clinical material from the treatment of children, adolescents, and their parents. The Scholarship application must be submitted by June 30.
Requirements: The Fellowship Programs are ideally suited for mental health clinicians who are in practice, recently graduated, or in training in their respective fields (psychiatry, psychology, social work, counseling, or nursing). It is also useful as preparation for clinicians who are considering more extensive training in psychoanalytic psychotherapy or psychoanalysis in the future.
Geographic Focus: All States
Date(s) Application is Due: Jun 30
Contact: Jean Lewis; (248) 851-3380; fax (248) 851-1806; jlewis@mpi-mpi.org
Internet: http://www.mpi-mps.org/joomla/index.php?option=com_content&task=view&id=62&Itemid=216
Sponsor: Michigan Psychoanalytic Institute and Society
32841 Middlebelt Road, Suite 411
Farmington Hills, MI 48334

Microsoft Research Cell Phone as a Platform for Healthcare Grants　　2548
Microsoft Research will support selected academic research with the goal of advancing the state-of-the-art in smart cell phones for health care applications. Successful projects must seek to increase the capacity of cell phone-based health care solutions in under-served rural and urban communities. They must also take into consideration the social context of application deployment. Topics to consider include: medical applications that are relevant, worldwide, for smart mobile phones (application and Web-enabled) in rural, and urban, communities; and appropriate services and infrastructures needed to provide affordable and accessible health care services. The total amount available under this request for proposals (RFP) is $1,000,000. Microsoft Research anticipates making approximately 10-12 awards averaging $80,000, with a maximum of $100,000 for any single award. Awards are made for the purpose of seed-funding larger initiatives, proofs of concept, or demonstrations of feasibility. It is important to understand that funding will continue after the first year only in exceptional circumstances, and that the principal investigators should therefore make every effort to leverage Microsoft Research funds as one component of a diverse funding base in a larger or longer-running project.
Requirements: All qualifying institutions are eligible without regard for geographic location. The proposing institution must be either: an accredited degree-granting college or university (or international equivalent) with non-profit status and awarding degrees at the baccalaureate level or above; or a research institution with non-profit status.
Restrictions: For all awards, payment of indirect costs (overhead) is not permitted.
Geographic Focus: All States
Date(s) Application is Due: Oct 29
Amount of Grant: Up to 100,000 USD
Contact: Rick Rashid; (800) 642-7676; fax (425) 936-7329; erpinq@microsoft.com

Internet: http://research.microsoft.com/en-us/um/redmond/about/collaboration/awards/cellphone-healthcare_awards.aspx
Sponsor: Microsoft Research
One Microsoft Way
Redmond, WA 98052-6399

Mid-Iowa Health Foundation Community Response Grants　　2549
Mid-Iowa Health Foundation awards grants to organizations working towards improving the health of people in greater Des Moines, Iowa. The Foundation is interested in work that affects specific health results and aligns with community-identified priorities. The focus for Community Response grants is the Greater Des Moines Health Safety Net System. The health care safety net provides appropriate, timely and affordable health services to people who experience barriers to accessing services from other providers due to financial, cultural, linguistic, or other issues. These core safety net providers offer care to patients in Greater Des Moines, Iowa regardless of their ability to pay for services, and primarily serve vulnerable, low-income patients who are uninsured, publicly insured or underinsured. Community Response grants average $10,000 to $30,000. The Foundation may consider partially funding a proposal, if acceptable to the grantee.
Requirements: Applicant organizations must: be tax-exempt, 501(c)3 and/or 509(a) status; serve the greater Des Moines, Iowa area (Polk, Warren, and/or Dallas Counties); and, offer health programs and services aligned with the Foundation's mission. The Foundation will consider proposals from core safety net providers for: preventive and primary safety net health services, including behavioral and oral health; critical elements of the safety net system such as coordinated outreach, system navigation, and culturally competent services; meeting increased demands on a safety net system with capacity, financial and workforce stressors. Mid-Iowa Health Foundation reviews Community Response proposals once annually. Proposals are due by noon on October 1; if the 1st falls on a weekend, proposals are due by noon on the preceding Friday.
Restrictions: The Mid-Iowa Health Foundation does not consider proposals for: individuals; scholarships; conference registration fees; programs that promote religious activities; general operations or special camps of disease- or condition-specific organizations; capital campaigns; endowment campaigns; debt reduction; or, fund raising events.
Geographic Focus: Iowa
Date(s) Application is Due: Oct 1
Amount of Grant: Up to 30,000 USD
Contact: Denise Swartz; (515) 277-6411; dswartz@midiowahealth.org
Internet: http://www.midiowahealth.org/grants.html
Sponsor: Mid-Iowa Health Foundation
3900 Ingersoll Avenue, Suite 104
Des Moines, IA 50312

Middlesex Savings Charitable Foundation Basic Human Needs Grants　　2550
Since June of 2000, the Middlesex Savings Charitable Foundation has provided more than $2 million in grants to over 200 non-profit organizations providing critical community services throughout Eastern and Central Massachusetts. Established as a nonprofit, private charitable foundation, the Foundation carries out the philanthropic mission of Middlesex Savings Bank, supporting nonprofit organizations, services and programs in a wide variety of fields, including education, public health and welfare, the arts, and community development. The Basic Human Needs Program funds projects and programs whose primary focus is on food, shelter, and clothing for low-and moderate-income and vulnerable populations. Food pantries may apply, but the request must be for a program or other initiative, with no more than 25% of grant proceeds used towards the purchase of food related to the broader initiative. Grant requests of up to $20,000 will be considered. Applicants may apply online. The annual deadlines are April 1 and August 1.
Requirements: Eastern Massachusetts 501(c)3 tax-exempt organizations serving one or more communities served by Middlesex Savings Bank, including Acton, Ashland, Ayer, Bedford, Bellingham, Berlin, Bolton, Boxborough, Carlisle, Chelmsford, Concord, Dover, Dunstable, Framingham, Franklin, Groton, Harvard, Holliston, Hopedale, Hopkinton, Hudson, Lexington, Lincoln, Littleton, Marlborough, Maynard, Medfield, Medway, Mendon, Milford, Millis, Natick, Needham, Newton, Norfolk, Northborough, Pepperell, Sherborn, Shirley, Southborough, Stow, Sudbury, Townsend, Tyngsborough, Upton, Walpole, Waltham, Wayland, Wellesley, Westborough, Westford, and Weston, are eligible. Projects for which support is requested should benefit people who live or work in the region.
Restrictions: The foundation will not fund political or sectarian activities.
Geographic Focus: Massachusetts
Date(s) Application is Due: Apr 1; Aug 1
Amount of Grant: Up to 20,000 USD
Contact: Mike Kuza, (508) 315-5361 or (508) 315-5360; mkuza@middlesexbank.com
Internet: https://www.middlesexbank.com/community-and-us/community-support/Pages/charitable-foundation.aspx
Sponsor: Middlesex Savings Charitable Foundation
P.O. Box 5210
Westborough, MA 01581-5210

Middlesex Savings Charitable Foundation Capacity Building Grants　　2551
Since June of 2000, the Middlesex Savings Charitable Foundation has provided more than $2 million in grants to over 200 non-profit organizations providing critical community services throughout Eastern and Central Massachusetts. Established as a nonprofit, private charitable foundation, the Foundation carries out the philanthropic mission of Middlesex Savings Bank, supporting nonprofit organizations, services and programs in a wide variety of fields, including education, public health and welfare, the arts, and community development. The Capacity Building Program funds initiatives designed to strengthen and increase the impact of local non-profits by improving their

organizational capacity. Desired outcomes for non-profits selected to receive grants through this program include one or more of the following: improved governance and leadership; improved staff skills; improved management systems and practices; completed strategic plans; improved, expanded, or additional services; and expanded strategic assets, including financial and human resources. Successful applicants will be able to describe how their enhanced capacity will ultimately benefit the communities that they serve. Grant requests of up to $20,000 will be considered. Applicants may apply online. The annual deadlines are April 1 and August 1.
Requirements: Eastern Massachusetts 501(c)3 tax-exempt organizations serving one or more communities served by Middlesex Savings Bank, including Acton, Ashland, Ayer, Bedford, Bellingham, Berlin, Bolton, Boxborough, Carlisle, Chelmsford, Concord, Dover, Dunstable, Framingham, Franklin, Groton, Harvard, Holliston, Hopedale, Hopkinton, Hudson, Lexington, Lincoln, Littleton, Marlborough, Maynard, Medfield, Medway, Mendon, Milford, Millis, Natick, Needham, Newton, Norfolk, Northborough, Pepperell, Sherborn, Shirley, Southborough, Stow, Sudbury, Townsend, Tyngsborough, Upton, Walpole, Waltham, Wayland, Wellesley, Westborough, Westford, and Weston, are eligible. Projects for which support is requested should benefit people who live or work in the region. Given the continuing economic challenges faced by MSCF communities, the Board will be focusing upon, and giving preference to, grant submissions from organizations providing basic human services such as food and shelter.
Restrictions: The Foundation will not fund political or sectarian activities.
Geographic Focus: Massachusetts
Date(s) Application is Due: Apr 1; Aug 1
Amount of Grant: Up to 20,000 USD
Contact: Mike Kuza, (508) 315-5361 or (508) 315-5360; mkuza@middlesexbank.com
Internet: https://www.middlesexbank.com/community-and-us/community-support/Pages/charitable-foundation.aspx
Sponsor: Middlesex Savings Charitable Foundation
P.O. Box 5210
Westborough, MA 01581-5210

Milagro Foundation Grants 2552
Started by musician Carlos Santana and his wife Deborah, the foundation supports educational efforts that help youths live healthy, literate, and culturally enriched lives. Priority is helping at-risk and vulnerable populations acquire the skills to succeed in life. The foundation is especially interested in arts education. Unsolicited completed applications are not accepted. Preliminary letters of inquiry are accepted at any time. The board meets in February, June, and October.
Requirements: Nonprofit organizations are eligible. In order of priority, the foundation has an interest in helping children in San Rafael, the remainder of California, the United States, and other nations in which Carlos Santana performs.
Geographic Focus: All States
Amount of Grant: 2,500 - 5,000 USD
Contact: Grants Administrator; (415) 460-9939; fax (415) 460-6802; info@milagrofoundation.org or apply@milagrofoundation.org
Internet: http://www.milagrofoundation.org/apply.asp
Sponsor: Milagro Foundation
P.O. Box 9125
San Rafael, CA 94912-9125

Miles of Hope Breast Cancer Foundation Grants 2553
The Fund provides programs and support services for women and families in the Hudson Valley, New York, affected by breast cancer. All funds raised for the Foundation are used in the Hudson Valley. The Miles of Hope Breast Cancer Foundation is currently offering grants for projects in the areas of breast health and breast cancer education, outreach, screening, treatment and support projects.
Requirements: Services must be provided in the eight counties served by the Miles of Hope Breast Cancer Foundation, which include: Columbia, Dutchess, Greene, Orange, Putnam, Rockland, Westchester and Ulster.
Restrictions: Project must be specific to breast health and/or breast cancer. Applicants must be a U.S. nonprofit (federally tax-exempt) organization. Nonprofit organizations, known educational institutions (i.e. college or university), government agencies, and Indian tribes are eligible. Equipment costs, if applicable, may not exceed 30% of direct costs and should be used exclusively on this project. Salaries, if requested, are for personnel related to this project only and not the general work of the employee. Funds will not be awarded to capital campaigns
Geographic Focus: New York
Contact: Administrator; (845) 527-6884 or (845) 264-2005; info@milesofhopebcf.org
Internet: http://www.milesofhopebcf.org/funds.html#grants
Sponsor: Miles of Hope Breast Cancer Foundation
P.O. Box 405
La Grangeville, NY 12540

Military Ex-Prisoners of War Foundation Grants 2554
The Military Ex-Prisoners of War Foundation was founded primarily to assist Military Ex-P.O.W Veterans, and to fund their National Educational Scholarship Program for qualified heirs. The mission of the Military Ex-Prisoners of War Foundation is to support educational programs designed to inform Americans about the P.O.W experience; offer scholarships to children and grandchildren of former P.O.Ws and other such activities as may be approved by the Foundation's board of directors.
Requirements: Children, grandchildren, and great-grandchildren of former prisoners-of-war (P.O.Ws) who served after December 7, 1941 are eligible to apply. Applicants must include the following: completed application (available at the Foundations website); short autobiographical statement (not to exceed one typed page); Transcript *Requirements:* copy of official transcripts of high school grades including SAT/ACT test scores, official transcripts of college grades; two letters of recommendation, one must be from a present/former teacher; short essay, limited to 1,000 words, on the impact of WWII on society today; statement about your grandparent's experience as a prisoner of war, 1,000 words or less. Completed applications should be mailed to: Dorris Livingstone, Recording Secretary, Military Ex-Prisoners of War Foundation, 1561 Glen Hollow Lane South, Dunedin, FL 34698. If you have any questions, email your inquiry to: Dorris2001@aol.com with Foundation Scholarship Application/Question typed into the Subject line.
Restrictions: No support for individuals not related to former prisoners of war, religious, political causes, or for organizations.
Geographic Focus: All States
Date(s) Application is Due: Apr 1
Amount of Grant: 2,000 USD
Contact: F. Paul Dallas; (910) 867-2775; fax (910) 867-0339; threatt273@aol.com
Internet: http://www.militarypowfoundation.org/072409-scholarships.html
Sponsor: Military Ex-Prisoners of War Foundation
916 Bingham Drive
Fayetteville, NC 27803

Milken Family Foundation Grants 2555
The purpose of the Milken Family Foundation is to discover and advance inventive and effective ways of helping people help themselves and those around them lead productive and satisfying lives. The Foundation advances this mission primarily through its work in education and medical research. In education, the Foundation is committed to: strengthening the profession by recognizing and rewarding outstanding educators, and by expanding their professional leadership and policy influence attracting, developing, motivating and retaining the best talent to the teaching profession by means of comprehensive, whole school reform; stimulating creativity and productivity among young people and adults through programs that encourage learning as a lifelong process; and building vibrant communities by involving people of all ages in programs that contribute to the revitalization of their community and to the well-being of its residents. In medical research, the Foundation is committed to: advancing and supporting basic and applied medical research, especially in the areas of prostate cancer and epilepsy, and recognizing and rewarding outstanding scientists in these areas; and supporting basic health care programs to assure the well-being of community members of all ages. Applicants may request funding at any time.
Requirements: Grants are made to 501(c)3 tax-exempt organizations. Grant recipients must have the financial potential to sustain the program for which funding is sought following the period of Foundation support. Preventive programs with long-range goals receive the closest consideration. Applicants should submit a brief written statement that includes: description of project, goals, procedure and personnel; brief background of organization, including number of years in operation, other areas of activity, applicant's qualifications for support, annual operating budget, and previous and current sources of funding; and a letter of exemption from the Internal Revenue Service.
Restrictions: Grants are not made directly to individuals.
Geographic Focus: All States
Contact: Richard Sandler; (310) 570-4800; fax (310) 570-4801; admin@mff.org
Internet: http://www.mff.org/about/about.taf?page=funding
Sponsor: Milken Family Foundation
1250 Fourth Street, 6th Floor
Santa Monica, CA 90401-1353

Miller Foundation Grants 2556
The Miller Foundation focuses on assisting local nonprofit, charitable organizations and governmental agencies with projects that provide the following for the Battle Creek, Michigan, area: economic development; education; health service; human service; neighborhood improvement; arts and culture; recreation and tourism; and leadership. The Miller Foundation Board of Trustees meets every other month to consider grant applications: January, March, May, July, September, and November. Applicants should submit grant applications by the 1st of the month for it to be considered at that month's Board meeting.
Requirements: Nonprofit 501(c)3 organizations located in and working to improve the Battle Creek community are eligible. Organizations should submit a preliminary letter of request, briefly describing their project, its estimated cost, amount requested, and funding from other sources. After reviewing the initial letter, the Foundation staff, if appropriate, will send a formal grant application to the requesting organization.
Restrictions: The Foundation seldom funds an entire project but rather joins with others as they work to improve the quality of life in the Battle Creek community. The Foundation does not make grants to individuals or for continuing operating funds of nonprofit organizations.
Geographic Focus: Michigan
Date(s) Application is Due: Jan 1; Mar 1; May 1; Jul 1; Sep 1; Nov 1
Contact: Sara Wallace, Executive Director; (269) 964-3542; fax (269) 964-8455
Internet: http://themillerfoundation.com/grants.htm
Sponsor: Miller Foundation
310 WahWahTaySee Way
Battle Creek, MI 49015

Mimi and Peter Haas Fund Grants 2557
The Mimi and Peter Haas Fund supports early childhood development. Their primary focus is for activities that provide San Francisco's young (ages 2-5), low-income children and their families with access to high-quality early childhood programs that are part of a comprehensive, coordinated system. The Fund recognizes the importance of connecting the work of its direct service grants to the ongoing discussions of public policy and seek

specific opportunities to collaborate with organizations to improve early childhood settings. The Fund will also continue trustee-initiated grantmaking to arts, education, public affairs, and health and human services organizations. Applicants should contact the trustee office to begin the application process. There are no particular application forms or deadlines with which to adhere.
Geographic Focus: California
Amount of Grant: Up to USD
Contact: Lynn Merz; (415) 296-9249; fax (415) 296-8842; mphf@mphf.org
Sponsor: Mimi and Peter Haas Fund
201 Filbert Street, 5th Floor
San Francisco, CA 94133-3238

Minneapolis Foundation Community Grants 2558
The foundation awards grants throughout Minnesota. Eligible activities include policy and systems change work in the following areas: affordable housing; economic opportunity; educational achievement; and the health and well-being of children, youth, and families. The foundation also focuses on systems and policy work that addresses the intersection of issues, such as housing and health. Types of support include program/project support, operating support, capital support (limited to the seven-county metro area); and some multiyear grants. Proposals are accepted throughout the year. Guidelines are available online.
Requirements: 501(c)3 nonprofit organizations located in the seven-county metropolitan area of Minneapolis and Saint Paul may apply.
Restrictions: The foundation does not fund individuals, organizations/activities outside of Minnesota, conference registration fees, memberships, direct religious activities, political organizations or candidates, direct fundraising activities, telephone solicitations, courtesy advertising, or financial deficits.
Geographic Focus: Minnesota
Date(s) Application is Due: Mar 15; Sep 15
Contact: Paul Verrette; (612) 672-3836; pverrette@mplsfoundation.org
Internet: http://www.mplsfoundation.org/grants/guidelines.htm
Sponsor: Minneapolis Foundation
800 IDS Center, 80 South Eighth Street
Minneapolis, MN 55402

MLA Continuing Education Grants 2559
The purpose of Medical Library Association (MLA)'s continuing education grants is to provide members with the opportunity to develop their knowledge of the theoretical, administrative, or technical aspects of librarianship. Members may submit applications for awards of $100 to $500 for either MLA courses or other Continuing Education (CE) activities. The application form and guidelines are available for download at the website. Applications and supporting documents must be received by MLA via email, fax, or mail by December 1. email applications will be accepted as PDF or MS Word files only. MLA may offer Continuing Education Grants more than once a year. Applicants are encouraged to visit the website or contact the association for additional information and deadlines.
Requirements: Applicants must satisfy the following eligibility *Requirements:* hold a graduate degree in library science; be a practicing health-science librarian with at least two years of professional experience; preferably be a regular member of the MLA; and be a citizen of or permanent resident in either the United States or Canada. (In exceptional cases, consideration will be given to an outstanding candidate not meeting the eligibility criteria outlined above.) Additionally, the continuing education activity must take place within the United States or Canada and the proposed course of study should be relevant to the library work in which the applicant is currently engaged. Grants will not be given to support work toward a degree or certificate, although course work that complements a candidate's program, yet is not part of the normal professional curriculum, will be considered.
Geographic Focus: All States, Canada
Date(s) Application is Due: Dec 1
Contact: Carla Funk, Executive Director; (312) 419-9094, ext. 14; fax (312) 419-8950; mlapd2@mlahq.org or grants@mlahq.org
Internet: http://www.mlanet.org/awards/grants/index.html
Sponsor: Medical Library Association
65 East Wacker Place, Suite 1900
Chicago, IL 60601-7246

MLA Cunningham Memorial International Fellowship 2560
The Medical Library Association (MLA) Cunningham Memorial International Fellowship was established in 1967 with a bequest from the estate of Eileen R. Cunningham. The fellowship supports the education and training of a health-science librarian from a country outside the United States or Canada and includes a stipend plus approved travel for four months in either country. The fellowship also pays for the cost of attending MLA's Annual Meeting. Guidelines and application materials are available at the website. All materials and letters of support must be received by MLA by December 1. Founded in 1898, the Medical Library Association (MLA) is a nonprofit, educational organization of more than 1,100 institutions and 3,600 individual members in the health sciences information field, committed to educating health-information professionals, supporting health-information research, promoting access to the world's health-sciences information, and working to ensure that the best health information is available to all.
Requirements: Candidates must have both an undergraduate degree and a master's-level library degree (latter requirement may be waived) and must either be working in or preparing to work in a medical library in their country. The fellowship recipient must agree to return to his or her country and work in a medical library for a period of two years. In addition, a satisfactory score must be achieved on the TOEFL English competency examination.
Restrictions: Fellowship is not available to U.S. or Canadian citizens. Travel arrangements to and from the United States or Canada must be paid by the fellow. Previous recipients of the Cunningham fellowship are ineligible.
Geographic Focus: All Countries
Date(s) Application is Due: Dec 1
Amount of Grant: 8,000 USD
Contact: Carla Funk, Program Contact; (312) 419-9094, ext. 14; fax (312) 419-8950; mlapd2@mlahq.org or grants@mlahq.org
Internet: http://www.mlanet.org/awards/grants/cunningham.html
Sponsor: Medical Library Association
65 East Wacker Place, Suite 1900
Chicago, IL 60601-7246

MLA David A. Kronick Traveling Fellowship 2561
Established in 2002, the David A. Kronick Traveling Fellowship grants one $2,000 award each year to cover the expenses involved in traveling to three or more medical libraries in the United States or Canada, for the purpose of studying a specific aspect of health information management. Activities that are operational in nature or have only local usefulness will be considered. Guidelines and application materials are available at the Medical Library Association (MLA) website. Applications and supporting documents must be received by MLA via email, fax, or mail by December 1. email applications and documents will be accepted as PDF or MS Word files only. Founded in 1898, the Medical Library Association (MLA) is a nonprofit, educational organization of more than 1,100 institutions and 3,600 individual members in the health sciences information field committed to educating health-information professionals, supporting health-information research, promoting access to the world's health-sciences information, and working to ensure that the best health information is available to all.
Requirements: Eligible candidates should satisfy the following *Requirements:* hold a graduate degree in library science; currently be a practicing health-sciences librarian with at least five years of professional experience; be an individual member of MLA; and be a citizen of or permanent resident in either the United States or Canada. Consideration will be given in exceptional cases to an outstanding candidate not meeting the criteria above. The award winner is required to submit a report to MLA on the results of the project, that may be published in appropriate health information journals.
Geographic Focus: All States
Date(s) Application is Due: Dec 1
Amount of Grant: 2,000 USD
Contact: Carla Funk, Executive Director; (312) 419-9094, ext. 14; fax (312) 419-8950; mlapd2@mlahq.org or grants@mlahq.org
Internet: http://www.mlanet.org/awards/grants/index.html
Sponsor: Medical Library Association
65 East Wacker Place, Suite 1900
Chicago, IL 60601-7246

MLA Donald A.B. Lindberg Research Fellowship 2562
The Medical Library Association (MLA) believes that access to high-quality information improves decision-making by health professionals, scientists, and consumers and is a major determinant in improved health and quality of care both nationally and internationally. MLA awards a $10,000 fellowship annually to support an individual whose research will extend the underlying knowledge-base of health-sciences information management or enhance the practice of the information professions, particularly health-sciences librarianship. The MLA's areas of interest include the organization, delivery, and use of information and knowledge and their impact on health-care access and delivery, public-health services, consumers' use of health information, biomedical research, and education for the health professions. The Lindberg Fellowship is an unrestricted grant that is awarded to the applicant, not to the sponsoring organization, and may be used for salaries, supplies, equipment, travel, fees, insurance, salaries for research assistants, and other research-related costs as specified in the grant application. The award may be used to supplement or extend other awards, including other private or government-supported fellowships, but is not contingent on receiving other awards. The award is not restricted to disbursement in a single year; funding may be disbursed over a period of up to five years depending on the needs of the research fellow. The application form and guidelines are available at the website. The application and supporting documents must be received by MLA via email, fax, or mail by November 15. Applications and documents sent via email will be accepted as PDF or MS Word files only. The Lindberg Research Fellowship is named in honor of Donald A.B. Lindberg, M.D., Director of the National Library of Medicine (NLM), in recognition of his significant national and international achievements at the NLM and its National Center for Biotechnology Information. The Lindberg Research Fellow program of MLA is administered through the Lindberg Research Fellow Jury.
Requirements: Health-sciences librarians, health professionals, researchers, educators, and administrators are eligible to apply. Applicants must be sponsored by an institution or organization; one or a combination of the following institutions or organizations may sponsor applicants: MLA's board, committees, sections, and chapters; graduate schools of library and information science; university deans or department chairs in health-professions schools or health-care organizations; library organizations; or scientific academies and societies. Applicants must have a bachelor's, master's, or doctor's degree or be enrolled in a program leading to such a degree and demonstrate a commitment to the health sciences; additionally applicants must be a citizen or permanent resident of either the United States or Canada, or have been lawfully admitted for permanent residence at the time of appointment.
Restrictions: The grant may not be used for institutional overhead, other indirect costs, income tax payments, or tuition. Acceptance of the grant may be subject to institutional rules and regulations and to all applicable tax laws. The fellowship is not designed to

support research for a doctoral dissertation or master's thesis. An applicant may resubmit an application to reapply for the Lindberg Fellowship for up to a two-year period.
Geographic Focus: All States, Canada
Date(s) Application is Due: Nov 15
Amount of Grant: 10,000 USD
Contact: Carla Funk, Executive Director; (312) 419-9094, ext. 14; fax (312) 419-8950; mlapd2@mlahq.org or grants@mlahq.org
Internet: http://www.mlanet.org/awards/grants/index.html
Sponsor: Medical Library Association
65 East Wacker Place, Suite 1900
Chicago, IL 60601-7246

MLA Estelle Brodman Academic Medical Librarian of the Year Award 2563
The Estelle Brodman Award was established in 1986 and is sponsored by a bequest of Irwin H. Pizer to honor Estelle Brodman's exemplary career as an educator, seminal thinker, able administrator, technological innovator, and skillful practitioner. The Brodman award recognizes an academic medical librarian at mid-career who demonstrates significant achievement, the potential for leadership, and continuing excellence. The recipient receives a certificate at the association's annual meeting and a cash award of $500 after the annual meeting. The recipient assumes all costs of attending the meeting and the ceremony at which the presentation is made. Guidelines and a nomination form are available at the website. All materials and letters of support must be received by MLA via email, fax, or mail (in order of preference) by November 1. The recipient will be notified in March before the annual meeting. Founded in 1898, the Medical Library Association (MLA) is a nonprofit, educational organization of more than 1,100 institutions and 3,600 individual members in the health sciences information field committed to educating health-information professionals, supporting health-information research, promoting access to the world's health-sciences information, and working to ensure that the best health information is available to all.
Requirements: Nominees must satisfy the following criteria: be members of MLA; be an academic health-sciences librarian at the time of the award; be in "mid-career" (defined as working at least five but nor more than fifteen years in an academic health-sciences library) at the time of the award; and have worked in an academic health-sciences library for each of the last five years immediately preceding the nomination. Nomination may be made for outstanding national or international contributions to academic health-sciences librarianship as demonstrated by excellence in performance (leadership), publications, research, service, or a combination of these four elements.
Restrictions: The nominee may not be a library director.
Geographic Focus: All States, All Countries
Date(s) Application is Due: Nov 1
Amount of Grant: 500 USD
Contact: Carla Funk, Program Contact; (312) 419-9094, ext. 14; fax (312) 419-8950; mlapd2@mlahq.org or awards@mlahq.org
Internet: http://www.mlanet.org/awards/honors/brodman.html
Sponsor: Medical Library Association
65 East Wacker Place, Suite 1900
Chicago, IL 60601-7246

MLA Graduate Scholarship 2564
The Medical Library Association (MLA) annually awards a $5,000 scholarship to a student who shows excellence in scholarship and potential for accomplishment in health-sciences librarianship. The scholarship is announced at the annual meeting of the association, where the recipient will also receive a one-year student membership in the Medical Library Association and free inclusive student registration at the association's annual meeting. The recipient will assume the other costs of attending the meeting and the ceremony at which the presentation is made. Guidelines and application materials are available at the website. All materials and letters of support must be received by MLA via email, fax, or mail by December 1. email applications and documents will be accepted as PDF or MS Word files only. The recipient will be notified in March before the annual meeting. Founded in 1898, the Medical Library Association is a nonprofit, educational organization of more than 1,100 institutions and 3,600 individual members in the health sciences information field, committed to educating health-information professionals, supporting health-information research, promoting access to the world's health-sciences information, and working to ensure that the best health information is available to all.
Requirements: To be eligible applicants must be either citizens or permanent residents of the United States or Canada. Additionally, applicants must either be entering or have completed no more than one-half of the academic requirements of a Masters program at an American Library Association (ALA)-accredited graduate school of library science at the time of the granting.
Restrictions: Past recipients of the MLA Scholarship or the MLA Scholarship for Minority Students are not eligible to apply.
Geographic Focus: All States, Canada
Date(s) Application is Due: Dec 1
Amount of Grant: Up to 5,000 USD
Contact: Carla Funk, Executive Director; (312) 419-9094, ext. 14; fax (312) 419-8950; mlapd2@mlahq.org or grants@mlahq.org
Internet: http://www.mlanet.org/awards/grants/scholar.html
Sponsor: Medical Library Association
65 East Wacker Place, Suite 1900
Chicago, IL 60601-7246

MLA Graduate Scholarship for Minority Students 2565
The Medical Library Association (MLA) annually awards a $5,000 scholarship to a student who shows excellence in scholarship and potential for accomplishment in health-sciences librarianship. The scholarship is announced at the annual meeting of the association, where the recipient will also receive a one-year student membership in the Medical Library Association and free inclusive student registration at the association's annual meeting. The recipient will assume the other costs of attending the meeting and the ceremony at which the presentation is made. Guidelines and application materials are available at the website. All materials and letters of support must be received by MLA via email, fax, or mail (in order of preference) by December 1. email applications and documents will be accepted as PDF or MS Word files only. The recipient will be notified in March before the annual meeting. Founded in 1898, the Medical Library Association (MLA) is a nonprofit, educational organization of more than 1,100 institutions and 3,600 individual members in the health sciences information field, committed to educating health-information professionals, supporting health-information research, promoting access to the world's health-sciences information, and working to ensure that the best health information is available to all.
Requirements: To be eligible the applicant must be a member of a minority group (Black or African-American, Hispanic or Latino, Asian, Aboriginal, North-American Indian or Alaskan Native, or Native Hawaiian or other Pacific Islander) and be a citizen or permanent resident of the United States or Canada. Additionally, the applicant must either be entering or have completed no more than one-half of the academic requirements of a Masters program at an American Library Association (ALA)-accredited graduate school of library science at the time of the granting.
Restrictions: Past recipients of the MLA Scholarship or the MLA Scholarship for Minority Students are not eligible to apply.
Geographic Focus: All States, Canada
Date(s) Application is Due: Dec 1
Amount of Grant: Up to 5,000 USD
Contact: Carla Funk, Program Contact; (312) 419-9094, ext. 14; fax (312) 419-8950; mlapd2@mlahq.org or grants@mlahq.org
Internet: http://www.mlanet.org/awards/grants/minstud.html
Sponsor: Medical Library Association
65 East Wacker Place, Suite 1900
Chicago, IL 60601-7246

MLA Ida and George Eliot Prize 2566
The Ida and George Eliot Prize was established by friends of the Medical Library Association (MLA), Ida and George Eliot owners of Eliot Health Sciences Books, Inc. The award is presented annually for a work published in the preceding calendar year which has been judged most effective in furthering medical librarianship. The award recipient receives a certificate at the association's annual meeting and a cash award of $200 after the annual meeting. The recipient assumes all costs of attending the meeting and the ceremony at which the presentation is made. Guidelines and a nomination form are available at the website. All materials and letters of support must be received by MLA via email, fax, or mail (in order of preference) by November 1. The recipient will be notified in March before the annual meeting. Founded in 1898, the Medical Library Association (MLA) is a nonprofit, educational organization of more than 1,100 institutions and 3,600 individual members in the health sciences information field committed to educating health-information professionals, supporting health-information research, promoting access to the world's health-sciences information, and working to ensure that the best health information is available to all.
Requirements: Nominations are accepted from the membership-at-large and from members of the Ida and George Eliot Prize Jury; the jury reviews publications other than those nominated and actively searches for publications, beginning the review process well before nominations may be received. It is preferred that nominees be members of MLA.
Geographic Focus: All States, All Countries
Date(s) Application is Due: Nov 1
Amount of Grant: 200 USD
Contact: Carla Funk, Program Contact; (312) 419-9094, ext. 14; fax (312) 419-8950; mlapd2@mlahq.org or awards@mlahq.org
Internet: http://www.mlanet.org/awards/honors/eliot.html
Sponsor: Medical Library Association
65 East Wacker Place, Suite 1900
Chicago, IL 60601-7246

MLA Janet Doe Lectureship Award 2567
The Janet Doe Lectureship was established in 1966 by an anonymous donation to support an annual lecture in honor of Janet Doe (1895-1985), former librarian of the New York Academy of Medicine, historical scholar, past president of the Medical Library Association (MLA), and editor of the first two editions of the Medical Library Practice handbook. The Janet Doe Lecturer is chosen for his or her unique perspective on the history or philosophy of medical librarianship. The lecturer receives a certificate at the annual meeting, a $250 honorarium, reimbursement for travel expenses to the site of the meeting, hotel for one night, and per diem for one day. Guidelines and a nomination form are available at the website. All materials must be received by MLA via email, fax, or mail (in order of preference) by November 1. The recipient will be notified in March, fourteen months before the lecture is to be given. Founded in 1898, the Medical Library Association (MLA) is a nonprofit, educational organization of more than 1,100 institutions and 3,600 individual members in the health sciences information field committed to educating health-information professionals, supporting health-information research, promoting access to the world's health-sciences information, and working to ensure that the best health information is available to all.

Requirements: Nominations are accepted from the MLA membership at large and from the members of the Janet Doe Lectureship Jury. Nominees should have been active in the profession a sufficient number of years to have acquired a broad perspective of medical librarianship and possess the qualities to communicate their experiences and ideas articulately. Nominees should also be regular members of the MLA and have made a substantial contribution to both the profession and to the MLA's work including leadership, development of or sustained work within a particular area within the association, dedicated service to medicine and the allied-health sciences, contributions to research or scholarship, effectiveness in administration/management, contributions to technology application and library practice, and/or effectiveness in teaching and training.
Geographic Focus: All States, All Countries
Date(s) Application is Due: Nov 1
Amount of Grant: 250 USD
Contact: Carla Funk, Program Contact; (312) 419-9094, ext. 14; fax (312) 419-8950; mlapd2@mlahq.org or awards@mlahq.org
Internet: http://www.mlanet.org/awards/honors/doe.html
Sponsor: Medical Library Association
65 East Wacker Place, Suite 1900
Chicago, IL 60601-7246

MLA Lois Ann Colaianni Award for Excellence and Achievement in Hospital Librarianship 2568

The Lois Ann Colaianni Award for Excellence and Achievement in Hospital Librarianship was established in 1989 to recognize a hospital librarian who has made significant contributions to the profession through overall distinction (as demonstrated by excellence and achievement) in service, advocacy, leadership, publications, presentations, teaching, research, technology, administration, special projects, or any combination of these areas. The award was renamed in 1999 to honor Ms. Colaianni, former Associate Director of the National Library of Medicine. The award recipient receives a certificate at the association's annual meeting and a cash award of $500 after the annual meeting. The recipient assumes all costs of attending the meeting and the ceremony at which the presentation is made. Guidelines and a nomination form are available at the website. In order for the nomination to be considered, all materials and letters of support must be received by MLA via email, fax, or mail (in order of preference) by November 1. The recipient will be notified in March before the annual meeting. Founded in 1898, the Medical Library Association (MLA) is a nonprofit, educational organization of more than 1,100 institutions and 3,600 individual members in the health sciences information field, committed to educating health-information professionals, supporting health-information research, promoting access to the world's health-sciences information, and working to ensure that the best health information is available to all.
Requirements: It is preferred that nominees be members of MLA. Nominees must be hospital librarians at the time of the award and must have worked in a hospital library for at least five (5) years immediately preceding the award.
Geographic Focus: All States, All Countries
Date(s) Application is Due: Nov 1
Amount of Grant: 500 USD
Contact: Carla Funk, Executive Director; (312) 419-9094, ext. 14; fax (312) 419-8950; mlapd2@mlahq.org or awards@mlahq.org
Internet: http://www.mlanet.org/awards/honors/colaianni.html
Sponsor: Medical Library Association
65 East Wacker Place, Suite 1900
Chicago, IL 60601-7246

MLA Louise Darling Medal for Distinguished Achievement in Collection Development in the Health Sciences 2569

Established in 1987 by Ballen Booksellers International, Inc., the Louise Darling Medal is presented annually by the Medical Library Association (MLA) in recognition of distinguished achievement in collection development in the health sciences. The medal honors Louise Darling's significant accomplishment in this professional specialty. The recipient receives a certificate at the association's annual meeting and a $1,000 cash award afterwards. The recipient assumes all costs of attending the meeting and the ceremony at which the presentation is made. Guidelines and a nomination form are available at the website. All materials and supporting documentation must be received by MLA via email, fax, or mail (in order of preference) by November 1. The recipient will be notified in March before the annual meeting. Founded in 1898, the Medical Library Association is a nonprofit, educational organization of more than 1,100 institutions and 3,600 individual members in the health sciences information field committed to educating health-information professionals, supporting health-information research, promoting access to the world's health-sciences information, and working to ensure that the best health information is available to all.
Requirements: Nominees may be individuals, institutions, or groups of individuals; it is preferred that nominees be members of MLA. Nominations are accepted from both the MLA membership at large and the members of the Louise Darling Medal Jury and may be made for overall distinction or leadership within the following areas: collection development; production of a definitive publication related to collection development; teaching of collection development; development of an extraordinary national information resource or collection in any format (e.g. printed materials, audiovisuals, electronic files, etc.); or for any other collection development activity deemed appropriate by the Board of Directors, the Awards Committee, and the Louise Darling Medal Jury.
Geographic Focus: All States, All Countries
Date(s) Application is Due: Nov 1
Amount of Grant: 1,000 USD
Contact: Carla Funk, Program Contact; (312) 419-9094, ext. 14; fax (312) 419-8950; mlapd2@mlahq.org or awards@mlahq.org
Internet: http://www.mlanet.org/awards/honors/index.html
Sponsor: Medical Library Association
65 East Wacker Place, Suite 1900
Chicago, IL 60601-7246

MLA Lucretia W. McClure Excellence in Education Award 2570

The Lucretia W. McClure Excellence in Education Award was established in 1998 to honor an outstanding practicing librarian or library educator in the field of health-sciences librarianship and informatics. The recipient receives a certificate at the Medical Library Association (MLA)'s annual meeting and a cash award of $500 afterwards. The recipient assumes all costs of attending the meeting and the ceremony at which the presentation is made. Guidelines and a nomination form are available at the website. All materials and letters of support must be received by MLA via email, fax, or mail (in order of preference) by November 1. The recipient will be notified in March before the annual meeting. Founded in 1898, the Medical Library Association is a nonprofit, educational organization of more than 1,100 institutions and 3,600 individual members in the health sciences information field committed to educating health-information professionals, supporting health-information research, promoting access to the world's health-sciences information, and working to ensure that the best health information is available to all.
Requirements: It is preferred that nominees be members of MLA. Nominees must be currently employed as a health-sciences librarian or educator at the time of the award and must have worked in such a position for at least five (5) years immediately preceding the award. The nomination may be made for contributions to education as demonstrated by excellence and achievement in the following areas: service; curriculum development; leadership at local, regional or national levels; publications; presentations; teaching; research; mentoring; special projects; or any combination of these.
Geographic Focus: All States, All Countries
Date(s) Application is Due: Nov 1
Amount of Grant: 500 USD
Contact: Carla Funk, Executive Director; (312) 419-9094, ext. 14; fax (312) 419-8950; mlapd2@mlahq.org or awards@mlahq.org
Internet: http://www.mlanet.org/awards/honors/index.html
Sponsor: Medical Library Association
65 East Wacker Place, Suite 1900
Chicago, IL 60601-7246

MLA Murray Gottlieb Prize Essay Award 2571

The Murray Gottlieb Prize was established in 1956 by Ralph and Jo Grimes of the Old Hickory Bookshop (Brinklow, Maryland) in memory of Murray Gottlieb, a New York antiquarian-book dealer. The purpose of the award is to stimulate the health-sciences librarian's interest in the history of medicine. Currently the award is sponsored by the Medical Library Association (MLA) History of the Health Sciences Section and is awarded annually for the best unpublished essay on the history of medicine and allied sciences written by a health-sciences librarian. The author of the winning essay receives complimentary registration to the annual meeting, a certificate at the association's annual meeting, and a cash award of $100 afterwards. The recipient assumes all costs of attending the meeting and the ceremony at which the presentation is made. Guidelines and a submission form are available at the website. All materials must be received by MLA via email, fax, or mail (in order of preference) by November 1. The recipient will be notified in March before the annual meeting. Founded in 1898, the Medical Library Association is a nonprofit, educational organization of more than 1,100 institutions and 3,600 individual members in the health sciences information field committed to educating health-information professionals, supporting health-information research, promoting access to the world's health-sciences information, and working to ensure that the best health information is available to all.
Requirements: Authors of papers submitted must be health-sciences librarians. The submitted paper must treat some aspect of the history of medicine or allied sciences. Entries will be judged according to quality of bibliographic research, quality of the experimental design or arguments developed to support a particular hypothesis, contribution to the study of the history of the health sciences, and originality and style (clarity, appearance, conciseness, and structure).
Restrictions: Papers may be under consideration for publication at the time of submission but cannot have been published.
Geographic Focus: All States, All Countries
Date(s) Application is Due: Nov 1
Amount of Grant: 100 USD
Contact: Carla Funk, Program Contact; (312) 419-9094, ext. 14; fax (312) 419-8950; mlapd2@mlahq.org or awards@mlahq.org
Internet: http://www.mlanet.org/awards/honors/gottlieb.html
Sponsor: Medical Library Association
65 East Wacker Place, Suite 1900
Chicago, IL 60601-7246

MLA Research, Development, and Demonstration Project Grant 2572

The purpose of the Medical Library Association (MLA)'s Research, Development, and Demonstration Project Grant is to provide support for research, development, or demonstration projects that will help to promote excellence in the field of health-sciences librarianship and information sciences. Grants range from $100 to $1000. The application form and guidelines are available for download at the website. Applications and supporting documents must be received by MLA via email, fax, or mail by December 1. emailed applications and documents will be accepted as PDF or MS Word files only. MLA may offer Research, Development, and Demonstration Project Grants more than once a year. Applicants are encouraged to visit the website or contact the association for information about additional deadlines. Founded in 1898, the Medical

Library Association (MLA) is a nonprofit, educational organization of more than 1,100 institutions and 3,600 individual members in the health sciences information field committed to educating health-information professionals, supporting health-information research, promoting access to the world's health-sciences information, and working to ensure that the best health information is available to all.
Requirements: Eligible applicants must satisfy the following criteria: hold a graduate degree in library science; be a practicing health-sciences librarian with at least two years of professional experience; be a citizen of or have permanent residence status in either the United States or Canada; and preferably be an individual member of the MLA. Consideration will be given in exceptional cases to an outstanding candidate(s) not meeting the criteria above.
Restrictions: Grants will not be given to support an activity that is operational in nature or that has only local usefulness.
Geographic Focus: All States, Canada
Date(s) Application is Due: Dec 1
Amount of Grant: 100 - 1,000 USD
Contact: Carla Funk, Executive Director; (312) 419-9094, ext. 14; fax (312) 419-8950; mlapd2@mlahq.org or grants@mlahq.org
Internet: http://www.mlanet.org/awards/grants/index.html
Sponsor: Medical Library Association
65 East Wacker Place, Suite 1900
Chicago, IL 60601-7246

MLA Rittenhouse Award 2573

The Rittenhouse Award, initiated in 1967 and sponsored by Rittenhouse Book Distributors, Inc., is presented annually by the Medical Library Association (MLA) for the best unpublished paper or web-based project on health-sciences librarianship or medical informatics submitted by a student in an ALA-accredited program of library and information studies or by an internship trainee in medical informatics or health-sciences librarianship. The author of the winning essay or project receives complimentary student registration to MLA's annual meeting, a certificate at the meeting, and a cash award of $500 after the meeting. Winning entries will be submitted to the Journal of the Medical Library Association for consideration. The recipient assumes all costs of attending the meeting and the ceremony at which the presentation is made. Guidelines and a submission form are available at the website. All materials must be received by MLA via email, fax, or mail (in order of preference) by November 1. The recipient will be notified in March before the annual meeting. Founded in 1898, the Medical Library Association is a nonprofit, educational organization of more than 1,100 institutions and 3,600 individual members in the health sciences information field committed to educating health-information professionals, supporting health-information research, promoting access to the world's health-sciences information, and working to ensure that the best health information is available to all.
Requirements: Authors of papers or web-based projects submitted for the Rittenhouse Award must be enrolled in a course for credit in an ALA-accredited graduate program of library and information studies or have recently graduated from such a program and be currently enrolled as an internship trainee in medical informatics or health-sciences librarianship. Submissions must treat some aspect of health-sciences librarianship or medical informatics and may be bibliographical, address a health-sciences issue or topic, or report the results of research. Authors of web-based projects should submit a needs assessment, information on development of the website, evaluation of the website, and a summary outlining the project plan and evaluation results. Entries will be judged according to the following standards: thoroughness and relevance of bibliographic research or needs assessment; importance of topic that would warrant a bibliographic treatment or web-based project; quality of experimental design or arguments developed to support a particular hypothesis or question; originality of argument and positions taken; methodological rigor; quality of website solution; impact on the field of health-sciences librarianship or medical informatics; and style (clarity, appearance, conciseness, organization, and structure).
Restrictions: Papers must have been written or web-based project completed during the 18 months preceding the submission deadline and during the student's course work toward a graduate degree in library and information studies or medical informatics. Papers that have been published or that are under consideration for publication are not eligible.
Geographic Focus: All States
Date(s) Application is Due: Nov 1
Amount of Grant: 500 USD
Contact: Carla Funk, Program Contact; (312) 419-9094, ext. 14; fax (312) 419-8950; mlapd2@mlahq.org or awards@mlahq.org
Internet: http://www.mlanet.org/awards/honors/ritten.html
Sponsor: Medical Library Association
65 East Wacker Place, Suite 1900
Chicago, IL 60601-7246

MLA Section Project of the Year Award 2574

The Medical Librarians Association (MLA) Section Project of the Year Award was established by the Board of Directors in 2010 and announced with a request for applications in 2011. The award was proposed as one of the ways to encourage health-sciences librarians to demonstrate their creativity, ingenuity, cooperation, and leadership within the framework of the mandate of the section. The recipient section will receive a certificate at the annual meeting and a cash award of $500 afterwards. The award will be presented to the recipient section's chair or other designated section representative. The recipient section or section representative assumes all costs of attending the meeting and the ceremony at which the presentation is made. Guidelines and an application form are available at the website. Applications and any supporting documentation must be received by MLA via email, fax, or mail (in order of preference) by November 1. The chair of the recipient section will be notified in March before the annual meeting. Founded in 1898, the Medical Library Association is a nonprofit, educational organization of more than 1,100 institutions and 3,600 individual members in the health sciences information field committed to educating health-information professionals, supporting health-information research, promoting access to the world's health-sciences information, and working to ensure that the best health information is available to all.
Requirements: Application is made for excellence in special projects or innovative operation-programming that demonstrate advocacy, leadership, service, technology, or innovations that contribute to the advancement of health-sciences librarianship. Applicants must satisfy the following criteria: be a section of the MLA as defined by the MLA bylaws; demonstrate section support for the application, as evidenced by an endorsement from the section chair on behalf of the executive committee or officiating body of the section; reference the project's goals, objectives, and outcomes or evaluations to provide evidence of the achievements and impact of the project on the profession and the members of the affected section; have completed the project within the three years preceding the date of the application or operate an ongoing program that accomplishes the goals set out for the project.
Restrictions: Projects or programs may be resubmitted as long as they continue to meet time frames and other eligibility criteria.
Geographic Focus: All States, All Countries
Date(s) Application is Due: Nov 1
Amount of Grant: 500 USD
Contact: Carla Funk, Executive Director; (312) 255-3666; fax (312) 419-8950; mlapd2@mlahq.org or awards@mlahq.org
Internet: http://www.mlanet.org/awards/honors/index.html
Sponsor: Medical Library Association
65 East Wacker Place, Suite 1900
Chicago, IL 60601-7246

MLA T. Mark Hodges International Service Award 2575

T. Mark Hodges, 1999 recipient of the Marcia C. Noyes Award from the Medical Library Association (MLA), was a lifelong believer in the importance of international connections between librarians. The T. Mark Hodges International Service Award (ISA) was established in 2007 to honor outstanding individual achievement in promoting, enabling, and/or delivering improvements in the quality of health information internationally through the development of health-information professionals, the improvement of libraries, or an increased use of health-information services. The recipient receives a certificate at the MLA annual meeting and the option of receiving a cash prize of $500 or a donation from the MLA in the amount of $500 to a charity of his or her choice. The recipient assumes all costs of attending the meeting and the ceremony at which the presentation is made. Guidelines and a nomination form are available at the website. All materials and letters of support must be received by MLA via email, fax, or mail (in order of preference) by November 1. The recipient will be notified in March before the annual meeting. Founded in 1898, the Medical Library Association (MLA) is a nonprofit, educational organization of more than 1,100 institutions and 3,600 individual members in the health sciences information field committed to educating health-information professionals, supporting health-information research, promoting access to the world's health-sciences information, and working to ensure that the best health information is available to all.
Requirements: The award is designed to enable MLA to recognize the widest range of achievement in the development of health-information services in the international context. Membership in MLA is not required. Nominees would usually hold a professional library- or information-science qualification but in the case of exceptional candidates this criterion may be waived. The nominee's achievement may cover the whole range of health-information services or a single aspect, and similarly, may be worldwide in impact or more narrow and intense in focus. MLA normally recognizes a professional contribution over a sustained period of time, but may also mark a single, outstanding achievement of global significance.
Geographic Focus: All Countries
Date(s) Application is Due: Nov 1
Amount of Grant: 500 USD
Contact: Carla Funk, Executive Director; (312) 419-9094, ext. 14; fax (312) 419-8950; mlapd2@mlahq.org or awards@mlahq.org
Internet: http://www.mlanet.org/awards/honors/index.html
Sponsor: Medical Library Association
65 East Wacker Place, Suite 1900
Chicago, IL 60601-7246

MLA Thomas Reuters/Frank Bradway Rogers Information Advancement Award 2576

The Thomson Reuters/Frank Bradway Rogers Information Advancement Award is presented annually by the Medical Library Association in recognition of outstanding contributions in the following areas: the use of technology to deliver health sciences information; the science of information; and facilitating the delivery of health-sciences information. Originally sponsored by the Institute for Scientific Information (ISI), the award was presented for the first time in 1983 to Frank Bradway Rogers, M.D., for whom it was subsequently renamed. The recipient receives a certificate at the association's annual meeting and a cash award of $500 after the annual meeting. Guidelines and a nomination form are available at the website. In order for the nomination to be considered, all materials and letters of support must be received by MLA via email, fax, or mail (in order of preference) by November 1. The recipient will be notified in March before the annual meeting. Founded in 1898, the Medical Library Association (MLA) is a nonprofit, educational organization of more than 1,100 institutions and 3,600 individual members in the health sciences information field, committed to educating health-information professionals, supporting health-information research, promoting access to the world's health-sciences information, and working to ensure that the best health information is available to all.
Requirements: It is preferred that an individual nominee be a member of MLA, or if a small group is nominated for a combined effort, that at least one member of the group

be a member of MLA. Nominations should be for innovative contributions to improved information handling and use. Preference will be given to pioneering efforts (i.e., new theories or applications).
Geographic Focus: All States, All Countries
Date(s) Application is Due: Nov 1
Amount of Grant: 500 USD
Contact: Carla Funk, Program Contact; (312) 419-9094, ext. 14; fax (312) 419-8950; mlapd2@mlahq.org or awards@mlahq.org
Internet: http://www.mlanet.org/awards/honors/rogers.html
Sponsor: Medical Library Association
65 East Wacker Place, Suite 1900
Chicago, IL 60601-7246

MMAAP Foundation Dermatology Fellowships 2577
The mission of the Milstein Medical Asian American Partnership Foundation (MMAAP Foundation) is to improve world health by developing mutually beneficial partnerships between the United States and China, as well as greater Asia. The Foundation invites the submission of applications from Chinese dermatologists and skin biologists for a Fellowship Award to support one-year training at a prominent sponsoring institution in the U.S. The Foundation will provide the sole support for these initiatives. The aim of the program is to build enduring partnership between the U.S. and Asia through training of future Chinese academic leaders and to encourage long-term collaborations between the two regions. The award will provide support for one fellow in the amount of $60,000 accompanied by a grant of $25,000 to the sponsoring U.S. institution. These funds will be distributed directly to the sponsoring U.S. institution. The annual application deadline is November 1.
Requirements: Chinese applicants should come from major medical centers with demonstrated excellence in dermatology and research; at the rank of Instructor or above by the application deadline; an MD or PhD within the specialty of dermatology; research experience of at least five years with a distinguished publication record; commitment to research and academics; demonstrated leadership skills and excellent English speaking proficiency. The ideal candidate should be coming from China at the beginning of the funding period by the MMAAP Foundation; while applicants who are working, studying or training in the U.S. or have plans to visit the U.S. under another program at the time of application may apply, they must disclose such information in the application. After application submission, applicants must provide updates to the MMAAP Foundation if there are any changes in their professional or personal situation.
Restrictions: It is not allowed to submit the same application for multiple awards in the same year. Indirect costs are not allowed.
Geographic Focus: All States, China
Date(s) Application is Due: Nov 1
Amount of Grant: 85,000 USD
Contact: Mia Wang, (212) 850-4505; fax (212) 889-4959; mia@mmaapf.org
Internet: http://www.mmaapf.org/en/grants/
Sponsor: Milstein Medical Asian American Partnership Foundation
335 Madison Avenue, 15th Floor
New York, NY 10017

MMAAP Foundation Dermatology Project Awards 2578
The mission of the Milstein Medical Asian American Partnership Foundation (MMAAP Foundation) is to improve world health by developing mutually beneficial partnerships between the United States and China, as well as greater Asia. The Foundation invites the submission of applications from Chinese dermatologists and skin biologists for a research project award which will have immediate impact on improving skin care in China. The Foundation will be providing the sole support for these initiatives. The aim of the program is to build enduring partnership between the U.S. and Asia through funding a research project and to encourage long term collaborations between the two regions. The award will provide support for the project to be conducted at a Chinese institution in the amount of $50,000 and support for a U.S. partner institution in the amount of $10,000. These funds will be distributed directly to the sponsoring Chinese and U.S. institutions. The annual deadline for applications is November 1.
Requirements: Funds may be used for support of a new or ongoing clinical research project with demonstrated potential for immediate impact on improving skin care in China. Knowledge gained should be applicable to dermatological clinical practice with clear generalizability. Applications should come from major Chinese medical centers with demonstrated excellence in dermatology with a prominent U.S. institution as the partner.
Geographic Focus: All States, China
Date(s) Application is Due: Nov 1
Amount of Grant: 60,000 USD
Contact: Mia Wang, (212) 850-4505; fax (212) 889-4959; mia@mmaapf.org
Internet: http://www.mmaapf.org/en/grants/
Sponsor: Milstein Medical Asian American Partnership Foundation
335 Madison Avenue, 15th Floor
New York, NY 10017

MMAAP Foundation Fellowship Award in Hematology 2579
The mission of the Milstein Medical Asian American Partnership Foundation (MMAAP Foundation) is to improve world health by developing mutually beneficial partnerships between the United States and China, as well as greater Asia. The Foundation invites the submission of applications from Chinese hematologists and blood researchers for a Fellowship Award to support one-year training at a prominent sponsoring institution in the U.S. The MMAAP Foundation will provide the sole support for these initiatives. The aim of the program is to build an enduring partnership between the U.S. and Asia through training of future Chinese academic leaders and to encourage long-term collaborations between the two regions. The award will provide support for one fellow in the amount of $60,000 accompanied by a grant of $25,000 to the sponsoring U.S. institution. These funds will be distributed directly to the sponsoring U.S. institution.
Requirements: Chinese applicants should come from major medical centers with demonstrated excellence in hematology and blood research; at the rank of Instructor or above by the application deadline; an MD or PhD within the specialty of hematology; research experience of at least five years with a distinguished publication record; commitment to research and academics; demonstrated leadership skills; and excellent English speaking proficiency. The ideal candidate should be coming from China at the beginning of the funding period by the MMAAP Foundation; while applicants who are working, studying or training in the U.S. or have plans to visit the U.S. under another program at the time of application may apply, they must disclose such information in the application. After application submission, applicants must provide updates to the MMAAP Foundation if there are any changes in their professional or personal situation.
Restrictions: It is not allowed to submit the same application for multiple awards in the same year. Indirect costs are not allowed.
Geographic Focus: All States, China
Amount of Grant: 85,000 USD
Contact: Mia Wang, (212) 850-4505; fax (212) 889-4959; mia@mmaapf.org
Internet: http://www.mmaapf.org/en/grants/
Sponsor: Milstein Medical Asian American Partnership Foundation
335 Madison Avenue, 15th Floor
New York, NY 10017

MMAAP Foundation Fellowship Award in Reproductive Medicine 2580
The mission of the Milstein Medical Asian American Partnership Foundation (MMAAP Foundation) is to improve world health by developing mutually beneficial partnerships between the United States and China, as well as greater Asia. The Foundation invites the submission of applications from Chinese investigators and researchers in reproductive medicine for a Fellowship Award to support one-year training at a prominent sponsoring institution in the U.S. The MMAAP Foundation will provide the sole support for these initiatives. The aim of the program is to build an enduring partnership between the U.S. and Asia through training of future Chinese academic leaders and to encourage long-term collaborations between the two regions. The award will provide support for one fellow in the amount of $60,000 accompanied by a grant of $25,000 to the sponsoring U.S. institution. These funds will be distributed directly to the sponsoring U.S. institution. The annual application deadline is November 1.
Requirements: Chinese applicants should come from major medical centers with demonstrated excellence in IVF research; at the rank of Instructor or above by the application deadline; an MD or PhD within the specialty of reproductive medicine; significant research experience with a distinguished publication record; commitment to research and academics; demonstrated leadership skills; and excellent English speaking proficiency. The ideal candidate should be coming from China at the beginning of the funding period by the MMAAP Foundation; while applicants who are working, studying or training in the U.S. or have plans to visit the U.S. under another program at the time of application may apply, they must disclose such information in the application. After application submission, applicants must provide updates to the MMAAP Foundation if there are any changes in their professional or personal situation.
Restrictions: It is not allowed to submit the same application for multiple awards in the same year. Indirect costs are not allowed.
Geographic Focus: All States, China
Date(s) Application is Due: Nov 1
Amount of Grant: 85,000 USD
Contact: Mia Wang, (212) 850-4505; fax (212) 889-4959; mia@mmaapf.org
Internet: http://www.mmaapf.org/en/grants/
Sponsor: Milstein Medical Asian American Partnership Foundation
335 Madison Avenue, 15th Floor
New York, NY 10017

MMAAP Foundation Fellowship Award in Translational Medicine 2581
The mission of the Milstein Medical Asian American Partnership Foundation (MMAAP Foundation) is to improve world health by developing mutually beneficial partnerships between the United States and China, as well as greater Asia. The Foundation invites the submission of applications from Chinese medical researchers for a Fellowship Award to support one-year training at a prominent sponsoring institution in the U.S. The MMAAP Foundation will provide the sole support for these initiatives. Translational Medicine refers to the process of converting scientific discoveries into drugs or medical devices to treat patients. The process includes preclinical studies, observational studies, testing treatment methods and the development of best practices of the treatment. The aim of the program is to build an enduring partnership between the U.S. and Asia through training of future Chinese academic leaders and to encourage long-term collaborations between the two regions. The award will provide support for one fellow in the amount of $60,000 accompanied by a grant of $25,000 to the sponsoring U.S. institution. These funds will be distributed directly to the sponsoring U.S. institution. The annual application deadline is November 1.
Requirements: Chinese applicants should come from major medical centers with demonstrated excellence in translational medicine; at the rank of Instructor or above by the application deadline; an MD or PhD within any specialty or medical sub-specialties; significant research experience with a distinguished publication record; commitment to research and academics; demonstrated leadership skills; and excellent English speaking proficiency. The ideal candidate should be coming from China at the beginning of the funding period by the MMAAP Foundation; while applicants who are working, studying or training in the U.S. or have plans to visit the U.S. under another program at the time

of application may apply, they must disclose such information in the application. After application submission, applicants must provide updates to the MMAAP Foundation if there are any changes in their professional or personal situation.
Restrictions: It is not allowed to submit the same application for multiple awards in the same year. Indirect costs are not allowed.
Geographic Focus: All States, China
Date(s) Application is Due: Nov 1
Amount of Grant: 85,000 USD
Contact: Mia Wang, (212) 850-4505; fax (212) 889-4959; mia@mmaapf.org
Internet: http://www.mmaapf.org/en/grants/
Sponsor: Milstein Medical Asian American Partnership Foundation
335 Madison Avenue, 15th Floor
New York, NY 10017

MMAAP Foundation Geriatric Project Awards 2582
The mission of the Milstein Medical Asian American Partnership Foundation (MMAAP Foundation) is to improve world health by developing mutually beneficial partnerships between the United States and China, as well as greater Asia. The Foundation invites the submission of applications from Chinese geriatricians for a research project award which will have immediate impact on improving skin care in China. The Foundation will be providing the sole support for these initiatives. The aim of the program is to build enduring partnership between the U.S. and Asia through funding a research project and to encourage long term collaborations between the two regions. The award will provide support for the project to be conducted at a Chinese institution in the amount of $50,000 and support for a U.S. partner institution in the amount of $10,000. These funds will be distributed directly to the sponsoring Chinese and U.S. institutions. The annual deadline for applications is November 1.
Requirements: Funds may be used for support of a new or ongoing clinical research project with demonstrated potential for immediate impact on improving senior health in China. Applications should come from major Chinese medical centers with demonstrated excellence in dermatology with a prominent U.S. institution as the partner.
Geographic Focus: All States, China
Date(s) Application is Due: Nov 1
Amount of Grant: 60,000 USD
Contact: Mia Wang, (212) 850-4505; fax (212) 889-4959; mia@mmaapf.org
Internet: http://www.mmaapf.org/en/grants/
Sponsor: Milstein Medical Asian American Partnership Foundation
335 Madison Avenue, 15th Floor
New York, NY 10017

MMAAP Foundation Hematology Project Awards 2583
The mission of the Milstein Medical Asian American Partnership Foundation (MMAAP Foundation) is to improve world health by developing mutually beneficial partnerships between the United States and China, as well as greater Asia. The Foundation invites the submission of applications from Chinese hematologists for a research project award which will have immediate impact on improving blood health in China. The Foundation will be providing the sole support for these initiatives. The aim of the program is to build enduring partnership between the U.S. and Asia through funding a research project and to encourage long term collaborations between the two regions. The award will provide support for the project to be conducted at a Chinese institution in the amount of $50,000 and support for a U.S. partner institution in the amount of $10,000. These funds will be distributed directly to the sponsoring Chinese and U.S. institutions. The annual deadline for applications is November 1.
Requirements: Funds may be used for support of a new or ongoing clinical research project with demonstrated potential for immediate impact on improving blood health in China. Knowledge gained should be applicable to hematological clinical practice with clear generalizability. Applications should come from major Chinese medical centers with demonstrated excellence in hematology with a prominent U.S. institution as the partner.
Geographic Focus: All States, China
Date(s) Application is Due: Nov 1
Amount of Grant: 60,000 USD
Contact: Mia Wang, (212) 850-4505; fax (212) 889-4959; mia@mmaapf.org
Internet: http://www.mmaapf.org/en/grants/
Sponsor: Milstein Medical Asian American Partnership Foundation
335 Madison Avenue, 15th Floor
New York, NY 10017

MMAAP Foundation Research Project Award in Reproductive Medicine 2584
The mission of the Milstein Medical Asian American Partnership Foundation (MMAAP Foundation) is to improve world health by developing mutually beneficial partnerships between the United States and China, as well as greater Asia. The Foundation invites the submission of applications from Chinese investigators in reproductive medicine for a research project award which will have immediate impact on n vitro fertilization (IVF) practices in China. The Foundation will be providing the sole support for these initiatives. The aim of the program is to build enduring partnership between the U.S. and Asia through funding a research project and to encourage long term collaborations between the two regions. The award will provide support for the project to be conducted at a Chinese institution in the amount of $50,000 and support for a U.S. partner institution in the amount of $10,000. These funds will be distributed directly to the sponsoring Chinese and U.S. institutions. The annual deadline for applications is November 1.
Requirements: Funds may be used for support of a new or ongoing clinical research project with demonstrated potential for immediate impact on improving IVF practices in China. Applications should come from major Chinese medical centers with demonstrated excellence in reproductive medicine with a prominent U.S. institution as the partner.
Geographic Focus: All States, China
Date(s) Application is Due: Nov 1
Amount of Grant: 60,000 USD
Contact: Mia Wang, (212) 850-4505; fax (212) 889-4959; mia@mmaapf.org
Internet: http://www.mmaapf.org/en/grants/
Sponsor: Milstein Medical Asian American Partnership Foundation
335 Madison Avenue, 15th Floor
New York, NY 10017

MMAAP Foundation Research Project Award in Translational Medicine 2585
The mission of the Milstein Medical Asian American Partnership Foundation (MMAAP Foundation) is to improve world health by developing mutually beneficial partnerships between the United States and China, as well as greater Asia. The Foundation invites the submission of applications from Chinese medical researchers for a research project award which will have immediate impact on improving health in China. The Foundation will be providing the sole support for these initiatives. Translational Medicine refers to the process of converting scientific discoveries into drugs or medical devices to treat patients. The process includes pre-clinical studies, observational studies, testing treatment methods and the development of best practices of the treatment. The aim of the program is to build enduring partnership between the U.S. and Asia through funding a research project and to encourage long term collaborations between the two regions. The award will provide support for the project to be conducted at a Chinese institution in the amount of $50,000 and support for a U.S. partner institution in the amount of $10,000. These funds will be distributed directly to the sponsoring Chinese and U.S. institutions. The annual deadline for applications is November 1.
Requirements: Funds may be used for support of a new or ongoing clinical research project with demonstrated potential for immediate impact on improving health in China. Applications should come from major Chinese medical centers with demonstrated excellence in translational medicine with a prominent U.S. institution as the partner.
Geographic Focus: All States, China
Date(s) Application is Due: Nov 1
Amount of Grant: 60,000 USD
Contact: Mia Wang, (212) 850-4505; fax (212) 889-4959; mia@mmaapf.org
Internet: http://www.mmaapf.org/en/grants/
Sponsor: Milstein Medical Asian American Partnership Foundation
335 Madison Avenue, 15th Floor
New York, NY 10017

MMAAP Foundation Senior Health Fellowship Award in Geriatrics Medicine and Aging Research 2586
The mission of the Milstein Medical Asian American Partnership Foundation (MMAAP Foundation) is to improve world health by developing mutually beneficial partnerships between the United States and China, as well as greater Asia. The Foundation invites the submission of applications from Chinese geriatricians for the Irma and Paul Milstein program for Senior Health Fellowship Award to support one year of training at a prominent sponsoring institution in the U.S. for two to four scholars in geriatrics from China. The Foundation will be providing the sole support for these initiatives. The aim of the program is to build enduring partnership between the U.S. and Asia through training of future Chinese academic leaders and to encourage long-term collaborations between the two regions. The award will provide support for the fellow in the amount of $60,000 accompanied by a grant of $25,000 to the sponsoring U.S. institution. These funds will be distributed directly to the sponsoring U.S. institution. The annual application deadline is November 1.
Requirements: Chinese applicants should come from major medical centers with demonstrated excellence in geriatrics and research; at the rank of Instructor or above by the application deadline; an MD or PhD within the specialty of geriatrics; research experience of at least five years with a distinguished publication record; commitment to research and academics; demonstrated leadership skills and excellent English speaking proficiency. The ideal candidate should be coming from China at the beginning of the funding period by the MMAAP Foundation; while applicants who are working, studying or training in the U.S. or have plans to visit the U.S. under another program at the time of application may apply, they must disclose such information in the application. After application submission, applicants must provide updates to the MMAAP Foundation if there are any changes in their professional or personal situation.
Restrictions: It is not allowed to submit the same application for multiple awards in the same year. Indirect costs are not allowed.
Geographic Focus: All States, China
Date(s) Application is Due: Nov 1
Amount of Grant: 85,000 USD
Contact: Mia Wang, (212) 850-4505; fax (212) 889-4959; mia@mmaapf.org
Internet: http://www.mmaapf.org/en/grants/
Sponsor: Milstein Medical Asian American Partnership Foundation
335 Madison Avenue, 15th Floor
New York, NY 10017

MMS and Alliance Charitable Foundation Grants for Community Action and Care for the Medically Uninsured 2587
The Massachusetts Medical Society and Alliance Charitable Foundation, through its Board of Directors, awards grants to nonprofit organizations and works with communities throughout Massachusetts to creatively address issues that affect the health, benefit, and welfare of the community. The Foundation supports physician-led volunteer initiatives to provide free care to uninsured/underinsured patients and increased access to care for the medically

underserved. Preference will be given to organizations working with interdisciplinary groups that address health care issues and where strong physician involvement exists.
Requirements: Given the spectrum of issues that influence health, the Foundation, as an organization of physicians and their families, focuses its efforts on programs that directly promote health in the community. Eligible programs may provide direct care services or target public health issues which impact the health care system and the health of communities. Programs applying for grants must address one or more of the following goals: (a) Affect the health and well-being of the community through community-based prevention, screening, early detection, health promotion, and/or increased access to medical care; (b) Promote healthy decision-making around behaviors and lifestyle choices by raising awareness, providing education, improving communication, and/or connecting community members with culturally appropriate programs and services; (c) Support physician-led volunteer initiatives to provide free care to uninsured/underinsured patients and/or increase access to care for the medically underserved; (d) Enable underserved, at-risk populations, to engage in activities that promote health and identify risky behaviors which contribute to diminished health. Applicants should submit a Letter of Inquiry (LOI) (Word Doc, 1 page) and accepted LOIs are invited to submit a full proposal. LOIs are due January 15, and proposals, if invited to submit, are due March 1.
Restrictions: The foundation does not provide funding support for: capital campaigns, endowments, building campaigns; for-profit organizations; fundraising drives and fundraising events; individuals (unless applying for International Health Studies Grant or scholarships through a directed giving program; private or parochial schools, colleges, or universities; government agencies (except in collaboration with community-based, nonprofit organizations which will lead the program and act as fiscal agent); organizations that advocate, support, or practice discrimination based on race, religion, age, national origin, language, sex, sexual preference, or physical handicap; religious organizations for religious purposes; research; or, political or lobbying activities.
Geographic Focus: Massachusetts
Date(s) Application is Due: Mar 1
Contact: Jennifer Day, Manager; (781) 434-7044; foundation@mms.org
Internet: http://www.massmed.org/Charitable_Foundation/Applying_for_Grants/Applying_for_Grants/#.U1gQtccRZlo
Sponsor: Massachusetts Medical Society and Alliance Charitable Foundation
c/o Massachusetts Medical Society
Waltham, MA 02451

MMS and Alliance Charitable Foundation International Health Studies Grants 2588
The primary goal of the International Health Studies Grants is to encourage international education, particularly focusing on underserved populations. This program is supported, in part, by an annual donation from the MMS. Additional monies have been raised through private donations to award a total of five grants annually. The Foundation Board of Directors will consider applications once a year. Applications must be submitted via the online application and be received by 4:00 pm on September 15.
Requirements: Medical students and resident physician members of the Massachusetts Medical Society (MMS) are eligible to apply for grants for up to $2,000 to defray the costs of study abroad. Preference will be given to projects providing health care related work and/or training of staff; and to applicants planning careers serving underprivileged populations in the world.
Restrictions: Research projects that do not involve direct clinical care or teaching will not be considered.
Geographic Focus: Massachusetts
Date(s) Application is Due: Sep 15
Amount of Grant: Up to 2,000 USD
Contact: Jennifer Day, Manager; (781) 434-7044; foundation@mms.org
Internet: http://www.massmed.org/Charitable_Foundation/Applying_for_Grants/International_Health_Studies_Program
Sponsor: Massachusetts Medical Society and Alliance Charitable Foundation
c/o Massachusetts Medical Society
Waltham, MA 02451

Mockingbird Foundation Grants 2589
The Mockingbird Foundation (Mockingbird) is a non-profit organization founded by Phish (the rock band) fans in 1996. Since then Mockingbird has distributed over $750,000 to support music-education programs for children. The foundation provides funding through its competitive grants, emergency-related grants, and tour-related grants. Emergency-related grants and tour-related grants are unsolicited grants. Emergency grants come from an Emergency Fund in which 3% of Mockingbird's gross revenues are designated for music-education programs affected by disasters (e.g., hurricanes and tornadoes). Tour-related grants support music-education programs in communities touched by Phish tours and are intended to inspire support for music and arts education and to generate positive press coverage. Competitive grants are awarded to schools and nonprofit organizations through a two-tiered grant-application process. Applicants must first complete an initial inquiry form at Mockingbird's website. If an applicant's project is selected for further consideration, the applicant will then be invited to submit a full and formal proposal. Initial inquiries may be submitted at any time. Mockingbird typically reviews inquiries in August and September, invites full proposals by Halloween, and announces new grants sometime between Christmas and the end of January. Mockingbird encourages projects with diverse or unusual musical styles, genres, forms, and philosophies; projects with unconventional outlets and forms of instruction as well as instruction in unconventional forms; projects that foster creative expression in any musical form (including composition, instrumentation, vocalization, or improvisation); and projects in which skills, as outcomes, are less assessable or even irrelevant. Projects should be experiential and directly engage students with creating and expressing music.

Preference is given to programs that benefit disenfranchised groups, including those with low skill levels, income, education, disabilities, or terminal illness, and/or those in foster-care homes, hospitals, and prisons. Prospective applicants should review the instructions and guidelines given at the website for complete details before making application.
Requirements: Schools and 501(c)3 nonprofit organizations in the United States are eligible. Applicants may apply through a sponsor who meets these qualifications. Mockingbird is particularly interested in organizations with low overhead, innovative approaches, and/or collaborative elements to their work. The foundation encourages geographic diversity and has funded forty-three states to date. Mockingbird is interested in targeting children 18 years or younger, but will consider projects that benefit college students, teachers/instructors, or adult students.
Restrictions: Grants are made on a one-time basis and are non-renewable and non-transferable. Mockingbird does not normally support individuals, fund-raising organizations or events, research, and programs that promote or engage in religious or political doctrine. It is hoped that applicants for Mockingbird grants hire staff and provide services without discriminating on the basis of race, religion, gender, sexual orientation, age, national origin, or disability. Mockingbird supports the provision of instruments, texts, and office materials, and the acquisition of learning space, practice space, performance space, and instructors/instruction. Mockingbird is particularly interested in projects that foster self-esteem and free expression, but does not fund music therapy which is neither education nor music appreciation which does not include participation.
Geographic Focus: All States
Amount of Grant: 100 - 5,000 USD
Samples: Connecticut Percussive Arts Society, Bridgeport, Connecticut, $5,000 - for scholarships to their Student to Student program; Rural Music Corps at Ethos, Portland, Oregon, $5,000 - for instruments; Jefferson Middle School, Albuquerque, New Mexico, $4,716 - for equipment for their Rock and Rhythm Band.
Contact: Ellis Godard, Executive Director; ellis@mbird.org
Kristin Godard, grants@mbird.org
Internet: http://mbird.org/funding/guidelines/
Sponsor: Mockingbird Foundation
6948 Luther Circle
Moorpark, CA 93021-2569

Moline Foundation Community Grants 2590
The resources and funds of the Moline Foundation are used throughout its community to assist others in times of need, plan for future workforce development, help its neighborhoods grow and prosper, and to better the quality of life with arts and culture. The Moline Foundation provides grants in the following designated charitable categories: Education; Health Care; Social Services; Arts and Humanities; and, Workforce and Economic Development. In general, grants are made for capital and programmatic purposes only, not for operating expenses. Grants are made to non-profit agencies located in the Moline Foundation service area which includes the Quad Cities region. Occasionally, grants may also be allocated to governmental entities such as public libraries and schools.
Requirements: The Moline Foundation grants are intended to support charitable projects that utilize companies, firms or vendors whose principal offices are located within the Moline Foundation area. Since funds that support the Moline Foundation Grantmaking Program are received from local donors, every effort should be made to return these grant dollars to the local community. To apply for a grant, you must be a recognized 501(c)3 not for profit organization as recognized by the Internal Revenue Service. Grants are also awarded on a limited basis to governmental entities such as libraries and schools. The Moline Foundation serves six counties in western Illinois (Henderson, Henry, Mercer, McDonough, Rock Island, and Warren) and Scott County in eastern Iowa including the urban area is known as the Quad Cities.
Restrictions: Grants are not generally available for those agencies and institutions that are funded primarily through tax support.
Geographic Focus: Illinois, Iowa
Date(s) Application is Due: Jan 31; Apr 15; Sep 30
Contact: Linda Martin, Director of Donor and Community Relations; (309) 764-4193
Internet: http://www.molinefoundation.org/Page/Our_Grantmaking.aspx?nt=1152
Sponsor: Moline Foundation
817 11th Avenue
Moline, IL 61265

Montgomery County Community Foundation Grants 2591
The Montgomery County Community Foundation Grants help fund non-profit organizations and agencies in Montgomery County, Indiana. Significant grants have been awarded to organizations such as the Crawfordsville District Public Library, Boys and Girls Club, and the Family Crisis Shelter. Organizations should contact the Foundation for a grant proposal application and other documentation required.
Requirements: A strong proposal will have several or all of the following characteristics: an estimate of who and how many will benefit; show long term potential; address a community problem of some significance for which funding is not covered by the regular budget; present an innovative and practical approach to solve a community problem or project; identify possible future funding, if needed; give evidence of the stability and qualifications of the organization applying; show cooperation within the organization and avoid duplication effort.
Restrictions: The Foundation will usually not fund any of the following: grants to individuals; programs which are religious or sectarian in nature, except when the program is open to the entire community; operating expenses such as salaries and utilities; parades, festivals and sporting events; endowment funds; any propaganda, political or otherwise, attempting to influence legislation or intervene in any political affairs or campaigns; an organization's past debts or existing obligations; or post-event or after-the-fact situations.

Geographic Focus: Indiana
Date(s) Application is Due: May 10
Amount of Grant: Up to 50,000 USD
Contact: Cheryl Keim; (765) 362-1267; fax (765) 361-0562; cheryl@mccf-in.org
Internet: http://www.mccf-in.org/Granthomenewpage.html
Sponsor: Montgomery County Community Foundation
119 East Main Street
Crawfordsville, IN 47933

Morehouse PHSI Project Imhotep Internships 2592
IMHOTEP is an eleven-week internship (May - August) managed by Public Health Sciences Institute (PHSI) and designed to increase the knowledge and skills of rising juniors and seniors and recent graduates of an undergraduate institution in biostatistics, epidemiology, and occupational safety and health. The internship begins with two weeks of intense educational training. The purpose of this training is to equip interns with the academic coursework and information necessary to complete the program. During the remaining nine weeks, interns conduct public health research with experts at the Centers for Disease Control and Prevention (C.D.C.), the National Institute for Medical Research (NIMR), Mexico Negro, and various other public health agencies. Since the inception of Project IMHOTEP, interns have been placed in various centers, institutes, and offices (CIOS) within the CD.C. and various domestic and international agencies, including, but not limited to: the National Center for Chronic Disease Prevention and Health Promotion (NCCDPHP); the National Center for Health Statistics (NCHS); the National Center for HIV, STD, and TB Prevention (NCHHSTP); the National Center for Infectious Diseases (now NCPD.C.ID, NCIRD, NCZVED, and NCHHSTP); the National Institute for Occupational Safety and Health (NIOSH); the NIMR; Mexico Negro; and Sister Love, Incorporated. Throughout the internship, students participate in a wide variety of seminars, workshops, and other educational initiatives and must complete a required number of community service hours. Interns receive research support and consultation in data analysis and other project requirements. At the conclusion of the internship, interns deliver an oral presentation and submit a written manuscript suitable for publication in a scientific journal. Interns receive a financial stipend, housing (at Morehouse and site location) and a travel allowance (travel allowance includes travel to and from Atlanta, site location, and home residence). Interns are required to pay for all expenses incurred during the two-week training (e.g., books) and other miscellaneous expenses (e.g., food). In FY 2005, this cost totaled approximately $300. A link is provided at the IMHOTEP website to apply online. Applicants are encouraged to check the internship website and/or contact the PHSI staff for complete details on deadlines and how to apply for the internship.
Requirements: Applicants must be Juniors, Seniors or Recent Graduates with a cumulative GPA of 2.7 or higher. The selection process is very rigorous and extremely competitive. Candidates should have a genuine interest in pursuing a career in public health with a particular emphasis in biostatistics, epidemiology, or occupational safety and health. Interns are required to sign a legal contract, adhering to all rules, regulations, and program requirements.
Geographic Focus: All States, All Countries
Date(s) Application is Due: Jan 31
Contact: Cynthia Trawick; (404) 681-2800; ctrawick@morehouse.edu
Internet: http://www.morehouse.edu/centers/phsi/imhotep/index.htm
Sponsor: Morehouse College
830 Westview Drive, S.W.
Atlanta, GA 30314

Morgan Adams Foundation Grants 2593
The Morgan Adams Foundation supports laboratory and clinical research in the area of pediatric cancer, with an emphasis on cancers of the brain and spine. The Foundation's particular interest is to provide seed and bridge grants for viable investigations not yet ripe enough for total funding by larger organizations. Additionally, the Foundation is able to partially fund translational studies, experimental therapeutics studies and/or Phase I and II clinical trials on a case by case basis. The Foundation is seeking out projects that are highly collaborative and/or multi-institutional in nature, hoping to facilitate through the strategic placement of funding, more rapid trials development in order to get the best possible treatment options into circulation with the greatest expediency possible. Through careful consideration and placement of its funding, the Foundation encourages and supports research intended to improve treatment effectiveness, improve treatment outcomes and improve the quality of life for children battling cancer.
Geographic Focus: All States
Contact: Joan Slaughter, Executive Director; (303) 758-2130; fax (303) 758-2134; joan@morganadamsfoundation.org or info@morganadamsfoundation.org
Internet: http://morganadamsfoundation.hotpressplatform.com/funding-research/how-funds-become-hope
Sponsor: Morgan Adams Foundation
5303 E. Evans Avenue, Suite 202
Denver, CO 80222

Morris and Gwendolyn Cafritz Foundation Grants 2594
The Morris & Gwendolyn Cafritz Foundation supports IRS-registered, tax-exempt, 501(c)3 organizations with a public charity status of 509(a)(1) or 509(a)(2) only. These organizations must serve residents in the District of Columbia, Prince George's and Montgomery Counties in Maryland, Arlington and Fairfax Counties, and the cities of Alexandria and Falls Church in Virginia. Grants are made in four program areas: Arts and Humanities, Community Services, Education and Health.
Requirements: Nonprofits serving residents in the District of Columbia, Prince George's and Montgomery Counties in Maryland, Arlington and Fairfax Counties, and the cities of Alexandria and Falls Church in Virginia may apply.
Restrictions: The Foundation does not generally fund the following projects: capital campaigns; endowments; multi-year grants; special events or tables for special events. Please also note, the Foundation does not fund: organizations that do not have 501(c)3 tax-exempt status with the IRS; private foundations; public charities with a non-private foundation status of 509(a)(3); individuals; organizations whose missions fall outside the Foundations funding priorities; organizations serving residents outside the Washington, D.C. metropolitan area.
Geographic Focus: District of Columbia, Maryland, Virginia
Date(s) Application is Due: Mar 1; Jul 1; Nov 1
Sample: Theater Downtown, Washington, D.C., $200,000 - for Staging Our Future Campaign; Prince Georges Child Resource Center, Largo, MD, $25,000 - for Family Literacy Program; Spanish Catholic Center, Washington, D.C., $75,000 - for oral health treatment and education for low-income, uninsured immigrants;
Contact: Rose Ann Cleveland, Executive Director; (202) 223-3100; fax (202) 296-7567; info@cafritzfoundation.org
Internet: http://www.cafritzfoundation.org/Applicant/app_guidelines.asp
Sponsor: Morris and Gwendolyn Cafritz Foundation
1825 K Sreet NW, Suite 1400
Washington, D.C. 20006

Morton K. and Jane Blaustein Foundation Health Grants 2595
The foundation makes grants in the United States and abroad. Support is provided in the program areas of educational opportunity, health, and human rights and social justice.Preference is given to programs in Baltimore, Maryland, New York City and Washington, D.C.. The goal of the Health program is to improve health care quality and health outcomes, especially for underserved populations. The Foundation's main emphasis is on strengthening primary care. Preference is given to programs with a public health approach of prevention and wellness, and those that are particularly responsive to the needs of underserved children, youth and families. The Foundation seeks to promote: effective and expanded use of nurse practitioners in primary care; integration of primary and mental health care; and links between primary care and public health. There are no annual deadlines, and the Foundation accepts applications on a rolling basis. Most recent awards have ranged from $25,000 to $230,000.
Requirements: 501(c)3 organizations in Maryland, Washington, D.C., and New York are eligible to apply.
Restrictions: No support is given for fundraising events, or direct mail solicitations. Grants are not made to individuals.
Geographic Focus: District of Columbia, Maryland, New York
Amount of Grant: 25,000 - 230,000 USD
Contact: Tanya C. Herbick, Senior Program Officer; (410) 347-7206 or (410) 347-7201; fax (410) 347-7210; info@blaufund.org
Internet: http://www.blaufund.org/foundations/mortonandjane_f.html#2
Sponsor: Morton K. and Jane Blaustein Foundation
One South Street, Suite 2900
Baltimore, MD 21202

Ms. Foundation for Women Ending Violence Grants 2596
The Ms. Foundation works in partnership with grassroots, state, and national organizations to end gender-based violence by transforming policies, beliefs, and behaviors that threaten the well-being of individuals, families and communities nationwide. The Foundation supports a range of community-based strategies to stop violence before it occurs--to prevent violence directed against women, girls, and LGBTQ individuals. The Foundation also supports a movement to advance a community-based, social justice approach to child sexual abuse.
Requirements: Applicants must be tax exempt nonprofits organizations. The application process, current deadlines, and previous grant recipients can be found on the Foundation website.
Geographic Focus: All States
Contact: Administrator; (212) 742-2300; fax (212) 742-1653; info@ms.foundation.org
Internet: http://ms.foundation.org/our_work/broad-change-areas/ending-violence
Sponsor: Ms. Foundation for Women
12 MetroTech Center, 26th Floor
Brooklyn, NY 11201

Ms. Foundation for Women Health Grants 2597
The Ms. Foundation for Women supports organizing at grassroots, state and national levels to promote equitable access to health care and education for women and youth. The Foundation delivers strategic funding, technical assistance, and networking support to organizations that are working in their communities and beyond to address the urgent priorities of those most affected by failed health policies, especially low-income women, women of color, immigrant women, women living with HIV/AIDS, and LGBTQ youth. The Foundation strives to build social movements and advance policy and culture change in key areas: reproductive health, rights and justice; sexuality education; and women and AIDS.
Requirements: Applicants must be tax exempt nonprofit organizations. The application process, current deadlines, and previous grant recipients can be found on the Foundation website.
Geographic Focus: All States
Contact: Contact; (212) 742-2300; fax (212) 742-1653; info@ms.foundation.org
Internet: http://ms.foundation.org/our_work/broad-change-areas/womens-health
Sponsor: Ms. Foundation for Women
12 MetroTech Center, 26th Floor
Brooklyn, NY 11201

Mt. Sinai Health Care Foundation Academic Medicine and Bioscience Grants 2598
The Foundation seeks to strengthen Cleveland's prominence as a national leader in medical education and biomedical science, perpetuating Mt. Sinai's leadership role in these areas. Emerging collaborations among Cleveland's academic medical centers present opportunities to build organizational capacity and advance progress. The Board of Directors of the Foundation meets on a quarterly basis to review proposals.
Requirements: 501(c)3 nonprofits serving greater Cleveland, Ohio may submit proposals for grant support. The foundation welcomes and encourages an informal conversation with program staff prior to the submission of a grant request.
Restrictions: In general, the foundation does not support general operating expenses, direct provision of health services, building or equipment expenses, fund-raising events, projects outside of greater Cleveland, endowment funds, lobbying, program advertising, grants for individuals, or scholarships.
Geographic Focus: Ohio
Date(s) Application is Due: Jan 1; Apr 1; Jul 1; Oct 1
Contact: Shelly Galvin; (216) 421-5500; fax (216) 421-5633; sgg2@case.edu
Internet: http://www.mtsinaifoundation.org/whatwefund_academic.html
Sponsor: Mount Sinai Health Care Foundation
11000 Euclid Avenue
Cleveland, OH 44106-1714

Mt. Sinai Health Care Foundation Health of the Jewish Community Grants 2599
The Mt. Sinai Health Care Foundation seeks to assist Greater Cleveland's organizations and leaders to improve the health and well-being of the Jewish and general communities now and for generations to come. The Foundation will support projects that build organizational capacity in those Jewish organizations that address these needs. The Board of Directors of the Foundation meets on a quarterly basis to review proposals.
Requirements: 501(c)3 nonprofits serving greater Cleveland, Ohio may submit proposals for grant support. The foundation welcomes and encourages an informal conversation with program staff prior to the submission of a grant request.
Restrictions: In general, the foundation does not support general operating expenses, direct provision of health services, building or equipment expenses, fund-raising events, projects outside of greater Cleveland, endowment funds, lobbying, program advertising, grants for individuals, or scholarships.
Geographic Focus: Ohio
Date(s) Application is Due: Jan 1; Apr 1; Jul 1; Oct 1
Contact: Ann Freimuth; (216) 421-5500; fax (216) 421-5633; aks17@case.edu
Internet: http://www.mtsinaifoundation.org/whatwefund_jewishcommunity.html
Sponsor: Mount Sinai Health Care Foundation
11000 Euclid Avenue
Cleveland, OH 44106-1714

Mt. Sinai Health Care Foundation Health of the Urban Community Grants 2600
In the tradition of The Mt. Sinai Medical Center, the Foundation is committed to improving the health of Greater Cleveland's most vulnerable individuals and families. To achieve impact in this area, scale is a significant factor. The Foundation seeks to support especially those projects focusing on health promotion and disease prevention that have the potential to access large populations through existing community infrastructure. To optimize impact in large populations, partnering with both public and private funding sources may be appropriate and necessary. Of particular interest are proposals in the areas of health-related early childhood development and health-related aging.
Requirements: 501(c)3 nonprofits serving greater Cleveland, Ohio may submit proposals for grant support. The foundation welcomes and encourages an informal conversation with program staff prior to the submission of a grant request.
Restrictions: In general, the foundation does not support general operating expenses, direct provision of health services, building or equipment expenses, fund-raising events, projects outside of greater Cleveland, endowment funds, lobbying, program advertising, grants for individuals, or scholarships.
Geographic Focus: Ohio
Date(s) Application is Due: Jan 1; Apr 1; Jul 1; Oct 1
Contact: Ann Freimuth; (216) 421-5500; fax (216) 421-5633; aks17@case.edu
Internet: http://www.mtsinaifoundation.org/whatwefund_urbancommunity.html
Sponsor: Mount Sinai Health Care Foundation
11000 Euclid Avenue
Cleveland, OH 44106-1714

Mt. Sinai Health Care Foundation Health Policy Grants 2601
The Mt. Sinai Health Care Foundation seeks to assist Greater Cleveland's organizations and leaders to improve the health and well-being of the Jewish and general communities now and for generations to come. Notwithstanding significant support from the private sector and philanthropy, government at all levels remains the single greatest financial contributor to the health of at-risk populations, including children, the elderly, and the poor. Through strategic initiatives in the area of health policy, the Foundation seeks to support projects that maximize the effectiveness of government in meeting its safety-net obligations and the obligations of the Affordable Care Act.
Requirements: 501(c)3 nonprofits serving greater Cleveland, Ohio may submit proposals for grant support. The foundation welcomes and encourages an informal conversation with program staff prior to the submission of a grant request.
Restrictions: In general, the foundation does not support general operating expenses, direct provision of health services, building or equipment expenses, fund-raising events, projects outside of greater Cleveland, endowment funds, lobbying, program advertising, grants for individuals, or scholarships.
Geographic Focus: Ohio

Date(s) Application is Due: Jan 1; Apr 1; Jul 1; Oct 1
Contact: Jodi Mitchell, Program Officer; (216) 421-5500; fax (216) 421-5633; Jodi.Mitchell@case.edu
Internet: http://www.mtsinaifoundation.org/whatwefund_policy.html
Sponsor: Mount Sinai Health Care Foundation
11000 Euclid Avenue
Cleveland, OH 44106-1714

Multiple Sclerosis Foundation Brighter Tomorrow Grants 2602
The goal of the grant is provide individuals with MS with goods or services (valued at up to $1000.00 per recipient) to improve their quality of life by enhancing safety, self-sufficiency, comfort, or well-being. Recipients have used the grant to purchase wheelchairs, walkers, eye glasses, computers, appliances, furniture, therapeutic equipment, hobby supplies, and for care repairs and various home modifications.
Requirements: To qualify, a person must be 18 years of age or older and diagnosed with MS, or the parent of a minor child diagnosed with MS, and be a permanent U.S. resident. They must not have any other means of fulfilling the need they express.
Geographic Focus: All States
Date(s) Application is Due: Oct 10
Amount of Grant: 1,000 USD
Contact: Program Services Department; (888) 673-6287 or (954) 776-6805; fax (954) 351-0630; support@msfocus.org
Internet: http://www.msfocus.org/Brighter-Tomorrow-Grant.aspx
Sponsor: Multiple Sclerosis Foundation
6350 North Andrews Avenue
Fort Lauderdale, FL 33309-2130

NAF Fellowships 2603
The foundation supports new and innovative studies that are relevant to the cause, pathogenesis, or treatment of the hereditary or sporadic ataxias. Postdoctoral fellowship awards serve as a bridge from postdoctoral positions to junior faculty positions. The award will permit individuals to spend an additional third year in a postdoctoral position and increase chances to establish an independent ataxia research program. Letters of Intent should arrive by August 15, with full proposals due by September 15. Application and guidelines are available online.
Requirements: Applicants should have completed at least one year of postdoctoral training, but not more than two at the time of application, and should have shown a commitment to research in the field of ataxia.
Restrictions: Indirect costs are not allowed.
Geographic Focus: All States
Date(s) Application is Due: Aug 15; Sep 15
Amount of Grant: 35,000 USD
Contact: Grants Administrator; (763) 553-0020; fax (763) 553-0167; naf@ataxia.org
Internet: http://www.ataxia.org/research/ataxia-research-grants.aspx
Sponsor: National Ataxia Foundation
2600 Fernbrook Lane, Suite 119
Minneapolis, MN 55447-4752

NAF Kyle Bryant Translational Research Award 2604
The Friedreich's Ataxia Research Alliance (FARA) and the National Ataxia Foundation (NAF) invite proposals, under a competitive Request for Applications (RFA) process, to award a grant focusing on pre-clinical investigations that will facilitate clinical trials for Friedreich's Ataxia. Funding will be for one year. The total award is limited to $120,000 (direct costs only). The Foundation is requesting that applicants submit Letters of Intent by July 15, and full applications by July 31.
Requirements: Each application must include a project objectives and lay summary that will be used as FARA and NAF wishes and should not contain confidential information. The other parts of the grant application are considered confidential and will only be released to review committee, FARA and NAF staff and Board of Directors. Applications will be accepted from for-profit organizations, non-profit organizations, public or private institutions, and foreign institutions.
Geographic Focus: All States
Date(s) Application is Due: Jul 15; Jul 31
Amount of Grant: Up to 120,000 USD
Contact: Susan Hagen, (763) 553-0020; susan@ataxia.org
Internet: http://www.ataxia.org/research/ataxia-research-grants.aspx
Sponsor: National Ataxia Foundation
2600 Fernbrook Lane, Suite 119
Minneapolis, MN 55447-4752

NAF Research Grants 2605
NAF research grants are for new and innovative studies that are relevant to the cause, pathogenesis or treatment of the hereditary or sporadic ataxias. Due to the larger availability of funding for Ataxia-Telangiectasia (A-T), those research proposals will receive a lower priority. However, a higher ranking will be given to those Ataxia-Telangiectasia research studies which lend themselves to an overall better understanding of the ataxia process. Research grants are offered primarily as seed monies to assist investigators in the early or pilot phase of their studies and as additional support for ongoing investigations on demonstration of need. It is hoped that these studies will be further developed to attract future funding from other sources. Grants are awarded for one year only. If funding for a second year is desired then another grant application should be submitted for that funding period. The average funds granted will be up to $15,000, but funding may be considered for up to $30,000 for projects deserving special consideration.

Restrictions: Indirect costs will not be funded.
Geographic Focus: All States
Date(s) Application is Due: Aug 15
Amount of Grant: 5,000 - 15,000 USD
Contact: Grants Administrator; (763) 553-0020; fax (763) 553-0167; naf@ataxia.org
Internet: http://www.ataxia.org/research/ataxia-research-grants.aspx
Sponsor: National Ataxia Foundation
2600 Fernbrook Lane, Suite 119
Minneapolis, MN 55447-4752

NAF Young Investigator Awards 2606
The purpose of this award is to encourage young clinical and scientific investigators to pursue a career in the field of ataxia research. It is the foundation's hope that ataxia research will be invigorated by the work of young, talented individuals supported by stable multi-year funding. This award will be made to the sponsoring institution and will be used as salary support or for direct research expenses. The sponsoring institution must agree in writing to the following provision: a clinical investigator (MD) must be free to allocate a minimum of half time for a project; a non-clinician investigator (PhD) must allocate a minimum of three-fourths time for the project. Periodic progress reports will be required. Applicants are asked to submit a brief email with the intentions to apply for NAF funding, including a full title of the research proposal, on or before August 1. Deadline for full application is September 1.
Requirements: Candidates must have attained the MD or PhD degree, and have an appointment as a junior faculty member. Clinicians must have finished their residency no more than five years prior to applying, and investigators must have completed their postdoctoral training no more than five years prior to applying.
Restrictions: Individuals at the associate professor level are not eligible. Awards are for direct costs only, not for indirect costs or institutional overhead.
Geographic Focus: All States
Date(s) Application is Due: Aug 1; Sep 1
Amount of Grant: 50,000 USD
Contact: Grants Administrator; (763) 553-0020; fax (763) 553-0167; naf@ataxia.org
Internet: http://www.ataxia.org/research/ataxia-research-grants.aspx
Sponsor: National Ataxia Foundation
2600 Fernbrook Lane, Suite 119
Minneapolis, MN 55447-4752

Nancy J. Pinnick Memorial Scholarship 2607
The Nancy J. Pinnick Memorial Scholarship was established by the Pinnick family to continually recognize her dedication and work as an educator in a meaningful way. A $1,000 scholarship will be awarded to a graduating senior from River Forest High School in Hobart, Indiana. The scholarship is designated for tuition and books and will be renewable for $1,000 for years two through four of the student's college studies. All college majors are eligible.
Requirements: The scholarship is awarded to a graduating high school senior with a cumulative GPA of at least 2.5. The student must be enrolled full time in a program leading to an associate or bachelor degree. The completed application package should include the application form, a current high school transcript, and answers, in essay form, to two questions asked on the application. A student will be selected based on his or her GPA, high school, and community activities.
Geographic Focus: Indiana
Date(s) Application is Due: Apr 15
Amount of Grant: 1,000 USD
Contact: Pinnick Scholarship Contact; (219) 962-7551; fax (219) 962-8338
Internet: http://www.legacyfoundationlakeco.org/scholarships/index.htm
Sponsor: Legacy Foundation
1000 East 80th Place, 302 South
Merrillville, IN 46410

Nancy R. Gelman Foundation Breast Cancer Seed Grants 2608
The program awards seed grants to fund projects aimed at improving outcomes for women with breast cancer. Proposed projects must: demonstrate a need for start-up, interim, or supplemental funding for a one-year period beginning April 1; not already be funded by other sources (projects that represent a clear new initiative extending beyond the scope of a currently funded project are permissible); and prospectively state an objective outcome measure that will be used to gauge project success. Types of projects seed grants might fund include: a mentored summer fellowship for a medical student to work in a laboratory, gaining experience in basic research on breast cancer mechanisms or therapies; community-based efforts to promote breast cancer awareness and early detection among populations at increased risk; development of information materials to encourage appropriate screening and to help guide newly diagnosed patients, or those whose cancer has recurred, through the challenging maze of treatment options; fledgling research studies to pave the way for further advances in breast cancer therapy; quality initiatives in cancer or emergency department nursing units to improve care for patients with breast cancer; and a summer research project studying end-of-life care for breast cancer patients and making recommendations to hospitals and hospice organizations to improve care. The application and guidelines are available online.
Geographic Focus: All States
Date(s) Application is Due: Aug 5
Amount of Grant: Up to 1,500 USD
Contact: Michael Gelman; (206) 441-7225; michael@nrgf.org or grants@nrgf.org
Internet: http://www.nrgf.org/grants.html
Sponsor: Nancy R. Gelman Foundation
1133 Broadway, Suite 706
New York, NY 10010

NAPNAP Foundation Elaine Gelman Scholarship 2609
The purpose of this annual scholarship is to support the education of a nurse practitioner student who demonstrates the ability to articulate and follow through an innovative solution through clinical competence, academic achievement and involvement in political activism relating to a pediatric health care issue. Preference is given to applicants in an accredited PNP program or those with interest or experience in health policy or advocacy. The recipient will receive an award up to $1000 if all criteria are met. The recipient will be recognized in The Pediatric Nurse Practitioner newsletter and again at the Annual NAPNAP Conference Awards Breakfast.
Requirements: Applicant must be a full or part time NP Student enrolled in an accredited NP program with an expected graduation date of two years or less. Applicants must submit two (2) letters of support from professional colleagues addressing the applicant's clinical competence, academic achievement and involvement in political activism relating to health care issues.
Geographic Focus: All States
Date(s) Application is Due: Jun 30
Amount of Grant: Up to 1,000 USD
Contact: Dolores Jones; (856) 857-9700; fax (856) 857-1600; djones@napnap.org
Internet: http://www.napnap.org/PNPResources/research/AvailGrantsScholarships.aspx
Sponsor: National Association of Pediatric Nurse Associates and Practitioners
20 Brace Road, Suite 200
Cherry Hill, NJ 08034-2634

NAPNAP Foundation Graduate Student Research Grant 2610
The program supports research by graduate students currently enrolled in a pediatric nurse practitioner program. The purposes of the award are to: encourage graduate students to participate in the research process; and stimulate interest in research regarding children and their families. The recipient of the award will be invited to participate at the research presentation session at the annual conference in the spring following the award. Application and guidelines are available online.
Requirements: The graduate student must be a current member of NAPNAP.
Geographic Focus: All States
Date(s) Application is Due: Apr 1
Amount of Grant: 1,000 USD
Samples: Cathy Woodward--for Routine Chest Radiographs Post Chest Tube Removal in Children, $1,000 (2006-07); Sheri Vlam--for ADHD Assessment Methods Used by APRNs, $500 (2003-04); Jacqueline Rychnovsky--for A Pediatric Perspective of Maternal Fatigue Across the First Ten Weeks Postpartum, $1,000 (2002-03).
Contact: Dolores Jones; (856) 857-9700; fax (856) 857-1600; djones@napnap.org
Internet: http://www.napnap.org/PNPResources/research/AvailGrantsScholarships.aspx
Sponsor: National Association of Pediatric Nurse Associates and Practitioners
20 Brace Road, Suite 200
Cherry Hill, NJ 08034-2634

NAPNAP Foundation Innovative Health Care Small Grant 2611
The Innovative Health Care Small Grant Program is designed to encourage NAPNAP members and chapters to develop and implement innovative, feasible, and cost-effective programs to improve children's health services. These projects are intended to improve the accessibility of health services to underserved children and those in need of health care; enhance family education regarding the importance of health care maintenance and preventive care; increase utilization rates for children without services or who are underserved; and encourage providers to pursue new mechanisms in delivery of health services to children in need. Guidelines are available online.
Requirements: Each application must be submitted by a national NAPNAP member (or group of members) or by a NAPNAP chapter; represent a local initiative that is community-based, family-centered, and culturally appropriate; describe the project's target population; offer innovative or improved access to health education and/or health care services for children; clearly state project goal; define measurable objectives of the project and identify those activities that lead to their achievement; develop a simple evaluation mechanism reflecting extent to which goal and objectives have been achieved; and include a budget for the project (monies may not be used for salary, consultation, or gifts).
Geographic Focus: All States
Date(s) Application is Due: Sep 15
Amount of Grant: Up to 4,000 USD
Contact: Dolores Jones; (856) 857-9700; fax (856) 857-1600; djones@napnap.org
Internet: http://www.napnap.org/PNPResources/research/AvailGrantsScholarships.aspx
Sponsor: National Association of Pediatric Nurse Associates and Practitioners
20 Brace Road, Suite 200
Cherry Hill, NJ 08034-2634

NAPNAP Foundation McNeil Grant-in-Aid 2612
The purpose of the NAPNAP Foundation is to support nursing research and clinical practice efforts that contribute to the improvement of the quality of life for children and their families. Through the generosity of grants from the McNeil Consumer Healthcare Division of McNeil-PPC, Inc., the NAPNAP Foundation offers up to $5,000 for members and one student to attend the annual conference.
Geographic Focus: All States
Date(s) Application is Due: Oct 4
Amount of Grant: 5,000 USD
Contact: Dolores Jones; (856) 857-9700; fax (856) 857-1600; djones@napnap.org
Internet: http://www.napnap.org/PNPResources/research/AvailGrantsScholarships.aspx
Sponsor: National Association of Pediatric Nurse Associates and Practitioners
20 Brace Road, Suite 200
Cherry Hill, NJ 08034-2634

NAPNAP Foundation McNeil PNP Scholarships 2613
NAPNAP will award two annual McNeil Scholarships to students enrolled in pediatric nurse practitioner programs. Scholarships for each academic year are awarded for both the fall and spring semesters/quarters. Individuals who are enrolled in a program within the current academic year will be reviewed. Students who begin a program in January may also apply the following May if clinical course work will continue into the next academic year. Scholarship recipients will be determined by NAPNAP and notified within two months of the application deadline. The recipients will be formally recognized at NAPNAP's annual conference. Airfare and conference registration fees will be paid by NAPNAP. Instructions and application are available online.
Requirements: Applicant must be a registered nurse with at least three years previous work experience in pediatrics; have documented acceptance at a recognized PNP program; have no previous formal pediatric nurse practitioner education; demonstrate financial need, based on submission of information regarding resources; and state rationale for seeking PNP education that is consistent with NAPNAP's mission statement. In addition, students must maintain the following: full-time status (nine or more credit hours/semester) while in PNP program; GPA of 3.0 or higher; NAPNAP membership; provide NAPNAP with ongoing accounts regarding student status, as requested.
Geographic Focus: All States
Date(s) Application is Due: Apr 30
Amount of Grant: 1,500 USD
Contact: Dolores Jones; (856) 857-9700; fax (856) 857-1600; djones@napnap.org
Internet: http://www.napnap.org/index.cfm?page=13&sec=102
Sponsor: National Association of Pediatric Nurse Associates and Practitioners
20 Brace Road, Suite 200
Cherry Hill, NJ 08034-2634

NAPNAP Foundation McNeil Rural and Underserved Scholarships 2614
Through the support of McNeil Consumer Healthcare Division of McNeil-PPC, Inc., NAPNAP members can apply for the McNeil Rural & Underserved Scholarship. This scholarship of up to $10,000 will cover a full academic year for a PNP student who will provide primary care to children and families in rural areas after graduating. The deadline for applications is June 15.
Requirements: To be eligible for the scholarship, applicants must meet the following criteria: plan on practicing in a rural or under served geographical region for the first two years after PNP education completed; be a registered nurse who has completed at least 2 semesters or quarters as defined by the university; show documented acceptance at a recognized Master's Degree program; have no previous formal pediatric nurse practitioner education; demonstrate financial need, based on submission of information regarding resources; and state rational for seeking PNP education that is consistent with NAPNAP's mission statement. In addition, students must maintain the following: full-time status (9 or more credit hours/semester) while in PNP program; G.P.A. of 3.0 or higher; NAPNAP membership; and provide NAPNAP with ongoing accounts regarding student status.
Geographic Focus: All States
Date(s) Application is Due: Jun 15
Amount of Grant: 10,000 USD
Contact: Dolores Jones; (856) 857-9700; fax (856) 857-1600; djones@napnap.org
Internet: http://www.napnap.org/PNPResources/research/AvailGrantsScholarships.aspx
Sponsor: National Association of Pediatric Nurse Associates and Practitioners
20 Brace Road, Suite 200
Cherry Hill, NJ 08034-2634

NAPNAP Foundation Nursing Research Grants 2615
The foundation supports nursing research efforts that contribute to the improvement of the quality of life for children and their families. The foundation is primarily interested in the following areas of research: impact of nurse practitioner role in healthcare delivery; development/implementation of innovative methods for delivery of health care services in the areas of school health, community health, behavior and development, adolescent health, and chronic disease management; health care problems of children and their families; advanced nursing practice problems and issues; exploratory/evaluative research of PNP education and credentialing; and educational programs and materials for healthcare providers and/or children and families. Up to two grants will be awarded annually.
Geographic Focus: All States
Date(s) Application is Due: Apr 1
Amount of Grant: 2,500 USD
Contact: Dolores Jones; (856) 857-9700; fax (856) 857-1600; djones@napnap.org
Internet: http://www.napnap.org/index.cfm?page=13&sec=102
Sponsor: National Association of Pediatric Nurse Associates and Practitioners
20 Brace Road, Suite 200
Cherry Hill, NJ 08034-2634

NAPNAP Foundation Reckitt Benckiser Student Scholarship 2616
Through a grant from Reckitt Benckiser Inc., the NAPNAP Foundation will award two grants of $1,600 to a Pediatric Nurse Practitioner Student to allow for attendance at the NAPNAP Annual Meeting in April. The postmark deadline for applications is November 2. The grant recipient will be determined by the NAPNAP Foundation and the recipient will be formally recognized at NAPNAP's Annual Conference.
Requirements: To be eligible for the grant, applicants must meet the following criteria: be a registered nurse and a graduate student who has completed at least 1 semester of graduate studies in a Pediatric Nurse Practitioner Program; provide documentation from the PNP Program Coordinator/Director that the student is enrolled in a recognized program of study leading to completion of qualifications for education and practice as a PNP; and be a member of NAPNAP.
Geographic Focus: All States
Date(s) Application is Due: Nov 2
Amount of Grant: 1,600 USD
Contact: Dolores Jones; (856) 857-9700; fax (856) 857-1600; djones@napnap.org
Internet: http://www.napnap.org/PNPResources/research/AvailGrantsScholarships.aspx
Sponsor: National Association of Pediatric Nurse Associates and Practitioners
20 Brace Road, Suite 200
Cherry Hill, NJ 08034-2634

NAPNAP Foundation Shourd Parks Immunization Project Small Grants 2617
The NAPNAP Foundation is pleased to offer the Shourd-Parks Immunization Project Small Grant, as part of a national effort to improve health care for infants through young adults. Biennially, a $500 grant will be awarded to a NAPNAP member or chapter. To keep grant activities consistent with the goals of NAPNAP and the NAPNAP Foundation, grant applications should focus on improving health care in the area of child, adolescent and/or young adult immunizations. Four copies of the proposal and the application should be mailed to the NAPNAP National Office and postmarked no later than September 15.
Requirements: Each NAPNAP Foundation application should meet at least one of the following: 1) Improve access to immunizations for children and/or adolescents in any health care setting. 2) Strengthen existing or new health education programs in the area of immunizations. 3) Strengthen existing or new methods in the delivery of immunizations. 4) Incorporates NAPNAP's and NAPNAP Foundation priorities. Each application must meet these specific *Requirements:* 1) Be submitted by a national NAPNAP member (or group of members) or by a NAPNAP Chapter. 2) Describe the project's target population. 3) Clearly state the project purpose, goal(s) or aim(s). 4) Define one measurable outcome for the project. 5) Develop an evaluation mechanism for the one outcome. 6) Include a brief budget.
Geographic Focus: All States
Date(s) Application is Due: Sep 15
Amount of Grant: 500 USD
Contact: Dolores Jones; (856) 857-9700; fax (856) 857-1600; djones@napnap.org
Internet: http://www.napnap.org/PNPResources/research/AvailGrantsScholarships.aspx
Sponsor: National Association of Pediatric Nurse Associates and Practitioners
20 Brace Road, Suite 200
Cherry Hill, NJ 08034-2634

NAPNAP Foundation Wyeth Pediatric Immunization Grant 2618
These grants will give NAPNAP members the opportunity to develop innovative programs to enhance the immunization rates in their communities. Examples of the ways in which a grant could be used include: parent education/awareness activities or materials; provider education workshops on current vaccines topics and true contra-indications to vaccinating; development of clinic policies to prevent missed opportunities; strategies for immunization registries; community outreach endeavors to assess and eliminate barriers to immunizations translation services for diverse communities; and evaluation research on effectiveness of vaccine programs to increase immunization rates. Two grants of $2,500 each and one grant of $7,500 are available.
Requirements: NAPNAP members are eligible to apply.
Geographic Focus: All States
Date(s) Application is Due: Dec 15
Amount of Grant: 2,500 - 7,500 USD
Contact: Dolores Jones; (856) 857-9700; fax (856) 857-1600; djones@napnap.org
Internet: http://www.napnap.org/PNPResources/research/AvailGrantsScholarships.aspx
Sponsor: National Association of Pediatric Nurse Associates and Practitioners
20 Brace Road, Suite 200
Cherry Hill, NJ 08034-2634

NARSAD Colvin Prize for Outstanding Achievement in Mood Disorders Research 2619
The Colvin Prize for Outstanding Achievement in Mood Disorders Research, is an award of $50,000 given to an outstanding scientist carrying out work on the causes, pathophysiology, treatment, or prevention of affective disorders. The scientist to be recognized is one who gives particular promise for advancing the understanding of affective illness or its basic brain mechanisms that will lead to new treatment approaches. The Prize is given at the annual New York City Awards Dinner held each October. Nominations are due by June 1.
Geographic Focus: All States
Date(s) Application is Due: Jun 1
Amount of Grant: 50,000 USD
Contact: Sho Tin Chen, Associate Manager, Research Grants; (516) 829-0091; fax (516) 487-6930; schen@bbrfoundation.org
Internet: http://bbrfoundation.org/outstanding-achievement-prizes#Lieber
Sponsor: Brain and Behavior Research Foundation
60 Cutter Mill Road, Suite 404
Great Neck, NY 11021

NARSAD Distinguished Investigator Grants 2620
The Brain and Behavior Research Foundation (formerly National Alliance for Research on Schizophrenia and Depression, or NARSAD) distributes funds for psychiatric brain and behavior disorder research. NARSAD Distinguished Investigator Grants provide support for experienced investigators (full professor or equivalent) conducting neurobiological and behavioral research. Areas of particular interest include patient populations with unique or unusual characteristics and central nervous system

developments. A one-year grant of up to $100,000 is provided for established scientists pursuing innovative projects in diverse areas of neurobiological research.
Requirements: Applicants must be a full professor (or equivalent), and maintain peer reviewed competitively funded scientific programs. Guidelines and application deadlines are at the Foundation's website.
Restrictions: Previous grantees may not receive the grant for a second time until five years have elapsed since the beginning date of the prior award. Grant requirements from all previous awards must be met. Only one applicant may apply per application. Applicants may submit only one application per cycle.
Geographic Focus: All States
Amount of Grant: Up to 100,000 USD
Contact: Sho Tin Chen, Associate Manager, Research Grants; (800) 829-8289 or (516) 829-5576; fax (516) 487-6930; grants@narsad.org
Internet: http://bbrfoundation.org/narsad-distinguished-investigator-grant
Sponsor: Brain and Behavior Research Foundation
60 Cutter Mill Road, Suite 404
Great Neck, NY 11021

NARSAD Goldman-Rakic Prize for Cognitive Neuroscience — 2621

The Goldman-Rakic Prize for Cognitive Neuroscience Research is given in recognition of a research scientist who has made distinguished contributions to the understanding of cognitive neuroscience. The Prize includes $40,000 and an honorary lecture at Yale University to honor the memory and accomplishments of Patricia Goldman-Rakic. The Prize is specifically for excellence in neurobiological research at the cellular, physiological, or behavioral levels that may lead to a greater understanding of underlying psychiatric or neurological disease. The Prize is given at the annual New York City Awards Dinner held each October. Nominations for the Award are due by June 1.
Geographic Focus: All States
Date(s) Application is Due: Jun 1
Amount of Grant: 40,000 USD
Contact: Sho Tin Chen, Associate Manager, Research Grants; (516) 829-0091; fax (516) 487-6930; schen@bbrfoundation.org
Internet: http://bbrfoundation.org/outstanding-achievement-prizes#Lieber
Sponsor: Brain and Behavior Research Foundation
60 Cutter Mill Road, Suite 404
Great Neck, NY 11021

NARSAD Independent Investigator Grants — 2622

The Brain and Behavior Research Foundation (formerly National Alliance for Research on Schizophrenia and Depression, or NARSAD) distributes funds for psychiatric brain and behavior disorder research. The NARSAD Independent Investigator Grants provide support for investigators during the critical period between the initiation of research and the receipt of sustained funding. Basic and/or clinical investigators are supported, but research must be relevant to schizophrenia, major affective disorders, or other serious mental illnesses. A two-year grant up to $100,000, or $50,000 per year, is provided to scientists at the associate professor level or equivalent, who are clearly independent and have won national competitive support as a principal investigator.
Requirements: Applicants must have a doctoral-level degree, be an associate professor (or equivalent), and have received, as Principal Investigator, competitive research support at a national level. Independent Investigators are eligible for a maximum of two grants. Previous awardees at the associate professor level are encouraged to apply for a second grant. Applications must be submitted through the online application system. Additional guidelines can be found at the Foundation's website.
Restrictions: The total budget request may not exceed $100,000. Awards cannot overlap in time, all prior grant requirements (including NARSAD Young Investigator Grants) must be met, and an applicant cannot submit more than one application per cycle. Investigators that have been promoted to associate professor, but have currently active NARSAD Young Investigator Grants, are not eligible.
Geographic Focus: All States
Date(s) Application is Due: Nov 15
Amount of Grant: Up to 50,000 USD
Contact: Sho Tin Chen, Associate Manager, Research Grants; (800) 829-8289 or (516) 829-5576; fax (516) 487-6930; grants@bbrfoundation.org
Internet: http://bbrfoundation.org/narsad-independent-investigator-grants
Sponsor: Brain and Behavior Research Foundation
60 Cutter Mill Road, Suite 404
Great Neck, NY 11021

NARSAD Lieber Prize for Schizophrenic Research — 2623

The NARSAD Lieber Prize for Schizophrenia Research is given in recognition of a research scientist who has made distinguished contributions to the understanding of schizophrenia. The $50,000 cash rewards past achievement and provides further incentive for an outstanding working scientist to continue to do exceptional research into the causes, prevention, and treatment of schizophrenia. The prize is bestowed at the annual New York City Awards Dinner held each October. The presentation of the prize helps build public understanding of the importance of schizophrenia research, and acts as a goal towards which researchers might strive. Nominations are due by June 1. The Lieber Prize winner will select the winner of the the Sidney R. Baer, Jr. Prize for Schizophrenia Research, worth $40,000. The winner is typically a NARSAD Young Investigator at the Lieber Prize winner's institution, conducting work in schizophrenia research.
Geographic Focus: All States
Date(s) Application is Due: Jun 1
Amount of Grant: 50,000 USD
Contact: Sho Tin Chen, Associate Manager, Research Grants; (516) 829-0091 or (800) 829-8289; fax (516) 487-6930; schen@bbrfoundation.org or info@bbrfoundation.org
Internet: http://bbrfoundation.org/outstanding-achievement-prizes#Lieber
Sponsor: Brain and Behavior Research Foundation
60 Cutter Mill Road, Suite 404
Great Neck, NY 11021

NARSAD Ruane Prize for Child and Adolescent Psychiatric Research — 2624

The Ruane Prize is an award of $50,000 given to an outstanding scientist carrying out research on the causes, pathophysiology, treatment, or prevention of severe mental illness in children. The scientist to be recognized is one who gives particular promise for advancing the understanding of psychotic, affective or other severe brain and behavior disorders having their onset in childhood or adolescence. Contributions may be for clinical research or relevant basic science. The Prize is given at the annual New York City Awards Dinner held each October. Nominations are due by June 1.
Geographic Focus: All States
Date(s) Application is Due: Jun 1
Amount of Grant: 50,000 USD
Contact: Sho Tin Chen, Associate Manager, Research Grants; (516) 829-0091; fax (516) 487-6930; schen@bbrfoundation.org
Internet: http://bbrfoundation.org/outstanding-achievement-prizes#Lieber
Sponsor: Brain and Behavior Research Foundation
60 Cutter Mill Road, Suite 404
Great Neck, NY 11021

NARSAD Young Investigator Grants — 2625

The Brain and Behavior Research Foundation (formerly National Alliance for Research on Schizophrenia and Depression, or NARSAD) distributes funds for psychiatric brain and behavior disorder research. The Foundation's NARSAD Young Investigator Grants offer up to $30,000 a year for up to two years to enable promising investigators to either extend their research fellowship training or to begin careers as independent research faculty. Grants are intended to facilitate innovative research opportunities and support basic, as well as translational and/or clinical investigators. Research must be relevant to the understanding, treatment and prevention of serious psychiatric disorders such as schizophrenia, bipolar, mood and anxiety disorders, or child and adolescent psychiatric disorders. Awardees become eligible for the Freedman Prize for Outstanding Research, with a cash prize of $1,000. They are also eligible for the Klerman Prize for Outstanding Clinical Research, also with a cash prize of $1,000. Young Investigator winners also become eligible for the Baer Prize for Schizophrenia Research, worth $40,000.
Requirements: Applicants must have a doctoral level degree and already be employed in research training, or be in a faculty or independent research position. Applicants must have an on-site mentor or senior collaborator who is an established investigator in areas relevant to psychiatric disorders. Applications must be submitted online from the Foundation's website.
Restrictions: Pre-doctoral students or investigators at the rank of associate professor or equivalent are not eligible. Institutional overhead is not allowable.
Geographic Focus: All States
Date(s) Application is Due: Sep 15
Amount of Grant: Up to 60,000 USD
Contact: Sho Tin Chen, Associate Manager, Research Grants; (800) 829-8289 or (516) 829-5576; fax (516) 487-6930; grants@bbrfoundation.org
Internet: http://bbrfoundation.org/narsad-young-investigator-grant
Sponsor: Brain and Behavior Research Foundation
60 Cutter Mill Road, Suite 404
Great Neck, NY 11021

Natalie W. Furniss Charitable Trust Grants — 2626

The Mission of the Furniss Foundation is to promote the humane treatment of animals by providing funding to societies for the prevention of cruelty to animals.
Requirements: Qualifying tax-exempt 501(c)3 organizations operating in the state of New Jersey are eligible to apply. Applications are accepted Ocotber 1 - December 1. Complete the Common Grant Application form available at the Wachovi website to apply for funding.
Restrictions: Grants are not made for political purposes, nor to organizations which discriminate on the basis of race, ethnic origin, sexual or religious preference, age or gender.
Geographic Focus: New Jersey
Date(s) Application is Due: Dec 1
Samples: Oasis Animal Sanctuary, Inc., $4,200--animal sterilization program; Woodford Cedar Run Wildlife Refuge, Inc., $4,5000--winter operating support for wildlife rehab; Society for the Prtevention of Cruelty to Animals, $5,000--humane education in Cumberland Co. School system; Greyhound Friends of New Jersey, Inc., $5,000--broken leg program.
Contact: Wachovia Bank, N.A., Trustee; grantinquiries6@wachovia.com
Internet: https://www.wachovia.com/foundation/v/index.jsp?vgnextoid=61078689fb0a a110VgnVCM1000004b0d1872RCRD&vgnextfmt=default
Sponsor: Natalie W. Furniss Foundation
100 North Main Street
Winston Salem, NC 27150

Nathan Cummings Foundation Grants — 2627

The Nathan Cummings Foundation is rooted in the Jewish tradition and committed to democratic values and social justice, including fairness, diversity, and community. It seeks to build a socially and economically just society that values nature and protects the ecological balance for future generations; promotes humane health care; and fosters arts

and culture that enriches communities. The Foundation's core programs include arts and culture; the environment; health; interprogram initiatives for social and economic justice; and the Jewish life and values/contemplative practice programs. Basic themes informing the Foundation's approach to grantmaking are: concern for the poor, disadvantaged, and underserved; respect for diversity; promotion of understanding across cultures; and empowerment of communities in need. The Board meets twice a year. Applicants should apply by January 15 to be considered for the spring Board meeting and by August 15 to be considered for the fall Board meeting.
Requirements: Eligible applicants must be 501(c)3 organizations. A two or three page letter of inquiry may be submitted with the following information: basic organizational information; contact person; grant purpose; key personnel; project budget and total organizational budget; amount requested and the length of time for which funds are being requested; and other funding sources. Projects that most closely fit with the Foundation's goals will be invited to submit a complete application.
Restrictions: The following is not funded: individuals; scholarships; sponsorships; capital or endowment campaigns; foreign-based organizations; specific diseases; local synagogues or institutions with local projects; Holocaust related projects; projects with no plans for replication; and general support for Jewish education.
Geographic Focus: All States
Samples: Alliance for Justice, Washington, D.C., $10,000 - general support; American Friends of the Heschel Center, Inc., Jersey City, New Jersey, $150,000 - envisioning a political agenda for 21st century Israel; and American Jewish World Service, Inc., New York, New York, $100,000 - group leadership training program.
Contact: Armanda Famiglietti, Director of Grants Management; (212) 787-7300; fax (212) 787-7377; info@nathancummings.org
Internet: http://www.nathancummings.org/programs/index.html
Sponsor: Nathan Cummings Foundation
475 10th Avenue, 14th Floor
New York, NY 10018

National Blood Foundation Research Grants 2628
The National Blood Foundation (NBF), established in 1983, has a history of supporting research and education that advances transfusion medicine and cellular therapies by funding scientific research that benefits patients and donors. Funds are raised annually from corporations, blood centers, foundations and individuals by the NBF for the National Blood Foundation Research and Education Trust Fund (NBFRET) and the NBF. The NBF and NBFRET, which are 501(c)3 organizations, provide grants for scientific research in the field of transfusion medicine and cellular therapies and support educational initiatives that benefit this community. The National Blood Foundation annually awards grants for one or two-year research projects, with a maximum award per grant of $75,000.
Requirements: NBF has also funded research and studies on critical issues that benefit patients and donors in the fields of transfusion medicine and cellular therapies, and are in addition to scientific research. Preference is given to new investigators in the field. There is a $150 application fee.
Geographic Focus: All States
Date(s) Application is Due: Dec 13
Amount of Grant: Up to 65,000 USD
Contact: Amy Quiggins, Manager; (301) 215-6552 or (301) 215-6551; fax (301) 215-5751; aquiggins@aabb.org or nbf@aabb.org
Internet: http://www.aabb.org/programs/nbf/Pages/default.aspx
Sponsor: National Blood Foundation
8101 Glenbrook Road
Bethesda, MD 20814-2749

National Center for Responsible Gaming Large Grants 2629
Large Grants provide up to $75,000 for up to two years for discrete, specified, circumscribed research projects related to gambling disorders. The proposed research investigation may focus on a broad range of research that develops and tests psychosocial or pharmacological approaches for prevention, intervention, treatment or relapse prevention of gambling disorders. The NCRG is especially interested in brief interventions targeted at underrepresented populations, such as minorities, young adults and persons with sub-clinical gambling disorders. Other priorities include the following topics: impact of Indian gaming; gambling and minorities; secondary data analysis; and technology and gambling. Applicants may request up to $75,000 in direct costs per year for a period not to exceed 24 months.
Requirements: Both public and private nonprofit organizations are eligible to apply. Profit-making organizations should contact Christine Reilly, senior research director (creilly@ncrg.org). Foreign institutions are required to collaborate with a U.S. institution. Applications involving a non-U.S. institution must have a principal investigator and fiscal agent based at a U.S. institution.
Geographic Focus: All States
Date(s) Application is Due: Aug 3
Amount of Grant: Up to 75,000 USD
Contact: Christine Reilly; (978) 338-6610; fax (978) 522-8452; creilly@ncrg.org
Internet: http://www.ncrg.org/research-center/apply-ncrg-funding/large-grants
Sponsor: National Center for Responsible Gaming
900 Cummings Center, Suite 216-U
Beverly, MA 01915

National Center for Responsible Gaming Postdoctoral Fellowships 2630
The National Center for Responsible Gaming Postdoctoral Fellowship is intended to help ensure that a pool of highly trained scientists is available to address the research needs of the field of gambling disorders. Eligible applicants are within five years of completing their terminal research degree or within five years of completing medical residency. The proposed research investigation may focus on a broad range of research that develops and tests psychosocial or pharmacological approaches for prevention, intervention, treatment or relapse prevention of gambling disorders. The NCRG is especially interested in brief interventions targeted at underrepresented populations, such as minorities, young adults and persons with sub-clinical gambling disorders. Other priorities include the impact of Indian gaming; gambling and minorities; secondary data analysis; and technology and gambling. Applicants may request up to $75,000 in direct costs for a period not to exceed twelve months.
Requirements: The Postdoctoral Fellowship is intended for individuals who are within five years of completing their terminal research degree or within five years of completing medical residency (or the equivalent). The candidate for this award may not have been a recipient of an NIH Career Development Award or an NCRG grant. Candidates who are not U.S. citizens but have an appointment at a U.S. institution are eligible to apply. Candidates based at foreign institutions are not eligible to apply.
Geographic Focus: All States
Date(s) Application is Due: Jun 3
Amount of Grant: Up to 75,000 USD
Contact: Christine Reilly; (978) 338-6610; fax (978) 522-8452; creilly@ncrg.org
Internet: http://www.ncrg.org/research-center/apply-ncrg-funding/addiction-fellowship
Sponsor: National Center for Responsible Gaming
900 Cummings Center, Suite 216-U
Beverly, MA 01915

National Center for Responsible Gaming Seed Grants 2631
National Center for Responsible Gaming Seed Grants support a variety of research activities, including: pilot and feasibility studies; secondary analysis of existing data; small, self-contained research projects; development of research methodology; development of new research technology. The proposed research investigation may focus on a broad range of research that develops and tests psychosocial or pharmacological approaches for prevention, intervention, treatment or relapse prevention of gambling disorders. The NCRG is especially interested in brief interventions targeted at underrepresented populations, such as minorities, young adults and persons with sub-clinical gambling disorders. Other priorities include the following topics: impact of Indian gaming, gambling and minorities, secondary data analysis and technology and gambling. Applicants may request up to $30,000 in direct costs for a period not to exceed 12 months. The annual deadline is September 1.
Requirements: Both public and private nonprofit organizations are eligible to receive grants from the NCRG. Applicants from profit-making companies should contact Christine Reilly, senior research director (creilly@ncrg.org). Foreign institutions are required to collaborate with a U.S. institution. Applications involving a non-U.S. institution must have a principal investigator and fiscal agent based at a U.S. institution.
Geographic Focus: All States
Date(s) Application is Due: Sep 1
Amount of Grant: Up to 30,000 USD
Contact: Christine Reilly; (978) 338-6610; fax (978) 522-8452; creilly@ncrg.org
Internet: http://www.ncrg.org/research-center/apply-ncrg-funding/seed-grants
Sponsor: National Center for Responsible Gaming
900 Cummings Center, Suite 216-U
Beverly, MA 01915

National Headache Foundation Research Grants 2632
The National Headache Foundation is interested in research protocols that are objectively sound and whose results can, when published in the medical literature, contribute to the better understanding and treatment of headache and pain. Grants may be used for the purchasing of supplies needed for the study (chemical or pharmacological reagents, laboratory animals, tissue culture materials. forms, etc.) and for costs related to the recruiting of patients, data analysis, interpreting or reading results, etc. Applications are available online, and are due by December 1.
Requirements: Funds for a research project approved by the National Headache Foundation are awarded to the individual who is the principal investigator of the study. The funds are sent to the institution that employs the PI but designated for the PI and the approved study.
Restrictions: Grants from the NHF will not cover costs including overhead, salaries (payment for technicians, others collecting data, administrative services, etc.), purchasing of equipment, rent or other indirect expenses.
Geographic Focus: All States
Date(s) Application is Due: Dec 1
Contact: Carolyn Smith, Executive Assistant; (312) 274-2652 or (312) 274-2650; fax (312) 640-9049; info@headaches.org
Internet: http://www.headaches.org/For_Professionals/Research_Grants
Sponsor: National Headache Foundation
820 N Orleans, Suite 217
Chicago, IL 60610-3132

National Headache Foundation Seymour Diamond Clinical Fellowship in Headache Education 2633
The one-year fellowship is intended to encourage, provide for, and engage physicians in the clinical management of headache patients. The program is an individualized hands-on approach at a headache clinic mutually selected by the committee and the applicant. The objectives are to develop competence in the management of common and complex headache problems, to publish a research paper keyed to the treatment of headache, and to prepare the applicant for a career in treating headache patients. The participating institution will be a specialized facility dedicated primarily to headache management, with facilities for inpatient and outpatient treatment. Up to 80 percent of the fellow's time will be clinical, 20 percent study/research.

Requirements: Applicant will be a graduate of an approved U.S. medical school and be four years or more postgraduate.
Geographic Focus: All States
Date(s) Application is Due: Feb 1
Amount of Grant: 40,000 USD
Contact: Selection Committee; (312) 274-2650; fax (312) 640-9049; info@headaches.org
Internet: http://www.headaches.org/For_Professionals/Fellowship_Grants
Sponsor: National Headache Foundation
820 N Orleans, Suite 217
Chicago, IL 60610-3132

National MPS Society Grants   2634
The society awards research grants and fellowships to qualified medical researchers to promote medical research in the fields of mucopolyssaccharidosis (MPS) and mucolipidosis (ML). The program supports MPS (I, II, III IV, V, and VI) research, ML II/III research, and general MPS research. In February of each year the Society announces the grants offered, and funding begins July 1. Occasionally a second grant offering is made later in the year. All applications are reviewed and ranked by members of the National MPS Society Scientific Advisory Board. The final decision of award recipients is determined by the Board of Directors of the National MPS Society. Funding for the grants comes from the annual National MPS Society 5k walk/runs in addition to donations and fundraisers. Grants and fellowships are typically awarded for a two-year period.
Requirements: Qualified medical researchers are eligible.
Geographic Focus: All States
Date(s) Application is Due: May 1
Amount of Grant: Up to 50,000 USD
Samples: Gustavo H.B. Maegawa, MD, PhD, Johns Hopkins School of Medicine, Department of Pediatrics, Baltimore, Maryland, $45,000 - two-year grant for induced-neuronal (iN) cells as tools to study the pathogenesis of neurological manifestations in MPS-II; Shunji Tomatsu, MD, PhD, Nemours Children's Clinic, Delaware Valley of the Nemours Foundation, Wilmington, Delaware, $40,000 - two-year grant for the development of Long Circulating Enzyme Replacement Therapy for MPS IVA.
Contact: Barbara Wedehase, Executive Director; (877) 677-1001 or (919) 806-0101; barbara@mpssociety.org or info@mpssociety.org
Internet: http://www.mpssociety.org/research/
Sponsor: National MPS Society
P.O. Box 14686
Durham, NC 27709-4686

National Parkinson Foundation Clinical Research Fund Grants   2635
The National Parkinson Foundation (NPF)'s Clinical Research Fund (CRF) is designed to support cutting-edge clinical studies of new therapeutic options for Parkinson's patients as well as research comparing the effectiveness of existing treatment options. Conducted by or in collaboration with NPF's Centers of Excellence, CRF-funded research aims to discover the best treatment and care options for people living with Parkinson's disease. Broad clinical-research areas funded by the CRF program are as follows: care quality; imaging; cognition; physical therapy; care models; exercise; and biomarkers. CRF Requests for Application are issued as funding becomes available, but no less than once per year. Applicants may request two-year funding (or less) for their projects. An example of the CRF application process is as follows: an RFP is released in August; NPF schedules pre-submission conference calls for September and October; applicants submit their proposals in December; peer-review is conducted January through March; awards are announced in April. Specific deadline dates will vary from year to year and will be detailed in the RFP. More detailed information is generally posted in a listserv to which all applicants are invited to subscribe (address given in the RFP). Applications must be submitted electronically through NPF's online grant submission system. Registration is required.
Requirements: Principal investigators must be clinical investigators affiliated with a currently-certified NPF Center of Excellence. Management of the grant budget and reporting must be done at the Center of Excellence. Further, each application must be cosigned by the institutional officer responsible for grants and contracts. Principal investigators must meet their institution's criteria for eligibility to receive grants and oversee expenditures. Individual investigators at non-certified institutions may apply as collaborators. Applicants must reasonably expect to have Institutional Review Board approval shortly after the award date.
Restrictions: There are no limits on the number of grants an institution may request but each grant is limited to no more than $250,000 over a two-year period. Postdoctoral fellows, fellows, and students cannot serve as a principal investigator. Grants cannot be used to purchase equipment or to support conference fees or travel (except for travel to a Center of Excellence collaborating on the research project). Indirect costs will not be funded. NPF no longer funds basic laboratory research.
Geographic Focus: All States, All Countries
Amount of Grant: Up to 250,000 USD
Samples: Antonio Strafella, M.D., Ph.D., F.R.C.P.C., Toronto Western Research Institute, Toronto, Ontario, Canada - "Early Detection of Cognitive Changes in the Brain"; Sherrilene Classen, Ph.D., M.P.H, O.T.R./L., University of Florida, Gainesville, Florida - "Visual Attention Deficits as an Early Sign of Cognitive Change"; David Houghton, M.D., M.P.H., University of Louisville, Louisville, Kentucky - "Comparing Physical Therapy Outcomes: LSVT-BIG© vs. Aquatic Methods".
Contact: Jorge Zamudio, (305) 243-4830; fax (305) 243-8144; jzamudio@parkinson.org or iiresearchgrants@parkinson.org
Internet: http://www.parkinson.org/Professionals/Grants---Research.aspx
Sponsor: National Parkinson's Foundation
1501 N.W. Ninth Avenue / Bob Hope Road
Miami, FL 33136-1494

National Psoriasis Foundation Research Grants   2636
The foundation will award up to five $50,000 one year pilot project grants to support advancement of both basic and clinical research into the cause and cure of psoriasis and psoriatic arthritis. The purpose of pilot project grants is to provide support for innovative early stage ideas. These grants are designed to allow the investigator to determine whether their idea is worth pursuing and then to secure longer-term funding.
Requirements: Applicants should use their NIH biographical sketches. Download the required application form from the website.
Geographic Focus: All States
Date(s) Application is Due: Oct 20
Amount of Grant: Up to 50,000 USD
Contact: Bruce F. Bebo, Jr., Director of Research and Medical Programs; (503) 244-7404 or (800) 723-9166; research@psoriasis.org
Internet: http://www.psoriasis.org/research/grants/index.php
Sponsor: National Psoriasis Foundation
6600 SW 92nd Avenue, Suite 300
Portland, OR 97223-7195

Nationwide Insurance Foundation Grants   2637
The Nationwide Insurance Foundation's mission is to improve the quality of life in communities in which a large number of Nationwide members, associates, agents and their families live and work. The foundation's grants fall into three categories: General Operating Support, Program and/or Project Support, and Capital Support. Funding priorities are then placed into one of four tiers. Tier 1-Emergency and basic needs: the foundation partners with organizations that provide life's necessities. Tier 2-Crisis stabilization: the foundation partners with organizations that provide resources to prevent crises or help pick up the pieces after one occurs. Tier 3-Personal and family empowerment: Nationwide helps at-risk youth and families in poverty situations who need tools and resources to advance their lives by partnering with organizations that assist individuals in becoming productive members of society. Tier 4-Community enrichment: the foundation partners with organizations that contribute to the overall quality of life in a community.
Requirements: In the following communities, the Nationwide Insurance Foundation will consider funding 501(c)3 organizations from all four tiers of funding priorities: Columbus, Ohio; Des Moines, Iowa; Scottsdale, Arizona. In the following communities, only Tiers 1 and 2 of the foundation's funding priorities will be considered: Sacramento, California; Denver, Colorado; Gainesville, Florida; Atlanta (Metro), Georgia; Baltimore, Maryland; Lincoln, Nebraska; Raleigh/Durham, North Carolina; Syracuse, New York; Canton, Ohio; Cleveland, Ohio; Harrisburg, Pennsylvania; Philadelphia (Metro), Pennsylvania; Nashville, Tennessee; Dallas (Metro), Texas; San Antonio, Texas; Lynchburg, Virginia; Richmond, Virginia; Wausau, Wisconsin.
Restrictions: The Nationwide Insurance Foundation generally does not fund national organizations (unless the applicant is a local branch or chapter providing direct services) or organizations located in areas with less than 100 Nationwide associates. Also, the foundation does not fund the following: Organizations that are not tax-exempt under paragraph 501(c)3 of the U.S. Internal Revenue Code; Fund-raising events such as walk-a-thons, telethons or sponsorships; Individuals for any purpose; Athletic events or teams, bands and choirs (including equipment and uniforms); Debt-reduction or retirement campaigns; Research; Public or private primary or secondary schools; Requests to support travel; Groups or organizations that will re-grant the foundation's gifts to other organizations or individuals (except United Way); Endowment campaigns; Veterans, labor, religious or fraternal groups (except when these groups provide needed services to the community at-large); Lobbying activities.
Geographic Focus: Arizona, California, Colorado, Florida, Georgia, Iowa, Maryland, Nebraska, New York, North Carolina, Ohio, Pennsylvania, Tennessee, Texas, Virginia, Wisconsin
Date(s) Application is Due: Sep 1
Amount of Grant: 5,000 - 50,000 USD
Contact: Chad Jester; (614) 249-4310 or (877) 669-6877; corpcit@nationwide.com
Internet: http://www.nationwide.com/about-us/nationwide-foundation.jsp
Sponsor: Nationwide Insurance Foundation
1 Nationwide Plaza, MD 1-22-05
Columbus, OH 43215-2220

NBME Stemmler Medical Education Research Fund Grants   2638
The purpose of the fund is to provide support for research or development of innovative evaluation methodology or techniques with the potential to advance assessment in medical education or practice, at eligible medical schools. Collaborative investigations within or among institutions are eligible, particularly as they strengthen the likelihood of the project's contribution and success. Proposed projects may request up to two years of support. Applicants for both new proposals and competing renewals may request up to $150,000 of NBME funding support for a project period of up to two years.
Requirements: Medical schools accredited by the Liaison Committee on Medical Education or the American Osteopathic Association are eligible.
Geographic Focus: All States
Date(s) Application is Due: Oct 3
Amount of Grant: Up to 150,000 USD
Contact: Officer of Research Support; (215) 590-9657; fax (215) 590-9488; stemmlerfund@nbme.org
Internet: http://www.nbme.org/research/stemmler/index.html
Sponsor: National Board of Medical Examiners
3750 Market Street
Philadelphia, PA 19104

NCCAM Exploratory Developmental Grants for Complementary and Alternative Medicine (CAM) Studies of Humans 2639

The National Center for Complementary and Alternative Medicine (NCCAM), the National Cancer Institute (NCI), the National Institute on Aging (NIA), and the Office of Dietary Supplements (ODS) invites high quality exploratory/developmental research grant applications of humans in all domains of complementary and alternative medicine (CAM), in order to obtain preliminary data that can be used as a foundation for a larger clinical study. NCCAM groups CAM practices into four domains: biologically-based practices; energy medicine; manipulative and body-based therapies, and mind-body medicine. In addition, NCCAM funds studies of whole medical systems that employ practices drawn from the four domains. The total project period for an application submitted in response to this funding opportunity may not exceed three years. Direct costs are limited to $400,000 over a three-year period, with no more than $250,000 in direct costs allowed in any single year.

Requirements: The following organizations and institutions may apply: public/state controlled institutions of higher education; private institutions of higher education; Hispanic-serving institutions; historically Black colleges and universities (HBCUs); tribally controlled colleges and universities (TCCUs); Alaska native and native Hawaiian serving institutions; nonprofits with 501(c)3 IRS status; nonprofits without 501(c)3 IRS status; small businesses; for-profit organizations; State governments; regional organizations; U.S. territories or possessions; Indian/Native American tribal governments; Indian/Native American tribally designated organizations; county governments; city or township governments; special district governments; independent school districts; public housing authorities/Indian housing authorities; eligible agencies of the Federal government; and faith-based or community based organizations.
Geographic Focus: All States
Date(s) Application is Due: Feb 16; Jun 16; Oct 16
Amount of Grant: Up to 400,000 USD
Contact: April Bower, (301) 451-3560; fax (301) 480-3621; bowera@mail.nih.gov
Internet: http://grants.nih.gov/grants/guide/pa-files/PAR-08-135.html#SectionVII
Sponsor: National Center for Complementary and Alternative Medicine
9000 Rockville Pike
Bethesda, MD 20892-5475

NCCAM Ruth L. Kirschstein National Research Service Awards for Postdoctoral Training in Complementary and Alternative Medicine 2640

The National Center for Complementary and Alternative Medicine (NCCAM) awards individual postdoctoral research training fellowships to promising applicants with the desire and potential to become productive, independent investigators in the field of complementary and alternative medicine (CAM) research. The primary objective of this funding opportunity is to help ensure that diverse pools of highly trained scientists will be available in adequate numbers and in appropriate research areas to carry out the Nation's biomedical, behavioral, and clinical CAM research agendas. Applicants may come from a range of career backgrounds, including biomedical research, conventional allopathic medicine, or CAM healing practices. The number of awards and the total amount of funding that NCCAM expects to award through this announcement will depend on the scientific merit of applications received, relevance to NCCAM's program priorities, and the availability of funds.

Requirements: Eligible organizations include for-profit and non-profit organizations, public or private institutions such as universities, colleges, hospitals and laboratories, units of State and local governments, eligible agencies and labs of the Federal government including NIH intramural labs, and domestic or foreign institutions/organizations. Applicant fellows must be citizens or non-citizen nationals of the United States, or have been lawfully admitted to the United States for permanent residence and have in their possession an Alien Registration Receipt Card.
Restrictions: Individuals on temporary or student visas are not eligible.
Geographic Focus: All States
Date(s) Application is Due: Apr 8; Aug 8; Dec 8
Contact: Nancy J. Pearson, (301) 594-0519; fax (301) 480-3621; pearsonn@mail.nih.gov
Internet: http://grants.nih.gov/grants/guide/pa-files/PAR-07-319.html
Sponsor: National Center for Complementary and Alternative Medicine
9000 Rockville Pike
Bethesda, MD 20892-5475

NCCAM Translational Tools for Clinical Studies of CAM Interventions Grants 2641

This Funding Opportunity Announcement (FOA) issued by the National Center for Complementary and Alternative Medicine (NCCAM) of the National Institutes of Health (NIH) encourages investigator(s)-initiated applications that propose to develop, enhance, and validate translational tools to facilitate rigorous study of complementary and alternative medicine (CAM) approaches that are in wide use by the public. Multidisciplinary studies and collaboration among investigators with expertise in appropriate disciplines are encouraged. NCCAM intends to award up to $3 million (total costs) per year, to support approximately 4 to 6 new grants. A maximum of $300,000 (direct costs) per year may be requested and the total project period for an application submitted in response to this FOA will be five years.

Requirements: The following organizations and institutions are eligible to apply: individuals; public/state controlled institutions of higher education; private institutions of higher education; Hispanic-serving institutions; historically Black colleges and universities (HBCUs); tribally controlled colleges and universities (TCCUs); Alaska Native and Native Hawaiian serving institutions; nonprofits with 501(c)3 IRS status; nonprofits without 501(c)3 IRS status; for-profit organizations; state governments; Indian/Native American tribal governments (federally recognized); Indian/Native American tribally designated organizations; county governments; city or township governments; special district governments; independent school districts; public housing authorities/Indian housing authorities; U.S. territory or possession; Indian/Native American tribal governments (other than federally recognized); regional organizations; non-domestic (non-U.S.) entities (foreign organizations); other eligible agencies of the federal government; and faith-based or community-based organizations.
Geographic Focus: All States
Date(s) Application is Due: Jul 17
Amount of Grant: 300,000 USD
Contact: Dr. Partap S. Khalsa, Division of Extramural Research Contact; (301) 594-3462; fax (301) 480-1587; khalsap@mail.nih.gov
Internet: http://grants.nih.gov/grants/guide/rfa-files/RFA-AT-09-002.html
Sponsor: National Center for Complementary and Alternative Medicine
9000 Rockville Pike
Bethesda, MD 20892-5475

NCHS National Center for Health Statistics Postdoctoral Research Appointments 2642

The National Center for Health Statistics (NCHS) is the Nation's principal health statistics agency and provides the information needed to develop the programs and policies that will improve the health of the American people. The objective of the NCHS Postdoctoral Research Program is to provide opportunities for postdoctoral candidates of unusual promise and ability to conduct research in problems of their choosing that are compatible with the interests of NCHS. A postdoctoral researcher is a temporary salaried employee of NCHS analogous to fellows or similar temporary researchers at the postdoctoral level in universities and other organizations. Appointments are two years in length, but extensions are possible with special consideration. Benefits include sick leave, annual leave, thrift savings, and health and life insurance. Flex time and alternate work schedules are available if desired, contingent on the supervisor's approval. Cost of relocation will be determined on an individual basis. An evaluation is conducted after one year to ensure that the postdoctoral researcher is making suitable progress. General areas of interest for research at NCHS include statistical theory, survey methodology, statistical computing, economics, demography, and social and behavioral science. Specific areas of interest are: confidentiality protection and related topics; data linkage; quality control, quality improvement, and its measurement; adjustment for nonresponse in sample surveys; telephone survey methods; data collection in a changing health care system; small-area estimation; cognitive processes and question response; data editing and imputation; statistical computing; sampling and estimation; questionnaire design and measurement research; human factors and usability research; and data mining. To apply, applicants must submit a statement of research interest, a curriculum vitae, official transcripts, and three reference letters. The statement of research interest should describe the purpose and context of the research project, the research approach and methods to be used, key activities to be undertaken during the fellowship period, and the expected contributions of the proposed research. For fullest consideration, applicants should submit the application by the deadline date to the address given. However, applications will be accepted throughout the year until positions are filled. Evaluation of proposals will be conducted annually in March or April by a special panel convened for this purpose. Notification of awards will be made beginning in mid-April. The date on which an appointment would begin is negotiable on an individual basis but is expected to occur between June 1 and December 31. Applicants can find complete information on the postdoctoral research program and on application submission requirements at the program website. Applicants are encouraged to contact NCHS to obtain specific information on current research and available technical facilities and to speak to NCHS researchers working in the specific area of interest.

Requirements: Awardees must hold a Ph.D. or other earned research degree recognized in the United States as equivalent to the Ph.D. or must present acceptable evidence of having completed all the formal academic requirements for the degree before appointment. Applicants must have demonstrated ability for creative research. An applicant's training and research experience may be in any appropriate discipline or combination of disciplines required for the proposed research. Postdoctoral researchers must devote their full-time effort to the research program and must be in residence at NCHS during the program. No period of tenure may be spent in residence at another agency or institution.
Restrictions: Applicants must be U.S. citizens or legal permanent residents with a work authorization. Postdoctoral research positions are awarded only to persons who have held doctorates for less than 3 years or are in the process of receiving doctorate degrees at the time of application. No additional monetary aid or other remuneration may be accepted from another appointment, fellowship, or similar grant during the period of the program.
Geographic Focus: All States
Date(s) Application is Due: Feb 28
Amount of Grant: 149,744 - 231,484 USD
Contact: Cynthia Link, (800) 232-4636 or (301) 458-4344; cynthia.link@cdc.hhs.gov
Internet: http://www.cdc.gov/nchs/about/postdoc.htm
Sponsor: National Center for Health Statistics
3311 Toledo Road
Hyattsville, MD 20782

NCI/NCCAM Quick-Trials for Novel Cancer Therapies 2643

This PA sponsored by NCI and NCCAM expands the initiative to all cancer sites and provides investigators with rapid access to support for pilot, phase I, and phase II cancer clinical trials testing new agents and patient monitoring and laboratory studies to ensure the timely development of new therapeutic approaches. QUICK-TRIAL will provide a new approach designed to simplify the grant application process and provide a rapid turnaround from application to funding. Features include a modular grant application and award process, inclusion of the clinical protocol within the grant application, and accelerated peer review with the goal of issuing new awards within five months of application receipt. Inclusion of the complete clinical protocol within the PHS 398

grant application is intended to simplify the application process by eliminating the need to duplicate protocol details in the research plan section. Investigators may apply for a maximum of two years of funding support using the exploratory/developmental (R21) grant mechanism for up to $250,000 direct costs per year. Therapeutic trials in human subjects employing new agents and therapeutic approaches, including CAM treatments, whether used as a single agent/modality or in combination, are appropriate. Preference will be given to clinical trials that include laboratory studies to validate mechanistic hypotheses or clinical correlates that can meaningfully guide further clinical development. A letter of intent is due one month prior to application receipt date. PA: PAR-04-155
Requirements: Applications may be submitted by foreign and domestic, for-profit and nonprofit organizations, both public and private; units of state and local governments; and eligible agencies of the federal government.
Geographic Focus: All States
Date(s) Application is Due: Apr 9; Aug 9; Dec 9
Amount of Grant: Up to 250,000 USD
Contact: Amy Connolly; (301) 496-8786; fax (301) 496-8601; amyconnolly@nih.gov
Internet: http://grants.nih.gov/grants/guide/pa-files/par-04-155.html
Sponsor: National Cancer Institute
6130 Executive Boulevard, Executive Plaza N, Suite 3157, MSC 7328
Bethesda, MD 20892-7328

NCI Academic-Industrial Partnerships for Development and Validation of In Vivo Imaging Systems and Methods for Cancer Investigations   2644

The overall goal of this Funding Opportunity Announcement (FOA) is to accelerate the transition of in vivo spectroscopic and imaging systems and methods into cancer research, clinical trials, and/or clinical practice. A related goal is to advance new technologies and methods for animal imaging. To promote these goals, the Cancer Imaging Program (CIP) of the National Cancer Institute (NCI) solicits applications from multi-institutional research partnerships formed by academic, industrial, and other investigators. The partners will establish an inter-disciplinary, multi-institutional research team for the purpose of developing imaging technology that is applicable to cancer research involving human subjects and/or animal models. The academic-industrial partnerships must be dedicated to the complete research and development cycle, culminating in the product prototype stage. The partnership model is expected to more readily overcome various barriers to the development of new imaging devices and technologies faced by either academia or industry working alone.
Requirements: Eligible institutions include: public and state controlled institutions of higher education; private institutions of higher education; nonprofits with 501(c)3 IRS status (other than institutions of higher education); nonprofits without 501(c)3 IRS status (other than institutions of higher education); small businesses; for-profit organizations (other than small businesses); state governments; U.S. territories or possessions; non-domestic (non-U.S.) entities (foreign organizations); and eligible agencies of the federal government.
Geographic Focus: All States
Date(s) Application is Due: Feb 5; Jun 5; Oct 5
Contact: Guoying Liu, (301) 496-9531; fax (301) 480-3507; guoyingl@mail.nih.gov
Internet: http://grants.nih.gov/grants/guide/pa-files/PAR-07-214.html
Sponsor: National Cancer Institute
6130 Executive Plaza, EPN Suite 6000, MSC 7412
Bethesda, MD 20892-7412

NCI Application and Use of Transformative Emerging Technologies in Cancer Research Grants   2645

This Funding Opportunity Announcement (FOA) solicits grant applications proposing exploratory research projects on evaluating the performance of emerging molecular and cellular analysis technologies utilizing an appropriate cancer-relevant biological system. Technologies proposed for evaluation may have broad applicability but must ultimately be suitable for cancer-relevant analyses such as the detection (at various molecular and cellular levels) of cancer-related characteristics and/or structural and/or functional alterations. Letters of intent are requested by January 23, April 27, and August 30; full proposal deadlines are February 23, May 27, and September 30. Annual deadline dates may vary; contact program staff for exact dates.
Requirements: Eligible institutions include: public and state controlled institutions of higher education; private institutions of higher education; nonprofits with 501(c)3 IRS status (other than institutions of higher education); nonprofits without 501(c)3 IRS status (other than institutions of higher education); small businesses; for-profit organizations (other than small businesses); state governments; U.S. territories or possessions; non-domestic (non-U.S.) entities (foreign organizations); and eligible agencies of the federal government.
Geographic Focus: All States
Date(s) Application is Due: Jan 23; Feb 23; Apr 27; May 27; Aug 30; Sep 30
Amount of Grant: Up to 275,000 USD
Contact: Richard Aragon, (301) 496-1550; fax (301) 496-7807; raragon@mail.nih.gov
Internet: http://grants.nih.gov/grants/guide/rfa-files/RFA-CA-09-006.html
Sponsor: National Cancer Institute
31 Center Drive, Building 31, Room 10A52, MSC 2580
Bethesda, MD 20892-2580

NCI Application of Metabolomics for Translational and Biological Research Grants   2646

The purpose of this Funding Opportunity Announcement (FOA), sponsored by multiple NIH Institutes/Centers (ICs), is to promote the use of metabolomic technologies in translational research in human health and disease for the purposes of enabling and improving disease detection, diagnosis, risk assessment, prognosis, and prediction of therapeutic responses. In addition to this overall goal, specific scopes and objectives of the participating ICs are presented below. Using the NIH Research Project Grant (R01) funding mechanism, this FOA focuses on well developed projects supported by preliminary data.
Requirements: Application can be made by the following institutions and organizations: public and state controlled institutions of higher education; private institutions of higher education; small businesses; for-profit organizations (other than small businesses); and non-domestic (non-U.S.) entities (foreign organizations).
Geographic Focus: All States
Date(s) Application is Due: Feb 5; Feb 16; Jun 5; Jun 16; Oct 5; Oct 16
Contact: Padma Maruvada, (301) 496-3893; fax (301) 402-8990; maruvadp@mail.nih.gov
Internet: http://grants.nih.gov/grants/guide/pa-files/PA-07-301.html
Sponsor: National Cancer Institute
6130 Executive Boulevard, Executive Plaza N, Suite 3144, MSC 7362
Bethesda, MD 20892-7362

NCI Basic Cancer Research in Cancer Health Disparities Grants   2647

Through this Funding Opportunity Announcement (FOA), the Center to Reduce Cancer Health Disparities (CRCHD) and the Division of Cancer Biology (D.C.B), at the National Cancer Institute (NCI), invite cooperative agreement research (U01) grant applications from investigators interested in conducting basic research studies into the causes and mechanisms of cancer health disparities. These awards will support pilot and feasibility studies, development and testing of new methodologies, secondary data analyses, and innovative mechanistic studies that investigate biological/genetic bases of cancer health disparities. This FOA is also designed to aid and facilitate the growth of a nationwide cohort of scientists with a high level of basic research expertise in cancer health disparities research who can develop resources and tools, such as biospecimens, cell lines and methods that are necessary to conduct basic research in cancer health disparities. The total project period for an application submitted in response to this funding opportunity may not exceed five years. Direct costs are limited to $250,000/year and may not exceed 5 years.
Requirements: The following organizations and institutions are eligible to apply: public/state controlled institutions of higher education; private institutions of higher education; Hispanic-serving institutions; historically Black colleges and universities (HBCUs); tribally controlled colleges and universities (TCCUs); Alaska Native and Native Hawaiian serving institutions; nonprofits with 501(c)3 IRS status; nonprofits without 501(c)3 IRS status; small businesses; for-profit organizations; state governments; Indian/Native American tribal governments (federally recognized); Indian/Native American tribally designated organizations; U.S. territories or possessions; Indian/Native American tribal governments (other than federally recognized); regional organizations; eligible agencies of the federal government; and units of local governments.
Geographic Focus: All States
Date(s) Application is Due: Jun 23; Nov 23
Amount of Grant: Up to 250,000 USD
Contact: Phillip J. Daschner, (301) 496-1951; fax (301) 496-2025; PD93U@nih.gov
Internet: http://grants.nih.gov/grants/guide/pa-files/PAR-09-161.html
Sponsor: National Cancer Institute
6130 Executive Boulevard, Room 5014, MSC 7344
Bethesda, MD 20892-7344

NCI Cancer Education and Career Development Program   2648

This Funding Opportunity Announcement (FOA) represents the continuation of the Cancer Education and Career Development Program (CECDP) established by the National Cancer Institute (NCI). The purpose of the CECDP is to support the development and implementation of institutional curriculum-dependent pre-doctoral and/or postdoctoral programs with specific core didactic and research requirements to train a cadre of scientists in interdisciplinary and collaborative cancer research settings. Examples of interdisciplinary/collaborative areas of cancer research that are particularly applicable to intent of the CECDP are cancer prevention and control, behavioral and population sciences, nutrition, imaging, and molecular diagnosis. This FOA makes an NCI-specific use of the NIH Education R25 grant mechanism.
Requirements: Applications may be submitted on behalf of the principal investigator by domestic, nonfederal, public or private organizations. Applications may include more than one institution to create a program through consortium agreements. Predoctoral and postdoctoral candidates must be citizens or noncitizen nationals of the US, or must have been lawfully admitted to the U.S. for permanent residence.
Restrictions: No application may request more than $500,000 in direct costs per annum without written approval from NCI.
Geographic Focus: All States
Amount of Grant: Up to 500,000 USD
Contact: Catherine Blount; (301) 496-3179; fax (301) 496-8601; blountc@mail.nih.gov
Internet: http://grants.nih.gov/grants/guide/pa-files/PAR-06-511.html
Sponsor: National Cancer Institute
6130 Executive Boulevard, Executive Plaza N, Suite 3144, MSC 7362
Bethesda, MD 20892-7362

NCI Centers of Excellence in Cancer Communications Research   2649

NCI invites applications for Centers of Excellence in Cancer Communications Research (CECCRs). The purpose is to identify areas of discovery that build upon important recent developments in knowledge and technology and that hold promise for making significant progress against all cancers. The centers must include three or more individual hypothesis-driven research projects, pilot or developmental research projects, shared resources, and a plan for career development. To be effective, the centers' research should integrate cancer communications appropriately into one or more contexts of the cancer continuum--from prevention through treatment to survivorship and end-of-life research. Communications research also is needed about challenging topics such as cancer information seeking,

decision making under uncertainty, and genetic testing. Centers' research also should provide insight into mechanisms underlying how people process information. Annual deadline dates may vary; contact program staff for exact dates. RFA: CA-03-007
Requirements: Applications may be submitted by domestic for-profit and nonprofit organizations, both public and private; units of state and local governments; and eligible agencies of the federal government.
Geographic Focus: All States
Amount of Grant: Up to 15,000,000 USD
Contact: Crystal Wolfrey, Grants Administration Branch; (301) 496-8634; fax (301) 496-8601; crystal.wolfrey@nih.gov or wolfreyc@mail.nih.gov
Internet: http://grants.nih.gov/grants/guide/rfa-files/RFA-CA-08-004.html
Sponsor: National Cancer Institute
6130 Executive Boulevard, Executive Plaza N, Suite 3144, MSC 7362
Bethesda, MD 20892-7362

NCI Diet, Epigenetic Events, and Cancer Prevention Grants **2650**
The purpose of this Funding Opportunity Announcement (FOA) is to promote innovative preclinical and clinical research to determine how diet and dietary factors, including dietary supplements, impact epigenetic processes involved in cancer prevention. Although much evidence exists that dietary components are linked to cancer prevention, the specific nutrients and sites of action remain elusive. Diet, in fact, has been implicated in many of the pathways of cancer, including apoptosis, cell cycle control, differentiation, inflammation, angiogenesis, DNA repair, and carcinogen metabolism. These are also processes that are likely regulated by epigenetic events to impact gene function and chromatin stability. Thus, research supported by this initiative could address one of the following issues: How bioactive food components regulate epigenetic events for cancer prevention, how bioactive food components might alter aberrant epigenetic patterns or events and restore gene function, and how these components might circumvent or compensate for genes and pathways that are altered by epigenetic events. Another important aim of this FOA is to encourage collaborations between nutrition scientists and experts in epigenetics to study bioactive food components with cancer-preventative properties and to examine key epigenetic events in cancer processes (e.g., carcinogen metabolism, cell division, differentiation, and apoptosis) in order to begin to establish linkages between epigenetic events and patterns, and tumor incidences/behaviors. NIDDK is particularly interested in receiving applications that address normal and neoplastic conditions focusing on the impact of epigenetics on cell division, differentiation, and/or apoptosis in the gastrointestinal (GI) tract and related organs. It is also expected that each application demonstrate experience in nutrition and cancer prevention as well as in epigenetic research. Because the nature and scope of the proposed research will vary from application to application, it is anticipated that the size and duration of each award will also vary.
Requirements: Application can be made by the following institutions and organizations: public/state controlled institutions of higher education; private institutions of higher education; nonprofits with 501(c)3 IRS status (other than institutions of higher education); nonprofits without 501(c)3 IRS status (other than institutions of higher education); small businesses; for-profit organizations (other than small businesses); state governments; U.S. territories or possessions; Indian/Native American tribal governments (federally recognized); Indian/Native American tribal governments (other than Federally recognized); Indian/Native American tribally designated organizations; non-domestic (non-U.S.) entities (foreign organizations); Hispanic-serving institutions; historically Black colleges and universities (HBCUs); tribally controlled colleges and universities (TCCUs); Alaska Native and Native Hawaiian serving institutions; regional organization; eligible agencies of the Federal government; and faith-based or community based organizations.
Geographic Focus: All States
Date(s) Application is Due: Feb 5; Jun 5; Oct 5
Contact: Vishnudutt Purohit, (301) 443-2689; fax (301) 594-0673; vpurohit@mail.nih.gov
Internet: http://grants.nih.gov/grants/guide/pa-files/PA-07-175.html
Sponsor: National Cancer Institute
6130 Executive Boulevard, Executive Plaza N, Suite 3157, MSC 7328
Bethesda, MD 20892-7328

NCI Exploratory Grants for Behavioral Research in Cancer Control **2651**
This FOA will support innovative pilot projects or feasibility studies, which will facilitate the growth of research science in the cancer control continuum from a behavioral perspective. This FOA includes and incorporates the research interests of the Behavioral Research program, the Office of Cancer Survivorship, and the Community Oncology and Prevention Trials Research Group. This FOA is appropriate for testing timely interventions in pilot studies for feasibility or using rigorous qualitative research methods to assess the potential efficacy of an intervention. It is also appropriate for the psychometric evaluation of new measures or culturally appropriate ones to be adapted for use in populations where measures have not yet been developed or validated. This FOA encourages applications that include small cross-disciplinary teams of investigators who bring perspectives from the behavioral and social sciences, as well as other fields of public health.
Requirements: Eligible institutions and agencies include: public/state controlled institutions of higher education; private institutions of higher education; nonprofits with 501(c)3 IRS status (other than institutions of higher education); nonprofits without 501(c)3 IRS status (other than institutions of higher education); small businesses; for-profit organizations (other than small businesses); state governments; U.S. territories or possessions; regional organizations; non-domestic (non-U.S.) entities (foreign organizations); eligible agencies of the federal government; and units of local governments.
Geographic Focus: All States
Date(s) Application is Due: Feb 16; Jun 16; Oct 16
Amount of Grant: Up to 275,000 USD
Contact: Sabra F. Woolley, (301) 435-4589; fax (301) 594-0673; woolleys@mail.nih.gov
Internet: http://grants.nih.gov/grants/guide/pa-files/PA-09-130.html
Sponsor: National Cancer Institute
6130 Executive Boulevard, Room 4084, MSC 7365
Bethesda, MD 20850-7365

NCI Improving Diet and Physical Activity Assessment Grants **2652**
This Funding Opportunity Announcement (FOA) is aimed at advancing the quality of measurements of dietary intake and physical activity pertinent to cancer and/or other pathologies through support of research on improved instruments, technologies, and/or statistical/analytical techniques. Studies proposed in the grant applications should be aimed at optimizing the combined use of objective and self-report measures of physical activity and/or dietary intake for testing in both general and diverse populations. Applicants responding to this FOA, which uses the R01 grant mechanism, should have already tested and validated new approaches and have obtained feasibility data in the context of cancer or other relevant diseases. Because the nature and scope of the proposed research will vary from application to application, it is anticipated that the size and duration of each award will also vary.
Requirements: Application can be made by the following institutions and organizations: public/state controlled institutions of higher education; private institutions of higher education; nonprofits with 501(c)3 IRS status (other than institutions of higher education); nonprofits without 501(c)3 IRS status (other than institutions of higher education); small businesses; for-profit organizations (other than small businesses); state governments; U.S. territories or possessions; Indian/Native American tribal governments (federally recognized); Indian/Native American tribal governments (other than Federally recognized); Indian/Native American tribally designated organizations; non-domestic (non-U.S.) entities (foreign organizations); Hispanic-serving institutions; historically Black colleges and universities (HBCUs); tribally controlled colleges and universities (TCCUs); Alaska Native and Native Hawaiian serving institutions; regional organization; eligible agencies of the Federal government; and faith-based or community based organizations.
Restrictions: Applicants without sufficiently extensive preliminary data or those who wish to explore the utility of new untested dietary or physical activity assessment methods in pilot studies are advised to consider submitting applications under the parallel FOA, PAR-06-103, which uses the Exploratory/Developmental Grant (R21) award mechanism.
Geographic Focus: All States
Date(s) Application is Due: Feb 5; Jun 5; Oct 5
Contact: Richard Troiano, (301) 435-6822; fax (301) 435-3710; troianor@mail.nih.gov
Internet: http://grants.nih.gov/grants/guide/pa-files/PAR-09-225.html
Sponsor: National Cancer Institute
6130 Executive Boulevard, Executive Plaza N, Room 4005, MSC 7344
Bethesda, MD 20892-7344

NCI Mentored Career Development Award to Promote Diversity **2653**
This mechanism establishes a pathway of recruiting, training, and retaining underrepresented investigators in research fields that address problems pertinent to the biology, etiology, pathogenesis, prevention, diagnosis, control, and treatment of human cancer and who can conduct independent competitive cancer research programs. The Mentored Career Development Award to Promote Diversity (K01) provides support for a sustained period of protected time for intensive research career development under the guidance of an experienced mentor, or sponsor, in the biomedical, behavioral or clinical sciences leading to research independence. The expectation is that through this sustained period of research career development and training, awardees will launch independent research careers and become competitive for new research project grant funding.
Requirements: The following organizations and institutions may apply: for-profit and non-profit organizations; public or private institutions, such as universities, colleges, hospitals, and laboratories; and domestic institutions.
Restrictions: Foreign institutions and organizations are not eligible to apply.
Geographic Focus: All States
Date(s) Application is Due: Feb 12; Mar 12; Jun 12; Jul 12; Oct 12; Nov 12
Amount of Grant: 75,000 USD
Contact: Belinda Locke, (301) 496-7344; fax (301) 402-4551; Lockeb@mail.nih.gov
Internet: http://grants.nih.gov/grants/guide/pa-files/PAR-06-220.html
Sponsor: National Cancer Institute
6116 Executive Boulevard, Suite 7031, MSC 8350
Bethesda, MD 20892-8350

NCI Ruth L. Kirschstein National Research Service Award Institutional Research Training Grants **2654**
The National Institutes of Health (NIH) will award Ruth L. Kirschstein National Research Service Award (NRSA) Institutional Research Training Grants (T32) to eligible institutions as the primary means of supporting predoctoral and postdoctoral research training to help ensure that a diverse and highly trained workforce is available to assume leadership roles related to the Nation's biomedical, behavioral and clinical research agenda. The primary objective of the T32 program is to prepare qualified individuals for careers that have a significant impact on the health-related research needs of the Nation. This program supports predoctoral, postdoctoral and short term research training programs at domestic institutions of higher education with the T32 funding mechanism. Stipends are provided as a subsistence allowance for trainees to help defray living expenses during the research training experience and are based on a 12-month appointment period.
Requirements: The following organizations are eligible to apply: public/state controlled institutions of higher education; private institutions of higher education; nonprofits with 501(c)3 IRS status (other than institutions of higher education); nonprofits without 501(c)3 IRS status (other than institutions of higher education); for-profit organizations (other than small businesses); state governments; U.S. territories or possessions; Indian/Native American tribal governments (Federally recognized); Hispanic-serving

institutions; historically Black colleges and universities (HBCUs); tribally controlled colleges and universities (TCCUs); Alaska Natives and Native Hawaiians serving institutions; regional organizations; eligible agencies of the federal government; faith-based or community based organizations.
Geographic Focus: All States
Date(s) Application is Due: Apr 8; Aug 8; Dec 8
Contact: Crystal Wolfrey, Grants Administration Branch, NCI; (301) 496-8634; fax (301) 496-8601; crystal.wolfrey@nih.gov or wolfreyc@mail.nih.gov
Internet: http://grants.nih.gov/grants/guide/pa-files/PA-08-226.html
Sponsor: National Cancer Institute
6120 Executive Boulevard, Room 243, MSC 7150
Bethesda, MD 20892-7150

NCI Stages of Breast Development: Normal to Metastatic Disease Grants   2655
The National Cancer Institute (NCI), National Institute of Child Health and Human Development (NICHD), National Institute of Diabetes and Digestive and Kidney Diseases (NIDDK), National Institute on Aging (NIA), and National Institute of Environmental Health Sciences (NIEHS) invite investigator-initiated research grant applications to study the molecular, cellular, endocrine, and other physiological influences on the development and maturation of the normal mammary gland and alterations involved in early malignant and metastatic breast cancer. Multi-disciplinary collaborations, for example between cell biologists, molecular endocrinologists, bioengineers, geneticists, and mammary pathologists, are encouraged. Appropriate studies include, but are not limited to, documenting the role of dynamic hormonal influences and determining the role of cell growth, apoptosis, and differentiation in mammary gland maturation; integrating knowledge of cell signaling in breast tissue with whole organ biology; developing models of breast differentiation; and studies focusing on the characteristics of breast tumor physiology particularly in relation to metastasis. It is expected that a complete understanding of mammary gland development will form critical underpinnings for continued advances in detecting, preventing and treating breast cancer.
Requirements: Applications may be submitted by domestic and foreign, for-profit and non-profit organizations, public and private, such as universities, colleges, hospitals, laboratories, units of State or local governments, and eligible agencies of the Federal government.
Geographic Focus: All States
Date(s) Application is Due: Feb 5; Jun 5; Oct 5
Amount of Grant: Up to 250,000 USD
Contact: Padma Maruvada, (301) 496-3893; fax (301) 402-8990; maruvadp@mail.nih.gov
Internet: http://grants.nih.gov/grants/guide/pa-files/PA-99-162.html
Sponsor: National Cancer Institute
6130 Executive Boulevard, Executive Plaza N, Suite 3144, MSC 7362
Bethesda, MD 20892-7362

NCI Technologies and Software to Support Integrative Cancer Biology Research (SBIR) Grants   2656
This Funding Opportunity Announcement (FOA), issued by the National Cancer Institute (NCI), invites Small Business Innovation Research (SBIR) grant applications from small business concerns (SBCs) that propose the development of commercial software tools, computational/mathematical methods, and technologies that will enable integrative cancer biology research. Integrative cancer biology focuses on understanding cancer as a complex biological system by utilizing both computational and experimental biology to integrate heterogeneous data sources and ultimately to generate predictive computational models of cancer processes. Phase I applications may request up to 2 years of support with total costs of up to $300,000 for the duration of the award (not to exceed $200,000 total costs in any one year). Budgetary requests of Phase II applications will conform to standard SBIR guidelines (up to $750,000 total costs for up to two years).
Requirements: Only United States small business concerns (SBCs) are eligible to submit SBIR applications.
Geographic Focus: All States
Date(s) Application is Due: Apr 5; Aug 5; Dec 5
Amount of Grant: 200,000 - 750,000 USD
Contact: Greg Evans, (301) 594-8807; fax (301) 480-4082; evansgl@mail.nih.gov
Internet: http://grants.nih.gov/grants/guide/pa-files/PA-09-188.html
Sponsor: National Cancer Institute
31 Center Drive, Building 31, Room 10A19, MSC 2580
Bethesda, MD 20892-2580

NCRR Networks and Pathways Collaborative Research Projects Grants   2657
This announcement solicits applications for research project grants that will leverage and complement the ongoing technology development being pursued in the National Technology Centers for Networks and Pathways (TCNPs), a program of the NIH Roadmap for Medical Research. These collaborative projects should focus either on addressing a challenging biological problem using the technology developed in one or more of the TCNPs, or on the development of technology that will complement that which is being developed in the centers. Applicants may request support for their own work as well as supplemental support for components pursued in the participating TCNP. While the nature and scope of these applications will vary, it is anticipated that the annual direct costs for these projects will typically range from $150,000 to $250,000.
Requirements: Application can be made by the following institutions and organizations: public/state controlled institutions of higher education; private institutions of higher education; nonprofits with 501(c)3 IRS status (other than institutions of higher education); nonprofits without 501(c)3 IRS status (other than institutions of higher education); small businesses; for-profit organizations (other than small businesses); state governments; U.S. territories or possessions; non-domestic (non-U.S.) entities (foreign organizations); historically Black colleges and universities (HBCUs); Tribally controlled colleges and universities (TCCUs); and other eligible agencies of the Federal government.
Geographic Focus: All States
Date(s) Application is Due: Feb 5; Jun 5; Oct 5
Amount of Grant: 150,000 - 250,000 USD
Contact: Douglas M. Sheeley, (301) 594-9762; sheeleyd@mail.nih.gov
Internet: http://grants.nih.gov/grants/guide/pa-files/PA-07-266.html
Sponsor: National Center for Research Resources
6701 Democracy Boulevard, MSC 4874
Bethesda, MD 20892-4874

NCRR Novel Approaches to Enhance Animal Stem Cell Research   2658
The purpose of this funding opportunity announcement (FOA) is to encourage the submission of applications for research to enhance animal stem cells as model biological systems. Innovative approaches to isolate, characterize and identify toti-potent and multi-potent stem cells from nonhuman biomedical research animal models, as well as to generate reagents and techniques to characterize and separate those stem cells from other cell types is encouraged.
Requirements: The following organizations and institutions may apply: public/state controlled institutions of higher education; private institutions of higher education; Hispanic-serving institutions; historically Black colleges and universities (HBCUs); tribally controlled colleges and universities (TCCUs); Alaska Native and Native Hawaiian serving institutions; nonprofits with 501(c)3 IRS status (other than institution of higher education); nonprofits without 501(c)3 IRS status (other than institution of higher education); non-domestic (non-U.S.) entities (foreign organizations); small businesses; for-profit organizations (other than small businesses).
Restrictions: Studies involving human subjects are not allowed.
Geographic Focus: All States
Date(s) Application is Due: Feb 5; Jun 5; Oct 5
Contact: John D. Harding, (301) 435-0744; fax (301) 480-3819; hardingj@mail.nih.gov
Internet: http://grants.nih.gov/grants/guide/pa-files/PA-07-303.html
Sponsor: National Center for Research Resources
6701 Democracy Boulevard, MSC 4874
Bethesda, MD 20892-4874

NEAVS Fellowsips in Women's Health and Sex Differences   2659
The American Fund for Alternatives to Animal Research (AFAAR) and NEAVS offer a one-year, $40,000 postdoctoral fellowship grant (with possible renewal) to a woman committed to developing, validating, or using alternatives to animal methods in the investigation of women's health or sex differences. Proposals are reviewed by an award committee with expertise in related fields of the project. The award is not limited to the U.S. – international applicants are welcome. The deadline for applications is December 15, with notice of awards sent by January 15.
Requirements: The award is available to postdoctoral female scientists researching women's health or sex differences whose research involves development, validation, or use of non-animal alternatives. Applicants must hold an interest in using or promoting non-animal alternatives in research. Applicants should send a cover letter explaining their interest in alternative research methods and their career goals, along with their CV, research proposal, and three letters of recommendation (including one from their mentor).
Geographic Focus: All States, All Countries
Date(s) Application is Due: Dec 15
Amount of Grant: 40,000 USD
Contact: Nina Farley; (617) 523-6020; fax (617) 523-7925; nfarley@neavs.org
Internet: http://alternativestoanimalresearch.org/afaar/programs
Sponsor: New England Anti-Vivisection Society
333 Washington Street, Suite 850
Boston, MA 02108

NEDA/AED Charron Family Research Grant   2660
Committed to promoting study into the causes, cures and prevention of eating disorders, NEDA first established the Research Grants Program in 2002 in partnership with the Academy for Eating Disorders (AED). The aim of these grants is to expand eating disorders research while drawing promising new scientists into the field. In 2004 Shelley Charron-Bahniuk died from complications of her eating disorder while she was in recovery. The Charron Family Research Grant was created as a direct result of this illness that impacted both her immediate and extended family.
Requirements: Individuals working in the field of eating disorders who have completed their training, and have the doctorate degree or beyond are eligible.
Geographic Focus: All States
Amount of Grant: 10,000 USD
Contact: Dee Christoff; (206) 382-3587, ext. 15; info@nationaleatingdisorders.org
Internet: http://www.nationaleatingdisorders.org/p.asp?WebPage_ID=712
Sponsor: National Eating Disorders Association / Academy for Eating Disorders
603 Stewart Street, Suite 803
Seattle, WA 98101

NEDA/AED Joan Wismer Research Grant   2661
Committed to promoting study into the causes, cures and prevention of eating disorders, NEDA first established the Research Grants Program in 2002 in partnership with the Academy for Eating Disorders (AED). The aim of these grants is to expand eating disorders research while drawing promising new scientists into the field.
Requirements: Individuals working in the field of eating disorders who have completed their training, and have the doctorate degree or beyond are eligible.

Geographic Focus: All States
Amount of Grant: 10,000 USD
Contact: Dee Christoff; (206) 382-3587, ext. 15; info@nationaleatingdisorders.org
Internet: http://www.nationaleatingdisorders.org/p.asp?WebPage_ID=818
Sponsor: National Eating Disorders Association / Academy for Eating Disorders
603 Stewart Street, Suite 803
Seattle, WA 98101

NEDA/AED Tampa Bay Eating Disorders Task Force Award 2662
The National Eating Disorders Association sponsors a small grants program to support research in the areas of eating disorders etiology, prevention, and treatment. Through the grants, the organization seeks to expand innovative eating disorders research while supporting investigators in the early stages of their careers in the eating disorders field. Support can be included for: salary for technicians, research assistants and other ancillary personnel; laboratory, radiological and pharmaceutical costs in clinical research projects; rating instruments and questionnaires; equipment and supplies; acquisition and maintenance of laboratory animals; subject payment; and advertising.
Requirements: Any junior investigator worldwide who works in the field of eating disorders and has completed their terminal graduate level degree and training is eligible.
Restrictions: Trainees (e.g. students, grad students, residents) are not eligible and investigators must be within seven years following the end of their training. Individuals who previously served or currently serve as a principal investigator on a substantial externally funded grant (e.g. RO1, RO3, or K award in the U.S.) are ineligible. Funds cannot be used for the following purposes: salary support for residents, fellows or other trainees; faculty or investigator salary; secretarial salary and benefits; tuition for residents, graduate students or post-doctoral fellows; publication costs or purchase of reprints or books; room and board for clinical subjects; studies that are also sponsored by pharmaceutical firms; computers; or travel.
Geographic Focus: All States
Amount of Grant: 10,000 USD
Contact: Dee Christoff; (206) 382-3587, ext. 15; info@nationaleatingdisorders.org
Internet: http://www.nationaleatingdisorders.org/p.asp?WebPage_ID=816
Sponsor: National Eating Disorders Association / Academy for Eating Disorders
603 Stewart Street, Suite 803
Seattle, WA 98101

NEI Clinical Study Planning Grant 2663
The National Eye Institute (NEI) supports large-scale clinical vision research projects, including randomized clinical trials and epidemiologic studies. At the time of submission, applications requesting support for these activities are expected to provide detailed information regarding the study's rationale, design, analytic techniques, protocols and procedures, facilities and environment, organizational structure, and collaborative arrangements. This information is best conveyed in a well-documented Manual of Procedures (MOP), the development of which represents a costly and time-consuming activity. The Clinical Study Planning Grant is designed to facilitate activities central to the refinement of a study's protocol and procedures and the development of a detailed MOP.
Requirements: Applications may be submitted by domestic for-profit and nonprofit organizations, both public and private; units of state and local governments; and eligible agencies of the federal government; domestic and foreign institutions.
Restrictions: Foreign institutions are not eligible.
Geographic Focus: All States
Amount of Grant: Up to 150,000 USD
Contact: Donald F. Everett, (301) 451-2020; fax (301) 402-0528; dfe@nei.nih.gov
Internet: http://grants.nih.gov/grants/guide/pa-files/PAR-05-115.html
Sponsor: National Eye Institute
5635 Fishers Lane, Suite 1300, MSC 9300
Bethesda, MD 20892-9300

NEI Clinical Vision Research Development Award 2664
This award is intended to strengthen interactions among clinicians, biostatisticians, epidemiologists, statistical geneticists, and other clinical trial specialists to facilitate the design and conduct of clinical research projects, such as the development of coordinating center capabilities. The intent of this projects is to advance the understanding, prevention, or clinical management of visual system disorders. A number of experienced investigators and research groups are needed to provide leadership in the design and implementation of such clinical vision research projects. Therefore, NEI wishes to help institutions with NEI-funded investigators augment staff expertise and acquire the resources necessary to enhance their clinical vision research programs. PA: PAR-00-050
Requirements: Applications may be submitted by domestic for-profit and non-profit organizations, public and private, such as universities, colleges, hospitals, laboratories, units of State and local governments, and eligible agencies of the Federal government.
Restrictions: Foreign institutions and institutions having current or past NEI Core Grant for Vision Research (P30) support for a biostatistics module are not eligible to apply.
Geographic Focus: All States
Amount of Grant: Up to 100,000 USD
Contact: Mary Frances Cotch, (301) 496-5983; fax (301) 402-0528; mfcotch@nei.nih.gov or deverett@nei.nih.gov
Internet: http://grants.nih.gov/grants/guide/pa-files/PAR-00-050.html
Sponsor: National Eye Institute
5635 Fishers Lane, Suite 1300, MSC 9300
Bethesda, MD 20892-9300

NEI Innovative Patient Outreach Programs & Ocular Screening Technologies To Improve Detection Of Diabetic Retinopathy Grants 2665
This Funding Opportunity Announcement (FOA) solicits Small Business Innovation Research (SBIR) grant applications from small business concerns (SBCs) that propose to 1) develop educational outreach programs to create a greater awareness of the blinding consequences of diabetes; and, 2) develop tools and systems to be used for increasing patient access to eye exams for detecting Diabetic Retinopathy (DR). This FOA will utilize the SBIR (R43/R44) grant mechanisms for Phase I, Phase II, and Fast-Track applications. The estimated amount of funds available for support of 2-4 projects awarded as a result of this announcement is $2 million for this fiscal year.
Requirements: Only United States small business concerns (SBCs) are eligible to submit SBIR applications. A small business concern is one that, at the time of award of Phase I and Phase II, meets all of the criteria listed. Any individual(s) with the skills, knowledge, and resources necessary to carry out the proposed research as the PD/PI is invited to work with his/her organization to develop an application for support. Individuals from underrepresented racial and ethnic groups as well as individuals with disabilities are always encouraged to apply for NIH support. See website for details.
Geographic Focus: All States
Date(s) Application is Due: Mar 23; Dec 23
Amount of Grant: 2,000,000 USD
Contact: Jerome R. Wujek, (301) 451-2020; fax (301) 402-0528; wujekjer@nei.nih.gov
Internet: http://grants.nih.gov/grants/guide/rfa-files/RFA-EY-09-001.html
Sponsor: National Eye Institute
5635 Fishers Lane, Suite 1300, MSC 9300
Bethesda, MD 20892-9300

NEI Mentored Clinical Scientist Development Program Award (K12) 2666
The NEI Mentored Clinician Scientist Development Award (K12) is an award to an educational institution or professional organization to facilitate and support career development experiences which lead clinician scientists to research independence. It is expected that scholars appointed to the program will subsequently apply for their own individual research support. The program should be designed to accommodate candidates with varying levels of research experience who require an individualized career development program ranging from two to five years. Applications for this award should propose a structured, phased development plan overseen by a Program Director and a Program Advisory Committee. A designated period of didactic training should be followed by a period of supervised research experience. Candidates should be exposed to both research that is typically undertaken with a few investigators and to research approaches that require a multidisciplinary team to address complex research questions as appropriate. It is expected that at the end of the career development period, appointees will transition successfully into positions as independent investigators. This award provides five years of potentially renewable support. The subsequent continuation will depend in part on the progress made by the Scholars in achieving independent research support. Renewal applications must contain a progress report documenting the accomplishments and current career activities of former Scholars. PA# PAR-09-083
Requirements: The following organizations and institutions are eligible to apply: public/state controlled institutions of higher education; private institutions of higher education; hispanic-serving Institutions; historically black colleges and universities; tribally controlled colleges and universities; Alaska Native and Native Hawaiian serving institutions; nonprofits with 501(c)3 IRS Status (other than institutions of higher education); nonprofits without 501(c)3 IRS Status (other than institutions of higher education). Any individual with the skills, knowledge, and resources necessary to carry out the proposed research is invited to work with his/her institution to develop an application for support. Individuals from underrepresented racial and ethnic groups as well as individuals with disabilities are always encouraged to apply for NIH support.
Restrictions: Foreign institutions are not eligible to apply.
Geographic Focus: All States
Date(s) Application is Due: May 13
Amount of Grant: 1,125,000 USD
Contact: Neeraj Agarwal, (301) 451-2020; fax (301) 402-0528; agarwalnee@nei.nih.gov
Internet: http://grants.nih.gov/grants/guide/pa-files/PAR-09-083.html
Sponsor: National Eye Institute
5635 Fishers Lane, Suite 1300, MSC 9300
Bethesda, MD 20892-9300

NEI Research Grant For Secondary Data Analysis Grants 2667
The NEI Research Grant for Secondary Data Analysis (R21) program is designed to provide investigators with the support necessary to conduct secondary data analyses utilizing existing database resources. Applications may be related to, but must be distinct from, the specific aims of the original data collection. The total project period for an application submitted in response to this funding opportunity may not exceed two years. Direct costs are limited to $275,000 over an R21 two-year period, with no more than $200,000 in direct costs allowed in any single year. PA # PAR-06-326
Requirements: Eligible organizations: for-profit and non-profit organizations; public and private institutions (such as universities, colleges, hospitals, and laboratories); units of state and local governments; eligible agencies of the federal government; units of state; local tribal governments; domestic and foreign institutions. Eligible Project Directors/Principal Investigators (PD/PIs): any individual with the skills, knowledge, and resources necessary to carry out the proposed research. Individuals from underrepresented racial and ethnic groups as well as individuals with disabilities are always encouraged to apply for NIH support.
Geographic Focus: All States
Date(s) Application is Due: May 8
Amount of Grant: 200,000 - 275,000 USD

Contact: Natalie Kurinij, (301) 451-2020; fax (301) 402-0528; kurinijn@mail.nih.gov
Internet: http://grants.nih.gov/grants/guide/pa-files/PAR-06-326.html
Sponsor: National Eye Institute
5635 Fishers Lane, Suite 1300, MSC 9300
Bethesda, MD 20892-9300

NEI Ruth L. Kirschstein National Research Service Award Short-Term Institutional Research Training Grants 2668

The National Institutes of Health (NIH) will award Ruth L. Kirschstein National Research Service Award (NRSA) Short-Term Institutional Research Training Grants (T35) to eligible institutions to develop or enhance research training opportunities for individuals interested in careers in biomedical, behavioral and clinical research. Many of the NIH Institutes and Centers (ICs) use this grant mechanism exclusively to support intensive, short-term research training experiences for students in health professional schools during the summer. In addition, the Short-Term Institutional Research Training Grant may be used to support other types of predoctoral and postdoctoral training in focused, often emerging scientific areas relevant to the mission of the funding IC. The proposed training must be in either basic, behavioral or clinical research aspects of the health-related sciences. This program is intended to encourage graduate and/or health professional students to pursue research careers by exposure to and short-term involvement in the health- related sciences. The training should be of sufficient depth to enable the trainees, upon completion of the program, to have a thorough exposure to the principles underlying the conduct of research. PA # PA-08-227
Requirements: The following organizations/institutions are eligible to apply: public/state controlled institutions of higher education; private institutions of higher education; hispanic-serving institutions; historically black colleges and universities (HBCUs); tribally controlled colleges and universities (TCCUs); Alaska Native and Native Hawaiian serving institutions; nonprofits with 501(c)3 IRS Status (Other than Institutions of Higher Education); nonprofits without 501(c)3 IRS Status (Other than institutions of higher education); indian/native american tribal governments (federally recognized); indian/native american tribally designated organizations; U.S. Territory or possession; indian/native american tribal governments (other than federally recognized); faith-based or community-based organizations. Only domestic, non-profit, private or public institutions may apply for grants to support National Research Service Award (NRSA) short-term research training programs. Foreign institutions are not eligible to apply. The applicant institution must have a strong and high quality research program in the area(s) proposed for the research training and must have the requisite staff and facilities on site to conduct the proposed research training program. Any individual with the skills, knowledge, and resources necessary to organize and implement a high-quality research training program is invited to work with their institution to develop an application for support. Individuals from underrepresented racial and ethnic groups, individuals with disabilities, and individuals from disadvantaged backgrounds are always encouraged to apply for NIH support.
Restrictions: Foreign institutions are not eligible to apply.
Geographic Focus: All States
Date(s) Application is Due: Sep 8
Contact: Neeraj Agarwal, (301) 451-2020; fax (301) 402-0528; agarwalnee@mail.nih.gov
Internet: http://grants.nih.gov/grants/guide/pa-files/PA-08-227.html
Sponsor: National Eye Institute
5635 Fishers Lane, Suite 1300, MSC 9300
Bethesda, MD 20892-9300

NEI Scholars Program 2669

This program provides an opportunity for outstanding individuals to obtain laboratory or clinical research training within NEI intramural environment and to facilitate the successful transition to continue their research career at an extramural institution as independent vision researchers. This is accomplished by first providing individuals with the necessary resources to receive high quality research training for three to four years at NEI. This is followed by providing extramural funding to support the research program for two years at the extramural institution to which NEI scholar is recruited. It is anticipated that NEI scholars will subsequently compete for independent funding to continue their research. During the intramural phase, the scholar is expected to spend full time on research. During the extramural phase, the scholar must spend a minimum of 75 percent of a full-time professional effort conducting research and engaging in research career development activities for the two years of the award. Applicants are encouraged to contact NEI extramural staff regarding their eligibility for this award and NEI intramural laboratory scientist regarding sponsorship information, prior to submitting an application. PA: PAR-98-107
Requirements: Individuals must have a research or health professional doctoral level degree or its equivalent and must have demonstrated the potential for a highly productive research career during predoctoral training and the immediate postdoctoral period, if applicable. Individuals at NEI or other NIH intramural laboratories who meet the eligibility requirements are eligible to apply. Before submitting an application for the program, the candidate must identify an individual in the NEI Division of Intramural Research who will serve as the sponsor and will be committed to supervise the training and research project. The sponsor must be an active investigator in the area of the proposed research and have research training experience and resources needed to support the NEI scholar.
Restrictions: Individuals who have had more than five years of postdoctoral research training at the time of application are not eligible to apply; however, clinical training does not count against the five years. Former principal investigators on NIH research project grants R01; SBIR/STTR awards R43, R44/R41, R42; subprojects of program project grants P01; center grants P30, P50; K08, K23, K24, or the equivalent are not eligible.
Geographic Focus: All States
Date(s) Application is Due: Feb 1; Jun 1; Oct 1
Amount of Grant: Up to 175,000 USD

Contact: Dr. Maria Giovanni, Division of Extramural Research; (301) 496-0484; fax (301) 402-0528; myg@nei.nih.gov
Internet: http://www.grants.nih.gov/grants/guide/pa-files/PAR-98-107.html
Sponsor: National Eye Institute
5635 Fishers Lane, Suite 1300, MSC 9300
Bethesda, MD 20892-9300

NEI Translational Research Program On Therapy For Visual Disorders 2670

This program continues support of collaborative, multidisciplinary research programs focused on new therapeutic approaches to restore or prevent the loss of function due to visual system diseases and disorders. The rapid and efficient translation of laboratory research findings into clinical development requires a comprehensive and highly integrated approach involving collaborative teams of scientists and clinicians with expertise in multiple disciplines. Such a collaborative approach would be particularly appropriate for research focused on pathways that will likely be targeted by biological intervention, such as gene therapy, cell-based therapy, pharmacological approaches, or development of appropriate delivery systems. The intention of this program is to make resources available to scientists from several disciplines to form research teams to address scientific and technical questions that would be beyond the capabilities of any one research group. It is anticipated that applications funded under this program will lead to therapies that can be tested in a clinical trial.
Requirements: Eligible Institutions, you may submit an application (s) if your organization has any of the following characteristics: for-profit organizations; non-profit organizations; public or private institutions, such as universities, colleges, hospitals, and laboratories; units of State government; units of local government; eligible agencies of the Federal government; domestic Institutions Foreign components to domestic applications will be allowed if the foreign component provides significant scientific input that is critical to the program and for which there is no similar resource in the United States. Eligible Individuals: any individual with the skills, knowledge, and resources necessary to carry out the proposed research is invited to work with their institution to develop an application for support. Individuals from underrepresented racial and ethnic groups as well as individuals with disabilities are always encouraged to apply for NIH programs.
Geographic Focus: All States
Amount of Grant: Up to 2,000,000 USD
Contact: Dr. Peter Dudley, (301) 451-2020; fax (301) 402-0528; pad@nei.nih.gov
Internet: http://grants.nih.gov/grants/guide/pa-files/PAR-05-110.html
Sponsor: National Eye Institute
5635 Fishers Lane, Suite 1300, MSC 9300
Bethesda, MD 20892-9300

NEI Vision Research Core Grants 2671

The primary objective of Core Grants for Vision Research is to provide groups of investigators who have achieved independent National Eye Institute (NEI) funding with additional, shared support to enhance their own and their institution's capability for conducting vision research. Secondary objectives of this program include facilitating collaborative studies and attracting other scientists to research on the visual system.
Requirements: Institutions applying for a Core Grant must hold, on the receipt date, a minimum of eight NEI research project grants (R01), Bioengineering Research Partnership Grants (R24), Mentored Clinical Scientist Development Awards (K08), Mentored Patient-Oriented Research Career Development Award (K23), or Midcareer Investigator Award in Patient-Oriented Research (K24), including any noncompeting extensions of these awards made with or without additional funds. No other mechanisms or source of research support will be considered in determining eligibility. Only one Core Grant will be made to any single applicant organization. For multicampus institutions, no more than one Core Grant will be made to each campus. Joint applications may be submitted by investigators at neighboring, independent institutions.
Geographic Focus: All States
Date(s) Application is Due: Sep 25
Amount of Grant: 2,000,000 - 2,500,000 USD
Contact: Hemin R. Chin, (301) 451-2020; fax (301) 402-0528; hemin@nei.nih.gov
Internet: http://www.nei.nih.gov/funding/p30.asp
Sponsor: National Eye Institute
5635 Fishers Lane, Suite 1300, MSC 9300
Bethesda, MD 20892-9300

Nelda C. and H.J. Lutcher Stark Foundation Grants 2672

The foundation awards grants in its areas of interest, including education, health and social services, medical and dental, community enrichment, arts and culture. Giving is primarily in Texas, with preference to programs that directly impact Southeast Texas and limited giving in Southwest Louisiana. There are no application forms. Submit a letter of proposal containing the following information: description of the project signed by the president or CEO; brief history, description, programs, and mission of the organization; project time line; specific amount requested; list of other funding sources and amounts; list of board of directors and occupations; form 990 for most recent year-end; most recent audited financial statements or organization prepared statements; and current copy of IRS determination letter certifying tax-exempt status. Guidelines are available online. All grants from the third quarter of 2008 and for 2009 have been suspended due to recovery efforts in which the Foundation must engage for its own operations as a result of damages from Hurricane Ike, together with the negative impact of the downturn in financial markets on the Foundation's corpus, the Stark Foundation will continue the prior suspension of the grant-making aspect of its operations through the entirety of Year 2009. Therefore, the Stark Foundation will not accept any grant applications for Year 2009, nor will be able to consider any grant applications that may be submitted notwithstanding this suspension.

Requirements: To be eligible for a grant, an organization must: be exempt from taxation under Section 501(c)3 of the Internal Revenue Code; not be a private foundation within the meaning of Section 509(a) of the Code. No grants are made for endowment purposes, to individuals, or to supporting organizations as described in Section 509(a)(3) of the Internal Revenue Code.
Geographic Focus: All States, Canada
Date(s) Application is Due: Mar 1; Jun 1; Oct 1
Contact: Grant Department
Internet: http://www.starkfoundation.org/Grants/Grant-Guidelines.aspx
Sponsor: Nelda C. and H.J. Lutcher Stark Foundation
P.O. Box 909
Orange, TX 77631-0909

Nell J. Redfield Foundation Grants — 2673

The foundation awards grants to eligible Nevada organizations in its areas of interest, including biomedical research, health care and health organizations, elderly, children with disabilities, education, human services, and religion. Types of support include building construction/renovation, capital campaigns, equipment acquisition, program development, and scholarship funds. Deadlines are at least ten days before a regularly scheduled quarterly meeting of the Board (March, June, September and December). Contact the office to request the application form and guidelines.
Requirements: Nevada nonprofit organizations are eligible with preferences for organizations in northern Nevada.
Geographic Focus: Nevada
Amount of Grant: Up to 100,000 USD
Contact: Gerald C. Smith; (775) 323-1373; redfieldfoundation@yahoo.com
Sponsor: Nell J. Redfield Foundation
P.O. Box 61
Reno, NV 89502

Nesbitt Medical Student Foundation Scholarship — 2674

The Nesbitt Medical Student Foundation provides scholarships to medical students who are in need of financial assistance in order to continue their medical education. These funds have been provided by the trust estate of Esther Mae Nesbit, who wished to assist needy medical students residing in Illinois, especially DeKalb County residents and women. Funds are given to encourage applicants to enter general practice in DeKalb County or in any county in Illinois having a population of less than 50,000 residents. The scholarship covers one academic year and is renewable; a recipient must reapply each year.
Requirements: Applicants must be U.S. citizens, residents of Illinois, and accepted for admission or already full-time students at approved medical schools. Preference is given to women and former residents of DeKalb County.
Geographic Focus: Illinois
Date(s) Application is Due: Jun 1
Amount of Grant: 500 - 2,000 USD
Contact: Darryl Coffman; (815) 754-7718 or (815) 895-2125; fax (815) 895-2575
Internet: https://www.banknbt.com/trust/wealthmanagement.html
Sponsor: Nesbitt Medical Student Foundation
230 West State Street, M-300
Sycamore, IL 60178

Nestle Foundation Large Research Grants — 2675

The Nestlé Foundation supports research in human nutrition with public health relevance in low-income and lower middle-income countries. The results of the research projects should ideally provide a basis for implementation and action which will lead to sustainable effects in the studied populations as generally applicable to the population at large. They should also enable institutional strengthening and capacity building in a sustainable manner in the host country and further cooperation and collaboration between institutions in developed and developing countries. Categories include: training grants, pilot grants, research grants (small and large), and re-entry grants. Large research grants include a full grant application of a complete research proposal. Applications are accepted all year and evaluated twice a year.
Requirements: The Foundation's work is primarily concerned with human nutrition research issues dealing with: maternal and child nutrition, including breastfeeding and complementary feeding; macro- and micronutrient deficiencies and imbalances; interactions between infection and nutrition; and nutrition education and health promotion. Interested scientists should first submit a letter of intent briefly describing the project and the estimated budget. Instructions are available on the Foundation's website. If approved applicants will receive an invitation to submit a full grant proposal. Guidelines and forms are available at the website. The letter of intent and the grant application should include detailed, evidence-based information about the public health relevance of the project as well as its immediate impact and sustainability. Funding is primarily based on the scientific quality, public health relevance in the short and long term, sustainability, capacity-building component, and budget considerations.
Restrictions: The Foundation expects research proposals to be primarily the initiative of local researchers from developing countries but considers applications jointly made by scientists from developed countries, provided it is clear that the initiative will result in capacity building and human resource development, and the bulk of the budget is spent in the developing country.
Geographic Focus: All Countries
Date(s) Application is Due: Jan 10; May 10
Contact: Catherine Loeb; +41 21 320 33 51; fax +41 21 320 33 92; nf@nestlefoundation.org
Internet: http://www.nestlefoundation.org/e/research.html
Sponsor: Nestle Foundation
4 Place de la Gare
Lausanne, CH-1001 Switzerland

Nestle Foundation Pilot Grants — 2676

The Nestlé Foundation supports research in human nutrition with public health relevance in low-income and lower middle-income countries. The results of the research projects should ideally provide a basis for implementation and action which will lead to sustainable effects in the studied populations as generally applicable to the population at large. They should also enable institutional strengthening and capacity building in a sustainable manner in the host country and further cooperation and collaboration between institutions in developed and developing countries. Categories include: training grants, pilot grants, research grants (small and large), and re-entry grants. Pilot grants provide support for research with a high potential to lead to a subsequent full research grant. Additional guidelines are available at the Foundation website. Applications are accepted all year and evaluated twice a year.
Requirements: The Foundation's work is primarily concerned with human nutrition research issues dealing with: maternal and child nutrition, including breastfeeding and complementary feeding; macro- and micronutrient deficiencies and imbalances; interactions between infection and nutrition; and nutrition education and health promotion. Interested scientists should first submit a letter of intent briefly describing the project and the estimated budget. Instructions are available on the Foundation's website. If approved, applicants will receive an invitation to submit a full grant proposal. Guidelines and forms are available at the website. The letter of intent and the grant application should include detailed, evidence-based information about the public health relevance of the project as well as its immediate impact and sustainability. Funding is primarily based on the scientific quality, public health relevance in the short and long term, sustainability, capacity-building component, and budget considerations.
Restrictions: The Foundation expects research proposals to be primarily the initiative of local researchers from developing countries but considers applications jointly made by scientists from developed countries with those from developing countries provided it is clear that the initiative will result in capacity building and human resource development in the latter and the bulk of the budget is spent in the developing country.
Geographic Focus: All Countries
Date(s) Application is Due: Jan 10; May 10
Amount of Grant: Up to 20,000 USD
Contact: Catherine Loeb; +41 21 320 33 51; fax +41 21 320 33 92; nf@nestlefoundation.org
Internet: http://www.nestlefoundation.org/e/research.html
Sponsor: Nestle Foundation
4 Place de la Gare
Lausanne, CH-1001 Switzerland

Nestle Foundation Re-entry Grants — 2677

The Nestlé Foundation supports research in human nutrition with public health relevance in low-income and lower middle-income countries. The results of the research projects should ideally provide a basis for implementation and action which will lead to sustainable effects in the studied populations as generally applicable to the population at large. They should also enable institutional strengthening and capacity building in a sustainable manner in the host country and further cooperation and collaboration between institutions in developed and developing countries. Categories include: training grants, pilot grants, research grants (small and large), and re-entry grants. Re-entry grants encourage the return and re-establishment of post-graduate students into careers in their own countries. Applications are accepted all year and evaluated twice a year.
Requirements: The Foundation's work is primarily concerned with human nutrition research issues dealing with: maternal and child nutrition, including breastfeeding and complementary feeding; macro- and micronutrient deficiencies and imbalances; interactions between infection and nutrition; and nutrition education and health promotion. Interested scientists should first submit a letter of intent briefly describing the project and the estimated budget. Instructions are available on the Foundation's website. If approved applicants will receive an invitation to submit a full grant proposal. Guidelines and forms are available at the website. The letter of intent and the grant application should include detailed, evidence-based information about the public health relevance of the project as well as its immediate impact and sustainability. Funding is primarily based on the scientific quality, public health relevance in the short and long term, sustainability, capacity-building component, and budget considerations.
Restrictions: The host institution will need to guarantee a post for the returnee and ensure career development within the host institution. Contribution of support to the eligible candidate from the host institution is essential, while support and collaboration from the overseas institution where the candidate trained is helpful.
Geographic Focus: All Countries
Date(s) Application is Due: Jan 10; May 10
Amount of Grant: Up to 50,000 USD
Contact: Catherine Loeb; +41 21 320 33 51; fax +41 21 320 33 92; nf@nestlefoundation.org
Internet: http://www.nestlefoundation.org/e/research.html
Sponsor: Nestle Foundation
4 Place de la Gare
Lausanne, CH-1001 Switzerland

Nestle Foundation Small Research Grants — 2678

The Nestlé Foundation supports research in human nutrition with public health relevance in low-income and lower middle-income countries. The results of the research projects should ideally provide a basis for implementation and action which will lead to sustainable effects in the studied populations as generally applicable to the population at large. They should also enable institutional strengthening and capacity building in a sustainable manner in the host country and further cooperation and collaboration between institutions in developed and developing countries. Categories include: training grants, pilot grants, research grants (small and large), and re-entry grants. Small research grants provide

support of a small research study that represents a continuation of a training grant or a pilot grant. Applications are accepted all year and evaluated twice a year.
Requirements: The Foundation's work is primarily concerned with human nutrition research issues dealing with: maternal and child nutrition, including breastfeeding and complementary feeding; macro- and micronutrient deficiencies and imbalances; interactions between infection and nutrition; and nutrition education and health promotion. Interested scientists should first submit a letter of intent briefly describing the project and the estimated budget. Instructions are available on the Foundation's website. If approved applicants will receive an invitation to submit a full grant proposal. Guidelines and forms are available at the website. The letter of intent and the grant application should include detailed, evidence-based information about the public health relevance of the project as well as its immediate impact and sustainability. Funding is primarily based on the scientific quality, public health relevance in the short and long term, sustainability, capacity-building component, and budget considerations.
Restrictions: The Foundation expects research proposals to be primarily the initiative of local researchers from developing countries but considers applications jointly made by scientists from developed countries, provided it is clear that the initiative will result in capacity building and human resource development, and the bulk of the budget is spent in the developing country.
Geographic Focus: All Countries
Date(s) Application is Due: Jan 10; May 10
Contact: Catherine Loeb; +41 21 320 33 51; fax +41 21 320 33 92; nf@nestlefoundation.org
Internet: http://www.nestlefoundation.org/e/research.html
Sponsor: Nestle Foundation
4 Place de la Gare
Lausanne, CH-1001 Switzerland

Nestle Foundation Training Grant 2679
The Nestlé Foundation supports research in human nutrition with public health relevance in low-income and lower middle-income countries. The results of the research projects should ideally provide a basis for implementation and action which will lead to sustainable effects in the studied populations as generally applicable to the population at large. They should also enable institutional strengthening and capacity building in a sustainable manner in the host country and further cooperation and collaboration between institutions in developed and developing countries. Categories include: training grants, pilot grants, research grants (small and large), and re-entry grants. Training grants support a small research project such as a master's degree in science, Ph.D. thesis project, or a training endeavor. Applications are accepted all year and evaluated twice a year.
Requirements: The Foundation's work is primarily concerned with human nutrition research issues dealing with: maternal and child nutrition, including breastfeeding and complementary feeding; macro- and micronutrient deficiencies and imbalances; interactions between infection and nutrition; and nutrition education and health promotion. Interested scientists should first submit a letter of intent briefly describing the project and the estimated budget. Instructions are available on the Foundation's website. If approved applicants will receive an invitation to submit a full grant proposal. Guidelines and forms are available on our website. The letter of intent and the grant application should include detailed, evidence-based information about the public health relevance of the project as well as its immediate impact and sustainability. Funding is primarily based on the scientific quality, public health relevance in the short and long term, sustainability, capacity-building component, and budget considerations.
Restrictions: The Foundation expects research proposals to be primarily the initiative of local researchers from developing countries but considers applications jointly made by scientists from developed countries with those from developing countries provided it is clear that the initiative will result in capacity building and human resource development in the latter and the bulk of the budget is spent in the developing country.
Geographic Focus: All Countries
Date(s) Application is Due: Jan 10; May 10
Contact: Catherine Loeb; +41 21 320 33 51; fax +41 21 320 33 92; nf@nestlefoundation.org
Internet: http://www.nestlefoundation.org/e/research.html
Sponsor: Nestle Foundation
4 Place de la Gare
Lausanne, CH-1001 Switzerland

New Jersey Osteopathic Education Foundation Scholarships 2680
Approximately five scholarships are awarded each year to deserving students entering their first year in an osteopathic college. They must be residents of New Jersey and have completed four years of pre-medical education. Selection of awardees is based on undergraduate academic achievement, financial need, motivation, and professional promise as an osteopathic physician.
Requirements: Student must be a resident of the state of New Jersey, must be osteopathically oriented, and must be a student member of the American Osteopathic Association and the New Jersey Association of Osteopathic Physicians and Surgeons throughout his/her medical education. Applicants should have a B average or a 3.0 grade point average or better on a 4.0 scale and be in the upper 25 percent in class standing. The required forms can be downloaded from the sponsor's website.
Geographic Focus: New Jersey
Date(s) Application is Due: Apr 30
Amount of Grant: 5,000 USD
Contact: Robert Bowen; (732) 940-9000; fax (732) 940-8899; njaops@njosteo.com
Internet: http://www.njosteo.com/displaycommon.cfm?an=1&subarticlenbr=68
Sponsor: New Jersey Osteopathic Education Foundation
One Distribution Way, Suite 201
Monmouth Junction, NJ 08852-3001

Newton County Community Foundation Grants 2681
The purpose of the Foundations is to assist donors in creating a source of assets to meet the ongoing needs and interests of the people living in Jasper and Newton County, Indiana communities. The Foundation welcomes grant applications from non-profit organizations whose programs benefit the residents of each county. Grant applications are available in the Foundation office or online at the Foundation's website.
Requirements: Favored grant requests for each county will generally reflect the following characteristics: projects which propose practical solutions to community needs; projects which promote cooperation among existing agencies without duplicating services; and projects that will become self-sustaining without requiring on-going Foundation funds. No restrictions are placed on gifts to these funds, allowing the Board of Directors to address community needs.
Restrictions: Grant are not normally made for budget deficits or sectarian religious activities.
Geographic Focus: Indiana
Date(s) Application is Due: Apr 1; Oct 1
Samples: Morocco Lions Club, Morocco, Indiana, $1,500 - funding for improvement of the Morocco Lions Club den and purchase of kitchen equipment; Newton/Jasper Community Band, $3,000 - funds to help purchase a tuba that will enable the tuba section to function on a higher level
Contact: Linda Harris, Chairperson; (219) 866-5899; fax (219) 866-0555; jasper@liljasper.com
Internet: http://www.jasperfdn.org/newtgrants.htm
Sponsor: Newton County Community Foundation
301 N. Van Rensselaer Street, P.O. Box 295
Rensselaer, IN 47978

Newton County Community Foundation Lilly Scholarships 2682
The Lilly Endowment Community Scholarships will provide full tuition, required fees, and a special allocation of up to $800 per year for required books and equipment for four years of undergraduate study leading to a baccalaureate degree at any Indiana public or private college or university. Any college major may apply.
Requirements: Any graduating senior who is a U.S. citizen and resident of Jasper or Newton County is eligible. The student must be an applicant for enrollment to an accredited Indiana university or college and pursuing a four year full time field of study. The student must demonstrate good citizenship qualities, high moral character, and be a positive role model to others. Also, he/she must have strong leadership skills and demonstrate their ability to be a good ambassador for the community. Along with the application, candidates must submit three letters of reference, and a privacy release statement. Candidates must then submit the application to their high school guidance counselor for scholastic information and their transcript, then forward the complete information to the Jasper Foundation for review. Candidates can contact the Jasper Foundation office for further information.
Geographic Focus: Indiana
Date(s) Application is Due: Jan 11
Contact: Rhonda Churchill; (219) 866-5899; fax (219) 866-0555; jasper@liljasper.com
Internet: http://www.jasperfdn.org/
Sponsor: Newton County Community Foundation
301 N. Van Rensselaer Street, P.O. Box 295
Rensselaer, IN 47978

Newton County Community Foundation Scholarships 2683
The purpose of the Jasper Foundation is to assist donors in creating a source of assets to meet the ongoing and changing charitable needs and interests of the people living in Jasper and Newton County, Indiana communities. The Jasper Foundation, Inc. and the Newton County Community Foundation support education as a lifelong pursuit. Through the generosity of their contributors, the Foundations have been able to provide financial assistance to many young people in Jasper and Newton counties seeking a post-secondary education. Scholarships are awarded by a separate advisory committee, either designated by the donor or Foundation committee. Selections are based upon established criteria and guidelines which vary with each scholarship. See website for further details and applications for each program offered.
Geographic Focus: Indiana
Date(s) Application is Due: Jan 8
Contact: Rhonda Churchill, Chairperson; (219) 866-5899; fax (219) 866-0555
Internet: http://www.jasperfdn.org/
Sponsor: Newton County Community Foundation
301 N. Van Rensselaer Street, P.O. Box 295
Rensselaer, IN 47978

New York University Steinhardt School of Education Fellowships 2684
The program awards fellowships for doctoral study in the field of education. Named fellowships include Founders Fellowships, Phyllis and Gerald LeBoff Fellowship in Media Ecology, and the Steinhardt Fellowship in Education and Jewish Studies. Fellowships provide up to three years of full-tuition support and annual stipends. Faculty in relevant disciplines are also invited to nominate qualified students for these awards for doctoral study in education, applied psychology, health, nursing, art, music, and media studies.
Requirements: Successful candidates are expected to have a history of collaborative and scholarly work with faculty, strong academic records, and an ability to draw on the resources of their program's faculty at the university.
Geographic Focus: All States
Date(s) Application is Due: Jan 15
Amount of Grant: 10,000 - 18,000 USD
Contact: Administrator; (212) 998-5030; ed.gradadmissions@nyu.edu

Internet: http://www.education.nyu.edu
Sponsor: New York University
82 Washington Square E, 2nd Floor
New York, NY 10003-6680

NFL Charities Medical Grants 2685
NFL Charities supports medical research and enhancing scientific knowledge in order to benefit all who are actively involved in competitive sports and recreational athletic activities. These grants support research that addresses some of the risk factors that exist for all athletes and citizens with active lifestyles. The NFL Charities Board considers Medical Grant proposals that place emphasis on scientific merit, clinical relevance and significance to the NFL, especially projects concerning concussion and traumatic brain injury, cardiovascular research and MRSA. Grants will be available in the end of June or early July and will be open for approximately 8 weeks.
Requirements: Funding is available for non-profit educational and research institutions within the United States. Applications must be submitted online via the NFL Charities Grant Application Management System (G.A.M.S.) website. Every field in the application must be populated with requested information. Indirect costs should not be included in the project's total budget, and only direct research costs are funded by the grant. Salaries for Principal Investigators (PI's) and Co-Principal Investigators (Co-PI's) (Ph.D's and M.D.'s) must follow NIH guidelines. PI's and Co-PI's must disclose all previous and/or current grant support and any potential overlap must be submitted. Postdoctoral fellows serving as PI or Co-PI must include assurances that they will remain at the same institution for the duration of the grant period; however, PI's and co-PI's may work at different institutions, but the grant must be submitted as associated with the institution of the primary PI.
Restrictions: For-profit groups are not eligible for funding. Only submission per PI or Co-PI will be funded, and only one submission per institution can be considered for NFL Charities Medical Grant funding. Prospective applicants first should be vetted through their institutional administrative offices in order to ensure that an institution only endorses and submits its top two proposals. Administrative contacts at the associated institution must sign off on the Compliance Certificate that accompanies each application(s) from a PI's institution, thereby ensuring that such vetting has occurred.
Geographic Focus: All States
Date(s) Application is Due: Jun 1
Amount of Grant: Up to 100,000 USD
Contact: Clare Graff, (212) 450-2435 or (917) 816-2885; clare.graff@nfl.com
Internet: http://www.nflcharities.org/grants/medical
Sponsor: NFL Charities
280 Park Avenue, 17th Floor
New York, NY 10017

NHGRI Mentored Research Scientist Development Award 2686
These awards are intended to foster the career development of individuals with expertise in scientific disciplines that would further technological developments critical to the success of the Human Genome Project and the understanding of the genetic basis of diseases. Up to three scientists may be selected and appointed to this program by the grantee institutions. Up to the Federal salary cap, plus fringe benefits, is available. Up to $20,000 of research support per year is also available.
Requirements: Eligibility is limited to individuals with degrees in computer sciences, mathematics, chemistry, engineering, physics, and closely related scientific disciplines, such as bioinformatics, computational biology, statistics, biomathematics, bioengineering. Applications will be accepted from domestic, nonfederal organizations of higher education that have strong, well-established research and training programs. The following organizations and institutions may apply: public/state controlled institutions of higher education; private institutions of higher education; Hispanic-serving institutions; historically Black colleges and universities (HBCUs); tribally controlled colleges and universities (TCCUs); Alaska native and native Hawaiian serving institutions; nonprofits with 501(c)3 IRS status; nonprofits without 501(c)3 IRS status; small businesses; for-profit organizations; State governments, regional organizations; U.S. territories or possessions; Indian/Native American tribal governments; Indian/Native American tribally designated organizations; eligible agencies of the Federal government; and faith-based or community based organizations.
Geographic Focus: All States
Date(s) Application is Due: Feb 12; Mar 12; Jun 12; Jul 12; Oct 12; Nov 12
Contact: Bettie J. Graham, (301) 496-7531; bettie_graham@nih.gov
Internet: http://grants.nih.gov/grants/guide/pa-files/PAR-98-062.html
Sponsor: National Human Genome Research Institute
5635 Fishers Lane, Suite 4076
Rockville, MD 20852

NHLBI Airway Smooth Muscle Function and Targeted Therapeutics in Human Asthma Grants 2687
The purpose of this funding opportunity announcement, issued by the NHLBI, NIH, is to solicit Research Project Grant applications from institutions and organizations that propose to investigate the complex role of airway smooth muscle (ASM) functions in the development of human asthma and identify novel therapeutic targets. Budgets for direct costs of up to $350,000 per year (exclusive of indirect costs associated with consortia) and a project duration of up to four years may be requested for a maximum of $1,400,000 direct costs over a four-year project period. Letters of Intent should be received by December 8, and full proposals are due by January 6.
Requirements: The following organizations and institutions are eligible to apply: public/state controlled institutions of higher education; private institutions of higher education; Hispanic-serving institutions; historically Black colleges and universities (HBCUs); tribally controlled colleges and universities (TCCUs); Alaska Native and Native Hawaiian serving institutions; nonprofits with 501(c)3 IRS status; nonprofits without 501(c)3 IRS status; for-profit organizations; state governments; Indian/Native American tribal governments (federally recognized); Indian/Native American tribally designated organizations; county governments; city or township governments; special district governments; independent school districts; public housing authorities/Indian housing authorities; U.S. territory or possession; Indian/Native American tribal governments (other than federally recognized); regional organizations; non-domestic (non-U.S.) entities (foreign organizations); other eligible agencies of the federal government; and faith-based or community-based organizations.
Geographic Focus: All States
Date(s) Application is Due: Jan 6; Dec 8
Amount of Grant: Up to 350,000 USD
Contact: Susan Banks-Schlegel, (301) 435-0202; Schleges@nhlbi.nih.gov
Internet: http://grants1.nih.gov/grants/guide/rfa-files/RFA-HL-09-007.html
Sponsor: National Institutes of Health
Building 31, Room 5A48
Bethesda, MD 20892

NHLBI Ancillary Studies in Clinical Trials 2688
This funding opportunity invites research grant applications to conduct time-sensitive ancillary studies related to heart, lung, and blood diseases and sleep disorders in conjunction with ongoing clinical trials and other large clinical studies supported by NIH or non-NIH entities. The program establishes an accelerated review/award process to support the crucial time frame in which these ancillary studies must be performed. Time-sensitive ancillary studies include those that require active longitudinal data collection and thus need to begin recruiting subjects as close as possible to the start of the parent study. The ancillary study can address any research questions related to the mission of NHLBI for which the parent study can provide participants, infrastructure, and data. The parent studies most often will be a clinical trial, but also can be an observational study or registry that can provide a sufficient cohort of well-characterized patients. Each ancillary study application must demonstrate the time-sensitive nature of the application and must explicitly address why an expedited review is essential to its feasibility. Applications are due May 24, September 24, and January 24 by 5:00 pm local time of applicant.
Requirements: The following may apply: public/state controlled and private institutions of higher education; nonprofits with or without 501(c)3 IRS status (other than institutions of higher education); small businesses; for-profit organizations (other than small businesses); state, county, city or township governments; special district governments; Indian/Native American Tribal governments; eligible agencies of the Federal government; U.S. territory or possession; independent school districts; public housing authorities/Indian housing authorities; faith-based or community-based organizations; regional organizations; non-domestic (non-U.S.) entities (Foreign Institutions). The following types of Higher Education Institutions are always encouraged to apply for NIH support as public or private institutions of higher education: Hispanic-serving institutions; historically Black colleges and universities; Tribally controlled colleges and universities; Alaska Native and Native Hawaiian serving institutions. Non-domestic (non-U.S.) components of U.S. Organizations are eligible to apply. Any individual(s) with the skills, knowledge, and resources necessary to carry out the proposed research as the Program Director/Principal Investigator (PD/PI) is invited to work with his/her organization to develop an application for support. Individuals from underrepresented racial and ethnic groups as well as individuals with disabilities are always encouraged to apply for NIH support. An application may request a budget for direct costs up to $250,000 (10 modules) per year, excluding subcontractor or consortium facilities and administrative (F&A) costs. The scope of the proposed project should determine the project period. The maximum project period is four years.
Geographic Focus: All States, Guam, Marshall Islands, Northern Mariana Islands, Puerto Rico, U.S. Virgin Islands, All Countries, American Samoa
Date(s) Application is Due: Jan 24; May 24; Sep 24
Contact: Suzanne Goldberg, R.N., M.S.N., (301) 435-0532; fax (301) 480-3667; goldbergsh@mail.nih.gov
Internet: http://grants.nih.gov/grants/guide/rfa-files/RFA-HL-14-004.html
Sponsor: National Institutes of Health
Building 31, Room 5A48
Bethesda, MD 20892

NHLBI Bioengineering and Obesity Grants 2689
This funding opportunity solicits Research Project Grant (R01) applications from institutions/organizations that propose to solicit applications to develop and validate new and innovative engineering approaches to address clinical problems related to energy balance, intake, and expenditure. Novel sensors, devices, imaging, and other technologies, including technologies to detect biochemical markers of energy balance are expected to be developed and evaluated by collaborating engineers, physical scientists, mathematicians, and scientists from other relevant disciplines with expertise in obesity and nutrition. Because the nature and scope of the proposed research will vary from application to application, it is anticipated that the size and duration of each award will also vary. The total amount awarded and the number of awards will depend upon the mechanism numbers, quality, duration, and costs of the applications received.
Requirements: The following are eligible to apply: public/state controlled and private institutions of higher education; nonprofit with or without 501(c)3 IRS status (other than institution of higher education); small business; for-profit organization (other than small business); state government; U.S. Territory or Possession; Indian/Native American tribal government (federally recognized); Indian/Native American tribal government (other than federally recognized); Indian/Native American tribally designated organization;

Non-domestic (non-U.S.) entity (foreign organization); Hispanic-serving institution; historically Black colleges and universities (HBCUs); tribally controlled colleges and universities (TCCUs); Alaska Native and Native Hawaiian serving institutions; regional organizations. Individuals with the skills, knowledge, and resources necessary to carry out the proposed research are invited to work with their institution/organization to develop an application for support. Individuals from underrepresented racial and ethnic groups as well as individuals with disabilities are always encouraged to apply for NIH support. More than one PD/PI, or multiple PDs/PIs, may be designated on the application. Applicants may submit more than one application, provided each application is scientifically distinct.
Geographic Focus: All States, Guam, Marshall Islands, Northern Mariana Islands, Puerto Rico, U.S. Virgin Islands, American Samoa
Date(s) Application is Due: Jan 25
Contact: Abby G. Ershow, (301) 435-0550; fax (301) 480-2858; ErshowA@mail.nih.gov
Internet: http://grants.nih.gov/grants/guide/pa-files/PA-07-354.html
Sponsor: National Institutes of Health
Building 31, Room 5A48
Bethesda, MD 20892

NHLBI Bioengineering Approaches to Energy Balance and Obesity Grants for SBIR 2690

This funding opportunity will develop and validate new and innovative bioengineering technology to address clinical problems related to energy balance, intake, and expenditure. Novel sensors, devices, imaging, and other approaches are expected to be developed and evaluated by collaborating engineers, physical scientists, and scientists from other relevant disciplines with expertise in obesity and nutrition. The goal is to increase the number of useful technologies and tools available to scientists to facilitate their research in energy balance and health. Eventually these research tools should facilitate therapeutic advances and behavioral changes to address such problems as weight control and obesity. The total amount awarded and the number of awards will depend upon the mechanism numbers, quality, duration, and costs of the applications received. This FOA will utilize the SBIR (R43/R44) grant mechanisms for Phase I, Phase II, and Fast-Track applications and runs in parallel with a FOA of identical scientific scope, PA-07-436, that solicits applications under the Small Business Technology Transfer (STTR) (R41/R42) grant mechanisms. The total amount awarded and the number of awards will depend upon the mechanism numbers, quality, duration, and costs of the applications received. Program Announcement (PA) Number: PA-07-435.
Requirements: Only United States small business concerns (SBCs) are eligible to submit SBIR applications. A small business concern is one that, at the time of award for both Phase I and Phase II SBIR awards, meets all of the following criteria: (1) is independently owned and operated, is not dominant in the field of operation in which it is proposing, has a place of business in the United States and operates primarily within the United States or makes a significant contribution to the U.S. economy, and is organized for profit; (2) is (a) at least 51% owned and controlled by one or more individuals who are citizens of, or permanent resident aliens in, the United States, or (b) for SBIR only, it must be a for-profit business concern that is at least 51% owned and controlled by another for-profit business concern that is at least 51% owned and controlled by one or more individuals who are citizens of, or permanent resident aliens in, the United States; (3) has, including its affiliates, an average number of employees for the preceding 12 months not exceeding. Any individual with the skills, knowledge, and resources necessary to carry out the proposed research is invited to work with his/her organization to develop an application for support. Individuals from underrepresented racial and ethnic groups as well as individuals with disabilities are always encouraged to apply for NIH support.
Geographic Focus: All States
Date(s) Application is Due: Apr 5; Aug 5; Dec 5
Contact: J. Timothy Baldwin, Ph.D., Biomedical Engineer; (301) 435-0513; fax (301) 480-1335; BaldwinT@mail.nih.gov
Internet: http://grants.nih.gov/grants/guide/pa-files/PA-07-435.html
Sponsor: National Institutes of Health
Building 31, Room 5A48
Bethesda, MD 20892

NHLBI Bioengineering Approaches to Energy Balance and Obesity Grants for STTR 2691

This funding opportunity will develop and validate new and innovative bioengineering technology to address clinical problems related to energy balance, intake, and expenditure. Novel sensors, devices, imaging, and other approaches are expected to be developed and evaluated by collaborating engineers, physical scientists, and scientists from other relevant disciplines with expertise in obesity and nutrition. The goal is to increase the number of useful technologies and tools available to scientists to facilitate their research in energy balance and health. Eventually these research tools should facilitate therapeutic advances and behavioral changes to address such problems as weight control and obesity. The total amount awarded and the number of awards will depend upon the mechanism numbers, quality, duration, and costs of the applications received. Program Announcement (PA) Number: PA-07-436.
Requirements: Only United States small business concerns (SBCs) are eligible to submit STTR applications. Individuals with the skills, knowledge, and resources necessary to carry out the proposed research are invited to work with their organization to develop an application for support. Individuals from underrepresented racial and ethnic groups as well as individuals with disabilities are always encouraged to apply for NIH support. On an STTR application, the PD/PI may be employed with the SBC or the participating non-profit research institution as long as he/she has a formal appointment with or commitment to the applicant SBC, which is characterized by an official relationship between the small business concern and that individual. Applicant SBCs may submit more than one application, provided each application is scientifically distinct. More than one PD/PI, or multiple PDs/PIs, may be designated on the application.
Geographic Focus: All States
Date(s) Application is Due: Apr 5; Aug 5; Dec 5
Contact: Abby G. Ershow, (301) 435-0550; fax (301) 480-1338; ErshowA@mail.nih.gov
Internet: http://grants.nih.gov/grants/guide/pa-files/PA-07-436.html
Sponsor: National Institutes of Health
Building 31, Room 5A48
Bethesda, MD 20892

NHLBI Biomedical Research Training Program for Individuals from Underrepresented Groups 2692

The NHLBI established the Biomedical Research Training Program for Individuals from Underrepresented Groups (BRTPUG) to offer opportunities for underrepresented post-baccalaureate individuals to receive training in basic, translational, and clinical research. The purpose of the program is to enhance career opportunities in biomedical sciences, including clinical and laboratory medicine, epidemiology, and biostatistics as applied to the etiology and treatment of heart, blood vessel, lung, and blood diseases. BRTPUG offers each participant the opportunity to work closely with leading research scientists in the Division of Intramural Research and extramural scientists in the Division of Cardiovascular Sciences -Prevention and Population Sciences Program. The program provides participants with hands-on training in a research environment, which will prepare them to continue their studies and advance their careers in clinical and basic research. The trainee's research internship is a one-time appointment of 1-2 year's beginning the summer of selection. Each trainee is assigned to a mentor who is responsible for designing a carefully planned training program.
Requirements: The program supports recently completed post-baccalaureates, who have completed academic training in course work relevant to biomedical, behavioral, or statistical research. Applicants must have a cumulative grade point average (GPA) or science course GPA of 3.3 or better on a 4.0 scale, or 4.3 or better on a 5.0 scale, and be a U.S. citizen or permanent resident. Underrepresented post-baccalaureate individuals planning to pursue advanced degrees in the biomedical sciences, who are planning to apply to graduate or professional (medical/dental/veterinary, pharmacy) school are eligible to apply.
Geographic Focus: All States
Date(s) Application is Due: Jan 31
Contact: Dr. Helena O. Mishoe, (301) 451-5081; mishoeh@nhlbi.nih.gov
Internet: http://www.nhlbi.nih.gov/funding/training/redbook/pbbrtu.htm
Sponsor: National Institutes of Health
Building 31, Room 5A48
Bethesda, MD 20892

NHLBI Cardiac Development Consortium Grants 2693

The National Heart, Lung, and Blood Institute (NHLBI) and the Canadian Institutes of Health Research (CIHR) invite applications to participate in a new NHLBI Cardiac Development Consortium, a cooperative investigative group that will drive an integrated approach to the investigation of cardiovascular development. The purpose of this initiative is to support basic collaborative research leading to a comprehensive understanding of the regulatory networks controlling cardiovascular development. The consortium will consist of up to four Research Centers, scientific Cores as necessary, and a Steering Committee. Research Center applicants will propose projects that will select key regulatory pathways, identify the components of the pathways and targets, and rapidly disseminate data to the scientific community. The Research Centers will collaborate to maximize interoperability of the data. This Consortium is part of a new NHLBI translational program in pediatric cardiovascular disease, which includes the companion NHLBI Pediatric Cardiac Genomics Consortium and Administrative Coordinating Center. These consortia will interact with each other and the NHLBI Pediatric Heart Network to encourage translation of results from basic science to clinical research, and to provide clinical input on pressing needs for basic research. Maximum allowable direct costs for the research centers are $160,000 in the Year 1 planning phase, and up to $1,070,000 per year in years two through six. Letters of Intent should be received by January 6, and full proposals are due by February 6.
Requirements: The following organizations and institutions are eligible to apply: public/state controlled institutions of higher education; private institutions of higher education; Hispanic-serving institutions; historically Black colleges and universities (HBCUs); tribally controlled colleges and universities (TCCUs); Alaska Native and Native Hawaiian serving institutions; nonprofits with 501(c)3 IRS status; nonprofits without 501(c)3 IRS status; small businesses; for-profit organizations; state governments; and non-domestic (non-U.S.) entities in North America.
Geographic Focus: All States
Date(s) Application is Due: Jan 6; Feb 6
Amount of Grant: Up to 5,510,000 USD
Contact: Charlene Schramm, (301) 435-0510; SchrammC@nhlbi.nih.gov
Internet: http://grants.nih.gov/grants/guide/rfa-files/RFA-HL-09-002.html
Sponsor: National Institutes of Health
Building 31, Room 5A48
Bethesda, MD 20892

NHLBI Career Transition Awards 2694

The purpose of the NHLBI Career Transition Award (K22) program is to provide highly qualified postdoctoral fellows with an opportunity to receive mentored research experience in the NHLBI Division of Intramural Research and then to provide them with funding to facilitate the transition of their research programs as new investigators to extramural institutions. To achieve these objectives, the NHLBI Career Transition

Award will support two phases of research: a mentored intramural phase (two years) and an extramural phase (three years), for a total of five years of combined support. Transition from the intramural phase of support to the extramural phase is not automatic. Approval of the transition will be based on the success of the awardees research program as determined by an NHLBI progress review, which will include an evaluation of a research plan to be carried out at the extramural institution. The total period of support is five years (two years intramural, plus three years extramural). Awards are not renewable. Applications are due by 5:00 pm local time of the applicant organization (see dates below).
Requirements: The following organizations are eligible to apply: public/state controlled and private institutions of higher education; Hispanic-serving institutions; historically black colleges and universities; tribally controlled colleges and universities; Alaska Native and Native Hawaiian serving institutions; nonprofits with or without 501(c)3 IRS status (other than institutions of higher education); small businesses; for-profit organizations (other than small businesses); state, county, and city/township governments; special district governments; Indian/Native American tribal governments (federally recognized and other than federally recognized); eligible agencies of the Federal government; U.S. territory or possession; independent school districts; public or Indiana housing authorities; Native American tribal organizations (other than federally recognized tribal governments); faith-based or community-based organizations; and regional organizations. Applicant organizations may submit more than one application, provided that each application is scientifically distinct. By the time of award, the individual must be a citizen or a non-citizen national of the United States or have been lawfully admitted for permanent residence. Candidates for this award must have earned a terminal clinical or research doctorate (including PhD, MD, DO, D.C., ND, DDS, DVM, ScD, DNS, PharmD., or equivalent doctoral degree), or a combined clinical and research doctoral degree. The candidate must have postdoctoral research experience, during which the potential for highly productive basic or clinical research was demonstrated. Individuals from underrepresented racial and ethnic groups as well as individuals with disabilities are always encouraged to apply for NIH support. During the intramural phase of the award, the candidate will spend full time on research. The required research experience must be completed in an intramural NIH laboratory. To obtain support for the extramural phase, at the time of the award candidates must have a full time formal tenure-track (or equivalent) appointment offer at the academic institution that is the applicant institution. Candidates who have VA appointments may not consider part of the VA effort toward satisfying the full time requirement at the applicant institution. Candidates with VA appointments should contact the NHLBI staff prior to preparing an application to discuss their eligibility. Support for the extramural phase will be provided to the extramural institution. Total direct costs for the extramural phase cannot exceed $249,000, including fringe benefits, per year. The total costs cannot exceed $747,000 for the three-year period.
Restrictions: NIH will not accept any application that is essentially the same as one already reviewed. An individual may not have two or more competing NIH career development applications pending review concurrently. Non-domestic (non-U.S.) entities (foreign institutions) and non-domestic (non-U.S.) components of U.S. organizations are not eligible to apply. Multiple Principal Investigators are not allowed.
Geographic Focus: All States, Guam, Marshall Islands, Northern Mariana Islands, Puerto Rico, U.S. Virgin Islands, American Samoa
Date(s) Application is Due: Feb 12; Jun 11; Oct 12
Amount of Grant: Up to 747,000 USD
Contact: Herbert M. Geller, Ph.D., Director, Office of Education; (301) 451-9440; fax (301) 594-8133; direducation@nhlbi.nih.gov
Internet: http://grants.nih.gov/grants/guide/pa-files/PAR-12-137.html
Sponsor: National Institutes of Health
Building 31, Room 5A48
Bethesda, MD 20892

NHLBI Characterizing the Blood Stem Cell Niche Grants   2695
This FOA issued by the National Heart, Lung, and Blood Institute, National Institutes of Health, solicits Research Project Grant applications in a specific area of stem cell research, the blood stem cell niche, an area critical to advancing stem cell biology and its applications to cellular therapeutics including hematopoietic stem cell transplantation. This FOA is being initiated to foster collaborative research projects and insightful approaches to dissect the cellular components and factors involved in the hematopoietic stem cell niche. Development of conditional genetic knock-out models to test the role of factors in specific cell lineages and imaging technology to facilitate following stem cell engraftment in the niche in vivo are integral to this initiative. An individual research grant application may not request more than $250,000 direct cost a year unless collaborative in vivo cell imaging studies are included, in which case up to $350,000 direct cost may be requested in a year. An applicant may request a project period of up to 4 years. Letters of Intent should be received by December 8, and full proposals are due by January 6.
Requirements: The following organizations and institutions are eligible to apply: public/state controlled institutions of higher education; private institutions of higher education; Hispanic-serving institutions; historically Black colleges and universities (HBCUs); tribally controlled colleges and universities (TCCUs); Alaska Native and Native Hawaiian serving institutions; nonprofits with 501(c)3 IRS status; nonprofits without 501(c)3 IRS status; for-profit organizations; state governments; Indian/Native American tribal governments (federally recognized); Indian/Native American tribally designated organizations; county governments; city or township governments; special district governments; independent school districts; public housing authorities/Indian housing authorities; U.S. territory or possession; Indian/Native American tribal governments (other than federally recognized); regional organizations; non-domestic (non-U.S.) entities (foreign organizations); other eligible agencies of the federal government; and faith-based or community-based organizations.
Geographic Focus: All States
Date(s) Application is Due: Jan 6; Dec 8
Amount of Grant: 250,000 - 350,000 USD
Contact: John Thomas, (301) 435-0065; fax (301) 451-5453; ThomasJ@NHLBI.NIH.Gov
Internet: http://grants.nih.gov/grants/guide/rfa-files/RFA-HL-09-010.html
Sponsor: National Institutes of Health
Building 31, Room 5A48
Bethesda, MD 20892

NHLBI Circadian-Coupled Cellular Function in Heart, Lung, and Blood Tissue Grants   2696
The purpose of this FOA is to stimulate Phase I translation (T1) of fundamental genome-based circadian biology to improve our understanding of heart, lung, and blood disease pathogenesis in non-neural tissues. For the purpose of this initiative, T1 research starts after gene discovery and has as its goal the development of molecular models with application to an improved understanding of human health and disease pathogenesis, predictive testing for disease susceptibility, prognostic testing, or the selection of therapeutic strategies. Budget for direct costs up to $325,000 per year for project duration of up to four years may be requested. Letters of Intent should be received by December 8, and full proposals are due by January 6.
Requirements: The following organizations and institutions are eligible to apply: public/state controlled institutions of higher education; private institutions of higher education; Hispanic-serving institutions; historically Black colleges and universities (HBCUs); tribally controlled colleges and universities (TCCUs); Alaska Native and Native Hawaiian serving institutions; nonprofits with 501(c)3 IRS status; nonprofits without 501(c)3 IRS status; for-profit organizations; state governments; Indian/Native American tribal governments (federally recognized); Indian/Native American tribally designated organizations; county governments; city or township governments; special district governments; independent school districts; public housing authorities/Indian housing authorities; U.S. territory or possession; Indian/Native American tribal governments (other than federally recognized); regional organizations; non-domestic (non-U.S.) entities (foreign organizations); other eligible agencies of the federal government; and faith-based or community-based organizations.
Geographic Focus: All States
Date(s) Application is Due: Jan 6; Dec 8
Amount of Grant: Up to 325,000 USD
Contact: Aaron Laposky; (301) 435-0199; fax (301) 480-3451; laposkya@nhlbi.nih.gov
Internet: http://grants.nih.gov/grants/guide/rfa-files/RFA-HL-09-012.html
Sponsor: National Institutes of Health
Building 31, Room 5A48
Bethesda, MD 20892

NHLBI Clinical Centers for the NHLBI Asthma Network Grants   2697
The purpose of this funding opportunity, issued by NHLBI, NIH, is to invite applications to participate in the NHLBI Asthma Network (AsthmaNet), a clinical research network that will develop and conduct multiple clinical trials to address the most important asthma management questions and new treatment approaches in pediatric and adult populations. AsthmaNet is designed to promote cooperation and coordination, facilitate scientific exchange, provides training opportunities, and leverage resources. AsthmaNet will include multiple Clinical Centers and one Data Coordinating Center. The protocols will include clinical trials to evaluate and/or compare existing or new therapeutic approaches to asthma management as well as a limited number of proof-of-concept studies to advance the development of novel therapies and studies that investigate the mechanistic bases for interventions examined in AsthmaNet. Letters of Intent should be received by December 30, and full proposals are due by January 30.
Requirements: The following organizations and institutions are eligible to apply: public/state controlled institutions of higher education; private institutions of higher education; Hispanic-serving institutions; historically Black colleges and universities (HBCUs); tribally controlled colleges and universities (TCCUs); Alaska Native and Native Hawaiian serving institutions; nonprofits with 501(c)3 IRS status; nonprofits without 501(c)3 IRS status; for-profit organizations; state governments; Indian/Native American tribal governments (federally recognized); Indian/Native American tribally designated organizations; county governments; city or township governments; special district governments; independent school districts; public housing authorities/Indian housing authorities; U.S. territory or possession; Indian/Native American tribal governments (other than federally recognized); regional organizations; non-domestic (non-U.S.) entities (foreign organizations); other eligible agencies of the federal government; and faith-based or community-based organizations.
Geographic Focus: All States
Date(s) Application is Due: Jan 30; Dec 30
Contact: Virginia S. Taggart, (301) 435-0202; taggartv@nhlbi.nih.gov
Internet: http://grants.nih.gov/grants/guide/rfa-files/RFA-HL-08-010.html
Sponsor: National Institutes of Health
Building 31, Room 5A48
Bethesda, MD 20892

NHLBI Developmental Origins of Altered Lung Physiology and Immune Function Grants   2698
This FOA issued by the NHLBI, National Institutes of Health, solicits Research Project Grant applications from institutions and organizations that propose to perform research that will enhance the understanding of how the pre- and postnatal environments affect the interplay of the lung and immune system during development resulting in sustained changes in lung physiology and immune function that compromise respiratory health and outcomes. Budgets for direct costs of up to $350,000 per year (exclusive of indirect costs associated with consortia) and a project duration of up to five years may be requested for

a maximum of $1,750,000 direct costs over a five-year project period. Letters of Intent should be received by September 19, and full proposals are due by October 21.
Requirements: The following organizations and institutions are eligible to apply: public/state controlled institutions of higher education; private institutions of higher education; Hispanic-serving institutions; historically Black colleges and universities (HBCUs); tribally controlled colleges and universities (TCCUs); Alaska Native and Native Hawaiian serving institutions; nonprofits with 501(c)3 IRS status; nonprofits without 501(c)3 IRS status; for-profit organizations; state governments; Indian/Native American tribal governments (federally recognized); Indian/Native American tribally designated organizations; county governments; city or township governments; special district governments; independent school districts; public housing authorities/Indian housing authorities; U.S. territory or possession; Indian/Native American tribal governments (other than federally recognized); regional organizations; non-domestic (non-U.S.) entities (foreign organizations); other eligible agencies of the federal government; and faith-based or community-based organizations.
Geographic Focus: All States
Date(s) Application is Due: Sep 19; Oct 21
Amount of Grant: Up to 350,000 USD
Contact: Patricia Noel, (301) 435-0202; noelp@nhlbi.nih.gov
Internet: http://grants.nih.gov/grants/guide/rfa-files/RFA-HL-08-009.html
Sponsor: National Institutes of Health
Building 31, Room 5A48
Bethesda, MD 20892

NHLBI Exploratory Studies in the Neurobiology of Pain in Sickle Cell Disease 2699
The goal of this initiative is to foster novel basic and translational research of the neurobiology of pain in sickle cell disease. Priority will be given to the application of investigational techniques that have been utilized in other pain syndromes in both human and nonhuman studies. A multidisciplinary collaboration, between neurobiologists, clinical specialists in pain, and hematologists will be necessary to maximize the research effort.
Requirements: The following organizations and institutions are eligible to apply: public/state controlled institutions of higher education; private institutions of higher education; Hispanic-serving institutions; historically Black colleges and universities (HBCUs); tribally controlled colleges and universities (TCCUs); Alaska Native and Native Hawaiian serving institutions; nonprofits with 501(c)3 IRS status; nonprofits without 501(c)3 IRS status; for-profit organizations; state governments; Indian/Native American tribal governments (federally recognized); Indian/Native American tribally designated organizations; county governments; city or township governments; special district governments; independent school districts; public housing authorities/Indian housing authorities; U.S. territory or possession; Indian/Native American tribal governments (other than federally recognized); regional organizations; non-domestic (non-U.S.) entities (foreign organizations); other eligible agencies of the federal government; and faith-based or community-based organizations.
Geographic Focus: All States
Date(s) Application is Due: Jan 5; Feb 3
Amount of Grant: Up to 300,000 USD
Contact: Harvey Luksenburg, (301) 435-0050; luksenburgh@mail.nih.gov
Internet: http://grants.nih.gov/grants/guide/rfa-files/RFA-HL-09-008.html
Sponsor: National Institutes of Health
Building 31, Room 5A48
Bethesda, MD 20892

NHLBI Immunomodulatory, Inflammatory, and Vasoregulatory Properties of Transfused Red Blood Cell Units as a Function of Preparation & Storage 2700
The goal of this Funding Opportunity Announcement (FOA) in blood banking and transfusion medicine is to solicit novel basic and translational research projects which will identify the molecular and cellular changes that occur during red blood cell unit preparation and storage; and evaluate the immunomodulatory, inflammatory, and vasoactive effects of storage lesion elements from red blood cell units on the blood vessel wall, host cells, and tissue oxygenation. Only research applications involving human red blood cell units (and their components) suitable for transfusion in the U.S. and that are collected and stored according to methods approved by the U.S. FDA will be considered responsive to this FOA. Applications are encouraged that propose collaborations between investigators from different disciplines such as hematology, blood banking, transfusion medicine, immunology, cell biology, rheology, nanotechnologies, molecular biology, and other relevant disciplines. This program will promote interactions and resource sharing among the award recipients to maximize their research efforts. Letters of Intent should be received by April 21 (1st cycle) or December 19 (2nd cycle), and full proposals are due by either May 21 (1st cycle) or January 21 (2nd cycle).
Requirements: The following organizations and institutions are eligible to apply: public/state controlled institutions of higher education; private institutions of higher education; Hispanic-serving institutions; historically Black colleges and universities (HBCUs); tribally controlled colleges and universities (TCCUs); Alaska Native and Native Hawaiian serving institutions; nonprofits with 501(c)3 IRS status; nonprofits without 501(c)3 IRS status; small businesses; for-profit organizations; state governments; Indian/Native American tribal governments; Indian/Native American tribally designated organizations; and U.S. territories or possessions.
Geographic Focus: All States
Date(s) Application is Due: Jan 21; Apr 21; May 21; Dec 19
Contact: Simone Glynn, (301) 435-0065; glynnsa@nhlbi.nih.gov
Internet: http://grants.nih.gov/grants/guide/rfa-files/RFA-HL-08-005.html
Sponsor: National Institutes of Health
Building 31, Room 5A48
Bethesda, MD 20892

NHLBI Independent Scientist Award 2701
The award is a special salary-only grant designed to provide protected time for newly independent scientists who currently have non-research obligations such as heavy teaching loads, clinical work, committee assignments, service, and administrative duties that prevent them from having a period of intensive research focus. The award is targeted to persons with doctoral degrees who have completed their research training, have independent peer-reviewed research support, and who need a period of protected research time in order to foster their research career development. New applications are due February 12, June 12, and October 12, while renewal applications are due March 12, July 12, and November 12. Up to $75,000 per year plus fringe benefits is available. No other research development support funds are provided.
Requirements: The following organizations and institutions are eligible to apply: public/state controlled institutions of higher education; private institutions of higher education; nonprofits with 501(c)3 IRS status; nonprofits without 501(c)3 IRS status; small businesses; for-profit organizations; State governments; U.S. territories or possessions; Indian/Native American tribal governments (Federally recognized and other than Federally recognized); Indian/Native American tribally designated organizations; Hispanic-serving institutions; historically Black colleges and universities (HBCUs); tribally controlled colleges and universities (TCCUs); Alaska Native and Native Hawaiian serving institutions; regional organizations; and faith-based or community based organizations.
Restrictions: The award is not intended for investigators who already have full time to perform research, or have substantial publication records or considerable research support indicating that they are well established in their fields. Foreign institutions are not eligible to apply.
Geographic Focus: All States
Date(s) Application is Due: Feb 12; Mar 12; Jun 12; Jul 12; Oct 12; Nov 12
Amount of Grant: 75,000 USD
Contact: Traci Heath Mondoro, (301) 435-0052; mondorot@nhlbi.nih.gov
Internet: http://grants.nih.gov/grants/guide/pa-files/PA-09-038.html
Sponsor: National Institutes of Health
Building 31, Room 5A48
Bethesda, MD 20892

NHLBI Intramural Research Training Awards 2702
The Office of Education of the Division of Intramural Research coordinates and assists the training program and provides administrative support and resources for all research training activities within the Division. Programs are available within the Laboratories and Branches within the Division to provide specialized research training for high school students, college students, graduate and medical students, postdoctoral fellows and medical residents and fellows. The Office of Education provides assistance to applicants at each of these levels. NHLBI is committed to improving the representation of underrepresented groups, including minorities, women, and scientists with disabilities in the mainstream of basic and clinical research.
Geographic Focus: All States
Contact: Herbert M. Geller, (301) 451-9440; direducation@nhlbi.nih.gov
Internet: http://www.training.nih.gov/student/pre-irta/previewpostbac.asp
Sponsor: National Heart, Lung, and Blood Institute
10 Center Drive, Building 10, Room 2N242, MSC 1754
Bethesda, MD 20892-1754

NHLBI Investigator Initiated Multi-Site Clinical Trials 2703
The support of multi-site clinical trials is one strategy NHLBI uses to improve the understanding of the clinical mechanisms of disease and to improve prevention, diagnosis, and treatment. The purpose of this opportunity is to provide a vehicle for submitting grant applications for investigator-initiated multi-site randomized controlled clinical trials. The trials may address any research question related to the mission and goals of the NHLBI and may test clinical or behavioral interventions. The funding opportunity is appropriate for applications to conduct phase II and phase III randomized clinical trials where participants are recruited from multiple sites. Large-scale pragmatic trials (such as comparative effectiveness trials) as well as trials designed to test efficacy of an intervention are appropriate. The trials may randomize at the individual (patient) level or at a group level (e.g., randomization of clinics, schools, worksites, etc.). Clinical trials involving NHLBI mission-related rare diseases that require coordination across multiple clinical sites are also suitable for submission to this funding opportunity. In any case, the trial should propose the most efficient study design to complete the specific aims. The maximum project period is 5 years.
Requirements: The following organizations are eligible to apply: public/state controlled and private institutions of higher education; Hispanic-serving institutions; historically black colleges and universities; tribally controlled colleges and universities; Alaska Native and Native Hawaiian serving institutions; Asian American Native American Pacific Islander serving institutions; nonprofits with or without 501(c)3 IRS status (other than institutions of higher education); small businesses; for-profit organizations (other than small businesses); state, county, and city/township governments; special district governments; Indian/Native American tribal governments (federally recognized and other than federally recognized); eligible agencies of the Federal government; U.S. territory or possession; independent school districts; public or Indiana housing authorities; Native American tribal organizations (other than federally recognized tribal governments); faith-based or community-based organizations; regional organizations; and non-domestic (non-U.S.) entities (foreign institutions) and non-domestic (non-U.S.) components of U.S. organizations. Clinical Coordination Center (CCC) and Data Coordination Center (D.C.C) applications are required to be submitted together when the proposed costs of the clinical trial exceed $500,000 (minus F&A for subcontracts) in any given year. Any individual(s) with the skills, knowledge, and resources necessary

to carry out the proposed research as the Program Director(s)/Principal Investigator(s) (PD(s)/PI(s)) is invited to work with his/her organization to develop an application for support. Individuals from underrepresented racial and ethnic groups as well as individuals with disabilities are always encouraged to apply for NIH support.
Geographic Focus: All States, All Countries
Date(s) Application is Due: Feb 12; Jun 12; Oct 12
Contact: David J. Gordon, (301) 435-0564; fax (301) 480-7971; gordond@nhlbi.nih.gov
Internet: http://grants.nih.gov/grants/guide/pa-files/PAR-13-128.html
Sponsor: National Institutes of Health
Building 31, Room 5A48
Bethesda, MD 20892

NHLBI Lymphatics in Health and Disease in the Digestive, Cardiovascular and Pulmonary Systems   2704

This program encourages Exploratory/Developmental Grant (R21) applications for research into aspects of lymphatic vessel physiology and pathophysiology related to health and disease of digestive, cardiovascular and pulmonary system organs and resolution of thromboembolic events, and inflammation and immune responses as they relate to these diseases. However, studies with the major focus on immune mechanisms will not be considered responsive. Studies to understand the factors that control local lymphatic vessel functional anatomy and physiology during health or disease in these organs and systems, and the mechanisms by which alterations of lymphatic vessel function affect organ function, are of interest. Specific areas of interests for the National Heart, Lung, and Blood Institute (NHLBI) include approaches that will identify the genetic, molecular, and cellular defects that contribute to congenital malformation of the lymphatic system; congenital-lymphatic-malformation-induced pulmonary dysfunction; whether and how lymphangiogenesis affects cardiovascular and pulmonary diseases; whether and how lymphatic drainage affects transplant of heart or lungs; whether and how lymphatic vessel hyperplasia in the dermal interstitium is involved in salt-sensitive hypertension; and whether and how the manipulation of platelet-lymphatic endothelial cell interaction may provide clinical benefit during thromboembolic events.
Requirements: The following organizations and institutions are eligible to apply: public/state controlled and private institutions of higher education; Hispanic-serving institutions; historically Black colleges and universities (HBCUs); Tribally controlled colleges and universities (TCCUs); Alaska Native and Native Hawaiian serving institutions; Nonprofits with or without 501(c)3 IRS status; small businesses; for-profit organizations; state, county, or city/township governments; special district governments; Indian/Native American tribal governments; Indian/Native American tribally designated organizations; independent school districts; public housing authorities/Indian housing authorities; U.S. territory or possession; Indian/Native American tribal governments (other than federally recognized); regional organizations; non-domestic (non-U.S.) entities (foreign organizations); other eligible agencies of the federal government; and faith-based or community-based organizations. Non-domestic (non-U.S.) components of U.S. Organizations are also eligible to apply. Applicant organizations may submit more than one application, provided that each application is scientifically distinct. Any individual(s) with the skills, knowledge, and resources necessary to carry out the proposed research as the Program Director(s)/Principal Investigator(s) (PD(s)/PI(s)) is invited to work with his/her organization to develop an application for support. Individuals from underrepresented racial and ethnic groups as well as individuals with disabilities are always encouraged to apply for NIH support.
Restrictions: NIH will not accept any application that is essentially the same as one already reviewed.
Geographic Focus: All States, Guam, Marshall Islands, Northern Mariana Islands, Puerto Rico, U.S. Virgin Islands, All Countries
Date(s) Application is Due: Feb 16; Jun 16; Oct 16
Contact: H. Eser Tolunay, Ph.D.; (301) 435-0560; Eser.Tolunay@nih.gov
Internet: http://grants.nih.gov/grants/guide/pa-files/PAR-12-260.html
Sponsor: National Institutes of Health
Building 31, Room 5A48
Bethesda, MD 20892

NHLBI Lymphatics in Health and Disease in the Digestive, Urinary, Cardiovascular and Pulmonary Systems   2705

This program is to encourage Research Project Grant (R01) applications for research into aspects of lymphatic vessel physiology and pathophysiology related to health and disease of digestive system and urinary tract organs, and cardiovascular and pulmonary systems; in resolution of thromboembolic events; and inflammation and immune responses as they relate to these diseases. However, studies with the major focus on immune mechanisms will not be considered responsive. Studies to understand the factors that control local lymphatic vessel functional anatomy and physiology during health or disease in these organs/systems, and the mechanisms by which alterations of lymphatic vessel function affect organ function, are of interest. Applications must be received by 5:00 PM local time of applicant organization.
Requirements: The following organizations and institutions are eligible to apply: public/state controlled and private institutions of higher education; Hispanic-serving institutions; historically Black colleges and universities (HBCUs); Tribally controlled colleges and universities (TCCUs); Alaska Native and Native Hawaiian serving institutions; Nonprofits with or without 501(c)3 IRS status; small businesses; for-profit organizations; state, county, or city/township governments; special district governments; Indian/Native American tribal governments; Indian/Native American tribally designated organizations; independent school districts; public housing authorities/Indian housing authorities; U.S. territory or possession; Indian/Native American tribal governments (other than federally recognized); regional organizations; non-domestic (non-U.S.) entities (foreign organizations); other eligible agencies of the federal government; and faith-based or community-based organizations. Non-domestic (non-U.S.) components of U.S. Organizations are also eligible to apply. Applicant organizations may submit more than one application, provided that each application is scientifically distinct. Any individual(s) with the skills, knowledge, and resources necessary to carry out the proposed research as the Program Director(s)/Principal Investigator(s) (PD(s)/PI(s)) is invited to work with his/her organization to develop an application for support. Individuals from underrepresented racial and ethnic groups as well as individuals with disabilities are always encouraged to apply for NIH support.
Restrictions: NIH will not accept any application that is essentially the same as one already reviewed.
Geographic Focus: All States, Guam, Marshall Islands, Northern Mariana Islands, Puerto Rico, U.S. Virgin Islands, All Countries
Date(s) Application is Due: Feb 5; Jun 5; Oct 5
Contact: H. Eser Tolunay, Ph.D.; (301) 435-0560; Eser.Tolunay@nih.gov
Internet: http://grants.nih.gov/grants/guide/pa-files/PAR-07-420.html
Sponsor: National Institutes of Health
Building 31, Room 5A48
Bethesda, MD 20892

NHLBI Mentored Career Award for Faculty at Institutions that Promote Diversity (K01)   2706

The National Institutes of Health (NIH) recognizes a unique and compelling need to promote diversity in the biomedical, behavioral, clinical and social sciences research workforce. The NIH expects efforts to diversify the workforce to lead to the recruitment of the most talented researchers from all groups, to improve the quality of the educational and training environment, to balance and broaden the perspective in setting research priorities, to improve the ability to recruit subjects from minority and other health disparity populations into clinical research protocols, and to improve the Nation's capacity to address and eliminate health disparities. The objectives of this program are to: (1) Advance the awardee's career development trajectory by strengthening research capacity, publishing and other scholarly activities; (2) Improve success and retention in a research career; (3) Promote scientific collaborations that lead to acquisition of new skills or research in other fields of scholarly interest; and (4) Increase the number of highly trained investigators at institutions that promote diversity whose basic and clinical research interests are grounded in the advanced methods and experimental approaches needed to solve problems related to cardiovascular, pulmonary, and hematologic diseases in general and in populations that suffer disproportionately from these conditions. An award made to a successful applicant to this FOA will be a three-to-five-year non-renewable career development award to support faculty at a non-research intensive institution with an institutional mission focused on serving diverse communities that are not well represented in NIH-funded research, or institutions that have been identified by federal legislation as having specially focused institutional missions. The candidates must have research experience and be committed to developing into independent biomedical investigators in research areas relevant to the mission of the NHLBI (i.e., cardiovascular, pulmonary, hematologic, or sleep disorders research).
Requirements: Higher Education Institutions (Public/State Controlled Institutions of Higher Education and Private Institutions of Higher Education) are eligible to apply. The following types of Higher Education Institutions are always encouraged to apply for NIH support as Public or Private Institutions of Higher Education: Hispanic-serving Institutions;Historically Black Colleges and Universities (HBCUs); Tribally Controlled Colleges and Universities (TCCUs); Alaska Native and Native Hawaiian Serving Institutions. The Institution must be a domestic college or university with an institutional mission focused on serving minority and other health disparity populations underrepresented in scientific research, or institutions that have been identified by federal legislation as having specifically focused institutional missions. To demonstrate need for research capacity development in heart, lung, and blood diseases and sleep disorders, applicant institutions should not have received over $1,515,000 in NIH funding (direct costs) from the NHLBI per year in any of the previous five years for research training and education grants in the above NHLBI mission areas. Applicant organizations may submit more than one application, provided that each application is scientifically distinct. Any candidate with the skills, knowledge, and resources necessary to carry out the proposed research as the Program Director/Principal Investigator (PD/PI) is invited to work with his/her mentor and organization to develop an application for support. Individuals from underrepresented racial and ethnic groups as well as individuals with disabilities are always encouraged to apply for NIH support. Multiple Principal Investigators are not allowed. By the time of award, the individual must be a citizen or a non-citizen national of the United States or have been lawfully admitted for permanent residence. Individuals on temporary or student visas are not eligible.
Restrictions: Non-domestic (non-U.S.) Entities (Foreign Institutions) and non-domestic (non-U.S.) components of U.S. Organizations are not eligible to apply. Foreign components are not allowed.
Geographic Focus: All States
Date(s) Application is Due: Feb 7
Contact: Henry Chang, M.D., (301) 435-0067; changh@nhlbi.nih.gov
Internet: http://www.nhlbi.nih.gov/funding/training/redbook/newrek01.htm
Sponsor: National Institutes of Health
Building 31, Room 5A48
Bethesda, MD 20892

NHLBI Mentored Career Award For Faculty At Minority Institutions   2707

The NHLBI Mentored Career Award for Faculty at Minority Institutions is a three-to-five-year non-renewable career development award made to faculty at a minority institution. The candidates are faculty members at a minority institution, must have research experience (length of time may vary), and be committed to developing into independent biomedical investigators in research areas relevant to the mission of the NHLBI (i.e, cardiovascular, pulmonary, hematologic, or sleep disorders research). The award will enable suitable faculty members holding doctoral degrees, such as the Ph.D., M.D., D.O., D.V.M., or an equivalent degree, to undertake special study and supervised research under a mentor who is an accomplished investigator in the research area proposed and has experience in developing independent investigators. Letters of Intent are due on June 21, while full proposals are due by July 19.

Requirements: Minority scientists and physicians with limited research experience needing guided coursework and supervised laboratory experiences, as well as minority faculty needing only an intensive research experience under the guidance of an established scientist, are eligible to apply. At the time of award, it is required that at least two years have elapsed since the receipt of the doctoral degree and that the candidate have at least one year of prior documented research experience.

Restrictions: Current or past principal investigators of an NIH grant or its equivalent, including the Clinical Investigator Award, Physician Scientist Award, Clinical Investigator Development Award, or Mentored Clinical Scientist Development Award are not eligible for the Mentored Research Scientist Development Award for Minority Faculty. Similarly, individuals serving as responsible investigators or project leaders on large grants, such as a Program Project Grant, are not eligible for the Mentored Research Scientist Development Award for Minority Faculty.

Geographic Focus: All States
Date(s) Application is Due: Jul 19
Amount of Grant: Up to 200,000 USD
Contact: Traci Heath Mondoro, (301) 435-0052; mondorot@nhlbi.nih.gov
Internet: http://grants.nih.gov/grants/guide/rfa-files/RFA-HL-05-016.html#SectionVII
Sponsor: National Institutes of Health
Building 31, Room 5A48
Bethesda, MD 20892

NHLBI Mentored Career Development Award to Promote Faculty Diversity/ Re-Entry in Biomedical Research — 2708

This program invites applications to increase the number of highly trained investigators, from diverse backgrounds underrepresented in research areas of interest to the NHLBI or from those individuals who wish to re-enter their research careers (e.g., after a hiatus due to family circumstances). Minorities are not the only group underrepresented in the scientific community. Individuals with disabilities are also underrepresented. It is geared toward individuals whose basic and clinical research interests are grounded in the advanced methods and experimental approaches needed to solve problems related to cardiovascular, pulmonary, and hematologic diseases in the general and health disparities populations. NHLBI encourages research training and career development crossing disciplinary boundaries (e.g., biophysics, biostatistics, bioinformatics, bioengineering) to develop a new interdisciplinary work force. Award budgets are composed of salary and other program-related expenses. The total project period may not exceed 5 years. Applications are due by the posted deadline date by 5:00 PM local time of applicant organization.

Requirements: The following organizations are eligible to apply: Public/state controlled and private institutions of higher education; nonprofits with and without 501(c)3 IRS Status (other than institutions of higher education); small businesses; for-profit organizations (other than small businesses); state governments; county governments; city or township governments; special district governments; Indian/Native American Tribal governments (federally recognized); Indian/Native American Tribal governments (other than federally recognized); U.S. territory or possession. The following types of higher education institutions are always encouraged to apply for NIH support as public or private institutions of higher education: Hispanic-serving institutions; historically Black colleges and universities; Tribally controlled colleges and universities; Alaska Native and Native Hawaiian serving institutions. Any candidate with the skills, knowledge, and resources necessary to carry out the proposed research as the Program Director/Principal Investigator (PD/PI) is invited to work with his/her mentor and organization to develop an application for support. Individuals from underrepresented racial and ethnic groups as well as individuals with disabilities are always encouraged to apply for NIH support. Multiple Principal Investigators are not allowed. By the time of award, the individual must be a citizen or a non-citizen national of the United States or have been lawfully admitted for permanent residence.

Restrictions: Non-domestic (non-U.S.) Entities (Foreign Institutions) are not eligible to apply. Non-domestic (non-U.S.) components of U.S. Organizations are not eligible to apply. Foreign components, as defined in the NIH Grants Policy Statement, are not allowed. Individuals on temporary or student visas are not eligible.

Geographic Focus: All States, Guam, Marshall Islands, Northern Mariana Islands, Puerto Rico, U.S. Virgin Islands, American Samoa
Date(s) Application is Due: Feb 7
Amount of Grant: Up to 625,000 USD
Contact: Lorraine M. Silsbee, M.H.S., (301) 435-0709; silsbeeL@nhlbi.nih.gov
Internet: http://www.nhlbi.nih.gov/funding/training/redbook/newsck01.htm
Sponsor: National Institutes of Health
Building 31, Room 5A48
Bethesda, MD 20892

NHLBI Mentored Clinical Scientist Research Career Development Awards — 2709

This award enables candidates holding professional degrees (e.g., M.D., D.O., D.V.M., or equivalent degrees) to undertake 3 to 5 years of special study and supervised research with the goal of becoming independent investigators. The award also allows awardees to pursue a research career development program suited to their experience and capabilities under a mentor who is competent to provide guidance in the chosen research area. Institutions may submit applications on behalf of candidates who hold professional degrees. At least 2 years must have elapsed since the health professional degree was granted. Candidates can have varying levels of clinical training and research experience. New applications are due February 12, June 12, and October 12, while resubmitted applications are due March 12, July 12, and November 12. Up to $75,000 per year plus fringe benefits is available, and up to $25,000 per year for research support for new awards.

Requirements: All candidates must be U.S. citizens, non-citizen nationals, or legal permanent residents of the U.S. Persons with temporary or student visas are not eligible. The grantee institution must have a strong, well-established research and research training program in the chosen area, accomplished faculty in the basic and clinical sciences, and a commitment to the candidate's research development. The proposed program should include an appropriate mentor.

Geographic Focus: All States
Date(s) Application is Due: Feb 12; Mar 12; Jun 12; Jul 12; Oct 12; Nov 12
Amount of Grant: Up to 100,000 USD
Contact: Lorraine M. Silsbee, (301) 435-0709; Lorraine_Silsbee@nih.gov
Internet: http://www.nhlbi.nih.gov/funding/training/redbook/newrek08.htm
Sponsor: National Institutes of Health
Building 31, Room 5A48
Bethesda, MD 20892

NHLBI Mentored Patient-Oriented Research Career Development Awards — 2710

This award supports the career development of investigators who are committed to patient-oriented research. It provides support for supervised study and research for clinically trained professionals who have the potential to develop into productive, clinical investigators focusing on patient-oriented research. This program provides research development opportunities for clinicians with varying levels of research experience. Support is provided for a minimum of 3 years and a maximum of 5 years. New applications are due February 12, June 12, and October 12. Resubmissions are due March 12, July 12, and November 12.

Requirements: The following organizations are eligible to apply: Public/state controlled and private institutions of higher education; nonprofits with and without 501(c)3 IRS Status (other than institutions of higher education); small businesses; for-profit organizations (other than small businesses); state governments; county governments; city or township governments; special district governments; Indian/Native American Tribal governments (federally recognized); Indian/Native American Tribal governments (other than federally recognized); U.S. territory or possession. The following types of higher education institutions are always encouraged to apply for NIH support as public or private institutions of higher education: Hispanic-serving institutions; historically Black colleges and universities; Tribally controlled colleges and universities; Alaska Native and Native Hawaiian serving institutions. Applicant organizations may submit more than one application, provided that each application is scientifically distinct. Any candidate with the skills, knowledge, and resources necessary to carry out the proposed research as the Program Director/Principal Investigator (PD/PI) is invited to work with his/her mentor and organization to develop an application for support. Individuals from underrepresented racial and ethnic groups as well as individuals with disabilities are always encouraged to apply for NIH support. Multiple Principal Investigators are not allowed. By the time of award, the individual must be a citizen or a non-citizen national of the United States or have been lawfully admitted for permanent residence. Candidates for this award must have a health-professional doctoral degree. Such degrees include but are not limited to the M.D., D.O., D.D.S., D.M.D., O.D., D.C., Pharm. D., N.D. (Doctor of Naturopathy), as well as a doctoral degree in nursing research or practice. Candidates with Ph.D. degrees are eligible for this award if the degree is in a clinical field and they usually perform clinical duties. Individuals with the Ph.D. or other doctoral degree in clinical disciplines such as clinical psychology, clinical genetics, speech-language pathology, audiology or rehabilitation are also eligible. Individuals holding the Ph.D. in a non-clinical discipline but who are certified to perform clinical duties should contact the appropriate Institute concerning their eligibility for a K23 award. Candidates also must have completed their clinical training, including specialty and, if applicable, subspecialty training prior to receiving an award. However, candidates may submit an application prior to the completion of clinical training.

Restrictions: Non-domestic (non-U.S.) Entities (Foreign Organizations) are not eligible to apply. Foreign (non-U.S.) components of U.S. Organizations are not allowed. Individuals on temporary or student visas are not eligible.

Geographic Focus: All States, Guam, Marshall Islands, Northern Mariana Islands, Puerto Rico, U.S. Virgin Islands, American Samoa
Date(s) Application is Due: Feb 12; Mar 12; Jun 12; Jul 12; Oct 12; Nov 12
Amount of Grant: Up to 620,000 USD
Contact: Sandra Colombini Hatch; (301) 435-0222; hatchs@nhlbi.nih.gov
Internet: http://www.nhlbi.nih.gov/funding/training/redbook/newk23.htm
Sponsor: National Institutes of Health
Building 31, Room 5A48
Bethesda, MD 20892

NHLBI Mentored Quantitative Research Career Development Awards — 2711

The goal of this program is to foster interdisciplinary collaboration in biomedical and behavioral research by supporting supervised research experiences for scientists with quantitative and engineering backgrounds. This award provides research and career development opportunities for scientists and engineers with little or no biomedical or behavioral research experience who are committed to establishing themselves in careers as independent biomedical or behavioral investigators. Examples of quantitative scientific and technical backgrounds outside of biology and medicine considered appropriate for this award include, but are not limited to: mathematics, statistics, computer science, informatics, physics, chemistry, and engineering. This mechanism is intended for research-oriented investigators from the postdoctoral level to the level of senior faculty. New applications are due February 12, June 12, October 12. Resubmissions are due

March 12, July 12, and November 12. Award budgets are composed of salary and other program-related expenses. The total project period may not exceed 5 years.
Requirements: Institutions may submit applications on behalf of candidates who hold an advanced degree in a quantitative area of science or engineering. The following organizations are eligible to apply: Public/state controlled and private institutions of higher education; nonprofits with and without 501(c)3 IRS Status (other than institutions of higher education); small businesses; for-profit organizations (other than small businesses); state governments; county governments; city or township governments; special district governments; Indian/Native American Tribal governments (federally recognized); Indian/Native American Tribal governments (other than federally recognized); U.S. territory or possession. The following types of higher education institutions are always encouraged to apply for NIH support as public or private institutions of higher education: Hispanic-serving institutions; historically Black colleges and universities; Tribally controlled colleges and universities; Alaska Native and Native Hawaiian serving institutions. Individuals with an advanced degree in a quantitative area of science or engineering are eligible candidates. Multiple Principal Investigators are not allowed. All candidates must be U.S. citizens, non-citizen nationals, or legal permanent residents of the U.S. The grantee institution must have a strong, well-established research and research training program in the chosen area, accomplished faculty in the basic and clinical sciences, and a commitment to the candidate's research development. The proposed program should include an appropriate mentor.
Restrictions: NHLBI has a 6 year limit of cumulative support on institutional and mentored Ks (e.g., K12 or KL2 plus the K25). A candidate for the K25 may not concurrently apply for or have an award pending for any other NIH career development award. Non-domestic (non-U.S.) Entities (Foreign Organizations) are not eligible to apply. Foreign (non-U.S.) components of U.S. Organizations are not allowed.
Geographic Focus: All States, Guam, Marshall Islands, Northern Mariana Islands, Puerto Rico, U.S. Virgin Islands, American Samoa
Date(s) Application is Due: Feb 12; Mar 12; Jun 12; Jul 12; Oct 12; Nov 12
Contact: Drew E. Carlson, Ph.D., Program Director, Office of Research Training and Career Development; (301) 435-0535; fax (301) 480-7404; carlsonde@nhlbi.nih.gov
Internet: http://www.nhlbi.nih.gov/funding/training/redbook/newk25.htm
Sponsor: National Institutes of Health
Building 31, Room 5A48
Bethesda, MD 20892

NHLBI Microbiome of the Lung and Respiratory Tract in HIV-Infected Individuals and HIV-Uninfected Controls 2712

The National Heart, Lung, and Blood Institute (NHLBI) solicits grant applications under this Funding Opportunity Announcement (FOA) to characterize the microbiome of the lung (the airways and airspaces below the glottis and the lung parenchyma) alone or in combination with the nasal and/or oropharyngeal cavities in HIV-infected individuals and matched HIV-uninfected controls, including normal healthy controls, using molecular techniques to identify bacteria and if possible other organisms, e.g., viruses, cell-wall deficient organisms, protozoa, and fungi. Investigators should use high-throughput technology platforms to create a data set of sufficient quality and depth to allow analysis of how changes of microbiota relate to HIV lung disease progression/complications. These data will be used to examine the impact of changes in the respiratory microbiome on the pathogenesis and progression of HIV disease, on HIV-related respiratory complications, and the effects of anti-HIV therapies. The NHLBI also invites applications for four Clinical/Sequencing research sites and one supporting Data Coordinating Center (D.C.C). Each application for a clinical/sequencing research site should propose approaches to characterize the lung microbiome and applicants are encouraged to include at least one hypothesis-driven mechanistic aim. Direct costs up to $525,000 per year for project duration of up to five years may be requested.
Requirements: The following organizations/institutions are eligible to apply: Public/State Controlled Institutions of Higher Education; Private Institutions of Higher Education; Hispanic-serving Institutions; Historically Black Colleges and Universities (HBCUs); Tribally Controlled Colleges and Universities (TCCUs); Alaska Native and Native Hawaiian Serving Institutions; Nonprofits with 501(c)3 IRS Status (Other than Institutions of Higher Education); Nonprofits without 501(c)3 IRS Status (Other than Institutions of Higher Education); Small Businesses; For-Profit Organizations (Other than Small Businesses); State Governments; Indian/Native American Tribal Governments (Federally Recognized); Indian/Native American Tribally Designated Organizations; County Governments; City or Township Governments; Special District Governments; Independent School Districts; Public Housing Authorities/Indian Housing Authorities; U.S. Territory or Possession; Indian/Native American Tribal Governments (Other than Federally Recognized); Regional Organizations; Other Eligible Agencies of the Federal Government; and, Faith-based or Community-based Organizations. Any individual with the skills, knowledge, and resources necessary to carry out the proposed research as the PD/PI is invited to work with his/her institution to develop an application for support. Individuals from underrepresented racial and ethnic groups as well as individuals with disabilities are always encouraged to apply for NIH support. More than one PD/PI, or multiple PDs/PIs, may be designated on the application for projects that require a "team science" approach and therefore clearly do not fit the single-PD/PI model. All PDs/PIs must be registered in the NIH eRA Commons prior to the submission of the application.
Restrictions: Foreign organizations are not eligible, however, domestic institutions may form consortia and enter into subcontracts with Non-domestic (non-U.S.) Entities (Foreign Organizations) if need can be demonstrated – foreign subcontract use the same criteria as are used for justifying foreign grants.
Geographic Focus: All States
Date(s) Application is Due: Feb 5; Jun 5; Oct 5

Amount of Grant: Up to 525,000 USD
Contact: Sandra Colombini Hatch, (301) 435-0222; Hatchs@nhlbi.nih.gov
Internet: http://grants.nih.gov/grants/guide/rfa-files/RFA-HL-09-006.html
Sponsor: National Institutes of Health
Building 31, Room 5A48
Bethesda, MD 20892

NHLBI Midcareer Investigator Award in Patient-Oriented Research 2713

This award provides protected time to mid-career clinical investigators who are typically at the Associate Professor level or the equivalent and who have their own independent peer-reviewed research support to provide mentoring to junior clinical investigators, particularly K23 grantees, in patient-oriented research and to stabilize the careers of these investigators so that they could continue to conduct patient-oriented research and be available as mentors in patient-oriented research. It is expected that the K24 recipients will obtain new or additional independent peer-reviewed funding for patient-oriented research as a Program Director/Principal Investigator and establish and assume leadership roles in collaborative patient-oriented research programs. In addition, it is expected that there will be an increased effort and commitment to act as a mentor to beginning clinician investigators in patient-oriented research to enhance the research productivity of both the K24 investigator and increase the pool of well-trained clinical researchers of the future. New applications are due on February 12, June 12 and October 12. Renewal applications are due March 12, July 12, and November 12.
Requirements: Candidates must be U.S. citizens, non citizen nationals, or legal permanent residents of the U.S. Candidates for this award must have a health-professional doctoral degree or its equivalent. Such degrees include but are not limited to the M.D., D.O., D.D.S., D.M.D., O.D., D.C., Pharm.D., N.D. (Doctor of Naturopathy), as well as a doctoral degree in nursing. Candidates with Ph.D. degrees are eligible for this award if the degree is in a clinical field or they perform patient-oriented research (P.O.R). This may include clinical psychologists, clinical geneticists, speech and language pathologists. Candidates should typically be in the midcareer stage at the Associate Professor level or functioning at that rank in an academic setting or equivalent non-academic setting and must have an established record of independent, peer-reviewed patient-oriented research grant funding including at the time of application and record of publications. This award is intended for individuals who have a record of supervising and mentoring patient-oriented researchers.
Geographic Focus: All States
Date(s) Application is Due: Feb 12; Mar 12; Jun 12; Jul 12; Oct 12; Nov 12
Contact: Sandra Colombini Hatch, M.D., Medical Officer; (301) 435-0222; fax (301) 480-3557; hatchs@nhlbi.nih.gov
Internet: http://www.nhlbi.nih.gov/funding/training/redbook/estk24.htm
Sponsor: National Institutes of Health
Building 31, Room 5A48
Bethesda, MD 20892

NHLBI National Research Service Award Programs in Cardiovascular Epidemiology and Biostatistics 2714

The National Research Service Award (NRSA) is one of many training programs sponsored by the National Heart, Lung, and Blood Institute (NHLBI). This program enables research institutions to support pre- and post-doctoral training in areas of particular interest to the Institute and to prepare individuals for careers in biomedical and behavioral research. The cardiovascular epidemiology and biostatistics research training programs offer opportunities in the fields of epidemiology of cardiovascular diseases and statistical analysis. These awards do not support study leading to the M.D., D.D.S. or other similar professional degrees, nor do they support residency training.
Requirements: Trainees are selected through review procedures established by the program director at the grantee institution. Students are required to pursue their research training full-time and must be U.S. citizens, noncitizen nationals or legal permanent residents of the United States. At the time of appointment, trainees must have received a baccalaureate degree and must be training at the post-baccalauareate level in a program leading to a Ph.D., Sc.D, or equivalent degree. Students who wish to interrupt their medical, dental, or other professional school studies to engage in full-time research training before completing their professional degrees are also eligible. Attention is given to recruiting individuals from minority groups that are under-represented nationally in the biomedical sciences (African Americans, Hispanics, Native Americans, Alaskan Natives, and Pacific Islanders).
Restrictions: Persons on temporary or student visas are not eligible.
Geographic Focus: All States
Date(s) Application is Due: Jan 25
Contact: Lorraine M. Silsbee, (301) 435-0709; fax (301) 480-1667; silsbeeL@nih.gov
Internet: http://www.nhlbi.nih.gov/funding/training/epi-bio/nrsa.htm
Sponsor: National Institutes of Health
Building 31, Room 5A48
Bethesda, MD 20892

NHLBI New Approaches to Arrhythmia Detection and Treatment Grants for SBIR 2715

The purpose of this initiative is to improve our ability to detect, treat and prevent cardiac arrhythmias using new or improved methods, tools, and technologies. Applications are invited for the development and significant improvements of innovative monitoring tools, diagnostic methods, and therapeutic approaches for arrhythmia detection, treatment and prevention.
Requirements: Only United States small business concerns (SBCs) are eligible to submit SBIR applications.
Geographic Focus: All States

Date(s) Application is Due: Apr 5; Aug 5; Dec 5
Contact: David A. Lathrop; (301) 435-0507; LathropD@NHLBI.NIH.GOV
Internet: http://grants.nih.gov/grants/guide/pa-files/PA-07-032.html
Sponsor: National Institutes of Health
Building 31, Room 5A48
Bethesda, MD 20892

NHLBI New Approaches to Arrhythmia Detection and Treatment Grants for STTR — 2716

The purpose of this initiative is to improve our ability to detect, treat and prevent cardiac arrhythmias using new or improved methods, tools, and technologies. Applications are invited for the development and significant improvements of innovative monitoring tools, diagnostic methods, and therapeutic approaches for arrhythmia detection, treatment and prevention.
Requirements: Only United States small business concerns (SBCs) are eligible to submit STTR applications.
Geographic Focus: All States
Date(s) Application is Due: Apr 5; Aug 5; Dec 5
Contact: David A. Lathrop; (301) 435-0507; LathropD@NHLBI.NIH.GOV
Internet: http://grants.nih.gov/grants/guide/pa-files/PA-07-031.html
Sponsor: National Institutes of Health
Building 31, Room 5A48
Bethesda, MD 20892

NHLBI Nutrition and Diet in the Causation, Prevention, and Management of Heart Failure Research Grants — 2717

The purpose of this Funding Opportunity Announcement (FOA) is to encourage submission of investigator-initiated research applications on the role of nutrition and diet in the causation, prevention, and treatment of cardiomyopathies and heart failure. Basic, translational, and applied interdisciplinary research applications with rigorous hypothesis-testing designs for projects in animals or humans are of interest. The overall goal is to develop a satisfactory science base for preventive approaches in high-risk individuals and for rational nutritional management of patients in various stages of heart failure.
Requirements: The following groups are eligible to apply: for-profit organizations on-profit organizations; public or private institutions, such as universities, colleges, hospitals, and laboratories; units of State government; units of local government; eligible agencies of the Federal government; foreign institutions; domestic institutions; faith-based or community-based organizations; units of State tribal government; and units of local Tribal government.
Geographic Focus: All States
Date(s) Application is Due: Feb 16; Jun 16; Oct 16
Amount of Grant: Up to 200,000 USD
Contact: Abby G. Ershow, (301) 435-0550; fax (301) 480-1338; ErshowA@mail.nih.gov
Internet: http://grants.nih.gov/grants/guide/pa-files/PA-06-136.html
Sponsor: National Institutes of Health
Building 31, Room 5A48
Bethesda, MD 20892

NHLBI Pathway to Independence Awards — 2718

The Pathway to Independence (PI) Award is designed to facilitate a timely transition from a mentored postdoctoral research position to a stable independent research position with independent NIH or other independent research support at an earlier stage than is currently the norm. The PI award will provide up to 5 years of support consisting of two phases. The initial phase will provide 1-2 years of mentored support for postdoctoral research scientists. This phase will be followed by up to 3 years of independent support contingent on securing an independent tenure-track or equivalent research position. New applications are due February 12, June 12, and October 12. Resubmitted applications are due March 12, July 12, and November 12.
Requirements: The following organizations are eligible to apply: Public/state controlled and private institutions of higher education; nonprofits with and without 501(c)3 IRS Status (other than institutions of higher education); small businesses; for-profit organizations (other than small businesses); state governments; county governments; city or township governments; special district governments; Indian/Native American Tribal governments (federally recognized); Indian/Native American Tribal governments (other than federally recognized); U.S. territory or possession. The following types of higher education institutions are always encouraged to apply for NIH support as public or private institutions of higher education: Hispanic-serving institutions; historically Black colleges and universities; Tribally controlled colleges and universities; Alaska Native and Native Hawaiian serving institutions. The applicant institution will be the mentored phase K99 institution. All institution/organization types listed above are eligible for both the mentored and independent phase, with one exception: eligible agencies of the Federal government, such as the NIH intramural program, are eligible only for the mentored phase. Any candidate with the skills, knowledge, and resources necessary to carry out the proposed research as the Program Director/Principal Investigator (PD/PI) is invited to work with his/her mentor and organization to develop an application for support. Individuals from underrepresented racial and ethnic groups as well as individuals with disabilities are always encouraged to apply for NIH support. Multiple Principal Investigators are not allowed. At the time of application submission (or resubmission) candidates for this award must have earned a terminal clinical or research doctorate (including Ph.D., M.D., D.O., D.C., N.D., D.D.S., D.V.M., Sc.D., D.N.S., Pharm. D., or equivalent doctoral degree, or a combined clinical and research doctoral degree); and have no more than 5 years (60 months) of research experience since completing the requirements of the doctoral degree (resubmissions must also comply with this requirement). Individuals whose terminal or research doctorate was awarded more than 60 months before submission of the K99/R00 application are strongly urged to contact one of the NHLBI Program staff designated below before submission to determine whether they are eligible to apply. This confirmation of eligibility should be documented in a cover letter to accompany the application. Applications from ineligible individuals will be returned without review. U.S. citizens and non-U.S. citizens are eligible.
Restrictions: Non-domestic (non-U.S.) Entities (Foreign Organizations) are not eligible to apply. Foreign (non-U.S.) components of U.S. Organizations are not allowed.
Geographic Focus: All States, Guam, Marshall Islands, Northern Mariana Islands, Puerto Rico, U.S. Virgin Islands, American Samoa
Date(s) Application is Due: Feb 12; Mar 12; Jun 12; Jul 12; Oct 12; Nov 12
Contact: Herbert Geller, Ph.D., Director, Office of Education; (301) 451-9440; fax (301) 594-8133; direducation@nhlbi.nih.gov
Internet: http://grants.nih.gov/grants/guide/pa-files/PA-11-197.html
Sponsor: National Institutes of Health
Building 31, Room 5A48
Bethesda, MD 20892

NHLBI Pediatric Cardiac Genomics Consortium Grants — 2719

The purpose of this initiative is to support collaborative genetic and genomic research leading to a comprehensive understanding of the causes and outcomes of congenital heart disease in the human population. The Consortium will consist of up to 6 Research Centers, Cores, and a Steering Committee. Research Center applicants will propose detailed investigation of the genetics and genomics of one or more human cardiac malformations, with the goal of identifying genetic influences on outcome, as well as causative genes. Individual Research Centers will collaborate in recruitment of individuals with congenital heart disease and their families for all Research Center projects. Maximum allowable direct costs for the Research Centers per year are $160,000 in Year 1, and $535,000 per year in years 2-6. Letters of Intent should be received by January 6, and full proposals are due by February 6.
Requirements: The following organizations and institutions are eligible to apply: public/state controlled institutions of higher education; private institutions of higher education; Hispanic-serving institutions; historically Black colleges and universities (HBCUs); tribally controlled colleges and universities (TCCUs); Alaska Native and Native Hawaiian serving institutions; nonprofits with 501(c)3 IRS status; nonprofits without 501(c)3 IRS status; for-profit organizations; small businesses; state governments; non-domestic (non-U.S.) entities in North America; and other eligible agencies of the Federal Government.
Geographic Focus: All States
Date(s) Application is Due: Jan 6; Feb 6
Amount of Grant: Up to 2,835,000 USD
Contact: Gail D. Pearson, (301) 435-0510; pearsong@mail.nih.gov
Internet: http://grants.nih.gov/grants/guide/rfa-files/RFA-HL-09-003.html
Sponsor: National Institutes of Health
Building 31, Room 5A48
Bethesda, MD 20892

NHLBI Phase II Clinical Trials of Novel Therapies for Lung Diseases — 2720

The purpose of this FOA is to solicit research grant applications to conduct Phase II clinical therapeutic trials that have the potential to advance development of novel therapies for a lung disease or a cardiopulmonary disorder of sleep. Each application will propose one Phase II interventional trial that will most likely use physiological or biochemical rather than clinical endpoints along with at least one smaller basic ancillary research study that is tightly related to the clinical question. Although definitive Phase III trials will not be supported, the proposed studies must provide proof of concept for a novel intervention that has high potential for modifying current treatments and could be disease modifying.
Requirements: The following organizations and institutions are eligible to apply: public/state controlled institutions of higher education; private institutions of higher education; Hispanic-serving institutions; historically Black colleges and universities (HBCUs); tribally controlled colleges and universities (TCCUs); Alaska Native and Native Hawaiian serving institutions; nonprofits with 501(c)3 IRS status; nonprofits without 501(c)3 IRS status; for-profit organizations; state governments; Indian/Native American tribal governments (federally recognized); Indian/Native American tribally designated organizations; county governments; city or township governments; special district governments; independent school districts; public housing authorities/Indian housing authorities; U.S. territories or possessions; Indian/Native American tribal governments (other than federally recognized); regional organizations; non-domestic (non-U.S.) entities (foreign organizations); other eligible agencies of the federal government; and faith-based or community-based organizations.
Geographic Focus: All States
Date(s) Application is Due: Jan 29; May 18; Jun 18; Aug 15; Sep 15; Dec 30
Amount of Grant: Up to 1,515,000 USD
Contact: Andrea L. Harabin, Ph.D; (301) 435-0222; harabina@nhlbi.nih.gov
Internet: http://grants.nih.gov/grants/guide/rfa-files/RFA-HL-10-003.html
Sponsor: National Institutes of Health
Building 31, Room 5A48
Bethesda, MD 20892

NHLBI Prematurity and Respiratory Outcomes Program (PROP) Grants — 2721

This RFA intends to fill the need for understanding the underlying pathophysiology by supporting the investigation of the molecular mechanisms that contribute to acute and chronic respiratory morbidity of the premature infant. The well phenotyped populations developed in this program will significantly increase the chance of studying groups with common pathophysiology and lay the groundwork for novel biomarker discovery. The program will use the cooperative agreement mechanism and require multidisciplinary

expertise including neonatologists/pulmonologists, and physiological, molecular, biological, or bioinformatics scientists. Participating sites will plan and coordinate their own research and will develop and implement shared protocols for respiratory phenotyping and respiratory outcomes of the premature infant NICU graduate. Letters of Intent should be received by May 4, and full proposals are due by June 2 each year.
Requirements: The following organizations and institutions are eligible to apply: public/state controlled institutions of higher education; private institutions of higher education; Hispanic-serving institutions; historically Black colleges and universities (HBCUs); tribally controlled colleges and universities (TCCUs); Alaska Native and Native Hawaiian serving institutions; nonprofits with 501(c)3 IRS status; nonprofits without 501(c)3 IRS status; for-profit organizations; state governments; Indian/Native American tribal governments (federally recognized); Indian/Native American tribally designated organizations; county governments; city or township governments; special district governments; independent school districts; public housing authorities/Indian housing authorities; U.S. territories or possessions; Indian/Native American tribal governments (other than federally recognized); regional organizations; non-domestic (non-U.S.) entities (foreign organizations); other eligible agencies of the federal government; and faith-based or community-based organizations.
Geographic Focus: All States
Date(s) Application is Due: May 4; Jun 2
Amount of Grant: 150,000 - 350,000 USD
Contact: Carol J. Blaisdell, (301) 435-0222; blaisdellcj@nhlbi.nih.gov
Internet: http://grants.nih.gov/grants/guide/rfa-files/RFA-HL-10-007.html
Sponsor: National Institutes of Health
Building 31, Room 5A48
Bethesda, MD 20892

NHLBI Progenitor Cell Biology Consortium Administrative Coordinating Center Grants 2722

The goal of the Consortium is to identify and characterize progenitor cell lineages, to direct the differentiation of stem and progenitor cells to desired cell fates, and to develop new strategies to address the unique challenges presented by the transplantation of these cells. The Consortium will assemble multiple independent research projects, each with a multi-disciplinary team of Principal Investigators, core research support facilities, and skills development components to establish synergistic virtual hubs focused on progenitor cell biology. The total project period for an application submitted in response to this FOA is 7 years. Maximum allowable direct costs are $600,000 per year for ACC operations. Letters of Intent are dur by April 1, while full applications should arrive by May 1.
Requirements: The following organizations and institutions are eligible to apply: public/state controlled institutions of higher education; private institutions of higher education; Hispanic-serving institutions; historically Black colleges and universities (HBCUs); tribally controlled colleges and universities (TCCUs); Alaska Native and Native Hawaiian serving institutions; nonprofits with 501(c)3 IRS status; nonprofits without 501(c)3 IRS status; small businesses; for-profit organizations; and state governments.
Geographic Focus: All States
Date(s) Application is Due: Apr 1; May 1
Amount of Grant: Up to 4,200,000 USD
Contact: Denis Buxton, (301) 435-0513; Buxtond@nhlbi.nih.gov
Internet: http://grants.nih.gov/grants/guide/rfa-files/RFA-HL-09-005.html
Sponsor: National Institutes of Health
Building 31, Room 5A48
Bethesda, MD 20892

NHLBI Protein Interactions Governing Membrane Transport in Pulmonary Health and Disease Grants 2723

This Funding Opportunity Announcement (FOA) issued by the National Heart, Lung, and Blood Institute, National Institutes of Health (NIH), solicits research grant applications to delineate the protein interactions and pathways governing membrane trafficking pathways operative in pulmonary health and disease and develop novel therapeutic interventions. Because the nature and scope of the proposed research will vary from application to application, it is anticipated that the size and duration of each award will also vary. The total amount awarded and the number of awards will depend upon the mechanism numbers, quality, duration, and costs of the applications received.
Requirements: The following organizations and institutions are eligible to apply: public/state controlled institutions of higher education; private institutions of higher education; Hispanic-serving institutions; historically Black colleges and universities (HBCUs); tribally controlled colleges and universities (TCCUs); Alaska Native and Native Hawaiian serving institutions; nonprofits with 501(c)3 IRS status; nonprofits without 501(c)3 IRS status; for-profit organizations; state governments; Indian/Native American tribal governments (federally recognized); Indian/Native American tribally designated organizations; county governments; city or township governments; special district governments; independent school districts; public housing authorities/Indian housing authorities; U.S. territory or possession; Indian/Native American tribal governments (other than federally recognized); regional organizations; non-domestic (non-U.S.) entities (foreign organizations); other eligible agencies of the federal government; and faith-based or community-based organizations.
Geographic Focus: All States
Date(s) Application is Due: Feb 5; Jun 5; Oct 5
Contact: Susan Banks-Schlegel, (301) 435-0202; Schleges@nhlbi.nih.gov
Internet: http://grants.nih.gov/grants/guide/pa-files/PA-07-137.html
Sponsor: National Institutes of Health
Building 31, Room 5A48
Bethesda, MD 20892

NHLBI Research Demonstration and Dissemination Grants 2724

The purpose of this Funding Opportunity Announcement (FOA) is to encourage the scientific community to conduct Demonstration and Dissemination (D&D) studies to test the effectiveness of interventions in children, adolescents, and/or adults to: promote healthful behaviors; reduce risk factors for heart, lung, and blood diseases, and sleep disorders; improve the prevention or management of heart, lung, and blood diseases, and sleep disorders, including the delivery of health care services; and enhance understanding of the processes of intervention implementation and diffusion, or sustainability in a defined population, or defined clinical or community setting. The total project period for an application submitted in response to this funding opportunity may not exceed 5 years. Direct costs are limited to less than $500,000 in any single year, unless pre-approval is obtained from the Institute.
Requirements: The following organizations and institutions are eligible to apply: public/state controlled institutions of higher education; private institutions of higher education; Hispanic-serving institutions; historically Black colleges and universities (HBCUs); tribally controlled colleges and universities (TCCUs); Alaska Native and Native Hawaiian serving institutions; nonprofits with 501(c)3 IRS status; nonprofits without 501(c)3 IRS status; for-profit organizations; state governments; Indian/Native American tribal governments (federally recognized); Indian/Native American tribally designated organizations; county governments; city or township governments; special district governments; independent school districts; public housing authorities/Indian housing authorities; U.S. territory or possession; Indian/Native American tribal governments (other than federally recognized); regional organizations; non-domestic (non-U.S.) entities (foreign organizations); other eligible agencies of the federal government; and faith-based or community-based organizations.
Geographic Focus: All States
Date(s) Application is Due: Jan 25; May 25; Sep 25
Amount of Grant: Up to 500,000 USD
Contact: Charlotte Pratt, (301) 435-0382; fax (301) 480-5158; PrattC@nhlbi.nih.gov
Internet: http://grants.nih.gov/grants/guide/pa-files/PA-07-017.html
Sponsor: National Institutes of Health
Building 31, Room 5A48
Bethesda, MD 20892

NHLBI Research Grants on the Relationship Between Hypertension and Inflammation 2725

This Funding Opportunity Announcement (FOA) solicits grant applications that propose the study of the sequence of events in which the vascular inflammatory state contributes to the development and maintenance of hypertension. Ample evidence suggests that inflammation may play a role in the pathogenesis of hypertension or that it may characterize a functional state of the vessel wall as a consequence of high blood pressure. Angiotensin II (Ang II), a widely recognized vasoconstrictor and anti-natriuretic involved in blood pressure regulation, also acts as a pro-inflammatory factor in the cardiovascular system. Ang II stimulates the expression of several inflammatory cytokines, which in turn affect blood pressure. A potential linkage between Ang II and immuno-cytokines is their shared ability to induce the level of reactive oxygen species (ROS), which serve as second messengers for many intracellular signaling pathways. The production of ROS not only decreases bioavailability of nitric oxide (NO), a vasodilator, but also initiates the functional and morphological alterations, such as remodeling, in the vascular wall that accompany the hypertensive state over time. This FOA would provide an opportunity to bring focus on the potential causal relationship between hypertension and inflammation in a cohesive, integrated manner. A new understanding of hypertension and inflammation would provide novel opportunities to prevent and treat the disease.
Requirements: The following organizations and institutions are eligible to apply: public/state controlled institutions of higher education; private institutions of higher education; Hispanic-serving institutions; historically Black colleges and universities (HBCUs); tribally controlled colleges and universities (TCCUs); Alaska Native and Native Hawaiian serving institutions; nonprofits with 501(c)3 IRS status; nonprofits without 501(c)3 IRS status; for-profit organizations; state governments; Indian/Native American tribal governments (federally recognized); Indian/Native American tribally designated organizations; county governments; city or township governments; special district governments; independent school districts; public housing authorities/Indian housing authorities; U.S. territory or possession; Indian/Native American tribal governments (other than federally recognized); regional organizations; non-domestic (non-U.S.) entities (foreign organizations); other eligible agencies of the federal government; and faith-based or community-based organizations.
Geographic Focus: All States
Date(s) Application is Due: Feb 5; Jun 5; Oct 5
Contact: Eser Tolunay, (301) 435-0560; tolunaye@nhlbi.nih.gov
Internet: http://grants.nih.gov/grants/guide/pa-files/PA-07-138.html
Sponsor: National Institutes of Health
Building 31, Room 5A48
Bethesda, MD 20892

NHLBI Research on the Role of Cardiomyocyte Mitochondria in Heart Disease: An Integrated Approach 2726

The National Heart, Lung, and Blood Institute (NHLBI) invites applications for collaborative research projects to develop an integrated understanding of cardiomyocyte mitochondria and its contributions to myocardial adaptations and heart disease progression by combining functional data with information derived from powerful new technologies. Budgets for direct costs of up to $500,000 per year and a project duration of up to four years may be requested for a maximum of $2,000,000 direct costs over a four-year project period. Because the nature and scope of the proposed research will vary from application to application, it is anticipated that the size and duration of each award will also vary.

Requirements: The following organizations/institutions are eligible to apply: Public/State Controlled Institutions of Higher Education; Private Institutions of Higher Education; Hispanic-serving Institutions; Historically Black Colleges and Universities (HBCUs); Tribally Controlled Colleges and Universities (TCCUs); Alaska Native and Native Hawaiian Serving Institutions; Nonprofits with 501(c)3 IRS Status (Other than Institutions of Higher Education); Nonprofits without 501(c)3 IRS Status (Other than Institutions of Higher Education); Small Businesses; For-Profit Organizations (Other than Small Businesses); State Governments; Indian/Native American Tribal Governments (Federally Recognized); Indian/Native American Tribally Designated Organizations; County Governments; City or Township Governments; Special District Governments; Independent School Districts; Public Housing Authorities/Indian Housing Authorities; U.S. Territory or Possession; Indian/Native American Tribal Governments (Other than Federally Recognized); Regional Organizations; Non-domestic (non-U.S.) Entities (Foreign Organizations); Other Eligible Agencies of the Federal Government; and, Faith-based or Community-based Organizations. Any individual(s) with the skills, knowledge, and resources necessary to carry out the proposed research as the PD/PI is invited to work with his/her organization to develop an application for support. Individuals from underrepresented racial and ethnic groups as well as individuals with disabilities are always encouraged to apply for NIH support. More than one PD/PI (i.e., multiple PDs/PIs) may be designated on the application for projects that require a "team science" approach and therefore clearly do not fit the single-PD/PI model. All PDs/PIs must be registered in the NIH electronic Research Administration (eRA) Commons prior to the submission of the application. Applicants may submit more than one application, provided each application is scientifically distinct. Request for Applications (RFA) Number: RFA-HL-10-002
Restrictions: Applicants are not permitted to submit a resubmission application. Renewals will not be allowed.
Geographic Focus: All States, Guam, Marshall Islands, Northern Mariana Islands, Puerto Rico, U.S. Virgin Islands, All Countries
Date(s) Application is Due: Feb 5; Jun 5; Oct 5
Amount of Grant: Up to 2,000,000 USD
Contact: Isabella Liang, (301) 435-0504; fax (301) 451-5458; Liangi@mail.nih.gov
Internet: http://grants.nih.gov/grants/guide/rfa-files/RFA-HL-10-002.html
Sponsor: National Institutes of Health
Building 31, Room 5A48
Bethesda, MD 20892

NHLBI Research Supplements to Promote Diversity in Health-Related Research 2727

This supplement enables principal investigators with eligible NHLBI research grants/contracts to include high school students in their projects. Priority is given to applications requesting support for individuals from underrepresented racial and ethnic groups, individuals with disabilities, and individuals from disadvantaged backgrounds. Nationally, underrepresented groups in biomedical research careers include but are not limited to African Americans, Hispanic Americans, American Indians/Alaska Natives, Native Hawaiians and Pacific Islanders. The research activity proposed for the student must be part of the approved research for the parent grant or contract. The student should be encouraged to participate in ongoing team discussions of research findings and directions. The purpose of this program is to provide high school students with an opportunity to obtain a meaningful experience in various aspects of health-related research to stimulate their interest in careers in biomedical, behavioral, biometric, clinical, nursing and social sciences. Students are expected to devote sufficient effort to the research project and related activities during the period of support to gain insight into the process of scientific discovery. Support for at least three months is encouraged during any one year. Principal Investigators are encouraged to seek high school students who will devote at least two years to this program (i.e., equivalent to two three-month, full-time periods). More than one high school research supplements can be awarded per research grant, subproject of a program project grant (P01), or contract. Applications will be accepted at any time, however, applications should arrive at least three months before the requested start date, to allow time for review.
Requirements: All principal investigators at U.S. institutions with eligible NHLBI research grants or contracts may apply. Grants/contracts with adequate time for a summer research experience or one year remaining at the time of award are eligible to apply for high school students. This announcement is for supplements to existing projects. To be eligible, the parent award must be active and the research proposed in the supplement must be accomplished within the competitive segment. All additional costs must be within the scope of the peer-reviewed and approved project. Before submitting an application for a research supplement, applicants are strongly encouraged to contact their program official to discuss the program. Supplemental awards under this program are limited to citizens or non-citizen nationals of the United States or to individuals who have been lawfully admitted for permanent residence in the United States (i.e., in possession of an Alien Registration Receipt Card or some other legal evidence of admission for permanent residence at the time of application). This program may not be used to provide technical support to NIH-supported investigators. Application budgets are limited to no more than the amount of the current parent award, and must reflect actual needs of the proposed project. Direct costs for individual administrative supplements vary from less than $5,000 to more than $100,000 depending on the career level of the candidate. Administrative supplements end with the competitive cycle of the parent grant.
Restrictions: Non-domestic (non-U.S.) Entities (Foreign Institutions) and non-domestic (non-U.S.) components of U.S. Organizations are not eligible to apply. Foreign components, as defined in the NIH Grants Policy Statement, are not allowed.
Geographic Focus: All States
Contact: Nara Gavini, (301) 451-5081; fax (301) 480-0862; gavininn@nhlbi.nih.gov
Internet: http://grants.nih.gov/grants/guide/pa-files/PA-12-149.html
Sponsor: National Institutes of Health
Building 31, Room 5A48
Bethesda, MD 20892

NHLBI Right Heart Function in Health and Chronic Lung Diseases Grants 2728

This program will promote research on cellular, molecular and physiological determinants of right ventricular function in health, and function/ dysfunction in association with chronic lung diseases. The goal is to encourage research which concentrates on building a foundation of basic knowledge of right ventricular physiology and the basic molecular and biomechanical factors contributing to right ventricular dysfunction associated with pulmonary diseases. This knowledge base will be used in determining the best methods of preventing, diagnosing and treating right heart failure. This FOA seeks to fill the current gaps that exist in our understanding of the right heart and how right heart function affects lung function in both health and disease. The emphasis on interaction of the heart and the lung during chronic lung disease in this program is intended to promote collaboration between the heart and lung research communities.
Requirements: The following organizations and institutions are eligible to apply: public/state controlled institutions of higher education; private institutions of higher education; Hispanic-serving institutions; historically Black colleges and universities (HBCUs); tribally controlled colleges and universities (TCCUs); Alaska Native and Native Hawaiian serving institutions; nonprofits with 501(c)3 IRS status; nonprofits without 501(c)3 IRS status; for-profit organizations; state governments; Indian/Native American tribal governments (federally recognized); Indian/Native American tribally designated organizations; county governments; city or township governments; special district governments; independent school districts; public housing authorities/Indian housing authorities; U.S. territory or possession; Indian/Native American tribal governments (other than federally recognized); regional organizations; non-domestic (non-U.S.) entities (foreign organizations); other eligible agencies of the federal government; and faith-based or community-based organizations.
Geographic Focus: All States
Date(s) Application is Due: Feb 5; Jun 5; Oct 5
Contact: Elizabeth M. Denholm, (301) 435-0222; denholme@mail.nih.gov
Internet: http://grants.nih.gov/grants/guide/pa-files/PA-07-043.html
Sponsor: National Institutes of Health
Building 31, Room 5A48
Bethesda, MD 20892

NHLBI Ruth L. Kirschstein National Research Service Awards for Individual Postdoctoral Fellows 2729

The National Heart, Lung, and Blood Institute is one of many Institutes in the National Institutes of Health that participate in the Ruth L. Kirschstein National Research Service Awards (NRSA) for Individual Postdoctoral Fellows. This program offers health scientists the opportunity to receive full-time research training for up to 3 years in areas that reflect the national need for biomedical, clinical and behavioral research in cardiovascular, pulmonary, hematologic, and sleep disorders. These grants are not intended for study leading to the M.D., D.O., D.D.S., or equivalent professional degrees, nor do they support residency training. The proposed postdoctoral training must offer an opportunity to enhance the applicant's understanding of the health-related sciences, and must be within the broad scope of biomedical, behavioral, or clinical research or other specific disciplines relevant to the research mission of the participating NIH Institutes and Centers. Applicants with a health professional doctoral degree may use the proposed postdoctoral training to satisfy a portion of the degree requirements for a master's degree, a research doctoral degree or any other advanced research degree program.
Requirements: Any applicant fellow with the skills, knowledge, and resources necessary to carry out the proposed research as the Project Director/Principal Investigator (PD/PI) is invited to work with his/her sponsor and organization to develop an application for support. Individuals from underrepresented racial and ethnic groups as well as individuals with disabilities are always encouraged to apply for NIH support. Applicants must be U.S. citizens, noncitizen nationals, or legal permanent residents of the U.S. Training can be conducted abroad if the site provides opportunities that are not available in this country. Applicants are cautioned that not all NIH Institutes and Centers (ICs) participate in this program, and that consultation with relevant IC staff prior to submission of an application is strongly encouraged. The participating ICs have different emphases and program requirements for this program. Before submitting a fellowship application, the applicant fellow must identify a sponsoring institution. The sponsoring institution must have staff and facilities available on site to provide a suitable environment for performing high-quality research. Eligible sponsoring institutions include: Public/State Controlled and Private Institutions of Higher Education; Nonprofits with or without 501(c)3 IRS Status (Other than Institutions of Higher Education); Small Businesses; For-Profit Organizations (Other than Small Businesses); Eligible Agencies of the Federal Government; and, Non-domestic (non-U.S.) Entities (Foreign Organizations). The following types of Higher Education Institutions are always encouraged to apply for NIH support as Public or Private Institutions of Higher Education: Hispanic-serving Institutions; Historically Black Colleges and Universities (HBCUs); Tribally Controlled Colleges and Universities (TCCUs); and, Alaska Native and Native Hawaiian Serving Institutions. When the fellowship begins, the applicant must have received a doctoral degree and have arranged to work with a sponsor affiliated with an institution that has the staff and facilities needed for the proposed training.
Restrictions: A Kirschstein-NRSA fellowship may not be used to support the clinical years of residency training. However, these awards are appropriate for the research fellowship years of a residency program. Research clinicians must devote full-time to their proposed research training and confine clinical duties to those activities that are part of the research training program.

Geographic Focus: All States, Guam, Marshall Islands, Northern Mariana Islands, Puerto Rico, U.S. Virgin Islands, All Countries
Date(s) Application is Due: Apr 8; Aug 8; Dec 8
Contact: Sandra Colombini Hatch, (301) 435-0222; hatchs@nhlbi.nih.gov
Internet: http://www.nhlbi.nih.gov/funding/training/redbook/phdf32.htm
Sponsor: National Institutes of Health
Building 31, Room 5A48
Bethesda, MD 20892

NHLBI Ruth L. Kirschstein National Research Service Awards for Individual Predoctoral Fellowships to Promote Diversity in Health-Related Research 2730

NHLBI is one of many Institutes in the NIH that participate in the Ruth L. Kirschstein National Research Service Awards for Individual Predoctoral Fellowships to Promote Diversity in Health-Related Research. This program encourages students from underrepresented racial and ethnic groups, individuals with disabilities, and individuals from disadvantaged backgrounds to seek research doctoral degrees in the biomedical and behavioral sciences to help increase the number of well-trained scientists from underrepresented groups. The fellowship provides up to 5 years of support for research training leading to the Ph.D. or equivalent research degree, the combined M.D./Ph.D. degree, or other combined degrees in the biomedical or behavioral sciences.
Requirements: Individuals with disabilities, or from underrepresented racial and ethnic groups, or individuals from disadvantaged backgrounds pursuing advanced degrees in the biomedical and behavioral sciences are eligible for fellowship awards. The Fellowship applicant must have a baccalaureate degree and be currently enrolled in an eligible doctoral program. At the time of appointment, students must be U.S. citizens, noncitizen nationals, or lawfully admitted to the U.S. for permanent residence. Individuals on temporary or student visas are not eligible. Before submitting a fellowship application, the applicant fellow must identify a sponsoring institution. The sponsoring institution must have staff and facilities available on site to provide a suitable environment for performing high-quality research. Eligible sponsoring institutions include Public/State Controlled and Private Institutions of Higher Education; Nonprofits with or without 501(c)3 IRS Status (Other than Institutions of Higher Education); Small Businesses; For-Profit Organizations (Other than Small Businesses); Eligible Agencies of the Federal Government; Non-domestic (non-U.S.) Entities (Foreign Organizations). The following types of Higher Education Institutions are always encouraged to apply for NIH support as Public or Private Institutions of Higher Education: Hispanic-serving Institutions; Historically Black Colleges and Universities (HBCUs); Tribally Controlled Colleges and Universities (TCCUs); Alaska Native and Native Hawaiian Serving Institutions.
Restrictions: NIH will not accept any application that is essentially the same as one already reviewed. An individual may not have two or more competing NIH fellowship applications pending review concurrently.
Geographic Focus: All States
Date(s) Application is Due: Apr 8; Aug 8; Dec 8
Contact: Xenia J. Tigno; (301) 435-0202; tignoxt@mail.nih.gov
Internet: http://www.nhlbi.nih.gov/funding/training/redbook/gradf31.htm
Sponsor: National Institutes of Health
Building 31, Room 5A48
Bethesda, MD 20892

NHLBI Ruth L. Kirschstein National Research Service Awards for Individual Predoctoral MD/PhD Fellows and Other Dual Degree Fellows 2731

The NHLBI is interested in supporting individual predoctoral fellowships for combined MD/PhD training in research areas relevant to the NHLBI mission. The purpose of the Ruth L. Kirschstein National Research Service Awards (Kirschstein-NRSA) is to provide support to individuals for combined MD/PhD and other dual doctoral degree training (e.g. DO/PhD, DDS/PhD, AuD/PhD). The participating Institutes award this Kirschstein-NRSA individual fellowship (F30) to qualified applicants with the potential to become productive, independent, highly trained physician-scientists and other clinician-scientists, including patient-oriented researchers in their scientific mission areas. This funding opportunity supports individual predoctoral F30 fellowships with the expectation that these training opportunities will increase the number of future investigators with both clinical knowledge and skills in basic, translational or clinical research. Award budgets are composed of stipends, tuition and fees, and institutional allowance.
Requirements: Domestic for-profit or non-profit institutions/organizations, or public or private institutions, such as universities, colleges, hospitals and laboratories, eligible agencies of the federal government, and NIH intramural laboratories are eligible to apply. The sponsoring institution must have staff and facilities available on site to provide a suitable environment for performing high-quality research training. The PhD phase of the program may be conducted outside the sponsoring institution, e.g. a Federal laboratory including the NIH intramural laboratories. This training, however, must be part of a combined MD/PhD program. Applications to the program will be accepted from students currently enrolled in a formally combined MD/PhD program. By the time of award, all candidates for the Kirschstein-NRSA F30 award must be citizens or non-citizen nationals of the United States, or must have been lawfully admitted to the United States for Permanent Residence.
Restrictions: Foreign institutions are not eligible to apply.
Geographic Focus: All States
Date(s) Application is Due: Apr 8; Aug 8; Dec 8
Contact: Drew E. Carlson; (301) 435-0535; carlsonde@nhlbi.nih.gov
Internet: http://www.nhlbi.nih.gov/funding/training/redbook/gradmdphdf30.htm
Sponsor: National Institutes of Health
Building 31, Room 5A48
Bethesda, MD 20892

NHLBI Ruth L. Kirschstein National Research Service Awards for Individual Senior Fellows 2732

The National Heart, Lung, and Blood Institute is one of many Institutes in the National Institutes of Health that participate in the Ruth L. Kirschstein National Research Service Awards (NRSA) for individual Senior Fellows (Parent F33). This program enables experienced scientists to change the direction of their research careers, broaden their scientific background, acquire new research capabilities, or enlarge their command of an allied research field. The purpose of the senior fellowship (F33) award is to provide senior fellowship support to experienced scientists who wish to make major changes in the direction of their research careers or who wish to broaden their scientific background by acquiring new research capabilities as independent research investigators in scientific health-related fields relevant to the missions of the participating NIH Institutes and Centers. These awards will enable individuals with at least seven years of research experience beyond the doctorate (individuals with PhD, MD, OD, associate professors, full professors, etc.), and who have progressed to the stage of independent investigator, to take time from regular professional responsibilities for the purpose of receiving training to increase their scientific capabilities. In most cases, this award is used to support sabbatical experiences for established independent scientists seeking support for retraining or additional career development. This program is not designed for postdoctoral level investigators seeking to enhance their research experience prior to independence. Award budgets are composed of stipends, tuition and fees, and institutional allowance.
Requirements: Applicants must be U.S. citizens, non-citizen nationals, or permanent residents of the U.S. When the fellowship begins, an applicant must have received the Ph.D., M.D., D.O., D.D.S., D.V.M., O.D., D.P.M., Sc.D., D.Eng., DNSc., or equivalent degree. The applicant must arrange to work full-time with a particular sponsor affiliated with an institution that has the staff and facilities needed for the proposed training. Training can be conducted abroad if the site provides opportunities not available in the U.S. The following are eligible sponsoring organizations: public/state controlled and private institutions of higher education; nonprofits with or without 501(c)3 IRS status (other than institutions of higher education); small businesses; for-profit organizations (other than small businesses); eligible agencies of the Federal government; non-domestic (non-U.S.) entities (foreign organizations). The following types of Higher Education Institutions are always encouraged to apply for NIH support as public or private institutions of higher education: Hispanic-serving institutions; historically Black colleges and universities; Tribally controlled colleges and universities; Alaska Native and Native Hawaiian serving institutions.
Restrictions: Multiple Principal Investigators are not allowed.
Geographic Focus: All States, Guam, Marshall Islands, Northern Mariana Islands, Puerto Rico, U.S. Virgin Islands, American Samoa
Date(s) Application is Due: Apr 8; Aug 8; Dec 8
Contact: Charlotte A. Pratt; (301) 435-0382; fax (301) 480-1864; prattc@nhlbi.nih.gov
Internet: http://www.nhlbi.nih.gov/funding/training/redbook/estf33.htm
Sponsor: National Institutes of Health
Building 31, Room 5A48
Bethesda, MD 20892

NHLBI Short-Term Research Education Program to Increase Diversity in Health-Related Research 2733

The National Heart, Lung, and Blood Institute, National Institutes of Health invites Research Education (R25) applications to promote diversity in undergraduate and health professional participant populations by providing short-term research education support to stimulate career development in cardiovascular, pulmonary, hematologic, and sleep disorders research. The overall goal of the program is to provide research opportunities for individuals from backgrounds underrepresented in biomedical science, including individuals from disadvantaged backgrounds, individuals from underrepresented racial and ethnic groups, and individuals with disabilities that will significantly contribute to a diverse research workforce in the future. The research opportunities should be of sufficient depth to enable the participants, upon completion of the program, to have a thorough exposure to the principles underlying the conduct of research, and help prepare participants interested in research to pursue competitive fellowships, or other research training or career development awards. NHLBI intends to fund three to five new awards corresponding to a total of $900,000 for new grants per fiscal year.
Requirements: The following types of organizations are eligible to apply: Higher Education Institutions (Public/State Controlled Institutions of Higher Education, Private Institutions of Higher Education); Nonprofits other than Institutions of Higher Education, with or without 501(c)3 IRS Status; For-Profit Organizations; Small Businesses; State Governments; County Governments; City or Township Governments; Special District Governments; Indian/Native American Tribal Governments (Federally Recognized); Indian/Native American Tribal Governments (Other than Federally Recognized); Eligible Agencies of the Federal Government; U.S. Territory or Possession; Independent School Districts; Public Housing Authorities/Indian Housing Authorities; Native American Tribal Organizations (other than Federally recognized tribal governments); Faith-based or Community-based Organizations; Regional Organizations. The following types of Higher Education Institutions are always encouraged to apply for NIH support as Public or Private Institutions of Higher Education: Hispanic-serving Institutions; Historically Black Colleges and Universities (HBCUs); Tribally Controlled Colleges and Universities (TCCUs); Alaska Native and Native Hawaiian Serving Institutions.Some institutions provide unique opportunities for access to participants and faculty from diverse backgrounds. These institutions can assist NIH in its efforts to recruit the most talented researchers from all groups, to improve the ability to recruit subjects from diverse backgrounds into clinical research protocols, and improve the Nation's capacity to address and eliminate health disparities. These institutions are therefore encouraged to apply for NIH support. The applicant

institution must have a strong research program in the area(s) proposed for research training and must have the requisite staff and facilities to carry out the proposed program. The applicant institution must have the available research facilities, personnel, and support for the program in the areas of cardiovascular, pulmonary, or hematologic diseases or sleep disorders. Institutions with adequate staff and resources in these areas are encouraged to apply. Although the size of the award may vary within the scope of the research education program proposed, it is expected that applications will stay within the following budgetary guidelines: the total institutional annual direct cost should not exceed $319,000. The total project period for an application submitted in response to this funding opportunity may not exceed 5 years.
Restrictions: Non-domestic (non-U.S.) Entities (Foreign Institutions) and non-domestic (non-U.S.) components of U.S. Organizations are not eligible to apply. Foreign components, as defined in the NIH Grants Policy Statement, are not allowed.
Geographic Focus: All States, Guam, Marshall Islands, Northern Mariana Islands, Puerto Rico, U.S. Virgin Islands
Date(s) Application is Due: Feb 7
Amount of Grant: Up to 2,700,000 USD
Contact: Drew Carlson, Ph.D., (301) 435-0535; carlsonde@nhlbi.nih.gov
Internet: http://grants.nih.gov/grants/guide/rfa-files/RFA-HL-13-020.html
Sponsor: National Institutes of Health
Building 31, Room 5A48
Bethesda, MD 20892

NHLBI Summer Institute for Training in Biostatistics II 2734
The National Heart, Lung, and Blood Institute (NHLBI) and the National Center for Research Resources (NCRR) invite applications for grants to develop, conduct, and evaluate summer courses in the basic principles and methods of biostatistics as employed in biomedical research. The courses would introduce participants to the field of biostatistics for the purpose of attracting new students into the field. The course will attract students from the entire USA and will cover the fundamental concepts of probability, statistical reasoning and inferential methods motivated, in part, by examples that include data collected in studies of heart, lung, blood, and sleep disorders. The course will be taught during the summer, with appropriate modifications or refinements following each of the first two summers' sessions. Programs may choose an area of emphasis, such as clinical trials, statistical genetics, bioinformatics or epidemiology or take a general approach. Recent college graduates as well as advanced undergraduates should be eligible to participate provided they have not participated in SIBS previously. Direct costs are limited to $248,000 per year for a three-year period. Letters of Intent should be received by December 5, and full proposals are due by January 6.
Requirements: The following organizations and institutions are eligible to apply: public/state controlled institutions of higher education; private institutions of higher education; Hispanic-serving institutions; historically Black colleges and universities (HBCUs); tribally controlled colleges and universities (TCCUs); Alaska Native and Native Hawaiian serving institutions; nonprofits with 501(c)3 IRS status; nonprofits without 501(c)3 IRS status; for-profit organizations; state governments; Indian/Native American tribal governments (federally recognized); Indian/Native American tribally designated organizations; county governments; city or township governments; special district governments; independent school districts; public housing authorities/Indian housing authorities; U.S. territory or possession; Indian/Native American tribal governments (other than federally recognized); regional organizations; non-domestic (non-U.S.) entities (foreign organizations); other eligible agencies of the federal government; and faith-based or community-based organizations.
Geographic Focus: All States
Date(s) Application is Due: Jan 6; Dec 5
Amount of Grant: Up to 248,000 USD
Contact: Song Yang, (301) 435-0431; fax (301) 480-1862; yangso@nhlbi.nih.gov
Internet: http://grants.nih.gov/grants/guide/rfa-files/RFA-HL-09-009.html
Sponsor: National Institutes of Health
Building 31, Room 5A48
Bethesda, MD 20892

NHLBI Summer Internship Program in Biomedical Research 2735
The goal of the Summer Internship Program in Biomedical Research at NHLBI is to expose students to research investigation in a highly enriched environment that is devoted exclusively to biomedical research and training. The Program is open to high school and college, graduate or Medical and Dental Students. Participants join a research laboratory for a minimum of ten weeks between June and August and conduct research in selected areas of investigation under the guidance of an NHLBI intramural research scientist. To augment the training experience, students participate in a Seminar Series at the NHLBI, as well as the National Institutes of Health Summer Seminar Series and Summer Research Program Poster Day. While applications may be filed between November 15 and March 1, early applications are encouraged.
Requirements: Applicants must be either U.S. citizens or permanent residents and be currently enrolled full-time in high school, college, professional or graduate school with a minimum GPA of 3.0 on a 4.0 scale. College graduates are eligible for the summer between graduation and beginning either graduate or professional school. High school students should be at least 16 years of age prior to June 1. Applications must be submitted online using the link for provided for Summer Internship Applicants.
Geographic Focus: All States
Date(s) Application is Due: Mar 1
Contact: Aurora Taylor, Summer Program Coordinator; (301) 451-9440; fax (301) 594-8133; taylora2@nhlbi.nih.gov or direducation@nhlbi.nih.gov
Internet: http://www.nhlbi.nih.gov/research/intramural/education/summer.htm

Sponsor: National Institutes of Health
Building 31, Room 5A48
Bethesda, MD 20892

NHLBI Targeted Approaches to Weight Control for Young Adults Grants 2736
The National Heart, Lung, and Blood Institute (NHLBI) of the National Institutes of Health (NIH) invites applications for cooperative agreements (U01) to conduct two-phase clinical research studies to develop, refine, and test innovative behavioral and/or environmental approaches for weight control in young adults at high risk for weight gain. Weight control interventions can address weight loss, prevention of weight gain, or prevention of excessive weight gain during pregnancy. The first phase will consist of formative research to refine the proposed intervention, recruitment, retention, and adherence strategies targeted to young adults. The second phase will consist of a randomized controlled trial to test the efficacy of the intervention. For the purpose of this FOA, young adults are defined as 18-35 years of age. The maximum award for the research component will be approximately $3,000,000 in direct costs, including direct costs of up to $500,000 for the first year, $650,000 for the second year, $650,000 for the third year, $650,000 for the fourth year, and $550,000 for the fifth year. Letters of Intent should be received by September 10, and full proposals are due by October 10.
Requirements: The following organizations and institutions are eligible to apply: public/state controlled institutions of higher education; private institutions of higher education; Hispanic-serving institutions; historically Black colleges and universities (HBCUs); tribally controlled colleges and universities (TCCUs); Alaska Native and Native Hawaiian serving institutions; nonprofits with 501(c)3 IRS status; nonprofits without 501(c)3 IRS status; for-profit organizations; Indian/Native American tribal governments; Indian/Native American Tribally Designated Organizations; eligible agencies of the federal government; faith-based or community-based organizations; and regional organizations.
Geographic Focus: All States
Date(s) Application is Due: Sep 10; Oct 10
Amount of Grant: 500,000 - 650,000 USD
Contact: Catherine (Cay) Loria, (301) 435-0702; loriac@nhlbi.nih.gov
Internet: http://grants.nih.gov/grants/guide/rfa-files/RFA-HL-08-007.html
Sponsor: National Institutes of Health
Building 31, Room 5A48
Bethesda, MD 20892

NHLBI Targeting Calcium Regulatory Molecules for Arrhythmia Prevention Grants 2737
The purpose of this Funding Opportunity Announcement (FOA) is to stimulate the development of therapeutic strategies for cardiac arrhythmias that target Ca2+ handling and Ca2+ signaling molecules. The overall goal is to exploit present knowledge of Ca2+ regulatory molecules and the role they play in arrhythmogenesis to develop novel therapeutic targets. Critical components of successful responses to this program announcement are: preclinical or early clinical testing of promising new therapeutic strategies to terminate and prevent cardiac arrhythmias in established models of human arrhythmogenic disease; and the feasibility to move positive findings to Phase l or ll clinical trials. Grantees must budget for and attend two Project Director/Principal Investigator (PD/PI) meetings in Bethesda, MD, usually scheduled during the second and fourth year of the award.
Requirements: The following organizations and institutions are eligible to apply: public/state controlled institutions of higher education; private institutions of higher education; Hispanic-serving institutions; historically Black colleges and universities (HBCUs); tribally controlled colleges and universities (TCCUs); Alaska Native and Native Hawaiian serving institutions; nonprofits with 501(c)3 IRS status; nonprofits without 501(c)3 IRS status; for-profit organizations; state governments; Indian/Native American tribal governments (federally recognized); Indian/Native American tribally designated organizations; county governments; city or township governments; special district governments; independent school districts; public housing authorities/Indian housing authorities; U.S. territory or possession; Indian/Native American tribal governments (other than federally recognized); regional organizations; non-domestic (non-U.S.) entities (foreign organizations); other eligible agencies of the federal government; and faith-based or community-based organizations.
Geographic Focus: All States
Date(s) Application is Due: Feb 5; Jun 5; Oct 5
Contact: Dennis A. Przywara; (301) 435-0506; przywarad@nhlbi.nih.gov
Internet: http://grants.nih.gov/grants/guide/pa-files/PA-07-101.html
Sponsor: National Institutes of Health
Building 31, Room 5A48
Bethesda, MD 20892

NHLBI Training Program Grants for Institutions that Promote Diversity 2738
The purpose of this program is to increase the participation of individuals from diverse backgrounds in cardiovascular, pulmonary, hematologic and sleep disorders research across the career development continuum. The NHLBI's T32 Training Program for Institutions That Promote Diversity is a Ruth L. Kirschstein National Research Service Award Program intended to support training of predoctoral and health professional students and individuals in postdoctoral training at non-research intensive institutions with an institutional mission focused on serving diverse communities that are not well represented in NIH-funded research, or identified federal legislation of same, with the potential to develop meritorious training programs in cardiovascular, pulmonary, hematologic, and sleep disorders. The Training Program is designed to expand the capability for biomedical research by providing grant support to institutions that have developed successful programs that promote diversity and that offer doctoral degrees

in the health professions or in health-related sciences. These institutions are uniquely positioned to engage health disparities populations in research and translation of research advances that impact health outcomes, as well as provide health care for these populations. The primary goals of the Training Program are to: (1) contribute to the expansion of the future pool of individuals from diverse backgrounds underrepresented in the biomedicine science enterprise, (2) enable trainees to increase their competitiveness for peer-review research funding, (3) strengthen publication record of trainees, and (4) foster institutional environments conducive to professional development in the biomedical science. The application is due by 5:00 pm local time of the applicant organization.

Requirements: The following organizations/institutions are eligible to apply: Public/State Controlled Institutions of Higher Education and Private Institutions of Higher Education. The following types of Higher Education Institutions are always encouraged to apply for NIH support as Public or Private Institutions of Higher Education: Hispanic-serving Institutions; Historically Black Colleges and Universities (HBCUs); Tribally Controlled Colleges and Universities (TCCUs); Alaska Native and Native Hawaiian Serving Institutions. The Institution must be a domestic college or university with an institutional mission focused on serving diverse communities or with federal recognition of the same. To demonstrate need for research capacity development in heart, lung, and blood diseases and sleep disorders, applicant institutions should not have received over $1,515,000 in funding (direct costs) from the NHLBI per year in each of the previous five years for research training and education grants in the above NHLBI mission areas. Applicant organizations may submit more than one application, provided that each application is programmatically distinct. Any individual(s) with the skills, knowledge, and resources necessary to carry out the proposed research as the Program Director/Principal Investigator (PD/PI) is invited to work with his/her organization to develop an application for support. Individuals from underrepresented racial and ethnic groups as well as individuals with disabilities are always encouraged to apply for NIH support. The PD/PI should be an established investigator in the scientific area in which the application is targeted and capable of providing both administrative and scientific leadership to the development and implementation of the proposed program. The PD/PI will be expected to monitor and assess the program and submit all documents and reports as required. Although a letter of intent is not required, is not binding, and does not enter into the review of a subsequent application, the information that it contains allows IC staff to estimate the potential review workload and plan the review.

Restrictions: Non-domestic (non-U.S.) Entities (Foreign Organizations) are not eligible to apply. Foreign (non-U.S.) components of U.S. Organizations are not allowed.
Geographic Focus: All States, Guam, Marshall Islands, Northern Mariana Islands, Puerto Rico, U.S. Virgin Islands
Date(s) Application is Due: Feb 7
Amount of Grant: Up to 250,000 USD
Contact: Sandra Colombini Hatch, M.D., (301) 435-0222; hatchs@nhlbi.nih.gov
Internet: http://grants.nih.gov/grants/guide/rfa-files/RFA-HL-12-032.html
Sponsor: National Institutes of Health
Building 31, Room 5A48
Bethesda, MD 20892

NHLBI Translating Basic Behavioral and Social Science Discoveries into Interventions to Reduce Obesity: Centers for Behavioral Intervention Development Grants 2739

This FOA solicits cooperative agreement (U01) applications from institutions/organizations that propose to translate findings from basic research on human behavior into more effective clinical, community, and population interventions to reduce obesity and improve obesity-related behaviors. This FOA will support Centers for Behavioral Intervention Development (CBIDs) in which interdisciplinary teams of basic and applied behavioral and social science researchers develop and refine novel interventions to reduce obesity and alter obesity-related health behaviors (e.g., diet, physical activity). The interventions to be developed include any of a wide range of strategies aimed at promoting weight loss and/or preventing weight gain. Interventions can be targeted to the individual, family, social network, community, environmental, clinical or population level, or combinations of these, and to any age group. Investigators at each CBID will conduct several types of studies during the funding period, including laboratory experimental studies, formative research, early phase trials and pilot/feasibility studies, in order to create promising new avenues for reducing obesity and improving obesity-related behaviors. The maximum award for the research component will be approximately $3,750,000 in direct costs across five years, with a limit of $750,000 in direct cost for any single year. Letters of Intent should be received by December 16, and full proposals are due by January 13.

Requirements: The following organizations and institutions are eligible to apply: public/state controlled institutions of higher education; private institutions of higher education; Hispanic-serving institutions; historically Black colleges and universities (HBCUs); tribally controlled colleges and universities (TCCUs); Alaska Native and Native Hawaiian serving institutions; nonprofits with 501(c)3 IRS status; nonprofits without 501(c)3 IRS status; for-profit organizations; state governments; Indian/Native American tribal governments (federally recognized); Indian/Native American tribally designated organizations; county governments; city or township governments; special district governments; independent school districts; public housing authorities/Indian housing authorities; U.S. territory or possession; Indian/Native American tribal governments (other than federally recognized); regional organizations; non-domestic (non-U.S.) entities (foreign organizations); other eligible agencies of the federal government; and faith-based or community-based organizations.
Geographic Focus: All States
Date(s) Application is Due: Jan 13; Dec 16
Amount of Grant: Up to 750,000 USD
Contact: Susan M. Czajkowski, (301) 435-0406; czajkows@mail.nih.gov
Internet: http://grants.nih.gov/grants/guide/rfa-files/RFA-HL-08-013.html
Sponsor: National Institutes of Health
Building 31, Room 5A48
Bethesda, MD 20892

NHLBI Translational Programs in Lung Diseases 2740

The purpose of this program is to encourage interdisciplinary approaches to lung translational research, allowing for potentially high-impact research, which will translate basic observations to applied clinical research. An application to NHLBI requires a minimum of three interrelated research projects that focus a number of scientific disciplines on investigations of a complex biomedical theme or research question. Applicants will be required to propose research that will become progressively more clinical and translational during the life of the program project, and they are expected to assemble interactive, multidisciplinary teams that have the combined expertise to formulate a plan for successful translation. Research teams are not required to have prior collaborative experience but must be able to demonstrate an integrated, practical approach that will result in the effective progression of mechanistic/basic concepts toward application in the clinic. This program allows for a maximum of two 5-year awards (a new application and one competing renewal). There are two application cycles. Letters of Intent should be received by April 25 and August 25 (respectively), and full proposals are due by May 25 and September 25 (respectively).

Requirements: The following organizations and institutions are eligible to apply: public/state controlled institutions of higher education; private institutions of higher education; Hispanic-serving institutions; historically Black colleges and universities (HBCUs); tribally controlled colleges and universities (TCCUs); Alaska Native and Native Hawaiian serving institutions; nonprofits with 501(c)3 IRS status; nonprofits without 501(c)3 IRS status; for-profit organizations; state governments; Indian/Native American tribal governments (federally recognized); Indian/Native American tribally designated organizations; county governments; city or township governments; special district governments; independent school districts; public housing authorities/Indian housing authorities; U.S. territories or possessions; Indian/Native American tribal governments (other than federally recognized); regional organizations; non-domestic (non-U.S.) entities (foreign organizations); other eligible agencies of the federal government; and faith-based or community-based organizations.
Geographic Focus: All States
Date(s) Application is Due: Apr 25; May 25; Aug 25; Sep 25
Contact: Patricia Noel, (301) 435-0202; noelp@nhlbi.nih.gov
Internet: http://grants.nih.gov/grants/guide/pa-files/PAR-09-185.html
Sponsor: National Institutes of Health
Building 31, Room 5A48
Bethesda, MD 20892

NHSCA Arts in Health Care Project Grants 2741

Arts in Health Care grants support arts activities, presentations and artist residencies that occur in health care facilities, rehabilitation centers and in centers serving the needs of the elderly. The overall goal of this grant category is to utilize the arts to enhance the quality of life and promote an environment conducive to healing for patients, residents, and/or clients. This grant category is in response to the Arts Council's commitment to meeting the needs of underserved populations, which can include the elderly, people with disabilities and people with health challenges. Requests may be made for $1,000 - $4,500.

Requirements: At a minimum, grants must be matched on a one-to-one basis. Health care facilities, hospitals, rehabilitation centers and facilities serving the elderly including: assisted living facilities, county and nonprofit nursing homes, veterans' homes, hospice care programs or visiting nurse associations may apply. Applicants must have 501(c)3 tax-exempt status from the IRS and not-for-profit incorporation in the State of New Hampshire and: make their programs accessible to people with disabilities; have submitted all required reports on past State Arts Council grants; and be in good standing with the N.H. Secretary of State's Office and the N.H. Attorney General's Office.
Restrictions: This grant does not fund: projects/activities that are not open to the general public; commercially viable for-profit publications, recordings or films; general operating expenses not directly related to the project; fundraising costs; projects already receiving funds from another State Arts Council grant category; or any cost item listed in the glossary under ineligible expenses.
Geographic Focus: New Hampshire
Date(s) Application is Due: Jun 2
Amount of Grant: 1,000 - 4,500 USD
Contact: Catherine O'Brian, Program Coordinator; (603) 271-0795 or (603) 271-2789; fax (603) 271-3584; Catherine.R.OBrian@dcr.nh.gov
Internet: http://www.nh.gov/nharts/grants/partners/artsinhealthcare.htm
Sponsor: New Hampshire State Council on the Arts
19 Pillsbury Street, 1st Floor
Concord, NH 03301

NIAAA Independent Scientist Award 2742

The award is a special salary-only grant designed to provide protected time for newly independent scientists who currently have non-research obligations such as heavy teaching loads, clinical work, committee assignments, service, and administrative duties that prevent them from having a period of intensive research focus. The award is targeted to persons with doctoral degrees who have completed their research training, have independent peer-reviewed research support, and who need a period of protected research time in order to foster their research career development. New applications are due February 12, June 12, and October 12, while renewal applications are due March 12, July 12, and November 12. Up to 75% per year of base salary plus fringe benefits is available. Research support is not provided

in most cases. Individuals engaged in predominantly theoretical work, such as modeling or computer simulation, may request up to $25,000 per year for research expenses.
Requirements: The following organizations and institutions are eligible to apply: public/state controlled institutions of higher education; private institutions of higher education; nonprofits with 501(c)3 IRS status; nonprofits without 501(c)3 IRS status; small businesses; for-profit organizations; State governments; U.S. territories or possessions; Indian/Native American tribal governments (Federally recognized and other than Federally recognized); Indian/Native American tribally designated organizations; Hispanic-serving institutions; historically Black colleges and universities (HBCUs); tribally controlled colleges and universities (TCCUs); Alaska Native and Native Hawaiian serving institutions; regional organizations; and faith-based or community based organizations. The candidate must have a doctoral degree and peer-reviewed, independent research support at the time the award is made. The candidate must spend a minimum of 75 percent effort conducting research during the period of the award.
Restrictions: The award is not intended for investigators who already have full time to perform research, or have substantial publication records or considerable research support indicating that they are well established in their fields. Foreign institutions are not eligible to apply.
Geographic Focus: All States
Date(s) Application is Due: Feb 12; Mar 12; Jun 12; Jul 12; Oct 12; Nov 12
Amount of Grant: Up to 100,000 USD
Contact: Michael Hilton, (301) 402-9402; mhilton@mail.nih.gov
Internet: http://grants.nih.gov/grants/guide/pa-files/PA-09-038.html
Sponsor: National Institute on Alcohol Abuse and Alcoholism
5635 Fishers Lane, MSC 9304
Bethesda, MD 20892-9304

NIAAA Mechanisms of Alcohol and Drug-Induced Pancreatitis Grants 2743
This Funding Opportunity Announcement (FOA) issued by the National Institute on Alcohol Abuse and Alcoholism (NIAAA), the National Institute of Diabetes and Digestive and Kidney Diseases (NIDDK), and the National Institute on Drug Abuse (NIDA), National Institutes of Health (NIH), solicits grant applications from institutions/organizations that propose to investigate the underlying molecular, biochemical, and cellular mechanisms by which long-term alcohol ingestion and drugs lead to the development of pancreatitis. Research is also encouraged to understand the role of various predisposing factors, including substance abuse, that make the pancreas susceptible to alcoholic injury. Understanding the mechanisms as well as the role of predisposing factors may help in developing strategies for the prevention or treatment of the disease. Because the nature and scope of the proposed research will vary from application to application, it is anticipated that the size and duration of each award will also vary.
Requirements: Application can be made by the following institutions and organizations: public/state controlled institutions of higher education; private institutions of higher education; nonprofits with 501(c)3 IRS status (other than institutions of higher education); nonprofits without 501(c)3 IRS status (other than institutions of higher education); small businesses; for-profit organizations (other than small businesses); state governments; U.S. territories or possessions; Indian/Native American tribal governments (federally recognized); Indian/Native American tribal governments (other than Federally recognized); Indian/Native American tribally designated organizations; non-domestic (non-U.S.) entities (foreign organizations); Hispanic-serving institutions; historically Black colleges and universities (HBCUs); tribally controlled colleges and universities (TCCUs); Alaska Native and Native Hawaiian serving institutions; regional organization; eligible agencies of the Federal government; and faith-based or community based organizations.
Geographic Focus: All States
Date(s) Application is Due: Feb 5; Jun 5; Oct 5
Contact: Vishnudutt Purohit, (301) 443-2689; fax (301) 443-0673; vpurohit@mail.nih.gov
Internet: http://grants.nih.gov/grants/guide/pa-files/PA-07-067.html
Sponsor: National Institute on Alcohol Abuse and Alcoholism
5635 Fishers Lane, MSC 9304
Bethesda, MD 20892-9304

NIAAA Mentored Clinical Scientist Research Career Development Awards 2744
This award enables candidates holding professional degrees (e.g., M.D., D.O., D.V.M., or equivalent degrees) to undertake 3 to 5 years of special study and supervised research with the goal of becoming independent investigators. The award also allows awardees to pursue a research career development program suited to their experience and capabilities under a mentor who is competent to provide guidance in the chosen research area. Institutions may submit applications on behalf of candidates who hold professional degrees. At least 2 years must have elapsed since the health professional degree was granted. Candidates can have varying levels of clinical training and research experience. New applications are due February 12, June 12, and October 12, while resubmitted applications are due March 12, July 12, and November 12. Up to Up to 75 percent of base salary plus fringe benefits is available, and up to $50,000 per year for research support for new awards.
Requirements: All candidates must be U.S. citizens, non-citizen nationals, or legal permanent residents of the U.S. Persons with temporary or student visas are not eligible. The grantee institution must have a strong, well-established research and research training program in the chosen area, accomplished faculty in the basic and clinical sciences, and a commitment to the candidate's research development. The proposed program should include an appropriate mentor.
Geographic Focus: All States
Date(s) Application is Due: Feb 12; Mar 12; Jun 12; Jul 12; Oct 12; Nov 12
Amount of Grant: 100,000 USD
Contact: Judith A. Arroyo, (301) 402-0717; fax (301) 443-6077; jarroyo@mail.nih.gov
Internet: http://grants.nih.gov/grants/guide/pa-files/pa-06-512.html
Sponsor: National Institute on Alcohol Abuse and Alcoholism
5635 Fishers Lane, MSC 9304
Bethesda, MD 20892-9304

NIA Aging, Oxidative Stress and Cell Death Grants 2745
The purpose of this Program Announcement is to encourage the submission of applications to support research on the relationship between oxidative stress and apoptosis, and how these biological processes are involved in aging and/or change with age. Although the National Institute of General Medical Sciences is not participating in this program announcement, that Institute continues to be interested in receiving applications on basic mechanisms of programmed cell death and oxidative stress when these applications do not focus on aging mechanisms. Modular Grant applications will request direct costs in $25,000 modules, up to a total direct cost request of $250,000 per year.
Requirements: The following organizations and institutions may apply: public/state controlled institutions of higher education; private institutions of higher education; Hispanic-serving institutions; historically Black colleges and universities (HBCUs); tribally controlled colleges and universities (TCCUs); Alaska native and native Hawaiian serving institutions; nonprofits with 501(c)3 IRS status; nonprofits without 501(c)3 IRS status; small businesses; for-profit organizations; State governments; regional organizations; U.S. territories or possessions; Indian/Native American tribal governments; Indian/Native American tribally designated organizations; county governments; city or township governments; special district governments; independent school districts; public housing authorities/Indian housing authorities; eligible agencies of the Federal government; and faith-based or community based organizations.
Geographic Focus: All States
Date(s) Application is Due: Feb 5; Jun 5; Oct 5
Amount of Grant: Up to 250,000 USD
Contact: Chyren Hunter, (301) 496-9322; fax (301) 402-0528; Hunterc@nia.nih.gov
Internet: http://grants.nih.gov/grants/guide/pa-files/PA-00-081.html
Sponsor: National Institute on Aging
Gateway Building, Room 2C218, 7201 Wisconsin Avenue, MSC 9205
Bethesda, MD 20892-9205

NIA Aging Research Dissertation Awards to Increase Diversity 2746
This Funding Opportunity Announcement (FOA) issued by the National Institute on Aging (NIA) of the National Institutes of Health (NIH) publicizes the re-issuance of an FOA that publicizes the availability of dissertation awards in aging research to increase the diversity of the research-on-aging workforce. These awards are available to all qualified Predoctoral students in accredited research doctoral programs in the United States (including Puerto Rico and other U.S. territories or possessions).
Requirements: Public or private institutions, such as universities, colleges, hospitals, and laboratories, that offer accredited research doctoral programs may submit an application.
Restrictions: Domestic institutions only are eligible.
Geographic Focus: All States
Date(s) Application is Due: Feb 16; Jun 16; Oct 16
Amount of Grant: 50,000 USD
Contact: Michael-David A.R.R. Kerns, (301) 402-7713; fax (301) 402-2945; michael-david.kerns@nih.hhs.gov
Internet: http://grants.nih.gov/grants/guide/pa-files/PAR-08-250.html
Sponsor: National Institute on Aging
Gateway Building, Room 2C218, 7201 Wisconsin Avenue, MSC 9205
Bethesda, MD 20892-9205

NIA AIDS and Aging: Behavioral Sciences Prevention Research Grants 2747
NIA, NIMH, and NINR invite qualified researchers to submit applications to investigate prevention issues relevant to AIDS in middle-aged and older populations. This announcement solicits AIDS prevention research proposals to study primary prevention of disease transmission as well as secondary and tertiary prevention of negative behavioral and social consequences of HIV/AIDS for persons with AIDS, their families, and communities. Thus, the primary goals are to identify social and behavioral factors associated with HIV transmission and disease progression in later life; examine behavioral and social consequences of HIV infection/AIDS across the life course; develop and evaluate age-appropriate behavioral and social interventions for preventing AIDS in middle-aged and older adults and/or ameliorating problems associated with older adult caregiver responsibilities and burdens; explore health care issues surrounding AIDS care; and strengthen existing research and evaluation methods.
Requirements: Applications may be submitted by domestic and foreign for-profit and non-profit organizations, public and private, such as universities, colleges, hospitals, laboratories, units of State and local governments, and eligible agencies of the Federal government. Applications may be submitted by single institutions or by a consortia of institutions.
Geographic Focus: All States
Date(s) Application is Due: Feb 5; Feb 16; Jun 5; Jun 16; Oct 5; Oct 16
Contact: Ying Tian, (301) 496-6761; fax (301) 402-1784; tiany@nia.nih.gov
Internet: http://www.grants.nih.gov/grants/guide/pa-files/PA-97-069.html
Sponsor: National Institute on Aging
Gateway Building, Suite 3C307, 7201 Wisconsin Avenue, MSC 9205
Bethesda, MD 20892-9205

NIA Alzheimer's Disease Core Centers Grants 2748
The National Institute on Aging (NIA) invites applications from qualified institutions for support of Alzheimer's Disease Core Centers (A.D.C.Cs). These centers are designed to support and conduct research on Alzheimer's disease (AD), to serve as shared research resources that will facilitate research in AD and related disorders, distinguish them from the

processes of normal brain aging and mild cognitive impairment (MCI), provide a platform for training, develop novel techniques and methodologies, and translate these research findings into better diagnostic, prevention and treatment strategies. New applications should request a project period of five years and a budget for direct costs of up to $750,000 per year. Competing renewal applications may request direct costs for all cores (both required and optional), and the other listed functions, i.e., satellites and pilot grants at a level not exceeding the combined direct costs of all funded activities awarded during the final year of the present funding period plus a 3% increase, or $750,000 per year, whichever is larger. Letters of Intent should arrive by the April 6 deadline, with full proposals due by May 5
Requirements: The following organizations and institutions are eligible to apply: public/state controlled institutions of higher education; private institutions of higher education; Hispanic-serving institutions; historically Black colleges and universities (HBCUs); tribally controlled colleges and universities (TCCUs); Alaska Native and Native Hawaiian serving institutions; nonprofits with 501(c)3 IRS status (other than institutions of higher education); and nonprofits without 501(c)3 IRS status (other than institutions of higher education).
Restrictions: At the time of award, the applicant institution cannot have another Alzheimer's Disease Center funded by the NIA.
Geographic Focus: All States
Date(s) Application is Due: Apr 6; May 5
Amount of Grant: 750,000 USD
Contact: Creighton H. Phelps, (301) 496-9350; fax (301) 496-1494; phelpsc@nia.nih.gov
Internet: http://grants.nih.gov/grants/guide/rfa-files/RFA-AG-10-001.html
Sponsor: National Institute on Aging
Gateway Building, Suite 350, 7201 Wisconsin Avenue, MSC 9205
Bethesda, MD 20892-9205

NIA Alzheimer's Disease Drug Development Program Grants 2749
The NIA and other NIH Institutes currently support research projects for the study of the epidemiology, etiology, diagnosis, and treatment of Alzheimer's disease (AD). Investigators in all relevant scientific disciplines and organizations are invited to develop and pre-clinically test compounds which can slow the disease progression, reverse the disease process, or delay the onset of or prevent AD, mild cognitive impairment (MCI), or age-related cognitive decline.
Requirements: The following organizations and institutions may apply: public/state controlled institutions of higher education; private institutions of higher education; nonprofits with 501(c)3 IRS status; nonprofits without 501(c)3 IRS status; small businesses; for-profit organizations; State governments; and non-domestic (non-U.S.) entities (foreign organizations).
Geographic Focus: All States
Date(s) Application is Due: Jan 2; Feb 2; Sep 1; Oct 0
Amount of Grant: 300,000 - 800,000 USD
Contact: Suzana Petanceska, (301) 496-9350; fax (301) 496-1494; petanceskas@nia.nih.gov
Internet: http://grants.nih.gov/grants/guide/pa-files/PAR-08-266.html
Sponsor: National Institute on Aging
Gateway Building, Suite 350, 7201 Wisconsin Avenue, MSC 9205
Bethesda, MD 20892-9205

NIA Alzheimer's Disease Pilot Clinical Trials 2750
The objective of the Alzheimer's Disease Pilot Clinical Trials initiative is to improve the quality of clinical research designed to evaluate interventions for the prevention and treatment of Alzheimer's disease (AD), mild cognitive impairment (MCI), and age-associated cognitive decline by stimulating applications for pilot clinical trials to test interventions aimed at delaying the onset of or preventing AD, MCI, and age-associated cognitive decline; slowing, halting, or, if possible, reversing the progressive decline in cognitive function; and modifying the cognitive and behavioral symptoms in AD and MCI. The goal is not to duplicate or compete with pharmaceutical companies but to encourage, complement, and accelerate the process of testing new, innovative, and effective treatments. The pilot research project application should directly address how it will inform and advance the design of a subsequent full-scale clinical trial. The application should also address the intrinsic scientific merit of the pilot trial itself. The application may include the following: studies to refine the intervention strategy (e.g. drug dosage, duration, delivery system; behavioral intensity and duration); studies to define and refine the target population and ensure adequate enrollment, protocol adherence and subject retention; collection of preliminary data for establishing measures of clinical efficacy including clinical/neuropsychological/behavioral measures, neuroimaging measures, and other biological measures in blood and cerebrospinal fluid; assessment of efficacy (including primary and secondary outcome and assessment measures for both patients and caregivers); and studies of safety.
Requirements: The following organizations and institutions may apply: public/state controlled institutions of higher education; private institutions of higher education; Hispanic-serving institutions; historically Black colleges and universities (HBCUs); tribally controlled colleges and universities (TCCUs); Alaska native and native Hawaiian serving institutions; nonprofits with 501(c)3 IRS status; nonprofits without 501(c)3 IRS status; small businesses; for-profit organizations; State governments; regional organizations; U.S. territories or possessions; Indian/Native American tribal governments; Indian/Native American tribally designated organizations; county governments; city or township governments; special district governments; independent school districts; public housing authorities/Indian housing authorities; eligible agencies of the Federal government; and faith-based or community based organizations.
Geographic Focus: All States
Date(s) Application is Due: Feb 5; Jun 5; Oct 5
Amount of Grant: 450,000 - 1,350,000 USD
Contact: Laurie M. Ryan, (301) 496-9350; fax (301) 496-1494; ryanl@nia.nih.gov

Internet: http://grants.nih.gov/grants/guide/pa-files/PAR-08-062.html
Sponsor: National Institute on Aging
Gateway Building, Suite 350, 7201 Wisconsin Avenue, MSC 9205
Bethesda, MD 20892-9205

NIA Alzheimer's Disease Research Centers Grants 2751
The National Institute on Aging (NIA) invites applications from qualified institutions for support of Alzheimer's Disease Research Centers (ADRCs). These centers are designed to support and conduct research on Alzheimer's disease (AD), to serve as shared research resources that will facilitate research in AD and related disorders, distinguish them from the processes of normal brain aging and mild cognitive impairment (MCI), provide a platform for training, develop novel techniques and methodologies, and translate these research findings into better diagnostic, prevention and treatment strategies. New applications should request a project period of five years and a budget for direct costs of up to $1 million per year. Letters of Intent are due by April 6 annually, with full applications received by May 5.
Requirements: The following organizations and institutions are eligible to apply: public/state controlled institutions of higher education; private institutions of higher education; Hispanic-serving institutions; historically Black colleges and universities (HBCUs); tribally controlled colleges and universities (TCCUs); Alaska native and native Hawaiian serving institutions; nonprofits with 501(c)3 IRS status (other than institutions of higher education); and nonprofits without 501(c)3 IRS status (other than institutions of higher education).
Geographic Focus: All States
Date(s) Application is Due: Apr 5; May 6
Amount of Grant: Up to 1,000,000 USD
Contact: Creighton H. Phelps, (301) 496-9350; fax (301) 496-1494; phelpsc@nia.nih.gov
Internet: http://grants.nih.gov/grants/guide/rfa-files/RFA-AG-10-002.html
Sponsor: National Institute on Aging
Gateway Building, Suite 350, 7201 Wisconsin Avenue, MSC 9205
Bethesda, MD 20892-9205

NIA Archiving and Development of Socialbehavioral Datasets in Aging Related Studies Grants 2752
This Funding Opportunity Announcement (FOA), issued by the National Institute on Aging (NIA) is seeking small grant applications to stimulate and facilitate data archiving and development related to cognitive psychology, behavioral interventions in the context of randomized controlled trials (RCTs), demography, economics, epidemiology, behavioral genetics and other behavioral research on aging for secondary analysis.
Requirements: The following organizations and institutions may apply: public/state controlled institutions of higher education; private institutions of higher education; Hispanic-serving institutions; historically Black colleges and universities (HBCUs); tribally controlled colleges and universities (TCCUs); Alaska native and native Hawaiian serving institutions; nonprofits with 501(c)3 IRS status; nonprofits without 501(c)3 IRS status; small businesses; for-profit organizations; State governments; regional organizations; U.S. territories or possessions; Indian/Native American tribal governments; Indian/Native American tribally designated organizations; county governments; city or township governments; special district governments; independent school districts; public housing authorities/Indian housing authorities; eligible agencies of the Federal government; and faith-based or community based organizations.
Geographic Focus: All States
Date(s) Application is Due: Feb 16; Jun 16; Oct 16
Amount of Grant: Up to 100,000 USD
Contact: Partha Bhattacharyya, (301) 496-3136; bhattacharyyap@mail.nih.gov
Internet: http://grants.nih.gov/grants/guide/pa-files/PA-08-252.html
Sponsor: National Institute on Aging
Gateway Building, Suite 2C231, 7201 Wisconsin Avenue, MSC 9205
Bethesda, MD 20892

NIA Awards to Support Research on the Biology of Aging in Invertebrates Grants 2753
This Funding Opportunity Announcement (FOA) issued by the National Institute on Aging (NIA), National Institutes of Health, invites applications that propose the identification and development of new invertebrate models for use in biomedical research that is highly relevant to the biology of aging. Applications should focus on the identification, development and characterization of new invertebrate models that have short lifespans. Invertebrate models with tractable genetics and genome sequence available or in progress, negligible or induced senescence, and tissue regeneration in adults are of particular interest to NIA. Examples of invertebrate models of interest include, but are not limited to planaria, hydra, rotifers, and tunicates. These invertebrate models are currently not well developed for aging research. Within the scope of this FOA, invertebrate models other than those mentioned above currently used by the biomedical research community, but not yet fully developed or realized for aging research, will be considered responsive to this FOA. In particular, applications that focus on the characterization of these novel models for aging research utilizing molecular, genetic, cellular or physiological tools and approaches to identify genes involved in longevity, senescence and cellular pathways in tissue homeostasis during aging are encouraged. Letters of Intent are due by September 30, with final applications due by October 30.
Requirements: The following organizations and institutions are eligible to apply for support: public/state controlled institutions of higher education; private institutions of higher education; Hispanic-serving institutions; historically Black colleges and universities (HBCUs); tribally controlled colleges and universities (TCCUs); Alaska Native and Native Hawaiian serving institutions; nonprofits with 501(c)3 IRS status (other than institutions of higher education); nonprofits without 501(c)3 IRS status (other

than institutions of higher education); Indian/Native American tribal governments (federally recognized); Indian/Native American tribally designated organizations; U.S. territories or possessions; and eligible agencies of the federal government.
Geographic Focus: All States
Date(s) Application is Due: Sep 30; Oct 30
Amount of Grant: Up to 250,000 USD
Contact: Anna M. McCormick, Ph.D; (301) 496-6402; mccormia@nia.nih.gov
Internet: http://grants.nih.gov/grants/guide/rfa-files/RFA-AG-10-004.html
Sponsor: National Institute on Aging
Gateway Building, Suite 2C231, 7201 Wisconsin Avenue, MSC 9205
Bethesda, MD 20892

NIA Behavioral and Social Research Grants on Disasters and Health 2754
The purpose of this FOA is to stimulate research in the behavioral and social sciences on the consequences of natural and man-made disasters for the health of children, the elderly, and vulnerable groups, with an ultimate goal of preventing and mitigating harmful consequences and health disparities. Disasters include severe weather-related events, earthquakes, large-scale attacks on civilian populations, technological catastrophes or perceived catastrophes, and influenza pandemics. For the elderly and for children and youth, the health outcomes of greatest interest include mortality, disability and resilience, severe distress and clinically significant morbidity (as opposed to mild or transient symptoms and dysphoria), and economic hardship sufficient to harm health. For children and youth, long-term effects on development are also of interest.
Requirements: The following organizations and institutions may apply: for-profit organizations; non-profit organizations; public or private institutions, such as universities, colleges, hospitals, and laboratories; units of State government; units of Local government; units of State Tribal government; units of Local Tribal government; eligible agencies of the Federal government; foreign institutions; domestic institutions; faith-based or community-based organizations; Indian/Native American tribal governments (federally recognized); Indian/Native American tribal governments (other than federally recognized); and Indian/Native American tribally designated organizations.
Geographic Focus: All States
Date(s) Application is Due: Feb 16; Jun 16; Oct 16
Contact: Rebecca L. Clark, (301) 496-1175; rclark@mail.nih.gov
Internet: http://grants.nih.gov/grants/guide/pa-files/PA-06-452.html
Sponsor: National Institute on Aging
Gateway Building, Room 533, 7201 Wisconsin Avenue, MSC 9205
Bethesda, MD 20892

NIA Bioenergetics, Fatigability, and Activity Limitations in Aging 2755
This FOA encourages Research Project Grant applications proposing to study bioenergetic factors underlying increased fatigability and activity limitations in aging. Increased fatigability is a significant cause of restricted physical and cognitive activity in older adults. Alterations in bioenergetics - the production and utilization of energy, and the regulation of these processes - may contribute significantly to increased fatigability. This FOA encourages applications that propose to: elucidate specific alterations in bioenergetics related to increased fatigability and activity limitations; develop and evaluate improved measures of fatigability related to bioenergetics; and evaluate interventions for increased fatigability and activity limitations that target alterations in bioenergetics and lead to improved quality of life.
Requirements: The following organizations and institutions may apply: public/state controlled institutions of higher education; private institutions of higher education; Hispanic-serving institutions; historically Black colleges and universities (HBCUs); tribally controlled colleges and universities (TCCUs); Alaska native and native Hawaiian serving institutions; nonprofits with 501(c)3 IRS status; nonprofits without 501(c)3 IRS status; small businesses; for-profit organizations; State governments; regional organizations; U.S. territories or possessions; Indian/Native American tribal governments; Indian/Native American tribally designated organizations; county governments; city or township governments; special district governments; independent school districts; public housing authorities/Indian housing authorities; eligible agencies of the Federal government; and faith-based or community based organizations.
Geographic Focus: All States
Date(s) Application is Due: Feb 5; Feb 16; Jun 5; Jun 16; Oct 5; Oct 16
Contact: Cheryl L. McDonald, (301) 435-0560; mcdonalc@nhlbi.nih.gov
Internet: http://grants.nih.gov/grants/guide/pa-files/PA-09-190.html
Sponsor: National Institute on Aging
Building 31, Room 5C27, 31 Center Drive, MSC 2292
Bethesda, MD 20892-7710

NIA Diversity in Medication Use and Outcomes in Aging Populations Grants 2756
The National Institute on Aging (NIA) in collaboration with the National Center for Complementary and Alternative Medicine, the Office of Research on Minority Health and the Office of Research on Women's Health invites qualified researchers to submit applications to investigate issues relevant to medication use and outcomes among older people. This program announcement solicits both basic and applied research proposals to investigate the multiple factors that influence medication use by the elderly. Goals of this announcement are to: identify the epidemiological, psychosocial, health care and clinical factors associated with medication use by older people, including the use of unconventional medical products; assess social, behavioral, psychological and cognitive factors that play a role in older people's understanding of and adherence to medication regimens; determine the role of medical and pharmaceutical professionals in facilitating or hindering proper use of prescribed and over-the-counter medications and unconventional medications including herbs and dietary supplements; refine methodologies to link these factors to medical or psychosocial outcomes; investigate interventions to improve medication adherence; and increase our understanding of biological factors which contribute to therapeutic outcomes in the use of medications by the elderly.
Requirements: Applications may be submitted by foreign and domestic for-profit and non-profit organizations, public and private, such as universities, colleges, hospitals, laboratories, units of state and local governments, and eligible agencies of the Federal government.
Geographic Focus: All States
Contact: Laurie M. Ryan, (301) 496-9350; fax (301) 496-1494; ryanl@nia.nih.gov
Internet: http://grants.nih.gov/grants/guide/pa-files/PA-99-097.html
Sponsor: National Institute on Aging
Gateway Building, Suite 350, 7201 Wisconsin Avenue, MSC 9205
Bethesda, MD 20892-9205

NIA Effects of Gene-Social Environment Interplay on Health and Behavior in Later Life Grants 2757
The National Institute on Aging (NIA) invites applications for the development of multidisciplinary collaborations among existing longitudinal twin and family studies, with a focus on social and behavioral factors associated with aging outcomes. This FOA is intended to lay the foundation for future studies of the role of gene-environment interplay in accounting for links between social experiences and physical health, functionality, and psychological well-being in midlife and older age. Of particular interest are applications that can embed this foundation within a lifespan perspective. Applicants are encouraged to use this FOA as a starting point for: developing collaborations among existing data sets; harmonizing social phenotypes across a number of twin and family studies; conducting analyses aimed at further refining social phenotypes for genetic analysis; using animal models of gene-social environment interplay to inform work using twin studies; and/or establishing preliminary data on the feasibility of genetic/genomic approaches that could be applied to questions of how social experiences are biologically embedded and influence middle and later life outcomes. Use of existing data is strongly encouraged; funds from this award are not intended for new data collection, but a small percentage of funds may be used for adding modules to ongoing data collection efforts to augment phenotypic or genetic data. NIA expects to make 1-2 awards totaling $750,000 total costs in the first year, and $3,750,000 in total costs over the 5 year project period. Direct costs are expected to range from $200,000 -$500,000 in the first year. Letters of Intent are due by October 9, with final applications due by November 9.
Requirements: The following organizations and institutions may apply: public/state controlled institutions of higher education; private institutions of higher education; Hispanic-serving institutions; historically Black colleges and universities (HBCUs); tribally controlled colleges and universities (TCCUs); Alaska native and native Hawaiian serving institutions; nonprofits with 501(c)3 IRS status; nonprofits without 501(c)3 IRS status; small businesses; for-profit organizations; State governments; regional organizations; U.S. territories or possessions; Indian/Native American tribal governments; Indian/Native American tribally designated organizations; county governments; city or township governments; special district governments; independent school districts; public housing authorities/Indian housing authorities; eligible agencies of the Federal government; and faith-based or community based organizations.
Geographic Focus: All States
Date(s) Application is Due: Oct 9; Nov 9
Amount of Grant: 200,000 - 500,000 USD
Contact: Erica L. Spotts, (301) 451-4503; fax (301) 402-0051; spottse@mail.nih.gov
Internet: http://grants.nih.gov/grants/guide/rfa-files/RFA-AG-10-006.html
Sponsor: National Institute on Aging
Gateway Building, Room 533, 7201 Wisconsin Avenue, MSC 9205
Bethesda, MD 20892

NIA Factors Affecting Cognitive Function in Adults with Down Syndrome Grants 2758
The purpose of this solicitation is to encourage investigator initiated research to focus on various factors that maximize and maintain cognitive function in adults (>21 years of age) with Down syndrome. This solicitation encourages applications from investigators from very diverse backgrounds and interdisciplinary and transdisciplinary proposals, as well as those that involve systematic longitudinal study rather than study of samples of convenience. These studies could serve to form an evidence base that will inform not only health care management and treatment, but also issues such as quality of life, access to services, education, and ultimately affect public policy. This solicitation encourages applications that would inform several areas of inquiry: the natural history of medical and mental health issues in adults with Down syndrome, their incidence and onset; health disparities that occur, and their prevention or effective treatment; the development of better diagnostic tools to assess functional and cognitive impairment in adults with Down syndrome; improved quality of life measures; community-based interventions that improve individual educational, occupational, and social outcomes; the impact of early entry to services, type of intervention, and/or family history on quality of life for adults with Down Syndrome; and the impact of family structure and ancestry on critical life transitions, co-occurring conditions, and developmental changes of adults with Down syndrome through longitudinal studies rather than studies of samples of convenience. The requested budget is limited to $500,000 total costs per year for a maximum of five years. Letters of Intent are due by May 29, and full applications should arrive by June 30.
Requirements: The following organizations and institutions are eligible to apply: public/state controlled institutions of higher education; private institutions of higher education; nonprofits with 501(c)3 IRS status (other than institutions of higher education); nonprofits without 501(c)3 IRS status (other than institutions of higher education); for-profit organizations (other than small businesses); State governments; county governments; eligible agencies of the Federal government; and faith-based or community-based organizations.

Restrictions: Foreign institutions may not submit applications, but collaborators at foreign institutions are allowed.
Geographic Focus: All States
Date(s) Application is Due: May 29; Jun 30
Amount of Grant: Up to 500,000 USD
Contact: Mary Lou Oster-Granite, (301) 435-6866; fax (301) 496-3791; mo96o@nih.gov
Internet: http://grants.nih.gov/grants/guide/rfa-files/RFA-HD-09-028.html
Sponsor: Eunice Kennedy Shriver National Institute of Child Health and Human Development
6100 Executive Boulevard, Room 4B05L, MSC 7510
Bethesda, MD 20892-7510

NIA Grants for Alzheimer's Disease Drug Discovery 2759
The objective of this solicitation is to stimulate preclinical research in the discovery, design, development and testing of novel compounds aimed at slowing, halting, or, if possible, reversing the progressive decline in cognitive function and modifying the behavioral symptoms in Alzheimer's disease as well as delaying the onset of or preventing AD. This initiative is intended to stimulate basic research and development efforts. The goal is not to duplicate or compete with pharmaceutical companies but to encourage, complement, and accelerate the process of discovering new, innovative, and effective compounds for the prevention and treatment of the cognitive impairment and behavioral symptoms associated with Alzheimer's disease. The development of compounds for ameliorating, modifying, or improving potential aberrations in neuronal cellular communication mechanisms is encouraged. These compounds should be designed to affect fundamental processes such as the neuronal dysfunction, death, and loss of connectivity associated with the disease by targeting molecules and mechanisms such as neurotransmitters, neuromodulators, and neurotrophins; receptors and ion channels; second and third messenger systems; all facets of relevant amyloid precursor protein, amyloid beta protein, and tau neurobiology; protein synthesis, aggregation, and degradation; energy utilization; and oxidative, immunological, and inflammatory mechanisms.
Requirements: The following organizations are eligible to apply: for-profit organizations; non-profit organizations; public or private institutions, such as universities, colleges, hospitals, and laboratories; units of State government; units of local government; eligible agencies of the Federal government; foreign institutions; and domestic institutions.
Geographic Focus: All States
Date(s) Application is Due: Feb 16; Jun 16; Oct 16
Contact: Neil S. Buckholtz, (301) 496-9350; fax (301) 496-1494; Buckholn@nia.nih.gov
Internet: http://grants.nih.gov/grants/guide/pa-files/PAS-06-261.html
Sponsor: National Institute on Aging
Gateway Building, Suite 350, 7201 Wisconsin Avenue, MSC 9205
Bethesda, MD 20892-9205

NIA Harmonization of Longitudinal Cross-National Surveys of Aging Grants 2760
The National Institute on Aging (NIA) will provide support through the R21 mechanism to facilitate the ex-ante and ex-post harmonization of international longitudinal surveys of aging populations to the Health and Retirement Study to maximize cross-national comparability of data. Facilitation will include, but is not limited to, meetings, pilots, and the development of methods to improve cross-national comparability of international aging surveys to the Health and Retirement Study and existing international data files that are harmonized with the HRS: the English Longitudinal Study on Ageing (ELSA); and the Survey of Health, Ageing, and Retirement in Europe (SHARE).
Requirements: The following organizations and institutions may apply: public/state controlled institutions of higher education; private institutions of higher education; nonprofits with 501(c)3 IRS status; nonprofits without 501(c)3 IRS status; for-profit organizations; State governments; and non-domestic (non-U.S.) entities (foreign organizations).
Geographic Focus: All States
Date(s) Application is Due: Feb 16; Jun 16; Oct 16
Contact: John W. R. Phillips, (301) 496-3138; fax (301) 402-0051; phillipj@mail.nih.gov
Internet: http://grants.nih.gov/grants/guide/pa-files/PAS-07-387.html
Sponsor: National Institute on Aging
Gateway Building, Room 533, 7201 Wisconsin Avenue, MSC 9205
Bethesda, MD 20892

NIA Health Behaviors and Aging Grants 2761
The National Institute on Aging (NIA) invites qualified researchers to submit applications for research and research training on those health-related behaviors and attitudes of older adults, their families, and significant others, that can affect health and functioning as people grow older. Studies are sought that extend scientific understanding of how older adults health behaviors and attitudes develop under varying social conditions; how they relate to health promotion and disease prevention, care and treatment of disease, rehabilitation or death; and how they can be modified as relevant new scientific knowledge is developed.
Requirements: Applications may be submitted by foreign and domestic, for-profit and non-profit organizations, public and private, such as universities, colleges, hospitals, laboratories, units of State and local governments, and eligible agencies of the Federal government. Applications from minority individuals and women are encouraged.
Geographic Focus: All States
Date(s) Application is Due: Feb 5; Jun 5; Oct 5
Amount of Grant: Up to 150,000 USD
Contact: Chyren Hunter, (301) 496-9322; fax (301) 402-0528; Hunterc@nia.nih.gov
Internet: http://grants.nih.gov/grants/guide/pa-files/PA-93-064.html
Sponsor: National Institute on Aging
Gateway Building, Room 2C218, 7201 Wisconsin Avenue, MSC 9205
Bethesda, MD 20892-9205

NIA Healthy Aging through Behavioral Economic Analyses of Situations Grants 2762
The National Institute on Aging (NIA), National Institutes of Health, solicits Research Project Grant applications that propose to translate basic findings from Behavioral Economics into behavior change interventions targeting health behaviors associated with chronic health conditions of mid-life and older age. Applications should propose small pilot clinical trials or demonstration projects, ideally based on collaborations between individuals with expertise in behavioral economics and psychologists, psychiatrists, clinicians, or others with expertise in aging or implementing behavioral interventions. Depending on the availability of funding, the NIA expects to make 4 or 5 awards totaling $1,250,000 total costs in the first year, and $3,750,000 over the 3 year project period. The direct cost amount for each individual award will be no greater than $525,000 in total for three years, and no more than $175,000 in any year. Letters of Intent are due by October 2, with final applications due by November 2.
Requirements: The following organizations and institutions may apply: public/state controlled institutions of higher education; private institutions of higher education; Hispanic-serving institutions; historically Black colleges and universities (HBCUs); tribally controlled colleges and universities (TCCUs); Alaska native and native Hawaiian serving institutions; nonprofits with 501(c)3 IRS status; nonprofits without 501(c)3 IRS status; small businesses; for-profit organizations; State governments; regional organizations; U.S. territories or possessions; Indian/Native American tribal governments; Indian/Native American tribally designated organizations; county governments; city or township governments; special district governments; independent school districts; public housing authorities/Indian housing authorities; eligible agencies of the Federal government; and faith-based or community based organizations.
Geographic Focus: All States
Date(s) Application is Due: Oct 2; Nov 2
Amount of Grant: Up to 175,000 USD
Contact: Dr. Jonathan W. King, (301) 402-4156; fax (301) 402-0051; kingjo@nia.nih.gov
Internet: http://grants.nih.gov/grants/guide/rfa-files/RFA-AG-10-008.html
Sponsor: National Institute on Aging
Gateway Building, Room 533, 7201 Wisconsin Avenue, MSC 9205
Bethesda, MD 20892

NIA Higher-Order Cognitive Functioning and Aging Grants 2763
The National Institute on Aging (NIA) invites qualified researchers to submit new applications for research projects that focus on adulthood and aging related changes in the higher-order processes and strategies required for judgment, decision-making, reasoning, problem-solving, and processing complex information. As in earlier periods of life older adults continue to make decisions related to everyday life, but with advanced age, new, and sometimes even more complex, decision-making is required of them. Recent research indicates that age-related limitations in cognitive processing resources (e.g., speed and working memory) may impact decision-making. Research also indicates that some older adults experience growth in specific areas of cognitive functioning (e.g., expertise, semantic knowledge, emotional regulation) and continue to use adaptive intelligence, demonstrating multi-directionality in adult cognitive change. It is generally recognized that research on higher-order processing is underdeveloped in the field of aging. Research proposals are needed that examine the actual processes that are engaged when older adults make important decisions, how these processes change with age and context, and what environmental supports, interventions, and training may be necessary for optimal functioning. Research may investigate either individual or collaborative, or social processes. Modular Grant applications will request direct costs in $25,000 modules, up to a total direct cost request of $250,000 per year.
Requirements: Applications may be submitted by foreign and domestic, for-profit and non-profit, public and private organizations such as universities, colleges, hospitals, laboratories, units of State and local governments, and eligible agencies of the Federal government. Racial/ethnic minority individuals, women, and persons with disabilities are encouraged to apply as principal investigators.
Geographic Focus: All States
Date(s) Application is Due: Feb 16; Jun 16; Oct 16
Amount of Grant: Up to 250,000 USD
Contact: Chyren Hunter, (301) 496-9322; fax (301) 402-0528; Hunterc@nia.nih.gov
Internet: http://grants.nih.gov/grants/guide/pa-files/PA-00-052.html
Sponsor: National Institute on Aging
Gateway Building, Room 2C218, 7201 Wisconsin Avenue, MSC 9205
Bethesda, MD 20892-9205

NIA Human Biospecimen Resources for Aging Research Grants 2764
The purpose of this announcement is to disseminate information about available human biospecimen resources for age-related studies, and promote independent use of existing human biospecimen resources through collaborative and other arrangements with the studies that collect biospecimens. Websites for the participating studies are accessed via the Virtual Repository. NIA is interested in aging-related research that integrates biomarker analyses with the study of normal aging, age-associated diseases, and behavioral and psychosocial data. Budgets for direct costs of up to $50,000 per year and a project duration of up to two years may be requested for a maximum of $100,000 direct costs over a two-year project period.
Requirements: The following organizations and institutions may apply: for-profit organizations; non-profit organizations; public or private institutions, such as universities, colleges, hospitals, and laboratories; units of State government; units of Local government; units of State Tribal government; units of Local Tribal government; eligible agencies of the Federal government; foreign institutions; domestic institutions; faith-based or community-based organizations; Indian/Native American tribal governments (federally recognized); Indian/Native American tribal governments (other than federally recognized); and Indian/Native American tribally designated organizations.

Geographic Focus: All States
Date(s) Application is Due: Feb 16; Jun 16; Oct 16
Amount of Grant: Up to 100,000 USD
Contact: Nancy L. Nadon, (301) 402-7744; fax (301) 402-0010; nadonn@nia.nih.gov
Internet: http://grants.nih.gov/grants/guide/pa-files/PA-06-443.html
Sponsor: National Institute on Aging
Gateway Building, Suite 2C231, 7201 Wisconsin Avenue, MSC 9205
Bethesda, MD 20892

NIAID Independent Scientist Award   2765

The award is a special salary-only grant designed to provide protected time for newly independent scientists who currently have non-research obligations such as heavy teaching loads, clinical work, committee assignments, service, and administrative duties that prevent them from having a period of intensive research focus. The award is targeted to persons with doctoral degrees who have completed their research training, have independent peer-reviewed research support, and who need a period of protected research time in order to foster their research career development. New applications are due February 12, June 12, and October 12, while renewal applications are due March 12, July 12, and November 12. Up to $75,000 plus fringe benefits per year is available. Additional research is not generally provided.

Requirements: The following organizations and institutions are eligible to apply: public/state controlled institutions of higher education; private institutions of higher education; nonprofits with 501(c)3 IRS status; nonprofits without 501(c)3 IRS status; small businesses; for-profit organizations; State governments; U.S. territories or possessions; Indian/Native American tribal governments (Federally recognized and other than Federally recognized); Indian/Native American tribally designated organizations; Hispanic-serving institutions; historically Black colleges and universities (HBCUs); tribally controlled colleges and universities (TCCUs); Alaska Native and Native Hawaiian serving institutions; regional organizations; and faith-based or community based organizations. The candidate must have a doctoral degree and peer-reviewed, independent research support at the time the award is made. The candidate must spend a minimum of 75 percent effort conducting research during the period of the award.
Restrictions: The award is not intended for investigators who already have full time to perform research, or have substantial publication records or considerable research support indicating that they are well established in their fields. Foreign institutions are not eligible to apply.
Geographic Focus: All States
Date(s) Application is Due: Feb 12; Mar 12; Jun 12; Jul 12; Oct 12; Nov 12
Amount of Grant: 75,000 USD
Contact: Milton J. Hernandez, (301) 496-3775; mh35c@nih.gov
Internet: http://grants.nih.gov/grants/guide/pa-files/PA-09-038.html
Sponsor: National Institute of Allergy and Infectious Diseases
6610 Rockledge Drive
Bethesda, MD 20892

NIA Independent Scientist Award   2766

The award is a special salary-only grant designed to provide protected time for newly independent scientists who currently have non-research obligations such as heavy teaching loads, clinical work, committee assignments, service, and administrative duties that prevent them from having a period of intensive research focus. The award is targeted to persons with doctoral degrees who have completed their research training, have independent peer-reviewed research support, and who need a period of protected research time in order to foster their research career development. New applications are due February 12, June 12, and October 12, while renewal applications are due March 12, July 12, and November 12. Up to 75% per year of base salary plus fringe benefits is available. No other research development support funds are provided.

Requirements: The following organizations and institutions are eligible to apply: public/state controlled institutions of higher education; private institutions of higher education; nonprofits with 501(c)3 IRS status; nonprofits without 501(c)3 IRS status; small businesses; for-profit organizations; State governments; U.S. territories or possessions; Indian/Native American tribal governments (Federally recognized and other than Federally recognized); Indian/Native American tribally designated organizations; Hispanic-serving institutions; historically Black colleges and universities (HBCUs); tribally controlled colleges and universities (TCCUs); Alaska Native and Native Hawaiian serving institutions; regional organizations; and faith-based or community based organizations. The candidate must have a doctoral degree and peer-reviewed, independent research support at the time the award is made. The candidate must spend a minimum of 75 percent effort conducting research during the period of the award.
Restrictions: The award is not intended for investigators who already have full time to perform research, or have substantial publication records or considerable research support indicating that they are well established in their fields. Foreign institutions are not eligible to apply.
Geographic Focus: All States
Date(s) Application is Due: Feb 12; Mar 12; Jun 12; Jul 12; Oct 12; Nov 12
Contact: Chyren Hunter, (301) 496-9322; fax (301) 402-0528; Hunterc@nia.nih.gov
Internet: http://grants.nih.gov/grants/guide/pa-files/PA-09-038.html
Sponsor: National Institute on Aging
Gateway Building, Room 2C218, 7201 Wisconsin Avenue, MSC 9205
Bethesda, MD 20892-9205

NIA Malnutrition in Older Persons Research Grants   2767

NIA continues to encourage research on the full spectrum of issues related to the causes, prevention, and treatment as well as sociobehavioral and economic aspects related to malnutrition in older persons. Support for this program will be through the research project grant and mentored research scientist development award. Applications will compete for available funds on the basis of scientific merit with all other applications assigned to the institute.

Requirements: Applications may be submitted by foreign and domestic, for-profit and non-profit organizations, public and private, such as universities, colleges, hospitals, laboratories, unit of State and local governments, and eligible agencies of the Federal government. Applications from minority individuals and women are encouraged.
Geographic Focus: All States
Date(s) Application is Due: Feb 12; Mar 12; Jun 12; Jul 12; Oct 12; Nov 12
Contact: Chyren Hunter, (301) 496-9322; fax (301) 402-0528; Hunterc@nia.nih.gov
Internet: http://grants.nih.gov/grants/guide/pa-files/PA-94-088.html
Sponsor: National Institute on Aging
Building 31, Room 5C27, 31 Center Drive, MSC 2292
Bethesda, MD 20892-7710

NIA Mechanisms, Measurement, and Management of Pain in Aging: from Molecular to Clinical   2768

This FOA encourages Research Project Grant applications from institutions and organizations that propose to: study biological, neurobiological, psychosocial, and clinical mechanisms and processes by which aging and/or age-related diseases affect the experience of pain; examine biological, neurobiological, psychosocial, and clinical factors that impact pain experience and prevalence in older people; evaluate existing pain assessment and/or management approaches in older adults; or develop new assessment methods and/or management strategies for pain with particular attention to the needs of older adults. Studies involving animal models or human subjects are appropriate under this program announcement.

Requirements: The following organizations and institutions may apply: public/state controlled institutions of higher education; private institutions of higher education; Hispanic-serving institutions; historically Black colleges and universities (HBCUs); tribally controlled colleges and universities (TCCUs); Alaska native and native Hawaiian serving institutions; nonprofits with 501(c)3 IRS status; nonprofits without 501(c)3 IRS status; small businesses; for-profit organizations; State governments; regional organizations; U.S. territories or possessions; Indian/Native American tribal governments; Indian/Native American tribally designated organizations; county governments; city or township governments; special district governments; independent school districts; public housing authorities/Indian housing authorities; eligible agencies of the Federal government; and faith-based or community based organizations.
Geographic Focus: All States
Date(s) Application is Due: Feb 5; Feb 16; Jun 5; Jun 16; Oct 5; Oct 16
Contact: Wen G. Chen, (301) 496-9350; fax (301) 496-1494; chenw@nia.nih.gov
Internet: http://grants.nih.gov/grants/guide/pa-files/PA-09-193.html
Sponsor: National Institute on Aging
Gateway Building, Suite 350, 7201 Wisconsin Avenue, MSC 9205
Bethesda, MD 20892-9205

NIA Mechanisms Underlying the Links between Psychosocial Stress, Aging, the Brain and the Body Grants   2769

This FOA encourages multidisciplinary and interdisciplinary research to elucidate the mechanistic links between psychosocial stress and health in aging, as well as how the aging process and age-related diseases affect the responses to psychosocial stressors. Generally, research should be focused on: aging and how neural mechanisms respond to psychosocial stress and affect other body systems; characterizing the behavioral, psychological and social mechanisms and pathways involved in transducing psychosocial stressors into health outcomes; how stressors modulate physiological process underlying life-span, immune mechanisms, and metabolism; and how psychosocial stress contributes to the development or progression of geriatric syndromes, chronic medical conditions, and disabilities in later life. Research is strongly encouraged that aims to identify appropriate targets for intervention, at any level of analysis, from societal to molecular. Research spanning multiple levels of analysis is particularly encouraged.

Requirements: The following organizations and institutions are eligible to apply: public/state controlled institutions of higher education; private institutions of higher education; Hispanic-serving institutions; historically Black colleges and universities (HBCUs); tribally controlled colleges and universities (TCCUs); Alaska native and native Hawaiian serving institutions; nonprofits with 501(c)3 IRS status (other than institutions of higher education); nonprofits without 501(c)3 IRS status (other than institutions of higher education); small businesses; for-profit organizations (other than small businesses); state governments; Indian/Native American tribal governments (federally recognized); Indian/Native American tribally designated organizations; county governments; city or township governments; special district governments; independent school districts; public housing authorities/Indian housing authorities; U.S. territories or possessions; Indian/native American tribal governments (other than federally recognized); regional organizations; and non-domestic (non-U.S.) entities (foreign organizations).
Restrictions: Research focused on oxidative stress or on environmental or physical stressors of a non-psychosocial nature is not appropriate for this FOA.
Geographic Focus: All States
Date(s) Application is Due: Mar 5; Jul 5; Nov 5
Contact: Suzana S. Petanceska, (301) 496-9350; petanceskas@mail.nih.gov
Internet: http://grants.nih.gov/grants/guide/pa-files/PA-09-216.html
Sponsor: National Institute on Aging
Gateway Building, Suite 350, 7201 Wisconsin Avenue, MSC 9205
Bethesda, MD 20892-9205

NIA Medical Management of Older Patients with HIV/AIDS Grants 2770

The National Institute on Aging (NIA), the National Institute of Allergy and Infectious Diseases (NIAID), the Center for Mental Health Research on AIDS of the National Institute of Mental Health (NIMH), and the National Institute of Nursing Research (NINR), invite Research Project Grant applications addressing clinical and translational health issues in the diagnosis and/or management of HIV infection and its consequences in older persons. The goal of this funding opportunity is to improve medical outcomes, functional status and quality of life in older patients with HIV/AIDS through improved understanding of interactions among aging processes, HIV viral infection, treatment effects and toxicities, and multiple morbidities commonly occurring in older persons.
Requirements: The following organizations and institutions may apply: public/state controlled institutions of higher education; private institutions of higher education; Hispanic-serving institutions; historically Black colleges and universities (HBCUs); tribally controlled colleges and universities (TCCUs); Alaska native and native Hawaiian serving institutions; nonprofits with 501(c)3 IRS status; nonprofits without 501(c)3 IRS status; small businesses; for-profit organizations; State governments; regional organizations; U.S. territories or possessions; Indian/Native American tribal governments; Indian/Native American tribally designated organizations; county governments; city or township governments; special district governments; independent school districts; public housing authorities/Indian housing authorities; eligible agencies of the Federal government; and faith-based or community based organizations.
Geographic Focus: All States
Date(s) Application is Due: Jan 7; May 7; Sep 7
Contact: Susan G. Nayfield, (301) 496-6761; fax (301) 402-1784; nayfiels@nia.nih.gov
Internet: http://grants.nih.gov/grants/guide/pa-files/PA-09-017.html
Sponsor: National Institute on Aging
Gateway Building, Suite 3C307, 7201 Wisconsin Avenue, MSC 9205
Bethesda, MD 20892-9205

NIA Mentored Clinical Scientist Research Career Development Awards 2771

The overall objective of the NIH Research Career Development Award program is to prepare qualified individuals for careers that have a significant impact on the health-related research needs of the Nation. The objective of the NIH Mentored Clinical Scientist Research Career Development Award (K08) program is to support didactic study and mentored research for individuals with clinical doctoral degrees (e.g., M.D., D.D.S., D.M.D., D.O., D.C., O.D., N.D., D.V.M., Pharm.D., or Ph.D. in clinical disciplines). The purpose of this Funding Opportunity Announcement (FOA) is to continue the long-standing NIH support of the K08 program. This award provides support and protected time for an intensive, mentored research career development experience in biomedical or behavioral research, including translational research. For the purpose of this award, translational research is defined as application of basic research discoveries toward the diagnosis, management, and prevention of human disease.
Requirements: The following organizations and institutions may apply: public/state controlled institutions of higher education; private institutions of higher education; Hispanic-serving institutions; historically Black colleges and universities (HBCUs); tribally controlled colleges and universities (TCCUs); Alaska native and native Hawaiian serving institutions; nonprofits with 501(c)3 IRS status; nonprofits without 501(c)3 IRS status; small businesses; for-profit organizations; State governments; regional organizations; U.S. territories or possessions; Indian/Native American tribal governments; Indian/Native American tribally designated organizations; county governments; city or township governments; special district governments; independent school districts; public housing authorities/Indian housing authorities; eligible agencies of the Federal government; and faith-based or community based organizations.
Restrictions: Individuals on temporary or student visas and organizations described in section 501(c)4 of the Internal Revenue Code that engage in lobbying are not eligible.
Geographic Focus: All States
Date(s) Application is Due: Feb 12; Mar 12; Jun 12; Jul 12; Oct 12; Nov 12
Amount of Grant: Up to 100,000 USD
Contact: Chyren Hunter, (301) 496-9322; fax (301) 402-0528; Hunterc@nia.nih.gov
Internet: http://grants.nih.gov/grants/guide/pa-files/PA-09-042.html
Sponsor: National Institute on Aging
Gateway Building, Room 2C218, 7201 Wisconsin Avenue, MSC 9205
Bethesda, MD 20892-9205

NIA Mentored Research Scientist Development Award 2772

The award is for research scientists who have established careers in biomedical, behavioral, or social research and wish to change career direction toward aging research; for more junior researchers with training in aging research who need additional periods of mentored research experience prior to becoming fully independent; or for researchers with training and experience in some aspect of aging research who wish to gain complementary training to expand their research interests in aging. The primary purpose of the program is to provide support and protected time (three, four, or five years) for an intensive, supervised career development experience in the biomedical, behavioral, or clinical sciences leading to research independence.
Requirements: The candidate must have a research or health-professional doctorate or its equivalent and must have demonstrated the capacity or potential for highly productive independent research in the period after the doctorate. The candidate must identify a mentor with extensive research experience and must spend a minimum of 75 percent of full-time professional effort conducting research for the period of the award. The following organizations and institutions may apply: public/state controlled institutions of higher education; private institutions of higher education; Hispanic-serving institutions; historically Black colleges and universities (HBCUs); tribally controlled colleges and universities (TCCUs); Alaska native and native Hawaiian serving institutions; nonprofits with 501(c)3 IRS status; nonprofits without 501(c)3 IRS status; small businesses; for-profit organizations; State governments; regional organizations; U.S. territories or possessions; Indian/Native American tribal governments; Indian/Native American tribally designated organizations; eligible agencies of the Federal government; and faith-based or community based organizations.
Geographic Focus: All States
Date(s) Application is Due: Feb 12; Mar 12; Jun 12; Jul 12; Oct 12; Nov 12
Contact: Chyren Hunter, (301) 496-9322; fax (301) 402-0528; Hunterc@nia.nih.gov
Internet: http://grants.nih.gov/grants/guide/pa-files/PAR-09-060.html
Sponsor: National Institute on Aging
Gateway Building, Room 2C218, 7201 Wisconsin Avenue, MSC 9205
Bethesda, MD 20892-9205

NIA Minority Dissertation Research Grants in Aging 2773

Small grants to support doctoral dissertation research will be available for underrepresented minority doctoral candidates. Grant support is designed to aid the research of such investigators and to encourage underrepresented minority individuals from a variety of academic disciplines and programs to conduct research related to aging. It is anticipated that a total of five to six new applications will be funded in each year. The maximum budget request should be limited to $25,000 in direct costs for the initial budget period. No more than $30,000 in total direct costs will be provided across the two year period.
Requirements: This research initiative is to provide underrepresented minority students assistance to complete their dissertation research on an aging-related topic and thereby increase their representation in aging research.
Geographic Focus: All States
Date(s) Application is Due: Mar 15; Nov 15
Contact: Laurie M. Ryan, (301) 496-9350; fax (301) 496-1494; ryanl@nia.nih.gov
Internet: http://grants.nih.gov/grants/guide/pa-files/PAR-98-110.html
Sponsor: National Institute on Aging
Gateway Building, Suite 350, 7201 Wisconsin Avenue, MSC 9205
Bethesda, MD 20892-9205

NIAMS Pilot and Feasibility Clinical Research Grants in Arthritis and Musculoskeletal and Skin Diseases 2774

The goal of this initiative is to encourage novel clinical research aimed at improving the understanding of the causes, treatment or prevention of arthritis and musculoskeletal and skin diseases. This Pilot and Feasibility Clinical Research Grants Program is designed to allow initiation of exploratory, short-term clinical studies, intended to facilitate the development of new ideas which may be investigated without stringent requirements for preliminary data. Research aims appropriate for an investigator-initiated R01 grant, such as understanding underlying biological mechanisms of disease, would not be expected in an application under this FOA. Proposed studies should focus on research questions that are likely to gather critical preliminary data in support of a future, planned clinical trial or to benefit clinical research and trials more broadly. Overall goals of these studies are intended to generate data critical to future clinical research or trials such as by demonstrating the feasibility of a recruitment target or approach, obtaining data in support of potential inclusion and exclusion criteria, or to gather preliminary evidence of efficacy, tolerability and/or toxicity of an available drug, biologic or device in a new population relevant to the NIAMS mission. Examples include testing of: new prevention strategies; a new application of existing pharmacologic or non-pharmacologic interventions or their unique combination; a new cost effective or adaptive trial design in a specific population or across populations; and studies intended to expand the science of clinical trials overall. Results generated from this program may lead to critical input into a new clinical trial; trials that inform decision-making in urgent or high priority clinical practice needs for arthritis and musculoskeletal and skin diseases; or preliminary data for trials in understudied and rare diseases. A high priority is the use of such studies to help stimulate the translation of promising research developments from the laboratory into clinical practice. The total project period for an application submitted in response to this funding opportunity may not exceed 2 years. Although the size of award may vary with the scope of research proposed, it is expected that applications will stay within the budgetary guidelines for an exploratory/developmental project; direct costs are limited to $275,000 over an R21 two-year period, with no more than $200,000 in direct costs allowed in any single year.
Requirements: The following organizations/institutions are eligible to apply: public/state controlled institutions of higher education; private institutions of higher education; Hispanic-serving institutions; historically Black colleges and universities; tribally controlled colleges and universities; Alaska native and native Hawaiian serving institutions; nonprofits with 501(c)3 IRS status (other than institutions of higher education); nonprofits without 501(c)3 IRS status (other than institutions of higher education); small businesses; for-profit organizations (other than small businesses); state governments; Indian/Native American tribal governments (federally recognized); Indian/Native American tribally designated organizations; county governments; city or township governments; special district governments; independent school districts; public housing authorities/Indian housing authorities; U.S. territory or possession; Indian/Native American tribal governments (other than federally recognized); regional organizations; non-domestic (non-U.S.) entities (foreign organizations); eligible agencies of the federal government; and faith-based or community-based organizations.
Geographic Focus: All States
Date(s) Application is Due: Mar 1; Jul 1; Nov 1
Amount of Grant: Up to 275,000 USD
Contact: James Witter (301) 594-1963; fax (301) 480-4543; witterj@mail.nih.gov
Internet: http://grants.nih.gov/grants/guide/pa-files/PAR-10-282.html#SectionVII
Sponsor: National Institute of Arthritis and Musculoskeletal and Skin Diseases
1 AMS Circle
Bethesda, MD 20892-3675

NIA Multicenter Study on Exceptional Survival in Families: The Long Life Family Study (LLFS) Grants 2775

This limited competition FOA is to continue the Long Life Family Study (LLFS). The LLFS was designed to determine the degree and patterns of familial transmission and aggregation of exceptional longevity and healthy survival to advanced age as characterized by a variety of phenotypic measures, with a further goal of also understanding potential genetic factors that contribute to exceptional survival. Five awards, to four study centers and one data management and coordinating center, were made in response to RFA-AG-03-004.
Geographic Focus: All States
Date(s) Application is Due: Aug 30; Sep 30
Contact: Winifred K. Rossi, (301) 496-3836; fax (301) 402-1784; winnie_rossi@nih.gov
Internet: http://grants.nih.gov/grants/guide/rfa-files/RFA-AG-10-005.html
Sponsor: National Institute on Aging
Gateway Building, Suite 3C307, 7201 Wisconsin Avenue, MSC 9205
Bethesda, MD 20892-9205

NIA Nathan Shock Centers of Excellence in Basic Biology of Aging Grants 2776

The NIA invites applications for support of centers, known as Nathan Shock Centers of Excellence in Basic Biology of Aging. These center grants will provide funding for research and training activities related to basic biology of aging. They are intended for institutions with a substantial investment in and commitment to aging research, but they are not intended to directly support clinical research or clinical trials. The goal of this program is to enhance the ability of institutions with well-developed research programs in basic research on aging to utilize state-of-the-art research resources to provide the strongest environment for the conduct of research on aging. Thus, this RFA is intended to enhance the quality of research in the basic biology of aging, facilitate the planning and coordination of aging research activities, provide support and a suitable environment for investigators new to aging research to acquire research skills and experience at institutions that have demonstrated commitment to, and expertise in, basic biology of aging research, and to develop potential regional and/or national resource centers. Annual deadline dates may vary.
Requirements: Applications may be submitted by domestic nonprofit organizations and institutions, state and local governments and their agencies, and authorized federal institutions. The applicant institution must currently support a minimum of 15 peer-reviewed, externally funded research projects on aging to be eligible. In the case of currently funded program projects (P01s), or similar multiproject research grants, each research project will be deemed to be a separate project. (Core components do not qualify.)
Geographic Focus: All States
Date(s) Application is Due: Apr 16; May 20
Contact: Dr. Huber Warner; (301) 496-4996; fax (301) 402-0010; warnerh@nia.nih.gov
Internet: http://grants.nih.gov/grants/guide/rfa-files/RFA-AG-04-010.html
Sponsor: National Institute on Aging
Gateway Building, Suite 2C231, 7201 Wisconsin Avenue, MSC 9205
Bethesda, MD 20892

NIA Network Infrastructure Support for Emerging Behavioral and Social Research Areas in Aging 2777

The purpose of this FOA is to provide infrastructure support in specific emerging interdisciplinary areas of behavioral and social research in aging using the NIH Resource-Related Research Project mechanism. The infrastructure support will facilitate research networks through meetings, conferences, small scale pilots, training, and dissemination to encourage growth and development in specified emerging areas and resources. Budgets for direct costs of up to $150,000 per year and project duration of 3-5 years may be requested with a maximum of $750,000 direct costs over a five-year project period. Letters of Intent are due by September 30, December 28, and April 26, respectively. Final applications due by October 30, January 25, and May 25, respectively..
Requirements: The following organizations and institutions may apply: public/state controlled institutions of higher education; private institutions of higher education; Hispanic-serving institutions; historically Black colleges and universities (HBCUs); tribally controlled colleges and universities (TCCUs); Alaska native and native Hawaiian serving institutions; nonprofits with 501(c)3 IRS status; nonprofits without 501(c)3 IRS status; small businesses; for-profit organizations; State governments; regional organizations; U.S. territories or possessions; Indian/Native American tribal governments; Indian/Native American tribally designated organizations; county governments; city or township governments; special district governments; independent school districts; public housing authorities/Indian housing authorities; eligible agencies of the Federal government; and faith-based or community based organizations.
Geographic Focus: All States
Date(s) Application is Due: Jan 25; Apr 26; May 25; Sep 30; Oct 30; Dec 28
Contact: Lis Nielsen, (301) 402-4156; fax (301) 402-0051; nielsenli@nia.nih.gov
Internet: http://grants.nih.gov/grants/guide/pa-files/PAR-09-233.html
Sponsor: National Institute on Aging
Building 31, Room 5C27, 31 Center Drive, MSC 2292
Bethesda, MD 20892-7710

NIA Pilot Research Grants 2778

NIA is seeking small grant applications in specific areas to stimulate and facilitate the entry of promising new investigators into aging research or encourage established investigators to enter new targeted, high-priority areas in this research field. This small grant program provides support for pilot research that is likely to lead to subsequent individual research project grants and/or a significant advancement of aging research. Annual deadline dates may vary; contact program staff for exact dates.
Requirements: Applications may be submitted by domestic for-profit and nonprofit organizations, both public or private; units of state and local governments; and eligible agencies of the federal government. The program supports new and established investigators. For a new investigator to be eligible, the individual should be in the first five years of his or her independent research career. If the applicant is in the final stages of training it is permissible to apply for an R03, but no award will be made to individuals who are still in training or fellowship status at the time of award. For an established investigator to be eligible, the individual must propose research that is unrelated to a currently funded research project in which the investigator participates.
Restrictions: Foreign organizations and institutions are not eligible.
Geographic Focus: All States
Amount of Grant: Up to 50,000 USD
Contact: Dr. David Finkelstein; (301) 496-6402; BAPquery@extramur.nia.nih.gov
Internet: http://grants.nih.gov/grants/guide/pa-files/PA-99-049.html
Sponsor: National Institute on Aging
Gateway Building, Suite 2C231, 7201 Wisconsin Avenue, MSC 9205
Bethesda, MD 20892

NIA Postdoctoral Training in Research on Aging in Canada 2779

The objective of the National Research Service Award (NRSA) F32 program is to provide support to promising postdoctoral applicants who have the potential to become productive and successful independent research investigators in scientific health-related fields relevant to the missions of the participating NIH Institutes and Centers (ICs). A primary purpose of this initiative is to offer training in research on aging that exposes developing American postdoctoral investigators to diverse combinations of expertise and resources singularly available in Canada through the Canadian Institutes of Health-Institute on Aging (CIHR-IA), combinations that also take into account the unique cultural and ethnic composition of Canada and its aging population.
Requirements: Applicants must choose Canadian institutions and mentors selected by CIHR-IA. These institutions and mentors can be located through the CIHR Funding Database. Any individual with the skills, knowledge, and resources necessary to carry out the proposed research training is invited to work with his/her sponsor and institution to develop an application for support. Individuals from underrepresented racial and ethnic groups as well as individuals with disabilities are always encouraged to apply for NIH support.
Restrictions: Candidates for the postdoctoral fellowship award must be citizens or non-citizen nationals of the United States, or must have been lawfully admitted to the United States for Permanent Residence.
Geographic Focus: All States
Date(s) Application is Due: Apr 8; Aug 8; Dec 8
Contact: Michael-David, (301) 402-7713; michael-david.kerns@nih.hhs.gov
Internet: http://grants.nih.gov/grants/guide/pa-files/PA-06-469.html
Sponsor: National Institute on Aging
Gateway Building, Room 2C218, 7201 Wisconsin Avenue, MSC 9205
Bethesda, MD 20892-9205

NIA Promoting Careers in Aging and Health Disparities Research Grants 2780

The Mentored Research Scientist Development Award provides support for a sustained period of protected time for intensive research career development under the guidance of an experienced mentor, or sponsor, in the biomedical, behavioral or clinical sciences leading to research independence. The expectation is that through this sustained period of research career development and training, awardees will launch independent research careers and become competitive for new research project grant funding.
Requirements: The following organizations and institutions may apply: public/state controlled institutions of higher education; private institutions of higher education; Hispanic-serving institutions; historically Black colleges and universities (HBCUs); tribally controlled colleges and universities (TCCUs); Alaska native and native Hawaiian serving institutions; nonprofits with 501(c)3 IRS status; nonprofits without 501(c)3 IRS status; small businesses; for-profit organizations; State governments; regional organizations; U.S. territories or possessions; Indian/Native American tribal governments; Indian/Native American tribally designated organizations; county governments; city or township governments; special district governments; independent school districts; public housing authorities/Indian housing authorities; eligible agencies of the Federal government; and faith-based or community based organizations.
Geographic Focus: All States
Date(s) Application is Due: Feb 12; Jun 12; Oct 12
Amount of Grant: 450,000 - 750,000 USD
Contact: Dr. J. Taylor Harden, (301) 496-0765; Taylor_Harden@nih.gov
Internet: http://grants.nih.gov/grants/guide/pa-files/PAR-09-136.html
Sponsor: National Institute on Aging
Building 31, Room 5C27, 31 Center Drive, MSC 2292
Bethesda, MD 20892-7710

NIA Renal Function and Chronic Kidney Disease in Aging Grants 2781

This Funding Opportunity Announcement (FOA) issued by the National Institute on Aging (NIA) and the National Institute of Diabetes and Digestive and Kidney Diseases (NIDDK), National Institutes of Health, invites applications that propose basic, clinical, and translational research on chronic kidney disease (CKD) and its consequences in aging and in older persons. Applications should focus on: biology and pathophysiology of CKD in animal models; etiology and pathophysiology of CKD in the elderly; epidemiology and risk factors for the development of CKD with advancing age; and/or diagnosis, medical management and clinical outcomes of CKD in this population. Research supported by this initiative should enhance knowledge of CKD and its consequences in the elderly and provide evidence-based guidance in the diagnosis, prevention, and treatment of CKD in older persons.

Requirements: The following organizations and institutions may apply: public/state controlled institutions of higher education; private institutions of higher education; Hispanic-serving institutions; historically Black colleges and universities (HBCUs); tribally controlled colleges and universities (TCCUs); Alaska native and native Hawaiian serving institutions; nonprofits with 501(c)3 IRS status; nonprofits without 501(c)3 IRS status; small businesses; for-profit organizations; State governments; regional organizations; U.S. territories or possessions; Indian/Native American tribal governments; Indian/Native American tribally designated organizations; county governments; city or township governments; special district governments; independent school districts; public housing authorities/Indian housing authorities; eligible agencies of the Federal government; and faith-based or community based organizations.
Geographic Focus: All States
Date(s) Application is Due: Feb 5; Feb 16; Jun 5; Jun 16; Oct 5; Oct 16
Contact: Susan G. Nayfield, (301) 496-6761; fax (301) 402-1784; nayfiels@nia.nih.gov
Internet: http://grants.nih.gov/grants/guide/pa-files/PA-09-165.html
Sponsor: National Institute on Aging
Gateway Building, Suite 3C307, 7201 Wisconsin Avenue, MSC 9205
Bethesda, MD 20892-9205

NIA Role of Apolipoprotein E, Lipoprotein Receptors and CNS Lipid Homeostasis in Brain Aging and Alzheimer's Disease 2782
This FOA encouarges multidisciplinary and interdisciplinary research to elucidate how Apolipoprotein E, lipoprotein receptors and CNS lipid homeostasis influence brain aging and the transition to neurodegeneration in Alzheimer's disease (AD). The ultimate goal is to gain an in depth understanding of the mechanisms by which the Apolipoprotein E e4 allele confers increased AD risk for the purpose of advancing the overall search for efficacious AD treatments and Apolipoprotein E e4-directed therapeutics in particular. To this end we encourage research spanning multiple levels of analysis in multiple species (from mice to man). Studies aimed at identifying new therapeutic targets for the treatment of AD and other age-related neurodegenerative conditions associated with lipid neurobiology are strongly encouraged. Also of great interest are projects that aim to use lipidomics, various types of imaging and other cutting edge technologies to identify and develop early biomarkers of neurodegeneration associated lipid dyshomeostasis. The estimated amount of funds available to support projects under this FOA is $2,500,000.
Requirements: The following organizations and institutions are eligible to apply: public/state controlled institutions of higher education; private institutions of higher education; Hispanic-serving institutions; historically Black colleges and universities (HBCUs); tribally controlled colleges and universities (TCCUs); Alaska native and native Hawaiian serving institutions; nonprofits with 501(c)3 IRS status (other than institutions of higher education); nonprofits without 501(c)3 IRS status (other than institutions of higher education); small businesses; for-profit organizations (other than small businesses); state governments; Indian/Native American tribal governments (federally recognized); Indian/Native American tribally designated organizations; county governments; city or township governments; special district governments; independent school districts; public housing authorities/Indian housing authorities; U.S. territories or possessions; Indian/native American tribal governments (other than federally recognized); regional organizations; and non-domestic (non-U.S.) entities (foreign organizations).
Geographic Focus: All States
Date(s) Application is Due: Mar 5; Jul 5; Nov 5
Contact: Suzana S. Petanceska, (301) 496-9350; petanceskas@mail.nih.gov
Internet: http://grants.nih.gov/grants/guide/pa-files/PA-09-217.html
Sponsor: National Institute on Aging
Gateway Building, Suite 350, 7201 Wisconsin Avenue, MSC 9205
Bethesda, MD 20892-9205

NIA Role of Nuclear Receptors in Tissue and Organismal Aging Grants 2783
This Funding Opportunity Announcement (FOA) solicits grant applications proposing to conduct research into underlying biologic mechanisms involving nuclear receptors, their co-regulators and intracellular signaling systems in the process of aging and the connections of the aging process with pathophysiology in middle and old age. The focus of the proposed research must be on processes of aging and/or age-related changes, i.e., research proposed in applications responding to this FOA must have clear relevance to aging. Potential topic areas of interest to the NIA include, but are not limited to, the following: age-changes in levels, sensitivity, and gene expression activation profiles of NRs and co-regulators, and their effects on processes of aging (e.g., age-changes in tissue function, inflammation, glucose and lipid metabolism, adipogenesis, muscle, and brain function) and connections with disease processes for diseases common in older individuals; modification of lifespan in model organisms (e.g., the NR homologue DAF-12 in the insulin-IGF-like pathway in the nematode, and mammalian homologues) and the health-protecting, life-extending effects of caloric restriction or other life-span altering genetic mutations or transgenic models in mammalian organisms; and age-changes in steroid metabolism or in tissue sensitivity and/or response to exogenous sex steroid (or tissue-selective sex steroid receptor modulators) therapy in middle- and old-age.
Requirements: The following organizations and institutions may apply: public/state controlled institutions of higher education; private institutions of higher education; Hispanic-serving institutions; historically Black colleges and universities (HBCUs); tribally controlled colleges and universities (TCCUs); Alaska native and native Hawaiian serving institutions; nonprofits with 501(c)3 IRS status; nonprofits without 501(c)3 IRS status; small businesses; for-profit organizations; State governments; regional organizations; U.S. territories or possessions; Indian/Native American tribal governments; Indian/Native American tribally designated organizations; eligible agencies of the Federal government; and faith-based or community based organizations.
Geographic Focus: All States
Date(s) Application is Due: Feb 5; Jun 5; Oct 5
Contact: Ronald Margolis, (301) 594-8819; fax (301) 435-6047; rm76f@nih.gov
Internet: http://grants.nih.gov/grants/guide/pa-files/PAS-07-267.html
Sponsor: National Institute on Aging
Gateway Building, Suite 2C231, 7201 Wisconsin Avenue, MSC 9205
Bethesda, MD 20892

NIA Support of Scientific Meetings as Cooperative Agreements 2784
The purpose of the program announcement is to inform the scientific community that the National Institute on Aging (NIA) will now support scientific meetings as cooperative agreements in addition to the current practice of supporting them through the traditional grant mechanism. This program provides guidelines for when it is appropriate to request support of a meeting as a cooperative agreement and explains procedures for preparing and submitting such applications. The Principal Investigator will have the primary authority and responsibility to define objectives and approaches; plan, publicize, and conduct the scientific meeting; and publish the results of the meeting.
Requirements: Applications may be submitted by U.S. institutions, including scientific or professional societies eligible to receive grants from Public Health Service (PHS) agencies. In the case of an international conference, the U.S. representative organization of an established international scientific or professional society is the eligible applicant.
Restrictions: Foreign institutions are not eligible.
Geographic Focus: All States
Contact: Chyren Hunter, (301) 496-9322; fax (301) 402-0528; Hunterc@nia.nih.gov
Internet: http://grants.nih.gov/grants/guide/pa-files/PA-00-128.html
Sponsor: National Institute on Aging
Gateway Building, Suite 2C231, 7201 Wisconsin Avenue, MSC 9205
Bethesda, MD 20892

NIA Thyroid in Aging Grants 2785
The purpose of this funding opportunity announcement (FOA) is to encourage submission of investigator-initiated research applications on the thyroid in aging. This FOA is intended to promote basic, translational, and clinical studies leading to increased understanding of the physiology of the aging thyroid and improved diagnosis and management of thyroid disease in the elderly.
Requirements: The following organizations and institutions may apply: public/state controlled institutions of higher education; private institutions of higher education; Hispanic-serving institutions; nonprofits with 501(c)3 IRS status; nonprofits without 501(c)3 IRS status; small businesses; for-profit organizations; State governments; regional organizations; U.S. territories or possessions; Indian/Native American tribal governments; Indian/Native American tribally designated organizations; non-domestic (non-U.S.) entities (foreign organizations); Hispanic-serving institutions; historically Black colleges and universities (HBCUs); tribally controlled colleges and universities (TCCUs); Alaska native and native Hawaiian serving institutions; and regional organizations.
Geographic Focus: All States
Date(s) Application is Due: Feb 5; Feb 16; Jun 5; Jun 16; Oct 5; Oct 16
Contact: Nancy J. Emenaker, PhD; (301) 496-0116; emenaken@mail.nih.gov
Internet: http://grants.nih.gov/grants/guide/pa-files/PA-08-037.html
Sponsor: National Institute on Aging
Gateway Building, Suite 3C307, 7201 Wisconsin Avenue, MSC 9205
Bethesda, MD 20892-9205

NIA Transdisciplinary Research on Fatigue and Fatigability in Aging Grants 2786
The purpose of this Funding Opportunity Announcement (FOA) is to encourage submission of research applications on fatigue and fatigability in aging. This FOA is intended to promote research studies employing transdisciplinary approaches that could lead to increased understanding of mechanisms contributing to, assessment of, or potential interventions for, increased fatigue or fatigability in older persons. Both animal models and humans are appropriate for study under this FOA.
Requirements: The following organizations and institutions may apply: public/state controlled institutions of higher education; private institutions of higher education; Hispanic-serving institutions; historically Black colleges and universities (HBCUs); tribally controlled colleges and universities (TCCUs); Alaska native and native Hawaiian serving institutions; nonprofits with 501(c)3 IRS status; nonprofits without 501(c)3 IRS status; small businesses; for-profit organizations; State governments; regional organizations; U.S. territories or possessions; Indian/Native American tribal governments; Indian/Native American tribally designated organizations; county governments; city or township governments; special district governments; independent school districts; public housing authorities/Indian housing authorities; eligible agencies of the Federal government; and faith-based or community based organizations.
Geographic Focus: All States
Date(s) Application is Due: Feb 5; Feb 16; Jun 5; Jun 16; Oct 5; Oct 16
Contact: Kathleen Jett, PhD; (301) 594-2154; fax (301) 480-8260; jettk@mail.nih.gov
Internet: http://grants.nih.gov/grants/guide/pa-files/PA-08-161.html
Sponsor: National Institute on Aging
Gateway Building, Suite 3C307, 7201 Wisconsin Avenue, MSC 9205
Bethesda, MD 20892-9205

NIA Translational Research at the Aging/Cancer Interface (TRACI) Grants 2787
The purpose of this Funding Opportunity Announcement (FOA) is to enhance translational research in the overlapping areas of human aging and cancer by: integrating knowledge of basic processes in cancer biology and aging into clinical care of older patients with cancer (bench to bedside); and exploring clinical observations from the patient care setting at more basic and molecular levels (bedside to bench). Ultimately, information

from the research supported by this initiative should lead to further improvements in prevention, diagnosis and disease management, improving the health and well-being of elderly patients at risk for, or diagnosed with, cancer and decreasing the functional impairment and morbidity associated with cancer in this population.
Requirements: The following organizations and institutions may apply: public/state controlled institutions of higher education; private institutions of higher education; Hispanic-serving institutions; historically Black colleges and universities (HBCUs); tribally controlled colleges and universities (TCCUs); Alaska native and native Hawaiian serving institutions; nonprofits with 501(c)3 IRS status; nonprofits without 501(c)3 IRS status; small businesses; for-profit organizations; State governments; regional organizations; U.S. territories or possessions; Indian/Native American tribal governments; Indian/Native American tribally designated organizations; county governments; city or township governments; special district governments; independent school districts; public housing authorities/Indian housing authorities; eligible agencies of the Federal government; and faith-based or community based organizations.
Geographic Focus: All States
Date(s) Application is Due: Feb 5; Feb 16; Jun 5; Jun 16; Oct 5; Oct 16
Contact: Susan G. Nayfield, (301) 496-6761; fax (301) 402-1784; nayfiels@nia.nih.gov
Internet: http://grants.nih.gov/grants/guide/pa-files/PA-08-230.html
Sponsor: National Institute on Aging
Gateway Building, Suite 3C307, 7201 Wisconsin Avenue, MSC 9205
Bethesda, MD 20892-9205

NIA Vulnerable Dendrites and Synapses in Aging and Alzheimer's Disease Grants 2788

The purpose of this initiative is to solicit research proposals that will investigate the factors regulating synaptic plasticity and dysfunction with a particular emphasis on the age-dependent changes in the functions of dendrites, spines and synapses of key cell types in regions of brain that are selectively vulnerable in certain neurodegenerative disorders such as AD, or with aging using, for example, in vitro and in vivo models.
Requirements: The following organizations and institutions may apply: public/state controlled institutions of higher education; private institutions of higher education; Hispanic-serving institutions; historically Black colleges and universities (HBCUs); tribally controlled colleges and universities (TCCUs); Alaska native and native Hawaiian serving institutions; nonprofits with 501(c)3 IRS status; nonprofits without 501(c)3 IRS status; small businesses; for-profit organizations; State governments; regional organizations; U.S. territories or possessions; Indian/Native American tribal governments; Indian/Native American tribally designated organizations; county governments; city or township governments; special district governments; independent school districts; public housing authorities/Indian housing authorities; eligible agencies of the Federal government; and faith-based or community based organizations.
Geographic Focus: All States
Date(s) Application is Due: Feb 5; Jun 5; Oct 5
Contact: D. Stephen Snyder, (301) 496-9350; fax (301) 496-1494; ss82f@nih.gov
Internet: http://grants.nih.gov/grants/guide/pa-files/PA-09-061.html
Sponsor: National Institute on Aging
Gateway Building, Suite 350, 7201 Wisconsin Avenue, MSC 9205
Bethesda, MD 20892-9205

NICHD Innovative Therapies and Clinical Studies for Screenable Disorders Grants 2789

This Funding Opportunity Announcement (FOA) issued by the National Institute of Child Health and Human Development (NICHD), the National Institute on Deafness and Other Communication Disorders (NID.C.D), and the National Institute of Diabetes and Digestive and Kidney Diseases (NIDDK), National Institutes of Health (NIH), solicit applications for research relevant to the basic understanding and development of therapeutic interventions for currently screened conditions and high priority genetic conditions for which screening could be possible in the near future. Because the nature and scope of the proposed research will vary from application to application, it is anticipated that the size and duration of each award will also vary. The total amount awarded and the number of awards will depend upon the numbers, quality, duration, and costs of the applications received.
Requirements: Application can be made by the following institutions and organizations: public/state controlled institutions of higher education; private institutions of higher education; nonprofits with 501(c)3 IRS status (other than institutions of higher education); nonprofits without 501(c)3 IRS status (other than institutions of higher education); small businesses; for-profit organizations (other than small businesses); state governments; U.S. territories or possessions; Indian/Native American tribal governments (federally recognized); Indian/Native American tribal governments (other than Federally recognized); Indian/Native American tribally designated organizations; non-domestic (non-U.S.) entities (foreign organizations); Hispanic-serving institutions; historically Black colleges and universities (HBCUs); tribally controlled colleges and universities (TCCUs); Alaska Native and Native Hawaiian serving institutions; regional organization; eligible agencies of the Federal government; and faith-based or community based organizations.
Geographic Focus: All States
Date(s) Application is Due: Feb 5; Mar 5; Jun 5; Jul 5; Oct 5; Nov 5
Contact: Gilian Engelson, (301) 451-2137; fax (301) 451-5512; engelsong@mail.nih.gov
Internet: http://grants.nih.gov/grants/guide/pa-files/PAR-07-184.html
Sponsor: National Institute of Child Health and Human Development
31 Center Drive, Building 31 Room 2A32, MSC 2425
Bethesda, MD 20892-2425

NICHD Ruth L. Kirschstein National Research Service Award Institutional Predoctoral Training Program in Systems Biology of Developmental Biology & Birth Defects 2790

The objective of this NRSA T32 program is to provide research training to predoctoral students interested in establishing research careers that use systems biology approaches to study developmental biology and the formation of structural birth defects. Stipends are provided as a subsistence allowance to help defray trainees' living expenses during the research training experience. Funds are provided to offset the cost of tuition and fees. The NIH provides 60% of the combined requested costs, up to $16,000 annually per Ph.D. student, and up to $21,000 annually for students in formal programs leading to combined professional and Ph.D. degrees. Letters of Intent are due by April 25, with full applications due May 25.
Requirements: The following organizations are eligible to apply: public/state controlled institutions of higher education; private institutions of higher education; nonprofits with 501(c)3 IRS status (other than institutions of higher education); nonprofits without 501(c)3 IRS status (other than institutions of higher education); for-profit organizations (other than small businesses); state governments; U.S. territories or possessions; Indian/Native American tribal governments (Federally recognized); Hispanic-serving institutions; historically Black colleges and universities (HBCUs); tribally controlled colleges and universities (TCCUs); Alaska Natives and Native Hawaiians serving institutions; regional organizations; eligible agencies of the federal government; faith-based or community based organizations.
Geographic Focus: All States
Date(s) Application is Due: Apr 25; May 25
Contact: James Coulombe, (301) 496-5541; fax (301) 480-0303; coulombej@mail.nih.gov
Internet: http://grants.nih.gov/grants/guide/pa-files/PAR-08-054.html
Sponsor: National Institute of Child Health and Human Development
31 Center Drive, Building 31 Room 2A32, MSC 2425
Bethesda, MD 20892-2425

Nicholas H. Noyes Jr. Memorial Foundation Grants 2791

The foundation awards grants to eligible Indiana nonprofit organizations in its areas of interest, including arts and culture, education from early childhood through higher education, disadvantaged, museums, social services, health, hospitals, family services, performing arts, and youth. Types of support include general operating support, endowment funds, and scholarship funds.
Requirements: Proposals are welcomed, and encouraged, to be submitted at any time after January 1st for the first funding cycle or after June 1st for the final funding cycle. Submit the following with your proposal: thirteen copies of the complete grant application (available at the foundations website or office) including attachments (i.e. applicable budget(s) as required). You may recreate the application form for ease in completing, but please limit your response to the space provided and use a minimum font of 12. It is also require that you use both sides of the paper i.e. side one is page 1 with page 2 on the flip-side etc; one copy of your organization's 501(c)3 tax determination letter from the IRS; one copy of your organization's most recent audited financial statement for all organizations requesting $25,000 or more.
Restrictions: Organizations may apply only one time in any calendar year. The Foundation does not make grants to individuals. The principle geographic region served by the Noyes Foundation is the greater Indianapolis area. If you intend to request a grant in excess of $50,000, please contact the Foundation's Program Officer before submitting your grant request.
Geographic Focus: Indiana
Date(s) Application is Due: Feb 1; Aug 29
Amount of Grant: Up to 150,000 USD
Contact: Kelly Mills; (317) 844-8009; fax (317) 844-8099; kmills@noyesfoundation.org
Internet: http://www.noyesfoundation.org
Sponsor: Nicholas H. Noyes Jr. Memorial Foundation
1950 E Greyhound Pass, #18
Carmel, IN 46033-7730

Nick Traina Foundation Grants 2792

The Foundation supports organizations involved in the diagnosis, research, treatment, and/or family support of manic-depression, and other forms of mental illness, suicide prevention, child abuse and children in jeopardy, and provides assistance to struggling musicians in the areas of mental illness. The Foundation may give special consideration to proposals that address manic-depression in children and young adults. There are no deadlines and there is no formal application form. The Board meets four times a year to review proposals.
Requirements: Applicants must be 501(c)3 organizations. Proposals should be kept to three pages in addition to attachments and should include: the organization's purpose, including a description of its history and mission and the population served; the reason for the grant request, including the amount requested and how the funds will be used; whether the request is for general organizational support or for a specific project (including the duration of a specific project); a copy of the IRS determination letter; and a copy of the most recent annual report and the last two years' financial statements.
Restrictions: Organizations may apply for one grant each calendar year. Requests from individuals are ineligible. Grant funding focuses on the San Francisco Bay area.
Geographic Focus: California
Contact: Danielle Steel, President; (415) 771-4224; info@nicktrainafoundation.org
Internet: http://nicktrainafoundation.com/grants.htm
Sponsor: Nick Traina Foundation
P.O. Box 470427
San Francisco, CA 94147-0427

NIDA Mentored Clinical Scientist Research Career Development Awards 2793

The overall objective of the NIH Research Career Development Award program is to prepare qualified individuals for careers that have a significant impact on the health-related research needs of the Nation. The objective of the NIH Mentored Clinical Scientist Research Career Development Award (K08) program is to support didactic study and mentored research for individuals with clinical doctoral degrees (e.g., M.D., D.D.S., D.M.D., D.O., D.C., O.D., N.D., D.V.M., Pharm.D., or Ph.D. in clinical disciplines). The purpose of this Funding Opportunity Announcement (FOA) is to continue the long-standing NIH support of the K08 program. This award provides support and protected time for an intensive, mentored research career development experience in biomedical or behavioral research, including translational research. For the purpose of this award, translational research is defined as application of basic research discoveries toward the diagnosis, management, and prevention of human disease.

Requirements: The following organizations and institutions may apply: public/state controlled institutions of higher education; private institutions of higher education; Hispanic-serving institutions; historically Black colleges and universities (HBCUs); tribally controlled colleges and universities (TCCUs); Alaska native and native Hawaiian serving institutions; nonprofits with 501(c)3 IRS status; nonprofits without 501(c)3 IRS status; small businesses; for-profit organizations; State governments; regional organizations; U.S. territories or possessions; Indian/Native American tribal governments; Indian/Native American tribally designated organizations; county governments; city or township governments; special district governments; independent school districts; public housing authorities/Indian housing authorities; eligible agencies of the Federal government; and faith-based or community based organizations.

Geographic Focus: All States
Date(s) Application is Due: Feb 12; Mar 12; Jun 12; Jul 12; Oct 12; Nov 12
Amount of Grant: Up to 140,000 USD
Contact: Mimi M. Ghim, (301) 443-6071; ghimm@mail.nih.gov
Internet: http://grants.nih.gov/grants/guide/pa-files/PA-09-042.html#SectionIII
Sponsor: National Institute on Drug Abuse
6001 Executive Boulevard, Room 5213, MSC 9561
Bethesda, MD 20892-9561

NIDA Pilot and Feasibility Studies in Preparation for Drug Abuse Prevention Trials 2794

This FOA for R34 applications seeks to support: pilot and/or feasibility testing of new, revised, or adapted preventive intervention approaches targeting the initiation of drug use, the progression to abuse or dependence, and the acquisition or transmission of HIV infection among diverse populations and settings; and pre-trial feasibility testing for prevention services and systems research.

Requirements: The following organizations and institutions may apply: public/state controlled institutions of higher education; private institutions of higher education; Hispanic-serving institutions; historically Black colleges and universities (HBCUs); tribally controlled colleges and universities (TCCUs); Alaska native and native Hawaiian serving institutions; nonprofits with 501(c)3 IRS status; nonprofits without 501(c)3 IRS status; small businesses; for-profit organizations; State governments; regional organizations; U.S. territories or possessions; Indian/Native American tribal governments; Indian/Native American tribally designated organizations; county governments; city or township governments; special district governments; independent school districts; public housing authorities/Indian housing authorities; eligible agencies of the Federal government; and faith-based or community based organizations.

Restrictions: The NIDA R34 mechanism does not support the development of intervention protocols, manuals, or the standardization of protocols.
Geographic Focus: All States
Date(s) Application is Due: Feb 16; Jun 16; Oct 16
Amount of Grant: 45,000 - 225,000 USD
Contact: Aleta Meyer; (301) 402-1725; fax (301) 443-2636; meyera2@nida.nih.gov
Internet: http://grants.nih.gov/grants/guide/pa-files/PA-09-146.html
Sponsor: National Institute on Drug Abuse
6001 Executive Boulevard, Room 5213, MSC 9561
Bethesda, MD 20892-9561

NIDCD Mentored Clinical Scientist Research Career Development Award 2795

This award enables candidates holding professional degrees (e.g., M.D., D.O., D.V.M., or equivalent degrees) to undertake 3 to 5 years of special study and supervised research with the goal of becoming independent investigators. The award also allows awardees to pursue a research career development program suited to their experience and capabilities under a mentor who is competent to provide guidance in the chosen research area. Institutions may submit applications on behalf of candidates who hold professional degrees. At least 2 years must have elapsed since the health professional degree was granted. Candidates can have varying levels of clinical training and research experience. New applications are due February 12, June 12, and October 12, while resubmitted applications are due March 12, July 12, and November 12. Up to $105,000 per year plus fringe benefits is available, and up to $85,000 per year for research support for new awards.

Requirements: A candidate must have a clinical degree or its equivalent, must have initiated postgraduate clinical training, must identify a mentor with extensive research experience, and must be willing to spend a minimum of 75 percent of full-time professional effort conducting research. Applications may be submitted on behalf of candidates by domestic, nonfederal organizations, public or private, such as medical, dental, or nursing schools or other institutions of higher education. Candidates must be U.S. citizens, noncitizen nationals, or permanent residents.

Geographic Focus: All States
Date(s) Application is Due: Feb 12; Mar 12; Jun 12; Jul 12; Oct 12; Nov 12
Amount of Grant: Up to 185,000 USD
Contact: Daniel Sklare, (301) 496-1804; fax (301) 402-6251; sklared@nidcd.nih.gov
Internet: http://grants.nih.gov/grants/guide/pa-files/pa-06-512.html
Sponsor: National Institute on Deafness and Other Communication Disorders
6120 Executive Boulevard, Executive Plaza S
Bethesda, MD 20892

NIDCD Mentored Research Scientist Development Award 2796

The NID.C.D utilizes the K01 Program for the retooling and transition of junior- and midcareer-level scientists and clinicians (generally below the rank of full professor, or the equivalent in nonacademic settings) into two research domains: translational research, i.e., the application of basic research discoveries toward advancing the diagnosis, management and prevention of deafness and other communication disorders; and clinical research. Junior-level clinically-trained individuals should apply for the Mentored Clinical Scientist Development Award or the Mentored Patient-Oriented Research Career Development Award, as appropriate.

Requirements: The candidate must have a research or health-professional doctorate or its equivalent, and must have demonstrated the capacity or potential for highly productive independent research in the period after the doctorate. The candidate must identify a mentor with extensive research experience, and must be willing to spend a minimum of 75 percent of full-time professional effort conducting research and research career development activities for the period of the award. Applications may be submitted on behalf of candidates by domestic, public or private, nonfederal organizations. Candidates must be U.S. citizens or noncitizen nationals.

Geographic Focus: All States
Date(s) Application is Due: Feb 12; Mar 12; Jun 12; Jul 12; Oct 12; Nov 12
Amount of Grant: Up to 105,000 USD
Contact: Daniel A. Sklare, (301) 496-1804; sklared@nidcd.nih.gov
Internet: http://www.grants.nih.gov/grants/guide/pa-files/PAR-99-027.html
Sponsor: National Institute on Deafness and Other Communication Disorders
6120 Executive Boulevard, Executive Plaza S
Bethesda, MD 20892

NIDDK Advances in Polycystic Kidney Disease Grants 2797

The National Institute of Diabetes and Digestive and Kidney Diseases (NIDDK) through its Division of Kidney, Urologic and Hematologic Diseases (DKUHD) invites experienced and new investigators to submit research grant applications to pursue basic and applied investigations in order to better understand the etiology and pathogenesis of Polycystic Kidney Disease (PKD), in both its autosomal dominant and autosomal recessive forms. Such applications may examine the genetic determinants, and cellular and molecular mechanisms, which disrupt normal kidney function; mechanisms of cyst formation and growth; development of experimental model systems; development of markers of disease progression; and the identification of innovative therapeutic interventions and gene targeted strategies to prevent progressive renal insufficiency due to this disorder. The intent of this funding opportunity is to intensify investigator-initiated research, to attract new investigators to the field, and to increase interdisciplinary research. The ultimate aim is to facilitate PKD-related research studies, which will provide the basis for new therapeutic approaches.

Requirements: The following organizations may apply: for-profit organizations; non-profit organizations; public or private institutions, such as universities, colleges, hospitals, and laboratories; units of State government; units of local government; eligible agencies of the Federal government; foreign institutions; domestic institutions; and faith-based or community-based organizations.

Geographic Focus: All States
Date(s) Application is Due: Feb 5; Jun 5; Oct 5
Contact: Catherine M. Meyers, (301) 594-7717; fax (301) 480-3510; cm420i@nih.gov
Internet: http://grants.nih.gov/grants/guide/pa-files/PA-06-158.html
Sponsor: National Institute of Diabetes and Digestive and Kidney Diseases
6707 Democracy Boulevard
Bethesda, MD 20892-5460

NIDDK Adverse Metabolic Side Effects of Second Generation Psychotropic Medications Leading to Obesity and Increased Diabetes Risk Grants 2798

This Funding Opportunity invites investigator-initiated research grant applications for studies examining the adverse metabolic effects (i.e., obesity and diabetes) of psychotropic medications in animal models and across the human lifespan (including pediatric, adult and geriatric populations). Applications responsive to this FOA should focus on: increasing the understanding of the nature, rates, and pathophysiology of adverse metabolic effects of psychotropic medications; elucidating biomedical and psychosocial risk factors for the development of metabolic adverse effects of psychiatric therapeutics; and develop interventions to prevent and/or mitigate metabolic adverse effects across the lifespan.

Requirements: The following organizations and institutions may apply: public/state controlled institutions of higher education; private institutions of higher education; Hispanic-serving institutions; historically Black colleges and universities (HBCUs); tribally controlled colleges and universities (TCCUs); Alaska native and native Hawaiian serving institutions; nonprofits with 501(c)3 IRS status; nonprofits without 501(c)3 IRS status; small businesses; for-profit organizations; State governments; regional organizations; U.S. territories or possessions; Indian/Native American tribal governments; Indian/Native American tribally designated organizations; eligible agencies of the Federal government; and faith-based or community based organizations.

Geographic Focus: All States
Date(s) Application is Due: Feb 22; Jun 22; Oct 22
Amount of Grant: Up to 500,000 USD

Contact: Christine Hunter, (301) 594-4728; hunterchristine@niddk.nih.gov
Internet: http://grants.nih.gov/grants/guide/pa-files/PAR-08-160.html
Sponsor: National Institute of Diabetes and Digestive and Kidney Diseases
6707 Democracy Boulevard
Bethesda, MD 20892-5460

NIDDK Ancillary Studies of Kidney Disease Accessing Information from Clinical Trials, Epidemiological Studies, and Databases Grants — 2799

The National Institute of Diabetes and Digestive and Kidney Diseases (NIDDK) invites investigator-initiated research project applications for ancillary studies to ongoing or completed clinical trials and epidemiological studies of kidney disease as well as clinical trials and epidemiological studies for other diseases or populations that lend themselves to the study of kidney disease. These studies may range from new analyses of existing datasets of completed studies to additional collection of data and biological specimens in ongoing investigations. The goal of these studies should be to extend our understanding of the risk factors for developing kidney disease and their associated co-morbid illnesses such as malnutrition and cardiovascular disease, factors associated with rapid decline in kidney function among persons with chronic kidney disease, and the impact of these diseases on quality of life and mental and physical functioning. Studies of biomarkers for patients with acute renal failure and other kidney diseases, for example, are also appropriate topics for further investigation. Studies ancillary to both government and non-government supported clinical trials and epidemiological studies are encouraged. Analysis of large public access databases and other databases is also encouraged.
Requirements: The following organizations and institutions may apply: public/state controlled institutions of higher education; private institutions of higher education; Hispanic-serving institutions; historically Black colleges and universities (HBCUs); tribally controlled colleges and universities (TCCUs); Alaska native and native Hawaiian serving institutions; nonprofits with 501(c)3 IRS status; nonprofits without 501(c)3 IRS status; small businesses; for-profit organizations; State governments; regional organizations; U.S. territories or possessions; Indian/Native American tribal governments; Indian/Native American tribally designated organizations; eligible agencies of the Federal government; and faith-based or community based organizations.
Geographic Focus: All States
Date(s) Application is Due: Feb 5; Jun 5; Oct 5
Contact: John W. Kusek, (301) 594-7735; fax (301) 480-3510; jk61x@nih.gov
Internet: http://grants.nih.gov/grants/guide/pa-files/PA-07-050.html
Sponsor: National Institute of Diabetes and Digestive and Kidney Diseases
6707 Democracy Boulevard
Bethesda, MD 20892-5460

NIDDK Ancillary Studies to Major Ongoing NIDDK and NHLBI Clinical Research Studies — 2800

This Funding Opportunity Announcement (FOA) solicits grant applications from qualified investigators to conduct ancillary studies of selected ongoing major clinical research studies, including clinical trials, epidemiological studies and disease databases, supported by the National Institute of Diabetes and Digestive and Kidney Diseases and the National Heart, Lung, and Blood Institute, National Institutes of Health (NIH). The objective is to introduce NIDDK-oriented ancillary studies into either NIDDK or NHLBI-funded parent studies, or NHLBI-oriented ancillary studies into NIDDK-funded parent studies.
Requirements: The following organizations and institutions may apply: public/state controlled institutions of higher education; private institutions of higher education; Hispanic-serving institutions; historically Black colleges and universities (HBCUs); tribally controlled colleges and universities (TCCUs); Alaska native and native Hawaiian serving institutions; nonprofits with 501(c)3 IRS status; nonprofits without 501(c)3 IRS status; small businesses; for-profit organizations; State governments; regional organizations; U.S. territories or possessions; Indian/Native American tribal governments; Indian/Native American tribally designated organizations; eligible agencies of the Federal government; and faith-based or community based organizations.
Geographic Focus: All States
Date(s) Application is Due: Feb 5; Jun 5; Oct 5
Contact: John W. Kusek, (301) 594-7735; fax (301) 480-3510; jk61x@nih.gov
Internet: http://grants.nih.gov/grants/guide/pa-files/PAR-07-024.html
Sponsor: National Institute of Diabetes and Digestive and Kidney Diseases
6707 Democracy Boulevard
Bethesda, MD 20892-5460

NIDDK Basic and Clinical Studies of Congenital Urinary Tract Obstruction Grants — 2801

The purpose of this funding opportunity is to address the numerous scientific and clinical uncertainties related to the development, treatment and prognosis of congenital obstructive uropathy, by encouraging and facilitating research in diverse areas. These areas include: the development of objective prognostic markers; the genetic determinants of this congenital disorder; the development of reliable animal models of the disorder; and, evaluation of the long-term effectiveness of various treatment strategies. Because the nature and scope of the proposed research will vary from application to application, it is anticipated that the size and duration of each award will also vary. The total amount awarded and the number of awards will depend upon the mechanism numbers, quality, duration, and costs of the applications received.
Requirements: The following organizations and institutions may apply: public/state controlled institutions of higher education; private institutions of higher education; Hispanic-serving institutions; historically Black colleges and universities (HBCUs); tribally controlled colleges and universities (TCCUs); Alaska native and native Hawaiian serving institutions; nonprofits with 501(c)3 IRS status; nonprofits without 501(c)3 IRS status; small businesses; for-profit organizations; State governments; regional organizations; U.S. territories or possessions; Indian/Native American tribal governments; Indian/Native American tribally designated organizations; eligible agencies of the Federal government; and faith-based or community based organizations.
Geographic Focus: All States
Date(s) Application is Due: Feb 5; Jun 5; Oct 5
Contact: Marva Moxey-Mims, (301) 594-7717; fax (301) 480-3510; mm726k@nih.gov
Internet: http://grants.nih.gov/grants/guide/pa-files/PA-07-059.html
Sponsor: National Institute of Diabetes and Digestive and Kidney Diseases
6707 Democracy Boulevard
Bethesda, MD 20892-5460

NIDDK Basic Research Grants in the Bladder and Lower Urinary Tract — 2802

This funding opportunity encourages basic cellular, molecular, developmental and genetic research relevant to the bladder and lower urinary tract. Basic research studies that address age and gender differences in bladder and lower urinary tract function are also encouraged. New and established investigators from related fields of study are encouraged to apply their expertise to these problem areas. Investigators with diverse basic science and clinical backgrounds are encouraged to develop collaborative research relationships. Discoveries resulting from these studies may serve as the basis for the development of new agents, techniques, and strategies for detecting, preventing, and treating diseases of the bladder and lower urinary tract, as well as dealing with bladder complications resulting from other diseases, such as diabetes.
Requirements: Application can be made by the following institutions and organizations: for-profit organizations; non-profit organizations; public or private institutions (such as universities, colleges, hospitals, and laboratories); units of State government; units of local government; eligible agencies of the Federal government; foreign institutions; and domestic institutions.
Geographic Focus: All States
Date(s) Application is Due: Feb 5; Jun 5; Oct 5
Contact: Chris Mullins, (301) 594-7717; fax (301) 480-3510; cm419z@nih.gov
Internet: http://grants.nih.gov/grants/guide/pa-files/PA-06-254.html
Sponsor: National Institute of Diabetes and Digestive and Kidney Diseases
6707 Democracy Boulevard
Bethesda, MD 20892-5460

NIDDK Calcium Oxalate Stone Diseases Grants — 2803

The purpose of this Funding Opportunity Announcement (FOA) is to increase investigator interest in research into the genetics and heritability of oxalate regulation and the oxalate stone diseases, and to develop new treatments for the disorder. This initiative solicits research as well as pilot and feasibility studies that utilize new and innovative approaches to study the diagnosis, treatment and prevention of these disorders. Because the nature and scope of the proposed research will vary from application to application, it is anticipated that the size and duration of each award will also vary. The total amount awarded and the number of awards will depend upon the mechanism numbers, quality, duration, and costs of the applications received.
Requirements: The following organizations and institutions may apply: public/state controlled institutions of higher education; private institutions of higher education; Hispanic-serving institutions; historically Black colleges and universities (HBCUs); tribally controlled colleges and universities (TCCUs); Alaska native and native Hawaiian serving institutions; nonprofits with 501(c)3 IRS status; nonprofits without 501(c)3 IRS status; small businesses; for-profit organizations; State governments; regional organizations; U.S. territories or possessions; Indian/Native American tribal governments; Indian/Native American tribally designated organizations; eligible agencies of the Federal government; and faith-based or community based organizations.
Geographic Focus: All States
Date(s) Application is Due: Feb 5; Jun 5; Oct 5
Contact: Rebekah S. Rasooly, (301) 594-6007; fax (301) 480-3510; rr185i@nih.gov
Internet: http://grants.nih.gov/grants/guide/pa-files/PA-07-051.html
Sponsor: National Institute of Diabetes and Digestive and Kidney Diseases
6707 Democracy Boulevard
Bethesda, MD 20892-5460

NIDDK Cell-Specific Delineation of Prostate & Genitourinary Development — 2804

This RFA is intended to encourage the development of new research tools and methods to study the development and biology of the prostate and genitourinary tract. Strategies to be supported include systematic assessment of gene expression in specific cell types in the developing prostate and genitourinary tract; methods to tag individual cell types for purification, analysis, and characterization; tools to manipulate gene expression in vivo in individual cell types; and development of clinically relevant cell lines. The major goal of this initiative is to ensure that research tools are available so that a broad range of innovative methods can be applied to studying prostate development. Annual deadline dates may vary; contact program staff for exact dates.
Requirements: Application can be made by the following institutions and organizations: public/state controlled institutions of higher education; private institutions of higher education; nonprofits with 501(c)3 IRS status (other than institutions of higher education); nonprofits without 501(c)3 IRS status (other than institutions of higher education); small businesses; for-profit organizations (other than small businesses); state governments; U.S. territories or possessions; Indian/Native American tribal governments (federally recognized); Indian/Native American tribal governments (other than Federally recognized); Indian/Native American tribally designated organizations; non-domestic (non-U.S.) entities (foreign organizations); Hispanic-serving institutions; historically Black colleges and universities (HBCUs); tribally controlled colleges and universities (TCCUs); Alaska

Native and Native Hawaiian serving institutions; regional organization; eligible agencies of the Federal government; and faith-based or community based organizations.
Geographic Focus: All States
Date(s) Application is Due: Feb 5; Mar 5; Jun 5; Jul 5; Oct 5; Nov 5
Amount of Grant: Up to 350,000 USD
Contact: Dr. Robert Star, (301) 594-7715; StarR@extra.niddk.nih.gov
Internet: http://grants.nih.gov/grants/guide/rfa-files/RFA-DK-00-015.html
Sponsor: National Institute of Diabetes and Digestive and Kidney Diseases
31 Center Drive, Room 9A-35, MSC 2560
Bethesda, MD 20892-2560

NIDDK Co-Activators and Co-Repressors in Gene Expression Grants   2805
This initiative stems from an NIDDK workshop in molecular endocrinology and is designed to stimulate research that addresses the fundamental underlying mechanisms by which nuclear accessory proteins mediate signaling through hormone receptors at the level of the regulation of gene expression. Staff contact prior to submission is highly recommended.
Requirements: Applications may be submitted by domestic and foreign, for-profit and nonprofit organizations, both public and private; units of state and local governments; and eligible agencies of the federal government.
Geographic Focus: All States
Date(s) Application is Due: Feb 16; Jun 16; Oct 16
Amount of Grant: 100,000 - 250,000 USD
Contact: Dr. Ronald Margolis, (301) 594-8819; fax (301) 435-6047; rm76f@nih.gov
Internet: http://grants.nih.gov/grants/guide/pa-files/PA-99-111.html
Sponsor: National Institutes of Health
6707 Democracy Boulevard, Building 2DEM, Room 693
Bethesda, MD 20892

NIDDK Collaborative Interdisciplinary Team Science in Diabetes, Endocrinology and Metabolic Diseases Grants   2806
The purpose of the Collaborative Interdisciplinary Team Science Program described in this announcement is to provide support to enable strong investigative teams to do inter- and/or trans-disciplinary research on a complex problem in biomedical science relevant to Diabetes, Endocrinology and Metabolic Diseases. It is anticipated that 1-3 projects will be funded per year, but because the nature and scope of the proposed research will vary from application to application, it is anticipated that the size and duration of each award will also vary. The total amount awarded and the number of awards will depend upon the quality, duration, and costs of the applications received, as well as on the availability of funds.
Requirements: The following organizations and institutions are eligible to apply: public/state controlled institutions of higher education; private institutions of higher education; Hispanic-serving institutions; historically Black colleges and universities (HBCUs); tribally controlled colleges and universities (TCCUs); Alaska Native and Native Hawaiian serving institutions; and non-domestic (non-U.S.) entities (foreign organizations).
Geographic Focus: All States
Date(s) Application is Due: Feb 24; Mar 24
Contact: Karen Salomon, (301) 594-7733; fax (301) 435-6047; ks495c@nih.gov
Internet: http://grants.nih.gov/grants/guide/pa-files/PAR-08-182.html
Sponsor: National Institute of Diabetes and Digestive and Kidney Diseases
6707 Democracy Boulevard
Bethesda, MD 20892-5460

NIDDK Developmental Biology and Regeneration of the Liver Grants   2807
The purpose of this FOA is to invite qualified scientific investigators to submit applications on liver development and regeneration to fully define the molecular and cellular mechanisms underlying these processes in health and disease and to apply these findings to developing improved therapies for liver disease. Because the nature and scope of the proposed research will vary from application to application, it is anticipated that the size and duration of each award will also vary. The total amount awarded and the number of awards will depend upon the mechanism numbers, quality, duration, and costs of the applications received.
Requirements: The following organizations and institutions may apply: public/state controlled institutions of higher education; private institutions of higher education; Hispanic-serving institutions; historically Black colleges and universities (HBCUs); tribally controlled colleges and universities (TCCUs); Alaska native and native Hawaiian serving institutions; nonprofits with 501(c)3 IRS status; nonprofits without 501(c)3 IRS status; small businesses; for-profit organizations; State governments; regional organizations; U.S. territories or possessions; Indian/Native American tribal governments; Indian/Native American tribally designated organizations; eligible agencies of the Federal government; and faith-based or community based organizations.
Geographic Focus: All States
Date(s) Application is Due: Feb 5; Jun 5; Oct 5
Contact: Samir Zakhari, (301) 443-0799; fax (301) 594-0673; szakhari@mail.nih.gov
Internet: http://grants.nih.gov/grants/guide/pa-files/PA-07-026.html
Sponsor: National Institute of Diabetes and Digestive and Kidney Diseases
6707 Democracy Boulevard
Bethesda, MD 20892-5460

NIDDK Development of Assays for High-Throughput Drug Screening Grants   2808
The overall goal of the Molecular Libraries Initiative, part of the NIH Roadmap for Medical Research, is to offer public sector researchers opportunities to screen small molecules with the high-throughput chemical screening (HTS) methods commonly used by the private sector to develop therapeutic agents. The purpose of this FOA is to support the development of innovative assays that may ultimately be adapted for automated screening. The assay should aim to identify new tools for basic research or promising new avenues for therapeutics development, especially in areas related to the missions of the participating institutes. Awards issued under this FOA are contingent upon the availability of funds and the submission of a sufficient number of meritorious applications. The total amount awarded and the number of awards will depend upon the quality and costs of the applications received.
Requirements: The following organizations and institutions may apply: public/state controlled institutions of higher education; private institutions of higher education; Hispanic-serving institutions; historically Black colleges and universities (HBCUs); tribally controlled colleges and universities (TCCUs); Alaska native and native Hawaiian serving institutions; nonprofits with 501(c)3 IRS status; nonprofits without 501(c)3 IRS status; small businesses; for-profit organizations; State governments; regional organizations; U.S. territories or possessions; Indian/Native American tribal governments; Indian/Native American tribally designated organizations; eligible agencies of the Federal government; and faith-based or community based organizations.
Geographic Focus: All States
Date(s) Application is Due: Feb 5; Jun 5; Oct 5
Contact: Laura K. Moen, (301) 594-4748; fax (301) 480-3510; lm232f@nih.gov
Internet: http://grants.nih.gov/grants/guide/pa-files/PA-07-320.html
Sponsor: National Institute of Diabetes and Digestive and Kidney Diseases
6707 Democracy Boulevard
Bethesda, MD 20892-5460

NIDDK Development of Disease Biomarkers Grants   2809
The goal of this Funding Opportunity Announcement (FOA) is to validate biomarkers for well-defined human diseases of liver, kidney, urological tract, digestive and hematologic systems, and endocrine and metabolic disorders, diabetes and its complications, and obesity, for which there are no or very few biomarkers, or for which standard biomarkers are currently prohibitively invasive or expensive. New biomarkers will stimulate bench to bedside translation by providing measures of the biological effects of potential new treatments. The ideal biomarker can be measured in a minimally invasive way, can be measured repeatedly over time, identifies early stages of disease, is indicative of disease prognosis, and correlates well with progression and response to therapy. Especially of interest would be studies designed to test the validity of candidate biomarkers or new technologies to monitor candidate biomarkers in patient tissue samples or small groups of well-characterized patients. For biomarkers already validated in human subjects, the larger effort might be to establish a reliable assay prior to pursuing larger patient studies. Priority will be given to those projects with high promise for improving clinical care in the relatively near future. Studies of potential biomarkers conducted in animal models of disease, and extended prospective clinical trials for marker validation against hard clinical endpoints, are outside the scope of this initiative. Because the nature and scope of the proposed research will vary from application to application, it is anticipated that the size and duration of each award will also vary.
Requirements: Application can be made by the following institutions and organizations: public/state controlled institutions of higher education; private institutions of higher education; nonprofits with 501(c)3 IRS status (other than institutions of higher education); nonprofits without 501(c)3 IRS status (other than institutions of higher education); small businesses; for-profit organizations (other than small businesses); state governments; U.S. territories or possessions; Indian/Native American tribal governments (federally recognized); Indian/Native American tribal governments (other than Federally recognized); Indian/Native American tribally designated organizations; non-domestic (non-U.S.) entities (foreign organizations); Hispanic-serving institutions; historically Black colleges and universities (HBCUs) and universities (TCCUs); Alaska Native and Native Hawaiian serving institutions; regional organization; eligible agencies of the Federal government; and faith-based or community based organizations.
Geographic Focus: All States
Date(s) Application is Due: Feb 5; Jun 5; Oct 5
Contact: Maren R. Laughlin, (301) 594-8802; fax (301) 480-0475; ml33q@nih.gov
Internet: http://grants.nih.gov/grants/guide/pa-files/PA-07-052.html
Sponsor: National Institute of Diabetes and Digestive and Kidney Diseases
6707 Democracy Boulevard
Bethesda, MD 20892-5460

NIDDK Diabetes, Endocrinology, and Metabolic Diseases NRSAs--Individual   2810
Individual National Research Service Awards (NRSAs) are given only at the postdoctoral level for research training in diabetes, endocrine, or metabolic diseases. To receive one, an individual must submit an application that describes a specific research project, which is guided and sponsored by a specific preceptor at a particular institution. The award may be for up to 36 months of training and consists of a stipend and an allowance to help meet expenses incurred during the course of fellowship.
Requirements: All awardees must be citizens or have been admitted to the United States for permanent residence. Individuals and public and private institutions, nonprofit and for-profit, that propose to establish, expand, and improve research activities in health sciences and related fields may apply for research grants. Individuals must be nominated and sponsored by public or nonprofit private institutions having staff and facilities appropriate to the proposed research training programs. SBIR grants can be awarded only to domestic small businesses. STTR grants can be awarded only to domestic small business concerns that partner with research institutions in cooperative research and development.
Geographic Focus: All States
Date(s) Application is Due: Apr 5; Aug 5; Dec 5
Amount of Grant: 16,600 - 100,000,000 USD
Contact: Dr. Judith Fradkin, Director; (301) 496-7348

Internet: http://grants.nih.gov/grants/guide/pa-files/PA-07-107.html
Sponsor: National Institute of Diabetes and Digestive and Kidney Diseases
45 Center Drive, Natcher Building
Bethesda, MD 20892-2560

NIDDK Diabetes Research Centers 2811

The mission of the Diabetes Centers is to serve as a key component of the NIDDK-supported research effort to develop new therapies and improve the health of Americans with, or at risk for, diabetes and related endocrine and metabolic disorders. The Centers promote new discoveries and enhance scientific progress through support of cutting-edge basic and clinical research related to the etiology and complications of diabetes, with the goal of rapidly translating research findings into novel strategies for the prevention, treatment and cure of diabetes and related conditions. This Funding Opportunity Announcement (FOA) solicits new and competing continuation applications for Diabetes Endocrinology Research Centers (DERCs) and Diabetes Research and Training Centers (DRTCs). Both types of Centers are designed to support and enhance the national research effort in diabetes and related endocrine and metabolic diseases. DERCs support three primary research-related activities: biomedical research cores, a Pilot and Feasibility (P&F) program, and an Enrichment program. DRTCs possess all elements of a DERC, with additional dedicated core services and P&F awards to support research in diabetes prevention and control. All activities pursued by Diabetes Centers are designed to enhance the efficiency, productivity, effectiveness and multidisciplinary nature of research in Diabetes Center topic areas.
Requirements: The following organizations and institutions are eligible to apply: public/state controlled institutions of higher education; private institutions of higher education; Hispanic-serving institutions; historically Black colleges and universities (HBCUs); tribally controlled colleges and universities (TCCUs); Alaska Native and Native Hawaiian serving institutions; nonprofits with 501(c)3 IRS status (other than institutions of higher education); nonprofits without 501(c)3 IRS status (other than institutions of higher education); for-profit organizations (other than small businesses); and state governments.
Geographic Focus: All States
Date(s) Application is Due: Jun 22; Jul 15
Contact: James F. Hyde, (301) 594-7692; fax (301) 480-3503; jh486z@nih.gov
Internet: http://grants.nih.gov/grants/guide/rfa-files/RFA-DK-08-008.html
Sponsor: National Institute of Diabetes and Digestive and Kidney Diseases
6707 Democracy Boulevard
Bethesda, MD 20892-5460

NIDDK Diet Composition and Energy Balance Grants 2812

The goal of this funding opportunity announcement (FOA) is to invite Research Project Grant applications investigating the role of diet composition in energy balance, including studies in both animals and humans. Both short and longer-term studies are encouraged, ranging from basic studies investigating the impact of micro-or macronutrient composition on appetite, metabolism, and energy expenditure through clinical studies evaluating the efficacy of diets differing in micro- or macronutrient composition, absorption, dietary variety, or energy density for weight loss or weight maintenance. Because the nature and scope of the proposed research will vary from application to application, it is anticipated that the size and duration of each award will also vary. The total amount awarded and the number of awards will depend upon the mechanism numbers, quality, duration, and costs of the applications received.
Requirements: The following organizations and institutions may apply: public/state controlled institutions of higher education; private institutions of higher education; Hispanic-serving institutions; historically Black colleges and universities (HBCUs); tribally controlled colleges and universities (TCCUs); Alaska native and native Hawaiian serving institutions; nonprofits with 501(c)3 IRS status; nonprofits without 501(c)3 IRS status; small businesses; for-profit organizations; State governments; regional organizations; U.S. territories or possessions; Indian/Native American tribal governments; Indian/Native American tribally designated organizations; eligible agencies of the Federal government; and faith-based or community based organizations.
Geographic Focus: All States
Date(s) Application is Due: Feb 5; Jun 5; Oct 5
Contact: Susan Z. Yanovski, (301) 594-8882; fax (301) 480-8300; sy29f@nih.gov
Internet: http://grants.nih.gov/grants/guide/pa-files/PA-07-218.html
Sponsor: National Institute of Diabetes and Digestive and Kidney Diseases
6707 Democracy Boulevard
Bethesda, MD 20892-5460

NIDDK Education Program Grants 2813

This funding opportunity announcement (FOA) solicits Research Education grant applications from applicant organizations that propose to create educational opportunities to attract undergraduate students, graduate students, and postdoctoral fellows to careers in areas of biomedical or behavioral research of particular interest to the NIDDK while fostering the career development of these students and fellows. The NIDDK is especially interested in attracting students and postdoctoral fellows from scientific disciplines underrepresented in disease-oriented biomedical research such as engineering, informatics, computer science, and computational sciences, to encourage them to apply their expertise to research relevant to diabetes and other endocrine and metabolic diseases, digestive and liver diseases, nutrition, obesity research and prevention, and kidney, urologic and hematologic diseases. The total project period for an application submitted in response to this funding opportunity may not exceed five years. Direct costs are limited to $100,000 per year.
Requirements: The following organizations and institutions may apply: public/state controlled institutions of higher education; private institutions of higher education; Hispanic-serving institutions; historically Black colleges and universities (HBCUs); tribally controlled colleges and universities (TCCUs); Alaska native and native Hawaiian serving institutions; nonprofits with 501(c)3 IRS status; nonprofits without 501(c)3 IRS status; and Indian/Native American tribally designated organizations.
Geographic Focus: All States
Date(s) Application is Due: Jan 25; May 25; Sep 25
Amount of Grant: Up to 100,000 USD
Contact: James F. Hyde, (301) 594-7692; fax (301) 480-3503; jh486z@nih.gov
Internet: http://grants.nih.gov/grants/guide/pa-files/PAR-06-554.html
Sponsor: National Institute of Diabetes and Digestive and Kidney Diseases
6707 Democracy Boulevard
Bethesda, MD 20892-5460

NIDDK Endoscopic Clinical Research Grants In Pancreatic And Biliary Diseases 2814

These small grants (R03) may be used as planning grants for full-scale multi-center clinical trials or for pilot studies that could lead to full-scale multi-center clinical trials designed to provide evidence for or against changes in the current standard of care. Pilot epidemiological studies are encouraged that could lead to more extended research that would provide evidence for or against changes in health policy, especially as related to disease and cancer prevention. It is expected that these R03 grants will serve as a basis for planning future multi-center research project grant applications (R01) or cooperative agreement (U01) awards. New and experienced investigators in relevant fields and disciplines may apply for these small grants. Investigators are encouraged to take advantage of recent endoscopic and laboratory developments. In addition, the small grant is a good mechanism for new and experienced investigators to become better equipped to perform clinical and epidemiological research. A project period of up to two years and a budget for direct costs of up to two $25,000 modules or $50,000 per year may be requested.
Requirements: The following organizations and institutions may apply: for-profit organizations; non-profit organizations; public or private institutions, such as universities, colleges, hospitals, and laboratories; units of State government; units of local government; eligible agencies of the Federal government; domestic institutions; and foreign institutions.
Geographic Focus: All States
Date(s) Application is Due: Feb 16; Jun 16; Oct 16
Contact: Jose Serrano, (301) 594-8871; fax (301) 480-8300; js362q@nih.gov
Internet: http://grants.nih.gov/grants/guide/pa-files/PAR-06-171.html
Sponsor: National Institute of Diabetes and Digestive and Kidney Diseases
6707 Democracy Boulevard
Bethesda, MD 20892-5460

NIDDK Enhancing Zebrafish Research with Research Tools and Techniques 2815

This FOA encourages investigator-initiated applications designed to exploit the power of the zebrafish as a vertebrate model for biomedical and behavioral research. Applications proposing to develop new research tools or techniques that are of high priority to the zebrafish community and that will advance the detection and characterization of genes, pathways, and phenotypes of interest in development and aging, organ formation, neural processes, behavior, sensory processing, physiological processes, and disease processes are welcome. Budgets for direct costs under $500,000 (direct costs) per year and a project duration of up to five years may be requested. Letters of Intent should arrive by August 17, with full proposals due by September 17.
Requirements: The following organizations and institutions may apply: public/state controlled institutions of higher education; private institutions of higher education; Hispanic-serving institutions; historically Black colleges and universities (HBCUs); tribally controlled colleges and universities (TCCUs); Alaska native and native Hawaiian serving institutions; nonprofits with 501(c)3 IRS status; nonprofits without 501(c)3 IRS status; small businesses; for-profit organizations; State governments; regional organizations; U.S. territories or possessions; Indian/Native American tribal governments; Indian/Native American tribally designated organizations; eligible agencies of the Federal government; and faith-based or community based organizations.
Geographic Focus: All States
Date(s) Application is Due: Aug 17; Sep 17
Contact: Dr. Rebekah S. Rasooly, (301) 594-6007; fax (301) 480-3510; rr185i@nih.gov
Internet: http://grants.nih.gov/grants/guide/pa-files/PAR-08-139.html
Sponsor: National Institute of Diabetes and Digestive and Kidney Diseases
6707 Democracy Boulevard
Bethesda, MD 20892-5460

NIDDK Erythroid Lineage Molecular Toolbox Grants 2816

The ultimate goals of this Funding Opportunity Announcement (FOA) will be to use erythroid cells at different stages of development and differentiation to assemble the complete collection of genes expressed, describe how they are expressed and elucidate physical parameters, including subcellular localization of the proteins translated during erythropoiesis. These reagents may then be used to discover the structure-function relationships that exist in erythroid cells with possible application to other cell types. Because the nature and scope of the proposed research will vary from application to application, it is anticipated that the size and duration of each award will also vary.
Requirements: Application can be made by the following institutions and organizations: public/state controlled institutions of higher education; private institutions of higher education; nonprofits with 501(c)3 IRS status (other than institutions of higher education); nonprofits without 501(c)3 IRS status (other than institutions of higher education); small businesses; for-profit organizations (other than small businesses); state governments; U.S. territories or possessions; Indian/Native American tribal governments (federally

recognized); Indian/Native American tribal governments (other than Federally recognized); Indian/Native American tribally designated organizations; non-domestic (non-U.S.) entities (foreign organizations); Hispanic-serving institutions; historically Black colleges and universities (HBCUs); tribally controlled colleges and universities (TCCUs); Alaska Native and Native Hawaiian serving institutions; regional organization; eligible agencies of the Federal government; and faith-based or community based organizations.
Geographic Focus: All States
Date(s) Application is Due: Feb 5; Jun 5; Oct 5
Contact: Terry Rogers Bishop, (301) 594-7726; tb232j@nih.gov
Internet: http://grants.nih.gov/grants/guide/pa-files/PA-07-011.html
Sponsor: National Institute of Diabetes and Digestive and Kidney Diseases
6707 Democracy Boulevard
Bethesda, MD 20892-5460

NIDDK Erythropoiesis: Components and Mechanisms Grants **2817**
This FOA issued by the National Institute of Diabetes and Digestive and Kidney Diseases (NIDDK) the National Institute of Aging (NIA), and the National Heart, Lung, and Blood Institute (NHLBI), National Institutes of Health, encourages investigator-initiated R01 applications that propose hypothesis-driven research using erythroid cells. The aim of this program is to support research efforts towards a complete description of the molecular and cellular components of erythropoiesis and how these components contribute to erythropoiesis. Components include genes that are expressed (transcriptome) in erythroid cells, either during development or during differentiation, and the proteins (proteome) that are translated in erythroid cells, especially with post-translational modifications or subcellular localizations that are unique to erythroid cells. A long range goal of this program is to generate a concise description of erythropoiesis that unifies genetics, molecular processes and cytokine determinants in the erythroid lineages so that new therapeutics may be developed to measure and combat anemia. Budgets for direct costs of up to $500,000 per year and a project duration of up to five years may be requested.
Requirements: The following organizations and institutions may apply: public/state controlled institutions of higher education; private institutions of higher education; Hispanic-serving institutions; historically Black colleges and universities (HBCUs); tribally controlled colleges and universities (TCCUs); Alaska native and native Hawaiian serving institutions; nonprofits with 501(c)3 IRS status; nonprofits without 501(c)3 IRS status; small businesses; for-profit organizations; State governments; regional organizations; U.S. territories or possessions; Indian/Native American tribal governments; Indian/Native American tribally designated organizations; county governments; city or township governments; special district governments; independent school districts; public housing authorities/Indian housing authorities; eligible agencies of the Federal government; and faith-based or community based organizations.
Geographic Focus: All States
Date(s) Application is Due: Feb 5; Jun 5; Oct 5
Contact: Terry Rogers Bishop, (301) 594-7726; tb232j@nih.gov
Internet: http://grants.nih.gov/grants/guide/pa-files/PA-09-255.html
Sponsor: National Institute of Diabetes and Digestive and Kidney Diseases
6707 Democracy Boulevard
Bethesda, MD 20892-5460

NIDDK Exploratory/Developmental Clinical Research Grants in Obesity **2818**
The goal of this initiative is to encourage exploratory/developmental clinical research that will accelerate the development of effective interventions for prevention or treatment of overweight or obesity in adults and/or children. The goal of this mechanism is to provide flexibility for initiating exploratory, short-term studies, thus allowing new ideas to be investigated in a more expeditious manner without stringent requirements for preliminary data. Such support is needed to encourage investigators to pursue new approaches, underdeveloped topics, or more creative avenues for research including new partnerships. Epidemiological research with a goal of informing translational/clinical research on prevention or treatment of obesity or overweight in adults and/or children is encouraged. The emphasis is thus on the development of exploratory clinical studies, pilot and feasibility studies, or small randomized clinical trials that will provide preliminary data for intervention and epidemiological studies that will inform translational/clinical research.
Requirements: The following organizations and institutions may apply: public/state controlled institutions of higher education; private institutions of higher education; Hispanic-serving institutions; historically Black colleges and universities (HBCUs); tribally controlled colleges and universities (TCCUs); Alaska native and native Hawaiian serving institutions; nonprofits with 501(c)3 IRS status; nonprofits without 501(c)3 IRS status; small businesses; for-profit organizations; State governments; regional organizations; U.S. territories or possessions; Indian/Native American tribal governments; Indian/Native American tribally designated organizations; eligible agencies of the Federal government; and faith-based or community based organizations.
Geographic Focus: All States
Date(s) Application is Due: Feb 16; Jun 16; Oct 16
Contact: Carolyn W. Miles, (301)451-3759; fax (301) 480-8300; cm294e@nih.gov
Internet: http://grants.nih.gov/grants/guide/pa-files/PA-09-124.html
Sponsor: National Institute of Diabetes and Digestive and Kidney Diseases
6707 Democracy Boulevard
Bethesda, MD 20892-5460

NIDDK Grants for Basic Research in Glomerular Diseases **2819**
This FOA invites applications from investigators with diverse scientific interests to apply their expertise in basic research to enhance the understanding of the pathogenesis of the various primary or secondary forms of glomerular disease. Recent observations regarding intrinsic glomerular cell biology, particularly in the podocyte, have provided exciting new insights into potential pathogenic mechanisms of human glomerular disease. Although both immune and nonimmune mechanisms of glomerular injury have been studied previously, experimental models of disease and recent techniques that provide tools for molecular and proteomic profiling show great promise for identifying glomerular disease biomarkers. Despite these recent advances, additional basic studies are needed to facilitate a better understanding of glomerular disease pathogenesis and ultimately impact on the treatment of human glomerulopathies. It is anticipated that applications submitted in response to this FOA could address a number of different aspects concerning the pathogenesis, natural history, therapy pre-emption or prevention of the various morphologic forms of experimental glomerular disease.
Requirements: The following organizations and institutions may apply: public/state controlled institutions of higher education; private institutions of higher education; Hispanic-serving institutions; historically Black colleges and universities (HBCUs); tribally controlled colleges and universities (TCCUs); Alaska native and native Hawaiian serving institutions; nonprofits with 501(c)3 IRS status; nonprofits without 501(c)3 IRS status; small businesses; for-profit organizations; State governments; regional organizations; U.S. territories or possessions; Indian/Native American tribal governments; Indian/Native American tribally designated organizations; eligible agencies of the Federal government; and faith-based or community based organizations.
Geographic Focus: All States
Date(s) Application is Due: Feb 5; Jun 5; Oct 5
Contact: Marva Moxey-Mims, (301) 594-7717; fax (301) 480-3510; mm726k@nih.gov
Internet: http://grants.nih.gov/grants/guide/pa-files/PA-07-367.html
Sponsor: National Institute of Diabetes and Digestive and Kidney Diseases
6707 Democracy Boulevard
Bethesda, MD 20892-5460

NIDDK Health Disparities in NIDDK Diseases **2820**
It is recognized that there are many diseases and disorders that disproportionately affect the health of racial and ethnic minority populations in the United States. It is evident that African-Americans, Hispanic Americans, American Indians, Alaska Natives, some Asian-Americans, and Native Hawaiians and other Pacific Islanders experience much higher risks and poorer health status than the general population. Several of the diseases that disproportionately afflict minorities are high priority research areas for NIDDK, including diabetes, obesity, nutrition-related disorders, hepatitis C, gallbladder disease, H. Pylori infection, sickle cell disease, kidney diseases, and metabolic, gastrointestinal, hepatic, and renal complications from infection with HIV. NIDDK encourages efforts health disparity research to reduce the human and economic costs resulting from inequities in health care and health outcomes. Because the nature and scope of the proposed research will vary from application to application, it is anticipated that the size and duration of each award will also vary. The total amount awarded and the number of awards will depend upon the mechanism numbers, quality, duration, and costs of the applications received.
Requirements: The following organizations and institutions may apply: public/state controlled institutions of higher education; private institutions of higher education; Hispanic-serving institutions; historically Black colleges and universities (HBCUs); tribally controlled colleges and universities (TCCUs); Alaska native and native Hawaiian serving institutions; nonprofits with 501(c)3 IRS status; nonprofits without 501(c)3 IRS status; small businesses; for-profit organizations; State governments; regional organizations; U.S. territories or possessions; Indian/Native American tribal governments; Indian/Native American tribally designated organizations; eligible agencies of the Federal government; and faith-based or community based organizations.
Geographic Focus: All States
Date(s) Application is Due: Feb 5; Jun 5; Oct 5
Contact: Myrlene Staten, (301) 402-7886; fax (301) 480-3503; ms808k@nih.gov
Internet: http://grants.nih.gov/grants/guide/pa-files/PA-07-027.html
Sponsor: National Institute of Diabetes and Digestive and Kidney Diseases
6707 Democracy Boulevard
Bethesda, MD 20892-5460

NIDDK Identifying and Reducing Diabetes and Obesity Related Health Disparities within Healthcare Systems **2821**
This FOA requests applications designed to identify or address factors or barriers that result in disparate outcomes within a healthcare system. All applications should measure the impact of identified factors or interventions on health outcomes. Research is sought that examines at least one of the following factors in a healthcare system and/or the interaction between these factors; healthcare professionals, healthcare organizations, and the patients and communities they serve.
Requirements: The following organizations and institutions may apply: public/state controlled institutions of higher education; private institutions of higher education; Hispanic-serving institutions; historically Black colleges and universities (HBCUs); tribally controlled colleges and universities (TCCUs); Alaska native and native Hawaiian serving institutions; nonprofits with 501(c)3 IRS status; nonprofits without 501(c)3 IRS status; small businesses; for-profit organizations; State governments; regional organizations; U.S. territories or possessions; Indian/Native American tribal governments; Indian/Native American tribally designated organizations; eligible agencies of the Federal government; and faith-based or community based organizations.
Geographic Focus: All States
Date(s) Application is Due: Feb 5; Jun 5; Oct 5
Contact: Christine Hunter, (301) 594-4728; fax (301) 480-3503; ch514c@.nih.gov
Internet: http://grants.nih.gov/grants/guide/pa-files/PA-07-388.html
Sponsor: National Institute of Diabetes and Digestive and Kidney Diseases
6707 Democracy Boulevard
Bethesda, MD 20892-5460

NIDDK Insulin Signaling and Receptor Cross Talk Grants — 2822

This Funding Opportunity Announcement (FOA) solicits investigator-initiated research projects that will define the mechanistic basis of crosstalk between the insulin receptor signaling pathway and other signaling pathways, with the goal of defining how insulin action is influenced by the complex microenvironment found in insulin responsive tissues.
Requirements: The following organizations and institutions may apply: public/state controlled institutions of higher education; private institutions of higher education; Hispanic-serving institutions; historically Black colleges and universities (HBCUs); tribally controlled colleges and universities (TCCUs); Alaska native and native Hawaiian serving institutions; nonprofits with 501(c)3 IRS status; nonprofits without 501(c)3 IRS status; small businesses; for-profit organizations; State governments; regional organizations; U.S. territories or possessions; Indian/Native American tribal governments; Indian/Native American tribally designated organizations; eligible agencies of the Federal government; and faith-based or community based organizations.
Geographic Focus: All States
Date(s) Application is Due: Feb 5; Jun 5; Oct 5
Contact: Kristin Abraham, (301) 451-8048; fax (301) 480-3503; ka136s@nih.gov
Internet: http://grants.nih.gov/grants/guide/pa-files/PA-07-058.html
Sponsor: National Institute of Diabetes and Digestive and Kidney Diseases
6707 Democracy Boulevard
Bethesda, MD 20892-5460

NIDDK Intestinal Failure, Short Gut Syndrome and Small Bowel Transplantation Grants — 2823

The National Institute of Diabetes and Digestive and Kidney Diseases (NIDDK) seeks grant applications to study the pathogenesis, natural history, treatment and complications of intestinal failure and its therapies, including parenteral nutrition and small bowel transplantation. The overall objective of this funding opportunity is to encourage basic and clinical research into intestinal failure, short gut syndrome and intestinal transplantation. Because the nature and scope of the proposed research will vary from application to application, it is anticipated that the size and duration of each award will also vary.
Requirements: Application can be made by the following institutions and organizations: public/state controlled institutions of higher education; private institutions of higher education; nonprofits with 501(c)3 IRS status (other than institutions of higher education); nonprofits without 501(c)3 IRS status (other than institutions of higher education); small businesses; for-profit organizations (other than small businesses); state governments; U.S. territories or possessions; Indian/Native American tribal governments (federally recognized); Indian/Native American tribal governments (other than Federally recognized); Indian/Native American tribally designated organizations; non-domestic (non-U.S.) entities (foreign organizations); Hispanic-serving institutions; historically Black colleges and universities (HBCUs); tribally controlled colleges and universities (TCCUs); Alaska Native and Native Hawaiian serving institutions; regional organization; eligible agencies of the Federal government; and faith-based or community based organizations.
Geographic Focus: All States
Date(s) Application is Due: Feb 5; Jun 5; Oct 5
Contact: Dr. Michael K. May, (301) 594-8884; fax (301) 480-8300; mm102i@nih.gov
Internet: http://grants.nih.gov/grants/guide/pa-files/PA-07-010.html
Sponsor: National Institute of Diabetes and Digestive and Kidney Diseases
6707 Democracy Boulevard
Bethesda, MD 20892-5460

NIDDK Intestinal Stem Cell Consortium Grants — 2824

This FOA invites new applications to participate in the Intestinal Stem Cell Consortium (ISCC). The goals of this initiative are: to establish a network of individual research projects and a Coordinating Center focused on research on stem cells of the small intestine; to isolate, culture, characterize, functionally validate, and compare stem cell populations from the small intestine epithelium in vivo and in vitro; to share data, biomaterials, models, reagents, resources and methods between projects in the ISCC; and make these publicly available through a web site to be developed by the Coordinating Center. The objective of support for this consortium is to accelerate research on stem cells of the intestinal epithelium in order to facilitate understanding of the biology and function of the intestine and aid development of therapies for intestinal diseases and conditions where damage and replacement of this epithelium are important components. Direct costs are limited to $250,000 per year for a maximum five-year period for research projects and $350,000 per year for a maximum five-year period for the Coordinating Center.
Requirements: Application can be made by the following institutions and organizations: public/state controlled institutions of higher education; private institutions of higher education; nonprofits with 501(c)3 IRS status (other than institutions of higher education); nonprofits without 501(c)3 IRS status (other than institutions of higher education); small businesses; for-profit organizations (other than small businesses); state governments; U.S. territories or possessions; Indian/Native American tribal governments (federally recognized); Indian/Native American tribal governments (other than Federally recognized); Indian/Native American tribally designated organizations; non-domestic (non-U.S.) entities (foreign organizations); Hispanic-serving institutions; historically Black colleges and universities (HBCUs); tribally controlled colleges and universities (TCCUs); Alaska Native and Native Hawaiian serving institutions; regional organization; eligible agencies of the Federal government; and faith-based or community based organizations.
Geographic Focus: All States
Date(s) Application is Due: Mar 18
Contact: Ronald Margolis, (301) 594-8819; fax (301) 435-6047; rm76f@nih.gov
Internet: http://grants.nih.gov/grants/guide/rfa-files/RFA-DK-08-010.html
Sponsor: National Institute of Diabetes and Digestive and Kidney Diseases
6707 Democracy Boulevard
Bethesda, MD 20892-5460

NIDDK Mentored Research Scientist Development Award — 2825

The overall objective of the NIH Research Career Development Award program is to prepare qualified individuals for careers that have a significant impact on the health-related research needs of the Nation. The objective of the NIDDK Mentored Research Scientist Development Award (K01) is to provide support for a sustained period of protected time for intensive research career development under the guidance of an experienced mentor, or sponsor, in the biomedical, behavioral or clinical sciences leading to research independence, thereby ensuring a future cadre of well-trained Ph.D. scientists working in research areas supported by NIDDK. The expectation is that through this sustained period of research career development and training, awardees will launch independent research careers and become competitive for new research project grant (R01) funding. By providing support for the critical transition period between postdoctoral training and independent R01 funding for non-clinical investigators, the NIDDK hopes to foster the careers of these investigators who are vital for the future excellence of the NIDDK research endeavor. Applicants must justify the need for a period of mentored research experience and provide a convincing case that the proposed period of support will substantially enhance their careers as independent investigators.
Requirements: The following organizations and institutions may apply: public/state controlled institutions of higher education; private institutions of higher education; Hispanic-serving institutions; historically Black colleges and universities (HBCUs); tribally controlled colleges and universities (TCCUs); Alaska native and native Hawaiian serving institutions; nonprofits with 501(c)3 IRS status; nonprofits without 501(c)3 IRS status; small businesses; for-profit organizations; State governments; regional organizations; U.S. territories or possessions; Indian/Native American tribal governments; Indian/Native American tribally designated organizations; eligible agencies of the Federal government; and faith-based or community based organizations.
Geographic Focus: All States
Date(s) Application is Due: Mar 12; Jul 12; Nov 12
Contact: James F. Hyde, (301) 594-7692; fax (301) 480-3503; jh486z@nih.gov
Internet: http://grants.nih.gov/grants/guide/pa-files/PAR-09-060.html
Sponsor: National Institute of Diabetes and Digestive and Kidney Diseases
6707 Democracy Boulevard
Bethesda, MD 20892-5460

NIDDK Multi-Center Clinical Study Cooperative Agreements — 2826

This PAR provides for grant applications for investigator-initiated, multi-center clinical studies. NIDDK supports investigator-initiated, multi-center clinical studies through a two-part process that includes an implementation planning (U34) grant. Beginning with the receipt date, NIDDK will accept, peer review, and consider for funding applications for investigator-initiated, multi-center clinical studies only from NIDDK Multi-Center Clinical Study Implementation Planning (U34) grant awardees. Because the nature and scope of the proposed research will vary from application to application, it is anticipated that the size and duration of each award will also vary. The total amount awarded and the number of awards will depend upon the mechanism numbers, quality, duration, and costs of the applications received.
Requirements: The following organizations and institutions may apply: public/state controlled institutions of higher education; private institutions of higher education; Hispanic-serving institutions; historically Black colleges and universities (HBCUs); tribally controlled colleges and universities (TCCUs); Alaska native and native Hawaiian serving institutions; nonprofits with 501(c)3 IRS status; nonprofits without 501(c)3 IRS status; small businesses; for-profit organizations; State governments; regional organizations; U.S. territories or possessions; Indian/Native American tribal governments; Indian/Native American tribally designated organizations; eligible agencies of the Federal government; and faith-based or community based organizations.
Geographic Focus: All States
Date(s) Application is Due: Feb 5; Jun 5; Oct 5
Contact: Catherine Meyers, (301) 451-4901; fax (301) 480-3510; meyersc@mail.nih.gov
Internet: http://grants.nih.gov/grants/guide/pa-files/PAR-08-058.html
Sponsor: National Institute of Diabetes and Digestive and Kidney Diseases
6707 Democracy Boulevard
Bethesda, MD 20892-5460

NIDDK Multi-Center Clinical Study Implementation Planning Grants — 2827

NIDDK supports investigator-initiated, multi-center clinical studies through a two-part process that includes an implementation planning grant. The planning grant is designed to: permit early peer review of the rationale for the proposed clinical study; permit assessment of the design/protocol of the proposed study; provide support for the development of a complete study protocol and associated documents including a manual of operations; and support the development of other essential elements required for the conduct of a clinical study. Completion of the required products of a U34 grant is a prerequisite for submission of a multi-center clinical study cooperative agreement application, which will support the actual conduct of the study. Because the nature and scope of the proposed research will vary from application to application, it is anticipated that the size and duration of each award will also vary. The total project period for an application submitted in response to this funding opportunity may not exceed two years. Applicants may request up to $250,000 in direct costs per year.
Requirements: The following organizations and institutions may apply: public/state controlled institutions of higher education; private institutions of higher education; Hispanic-serving institutions; historically Black colleges and universities (HBCUs); tribally controlled colleges and universities (TCCUs); Alaska native and native Hawaiian serving institutions; nonprofits with 501(c)3 IRS status; nonprofits without 501(c)3 IRS status; small businesses; for-profit organizations; State governments; regional organizations; U.S. territories or possessions; Indian/Native American tribal

governments; Indian/Native American tribally designated organizations; eligible agencies of the Federal government; and faith-based or community based organizations.
Geographic Focus: All States
Date(s) Application is Due: Mar 18; Jun 23
Contact: Catherine Meyers, (301) 451-4901; fax (301) 480-3510; meyersc@mail.nih.gov
Internet: http://grants.nih.gov/grants/guide/pa-files/PAR-08-057.html
Sponsor: National Institute of Diabetes and Digestive and Kidney Diseases
6707 Democracy Boulevard
Bethesda, MD 20892-5460

NIDDK Neurobiology of Diabetic Complications Grants 2828
NIDDK invites investigator-initiated research grant applications to study the mechanisms by which diabetes results in painful and disabling neuropathies and other neurological complications and to apply this information to the development of interventions to prevent, limit, or reverse these complications. The intent of this RFA is to attract basic neuroscientists to the study of diabetic neuropathy and neurobiology relevant to diabetes, and enhance interdisciplinary approaches to research in this area. Annual deadline dates may vary; contact the appropriate NIH institute for exact dates. Letters of intent are due by March 24, with full applications due by April 25 each year.
Requirements: Applications may be submitted by domestic and foreign, for-profit and nonprofit organizations, both public and private; units of state and local governments; and eligible agencies of the federal government.
Geographic Focus: All States
Date(s) Application is Due: Mar 24; Apr 25
Amount of Grant: Up to 3,000,000 USD
Contact: Dr. Paul Nichols, (301) 496-9964; fax (301) 402-2060; pn13w@nih.gov
Internet: http://grants.nih.gov/grants/guide/rfa-files/RFA-NS-00-002.html
Sponsor: National Institute of Diabetes and Digestive and Kidney Diseases
45 Center Drive, MSC 6600
Bethesda, MD 20892-6600

NIDDK Neuroimaging in Obesity Research Grants 2829
This FOA issued by the National Institute of Diabetes and Digestive and Kidney Diseases, the National Institute on Drug Abuse, and the National Institute of Biomedical Imaging and Bioengineering, of the National Institutes of Health, solicits Research Project Grant (R01) applications from institutions/ organizations that propose to use neuroimaging approaches in obesity research in human subjects and animal models. Many areas of the brain interact or communicate with other organs to control eating behavior, physical activity and energy metabolism, and functional neuroimaging holds enormous promise for expanding our understanding of how food intake and energy expenditure are mismatched in a setting of abundantly available nutrients, leading to excessive fat storage. The total amount of funding that the ICs expect to award through this announcement is up to $4.5 Million, to fund approximately 9-15 awards. The duration and requested direct costs will vary depending on the scope of the proposed research, but cannot exceed 5 years and $500,000 per year. Letters of intent are due by February 18, with full applications due by 5:00 p.m. on March 18.
Requirements: Application can be made by the following institutions and organizations: public/state controlled institutions of higher education; private institutions of higher education; nonprofits with 501(c)3 IRS status (other than institutions of higher education); nonprofits without 501(c)3 IRS status (other than institutions of higher education); small businesses; for-profit organizations (other than small businesses); state governments; U.S. territories or possessions; Indian/Native American tribal governments (federally recognized); Indian/Native American tribal governments (other than Federally recognized); Indian/Native American tribally designated organizations; non-domestic (non-U.S.) entities (foreign organizations); Hispanic-serving institutions; historically Black colleges and universities (HBCUs); tribally controlled colleges and universities (TCCUs); Alaska Native and Native Hawaiian serving institutions; regional organization; eligible agencies of the Federal government; and faith-based or community based organizations.
Geographic Focus: All States
Date(s) Application is Due: Feb 18; Mar 18
Amount of Grant: Up to 500,000 USD
Contact: Maren R. Laughlin, (301) 594-8802; fax (301) 480-0475; ml33q@nih.gov
Internet: http://grants.nih.gov/grants/guide/rfa-files/RFA-DK-08-009.html
Sponsor: National Institute of Diabetes and Digestive and Kidney Diseases
6707 Democracy Boulevard
Bethesda, MD 20892-5460

NIDDK New Technologies for Liver Disease SBIR 2830
The purpose of this Funding Opportunity Announcement (FOA) is to solicit Small Business Innovation Research (SBIR) grant applications from small business concerns (SBCs) that propose to develop resources, research tools, instrumentations, biomarkers, devices, drugs or new and innovative approaches to diagnosis, monitoring, management, treatment and prevention of liver diseases. Areas of interest include development of reliable and practical means of diagnosis of liver diseases; biomarkers for disease activity and stage; noninvasive tests for inflammation, fibrosis and fat in the liver; and drugs, complementary and alternative modalities, biologics or molecular reagents for the therapy or prevention of liver diseases. The goal of this announcement is to enlist members of the small business research community in advancing means of diagnosis, treatment and prevention of liver disease and facilitate the goals outlined in the trans-NIH Action Plan for Liver Disease Research.
Requirements: Only United States small business concerns (SBCs) are eligible to submit SBIR applications.
Geographic Focus: All States
Date(s) Application is Due: Apr 5; Aug 5; Dec 5
Amount of Grant: 100,000 - 750,000 USD
Contact: Christine L. Densmore, (301) 402-8714; fax (301) 480-8300; cd121z@nih.gov
Internet: http://grants.nih.gov/grants/guide/pa-files/PA-09-095.html
Sponsor: National Institute of Diabetes and Digestive and Kidney Diseases
6707 Democracy Boulevard
Bethesda, MD 20892-5460

NIDDK New Technologies for Liver Disease STTR 2831
The objective of this FOA is to encourage and enable scientists at small businesses to develop and evaluate new technologies, drugs, devices, and approaches to diagnosis, management, and prevention of liver disease. Development of new technologies as well as application of existing technologies may be proposed. Studies may include use of animal models or human participants or both. If appropriate, plans for manufacturing and clinical evaluation of developed technologies, drugs, devices and innovative approaches should be included in the application. However, clinical trials beyond Phase I studies will not be considered appropriate to this announcement.
Requirements: Only United States small business concerns (SBCs) are eligible to submit STTR applications.
Geographic Focus: All States
Date(s) Application is Due: Apr 5; Aug 5; Dec 5
Amount of Grant: 100,000 - 750,000 USD
Contact: Christine L. Densmore, (301) 402-8714; fax (301) 480-8300; cd121z@nih.gov
Internet: http://grants.nih.gov/grants/guide/pa-files/PA-09-094.html
Sponsor: National Institute of Diabetes and Digestive and Kidney Diseases
6707 Democracy Boulevard
Bethesda, MD 20892-5460

NIDDK Non-Invasive Methods for Diagnosis and Progression of Diabetes, Kidney, Urological, Hematological and Digestive Diseases 2832
This funding opportunity is a call for the application of imaging and other non- or minimally-invasive technologies to detect, characterize, diagnose, identify persons with predisposition to, or monitor treatment of diseases of interest to the National Institute of Diabetes and Digestive and Kidney Diseases. Also needed are new, robust surrogate markers for clinical trial endpoints, and new ways to characterize normal and pathological tissues in vivo. Diseases of interest include type 1 and 2 diabetes, obesity, and kidney, liver, urologic, hematologic, digestive, endocrine and metabolic diseases and their complications. Applicable techniques include molecular imaging and functional imaging approaches, imaging methods with high spatial, chemical or time resolution, and new spectroscopic or sensor array technologies for monitoring metabolic or physiological events.
Requirements: The following organizations and institutions may apply: for-profit organizations; non-profit organizations; public or private institutions, such as universities, colleges, hospitals, and laboratories; units of State government; units of local government; eligible agencies of the Federal government; foreign institutions; domestic institutions; and faith-based or community-based organizations.
Geographic Focus: All States
Date(s) Application is Due: Feb 5; Jun 5; Oct 5
Contact: Maren R. Laughlin, (301) 594-8802; fax (301) 480-0475; ml33q@nih.gov
Internet: http://grants.nih.gov/grants/guide/pa-files/PA-06-143.html
Sponsor: National Institute of Diabetes and Digestive and Kidney Diseases
6707 Democracy Boulevard
Bethesda, MD 20892-5460

NIDDK Pancreatic Development and Regeneration Grants: Toward Cellular Therapies for Diabetes 2833
This Funding Opportunity Announcement (FOA) is intended to promote basic research in pancreatic development, stem cell biology and regeneration that will provide the foundation for the rationale design of future cell replacement therapies for diabetes. Appropriate topics for investigation include, but are not limited to: prospective isolation, purification and characterization of pancreatic stem/progenitor cells; identification of the signals, signaling pathway components and transcriptional factors that could direct the formation of pancreatic endoderm and progenitors, islet cell differentiation, morphogenesis, growth, function and survival in vitro; use of existing NIH-approved human ES cell lines or mouse embryonic cells, pancreatic stem cells, non-pancreatic adult stem cells such as those from liver or gut, human amniotic cells or bone marrow-derived cells to efficiently generate endoderm, pancreatic progenitor cells, or differentiated beta cells in culture; development of methods and reagents to assess quantitatively, the survival, migration, fate and function of transplanted animal or human pancreatic progenitors and their progeny in the living animal; determination of the factors important for beta cell regeneration and designing strategies to foster regeneration; development of new animal models that can used to identify exogenous or endogenous factors that might promote beta cell regeneration or neogenesis; characterization the lineage of cells that may contribute or participate in beta cell regeneration by genetic marking; and development of assays to screen chemical libraries for modifiers of beta cell regeneration, differentiation, and or expansion.
Requirements: The following organizations and institutions may apply: public/state controlled institutions of higher education; private institutions of higher education; Hispanic-serving institutions; historically Black colleges and universities (HBCUs); tribally controlled colleges and universities (TCCUs); Alaska native and native Hawaiian serving institutions; nonprofits with 501(c)3 IRS status; nonprofits without 501(c)3 IRS status; small businesses; for-profit organizations; State governments; regional organizations; U.S. territories or possessions; Indian/Native American tribal governments; Indian/Native American tribally designated organizations; eligible agencies of the Federal government; and faith-based or community based organizations.
Geographic Focus: All States

Date(s) Application is Due: Feb 5; Jun 5; Oct 5
Contact: Sheryl M. Sato, (301) 594-8811; fax (301) 480-0475; ss68z@nih.gov
Internet: http://grants.nih.gov/grants/guide/pa-files/PA-07-056.html
Sponsor: National Institute of Diabetes and Digestive and Kidney Diseases
6707 Democracy Boulevard
Bethesda, MD 20892-5460

NIDDK Pilot and Feasibility Clinical Research Grants in Diabetes, Endocrine and Metabolic Diseases — 2834

The goal of this initiative is to encourage exploratory/developmental clinical research related to the prevention or treatment of diabetes, obesity and endocrine and genetic metabolic diseases. The Pilot and Feasibility Clinical Research Grants Program is designed to allow initiation of exploratory, short-term clinical studies, so that new ideas may be investigated without stringent requirements for preliminary data. Such support can be used by experienced investigators, as well as new investigators, to pursue new approaches and investigate underdeveloped research areas. The short-term studies should focus on research questions that are likely to have high clinical impact. They can include testing a new prevention strategy, a new intervention, or unique combinations of therapies. A high priority is the use of such studies to help stimulate the translation of promising research developments from the laboratory into clinical practice in diabetes, endocrine diseases and genetic metabolic diseases, including cystic fibrosis.
Requirements: The following organizations and institutions may apply: public/state controlled institutions of higher education; private institutions of higher education; Hispanic-serving institutions; historically Black colleges and universities (HBCUs); tribally controlled colleges and universities (TCCUs); Alaska native and native Hawaiian serving institutions; nonprofits with 501(c)3 IRS status; nonprofits without 501(c)3 IRS status; small businesses; for-profit organizations; State governments; regional organizations; U.S. territories or possessions; Indian/Native American tribal governments; Indian/Native American tribally designated organizations; eligible agencies of the Federal government; and faith-based or community based organizations.
Geographic Focus: All States
Date(s) Application is Due: Feb 16; Jun 16; Oct 16
Amount of Grant: 200,000 - 275,000 USD
Contact: Karen Salomon, (301) 594-7733; fax (301) 435-6047; ks495c@nih.gov
Internet: http://grants.nih.gov/grants/guide/pa-files/PA-09-133.html
Sponsor: National Institute of Diabetes and Digestive and Kidney Diseases
6707 Democracy Boulevard
Bethesda, MD 20892-5460

NIDDK Pilot and Feasibility Clinical Research Grants in Kidney or Urologic Diseases — 2835

This funding opportunity announcement (FOA) is a re-issuance of PAR-06-113 and specifically encourages the submission of applications for small scale or pilot and feasibility clinical and translational research studies, including epidemiological studies or clinical trials related to kidney or urologic disease research that address important clinical and translational questions and are potentially of high impact. It is anticipated that some applications for pilot and feasibility studies may lead to full-scale clinical studies including evaluation of diagnostic strategies, epidemiologic studies, or trials in the diagnosis, prevention, or treatment of kidney or urologic diseases. These grants may be used to plan, evaluate the feasibility, or implement clinical trials that assess pharmacological, dietary, surgical, or behavioral interventions for the prevention or treatment of kidney or urologic disease. Small scale or pilot epidemiological studies are also encouraged.
Requirements: The following organizations and institutions may apply: public/state controlled institutions of higher education; private institutions of higher education; Hispanic-serving institutions; historically Black colleges and universities (HBCUs); tribally controlled colleges and universities (TCCUs); Alaska native and native Hawaiian serving institutions; nonprofits with 501(c)3 IRS status; nonprofits without 501(c)3 IRS status; small businesses; for-profit organizations; State governments; regional organizations; U.S. territories or possessions; Indian/Native American tribal governments; Indian/Native American tribally designated organizations; eligible agencies of the Federal government; and faith-based or community based organizations.
Geographic Focus: All States
Date(s) Application is Due: Feb 16; Jun 16; Oct 16
Contact: Catherine M. Meyers, (301) 594-7717; fax (301) 480-3510; cm420i@nih.gov
Internet: http://grants.nih.gov/grants/guide/pa-files/PAR-09-077.html
Sponsor: National Institute of Diabetes and Digestive and Kidney Diseases
6707 Democracy Boulevard
Bethesda, MD 20892-5460

NIDDK Pilot and Feasibility Clinical Research Studies in Digestive Diseases and Nutrition — 2836

The goal of this pilot and feasibility mechanism is to provide flexibility for initiating preliminary, short-term clinical studies, thus allowing new ideas to be investigated in a more expeditious manner without stringent requirements for preliminary data. Such support is needed to encourage experienced investigators as well as new investigators to pursue new approaches, underdeveloped topics, or more creative avenues of research. If successful, these awards should lead to significant scientific advances in digestive disease research and should facilitate translation into clinical practice to improve patient outcomes.
Requirements: The following organizations and institutions are eligible to apply: public/state controlled institutions of higher education; private institutions of higher education; Hispanic-serving institutions; historically Black colleges and universities (HBCUs); tribally controlled colleges and universities (TCCUs); Alaska Native and Native Hawaiian serving institutions; nonprofits with 501(c)3 IRS status (other than institutions of higher education); nonprofits without 501(c)3 IRS status (other than institutions of higher education); small businesses; for-profit organizations (other than small businesses); state governments; Indian/Native American tribal governments (federally recognized); Indian/Native American tribally designated organizations; county governments; city or township governments; special district governments; independent school districts; public housing authorities; Indian housing authorities; U.S. territory or possession; Indian/Native American tribal governments (other than federally recognized); regional organizations; non-domestic (non-U.S.) entities (foreign organizations); other eligible agencies of the federal government; and faith-based or community-based organizations.
Geographic Focus: All States
Date(s) Application is Due: Feb 16; Jun 16; Oct 16
Contact: Mary Evans, (301) 594-4578; fax (301) 480-8300; me189d@nih.gov
Internet: http://grants.nih.gov/grants/guide/pa-files/PA-09-151.html
Sponsor: National Institute of Diabetes and Digestive and Kidney Diseases
6707 Democracy Boulevard
Bethesda, MD 20892-5460

NIDDK Planning Grants For Translational Research For The Prevention and Control Of Diabetes And Obesity — 2837

The National Institute of Diabetes and Digestive and Kidney Diseases (NIDDK) seeks to develop cost effective and sustainable interventions, that can be adopted in real world settings, for the prevention and control of diabetes and obesity. Research should be based on interventions already proven efficacious in clinical trials to prevent and reverse obesity and type 2 diabetes, to improve care of type 1 and type 2 diabetes and to prevent or delay its complications. The interventions proposed under this solicitation should have the potential to be widely disseminated to clinical practice, individuals and communities at risk. Relevant topics include: methods to improve health care delivery to patients with or at risk of diabetes; to enhance diabetes self management; methods to develop strategies to promote healthy lifestyles to reduce the risk of diabetes; and methods to promote lifestyle change that will prevent or reverse overweight and obesity, and 5) new cost effective ways to identify people with pre-diabetes and undiagnosed diabetes. This FOA will use the National Institutes of Health (NIH) clinical trial planning grant (R34) award mechanism, which will provide up to $150,000 in direct costs per year.
Requirements: Applications may be submitted by: for-profit organizations; non-profit organizations; public or private institutions, such as universities, colleges, hospitals, and laboratories; units of State government; units of local government; eligible agencies of the Federal government; domestic institutions; faith-based or community-based organizations; Indian/Native American tribal governments; and Indian/Native American tribally designated organizations.
Geographic Focus: All States
Date(s) Application is Due: Feb 16; Jun 16; Oct 16
Contact: Sanford Garfield, (301) 594-8803; fax (301) 480-3503; sg50o@nih.gov
Internet: http://grants.nih.gov/grants/guide/pa-files/PAR-06-358.html
Sponsor: National Institute of Diabetes and Digestive and Kidney Diseases
6707 Democracy Boulevard
Bethesda, MD 20892-5460

NIDDK Proteomics: Diabetes, Obesity, And Endocrine, Digestive, Kidney, Urologic, And Hematologic Diseases — 2838

This Funding Opportunity Announcement (FOA) issued by the National Institute of Diabetes and Digestive and Kidney Diseases (NIDDK), National Institutes of Health (NIH), solicits projects that advance research to identify and quantitate protein expression patterns, post-translational modification of proteins, and protein-protein interactions on cells, tissues, organ systems relevant to diabetes, obesity, endocrine and metabolic diseases, nutritional function and diseases of the alimentary tract, exocrine pancreas, liver, kidney, bladder and prostate and normal biological processes related to the function of these systems. The development and improvement of innovative proteomic technologies is also encouraged through their application to relevant biological questions related to the pathophysiology of endocrine glands, gastrointestinal tract, liver and kidney, bladder and prostate.
Requirements: The following organizations and institutions may apply: public/state controlled institutions of higher education; private institutions of higher education; Hispanic-serving institutions; historically Black colleges and universities (HBCUs); tribally controlled colleges and universities (TCCUs); Alaska native and native Hawaiian serving institutions; nonprofits with 501(c)3 IRS status; nonprofits without 501(c)3 IRS status; small businesses; for-profit organizations; State governments; regional organizations; U.S. territories or possessions; Indian/Native American tribal governments; Indian/Native American tribally designated organizations; eligible agencies of the Federal government; and faith-based or community based organizations.
Geographic Focus: All States
Date(s) Application is Due: Feb 5; Jun 5; Oct 5
Contact: Salvatore Sechi, (301) 594-8814; fax (301) 480-2688; ss24q@nih.gov
Internet: http://grants.nih.gov/grants/guide/pa-files/PA-07-016.html
Sponsor: National Institute of Diabetes and Digestive and Kidney Diseases
6707 Democracy Boulevard
Bethesda, MD 20892-5460

NIDDK Research Grants for Studies of Hepatitis C in the Setting of Renal Disease — 2839

It is the intent of this Funding Opportunity Announcement (FOA) to invite applications from investigators with diverse scientific interests, who wish to apply their expertise into basic and applied research to enhance the understanding of the pathogenesis, natural history, therapy, and prevention of hepatitis C in the setting of renal disease or renal transplantation. Because the nature and scope of the proposed research will vary from

application to application, it is anticipated that the size and duration of each award will also vary. The total amount awarded and the number of awards will depend upon the mechanism numbers, quality, duration, and costs of the applications received.
Requirements: The following organizations and institutions may apply: public/state controlled institutions of higher education; private institutions of higher education; Hispanic-serving institutions; historically Black colleges and universities (HBCUs); tribally controlled colleges and universities (TCCUs); Alaska native and native Hawaiian serving institutions; nonprofits with 501(c)3 IRS status; nonprofits without 501(c)3 IRS status; small businesses; for-profit organizations; State governments; regional organizations; U.S. territories or possessions; Indian/Native American tribal governments; Indian/Native American tribally designated organizations; eligible agencies of the Federal government; and faith-based or community based organizations.
Geographic Focus: All States
Date(s) Application is Due: Feb 5; Jun 5; Oct 5
Contact: Catherine M. Meyers, (301) 594-7717; fax (301) 480-3510; cm420i@nih.gov
Internet: http://grants.nih.gov/grants/guide/pa-files/PA-07-015.html
Sponsor: National Institute of Diabetes and Digestive and Kidney Diseases
6707 Democracy Boulevard
Bethesda, MD 20892-5460

NIDDK Research Grants on Improving Health Care for Obese Patients 2840
This funding opportunity announcement (FOA) solicits Research Project Grant applications from institutions and organizations that propose to conduct research to determine the barriers to optimal health care for obese patients, and to test innovations or modifications in care delivery to improve health outcomes for obese patients independent of weight loss. Awards issued under this FOA are contingent upon the availability of funds and the submission of a sufficient number of meritorious applications. Applications can be renewed by competing for additional project periods.
Requirements: The following organizations and institutions may apply: public/state controlled institutions of higher education; private institutions of higher education; Hispanic-serving institutions; historically Black colleges and universities (HBCUs); tribally controlled colleges and universities (TCCUs); Alaska native and native Hawaiian serving institutions; nonprofits with 501(c)3 IRS status; nonprofits without 501(c)3 IRS status; small businesses; for-profit organizations; State governments; regional organizations; U.S. territories or possessions; Indian/Native American tribal governments; Indian/Native American tribally designated organizations; eligible agencies of the Federal government; and faith-based or community based organizations.
Geographic Focus: All States
Date(s) Application is Due: Feb 5; Jun 5; Oct 5
Contact: Susan Z. Yanovski, (301) 594-8882; fax (301) 480-8300; sy29f@nih.gov
Internet: http://grants.nih.gov/grants/guide/pa-files/PA-07-013.html
Sponsor: National Institute of Diabetes and Digestive and Kidney Diseases
6707 Democracy Boulevard
Bethesda, MD 20892-5460

NIDDK Role of Gastrointestinal Surgical Procedures in Amelioration of Obesity-Related Insulin Resistance & Diabetes Independent of Weight Loss 2841
This Funding Opportunity Announcement (FOA) issued by the National Institute of Diabetes and Digestive and Kidney Diseases, National Institutes of Health encourages Research Project Grant (R01) applications from institutions and organizations that propose to explain the underlying mechanism(s) by which various gastrointestinal surgical procedures ameliorate obesity-related insulin resistance and diabetes independent of the resultant weight loss. Awards issued under this FOA are contingent upon the availability of funds and the submission of a sufficient number of meritorious applications.
Requirements: The following organizations and institutions may apply: public/state controlled institutions of higher education; private institutions of higher education; Hispanic-serving institutions; historically Black colleges and universities (HBCUs); tribally controlled colleges and universities (TCCUs); Alaska native and native Hawaiian serving institutions; nonprofits with 501(c)3 IRS status; nonprofits without 501(c)3 IRS status; small businesses; for-profit organizations; State governments; regional organizations; U.S. territories or possessions; Indian/Native American tribal governments; Indian/Native American tribally designated organizations; eligible agencies of the Federal government; and faith-based or community based organizations.
Geographic Focus: All States
Date(s) Application is Due: Feb 5; Jun 5; Oct 5
Contact: Carolyn W. Miles, (301) 451-3759; fax (301) 480-8300; cm294e@nih.gov
Internet: http://grants.nih.gov/grants/guide/pa-files/PA-08-014.html
Sponsor: National Institute of Diabetes and Digestive and Kidney Diseases
6707 Democracy Boulevard
Bethesda, MD 20892-5460

NIDDK Secondary Analyses in Obesity, Diabetes and Digestive and Kidney Diseases Grants 2842
The specific objectives of this announcement on Secondary Analyses In Obesity, Diabetes, Digestive and Kidney Diseases are to support the following: secondary analyses of data related to the epidemiology of disease areas of NIDDK; important and/or innovative hypotheses explored through analysis of existing data sets; secondary analyses designed to inform and support subsequent applications for individual research awards; rapid analyses of new databases and experimental modules to inform the design and content of future studies; the archiving of datasets to be made publicly available for research purposes related to disease areas of NIDDK, including both epidemiological studies and multi-center clinical trials. Research that employs analytic techniques that demonstrate or promote methodological advances in patient oriented and epidemiologic research is also of interest. International comparative analyses are encouraged. Applications that are innovative and high risk with the likelihood for high impact are especially encouraged.
Requirements: The following organizations and institutions may apply: public/state controlled institutions of higher education; private institutions of higher education; Hispanic-serving institutions; historically Black colleges and universities (HBCUs); tribally controlled colleges and universities (TCCUs); Alaska native and native Hawaiian serving institutions; nonprofits with 501(c)3 IRS status; nonprofits without 501(c)3 IRS status; small businesses; for-profit organizations; State governments; regional organizations; U.S. territories or possessions; Indian/Native American tribal governments; Indian/Native American tribally designated organizations; eligible agencies of the Federal government; and faith-based or community based organizations.
Geographic Focus: All States
Date(s) Application is Due: Feb 16; Jun 16; Oct 16
Amount of Grant: 200,000 - 275,000 USD
Contact: James E. Everhart, (301) 594-8878; fax (301) 480-8300; je17g@nih.gov
Internet: http://grants.nih.gov/grants/guide/pa-files/PA-09-131.html
Sponsor: National Institute of Diabetes and Digestive and Kidney Diseases
6707 Democracy Boulevard
Bethesda, MD 20892-5460

NIDDK Seeding Collaborative Interdisciplinary Team Science in Diabetes, Endocrinology and Metabolic Diseases Grants 2843
The purpose of the Seeding Collaborative Interdisciplinary Team Science Program described in this announcement is to provide initial support to enable strong new investigative teams to form, and to foster preliminary research activities. It is anticipated that research teams receiving seeding support under this program will be well positioned to compete for funds through the parent NIDDK Collaborative Interdisciplinary Research Program in Diabetes, Endocrinology and Metabolic Diseases solicitation. It is anticipated that 5-10 projects will be funded, but because the nature and scope of the proposed research will vary from application to application, it is anticipated that the size and duration of each award will also vary. Budgets for direct costs of up to $300,000 per year with a project duration of up to one year may be requested.
Requirements: The following organizations and institutions are eligible to apply: public/state controlled institutions of higher education; private institutions of higher education; Hispanic-serving institutions; historically Black colleges and universities (HBCUs); tribally controlled colleges and universities (TCCUs); Alaska Native and Native Hawaiian serving institutions; and non-domestic (non-U.S.) entities (foreign organizations).
Geographic Focus: All States
Date(s) Application is Due: Jan 25; May 25; Sep 25
Amount of Grant: Up to 375,000 USD
Contact: Karen Salomon, (301) 594-7733; fax (301) 435-6047; ks495c@nih.gov
Internet: http://grants.nih.gov/grants/guide/pa-files/PAR-08-181.html
Sponsor: National Institute of Diabetes and Digestive and Kidney Diseases
6707 Democracy Boulevard
Bethesda, MD 20892-5460

NIDDK Silvio O. Conte Digestive Diseases Research Core Centers Grants 2844
The objective of the Silvio O. Conte Digestive Diseases Research Core Centers is to bring together, on a cooperative basis, basic science and clinical investigators to enhance the effectiveness of their research related to digestive and/or liver diseases and their complications. A core center must be an identifiable unit within a single university medical center or a consortium of cooperating institutions, including an affiliated university. An existing program of excellence in biomedical research in the area of digestive and/or liver diseases is a prerequisite for applying. This research must be in the form of NIH research projects, program projects, or other peer-reviewed research that is already funded at the time of submission of a center grant application. Close cooperation, communication, and collaboration among all involved personnel of various professional disciplines are the ultimate objectives.
Requirements: The following organizations and institutions are eligible to apply: public/state controlled institutions of higher education; private institutions of higher education; nonprofits with 501(c)3 IRS status (other than institutions of higher education); nonprofits without 501(c)3 IRS status (other than institutions of higher education); and for-profit organizations (other than small businesses).
Geographic Focus: All States
Date(s) Application is Due: Oct 12; Nov 9
Amount of Grant: 750,000 USD
Contact: Judith Podskalny, (301) 594-8876; fax (301) 480-8300; jp53s@nih.gov
Internet: http://grants.nih.gov/grants/guide/rfa-files/RFA-DK-09-005.html
Sponsor: National Institute of Diabetes and Digestive and Kidney Diseases
6707 Democracy Boulevard
Bethesda, MD 20892-5460

NIDDK Small Business Innovation Research to Develop New Therapeutics & Monitoring Technologies for Type 1 Diabetes (T1D)--Towards an Artificial Pancreas (SBIR) 2845
A promising therapeutic option for the treatment of diabetes is a system (termed an artificial pancreas or closed-loop) that can mimic normal pancreatic beta cell function thereby restoring normal metabolic homeostasis without causing hypoglycemia. The design of any system capable of achieving this goal is complex and raises scientific, clinical and regulatory challenges. This FOA solicits Small Business Innovation Research (SBIR) grant applications from small business concerns (SBCs) for funding to perform research leading to the development of innovative technologies that may advance progress toward

integrated long term glucose regulated insulin delivery system (artificial pancreas). Letters of Intent are due by March 16, with full proposals to be received by April 15.
Requirements: Only United States small business concerns (SBCs) are eligible to submit SBIR applications.
Geographic Focus: All States
Date(s) Application is Due: Mar 16; Apr 15
Amount of Grant: 500,000 - 1,000,000 USD
Contact: Dr. Guillermo Arreaza, (301) 594-4724; fax (301) 480-3503; ga96b@nih.gov
Internet: http://grants.nih.gov/grants/guide/rfa-files/RFA-DK-09-001.html
Sponsor: National Institute of Diabetes and Digestive and Kidney Diseases
6707 Democracy Boulevard
Bethesda, MD 20892-5460

NIDDK Small Grant Program for K08/K23 Recipients — 2846

The NIDDK invites recipients of its K08/K23 awards to apply for grant support during the final two years of their awards through the NIDDK Small Grant Program. This additional support will enable K awardees to either expand their current research objectives or to branch out to a closely related pilot study, thus demonstrating their growth as investigators and their independence. This should facilitate the transition to fully independent investigator status. The added grant support in the latter years of an award is expected to have the following benefits: increased fiscal independence for the K08/K23 award recipient as a precursor to complete independence; an opportunity for the recipient to generate additional publications and data to form the basis for an R01 application; an opportunity for a Scientific Review Group (SRG) to evaluate accomplishments made during the first two years of the K08/K23 award; an opportunity for the applicant to provide more detailed research plans for the last two years of the K award; and an opportunity for the applicant to demonstrate additional success in the peer review process during the course of their K08/K23 award. A project period of up to two years and a budget for direct costs of up to two $25,000 modules or $50,000 per year may be requested.
Requirements: This award is available only to those currently holding the K08 award from NIDDK. Details of the K08 award are given in NIH Program Announcement PA-00-003. Applications may be submitted on behalf of candidates by domestic non-federal nonprofit organizations, public or private.
Geographic Focus: All States
Date(s) Application is Due: Feb 16; Jun 16; Oct 16
Amount of Grant: 50,000 USD
Contact: Dr. Judith Podskalny, (301) 594-8876; fax (301) 480-8300; jp53s@nih.gov
Internet: http://grants.nih.gov/grants/guide/pa-files/PAR-06-172.html
Sponsor: National Institute of Diabetes and Digestive and Kidney Diseases
6707 Democracy Boulevard
Bethesda, MD 20892-5460

NIDDK Small Grants for Underrepresented Minority Scientists in Diabetes and Digestive and Kidney Diseases — 2847

The goal of this program is to enable the applicant to accept a tenure-earning position, gain additional research experience and obtain preliminary data on which to base a subsequent research grant application in an area of diabetes, endocrinology, metabolism, digestive diseases, and nutrition, kidney, urology, or hematology. Annual deadline dates may vary; contact program staff for exact dates.
Requirements: Awards will be limited to minority citizens or non-citizen nationals of the United States or to individuals who have been lawfully admitted for permanent residence at the time of application. Applicants must also have a doctoral degree.
Restrictions: Applicants may not hold, nor apply concurrently for any other PHS research project grant at the time of this application.
Geographic Focus: All States
Date(s) Application is Due: Mar 22
Amount of Grant: Up to 250,000 USD
Contact: Dr. Charles H. Rodgers, (301) 594-7717; rodgersc@extra.niddk.nih.gov
Internet: http://grants.nih.gov/grants/guide/rfa-files/RFA-DK-00-007.html
Sponsor: National Institute of Diabetes and Digestive and Kidney Diseases
45 Center Drive, MSC 6600
Bethesda, MD 20892-6600

NIDDK Traditional Conference Grants — 2848

Conference Grants are awarded to institutions and organizations (not individuals) to provide partial support for international or domestic meetings, conferences, and workshops to coordinate, exchange, and disseminate research information. NIDDK is especially interested in conference grants used to cover travel expenses for young and minority investigators, although meeting publications, salaries, consultant services, equipment rental, and supplies may be requested. Indirect costs normally are not allowed.
Requirements: Applicants must submit a letter of intent to apply at least 8 weeks before the application receipt date. This letter should include the names of the organizing committee, a preliminary agenda, a list of accepted or invited speakers, a proposed budget and other planned sources of funding, and a description of how the conference will address each of the above criteria.
Geographic Focus: All States
Contact: Dr. M. Ken May, Nutrient Metabolism Program Director; (301) 594-8884; fax (301) 480-8300; maym@mail.nih.gov
Internet: http://www2.niddk.nih.gov/Funding/Grants/ApplicantGuidelines/ConferenceSupport.htm
Sponsor: National Institute of Diabetes and Digestive and Kidney Diseases
6707 Democracy Boulevard
Bethesda, MD 20892-5460

NIDDK Training in Clinical Investigation in Kidney and Urology Grants — 2849

The Division of Kidney, Urology, and Hematology (DKUH) has long been interested in stimulating more training in clinical research methods in programs related to nephrology and urology. At the same time, there is a dearth of clinicians that have advanced training in clinical trial methodology. The purpose of this program announcement is to encourage applications for new training programs and to encourage program directors of existing Institutional Training Grants supported by DKUH to collaborate with schools of public health and departments of epidemiology, statistics, and biostatistics to offer didactic and practical training in clinical research methodologies. The goal of this program announcement is to increase the number of well-trained investigators who are capable of conceiving, designing, conducting, analyzing, and presenting the results of clinical research in areas pertaining to nephrology or urology.
Requirements: For new training programs, only domestic, nonprofit, private or public institutions may apply. The applicant institution must have the staff and facilities required for the proposed program. The research training program director at the institution will be responsible for the selection and appointment of trainees to receive National Research Service Award (NRSA) support and for the overall direction of the program.
Restrictions: Positions on NRSA institutional grants may not be used for study leading to the MD, DDS or other clinical health professional degrees except when those studies are a part of a formal combined research degree program such as the MD/PhD. Similarly, trainees may not accept NRSA support for studies which are a part of residency training leading to a medical specialty or subspecialty except when the residency program credits a period of full-time, postdoctoral research training toward board certification and the trainee intends to pursue a research career.
Geographic Focus: All States
Date(s) Application is Due: Mar 10
Amount of Grant: 2,060 USD
Contact: Charles Rodgers, (301) 594-7717; rodgersc@extra.niddk.nih.gov
Internet: http://grants.nih.gov/grants/guide/pa-files/PAR-99-012.html
Sponsor: National Institute of Diabetes and Digestive and Kidney Diseases
45 Center Drive, MSC 6600
Bethesda, MD 20892-6600

NIDDK Translational Research for the Prevention and Control of Diabetes and Obesity — 2850

The National Institute of Diabetes and Digestive and Kidney Diseases (NIDDK), the National Institute of Nursing Research and the Office of Behavior and Social Sciences Research seek to develop cost effective and sustainable interventions that can be adopted in real world settings, for the prevention and control of diabetes and obesity. Research should be based on interventions already proven efficacious in clinical trials to prevent and reverse obesity and type 2 diabetes, to improve care of type 1 and type 2 diabetes and to prevent or delay its complications. The interventions proposed under this solicitation should have the potential to be widely disseminated to clinical practice, individuals and communities at risk. Relevant topics include: methods to improve health care delivery to patients with or at risk of diabetes; strategies to enhance diabetes self management; methods to develop strategies to promote healthy lifestyles to reduce the risk of diabetes; methods to promote lifestyle change that will prevent or reverse overweight and obesity; and new cost effective ways to identify people with pre-diabetes and undiagnosed diabetes.
Requirements: The following organizations and institutions may apply: public/state controlled institutions of higher education; private institutions of higher education; Hispanic-serving institutions; historically Black colleges and universities (HBCUs); tribally controlled colleges and universities (TCCUs); Alaska native and native Hawaiian serving institutions; nonprofits with 501(c)3 IRS status; nonprofits without 501(c)3 IRS status; small businesses; for-profit organizations; State governments; regional organizations; U.S. territories or possessions; Indian/Native American tribal governments; Indian/Native American tribally designated organizations; eligible agencies of the Federal government; and faith-based or community based organizations.
Geographic Focus: All States
Date(s) Application is Due: Jan 25; May 25; Sep 25
Contact: Sanford Garfield, (301) 594-8803; fax (301) 480-3503; sg50o@nih.gov
Internet: http://grants.nih.gov/grants/guide/pa-files/PAR-06-532.html
Sponsor: National Institute of Diabetes and Digestive and Kidney Diseases
6707 Democracy Boulevard
Bethesda, MD 20892-5460

NIDDK Transmission of Human Immunodeficiency Virus (HIV) In Semen — 2851

The National Institute of Diabetes and Digestive and Kidney Diseases (NIDDK) through its Division of Kidney, Urologic and Hematologic Diseases (DKUHD) and the National Institute of Child Health and Human Development, Center for Population Research invite investigators to submit research grant applications that will increase the basic and clinical knowledge of the biology of HIV in semen. Direct exposure to semen of HIV seropositive men is a major route for transmission of HIV type I. This Funding Opportunity Announcement (FOA) focuses on studies that will elucidate the factors that determine HIV shedding in the male genital tract. This includes studies that elucidate the infectivity of HIV in semen fractions, the effect of antiretroviral therapy on HIV infectivity in semen fractions, the relationship between the immunobiology of the male genital tract and HIV replication and infectivity, and factors such as genital tract inflammation which influence HIV transmission through semen. Of particular interest are studies of the anatomic sites of virus infected cells and virus production in the male genital tract. The intent of this Funding opportunity is to encourage investigator-initiated research, to attract new investigators to the field, and to increase interdisciplinary research to enhance the scope and effectiveness of research in this area. Because the nature and scope of the proposed research will vary from application

to application, it is anticipated that the size and duration of each award will also vary. The total amount awarded and the number of awards will depend upon the mechanism numbers, quality, duration, and costs of the applications received.
Requirements: The following organizations and institutions are eligible to apply: public/state controlled institutions of higher education; private institutions of higher education; Hispanic-serving institutions; historically Black colleges and universities (HBCUs); tribally controlled colleges and universities (TCCUs); Alaska Native and Native Hawaiian serving institutions; nonprofits with 501(c)3 IRS status (other than institutions of higher education); nonprofits without 501(c)3 IRS status (other than institutions of higher education); small businesses; for-profit organizations (other than small businesses); state governments; Indian/Native American tribal governments (federally recognized); Indian/Native American tribally designated organizations; county governments; city or township governments; special district governments; independent school districts; public housing authorities; Indian housing authorities; U.S. territory or possession; Indian/Native American tribal governments (other than federally recognized); regional organizations; non-domestic (non-U.S.) entities (foreign organizations); other eligible agencies of the federal government; and faith-based or community-based organizations.
Geographic Focus: All States
Date(s) Application is Due: Feb 5; Jun 5; Oct 5
Contact: Leroy M. Nyberg, Jr., (301) 594-7717; ln10f@nih.gov
Internet: http://grants.nih.gov/grants/guide/pa-files/PA-07-044.html
Sponsor: National Institute of Diabetes and Digestive and Kidney Diseases
6707 Democracy Boulevard
Bethesda, MD 20892-5460

NIDDK Type 2 Diabetes in the Pediatric Population Research Grants   2852
NIDDK and NICHD invite investigator-initiated research grant applications to study the epidemiology, natural history, pathophysiology, prevention, and treatment of type 2 diabetes in children in the United States. Type 2 diabetes has traditionally been viewed as a disease of adults. More recently, however, it has become apparent that an increasing number of cases of type 2 diabetes are being reported in children. New hypothesis-driven studies are needed to describe the epidemiology, refine the diagnosis, define the metabolic abnormalities, and formulate treatment options for type 2 diabetes in children. Annual deadline dates may vary; contact program staff for exact dates.
Requirements: Applications may be submitted by domestic for-profit and nonprofit organizations, both public and private; units of state and local governments; and eligible agencies of the federal government.
Geographic Focus: All States
Contact: Dr. Gilman Grave, (301) 496-5593; fax (301) 480-9791
Internet: http://www2.niddk.nih.gov/Research/ScientificAreas/Diabetes/Type2Diabetes/DM2K.htm
Sponsor: National Institute of Diabetes and Digestive and Kidney Diseases
6707 Democracy Boulevard
Bethesda, MD 20892-5460

NIDRR Field-Initiated Projects   2853
The goals of this project include: developing methods and technologies that maximize the integration of individuals with disabilities into society; employment and family environments; and improving the effectiveness of services authorized under the Rehabilitation Act. Applicants who receive awards under this program may carry out either research or development activities, and applications for each type of project will be reviewed separately. The applicant is responsible for indicating which type of project is being proposed. Projects may last up to three years. Annual deadline dates may vary; contact program staff for exact dates.
Requirements: Proposals will be accepted from any public or private agency or organization.
Geographic Focus: All States
Contact: Donna Nangle, (202) 245-7462; fax (202) 245-7323; Donna.Nangle@ed.gov
Internet: http://www.ed.gov/fund/grant/apply/nidrr/index.html
Sponsor: National Institute on Disability and Rehabilitation Research
550 12th Street SW, 6th Floor, Switzer Building
Washington, D.C. 20202-6510

NIEHS Hazardous Materials Worker Health and Safety Training Grants   2854
This FOA issued by the National Institute of Environmental Health Sciences (NIEHS) invites applications for cooperative agreements to support the development of model programs for the training and education of workers engaged in activities related to hazardous materials and waste generation, removal, containment, transportation and emergency response. The major objective of this solicitation is to prevent work-related harm by assisting in the training of workers in how best to protect themselves and their communities from exposure to hazardous materials encountered during hazardous waste operations, hazardous materials transportation, environmental restoration of contaminated facilities or chemical emergency response. A variety of sites, such as those involved with chemical waste clean up and remedial action and transportation-related chemical emergency response may pose severe health and safety concerns to workers and the surrounding communities. These sites contain a multiplicity of hazardous substances, sometimes unknown substances, and often the site is uncontrolled. A major goal of the Worker Education and Training Program (WETP) is to provide assistance to organizations in developing their institutional competency to provide appropriate model training and education programs. The total project period for an application submitted in response to this funding opportunity must be 5 years. A new applicant may request a budget for direct costs of up to $700,000 for the first year. Letters of Intent (LOIs) are due by October 23, with full proposals due by November 23.
Requirements: The following organizations and institutions are eligible to apply: public/state controlled institutions of higher education; private institutions of higher education; Hispanic-serving institutions; historically Black colleges and universities (HBCUs); tribally controlled colleges and universities (TCCUs); Alaska Native and Native Hawaiian serving institutions; and nonprofits with 501(c)3 IRS status (other than institutions of higher education).
Geographic Focus: All States
Date(s) Application is Due: Oct 23; Nov 23
Amount of Grant: Up to 700,000 USD
Contact: Joseph Hughes; (919) 541-0217; fax (919) 541-0462; hughes3@niehs.nih.gov
Internet: http://grants.nih.gov/grants/guide/rfa-files/RFA-ES-09-004.html
Sponsor: National Institute of Environmental Health Sciences
Keystone Building, Room 3039, P.O. Box 12233, MD K3-14
Research Triangle Park, NC 27709

NIEHS Hazmat Training at Doe Nuclear Weapons Complex Grants   2855
The National Institute of Environmental Health Sciences (NIEHS) invites applications for cooperative agreements to support the development of model programs for the training and education of workers engaged in activities related to hazardous materials and waste generation, removal, containment, transportation and emergency response within the Department of Energy (DOE) Nuclear Weapons Complex. The major objective of this solicitation is to prevent work related harm by assisting in the training and education of workers in the DOE nuclear weapons complex. Safety and health training will transmit skills and knowledge to workers in how best to protect themselves and their communities from exposure to hazardous materials encountered during hazardous waste operations, facility decommissioning and decontamination, hazardous materials transportation, environmental restoration of contaminated facilities or chemical emergency response. The total project period for an application submitted in response to this funding opportunity must be 5 years. A new applicant may request a budget for direct costs of up to $700,000 for the first year. Letters of Intent (LOIs) are due by October 23, with full proposals due by November 23.
Requirements: The following organizations and institutions are eligible to apply: public/state controlled institutions of higher education; private institutions of higher education; Hispanic-serving institutions; historically Black colleges and universities (HBCUs); tribally controlled colleges and universities (TCCUs); Alaska Native and Native Hawaiian serving institutions; and nonprofits with 501(c)3 IRS status (other than institutions of higher education).
Geographic Focus: All States
Date(s) Application is Due: Oct 23; Nov 23
Amount of Grant: Up to 700,000 USD
Contact: Joseph Hughes; (919) 541-0217; fax (919) 541-0462; hughes3@niehs.nih.gov
Internet: http://grants.nih.gov/grants/guide/rfa-files/RFA-ES-09-003.html
Sponsor: National Institute of Environmental Health Sciences
Keystone Building, Room 3039, P.O. Box 12233, MD K3-14
Research Triangle Park, NC 27709

NIEHS Independent Scientist Award   2856
The award is a special salary-only grant designed to provide protected time for newly independent scientists who currently have non-research obligations such as heavy teaching loads, clinical work, committee assignments, service, and administrative duties that prevent them from having a period of intensive research focus. The award is targeted to persons with doctoral degrees who have completed their research training, have independent peer-reviewed research support, and who need a period of protected research time in order to foster their research career development. Strong preference is given to candidates with active research project grant support from the NIEHS. New applications are due February 12, June 12, and October 12, while renewal applications are due March 12, July 12, and November 12. Up to $75,000 per year of base salary plus fringe benefits is available. Research support of up to $25,000 is provided in some cases. Individuals engaged in predominantly theoretical work, such as modeling or computer simulation, may request up to $25,000 per year for research expenses.
Requirements: The following organizations and institutions are eligible to apply: public/state controlled institutions of higher education; private institutions of higher education; nonprofits with 501(c)3 IRS status; nonprofits without 501(c)3 IRS status; small businesses; for-profit organizations; State governments; U.S. territories or possessions; Indian/Native American tribal governments (Federally recognized and other than Federally recognized); Indian/Native American tribally designated organizations; Hispanic-serving institutions; historically Black colleges and universities (HBCUs); tribally controlled colleges and universities (TCCUs); Alaska Native and Native Hawaiian serving institutions; regional organizations; and faith-based or community based organizations. The candidate must have a doctoral degree and peer-reviewed, independent research support at the time the award is made. The candidate must spend a minimum of 75 percent effort conducting research during the period of the award.
Restrictions: The award is not intended for investigators who already have full time to perform research, or have substantial publication records or considerable research support indicating that they are well established in their fields. Foreign institutions are not eligible to apply.
Geographic Focus: All States
Contact: Carol Shreffler, (919) 541-1445; fax (919) 541-5064; shreffl1@niehs.nih.gov
Internet: http://grants.nih.gov/grants/guide/pa-files/PA-09-038.html
Sponsor: National Institute of Environmental Health Sciences
Keystone Building, Room 3039, P.O. Box 12233, MD K3-14
Research Triangle Park, NC 27709

NIEHS Mentored Clinical Scientist Research Career Development Awards   2857

This award enables candidates holding professional degrees (e.g., M.D., D.O., D.V.M., or equivalent degrees) to undertake 3 to 5 years of special study and supervised research with the goal of becoming independent investigators. The award also allows awardees to pursue a research career development program suited to their experience and capabilities under a mentor who is competent to provide guidance in the chosen research area. Institutions may submit applications on behalf of candidates who hold professional degrees. At least 2 years must have elapsed since the health professional degree was granted. Candidates can have varying levels of clinical training and research experience. New applications are due February 12, June 12, and October 12, while resubmitted applications are due March 12, July 12, and November 12. Up to $75,000 per year plus fringe benefits is available, and up to $25,000 per year for research support for new awards.
Requirements: All candidates must be U.S. citizens, non-citizen nationals, or legal permanent residents of the U.S. Persons with temporary or student visas are not eligible. The grantee institution must have a strong, well-established research and research training program in the chosen area, accomplished faculty in the basic and clinical sciences, and a commitment to the candidate's research development. The proposed program should include an appropriate mentor.
Geographic Focus: All States
Date(s) Application is Due: Feb 12; Mar 12; Jun 12; Jul 12; Oct 12; Nov 12
Amount of Grant: Up to 100,000 USD
Contact: Carol Shreffler, (919) 541-1445; fax (919) 541-5064; shreffl1@niehs.nih.gov
Internet: http://grants.nih.gov/grants/guide/contacts/pa-06-512_contacts.htm
Sponsor: National Institute of Environmental Health Sciences
Keystone Building, Room 3039, P.O. Box 12233, MD K3-14
Research Triangle Park, NC 27709

NIEHS Mentored Research Scientist Development Award   2858

The MRSDA is for a research scientist who needs an additional period of sponsored research experience as a way to gain expertise in a research area new to the candidate or in an area that would demonstrably enhance the candidate's scientific career. Research projects proposed in response to this Announcement will be expected to have a defined impact on the environmental health sciences and be directly responsive to the mission of the NIEHS, which is distinguished from that of other Institutes by its focus on research programs seeking to link the effects of environmental exposures to the cause, mechanisms, moderation, or prevention of a human disease or disorder or relevant pathophysiologic process. Applications must include as part of the justification the specific human disease, dysfunction, pathophysiologic condition, or relevant human biologic process altered, and the specific environmentally relevant toxicant. Examples of environmentally relevant toxicants include industrial chemicals or manufacturing byproducts, metals, pesticides, herbicides, air pollutants and other inhaled toxicants, particulates or fibers, fungal, bacterial, or biologically derived toxins. Agents considered non-responsive to this announcement include, but are not limited to: alcohol, chemotherapeutic agents, radiation which is not the result of an ambient environmental exposure, drugs of abuse, and infectious or parasitic agents. Ecologic, biomonitoring, biotransformation or biodegradation studies are also not responsive. Applicants must demonstrate a significant career shift or career enhancement activity which will achieve research independence upon completion of the program. Preference will be given to candidates whose mentor has NEIHS research grant support.
Requirements: The candidate must have a research or health-professional doctorate or its equivalent and must have demonstrated the capacity or potential for highly productive independent research in the period after the doctorate. The candidate must identify a mentor with extensive research experience and must spend a minimum of 75 percent of full-time professional effort conducting research for the period of the award. Applications may be submitted on behalf of candidates by domestic, nonfederal organizations, public or private, such as medical, dental, or nursing schools or other institutions of higher education. Candidates must be U.S. citizens, noncitizen nationals, or permanent residents.
Geographic Focus: All States
Date(s) Application is Due: Feb 12; Mar 12; Jun 12; Jul 12; Oct 12; Nov 12
Amount of Grant: 75,000 USD
Contact: Carol Shreffler, (919) 541-1445; fax (919) 541-5064; shreffl@niehs.nih.gov
Internet: http://grants.nih.gov/grants/guide/pa-files/PA-09-040.html
Sponsor: National Institute of Environmental Health Sciences
Keystone Building, Room 3039, P.O. Box 12233, MD K3-14
Research Triangle Park, NC 27709

NIEHS SBIR E-Learning for HAZMAT and Emergency Response Grants   2859

This funding opportunity announcement (FOA) solicits Small Business Innovation Research (SBIR) grant applications from small business concerns (SBCs) that propose to further the development of Advanced Technology Training (ATT) Products for the health and safety training of hazardous materials (HAZMAT) workers, emergency responders, and skilled support personnel. These products would complement the goals and objectives of the Worker Education and Training Program (WETP). The major objective of the NIEHS/WETP is to prevent work related harm by assisting in the training of workers in how best to protect themselves and their communities from exposure to hazardous materials. There is a need to ensure that learning and training technologies are further developed, field tested and applied to real world situations. It is the intent of this solicitation to support the development of products to support e-collaboration, e-teaching, e-certification, and e-learning in safety and health training for workers engaged in hazardous materials response. A new applicant may request a budget for direct costs of up to $300,000 for the first year. Letters of Intent (LOIs) are not required, and full proposals are due by August 3.
Requirements: Only United States small business concerns (SBCs) are eligible to submit SBIR applications. A small business concern is one that, at the time of award of SBIR Phase I and Phase II, meets all of the following criteria: is organized for profit, with a place of business located in the United States, which operates primarily within the United States or which makes a significant contribution to the United States economy through payment of taxes or use of American products, materials or labor; is in the legal form of an individual proprietorship, partnership, limited liability company, corporation, joint venture, association, trust or cooperative, except that where the form is a joint venture, there can be no more than 49 percent participation by foreign business entities in the joint venture; is at least 51 percent owned and controlled by one or more individuals who are citizens of, or permanent resident aliens in, the United States, or it must be a for-profit business concern that is at least 51% owned and controlled by another for-profit business concern that is at least 51% owned and controlled by one or more individuals who are citizens of, or permanent resident aliens in, the United States, except in the case of a joint venture, where each entity to the venture must be 51 percent owned and controlled by one or more individuals who are citizens of, or permanent resident aliens in, the United States; and; has, including its affiliates, not more than 500 employees.
Geographic Focus: All States
Date(s) Application is Due: Aug 3
Amount of Grant: Up to 300,000 USD
Contact: Ted Outwater, (919) 541-2972; fax (919) 541-0462; outwater@niehs.nih.gov
Internet: http://grants.nih.gov/grants/guide/rfa-files/RFA-ES-09-006.html
Sponsor: National Institute of Environmental Health Sciences
Keystone Building, Room 3039, P.O. Box 12233, MD K3-14
Research Triangle Park, NC 27709

NIEHS Small Business Innovation Research (SBIR) Program Grants   2860

The focus of these grants is on understanding how chemical and physical agents cause pathological changes in molecules, cells, tissues, and organs and become manifested as respiratory disease; neurological, behavioral, and developmental abnormalities; cancer; and other disorders. By understanding the relationship between environmental exposures and the subsequent development of disease or biological injury, human health may be protected. SBIR program objectives are to stimulate technological innovation, use small businesses to meet federal research and development needs, increase private sector commercialization of innovations derived from federal research and development, and foster and encourage participation by minority and disadvantaged persons in technological innovation. Phase I awards, generally for six months, establish the technical merit and feasibility of proposed research efforts that may lead to commercial products or processes. Phase II awards are for the continuation of research initiated in Phase I.
Requirements: SBIR grants can be awarded only to domestic small businesses (entities that are independently owned and operated for profit, are not dominant in the field in which research is proposed, and have no more than 500 employees). Primary employment (more than one-half) of the principal investigator must be with the small business at the time of award and during the conducting of the proposed research. In both Phase I and II, the research must be performed in the United States.
Geographic Focus: All States
Date(s) Application is Due: Apr 15; Aug 15; Dec 15
Amount of Grant: 100,000 - 750,000 USD
Contact: Jo Anne Goodnight, SBIR/STTR; (301) 435-2688; fax (301) 480-0146; jg128w@nih.gov or sbir@od.nih.gov
Internet: http://grants.nih.gov/grants/funding/sbir.htm#sbir
Sponsor: National Institute of Environmental Health Sciences
Keystone Building, Room 3039, P.O. Box 12233, MD K3-14
Research Triangle Park, NC 27709

NIGMS Advancing Basic Behavioral and Social Research on Resilience: an Integrative Science Approach   2861

This FOA solicits applications that will elucidate mechanisms and processes of resilience within a general framework that emphasizes its dynamics and interactions across both time and scale, multiple contexts, multiple outcomes, and multiple time frames. This framework can be applied to multiple contexts (e.g., acute or chronic stress exposure, including disease, disaster, unemployment, divorce, etc.), multiple outcomes (health, function, psychological well-being, etc.), and multiple time frames (seconds, minutes, days, years). The framework has four features: assessment of a baseline level prior to a challenge; characterization of a specific challenge (acute or chronic); post-challenge measures of outcomes that characterize the response over time; and predictors of outcomes (including predisposing factors at the individual level and environmental moderators). Measurement of response across multiple domains (physiological, psychological etc.) is critical to this holistic framework. Consistent application of this general framework has the potential to reveal underlying principles that describe dynamic trajectories of adaptation to a challenge and to identify potential malleable processes or mechanisms that shape those dynamic response patterns. Typically, it will require repeated measures of outcomes to differentiate resilient patterns of response from non-resilient trajectories. It may also offer insights into individual differences in reactivity and recovery that characterize resilient phenotypes. Previous investigations have identified several genetic, biological and psychological factors that are associated with individuals more successful at adapting to an adverse experience. However, most prior research on resilience has examined processes of adaptation for a single or a limited number of outcome parameters and has addressed the phenomenon from the perspective of a single aspect of functioning, e.g. cardiovascular, mental health, disease progression, substance abuse, etc. In addition, the focus of the majority of the current research has been on the individual; only a few studies have assessed resilience at the level of the family or community. This fragmentation makes it difficult to fully test hypotheses derived from holistic dynamic models of adaptation and recovery. A full model of resilience will require measuring outcomes, processes, and moderators at multiple levels (e.g. genetic, neurobiological, physiological, psychological, behavioral, social, and environmental) as these unfold

at different timescales. This poses complex research challenges and may require the development or application of novel methods from other fields, such as engineering and complex systems science. Large sample sizes will be required to address some questions, and integrated expertise is essential. Computational approaches may augment traditional methods of research. This initiative is designed to develop the infrastructure and protocols needed to support well-designed studies of resilience in human subjects (individuals or communities) to examine patterns of response (pre-, during, and post-) to a well characterized challenge as they evolve over time. In order to develop an integrated framework for the concept of resilience, it is critical that investigators examine the concept at multiple levels of analysis and examine interactions, mediators, moderators and potential mechanisms. The goal is to develop comprehensive models of individual and/or social resilience in humans, and to shed light on the mechanisms and processes that account for adaptive or "resilient" response profiles, by leveraging the expertise of investigators from multiple disciplines working in a cooperative and integrated fashion. This FOA seeks teams of investigators who will develop, in the first phase (UG3), the infrastructure and research design for a project that will support testing of hypotheses derived from dynamic models of resilience, incorporating the four key aspects of the general framework articulated above. This approach is intended to address the acknowledged gaps in resilience research in the behavioral and social sciences. Upon completion of the planning phase (UG3), progress will be administratively reviewed for consideration of awarding a second phase (UH3). Letters of intent (not required) are due by November 1, with the application deadline being December 1. UG3 phase support is limited to $250,000 per year with the exception of studies incorporating well-justified pilot studies, in which case there is no budget limit. The budget for the UH3 phase is not limited but needs to reflect the actual needs of the proposed project.
Requirements: Eligible organizations include: higher education institutes; nonprofits other than institutions of higher education, either with or without 501(c)3 status; for-profit organizations; governments, including state, county, local, Native American Tribal, and U.S. territories or possessions; independent school districts; public and Indian housing authorities; Native American Tribal organizations; faith-based organizations; community-based organizations; regional organizations; and non-domestic (foreign) entities.
Geographic Focus: All States, Guam, Marshall Islands, Northern Mariana Islands, Puerto Rico, U.S. Virgin Islands, All Countries, American Samoa
Date(s) Application is Due: Dec 1
Contact: Stephen Marcus, Program Officer; (301) 451-6446 or (301) 496–7301; marcusst@mail.nih.gov or info@nigms.nih.gov
Internet: http://grants.nih.gov/grants/guide/pa-files/PAR-16-326.html
Sponsor: National Institutes of Health
45 Center Drive MSC 6200
Bethesda, MD 20892-6200

NIGMS Enabling Resources for Pharmacogenomics Grants   2862
The purpose of this funding opportunity announcement is to support critical enabling resources that will accelerate new research discoveries and/or implementation of research discoveries in pharmacogenomics. A proposed resource must meet an ascertained community demand and benefit the entire scientific field of users. The FOA will support activities that can be clearly and specifically defined, are optimally designed, have evaluative measures built-in, are judiciously staffed, have formed partnerships where appropriate, and ideally have a proven track record and a finite lifetime. The outcome of an enabling resource must be highly impactful in a demonstrable way. Advance consultation with Scientific/Research staff to ensure that a proposed resource fits well with this opportunity is highly encouraged. The current closing date is September 25.
Requirements: Eligible applicants include: Native American tribal governments (Federally recognized); private institutions of higher education; special district governments; public and state controlled institutions of higher education; county governments; nonprofits that do not have a 501(c)3 status with the IRS, other than institutions of higher education; for profit organizations other than small businesses; nonprofits having a 501(c)3 status with the IRS, other than institutions of higher education; state governments; small businesses; city or township governments; public housing authorities; Indian housing authorities; independent school districts, Native American tribal organizations (other than Federally recognized tribal governments); Alaska Native and Native Hawaiian serving institutions; Asian American and Native American Pacific Islander serving institutions; eligible agencies of the federal government; faith-based organizations; community-based organizations; Hispanic-serving institutions; historically Black colleges and universities; Indian/Native American tribal governments (other than federally recognized); regional organizations; Tribally controlled colleges and universities (TCCUs); and U.S. territories or possessions.
Restrictions: Non-domestic (non-U.S.) entities (foreign institutions) are not eligible.
Geographic Focus: All States, District of Columbia, Marshall Islands, Northern Mariana Islands, Puerto Rico, U.S. Virgin Islands, American Samoa
Date(s) Application is Due: Sep 25
Contact: Rochelle M. Long, Research Contact; (301) 594-3827 or (301) 496–7301; rochelle.long@nih.gov or info@nigms.nih.gov
Internet: http://grants.nih.gov/grants/guide/pa-files/PAR-14-185.html
Sponsor: National Institutes of Health
45 Center Drive MSC 6200
Bethesda, MD 20892-6200

NIGMS Ruth L. Kirschstein National Research Service Awards for Individual Predoctoral Fellows in PharmD/PhD Grants   2863
NIGMS is committed to increasing the number of pharmacy students who wish to pursue doctoral research in the pharmaceutical sciences. It recognizes that current career paths for Doctor of Pharmacy graduates have reduced the number of individuals who pursue academic research training. NIGMS's intent in issuing this Funding Opportunity Announcement (FOA) is to provide individual research training fellowships to promising predoctoral applicants who are enrolled in PharmD/PhD programs in Schools of Pharmacy who desire to become productive and successful independent investigators whose research meets the mission of NIGMS. This fellowship program will provide support for the dissertation stage of predoctoral students enrolled in joint PharmD/PhD programs. It is expected that the length of time for an award will be 3 years or less.
Requirements: The following organizations and institutions are eligible to apply: for-profit organizations; non-profit organizations; domestic institutions; public or private institutions, such as universities, colleges, and hospitals; units of State and local governments; eligible agencies of the Federal government.
Restrictions: Individuals on temporary or student visas are not eligible.
Geographic Focus: All States
Date(s) Application is Due: Apr 8; Aug 8; Dec 8
Contact: Richard Okita, (301) 594-3827; fax (301) 480-2802; okitar@nigms.nih.gov
Internet: http://grants.nih.gov/grants/guide/pa-files/PA-09-029.html
Sponsor: National Institute of General Medical Sciences - 4
45 Center Drive, Natcher Building, Room 2AS-49, MSC 6200
Bethesda, MD 20892-6200

NIH Academic Research Enhancement Awards (AREA)   2864
NIH makes a special effort to stimulate research in educational institutions that provide baccalaureate training for a significant number of the nation's research scientists but that historically have not been major recipients of NIH support. The AREA funds are intended to support new research projects or expand ongoing research activities proposed by faculty members of these institutions in areas related to the health sciences. Funding decisions will be based on the proposed research project's scientific merit and relevance to NIH programs and the institution's contribution to the undergraduate preparation of doctoral-level health professionals. Annual deadline dates vary; contact program staff for exact dates. PA: PA-99-062
Requirements: All health professional schools/colleges and other academic components of domestic institutions offering baccalaureate or advanced degrees in the sciences related to health are eligible to apply.
Restrictions: Applicant principal investigators must not have active research grant support from either NIH or ADAMHA at the applicant institution at the time of award of an AREA grant; may not submit a regular NIH or ADAMHA research grant application for essentially the same project as a pending AREA application; are expected to conduct the majority of their research at their own institutions, although limited access to special facilities or equipment other institutions is permitted; and may not be awarded more than one AREA grant at a time nor be awarded a second AREA grant to continue research initiated under the first AREA grant.
Geographic Focus: All States
Date(s) Application is Due: May 7; Jun 25; Sep 7
Amount of Grant: Up to 150,000 USD
Contact: Denise Russo, Office of Extramural Research; (301) 451-7972; fax (301) 480-0146; drusso1@mail.nih.gov
Internet: http://grants.nih.gov/grants/guide/pa-files/PA-06-042.html
Sponsor: National Institutes of Health
6705 Rockledge Drive, Room 3524
Bethesda, MD 20892

NIH Biobehavioral Research for Effective Sleep   2865
The goal of this program announcement is to stimulate clinical and applied research on behavioral, psychosocial, and physiological consequences of acute and chronic partial sleep deprivation in either chronically ill or healthy individuals and to develop environmental, clinical management, and other interventions with the potential to reduce sleep disturbances and significantly improve the health of large numbers of people. Although sleep disorders are a cause of sleep loss in affected individuals, the questions to be addressed under this solicitation should focus on causes and consequences of sleep deprivation, apart from any sleep pathology. Inquiries may be directed to program staff at the appropriate NIH institute.
Requirements: Applications may be submitted by domestic and foreign, for-profit and non-profit organizations, public and private, such as universities, colleges, hospitals, laboratories, units of State and local governments, and eligible agencies of the Federal government. Racial/ethnic minority individuals, women, and persons with disabilities are encouraged to apply as Principal Investigators.
Geographic Focus: All States
Date(s) Application is Due: Feb 16; Jun 16; Oct 16
Amount of Grant: Up to 250,000 USD
Contact: Susan Grove; (301) 435-0714; sg16d@nih.gov or grantsinfo@od.nih.gov
Internet: http://grants.nih.gov/grants/guide/pa-files/PA-00-046.html
Sponsor: National Institutes of Health
6701 Rockledge Drive
Bethesda, MD 20892

NIH Biology, Development, and Progression of Malignant Prostate Disease Research Grants   2866
Investigator-initiated research grants will be made to examine a range of fundamental biological issues considered critical for progress in defeating prostate cancer. The purpose of funding is to encourage new projects focusing on the biology that underlies the development and progression of malignant prostatic disease. Multidisciplinary collaborations between basic and clinical scientists in projects ranging from exploratory basic biology studies to disease progression in appropriate biological systems or models are encouraged.
Requirements: Applications may be submitted by foreign or domestic for-profit and nonprofit organizations, both public and private; units of state and local government; and eligible agencies of the federal government.

Geographic Focus: All States
Date(s) Application is Due: Feb 5; Feb 16; Jun 5; Jun 16; Oct 5; Oct 16
Amount of Grant: Up to 250,000 USD
Contact: Frank Bellino, (301) 496-6402; fax (301) 402-0010; fb12a@nih.gov
Internet: http://grants.nih.gov/grants/guide/pa-files/PA-99-081.html
Sponsor: National Institutes of Health
Gateway Building, 7201 Wisconsin Avenue, Suite 2C231
Bethesda, MD 20892

NIH Biomedical Research on the International Space Station   2867
The National Institutes of Health (NIH) and the National Aeronautics and Space Administration (NASA) are cooperating to facilitate biomedical research in space for better understanding of human physiology and human health on Earth. Applications to this FOA should propose innovative biomedical research on the molecular or cellular level that is directly relevant to the NIH mission and can be carried out on the ISS. Awards made through this FOA will initially support milestone-driven, ground based preparatory studies (UH2 ground feasibility phase), with possible rapid transition to the second, ISS-based research phase (UH3 ISS experimental phase). The ground feasibility phase (UH2) will allow investigators to focus on ground-based preparatory work to meet scientific milestones and technical requirements leading to the ISS experimental phase (UH3). The UH3 phase will include preparing the experiments for launch, conducting them on the ISS, and the subsequent data analyses on Earth. UH3s will be awarded after administrative review of the eligible UH2s that have met the scientific milestones and feasibility requirements necessary to conduct research on the ISS. Applicants should contact individuals at the specific departments listed on the web site. Letters of Intent are due by August 31, and full applications should arrive by September 30.
Requirements: The UH2/UH3 application must be submitted as a single application, and applicants should note specific instructions for each phase in this FOA. The following organizations and institutions may apply: public/state controlled institutions of higher education; private institutions of higher education; Hispanic-serving institutions; historically Black colleges and universities (HBCUs); tribally controlled colleges and universities (TCCUs); Alaska native and native Hawaiian serving institutions; nonprofits with 501(c)3 IRS status; nonprofits without 501(c)3 IRS status; small businesses; for-profit organizations; State governments; regional organizations; U.S. territories or possessions; Indian/Native American tribal governments; Indian/Native American tribally designated organizations; county governments; city or township governments; special district governments; independent school districts; public housing authorities/Indian housing authorities; eligible agencies of the Federal government; and faith-based or community based organizations.
Geographic Focus: All States
Date(s) Application is Due: Aug 31; Sep 30
Amount of Grant: 300,000 USD
Contact: Susan Grove; (301) 435-0714; sg16d@nih.gov or grantsinfo@od.nih.gov
Internet: http://grants.nih.gov/grants/guide/pa-files/PAR-09-120.html
Sponsor: National Institutes of Health
6701 Rockledge Drive
Bethesda, MD 20892

NIH Bone and the Hematopoietic and Immune Systems Research Grants   2868
The participating institutes seek investigator-initiated projects that have the potential to illuminate functional interactions between bone and the hematopoietic and immune systems. Recent observations underscore the linkage between endochondral bone formation and the establishment of hematopoietic marrow, and suggest that interactions between bone, marrow, and the immune system persist in the mature skeleton. Marrow stromal cells include the precursors of the osteochondrogenic lineage, exert important influences on osteoclastogenesis and lymphopoiesis, and mediate the effects of some systemic factors on bone turnover. Recent evidence indicates that hematopoietic cells can influence the differentiation of osteogenic cells, and suggests that mature lymphocytes can influence osteoclastic and osteoblastic functions. In order to explore the mechanisms that underlie these interactions, the participating institutes support outstanding projects that have the potential to either clarify the importance of specific cell types and effector molecules or identify previously unrecognized cellular and molecular agents that influence bone physiology. Collaborations among bone biologists, hematologists, and immunologists, and between basic scientists and clinical investigators, are particularly encouraged. Inquiries may be directed to the staff at the appropriate NIH institute: Dr. William Sharrock, Musculoskeletal Diseases Branch, NIAMS, 45 Center Drive, Bethesda, MD 20892; Dr. Frank Bellino, (301) 496-6402, Biology of Aging Program, NIA, Gateway Building, Bethesda, MD 20892; Dr. Kenneth Gruber, (301) 584-4836, Chronic and Disabling Diseases Branch, NID.C.R, 45 Center Drive, Bethesda, MD 20892; and Dr. Charles Peterson, (301) 435-0050, Division of Blood Diseases and Resources, NHLBI, 6701 Rockledge Drive, Bethesda, MD 20892. PA: PA-99-085
Requirements: Applications may be submitted by domestic and foreign, for-profit and nonprofit, public and private organizations, units of state and local governments, and eligible agencies of the federal government.
Restrictions: Foreign institutions are not eligible for program project grant support.
Geographic Focus: All States
Amount of Grant: Up to 250,000 USD
Contact: Frank Bellino, (301) 496-6402; fax (301) 402-0010; fb12a@nih.gov
Internet: http://grants.nih.gov/grants/guide/pa-files/PA-99-085.html
Sponsor: National Institutes of Health
Gateway Building, 7201 Wisconsin Avenue, Suite 2C231
Bethesda, MD 20892

NIH Building Interdisciplinary Research Careers in Women's Health Grants   2869
The NIH Office of Research on Women's Health (ORWH) and its cosponsors invite institutional career development award applications for Building Interdisciplinary Research Careers in Women's Health (BIRCWH) Career Development Programs, hereafter termed Programs. Programs will support mentored research career development of junior faculty members, known as BIRCWH Scholars, who have recently completed clinical training or postdoctoral fellowships, and who will be engaged in interdisciplinary basic, translational, behavioral, clinical, and/or health services research relevant to women's health or sex/gender factors. Direct costs are limited to $500,000 total costs for a five-year period. Letters of Intent are due by September 23, with final applications due by October 22.
Requirements: The following organizations and institutions are eligible to apply for support: public/state controlled institutions of higher education; private institutions of higher education; Hispanic-serving institutions; historically Black colleges and universities (HBCUs); tribally controlled colleges and universities (TCCUs); Alaska Native and Native Hawaiian serving institutions; nonprofits with 501(c)3 IRS status (other than institutions of higher education); nonprofits without 501(c)3 IRS status (other than institutions of higher education); Indian/Native American tribal governments (federally recognized); Indian/Native American tribally designated organizations; U.S. territories or possessions; and eligible agencies of the federal government.
Geographic Focus: All States
Date(s) Application is Due: Sep 23; Oct 22
Amount of Grant: Up to 500,000 USD
Contact: Joan Davis Nagel, (301) 496-9186; fax (301) 402-1798; joandav@mail.nih.gov
Internet: http://grants.nih.gov/grants/guide/rfa-files/RFA-OD-09-006.html
Sponsor: National Institutes of Health
6707 Democracy Boulevard, Suite 400
Bethesda, MD 20892

NIHCM Foundation Health Care Print Journalism Awards   2870
To encourage outstanding work from journalists furthering innovation in health care policy and management, The National Institute for Health Care Management Research and Educational Foundation (NIHCM Foundation) recognizes excellence in health care reporting and writing on the financing and delivery of health care and the impact of health care policy with two $10,000 awards. NIHCM Foundation awards prizes to the winning entry in the following two categories, Health Care Article or Series from General Circulation Publications and Health Care Article or Series from Trade Publications. The award prize will be presented to the winner at a dinner in Washington, D.C. in May.
Requirements: Articles must originally have been published during calendar year 2009. Entries will be accepted from all staff reporters and editors, as well as freelance writers for articles that appeared in a General Circulation or Trade Publication. Online articles will be accepted if produced and disseminated originally on the web in affiliation with a General Circulation or Trade Publication. Articles should be timely and produce new insights into health care financing and delivery or health care policy.
Geographic Focus: All States
Date(s) Application is Due: Feb 26
Amount of Grant: 10,000 USD
Contact: Nancy Chockley; (202) 296-4426; fax (202) 296-4319; nihcm@nihem.org
Internet: http://nihcm.org/awards/journalism
Sponsor: National Institute for Health Care Management Foundation
1225 19th Street NW, Suite 710
Washington, D.C. 20036

NIHCM Foundation Health Care Research Awards   2871
To encourage outstanding work from researchers furthering innovation in health care policy and management, the National Institute for Health Care Management (NIHCM) Research and Educational Foundation recognizes excellence in original and creative research with a $10,000 award. Selection criteria for the award includes the originality and creativity of the research. Impact that the research may have in advancing the public policy debate, improving health care delivery, management and the relevance to one or more of the subject areas. The award will be presented to the winner at a dinner in Washington, D.C. in May.
Requirements: Articles must have been published in a peer-reviewed journal or similar quality publication during calendar year 2009. Applications are encouraged from authors in diverse fields, including economics, health policy, political science, public health, etc. and from individuals in non-academic settings, such as research firms and policy organizations. All research articles must address some aspect of health care financing, delivery, organization or the implementation of health care policy.
Geographic Focus: All States
Date(s) Application is Due: Feb 26
Amount of Grant: 10,000 USD
Contact: Nancy Chockley; (202) 296-4426; fax (202) 296-4319; nihcm@nihem.org
Internet: http://nihcm.org/awards/research
Sponsor: National Institute for Health Care Management Foundation
1225 19th Street NW, Suite 710
Washington, D.C. 20036

NIHCM Foundation Health Care Television and Radio Journalism Awards   2872
The Third Annual NIHCM Foundation Health Care Television and Radio Journalism Award recognizes excellence in television and radio reporting on health care issues and policy. Selection criteria will include, new insights generated by the reporting, impact on health care policy and quality of writing and production. A $10,000 prize will be presented to the winner at a dinner in Washington, D.C. in May.
Requirements: Entries must originally have aired during calendar year 2009. Entries will be accepted from staff employees of national or local news organizations, nationally-

distributed cable programs, or from freelance journalists who have produced reports disseminated by same. Online stories will be accepted if produced for and disseminated originally on the web in affiliation with a television or radio news organization. Material repurposed from other media will not be accepted.
Geographic Focus: All States
Date(s) Application is Due: Feb 26
Amount of Grant: 10,000 USD
Contact: Nancy Chockley; (202) 296-4426; fax (202) 296-4319; nihcm@nihem.org
Internet: http://nihcm.org/awards/televison_and_radio_journalism
Sponsor: National Institute for Health Care Management Foundation
1225 19th Street NW, Suite 710
Washington, D.C. 20036

NIH Dietary Supplement Research Centers: Botanicals Grants 2873
ODS, NCCAM and NCI invite new and renewal applications to support research centers which will: promote collaborative integrated interdisciplinary study of botanicals, particularly those found as ingredients in dietary supplements; and conduct research of high potential for being translated into practical benefits for human health. This initiative is intended to advance the spectrum of botanical research activities, ranging from plant identification and characterization to early phase clinical studies. Preclinical research that will inform future clinical studies is encouraged as the primary research focus of this program. ODS has a particular interest in botanicals as part of health care for conditions relating to health maintenance or primary prevention. NCCAM is interested in the study of botanicals broadly used by the American public. The NCI Division of Cancer Prevention is interested in supporting research focused on mechanisms by which botanically derived bioactive food components might influence cancer risk and tumor behavior. An applicant may request a project period of up to five years; an annual budget in direct costs up to $1.0 million.
Requirements: The following organizations and institutions are eligible to apply: individuals; public/state controlled institutions of higher education; private institutions of higher education; Hispanic-serving institutions; historically Black colleges and universities (HBCUs); tribally controlled colleges and universities (TCCUs); Alaska Native and Native Hawaiian serving institutions; nonprofits with 501(c)3 IRS status; nonprofits without 501(c)3 IRS status; for-profit organizations; state governments; Indian/Native American tribal governments (federally recognized); Indian/Native American tribally designated organizations; county governments; city or township governments; special district governments; independent school districts; public housing authorities/Indian housing authorities; U.S. territory or possession; Indian/Native American tribal governments (other than federally recognized); regional organizations; non-domestic (non-U.S.) entities (foreign organizations); other eligible agencies of the federal government; and faith-based or community-based organizations.
Geographic Focus: All States
Date(s) Application is Due: Dec 3
Amount of Grant: 1,000,000 USD
Contact: Christine A. Swanson, Director, Botanical Research Centers Program; (301) 435-2920; fax (301) 480-1845; Swansonc@od.nih.gov
Internet: http://grants.nih.gov/grants/guide/pa-files/PAR-09-091.html
Sponsor: National Institutes of Health
6100 Executive Boulevard, Room 3B01, MSC 7517
Bethesda, MD 20892

NIH Earth-Based Research Relevant to the Space Environment 2874
The purpose of this Program Announcement (PA) is to stimulate ground-based research on basic, applied, and clinical biomedical and behavioral problems that are relevant to human space flight or that could use the space environment as a laboratory. Although none of the research supported under this initiative would be conducted in space, it is anticipated that it would form a basis for future competitively reviewed studies which could be conducted on the International Space Station, or other space flight opportunities, by skilled on-board specialists. Modular Grant applications will request direct costs in $25,000 modules, up to a total direct cost request of $250,000 per year.
Requirements: The following organizations and institutions may apply: public/state controlled institutions of higher education; private institutions of higher education; Hispanic-serving institutions; historically Black colleges and universities (HBCUs); tribally controlled colleges and universities (TCCUs); Alaska native and native Hawaiian serving institutions; nonprofits with 501(c)3 IRS status; nonprofits without 501(c)3 IRS status; small businesses; for-profit organizations; State governments; regional organizations; U.S. territories or possessions; Indian/Native American tribal governments; Indian/Native American tribally designated organizations; county governments; city or township governments; special district governments; independent school districts; public housing authorities/Indian housing authorities; eligible agencies of the Federal government; and faith-based or community based organizations.
Geographic Focus: All States
Date(s) Application is Due: Feb 5; Jun 5; Oct 5
Amount of Grant: Up to 250,000 USD
Contact: Susan Grove; (301) 435-0714; sg16d@nih.gov or grantsinfo@od.nih.gov
Internet: http://grants.nih.gov/grants/guide/pa-files/PA-00-088.html
Sponsor: National Institutes of Health
6701 Rockledge Drive
Bethesda, MD 20892

NIH Enhancing Adherence to Diabetes Self-Management Behaviors 2875
This PA solicits applications for investigator-initiated research related to sociocultural, environmental, and behavioral mechanisms and biological/technological factors that contribute to successful and ongoing self-management in diabetes. Self-management is defined as client strategies and behaviors that contribute to blood glucose normalization, improved health, and prevention or reduction of complications. It is broader than adherence to specific regimen components and incorporates deliberate problem solving and decision-making processes. Applications are encouraged for both type 1 and type 2 diabetes; representative and minority populations; and all age groups. Inquiries may be directed to program staff at the appropriate NIH institute. Modular Grant applications will request direct costs in $25,000 modules, up to a total direct cost request of $250,000 per year.
Requirements: Applications may be submitted by domestic and foreign, for-profit and non-profit organizations, public and private, such as universities, colleges, hospitals, laboratories, units of State and local governments, and eligible agencies of the Federal government. Racial/ethnic minority individuals, women, and persons with disabilities are encouraged to apply as principal investigators.
Geographic Focus: All States
Date(s) Application is Due: Feb 16; Jun 16; Oct 16
Amount of Grant: Up to 250,000 USD
Contact: Susan Grove; (301) 435-0714; sg16d@nih.gov or grantsinfo@od.nih.gov
Internet: http://grants.nih.gov/grants/guide/pa-files/PA-00-049.html
Sponsor: National Institutes of Health
6701 Rockledge Drive
Bethesda, MD 20892

NIH Exceptional, Unconventional Research Enabling Knowledge Acceleration (EUREKA) Grants 2876
The purpose of the EUREKA (Exceptional Unconventional Research Enabling Knowledge Acceleration) initiative is to foster exceptionally innovative research that, if successful, will have an unusually high impact on the areas of science that are germane to the mission of one or more of the participating NIH Institutes. Before submitting an application, it is extremely important to verify that the proposed research is of interest to at least one of the NIH Institutes that is participating in the EUREKA FOA, since applications that are not germane to the mission of one or more of the participating Institutes will be withdrawn without review.
Requirements: The following organizations and institutions are eligible to apply: public/state controlled institutions of higher education; private institutions of higher education; Hispanic-serving institutions; historically Black colleges and universities (HBCUs); tribally controlled colleges and universities (TCCUs); Alaska Native and Native Hawaiian serving institutions; nonprofits with 501(c)3 IRS status (other than institutions of higher education); nonprofits without 501(c)3 IRS status (other than institutions of higher education); small businesses; for-profit organizations (other than small businesses); non-domestic (non-U.S.) entities (foreign organizations).
Restrictions: EUREKA is for new projects, not for continuation of existing projects. Nor is EUREKA for support of pilot projects, i.e., projects of limited scope that are designed primarily to generate data that will enable the PI to seek other funding opportunities.
Geographic Focus: All States
Date(s) Application is Due: Nov 24
Amount of Grant: Up to 800,000 USD
Contact: Laurie Tompkins, (301) 594-0943; tompkinL@nigms.nih.gov
Internet: http://grants.nih.gov/grants/guide/rfa-files/RFA-GM-10-009.html
Sponsor: National Institutes of Health
45 Center Drive, MSC 6200
Bethesda, MD 20892-6200

NIH Exploratory/Developmental Research Grant 2877
The Exploratory/Developmental Grant (R21) mechanism is intended to encourage exploratory and developmental research projects by providing support for the early and conceptual stages of these projects. These studies may involve considerable risk but may lead to a breakthrough in a particular area, or to the development of novel techniques, agents, methodologies, models, or applications that could have a major impact on a field of biomedical, behavioral, or clinical research.
Requirements: Eligible organizations: for profit organizations; non-profit organizations; public or private institutions, such as universities, colleges, hospitals and laboratories; units of state government; units of local government; eligible institutions of the federal government; domestic institutions; foreign institutions; faith-based or community-based organizations; units of state tribal government; and units of local tribal government. Eligible Project Directors/Principal Investigators: any individual with the skills, knowledge and resources necessary to carry out the proposed research, is invited to work with his/her institution to develop an application for support; individuals from underrepresented racial and ethnic groups; individuals with disabilities
Geographic Focus: All States
Date(s) Application is Due: Feb 16; Jun 16; Oct 16
Amount of Grant: Up to 275,000 USD
Contact: Ellen S. Liberman, (301) 451-2020; esl@nei.nih.gov
Internet: http://grants.nih.gov/grants/guide/pa-files/PA-06-181.html
Sponsor: National Eye Institute
5635 Fishers Lane, Suite 1300, MSC 9300
Bethesda, MD 20892-9300

NIH Exploratory Innovations in Biomedical Computational Science and Technology (BISTI) Grants 2878
The NIH is interested in promoting research and developments in computational science and technology that will support rapid progress in areas of scientific opportunity in biomedical research. As defined here, biomedical computing or biomedical information science and technology includes database design, graphical interfaces, querying approaches, data retrieval, data visualization and manipulation, data integration through

the development of integrated analytical tools, and tools for electronic collaboration, as well as computational and mathematical research including the development of structural, functional, integrative, and analytical models and simulations. This particular FOA is intended to support exploratory biomedical informatics and computational biology research - applications should be innovative, with high risk/high impact in new areas that are lacking preliminary data or development. Because the nature and scope of the proposed research will vary from application to application, it is anticipated that the size and duration of each award will also vary.

Requirements: Application can be made by the following institutions and organizations: for-profit organizations; non-profit organizations; public or private institutions (such as universities, colleges, hospitals, and laboratories); units of state government; units of local government; eligible agencies of the Federal government; foreign institutions; domestic institutions; faith-based or community-based organizations; Indian/Native American tribal governments (federally recognized); Indian/Native American tribal governments (other than federally recognized); and Indian/Native American tribally designated organizations.
Geographic Focus: All States
Date(s) Application is Due: Feb 16; Jun 16; Oct 16
Contact: Peter Lyster, (301) 451-6446; lysterp@mail.nih.gov
Internet: http://grants.nih.gov/grants/guide/pa-files/PAR-06-411.html
Sponsor: National Institutes of Health
45 Center Drive, Room 2AS.55k, MSC 6200
Bethesda, MD 20892-6200

NIH Fine Mapping and Function of Genes for Type 1 Diabetes Grants 2879
The purpose of this initiative is to bring together investigators with experience in genetics, immunology, and biochemistry to perform fine mapping of the loci discovered and study the function of the genes being identified and to identify the mechanisms by which newly discovered genes predispose to T1D. This would lead to elucidation of the mechanisms whereby changes in the function or regulation of these genes are likely to provide crucial new insights into disease pathogenesis. The discovery of the genes would be relevant to developing a predictive strategy for individuals who may develop diabetes. In the future, genetic testing may lead to personalized treatment regimens by identifying the most appropriate class of drugs for particular patients. Total funding available for this five year program is $20 million to fund 4-10 applications, contingent upon the availability of funds and the submission of a sufficient number of meritorious applications. The maximum direct costs are $5 million that can be spent on a project period of up to 5 years. Letters of Intent are due by February 27, with final applications due by March 30. The office strongly urges communication from applicants via email.
Requirements: The following organizations and institutions are eligible to apply: public/state controlled institutions of higher education; private institutions of higher education; Hispanic-serving institutions; historically Black colleges and universities (HBCUs); tribally controlled colleges and universities (TCCUs); Alaska Native and Native Hawaiian serving institutions; nonprofits with 501(c)3 IRS status (other than institutions of higher education); nonprofits without 501(c)3 IRS status (other than institutions of higher education); small businesses; for-profit organizations (other than small businesses); non-domestic (non-U.S.) entities (foreign organizations); and other eligible agencies of the federal government.
Geographic Focus: All States
Date(s) Application is Due: Feb 27; Mar 30
Amount of Grant: 1,500,000 - 5,000,000 USD
Contact: Beena Akolkar, PhD; (301) 594-8812; ba92i@nih.gov
Internet: http://grants.nih.gov/grants/guide/rfa-files/rfa-dk-08-006.html
Sponsor: National Institutes of Health
6707 Democracy Boulevard, Room 6105
Bethesda, MD 20892-5460

NIH Health Promotion Among Racial and Ethnic Minority Males 2880
This Funding Opportunity Announcement solicits Exploratory/Developmental grant applications from applicants that propose to stimulate and expand research in the health of minority men. Specifically, this initiative is intended to: enhance our understanding of the numerous factors (e.g., sociodemographic, community, societal, personal) influencing the health promoting behaviors of racial and ethnic minority males and their sub-populations across the life cycle; and solicit applications focusing on the development and testing of culturally and linguistically appropriate health-promoting interventions designed to reduce health disparities among racially and ethnically diverse males and their sub-populations age 21 and older. The total project period for an application submitted in response to this funding opportunity may not exceed two years. Direct costs are limited to $275,000 over an R21 two-year period, with no more than $200,000 in direct costs allowed in any single year.
Requirements: Application can be made by the following institutions and organizations: public/state controlled institutions of higher education; private institutions of higher education; nonprofits with 501(c)3 IRS status (other than institutions of higher education); nonprofits without 501(c)3 IRS status (other than institutions of higher education); small businesses; for-profit organizations (other than small businesses); state governments; U.S. territories or possessions; Indian/Native American tribal governments (federally recognized); Indian/Native American tribal governments (other than Federally recognized); Indian/Native American tribally designated organizations; non-domestic (non-U.S.) entities (foreign organizations); Hispanic-serving institutions; historically Black colleges and universities (HBCUs); tribally controlled colleges and universities (TCCUs); Alaska Native and Native Hawaiian serving institutions; regional organization; eligible agencies of the Federal government; and faith-based or community based organizations.
Geographic Focus: All States
Date(s) Application is Due: Feb 16; Jun 16; Oct 16
Amount of Grant: 200,000 - 275,000 USD

Contact: Dr. Charlotte Pratt; (301) 435-0382; fax (301) 480-5158; prattc@nhlbi.nih.gov
Internet: http://grants.nih.gov/grants/guide/pa-files/PA-07-421.html
Sponsor: National Institutes of Health
6701 Rockledge Drive, Suite 10118, MSC 7936
Bethesda, MD 20892-7936

NIH Human Connectome Project Grants 2881
This Funding Opportunity Announcement is issued as an initiative of the NIH Blueprint for Neuroscience Research. The Neuroscience Blueprint is a collaborative framework through which 16 NIH Institutes, Centers and Offices jointly support neuroscience-related research, with the aim of accelerating discoveries and reducing the burden of nervous system disorders. The overall purpose of this five year Human Connectome Project (HCP) is to develop and share knowledge about the structural and functional connectivity of the human brain. This purpose will be pursued through the following specific efforts: existing, but cutting-edge, non-invasive imaging technologies will be optimized and combined to acquire structural and functional in vivo data about axonal projections and neural connections from brains of hundreds of healthy adults. Demographic data and data regarding sensory, motor, cognitive, emotional, and social function will also be collected for each subject, as will DNA samples and blood (to establish cell lines). Models to better understand and use these data will be developed. Connectivity patterns will be linked to existing architectonic data. Data and models will be made available to the research community immediately via a user-friendly system to include tools to query, organize, visualize and analyze data. Outreach activities will be conducted to engage and educate the research community about the imaging tools, data, models, and informatics tools. After five years, these specific efforts are expected to deliver: a set of integrated, non-invasive imaging tools to obtain connectivity data from humans in vivo; a high quality and well characterized, quantitative set of human connectivity data linked to behavioral and genetic data as well as to general, existing architectonic data, and associated models, from up to hundreds of healthy adult female and male subjects; and rapid, user-friendly dissemination of connectivity data, models, and tools to the research community via outreach activities and an informatics platform. Letters of Intent are due by October 24, with final applications due by November 24.
Requirements: The following organizations and institutions are eligible to apply for support: public/state controlled institutions of higher education; private institutions of higher education; Hispanic-serving institutions; historically Black colleges and universities (HBCUs); tribally controlled colleges and universities (TCCUs); Alaska Native and Native Hawaiian serving institutions; nonprofits with 501(c)3 IRS status (other than institutions of higher education); nonprofits without 501(c)3 IRS status (other than institutions of higher education); Indian/Native American tribal governments (federally recognized); Indian/Native American tribally designated organizations; U.S. territories or possessions; and eligible agencies of the federal government.
Geographic Focus: All States
Date(s) Application is Due: Oct 24; Nov 24
Amount of Grant: Up to 6,000,000 USD
Contact: Michael F. Huerta, (301) 443-1815; fax (301) 443-1731; mhuert1@mail.nih.gov
Internet: http://grants.nih.gov/grants/guide/rfa-files/RFA-MH-10-020.html
Sponsor: National Institutes of Health
6001 Executive Boulevard, Room 7202, MSC 9645
Bethesda, MD 20892-9645

NIH Independent Scientist Award 2882
The award is a special salary-only grant designed to provide protected time for newly independent scientists who currently have non-research obligations such as heavy teaching loads, clinical work, committee assignments, service, and administrative duties that prevent them from having a period of intensive research focus. The award is targeted to persons with doctoral degrees who have completed their research training, have independent peer-reviewed research support, and who need a period of protected research time in order to foster their research career development. New applications are due February 12, June 12, and October 12, while renewal applications are due March 12, July 12, and November 12.
Requirements: The following organizations and institutions are eligible to apply: public/state controlled institutions of higher education; private institutions of higher education; nonprofits with 501(c)3 IRS status; nonprofits without 501(c)3 IRS status; small businesses; for-profit organizations; State governments; U.S. territories or possessions; Indian/Native American tribal governments (Federally recognized and other than Federally recognized); Indian/Native American tribally designated organizations; Hispanic-serving institutions; historically Black colleges and universities (HBCUs); tribally controlled colleges and universities (TCCUs); Alaska Native and Native Hawaiian serving institutions; regional organizations; and faith-based or community based organizations. The candidate must have a doctoral degree and peer-reviewed, independent research support at the time the award is made. The candidate must spend a minimum of 75 percent effort conducting research during the period of the award.
Restrictions: The award is not intended for investigators who already have full time to perform research, or have substantial publication records or considerable research support indicating that they are well established in their fields. Foreign institutions are not eligible to apply.
Geographic Focus: All States
Date(s) Application is Due: Feb 12; Mar 12; Jun 12; Jul 12; Oct 12; Nov 12
Amount of Grant: 75,000 - 100,000 USD
Contact: Susan Grove; (301) 435-0714; sg16d@nih.gov or grantsinfo@od.nih.gov
Internet: http://grants.nih.gov/grants/guide/contacts/PA-09-038_contacts.html
Sponsor: National Institutes of Health
6701 Rockledge Drive
Bethesda, MD 20892

NIH Innovations in Biomedical Computational Science & Technology Grants 2883
The NIH is interested in promoting research and developments in computational science and technology that will support rapid progress in areas of scientific opportunity in biomedical research. As defined here, biomedical computing or biomedical information science and technology includes database design, graphical interfaces, querying approaches, data retrieval, data visualization and manipulation, data integration through the development of integrated analytical tools, and tools for electronic collaboration, as well as computational and mathematical research including the development of structural, functional, integrative, and analytical models and simulations. Because the nature and scope of the proposed research will vary from application to application, it is anticipated that the size and duration of each award will also vary.
Requirements: Application can be made by the following institutions and organizations: for-profit organizations; non-profit organizations; public or private institutions (such as universities, colleges, hospitals, and laboratories); units of state government; units of local government; eligible agencies of the Federal government; foreign institutions; domestic institutions; faith-based or community-based organizations; Indian/Native American tribal governments (federally recognized); Indian/Native American tribal governments (other than federally recognized); and Indian/Native American tribally designated organizations.
Geographic Focus: All States
Date(s) Application is Due: Feb 5; Jun 5; Oct 5
Amount of Grant: Up to 500,000 USD
Contact: Peter Lyster, (301) 451-6446; lysterp@mail.nih.gov
Internet: http://grants.nih.gov/grants/guide/pa-files/PAR-09-218.html
Sponsor: National Institutes of Health
45 Center Drive, Room 2AS.55k, MSC 6200
Bethesda, MD 20892-6200

NIH Innovations in Biomedical Computational Science and Technology Initiative Grants for SBIR 2884
This funding opportunity announcement (FOA) solicits Small Business Innovation Research (SBIR) grant applications from small business concerns (SBCs) that propose innovative research in biomedical computational science and technology to promote the progress of biomedical research. There exists an expanding need to speed the progress of biomedical research through the power of computing to manage and analyze data and to model biological processes. The NIH is interested in promoting research and developments in biomedical computational science and technology that will support rapid progress in areas of scientific opportunity in biomedical research. As defined here biomedical computing or biomedical information science and technology includes database design, graphical interfaces, querying approaches, data retrieval, data visualization and manipulation, data integration through the development of integrated analytical tools, and tools for electronic collaboration, as well as computational research including the development of structural, functional, integrative, and analytical models and simulations.
Requirements: Only United States small business concerns (SBCs) are eligible to submit SBIR applications.
Geographic Focus: All States
Date(s) Application is Due: Apr 5; Aug 5; Dec 5
Amount of Grant: 150,000 - 750,000 USD
Contact: Peter Lyster, (301) 451-6446; lysterp@mail.nih.gov
Internet: http://grants.nih.gov/grants/guide/pa-files/PAR-07-160.html
Sponsor: National Institutes of Health
45 Center Drive, Room 2AS.55k, MSC 6200
Bethesda, MD 20892-6200

NIH Innovations in Biomedical Computational Science and Technology Initiative Grants for STTR 2885
This funding opportunity announcement (FOA) solicits Small Business Technology Transfer (STTR) grant applications from small business concerns (SBCs) that propose innovative research in biomedical computational science and technology to promote the progress of biomedical research. There exists an expanding need to speed the progress of biomedical research through the power of computing to manage and analyze data and to model biological processes. The NIH is interested in promoting research and developments in biomedical computational science and technology that will support rapid progress in areas of scientific opportunity in biomedical research. As defined here biomedical computing or biomedical information science and technology includes database design, graphical interfaces, querying approaches, data retrieval, data visualization and manipulation, data integration through the development of integrated analytical tools, and tools for electronic collaboration, as well as computational research including the development of structural, functional, integrative, and analytical models and simulations.
Requirements: Only United States small business concerns (SBCs) are eligible to submit STTR applications.
Geographic Focus: All States
Date(s) Application is Due: Apr 5; Aug 5; Dec 5
Amount of Grant: 150,000 - 750,000 USD
Contact: Peter Lyster, (301) 451-6446; lysterp@mail.nih.gov
Internet: http://grants.nih.gov/grants/guide/pa-files/PAR-07-161.html
Sponsor: National Institutes of Health
45 Center Drive, Room 2AS.55k, MSC 6200
Bethesda, MD 20892-6200

NIH Mentored Clinical Scientist Development Award 2886
This award enables candidates holding professional degrees (e.g., M.D., D.O., D.V.M., or equivalent degrees) to undertake 3 to 5 years of special study and supervised research with the goal of becoming independent investigators. The award also allows awardees to pursue a research career development program suited to their experience and capabilities under a mentor who is competent to provide guidance in the chosen research area. Institutions may submit applications on behalf of candidates who hold professional degrees. At least 2 years must have elapsed since the health professional degree was granted. Candidates can have varying levels of clinical training and research experience. New applications are due February 12, June 12, and October 12, while resubmitted applications are due March 12, July 12, and November 12.
Requirements: All candidates must be U.S. citizens, non-citizen nationals, or legal permanent residents of the U.S. Persons with temporary or student visas are not eligible. The grantee institution must have a strong, well-established research and research training program in the chosen area, accomplished faculty in the basic and clinical sciences, and a commitment to the candidate's research development. The proposed program should include an appropriate mentor.
Geographic Focus: All States
Date(s) Application is Due: Feb 12; Mar 12; Jun 12; Jul 12; Oct 12; Nov 12
Contact: Susan Grove; (301) 435-0714; sg16d@nih.gov or grantsinfo@od.nih.gov
Internet: http://grants.nih.gov/grants/guide/pa-files/pa-06-512.html
Sponsor: National Institutes of Health
6701 Rockledge Drive
Bethesda, MD 20892

NIH Mentored Clinical Scientist Research Career Development Award 2887
This award enables candidates holding professional degrees (e.g., M.D., D.O., D.V.M., or equivalent degrees) to undertake three to five years of special study and supervised research with the goal of becoming independent investigators. The award also allows awardees to pursue a research career development program suited to their experience and capabilities under a mentor who is competent to provide guidance in the chosen research area. Institutions may submit applications on behalf of candidates who hold professional degrees. At least 2 years must have elapsed since the health professional degree was granted. Candidates can have varying levels of clinical training and research experience. Up to $100,000 per year plus fringe benefits is available, and up to $25,000 per year for research support for new awards.
Requirements: Due to the difference in individual Institute and Center program requirements for this FOA, prospective applications should consult the Table of IC-Specific Information, Requirements and Staff Contacts, to make sure that their application is responsive to the requirements of one of the participating NIH Institute and Center programs. The candidate must have an MD degree or its equivalent, must have completed postgraduate clinical training and have secured a faculty appointment in an appropriate research-intensive environment, must identify a mentor with extensive research experience, and must be willing to spend a minimum of 75 percent of full-time professional effort conducting research and research career development. Applications may be submitted on behalf of candidates by domestic, non-federal organizations, public or private, such as medical schools. Minorities and women are encouraged to apply. Candidates must be U.S. citizens or non-citizen nationals, or must have been lawfully admitted for permanent residence and possess Alien Registration Receipt Card (I-551) or some other verification of legal admission as permanent residents.
Restrictions: Former principal investigators on NIH research project (R01), sub-projects of program project (P01), or center grants (P50), or the equivalent, are not eligible. A candidate for the award may not concurrently apply for any other PHS award that duplicates the provisions of this award nor have another application pending award. K08 recipients are encouraged to apply for independent research grant support during the period of this award. The K08 recipient would be allowed to maintain the award if other PHS support is procured, as long as the new support does not interfere with the ability to meet the MCSDA requirements.
Geographic Focus: All States, District of Columbia, Guam, Marshall Islands, Northern Mariana Islands, U.S. Virgin Islands, American Samoa
Amount of Grant: Up to 125,000 USD
Contact: Alison Cole, Scientific Contact; (301) 594-3827 or (301) 496–7301; fax (301) 480-2802; colea@nigms.nih.gov
Internet: http://grants.nih.gov/grants/guide/pa-files/PA-16-191.html
Sponsor: National Institutes of Health
9000 Rockville Pike
Bethesda, MD 20892

NIH Mentored Patient-Oriented Research Career Development Award 2888
The purpose of this program is to support the career development of investigators who have made a commitment to focus their research endeavors on patient-oriented research. This mechanism provides support for three to five years of supervised study and research for clinically trained professionals who have the potential to develop into productive, clinical investigators focusing on patient-oriented research. Inquiries may be directed to program staff at the appropriate NIH institute. See the following web page for specific contact information: http://grants.nih.gov/grants/guide/contacts/PA-09-043_contacts.html
Requirements: Applicants must have a clinical doctoral degree or its equivalent and must have completed their clinical training, specialty and, if applicable, sub-specialty training prior to receiving an award. However, candidates may submit an application prior to the completion of clinical training. Candidates must identify a mentor with extensive research experience, and must be willing to spend a minimum of 75 percent of full-time professional effort conducting research career development and clinical research.
Restrictions: Current and former principal investigators on NIH research project (R01), FIRST Awards (R29), comparable career development awards (K01, K07, or K08), sub-projects of program project (P01) or center grants (P50), and the equivalent are not eligible for funding under this grant.
Geographic Focus: All States

Contact: Susan Grove; (301) 435-0714; sg16d@nih.gov or grantsinfo@od.nih.gov
Internet: http://grants.nih.gov/grants/guide/pa-files/PA-09-043.html#SectionVII
Sponsor: National Institutes of Health
6701 Rockledge Drive
Bethesda, MD 20892

NIH Mentored Research Scientist Development Awards 2889
The overall objective of the NIH Research Career Development Award program is to prepare qualified individuals for careers that have a significant impact on the health-related research needs of the Nation. The objective of the NIH Mentored Research Scientist Development Award is to provide support for a sustained period of protected time for intensive research career development under the guidance of an experienced mentor, or sponsor, in the biomedical, behavioral or clinical sciences leading to research independence. The expectation is that through this sustained period of research career development and training, awardees will launch independent research careers and become competitive for new research project grant funding. Inquiries should be directed to program staff at the appropriate NIH institute. See the website for specific contact information.
Requirements: Applications may be submitted, on behalf of candidates, by domestic, non-federal organizations, public or private, such as medical, dental, or nursing schools or other institutions of higher education. Candidates must be U.S. citizens or non-citizen nationals, or must have been lawfully admitted for permanent residence by the time of award. Candidates must have a research or a health-professional doctorate or its equivalent; have demonstrated the capacity or potential for highly productive independent research in the period after the doctorate; must identify a mentor with extensive research experience; and must be willing to spend a minimum of 75 percent of full-time professional effort conducting research and research career development during the entire award period.
Restrictions: Individuals on temporary or student visas are not eligible for this award.
Geographic Focus: All States
Date(s) Application is Due: Feb 12; Mar 12; Jun 12; Jul 12; Oct 12; Nov 12
Amount of Grant: 75,000 - 100,000 USD
Contact: Susan Grove; (301) 435-0714; sg16d@nih.gov or grantsinfo@od.nih.gov
Internet: http://grants.nih.gov/grants/guide/pa-files/PA-09-040.html
Sponsor: National Institutes of Health
6701 Rockledge Drive
Bethesda, MD 20892

NIH Nonhuman Primate Immune Tolerance Cooperative Study Group Grants 2890
NIAID and NIDDK invite applications from single institutions or consortia of institutions to participate in a multi-center, cooperative research program to evaluate existing and new tolerance induction treatment regimens and to elucidate the underlying mechanisms of the induction, maintenance and/or loss of tolerance in nonhuman primate models of kidney and islet transplantation. The goals of this research program are to evaluate further the safety, toxicity, and efficacy of existing tolerance induction regimens; define the underlying mechanisms of action of the therapeutic approaches under investigation; and develop and validate immune and/or surrogate markers of the induction, maintenance, and loss of tolerance, graft function, graft acceptance, and graft survival. The deadline for Letters of Intent is August 22, with full applications due September 22.
Requirements: Applications may be submitted by domestic for-profit and nonprofit organizations and public and private institutions.
Restrictions: Foreign institutions are not eligible to apply. This RFA will not support: studies in animal models other than non-human primates; human studies or trials; xenotransplantation; preliminary development of a non-human primate transplantation model; studies to improve the isolation, preservation, or supply of organs, tissues or cells; hematopoietic stem cell transplantation, unless in the context of kidney or islet transplantation (i.e. bone marrow chimerism experiments); heart or lung transplantation models; or studies of non-immunological side effects of an immuno-modulatory agent.
Geographic Focus: All States
Date(s) Application is Due: Aug 22; Sep 22
Amount of Grant: Up to 67,000,000 USD
Contact: Kristy Kraemer, Division of Allergy, Immunology and Transplantation; (301) 496-5598; fax (301) 480-0693; kkraemer@niaid.nih.gov
Internet: http://grants.nih.gov/grants/guide/rfa-files/RFA-AI-06-018.html
Sponsor: National Institutes of Health
6610 Rockledge Drive, Room 3043, MSC-6601
Bethesda, MD 20892

NIH Receptors and Signaling in Bone in Health and Disease 2891
The objective of this initiative is to elicit grant submissions that focus on systemic hormones, local growth factors, and bone-active cytokines, their receptors and mechanisms of signaling in bone. While the primary focus is on basic research, the long-term emphasis is on identifying mechanisms or processes related to hormone action with potential applicability as targets for therapeutic agents that may have efficacy in the treatment of diseases that adversely affect bone, such as osteoporosis and primary hyperparathyroidism. Modular Grant applications will request direct costs in $25,000 modules, up to a total direct cost request of $250,000 per year.
Requirements: Applications may be submitted by domestic and foreign for-profit and nonprofit organizations, public and private, such as universities, colleges, hospitals, laboratories, units of State and local governments, and eligible agencies of the Federal Government. Racial/ethnic minority individuals, women, and persons with disabilities are encouraged to apply as principal investigators.
Geographic Focus: All States
Date(s) Application is Due: Feb 5; Jun 5; Oct 5
Amount of Grant: Up to 250,000 USD
Contact: Susan Grove; (301) 435-0714; sg16d@nih.gov or grantsinfo@od.nih.gov
Internet: http://grants.nih.gov/grants/guide/pa-files/PA-00-017.html
Sponsor: National Institutes of Health
6701 Rockledge Drive
Bethesda, MD 20892

NIH Recovery Act Limited Competition: Academic Research Enhancement Awards 2892
The purpose of the Academic Research Enhancement Award (AREA) program is to stimulate research in educational institutions that provide baccalaureate or advanced degrees for a significant number of the Nation's research scientists, but that have not been major recipients of NIH support. These AREA grants create opportunities for scientists and institutions otherwise unlikely to participate extensively in NIH programs, to contribute to the Nation's biomedical and behavioral research effort. AREA grants are intended to support small-scale health-related research projects proposed by faculty members of eligible, domestic institutions. The requested budget is limited to $300,000 total costs for the entire three-year project (maximum time period). Applicants should contact individuals at the specific departments listed on the web site. No Letters of Intent are necessary, though full applications should arrive by September 24.
Requirements: The following institutions and organizations may apply: public/state controlled institutions of higher education; private institutions of higher education; Hispanic-serving institutions; historically Black colleges and universities (HBCUs); tribally controlled colleges and universities (TCCUs); Alaska Native and Native Hawaiian serving institutions; public or private institutions and components of institutions defined below as health professional schools or colleges; and other academic components.
Restrictions: Foreign organizations/institutions are not permitted as the applicant organization.
Geographic Focus: All States
Date(s) Application is Due: Sep 24
Amount of Grant: Up to 300,000 USD
Contact: Susan Grove; (301) 435-0714; sg16d@nih.gov or grantsinfo@od.nih.gov
Internet: http://grants.nih.gov/grants/guide/rfa-files/RFA-OD-09-007.html
Sponsor: National Institutes of Health
6701 Rockledge Drive
Bethesda, MD 20892

NIH Recovery Act Limited Competition: Biomedical Research, Development, and Growth to Spur the Acceleration of New Technologies Pilot Grants 2893
This Funding Opportunity Announcement (FOA), supported by funds provided to the NIH under the American Recovery & Reinvestment Act of 2009 (Recovery Act or ARRA), Public Law 111-5, solicits grant applications for a new initiative called Biomedical Research, Development, and Growth to Spur the Acceleration of New Technologies (BRDG-SPAN) Pilot Program. The purpose of this pilot program is to address the funding gap between promising research and development (R&D) and transitioning to the market -- often called the Valley of Death -- by contributing to the critical funding needed by applicants to pursue the next appropriate milestone(s) toward ultimate commercialization; i.e., to carry out later stage research activities necessary to that end. This program aims to accelerate the transition of research innovations and technologies toward the development of products or services that will improve human health, help advance the mission of NIH and its Institutes and Centers (ICs), and create significant value and economic stimulus. This program also aims to foster partnerships among a variety of research and development (R&D) collaborators working toward these aims. The requested budget is limited to $1 million total costs per year for a maximum of three years. Applicants should contact individuals at the specific departments listed on the web site. Letters of Intent are due by August 3, and full applications should arrive by September 1.
Requirements: The applicant United States institution and organization must be located in the 50 states, territories and possessions of the U.S., Commonwealth of Puerto Rico, Trust Territory of the Pacific Islands, or District of Columbia.
Restrictions: Foreign organizations/institutions are not permitted as to apply.
Geographic Focus: All States
Date(s) Application is Due: Aug 3; Sep 1
Amount of Grant: Up to 1,000,000 USD
Contact: Susan Grove; (301) 435-0714; sg16d@nih.gov or grantsinfo@od.nih.gov
Internet: http://grants.nih.gov/grants/guide/rfa-files/RFA-OD-09-008.html
Sponsor: National Institutes of Health
6701 Rockledge Drive
Bethesda, MD 20892

NIH Recovery Act Limited Competition: Building Sustainable Community Linked Infrastructure to Enable Health Science Research 2894
Applications are invited from domestic (United States) institutions/organizations proposing to develop or expand needed infrastructures that will fundamentally transform collaboration and communication between academic health centers and local communities. Such collaborative infrastructures are essential to advance the health science research enterprise while ensuring that important research findings are effectively disseminated and implemented to improve public health and advance health care delivery. The primary intent is to fund projects to develop infrastructure for productive and sustainable academic-community research partnerships that can be leveraged in the future for efficiently conducting research that includes and is relevant to affected communities, and through which research findings can be disseminated in a manner that maximizes impact on public health. It is expected that each grantee will be able to leverage this Recovery Act funding into future research grants using the infrastructure partnership between itself and the participating community. Letters of Intent are due by November 12, with final applications due by December 11.

Requirements: The following organizations and institutions may apply: public/state controlled institutions of higher education; private institutions of higher education; Hispanic-serving institutions; historically Black colleges and universities (HBCUs); tribally controlled colleges and universities (TCCUs); Alaska native and native Hawaiian serving institutions; nonprofits with 501(c)3 IRS status; nonprofits without 501(c)3 IRS status; small businesses; for-profit organizations; State governments; regional organizations; U.S. territories or possessions; Indian/Native American tribal governments; Indian/Native American tribally designated organizations; county governments; city or township governments; special district governments; independent school districts; public housing authorities/Indian housing authorities; eligible agencies of the Federal government; and faith-based or community based organizations.
Restrictions: This RFA is not intended to fund research or evaluation projects, clinical trials, or public health campaigns. Foreign organizations/institutions are not permitted as the applicant organization.
Geographic Focus: All States
Date(s) Application is Due: Nov 12; Dec 11
Contact: Dr. Suzanne Heurtin-Roberts, (301) 402-2277; sheurtin@mail.nih.gov
Internet: http://grants.nih.gov/grants/guide/rfa-files/RFA-OD-09-010.html
Sponsor: National Institutes of Health
1 Center Drive, Room B3-11
Bethesda, MD 20892

NIH Recovery Act Limited Competition: Small Business Catalyst Awards for Accelerating Innovative Research Grants 2895

This NIH Funding Opportunity Announcement, supported by funds provided to the NIH under the American Recovery and Reinvestment Act of 2009 (Recovery Act or ARRA), Public Law 111-5, invites grant applications from small business concerns that propose to accelerate innovation through high risk, high reward research and development (R&D) that has commercial potential and is relevant to the mission of the NIH. The Small Business Catalyst Award is further expected to support entrepreneurs of exceptional creativity, drawn from scientific and technological environments beyond NIH, who propose pioneering and possibly transformative approaches to addressing major biomedical or behavioral challenges with the potential for downstream commercial development. The Small Business Catalyst Award for Accelerating Innovative Research funding opportunity seeks to encourage fresh research perspectives and approaches to serve the mission of NIH. In particular, applications from small business concerns without a history of NIH Small Business Innovation Research (SBIR) or Small Business Technology Transfer (STTR) support may receive funding priority. Solicited are applications for support for projects that have the potential to generate high impact results (e.g., products, processes or services) and/or innovative research applications, research tools, techniques, devices, inventions, or methodologies. The outcomes of the research supported should have potential to lead to products that will improve public health and create significant value and economic stimulus. This FOA solicits early-stage ideas that promise to lead to major leaps forward in capabilities important to serving the mission of NIH rather than incremental improvements of existing technologies. In accord with the funding priority of this initiative to attract applicants without a history of SBIR/STTR support from NIH, the focus of the projects solicited by this FOA is on early stage technology development. Budget requests are limited to $200,000 total costs for a maximum project period of one year. Applicants should contact individuals at the specific departments listed on the web site. Letters of Intent are due by August 3, and full applications should arrive by September 1.
Requirements: Only United States small business concerns (SBCs) are eligible to submit SBIR applications. A small business concern is one that, at the time of award of SBIR Phase I and Phase II, meets all of the following criteria: is organized for profit, with a place of business located in the United States, which operates primarily within the United States or which makes a significant contribution to the United States economy through payment of taxes or use of American products, materials or labor; is in the legal form of an individual proprietorship, partnership, limited liability company, corporation, joint venture, association, trust or cooperative, except that where the form is a joint venture, there can be no more than 49 percent participation by foreign business entities in the joint venture; is at least 51 percent owned and controlled by one or more individuals who are citizens of, or permanent resident aliens in, the United States, or it must be a for-profit business concern that is at least 51% owned and controlled by another for-profit business concern that is at least 51% owned and controlled by one or more individuals who are citizens of, or permanent resident aliens in, the United States, except in the case of a joint venture, where each entity to the venture must be 51 percent owned and controlled by one or more individuals who are citizens of, or permanent resident aliens in, the United States; and has, including its affiliates, not more than 500 employees.
Geographic Focus: All States
Date(s) Application is Due: Aug 3; Sep 1
Amount of Grant: 200,000 USD
Contact: Susan Grove; (301) 435-0714; sg16d@nih.gov or grantsinfo@od.nih.gov
Internet: http://grants.nih.gov/grants/guide/rfa-files/RFA-OD-09-009.html
Sponsor: National Institutes of Health
6701 Rockledge Drive
Bethesda, MD 20892

NIH Recovery Act Limited Competition: Supporting New Faculty Recruitment to Enhance Research Resources through Biomedical Research Core Centers Grants 2896

This initiative, which is supported by funds provided to the NIH under the Recovery Act, is designed to provide the necessary resources for U.S. academic institutions/organizations to enhance their biomedical research efforts through the development of Biomedical Research Core Centers. For this announcement, Biomedical Core Centers are defined as a community of multidisciplinary researchers focusing on areas of biomedical research relevant to NIH, such as centers, departments, programs, and/or trans-departmental collaborations or consortia. These Core Centers are designed to provide scientific and programmatic support for promising research faculty and their areas of research. The NIH invites applications that include plans to recruit and hire investigators to conduct biomedical research in all scientific disciplines, including the field of bioethics. Applicants should contact individuals at the specific departments listed on the web site. Letters of Intent are due by April 29, and full applications should arrive by May 29.
Requirements: The following institutions and organizations may apply: public/state controlled institutions of higher education; private institutions of higher education; Hispanic-serving institutions; historically Black colleges and universities (HBCUs); tribally controlled colleges and universities (TCCUs); Alaska Native and Native Hawaiian serving institutions; nonprofits with 501(c)3 IRS status (other than institutions of higher education); nonprofits without 501(c)3 IRS status (other than institutions of higher education); and Indian/Native American tribally designated organizations .
Restrictions: Foreign organizations/institutions are not permitted as the applicant organization.
Geographic Focus: All States
Date(s) Application is Due: Apr 29; May 29
Contact: Susan Grove; (301) 435-0714; sg16d@nih.gov or grantsinfo@od.nih.gov
Internet: http://grants.nih.gov/grants/guide/rfa-files/RFA-OD-09-005.html
Sponsor: National Institutes of Health
6701 Rockledge Drive
Bethesda, MD 20892

NIH Research On Ethical Issues In Human Subjects Research 2897

The purpose of this funding opportunity announcement is to solicit research addressing the ethical challenges of human subjects research in order to optimize the protection of human subjects and enhance the ethical conduct of human subjects research. The research design for studies on ethical issues in human subjects research should be appropriate to the nature of the project(s) proposed and the disciplines involved. Given the conceptual and methodological complexity of many of these research questions, interdisciplinary and collaborative projects are encouraged, particularly those involving clinical researchers, ethicists, and behavioral/social scientists.
Requirements: The following organizations and institutions may apply: public/state controlled institutions of higher education; private institutions of higher education; nonprofits with 501(c)3 IRS status (other than institution of higher education); nonprofits without 501(c)3 IRS status (other than institution of higher education); small businesses; for-profit organizations (other than small business); State governments; U.S. territories or possessions; Indian/Native American tribal governments (Federally recognized); Indian/Native American tribal governments (other than Federally recognized); Indian/Native American tribally designated organizations; non-domestic (non-U.S.) entities (foreign organization); eligible agencies of the Federal government; and faith-based or community based organizations.
Geographic Focus: All States
Date(s) Application is Due: Feb 5; Jun 5; Oct 5
Contact: Kim Witherspoon, (301) 496-8866; withersk@ctep.nci.nih.gov
Internet: http://grants.nih.gov/grants/guide/pa-files/PA-07-277.html
Sponsor: National Institutes of Health
6130 Executive Boulevard, Executive Plaza N, Room 7009, MSC 7432
Bethesda, MD 20892-7432

NIH Research on Sleep and Sleep Disorders 2898

This Funding Opportunity Announcement (FOA) solicits grant applications proposing research to advance biomedical knowledge related to sleep or sleep disorders, improve understanding of the neurobiology or functions of sleep over the life-span, enhance timely diagnosis and effective treatment for individuals affected by sleep-related disorders, or implement and evaluate innovative community-based public health education and intervention programs. Because the nature and scope of the proposed research will vary from application to application, it is anticipated that the size and duration of each award will also vary. The total amount awarded and the number of awards will depend upon the mechanism numbers, quality, duration, and costs of the applications received.
Requirements: The following organizations and institutions may apply: public/state controlled institutions of higher education; private institutions of higher education; nonprofits with 501(c)3 IRS status; nonprofits without 501(c)3 IRS status; small businesses; State governments; non-domestic (non-U.S.) entities (foreign organizations); eligible agencies of the Federal government; and faith-based or community based organizations.
Geographic Focus: All States
Date(s) Application is Due: Feb 5; Jun 5; Oct 5
Contact: Susan Grove; (301) 435-0714; sg16d@nih.gov or grantsinfo@od.nih.gov
Internet: http://grants.nih.gov/grants/guide/pa-files/pa-07-140.html
Sponsor: National Institutes of Health
6701 Rockledge Drive
Bethesda, MD 20892

NIH Research Project Grants 2899

The Research Project Grant (R01) is an award to support a discrete, specified, circumscribed project to be performed by named Project Directors/Principal Investigators (PDs/PIs) in areas representing the investigators' specific interests and competencies, based on the mission of the NIH. The R01 is the original and historically the oldest grant mechanism used by the NIH to support health-related research and development. The NIH awards R01 grants to institutions/organizations of all types. This mechanism allows the PDs/PIs to define the scientific focus or objective of the research based on particular areas of interest and competence. Although the PDs/PIs write the grant application and are responsible for conducting and supervising the research, the actual applicant is the research institution/organization. Research grant applications are assigned to an NIH IC based on receipt and referral guidelines and many applications are assigned to multiple ICs with related research interests. Each IC maintains a Web site with funding opportunities and areas of interest. Contact with an IC representative may help focus the research plan based on an understanding of the mission of the IC. For specific information about the mission of each NIH IC, see http://www.nih.gov/icd, which provides a brief summary of the research interests in each IC and access to individual IC home pages.

Requirements: The following organizations and institutions may apply: public/state controlled institutions of higher education; private institutions of higher education; Hispanic-serving institutions; historically Black colleges and universities (HBCUs); tribally controlled colleges and universities (TCCUs); Alaska native and native Hawaiian serving institutions; nonprofits with 501(c)3 IRS status; nonprofits without 501(c)3 IRS status; small businesses; for-profit organizations; State governments; regional organizations; U.S. territories or possessions; Indian/Native American tribal governments; Indian/Native American tribally designated organizations; eligible agencies of the Federal government; and faith-based or community based organizations.

Geographic Focus: All States
Date(s) Application is Due: Feb 5; Jun 5; Oct 5
Contact: Grants Coordinator; (301) 435-0714; grantsinfo@od.nih.gov
Internet: http://grants.nih.gov/grants/guide/pa-files/PA-07-070.html
Sponsor: National Institutes of Health
6100 Executive Boulevard
Bethesda, MD 20892

NIH Ruth L. Kirschstein National Research Service Award Institutional Research Training Grants 2900

This award enables research institutions to support predoctoral research training in specific areas and fields of shortage. Selection of institutions is by national competition; trainees are selected through local review procedures established by the program director at the grantee institution. Trainees are required to pursue their research training full-time, and trainees in clinical areas are expected to confine their clinical duties to those that are part of their research training. Institutional grants are renewable, through competition, for up to 5 years. The NHLBI will accept all types of competing T32 applications (new, renewal/competing continuation, resubmission/amended) until January 25. Only resubmission/amended T32 applications will be accepted. Because of the difference in individual Institute and Center program requirements for this FOA, prospective applications must consult the Table of IC-Specific Information, Requirements, and Staff Contacts, to make sure that their application is responsive to the requirements of one of the participating NIH Institute and Centers.

Requirements: Trainees must be U.S. citizens, noncitizen nationals, or legal permanent residents of the U.S. Persons on temporary or student visas are not eligible. At the time of appointment, trainees must have received a baccalaureate degree and must be training at the graduate level in a program leading to the Ph.D., Sc.D., or other equivalent degree. Trainees who wish to interrupt their medical, dental, veterinary, or other professional school studies to engage in full-time research training before completing their professional degrees are also eligible. Applications requesting $500,000 or more in direct costs in any given year must receive prior approval from the NHLBI before submitting an application.

Restrictions: These awards do not support study leading to the M.D., D.O., D.D.S., or other similar professional degrees, nor do they support residency training.

Geographic Focus: All States, District of Columbia, Guam, Marshall Islands, Northern Mariana Islands, Puerto Rico, U.S. Virgin Islands, American Samoa
Contact: Jane Scott, Scientific Program Contact; (301) 435-0535 or (301) 592-8573; scottj2@nhlbi.nih.gov or nhlbiinfo@nhlbi.nih.gov
Internet: http://grants.nih.gov/grants/guide/pa-files/PA-16-152.html
Sponsor: National Institutes of Health
9000 Rockville Pike
Bethesda, MD 20892

NIH Ruth L. Kirschstein National Research Service Awards (NRSA) for Individual Senior Fellows 2901

The objective of the National Research Service Award (NRSA) F33 program is to provide senior fellowship support to experienced scientists who wish to make major changes in the direction of their research careers or who wish to broaden their scientific background by acquiring new research capabilities as independent research investigators in scientific health-related fields relevant to the missions of the participating NIH Institutes and Centers (ICs). Senior fellowship applicants for the F33 award must include a research training proposal that offers an opportunity for individuals to broaden their scientific background or to extend their potential for research in health-related areas as independent researchers. These awards will enable individuals with at least seven years of research experience beyond the doctorate, and who have progressed to the stage of independent investigator, to take time from regular professional responsibilities for the purpose of receiving training to increase their scientific capabilities. In most cases, this award is used to support sabbatical experiences for established independent scientists seeking support for retraining or additional career development.

Requirements: The following organizations are eligible to apply: for-profit organizations; non-profit organizations; public or private institutions, such as universities, colleges, hospitals, and laboratories; eligible agencies and labs of the Federal government, including NIH intramural labs; domestic institutions; and foreign institutions.

Restrictions: This program is not designed for postdoctoral level investigators seeking to enhance their research experience prior to independence.

Geographic Focus: All States
Date(s) Application is Due: Apr 8; Aug 8; Dec 8
Contact: Nancy C. Lohrey, (301) 496-8580; fax (301) 402-4472; lohreyn@mail.nih.gov
Internet: http://grants.nih.gov/grants/guide/pa-files/PA-07-172.html
Sponsor: National Institutes of Health
6116 Executive Boulevard, Suite 7019, MSC 8346
Bethesda, MD 20892-8346

NIH Ruth L. Kirschstein National Research Service Awards for Individual Postdoctoral Fellowships 2902

The purpose of the National Research Service Award Act (NRSA) is to help ensure that diverse pools of highly trained scientists will be available in adequate numbers and in appropriate research areas to carry out the Nation's biomedical, behavioral and clinical research agendas. The National Institutes of Health (NIH) awards individual postdoctoral fellowships (F32) to promising applicants who have the potential to become productive, independent investigators in fields related to the mission of participating NIH Institutes and Centers. The proposed postdoctoral training must be within the broad scope of biomedical, behavioral, or clinical research or other specific disciplines relevant to the research mission of the participating NIH Institutes and Centers. The proposed training must offer an opportunity to enhance the fellow's understanding of the health-related sciences and extend his/her potential for a productive research career. Because of the difference in individual Institute and Center program requirements for this FOA, prospective applications must consult the Table of IC-Specific Information, Requirements, and Staff Contacts, to make sure that their application is responsive to the requirements of one of the participating NIH Institute and Centers.

Requirements: Eligible applicants include: county governments; independent school districts; public and state controlled institutions of higher education; for profit organizations other than small businesses; nonprofits that do not have a 501(c)3 status with the IRS, other than institutions of higher education; public housing authorities; Indian housing authorities; private institutions of higher education; city or township governments; special district governments; small businesses; nonprofits having a 501(c)3 status with the IRS, other than institutions of higher education; state governments; Native American tribal governments (federally recognized); Native American tribal organizations (other than Federally recognized); Alaska Native and Native Hawaiian serving institutions; Asian American Native American Pacific Islander serving institutions; eligible agencies of the federal government; faith-based organizations; community-based organizations; Hispanic-serving institutions; Historically Black Colleges and Universities; Indian/Native American tribal governments (other than federally recognized); non-domestic (non-U.S.) entities (foreign organizations); regional organizations; Tribally controlled colleges and universities; and U.S. territories or possessions.

Geographic Focus: All States, District of Columbia, Guam, Marshall Islands, Northern Mariana Islands, Puerto Rico, U.S. Virgin Islands, American Samoa
Contact: Leslie A. Frieden, Scientific Program Contact; (240) 276-5630 or (800) 422-6237; jakowles@mail.nih.gov
Internet: http://grants.nih.gov/grants/guide/pa-files/PA-16-307.html
Sponsor: National Institutes of Health
9000 Rockville Pike
Bethesda, MD 20892

NIH Ruth L. Kirschstein National Research Service Awards for Individual Predoctoral Fellows 2903

The objective of this funding opportunity announcement is to help ensure that highly trained scientists will be available in adequate numbers and in appropriate research areas to carry out the Nation's biomedical, behavioral, and clinical research agenda. The participating Institutes of the National Institutes of Health (NIH) provide individual predoctoral research training fellowship awards to promising doctoral candidates who have the potential to become productive, independent investigators in research fields relevant to the missions of these participating NIH Institutes and Centers. Each participating NIH Institute has a unique scientific purview and different program goals and initiatives that evolve over time. Prior to preparing an application, it is critical that all applicants consult the appropriate Institute website for details of research areas supported by that Institute. Applicants should also contact the appropriate Institute scientific/research contact to obtain current information about specific program priorities and policies. Because of the difference in individual Institute and Center program requirements for this FOA, prospective applications must consult the Table of IC-Specific Information, Requirements, and Staff Contacts, to make sure that their application is responsive to the requirements of one of the participating NIH Institute and Centers.

Requirements: Eligible applicants include: county governments; independent school districts; public and state controlled institutions of higher education; for profit organizations other than small businesses; nonprofits that do not have a 501(c)3 status with the IRS, other than institutions of higher education; public housing authorities; Indian housing authorities; private institutions of higher education; city or township governments; special district governments; small businesses; nonprofits having a 501(c)3 status with the IRS, other than institutions of higher education; state governments; Native American tribal governments (federally recognized); Native American tribal organizations (other than Federally recognized); Alaska Native and Native Hawaiian serving institutions; Asian American Native American Pacific Islander serving institutions; eligible agencies of the federal government; faith-based organizations; community-based organizations; Hispanic-serving institutions; Historically Black Colleges and Universities; Indian/Native American tribal governments (other than federally recognized); non-domestic (non-U.S.) entities (foreign organizations); regional organizations; Tribally controlled colleges and universities; and U.S. territories or possessions.

Geographic Focus: All States, District of Columbia, Guam, Marshall Islands, Northern Mariana Islands, Puerto Rico, U.S. Virgin Islands, American Samoa
Contact: John Williamson; (301) 496-2583 or (888) 644-6226; john.williamson@nih.gov

Internet: http://grants.nih.gov/grants/guide/pa-files/PA-16-309.html
Sponsor: National Institutes of Health
9000 Rockville Pike
Bethesda, MD 20892

NIH Ruth L. Kirschstein National Research Service Awards for Individual Predoctoral Fellowships to Promote Diversity in Health-Related Research   2904

The purpose of the F31 predoctoral fellowship to promote diversity in health-related research is to provide up to five years of support for research training leading to the PhD or equivalent research degree, the combined MD/PhD degree; or another formally combined professional degree and research doctoral degree in biomedical, behavioral, health services, or clinical sciences. These fellowships will enhance the diversity of the biomedical, behavioral, health services, and clinical research labor force in the United States by providing opportunities for academic institutions to identify and recruit students from diverse population groups to seek graduate degrees in health-related research and apply for this fellowship. The goal of this program is to increase the number of scientists from diverse population groups who are prepared to pursue careers in biomedical, behavioral, social, clinical, or health services research.

Requirements: The following organizations and institutions are eligible to apply: public/state controlled institutions of higher education; private institutions of higher education; Hispanic-serving institutions; historically Black colleges and universities (HBCUs); tribally controlled colleges and universities (TCCUs); Alaska Native and Native Hawaiian serving institutions; nonprofits with 501(c)3 IRS status (other than institutions of higher education); nonprofits without 501(c)3 IRS status (other than institutions of higher education); for-profit organizations (other than small businesses); non-domestic (non-U.S.) entities (foreign organizations); and other eligible agencies of the Federal Government.
Geographic Focus: All States
Date(s) Application is Due: Apr 8; Aug 8; Dec 8
Contact: Susan Grove; (301) 435-0714; sg16d@nih.gov or grantsinfo@od.nih.gov
Internet: http://grants.nih.gov/grants/guide/pa-files/PA-09-209.html
Sponsor: National Institutes of Health
6701 Rockledge Drive
Bethesda, MD 20892

NIH Ruth L. Kirschstein National Research Service Awards for Individual Predoctoral MD/PhD and Other Dual Doctoral Degree Fellows   2905

The purpose of the Ruth L. Kirschstein National Research Service Awards (Kirschstein-NRSA) is to provide support to individuals for combined MD/PhD and other dual doctoral degree training (e.g. DO/PhD, DDS/PhD, AuD/PhD). The participating Institutes award this Kirschstein-NRSA individual fellowship to qualified applicants with the potential to become productive, independent, highly trained physician-scientists and other clinician-scientists, including patient-oriented researchers in their scientific mission areas. This funding opportunity supports individual predoctoral F30 fellowships with the expectation that these training opportunities will increase the number of future investigators with both clinical knowledge and skills in basic, translational or clinical research.

Requirements: The following organizations and institutions are eligible to apply: public/state controlled institutions of higher education; private institutions of higher education; Hispanic-serving institutions; historically Black colleges and universities (HBCUs); tribally controlled colleges and universities (TCCUs); Alaska Native and Native Hawaiian serving institutions; nonprofits with 501(c)3 IRS status (other than institutions of higher education); nonprofits without 501(c)3 IRS status (other than institutions of higher education); for-profit organizations (other than small businesses); and other eligible agencies of the Federal Government.
Restrictions: Non-domestic (non-U.S.) entities (foreign organizations) are not eligible.
Geographic Focus: All States
Date(s) Application is Due: Apr 8; Aug 8; Dec 8
Contact: Nancy C. Lohrey, (301) 496-8580; fax (301) 402-4472; lohreyn@mail.nih.gov
Internet: http://grants.nih.gov/grants/guide/pa-files/PA-09-207.html
Sponsor: National Institutes of Health
6116 Executive Boulevard, Suite 7019, MSC 8346
Bethesda, MD 20892-8346

NIH Ruth L. Kirschstein National Research Service Award Short-Term Institutional Research Training Grants   2906

This program provides funds to research institutions to make awards to individuals in health professional schools for research opportunities that would not be available through their regular course of study. Awards are made to training institutions by national competition. Many of the NIH Institutes and Centers (ICs) use this grant mechanism exclusively to support intensive, short-term research training experiences for students in health professional schools during the summer. In addition, the Short-Term Institutional Research Training Grant may be used to support other types of predoctoral and postdoctoral training in focused, often emerging scientific areas relevant to the mission of the funding IC. Grants may be for project periods up to five years in duration and are renewable. Trainees selected for short-term training are required to pursue research training for 2-3 months on a full-time basis. Because of the difference in individual Institute and Center program requirements for this FOA, prospective applications must consult the Table of IC-Specific Information, Requirements, and Staff Contacts, to make sure that their application is responsive to the requirements of one of the participating NIH Institute and Centers.

Requirements: Public and private institutions of higher education; nonprofits with or without 501(c)3 IRS status (other than Institutions of Higher Education); Indian/Native American Tribal Governments; U.S. Territories or Possessions; Native American tribal organizations; and, Faith-based or Community-based Organizations are eligible to apply.

The following types of Higher Education Institutions are always encouraged to apply for NIH support as Public or Private Institutions of Higher Education: Hispanic-serving Institutions; Historically Black Colleges and Universities (HBCUs); Tribally Controlled Colleges and Universities (TCCUs); and, Alaska Native and Native Hawaiian Serving Institutions. The proposed training must be in basic, behavioral or clinical research aspects of the health-related sciences. This program is intended to encourage graduate and/or health professional students to pursue research careers by exposure to and short-term involvement in the health-related sciences. The training should be of sufficient depth to enable the trainees, upon completion of the program, to have a thorough exposure to the principles underlying the conduct of research. Trainees are selected by the grantee institution and must be U.S. citizens, noncitizen nationals, or legal permanent residents of the U.S. Trainees should have successfully completed at least one semester at an accredited school of medicine, optometry, osteopathy, dentistry, veterinary medicine, pharmacy, or public health before entering the program. The award can be used to support individuals who already have an M.S. or Ph.D. and have been accepted to health professional schools.
Restrictions: The award cannot be used to support courses that are required for the M.D., D.O., D.D.S., D.V.M., or similar professional degrees. Non-domestic (non-U.S.) Entities (Foreign Organizations) are not eligible to apply. Foreign (non-U.S.) components of U.S. Organizations are not allowed.
Geographic Focus: All States, Guam, Marshall Islands, Northern Mariana Islands, Puerto Rico, U.S. Virgin Islands, American Samoa
Contact: Charlotte Pratt, Ph.D., RD, Scientific Program Contact; (301) 435-0382 or (301) 592-8573; fax (301) 480-1864; prattc@nhlbi.nih.gov or nhlbiinfo@nhlbi.nih.gov
Internet: http://grants.nih.gov/grants/guide/pa-files/PA-16-151.html
Sponsor: National Institutes of Health
9000 Rockville Pike
Bethesda, MD 20892

NIH Sarcoidosis: Research into the Cause of Multi-Organ Disease and Clinical Strategies for Therapy Grants   2907

The purpose of this Funding Opportunity Announcement (FOA) is to stimulate research on the etiology and management of sarcoidosis, an immune-mediated granulomatous inflammatory disorder. Studies supported by this FOA would include investigations to find the etiology(ies) and related host factors that might enhance susceptibility to sarcoidosis, especially development of symptomatic multi-organ disease that has a propensity to involve critical organs (such as lungs, heart, eyes, central/peripheral nervous system, liver, kidneys and other abdominal viscera) that create serious illness. Study of the still problematic management and the biological, behavioral, or psychosocial burden on individuals, families, and community is encouraged. Moreover, there is interest for approaches to risk reduction, psychological coping, and management of complications and side effects of treatment.

Requirements: The following organizations and institutions may apply: public/state controlled institutions of higher education; private institutions of higher education; Hispanic-serving institutions; historically Black colleges and universities (HBCUs); tribally controlled colleges and universities (TCCUs); Alaska native and native Hawaiian serving institutions; nonprofits with 501(c)3 IRS status; nonprofits without 501(c)3 IRS status; small businesses; for-profit organizations; State governments; regional organizations; U.S. territories or possessions; Indian/Native American tribal governments; Indian/Native American tribally designated organizations; eligible agencies of the Federal government; and faith-based or community based organizations.
Geographic Focus: All States
Date(s) Application is Due: Feb 5; Jun 5; Oct 5
Contact: Herbert Y. Reynolds, (301) 435-0222; Reynoldh@nhlbi.nih.gov
Internet: http://grants.nih.gov/grants/guide/pa-files/PA-07-136.html
Sponsor: National Institutes of Health
6701 Rockledge Drive, Room 10018, MSC 7936
Bethesda, MD 20892-7952

NIH School-based Interventions to Prevent Obesity   2908

This Funding Opportunity Announcement (FOA) encourages the formation of partnerships between academic institutions and school systems in order to develop and implement controlled, school-based intervention strategies designed to reduce the prevalence of obesity in childhood. This FOA also encourages evaluative comparisons of different intervention strategies, as well as the use of methods to detect synergistic interactions between different types of interventions. Because the nature and scope of the proposed research will vary from application to application, it is anticipated that the size and duration of each award will also vary. The total amount awarded and the number of awards will depend upon the mechanism numbers, quality, duration, and costs of the applications received.

Requirements: The following organizations and institutions may apply: public/state controlled institutions of higher education; private institutions of higher education; Hispanic-serving institutions; historically Black colleges and universities (HBCUs); tribally controlled colleges and universities (TCCUs); Alaska native and native Hawaiian serving institutions; nonprofits with 501(c)3 IRS status; nonprofits without 501(c)3 IRS status; small businesses; for-profit organizations; State governments; regional organizations; U.S. territories or possessions; Indian/Native American tribal governments; Indian/Native American tribally designated organizations; eligible agencies of the Federal government; and faith-based or community based organizations.
Geographic Focus: All States
Date(s) Application is Due: Feb 5; Jun 5; Oct 5
Contact: Gilman Grave, (301) 496-5593; fax (301) 480-9791; graveg@mail.nih.gov
Internet: http://grants.nih.gov/grants/guide/pa-files/PA-07-180.html
Sponsor: National Institutes of Health
6100 Executive Boulevard, 4B-11, MSC 7510
Bethesda, MD 20892-7510

NIH Self-Management Strategies Across Chronic Diseases Grants 2909

The purpose of this Program Announcement (PA) is to solicit applications to expand research on established self-management interventions to multiple chronic diseases across the life-course. Interventions aimed at chronic disease self-management are numerous and many are well described in the literature. They are often presented as specific to a particular chronic disease. This PA encourages applicants to investigate the applicability of effective self-management interventions to a broader spectrum of chronic diseases. Chronic disease, for this announcement, is defined as illnesses that are prolonged, are rarely cured completely, and require self-management behaviors by affected individuals and/or their caretakers. Modular Grant applications will request direct costs in $25,000 modules, up to a total direct cost request of $250,000 per year.
Requirements: The following organizations and institutions may apply: public/state controlled institutions of higher education; private institutions of higher education; Hispanic-serving institutions; historically Black colleges and universities (HBCUs); tribally controlled colleges and universities (TCCUs); Alaska native and native Hawaiian serving institutions; nonprofits with 501(c)3 IRS status; nonprofits without 501(c)3 IRS status; small businesses; for-profit organizations; State governments; regional organizations; U.S. territories or possessions; Indian/Native American tribal governments; Indian/Native American tribally designated organizations; county governments; city or township governments; special district governments; independent school districts; public housing authorities/Indian housing authorities; eligible agencies of the Federal government; and faith-based or community based organizations.
Geographic Focus: All States
Date(s) Application is Due: Feb 5; Jun 5; Oct 5
Amount of Grant: Up to 250,000 USD
Contact: Susan Grove; (301) 435-0714; sg16d@nih.gov or grantsinfo@od.nih.gov
Internet: http://grants.nih.gov/grants/guide/pa-files/PA-00-109.html
Sponsor: National Institutes of Health
6701 Rockledge Drive
Bethesda, MD 20892

NIH Solicitation of Assays for High Throughput Screening (HTS) in the Molecular Libraries Probe Production Centers Network (MLPCN) 2910

he NIH Molecular Libraries Roadmap Initiative wishes to encourage HTS assay applications from investigators who have the interest and capability to work with the Molecular Libraries Probe Production Centers Network (MLPCN) for chemical probe development. This Funding Opportunity Announcement (FOA) promotes discovery and development of new chemical probes as research tools for use by scientists in both the public and private sectors to advance the understanding of biological functions and disease mechanisms. This initiative is one of the integrated components of the NIH Molecular Libraries Roadmap initiative that offers biomedical researchers access to large-scale automated high throughput screening (HTS) centers in the MLPCN, diverse compound libraries in the Small Molecule Repository (MLSMR) and information on biological activities of small molecules in the PubChem BioAssay public database. Letters of Intent (LOIs) are due on December 4, April 4, and August 3 respectively. Full application must be received by January 4, May 4, and September 3 respectively.
Requirements: The following organizations and institutions are eligible to apply: public/state controlled institutions of higher education; private institutions of higher education; Hispanic-serving institutions; historically Black colleges and universities (HBCUs); tribally controlled colleges and universities (TCCUs); Alaska native and native Hawaiian serving institutions; nonprofits with 501(c)3 IRS status (other than institutions of higher education); nonprofits without 501(c)3 IRS status (other than institutions of higher education); small businesses; for-profit organizations (other than small businesses); state governments; Indian/Native American tribal governments (federally recognized); Indian/Native American tribally designated organizations; county governments; city or township governments; special district governments; independent school districts; public housing authorities/Indian housing authorities; U.S. territories or possessions; Indian/native American tribal governments (other than federally recognized); regional organizations; non-domestic (non-U.S.) entities (foreign organizations); eligible agencies of the federal government; and faith-based or community-based organizations.
Geographic Focus: All States
Date(s) Application is Due: Jan 4; May 4; Sep 3
Amount of Grant: Up to 50,000 USD
Contact: Yong Yao, (301) 443-6102; yyao@mail.nih.gov
Internet: http://grants.nih.gov/grants/guide/pa-files/PAR-09-129.html#SectionIII
Sponsor: National Institutes of Health
6001 Executive Boulevard, Room 7175, MSC 9641
Bethesda, MD 20892-9641

NIH Structural Biology of Membrane Proteins Grants 2911

This PA solicits applications to develop research and methods to enhance the rate of membrane protein structure determination and to determine specific membrane protein structures. Innovative methods for expression, oligomerization, solubilization, stabilization, purification, characterization, crystallization, isotopic labeling, and structure determination of unique and biologically significant membrane proteins by x-ray diffraction, nuclear magnetic resonance (NMR), electron microscopic, mass spectrometry, and other biophysical techniques are encouraged. Projects that will lead in the near term to determining the structures of biologically important membrane proteins are also encouraged.
Requirements: Applicant organizations must be: for-profit organizations; non-profit organizations; public or private institutions, such as universities, colleges, hospitals, and laboratories; units of State government; units of local government; eligible agencies of the Federal government; foreign institutions; domestic institutions; faith-based agencies; or community-based organizations.
Geographic Focus: All States
Date(s) Application is Due: Feb 5; Jun 5; Oct 5
Contact: Peter C. Preusch, (301) 594-3827; preuschp@nigms.nih.gov
Internet: http://grants.nih.gov/grants/guide/pa-files/PA-06-119.html
Sponsor: National Institutes of Health
45 Center Drive, Building 45, 2AS.55E, MSC 6200
Bethesda, MD 20892

NIH Summer Internship Program in Biomedical Research 2912

Summer programs at the National Institutes of Health (NIH) provide an opportunity to spend a summer working side-by-side with some of the leading scientists in the world, in an environment devoted exclusively to biomedical research. The NIH consists of the 240-bed Mark O. Hatfield Clinical Research Center and more than 1,200 laboratories located on the main campus in Bethesda, Maryland, as well as in Baltimore and Frederick, Maryland; Research Triangle Park, North Carolina; Phoenix, Arizona; Hamilton, Momtana; and Detroit, Michigan. Awards cover a minimum of eight weeks, with students generally arriving at the NIH in May or June. The stipends for trainees are adjusted yearly, with supplements for prior experience. Prospective candidates must apply online. The application is available from mid-November to March 1.
Requirements: The Summer Internship Program is for students who will be sixteen years of age or older at the time they begin the program and who are currently enrolled at least half-time in high school or an accredited U.S. college or university. Students who have been accepted into a college or university program may also apply.
Restrictions: To be eligible, candidates must be U.S. citizens or permanent residents.
Geographic Focus: All States
Date(s) Application is Due: Mar 1
Contact: Arnetta Courtney, (301) 827-3724; arnetta.courtney@fda.hhs.gov
Internet: http://www.training.nih.gov/student/sip/info.asp
Sponsor: National Institutes of Health
31 Center Drive, Building 31, Room 9A06, MSC 2560
Bethesda, MD 20892-2560

NIH Support of Competitive Research (SCORE) Pilot Project Awards 2913

The objective of the SCORE program is to foster the development of faculty at minority serving institutions (MSIs) in order to increase their research competitiveness and promote their transition to non-SCORE external sources of funding. This objective is expected to translate into an increase in the number of individuals from groups underrepresented in biomedical and behavioral research professionally engaged in these areas of research, and an enhancement of an institution's research base. SCORE SC grants are offered to eligible MSIs. For the purposes of this program, eligible MSIs are those with more than 50% student enrollment of individuals from groups underrepresented in biomedical and behavioral research. The SC2 awards may not exceed $300,000 (direct costs) for the entire length of the award which may be one to three years maximum.
Requirements: Eligible applicants include: public or private post secondary educational institution, such as a university or college awarding associate, undergraduate or graduate degrees located in the United States of America or its territories including Puerto Rico, Guam and Virgin Islands with more than 50 percent student enrollment from groups underrepresented in the biomedical and behavioral sciences (such as African American, Hispanic American, Native American, Alaska Natives or natives of the U.S. Pacific Islands); an Indian tribe that has a recognized governing body and that performs substantial governmental functions, or an Alaska Regional Corporation (ARC), as defined in the Alaska Native Claims Settlement Act (43 U.S.C. 1601 et. Seq.); or an institution that has a significantly high student enrollment from groups underrepresented in the biomedical or behavioral sciences but less than 50% (if the Secretary of the Department of Health and Human Services, through the MBRS Chief, determines that the institution has demonstrated special commitment to the retention and graduation of students from groups underrepresented in the biomedical and behavioral sciences and to the hiring and retention of science faculty from underrepresented groups).
Restrictions: Foreign institutions are not eligible to apply.
Geographic Focus: All States, District of Columbia, Guam, Marshall Islands, Northern Mariana Islands, Puerto Rico, U.S. Virgin Islands, American Samoa
Date(s) Application is Due: Jan 25; May 25; Sep 7
Amount of Grant: 50,000 - 100,000 USD
Contact: Hinda Zlotnik, Research Contact; (301) 594-5132 or (301) 496–7301; fax (301) 480-2753; zlotnikh@nigms.nih.gov
Internet: http://grants.nih.gov/grants/guide/pa-files/PAR-13-069.html
Sponsor: National Institutes of Health
9000 Rockville Pike
Bethesda, MD 20892

NIH Support of Competitive Research Research Advancement Awards 2914

The objective of the SCORE program is to foster the development of faculty at minority serving institutions (MSIs) in order to increase their research competitiveness and promote their transition to non-SCORE external sources of funding. This objective is expected to translate into an increase in the number of individuals from groups underrepresented in biomedical and behavioral research professionally engaged in these areas of research, and an enhancement of an institution's research base. SCORE SC grants are offered to eligible MSIs. For the purposes of this program, eligible MSIs are those with more than 50% student enrollment of individuals from groups underrepresented in biomedical and behavioral research. The anticipated amount of individual SC1 awards is expected to range between $125,000 to $250,000 maximum direct costs per year.
Requirements: Eligible applicants include: public or private post secondary educational institution, such as a university or college awarding associate, undergraduate or graduate

degrees located in the United States of America or its territories including Puerto Rico, Guam and Virgin Islands with more than 50 percent student enrollment from groups underrepresented in the biomedical and behavioral sciences (such as African American, Hispanic American, Native American, Alaska Natives or natives of the U.S. Pacific Islands); an Indian tribe that has a recognized governing body and that performs substantial governmental functions, or an Alaska Regional Corporation (ARC), as defined in the Alaska Native Claims Settlement Act (43 U.S.C. 1601 et. Seq.); or an institution that has a significantly high student enrollment from groups underrepresented in the biomedical or behavioral sciences but less than 50% (if the Secretary of the Department of Health and Human Services, through the MBRS Chief, determines that the institution has demonstrated special commitment to the retention and graduation of students from groups underrepresented in the biomedical and behavioral sciences and to the hiring and retention of science faculty from underrepresented groups).
Restrictions: Foreign institutions are not eligible to apply.
Geographic Focus: All States, District of Columbia, Guam, Marshall Islands, Northern Mariana Islands, Puerto Rico, U.S. Virgin Islands, American Samoa
Date(s) Application is Due: Mar 4; May 25; Sep 25
Amount of Grant: 125,000 - 250,000 USD
Contact: Hinda Zlotnik, Research Contact; (301) 594-5132 or (301) 496–7301; fax (301) 480-2753; zlotnikh@nigms.nih.gov or info@nigms.nih.gov
Internet: http://grants.nih.gov/grants/guide/pa-files/PAR-14-019.html
Sponsor: National Institutes of Health
9000 Rockville Pike
Bethesda, MD 20892

NIMH AIDS and Aging: Behavioral Sciences Prevention Research Grants 2915
NIA, NIMH, and NINR invite qualified researchers to submit applications to investigate prevention issues relevant to AIDS in middle-aged and older populations. This announcement solicits AIDS prevention research proposals to study primary prevention of disease transmission as well as secondary and tertiary prevention of negative behavioral and social consequences of HIV/AIDS for persons with AIDS, their families, and communities. Thus, the primary goals are to identify social and behavioral factors associated with HIV transmission and disease progression in later life; examine behavioral and social consequences of HIV infection/AIDS across the life course; develop and evaluate age-appropriate behavioral and social interventions for preventing AIDS in middle-aged and older adults and/or ameliorating problems associated with older adult caregiver responsibilities and burdens; explore health care issues surrounding AIDS care; and strengthen existing research and evaluation methods.
Requirements: Applications may be submitted by domestic and foreign for-profit and non-profit organizations, public and private, such as universities, colleges, hospitals, laboratories, units of State and local governments, and eligible agencies of the Federal government. Applications may be submitted by single institutions or by a consortia of institutions.
Geographic Focus: All States
Date(s) Application is Due: Feb 5; Feb 16; Jun 5; Jun 16; Oct 5; Oct 16
Contact: Dr. Willo Pequegnat, Division of Mental Disorders, Behavioral Research and AIDS; (301) 443-6100; fax (301) 443-9719; wpequegn@nih.gov
Internet: http://www.grants.nih.gov/grants/guide/pa-files/PA-97-069.html
Sponsor: National Institute of Mental Health
5600 Fishers Lane, Parklawn Building
Rockville, MD 20857

NIMH Basic and Translational Research in Emotion Grants 2916
This Funding Opportunity Announcement encourages Research Project Grant applications to expand basic and translational research on the processes and mechanisms involved in the experience, expression, and regulation of emotion. The study of emotion encompasses a wide range of physiological, psychological, social, cognitive, and developmental phenomena. Emotional processes can be studied in human or animal subjects. Important objects of study include, but are not limited to: central and peripheral nervous system activity in the origins, expression, regulation and modulation of emotion; the contribution of emotional and motivational systems to cognitive faculties such as perception, attention, learning, and memory; investigations of overt behaviors, interpersonal relationships, communication and decision making; and the environmental circumstances and experiences that evoke and modulate different emotions.
Requirements: The following organizations and institutions may apply: public/state controlled institutions of higher education; private institutions of higher education; Hispanic-serving institutions; historically Black colleges and universities (HBCUs); tribally controlled colleges and universities (TCCUs); Alaska native and native Hawaiian serving institutions; nonprofits with 501(c)3 IRS status; nonprofits without 501(c)3 IRS status; small businesses; for-profit organizations; State governments; regional organizations; U.S. territories or possessions; Indian/Native American tribal governments; Indian/Native American tribally designated organizations; county governments; city or township governments; special district governments; independent school districts; public housing authorities/Indian housing authorities; eligible agencies of the Federal government; and faith-based or community based organizations.
Geographic Focus: All States
Date(s) Application is Due: Feb 5; Jun 5; Oct 5
Contact: Wendy Nelson, (301) 435 4590; fax (301) 435 7547; nelsonw@mail.nih.gov
Internet: http://grants.nih.gov/grants/guide/pa-files/PA-09-137.html
Sponsor: National Institute of Mental Health
6001 Executive Boulevard, Room 7179, MSC 9637
Bethesda, MD 20892-9637

NIMH Curriculum Development Award in Neuroinformatics Research and Analysis 2917
The purpose of this program is to encourage and support applications from individuals, with the requisite scientific expertise and leadership, for the development of courses and curricula designed to train interdisciplinary neuroinformatics scientists at U.S. educational institutions. The field of neuroinformatics combines neuroscience research with informatics research developed from the computer sciences, mathematics, physics, engineering or closely related sciences. It is anticipated that these courses or curricula would be useful to students and scientists who wish: to develop new conceptual approaches to basic and/or clinical neuroscientific research and analysis; or to acquire, store, retrieve, organize, manage, analyze, visualize, manipulate, integrate, synthesize, disseminate, and share data about the brain and behavior. Development of courses at the graduate and undergraduate level is encouraged. As part of this program, awardees will be expected to develop and implement the courses or curricula in their institution. It is expected that such courses and curricula will be models that could be transferable to other institutions in whole or in part.
Requirements: The principal investigator must be engaged in neuroscience research or research in computer science, mathematics, physics, engineering, or a related informatics field. Collaborator(s) who will contribute to the interdisciplinary nature of the courses or curricula must be identified. The principal investigator must be willing to spend at least 20 percent of full-time professional effort on course(s) and curricula development during the period of the award. The principal investigator must also identify appropriate researcher(s) who will agree to collaborate on the development of course(s) and curricula.
Geographic Focus: All States
Date(s) Application is Due: Feb 16; Jun 16; Oct 16
Contact: Stephen H. Koslow, (301) 443-1815; fax (301) 443-1867; koz@helix.nih.gov
Internet: http://grants.nih.gov/grants/guide/pa-files/PAR-99-135.html
Sponsor: National Institute of Mental Health
6001 Executive Boulevard, Room 6167, MSC 9613
Bethesda, MD 20892

NIMH Early Identification and Treatment of Mental Disorders in Children & Adolescents Grants 2918
This program supports research on early identification and treatment of mental disorders in children and adolescents. In particular, this announcement intends to encourage research on disorders such as schizophrenia, schizoaffective disorder, bipolar disorder, major depression, obsessive-compulsive disorder, and anorexia nervosa, alone or comorbid with other common mental or substance abuse disorders. PA: PA-00-094
Requirements: Applications may be submitted by domestic for-profit and nonprofit organizations, both public and private; units of state or local governments; and eligible agencies of the federal government.
Geographic Focus: All States
Contact: Shelli Avenevoli, (301) 443-5944; avenevos@mail.nih.gov
Internet: http://grants.nih.gov/grants/guide/notice-files/NOT-MH-08-006.html
Sponsor: National Institute of Mental Health
6001 Executive Boulevard
Bethesda, MD 20892

NIMH Jointly Sponsored Ruth L. Kirschstein National Research Service Award Institutional Predoctoral Training Program in the Neurosciences 2919
The Jointly Sponsored NIH Predoctoral Training Program in the Neurosciences supports broad and fundamental, early-stage graduate research training in the neurosciences via institutional NRSA research training grants (T32) at domestic institutions of higher education. Trainees are supported during years 1 and 2 of their graduate training when they are typically not committed to a dissertation laboratory. The primary objective is to prepare qualified individuals for careers in neuroscience that have a significant impact on the health-related research needs of the Nation. Because the nature and scope of the proposed research training will vary from application to application, it is anticipated that the size and duration of each award will also vary. The total amount awarded and the number of awards will depend upon the number, quality, duration, and costs of the applications received.
Requirements: The following organizations are eligible to apply: non-profit organizations; public or private institutions, such as universities, colleges, hospitals, and laboratories; and domestic institutions.
Geographic Focus: All States
Date(s) Application is Due: Apr 25; May 25
Contact: Lindsey Grandison, (301) 443-0606; fax (301) 443-1650; lgrandis@mail.nih.gov
Internet: http://grants.nih.gov/grants/guide/pa-files/PAR-08-101.html
Sponsor: National Institute of Mental Health
5635 Fishers Lane, MSC 9304
Bethesda, MD 20892-9304

NIMH Short Courses in Neuroinformatics 2920
The purpose of this program is to encourage and support short-term research for the development of short courses, seminars, and workshops on interdisciplinary neuroinformatics education . This short-term training will be provided to scientists seeking to combine knowledge about the various subdisciplines of neuroscience and behavioral science research with expertise in informatics research. It is anticipated that these short courses will allow the participants to acquire new conceptual approaches to basic neuroscience research and analyses; and learn to develop unique strategies for acquiring, storing, retrieving, organizing, managing, analyzing, visualizing, manipulating, integrating, synthesizing, disseminating, and sharing data about the brain and behavior.
Requirements: Any nonprofit or for-profit organization engaged in health-related education or research and located in the United States, its possessions, or territories is eligible to apply.

Geographic Focus: All States
Date(s) Application is Due: Feb 16; Jun 16; Oct 16
Amount of Grant: Up to 150,000 USD
Contact: Stephen H. Koslow, (301) 443-1815; fax (301) 443-1867; koz@helix.nih.gov
Internet: http://grants.nih.gov/grants/guide/pa-files/PAR-99-137.html
Sponsor: National Institute of Mental Health
6001 Executive Boulevard, Room 6167, MSC 9613
Bethesda, MD 20892

Nina Mason Pulliam Charitable Trust Grants 2921

The Nina Mason Pulliam Charitable Trust provides grants that focus on the areas of service Nina Pulliam supported during her lifetime: helping people in need; protecting animals and nature; enriching community life. The Trust awards grants for program projects and capital needs, and provides application opportunities three times during the calendar year. The Trust prefers to disperse funds as a one-year grant, but will consider projects of up to three years.
Requirements: Primary consideration is given to 501(c)3 charitable organizations that serve metropolitan Indianapolis and Phoenix. Secondary consideration is given to the states of Indiana and Arizona. National organizations whose programs benefit these priority areas and/or benefit society as a whole are occasionally considered.
Restrictions: Grants are not awarded for international purposes or academic research, Grants are not awarded to individuals, sectarian organizations for religious purposes, or to non-operating private foundations except in extraordinary circumstances.
Geographic Focus: Arizona, Indiana
Date(s) Application is Due: Jan 4; May 4; Sep 4
Samples: Chrysalis Academy of Life and Learning, $75,000 for a residential program to develop self-sufficiency among young men ages 18-24; Indiana Legal Services, to continue a project that helps individuals avoid losing their homes in inner-city Indianapolis neighborhoods because of abusive lending practices, $35,000; Camptown, Inc., to send 100 underprivileged youths on wilderness trips, $28,000; St. Francis Healthcare Foundation, Inc., $35,000 to hire an additional social worker to serve residents in the Garfield Park area.
Contact: David Hillman, (317) 231-6075; fax (317) 231-9208
Internet: http://www.ninapulliamtrust.org
Sponsor: Nina Mason Pulliam Charitable Trust
135 N Pennsylvania Street, Suite 1200
Indianapolis, IN 46204

NINDS Optimization of Small Molecule Probes for the Nervous System (SBIR) Grants 2922

This Funding Opportunity Announcement (FOA) encourages Small Business Innovation Research (SBIR) grant applications from small business concerns (SBCs) that propose to develop new small molecule probes for investigating biological function in the nervous system via the application of advanced medicinal chemistry and the biological testing of compounds. Eligible SBCs will have identified probe candidates via screening of small molecule collections, using in vitro assays of biological activity developed to interrogate these collections, and be able to show that the structural features of these small molecules are related to their biological activity. Applications should nominate small molecule probe candidates from distinct structural series for the further, iterative design and testing of analogues in structure-activity relationship studies, using in vitro assays of biological function adapted to the medium throughput screening requirements of this work. These studies should have the goal of developing a small molecule probe possessing the attributes (eg: affinity, selectivity, activity) required for its use in future pharmacological studies proposed by the SBC. Applicants are strongly encouraged to utilize publicly available cheminformatic capabilities for the acquisition of compounds, and semi-custom synthesis of analogues.
Requirements: Only United States small business concerns (SBCs) are eligible to submit SBIR applications.
Geographic Focus: All States
Date(s) Application is Due: Apr 5; Aug 5; Dec 5
Amount of Grant: Up to 150,000 USD
Contact: Mark Scheideler, (301) 496-1779; scheiderm@ninds.nih.gov
Internet: http://grants.nih.gov/grants/guide/pa-files/PAR-09-260.html
Sponsor: National Institute of Neurological Disorders and Stroke
6001 Executive Boulevard, NSC 2107
Bethesda, MD 20892-9527

NINDS Optimization of Small Molecule Probes for the Nervous System (STTR) Grants 2923

This Funding Opportunity Announcement (FOA) encourages Small Business Technology Transfer (STTR) grant applications from small business concerns (SBCs) that propose to develop new small molecule probes for investigating biological function in the nervous system via the application of advanced medicinal chemistry and the biological testing of compounds. Eligible SBCs will have identified probe candidates via screening of small molecule collections, using in vitro assays of biological activity developed to interrogate these collections, and be able to show that the structural features of these small molecules are related to their biological activity. Applications should nominate small molecule probe candidates from distinct structural series for the further, iterative design and testing of analogues in structure-activity relationship studies, using in vitro assays of biological function adapted to the medium throughput screening requirements of this work. These studies should have the goal of developing a small molecule probe possessing the attributes (eg: affinity, selectivity, activity) required for its use in future pharmacological studies proposed by the SBC. Applicants are strongly encouraged to utilize publicly available cheminformatic capabilities for the acquisition of compounds, and semi-custom synthesis of analogues, which is required of these studies.
Requirements: Only United States small business concerns (SBCs) are eligible to submit STTR applications.
Geographic Focus: All States
Date(s) Application is Due: Apr 5; Aug 5; Dec 5
Amount of Grant: Up to 150,000 USD
Contact: Mark Scheideler, (301) 496-1779; scheiderm@ninds.nih.gov
Internet: http://grants.nih.gov/grants/guide/pa-files/PAR-09-260.html
Sponsor: National Institute of Neurological Disorders and Stroke
6001 Executive Boulevard, NSC 2107
Bethesda, MD 20892-9527

NINDS Optimization of Small Molecule Probes for the Nervous System Grants 2924

This FOA issued by participating institutes of the National Institutes of Health, encourages research grant applications from institutions/ organizations that propose to develop new small molecule probes for investigating biological function in the nervous system via the application of advanced medicinal chemistry and the biological testing of compounds. Eligible investigators will have identified probe candidates via screening of small molecule collections, using in vitro assays of biological activity developed to interrogate these collections, and be able to show that the structural features of these small molecules are related to their biological activity. Proposals should nominate small molecule probe candidates from distinct structural series for the further, iterative design and testing of analogues in structure-activity relationship studies, using in vitro assays of biological function adapted to the medium throughput screening requirements of this work. These studies should have the goal of developing a small molecule probe possessing the attributes (eg: affinity, selectivity, activity) required for its use in future pharmacological studies proposed by the investigator. Applicants are strongly encouraged to utilize publicly available cheminformatic capabilities for the acquisition of compounds, and semi-custom synthesis of analogues.
Requirements: The following organizations and institutions are eligible to apply: public/state controlled institutions of higher education; private institutions of higher education; Hispanic-serving institutions; historically Black colleges and universities (HBCUs); tribally controlled colleges and universities (TCCUs); Alaska native and native Hawaiian serving institutions; nonprofits with 501(c)3 IRS status (other than institutions of higher education); nonprofits without 501(c)3 IRS status (other than institutions of higher education); small businesses; for-profit organizations (other than small businesses); state governments; Indian/Native American tribal governments (federally recognized); Indian/Native American tribally designated organizations; county governments; city or township governments; special district governments; independent school districts; public housing authorities/Indian housing authorities; U.S. territories or possessions; Indian/native American tribal governments (other than federally recognized); regional organizations; non-domestic (non-U.S.) entities (foreign organizations); eligible agencies of the federal government; and faith-based or community-based organizations.
Geographic Focus: All States
Date(s) Application is Due: Feb 16; Jun 16; Oct 16
Amount of Grant: Up to 150,000 USD
Contact: Mark Scheideler, (301) 496-1779; scheiderm@ninds.nih.gov
Internet: http://grants.nih.gov/grants/guide/pa-files/PAR-09-251.html
Sponsor: National Institute of Neurological Disorders and Stroke
6001 Executive Boulevard, NSC 2107
Bethesda, MD 20892-9527

NINDS Support of Scientific Meetings as Cooperative Agreements 2925

The purpose of the program announcement is to inform the scientific community that the National Institute of Neurological Disorders and Stroke (NINDS) will now support scientific meetings as cooperative agreements in addition to the current practice of supporting them through the traditional grant mechanism. This program provides guidelines for when it is appropriate to request support of a meeting as a cooperative agreement and explains procedures for preparing and submitting such applications. The Principal Investigator will have the primary authority and responsibility to define objectives and approaches; plan, publicize, and conduct the scientific meeting; and publish the results of the meeting.
Requirements: Applications may be submitted by U.S. institutions, including scientific or professional societies eligible to receive grants from Public Health Service (PHS) agencies. In the case of an international conference, the U.S. representative organization of an established international scientific or professional society is the eligible applicant.
Restrictions: Foreign institutions are not eligible.
Geographic Focus: All States
Contact: Danilo A. Tagle, (301) 496-5745; fax (301) 402-2060; tagled@ninds.nih.gov
Internet: http://grants.nih.gov/grants/guide/notice-files/NOT-NS-02-003.html
Sponsor: National Institute of Neurological Disorders and Stroke
6001 Executive Boulevard, Neuroscience Center, Room 2114
Bethesda, MD 20892-9525

NINR Acute & Chronic Care During Mechanical Ventilation Research Grants 2926

NINR seeks research applications for investigator-initiated research to improve the clinical management and care delivery of patients who require mechanical ventilation. For the purpose of this program announcement, the time course of mechanical ventilation is defined as beginning with endotracheal or tracheal tube intubation, maintenance of full and/or partial ventilatory support, discontinuation of mechanical ventilation via weaning techniques, and concluding with care of patients who require chronic mechanical ventilation. Two different, but related, types of basic science and/or clinical research on acute and/or chronic mechanical ventilation are encouraged: studies examining the effects of physiological, environmental, and/or psychological approaches to facilitate the

care of patients during any phase of mechanical ventilation, and studies directed toward examining the physical and psychological ramifications of technology dependency on patients and their families/significant others. Priority research areas include optimum ways to deliver patient care during mechanical ventilatory support; ethical issues related to initiation, maintenance, and withdrawal of life support technology; emotional support and communication during mechanical ventilation; methods of optimally managing the artificial airway; impact of care on family dynamics; and decision making regarding withdrawal of ventilatory support. Research is also needed to describe and intervene with care-giving resources, family functioning, and quality of life for patients on mechanical ventilation particularly chronically ill children and adults being maintained on home ventilators; optimum methods and techniques to deliver nursing care while critically ill patients receive mechanical ventilation; and determine optimal measures for promoting recovery as patients move through levels of care in acute, intermediate, and long-term care. This PA covers a wide range of topics relevant to clinical management and clinical problem solving issues for patients who require mechanical ventilation.
Requirements: Applications may be submitted by public and private, domestic and foreign, for-profit and nonprofit organizations; units of state and local governments; and eligible agencies of the federal government.
Restrictions: The total project period for an application may not exceed five years.
Geographic Focus: All States
Date(s) Application is Due: Feb 5; Feb 16; Jun 5; Jun 16; Oct 5; Oct 16
Contact: Josephine Boyington, (301) 594-2542; boyingtonje@mail.nih.gov
Internet: http://grants.nih.gov/grants/guide/pa-files/PA-99-003.html
Sponsor: National Institute of Nursing Research
31 Center Drive, Room 5B10
Bethesda, MD 20892-2178

NINR AIDS and Aging: Behavioral Sciences Prevention Research Grants 2927
NIA, NIMH, and NINR invite qualified researchers to submit applications to investigate prevention issues relevant to AIDS in middle-aged and older populations. This announcement solicits AIDS prevention research proposals to study primary prevention of disease transmission as well as secondary and tertiary prevention of negative behavioral and social consequences of HIV/AIDS for persons with AIDS, their families, and communities. Thus, the primary goals are to identify social and behavioral factors associated with HIV transmission and disease progression in later life; examine behavioral and social consequences of HIV infection/AIDS across the life course; develop and evaluate age-appropriate behavioral and social interventions for preventing AIDS in middle-aged and older adults and/or ameliorating problems associated with older adult caregiver responsibilities and burdens; explore health care issues surrounding AIDS care; and strengthen existing research and evaluation methods.
Requirements: Applications may be submitted by domestic and foreign for-profit and non-profit organizations, public and private, such as universities, colleges, hospitals, laboratories, units of State and local governments, and eligible agencies of the Federal government. Applications may be submitted by single institutions or by a consortia of institutions.
Geographic Focus: All States
Date(s) Application is Due: Feb 5; Feb 16; Jun 5; Jun 16; Oct 5; Oct 16
Contact: Josephine Boyington, (301) 594-2542; boyingtonje@mail.nih.gov
Internet: http://www.grants.nih.gov/grants/guide/pa-files/PA-97-069.html
Sponsor: National Institute of Nursing Research
31 Center Drive, Room 5B10
Bethesda, MD 20892-2178

NINR Chronic Illness Self-Management in Children and Adolescents Grants 2928
The purpose of this Funding Opportunity Announcement (FOA) issued by the NINR, NICHD, NIDDK, NCI, and the ODS is to solicit research to improve self-management and quality of life in children and adolescents with chronic illnesses. Biobehavioral studies of children within the context of family and family-community dynamics are encouraged. Children with a chronic disease and their families have a long-term responsibility for self-management. The child with the illness will have a life-long responsibility to maintain and promote health and prevent complications. Research related to sociocultural, environmental, and behavioral mechanisms as well as biological/technological factors that contribute to successful and ongoing self-management of particular chronic illnesses in children is encouraged. Because the nature and scope of the proposed research will vary from application to application, it is anticipated that the size and duration of each award will also vary.
Requirements: Application can be made by the following institutions and organizations: public/state controlled institutions of higher education; private institutions of higher education; nonprofits with 501(c)3 IRS status (other than institutions of higher education); nonprofits without 501(c)3 IRS status (other than institutions of higher education); small businesses; for-profit organizations (other than small businesses); state governments; U.S. territories or possessions; Indian/Native American tribal governments (federally recognized); Indian/Native American tribal governments (other than Federally recognized); Indian/Native American tribally designated organizations; non-domestic (non-U.S.) entities (foreign organizations); Hispanic-serving institutions; historically Black colleges and universities (HBCUs); tribally controlled colleges and universities (TCCUs); Alaska Native and Native Hawaiian serving institutions; regional organization; eligible agencies of the Federal government; and faith-based or community based organizations.
Restrictions: This FOA is restricted to studies of chronic illnesses in children and adolescents ages 8 to 21 grouped by developmental stages according to the discretion of the investigator. Studies of chronic mental illness or serious cognitive disability are beyond the scope of this FOA.
Geographic Focus: All States
Date(s) Application is Due: Feb 5; Jun 5; Oct 5

Contact: Dr. Linda S. Weglicki, (301) 594-6908; weglickils@mail.nih.gov
Internet: http://grants.nih.gov/grants/guide/pa-files/PA-07-097.html
Sponsor: National Institute of Nursing Research
31 Center Drive, Room 5B10
Bethesda, MD 20892-2178

NINR Clinical Interventions for Managing the Symptoms of Stroke Research Grants 2929
NINR, NINDS, NIA, NID.C.D, and NICHD seek research applications that address effective nonpharmacological approaches to managing the initial events and subsequent symptoms of a stroke. This research would focus on reducing the impairments, preventing secondary complications, and improving the functional independence and the quality of life for the individual following a stroke. Research areas that need to be explored include, but are not limited to: techniques to improve assessment during the initial stroke event and at each stage of recovery to guide nonpharmacologic treatment decisions and monitor patient progress; interventions to manage common symptoms, such as dysphagia, aphasia, bowel and bladder problems, sensorimotor deficits, depression, cognitive and perceptual deficits as well as difficulties in nutrition, hydration, skin integrity, and sleep; strategies to determine optimal timing, intensity, and duration of nonpharmacologic treatments; methods to prevent secondary complications, improve function, and address the sequelae of common comorbid conditions; strategies to promote self-care in both activities of daily living and instrumental activities of daily living and to enhance or maintain health promotion activities; measures to enhance family caregiving during all stages of recovery; methods to improve care during transition from one setting to another and care in specialized nurse managed units; and studies on the extent to which age and pre-existing conditions, such as specific diseases, sensory, motor, and cognitive declines, influence treatment outcome and management after stroke.
Requirements: Applications may be submitted by domestic and foreign, for-profit and nonprofit organizations, public and private.
Geographic Focus: All States
Amount of Grant: Up to 250,000 USD
Contact: Karin F. Helmers, (301) 594-2177; karin_helmers@nih.gov
Internet: http://www.grants.nih.gov/grants/guide/pa-files/PA-99-088.html
Sponsor: National Institute of Nursing Research
31 Center Drive, Room 5B10
Bethesda, MD 20892-2178

NINR Diabetes Self-Management in Minority Populations Grants 2930
This Program Announcement (PA) solicits applications for investigator-initiated research related to sociocultural, environmental, and behavioral mechanisms and biological/technological factors that contribute to successful and ongoing self-management of diabetes in minority populations. Applications that expand accepted intervention strategies in majority populations to minority populations are encouraged. Testing new interventions designed to promote self-management in minority diabetes populations will also be responsive to the PA. Self-management is defined as client strategies and behaviors that contribute to blood glucose normalization, improved health, and prevention or reduction of complications. The concept is broader than adherence to specific regimen components and incorporates deliberate problem solving and decision making processes. Applications are encouraged for both type 1 and type 2 diabetes and all age groups. Modular Grant applications will request direct costs in $25,000 modules, up to a total direct cost request of $250,000 per year.
Requirements: Applications may be submitted by domestic and foreign, for-profit and non-profit organizations, public and private, such as universities, colleges, hospitals, laboratories, units of State and local governments, and eligible agencies of the Federal government. Racial/ethnic minority individuals, women, and persons with disabilities are encouraged to apply as principal investigators.
Geographic Focus: All States
Date(s) Application is Due: Feb 5; Jun 5; Oct 5
Amount of Grant: Up to 250,000 USD
Contact: Dr. Karen Huss, (301) 594-5970; fax (301) 451-5649; husk@mail.nih.gov
Internet: http://grants.nih.gov/grants/guide/pa-files/PA-00-113.html
Sponsor: National Institute of Nursing Research
31 Center Drive, Room 5B10
Bethesda, MD 20892-2178

NINR Mechanisms, Models, Measurement, & Management in Pain Research Grants 2931
The purpose of this Funding Opportunity Announcement (FOA), Mechanisms, Models, Measurement, and Management in Pain Research, is to inform the scientific community of the pain research interests of the various Institutes and Centers (ICs) at the National Institutes of Health (NIH) and to stimulate and foster a wide range of basic, clinical, and translational studies on pain as they relate to the missions of these ICs. New advances are needed in every area of pain research, from the micro perspective of molecular sciences to the macro perspective of behavioral and social sciences. Although great strides have been made in some areas, such as the identification of neural pathways of pain, the experience of pain and the challenge of treatment have remained uniquely individual and unsolved. Furthermore, our understanding of how and why individuals transition to a chronic pain state after an acute insult is limited. Research to address these issues conducted by interdisciplinary and multidisciplinary research teams is strongly encouraged, as is research from underrepresented, minority, disabled, or female investigators. Direct costs are limited to $275,000 over a two-year period, with no more than $200,000 in direct costs allowed in any single year.
Requirements: The following organizations and institutions are eligible to apply: public/state controlled institutions of higher education; private institutions of higher education;

Hispanic-serving institutions; historically Black colleges and universities (HBCUs); tribally controlled colleges and universities (TCCUs); Alaska native and native Hawaiian serving institutions; nonprofits with 501(c)3 IRS status (other than institutions of higher education); nonprofits without 501(c)3 IRS status (other than institutions of higher education); small businesses; for-profit organizations (other than small businesses); state governments; Indian/Native American tribal governments (federally recognized); Indian/Native American tribally designated organizations; county governments; city or township governments; special district governments; independent school districts; public housing authorities/Indian housing authorities; U.S. territories or possessions; Indian/native American tribal governments (other than federally recognized); regional organizations; non-domestic (non-U.S.) entities (foreign organizations); eligible agencies of the federal government; and faith-based or community-based organizations.
Geographic Focus: All States
Date(s) Application is Due: Feb 16; Jun 16; Oct 16
Amount of Grant: Up to 275,000 USD
Contact: Susan Marden, (301) 496-9623; fax (301) 480-8260; mardens@mail.nih.gov
Internet: http://grants.nih.gov/grants/guide/pa-files/PA-10-007.html
Sponsor: National Institute of Nursing Research
31 Center Drive, Room 5B10
Bethesda, MD 20892-2178

NINR Mentored Research Scientist Development Award   2932
The objective of the NIH Mentored Research Scientist Development Award is to provide support for a sustained period of protected time for intensive research career development under the guidance of an experienced mentor, or sponsor, in the biomedical, behavioral or clinical sciences leading to research independence. The expectation is that through this sustained period of research career development and training, awardees will launch independent research careers and become competitive for new research project grant funding. NINR encourages mentored research scientist development in areas of symptom management, pulmonary, critical care, trauma, reproductive health, end-of-life and palliative care. NINR limits the length of the K01 award to a 3-year period and generally considers career development applications only from doctorally prepared applicants who will advance the science of nursing.
Requirements: The candidate must have a research or health-professional doctorate or its equivalent and must have demonstrated the capacity or potential for highly productive independent research in the period after the doctorate. The candidate must identify a mentor with extensive research experience and must spend a minimum of 75 percent of full-time professional effort conducting research for the period of the award. Applications may be submitted on behalf of candidates by domestic, nonfederal organizations, public or private, such as medical, dental, or nursing schools or other institutions of higher education. Candidates must be U.S. citizens, noncitizen nationals, or permanent residents.
Geographic Focus: All States
Date(s) Application is Due: Feb 12; Mar 12; Jun 12; Jul 12; Oct 12; Nov 12
Amount of Grant: 50,000 USD
Contact: David Banks, (301) 496-9558; fax (301) 480-8260; banksd@mail.nih.gov
Internet: http://grants.nih.gov/grants/guide/pa-files/PA-09-040.html
Sponsor: National Institute of Nursing Research
31 Center Drive, Room 5B10
Bethesda, MD 20892-2178

NINR Quality of Life for Individuals at the End of Life Grants   2933
The National Institute of Nursing Research (NINR) and 6 other ICs seek research grant applications that will generate scientific knowledge to improve the quality of life for individuals who are facing end-of-life issues and for their families. Research applications may include basic, clinical or care delivery studies focused on management of physical and psychological symptoms, patient-provider and patient-family communication, ethics and clinical decision-making, caregiver support, or the context of care delivery for those facing life-limiting illnesses. In a broad sense the purpose of this program announcement is to enhance the quality of life remaining for individuals who are nearing the end of their lives. Modular Grant applications will request direct costs in $25,000 modules, up to a total direct cost request of $250,000 per year.
Requirements: Applications may be submitted by domestic and foreign, for-profit and non-profit organizations, public and private, such as universities, colleges, hospitals, laboratories, units of State and local governments, and eligible agencies of the Federal government. Racial/ethnic minority individuals, women, and persons with disabilities are encouraged to apply as principal investigators.
Geographic Focus: All States
Date(s) Application is Due: Feb 5; Jun 5; Oct 5
Amount of Grant: Up to 250,000 USD
Contact: Nancy C. Lohrey, (301) 496-8580; fax (301) 402-4472; lohreyn@mail.nih.gov
Internet: http://grants.nih.gov/grants/guide/pa-files/PA-00-127.html
Sponsor: National Institute of Nursing Research
31 Center Drive, Room 5B10
Bethesda, MD 20892-2178

NINR Ruth L. Kirschstein National Research Service Award for Individual Predoctoral Fellows in Nursing Research   2934
This grant program will provide predoctoral training support for doctoral students. The applicant must propose a research training program and dissertation research that is consistent with the scientific mission of the NINR. Research topics and skills that will serve as a foundation for an ongoing program of research are of particular interest. The research training experience must enhance the applicant's conceptualization of research problems and research skills, under the guidance and supervision of a committed mentor who is an active and established investigator in the area of the applicant's proposed research. The research training program should be carried out in a research environment that includes appropriate human and technical resources and is demonstrably committed to the research training of the applicant in the program he/she proposes in the application. Up to five years of aggregate NRSA support may be provided. Fellowship awardees are required to pursue their research training on a full-time basis, devoting at least 40 hours per week to the training program.
Requirements: The following organizations are eligible to apply: for-profit organizations; non-profit organizations; public or private institutions, such as universities, colleges, hospitals, and laboratories; and domestic institutions.
Geographic Focus: All States
Date(s) Application is Due: Apr 8; Aug 8; Dec 8
Contact: David Banks, (301) 496-9558; fax (301) 480-8260; banksd@mail.nih.gov
Internet: http://grants.nih.gov/grants/guide/pa-files/PAR-09-227.html
Sponsor: National Institute of Nursing Research
31 Center Drive, Room 5B10
Bethesda, MD 20892-2178

NKF Clinical Scientist Awards   2935
The award supports investigators who have demonstrated outstanding clinical or basic research potential to promote their continued success as independent investigators. Scientists are expected to devote a minimum of 75 percent of their time to research. This undertaking includes the establishment of a new Study Section which will review all proposals involving clinical research for all funding mechanisms, based on the following definition from the Nathan Report, which describes three categories of clinical research: Patient-Oriented Research, defined as research conducted with human subjects (or on material of human origin such as tissues, specimens and cognitive phenomena) for which an investigator (or colleague) directly interacts with human subjects in either an outpatient or inpatient setting; Epidemiologic and Biobehavioral Studies; and Outcomes Research and Health Services Research. Applications must be submitted electronically no later than November 30.
Requirements: Applicants must be citizens or permanent residents of the United States at the time of application. Applicants must hold an M.D. or equivalent domestic or foreign degree and must be full-time staff members of a department within an institution in the United States at the time of the award. Applicants will generally have at least five, but not more than 10 years of postdoctoral research experience in nephrology by the beginning of the award date.
Geographic Focus: All States
Date(s) Application is Due: Nov 30
Amount of Grant: 50,000 USD
Contact: Emily B. Newell, Scientific Communications Director; (212) 889-2210, ext. 288 or (800) 622-9010; fax (212) 689-9261; emilyn@kidney.org or research@kidney.org
Internet: http://www.kidney.org/professionals/research/index.cfm
Sponsor: National Kidney Foundation
30 East 33rd Street
New York, NY 10016

NKF Franklin McDonald/Fresenius Medical Care Clinical Research Grants   2936
The National Kidney Foundation is soliciting applications for the annual Franklin McDonald/Fresenius Medical Care Clinical Research Grant. Applicants should be able to spend a minimum of 50% of their time conducting research and 25% of effort in the direct care of maintenance dialysis patients and/or in teaching medical students, residents or nephrology fellows the pathophysiology and treatment of CKD, ESRD, or maintenance dialysis. Proposals are due on December 1.
Requirements: Applicants must hold a full-time faculty appointment at a university or an equivalent position as a scientist on the staff of a research organization.
Geographic Focus: All States
Date(s) Application is Due: Dec 1
Contact: Emily B. Newell, Scientific Communications Director; (212) 889-2210, ext. 288 or (800) 622-9010; fax (212) 689-9261; emilyn@kidney.org or research@kidney.org
Internet: http://www.kidney.org/professionals/research/index.cfm
Sponsor: National Kidney Foundation
30 East 33rd Street
New York, NY 10016

NKF Professional Councils Research Grants   2937
The purpose of the CNNT Research Grants Program is to further knowledge of nursing and technician issues in the management of kidney failure. The grants may cover basic or applied research on nursing and technical issues in the area of kidney failure; early intervention and treatment of chronic kidney disease; development and evaluation of education programs to enhance patient/family understanding of kidney failure treatment; demonstration projects related to kidney failure and rehabilitation. Letter of intent due on October 15, and full proposals must be received by December 1.
Requirements: Applicants must hold a regular membership in CNNT, a minimum of two years work experience as a nephrology nurse or technician, residence in the U.S. or its territories, written approval from the department head or facility director at the institution/facility where the research is to be conducted, and have prior research training or evidence of support from an individual with research experience.
Restrictions: Project support and stipend not to exceed 20% of requested amount; consultant fees not to exceed 15% of total requested funds.
Geographic Focus: All States
Date(s) Application is Due: Oct 15; Dec 1
Amount of Grant: Up to 20,000 USD
Contact: Beverly Sneed, Grant Contact; (214) 514-3846 or (800) 622-9010; fax (212) 689-9261; bevyann1@sbcglobal.net or research@kidney.org
Internet: http://www.kidney.org/professionals/research/profcouncil.cfm

Sponsor: National Kidney Foundation
30 East 33rd Street
New York, NY 10016

NKF Research Fellowships 2938
The purpose of National Kidney Foundation (NKF) Research Fellowships is: to foster training of young and new investigators with the potential of making contributions to the understanding and cure of kidney diseases; to foster, through training fellows, kidney-related research that is of high merit; and to encourage high quality applicants who want to make a career change into academic Nephrology. The fellowships are open to qualified investigators of any nationality who are interested in a career in kidney research and who will not have had more than four and one-half years of research training prior to the commencement of the award. The one-year awards may be renewed.
Requirements: Applicants must have completed no more than 4.5 years of post-doctoral research training (after receipt of M.D., Ph.D, D.O., or equivalent degree) at time of the activation of the award. For those applying for a second year (competitive renewal), applicants must have completed no more than 5.5 years of post-doctoral training.
Restrictions: Anyone who has ever held a faculty appointment at the level of Assistant Professor is ineligible.
Geographic Focus: All States
Date(s) Application is Due: Nov 30
Amount of Grant: 50,000 USD
Contact: Emily B. Newell, Scientific Communications Director; (212) 889-2210, ext. 288 or (800) 622-9010; fax (212) 689-9261; emilyn@kidney.org or research@kidney.org
Internet: http://www.kidney.org/professionals/research/awards.cfm
Sponsor: National Kidney Foundation
30 East 33rd Street
New York, NY 10016

NKF Young Investigator Grants 2939
The purpose of this program is to support research in the fields of nephrology, urology, and related disciplines by junior faculty now starting their investigative work, who are often hampered in their progress by difficulty in obtaining suitable financial support for their research during the formative periods of their careers. The applicant's research and career goals must be directed to the study of normal or abnormal kidney function or of diseases of the kidney or urinary tract.
Requirements: Applicants must have completed research fellowship training in nephrology, urology or closely related field prior to the start of the grant award, intend to pursue directly related research, and hold a full-time appointment to a faculty position at a university or an equivalent position as a scientist on the staff of a research oriented institution for no more than four years at the start of the award. Recipients typically hold an appointment as Assistant Professor.
Geographic Focus: All States
Date(s) Application is Due: Nov 30
Amount of Grant: Up to 50,000 USD
Contact: Emily B. Newell, Scientific Communications Director; (212) 889-2210, ext. 288 or (800) 622-9010; fax (212) 689-9261; emilyn@kidney.org or research@kidney.org
Internet: http://www.kidney.org/professionals/research/awards.cfm
Sponsor: National Kidney Foundation
30 East 33rd Street
New York, NY 10016

NLM Academic Research Enhancement Awards (AREA) 2940
The purpose of the Academic Research Enhancement Award (AREA) program is to stimulate research in educational institutions that provide baccalaureate or advanced degrees for a significant number of the Nation's research scientists, but that have not been major recipients of support. NLM's participation is confined to applications that address research in biomedical informatics and bioinformatics. Only electronic applications are accepted. Deadlines for new applications are February 25, June 25 and October 25 each year.
Geographic Focus: All States
Date(s) Application is Due: Feb 25; Jun 25; Oct 25
Contact: Dr. Hua-Chuan Sim, Clinical and Public Health Informatics; (301) 594-5983; fax (301) 402-1384; simh@mail.nih.gov
Internet: http://www.nlm.nih.gov/ep/area.html
Sponsor: U.S. National Library of Medicine
8600 Rockville Pike
Bethesda, MD 20894

NLM Exceptional, Unconventional Research Grants Enabling Knowledge Acceleration 2941
This RFA is for exceptionally innovative research on novel hypotheses or difficult problems, solutions to which would have an extremely high impact on biomedical or biobehavioral research that is germane to the mission of the National Library of Medicine. The RFA supports new projects, not continuation of projects that have already been initiated. For the EUREKA program, NLM seeks innovative applications in the following two discovery science areas: integrated discovery mining for biology and medicine; and integrated hypothesis testing for biology and medicine. Only electronic applications are accepted. The deadline for Letter of Intent is September 29, and the deadline for new applications is October 28.
Restrictions: This grant does not support pilot projects, i.e., projects of limited scope that are designed primarily to generate data that will enable the PI to seek other funding opportunities.
Geographic Focus: All States
Date(s) Application is Due: Sep 29; Oct 28

Contact: Dr. Valerie Florance, (301) 594-5983; florancev@mail.nih.gov
Internet: http://www.nlm.nih.gov/ep/eureka.html
Sponsor: U.S. National Library of Medicine
8600 Rockville Pike
Bethesda, MD 20894

NLM Exploratory/Developmental Grants 2942
The purpose of this program is to support for early stage conceptual work and feasibility tests in biomedical informatics and bioinformatics. These studies may involve considerable risk but may lead to a breakthrough in a particular area, or to the development of novel techniques or approaches that could have major impact on a field of biomedical, behavioral, or clinical research. Preliminary data are not required, but applicants are urged to provide milestones by which progress on the project can be measured. Only electronic applications are accepted. Deadlines for new applications are February 16, June 16 and October 16 each year. Deadlines for revised applications are March 16, July 16 and November 16 each year.
Geographic Focus: All States
Date(s) Application is Due: Feb 16; Mar 16; Jun 16; Jul 16; Oct 16; Nov 16
Contact: Dr. Jane Ye, Bioinformatics and Computational Biology; (301) 594-5983; fax (301) 402-1384; yej@mail.nih.gov
Internet: http://www.nlm.nih.gov/ep/GrantExDev.html
Sponsor: U.S. National Library of Medicine
8600 Rockville Pike
Bethesda, MD 20894

NLM Express Research Grants in Biomedical Informatics 2943
The National Library of Medicine (NLM) offers support for innovative research in biomedical informatics. The scope of NLM's interest in the research domain of informatics is interdisciplinary, encompassing informatics problem areas in the application domains of health care, public health, basic biomedical research, bioinformatics, biological modeling, translational research and health information management in disasters. NLM defines biomedical informatics as the science of optimal organization, management, presentation and utilization of information relevant to human health and biology. Informatics research produces concepts, tools and approaches that advance what is known in the field and have the capacity to improve human health. The number of awards is contingent upon NIH appropriations and the submission of a sufficient number of meritorious applications. The NLM Express Research Grant has a limit of $250,000 per year in direct costs. The scope of the proposed project should determine the project period. The maximum project period is 4 years. (R01; FOA# PAR-13-300)
Requirements: Any individual(s) with the skills, knowledge, and resources necessary to carry out the proposed research as the Program Director(s)/Principal Investigator(s) (PD(s)/PI(s)) is invited to work with his/her organization to develop an application for support. Individuals from underrepresented racial and ethnic groups as well as individuals with disabilities are always encouraged to apply for NIH support. Eligible organizations include: higher education institutions (public/state-controlled institutions of higher education, private institutions of higher education); nonprofits other than institutions of higher education (nonprofits with or without 501(c)3 IRS status); for-profit organizations (small businesses, for-profit organizations other than small businesses); governments (state governments, county governments, city or township governments, special district governments, Indian/Native American Tribal Governments (federally recognized), Indian/Native American Tribal Governments (other than federally recognized), eligible agencies of the Federal Government, U.S. Territory or Possession); other (independent school districts, public housing authorities/Indian housing authorities, Native American Tribal Organizations (other than Federally recognized tribal governments), faith-based or community-based organizations, regional organizations, non-domestic (non-U.S.) entities; and, foreign institutions. The following types of Higher Education Institutions are always encouraged to apply for NIH support as public or private institutions of higher education: Hispanic-serving institutions; Historically Black Colleges and Universities (HBCUs); Tribally Controlled Colleges and Universities (TCCUs); Alaska Native and Native Hawaiian Serving Institutions; Asian American Native American Pacific Islander Serving Institutions (AANAPISIs). Applicant organizations must complete and maintain the following registrations as described in the SF 424 (R&R) Application Guide to be eligible to apply for or receive an award: DUNS Number; System for Award Management (SAM); NATP Commercial and Government Entity (NCAGE) Code; eRA Commons; and Grants.gov. All registrations must be completed prior to the application being submitted. Registration can take 6 weeks or more, so applicants should begin the registration process as soon as possible. Standard NIH deadline dates apply, by 5:00 pm local time of applicant organization.
Geographic Focus: All States, All Countries
Date(s) Application is Due: Feb 5; Jun 5; Oct 5
Amount of Grant: Up to 1,000,000 USD
Contact: Dr. Hua-Chuan Sim, Clinical and Public Health Informatics; (301) 594-4882; hua-chuan.sim@nih.gov
Internet: http://www.nlm.nih.gov/ep/GrantResearch.html
Sponsor: U.S. National Library of Medicine
8600 Rockville Pike
Bethesda, MD 20894

NLM Grants for Scholarly Works in Biomedicine and Health 2944
Small grants are awarded for up to three years to provide short-term assistance for the preparation of book-length manuscripts and, in some cases, the publication of important scientific information needed by U.S. health professionals. Grants are awarded for major critical reviews and analyses of current developments in important areas of the health sciences, historical studies, works about health sciences informatics, librarianship,

and certain kinds of secondary information and literature tools in the health sciences. Publication may be in formats other than print-on-paper (e.g., electronic, film, etc.), and may involve new and innovative ways of organizing and presenting information. Only electronic applications are accepted.
Requirements: Grants may be made to public or private, nonprofit institutions on behalf of a principal investigator; or to unaffiliated individuals to support salaries, consultant fees, equipment and supplies, travel, and other justified costs.
Restrictions: The program does not support textbooks, the production of curriculum materials, the initial reporting of original biomedical scientific research findings, proceedings of annual meetings, projects highly likely to have substantial commercial sales (best-seller type publications), or projects of local interest only. The program does not provide support for the free distribution of publications resulting from grants.
Geographic Focus: All States
Date(s) Application is Due: Feb 2
Amount of Grant: 50,000 - 150,000 USD
Contact: Dr. Hua-Chuan Sim, Clinical and Public Health Informatics; (301) 594-5983; fax (301) 402-1384; simh@mail.nih.gov
Internet: http://www.nlm.nih.gov/ep/GrantPubs.html
Sponsor: U.S. National Library of Medicine
8600 Rockville Pike
Bethesda, MD 20894

NLM Informatics Conference Grants 2945
The purpose of the program is to provide support for scientific meetings, conferences, and workshops that are relevant to scientific development in biomedical informatics, bioinformatics, computational biology, and information sciences. The typical award is $10,000 per year in direct costs. The maximum amount that can be requested for a single year is $20,000 in direct costs. Only electronic applications are accepted.
Restrictions: NLM will not support the annual meeting for an organization.
Geographic Focus: All States
Date(s) Application is Due: Apr 12; Aug 12; Dec 12
Amount of Grant: Up to 20,000 USD
Contact: Dr. Jane Ye, Bioinformatics and Computational Biology; (301) 594-5983; fax (301) 402-1384; yej@mail.nih.gov
Internet: http://www.nlm.nih.gov/ep/GrantConf.html
Sponsor: U.S. National Library of Medicine
8600 Rockville Pike
Bethesda, MD 20894

NLM Internet Connection for Medical Institutions Grants 2946
NLM encourages the development of a communications infrastructure to promote the rapid interchange of medical information nationally and throughout the world. Internet access provides health professionals engaged in education, research, clinical care, and administration with a means of accessing remote databases and libraries, transferring files and images, and interacting with colleagues throughout the world. To accelerate the pace with which health-related institutions become part of the electronic information web, NLM offers grants to support Internet connections. Funds may be used for the purchase and installation of a gateway system and associated connection hardware. Funds may also be used to defray the cost of installation and leasing of communication circuits to connect the network. Annual deadline dates may vary; contact the program office for exact dates. RFA: LM-00-001
Requirements: Domestic public and private nonprofit institutions working in health sciences administration, education research, and/or clinical care services can apply for these grants. Groups that are health-related institutions are also encouraged to apply.
Geographic Focus: All States
Amount of Grant: 30,000 - 50,000 USD
Contact: Dr. Jane Ye, Bioinformatics and Computational Biology; (301) 594-5983; fax (301) 402-1384; yej@mail.nih.gov
Internet: http://grants.nih.gov/grants/guide/rfa-files/RFA-LM-00-001.html
Sponsor: National Library of Medicine
8600 Rockville Pike, Building 38, Room B1E08
Bethesda, MD 20894

NLM Limited Competition for Continuation of Biomedical Informatics Bioinformatics Resource Grants 2947
This limited competition funding opportunity announcement (FOA) solicits competitive renewal applications from investigators currently supported under PAR-04-142, NLM Resource Grant in Biomedical Informatics/Bioinformatics (P41) or its predecessors. NLM's Biomedical Informatics/Bioinformatics Resource Grant program provides support for five (5) resource centers that support activities, databases or software tools that advance research or practice in the biomedical sciences, clinical medicine or health services research. Direct costs for the first year may not exceed 5% above the current award of the active year of NLM funding at the time of application, or a maximum of $500,000, whichever is smaller. Applicants may include cost-of-living increases of up to 3% per year for direct costs in future years. The deadline for Letters of Intent is July 28, while the deadline for Renewal Applications is August 27.
Geographic Focus: All States
Date(s) Application is Due: Jul 28; Aug 27
Contact: Dr. Valerie Florance, (301) 594-5983; florancev@mail.nih.gov
Internet: http://www.nlm.nih.gov/ep/grantp41.html
Sponsor: U.S. National Library of Medicine
8600 Rockville Pike
Bethesda, MD 20894

NLM Manufacturing Processes of Medical, Dental, and Biological Technologies (SBIR) Grants 2948
NLM is encouraging eligible United States small business concerns to submit SBIR and STTR Phase I, Phase II, and Fast-Track grant applications whose biomedical research is related to advanced processing, manufacturing processes, equipment and systems, and manufacturing workforce skills and protection. Only electronic applications are accepted.
Geographic Focus: All States
Date(s) Application is Due: Apr 5; Aug 5; Dec 5
Contact: Dr. Jane Ye, Bioinformatics and Computational Biology; (301) 594-5983; fax (301) 402-1384; yej@mail.nih.gov
Internet: http://www.nlm.nih.gov/ep/grantsbmanu.html
Sponsor: U.S. National Library of Medicine
8600 Rockville Pike
Bethesda, MD 20894

NLM New Technologies for Liver Disease Grants 2949
The goal of this announcement is to enlist members of the small business research community in advancing means of diagnosis, treatment and prevention of liver disease and facilitate the goals outlined in the trans-NIH Action Plan for Liver Disease Research. Annual deadline dates are April 5, August 5 and December 5.
Geographic Focus: All States
Date(s) Application is Due: Apr 5; Aug 5; Dec 5
Contact: Dr. Jane Ye, Bioinformatics and Computational Biology; (301) 594-5983; fax (301) 402-1384; yej@mail.nih.gov
Internet: http://www.nlm.nih.gov/ep/grantsbliver.html
Sponsor: U.S. National Library of Medicine
8600 Rockville Pike
Bethesda, MD 20894

NLM Predictive Multiscale Models of the Physiome in Health and Disease Grants 2950
The goal of this funding opportunity announcement (FOA) is to move the field of biomedical computational modeling forward through the development of more realistic and predictive models of health and disease. NIH recognizes the need for sophisticated, predictive, computational models of development and disease that encompass multiple biological scales. These models may be designed to uncover biological mechanisms or to make predictions about clinical outcome and may draw on a variety of data sources including relevant clinical data. Ultimately the models and the information derived from their use will enable researchers and clinicians to better understand, prevent, diagnose and treat the diseases or aberrations in normal development. Specifically this FOA solicits the development of predictive multiscale models of health and disease states that must include higher scales of the physiome. The specific objectives are to develop multiscale models that are physiologically mechanistic and biomedically relevant, to bring together modeling and biomedical expertise to collaborate on building models, to validate and test models with standard datasets, and to develop models that can be explicitly shared with other modelers.
Geographic Focus: All States
Date(s) Application is Due: Jan 14; May 14; Sep 15
Contact: Dr. Jane Ye, Bioinformatics and Computational Biology; (301) 594-5983; fax (301) 402-1384; yej@mail.nih.gov
Internet: http://www.nlm.nih.gov/ep/predictive.html
Sponsor: U.S. National Library of Medicine
8600 Rockville Pike
Bethesda, MD 20894

NLM Research Project Grants 2951
The primary purpose of this program is to provide support for rigorous scientific research in biomedical informatics and bioinformatics. Medical informatics is an interdisciplinary field that combines health sciences specialty areas with the information and computer sciences, computational linguistics, and decision theory, among others. NLM's program emphasizes knowledge issues without restriction as to specific categorical diseases. Proposed research must be clearly health relevant. Areas of interest include knowledge representation, more effective human-machine interfaces, and more efficient information storage and retrieval. The cognitive aspects of information-related behaviors are of particular interest. This area includes problems of medical decision analysis, medical decision support systems, and retrieval from image databases. Deadlines for new applications are February 5, June 5 and October 5 each year. Deadlines for revised applications are March 5, July 5 and November 5 each year.
Requirements: Institutions or organizations with research capabilities in the health information fields or in medical informatics are eligible to apply.
Geographic Focus: All States
Date(s) Application is Due: Feb 5; Mar 5; Jun 5; Jul 5; Oct 5; Nov 5
Amount of Grant: 25,000 - 2,000,000 USD
Contact: Dr. Jane Ye; (301) 594-5983; fax (301) 402-1384; yej@mail.nih.gov
Internet: http://www.nlm.nih.gov/ep/GrantResearchParent.html
Sponsor: U.S. National Library of Medicine
8600 Rockville Pike
Bethesda, MD 20894

NLM Research Supplements to Promote Reentry in Health-Related Research 2952
Principal Investigators holding NIH research grants of the specified types (see announcement) can apply for administrative supplements to support individuals with high potential to reenter an active research career after taking time off to care for children or attend to other family responsibilities. It is anticipated that at the completion of the

supplement, the reentry scientist will be in a position to apply for a career development award, a research award, or some other form of independent research support. The grantee institution must submit the application for supplemental funds directly to the awarding component that supports the parent grant. The application must not be submitted to the NIH Center for Scientific Review.
Geographic Focus: All States
Contact: Hua-Chuan Sim; (301) 594-5983; fax (301) 402-1384; simh@mail.nih.gov
Internet: http://www.nlm.nih.gov/ep/grantsupreentry.html
Sponsor: U.S. National Library of Medicine
8600 Rockville Pike
Bethesda, MD 20894

NLM Understanding and Promoting Health Literacy Research Grants 2953
The goal of these Program Announcements (Understanding and Promoting Health Literacy Research Grants and Understanding and Promoting Health Literacy Small Project Grants) is to increase scientific understanding of the nature of health literacy and its relationship to healthy behaviors, illness prevention and treatment, chronic disease management, health disparities, risk assessment of environmental factors, and health outcomes including mental and oral health.
Geographic Focus: All States
Date(s) Application is Due: Jan 25; May 25
Contact: Hua-Chuan Sim; (301) 594-5983; fax (301) 402-1384; simh@mail.nih.gov
Internet: http://www.nlm.nih.gov/ep/healthlit.html
Sponsor: National Library of Medicine
8600 Rockville Pike, Building 38, Room B1E08
Bethesda, MD 20894

NMF Aura E. Severinghaus Award 2954
One award is presented annually to an underrepresented minority (African American, Mexican American, mainland Puerto Rican, or Native American) student attending Columbia University's College of Physicians and Surgeons in recognition of outstanding academic achievement, leadership, and community service. A committee of faculty and administrators at the college selects the student most deserving of this award and presents the candidate's dossier to NMF for final approval.
Requirements: This award is open only to underrepresented minority students enrolled in their senior year at the College of Physicians and Surgeons at Columbia University.
Geographic Focus: New York
Amount of Grant: 2,000 USD
Contact: Coordinator; (212) 483-8880; fax (212) 483-8897; nmf1@nmfonline.org
Internet: http://www.nmfonline.org/index.html
Sponsor: National Medical Fellowships
5 Hanover Square
New York, NY 10004

NMF Franklin C. McLean Award 2955
One award is presented annually to a senior underrepresented minority (African American, Mexican American, mainland Puerto Rican, and Native American) student enrolled in an accredited U.S. medical school in recognition of outstanding academic achievement, leadership, and community service. Annual deadline dates may vary; contact program staff for exact dates.
Requirements: Candidates must be nominated by medical school deans. Nominations are requested in June. Schools are required to submit letters of recommendation and official academic transcripts for each candidate.
Geographic Focus: All States
Amount of Grant: 3,000 USD
Contact: Coordinator; (212) 483-8880; fax (212) 483-8897; nmf1@nmfonline.org
Internet: http://www.nmf-online.org/Programs/MeritAwards/McLean/Overview.htm
Sponsor: National Medical Fellowships
5 Hanover Square
New York, NY 10004

NMF General Electric Medical Scholars Fellowships 2956
The sponsor has established a partnership with General Electric (GE) to provide need-based scholarship support to historically underrepresented minority medical students, and to allow them to engage in clinical work in underserved communities. The program also offers eligible students an opportunity to complete a two-month elective in Ghana, West Africa focusing on areas of critical need such as HIV/AIDS, diabetes, tuberculosis, heart disease, cancer and infant mortality. Five fellowships are awarded annually. Students will be matched with a mentor for the period of the program. Travel, hosing and a stipend will be provided for students.
Requirements: Students must be nominated by their medical school deans. Eligible candidates must be underrepresented minority students enrolled in their fourth year in an accredited U.S. medical school. Candidates must demonstrate outstanding academic achievement, leadership, and potential for community-based work or initiation of innovative projects in school or community. Additionally, candidates must also demonstrate potential for distinguished contributions to medicine. Complete application requirements can be found at the sponsor's website.
Geographic Focus: All States
Contact: Dr. Stephen Keith, President; (240) 243-8000; fax (240) 465-0450
Internet: http://www.nmfonline.org/index.html
Sponsor: National Medical Fellowships
5 Hanover Square
New York, NY 10004

NMF Henry G. Halladay Awards 2957
Five supplemental scholarships are presented annually to African American men who have been accepted into the first-year classes of accredited U.S. medical schools and who have overcome significant obstacles to obtain a medical education. Staff members review NMF first-year general scholarship applicants and select five students who, on the basis of recommendations, personal statements, and financial need, are most deserving of these awards. Annual deadline dates may vary; contact program staff for exact dates.
Requirements: Applicants must have been accepted to AAMC or AOA-accredited U.S. medical schools for study leading to MD or DO degrees.
Geographic Focus: All States
Amount of Grant: 760 USD
Contact: Coordinator; (212) 483-8880; fax (212) 239-9718; nmf2@nmfonline.org
Internet: http://www.nmfonline.org/index.html
Sponsor: National Medical Fellowships
5 Hanover Square
New York, NY 10004

NMF Hugh J. Andersen Memorial Scholarship 2958
This scholarship program was established by the family of the late Hugh J. Andersen. Need-based merit scholarships are annually presented to Minnesota residents enrolled in any accredited U.S. medical school or students attending medical schools in Minnesota. Up to 5 scholarships of $2,500 are presented annually.
Requirements: Eligible candidates must be underrepresented minority students enrolled beyond their first year in an accredited U.S. medical school. Candidates must demonstrate outstanding academic achievement as indicated in academic transcripts, faculty evaluations, by receipt of special academic honors, fellowships, awards or induction into national medical honor societies, leadership (as indicated by active participation in community-based work or initiation of innovative projects in school or community), and potential for distinguished contributions to medicine as indicated by participation in research, publications or unique clerkships. Students must also provide documented proof of financial need. Students must be nominated by medical school deans. Schools are required to submit letters of recommendations and official academic transcripts. Applicants must submit scholarship applications and include documented proof of financial need. See the sponsor's website for the required application form and further instructions.
Geographic Focus: Minnesota
Amount of Grant: 2,500 USD
Contact: Coordinator; (212) 714-0933; fax (212) 239-9718; nmf1@nmfonline.org
Internet: http://www.nmfonline.org/index.html
Sponsor: National Medical Fellowships
5 Hanover Square
New York, NY 10004

NMF Irving Graef Memorial Scholarship 2959
This two-year scholarship is presented annually to a third-year student and recognizes outstanding academic achievement, leadership, and community service. The scholarship is renewable in the fourth year if the award recipient continues in good academic standing. The scholarship is open only to rising third-year minority medical students who received NMF financial assistance during their second year. One new scholarship is awarded each year. Annual deadline date may vary; contact program staff for exact dates.
Requirements: Students must be nominated by medical school deans. Schools are required to submit letters of recommendation as well as official academic transcripts. Candidates must demonstrate outstanding academic achievement, submit the scholarship application (excluding proof of financial need), personal essay and current cirriculum vitae.
Geographic Focus: All States
Amount of Grant: 2,000 USD
Contact: Coordinator; (212) 483-8880; fax (212) 483-8897; nmf1@nmfonline.org
Internet: http://www.nmfonline.org/index.html
Sponsor: National Medical Fellowships
5 Hanover Square
New York, NY 10004

NMF Mary Ball Carrera Scholarship 2960
This need-based merit scholarship is annually presented to a Native American woman enrolled in any accredited U.S. medical school and recognizes outstanding academic achievement, leadership and community services. Recipients receive a one-time award of $2,500. Scholarships are granted on the basis of financial need as determined by the student's total resources (including parental and spousal support), education costs, and receipt of other scholarships and grants.
Requirements: Eligible candidates must be first- or second-year medical students. Students must provide proof of admission by an accredited medical school. Applicants must demonstrate and document financial need and submit a current financial aid award letter.
Restrictions: All applicants must reside within the 50 U.S.A. states and provide proof of citizenship.
Geographic Focus: All States
Amount of Grant: 2,500 USD
Contact: Coordinator; (212) 483-8880; fax (212) 483-8897; nmf1@nmfonline.org
Internet: http://www.nmfonline.org/index.html
Sponsor: National Medical Fellowships
5 Hanover Square
New York, NY 10004

NMF Metropolitan Life Foundation Awards Program for Academic Excellence in Medicine — 2961

Need-based scholarships are awarded annually to third- and fourth-year underrepresented minority medical students who have demonstrated academic achievement and leadership. Applicants must attend medical schools or have legal residence in cities, designated each year; contact office for current eligible area. Nominations are requested in August. Annual deadline dates may vary; contact program staff for exact dates.
Requirements: Students must be nominated by medical school deans. Schools are required to submit letters of recommendation as well as official academic transcripts. Candidates are required to submit scholarship applications including personal statements and verification of financial need.
Geographic Focus: All States
Amount of Grant: 4,000 USD
Contact: Coordinator; (212) 483-8880; fax (212) 483-8897; nmf1@nmfonline.org
Internet: http://www.nmfonline.org/index.html
Sponsor: National Medical Fellowships
5 Hanover Square
New York, NY 10004

NMF National Medical Association Awards for Medical Journalism — 2962

In general, these awards recognize academic achievement, leadership, and potential for distinguished contributions to medicine. Specifically, these awards recognize demonstrated skills in journalism. Deadlines vary - contact the sponsor for dates.
Requirements: Eligibility is limited to African-American medical students. Each medical school may submit up to two nominations. Candidates must be in the third or fourth year of medical school and must have published articles and photographs in, or been writers, editors, photographers on the staffs of: medical school newspapers; medical students' journals; recognized professional journals; or, other accredited scientific journal. Students who have written, produced or directed health-related films, commercials or videos are also encouraged to submit their applications and samples of their work. Applicants must also submit samples of their journalistic work (articles, photographs, audio tapes, video tapes, etc.) for review by the selection committee. Entries must have been published or produced within the last three years. The application form and more detailed guidelines can be found at the sponsor's website.
Geographic Focus: All States
Amount of Grant: 2,500 USD
Contact: Coordinator; (212) 483-8880; fax (212) 239-9718; nmf2@nmfonline.org
Internet: http://www.nmfonline.org/index.html
Sponsor: National Medical Fellowships
5 Hanover Square
New York, NY 10004

NMF National Medical Association Emerging Scholar Awards — 2963

The National Medical Association and NMF sponsor need-based award programs that recognize and reward African American medical students for special achievements, academic excellence, leadership, and potential for outstanding contributions to medicine. The NMA Emerging Scholar Awards are presented to current first, second, and third-year medical students.
Requirements: Eligibility for these awards is limited to African-American medical students. Each school may submit two nominations. Applicants must demonstrate and document financial need. Deadlines vary - contact the sponsor for specific dates.
Geographic Focus: All States
Amount of Grant: 2,250 USD
Contact: Coordinator; (212) 483-8880; fax (212) 483-8897; nmf1@nmfonline.org
Internet: http://www.nmfonline.org/index.html
Sponsor: National Medical Fellowships
5 Hanover Square
New York, NY 10004

NMF National Medical Association Patti LaBelle Award — 2964

This award is one of the highest honors presented to students by the National Medical Association. It recognizes academic achievement, leadership and potential for distinguished contributions to medicine. This award is presented to current first-, second-, and third-year medical students with demonstrated financial need.
Requirements: Eligibility is limited to African-American medical students. Application materials must include a letter of recommendation that fully discusses the candidate's academic performance, extracurricular activities and other qualifications; a complete scholarship application form; an essay of at least 500 words discussing the applicant's motivation for a career in medicine and plans over the next ten years; and, an official academic transcript.
Geographic Focus: All States
Amount of Grant: 5,000 USD
Contact: Coordinator; (212) 483-8880; fax (212) 483-8897; nmf1@nmfonline.org
Internet: http://www.nmfonline.org/index.html
Sponsor: National Medical Fellowships
5 Hanover Square
New York, NY 10004

NMF Need-Based Scholarships — 2965

The program provides financial assistance to United States citizens from groups currently underrepresetned in the medical profession; specifically, African-Americans, Mexican-Americans, Native Americans, Alaska Natives, Native Hawaiians, and mainland Puerto Ricans who permanently reside within the 50 U.S. states. These awards are offered to first- and second-year medical students. Scholarships are granted on the basis of financial need as determined by the student's total resources (including parental and spousal support), cost of education, and receipt of other scholarships and grants. Awards have ranged from $500 to $10,000.
Requirements: Applicants must be U.S. citizens by the application deadline and must have been accepted to AAMC or AOA-accredited U.S. medical schools for study leading to M.D. or D.O. degrees. The application form and FAQs can be found at the sponsor's website. Applicants must demonstrate and document financial need by submitted complete copies of their parents', spouse's and their own most recent 1040, 1040A or 1040EZ tax forms. All nontaxable income (e.g. AFD.C., AD.C., Social Security benefits, etc.) must also be documented by the appropriate agency. Applicants with neither taxable nor nontaxable income must provide verification of means of support. All parental financial data must be submitted in order to evaluate dependent status.
Geographic Focus: All States
Date(s) Application is Due: Aug 31
Amount of Grant: 500 - 10,000 USD
Contact: Programs Department; (212) 483-8880, ext. 304; fax (212) 483-8897; NMF1@NMFonline.org or info@NMFonline.org
Internet: http://www.nmfonline.org/programs.html
Sponsor: National Medical Fellowships
5 Hanover Square
New York, NY 10004

NMF Ralph W. Ellison Prize — 2966

The prize is presented to a graduating underrepresented medical student who has demonstrated outstanding academic achievement, leadership, and potential to make significant contributions to medicine. One prize is presented each year (nonrenewable) and includes a certificate of merit and a stipend. Annual deadline date may vary; contact program staff for exact dates.
Requirements: African-Americans, mainland Puerto Ricans, Mexican-Americans, Native Hawaiians, Alaska Natives, and American Indians who are U.S. citizens attending M.D. degree-granting institutions accredited by the Liaison Committee on Medical Education of the Association of American Medical Colleges, or in D.O. degree-granting colleges of osteopathic medicine accredited by the Bureau of Professional Education of the American Osteopathic Association are eligible.
Geographic Focus: All States
Amount of Grant: 500 USD
Contact: Coordinator; (212) 483-8880; fax (212) 483-8897; nmf1@nmfonline.org
Internet: http://www.nmfonline.org/index.html
Sponsor: National Medical Fellowships
5 Hanover Square
New York, NY 10004

NMF William and Charlotte Cadbury Award — 2967

This award is presented annually to a senior, underrepresented student enrolled in an accredited U.S. medical school. Candidates must demonstrate outstanding academic achievement, leadership, and community service. A certificate of merit is included in the award. Annual deadline dates may vary; contact program staff for exact dates.
Requirements: Students must be nominated by medical school deans. Schools are required to submit letters of recommendation and official academic transcripts for each candidate. Competition is open to senior, underrepresented minority students enrolled in accredited U.S. medical schools. Candidates must demonstrate outstanding academic achievement and leadership.
Geographic Focus: All States
Amount of Grant: 2,000 USD
Contact: Coordinator; (212) 483-8880; fax (212) 483-8897; nmf1@nmfonline.org
Internet: http://www.nmfonline.org/index.html
Sponsor: National Medical Fellowships
5 Hanover Square
New York, NY 10004

NMSS Scholarships — 2968

The society awards unrestricted scholarships to defray expenses related to post-high school education. Decisions are based on financial need, an academic record that shows the applicant is able to succeed in his or her chosen area, and on a personal essay discussing the impact of multiple sclerosis (MS) on the applicant's life.
Requirements: Application is open to high school seniors who have MS; high school seniors who have a parent or guardian with MS; and adults with MS who have never enrolled in a postsecondary institution.
Geographic Focus: All States
Date(s) Application is Due: Mar 1
Amount of Grant: 2,000 - 3,000 USD
Contact: Coordinator; (800) 344-4467 or (510) 268-0572; info@msconnection.org
Internet: http://www.nationalmssociety.org/can/event/event_detail.asp?e=8108#scholarship
Sponsor: National Multiple Sclerosis Society
733 Third Avenue
New York, NY 10017-3288

Noble County Community Foundation - Arthur A. and Hazel S. Auer Scholarship — 2969

The Noble County Community Foundation offers 40 scholarships with a variety of requirements and specific criteria for eligible Noble County residents. Applicants can review the scholarship list and choose to apply for several scholarships through one application. The Arthur A. and Hazel S. Auer Scholarship helps fund two East Noble high school seniors preparing to enter a four year bachelor's program at any accredited

college or university. The scholarship pays $500 per semester for the first two years for a total of $2,000. Any college major is eligible.
Requirements: Applicants must have a GPA of 3.7 or above. Candidates must apply through the Foundation's online application and follow the instructions for submission listing detailed information about their school performance, community service and volunteer activities, work experience, expected family contribution from the FAFSA, and three references.
Geographic Focus: Indiana
Date(s) Application is Due: Jan 28
Amount of Grant: 2,000 USD
Contact: Jennifer Shultz, Scholarship Manager; (260) 894-3335; fax (260) 894-9020; jennifer@noblecountycf.org or info@noblecountrycf.org
Internet: http://www.noblecountycf.org/body.cfm?lvl1=servic&lvl2=schola&lvl3=availa
Sponsor: Noble County Community Foundation
1599 Lincolnway South
Ligonier, IN 46767

Noble County Community Foundation - Art Hutsell Scholarship 2970
The Noble County Community Foundation offers 40 scholarships with a variety of requirements and specific criteria for eligible Noble County residents. Applicants can review the scholarship list and choose to apply for several scholarships through one application. The Art Hutsell Scholarship helps fund a West Noble High School senior pursing advanced education with a one-time payment of $300.
Requirements: Applicants must be pursuing post-secondary education in emergency services involving fire, police, and/or EMS. Candidates must apply through the Foundation's online application and follow the instructions for submission by listing detailed information about their school performance, community service and volunteer activities, work experience, expected family contribution from the FAFSA, and three references.
Geographic Focus: Indiana
Date(s) Application is Due: Jan 28
Amount of Grant: 300 USD
Contact: Jennifer Shultz, Scholarship Manager; (260) 894-3335; fax (260) 894-9020; jennifer@noblecountycf.org or info@noblecountrycf.org
Internet: http://www.noblecountycf.org/body.cfm?lvl1=servic&lvl2=schola&lvl3=availa
Sponsor: Noble County Community Foundation
1599 Lincolnway South
Ligonier, IN 46767

Noble County Community Foundation-Delta Theta Tau Sorority IOTA IOTA Chapter Riecke Scholarship 2971
The Noble County Community Foundation offers 40 scholarships with a variety of requirements and specific criteria for eligibility for Noble County residents. Applicants can review the scholarship list and choose to apply for several scholarships through one application. The Delta Theta Tau Sorority Tammy Riecke Scholarship offers funding for an East Noble high school senior who has worked to promote philanthropy and charity for the community. The scholarship amount is to be determined but is paid during the fall and spring semester of the first year.
Requirements: Candidates must apply through the Foundation's online application and follow the instructions for submission listing detailed information about their school performance, community service and volunteer activities, work experience, expected family contribution from the FAFSA, and three references. The application section for this scholarship includes an essay question, "In your own words, what is a humanitarian? List anything you have done that you feel is humanitarian related."
Geographic Focus: Indiana
Date(s) Application is Due: Jan 28
Contact: Jennifer Shultz, Scholarship Manager; (260) 894-3335; fax (260) 894-9020; jennifer@noblecountycf.org or info@noblecountrycf.org
Internet: http://www.noblecountycf.org/body.cfm?lvl1=servic&lvl2=schola&lvl3=availa
Sponsor: Noble County Community Foundation
1599 Lincolnway South
Ligonier, IN 46767

Noble County Community Foundation - Democrat Central Committee Scholarships 2972
The Noble County Community Foundation offers 40 scholarships with a variety of requirements and specific criteria for eligible Noble County residents. Applicants can review the scholarship list and choose to apply for several scholarships through one application. The Democrat Central Committee Scholarships are offered to several Noble County residents ad/or high school seniors pursuing post-high school education.
Requirements: Preference is given to current graduating high school seniors. Candidates must apply through the Foundation's online application and follow the instructions for submission by listing detailed information about their school performance, community service and volunteer activities, work experience, expected family contribution from the FAFSA, and three references.
Geographic Focus: Indiana
Date(s) Application is Due: Jan 28
Amount of Grant: 400 USD
Contact: Jennifer Shultz, Scholarship Manager; (260) 894-3335; fax (260) 894-9020; jennifer@noblecountycf.org or info@noblecountrycf.org
Internet: http://www.noblecountycf.org/body.cfm?lvl1=servic&lvl2=schola&lvl3=availa
Sponsor: Noble County Community Foundation
1599 Lincolnway South
Ligonier, IN 46767

Noble County Community Foundation - Kathleen June Earley Memorial Scholarship 2973
The Noble County Community Foundation offers 40 scholarships with a variety of requirements and specific criteria for eligible Noble County residents pursuing higher education. Applicants can review the scholarship list and choose to apply for several scholarships through one application. The Kathleen June Earley Scholarship funds an East Noble high school senior planning to study at Ball State University with a one-time payment of $250. Any college major is eligible.
Requirements: Candidates must apply through the Foundation's online application and follow the instructions for submission listing detailed information about their school performance, community service and volunteer activities, work experience, expected family contribution from the FAFSA, and three references.
Geographic Focus: Indiana
Date(s) Application is Due: Jan 28
Contact: Jennifer Shultz, Scholarship Manager; (260) 894-3335; fax (260) 894-9020; jennifer@noblecountycf.org or info@noblecountrycf.org
Internet: http://www.noblecountycf.org/body.cfm?lvl1=servic&lvl2=schola&lvl3=availa
Sponsor: Noble County Community Foundation
1599 Lincolnway South
Ligonier, IN 46767

Noble County Community Foundation - Lolita J. Hornett Memorial Nursing Scholarship 2974
The Noble County Community Foundation offers 40 scholarships with a variety of requirements and specific criteria for eligible Noble County residents. Applicants can review the scholarship list and choose to apply for several scholarships through one application. The Lolita Hornett Nursing Scholarship helps fund Noble County residents entering their first year in post-secondary education. A one-time payment of $500 is given at the completion of the first semester.
Requirements: Applicants must be East Noble High School seniors or graduates entering a nursing degree program as a freshman. They must have also maintained a 3.0 GPA in high school. Candidates must apply through the Foundation's online application and follow the instructions for submission by listing detailed information about their school performance, community service and volunteer activities, work experience, expected family contribution from the FAFSA, and three references.
Geographic Focus: Indiana
Date(s) Application is Due: Jan 28
Contact: Jennifer Shultz, Scholarship Manager; (260) 894-3335; fax (260) 894-9020; jennifer@noblecountycf.org or info@noblecountrycf.org
Internet: http://www.noblecountycf.org/body.cfm?lvl1=servic&lvl2=schola&lvl3=availa
Sponsor: Noble County Community Foundation
1599 Lincolnway South
Ligonier, IN 46767

Noble County Community Foundation Grants 2975
The Noble County Community Foundation makes grants for innovative, creative projects and programs responsive to changing community needs in the areas of health and human services, education, arts and culture, and civic affairs. The funding is not limited to these areas and grant seekers are encouraged to respond to emerging community needs. The Foundation considers only not-for-profit projects that: promote cooperation among organizations without duplicating services; promote volunteer involvement; demonstrate practical approaches to current community issues; enhance or improve an organization's self-sufficiency and effectiveness; emphasize prevention. In addition, the Foundation considers projects that: affect a broad segment of the population; are pilot programs clearly replicable in their design and have reasonable prospects for future support; serve people whose needs are not being met by existing services and which encourage independence; move the community to a higher cultural awareness. The Foundation offers grants in April, June, August, and December, with decisions made 60 days after each deadline. Applicants are encouraged to contact the Foundation before submitting a proposal to find out if the project is appropriate for funding. Applicants should refer to the website for further information.
Requirements: In applying for grants, the following points must be addressed (with explanation if information is not available): the specific purpose and the need for the funds requested; the need for the program/project in the community; the amount requested; a detailed description of how the money would be spent; a statement of how the project will improve life in Noble County and how the outcomes will be measured; recent grants received and applications pending for this program/project; a listing of the current board and/or project organizers and their contact information; a copy of the organization's IRS tax-exempt letter stating 501(c)3 status; and for schools applying, a letter of support from the organization's superintendent/principal. After the organization's request has been received, it may be contacted by a Foundation representative requesting a site visit or additional information.
Restrictions: Discretionary funds will not be used under any circumstances to support: deficit spending; political purposes; annual fund campaigns; lobbying; organizations whose primary function is to allocate funds to other charitable organizations or projects; projects that do not serve residents of Noble County; travel; augmenting endowments; underwriting for fundraising events; loans.
Geographic Focus: Indiana
Date(s) Application is Due: Mar 2; May 2; Jul 2; Nov 2
Contact: Linda Speakman-Yerick, Executive Director; (260) 894-3335; fax (260) 894-9020; linda@noblecountycf.org or info@noblecountrycf.org
Internet: http://noblecountycf.org/noble-county-community-foundation-inc-grants/
Sponsor: Noble County Community Foundation
1599 Lincolnway South
Ligonier, IN 46767

Noble County Community Foundation Lilly Endowment Scholarship 2976
The Noble County Community Foundation Lilly Endowment Scholarship awards four years of full tuition and required fees and books to a Noble County high school student pursuing a baccalaureate degree at an Indiana public or private college or university. All college majors are eligible.
Requirements: Application criteria includes community service and at least a 3.5 high school GPA. Candidates may apply for the scholarship through the online joint scholarship application on the Foundation's website.
Restrictions: Applicants must be U.S. citizens and residents of Noble County for at least one year.
Geographic Focus: Indiana
Contact: Jennifer Shultz, Scholarship Coordinator; (260) 894-3335; fax (260) 894-9020; jennifer@noblecountycf.org or info@noblecountycf.org
Internet: http://www.noblecountycf.org/body.cfm?lvl1=servic&lvl2=schola&lvl3=availa
Sponsor: Noble County Community Foundation
1599 Lincolnway South
Ligonier, IN 46767

Noble County Community Foundation Scholarships 2977
The Noble County Community Foundation helps build an enduring source of charitable assets to meet the changing needs of the people in Noble County, Indiana. The Foundation administers separate scholarship funds established by individuals, families, corporations, and organizations to assist local students in pursuit of advanced education. Eligibility, criteria, and supplemental requirements are different for each scholarship. See the Noble County Community Foundation website for a description of each scholarship offered. A joint scholarship application is available online so that applicants may apply for several scholarships at one time. Candidates are encouraged to apply as soon as possible, but may apply from mid-November through the following January.
Requirements: Applicants must submit 7 copies of the application to the Foundation office. Applicants are also required to agree to several criteria listed on the application introduction. Applicants are required to list detailed information about their school performance, community service activities, volunteering, work experience, expected family financial contribution to their college studies, and three references.
Geographic Focus: Indiana
Date(s) Application is Due: Jan 28
Contact: Jennifer Shultz, Scholarship Manager; (260) 894-3335; fax (260) 894-3335; jennifer@noblecountycf.org or info@noblecountycf.org
Internet: http://www.noblecountycf.org/body.cfm?lvl1=servic&lvl2=schola&lvl3=availa
Sponsor: Noble County Community Foundation
1599 Lincolnway South
Ligonier, IN 46767

NOCSAE Research Grants 2978
Due to the diversity and complexity of potential research proposals, NOCSAE has instituted a two-phase application procedure. Those interested in seeking funding are required to first submit a Preliminary Grant Application. These brief, one page proposals are reviewed by the NOCSAE Board of Directors. The Board votes to invite Full Grant Applications from these preliminary proposals after consideration of the current goals and the available funds. The invited full proposals are assigned and reviewed by an external scientific study section. Applications should describe projects based upon basic and/or applied research bearing a rational relationship toward increasing our understanding of sports injury mechanisms and injury prevention through the use of protective sports equipment. Priority will be given to projects that demonstrate evidence of recurring injury where the injury is either catastrophic, serious, or costly.
Requirements: Applications should describe projects based upon basic and/or applied research bearing a rational relationship towards increasing our understanding of sports injury mechanisms and injury prevention through the use of protective sports equipment. Projects related specifically to equipment design are not appropriate. Projects analyzing athletic performance are not appropriate. Any individual and institution residing in North America is eligible. All applications regardless of origin will be evaluated by the same strict scientific guidelines.
Restrictions: Projects related specifically to equipment design and development are not appropriate and will not be accepted. Projects analyzing athletic performance are also not appropriate.
Geographic Focus: All States
Date(s) Application is Due: Oct 1
Amount of Grant: 50,000 USD
Contact: Mike Oliver; (913) 888-1340; fax (913) 498-8817; mike.oliver@nocsae.org
Internet: http://www.nocsae.org/research/funding.html
Sponsor: National Operating Committee on Standards for Athletic Equiipment
11020 King Street, Suite 215
Overland, KS 66210

Norcliffe Foundation Grants 2979
The Norcliffe Foundation is a private nonprofit family foundation established in 1952 by Paul Pigott for the purpose of improving the quality of life for all people in the Puget Sound region. The foundation provides grants in the areas of health, education, social services, civic improvement, religion, culture and the arts, the environment, historic preservation and youth programs. Foundation funding types include capital campaigns for building and equipment, certain operating budgets, endowments, challenge/matching grants, land acquisition, new projects, start-up funds, renovation, and research. Requests for publication, videos/films, and website production may be considered as may scholarships, fellowships / chairs, conferences / seminars, social enterprise development, and technical assistance. Multi-year and renewable funding may be considered. Applications are accepted year-round. One copy of a letter proposal and/or common grant application form should be directed to the President in care of the Foundation Manager at the contact information given. An initial phone call is optional. Guidelines, instructions, and a copy of the common grant proposal form are provided at the website. Funding decisions and notification generally occur three to six months after receipt of request.
Requirements: 501(c)3 organizations in the Puget Sound region in and around Seattle, Washington are eligible to apply.
Restrictions: The Norcliffe Foundation does not provide the following types of assistance: deficit financing, emergency funds, grants to individuals, matches to employee-giving, Program-Related Investments / Loans (PRIs), in-kind services, volunteer / loaned executive, and internships. Applicants may only submit one request per year from date of funding or denial. Applicants who have previously received grants of $50,000 or more from the Foundation must wait two years from final payment before submitting a new application.
Geographic Focus: Washington
Amount of Grant: 1,000 - 25,000 USD
Contact: Arline Hefferline, Foundation Manager; (206) 682-4820; fax (206) 682-4821; arline@thenorcliffefoundation.com
Internet: http://www.thenorcliffefoundation.com/
Sponsor: Norcliffe Foundation
999 Third Avenue, Suite 1006
Seattle, WA 98104-4001

Nord Family Foundation Grants 2980
The Nord Family Foundation, in the tradition of its founders, Walter and Virginia Nord, endeavors to build community through support of projects that bring opportunity to the disadvantaged, strengthen the bond of families, and improve the quality of people's lives. Grants are awarded in the fields of arts and culture, civic affairs, education and health and social services. Most grants are made for program related activities, with some grants to support capital improvements and capital campaigns when special criteria are met. Projects that are directed at the root causes of social problems are of special interest. Grants range from $2,000 to $50,000.
Requirements: Applicants must be 501(c)3 non-profit organizations or selected public sector activities. Awards are made primarily in the environs of Lorain County, Ohio, with additional support in Cuyahoga County, Ohio. Particular attention is given to Cuyahoga County non-profits that provide services to Lorain County residents. Occasionally awards are made outside of this area, to organizations located in areas where voting members of the Foundation reside or to organizations with a national or international focus. Application is made online from the Foundation's website.
Restrictions: The Foundation does not support debt reduction, research projects, and tickets or advertising for fundraising activities.
Geographic Focus: Ohio
Date(s) Application is Due: Apr 1; Aug 1; Dec 1
Samples: Augustana Arts, Denver, Colorado, $10,000, support of the City Strings youth outreach; Amherst Downtown Betterment Association, Amherst, Ohio, commercial revitalization efforts; and Boys and Girls Clubs of Loraine County, Elyria, Ohio, $50,000, capital improvement.
Contact: John Mullaney, Executive Director; (800) 745-8946 or (440) 984-3939; fax (440) 984-3934; execdir@nordff.org or info@nordff.org
Internet: http://www.nordff.org
Sponsor: Nord Family Foundation
747 Milan Avenue
Amherst, OH 44001

Nordson Corporation Foundation Grants 2981
The corporate foundation awards grants to nonprofit organizations in geographic areas where Nordson facilities and employees are located: Cuyahoga and Lorain counties in Ohio; the Greater Atlanta, Georgia area; California, Rhode Island and Southeastern Massachusetts. The foundation provides a source of stable funding for community programs and projects in the areas of education, human welfare, civic, and arts and culture. Educational support is generally limited to improving elementary and secondary schools and certain programs for public and private higher education. Human welfare grants focus on children and youth. The foundation considers other funding areas, including urban affairs, volunteerism, public policy, health and health organizations, and literacy. Grants are awarded for general operating support, continuing support, annual campaigns, capital campaigns, building construction/renovation, equipment acquisition, emergency funds, seed money, scholarship funds, employee matching gifts, and technical assistance. The board meets four times each year. Deadlines vary by state; contact program staff for appropriate deadlines. Application forms are available on the website and may only be submitted online.
Requirements: Private, nonprofit organizations in California, Georgia, Ohio, Rhode Island, and southeastern Massachusetts may apply.
Restrictions: Funding is not provided for organizations whose services are not provided within the foundation's geographic areas of interest; direct grants or scholarships to individuals; organizations not eligible for tax-deductible support; organizations not exempt under Section 501(c)3 of the Internal Revenue Code; political causes, candidates, organizations or campaigns; organizations that discriminate on the basis of race, sex, or religion; or special occasion, goodwill advertising, i.e., journals or dinner programs.
Geographic Focus: California, Georgia, Massachusetts, Ohio, Rhode Island
Date(s) Application is Due: Feb 15; May 15; Aug 15; Nov 15
Amount of Grant: Up to 15,000,000 USD

Samples: Ctr for Leadership in Education (Lorain, OH)--for general operating support, $250,000; Salvation Army of Lorain (Lorain, OH)--for the capital campaign, $50,000; Neighborhood House Assoc of Lorain County (Lorain, OH)--for rejuvenation of the Cityview Ctr, $39,625.
Contact: Cecilia H. Render; (440) 892-1580, ext. 5172; crender@nordson.com
Internet: http://www.nordson.com/Corporate/Community/Foundation
Sponsor: Nordson Corporation Foundation
28601 Clemens Road
Westlake, OH 44145-1119

Norfolk Southern Foundation Grants 2982
The Norfolk Southern Foundation awards grants to nonprofits in the territory served by Norfolk Southern Railway Company. The foundation offers grants in four principal areas: educational programs, primarily at the post-secondary level; community enrichment focusing on cultural and artistic organizations; environmental programs; and health and human services (primarily food banks, homeless programs, and free clinics). Decision are made by December 31 for the following year's discretionary funding program, and applicants are notified early in the following calendar year.
Requirements: Applicants should include the following in their information packet: a valid 501(c)3 or 170(c)1 letter from the IRS; a short request outlining objectives of the organization, project, and intended use of funds; a listing of the applicant's board of directors; current and potential sources of funding; audited financial statements from the last three years; and the applicant's contact name, phone number, and email address.
Restrictions: Grant requests are only accepted between July 15 and September 30 for funding in the following calendar year. Applications will not be accepted from: organizations not in the NS territory (must be served by Norfolk Southern to be considered); organizations that do not have a 501(c)3 or 170(c)(1) IRS letter; individuals or organizations established to help individuals; religious, fraternal, social or veterans organizations; political or lobbying organizations; public or private elementary and secondary schools; fundraising events, telethons, races or benefits; sports or athletic organizations or activities; community or private foundations, or other organizations that merely redistribute to other eligible organizations aggregated contributions; disease-related organizations and hospitals; mentoring programs; Boys and Girls Scout programs or similar organizations; animal organizations; non-U.S. based charities; and organizations whose programs have national scope
Geographic Focus: Alabama, Connecticut, Delaware, District of Columbia, Florida, Georgia, Illinois, Indiana, Iowa, Kentucky, Louisiana, Maine, Maryland, Massachusetts, Michigan, Mississippi, Missouri, New Hampshire, New Jersey, New York, North Carolina, Ohio, Pennsylvania, Rhode Island, South Carolina, Tennessee, Texas, Vermont, Virginia, West Virginia, Wisconsin
Date(s) Application is Due: Sep 30
Amount of Grant: Up to USD
Contact: Katie Fletcher; (757) 629-2881; fax (757) 629-2361; katie.fletcher@nscorp.com
Internet: http://www.nscorp.com/nscportal/nscorp/Community/NS%20Foundation/
Sponsor: Norfolk Southern Foundation
P.O. Box 3040
Norfolk, VA 23514-3040

Norman Foundation Grants 2983
The Norman Foundation supports efforts that strengthen the ability of communities to determine their own economic, environmental and social well-being, and that help people control those forces that affect their lives. These efforts may: promote economic justice and development through community organizing, coalition building and policy reform efforts; work to prevent the disposal of toxics in communities, and to link environmental issues with economic and social justice; link community-based economic and environmental justice organizing to national and international reform efforts. The following is considered when evaluating grant proposals: does the project arise from the hopes and efforts of those whose survival, well-being and liberation are directly at stake; does it further ethnic, gender and other forms of equity; is it rooted in organized, practical undertakings; and is it likely to achieve systemic change. In pursuing systemic change, the Foundation would hope that: the proposed action may serve as a model; the spread of the model may create institutions that can survive on their own; their establishment and success may generate beneficial adaptations by other political, social and economic institutions and structures. The Foundation provides grants for general support, projects, and collaborative efforts. We also welcome innovative proposals designed to build the capacity of social change organizations working in our areas of interest. Priority is given to organizations with annual budgets of under $1 million.
Requirements: Programs must be 501(c)3 tax-exempt organizations that focused on domestic United States issues. Prospective grantees should initiate the application process by sending a short two or three page letter of inquiry to the Program Director (fax, email or regular mail). There are no set deadlines, and letters of inquiry are reviewed throughout the year. The Foundation only accepts full proposals upon positive response to the letter of inquiry. The letter of inquiry should briefly explain: the scope and significance of the problem to be addressed; the organization's proposed response and (if appropriate) how this strategy builds upon the organization's past work; the specific demonstrable effects the project would have if successful, especially its potential to effect systemic (fundamental, institutional and significant) change; how the project promotes change on a national level and is otherwise related to the foundation's guidelines; the size of the organization's budget. All inquiries will be acknowledged and, if deemed promising, the Foundation will request a full proposal.
Restrictions: The Foundation does not make grants to individuals or universities; or to support conferences, scholarships, research, films, media and arts projects; or to capital funding projects, fundraising drives or direct social service programs, such as shelters or community health programs. The Foundation's grant making is restricted to U.S.-based organizations.
Geographic Focus: All States
Amount of Grant: 5,000 - 30,000 USD
Samples: Direct Action Welfare Group, Charleston, West Virginia, $20,000 - renewed support of statewide organizing and empowerment of people living in poverty in West Virginia; Interfaith Action of Southwest Florida, Immokalee, Florida, $20,000 - renewed support for joint work with Coalition of Immokalee Workers in support of low-wage farmworkers in Florida; Alaska Community Action on Toxics, Anchorage, Alaska, $30,000 - renewed support for work with tribal villages and other communities to eliminate environmental contaminants.
Contact: June Makela; (212) 230-9830; fax (212) 230-9849; norman@normanfdn.org
Internet: http://www.normanfdn.org/
Sponsor: Norman Foundation
147 East 48th Street
New York, NY 10017

North Carolina Biotechnology Center Event Sponsorship Grants 2984
This popular Biotechnology Center grant promotes and supports events advancing the understanding or application of biotechnology to benefit North Carolina. Events must promote information sharing and personal interaction focused on biotechnology research, education, or business. The highest amount awarded is $3,000. Awards typically cover costs such as speaker fees, travel expenses, event-site rental and publicity.
Requirements: All award money is disbursed directly to the applicant's organization. The money must be used solely to support the event outlined in this application.
Restrictions: This sponsorship does not provide money either to promote a specific product to benefit a few companies or individuals. This sponsorship does not cover food or refreshments for event attendees.
Geographic Focus: North Carolina
Amount of Grant: Up to 3,000 USD
Contact: Ginny DeLuca, (919) 549-8842; virginia_deluca@ncbiotech.org
Internet: http://www.ncbiotech.org/services_and_programs/grants_and_loans/biotechnology_event_sponsorship/index.html
Sponsor: North Carolina Biotechnology Center
15 T.W. Alexander Drive, P.O. Box 13547
Research Triangle Park, NC 27709-3547

North Carolina Biotechnology Center Grantsmanship Training Grants 2985
The Grantsmanship Training Grant (GTG) program is a one-time funding opportunity that provides financial support for grant writing training activities for faculty and staff of North Carolina colleges and universities and other non-profits who are involved in research or programs related to the life sciences. Funds can be requested to support a variety of grantsmanship training activities, including workshops and other training events. The following individuals and groups would be eligible for training support under the GTG program: academic faculty members who seek funding related to life science research or life science-related economic development; research administration staff affiliated with departments or institutions that regularly submit proposals in life science research or life science-related economic development; and staff of non-profit organizations involved in activities related to biotechnology or the life sciences. A maximum of $5,000 may be requested based on the type of training activity proposed.
Requirements: Proposed programs and activities must clearly relate to biotechnology research, development, commercialization, education, or training. Priority will be giving to proposals requesting support for innovative programs or activities in biotechnology. Applicants are required to contribute a cash match equaling the requested amount (dollar-for-dollar).
Restrictions: Awards are not made to individuals.
Geographic Focus: North Carolina
Amount of Grant: Up to 5,000 USD
Contact: Deborah De; (919) 549-8845; fax (919) 549-9710; deborah_de@ncbiotech.org
Internet: http://www.ncbiotech.org/gtg/
Sponsor: North Carolina Biotechnology Center
15 T.W. Alexander Drive, P.O. Box 13547
Research Triangle Park, NC 27709-3547

North Carolina Biotechnology Center Meeting Grants 2986
This grant promotes and supports national and international meetings which advance the understanding or application of biotechnology and focus national and international attention on the North Carolina scientific community. Meetings must promote information sharing and personal interaction focused on biotechnology research, education, or business. The highest amount awarded is $10,000. Awards typically cover costs such as speaker fees, travel expenses, event-site rental and publicity.
Requirements: All award money is given to the applicant's organization. The money must be used solely to support the event outlined in this application.
Restrictions: This sponsorship does not provide money either to promote a specific product to benefit a few companies or individuals. This sponsorship does not cover food or refreshments for event attendees.
Geographic Focus: North Carolina
Amount of Grant: Up to 10,000 USD
Contact: Ginny DeLuca, (919) 549-8842; virginia_deluca@ncbiotech.org
Internet: http://www.ncbiotech.org/services_and_programs/grants_and_loans/BiotechnologyMeetingGrant.html
Sponsor: North Carolina Biotechnology Center
15 T.W. Alexander Drive, P.O. Box 13547
Research Triangle Park, NC 27709-3547

North Carolina Biotechnology Center Regional Development Grants   2987
The goal of the Regional Development Grant Program (PDF) is to build capacity through collaborative projects, providing a foundational resource for biotechnology development in the community that was not there previously and that the community could not have achieved without Biotechnology Center funding. Preproposals are required: cycle 1 preproposals are due no later than January 28, while cycle 2 preproposals should be received by August 20. The deadlines for final proposals are April 1 (cycle 1) and October 15 (cycle 2).
Requirements: Proposed programs and activities must clearly relate to biotechnology research, development, commercialization, education, or training. Priority will be giving to proposals requesting support for innovative programs or activities in biotechnology. Only North Carolina academic institutions, non profit research organizations, or businesses may apply.
Restrictions: Awards are not made to individuals.
Geographic Focus: North Carolina
Date(s) Application is Due: Jan 28; Apr 1; Aug 20; Oct 15
Amount of Grant: Up to 75,000 USD
Contact: Deborah De; (919) 549-8845; fax (919) 549-9710; deborah_de@ncbiotech.org
Internet: http://www.ncbiotech.org/services_and_programs/grants_and_loans/regional_development/index.html
Sponsor: North Carolina Biotechnology Center
15 T.W. Alexander Drive, P.O. Box 13547
Research Triangle Park, NC 27709-3547

North Carolina Biotechnology Centers of Innovation Grants   2988
The Centers of Innovation Program is bringing together North Carolina's best scientific and technical minds in the life sciences. This program is designed to focus the state's efforts in biotechnology research, development and commercialization in targeted industrial sectors important to economic development and job creation. The North Carolina Biotechnology Center (NCBC) will work with university researchers, technology transfer offices, industrial partners, non-profit stake holders as well as regional and state-wide community leaders to establish nine Centers of Innovation (COI). Initial Centers of Innovation will complement efforts already under way in the state to align academic and industrial resources. The Phase I award is a 12-month $100,000 planning grant that will allow the development of a business plan and organization plan. The Phase II award is funding for the implementation of the business plan. Only invited applicants may apply. Potential applicants will need to demonstrate that they have established a cohesive academic-industry consortium in order to receive an invitation to submit an application.
Geographic Focus: North Carolina
Amount of Grant: Up to 2,400,000 USD
Contact: Mary Beth Thomas, (919) 541-9366; marybeth_thomas@ncbiotech.org
Internet: http://www.ncbiotech.org/services_and_programs/COI/index.html
Sponsor: North Carolina Biotechnology Center
15 T.W. Alexander Drive, P.O. Box 13547
Research Triangle Park, NC 27709-3547

North Carolina Biotechnology Center Technology Enhancement Grants   2989
The Technology Enhancement Grant (TEG) Program provides funding to North Carolina Universities or other research institutions in the state through their respective technology transfer offices. Under this program, awards of up to $50,000 are available to fund research studies determined to be critical to enhancing the university's licensing position for specified technologies with commercial opportunity. Eligible technologies are those for which intellectual property rights are assigned to the university or institution and have not been committed through license, option or letter-of-intent to license to any third party commercial interest at the time of award. Funds are intended to support an approved research project designed to yield a significant technical milestone in the evolution of a technology from basic science to licensable commercial product or service. Examples of such projects include proof-of-concept, validation, lead identification and optimization using a proprietary platform technology, etc. Due to legal constraints, funds provided under the TEG program can only be used to offset direct costs associated with the project.
Requirements: Technology transfer offices associated with any North Carolina university or not-for-profit research institution are eligible and encouraged to apply.
Restrictions: Prohibited charges include (but are not limited to): purchase or lease of real estate, building construction or renovations; salaries for administrative personnel; overhead or indirect costs; and atenting costs.
Geographic Focus: North Carolina
Amount of Grant: Up to 50,000 USD
Contact: Dr. Rob Lindberg, (919) 541-9366; rob_lindberg@ncbiotech.org
Internet: http://www.ncbiotech.org/services_and_programs/grants_and_loans/technology_enhancement/index.html
Sponsor: North Carolina Biotechnology Center
15 T.W. Alexander Drive, P.O. Box 13547
Research Triangle Park, NC 27709-3547

North Carolina Community Foundation Grants   2990
The community foundation exists to meet the needs of nonprofit organizations and improve the lives of citizens of North Carolina counties, including Allegheny, Caldwell, Cherokee, Clay, Craven, Eastern Bank of Cherokee Indians, Franklin Community, Franklin, Graham, Harnett, Haywood, Jackson, Johnston, Madison, Montgomery, Moore, Pender, Pitt, Randolph, Rockingham, Swain, Wake, Watauga, and Wilkes. Support is available for programs that enrich the quality of life in the areas of civic affairs, education, health, religion, social services, arts, and the conservation and preservation of environmental, historical preservation, and cultural resources. Grants will be awarded to support programs and projects, endowments, conferences and seminars, and consulting services. Grants are awarded in local communities; use the Affiliate Webpage link to contact the office serving your area.
Requirements: Nonprofits throughout North Carolina are eligible to apply.
Geographic Focus: North Carolina
Contact: Jennifer Tolle Whiteside, President; (800) 201-9533 or (919) 828-4387; fax (919) 828-5495; jtwhiteside@nccommunityfoundation.org
Internet: http://nccommunityfoundation.org/05_grants_available.php
Sponsor: North Carolina Community Foundation
4601 Six Forks Road, Suite 524
Raleigh, NC 27609

North Carolina GlaxoSmithKline Foundation Grants   2991
The foundation supports activities primarily in North Carolina that help meet the needs of today's society and future generations by funding programs that emphasize the understanding and application of health, science, and mathematics at all educational and professional levels. Although providing seed funds for new and worthwhile educational programs is the foundation's primary focus, requests for funding of ongoing projects will also receive consideration. Types of support include operating budgets, professorships, and conferences and seminars. Proposals may be submitted for either one-year or multi-year funding with a maximum of five years' duration. The foundation's board of directors meets four times a year to consider and award grants. Completed applications must be received by the listed deadline dates. If the stated deadline falls on a weekend or holiday, the next business day serves as the official deadline. All applicants will be notified of the board's decisions within 15 days of the board meeting.
Requirements: The foundation makes grants only to North Carolina 501(c)3 tax-exempt organizations and institutions or to governmental agencies.
Restrictions: Grants are not made to individuals for construction or restoration projects, or for international programs unless specifically exempted by the board. Funds are not ordinarily provided to programs that benefit a limited geographic region.
Geographic Focus: North Carolina
Date(s) Application is Due: Jan 1; Apr 1; Jul 1; Oct 1
Amount of Grant: 25,000 - 1,000,000 USD
Contact: Marilyn Foote-Hudson, Executive Director; (919) 483-2140; fax (919) 315-3015; community.partnership@gsk.com
Internet: http://us.gsk.com/html/community/community-grants-foundation.html
Sponsor: North Carolina GlaxoSmithKline Foundation
5 Moore Drive, P.O. Box 13398
Research Triangle Park, NC 27709

North Central Health Services Grants   2992
North Central Health Services is committed to addressing a wide range of health issues, and to enhancing the quality of life for individuals, families, and communities in our eight county Service Area. NCHS will support not-for-profit organizations and agencies that share our commitment to Health and Healthy Communities, primarily through grants for capital projects. NCHS prefers to fund projects that have a significant potential for positive impact on the community. NCHS encourages collaboration among agencies and organizations that efficiently utilize resources, and provide programs and services that are not duplicative. The first step for an organization to apply for a grant is to submit a written letter of inquiry. This letter should include a concise description of the organization, a statement of the problem or need being addressed, and an explanation of the project, estimated total costs, and the amount being requested from NCHS. Grants have ranged from $600 to $5,000,000.
Requirements: Applicants should be 501(c)3 not-for-profit organizations who serve Benton, Carroll, Clinton, Fountain, Montgomery, Tippecanoe, Warren, and White Counties.
Geographic Focus: Indiana
Amount of Grant: Up to 5,000,000 USD
Samples: Lyn Treece Boys and Girls Club, Lafayette, Indiana, $150,000 - to expand program space at Beck Lane site Boys & Girls Club to serve increasing numbers of kids who are attending the club; Tippecanoe Arts Federation, Lafayette, Indiana, $300,000 - to administer a distribution program for 2012 capital projects funding to arts and cultural member organizations within the eight county NCHS service area; Montgomery County Youth Camps, Crawfordsville, Indiana, $19,000 - to add bathrooms and septic system to meet increasing utilization of Camp Rotary.
Contact: Rita Smith, Program Director, (765) 423-1604
Internet: http://www.nchsi.com/grantmaking.cfm
Sponsor: North Central Health Services
P.O. Box 528
Lafayette, IN 47902

North Central Health Services Grants   2993
North Central Health Services (NCHS) was created in 1984 to serve as the parent company of a family of corporations which included Lafayette Home Hospital, Home Hospital Foundation, and Service Frontiers Incorporated. Today, NCHS is a medical services company whose primary purpose is to operate an ambulatory surgery center, where sterilization services are made available to the Greater Lafayette community. NCHS also is a grant making organization providing primarily capital grants to 501(c)3 organizations serving the citizens of Benton, Carroll, Clinton, Fountain, Montgomery, Tippecanoe, Warren, and White counties in Indiana. Grants are awarded to organizations for projects that relate to Health, and Healthy Communities. NCHS prefers to fund projects that have a significant potential for positive impact on the community. NCHS encourages collaboration among agencies and organizations that efficiently utilize resources, and provide services that are not duplicative of existing programs. NCHS provides grants that will: advance technology and scientific development in health care; develop and promote community health education; provide direct, individual social welfare and human service; promote a Healthy Community; improve the quality of life for all the communities people.

Requirements: The first step for an organization to apply for a grant is to submit a written letter of inquiry. This letter should include a concise description of the organization, a statement of the problem or need being addressed, and an explanation of the project, estimated total costs, and the amount being requested from NCHS. The letter should be addressed to: North Central Health Services, Inc., Attn: Rita Smith, Program Director, P.O. Box 528, Lafayette, In. 47902. Upon receipt of the letter of inquiry, initial review will determine whether or not the request meets the funding focus and priorities of NCHS. You will be promptly notified in writing of the initial determination, and informed of application opportunities.
Restrictions: Beginning in 2008, NCHS implemented the following grant categories and cycles. Grant requests and cycles: letter of inquiry requesting funds in excess of $250,000, accepted May 1 to August 15 of each calendar year. If approved, funding would be available for distribution after January 1 of the following year; letter of inquiry requesting funds of $100,000 to $250,000 will be accepted throughout the calendar year. If approved, funding would generally be available for distribution within 90 days of board approval; letter of inquiry requesting funds of $25,000 to $100,000 will be accepted throughout the calendar year. If approved, funding would generally be available within 60 days of board approval; letter of inquiry requesting funds of less than $25,000 will be accepted throughout the calendar year. If approved, funding would generally be available within 60 days of approval.
Geographic Focus: Indiana
Samples: Lauramie Township EMS - Purchase Zoll E series AED to update equipment for EMS basic life support ambulance service, $8,693.00; St. Elizabeth Regional Health - Gifts for hospitalized children, $4,155.00; Indiana Regular Baptist Youth Camp, Inc. - Build a new Health Center for Twin Lakes Camp & Confererence Center, $150,000.00.
Contact: Rita Smith, Program Director; (765) 423-1604
Internet: http://www.nchsi.com/grantmaking.cfm
Sponsor: North Central Health Services
201 Main Street, Suite 606, P.O. Box 528
Lafayette, IN 47902

North Dakota Community Foundation Grants 2994
The Foundation serves North Dakota communities statewide with the goal of improving the quality of life for the state's citizens. The Foundation does not have a narrow area of focus. Each project is reviewed on its own merits with an emphasis on helping applicants who have limited access to other sources of funding. Applicants may submit a concise letter of request not to exceed two pages describing the organization, the project, the approximate project cost, and the amount requested from the Foundation. If the Board is interested in additional information, formal application materials will be sent.
Requirements: Eligible applicants must be 501(c)3 organizations in North Dakota.
Restrictions: The Foundation accepts applications by mail but not by fax or email. Only one request per agency per year may be submitted. The following is not funded: grants to organizations and projects that exist to influence legislation, carry on propaganda, participate in political campaigns, or which threaten to cause significant controversy or divisiveness; grants to individuals; and multi-year grant commitments. Low priority is given to the following: projects already substantially supported by government, or which in the opinion of the Board, can and should be provided for by taxes; grants for sectarian projects; grants to national organizations; and organizations which field substantial fund-raising each year with paid and volunteer staff.
Geographic Focus: North Dakota
Date(s) Application is Due: Aug 15
Contact: Kevin Dvorak; (701) 222-8349; fax (701) 222-8349; kdvorak@ndcf.net
Internet: http://www.ndcf.net/Information/GrantGuidelines.asp
Sponsor: North Dakota Community Foundation
309 North Mandan Street, Suite 2, P.O. Box 387
Bismarck, ND 58502-0387

Northern Chautauqua Community Foundation Community Grants 2995
The foundation awards grants to nonprofit, tax-exempt organizations located in Northern Chautauqua County, New York. Priority will be given to programs representing innovative and efficient approaches to serving community needs and opportunities, projects that assist citizens whose needs are not met by existing services, projects that expect to test or demonstrate new approaches and techniques in the solutions of community problems, and projects that promote volunteer participation and citizen involvement in the community. Consideration will be given to the potential impact of the request and the number of people who will benefit. Seed grants will be awarded to initiate promising new programs in the foundation's field of interest as well as challenge grants to encourage matching gifts. Except in unusual circumstances, grants are approved for one year at a time. Contact the foundation office for an application.
Requirements: 501(c)3 organizations in Northern Chautauqua County, New York, are eligible.
Restrictions: Areas generally not funded are capital campaigns to establish or add to endowment funds, general operating budgets for existing organizations, publication of books, conferences, or annual fund-raising campaigns.
Geographic Focus: New York
Date(s) Application is Due: Mar 23; Sep 21
Amount of Grant: 12,500 USD
Contact: Diane E. Hannum, Executive Director; (716) 366-4892; fax (716) 366-3905; dhannum@nccfoundation.org or info@nccfoundation.org
Internet: http://www.nccfoundation.org/Grantseekers/ApplyforaGrant/tabid/256/Default.aspx
Sponsor: Northern Chautauqua Community Foundation
212 Lake Shore Drive West
Dunkirk, NY 14048

Northern New York Community Foundation Grants 2996
The foundation is looking for innovative programs that address problems to be solved, or opportunities to be seized, in the northern New York area. Grants are made only for capital items and for seed money for new agencies, or new projects by established agencies.
Requirements: Grants are made to nonprofit, tax-exempt organizations in or serving Jefferson and/or Lewis counties in New York State. Governmental units are also eligible provided that the purpose of the grant goes beyond the expected limits of government service. The foundation does not make grants to churches and religious organizations except in cases where projects clearly benefit the entire community. Applicants should contact the office to provide a brief summary of the project and needs. The applicant will be notified of grant award or denial by letter after the board of directors meets, which is generally one month following application deadline dates. Grant requests under $5,000 may be approved at the committee level. Larger grant requests are referred to the board of directors with a committee recommendation.
Restrictions: Individuals are ineligible. The foundation does not fund operations or pay off deficits.
Geographic Focus: New York
Date(s) Application is Due: Jan 30; Apr 24; Aug 28; Oct 23
Samples: 4th Qtr 2008: The Handweaving Museum and Arts Center - $35,000 to help fund a new Director of Development position; Northern Area Health Education Center (NAHEC) - $3,575 for student internships in technical fields; Ogdensburg Command Performances - $7,000 to match a Land Trust Alliance grant to develop a Point Peninsula conservation plan; Pendragon Theatre - $2,784 for three Shakespeare performances at Jefferson Community College; Trinity Episcopal Church, Watertown - $2,500 for a Trinity Concert Series performance of Handel's Messiah; Watertown Housing Authority - $5,000 to help fund a new playground for East Hills and Meadowbrook family housing projects; American Red Cross of Northern New York - $5,306 for access to donor management software.
Contact: Alex Velto; (315) 782-7110; fax (315) 782-0047; info@nnycf.org
Internet: http://www.nnycf.org/grantguidelines.asp?mm=5
Sponsor: Northern New York Community Foundation
120 Washington Street, Suite 400
Watertown, NY 13601

Northrop Grumman Corporation Grants 2997
Northrop Grumman is committed to supporting communities throughout the U.S., especially those where its employees live and work. The Contributions Program seeks to address critical issues and needs by providing financial assistance to accredited schools and 501(c)3 nonprofit organizations. The majority of these contributions address education, services for veterans and the military, health and human services, and the environment.
Requirements: The following information must be included in the grant request: Tax Exempt Number; non-profit contact name, title, address, phone number, fax number; brief history of the organization, mission statement, goals and objectives; type and scope of services offered, and the geographical area served; specific details as to how requested funds would be used; project budget and amount requested; demographic impact; current operating budget including latest financial statement; list of other corporate funders; list of directors and/or officers, and their affiliations; Northrop Grumman employee sponsor, if applicable (name and phone number); and contact information.
Restrictions: Grants are not awarded to: religious organizations; political groups; fraternal organizations; individuals; athletic groups or activities, including charity-benefit sporting events; charter schools, unless they have open enrollment and hold the same standards as public schools; bands or choirs; capital campaigns; organizations providing services primarily to animals; communities outside of the United States; or organizations whose programs discriminate based on race, color, age, sex, religion, national origin, sexual orientation, disability, veteran status or any other characteristic protected by law.
Geographic Focus: All States
Date(s) Application is Due: Apr 30; Sep 30
Contact: Cheryl Horn, cheryl.horn@ngc.com
Internet: http://www.northropgrumman.com/corporate-responsibility/corporate-citizenship/contribution-guidelines.html
Sponsor: Northrop Grumman Corporation
1840 Century Park East
Los Angeles, CA 90067

Northwest Airlines KidCares Medical Travel Assistance 2998
The program provides air travel to children age 18 and younger who are unable to receive treatment in their home area. The KidCares program is fueled by generous mileage donations of Northwest's WorldPerks members. Availability of the program is based on that donated WorldPerks mileage, and consideration of each request is based on the guidelines below. Each application received must be evaluated on a case-by-case basis.
Requirements: Guidelines are as follows: (1) Only one round-trip per family is permitted. (2) A letter from the child's physician must accompany the application, stating: (a) patient's diagnosis; (b) confirmation of the need for travel; (c) the patient has medical clearance for air travel; (d) dates of appointments/surgery; and (e) length of stay. (3) The patient must be 18 years old or younger at the time the travel request is made. (4) The medical treatment for which the child is traveling must be unavailable in the child's home location. (5) KidCares travel can only be provided by Northwest Airlines or one of its airline partners. (6) Families demonstrating financial need are given priority consideration for KidCares travel. (7) One adult companion may accompany the child on KidCares travel at no additional cost. Applications may be found on Northwest's Web site, nwa.com at the following location http://www.nwa.com/corpinfo/aircares/about/KidCares_Application.pdf. You may also call (612) 726-4206 and leave detailed information regarding where the application should be sent and to whom.

Restrictions: KidCares travel is not available to families with insurance coverage that covers the cost of travel for medical treatment. KidCares travel is not eligible for Northwest Airlines WorldPerks bonus mileage credit. KidCares travel is not provided to St. Jude Children's Research Hospital, which is covered by a separate Northwest Airlines travel program.
Geographic Focus: All States
Contact: Manager; (612) 726-4206; fax (612) 726-3942; aircares@nwa.com
Internet: http://www.nwa.com/corpinfo/aircares/feature/
Sponsor: Northwest Airlines
2700 Lone Oak Parkway, Department A1300
St. Paul, MN 55121

Northwestern Mutual Foundation Grants 2999
The people of Northwestern Mutual have a long history of community involvement. To carry out its corporate contributions program, Northwestern Mutual created a private foundation, The Northwestern Mutual Foundation. The Northwestern Mutual Foundation focuses its funding in Milwaukee, where the company is headquartered. Therefore, applying for a grant from the Northwestern Mutual Foundation is limited to Milwaukee-area nonprofit organizations with programs centered on Education, Health & Human Services or Arts & Culture.
Requirements: Northwestern Mutual Foundation does not accept unsolicited grant requests. Groups new to the Foundation are asked to submit a written letter of intent, describing the organization, funding request, and provide information about the program for which you are seeking support. The Foundations team will review your letter of intent and, if they are interested in learning more about your project, you will receive an application form. Groups that have an active relationship with the Foundation and currently receive funding are allowed to apply annually for support. Once an organization submits an application, the review process takes about 90 days. The Foundation review's Milwaukee-area grants on a monthly basis.
Restrictions: Milwaukee-area nonprofits are eligible to apply. The Foundation defines the Milwaukee-area as Milwaukee County, with limited programmatic grants in Waukesha, Racine and Ozaukee Counties.
Geographic Focus: Wisconsin
Amount of Grant: 10,000 - 800,000 USD
Samples: Boys and Girls Clubs of Greater Milwaukee, Milwaukee, WI, $800,000; United Performing Arts Fund, Milwaukee, WI, $605,000; Scholarship America, Saint Peter, MN, $205,425;
Contact: Edward J. Zore, President; (414) 271-1444
Internet: http://www.nmfn.com/tn/aboutus--fd_intro
Sponsor: Northwestern Mutual Foundation
720 East Wisconsin Avenue
Milwaukee, WI 53202-4797

Northwest Minnesota Foundation Women's Fund Grants 3000
Formed in 1997 as a tax exempt charitable component of the Northwest Minnesota Foundation (NMF), the Women's Fund serves as a catalyst for improving the quality of life for women and girls by promoting ideas and supporting programs for the continued strengthening and empowerment of all women. The Women's Fund will consider grant requests of up to $2,500 addressing one or more of the following areas meeting the needs of females: cultural interactions; economic development; education; healthcare; and housing.
Requirements: Minnesota counties served by NMF are Beltrami, Clearwater, Hubbard, Kittson, Lake of the Woods, Mahnomen, Marshall, Norman, Pennington, Polk, Red Lake, and Roseau. Priority is given to applications supporting the following goals: developing entrepreneurial and economic opportunities for women; supporting programs that lead to a safer environment for women and girls; building networks of women in leadership positions throughout the region; and encouraging and equipping women and girls to achieve their full potential.
Geographic Focus: Minnesota
Date(s) Application is Due: May 1
Amount of Grant: 2,500 USD
Contact: Lisa Peterson, Communications Director; (800) 659-7859; lisap@nwmf.org
Internet: http://www.nwmf.org/component-funds/community-fund-websites/nw-mn-womens-fund/womens_fund_grants_and_programs.html
Sponsor: Northwest Minnesota Foundation
4225 Technology Drive NW
Bemidji, MN 56601

Norwin S. and Elizabeth N. Bean Foundation Grants 3001
The Norwin S. and Elizabeth N. Bean Foundation was established in 1967 as a general purpose foundation to serve the communities of Manchester and Amherst and, consistent with the wishes of the founders, grants are made in the fields of arts and humanities, education, environment, health, human services, and public/society benefit.
Requirements: Applications are accepted from nonprofit 501(c)3 organizations and municipal and public agencies serving the communities of Manchester and Amherst, New Hampshire. Priority consideration is given to organizations operating primarily in those two communities. However, the Foundation will consider applications from statewide or regional organizations which provide a substantial and documented level of service to Manchester and Amherst.
Restrictions: The foundation does not make grants to individuals or provide scholarship aid.
Geographic Focus: New Hampshire
Date(s) Application is Due: Apr 1; Sep 1; Dec 1
Contact: Kathleen Cook, Manager; (603) 493-7257; KCook@BeanFoundation.org
Internet: http://www.beanfoundation.org/grants/bean-foundation-grant/application-criteria.aspx
Sponsor: Norwin S. and Elizabeth N. Bean Foundation
40 Stark Street
Manchester, NH 03301

Notsew Orm Sands Foundation Grants 3002
The Notsew Orm Sands Foundation was established in 1995 via an initial donation from Charles Burnett, III, a British-born race car driver who set the steam-powered land speed record in 2009. Giving is on a national and international basis, with some emphasis on Houston, Texas, and the United Kingdom. Primary fields of interest are: higher education, hospitals, human services, medical research, and Protestant agencies and churches. There are no specific deadlines, and applicants should submit a detailed description of project and amount of funding requested, along with a copy of current year's organizational budget and/or project budget.
Restrictions: No grants are given to individuals.
Geographic Focus: All States
Amount of Grant: 250 - 75,000 USD
Samples: Baylor College of Medicine, Houston, Texas, $75,000; University of Saint Thomas, Houston, Texas, $61,000; Yale University, New Haven, Connecticut, $55,636; Centurion Ministries, Princeton, New Jersey, $20,000.
Contact: Charles Burnett, III, President and Secretary; (281) 497-0744
Sponsor: Notsew Orm Sands Foundation
2470 S. Dairy Ashford Street, Suite 802
Houston, TX 77077-5716

NSERC Brockhouse Canada Prize for Interdisciplinary Research in Science 3003
and Engineering Grant
The Brockhouse Canada Prize for Interdisciplinary Research in Science and Engineering recognizes outstanding Canadian teams of researchers from different disciplines who have combined their expertise to produce achievements of outstanding international significance in the natural sciences and engineering. The prize, accompanied by a team research grant of $250,000, reflects NSERC's commitment to supporting Canadian research through strategic investments in people, discovery and innovation. The grant supports the direct costs of university-based research and/or the enhancement of research facilities. The grant may be distributed in one lump sum or up to five instalments, depending on the needs of the recipients.
Requirements: Research teams nominated for the Brockhouse Canada Prize must have at least two members who are independent researchers, one of whom must hold an NSERC grant. The team can be part of an international effort, but the majority of the nominated team members must be employed at a Canadian university or public or private organization. NSERC recognizes that teams may change between the time of the specific research achievements and the time of nomination. Nominations will be accepted when changes have occurred but only as long as the core of the team remains intact. Contributions must be primarily in the natural sciences and engineering, and of an interdisciplinary and collaborative nature.
Restrictions: Any Canadian citizen may nominate a research team for the Brockhouse Canada Prize. Self nominations will not be accepted. Current NSERC Council members cannot be nominated. Since the prize is for interdisciplinary research, nominators should consult NSERC's Guidelines for the Preparation and Review of Applications in Interdisciplinary Research. Nominators should also consult NSERC's Guidelines for the Preparation and Review of Applications in Engineering and the Applied Sciences. See website for deadline & hyper links to above mentioned guidelines.
Geographic Focus: All States, Canada
Date(s) Application is Due: Jun 1
Amount of Grant: 250,000 USD
Contact: Administrator; (613) 995-5829; fax (613) 992-5337; brockhouse@nserc-crsng.gc.ca
Internet: http://www.nserc-crsng.gc.ca/Prizes-Prix/Brockhouse-Brockhouse/Index-Index_eng.asp
Sponsor: Natural Sciences and Engineering Research Council of Canada
350 Albert Street
Ottawa, ON K1A 1H5 Canada

NSF-NIST Interaction in Chemistry, Materials Research, Molecular 3004
Biosciences, Bioengineering, and Chemical Engineering
This program is intended to facilitate interactions between faculty and students supported by the National Science Foundation (NSF) and scientists at the National Institute of Standards and Technology's (NIST) Chemical Science and Technology Laboratory (CSTL) and Materials Science and Engineering Laboratory (MSEL), including the NIST Center for Neutron Research (NCNR). Chemistry, materials research, molecular biology, bioengineering, and chemical engineering are centralized at NIST in these laboratories. Support may be requested for supplements to existing NSF awards to provide the opportunity for faculty and students to participate in research at NIST facilities. Support may be requested as supplements to existing NSF awards for travel expenses and per diem associated with work at these NIST laboratories for faculty, students and other personnel associated with the NSF/NIST activity. It is estimated that 10-20 supplements will be made annually, up to $20,000 each.
Requirements: Support may be requested only by organizations that have currently active awards in any of the participating NSF Divisions, viz. the Division of Bioengineering and Environmental Systems (BES), the Division of Chemical and Transport Systems (CTS), the Division of Chemistry (CHE), the Division of Materials Research (DMR), and the Division of Molecular and Cellular Biosciences (MCB). Only faculty who are principal investigators on current NSF awards from BES, CHE, CTS, DMR, or MCB are eligible to apply for supplements. Standard NSF Grant Proposal Guidelines apply (see http://www.nsf.gov/publications/pub_summ.jsp?ods_key=gpg).

GRANT PROGRAMS | 485

Restrictions: Indirect costs (F&A) are not allowed.
Geographic Focus: All States
Amount of Grant: Up to 20,000 USD
Contact: Dr. William F. Koch, Deputy Director; (301) 975-8301; fax (301) 975-3845; william.koch@nist.gov
Internet: http://www.nsf.gov/publications/pub_summ.jsp?ods_key=nsf03568
Sponsor: National Science Foundation
4201 Wilson Boulevard
Arlington, VA 22230

NSF Accelerating Innovation Research 3005
To accelerate the process of innovation, NSF is undertaking two related, new activities. The first will encourage the translation of the numerous, technologically-promising, fundamental discoveries made by NSF researchers, while drawing upon and building the entrepreneurial spirit of the researchers and students. The second activity will foster connections between an existing NSF innovation research alliance (including consortia such as Engineering Research Centers (ERC), Industry University Cooperative Research Centers (I/UCRC), Partnerships for Innovation (PFI), Science and Technology Centers (STC), Nanoscale Science and Engineering Centers (NSEC), Materials Research Science and Engineering Centers (MRSEC) grantees) and other institutions, whose complementary focus will spur the development of discoveries into innovative technologies through collaboration. Both of these activities are designed to strengthen the U.S. innovation ecosystem.
Requirements: For the Technology Translation Competition, the Principal Investigator or a co-PI must be a current or prior NSF awardee and a faculty member at a U.S. college or university at the time of award in the current competition. Subject technology must be derived from a discovery research project already conducted or initiated by the National Science Foundation. For the Research Alliance Competition, one of the partners must be an NSF funded innovation research alliance (including Centers for Analysis and Synthesis, Centers for Chemical Innovation, Engineering Research Centers, Industry/University Collaborative Research Centers, Materials Research Science and Engineering Centers, Nanoscale Science and Engineering Centers, Science and Technology Centers, and Science of Learning Centers). The research alliance must be active at the time of award in the current competition. The other partner(s) may be another research entity (either NSF-funded, other government agency funded, or privately funded), a small business consortium, or a local or regional innovation entity.
Geographic Focus: All States
Date(s) Application is Due: Feb 1
Amount of Grant: 350,000 - 1,000,000 USD
Contact: Rathindra DasGupta; (703) 292-8353; fax (703) 292-9057; rdasgupt@nsf.gov
Internet: http://www.nsf.gov/funding/pgm_summ.jsp?pims_id=503553
Sponsor: National Science Foundation
4201 Wilson Boulevard
Arlington, VA 22230

NSF Alan T. Waterman Award 3006
The annual award recognizes an outstanding young researcher in any field of science or engineering supported by the National Science Foundation. In addition to a medal, the awardee receives a grant of $1,000,000 over a five year period for scientific research or advanced study in the mathematical, physical, biological, engineering, social, or other sciences at the institution of the recipient's choice.
Requirements: Institutions may nominate an unlimited number of individuals. Nomination packages consist of a nomination and four letters of reference submitted via FastLane. Candidates must be U.S. citizens, normally 35 years of age or younger, or not more than seven years beyond receipt of the PhD degree by December 31 of the year nominated. Candidates should have demonstrated exceptional individual achievements in scientific or engineering research of sufficient quality to place them at the forefront of their peers. Criteria include originality, innovation, and significant impact on the field.
Geographic Focus: All States
Date(s) Application is Due: Dec 31
Amount of Grant: 1,000,000 USD
Samples: David Charbonneau, Professor of Astronomy, Harvard University; Terence Tao, Professor of Mathematics, University of California, Los Angeles; Peidong Yang, Associate Professor of Chemistry, University of California, Berkeley.
Contact: Mayra N. Montrose; (703) 292-8040; fax (703) 292-9040; mmontros@nsf
Internet: http://www.nsf.gov/od/waterman/waterman.jsp
Sponsor: National Science Foundation
4201 Wilson Boulevard
Arlington, VA 22230

NSF Animal Developmental Mechanisms Grants 3007
Animal Developmental Systems programmatic area supports research that seeks to understand the processes that result in the complex phenotype of animals. Because different organisms may be more amenable to certain approaches than others, analyses of development in a wide range of different species are encouraged. The Developmental Systems Cluster is also particularly interested in understanding how emergent properties result in the development of complex phenotypes and lead to the evolution of developmental mechanisms. Proposals may be submitted at any time; however, receipt by target dates will ensure review at the regular panel meeting. Full proposals should be submitted via the FastLane system at NSF or Grants.gov.
Requirements: The most frequent recipients of support are academic institutions and nonprofit research groups. In special circumstances, grants also are awarded to other types of institutions and to individuals.

Geographic Focus: All States
Date(s) Application is Due: Jan 12; Jul 12
Contact: James Deshler, Program Director; (703) 292-8417; jdeshler@nsf.gov
Internet: http://www.nsf.gov/funding/pgm_summ.jsp?pims_id=501087
Sponsor: National Science Foundation
4201 Wilson Boulevard
Arlington, VA 22230

NSF Biological Physics (BP) 3008
The program supports projects in which the analytical and experimental tools of physics are applied to the study of problems originating in the living world. Both experimental and theoretical projects will be considered, although the main focus of the program is in the experimental area. Of particular interest are projects in which new experimental approaches are brought to bear on a well-identified problem. These approaches should at the same time have the potential for broad applicability to a set of similar problems, thereby adding to the set of tools the scientist has for addressing biological problems in general. While the problems under study must be important to advancing understanding of the living world in a meaningful way, particular emphasis will be placed on those projects in which the lessons learned from the application serve to foster new concepts and ideas that expand the intellectual basis of physics. The program funds individual investigators, although collaborative proposals between physicists and biologists are welcome.
Requirements: Unrestricted (i.e., open to any type of entity)
Geographic Focus: All States
Date(s) Application is Due: Jul 31
Amount of Grant: 135,000 - 500,000 USD
Samples: Trustees of Boston University - ($195,000) Electronic Recognition of Gene Regulatory Proteins Bound to DNA; University of California-Berkeley - ($300,000) In vitro and in vivo Single Molecular Experiments of Biological Systems; University of Illinois at Chicago - ($180,000) Study of the Dynamics of Protein-DNA Interactions to Probe Site-Specific Recognition; University of California-Davis - ($135,000) Computational study of DNA repair enzyme photolyase.
Contact: Ramona Winkelbauer, Program Contact; (703) 292-7390; winkelb@nsf.gov
Internet: http://nsf.gov/funding/pgm_summ.jsp?pims_id=6673&org=NSF&sel_org=NSF&from=fund
Sponsor: National Science Foundation
4201 Wilson Boulevard
Arlington, VA 22230

NSF Biomedical Engineering and Engineering Healthcare Grants 3009
The mission of the Biomedical Engineering and Engineering Healthcare cluster is to provide opportunities to develop novel ideas into projects that integrate engineering and life science principles in solving biomedical problems that serve humanity. The cluster focuses on high impact transforming technologies for deriving information from cells, tissues, organs, and organ systems, extraction of useful information from complex biomedical signals, new approaches to the design of structures and materials for eventual medical use, biophotonics, and new methods of controlling living systems. This cluster is also directed toward the characterization, restoration, and/or substitution of normal functions in humans.
Geographic Focus: All States
Contact: Marcia Rawlings; (703) 292-7956; fax (703) 292-9013; mrawling@nsf.gov
Internet: http://www.nsf.gov/funding/pgm_summ.jsp?pims_id=501032
Sponsor: National Science Foundation
4201 Wilson Boulevard
Arlington, VA 22230

NSF Biomolecular Systems Cluster Grants 3010
The Biomolecular Systems Cluster emphasizes the structure, function, dynamics, interactions, and interconversions of biological molecules. The context for such studies can range from investigations of individual macromolecules to the large-scale integration of metabolic and energetic processes. Research supported by this cluster includes development of cutting-edge technologies integrating theoretical, computational, and experimental approaches to the study of biological molecules and their functional complexes; mechanistic studies of the regulation and catalysis of enzymes and RNA, and higher-order characterization of the biochemical processes by which all organisms acquire, transform, and utilize energy from substrates. This cluster emphasizes the importance of multi-disciplinary research carried out at the interfaces of biology, physics, chemistry, mathematics and computer science, and engineering.
Requirements: Standard NSF Grant Proposal Guidelines apply (see http://www.nsf.gov/publications/pub_summ.jsp?ods_key=gpg).
Geographic Focus: All States
Date(s) Application is Due: Jan 12; Jul 12
Contact: Kamal Shukla; (703) 292-7131; fax (703) 292-9061; kshukla@nsf.gov
Internet: http://www.nsf.gov/funding/pgm_summ.jsp?pims_id=12771
Sponsor: National Science Foundation
4201 Wilson Boulevard
Arlington, VA 22230

NSF Cell Systems Cluster Grants 3011
The Cellular Systems Cluster, one of three thematic areas within the Division of Molecular and Cellular Biosciences, supports research, across all taxa, into the structure and organization of cells and the dynamics of cellular processes. The Cellular Systems cluster is interested not only in traditional areas of cell biology (such as the organization, function, and dynamics of membranes, organelles and other subcellular compartments, and intracellular and transmembrane signal transduction mechanisms and cell-

cell signaling processes) but also in the development of quantitative, theory-driven approaches to cell biology that integrate experimental studies at the molecular genetic, biochemical, biophysical, transcriptomic and proteomic levels. Network theory (e.g., as applied to signal transduction) and molecular dynamic modeling (e.g., as applied to the structure/function relationships of cellular structures) are also of particular interest.
Requirements: While proposals using approaches and model systems traditional in the field of cell biology are welcome, studies focused on novel, unique approaches and on non-traditional model organisms are encouraged.
Geographic Focus: All States
Date(s) Application is Due: Jan 12; Jul 12
Contact: LaJoyce Debro, Program Director; (703) 292-7135; ldebro@nsf.gov
Internet: http://www.nsf.gov/funding/pgm_summ.jsp?pims_id=12772
Sponsor: National Science Foundation
4201 Wilson Boulevard
Arlington, VA 22230

NSF Chemistry Research Experiences for Undergraduates (REU) 3012
The program supports active research participation by undergraduate students in any of the areas of research funded by the National Science Foundation. The NSF Division of Chemistry funds about 20-25 REU Sites a year (depending on the availability of funds) as part of the NSF-wide REU activity. Projects involve students in meaningful ways in ongoing research programs or in research projects specifically designed for the program. Two mechanisms are featured for support of student research: (1) REU Sites are based on independent proposals to initiate and conduct projects that engage a number of students in research. REU Sites may be based in a single discipline or academic department, or on interdisciplinary or multi-department research opportunities with a coherent intellectual theme. Proposals with an international dimension are welcome. A partnership with the Department of Defense supports REU Sites in DoD-relevant research areas. (2) REU Supplements may be requested for ongoing NSF-funded research projects or may be included as a component of proposals for new or renewal NSF grants or cooperative agreements. NSF 07-569.
Requirements: For REU Site proposals, a single individual should be designated as the Principal Investigator. This individual will be responsible for overseeing all aspects of the award. However, one additional person may be designated as Co-Principal Investigator, if developing and operating the REU Site would involve such shared responsibility. Other anticipated research supervisors should be listed as Non-Co-PI Senior Personnel. Undergraduate student participants in either Sites or Supplements must be citizens or permanent residents of the United States or its possessions. An undergraduate student is a student who is enrolled in a degree program (part-time or full-time) leading to a baccalaureate or associate degree. Students who are transferring from one college or university to another and are enrolled at neither institution during the intervening summer may participate. High school graduates who have been accepted at an undergraduate institution but who have not yet started their undergraduate study are also eligible to participate. Students who have received their bachelor's degrees and are no longer enrolled as undergraduates are generally not eligible to participate. For REU Sites, a significant fraction of the student participants should come from outside the host institution or organization. Some NSF directorates encourage inclusion in the REU program of K-12 teachers of science, technology, engineering, and mathematics. Please contact the appropriate disciplinary program officer for guidance.
Restrictions: Students may not apply to NSF to participate in REU activities. Students apply directly to REU Sites and should consult the directory of active REU Sites.
Geographic Focus: All States
Contact: Wilfredo Colon; (703) 292-8171; fax (703) 292-9061; wcolon@nsf.gov
Internet: http://www.nsf.gov/funding/pgm_summ.jsp?pims_id=503210
Sponsor: National Science Foundation
4201 Wilson Boulevard
Arlington, VA 22230

NSF Doctoral Dissertation Improvement Grants in the Directorate for Biological Sciences (DDIG) 3013
This program awards grants in selected areas of the biological sciences. These grants provide partial support of doctoral dissertation research to improve the overall quality of research. Allowed are costs for doctoral candidates to participate in scientific meetings, to conduct research in specialized facilities or field settings, and to expand an existing body of dissertation research. NSF 08-564.
Requirements: U.S. institutions and organizations that are eligible for awards from the National Science Foundation, including colleges, universities, and other nonprofit research organizations such as botanical gardens, marine and freshwater institutes, and natural history museums may submit proposals. See Chapter I, Section E of the NSF Grant Proposal Guide (GPG) for specific definitions of these categories of proposers. The NSF encourages collaborations with scientists at foreign organizations; however, primary support for any foreign participants' activities must be secured through their own national sources. A student must have advanced to candidacy for a Ph.D. degree before the submission deadline to be eligible to submit a proposal. A statement that the student has advanced to candidacy for a Ph.D., signed and dated by the department chairperson, graduate dean, or similar administrative official is required. The proposal must be submitted through regular organizational channels by the dissertation advisor(s) on behalf of a graduate student who is at the point of initiating or is already conducting dissertation research. The student must be enrolled at a U.S. institution, but need not be a U.S. citizen. Organizations should limit applications to outstanding dissertation proposals with unusual financial requirements that cannot be met otherwise. Preference may be given to projects that are underway and for which feasibility is demonstrated.
Restrictions: Indirect costs not allowed.
Geographic Focus: All States
Date(s) Application is Due: Nov 20
Amount of Grant: Up to 15,000 USD
Contact: IOS Program Officer; (703) 292-8423; ddig-ios@nsf.gov
Internet: http://www.nsf.gov/funding/pgm_summ.jsp?pims_id=5234&org=BIO&sel_org=BIO&from=fund
Sponsor: National Science Foundation
4201 Wilson Boulevard
Arlington, VA 22230

NSF Emerging Frontiers in Research and Innovation (EFRI) Grants 3014
The Directorate for Engineering at the National Science Foundation has established the Office of Emerging Frontiers in Research and Innovation (EFRI) to serve a critical role in focusing on important emerging areas in a timely manner. The EFRI Office is launching a new funding opportunity for interdisciplinary teams of researchers to embark on rapidly advancing frontiers of fundamental engineering research. For this program, they will consider proposals that aim to investigate emerging frontiers in the following two specific research areas: (1) BioSensing & BioActuation: Interface of Living and Engineered Systems (BSBA), and (2) Hydrocarbons from Biomass (HyBi). EFRI seeks proposals with transformative ideas that represent an opportunity for a significant shift in fundamental engineering knowledge with a strong potential for long term impact on national needs or a grand challenge. It is anticipated that 11 or more standard grants will be made (4-year awards).
Requirements: Proposals may be submitted by a single organization or a group of organizations consisting of a lead organization in partnership with one or more partner organizations. Only U.S. academic institutions with significant research and degree-granting education programs in disciplines normally supported by NSF are eligible to be the lead organization. Principal investigators are encouraged to form synergistic collaborations among researchers and with private and public sector organizations, government laboratories, and scientists and engineers at foreign organizations where appropriate, though no NSF funds will be provided to those organizations. Submission of Letters of Intent is required, due by October 14. Submission of Preliminary Proposals is required, due by December 2. Full proposals will be due by April 30. Each project team may receive support of up to a total of $500,000 per year for up to four years, pending the availability of funds. It is not expected that all awards will receive the maximum amount; the size of awards will depend upon the type of research program proposed. See the website for the detailed solicitation, which includes guidelines, requirements, and application instructions.
Restrictions: The principal investigator and co-principal investigators may participate in one proposal submitted to this program.
Geographic Focus: All States
Date(s) Application is Due: Apr 30; Oct 14; Dec 2
Amount of Grant: Up to 500,000 USD
Contact: Sohi Rastegar, Director, Office of Emerging Frontiers in Research and Innovation; (703) 292-8305; srastega@nsf.gov
Internet: http://www.nsf.gov/funding/pgm_summ.jsp?pims_id=13708
Sponsor: National Science Foundation
4201 Wilson Boulevard
Arlington, VA 22230

NSF Genes and Genome Systems Cluster Grants 3015
Grants support studies on genomes and genetic mechanisms in all organisms, whether prokaryote, eukaryote, phage, or virus. Proposals on the structure, maintenance, expression, transfer, and stability of genetic information in DNA, RNA, and proteins and how those processes are regulated are appropriate. Areas of interest include genome organization, molecular and cellular evolution, replication, recombination, repair, and vertical and lateral transmission of heritable information. Of equal interest are the processes that mediate and regulate gene expression, such as chromatin structure, epigenetic phenomena, transcription, RNA processing, editing and degradation, and translation. The use of innovative in vivo and/or in vitro approaches, including biochemical, physiological, genetic, genomic, and/or computational methods, is encouraged, as is research at the interfaces of biology, physics, chemistry, mathematics and computer science, and engineering.
Requirements: Standard NSF Grant Proposal Guidelines apply (see http://www.nsf.gov/publications/pub_summ.jsp?ods_key=gpg).
Geographic Focus: All States
Date(s) Application is Due: Jan 12; Jul 12
Contact: Patrick P. Dennis; (703) 292-7145; fax (703) 292-9061; pdennis@nsf.gov
Internet: http://www.nsf.gov/funding/pgm_summ.jsp?pims_id=12780
Sponsor: National Science Foundation
4201 Wilson Boulevard
Arlington, VA 22230

NSF Grant Opportunities for Academic Liaison with Industry (GOALI) 3016
Grant Opportunities for Academic Liaison with Industry (GOALI) promotes university-industry partnerships by making project funds or fellowships/traineeships available to support an eclectic mix of industry-university linkages. Special interest is focused on affording the opportunity for: faculty, postdoctoral fellows, and students to conduct research and gain experience in an industrial setting; industrial scientists and engineers to bring industry's perspective and integrative skills to academe; and interdisciplinary university-industry teams to conduct research projects. This solicitation targets high-risk/high-gain research with a focus on fundamental research, new approaches to solving generic problems, development of innovative collaborative industry-university educational programs, and direct transfer of new knowledge between academe and industry. GOALI seeks to fund transformative research that lies beyond that which industry would normally fund.
Geographic Focus: All States

Contact: Donald Senich; (703) 292-7082; fax (703) 292-9056; dsenich@nsf.gov
Internet: http://www.nsf.gov/funding/pgm_summ.jsp?pims_id=13706
Sponsor: National Science Foundation
4201 Wilson Boulevard
Arlington, VA 22230

NSF Instrument Development for Biological Research (IDBR) 3017
The IDBR program provides support for development of the following: (1) New instrumentation that addresses emerging biological research needs with the capacity to transform biological research; (2) Novel instrument configurations (hybrids, etc.) that demonstrably address a fundamental biological research challenge; (3) Instrumentation that leverages advances in micro- and nano-fabrication and device design to provide new capabilities in measurement and observation of biological phenomena; (4) Concept and proof-of-concept of novel instruments for biological research; (5) New instruments that provide new capabilities for detection, measurement, and/or observation of biological phenomena, or that greatly extend currently achievable sensitivity, accuracy or resolution; (6) Sensors that meet emerging biological research needs and that have the potential to transform biological research at any level of organization (7) Data acquisition and analysis tools that, through the development of novel devices/instruments, meet biological research needs. The program is not limited to the above mentioned activities. Any proposal that is designed to meet the goals of the Program will be considered. Proposals must, in all cases, demonstrate a strong connection to biological research needs through, for example, collaboration with biological researchers as well as demonstrate that the need for the instrumentation is firmly based on biological research drivers. NSF Program Solicitation 08-566.
Requirements: The program encourages proposals which conduct collaborative and planning activities such as workshops and the development of virtual organization frameworks. Those activities which promote interactions between the instrument development community and biological researchers as well as innovative networking strategies that foster research collaborations or enable new instrument development directions are especially encouraged. Activities which increase participation of colleagues at smaller institutions, minority-serving institutions, community colleges, and K-12 students and teachers are also recommended. Recognizing that the development and use of novel instrumentation have become increasingly integral to activities supported by all BIO programs, the IDBR program will place a higher priority on proposals that create novel instrumentation that enables biological research in a new and potentially transformative way and that serves a broad user community. Check the NSF Grant Proposal Guide (GPG) available electronically on the NSF website at: http://www.nsf.gov/publications/pub_summ.jsp?ods_key=gpg for other general eligibility guidelines.
Restrictions: The program does not support research or technique development activities, except to the extent these are required as part of the development of the new or improved instrument, or for the testing of its utility. Projects emphasizing the development of new research techniques should be addressed to an appropriate research program. The anticipated uses of the instrumentation to be developed or improved should include areas of research that fall within the scope of the Directorate for Biological Sciences (see BIO Home Page at http://www.nsf.gov/bio). Projects aimed at instrumentation whose primary use will be in studies of the etiology, diagnosis or treatment of physical or mental disease, abnormality, or malfunction in human beings or animals, is not supported by IDBR. Similarly, the development or testing of drugs or of instruments whose primary application is in pharmaceutical chemistry are not eligible for support. Such projects should be addressed to an appropriate program in another NSF Directorate or to another agency. Projects in which the main portion of the instrument development activity will be subcontracted to a Federally Funded Research and Development Center (FFRD.C.) or a commercial (for profit) organization are also not eligible for support by IDBR, and should be addressed to an appropriate NSF program or to another agency.
Geographic Focus: All States
Date(s) Application is Due: Aug 28
Contact: Robyn Hannigan; (703) 292-8470; fax (703) 292-9063; dbiiid@nsf.gov
Internet: http://www.nsf.gov/publications/pub_summ.jsp?ods_key=nsf08566&org=NSF
Sponsor: National Science Foundation
4201 Wilson Boulevard
Arlington, VA 22230

NSF Major Research Instrumentation Program (MRI) Grants 3018
The MRI program assists in the acquisition or development of major research instrumentation that is, in general, too costly for support through other NSF programs. This program seeks to improve the quality and expand the scope of research and research training in science and engineering, and to foster the integration of research and education by providing instrumentation for research-intensive learning environments. For proposals over $2 million, requests must be for the acquisition of a single instrument. For proposals requesting $2 million or less, investigators may seek support for instrument development or for acquisition of a single instrument, a large system of instruments, or multiple instruments that share a common or specific research focus. Submission of Letters of Intent is required (only for acquisition requests between $2 million and $4 million). NSF 08-503
Requirements: Proposals may only be submitted by the following: (1) U.S. colleges, universities and organizations of higher education located in the US, its territories and possessions; (2) U.S. independent research museums located in the US, its territories and possessions; (3) U.S. independent nonprofit research organizations located in the US, its territories and possessions, including consortia whose members consist only of organizations described in items (1) and (2). (Requests for instrumentation that will be located at a Federally Funded Research and Development Center (FFRD.C.) must be submitted as consortium proposals. This is the only mechanism by which instrumentation, whether purchased or developed, can be placed at an FFRD.C..); (4) U.S. small businesses located in the US, its territories and possessions are eligible for instrument development support as private sector partners with submitting organizations; they may not submit proposals as a lead organization. The deadline is the fourth Thursday in January. Proposers are required to prepare and submit all proposals for this program solicitation through use of the NSF FastLane system.
Restrictions: The MRI program will NOT support proposal requests for: Computer networks used for general-purposes; General laboratory equipment or assorted instruments that do not share a common or specific research or research training focus; Instrumentation used primarily for standard science and engineering courses; Renovation or modernization of research facilities, fixed equipment, or facilities such as research vessels, airplanes, large telescopes, and supercomputing centers. The term 'research facilities' refers to the bricks-and-mortar physical plant in which sponsored or unsponsored research activities (including research training) take place, including related infrastructure, systems (e.g., HVAC and power systems, toxic waste removal systems), and fixed equipment. The term 'fixed equipment' refers to the permanent components of a research facility that are integral (i.e., built in, rather than affixed) to the facility (e.g., clean rooms, fume hoods, elevators, laboratory casework); their removal would affect the integrity or basic operation of the facility.
Geographic Focus: All States
Date(s) Application is Due: Jan 22
Amount of Grant: Up to 4,000,000 USD
Contact: Randy Phelps, (703) 292-8040; rphelps@nsf.gov
Internet: http://nsf.gov/funding/pgm_summ.jsp?pims_id=5260
Sponsor: National Science Foundation
4201 Wilson Boulevard
Arlington, VA 22230

NSF Perception, Action & Cognition Research Grants 3019
This program supports research on perception, action and cognition including the development of these capacities. Emphasis is on research strongly grounded in theory. Research topics include vision, audition, haptics, attention, memory, reasoning, written and spoken discourse, motor control, and developmental issues in all topic areas. The program encompasses a wide range of theoretical perspectives, such as symbolic computation, connectionism, ecological, nonlinear dynamics, and complex systems, and a variety of methodologies including both experimental studies and modeling. Research involving acquired or developmental deficits is appropriate if the results speak to basic issues of perception, action, and cognition.
Geographic Focus: All States
Date(s) Application is Due: Feb 1; Aug 1
Amount of Grant: 1,100 - 49,000,000 USD
Contact: Vincent R. Brown, Program Director; (703) 292-7305; vrbrown@nsf.gov
Internet: http://www.nsf.gov/funding/pgm_summ.jsp?pims_id=5686
Sponsor: National Science Foundation
4201 Wilson Boulevard
Arlington, VA 22230

NSF Postdoctoral Research Fellowships in Biology (PRFB) 3020
The Directorate for Biological Sciences (BIO) awards Postdoctoral Research Fellowships in Biology to recent recipients of the doctoral degree for research and training in selected areas of biology supported by BIO to encourage independence early in their research careers and to permit them to pursue their research and training goals in the most appropriate research locations regardless of the availability of funding for the Fellows at that site. The fellowships are also designed to provide active mentoring of the Fellows by the sponsoring scientists who will benefit from having additional members in their research groups. The research and training plan of each fellowship must address important scientific questions in contemporary biology within the scope of the BIO Directorate and the specific guidelines in this fellowship program solicitation. Because the fellowships are offered only to postdoctoral scientists early in their careers, doctoral advisors are encouraged to discuss the availability of BIO fellowships with their graduate students early in their doctoral programs. Fellowships are awards to individuals, not institutions, and are administered by the Fellows. Fellowship stipends are being adjusted on a yearly basis. NSF 07-580.
Requirements: Only individuals may apply. NSF postdoctoral fellowships are awards to individuals, and applications are submitted directly by applicants to NSF. However, applications must include sponsoring scientists' statements and the applicants must affiliate with institutions (e.g., colleges and universities, and privately-sponsored nonprofit institutes and museums, government agencies and laboratories, and, under special conditions, for-profit organizations) anywhere in the world. Applicants must: (1) be U.S. citizens (or nationals) or legally admitted permanent residents of the United States (i.e., have a 'green card') at the time of application; (2) earn or plan to earn the doctoral degree in a scientific or engineering field prior to the requested start date of the fellowship; (3) either currently be a graduate student or, at the deadline date, have served in a position requiring the doctoral degree for no more than 12 full time months since earning the degree. There is a one time exception to this criterion for Biological Informatics in November, 2007 only, as described below; (4) must present a research and training plan that falls within the purview of BIO and includes the required information for the specific competition as described below; (5) select a host institution and sponsoring scientist different from the doctoral degree and current position or provide compelling justification for why such a change is not being proposed; (6) not have received Federal funding of more than $20,000 as PI or co-PI (except graduate fellowships and doctoral dissertation improvement grants); (7) not have submitted concurrently the same project to another NSF program; (8) not be a named participant on any other proposal submitted to NSF, including regular research proposals, concurrent with the fellowship application, regardless of who is the named principal investigator; AND (9) meet other limitations on eligibility imposed within the specific competitive areas, if any. There is no limit on the number of applicants that an institution may host. For now, submission can only be made

using NSF's FastLane system, given that Grants.gov does not support all required forms. Applications are due the first Monday in November by 5 p.m. proposer's local time.
Geographic Focus: All States
Date(s) Application is Due: Nov 0
Amount of Grant: 60,000 - 66,000 USD
Contact: Peter McCartney, Program Director; (703) 292-8470; pmccartn@nsf.gov
Internet: http://www.nsf.gov/publications/pub_summ.jsp?ods_key=nsf07580&org=BIO
Sponsor: National Science Foundation
4201 Wilson Boulevard
Arlington, VA 22230

NSF Presidential Early Career Awards for Scientists and Engineers Grants   3021
Each year NSF selects nominees for the Presidential Early Career Awards for Scientists and Engineers (PECASE) from among the most meritorious new CAREER awardees. Selection for this award is based on two important criteria: innovative research at the frontiers of science and technology that is relevant to the mission of the sponsoring organization or agency, and community service demonstrated through scientific leadership, education or community outreach. These awards foster innovative developments in science and technology, increase awareness of careers in science and engineering, give recognition to the scientific missions of the participating agencies, enhance connections between fundamental research and national goals, and highlight the importance of science and technology for the Nation's future. Individuals cannot apply for PECASE. These awards are initiated by the participating federal agencies. At NSF, up to twenty nominees for this award are selected each year from among the PECASE-eligible CAREER awardees who are most likely to become the leaders of academic research and education in the twenty-first century. The White House Office of Science and Technology Policy makes the final selection and announcement of the awardees.
Requirements: Applicant must be employed at an institution in the United States, its territories or possessions, or Puerto Rico, that awards a baccalaureate or advanced degree in a field supported by NSF; be in their first or second full-time tenure-track academic appointment and have begun the first tenure-track or appointment (at any institution) on or after October 1, 1996, and before October 1, 2001; not be tenured or have held tenure on or before October 1, 2001; and not be a previous recipient of a CAREER or PECASE award. Contact program officer for exact requirements.
Restrictions: No salary support for other Senior Personnel is permitted, in either the primary budget or in any sub awards.
Geographic Focus: All States
Date(s) Application is Due: Jul 21; Jul 22; Jul 23
Amount of Grant: 400,000 - 500,000 USD
Contact: Elizabeth L. Rom; (703) 292-7709; fax (703) 292-9085; elrom@nsf.gov
Internet: http://www.nsf.gov/funding/pgm_summ.jsp?pims_id=5262
Sponsor: National Science Foundation
4201 Wilson Boulevard
Arlington, VA 22230

NSF Small Business Innovation Research (SBIR) Grants   3022
The Small Business Innovation Research (SBIR) Program stimulates technological innovation in the private sector by strengthening the role of small business concerns in meeting Federal research and development needs, increasing the commercial application of federally supported research results, and fostering and encouraging participation by socially and economically disadvantaged and women-owned small businesses.
Requirements: Eligible to apply are small business firms--those organized for profit, individually owned or operated, not dominant in the field in which they are bidding, and with an average of no more than 500 employees in all affiliated firms. In addition, the primary employment of the principal investigator must be with the small business firm at the time of the award.
Geographic Focus: All States
Amount of Grant: 5,000 - 4,000,000 USD
Samples: Inrad Inc (Northvale, NJ)--Growth of potassium iodate single crystals for nonlinear applications (Phase I SBIR), $65,000; Metcal Inc (Menlo Park, CA)--Computer design of a new family of magnetic alloys (Phase I SBIR), $65,000.
Contact: Gregory T. Baxter; (703) 292-7795; fax (703) 292-9057; gbaxter@nsf.gov
Internet: http://www.nsf.gov/funding/pgm_summ.jsp?pims_id=503361
Sponsor: National Science Foundation
4201 Wilson Boulevard
Arlington, VA 22230

NSF Social Psychology Research Grants   3023
The Social Psychology Program at NSF supports basic research on human social behavior, including cultural differences and development over the life span. Among the many research topics supported are: attitude formation and change, social cognition, personality processes, interpersonal relations and group processes, the self, emotion, social comparison and social influence, and the psychophysiological and neurophysiological bases of social behavior. The scientific merit of a proposal depends on four important factors: the problems investigated must be theoretically grounded; the research should be based on empirical observation or be subject to empirical validation; the research design must be appropriate to the questions asked; and the proposed research must advance basic understanding of social behavior. Annual full proposal target dates are July 15 and January 15.
Requirements: Except where a program solicitation establishes more restrictive eligibility criteria, individuals and organizations in the following categories may submit proposals: universities and colleges; non-profit, non-academic organizations; for-profit organizations; state and local governments; unaffiliated individuals; foreign organizations; and other Federal agencies.
Geographic Focus: All States, All Countries
Date(s) Application is Due: Jan 15; Jul 15
Contact: Sally Dickerson; (703) 292-7277; fax (703) 292-9068; sdickers@nsf.gov
Internet: http://www.nsf.gov/funding/pgm_summ.jsp?pims_id=5712
Sponsor: National Science Foundation
4201 Wilson Boulevard
Arlington, VA 22230

NSF Systematic Biology and Biodiversity Inventories Grants   3024
The Systematic Biology and Biodiversity Inventories Cluster supports research in taxonomy and systematics that contributes to: 1) using phylogenetic methods to understand the evolution of life in time and space, 2) discovery, description, and cataloguing global species diversity, and 3) organizing information from the above in efficiently retrievable forms that best meet the needs of science and society. The Systematic Biology and Biodiversity Inventories Cluster funds projects within the two Programs, Systematic Biology and Biodiversity Inventories, in addition to the PEET and PBI solicitations listed below. In addition, the cluster participates in AToL and other related funding opportunities. The SBBI Cluster continues to encourage and support studies that seek to synthesize available and new species-level taxonomic information in the context of providing revisionary treatments and predictive classifications for particular groups of organisms. NSF PD 047374.
Requirements: The most frequent recipients are academic institutions and nonprofit research groups. In special circumstances, grants also are awarded to other types of institutions and to individuals. Standard NSF Grant Proposal Guidelines apply (see http://www.nsf.gov/publications/pub_summ.jsp?ods_key=gpg).
Geographic Focus: All States
Date(s) Application is Due: Jan 9; Jul 9
Contact: Maureen Kearney, Program Director; (703) 292-7187; mkearney@nsf.gov
Internet: http://www.nsf.gov/publications/pub_summ.jsp?ods_key=pd047374&org=NSF
Sponsor: National Science Foundation
4201 Wilson Boulevard
Arlington, VA 22230

NSF Undergraduate Research and Mentoring in the Biological Sciences   3025
This program will fund projects that have strong research and mentoring activities designed to prepare students for successful entry into graduate programs. URM will support projects involving the recruitment, retention and development of undergraduate students, especially those from underrepresented groups, for the purpose of preparing them for graduate study in the biological sciences. Proposed projects are expected to create a URM program that will actively engage students in interesting and exciting research ideas, provide hands-on research experience, and develop their academic skills. URM will enable institutions to create innovative programs that will increase the number and diversity of students who enter graduate research programs in the biological sciences. URM will support well-defined research and mentoring activities that will enable students, especially underrepresented minorities, to become independent thinkers and effective communicators, as well as professional development activities that will expose students to exciting careers in biology. Students are expected to be recruited early in their academic career. The project should frame all student activities around an effective mentoring strategy. Factors that are known to increase the likelihood of success for students to enter, and complete, graduate research programs should be emphasized. Approximately $4,000,000 will be available for this program, subject to the availability of funds. Under the solicitation, requests may be submitted for funding amounts up to a total of $1,000,000 for up to 5 years. However, funding beyond the first year is contingent upon satisfactory progress in the program. The program expects to make at least eight awards, depending upon the quality of submissions and the availability of funds. NSF 06-591
Requirements: Preliminary Proposals (deadline: Sep. 18) are required prior to submitting a Full Proposal (deadline: Mar. 4). Proposals may only be submitted by academic institutions located in the United States. Collaborative Proposals from two or more institutions are allowed. The proposing institution is expected to foster an enriched and intellectually stimulating research and educational environment. URM proposals should involve year-round mentoring and include a major emphasis on direct student participation in research leading to publishable data. URM projects may be based in a single discipline or academic department, or on interdisciplinary or multi-department research opportunities. A proposal should reflect the unique combination of the proposing organization's interests and capabilities and those of any partnering organizations. Partnerships between minority-serving institutions, especially Historically Black Colleges and Universities (HBCU), Tribal Colleges and Universities (TCU), and Hispanic-Serving Institutions (HSI), community colleges, and research intensive institutions are encouraged, if such partnerships will result in a diverse pool of qualified participants as well as an enriched research environment for the students. Undergraduate student participants supported with NSF funds must be citizens or permanent residents of the United States or its possessions. An undergraduate student is one who is enrolled in a degree program (part-time or full-time) leading to a baccalaureate or associates degree. High school graduates who have not yet enrolled, and students who have received their bachelor's degrees and are no longer enrolled as undergraduates, are not eligible.
Restrictions: All collaborative proposals submitted as separate submissions from multiple organizations must be submitted via the NSF FastLane system.
Geographic Focus: All States
Date(s) Application is Due: Mar 4; Sep 18
Amount of Grant: Up to 1,000,000 USD
Samples: University of Arkansas (AR) $149,233 - RIG: Biophysical Characterization of Cdc42Hs Forms Associated with Cell Transformation; University of Montana (MT) $170,382 - Collaborative UMEB: Training American Indians in Environmental Biology;

Salish Kootenai College (MT) $132,715 - Collaborative UMEB: Training American Indians in Environmental Biology; University of Maryland Eastern Shore (MD) $150,000 - Discovery of Novel Transcription Factor III Proteins in Trypanosomes
Contact: URM Program Director; (703) 292-8470; biourm@nsf.gov
Internet: http://www.nsf.gov/funding/pgm_summ.jsp?pims_id=500036
Sponsor: National Science Foundation
4201 Wilson Boulevard
Arlington, VA 22230

Nuffield Foundation Africa Program Grants 3026
The Commission for Africa highlighted the importance of strengthening Africa's science capacity, recommending that donors develop incentives for research and development in health that meet Africa's needs, and increase funding to African-led research. With that in mind, the Foundation has recently replaced its Commonwealth Program with Africa Program grants, which aims to improve services in health, education and civil justice in Southern and Eastern Africa through the development of the expertise and experience of practitioners and policy makers.
Geographic Focus: All States, Algeria, Angola, Benin, Botswana, Burkina Faso, Burundi, Cameroon, Cape Verde, Central African Republic, Chad, Comoros, Congo, Congo, Democratic Republic of, Cote d' Ivoire (Ivory Coast), Djibouti, Egypt, Equatorial Guinea, Eritrea, Ethiopia, Gabon, Gambia, Ghana, Guinea, Guinea-Bissau, Kenya, Lesotho, Liberia, Libya, Madagascar, Malawi, Mali, Mauritania, Mauritius, Morocco, Mozambique, Namibia, Niger, Nigeria, Rwanda, Sao Tome & Principe, Senegal, Seychelles, Sierra Leone, Somalia, South Africa, Sudan, Swaziland, United Kingdom
Contact: Sarah Lock, 44-171-631-0566; africaprogram@nuffieldfoundation.org
Internet: http://www.nuffieldfoundation.org/go/grants/commonwealth/page_100.html
Sponsor: Nuffield Foundation
28 Bedford Square
London, WC1B 3JS England

Nuffield Foundation Children and Families Grants 3027
The foundation was founded for the advancement of health through medical research and teaching and for the care of the elderly through scientific research and education. The Children and Families program supports work to help ensure that the legal and institutional framework is best adapted to meet the needs of children and families. At present, particular interests include (but are not limited to): work that links education and child development, either in the case of adolescent mental health or younger children; work that considers policies relevant to child welfare in a broader institutional context: parents' paid working patterns; childcare and early years provision; work that considers especially the well-being of children growing up in adverse conditions, and what institutional responses may be appropriate; work in family law, including cohabitation, child contact, child support; and work in child protection and placement (adoption and fostering) but only when it raises significant issues. Where a proposal is for a research study, Trustees are interested in the dispassionate examination of evidence. It notes that evidence is likely to be different in different cases, for different types of children and families, and are more likely to support work that takes this approach.
Requirements: Usually grants are made only to United Kingdom organizations, and support work that will be mainly based in the UK, although the Trustees welcome proposals for collaborative projects involving partners in European or Commonwealth countries.
Restrictions: The Foundation does not make grants for the running costs of voluntary bodies but will consider making a contribution to voluntary sector overheads on funded projects.
Geographic Focus: All States, United Kingdom
Date(s) Application is Due: Jan 8; May 7; Sep 3
Amount of Grant: 5,000 - 150,000 GBP
Samples: Gillian Douglas, Cardiff Law School--for Attitudes towards the law of inheritance in the context of changing family structures, 193,542 GBP; Ingrid Schoon, Department of Quantitative Social Sciences, Institute of Education, 152,771 GBP.
Contact: Sharon Witherspoon, 44-171-631-0566; fax 44-171-323-4877; sfwpa@nuffieldfoundation.org or info@nuffieldfoundation.org
Internet: http://www.nuffieldfoundation.org/go/grants/cplfj/page_30.html
Sponsor: Nuffield Foundation
28 Bedford Square
London, WC1B 3JS England

Nuffield Foundation Law and Society Grants 3028
The foundation was founded for the advancement of health through medical research and teaching and for the care of the elderly through scientific research and education. The Foundation has long had an interest in the area of law. Its current objectives in this area are: to promote developments in the legal system that will improve its accessibility to all people; to promote wider access to legal services and advice and a better understanding of the obstacles to access to justice; to fund research and promote developments in alternative dispute resolution; to help promote a greater knowledge of the rights and duties of the individual, including those of the European citizen; to examine the implications of new human rights' obligations on civil (not criminal) justice; and to help promote a greater public understanding of the role of law in society and of the legal system.
Requirements: Usually grants are made only to United Kingdom organizations, and support work that will be mainly based in the UK, although the Trustees welcome proposals for collaborative projects involving partners in European or Commonwealth countries.
Geographic Focus: All States, United Kingdom
Date(s) Application is Due: Jan 8; May 7; Sep 3
Amount of Grant: 5,000 - 150,000 GBP
Contact: Sharon Witherspoon, 44-171-631-0566; fax 44-171-323-4877; sfwpa@nuffieldfoundation.org or info@nuffieldfoundation.org
Internet: http://www.nuffieldfoundation.org/go/grants/accesstojustice/page_57.html
Sponsor: Nuffield Foundation
28 Bedford Square
London, WC1B 3JS England

Nuffield Foundation Oliver Bird Rheumatism Program Grants 3029
The Fund will offer grants to up to five United Kingdom Higher Education Institutions where students of outstanding potential will be recruited into doctoral training. The institutions will provide multidisciplinary environments for researching problems in rheumatic disease, and be able to develop both the students' awareness of the broader context of these diseases and the professional skills required of future scientists. For each studentship the grant will provide: a stipend of not less than 13,500GBP per annum plus United Kingdom fees; research expenses at a rate of 8,000GBP per annum; and a meeting and travel allowance of 1,000GBP per annum. In addition the Fund will provide a grant to each institution of 25,000GBP over the five years of the program to support such activities as: meetings within the Collaborative Center to promote the field and assist the recruitment and training of doctoral students; the hosting of a program meeting to be attended by all program participants from the other Collaborative Centers; and occasional bursaries for interested undergraduates to work alongside the doctoral students and potential supervisors or to attend meetings in the Collaborative Centers.
Geographic Focus: All States, United Kingdom
Date(s) Application is Due: Feb 28
Amount of Grant: 27,500 GBP
Contact: Vicki Hughes, 44-171-631-0566; fax 44-171-323-4877; oliverbird@nuffieldfoundation.org
Internet: http://www.nuffieldfoundation.org/go/grants/oliverbird/page_86.html
Sponsor: Nuffield Foundation
28 Bedford Square
London, WC1B 3JS England

Nuffield Foundation Open Door Grants 3030
The Foundation keeps an open door to proposals of exceptional merit for research projects or practical innovations that lie outside our main program areas, but that meet Trustees' wider interests. These must have some bearing on the Foundation's widest charitable object. Particular interest is given to projects which identify change or interventions which will have a practical implications for policy or practice, or that will improve the quality of research evidence in areas of public debate. Through the Open Door, the Foundation may also identify emerging areas that justify more sustained attention.
Requirements: Usually grants are made only to United Kingdom organizations, and support work that will be mainly based in the UK, although the Trustees welcome proposals for collaborative projects involving partners in European or Commonwealth countries.
Geographic Focus: All States, United Kingdom
Date(s) Application is Due: Jan 8; May 7; Sep 3
Amount of Grant: 5,000 - 150,000 GBP
Samples: Dr Alex Bryson, National Institute of Economic & Social Research--for British Workplaces in the shadow of recession, 219,167 GBP; Colin Aitken, School of Mathematics, University of Edinburgh, 75,869 GBP.
Contact: Sharon Witherspoon, 44-171-631-0566; fax 44-171-323-4877; sfwpa@nuffieldfoundation.org or info@nuffieldfoundation.org
Internet: http://www.nuffieldfoundation.org/go/grants/opendoor/page_115.html
Sponsor: Nuffield Foundation
28 Bedford Square
London, WC1B 3JS England

Nuffield Foundation Small Grants 3031
Most recently, the Foundation has decided that it will no longer try to fund across the full range of all social sciences, but will focus on work that more closely matches the wider interests of the Foundation. As a consequence, the budgetary ceiling for individual grants will be raised from the current maximum of 12,000GBP.
Geographic Focus: All States, United Kingdom
Date(s) Application is Due: Jul 17
Amount of Grant: 12,000 GBP
Contact: Sarah Lock, 44-171-631-0566; smallgrants@nuffieldfoundation.org
Internet: http://www.nuffieldfoundation.org/go/grants/smallgrants/page_123.html
Sponsor: Nuffield Foundation
28 Bedford Square
London, WC1B 3JS England

NWHF Community-Based Participatory Research Grants 3032
The Northwest Health Foundation supports community-based participatory research as a tool to generate meaningful information about community health and build the capacity of groups united by common challenges to organize and advocate for change. The RFP will seek to support community-based participatory research (CBPR) projects that: build community capacity for policy advocacy; and apply research findings to support policy change that improves community health.
Requirements: 501(c)3 educational institutions and governmental entities are eligible. Generally speaking, foundation programs serve the entire state of Oregon, and the following Washington counties: Clark, Cowlitz, Pacific, Skamania, and Wahkiakum. Specific programs and funds may have additional geographic restrictions. Applying organizations must sign a statement of nondiscrimination in leadership, staffing, and services.
Restrictions: Support will not be considered for basic operations, or for salaries or other expenses associated with ongoing programs.
Geographic Focus: Oregon, Washington

Date(s) Application is Due: Sep 8
Contact: Judith Woodruff; (503) 220-1955; fax (503) 220-1355; judith@nwhf.org
Internet: http://www.nwhf.org/index.php?/apply/cbpr
Sponsor: Northwest Health Foundation
221 NW Second Avenue, Suite 300
Portland, OR 97209

NWHF Health Advocacy Small Grants 3033
Awards of up to $10,000 will be available for 15-20 projects dedicated to one of the following areas: grassroots organizing; coalition building; communications; and policy analysis. For example, organizations may request funds to develop communications materials or enhance their Web site, dedicate staff time to keep abreast of health-reform related activity in Salem during the legislative session, or build their organizational capacity to develop a coalition focused on health reform. In addition to grant dollars, organizations can apply for approximately two days of technical assistance from the Consumer Voices for Coverage team in one or more of the areas listed above.
Requirements: 501(c)3 educational institutions and governmental entities are eligible. Generally speaking, foundation programs serve the entire state of Oregon, and the following Washington counties: Clark, Cowlitz, Pacific, Skamania, and Wahkiakum. Specific programs and funds may have additional geographic restrictions. Applying organizations must sign a statement of nondiscrimination in leadership, staffing, and services.
Restrictions: Support will not be considered for basic operations, or for salaries or other expenses associated with ongoing programs.
Geographic Focus: Idaho, Oregon, Washington
Date(s) Application is Due: Feb 4
Amount of Grant: Up to 10,000 USD
Contact: Chris DeMars; (971) 230-1292 or (503) 220-1955; cdemars@nwhf.org
Internet: http://www.nwhf.org/index.php?/apply/access
Sponsor: Northwest Health Foundation
221 NW Second Avenue, Suite 300
Portland, OR 97209

NWHF Kaiser Permanente Community Fund Grants 3034
The Kaiser Permanente Community Fund (KPCF) at Northwest Health Foundation was established in late 2004 to advance the health of the communities served by Kaiser Permanente Northwest. The Fund intends to achieve this goal by addressing those factors in the social, policy, and physical environment that impact community health. Often referred to as the social determinants of health, these factors have been shown to play a major role in the development of health disparities based on race, ethnicity, and socio-economic status.
Requirements: To be eligible to apply for a grant from the Kaiser Permanente Community Fund, the service area of your project must fall within the geographic region roughly spanning from Longview, Washington to Salem, Oregon and portions of the Willamette Valley.
Geographic Focus: Oregon, Washington
Date(s) Application is Due: Jul 2
Contact: Judith Woodruff; (503) 220-1955; fax (503) 220-1355; judith@nwhf.org
Internet: http://www.nwhf.org/index.php?/apply/kaiser
Sponsor: Northwest Health Foundation
221 NW Second Avenue, Suite 300
Portland, OR 97209

NWHF Mark O. Hatfield Research Fellowship 3035
The Mark O. Hatfield Research Fellowship Program was established in 2005 to honor the legacy of Senator Mark O. Hatfield, founding Board Chairman of the Northwest Health Foundation (NWHF). The purpose of the Fellowship program is to advance the research career of health practitioners who serve communities in Oregon and southwest Washington. The fellowship is open to health professionals from a variety of disciplines who wish to launch, renew, or expand their research activities in biomedical, applied, or translational research. A secondary purpose is to increase the intellectual resources of research-based institutions. One two-year fellowship of up to $150,000 will be awarded. Funds will be distributed to the fellow's sponsoring institution in two annual disbursements of half of the amount awarded.
Requirements: Eligible candidates must: be a clinical, allied health, public health, or social service provider (early-career researchers specializing in translational research are also encouraged to apply); possess (or be in the process of completing) a doctoral degree or equivalent in an appropriate field related to medicine, nursing, health care, public health, or biomedical sciences; obtain sponsorship from a research-based institution that can act as fiscal manager of the award; and demonstrate a commitment to communities in Oregon and southwest Washington.
Geographic Focus: All States
Date(s) Application is Due: Mar 16
Amount of Grant: Up to 150,000 USD
Contact: R. David Rebanal, Program Officer; (503) 220-1955; drebanal@nwhf.org
Internet: http://www.nwhf.org/index.php?/apply/hatfield
Sponsor: Northwest Health Foundation
221 NW Second Avenue, Suite 300
Portland, OR 97209

NWHF Partners Investing in Nursing's Future Grants 3036
Partners Investing in Nursing's Future is a collaborative initiative of the RWJ Foundation and Northwest Health Foundation that will address nursing issues at the local level through funding partnerships with community and regional foundations. The goal of this initiative is to enable local foundations to act as catalysts in developing the comprehensive strategies that are vital to establishing a stable, adequate nursing workforce. Grants support the capacity, involvement, and leadership of local foundations to advance nursing workforce solutions in their communities. Local foundations must match funds with at least $1 for every $1 provided by the program. Proposals must be submitted online. Guidelines are available online. Letters of Intent are due by October 2, and invited proposals are due on December 19.
Requirements: U.S. local or regional private, family, or community foundations are eligible. Eligible foundations are those classified under IRS codes as 501(c)3 tax exempt, a nonexempt charitable trust treated as a private foundation under 4947(a)1, or organizations claiming status as a private operating foundation under 4942(j)3 or 5. Government entities, corporations, or corporate grantmakers may participate in funding collaboratives but may not serve as the applicant organization.
Geographic Focus: All States
Date(s) Application is Due: Oct 2; Dec 19
Amount of Grant: Up to 250,000 USD
Contact: Judith Woodruff; (503) 220-1955; fax (503) 220-1355; judith@nwhf.org
Internet: http://www.partnersinnursing.org/grants.html
Sponsor: Northwest Health Foundation
221 NW Second Avenue, Suite 300
Portland, OR 97209

NWHF Physical Activity and Nutrition Grants 3037
The Foundation is soliciting applications from organizations engaged in creating community environments and policies that support healthy lifestyles, specifically with regards to healthy food choices and opportunities for regular physical activity. It is looking to fund initiatives that: identify the social, policy or environmental factors to be addressed, and how they will support healthy physical activity and nutrition behaviors; articulate goals, strategies and priorities that have been developed in full partnership with the community to be served; build and mobilize constituencies to advocate for improved physical activity and nutrition policies, practices and environments; reduce health disparities related to unequal access to healthy food and opportunities for daily physical activity; involve collaboration among different organizations or agencies, especially across different sectors, to achieve common goals; and implement recommendations proposed in peer-reviewed national and state reports on physical activity and nutrition. The Foundation's intent is to award a portfolio of five to eight grants of various sizes in the $25,000 to $100,000 range (total award amount - not per year). Projects will be funded for up to three years.
Requirements: The Northwest Health Foundation invites proposals from non-profit organizations, local government agencies and/or tribal organizations
Restrictions: For this RFP, the Foundation is not interested in funding curriculum development, research studies, clinical interventions, exercise or cooking classes, educational programs, capital construction costs or general operating support.
Geographic Focus: Oregon, Washington
Date(s) Application is Due: Mar 16
Amount of Grant: 25,000 - 100,000 USD
Contact: Judith Woodruff; (503) 220-1955; fax (503) 220-1355; judith@nwhf.org
Internet: http://www.nwhf.org/index.php?apply/obesity
Sponsor: Northwest Health Foundation
221 NW Second Avenue, Suite 300
Portland, OR 97209

NYAM Brock Lecture, Award and Visiting Professorship in Pediatrics 3038
The New York Academy of Medicine invites pediatric program directors from New York area medical schools and medical centers to nominate candidates for the annual fall Millie and Richard Brock Lecture, Award and Visiting Professorship in Pediatrics. The Brock Lecture, Award and Visiting Professorship in Pediatrics was established by Millie and Richard Brock in 1995 on the 100th anniversary of The Academy's Section on Pediatrics to sponsor a nationally recognized leader in pediatrics to engage in a one to two day visiting professorship at a New York-area pediatrics training program, to deliver the annual Brock Lecture, and to receive the Brock Award for distinguished contributions to pediatrics. The one to two day visiting professorship to a New York area pediatric training program should be part of a program which include a grand rounds lecture as well as teaching interactions with medical students, pediatric residents, fellows, and members of the attending staff. The Brock Lecture held at the Academy must address issues concerned with providing care for underserved children. Suggested topics for discussion include, but are not restricted to: childhood nutrition, HIV and AIDS, substance abuse, child abuse, asthma, violence, and environmental issues such as lead poisoning. The Brock program includes an honorarium of $1,000 and all expenses for travel, lodging, and meals. Letters of nomination, including an outline of the proposed plan for the visit and the topic for the Brock Lecture, as well as the curriculum vitae of the nominee, should be forwarded via email to: brock@nyam.org.
Geographic Focus: All States
Date(s) Application is Due: Mar 1
Amount of Grant: 1,000 USD
Contact: Program Coordinator; (212) 822-7204; fax (212) 822-7338; brock@nyam.org
Internet: https://www.nyam.org/grants/brock.shtml
Sponsor: New York Academy of Medicine
1216 Fifth Avenue
New York, NY 10029-5202

NYAM Charles A. Elsberg Fellowship 3039
The Fellowship in Neurological Surgery was established to support research training in the specialty of neurological surgery for individuals who have completed, or will shortly complete, accredited residency training in neurological surgery, and who intend to use research training for continued development of academic careers in this field. The Elsberg

Fellowship program seeks to support individuals in supervised programs that will develop the candidate's capacity to perform independent clinical or laboratory research.
Requirements: Candidates who have, or will have completed an accredited neurological surgery residency program by June 30, and whose career direction suggests eligibility for subsequent research support are invited to apply. In general, applicants are expected to be at the start of their academic careers, having completed their residency training within the past year or two. Candidates must be board certified or eligible for certification by the beginning of the grant period. The proposed research must be conducted at an institution located in the United States under the mentorship of a senior researcher.
Geographic Focus: All States
Amount of Grant: 50,000 USD
Samples: Aaron S. Dumont, MD, University of Virginia School of Medicine (2008-09); Ziv Williams, MD, Massachusetts General HospitalJason (2007-08); H. Huang, MD, University of Rochester School of Medicine and Dentistry (2006-07).
Contact: Administrator; (212) 822-7204; fax (212) 822-7338; elsberg@nyam.org
Internet: http://www.nyam.org/grants/elsberg.shtml
Sponsor: New York Academy of Medicine
1216 Fifth Avenue
New York, NY 10029-5202

NYAM David E. Rogers Fellowships 3040
The Fellowship is meant to enrich the educational experiences of medical and dental students through projects that bear on medicine and dentistry as social enterprises, such as enterprises devoted to the capacity of these professions in any and all of their expressions to serve human needs. Particularly, the focus is on the needs of underserved or disadvantaged patients or populations. The content of the Fellowship might include clinical investigation, health policy analysis, activities linking biomedicine, the social infrastructure and human need, or community activities.
Requirements: Available to first-year medical and dental students for support of projects to be executed during the summer between the first and second years of medical or dental school.
Geographic Focus: All States
Amount of Grant: 3,500 USD
Samples: Abdulrahman Mohamed El-Sayed, University of Michigan Medical School; Patricia Kuan-Pei Foo, Stanford University School of Medicine; Christina Marie Hupman, Medical College of Georgia.
Contact: Academic Affairs; (212) 822-7204; fax (212) 996-7338; rogers@nyam.org
Internet: http://www.nyam.org/grants/rogers.shtml
Sponsor: New York Academy of Medicine
1216 Fifth Avenue
New York, NY 10029-5202

NYAM Edward N. Gibbs Memorial Lecture and Award in Nephrology 3041
The New York Academy of Medicine seeks nominations of physician scientists who have dedicated their careers to advances in nephrology or are presently making cutting-edge discoveries in the field. The distinguished recipient of this honor will present his or her research at a lecture before a broad audience of scientists and clinicians, to be held at The New York Academy of Medicine in the fall. The award recipient will receive the Edward N. Gibbs Memorial Medal and an honorarium of $7,500; all travel expenses associated with the lecture will be paid by the Edward N. Gibbs Memorial Endowment. It is expected that the manuscript from this lecture will be submitted to a scholarly journal for publication.
Requirements: Candidates must hold an MD degree and be citizens of the United States to be considered for this award.
Geographic Focus: All States
Date(s) Application is Due: Mar 1
Amount of Grant: 7,500 USD
Contact: Program Coordinator; (212) 822-7204; fax (212) 822-7338; gibbs@nyam.org
Internet: https://www.nyam.org/grants/gibbs.shtml
Sponsor: New York Academy of Medicine
1216 Fifth Avenue
New York, NY 10029-5202

NYAM Edwin Beer Research Fellowship 3042
The Edwin Beer Research Fellowship seeks to increase the number of investigators in urology and urology-related fields by providing transitional funding in support of research conducted by qualified candidates. The program provides two-year grants of up to $80,000 for research projects commencing July of the award year. The grants are made to the awardees' institutions for the direct support of their salary and research activities. Fellows are required to devote at least fifty percent of their time to the research funded by the program.
Requirements: Candidates must hold an MD, PhD, or equivalent degree and are expected to conduct research in a supervised program in the United States. Applicants should have completed residency and fellowship or postdoctoral training, having demonstrated their ability to perform and direct research, and should currently be under the mentorship of a senior researcher. Fellowships are awarded to qualified candidates who are at a stage of transition prior to receiving major independent funding. Because the program seeks to increase the number of investigators in urology and urology-related fields, applicants must have a direct association with the department of urology at their sponsoring institution. Candidates are not required to be United States citizens; however, non-citizen applicants are required to provide assurances of employment eligibility to perform the proposed research.
Geographic Focus: All States
Date(s) Application is Due: Feb 2
Amount of Grant: Up to 80,000 USD
Samples: David C. Miller, MD, University of Michigan Medical School; Aria F. Olumi, MD, Massachusetts General Hospital; Jayoung Kim, PhD, Harvard Medical School.
Contact: Academic Affairs; (212) 822-7204; fax (212) 822-7338; beer@nyam.org
Internet: http://www.nyam.org/grants/beer.shtml
Sponsor: New York Academy of Medicine
1216 Fifth Avenue
New York, NY 10029-5202

NYAM Ferdinand C. Valentine Fellowship 3043
The Fellowship offers one-year, $50,000 fellowships in support of research by individuals who have completed, or will shortly complete, residency training acceptable to the American Board of Urology, and who intend to pursue academic research careers in urology.
Requirements: Fellows must commit no less than fifty percent of their time to the supported research and preference is given to investigators conducting their research at institutions in the greater New York area.
Restrictions: Indirect costs and fringe benefits will not be paid.
Geographic Focus: All States
Date(s) Application is Due: Feb 2
Amount of Grant: 50,000 USD
Samples: Gerald J. Wang, MD, New York Presbyterian Hospital, Weill Cornell Medical Center (2009-10); Maurico Davalos, MD, New York Medical College (2008-09); Gerald Y. Tan, MD, Weill Medical College of Cornell University (2009-09).
Contact: Administrator; (212) 822-7204; fax (212) 822-7338; valentine@nyam.org
Internet: http://www.nyam.org/grants/valentine.shtml
Sponsor: New York Academy of Medicine
1216 Fifth Avenue
New York, NY 10029-5202

NYAM Lewis Rudin Glaucoma Grants 3044
This $50,000 prize is awarded for the most outstanding scholarly article on glaucoma published in a peer-reviewed journal during the previous calendar year. Nominations are solicited from leaders of the ophthalmologic community and reviewed in the spring of the award year. One prize winner is recommended for ratification by the Academy's Board of Trustees by the Lewis Rudin Glaucoma Prize Selection Committee, a group of nationally recognized experts in glaucoma research chaired by David H. Abramson, MD of Memorial Sloan-Kettering Cancer Center and Weill Medical College of Cornell University. The winner is announced in July of the award year.
Requirements: Nominees for this award must be the first or last author of the published work and must hold primary responsibility for the research. Authors may nominate themselves but may not submit multiple articles and must choose only one article to submit for consideration by the committee. All authors of the article receive recognition, however the monetary prize is granted solely to the primary researcher specified in the nomination. Copies of the published article must accompany the completed nomination form.
Geographic Focus: All States
Date(s) Application is Due: May 1
Amount of Grant: 50,000 USD
Samples: Oscar A. Candia, MD, Mount Sinai School of Medicine; Claude Burgoyne, MD, Devers Eye Institute; John Danias, MD, PhD, Mount Sinai School of Medicine; Cynthia Lee Grosskreutz, MD, PhD, Harvard Medical School and Massachusetts Eye and Ear Infirmary.
Contact: Coordinator; (212) 822-7204; fax (212) 822-7338; rudinglaucoma@nyam.org
Internet: https://www.nyam.org/grants/rudin.shtml
Sponsor: New York Academy of Medicine
1216 Fifth Avenue
New York, NY 10029-5202

NYAM Mary and David Hoar Fellowship 3045
The Mary and David Hoar Fund was established to promote research in the prevention and treatment of hip fractures. The Fund provides a two-year grant of $100,000 to support pilot programs in clinical, epidemiologic and health services research in the following areas related to hip fractures: clinical care and treatment; prevention; rehabilitation; and model program development.
Requirements: Candidates must hold an MD, PhD, or equivalent degree and are expected to conduct research in a supervised program in the greater New York area. Preferential consideration will be given to new investigators for whom this award is likely to provide interim support pending future funding, or to more senior investigators who are making a change in career or research direction for whom this award will facilitate the transition. Candidates are not required to be United States citizens; however, non-citizen applicants are required to provide assurances, such as permanent resident status, of legal eligibility for employment to perform the proposed research throughout the award period, regardless of the disposition of this or any other pending grants.
Geographic Focus: All States
Amount of Grant: 100,000 USD
Samples: Amanda S. Carmel, Weill Medical College of Cornell University (2008-10); Marcella D. Walker, Columbia University College of Physicians and Surgeons (2007-09); Knox H. Todd, Beth Israel Medical Center (2006-08).
Contact: Program Administrator; (212) 822-7204; fax (212) 822-7338; hoar@nyam.org
Internet: http://www.nyam.org/grants/hoar.shtml
Sponsor: New York Academy of Medicine
1216 Fifth Avenue
New York, NY 10029-5202

NYAM Student Essay Grants 3046
The New York Academy of Medicine invites entries for the New York Academy of Medicine Student Essay Prize, awarded to the best unpublished essay by a graduate student in a medical, public health, pharmacy, or nursing program in the United States. Essays should address topics in the history of public health or medicine as they relate to urban health issues; social or environmental factors in the health of urban populations, institutional histories, or specific diseases may be considered. The winner will receive $500, and the winning essay will receive expedited review for possible publication in the Journal of Urban Health. Honorable Mention prizes may also be awarded at the discretion of the Prize Committee.
Requirements: The contest is open to students in accredited professional degree programs in medicine, nursing, pharmacy, and public health. The writer must have been a student at the time the essay was written.
Geographic Focus: All States
Amount of Grant: 500 USD
Contact: Coordinator; (212) 822-7314; fax (212) 822-7338; historyessay@nyam.org
Internet: https://www.nyam.org/grants/studentessay.shtml
Sponsor: New York Academy of Medicine
1216 Fifth Avenue
New York, NY 10029-5202

NYC Managed Care Consumer Assistance Workshop Re-Design Grants 3047
The New York City Managed Care Consumer Assistance Program (NYC MCCAP), a program of the Community Service Society of NY invites you to submit a proposal for the workshop re-design project. We are seeking individuals or agencies that have interest and knowledge around training, organizational structuring and curriculum development expertise. The goal of the workshop project is to create a more flexible workshop structure that can be easily adapted to different trainers, audiences and settings ranging from busy hospitals and clinics to classrooms and health fairs.
Requirements: 501(c)3 social service agencies with a history of providing education and counseling for low-income communities, immigrants, and populations with special health care needs are encouraged to apply.
Geographic Focus: All States
Date(s) Application is Due: May 21
Contact: Priya Mendon, (212) 614-5400; fax (212) 614-5305; pmendon@cssny.org
Internet: http://www.nycmccap.org/newsitems.html#rfp
Sponsor: Community Service Society of New York
105 East 22nd Street, 8th Floor
New York, NY 10010

NYCT AIDS/HIV Grants 3048
The Trust supports projects that strengthen preventive health care, improve access to services, promote the efficient use of health resources, and develop the skills and independence of people with special needs. The goal of this program is to address the complex social, medical, and legal problems of people with HIV. Since this issue cuts across a number of program areas, projects may fall within a number of funding categories. A limited amount of support is available specifically for policy research and advocacy efforts that: increase public understanding of AIDS and HIV infection; improve the funding and delivery of services; and improve coordination among service organizations. The grant review process takes from two to six months. Grants range from $5,000 to $100,000; an average grant is around $60,000.
Requirements: Grants are made primarily to nonprofit organizations located in the five boroughs of New York City. Grants for programs outside the area are generally from funds designated for specific charities or have been made at the suggestion of donors.
Restrictions: The trust does not make grants to individuals, offer general or capital funding, endowments, building construction/renovation, deficit financing, films, or religion.
Geographic Focus: New York
Date(s) Application is Due: Sep 14
Amount of Grant: 5,000 - 100,000 USD
Contact: Joyce M. Bove, Senior Vice President; (212) 686-0010, ext. 552; fax (212) 534-8528; info@nycommunitytrust.org or grants@nycommunitytrust.org
Internet: http://www.nycommunitytrust.org/HowtoApply/RequestsforProposals/NewYorkCityAIDSFund/tabid/408/Default.aspx
Sponsor: New York Community Trust
909 Third Avenue, 22nd Floor
New York, NY 10022

NYCT Biomedical Research Grants 3049
The Trust supports projects that strengthen preventive health care, improve access to services, promote the efficient use of health resources, and develop the skills and independence of people with special needs. The Trust provides funding for biomedical research projects in four areas: blood diseases, cancer, leprosy and tuberculosis. Grants for these projects are awarded in response to requests for proposals.
Requirements: Grants are made primarily to colleges, universities, and research organizations located in the five boroughs of New York City. Grants for programs outside the area are generally from funds designated for specific charities or have been made at the suggestion of donors.
Restrictions: The trust does not make grants to individuals, offer general or capital funding, endowments, building construction/renovation, deficit financing, films, or religion.
Geographic Focus: New York
Amount of Grant: 5,000 - 100,000 USD
Contact: Joyce M. Bove, Senior Vice President; (212) 686-0010, ext. 552; fax (212) 534-8528; info@nycommunitytrust.org or grants@nycommunitytrust.org
Internet: http://www.nycommunitytrust.org/ForGrantSeekers/GrantmakingGuidelines/HealthandPeoplewithSpecialNeeds/tabid/207/Default.aspx
Sponsor: New York Community Trust
909 Third Avenue, 22nd Floor
New York, NY 10022

NYCT Blindness and Visual Disabilities Grants 3050
The Trust supports projects that strengthen preventive health care, improve access to services, promote the efficient use of health resources, and develop the skills and independence of people with special needs. The goals of this program are to support program innovation and reform and eliminate service gaps. Specifically, grants are made to projects that: improve services to those people with visual disabilities who are presently underserved, such as the elderly, minorities, people with multiple disabilities, people in institutions, and youth in transition from school to work (projects need not be limited to direct service, and may include advocacy and organizational improvement); involve people with visual disabilities more fully in community activities; expand programs that identify people with vision problems at an early stage and link them with appropriate resources; empower persons with visual disabilities by enabling them to participate in planning programs that affect them; and support research in the prevention and treatment of blindness (this may include clinical, epidemiological, and applied studies). The grant review process takes from two to six months. Grants range from $5,000 to $100,000; an average grant is around $60,000.
Requirements: Grants are made primarily to nonprofit organizations located in the five boroughs of New York City. Grants for programs outside the area are generally from funds designated for specific charities or have been made at the suggestion of donors.
Restrictions: The trust does not make grants to individuals, offer general or capital funding, endowments, building construction/renovation, deficit financing, films, or religion.
Geographic Focus: New York
Amount of Grant: 5,000 - 100,000 USD
Contact: Joyce M. Bove, Senior Vice President; (212) 686-0010, ext. 552; fax (212) 534-8528; info@nycommunitytrust.org or grants@nycommunitytrust.org
Internet: http://www.nycommunitytrust.org/ForGrantSeekers/GrantmakingGuidelines/HealthandPeoplewithSpecialNeeds/tabid/207/Default.aspx
Sponsor: New York Community Trust
909 Third Avenue, 22nd Floor
New York, NY 10022

NYCT Blood Disease Research Grants 3051
The New York Community Trust is a charitable foundation concerned with the needs of the Greater New York area. The NYCT has a special fund to be used in support of research in the field of blood diseases. A panel of recognized hematologists will convene to review proposals solicited from the major medical research institutions on a competitive basis. It is estimated that about $700,000 will be available this year to support such research projects. A total of no more than two projects may be submitted from any one organization. Basic laboratory research. Candidates may apply for support totaling $100,000 for up to 2 years, but funds for any one project year may not exceed $50,000. Second-year funding is contingent upon submission of a satisfactory progress report at the end of the first year. Research that may have application to New York City residents is of particular interest to The Trust. Young investigators will be given special consideration.
Requirements: Proposals are due no later than 5:00 p.m. on September 25th. Additional guidelines and forms are available online at the NYCT website.
Geographic Focus: New York
Date(s) Application is Due: Sep 25
Amount of Grant: 100,000 USD
Contact: Joyce M. Bove, Senior Vice President; (212) 686-0010, ext. 552; fax (212) 534-8528; info@nycommunitytrust.org or grants@nycommunitytrust.org
Internet: http://www.nycommunitytrust.org/BloodDiseaseResearch/tabid/504/Default.aspx
Sponsor: New York Community Trust
909 Third Avenue, 22nd Floor
New York, NY 10022

NYCT Children and Youth with Disabilities Grants 3052
The Trust supports projects that strengthen preventive health care, improve access to services, promote the efficient use of health resources, and develop the skills and independence of people with special needs. The goals of this program are to stimulate policymakers and service providers to improve existing services for children with disabilities and to encourage a service approach that emphasizes independence and the development of full potential. Specifically, grants are made to projects that: foster integration into community life, independent living, and improved self-image; improve early identification of disability, encourage early intervention, and increase access to early childhood education; provide comprehensive treatment, planning, and referral programs for children and their families; and assess needs, develop policy, and advocate to improve the delivery and coordination of services. The grant review process takes from two to six months. Grants range from $5,000 to $100,000; an average grant is around $60,000.
Requirements: Grants are made primarily to nonprofit organizations located in the five boroughs of New York City. Grants for programs outside the area are generally from funds designated for specific charities or have been made at the suggestion of donors.
Restrictions: The trust does not make grants to individuals, offer general or capital funding, endowments, building construction/renovation, deficit financing, films, or religion.
Geographic Focus: New York
Amount of Grant: 5,000 - 100,000 USD
Contact: Joyce M. Bove, Senior Vice President; (212) 686-0010, ext. 552; fax (212) 534-8528; info@nycommunitytrust.org or grants@nycommunitytrust.org

Internet: http://www.nycommunitytrust.org/ForGrantSeekers/GrantmakingGuidelines
/HealthandPeoplewithSpecialNeeds/tabid/207/Default.aspx
Sponsor: New York Community Trust
909 Third Avenue, 22nd Floor
New York, NY 10022

NYCT Girls and Young Women Grants 3053
The Trust works in partnership with government and private agencies to develop the strengths of families and young people; to improve their living and working conditions; to improve family and child welfare services; and to advance social work practice. The goal of this Program is to improve conditions and opportunities for those who are poor and disadvantaged, particularly minorities and young, single mothers. Specifically, the program funds projects that: prepare girls and young women for economic self-sufficiency by providing the economic and personal support necessary to gain access to jobs and promoting innovative ways to include indigent and vulnerable young women in the workplace; and reduce the effects of health and social problems that threaten the lives and well-being of girls and young women through direct services and policy and advocacy efforts. Particular emphasis is placed on projects addressing the issues of teen pregnancy, homelessness, and AIDS. The grant review process takes from two to six months. Grants range from $5,000 to $100,000; an average grant is around $60,000.
Requirements: Grants are made primarily to nonprofit organizations located in the five boroughs of New York City. Grants for programs outside the area are generally from funds designated for specific charities or have been made at the suggestion of donors.
Restrictions: The trust does not make grants to individuals, offer general or capital funding, endowments, building construction/renovation, deficit financing, films, or religion.
Geographic Focus: New York
Amount of Grant: 5,000 - 100,000 USD
Contact: Joyce M. Bove, Senior Vice President; (212) 686-0010, ext. 552; fax (212) 534-8528; info@nycommunitytrust.org or grants@nycommunitytrust.org
Internet: http://www.nycommunitytrust.org/ForGrantSeekers/GrantmakingGuidelines
/ChildrenYouthandFamilies/tabid/205/Default.aspx
Sponsor: New York Community Trust
909 Third Avenue, 22nd Floor
New York, NY 10022

NYCT Grants for the Elderly 3054
The Trust supports projects that strengthen preventive health care, improve access to services, promote the efficient use of health resources, and develop the skills and independence of people with special needs. The goals of this program are to enable elderly people to remain active in their communities and to meet the basic needs of those who are vulnerable and dependent. Resources are primarily targeted toward the most underserved elderly and those whose needs are most acute, including members of racial and ethnic minorities, the poor, and those with chronic illnesses or mental or functional disabilities. Specifically, grants are made to projects that: increase the number of elderly who are able to participate in community activities, particularly those involving contact with young people; enable the elderly to assume a leadership role in planning, influencing, or changing the programs that serve them; improve the management, capacity, and resources of government and voluntary agencies serving the elderly, and encourage collaboration among agencies; and assess need, develop policy, and advocate to improve the delivery and coordination of services. The grant review process takes from two to six months. Grants range from $5,000 to $100,000; an average grant is around $60,000.
Requirements: Grants are made primarily to nonprofit organizations located in the five boroughs of New York City. Grants for programs outside the area are generally from funds designated for specific charities or have been made at the suggestion of donors.
Restrictions: The trust does not make grants to individuals, offer general or capital funding, endowments, building construction/renovation, deficit financing, films, or religion.
Geographic Focus: New York
Amount of Grant: 5,000 - 100,000 USD
Contact: Joyce M. Bove, Senior Vice President; (212) 686-0010, ext. 552; fax (212) 534-8528; info@nycommunitytrust.org or grants@nycommunitytrust.org
Internet: http://www.nycommunitytrust.org/ForGrantSeekers/GrantmakingGuidelines
/HealthandPeoplewithSpecialNeeds/tabid/207/Default.aspx
Sponsor: New York Community Trust
909 Third Avenue, 22nd Floor
New York, NY 10022

NYCT Health Services, Systems, and Policies Grants 3055
The Trust supports projects that strengthen preventive health care, improve access to services, promote the efficient use of health resources, and develop the skills and independence of people with special needs. The primary goal of this program is to improve the effectiveness, responsiveness, and equity of health care in New York City. Specifically, the program makes grants: for services, policy research, advocacy, and technical assistance that promote the accessibility of basic health services, especially in minority and immigrant communities; to strengthen health service providers, especially those serving the city's poorest residents; and to promote healthy lifestyles. The grant review process takes from two to six months. Grants range from $5,000 to $100,000; an average grant is around $60,000.
Requirements: Grants are made primarily to nonprofit organizations located in the five boroughs of New York City. Grants for programs outside the area are generally from funds designated for specific charities or have been made at the suggestion of donors.
Restrictions: The trust does not make grants to individuals, offer general or capital funding, endowments, building construction/renovation, deficit financing, films, or religion.
Geographic Focus: New York
Amount of Grant: 5,000 - 100,000 USD
Contact: Joyce M. Bove, Senior Vice President; (212) 686-0010, ext. 552; fax (212) 534-8528; info@nycommunitytrust.org or grants@nycommunitytrust.org
Internet: http://www.nycommunitytrust.org/ForGrantSeekers/GrantmakingGuidelines
/HealthandPeoplewithSpecialNeeds/tabid/207/Default.aspx
Sponsor: New York Community Trust
909 Third Avenue, 22nd Floor
New York, NY 10022

NYCT Mental Health and Mental Retardation Grants 3056
The Trust supports projects that strengthen preventive health care, improve access to services, promote the efficient use of health resources, and develop the skills and independence of people with special needs. The goals of this program are to foster the independence of people with mental illness and mental retardation, and to encourage a community-based system of care. Specifically, grants are made to projects that: improve the quality and availability of community housing and services for people with chronic mental illnesses; improve and expand the delivery of mental health services to children and adolescents, especially those that stress the early identification and remediation of problems; strengthen advocacy groups that promote reimbursement practices and allocation of mental health resources that correspond to community needs; and direct the attention of service providers, policy makers, and the general public to the mental health concerns of racial and ethnic minorities. The grant review process takes from two to six months. Grants range from $5,000 to $100,000; an average grant is around $60,000.
Requirements: Grants are made primarily to nonprofit organizations located in the five boroughs of New York City. Grants for programs outside the area are generally from funds designated for specific charities or have been made at the suggestion of donors.
Restrictions: The trust does not make grants to individuals, offer general or capital funding, endowments, building construction/renovation, deficit financing, films, or religion.
Geographic Focus: New York
Amount of Grant: 5,000 - 100,000 USD
Contact: Joyce M. Bove, Senior Vice President; (212) 686-0010, ext. 552; fax (212) 534-8528; info@nycommunitytrust.org or grants@nycommunitytrust.org
Internet: http://www.nycommunitytrust.org/ForGrantSeekers/GrantmakingGuidelines
/HealthandPeoplewithSpecialNeeds/tabid/207/Default.aspx
Sponsor: New York Community Trust
909 Third Avenue, 22nd Floor
New York, NY 10022

NYCT Substance Abuse Grants 3057
The Trust works in partnership with government and private agencies to develop the strengths of families and young people; to improve their living and working conditions; to improve family and child welfare services; and to advance social work practice. The goal of this program is to address the treatment needs of the most under-served, with a focus on women, ex-offenders, and adolescents. Specifically, grants are made to projects that: promote coordinated drug treatment, particularly between treatment providers and other systems such as mental health and child welfare; expand treatment by building the capacity of mid-sized drug treatment programs. We will help groups strengthen and increase existing services and develop new capacity to accommodate people not being treated; and promote strategies that reduce or prevent substance abuse among elders. The grant review process takes from two to six months. Grants range from $5,000 to $100,000; an average grant is around $60,000.
Requirements: Grants are made primarily to nonprofit organizations located in the five boroughs of New York City. Grants for programs outside the area are generally from funds designated for specific charities or have been made at the suggestion of donors.
Restrictions: The trust does not make grants to individuals, offer general or capital funding, endowments, building construction/renovation, deficit financing, films, or religion.
Geographic Focus: New York
Amount of Grant: 5,000 - 100,000 USD
Contact: Joyce M. Bove, Senior Vice President; (212) 686-0010, ext. 552; fax (212) 534-8528; info@nycommunitytrust.org or grants@nycommunitytrust.org
Internet: http://www.nycommunitytrust.org/ForGrantSeekers/GrantmakingGuidelines
/ChildrenYouthandFamilies/tabid/205/Default.aspx
Sponsor: New York Community Trust
909 Third Avenue, 22nd Floor
New York, NY 10022

NYCT Technical Assistance Grants 3058
The Trust focuses on relieving New York's chronic shortage of affordable housing, strengthening the local economy, and protecting the environment. It supports community-based agencies working on these issues at the neighborhood level, and government and nonprofit institutions developing strategies for the City as a whole. The goal of this program is to improve the management capacity of nonprofits and strengthen the nonprofit sector. Specifically, The Trust provides grants: up to $10,000 to current and prospective grantees to hire consultants to help with specific management needs; to advance public and nonprofit service by developing skills and expertise of professionals in the field; and to support service and umbrella organizations providing technical assistance to groups of nonprofits. The grant review process takes from two to six months. Grants range from $5,000 to $100,000; an average grant is around $60,000.
Requirements: Grants are made primarily to nonprofit organizations located in the five boroughs of New York City. Grants for programs outside the area are generally from funds designated for specific charities or have been made at the suggestion of donors.
Restrictions: The trust does not make grants to individuals, offer general or capital funding, endowments, building construction/renovation, deficit financing, films, or religion.

Geographic Focus: New York
Amount of Grant: 5,000 - 100,000 USD
Contact: Joyce M. Bove, Senior Vice President; (212) 686-0010, ext. 552; fax (212) 534-8528; info@nycommunitytrust.org or grants@nycommunitytrust.org
Internet: http://www.nycommunitytrust.org/ForGrantSeekers/GrantmakingGuidelines/CommunityDevelopmentandtheEnvironment/tabid/204/Default.aspx
Sponsor: New York Community Trust
909 Third Avenue, 22nd Floor
New York, NY 10022

Oak Foundation Child Abuse Grants 3059
The foundation awards grants worldwide to address international social and environmental issues, particularly those that have a major impact on the lives of the disadvantaged. In the Child Abuse Program, the Foundation envisions a world in which all children are protected from sexual abuse and sexual exploitation. Recognising that for many children these forms of abuse do not exist in isolation from other forms of abuse and violence, Oak supports initiatives that: directly address sexual abuse and sexual exploitation; and/or diminish other forms of abuse and violence that are related to or impact upon sexual abuse and sexual exploitation. Oak has a particular interest in promoting and supporting learning from the work of partners. This is done through the identification of learning opportunities within its existing partnerships, as well as through new partnerships specifically designed to drive learning forward across the sector.
Requirements: Within this program, Oak funds organizations in the United States, Canada, Brazil, Bulgaria, Ethiopia, Latvia, Mexico, Moldova, Netherlands, South Africa, Switzerland, Tanzania, and Uganda.
Restrictions: Oak does not provide support to individuals, and does not provide funding for scholarships or tuition assistance for undergraduate or postgraduate studies. The Foundation also does not fund religious organizations for religious purposes, election campaigns, or general fund-raising drives.
Geographic Focus: All States, Brazil, Bulgaria, Canada, Ethiopia, Latvia, Mexico, Moldova, Netherlands, South Africa, Switzerland, Tanzania, Uganda, United Kingdom
Amount of Grant: 25,000 - 2,000,000 USD
Samples: Anti-Slavery International, London, England, $137,008 - safe house for victims of domestic violence; Firelight Foundation, Santa Cruz, California, $844,955 - generating learning to enhance community-based child protection mechanisms; Ethiopian Sociology Social Anthropology and Social Work Association, Addis Ababa, Ethiopia, $167,023 - enhancing competences of professionals to prevent child sexual abuse and exploitation.
Contact: Florence Bruce, Director, Child Abuse Programs; cap@oakfnd.ch
Internet: http://www.oakfnd.org/node/1296
Sponsor: Oak Foundation
Case Postale 115 58, Avenue Louis Casai
Geneva, Cointrin 1216 Switzerland

Obesity Society Grants 3060
The Obesity Society's Grants Program was founded to promote, reward, and encourage research in the field of obesity. Pilot grants of up to $25,000 for a 1-year period will be offered to foster and stimulate new research ideas in any area of investigation related to obesity.
Requirements: Members of The Obesity Society (domestic and international) are eligible to apply. Individuals must be qualified by training and experience to conduct his or her responsibilities on the research project.
Geographic Focus: All States
Amount of Grant: 25,000 USD
Contact: Sadie Campbell, (301) 563-6526; fax (301) 563-6595; scampbell@obesity.org
Internet: http://www.obesity.org/about-us/obesity-society-grants.htm
Sponsor: Obesity Society
8757 Georgia Avenue, Suite 1320
Silver Spring, MD 20910

OBSSR Behavioral and Social Science Research on Understanding and Reducing Health Disparities Grants 3061
The purpose of this program is to encourage behavioral and social science research on the causes and solutions to health and disabilities disparities in the U. S. population. Health disparities between, on the one hand, racial/ethnic populations, lower socioeconomic classes, and rural residents and, on the other hand, the overall U.S. population are major public health concerns. Emphasis is placed on research in and among three broad areas of action: public policy; health care; and disease/disability prevention. Particular attention is given to reducing health gaps among groups. Proposals that utilize an interdisciplinary approach, investigate multiple levels of analysis, incorporate a life-course perspective, and/or employ innovative methods such as system science or community-based participatory research are particularly encouraged. The NIH anticipates supporting 20 to 30 awards. Because the nature and scope of the proposed research will vary from application to application, it is anticipated that the size and duration of each award will also vary.
Requirements: Application can be made by the following institutions and organizations: public/state controlled institutions of higher education; private institutions of higher education; nonprofits with 501(c)3 IRS status (other than institutions of higher education); nonprofits without 501(c)3 IRS status (other than institutions of higher education); small businesses; for-profit organizations (other than small businesses); state governments; U.S. territories or possessions; Indian/Native American tribal governments (federally recognized); Indian/Native American tribal governments (other than Federally recognized); Indian/Native American tribally designated organizations; non-domestic (non-U.S.) entities (foreign organizations); Hispanic-serving institutions; historically Black colleges and universities (HBCUs); tribally controlled colleges and universities (TCCUs); Alaska Native and Native Hawaiian serving institutions; regional organization; eligible agencies of the Federal government; and faith-based or community based organizations.
Geographic Focus: All States
Date(s) Application is Due: Feb 5; Jun 5; Oct 5
Contact: Ronald P. Abeles, (301) 496-7859; fax (301) 435-8779; abeles@nih.gov
Internet: http://grants.nih.gov/grants/guide/pa-files/PAR-07-379.html
Sponsor: Office of Behavioral and Social Sciences Research
31 Center Drive, Building 31, Room B1C19
Bethesda, MD 20892-2027

OceanFirst Foundation Major Grants 3062
The OceanFirst Foundation concentrates its grantmaking around four core priority areas: health and wellness; housing; improving quality of life; and youth development and education. In addition, grants are made to support emerging community needs and special initiatives consistent with the priorities of the Foundation. Requests for Major Grants are defined as those that are more than $5,000 for program or project support. Such requests must address one of the Foundation's priorities, and applicants must be able to demonstrate a significant impact within the OceanFirst market area. Potential applicants are strongly encouraged to review the application and requirement checklist prior to completing a request to ensure that all qualifications can be met.
Requirements: Organizations may request one major grant per year and if the request is declined, the organization must wait one year before reapplying to the Foundation.
Restrictions: The Foundation, in general, does not fund outside these specific strategic areas or provide support to organizations that cannot demonstrate a significant level of service within the OceanFirst market area. The Foundation cannot provide funding for the following: individuals; research; organizations not exempt under Section 501(c)3 of the Internal Revenue Code; religious congregations; political causes, candidates, organizations or campaigns; organizations whose primary purpose is to influence legislation; or sports leagues and teams.
Geographic Focus: New Jersey
Amount of Grant: 5,000 - 100,000 USD
Contact: Katherine Durante, Executive Director; (732) 341-4676; fax (732) 473-9641; kdurante@oceanfirstfdn.org or info@oceanfirstfdn.org
Internet: http://www.oceanfirstfdn.org/major-grants.php
Sponsor: OceanFirst Foundation
1415 Hooper Avenue, Suite 304
Toms River, NJ 08753

Oceanside Charitable Foundation Grants 3063
The Oceanside Charitable Foundation will focus on supporting projects that inspire and strengthen Civil Society in its communities; in particular, projects that foster community participation, consensus-building and collective problem solving in the field of children and youth programs. The Oceanside Charitable Foundation will fund Civil Society projects for school-aged (K-12) children and youth that are managed by non-profit organizations or government agencies.
Requirements: Projects for which funding is requested must serve children and/or youth in the geographical limits of the City of Oceanside.
Restrictions: The Oceanside Charitable Foundation does not make grants for: annual campaigns and fund raising events for non-specific purposes; capital campaigns for buildings or facilities; stipends for attendance at conferences; endowments or chairs; for-profit organizations or enterprises; individuals unaffiliated with a qualified fiscal sponsor; projects that promote religious or political doctrine; research projects (medical or otherwise); scholarships; or existing obligations or debt.
Geographic Focus: California
Date(s) Application is Due: Dec 18
Contact: Trudy Amstrong; (619) 814-1384; trudy@sdfoundation.org
Internet: http://www.endowoceanside.org/grants.html
Sponsor: Oceanside Charitable Foundation / San Diego Foundation
2508 Historic Decatur Road, Suite 200
San Diego, CA 92106

OECD Sexual Health and Rights Project Grants 3064
The Open Society Public Health Program supports marginalized populations to fight discrimination and protect their fundamental rights. The program aims to build societies committed to inclusion, human rights, and justice, in which health-related policies and practices reflect these values and are based on evidence. The Sexual Health and Rights Project promotes human rights-based approaches to advancing the health of sex workers and transgender individuals. The project often aims to develop, pilot, and disseminate information on innovative approaches to improving the health of sex workers and transgender individuals in order to integrate these approaches into wide-scale health programs and plans.
Requirements: The Initiative has an open pre-application process. To be considered for an invitation to submit a full proposal, organizations may write a one page letter containing the following information: its purpose and goals; full description of the project and amount requested; organization's total income for the past fiscal year; and biographical information about the organization's leadership.
Geographic Focus: All States, All Countries
Contact: Maureen Aung-Thwin, Director; (212) 548-6919; fax (212) 548-4676; maureen.aung-thwin@opensocietyfoundations.org
Internet: https://www.opensocietyfoundations.org/about/programs/public-health-program
Sponsor: Open Society Foundations
224 West 57th Street
New York, NY 10019

Office Depot Corporation Community Relations Grants 3065
The corporate giving program awards grants to nonprofit organizations aligned with Office Depot's mission to directly impact the health, education, and welfare of children. Office Depot supports nonprofit organizations at the local level with donations of products, contributions of funds, and efforts to encourage employees and customers to become involved. To request a monetary donation, provide a brief description of the organization, the federal tax-ID number, an explanation of what is being requested, and the rationale based on Office Depot's charitable giving guidelines. The request should be submitted on organization letterhead and sent to the office.
Requirements: 501(c)3 nonprofit organizations are eligible.
Restrictions: Grants do not support individuals, advertising, athletic teams or events, fashion shows, project graduation, capital campaigns, individual or group travel, political causes, film/video projects, or nonprofit organizations that spend more than 25 percent of their revenue on management overhead and fundraising expenses.
Geographic Focus: All States
Contact: Melissa Perlman; (561) 438-0704; Melissa.Perlman@officedepot.com
Internet: http://www.community.officedepot.com/local.asp
Sponsor: Office Depot Corporation
2200 Old Germantown Road
Delray Beach, FL 33445

Ogden Codman Trust Grants 3066
The trust awards grants to eligible Massachusetts nonprofit organizations in its areas of interest, including environmental conservation, historic preservation, cultural activities, health care, and social services. Grants support special programs intended to improve the quality of life. Matching funds are sometimes requested
Requirements: Lincoln, MA, nonprofit organizations are eligible to apply.
Restrictions: General operating grants are rarely awarded.
Geographic Focus: All States
Contact: Susan Monahan; (617) 951-1108; fax (617) 542-7437; smonahan@rackemann.com
Sponsor: Ogden Codman Trust
160 Federal Street, 13th Floor
Boston, MA 02110-1700

Ohio County Community Foundation Board of Directors Grants 3067
The Ohio County Community Foundation may, on occasion, find it necessary to issue discretionary small grants from the unrestricted funds outside the Grants Committee recommendations and full Board approval. These grants are on a first come, first serve basis until allotted funding for the current calendar year has been exhausted. Each grant application will be reviewed after submission to ensure the application is complete and the organization is eligible to make application. Specifically, Board of Directors Grants are given in an amount not to exceed $300 per application with a maximum of $600 per year per organization and a maximum of four Board Grants per year.
Requirements: Grants will only be awarded for projects and programs that benefit the residents of Ohio County. Applicants must qualify as an exempt organization under the IRS Code 501(c), or be sponsored by such organizations, or qualify as a governmental or educational entity or possess similar attributes per IRS Code Section 509(a). Grants applied for brick and mortar projects must be only for charitable purposes, to further the mission of a public charity.
Restrictions: No grants will be made solely to individuals but can be made for the benefit of certain individuals for such purposes as scholarships and special programs through educational institutions and other sponsoring recipient organizations. In addition, no grants will be made specifically for sectarian religious purposes but can be made to religious organizations for general community programs.
Geographic Focus: Indiana
Contact: Stephanie Scott; (812) 438-9401; fax (812) 438-9488; sscott@occfrisingsun.com
Internet: http://www.occfrisingsun.com/CombineApplicationSmallGrants.htm
Sponsor: Ohio County Community Foundation
591 Smart Drive, P.O. Box 170
Rising Sun, IN 47040

Ohio County Community Foundation Grants 3068
The Ohio County Community Foundation is charged with assisting donors in building, managing, and distributing a lasting source of charitable funds for the good of Ohio County. The Foundation funds projects and programs for economic development, education, human services, cultural affairs, and health. The Foundation will: offer grant awards that strive to anticipate the changing needs of the community and be flexible in responding to them; focus on those types of grants which will have the greatest benefit per dollar granted; encourage the participation of other contributions by using matching, challenge and other grant techniques; offer funding that closely relates and coordinates with the programs of other sources for funding, such as the government, other foundations, and associations; induces grant recipients to achieve certain objectives such as becoming more efficient, increasing fundraising capabilities, and delivering better products; and consider grants in the form of technical assistance and staff assisted special projects which are intended to respond to a variety of needs in the county.
Requirements: Grants will be made only to organizations whose programs benefit the residents of Ohio County. The Foundation uses the following criteria when evaluating proposals: is there an established need for the program or project; are there other more compatible sources for potential funding; does the Foundation have adequate sources to respond: and does the grant support a charitable purpose.
Restrictions: Funding is not available to individuals. Grants are not made to enable individuals or groups to take trips except where there are special circumstances which will benefit the larger community.
Geographic Focus: Indiana
Date(s) Application is Due: Apr 15; Oct 15
Amount of Grant: Up to 3,000 USD
Contact: Stephanie Scott; (812) 438-9401; fax (812) 438-9488; sscott@occfrisingsun.com
Internet: http://www.occfrisingsun.com/GrantApplicationNew.htm
Sponsor: Ohio County Community Foundation
591 Smart Drive, P.O. Box 170
Rising Sun, IN 47040

Ohio County Community Foundation Mini-Grants 3069
The Ohio County Community Foundation assists donors in building, managing and distributing a lasting source of charitable funds for the good of Ohio County. The Foundation Mini-Grants fund projects and programs for economic development, education, human services, cultural affairs, and health. Amounts do not exceed $100 per application with a maximum of $200 per year per organization and a maximum of four mini-grants per year. These grants are on a first come, first serve basis until allotted funding for the current calendar year has been exhausted.
Requirements: Grants will be made only to organizations whose programs benefit the residents of Ohio County. The Foundation uses the following criteria when evaluating proposals: is there an established need for the program or project; are there other more compatible sources for potential funding; does the Foundation have adequate sources to respond; and does the grant support a charitable purpose.
Restrictions: Funding is not available to individuals. Grants are not made to enable individuals or groups to take trips except where there are special circumstances which will benefit the larger community.
Geographic Focus: Indiana
Amount of Grant: Up to 100 USD
Contact: Stephanie Scott; (812) 438-9401; fax (812) 438-9488; sscott@occfrisingsun.com
Internet: http://www.occfrisingsun.com/CombineApplicationSmallGrants.htm
Sponsor: Ohio County Community Foundation
591 Smart Drive, P.O. Box 170
Rising Sun, IN 47040

Ohio Learning Network Grants 3070
The network funds programs to help Ohioans increase work-related knowledge and skills through college credits; assist faculty in teaching using emerging technologies; and better understand the role of technology in learning. Grants support emerging needs, learning communities, distance learning research, partnership grants, and technology initiatives/innovations.
Geographic Focus: Ohio
Amount of Grant: Up to 100,000 USD
Contact: Judith M. Leach; (614) 485-6738 or (614) 995-3240; jleach@oln.org
Internet: http://www.oln.org/about_oln/grants.php
Sponsor: Ohio Learning Network
35 E Chestnut Street, 8th Floor
Columbus, OH 43215-2541

Oleonda Jameson Trust Grants 3071
The trust awards grants to New Hampshire nonprofit organizations in its areas of interest, including arts/culture, community foundations, children and youth, health care, housing, social sciences, human services, federated giving programs and, social services. Types of support include capital campaigns, building construction/renovation, equipment acquisition, emergency funds, and scholarship funds. There are no application forms or deadlines. The board meets in March, June, September, and December.
Requirements: New Hampshire nonprofit organizations are eligible. Requests from Concord nonprofits receive preference.
Restrictions: Grants do not support endowments or general operating expenses.
Geographic Focus: New Hampshire
Amount of Grant: 1,000 - 60,000 USD
Contact: Grants Management Officer; (603) 226-0400
Sponsor: Oleonda Jameson Trust
11 South Main Street, Suite 500
Concord, NH 03301-4945

Olive Higgins Prouty Foundation Grants 3072
The Olive Higgins Prouty Foundation was established in Massachusetts in 1952, in honor of an American novelist and poet, best known for her 1922 novel Stella Dallas and her pioneering consideration of psychotherapy in her 1941 novel Now, Voyager. The Foundation's primary fields of interest continue to be: arts and culture; child welfare; higher education; hospital care; music, and secondary education. Giving is generally centered around New Bedford and Worcester, Massachusetts, although it sometimes is given out-of-state. Support is typically given for annual campaigns, capital campaigns, and general operations. Amounts range from $1,000 to $50,000, with approximately forty awards given annually. Applications should be submitted in letter form, along with details of the proposal and budgetary needs. The annual deadline for submission is September 30.
Requirements: Any 501(c)3 serving the residents of New Bedford and Worcester, Massachusetts, are eligible to apply, along with a selection of out-of-state organizations.
Geographic Focus: All States
Amount of Grant: 1,000 - 50,000 USD
Contact: Charlene Teja; (617) 434-6565 or (401) 278-6058
Sponsor: Olive Higgins Prouty Foundation
P.O. Box 1802, One Financial Plaza
Providence, RI 02901-1802

Olympus Corporation of Americas Corporate Giving Grants 3073

The Olympus Corporation of Americas (OCA) targets corporate giving initiatives designed to benefit and sustain society, making positive contributions. Globally, Olympus strives to play an integral role in helping people around the world to lead safer, healthier and more fulfilling lives. In addition to offering knowledge, expertise and world-renowned products, its employees go further to fulfill the corporate responsibility to its neighbors and to the planet. As a leading medical device company with roots in healthcare going back nearly 100 years, our corporate giving philosophy is aligned with industry guidance and best practices for healthcare companies. Ever cognizant of our duty to comply with laws governing interactions with health care professionals, we adhere to the charitable contributions requirements of the Advanced Medical Technology Association's (AdvaMed) Code of Ethics on Interactions with Health Care Professionals.
Requirements: Olympus identifies 501(c)3 charitable organizations* that support the above advancements toward building a sustainable society.
Restrictions: Olympus will not make contributions to (a) organizations that discriminate on the basis of race, creed, color, gender, or national origin or (b) individuals. Olympus will not make contributions to denominational religious or political organizations, or similar limited-purpose groups, unless the funds are for purposes that will benefit the community as a whole or further the interests of Olympus.
Geographic Focus: All States, Argentina, Bolivia, Brazil, Canada, Caribbean, Chile, Colombia, Costa Rica, Ecuador, El Salvador, Guatemala, Honduras, Mexico, Netherlands Antilles, Nicaragua, Panama, Paraguay, Peru, Uruguay, Venezuela
Contact: Manager of Corporate Giving; (484) 896-5000
Internet: http://www.olympusamerica.com/corporate/social_responsibility/community_involvement.asp
Sponsor: Olympus Corporation of Americas
3500 Corporate Parkway
Center Valley, PA 18034

Olympus Corporation of Americas Fellowships 3074

The Olympus Fellows program is a full-time, benefits-eligible leadership development program designed for new college graduates who have strong desires to pursue careers with a company whose products and services can help save lives and capture life's most precious moments. It is designed to be educational and challenging while allowing Olympus Fellows to apply analytical and creative skills to make valuable contributions to the business. Based on prior studies and work experience, Olympus has Fellows opportunities in its Medical Systems, Imaging, Scientific Equipment or Corporate Groups. The Fellowship accepts all majors from any Bachelor's or Master's degree program. Graduate and current Fellows have a range of majors, including marketing, finance, engineering, psychology, biochemistry, international relations, supply chain management and information systems, economics, kinesiology, business administration and more. The amount of travel a Fellow can expect will depend on each rotational assignment. Although there is no set deadline, the program encourages students to apply as early in the academic year as possible. Once the next class of Fellows has been identified, the position will be removed from the job posting website until the beginning of the next academic year.
Geographic Focus: All States
Contact: Manager of Fellowships; (484) 896-5000
Internet: http://www.olympusamerica.com/corporate/recruiting/index.asp
Sponsor: Olympus Corporation of Americas
3500 Corporate Parkway
Center Valley, PA 18034

OneFamily Foundation Grants 3075

The foundation awards grants to eligible Washington nonprofit organizations working to improve the lives of women living in poverty and at-risk youth, for support services for abused women, and for efforts to end sexual abuse against women and children. Consideration is given to programs that provide training and skills development to low-income women and services providing basic needs such as shelter, counseling, food, and childcare; educational and mentoring projects to help prevent teen pregnancy; job-training programs for youth and school/community-based programs to help low-income youth complete their education; parenting, training, and education programs to help break the cycle of family violence; shelters and services to support abused and neglected children and women; and hands-on programs to encourage philanthropy among children and youth. Grants will support operating expenses, special projects, and minor capital costs necessary to assure the success of a funded project. For general grants, two-page preapplication letters are due on the third Friday in March, July, and November and final proposals are due the second Friday in January, May, and September. Annual deadlines may vary; contact program staff for exact dates.
Requirements: Nonprofit 501(c)3 organizations based in King County, Snohomish County, or the Olympic Peninsula of Washington State are eligible to apply for funding.
Restrictions: Grants are not made to: individuals, scholarships, schools, research, summer camps, athletic events, video or film projects, website development, book publications. No multi-year requests are considered. Groups who have been declined three times are ineligible to reapply.
Geographic Focus: Washington
Amount of Grant: 5,000 - 12,000 USD
Contact: Therese Ogle; (206) 781-3472; fax (206) 784-5987; Oglefounds@aol.com
Internet: http://fdncenter.org/grantmaker/onefamily
Sponsor: OneFamily Foundation
6723 Sycamore Avenue NW
Seattle, WA 98117

OneSight Research Foundation Block Grants 3076

Formerly known as the Pearle Vision Foundation, the OneSight Research Foundation is dedicated to sight preservation through research and education. A significant portion of their grant dollars will be awarded to fund research and treatment for diabetic eye diseases, hoping to find a cure for diabetic retinopathy, the number one cause of adult blindness in the United States.
Requirements: U.S. 501(c)3 nonprofits are eligible for grant support. Requests for funding should cover a period of time not to exceed one year. Exceptions to this policy will be considered on an individual basis.
Restrictions: Grants will not be considered for endowments or operating expenses.
Geographic Focus: All States
Date(s) Application is Due: Jun 30; Dec 31
Amount of Grant: 5,000 - 50,000 USD
Contact: Trina Parasiliti; (972) 277-6191; fax (972) 277-6414; tparasil@onesight.org
Internet: http://www.onesight.org/northamerica/na/
Sponsor: OneSight Research Foundation
2465 Joe Field Road
Dallas, TX 75229

OneStar Foundation AmeriCorps Grants 3077

OneStar NSC seeks to fund programs that engage members to address health, public safety, homeland security, education, or human service needs in Texas communities. Funds support the state priority grant, professional corps grant, and education award grant. AmeriCorps education awards for members who successfully complete their full term of service will also be supported. Guidelines are available online.
Requirements: Local or statewide nonprofit organizations, state and local units of government, and institutions of higher education are eligible.
Geographic Focus: All States
Contact: Administrator; (512) 287-2000; fax (512) 473-8228; onestar@onestarfoundation.org
Internet: http://www.onestarfoundation.org/page/americorpstexas
Sponsor: OneStar Foundation
816 Congress, Stuite 900
Austin, TX 78701

Oppenstein Brothers Foundation Grants 3078

Grants are awarded in the metropolitan area of Kansas, City, Missouri, primarily for social services and early childhood, elementary, secondary, adult-basic, vocational, and higher education; family planning and services; social services and welfare agencies; Jewish welfare organizations; and programs for youth, the handicapped, disadvantaged, mentally ill, homeless, minorities, and elderly. Additional areas of interest include arts/cultural programs, the performing arts, museums, health care and health organizations, and AIDS research. The foundation considers requests for building and renovation, capital campaigns, curriculum development, emergency funds, equipment, general operating support, program development, seed money, technical support, conferences and seminars, consulting services, and matching funds. Application guidelines are available on request. The board meets every other month. Deadlines are generally three weeks prior to board meetings. Notification of award will take place within two to four months.
Requirements: Nonprofit organizations serving the metropolitan area of Kansas City, Missouri are eligible to apply.
Restrictions: The foundation primarily supports 501(c)3 nonprofit organizations serving the metropolitan area of Kansas City, Missouri. Grants are not awarded to support individuals or for annual campaigns.
Geographic Focus: Missouri
Contact: Beth Radtke, Program Officer; (816) 234-2577
Sponsor: Oppenstein Brothers Foundation
922 Walnut Street, Suite 200
Kansas City, MO 64106-1809

Orange County Community Foundation Grants 3079

The Foundation's mission is to encourage, support and facilitate philanthropy in Orange County, California. One of our most important strategic priorities is strengthening the capacity of Orange County's nonprofit sector. Foundation Grants are made possible by the income earned from unrestricted, field-of-interest endowment funds and legacy funds. Grants funding is determined through research of community needs and approved by the Board. In addition, Grants are periodically administered in partnership with statewide and national foundations as well as Foundation donors to advance special projects and initiatives in Orange County. Fields of interest include arts and culture, environment, education, health, an human services.
Requirements: Nonprofits agencies in Orange County, California, are eligible. Unsolicited grant proposals are not accepted. Nonprofit agencies wishing to apply should pay special attention to the website for requests for proposal and deadlines. Every grant has a formal application and its own annual cycle.
Geographic Focus: California
Contact: Patricia Benevenia; (949) 553-4202, ext. 37; penevenia@oc-cf.org
Internet: http://www.oc-cf.org/Page.aspx?pid=496
Sponsor: Orange County Community Foundation
4041 MacArthur Boulevard, Suite 510
Newport Beach, CA 92660

Ordean Foundation Grants 3080

The Foundation, established in 1933, supports organizations involved with alcoholism, children/youth, services, crime/violence prevention, youth, education, family services, food services, health care, homeless, human services, housing/shelter, human services, medical care,

rehabilitation, mental health/crisis services, substance abuse, services, YM/YWCAs & YM/YWHAs, aging, disabilities, economically disadvantaged. Grants will be awarded for up to three years. Applications are reviewed on an ongoing basis; the deadline is the 15th of each month.
Requirements: Non-profits in the Duluth and contiguous cities and townships in St. Louis County, Minnesota are eligible. The Foundation accepts the Minnesota Common Grant Application Form.
Restrictions: No support for direct religious purposes, or for political campaigns or lobbying activities. No grants to individuals (directly), or for endowment funds, travel, conferences, seminars or workshops, telephone solicitations, benefits, dinners, research, including biomedical research, deficit financing, national fund raising campaigns, or to supplant government funding.
Geographic Focus: Minnesota
Amount of Grant: 1,000 - 290,000 USD
Contact: Stephen A. Mangan; (218) 726-4785; ordean@computerpro.com
Sponsor: Ordean Foundation
501 Ordean Building
Duluth, MN 55802-4725

Oregon Community Foundation Better Nursing Home Care Grants 3081
The Better Nursing Home Care Fund was created at The Oregon Community Foundation in 1986 by the U.S. District Court, as partial restitution in a court case involving the fraudulent operation of nursing homes. The U.S. Department of Justice added to the fund in 1992, via a settlement in connection with a separate court case. In addition, private donors contribute to the fund as a way of supporting improvements in nursing home care. This fund generally supports two or three projects each year that focus on improving care for nursing home residents in Oregon. The Foundation welcomes proposals from outside the Willamette Valley. About $25,000 is available for these grants, which are awarded each spring. The application deadline is February 7. Applicants are notified by May 31 of funding decisions.
Requirements: Projects may address issues or needs of both nonprofit and for-profit nursing homes. However, applications are accepted only from nonprofits with tax-exempt status and public agencies that operate within Oregon. At this time the fund is focused only on nursing homes and is not available for assisted living facilities.
Restrictions: Grants will not be made for equipment or capital projects, or to replace other funding where it is not demonstrated that OCF support will make a significant difference.
Geographic Focus: Oregon
Date(s) Application is Due: Feb 7
Amount of Grant: Up to 25,000 USD
Contact: Melissa Hansen, Program Officer; (503) 227-6846 or (503) 944-2108; fax (503) 274-7771; mhansen@oregoncf.org
Internet: http://www.oregoncf.org/grants-scholarships/grants/ocf-funds/better-nursing-home-care
Sponsor: Oregon Community Foundation
1221 SW Yamhill Street, Suite 100
Portland, OR 97205

Oregon Community Foundation Community Grants 3082
The Oregon Community Foundation's Community Grant program addresses community needs and fosters civic leadership and engagement throughout the state. The program awards about 220 to 240 grants each year, mostly to small- and moderate-size nonprofits. The average grant award is $20,000. OCF typically receives 300 to 350 proposals per grant cycle and funds 110 to 120 of these. Concerns central to OCF's evaluation of proposed projects include: the strength of local support for the project; the strength of the applicant organization; and whether the project addresses a significant community need. Four different application areas are considered: health and well being of vulnerable populations (30 to 40 percent of grants); educational opportunities and achievement (30 to 40 percent of grants); arts and cultural organizations (15 to 25 percent of grants); and community livability, environment and citizen engagement (10 to 20 percent of grants). The two annual deadlines for application submission are January 15 and July 15.
Requirements: Applicants must: have 501(c)3 tax-exempt status and be classified as a public entity rather than a private foundation as defined by section 509(a) of the Internal Revenue Code (alternatively, they must have a qualified fiscal sponsor (i.e., a sponsoring tax-exempt organization that is tax-exempt); and have submitted required evaluation reports for all prior grants from the Foundation.
Restrictions: Organizations are not eligible to apply who either currently have an active Community Grant, or who are requesting funding for a program or project that has been previously funded by a Community Grant. Generally speaking, applicants may submit only one Community Grant application per 12-month period. Activities typically not eligible for funding include: annual fund appeals or endowment funds; sponsorship of one-time events or performances; sponsorship of regular events or performances (e.g., a season); projects in individual schools; grants to scholarship or regranting programs; subsidies to allow individuals to participate in conferences; capital projects that will not clearly benefit the community; purchases or activities that occur prior to grant decisions; deficit funding; replacement of government funding; lobbying to influence legislation; scientific research; religious activities; or operating support (except where a grant may have strategic impact on the long-term viability of programs of high priority).
Geographic Focus: Oregon
Date(s) Application is Due: Jan 15; Jul 15
Amount of Grant: Up to 40,000 USD
Contact: Megan Schumaker, Senior Program Officer for Community Grants; (503) 227-6846; fax (503) 274-7771; mschumaker@oregoncf.org or grants@oregoncf.org
Internet: http://www.oregoncf.org/grants-scholarships/grants/community-grants
Sponsor: Oregon Community Foundation
1221 SW Yamhill Street, Suite 100
Portland, OR 97205

Orthopaedic Trauma Association Kathy Cramer Young Clinician Memorial Fellowships 3083
The AAOS/OREF/ORS Clinician Scholar Career Development Program (CSCDP) is an annual program seeking applicants in their PGY2-PGY 5 residency years, in fellowships, and Junior Faculty through year three who have the potential/desire to become orthopaedic clinician scientists. The program supports two scholarships for the New Investigators Workshop, as well as two scholarships for the annual Clinician Scholar Career Development program.
Requirements: Kathy Cramer Young Clinician Scholarship Award applicants for the AAOS/OREF/ORS Clinician Scholar Career Development Program must be an OTA member in North America ideally in their PGY2-PGY5 residency years, in fellowships and junior faculty through year three who have the potential/desire to become orthopaedic clinician scientists.
Geographic Focus: All States
Contact: Melanie Hopkins, Fellowship Coordinator; (847) 698-1631; fax (847) 823-0536; hopkins@aaos.org or OTA@aaos.org
Internet: http://ota.org/research/research-studies/
Sponsor: Orthopaedic Trauma Association
9400 W. Higgins Road, Suite 305
Rosemont, IL 60018-4975

Orthopaedic Trauma Association Research Grants 3084
The Orthopaedic Trauma Association is accepting pre-proposals for clinical research applications from OTA International Members residing and conducting research. Depending on the year, directed research grants will be given priority evaluation by the research committee. There is no guarantee that these directed grant applications (topics chosen by the OTA Board of Directors) will receive funding, as they will be graded on scientific merit compared to all other grants received that year. The following grants are available for any research issue related to musculoskeletal trauma: Clinical Research Grants ($40,000 per year maximum, for 2 year grant cycle); Basic Research Grants ($50,000 maximum, for 2 year grant cycle); and Resident Research Grant ($20,000 maximum). Full proposal submission deadlines fall during the first two weeks of June.
Geographic Focus: All States
Date(s) Application is Due: Jun 8
Amount of Grant: Up to 50,000 USD
Contact: Melanie Hopkins, Fellowship Coordinator; (847) 698-1631; fax (847) 823-0536; hopkins@aaos.org or OTA@aaos.org
Internet: http://ota.org/research/research-studies/
Sponsor: Orthopaedic Trauma Association
9400 W. Higgins Road, Suite 305
Rosemont, IL 60018-4975

Oscar Rennebohm Foundation Grants 3085
The foundation awards grants to nonprofit organizations in the Madison, WI area. Areas of interest, include: higher education, the arts, environmental conservation, health and social service agencies. Types of support include building construction/renovation, equipment acquisition, and research. There are no application forms or deadlines.
Geographic Focus: Wisconsin
Contact: Grants Administrator; (608) 274-5991
Sponsor: Oscar Rennebohm Foundation
P.O. Box 5187
Madison, WI 53705

OSF Access to Essential Medicines Initiative 3086
The Open Society Public Health Program supports marginalized populations to fight discrimination and protect their fundamental rights. The program aims to build societies committed to inclusion, human rights, and justice, in which health-related policies and practices reflect these values and are based on evidence. The Access to Essential Medicines Initiative engages in grantmaking, capacity building, and advocacy on three areas: promoting national and international laws, policies, and practices that safeguard access to medicines; fostering new thinking and action on needs-driven pharmaceutical innovation; and using existing legal and policy frameworks to improve access to medicines at the country level.
Requirements: The Initiative has an open pre-application process. To be considered for an invitation to submit a full proposal, organizations may write a one page letter containing the following information: its purpose and goals; full description of the project and amount requested; organization's total income for the past fiscal year; and biographical information about the organization's leadership.
Geographic Focus: All States, All Countries
Contact: Maureen Aung-Thwin, Director; (212) 548-6919; fax (212) 548-4676; maureen.aung-thwin@opensocietyfoundations.org
Internet: https://www.opensocietyfoundations.org/about/programs/public-health-program
Sponsor: Open Society Foundations
224 West 57th Street
New York, NY 10019

OSF Accountability and Monitoring in Health Initiative Grants 3087
The Open Society Public Health Program supports marginalized populations to fight discrimination and protect their fundamental rights. The program aims to build societies committed to inclusion, human rights, and justice, in which health-related policies and practices reflect these values and are based on evidence. The Accountability and Monitoring in Health Initiative seeks to strengthen meaningful and sustained engagement by affected communities in the development, implementation, and monitoring of health

budgets, policies, programs and practices; promote government accountability to citizens; and foster an informed and open dialogue about the governance of public health systems, provision of health services, and advancement of health and rights.
Requirements: The Initiative has an open pre-application process. To be considered for an invitation to submit a full proposal, organizations may write a one page letter containing the following information: its purpose and goals; full description of the project and amount requested; organization's total income for the past fiscal year; and biographical information about the organization's leadership.
Geographic Focus: All States, All Countries
Contact: Maureen Aung-Thwin, Director; (212) 548-6919; fax (212) 548-4676; maureen.aung-thwin@opensocietyfoundations.org
Internet: https://www.opensocietyfoundations.org/about/programs/public-health-program
Sponsor: Open Society Foundations
224 West 57th Street
New York, NY 10019

OSF Global Health Financing Initiative Grants 3088
The Open Society Public Health Program supports marginalized populations to fight discrimination and protect their fundamental rights. The program aims to build societies committed to inclusion, human rights, and justice, in which health-related policies and practices reflect these values and are based on evidence. The Global Health Financing Initiative works to ensure that global funding for health is raised, allocated, and used in ways that meet the health needs of marginalized persons, strengthen civil society engagement in decision-making, promote and respect human rights, and lead to greater accountability and transparency.
Requirements: The Initiative has an open pre-application process. To be considered for an invitation to submit a full proposal, organizations may write a one page letter containing the following information: its purpose and goals; full description of the project and amount requested; organization's total income for the past fiscal year; and biographical information about the organization's leadership.
Geographic Focus: All States, All Countries
Contact: Maureen Aung-Thwin, Director; (212) 548-6919; fax (212) 548-4676; maureen.aung-thwin@opensocietyfoundations.org
Internet: https://www.opensocietyfoundations.org/about/programs/public-health-program
Sponsor: Open Society Foundations
224 West 57th Street
New York, NY 10019

OSF Health Media Initiative Grants 3089
The Open Society Public Health Program supports marginalized populations to fight discrimination and protect their fundamental rights. The program aims to build societies committed to inclusion, human rights, and justice, in which health-related policies and practices reflect these values and are based on evidence. The Health Media Initiative seeks to build the capacity of civil society organizations to effectively use media to advocate on key health issues that impact marginalized populations. Through ongoing technical assistance and training workshops, the initiative provides organizations with the tools they need to strengthen their capacity to use media, including the press, social media, and new communications technologies.
Requirements: The Initiative has an open pre-application process. To be considered for an invitation to submit a full proposal, organizations may write a one page letter containing the following information: its purpose and goals; full description of the project and amount requested; organization's total income for the past fiscal year; and biographical information about the organization's leadership.
Geographic Focus: All States, All Countries
Contact: Maureen Aung-Thwin, Director; (212) 548-6919; fax (212) 548-4676; maureen.aung-thwin@opensocietyfoundations.org
Internet: https://www.opensocietyfoundations.org/about/programs/public-health-program
Sponsor: Open Society Foundations
224 West 57th Street
New York, NY 10019

OSF International Harm Reduction Development Program Grants 3090
The Open Society Public Health Program supports marginalized populations to fight discrimination and protect their fundamental rights. The program aims to build societies committed to inclusion, human rights, and justice, in which health-related policies and practices reflect these values and are based on evidence. The International Harm Reduction Development Program works to advance the health and human rights of people who use drugs. Through grant making, capacity building, and advocacy, the program aims to reduce HIV, fatal overdose, and other drug-related harms; to decrease abuse by police and in places of detention; and to improve the quality of health services. The program supports community monitoring and advocacy, legal empowerment, and strategic litigation.
Requirements: The Initiative has an open pre-application process. To be considered for an invitation to submit a full proposal, organizations may write a one page letter containing the following information: its purpose and goals; full description of the project and amount requested; organization's total income for the past fiscal year; and biographical information about the organization's leadership.
Geographic Focus: All States, All Countries
Contact: Maureen Aung-Thwin, Director; (212) 548-6919; fax (212) 548-4676; maureen.aung-thwin@opensocietyfoundations.org
Internet: https://www.opensocietyfoundations.org/about/programs/public-health-program
Sponsor: Open Society Foundations
224 West 57th Street
New York, NY 10019

OSF International Palliative Care Initiative Grants 3091
The Open Society Public Health Program supports marginalized populations to fight discrimination and protect their fundamental rights. The program aims to build societies committed to inclusion, human rights, and justice, in which health-related policies and practices reflect these values and are based on evidence. The International Palliative Care Initiative works to improve end-of-life care for patients and their families, with a special focus on vulnerable populations including the elderly, children, and patients with cancer or AIDS. The initiative aims to integrate palliative care into national health care plans, policies, and systems of care. It also supports palliative care education for health care professionals and advocates for increasing access to essential medicines for pain and symptom management.
Requirements: The Initiative has an open pre-application process. To be considered for an invitation to submit a full proposal, organizations may write a one page letter containing the following information: its purpose and goals; full description of the project and amount requested; organization's total income for the past fiscal year; and biographical information about the organization's leadership.
Geographic Focus: All States, All Countries
Contact: Maureen Aung-Thwin, Director; (212) 548-6919; fax (212) 548-4676; maureen.aung-thwin@opensocietyfoundations.org
Internet: https://www.opensocietyfoundations.org/about/programs/public-health-program
Sponsor: Open Society Foundations
224 West 57th Street
New York, NY 10019

OSF Law and Health Initiative Grants 3092
The Open Society Public Health Program supports marginalized populations to fight discrimination and protect their fundamental rights. The program aims to build societies committed to inclusion, human rights, and justice, in which health-related policies and practices reflect these values and are based on evidence. The Law and Health Initiative supports legal strategies to advance the health and human rights of marginalized and vulnerable groups worldwide. The initiative works to develop individual and organizational leadership in the field of health and human rights, pilot innovative access to justice tools as health interventions, advocate for rights-based legal protections that improve health, and leverage sustainable funding for other health and human rights efforts.
Requirements: The Initiative has an open pre-application process. To be considered for an invitation to submit a full proposal, organizations may write a one page letter containing the following information: its purpose and goals; full description of the project and amount requested; organization's total income for the past fiscal year; and biographical information about the organization's leadership.
Geographic Focus: All States, All Countries
Contact: Maureen Aung-Thwin, Director; (212) 548-6919; fax (212) 548-4676; maureen.aung-thwin@opensocietyfoundations.org
Internet: https://www.opensocietyfoundations.org/about/programs/public-health-program
Sponsor: Open Society Foundations
224 West 57th Street
New York, NY 10019

OSF Mental Health Initiative Grants 3093
The Mental Health Initiative provides grants to projects that stimulate the reform of national health, social welfare, education, and employment policies. The initiative also provides technical assistance and training in substantive areas to its grantees. Many grantees provide high-quality, community-based services which demonstrate that people with intellectual disabilities can live in their communities when they receive appropriate support. The Initiative supports projects that include community-based housing, early intervention, inclusive education, and supported employment for people with intellectual disabilities. The Initiative also provides support for organizations working on policy-based advocacy at local or national levels with the aim of promoting community living for people with intellectual disabilities.
Requirements: The Mental Health Initiative provides funding to non-governmental organizations in Central and Eastern Europe and the former Soviet Union, or to organizations based in other countries that focus their activities in this region. The Initiative has an open pre-application process. To be considered for an invitation to submit a full proposal, organizations may write a one page letter containing the following information: its purpose and goals; full description of the project and amount requested; organization's total income for the past fiscal year; and biographical information about the organization's leadership.
Restrictions: The Mental Health Initiative does not fund projects which are connected to increasing the capacity of or to improving residential institutions for people with disabilities. This includes renovations or any other upgrades, equipment, charitable contributions or humanitarian aid, events organized within an institution, and any other form of core support to residential institutions.
Geographic Focus: Albania, Bosnia & Herzegovina, Bulgaria, Croatia, Czech Republic, Estonia, Germany, Hungary, Kosovo, Latvia, Lithuania, Macedonia, Montenegro, Poland, Romania, Russia, Serbia, Slovakia, Slovenia
Contact: Maureen Aung-Thwin, Director; (212) 547-6919; fax (212) 548-4676; maureen.aung-thwin@opensocietyfoundations.org or mhi@osi.hu
Internet: http://www.opensocietyfoundations.org/grants/mental-health-initiative
Sponsor: Open Society Foundations
224 West 57th Street
New York, NY 10019

OSF Public Health Program Grants in Kyrgyzstan 3094
The Public Health Program will promote initiatives aimed at ensuring equal rights to public health services for vulnerable and marginalized segments of the population. Major activities include: support of initiatives for the institutionalized development of pilot projects

providing medical, psychosocial, and legal services to vulnerable groups; and implementation of advocacy projects aimed at reforming existing health care policies and practices in order to ensure that marginalized and vulnerable groups have equal rights of access to health and legal services. Potential partners include: governmental and non-governmental organizations, international partner organizations working with marginalized, vulnerable groups. The program operates through grant-based and operational mechanisms.
Geographic Focus: All States, Kyrgyzstan
Contact: Aybek Mukambetov, Public Health Program Director; +962-6-5929994, ext. 135; fax +962-6-5929996; amukambetov@soros.kg
Internet: http://soros.kg/en/programs/public-health-program
Sponsor: Open Society Foundations
55-a Logvinenko Street
Bishkek, 720040 Kyrgyzstan

OSF Roma Health Project Grants 3095
The Open Society Public Health Program supports marginalized populations to fight discrimination and protect their fundamental rights. The program aims to build societies committed to inclusion, human rights, and justice, in which health-related policies and practices reflect these values and are based on evidence. The Roma Health Project works to advance the health and human rights of the Roma in Central and Eastern Europe. The project focuses mainly on grantmaking and advocacy efforts, including strengthening the capacity of Roma organizations and activists. The project does not fund health services or health education activities.
Requirements: The Initiative has an open pre-application process. To be considered for an invitation to submit a full proposal, organizations may write a one page letter containing the following information: its purpose and goals; full description of the project and amount requested; organization's total income for the past fiscal year; and biographical information about the organization's leadership.
Geographic Focus: All States, Albania, Belarus, Bosnia & Herzegovina, Croatia, Czech Republic, Hungary, Kosovo, Macedonia, Moldova, Montenegro, Poland, Romania, Russia, Serbia, Slovakia, Turkey, Ukraine
Contact: Maureen Aung-Thwin, Director; (212) 548-6919; fax (212) 548-4676; maureen.aung-thwin@opensocietyfoundations.org
Internet: https://www.opensocietyfoundations.org/about/programs/public-health-program
Sponsor: Open Society Foundations
224 West 57th Street
New York, NY 10019

OSF Tackling Drug Addiction Grants in Baltimore 3096
The Tackling Drug Addiction Treatment Initiative of the Open Society Foundations–Baltimore seeks to ensure universal access to treatment services for all in need regardless of income or insurance status. The Tackling Drug Addiction Treatment Initiative funds grantees who focus on the following three priorities: using the opportunity of health care reform to help Baltimore City and Maryland as a whole reach nearly universal access to a comprehensive, high-quality public treatment system; ensuring access to high quality public substance use disorder services for those that remain uninsured after the implementation of health care reform; and facilitating the creation or and help to sustain a strong, diverse addition treatment advocacy community, inclusive of those most affected by substances use disorder services policies. Applications are accepted at any time. Applications are accepted on an ongoing basis.
Requirements: Applicants should submit a copy of the IRS letter stating tax-exempt status and a letter of inquiry not exceeding three pages. If OSF-Baltimore determines that the proposal is of interest, the applicant is invited to submit a full proposal. Additional guidelines are available at the website.
Geographic Focus: Maryland
Contact: Scott Nolen, Director, Drug Addiction Treatment; (410) 234-1091; fax (410) 234-2816; scott.nolen@opensocietyfoundations.org or osi.baltimore@opensocietyfoundations.org
Internet: http://www.opensocietyfoundations.org/grants/tackling-drug-addiction
Sponsor: Open Society Foundations
201 North Charles Street, Suite 1300
Baltimore, MD 21201

Osteosynthesis and Trauma Care Foundation European Visiting Fellowships 3097
The Osteosynthesis and Trauma Care Foundation offers Visiting Fellowships in the field of trauma care and orthopaedic surgery. They receive a Travel Grant to visit one of the Host Centers participating in the program. Eligible are surgeons in training or practice in Europe. Objectives of the program are to: enhance professional skills through continuous education; provide experience on latest surgical intervention approaches; and stimulate inter-country exchange between surgeons in Europe. The Visiting Fellowship allows for knowledge acquisition in a variety of fields such as: surgical as well as non-surgical management of musculoskeletal injuries; clinical and radiological musculoskeletal evaluation of the injured patient; algorithmic decision-making with respect to timing and sequencing of multiple injury management; pathophysiology of the multiple-injured patient, including the prophylaxis and/or treatment of complications of musculoskeletal injuries; understanding of advanced technology and instrumentation in treating trauma patients; and ethical, economic, administrative and legal issues as they pertain to orthopaedic trauma care.
Geographic Focus: All States
Contact: Lee-Ann Finno; (212) 305-6392; education@otcfoundation.org
Internet: http://www.otcfoundation.org/fellowships/european-visiting-fellowships-program/
Sponsor: Osteosynthesis and Trauma Care Foundation
622 West 168th Street, 11th Floor
New York, NY 10032

Oticon Focus on People Awards 3098
Oticon Focus on People Awards, a national awards program now past its 14th year, honors hearing-impaired students, adults, and advocacy volunteers who have demonstrated through their accomplishments that hearing loss does not limit a person's ability to make a difference in their families, communities and the world. By spotlighting people with hearing loss and their contributions, Oticon aims to change outdated stereotypes that discourage people from seeking professional help for their hearing loss. Focus on People Awards are offered in four categories: Student (for young people with hearing loss, ages 6-21, who are full-time students); Adults (for people with hearing loss, ages 21 and above); Advocacy (for adults with hearing loss, ages 21 and above who actively volunteer their time in advocacy or support efforts for the hearing-impaired and deaf community); and Practitioner (a special award for hearing-care professionals who go "above and beyond"). First-place category winners will receive a $1,000 award and a $1,000 donation by Oticon to a not-for-profit cause of their choice. First-place winners in the Student, Adult, and Advocacy categories will also receive a set of advanced-technology Oticon hearing solutions. Second place winners in each category will receive a $500 award and third-place winners will receive a $250 award. Guidelines and a nomination form are available online during the annual nomination period. The deadline for nomination may vary from year to year. Applicants are encouraged to visit the grant website to verify the current deadline date.
Requirements: Anyone may nominate themselves or another individual with a hearing loss. Nominees in the Practitioner category are not required to have a hearing loss to qualify. Full-time students who volunteer their time in advocacy should apply for the Student category.
Geographic Focus: All States
Date(s) Application is Due: Jul 27
Amount of Grant: 250 - 2,000 USD
Contact: Peer Lauritsen; (732) 560-1220; fax (732) 560-0029; peoplefirst@oticonusa.com
Internet: http://www.oticonusa.com/Oticon/Professionals/FocusOnPeople.html
Sponsor: Oticon Corporation
29 Schoolhouse Road
Somerset, NJ 08873

Otto Bremer Foundation Grants 3099
The foundation concentrates its grantmaking activity in communities served by Bremer-affiliated banks and provides financial assistance to nonprofit organizations whose work contributes to the well-being of these towns. The foundation looks to support programs that promote civil and political rights, including freedom of assembly, speech, and religion; economic and social rights, including the right to education, food, health care, and shelter; and cultural and environmental rights, including the right to live in a clean environment and participate in the cultural and political events of one's community. Types of support include program development, operating support, capital (including building and equipment), matching or challenge grants, and internships. Applications are accepted throughout the year. Most grants are given for a one-year period, although some multiyear grants are awarded. The foundation encourages initial telephone or written inquiries concerning its interest in a particular project. Applicants are also encouraged to contact foundation staff for assistance in the development of a proposal.
Requirements: Private nonprofit or public 501(c)3 tax-exempt organizations whose beneficiaries are residents of Minnesota, North Dakota, or Wisconsin with priority given to those communities or regions served by Bremer affiliates.
Restrictions: Requests for the following types of projects are discouraged: annual fund drives; benefit events; camps; commercial and business development; K-12 education; medical research; sporting activities; building endowments other than for the development of community foundations; capital requests for hospitals and nursing homes; theatrical productions, including motion pictures, books, and other artistic or media projects; municipal and government services; or historical preservation, museums, and interpretive centers.
Geographic Focus: Minnesota, North Dakota, Wisconsin
Amount of Grant: 552 - 259,235 USD
Contact: Danielle Cheslog, Grants Manager; (651) 312-3717 or (651) 227-8036; fax (651) 312-3665; danielle@ottobremer.org or obf@ottobremer.org
Sponsor: Otto Bremer Foundation
445 Minnesota Stret, Suite 2250
Saint Paul, MN 55101-2107

Owen County Community Foundation Grants 3100
The Owen County Community Foundation was established in 1996 and serves the Owen County, Indiana area. The Foundation connects donors with the causes they care about. The Foundation provides support for organizations involved with arts and culture, education, animal welfare, health care, human services, and community development.
Requirements: A letter of intent is required prior to submitting a grant request, in addition to a tax-exempt verification. Applicants will be contacted within 30 days as to whether they should submit a full proposal. A full proposal should include the following: grant application and agreement page; the organization's background, including history, mission statement, number of staff, and number of persons served by agency or program; proposed program. Submitted information should also include desired outcomes; program methods; evaluation; funding the program; timetable; project budget; agency budget; governing organization; and financial statement and balance sheet.
Restrictions: Giving is limited to Owen County, Indiana. No grants for endowments, deficit funding, conferences, publications, films, television, or radio programs, travel, annual appeals, religious purposes or membership contributions.
Geographic Focus: Indiana
Date(s) Application is Due: Sep 15
Amount of Grant: Up to 30,000 USD

Contact: Marilyn Hart; (812) 829-1725; marilyn@owencountycf.org
Internet: http://www.owencountycf.org/Grants.asp
Sponsor: Owen County Community Foundation
201 W. Morgan Street, Suite 202, P.O. Box 503
Spencer, IN 47460

PACCAR Foundation Grants 3101

The PACCAR Foundation is a private foundation which usually directs its grants to organizations in those communities within the service area of a significant PACCAR presence, such as a factory or a major office. Grant recipients in locations where there is a significant PACCAR presence usually include United Way, universities, hospitals and programs for the arts and economic education. The balance of the Foundation's grants is normally reserved for capital campaigns involving acquisition or improvement of facilities used for social and health services, education and cultural affairs.
Requirements: Organizations submitting a proposal should include the following information: complete name, mailing address and website of the organization; contact name, title, phone number and email address; information on the organization's history and goals; background data on the organization, personnel and Board of Directors or Board of Trustees; most recent annual report and audited financial statements; a list of sources for ongoing operational support; evidence of 501(c)3 tax-exempt status specific amount requested; description of the specific project; an overall budget for the project to be funded and a total annual budget; a time schedule for project commencement and completion; and other project funding sources.
Restrictions: Proposals for program funds, support of operation budgets, and fundraising events are seldom funded. Although applications are taken anytime, the Foundation does not accept more than one request per organization per year; provide financial support by purchasing sponsorships, dinner or event tickets; purchase advertising space for charitable causes in yearbooks, programs or other publications; accept telephone solicitations; support specific churches for the purpose of religious advocacy; automatically renew a contribution from year to year; award scholarships or other grants for individuals; or grant personal interviews with trustees.
Geographic Focus: Oklahoma, Texas, Washington, Australia, Canada, Mexico, Netherlands, United Kingdom
Contact: Ken Hastings; (425) 468-7400; fax (425) 468-8216; ken.hastings@paccar.com
Internet: http://www.paccar.com/company/foundation.asp
Sponsor: PACCAR Foundation
777 106th Avenue N.E.
Bellevue, WA 98009

PacifiCare Foundation Grants 3102

The foundation's mission is to improve the quality of life for residents of areas where PacifiCare Health Systems does business. The foundation's focus areas are: child/youth, including child care, youth activity programs, at-risk youth, and counseling programs; education, including school programs that promote self-esteem, encourage academic achievement and the development of specific skills, literacy programs, training programs, and programs that improve the effectiveness of the educational system; health, including prevention, health education, access to health care, and improved quality of health care of targeted populations; human/social services, including housing, shelters, education, protection, community development, crime prevention, food, transportation, and other social services for targeted populations; and senior, including social services, nutrition, education, volunteer, and and adult day care. Preference will be given to proposals for specific projects. Requests for operating costs will be considered if the request is very specific and clearly defined. Seed grant requests also receive consideration. Organizations funded by the foundation are welcome to reapply annually. Application forms are available on the foundation's website.
Requirements: IRS 501(c)3 nonprofit organizations serving residents of PacifiCare regions in Arizona, California, Colorado, Nevada, Oklahoma, Oregon, Texas, and Washington are eligible. The proposal must include two copies of the following: application form; checklist form; cover letter accompanying the proposal, signed by either the CEO or appointee of the organization, summarizing the proposed project, the problem addressed, the amount requested, and the name and phone number of the contact person; the written proposal, which should not exceed 2-5 pages in length and should include background information, description of the problem, need or issue being addressed, and a complete description of the proposed project; most recent audited financial statement and 990; current operating budget and line item budget for the specific project; list of major funders and amounts; list of board of directors; and one paragraph summary of previous support from the PacifiCare Foundation.
Restrictions: The foundation will not consider grants for arts/cultural programs, associations, annual campaigns, associations (professional/technical), capital campaigns, challenge/matching grants, hosting/supporting conferences, individual support, private foundations, programs that promote religious doctrine, research, scholarships, or sponsorship of special events.
Geographic Focus: Arizona, California, Colorado, Nevada, Oklahoma, Oregon, Texas, Washington
Date(s) Application is Due: Jan 1; Jul 1
Samples: Casa de Esperanza, Green Valley, Arizona, $10,000 - to provide intergenerational community programming for children and the elderly; Escape Family Resource Center, Houston, Texas, $10,000 - for a school-based child abuse prevention and family support program.
Contact: Riva Gebel, Director; (714) 825-5233
Internet: http://www.pacificare.com/vgn/images/portal/cit_60701/127503Guidelines_for_Charitable_Giving.pdf
Sponsor: PacifiCare Foundation
P.O. Box 25186, MS LC03-159
Santa Ana, CA 92799

Pacific Life Foundation Grants 3103

Pacific Life has long recognized the importance of helping communities where their employees reside and work and has a record of community involvement that spans the history of the company. The Foundation accepts proposals from agencies seeking funds for programs and projects in the areas of health and human services; education; arts and culture; and civic, community, and environment. Grant proposals are generally accepted from July 15 through August 15. See the Foundation's website to verify deadlines.
Requirements: Funding is made primarily in areas with large concentrations of Pacific Life employees: generally, the greater Orange County, California, area and other areas, such as Omaha, Nebraska. Ideally, agencies should serve a large area, usually including more than one city or community. Some California statewide and national organizations also receive support. General grants range from $5,000 to $10,000 for a one-year period of funding and are given to support programs, operating expenses, or collaborative programs with other agencies. Capital grants range from $10,000 to $100,000 and are paid over multiple years. Capital grants are generally given to an agency with an organized campaign already under way to raise substantial funds. More than fifty percent of the campaign goal (excluding in-kind donations, anonymous gifts and loans) must be pledged prior to consideration by the Foundation.
Restrictions: The following is not funded: individuals; political parties, candidates, or partisan political organizations; labor organizations, fraternal organizations, athletic clubs, or social clubs; K-12 schools, school districts, or school foundations; sectarian or denominational religious organizations, except for programs that are broadly promoted, available to anyone, and free from religious orientation; fundraising events; sports leagues or teams; and advertising sponsorship or conference underwriting/sponsorship.
Geographic Focus: All States
Samples: Aquarium of the Pacific, Long Beach, California, $258,503 - civic, community and environment support; Greater Santa Ana Vitality Foundation, Santa Ana, California, $10,000 - education; and Southern California Public Radio, Pasadena, California, $40,000 - arts and culture.
Contact: Brenda Hardwig, Program Contact; (949) 219-3787; PLFoundation@PacificLife.com
Internet: http://www.pacificlife.com/About+Pacific+Life/Foundation+or+Community
Sponsor: Pacific Life Foundation
700 Newport Center Drive
Newport Beach, CA 92660

Packard Foundation Children, Families, and Communities Grants 3104

The Foundation's Children, Families, and Communities Grants strive to ensure that all children have the opportunity to reach their full potential. Strategies address two interrelated and fundamental needs that must be met for children to thrive: health and education. The Foundation's grants have three goals. The central goal is to create publicly supported, high-quality preschool opportunities for all three- and four-year-olds. This is funded through Preschool for California's Children Grants. A second goal is to ensure that all children receive appropriate health care by creating nationwide systems that provide access to health insurance for all children. Work is done in over a dozen states and at the federal level through Children's Health Insurance Grants. A third goal is to strengthen California's public commitment to school-based, after-school programs, while also spurring the expansion of these programs into summer. Such expanded learning opportunities that are aligned with the school day promote positive youth development for elementary and middle school children. The Foundation emphasizes literacy, good nutrition, and out-of-door experiences in these expanded learning settings. Work is supported at the federal level through After School and Summer Enrichment Grants. Samples of previously funded projects are discussed on the Foundation website.
Requirements: The Foundation provides grants for charitable, educational, or scientific purposes, primarily from tax-exempt, charitable organizations. See the Foundation's website for current opportunities.
Geographic Focus: All States
Contact: Meera Mani, Director; (650) 948-7658; cfc@packard.org
Internet: http://www.packard.org/what-we-fund/children-families-and-communities/
Sponsor: David and Lucile Packard Foundation
343 Second Street
Los Altos, CA 94022

Packard Foundation Local Grants 3105

The Foundation supports an array of nonprofit organizations in geographic areas that are significant to the Packard family. These include the five California counties that surround the Foundation's headquarters in Los Altos, California (San Mateo, Santa Clara, Santa Cruz, Monterey, and San Benito) as well as Pueblo, Colorado, the birthplace of David Packard. The goal in supporting these communities is to help make them stronger and more vibrant places where all families can thrive and reach their potential. To achieve this goal, Local Grants focus resources on addressing five fundamental issue areas: arts; children and youth; conservation and science; food and shelter; and population and reproductive health. There are no deadlines. Samples of previously funded projects are discussed at the Foundation website.
Requirements: The Foundation accepts grant proposals only for charitable, educational, or scientific purposes, primarily from tax-exempt, charitable organizations. An online letter of inquiry form is available at the Foundation's website. Applicants typically receive a response within three to six weeks. If accepted, further details about completing a full proposal will be provided.
Restrictions: The following is not funded: public policy work, capital campaigns, specific performances or productions, one-time events, event sponsorships, religious or business organizations, and individuals. Requests should generally not exceed 25% of the organization's operating budget.

Geographic Focus: California, Colorado
Amount of Grant: 15,000 - 150,000 USD
Samples: TC Hoffman and Associates LLC, Oakland, California, $149,900 - conservation and science; The Legal Aid Society, Employment Law Center, San Francisco, California $10,000 - children, families and communities; and Silicon Valley Community Foundation, Mountain View, California, $25,000 - special opportunities for children, conservation and population.
Contact: Linda Schuurmann Baker, Program Officer, Santa Cruz and Monterey Counties; (650) 917-7238; fax (650) 948-1361; local@packard.org
Curt Riffle, Program Officer; (650) 948-7658; fax (650) 948-1361
Internet: http://www.packard.org/what-we-fund/local-grantmaking/
Sponsor: David and Lucile Packard Foundation
343 Second Street
Los Altos, CA 94022

Packard Foundation Population and Reproductive Health Grants 3106

The Foundation's Population and Reproductive Health Grants fund innovative work that addresses population growth and promotes positive reproductive health. The goals are to slow population growth rates in high-fertility areas, and to ensure individual reproductive health and rights in order to improve the quality of life for more people. Currently work is focused at the global level, regionally on South Asia and Sub-Saharan Africa, and on selected initiatives within the United States. Samples of previously funded projects are discussed at the Foundation website.
Requirements: Interested applicants may submit a letter of inquiry not to exceed three pages. Only letters that clearly support a particular subprogram strategy will be considered. Letters should include a brief overview of the applicant organization and its mission; a description of the project for which funding is being requested, including goals, major strategies, and rationale; and monitoring and evaluation plans. Every attempt is made to review and respond within four to six weeks. If accepted, applicants will be invited to submit a full grant proposal for further review.
Geographic Focus: All States
Samples: Walta Information Center, Addis Ababa, Ethiopia, $250,000 - population; Women's Health Care Foundation, Quezon City, Philippines, $100,000 - population; and Women's Health and Action Research Centre, Benin City, Nigeria, $120,912 - population.
Contact: Amy Gavin; (650) 948-7658; fax (650) 948-1361; population@packard.org
Internet: http://www.packard.org/what-we-fund/population-reproductive-health/
Sponsor: David and Lucile Packard Foundation
343 Second Street
Los Altos, CA 94022

PAHO-ASM International Professorship for Latin America 3107

The program provides funding support to an ASM member from the U.S. who is scientifically recognized in his/her area to teach a hands-on, highly interactive short course in the area of antimicrobial resistance at an institution of higher learning in Bolivia, Colombia, Costa Rica, Dominican Republic, Ecuador, El Salvador, Guatemala, Honduras, Nicaragua, Panama, Paraguay, or Peru. One Teaching Professorship offered per program year.
Requirements: The application must be made jointly between a hosting institution and the visiting professor. Eligibility for Host Institution: institution with graduate students enrolled in a masters, doctoral or equivalent program, post-doctoral fellows or residents, and teaching faculty; institution where at least 12 students will be enrolled full-time in the short course; commitment to maximizing use of the course, as demonstrated by the applicability of the course's contents to existing programs at the institution or to the field of research of the Research Professor. demonstrated commitment to international collaborations and partnerships; demonstrated commitment to advancing the microbiological sciences through interactions with professional organizations; institutions in developed nations, as defined by the UN high HDI group, are ineligible. Eligibility for Visiting Teaching Professor: scientifically recognized their area of microbiological expertise; actively engaged in teaching at the post-secondary level; commitment to international collaborations and partnerships; must be an ASM member during the term of the Professorship. Preference will be given to: applicants who will be able to make maximal use of the course, as demonstrated either by the applicability of the course's contents to existing programs at the host institution or to the field of research of the visiting Professor; applications that demonstrate a commitment to building an ongoing institutional relationship. It is strongly recommended that the course include a hands-on component, such as a wet lab or other practical activity.
Geographic Focus: All States
Date(s) Application is Due: Apr 15; Oct 15
Amount of Grant: 4,000 USD
Contact: Supervisor; (202) 942-9368; fax (202) 942-9328; international@asmusa.org
Internet: http://www.asm.org/International/index.asp?bid=57785
Sponsor: American Society for Microbiology
1752 N Street, N.W.
Washington, D.C. 20036-2904

Pajaro Valley Community Health Health Trust Insurance/Coverage & Education on Using the System Grants 3108

The Trust provides grants for projects that advance our mission to improve the health and quality of life for all people of the greater Pajaro Valley. The primary goals of the Health Insurance/Coverage & Education on Using the System Initiative are to increase the number of Pajaro Valley residents with health insurance, increase residents understanding and appropriate use of a medical home (family practitioner), and decrease inappropriate use of the emergency department. The Trust will support programs that increase the number of Pajaro Valley residents that have health insurance, as well as programs that improve access to health care for our community's more vulnerable populations. Additionally, the Trust will look at community-wide solutions to these issues.
Requirements: Applicant organizations must provide or plan to provide programs/services benefiting the health of residents in the Trust's primary geographic service area. Communities within this service area include Watsonville, Pajaro, Freedom, and Aromas. The home office of the applicant organization need not be located in the Pajaro Valley, but the applicant organization must demonstrate that it provides or plans to provide services that directly benefit residents of the Pajaro Valley. The applicant organization must be a nonprofit, 501(c)3 tax-exempt organization; a school-based health program; or have a 501(c)3, tax-exempt organization as a fiscal sponsor.
Restrictions: In general, the Trust's Board of Directors prefers not to fund programs or projects administered by a city, county, state, or federal government with the exception of school-based health programs. In general, the Trust does not give grants to: projects that do not substantially benefit residents of the Pajaro Valley; projects and proposals unrelated to the Trust's mission, eligibility requirements, and current strategic plan funding priorities and objectives; individuals, with the exception of the Trust's scholarship programs; religious organizations for secular purposes; endowments, building campaigns, annual fund appeals, fundraising events, or celebrations; or commercial ventures.
Geographic Focus: California
Samples: La Manzana Community Resources, La Manzana, California, $12,000 - to assist low-income families in accessing public health benefits; Healthy Kids of Santa Cruz County, Santa Cruz, California, $15,000 - to provide access to health insurance to children in the Pajaro Valley who have incomes at or below 300% of the federal poverty level; Santa Cruz Community Counseling Center, Santa Cruz, $5,000 - to launch a health literacy campaign focusing on low-income families with children birth to five years old.
Contact: Raquel Ramirez Ruiz, Director of Programs; (831) 763-6456 or (831) 761-5639; fax (831) 763-6084; info@pvhealthtrust.org or raquel_dhc@pvhealthtrust.org
Internet: http://www.pvhealthtrust.org/grants_core.html
Sponsor: Pajaro Valley Community Health Trust
85 Nielson Street
Watsonville, CA 95076

Pajaro Valley Community Health Trust Diabetes and Contributing Factors Grants 3109

The Trust provides grants for projects that advance our mission to improve the health and quality of life for all people of the greater Pajaro Valley. The primary goals of the Diabetes and Contributing Factors Initiative are to reduce the risk factors associated with diabetes, reduce complications related to diabetes, and decrease the prevalence of childhood and adult obesity in the Pajaro Valley. The Trust will mobilize communities in the tri-county area to prevent the increase of type-2 diabetes in youth and young adult populations; teach diabetes self-management, and provide medical nutrition therapy to people living with diabetes thereby preventing the life-threatening complications associated with diabetes. Further, the Trust will promote "best practices" in clinical management of diabetes throughout the region. The Trust will seek to minimize factors that contribute to diabetes, including obesity, poor nutrition, and lack of physical activity.
Requirements: Applicant organizations must provide or plan to provide programs/services benefiting the health of residents in the Trust's primary geographic service area. Communities within this service area include Watsonville, Pajaro, Freedom, and Aromas. The home office of the applicant organization need not be located in the Pajaro Valley, but the applicant organization must demonstrate that it provides or plans to provide services that directly benefit residents of the Pajaro Valley. The applicant organization must be a nonprofit, 501(c)3 tax-exempt organization; a school-based health program; or have a 501(c)3, tax-exempt organization as a fiscal sponsor.
Restrictions: In general, the Trust's Board of Directors prefers not to fund programs or projects administered by a city, county, state, or federal government with the exception of school-based health programs. In general, the Trust does not give grants to: projects that do not substantially benefit residents of the Pajaro Valley; projects and proposals unrelated to the Trust's mission, eligibility requirements, and current strategic plan funding priorities and objectives; individuals, with the exception of the Trust's scholarship programs; religious organizations for secular purposes; endowments, building campaigns, annual fund appeals, fundraising events, or celebrations; or commercial ventures.
Geographic Focus: California
Amount of Grant: 5,000 - 30,000 USD
Samples: Community Action Board of Santa Cruz County, Santa Cruz, California, $10,000 - to support the REAL for Diabetes Prevention project; Ecology Action, Willits, California, $9,000 - to support the Boltage program, a daily student biking and walking tracking program; Mesa Verde Gardens, Mesa Verde, California, $15,000 - to support the start-up of a second Community Garden.
Contact: Raquel Ramirez Ruiz, Director of Programs; (831) 763-6456 or (831) 761-5639; fax (831) 763-6084; info@pvhealthtrust.org or raquel_dhc@pvhealthtrust.org
Internet: http://www.pvhealthtrust.org/grants_core.html
Sponsor: Pajaro Valley Community Health Trust
85 Nielson Street
Watsonville, CA 95076

Pajaro Valley Community Health Trust Oral Health: Prevention & Access Grants 3110

The Trust provides grants for projects that advance our mission to improve the health and quality of life for all people of the greater Pajaro Valley. The primary goals of the Oral Health Initiative are to reduce the risk factors associated with oral health disease, increase the number of Pajaro Valley residents with ready access to comprehensive dental care, and decrease the prevalence of dental disease among residents of the Pajaro Valley. The Trust's goals include improving access to dental treatment and preventing dental disease. Through this initiative, the

Trust will look at systematic issues facing oral health care, particularly in the areas of prevention and access to care, and work with others in the community to remove these barriers.
Requirements: Applicant organizations must provide or plan to provide programs/services benefiting the health of residents in the Trust's primary geographic service area. Communities within this service area include Watsonville, Pajaro, Freedom, and Aromas. The home office of the applicant organization need not be located in the Pajaro Valley, but the applicant organization must demonstrate that it provides or plans to provide services that directly benefit residents of the Pajaro Valley. The applicant organization must be a nonprofit, 501(c)3 tax-exempt organization; a school-based health program; or have a 501(c)3, tax-exempt organization as a fiscal sponsor.
Restrictions: In general, the Trust's Board of Directors prefers not to fund programs or projects administered by a city, county, state, or federal government with the exception of school-based health programs. In general, the Trust does not give grants to: projects that do not substantially benefit residents of the Pajaro Valley; projects and proposals unrelated to the Trust's mission, eligibility requirements, and current strategic plan funding priorities and objectives; individuals, with the exception of the Trust's scholarship programs; religious organizations for secular purposes; endowments, building campaigns, annual fund appeals, fundraising events, or celebrations; or commercial ventures.
Geographic Focus: California
Amount of Grant: 5,000 - 30,000 USD
Samples: Dientes Community Dental Care, Santa Cruz, California, $10,000 - to support the delivery of dental services to low-income, uninsured adults from the Pajaro Valley; Salud Para La Gente, Watsonville, California, $10,000 - to support the delivery of dental services to low-income, uninsured adults from the Pajaro Valley.
Contact: Raquel Ramirez Ruiz, Director of Programs; (831) 763-6456 or (831) 761-5639; fax (831) 763-6084; info@pvhealthtrust.org or raquel_dhc@pvhealthtrust.org
Internet: http://www.pvhealthtrust.org/grants_core.html
Sponsor: Pajaro Valley Community Health Trust
85 Nielson Street
Watsonville, CA 95076

Pajaro Valley Community Health Trust Promoting Entry & Advancement in the Health Professions Grants 3111

The Trust provides grants for projects that advance our mission to improve the health and quality of life for all people of the greater Pajaro Valley. The primary goals of the Promoting Entry & Advancement in the Health Professions Initiative are to increase the number of culturally competent healthcare workers in the Pajaro Valley, increase the proportion of bilingual/bicultural healthcare workers in the Pajaro Valley, and reduce the number of unfilled healthcare position in the Pajaro Valley. The Trust will support and promote programs that encourage individuals to choose the health professions as a career/professional path. It is a priority for the Trust to encourage all individuals to enter the health professions, particularly bilingual/bicultural individuals that will return to the Pajaro Valley to serve this community's healthcare needs.
Requirements: Applicant organizations must provide or plan to provide programs/services benefiting the health of residents in the Trust's primary geographic service area. Communities within this service area include Watsonville, Pajaro, Freedom, and Aromas. The home office of the applicant organization need not be located in the Pajaro Valley, but the applicant organization must demonstrate that it provides or plans to provide services that directly benefit residents of the Pajaro Valley. The applicant organization must be a nonprofit, 501(c)3 tax-exempt organization; a school-based health program; or have a 501(c)3, tax-exempt organization as a fiscal sponsor.
Restrictions: In general, the Trust's Board of Directors prefers not to fund programs or projects administered by a city, county, state, or federal government with the exception of school-based health programs. In general, the Trust does not give grants to: projects that do not substantially benefit residents of the Pajaro Valley; projects and proposals unrelated to the Trust's mission, eligibility requirements, and current strategic plan funding priorities and objectives; individuals, with the exception of the Trust's scholarship programs; religious organizations for secular purposes; endowments, building campaigns, annual fund appeals, fundraising events, or celebrations; or commercial ventures.
Geographic Focus: California
Amount of Grant: 2,000 - 10,000 USD
Samples: Aptos Adult Education, Watsonville, California, $5,000 - to support scholarship funding to train individuals as Certified Nurse Assistants; Watsonville High School Health Careers Academy, Watsonville, California, $6,000 - for enhancement of its program that provides a specialized course of study for high school students interested in entering the health professions.
Contact: Raquel Ramirez Ruiz, Director of Programs; (831) 763-6456 or (831) 761-5639; fax (831) 763-6084; info@pvhealthtrust.org or raquel_dhc@pvhealthtrust.org
Internet: http://www.pvhealthtrust.org/grants_core.html
Sponsor: Pajaro Valley Community Health Trust
85 Nielson Street
Watsonville, CA 95076

Palmer Foundation Grants 3112

The Palmer Foundation was founded by Rogers Palmer and his wife, Mary, in 1990. Today, the Foundation considers proposals that empower young people up to the age of 25. General categories include arts and culture, education, health, and human services. In making awards, the foundation places the highest priority on applications that demonstrate a level of cooperation with other organizations, including leveraging financial and in-kind support from other groups; clearly avoids duplication of existing services; where applicable, demonstrates potential for continued funding from internal or other sources after the grant period; and offer challenge grants in order to stimulate support of the project by other organizations or individuals. The board will review Letters of Intent each month through the listed deadlines; full proposals are by invitation. Letters of intent may be submitted online. Guidelines are available online.
Requirements: 501(c)3 tax-exempt organizations are eligible. The foundation's geographical region is usually limited to the Mid West states, the Mid Atlantic states. Guatemala, and El Salvador.
Restrictions: The foundation does not usually support the following types of activities: grants to individuals or scholarships; endowment drives or capital campaigns; general operating expenses for an existing project, except to support an innovative program during the initial years of operation; grants for political activities, sectarian religious purposes, or individual medical or scientific research projects; or repeated funding to the same program or organization for the same purposes on an annual or ongoing basis.
Geographic Focus: All States
Samples: Art Institute and Gallery, Salisbury, Maryland - general operating support; Carolina Philharmonic, Pinehurst, North Carolina - to bring Carnegie Hall's Link-up educational program to the elementary school children of Moore County; Crohn's and Colitis Foundation of America, New York, New York - support basic and clinical research as well as educational and support programs designed for patients, caregivers, and health-care professionals.
Contact: Charlly Enroth; (202) 595-1020; admin@thepalmerfoundation.org
Internet: http://www.thepalmerfoundation.org
Sponsor: Palmer Foundation
1201 Connecticut Avenue, NW, Suite 300
Washington, D.C. 20036

Pancreatic Cancer Action Network-AACR Fellowship 3113

The Pancreatic Cancer Action Network-AACR Fellowship represents a joint effort to attract young scientist to careers in pancreatic cancer research. The Fellowship is open to Postdoctoral Fellows and Clinical Research Fellows working at an academic, medical, or research institution who will be in the 1st, 2nd, or 3rd year of their postdoctoral training at the start of the grant term on July 1. The Fellowship provides a one-year grant of $45,000 to support the salary and benefits of the Fellow while working on a mentored pancreatic cancer research project. A partial amount of funds, up to 25% of the total grant, may be designated for direct research support. The research proposed for funding may be basic, translational, clinical or epidemiological in nature and must have direct application and relevance to pancreatic cancer. It is anticipated that one Fellowship will be funded. AACR requires applicants submit both an online and a paper application. The online application must be submitted by 12:00 noon (United Sates Eastern Time) on October 28, using the proposalCENTRAL website at http://proposalcentral.altum.com. The paper copy with original signatures and all required documents must be postmarked and sent no later then October 30.
Requirements: Applicants must have acquired a doctoral degree (including Ph.D., M.D., D.O., D.C., N.D., D.D.S., D.V.M., Sc.D., D.N.S., Pharm.D., or equivalent doctoral degree, or a combined clinical and research doctoral degree) in a related field and may not currently be candidate for another degree. Applicants must be in the 1st, 2nd, or 3rd year of Postdoctoral or Clinical Research Fellowship at an academic, medical, or research institution within the United States at the start of the Fellowship term on July 1.
Restrictions: Scientific investigators or health professionals who are funded by the tobacco industry for any project, or whose named mentors in the case of mentored grants are funded by the tobacco industry for any project, may not apply and will not be eligible for AACR grants.
Geographic Focus: All States
Date(s) Application is Due: Oct 28; Oct 30
Amount of Grant: 45,000 USD
Contact: Hanna Hopfinger; (267) 646-0665; hanna.hopfinger@aacr.org
Internet: http://www.aacr.org/home/scientists/research-funding--fellowships/postdoctoral-fellowships.aspx
Sponsor: American Association for Cancer Research
615 Chestnut Street, 17th Floor
Philadelphia, PA 19106-4404

Pancreatic Cancer Action Network-AACR Pathway to Leadership Grants 3114

The Pancreatic Cancer Action Network-AACR Pathway to Leadership Grant represents a joint effort to ensure the future leadership of pancreatic cancer research by supporting outstanding early career investigators beginning in their postdoctoral research positions and continuing through their successful transition to independence. Applicants must be in the first five years of their postdoctoral or clinical research fellowships at the time of application and not already have a full-time tenure track assistant professor position (or equivalent). The Pathway to Leadership Grant will provide up to five years of support, for a total of $600,000, consisting of two phases. The initial Mentored Phase lasts up up two years, during which time recipients will receive $75,000 per year and are expected to work closely with mentors to develop their research. During the subsequent three years, recipients are expected to be independent research positions and will be funded at $150,000 per year. AACR requires applicants to submit both an online and a paper application. Online applications must be completed by 12:00 noon (United States Eastern Daylight Time) on October 28, using the proposalCENTRAL website. The paper copy with original signatures and all supporting documents must be postmarked and sent no later than October 30.
Requirements: Applicants must be in mentored training positions at an academic, medical, or research institution within the United States. There are no citizenship requirements. The applicant institution will be the mentored phase institution. Employees or subcontractors of a government entity or for-profit private industry are not eligible. Exceptions may apply if an applicant holds a full-time position at a Veterans' Hospital or national laboratory in the U.S. Contact AACR before submitting an application to determine your eligibility. Applicants must be Postdoctoral or Clinical Research Fellows who completed all doctoral degree requirements no more then five years prior to the start of the grant term.

Geographic Focus: All States
Date(s) Application is Due: Oct 28; Oct 30
Amount of Grant: 600,000 USD
Contact: Elizabeth Martin; (267) 646-0664; fax (215) 440-9372; grants@aacr.org
Internet: http://www.aacr.org/home/scientists/research-funding--fellowships/postdoctoral-fellowships.aspx
Sponsor: American Association for Cancer Research
615 Chestnut Street, 17th Floor
Philadelphia, PA 19106-4404

Parke County Community Foundation Grants 3115
The Parke County Community Foundation Grants fund fields of interest which are community-enhancing such as agricultural interests, family support, fine arts/culture, handicapped persons, historic preservation, individual township interests, religion, scholarship/education, and youth/recreation. The Foundation accepts grant applications year-round.
Requirements: Applications for grants are accepted from any new or existing charitable organization or community agency with a charitable purpose in Parke County. Organizations should fill out the online application with the following information: a detailed description of their organization, project, and budget required; other organizations who might partner with them with a similar project. For requests of $1,000 or less, a one-page letter or email is acceptable. This letter should describe the project, sharing the organization's perception of need. It should also describe who and how many are likely to benefit and the factors the organization will use to evaluate its success. Applicants are encouraged to contact the Foundation to discuss if their project is appropriate for funding before submitting the application.
Restrictions: The Foundation usually funds only to nonprofit organizations, but may fund other organizations if the project is designed to assist the needy and promote well-being in Parke County.
Geographic Focus: Indiana
Contact: Brad Bumgardner, Executive Director; (765) 569-7223; fax (765) 569-5383; bradbum@yahoo.com or parkeccf@yahoo.com
Internet: http://www.parkeccf.org/Grants.html
Sponsor: Parke County Community Foundation
115 North Market Street
Rockville, IN 47872

Parkersburg Area Community Foundation Action Grants 3116
Community Action Grants help organizations by supporting vital projects in the fields of arts and culture, education, health and human services, community and economic development, youth and family services, and recreation. The Foundation accepts applications for the spring deadline on March 1 and for the fall deadlines on September 1. Applicants are encouraged to contact the Foundation with any questions about the grants or the application process.
Requirements: Grants are awarded to 501(c)3 tax-exempt nonprofit organizations in specific counties of West Virginia (Calhoun; Doddridge; Gilmer; Jackson; Pleasants; Mason; Ritchie; Roane; Wirt; and Wood) and Washington County, Ohio.
Restrictions: Grants do not support religious purposes, travel, meetings, seminars, student exchange programs, annual campaigns, endowment funds, operating budgets, or debt reduction.
Geographic Focus: Ohio, West Virginia
Date(s) Application is Due: Mar 1; Sep 1
Amount of Grant: 1,000 - 10,000 USD
Contact: Marian Clowes; (866) 428-4438 or (304) 428-4438; info@pacfwv.com
Internet: http://www.pacfwv.com
Sponsor: Parkersburg Area Community Foundation
501 Avenuery Street, P.O. Box 1762
Parkersburg, WV 26102-1762

Partnership for Cures Charles E. Culpeper Scholarships in Medical Science 3117
The program awards scholarships in medical science to support the career development of academic physicians. In direct costs, $100,000 per year to the sponsoring institution will be committed for up to three consecutive years. It is expected that a significant portion of the award will be used as a contribution to the Scholar's salary. Up to $50,000 (plus fringe benefits) of this sum may be used for this purpose; the rest for support of research including personnel and some travel costs. Eight percent will be added to each award for indirect costs. All scientific research relevant to human health is eligible for consideration.
Requirements: Candidates must: be U.S. citizens or aliens who have been granted permanent U.S. residence (proof required); hold the M.D. degree from a U.S. medical school or the equivalent of an M.D. degree from an educational institution equivalent to a United States medical school; and be judged worthy of support by virtue of the quality of their research proposals and their potential for successful careers in academic medicine. Applicants must have at least one year of post-doctoral clinical training.
Restrictions: Individuals who have achieved a rank above assistant professor are ineligible for this program, and an awardee who is promoted above the rank of assistant professor before the initiation of the scholarship will be ineligible to receive the award.
Geographic Focus: All States
Date(s) Application is Due: Aug 16
Amount of Grant: 108,000 USD
Contact: Dr. Bruce E. Bloom; (312) 696-1366; fax (312) 263-0939; Bruce@4cures.org
Internet: http://www.4cures.org/home/culpeper_scholar_program_research
Sponsor: Partnership for Cures
70 West Madison Street, Suite 1500
Chicago, IL 60602

Partnership for Cures Two Years To Cures Grants 3118
The Initiative is intended to bring researchers, funders and institutions together to produce better treatments and cures for patients with disease right now. The Program works to quickly bring together: breakthrough scientific discoveries about diseases and treatments; FDA approved drugs and other medical therapies (Eastern and Western) with new potential uses; research teams at top institutions ready to proceed with urgency and focus; and independent validation, compressed time lines, and financial leverage. Projects must be able to create a direct patient impact in 24 months from the date of initiation of the project. Budgets can range from $35,000 to $500,000. The Program is accepting research projects in any catastrophic disease, but has special interest in the following diseases: lupus, Crohns disease, MS, ALS, protein intolerances in children, ischemic colitis, autism, type I diabetes, lung cancer, osteosarcoma, Fragile X syndrome, Jewish genetic diseases, and Parkinsons disease. There is no deadline for submissions, but funders are looking for projects right now.
Requirements: Each project must come with a minimum of 30% matching funds. Those funds can come from the institution itself, from government or other public funders, or from individuals, foundations or other private funders.
Restrictions: No more than 10% of the budget can be allocated to overhead, travel and publications.
Geographic Focus: All States
Amount of Grant: 35,000 - 500,000 USD
Contact: Dr. Bruce E. Bloom; (312) 696-1366; fax (312) 263-0939; Bruce@4cures.org
Internet: http://www.4cures.org/home/funding_opportunities
Sponsor: Partnership for Cures
70 West Madison Street, Suite 1500
Chicago, IL 60602

Paso del Norte Health Foundation Grants 3119
The Paso del Norte Health Foundation (PdNHF) is one of the largest private foundations on the U.S. - Mexico border. It was established in 1995 from the sale of Providence Memorial Hospital to Tenet Healthcare Corporation for $130 million. The purpose of the Foundation is to improve the health and promote the wellness of the people living in West Texas, Southern New Mexico, and Ciudad Juárez, Mexico through education and prevention. The foundation currently issues RFPs for the following initiatives: physical activity and balanced nutrition; tobacco, alcohol, and illicit drug use; health care and mental health services; healthy families and social environments; and leadership. Interested organizations can sign up for the foundation's RFP mailing list at the grant website. The foundation does not accept unsolicited proposals but offers workshops when new RFPs are issued and also encourages organizations to contact the Program Officer with suggestions and ideas for new programs, especially as they relate to the foundation's initiatives.
Requirements: Proposals must be solicited, and must be linked to an established initiative. Applicants must be 501(c)3 or equivalent organizations operated for charitable, educational, or religious purposes. Funding is principally for in-field intervention projects. Research, studies, or planning activities may be considered only if they directly assist in the implementation of a project. Ineligible expenses include general operating and overhead expenses, capital acquisition or construction (such as property or buildings), purchase of motorized vehicles, and the direct provision of medical services and medicines for acute care.
Restrictions: Funding is not provided to individuals. Proposals are only considered from the Paso del Norte region which encompasses El Paso and Hudspeth Counties in West Texas; Doña Ana, Luna, and Otero Counties in Southern New Mexico; and Ciudad Juárez and Chihuahua in Northern Mexico.
Geographic Focus: Texas, Mexico
Contact: Enrique Mata, Senior Program Officer; (915) 544-7636, ext. 1918; fax (915) 544-7713; emata@pdnhf.org or health@pdnhf.org
Internet: http://www.pdnhf.org/index.php?option=com_content&view=article&id=91&Itemid=68&lang=us
Sponsor: Paso del Norte Health Foundation
221 N. Kansas, Suite 1900
El Paso, TX 79901

Patrick and Anna M. Cudahy Fund Grants 3120
The Patrick and Anna Cudahy Fund is a general purpose foundation which primarily supports organizations in Wisconsin and the metropolitan Chicago area. The principal areas of interest are social service, youth, and education with some giving for the arts, and other areas. Some support is also given for local and national programs concerned with public interest and environmental issues. A few grants are given for international programs but only to those which are represented by a United States based organization.
Requirements: Nonprofits organizations in Wisconsin and the metropolitan Chicago area are eligible. See Foundations website for additional guidelines and application form: http://www.cudahyfund.org/Guideliines.htm
Restrictions: Organizations may submit only one proposal during any calendar year. Requests are not considered for: organizations and projects primarily serving a local constituency outside of Wisconsin and the Chicago metropolitan area; organizations outside the United States who are not represented by a United States based 501 [c]3 organization; grants to individuals; loans or endowments.
Geographic Focus: Illinois, Wisconsin
Date(s) Application is Due: Jan 5; Apr 5; Jul 5; Oct 5
Amount of Grant: 5,000 - 25,000 USD
Contact: Janet S. Cudahy, President; (847) 866-0760; fax (847) 475-0679
Internet: http://www.cudahyfund.org/Guideliines.htm
Sponsor: 1609 Sherman Avenue, #207
Evanston, IL 60201

Patron Saints Foundation Grants 3121
The Patron Saints Foundation is a private foundation that provides grants to public charities that improve the health of individuals residing in the West San Gabriel Valley through health care programs which are consistent with the moral and religious teachings of the Roman Catholic Church. Grants will be made to qualified public charities to sponsor charitable, scientific or educational health care programs which are not inconsistent with the moral and religious teachings of the Roman Catholic Church, in the following health care categories: community health services (direct health care services); community health care education; capital expenditures; equipment and supplies; medical research. Grants typically range from $5,000 to $15,000. Larger or smaller grants are at the discretion of The Patron Saints Foundation Board of Directors based on the merits of the project.
Requirements: California 501(c)3 tax-exempt organizations serving the West San Gabriel Valley are eligible, including: the cities of Alhambra, Arcadia, Duarte, El Monte, La Canada Flintridge, Monrovia, Monterey Park, Pasadena, Rosemead, San Gabriel, San Marino, Sierra Madre, South El Monte, South Pasadena, Temple City and the unincorporated areas of Los Angeles County known as Altadena and South San Gabriel along with the unincorporated portions of Los Angeles County within the West San Gabriel Valley.
Restrictions: Grants are not awarded for endowment funds, political activities, travel, surveys or fund raising activities.
Geographic Focus: California
Date(s) Application is Due: Mar 1; May 1
Amount of Grant: 5,000 - 15,000 USD
Contact: Kathleen T. Shannon, Executive Director; (626) 564-0444; fax (626) 564-0444; patronsaintsfdn@sbcglobal.net
Internet: http://www.patronsaintsfoundation.org/index.php?nav=Grant%20Guidelines
Sponsor: Patron Saints Foundation
260 S Los Robles Avenue, Suite 201
Pasadena, CA 91101-3614

Paul Balint Charitable Trust Grants 3122
The trust operates in the United Kingdom and internationally and awards grants to nonprofit charities in the fields of Jewish charities, medicine, disability, with a specific interest in aging and elderly and improving the daily lives of those with disabilities.
Geographic Focus: All States
Contact: Andrew Balint, 020 7624 2098
Sponsor: Paul Balint Charitable Trust
26 Church Crescent
London, N20 0JP England

Paul Marks Prizes for Cancer Research 3123
The Paul Marks Prizes for Cancer Research recognize outstanding young investigators who have made significant contributions to increase the understanding of cancer or improve the treatment of the disease through basic or clinical research. The winners will present their work at the center, be honored at a dinner, and share a cash award of $150,000. Nomination packets must include a letter from the nominator outlining the significance of the accomplishments for which the candidate should be recognized; a one-page scientific biography of the candidate; a list of up to eight of the candidate's significant published papers with a brief (fewer than 100 words) explanation of the importance of each one; and the candidate's curriculum vita. Up to three supporting letters may also be submitted. The bi-annual deadline date is April 30 of odd years.
Requirements: Nominees are required to be age 45 or younger at the time of the submission deadline.
Geographic Focus: All States
Date(s) Application is Due: Apr 30
Amount of Grant: 150,000 USD
Contact: Program Contact; (212) 639-6561 or (212) 639-2000
Internet: http://www.mskcc.org/mskcc/html/53983.cfm
Sponsor: Memorial Sloan-Kettering Cancer Center
1275 York Avenue
New York, NY 10021

Paul Ogle Foundation Grants 3124
The Paul Ogle Foundation, Inc., was incorporated as an Indiana non-profit corporation in 1979 and is qualified under the IRS 501(c)3 as a private foundation. The Foundation's founder and original benefactor was Paul W. Ogle, who lived in Clark County, Indiana. The Foundation is headquartered in Jeffersonville, Indiana, and provides grants to deserving IRS 501(c)3 organizations in Clark, Floyd, Harrison, Switzerland, Scott, and Washington Counties in Indiana and Jefferson County, Kentucky. The Board of Directors has established guidelines for programs which they will consider for grant funding of capital projects and endowments. The guidelines are published on the Foundations website and make up the current basis for recognized charities to apply for grant funding.
Requirements: First read guidelines prior to contacting the foundations office, then send a letter of intent to the office. The letter should be typed on your institutions letterhead and should not exceed for typewritten pages in length. It should contain the following information about the proposed project: a brief description of the need that will be addressed by the proposal; a statement of the project's principal objectives; how the need will be met; the total estimated project budget and other sources of support that may be forthcoming, including a list of donors and amounts; the staff that will carry out the project and their qualifications; a timetable for the project; brief background on the organization; list of board members; the name of the primary contact person for follow-up; copy of IRS 501(c)3 exemption letter. Based on review of your letter of intent, the foundation may request a full proposal. If a proposal is requested, you will be provided with a grant form along with specific instructions. All projects are funded only on the authorization of the Foundation's Board of Director, this may take up to six months. Formal written communications from the foundation is the ONLY way you will be advised of a favorable funding decision.
Restrictions: Grants are not made to: private foundations; religious organizations for religious purposes; individuals; primary or secondary schools; political entities; national organizations, even if for local projects; annual operating support or debt reduction.
Geographic Focus: Indiana, Kentucky
Contact: Robert W. Lanum, Chair; (812) 284-5519; klanum@ogle-fdn.org
Internet: http://www.ogle-fdn.org/Guidelines.htm
Sponsor: Paul Ogle Foundation
321 East Court Avenue, P.O. Box 845
Jeffersonville, IN 47131

Paul Rapoport Foundation Grants 3125
The Paul Rapoport Foundation was established in 1987 with funds from the estate of Paul Rapoport, a founder of both New York City's LGBT Community Services Center and GMHC. For its final years of grantmaking the Foundation's focus will be on three populations of low or no income: transgender communities of color; LGTBQ youth of color, ages 24 and under; and LGTB seniors of color aged 60 and over. The Foundation will also consider funding programs of organizations not focused exclusively on the LGTB community if the number of LGTBQ clients of color served by the program is at least 50%. The foundation awards grants to nonprofits in the metropolitan area of New York, New York. Types of support include general operating support, continuing support, building construction and renovation, program development, conferences and seminars, publications, seed money, technical assistance, and matching funds. There are no application forms. The board meets in February, June, and October to consider requests. Grants of $50,000 and higher, per year are awarded.
Requirements: The Foundation funds only non-profit, charitable organizations as defined by the Internal Revenue Service Code Section 509(a). The Foundation funds primarily within the five boroughs of New York City, as well as on Long Island, in Westchester and nearby New Jersey. It will only fund national organizations when they request funding for programs specific to the New York metropolitan area.
Restrictions: The Foundation will no longer support start-up organizations. The Foundation does not support medical research, cultural and artistic activities, major building campaigns, endowments, grants or scholar-ships to individuals or to other foundations. The Foundation does not make grants for purposes of influencing elections or legislation, or for any other activity that may jeopardize the Foundation's tax-exempt status.
Geographic Focus: New York
Date(s) Application is Due: Feb 1; Jun 1; Oct 1
Samples: Ali Forney Center (AFC), New York, New York, $150,000 (1st year of 3-year grant totaling $450,000) - creation of an In-house Development Department; Friends of Green Chimneys, New York, New York, $230,000 (1st year of 4-year grant totaling $665,000) - (a) Career Development Program for LGTB Foster Care and Runaway and Homeless Youth, (b) Infrastructure Development for NYC Division of Green Chimneys, (c) Engineering Study of Basement of Gramercy Building.
Contact: Jane D. Schwartz, Executive Director; (212) 888-6578; fax (212) 980-0867
Ona M. Winet, Program Director; (212) 888-6578; fax (212) 980-0867
Internet: http://www.paulrapoportfoundation.org/guide.html
Sponsor: Paul Rapoport Foundation
220 E 60th Street, Suite 3H
New York, NY 10022

PC Fred H. Bixby Fellowships 3126
The Population Council offers a Fellowship program to expand training opportunities for social scientists and biomedical researchers in the health and population fields. This Fellowship is geared toward developing-country nationals in the early stages of their careers, and to those with a demonstrated commitment to remaining in their home countries to build capacity in local institutions or returning home after working/studying abroad. The Council's Fred H. Bixby Fellowship is a ten-year program which offers a limited number of Fellowships each year. The Bixby Fellowship allows candidates to work with experienced mentors in the Council's network of offices all over the world. Fellows work on projects in the areas of HIV and AIDS, poverty, gender, and youth, and reproductive health. The term of appointment is two years, with the second year contingent upon a successful evaluation and review of performance in the first year. The Fellowship is a full-time commitment, meaning no additional research, studies, or work may be pursued for the duration of the Fellowship. Awards consist of a monthly stipend, health insurance, an allowance for relocation expenses, and attendance and travel to one international professional meeting per year. The Fellow is responsible for all local accommodations and living costs. Fellows will be provided with office space in the Council's office, plus access to office support facilities and a notebook computer.
Requirements: Candidates must have recently completed a Ph.D. or equivalent degree within the last five years in the social sciences, public health, or biomedical sciences. All applicants should have previous direct experience with either biomedical research, program research, or policy-relevant social science research. Applicants must be legal citizens of a developing country and proficient in English. Candidates interested in applying for a Fred H. Bixby Fellowship must first email the Fellowship Coordinator with a letter indicating professional objectives and goals, a short statement about research interests and plans for the proposed Fellowship period, as well as the name(s) of the mentor(s) (up to three) with whom they would like to work. They should also include a copy of their curriculum vitae. The Fellowship Coordinator will then initiate contact with the appropriate mentor(s) and instruct each candidate on how to proceed. Additional information about the application process is available at the Council website.

Restrictions: Candidates with master's-level degrees in public health or a related field may be eligible only if they have considerable skills, background, experience, and peer-reviewed publications, including 2–3 years of relevant job experience post-master's.
Geographic Focus: All Countries
Date(s) Application is Due: Jan 31
Contact: Coordinator; (212) 339-0500; fax (212) 755-6052; bixbyfellowship@popcouncil.org
Internet: http://www.popcouncil.org/what/bixby.asp
Sponsor: Population Council
One Dag Hammarskjold Plaza
New York, NY 10017

PDF-AANF Clinician-Scientist Development Awards 3127
The Parkinson's Disease Foundation (PDF) and the American Academy of Neurology Foundation (AANF) collaborated in 2007 to create the Clinician Scientist Development Award to support our mutual interest in training outstanding residency-level clinicians in the field of Parkinson's disease. Successful applicants will receive a three-year award of $85,000. This award supports three years of research training in an environment where talented young clinicians address problems in Parkinson's disease with the most current scientific tools. It is expected that upon completion of the program, participants will be committed to a research or combined clinical/research career in PD and will be in line to direct robust research programs relevant to PD. This award is given every three years.
Requirements: Eligibility & Guidelines: applicants must hold an M.D., D.O., or equivalent clinical degree from an accredited institution; applicants must have completed residency training but be less than seven years from completion of residency when funding begins; there is no citizenship requirement; however, the individual applying for the award must be licensed to practice medicine in the United States at the time of application; may only submit one application per department. The department is considered to be the one at which the applicant's training and research program will occur; each applicant must agree to a phone interview with a member of the PDF-AANF Advisory Committee, if necessary, during the application review period.
Geographic Focus: All States
Amount of Grant: 85,000 USD
Contact: Terry Heinz, Grants Administrator; (651) 695-2746; theinz@aan.com
Internet: http://www.pdf.org/en/grant_funding_fellow#pdflabres
Sponsor: Parkinson's Disease Foundation
1359 Broadway, Suite 1509
New York, NY 10018

PDF International Research Grants 3128
The International Research Grants Program (IRGP) is designed to promote innovative research that has a high potential to significantly advance the knowledge of Parkinson's disease (PD) but little likelihood of securing funding through more traditional sources. By supporting novel, high risk/high reward research, the goal of the IRGP is to enable investigators to demonstrate the feasibility of their ideas while generating preliminary data necessary for the support of future funding. At its core, research supported by the IRGP must be directly relevant to the treatment and/or understanding of Parkinson's disease. Successful IRGP projects are: novel research hypotheses of Parkinson's disease; inventive in terms of methodology or approach; and clinical, preclinical, or basic research proposals that will directly impact Parkinson's disease or its treatment. IRGP awards will now reach a maximum of $75,000 per year for two years (compared to the previous maximum of $50,000 per year for one year of research), subject to review of first year progress. Research grants are now also eligible for an additional ten percent in indirect costs. All applications must be submitted online through the PDF website; no paper copies will be accepted. Visit grants.pdf.org, after Monday, November 2, to create an account, see more detailed instructions and submit your application. The deadline for submission of grant applications will be 11:59 PM EST on Tuesday, February 2.
Requirements: The IRGP is open to applicants both nationally and internationally and basic, translational and clinical research proposals are eligible for support. All applicants must possess a Ph.D. or a M.D. (or equivalent). Preference will be given to scientists who are at an early stage in their professional careers.
Restrictions: Investigators who receive funding from PDF may not receive funding for other projects supported by PDF during the same year, nor receive funding from other foundations or institutions for the same project during that year.
Geographic Focus: All States
Date(s) Application is Due: Feb 2
Amount of Grant: 75,000 USD
Contact: Robin Anthony Elliott; (212) 923-4700; fax (212) 923-4778; info@pdf.org
Internet: http://www.pdf.org/en/grant_funding_irg#focus
Sponsor: Parkinson's Disease Foundation
1359 Broadway, Suite 1509
New York, NY 10018

PDF Postdoctoral Fellowships for Basic Scientists 3129
The Postdoctoral Fellowships for Basic Scientists are one-year fellowships for young scientists, fresh from their Ph.D. training, to study at major research institutions. The programs grants funding in the amount of $40,000 for one year as well as a research allowance of $5,000 to be used at the discretion of the Fellow, with approval of the sponsor/mentor, to pay for such items as books, training courses, travel costs (up to $2,000) or a computer. There is no provision for the deduction of postdoctoral taxes, institutional overhead or fees. However, if the Fellow submits a written request, $3,000 of the research allowance may be used to defray the cost of premiums for health insurance. Fellows may competitively reapply for continued funding after their first year for a total of three years of support. In these instances, a progress report of the current research would be due along with the new application. All applications must be submitted online through the PDF website; no paper copies will be accepted. The deadline for submission of grant applications will be 11:59 PM ET on Tuesday, February 2. Visit grants.pdf.org after Monday, November 2, to create an account, see more detailed instructions and submit your application. Final grant decisions are expected by mid-May.
Requirements: The applicant must have completed a Ph.D. or M.D. and must identify a research mentor who will serve as mentor for the fellow and supervisor of the research.
Geographic Focus: All States
Date(s) Application is Due: Feb 2
Amount of Grant: 45,000 USD
Contact: Robin Anthony Elliott; (212) 923-4700; fax (212) 923-4778; info@pdf.org
Internet: http://www.pdf.org/en/grant_funding_fellow#pdflabres
Sponsor: Parkinson's Disease Foundation
1359 Broadway, Suite 1509
New York, NY 10018

PDF Postdoctoral Fellowships for Neurologists 3130
For young clinicians who have completed their neurology residency and are seeking clinical research experience, PDF offers the Postdoctoral Fellowships for Neurologists, one-year awards in the amount of $55,000. A research allowance of $5,000 may be used at the discretion of the Fellow, with approval of the mentor, to pay for such items as books, training courses, travel costs (up to $2,000) or a computer. There is no provision for the deduction of postdoctoral taxes, institutional overhead or fees. However, if the Fellow submits a written request, $3,000 of the research allowance may be used to defray the cost of premiums for health insurance. Fellows may competitively reapply for continued funding after their first year for a total of three years of support. In these instances, a progress report of the current research would be due along with the new application. All applications must be submitted online through the PDF website; no paper copies will be accepted. The deadline for submission of grant applications will be 11:59 PM EST on Tuesday, February 2. Visit grants.pdf.org, after Monday, November 2, to create an account, see more detailed instructions and submit your application. Final decisions are expected by mid-May.
Requirements: Applicants seeking a Postdoctoral Fellowship for Neurologists must have an M.D. and be completing or have completed a neurology residency.
Restrictions: Applicants may not have their own lab.
Geographic Focus: All States
Date(s) Application is Due: Feb 2
Amount of Grant: 55,000 USD
Contact: Robin Anthony Elliott; (212) 923-4700; fax (212) 923-4778; info@pdf.org
Internet: http://www.pdf.org/en/grant_funding_fellow#pdflabres
Sponsor: Parkinson's Disease Foundation
1359 Broadway, Suite 1509
New York, NY 10018

PDF Summer Fellowships 3131
PDF's Summer Fellowship Program is used to support students from advanced undergraduates to graduate and medical students in their pursuit of Parkinson's-related summer research projects. The goal of the Summer Fellowship is to cultivate an early interest in Fellows into the cause(s) and possible treatments for Parkinson's disease. Fellows work under the close supervision of a sponsor who is an expert in the Parkinson's community and oversees the project. Typically, summer fellowships are offered for 10 weeks of laboratory work with an award of $3,000. The application period for Summer Student Fellowships will open on Monday, November 2, apply by 11:59 PM ET on Friday, January 15.
Requirements: Undergraduate students, graduate students and medical students are all eligible for summer fellowships. Each applicant must identify a mentor, with whom he or she will conduct the proposed project.
Geographic Focus: All States
Date(s) Application is Due: Jan 15
Amount of Grant: 3,000 USD
Contact: Robin Anthony Elliott; (212) 923-4700; fax (212) 923-4778; info@pdf.org
Internet: http://www.pdf.org/en/grant_funding_fellow#pdflabres
Sponsor: Parkinson's Disease Foundation
1359 Broadway, Suite 1509
New York, NY 10018

Peabody Foundation Grants 3132
The Peabody Foundation was established in 1894 in Massachusetts. Grants provide care, treatment, rehabilitation, education, and assistance to children with physical disabilities as well as encouraging and supporting medical research in the causes of crippling disease, particularly in children. The majority of recipient organizations will be located in the Boston, Massachusetts area. Proposals may be submitted at any time and are reviewed in the spring. Funding ranges from $20,000 to $200,000. Interested organizations should contact the Administrative Director for application procedures and eligibility information.
Restrictions: Grants are not made to individuals. Giving is limited to Massachusetts, with emphasis on the Boston area.
Geographic Focus: Massachusetts
Date(s) Application is Due: Feb 1
Amount of Grant: 20,000 - 200,000 USD
Contact: Judi Mullen; (508) 728-8780; jemullen12@comcast.net
Jonathan Bashein, Office Administrator; (617) 345-1000 or (617) 310-4100; fax (617) 345-1300; jbashein@nixonpeabody.com
Sponsor: Peabody Foundation
100 Summer Street
Boston, MA 02110-2131

Peacock Foundation Grants 3133

Established by Henry B. Peacock, Jr. in 1947, the mission of Peacock Foundation, Inc. is to enhance and promote the good health and well being of children, families, and underprivileged persons in Southeast Florida, through contributions, gifts, and grants to eligible nonprofit organizations. The priorities of Peacock Foundation, Inc. include: making grants to human services providers that promote youth development, assist abused or neglected children, women, and the elderly, and seek to reduce abuse, prevent homelessness, and end hunger in our community; supporting educational programs in the arts and the environment, as well as special education for disabled persons; contributing to medical research, health care organizations, and hospitals.
Requirements: All applicants must be IRS recognized 501(c)3 public charities classified as not a private foundation, registered with the Department of Agriculture to solicit funds in Florida, when applicable, and located in and/or of significant benefit to residents to the Southeast Florida counties of Miami-Dade, Broward, or Monroe. In order for a proposal to be considered for funding, the applicant first must send a brief letter of inquiry. See Foundations website for letter of inquiry guidelines: http://www.peacockfoundationinc.org/review_progress.html.
Restrictions: Peacock Foundation, Inc. does not fund: capital campaigns, construction, or renovation projects; deficit financing or debt reduction; conferences or festivals; fundraising events or advertising; special events or athletic events; individuals; lobbying to influence legislation; religious organizations, unless engaged in a significant project benefiting the entire community.
Geographic Focus: Florida
Contact: Joelle Allen, Executive Director; (305) 373-1386
Internet: http://www.peacockfoundationinc.org/eligibility.html
Sponsor: Peacock Foundation
100 SE Second Street, Suite 2370
Miami, FL 33131

Pediatric Brain Tumor Foundation Research Grants 3134

The Pediatric Brain Tumor Foundation (PBTF) is a 501(c)3 nonprofit charitable organization whose mission is as follows: to find the cause of and cure for childhood brain tumors by supporting medical research; to increase public awareness about the severity and prevalence of childhood brain tumors; to aid in the early detection and treatment of childhood brain tumors; to support a national database on all primary brain tumors; and to provide educational and emotional support for children and families affected by this life-threatening disease. PBTF grants are meant to expand scientists' knowledge of the causes of pediatric brain tumors and should demonstrate the potential to advance discoveries into new treatment methodologies. Each year PBTF issues a Request for Application for Basic or Translational Research support. The foundation accepts requests for seed grants (start-up research programs) and for fellowships. Guidelines and application are available at the foundation website.
Requirements: Research projects must be hypothesis-driven basic or translational scientific research on pediatric brain tumors. Applicants with MDs must have no more than five years of post-residency training as of September of the current year of application. Doctoral applicants must have no more than five years postdoctoral laboratory experience as of current year of application. Applicant organizations must be 501(c)3 institutions located in the U.S., Canada, and Australia.
Restrictions: Grant requests must not exceed $50,000 over the period of one year.
Geographic Focus: All States, Australia, Canada
Date(s) Application is Due: Sep 7
Amount of Grant: Up to 50,000 USD
Samples: Fredrik Johansson Swartling, P.D., University of California, San Franscisco, $50,000 - "Characterizing and Treating Cancer Stem Cells Isolated from a MYCN Medulloblastoma Model" (2007-2010); Uri Tabori, M.D., Hospital for Sick Children, Toronto, Ontario, Canada, $100,000 - "Predictors of Functional and Neuro-Cognitive Outcomes in Long-Term Survivors of Pediatric Low-Grade Gliomas" (2009-2010).
Contact: Dianne Traynor, President and Chairman of the Board; (800) 253-6530 or (828) 665-6891; fax (828) 665-6894; dtraynor@pbtfus.org or pbtfus@pbtfus.org
Internet: http://www.pbtfus.org/medcomm/research/requests-for-applications.html
Sponsor: Pediatric Brain Tumor Foundation of the United States
302 Ridgefield Court
Asheville, NC 28806

Pediatric Cancer Research Foundation Grants 3135

Since its establishment in 1982 as a grass-roots organization, the Pediatric Cancer Research Foundation (PCRF) has focused its efforts upon improving the care, quality of life, and survival rate of children with malignant diseases. With these ends in mind, PCRF concentrates on laboratory research that will translate into immediate treatment for children with cancer. PCRF primarily funds research that brings innovative new drugs and treatment regimens to children. Proposals should be based on molecular, cellular, or integrated systems; be conceptually innovative; and have a clear plan for study and future clinical use of the factual data obtained. Current areas of foundation funding include stem-cell transplantation, stem-cell biology, molecular oncology, and molecular and cellular genetics. Awards will generally be made for a one- to three-year period at the Board's discretion. Renewals for additional years may be made for promising projects as recognized by the Scientific Review Committee. Prospective applicants can download the application form at the PCRF website. Completed applications must be received at the PCRF office by 5:00 PM on May 1st. If a deadline falls on a Saturday or Sunday, the applications are due the next business day.
Requirements: Applications may be submitted by individuals holding a M.D., Ph.D., or equivalent degree and working in domestic non-profit organizations such as universities, colleges, hospitals, or laboratories (the "sponsoring institution"). Applications may be multi-institutional in nature. Applicants currently holding a grant may re-apply for extension of similar research by submitting a Renewal Application.
Restrictions: No overhead or indirect costs to institutions will be funded.
Geographic Focus: All States
Date(s) Application is Due: May 1
Contact: Joseph M. Galosic, Director of Scientific Affairs; (949) 859-6312; fax (949) 859-6323; admin@pcrf-kids.org
Internet: http://www.pcrf-kids.org/
Sponsor: Pediatric Cancer Research Foundation
9272 Jeronimo Road, Suite 122
Irvine, CA 92618

Penny Severns Breast, Cervical and Ovarian Cancer Research Grants 3136

The Penny Severns Breast, Cervical and Ovarian Cancer Research Fund is a special fund within the state treasury that is used for breast, cervical and ovarian cancer research grants. Revenue sources include general revenue funds, income tax contributions and gifts, as well as grants and awards from private foundations, nonprofit organizations and other governmental entities or persons. Grants support research in areas related to breast, cervical and ovarian cancer prevention, etiology, pathogenesis, early detection, treatment and behavioral sciences. Research also may include clinical trials. One-year grants are available with the possibility of two subsequent 12-month renewals. Standard research grants are intended to develop and advance the understanding, techniques and modalities effective in early detection, prevention, cure, screening and treatment of breast, cervical and ovarian cancers and may include clinical trials. The maximum award for standard research grants will not exceed $75,000. Fellowship research grants support supervised post-doctoral research training. These grants are intended to further develop the skills necessary for a career in breast, cervical and ovarian cancer research. Fellowship research grants will receive maximum annual funding of $35,000. Special consideration is given to single-year applications and to applications for pilot projects that have the potential for subsequent funding from other sources. Continuation grants, with a maximum of two additional years, are available for those grantees that upon peer review demonstrate progress toward stated goals.
Requirements: Although a majority of the applications submitted are bio-medical in nature, researchers in the fields of behavioral and social sciences also are encouraged to apply. Funding is granted to institutions, not individuals. Research is conducted by an individual(s) under the authority of an institution.
Geographic Focus: Illinois
Amount of Grant: 20,000 - 75,000 USD
Contact: Coordinator; 217-524-6088; fax 217-557-3326; dph.owhline@illinois.gov
Internet: http://www.idph.state.il.us/about/womenshealth/grants/penny.htm
Sponsor: Illinois Department of Public Health
535 W. Jefferson Street, First Floor
Springfield, IL 62761-0001

PepsiCo Foundation Grants 3137

The foundation's three focus areas include health (food security, improved and optimum nutrition, and energy balance), environment (water security, sustainable agriculture, and adaptive approaches to climate change), and education (access to education, dropout prevention, women's empowerment, and skills training for the under-served). Additionally the foundation provides financial assistance, in-kind product donations, and human-resource contributions to help respond to people and communities affected by major disasters. The foundation divides its grants into two categories: major requests (over $100,000) and other requests ($100,000 and under). Foundation staff must solicit proposals for all major grants. All other requests must be submitted as a Letter of Interest (LOI) to the email address given. Specific guidelines for LOI contents are available at the website. LOIs are reviewed on a rolling basis. Consideration regularly takes several months, especially during peak periods. PepsiCo Foundation was established in 1962 for charitable and educational purposes. In the 1970s, the Foundation began to support fitness research, and by the 1980s had established a focus on preventive medicine. Later, the Foundation's focus was expanded to funding fitness education for youth. Today the foundation has evolved its goals to reflect the needs of under-served populations and has extended its grant-making to the global community.
Requirements: Eligible organizations must have official tax-exempt status under Section 501(c)3 of the Internal Revenue Code (or the equivalent of such status) and have a primary focus in the areas of health, environment, or education. In evaluating requests, the foundation will consider the following criteria: the extent to which the request addresses specific goals, methodologies, and approaches; the degree to which the request advances or fulfills PepsiCo Foundation's stated goals and priorities; evidence of proven success in the field or scope of work specific to the request; and a method by which to measure and track impact and progress.
Restrictions: PepsiCo Foundation does not fund the following entities (or causes): individuals; private charities or foundations; organizations not exempt under Section 501(c)3 of the Internal Revenue Code and not eligible for tax-deductible support; religious organizations; political causes, candidates, organizations, or campaigns; organizations that discriminate on the basis of age, race, citizenship or national origin, disability or disabled-veteran status, gender, religion, marital status, sexual orientation, military service or status, or Vietnam-era veteran status; organizations whose primary purpose is to influence legislation; endowments or capital campaigns; playgrounds, sports fields, or equipment; film, music, TV, video, and media production companies; sports sponsorships; performing arts tours; and association memberships.
Geographic Focus: All States, All Countries
Amount of Grant: Up to 100,000 USD
Samples: Save the Children, Westport Connecticut, $3 million - to help promote the survival and well-being of children living in rural India and Bangladesh; Safe Water Network, Westport, Connecticut, $3.35 million - to implement safe water initiatives for village water systems in Ghana, India, and Kenya, as well as rainwater-harvesting

systems in India; Diplomas Now, Baltimore, Maryland, $6 million - to invest in a proven approach to helping the toughest schools in America's largest cities succeed in helping every student graduate, ready for college or a career.
Contact: Maura Smith, President; (914) 253-3153 or (914) 253-2000; fax (914) 253-2788; pepsico.foundation@pepsico.com
Internet: http://www.pepsico.com/Purpose/PepsiCo-Foundation/Grants.html
Sponsor: PepsiCo Foundation
700 Anderson Hill Road
Purchase, NY 10577

Percy B. Ferebee Endowment Grants 3138
Grants from the Percy B. Ferebee Endowment are awarded to support charitable, scientific and literary projects, and in particular, governmental and civic projects designed to further the cultural, social, economic and physical well-being of residents of Cherokee, Clay, Graham, Jackson, Macon and Swain Counties of North Carolina and the Cherokee Indian Reservation. Grants are also awarded in the form of scholarships to assist worthy and talented young men and women who reside in said counties in pursuing their college/university degree education within the state of North Carolina. The deadline for the submission of a grant application is September 30. The deadline for scholarship applications is January 31. Application forms are available online. Applicants will receive notice acknowledging receipt of the grant request, and subsequently be notified of the grant declination or approval.
Requirements: 501(c)3 non-profits in the Cherokee, Clay, Graham, Jackson, Macon and Swain Counties of North Carolina are eligible to apply for grants. Proposals should be submitted in the following format: completed Common Grant Application Form; an original Proposal Statement; an audited financial report and a current year operating budget; a copy of your official IRS Letter with your tax determination; a listing of your Board of Directors. Proposal Statements (second item in the above Format) should answer these questions: what are the objectives and expected outcomes of this program/project/request; what strategies will be used to accomplish your objective; what is the timeline for completion; if this is part of an on-going program, how long has it been in operation; what criteria will you use to measure success; if the request is not fully funded, what other sources can you engage; an Itemized budget should be included; please describe any collaborative ventures. Prior to the distribution of funds, all approved grantees must sign and return a Grant Agreement Form, stating that the funds will be used for the purpose intended. Progress reports and Completion reports must also be filed as required for your specific grant. All current grantees must be in good standing with required documentation prior to submitting new proposals to any foundation. Scholarship recipients must be a resident of these areas and must attend a college or university in the state of North Carolina. Contact the Foundation for additional application requirements.
Restrictions: Grants are not made for political purposes, nor to organizations which discriminate on the basis of race, ethnic origin, sexual or religious preference, age or gender.
Geographic Focus: North Carolina
Date(s) Application is Due: Jan 31; Sep 30
Amount of Grant: 3,000 - 15,000 USD
Samples: Nantahala Regional Library, $15,000--general operating support; Macon County Historical Society, $7,500--general operating expenses; Reach, Inc. Resources Education Assistance Counseling & Housing, $4,000--general operating support.
Contact: Wachovia Bank, N.A., Trustee; grantinquiries6@wachovia.com
Internet: https://www.wachovia.com/foundation/v/index.jsp?vgnextoid=5d6852199c0a a110VgnVCM1000004b0d1872RCRD&vgnextfmt=default
Sponsor: Percy B. Ferebee Endowment
Wachovia Bank, NC6732, 100 North Main Street
Winston Salem, NC 27150

Perkin Fund Grants 3139
The Perkin Fund supports projects and programs in the fields of astronomy, medicine, and scientific research, as well as limited giving to leading organizations in the arts, education, and social services. Most grants are awarded to well-established institutions. However, small and medium-sized institutions doing significant work in the fields of interest also are encouraged to apply. Deadlines are March 15 and September 15, with board meetings in May and November. Final notification of funding is usually two to four weeks.
Requirements: U.S. nonprofit institutions and organizations are eligible. Applicants should submit a letter with a detailed description of their project, the amount of funding requested, and a copy of their IRS determination letter.
Restrictions: The Fund does not grant to individuals or institutions outside the United States. Funding is limited to Connecticut, Massachusetts, and New York.
Geographic Focus: Connecticut, Massachusetts, New York
Date(s) Application is Due: Mar 15; Sep 15
Amount of Grant: 20,000 - 100,000 USD
Contact: Winifred Gray, Treasurer; (978) 468-2266; theperkinfund@verizon.net
Sponsor: Perkin Fund
176 Bay Road, P.O. Box 2220
South Hamilton, MA 01982-2232

Perkins Charitable Foundation Grants 3140
The Perkins Charitable Foundation Trust was established in 1950, by members of the Perkins family. The Foundation gives primarily for education, the arts, environmental conservation, animals, wildlife, health and medical care, and children, youth and social services. They offer funding on a national basis, with some emphasis on Ohio and Vermont.
Requirements: Contact the Foundation's office by letter or phone prior to your proposal. No Application form is required.
Restrictions: No grants to individuals.
Geographic Focus: All States
Contact: Marilyn Best, Secretary; (216) 621-0465
Sponsor: Perkins Charitable Foundation
1030 Hanna Building, 1422 Euclid Avenue
Cleveland, OH 44115-2001

Perpetual Trust for Charitable Giving Grants 3141
The Perpetual Trust for Charitable Giving was established in 1957 to support and promote quality educational, human-services, and health-care programming for underserved populations. Special consideration is given to medical aid and medical research organizations, as well as to institutions of higher learning. Grant requests for general operating support are strongly encouraged. Program support will also be considered. Small, program-related capital expenses may be included in general operating or program requests. To better support the capacity of nonprofit organizations, multi-year funding requests are strongly encouraged. Applicants must apply online at the grant website. Applicants are strongly encouraged to do the following before applying: review the downloadable state application procedures for additional helpful information and clarifications; review the downloadable online-application guidelines at the grant website; review the trust's funding history (link is available from the grant website); review the online application questions in advance; and review the list of required attachments. These will generally include: a list of board members, financial statements (audited, reviewed, or compiled by independent auditor); an organization summary; a list of other funding sources; an IRS Determination letter; and other required documents. All attachments must be uploaded in the online application as PDF, Word, or Excel files. The application deadline for the Perpetual Trust is 11:59 p.m. on September 1. Applicants will be notified of grant decisions before November 30.
Restrictions: The trust does not support requests from individuals, organizations attempting to influence policy through direct lobbying, or any political campaigns.
Geographic Focus: Massachusetts
Date(s) Application is Due: Sep 1
Samples: Forsyth Institute's Forsyth Dental Infirmary for Children, Boston, Massachusetts, $60,000, ForsythKids Program; Bridge Over Troubled Waters, Boston, Massachusetts, $30,000, general operating support; Codman Academy Foundation, Dorchester, Massachusetts, $25,000, general operating support.
Contact: Miki C. Akimoto, Vice President; (866) 778-6859; miki.akimoto@baml.com
Internet: https://www.bankofamerica.com/philanthropic/fn_search.action
Sponsor: Perpetual Trust for Charitable Giving
225 Franklin Street, 4th Floor, MA1-225-04-02
Boston, MA 02110

Perry County Community Foundation Grants 3142
The Perry County Community Foundation is a nonprofit, public charity created for the people of Perry County, Indiana. We connect donors with the causes they care about. The Foundation will accept grant proposals from all charitable organizations seeking funding; however, special attention will be given to projects that promote healthy living and that offer training and/or support to potential small business entrepreneurship. Funding priority areas include arts and culture; community development; education; health; human services; environment; recreation; and youth development. The Foundation considers proposals for grants on a yearly cycle which begins each July. At the start of each cycle, a notice is mailed to nonprofit organizations that have applied for or received grants in the past, or have otherwise requested notification of the start of each cycle. The Foundation's grant cycle runs from July through November each year. Proposals are accepted from July through September. The grant committee makes its recommendations on funding to the Foundation's Board of Trustees, which will make final funding recommendations to the board of directors of the Community Foundation Alliance. All organizations who submit proposals are notified committee's decision no later than December 1. The application is available at the Foundation's website.
Requirements: The Foundation welcomes proposals from nonprofit organizations that are 501(c)3 and 509(a) tax-exempt under the IRS code and from governmental agencies serving the county. Proposals from nonprofit organizations not classified as a 501(c)3 public charity may be considered if the project is charitable and supports a community need. Proposals submitted by an entity under the auspices of another agency must include a written statement signed by the agency's board president on behalf of the board of directors agreeing to act as the entity's fiscal sponsor, to receive grant monies if awarded, and to oversee the proposed project.
Restrictions: Project areas not considered for funding: religious organizations for religious purposes; political parties or campaigns; endowment creation or debt reduction; operating costs; capital campaigns; annual appeals or membership contributions; travel requests for groups or individuals such as bands, sports teams, or classes.
Geographic Focus: Indiana
Contact: Renate Warner; (812) 547-3176; renate@perrycommunityfoundation.org
Internet: http://www.perrycommunityfoundation.org/disc-grants-program
Sponsor: Perry County Community Foundation
817 12th Street, P.O. Box 13
Tell City, IN 47586

Pet Care Trust Sue Busch Memorial Award 3143
Incorporated in Washington, D.C. in 1990, The Pet Care Trust is a non-profit, charitable, public foundation. The purpose of The Pet Care Trust is to help promote public understanding regarding the value of and right to enjoy companion animals, to enhance knowledge about companion animals through research and education, and to promote professionalism among members of the companion animal community. The Pet Care Trust invites you submit a letter of recommendation for a graduating veterinary technician

student who has been selected by your faculty as having a keen interest in companion animals (dogs, cats, small pet mammals, birds, reptiles, amphibians or pond and aquarium fish) as well as dedicating time to community service related to animals while maintaining a high academic standard. Please include instances relating to the above mentioned criteria. If selected, the student you have recommended will receive one of three $500 Sue Busch Memorial Awards for the year. Our goal is not only to bring recognition to a fine student but recognition as well to the institution instrumental in educating that student. Please submit your recommendations on or before Feb 1 You will be notified on or before April 30 in order for the Check and Certificate to be presented at your graduation ceremony.
Geographic Focus: All States
Date(s) Application is Due: Feb 1
Amount of Grant: 500 USD
Contact: Steve Hellem, Executive Director; (202) 530-5910; shellem@navista.net
Internet: http://www.petcaretrust.org/i4a/pages/index.cfm?pageid=3297
Sponsor: Pet Care Trust
1155 Fifteenth Street NW, Suite 500
Washington, D.C. 20005

Peter and Elizabeth C. Tower Foundation Annual Intellectual Disabilities Grants 3144

Elizabeth Nelson Clarke was born in Mt. Vernon, New York, in 1920, and Peter Tower in Niagara Falls, New York, in 1921. They met while attending Cornell University and married in the summer of 1942. Two daughters, Mollie and Cynthia, were born in 1944 and 1947. Peter entered the Army Air Force in January 1943. After service in Texas and Europe, he hoped to find work in the fledgling air transport business which he expected to thrive. Taking a "temporary" clerk job for the family's customhouse broker business C. J. Tower and Sons, he stayed on to see the partnership evolve into a corporation. In 1986 it processed a total of $45 billion worth of merchandise and Peter then sold the business to McGraw-Hill Inc. who later sold it to the Federal Express Company, which utilizes Tower skills and systems worldwide. Meanwhile, Liz, who had studied art at Cornell, had become a notable artist, working mostly in oils. True to other personal philosophies, Peter and Liz knew that with prosperity came responsibility. Their desire was to assure that the resources they had acquired over the years were put to good use and to see the benefits spread among many. Formed December 31, 1990, the Peter and Elizabeth C. Tower Foundation seeks to support community programming that will help children, adolescents, and young adults affected by substance abuse, learning disabilities, mental illness, and intellectual disabilities achieve their full potential. The foundation's funding objective in the intellectual disabilities category is to improve service delivery for children, adolescents, and young adults to age 26 with intellectual disabilities. The foundation defines an intellectual disability as a disability characterized by significant limitations both in intellectual functioning and adaptive behavior, which covers many everyday social and practical skills. This disability originates before the age of 18. The Foundation will give preference to projects addressing one or more of the following priority areas: reducing obstacles to seeking services/treatment; stigma reduction; transitional services; early identification and linkage to services; effective treatment/programming; co-occurring disorders; and provider workforce shortage and readiness (capacity building). The foundation will also consider other project ideas that have the potential to advance its objective in the category. Typical funding range is $25,000 - $75,000 per year. Multi-year grants are encouraged; projects should be sustainable after grant funding ends. Preference is given to projects where the majority of costs are both new to the organization and directly related to the proposed initiative. Interested organizations must submit a letter of inquiry by 5:00 p.m. June 8 to the Chief Program Officer. Guidelines for the letter of inquiry are provided at the grant website. Applicants will be notified by July 27 of the result of their letters of inquiry. Selected applicants must submit a full grant application by June 22. Final grantees will be informed by September 14. Deadline dates may vary from year to year. Interested organizations should visit the grant website to verify current deadline dates.
Requirements: Letters of inquiry must be mailed or hand-delivered. emailed and faxed copies will not be accepted. Signature of organization's Executive Director, Superintendent or Headmaster is required; designee signatures are not acceptable.
Restrictions: The Foundation accepts only one letter of inquiry per applicant. The Foundation makes grants only to: tax-exempt organizations with 501(c)3 tax-exempt status from the Internal Revenue Service that are neither private foundations nor described as 509(a)3 organizations; diocesan and public school districts; charter schools; and to nonprofit public benefit corporations. Organizations must be located in and primarily serve residents of one of the following geographic areas: Barnstable County, Massachusetts; Dukes County, Massachusetts; Essex County, Massachusetts; Nantucket County, Massachusetts; Erie County, New York; or Niagara County, New York. The Foundation does not provide funds that: may be used for the private benefit of any grant recipient or affiliated person; attempt to influence legislation; attempt to influence or intervene in any political campaign; support capital campaigns or improvements; support individual scholarships; provide general operating support; or that subsidize individuals for the cost of care.
Geographic Focus: Massachusetts, New York
Date(s) Application is Due: Sep 14
Amount of Grant: 25,000 - 75,000 USD
Samples: Community Services for the Developmentally Disabled, Buffalo, New York, $189,071, "Children's Admission Services"; League for the Handicapped, Inc., Springville, New York, $31,424, to provide staff training in Brain Gym and Ooey Gooey; Sandwich Public School District, Sandwich, Massachusetts, $143,060, to provide early intervention staff with training and consultation in the Center for Autism and Related Disorders (CARD) model.
Contact: Tracy A. Sawicki, Executive Director; (716) 689-0370; fax (716) 689-3716; info@thetowerfoundation.org or tas@thetowerfoundation.org
Internet: http://www.thetowerfoundation.com/WhatWeFund/ID/IDGrantGuidelines
Sponsor: Peter and Elizabeth C. Tower Foundation
2351 North Forest Road
Getzville, NY 14068-1225

Peter and Elizabeth C. Tower Foundation Annual Mental Health Grants 3145

Elizabeth Nelson Clarke was born in Mt. Vernon, New York, in 1920, and Peter Tower in Niagara Falls, New York, in 1921. They met while attending Cornell University and married in the summer of 1942. Two daughters, Mollie and Cynthia, were born in 1944 and 1947. Peter entered the Army Air Force in January 1943. After service in Texas and Europe, he hoped to find work in the fledgling air transport business which he expected to thrive. Taking a "temporary" clerk job for the family's customhouse broker business C. J. Tower and Sons, he stayed on to see the partnership evolve into a corporation. In 1986 it processed a total of $45 billion worth of merchandise and Peter then sold the business to McGraw-Hill Inc. who later sold it to the Federal Express Company, which utilizes Tower skills and systems worldwide. Meanwhile, Liz, who had studied art at Cornell, had become a notable artist, working mostly in oils. True to other personal philosophies, Peter and Liz knew that with prosperity came responsibility. Their desire was to assure that the resources they had acquired over the years were put to good use and to see the benefits spread among many. Formed December 31, 1990, the Peter and Elizabeth C. Tower Foundation seeks to support community programming that will help children, adolescents, and young adults affected by substance abuse, learning disabilities, mental illness, and intellectual disabilities achieve their full potential. The foundation has a particular interest in serious mental illnesses, including major depression, schizophrenia, bipolar disorder, obsessive-compulsive disorder (OCD), panic disorder, post-traumatic stress disorder (PTSD) and borderline personality disorder. The foundation's annual mental health grants seek to prevent or alleviate psychological disorders in children, adolescents, and young adults to age 26. The foundation will give preference to projects involving one or more of the following priorities: providing direct benefit to individuals with serious mental illness; reducing barriers to seeking treatment or services; reducing stigma and prejudice often associated with mental illness; fostering social and emotional development; providing early identification and linkage to services; treating co-occurring disorders; offering life-skills programming for persons with serious mental illness; building provider workforce readiness (capacity building); providing indicated or selected prevention programming; and providing effective treatment/programming. The foundation will also consider other project ideas that have the potential to advance its objective in the category. Typical funding range is $25,000 - $75,000 per year. Multi-year grants are encouraged; projects should be sustainable after grant funding ends. Preference is given to projects where the majority of costs are both new to the organization and directly related to the proposed initiative. Interested organizations must submit a letter of inquiry by 5:00 p.m. March 30 to the Chief Program Officer. Guidelines for the letter of inquiry are provided at the grant website. Applicants will be notified by May 11 of the result of their letters of inquiry. Selected applicants must submit a full grant application by June 22. Final grantees will be informed by September 7. Deadline dates may vary from year to year. Interested organizations should visit the grant website to verify current deadline dates.
Requirements: To be eligible to apply, organizations must operate programs that provide services for children, adolescents, or young adults to age 26 with psychological disorders; and/or offer programming to prevent the onset of psychological disorders in children, adolescents, and young adults to age 26 who, based on a range of socio-economic factors, have a higher-than-average likelihood of developing these conditions. Letters of inquiry must be mailed or hand-delivered. emailed and faxed copies will not be accepted. Signature of organization's Executive Director, Superintendent or Headmaster is required; designee signatures are not acceptable.
Restrictions: The Foundation accepts only one letter of inquiry per applicant. Universal programs (programs applied to general population groups without reference to or identification of those at particular risk) will not be considered for funding. Certain general restrictions apply to all grant-seeking organizations. The Foundation makes grants only to: tax-exempt organizations with 501(c)3 tax-exempt status from the Internal Revenue Service that are neither private foundations nor described as 509(a)3 organizations; diocesan and public school districts; charter schools; and to nonprofit public benefit corporations. Organizations must be located in and primarily serve residents of one of the following geographic areas: Barnstable County, Massachusetts; Dukes County, Massachusetts; Essex County, Massachusetts; Nantucket County, Massachusetts; Erie County, New York; or Niagara County, New York. The Foundation does not provide funds that: may be used for the private benefit of any grant recipient or affiliated person; attempt to influence legislation; attempt to influence or intervene in any political campaign; support capital campaigns or improvements; support individual scholarships; provide general operating support; or that subsidize individuals for the cost of care.
Geographic Focus: Massachusetts, New York
Amount of Grant: 25,000 - 75,000 USD
Samples: Lynn Public Schools, Lynn, Massachusetts, $276,230, "Paving the Way to Independence, One Step at a Time"; Child and Family Services of Erie County, Buffalo, New York, $191,930, a two-year training program for clinicians and counselors who work with children and adolescents that have experienced trauma in their lives; Catholic Charities of Buffalo, Buffalo, New York, $200,033, a comprehensive training program for clinicians and therapists who work for managing behavioral and emotional problems stemming from the experience of personal trauma.
Contact: Tracy A. Sawicki, Executive Director; (716) 689-0370; fax (716) 689-3716; info@thetowerfoundation.com or tas@thetowerfoundation.org
Internet: http://www.thetowerfoundation.com/WhatWeFund/MentalHealth/AnnualMentalHealthGrantGuidelines
Sponsor: Peter and Elizabeth C. Tower Foundation
2351 North Forest Road
Getzville, NY 14068-1225

Peter and Elizabeth C. Tower Foundation Learning Disability Grants 3146
Elizabeth Nelson Clarke was born in Mt. Vernon, New York, in 1920, and Peter Tower in Niagara Falls, New York, in 1921. They met while attending Cornell University and married in the summer of 1942. Peter entered the Army Air Force in January 1943. After serving in Texas and Europe, he took a "temporary" clerk job for the family's customhouse broker business C. J. Tower and Sons. Two daughters, Mollie and Cynthia, were born in 1944 and 1947. Although he had hoped to find work in the fledgling air transport business which he expected to thrive, he stayed with Tower and Sons to see the partnership evolve into a corporation. In 1986 it processed a total of $45 billion worth of merchandise and Peter then sold the business to McGraw-Hill Inc. who later sold it to the Federal Express Company, which utilizes Tower skills and systems worldwide. Meanwhile, Liz, who had studied art at Cornell, had become a notable artist, working mostly in oils. True to other personal philosophies, Peter and Liz knew that with prosperity came responsibility. Their desire was to assure that the resources they had acquired over the years were put to good use and that the benefits were spread among many. Formed December 31, 1990, the Peter and Elizabeth C. Tower Foundation supports community programming that helps children, adolescents, and young adults affected by substance abuse, learning disabilities, mental illness, and intellectual disabilities achieve their full potential. Learning disabilities are defined as neurological disorders affecting the brain's ability to receive, process, store, and respond to information. These constitute disorders in one or more of the basic psychological processes involved in understanding or using language, spoken or written, and may manifest themselves in the imperfect ability to listen, think, speak, read, write, spell or do mathematical calculations. These disorders do not include learning problems that are primarily the result of visual, hearing, or motor abilities, of mental retardation, of emotional disturbance, of traumatic brain injury, or of environmental, cultural, or economic disadvantage. The Tower Foundation is conducting research on the current state of the field. The foundation plans to offer grant opportunities to organizations that offer programming to help the learning disabled after it completes the study. These organizations are encouraged to check the grant website in 2012.
Requirements: Eligible organizations must focus on providing services and programming that will help children, adolescents, and young adults affected by substance abuse, learning disabilities, mental illness, and intellectual disabilities achieve their full potential.
Restrictions: Certain general restrictions apply to all grant-seeking organizations. The Foundation makes grants only to the following: tax-exempt organizations with 501(c)3 tax-exempt status that are neither private foundations nor described as 509(a)3 organizations; diocesan and public school districts; charter schools; and nonprofit public benefit corporations. Organizations must be located in and primarily serve residents of one of the following geographic areas: Barnstable County, Massachusetts; Dukes County, Massachusetts; Essex County, Massachusetts; Nantucket County, Massachusetts; Erie County, New York; or Niagara County, New York. The Foundation does not provide funds for the following purposes: the private benefit of any grant recipient or affiliated person; attempts to influence legislation; attempts to influence or intervene in any political campaign; capital campaigns or improvements; individual scholarships; general operating support; or subsidy of individuals for the cost of care.
Geographic Focus: Massachusetts, New York
Contact: Tracy A. Sawicki, Executive Director; (716) 689-0370; fax (716) 689-3716; info@thetowerfoundation.org or tas@thetowerfoundation.org
Internet: http://www.thetowerfoundation.com/WhatWeFund/LearningDisabilities
Sponsor: Peter and Elizabeth C. Tower Foundation
2351 North Forest Road
Getzville, NY 14068-1225

Peter and Elizabeth C. Tower Foundation Mental Health Reference and Resource Materials Mini-Grants 3147
Elizabeth Nelson Clarke was born in Mt. Vernon, New York, in 1920, and Peter Tower in Niagara Falls, New York, in 1921. They met while attending Cornell University and married in the summer of 1942. Two daughters, Mollie and Cynthia, were born in 1944 and 1947. Peter entered the Army Air Force in January 1943. After service in Texas and Europe, he hoped to find work in the fledgling air transport business which he expected to thrive. Taking a "temporary" clerk job for the family's customhouse broker business C. J. Tower and Sons, he stayed on to see the partnership evolve into a corporation. In 1986 it processed a total of $45 billion worth of merchandise and Peter then sold the business to McGraw-Hill Inc. who later sold it to the Federal Express Company, which utilizes Tower skills and systems worldwide. Meanwhile, Liz, who had studied art at Cornell, had become a notable artist, working mostly in oils. True to other personal philosophies, Peter and Liz knew that with prosperity came responsibility. Their desire was to assure that the resources they had acquired over the years were put to good use and to see the benefits spread among many. Formed December 31, 1990, the Peter and Elizabeth C. Tower Foundation seeks to support community programming that will help children, adolescents, and young adults affected by substance abuse, learning disabilities, mental illness, and intellectual disabilities achieve their full potential. The Mental Health Mini-Grant initiative seeks to prevent or alleviate psychological disorders in children, adolescents, and young adults up to age 26, as well as to build mental health providers' capacity. The mini-grants will provide funds for the purchase of clinical reference and resource materials for these purposes. Grant requests of up to $7,500 will be considered for mental health providers requesting reference materials for multiple sites within an agency. Grant requests of up to $3,000 will be considered for mental health providers requesting reference materials for one site. The Foundation anticipates making 30-45 awards. Applicants must mail or hand-deliver a letter of application (along with required supporting materials) by October 31. Guidelines are available at the grant website. Questions may be directed to the Program Officer. A FAQ is also available at the grant website. Award checks are distributed November 30.
Requirements: Letters of application from community-based organizations must have the Executive Director/CEO's original signature. Requests from school districts, private schools, or charter schools must have the Superintendent/School Headmaster's original signature.
Restrictions: Only one request per community-based organization, charter school, private school or school district will be accepted. Mini-grant monies may not be used for: the purchase of office equipment, furniture or supplies; computers or smart boards; non-mental health reference materials or consumable items; screening or assessment tools; subscriptions to journals/newsletters; professional development or staff training; staff salaries or other general operating expense; or for capital improvements. Certain general restrictions apply to all grant-seeking organizations. The Foundation makes grants only to: tax-exempt organizations with 501(c)3 tax-exempt status from the Internal Revenue Service that are neither private foundations nor described as 509(a)3 organizations; diocesan and public school districts; charter schools; and to nonprofit public benefit corporations. Organizations must be located in and primarily serve residents of one of the following geographic areas: Barnstable County, Massachusetts; Dukes County, Massachusetts; Essex County, Massachusetts; Nantucket County, Massachusetts; Erie County, New York; or Niagara County, New York. The Foundation does not provide funds that: may be used for the private benefit of any grant recipient or affiliated person; attempt to influence legislation; attempt to influence or intervene in any political campaign; support capital campaigns or improvements; support individual scholarships; provide general operating support; or that subsidize individuals for the cost of care.
Geographic Focus: Massachusetts, New York
Date(s) Application is Due: Oct 31
Amount of Grant: 3,000 - 7,500 USD
Contact: Tracy A. Sawicki, Executive Director; (716) 689-0370; fax (716) 689-3716; info@thetowerfoundation.org or tas@thetowerfoundation.org
Internet: http://www.thetowerfoundation.com/WhatWeFund/MentalHealth/MH-RRMaterialsMiniGrant
Sponsor: Peter and Elizabeth C. Tower Foundation
2351 North Forest Road
Getzville, NY 14068-1225

Peter and Elizabeth C. Tower Foundation Organizational Scholarships 3148
Elizabeth Nelson Clarke was born in Mt. Vernon, New York, in 1920, and Peter Tower in Niagara Falls, New York, in 1921. They met while attending Cornell University and married in the summer of 1942. Peter entered the Army Air Force in January 1943. After serving in Texas and Europe, he took a "temporary" clerk job for the family's customhouse broker business C. J. Tower and Sons. Two daughters, Mollie and Cynthia, were born in 1944 and 1947. Although he had hoped to find work in the fledgling air transport business which he expected to thrive, he stayed with Tower and Sons to see the partnership evolve into a corporation. In 1986 it processed a total of $45 billion worth of merchandise and Peter then sold the business to McGraw-Hill Inc. who later sold it to the Federal Express Company, which utilizes Tower skills and systems worldwide. Meanwhile, Liz, who had studied art at Cornell, had become a notable artist, working mostly in oils. True to other personal philosophies, Peter and Liz knew that with prosperity came responsibility. Their desire was to assure that the resources they had acquired over the years were put to good use and that the benefits were spread among many. Formed December 31, 1990, the Peter and Elizabeth C. Tower Foundation seeks to support community programming that will help children, adolescents, and young adults affected by substance abuse, learning disabilities, mental illness, and intellectual disabilities achieve their full potential. Recognizing the benefits of using existing community programming aimed at building organizational infrastructure, the Tower Foundation partners with local providers to offer organizational scholarships for selected programs at the Canisius College Center for Professional Development, the University at Buffalo Institute for Nonprofit Agencies, the Boston University Institute for Nonprofit Management and Leadership, the Merrimack College's Non-Profit Certificate Program, and the Martha's Vineyard Donors Collaborative. More information is available at the grant website.
Restrictions: Funding is limited to organizations whose service recipients are children, adolescents, and young adults to age 26 affected by mental illness, intellectual disabilities, learning disabilities, and/or substance abuse. The Foundation makes grants only to 501(c)3 organizations that are neither private foundations nor described as 509(a)3 organizations, diocesan and public school districts, charter schools, and nonprofit public benefit corporations. Organizations must be located in and primarily serve residents of one of the following geographic areas: Barnstable County, Massachusetts; Dukes County, Massachusetts; Essex County, Massachusetts; Nantucket County, Massachusetts; Erie County, New York; or Niagara County, New York.
Geographic Focus: Massachusetts, New York
Contact: Tracy A. Sawicki, Executive Director; (716) 689-0370; fax (716) 689-3716; info@thetowerfoundation.org or tas@thetowerfoundation.org
Internet: http://www.thetowerfoundation.com/WhatWeFund/OrganizationalCapacity-Building/OrganizationalScholarships
Sponsor: Peter and Elizabeth C. Tower Foundation
2351 North Forest Road
Getzville, NY 14068-1225

Peter & Elizabeth C. Tower Foundation Phase II Technology Initiative Grants 3149
The Peter and Elizabeth C. Tower Technology Initiative program connects not-for-profit organizations to technological expertise in order to develop and execute a strategic technology plan. A strategic technology plan aligns an organization's administrative and business needs (as outlined in the organization's strategic plan) with its technology needs thus allowing it to function more efficiently and effectively. The application process for the technology initiative grants occurs in two phases: in Phase I, the foundation provides funds to hire a technology consultant to conduct a technology inventory and needs assessment

and to assist an organization in developing a three-to-five-year technology plan; in Phase II, the foundation provides a dollar-for-dollar match (up to $125,000) to organizations wishing to implement their technology plans. The Tower Foundation anticipates making multiple Phase II awards in each of the geographic areas it serves. Each matching grant will be for a time period of up to three years; funds will be disbursed each year, not as a single lump sum. Grant awards will vary. The foundation does not accept unsolicited proposals. Applicants must first call the foundation to clarify the intent, scope, and details of the technology initiative; establish the organization's eligibility for the initiative; and to discuss their capacity and readiness to implement their strategic technology plan. Based on this telephone call, the foundation will determine whether or not it wishes to request a full application from the organization. Arrangements for this pre-screening telephone call must be made prior to January 6 for the summer cycle, or July 6 for the winter cycle. The deadline for full proposals is March 14 for the summer cycle and September 12 for the winter cycle. Successful applicants will be notified by June 15 or December 15. More thorough information on the application process is available at the grant website. Details on Phase I of the application process are also available at the website.

Requirements: Organizations must be located in and primarily serve residents of one of the following geographic areas: Barnstable County, Massachusetts; Dukes County, Massachusetts; Essex County, Massachusetts; Nantucket County, Massachusetts; Erie County, New York; or Niagara County, New York. Organizations must focus on the following: providing services for children, adolescents and young adults to age 26 with intellectual disabilities; treating mental illness among children, adolescents and young adults to age 26; or preventing or treating substance abuse among children, adolescents and young adults to age 26. To be eligible for Phase II technology initiative grants, organizations must have already developed a strategic technology plan (based on an up-to-date strategic plan) that recommends specific technologies/policies/practices, identifies their relationship to administrative needs, and provides an estimated implementation budget for each recommendation. Preference will be given to technology plans that also do the following: identify the current state of the organization's technology (including network infrastructure, desktop/server hardware and software, staff skills and training needs, end-user support, periodic systems maintenance, and current practices, policies, procedures, and documentation); place a priority on each recommendation, as well as identify benefits associated with each recommendation and consequences of failing to implement recommendations; and propose an implementation timeline and replacement schedule. Organizations funded through this initiative are expected to provide dollar for dollar matching funds for the Tower Foundation's award. These funds may be obtained through any source of unrestricted funds or awards designated for the technology implementation project specifically.

Restrictions: The foundation makes technology initiative grants only to private and charter schools, nonprofit public benefit corporations, and organizations with 501(c)3 tax-exempt status that are neither private foundations nor described as 509(a)3 organizations. Diocesan and public school districts are not eligible to apply for technology initiative grants. Phase II grant money may not used for the following: the private benefit of any grant recipient or affiliated person; endowments; non-technology capital projects or campaigns; the development of an organization's strategic plan; the development of a strategic technology plan; the development of custom software applications or websites; advanced information technology training or certifications; general operating support; or individual scholarships. Please note that the RFP does not provide funds for hiring new staff nor does it provide funds for existing staff to conduct Phase II technology-initiative-grant activities.

Geographic Focus: Massachusetts, New York

Date(s) Application is Due: Mar 14; Sep 12

Amount of Grant: Up to 125,000 USD

Contact: Tracy A. Sawicki, Executive Director; (716) 689-0370; fax (716) 689-3716; info@thetowerfoundation.org or tas@thetowerfoundation.org

Internet: http://www.thetowerfoundation.com/WhatWeFund/OrganizationalCapacity-Building/TechnologyInitiative/PhaseII2012

Sponsor: Peter and Elizabeth C. Tower Foundation
2351 North Forest Road
Getzville, NY 14068-1225

Peter and Elizabeth C. Tower Foundation Phase I Technology Initiative Grants **3150**

The Peter and Elizabeth C. Tower Technology Initiative program connects not-for-profit organizations to technological expertise in order to develop and execute a strategic technology plan. A strategic technology plan aligns an organization's administrative and business needs (as outlined in the organization's strategic plan) with its technology needs thus allowing it to function more efficiently and effectively. The application process for the technology initiative grants occurs in two phases: in Phase I, the foundation provides funds to hire a technology consultant to conduct a technology inventory and needs assessment and to assist an organization in developing a three-to-five-year technology plan; in Phase II, the foundation provides a dollar-for-dollar match (up to $125,000) to organizations wishing to implement their technology plans. The Tower Foundation anticipates making multiple Phase I awards in each of the geographic areas it serves. Each grant will be for a one-year time period. Grant awards will range from $10,000 to $50,000 depending on the size of the organization and the complexity of its business needs. The foundation does not accept unsolicited proposals. Applicants must first call the foundation to clarify the intent, scope, and details of the technology initiative; establish the organization's eligibility for the initiative; and to discuss their capacity and readiness to undertake the technology planning process. Based on this telephone call, the foundation will determine whether or not it wishes to request a full application from the organization. Arrangements for this pre-screening telephone call must be made prior to January 6 for the summer funding cycle, or July 6 for the winter funding cycle. The deadline for full proposals is March 14 for the summer cycle and September 12 for the winter cycle. Successful applicants will be notified by June 15 or December 15. More information on the application process is available at the grant website. Details on Phase II of the application process are also available at the website.

Requirements: Organizations must be located in and primarily serve residents of one of the following geographic areas: Barnstable County, Massachusetts; Dukes County, Massachusetts; Essex County, Massachusetts; Nantucket County, Massachusetts; Erie County, New York; or Niagara County, New York. Organizations must focus on the following: providing services for children, adolescents and young adults to age 26 with intellectual disabilities; preventing or treating mental illness among children, adolescents and young adults to age 26; or preventing or treating substance abuse among children, adolescents and young adults to age 26. Additionally, organizations must have completed a strategic planning document within the past 36 months explicitly identifying technology as a focal area.

Restrictions: The foundation makes technology initiative grants only to private and charter schools, nonprofit public benefit corporations, and organizations with 501(c)3 tax-exempt status that are neither private foundations nor described as 509(a)3 organizations. Diocesan and public school districts are not eligible to apply for technology initiative grants. Phase I grant money may not used for the following: the private benefit of any grant recipient or affiliated person; endowments; capital projects or campaigns; staffing costs associated with technology planning or implementation; the development of an agency's strategic plan; general operating support; individual scholarships; or the purchase of computer hardware or software.

Geographic Focus: Massachusetts, New York

Date(s) Application is Due: Mar 14; Sep 12

Amount of Grant: 10,000 - 50,000 USD

Contact: Tracy A. Sawicki, Executive Director; (716) 689-0370 x206; fax (716) 689-3716; info@thetowerfoundation.org or tas@thetowerfoundation.org

Internet: http://www.thetowerfoundation.com/WhatWeFund/OrganizationalCapacity-Building/TechnologyInitiative/PhaseI-2012

Sponsor: Peter and Elizabeth C. Tower Foundation
2351 North Forest Road
Getzville, NY 14068-1225

Peter and Elizabeth C. Tower Foundation Social and Emotional Preschool Curriculum Grants **3151**

Elizabeth Nelson Clarke was born in Mt. Vernon, New York, in 1920, and Peter Tower in Niagara Falls, New York, in 1921. They met while attending Cornell University and married in the summer of 1942. Two daughters, Mollie and Cynthia, were born in 1944 and 1947. Peter entered the Army Air Force in January 1943. After service in Texas and Europe, he hoped to find work in the fledgling air transport business which he expected to thrive. Taking a "temporary" clerk job for the family's customhouse broker business C. J. Tower and Sons, he stayed on to see the partnership evolve into a corporation. In 1986 it processed a total of $45 billion worth of merchandise and Peter then sold the business to McGraw-Hill Inc. who later sold it to the Federal Express Company, which utilizes Tower skills and systems worldwide. Meanwhile, Liz, who had studied art at Cornell, had become a notable artist, working mostly in oils. True to other personal philosophies, Peter and Liz knew that with prosperity came responsibility. Their desire was to assure that the resources they had acquired over the years were put to good use and to see the benefits spread among many. Formed December 31, 1990, the Peter and Elizabeth C. Tower Foundation seeks to support community programming that will help children, adolescents, and young adults affected by substance abuse, learning disabilities, mental illness, and intellectual disabilities achieve their full potential. The foundation's Social and Emotional Preschool Curriculum Grants provide funds to implement selected preschool curricula to enhance the social-emotional development of preschool children. Eligible social-emotional curricula include: "Al's Pals: Kids Making Healthy Choices"; "The Incredible Years: Dina Dinosaur Classroom Curriculum Preschool"; "PATHS Preschool (Promoting Alternative-Thinking Strategies)"; "Second Step"; and "Tools of the Mind." Grants awards will be for two or three years with grant payments made annually. Funding levels will depend on the specific curriculum selected. In general funds may be used for the cost of training and technical assistance from program developers or publishers, including travel expense; direct program expense including materials and equipment supporting curriculum implementation; and substitute teacher or stipends for out-of-school time training expenses, if applicable. The Social and Emotional Preschool Curriculum Grants have a two-step application process. Applicants must submit a pre-application by 5:00 p.m. on December 1; selected applicants must submit a full proposal by February 9; grantees will be notified by May 6. Further information on the grant application process, including downloadable forms, are available at the grant website. Deadline dates may vary from cycle to cycle. Interested organizations are encouraged to visit the website to verify current deadline dates.

Requirements: The pre-application must be mailed or hand-delivered to the Chief Program Officer. The foundation requires organizations to commit to the highest level of staff training available. However, only those costs associated with the specific training and technical assistance recommended and provided by program developers or publishers will be permitted.

Restrictions: emailed or faxed proposals are not accepted. Social and Emotional Preschool Curriculum Grants do not provide funds for existing personnel expense or to hire new staff. The foundation will accept only one request for a single curriculum from each eligible organization. Certain general restrictions apply to all grant-seeking organizations. The Foundation makes grants only to: tax-exempt organizations with 501(c)3 tax-exempt status from the Internal Revenue Service that are neither private foundations nor described as 509(a)3 organizations; diocesan and public school districts; charter schools; and to nonprofit public benefit corporations. Organizations must be located in and primarily serve residents of one of the following geographic areas: Barnstable County, Massachusetts; Dukes County, Massachusetts; Essex County, Massachusetts; Nantucket County, Massachusetts; Erie County, New York; or Niagara

County, New York. The Foundation does not provide funds that: may be used for the private benefit of any grant recipient or affiliated person; attempt to influence legislation; attempt to influence or intervene in any political campaign; support capital campaigns or improvements; support individual scholarships; provide general operating support; or that subsidize individuals for the cost of care.
Geographic Focus: Massachusetts, New York
Date(s) Application is Due: Feb 9
Samples: Cape Cod Child Development, Cape Cod, Massachusetts, $36,034, "Promoting Alternative Thinking Strategies (PATHS) Preschool Curriculum"; Niagara Wheatfield Central School District, Niagara Falls, New York, $28,514, "PATHS (Promoting Alternative Thinking Strategies)"; YMCA of the North Shore, Boston, Massachusetts, $25,829, implementation of the Second Step curriculum in the six preschool programs in five Essex County communities.
Contact: Tracy A. Sawicki, Executive Director; (716) 689-0370; fax (716) 689-3716; info@thetowerfoundation.org or tas@thetowerfoundation.org
Internet: http://www.thetowerfoundation.com/WhatWeFund/MentalHealth/2011SocialandEmotionalPreschoolRFPs
Sponsor: Peter and Elizabeth C. Tower Foundation
2351 North Forest Road
Getzville, NY 14068-1225

Peter and Elizabeth C. Tower Foundation Substance Abuse Grants 3152
Elizabeth Nelson Clarke was born in Mt. Vernon, New York, in 1920, and Peter Tower in Niagara Falls, New York, in 1921. They met while attending Cornell University and married in the summer of 1942. Peter entered the Army Air Force in January 1943. After serving in Texas and Europe, he took a "temporary" clerk job for the family's customhouse broker business C. J. Tower and Sons. Two daughters, Mollie and Cynthia, were born in 1944 and 1947. Although he had hoped to find work in the fledgling air transport business which he expected to thrive, he stayed with Tower and Sons to see the partnership evolve into a corporation. In 1986 it processed a total of $45 billion worth of merchandise and Peter then sold the business to McGraw-Hill Inc. who later sold it to the Federal Express Company, which utilizes Tower skills and systems worldwide. Meanwhile, Liz, who had studied art at Cornell, had become a notable artist, working mostly in oils. True to other personal philosophies, Peter and Liz knew that with prosperity came responsibility. Their desire was to assure that the resources they had acquired over the years were put to good use and that the benefits were spread among many. Formed December 31, 1990, the Peter and Elizabeth C. Tower Foundation supports community programming that helps children, adolescents, and young adults affected by substance abuse, learning disabilities, mental illness, and intellectual disabilities achieve their full potential. Substance Abuse is defined as the use of illegal drugs or the use of prescription or over-the-counter drugs or alcohol for purposes other than those prescribed, or in excessive amounts. Substance abuse may lead to social, physical, emotional, and job-related problems. The Tower Foundation is conducting research on the current state of the field. The foundation plans to offer grant opportunities to organizations that offer treatment/programming for substance abusers after it completes the study. These organizations are encouraged to check the grant website in 2012.
Requirements: Eligible organizations must focus on providing services and programming that help children, adolescents, and young adults affected by substance abuse, learning disabilities, mental illness, and intellectual disabilities to achieve their full potential.
Restrictions: Certain general restrictions apply to all grant-seeking organizations. The Foundation makes grants only to: tax-exempt organizations with 501(c)3 tax-exempt status that are neither private foundations nor described as 509(a)3 organizations; diocesan and public school districts; charter schools; and to nonprofit public benefit corporations. Organizations must be located in and primarily serve residents of one of the following geographic areas: Barnstable County, Massachusetts; Dukes County, Massachusetts; Essex County, Massachusetts; Nantucket County, Massachusetts; Erie County, New York; or Niagara County, New York. The Foundation does not provide funds for the following purposes: the private benefit of any grant recipient or affiliated person; attempts to influence legislation; attempts to influence or intervene in any political campaign; capital campaigns or improvements; individual scholarships; general operating support; or subsidy of individuals for the cost of care.
Geographic Focus: Massachusetts, New York
Contact: Tracy A. Sawicki, Executive Director; (716) 689-0370; fax (716) 689-3716; info@thetowerfoundation.org or tas@thetowerfoundation.org
Internet: http://www.thetowerfoundation.com/WhatWeFund/SubstanceAbuse
Sponsor: Peter and Elizabeth C. Tower Foundation
2351 North Forest Road
Getzville, NY 14068-1225

Peter F. McManus Charitable Trust Grants 3153
The Peter F. McManus Charitable Trust was established in Pennsylvania in 2000, and designated to support mental health research on the causes of alcoholism and other substance abuses. Proposed projects may examine basic, clinical, social, and environmental causes of substance abuse. Grants may be requested in an amount up to a maximum of $50,000. Since a formal application is not required, interested parties should forward a brief summary proposal describing the project and the amount of funding requested, along with investigator's biosketch, proposed budget for the project, and a copy of the organization's IRS determination letter. The Board meets once annually, in either October or November. The annual application submission deadline is August 31
Requirements: The Trust gives primarily to 501(c)3 organizations in California, Connecticut, Maryland, and Massachusetts.
Restrictions: No funding is available for treatment of alcohol or substance abuse. No more than 10% of the grant amount may be used for indirect costs.
Geographic Focus: California, Connecticut, Maryland, Massachusetts
Date(s) Application is Due: Aug 31
Amount of Grant: Up to 50,000 USD
Samples: John Hopkins University, Baltimore, Maryland, $50,000 - dopamine system research project (2014); Boston University School of Medicine, Boston, Massachusetts, $50,000 - neurological basis of alcohol addiction (2014); Stanford University, Stanford, California, $50,000 - exploring physician opioid (2014).
Contact: Katharine G. Lidz, Trust Administrator; (610) 647-4974; fax (610) 647-8316
Sponsor: Peter F. McManus Charitable Trust
31 Independence Place
Chesterbrook, PA 19087-5824

Peter Kiewit Foundation General Grants 3154
The foundation supports nonprofits and individuals in designated geographic areas for arts and cultural programs, higher and other education, health care, human services, youth services, rural development, community development, and government/public administration. In the general purpose grants program, there are no limitations on the size or duration of the grants that may be requested. Any applicant may submit a total of up to two applications, for two separate projects, in any 12 month period. All Peter Kiewit Foundation grants are awarded on a matching funds basis.
Requirements: Grant application guidelines and application forms are required to submit a funding request and these materials are available through the Foundation office only. Potential applicants should contact the Foundation to establish an organization's eligibility to apply and to discuss the proposed project. Nonprofit 501(c)3 organizations in Rancho Mirage, California; western Iowa; Nebraska; and Sheridan, Wyoming are eligible for a maximum of 50% of the total project cost. Units of government (tax supported) in Rancho Mirage, California; western Iowa; Nebraska; and Sheridan, Wyoming may apply for a maximum of 25% of the total project cost.
Restrictions: Grants are not awarded to support elementary or secondary schools, churches, or religious groups. Grants are not awarded to individuals (except for scholarships), or for endowment funds or annual campaigns.
Geographic Focus: California, Iowa, Nebraska, Wyoming
Date(s) Application is Due: Jan 15; Apr 15; Jul 15; Oct 15
Amount of Grant: 10,000 - 500,000 USD
Samples: Camp Fire USA, Omaha, Nebraska, $40,000 - program support; Bemis Center for Contemporary Arts, Omaha, Nebraska, $450,000 - capital improvements; Fremont Opera House, Fremont, Nebraska, $75,000 - capital improvements.
Contact: Lynn Wallin Ziegenbein; (402) 344-7890; fax (402) 344-8099
Internet: http://www.peterkiewitfoundation.org/page.aspx?id=13&pid=3
Sponsor: Peter Kiewit Foundation
8805 Indian Hills Drive, Suite 225
Omaha, NE 68114-4096

Peter Kiewit Foundation Small Grants 3155
The foundation supports nonprofits and individuals in designated geographic areas for arts and cultural programs, higher and other education, health care, human services, youth services, rural development, community development, and government/public administration. The small grants program allows the Trustees to assist a large number of worthy organizations with a broad array of small projects which are limited in scope but significant for the organization. Small grants are rarely awarded to large organizations. Small grants range in size from $500 to $10,000. The Trustees created this category of grants to support small, defined projects; not to contribute small amounts to much larger budgets.
Requirements: Grant application guidelines and application forms are required to submit a funding request and these materials are available through the Foundation office only. Potential applicants should contact the Foundation to establish an organization's eligibility to apply and to discuss the proposed project. Nonprofit 501(c)3 organizations in Rancho Mirage, California; western Iowa; Nebraska; and Sheridan, Wyoming are eligible for a maximum of 50% of the total project cost. Units of government (tax supported) in Rancho Mirage, California; western Iowa; Nebraska; and Sheridan, Wyoming may apply for a maximum of 25% of the total project cost.
Restrictions: Grants are not awarded to support elementary or secondary schools, churches, or religious groups. Grants are not awarded to individuals (except for scholarships), or for endowment funds or annual campaigns.
Geographic Focus: California, Iowa, Nebraska, Wyoming
Date(s) Application is Due: Jan 15; Apr 15; Jul 15; Oct 15
Amount of Grant: 500 - 10,000 USD
Samples: Bone Creek Art Museum, David City, Nebraska, $10,000 - program support; Brownsville Fine Arts Association, Brownsville, Nebraska, $5,000 - program support; Joslyn Art Museum, Omaha, Nebraska, $8,500 - program support.
Contact: Lynn Wallin Ziegenbein; (402) 344-7890; fax (402) 344-8099
Internet: http://www.peterkiewitfoundation.org/page.aspx?id=14&pid=3
Sponsor: Peter Kiewit Foundation
8805 Indian Hills Drive, Suite 225
Omaha, NE 68114-4096

Petersburg Community Foundation Grants 3156
The Petersburg Community Foundation has a competitive award process for unrestricted grants. Applications are available online. Most recently, total grant funds allocated equaled a combined $11,000, with awards ranging from $1,000 to $3,000. Specific areas of interest include: arts and culture; education; literacy; library support; community development; animals; the environment; medical supplies; medical care and access; and service delivery programs. Preference will be given to applications which have the potential to impact a broad range of Petersburg area residents. Applications should detail measurable and achievable outcomes and demonstrate other sources of support,

collaboration and/or cooperation. Applications should also address the sustainability of the proposed program or project for which funding is desired. Awarded grant proposals must be completed within one year. The annual deadline for applications is April 17.
Requirements: The Foundation seeks applications from qualified tax-exempt 501(c)3 organizations that support the organizations and programs in the Petersburg area and serve the people's needs in such areas as health, education, community heritage, the arts, vulnerable populations, recreation, safety, and community and economic development.
Restrictions: Individuals, for-profit, and 501(c)4 or 501(c)6 organizations, non-Alaska based organizations and state or federal government agencies are not eligible for competitive grants. Applications for religious indoctrination or other religious activities, endowment building, deficit financing, fundraising, lobbying, electioneering and activities of political nature will not be considered, nor will proposals for ads, sponsorships, or special event and any proposals which discriminate as to race, gender, marital status, sexual orientation, age, disability, creed or ethnicity.
Geographic Focus: Alaska
Date(s) Application is Due: Apr 17
Amount of Grant: 1,000 - 3,000 USD
Contact: Ricardo Lopez, Affiliate Program Officer; (907) 274-6707 or (907) 249-6609; fax (907) 334-5780; rlopez@alaskacf.org or petersburg@alaskacf.org
Internet: http://petersburgcf.org/projects/
Sponsor: Petersburg Community Foundation
P.O. Box 1024
Petersburg, AK 99833

Pew Charitable Trusts Biomedical Research Grants 3157
Pew invests in an array of programs related to science that are aimed at improving the quality of scientific research as well as making data widely available. These projects are working toward solutions to environmental, health and safety dilemmas. Pew also has a decades-long commitment to support groundbreaking research by biomedical researchers early in their careers. All of these initiatives are part of our broader effort to study and promote nonpartisan policy solutions for pressing and emerging problems affecting the American public and the global community. Pew's experts partner with leading authorities in science, the environment and associated fields to conduct research and advance fact-based solutions to compelling problems. Pew targets the frontiers of scientific knowledge where it can make a difference.
Requirements: Grants are made only to colleges and universities, as well as 501(c)3 tax-exempt organizations that are not private foundations.
Restrictions: Grants are not made to individuals or for endowments, capital campaigns, unsolicited construction requests, debt reduction, or scholarships or fellowships that are not part of a program initiated by the Trusts.
Geographic Focus: All States
Contact: Susan A. Magill, Managing Director; (202) 552-2129 or (202) 552-2000; fax (202) 552-2299; smagill@pewtrusts.org or info@pewtrusts.com
Internet: http://www.pewtrusts.org/our_work_category.aspx?id=336
Sponsor: Pew Charitable Trusts
2005 Market Street, Suite 1700
Philadelphia, PA 19103-7077

Pew Charitable Trusts Children and Youth Grants 3158
The Pew Charitable Trusts seeks to study and promote nonpartisan policy solutions for pressing and emerging problems that affect the next generation. Pew supports initiatives, grounded in research and evidence, that aim to help children and youth become active, contributing members of society both in Philadelphia and around the country. At the national level, Pew supports efforts to prevent children from languishing in foster care without safe, permanent families. Pew works to provide access to high-quality preschool for all three- and four-year-olds, and supports a public health initiative that seeks to reduce young people's exposure to alcohol advertising. Locally, the Pew Fund for Health and Human Services in Philadelphia offers operating and project-specific support to a number of nonprofits that aid youth in the city and four nearby counties.
Requirements: Grants are made only to 501(c)3 tax-exempt organizations that are not private foundations.
Restrictions: Grants are not made to individuals or for endowments, capital campaigns, unsolicited construction requests, debt reduction, or scholarships or fellowships that are not part of a program initiated by the Trusts.
Geographic Focus: All States
Contact: Susan A. Magill, Managing Director; (202) 552-2129 or (202) 552-2000; fax (202) 552-2299; smagill@pewtrusts.org or info@pewtrusts.com
Internet: http://www.pewtrusts.org/our_work_category.aspx?id=4
Sponsor: Pew Charitable Trusts
2005 Market Street, Suite 1700
Philadelphia, PA 19103-7077

PeyBack Foundation Grants 3159
The PeyBack Foundation was established by NFL quarterback Peyton Manning with the purpose of promoting the future success of disadvantaged youth by assisting programs that provide leadership and growth opportunities for children at risk (ages 6-18). The nature of the programs and their immediate long-term benefit shall be guiding considerations in funding grants. Although the foundation does not have a dollar limit on grant requests, most grant amounts range between $1,500 and $10,000. The deadline to submit a PeyBack Foundation Grant Application is February 1 each year. In order to be considered, all applications must be submitted in entirety to the PeyBack Foundation by this date.
Requirements: Due to the close association of Peyton Manning with the Indiana, Tennessee, Denver, and the New Orleans Metropolitan area, programs and projects related to the youth in these areas are of primary concern to the foundation. Proposals will be only considered from organizations that have tax-exempt status under Section 501(c)3 of the Internal Revenue Code. A proposal asking to consider providing a portion of the support for a project will generally receive greater preference than one seeking exclusive funding. Download the required application form at the website.
Restrictions: The following are not areas that the PeyBack Foundation supports: organizations without 501(c)3 tax-exempt status will immediately be eliminated; Fundraising and sponsorship events (e.g., golf tournaments, telethons, banquets); Groups outside of Indiana, Tennessee and New Orleans, LA; Projects/groups benefiting an individual or just a few persons; Building/renovating expenses of any kind; To defray meeting, conferences, workshops or seminars expenses; Payment of travel of individuals or groups; Re-granting organizations; Post-event fundraising; Multi-year gifts.
Geographic Focus: Colorado, Indiana, Louisiana, Tennessee
Date(s) Application is Due: Feb 1
Amount of Grant: 1,500 - 10,000 USD
Contact: Elizabeth Ellis; (877) 873-9225; PeyBack@PeytonManning.com
Internet: http://www.peytonmanning.com/peyback-foundation/requests/funding-requests
Sponsor: PeyBack Foundation
6325 North Guilford, Suite 201
Indianapolis, IN 46220

Peyton Anderson Foundation Grants 3160
The Peyton Anderson Foundation initiates projects to meet needs in the community and reacts to requests from charitable organizations in Macon, Bibb County, and Middle Georgia.
Requirements: Applicants must serve Macon and/or Bibb County Georgia in order to be considered for funding. Preference is given to organizations that have a substantial presence in Bibb County. Applications can be downloaded from the Foundation's website and must contain the original application plus five (5) copies, including copies of all required application items. In addition, organizations must submit one (1) copy of the organization's current annual operating budget, IRS determination letter, and most recent annual financial statement.
Restrictions: Grants may not be made to private foundations, individuals, private schools, endowments, churches, or for festivals and trips. Grants are also only awarded to organizations with current 501(c)3 status that benefit Macon and Bibb County, Georgia. DO NOT include any forms or information other than what is required above.
Geographic Focus: Georgia
Date(s) Application is Due: Apr 1; Aug 1
Amount of Grant: 1,000 - 500,000 USD
Contact: Juanita T. Jordan, President; (478) 743-5359; fax (478) 742-5201; jtjordan@pafdn.org or grants@pafdn.org
Internet: http://www.peytonanderson.org
Sponsor: Peyton Anderson Foundation
577 Mulberry Street, Suite 830
Macon, GA 31201

Pezcoller Foundation-AACR International Award for Cancer Research 3161
The Award for Cancer Research was established in 1997 to annually recognize a scientist: who has made a major scientific discovery in basic cancer research OR who has made significant contributions to translational cancer research; who continues to be active in cancer research and has a record of recent, noteworthy publications; and whose ongoing work holds promise for continued substantive contributions to progress in the field of cancer. More than one scientist may be co-nominated and selected to share the Award when their investigations are closely related in subject matter and have resulted in work that is worthy of the Award. The Award consists of an unrestricted grant of Euro 75,000 and a commemorative plaque.
Requirements: Eligible candidates are cancer researchers affiliated with institutions in academia, industry, or government that are involved in cancer research, cancer medicine, or cancer-related biomedical science anywhere in the world.
Restrictions: Institutions or organizations are not eligible for the Award.
Geographic Focus: All States
Date(s) Application is Due: Sep 15
Amount of Grant: 75,000 EUR
Contact: Monique P. Eversley, Staff Associate; (267) 646-0576 or (215) 440-9300, ext. 1400; fax (215) 440-9372; monique.eversley@aacr.org or awards@aacr.org
Internet: http://www.aacr.org/home/scientists/scientific-achievement-awards/pezcoller-aacr-international-award.aspx
Sponsor: Pezcoller Foundation / American Association for Cancer Research
615 Chestnut Street, 17th Floor
Philadelphia, PA 19106-4404

Pfizer/AAFP Foundation Visiting Professorship Program in Family Medicine 3162
This program is designed to bring educational benefits to U.S. medical schools, teaching hospitals, and community hospitals. Recipient institutions may select and invite prominent physician scientists for three days of teaching and interaction with the faculty, students, residents, research fellows, and physicians involved in family medicine at the medical center. During the visit, the guest faculty may give lectures and participate in rounds, seminars, and conferences. Awards are intended to cover the Visiting Professor's honorarium, travel expenses, and other direct expenses incurred by the host institution in conducting program activities. Amounts allotted for each item are at the discretion of the host institution.
Requirements: This program is open to any family medicine department within a U.S. medical school or any accredited family medicine residency program. Applications from community-based programs are encouraged. Each hospital may submit only one application. Submissions must originate from the Chair of the Family Medicine Department or the Residency Director.

Restrictions: Only one visit per year per visiting professor will be funded.
Geographic Focus: All States
Date(s) Application is Due: Mar 28
Amount of Grant: 7,500 USD
Contact: Susie Morantz; (800) 274-2237, ext. 4470; smorantz@aafp.org
Internet: http://www.aafpfoundation.org/x500.xml
Sponsor: American Academy of Family Physicians Foundation
11400 Tomahawk Creek Parkway, Suite 440
Leawood, KS 66211-2672

Pfizer Advancing Research in Transplantation Science Research Awards   3163

The mission of the Pfizer ARTS award program is to support basic and clinical science through a competitive, independent, peer-reviewed program that advances the understanding and medical knowledge of mTOR pathway inhibition in transplantation and related areas. Pfizer will support clinical and basic science research exploring mTOR pathway inhibition in the following areas: mechanisms of alloimmunity and allograft injury; mechanisms of cellular and molecular responses; pharmacogenomics and pharmacogenetics in transplantation; mechanisms, incidence, and treatment of adverse events; patient outcomes in transplantation; and novel applications in areas of unmet medical need. Projects are expected to yield results that will advance the medical knowledge in the research areas outlined above and that will be presented at scientific meetings and published in peer-reviewed journals. Proposed projects should last no longer than two years in duration. Award funding is limited to $250,000 per award for the entirety of the project. Funds awarded may be used to cover the awardees' salary and fringe benefits, as well as other direct expenses and indirect expenses incurred during the research project. Direct expenses may include the salary of technical associates; the purchase of relevant laboratory supplies; direct expenses necessary for proper conduct of research (e.g., subject-related costs, study-related personnel costs, diagnostic fees/services, data management expenses). Indirect expenses may include travel expenses to scientific meetings, software licensing fees, publication costs, etc. The selection of research proposals will be performed by an independent, external review committee comprised of academic clinicians/scientists who are experts in the field. Applicants must submit their proposals through the Integrated System for Pfizer Investigator Initiated Research (INSPIIRE) web portal, the link to which is available from the competitive grant announcement web page. Applications must be received by 11:59 p.m. eastern daylight time on the deadline date. Due dates may vary from year to year. Interested applicants are advised to verify current deadlines at the website or by contacting Pfizer. Pfizer, Inc. is an American multinational research-based pharmaceutical company that was founded in 1849 by two cousins Charles Pfizer and Charles Erhart. It topped the 2009 list of America's most generous companies, giving $2.3 billion in products and cash. Pfizer's grants and contributions include the following types of support: support for medical, scientific, and patient organizations; research; lobbying and political contributions; medical education grants; medical and academic partnerships; healthcare charitables; sponsorships; and special events.
Requirements: Applicants must have a professional degree (M.D., Ph.D., Pharm.D., or equivalent) and reside in the United States. Proposals will be reviewed on the following criteria: relevance to clinical transplantation; scientific excellence; originality; and realistic expectations for near term (3-5 years) clinical application.
Restrictions: Other support (funding and/or drug) provided by governmental agencies, non-governmental entities, or other pharmaceutical companies may be used to support the project; however, there should not be any overlap between the components of the research project supported by the Pfizer ARTS award program and those supported by another source.
Geographic Focus: All States
Date(s) Application is Due: Apr 30
Amount of Grant: Up to 250,000 USD
Samples: Reza Abdi, M.D., Brigham and Women's Hospital, Boston, Massachusetts - The Synergistic Role of P14K Gamma and mTOR Signaling Inhibition in Allotransplanation; David Briscoe, M.D., Children's Hospital, Boston, Massachusetts - DEPTOR: A Novel Cell Intrinsic Protein Regulating mTOR/Akt Signaling In vitro and Allograft Rejection In vivo; Mandy Ford, Ph.D., Emory University, Atlanta, Georgia - Mechanisms Underlying the Differential Impact of mTOR Inhibition on Graft- vs. Pathogen-specific T cell Responses.
Contact: Mikael Dolssten, President; (212) 733-2323
Pfizer Medical Representative; (212) 733-2323; Questions.PfizerARTS@pfizer.comQuestions.PfizerART
Internet: http://www.pfizerarts.com/
Sponsor: Pfizer
235 East 42nd Street
New York, NY 10017-5755

Pfizer Anti Infective Research EU Grants   3164

Pfizer's Anti-Infectives Research Foundation seeks high-quality and innovative anti-infectives research which may result in excellence in patient care. The foundation anticipates making non-renewable awards of approximately £50,000 - £100,000 each for research to be undertaken within a twelve-month period. Downloadable application guidelines are available from the grant web page. Applications must be submitted to Pfizer through the company's global investigator-initiated research (INSPIIRE) web portal. First time applicants must "Create an Account". Applications must be received by the deadline date. Due dates and focus areas may vary from year to year; applicants are encouraged to verify current submission requirements at the website or by contacting Pfizer. Pfizer, Inc. is an American multinational research-based pharmaceutical company that was founded in 1849 by two cousins Charles Pfizer and Charles Erhart. It topped the 2009 list of America's most generous companies, giving $2.3 billion in products and cash. Pfizer's grants and contributions include the following types of support: support for medical, scientific, and patient organizations; research; lobbying and political contributions; medical education grants; medical and academic partnerships; healthcare charitables; sponsorships; and special events.
Requirements: Applications for funding are open to all clinical practitioners and scientists working in the UK who hold medical or pharmacy degrees or Ph.D.s and who do not work for Pfizer. Applications should focus on translational (clinically-oriented) research that will impact patient outcomes. Applications regarding research in the late-laboratory phase will also be considered. Proposed projects should be in the following areas of Pfizer's therapeutic focus (however these may change from year to year): clinical diagnostics for invasive aspergillosis; epidemiology and management of serious fungal infections (Aspergillus and Candida) in the UK; cost-effective approaches to management of Complicated Skin and Soft Tissue Infections (cSSTI)s involving Multi-Drug-Resistant (MDR) organisms; and antibiotic stewardship programmes looking at cost and outcomes where the aim is to drive down resistant pathogens in the hospital environment. In exceptional circumstances, applications outside of the stated criteria, such as proposals of high scientific merit that will advance current clinical knowledge in the field of anti-infectives, may also be considered.
Restrictions: Pfizer reserves the right to not support projects that fall outside of Pfizer's internal compliance procedures or those that replicate existing research.
Geographic Focus: United Kingdom
Date(s) Application is Due: Mar 31
Amount of Grant: 50,000 - 100,000 GBP
Contact: Mikael Dolssten, President; (212) 733-2323
Coordinator; (212) 733-2323 or (203) 316-9059; Air.Foundation@Pfizer.com
Internet: http://www.pfizer.com/research/investigator/competitive_grant_programs.jsp
Sponsor: Pfizer
235 East 42nd Street
New York, NY 10017-5755

Pfizer ASPIRE EU Antifungal Research Awards   3165

Pfizer invites investigators to submit research proposals that evaluate early diagnosis of invasive mould infections. Pfizer will fund 5 awards up to € 50,000 each for a one-year period. The grants will be awarded to the host institutions on behalf of the awardees. Each award amount includes direct costs (labor and study costs), institutional overhead costs, and indirect costs (additional expenses such as publication). Prior to submitting an application, applicants are requested to have a "medical to medical conversation" with their Pfizer representative. Applicants can email iir@pfizer.com to obtain the appropriate Pfizer contact information for their region. Downloadable application guidelines and online submission instructions are available at the grant website. Applications must be submitted to Pfizer through an online submission website. Applicants should visit http://www.europeaspire.org and click on "apply". From there they will be routed to Pfizer's global investigator-initiated research (INSPIIRE) web portal where they should click on "Submit an IIR Request" and follow the online instructions. First-time applicants must "Create an Account". Applications must be received by the deadline date. (Due dates may vary from year to year; applicants are encouraged to verify current deadlines at the website or by contacting Pfizer). Pfizer, Inc. is an American multinational research-based pharmaceutical company that was founded in 1849 by two cousins Charles Pfizer and Charles Erhart. It topped the 2009 list of America's most generous companies, giving $2.3 billion in products and cash. Pfizer's grants and contributions include the following types of support: support for medical, scientific, and patient organizations; lobbying and political contributions; medical education grants; medical and academic partnerships; healthcare charitables; sponsorships; and special events. Particular to the grant opportunity listed on this page, Pfizer offers a competitive research grants program ASPIRE (Advancing Science Through Pfizer-Investigator Research Exchange). ASPIRE grant opportunities are available by geographic area, specifically: worldwide, North America, Europe, Asia, and Australia. Notices of opportunities are posted on Pfizer's competitive grants and regional ASPIRE websites. Prospective applicants may fill out a web-based form at the main ASPIRE website (http://www.aspireresearch.org/) to be notified of future competitive grant opportunities via email.
Requirements: Applicants must have a professional degree (M.D., Ph.D., Pharm.D., or equivalent) and reside in one of the EU 16 countries: Austria, Belgium, Denmark, Finland, France, Germany, Greece, Ireland, Italy, Netherlands, Norway, Portugal, Spain, Sweden, Switzerland, or the United Kingdom. Applicants must have a strong academic career interest in the research focus area. Investigators are expected to generate epidemiological, microbiological, and clinical data to better understand how non-culture-based diagnostic methods (e.g., imaging or detection of galactomannan, ß-D-glucan, or nucleic acid) can contribute to early diagnosis of invasive mould disease. Proposals should be clinical in design. Prospective evaluations are preferred however, well-designed retrospective analyses may also be considered for inclusion into the overall proposal. An important criterion in protocol assessment will be how the research contributes to improvement in the quality of care. It is expected that results will be presented at scientific meetings and published in peer reviewed journals. All applications will be formally reviewed by an independent committee of European medical experts in relevant fields. Committee members will evaluate proposals using the following criteria: scientific merit of the research proposal; qualifications of the applicant; relevance of proposed research to the program's mission; and evidence of a suitable research environment.
Restrictions: No other government, non-governmental, or industry-sponsored projects may cover the same work scope as the grant application to the EUROPE ASPIRE Program. However, a EUROPE ASPIRE grant may be related to other funding from foundations or government agencies, as long as there is no direct overlap. The institutional overhead costs cannot exceed 28% of the direct costs. Pfizer does not pay overhead on indirect costs.
Geographic Focus: Austria, Belgium, Denmark, Finland, France, Germany, Greece, Ireland, Italy, Netherlands, Norway, Portugal, Spain, Sweden, Switzerland, United Kingdom

Date(s) Application is Due: Mar 11
Amount of Grant: Up to 50,000 USD
Contact: Mikael Dolssten, President; (212) 733-2323
Internet: http://www.europeaspire.com/antifungals/index.html
Sponsor: Pfizer
235 East 42nd Street
New York, NY 10017-5755

Pfizer ASPIRE EU Dupuytren's Contracture Research Awards 3166

Current gold standard treatment for the management of Dupuytren's contracture is surgical excision involving removal of the affected palmar fascial tissue. There is however growing interest in nonsurgical, outpatient-based treatments that could provide reduced morbidity, show a decreased rate of recurrence, and give patients with Dupuytren's contracture an improved quality of life when compared with traditional surgical management. Pfizer is funding three awards up to € 60,000 each, for one year, to conduct clinical research into surgical alternatives to treating Dupuytren's contracture. Prior to submitting an application, applicants are requested to have a "medical to medical conversation" with their Pfizer representative. Applicants can email iir@pfizer.com to obtain the appropriate contact information for their region. Downloadable application guidelines and online submission instructions are available at the grant website. Applications must be submitted to Pfizer through an online submission website. Applicants should visit http://www.europeaspire.org and click on "apply". From there they will be routed to Pfizer's global investigator-initiated research (INSPIIRE) web portal where they should click on "Submit an IIR Request" and follow the online instructions. First-time applicants must "Create an Account". Applications must be received by the deadline date. (Due dates may vary from year to year; applicants are encouraged to verify current deadlines at the website or by contacting Pfizer.) Pfizer, Inc. is an American multinational research-based pharmaceutical company that was founded in 1849 by two cousins Charles Pfizer and Charles Erhart. It topped the 2009 list of America's most generous companies, giving $2.3 billion in products and cash. Pfizer's grants and contributions include the following types of support: support for medical, scientific, and patient organizations; lobbying and political contributions; medical education grants; medical and academic partnerships; healthcare charitables; sponsorships; and special events. Particular to the grant opportunity listed on this page, Pfizer offers a competitive research grants program ASPIRE (Advancing Science Through Pfizer-Investigator Research Exchange). ASPIRE grant opportunities are available by geographic area, specifically: worldwide, North America, Europe, Asia, and Australia. Notices of opportunities are posted on Pfizer's competitive grants and regional ASPIRE websites. Prospective applicants may fill out a web-based form at the main ASPIRE website (http://www.aspireresearch.org/) to be notified of future competitive grant opportunities via email.

Requirements: Applicants must have a professional degree (M.D., Ph.D., Pharm.D., or equivalent) and reside in one of the EU 16 countries: Austria, Belgium, Denmark, Finland, France, Germany, Greece, Ireland, Italy, Netherlands, Norway, Portugal, Spain, Sweden, Switzerland, or the United Kingdom. Applicants must have a strong interest in use of Collagenase Clostridium Histcolytium (CCH) in relation to the following areas of interest: various injection techniques; treatment of various cords; timing of finger extension; splinting techniques and their outcomes after treatment; Pre- and Post-treatment pain management; studies looking at Mode of Action using imaging techniques; studies splitting the volume of the reconstituted product into multiple injection sites in the same cord (0.58 mg dose stays the same); and studies assessing treatments aiming at decreasing the extent of local inflammatory reaction (e.g. use of ice packs after injection). Non-drug studies can include the development of hand functionality instruments. It is expected that the results will be presented at scientific meetings and published in peer reviewed journals. An independent committee of European medical experts in the field of hand surgery will review the research proposals and select the grant recipients. In their evaluation, committee members will consider the following criteria: scientific merit of the research proposal; qualifications of the applicant; relevance of proposed research to the program's mission; and evidence of a suitable research environment.

Restrictions: No other government, non-governmental, or industry-sponsored projects may cover the same work scope as the grant application to the EUROPE ASPIRE Program. However, a EUROPE ASPIRE Program grant may be related to other funding from foundations or government agencies, as long as there is no direct overlap. Due to the competitive nature of these awards, Pfizer cannot provide any additional funding and/or drug support beyond what has been requested and approved by the external review panel.

Geographic Focus: Austria, Belgium, Denmark, Finland, France, Germany, Greece, Ireland, Italy, Netherlands, Norway, Portugal, Spain, Sweden, Switzerland, United Kingdom
Date(s) Application is Due: Jun 15
Amount of Grant: Up to 60,000 EUR
Contact: Mikael Dolssten, President; (212) 733-2323
Internet: http://www.europeaspire.org/Dupuytrens/index.html
Sponsor: Pfizer
235 East 42nd Street
New York, NY 10017-5755

Pfizer ASPIRE EU Emerging Mechanisms of Resistance Antibacterial Research Awards 3167

Pfizer invites investigators to submit research proposals with the primary objective of further understanding the emerging mechanisms of resistance of selected Gram-negative bacteria and the susceptibility of these pathogens to antimicrobial agents. Pfizer will fund five proposals up to 50,000 Euros each for a one-year period. Funds can be used to cover the awardees' salary and fringe benefits, as well as other direct expenses incurred during the research project. Direct expenses may include the salary of technical associates, the purchase of relevant laboratory supplies, and direct expenses necessary for proper conduct of research (e.g. subject-related costs, study-related personnel costs, diagnostic fees/services, data management expenses). Indirect costs may include travel expenses to scientific meetings, software licensing fees, publication costs, etc. Prior to submitting an application, applicants are requested to have a "medical to medical conversation" with their Pfizer representative. Applicants can email iir@pfizer.com to obtain the appropriate Pfizer contact information for their region. Downloadable application guidelines and online submission instructions are available at the grant website. Applications must be submitted to Pfizer through an online submission website. Applicants should visit http://www.europeaspire.org and click on "apply". From there they will be routed to Pfizer's global investigator-initiated research (INSPIIRE) web portal where they should click on "Submit an IIR Request" and follow the online instructions. First-time applicants must "Create an Account". Applications must be received by the deadline date. (Due dates may vary from year to year; applicants are encouraged to verify current deadlines at the website or by contacting Pfizer.) Pfizer, Inc. is an American multinational research-based pharmaceutical company that was founded in 1849 by two cousins Charles Pfizer and Charles Erhart. It topped the 2009 list of America's most generous companies, giving $2.3 billion in products and cash. Pfizer's grants and contributions include the following types of support: support for medical, scientific, and patient organizations; lobbying and political contributions; medical education grants; medical and academic partnerships; healthcare charitables; sponsorships; and special events. Particular to the grant opportunity listed on this page, Pfizer offers a competitive research grants program ASPIRE (Advancing Science Through Pfizer-Investigator Research Exchange). ASPIRE grant opportunities are available by geographic area, specifically: worldwide, North America, Europe, Asia, and Australia. Notices of opportunities are posted on Pfizer's competitive grants and regional ASPIRE websites. Prospective applicants may fill out a web-based form at the main ASPIRE website (http://www.aspireresearch.org/) to be notified of future competitive grant opportunities via email.

Requirements: Applicants must have a professional degree (M.D., Ph.D., Pharm.D., or equivalent) and reside in one of the EU 16 countries: Austria, Belgium, Denmark, Finland, France, Germany, Greece, Ireland, Italy, Netherlands, Norway, Portugal, Spain, Sweden, Switzerland, or the United Kingdom. Applicants must have a strong academic career interest in the research focus area. Research proposals should focus on the activity of tigecycline alone or in combination with other antibiotics on Enterobacteriaceae (mainly Escherichia coli and Klebsiella pneumoniae) showing multi-drug or extreme-drug resistance to other antibiotics. This includes epidemiological studies, microbiological in-vitro susceptibility studies; In-vivo studies in animal model; and studies for determination of antimicrobial mechanisms of resistance. It is expected that results will be presented at scientific meetings and published in peer reviewed journals. All applications will be formally reviewed by an independent committee of European medical experts in relevant fields. Committee members will evaluate proposals using the following criteria: scientific merit of the research proposal; qualifications of the applicant; relevance of proposed research to the program's mission; and evidence of a suitable research environment.

Restrictions: No other government, non-governmental, or industry-sponsored projects may cover the same work scope as the grant application to the EUROPE ASPIRE Program. However, a EUROPE ASPIRE grant may be related to other funding from foundations or government agencies, as long as there is no direct overlap. No more than 28% of the funds may be used for institutional overhead costs. The following topics fall outside of the scope of the EUROPE ASPIRE Program: general education and / or training; public health; and support for ongoing clinical programs that are part of an organization's routine operations.

Geographic Focus: Austria, Belgium, Denmark, Finland, France, Germany, Greece, Ireland, Italy, Netherlands, Norway, Portugal, Spain, Sweden, Switzerland, United Kingdom
Date(s) Application is Due: Mar 11
Amount of Grant: Up to 50,000 USD
Contact: Mikael Dolssten, President; (212) 733-2323
Internet: http://www.europeaspire.com/Tigecycline/index.html
Sponsor: Pfizer
235 East 42nd Street
New York, NY 10017-5755

Pfizer ASPIRE EU MRSA Nosocomial Pneumonia & MRSA Complicated Skin & Soft Tissue Infections Antibacterial Research Awards 3168

Pfizer invites investigators in the EU 16 to submit proposals to conduct clinical research generating data that will lead to a better understanding of the safety and efficacy of linezolid to enhance the clinical care of patients with MRSA nosocomial pneumonia and MRSA complicated skin and soft tissue infections. Pfizer will fund five proposals up to 50,000 Euros each for a one-year period. Funds can be used to cover the awardees' salary and fringe benefits, as well as other direct expenses incurred during the research project. Direct expenses may include the salary of technical associates, the purchase of relevant laboratory supplies, and direct expenses necessary for proper conduct of research (e.g. subject-related costs, study-related personnel costs, diagnostic fees/services, data management expenses). Indirect costs may include travel expenses to scientific meetings, software licensing fees, and publication costs, etc. Prior to submitting an application, applicants are requested to have a "medical to medical conversation" with their Pfizer representative. Applicants can email iir@pfizer.com to obtain the appropriate Pfizer contact information for their region. Downloadable application guidelines and online submission instructions are available at the grant website. Applications must be submitted to Pfizer through an online submission website. Applicants should visit http://www.europeaspire.org and click on "apply". From there they will be routed to Pfizer's global investigator-initiated research (INSPIIRE) web portal where they should click on "Submit an IIR Request" and follow the online instructions. First-time applicants must "Create an Account". Applications must be received by the deadline date. (Due dates may vary from year to year; applicants are encouraged to verify current due dates at the website or by contacting Pfizer.) Pfizer, Inc. is an American multinational research-based pharmaceutical company that was founded in 1849 by two cousins Charles Pfizer and Charles Erhart. It topped the 2009 list of America's most generous companies, giving $2.3 billion in products and cash. Pfizer's grants and contributions include the following types of support: support for medical, scientific, and patient organizations; lobbying and political contributions; medical education grants; medical and academic partnerships; healthcare charitables; sponsorships; and special events. Particular to the grant opportunity listed on this page, Pfizer offers a competitive research grants program ASPIRE (Advancing Science Through Pfizer-Investigator Research Exchange). ASPIRE grant opportunities are available by geographic area, specifically: worldwide, North America, Europe, Asia, and Australia. Notices of opportunities are posted on Pfizer's competitive grants and regional ASPIRE websites. Prospective applicants may fill out a web-based form at the main ASPIRE website (http://www.aspireresearch.org/) to be notified of future competitive grant opportunities via email.

Requirements: Applicants must have a professional degree (M.D., Ph.D., Pharm.D., or equivalent) and reside in one of the EU 16 countries: Austria, Belgium, Denmark, Finland, France, Germany, Greece, Ireland, Italy, Netherlands, Norway, Portugal, Spain, Sweden, Switzerland, or the United Kingdom. Applicants must have a strong academic career interest in the research focus area and propose innovative clinical studies which clinically evaluate linezolid in treatment of new or recurrent infections due to MRSA in nosocomial pneumonia or in complicated skin and soft tissue infections in patient populations with diabetes, renal insufficiency, morbid obesity, vascular disease, and pulmonary comorbidities. Proposals should be clinical in design. Prospective evaluations are preferred however well-designed retrospective analyses may also be considered for inclusion into the overall proposal to generate and test hypotheses. It is expected that results will be presented at scientific meetings and published in peer reviewed journals. All applications will be formally reviewed by an independent committee of European medical experts in relevant fields. Committee members will evaluate proposals using the following criteria: scientific merit of the research proposal; qualifications of the applicant; relevance of proposed research to the program's mission; and evidence of a suitable research environment.

Restrictions: No other government, non-governmental, or industry-sponsored projects may cover the same work scope as the grant application to the EUROPE ASPIRE Program. However, a EUROPE ASPIRE grant may be related to other funding from foundations or government agencies, as long as there is no direct overlap. Due to the competitive nature of these awards, Pfizer cannot provide any additional funding and/or drug support beyond what has been requested and approved by the external review panel. No more than 28% of the funds may be used for institutional overhead costs. The following topics fall outside of the scope of the EUROPE ASPIRE Program: general education and / or training; public health; and support for ongoing clinical programs that are part of an organization's routine operations.

Geographic Focus: Austria, Belgium, Denmark, Finland, France, Germany, Greece, Ireland, Italy, Netherlands, Norway, Portugal, Spain, Sweden, Switzerland, United Kingdom

Date(s) Application is Due: Mar 11

Amount of Grant: Up to 50,000 EUR

Samples: Dr. Saeed, United Kingdom - The prevalence, virulence, and risk factors of OS-MRSA nosocomial pneumonia, bacteraemia, and skin and soft tissue infections: is it time to switch empirical treatment from flucloxacillin to linezolid; Dr. Blot, Belgium - Pharmacokinetic and pharmacodynamic evaluation of linezolid administered intravenously in morbidly obese patients with MRSA pneumonia; Dr. Maynar, Spain - Pharmacokinetic pharmacodynamic study of linezolid in critically ill septic patients undergoing continuous haemodiafiltration.

Contact: Mikael Dolssten, President; (212) 733-2323

Internet: http://www.europeaspire.com/Linezolid2012/eligibility.html

Sponsor: Pfizer
235 East 42nd Street
New York, NY 10017-5755

Pfizer ASPIRE North America Broad Spectrum Antibiotics for the Treatment of Gram-Negative or Polymicrobial Infections Research Awards 3169

While considerable clinical trial data exists for tygecycline in a variety of patients with known infections, most of this information is from controlled clinical trials which leave patients with certain disease complexities underrepresented. Pfizer invites investigators to submit innovative clinical research proposals that evaluate tigecycline's role in the treatment of infections within approved indications. Studies with well-designed retrospective analyses are preferred, however small pilot studies that attempt to test hypotheses will be considered as long as the study can be completed within 1 year. Funds may be used to cover the awardees' salary and fringe benefits, as well as other direct expenses incurred during the research project. Direct expenses may include the salary of technical associates; the purchase of relevant laboratory supplies; direct expenses necessary for proper conduct of research (e.g. subject-related costs, study-related personnel costs, diagnostic fees/services, and data management expenses). Indirect costs may include travel expenses to scientific meetings, software licensing fees, publication costs, etc. Downloadable application guidelines are available at the grant website. Applicants must submit their proposals through the Integrated System for Pfizer Investigator Initiated Research (INSPIIRE) web portal, the link to which is available from the competitive grant announcement on the ASPIRE website. Applications must be received by 11:59 p.m. eastern daylight time on the deadline date. Due dates may vary from year to year. Applicants are encouraged to verify current deadlines at the website or by contacting Pfizer. Pfizer, Inc. is an American multinational research-based pharmaceutical company that was founded in 1849 by two cousins Charles Pfizer and Charles Erhart. It topped the 2009 list of America's most generous companies, giving $2.3 billion in products and cash. Pfizer's grants and contributions include the following types of support: support for medical, scientific, and patient organizations; lobbying and political contributions; medical education grants; medical and academic partnerships; healthcare charitables; sponsorships; and special events. Particular to the grant opportunity listed on this page, Pfizer offers a competitive research grants program ASPIRE (Advancing Science Through Pfizer-Investigator Research Exchange). ASPIRE grant opportunities are available by geographic area, specifically: worldwide, North America, Europe, Asia, and Australia. Notices of opportunities are posted on Pfizer's competitive grants and regional ASPIRE websites. Prospective applicants may fill out a web-based form at the ASPIRE website to by notified of future competitive grant opportunities via email.

Requirements: Applicants must have a professional degree (M.D., Ph.D., Pharm.D., or equivalent) and reside in the United States. Applications submitted by junior investigators (within 5 years of terminal training) will be afforded preferential consideration. Applicants must have a strong academic career interest in clinical evaluations in the following focus areas and patient populations: the disease states of complicated intra-abdominal infections, complicated skin and skin-structure infection, and community-acquired bacterial pneumonia; immunocompromised patient populations; patient populations older than fifty-five years of age; patient populations suffering from diabetes, renal insufficiency, pulmonary comorbidities, vascular disease, obesity, or thermal injuries; and patient populations that have transplants or that are critically ill. Within the specified patient populations with infection types of interest, investigators are expected to generate exploratory clinical data in patients receiving tigecycline to better understand disease pathogenesis, risk factors, host response and treatment outcomes. It is expected that results will be presented at scientific meetings and published in peer-reviewed journals. ASPIRE applications will be reviewed by an independent, external review committee comprised of medical and scientific experts. Review committee members will consider the following criteria in evaluating proposals: scientific merit of the research proposal; qualifications of the applicant; relevance of the proposal to the ASPIRE program's mission; evidence of the applicant's commitment to an academic research career, and evidence of a suitable research environment.

Restrictions: No other government, non-governmental, or industry-sponsored funding may cover the same work scope as the grant application to the ASPIRE program; however the proposal may be related to other funding as long as there is no direct overlap. The following topics fall outside of the scope of the ASPIRE Antibacterial Research Program: general education and/or training; public health; support for ongoing clinical programs that are part of an organization's routine operations; and proposals that do not evaluate tigecycline. Investigators may consider evaluating additional clinical measures (e.g. length of stay, readmission, ventilator days, etc.). However, proposals seeking to financially quantify outcomes are outside the scope of the program. Studies designed to address how patients / payers use evidence about Pfizer products to make coverage and reimbursement decisions are prohibited.

Geographic Focus: All States

Date(s) Application is Due: Apr 24

Samples: Vincent B. Young, M.D., Ph.D., University of Michigan, Ann Arbor, Michigan - Effects of tigecycline on the murine intestinal microbiome and experimental Clostridium difficile infection; Amee R. Manges, M.P.H., Ph.D., McGill University, Montreal, Quebec, Canada - Antimicrobial use, intestinal microbiota and Clostridium difficile colonization and infection in hospitalized patients; Michael Aldape, Ph.D., Veteran's Administration Medical Center, Boise, Idaho - Targeting the inflammatory response during Clostridium difficile infection with tigecycline.

Contact: Mikael Dolssten, President; (212) 733-2323

Internet: https://www.aspireresearch.org/ASPIRE-AR_US/ASPIRE%20AR%20Web site%20Tab%20Send/tigecycline/index.html

Sponsor: Pfizer
235 East 42nd Street
New York, NY 10017-5755

Pfizer ASPIRE North America Hemophilia Research Awards 3170

Pfizer's ASPIRE Hemophilia Research awards are intended to support basic-science, translational, and clinical research that advances medical knowledge, pathogenesis, and treatment of hemophilia and to support academic research as well as the career development of promising young and established scientists. Projects are expected to yield results that will advance the medical knowledge of Hemophilia A or Hemophilia B. It is expected that results will be presented at scientific meetings and published in peer reviewed journals. Proposed projects should last 1-2 years and cost approximately $125,000/year, inclusive of overhead costs. Funds may be used to cover the awardees' salary and fringe benefits, as well as other direct expenses incurred during the research project. Direct expenses may include the salary of technical associates; the purchase of relevant laboratory supplies; direct expenses necessary for proper conduct of research (e.g. subject-related costs, study-related personnel costs, diagnostic fees/services, data management expenses). Indirect costs may include travel expenses to scientific meetings, software licensing fees, publication costs, etc. Research proposals will be evaluated and selected by an independent, external expert panel comprised of nationally known academic clinicians. Applicants must submit their proposals through the Integrated System for Pfizer Investigator Initiated Research (INSPIIRE) web portal, the link to which is available from the competitive grant announcement on the ASPIRE website (http://www.aspireresearch.org/). Applications must be received by 11:59 p.m. eastern daylight time on the deadline date. Deadlines may vary from year to year. Interested applicants are advised to verify the current deadline date at the website or by contacting Pfizer. Pfizer, Inc. is an American multinational research-based pharmaceutical company that was founded in 1849 by two cousins Charles Pfizer and Charles Erhart. It topped the 2009 list of America's most generous companies, giving $2.3 billion in products and cash. Pfizers grants and contributions include the following types of support: support for medical, scientific, and patient organizations; lobbying and political contributions; medical education grants; medical and academic partnerships; healthcare charitables;

sponsorships; and special events. Particular to the grant opportunity listed on this page, Pfizer offers a competitive research grants program ASPIRE (Advancing Science Through Pfizer-Investigator Research Exchange). ASPIRE grant opportunities are available by geographic area, specifically: worldwide, North America, Europe, Asia, and Australia. Notices of opportunities are posted on Pfizer's competitive grants and regional ASPIRE websites. Prospective applicants may fill out a web-based form at the ASPIRE website to by notified of future competitive grant opportunities via email.
Requirements: Applicants must have a professional degree (M.D., Ph.D., Pharm.D., or equivalent) and reside in the United States. Applicants should consider the following components when developing their research proposals: in vitro, in vivo, and clinical investigations; evaluation of disease pathogenesis and host response; and comparison of host response and clinical impact. Proposals will be evaluated on the following criteria: scientific merit of the research proposal; qualifications of the applicant; relevance of the proposed research to Pfizer's hematology program mission; evidence of the applicant's commitment to an academic research career; and evidence of a suitable research environment.
Restrictions: ASPIRE Hemophilia award winners may apply for awards in the future; awards are based upon scientific merit commensurate with funding that is available. No other government, non-governmental, or industry-sponsored funding may cover the same work scope as the grant application to the ASPIRE program; however the proposal may be related to other funding as long as there is no direct overlap.
Geographic Focus: All States
Date(s) Application is Due: May 14
Amount of Grant: 100,000 - 200,000 USD
Contact: Mikael Dolssten, President; (212) 733-2323
Internet: https://www.aspireresearch.org/ASPIRE-Hemophilia/index.html
Sponsor: Pfizer
235 East 42nd Street
New York, NY 10017-5755

Pfizer ASPIRE North America Narrow Spectrum Antibiotics for Treatment of MRSA Research Awards 3171
While considerable clinical trial data exists for linezolid in a variety of patients with known infections, most of this information is from controlled clinical trials which leave patients with certain disease complexities underrepresented. Pfizer invites investigators to submit innovative clinical research proposals that evaluate linezolid's role in the treatment of infections in complex adult patients within approved indications. Studies with well-designed retrospective analyses are preferred, however small pilot studies that attempt to test hypotheses will be considered as long as the study can be completed within 1 year. Funds may be used to cover the awardees' salary and fringe benefits, as well as other direct expenses incurred during the research project. Direct expenses may include the salary of technical associates; the purchase of relevant laboratory supplies; direct expenses necessary for proper conduct of research (e.g. subject-related costs, study-related personnel costs, diagnostic fees/services, and data management expenses). Indirect costs may include travel expenses to scientific meetings, software licensing fees, publication costs, etc. Downloadable application guidelines are available at the grant website. Applicants must submit their proposals through the Integrated System for Pfizer Investigator Initiated Research (INSPIIRE) web portal, the link to which is available from the competitive grant announcement on the ASPIRE website. Applications must be received by 11:59 p.m. eastern daylight time on the deadline date. Due dates may vary from year to year. Applicants are encouraged to verify current deadlines at the website or by contacting Pfizer. Pfizer, Inc. is an American multinational research-based pharmaceutical company that was founded in 1849 by two cousins Charles Pfizer and Charles Erhart. It topped the 2009 list of America's most generous companies, giving $2.3 billion in products and cash. Pfizer's grants and contributions include the following types of support: support for medical, scientific, and patient organizations; lobbying and political contributions; medical education grants; medical and academic partnerships; healthcare charitables; sponsorships; and special events. Particular to the grant opportunity listed on this page, Pfizer offers a competitive research grants program ASPIRE (Advancing Science Through Pfizer-Investigator Research Exchange). ASPIRE grant opportunities are available by geographic area, specifically: worldwide, North America, Europe, Asia, and Australia. Notices of opportunities are posted on Pfizer's competitive grants and regional ASPIRE websites. Prospective applicants may fill out a web-based form at the ASPIRE website to by notified of future competitive grant opportunities via email.
Requirements: Applicants must have a professional degree (M.D., Ph.D., Pharm.D., or equivalent) and reside in the United States. Applications submitted by junior investigators (within 5 years of terminal training) will be afforded preferential consideration. Applicants must have a strong academic career interest in clinical evaluations of linezolid in the following focus areas and populations: the disease states of nosocomial pneumonia (e.g. HAP, VAP, HCAP), skin and soft tissue infection (e.g. surgical site infection, wound, and complicated cellulitis), and diabetic foot infection without osteomyelitis; patient populations with diabetes, renal insufficiency, pulmonary dysfunction (e.g. COPD), peripheral vascular disease, morbid obesity, and immunosuppression; older adults stratified by age (e.g. 60-70 years, 79-80 years, >80 years); adults stratified by age (e.g. 18-35 years, 36-55 years, 56-72 years, >72 years; the critically ill; pharmakinetics, host immune response; and rate of resolution of signs/symptoms of infection. Within the specified patient populations with infection types of interest, investigators are expected to generate exploratory clinical data to better understand the disease pathogenesis, risk factors, host response, and treatment outcomes. It is expected that results will be presented at scientific meetings and published in peer reviewed journals. ASPIRE applications will be reviewed by an independent, external review committee comprised of medical and scientific experts. Review committee members will consider the following criteria in evaluating proposals: scientific merit of the research proposal; qualifications of the applicant; relevance of the proposal to the ASPIRE program's mission; evidence of the applicant's commitment to an academic research career, and evidence of a suitable research environment.

Restrictions: No other government, non-governmental, or industry-sponsored funding may cover the same work scope as the grant application to the ASPIRE program; however the proposal may be related to other funding as long as there is no direct overlap. The following topics fall outside of the scope of the ASPIRE Antibacterial Research Program: general education and/or training; public health; support for ongoing clinical programs that are part of an organization's routine operations; and proposals that do not evaluate linezolid. Investigators may consider evaluating additional clinical measures (e.g. length of stay, readmission, ventilator days, etc.). However, proposals seeking to financially quantify outcomes are outside the scope of the program. Studies designed to address how patients / payers use evidence about Pfizer products to make coverage and reimbursement decisions are prohibited.
Geographic Focus: All States
Date(s) Application is Due: May 1
Amount of Grant: 50,000 - 100,000 USD
Samples: Ganapathi Parameswaran, M.B., B.S., Veteran's Administration Western New York Healthcare System, Batavia, New York - Methicillin resistant Staphylococcus aureus infection in COPD and effects of linezolid treatment on inflammation; Delores Tseng, Ph.D., Cedars-Sinai Medical Center, Los Angeles, California - Effect of linezolid versus vancomycin on treatment of MRSA sepsis in aged hosts; Michael David, M.D., Ph.D., University of Chicago, Chicago, Illinois - Asymptomatic colonization with S. aureus after therapy with linezolid or clindamycin for acute S. aurus skin and skin structure infections in patients with comorbid conditions: A randomized trial.
Contact: Mikael Dolssten, President; (212) 733-2323
Internet: https://www.aspireresearch.org/ASPIRE-AR_US/ASPIRE%20AR%20Web site%20Tab%20Send/linezolid/index.html
Sponsor: Pfizer
235 East 42nd Street
New York, NY 10017-5755

Pfizer ASPIRE North America Rheumatology Research Awards 3172
Pfizer's ASPIRE Rheumatology Research awards are designed to increase the understanding of health disparities as they relate to rheumatoid arthritis in Hispanic and other underrepresented minority populations. Specifically, proposals should address the following areas: epidemiology and current trends in diagnosis and management; clinical phenotypic differences; and genomic research into the heterogeneity of and within the targeted population. Pfizer will provide $100,000 to fund research conducted over a two-year period. Applicants must submit their proposals through the Integrated System for Pfizer Investigator Initiated Research (INSPIIRE) web portal, the link to which is available from the competitive grant announcement on the ASPIRE website (http://www.aspireresearch.org/). Applications must be received by 11:59 p.m. eastern daylight time on July 10. Pfizer, Inc. is an American multinational research-based pharmaceutical company that was founded in 1849 by two cousins Charles Pfizer and Charles Erhart. It topped the 2009 list of America's most generous companies, giving $2.3 billion in products and cash. Pfizer's grants and contributions include the following types of support: support for medical, scientific, and patient organizations; lobbying and political contributions; medical education grants; medical and academic partnerships; healthcare charitables; sponsorships; and special events. Particular to the grant opportunity listed on this page, Pfizer offers a competitive research grants program ASPIRE (Advancing Science Through Pfizer-Investigator Research Exchange). ASPIRE grant opportunities are available by geographic area, specifically: worldwide, North America, Europe, Asia, and Australia. Notices of opportunities are posted on Pfizer's competitive grants and regional ASPIRE websites. Prospective applicants may fill out a web-based form at the ASPIRE website to by notified of future competitive grant opportunities via email.
Requirements: Applicants must have a professional degree (M.D., Ph.D., Pharm.D., or equivalent) and reside in the United States. Investigators are expected to generate exploratory data to better understand the variance in clinical presentation, severity of illness, response to therapy, and predictors of efficacy and tolerability in the targeted population. ASPIRE review committee members will consider the following criteria in evaluating proposals: scientific merit of the research proposal; qualifications of the applicant; relevance of the proposal to the ASPIRE program's mission; evidence of the applicant's commitment to an academic research career, evidence of a suitable research environment; and evidence of a mentorship program.
Restrictions: ASPIRE Rheumatology Research awards are limited to one award per investigator per lifetime. No other government, non-governmental, or industry-sponsored funding may cover the same work scope as the grant application to the ASPIRE program; however the proposal may be related to other funding as long as there is no direct overlap.
Geographic Focus: All States
Date(s) Application is Due: Jul 10
Amount of Grant: Up to 100,000 USD
Contact: Mikael Dolssten, President; (212) 733-2323
Internet: http://www.pfizer.com/research/investigator/competitive_grant_programs.jsp
Sponsor: Pfizer
235 East 42nd Street
New York, NY 10017-5755

Pfizer ASPIRE Worldwide Endocrine Young Investigator Grants 3173
The mission of the ASPIRE Young Investigator Awards in Endocrine Research is to advance the medical knowledge of the fundamental mechanisms of endocrine disease and the mechanisms of action of current and potential future treatments to improve the care of patients with endocrine disorders. Pfizer invites basic-science, innovative-translational, and clinical-research proposals that evaluate endocrine disease, disease mechanisms, and outcomes, primarily in the growth-hormone insulin-like growth-

factor field. Funding will be provided to research proposals that address understanding disease mechanisms, genetic factors involved in disease development, factors affecting response to therapeutic agents, identification of novel biomarkers, and identification of novel therapeutic targets or therapeutic approaches. Pfizer will fund up to three awards of $50,000 each for one year. Applicants must submit their proposals through the Integrated System for Pfizer Investigator Initiated Research (INSPIIRE) web portal, the link to which is available from the competitive grant announcement on the ASPIRE website (http://www.aspireresearch.org/). Applications must be received by 11:59 p.m. eastern daylight time on June 15. Interested applicants can download an application guide from this location for more detailed information about application requirements. Pfizer, Inc. is an American multinational research-based pharmaceutical company that was founded in 1849 by two cousins Charles Pfizer and Charles Erhart. It topped the 2009 list of America's most generous companies, giving $2.3 billion in products and cash. Pfizer's grants and contributions include the following types of support: support for medical, scientific, and patient organizations; lobbying and political contributions; medical education grants; medical and academic partnerships; healthcare charitables; sponsorships; and special events. Particular to the grant opportunity listed on this page, Pfizer offers a competitive research grants program ASPIRE (Advancing Science Through Pfizer-Investigator Research Exchange). ASPIRE grant opportunities are available by geographic area, specifically: worldwide, North America, Europe, Asia, and Australia. Notices of opportunities are posted on Pfizer's competitive grants and regional ASPIRE websites. Prospective applicants may fill out a web-based form at the ASPIRE website to by notified of future competitive grant opportunities via email.
Requirements: Applicants must hold a professional degree (M.D., Ph.D., Pharm.D., or equivalent), be qualified to carry out the proposed investigation, and still be in the early stages of their career (preferably within ten years of having completed a post-graduate degree). ASPIRE review committee members will consider the following criteria in evaluating proposals: scientific merit of the research proposal; qualifications of the applicant; relevance of the proposal to the ASPIRE program's mission; evidence of the applicant's commitment to an academic research career, evidence of a suitable research environment; and evidence of suitable mentors.
Restrictions: No other government, non-governmental, or industry-sponsored funding may cover the same work scope as the grant application to the ASPIRE program; however the proposal may be related to other funding as long as there is no direct overlap. Due to the Foreign Corrupt Practices Act (FCPA), submissions selected for awards must go through an internal review at Pfizer to ensure provision of the grant is appropriate and that relevant information regarding potentially influencing government official status, as well as beneficiaries and controllers information, is captured, when applicable.
Geographic Focus: All States, All Countries
Date(s) Application is Due: Jun 15
Amount of Grant: 50,000 USD
Samples: Interested applicants may download alphabetical listings of grant recipients per year from Pfizer's Transparency In Grants webpage at http://www.pfizer.com.
Contact: Mikael Dolssten, President; (212) 733-2323
Internet: http://www.pfizer.com/research/investigator/competitive_grant_programs.jsp
Sponsor: Pfizer
235 East 42nd Street
New York, NY 10017-5755

Pfizer Australia Cancer Research Grants 3174
Pfizer Australia Cancer Research Grants were established in 2007 to support and advance clinical research in the field of oncology. Grants are available up to AUD $55,000 (including the Australian Goods and Services Tax) and can be used to cover part salaries, equipment, research materials, and expenses. Proposals should involve clinical or translational research. Translation applications are required to describe how the outcomes from the proposed research project will potentially impact clinical practice, policy, or further research in cancer control. Applications must be submitted via an online application form which can be accessed at the website from January to May. Submission deadlines may vary from year to year. Interested applicants are encouraged to verify current due dates at the website or by contacting Pfizer Australia. All applications are reviewed by two expert referees and Pfizer (for proposals involving Pfizer compounds); final decisions are then made by the Australian Physicians Independent Committee (APIC), composed of eight leading specialists. Pfizer Australia awards up to $1.5 million in grants to support medical graduates conducting research in the areas of neuroscience, oncology, paediatric endocrinology, and related areas. Grants are awarded for 12 months and assist researchers to establish a basis for further research funding from granting bodies such as the National Health and Medical Research Council.
Requirements: Eligible applicants will have entered the field of research, (or returned after an appropriate break) within the last five years of having attained post-graduate research or clinical qualifications (e.g., Ph.D./M.D. or College Fellowship). The principal applicant must be a citizen or permanent resident of Australia. Grant recipients must conduct the majority of the research within Australia. Pfizer Australia Cancer Research Grants are awarded with the expectation that timely submissions for presentation will be made at appropriate medical meetings or research forums and that papers will be submitted for publication in appropriate peer-reviewed medical journals. The grant recipient will be required to provide Pfizer with an opportunity to review any proposed publications.
Restrictions: Previous Cancer Research Grant recipients are ineligible. Only one Pfizer Australia Cancer Research Grant will be allocated to a single applicant in a given year although applications will be accepted and reviewed for more than one project. Applicants are not eligible to receive funding from more than one Pfizer Australia Research Grant program in the same financial year.
Geographic Focus: Australia
Date(s) Application is Due: May 11
Amount of Grant: Up to 55,000 AUD
Samples: Dr. Arun Asad, Peter MacCallum Cancer Centre, Melbourne, Victoria - Genetic targeting of the DNA damage response using a novel siRNA delivery technology; Dr. Fiona Chionh, Austin Hospital, Heidelberg, Victoria - Understanding the molecular basis of inherent and acquired resistance to targeted therapies in metastatic colorectal cancer; Dr. Andrew Moore, Queensland Children's Medical Research Institute, Brisbane, Queensland - The role Survivin and XIAP (X-linked inhibitor of apoptosis protein) as biomarkers and therapeutic targets in paediatric acute myeloid leukaemia.
Contact: Trudy Snape, Grants Coordinator; +61 2 9850 3973
Internet: https://www.pfizergrants.com.au/Grants/CancerResearch.aspx
Sponsor: Pfizer
235 East 42nd Street
New York, NY 10017-5755

Pfizer Australia Neuroscience Research Grants 3175
The Pfizer Australia Neuroscience Research Grants were established in 2002 to support and advance basic and clinical research and to encourage young researchers to enter or continue in the field of neuroscience. Grants are available up to AUD $44,000 (including the Goods and Services Tax) and can be used to cover part salaries, equipment, research materials, and expenses. Applications must be submitted via an online application form which can be accessed on the website between January 1 and May 11. Submission deadlines may vary from year to year; interested applicants are encouraged to verify current due dates by visiting the website or by contacting Pfizer Australia. Applications are reviewed by two expert referees and by Pfizer (for proposals involving Pfizer compounds); final decisions are then made by the Australian Physicians Independent Committee (APIC), composed of eight leading specialists. Pfizer Australia awards up to $1.5 million in grants to support medical graduates conducting research in the areas of neuroscience, oncology, paediatric endocrinology, and related areas. Grants are awarded for 12 months and assist researchers to establish a basis for further research funding from granting bodies such as the National Health and Medical Research Council.
Requirements: The interested applicant must meet the following eligibility criteria: be an Australian citizen or permanent resident; have entered or returned to the field of research within the last five years; and be a medical graduate who has obtained or is in process of obtaining specialist qualifications or a doctorate or who is currently in advanced training. Applications are welcome for research into any aspect of cognitive decline, mood disorders or pain. Proposals should involve clinical research or basic research involving humans. The majority of the research must be conducted within Australia. Pfizer grants are awarded with the expectation that timely submissions for presentation will be made at appropriate medical meetings or research forums and that papers will be submitted for publication in appropriate peer-reviewed medical journals. The grant recipient will be required to provide Pfizer with an opportunity to review any proposed publications.
Restrictions: Applications will not be accepted from established researchers including professors, associate professors, or CIAs with any competitive grant of $50,000 or above, or in receipt of a fellowship of $100,000 or more. Previous successful applicants are not eligible. Only one Neuroscience Research Grant will be allocated to a single applicant in a given year although applications will be accepted and reviewed for more than one project. Applicants will not be eligible to receive funding from more than one Pfizer Australia Research Program in the same financial year.
Geographic Focus: Australia
Date(s) Application is Due: May 11
Amount of Grant: Up to 44,000 AUD
Samples: Dr. Natalie Mills, Queensland Institute of Medical Research, Herston, Queensland - The Role of Cytokines in Depression and Cognition in Adolescents; Dr. Houman Ebrahimi, John Hunter Hospital, New Lambton Heights, New South Wales - The Olfactory Stress Test: association with antecedents of Alzheimers disease in a cognitively normal sample; Dr. Chris Moran, Monash Medical Centre, Melbourne, Victoria - Type 2 Diabetes and pre-clinical dementia: an FDG-PET co-twin study.
Contact: Trudy Snape, Grants Coordinator; +61 2 9850 3973; trudy.snape@pfizer.com
Internet: https://www.pfizergrants.com.au/Grants/NeuroscienceResearch.aspx
Sponsor: Pfizer
235 East 42nd Street
New York, NY 10017-5755

Pfizer Australia Paediatric Endocrine Care Research Grants 3176
The Australia Paediatric Endocrine Care (APEC) research grants were established in 2007 to support and advance clinical research in the field of Paediatric Endocrinology Care and to encourage young researchers to enter or continue in the field. Grants are available up to AUD $55,000 (including the Australian Goods and Services Tax) and can be used to cover part salaries, equipment, research materials, and expenses. Applications must be submitted via an online application form which can be accessed on the website from January 1 - March 23. Submission deadlines may vary from year to year. Interested applicants are encouraged to verify current due dates by visiting the website or by contacting Pfizer Australia. Applications are reviewed by two expert referees and by Pfizer (for proposals involving Pfizer compounds); final decisions are then made by the Australian Physicians Independent Committee (APIC), composed of eight leading specialists. Pfizer Australia awards up to $1.5 million in grants to support medical graduates conducting research in the areas of neuroscience, oncology, paediatric endocrinology, and related areas. Grants are awarded for 12 months and assist researchers to establish a basis for further research funding from granting bodies such as the National Health and Medical Research Council.
Requirements: The interested applicant must meet the following criteria to be eligible to apply: be a medical graduate currently residing, working, and conducting research within Australasia; be registered to practice medicine in Australasia; be under the age of 40 at time of application; and be a current trainee in Paediatric Endocrinology in

Australia or within five years of conferring of FRACP (Fellow of the Royal Australasian College of Physicians) or within five years of accreditation of Paediatric Endocrinology training from the Endocrinology SAC (Specialist Advisory Committee) of the Royal Australian College of Physicians. Pfizer grants are awarded with the expectation that timely submissions for presentation will be made at appropriate medical meetings or research forums and that papers will be submitted for publication in appropriate peer-reviewed medical journals. The grant recipient will be required to provide Pfizer with an opportunity to review any proposed publications.
Restrictions: Awardees must remain in Australasia for the duration of the study.
Geographic Focus: All Countries
Date(s) Application is Due: Mar 23
Amount of Grant: Up to 55,000 AUD
Samples: Dr. Andrew Biggin, Children's Hospital, Westmead, New South Wales - The mutational events leading to osteogenesis imperfecta, its genotype-phenotype correlations and response to biphosphonate therapy in the paediatric population; Dr. Vinutha Shetty, Princess Margaret Hospital for Children, Perth, Western Australia - Improving guidelines to allow children with type 1 diabetes to exercise safely; Dr. Stephanie Johnson, Mater Children's Hospital, Brisbane, Queensland - Evaluating the efficacy of Next Generation DNA sequencing in the diagnosis of disorders of beta cell function.
Contact: Trudy Snape, Grants Coordinator; +61 2 9850 3973
Internet: https://www.pfizergrants.com.au/grants/PaediatricEndocrine.aspx
Sponsor: Pfizer
235 East 42nd Street
New York, NY 10017-5755

PFizer Compound Transfer Agreements   3177
Pfizer has a pure-powder-transfer program encompassing all of the company's publicly-known compounds and routinely enters Compound Transfer Agreements (CTAs) that provide compounds free of charge to third parties to conduct agreed-upon studies. To request a Pfizer pure-substance compound, applicants should complete the submission form located on the Integrated System for Pfizer Investigator Initiated Research (INSPIIRE) web portal, the link to which is available on the Compound-Transfer-Program web page. Processing time from initial submission to shipment of the approved compound(s) takes approximately 12 weeks. Applicants will receive an auto-reply email confirming their submission has been completed and can track the status of their submissions on the INSPIIRE portal. Pfizer, Inc. is an American multinational research-based pharmaceutical company that was founded in 1849 by two cousins Charles Pfizer and Charles Erhart. It topped the 2009 list of America's most generous companies, giving $2.3 billion in products and cash. Most of Pfizer's philanthropic activities are managed by its Corporate Responsibility Department through two foundations, the Pfizer Foundation and the Pfizer Patient Assistance Foundation. Pfizer's philanthropic activities are generally divided among its global health programs, its various grants and contributions programs, and its work with health care professionals. Pfizer's grants and contributions include the following types of support: support for medical, scientific, and patient organizations; lobbying and political contributions; medical education grants; medical and academic partnerships; healthcare charitables; sponsorships; and special events.
Requirements: An attachment with a full description of the specific research study proposed should be included with the applicant's on-line submission. Pfizer reviews each request for content, value, and consistency with Pfizer's direction for the requested compound and reserves the right to deny any request the reviewers feel does not advance the compound(s) or add value to human or animal health. Pfizer requires a manuscript or report of the conclusions to bring closure to the study.
Restrictions: Studies involving humans or requesting funding support should be submitted through Pfizer's Investigator Initiated Research program.
Geographic Focus: All States, All Countries
Samples: Interested applicants may download alphabetical listings of grant recipients per year from Pfizer's Transparency In Grants webpage at http://www.pfizer.com.
Contact: Mikael Dolssten, President; (212) 733-2323
Internet: http://www.pfizer.com/research/rd_works/compound_transfer_program.jsp
Sponsor: Pfizer
235 East 42nd Street
New York, NY 10017-5755

Pfizer Global Investigator-Initiated Research Grants   3178
The mission and purpose of Pfizer's Investigator-Initiated Research (IIR) program is to provide support for investigator-initiated research that advances medical and scientific knowledge about Pfizer products and that generates promising medical interventions. This global program is open to all researchers who are interested in conducting their own research. Pfizer support is typically provided in the form of funding and/or drugs (either formulated or pure compound). Pfizer accepts concept submissions for IIR grant requests. If a concept submission is of interest, Pfizer will issue the applicant a follow-up request for a full submission. Requests are submitted to Pfizer through the Integrated System for Pfizer Investigator Initiated Research (INSPIIRE) web portal, the link to which is available on the grant web page. Prior to submitting a grant request, applicants should contact a Pfizer Regional Medical Research Specialist (RMRS), a Pfizer Country Medical Representative, or other Pfizer Medical personnel for more information. Applicants may use the email address provided in the contact section of this page to request the appropriate Pfizer contact information. Interested applicants can download a program brochure from the grant web page to obtain detailed information about application requirements. Pfizer, Inc. is an American multinational research-based pharmaceutical company that was founded in 1849 by two cousins Charles Pfizer and Charles Erhart. It topped the 2009 list of America's most generous companies, giving $2.3 billion in products and cash. Most of Pfizer's philanthropic activities are managed by its Corporate Responsibility Department through two foundations, the Pfizer Foundation and the Pfizer Patient Assistance Foundation. Pfizer's philanthropic activities are generally divided among its global health programs, its various grants and contributions programs, and its work with health care professionals. Pfizer's grants and contributions include the following types of support: support for medical, scientific, and patient organizations; lobbying and political contributions; medical education grants; medical and academic partnerships; healthcare charitables; sponsorships; and special events.
Requirements: Following are types of research eligible for support: clinical studies of approved and unapproved uses involving approved or unapproved Pfizer drugs; observational studies, such as epidemiology studies and certain outcomes research studies where the primary focus is the scientific understanding of disease; other types of independent research on disease states, including novel diagnostic screening tools and surveys where Pfizer has no direct commercial interest; and in vitro or animal studies which include a request for funding. Applicants should note that Pfizer requires the following documents before it can initiate support (either drug and/or funding) of an IIR request: internal review board (IRB)/ ethics committee approval; regulatory response documentation - Investigational New Drug (IND) or Compound Transfer Agreement (CTA) documentation if applicable, a fully executed IIR agreement, and a final study protocol.
Restrictions: Pfizer will not support the following types of requests: requests for ongoing or new research without an associated study protocol or synopsis; general education and training activities; support for ongoing clinical programs that are part of an organization's routine operations; start-up funds to establish new clinical or research programs or to expand existing programs; purchases of capital equipment unrelated to the study or that would generate revenue; construction funds to build new facilities; and hiring of staff that are not dedicated to the study.
Geographic Focus: All States, All Countries
Samples: Interested applicants may download alphabetical listings of grant recipients per year from Pfizer's Transparency In Grants webpage at http://www.pfizer.com.
Contact: Mikael Dolssten, President; (212) 733-2323
Internet: http://www.pfizer.com/research/investigator/investigator_initiated_research.jsp
Sponsor: Pfizer
235 East 42nd Street
New York, NY 10017-5755

Pfizer Global Research Awards for Nicotine Independence   3179
The mission of Pfizer's Global Research Awards for Nicotene Independence (GRAND) Program is to advance the understanding of the mechanisms of tobacco and nicotine dependence and its treatment. Projects should yield results that merit first submission as abstracts to scientific meetings and, subsequently, publication in peer-reviewed journals. Pfizer intends to fund at least five grants in amounts between $50,000 and $200,000. Funds from the GRAND program may be used to support the applicant's salary and fringe benefits, technical salaries, and supplies. Applicants must submit their proposals through the Integrated System for Pfizer Investigator Initiated Research (INSPIIRE) web portal, the link to which is available from the competitive-grant-announcement web page. Applications must be received by 11:59 p.m. eastern daylight time on June 15. Interested applicants can also download an application guide from the announcement web page for more detailed information about application requirements. Pfizer, Inc. is an American multinational research-based pharmaceutical company that was founded in 1849 by two cousins Charles Pfizer and Charles Erhart. It topped the 2009 list of America's most generous companies, giving $2.3 billion in products and cash. Pfizer's grants and contributions include the following types of support: support for medical, scientific, and patient organizations; research; lobbying and political contributions; medical education grants; medical and academic partnerships; healthcare charitables; sponsorships; and special events.
Requirements: Applicants holding an M.D., a Ph.D., or equivalent degree are eligible to apply. Research projects should aim to provide information that could directly improve the use of pharmacotherapy in clinical practice. The GRAND review committee will consider the following criteria in evaluating proposals: scientific merit of the research proposal; qualifications of the applicant; relevance of the proposed research to the advancement of the treatment of nicotine dependence; evidence of a suitable research environment; and feasibility of the project. Prior to submitting an application via the INSPIIRE portal, applicants should notify the GRAND coordinator via the current email address given in the guidelines (at this writing, applications@grand2012.org) so that the GRAND coordinator can expedite and track the processing of the application.
Restrictions: The following topics fall outside of the scope of the GRAND Program in the current year: animal studies; genetics; epidemiology (unrelated to pharmacotherapy, e.g. harm and prevalence of dependence); tobacco policy; public health; community interventions; economic and social factors; and educational programs. No other government, non-governmental, or industry-sponsored funding may cover the same work scope as the grant application to the GRAND program; however the proposal may be related to other funding as long as there is no direct overlap. If the conduct of the research results in any invention or discovery by the awardee that relates to a Pfizer product, the awardee will grant to Pfizer a perpetual, royalty-free worldwide, non-exclusive license to each such invention.
Geographic Focus: All States, All Countries
Date(s) Application is Due: Jun 15
Amount of Grant: 50,000 - 200,000 USD
Contact: Mikael Dolssten, President; (212) 733-2323
Internet: http://www.pfizer.com/research/investigator/competitive_grant_programs.jsp
Sponsor: Pfizer
235 East 42nd Street
New York, NY 10017-5755

Pfizer Healthcare Charitable Contributions 3180
Pfizer's healthcare charitables contributions support the following types of requests/projects: patient education (including health screening); patient advocacy for disease awareness; and improving patent access to care. Funding is currently available in the following clinical areas to support healthcare charitable contributions: arthritis & pain management; Gaucher disease; growth disorders; hemophilia; infectious disease (bacterial, pneumococcal disease prevention); neurology; amyloidosis; dementia; diabetic peripheral neuropathy; fibromyalgia; multiple sclerosis; oncology; breast cancer; gastrointestinal stromal tumors; leukemia (CML and ALL); lung cancer; non-Hodgkin's lymphoma; pancreatic neuroendocrine tumors; renal cell carcinoma; pulmonary arterial hypertension; respiratory (COPD, smoking cessation); rheumatoid arthritis; transplantation (kidney transplant); urology (overactive bladder); and women's health (menopause, cardiovascular risk, depression, bladder health, and sexual health). Applicants have four opportunities during the year to apply for Pfizer's Healthcare Charitable Contributions: from December 1 through January 15 for activities starting in April; March 1 through April 15 for activities starting in July; June 1 through July 15 for activities starting in October; and September 1 through October 15 for activities starting in January. Interested applicants must apply through the Pfizer Grants Application System. Applicants can register and log into the system from the grant webpage. Pfizer, Inc. is an American multinational research-based pharmaceutical company that was founded in 1849 by two cousins Charles Pfizer and Charles Erhart. It topped the 2009 list of America's most generous companies, giving $2.3 billion in products and cash. Most of Pfizer's philanthropic activities are managed by its Corporate Responsibility Department through two foundations, the Pfizer Foundation and the Pfizer Patient Assistance Foundation. Pfizer's philanthropic activities are generally divided among its global health programs, its various grants and contributions programs, and its work with health care professionals. Pfizer's grants and contributions include the following types of support: support for medical, scientific, and patient organizations; lobbying and political contributions; medical education grants; medical and academic partnerships; healthcare charitables; sponsorships; and special events.
Requirements: 501(c)3 organizations are eligible to apply.
Restrictions: Pfizer must not receive any significant value in terms of good or services in return for their donation.
Geographic Focus: All States, All Countries
Date(s) Application is Due: Jan 15; Apr 15; Jul 15; Oct 15
Samples: Interested applicants may download alphabetical listings of grant recipients per year from Pfizer's Transparency In Grants webpage at http://www.pfizer.com.
Contact: Caroline Roan; (866) 634-4647 or (212) 209-8997; healthcharitables@pfizer.com
Internet: https://www.pfizerhealthcharitables.com/
Sponsor: Pfizer
235 East 42nd Street
New York, NY 10017-5755

Pfizer Inflammation Competitive Research Awards (UK) 3181
The Pfizer Inflammation Competitive Research Programme (I-CRP) seeks innovative research proposals in the field of inflammation which may further the understanding of certain rheumatology and dermatology conditions and contribute to excellence in patient care. Pfizer anticipates making non-renewable awards of approximately £50,000 - £100,000 each for research to be undertaken within a time period of 12-18 months. Prior to submitting an application, applicants are requested to have a "medical to medical conversation" with their Pfizer representative. Applicants can email iir@pfizer.com to obtain the appropriate Pfizer contact information for their region. Downloadable application guidelines are available at the grant website. Applications must be submitted to Pfizer through the company's global investigator-initiated research (INSPIIRE) web portal. Applicants should click on "Submit an IIR Request" at that site and follow the online instructions. First-time applicants must "Create an Account". Applications must be received by the deadline date. (Due dates may vary from year to year; applicants are encouraged to verify current due dates at the website or by contacting Pfizer.) Pfizer, Inc. is an American multinational research-based pharmaceutical company that was founded in 1849 by two cousins Charles Pfizer and Charles Erhart. It topped the 2009 list of America's most generous companies, giving $2.3 billion in products and cash. Pfizer's grants and contributions include the following types of support: support for medical, scientific, and patient organizations; research; lobbying and political contributions; medical education grants; medical and academic partnerships; healthcare charitables; sponsorships; and special events.
Requirements: Clinical practitioners and scientists working in the United Kingdom who hold a medical or pharmacy degree or a Ph.D., do not directly work for Pfizer or any other pharmaceutical company, and are not members of the current I-CRP Grant Awarding Committee are eligible to apply. Application should focus on translational (clinically-oriented) research that will impact patient outcomes and relate to an area aligned to Pfizer's areas of therapeutic focus: Rheumatoid Arthritis, Psoriatic Arthritis, Anklylosing Spondylitis, Psoriasis, or Juvenile Idiopathic Arthritis. (Proposals outside these high-interest areas will also be considered but only supported if deemed of exceptional scientific merit.) Studies are not necessarily expected to involve treatment with a Pfizer medication. It is expected that results will be presented at scientific meetings and published in peer-reviewed journals. All applications will be formally reviewed by an independent committee of European medical experts in rheumatology and dermatology. Projects that are awarded funding through I-CRP may also be eligible for National Institute for Health Research (NIHR)'s Clincial Research Network (CRN) support.
Restrictions: Pfizer reserves the right not to support projects that fall outside of Pfizer's internal compliance policies and procedures or those that replicate existing research. Pfizer is unable to fund direct comparative-treatment studies primarily involving competitor products, studies where the risk/benefit ratio to patients is not obviously positive, or studies that involve use of compounds in development. Due to the competitive nature of the awards, Pfizer cannot provide any additional funding and/or drug support beyond what has been requested and approved by the external review panel. Current Grant Awarding Committee members cannot be an applicant or co-applicant on a proposal submitted for the programme, nor be an author on any publication ensuing from the award. Committee members will not participate in the review of any proposal that poses a conflict of interest. Proposals received from current Grant Awarding Committee members' institutions will still be allowed but the scoring and review decisions of these proposals will be made by the committee excluding the potentially conflicted member.
Geographic Focus: United Kingdom
Date(s) Application is Due: Apr 12
Amount of Grant: 50,000 - 100,000 GBP
Contact: Mikael Dolssten, President; (212) 733-2323
Internet: http://www.pfizer.com/research/investigator/competitive_grant_programs.jsp
Sponsor: Pfizer
235 East 42nd Street
New York, NY 10017-5755

Pfizer Medical Education Track One Grants 3182
The mission of the Pfizer medical-education-grants program is to cooperate with healthcare delivery organizations and professional associations to narrow professional-practice gaps through support of continuing education, learning, and continuous-improvement strategies that result in measurable improvement in competence, performance, or patient outcomes. Applicants may apply for two types of medical-education grants: healthcare quality improvement (track one) grants; and annual meetings for purposes of emerging science/knowledge exchange (track two) grants. For track one the Pfizer Medical Education Group (MEG) will identify, on an annual basis, public health concerns (for which education is likely to improve patient care) and issue Requests for Proposals (RFPs) for training, continuing education, and/or certification in these areas. (As of this update, identified priorities are vaccines, oncology, smoking cessation, pain and inflammation, infectious disease, and women's health.) MEG will post the RFPs at the website and also disseminate them through email to all registered organizations. Application must be made through MEG's online Grant Management System (GMS) which is available at the website. To be considered, applicants must first register with the GMS. Only applicants whose registration has been approved by MEG will be able to submit a letter of intent (LOI) in response to an RFP. From the submitted LOIs, MEG will select semi-finalists who will be invited to submit full proposals. Each published RFP will include submission timelines and a Letter of Intent template. Interested applicants are encouraged to visit the Pfizer website and register with the GMS to receive notice of new RFPs. Pfizer, Inc. is an American multinational research-based pharmaceutical company that was founded in 1849 by two cousins Charles Pfizer and Charles Erhart. It topped the 2009 list of America's most generous companies, giving $2.3 billion in products and cash. Most of Pfizer's philanthropic activities are managed by its Corporate Responsibility Department through two foundations, the Pfizer Foundation and the Pfizer Patient Assistance Foundation. Pfizer's philanthropic activities are generally divided among its global health programs, its various grants and contributions programs, and its work with health care professionals. Pfizer's grants and contributions include the following types of support: support for medical, scientific, and patient organizations; lobbying and political contributions; medical education grants; medical and academic partnerships; healthcare charitables; sponsorships; and special events.
Requirements: Types of organizations eligible to apply for grants include hospitals, academic medical centers, schools of nursing or pharmacy, and professional societies and associations.
Geographic Focus: All States, All Countries
Samples: Interested applicants may download alphabetical listings of grant recipients per year from Pfizer's Transparency In Grants webpage at http://www.pfizer.com.
Contact: Caroline Roan; (866) 634-4647 or (212) 209-8997; healthcharitables@pfizer.com
Internet: www.pfizermededgrants.com
Sponsor: Pfizer
235 East 42nd Street
New York, NY 10017-5755

Pfizer Medical Education Track Two Grants 3183
The mission of the Pfizer medical-education-grants program is to cooperate with healthcare delivery organizations and professional associations to narrow professional-practice gaps through support of continuing education, learning, and continuous-improvement strategies that result in measurable improvement in competence, performance, or patient outcomes. Applicants may apply for two types of medical-education grants: healthcare quality improvement (track one) grants; and annual meetings for purposes of emerging science/knowledge exchange (track two) grants. The purpose of track-two grants is to support the live presentation of original research and the exchange of emerging clinical information. Eligible providers can request funding to support live (fact-to-face) activities at major national and regional conferences and congresses in the following clinical areas: amyloidosis (familial amyloid polyneuropathy); cardiovascular metabolic risk (lipids, diabetes); chronic obstructive pulmonary disease; cough and cold; diabetic peripheral neuropathy; dietary supplements; fibromyalgia; Gaucher disease; growth disorders; hemophilia; infectious disease; multiple sclerosis; oncology (breast cancer, gastrointestinal stromal tumors; leukemia, lung cancer, non-Hodgkin's lymphoma, pancreatic neuroendocrine tumors, and renal cell carcinoma); pain; pulmonary arterial hypertension; rheumatoid arthritis; smoking cessation; transplantation (kidney); urology (overactive bladder); women's health (menopause); and support for the Continuing Medical Education (CME)/Continuing Education (CE)/Continuing Professional Development (CPD) profession. Track two grants have five submission cycles annually: January 1 through February 15 for activities starting in April; March 1 through April 15 for activities starting in June; May 1 through June 15 for activities starting in August; July 1 through August 15 for activities starting in October; and September 1 through October

15 for activities starting in December. Regional meeting requests are capped at $25,000 and national meeting requests have a $50,000 cap. Application must be made through the Medical Education Group's online Grant Management System (GMS) which is also available at the website. Applicants must first register themselves in the GMS to create an account. Pfizer, Inc. is an American multinational research-based pharmaceutical company that was founded in 1849 by two cousins Charles Pfizer and Charles Erhart. It topped the 2009 list of America's most generous companies, giving $2.3 billion in products and cash. Most of Pfizer's philanthropic activities are managed by its Corporate Responsibility Department through two foundations, the Pfizer Foundation and the Pfizer Patient Assistance Foundation. Pfizer's philanthropic activities are generally divided among its global health programs, its various grants and contributions programs, and its work with health care professionals. Pfizer's grants and contributions include the following types of support: support for medical, scientific, and patient organizations; lobbying and political contributions; medical education grants; medical and academic partnerships; healthcare charitables; sponsorships; and special events.
Requirements: Eligibility is based on the following criteria: whether or not the activity aligns with Pfizer's posted areas of interest (see Description); whether the activity is a live (face-to-face) annual activity which serves as a platform for the exchange of new clinical and scientific information and reaches a national or regional audience; whether the activity was developed by the applicant organization specifically for its professional members or constituents; whether the activity is an established part of the applicant organization's ongoing educational program; and whether the activity (if new) is clearly based on an assessment of the educational needs of the applicant's target audience.
Restrictions: Only one request per annual meeting will be accepted.
Geographic Focus: All States, All Countries
Date(s) Application is Due: Feb 15; Apr 15; Jun 15; Aug 15; Oct 15
Amount of Grant: 25,000 - 50,000 USD
Samples: Interested applicants may download alphabetical listings of grant recipients per year from Pfizer's Transparency In Grants webpage at http://www.pfizer.com.
Contact: Caroline Roan; (866) 634-4647 or (212) 209-8997; healthcharitables@pfizer.com
Internet: http://www.pfizermededgrants.com
Sponsor: Pfizer
235 East 42nd Street
New York, NY 10017-5755

Pfizer Research Initiative Dermatology Grants (Germany) 3184
Pfizer GmbH seeks proposals to study the effectiveness and safety of treating dermatologic diseases with TNF-blockade. Downloadable application guidelines (in German) are available from the grant web page. Interested applicants are encouraged to contact the Medical Science Liason for more information. Pfizer, Inc. is an American multinational research-based pharmaceutical company that was founded in 1849 by two cousins Charles Pfizer and Charles Erhart. It topped the 2009 list of America's most generous companies, giving $2.3 billion in products and cash. Pfizer's grants and contributions include the following types of support: support for medical, scientific, and patient organizations; research; lobbying and political contributions; medical education grants; medical and academic partnerships; healthcare charitables; sponsorships; and special events. Pfizer, Inc. is an American multinational research-based pharmaceutical company that was founded in 1849 by two cousins Charles Pfizer and Charles Erhart. It topped the 2009 list of America's most generous companies, giving $2.3 billion in products and cash. Pfizer's grants and contributions include the following types of support: support for medical, scientific, and patient organizations; research; lobbying and political contributions; medical education grants; medical and academic partnerships; healthcare charitables; sponsorships; and special events.
Geographic Focus: Germany
Date(s) Application is Due: Aug 1
Amount of Grant: Up to 60,000 EUR
Contact: Dr. Ekkehard Lange; +49 (0) 30 550055-52975; ekkehard.lange@pfizer.com
Internet: http://www.pfizer.com/research/investigator/competitive_grant_programs.jsp
Sponsor: Pfizer
235 East 42nd Street
New York, NY 10017-5755

Pfizer Research Initiative Rheumatology Grants (Germany) 3185
Pfizer GmbH seeks proposals to study significant or pathophysiological issues in treating rheumatic diseases with TNF-alpha blockade. Downloadable application guidelines (in German) are available from the grant web page. Interested applicants are encouraged to contact the Medical Science Liason for more information. Pfizer, Inc. is an American multinational research-based pharmaceutical company that was founded in 1849 by two cousins Charles Pfizer and Charles Erhart. It topped the 2009 list of America's most generous companies, giving $2.3 billion in products and cash. Pfizer's grants and contributions include the following types of support: support for medical, scientific, and patient organizations; research; lobbying and political contributions; medical education grants; medical and academic partnerships; healthcare charitables; sponsorships; and special events. Pfizer, Inc. is an American multinational research-based pharmaceutical company that was founded in 1849 by two cousins Charles Pfizer and Charles Erhart. It topped the 2009 list of America's most generous companies, giving $2.3 billion in products and cash. Pfizer's grants and contributions include the following types of support: support for medical, scientific, and patient organizations; research; lobbying and political contributions; medical education grants; medical and academic partnerships; healthcare charitables; sponsorships; and special events.
Geographic Focus: Germany
Date(s) Application is Due: Aug 1
Amount of Grant: Up to 60,000 EUR
Contact: Dr. Ekkehard Lange; +49 (0) 30 550055-52975; ekkehard.lange@pfizer.com
Internet: http://www.pfizer.com/research/investigator/competitive_grant_programs.jsp
Sponsor: Pfizer
235 East 42nd Street
New York, NY 10017-5755

Pfizer Special Events Grants 3186
Pfizer, Inc. provides limited support to external, independent, not-for-profit organizations for special events, e.g. fundraising dinners, walks, biking and golf events, galas, awards ceremonies, and other similar events that do not provide Pfizer with a tangible benefit. Requests must be submitted at least sixty days prior to the event. Applicants must submit their requests via email to the contact information given. Interested applicants can download a special event brochure from the grant webpage to obtain detailed information about application requirements, processes, and materials. Pfizer, Inc. is an American multinational research-based pharmaceutical company that was founded in 1849 by two cousins Charles Pfizer and Charles Erhart. It topped the 2009 list of America's most generous companies, giving $2.3 billion in products and cash. Most of Pfizer's philanthropic activities are managed by its Corporate Responsibility Department through two foundations, the Pfizer Foundation and the Pfizer Patient Assistance Foundation. Pfizer's philanthropic activities are generally divided among its global health programs, its various grants and contributions programs, and its work with health care professionals. Pfizer's grants and contributions include the following types of support: support for medical, scientific, and patient organizations; lobbying and political contributions; medical education grants; medical and academic partnerships; healthcare charitables; sponsorships; and special events.
Requirements: External, independent, not-for-profit organizations - e.g. patient advocacy groups, professional medical associations, trade associations and other charitable organizations with a 501(c)3 or similar status, are eligible to apply. Events must provide broad public benefit and primarily benefit patient care, advance medical science, or otherwise align with Pfizer's policy or business goals.
Restrictions: Pfizer will not provide funding for the following types of activities: special events that have already occurred; activities aimed at improperly influencing any entity, including but not limited to healthcare professionals and government officials; activities to improperly influence prescribing, formulary positioning, or recommendation of Pfizer products; activities that may result in off-label promotion of Pfizer products; activities that would undermine in any way an organization's independence; activities offered as quid pro quo for service or support; capital support for an organization, including building costs or other similar requests related to "start up" costs; proposals from individual persons; and non-research proposals from individual healthcare professionals or group practices.
Geographic Focus: All States, All Countries
Contact: Sally Susman, Executive Vice President, Policy, External Affairs, and Communications; (212) 733-2323; publicaffairssupport@pfizer.com
Internet: http://www.pfizer.com/responsibility/grants_contributions/special_events.jsp
Sponsor: Pfizer
235 East 42nd Street
New York, NY 10017-5755

Pharmacia-ASPET Award for Experimental Therapeutics 3187
The Pharmacia-ASPET Award in Experimental Therapeutics is given annually to recognize and stimulate outstanding research in pharmacology and experimental therapeutics - basic laboratory or clinical research that has had, or potentially will have, a major impact on the pharmacological treatment of disease. The winner will receive a $2,500 honorarium, a plaque, hotel and economy airfare for the winner and spouse to the award ceremony at the ASPET annual meeting.
Requirements: Nominations must be made by ASPET members; however, the nominee need not be a member. No restrictions exist as to age or institutional affiliation of nominee.
Geographic Focus: All States
Date(s) Application is Due: Sep 15
Amount of Grant: 2,500 USD
Contact: Christine Carrico; (301) 634-7060; fax (301) 634-7061; ccarrico@aspet.org
Internet: http://www.aspet.org/public/awards/exp_ther_award.html
Sponsor: Pfizer / American Society for Pharmacology and Experimental Therapeutics
9650 Rockville Pike
Bethesda, MD 20814-3995

Phelps County Community Foundation Grants 3188
The Phelps County Community Foundation's grants program provides a means by which not-for-profit charitable organizations may secure financial assistance for projects and programs which will enhance the quality of life for residents of Phelps County, Nebraska. The foundation awards grants in the areas of education, culture, human services, health and recreation, and community. Priority is given to seed grants to initiate promising new projects or programs, programs representing innovative and efficient approaches to serving community needs and opportunities, challenge grants, organizations that work cooperatively with other community agencies, projects or programs where a moderate amount of grant money can effect a significant result, and projects or programs that enlist volunteer participation and citizen involvement. Types of support include general operating grants, continuing support grants, building construction/renovation, equipment acquisition, program development, publication, seed money, scholarship funds, and matching grants. Applicants must have a plan for future funding and support from other sources.
Requirements: Grants are made to 501(c)3 nonprofit organizations in Phelps County, Nebraska, and sometimes to governmental agencies for capital expenditures and/or capital improvements within Phelps County.

Restrictions: Grants are not made to individuals, to support political activities, to support operating expenses of well-established organizations or public service agencies, to establish new endowment funds, for travel or related expenses for individuals or groups, for operating support of governmental agencies, to religious groups for religious purposes, to profit-making enterprises, or to agencies serving a populace outside of Phelps County. In addition, grants are not made to support annual fund drives or to eliminate previously incurred deficits.
Geographic Focus: Nebraska
Date(s) Application is Due: Apr 1; Oct 1
Amount of Grant: Up to 25,000 USD
Samples: American Legion Baseball, Holdrege, Nebraska, $22,000 - new concession stand; Bertrand Nursing Home, Bertrand, Nebraska, $7,632 - operating expenses.
Contact: Vickie Klein; (308) 995-6847; fax (308) 995-2146; vlpccf@phelpsfoundation.org
Internet: http://www.phelpsfoundation.org/grants.html
Sponsor: Phelps County Community Foundation
504 4th Avenue
Holdrege, NE 68949

Philadelphia Foundation Organizational Effectiveness Grants 3189
Organizational Effectiveness Grants are designed to strengthen and improve the business and operational practices of nonprofit organizations in Bucks, Chester, Delaware, Montgomery and Philadelphia counties. As in business, these practices are critical to the success of nonprofits' ability to lead, adapt, manage and effectively operate their organizations. These core competencies are critical to nonprofits at all stages of development and size for the effective delivery of services, implementation of best practices, utilization of data to drive decision-making and the ability to adapt to new demands, risks and opportunities. new applications for General Operating Support or Organizational Effectiveness cannot be accepted until January 1st.
Requirements: IRS 501(c)3 tax-exempt organizations located in Bucks, Chester, Delaware, Montgomery, and Philadelphia Counties of Pennsylvania are eligible.
Restrictions: Grants are rarely made to affiliates of national or international organizations, government agencies, organizations not located in Southeastern Pennsylvania, organizations with budgets of more than $1.5 million, private schools, or umbrella-funding organizations. Requests usually are denied for capital campaigns, conferences, deficit financing, endowments, publications, research projects, tours, and trips. Individuals are ineligible.
Geographic Focus: Pennsylvania
Amount of Grant: 3,000 - 30,000 USD
Contact: Libby Walsh, Program Associate; (215) 563-6417; fax (215) 563-6882; lwalsh@philafound.org or oeapplications@philafound.org
Internet: https://www.philafound.org/ForNonprofits/DiscretionaryGrantmaking/OrganizationalEffectiveness/tabid/238/Default.aspx
Sponsor: Philadelphia Foundation
1234 Market Street, Suite 1800
Philadelphia, PA 19107

Philip L. Graham Fund Health and Human Services Grants 3190
The Philip L. Graham Fund awards Health and Human Services grants to organizations in the Washington, D.C., metropolitan area, including Maryland and Virginia. The Health and Human Services segment of the Fund's giving portfolio focuses on organizations serving those in greatest need in the community. Organizations providing shelter, food, medical care, and counseling to low income members of our community remain a high priority for the Fund. Interested partied must submit a Letter of Inquiry (LOI) online prior to each submission deadline. The annual LOI deadlines are March 16, June 29, and December 2. Recent awards have ranged from $10,000 to $200,000.
Requirements: Applicants are required to submit a letter of inquiry through the Fund's online application system before one of three deadline dates. Organizations must be a tax-exempt 501(c)3 organization to apply and located within the greater Washington D.C. metropolitan area.
Restrictions: The Fund does not accept proposals from hospitals. Proposals for the following purposes are not considered: advocacy or litigation; research; endowments; special events, conferences workshops or seminars; travel expenses; annual giving campaigns, benefits or sponsorships; courtesy advertising; and production of films or publications. Also, independent schools, institutions of post-secondary education, national or international organizations, and hospitals are not eligible to apply. Grants are also not made to: individuals; religious, political or lobbying activities; to membership organizations; or to any organization that has received a grant from the Fund within the previous thirty-six months.
Geographic Focus: District of Columbia, Maryland, Virginia
Date(s) Application is Due: Apr 1; Jul 15; Dec 15
Amount of Grant: 10,000 - 200,000 USD
Contact: Eileen F. Daly; (202) 334-6640; fax (202) 334-4498; plgfund@ghco.com
Internet: http://www.plgrahamfund.org/content/interest-areas/health-human-services
Sponsor: Philip L. Graham Fund
1300 North 17th Street, Suite 1700
Arlington, VA 22209

Phi Upsilon Omicron Alumni Research Grant 3191
Phi Upsilon Omicron National Honor Society was founded in 1910 at the University of Minnesota in the area of Family and Consumer Sciences. The purpose of the society's Alumni Research Grant is to support post-graduate research in family and consumer sciences. Proposed research should contribute to the knowledge base of family and consumer sciences or further the purposes of Phi Upsilon Omicron, especially in regard to academic excellence, leadership development, and/or professional service. The grant is awarded on even-numbered years. Application forms may be obtained from the Society's National Office (see contact section on this page). The application packet must be postmarked no later than November 1. Applicants will be notified of the disposition of their application following the meeting of the Alumni Research Grant Advisory Committee.
Requirements: One or more of the following entities are eligible to apply: individual society alumni; society alumni committees or task forces; and alumni chapters.
Geographic Focus: All States
Date(s) Application is Due: Nov 1
Amount of Grant: 2,500 USD
Contact: Susan Rickards, Executive Director; (304) 368-0612; rickards@phiu.org or info@phiu.org
Internet: http://phiu.org/awards.htm
Sponsor: Phi Upsilon Omicron, Inc.
P.O. Box 329
Fairmont, WV 26555

Phi Upsilon Omicron Florence Fallgatter Distinguished Service Award 3192
Phi Upsilon Omicron National Honor Society was founded in 1910 at the University of Minnesota in the area of Family and Consumer Sciences. The society's Florence Fallgatter Distinguished Service Award is granted biennially to a qualified alumni member of the society for outstanding achievement in family and consumer sciences. The recipient's expenses to attend the society's biennial meeting will be paid, and the member will be recognized with a plaque at the meeting. The nomination deadline is November 1. Interested applicants should contact the society for information on nomination procedures.
Requirements: A nominee must exhibit the following characteristics: be an alumni member of the society; have finished his or her bachelor's degree no less than ten years ago; have demonstrated excellence in business, dietetics/nutrition, education, cooperative extension; journalism; research; social services, or volunteer services; and have had a significant positive impact on the family and consumer sciences profession.
Geographic Focus: All States
Date(s) Application is Due: Nov 1
Contact: Susan Rickards; (304) 368-0612; rickards@phiu.org or info@phiu.org
Internet: http://phiu.org/awards.htm
Sponsor: Phi Upsilon Omicron, Inc.
P.O. Box 329
Fairmont, WV 26555

Phi Upsilon Omicron Frances Morton Holbrook Alumni Award 3193
Phi Upsilon Omicron National Honor Society was founded in 1910 at the University of Minnesota in the area of Family and Consumer Sciences. The society grants its Frances Morton Holbrook Alumni Award biennially to a qualified alumni member of the society for fulfilling personal and professional goals which promote the purposes of family and consumer sciences. Criteria includes demonstrated excellence in one of the areas of family and consumer sciences. The recipient's expenses to attend the society's biennial meeting will be paid, and the member will be recognized with a plaque at the meeting. The nomination deadline is November 1. Interested applicants should contact the society for information on nomination procedures.
Requirements: Nominees must be alumni members of the society to be eligible for the award.
Geographic Focus: All States
Date(s) Application is Due: Nov 1
Contact: Susan Rickards; (304) 368-0612; rickards@phiu.org or info@phiu.org
Internet: http://phiu.org/awards.htm
Sponsor: Phi Upsilon Omicron, Inc.
P.O. Box 329
Fairmont, WV 26555

Phi Upsilon Omicron Geraldine Clewell Senior Awards 3194
Phi Upsilon Omicron National Honor Society was founded in 1910 at the University of Minnesota in the area of Family and Consumer Sciences. The society annually awards its Geraldine Clewell Senior Awards to its graduating seniors for exemplary service to their local chapters. Prospective applicants can download complete application guidelines from the award website. The application packet must be postmarked no later than February 1 and must include the application (signed by the applicant's local chapter advisor) and three letters of recommendation (one of which must be from the local chapter advisor). Phi Upsilon Omicron offers various collegiate awards; applicants may apply for more than one award with the same application packet and should list the name of the award(s) for which application is being made.
Requirements: Applicants must be graduating seniors and members of Phi Upsilon Omicron.
Geographic Focus: All States
Date(s) Application is Due: Feb 1
Amount of Grant: 250 - 300 USD
Contact: Susan Rickards; (304) 368-0612; rickards@phiu.org or info@phiu.org
Internet: http://phiu.org/awards.htm
Sponsor: Phi Upsilon Omicron, Inc.
P.O. Box 329
Fairmont, WV 26555

Phi Upsilon Omicron Janice Cory Bullock Collegiate Award 3195
Phi Upsilon Omicron National Honor Society was founded in 1910 at the University of Minnesota in the area of Family and Consumer Sciences. The society awards its Janice Cory Bullock Award annually to a non-traditional student who desires to continue education in family and consumer sciences or a related area which will enable gainful employment. Detailed application guidelines can be downloaded from the award website. The application packet must be postmarked no later than February 1 and must

include the application (signed by the applicant's chapter advisor) and three letters of recommendation. The society offers various collegiate awards; applicants may apply for more than one award with the same application packet and should list the name of the award(s) for which application is being made.
Requirements: Applicants must be student members of Phi Upsilon Omicron.
Geographic Focus: All States
Date(s) Application is Due: Feb 1
Amount of Grant: 250 - 300 USD
Contact: Susan Rickards; (304) 368-0612; rickards@phiu.org or info@phiu.org
Internet: http://phiu.org/awards.htm
Sponsor: Phi Upsilon Omicron, Inc.
P.O. Box 329
Fairmont, WV 26555

Phi Upsilon Omicron Lillian P. Schoephoerster Award 3196
Phi Upsilon Omicron National Honor Society was founded in 1910 at the University of Minnesota in the area of Family and Consumer Sciences. The society awards its Lillian P. Schoephoerster Award to a part-time, nontraditional, junior or senior undergraduate. Prospective applicants can download complete application guidelines from the award website. The application packet must be postmarked no later than February 1 and must include the application (signed by the applicant's chapter advisor) and three letters of recommendation. Phi Upsilon Omicron offers various collegiate awards; applicants may apply for more than one award with the same application packet and should list the name of the award(s) for which application is being made.
Requirements: Applicants must be nontraditional, part-time undergraduates who have attained at least junior status and who are members of Phi Upsilon Omicron.
Geographic Focus: All States
Date(s) Application is Due: Feb 1
Amount of Grant: 250 - 300 USD
Contact: Susan Rickards; (304) 368-0612; rickards@phiu.org or info@phiu.org
Internet: http://phiu.org/awards.htm
Sponsor: Phi Upsilon Omicron, Inc.
P.O. Box 329
Fairmont, WV 26555

Phi Upsilon Omicron Margaret Jerome Sampson Scholarships 3197
Phi Upsilon Omicron National Honor Society was founded in 1910 at the University of Minnesota in the area of Family and Consumer Sciences. The society annually awards its Margaret Jerome Sampson Scholarships to five undergraduate students pursuing a baccalaureate degree in family and consumer sciences or a related area. Detailed application guidelines can be downloaded from the scholarship website. Applicants should note that one of the three required recommendations must come from their chapter advisor. The application packet must be postmarked no later than February 1. The society offers various scholarship opportunities; applicants may apply for more than one scholarship with the same application packet and should check the name of the scholarship(s) for which application is being made on the required cover sheet (also available for download at the website).
Requirements: Applicants must be members of Phi Upsilon Omicron and be enrolled as full-time students pursuing a baccalaureate degree in consumer and family sciences (or a related area) for the full academic year in which scholarship money is received. Preference is given to majors in dietetics or food and nutrition. Selection will be based on the applicant's financial statement, statement of need, scholastic record, participation in Phi U, and professional aims and goals.
Geographic Focus: All States
Date(s) Application is Due: Feb 1
Amount of Grant: 4,000 USD
Contact: Susan Rickards; (304) 368-0612; rickards@phiu.org or info@phiu.org
Internet: http://www.phiu.org/scholarships.htm
Sponsor: Phi Upsilon Omicron, Inc.
P.O. Box 329
Fairmont, WV 26555

Phi Upsilon Omicron Orinne Johnson Writing Award 3198
Phi Upsilon Omicron National Honor Society was founded in 1910 at the University of Minnesota in the area of Family and Consumer Sciences. The society annually awards its Orinne Johnson Writing Award to a collegiate member who submits the best article for publication in the Phi Upsilon Omicron publication "The Candle." The article should be related to Phi Upsilon's three main purposes: to recognize and promote academic excellence; to enhance qualities of leadership by proving opportunities for service; and to encourage lifelong learning and commitment to advance family and consumer sciences and related areas. Detailed application and article guidelines can be downloaded from the award website. The application packet must be postmarked no later than February 1 and must include the application (signed by the applicant's chapter advisor) and the article.
Requirements: Applicants must be graduate or undergraduate students and members of Phi Upsilon Omicron.
Geographic Focus: All States
Date(s) Application is Due: Feb 1
Amount of Grant: 250 - 300 USD
Contact: Susan Rickards; (304) 368-0612; rickards@phiu.org or info@phiu.org
Internet: http://phiu.org/awards.htm
Sponsor: Phi Upsilon Omicron, Inc.
P.O. Box 329
Fairmont, WV 26555

Phi Upsilon Omicron Sarah Thorniley Phillips Leader Awards 3199
Phi Upsilon Omicron National Honor Society was founded in 1910 at the University of Minnesota in the area of Family and Consumer Sciences. The society annually awards its Sarah Thorniley Phillips Leadership Awards to its undergraduate members for outstanding leadership and participation in collegiate and community programs. Application packets must be postmarked no later than February 1 and must include the application (signed by the applicant's chapter advisor) and three letters of recommendation. The society offers various collegiate awards; applicants may apply for more than one award with the same application packet and should list the name of the awards(s) for which application is being made.
Requirements: Applicants must be undergraduate members of Phi Upsilon Omicron.
Geographic Focus: All States
Date(s) Application is Due: Feb 1
Amount of Grant: 250 - 300 USD
Contact: Susan Rickards; (304) 368-0612; rickards@phiu.org or info@phiu.org
Internet: http://phiu.org/awards.htm
Sponsor: Phi Upsilon Omicron, Inc.
P.O. Box 329
Fairmont, WV 26555

Phi Upsilon Omicron Undergraduate Karen P. Goebel Conclave Award 3200
Phi Upsilon Omicron National Honor Society was founded in 1910 at the University of Minnesota in the area of Family and Consumer Sciences. The society awards its Undergraduate Karen P. Goebel Conclave Award in even-numbered years to help one student attend the society's biennial Conclave event. Prospective applicants can download complete application guidelines from the award website. The application packet must be postmarked no later than February 1 and must include the application (signed by the applicant's chapter advisor) and three letters of recommendation, two of which must come from a current or previous academic advisor and a current or previous professor. Phi Upsilon Omicron offers various collegiate awards; applicants may apply for more than one award with the same application packet and should list the name of the award(s) for which application is being made.
Requirements: This award is available to Phi Upsilon Omicron members to be enrolled during the full academic year (2 semesters or 3 quarters remaining in school after the Conclave). The total amount of the award is $500. The recipient will be awarded $300 at the Conclave. To receive the balance of the award money, the recipient must then turn the application into an article to be submitted for The Candle by December 1 of the Conclave year. The article should describe the recipient's overall plan and fundraising efforts for attending the Conclave and the actual experience and benefits gained from attending. Upon receipt of this article, the society will send the student the additional $200 of the award.
Geographic Focus: All States
Date(s) Application is Due: Feb 1
Amount of Grant: 500 USD
Contact: Susan Rickards; (304) 368-0612; rickards@phiu.org or info@phiu.org
Internet: http://phiu.org/awards.htm
Sponsor: Phi Upsilon Omicron, Inc.
P.O. Box 329
Fairmont, WV 26555

Phoenix Suns Charities Grants 3201
Ranging in size from $1,000 to $10,000, Phoenix Suns Charities Program Grants are intended for Arizona non-profit organization whose programs and activities focus on helping children and families maximize their potential. The foundation's largest annual gift, the Playmaker Award, is a one-time $100,000 grant which can be used for capital or programs, or a combination of both. For this grant, Suns Charities looks favorably on collaborative ideas and naming or branding opportunities.
Requirements: Applications and supporting documents must be submitted electronically through ZoomGrants at the Suns website. Organizations must consider the prerequisites, answer all questions and carefully follow directions. Applications are available in November, evaluated by Suns board members in April and May, with funding in June.
Geographic Focus: Arizona
Date(s) Application is Due: Apr 1
Amount of Grant: 1,000 - 10,000 USD
Contact: Kathryn Pidgeon; (602) 379-7948; fax (602) 379-7990; kpidgeon@suns.com
Internet: http://www.nba.com/suns/charities.html
Sponsor: Phoenix Suns
201 East Jefferson Street
Phoenix, AZ 85004

PhRMA Foundation Health Outcomes Post Doctoral Fellowships 3202
The program provides stipend support for individuals engaged in a research training program that will create or extend their credentials in health outcomes. The intent of this program is to support post doctoral career development activities of individuals prepared (or preparing) to engage in research that will strengthen representation of health outcomes in schools of pharmacy, medicine, nursing and public health. To accomplish these goals, support will be provided for a two-year period to selected individuals who are beginning careers in health outcomes research and who give promise of outstanding development as researchers. The award, consisting of a $40,000 annual stipend, is made to the institution on behalf of the fellow. The award is intended solely as a stipend and may not be used otherwise.
Requirements: Eligible applicants include well-trained graduates from Pharm. D., M.D., and Ph.D. programs who seek to further develop and refine their research skills through formal postdoctoral training. They must also have a firm commitment from an accredited U.S. university and be a U.S. citizen or permanent resident. The application must include a research plan written by the applicant, the mentor's research record, and a description of

how the mentored experience will enhance the applicant's career development in health outcomes research. The sponsor (mentor) of the post doctoral program must describe how the goals of the research training program will be accomplished and provide assurance that key collaborating mentors endorse and are willing to support the training plan.
Restrictions: Applications, however good, which do not meet the aims of the program will be disapproved.
Geographic Focus: All States
Amount of Grant: 40,000 USD
Contact: Eileen M. Cannon; (202) 572-7756; fax (202) 572-7799; info@phrmafoundation.org
Internet: http://www.phrmafoundation.org/awards/outcomes/postdoc.php
Sponsor: Pharmaceutical Research and Manufacturers of America Foundation
950 F Street NW, Suite 300
Washington, D.C. 20004

PhRMA Foundation Health Outcomes Pre-Doctoral Fellowships   3203
To provide some assistance in this training sequence, the PhRMA Foundation program aims at supporting promising students during their thesis research. The program provides a stipend and funds to cover costs incidental to the training. Applications will be accepted for a minimum of one year and a maximum of two years. The program is designed for candidates who expect to complete the requirements for the Ph.D. in the field of health outcomes in two years or less from the time the fellowship begins. The program assumes that the fellows will devote full-time (including summers) to their research. The award is made to the university on behalf of the fellow. The fellowship provides a stipend of $20,000 a year payable monthly for a minimum of one year and a maximum of two years. Of the $20,000 awarded annually, $500 a year may be used for incidentals directly associated with the thesis research preparation (e.g., secretarial help, artwork, books, travel, etc.). The second year of a fellowship is contingent upon certification by the thesis advisor that satisfactory progress is being achieved.
Requirements: The fellowship program of pre-doctoral support is designed to assist full-time, in-residence Ph.D. candidates in the fields of health outcomes who are enrolled in U.S. schools of medicine, pharmacy, dentistry or schools of public health. The program seeks to support advanced students who will have completed the bulk of their pre-thesis requirements (two years of study) and are starting their thesis research by the time the award is activated. Students just starting in graduate school should not apply. A candidate enrolled in an M.D./Ph.D. program should not be taking required clinical course work or clerkships during the tenure of the fellowship. Before an individual is eligible to apply for a PhRMA Foundation award, the applicant must first have a firm commitment from a U.S. university. Applications must be submitted by an accredited U.S. school, and all applicants must be U.S. citizens or permanent residents.
Restrictions: The fellowship may not be used as a supplement to funds from other fellowships, traineeships or assistantships. The focus of the program is to assist in the pre doctoral training of the candidate. It should not be viewed as a program to fund a research project. An individual may not simultaneously hold or interrupt any other fellowship providing stipend support during the PhRMA Foundation fellowship.
Geographic Focus: All States
Amount of Grant: 20,000 USD
Contact: Eileen M. Cannon; (202) 572-7756; fax (202) 572-7799; info@phrmafoundation.org
Internet: http://www.phrmafoundation.org/awards/outcomes/predoc.php
Sponsor: Pharmaceutical Research and Manufacturers of America Foundation
950 F Street NW, Suite 300
Washington, D.C. 20004

PhRMA Foundation Health Outcomes Research Starter Grants   3204
The purpose of the program is to offer financial support to individuals beginning their independent research careers at the faculty level. The program provides a research grant of $30,000 per year for up to two years. This program supports individuals beginning independent research careers in academia who do not have other substantial sources of research.
Requirements: Those holding the academic rank of instructor or assistant professor and investigators at the doctoral level with equivalent positions are eligible to apply, providing their proposed research is neither directly nor indirectly subsidized to any significant degree by an extramural support mechanism. The program is not intended for those in postdoctoral training programs. However, individuals in postdoctoral training scheduled to conclude and who will hold an academic appointment by January 1 may apply. Applicants must be sponsored by the department or unit within which the proposed research is to be undertaken. The grant is made to the university on behalf of the applicant and with the understanding that the university will administer the funds. Schools of medicine, pharmacy, public health, nursing, dentistry and schools of other areas where appropriate are eligible for this award. Applications must be submitted by an accredited school in the U.S., and all applicants must be U.S. citizens or permanent residents. Before an individual is eligible to apply for a PhRMA Foundation award, the applicant must first have a firm commitment from a U.S. university.
Restrictions: This program is not offered as a means to augment substantially funded research efforts. It is intended to offer support for researchers who are starting their independent research efforts. This award is granted in part based on need. If an individual currently has or is guaranteed substantial funding, they should not apply.
Geographic Focus: All States
Amount of Grant: 30,000 USD
Contact: Eileen M. Cannon; (202) 572-7756; fax (202) 572-7799; info@phrmafoundation.org
Internet: http://www.phrmafoundation.org/awards/outcomes/starter.php
Sponsor: Pharmaceutical Research and Manufacturers of America Foundation
950 F Street NW, Suite 300
Washington, D.C. 20004

PhRMA Foundation Health Outcomes Sabbatical Fellowships   3205
The fellowship provides stipend support for individuals engaged in a multidisciplinary research training program that will create or extend their credentials in health outcomes. The program enables faculty with active research programs to work outside of their home institution for periods of six months to one year to learn new skills or develop new collaborations that will enhance their research and research training capabilities in health outcomes. The applicant and mentor of the program must describe how the multidisciplinary goals of the research experiential program will be accomplished and provide assurance that key collaborating mentors endorse and are willing to support the plan. Matching funds must be provided by the home institution. Awards may be activated beginning January 1 or on the first day of any month thereafter, up to and including December 1. Guidelines and application are available online.
Requirements: Eligible applicants must (1) hold a PhD, Pharm.D., M.D. or Sc.D. degree in a field of study logically or functionally related to the proposed post doctoral activities, (2) hold a faculty appointment that imparts eligibility for a sabbatical leave from their home institution, (3) have institutional approval of a sabbatical plan that includes partial salary that matches the PhRMA stipend, (4) hold an endorsement from a mentor who agrees to sponsor the applicant's visiting scientist activity, and (5) be a U.S. citizen or permanent resident. Matching funds must be provided by the home institution.
Restrictions: The program provides no other subsidies (travel, tuition, fringe benefit costs, etc.), and indirect costs to the institution are not provided.
Geographic Focus: All States
Amount of Grant: 40,000 USD
Contact: Eileen M. Cannon; (202) 572-7756; fax (202) 572-7799; info@phrmafoundation.org
Internet: http://www.phrmafoundation.org/awards/outcomes/sabbatical.php
Sponsor: Pharmaceutical Research and Manufacturers of America Foundation
950 F Street NW, Suite 300
Washington, D.C. 20004

PhRMA Foundation Informatics Post Doctoral Fellowships   3206
The postdoctoral program provides stipend support for individuals engaged in a multidisciplinary research training program that will create or extend their credentials in informatics. The intent of this program is to support postdoctoral career development activities of individuals preparing to engage in research that will bridge the gap between experimental and computational approaches in genomic and biomedical studies. It is anticipated that this research training will be accomplished in academic and/or industrial laboratory settings where multidisciplinary teams are organized to address problems that span the range of biological complexity rather than focus on the application of single technologies. The award, consisting of a $40,000 annual stipend, is made to the institution on behalf of the fellow. The award is intended solely as a stipend and may not be used otherwise.
Requirements: Eligible applicants must either hold a PhD degree in a field of study logically or functionally related to the proposed postdoctoral activities, or expect to receive the PhD before activating the award. They must also have a firm commitment from an accredited U.S. university and be a U.S. citizen or permanent resident. The sponsor (mentor) of the postdoctoral program must describe how the multidisciplinary goals of the research training program will be accomplished and provide assurance that key collaborating mentors endorse and are willing to support the training plan.
Restrictions: The program provides no other subsidies (travel, tuition, fringe benefit costs, etc.) and indirect costs to the institution are not provided. The second year of this award is contingent upon a progress report approved by the Foundation and submission of a financial report.
Geographic Focus: All States
Amount of Grant: 40,000 USD
Contact: Eileen M. Cannon; (202) 572-7756; fax (202) 572-7799; info@phrmafoundation.org
Internet: http://www.phrmafoundation.org/awards/informatics/postdoc.php
Sponsor: Pharmaceutical Research and Manufacturers of America Foundation
950 F Street NW, Suite 300
Washington, D.C. 20004

PhRMA Foundation Informatics Pre Doctoral Fellowships   3207
The program is designed for candidates who expect to complete the requirements for the Ph.D. in the fields of informatics in two years or less from the time the fellowship begins. It aims at supporting promising students during their thesis research. The award is made to the university on behalf of the fellow. The fellowship provides a stipend of $20,000 a year payable monthly for a minimum of one year and a maximum of two years. Of the $20,000 awarded annually, up to $500 a year may be used for incidentals directly associated with the thesis research preparation (e.g., secretarial help, artwork, books, travel, etc.). Applications will be accepted for a minimum of one year and a maximum of two years. The program assumes that the fellows will devote full-time (including summers) to their research.
Requirements: The fellowship program of pre doctoral support is designed to assist full-time, in-residence Ph.D. candidates in the fields of informatics who are enrolled in USA schools of medicine, pharmacy, dentistry or veterinary medicine. The program supports full-time advanced students who will have completed the bulk of their pre-thesis requirements (at least two years of study) and are engaged in thesis research as Ph.D. Candidates by the time the award is activated. Students just starting in graduate school should not apply. Highest priority will be given to applications who support the goal of the Informatics Programs. The goal of the program is to promote the use of informatics in an integrative approach to the understanding of biological and disease processes. Informatics awards will support career development of scientists engaged in computational and experimental research to integrate cutting-edge information technology with advanced biological, chemical, and pharmacological sciences. Applications must be submitted by an accredited U.S. school, and all applicants must be U.S. citizens or permanent residents.

Restrictions: The fellowship may not be used to supplement funds from other fellowships, traineeships or assistantships, unless necessary to make stipend levels compliant with institutional policy. The focus of the program is to assist in the predoctoral training of the candidate. It should not be viewed as a program to fund a research project. Indirect costs are not provided to the institution and PhRMA Foundation grant funds may not be used for this purpose.
Geographic Focus: All States
Amount of Grant: 20,000 USD
Contact: Eileen M. Cannon; (202) 572-7756; fax (202) 572-7799; info@phrmafoundation.org
Internet: http://www.phrmafoundation.org/awards/informatics/predoc.php
Sponsor: Pharmaceutical Research and Manufacturers of America Foundation
950 F Street NW, Suite 300
Washington, D.C. 20004

PhRMA Foundation Informatics Research Starter Grants — 3208

This program supports individuals beginning independent research careers in academia who do not have other substantial sources of research. The areas of interest within this program consists of research that supports career development of scientists engaged in computational and experimental research to integrate cutting edge-edge information technology with advanced biological, chemical, and pharmacological sciences in: Genetics (Molecular, Medical-human, Pharmaco, Population); Genomics (Function, Structural, Toxico, Pharmaco, Comparative); Proteomics; Biological pathways. The program provides a research grant of $30,000 per year for up to two years. The 'starter' aspect of the program strives to assist those individuals who are establishing careers as independent investigators in the field of informatics. The program is not offered as a means to augment an ongoing research effort or is the grant intended to be used for any direct effort to obtain further extramural funding. The funds are to be used to conduct the proposed research.
Requirements: Those holding the academic rank of instructor or assistant professor and investigators at the doctoral level with equivalent positions are eligible to apply for these research starter grants, providing their proposed research is neither directly nor indirectly subsidized to any significant degree by an extramural support mechanism. The program is not intended for those in postdoctoral training programs. However, individuals in postdoctoral training scheduled to conclude and who will hold an academic appointment by January 1 may apply. Applicants must be sponsored by the department or unit in which the proposed research is to be undertaken. The grant is made to the university on behalf of the applicant and with the understanding that the university will administer the funds. Applications must be submitted by an accredited U.S. school, and all applicants must be U.S. citizens or permanent residents.
Restrictions: Research projects that extend or develop the proprietary value of specific drug products are not acceptable in this program. This exclusion does not preclude research in which specific drug products are used to test hypotheses that have a general applicability. The funds may not be used for salary support of the grantee, but may be used to support technical assistance. No more than $500 a year may be used for travel to professional meetings by the grantee. Indirect costs are not provided to the institution, and grant funds may not be used for this purpose.
Geographic Focus: All States
Amount of Grant: 30,000 USD
Contact: Eileen M. Cannon; (202) 572-7756; fax (202) 572-7799; info@phrmafoundation.org
Internet: http://www.phrmafoundation.org/awards/informatics/starter.php
Sponsor: Pharmaceutical Research and Manufacturers of America Foundation
950 F Street NW, Suite 300
Washington, D.C. 20004

PhRMA Foundation Informatics Sabbatical Fellowships — 3209

The fellowship supports individuals engaged in a multidisciplinary research training program that will create or extend their credentials in informatics. The program enables faculty with active research programs to work outside of their home institution for periods of six months to one year to learn new skills or develop new collaborations that will enhance their research and research training capabilities in informatics. Applicants are expected to engage in research that supports career development of scientists engaged in computational and experimental research to integrate cutting edge information technology with advanced biological, chemical, and pharmacological sciences in the fields of genetics, genomics, and proteomics biological pathways. Emphasis will be placed on the development of new informatics technologies that demonstrate the translation of genomic data into an elucidation and understanding of biological and disease processes. The applicant and mentor of the program must describe how the multidisciplinary goals of the research experiential program will be accomplished and provide assurance that key collaborating mentors endorse and are willing to support the plan. The award provides up to $40,000 stipend support for mid-career scientists to engage in an academic year or calendar year experiential program intended to redirect their core research focus to an area of emerging importance to pharmaceutical research and development.
Requirements: Eligible applicants must hold a PhD or MD degree in a field of study logically or functionally related to the proposed postdoctoral activities; hold a faculty appointment that imparts eligibility for a sabbatical leave from their home institution; have institutional approval of a sabbatical plan that includes partial salary that matches the PhRMA stipend; and be a U.S. citizen or permanent resident. Preference will be given to those individuals whose research combines novel computational methods with experimental validation. Emphasis will be placed on the development of new informatics technologies that demonstrate the translation of genomic data into an elucidation and understanding of biological and disease processes. Matching funds must be provided by the home institution.
Restrictions: The program provides no other subsidies (travel, tuition, fringe benefit costs, etc.), and indirect costs to the institution are not provided.
Geographic Focus: All States
Amount of Grant: 40,000 USD
Contact: Eileen M. Cannon; (202) 572-7756; fax (202) 572-7799; info@phrmafoundation.org
Internet: http://www.phrmafoundation.org/awards/informatics/sabbatical.php
Sponsor: Pharmaceutical Research and Manufacturers of America Foundation
950 F Street NW, Suite 300
Washington, D.C. 20004

PhRMA Foundation Paul Calabresi Medical Student Research Fellowship — 3210

The fellowship is offered to medical or dental students who have substantial interests in research and teaching careers in pharmacology - clinical pharmacology and who are willing to spend full-time in a specific research effort within a pharmacology or clinical pharmacology unit. Fellowships are available for a minimum period of three months or any period of time up to 24 months. The commitment must be full-time. The student may undertake this investigative effort at his/her own school or at another institution. Requests for a fellowship are to be submitted online by the appropriate representative of the school or university. The fellowship is made to the institution on behalf of the student and the institution must agree to administer the funds. The program offers funding of up to $1,500 per month. It is expected that the funds will be distributed to the fellow in regular payments over the period of the fellowship by the recipient school. The funds may be used towards the stipend, tuition or affiliated fees for the fellow. Fellowships are available for a minimum period of six months or any period of time up to 24 months. However, if the student's award period is 24 months, the maximum stipend is still $18,000.
Requirements: A candidate must be enrolled in a U.S. medical/dental school and have finished at least one year of the school curriculum. Priority consideration will be given to those candidates who project strong commitments to careers in the field of clinical pharmacology. The applicant must be sponsored by the pharmacology or clinical pharmacology program in which the investigative project is to be undertaken. The sponsoring unit must have a demonstrated commitment to training in the fields of pharmacology or clinical pharmacology. Applications must be submitted by an accredited U.S. school, and all applicants must be U.S. citizens or permanent residents.
Restrictions: There are no indirect costs provided to the institution nor may any of the funds be used for this purpose. Once an individual has received a Medical Student Fellowship, that individual may not reapply.
Geographic Focus: All States
Amount of Grant: Up to 18,000 USD
Contact: Eileen M. Cannon; (202) 572-7756; fax (202) 572-7799; info@phrmafoundation.org
Internet: http://www.phrmafoundation.org/awards/clinical/medstudent.php
Sponsor: Pharmaceutical Research and Manufacturers of America Foundation
950 F Street NW, Suite 300
Washington, D.C. 20004

PhRMA Foundation Pharmaceutics Postdoctoral Fellowships — 3211

The purpose of the fellowships is to encourage graduates from PhD programs in pharmaceutics to continue to refine their research skills through formal postdoctoral training. Applicants should indicate a strong determination to continue their research careers in pharmaceutics following completion of the fellowship. Applications will be accepted for postdoctoral research training extending over a minimum of one year and a maximum of two years. Each fellowship carries an annual stipend of $40,000 which can be supplemented by the sponsor and/or his/her institution.
Requirements: To be eligible, applicants must either hold a PhD degree in pharmaceutics from a school of pharmacy accredited by the American Council on Pharmaceutical Education or expect to receive such a degree before activating the fellowship. The mentor of the fellow must be in an accredited school of pharmacy. Alternatively, a student with a Ph.D. from a school of pharmacy may use the fellowship to study with a mentor in any science or engineering department. An equally important aspect of eligibility is that suitable facilities for the necessary training and research must be available to the applicant. The program with which the applicant is, or will be, associated must provide an environment where the applicant's potential can be developed to the fullest extent possible.
Restrictions: The award is intended solely as a stipend and may not be used otherwise. If necessary, the institution may supplement the award to a level that is consistent with other postdoctoral fellowships it currently offers. Research projects that extend or develop the proprietary value of specific drug products are not acceptable in this program. The program provides no other subsidies (travel, tuition, fringe benefit costs, etc.) and indirect costs to the institution are not provided.
Geographic Focus: All States
Amount of Grant: 40,000 USD
Contact: Eileen M. Cannon; (202) 572-7756; fax (202) 572-7799; info@phrmafoundation.org
Internet: http://www.phrmafoundation.org/awards/pharmaceutics/postdoc.php
Sponsor: Pharmaceutical Research and Manufacturers of America Foundation
950 F Street NW, Suite 300
Washington, D.C. 20004

PhRMA Foundation Pharmaceutics Pre Doctoral Fellowships — 3212

To encourage the entry of more qualified individuals into this area, the program aims at supporting promising students during their thesis research by providing assistance in the form of stipend and funds to cover costs incidental to the training. Applications will be accepted for a minimum of one year and a maximum of two years. The program is designed for applicants who expect to complete the requirements for the Ph.D. in pharmaceutics in two years or less from the time the fellowship begins. For the purposes of this program, pharmaceutics includes basic pharmaceutics, biopharmaceutics and pharmaceutical technology. The fundamental aspects of pharmacokinetics are not included since these are covered by PhRMA Foundation awards in pharmacology and toxicology.
Requirements: To be eligible, applicants must hold a B.S., M.S., or Pharm.D. degree in pharmacy or a related area such as chemistry or biology from an accredited school in the U.S. A candidate enrolled in a Pharm.D./Ph.D. program should not be taking required

clinical course work or clinical clerkships during the tenure of the fellowship. Also, before an individual is eligible to apply for a PhRMA Foundation award, the applicant must first have a firm commitment from a U.S. university, and must be a U.S. citizen or permanent resident. The program assumes that the fellows will devote full-time (including summers) to their research. The fellowship may not be used as a supplement to funds from other fellowships or assistantships. However, this does not preclude institutions supplementing a fellowship with their own funds in order to bring the stipend within the normal range of the institution.
Restrictions: Research projects which extend or develop the proprietary value of specific drug products are not acceptable in this program. This exclusion does not preclude research in which specific drug products are used to test hypotheses which have a general applicability. Applications, however good, which do not meet the aims of the program will be disapproved.
Geographic Focus: All States
Contact: Eileen M. Cannon; (202) 572-7756; fax (202) 572-7799; info@phrmafoundation.org
Internet: http://www.phrmafoundation.org/awards/pharmaceutics/predoc.php
Sponsor: Pharmaceutical Research and Manufacturers of America Foundation
950 F Street NW, Suite 300
Washington, D.C. 20004

PhRMA Foundation Pharmaceutics Research Starter Grants   3213
The purpose of the starter grants is to offer financial support to individuals beginning their independent research careers at the faculty level. The areas of interest within this program are research efforts in the fields of pharmaceutics. The program provides a research grant of $30,000 per year for up to two years. For the purposes of this program, pharmaceutics includes basic pharmaceutics, biopharmaceutics, pharmaceutical technology and pharmaceutical biotechnology. The fundamental aspects of pharmacokinetics are not included since they are covered by PhRMA Foundation programs in pharmacology and toxicology. The 'starter' aspect of the program strives to assist those individuals who are establishing careers as independent investigators in the fields listed above.
Requirements: Those holding the academic, tenure track, rank of instructor or assistant professor and investigators at the doctoral level with equivalent positions are eligible to apply for these research starter grants, providing their proposed research is neither directly nor indirectly subsidized to any significant degree by an extramural support mechanism. It is anticipated that this research experience will occur in an academic and/or industrial laboratory setting. Applications must be submitted by a school of pharmacy accredited by the American Council on Pharmaceutical Education, and all applicants must be U.S. citizens or permanent residents. The grant is made to the university on behalf of the applicant and with the understanding that the university will administer the funds. Only schools of pharmacy are eligible for this award.
Restrictions: The program is not offered as a means to augment an ongoing research effort. Nor is the grant intended to be used for any direct effort to obtain further extramural funding. The funds are to be used to conduct the proposed research. The funds may not be used for salary support of the grantee, but may be used to support technical assistance. No more than $500 a year may be used for travel to professional meetings by the grantee. Indirect costs are not provided to the institution, and grant funds may not be used for this purpose.
Geographic Focus: All States
Amount of Grant: 30,000 USD
Contact: Eileen M. Cannon; (202) 572-7756; fax (202) 572-7799; info@phrmafoundation.org
Internet: http://www.phrmafoundation.org/awards/pharmaceutics/starter.php
Sponsor: Pharmaceutical Research and Manufacturers of America Foundation
950 F Street NW, Suite 300
Washington, D.C. 20004

PhRMA Foundation Pharmaceutics Sabbatical Fellowships   3214
The purpose of this program is to enable faculty with active research programs to work outside of their home institution for periods of 6 months to one year to learn new skills or develop new collaborations that will enhance their research and research training capabilities in pharmaceutics. For the purposes of this program, pharmaceutics includes basic pharmaceutics, biopharmaceutics, pharmaceutical technology and pharmaceutical biotechnology. The fundamental aspects of pharmacokinetics are not included since they are covered by PhRMA Foundation programs in pharmacology and toxicology. It is anticipated that this research experience will occur in an academic and/or industrial laboratory setting. The award provides up to $40,000 stipend support for mid-career scientists to engage in an academic year or calendar year experiential program intended to redirect their core research focus to an area of emerging importance to pharmaceutical research and development.
Requirements: Eligible applicants must hold a PhD degree in pharmaceutics from a School of Pharmacy accredited by the American Council on Pharmaceutical Education; hold a faculty appointment that imparts eligibility for a sabbatical leave from their home institution; have institutional approval of a sabbatical plan that includes partial salary that matches the PhRMA stipend; hold an endorsement from a mentor who agrees to sponsor the applicants visiting scientist activity; and be a U.S. citizen or permanent resident. The applicant and mentor of the program must describe how the multidisciplinary goals of the research experiential program will be accomplished and provide assurance that key collaborating mentors endorse and are willing to support the plan. Matching funds must be provided by the home institution.
Restrictions: The program provides no other subsidies (travel, tuition, fringe benefit costs, etc.) and indirect costs to the institution are not provided.
Geographic Focus: All States
Amount of Grant: 40,000 USD
Contact: Eileen M. Cannon; (202) 572-7756; fax (202) 572-7799; info@phrmafoundation.org
Internet: http://www.phrmafoundation.org/awards/pharmaceutics/sabbatical.php
Sponsor: Pharmaceutical Research and Manufacturers of America Foundation
950 F Street NW, Suite 300
Washington, D.C. 20004

PhRMA Foundation Pharmacology/Toxicology Post Doctoral Fellowships   3215
The postdoctoral program provides stipend support for individuals engaged in a multidisciplinary research training program that will create or extend their credentials in pharmacology or toxicology. The intent of this program is to support postdoctoral career development activities of individuals prepared (or preparing) to engage in research that integrates information on molecular or cellular mechanisms of action with information on the effect of an agent in the intact organism. It is anticipated that this research training will be accomplished in academic and/or industrial laboratory settings where multidisciplinary teams are organized to integrate informatics, molecular, cell, and systems biology with pharmacology/toxicology research. Funding is for up to 2 years but not beyond the 3rd year of the existing Post Doc.
Requirements: Eligible applicants must either hold a PhD degree or appropriate terminal research doctorate in a field of study logically or functionally related to the proposed post doctoral activities; or expect to receive the PhD before activating the award. They must also have a firm commitment from an accredited U.S. university and be a U.S. citizen or permanent resident. The sponsor (mentor) must describe how the multidisciplinary goals of the research training program will be accomplished and provide assurance that key collaborating mentors endorse and are willing to support the training plan. The award, consisting of a $40,000 annual stipend, is made to the institution on behalf of the fellow. The award is intended solely as a stipend and may not be used otherwise. If necessary, the institution may supplement the award to a level that is consistent with other postdoctoral fellowships it currently offers.
Restrictions: The program provides no other subsidies (travel, tuition, fringe benefit costs, etc.) and indirect costs to the institution are not provided. The second year of this award is contingent upon a progress report approved by the foundation and submission of a financial report.
Geographic Focus: All States
Amount of Grant: 40,000 USD
Contact: Eileen M. Cannon; (202) 572-7756; fax (202) 572-7799; info@phrmafoundation.org
Internet: http://www.phrmafoundation.org/awards/pharmacology/postdoc.php
Sponsor: Pharmaceutical Research and Manufacturers of America Foundation
950 F Street NW, Suite 300
Washington, D.C. 20004

PhRMA Foundation Pharmacology/Toxicology Pre Doctoral Fellowships   3216
The program is designed for candidates who expect to complete the requirements for the Ph.D. in the fields of pharmacology or toxicology in two years or less from the time the fellowship begins. The award is made to the university on behalf of the fellow. The fellowship provides a stipend of $20,000 a year payable monthly for a minimum of one year and a maximum of two years. Of the $20,000 awarded annually, up to $500 a year may be used for incidentals directly associated with the thesis research preparation (e.g., secretarial help, artwork, books, travel, etc.). Awards may be activated beginning January 1 or on the first day of any month thereafter, up to and including August 1. The second year of a fellowship is contingent upon certification by the thesis advisor that satisfactory progress is being achieved.
Requirements: The fellowship program of pre doctoral support is designed to assist full-time, in-residence Ph.D. candidates in the fields of pharmacology or toxicology who are enrolled in USA schools of medicine, pharmacy, dentistry or veterinary medicine. The program supports full-time advanced students who will have completed the bulk of their pre-thesis requirements (at least two years of study) and are engaged in thesis research as Ph.D. Candidates by the time the award is activated. Students just starting in graduate school should not apply. In the field of pharmacology or toxicology, highest priority will be given to applications for research that attempts to integrate information on the mechanism of action of a drug or chemical at a molecular or cellular level with the drug's effects in the intact laboratory animal or human, encompassing the potential influences of biochemical, physiological or behavioral systems. The pharmacology/toxicology research and teaching mission of the unit awarding the degree must be apparent as a core mission component on their web site. Applications must be submitted by an accredited U.S. school, and all applicants must be U.S. citizens or permanent residents. The program assumes that the fellows will devote full-time (including summers) to their research. The PhRMA Foundation fellowship may not be used to supplement funds from other fellowships, traineeships or assistantships, unless necessary to make stipend levels compliant with institutional policy. The focus of the program is to assist in the predoctoral training of the candidate. It should not be viewed as a program to fund a research project.
Restrictions: A candidate enrolled in an M.D./Ph.D. program should not be taking required clinical course work or clerkships during the tenure of the fellowship. The foundation will accept only one application per academic institution. Exceptions may be considered for institutions with multiple campuses. Indirect costs are not provided to the institution and PhRMA Foundation grant funds may not be used for this purpose.
Geographic Focus: All States
Amount of Grant: 20,000 USD
Contact: Eileen M. Cannon; (202) 572-7756; fax (202) 572-7799; info@phrmafoundation.org
Internet: http://www.phrmafoundation.org/awards/pharmacology/predoc.php
Sponsor: Pharmaceutical Research and Manufacturers of America Foundation
950 F Street NW, Suite 300
Washington, D.C. 20004

PhRMA Foundation Pharmacology/Toxicology Research Starter Grants   3217
The purpose of the program is to offer financial support to individuals beginning their independent research careers at the faculty level. The areas of interest within this program are research efforts in the fields of pharmacology and drug toxicology. The program provides a research grant of $30,000 per year for up to two years. The 'starter' aspect of the program strives to assist those individuals who are establishing careers as independent investigators in the fields listed above. The program is not offered as a means to augment an ongoing research effort. The funds are to be used to conduct the proposed research.

This program supports individuals beginning independent research careers in academia who do not have other substantial sources of research. Applicants will be judged on the scientific merit of the proposed research, and on the degree of financial need.
Requirements: Those holding the academic rank of instructor or assistant professor and investigators at the doctoral level with equivalent positions are eligible to apply for these research starter grants, providing their proposed research is neither directly nor indirectly subsidized to any significant degree by a competitive extramural grant. The program is not intended for those in postdoctoral training programs. However, individuals in postdoctoral training scheduled to conclude and who will hold an academic appointment by January 1 may apply. Applicants must be sponsored by the department or unit in which the proposed research is to be undertaken. The sponsoring unit must have responsibility for pharmacology/toxicology teaching and research as part of its core mission. If the pharmacology/toxicology mission is not apparent on the sponsoring unit's web site, the sponsor letter must describe how the academic appointment will support the applicant's career development in pharmacology/toxicology. The grant is made to the university on behalf of the applicant, with the understanding that the university will administer the funds. Schools of medicine (human or veterinary), pharmacy and dentistry are eligible for this award. In the field of pharmacology or toxicology, highest priority will be given to applications for research that attempts to integrate information on the mechanism of action of a drug or chemical at a molecular or cellular level with the drug's effects in the intact laboratory animal or human, encompassing the potential influences of biochemical, physiological or behavioral systems. Applications must be submitted by an accredited U.S. school, and all applicants must be U.S. citizens or permanent residents.
Restrictions: Research projects that extend or develop the proprietary value of specific drug products are not acceptable in this program. This exclusion does not preclude research in which specific drug products are used to test hypotheses that have a general applicability. The funds may not be used for salary support of the grantee, but may be used to support technical assistance. No more than $500 a year may be used for travel to professional meetings by the grantee. Indirect costs are not provided to the institution, and grant funds may not be used for this purpose. Applicants holding multi-year national competitive awards or grants supporting their independent research will not be considered. If a national competitive grant is obtained after the application has been submitted, the Foundation must be so informed.
Geographic Focus: All States
Amount of Grant: 30,000 USD
Contact: Eileen M. Cannon; (202) 572-7756; fax (202) 572-7799; info@phrmafoundation.org
Internet: http://www.phrmafoundation.org/awards/pharmacology/starter.php
Sponsor: Pharmaceutical Research and Manufacturers of America Foundation
950 F Street NW, Suite 300
Washington, D.C. 20004

PhRMA Foundation Pharmacology/Toxicology Sabbatical Fellowships   3218
This program provides stipend funding to enable faculty members at all levels with active research programs an opportunity to work at other institutions for periods of six months to one year to learn new skills or develop new collaborations that will enhance their research and research training activities in pharmacology/toxicology. Applicants are expected to engage in multidisciplinary research that integrates information on molecular or cellular mechanisms of action with information on the effect of an agent in the intact organism. It is anticipated that this research experience will occur in academic and/or industrial laboratory settings as part of a multidisciplinary team organized to integrate informatics, molecular, cell, and systems biology with pharmacology/toxicology research. The award provides up to $40,000 stipend support for mid-career scientists to engage in an academic year or calendar year experiential program intended to redirect their core research focus to an area of emerging importance to pharmaceutical research and development.
Requirements: Eligible applicants must hold a PhD degree or appropriate terminal doctorate and record of research accomplishment in a field of study logically or functionally related to the proposed post doctoral activities; hold a faculty appointment that imparts eligibility for a sabbatical leave from their home institution; have institutional approval of a sabbatical plan that includes partial salary that matches the PhRMA stipend; hold an endorsement from a mentor who agrees to sponsor the applicants visiting scientist activity; and be a U.S. citizen or permanent resident. The applicant and mentor of the program must describe how the multidisciplinary goals of the research experiential program will be accomplished and provide assurance that key collaborating mentors endorse and are willing to support the plan. Matching funds must be provided through the university.
Restrictions: The program provides no other subsidies (travel, tuition, fringe benefit costs, etc.) and indirect costs to the institution are not provided.
Geographic Focus: All States
Amount of Grant: 40,000 USD
Contact: Eileen M. Cannon; (202) 572-7756; fax (202) 572-7799; info@phrmafoundation.org
Internet: http://www.phrmafoundation.org/awards/pharmacology/sabbatical.php
Sponsor: Pharmaceutical Research and Manufacturers of America Foundation
950 F Street NW, Suite 300
Washington, D.C. 20004

Piedmont Health Foundation Grants   3219
The Piedmont Health Care Foundation was established in 1985 through the sale of the first HMO in Greenville County, South Carolina. During the past 25 years, the foundation has invested more than $3.4 million in dozens of nonprofit organizations in the Greenville, South Carolina area. The Piedmont Health Care Foundation has played an important role in catalyzing and providing seed funding to critical projects, and it has provided operating and programmatic funds needed by local health-service organizations. In looking ahead, the foundation recognizes that much of what makes a healthy community takes place outside of the health-care system. So for its 25th anniversary, the foundation decided to change its name to the Piedmont Health Foundation. The foundation currently focuses on the area of policy, system, and environmental change to reduce childhood obesity rates. Applications are accepted on a quarterly basis. Downloadable guidelines and editable forms are provided at the foundation website. Applicants should email their completed forms to the address given by midnight of the deadline date.
Requirements: Nonprofits that serve Greenville County are eligible to apply.
Geographic Focus: South Carolina
Date(s) Application is Due: Jan 31; Apr 10; Jul 10; Oct 10
Amount of Grant: 1,000 - 15,000 USD
Contact: Katy Smith, Executive Director; (864) 370-0212; fax (864) 370-0212; katypughsmith@bellsouth.net or katysmith@piedmonthealthfoundation.org
Internet: http://www.phcfdn.org/grantmaking.php
Sponsor: Piedmont Health Foundation
P.O. Box 9303
Greenville, SC 29604

Piedmont Natural Gas Corporate and Charitable Contributions   3220
Piedmont Natural Gas supports local non-profit organizations sponsored by its employees through matching gifts, financial assistance and volunteer support.
Requirements: Piedmont supports 501(c)3 organizations that are sponsored or assisted by its employees. Eligible organizations with employee sponsors are also able to apply for Employee Matching Gifts. Applications are available at Piedmont's website.
Restrictions: As a general rule, grants do not support: individuals or non-501(c)3 organizations; pre-college level private schools except through employee matching gifts; travel and conferences; third-party professional fund-raising organizations; controversial social causes; fraternal and veterans organizations or private clubs; religious organizations with programs limited to or expressly for their membership only; agencies already receiving corporate support through United Way or a united arts drive, with the exception of approved capital campaigns; or athletic events and programs.
Geographic Focus: North Carolina, South Carolina, Tennessee
Contact: George Baldwin, Managing Director; (704) 731-4063; fax (704) 731-4086; george.baldwin@piedmontng.com
Internet: http://www.piedmontng.com/ourcommunity/communityoutreach.aspx
Sponsor: Piedmont Natural Gas Corporation
4720 Piedmont Row Drive
Charlotte, NC 28210

Piedmont Natural Gas Foundation Health and Human Services Grants   3221
Piedmont Natural Gas Foundation's Health and Human Services grant focuses funding toward: organizations providing outreach services to community members with basic needs, including shelter, food and clothing; substance abuse or mental illness; organizations providing services or programs for a range of human service needs including youth engagement and mentoring, special needs and disability assistance, substance abuse or mental illnesses, transitional housing and situational homelessness support, and gang violence prevention; organizations providing emergency/disaster relief; increased access to critical healthcare services and comprehensive medical treatment including preventative care, prescription medication, medical exams, screenings, immunizations and dental care; and increased access to mental health services.
Requirements: Organizations must be a 501(c)3 non-profit organization or a qualified government entity. All grant requests must be submitted through Piedmont's online grant application form located at Piedmont's website.
Restrictions: Piedmont Natural Gas Foundation will not fund: religious, fraternal, political or athletic groups; four-year colleges and universities; private foundations; or social or veterans' organizations. In addition to these restrictions, contributions will generally not be made to or for: individuals; pre-college level private schools, except through the Employee Matching Gifts program; travel and conferences; third-party professional fundraising organizations; controversial social causes; religious organizations with programs limited to or expressly for their membership only; athletic events and programs; agencies already receiving Piedmont support through United Way or a united arts drive, with the exception of an approved capital campaign; or any proposal outside of the geographic area where Piedmont Natural Gas does business.
Geographic Focus: North Carolina, South Carolina, Tennessee
Amount of Grant: 500 - 30,000 USD
Contact: George Baldwin; (704) 731-4063; george.baldwin@piedmontng.com
Internet: http://www.piedmontng.com/ourcommunity/ourfoundation.aspx#guidelines
Sponsor: Piedmont Natural Gas Foundation
4720 Piedmont Row Drive
Charlotte, NC 28210

Pike County Community Foundation Grants   3222
The Pike County Community Foundation is a charitable organization formed to strengthen the Pike County, Indiana community by awarding grants to local nonprofits, by bringing individuals together to address community needs, and by offering personalized charitable gift planning services to donors. At the start of each cycle, a notice is mailed to nonprofit organizations that have previously applied for and received grants, or have otherwise requested notification of each cycle. Proposals are accepted from July through the September deadline. The grants committee will make its recommendations on funding to the Foundation's Board of Trustees, which will make final funding recommendations to the board of directors of the Community Foundation Alliance. All organizations that have submitted grant proposals are notified of the final outcome no later than December 1. Application instructions and samples of previously funded projects are available at the website.

Requirements: The Foundation welcomes proposals from nonprofit organizations that are deemed tax-exempt under sections 501(c)3 and 509(a) of the Internal Revenue Code and from governmental agencies serving Pike County. Proposals from nonprofit organizations not classified as a 501(c)3 public charity may be considered provided the project is charitable and supports a community need. Proposals submitted by an entity under the auspices of another agency must include a written statement signed by the agency's board president on behalf of the board of directors agreeing to act as the entity's fiscal sponsor, to receive grant monies if awarded, and to oversee the proposed project.
Restrictions: Project areas NOT considered for funding: religious organizations for religious purposes; political parties or campaigns; endowment creation or debt reduction; operating costs; capital campaigns; annual appeals or membership contributions; travel requests for groups or individuals such as bands, sports teams, or classes.
Geographic Focus: Indiana
Date(s) Application is Due: Nov 1
Contact: Director; (812) 354-6797; director@pikecommunityfoundation.org
Internet: http://www.pikecommunityfoundation.org/disc-grants
Sponsor: Pike County Community Foundation
714 Main Street, P.O. Box 587
Petersburg, IN 47567

Pinellas County Grants   3223
The Pinellas Community Foundation was established in 1969, it distributes grants twice annually to a wide variety of non-profit agencies, organizations and programs that enhance and support the quality of life in Pinellas County, Florida. Areas of interest are: art, culture, health care, environment, community development, employment opportunities, the underserved, and social services.
Requirements: Eligibility: a non-profit, 501(c)3; be headquartered in Pinellas County; provide social services to people in Pinellas County; not have a large endowment or fund raising staff; provide recent audited financial statements. For Grant Applications call Pinellas Community Foundation: (727)531-0058.
Geographic Focus: Florida
Date(s) Application is Due: Jun 15; Oct 1
Contact: Julie Scales; (727) 531-0058; fax (727) 531-0053; info@pinellasccf.org
Internet: http://www.pinellasccf.org/pinellas-community-foundation-pcf-clearwater-fl-our-grants-programs.htm
Sponsor: Pinellas County Community Foundation
5200 East Bay Drive, Suite 202
Clearwater, FL 33764

Pinkerton Foundation Grants   3224
The Pinkerton Foundation is an independent grantmaking foundation established in 1966 by Robert Allan Pinkerton with the broad directive to reduce the incidence of crime and to prevent juvenile delinquency. The Foundation's principal program interests are focused on economically disadvantaged children, youth and families, and severely learning disabled children and adults of borderline intelligence. The Foundation supports efforts to strengthen and expand community-based programs for children, youth and families in New York City. The Foundation also occasionally funds research, demonstration and evaluation projects in its principal program areas. While grants for direct service projects are usually limited to New York City, those with potential for national impact or replication may go beyond this geographic limitation. The Foundation's Board of Directors have two grantmaking meetings per year, in May and in December. Letters of inquiry are welcome throughout the year. Additional guidelines are available at the Foundation's website.
Requirements: Grants are awarded primarily to New York City 501(c)3 nonprofit public charitable organizations.
Restrictions: The foundation does not grant requests for emergencies, medical research, direct provision of health care, religious education, conferences, publications, or capital projects.
Geographic Focus: All States
Date(s) Application is Due: Feb 1; Sep 1
Amount of Grant: 25,000 - 250,000 USD
Samples: Reel Works Teen Filmmaking, Brooklyn, NY, $35,000--operating support for filmmaking program that serves up to 180 high school students; Madison Square Boys & Girls Club (New York, NY) $250,000--for after school and summer program operations at Madison's clubhouses; New York Cares, Inc., New York, NY, $20,000--in support of their SAT Prep Program.
Contact: Joan Colello; (212) 332-3385; fax (212) 332-3399; pinkfdn@pinkertonfdn.org
Internet: http://fdncenter.org/grantmaker/pinkerton
Sponsor: Pinkerton Foundation
610 Fifth Avenue, Suite 316
New York, NY 10020

Pinnacle Foundation Grants   3225
The Pinnacle Foundation was established in the State of Washington with a primary interest in family, military family support, critical housing, and homeless programs. Although there are no specified annual deadlines, a formal application is required. The standard application should include contact information, a description of the charitable function, and an overview of the project.
Geographic Focus: All States
Contact: Stanley J. Harrelson, President; (206) 215-9700 or (206) 215-9747
Sponsor: Pinnacle Foundation
2801 Alaskan Way, Suite 310
Seattle, WA 98121-1136

Pioneer Hi-Bred Society Fellowships   3226
The Pioneer Hi-Bred Society Fellowship is designed to contribute to agricultural research and education by providing financial support for outstanding students studying for graduate level degrees in plant breeding. Leading societies that support plant improvement graduate programs in crops of interest to Pioneer have been selected by the Fellowship Sub-committee. The institutions are responsible for student selection and administering the program. Funded societies include: Agronomic Science Foundation; American Phytopathological Society; American Society of Plant Biologists; the Consultative Group on International Agricultural Research (CGIAR); and the Entomological Foundation.
Requirements: Pioneer prefers this grant to be used solely to fund a student(s) progressing towards a graduate degree in plant breeding from the funded institution. No university overhead will be paid from these funds.
Geographic Focus: All States
Contact: David Harwood, (515) 270-3200 or (800) 247-6803, ext. 3915; fax (515) 334-4415; David.Harwood@pioneer.com
Internet: http://www.pioneer.com/web/site/portal/menuitem.07196fe7964ae318bc0c0a03d10093a0/
Sponsor: DuPont Pioneer
P.O. Box 1000
Johnston, IA 50131-0184

Piper Jaffray Foundation Communities Giving Grants   3227
The foundation supports organizations and programs that enhance the lives of people living and working in communities in which the company has offices. Of primary interest is support for organizations that increase opportunities for individuals to improve their lives and help themselves. Highest priority is given to family stability programs (including housing, family violence, responsible parenting), early childhood development, job training/career development, youth development, and adult education services. The foundation will also consider requests from organizations that work to increase citizen understanding or involvement in civic affairs or that enhance the artistic and cultural life of the community. Requests for general operating support from proven nonprofit organizations will be considered. Requests for project and capital support will be considered on a very selective basis. Support for higher education and K-12 public and private schools is provided primarily through the company gift-matching program. Contact the foundation for deadline dates.
Requirements: IRS 501(c)3 organizations located in the Minneapolis/Saint Paul metropolitan area should submit requests directly to the foundation. Organizations located outside the Minneapolis/Saint Paul metropolitan area should submit requests to the nearest Piper Jaffray office for forwarding to the foundation. Offices are located in communities in Arizona, California, Colorado, Idaho, Illinois, Iowa, Kansas, Kentucky, Minnesota, Missouri, Montana, Nebraska, Nevada, North Dakota, Ohio, Oregon, South Dakota, Tennessee, Utah, Washington, Wisconsin, and Wyoming.
Restrictions: Requests will not be considered from newly formed nonprofit organizations; individuals; teams; religious, political, veterans, or fraternal organizations; or organizations working to treat or eliminate specific diseases. Support is not available for basic or applied research, travel, event sponsorship, benefits or tickets, or to eliminate an organization's operating deficit.
Geographic Focus: Arizona, Arkansas, California, Colorado, Idaho, Illinois, Iowa, Kansas, Kentucky, Minnesota, Missouri, Montana, Nebraska, Nevada, North Dakota, Ohio, Oregon, South Dakota, Tennessee, Utah, Washington, Wisconsin, Wyoming
Date(s) Application is Due: Mar 18
Amount of Grant: 1,000 - 5,000 USD
Contact: Connie McCuskey, Vice President; (612) 303-1309; fax (612) 342-6085; communityrelations@pjc.com
Internet: http://www.piperjaffray.com/2col_largeright.aspx?id=127
Sponsor: Piper Jaffray Foundation
800 Nicollet Mall, Suite 800
Minneapolis, MN 55402

Piper Trust Healthcare and Medical Research Grants   3228
The Piper Trust's grantmaking focuses on Virginia Galvin Piper's commitment to improving the quality of life for residents of Maricopa County. Piper Trust's particular interest lies with projects that benefit young children, adolescents and older adults in Maricopa County. The Trust makes grants to faith-based organizations that serve these target populations in a manner consistent with program guidelines. For Healthcare and Medical Research grants, the trust is most interested in proposals that address improved facilities for children, adolescents and older adults; better trained healthcare workforce; increased access to basic healthcare; and, centers for advancement in personalized medicine.
Requirements: Piper Trust makes grants to actively operating Section 501(c)3 organizations in Maricopa County. These organizations must have been in operation for at least three years from the effective date of their IRS ruling. Special rules apply to private foundations and 509(a)3 (Type III) organizations. There are no deadlines on initial proposals, and letters of inquiry throughout the year are reviewed throughout the year. If the Trust asks for a full proposal, its disposition depends on its completeness and the meeting schedule of the Piper trustees.
Geographic Focus: Arizona
Contact: Terri Leon, Program Officer; (480) 556-7121; tleon@pipertrust.org
Internet: http://pipertrust.org/our-grants/healthcare-medical-research/
Sponsor: Virginia G. Piper Charitable Trust
1202 East Missouri Avenue
Phoenix, AZ 85014

Piper Trust Older Adults Grants 3229
The Piper Trust's grantmaking focuses on Virginia Galvin Piper's commitment to improving the quality of life for residents of Maricopa County. Piper Trust's particular interest lies with projects that benefit young children, adolescents and older adults in Maricopa County. The Trust makes grants to faith-based organizations that serve these target populations in a manner consistent with program guidelines. For Older Adults grants, the trust is most interested in proposals that address disease and disability prevention; assistance for older adults to remain independent; and, volunteerism, "recareering" and community engagement.
Requirements: Piper Trust makes grants to actively operating Section 501(c)3 organizations in Maricopa County. These organizations must have been in operation for at least three years from the effective date of their IRS ruling. Special rules apply to private foundations and 509(a)3 (Type III) organizations. There are no deadlines on initial proposals, and letters of inquiry throughout the year are reviewed throughout the year. If the Trust asks for a full proposal, its disposition depends on its completeness and the meeting schedule of the Piper trustees.
Geographic Focus: Arizona
Contact: Terri Leon, Program Officer; (480) 556-7121; tleon@pipertrust.org
Internet: http://pipertrust.org/our-grants/older-adults/
Sponsor: Virginia G. Piper Charitable Trust
1202 East Missouri Avenue
Phoenix, AZ 85014

Pittsburgh Foundation Community Fund Grants 3230
The Pittsburgh Foundation, established in 1945, is the 14th largest community foundation in the county. The foundation awards grants to organizations operating primarily in the Pittsburgh area and Allegheny County, PA. The Foundation supports five targeted areas of impact: education; economic development; families, children and youth; health care; arts.
Requirements: Grants in the five targeted areas for impact are awarded to nonprofit organizations that are defined as tax-exempt under Section 501(c)3 of the Internal Revenue Code and are located within Allegheny County, or can demonstrate that a significant majority of their population served is from Allegheny County.
Geographic Focus: Pennsylvania
Amount of Grant: 30,000 - 150,000 USD
Contact: Jane Downing; (412) 391-5122; fax (412) 391-7259; downingj@pghfdn.org
Internet: http://www.pittsburghfoundation.org/page8894.cfm
Sponsor: Pittsburgh Foundation
5 PPG Place, Suite 250
Pittsburgh, PA 15222-5414

Pittsburgh Foundation Medical Research Grants 3231
Through the generosity of individuals and families who have established medical research funds with The Pittsburgh Foundation, the Foundation provides support for this cutting edge research. Based on the donors' fields of interest, the Foundation provides grants for the following: Richard S. Caliguiri Fund - for research in the field of amyloidosis; The Walter L. Copeland Fund - for cranial research (disease of the brain/neurological disease), to be conducted within the Department of Neurological Surgery at the University of Pittsburgh; John F. and Nancy A. Emmerling Fund - for research in the field of neuropsychobiology (with emphasis on biochemistry) of mental disorders, with the goal of discovering the causes and cure, of amelioration, of mental disorders. The principal investigator must have full time status at the University of Pittsburgh; Albert B. Ferguson, M.D. Orthopaedic Fund - for orthopaedic research; GLBT Health and Wellness Fund - to assess the incidence, cause or cure of certain diseases (i.e., cancer, heart disease, diabetes, arthritis, AIDS and Alzheimer's disease) in the Gay, Lesbian, Bisexual and Transgender (GLBT) community; Medical Research Funding - for research in the field of cancer, heart disease, arthritis, diabetes, multiple sclerosis and geriatric diseases (with focuses particularly on Alzheimer's disease and other related diseases of the elderly).
Requirements: Pennsylvania not-for-profit institutions with 501(c)3 tax-exempt status are invited to apply for funding. Support is given to investigators in Allegheny County with a proven record of performance in the area under investigation. Guidelines and applications vary, see: http://www.pittsburghfoundation.org/page32428.cfm for a complete listing of each grant.
Geographic Focus: Pennsylvania
Contact: Judy Powell, Administrative Support; (412) 394-4294; powellj@pghfdn.org
Internet: http://www.pittsburghfoundation.org/page32428.cfm
Sponsor: Pittsburgh Foundation
5 PPG Place, Suite 250
Pittsburgh, PA 15222-5414

PKD Foundation Lillian Kaplan International Prize for the Advancement in the Understanding of Polycystic Kidney Disease 3232
The prize is a joint venture between the foundation and the International Society of Nephrology. The winning nominee of the initial award received the prize in Berlin; it has been awarded every other year since. Each award consists of a monetary award, a framed certificate with an appropriate citation, and a designated sculpture to clinicians and scientists who have made outstanding discoveries in basic science or who have made outstanding clinical contributions to the care of individuals. This is the largest cash prize for a clinical medical specialty in the world signaling the prestige and importance it carries. PKD experts from both organizations judge nominations.
Requirements: Awards are made to residents of any country without restriction to gender, race, religion, creed, or nationality. Nomination letters should describe the key role of the nominee(s) in a biomedical advance relevant to PKD and should be accompanied by a full curriculum vita with a complete mailing address and at least two supporting letters that detail a major discovery or sustained advance. If two individuals are nominated to share the prize, the reasons for why the prize would be shared must be clearly delineated.
Geographic Focus: All States
Date(s) Application is Due: Feb 1
Amount of Grant: 50,000 USD
Samples: Dr. Corinne Antignac, Necker Hospital, Paris, France, $50,000; Dr. Lisa Guay-Woodford, University of Alabama at Birmingham, $50,000; Dr. Friedhelm Hildebrandt, University of Michigan, $50,000.
Contact: Dan Lara, Government Relations Manager; (816) 268-8480; fax (816) 931-8655; larad@pkdcure.org or pkdcure@pkdcure.org
Internet: http://www.pkdcure.org
Sponsor: PKD Foundation
9221 Ward Parkway, Suite 400
Kansas City, MO 64114-3367

PKD Foundation Research Grants 3233
The PKD Foundation, established in 1982 by Joseph H. Bruening and Jared J. Grantham, leads the fight against polycistic kidney disease through vital research funding and patient education. Annually, the PKD Foundation funds more than $4.2 million in vital PKD research. Most recently, clinical trials in cardiovascular health, Rapamycin, Somatostatin, Tolvaptan, and more have offered help and hope to the 12.5 million worldwide with the disease.
Geographic Focus: All States
Amount of Grant: 25,000 - 1,000,000 USD
Contact: Dan Lara, Government Relations Manager; (816) 268-8480; fax (816) 931-8655; larad@pkdcure.org or pkdcure@pkdcure.org
Internet: http://www.pkdcure.org/Research/tabid/82/Default.aspx
Sponsor: PKD Foundation
9221 Ward Parkway, Suite 400
Kansas City, MO 64114-3367

Plastic Surgery Educational Foundation/AAO-HNSF Combined Grant 3234
The PSEF/AAO-HNSF Combined Grant is designed to promote interdisciplinary research by encouraging a collaborative research effort between the specialties of Plastic Surgery and Otolaryngology. Recipients of this grant must demonstrate a potential for excellence in research and training and a commitment to an academic research career in either Plastic Surgery or Otolaryngology.
Requirements: Applicants must have demonstrated a potential for excellence in research and teaching and serious commitment to an academic research career in otolaryngology or plastic surgery. Priority will be given to senior residents, fellows or faculty who have completed residencies or fellowships within four years of the application receipt date.
Geographic Focus: All States
Date(s) Application is Due: Dec 15
Amount of Grant: Up to 20,000 USD
Contact: Carol V. Wargo, Director of Corporate Relations; (847) 228-3358 or (847) 228-9900; fax (847) 228-9131; cvw@plasticsurgery.org or cschmieden@plasticsurgery.org
Internet: http://www.plasticsurgery.org/medical_professionals/research/Grant-Applications.cfm
Sponsor: Plastic Surgery Educational Foundation
444 E Algonquin Road
Arlington Heights, IL 60005

Plastic Surgery Educational Foundation Basic Research Grants 3235
The Plastic Surgery Educational Foundation and the American Society for Aesthetic Plastic Surgery offer annual research grants to provide seed money to U.S. and Canadian surgeons for research projects in the field of plastic and reconstructive surgery. Nonmembers of PSEF must submit a letter of sponsorship from a foundation member.
Requirements: Applicant must be a plastic surgeon, MD, or PhD working in plastic surgery.
Geographic Focus: All States
Date(s) Application is Due: Jan 17
Contact: Carol V. Wargo, Director of Corporate Relations; (847) 228-3358 or (847) 228-9900; fax (847) 228-9131; cvw@plasticsurgery.org or cschmieden@plasticsurgery.org
Internet: http://www.plasticsurgery.org/medical_professionals/research/Grant-Applications.cfm
Sponsor: Plastic Surgery Educational Foundation
444 E Algonquin Road
Arlington Heights, IL 60005

Plastic Surgery Educational Foundation Research Fellowship Grants 3236
The Research Fellowship Grants are awarded for the purpose of encouraging research and academic career development in Plastic Surgery. The PSEF Research Fellowship is for training in any area of Plastic Surgery and is designed for those who wish to supplement one year of clinical training for a research experience. The PSEF/KCI Wound Care Research Fellowship is open to plastic surgery residents and recent plastic surgery resident graduates focusing on a project in wound care management.
Geographic Focus: All States
Contact: Carol V. Wargo, Director of Corporate Relations; (847) 228-3358 or (847) 228-9900; fax (847) 228-9131; cvw@plasticsurgery.org or cschmieden@plasticsurgery.org
Internet: http://www.plasticsurgery.org/medical_professionals/research/Grant-Applications.cfm
Sponsor: Plastic Surgery Educational Foundation
444 E Algonquin Road
Arlington Heights, IL 60005

Playboy Foundation Grants 3237
The Playboy Foundation seeks to foster social change by confining its grants and other support to projects of national impact and scope involved in fostering open communication about, and research into, human sexuality, reproductive health and rights; protecting and fostering civil rights and civil liberties in the United States for all people, including women, people affected and impacted by HIV/AIDS, gays and lesbians, racial minorities, the poor and the disadvantaged; and eliminating censorship and protecting freedom of expression and First Amendment rights. The Foundation does not accept unsolicited proposals, but welcomes letters of inquiry via post for the areas of interest noted above. Grants awarded by the Foundation are typically up to $10,000.
Restrictions: The foundation will not consider religious programs, individual needs, capital campaigns, endowments, scholarships, or fellowships; social services, including residential care, clinics, treatment, or recreation programs; national health, welfare, educational, or cultural organizations, or their state affiliates; or government agencies or projects.
Geographic Focus: All States
Amount of Grant: 5,000 - 10,000 USD
Contact: Executive Director; (312) 373-2437 or (312) 751-8000; fax (312) 751-2818; giving@playboy.com
Internet: http://www.playboyenterprises.com/foundation
Sponsor: Playboy Foundation
680 North Lake Shore Drive
Chicago, IL 60611

Plough Foundation Grants 3238
The foundation awards grants to nonprofit organizations in Shelby County, Tennessee, (emphasis on Memphis) for charitable purposes that will benefit the greatest number of people in the state. Areas of interest include early childhood and elementary education, crime, health care, economic development, social service agencies, housing and homelessness, and the arts. Types of support include building/renovation, capital campaigns, endowments, equipment, land acquisition, management development/capacity building, matching/challenge support, professorships, program-related investments/loans, program development, program evaluation, and seed money. Applicants should submit a brief letter of three pages or less explaining the specific project for which funds are needed and the results anticipated if the project is funded. Organizations whose projects are within the focus area will be invited to submit full proposals. The board meets in February, May, August, and November to review requests.
Requirements: Nonprofit organizations in Shelby County, TN, with particular emphasis on Memphis, may submit letters of request.
Restrictions: Funding requests are denied for annual operating expenses, individuals, and projects outside the Memphis, TN, area.
Geographic Focus: Tennessee
Date(s) Application is Due: Jan 10; Apr 10; Jul 10; Oct 10
Amount of Grant: 1,000 - 1,250,000 USD
Contact: Scott McCormick, Executive Director; (901) 529-4063; mail@plough.org
Sponsor: Plough Foundation
62 North Main Street, Suite 201
Memphis, TN 38103

PMI Foundation Grants 3239
The PMI Foundation awards grants nationally, with emphasis on California, to a wide range of organizations with the goal of expanding homeownership. The foundation also contributes generously to deserving causes and charities in the areas of arts and culture, health and human services, education, civic organizations, and community development. Application guidelines are available for download from the PMI website. There are no application deadlines. Check with foundation staff to verify whether they are currently accepting applications.
Requirements: 501(c)3 organizations are eligible. Requests must target the disadvantaged, the poor, and distressed populations. Requests must either focus on increasing affordable housing opportunities or directly contribute to the quality of life in under-served communities.
Restrictions: The PMI Foundation does not accept requests for the following purposes: individuals; fraternal, veteran, labor, athletic or religious organizations serving a limited constituency; political or lobbying organizations, or those supporting the candidacy of a particular individual; travel funds; and films, videotapes or audio productions.
Geographic Focus: All States
Amount of Grant: 200 - 200,000 USD
Contact: Laura Kinney, Foundation Grant Administrator; (925) 658-6562
Internet: http://www.pmifoundation.org/index.html
Sponsor: PMI Foundation
3003 Oak Road
Walnut Creek, CA 94597-2098

PNC Foundation Community Services Grants 3240
The Foundation supports a variety of nonprofit organizations with a special emphasis on those that work to achieve sustainability and touch a diverse population, in particular, those that support early childhood education and/or economic development. Through this program, support is given to social services organizations that benefit the health, education, quality of life or provide essential services for low-and moderate-income individuals and families. The Foundation supports job training programs and organizations that provide essential services for their families. PNC provides support for early learning and educational enrichment programs for children in low-and moderate-income families as well as for the construction of community facilities that benefit low-and moderate-income communities.
Requirements: Organizations receiving support from the PNC Foundation must have a current Internal Revenue Service tax-exempt designation and be eligible to receive charitable contributions. In addition, the proposed activity must occur in a community where PNC has a significant presence. This includes select counties in the following states: Alabama; Delaware; District of Columbia; Florida; Georgia; Illinois; Indiana; Kentucky; Maryland; Michigan; Missouri; New Jersey; North Carolina; Ohio; Pennsylvania; South Carolina; Virginia; and Wisconsin.
Restrictions: The Foundation will not provide support for: organizations that discriminate by race, color, creed, gender or national origin; religious organizations, except for non-sectarian activities; advocacy groups; operating funds for agencies that receive funds through PNC United Way allocation; individuals or private foundations; annual funds of hospitals or colleges and universities; conferences and seminars; or tickets and goodwill advertising.
Geographic Focus: Alabama, Delaware, District of Columbia, Florida, Georgia, Illinois, Indiana, Kentucky, Maryland, Michigan, Missouri, New Jersey, North Carolina, Ohio, Pennsylvania, South Carolina, Virginia, Wisconsin
Contact: Sally McCrady, Chairwoman; (412) 768-8371 or (888) 762-2265; fax (412) 705-3584; sally.mccrady@pnc.com
Internet: https://www1.pnc.com/pncfoundation/foundation_grantProcess.html
Sponsor: PNC Foundation
630 Liberty Avenue, Tenth Floor
Pittsburg, PA 15222-2705

Pohlad Family Foundation 3241
The Pohlad Foundation concentrates their funding on the arts, health and human services, capital grants, and youth programs. The Foundation strives to support programs that: support all types and sizes of art programs of interest to diverse audiences; provide essential and effective human services to disadvantaged children and families, fund a range of services from emergency needs to workforce development, and have a positive multi-year impact;
Requirements: Applications are considered only in response to a request for proposal or an invitation to apply. Organizations must be 501(c)3 non-profit organizations. Priority is given to organizations that demonstrate a commitment to their communities by ensuring that their governing bodies include representatives from within the community and projects that directly benefit communities in need. Organizations may view sample grant recipients and specific instructions on how to apply on the website.
Restrictions: Grants are focused in Minnesota, primarily St. Paul and Minneapolis. Funding is not available for: direct funding to individuals; health or housing-related emergency assistance to individuals; benefits, fundraisers, walk-a-thons, telethons, galas, or other revenue generating events; advertising; organizations that discriminate on the basis of race, gender, religion, culture, age, physical ability or disability, sexual orientation, gender identity, status as a military veteran or genetic information; veterans' and fraternal organizations; political or lobbying organizations; replacement of government funding.
Geographic Focus: Minnesota
Amount of Grant: 5,000 - 20,000 USD
Contact: Rose Peterson, Program Manager; (612) 661-3903; fax (612) 661-3715; rpeterson@pohladfamilygiving.org
Internet: http://pohladfoundation.org/giving/grant-guidelines.html
Sponsor: Pohlad Family Foundation
60 South Sixth Street, Suite 3900
Minneapolis, MN 55402

Pokagon Fund Grants 3242
The mission of the Pokagon Fund is to enhance the lives of the residents in the New Buffalo region of southwest Michigan through the financial support of local governments, nonprofits, charities and other organizations. Funding supports initiatives in the areas of health and human services, education, arts and culture, recreation, and environment. A number of distinct application forms are available at the website, including: a discretionary application; a band application; a municipal application; and a bus application. Annual deadlines for all application submissions are January 15, April 15, July 15, and October 15.
Requirements: 501(c)3 organizations throughout the Pokagon service area are eligible.
Geographic Focus: Michigan
Date(s) Application is Due: Jan 15; Apr 15; Jul 15; Oct 15
Contact: Janet Cocciarelli, Executive Director; (269) 469-9322; jcocciarelli@pokagonfund.org or grants@pokagonfund.org
Internet: http://www.pokagonfund.org/How.asp
Sponsor: Pokagon Fund
821 E. Buffalo Street
New Buffalo, MI 49117

Polk Bros. Foundation Grants 3243
The primary focus of the Foundation is programs that work with populations of need, particularly children, youth, and families in underserved Chicago communities. Very few awards are made to organizations located outside the city of Chicago. Grants are made for both new and ongoing initiatives in four program areas: social service, education, culture and health care. In all areas, proposals should address increased access to services and improvement of the quality of life for area residents. Grants are seldom made for capital support.
Requirements: Illinois 501(c)3 nonprofit organizations are eligible. Preference is given to requests from Chicago. An organization that has not previously received a grant from the Foundation should first call the Foundation office or complete the pre-application form available on the Foundation website. The Foundation will then mail an application form or contact the organization with further questions.
Restrictions: The Polk Bros. Foundation will not support: organizations that devote a substantial portion of their activities to attempting to influence legislation or to participating in campaigns on behalf of candidates for public office; religious institutions seeking support for programs whose participants are restricted by religious affiliation or whose services promote a particular creed; purchase of dinner or raffle tickets or

advertising in dinner programs; medical, scientific, or academic research; grants to individuals; tax-generating entities (municipalities, school districts, etc.) for services within their normal responsibilities. The Foundation will not consider more than one request from an organization or its affiliates in a 12-month period, nor will it generally fund more than eight percent of an organization's operating budget.
Geographic Focus: Illinois
Amount of Grant: Up to 15,000 USD
Samples: Chicago High School for the Arts, $750,000 - general operating; Advocate Illinois Masonic Medical Center, $90,000 - school-based health centers at Amundsen and Lake View High Schools; Albany Park Theater Project, $80,000 - youth development through theater program; America Scores Chicago, $25,000 - salary support for the education and soccer directors;
Contact: Suzanne Doombos Kerbow; (312) 527-4684; questions@polkbrosfdn.org
Internet: http://www.polkbrosfdn.org/guidelines.htm
Sponsor: Polk Brothers Foundation
20 West Kinzie Street, Suite 1110
Chicago, IL 60611

Pollock Foundation Grants 3244
The foundation awards grants to Texas nonprofit organizations in its areas of interest, including cultural programs, dental education and schools, health care and health organizations, Jewish organizations and temples, libraries and library science, nursing, public health education and schools, social services, and youth development. Grants support program develoment and general operating expenses. There are no application deadlines. Contact the office for application materials.
Requirements: Texas nonprofits are eligible. Preference is given to requests from Dallas.
Geographic Focus: Texas
Amount of Grant: 1,000 - 300,000 USD
Samples: Reading and Radio Resource, Dallas, TX, $6,000; Dallas Symphony Association, Dallas, TX, $300,000; Temple Emanu-El, Dallas, TX, $50,000.
Contact: Robert Pollock, Trustee; (214) 871-7155; fax (214) 871-8158
Sponsor: Pollock Foundation
2626 Howell Street, Suite 895
Dallas, TX 75204

Porter County Foundation Lilly Endowment Community Scholarships 3245
The Lilly Endowment Community Scholarship Program is designed to raise the level of educational attainment and improve the quality of life for Indiana residents. To help further this goal, Lilly Endowment is providing funding to enable Indiana community foundations to offer the scholarship program. Three full tuition scholarships will be awarded to residents of Porter County who have graduated from an accredited Indiana high school. Students must pursue a baccalaureate degree at any accredited public or private Indiana college or university. The scholarship covers a four-year degree program including full tuition, required fees, and a special allocation of up to $800 per year for required books and equipment. Students will be evaluated on financial need, academic accomplishments, ability to communicate, motivation and desire to succeed, along with activities and leadership capabilities. The scholarship will be awarded to those who demonstrate that the scholarship will make a difference in their lives. Scholarship recipients may transfer from one Indiana college or university to another during the four-year period.
Requirements: Applicants must be U.S. citizens residing in Porter County, and graduates of an accredited Indiana high school. They must also be accepted to begin a full time four year study at an accredited public of private Indiana college or university. All college majors are eligible. In addition to the application, students must submit two letters of recommendation; an official high school transcript; copy of notification of SAT/ACT score; their completed financial information sheet; and a one page essay addressing the importance of the scholarships in helping them achieve their future goals. Applicants should refer to the Porter County website for current applications and deadlines. The application and supporting materials should be submitted to the candidate's individual high school counselor.
Restrictions: The scholarship will not extend beyond the four-year period.
Geographic Focus: Indiana
Contact: submit application materials to Individual Guidance Counselor
Internet: http://www.portercountyfoundation.org/scholarships.html
Sponsor: Porter County Community Foundation
57 South Franklin Street, Suite 207, P.O. Box 302
Valparaiso, IN 46384

Porter County Health and Wellness Grant 3246
The Porter County Health and Wellness Fund awards grants to nonprofit organizations that promote, support, and/or advance health care in Porter County. Funding priorities include: increasing health care access for the underserved; improving and promoting healthy lifestyles for youth; and improving the nonprofit's operational capabilities to provide health care services. The maximum grant amount is $25,000.
Requirements: In addition to the application, all grant application packets must include the following information: a grant request cover page; a grant narrative; a project budget; a current operating budget and financial statement; the names and principal occupations of the organization's Board of Directors; the organization's grant application approval by their Board of Directors; and a copy of the organization's 501(c)3 tax exemption ruling.
Restrictions: Grants will not be made to: individuals; membership contributions; event sponsorships; programs that are sectarian or religious in nature; political organizations or candidates; contributions to endowment campaigns; campaigns to reduce previously incurred debt; and programs already completed.
Geographic Focus: Indiana
Date(s) Application is Due: Jun 15
Amount of Grant: 25,000 USD
Contact: Brenda Sheetz, Health/Wellness Fund Contact; (219) 465-0294; fax (219) 464-2733; bsheetz@portercountyfoundation.org
Internet: http://www.portercountyfoundation.org/grantprograms.html
Sponsor: Porter County Community Foundation
57 South Franklin Street, Suite 207, P.O. Box 302
Valparaiso, IN 46384

Porter County Health Occupations Scholarship 3247
The Porter County Health Occupations Scholarship provides funding for those wishing to continue their studies in a health-related field at a 2 or 4 year college or university in Indiana. The scholarship is evaluated on the student's financial need, academics, and capacity to succeed.
Requirements: Applicants must be U.S. citizens and residents of Porter County, and must have already completed one year of study. They must be in good standing at an Indiana accredited university or college, and enrolled in a health-related field of study. Along with the application, candidates should include their college transcript, a copy of their current Student Aid Report from FAFSA, a letter of recommendation from a professor, employer, or other professional affiliation, a one page statement describing why they chose a health occupation, and where they see themselves in ten years.
Geographic Focus: Indiana
Date(s) Application is Due: May 31
Contact: Scholarships; (219) 465-0294; fax (219) 464-2733; info@portercountyfoundation.org
Internet: http://www.portercountyfoundation.org/scholarships.html
Sponsor: Porter County Community Foundation
57 South Franklin Street, Suite 207, P.O. Box 302
Valparaiso, IN 46384

Porter County Women's Grant 3248
The Women's Grant of Porter County seeks to improve the quality of life for women and children in Porter County by collectively funding high impact grants for charitable initiatives with the same purpose. The group will award a $45,000 grant to a nonprofit organization in support of a project or program that addresses the issues of women and children and is sustainable. Priority will be given to innovative programs that demonstrate positive outcomes in one of the following areas: education and training to promote economic security and self-sufficiency; leadership development and programs designed to build self-esteem; access to women's health services and healthy lifestyles; safe environments and freedom from violence; and access to affordable daycare services that will expand hours of service, increase the number served on a sustainable basis and/or improve quality.
Requirements: In addition to the online application form, all application packets must include the following information: a grant request cover page; a grant narrative (see application for narrative details); a project budget; a copy of the organization's current year operating budget; the organization's most recent financial statement; names and principal occupations of the Board of Directors; evidence of Board approval for this application; a copy of the organization's 501(c)3 tax exemption.
Restrictions: The grant will not fund the following: projects or programs that do not address issues facing women and/or children; scholarship programs including daycare and program participation fees; annual appeals or membership contributions; event sponsorships; programs that are sectarian or religious in nature; political organizations or candidates; contributions to endowment campaigns; campaigns to reduce previously incurred debt; individuals; programs already completed and/or equipment already contracted for; and travel for bands, sports teams and similar groups.
Geographic Focus: Indiana
Date(s) Application is Due: Apr 15
Contact: Brenda Sheetz; (219) 465-0294; bsheetz@portercountyfoundation.org
Internet: http://www.portercountyfoundation.org/grantprograms.html
Sponsor: Porter County Community Foundation
57 South Franklin Street, Suite 207, P.O. Box 302
Valparaiso, IN 46384

Portland Foundation - Women's Giving Circle Grant 3249
The mission of The Portland Foundation Women's Giving Circle is to build a community of women philanthropists through the pooling of knowledge and resources for the purpose of providing grants to Jay County organizations and initiatives that address mutually-agreeable issues. The Circle awards grants which focus on enhancing the capacity of Jay County's not-for-profit organizations to support programming that addresses needs for youth and families. The Circle strives to fund innovative endeavors to benefit Jay County residents from toddlers to senior citizens.
Requirements: Organizations are encouraged to contact the Foundation to discuss whether their project is appropriate for funding. The application is available at the Foundation site. In addition to the application, proposals should include 10 copies of the following information: a copy of the IRS determination letter confirming tax-exempt status; organization's most recent financial statement, including budget and year-to-date income and expenses; and a detailed list of the Board of Directors.
Restrictions: The following are excluded from funding: organizations for religious or sectarian purposes; make-up of operating deficits, post-event or after-the-fact situations; endowment or capital projects and campaigns; for any propaganda, political or otherwise, attempting to influence legislation or intervene in any political affairs or campaigns; and dinner galas, advertising or other special fundraising events.
Geographic Focus: Indiana
Amount of Grant: 200 - 1,000 USD
Samples: Arts Place, Inc., Arts in the Parks Clay Camp and Family Pottery Nights, $525,; Jay County Public Library, 1000 Books Before Kindergarten, $525; State of the Heart Hospice, Camp BEARable, $200.

Contact: Douglas Inman; (260) 726-4260; fax (260) 726-4273; tpf@portlandfoundation.org
Internet: http://www.portlandfoundation.org/womens-giving-circle-grant
Sponsor: Portland Foundation
112 East Main Street
Portland, IN 47371

Portland General Electric Foundation Grants 3250
Portland General Electric is committed to improving the quality of life for all Oregonians. The Foundation focuses giving in three areas: education, arts and culture, and healthy families. Education funding dedicates awards to scholarships, innovation in classroom instruction, transitional bridges between grade levels, career readiness and at-risk youth programs. Arts and Culture awards support youth programs and adult cultural programs that enhances understanding in communities. Healthy Families funding promotes access and services in areas of health, domestic violence, parenting, foster care, and other services that benefit families.
Requirements: 501(c)3 nonprofits in Oregon that address issues in the three focus areas of the Foundation are eligible for funding. Applicants must first submit a Letter of Inquiry to the Foundation; applicants may then be invited to submit a full application. Both of these steps must be completed online.
Restrictions: The foundation does not fund: bridge grants, debt retirement or operational deficits; endowment funds; general fund drives or annual appeals; political entities, ballot measure campaigns or candidates for political office; organizations that discriminate against individuals on the basis of creed, color, gender, sexual orientation, age, religion or national origin; fraternal, sectarian and religious organizations; individuals; travel expenses; or conferences, symposiums, festivals, events, team sponsorships or user fees. The Foundation also does not directly fund public K-12 education.
Geographic Focus: Oregon
Date(s) Application is Due: Jan 11; Apr 5; Jul 5; Nov 1
Amount of Grant: 2,500 - 10,000 USD
Samples: Artists Repertory Theatre, Oregon, $7,500 - for educational programming; Chess for Success, Oregon, $5,000 - after-school chess programs and tournaments for elementary schools; Lifeworks NW, Oregon, $10,000 - vocational assistance to adults living with severe mental illness.
Contact: Paige Haxton; (503) 464-8818; fax (503) 464-2929; pgefoundation@pgn.com
Internet: http://www.pgefoundation.org/how_we_fund.html
Sponsor: Portland General Electric Foundation
121 SW Salmon Street, One World Trade Center, 3rd Floor
Portland, OR 97204

Posey Community Foundation Women's Fund Grants 3251
The Posey County Community Foundation is a nonprofit, public charity created by and for the people of Posey County, Indiana. The Posey County's Women's Fund makes yearly grants to support a variety of projects or programs serving women and girls in Posey County, Indiana. These programs include those that prevent domestic violence, secure family-supporting jobs, promote health and education, and develop confidence. Grant proposals are accepted once each year. Grants are normally given as one time support of a project but may be considered for additional support for expansions or outgrowths of an initial project. The application form and examples of previously funded projects available at the website.
Requirements: Projects must address the needs which support the Fund's mission by providing opportunities, encouragement, knowledge, information, and hope for the community's women and girls. The Women's Fund welcomes proposals from non-profit organizations that are tax-exempt under sections 501(c)3 and 509(a) of the Internal Revenue Code and from governmental agencies serving Posey County women and girls. Proposals from other non-profit organizations that address issues facing women and girls in the county may be accepted. Proposals submitted by an entity under the auspices of another agency must include a written statement signed by the agency's board president on behalf of the board of directors agreeing to act as the entity's fiscal sponsor, to receive grant monies if awarded, and to oversee the proposed project.
Geographic Focus: Indiana
Date(s) Application is Due: Jul 5
Contact: Johnna Denning, Director; (812) 838-0288; fax (812) 838-8009; johnna@poseycommunityfoundation.org
Internet: http://www.poseycommunityfoundation.org/wf-grantmaking
Sponsor: Posey County Community Foundation
402 Main Street, P.O. Box 746
Mt. Vernon, IN 47620

Posey County Community Foundation Grants 3252
The Posey County Community Foundation is a nonprofit, public charity created by and for the people of Posey County, Indiana. Grant proposals are accepted once each year as a one-time project support. At the beginning of each cycle in January, notices are sent to nonprofit organizations that have previously applied for grants, have received grants in the past, or have otherwise requested notification. Applicants are encouraged to schedule a meeting with the Foundation's director to receive an overview of the grant process. Proposals are accepted from January through March. The grant application is available at the website.
Requirements: The Foundation welcomes funding requests from nonprofit organizations that are 501(c)3 tax exempt. For those organizations not tax exempt, requests may be considered if the project is charitable and supports a community need.
Restrictions: Funding is not available for religious organizations for religious purposes; political parties or campaigns; endowment creation or debt reduction; operating costs; capital campaigns; annual appeals or membership contributions; or travel requests for groups or individuals such as bands, sports teams, or classes.
Geographic Focus: Indiana

Date(s) Application is Due: Mar 6
Amount of Grant: Up to 5,000 USD
Contact: Johnna Denning, Director; (812) 838-0288; fax (812) 838-8009; johnna@poseycommunityfoundation.org
Internet: http://www.poseycommunityfoundation.org/disc-grants
Sponsor: Posey County Community Foundation
402 Main Street, P.O. Box 746
Mt. Vernon, IN 47620

Pott Foundation Grants 3253
Established in 1963 by Herman T. Pott, The Pott Foundation is a non-profit organization supporting children, education and health and human services. Most of the grant recipients are located in Herman Pott's adopted city of St. Louis, Missouri. The Foundation distributes approximately $1.5 million annually to over 100 charities.
Requirements: Non-profit 501(c)3 organization are eligible to apply in Missouri. Mail grant applications to: The Pott Foundation, C/O U.S. Bank, N.A., The Private Client Reserve, Attn: Carol Eaves, Mail Loc: SL-MO-CTCS, 10 North Hanley, Clayton, MO 63105.
Geographic Focus: Missouri
Date(s) Application is Due: Apr 1
Contact: Carol Eaves, (314) 418-8317
Sponsor: Pott Foundation
10 North Hanley
Clayton, MO 63105

Potts Memorial Foundation Grants 3254
The Potts Memorial Foundation is a private foundation, incorporated in 1922 in New York. The Foundation gives on a national and international basis. The foundation was established to provide for the care, treatment, and rehabilitation of persons afflicted with tuberculosis; support for tuberculosis eradication, including fellowship programs for physicians. Grants given for tuberculosis and related disorders.
Requirements: Application form not required, Applicants should submit a detailed description of project and amount of funding requested. Proposal must be submitted via mail.
Restrictions: No grants to individuals, or for endowment funds or matching gifts; no loans.
Geographic Focus: All States
Date(s) Application is Due: Apr 15; Sep 15
Amount of Grant: 15,000 - 45,000 USD
Samples: Johns Hopkins University, Baltimore, Maryland, $41,800; Columbia Memorial Hospital, Hudson, New York, $20,000;
Contact: Charles Inman, Secretary; (518) 828-3365
Sponsor: Potts Memorial Foundation
444 Warren Street
Hudson, NY 12534-2415

Powell Foundation Grants 3255
The purpose of the Powell Foundation is to distribute funds for public charitable purposes, principally for the support, encouragement and assistance to education, health, conservation, and the arts with a direct impact within the Foundation's geographic zone of interest The Foundation places priority on organizations and programs that serve residents in Harris, Travis and Walker counties, Texas, principally in the fields of education, the arts, health and conservation. The Foundation's current emphasis is in the field of public education in the broadest sense. Other areas of interest continue to be community service projects focused on the needs of children, the disadvantaged, the urban environment, and the visual and performing arts, especially in the Greater Houston, Texas area. The Foundation operates on a calendar year and its Board meets twice a year in the spring and in the fall. Submission of proposals is required at least two months prior to a meeting for consideration at that meeting. To allow for optimum consideration and due diligence, those seeking grants are encouraged to apply to the foundation on an ongoing basis. Grants that do not make the deadline for one meeting will be carried forward to the next meeting. Each request must be in writing and should be accompanied by the proposal summary and the required list of attachments. See the foundation's website for additional guidelines.
Requirements: Texas IRS 501(c)3 tax-exempt organizations serving Harris, Walker, and Travis counties are eligible to apply.
Restrictions: Normally, the Foundation will not support: requests for building funds or grant commitments extending into successive calendar years; grants to religious organizations for religious purposes; fund raising events or advertising; grants to other private foundations; grants to cover past operating deficits or debt retirement; grants for support to individuals; grants that impose the exercise of responsibility upon the Foundation. For example: private operating foundations or certain supporting organizations.
Geographic Focus: Texas
Amount of Grant: 1,000 - 20,000 USD
Samples: Alley Theatre, Houston, Texas - educational outreach support; Travis Audubon Society, Houston, Texas - operational support; Great Expectations Foundation, Tahlequah, Oklahoma - professional development and mentoring for teachers.
Contact: Caroline J. Sabin, Executive Director; (713) 523-7557; fax (713) 523-7553; info@powellfoundation.org
Internet: http://www.powellfoundation.org/powellguide.htm
Sponsor: Powell Foundation of Houston
2121 San Felipe, Suite 110
Houston, TX 77019-5600

PPCF Community Grants 3256

The Pikes Peak Community Foundation gives priority to high-impact initiatives that provide maximum benefit for the community. In addition, the Foundation prefers to fund specific programs and projects that demonstrate measurable results. Primary areas of funding include: civic improvement; community development; education; the environment; health; and human services. Interested applicants should first forward an online letter of inquiry (LOI), detailing their proposal. If approved, an email with a link to the online application will be provided. There are three annual deadlines for completed full applications, including: March 1, July 1, and November 1.
Requirements: Any 501(c)3 organization serving the residents of the Pikes Peak region, defined as El Paso County, Teller County, and adjacent communities, is eligible to apply.
Restrictions: The Foundation generally does not fund: organizations that do not have an active 501(c)3 tax status; other foundations or organizations that distribute money to nonprofit recipients of its own selection; debt retirement, endowments or other reserve funds; individuals; medical, scientific, or academic research; political or religious doctrine; sponsorships; camperships; travel; vehicle purchases; conference fees; symposium fees; workshop fees; writing, publications, or distribution of books, articles, newsletters, and electronic media; annual memberships; or dinners.
Geographic Focus: Colorado
Date(s) Application is Due: Mar 1; Jul 1; Nov 1
Contact: Whitney Calhoun; (719) 389-1251, ext. 115; wcalhoun@ppcf.org
Internet: https://www.ppcf.org/community-grants/
Sponsor: Pikes Peak Community Foundation
730 North Nevada Avenue
Colorado Springs, CO 80903

PPG Industries Foundation Grants 3257

Funding requests for a variety of project proposals that advance the foundation's interests are eligible for consideration. These may include capital projects, operating grants and special projects. In general, the foundation gives priority to applications from organizations dedicated to enhancing the welfare of communities in which PPG is a resident. Each grant application is reviewed with regard to the: compatibility of the applicant's goals with the foundation's priorities and available resources; financial needs of the organization; past practices of the foundation with respect to that organization; capability and reputation of the applicant; funds available to the applicant from other sources; extent to which the work of the applicant duplicates that of other organizations; public scope and impact of the applicant's proposal; and the interest of other corporate foundations with respect to the applicant. Historically the foundation has supported nonprofits in the areas of human services, health and safety, civic and community affairs, education, and cultural and arts. Requests for funding are accepted year-round. Determinations are made by the foundation's screening committee and board of directors.
Requirements: Applicants must use PPG's online grant making system to apply. The link is on the website. PPG Industries Foundation will review applications on a regular basis and will contact all grantseekers with proposals of interest. Organizations located in the Pittsburgh area and organizations of national scope should direct any questions to the executive director of the foundation. Organizations serving communities where PPG facilities are located should direct any questions to the local PPG Industries Foundation agent in their area. A list of these may be found at the PPG Foundation website under the Foundation Governance link.
Restrictions: The foundation will not award grants for: advertising or sponsorships; endowments; political or religious purposes; projects which would directly benefit PPG Industries, Inc; or special events and telephone solicitations. Operating grants are not made to United Way agencies.
Geographic Focus: All States, Pennsylvania
Samples: YMCA of Metropolitan Milwaukee, South Shore Center, Milwaukee, Wisconsin, $3,000 - funding supports SPLASH swimming program for second-graders; Carlisle Regional Performing Arts Center, location unspecified, $3,000 - funding supports maintenance, technology upgrade costs; Robert Morris University, Moon Township, Pennsylvania, $350,000 - funding supports renovation of the Career and Leadership Development Center.
Contact: Sue Sloan; (412) 434-2453; fax (412) 434-4666; foundation@ppg.com
Internet: http://www.ppg.com/en/ppgfoundation/Pages/Grant_Policies.aspx
Sponsor: PPG Industries Foundation
One PPG Place
Pittsburgh, PA 15272

Premera Blue Cross Grants 3258

Premera is committed not only to improving the health of its members, but is also dedicated to making a difference in the communities it serves through its corporate philanthropy and community service program. Premera supports community programs, events and charities each year with financial and in-kind donations as well as volunteer time. The company invites all eligible charitable organizations to submit requests for funding. It focuses giving criteria on key areas that have an impact on the health and well-being of people in its communities. In order to address the common risk factors related to many of the major diseases and health conditions that have a large impact on the residents in our communities, Premera supports nonprofit organizations and programs that address wellness, with a specific focus on exercise and stress management. Applications are accepted online only.
Requirements: To be eligible for consideration, proposals for donations must: be from organizations that are defined as exempt from taxation under Section 501(a) as organizations described in Section 501(c)3 of the Internal Revenue Code; or a government agency or program, a public school or school district (but only if the contribution or gift is made for exclusively public purposes). Applicants must assist residents in the areas it serves in Washington.
Restrictions: Premera generally does not make contributions to capital campaigns. Premera funds are not used for charitable grants or contributions to: individuals; organizations that are for profit; arts organizations; religious organizations, unless the gift is designated to an ongoing secular community service program sponsored by these organizations and does not propagate a belief in a specific faith; or organizations that discriminate against individuals based on age, gender, race, religion, ethnicity, disability or sexual orientation, or that discriminate against individuals on any impermissible basis, or which advocate a position contrary to established public policy.
Geographic Focus: Washington
Contact: Stefanie Bruno; (509) 252-7431 or (425) 918-5933; stefanie.bruno@premera.com
Internet: https://www.premera.com/wa/visitor/about-premera/corporate-citizenship/giving-guidelines/
Sponsor: Premera Blue Cross
7001 220th Street SW, Building 1
Mountlake Terrace, WA 98043-2160

Presbyterian Health Foundation Bridge, Seed and Equipment Grants 3259

The primary objective of bridge grants is to provide funding of limited duration to enhance faculty competitiveness for national extramural funding. The seed grant program provides start-up funding for young investigators or for investigators launching into a new research direction. The equipment grants program provides the funding for laboratory equipment needed to advance research.
Requirements: Oklahoma nonprofit organizations are eligible. Submit a letter of application. Applications are due the last week in March, June, September, and December.
Restrictions: Grants are not made to individuals, private foundations, for-profit organizations or for operating funds of organizations.
Geographic Focus: Oklahoma
Amount of Grant: 25,000 - 250,000 USD
Samples: U of Oklahoma Health Sciences Ctr (OK)--for research on genetics, $1.2 million over two years; U of Oklahoma Health Sciences Ctr (OK)--to build the Stanton L. Young Research Center, $5 million.
Contact: Michael D. Anderson; (405) 319-8150; manderson@phfokc.com
Internet: http://www.phfokc.com/bridge.htm
Sponsor: Presbyterian Health Foundation
655 Research Park Way, Suite 500
Oklahoma City, OK 73104-3603

Presbyterian Patient Assistance Program 3260

The goal program is to be able to provide temporary assistance in times of urgent and emergency need to patients and clinical clients and their immediate families. Funding can be used to assist with temporary medication, food, utility and housing assistance, and/or clinically-related travel. Costs, expenses and other related expenses are sometimes covered by these funds.
Requirements: Requests for funding come to the foundation via their Chaplains, social workers, case workers, and other related individuals. Each case is evaluated on an individual basis and applicants must be able to provide documentation that verifies the needs for which assistance is requested. Those having the most devastating direct impact on an patient will take precedence.
Geographic Focus: All States
Contact: Program Coordinator; (505) 724-6580
Internet: http://www.phs.org/foundation/assistance.shtml
Sponsor: Presbyterian Healthcare Foundation
P.O. Box 26666
Albuquerque, NM 87125-6666

Price Chopper's Golub Foundation Grants 3261

Price Chopper's Golub Foundation provides financial support to eligible charitable organizations with a current 501(c)3 tax exempt status. Contributions are made through planned, continued giving programs in the areas of health and human services, arts, culture, education, and youth activities, within Price Chopper marketing areas. To be considered for funding, mail a written request, on letterhead for the organization seeking the donation, six to eight weeks prior to needed support or response deadlines. The Foundation reviews capital campaign requests quarterly, so please allow three to four months for a response.
Requirements: The Foundation's six state marketing area includes a specific mile radius around its stores in New York (Albany, Broome, Cayuga, Chenango, Clinton, Columbia, Cortland, Delaware, Dutchess, Essex, Franklin, Fulton, Greene, Hamilton, Herkimer, Jefferson, Lewis, Madison, Montgomery, Oneida, Onondaga, Orange, Oswego, Otsego, Rensselaer, St. Lawrence, Saratoga, Schenectady, Schoharie, Sullivan, Tioga, Tompkins, Ulster, Warren, and Washington counties), Massachusetts (Berkshire, Hampden, Hampshire, Middlesex, and Worcester counties), Vermont (Addison, Bennington, Caledonia, Chittenden, Essex, Franklin, Grand Isle, Lamoille, Orange, Orleans, Rutland, Washington, Windham, and Windsor counties), Pennsylvania (Lackawanna, Luzerne, Pike, Susquehanna, Wayne, and Wyoming counties), Connecticut (Hartford, Litchfield, New Haven, Tolland, and Windham counties) and New Hampshire (Cheshire, Grafton, and Sullivan counties).
Restrictions: The Foundation does not support: individuals; annual meetings; endowments; film and video projects; program advertising; funding for travel; organizations or events outside of its marketing area; events to raise funds for groups outside of its local community; conferences, conventions, or symposiums; publishing; operating expenses; scholarship programs outside of its own; or capital campaigns of national, religious or political organizations.
Geographic Focus: Connecticut, Massachusetts, New Hampshire, New York, Pennsylvania, Vermont
Contact: Deborah Tanski; (518) 356-9450 or (518) 379-1270; fax (518) 374-4259

Internet: http://www.pricechopper.com/GolubFoundation/GolubFoundation_S.las
Sponsor: Price Chopper's Golub Foundation
P.O. Box 1074
Schenectady, NY 12301

Price Family Charitable Fund Grants 3262

Established in 1983, the Price Family Charitable Fund serves the San Diego, Carlsbad and, San Marcos, California region. The foundation is interested in supporting the economically disadvantaged, giving primarily for education and philanthropy purposes. The types of support include: annual campaigns; fellowships; program evaluation; scholarship funds; scholarships to individuals and; general operating support. There's no formal application to submit. Potential grantees much first submit a letter of Inquiry including: project goals, objectives and expected results (maximum of one page); project narrative (maximum of five pages); project budget (maximum of one page); a list of grants, if any, received by organization in the last 12 months for this program/project (sources and amounts); a list of the organization's current board of directors, including each member's name, profession, and office help on the board, if any.
Requirements: Non-profit organizations with 501(c)3 status or governmental units such as public schools or city departments in are available for funding. The Foundation gives primarily in the following region of: San Diego; Carlsbad; San Marcos, California.
Restrictions: Grants are not made: to organizations whose primary purpose is religious, or for propagandizing, influencing legislation and/or elections, promoting voting registration, for political candidates, political campaigns or organizations engaged in political activities; or to federal appeals or to organizations the collect funds for redistribution to other non-profit groups. Unsolicited requests for funds are not accepted.
Geographic Focus: California
Contact: Terry Malavenda, (858) 551-2330
Sponsor: Price Family Charitable Fund
7979 Ivanhoe Avenue, Suite 520
La Jolla, CA 92037-4513

Priddy Foundation Capital Grants 3263

The Priddy Foundation is a general purpose foundation, interested primarily in programs that have the potential for lasting and favorable impact on individuals and organizations. Considerations for funding include the geographic area served by the project, the individuals and groups served, the problem being addressed, the availability of existing resources and the degree of need. The Foundation board will consider capital projects for buildings and major items of equipment. Approval is more likely if the project has broad support from organizations and individuals to the extent that Foundation's requested share of the project does not exceed 20% of the total project budget. Before a capital grant is funded, the organization must attain the project fund-raising goal and document that funds raised are sufficient to complete the project as presented in the grant application. Deadlines for preliminary applications are February 1 and August 1, while final applications are due March 1 and September 1.
Requirements: 501(c)3 Texas and Oklahoma nonprofit organizations are eligible. The foundation considers grant applications from organizations in the Wichita Falls, Texas area. In Texas, this includes the following counties: Archer, Baylor, Childress, Clay, Cottle, Foard, Hardeman, Haskell, Jack, King, Knox, Montague, Stonewall, Throckmorton, Wichita, Wilbarger, Wise, and Young. In Oklahoma, it includes the following counties: Comanche, Cotton, Jackson, Jefferson, Stephens, and Tillman.
Restrictions: The Priddy Foundation does not normally make grants for the following purposes: operating deficits; endowments; debt retirement; organizations that make grants to others; charities operated by service clubs; a request for capital funds for a project previously supported; any grant that would tend to obligate the foundation to future funding; fund raising programs and events; grants that impose expenditure responsibility on the foundation; grants to individuals, including individual scholarship awards; start-up funding for new organizations; individual public elementary or secondary schools (K-12); religious institutions except for non-sectarian, human service programs offered on a non-discriminatory basis; basic or applied research; media productions or publications; school trips; conferences or other educational events except through an organizational development grant; or direct grants to volunteer fire departments.
Geographic Focus: Oklahoma, Texas
Date(s) Application is Due: Mar 1; Sep 1
Amount of Grant: 30,000 - 1,500,000 USD
Contact: Debbie C. White; (940) 723-8720; fax (940) 723-8656; debbiecw@priddyfdn.org
Internet: https://priddyfdn.org/policy/
Sponsor: Priddy Foundation
807 Eighth Street, Suite 1010
Wichita Falls, TX 76301-3310

Priddy Foundation Operating Grants 3264

The Priddy Foundation is a general purpose foundation, interested primarily in programs that have the potential for lasting and favorable impact on individuals and organizations. Considerations for funding include the geographic area served by the project, the individuals and groups served, the problem being addressed, the availability of existing resources and the degree of need. Although the Foundation is wary of fostering annual budget dependency on the part of a grantee agency, its board recognizes that there are circumstances in which a grant for general operating purposes might be critical to an organization's success or viability. Such grants would be for a limited period of time. Among other conditions which might be imposed, based on a specific organization's application, a grantee organization will be required to present a practicable plan to achieve self-sufficiency without additional foundation funding. During the term of the grant the grantee organization might also be required to enter into a formal consulting arrangement with a Center for Non-profit Management, or a similar organization, also with the objective of becoming self-sufficient. Deadlines for preliminary applications are February 1 and August 1, while final applications are due March 1 and September 1.
Requirements: 501(c)3 Texas and Oklahoma nonprofit organizations are eligible. The foundation considers grant applications from organizations in the Wichita Falls, Texas area. In Texas, this includes the following counties: Archer, Baylor, Childress, Clay, Cottle, Foard, Hardeman, Haskell, Jack, King, Knox, Montague, Stonewall, Throckmorton, Wichita, Wilbarger, Wise, and Young. In Oklahoma, it includes the following counties: Comanche, Cotton, Jackson, Jefferson, Stephens, and Tillman.
Restrictions: The Priddy Foundation does not normally make grants for the following purposes: operating deficits; endowments; debt retirement; organizations that make grants to others; charities operated by service clubs; a request for capital funds for a project previously supported; any grant that would tend to obligate the foundation to future funding; fund raising programs and events; grants that impose expenditure responsibility on the foundation; grants to individuals, including individual scholarship awards; start-up funding for new organizations; individual public elementary or secondary schools (K-12); religious institutions except for non-sectarian, human service programs offered on a non-discriminatory basis; basic or applied research; media productions or publications; school trips; conferences or other educational events except through an organizational development grant; or direct grants to volunteer fire departments.
Geographic Focus: Oklahoma, Texas
Date(s) Application is Due: Mar 1; Sep 1
Amount of Grant: 20,000 - 120,000 USD
Samples: Communities In Schools, Wichita Falls, Texas, $120,00 - program operations; Wichita Falls Alliance for the Mentally Ill, Wichita Falls, Texas, $20,000 - operating support; Wichita-Archer-Clay Christian Womens Job Corps, Wichita Falls, Texas, $30,000 - operating support.
Contact: Debbie C. White; (940) 723-8720; fax (940) 723-8656; debbiecw@priddyfdn.org
Internet: https://priddyfdn.org/policy/
Sponsor: Priddy Foundation
807 Eighth Street, Suite 1010
Wichita Falls, TX 76301-3310

Priddy Foundation Organizational Development Grants 3265

The Priddy Foundation is a general purpose foundation, interested primarily in programs that have the potential for lasting and favorable impact on individuals and organizations. Considerations for funding include the geographic area served by the project, the individuals and groups served, the problem being addressed, the availability of existing resources and the degree of need. In the area of Organizational Development, the Foundation will consider grants to organizations for such things as board and staff development, planning initiatives, technical assistance, technology enhancements, and capital projects. Organizational development grants will be dependent on a comprehensive plan supported by the organization's board, outside professional assistance (e.g., Center for Non-profit Management), if appropriate, and absolute linkage between the development plan and the ability of the organization to achieve its mission more effectively. Deadlines for preliminary applications are February 1 and August 1, while final applications are due March 1 and September 1.
Requirements: 501(c)3 Texas and Oklahoma nonprofit organizations are eligible. The foundation considers grant applications from organizations in the Wichita Falls, Texas area. In Texas, this includes the following counties: Archer, Baylor, Childress, Clay, Cottle, Foard, Hardeman, Haskell, Jack, King, Knox, Montague, Stonewall, Throckmorton, Wichita, Wilbarger, Wise, and Young. In Oklahoma, it includes the following counties: Comanche, Cotton, Jackson, Jefferson, Stephens, and Tillman.
Restrictions: The Priddy Foundation does not normally make grants for the following purposes: operating deficits; endowments; debt retirement; organizations that make grants to others; charities operated by service clubs; a request for capital funds for a project previously supported; any grant that would tend to obligate the foundation to future funding; fund raising programs and events; grants that impose expenditure responsibility on the foundation; grants to individuals, including individual scholarship awards; start-up funding for new organizations; individual public elementary or secondary schools (K-12); religious institutions except for non-sectarian, human service programs offered on a non-discriminatory basis; basic or applied research; media productions or publications; school trips; conferences or other educational events except through an organizational development grant; or direct grants to volunteer fire departments.
Geographic Focus: Oklahoma, Texas
Date(s) Application is Due: Mar 1; Sep 1
Amount of Grant: 1,500 - 120,000 USD
Contact: Debbie C. White; (940) 723-8720; fax (940) 723-8656; debbiecw@priddyfdn.org
Internet: https://priddyfdn.org/policy/
Sponsor: Priddy Foundation
807 Eighth Street, Suite 1010
Wichita Falls, TX 76301-3310

Priddy Foundation Program Grants 3266

The Priddy Foundation is a general purpose foundation, interested primarily in programs that have the potential for lasting and favorable impact on individuals and organizations. Considerations for funding include the geographic area served by the project, the individuals and groups served, the problem being addressed, the availability of existing resources and the degree of need. In the area of Program Grants, the Foundation gives highest priority to organizations seeking funds for service extension or implementation of new services. Projects should make a difference in the lives of people served by dealing effectively with known problems or opportunities. Results should be capable of evaluation against defined standards of measurement. Proposals should be realistic concerning the

ability of the organization to conduct the program and to sustain the program beyond the period a grant from the Foundation may cover. Deadlines for preliminary applications are February 1 and August 1, while final applications are due March 1 and September 1.
Requirements: 501(c)3 Texas and Oklahoma nonprofit organizations are eligible. The foundation considers grant applications from organizations in the Wichita Falls, Texas area. In Texas, this includes the following counties: Archer, Baylor, Childress, Clay, Cottle, Foard, Hardeman, Haskell, Jack, King, Knox, Montague, Stonewall, Throckmorton, Wichita, Wilbarger, Wise, and Young. In Oklahoma, it includes the following counties: Comanche, Cotton, Jackson, Jefferson, Stephens, and Tillman.
Restrictions: The Priddy Foundation does not normally make grants for the following purposes: operating deficits; endowments; debt retirement; organizations that make grants to others; charities operated by service clubs; a request for capital funds for a project previously supported; any grant that would tend to obligate the foundation to future funding; fund raising programs and events; grants that impose expenditure responsibility on the foundation; grants to individuals, including individual scholarship awards; start-up funding for new organizations; individual public elementary or secondary schools (K-12); religious institutions except for non-sectarian, human service programs offered on a non-discriminatory basis; basic or applied research; media productions or publications; school trips; conferences or other educational events except through an organizational development grant; or direct grants to volunteer fire departments.
Geographic Focus: Oklahoma, Texas
Date(s) Application is Due: Mar 1; Sep 1
Amount of Grant: 1,500 - 1,500,000 USD
Samples: Wichita Adult Literacy Council, Wichita Falls, Texas, $25,000 - basic program funding; Kairos Prison Ministry International, Winter Park, Florida, $10,000 - Wichita Falls Area Kairos Outside Program; Presbyterian Children's Homes and Services, Wichita Falls, Texas, $90,000 - child and foster care program.
Contact: Debbie C. White; (940) 723-8720; fax (940) 723-8656; debbiecw@priddyfdn.org
Internet: https://priddyfdn.org/policy/
Sponsor: Priddy Foundation
807 Eighth Street, Suite 1010
Wichita Falls, TX 76301-3310

Pride Foundation Fellowships 3267
The Pride Foundation Fellowship Program seeks to cultivate leaders and strengthen the Pacific Northwest LGBTQ community. By matching exceptional Pride Foundation scholarship recipients and other LGBTQ and ally students with Pride Foundation grantees for a substantial project-based fellowship, this fellowship experience will provide an opportunity for professional development as well as an introduction to the work of community leadership organizations. There are six internship opportunities, with four in Seattle, one in Sedro Wolley, Washington, and one in Portland, Oregon. Applicants will be asked to rank their top choices to help the selection committee determine eligible matches. Fellows are paid a $3,500 stipend at the end of the fellowship for their work. Fellows are expected to commit to 200 hours of work at their internship site (20 hours per week for 10 weeks), and must be available from mid-June through mid-August for their assignments.
Requirements: This program is open to current college students only (this includes students enrolled in community college, four year public or private colleges/universities, graduate school, professional programs, vocational or trade programs) pursuing any degree or major. Along with the online application, candidates are asked to provide basic contact information, summer availability, a list of applicable skills, and areas of potential development. Two references are also required.
Geographic Focus: All States
Date(s) Application is Due: Jan 31
Amount of Grant: 3,500 USD
Contact: Coordinator; (800) 735-7287 or (206) 323-3318; fellowship@pridefoundation.org
Internet: http://www.pridefoundation.org/scholarships/fellowship/
Sponsor: Pride Foundation
1122 East Pike, PMB 1001
Seattle, WA 98122

Pride Foundation Grants 3268
The Pride Foundation works to strengthen the lesbian, gay, transgender, and bisexual community primarily in Washington state and extending to the four neighboring states of Alaska, Idaho, Montana, and Oregon. The Foundation awards grants to projects in arts and recreation; education, advocacy, and outreach; health and community service; HIV/AIDS service delivery and prevention; lesbian health; and youth and family services. Applicants will first submit the letter of inquiry online applications by August 19. If organizations are invited to submit the full application, they will be notified by September 23. Funds will be available in December.
Requirements: IRS 501(c)3 tax-exempt organizations or organizations affiliated with tax-exempt organizations are eligible. Projects or programs must directly benefit the lesbian, gay, bisexual, and transgender community; people affected by HIV/AIDS; and/or their friends and families.
Restrictions: Grants to individuals cannot be considered.
Geographic Focus: Alaska, Idaho, Montana, Oregon, Washington
Date(s) Application is Due: Sep 20
Amount of Grant: Up to 5,000 USD
Contact: Jeff Hedgepeth; (800) 735-7287 or (206) 323-3318; jeff@pridefoundation.org
Internet: http://www.pridefoundation.org/grants/overview/
Sponsor: Pride Foundation
1122 East Pike, PMB 1001
Seattle, WA 98122

Pride Foundation Scholarships 3269
Pride Foundation provides post-secondary educational scholarships to lesbian, gay, bisexual, transgender, queer, and straight-ally leaders and role models from Alaska, Oregon, Idaho, Montana, and Washington. The Foundation offers one of the largest programs of its kind in the U.S. with nearly 50 specific scholarships for: students from various geographic areas in the Northwest; students of color; and students raised by LGBT parents. Scholarships cover nearly any accredited post-secondary school or program, including: community colleges; 4-year public or private colleges and universities; trade or vocational training; creative studies programs; certificate programs; medical or law schools; graduate studies; or other accredited degree programs. Candidates may apply scholarships to any of the following: tuition; course, program, or student fees; on campus room and board charges; school organized study abroad programs; books or course supplies; or graduate research related expenses. The application is available online.
Requirements: Requirements and scholarship amounts vary for each scholarship, but only one application is necessary for all scholarships. Applicants may contact the Foundation for more information.
Restrictions: Candidates outside the five state area are not eligible. Scholarship funds cannot be applied to off-campus housing, personal expenses, transportation costs, dependent care, computer or software purchases (unless required by the program); or other expenses not posted on the Foundation website.
Geographic Focus: Alaska, Idaho, Montana, Oregon, Washington
Date(s) Application is Due: Jan 31
Amount of Grant: Up to 10,000 USD
Contact: Anthony Papini, Director of Program Strategies; (800) 735-7287 or (206) 323-3318; anthony@pridefoundation.org
Internet: http://www.pridefoundation.org/scholarships/overview/
Sponsor: Pride Foundation
1122 East Pike, PMB 1001
Seattle, WA 98122

Prince Charitable Trusts Chicago Grants 3270
The trusts awards grants to eligible Chicago nonprofit organizations in its areas of interest: arts and culture, education, environment, health, and social services.
Requirements: The Trusts Chicago program only funds organizations within the city limits of Chicago (with the exception of grants made through the MacArthur Fund for Arts and Culture at Prince). The Trusts make grants only to charitable organizations that are exempt from federal income tax under Section 501(c)3 of the Internal Revenue Code and are classified as public charities under Sections 509(a)(1) or 509(a)(2). All grant applications must include a Prince Charitable Trusts cover sheet, see the Trusts website, http://foundationcenter.org/grantmaker/prince/chi_app.html for proper form and additional guidelines.
Restrictions: The Trusts do not fund projects that promote or proselytize any religion. While the Trusts do fund the projects of faith-based organizations, those projects must be secular in nature. The Trusts do not fund organizations that discriminate on the basis of ethnicity, race, color, creed, religion, gender, national origin, age, disability, marital status, sexual orientation, gender identity, or any veteran's status.
Geographic Focus: Illinois
Date(s) Application is Due: Jan 13; May 1; Jun 1
Contact: Sharon Robison; (312) 419-8700; fax (312) 419-8558; srobison@prince-trusts.org
Internet: http://www.fdncenter.org/grantmaker/prince/chicago.html
Sponsor: Prince Charitable Trusts
303 West Madison Street, Suite 1900
Chicago, IL 60606

Prince Charitable Trusts District of Columbia Grants 3271
The trusts awards grants to eligible Washington D.C. nonprofit organizations in its areas of interest: arts and culture, community, environment, health, emergency services, youth and provide a limited number of capital grants each year.
Requirements: The Trusts make grants only to charitable organizations that are exempt from federal income tax under Section 501(c)3 of the Internal Revenue Code and are classified as public charities under Sections 509(a)(1) or 509(a)(2). Electronic proposals are preferred. Attachments may be mailed separately. Proposals should include the Prince Charitable Trust Grant Application Cover Sheet and the Common Grant Application Format of Washington Grantmakers. These forms and additional guidelines may be obtained at the Trusts website, http://foundationcenter.org/grantmaker/prince/dc_app.html.
Restrictions: The Trusts do not make grants to individuals, nor does it fund projects that promote or proselytize any religion. While the Trusts do fund the projects of faith-based organizations, those projects must be secular in nature.
Geographic Focus: District of Columbia
Date(s) Application is Due: Feb 1; Aug 10; Sep 1
Amount of Grant: 10,000 - 30,000 USD
Contact: Kristin Pauly; (202) 728-0646; fax (202) 466-4726; kpauly@princetrusts.org
Internet: http://www.fdncenter.org/grantmaker/prince/dc_interest.html
Sponsor: Prince Charitable Trusts
816 Connecticut Avenue NW
Washington, D.C. 20006

Princeton Area Community Foundation Greater Mercer Grants 3272
The Greater Mercer Grants Program is a competitive program for projects benefiting residents of Mercer County, New Jersey, and the immediately adjoining areas of surrounding counties. There are three distinct categories. In addressing the needs of low-income individuals and families (Category 1), grants up to $15,000 are made to nonprofit organizations only. In support of community organizing in low-income neighborhoods

(Category 2), grants of $500 to $20,000 are made to 501(c)3 nonprofit organizations, neighborhood associations, and groups of residents forming a neighborhood association to work on a project. And in support of building community within and among municipalities (Category 3), grants generally range from $10,000 to $35,000, but may be as much as $50,000, made to nonprofit organizations only.
Requirements: Nonprofit organizations that have tax-exempt status under Section 501(c)3 of the Internal Revenue Code are eligible. They must be registered with the State of New Jersey as a charity, unless they are religious organizations, or schools that file their curricula with the Department of Education and are exempt from the provisions of the New Jersey Charitable Registration and Investigation Act.
Restrictions: Nonprofits may submit one application per calendar year in each category, for a maximum of three. Neighborhood associations or groups of residents who will form a neighborhood association may apply only once annually for a Community Organizing in Low-Income Neighborhoods grant.
Geographic Focus: New Jersey
Contact: Deborah Thomas; (609) 219-1800; fax (609) 219-1850; daubert-thomas@pacf.org
Internet: http://www.pacf.org/grants/polCalendarEvent.cfm?Program_Code=12
Sponsor: Princeton Area Community Foundation
15 Princess Road
Lawrenceville, NJ 08648

Principal Financial Group Foundation Grants 3273
The foundation addresses concerns in the areas of health and human services, education, arts and culture, environment, and recreation and tourism. The primary objective is to support, through charitable contributions, selected nonprofit organizations primarily located in the greater Des Moines, IA, area. The foundation also will consider requests from organizations located in areas where the corporation has offices, including Des Moines, Mason City and Cedar Falls, IA; Grand Island, NE; Spokane, WA; Wilmington, DE; Appleton, WI; and Phoenix, AZ. The objectives, priorities, and programs seek to reflect the needs and concerns of communities in which the corporation operates. Support is given for annual campaigns, building funds, capital campaigns, continuing support, employee matching gifts, in-kind gifts, performances and exhibitions, conferences and workshops, adult basic education, vocational programs, operating budgets, internships, demonstration grants, matching grants, and seed grants. Contribution requests are considered on a quarterly basis, in accordance with the following schedule: health and human services, March 1; education, June 1; arts and culture, September 1; and environment, recreation, and tourism, December 1.
Requirements: 501(c)3 tax-exempt organizations in company operating locations may apply.
Restrictions: Proposals for athletic groups, conferences, endowments, fellowships, festivals, fraternal organizations, health care facility fund drives, libraries, or religious groups are denied.
Geographic Focus: Delaware, Iowa, Nebraska, Washington
Date(s) Application is Due: Mar 1; Jun 1; Sep 1; Dec 1
Contact: Laura Sauser; (515) 247-7227; fax (515) 246-5475
Internet: http://www.principal.com/about/giving/grant.htm
Sponsor: Principal Financial Group Foundation
711 High Street
Des Moines, IA 50392-0150

Procter and Gamble Fund Grants 3274
The fund supports nonprofit organizations in company-operating locations in the areas of education, health and social services, civic projects, cultural organizations, disaster relief, and environmental efforts. Grants are awarded to education initiatives in local communities, such as teacher training efforts, with a focus on economic teaching; and other efforts by public policy, research, and economic education organizations. Employee voluntarism is prevalent in many K-12 initiatives. Most health and human services funding supports the United Way. The Salvation Army, Red Cross, hospitals, food banks, and other social service organizations receive support. Community support is awarded through grants that bolster economic growth and enrichment, including support for a youth jobs program, libraries, zoos, and local chambers of commerce. Support is awarded to a variety of arts organizations, including theater, dance, music, and visual arts. Major environmental groups also receive support. Grant Application Cycles are July 1 through September 30 and December 1 through February 28, grant requests are only accepted during those times.
Requirements: 501(c)3 organizations in communities where Procter and Gamble Company manufacturing plants are located are eligible.
Geographic Focus: All States
Contact: Brenda Ratliff; (513) 945-8454; fax (513) 945-5211; pgfund.im@pg.com
Internet: http://www.pg.com/company/our_commitment/grant_application_guidelines.shtml
Sponsor: Procter and Gamble Fund
P.O. Box 599
Cincinnati, OH 45201

Prospect Burma Scholarships 3275
By funding education for young Burmese, Prospect Burma seeks to help create a cadre of Burmese who will be competent to run the country once democracy returns. Its aim is to educate those who will bring their knowledge and skills back to Burma for its future development. Preference is given to applicants studying one of the following subjects: Agriculture; Civil Society; Development; Ecology/Conservation; Economics; Education/Teacher Training; Engineering; Human Rights; Nursing; Public Administration; Public Health; or Science subjects. Preference is also given to students at educational institutions in South-East Asia or South Asia.
Requirements: You may qualify for an award if you are of Burmese origin and have been accepted for, or enrolled in a degree course at an accredited college or university, and are: accepted for and/or enrolled in a first degree (undergraduate) course at a college or university; already studying, or accepted for, a Master's degree course; or a postgraduate student who has already started on a Doctorate, or has a confirmed offer of a place to read for a Doctorate.
Restrictions: The Program does not contribute to international travel, computer equipment or debts. Furthermore, it does not make a contribution to the costs of dependants (spouses, partners, children).
Geographic Focus: All States
Date(s) Application is Due: Mar 31
Contact: Shona Kirkwood, Education Project Coordinator; bpeoshona@gmail.com
Internet: http://www.prospectburma.org/index.php?option=com_content&task=category§ionid=5&id=18&Itemid=39
Sponsor: Prospect Burma
Rivermead Court, Ranelagh Gardens
London, SW6 3SF England

Prudential Foundation Education Grants 3276
The Prudential Foundation's areas of interest are ready-to-learn programs, ready-to-work programs, and ready-to-live programs. In order to promote sustainable communities and improve social outcomes for community residents, the Foundation focuses its strategy in the following educational areas: education leadership to support reform in public education by increasing the capacity of educators, parents, and community residents to implement public school reform; and youth development to build skills and competencies needed for young people to be productive citizens (this includes expanding arts education opportunities and supporting effective out-of-school-time programs for young people). Finally, the Foundation also funds organizations whose efforts influence policy that adapts promising practices and evidence-based approaches to instruction and learning in schools. Types of support include operating support, continuing support, annual campaigns, seed money, matching funds and employee matching gifts, consulting services, technical assistance, employee-related scholarships, research, capital campaigns, conferences and seminars, and projects/programs. The Foundation is especially interested in proposals that anticipate and address potential major problems. Funds are targeted to areas where Prudential has a strong presence. Applicant should make initial contact with a brief letter to determine whether a more detailed proposal would be acceptable.
Requirements: The Prudential Foundation supports nonprofit, charitable organizations, and programs whose mission and operations are broad and non-discriminatory. The Foundation focuses its resources to support organizations whose activities address social needs or benefit underserved groups and communities. Priority in order of preference goes to programs in Newark, New Jersey, and surrounding communities; Los Angeles, California; Hartford, Connecticut; New York, New York; Chicago, Illinois; Jacksonville, Florida; Atlanta, Georgia; Minneapolis, Minnesota; Philadelphia and Scranton, Pennsylvania; Houston and Dallas, Texas; Dubuque, Iowa; Phoenix, Arizona; and New Orleans, Louisiana.
Restrictions: The Foundation does not fund: organizations that are not tax-exempt under paragraph 501(c)3 of the U.S. Internal Revenue Code; labor, religious, political, lobbying, or fraternal groups—except when these groups provide needed services to the community at large; direct grants or scholarships to individuals; support for single-disease health groups; or good will advertising.
Geographic Focus: Arizona, California, Connecticut, Florida, Georgia, Illinois, Iowa, Louisiana, Minnesota, New Jersey, New York, Pennsylvania, Texas
Contact: Lata Reddy; (973) 802-4791; community.resources@prudential.com
Internet: http://www.prudential.com/view/page/public/12373
Sponsor: Prudential Foundation
751 Broad Street, 15th Floor
Newark, NJ 07102-3777

PSG Mentored Clinical Research Awards 3277
The Mentored Clinical Research Award (MCRA) for new investigators is funded by a grant from the Parkinson's Disease Foundation (PDF) to the Parkinson Study Group (PSG). The PDF has partnered with the PSG to encourage the professional and scientific development of young investigators on their path to independence. To this end, this grant will support a new investigator for a one year project in patient oriented research in Parkinson's disease or other parkinsonian disorders under the mentorship of an experienced investigator. The training should lead a junior investigator to gain skills in clinical research. The research plan should address unmet needs of people living with PD, have the potential for broad application among the PD community, and lead to advances in clinically relevant treatment options. An award of $75,000 will be available this year. An electronic copy of the proposal in Microsoft Word or pdf-format must be received by the PSG on or before Friday, March 26. Applications should be sent by email to: Roseanna.Battista@ctcc.rochester.edu with a cover note that includes the candidate's name and the title of proposal. The candidate's proposal should be formatted according to the guidelines indicated in the toolkit available on the PSG website http://www.parkinson-study-group.org/PSGToolkit.asp. Applicants will be notified of the results of these reviews by May 21.
Requirements: Appropriate applicants for the MCRA are clinicians and scientists who are within 5 years of having completed formal training (this includes PhD professionals working in the field of Parkinson's disease). Fellows may apply. See website for additional guidelines.
Geographic Focus: All States
Date(s) Application is Due: Mar 26
Contact: Roseanna Battista; roseanna.battista@ctcc.rochester.edu
Internet: http://www.parkinson-study-group.org/Resources/PSG%20SRC%20Mentored%20Clinical%20Research%20Award%20RFP%202010-2011%20FINAL.pdf
Sponsor: Parkinson Study Group
University of Rochester, 1351 Mt. Hope Avenue, Suite 223
New York, NY 14620

Puerto Rico Community Foundation Grants 3278

The Foundation wishes to develop the capacities of communities in Puerto Rico so that they may achieve social transformation and economic self-sufficiency, by stimulating investment in communities and maximizing the impact and yield of each contribution. Grants are awarded in the areas of: education; community development; financial development; development of social interest housing; and philanthropy. Types of support include: general operating support; emergency funds; conferences and seminars; professorships; publications; research; technical assistance; consulting services; and matching funds. There are no application deadlines; the board meets in March, June, September, and December to consider requests.

Requirements: Organizations applying for grants must comply with the conditions below in order to demonstrate eligibility: be duly incorporated and registered as a nonprofit organization, according to the laws of the Commonwealth of Puerto Rico; be located and offer services in Puerto Rico. Present a copy of the following documents: certificate of good standing from the State Department; certificate from the Treasury Department; statement of organization's total budget for the year for which funds are solicited; financial statements; list of current members of Board of Directors. Should include each member's address and phone number; resume or curriculum vitae of the Project Director.

Restrictions: The foundation does not make grants to support individuals, annual campaigns, seed money, endowments, deficit financing, scholarships, or building funds.
Geographic Focus: Puerto Rico
Amount of Grant: 1,000 - 40,000 USD
Contact: Grants Administrator; (787) 721-1037; fax (787) 721-1673; fcpr@fcpr.org
Internet: http://www.fcpr.org/
Sponsor: Puerto Rico Community Foundation
P.O. Box 70362
San Juan, PR 00936-8362
Puerto Rico

Pulaski County Community Foundation Grants 3279

The Pulaski County Community Foundation was established to serve the citizens of Pulaski County, Indiana. The Foundation welcomes grant requests from any nonprofit organization in Pulaski County. The Foundation also invites applications to help fund new organizations who meet demonstrated needs or benefit the community through creative and innovative projects and programs. Grant seekers may access the online application but are also encouraged to discuss their project with the Foundation office before submitting a grant proposal.

Requirements: The Foundation favors grant requests which: impact a substantial number of people in the Pulaski community; propose practical solutions to current problems or address a current community interest; examine and address underlying causes of local needs; encourage cooperation and elimination of duplicate services; build the capacity of the applying organization; are from established non-profit organizations.

Restrictions: Grants are made only to organizations that serve the Pulaski county area. As a general rule, the Foundation does not make grants from its discretionary funds for the following: ongoing operating expenses or annual fund raising drives; existing obligations, debt reduction or building campaigns; individuals or travel expenses; loans or endowments; political purposes. Grantees are required to complete a Final Report (program and financial) detailing how the grant funds were spent.
Geographic Focus: Indiana
Contact: Wendy Rose; (574) 946-0906; fax (574) 946-0971; wrose@pulaskionline.org
Internet: http://www.pulaskionline.org/content/view/97/432/
Sponsor: Pulaski County Community Foundation
127 E. Pearl Street, P.O.Box 407
Winamac, IN 46996

Pulaski County Community Foundation Lilly Endowment Community Scholarships 3280

The Lilly Endowment Community Scholarships help to raise the level of educational access in Indiana and enhance the quality of life for the state's residents. It provides full tuition and an $800 book stipend for four years of undergraduate study on a full-time basis at any accredited Indiana public or private college or university. All majors are eligible to apply.
Requirements: Applicants must be graduating high school seniors from Pulaski county. Candidates should contact the Foundation office or their guidance office for an application and further information about application requirements.
Geographic Focus: Indiana
Contact: Kim Krause, Scholarship Coordinator; (574) 946-0906; fax (574) 946-0971
Internet: http://www.pulaskionline.org/content/view/109/436/
Sponsor: Pulaski County Community Foundation
127 E. Pearl Street, P.O.Box 407
Winamac, IN 46996

Pulaski County Community Foundation Scholarships 3281

The Pulaski County Community Foundation offers nearly thirty scholarships to provide financial support to local students each year. Each scholarship has specific criteria and deadlines for application. All college majors may apply. Students should contact the Foundation or their guidance office for further information.
Geographic Focus: Indiana
Contact: Kim Krause, Scholarship Coordinator; (574) 946-0906; fax (574) 946-0971
Internet: http://www.pulaskionline.org/content/view/151/435/
Sponsor: Pulaski County Community Foundation
127 E. Pearl Street, P.O.Box 407
Winamac, IN 46996

Pulido Walker Foundation 3282

Established in 1996, the Pulido Walker Foundation offers grant support throughout California and South Carolina (though awards are occasionally given in other states). Its primary fields of interest include: the arts; general education; health and health care organizations; higher education; and a variety of human services. Typically, awards are given for general operating support. A formal application should be secured from the Foundation office, though no annual application deadlines have been identified. Most recent grants have ranged from $100 to $10,000.
Requirements: 501(c)3 organizations either located in, or serving residents of, California and South Carolina are eligible to apply.
Geographic Focus: California, South Carolina
Amount of Grant: 100 - 10,000 USD
Samples: American Diabetes Association, San Diego, California, $1,000 - general operating support; Classics for Kids Foundation, Holliston, Massachusetts, $10,000 - general operating support; Ranch Santa Fe, California, $1,000 - general operating support.
Contact: Donna J. Walker, President; (858) 756-6150 or (858) 558-9200
Sponsor: Pulido Walker Foundation
P.O. Box 1334
Rancho Santa Fe, CA 92067-1334

Putnam County Community Foundation Grants 3283

The Putnam County Community Foundation is a nonprofit public charity established to administer funds, award grants and provide leadership, enriching the quality of life and strengthening community in Putnam County. The Foundation makes grants to qualified non-profit organizations seeking to make a different in Putnam County and its residents. Grants are made in the following areas: animal welfare; arts and culture; civic and community; economic development; education; environment; health and human services; recreation; and youth. The application and samples of previously funded grants are available at the website.
Requirements: To be considered for funding, organizations must first submit a preliminary grant application form. The Grants Committee will review all preliminary applications to determine who will be invited to submit a full grant application.
Restrictions: Funding is not allowed for the following: individuals; ongoing operational expenses, i.e. salaries, rent, and utilities; projects that do not serve Putnam County citizens; projects normally fully funded by units of government; programs to build or fund an endowment; religious activities or programs that appear to serve one denomination and not the community at large; political organizations or campaigns; national and state-wide fund raising projects; for-profit companies; or projects requesting retroactive funding.
Geographic Focus: Indiana
Date(s) Application is Due: Feb 1; Mar 9; Aug 1; Sep 9
Contact: M. Elaine Peck, Executive Director; (765) 653-4978; fax (765) 653-6385; epeck@pcfoundation.org or info@pcfountation.org
Internet: http://www.pcfoundation.org/grant_what_we_fund.html
Sponsor: Putnam County Community Foundation
2 South Jackson Street, P.O. Box 514
Greencastle, IN 46135

PVA Education Foundation Grants 3284

The PVA Education Foundation helps develop tools that share spinal cord injury and disease (SCI/D) knowledge and improve the lives of those with SCI/D. The Foundation provides funding in five project categories: consumer and community education to improve the health, independence and quality of life for individuals with SCI/D; professional development and education to improve the knowledge and competencies of health professionals who serve the SCI/D community, including fellowship and traineeship programs; research utilization and dissemination, which translates findings into practice; assistive technology—Development of teaching tools or pilot programs that demonstrate innovative approaches to the use of assistive devices; and conferences and symposia that provide education and collaboration opportunities for members of the SCI/D community. The Foundation supports one-year (12 months) or two- year (24 months) projects for a maximum of $50,000 per year.
Requirements: Eligible applicants should be members of academic institutions, health care providers and organizations, or consumer advocates and organizations. Grantee institutions must be located in the United States or Canada. However, project directors and fellows are not required to be U.S. or Canadian citizens. All applications must be submitted in the name of the project director by fiscally responsible organizational entities. Each application must be endorsed by the organization's signing official (the person responsible for organizational approval) and the financial officer (the person responsible for financial reporting).
Geographic Focus: All States, Canada
Date(s) Application is Due: Feb 1
Amount of Grant: Up to 50,000 USD
Contact: Barbara Zupnik, Grants Portfolio Manager; (202) 416-7652 or (800) 424-8200; fax (202) 416-7641; barbaraz@pva.org or foundations@pva.org
Internet: http://www.pva.org/site/c.ajIRK9NJLcJ2E/b.6305829/k.6E40/PVA_Education_Foundation.htm
Sponsor: Paralyzed Veterans of America Education Foundation
801 Eighteenth Street, NW
Washington, D.C. 20006-3517

PVA Research Foundation Grants 3285

The foundation, established by the Paralyzed Veterans of America, supports innovative research and fellowships that improve the lives of those with spinal cord injury and disease (SCI/D). The Foundation funds five research categories: laboratory research in the basic sciences to find a cure for SCI/D; clinical and functional studies of the medical,

psychosocial and economic effects of SCI/D, and interventions to alleviate these effects; design and development of assistive technology for people with SCI/D, which includes improving the identification, selection and utilization of these devices; fellowships for postdoctoral scientists, clinicians and engineers to encourage training and specialization in the field of spinal cord research; and conferences and symposia that provide opportunities for collaboration and interaction among scientists, health care providers and others involved in the SCI/D research community. The PVA Research Foundation online application site will open on May 1 to receive new applications. The annual deadline is September 1.
Requirements: Eligible grantee institutions must be located in the United States or Canada. However, investigators and fellows are not required to be U.S. or Canadian citizens. All grant applicants must have a professional degree: Ph.D. or M.D. preferred, master's degree acceptable under some circumstances. Senior fellows are encouraged to apply as principal investigators. Post-doctoral individuals are eligible to apply for fellowship support within four years of receiving a Ph.D. or completing M.D. residency. Graduate students can participate in Foundation-related research and be paid from a Foundation award.
Restrictions: Graduate students cannot apply for a Foundation grant either as a fellow or as a principal investigator.
Geographic Focus: All States, Canada
Date(s) Application is Due: Sep 1
Amount of Grant: 50,000 - 75,000 USD
Contact: Barbara Zupnik, Grants Portfolio Manager; (202) 416-7652 or (800) 424-8200; fax (202) 416-7641; barbaraz@pva.org or foundations@pva.org
Internet: http://www.pva.org/site/c.ajIRK9NJLcJ2E/b.6305827/k.7268/PVA_Research_Foundation.htm
Sponsor: Paralyzed Veterans of America Research Foundation
801 Eighteenth Street, NW
Washington, D.C. 20006-3517

Quaker Chemical Foundation Grants 3286
The Quaker Chemical Foundation supports nonprofit organizations in company-operating areas whose activities may be described as cultural, educational, health and welfare, and civic and community. Types of support include: program grants; matching gifts; and scholarships. As much as possible, a grant application should contain a project description with a pro-forma budget; a current agency annual operating budget and audited financial statements; a list of funding sources for the organization, including past major contributors with amounts, recent applications with results, and anticipated future funding sources; and a list of board of directors and officers. Of particular interest are programs and activities that promote education and science, especially chemistry. However, the Foundation also accepts grant requests from organizations operating in the areas of health and welfare programs, civic and community concerns, and cultural activities (i.e. theater, music, museums). The foundation chooses to make smaller grants to a large number of organizations rather than larger grants to a small number of organizations. Grants are typically awarded once each year (August) in the range of $1,000 to $5,000.
Requirements: Submitting organizations must have tax-exempt 501(c)3 status as defined by the Internal Revenue Code. Programs and activities outlined in the grant request must directly benefit the local communities in which Quaker operates in the U.S., which includes: Santa Fe Springs, California; Aurora, Illinois; Downers Creek, Illinois; Bingham Farms, Michigan; Detroit, Michigan; Batavia, New York; Dayton, Ohio; Middletown, Ohio; and Conshohocken, Pennsylvania. Also considered are submissions from organizations outside local areas of operations in which Quaker associates are actively involved on a regular basis.
Restrictions: The foundation generally does not support national organizations or bricks and mortar projects.
Geographic Focus: All States
Date(s) Application is Due: Apr 30
Amount of Grant: 1,000 - 5,000 USD
Contact: Amy O'Neill, Secretary; (610) 832-4301 or (610) 832-4127
Internet: http://www.quakerchem.com/about_us/foundation_guidelines.pdf
Sponsor: Quaker Chemical Foundation
One Quaker Pike, 901 East Hector Street
Conshohocken, PA 19428-2380

Qualcomm Grants 3287
The philanthropic endeavors of Qualcomm develop and strengthen communities worldwide. Qualcomm invests human and financial resources in inspirational, innovative programs that serve diverse populations. Specifically their goal is to create educated, healthy, sustainable, culturally vibrant communities and to support employees' commitment to global communities through various programs. The company focuses primarily in geographic regions where they have a business presence. There are three focus areas. First is educated communities. The company is committed to improving science, technology, engineering and math education for students during their primary, secondary, and higher education years, and to expanding educational opportunities for under-represented students. Second is healthy, sustainable communities. The company strives to better the lives of underserved populations by providing basic human needs, with a focus on enhancing the welfare of children. They are also committed to protecting and enhancing our global environment. Third is culturally vibrant communities. Through their support of arts education and outreach programs, they help young people develop innovative minds and expand cultural enrichment opportunities to in-need populations. Applicants may submit a letter of inquiry form online and some organizations will be invited to submit a proposal for funding consideration. The submission deadlines are based on the grantmaking focus areas. The schedule can be found on the website.
Requirements: Eligible applicants must be 501(c)3 organizations.
Restrictions: The following are not eligible: individuals; sporting events without a charitable beneficiary; sectarian or denominational religious groups; faith-based schools unless the school accepts students from all religious and non-religious backgrounds and the students are not required to adhere to or convert to any religious doctrine; faith-based organizations unless the programs are broadly promoted and the program's beneficiaries are not encouraged or required to learn about, adhere, or convert to any religious doctrine; organizations that advocate, support, or practice activities inconsistent with Qualcomm's non-discrimination policies, whether based on race, religion, color, national origin, ancestry, mental or physical disability, age, gender, gender identity and/or expression, sexual orientation, veteran status, pregnancy, medical condition, marital status, or other basis protected by law; primary and secondary schools (note, however, these entities may be eligible for employee matching grants as long as they are not deemed ineligible by other exclusions and restrictions); and political contributions.
Geographic Focus: California, Colorado, Georgia, New Jersey, North Carolina, Texas
Contact: Administrator; (858) 651-3200; fax (858) 651-3255; giving@qualcomm.com
Internet: http://www.qualcomm.com/citizenship/global-social-responsibility/philanthropy/guidelines
Sponsor: Qualcomm
5775 Morehouse Drive
San Diego, CA 92121

Quality Health Foundation Grants 3288
The Quality Health Foundation (QHF) awards grants to eligible organizations that work to improve healthcare for individuals and communities through measurable outcome improvement projects. The Foundation is the mission arm of Quality Health Strategies, a group of companies with 40 years of experience in conducting and evaluating health care quality improvement initiatives. QHF will fund various projects, including service, demonstration, education and clinical programs producing high impact results on health outcomes. Funding will be prioritized based on a project's potential impact on healthcare improvements and access for individuals and communities, particularly the uninsured and under-served population, and/or projects that can be replicated. Customarily, grant awards are for one year with the potential for additional funding in subsequent years. These awards may be up to $50,000.
Requirements: Applicants must be from Maryland or the District of Columbia. All organizations focused on improving the health of individuals at the community level are encouraged to apply. Priority areas include: improved treatment through the use of "best practices"; improved access to health care services; improved understanding of health issues.
Restrictions: QHF will not fund: reimbursable direct patient care services; facility construction/remodeling of facilities; lobbying; or fundraising activities.
Geographic Focus: District of Columbia, Maryland
Date(s) Application is Due: Jan 16
Amount of Grant: Up to 50,000 USD
Contact: Glennda Moragne El; (410) 872-9632; moragneelg@dfmc.org
Internet: http://www.qualityhealthfoundation.org/funding
Sponsor: Quality Health Foundation
9240 Centreville Road
Easton, MD 21601

Quantum Corporation Snap Server Grants 3289
The Snap Server Donation program focuses its charitable efforts on increasing access to storage technology by supporting organizations in the areas of human services and civic development. Quantum makes grants of storage appliances to nonprofit organizations worldwide. Its goal is to help bring the benefits of storage technology to people and their communities, to provide support to the communities in which its employees live, and to support its employees by taking an active role in their community. Snap Server donations are focused on youth organizations, local community programs, science and technology organizations, and humanitarian causes.
Geographic Focus: California
Contact: Sean Lamb; (408) 879-8776; fax (408) 371-1783; sean.lamb@quantum.com
Internet: http://www.quantum.com
Sponsor: Quantum Corporation
2001 Logic Drive
San Jose, CA 95035

Quantum Foundation Grants 3290
The foundation makes grants to approved charitable organizations, as well as state and local governmental entities, serving Palm Beach County, Florida, in the areas of health--improving access to insurance, health care, and prevention programs and reforming the health care delivery system; education--early childhood development, along with academic and career achievement for K-12 students; and community betterment (by invitation only)--promoting diversity, helping special needs populations, reducing family violence, and reducing child abuse and neglect. Types of support include capital grants, challenge/matching grants, program development grants, and seed money grants. Submit concept papers before full proposals.
Requirements: Nonprofits serving West Palm Beach, Florida, are eligible.
Geographic Focus: Florida
Contact: Christine Koehn; (561) 832-7497; fax (561) 832-5794; chrisk@quantumfnd.org
Internet: http://www.quantumfnd.org
Sponsor: Quantum Foundation
2701 North Australian Avenue, Suite 200
West Palm Bach, FL 33407

Questar Corporate Contributions Grants 3291
The corporation funds health and social services in the West, including some parts of Colorado, Oklahoma, Utah, and Wyoming. Grants have been awarded to hospitals, especially to build emergency rooms; substance abuse prevention programs, especially in the schools; child abuse prevention programs; and HIV/AIDS prevention and care. Questar determines the size of the grants based on the number of its employees in its facilities in the area, as well as how many other organizations of the same type it has funded nearby. Requests should be submitted in writing and describe the problem, outline the project, and state the amount sought. Applications are accepted at any time.
Requirements: Nonprofits in the West are eligible.
Geographic Focus: Colorado, Oklahoma, Utah, Wyoming
Amount of Grant: 5,000 - 20,000 USD
Contact: Jan Bates; (801) 324-5132 or (801) 324-5202; Jan.Bates@Questar.com
Internet: http://www.questar.com/about_us/community/contributions.html
Sponsor: Questar Corporation
180 East 100 South, P.O. Box 45433, Mailstop QB 811
Salt Lake City, UT 84145-0433

R.C. Baker Foundation Grants 3292
The foundation makes grants to U.S. nonprofit organizations for projects and programs that support social services for youth and the elderly, crime prevention, education, religion (Christian, Episcopal, Friends, Jewish, Methodist, and Presbyterian), scientific research, culture, and health. Support will be provided for fellowships and scholarships, general operating grants, challenge/matching grants, emergency funds, building funds, equipment, continuing support, annual campaigns, capital campaigns, renovation projects, and special projects. Submit cover letter with proposal. The board meets in June and November to consider requests.
Requirements: Nonprofit organizations are eligible. $25,000 to
Restrictions: Anaheim Memorial Medical Center, Anaheim, CA - $40,000; Harvey Mudd College, Claremont, CA - $25,000;
Date(s) Application is Due: May 1; Oct 1
Amount of Grant: 1,000 - 280,000 USD
Contact: Frank Scott, Chairman; (714) 750-8987
Sponsor: R.C. Baker Foundation
P.O. Box 6150
Orange, CA 92863-6150

R.S. Gernon Trust Grants 3293
The R.S. Gernon Trust was established in 1975 to support and promote quality educational, human-services, and health-care programming for underserved populations. Grants from the R.S. Gernon Trust are primarily one year in duration. On occasion, multi-year support is awarded. Applicants must apply online at the grant website. Applicants are strongly encouraged to do the following before applying: review the downloadable state application procedures for additional helpful information and clarifications; review the downloadable online-application guidelines at the grant website; review the trust's funding history (link is available from the grant website); review the online application questions in advance; and review the list of required attachments. These will generally include: a list of board members, financial statements (audited, reviewed, or compiled by independent auditor); an organization summary; a list of other funding sources; an IRS Determination letter; and other required documents. All attachments must be uploaded in the online application as PDF, Word, or Excel files. The R. S. Gernon Trust has biannual deadlines of February 15 and August 15. Applications must be submitted by 11:59 p.m. on the deadline dates. Applicants will be notified of grant decisions by letter within two to three months after each respective proposal deadline.
Requirements: Applicants must have 501(c)3 tax-exempt status and serve the people of Norwich, Connecticut.
Restrictions: Grant requests for capital projects are generally not considered. Applicants will not be awarded a grant for more than 3 consecutive years. The trust does not support requests from individuals, organizations attempting to influence policy through direct lobbying, or any political campaigns.
Geographic Focus: Connecticut
Date(s) Application is Due: Feb 15; Aug 15
Contact: Kate Kerchaert; (860) 657-7016; kate.kerchaert@baml.com
Internet: https://www.bankofamerica.com/philanthropic/fn_search.action
Sponsor: R.S. Gernon Trust
200 Glastonbury Boulevard, Suite # 200, CT2-545-02-05
Glastonbury, CT 06033-4056

R.T. Vanderbilt Trust Grants 3294
The trust awards grants in its areas of interest, including the arts, education, environmental programs, health care, historic preservation and societies, hospitals, and human services. Types of support include building construction/renovation, endowments, general operating support, and program development. The board meets in April, June, September, and December.
Requirements: There are no applications to submit. Proposals should be submitted in writing.
Restrictions: Applications are not accepted. Gives primarily to Connecticut, with some giving in Maine and New York. The trust does not give grants to individuals.
Geographic Focus: Connecticut, Maine, New York
Amount of Grant: Up to USD
Contact: Gloria Kallas, Trustee; (908) 598-3582; fax (336) 732-2024/(336) 747-8722
Sponsor: R.T. Vanderbilt Trust
1525 West WT Harris Boulevard
Charlotte, NC 28288-5709

RACGP Cardiovascular Research Grants in General Practice 3295
Individual grants of up to $25,000 will be offered for a period of one year to fund projects investigating various aspects of cardiovascular disease including hypertension, stroke, lipid disorder, health promotion and lifestyle changes. The Cardiovascular Research Grants (CVRGs) in General Practice are a joint initiative of the RACGP and the Australian Association of Academic General Practice (AAAGP).
Requirements: To be eligible for the Cardiovascular Research Grants, the chief investigator must be a GP or GP registrar who is a member of the RACGP or the AAAGP. New and emerging researchers are particularly encouraged to apply.
Restrictions: Unless otherwise indicated, Grants must be used in the year for which they are awarded. Grants will be paid only to incorporated bodies that have an ABN and ACN and not to individuals. By applying for a Grant each applicant will be taken to have consented to permit the RACGP to publicise their name, institution, and title of their research project.
Geographic Focus: All States, Australia
Date(s) Application is Due: May 8
Amount of Grant: Up to 25,000 USD
Contact: Samantha Fernandes, Research and Grants Program Administrator; 03-8699-0496; fax 03-8699-0400; research@racgp.org.au
Internet: http://www.racgp.org.au/researchfoundation/awards/cvrg
Sponsor: Royal Australian College of General Practitioners
College House, 1 Palmerston Crescent
South Melbourne, VIC 3205 Australia

RACGP National Asthma Council Research Award 3296
The Award is awarded annually by the National Asthma Council to the best presenter of an asthma related abstract or poster at the RACGP Annual Scientific Convention. The award is to the value of $500 and is presented at the closing ceremony.
Requirements: To be eligible for the award, an individual must be: a GP or GP registrar; a member of the RACGP; in the early stages of their research career; presenting an asthma-related paper or poster at the RACGP Annual Scientific Conference, and; a major contributor to the research study being presented.
Restrictions: By applying for the award each applicant will be taken to have consented to permit RACGP to publicise their name, their institution, and the title of their research project.
Geographic Focus: All States, Australia
Date(s) Application is Due: Sep 1
Amount of Grant: 500 USD
Contact: Samantha Fernandes, Research and Grants Program Administrator; 03-8699-0496; fax 03-8699-0400; research@racgp.org.au
Internet: http://www.racgp.org.au/researchfoundation/awards/nac
Sponsor: Royal Australian College of General Practitioners
College House, 1 Palmerston Crescent
South Melbourne, VIC 3205 Australia

RACGP Vicki Kotsirilos Integrative Medicine Grants 3297
The Vicki Kotsirilos Integrative Medicine Grant is to encourage research that assists the integration of safe and ethical forms of natural and complementary medicines or therapies into mainstream general practice. The Grant is designed to encourage research that assists the integration of safe and ethical forms of natural and complementary medicines or therapies into mainstream general practice. A Grant of up $5,000 is offered to applicants conducting general practice research in the following areas: clinical research that contributes to the knowledge base concerning the efficacy of natural and complementary medicines or therapies; any project that supports research in behavioural and lifestyle factors impacting on health eg. diet, exercise, stress management, and; research that contributes to the efficient dissemination of the evidence regarding integrative medicine.
Requirements: Applicants must be: a General Practitioner or General Practice Registrar, and; a member of the RACGP.
Restrictions: Unless otherwise indicated, the Grant must be used in the year for which it is awarded. The Grant will be paid only to incorporated bodies that have an ABN and ACN and not to individuals. By applying for the Grant each applicant will be taken to have consented to permit the RACGP to publicise their name, institution, and title of their research project.
Geographic Focus: All States, Australia
Date(s) Application is Due: May 8
Amount of Grant: 5,000 USD
Contact: Samantha Fernandes, Research and Grants Program Administrator; 03-8699-0496; fax 03-8699-0400; research@racgp.org.au
Internet: http://www.racgp.org.au/researchfoundation/awards/kotsirilos
Sponsor: Royal Australian College of General Practitioners
College House, 1 Palmerston Crescent
South Melbourne, VIC 3205 Australia

Radcliffe Institute Carol K. Pforzheimer Student Fellowships 3298
The Arthur and Elizabeth Schlesinger Library on the History of Women in America invites Harvard undergraduates to make use of the library's collections with competitive awards of amounts up to $2,500 for relevant research projects. Preference will be given to applicants pursuing research in the history of work and the family, community service and volunteerism, the culinary arts, or women's health. The research may be, but is not required to be, in connection with a project for academic credit. Applications will be evaluated on the significance of the research and the project's potential contribution to the advancement of knowledge as well as its creativity in drawing on the library's holdings. The awards may be used to cover photocopying, other incidental research expenses, or living expenses in lieu of term time or summer employment. See website for current deadlines and application.
Requirements: Each application must include: the Schlesinger Library Grants and Fellowships cover page; a project description (no longer than four double-spaced pages in twelve-point font) indicating the purpose of the research, the Schlesinger Library holdings

to be consulted, and the significance of these holdings to the project overall; a one-page bibliography of principal secondary sources consulted in designing the research; a résumé no longer than two pages; the name of one reference who has agreed to send a supporting letter; and a proposed budget indicating how the funds requested will be spent.
Restrictions: Only Harvard undergraduates are eligible. Funding does not cover the purchase of durable equipment.
Geographic Focus: All States
Amount of Grant: Up to 2,500 USD
Contact: (617) 496-3048; science@radcliffe.edu
Internet: http://www.radcliffe.edu/schles/pforzheimer_grant.aspx
Sponsor: Radcliffe Institute for Advanced Study
10 Garden Street
Cambridge, MA 02138

Radcliffe Institute Individual Residential Fellowships 3299
The Radcliffe Institute Fellowship Program is a scholarly community where individuals pursue advanced work across a wide range of academic disciplines, professions, and creative arts. Radcliffe Institute fellowships are designed to support scholars, scientists, artists, and writers of exceptional promise and demonstrated accomplishment who wish to pursue work in academic and professional fields and in the creative arts. In recognition of Radcliffe's historic contributions to the education of women and to the study of issues related to women, the Radcliffe Institute sustains a continuing commitment to the study of women, gender, and society. Applicants need not focus on gender, however. Women and men from across the United States and throughout the world are encouraged to apply. email fellowships@radcliffe.edu for questions regarding humanities, social sciences, or creative arts fellowships. email sciencefellowships@radcliffe.edu for questions regarding natural science and mathematics fellowships.
Requirements: Scholars in any field with a doctorate or appropriate terminal degree at least two years prior to appointment (by December of the prior year) in the area of the proposed project are eligible to apply. Only scholars who have published at least two articles or monographs are eligible to apply. Artists and writers need not have a PHD or an MFA to apply; however, they must meet other specific eligibility requirements. Fellows are expected to be free of their regular commitments so they may devote themselves full time to the work. Applicants must reside in the Boston area, with their primary office at the Institute. Applications are available online, but applicants should refer to website for current deadlines and application requirements for each academic discipline.
Restrictions: Former fellows of the Radcliffe Institute Fellowship Program (1999 to present) are not eligible to apply.
Geographic Focus: All States, All Countries
Amount of Grant: 35,000 - 70,000 USD
Contact: (617) 496-1324
Internet: http://www.radcliffe.edu/fellowships/apply.asp
Sponsor: Radcliffe Institute for Advanced Study
10 Garden Street
Cambridge, MA 02138

Rainbow Endowment Grants 3300
Once a year, the Rainbow Endowment invites non-profit organizations having a national impact on the LGBT community to submit a request for funding. The proposals are reviewed by the Endowment's Grant Committee, which makes funding recommendations to the Board of Directors. The endowment supports national efforts to encourage positive physical and mental health; promote increased visibility; and advance full participation and access to social, cultural, and civic life for the gay and lesbian community. The Endowment seeks to support LGBT organizations whose efforts are national in scope in the areas of health and community. Priority is given to efforts that have practical implications or applications. In the area of health, grants are made to promote awareness of gay and lesbian mental and physical health issues among consumers and health care providers; to advocate for and develop public policy for equal health care treatment or equal access to health care treatment; and to advocate for and develop public policy to prevent further HIV/AIDS infection and to improve the lives of people with HIV/AIDS. In the area of community, grants are made to protect lesbian and gay rights; promote coalitions that work to strengthen advocacy efforts to improve conditions for lesbians and gays; and to develop policies that benefit lesbian and gay youth and the children of lesbian and gay families. Proposals can be submitted at anytime; however, they are reviewed only once per year. The deadline is the third Friday in May.
Requirements: U.S. nonprofit 501(c)3 organizations are eligible to apply.
Restrictions: The endowment does not support direct services, state or local projects, government agencies, individuals, regularly held meetings, fund-raising events, K-12 schools, religious purposes, endowments, or capital projects. Organizations may not apply more than once each calendar year.
Geographic Focus: All States
Amount of Grant: 5,000 - 20,000 USD
Contact: Jean E. Bochnowski; (215) 241-7280; fax (215) 241-7278; jeb35@aol.com
Internet: http://www.rainbowendowment.org/what.html
Sponsor: Rainbow Endowment
1501 Cherry Street
Philadelphia, PA 19102-1403

Rainbow Fund Grants 3301
The Trust was established in 1954 in Georgia. Giving is primarily focused on theological education and other Christian endeavors. Support is also offered for substance abuse treatment and music education through mentoring high school students and conservatory musicians. There are no specific guidelines or application forms, and potential applicants are advised to contact the office directly.
Requirements: Eligible applicants from Florida, Georgia, Kentucky, Mississippi, and Texas can apply.
Geographic Focus: Florida, Georgia, Kentucky, Mississippi, Texas
Amount of Grant: 25,000 - 2,000,000 USD
Samples: Wesley Biblical Center, Jackson, Mississippi, $2,000,000--for theology instruction; Champions for Life, Duncanville, Texas, $50,000--for prison ministry.
Contact: Burton S. Luce; (954) 764-7724; fax (954) 764-3603; lluce@drmail.com
Sponsor: Rainbow Fund
2408 Sunrise Key Boulevard
Fort Lauderdale, FL 33304

Rajiv Gandhi Foundation Grants 3302
The foundation conducts activities in India in the areas of the arts and humanities, conservation and the environment, economic affairs, education, international affairs, law and human rights, medicine and health, science and technology, and social welfare. Main projects are concerned with literacy and primary education, empowering women and children, aid for the disabled, health care services in rural and poor urban areas, and decentralized government. Types of support include research grants to institutions and scholarships, fellowships, and prizes to individuals.
Geographic Focus: All States
Contact: Jawahar Bhawan, Director; (091-11) 23755117 or (091-11) 23312456; fax (091-11) 23755119; info@rgfindia.com
Internet: http://www.rgfindia.com
Sponsor: Rajiv Gandhi Foundation
Dr. Rajendra Prasad Road
New Delhi, 110 001 India

Ralph F. Hirschmann Award in Peptide Chemistry 3303
This award, supported by Merck Research Laboratories, is given annually to recognize and encourage outstanding achievements in the chemistry, biochemistry, and biophysics of peptides. The award is granted without regard to age or nationality of the recipient. The award consists of $5,000 and a certificate. Up to $2,500 for travel expenses to the meeting at which the award will be presented will be reimbursed.
Requirements: Any individual, except a member of the award committee, may submit one nomination or seconding letter for each award in any given year. Nominating documents consist of a letter of not more than 1000 words containing an evaluation of the nominee's accomplishments and a specific identification of the work to be recognized, a biographical sketch including date of birth, and a list of publications and patents authored by the nominee.
Restrictions: Self-nominations are not accepted.
Geographic Focus: All States
Date(s) Application is Due: Nov 1
Amount of Grant: 5,000 USD
Samples: Morten P. Meldal--award winner, $5,000; William F. DeGrado--award winner, $5,000; Samuel H. Gellman--award winner, $5,000.
Contact: Felicia Dixon, Awards Administrator; (800) 227-5558 or (202) 872-4408; fax (202) 776-8008; f_dixon@acs.org or awards@acs.org
Internet: http://portal.acs.org/portal/acs/corg/content?_nfpb=true&_pageLabel=PP_ARTICLEMAIN&node_id=1319&content_id=CTP_004547&use_sec=true&sec_url_var=region1
Sponsor: American Chemical Society
1155 Sixteenth Street, NW
Washington, D.C. 20036-4801

Ralph M. Parsons Foundation Grants 3304
The Ralph M. Parsons Foundation strives to support and facilitate the work of the region's best nonprofit organizations, recognizing that many of those in need today will go on to shape the future of Southern California, to define it, redefine it, and help it set and achieve new goals. The Foundation focuses on four areas: social impact, civic and cultural programs, health, and higher education. Applicants may submit a letter of inquiry, and there are no deadlines. If the Foundation decides to explore specifics of the request in more detail, applicants will be asked to submit a full proposal. Guidance will be provided in writing. Approximately 200 awards are made per year ranging from $25,000 to $50,000.
Requirements: Eligible applicants are 501(c)3 organizations located in Los Angeles County. Some occasional exceptions are made in the area of higher education. Excellence, access for disadvantaged populations, and the active participation of volunteers, board and staff are key characteristics the Foundation seeks in its applicants. Funding of direct services is a priority.
Restrictions: The following is ineligible: fundraising events, dinners and mass mailings; direct aid to individuals; conferences, seminars, workshops, etc; sectarian, religious or fraternal purposes; federated fundraising appeals; support of candidates for political office or to influence legislation; for-profit organizations or businesses; organizations outside of Los Angeles County (with occasional exceptions in the area of higher education); animal welfare; environment; documentary filmmaking; and scientific and/or medical research. Scholarship support is provided only to nonprofit institutions; individuals are not eligible to apply.
Geographic Focus: California
Contact: Wendy Garen, President and Chief Executive Officer; (213) 362-7600
Internet: http://www.rmpf.org
Sponsor: Ralph M. Parsons Foundation
1888 West Sixth Street, Suite 700
Los Angeles, CA 90017

Ralphs Food 4 Less Foundation Grants 3305
The foundation supports nonprofits primarily in areas of company operations in southern California, from Santa Barbara to San Diego, for programs and activities to improve the well-being of youth through education, recreation, and health-related programs; expand cultural awareness and appreciation of the arts; strengthen neighborhoods; and assist communities in the aftermath of local disasters. Types of support include special projects and general operating expenses. There are no application deadlines. Applicants should submit a letter of application.
Requirements: To be eligible, your organization must be: a 501(3) tax exempt nonprofit; located in Southern California; working in one of the focus areas. Requests for funding must arrive eight (8) weeks prior to the event date or date of need.
Restrictions: Funding is not available to: individuals; capital campaigns; travel expenses; projects of sectarian or religious organizations whose principal benefit is for their own members or adherents; organizations that discriminate on the basis of sex, race, religion, sexual orientation or national origin; third party giving.
Geographic Focus: California
Samples: Los Angeles Rescue Mission, Los Angeles, CA - $10,000; Community Environmental Council, Santa Barbara, CA - to support its annual Earth Day Festival $1,250; The Museum of Tolerance (MOT), Los Angeles, CA - to fund the MOT's student programs $10,000;
Contact: Michelle Williams, Executive Director; (310) 884-6205 or (310) 900-3522
Internet: http://www.ralphs.com/corpnewsinfo_charitablegiving_art5.htm
Sponsor: Ralph's-Food 4 Less Foundation
P.O. Box 54143
Los Angeles, CA 90054

Raskob Foundation for Catholic Activities Grants 3306
The Raskob Foundation is an independent private Catholic family foundation that makes grants worldwide for projects and programs associated with the Catholic Church. Grants support elementary and secondary education, community action and development, missionary activities, ministries (including youth and parish), health care, social concerns, AIDS victims, finance and development, and relief services. Types of support include operating budgets, seed money, emergency funds, equipment, land acquisition, conferences and seminars, program-related investments, renovation projects, special projects, and matching funds. Deadlines are June 8 and August 8 for the fall meeting; and December 8 and February 8 for the spring meeting.
Requirements: Roman Catholic organizations listed in the Kenedy Directory of Official Catholic Organizations may apply. Organizations should refer to the application guidelines for specific instruction on how to apply and information to submit.
Restrictions: The Foundation does not accept applications for the following purposes: tuition, scholarships or fellowships; reduction of debt; endowment funds; grants made by other grantmaking organizations; individual scholarly research; lobbying or legislation; or projects completed prior to our board meetings (mid-May and late November).
Geographic Focus: All States, All Countries
Amount of Grant: 5,000 - 15,000 USD
Contact: Maureen Horner; (302) 655-4440; fax (302) 655-3223; info@rfca.org
Internet: http://www.rfca.org/en/Grantmaking/tabid/63/Default.aspx
Sponsor: Raskob Foundation for Catholic Activities
P.O. Box 4019
Wilmington, DE 19807

Rasmuson Foundation Tier One Grants 3307
The Rasmuson Foundation supports non-profit organizations which strive to improve the quality of life for people throughout the state of Alaska. Tier One Grants provide funding of up to $25,000 for capital projects, technology updates, capacity building, program expansion, and creative works. The Foundation encourages applicants to discuss proposals prior to submission. Applications are available on the website and are accepted at any time. The Foundation accepts Tier 1 grant applications year-round, with review and awards handled on a rolling basis throughout the year. Organizations will be notified via email when an application has been received. If an applicant organization is successful, a formal grant agreement and grant check will be sent.
Requirements: Alaskan organizations that have received 501(c)3 status and are classified as not a private foundation under section 509(a) of the Internal Revenue Service Code, units of government, and federally-recognized tribes are eligible.
Restrictions: For religious organizations, only projects with a broad community impact are considered. For units of government and tribes, only projects with a broad community impact beyond traditional government functions are considered. The following is not eligible: general operations, administrative, indirect, or overhead costs; deficits or debt reduction; endowments; scholarships; fundraising events or sponsorships; in general, K-12 education; reimbursement for items already purchased; and electronic health records and other emerging technologies.
Geographic Focus: Alaska
Amount of Grant: Up to 25,000 USD
Samples: Alaska Association for Historic Preservation, Anchorage, Alaska, $20,000 - to restore launch control building on Nike Site Summit in Anchorage (2015); Yakutat Tlingit Tribe, Yakutat, Alaska, $18,837 - to upgrade building infrastructure and purchase equipment for a culture camp in Yakutat (2015); Valley Residential Services, Wasilla, Alaska, $23,845 - for technology upgrades (2015).
Contact: Barbara Bach, Director of Grant Management; (907) 297-2825 or (907) 297-2700; fax (907) 297-2770; bbach@rasmuson.org or rasmusonfdn@rasmuson.org
Internet: http://www.rasmuson.org/index.php?switch=viewpage&pageid=32
Sponsor: Rasmuson Foundation
301 West Northern Lights Boulevard, Suite 400
Anchorage, AK 99503

Rasmuson Foundation Tier Two Grants 3308
The Rasmuson Foundation supports non-profit organizations which strive to improve the quality of life for people throughout the state of Alaska. Tier Two Grants are available of more than $25,000 for large capital (building) projects, projects of demonstrable strategic importance or innovative nature, or the expansion or start-up of innovative programs that address issues of broad community or statewide significance. Tier 2 grants may also support technology updates and creative works. The project must demonstrate long-term benefits or impacts, and be initiated by an established organization(s) with a history of accomplishment. Applying for a Tier 2 grant is a two-step process. The first step is to prepare and submit a Letter of Inquiry. If the Foundation is interested in the project, then the organization will be invited to submit a full Tier 2 proposal. Tier 2 Letter of Inquiry reviews can take up to 90 days. Tier 2 grants are awarded during the biannual board meetings, which generally take place in early July and early December.
Requirements: Alaskan organizations that have received 501(c)3 status and are classified as not a private foundation under section 509(a) of the Internal Revenue Service Code, units of government, and federally-recognized tribes are eligible.
Restrictions: For religious organizations, only projects with a broad community impact are considered. For units of government and tribes, only projects with a broad community impact beyond traditional government functions are considered. The following is not eligible: general operations, administrative, indirect, or overhead costs; deficits or debt reduction; endowments; scholarships; fundraising events or sponsorships; in general, K-12 education; reimbursement for items already purchased; and electronic health records and other emerging technologies.
Geographic Focus: Alaska
Amount of Grant: 25,000 - 3,000,000 USD
Samples: Southcentral Foundation, Anchorage, Alaska, $2,605,000 - construction of NUKA Institute and start-up operations (2014); Anchorage Museum Association, Anchorage, Alaska, $495,000 - implementation of a Polar Lab (2014); Alaska Pacific University, Anchorage, Alaska, $500,000 - initiatives to enhance sustainability (2014).
Contact: Barbara Bach, Director of Grant Management; (907) 297-2825 or (907) 297-2700; fax (907) 297-2770; bbach@rasmuson.org or rasmusonfdn@rasmuson.org
Sponsor: Rasmuson Foundation
301 West Northern Lights Boulevard, Suite 400
Anchorage, AK 99503

Rathmann Family Foundation Grants 3309
The foundation's main funding areas are education, with priority given to science and math; the arts; children and youth health organizations; and preservation of the environment. Types of support include general operating support, continuing support, capital campaigns, equipment acquisition, endowment funds, program development, conferences and seminars, seed grants, curriculum development, fellowships, internships, scholarship funds, research, and matching funds. There are no specific deadlines or application forms, and interested parties should begin by contacting the Foundation directly.
Requirements: Grants are awarded to organizations in the San Francisco Bay, California area; the Annapolis, Maryland area; the Seattle, Washington area; the Philadelphia, Pennsylvania area: and metropolitan Minneapolis/Saint Paul, Minnesota.
Restrictions: Grants are not awarded to/for private foundations; religious organizations for religious activities; civil rights; social action; advocacy organizations; fraternal groups; political purposes; mental health counseling; individuals; or fundraisers, media events, public relations, annual appeals, or propaganda.
Geographic Focus: California, Maryland, Minnesota, Pennsylvania, Washington
Amount of Grant: 1,000 - 100,000 USD
Contact: Rick Rathmann, (410) 349-2376; fax (410) 349-2377
Sponsor: Rathmann Family Foundation
1290 Bay Dale Drive, P.O. Box 352
Arnold, MD 21012

Raymond John Wean Foundation Grants 3310
The foundation awards grants to children's nonprofits in northeast Ohio and southwest Pennsylvania in the areas of education and human services. In education, the foundation supports programs helping teachers and schools prepare students for life and work, especially projects improving educational opportunities, standards, performance, and achievement. In the human services category, the foundation prefers programs strengthening families, providing opportunities for at-risk youth, and helping disadvantaged people. The foundation also is interested in improving healthcare access.
Requirements: Nonprofit organizations serving northeast Ohio and southwest Pennsylvania are eligible.
Geographic Focus: Ohio, Pennsylvania
Date(s) Application is Due: Mar 1; Jun 1; Sep 1; Dec 1
Amount of Grant: 1,000 - 5,000 USD
Contact: Administrator; (330) 394-5600; info@rjweanfoundation.org
Internet: http://www.rjweanfoundation.org
Sponsor: Raymond John Wean Foundation
108 Main Avenue SW, Suite 1005
Warren, OH 44481-1058

Rayonier Foundation Grants 3311
The foundation awards grants to eligible nonprofit organizations in its areas of interest, including children and youth, community development, disadvantaged (economically), education, engineering, environmental programs, families, health care and hospitals, libraries, minorities, performing arts, recreation science, social services, technology, and volunteerism. Types of support include annual campaigns, building construction/renovation, continuing support, employee matching gifts, employee-related scholarships,

endowments, equipment acquisition, general operating support, matching/challenge grants, performing arts, program development, recreation, science, research grants, scholarship funds, and seed grants.
Requirements: Nonprofits in Florida, Georgia, and Washington are eligible.
Geographic Focus: Florida, Georgia, Washington
Amount of Grant: 250 - 25,000 USD
Samples: United Way of Clallam County, Port Angeles, WA - $4,243; Florida Community College at Jacksonville Foundation, Jacksonville, FL - $25,000; University of Virginia, Charlottesville, VA - $7,500;
Contact: Charles H. Hood; (904) 357-9100; fax (904) 357-9101; info@rayonier.com
Sponsor: Rayonier Foundation
50 North Laura Street, Stuite 1900
Jacksonville, FL 32202

RCF General Community Grants 3312
Six times a year, the Richland County Foundation awards General Community Grants to nonprofit organizations through a competitive process. In doing so, the Foundation looks to partner with nonprofit organizations to respond to current community needs in the following areas: education; health services; arts and culture; community services; children, youth and families; human services; environment; and employment and economic development. Application deadlines are 5:00 p.m. on the first Friday of January, March, May, July, September, and November. Applications must be at the Foundation Office by 5:00 p.m. on the due date to meet the deadline. Final decisions on all grant applications are made by the Board of Trustees approximately 6 – 8 weeks following the grant deadline.
Requirements: 501(c)3 public charities, government entities, schools, and nonprofit medical facilities serving residents of Richland County are eligible to apply. The application procedure should begin with a telephone call to the Program Officer to schedule an initial meeting to discuss the project. The Foundation typically looks for several of the following key elements in submitted applications: a one-time grant, especially for a pilot project which can serve as a model or be replicated; a project in which the Foundation is a funding partner, rather than the sole funder; a project or program which promotes volunteer involvement; an organization which can demonstrate the ability to sustain the project in the future when Foundation grant dollars end; projects or programs which are a collaborative effort(s) among nonprofit organizations in the community which eliminate duplication of services; a project which is likely to make a clear difference in the quality of life of a substantial number of people; an organization which is proposing a practical approach to a solution of a current community problem; a project or program which is focusing on prevention; and a worthy community project for which a grant from the Foundation will most likely leverage additional financial support. Applicants who have a program or project that meets these criteria are encouraged to contact the Foundation to discuss submitting an application.
Restrictions: Community grants are awarded from endowed, unrestricted and field of interest funds. Richland County Foundation typically does not provide funding from unrestricted funds for the following: sectarian activities of religious organizations; operating expenses for annual drives or to eliminate debt; medical, scientific or academic research; individuals other than for college scholarships; travel to or in support of conferences, or travel for groups such as bands, sports teams and classes (unless through special grant programs such as Summertime Kids or the Teacher Assistance Program); capital improvements to building and property not owned by the organization or covered by a long term lease; computer systems; projects that taxpayers support or expected to support; and political issues.
Geographic Focus: Ohio
Samples: Ashland University, Ashland, Ohio, $250,000 - this grant is part of a $15.5 million campaign to build a College of Nursing facility in Richland County; Richland Community Development Group, Mansfield, Ohio, $50,000 - this supports phase two of a pilot project to coordinate a county-wide collaborative effort for both economic and community development.
Contact: Bradford Groves; (419) 525-3020; fax (419) 525-1590; bgroves@rcfoundation.org
Internet: http://www.richlandcountyfoundation.org/grant-information/types-of-grants/community
Sponsor: Richland County Foundation
24 West Third Street, Suite 100
Mansfield, OH 44902-1209

RCF Individual Assistance Grants 3313
Twice a year, the Richland County Foundation awards Individual Assistance Grants to nonprofit organizations which operate programs that assist the needy. Grant dollars may be used for emergency assistance, basic human needs and sustaining basic health. Application deadlines are 5:00 p.m. on the first Friday of March and September. Applications must be at the Foundation Office by 5:00 p.m. on the due date to meet the deadline. Final decisions on all grant applications are made by the Board of Trustees approximately 6 – 8 weeks following the grant deadline. Community grants are awarded from endowed, unrestricted and field of interest funds.
Requirements: 501(c)3 public charities, government entities, schools, and nonprofit medical facilities providing emergency assistance, basic human needs, and basic health programs for residents of Richland County are eligible to apply. The application procedure should begin with a telephone call to the Program Officer to schedule an initial meeting to discuss the project.
Geographic Focus: Ohio
Contact: Bradford Groves; (419) 525-3020; fax (419) 525-1590; bsmith@rcfoundation.org
Internet: http://www.richlandcountyfoundation.org/grant-information/types-of-grants/community
Sponsor: Richland County Foundation
24 West Third Street, Suite 100
Mansfield, OH 44902-1209

RCF Summertime Kids Grants 3314
Richland County Foundation's Summertime Kids Grants provide funding to nonprofit organizations that develop creative, educational and fun-filled activities for Richland County children throughout the summer months. Proposals are due the second Friday in February and must be at the Foundation Office by 5:00 p.m. on the due date to meet the deadline. The application with original signatures plus 18 copies must be submitted (the Foundation requests that applicants refrain from attachments). Awards will be announced in early April.
Requirements: 501(c)3 organizations, schools, churches, government entities and health service organizations serving residents of Richland County are eligible to apply. Collaborating applicants should submit only one application and note on it the lead agency.
Restrictions: Grant dollars should be used for direct programming expenses rather than operational expenses.
Geographic Focus: Ohio
Amount of Grant: Up to 2,500 USD
Samples: City of Mansfield, Mansfield, Ohio - "Hooked on Fishing"; Ohio Bird Sanctuary, Mansfield, Ohio - Nature Camp; Raemelton Therapeutic Equestrian Center, Mansfield, Ohio - Summer Horse Camp.
Contact: Becky Smith; (419) 525-3020; fax (419) 525-1590; bsmith@rcfoundation.org
Internet: http://richlandcountyfoundation.spirecms.com/grant-information/types-of-grants/summertime-kids
Sponsor: Richland County Foundation
24 West Third Street, Suite 100
Mansfield, OH 44902-1209

RCF The Women's Fund Grants 3315
Richland County Foundation makes grants annually from "The Women's Fund" a permanent endowment established in 1996 to promote the physical, intellectual, emotional, social, economic, and cultural growth of women of all ages. While the current grant cycle is open to any nonprofit organization for programs which benefit women and girls in Richland County, a preference will be given to programs addressing childhood obesity, which exclusively target girls up to the age of (18) eighteen. The Foundation offers a workshop in August for interested applicants. Details are available at the grant website. Grant applications will be available at the workshop and will be made available for download from the website afterwards. Awards will be announced at the annual Women's Fund Luncheon in November.
Requirements: 501(c)3 public charities, government entities, schools, and nonprofit medical facilities serving residents of Richland County are eligible to apply. Deadline dates may vary from year to year. Interested applicants are encouraged to check the website, call the Program Officer with any questions and attend the Foundation's annual workshop for this grant.
Geographic Focus: Ohio
Date(s) Application is Due: Sep 16
Amount of Grant: 250 - 10,000 USD
Samples: Center for Individual and Family Services, Mansfield, Ohio - "Ladies in Recovery"; Emergency Pregnancy Contact, Mansfield, Ohio - "Helping Mom Succeed III"; Eastern Elementary School, Lexington, Ohio - "Women in Science Day"
Contact: Bradford Groves; (419) 525-3020; fax (419) 525-1590; bgroves@rcfoundation.org
Internet: http://richlandcountyfoundation.spirecms.com/womens-fund/womens-fund-application-process
Sponsor: Richland County Foundation
24 West Third Street, Suite 100
Mansfield, OH 44902-1209

RCPSC/AMS CanMEDs Research and Development Grants 3316
The primary objective of the AMS CanMEDs Research and Development grant is to support research, development, and/or implementation of projects to enhance specialty education in the promotion of the key roles for specialists as defined by CanMEDS. Applications may be individual or joint, involving one or more departments, disciplines, or faculties of medicine. It is not necessary to be a Fellow of the RCPSC or a physician; however, a Fellow of the Royal College must contribute to the project. Eligible projects may include faculty development, course development and/or medical education research that both apply and promote the CanMEDS roles. Budgets may include requests to defray the costs for travel, accommodation and meals for investigators to disseminate the project findings. One or more grants of up to $25,000 will be awarded each year. The annual deadline for applications is March 3.
Requirements: Applicants must be involved in teaching and/or medical education research in Canada. Applications that include research must meet contemporary standards of educational research, which include the relevance of the research question, the quality of the research design, ethical review of the submission, accountability for funds received, and the ability to complete the project during the funding period. Applications that include project development must include an evaluative component assessing the impact of the project.
Restrictions: Budgets should not include salary or honoraria for investigators. Ineligible costs include tuition fees, capital equipment costs (e.g. computer hardware, office equipment), overhead costs, university or other organization administrative fees.
Geographic Focus: All States
Date(s) Application is Due: Mar 3
Amount of Grant: Up to 25,000 USD
Contact: Allison Montpetit, Grant Administrator; (800) 668-3740, ext. 291 or (613) 730-8177; fax (613) 730-8830; researchgrants@royalcollege.ca or awards@royalcollege.ca
Internet: http://www.royalcollege.ca/portal/page/portal/rc/awards/grants/medical_education/ams_grant
Sponsor: Royal College of Physicians and Surgeons of Canada
774 Echo Drive
Ottawa, ON K1S 5N8 Canada

RCPSC/AMS Donald Richards Wilson Award 3317
The Royal College of Physicians and Surgeons of Canada, in collaboration with Associated Medical Services, has established an award to honor and acknowledge the contribution of Dr. Donald R. Wilson (President, RCPSC, 1988-1990) to medical education. This award will be given annually to a medical educator (not necessarily a Fellow) or an identified leader of a team, program, or department who has demonstrated excellence in integrating the CanMEDS roles into a Royal College or other health related training programs. Nominations will be received from Deans, Department or program heads, peers, or trainees. The deadline for applications is September 2 each year. The successful applicant(s) will receive an award of $2,000 and a suitably engraved recognition.
Restrictions: Self-applications are ineligible.
Geographic Focus: Canada
Date(s) Application is Due: Sep 2
Amount of Grant: 2,000 USD
Contact: Kimberly Ross, Grant Administrator; (800) 668-3740, ext. 355 or (613) 730-8177; fax (613) 730-8830; kross@royalcollege.ca or awards@royalcollege.ca
Internet: http://www.royalcollege.ca/portal/page/portal/rc/awards/awards/personal_achievement/wilson_award
Sponsor: Royal College of Physicians and Surgeons of Canada
774 Echo Drive
Ottawa, ON K1S 5N8 Canada

RCPSC Balfour M. Mount Visiting Professorship in Palliative Medicine 3318
The Royal College of Physicians and Surgeons of Canada, in collaboration with the Robert Pope Foundation of Nova Scotia, has established this professorship in order to bring to the health science faculties in Canada the most distinguished international lecturers in the palliative care community to engage in discussion about, and to lecture on, the physical, psycho-social and spiritual needs of patients requiring palliative care. The professorship is meant to facilitate the work of the multidisciplinary team that contributes to palliative care expertise in the community and university being visited. The professor will be appointed in alternate years and will visit the health science faculties of at least two universities in Canada. Using a variety of teaching tools, the professor will bring to the host faculties knowledge and research related to the physical, psycho-social, and spiritual needs of patients requiring palliative care. The annual deadline for applications is October 15.
Geographic Focus: Canada
Date(s) Application is Due: Oct 15
Contact: Kimberly Ross, Grant Administrator; (800) 668-3740, ext. 355 or (613) 730-8177; fax (613) 730-8830; kross@royalcollege.ca or awards@royalcollege.ca
Internet: http://www.royalcollege.ca/portal/page/portal/rc/awards/awards/visiting_professorships/balfour_mount_award
Sponsor: Royal College of Physicians and Surgeons of Canada
774 Echo Drive
Ottawa, ON K1S 5N8 Canada

RCPSC Continuing Proffesional Development Grants 3319
Continuing Professional Development grants are provided to physician organizations to assist in the development of group learning events. These group learning events must be developed to meet the educational and ethical standards established by the Royal College for accredited group learning within Section 1 of the Maintenance of Certification program. All successful grant applicants will receive a maximum of $1,000. The amount awarded will be based on the following criteria: the total amount of industry funding; and the anticipated expenses in relation to anticipated revenue.
Requirements: Each applicant must be a regional or provincial organization that meets the Royal College definition of a physician organization.
Restrictions: National organizations are not eligible to apply for this grant.
Geographic Focus: Canada
Amount of Grant: Up to 1,000 USD
Contact: Melanie Blackburn, Grant Administrator; (800) 668-3740, ext. 355 or (613) 730-8177; fax (613) 730-8830; awards@rcpsc.edu
Internet: http://www.royalcollege.ca/portal/page/portal/rc/awards/grants/cpd/cpd_activity_grant
Sponsor: Royal College of Physicians and Surgeons of Canada
774 Echo Drive
Ottawa, ON K1S 5N8 Canada

RCPSC Detweiler Traveling Fellowships 3320
Fellowships are made to improve the quality of medical and surgical practice in Canada. The Detweiler Traveling Fellowship will enable the recipients to visit medical centers in Canada or abroad to study or gain experience in the use or application of new knowledge or techniques in their fields or to further the pursuit of a fundamental or clinical research problem. Five or more fellowships offered annually to a maximum of $21,000 will be reserved for junior applicants. Two additional fellowships offered annually will be reserved for senior applicants: one will be granted to a Fellow who has held Royal College certification for five years or more and has been in private practice for at least three years, and one will be granted to a Fellow who has held Royal College certification for five years or more and has been in academic practice for at least three years. Each award will be to a maximum of $10,500.
Requirements: Applicants must be Fellows of the Royal College in good standing, or residents in the final year of their training who will be admitted to Fellowship in the Royal College prior to December 31st.
Restrictions: Travel fellowships will not be awarded retroactively.
Geographic Focus: Canada
Date(s) Application is Due: Sep 30
Amount of Grant: 1,750 - 21,000 USD
Contact: Kimberly Ross, Awards and Grants Section; (800) 668-3740, ext. 355 or (613) 730-8177; fax (613) 730-8830; kross@royalcollege.ca or awards@royalcollege.ca
Internet: http://www.royalcollege.ca/portal/page/portal/rc/awards/grants/travelling_fellowships/detweiler_travelling_grant
Sponsor: Royal College of Physicians and Surgeons of Canada
774 Echo Drive
Ottawa, ON K1S 5N8 Canada

RCPSC Dr. Thomas Dignan Indigenous Health Award 3321
The Royal College Dr. Thomas Dignan Indigenous Health Award was founded in 2014 in honour of Dr. Thomas Dignan, OOnt, MD, chair of the Royal College Indigenous Health Advisory Committee. Dr. Dignan of Thunder Bay, Ontario, a Mohawk from Six Nations of the Grand River Territory is a tireless advocate for eradicating disparities in health outcomes and inequities in the quality of health care facing Indigenous people. Dr. Dignan is a former Royal College council member, co-founder of the Native Physicians Association of Canada and a founding member and first president of the Native Nurses Association of Canada. Dr. Dignan has dedicated his life to improving the health of Indigenous Peoples in Canada, particularly drawing attention to issues of institutional racism. The recipient will receive an award of $1,000, and will be presented with an engraved memento at a Royal College Annual Conference or other suitable event where Indigenous culture is recognized. Travel and related expenses of the award winner and a companion will be paid in accordance with Royal College guidelines. A maximum of one award recipient will be named in any given year. At the Awards Committee's discretion, the award may not necessarily be bestowed every year. The annual deadline for application is September 2.
Geographic Focus: Canada
Date(s) Application is Due: Sep 2
Amount of Grant: 1,000 USD
Contact: Kimberly Ross, Grant Administrator; (800) 668-3740, ext. 355 or (613) 730-8177; fax (613) 730-8830; kross@royalcollege.ca or awards@royalcollege.ca
Internet: http://www.royalcollege.ca/portal/page/portal/rc/awards/awards/personal_achievement/thomas_dignan_award
Sponsor: Royal College of Physicians and Surgeons of Canada
774 Echo Drive
Ottawa, ON K1S 5N8 Canada

RCPSC Duncan Graham Award 3322
One of the notable and outstanding awards that the Royal College of Physicians and Surgeons of Canada may bestow upon an individual is the Duncan Graham Award. The Duncan Graham Award was introduced in 1969 in honor of the late Dr. Duncan Graham, for many years a professor of medicine at the University of Toronto. The Award is conferred upon any individual, whether physician or not, in recognition of outstanding lifelong contribution to medical education. The award is in the amount of $1,000 accompanied by a suitably engraved memento. It is usually conferred each year at the Convocation of the Royal College Annual Meeting. Nominations are solicited from the Deans of the Faculties of medicine. The Awards Committee of the Royal College selects the recipient of the Duncan Graham Award. At its discretion, the Awards Committee may add to the list of nominations. The nomination deadline is October 15, and applications must be submitted electronically. The award includes $1,000 and is supported financially by donations to the Royal College's Educational Fund from former students and friends of Dr. Graham.
Geographic Focus: Canada
Date(s) Application is Due: Oct 15
Amount of Grant: 1,000 USD
Contact: Kimberly Ross, Grant Administrator; (800) 668-3740, ext. 355 or (613) 730-8177; fax (613) 730-8830; kross@royalcollege.ca or awards@royalcollege.ca
Internet: http://www.royalcollege.ca/portal/page/portal/rc/awards/awards/personal_achievement/duncan_graham_award
Sponsor: Royal College of Physicians and Surgeons of Canada
774 Echo Drive
Ottawa, ON K1S 5N8 Canada

RCPSC Harry S. Morton Lectureship in Surgery 3323
The Harry S. Morton Lectureship in Surgery, established by the Royal College of Physicians and Surgeons of Canada in 2013, is funded from the interest income of the H.S. Morton Exchange Fellowship Fund of Canada. Dr. Morton obtained his medical degrees from the Royal London Hospital Medical School in England and joined the Royal Canadian Navy in 1938. Following the Second World War, he taught at McGill and practiced as Chief Surgeon at Queen Mary's Veterans Hospital in Montreal. He was made a Patron of the Royal College of Surgeons in 1999. Dr. Morton died in Halifax in December 2001. The purpose of this lectureship is to advance and to promote study of and research in surgical training and the practice of surgery, by providing support for a nominated surgeon from the United Kingdom to travel for not less than three days to centres in Canada, timed to be held in conjunction with the annual Surgical Forum other academic surgical conference, or other meeting of a National Specialty Society of a surgical discipline. Specifically, the purpose of the Morton Lectureship is to deliver opportunities for interactive discourse on emerging topics or innovative practices in a surgical specialty or sub-specialty recognized by the Royal College. The lectureship will also promote opportunities for the travelling surgeon to learn from the centres visited, and bring back knowledge and techniques. The lectureship will provide up to $7,000.00 to support travel and accommodation for the visiting lecturer and to cover expenses incurred by the host organization related to the lectureship. Up to 75% of the award will be given prior to the lectureship, and the remainder upon receipt of a final report at the end of the lectureship. One or more Lectureships shall be offered annually. The number shall be determined by the Awards Committee based on available funding, but shall not exceed five. The annual deadline for applications is October 15.

Requirements: Nominations shall be solicited from the surgical National Specialty Societies and from potential host academic centres. The nomination shall include a completed application form and an outline of the proposed itinerary/program of activity for the visiting lecturer. The successful host organization shall be responsible for the arrangements and itinerary for the Harry S. Morton Lectureship.
Geographic Focus: Canada
Date(s) Application is Due: Oct 15
Amount of Grant: Up to 7,000 USD
Contact: Kimberly Ross, Grant Administrator; (800) 668-3740, ext. 355 or (613) 730-8177; fax (613) 730-8830; kross@royalcollege.ca or awards@royalcollege.ca
Internet: http://www.royalcollege.ca/portal/page/portal/rc/awards/awards/visiting_professorships/morton_award
Sponsor: Royal College of Physicians and Surgeons of Canada
774 Echo Drive
Ottawa, ON K1S 5N8 Canada

RCPSC Harry S. Morton Traveling Fellowship in Surgery　　3324
The Harry S. Morton Traveling Fellowship in Surgery, established by The Royal College of Physicians and Surgeons of Canada in 2013, is funded from the investment income of the H. S. Morton Exchange Fellowship Fund of Canada. Dr. Morton obtained his medical degrees from the Royal London Hospital Medical School in England and joined the Royal Canadian Navy in 1938. Following the Second World War, he taught at McGill and practiced as Chief Surgeon at Queen Mary's Veterans Hospital in Montreal. He was made a Patron of the Royal College of Surgeons in 1999. The purpose of this Fellowship is to advance and to promote the study of and research in surgical training, and the practice of surgery, and to support the building of capacity and additional expertise within Canada's academic surgical programs. The Fellowship provides funds for Canadian surgeons and surgical residents travelling to the United Kingdom and may be used to support surgery-related clinical, scientific, or medical education studies or research. The Travelling Fellowship will provide extended training support (3 months or more) for individuals with academic surgical career plans when they return to practice in Canada. Fellowships are supported by a grant of $50,000 per year, pro-rated monthly, and are contingent upon a financial commitment from a University/Department of Surgery to provide funding support for salary/living expenses (amount should be reflected on application form). The annual deadline for applications is September 2.
Requirements: The Traveling Fellowship is open to Royal College Fellows certified in any surgical discipline recognized by the Royal College, and to residents in the final year of their surgical training who will be admitted to Fellowship in the Royal College prior to December 31 of the year following receipt of the Fellowship.
Geographic Focus: Canada
Date(s) Application is Due: Sep 2
Amount of Grant: Up to 50,000 USD
Contact: Kimberly Ross, Grant Administrator; (800) 668-3740, ext. 355 or (613) 730-8177; fax (613) 730-8830; kross@royalcollege.ca or awards@royalcollege.ca
Internet: http://www.royalcollege.ca/portal/page/portal/rc/awards/grants/travelling_fellowships/morton_travelling_fellowship
Sponsor: Royal College of Physicians and Surgeons of Canada
774 Echo Drive
Ottawa, ON K1S 5N8 Canada

RCPSC International Residency Educator of the Year Award　　3325
The International Residency Educator of the Year Award is given annually to an international residency educator who has demonstrated a commitment to enhancing residency education as evidenced by innovation and impact beyond his/her program. Nominations may be submitted by program directors, deans, postgraduate deans, department chairs/division director, peer teachers or trainees. Applications for this award are adjudicated by a sub-committee of the Awards Committee of the Royal College of Physicians and Surgeons of Canada. One award will be presented annually. Award winner will receive a plaque recognizing their contribution as well as travel and registration to the ICRE conference. The annual deadline for application is April 4.
Requirements: Nominations must be submitted electronically by completing the official award application form. The criteria includes: demonstrated commitment to innovation in residency education; a lasting impact on the quality of residency education and measurable outcomes beyond their program; evidence of effective leadership and mentorship ability; embodies/integrates the CanMEDS Roles: Professional, Communicator, Collaborator, Manager, Health Advocacy, Scholar, Medical Expert; and a strong track record for promoting ethics, reflection and humanism in residency education.
Restrictions: Self-applications are ineligible. Individuals may be re-nominated in subsequent years; however, previous winners of the award will not be eligible to win again.
Geographic Focus: Canada
Contact: Azura Fennell, Office of Specialty Education; (800) 668-3740, ext. 170 or (613) 730-8177; fax (613) 730-8830; icreawards@royalcollege.ca
Internet: http://www.royalcollege.ca/portal/page/portal/rc/awards/awards/personal_achievement/intl_res_ed_year
Sponsor: Royal College of Physicians and Surgeons of Canada
774 Echo Drive
Ottawa, ON K1S 5N8 Canada

RCPSC International Resident Leadership Award　　3326
The International Resident Leadership Award is given annually to an international resident who has demonstrated leadership in specialty education and encourages the development of future leaders in medicine. Nominations may be submitted by postgraduate deans, program directors, supervisors, department chairs/division directors, or resident peers (current or recently graduated). Applications for this award are adjudicated by a sub-committee of the Awards Committee of the Royal College of Physicians and Surgeons of Canada. One award will be presented annually. Award winner will receive a plaque recognizing their contribution; and registration and travel to attend the International Resident Leadership Summit. The annual deadline for application is April 4.
Requirements: Nominations must be submitted electronically by completing the official award application form
Restrictions: Self-applications are ineligible. Individuals may be re-nominated in subsequent years; however, previous winners of the award will not be eligible to win again.
Geographic Focus: Canada
Date(s) Application is Due: Apr 4
Contact: Azura Fennell, Office of Specialty Education; (800) 668-3740, ext. 170 or (613) 730-8177; fax (613) 730-8830; icreawards@royalcollege.ca
Internet: http://www.royalcollege.ca/portal/page/portal/rc/awards/awards/personal_achievement/intl_resident_leadership_award
Sponsor: Royal College of Physicians and Surgeons of Canada
774 Echo Drive
Ottawa, ON K1S 5N8 Canada

RCPSC James H. Graham Award of Merit　　3327
In 1987, the Council of The Royal College of Physicians and Surgeons of Canada recommended that an award of merit be given to a person whose outstanding achievements reflect the aims and objectives of the Royal College. This person need not be a physician. The Awards Committee will be the selection committee for the award of merit. The selection committee will receive nominations (with curriculum vitae) from the RCPSC Council (limit of two), and the Regional Advisory Committees (one per RAC). The nominations are submitted electronically, and the annual deadline is October 15. The presentation for the award of merit will be made at the Annual Meeting of the Royal College. The winner will receive a suitably engraved memento. The travel and maintenance expenses of the award winner and spouse will be paid in accordance with Royal College guidelines.
Geographic Focus: Canada
Date(s) Application is Due: Oct 15
Contact: Kimberly Ross, Grant Administrator; (800) 668-3740, ext. 355 or (613) 730-8177; fax (613) 730-8830; kross@royalcollege.ca or awards@royalcollege.ca
Internet: http://www.royalcollege.ca/portal/page/portal/rc/awards/awards/personal_achievement/james_graham_award
Sponsor: Royal College of Physicians and Surgeons of Canada
774 Echo Drive
Ottawa, ON K1S 5N8 Canada

RCPSC Janes Visiting Professorship in Surgery　　3328
The Janes Visiting Professorship in Surgery, established by The Royal College of Physicians and Surgeons of Canada, in cooperation with the Janes Surgical Society, is funded from the interest income of the Janes Fund. The Fund was created by former students of Dr. Robert M. Janes, Professor of Surgery at the University of Toronto from 1947 to 1957. The purpose of the Professorship is to stimulate and increase an interest in the latest developments in surgery and surgical techniques. The visiting professor, preferably a Canadian surgeon of national stature, spends up to one week visiting one to three medical schools according to a mutually agreed schedule. The professor participates in undergraduate and postgraduate teaching and in the exchange of ideas with practicing surgeons, residents and students. The professorship is supported by a grant of $7,000 and is usually selected every second year. The annual deadline for applications is September 30.
Geographic Focus: Canada
Date(s) Application is Due: Sep 30
Amount of Grant: 7,000 USD
Contact: Kimberly Ross, Grant Administrator; (800) 668-3740, ext. 355 or (613) 730-8177; fax (613) 730-8830; kross@royalcollege.ca or awards@royalcollege.ca
Internet: http://www.royalcollege.ca/portal/page/portal/rc/awards/awards/visiting_professorships/janes_award
Sponsor: Royal College of Physicians and Surgeons of Canada
774 Echo Drive
Ottawa, ON K1S 5N8 Canada

**RCPSC John G. Wade Visiting Professorship in Patient Safety and Simulation-　　3329
Based Medical Education**
The purpose of the John G. Wade Visiting Professorship is to provide support for a distinguished educator to visit a Royal College accredited simulation centre in Canada each year, in order to promote the use of simulation in medical education, and its applications to improving patient safety. Ideally, the visiting professor will interact with a wide variety of medical professionals: students, staff, and other members of the interdisciplinary health care team. The visiting professor is not required to restrict himself/herself to speaking only at the designated simulation center. Nominations will be sought from Canadian Royal College accredited simulation programs. Nominators are asked to summarize the reasons for their nomination, put forward a proposed itinerary for the visiting professor, and attach the nominee's most recent curriculum vitae. Nominees need not be physicians or Fellows of the Royal College. International nominees are welcome. The travel and accommodation expenses of the visiting professor and a $1,000.00 honorarium (to a total maximum of C$4,000) will be paid in accordance with Royal College guidelines. The annual deadline for applications is September 15.
Geographic Focus: Canada
Date(s) Application is Due: Sep 15
Amount of Grant: Up to 4,000 CAD

Contact: Kimberly Ross, (800) 668-3740, ext. 355 or (613) 730-8177; fax (613) 730-8830; kross@royalcollege.ca or awards@royalcollege.ca
Internet: http://www.royalcollege.ca/portal/page/portal/rc/awards/awards/visiting_professorships/wade_award
Sponsor: Royal College of Physicians and Surgeons of Canada
774 Echo Drive
Ottawa, ON K1S 5N8 Canada

RCPSC KJR Wightman Award for Scholarship in Ethics   3330
The Royal College of Physicians and Surgeons of Canada offers the KJR Wightman Award for Scholarship in Ethics for the best scholarly paper presented by a resident registered in a postgraduate program accredited by the Royal College. Papers arising from a clinical case, addressing ethical issues pertinent to the CanMEDS roles (medical expert, communicator, collaborator, manager, health advocate, scholar, professional), or on the topic of equity in any aspect relevant to the work of the College are welcome. The recipient will receive $1,000 as well as a certificate. The award will be presented during the Convocation Ceremony at the Annual Conference. The travel expenses of the recipient and guest to the Annual Conference will be paid in accordance with Royal College guidelines. The annual deadline for application is May 5.
Requirements: Submissions for the award must be accompanied by a curriculum vitae and submitted electronically. The manuscript must be between 2500 and 5000 words, excluding the bibliography.
Geographic Focus: Canada
Date(s) Application is Due: May 5
Amount of Grant: 1,000 CAD
Contact: Kimberly Ross, Grant Administrator; (800) 668-3740, ext. 355 or (613) 730-8177; fax (613) 730-8830; kross@royalcollege.ca or awards@royalcollege.ca
Internet: http://www.royalcollege.ca/portal/page/portal/rc/awards/awards/original_research/wightman_ethics_award
Sponsor: Royal College of Physicians and Surgeons of Canada
774 Echo Drive
Ottawa, ON K1S 5N8 Canada

RCPSC KJR Wightman Visiting Professorship in Medicine   3331
The KJR Wightman Visiting Professor in Medicine, established by The Royal College of Physicians and Surgeons of Canada, in cooperation with the Wightman Club, is funded from the interest income of the Wightman Fund. The Fund was created by former students of Dr. K.J.R. Wightman, former Sir John and Lady Eaton Professor of Medicine, and Chairman of the Department of Medicine of the University of Toronto from 1960 to 1970. The Visiting Professor participates in undergraduate and postgraduate teaching and in discussions with peers, both in internal medicine and in the professor's particular area of academic interest. The Professor spends one or two weeks visiting two or three Canadian faculties of medicine. The position is supported by a grant of $7,000. The annual deadline for applications is September 30.
Geographic Focus: Canada
Date(s) Application is Due: Sep 30
Amount of Grant: 7,000 USD
Contact: Kimberly Ross, Grant Administrator; (800) 668-3740, ext. 355 or (613) 730-8177; fax (613) 730-8830; kross@royalcollege.ca or awards@royalcollege.ca
Internet: http://www.royalcollege.ca/portal/page/portal/rc/awards/awards/visiting_professorships/wightman_medicine_award
Sponsor: Royal College of Physicians and Surgeons of Canada
774 Echo Drive
Ottawa, ON K1S 5N8 Canada

RCPSC Kristin Sivertz Resident Leadership Award   3332
The Kristin Sivertz Resident Leadership Award is given annually to a resident who has demonstrated leadership in Canadian specialty education and encourages the development of future leaders in medicine. Nominations may be submitted by postgraduate deans, program directors, department chairs/division directors, or resident peers (current or recently graduated). Applications for this award are adjudicated by a sub-committee of the Awards Committee of the Royal College of Physicians and Surgeons of Canada. Up to two awards will be presented annually. Award winners will receive a plaque recognizing their contribution; and registration and travel to attend the International Resident Leadership Summit. The annual deadline for applications is April 4.
Requirements: Nominations must be submitted electronically.
Restrictions: Self-applications are ineligible. Individuals may be re-nominated in subsequent years; however, previous winners of the award will not be eligible to win again.
Geographic Focus: Canada
Date(s) Application is Due: Apr 4
Contact: Azura Fennell, Office of Specialty Education; (800) 668-3740, ext. 170 or (613) 730-8177; fax (613) 730-8830; icreawards@royalcollege.ca
Internet: http://www.royalcollege.ca/portal/page/portal/rc/awards/awards/personal_achievement/resident_leadership_award
Sponsor: Royal College of Physicians and Surgeons of Canada
774 Echo Drive
Ottawa, ON K1S 5N8 Canada

RCPSC McLaughlin-Gallie Visiting Professorship   3333
The Royal College of Physicians and Surgeons of Canada established the McLaughlin-Gallie Visiting Professorship in 1960. The professorship commemorates the outstanding contributions to surgical science of Dr. William Edward Gallie of Toronto. McLaughlin-Gallie Visiting Professors may be selected from any country in the world and may represent any branch of medicine. They spend up to one week at each of two Canadian faculties of medicine. They participate in undergraduate and postgraduate teaching, and in the exchange of ideas with the staff, the researchers and the undergraduate and postgraduate students. The professorship is supported each year by a $10,000 grant made to the Royal College Educational Fund from the R. Samuel McLaughlin Foundation. The visiting professor will spend up to one week at each of two Canadian faculties of medicine. They participate in undergraduate and postgraduate teaching, and in the exchange of ideas with the staff, the researchers and the undergraduate and postgraduate students. The annual deadline for applications is May 30.
Requirements: McLaughlin-Gallie Visiting Professors may be selected from any country in the world and may represent any branch of medicine.
Geographic Focus: All States, All Countries
Date(s) Application is Due: May 30
Amount of Grant: 10,000 USD
Contact: Kimberly Ross, Grant Administrator; (800) 668-3740, ext. 355 or (613) 730-8177; fax (613) 730-8830; kross@royalcollege.ca or awards@royalcollege.ca
Internet: http://www.royalcollege.ca/portal/page/portal/rc/awards/awards/visiting_professorships/mclaughlin_gallie_award
Sponsor: Royal College of Physicians and Surgeons of Canada
774 Echo Drive
Ottawa, ON K1S 5N8 Canada

RCPSC Medical Education Research Grants   3334
The primary objective of the Medical Education Research Grant is to support quality research in Canada relevant to the field of postgraduate medical education or continuing professional development. Applications may be individual or joint, involving one or more departments, disciplines, or faculties of medicine. It is not necessary to be a Fellow of the RCPSC or a physician; however, a Fellow of the Royal College must contribute to the research project. The following areas of research are encouraged: curriculum design and implementation; instructional methods; assessment; program evaluation; continuing education; professional development; and faculty development. Up to $25,000 per year of study to a maximum of $50,000 per project is available. The annual deadline for applications is March 2.
Requirements: Applicants must be involved in teaching and/or medical education research in Canada. Applications must meet contemporary standards of educational research, which include the relevance of the research question, the quality of the research design, ethical review of the submission, accountability for funds received, and the ability to complete the project during the funding period.
Restrictions: Ineligible costs include tuition fees, capital equipment costs (e.g. computer hardware, office equipment), overhead costs, university or other organization administrative fees.
Geographic Focus: All States, Canada
Date(s) Application is Due: Mar 2
Contact: Allison Montpetit, Grant Administrator; (800) 668-3740, ext. 291 or (613) 730-8177; fax (613) 730-8830; researchgrants@royalcollege.ca or awards@royalcollege.ca
Internet: http://www.royalcollege.ca/portal/page/portal/rc/awards/grants/medical_education/meded_research_grant
Sponsor: Royal College of Physicians and Surgeons of Canada
774 Echo Drive
Ottawa, ON K1S 5N8 Canada

RCPSC Mentor of the Year Award   3335
The Mentor of the Year award was established to recognize Fellows of the Royal College in good standing who have had a significant impact on the career development of students, residents and/or Fellows. The nominee must have demonstrated their ability to be an excellent role model in demonstrating the qualities or competencies of Manager, Scholar and Professional as described in the CanMEDS framework.
Requirements: The nominee must be a Fellow of the Royal College in good standing. A minimum of three and a maximum of five nominations (including the lead Nominator) are required to support each nominee.
Restrictions: Previous recipients are not eligible to be nominated in the future. Self-nominations will not be accepted.
Geographic Focus: Canada
Date(s) Application is Due: Sep 2
Contact: Kora McNulty, Administrative Coordinator; (800) 668-3740, ext. 173 or (613) 730-8177; fax (613) 260-4173; kmcnulty@royalcollege.ca
Internet: http://www.royalcollege.ca/portal/page/portal/rc/awards/awards/personal_achievement/mentor_award
Sponsor: Royal College of Physicians and Surgeons of Canada
774 Echo Drive
Ottawa, ON K1S 5N8 Canada

RCPSC Peter Warren Traveling Scholarship   3336
Dr. Charles Peter Warren was the inaugural Chair of the Royal College's History and Heritage Advisory Committee, and provided leadership to the Royal College's history and heritage activities until his sudden passing in May of 2011. This traveling scholarship was introduced in 2013 in recognition of Dr. Warren's significant contributions to the History of Medicine in Canada. It is intended to encourage post-graduate trainees in medicine to undertake research projects in the history of specialty medicine, the history of the Royal College of Physicians and Surgeons of Canada, and/or the history of Post-Graduate Medical Education in Canada; and to encourage the use of the Royal College historical and archival collections as a resource for scholarly work within the field of the history of medicine. Up to $1,500, based on the budget provided with the application package, is available. Applications should be submitted electronically by the annual September 2 deadline.

Requirements: The Peter Warren Traveling Scholarship is open to applicants from all medical and surgical disciplines recognized by the Royal College. Applicants must be post-graduate trainees in a Canadian medical school who are undertaking a clearly defined project in the history of specialty medicine including but not limited to the history of the Royal College and/or the history of Post-Graduate Medical Education in Canada.
Geographic Focus: Canada
Date(s) Application is Due: Sep 2
Amount of Grant: Up to 1,500 CAD
Contact: Kimberly Ross, Grant Administrator; (800) 668-3740, ext. 355 or (613) 730-8177; fax (613) 730-8830; kross@royalcollege.ca or awards@royalcollege.ca
Internet: http://www.royalcollege.ca/portal/page/portal/rc/awards/grants/travelling_fellowships/warren_scholarship
Sponsor: Royal College of Physicians and Surgeons of Canada
774 Echo Drive
Ottawa, ON K1S 5N8 Canada

RCPSC Prix D'excellence (Specialist of the Year) Award 3337
The Prix D'excellence (Specialist of the Year) award was established to recognize Fellows of the Royal College in good standing who have made significant contributions in providing outstanding patient care to their patients and the community in which they practice. The nominee should be a role model for excellence in patient care. There will only be one recipient of the Prix D'excellence (Specialist of the Year) per region. A sub-committee of the Regional Advisory Committee (RAC) will be created each year to adjudicate the nominations and select the recipients.
Requirements: The nominee must be a Fellow of the Royal College in good standing. For each nominee there must be at least three and a maximum of five nominations (including the lead Nominator), with at least two of the nominators being Fellows of the Royal College in good standing.
Restrictions: Previous recipients are not eligible to be nominated in the future. Self-nominations will not be accepted.
Geographic Focus: Canada
Contact: Kora McNulty, Administrative Coordinator; (800) 668-3740, ext. 173 or (613) 260-4173; fax (613) 730-8830; kmcnulty@royalcollege.ca
Internet: http://www.royalcollege.ca/portal/page/portal/rc/awards/awards/personal_achievement/prix_dexcellence_award
Sponsor: Royal College of Physicians and Surgeons of Canada
774 Echo Drive
Ottawa, ON K1S 5N8 Canada

RCPSC Professional Development Grants 3338
Professional Development Grants are self-learning activities that are prospectively planned by individual Fellows in collaboration with a mentor, supervisor or instructor. These self-learning activities are planned to achieve a defined set of learning objectives and qualify as planned learning projects within Section 2 – Planned Learning (2 credits per hour) of the Royal College's Maintenance of Certification (MOC) program. Professional Development grants can be developed as individual traineeships to learn a skill or expand an area of competence relevant to their scope of practice. Professional Development grants can equally support participation in formal courses offered by a university, college or institute. All successful applicants will receive up to $1,000 per week (up to $200 per day) up to a maximum of $4,000.
Requirements: Only Fellows are eligible to apply for funding for a Professional Development grant. All Professional Development grants must be structured to meet the self-learning requirements of traineeships or formal courses as defined by the Royal College's MOC Program. Each applicant must be practicing in the region at the time of application and intending to return to practicing in the region following the completion of the professional development activity. Each applicant is eligible to receive a Professional Development grant once every 5 years.
Restrictions: This grant is not applicable to observerships (due to the lack of formal, structured feedback and evaluation). Professional Development grants are not open to individuals registered/registering in a postgraduate program accredited by the Royal College.
Geographic Focus: Canada
Amount of Grant: Up to 4,000 CAD
Contact: Melanie Blackburn, Grant Administrator; (800) 668-3740, ext. 355 or (613) 730-8177; fax (613) 730-8830; awards@rcpsc.edu
Internet: http://www.royalcollege.ca/portal/page/portal/rc/awards/grants/cpd/professional_dev_grant
Sponsor: Royal College of Physicians and Surgeons of Canada
774 Echo Drive
Ottawa, ON K1S 5N8 Canada

RCPSC Program Administrator Award for Innovation and Excellence 3339
The Program Administrator Award for Innovation and Excellence is given annually to a Residency Program Administrator (PA) who has demonstrated a commitment to enhancing residency education as evidenced by innovation and excellence beyond his/her program. Nominations can be submitted by a Program Director or other members of the Faculty of Medicine, Residents and/or Fellows currently registered in the Faculty. Such nominations can be made by individuals, or groups of individuals. Applications for this award are adjudicated by a sub-committee of the Awards Committee of the Royal College of Physicians and Surgeons of Canada. One award will be presented annually at the (ICRE) Program Administrators' Conference. The winner will receive a Royal College certificate recognizing their contribution as well as travel compensation and complimentary registration to the ICRE Program Administrators' Conference, the ICRE conference and an invitation to attend the ICRE Residency Awards dinner. The annual deadline for application is April 17.

Requirements: All applicants should be a current postgraduate specialty residency program administrator for an accredited Royal College residency program. Nominations must be submitted electronically by completing the official award application form.
Restrictions: Self-applications are ineligible. Current members of the Royal College Program Administrator Conference steering committee are not eligible. Individuals may be re-nominated in subsequent years; however, previous winners of the award will not be eligible to win again.
Geographic Focus: Canada
Date(s) Application is Due: Apr 17
Contact: Diane Cyr, Office of Specialty Education; (800) 668-3740, ext. 548 or (613) 730-8177; fax (613) 730-8830; paaward@royalcollege.ca
Internet: http://www.royalcollege.ca/portal/page/portal/rc/awards/awards/personal_achievement/program_admin_award
Sponsor: Royal College of Physicians and Surgeons of Canada
774 Echo Drive
Ottawa, ON K1S 5N8 Canada

RCPSC Program Director of the Year Award 3340
The Program Director of the Year Award is given annually to a program director who has demonstrated a commitment to enhancing residency education as evidenced by innovation and impact beyond his/her program. Nominations may be submitted by program directors, deans, postgraduate deans, or department chairs/division director. Up to two awards will be presented annually. Award winners will receive a plaque recognizing their contribution as well as travel and registration to the ICRE conference. The annual deadline for application submission is April 4.
Requirements: Any current program director of a residency program accredited by the Royal College is eligible.
Restrictions: Self-applications are ineligible. Individuals may be re-nominated in subsequent years; however, previous winners of the award will not be eligible to win again.
Geographic Focus: Canada
Date(s) Application is Due: Apr 4
Contact: Azura Fennell, Office of Specialty Education; (800) 668-3740, ext. 170 or (613) 730-8177; fax (613) 730-8830; icreawards@royalcollege.ca
Internet: http://www.royalcollege.ca/portal/page/portal/rc/awards/awards/personal_achievement/program_director_award
Sponsor: Royal College of Physicians and Surgeons of Canada
774 Echo Drive
Ottawa, ON K1S 5N8 Canada

RCPSC Robert Maudsley Fellowship for Studies in Medical Education 3341
Dr. Robert Maudsley, MD, FRCSC was a passionate mentor and innovator in the field of medical education. He was instrumental in the development of the original CanMEDS Framework, and at the time of his passing was Associate Director, Internationally Educated Health Professionals at the Royal College. Dr. Maudsley passed away suddenly in October 2012. he goal of this one-year fellowship is to increase the number and quality of professionally trained medical educators in Canada by providing training in the science of medical education to selected promising candidates. The program is to help specialists acquire knowledge and skills in the field of medical education in order to develop educational programs, evaluation methods and medical education research that apply to any phase of the continuum of medical education. Up to $40,000 for residents and Fellows per year of study is available. The annual deadline for applications is September 2.
Requirements: Applicants must: register in a Masters- or PhD-level university program leading to a degree in Education, or a subject field/program of study closely related to medical education; be registered and attending university as a full-time or part-time student, either within Canada or abroad; be Canadian citizens or permanent residents; be Fellows of the Royal College in good standing or residents who, at the time of the application, are registered in a program accredited by the Royal College; and provide a letter of support from a Canadian Faculty of Medicine (residents must also provide a letter of support from their RCPSC Residency Program Director indicating that the Program Director has approved the resident's intended course of study in education).
Geographic Focus: Canada
Date(s) Application is Due: Sep 2
Amount of Grant: Up to 40,000 CAD
Contact: Allison Montpetit, Grant Administrator; (800) 668-3740, ext. 291 or (613) 730-8177; fax (613) 730-8830; researchgrants@royalcollege.ca
Internet: http://www.royalcollege.ca/portal/page/portal/rc/awards/grants/medical_education/fellowship_grant
Sponsor: Royal College of Physicians and Surgeons of Canada
774 Echo Drive
Ottawa, ON K1S 5N8 Canada

RCPSC Royal College Medal Award in Medicine 3342
The Royal College of Physicians and Surgeons of Canada offers annually a competition for the RCPSC Medal in Medicine. This award provides national recognition for original work by a clinical investigator who has completed his/her training within the past 10 years. Any original work in clinical investigation or in the basic sciences relating to medicine is acceptable. The deadline for receipt of the manuscript is November 28. The recipient will receive a gold medal and $5,000.
Requirements: The candidate must be a Fellow of the Royal College and the work must have been done mainly in Canada. Only applicants who have not previously won the Medal are eligible.
Restrictions: Work which is directed or overseen by a supervisor such as in a graduate or post-doctoral training program is not eligible.

Geographic Focus: Canada
Date(s) Application is Due: Nov 28
Amount of Grant: 5,000 USD
Contact: Kimberly Ross, Grant Administrator; (800) 668-3740, ext. 355 or (613) 730-8177; fax (613) 730-8830; kross@royalcollege.ca or awards@royalcollege.ca
Internet: http://www.royalcollege.ca/portal/page/portal/rc/awards/awards/original_research/medal_awards
Sponsor: Royal College of Physicians and Surgeons of Canada
774 Echo Drive
Ottawa, ON K1S 5N8 Canada

RCPSC Royal College Medal Award in Surgery   3343
The Royal College of Physicians and Surgeons of Canada offers annually a competition for the RCPSC Medal in Surgery. This award provides national recognition for original work by a clinical investigator who has completed his/her training within the past 10 years. Any original work in clinical investigation or in the basic sciences relating to surgery is acceptable. The deadline for receipt of the manuscript is November 30. The recipient will receive a gold medal and $5,000.
Requirements: The candidate must be a Fellow of the Royal College and the work must have been done mainly in Canada. Only applicants who have not previously won the Medal are eligible.
Restrictions: Work which is directed or overseen by a supervisor such as in a graduate or post-doctoral training program is not eligible.
Geographic Focus: Canada
Date(s) Application is Due: Nov 30
Amount of Grant: 5,000 USD
Contact: Kimberly Ross, Grant Administrator; (800) 668-3740, ext. 355 or (613) 730-8177; fax (613) 730-8830; kross@royalcollege.ca or awards@royalcollege.ca
Internet: http://www.royalcollege.ca/portal/page/portal/rc/awards/awards/original_research/medal_awards
Sponsor: Royal College of Physicians and Surgeons of Canada
774 Echo Drive
Ottawa, ON K1S 5N8 Canada

RCPSC Teasdale-Corti Humanitarian Award   3344
The purpose of the Teasdale-Corti Humanitarian Award is to acknowledge and celebrate Canadian physicians who, while providing health care or emergency medical services, go beyond the accepted norms of routine practice, which may include exposure to personal risk. The recipient's action will exemplify altruism and integrity, courage and perseverance in the alleviation of human suffering. The award may be bestowed upon Canadian physicians whose current practice reflects the above description or upon those who have made such a contribution in the past. The Award will be open to Canadian physicians worldwide. Nominations will be widely solicited and may be submitted by any Fellow of the Royal College. The recipient will be presented with a suitably engraved memento that will be presented at the Royal College Annual Conference or other suitable venue. The travel and maintenance expenses of the award winner and companion will be paid in accordance with Royal College guidelines. The Royal College may make a donation to the relief organization with which the recipient is affiliated. The application deadline is September 2.
Requirements: Canadian physicians whose current practice reflects altruism and integrity, courage and perseverance in the alleviation of human suffering are eligible; or those who have made such a contribution in the past.
Geographic Focus: Canada
Date(s) Application is Due: Sep 2
Contact: Kimberly Ross, Grant Administrator; (800) 668-3740, ext. 355 or (613) 730-8177; fax (613) 730-8830; kross@royalcollege.ca or awards@royalcollege.ca
Internet: http://www.royalcollege.ca/portal/page/portal/rc/awards/awards/personal_achievement/teasdale_corti_award
Sponsor: Royal College of Physicians and Surgeons of Canada
774 Echo Drive
Ottawa, ON K1S 5N8 Canada

Reader's Digest Partners for Sight Foundation Grants   3345
The Reader's Digest Partners for Sight (PFS) Foundation welcomes proposals for initiatives that further its charge to devote resources to assist the visually impaired and blind. The foundation is seeking to fund projects or programs with broad, practical applications and measurable outcomes as opposed to outcomes that are pure research, narrow in geographical scope, or otherwise very limited i potential reach and scale. The preference is for funding new initiatives, but in select cases, the foundation my support projects in the development phase that require additional funding in order to bring them to completion. The initial grant will be up to $300,000 in year one, and up to $200,000 for two to three subsequent years based on proven results.
Requirements: 501(c)3 nonprofits are eligible, however other options may be considered.
Restrictions: PFS will not fund individuals, lobbying organizations, medical research, endowments, charities operating outside the United States, and/or organizations or entities whose primary function is fundraising activities.
Geographic Focus: All States
Date(s) Application is Due: Sep 30
Contact: Dianna Kelly, Grant Program Manager; (914) 244-5830 or (800) 877-5293; fax (914) 244-7481; dianna@partnersforsight.org or susan@partnersforsight.org
Internet: http://www.partnersforsight.org/grants.shtml
Sponsor: Reader's Digest Partners for Sight Foundation
100 South Bedford Road, Suite 340
Mount Kisco, NY 10549

Regence Foundation Access to Health Care Grants   3346
Grants in this category support the health care safety-net and insurance access programs, including efforts to: increase access to health and medical care for low-income, uninsured or underinsured individuals; provide a medical home to those who would otherwise go without; improve medical outcomes; and help eligible individuals and families navigate the process of accessing available health insurance coverage.
Requirements: The Foundation funds: organizations that are nonprofit and tax-exempt under 501(c)3 of the Internal Revenue Service Code (IRC) and defined as a public charity under 509(a) 1, 2, or 3 (types I, II, or a functionally integrated type III); organizations that serve residents of Idaho, Oregon, Washington and/or Utah; accredited schools or universities; units of government and public agencies; and programs and services that align with our mission and program areas.
Restrictions: The Foundation does not fund: capital construction; award dinners, athletic events, competitions, special events or tournaments; conferences or seminars; religious organizations for religious purposes; political causes, candidates, organizations or campaigns; grants to individuals; grants to 509(a)3, type III supporting organizations that are not functionally integrated; or grants to organizations that are not doing significant work in Idaho, Oregon, Washington and/or Utah.
Geographic Focus: Idaho, Oregon, Utah, Washington
Amount of Grant: 50,000 - 115,000 USD
Contact: Monique Barton; (503) 276-1965; mxbarto@regence.com
Internet: http://www.regencefoundation.org/programs/transformation.html
Sponsor: Regence Foundation
100 SW Market Street, MS E-8T
Portland, OR 97201

Regence Foundation Health Care Community Awareness and Engagement Grants   3347
At the core of this program area is a laser-focus on increasing the accessibility and usability of information for consumers to make informed health care decisions. Grants in this category support efforts that encourage consumers to seek information about the value and quality of their health care, and develop tools to access that information, such as: community collaborative pilot projects; campaigns promoting consumer education and engagement; and research on consumer use of health care information.
Requirements: The Foundation funds: organizations that are nonprofit and tax-exempt under 501(c)3 of the Internal Revenue Service Code (IRC) and defined as a public charity under 509(a) 1, 2, or 3 (types I, II, or a functionally integrated type III); organizations that serve residents of Idaho, Oregon, Washington and/or Utah; accredited schools or universities; units of government and public agencies; and programs and services that align with our mission and program areas.
Restrictions: The Foundation does not fund: capital construction; award dinners, athletic events, competitions, special events or tournaments; conferences or seminars; religious organizations for religious purposes; political causes, candidates, organizations or campaigns; grants to individuals; grants to 509(a)3, type III supporting organizations that are not functionally integrated; or grants to organizations that are not doing significant work in Idaho, Oregon, Washington and/or Utah.
Geographic Focus: Idaho, Oregon, Utah, Washington
Amount of Grant: 45,000 - 150,000 USD
Contact: Monique Barton, Executive Director; (503) 276-1965; mxbarto@regence.com or RegenceFoundation@regence.com
Internet: http://www.regencefoundation.org/programs/transparency.html
Sponsor: Regence Foundation
100 SW Market Street, MS E-8T
Portland, OR 97201

Regence Foundation Health Care Connections Grants   3348
Grants in this category support general health programs that are innovative and solve community-identified problems, including efforts to: manage chronic conditions; make healthy lifestyle choices; and serve a target population whose health care needs may be unmet and/or who may be experiencing health disparities.
Requirements: The Foundation funds: organizations that are nonprofit and tax-exempt under 501(c)3 of the Internal Revenue Service Code (IRC) and defined as a public charity under 509(a) 1, 2, or 3 (types I, II, or a functionally integrated type III); organizations that serve residents of Idaho, Oregon, Washington and/or Utah; accredited schools or universities; units of government and public agencies; and programs and services that align with our mission and program areas.
Restrictions: The Foundation does not fund: capital construction; award dinners, athletic events, competitions, special events or tournaments; conferences or seminars; religious organizations for religious purposes; political causes, candidates, organizations or campaigns; grants to individuals; grants to 509(a)3, type III supporting organizations that are not functionally integrated; or grants to organizations that are not doing significant work in Idaho, Oregon, Washington and/or Utah.
Geographic Focus: Idaho, Oregon, Utah, Washington
Amount of Grant: 15,000 - 50,000 USD
Contact: Monique Barton, Executive Director; (503) 276-1965; mxbarto@regence.com or RegenceFoundation@regence.com
Internet: http://www.regencefoundation.org/programs/transformation.html
Sponsor: Regence Foundation
100 SW Market Street, MS E-8T
Portland, OR 97201

Regence Foundation Improving End-of-Life Grants 3349
The Regence Foundation works to transform health care and address core problems in our health care system with innovative solutions. The Foundation supports a transformational health care system where consumers partner with their physicians and other providers to make health care decisions based on what's valuable to them - such as end-of-life decisions - just as they do when making other economic decisions.
Requirements: The Foundation funds: organizations that are nonprofit and tax-exempt under 501(c)3 of the Internal Revenue Service Code (IRC) and defined as a public charity under 509(a) 1, 2, or 3 (types I, II, or a functionally integrated type III); organizations that serve residents of Idaho, Oregon, Washington and/or Utah; accredited schools or universities; units of government and public agencies; and programs and services that align with our mission and program areas.
Restrictions: The Foundation does not fund: capital construction; award dinners, athletic events, competitions, special events or tournaments; conferences or seminars; religious organizations for religious purposes; political causes, candidates, organizations or campaigns; grants to individuals; grants to 509(a)3, type III supporting organizations that are not functionally integrated; or grants to organizations that are not doing significant work in Idaho, Oregon, Washington and/or Utah.
Geographic Focus: Idaho, Oregon, Utah, Washington
Contact: Monique Barton, Executive Director; (503) 276-1965; mxbarto@regence.com or RegenceFoundation@regence.com
Internet: http://www.regencefoundation.org/programs/endOfLife.html
Sponsor: Regence Foundation
100 SW Market Street, MS E-8T
Portland, OR 97201

Regence Foundation Tools and Technology Grants 3350
At the core of this program area is a laser-focus on increasing the accessibility and usability of information for consumers to make informed health care decisions. Grants in this category support innovative technology solutions that work to overcome practical barriers and improve the quality and safety of health care. Projects might include: multi-stakeholder technology collaborations; not-for-profit technology initiatives that include providers in development and testing; research on the usage and effects of technology collaborations in health care; and approaches that address patients who receive services from multiple providers across the spectrum.
Requirements: The Foundation funds: organizations that are nonprofit and tax-exempt under 501(c)3 of the Internal Revenue Service Code (IRC) and defined as a public charity under 509(a) 1, 2, or 3 (types I, II, or a functionally integrated type III); organizations that serve residents of Idaho, Oregon, Washington and/or Utah; accredited schools or universities; units of government and public agencies; and programs and services that align with our mission and program areas.
Restrictions: The Foundation does not fund: capital construction; award dinners, athletic events, competitions, special events or tournaments; conferences or seminars; religious organizations for religious purposes; political causes, candidates, organizations or campaigns; grants to individuals; grants to 509(a)3, type III supporting organizations that are not functionally integrated; or grants to organizations that are not doing significant work in Idaho, Oregon, Washington and/or Utah.
Geographic Focus: Idaho, Oregon, Utah, Washington
Contact: Monique Barton, Executive Director; (503) 276-1965; mxbarto@regence.com or RegenceFoundation@regence.com
Internet: http://www.regencefoundation.org/programs/transparency.html
Sponsor: Regence Foundation
100 SW Market Street, MS E-8T
Portland, OR 97201

Regenstrief General Internal Medicine - General Pediatrics Research Fellowship Program 3351
The Regenstrief Institute and the Division of General Internal Medicine offers a 2-year research oriented fellowship (with an optional third year) to prepare physicians for academic careers in general internal medicine and general pediatrics. Fellows receive formal didactic training through classes and seminars taught by faculty from both the Institute as well as Indiana University.
Requirements: Candidates for the Fellowship must hold the M.D. or equivalent degree, and must be U.S. citizens or permanent residents. Minority applicants are encouraged to apply.
Geographic Focus: All States
Contact: Kurt Kroenke; (317) 630-7447; fax (317) 630-6611; KKroenke@regenstrief.org
Internet: http://www.regenstrief.org/training/general-internal-medicine-pediatric-research-fellowship-program
Sponsor: Regenstrief Institute
410 West 10th Street, Suite 2000
Indianapolis, IN 46202-3012

Regenstrief Geriatrics Fellowship 3352
The Department of Medicine and the Division of General Internal Medicine and Geriatrics offer a 2-year fellowship in Geriatric Medicine (with an optional third year). Residents who have completed three years of training in internal medicine and family practice and who are board eligible or board certified may apply for the fellowship. Fellows will be trained in clinical geriatrics, research, teaching, and/or administration in preparation for a successful career in academic geriatric medicine. Formal instruction will be provided by a structured curriculum with goals and objectives that will be met by clinical experiences in a variety of settings, didactic seminars to augment the clinical experiences, and seminars in research, teaching and administration. For further information on the Geriatric Fellowship contact the Program Director.
Geographic Focus: All States
Contact: Glenda Westmorland, Program Director; (317) 630-6906; fax (317) 630-2667; GWestmorland@regenstrief.org
Internet: http://www.regenstrief.org/training/geriatrics-fellowship-program
Sponsor: Regenstrief Institute
410 West 10th Street, Suite 2000
Indianapolis, IN 46202-3012

Regenstrief Master Of Science In Clinical Research/CITE Program 3353
Indiana University has received funding from the National Institute of Health through a K-30 grant to develop and implement the Clinical Investigation and Translational Education (CITE) program. The purpose of this program is to prepare health care professionals for a career in clinical research. Following completion of the program, graduates can embark on a career in clinical research with the skills necessary to successfully compete for grant funding, conduct and analyze research findings, and publish their work in scientific journals. By participating in the program, CITE trainees will complete a two-year formal clinical research curriculum, at the end of which they will receive a Master of Science in Clinical Research degree. The program is designed for participants that already have or are in training for a doctoral degree (e.g., MD, PhD, DNS, DDS, DPT).
Requirements: Most participants will already have a doctoral degree (e.g., MD, PhD, DNS, DDS, or DPT) or be in training for a doctoral degree. Potential applicants include, but are not limited to, the following: fellows or other health care professional trainees who have substantial protected time for clinical research; junior faculty who have career awards or support from their division or department head for participation in the CITE program; other faculty who previously have focused on basic research or other academic activities but who are now committed to pursuing a career in clinical research CITE constitutes the formal didactic requirements for certain types of federal training grants (such as K-23's) and other career awards. Individuals in these programs would be expected to participate in CITE unless they were enrolled in some other type of formal degree program.
Geographic Focus: All States
Contact: Kurt Kroenke, Program Director; (317) 630-7447; fax (317) 630-6611; KKroenke@regenstrief.org
Internet: https://www.indianactsi.org/cite/citeapplication
Sponsor: Regenstrief Institute
1050 Wishard Boulevard, RG-6
Indianapolis, IN 46202-2872

Rehabilitation Nursing Foundation Research Grants 3354
The Foundation makes available research grants totaling $44,500. RNF offers funding of up to $30,000 for rehabilitation nursing research which is awarded in the form of multiple grants. A New Investigator Award, offered to encourage nurses who are novice researchers, will be awarded for up to $10,000. Up to two grants from the remaining amount will be awarded to recipients, who will be designated RNF Research Fellows. Research proposals that address clinical practice or educational or administrative dimensions of rehabilitation nursing are accepted. Quantitative and qualitative research projects will be reviewed. Funding may be awarded in the form of multiple grants. The Research Grant Committee reviews proposals once a year. Proposals must be postmarked by February 1. Grant recipients will be notified in September, and funding begins January of the following year for the selected proposals.
Requirements: The principal investigator for the research project must be a registered nurse who is active in rehabilitation or who demonstrates interest in and significant contributions to rehabilitation nursing. Membership in the Association of Rehabilitation Nurses is not required. Graduate student researchers are eligible for funding.
Geographic Focus: All States
Date(s) Application is Due: Feb 1
Amount of Grant: Up to 44,500 USD
Contact: Program Director; (847) 375-4710 or (800) 229-7530; info@rehabnurse.org
Internet: http://www.rehabnurse.org/research/researchgrants.html
Sponsor: Rehabilitation Nursing Foundation
4700 W Lake Avenue
Glenview, IL 60025-1485

Rehabilitation Nursing Foundation Scholarships 3355
ARN offers financial assistance in the form of two distinct types of scholarships: a $1,000 award to nurses working toward a Bachelor of Science in Nursing (BSN) degree; and the Mary Ann Mikulic Scholarship, which covers one full tuition to the Professional Rehabilitation Nursing Course. Each award has a deadline of June 1.
Requirements: The BSN Scholarship eligibility requirements include: membership and involvement in ARN; enrollment in a Bachelor of Science in Nursing (BSN) program in good standing and successful completion of at least one course; current practice in rehabilitation nursing; and a minimum of 2 years experience in rehabilitation nursing. The Mary Ann Mikulic Scholarship eligibility requirements include: being a registered nurse with a current license; currently practicing in the specialty of rehabilitation nursing; and the ability to meet all other financial responsibilities incurred by participating in the course.
Geographic Focus: All States
Date(s) Application is Due: Jun 1
Amount of Grant: 1,000 USD
Contact: Gayle Elliott, (847) 375-4710 or (800) 229-7530; gelliott@connect2amc.com
Internet: http://www.rehabnurse.org/awards/allscholar.html
Sponsor: Rehabilitation Nursing Foundation
4700 W Lake Avenue
Glenview, IL 60025-1485

Rehab Therapy Foundation Grants   3356
The foundation awards grants to North Carolina nonprofit organizations. Areas of priority include organizations that serve children with developmental disabilities and their families; and impact inadequately insured or underserved children with developmental disabilities and their families. Grant applications are available year round at the Foundations website however, applications will only be accepted once a year beginning in January and must be received by February 28th. Award recipients will be notified by mail on or about May 15th.
Requirements: Applicant must be a 501(c)3 organization or public school in North Carolina.
Restrictions: Proposals benefiting religious organizations, university level education, research, events, individuals, trusts, or political causes will not be funded.
Geographic Focus: North Carolina
Date(s) Application is Due: Feb 28
Amount of Grant: Up to 50,000 USD
Contact: Kimberly D. Reilly; info@rehabtherapyfoundation.org
Internet: http://www.rehabtherapyfoundation.org/grant.htm
Sponsor: Rehab Therapy Foundation
150 Fayetteville Street, Seventh Floor
Raleigh, NC 27602

Reinberger Foundation Grants   3357
Clarence T. Reinberger was born in 1894 on Cleveland's west side, and began his business career in the 1920's as a pioneer in the automobile replacement parts field. Starting as a clerk in the Cleveland retail store of the National Automotive Parts Association's (NAPA) Automotive Parts Company, he became president of that company in 1948. In the 1960's the Automotive Parts Company merged with the Genuine Parts Company of Atlanta. Mr. Reinberger held the position of Chairman of the Board of Genuine Parts Company until his death in 1968. Louise Fischer Reinberger was born in Germany. After graduation from high school in the United States, she was employed by the Halle Brothers Company, a large Cleveland department store. The Reinberger Foundation was established by Mr. and Mrs. Reinberger in 1966. Mr. Reinberger left a substantial bequest to the Foundation at his death in 1968. Upon Mrs. Reinberger's death in 1984, the major portion of her estate was also bequeathed to the Foundation. Although the Reinbergers had no children, the foundation continues to be managed by several generations of Mr. Reinberger's family. Since its inception, the foundation has distributed over $91,000,000 to the non-profit community. The foundation divides its support among the following program areas: Arts, Culture, and Humanities; Education; Human Service - Health; and Human Service - Other. Categories supported under Arts, Culture and Humanities include museums, visual arts, performing arts, media and communication, arts education, zoos, and public recreation. Categories supported under Education include K-12 schools, early childhood education, adult education/literacy, libraries, and workforce development. Categories supported under Human Service - Health include hospitals and clinics, substance abuse prevention/treatment, medical and disease research, disease prevention, and speech and hearing. Categories supported under Human Service - Other include children and youth services, residential and home care, emergency food programs, youth development, domestic violence, and temporary housing/homeless shelters. Letters of Inquiry for grants of any type may be submitted according to the following program-area schedule: Education - March 1; Human Service (Health) - June 1; Human Service (Other) - September 1; and Arts, Culture, and Humanities - December 1. Letters of Inquiry may either be sent by U.S. Mail or emailed and are due in the foundation office by the deadline date. Full proposals are accepted only at the request of the foundation.
Requirements: Applicants must be 501(c)3 organizations. Preferential consideration is given to organizations serving Northeast Ohio, or the greater-Columbus area.
Restrictions: No loans are made, nor are grants given to individuals. The Reinberger Foundation does not make more than one grant to a particular organization during a given calendar year, nor will new proposals be considered until existing multi-year commitments have been paid.
Geographic Focus: Ohio
Date(s) Application is Due: Mar 1; Jun 1; Sep 1; Dec 1
Contact: Karen R. Hooser, President; (216) 292-2790; fax (216) 292-4466; info@reinbergerfoundation.org
Internet: http://reinbergerfoundation.org/apply.html
Sponsor: Reinberger Foundation
30000 Chagrin Boulevard #300
Cleveland, OH 44122

Religion and Health: Effects, Mechanisms, and Interpretation RFP   3358
The Center seeks proposals to elucidate how religious involvement (religious attendance, worship, altruistic and caring activities) influences individual and community health. The goals of this research program are to: quantitatively document effects on health (where health is broadly defined); clarify the biological, social, psychological mechanisms involved; and interpret what the findings mean for individual, congregational, and community health through trans-disciplinary collaboration and deliberation. This RFP focuses on research that examines the contribution of context and caring as part of the mechanism that accounts for the effects of religion on health, in particular the effects of membership and practices within a faith community. Letters of Intent are due July 15. Thirty finalists will be invited to submit full proposals; seven $200,000 grants will be awarded to outstanding research proposals. Finalists will be selected by Aug 5, with full proposals due by November 1, and awards announced on January 1.
Geographic Focus: All States
Date(s) Application is Due: Jul 15
Amount of Grant: 200,000 USD
Contact: Catherine Craver, Assistant Director; (919) 660-7578; fax (919) 684-8569; catherine.craver@duke.edu
Internet: http://www.dukespiritualityandhealth.org/proposals/
Sponsor: Center for Spirituality, Theology and Health at Duke University Med Center
Box 3825, Duke University Medical Center
Durham, NC 27710

Retirement Research Foundation General Program Grants   3359
The Retirement Research Foundation, based in Chicago, was established in the 1950s and endowed in 1978 by the late John D. MacArthur. The Foundation is devoted solely to serving the needs of older persons in the U.S. and enhancing their quality of life. The Foundation supports a range of programs and special initiatives designed to: improve access to and quality of community-based and residential health and long-term care; promote economic security for all older adults by strengthening social insurance, pension, and personal savings programs; and support adequate training of, and compensation for, those already working directly with older persons and their families to bring higher quality of care to larger numbers of older adults. The Foundation's historic interest in innovative projects continues. RRF also has a strong interest in projects that implement or adapt proven models that address clearly identified needs and gaps. Requests for support of projects focusing on advocacy, research, or education and training will be considered from anywhere in the United States. Direct service requests will be considered only from organizations in these seven states: Illinois, Indiana, Iowa, Kentucky, Missouri, Wisconsin, and Florida.
Requirements: The Retirement Research Foundation does not use a standard application form, proposals for funds should include the following elements: cover page; Summary; project significance; statement of objectives; description of methodology; dissemination; budget and timetable; plans for continued support; list of personnel; applicant organization; tax exempt status. Application submission must include: the signature of the chief executive officer of the applicant organization; the proposal and attachments must be submitted in triplicate; only one copy of the annual report and financial/audit are needed.
Restrictions: Funding for vans is limited to applicants from Illinois and Florida. Funding is not generally available for: biomedical research; computer equipment; conferences, publications or travel unless they are components of other larger Foundation-funded projects; construction of facilities; dissertation research; endowment or developmental campaigns; general operating expenses of established organizations; grants or scholarships to individuals; production of films and videos; projects of governmental organizations except for state universities, area agencies on aging, and programs of the Veterans Administration; projects outside the United States
Geographic Focus: All States
Date(s) Application is Due: Feb 1; May 1; Aug 1
Contact: Irene Frye; (773) 714-8080; fax (773) 714-8089; info@rrf.org
Internet: http://www.rrf.org/generalProgram.htm
Sponsor: Retirement Research Foundation
8765 W Higgins Road, Suite 430
Chicago, IL 60631-4170

Reynolds and Reynolds Associate Foundation Grants   3360
Reynolds has a long-standing reputation as a leading corporate citizen within the communities where it has a large presence. The company encourages its associates to roll up their sleeves and get involved in organizations they believe in, and in which they often are involved personally. The mission of The Foundation is to provide a vehicle for Dayton associates to improve health and human services in the local communities. Since 1956, the Foundation has provided financial support to area health and human service agencies throughout an eight-county area. Focus areas include: health and human services with emphasis on domestic violence and child abuse; youth-at-risk; hunger; issues of the elderly; homelessness; literacy; and life-threatening illnesses.
Requirements: Giving is limited to the eight-county region in and around Dayton, Ohio.
Restrictions: There is no support available for non 501(c)3 organizations, sectarian organizations with a predominately religious purpose, fraternal or veterans' organizations, individual primary or secondary schools (except for occasional special projects), or tax-supported universities and colleges (except for occasional special projects) No grants are offered to individuals, or for endowments, courtesy advertising, fund raising events, deficit or debt retirement, or capital campaigns.
Geographic Focus: Ohio
Date(s) Application is Due: Jan 7; Mar 11; Jun 10; Sep 9
Amount of Grant: Up to 40,000 USD
Contact: Alice Davisson; (937) 485-4409 or (937) 485-8138; alice_davisson@reyrey.com
Internet: http://www.reyrey.com/company/community/associate_foundation.asp
Sponsor: Reynolds and Reynolds Associate Foundation
P.O. Box 2608
Dayton, OH 45401-2608

RGk Foundation Grants   3361
RGK Foundation awards grants in the broad areas of Education, Community, and Medicine/Health. The foundation's primary interests within Education include programs that focus on formal K-12 education (particularly mathematics, science and reading), teacher development, literacy, and higher education. Within Community, the foundation supports a broad range of human services, community improvement, abuse prevention, and youth development programs. Human service programs of particular interest to the foundation include children and family services, early childhood development, and parenting education. The foundation supports a variety of Community Improvement programs including those that enhance non-profit management and promote philanthropy and voluntarism. Youth development programs supported by the foundation typically include after-school educational enrichment programs that

supplement and enhance formal education systems to increase the chances for successful outcomes in school and life. The foundation is also interested in programs that attract female and minority students into the fields of mathematics, science, and technology. The foundation's current interests in the area of Medicine/Health include programs that promote the health and well-being of children, programs that promote access to health services, and Foundation-initiated programs focusing on ALS. Annual deadlines for applicants are March 6, June 12, and September 16.
Requirements: Although there are no geographic restrictions to the foundation's grantmaking program, the foundation no longer accepts unsolicited requests for international agencies or programs. All applicants must complete an electronic Letter of Inquiry (LOI) from the website as the first step. RGK Foundation will entertain one electronic LOI per organization in a twelve-month period. While the foundation occasionally awards grants for operating expenses, capital campaigns, endowments, and international projects, such grants are infrequent and usually initiated by the foundation. Grants are made only to nonprofit organizations certified as tax exempt under Sections 501(c)3 or 170(c) of the IRS code. Hospitals, educational institutions, and governmental institutions meeting these requirements are eligible to apply.
Restrictions: The foundation refrains from funding annual funds, galas or other special-event fundraising activities; debt reduction; emergency or disaster relief efforts; dissertations or student research projects; indirect and/or administrative costs; sectarian religious activities, political lobbying, or legislative activities; institutions that discriminate on the basis of race, creed, gender, or sexual orientation in policy or in practice; or loans, scholarships, fellowships, or grants to individuals; unsolicited requests for international organizations or programs or for ALS research projects.
Geographic Focus: All States
Amount of Grant: 1,000 - 200,000 USD
Samples: Chicago Youth Programs, Chicago, Illinois, $15,000 - support to expand early literacy services in four high-need Chicago neighborhoods (2015); Menninger Clinic Foundation, Houston, Texas, $13,860 - support for genetic analysis of adolescent blood samples for the McNair Initiative for Neuroscience Discovery (2015); Colorado Humanities, Greenwood Village, Colorado, $25,000 - support for Motheread/Fatheread Early Childhood and Family Literacy Program (2015).
Contact: Suzanne Haffey; (512) 474-9298; fax (512) 474-7281; shaffey@rgkfoundation.org
Internet: http://www.rgkfoundation.org/public/guidelines
Sponsor: RGk Foundation
1301 West 25th Street, Suite 300
Austin, TX 78705-4236

Rhode Island Foundation Grants 3362
The foundation seeks to promote philanthropic activities that will improve the living conditions and well-being of the inhabitants of Rhode Island. Grants for capital and operating purposes principally to agencies working in the fields of education, health care, the arts and cultural affairs, youth, the aged, social services, urban affairs, historic preservation, and the environment.
Requirements: Rhode Island nonprofits organizations may apply. See Foundations website for additional guidelines.
Restrictions: The foundation does not make grants for endowments, research, religious groups for religious purposes, hospital equipment, capital needs of health organizations, or to educational institutions for general operating expenses.
Geographic Focus: Rhode Island
Samples: Ocean State Action Fund, Cranston, RI, $50,000 - for continued funding for grassroots organizing, coalition work, and public communications in support of affordable, quality health care for all Rhode Islanders; Community Housing Land Trust of Rhode Island, Providence, RI, $50,000 - for new program targeting foreclosed properties or those close to being foreclosed upon; Sakonnet Preservation Association, Little Compton, RI, $25,000 - for three-community partnership to preserve drinking water quality and open space in Newport County;
Contact: Owen Heleen, Vice President for Grant Programs; (401) 427-4009 or (401) 274-4564; fax (401) 331-8085; oheleen@rifoundation.org
Internet: http://www.rifoundation.org/Nonprofits/GrantOpportunities/tabid/175/Default.aspx
Sponsor: Rhode Island Foundation
1 Union Station
Providence, RI 02903

Rice Foundation Grants 3363
The foundation awards grants to Illinois nonprofit organizations in its areas of interest, including civic affairs, higher education, hospitals, libraries, medical education, and youth programs. Project and operating support are available. There are no application forms or deadlines.
Requirements: Applicants should submit a proposal with directly to the Foundation including the following: statement of problem project will address; copy of IRS Determination Letter; brief history of organization and description of its mission; detailed description of project and amount of funding requested.
Geographic Focus: Illinois
Amount of Grant: 25,000 - 1,000,000 USD
Samples: John G. Shedd Aquarium, Chicago, IL $1,000,000 - for general support; Northwestern University, Evanston, IL $386,700 - for general support; Big Shoulders Fund, Chicago, IL $30,000 - for general support;
Contact: Peter Nolan, President; (847) 581-9999
Sponsor: Daniel F. and Ada L. Rice Foundation
8600 Gross Point Road
Skokie, IL 60077-2151

Richard and Helen DeVos Foundation Grants 3364
The foundation supports nonprofit organizations primarily in western Michigan and central Florida in its areas of interest, including religious agencies and churches and education and outreach, social services, the arts, public policy, and health care. Types of support include general operating support, continuing support, annual campaigns, capital campaigns, building construction/renovation, program development, seed grants, and matching funds.
Requirements: Application forms are not required. Applicants should submit the following: copy of IRS Determination Letter; copy of current year's organizational budget and/or project budget. The Board meets every 3 months, submit your proposal two weeks prior to the review (contact Foundation for deadline dates). Mail applications to: 126 Ottawa Avenue, N.W., Suite 500, Grand Rapids, MI 49503
Restrictions: No grants to individuals.
Geographic Focus: Florida, Michigan
Amount of Grant: 300 - 5,000,000 USD
Samples: Michigan State University, East Lansing, Michigan, $5,000,000 - For general support, payable over 1 year); Salvation Army of West Palm Beach, West Palm Beach, Florida, $50,000 - For general support, payable over 1 year;
Contact: Ginny Vander Hart; (616) 643-4700; fax (616) 774-0116; virginiav@rdvcorp.com
Sponsor: Richard and Helen DeVos Foundation Grants
P.O. Box 230257
Grand Rapids, MI 49523-0257

Richard and Susan Smith Family Foundation Grants 3365
The foundation awards grants to Massachusetts nonprofit organizations in its areas of interest, including arts and culture, children and youth, disadvantaged (economically), education (early childhood, elementary, and higher), minorities, biomedical research, health care and social services delivery. Types of support include annual campaigns, building construction/renovation, capital campaigns, curriculum development, general operating grants, program/project support, seed grants, and service delivery programs. There are no application forms.
Requirements: Massachusetts nonprofits serving the greater Boston area are eligible.
Restrictions: Grants do not support political activities; religious activities; individuals; or requests for deficit financing or endowment funds.
Geographic Focus: Massachusetts
Date(s) Application is Due: Mar 15; Aug 15
Amount of Grant: 5,000 - 238,000 USD
Contact: David Ford; (617) 278-5200; fax (617) 278-5250; dford@smithfamilyfoundation.net
Internet: http://www.smithfamilyfoundation.net
Sponsor: Richard and Susan Smith Family Foundation
1280 Boylston Street, Suite 100
Chestnut Hill, MA 02467

Richard D. Bass Foundation Grants 3366
The Richard D. Bass Foundation's area of interest include: arts/cultural-programs, including music and dance companies; Catholic/Protestant agencies & churches; community/economic development; education; health organizations, association. The type of support available include: annual campaigns; building/renovation; capital campaigns; general/operating support. Giving primarily in the metropolitan Dallas, Texas area. Grants range from $500 - $15,000. Application deadlines are March 31 and September 30 annually.
Requirements: IRS 501(c)3 nonprofit organizations are available to apply. The Foundations gives primarily to the Dallas, Texas region of the United States but funding is not limited to Texas. Grant proposals are available through out the U.S., contact the Foundation directly before submitting a proposal to access the likely hood of funding for your project. There is no application form required when applying for funding. Submit the proposal with one copy of the organizations IRS Determination Letter.
Restrictions: Funding is not available for: private foundations; individuals.
Geographic Focus: All States
Date(s) Application is Due: Mar 31; Sep 30
Amount of Grant: 500 - 10,000 USD
Samples: Dallas Opera, Dallas, TX, $5,000--arts grant; Metro Dallas Homeless Alliance, Dallas, TX, $2,500--community grant; Yale University, New haven, CT, $5,000--education grant; First Baptist Church, Valley Mills, TX, $500--religious grant; American Foundation for the Blind, Dallas, TX, $2,500--health grant.
Contact: Barbara B. Moroney, Treasurer; (214) 351-6994
Sponsor: Richard D. Bass Foundation
4516 Wildwood Road
Dallas, TX 75209-1926

Richard Davoud Donchian Foundation Grants 3367
The Richard Davoud Donchian Foundation was founded after Mr. Donchian's death in April of 1993. Extending Mr. Donchian's lifetime passion, its mission is to help others meet their potential and achieve high degrees of personal and professional success. The Foundation's Board of Directors is continuous in its efforts to preserve the Donchian legacy of leadership and integrity, as well as its commitment to learning and personal growth. Consequently, the Foundation concentrates its primary giving activities in the areas of ethical leadership in business and community affairs; education, personal development and literacy; and moral, ethical and spiritual advancement in all areas of life. Applications are accepted throughout the year and are reviewed in the order they are received. The Foundation encourages pilot initiatives that test new program models.
Requirements: The majority of the Foundation's grantmaking is focused in the northeastern United States, although, occasionally, grants may be made in other regions of the country and/or abroad. All applicants must have tax-exempt 501(c)3 status as a non-profit organization as defined by the Internal Revenue Service. The applicant must have an active board of

directors with policy-making authority. Grants may range from a few thousand dollars up to $50,000. In unique circumstances, the Foundation does consider a more significant grant for a program having a major impact in one or more of our areas of interest. Of particular interest to the Foundation are organizations that promote partnerships and collaborative efforts among multiple groups and organizations. Priority will be given to requests that show specific plans for funding beyond the present and program innovation and tangible outcomes, with an emphasis on opportunities for significant and lasting social improvement.
Restrictions: The Foundation generally will not provide grants for the following: organizations not determined to be tax-exempt under section 501(c)3 of the Internal Revenue Code; individuals; general fundraising drives; endowments; government agencies; or organizations that subsist mainly on third party funding and have demonstrated no ability or expended little effort to attract private funding.
Geographic Focus: Connecticut, Delaware, District of Columbia, Florida, Maine, Maryland, New Jersey, New York, North Carolina, South Carolina, Vermont, Virginia
Amount of Grant: Up to 50,000 USD
Contact: Donchian Administrator, (203) 629-8552; fax (203) 547-6112; rdd@fsllc.net
Internet: http://www.foundationservices.cc/RDD2/grantrequests.htm#Guidelines
Sponsor: Richard Davoud Donchian Foundation
640 W. Putnam Avenue, 3rd Floor
Greenwich, CT 06830

Richard E. Griffin Family Foundation Grants 3368
The Foundation, established in 1997, is dedicated to supporting higher education, human and social services, and federated giving programs. Funding most often comes in the form of general operating support. Though there are no specific application forms or deadlines with which to adhere, applicants should make an initial approach by letter outlining the project and budgetary need.
Geographic Focus: All States
Amount of Grant: 1,000 - 100,000 USD
Samples: Nantucket Historical Association, Nantucket, Massachusetts, $36,000--for general operations; Massachusetts College of Pharmacy and Health Sciences, Boston, Massachusetts, $55,000--for general operations.
Contact: Richard E. Griffin, (603) 472-4652
Sponsor: Richard E. Griffin Family Foundation
75 Federal Street, Suite 413
Boston, MA 02110-1904

Richard King Mellon Foundation Grants 3369
On a national basis, the foundation makes grants to acquire and preserve key tracts of land in danger of being lost to urban growth and environmentally insensitive development. Mellon gives to nonprofits in Pittsburgh and throughout southwestern Pennsylvania to improve human services, education, medical care, civic affairs, and cultural activities. Types of support include capital grants, challenge/matching grants, general operating grants, project grants, and seed money grants. Application and guidelines are available online.
Requirements: Projects originating in Pittsburgh and southwestern Pennsylvania are given special priority.
Restrictions: The Foundation gives priority to projects and programs that have clearly defined outcomes and an evaluation component. It does not consider requests on behalf of individuals, or from outside the Untied States. The Foundation does not encourage requests from outside Pennsylvania.
Geographic Focus: Pennsylvania
Contact: Michael Watson, Senior Vice President; (412) 392-2800; fax (412) 392-2837
Internet: http://foundationcenter.org/grantmaker/rkmellon/
Sponsor: Richard King Mellon Foundation
500 Grant Street, Suite 4106
Pittsburgh, PA 15219-2502

Richard M. Fairbanks Foundation Grants 3370
The Richard M. Fairbanks Foundation, Inc. was established in 1986 by Richard M. Fairbanks, founder and owner of Fairbanks Communications. An independent private foundation granting funds to qualifying nonprofit organizations, programs, and projects in the greater Indianapolis, Indiana area. Exceptions to this geographic limitation are normally made only for organizations historically supported by the Fairbanks family or for national disasters. If you are interested in learning if your organization and project/program matches the Richard M. Fairbanks Foundation's areas of interest and funding guidelines, you may contact the foundation by telephone, email or brief written letter of inquiry. Please note that the Fairbanks Foundation does not accept unsolicited proposals.
Requirements: The Foundation: makes grants only to tax exempt public charities as defined in Sections 501(c)3 & 509(a)(1)(2)(3) of the Internal Revenue Code, except as prohibited for 509(a)(3) Type III organizations; will not consider requests for loans and grants to individuals; does not make grants or give support to conferences, seminars, media events, or workshops unless they are an integral part of a broader program; ordinarily considers grant requests only from organizations located in Greater Indianapolis, Indiana.
Geographic Focus: Indiana
Samples: Prevent Blindness Indiana - in support of General Operations, $15,000; Indiana University School of Nursing - in support of School of Nursing Learning Consortium for Faculty Development in Simulation Technology, $167,887.
Contact: Claire Fiddian-Green, Grants Officer; (317) 663-4189 or (317) 846-7111; fax (317) 844-0167; Fiddiangreen@rmfairbanksfoundation.org
Internet: http://www.rmfairbanksfoundation.org/default.asp?p=3
Sponsor: Richard M. Fairbanks Foundation
9292 North Meridian Street, Suite 304
Indianapolis, IN 46260

Richland County Bank Grants 3371
The Richland County Bank has had a long tradition of community involvement, dedication and volunteerism. The Bank is committed to supporting the community in which it serves. In addition to monetary contributions, the bank encourages its employees to take an active role in our community. As a corporate sponsor, we support various organizations in Richland and surrounding counties, including: Vernon, Crawford, Grant, Iowa, and Sauk counties. Areas of interest include: health care; higher education; agricultural agencies; community development; children and youth programs; and scholarship funds.
Requirements: Applicants must be 501(c)3 organizations serving the Wisconsin counties of Richland, Vernon, Crawford, Grant, Iowa, or Sauk.
Geographic Focus: Wisconsin
Contact: Gail Surrem, Vice President; (608) 647-6306
Internet: http://www.richlandcountybank.com/aboutInvolvement.cfm
Sponsor: Richland County Bank
195 West Court Street
Richland Center, WI 53581

Richmond Eye and Ear Fund Grants 3372
The fund is committed to serving the city of Richmond, Colonial Heights, Hopewell and Petersburg, and the counties of Chesterfield, Hanover, Henrico, Goochland, New Kent, Charles City, Lancaster, Northumberland, Richmond, Westmoreland and Powhatan in the fields of ophthalmology, otolaryngology, and oral maxillofacial surgery for qualified organizations. Grantmaking focuses on underserved, uninsured, or medically indigent children and adults in need of ophthalmological (eye), otolaryngological (ear, nose, and throat), and/or oral maxillofacial services (conditions, defects, injuries, and aesthetic aspects of the mouth, teeth, and jaws); improving access and outreach as it relates to the stated services mentioned; and healthcare education/preventive care consistent with purposes of this fund. Guidelines are available online. The Fund???s grantmaking focuses on: Underserved, uninsured or medically indigent children and adults in need of ophthalmological (eye), otolaryngological (ear, nose and throat), and/or oral maxillofacial services (conditions, defects, injuries, and aesthetic aspects of the mouth, teeth, and jaws). Improving Access and Outreach as it relates to the services mentioned above. Healthcare Education/preventive care consistent with purposes of this fund.
Requirements: Proposals will be accepted from charitable organizations, which serve the residents of the focus area.
Geographic Focus: Virginia
Date(s) Application is Due: May 5; Nov 5
Amount of Grant: Up to 50,000 USD
Contact: Elaine Summerfield, Program Officer; (804) 330-7400; fax (804) 330-5992; esummerfield@tcfrichmond.org
Internet: http://www.tcfrichmond.org/Page2954.cfm#ear
Sponsor: Community Foundation
7501 Boulders View Drive, Suite 110
Richmond, VA 23225

Ricks Family Charitable Trust Grants 3373
The Ricks Family Charitable Trust (formerly known as the 1104 Foundation) offers support to religious programs, health, education, children and youth, and community service organizations. Most recently, amounts have ranged from $100 to $1,000. There are no application forms or deadlines to which applicants must adhere, and interested parties should contact the Trust prior to submitting a proposal. Though giving is primarily restricted to the Charlotte, North Carolina, region, grants are also awarded to state-wide organizations which are aligned with the foundation's interests.
Geographic Focus: North Carolina
Samples: Sharon United Methodist Church, Charlotte, North Carolina, $250 - general operating support (2014); First Presbyterian Church, Charlotte, North Carolina, $1,000 - general operating support (2014); Alzheimers Association, Charlotte, North Carolina, $500 - general operating support (2014).
Contact: Charles V. Ricks, Trustee; (704) 537-0526
Sponsor: Ricks Family Charitable Trust
6000 Monroe Road, Suite 100
Charlotte, NC 28212-6119

Riley Foundation Grants 3374
The foundation's purpose is to make grants, distributions, and/or loans to charitable and governmental organizations for charitable purposes and to provide financial resources and assistance for community wide projects and programs in health care; education; and the betterment of cultural, environmental, and economic conditions for the people of Meridian and Lauderdale County, Mississippi. Applications are reviewed at quarterly meetings in January, April, July, and October and must be in by noon on the listed deadline days.
Requirements: Mississippi governmental agencies and 501(c)3 organizations that are not private foundations as defined under Section 509(a) are eligible.
Restrictions: Best of Both Worlds Addiction Center, Inc. $75,000 - Operating Support; Meridian Community College $5,000 - Lucile Reisman Rosenbaum Memorial Scholarship Endowmen;
Geographic Focus: Mississippi
Date(s) Application is Due: Feb 15; May 15; Aug 15; Nov 15
Amount of Grant: 2,000 - 300,000 USD
Contact: Becky Lewis, (601) 481-1430; fax (601) 481-1434; info@rileyfoundation.org
Internet: http://www.rileyfoundation.org/guide.htm?46,10
Sponsor: Riley Foundation
4518 Poplar Springs Drive
Meridian, MS 39305

Ringing Rocks Foundation Discretionary Fund Grants 3375
The foundation's mission is to explore, document, and preserve indigenous cultures and healing practices. Grants are awarded to organizations throughout the world that promote indigenous healing, work with indigenous cultures, and educate the public at large about these topics. Grants can be used for start-up costs, program development, and general operating expenses. Interested organizations should be able to show substantial leadership from the community they serve at both the board and staff levels. Submit a completed eligibility statement along with a one-page letter of intent in English briefly describing the organization along with type and amount of funding sought. Application materials are available online. The listed application deadline is for letters of request; full proposals are by invitation.
Requirements: Applicant organizations must be recognized as a charity by their governments. The project must have support from the indigenous community it serves.
Restrictions: Faxed letters of Intent, eligibility forms, and applications cannot be accepted.
Geographic Focus: All States
Date(s) Application is Due: Feb 2
Amount of Grant: 500 - 5,000 USD
Contact: Administrator; (928) 282-1298; fax (928) 282-1327; grants@ringingrocks.org
Internet: http://www.ringingrocks.org/www/index.php?grants_fundProgram
Sponsor: Ringing Rocks Foundation
1890 West Highway 89A, Suite B-1
Sedona, AZ 86336-5571

Ripley County Community Foundation Grants 3376
The Ripley County Community Foundation (RCCF) was established to improve the quality of life for Ripley County, Indiana residents. The mission of the Ripley County Community Foundation is to assist donors in building an enduring source of charitable assets to benefit the citizens and qualified organizations of Ripley County. The RCCF strives to provide responsible stewardship of the gifts donated; to promote leadership in addressing Ripley County's issues; and to make grants in the fields of community service, social service, education, health, environment, and the arts. The application and samples of previously funded projects are available at the website.
Requirements: Grant applicants must qualify as 501(c)3 or 509(a) tax exempt organizations or hold sponsorship with such organizations. Because the grant guidelines and policies are brief and do not address every aspect of the RCCF granting program, the most effective means of making initial contact with the RCCF is through a letter or phone call of inquiry to the RCCF.
Restrictions: Funding is not available for the following: individuals; travel or lodging expenses to enable individuals or groups to attend seminars or take trips; endowment purposes of recipient organizations; programs funded prior to the RCCF date for grant decisions; to repay acquisition costs for equipment already purchased or paid for; for acquisition of weapons, firearms or destructive devices; sectarian religious purposes; to attempt to influence legislation or to intervene in any political campaign. RCCF reserves the right to refuse any and all grant applications.
Geographic Focus: Indiana
Date(s) Application is Due: Sep 14
Amount of Grant: Up to 2,500 USD
Contact: Jane Deiwert, Program Officer; (877) 234-5220 or (812) 933-1098; fax (812) 933-0096; jdeiwert@rccfonline.org
Internet: http://www.rccfonline.org/grants.asp
Sponsor: Ripley County Community Foundation
4 South Park, Suite 210
Batesville, IN 47006

Ripley County Community Foundation Small Project Grants 3377
Small Project Grants are available throughout the year (not just during the traditional Fall Granting Cycle) for amounts up to $500. The projects must meet the Foundation's charitable guidelines for traditional grants. Organizations may only apply for one small project grant each year; they may also apply for a grant during the traditional fall granting cycle if they receive a small project grant. Organizations may only apply for one larger grant in the fall. Applications will be accepted anytime during the year but decisions will only be made by the 30th of April, June, August, and October.
Requirements: To be considered, application must be made by the second Friday of each of these months. Applications not received by the second Friday will be held for the next Small Project Grant period.
Geographic Focus: Indiana
Date(s) Application is Due: Apr 30; Jun 30; Aug 30; Oct 30
Amount of Grant: Up to 500 USD
Contact: Jane Deiwert, Program Officer; (877) 234-5220 or (812) 933-1098; fax (812) 933-0096; jdeiwert@rccfonline.org
Internet: http://www.rccfonline.org/grants.asp
Sponsor: Ripley County Community Foundation
4 South Park, Suite 210
Batesville, IN 47006

Robbins Charitable Foundation Grants 3378
The Robbins Charitable Foundation, based in Brookline, Massachusetts, provides support throughout the State of Massachusetts in its primary fields of interest. These interest areas include: arts and culture; the environment; health organizations; and religious welfare programs. Typically, awards are given for general operations, and most recent grants have ranged from $50 to $800. There is no formal application required, and no specified annual deadlines for submission. Interested parties should begin by contacting the Foundation office directly.
Requirements: 501(c)3 organizations in the State of Massachusetts are eligible to apply.
Restrictions: No grants are given directly to individuals.
Geographic Focus: Massachusetts
Amount of Grant: 50 - 800 USD
Samples: Dana Farber Jimmy Fund, Brookline, Massachusetts, $320 - general operating support (2014); Perkins Institute for the Blind, Watertown, Massachusetts, $200 - general operating support (2014); Museum of Fine Arts, Boston, Massachusetts, $500 - general operating support (2014).
Contact: Phillis Robbins, Trustee; (617) 566-4919
Sponsor: Robbins Charitable Foundation
77 Marion Street
Brookline, MA 02246

Robert and Joan Dircks Foundation Grants 3379
The foundation supports non-profit organizations that enrich and improve the quality of life for individuals primarily located in the New Jersey area. The focus is to encourage innovative programs and projects that benefit and improve the lives of children and individuals who are physically mentally or economically disadvantaged.
Requirements: Submit your inquiry online, using the Grant Request Information form provided on the Robert and Joan Dircks Foundation website: http://www.dircksfoundation.org/application2.asp
Restrictions: The Robert and Joan Dircks Foundation does not award grants for political or lobbying activities, environmental or cultural projects, capital or annual campaigns, endowments, operating budgets, deficit or debt reduction, loans or housing projects. Also, grants are not made to programs of national or international scope.
Geographic Focus: Connecticut, Maine, Massachusetts, New Hampshire, Rhode Island, Vermont
Amount of Grant: 1,000 - 15,000 USD
Contact: Grants Administrator; grants@dircksfoundation.org
Internet: http://www.dircksfoundation.org/guidelines.html
Sponsor: Robert and Joan Dircks Foundation
P.O. Box 6
Mountain Lakes, NJ 07046

Robert B. Adams Foundation Scholarships 3380
Established in Alabama in 1997, the Robert B. Adams Foundation offers support for higher education in Alabama, including scholarship awards paid directly to the college or university for students pursuing a degree in the field of clinical laboratory sciences. Funding can be used for tuition, supplies, and other educational support, as well as the purchase of equipment. Since there is no formal application, interested parties should contact the Foundation office in writing. There are no annual deadlines specified.
Geographic Focus: Alabama
Contact: Kathy Jones, Secretary; (334) 244-3606 or (334) 244-3254
Sponsor: Robert B. Adams Foundation
P.O. Box 244023
Montgomery, AL 36124-4023

Robert B McMillen Foundation Grants 3381
The Robert B McMillen Foundation is a non-profit charitable foundation. The Foundation offers funding in the states of Washington and Alaska. The two areas of intent are: medical and social enhancement. Fifty percent of the annual funding is earmarked for medical research. The Foundation will consider making grants to non-profit organizations involved in researching cardiology, lipid and organ transplants. Twenty-five percent of annual giving is earmarked for social areas including, but not limited to, Goodwill, Salvation Army & United Way. Preference is given to organizations and/or programs that use art as the vehicle to impact communities and change individual lives.
Requirements: The Foundation does not accept uninvited medical or social proposals. Contact the Foundations office for information on, how to be invited to make a medical or social proposal to the McMillen Foundation.
Restrictions: The Foundation do not make grants to: religious, fraternal or other organizations primarily benefiting their own members; other private foundations nor political organizations; for-profit entities; other than scholarships, the foundation will not make multi-year grants.
Geographic Focus: Alaska, Washington
Contact: Cassandra Town, President; (425) 313-5711; fax (425) 313-8955; cassandra@mcmillenfoundation.org
Internet: http://www.mcmillenfoundation.org/mission.htm
Sponsor: Robert B McMillen Foundation
55 1st Place NW, Suite 2, P.O. Box 1523
Issaquah, WA 98027

Robert F. Stoico/FIRSTFED Charitable Foundation Grants 3382
The Robert F. Stoico/FIRSTFED Charitable Foundation was established to support the local communities of southeastern Massachusetts and Rhode Island. The Foundation's main areas of interest are: affordable housing; job development; education; accessibility to arts and cultural programs; and accessible to health care. Small grants (under $5,001) are reviewed throughout the year on a rolling basis. Large grant requests are reviewed four times a year, with the following quarterly deadlines being identified: February 28, May 31, August 31, and November 30.
Requirements: Massachusetts and Rhode Island 501(c)3 nonprofit organizations in the counties where FIRSTFED America Bancorp had business involvement are eligible.
Restrictions: The foundation does not support sports programs, political organizations, individual schools, individuals, or organizations that are not classified as 501(c)3 by the IRS.
Geographic Focus: Massachusetts, Rhode Island
Date(s) Application is Due: Feb 28; May 31; Aug 31; Nov 30
Amount of Grant: Up to 20,000 USD

Contact: Cecilia Viveiros, Executive Director; (508) 235-1368; fax (508) 300-2588; Cecilia@stoicofirstfed.org
Internet: http://www.stoicofirstfed.org/apply.html
Sponsor: Robert F. Stoico/FIRSTFED Charitable Foundation
P.O. Box 438
Swansea, MA 02777

Robert Leet and Clara Guthrie Patterson Trust Grants 3383

The Robert E. Leet and Clara Guthrie Patterson Trust was established in 1980 to support and promote the investigation of human disease and the alleviation of human suffering through improved treatment. While the funding history shows that the trust has recently awarded grants, the grant website states that the trust is not currently accepting applications. Individuals desiring more information about funding opportunities from the Robert E. Leet and Clara Guthrie Patterson Trust are advised to contact the Senior Foundation Officer for further information.
Geographic Focus: Connecticut, New Jersey, New York
Amount of Grant: 5,000 - 500,000 USD
Samples: Damon Runyon Cancer Research Foundation, New York, New York, $184,000 - research (2014); Yale University, New Haven, Connecticut, $500,000 - research (2015); Kessler Foundation, West Orange, New Jersey, $75,000 - research (2015).
Contact: Carmen Britt; (860) 657-7019; carmen.britt@baml.com
Internet: https://www.bankofamerica.com/philanthropic/foundation.go?fnId=106
Sponsor: Robert Leet and Clara Guthrie Patterson Trust
200 Glastonbury Boulevard, Suite # 200, CT2-545-02-05
Glastonbury, CT 06033-4056

Robert P. & Clara I. Milton Fund for Senior Housing 3384

The purpose of this fund is to support innovative programs focusing on issues related to senior housing. Projects will be considered that make a meaningful impact while promoting dignity, independence and quality of life. This fund strives to make St. Joseph County a regional model for senior housing by acting as a catalyst through the funding of innovative approaches. Senior housing may range from institutional care to in-home care, as well as the various services related to this issue. Special consideration will be given to projects that serve the most vulnerable of the aging population. Priorities for funding include education/training; home and community based services; programs and project serving vulnerable seniors; and public policy issues. Grants are judged by community impact, concept/idea, improved solutions to current and projected needs, the ability to implement the project, the quality of need, and its sustainability.
Requirements: In addition to the online application, the following attachments are required: up to a two page proposal narrative; a detailed project budget; current board roster with officers identified; and proof of nonprofit status. Detailed guidelines and application are available at the website.
Restrictions: Annual appeals or membership contributions, grants to individuals directly or special event underwriting will not be funded.
Geographic Focus: Indiana
Date(s) Application is Due: May 1; Nov 1
Samples: Housing Authority, South Bend, Indiana: salary of the Senior Assessment Coordinator, $30,000: Logan Center, South Bend, Indiana: renovation of their senior day program facility, $22,000.
Contact: Christopher Nanni; (574) 232-0041; fax (574) 233-1906; chris@cfsjc.org
Internet: http://www.cfsjc.org/initiatives/senior/milton_grants.html
Sponsor: Community Foundation of St. Joseph County
205 W Jefferson Boulevard, P.O. Box 837
South Bend, IN 46624

Robert R. McCormick Tribune Foundation Community Grants 3385

The Communities Program helps to transform communities by giving under-served people access to programs which improve their lives. The McCormick Foundation partners with media outlets, sports teams and philanthropic organizations to raise money for local needs and provides matching funds to increase the impact of charitable giving. Through the partnership, grants are made to qualified nonprofit organization with programs that help transition low-income children, adults and families to self-sufficiency. To achieve greater impact, the Foundation focuses on programs for children and youth in education, literacy, health & wellness and abuse prevention; hunger & housing; and adult workforce development and literacy. The McCormick Foundation is committed to measurable change with these projects and monitors the impact of all grants made. Community grants are made on the basis of requests received from 501(c)3 organizations in each local community where the Foundation has fundraising partners (Chicago, Denver, Fort Lauderdale, Los Angeles, Orlando, Washington, D.C., and Long Island, New York). Each fund partner has a unique set of guidelines that emphasize the needs of the particular community. Areas of focus include: child abuse prevention and treatment; child and youth education; health and wellness; housing; hunger; literacy; workforce development; and youth sports.
Requirements: Organizations must apply through the individual funding partners listed on the Foundation website. Specific information, including contacts, is included at each website.
Restrictions: Funding is not available for individuals or for-profit organizations.
Geographic Focus: California, Colorado, District of Columbia, Florida, Illinois, New York
Contact: Lesley Kennedy, Communities Program Officer; (312) 445-5000; info@mccormickfoundation.org
Internet: http://www.mccormickfoundation.org/page.aspx?pid=594
Sponsor: Robert R. McCormick Tribune Foundation
205 North Michigan Avenue, Suite 4300
Chicago, IL 60601

Robert R. McCormick Tribune Veterans Initiative Grants 3386

The mission of the McCormick Foundation Veterans Initiative is to create welcoming and inclusive communities for returning military and their families where each is able to reach their maximum potential. Current initiatives include workforce development programs to create pathways to careers for new veterans; a peer to peer model to assist veterans in finding pathways to information, services, and connection with other veterans; and strategic partnerships with medical centers across the country, local VA medical centers, and research institutions to explore innovative strategies to engage veterans and their families in mental health services that are user friendly, free of stigma, and readily accessible. A detailed list of previously funded project is available at the Foundation website. The application process and deadlines vary according to individual program.
Requirements: Grantees are solicited through a request for proposal (RFP) process at the Foundation website.
Restrictions: Support is concentrated in the Chicago area, but other locations are considered for funding.
Geographic Focus: Illinois
Samples: Albany Park Community Center, Chicago, Illinois, Veterans Transition and Workforce Development Program, $300,000 (2011-2013); Health and Disability Advocates, Chicago, Illinois, for the Returning National Guard Project and Warrior to Warrior in Illinois - Connecting Soldiers to Wellness, $475,000 (2011-2013); Cornell University, Ithaca, New York, Program for Anxiety and Traumatic Stress Studies, $200,000.
Contact: Anna LauBach, Director of Veterans Initiatives; (312) 445-5000; fax (312) 445-5001; ALauBach@mccormickfoundation.org
Internet: http://donate.mccormickfoundation.org/page.aspx?pid=627
Sponsor: Robert R. McCormick Tribune Foundation
205 North Michigan Avenue, Suite 4300
Chicago, IL 60601

Robert R. Meyer Foundation Grants 3387

The Robert R. Meyer Foundation is a private foundation established in 1949 by Mr. Robert R. Meyer and further funded by bequests from the wills of Robert R. Meyer and John Meyer. Mr. Meyer desired that assets from his foundation be used to address needs in Birmingham and its vicinity. The foundation has made awards in the areas of arts and culture, education, environment, health, human services, and public/society benefit. The foundation meets twice a year in the spring and fall to review proposals. Applicants should contact the Trustee for application forms and guidelines.
Requirements: Giving is limited to 501(c)3 organizations in the metropolitan Birmingham, Alabama area. All applicant organizations are encouraged (but not required) to join the Alabama Association of Nonprofits.
Geographic Focus: Alabama
Date(s) Application is Due: Mar 1; Sep 1
Amount of Grant: 5,000 - 100,000 USD
Samples: A+ College Ready, Birmingham, Alabama, $50,000—to increase participation and performance of high school students in rigorous, college-level advancement placement courses in math, science, and English; Alabama Ballet, Birmingham, Alabama, $25,000—to help fund a new full-length ballet, Alice in Wonderland, in collaboration with the Alabama Symphony; Alabama Ear Institute, Mountain Brook, Alabama, $12,000—to help fund the Auditory-Verbal Mentoring Program.
Contact: Carla B. Gale, Vice President and Trust Officer; (205) 326-5382
Sponsor: Robert R. Meyer Foundation
P.O. Box 11647
Birmingham, AL 35202-1647

Robert Sterling Clark Foundation Reproductive Rights and Health Grants 3388

The objective of the Foundation's work in this field is to protect and expand women's access to reproductive health services in the belief that the ability to control one's fertility and prevent unintended pregnancy is fundamental to the advancement of women's opportunities. The strategy employed by the Foundation to achieve this objective is to support policy analysis, advocacy, litigation, research, message development, and/or organizing aimed at: promoting the implementation of laws, policies and practices that will enable all women to have access to comprehensive reproductive health information and services, including emergency contraception and abortion; challenging laws and legal decisions that undermine reproductive rights and developing legal theory in support of these rights; mobilizing an informed grassroots constituency; developing new pro-choice messages, based on current opinion research, that resonate with a broad public; informing the development of judicial selection criteria that will produce non-ideological courts; and promoting the use of comprehensive sexuality education curricula in the schools.
Requirements: The Foundation makes support available to national and regional organizations working on these issues.
Geographic Focus: All States
Contact: Margaret C. Ayers; (212) 288-8900; fax (212) 288-1033; rcsf@rsclark.org
Internet: http://www.rsclark.org/index.php?page=reproductive-right-and-health
Sponsor: Robert Suiterling Clark Foundation
135 East 64th Street
New York, NY 10065-7045

Robert W. Woodruff Foundation Grants 3389

The foundation is interested in supporting programs and projects in the areas of elementary, secondary, and higher education; health care and education; Human services, particularly for children and youth; economic development and civic affairs; art and cultural activities; and conservation of natural resources and environmental education. Preference is given to one-time capital projects, matching/challenge grants, specific projects, construction, land acquisition, and equipment purchases.

Requirements: Most grants are awarded to tax-exempt organizations in Georgia. Grants to qualified public charities headquartered outside Georgia are considered when it is demonstrated that the proposed project or program will have particular impact in Georgia. Application forms are not required. Proposals should be made in letter form and describe the organization, including its purposes, staffing, and governing board; latest financial statement and most recent audit report; description of the proposed project and justification for funding; itemized project budget, including other sources of support; and evidence of tax-exempt status. The board meets in April and November to consider requests.
Restrictions: Grants are not awarded to support individuals, annual operating support, festivals and performances, conferences, films and documentaries, start-up funding or seed money, churches or their denominational programs, or youth services outside Atlanta.
Geographic Focus: Georgia
Date(s) Application is Due: Feb 1; Sep 1
Amount of Grant: 50,000 - 750,000 USD
Samples: The Atlanta Lyric Theater, $75,000 for costs associated with the Lyric's move to the Strand Theater in Cobb County; Georgia Tech Foundation, Inc., $2,500,000 for construction of the Innovative Learning Resource Center; Arch Foundation, $250,000 for construction of a new horticulture complex at the State Botanical Garden of Georgia; Bobby Dodd Institute, $500,000 for establishment of an Enterprise Fund to provide working capital for Bobby Dodd's social enterprises.
Contact: P. Russell Hardin; (404) 522-6755; fax (404) 522-7026; fdns@woodruff.org
Internet: http://www.woodruff.org/appGuidelines_rww.aspx
Sponsor: Robert W. Woodruff Foundation
50 Hurt Plaza, Suite 1200
Atlanta, GA 30303

Robins Foundation Grants 3390
The foundation awards grants to nonprofit organizations in the Richmond, VA, area. Areas of interest include general grants--including cultural, charitable, scientific, environmental, and educational programs, and at-risk youth; and early childhood/quality improvement grants-- for organizations that devote a major portion of their resources to young children and their families. The goal is to help improve the quality of services by providing funding for accreditation, staff training, facilities improvements, and similar initiatives. The foundation seeks projects that meet well-defined community needs, use effective approaches that build on proven programs, develop models with potential for wider application, foster self-reliance and/or end dependency, and focus on prevention as well as treatment. Types of support include capital grants and endowments.
Requirements: 501(c)3 organizations based in Virginia are eligible. Organizations based in Virginia that have or support programs outside Virginia or the United States also are eligible.
Restrictions: In general, the foundation does not make grants to support annual operating funds or budgets, special events or fundraising benefits, or religious purposes unless they are otherwise compatible with the objectives of the foundation.
Geographic Focus: Virginia
Amount of Grant: 10,000 - 100,000 USD
Contact: William Roberts Jr.; (804) 697-6917; wlrjr@robins-foundation.org
Internet: http://www.robins-foundation.org
Sponsor: Robins Foundation
Capitol Station, P.O. Box 1124
Richmond, VA 23218-1124

Rochester Area Community Foundation Grants 3391
The foundation seeks to improve the quality of life in the community and awards grants in the areas of child development, education, the environment, arts and cultural programs, health services, community development, and social services and general charitable giving. Types of support include general operating support, building construction/renovation, equipment acquisition, program development, conferences and seminars, publication, seed grants, scholarship funds, technical assistance, consulting services, and scholarships to individuals. There are no application deadlines. The board meets in January, February, March, May, June, July, October, and November.
Requirements: Nonprofits in Monroe, Livingston, Ontario, Orleans, Genessee, and Wayne Counties, NY, are eligible.
Restrictions: Grants do not support partisan political organizations or religious projects, individuals, annual campaigns, deficit financing, land acquisition, endowments, or emergency funds.
Geographic Focus: New York
Contact: Marlene Cole; (585) 341-4333; fax (585) 271-4292; mcole@racf.org
Internet: http://www.racf.org/page10000903.cfm
Sponsor: Rochester Area Community Foundation
500 East Avenue
Rochester, NY 14607-1912

Rochester Area Foundation Grants 3392
The foundation makes grants in the fields of arts and culture, community development, education, human services and recreation in the greater Rochester, Minnesota area.
Requirements: Organizations must be one of the following to be eligible to receive grant funding: a tax-exempt 501(c)3 organization, a government unit (city, township, county), or a government-created organization such as a public agency. Download the required pre-application form from the website. If your pre-application is approved, you will be notified and asked to submit a full application to the foundation. (The required full application form is also available at the website.)
Geographic Focus: Minnesota
Date(s) Application is Due: Jan 2; May 1; Sep 1
Amount of Grant: 1,000 - 25,000 USD
Contact: Steve Thornton, Executive Director; (507) 424-2400 or (507) 424-3755, ext. 102; fax (507) 282-4938; Steve@RochesterArea.org
Internet: http://www.rochesterarea.org/grant-resources/index.html
Sponsor: Rochester Area Foundation
400 South Broadway, Suite 300
Rochester, MN 55904

Rockefeller Archive Center Research Grants 3393
The foundation awards grants for travel and research at the Rockefeller Archive Center, which contains records of the Rockefeller Foundation, Rockefeller University, and other affiliated philanthropic organizations and individuals. The center also will award grants to support research on the history of basic medical research, the life of the Nobel laureate Paul Ehrlich, and on postcolonial studies.
Requirements: Scholarly reseearchers, particularly students preparing dissertations, are eligible.
Restrictions: The Archive Center's programs do not support research at other institutions, and they do not provide general tuition support.
Geographic Focus: All States
Contact: Camilla Harris, Grants Administrator; (914) 366-6311 or (914) 631-4505; fax (914) 631-6017; charris@rockarch.org or archive@rockarch.org
Internet: http://www.rockarch.org/grants/
Sponsor: Rockefeller Foundation
15 Dayton Avenue
Sleepy Hollow, NY 10591-1598

Rockefeller Brothers Fund Peace and Security Grants 3394
The Fund's Peace and Security Program focuses on two factors that may be key to advancing or undermining global problem solving: the content and style of U.S. global engagement in the face of new perils and opportunities; and the strength and quality of relationships between Muslim and Western societies. In addition, peace and security is a theme that may be identified for attention in one or more of the Fund's pivotal places. Two specific goals have emerged: to advance U.S. policies and behaviors that reflect a broadly shared vision of constructive, cooperative, principled, farsighted, and effective global engagement; and to reduce the divisive and destabilizing tensions that exist between much of the Islamic world and the West, particularly the United States, and to increase the potential for collaboration among Muslim and Western societies on behalf of a better, safer world.
Requirements: A prospective grantee in the United States or foreign counterpart must be either a tax-exempt organization or an organization seeking support for a project that would qualify as educational or charitable.
Restrictions: The fund does not make grants to individuals, nor does it as a general rule support research, graduate study, or the writing of books or dissertations by individuals.
Geographic Focus: All States
Amount of Grant: 25,000 - 300,000 USD
Contact: Benjamin R. Shute; (212) 812-4200; fax (212) 812-4299; info@rbf.org
Internet: http://www.rbf.org/programs/
Sponsor: Rockefeller Brothers Fund
437 Madison Avenue, 37th Floor
New York, NY 10022-7001

Rockefeller Brothers Fund Pivotal Places Grants: South Africa 3395
The Fund will support human advancement through its work to improve basic education and to meet the developmental needs of orphans and vulnerable children. In addition, the RBF is exploring linkages between HIV/AIDS and the Fund's other substantive concerns: democratic practice, sustainable development, and peace and security. While the Fund recognizes the critical importance of HIV/AIDS education, prevention, and treatment, many other donors are active in these fields. The RBF seeks to complement these efforts by contributing to the development of integrated responses to the broad societal impacts of the pandemic. Specifically, the Fund seeks to: improve the quality and accessibility of basic education for children in the areas of early childhood development and primary learning; assist orphans and vulnerable children in achieving their full potentials individuals and as members of society; and strengthen understanding of the linkages between HIV/AIDS and democratic practice, sustainable development, and peace and security issues in South Africa, and to promote integrated responses to the broad societal impacts of the HIV/AIDS pandemic.
Requirements: A prospective grantee in the United States or foreign counterpart must be either a tax-exempt organization or an organization seeking support for a project that would qualify as educational or charitable.
Restrictions: The fund does not make grants to individuals, nor does it as a general rule support research, graduate study, or the writing of books or dissertations by individuals.
Geographic Focus: All States, Algeria, Angola, Benin, Botswana, Burkina Faso, Burundi, Cameroon, Cape Verde, Central African Republic, Chad, Comoros, Congo, Congo, Democratic Republic of, Cote d' Ivoire (Ivory Coast), Djibouti, Egypt, Equatorial Guinea, Eritrea, Ethiopia, Gabon, Gambia, Ghana, Guinea, Guinea-Bissau, Kenya, Lesotho, Liberia, Libya, Madagascar, Malawi, Mali, Mauritania, Mauritius, Morocco, Mozambique, Namibia, Niger, Nigeria, Rwanda, Sao Tome & Principe, Senegal, Seychelles, Sierra Leone, Somalia, South Africa, Sudan, Swaziland
Contact: Nancy L. Muirhead, Program Director; (212) 812-4200; fax (212) 812-4299; nmuirhead@rbf.org or info@rbf.org
Internet: http://www.rbf.org/programs/
Sponsor: Rockefeller Brothers Fund
437 Madison Avenue, 37th Floor
New York, NY 10022-7001

Rockefeller Foundation Grants 3396

The foundation provides grants to institutions and individuals seeking to improve the lives of poor people with a focus on the issues and region where the foundation works. The foundation works globally but provides the majority of its grants to organizations whose work is focused in Southern and Eastern Africa, Southeast Asia, and North America through programs that address agriculture, health, employment, housing, education, arts and culture, and global policy. The foundation also operates special programs, including a conference and study program at its Bellagio Center, the Program Venture Experiment, and the Philanthropy Workshop.
Requirements: Nonprofit organizations are eligible.
Restrictions: The foundation does not give or lend money for personal aid to individuals, support attempts to influence legislation, or, except in rare cases, provide general institutional support, fund endowments, or contribute to building and operating funds.
Geographic Focus: All States
Samples: Criterion Ventures, Haddam, CT--toward the start-up costs of a new mission-driven venture that will provide healthcare financial services to groups of American workers who are struggling with the negative economic and health impacts of paying cash for health services, $500,000; Water and Sanitation for the Urban Poor, London, England--toward the costs of building the capacity of community and local service groups and of testing innovative water and sanitation delivery systems in Gatwekera, Nairobi and the surrounding municipality to provide sustainable, equitable and affordable water and sanitation services, $249,900
Contact: Grants Administrator; (212) 869-8500; fax (212) 764-3468
Internet: http://www.rockfound.org/grants/grants.shtml
Sponsor: Rockefeller Foundation
420 Fifth Avenue
New York, NY 10018-2702

Rockwell Collins Charitable Corporation Grants 3397

The Rockwell Collins Charitable Corporation has identified two distinct funding priorities, and special consideration will be given to proposals that integrate these issues: education and youth development, with emphasis in math, science and engineering; and culture and the arts, with emphasis on youth educational programs. Rockwell Collins also contributes to health, human services, and civic organizations. Special consideration is given to qualifying organizations where employees volunteer; however, the majority of gifts to these organizations are made through Rockwell Collins United Way corporate contributions or employee campaigns. A list of contacts by region is detailed on the web site.
Requirements: Applying organizations must be tax exempt under the federal code; able to provide current full, certified, audited financial statements; and not be private foundations. Non-U.S. organizations must provide nongovernmental organization (NGO) documentation.
Restrictions: Funding will not be considered for general endowments, deficit reduction, grants to individuals, federated campaigns, religious organizations for religious purposes, or fraternal or social organizations.
Geographic Focus: All States
Amount of Grant: Up to 5,000 USD
Contact: Jenny Becker, Manager Community Relations; (319) 295-7444; fax (319) 295-9374; jlbecker@rockwellcollins.com or communityrelations@rockwellcollins.com
Internet: http://www.rockwellcollins.com/Our_Company/Corporate_Responsibility/Community_Overview/Charitable_Giving.aspx
Sponsor: Rockwell Collins Corporation
400 Collins Road NE
Cedar Rapids, IA 52498-0001

Rockwell Fund, Inc. Grants 3398

Rockwell Fund, Inc. was founded in 1931 from the estate of James M. Rockwell. Rockwell's grant making is concentrated on the following issue areas: community health, concentrating on health issues that affect the broader community, including mental and behavioral health; education, concentrating on dropout prevention strategies that target the intermediate and middle school years; employment, concentrating on training/placement and jobs creation/enterprise development opportunities; and supportive housing, concentrating on affordable housing coupled with onsite, sustained services to support individuals and families in achieving self-sufficiency.
Requirements: Eligible applicants must be nonprofit tax exempt organizations. To be eligible, an organization must: have a determination letter from the Internal Revenue Service indicating that it is an organization described in Section 501(c)3 of the Internal Revenue Code or be a church or political subdivision that is not required to obtain a Section 501(c)3 designation in order to be a permitted donee of a private foundation. Funding is provided for the Houston area or in some cases for a purpose that will benefit a Houston cause. Interested applicants must first complete an on-line inquiry form. Applicants will then either be invited to submit a full grant application for further consideration or declined. Inquiries are accepted throughout the year.
Restrictions: The following projects or programs are ineligible: feasibility studies; annual fund drives; direct mail or other mass solicitations; grants that impose the exercise of "expenditure responsibility" upon Rockwell Fund, for example, private operating foundations or certain support organizations; houses of worship; individuals; medical or scientific research projects or organizations that target a specific disease; parochial or private primary and secondary schools; and underwriting for benefits, dinners, galas, golf tournaments or other fundraising or special events.
Geographic Focus: Texas
Contact: Judy Ahlgrim; (713) 341-5338; fax (713) 629-7702; jahlgrim@rockfund.org
Internet: http://www.rockfund.org/giving/howToApply.shtml
Sponsor: Rockwell Fund, Inc.
770 S. Post Oak Lane, Suite 525
Houston, TX 77056

Rockwell International Corporate Trust Grants 3399

Grants are awarded in the areas of education and youth development, with emphasis in math, science, and engineering; and culture and the arts, with emphasis on youth educational programs. Rockwell also contributes to health, human services, and civic organizations and also gives special consideration to qualifying organizations in which employees are involved as volunteers. Types of support include capital grants, challenge/matching grants, conferences and seminars, development grants, endowments, research grants, scholarship funds, and general operating grants. Awards include single-year, multiple-year, and provisional continuing support of specific programs. Applicants should submit full, detailed proposals. Proposals are accepted at any time. At the Web site, click on About Us, then Corporate Citizenship, and then Rockwell Automation Corporation Trust.
Restrictions: Organizations are not eligible if they have not received a permanent, tax-exempt ruling determination from the federal government; if they cannot provide current full, certified, audited financial statements; or if they are private foundations. Funding will not be considered for the following purposes: general endowments, deficit reduction, grants to individuals, federated campaigns, organizations or projects outside the United States, religious organizations for religious purposes, or fraternal or social organizations.
Geographic Focus: All States
Amount of Grant: 1,000 - 50,000 USD
Contact: Rockwell International Corporation General Information; (562) 797-3311
Internet: http://www.rockwell.com
Sponsor: Rockwell International Corporate Trust
1201 S Second Street
Milwaukee, WI 53204

Roeher Institute Research Grants 3400

The institute, through the support of the Scottish Rite Charitable Foundation of Canada, offers research grants to associates, associations, and agencies in a broad range of fields relating to human services and intellectual disability. The renewable grants are offered for one year. Research projects must be under the direction of the Roeher Institute academic associates or other university faculty or under researchers approved by the Roeher Institute. Areas constituting research priorities for the Roeher Institute are issues affecting people with intellectual disabilities; integration of people who have an intellectual disability into society; prevention, early identification, and minimization of disabling conditions; and strategies for social change that improve the quality of life of persons with intellectual disabilities. Applicants are asked to submit a summary of their research project, budget, and financial requirements.
Requirements: Applications are invited from university faculty, department heads, supervisors, Roeher Institute associates, and/or consultants.
Geographic Focus: All States
Amount of Grant: Up to 10,000 CAD
Contact: Cameron Crawford, President; (416) 661-9611; fax (416) 661-5701; cameronc@roeher.ca or cameronc@worldchat.com
Internet: http://www.roeher.ca/english/about/about.htm
Sponsor: Roeher Institute
4700 Keele Street, York University, Kinsmen Building.
North York, ON M3J 1P3 Canada

Rogers Family Foundation Grants 3401

The Rogers Family Foundation funds nonprofit organizations that provide educational, medical, artistic, and religious services within Massachusetts' Merrimack Valley and North Shore and southeastern New Hampshire. See website for specific areas served.
Requirements: Applicants must be tax exempt under section 501(c)3 of the Internal Revenue Code and classified as "not a private foundation" under Section 509(a) of the Code. All applications are submitted on-line. The online application follows the Common Proposal designed and published by Associated Grant Makers, Inc. See website for detailed online application and instructions.
Restrictions: Traditionally the Foundation awards grants to organizations located in Massachusetts' Merrimack Valley and North Shore and southeastern New Hampshire. From time to time grants may be made outside of this normal geographic area.
Geographic Focus: Massachusetts, New Hampshire
Date(s) Application is Due: Mar 1; Sep 1
Amount of Grant: 25,000 USD
Contact: Susan Haff, (617) 426-7080; fax (617) 426-7087
Internet: http://www.rogersfamilyfoundation.com/app/
Sponsor: Rogers Family Foundation
c/o GMA Foundations
Boston, MA 02110

Roget Begnoche Scholarship 3402

The Roger Begnoche Scholarship was established to encourage young men to pursue nursing or education careers. Scholarships of $4,000 per year for a maximum of 4 years will be awarded. A student is eligible for the renewable scholarship if they remain in good academic standing, continue toward a nursing or education degree, and submit the appropriate paperwork to the Legacy Foundation in a timely manner.
Requirements: Applicants are eligible to apply if they meet the following criteria: they are a U.S. citizen and resident of Indiana; they are a current high school graduate; they are male and have been accepted as a full time student at Purdue University Calumet, Indiana University Northwest, Calumet College of St. Joseph, or Sisters of St. Francis nursing program; they have been accepted into the School of Nursing or are pursuing a degree in education; and they have an SAT score above 900 for a nursing student or 850 for an education student. Selection of scholarship recipients will be based on: academics,

school activities, community or volunteer involvement; financial need; work experience; a recommendation letter from an adult; and an original essay.
Restrictions: Only specific colleges in Indiana will meet the selection criteria.
Geographic Focus: Indiana
Date(s) Application is Due: May 15
Amount of Grant: 4,000 USD
Contact: Begnoche Scholarship Contact; (219) 736-1880; fax (219) 736-1940; legacy@legacyfoundationlakeco.org
Internet: http://www.legacyfoundationlakeco.org/scholarships/index.htm
Sponsor: Legacy Foundation
1000 East 80th Place, 302 South
Merrillville, IN 46410

Rohm and Haas Company Grants 3403
The company awards grants in five key philanthropic categories: education, environment, civic and community, health and human services, and arts and culture. Education grants support programs linking education in science, technology, and math to workplace and career opportunities; and educational enrichment for students, particularly in the areas of math, science, technology, and the environment either during or after school hours (including weekends, summer, and other times when school is not in session). Environment grants support after-school programs that build environmental awareness and understanding; sustainability programs that educate and promote development that meets the needs of the present society without compromising the ability of future generations to meet their own needs; pollution prevention and waste management, environmental conservation and biodiversity, and energy and water conservation; and community beautification. Civic and community grants support regional efforts that build local competitiveness and strengthen the economic and social base in Rohm and Haas host communities; programs that cut across multiple giving categories and/or that impact key stakeholders in Rohm and Haas communities; and programs that focus on volunteerism. Health and human services grants support public safety awareness; disaster preparedness; structured youth development such as leadership training; and structured youth activities and recreation. Arts and culture grants support organizational capacity building in the form of plan implementation--technology implementation and/or enhancement, communications and marketing, outreach and audience development, financial management and fundraising, board development, strategic development, professional development, and educational program development. Application and guidelines are available online.
Requirements: Eligible national and international organizations include performing and visual arts, humanities and historical societies, and museums. Preference is given to requests from Chicago, IL, and the greater Philadelphia area.
Restrictions: Grants are not awarded to individuals or for fundraising, advertising, or testimonials.
Geographic Focus: All States
Amount of Grant: 250 - 50,000 USD
Contact: Alexandra Samuels, (215) 592-3644; alexandra_samuels@rohmhaas.com
Internet: http://www.rohmhaas.com/community/giving/guidelines/guidelines.html
Sponsor: Rohm and Haas Company
100 Independence Mall W
Philadelphia, PA 19106-2399

Rollins-Luetkemeyer Foundation Grants 3404
The Foundation awards grants to eligible Maryland nonprofit organizations in its areas of interest, including education (early childhood, elementary school, and higher education), health care and health organizations, historic preservation/societies, and social services. Types of support include annual campaigns, building construction/renovation, general operating support, and project support.
Requirements: Applications are not required. Letters of intent are accepted at any time.
Restrictions: Maryland nonprofit organizations, with preference given to the Baltimore area, are eligible. No grants to individuals.
Geographic Focus: Maryland
Amount of Grant: 2,000 - 2,000,000 USD
Contact: John A. Luetkemeyer, Jr., President; (443) 921-4358
Sponsor: Rollins-Luetkemeyer Foundation
1427 Clarkview Road, Suite 500
Baltimore, MD 21209

Ronald McDonald House Charities Grants 3405
Ronald McDonald House Charities provides grants to nonprofit, tax-exempt organizations whose national or global programs help improve the health and well being of children under 21. Types of support include capital grants, challenge/matching grants, formula grants, professorships, program grants, research grants, scholarships, seed money grants, and visiting scholars. The board meets quarterly to review, select, and award grants to organizations that have demonstrated an ability to respond to the needs of specific groups of children in a definitive, hands-on manner that yields measurable results.
Requirements: To be considered for funding, an applicant must be designated a not-for-profit, tax-exempt charitable organization. U.S. charities must have a current 501(c)3 tax-exempt status letter on file with the Internal Revenue Service. Organizations seeking funding should have: a broad base of funding support; management and staff capacity to effectively execute the project; a longer-term organizational strategic plan (evidence to ensure program and organization sustainability); and a clear, concise plan for project evaluation with outcome measurement that shows potential for meaningful change. Organizations seeking funding should have a specific program that: directly improves the health and well being of children; addresses a significant funding gap or critical opportunity; has long-term impact in terms of replication or reach; produces measurable results; and is sustainable without relying on RMHC funding. Prior to submitting a full proposal, organizations must submit a letter of inquiry which will be reviewed by the Board. If interested, the Board will invite organizations to submit a formal grant proposal.
Restrictions: The charity does not fund to projects/programs that are local in scope and that do not benefit children under 21; to reduce debt; for annual fund appeals; for medical research; for for-profit organizations; for fundraising sponsorships; for programs/projects administered by activist groups; for projects that have already been completed; for sectarian or religious purposes; for scholarships and fellowships to individuals; to propagandize or influence elections or legislation; for advertising or fundraising drives; to intermediary funding agencies; for endowment campaigns; for individuals.
Geographic Focus: All States
Date(s) Application is Due: Mar 5; Sep 7
Contact: Michael Singer; (630) 623-7048; fax (630) 623-7488; grants@rmhc.org
Internet: http://rmhc.org/what-we-do/grants/how-to-apply/
Sponsor: Ronald McDonald House Charities
1 Kroc Drive
Oak Brook, IL 60523

Rosalinde and Arthur Gilbert Foundation/AFAR New Investigator Awards in 3406
Alzheimer's Disease
The major goal of this program is to support important research in areas in which more scientific investigation is needed to improve the prevention, diagnosis, and treatment of Alzheimer's Disease. The program will also serve to encourage junior investigators in the United States and Israel to pursue research and academic careers in the neurosciences, and Alzheimer's disease in particular. Projects in basic and translational research related to Alzheimer's disease (AD) that are clinically relevant, will be considered. Examples of promising areas of research include, but are not limited to: basic mechanisms of aging in the central nervous system; genetics of AD; neuroimaging and precursors of AD; cellular and Molecular pathways of AD; biological Markers of AD; exercise and dietary factors; neurogenesis and AD; impact of environmental agents in CNS aging and early AD; inflammation; cardiovascular and cerebrovascular factors; drug discovery. It is anticipated that 5 grants of up to $75,000 each will be awarded. Applicants may propose to use the award over the course of one or two years as justified by the proposed research. Funds may not be requested for overhead or indirect costs. Funding will begin July 1.
Requirements: The applicant must be an independent investigator with assigned independent space and must have received a junior faculty appointment (instructor, assistant professor or equivalent) no sooner then July 1st. The proposed research must be conducted at any type of not-for-profit setting in the United States or Israel. Five criteria are used to determine the merit of an application: qualifications of the applicant; quality of the proposed research; relevance of the proposal to how mechanisms of aging may lead to AD; excellence of the research environment; likelihood that the project will advance the applicant's career in basic research on the mechanisms of aging and AD. All candidates must submit applications endorsed by their institution. The deadline for receipt of applications and supporting materials from candidates based in the U.S. is December 16th, at 5:00 p.m. EST. The deadline for receipt of applications and supporting materials from candidates based in Israel is December 30th. See website for instruction sheet and application for complete application procedures. Incomplete applications cannot be considered.
Restrictions: The Gilbert/AFAR Research Grant Program does not provide support for: Postdoctoral fellows in the laboratory of a senior investigator; investigators who have already received major extramural funding for research on Alzheimer's disease and/or aging (such as an R01 grant); Senior faculty, i.e. at the rank of Associate Professor level or higher; NIH Intramural program employees. Applicants cannot apply for the AFAR Research Grant Program or Glenn/AFAR BIG Program.
Geographic Focus: All States, Israel
Date(s) Application is Due: Dec 16; Dec 30
Contact: Grants Manager; (212) 703-9977 or (888) 582-2327; fax (212) 997-0330; grants@afar.org or info@afar.org
Internet: http://afar.org/Gilbert.html
Sponsor: American Federation for Aging Research
55 West 39th Street, 16th Floor
New York, NY 10018

Rosalynn Carter Institute Georgia Caregiver of the Year Awards 3407
To honor their work and to focus public attention on the contributions of caregivers throughout the State of Georgia, the Rosalynn Carter Institute on Caregiving in cooperation with the Georgia CARE-NETS initiated the Georgia Caregiver of the Year Awards in 2007. Awardess are presented with a gilded rose and a check for $1,000.
Requirements: Nominees must be caregivers from the State of Georgia.
Geographic Focus: Georgia
Amount of Grant: 1,000 USD
Contact: Laura Bauer Granberry; (229) 931-2034; fax (229) 931-2663; lbgran@gsw.edu
Internet: http://www.rosalynncarter.org/caregiver2/
Sponsor: Rosalynn Carter Institute for Caregiving
800 GSW Drive
Americus, GA 31709-4379

Rosalynn Carter Institute John and Betty Pope Fellowships 3408
The John and Betty Pope Fellowship in Caregiving is a unique program of the Rosalynn Carter Institute for Caregiving at Georgia Southwestern State University. The fellowship provides financial support for outstanding individuals pursuing training and careers in fields related to caregiving, and provides a structure and course of study intended to prepare them for leadership in the field. The certificate program of study includes courses in psychology, nursing, sociology and special education taken over up to four years of study at GSW. The amount of the award is determined annually and is currently set at $3750 per academic year.

Requirements: Award of Pope Fellowships is currently limited to students pursuing a degree program at Georgia Southwestern State University and who are or will be enrolled in the Caregiver Specialist Certificate at GSW. Applicants must have a grade point average of 3.0 as well as an established interest in academic study related to caregiving or a record of outstanding professional or informal caregiving. In addition, Pope Fellows must: enroll for minimum of 12 credit hours each semester (undergraduate) or 9 credit hours (graduate) for a full award; attend Pope Fellowship orientation and other required meetings; and make steady progress in fulfilling requirements of the Caregiver Specialist Certificate.
Geographic Focus: All States
Date(s) Application is Due: Feb 15
Amount of Grant: 6,000 - 24,000 USD
Contact: Leisa Easom; (229) 928-1234; fax (229) 931-2663; leasom@gsw.edu
Internet: http://www.rosalynncarter.org/PopeFellowship/
Sponsor: Georgia Southwestern State University
800 Wheatley Street
Americus, GA 31709-4379

Rosalynn Carter Institute Mattie J. T. Stepanek Caregiving Scholarships 3409
The Mattie Stepanek Caregiving Scholarship was created in memory of renowned poet and peacemaker Mattie J.T. Stepanek. The purpose of the Scholarship is to provide financial assistance to family, professional, or paraprofessional caregivers who are seeking training or education in specific skills, procedures and strategies that lead to more effective care, while also protecting the health and well-being of the caregiver.
Geographic Focus: All States
Contact: Laura Bauer Granberry; (229) 931-2034; fax (229) 931-2663; lbgran@gsw.edu
Internet: http://www.rosalynncarter.org/08mattie/
Sponsor: Rosalynn Carter Institute for Caregiving
800 GSW Drive
Americus, GA 31709-4379

Rose Community Foundation Aging Grants 3410
In its Aging program area, Rose Community Foundation promotes change in how communities organize care and support for both seniors and caregivers, with particular attention to the needs of low- and moderate-income seniors. In addition to funding community-based programs, the Foundation plays a leadership role in strengthening the existing network of aging resources, including initiatives that bring together community and government partners to address key issues in aging. The Foundation is especially interested in projects that address the following priorities: direct in-home and community-based services; transportation; and end-of-life care. There are no annual deadlines.
Requirements: Colorado 501(c)3 tax-exempt organizations serving the residents of Adams, Arapahoe, Boulder, Denver, Douglas, and Jefferson Counties are eligible.
Restrictions: The Foundation will generally not support: grants to individuals or endowments, including academic chairs and scholarships; grants to one organization to be passed to another; annual appeals or membership drives; fundraisers and other one-time events; or financial support for political candidates. An application will not be accepted if: reports for a prior grant to the organization/program in question are past due; or the same applicant has other applications pending.
Geographic Focus: Colorado
Contact: Therese Ellery; (303) 398-7413 or (303) 398-7400; tellery@rcfdenver.org
Internet: http://www.rcfdenver.org/content/aging
Sponsor: Rose Community Foundation
600 South Cherry Street, Suite 1200
Denver, CO 80246-1712

Rose Community Foundation Health Grants 3411
The Rose Community Foundation recognizes that improving access to quality care requires well-informed, visionary leaders. For this reason, the Foundation promotes initiatives that develop health-policy and public-health leadership. While the Foundation invests in health leadership that can improve access to quality care over the long term, it also works to effect more immediate change. Committed to improving access to care for low-income children, youth and families, the Foundation continues to support efforts to enroll them in programs such as Child Health Plan Plus (CHP+), a publicly supported health insurance program. Because at least half of all premature deaths are the result of lifestyle choices that individuals may be able to change, the Foundation also supports health promotion and disease prevention programs that encourage healthy choices and discourage behaviors that lead to illness and injury. In addition, the Foundation funds community organizations that provide information about the steps individuals can take to prevent such conditions as diabetes, teen pregnancy, HIV, heart disease and cancer. There are no annual deadlines.
Requirements: Colorado 501(c)3 tax-exempt organizations serving the residents of Adams, Arapahoe, Boulder, Denver, Douglas, and Jefferson Counties are eligible.
Restrictions: The Foundation will generally not support: grants to individuals or endowments, including academic chairs and scholarships; grants to one organization to be passed to another; annual appeals or membership drives; fundraisers and other one-time events; or financial support for political candidates. An application will not be accepted if: reports for a prior grant to the organization/program in question are past due; or the same applicant has other applications pending.
Geographic Focus: Colorado
Contact: Whitney Gustin Connor, Program Officer; (303) 398-7410 or (303) 398-7400; fax (303) 398-7430; wconnor@rcfdenver.org
Internet: http://www.rcfdenver.org/content/health
Sponsor: Rose Community Foundation
600 South Cherry Street, Suite 1200
Denver, CO 80246-1712

ROSE Fund Grants 3412
The ROSE Fund (Regaining One's Self Esteem) is a New England non-profit committed to recognizing, assisting, and empowering women who have broken the cycle of domestic violence. In partnership with our medical affiliates the Fund provides female survivors of domestic abuse with access to medical and dental reconstructive procedures to help them to regain their self-esteem. It also is working to actively break the silence and the NIMBY (Not in My Backyard) myths associated with these issues on a high school by high school basis across New England. The Fund exposes these issues by localizing them. It hopes to educate, energize and empower groups of concerned parents and principals to become an active and effective part of the solution. The bulk of grants are made to small organizations with annual budgets under $500,000. Priority is given to organizations in the Northeast. Grants are made for one year. Guidelines are available online.
Requirements: 501(c)3 tax-exempt organizations are eligible.
Geographic Focus: Connecticut, Maine, Massachusetts, New Hampshire, Rhode Island, Vermont
Date(s) Application is Due: Oct 1
Amount of Grant: Up to 10,000 USD
Contact: Gail Walsh; (617) 482-5400; fax (617) 482-3443; gwalsh@rosefund.org
Internet: http://www.rosefund.org/
Sponsor: ROSE Fund
200 Harvard Mill Square, Suite 310
Wakefield, MA 01880

Rose Hills Foundation Grants 3413
The Rose Hills Foundation supports organizations that promote the welfare of humankind, including but not limited to: arts and culture; civic and community services; education; community-based health programs; youth activities; and the advancement of knowledge. Giving is centered in southern California, with emphasis on the San Gabriel Valley and East Los Angeles area. The Foundation accepts and processes applications throughout the year with the expectation of grant distributions every six months. Depending on timing, requests are reviewed at six annual Board Meetings. At times, there may be an approximate wait of up to six months prior to a request being reviewed. Grants range from $5,000 to million dollar commitments. Foundation directors may opt to grant less than the amount requested, depending upon resources available, or spread payments over more than one year.
Requirements: Preferential attention is given to organizations that exhibit the following criteria: a history of achievement, good management, and a stable financial condition; self-sustaining programs that are unlikely to depend on future Foundation funding; significant programs that make a measurable impact; funding that is matched or multiplied by other sources; projects or programs that benefit people of southern California; and programs that reach the greatest number of people at the most reasonable cost. Organizations should send a two page letter of introduction (LOI) addressed to the Foundation's President. Initial correspondence should include the following information: brief purpose and history of organization; brief outline of program/project for which funds are being sought; program/project budget; specific amount being requested from Foundation; geographic area, demographics of population, and the number of individuals served annually; list of Board of Directors; current operating budget and most recent audited financials (if not available, please provide most recent financial statement); copy of IRS determination letter; and a detailed funding history.
Restrictions: Funding is not available for propagandizing, influencing legislation and/or elections or promoting voter registration; political candidates, political campaigns or organizations engaging in political activities; programs which promote religious doctrine; individuals (except as permitted by the IRS); governmental agencies; or endowments.
Geographic Focus: California
Amount of Grant: 5,000 - 1,000,000 USD
Contact: Victoria B. Rogers, President; (626) 696-2220; fax (626) 696-2210; vbrogers@rosehillsfoundation.org
Internet: http://www.rosehillsfoundation.org/AppProcedures.htm
Sponsor: Rose Hills Foundation
225 South Lake Avenue, Suite 1250
Pasadena, CA 91101

Royal Norwegian Embassy Kavli Prizes 3414
The Kavli Prize is awarded every two years and includes three separate prizes awarded for outstanding scientific research in the fields of astrophysics, neuroscience and nanoscience. The Prize, which consists of a cash award, a scroll, and a medal, is intended not only as an accolade for the individual scientists concerned, but also as a recognition of important research efforts. It aims to promote international cooperation between researchers and to increase awareness among the general public of the importance of research. The Norwegian Academy of Science and Letters has full responsibility for selecting the prize winners to ensure that this process is independent of the funding organization. The Academy has set up three selection committees made up of top international scientists within the fields for which the prize is awarded. Four of the committee members are Nobel Laureates. The committee members have been chosen in close cooperation with the corresponding institutions in the USA, China, France, the UK and Germany.
Requirements: The nomination process is open to all who wish to nominate candidates. The prize can be awarded to a single person for profound scientific achievements or shared for closely related fundamental contributions. Specific guidelines and a current nomination form, which must be completed online in one session, are available at the website.
Restrictions: Self nominations are not accepted, and deceased persons may not be nominated.
Geographic Focus: All States, All Countries
Date(s) Application is Due: Dec 1
Amount of Grant: 1,000,000 USD
Contact: Eirik Lislerud; +47 221236640; eirik.lislerud@dnva.no
Internet: http://www.norway.org/studywork/research/prizes/Kavli

Sponsor: Royal Norwegian Embassy in Washington
2720 34th Street NW
Washington, D.C. 20008

Roy and Christine Sturgis Charitable Trust Grants 3415
The Roy and Christine Sturgis Charitable Trust was established in 1981 to support and promote quality educational, cultural, human-services, and health-care programming for all people. Roy Sturgis was one of ten children of an Arkansas farmer and homemaker. He dropped out of school after the tenth grade to join the Navy during World War I. Sturgis returned to his family home in southern Arkansas after the war and went to work in the local sawmills. In 1933, he married Texas native Christine Johns. They became very successful in the timber, lumber, and sawmill industries in Arkansas, owned other prosperous business enterprises, and had notable success managing their investments. The Sturgis' spent most of their lives in Arkansas and Dallas, Texas. They did not have children, but were particularly interested in educational opportunities for young people. In addition, they supported organizations working in the areas of health, social services, and the arts. The Sturgis Charitable Trust encourages requests for the following types of grants: capital, project-related, medical research, and endowment campaign. Funding for start-up programs and limited general-operating requests will also be considered. Approximately 65% of the trust's annual distributions are made within the state of Arkansas. The remaining 35% of grants are distributed within the state of Texas with strong preference given to organizations located in the Dallas area. The majority of grants from the Sturgis Charitable Trust are one year in duration; on occasion, multi-year support is awarded. Applicants should apply online at the grant website, and are strongly encouraged to do the following before applying: review the downloadable state application procedures for additional helpful information and clarifications; review the downloadable online-application guidelines at the grant website; review the trust's funding history (link is available from the grant website); review the online application questions in advance; and review the list of required attachments. These will generally include: a list of board members, financial statements (audited, reviewed, or compiled by independent auditor); an organization summary; a list of other funding sources; an IRS Determination letter; and other required documents. Most recent awards have ranged from $150,000 to $200,000. The annual application deadline is March 1.
Requirements: 501(c)3 nonprofit organizations in Texas and Arkansas are eligible.
Restrictions: Former grantees must skip a year before submitting a subsequent application. The trust does not support requests from individuals, organizations attempting to influence policy through direct lobbying, or any political campaigns.
Geographic Focus: Arkansas, Texas
Date(s) Application is Due: Mar 1
Amount of Grant: 150,000 - 200,000 USD
Contact: Robert Fox, Senior Philanthropic Relationship Manager; (214) 209-1965 or (214) 209-1370; tx.philanthropic@ustrust.com
Internet: https://www.bankofamerica.com/philanthropic/foundation.go?fnId=88
Sponsor: Roy and Christine Sturgis Charitable Trust
901 Main Street, 19th Floor
Dallas, TX 75202-3714

Roy J. Carver Charitable Trust Medical and Science Research Grants 3416
The trust supports Iowa and western Illinois nonprofit organizations and projects in the area of medical and scientific research. The goal of the program in medical and scientific research is to provide support for innovative investigation that may hold great promise for advancing scientific knowledge and improving human health. In this context, the Trust has recognized a variety of university-based scientific endeavors, most notably in the fields of medicine, engineering and the natural sciences. Although grants are occasionally awarded to address institutional capital needs, the focus of this program area is typically on the work of specific laboratories, often emphasizing multidisciplinary approaches and the development of untried, inherently risky research directions. Applicants are encouraged to contact staff prior to the development of a proposal. The board considers requests at quarterly meetings in January, April, July, and October. Approved grant payments are made semiannually in April and October.
Requirements: Grants are made to 501(c)3 tax-exempt scientific organizations in Iowa and western Illinois.
Restrictions: The trust does not support annual campaigns or ongoing operations, direct grants to individuals, religious activities, organizations without 501(c)3 status, fund-raising benefits or program advertising, or political parties or candidates.
Geographic Focus: Illinois, Iowa
Date(s) Application is Due: Feb 15; May 15; Aug 15; Nov 15
Amount of Grant: 10,000 - 150,000 USD
Samples: Iowa State University, Ames IA--A study of high-intensity ultrasound technology for clinical applications, $166,458; University of Illinois, Urbana, IL--Enhance bioinformatics & mass spectrometry resources, $1,804,293; University of Iowa Foundation, Iowa City, IA--Continuing support for program, endowment and capital needs at the Carver College of Medicine, $6,104,365.
Contact: Dr. Troy Ross; (563) 263-4010; fax (563) 263-1547; info@carvertrust.org
Internet: https://www.carvertrust.org/index.php?page=56
Sponsor: Roy J. Carver Charitable Trust
202 Iowa Avenue
Muscatine, IA 52761-3733

RRF General Program Grants 3417
The program funds service, education, research, and advocacy projects. The foundation is particularly interested in innovative projects that have the potential to change practice, policy, or delivery systems. The foundation's programs seek to improve the availability and quality of community-based and institutional long-term care programs; expand opportunities for older persons to play meaningful roles in society; support selected basic, applied, and policy research into the causes and solutions of significant problems of the aged; and increase the number of professionals and paraprofessionals adequately prepared to serve the elderly. Proposals fall into three general categories: research, model projects and service, and education and training. Guidelines are available online.
Requirements: Direct service projects that seek to improve the availability and quality of community-based and institutional long-term care programs; expand opportunities for older persons to play meaningful roles in society; support selected applied and policy research into the causes and solutions of significant problems of the aged; increase the number of professionals and paraprofessionals adequately prepared to serve the elderly.
Restrictions: The foundation will only consider proposals from applicants in Illinois, Indiana, Iowa, Kentucky, Missouri, Wisconsin, and Florida. Priority is given to nonprofit organizations serving the Chicago metropolitan area.
Geographic Focus: Florida, Illinois, Indiana, Iowa, Kentucky, Missouri, Wisconsin
Date(s) Application is Due: Feb 1; May 1; Aug 1
Amount of Grant: Up to 15,000,000 USD
Contact: Marilyn Hennessy; (773) 714-8080; fax (773) 714-8089; info@rrf.org
Internet: http://www.rrf.org/generalProgram.htm
Sponsor: Retirement Research Foundation
8765 W Higgins Road, Suite 430
Chicago, IL 60631-4170

RRF Organizational Capacity Building Grants 3418
The foundation provides grants for capacity-building activities that will help a nonprofit organization improve its management or governance to better serve the elderly. The following are examples of activities that will be supported but are not meant to exclude other ideas: planning (strategic, business, and financial plans) to steer an organization through changes; organizational assessments to identify problem areas that need to be strengthened; evaluation to develop outcome measures and determine progress in order to strengthen accountability to constituents and funding sources; public relations, communications, and marketing to increase awareness and utilization of services; financial management and resource development to improve financial systems, explore revenue options, and develop a stronger funding base; human resources management to enhance professional development, do team building, and improve the working environment; information systems management to identify and incorporate technology to improve efficiency; restructuring and building relations among organizations to strengthen service delivery, reduce costs, share resources, develop joint evaluations, or achieve other benefits; and board assessment, recruitment, training and structuring to strengthen the governance of the organization. All organizations that receive a grant are eligible for additional funding over the course of the grant, to be used flexibly for technical assistance opportunities that arise during the course of the grant. These may include seminars, workshops, short-term courses, publications, or other training opportunities related to organizational capacity building.
Requirements: For an organization to be eligible, it must meet the following criteria: be a nonprofit organization, but not a unit of government; be located in Cook, DuPage Kane, Kendall, Lake, or McHenry County, IL; provide services to the elderly; and intentionally give a high priority to the elderly.
Restrictions: The foundation will not fund building, renovation or capital improvements; emergency needs; deficits; or one-time events that do not build long-term capacity, i.e. fundraising events or one-time conferences.
Geographic Focus: Illinois
Date(s) Application is Due: Feb 1; May 1; Aug 1
Amount of Grant: 30,000 - 75,000 USD
Contact: Mary O'Donnell, Coordinator; (773) 714-8080; info@rrf.org
Internet: http://www.rrf.org/organizationalCapacityBuildingProgram.htm
Sponsor: Retirement Research Foundation
8765 W Higgins Road, Suite 430
Chicago, IL 60631-4170

RSC Abbyann D. Lynch Medal in Bioethics 3419
The award was created to honour bioethicist and Past-President of AMS, Professor Abbyann D. Lynch, C.M., Ph.D., for her substantial achievements and support of biomedical ethics in Canada. Its purpose is to raise the profile of ethics in medicine and science by honouring leading Canadian exponents in these fields and bring recognition to their work. The bronze medal and a cash amount of $2,000 are offered every year if there is a suitable candidate.
Requirements: The medal is awarded for a major contribution in bioethics by a Canadian. The contribution may be a book, a report, a scholarly article, a monograph or series of articles, which will normally be published in the two years preceding the nomination. Candidates shall be Canadian citizens or persons having permanent resident status, and who have lived in Canada during the three years preceding the date of nomination. Nominations shall be made by three persons with the initiator being a Fellow of the Society. The form for nomination can be downloaded from the sponsor's website.
Restrictions: Self-nomination for an award is not permitted.
Geographic Focus: All States, Canada
Date(s) Application is Due: Dec 1
Amount of Grant: 2,000 CAD
Contact: Erin Hearty; (613) 991-6990; fax (613) 991-6996; fellowship@rsc-src.ca
Internet: http://www.rsc.ca/index.php?page_id=61&lang_id=1&award_id=16
Sponsor: Royal Society of Canada
170 Waller Street
Ottawa, ON K1N 9B9 Canada

RSC Jason A. Hannah Medal 3420

The Jason A. Hannah Medal was established in 1976 by RSC with the assistance of Associated Medical Services Incorporated, through its Hannah Institute for the History of Medicine. The medal was established to honor the late President and Managing Director of AMS Inc., Dr. Jason A. Hannah, and to bring recognition to the work of Canadian research in the history of medicine. The award is normally made for a book published in the two years preceding its nomination. The bronze medal and a cash amount are offered every year if there is a suitable nomination.
Requirements: The medal is awarded for an important Canadian publication in the history of medicine. In this context, the word 'Canadian' refers to the citizenship or residence of the author, or to a content clearly relevant to Canadian medicine and health care. Nominations shall be made by three persons with the initiator being a Fellow of the Society.
Geographic Focus: All States, Canada
Date(s) Application is Due: Mar 1
Amount of Grant: 1,500 CAD
Samples: Michael Bliss, 'Harvey Cushing: A Life in Surgery'. James Opp, 'Home, Work, and Play: Situating Canadian Social History, 1840-1980'. David and Rosemary R. Gagan, 'or Patients of Moderate Means: A Social History of the Voluntary Public General Hospital in Canada, 1890-1950'. F. J. Paul Hackett, 'A Very Remarkable Sickness: Disease and Epidemics in the Petit Nord, 1670-1846'.
Contact: Erin Hearty; (613) 991-6990; fax (613) 991-6996; fellowship@rsc-src.ca
Internet: http://www.rsc.ca/index.php?page_id=61&lang_id=1&award_id=14
Sponsor: Royal Society of Canada
170 Waller Street
Ottawa, ON K1N 9B9 Canada

RSC McLaughlin Medal 3421

This medal and cash award are given annually for important research of sustained excellence in any branch of medical sciences. Nominations must be received by the listed application deadline.
Requirements: Candidates are to be Canadian citizens or persons who have been Canadian residents for the five years prior to award date. Nominations must be made by three persons, with the initiator being a fellow of the society, and consist of citation of not more than 250 words stating concisely why the nominee should be considered, curriculum vita, list of publications, and detailed appraisal of nominee's work. The nomination dossier will remain valid for three years from the year of first nomination.
Restrictions: Self-nomination for an award is not permitted.
Geographic Focus: All States, Canada
Date(s) Application is Due: Dec 1
Amount of Grant: 2,500 CAD
Samples: 2007 - John R. G. Challis. 2006 - Michael D. Tyers. 2005 - Robert E. W. Hancock. 2004 - John J.M. Bergeron. 2003 - Robert G. Korneluk.
Contact: Erin Hearty; (613) 991-6990; fax (613) 991-6996; fellowship@rsc-src.ca
Internet: http://www.rsc.ca/index.php?page_id=61&lang_id=1&award_id=17
Sponsor: Royal Society of Canada
170 Waller Street
Ottawa, ON K1N 9B9 Canada

RSC McNeil Medal for the Public Awareness of Science 3422

The McNeil Medal for the Public Awareness of Science was established in 1991 through McNeil Consumer Healthcare. It is intended to highlight the important role that science plays within our society and to encourage the communication of science to students and the public. The bronze medal and a cash amount are awarded every year if there is a suitable candidate.
Requirements: The medal is awarded to a candidate who has demonstrated outstanding ability to promote and communicate science to students and the public within Canada (the term public is defined in its broadest sense). Candidates shall be Canadian citizens or persons having permanent resident status, and who have lived in Canada during the three years preceding the date of nomination. Nominations shall be made by three persons with the initiator being a Fellow of the Society.
Restrictions: Self-nomination for an award is not permitted.
Geographic Focus: All States, Canada
Date(s) Application is Due: Dec 1
Amount of Grant: 1,500 CAD
Samples: Mary Anne White - Her contributions have been made through helping establish an interactive science center, newspaper columns, encyclopedia contributions and science booklets, and through public lectures, innovative courses, radio, television, and many public venues, including libraries, shopping malls, and Parliament Hill. Reginald Harry Mitchell - In the area of awareness of science, his principal audience is young people and his Chemistry 'magic show', experienced by more than 30,000 BC students to date, is legendary.
Contact: Erin Hearty; (613) 991-6990; fax (613) 991-6996; fellowship@rsc-src.ca
Internet: http://www.rsc.ca/index.php?page_id=61&lang_id=1&award_id=18
Sponsor: Royal Society of Canada
170 Waller Street
Ottawa, ON K1N 9B9 Canada

RSC Thomas W. Eadie Medal 3423

The medal is awarded annually in recognition of major contributions to any field through engineering or applied science, with preference given to those having an impact on communications, in particular the development of the Internet. Nominations must be received by the listed application deadline. The award includes a bronze medal and a cash amount.
Requirements: Candidates are to be Canadian citizens or persons who have been Canadian residents for the five years prior to award date. Nominations must be made by three persons, with the initiator being a fellow of the society, and consist of citation of not more than 250 words stating concisely why the nominee should be considered, curriculum vita, list of publications, and detailed appraisal of nominee's work. The nomination dossier will remain valid for five years from the year of first nomination.
Restrictions: Self-nomination for an award is not permitted.
Geographic Focus: All States, Canada
Date(s) Application is Due: Dec 1
Amount of Grant: 3,000 CAD
Samples: 2007 - Hussein T. Mouftah; 2006 - Alberto Leon-Garcia; 2005 - Norman C. Beaulieu; 2004 - Vijay K. Bhargava; 2003 - Morrel P. Bachynski.
Contact: Erin Hearty; (613) 991-6990; fax (613) 991-6996; fellowship@rsc-src.ca
Internet: http://www.rsc.ca/index.php?page_id=61&lang_id=1&award_id=5
Sponsor: Royal Society of Canada
170 Waller Street
Ottawa, ON K1N 9B9 Canada

Rush County Community Foundation Grants 3424

The Rush County Community Foundation, is a nonprofit public charity established in 1991 to serve donors, award grants and scholarships, and provide leadership to enrich and enhance the quality of life in Rush County, Indiana. The Rush County Community Foundation now holds over 130 funds, and have permanent endowment assets of over $6.5 million. As a public foundation, it helps donors provide grant making dollars for not-for-profit organizations that serve Rush County citizens. For additional information contact the Foundations office.
Requirements: Funding projects must serve the Rush County citizens.
Geographic Focus: Indiana
Contact: Garry Cooley; (765) 938-1177; fax (765) 938-1719; garryc@rushcountyfoundation.org
Internet: http://www.rushcountyfoundation.org/funds.php
Sponsor: Rush County Community Foundation
117 North Main Street
Rushville, IN 46173

Ruth Anderson Foundation Grants 3425

The Ruth Anderson Foundation was established in Florida in 1989, and currently awards grants to Florida nonprofit organizations throughout Dade County. The Foundation's funding interests include: AIDS research; children and youth services; homeless/housing shelters; human services; and substance abuse services. There are no application forms or annual deadlines. The initial approach should consist of a brief exploratory letter; full proposals will be by invitation only. The board meets throughout the year to consider requests for funding. Most recent awards have ranged from $1,500 to $10,000, and between ten and fifteen awards are given each year.
Requirements: Miami Dade County 501(c)3 tax-exempt organizations are eligible.
Restrictions: Grants are not awarded to individuals.
Geographic Focus: Florida
Amount of Grant: 1,500 - 10,000 USD
Samples: Sunrise Community, Miami, Florida, $10,000 - services support for the developmentally challenged (2015); Camillus House, Miami, Florida, $10,000 - support for shelter programs for homeless men (2015); Dade Heritage Trust, Miami, Florida, $10,000 - preservation and restoration of historical buildings (2015).
Contact: Ruth Admire, Administrator; (305) 444-6121; fax (305) 444-5508; info@sullivanadmire.com or ruth.admire@sullivanadmire.com
Internet: http://www.sullivanadmire.com/charitable.html
Sponsor: Ruth Anderson Foundation
255 Ponce de Leon Boulevard, Suite 320
Coral Gables, FL 33134

Ruth and Vernon Taylor Foundation Grants 3426

The foundation awards grants to nonprofit organizations in the areas of arts and humanities, civic and public affairs, secondary schools, higher education, environment, hospitals, human services, health, youth services, and social services. Types of support include general operating support, building construction and renovation, endowment funds, and research. The foundation suggests that initial contact be made in writing, since unsolicited requests for funds are not accepted. The Board meets in May and September.
Requirements: Organizations located in Colorado, Illinois, Montana, New Jersey, New York, Pennsylvania, Texas, or Wyoming are eligible.
Restrictions: Grants are not awarded to individuals.
Geographic Focus: Colorado, Illinois, Montana, New Jersey, New York, Pennsylvania, Texas, Wyoming
Amount of Grant: 1,000 - 20,000 USD
Contact: Douglas Taylor, Trustee; (303) 893-5284; fax (303)893-8263
Sponsor: Ruth and Vernon Taylor Foundation
518 17th Street, Suite 1670
Denver, CO 80202

Ruth Eleanor Bamberger and John Ernest Bamberger Memorial Foundation Grants 3427

The Foundation is dedicated to fulfilling the Founders' desires to help people reach their individual potential. The Foundation assists people of all ages, but especially children and young people through educational opportunities and scholarships, supporting crisis care and protective services, dental aid, after school programs, etc.
Requirements: Only residents of Utah may apply.
Restrictions: Grants and scholarships are given only to organizations, not to individuals.
Geographic Focus: Utah
Date(s) Application is Due: Mar 16; Sep 28
Amount of Grant: 1,000 - 20,000 USD

Contact: Eleanor Roser; (801) 364-2045; bambergermemfdn@qwestoffice.net
Internet: http://www.ruthandjohnbambergermemorialfdn.org/
Sponsor: Ruth Eleanor Bamberger and John Ernest Bamberger Memorial Foundation
136 S Main, Suite 418
Salt Lake City, UT 84101-1690

Ruth Estrin Goldberg Memorial for Cancer Research Grants   3428
The Ruth Estrin Goldberg Memorial for Cancer Reearch Grants are intended to support cancer research projects being conducted by an investigator at a recognized hospital, university, or other research institution. Grants are awarded for one year with possible renewal. Application forms and guidelines are available upon request.
Requirements: The principal investigator must be affiliated with an accredited institution.
Restrictions: Focus is limited to New Jersey, New York, and Pennsylvania.
Geographic Focus: New Jersey, New York, Pennsylvania
Amount of Grant: 10,000 - 15,000 USD
Contact: Rhoda Goodman, Grants Contact; (908) 686-5508 or (973) 467-3838
Sponsor: Ruth Estrin Goldberg Memorial for Cancer Research
c/o Greenberg and Company
Springfield, NJ 07081

Ruth H. and Warren A. Ellsworth Foundation Grants   3429
The foundation awards grants to nonprofits in Massachusetts in support of the arts, children and youth, community development, education and higher education, and health care and hospitals. Types of support include general operating support, continuing support, annual campaigns, building construction/renovation, equipment acquisition, emergency funds, and seed grants. There are no application forms.
Requirements: Nonprofits in the Worcester, MA, area are eligible.
Restrictions: Grants are not awarded to individuals or for endowment funds, scholarships, fellowships, research, publications, conferences, matching gifts, or loans.
Geographic Focus: Massachusetts
Amount of Grant: 500 - 100,000 USD
Contact: Sumner Tilton Jr., Trustee; (508) 798-8621
Sponsor: Ruth H. and Warren A. Ellsworth Foundation
370 Main Street, 12th Floor, Suite 1250
Worcester, MA 01608-1723

RWJF Changes in Health Care Financing and Organization Grants   3430
The Robert Wood Johnson Foundation, through its Changes in Health Care Financing and Organization (HCFO) initiative, supports research, policy analysis and evaluation projects that provide policy leaders timely information on health care policy, financing and organization issues. Supported projects include: examining significant issues and interventions related to health care financing and organization and their effects on health care costs, quality and access; and exploring or testing major new ways to finance and organize health care that have the potential to improve access to more affordable and higher quality health services. Grants will be awarded in two categories: small grants for projects requiring $100,000 or less and projected to take up to twelve months or less; and large grants for projects requiring more than $100,000 and/or projected to take longer than twelve months.
Requirements: Researchers, as well as practitioners and public and private policy-makers working with researchers, are eligible to submit proposals through their organizations. Projects may be initiated from within many disciplines, including health services research, economics, sociology, political science, public policy, public health, public administration, law and business administration. RWJF encourages proposals from organizations on behalf of researchers who are just beginning their careers, who can serve either individually as principal investigators or as part of a project team comprising researchers or other collaborators with more experience. Only organizations and government entities are eligible to receive funding under this program. Preference will be given to applicants that are either public entities or nonprofit organizations that are tax-exempt under Section 501(c)3 of the Internal Revenue Code and are not private foundations as defined under Section 509(a). Proposals for this solicitation must be submitted via the RWJF online system. Applicants should direct inquiries to the national program office.
Restrictions: Grants are not available for ongoing general operating expenses or existing deficits; endowment or capital costs, including construction, renovation, or equipment purchases; basic biomedical research; conferences, symposia, publications, or media projects, unless they are clearly related to the foundation's goals; research on unapproved drug therapies or devices; international programs and institutions; or direct support to individuals.
Geographic Focus: All States
Amount of Grant: Up to 500,000 USD
Contact: Bonnie J. Austin, HCFO Deputy Director; (202) 292-6700 or (202) 292-6756; fax (202) 292-6800; bonnie.austin@academyhealth.org or hcfoproposals@academyhealth.org
Internet: http://www.hcfo.org/funding
Sponsor: Robert Wood Johnson Foundation
Route 1 and College Road East, P.O. Box 2316
Princeton, NJ 08543-2316

RWJF Childhood Obesity Grants   3431
Through its Childhood Obesity Grant program, RWJF has developed three integrated strategies to reverse the childhood obesity epidemic: funding local organizations to make positive changes at the community level, advocating for healthier policies in the public and private sectors, and providing grants to researchers and evaluators to strengthen the evidence about what works. The Foundation's grant making is centered on advancing these strategies in ways that make it easier for all children to lead healthy lives. The program is intended to improve understanding of school, community, state and national policies and environmental factors affecting youth diet, physical activity, obesity, and tobacco, alcohol and drug use, and to evaluate the effectiveness of interventions to prevent youth obesity.
Requirements: Eligible applicants for research grants are scholars associated with educational institutions, research organizations, health care providers or other public or nonprofit organizations. Preference will be given to public entities or nonprofit organizations tax exempt under Section 501(c)3 of the Internal Revenue Code. Applicants may be independent of or associated with organizations currently receiving grants.
Geographic Focus: All States
Contact: C. Tracy Orleans, Senior Program Officer/Senior Scientist; (609) 627-5962 or (877) 843-7953; cto@rwjf.org
Internet: http://www.rwjf.org/en/about-rwjf/program-areas/childhood-obesity/programs-and-grants.html
Sponsor: Robert Wood Johnson Foundation
Route 1 and College Road East, P.O. Box 2316
Princeton, NJ 08543-2316

RWJF Community Health Leaders Awards   3432
Each year, the RWJF Community Health Leaders Award program honors outstanding individuals who overcome daunting odds to expand access to health care and social services to under-served and isolated populations in communities across the United States. Awards include a personal stipend and a grant for program enhancement. Potential nominators must submit a letter of intent; if a letter of intent is approved, a packet is sent to the nominator. For more information, contact the program office directly.
Requirements: Candidates must be residents of the United States, be affiliated with a 501(c)3 tax-exempt organization, have a five- to 15-year record of work in the community-health field, have experience as a direct provider of health-care services or be involved in the improvement of health-care access and delivery, demonstrate a full-time commitment to working with needy people, and have received no significant national recognition.
Restrictions: Nominations are not accepted from development and public relations departments or professional grant writers. Self-nominations are not accepted.
Geographic Focus: All States
Date(s) Application is Due: Oct 15
Amount of Grant: 125,000 USD
Contact: Janice Ford Griffin, National Program Director; (832) 319-7380; fax (832) 319-7385; info@communityhealthleaders.org
Internet: http://www.communityhealthleaders.org/
Sponsor: Robert Wood Johnson Foundation
Route 1 and College Road East, P.O. Box 2316
Princeton, NJ 08543-2316

RWJF Disparities Research for Change Grants   3433
The RWJF Disparities Research for Change program was created to support rigorous evaluations of practical and replicable solutions designed to reduce and eliminate disease specific racial and ethnic health care disparities. It seeks to improve the quality of health care provided to patients affected by racial and ethnic disparities by: Granting funds to discover and evaluate practical and replicable solutions designed to reduce and eliminate disease specific racial and ethnic health care disparities; Focusing on interventions aimed at health care delivery for one or more of the following health concerns: cardiovascular disease, depression, or diabetes; Conducting systematic reviews of the literature regarding racial and ethnic health care disparities interventions; and Disseminating results from these research efforts and systematic reviews to encourage health care systems to address racial and ethnic gaps in care.
Requirements: For-profit organizations are eligible to apply, however preference will be given to applicants that are public agencies or are tax-exempt under Section 501(c)3 of the Internal Revenue Code and are not private foundations as defined under Section 509(a). Proposals must focus on innovative interventions that reduce racial and ethnic disparities in the treatment of cardiovascular disease, depression, and diabetes. There are no age restrictions for the target populations of the interventions proposed for evaluation using program funds. Although there are no restrictions on the age of target populations, proposals targeting pediatric and geriatric populations are encouraged. Applications are accepted in two phases. The first phase is a brief proposal submission open to all interested research teams. The second phase is a full proposal submission and will be accepted by invitation only. All applications, both the brief and full proposals, must be submitted through the Robert Wood Johnson Foundation's Grantmaking Online system.
Restrictions: Individuals are not eligible to apply. Stand-alone prevention interventions such as tobacco cessation or exercise programs are not eligible unless they are one part of a larger, more comprehensive, intervention. The program will only directly fund health care delivery institutions. However, eligible institutions may partner/sub-contract with academic or research groups.
Geographic Focus: All States
Amount of Grant: 100,000 - 275,000 USD
Contact: Marshall H. Chin, Director; (866) 344-9800 or (773) 702-4769; fax (773) 702-4620; mchin@medicine.bsd.uchicago.edu or info@solvingdisparities.org
Internet: http://www.solvingdisparities.org/
Sponsor: Robert Wood Johnson Foundation
Route 1 and College Road East, P.O. Box 2316
Princeton, NJ 08543-2316

RWJF Harold Amos Medical Faculty Development Research Grants   3434
The Harold Amos Medical Faculty Development program was established to increase the number of faculty from historically disadvantaged backgrounds who can achieve senior rank in academic medicine and dentistry and who will encourage and foster the development of succeeding classes of such physicians and dentists. Four-year postdoctoral research awards are offered to universities, schools of medicine and dentistry and research

institutions to support the research and career development of physicians and dentists from historically disadvantaged backgrounds who are committed to developing careers in academic medicine and dentistry and to serving as role models for students and faculty of similar background. The program defines the term historically disadvantaged to mean the challenges facing individuals because of their race, ethnicity, socioeconomic status or similar factors. The program will fund up to ten four-year awards up to $420,000 each, with scholars to receive an annual stipend of up to $75,000 each, complemented by a $30,000 annual grant toward support of research activities.
Requirements: Eligible to apply are minority physicians who are U.S. citizens at the time of application, have excelled in their post-secondary education within the United States, are now completing or will have completed formal clinical training, and are committed to academic careers.
Geographic Focus: All States
Date(s) Application is Due: Mar 15
Amount of Grant: 75,000 - 420,000 USD
Contact: Nina Ardery, Deputy Director; (317) 278-0500; amfdp@starpower.net
Internet: http://www.rwjf.org/en/library/funding-opportunities/2016/harold-amos-medical-faculty-development-program--amfdp--.html
Sponsor: Robert Wood Johnson Foundation
Route 1 and College Road East, P.O. Box 2316
Princeton, NJ 08543-2316

RWJF Harold Amos Medical Faculty Development Scholars 3435
The Harold Amos Medical Faculty Development Program supports basic/biomedical, clinical, and health services/epidemiology research. The purpose of the AMFDP award is to facilitate the transition of the newly-trained clinician who wishes to develop into an independent investigator. A completed application consists of two parts: the completion of the online proposal process, and submission of other (paper) supporting documents. The maximum amount of support is $420,000 over a four-year period. This includes an annual $75,000 stipend, plus $30,000 per year for research support.
Requirements: To be eligible, applicants must be physicians from historically disadvantaged backgrounds (ethnic, financial, or educational) who are U.S. citizens or permanent residents at the time of application deadline; have excelled in their education; are now completing or have completed their formal clinical training; are prepared to devote four consecutive years to research; are committed to pursuing academic careers; and are committed to improving the health status of the under-served, decreasing health disparities, or serving as role models for students and faculty from historically disadvantaged backgrounds Preference will be given to physicians who have recently completed their formal clinical training. In order to pursue the advanced research training required by this program, applicants must identify faculty mentors with whom a research plan will be developed. The mentor's experience in the supervision of trainees and the adequacy of the mentor's research environment and support will be strongly considered.
Geographic Focus: All States
Date(s) Application is Due: Mar 15
Amount of Grant: Up to 105,000 USD
Contact: Nina Ardery, Deputy Director; (301) 565-4080 or (317) 278-0500; fax (301) 565-4088; nardery@erols.com or amfdp@starpower.net
Internet: http://www.amfdp.org/for-applicants
Sponsor: Robert Wood Johnson Foundation
Route 1 and College Road East, P.O. Box 2316
Princeton, NJ 08543-2316

RWJF Health and Society Scholars 3436
The RWJF Health and Society Scholars program is a two-year fellowship designed to build the nation's capacity for research, leadership, and action to address the broad range of factors affecting health. The goal of the program is to improve health by training scholars to rigorously investigate the connections between biological, behavioral, environmental, economic, and social determinants of health; and develop, evaluate, and disseminate knowledge and interventions based on integration of these determinants. Each year the program will enable up to twelve outstanding individuals who have completed doctoral training to engage in an intensive two-year program at one of four nationally prominent universities. Scholars will have access to a full range of university resources and will receive an annual stipend of $80,000, plus health insurance from their university site. Scholars will also have access at each university to financial support for research-related expenses, training workshops and travel to professional meetings.
Requirements: Individuals must apply only through the foundation's online system. To be eligible, scholars must: have completed doctoral training by the time of entry into the program (August or September) in one of a variety of fields including, but not limited to, the behavioral and social sciences, the biological and natural sciences, health professions, public policy, public health, history and ethics; have significant research experience; clearly connect their research interests to substantive population health concerns; and be citizens or permanent residents of the United States or its territories. Applications from candidates with diverse backgrounds are encouraged.
Restrictions: Individuals placed in the program must be willing to relocate to the institution designated during the match process for the duration of the two-year program.
Geographic Focus: All States
Date(s) Application is Due: Sep 20
Amount of Grant: 89,000 USD
Contact: Gerard P. Lebeda; (212) 419-3566; fax (212) 419-3569; hss@nyam.org
Internet: http://www.healthandsocietyscholars.org
Sponsor: Robert Wood Johnson Foundation
Route 1 and College Road East, P.O. Box 2316
Princeton, NJ 08543-2316

RWJF Healthy Eating Research Grants 3437
Healthy Eating Research is a national program that supports research to identify, analyze, and evaluate environmental and policy strategies that can promote healthy eating among children and prevent childhood obesity. Special emphasis is given to research projects that benefit children in the low-income and racial/ethnic populations at highest risk for obesity. The program funds two types of research grants: studies to identify and/or evaluate promising school food environment and policy changes (12- to 18-month awards up to $100,000 each, and 18- to 36-month awards up to $400,000 each); and analyses of macro-level policy or system determinants of school food environments and policies (12- to 18-month awards up to $75,000 each). Applications must be submitted through the online process in response to future calls for proposals. Calls for Proposals and their guidelines are available online.
Requirements: Preference will be given to those applicants that may be either public entities or nonprofit organizations that are tax-exempt under Section 501(c)3 of the Internal Revenue Code. Applicant organizations must be based in the United States or U.S. Territories. The focus of this program is the United States; studies of policies in other countries will be considered only to the extent that they may directly inform U.S. policy. The experience and qualifications of the research team is one of the primary criteria for proposal review. A doctorate or terminal degree is preferred for the principal investigator.
Restrictions: This program does not fund demonstration projects. The foundation does not award grants to private individuals or applicants from outside the U.S. or U.S. Territories. No faxed, emailed or mailed proposals will be accepted.
Geographic Focus: All States
Amount of Grant: Up to 400,000 USD
Contact: Kathy Kosiak; (800) 578-8636; fax (612) 624-9328; healthyeating@umn.edu
Internet: http://www.healthyeatingresearch.org/
Sponsor: Robert Wood Johnson Foundation
Route 1 and College Road East, P.O. Box 2316
Princeton, NJ 08543-2316

RWJF Interdisciplinary Nursing Quality Research Initiative Grants 3438
The Interdisciplinary Nursing Quality Research Initiative has most recently awarded two-year grants to support the generation, dissemination, and translation of research to improve the contributions of nursing to the quality of hospital care. It will continue to support interdisciplinary teams of nurse scholars and scholars from other disciplines to address gaps in knowledge about the relationship between nursing and health care quality. Topic areas may include: identifying, developing, and testing new measures to advance quality nursing research; and testing the effects of implementing the nursing-sensitive performance measures endorsed by the National Quality Forum. Investigator-initiated ideas that are consistent with the broader set of program goals also will be considered. The two-phase application process begins with online submittal of a brief proposal; full proposals are by invitation, and are due approximately four months after invitations to submit are given. All current and future calls for proposals are detailed at the web site.
Requirements: Teams of researchers who are U.S. citizens or permanent residents are eligible. Each team must have at least one nursing researcher and at least one researcher from another discipline. Preference will be given to 501(c)3 applicants that are not 509(a) private foundations.
Geographic Focus: All States
Amount of Grant: Up to 300,000 USD
Samples: University of North Carolina at Chapel Hill, Chapel Hill, North Carolina - Evaluating the Business Case for Nurse Residency Programs; Institute of Health Professions, Boston, Massachusetts - Scope of Practice Influences on Workforce, Workplace, and Outcomes of Care; Pennsylvania School of Nursing, Philadelphia, Pennsylvania - Nurse Staffing Public Reporting Laws: Effects on Staffing and Patient Safety.
Contact: Lori A. Melichar, Senior Program Officer; (301) 571-8161 or (800) 978-8309; INQRIhelpdesk@constellagroup.com
Internet: http://www.inqri.org/grants
Sponsor: Robert Wood Johnson Foundation
Route 1 and College Road East, P.O. Box 2316
Princeton, NJ 08543-2316

RWJF New Careers in Nursing Scholarships 3439
The Robert Wood Johnson Foundation New Careers in Nursing is a scholarship program to help alleviate the nursing shortage and increase the diversity of nursing professionals. Through grants to schools of nursing, the program will provide scholarships to college graduates with degrees in other fields who are enrolled in accelerated baccalaureate and master's degree nursing programs. A school of nursing may apply for between five and 30 scholarships per year to be awarded to students from underrepresented groups in nursing or disadvantaged backgrounds. Up to 400 scholarship awards will be made during this funding cycle. The deadline for receipt of applications and supporting documents submitted online is December 15.
Requirements: To be eligible, applicant institutions must: offer an entry-level accelerated baccalaureate nursing program or master's nursing program for non-nursing college graduates; and be accredited by a nursing accrediting agency recognized by the U.S. Department of Education.
Geographic Focus: All States
Date(s) Application is Due: Jan 9
Amount of Grant: 10,000 USD
Contact: Vernell DeWitty, Deputy Director; (202) 463-6930, ext. 265; fax (202) 785-8320; vdewitty@aacn.nche.edu or ncin@aacn.nche.edu
Internet: http://www.newcareersinnursing.org/about-ncin/funding-information
Sponsor: Robert Wood Johnson Foundation
Route 1 and College Road East, P.O. Box 2316
Princeton, NJ 08543-2316

RWJF New Jersey Health Initiatives Grants 3440
The RWJF New Jersey Health Initiatives program funds grants that promote the development of innovative, community-based health services in New Jersey. Special emphasis is placed on projects that improve access to primary health care, address the complex service needs of people with chronic health conditions, and reduce substance abuse. Grant funds may be used for staff salaries, project-related travel, data processing, supplies, a limited amount of equipment, and other expenses essential to the project. The primary emphasis of the program is the demonstration of new service delivery approaches, although a limited amount of funding will be available for planning, feasibility, and program development. The program will also support a small number of health services research or evaluation projects in the goal areas. Projects funded through NJHI are generally three years in duration with proposals submitted annually through calls for proposals issued by the Robert Wood Johnson Foundation.
Requirements: Public entities or IRS 501(c)3 organizations located in New Jersey are eligible for funding. Applicants are asked to propose projects that respond to one of the following RWJF interest areas: addiction prevention and treatment; childhood obesity; or vulnerable populations. Projects must address all of the following: reach people who are not served by traditional health and social services; focus on meeting the needs of individuals and families, touching their lives directly; and address the lack of policy, financing and service integration among state and local agencies that creates barriers to helping people.
Restrictions: Private 509(a) foundations are not eligible. Grant funds may not be used for capital costs (including construction, renovation, and most equipment purchases), to support ongoing general operating expenses or existing deficits, or to substitute for funds currently supporting similar services. No hard-copy brief or full proposals will be accepted.
Geographic Focus: New Jersey
Date(s) Application is Due: Jan 15
Amount of Grant: 50,000 - 150,000 USD
Contact: Bob Atkins, Director; (856) 225-6734 or (856) 225-6733; fax (856) 225-6726; info@njhi.org or atkins@njhi.org
Internet: http://www.njhi.org/our-grants/
Sponsor: Robert Wood Johnson Foundation
Route 1 and College Road East, P.O. Box 2316
Princeton, NJ 08543-2316

RWJF Nurse Faculty Scholars 3441
The goal of the Robert Wood Johnson Foundation Nurse Faculty Scholars (NFS) program is to develop the next generation of national leaders in academic nursing through career development awards for outstanding junior nursing faculty. The program aims to strengthen the academic productivity and overall excellence of nursing schools by providing mentorship, leadership training, and salary and research support to junior faculty. The Nurse Faculty Scholars program is part of the Foundation's Human Capital Portfolio, which aims to ensure that the nation has a diverse, well-trained workforce to meet the health care needs of all Americans. Up to 12 awards of up to $350,000 each over three years will be available. February 12 is the deadline for receipt of applications.
Requirements: Accredited colleges and universities within the United States are eligible to apply. To be eligible, candidates must meet the following criteria: be a registered nurse with a research doctorate in nursing or a related discipline; be a junior faculty member in an accredited school of nursing with an academic position that could lead to tenure; after completing the doctorate, be in an academic position that could lead to tenure for at least two years and no more than five years at the start of the program; identify at least one senior leader in the school of nursing to serve as a primary (nursing) mentor for academic career planning, and to provide access to organizations, programs and colleagues helpful to the candidate's work as a scholar; identify at least one senior researcher in the university with similar or complementary research interests to serve as a research mentor; and be a U.S citizen or permanent resident at the time of application.
Geographic Focus: All States
Date(s) Application is Due: Feb 12
Amount of Grant: Up to 350,000 USD
Contact: Maryjoan D. Ladden; (877) 738-0737; fax (410) 614-8285; rwjfnfs@jhu.edu
Internet: http://www.rwjf.org/applications/solicited/npo.jsp?FUND_ID=55121
Sponsor: Robert Wood Johnson Foundation
Route 1 and College Road East, P.O. Box 2316
Princeton, NJ 08543-2316

RWJF Pioneer Portfolio Grants 3442
The Pioneer Portfolio Grant program aims to support novel, high-return ideas that may have far-reaching impact on people's health, the quality of care they receive and the systems through which that care is provided. It seeks ideas not only from the mainstream of health and health care but also looks to sources outside of these fields for innovations that might have transformative impact. Several projects apply approaches from diverse sectors such as finance, design and entertainment to forge new solutions in health and health care arenas.
Requirements: Preference is given to public agencies, public charities or are tax-exempt under section 501(c)3 of the Internal Revenue Code. Organizations must be located in the United States or one of its territories.
Restrictions: The organization does not fund any of the following: Ongoing general operating expenses or existing deficits; Endowment or capital costs; Basic biomedical research; Research on drug therapies or devices; International programs; Direct support of individuals; Lobbying of any kind. Additionally, RWJF no longer funds projects in the following areas: End-of-life care; Long-term care; Specific chronic conditions (that are not part of the strategies of other Program Areas); Physical activity for adults age 50 or older.
Geographic Focus: All States
Contact: Deborah H. Bae; (609) 627-5812 or (877) 843-7953; dbae@rwjf.org
Internet: http://www.rwjf.org/en/grants/what-we-fund.html
Sponsor: Robert Wood Johnson Foundation
Route 1 and College Road East, P.O. Box 2316
Princeton, NJ 08543-2316

RWJF Policies for Action: Policy and Law Research to Build a Culture of Health Grants 3443
Policies for Action: Policy and Law Research to Build a Culture of Health (P4A) was created to help build an evidence base for policies that can lead to a Culture of Health. P4A seeks to engage long-standing health and health care researchers, as well as experts in fields like housing, education, transportation, and the built environment, to name a few, who have not worked in health before. The goal is to develop research that generates actionable evidence—the data and information that can guide legislators and other policymakers, public agencies, educators, advocates, community groups, and individuals. The research may examine established laws, regulations, and policies as well as potential new policies and approaches. The research funded under this call for proposals (CFP) should inform the significant gaps in our knowledge regarding what policies can serve as levers to improve population health and well-being, and achieve greater levels of health equity. Approximately $1.5 million will be awarded through this CFP. Each grant will award up to $250,000 for a maximum funding period of 24 months. The deadline for required Letters of Intent is March 15, with full invited proposals due June 17.
Requirements: Applicants may be either 501(c)3 public entities or nonprofit organizations and not classified as a private foundation under Section 509(a).
Restrictions: In-kind services and money to pay for capital costs or renovation may not be used to match funds.
Geographic Focus: All States
Date(s) Application is Due: Jun 17
Amount of Grant: Up to 250,000 USD
Contact: Bethany Saxon, Director of Communications; (215) 204-2134
Internet: http://www.rwjf.org/en/library/funding-opportunities/2016/policies-for-action--policy-and-law-research-to-build-a-culture-.html
Sponsor: Robert Wood Johnson Foundation
Route 1 and College Road East, P.O. Box 2316
Princeton, NJ 08543-2316

RWJF Vulnerable Populations Portfolio Grants 3444
Programs within the Vulnerable Populations Portfolio have four elements in common: they offer an opportunity to improve health by taking a fresh approach to a long standing problem; they address poor health status in the context of other factors like housing, education and poverty; they make fundamental changes in how services are organized and delivered; and they address the lack of policy, financing, or service integration among local service providers and state and federal agencies. While RWJF awards most of its grants in response to calls for proposals, it also award grants in response to unsolicited proposals in the Vulnerable Populations Portfolio.
Requirements: Preference is given to public agencies, public charities or are tax-exempt under section 501(c)3 of the Internal Revenue Code. Organizations must be located in the United States or one of its territories.
Restrictions: This program will not provide funding for efforts that do not incorporate the social factors that drive health status as part of their proposed model. Additionally, the program does not fund documentaries, research, programs that address a single medical condition, or provide core support for free or safety net clinics, disease management models, or well-tested models whose effectiveness have already been established and that have been widely disseminated.
Geographic Focus: All States
Contact: Kristin B. Schubert; (888) 631-9989; kschubert@rwjf.org
Internet: http://www.rwjf.org/en/about-rwjf/program-areas/vulnerable-populations/Programs-and-Grants.html
Sponsor: Robert Wood Johnson Foundation
Route 1 and College Road East, P.O. Box 2316
Princeton, NJ 08543-2316

Ryan Gibson Foundation Grants 3445
The Foundation welcomes any and all research grant requests from around the country. The main focus is on blood related cancers. Applicants should submit a one to two page proposal with the following information: description of the procedure and or drug that is to be tested/developed, and what the desired/expected result will be; some background on the researcher and the facility that will be used (past projects success and or failures should be included); and a budget of anticipated costs. Mail all requests to the following address: The Ryan Gibson Foundation, Attn: Research Proposal, 18111 Preston Road, Suite 650, Dallas, TX 75252
Geographic Focus: All States
Amount of Grant: 50,000 USD
Contact: National Director; (214) 987-0008; fax (214) 365-9293; mail@trgf.org
Internet: http://www.trgf.org/grants.shtml
Sponsor: Ryan Gibson Foundation
5910 North Central Expressway, Suite 770
Dallas, TX 75206

S. D. Bechtel, Jr. Foundation / Stephen Bechtel Fund Preventive Healthcare and Selected Research Grants 3446
The Foundation?fund believe that access to quality care, preventive health care and research programs can help individuals of all ages increase life expectancy and improve their quality of life. The directors have a particular interest in promoting nutritional health, fitness, and wellness in order to increase positive health outcomes. The Foundation/Fund support organizations working to achieve the following objectives: enabling San Francisco Bay Area

community clinics and school-based health centers to improve the quality and delivery of medical care; enhancing and expand school-based efforts in the San Francisco Bay Area that improve the nutritional health and fitness of children; and advancing research related to Alzheimer's, aging, nutrition or other Board-determined specific research initiatives.
Requirements: The primary geographic focus of the Foundation/Fund is the San Francisco Bay area. The Foundation/Fund: support non-profit organizations providing quality programs in science, technology, engineering and math (STEM) education, environment, environmental education, character and citizenship development and preventive healthcare and selected research; provide capital support; provide operational support; and provide project support.
Restrictions: The Foundation/Fund do not provide endowment funding, international grants, or grants for individuals.
Geographic Focus: California
Date(s) Application is Due: Oct 1
Amount of Grant: 5,000 - 25,000 USD
Contact: Coordinator; (415) 284-8675; fax (415) 284-8571; sdbjr@sdbjrfoundation.org
Internet: http://www.sdbjrfoundation.org/program_areas.htm#health
Sponsor: S. D. Bechtel, Jr. Foundation / Stephen Bechtel Fund
P.O. Box 193809
San Francisco, CA 94119-3809

S.H. Cowell Foundation Grants 3447
The foundation awards grants for a wide variety of causes to nonprofit organizations in northern California. Projects of interest include those that support children, youth, and families; education; housing; alcohol abuse prevention; religious organizations, education, and welfare; school-to-work employment training; population and environment; family planning; and conventional arms control. Applicants are encouraged to obtain most of their operating and project funding from other sources. Matching and challenge grants will be awarded under appropriate circumstances. The foundation prefers to award grants for one-time capital needs or for specific projects that are time-definite in nature and likely to become self-sufficient within several years. The application process should begin with a phone call. If interested, the foundation will request a short letter and then a formal proposal.
Requirements: Grants are made only to 501(c)3 tax-exempt organizations primarily in northern California.
Restrictions: The foundation does not normally make grants to individuals, for start-up of new organizations, for academic or other research, for general support, for annual fund-raising, to governmental agencies, to churches for religious support, to hospitals for medical research or treatment, for conferences, for media projects, or for political lobbying.
Geographic Focus: California
Amount of Grant: Up to 500,000 USD
Samples: Adventure Risk Challenge, Truckee, CA, $25,000 - to launch year-round expansion of the Adventure Risk Challenge (ARC) summer program; Boys & Girls Clubs of the North Valley Chico, CA, $500,000 - to launch a Boys & Girls Club in South Oroville; East Bay Asian Local Development Corporation Oakland, CA, $375,000 - to complete the budget to install a park at Lion Creek Crossings;
Contact: Susan Vandiver; (415) 397-0285; fax (415) 986-6786; info@shcowell.org
Internet: http://www.shcowell.org/grant/grant.php
Sponsor: S.H. Cowell Foundation
120 Montgomery Street, Suite 2570
San Francisco, CA 94104

S. Livingston Mather Charitable Trust Grants 3448
The S. Livingston Mather Charitable Trust was established in 1953 in Ohio by Cleveland-Cliffs vice-president Samuel Livingston Mather. The Trust's primary areas of interest include cultural programs, education, child welfare, and social services. Support is also available for youth programs, the environment, and natural resources. Giving is primarily restricted to the northeastern Ohio area. Applicants should contact the Trust prior to submitting an application.
Requirements: Unsolicited requests for funds not accepted, contact the Trust before sending a proposal.
Restrictions: No support is available for: endowments; science; medical research programs; in areas appropriately supported by the government and/or the United Way; individuals; or deficit financing.
Geographic Focus: Ohio
Amount of Grant: 100 - 100,000 USD
Samples: Cleveland Museum of Art, Cleveland, OH, $20,000; Cuyahoga Valley Countryside Conservancy, Peninsula, OH, $7,500.
Contact: Janet W. Havener, Director, c/o Glenmede Trust Company; (215) 419-6000
Sponsor: S. Livingston Mather Charitable Trust
1 Corporate Exchange, 25825 Science Park Drive, Suite 110
Beachwood, OH 44122

S. Mark Taper Foundation Grants 3449
The S. Mark Taper Foundation, founded in 1989, is a private family foundation dedicated to enhancing the quality of people's lives by supporting nonprofit organizations and their work in the Los Angeles, Long Beach, Santa Ana, California communities. Areas of interest are broad and include but are not limited to education, the environment, independent living for the disabled, abused women, immigrant health care, children, hunger, housing, AIDS, teenage pregnancy prevention, job creation and economic revitalization, individuals with visual impairments, and the arts. Types of support include capital grants, challenge/matching grants, general operating grants, research grants, program/project grants, seed money grants, and scholarships. Grants are made generally for one year.
Requirements: 501(c)3 nonprofit organizations in California are eligible for grant support. Application forms are required therefore your initial approach should be a Letter of Inquiry containing one copy of the proposal and the following: brief history of organization and description of its mission; copy of most recent annual report/audited financial statement/990; detailed description of project and amount of funding requested; list of source(s) of last three years of funding for the specific project (if any) and, the organization.
Geographic Focus: California
Date(s) Application is Due: Dec 1
Amount of Grant: 5,000 - 500,000 USD
Samples: Korean Health Education Information and Research Center, Los Angeles, CA - $500,000; Santa Clarita Child and Family Development Center, Santa Clarita, CA - $100,000; Los Angeles Unified School District Education Foundation, Los Angeles, CA - $200,000;
Contact: Raymond Reister, Executive Director; (310) 476-5413; fax (310) 471-4993; rreisler@smtfoundation.org or info@smtfoundation.org
Internet: http://www.smtfoundation.org/
Sponsor: S. Mark Taper Foundation
12011 San Vicente Boulavard, Suite 400
Los Angeles, CA 90049

Saginaw Community Foundation Senior Citizen Enrichment Fund 3450
The Senior Citizens Enrichment Fund assists programs supporting charitable purposes that benefit the elderly in Saginaw County. The priority of this grant is to assist programs and projects that enhance the quality of life for seniors maintaining an independent residence.
Requirements: Applicants must first call the Foundation to discuss the project with the staff. To be eligible for funding, organizations must have a nonprofit status and directly benefit Saginaw County. Grants are ordinarily only made for one year. Applications can be downloaded at the website but must be received by mail or dropped off at the agency.
Restrictions: The Saginaw Community Foundation does not support: operating budgets; basic municipal services; basic educational functions; endowment campaigns; previously incurred debt; or sectarian religious programs.
Geographic Focus: Michigan
Date(s) Application is Due: Feb 1; May 1; Aug 1; Nov 1
Amount of Grant: Up to 10,000 USD
Contact: Kendra Kempf, Program Associate/FORCE Coordinator; (989) 755-0545; fax (989) 755-6524; kendra@saginawfoundation.org
Internet: http://www.saginawfoundation.org/grants_and_scholarships/grants/
Sponsor: Saginaw Community Foundation
100 South Jefferson, Suite 201
Saginaw, MI 48602

Saigh Foundation Grants 3451
Fred M. Saigh was born in Springfield, Illinois on June 27, 1905, the son of Lebanese immigrants who owned a chain of grocery stores. He passed away in 1999. Although he is best known as the owner of the St. Louis Cardinals (1948 - 1953), many of his friends and acquaintances will always remember him as a perceptive and caring benefactor - "a one-man charity fund," in the words of sportswriter Mike Eisenbath. The Saigh Foundation continues the important work begun by Mr. Saigh by assisting St. Louis-area organizations that benefit children and youth, particularly in the areas of education and healthcare. The foundation is particularly interested in stimulating the development of new ventures, as well as in supporting organizations that feature innovative approaches or programs. Like Mr. Saigh, the foundation is especially dedicated to aiding those who might not otherwise receive assistance. Proposals are requested at least three months before any quarterly meeting. These are normally held in January, April, July, and October. The foundation uses a customized version of the Missouri Common Grant Application (CGA). It is available at the grant website along with guidelines and budget templates and should be submitted via mail or fax to the address given.
Requirements: Saint Louis-area nonprofit organizations are eligible.
Restrictions: The foundation does not participate in annual appeals, dinner functions, and fundraising events; capital campaigns; loans and deficits; grants for films and travel; or nonprofit organizations outside of the metropolitan Saint Louis area.
Geographic Focus: Missouri
Contact: JoAnn Hejna; (314) 862-3055; fax (314) 862-9288; saigh@thesaighfoundation.org
Internet: http://www.thesaighfoundation.org/grant_guide.html
Sponsor: Saigh Foundation
7777 Bonhomme Avenue, Suite 2007
Saint Louis, MO 63105

Sain-Orr and Royak-DeForest Steadman Foundation Grants 3452
The Sain-Orr, Royak, DeForest, Steadman Foundation (formerly Sain-Orr Foundation) was established in Florida in 2002. Its primary fields of interest include: arts and culture; elementary and secondary education; health care services; higher education; and human services. Though unsolicited requests are not accepted, interested parties should contact the Foundation managers with a concept in writing. There are no annual deadlines for invited applicants. Between twenty and twenty-five grants are awarded each year, ranging from $750 to $10,000.
Requirements: 501(c)3 organizations serving the residents of Dade County, Florida, are eligible to apply.
Geographic Focus: Florida
Amount of Grant: 750 - 10,000 USD
Contact: Ruth Admire, Administrator; (305) 444-6121; fax (305) 444-5508; info@sullivanadmire.com or ruth.admire@sullivanadmire.com
Internet: http://www.sullivanadmire.com/charitable.html
Sponsor: Sain-Orr and Royak-DeForest Steadman Foundation
255 Ponce de Leon Boulevard, Suite 320
Coral Gables, FL 33134

Saint Louis Rams Foundation Community Donations 3453
The foundation supports nonprofits that help inspire positive change for youth in the Saint Louis area. Programs that impact youth in the general fields of education, literacy, health, and recreation will be considered. Annually, the Rams provide to charitable groups more than 3,500 items, helping recipient organizations raise thousands of dollars through raffles, auctions and other fundraising endeavors. Other types of financial support include program development grants and general operating grants. The foundation does not accept unsolicited requests, but initial information may be sent for the office to keep on file for future opportunities.
Requirements: Nonprofits in the metropolitan Saint Louis, Missouri, area, including southern Illinois and eastern Missouri, are eligible. Preference is given to organizations that partner with other local nonprofits and offer creative approaches for more than grants (i.e., personnel involvement or in-kind support) and ways the Rams can participate.
Restrictions: The Rams do not provide monetary contributions or merchandise donations for the following: businesses, retail and otherwise; capital campaigns/start-up funding for new businesses; on-line auctions; chamber of commerce/city/neighborhood festivals such as homecoming celebrations and carnivals that do not directly benefit a charitable organization; class reunions; family reunions; pageant contestants (beauty and otherwise); student ambassador/exchange programs; or non-charity events and organizations such as company picnics, employee golf tournaments, employee recognition/incentive programs, card clubs, car shows, "poker runs", and organized adult leisure sports teams.
Geographic Focus: Missouri
Date(s) Application is Due: Jan 1; Jul 1
Contact: Coordinator; (314) 516-8788 or (314) 982-7267; fax (314) 770-0392
Internet: http://www.stlouisrams.com/community/donations.html
Sponsor: Saint Louis Rams Foundation
1 Rams Way
Saint Louis, MO 63045

Saint Luke's Foundation Grants 3454
The foundation exists to improve the health and well being of individuals and families living in Greater Cleveland. Proposals are invited for activities intended to enhance community involvement and ownership; promote healthy behaviors and life styles by education and outreach programs aimed at increasing the capacity of individuals and families to protect and improve their own health; increase and improve health care by improving access to affordable, high-quality, comprehensive, culturally competent and appropriate health care; educate health care professionals to increase their capacity in serving the health and health care needs of inner-city residents; and increase knowledge by furthering research on: the determinants of health; innovative systems, strategies, and collaborations designed to improve health and well being in the community; and the underlying causes of diseases and disabilities particularly prevalent in the communities.
Requirements: 501(c)3 tax-exempt organizations and governmental units/agencies are eligible. The grant application is a 3 step process: step 1: Letter of Inquiry; step 2: site visit; step 3: proposal submission. See website for detailed description of each step. The deadlines posted are for the Letter of Inquiry.
Restrictions: The Foundation does not fund: individuals; religious organizations for religious or evangelical purposes; projects outside Greater Cleveland that do not directly benefit Cleveland residents; fundraising events; endowment funds; biomedical research; debt retirement; lobbying.
Geographic Focus: Ohio
Date(s) Application is Due: Jan 5; Apr 1; Jul 1; Aug 1; Oct 1
Contact: Peg Butler; (216) 431-8010; fax (216) 431-8015; pbutler@saintlukesfoundation.org
Internet: http://www.saintlukesfoundation.org/grants/community-grants.html
Sponsor: Saint Luke's Foundation of Cleveland, Ohio
4208 Prospect Avenue
Cleveland, OH 44103

Saint Luke's Health Initiatives Grants 3455
The trust was established for the purpose of addressing critical health needs in Maricopa County, AZ. Community grants support innovative approaches that address community health from a broad perspective, including funding for such projects as professional education and training, public education/advocacy demonstration projects, program planning and expansion, and equipment to address the health needs of populations that are considered underserved and vulnerable. Community grants are awarded for up to two years. Bridge grants support small, emerging organizations or new programs of larger organizations that have infrastructure or developmental needs. Grants support strategic planning, technical training, purchasing equipment, or convening discussion groups on community health issues. Bridge grants are awarded for one year. Inquiries are encouraged prior to proposal submission.
Requirements: To be eligible for a community partnership grant, a prospective organization must: demonstrate tax-exempt status under Section 501(c)3 of the Internal Revenue Service Code or be recognized as an instrumentality of state/local government; be located in Arizona. Preference is given to organizations within Maricopa County, although a statewide program with impact on Maricopa County will be considered.
Restrictions: The trust does not fund individuals, participate in large capital campaigns or endowment drives, provide grants for ongoing annual operating support, or fund projects of sectarian or religious organizations.
Geographic Focus: Arizona
Date(s) Application is Due: Feb 12; Aug 14
Amount of Grant: 10,000 - 100,000 USD
Contact: Jane Pearson; (602) 385-6500; fax (602) 385-6510; Jane.Pearson@slhi.org
Internet: http://www.slhi.org/community_grants/how_to_apply.shtml
Sponsor: Saint Luke's Health Initiatives
2929 North Central Avenue, Suite 1550
Phoenix, AZ 85012

Salem Foundation Charitable Trust Grants 3456
The Foundation provides funding for a large variety of community based programs that serve the greater Salem area. In partnership with donors, the Foundation supports nonprofit and other community organizations with funds for health and human services, affordable housing, early childhood development, community arts, culture and other important areas of need. Grant making decisions are made by the Distribution Committee, comprised of five Salem area residents.
Requirements: Must be a community-based organization that serves the greater Salem region. Grants may be made only to organizations which have received appropriate determination letters from the Internal Revenue Service that they are exempt under the Internal Revenue Code.
Geographic Focus: Oregon
Date(s) Application is Due: May 1; Dec 1
Samples: Camerata Musica--to support 2007-2008 concerts at the Salem Public Library; Family Building Blocks--for staff members to attend National Conference on Child Abuse and Neglect; Lord & Schryver Conservancy--Garden rehabilitation at Historic Deepwood Estate.
Contact: Trustee; (503) 363-3136; salemfoundation@pioneertrustbank.com
Internet: http://www.pioneertrustbank.com/salemfoundation/index.html
Sponsor: Salem Foundation Charitable Trust
109 Commercial Street NE, P.O. Box 2305
Salem, OR 97308

Salem Foundation Grants 3457
The primary purpose of the Foundation is support of higher education, cultural activities, health services, hospital building funds, and Christian giving. Its fields of interest include: arts; Christian agencies and churches; the disabled; education (secondary and higher); the environment; family services; federated giving programs; health care (particularly heart and circulatory disease research); hospitals; human services; museums; and Protestant federated giving programs. Types of support include: annual campaigns, building/renovation projects, capital campaigns, general operating support, and program development. There is no formal application process and no deadlines. Please contact via letter.
Restrictions: Giving primarily in Florida, Maine and Minnesota. No grants to individuals.
Geographic Focus: Florida, Maine, Minnesota
Amount of Grant: Up to 10,000 USD
Contact: Robert S. Parish, President; (952) 476-6292
Sponsor: Salem Foundation
2181 Springwood Road
Wayzata, MN 55391-2254

Salmon Foundation Grants 3458
Established in New York in 1991, the Salmon Foundation provides funding to nonprofit organizations for programs providing for children and youth programs, family services, and education. Grants are awarded for projects, scholarship support, operating support, and capital funding. Funding ranges from $2,000 to $40,000.
Requirements: Funding provided to nonprofit organizations. Organizations tend to be on the east coast and include Connecticut, New Hampshire, Vermont, Ohio, Pennsylvania, Alabama, Tennessee, Virginia, California, Maryland, Colorado and District of Columbia.
Restrictions: No individual scholarships are made.
Geographic Focus: Alabama, California, Colorado, Connecticut, District of Columbia, Maryland, New Hampshire, Pennsylvania, Tennessee, Virginia
Amount of Grant: 2,000 - 40,000 USD
Samples: Aspen Education Foundation, Aspen, Colorado, $7,500 - elementary math and literacy program support; Baltimore Symphony Orchestra, Baltimore, Maryland, $15,000 - support for the OrchKids program; The Family Place, Norwich, Vermont, $30,000 - for the Families Learning Together Program.
Contact: Emily Grand, Administrator; (212) 708-9316 or (212) 812-4362
Sponsor: Salmon Foundation, Inc.
6 West 48th Street, 10th Floor
New York, NY 10036-1802

Salt River Project Health and Human Services Grants 3459
Salt River Project (SRP) is an energy/utilities company serving electric customers and water shareholders in the Phoenix metropolitan area. SRP provides funding to nonprofit organizations that address critical needs within its service communities. SRP is committed to safe and healthy communities. Health and Human Services Grants support programs that reach out to underserved communities to promote the individual's ability to overcome barriers and be self-sufficient; increase the community's ability to care for individuals who are in need of food, shelter, and safety from violent or crisis situations; increase the ability of children to participate in youth programs which promote personal development and positive life choices; support increasing underserved communities' access to hospitals and medical care as an integral part of a thriving community; and sponsor programs that seek to highlight the never ceasing need for water and electric safety.
Requirements: Eligible applicants must be 501(c)3 nonprofit, organizations within SRP's service area. SRP's service area is central Arizona and includes the following cities and towns: Phoenix, Mesa, Tempe, Paradise Valley, Fountain Hills, Scottsdale, Apache Junction, Peoria, Queen Creek, Avondale, Chandler, Gilbert, Glendale, Guadalupe, and Tolleson. There are no specific grant deadlines. Requests are reviewed in an on-going process which typically takes eight weeks.
Restrictions: The following are ineligible: individuals, including support for specific students, researchers, travel expenses, conference fees; organizations that discriminate on the basis of race, creed, color, sex, or national origin; endowment programs; medical

research projects or medical procedures for individuals; professional schools of art, academic art programs, individual high school or college performing groups; political or lobbying groups or campaigns; fraternal organizations, veterans' organizations, professional associations, and similar membership groups; public or commercial broadcasting programs; religious activities or church-sponsored programs limited to church membership; and debt-reduction campaigns. SRP does not donate services, including water or electricity, or equipment for which a fee is normally charged.
Geographic Focus: Arizona
Contact: Contributions Administrator; (602) 236-5900; webmstr@srpnet.com
Internet: http://www.srpnet.com/community/contributions/guidelines.aspx
Sponsor: Salt River Project
1521 North Project Drive
Tempe, AZ 85281-1298

SAMHSA Campus Suicide Prevention Grants 3460
The purpose of this program is to facilitate a comprehensive approach to preventing suicide in institutions of higher education. This program is designed to assist colleges and universities in their efforts to prevent suicide attempts and completions and to enhance services for students with mental and behavioral health problems, such as depression and substance abuse, which put them at risk for suicide and suicide attempts. Applications are due by November 25.
Requirements: Eligibility for SAMHSA's Campus Suicide Prevention Grant program is limited to institutions of higher education. Applicants from both public and private institutions may apply, including State universities, private four-year colleges and universities (including those with religious affiliations), Minority Serving Institutions of higher learning, and community colleges.
Geographic Focus: All States
Date(s) Application is Due: Nov 25
Amount of Grant: 100,000 USD
Contact: Scott J. Salvatore; (240) 276-1866; Scott.Salvatore@samhsa.hhs.gov
Internet: http://www.samhsa.gov/Grants/2009/sm_09_001.aspx
Sponsor: Substance Abuse and Mental Health Services Administration
1 Choke Cherry Road, Room 7-1085
Rockville, MD 20857

SAMHSA Child Mental Health Initiative (CMHI) Grants 3461
The purpose of this program is to support States, political subdivisions within States, the District of Columbia, Territories, Native American Tribes and tribal organizations, in developing integrated home and community-based services and supports for children and youth with serious emotional disturbances and their families by encouraging the development and expansion of effective and enduring systems of care. Applications are due by January 15.
Requirements: Eligibility for this program is statutorily limited to public entities such as: state governments; Indian or tribal organizations (as defined in Section 4[b] and Section 4[c] of the Indian Self-Determination and Education Assistance Act); governmental units within political subdivisions of a State, such as a county, city or town; District of Columbia government; and Commonwealth of Puerto Rico, Northern Mariana Islands, Virgin Islands, American Samoa and Trust Territory of the Pacific Islands (now Palau, Micronesia and the Marshall Islands).
Restrictions: Proposed budgets cannot exceed the allowable amount in total costs (direct and indirect) in any year of the proposed project. Annual continuation awards will depend on the availability of funds, grantee progress in meeting project goals and objectives, timely submission of required data and reports and compliance with all terms and conditions of award.
Geographic Focus: All States
Date(s) Application is Due: Jan 15
Amount of Grant: Up to 1,000,000 USD
Contact: Gwendolyn Simpson; (240) 276-1408; gwendolyn.simpson@samhsa.hhs.gov
Internet: http://www.samhsa.gov/Grants/2009/sm_09_002.aspx
Sponsor: Substance Abuse and Mental Health Services Administration
1 Choke Cherry Road, Room 7-1085
Rockville, MD 20857

SAMHSA Conference Grants 3462
The purpose of the Conference Grant program is to disseminate knowledge about practices within the mental health services and substance abuse prevention and treatment fields and to integrate that knowledge into real-world practice as effectively and efficiently as possible. The maximum Conference Grant award is $50,000 for a 12-month project period. Only direct costs will be funded under this program. Applications are due on the recurring dates of March 31 and September 30 each year.
Requirements: Applications may be submitted by public and domestic private nonprofit and for-profit entities.
Geographic Focus: All States
Date(s) Application is Due: Mar 31; Sep 30
Amount of Grant: Up to 50,000 USD
Contact: Gwendolyn Simpson; (240) 276-1408; gwendolyn.simpson@samhsa.hhs.gov
Internet: http://www.samhsa.gov/Grants/2008/OA_08_002cmhs.aspx
Sponsor: Substance Abuse and Mental Health Services Administration
1 Choke Cherry Road, Room 7-1085
Rockville, MD 20857

SAMHSA Drug Free Communities Support Program Grants 3463
This Program is a collaborative initiative sponsored by OND.C.P in partnership with SAMHSA in order to achieve two major goals: establish and strengthen collaboration among communities, private nonprofit agencies, and Federal, State, local, and tribal governments to support the efforts of community coalitions to prevent and reduce substance abuse among youth; and reduce substance abuse among youth and, over time, among adults by addressing the factors in a community that increase the risk of substance abuse and promoting the factors that minimize the risk of substance abuse. Substances include, but are not limited to, narcotics, depressants, stimulants, hallucinogens, cannabis, inhalants, alcohol, and tobacco, where their use is prohibited by Federal, State, or local law. Approximately $17 million for 130 FY 2009 DFC grants will be awarded through this RFA. DFC grants will be available to eligible coalitions in amounts of up to $125,000 per year over a five-year period. Applications are due by March 20.
Requirements: Grant funds are intended to support eligible community-based coalitions.
Geographic Focus: All States
Date(s) Application is Due: Mar 30
Amount of Grant: Up to 125,000 USD
Contact: Dan Fletcher, (240) 276-1270; dfcnew2009@samhsa.hhs.gov
Internet: http://www.samhsa.gov/grants/2009/sp_09_002.aspx
Sponsor: Substance Abuse and Mental Health Services Administration
1 Choke Cherry Road, Room 7-1085
Rockville, MD 20857

SAMHSA Strategic Prevention Framework State Incentive Grants 3464
The purpose of the SPF SIG program is to provide funding to States, Federally recognized Tribes and U.S. Territories in order to: prevent the onset and reduce the progression of substance abuse, including childhood and underage drinking; reduce substance abuse-related problems; and build prevention capacity and infrastructure at the State, tribal, territorial and community-levels. Applications are due by November 7.
Requirements: Eligible applicants are the immediate Office of the Chief Executive (e.g., Governor) in the States, U.S. Territories or District of Columbia; and Federally recognized Tribes. The Chief Executive of the State, Territory, or District of Columbia; or of the Federally recognized Tribe must sign the application.
Geographic Focus: All States
Date(s) Application is Due: Nov 7
Amount of Grant: Up to 23,000,000 USD
Contact: Mike Lowther; (240) 276-2581; mike.lowther@samhsa.hhs.gov
Internet: http://www.samhsa.gov/Grants/2009/sp_09_001.aspx
Sponsor: Substance Abuse and Mental Health Services Administration
1 Choke Cherry Road, Room 7-1085
Rockville, MD 20857

Samueli Foundation Health Grants 3465
The Samueli Foundation considers grants to agencies, primarily in Orange County, California, that serve the community, and whose programs meet the guidelines listed. Grants are usually approved for a defined period of time, but may be paid over a multi-year period. In the area of health, its goals are to: support programs promoting alternative and integrative medicine; and utilize creative and innovative approaches to education, prevention, treatment and research of disease and illness.
Requirements: 501(c)3 tax-exempt organizations throughout the U.S. are eligible .
Restrictions: The Foundation does not fund umbrella fund raising organizations, political campaigns, or grants to individuals.
Geographic Focus: All States
Amount of Grant: 1,000 - 50,000 USD
Contact: Gerald R. Solomon; (949) 760-4400; fax (949) 759-5707; Info@samueli.org
Internet: http://www.samueli.org/health.aspx
Sponsor: Samueli Foundation
2101 East Coast Highway, 3rd Floor
Corona del Mar, CA 92625

Samueli Institute Scientific Research Grants 3466
The Institute operates in conjunction with the Uniformed Services University of Health Sciences and Universitatsklinikum Freiburg Institute. Its mission is is to transform health care through the scientific exploration of healing. The Institute is one of an elite group of organizations in the nation with a track record in complementary and integrative medicine, healing relationships and military medical research. The major purpose of funding is to investigate healing processes and their application in promoting health and wellness, preventing illness, and treating disease. Researchers should respond to Requests for Applications or Program Announcements.
Geographic Focus: All States
Amount of Grant: 2,000 - 600,000 USD
Contact: Matt Fritts, (703) 299-4800; fax (703) 299-4800; mfritts@SamueliInstitute.org
Internet: http://www.siib.org/funding/investigator-home/218-SIIB.html
Sponsor: Samueli Institute
1737 King Street, Suite 600
Alexandria, VA 22314

Samuel N. and Mary Castle Foundation Grants 3467
The Samuel N. and Mary Castle Foundation is committed to providing resources to improve the life of Hawaii's children and families by improving the quality and quantity of early education. Efforts are concentrated on creating greater social equality and opportunity through improving access to high quality pre-K education. Secondarily, the Foundation provides limited support for independent K-12 education, the arts, health, and historical and cultural projects. Grants generally range from $5,000 to $25,000. Grants for major capital improvements typically range from $10,000 to $100,000. The deadlines are January 15 for the April meeting, June 1 for the August/September meeting, and September 15 for the December meeting.

Requirements: Eligible organizations must be tax exempt, publicly supported and charitable as determined by the Internal Revenue Service. Grants are primarily awarded to organizations located within the state of Hawaii, for programs and projects benefiting the people of Hawaii. Proposed programs or projects must be in response to a documented community need, and not solely an organizational need. Grants may be awarded for innovative programs, demonstration projects and "start-up" funding. Program and project support does not generally exceed three years, and funding must be applied for on a yearly basis. Applicants must contact the Foundation's Executive Director by letter, email, phone or a personal visit before making fund application. A site visit may be required if the organization has not applied before or in many years.
Restrictions: The following are ineligible: charter schools; endowments; regular operating costs such as salaries, rents, or maintenance; more than forty to fifty percent of total project costs; programs not open to all racial and ethnic groups; projects in which parents and the community have not been appropriately involved in planning and funding; publication projects; general student scholarships for tuition, travel, or conferences; video projects; and annual fund drives or sponsorships.
Geographic Focus: Hawaii
Date(s) Application is Due: Jan 15; Jun 1; Sep 15
Samples: Chaminade University of Honolulu, Honolulu, Hawaii, $85,500 - support of the Castle Colleagues Pre-School Directors Management Training program (2015); Christian Liberty Academy, Keaau, Hawaii, $25,000 - support of pre-school expansion (2015); Maui Arts and Cultural Center, Kahului, Hawaii, $10,000 - support of an arts program for 4-7 year-old children (2015).
Contact: Alfred L. Castle, Executive Director; (808) 522-1101; fax (808) 522-1103; acastle@aloha.net or snandmarycastle@hawaii.rr.com
Internet: http://fdnweb.org/castle/grantmaking/
Sponsor: Samuel N. and Mary Castle Foundation
733 Bishop Street, Suite 1275
Honolulu, HI 96813

Samuel S. Johnson Foundation Grants 3468
The Samuel S. Johnson Foundation was incorporated in 1948 and supports organizations primarily in the Oregon and Clark County, Washington, region. The Foundation gives to: formal education programs leading to an R.N. status or baccalaureate or higher college/university degree in nursing; vocational education programs targeting high school drop-outs and high school grads who are not able to pursue junior college or higher formal education and which offer them job-specific technical training, mentoring or coaching; emergency food assistance programs; rural mobile health screening/care projects benefiting the uninsured medically needy; environmental programs, coastal & marine ecosystems, sustainable agriculture and communities. Most recent awards have range from $1,000 to $30,000. Though there are no specified deadlines, the board meets in July and November to make grant making decisions.
Requirements: Grants are awarded to non-profit organizations in Oregon and Clark County, Washington. Contact the Foundation for current focus and guidelines with a phone call before submitting a proposal. No Application form is required, however you must include the following in your proposal: copy of IRS Determination Letter; brief history of organization and description of its mission; copy of most recent annual report/audited financial statement/990; listing of board of directors, trustees, officers and other key people and their affiliations; detailed description of project and amount of funding requested; contact person; copy of current year's organizational budget and/or project budget; listing of additional sources and amount of support. Include one copy of the proposal. The board meets twice a year, in May and October, with no deadline date for the submitting of proposals. If your proposal is accepted, you will receive notification within 2 - 3 weeks after the board meets.
Restrictions: No support for foreign organizations. No grants or scholarships to individuals, or for leadership training or staff development, campaigns to retire debt, annual campaigns, deficit financing, construction, sole underwriting of major proposals or projects, demolition or endowments.
Geographic Focus: Oregon, Washington
Amount of Grant: 500 - 26,000 USD
Contact: Mary A. Krenowicz; (541) 548-8104; fax (541) 548-2014; mary@ssjf.org
Sponsor: Samuel S. Johnson Foundation
P.O. Box 356
Redmond, OR 97756-0079

San Antonio Area Foundation Grants 3469
Grants are usually awarded in the areas of health care and biomedical research, community and social services, arts and culture, education (early childhood education, higher education, medical schools, nursing schools, adult basic education and literacy), and animal services. Types of support include operating budgets, continuing support, annual campaigns, seed grants, emergency funds, equipment acquisition, matching funds, scholarship funds, research, lectureships, professorships, and building renovations. Applicants must submit a letter of intent to apply; if the letter is approved, the foundation will send a request for a proposal to the applicant. The foundation only reviews full proposals from applicants whose letters of intent have been approved. If asked to submit a proposal, the applicant will be sent an application package with a letter of notification in early February.
Requirements: Grants are made to organizations in the San Antonio, Texas, area.
Geographic Focus: Texas
Date(s) Application is Due: Nov 15
Amount of Grant: 250 - 250,000 USD
Samples: San Antonio Symphony (TX)--for the Young People's Concert series, $25,000; Southwest Foundation for Biomedical Research (San Antonio, TX)--to purchase equipment, $15,000.
Contact: Lydia R. Saldana, Program Officer; (210) 228-3753 or (210) 225-2243; fax (210) 225-1980; lsaldana@saafdn.org or info@saafdn.org
Internet: http://www.saafdn.org/NetCommunity/Page.aspx?pid=254
Sponsor: San Antonio Area Foundation
110 Broadway Street, Suite 230
San Antonio, TX 78205-1948

San Diego Foundation for Change Grants 3470
The foundation seeks funding proposals from grassroots community organizations who work to achieve social, economic, and environmental justice in San Diego County. Foundation concerns cover the range of progressive social issues, including: ending discrimination in any form; halting environmental pollution; promoting peace; expanding immigrant rights; improving San Diego's neighborhoods and workplaces; working toward racial equality; fighting for women's rights; developing youth projects; addressing domestic violence; and preventing violence in the community. Priority goes to projects with limited access to other sources of funding. In addition to the general funds, the foundation has the following special grants: the Danzig Award; and the James Mitsuo Cua Award. Organizations submitting a general application will be considered for these special grants with no additional application necessary. Letters of Intent are due on February 5, while full proposals are due by April 2.
Requirements: California nonprofit organizations in San Diego are eligible.
Restrictions: Grants do not support social or human service organizations that do not have a strong community organizing component; individual efforts; national or statewide projects unless there is a strong local focus; direct union organizing; private businesses or profit-making organizations; or electoral campaigns or candidates.
Geographic Focus: All States
Date(s) Application is Due: Feb 5; Apr 2
Contact: John Fanestil, Executive Director; (619) 692-0527; fax (619) 255-3640; john@foundation4change.org or info@foundation4change.org
Internet: http://www.foundation4change.org/forapplicants.php
Sponsor: San Diego Foundation for Change
3758 30th Street
San Diego, CA 92104

San Diego Foundation Health & Human Services Grants 3471
The San Diego Foundation's Health and Human Services Working Group was established with the vision to improve the health, well-being and self-sufficiency of individuals, families and communities in the San Diego region. The Health and Human Services Working Group's research, investigation and grant awards to date indicate that families who are vulnerable to episodic homelessness benefit from financial education that leads to budget management, traditional banking relationships, reduction of debt improved credit, and increased savings. This initiative will support financial education and asset building programs of community-based organizations that serve the residents of the San Diego region.
Requirements: To be eligible, organizations must be providing services in San Diego County. The organizations must have a 501(c)3 IRS tax exempt status. An organization may serve as a fiscal sponsor for a charitable organization that does not have a 501(c)3 status if a cooperative relationship between the two can clearly be demonstrated. The fiscal sponsor must be willing to administer the grant if awarded.
Restrictions: Generally, the Foundation does not make grants for: organizations that have previously received funding from the Foundation but have not submitted required final reports; major building/capital campaigns; scholarships; endowments; for-profit organizations; projects that promote religious doctrine; individuals; organizations outside San Diego County; marketing and/or promotional materials (annual reports, brochures, video productions); re-granting dollars to other nonprofit organizations or individuals; short-term, annual or one time events, including festivals, performances and conferences; travel outside of the San Diego region, or; existing obligations or debt.
Geographic Focus: California
Date(s) Application is Due: Jan 16
Amount of Grant: 30,000 USD
Contact: Shelley Lyford; (619) 235-2300; Shelley@sdfoundation.org
Internet: http://www.sdfoundation.org/communityimpact/cycle2006.html#ac
Sponsor: San Diego Foundation
2508 Historic Decatur Road, Suite 200
San Diego, CA 92106

San Diego Foundation Paradise Valley Hospital Community Fund Grants 3472
The Foundation has created a joint process managed by the San Diego Foundation and Alliance Healthcare Foundation for the distribution of funds to tax-exempt community-based health and social service organizations. Grant making decisions and oversight of funded projects will be provided by an advisory committee comprised of diverse individuals representing stakeholders in the PVH service area, as well as members of The San Diego Foundation's Health & Human Services Working Group.
Requirements: To be eligible, organizations must be providing services in San Diego County. The organizations must have a 501(c)3 IRS tax exempt status. An organization may serve as a fiscal sponsor for a charitable organization that does not have a 501(c)3 status if a cooperative relationship between the two can clearly be demonstrated. The fiscal sponsor must be willing to administer the grant if awarded.
Geographic Focus: California
Contact: Anahid Brakke; fax (619) 235-2300; anahid@sdfoundation.org
Internet: http://www.sdfoundation.org/communityimpact/healthservices/pvhfund.html
Sponsor: San Diego Foundation
2508 Historic Decatur Road, Suite 200
San Diego, CA 92106

San Diego HIV Funding Collaborative (SDHFC) Grants 3473

The mission of the Collaborative is to raise, leverage and collaboratively allocate private funds to reduce the impact of HIV in the San Diego region. Donations are pooled and re-granted to local service organizations to help people affected by HIV/AIDS. Funds from the local community are used to make grants to local organizations and assist grantees with technical support. All grant applicants must start by submitting a Letter of Intent.
Requirements: The Foundation does not fund projects and programs outside San Diego and Imperial Counties.
Restrictions: The Foundation does not fund: research; lobbying; underwriting of medical expenses; general operating expenses we deem to be excessive; construction or renovation; the purchase of costly equipment; development activities, such as fund raising events, capital campaigns or annual fund drives; projects or proposals from individuals; or organizations that do not have 501(c)3 status.
Geographic Focus: California
Contact: Hamse Warfa, Program Officer; (858) 614-4892 or (858) 874-3788; fax (858) 874-3656; hwarfa@alliancehf.org
Internet: http://www.alliancehf.org/sdhiv/who_we_are.html
Sponsor: Alliance Healthcare Foundation
9325 Sky Park Court, Suite 350
San Diego, CA 92123

San Diego Women's Foundation Grants 3474

The San Diego Women's Foundation (SDWF) strengthens and improves women's capacities to engage in significant philanthropy in the San Diego region. The foundation chooses one focus area of grant making each year. Its primary focus areas include: education, arts and culture, environment, employment and economic development, civil society, and health and human services. Check the web site each September for that year's focus guidelines.
Requirements: Only 501(c)3 organization within San Diego County should apply.
Geographic Focus: California
Date(s) Application is Due: Dec 6
Amount of Grant: 25,000 USD
Contact: Tracy Johnson; (619) 814-1374 or (619) 235-2300; tracy@sdfoundation.org
Internet: http://www.sdwomensfoundation.org/
Sponsor: San Diego Women's Foundation
2508 Historic Decatur Road, Suite 200
San Diego, CA 92106

Sandler Program for Asthma Research Grants 3475

The mission of the foundation is to develop important new pathways of investigation in basic research regarding asthma. The program, which is open to investigators from all fields, particularly encourages applications from investigators not currently studying asthma. Innovation and risk are strongly encouraged. The program especially (but not exclusively) seeks applications in the following areas: development of new animal models for asthma; studies in humans or of human materials (excluding clinical trials); genetics of asthma; and technology development relevant to asthma. Three-year senior investigator awards and three-year junior investigator awards will be made. Awards to junior investigators are intended to support investigators who have demonstrated exceptional early accomplishment as independent investigators, permitting the expansion of their work into the field of asthma; they are not start-up funds. The proposed work should be directed toward uncovering basic mechanisms in the pathogenesis of asthma.
Requirements: Senior investigators will have well-established research programs and an international reputation for their research. They will usually hold a full-time academic appointment as professor, tenured associate professor, or the equivalent. Junior investigators will have already established an independent research program. They will usually hold a full-time academic appointment as an assistant professor, early associate professor, or the equivalent. Investigators may apply from any U.S. or Canadian nonprofit organization that can and will provide the necessary facilities and infrastructure for the research, and that will assure compliance with guidelines for animal and human studies established by the U.S. Department of Health and Human Services. Applicants from Canadian institutions must also provide evidence that the institution has been designated as a public charity by the U.S. Internal Revenue Service.
Geographic Focus: All States
Date(s) Application is Due: Feb 8
Amount of Grant: 150,000 - 250,000 USD
Contact: William E. Seaman; (415) 514-0730; seamanb@sandlerresearch.org
Internet: http://www.sandlerresearch.org
Sponsor: Sandler Foundation
Box 0509, UCSF
San Francisco, CA 94143-0509

Sands Foundation Grants 3476

The Venetian Foundation (now the Sands Foundation) was formed December 7, 2000, by the Venetian Casino Resort. Today, the Las Vegas Sands Corporation's primary philanthropic initiative is pursued through Sands Foundation, a non-profit 501(c)3 organization. Sands Foundation pursues a mission of supporting charitable organizations and endeavors that concentrate on assisting youth, promoting health, and expanding educational opportunities within the local communities. The Foundation also supports causes that empower minority communities and improve underprivileged areas, as well as other valuable charitable and philanthropic activities permitted under relevant tax-exempt laws. Sands Foundation pursues a mission of supporting charitable organizations and endeavors that concentrate on assisting youth, promoting health, and expanding educational opportunities within our local communities. The Foundation also supports causes that empower minority communities and improve underprivileged areas, as well as other valuable charitable and philanthropic activities permitted under relevant tax-exempt laws. Charitable requests along with supporting documents may either be faxed or mailed.
Requirements: All charitable requests must be submitted in writing. Written requests should include the following: agency/organization information (brochures, information packet, list of the board of directors, history, background, or other helpful information); 501(c)3 tax identification number; contact person; mailing address and telephone number; overview of project or event at hand; date, time, location for event requests; purpose of request; very specific information about the amount/item(s) requested; and target population which will benefit from support.
Geographic Focus: Nevada, Pennsylvania, Macau, Singapore
Amount of Grant: 5,000 - 100,000 USD
Contact: Community Development; (702) 607-1677; foundation@venetian.com
Internet: http://www.lasvegassands.com/LasVegasSands/Sands_Foundation/Donation_Request.aspx
Sponsor: Sands Foundation
3355 Las Vegas Boulevard South
Las Vegas, NV 89109-8941

Sandy Hill Foundation Grants 3477

The Sandy Hill Foundation offers awards to eligible nonprofit organizations in the areas of education, health care, and social services. The foundation gives primarily to the arts and culture, higher education, hospitals, health associations, social services, and federated giving programs. Their areas of interest include the arts; child and youth services; community and economic development; health organizations and associations; higher education; hospitals; human services; Protestant agencies and churches; recreation camps; and United Way and Federated Giving Programs. It also offers college scholarships for designated local area schools.
Requirements: There is no application or specific deadline for nonprofit giving. See contact information for current scholarship application due April 1.
Restrictions: The foundation gives primarily to the greater Hudson Falls, NY area. No grants are given to individuals.
Geographic Focus: New York
Date(s) Application is Due: Apr 1
Amount of Grant: Up to USD
Contact: Nancy Juckett Brown, Trustee; (518) 791-3490
Sponsor: Sandy Hill Foundation
P.O. Box 30
Hudson Falls, NY 12839-0030

San Francisco Foundation Community Health Grants 3478

The San Francisco Foundation welcomes unsolicited applications twice a year. The Community Health Program works to expand access to health services and promote health prevention in the five counties we serve. Community Health's efforts focus on improving health on a community level and providing the social supports for communities to improve health. The Bay Area counties included are: Martin, Contra Costa, San Francisco, Alameda and San Mateo.
Requirements: Grants are made to nonprofit organizations in Alameda, Contra Costa, Marin, San Francisco, and San Mateo Counties of California.
Restrictions: The foundation generally does not fund projects outside the Bay Area community; long-term operating support; medical, academic, or scientific research; religious activities (religious institutions may apply for nonsectarian programs); direct assistance to individuals; or conferences or one-time events.
Geographic Focus: California
Date(s) Application is Due: Jan 16; Jul 1
Amount of Grant: 15,000 - 80,000 USD
Samples: California Pan-Ethnic Health Network, Oakland, CA--to provide core operating support for public policy work and building relationships with primary constituents, $50,000;Latina Breast Cancer Agency, San Francisco, CA--to perform a needs and community assessment around the California Every Women Counts Cancer Detection Program, $25,000.
Contact: Denise Martin; (415) 733-8500 or (415) 733-8569; dkm@sff.org
Internet: http://www.sff.org/programs/community-health
Sponsor: San Francisco Foundation
225 Bush Street, Suite 500
San Francisco, CA 94104-4224

San Francisco Foundation Disability Rights Advocate Fund Emergency Grants 3479

The program awards emergency grants to positively impact large populations of people with disabilities in the Bay Area and northern California. Grants are available to organizations to assist mobilizing people with disabilities to act upon time-sensitive and urgent matters affecting the disability community. For example, funds may be used to cover the transportation and/or personal assistant costs associated with a community mobilization effort. Requests are accepted throughout the year; decisions are made, based on availability of funds, within four to six weeks of submission. Requests should be submitted via electronic mail. Guidelines are available online.
Requirements: Organizations located in the Bay Area and northern California are eligible.
Geographic Focus: California
Amount of Grant: 10,000 USD
Contact: Denise Martin; (415) 733-8500 or (415) 733-8569; dkm@sff.org
Internet: http://www.sff.org
Sponsor: San Francisco Foundation
225 Bush Street, Suite 500
San Francisco, CA 94104-4224

San Juan Island Community Foundation Grants 3480

The foundation's mission is to enhance the quality of life on San Juan Island and make a positive difference in the community. The foundation awards grants to tax-exempt organizations in its areas of interest, including arts and culture, health and wellness, local economy, community infrastructure, environment, education, and basic social needs.
Requirements: The Foundation only gives grants to tax-exempt organizations which include 501(c)3's, local non-profit branches of 501(c)1's, government agencies, non-profit schools and religious organizations (but only for non-religious purposes). Faith-based organizations are eligible but only for non-religious and unrestricted public service projects. Foundation grants, by policy, are targeted to the local community and would usually only be given to an outside nonprofit for work that affected the local community. A meeting with Foundation representatives is required before a grant application is submitted. Call the office to request a "pre-grant application 1-on-1 meeting". Partner funding will be a critical component of each required, pre-submission meeting including careful identification of the resources that will be allocated by the applying organization itself. The required grant application can be downloaded from the website or requested by phone. Applications may be submitted at anytime following the required online registration and the pre-submission 1:1 meeting with Foundation representatives.
Restrictions: Grants do not support religious organizations where the funds would be used to further the organization's religious purposes; individuals; other endowments; political purposes; or any purpose that discriminates as to race, creed, ethnic group, or gender.
Geographic Focus: Washington
Date(s) Application is Due: Apr 28; Jul 22; Oct 27; Dec 30
Amount of Grant: 500 USD
Contact: Jeanne Peihl; (360) 378-1001; info@sjicf.org
Internet: http://sjicf.org/for-nonprofits/
Sponsor: San Juan Island Community Foundation
P.O. Box 1352
Friday Harbor, WA 98250

Santa Barbara Foundation Strategy Grants - Core Support 3481

As the community foundation for all of Santa Barbara county, the foundation funds a wide range of initiatives and projects that address community needs, strengthen the nonprofit sector, develop community leadership, and encourage collaboration. Strategy Grants are distributed annually and aim to help nonprofit organizations fulfill their important missions so that community needs and affect positive change can be addressed. Core Support Grants are available to nonprofit organizations looking to sustain their organizational infrastructure. Core support is defined as unrestricted funding that enables an organization to carry out its mission. A Core Support Grant can be used to underwrite administrative infrastructure and/or to maintain core programs and essential staff. The foundation's philanthropic purpose for offering core support at this time is to maintain organizations that provide safety net services for the poor and underserved members of our community, and to support increased demand for services whether due to sustained need and/or significant reductions in funding that impact the delivery of core programs and services.
Requirements: The foundation provides grants to nonprofit organizations serving Santa Barbara County. Organizations must have 501(c)3 tax-exempt status or operate under a fiscal agent. Organizations may submit one Strategy Grant application (Core Support, Capital, or Innovation), as well as be part of one or more Collaborative Strategy Grant applications. The organization must be addressing issues of hunger or shelter, or providing primary or behavioral health care for vulnerable populations. The organization must demonstrate increasing demand for services and/or significant reductions in funding over the past three years. Priority for Core Support Grants will be given to organizations that are providing direct services to address hunger, shelter, and primary and/or behavioral health care; are established, well-managed, financially viable, and operate effective programs that primarily serve the needs of poor and underserved communities; and, have developed short- and long-term strategies for addressing identified organizational needs. Applications must be received by the foundation by 5:00 pm of the deadline date.
Restrictions: The foundation does not make grants for the following purposes or activities: debt; endowment; fundraising events; individuals; religious organizations for religious purposes; government entities (including schools) for basic services or capital needs; projects that discriminate on the basis of ethnicity, race, color, creed, religion, gender, national origin, age, disability, marital status, sexual orientation, gender identity, gender expression, or any veteran status; or, activities that occurred prior to the beginning date of the grant.
Geographic Focus: California
Amount of Grant: Up to 50,000 USD
Contact: Jack Azar; (805) 963-1873; fax (805) 966-2345; jazar@sbfoundation.org
Internet: http://www.sbfoundation.org/page.aspx?pid=778
Sponsor: Santa Barbara Foundation
1111 Chapala Street, Suite 200
Santa Barbara, CA 93101

Santa Fe Community Foundation Seasonal Grants-Fall Cycle 3482

Through its outreach to nonprofits, donors and community leaders, the Foundation organizes its annual grants cycle into a two-season grants program. Each season (Spring and Fall) focuses on its own specific goals and strategies. The Foundation is devoted to building healthy and vital communities in the region where: racial, cultural or economic differences do not limit access to health, education or employment; diverse audiences enjoy the many arts and cultural heritages of our region; and, all sectors of its community take responsibility for ensuring a healthy environment. The areas of interest for the Fall Cycle are Arts, Animal Welfare, and Health and Human Services. For Arts proposals, projects should: increase public engagement in the arts; and, support public policy, community organizing or public information to strengthen the arts segment of the creative economy locally. For Health and Human Services proposals, projects should: improve the health of underserved residents of the Santa Fe region; improve access to affordable healthy food; strengthen the delivery of homelessness services; improve safety for children, women, families, sexual minorities and the elderly; and/or, support public policy, civic engagement, community organizing or public information to improve the health and well-being of local residents. For Animal Welfare proposals, the Foundation has approximately $25,000 available for animal welfare-related grants, and will include summaries of all animal welfare proposals (that meet basic due diligence) in the 'Giving Together' catalogue that accompanies the Fall Community Grant Cycle. The Giving Together catalogue is then shared with the Foundation's fundholders who are invited to make grants toward any proposal in the catalogue.
Requirements: Applications will be accepted from organizations that: are located in or serve the people of Santa Fe, Rio Arriba, Taos, Los Alamos, San Miguel or Mora Counties; are tax-exempt under Section 501(c)3 of the Internal Revenue Code or are a public or governmental agency or a federally recognized tribe in the state of New Mexico, or that have a fiscal sponsor; employ staff and provide services without discrimination on the basis of race, religion, sex, age, national origin, disability, or sexual orientation; and, are at least three years old. Each nonprofit entity may only apply for funding once per year. All grants will be $5,000, $10,000 or $15,000, depending on your annual budget. For organizations whose annual budget is under $150,000, you may apply for a $5,000 grant; for organizations whose annual budget is between $150,000 and $500,000, you may apply for a $10,000 grant; for organizations with an annual budget over $500,000, you may apply for a $15,000 grant. Grant applications will be accepted online only. Applications must be received by 5:00 pm of the deadline date.
Restrictions: The foundation does not award grants for religious purposes, capital campaigns or endowments, scholarships, or individuals. Organizations that received a community grant from SFCF in the last calendar year are not eligible to apply for a community grant in the current calendar year.
Geographic Focus: New Mexico
Date(s) Application is Due: Aug 26
Amount of Grant: 5,000 - 15,000 USD
Contact: Christa Coggins; (505) 988-9715 x 7002; ccoggins@santafecf.org
Internet: http://www.santafecf.org/nonprofits/grantseekers/general-grant-information
Sponsor: Santa Fe Community Foundation
501 Halona Street
Santa Fe, NM 87505

Sara Elizabeth O'Brien Trust Grants 3483

The Sara Elizabeth O'Brien Trust was established in 1981 to support charitable organizations that provide treatment or care to those who are blind or have cancer. The O'Brien Trust also supports medical research in the areas of blindness or cancer. The majority of grants from the O'Brien Trust are one year in duration; on occasion, multi-year support is awarded. Applicants must apply online at the grant website. Applicants are strongly encouraged to do the following before applying: review the downloadable state application procedures for additional helpful information and clarifications; review the downloadable online-application guidelines at the grant website; review the trust's funding history (link is available from the grant website); review the online application questions in advance; and review the list of required attachments. These will generally include: a list of board members, financial statements (audited, reviewed, or compiled by independent auditor); an organization summary; a list of other funding sources; an IRS Determination letter; and other required documents. All attachments must be uploaded in the online application as PDF, Word, or Excel files. The application deadline for the Sara Elizabeth O'Brien Trust is 11:59 p.m. on October 15. Applicants will be notified of grant decisions before December 31.
Requirements: Applicants must have 501(c)3 tax-exempt status.
Restrictions: The trust does not support requests from individuals, organizations attempting to influence policy through direct lobbying, or any political campaigns.
Geographic Focus: Massachusetts
Date(s) Application is Due: Oct 15
Samples: Foundation of the Massachusetts Eye and Ear Infirmary, Boston, Massachusetts, $60,000, to support research aimed at discovering genes associated with diabetic retinopathy; Dana-Farber Cancer Institute, Boston, Massachusetts, $33,000, fellowships in pediatric palliative care; MAB Community Services, Brookline, Massachusetts, $47,000, Seeing is Not Falling Project.
Contact: Michealle Larkins; (866) 778-6859; michealle.larkins@baml.com
Internet: https://www.bankofamerica.com/philanthropic/fn_search.action
Sponsor: Sara Elizabeth O'Brien Trust
225 Franklin Street, 4th Floor, MA1-225-04-02
Boston, MA 02110

Sarasota Memorial Healthcare Foundation Grants 3484

The Foundation is an organization with the mission to improve the delivery of health care in the Sarasota area through the acquisition and utilization of philanthropic funds. Although the Foundation primarily supports Sarasota Memorial Health Care System, it also awards grants to other area nonprofit organizations to meet the community's changing health care needs. Grants totaling more than $32,000,000 have been provided since the Foundation's establishment in 1976. Current fields of interest include: health care; medical care; and community health systems.
Requirements: Health care programs and organization in the Sarasota, Florida, region are eligible to apply.
Geographic Focus: Florida
Contact: Alexandra Quarles, President; (941) 917-1286; alex-quarles@smh.com
Internet: http://www.smhf.org
Sponsor: Sarasota Memorial Healthcare Foundation
1515 S Osprey Avenue, Suite B4
Sarasota, FL 34239

Sarkeys Foundation Grants   3485

Governed by a dedicated Board of Trustees, the foundation that bears SJ Sarkeys' name is deeply committed to furthering his vision to improve the quality of life in Oklahoma. The Foundation provides grants to a diverse group of nonprofit organizations and institutions, almost all of which are located in Oklahoma. Major areas of foundation support include education, arts and cultural endeavors, scientific research, animal welfare, social service and human service needs, and cultural and humanitarian programs of regional significance. Grant proposals are considered at the April and October meetings of the board of trustees.
Requirements: Most organizations classified by the IRS as being a 501(c)3 that is not a private foundation and is headquartered and offering services in Oklahoma may apply. Preference is given to organizations that have been in operation at least 3 years. Organizations are required to submit a Letter of Inquiry to determine whether they meet the criteria and priorities for funding. Representatives are encouraged to speak with a program officer for more information and to ask any questions about the process. An organization may submit one request in a twelve month period.
Restrictions: The Sarkeys Foundation will not fund: local programs appropriately financed within the community; direct mail solicitations and annual campaigns; out of state institutions; hospitals; operating expenses; purchase of vehicles; grants to individuals; responsibility for permanent financing of a program; programs whose ultimate intent is to be profit making; start-up funding for new organizations; feasibility studies; grants which trigger expenditure responsibility by Sarkeys Foundation; direct support to government agencies; individual public or private elementary or secondary schools, unless they are serving the needs of a special population which are not being met elsewhere; and religious institutions and their subsidiaries.
Geographic Focus: Oklahoma
Date(s) Application is Due: Feb 3; Aug 1
Amount of Grant: Up to 50,000 USD
Contact: Susan Frantz; (405) 364-3703; susan@sarkeys.org or sarkeys@sarkeys.org
Internet: http://www.sarkeys.org/grant_guidelines.html
Sponsor: Sarkeys Foundation
530 East Main Street
Norman, OK 73071

Sartain Lanier Family Foundation Grants   3486

The Sartain Lanier Family Foundation awards grants to Georgia nonprofits in support of education, health and human services, arts, environment, and community development, with the majority of new grantmaking in the area of education. Types of support include building and renovation; capital campaigns; endowments; general operating support; program development; and program-related investments and loans. The foundation's board meets in May and December of each year to consider grant requests, which will be by invitation only. Interested applicants should provide an organizational overview for consideration purposes. Prior to submitting a full proposal, interested parties should submit a letter limited to two pages summarizing the request. The letter should include a brief description of the organization and its purpose, the project for which funding is requested, the total cost of the project, and the amount being requested.
Requirements: Nonprofit organizations in the southeastern United States are eligible to apply if invited to do so; however, the majority of recipients are located in Georgia and specifically the Atlanta metro area.
Restrictions: The foundation does not make grants for individuals; churches or religious organizations for projects that primarily benefit their own members; partisan political purposes; tickets to charitable events or dinners, or to sponsor special events or fundraisers.
Geographic Focus: Georgia
Amount of Grant: 5,000 - 1,000,000 USD
Contact: Patricia E. Lummus, Executive Director; (404) 564-1259; fax (404) 564-1251; plummus@lanierfamilyfoundation.org
Internet: http://www.lanierfamilyfoundation.org/funding-priorities/
Sponsor: Sartain Lanier Family Foundation
25 Puritan Mill, 950 Lowery Boulevard NW
Atlanta, GA 30318

Sasco Foundation Grants   3487

The Foundation focus of support is on the arts, education, the environment, health, and human services. Although the main geographic areas are limited to Connecticut, Maine, and New York, the Foundation also occasionally supports programs of national importance. There are no deadlines and no formal applications. Please contact the Foundation for further details.
Requirements: Must be a 501(c)3 organization. Grants are not given to individuals.
Geographic Focus: Connecticut, Maine, New York
Contact: Uwe Linder, J.P. Morgan Private Bank Trustee; (800) 576-6209
Sponsor: Sasco Foundation
345 Park Avenue, 6th Floor
New York, NY 10154-1002

Savoy Foundation Post-Doctoral and Clinical Research Fellowships   3488

Fellowships of the Savoy Foundation are available to Canadian researchers and to foreign nationals or for projects conducted in Canada. Post-doctoral and clinical research fellowships of $30,000 will be awarded to scientists or medical specialists (Ph.D. or M.D.) wishing to carry out a full time research project in the field of epilepsy. The grant will be for one (1) year, but will be renewable, exceptionally, once upon request. As a rule, the modalities of attribution and amounts awarded will be similar to those of governmental funding agencies (MRC or FRSQ). The fellow with the highest mark will received the prestigious Steriade-Savoy award of an additional $1,500.
Restrictions: Fellows funded by the Savoy Foundation must not cumulate two major awards.
Geographic Focus: All States, Canada
Date(s) Application is Due: Jan 15
Amount of Grant: 30,000 USD
Contact: Coordinator; (450) 358-9779; fax (450) 346-1045; epilepsy@savoy-foundation.ca
Internet: http://www.savoy-foundation.ca/eng/mission/defaut.htm
Sponsor: Savoy Foundation
230 Foch Street
St Jean Sur Richelieu, QC J3B 6B2 Canada

Savoy Foundation Research Grants   3489

A small number of research grants are available each year to clinicians and/or established scientists working on epilepsy or related subjects. These grants will be attributed for the following purposes: launching of a new research project, preliminary studies in preparation of a more substantial request to be addressed to another agency, continuation or completion of a project, sharing of expenses, or a limited project not requiring funding in excess of the limit. As a rule, these grants are not renewable. Contact the foundation for application forms.
Requirements: Research grants are available to Canadian citizens or for projects conducted in Canada.
Geographic Focus: All States, Canada
Date(s) Application is Due: Jan 15
Amount of Grant: 25,000 USD
Contact: Administrator; (450) 358-9779; fax (450) 346-1045; epilepsy@savoy-foundation.ca
Internet: http://www.savoy-foundation.ca/eng/mission/defaut.htm
Sponsor: Savoy Foundation
230 Foch Street
St Jean Sur Richelieu, QC J3B 6B2 Canada

Savoy Foundation Studentships   3490

Studentships of the Savoy Foundation are available to Canadian researchers and to foreign nationals or for projects conducted in Canada. Studentships will be awarded to meritorious applicants wishing to acquire training and pursue research in a biomedical discipline, the health sciences or social sciences related to epilepsy. These studentships last one year, but will be renewable three times, upon request, for a maximum duration of four years. The stipend will be $15,000 per year. The student with the highest mark will receive the prestigious Van Gelder-Savoy award of an additional $1,500. An annual sum of $1,000 will be allocated to the laboratory or institution as additional support for the research project.
Requirements: To be eligible, the candidate must have a good university record (B. Sc., M.D. or equivalent diploma) and have ensured that a qualified researcher affiliated to a university and/or hospital, will supervise his/her work. Concomitant registration in a graduate program (M. Sc. or Ph. D.) is encouraged.
Restrictions: Students funded by the Foundation must not cumulate two studentship awards.
Geographic Focus: All States, Canada
Date(s) Application is Due: Jan 15
Amount of Grant: 15,000 USD
Contact: Caroline Savoy, (450) 358-9779; fax (450) 346-1045; epileps@savoy-foundation.ca
Internet: http://www.savoy-foundation.ca/eng/mission/defaut.htm
Sponsor: Savoy Foundation
230 Foch Street
St Jean Sur Richelieu, QC J3B 6B2 Canada

Schering-Plough Foundation Community Initiatives Grants   3491

The Schering-Plough Foundation, a non-profit membership corporation established in 1955, is dedicated to working with the citizens of our communities to help them realize their full potential and enhance their quality of life. Support for community development takes many forms and allows the Foundation to reach out to numerous, highly diverse groups within its communities. The Foundation continues to support organizations that promote culture and the arts, environmental issues, legal services, etc. and are always seeking out new and innovative ways to serve the citizens of its communities.
Requirements: The Foundation considers requests from tax-exempt, 501(c)3 non-profit organizations located in the United States, or its possessions, whose goals and activities fall within its stated objectives and areas of interest. All requests for funding must be made online. National organizations are eligible to apply.
Restrictions: Grants are not made to individuals.
Geographic Focus: All States
Amount of Grant: 5,000 - 200,000 USD
Contact: Christine Fahey, Assistant Secretary; (908) 298-7232; fax (908) 298-7349
Internet: http://www.schering-plough.com/company/foundation.aspx
Sponsor: Schering-Plough Foundation
2000 Galloping Hill Road
Kennilworth, NJ 07033-0530

Schering-Plough Foundation Health Grants   3492

The Schering-Plough Foundation, a non-profit membership corporation established in 1955, is dedicated to working with the citizens of our communities to help them realize their full potential and enhance their quality of life. The Foundation's interest in improving the health and well-being of its communities remains a top priority and is reflected in the numerous initiatives that have been developed to benefit these communities. The Foundation invests in the future of those communities by building meaningful partnerships that enhance the quality of life. Currently, the Foundation is involved in several partnerships that seek to expand the quality and availability of health care in its underserved communities.
Requirements: The Foundation considers requests from tax-exempt, 501(c)3 non-profit organizations located in the United States, or its possessions, whose goals and activities

fall within its stated objectives and areas of interest. All requests for funding must be made online. National organizations are eligible to apply.
Restrictions: Grants are not made to individuals.
Geographic Focus: All States
Amount of Grant: 5,000 - 200,000 USD
Contact: Christine Fahey, Assistant Secretary; (908) 298-7232; fax (908) 298-7349
Internet: http://www.schering-plough.com/company/foundation.aspx
Sponsor: Schering-Plough Foundation
2000 Galloping Hill Road
Kennilworth, NJ 07033-0530

Schlessman Family Foundation Grants 3493
The foundation awards grants to Colorado nonprofits in its areas of interest: education, disadvantaged youth programs and services, elderly/senior programs, special needs groups, and established cultural institutions (such as museums, libraries and zoos). Performing arts grants are available but are very limited.
Requirements: Grants are limited to Colorado charities, primarily greater metro-Denver organizations. The Foundation accepts, but does not require, the Colorado Common Grant Application. All requests must be in writing and discourages lengthy proposals with multiple attachments. If additional information is required, you will be contacted. Proposals are accepted throughout the year, however they must be postmarked on or before the deadline date if they are to be considered in time for the once-a-year distributions on March 31.
Restrictions: The following are ineligible: individuals, start-ups, support for benefits or conferences, public/private/charter schools.
Geographic Focus: Colorado
Date(s) Application is Due: Dec 31
Contact: Patricia Middendorf, (303) 831-5683; contact@schlessmanfoundation.org
Internet: http://www.schlessmanfoundation.org
Sponsor: Schlessman Family Foundation
1555 Blake Street, Suite 400
Denver, CO 80202

Schrage Family Foundation Scholarships 3494
The Centier Bank-Schrage Family Foundation Scholarship rewards academic achievement, as well as school and community leadership. The $2,000 award is distributed for four years providing that the student achieves a GPA of 3.0 each semester. Centier Bank communities offering the Schrage Scholarship include Chesterton, Crown Point, Dyer, East Chicago, Gary, Griffith, Hammond, Hanover, Highland, Hobart, LaPorte, Lowell, Merrillville, Munster, Plymouth, Portage, Schererville, St. John, Valparaiso, and Whiting. Students should apply to their corresponding bank location.
Requirements: To be considered for the Schrage Scholarship, applicants must: be in the top 20% of their graduating class; obtain high level SAT or ACT scores; occupy leadership roles in school and/or community activities; and be willing to sign a publicity release form. Any college major is eligible for the scholarship, with preference given to Business majors. Students will be required to submit their grades to Centier Bank at the end of each school year and communicate openly to Centier Bank regarding any changes in their scholastic involvement (i.e. switching schools, internship programs, etc.).
Geographic Focus: Indiana
Amount of Grant: 2,000 USD
Contact: Vicki Bukowski; (219) 755-6120, ext. 1815; vbukowsk@centier.com
Internet: https://www.centier.com/community/schrage.asp
Sponsor: Legacy Foundation
1000 East 80th Place, 302 South
Merrillville, IN 46410

Schramm Foundation Grants 3495
The foundation awards grants to Colorado nonprofit organizations in its areas of interest, including arts and culture, civic affairs, community development, education (elementary, secondary, and higher), health care, housing, humanities, medical research, science, social services delivery, technology, women's issues, and youth. Types of support include building construction/renovation, continuing support, equipment acquisition, general operating support, matching/challenge grants, program development, and scholarship funds. Applications are accepted from July 1 through August 31 (postmarked).
Requirements: Colorado 501(c)3 nonprofit organizations are eligible. Preference is given to requests from the Denver area. Applications must clearly express the reason(s) for the request, attach financial statements and copy of exemption letter.
Restrictions: Grants do not support advertising, advocacy organizations, individuals, international organizations, political organizations, religious organizations, school districts, special events, or veterans organizations.
Geographic Focus: Colorado
Date(s) Application is Due: Aug 31
Contact: Gary Kring, President; (303) 861-8291
Sponsor: Schramm Foundation
800 Grant Street, Suite 330
Denver, CO 80203-2944

Schurz Communications Foundation Grants 3496
The Schurz Communications Foundation, Inc. is a company sponsored foundation which was incorporated in 1940. The Foundations donors are: Schurz Communications, Incorporated; South Bend Tribune Corporation; WSBT, Incorporated. The Foundation supports community service clubs and organizations involved with arts and culture, higher education, youth development and disabled people.
Requirements: Giving is limited to the South Bend Indiana area.
Geographic Focus: Indiana
Amount of Grant: 1,000 - 40,000 USD
Contact: David C. Ray, President; (574) 235-6241
Sponsor: Schurz Communications Foundation
225 West Colfax Avenue
South Bend, IN 46624

Scleroderma Foundation Established Investigator Grants 3497
The Scleroderma Foundation is seeking applications from promising established investigators both inside and outside the field of SSc who wish to propose pilot studies to obtain preliminary data dealing with a highly innovative and/or highly relevant theme related to SSc. Applications along with the instructions are available at the SF website.
Requirements: Applications may be submitted by domestic and foreign nonprofit organizations, public and private, such as universities, colleges, hospitals, and laboratories. Applicants must have a doctorate degree in medicine, osteopathy, veterinary medicine, or one of the sciences; must have completed a postdoctoral fellowship by the grant award date; and have been principal investigator on grants from the foundation or other national, private, or government agencies. Applicants may request up to $75,000. Up to 8% of the direct costs per year of award may be set aside as indirect costs. The indirect cost amount will be subtracted from the yearly total [up to $75,000) of the award) per year for up to two years (total for 2 years not to exceed $150,000).
Restrictions: Funding are not renewable. Before completion of this grant, investigators are encouraged to seek more substantial continuing support for research through other grant mechanisms through private or government agencies. Replacement of the principal investigator on this award is not permitted. There will be no routine escalation for future years. Awards are contingent on the availability of funds and the receipt of sufficiently meritorious applications meeting the stated eligibility requirements.
Geographic Focus: All States, All Countries
Date(s) Application is Due: Sep 15
Amount of Grant: Up to 75,000 USD
Contact: Tracey Sperry, Director of Development and Research; (800) 722-4673 or (978) 463-5843; fax (978) 463-5809; tsperry@scleroderma.org
Internet: http://www.scleroderma.org/site/PageServer?pagename=prof_research_types#.UCz0vXAZzbg
Sponsor: Scleroderma Foundation
300 Rosewood Drive, Suite 105
Danvers, MA 01923

Scleroderma Foundation New Investigator Grants 3498
The Scleroderma Foundation is seeking applications from promising new investigators who hold faculty or equivalent positions and who wish to pursue a career in research related to systemic sclerosis (scleroderma). This grant will support promising research that is likely to lead to individual research project grants. Awards are contingent on the availability of funds and the receipt of sufficiently meritorious applications meeting the state eligibility requirements. The application is available at the SF website.
Requirements: Applications may be submitted by domestic non-profit organizations, public and private such as universities, colleges, hospitals and laboratories. Applicants must have a doctoral degree in Medicine, Osteopathy, Veterinary Medicine or one of the sciences and must have completed a postdoctoral fellowship by the grant award date. Applicants may request up to $50,000. Up to 8% of the direct costs per year of award may be set aside as indirect costs. The indirect costs will be subtracted from the yearly total (up to $50,000 of the award) per year for up to three years (total for three years not to exceed $150,000).
Restrictions: Foreign organizations and institutions are not eligible. These new investigator grants may not be used to support thesis or dissertation research or fellowship training. Applicants who have been a Principal Investigator on grants from the Scleroderma Foundation or other national, private or government agencies other than fellowship grants are not eligible for this award. Awards are not renewable. Before completion of this grant, investigators are encouraged to seek continuing support for research through other grant mechanisms through private or government agencies. Replacement of the Principal Investigator on this award is not permitted. There will be no routine escalation for future years.
Geographic Focus: All States
Date(s) Application is Due: Sep 15
Amount of Grant: Up to 50,000 USD
Contact: Tracey Sperry, Director of Development and Research;)978) 463-5843; fax (978) 463-5809; tsperry@scleroderma.org
Internet: http://www.scleroderma.org/site/PageServer?pagename=prof_research_types#.UCzz-3AZzbg
Sponsor: Scleroderma Foundation
300 Rosewood Drive, Suite 105
Danvers, MA 01923

Scott B. and Annie P. Appleby Charitable Trust Grants 3499
The trust supports programs and projects of nonprofit organizations in the categories of higher education, cultural programs, and child welfare. Types of support include general operating support, continuing support, capital campaigns, building construction and renovation, research, and scholarship funds. An application form is required, although there are no specified deadlines. Most recent awards have ranged from $1,000 to $100,000.
Requirements: The foundation awards grants to nonprofit organizations in the United States. There are no deadlines. Interested applicants are encouraged to submit a letter describing the intent and purpose of the organization with a specific proposal for allocation of funds.
Geographic Focus: All States
Amount of Grant: 1,000 - 100,000 USD

Samples: Asheville Art Museum, Asheville, North Carolina, $100,000 - general operations; Medical Foundation of North Carolina, Chapel Hill, North Carolina, $25,000 - general operations; Regents of the University of California, Oakland, California, $40,000 - general operations.
Contact: Benjamin N. Colby, Co-Trustee; (941) 329-2628; bncolby@uci.edu
Sponsor: Scott B. and Annie P. Appleby Charitable Trust
c/o The Northern Trust Company
Sarasota, FL 34236

Scott County Community Foundation Grants 3500
The purpose of the Scott County Community Foundation is to improve the quality of life of the residents of Scott County, Indiana. Grant funding will focus on the encouragement of programs that enhance cooperating and collaboration among institutions within the Scott County community. Funding focuses on effectiveness of special non-recurring projects which enrich health, education, cultural or recreational situations in Scott County. Specific areas of interest include community service; social service; education; the arts; environment; and entrepreneurship. Special consideration is given to agencies who partner with other organizations to complete a unified project. Funding will seek to offer leverage funds through the use of seed money, match and challenge grants. Applications are typically accepted in the spring and recipients announced during the summer. The application and additional guidelines are available at the Foundation website.
Restrictions: The Foundation does not typically fund: on going operational expenses; existing obligations; services primarily supported by tax dollars or responsibility of a public agency; individuals or travel expenses; multi-year grants or repeat funding; advocacy or political purposes; religious or sectarian purposes; and loans or endowments.
Geographic Focus: Indiana
Contact: Jaime Toppe; (812) 752-2057; info@scottcountyfoundation.org
Internet: http://www.scottcountyfoundation.org/grants.htm
Sponsor: Scott County Community Foundation
60 North Main Street, P.O. Box 25
Scottsburg, IN 47170

Seabury Foundation Grants 3501
In general, The Seabury Foundation makes one-time, special project grants, rather than general operating support. Areas of interest of the Foundation include: Arts, Culture and Humanities; Human Capacity-building; Health; Education, Conservation and Environment; Children, Youth and Families. On occasion, the Foundation may invite grant renewal requests for up to two consecutive succeeding years. After that period, the Foundation will not consider a request for funding from that organization until a minimum of two full years have passed since the final grant was awarded. Please note that all grants awarded are within the City of Chicago with the exception of projects personally sponsored by family members.
Requirements: Chicago, IL, nonprofit organizations are eligible. Initial contact should be through a Letter of Inquiry, letters need only be 1-2 pages in length. Deadline dates for Letter of Inquiry are: September 15; January 15; June 15. If after reviewing the Letter of Inquiry, if your organization is invited to submit a full proposal, the deadline dates are: November 1; March 1; August 1.
Restrictions: The Seabury Foundation do not entertain grants in support of gala events, annual fundraising events, or annual fundraising drives.
Geographic Focus: Illinois
Date(s) Application is Due: Jan 15; Mar 1; Jun 15; Aug 1; Sep 15; Nov 1
Contact: Boyd McDowell III, Director; (312) 587-7146; fax (312) 587-7332; bmcdowell@seaburyfoundation.org
Internet: http://www.seaburyfoundation.org/apply.htm
Sponsor: Seabury Foundation
1111 North Wells Street, Suite 503
Chicago, IL 60610-7633

Seagate Technology Corporation Capacity to Care Grants 3502
Seagate's community engagement program addresses the varied interests of employees and the growing needs of its local communities. Known as Capacity to Care, the program aligns Seagate's giving with employees' localized interests, leveraging the impact of combined financial, product and volunteer contributions. Seagate focuses its giving on K-12 science and technology education. Innovative programs that enhance creativity, encourage hands-on learning, and reach all populations (specifically low-income and minority communities where fewer opportunities exist) are sought. Programs that support and assist people in extreme need of shelter, food, healthcare, and support services are given preference. Arts and culture, civic, environmental, and diversity-focused programs are valuable to building a well-balanced, healthy community. Volunteerism is an important component of Seagate's philanthropic program, and consideration will be given to proposals that offer employee involvement opportunities. The company offers support through donations of cash; excess furniture, computer and office equipment; in-kind donations of disc drives, tape drives and software; and by encouraging employee volunteerism. There are no application deadlines. Submit requests on organization letterhead.
Requirements: Nonprofit organizations must: maintain program expenses at or below 20 percent; demonstrate a clear objective and have tools in place to measure, track, and report the impact of a contribution; and operate within 50 miles of those communities where Seagate maintains a significant business presence. This includes: Silicon Valley, California; Longmont, Colorado; Bloomington and Normandale, Minnesota; Oklahoma City, Oklahoma; Shrewbury, Massachusetts; Londondary, Northern Ireland; Singapore; China; Indonesia; Malaysia; Mexico; and Thailand.
Restrictions: Grants do not support organizations that raise funds for redistribution to other organizations; organizations, or events sponsored by organizations, that discriminate by race, creed, gender, disability, sexual orientation, age, religion or nationality; political parties, religious, or labor organizations (exceptions occasionally made for programs sponsored for the direct benefit of the community); luncheons, banquets or other forms of indirect support; private foundations; endowment funds or capital fund drives; fundraising organizations; special interest groups; athletic teams or events; individuals; travel funds; or courtesy advertising-related projects.
Geographic Focus: California, Colorado, Massachusetts, Minnesota, Oklahoma, China, Ireland, Malaysia, Mexico, Singapore, Thailand
Contact: Director; (405) 324-4700 or (800) 732-4283; community.involvement@seagate.com
Internet: http://www.seagate.com/about/global-citizenship/?navtab=community-involvement
Sponsor: Seagate Tecdhnologies Corporation
10200 South De Anza Boulevard
Cupertino, CA 95014

Searle Scholars Program Grants 3503
The aim of the Searle Scholars Program (administered by the Kinship Foundation) is to identify junior faculty members who have demonstrated a potential for doing innovative research in medicine, chemistry, and the biological sciences. Approximately 15 scholars are selected each year to receive the award. Each scholar receives $100,000 per year to support their research for three years. Submissions are restricted to selected candidates from invited institutions only and are not open to individuals who have not been selected by their institution to participate in the competition. The list of invited institutions can be found on the Participating Institutions page on the Program's website. Individuals from an invited institution who are considering submitting an application should consult with the individual at their institution who handles outside awards to determine how many applications their institution has been allowed to submit, as well as how selections will be made within their institution.
Requirements: Applicants will be expected to be pursuing independent research careers in biochemistry, cell biology, genetics, immunology, neuroscience, pharmacology, related areas in chemistry, medicine, and the biological sciences. Candidates should be in their first or second year at the assistant professor level. This appointment must be a tenure-track position and must be in an academic department of an invited, degree-granting institution. Potential applicants whose institutions do not have tenure-track appointments should consult with the scientific director of the program prior to preparing an application.
Restrictions: Applications from any institution not invited to participate will not be considered. No more than two applications will be accepted from any one institution.
Geographic Focus: All States
Date(s) Application is Due: Sep 28
Amount of Grant: 100,000 USD
Contact: Jennifer Nesler, Administrative and Grants Coordinator; (312) 803-6200; fax (312) 803-6201; jennifer.nesler@kinshipfoundation.org
Internet: http://www.searlescholars.net/go.php?id=5
Sponsor: Searle Scholars Program
303 West Madison Street, Suite 1800
Chicago, IL 60606

Seattle Foundation Benjamin N. Phillips Memorial Fund Grants 3504
The Benjamin N. Phillips Memorial Fund was established by the estate of Joy Phillips to honor her late husband in 2006 as an area of interest fund of The Seattle Foundation. The goal of the Fund is to make grants to organizations improving the lives of Clallam County, Washington, residents. The Benjamin N. Phillips Memorial Fund is interested in supporting organizations that have: a mission statement that clearly defines the organization's purpose and reflects its understanding of the communities they serve; a clear articulation of why it believed what it is doing is important and that it will be effective and produce desired results; clearly defined priorities, goals and measurable outcomes; experienced and highly qualified staff and volunteer leadership; a skilled governing board whose knowledge includes management, fundraising and the community served; a funding plan appropriate to agency size and developmental state-guiding development efforts; sound financial management practices; support in the community and constituent involvement; and proven ability to mobilize financial and in-kind support, including volunteers. Grants are predominately made for one year, with no implied renewal funding. However, a two-year grant will be considered if a case is made for why funding is required for a longer period. An example of this exception is a planning or capacity-building process occurring over a two-year period of time. Approximately $250,000 will be distributed annually, with grants ranging in size from $1,000 to $25,000; the average grant size is $11,000.
Requirements: To qualify for a grant from the Foundation, an organization must: be a 501(c)3 tax-exempt nonprofit organization serving residents of Clallam County, Washington.
Geographic Focus: Washington
Date(s) Application is Due: Jul 1
Amount of Grant: 1,000 - 25,000 USD
Samples: Compassion and Choices of Washington, Seattle, Washington, $10,000 - support two years of added outreach, partnership development and fund development activities in Clallam County; First Book of Clallam County, Seattle, Washington, $1,000 - to support the purchase of books for low-income children; Juan de Fuca Festival of the Arts, Port Angeles, Washington, $8,000 - support sponsorship of Baka Beyond workshops and concerts.
Contact: Ceil Erickson, Grantmaking Director; (206) 515-2131 or (206) 515-2109; fax (206) 622-7673; c.erickson@seattlefoundation.org or phillips@seattlefoundation.org
Internet: http://www.seattlefoundation.org/nonprofits/phillips/Pages/benjaminphillipsmemorialfund.aspx
Sponsor: Seattle Foundation
1200 Fifth Avenue, Suite 1300
Seattle, WA 98101-3151

Seattle Foundation Doyne M. Green Scholarships 3505
Doyne M. Green Scholarships are intended to support and honor female graduate students of medicine, law or social and public services. This scholarship is available to students who have completed at least one year of graduate work. The scholarship amount is $4,000, and two awards are given each year. The annual application deadline is March 1.
Geographic Focus: Washington
Date(s) Application is Due: Mar 1
Amount of Grant: 4,000 USD
Contact: Ceil Erickson; (206) 515-2131 or (206) 515-2109; c.erickson@seattlefoundation.org
Internet: https://fortress.wa.gov/hecb/thewashboard/ScholarshipDetails/The+Seattle+Foundation/2012-2013/Doyne+M+Green+Scholarship
Sponsor: Seattle Foundation
1200 Fifth Avenue, Suite 1300
Seattle, WA 98101-3151

Seattle Foundation Hawkins Memorial Scholarship 3506
he purpose of the Hawkins Memorial Scholarship is to honor and to give financial support to graduating seniors currently enrolled at Lake Stevens High School who are pursuing a course of study in life sciences, medicine or pharmacy. Preference will be given to students attending the University of Washington. The scholarship amount is $1,500, and one award is given each year. The annual application deadline is March 16.
Geographic Focus: Washington
Date(s) Application is Due: Mar 16
Amount of Grant: 1,500 USD
Contact: Ceil Erickson; (206) 515-2131 or (206) 515-2109; c.erickson@seattlefoundation.org
Internet: https://fortress.wa.gov/hecb/thewashboard/ScholarshipDetails/The+Seattle+Foundation/2012-2013/Hawkins+Memorial+Scholarship
Sponsor: Seattle Foundation
1200 Fifth Avenue, Suite 1300
Seattle, WA 98101-3151

Seattle Foundation Health and Wellness Grants 3507
The Seattle Foundation is committed to fostering health and wellness for people throughout the King County, Washington, region. The Foundation awards grants to nonprofits in all fields that improve the quality of life for county residents, and is focused on making sure that everyone in the county has access to quality care, including physical and dental health, cognitive, emotional, and mental health. Goals within the Health and Wellness element are to: improve access to: basic healthcare, especially services and treatment for those who are low-income, uninsured and/or underinsured; support efforts designed to reduce and/or eliminate disparities in health status due to poverty and/or race; foster efforts to strengthen the ability and capacity of providers to deliver quality services; and support efforts that protect the safety net for the vulnerable in our community through case management, treatment, and counseling services. The fields of interest and populations captured in this element include: healthcare, dental care, mental health, domestic violence, developmentally disabled, physically disabled, seniors, and birth to three programs, substance abuse programs and child welfare programs. The deadline for Health and Wellness is May 1.
Requirements: To qualify for a grant from the Seattle Foundation's Grantmaking Program, an organization must: be a 501(c)3 tax-exempt nonprofit organization; serve residents of King County; and provide programming in Health and Wellness that aligns with one of our three strategies.
Restrictions: The Foundation does not support disease-specific organizations and does not support health research projects. Multi-year grants are not considered.
Geographic Focus: Washington
Date(s) Application is Due: May 1
Contact: Ceil Erickson; (206) 515-2131 or (206) 515-2109; c.erickson@seattlefoundation.org
Internet: http://www.seattlefoundation.org/nonprofits/grantmaking/healthwellness/Pages/HealthWellness.aspx
Sponsor: Seattle Foundation
1200 Fifth Avenue, Suite 1300
Seattle, WA 98101-3151

Seattle Foundation Medical Funds Grants 3508
The Seattle Foundation administers the Medical Funds program to support medical research of potential benefit to the community and to address specific healthcare needs. In the area of medical research, grants are available in the fields of cancer, cardio-pulmonary disease, multiple sclerosis and diabetes. Special consideration will be given to research projects related to immunology, oncology, neurology, molecular biology and genetics. Grants will also be given to requests for specific equipment. Grants are also available to organizations that are administering projects addressing the healthcare needs of low-income children. In this area, grants will be given to requests for specific equipment and support for capital campaigns or facility renovation projects. A total amount of $200,000 is available for grants annually in this entire program. Typically no more than $50,000 is disbursed to any one organization.
Requirements: Preference will be given to organizations/institutions with the capacity to disseminate research findings and who receive regular support from other recognized funding sources (including the federal government).
Restrictions: These funds are not to be used for patient care. Research activities can include patient care, but the primary purpose of these grants is to support the purchase of equipment used in medical research.
Geographic Focus: Washington
Contact: Ceil Erickson; (206) 515-2131 or (206) 515-2109; c.erickson@seattlefoundation.org
Internet: http://www.seattlefoundation.org/nonprofits/medicalfunds/Pages/Medical Funds.aspx

Sponsor: Seattle Foundation
1200 Fifth Avenue, Suite 1300
Seattle, WA 98101-3151

Seaver Institute Grants 3509
The Institute's focuses its giving in four primary areas: scientific and medical research, education, public affairs, and the cultural arts. Seed grants are awarded to nonprofit organizations for projects that offer the potential for a significant advancement in their fields. The institute seeks proposals that incorporate state-of-the-art scientific research, exceptional educational development, unique approaches to the creative arts, and the ramifications of shifting societal realities. There are no application deadlines.
Requirements: U.S. 501(c)3 nonprofits are eligible. Interested applicants should submit an inquiry letter of two or three pages and include the goals of the project, a preliminary budget, and the qualifications of the organization and principal investigator. If interested, the institute will invite a full proposal.
Restrictions: Grants are not made for construction, endowments, ongoing projects, operating budgets, deficits, or individual scholarships.
Geographic Focus: All States
Amount of Grant: 1,000 USD
Contact: Victoria Seaver Dean, President; (310) 979-0298; vsd@theseaverinstitute.org
Sponsor: Seaver Institute
12400 Wilshire Boulevard
Los Angeles, CA 90025

Self Foundation Grants 3510
The Self Family Foundation primarily serves the Greenwood, South Carolina and surrounding counties. The Self Family Foundation will consider requests from other regions of South Carolina, if they have statewide impact. The Foundation supports non-profit organizations in the following areas of interest: Education - enhancing children's school readiness and ability to achieve educational goals with success; working together with schools, families, youth-serving organizations and other community organizations to support the intellectual and social development of young people; Arts, Culture & History - increasing awareness of and access to activities that broaden the horizons and knowledge of all residents, within Greenwood, South Carolina; Civic & Community - improving neighborhood resources and community attitudes and services, emphasizing the utilization of existing facilities, skills and talents to create a stronger, more unified community through collaboration with organizations, businesses, public institutions and residents; Health and Human Services - promoting individual and community wellness, prevention and literacy.
Requirements: IRS 501(c)3 tax-exempt or 509(a) nonprofit organizations in South Carolina are eligible. Letters of Inquiry or requests for information may be emailed to: application@selffoundation.org or mailed to Foundation. Contact Foundation Program Officer by email or phone prior to submitting proposal. All applicants must have the financial potential to sustain the project on a continuing basis after any funding is approved.
Restrictions: No grants to individuals and no loans available.
Geographic Focus: South Carolina
Date(s) Application is Due: Feb 15; May 15; Aug 15; Nov 15
Amount of Grant: 1,500 - 50,000 USD
Contact: Administrator; (864) 953-2441 or (864) 941-4036; info@selffoundation.org
Internet: http://www.selffoundation.org/grantmaking4.htm
Sponsor: Self Family Foundation
P.O. Box 1017
Greenwood, SC 29648-1017

Seneca Foods Foundation Grants 3511
The Seneca Foods Foundation was established in 1988 in New York and is sponsored by the Seneca Foods Corporation. The foundation supports fire departments and organizations involved with education, health, Down Syndrome, agriculture and food, recreation, and youth. Funding is primarily in areas of company processing facilities, with emphasis on New York.
Requirements: Nonprofit organizations in Seneca Foods operating locations are eligible. Contact foundation for application form.
Restrictions: Grants are not made to individuals.
Geographic Focus: All States
Contact: Kraig H. Kayser, President; (315) 926-8100; foundation@senecafoods.com
Sponsor: Seneca Foods Corporation
3736 South Main Street
Marion, NY 14505-9751

Sengupta Family Scholarship 3512
The Sengupta Family Scholarship is available to the valedictorian of Lake Central High School in St. John, Indiana. The award is $1,000 per year for four years. The applicant should contact the Lake Central High School Guidance Office for the current application, current deadline, and the Dollars for Scholars application. All college majors are eligible.
Geographic Focus: Indiana
Date(s) Application is Due: Jan 1
Amount of Grant: 1,000 USD
Contact: Sengupta Scholarship Contact; (219) 365-8551; fax (219) 365-7156
Internet: http://www.legacyfoundationlakeco.org/scholarships/index.htm
Sponsor: Legacy Foundation
1000 East 80th Place, 302 South
Merrillville, IN 46410

Sensient Technologies Foundation Grants 3513
The foundation supports organizations involved with arts and culture; children/youth, services; community/economic development; education; education, fund raising/fund distribution; education, research; family services; food services; general charitable giving; health care; higher education; homeless, human services; hospitals (general); human services; medical research, institute; mental health/crisis services; nutrition; performing arts; residential/custodial care, hospices; United Ways and Federated Giving Programs; Urban/community development; and voluntarism promotion. The foundation's types of support include: annual campaigns; capital campaigns; emergency funds; endowments; general operating support; matching/challenge support; program development; research; and scholarship funds.
Requirements: An application form is not required. Submit a letter of inquiry as an initial approach. The advisory board meets in January and June/July or as needed, with deadline for proposal review one month prior to board meetings.
Restrictions: The foundation gives primarily to its areas of interest in Indianapolis, Indiana; St. Louis, Missouri; and Milwaukee, Wisconsin. It does not support sectarian religious, fraternal, or partisan political organizations. It does not give grants to individuals.
Geographic Focus: Indiana, Missouri, Wisconsin
Contact: Douglas L. Arnold, (414) 271-6755; fax (414) 347-4783
Sponsor: Sensient Technologies Foundation
777 E Wisconsin Avenue
Milwaukee, WI 53202-5304

Seward Community Foundation Grants 3514
As an organization, the Seward Community Foundation goal is to support projects that enhance the quality of life for Seward and Moose Pass residents, addressing immediate needs while working toward long-term improvements. The Foundation is continually listening and learning about what is important to its residents. Currently, grantmaking supports organizations and programs that serve the needs of people of all ages in such areas as health, education, recreation, social services, arts and culture, the environment, and community development. The Foundation seeks projects that offer maximum impacts for its community and shows collaboration with other organizations. Typically, awards have ranged from $1,000 to $5,000, though larger grants are considered. The annual application cycle begins on January 1 and runs through the deadline of March 1.
Requirements: Applications are accepted from qualified 501(c)3 nonprofit organizations, or equivalent organizations located in the state of Alaska and serving the Seward/Moose Valley region. Equivalent organizations may include tribes, local or state governments, schools, or Regional Educational Attendance Areas.
Geographic Focus: Alaska
Date(s) Application is Due: Mar 1
Amount of Grant: 500 - 25,000 USD
Samples: Qutekcak Native Tribe, Seward, Alaska, $1,000 - support for the Seward Native Youth Olympics (NYO) State Meet (2015); Independent Living Center, Seward, Alaska, $$6,750 - support for the TRAILS program (2015); Seward Little League, Seward, Alaska, $2,500 - purchase of a pitching machine for youth team (2015).
Contact: Mariko Sarafin, Senior Program Associate; (907) 249-6609 or (907) 334-6700; fax (907) 334-5780; msarafin@alaskacf.org
Internet: http://sewardcf.org/projects/
Sponsor: Seward Community Foundation
P.O. Box 933
Seward, AK 99664

Seward Community Foundation Mini-Grants 3515
The Seward Community Foundation Mini-Grant program is available throughout the year. These are a quick method to assist nonprofit organizations and projects in the Seward/Moose Pass communities. The maximum grant award is up to $1,000 and can be applied for easily with the Foundation's online application. The Foundation's advisory board will review applications after they are submitted, generally at their next scheduled meeting. These meeting are held on the 3rd Wednesday of every month.
Requirements: Applications are accepted from qualified 501(c)3 nonprofit organizations, or equivalent organizations located in the state of Alaska and serving the Seward/Moose Valley region. Equivalent organizations may include tribes, local or state governments, schools, or Regional Educational Attendance Areas.
Geographic Focus: Alaska
Amount of Grant: Up to 1,000 USD
Contact: Mariko Sarafin, Senior Program Associate; (907) 249-6609 or (907) 334-6700; fax (907) 334-5780; msarafin@alaskacf.org
Internet: http://sewardcf.org/projects/
Sponsor: Seward Community Foundation
P.O. Box 933
Seward, AK 99664

Sexually Transmitted Diseases Postdoctoral Fellowships 3516
The ASHA Research Fund fellowships program is intended to expand the number of trained scientists working in the field of sexually transmissible diseases research by providing postdoctoral candidates with the opportunity to spend two years in a challenging, stimulating, and instructive STD research environment under the sponsorship of an established researcher. Applications for research training relative to any aspect of the sexually transmissible diseases are welcome. ASHA has supported projects in chancroid, human papilloma virus (HPV), group B streptococcal infection, herpes, gonorrhea, chlamydia, HIV, pelvic inflammatory disease (PID), and syphilis. In funding HIV-related research, ASHA gives priority to researchers exploring the relationship between HIV and other STDs. Although applications covering fellowship training in basic investigations are not discouraged, it is the intent of this program to focus on research of short- or medium-term duration. Requests for applications are usually issued in May of odd-numbered years, with an application deadline of October 15, and the award being given the following spring for a fellowship period commencing on July 1.
Requirements: Applications are requested from MDs, PhDs, DSc who will be conducting two-year research projects under the sponsorship of established STD investigators. Applicants are required to study at established U.S. institutions.
Restrictions: ASHA fellowships may not be used in support of faculty positions unless approved by ASHA's management. Current ASHA fellows are not eligible to apply for other ASHA fellowships.
Geographic Focus: All States
Date(s) Application is Due: Nov 15
Amount of Grant: 35,500 - 36,000 USD
Contact: Grant Administrator; (919) 361-8400; fax (919) 361-8425
Internet: http://www.ashastd.org/programs/arf.html
Sponsor: American Social Health Association
P.O. Box 13827
Research Triangle Park, NC 27709

SfN Award for Education in Neuroscience 3517
The Award for Education in Neuroscience recognizes individuals who have made outstanding contributions to neuroscience education and training. The recipient receives $5,000, along with complimentary transportation, hotel, and registration for the SfN annual meeting. The Award is presented at the Neuroscience Departments and Programs Reception. All nomination materials should be submitted using SfN's award nomination site. Nominations open on March 7 and close on May 27.
Requirements: Nominees must: have notable contributions to educating and training others about neuroscience; and be made or endorsed by an SfN member (although nominees do not need to be SfN members).
Restrictions: Self nomination is not permitted, and no person may nominate more than one candidate. Current SfN officers and councilors are ineligible for nomination.
Geographic Focus: All States, Puerto Rico, Argentina, Australia, Brazil, Canada, China, Denmark, India, Israel, Malaysia, Mexico, New Zealand, Nigeria, Norway, Singapore, South Korea, Spain, Switzerland, Turkey, Ukraine, United Kingdom
Date(s) Application is Due: May 27
Amount of Grant: 5,000 USD
Contact: Ronald L Calabrese, (404) 727-0319; ronald.calabrese@emory.edu
Internet: http://www.sfn.org/Awards-and-Funding/Individual-Prizes-and-Fellowships/Science-Education-and-Outreach/Award-for-Education-in-Neuroscience
Sponsor: Society for Neuroscience
1121 14th Street, NW, Suite 1010
Washington, D.C. 20005

SfN Bernice Grafstein Award for Outstanding Accomplishments in Mentoring 3518
The Bernice Grafstein Award for Outstanding Accomplishments in Mentoring recognizes individuals who are dedicated to promoting women's advancement in the field of neuroscience, and who have made outstanding accomplishments in mentoring. The recipient receives a $2,000 award, in addition to complimentary transportation, hotel, and registration for the SfN annual meeting. All nomination materials will be submitted using SfN's award nomination site. Nominations open on March 7 and close on June 10.
Requirements: Nominees must have shown a dedication to, and success in, mentoring women neuroscientists, and in facilitating their entry into or retention within the field of neuroscience. Nominees may also be recognized for related efforts and achievements that promote women's advancement in neuroscience. Nominations must be made or endorsed by Society members, although nominees do not need to be Society members.
Restrictions: Self nomination is not permitted, and no person may nominate more than one candidate. Current SfN officers and councilors are not eligible for nomination.
Geographic Focus: All States, Puerto Rico, Argentina, Australia, Brazil, Canada, China, Denmark, India, Israel, Malaysia, Mexico, New Zealand, Nigeria, Norway, Singapore, South Korea, Spain, Switzerland, Turkey, Ukraine, United Kingdom
Date(s) Application is Due: Jun 10
Amount of Grant: 2,000 USD
Contact: Nancy Y Ip, Ph.D.; 85-223-587-267; fax 85-223-581-464; boip@ust.hk
Internet: http://www.sfn.org/Awards-and-Funding/Individual-Prizes-and-Fellowships/Promotion-and-Mentoring-of-Women-in-Neuroscience/Bernice-Grafstein-Award-for-Outstanding-Accomplishments-in-Mentoring
Sponsor: Society for Neuroscience
1121 14th Street, NW, Suite 1010
Washington, D.C. 20005

SfN Chapter-of-the-Year Award 3519
The Chapter-of-the-Year Award recognizes outstanding chapters for their efforts and accomplishments across a broad range of activities that are in line with the mission and strategic initiatives of SfN. Recipients receive a $1,000 grant to support chapter goals and programs. Awards are presented during the Chapters Workshop at the SfN annual meeting.
Requirements: For award consideration, chapters must: submit an annual chapter report; and meet reporting requirements for all other programs through which funding from SfN was received (e.g., chapter grants). Additionally, a representative from the chapter leadership must be present at the Chapter's Workshop to accept the award. If requested, the chapter will give a presentation on a topic selected by SfN at the Chapter's Workshop.
Geographic Focus: All States, All Countries
Date(s) Application is Due: Jun 10
Amount of Grant: 1,000 USD
Contact: Program Director; (202) 962-4000; chapters@sfn.org

Internet: http://www.sfn.org/Awards-and-Funding/Chapter-Grants-and-Awards/Chapter-of-the-Year-Award
Sponsor: Society for Neuroscience
1121 14th Street, NW, Suite 1010
Washington, D.C. 20005

SfN Chapter Grants 3520
SfN's offers annual Chapter Grants of $500 to $2,000 to support chapter activities that advance SfN's mission, including outreach, advocacy, and professional development programs. For award consideration, chapters must have submitted at least one annual report during the previous two calendar years and must have provided reports from all other programs through which funding was received. Previous grants have funded activities such as: student-oriented regional conferences; beginning, continuing, or expanding Brain Awareness Week activities, including increased community involvement or the purchase of educational supplies; hosting a local political representative to promote neuroscience; hosting a lecturer from outside the chapter's geographic region (previously funded by the Distinguished Traveling Scientists program); collaborating with local organizations on professional development initiatives; and supporting a chapter annual meeting.
Requirements: Chapters must spend grant money on the approved budget (submitted with the application) within one year of being awarded. A final financial and project report, with return of unused funds, is due to SfN 30 days after the event.
Geographic Focus: All States, All Countries
Amount of Grant: 500 - 2,000 USD
Contact: Program Coordinator; (202) 962-4000; chapters@sfn.org
Internet: http://www.sfn.org/Awards-and-Funding/Chapter-Grants-and-Awards/Chapter-Grants
Sponsor: Society for Neuroscience
1121 14th Street, NW, Suite 1010
Washington, D.C. 20005

SfN Donald B. Lindsley Prize in Behavioral Neuroscience 3521
The Donald B. Lindsley Prize in Behavioral Neuroscience recognizes a young neuroscientist's outstanding Ph.D. thesis in the general area of behavioral neuroscience (defined as neuroscience research involving behavioral variables or oriented toward the solution of behavioral problems). The recipient receives a $2,500 prize; and air or ground transportation, two nights hotel, and complimentary registration for the Society's annual meeting. The Society's President presents the prize at a lecture at the annual meeting. All nomination materials must be submitted through the Society's award nomination link at the Society's website. Nominations open on March 7 and close on May 20.
Requirements: Nominees must have a thesis submitted and approved between January 1 and December 31 of the preceding calendar year. Nominations must be made or endorsed by SFN members, although nominees do not need to be SFN members.
Restrictions: Self nomination is not permitted. No person may nominate more than one candidate. Current Society officers and councilors are ineligible for nomination.
Geographic Focus: All States, Puerto Rico, Argentina, Australia, Brazil, Canada, China, Denmark, India, Israel, Malaysia, Mexico, New Zealand, Nigeria, Norway, Singapore, South Korea, Spain, Switzerland, Turkey, Ukraine, United Kingdom
Date(s) Application is Due: May 20
Contact: Steven E Petersen; (314) 362-3317; fax (314) 362-2186; sep@npg.wustl.edu
Internet: http://www.sfn.org/Awards-and-Funding/Individual-Prizes-and-Fellowships/Young-Scientists-Achievements-and-Research/Donald-B-Lindsley-Prize-in-Behavioral-Neuroscience
Sponsor: Society for Neuroscience
1121 14th Street, NW, Suite 1010
Washington, D.C. 20005

SfN Federation of European Neuroscience Societies Forum Travel Awards 3522
The SfN offers up to fifteen (15) travel awards of $2,000 each to support the participation of graduate students from the U.S., Canada, and Mexico at the Federation of European Neuroscience Societies (FENS) Forum, held every two years. Recipients of the travel awards, selected by the SfN International Affairs Committee, are based on the scientific merit of their abstract submitted to the Forum, and the strength of related application materials. Applicants will be emailed about the status of their application. Recipients must attend the Forum to receive their award checks.
Requirements: The applicant must be a U.S., Canadian, or Mexican citizen or permanent resident and graduate student who has advanced to candidacy for a Ph.D. The nominee must be the first author of an abstract to be presented at the Forum. Both the applicant and his or her advisor must be SfN members, and the applicant may not be a recipient of another SfN travel award in the same year. In addition to the online application form, applicants must submit a paragraph describing what they hope to achieve by attending the Forum; a copy of the abstract that would be submitted for the Forum; a one page letter of recommendation from their advisor; and a one page CV which includes their education, honors/awards, and publications (abstracts and manuscripts).
Geographic Focus: All States, Canada, Mexico
Date(s) Application is Due: Feb 2
Amount of Grant: 2,000 USD
Contact: Kelly McCabe, Global Programs Assistant; (202) 962-4000; fax (202) 962-4941; kmccabe@sfn.org or globalaffairs@sfn.org
Internet: http://www.sfn.org/Awards-and-Funding/Individual-Prizes-and-Fellowships/Professional-Development-Awards/FENS-Forum-Travel-Awards
Sponsor: Society for Neuroscience
1121 14th Street, NW, Suite 1010
Washington, D.C. 20005

SfN Jacob P. Waletzky Award 3523
The Jacob P. Waletzky Award is presented to a scientist who has conducted research or plans to conduct research in the area of substance abuse and the brain and nervous system. Recipients receive a $25,000 prize and complimentary registration, transportation (economy air or ground), and two nights hotel accommodations for the SfN annual meeting. The SfN president presents the prize at a lecture at the meeting. Nominations open on March 7 and close on June 3.
Requirements: Nominations must be made or endorsed by a Society member, although nominees do not need to be Society members. All materials are submitted using the Society's award nomination process located at the Society's website.
Geographic Focus: All States, Puerto Rico, Argentina, Australia, Brazil, Canada, Denmark, India, Israel, Malaysia, Mexico, New Zealand, Nigeria, Norway, Singapore, South Korea, Spain, Switzerland, Turkey, Ukraine, United Kingdom
Date(s) Application is Due: Jun 3
Amount of Grant: 25,000 USD
Contact: Daniel H Geschwind; (310) 206-6814; dhg@mednet.ucla.edu
Internet: http://www.sfn.org/Awards-and-Funding/Individual-Prizes-and-Fellowships/Outstanding-Research-and-Career-Awards/Jacob-P-Waletzky-Award
Sponsor: Society for Neuroscience
1121 14th Street, NW, Suite 1010
Washington, D.C. 20005

SfN Janett Rosenberg Trubatch Career Development Award 3524
The Janett Rosenberg Trubatch Award recognizes individuals who have demonstrated originality and creativity in research. The Award also promotes success during academic transitions prior to tenure. The recipient receives a $2,000 award and complimentary registration to the Society's annual meeting. Two recipients are selected each year. The Awards are presented during the Society's annual meeting. All materials must be submitted using the Society's online nomination site. Nominations open on March 7 and close on May 20.
Requirements: Nominees with a Ph.D. or M.D. (or equivalent) are eligible until they have completed ten years of full-time work. Candidates working in non-academic environments are eligible if their work is published and meets academic standards. Nominees must not have received university tenure by the nomination deadline. Nominations must be made or endorsed by a Society member, although nominees do not need to be Society members.
Restrictions: Self-nomination is not permitted. No person may nominate more than one candidate. Current Society officers and councilors are ineligible for nomination.
Geographic Focus: All States, Puerto Rico, Argentina, Australia, Brazil, Canada, China, Denmark, India, Israel, Malaysia, Mexico, New Zealand, Nigeria, Norway, Singapore, South Korea, Spain, Switzerland, Turkey, Ukraine, United Kingdom
Date(s) Application is Due: May 20
Amount of Grant: 2,000 USD
Contact: Nancy Y Ip, Ph.D.; 85-223-587-267; fax 85-223-581-464; boip@ust.hk
Internet: http://www.sfn.org/Awards-and-Funding/Individual-Prizes-and-Fellowships/Young-Scientists-Achievements-and-Research/Janett-Rosenberg-Trubatch-Career-Development-Award
Sponsor: Society for Neuroscience
1121 14th Street, NW, Suite 1010
Washington, D.C. 20005

SfN Japan Neuroscience Society Meeting Travel Awards 3525
The Society for Neuroscience (SfN) and the Japan Neuroscience Society (JNS) support a joint travel award exchange program for graduate students or postdoctoral investigators to their respective annual meetings. The Society sponsors five eligible members from Canada, Mexico, and the U.S. who intend on participating in the JNS meeting in Japan. The JNS will also sponsor five members to attend the SfN annual meeting in the U.S. (locations vary). Recipients are selected on the merit of their abstract, CV, and letters of recommendation. Students are selected by the International Affairs Committee. Award checks are distributed to selectees following the meeting. Awardees will need to provide a certificate of attendance and airline boarding passes to receive their award.
Requirements: Applicants must be Ph.D. candidates or postdoctoral scientists studying or working in Canada, Mexico, or the U.S. They must also be citizens or permanent residents of those locations. Candidates must be SfN members at the time of application and first author on an abstract to be presented at the JNS meeting. Candidates may not have been selected for another SfN travel award within the last calendar year. Preference is given to those who have not previously had extensive experience living or working in Japan. In addition to the online application, candidates must submit a paragraph describing what the applicant hopes to achieve by attending the JNS meeting; a copy of the abstract to be submitted; a one page recommendation from their advisor; and the applicant's CV including information on education, date of Ph.D. candidacy, honors/awards, and publications.
Geographic Focus: All States, Canada, Japan, Mexico
Date(s) Application is Due: Feb 3
Amount of Grant: 2,000 USD
Contact: Kelly McCabe, Global Programs Assistant; (202) 962-4000; fax (202) 962-4941; kmccabe@sfn.org or globalaffairs@sfn.org
Internet: http://www.sfn.org/Awards-and-Funding/Individual-Prizes-and-Fellowships/Professional-Development-Awards/JNS-Meeting-Travel-Awards
Sponsor: Society for Neuroscience
1121 14th Street, NW, Suite 1010
Washington, D.C. 20005

574 | Grant Programs

SfN Julius Axelrod Prize   3526

The Julius Axelrod Prize, supported by the Eli Lilly and Company Foundation, was established to honor a scientist with distinguished achievements in the broad field of neuropharmacology (or a related area) and exemplary efforts in mentoring young scientists. The recipient receives a $25,000 prize. They also attend the SfN annual meeting with complimentary air or ground transportation, hotel, and registration. The SfN president presents the prize at a lecture at the meeting. Nominations open on March 7 and close on June 3.
Requirements: Nominees must have accomplishments in the field of neuropharmacology or a related area. They also need to have shown dedication to mentoring young scientists. Nominations are made or endorsed by SfN members, although nominees do not need to be SfN members. All nomination materials are submitted using the Society's award nomination site.
Restrictions: Self nomination is not permitted, and no person may nominate more than one candidate. Current SfN officers and councilors are ineligible for nomination.
Geographic Focus: All States, Puerto Rico, Argentina, Australia, Brazil, Canada, China, Denmark, India, Israel, Malaysia, Mexico, New Zealand, Nigeria, Norway, Singapore, South Korea, Spain, Switzerland, Turkey, Ukraine, United Kingdom
Date(s) Application is Due: Jun 3
Amount of Grant: 25,000 USD
Contact: Guoping Feng; (617) 715-4898; fax (617) 324-6752; fengg@mit.edu
Internet: http://www.sfn.org/Awards-and-Funding/Individual-Prizes-and-Fellowships/Outstanding-Research-and-Career-Awards/Julius-Axelrod-Prize
Sponsor: Society for Neuroscience
1121 14th Street, NW, Suite 1010
Washington, D.C. 20005

SfN Louise Hanson Marshall Specific Recognition Award   3527

The Louise Hanson Marshall Special Recognition Award recognizes an individual who has significantly promoted the professional development of women in neuroscience through teaching, organizational leadership, public advocacy, or other efforts that are not necessarily related to research. The recipient receives the Marshall Award, along with complimentary transportation, hotel, and registration for the SfN annual meeting. The Award is presented at the Celebration of Women in Neuroscience luncheon held during the SfN annual meeting. All nominating materials should be submitted using SfN's award nomination site. Nominations for the award open on March 7 and close on June 10.
Requirements: Nominees must have served the professional through SfN and/or related organizations. They must also have attained recognition at national or international levels as a scientist, educator, businessperson, or administrator in neuroscience. In addition, nominees must have demonstrated a high degree of imagination, innovation, and initiative in the pursuit of neuroscience, and facilitated mentoring or the advancement of young women in neuroscience. Nominees may only be made by SfN members, but the nominees do not need to be SfN members.
Restrictions: Self nomination is not permitted. No one may nominate more than one candidate.
Geographic Focus: All States, Puerto Rico, Argentina, Australia, Brazil, Canada, China, Denmark, India, Israel, Malaysia, Mexico, New Zealand, Nigeria, Norway, Singapore, South Korea, Spain, Switzerland, Turkey, Ukraine, United Kingdom
Date(s) Application is Due: Jun 10
Contact: Nancy Y Ip, Ph.D.; 85-223-587-267; fax 85-223-581-464; boip@ust.hk
Internet: http://www.sfn.org/Awards-and-Funding/Individual-Prizes-and-Fellowships/Promotion-and-Mentoring-of-Women-in-Neuroscience/Louise-Hanson-Marshall-Special-Recognition-Award
Sponsor: Society for Neuroscience
1121 14th Street, NW, Suite 1010
Washington, D.C. 20005

SfN Mika Salpeter Lifetime Achievement Award   3528

The Mika Salpeter Lifetime Achievement Award recognizes an individual with outstanding career achievements in neuroscience who has also significantly promoted the professional advancement of women in neuroscience. The recipient receives a $5,000 award, in addition to complimentary air or ground transportation, hotel, and registration to the SfN annual meeting. The nominee must have exhibited a dedication to facilitating the mentoring and entry of young women into neuroscience or to the advancement of women in neuroscience; sustained exceptional achievements in neuroscience as evidenced by publications, inventions, and/or awards; served the profession through SfN and/or related organizations; recognized at national or international levels as a scientist, educator, businessperson, or administrator in neuroscience; and demonstrated a high degree of imagination, innovation, and initiative in the pursuit of neuroscience. Nominations for the award open on March 7 and close on June 10.
Requirements: Nominations must be made or endorsed by a Society member, although nominees do not need to be Society members. All nomination materials are submitted using SfN's award nomination site.
Restrictions: Self nomination is not permitted, and no person may nominate more than one candidate. Current SfN officers and councilors are ineligible for nomination.
Geographic Focus: All States, Puerto Rico, Argentina, Australia, Brazil, Canada, China, Denmark, India, Israel, Malaysia, Mexico, New Zealand, Nigeria, Norway, Singapore, South Korea, Spain, Switzerland, Turkey, Ukraine, United Kingdom
Date(s) Application is Due: Jun 10
Amount of Grant: 5,000 USD
Contact: Nancy Y Ip, Ph.D.; 85-223-587-267; fax 85-223-581-464; boip@ust.hk
Internet: http://www.sfn.org/Awards-and-Funding/Individual-Prizes-and-Fellowships/Outstanding-Research-and-Career-Awards/Mika-Salpeter-Lifetime-Achievement-Award
Sponsor: Society for Neuroscience
1121 14th Street, NW, Suite 1010
Washington, D.C. 20005

SfN Nemko Prize in Cellular or Molecular Neuroscience   3529

The Nemko Prize in Cellular or Molecular Neuroscience, supported by The Nemko Family, recognizes a young neuroscientist's outstanding Ph.D. thesis advancing understanding of molecular, genetic, or cellular mechanisms underlying brain function, including higher function and cognition. Recipients receive a $2,500 prize and complimentary registration, transportation (economy air or ground), and two nights hotel accommodations for the SfN annual meeting. The SfN president presents the prize at a lecture at the meeting. Nominations for the award open on March 7 and close on May 20.
Requirements: Nominees must have a thesis submitted and approved between January 1 and December 31 of the preceding calendar year.
Geographic Focus: All States
Date(s) Application is Due: May 20
Amount of Grant: 2,500 USD
Contact: Guoping Feng, Ph.D.; (617) 715-4898; fax (617) 324-6752; fengg@mit.edu
Internet: http://www.sfn.org/Awards-and-Funding/Individual-Prizes-and-Fellowships/Young-Scientists-Achievements-and-Research/Nemko-Prize-in-Cellular-or-Molecular-Neuroscience
Sponsor: Society for Neuroscience
1121 14th Street, NW, Suite 1010
Washington, D.C. 20005

SfN Neuroscience Fellowships   3530

The Neuroscience Scholars Program (NSP) is a two-year fellowship administered by the Society for Neuroscience to enhance career development and professional networking opportunities for underrepresented diverse undergraduate and graduate students, and postdoctoral fellows in neuroscience. The program, funded by the National Institute of Neurological Disorders and Stroke, provides scholars with: annual stipend for enrichment activities outside the scholar's home institution; support for and annual meeting travel expenses; access to annual meeting workshops, courses, and events; complimentary SfN meeting registration and abstract fee waivers; complimentary SfN membership and online subscription to The Journal of Neuroscience; and mentoring. Other benefits include opportunities to network, expand professional contacts, and acquire professional skills. The application and samples of the biosketch required are located at the SfN website.
Requirements: Applicants must be citizens or permanent residents of the U.S., and enrolled in a degree-granting program or postdoctoral fellowship. Candidates must be from a group recognized as underrepresented in the biomedical, behavioral, clinical, and social sciences. Additional information is available at the website.
Restrictions: Previous fellowship recipients are ineligible.
Geographic Focus: All States
Contact: Mona Miller, Director of Programs; (202) 962-4000; mmiller@sfn.org
Internet: http://www.sfn.org/Awards-and-Funding/Individual-Prizes-and-Fellowships/Fellowships/Neuroscience-Scholars-Program
Sponsor: Society for Neuroscience
1121 14th Street, NW, Suite 1010
Washington, D.C. 20005

SfN Next Generation Award   3531

The SfN Next Generation Award recognizes Society chapter members who have made outstanding contributions to public communication, outreach, and education about neuroscience. Each year, one award is given at the predoctoral/postdoctoral level and one at the junior faculty level. The recipients receive a $750 travel award to help defray the cost of attending the Society's annual meeting, plus a $300 honorarium. The recipient's chapter receives a $2,000 chapter grant to be used to continue the chapter's outreach efforts for the following year. All nominating materials must be submitted through the Society's award nomination site. Nominations for the award open on March 7 and close on May 27.
Requirements: Nominating chapters must be active, and must have submitted an annual chapter report within the past year. Predoctoral/postdoctoral nominees must be current pre or post-doc students. In cases where cooperation was essential for the success of the activity, a group of up to four may be nominated at those levels by one chapter. Junior faculty nominees must be scientists within no more than ten years of receiving an advanced degree. Nominees must have made outstanding contributions to outreach and science education activities for SfN or its local chapters. The nominee may not be a recipient of another SfN travel award in the same year. The following year, award recipients must take leadership roles designing and/or executing chapter outreach efforts which the annual chapter report must document.
Geographic Focus: All States, Puerto Rico, Argentina, Australia, Brazil, Canada, China, Denmark, India, Israel, Malaysia, Mexico, New Zealand, Nigeria, Norway, Singapore, South Korea, Spain, Switzerland, Turkey, Ukraine, United Kingdom
Date(s) Application is Due: May 27
Amount of Grant: 3,050 USD
Contact: Jerold J.M. Chun, Chairperson; (858) 784-8410; jchun@scripps.edu
Internet: http://www.sfn.org/Awards-and-Funding/Individual-Prizes-and-Fellowships/Science-Education-and-Outreach/Next-Generation-Award
Sponsor: Society for Neuroscience
1121 14th Street, NW, Suite 1010
Washington, D.C. 20005

SfN Patricia Goldman-Rakic Hall of Honor   3532

The Patricia Goldman-Rakic Hall of Honor is a posthumous award for a neuroscientist who pursued career excellence and exhibited dedication to the advancement of women in neuroscience. The recipient is recognized at the Celebration of Women in Neuroscience Luncheon held during the SfN annual meeting and the family of the deceased honoree

receives an engraved Tiffany and Company crystal bowl. Nominations for the award open on March 7 and close on June 10.
Requirements: Nominees must have: exhibited dedication to facilitating the mentoring and entry of young women into neuroscience or to the advancement of women in neuroscience; sustained exceptional achievements in neuroscience as evidenced by publications, inventions, and/or awards; served the profession through SfN and/or related organizations; been recognized at a national or international level as a scientist, educator, businessperson, or administrator in neuroscience; and demonstrated a high degree of imagination, innovation, and initiative in the pursuit of neuroscience.
Geographic Focus: All States, All Countries
Date(s) Application is Due: Jun 10
Contact: Nancy Y Ip, Ph.D.; 85-223-587-267; fax 85-223-581-464; boip@ust.hk
Internet: http://www.sfn.org/Awards-and-Funding/Individual-Prizes-and-Fellowships/Promotion-and-Mentoring-of-Women-in-Neuroscience/Patricia-Goldman-Rakic-Hall-of-Honor
Sponsor: Society for Neuroscience
1121 14th Street, NW, Suite 1010
Washington, D.C. 20005

SfN Peter and Patricia Gruber International Research Award 3533
The Peter and Patricia Gruber International Research Award in Neuroscience recognizes two young neuroscientists for outstanding research and educational pursuit in an international setting. If the nominee is a U.S. citizen or permanent resident, the nominee must be studying or working at an institution located outside the U.S; if a non-U.S. citizen or permanent resident, the nominee must be studying or working at an institution that is located in the U.S. Nominees must be active in research at the time the award is given. Successful candidates are graduate students, postdoctoral fellows, or research associates. Each recipient receives a $25,000 prize, in addition to air or ground transportation to the SfN annual meeting with two nights hotel accommodations, and complimentary registration. All nominating materials must be submitted using the SfN's award nomination site. Nominations for the award open on March 7 and close on May 20.
Requirements: Although nominees do not need to be SfN members, they must be nominated or endorsed by an SfN member.
Restrictions: Self-nomination is not permitted. No person may nominate more than one candidate. Current SfN officers and councilors are ineligible for nomination.
Geographic Focus: All States, Puerto Rico, Argentina, Australia, Brazil, Canada, China, Denmark, India, Israel, Malaysia, Mexico, New Zealand, Nigeria, Norway, Singapore, South Korea, Spain, Switzerland, Turkey, Ukraine, United Kingdom
Date(s) Application is Due: May 20
Amount of Grant: 25,000 USD
Contact: Roberta Diaz Brinton, Ph.D., Chairperson; (323) 442-1430; fax (323) 442-1740; awards@sfn.org or rbrinton@usc.edu
Internet: http://www.sfn.org/Awards-and-Funding/Individual-Prizes-and-Fellowships/Young-Scientists-Achievements-and-Research/Peter-and-Patricia-Gruber-International-Research-Award
Sponsor: Society for Neuroscience
1121 14th Street, NW, Suite 1010
Washington, D.C. 20005

SfN Ralph W. Gerard Prize in Neuroscience 3534
The Ralph W. Gerard Prize in Neuroscience honors an outstanding scientist who has made significant contributions to neuroscience throughout his or her career. The Gerard Prize is the highest recognition conferred by the Society, and open to all scientists with contributions to the field of neuroscience. The recipient receives $25,000, in addition to complimentary air or ground transportation, hotel, and registration for the SfN annual meeting. Nominations for the award open on March 7 and close on June 3.
Requirements: Nominations must be made or endorsed by Society member, although nominees do not need to be Society members. All nomination materials need to be submitted using SfN's award nomination site.
Restrictions: Self nominations are not permitted, and no person may nominate more than one candidate. Current SfN officers and councilors are ineligible for nomination.
Geographic Focus: All States, Puerto Rico, Argentina, Australia, Brazil, Canada, China, Denmark, India, Israel, Malaysia, Mexico, New Zealand, Nigeria, Norway, Singapore, Spain, Switzerland, Turkey, Ukraine, United Kingdom
Date(s) Application is Due: Jun 3
Amount of Grant: 25,000 USD
Contact: Nicholas C Spitzer; (858) 534-3896; fax (858) 534-7309; nspitzer@ucsd.edu
Internet: http://www.sfn.org/Awards-and-Funding/Individual-Prizes-and-Fellowships/Outstanding-Research-and-Career-Awards/Ralph-W-Gerard-Prize-in-Neuroscience
Sponsor: Society for Neuroscience
1121 14th Street, NW, Suite 1010
Washington, D.C. 20005

SfN Science Educator Award 3535
The Science Educator Award honors an outstanding neuroscientist who has made significant contributions to educating the public about neuroscience. The recipient receives a $5,000 honorarium and the opportunity to write a feature commentary on science education in the Journal of Neuroscience. He or she also receives complimentary transportation, hotel, and registration to attend the SfN annual meeting where they will receive the Award. Nominee must be actively involved in teaching and/or outreach activities at the time the prize is given. Types of activities considered include: programs for professional development of teachers and for student research mentorships; development of educational resources, books, magazines, and newspaper articles; broadcasting; lectures; exhibit design; and website development and other public education and outreach activities about neuroscience. Nominators are encouraged to identify candidates whose contributions reach broad audiences that include children, K-12 teachers, women, minorities, and persons with disabilities. Nomination materials should be submitted using SfN's award nomination site. Nominations for the award open on March 7 and close on May 27.
Requirements: Nominations must be made or endorsed by an SfN member, although nominees do not need to be SfN members.
Restrictions: Self nomination is not permitted, and no person may nominate more than one candidate. Current SfN officers and councilors are ineligible for nomination.
Geographic Focus: All States, Puerto Rico, Argentina, Australia, Brazil, Canada, China, Denmark, India, Israel, Malaysia, Mexico, New Zealand, Nigeria, Norway, Singapore, South Korea, Spain, Switzerland, Turkey, Ukraine, United Kingdom
Date(s) Application is Due: May 27
Amount of Grant: 5,000 USD
Contact: Jerold J.M. Chun; (858) 784-8410; fax (858) 534-7309; jchun@scripps.edu
Internet: http://www.sfn.org/Awards-and-Funding/Individual-Prizes-and-Fellowships/Science-Education-and-Outreach/Science-Educator-Award
Sponsor: Society for Neuroscience
1121 14th Street, NW, Suite 1010
Washington, D.C. 20005

SfN Science Journalism Student Awards 3536
The Society for Neuroscience Science Journalism Student Awards give students pursuing a science or medical journalism career a chance to attend the Society's annual meeting – the world's largest source of emerging news about brain science and health – and mentor with a practicing journalist writing about neuroscience issues. Two awards are given each year and include complimentary meeting registration, lodging for four nights, and $750 to help defray the expenses of attending. Every effort is made to pair awardees with a mentor based on mutual interests in print or broadcast journalism. The awardees also receive an orientation with SfN Public Information staff on navigating and reporting on the meeting as a journalist.
Requirements: The award is open to undergraduate and graduate students who are enrolled in or have recently completed formal education in journalism and can demonstrate the intent to cover science or medicine. Students may also apply if they are enrolled in or have recently completed formal education in a scientific or medical field and can demonstrate the intent to cover science and medicine. To apply, candidates send a cover letter describing their career goals and how attending the meeting will help achieve them; a resume or CV; two published clips, if available; and a letter of recommendation from an instructor or professor.
Geographic Focus: All States
Date(s) Application is Due: Sep 11
Amount of Grant: 750 USD
Contact: Award Coordinator; (202) 962-4000; fax (202) 962-4941; pidaward@sfn.org
Internet: http://www.sfn.org/Awards-and-Funding/Individual-Prizes-and-Fellowships/Professional-Development-Awards/Science-Journalism-Student-Award
Sponsor: Society for Neuroscience
1121 14th Street, NW, Suite 1010
Washington, D.C. 20005

SfN Swartz Prize for Theoretical and Computational Neuroscience 3537
The Swartz Prize is presented annually to an individual whose activities have produced a significant cumulative contribution to theoretical models or computational methods in neuroscience, or who has made a noteworthy advance in theoretical or computational neuroscience. The recipient receives $25,000, in addition to complimentary transportation, hotel, and registration to the SfN annual meeting, where the Prize is presented. The recipient also receives a lecture slot at the Dynamical Neuroscience Satellite Event, sponsored by the NIMH. All nomination materials must be submitted using SfN's award nomination site. Nominations for the award open on March 7 and close on June 3.
Requirements: Nominations are open to all individuals who have contributed to theoretical or computational neuroscience. Nominations should be made or endorsed by SfN members, although nominees do not need to be SfN members.
Restrictions: Self nomination is not permitted, and no person may nominate or write a letter of recommendation for more than one candidate. Current SfN officers and councilors are not eligible for nomination.
Geographic Focus: All States, Puerto Rico, Argentina, Australia, Brazil, Canada, China, Denmark, India, Israel, Malaysia, Mexico, New Zealand, Nigeria, Norway, Singapore, South Korea, Spain, Switzerland, Turkey, Ukraine, United Kingdom
Date(s) Application is Due: Jun 3
Amount of Grant: 25,000 USD
Contact: Thomas D. Albright; (858) 453-4100; fax (858) 546-8526; tom@salk.edu
Internet: http://www.sfn.org/Awards-and-Funding/Individual-Prizes-and-Fellowships/Outstanding-Research-and-Career-Awards/Swartz-Prize-for-Theoretical-and-Computational-Neuroscience
Sponsor: Society for Neuroscience
1121 14th Street, NW, Suite 1010
Washington, D.C. 20005

SfN Trainee Professional Development Awards 3538
Trainee Professional Development Awards are presented to deserving young scientists who demonstrate scientific merit and excellence in their research. These awards provide those in the early stages of their career the chance to present a scientific abstract and to meet and network with senior scientists at the SfN annual meeting. The aim of these awards is to promote the advancement of career training to neuroscientists from a wide range of institutions. SfN seeks to promote gender equality and increase diversity, in all

of its forms, throughout its programs. Awards are distributed based on the merit of the trainee's application. Recipients of the Trainee Professional Development Award will receive complimentary registration to attend the annual meeting. An award in the amount of $1,000 will be given to recipients based at institutions within North America. Recipients based at international institutions will receive an award in the amount of $2,000.
Requirements: Applicants must be the first author of an abstract to be presented at the annual meeting, and a postdoctoral fellow. The applicant and his or her advisor must be members of SfN. Nominee is not a recipient of this or another SfN travel award in the same or previous year. Awardees are required to attend at least two professional development workshops at the annual meeting, and to complete a post event survey (to be provided digitally).
Geographic Focus: All States, All Countries
Amount of Grant: 1,000 - 2,000 USD
Contact: Joanne E Berger-Sweeney, Chairperson; (617) 627-3864; fax (617) 627-3703; Joanne.Berger-Sweeney@trincoll.edu
Internet: http://www.sfn.org/Awards-and-Funding/Individual-Prizes-and-Fellowships/Professional-Development-Awards/Trainee-Professional-Development-Awards
Sponsor: Society for Neuroscience
1121 14th Street, NW, Suite 1010
Washington, D.C. 20005

SfN Travel Awards for the International Brain Research Organization World Congress 3539

The Society for Neuroscience (SfN) offers up to 15 travel awards of $2,000 each to support the participation of U.S., Canadian, and Mexican graduate students at the IBRO World Congress held every four years. Recipients of the travel awards are selected by the SfN International Affairs Committee, based on the scientific merit of the abstract submitted to the Congress and the strength of related application materials. Award checks are presented at the conference. Applicants are notified by email of the status of their application. Deadlines vary according to the date of the World Congress.
Requirements: Applicants must be U.S., Canadian, or Mexican citizens or permanent residents, and graduate students who have advanced to candidacy for a Ph.D. Nominees must be first authors of an abstract to be presented at the conference. Applicants and their advisors must be SfN members. Applicants may not be recipients of another SfN travel award in the same award year. In addition to the online application form, applicants must submit: a paragraph describing what they hope to achieve by attending the conference; a copy of the abstract submitted for the conference; a one page letter of recommendation from their advisor; and a one page CV which includes their education, honors/awards, and publications (abstracts and manuscripts).
Geographic Focus: All States, Canada, Mexico
Amount of Grant: 2,000 USD
Contact: Chairperson; (202) 962-4000; globalaffairs@sfn.org
Internet: http://www.sfn.org/Awards-and-Funding/Individual-Prizes-and-Fellowships/Professional-Development-Awards/IBRO-World-Congress-Travel-Awards
Sponsor: Society for Neuroscience
1121 14th Street, NW, Suite 1010
Washington, D.C. 20005

SfN Undergraduate Brain Awareness Travel Award 3540

The Society for Neuroscience and the Faculty for Undergraduate Neuroscience (FUN) cosponsor a travel award to honor an outstanding undergraduate student active in neuroscience education and outreach. The award includes complimentary registration to the SfN annual meeting and a $750 stipend to offset travel expenses. Applications are evaluated on: the importance, quality, and originality of the Brain Awareness work; the student's contribution to the Brain Awareness project; and the candidate's potential as a scientist or a science educator. The online application must be completed by June 26.
Requirements: The student should be an author on a Theme H poster, which should present the results from a Brain Awareness project done while the candidate was an undergraduate. All applications must be sponsored by a dues-paying member of the Faculty for Undergraduate Neuroscience.
Geographic Focus: All States
Date(s) Application is Due: Jun 26
Amount of Grant: 750 USD
Contact: Lisa A. Gabel; (610) 330-5296 or (610) 330-5000; gabell@lafayette.edu
Internet: http://www.funfaculty.org/drupal/BrainAwarenessTravelAward
Sponsor: Society for Neuroscience
1121 14th Street, NW, Suite 1010
Washington, D.C. 20005

SfN Young Investigator Award 3541

The Young Investigator Award recognizes the outstanding achievements and contributions by a young neuroscientist who has recently received his or her advanced professional degree. The recipient receives a $15,000 award, along with complimentary ground or air transportation, hotel, and registration for the SfN annual meeting, where the awards are presented. Nominations must be submitted using SfN's award nomination site. Nominations for the award open on March 7 and close on June 3.
Requirements: Nominees must be active in research at the time the award is given, and have received a Ph.D. or M.D. (or equivalent) within the past ten years. Nominations must be made or endorsed by an SfN members, but nominees do not need to be SfN members.
Restrictions: Self-nomination is not permitted. No person may nominate more than one candidate. Current SfN officers and councilors are ineligible for nomination.
Geographic Focus: All States, Puerto Rico, Argentina, Australia, Brazil, Canada, China, Denmark, India, Israel, Malaysia, Mexico, New Zealand, Nigeria, Norway, Singapore, South Korea, Spain, Switzerland, Turkey, Ukraine, United Kingdom

Date(s) Application is Due: Jun 3
Amount of Grant: 15,000 USD
Contact: Catherine G Dulac; (617) 495-7893; fax (617) 495-1819; dulac@fas.harvard.edu
Internet: http://www.sfn.org/Awards-and-Funding/Individual-Prizes-and-Fellowships/Young-Scientists-Achievements-and-Research/Young-Investigator-Award
Sponsor: Society for Neuroscience
1121 14th Street, NW, Suite 1010
Washington, D.C. 20005

Shaw's Supermarkets Donations 3542

Shaw's focuses its support on not-for-profit 501(c)3 organizations that help create healthy, thriving communities. Shaw's provides financial support, volunteer support and in-kind product donations to organizations that benefit communities in areas where they operate, and meet the following focus areas: (1) Hunger Relief - Ending hunger in the stores' local communities via both product and financial support to hunger-relief organizations. (2) Nutrition Education - support for organizations that educate and promote healthy lifestyles and nutrition by emphasizing disease management and prevention through diet. (3) Environmental Stewardship - environmental stewardship and sustainable operations and that will continually work to use energy more efficiently and reduce waste. Shaw's supports local efforts towards sustainability. (4) Community Connections - select local events that build strong communities by broadly engaging community pride and spirit.
Requirements: 501(c)3 tax-exempt organizations in Connecticut, Maine, Massachusetts, New Hampshire, Rhode Island, and Vermont are eligible. If the grant request involves an event, the request should be sent at least 8 weeks prior to event date. Download the application form from the website.
Restrictions: Shaw's will generally not support: individuals; travel or research expenses; fees for participation in competitive programs; veteran, fraternal or labor organizations; lobbying, political or religious programs; or, organizations that are not tax-exempt.
Geographic Focus: Connecticut, Maine, Massachusetts, New Hampshire, Rhode Island, Vermont
Contact: Donation Committee; shaws.donations@shaws.com
Internet: http://www.shaws.com/pages/community/donationRequests.php
Sponsor: Shaw's Supermarkets
P.O. Box 600
East Bridgewater, MA 02333

Shell Deer Park Grants 3543

The company supports organizations that work to improve the quality of life and the general welfare of its employees and the citizens of their home communities. Contributions are made in the areas of education and workforce development, health and human services, the environment, civic and community betterment, and arts and culture. In addition to local community funds, support may be given to institutions for the handicapped, associations for the blind, or other well-established organizations that enjoy wide community support and receive no support from local United Funds. Financial aid to education is directed primarily to accredited colleges and universities where the company has significant operations, where it recruits, or where research of interest or applicability to its business is being conducted. Approximately one-half of Pennzoil's corporate giving is directed toward educational support. Corporate contributions to hospitals, clinics, rehabilitation centers, and other health-related organizations are restricted to building programs, equipment additions, and unusual research programs. Contributions to campaigns conducted for the more prevalent diseases or disabilities may also be considered. Contributions are made to organizations engaged in the performing or visual arts, museums, zoos, libraries, and other cultural endeavors. Support will be given to selected civic organizations working to improve the quality of life in local Pennzoil communities. There may be causes outside the preceding classifications that are deserving of support and to which consideration will be given on an individual basis.
Requirements: Shell Deer Park will place priority on charitable contributions to eligible nonprofit organizations in nearby Texas communities, including Deer Park, Pasadena, La Porte, and North Channel.
Restrictions: Shell does not generally support: individuals; private foundations; non-profit organizations without a current 501(c)3 exempt status; conferences, workshops, or seminars not directly related to Shell business interests; religious organizations that do not serve the general public on a non-denominational basis; or organizations located in or benefiting nations other than U.S. and its territories.
Geographic Focus: Texas
Contact: Janet Noble; (713) 246-7301 or (713) 246-7137; fax (713) 246-7800
Internet: http://www.shell.us/about-us/projects-and-locations/deer-park-manufacturing-site/social-investment-and-grants.html
Sponsor: Shell Deer Park
5900 Highway 225, P.O. Box 100
Deer Park, TX 77536

Shell Oil Company Foundation Community Development Grants 3544

The mission of the Shell Oil Company Foundation is to help foster the general well being of communities where Shell Oil Company employees live and work and to provide educational opportunities that prepare students and faculty to succeed while meeting the needs of the ever-changing workplace. Focus areas for funding include: community development; education; and the environment. In the area of community development, the focus is on civic and human needs in the community while promoting healthy lifestyles, major and cultural arts that promote access to under-served students and communities, and disaster relief efforts. The Foundation funds a broad array of community outreach projects, particularly in areas where employees work and live. These projects range from local neighborhood improvement efforts to regional non-profit

organizations. The Foundation is especially interested in supporting groups that reflect the diversity and inclusiveness of its communities, which is a Shell core value.
Requirements: Shell will consider charitable contributions to eligible nonprofit organizations with priority consideration given to organizations serving in or near U.S. communities where Shell has a major presence.
Restrictions: Shell Oil Company and the Shell Oil Company Foundation will not consider contributions for the following purposes: individuals; private foundations; non-profit organizations without a current 501(c)3 exempt status; conferences or symposia; endowment funds; fraternal and labor organizations; capital campaigns; conferences, workshops, or seminars not directly related to Shell business interests; religious organizations that do not serve the general public on a non-denominational basis; organizations located in or benefiting nations other than U.S. and its territories; or organizational operating expenses.
Geographic Focus: All States
Amount of Grant: 10,000 - 50,000 USD
Contact: Hasting Stewart, Social Investment Manager; (713) 241-0663 or (713) 241-4544; fax (713) 241-3329; hasting.stewart@shell.com
Internet: http://www.shell.us/sustainability/request-for-a-grant-from-shell.html
Sponsor: Shell Oil Company Foundation
One Shell Plaza 910 Louisiana, Suite 4478A
Houston, TX 77002

Sheltering Arms Fund Grants   3545
The purpose of the fund is to complement and extend Sheltering Arms Hospital's mission by funding health related initiatives that serve uninsured and underinsured adults who are experiencing physical or cognitive disabilities or who are at risk for developing functional limitations. Emphasis will be placed on responding to the needs of individuals with a Rehab Impairment Category diagnosis as defined by the Centers for Medicaid and Medicare (CMS)
Requirements: Proposals will be accepted from charitable organizations, which serve the residents of the metropolitan Richmond and Central Virginia. Proposed programs should align with the goals of the fund: Improve access to health and related support services such as Primary Care, Specialty clinics (Diabetes, Podiatry, Dental, Women's Health, etc.), and Nutritional Support; Increase functional independence (Home care, Day care, Transportation, Use of assistive devices and technology, Facility renovations to improve handicapped access); Promote disability prevention for at risk populations through education and/or public health initiatives.
Geographic Focus: Virginia
Date(s) Application is Due: May 5
Amount of Grant: Up to 100,000 USD
Contact: Susan Hallett; (804) 330-7400; fax (804) 330-5992; shallett@tcfrichmond.org
Internet: http://www.tcfrichmond.org/page2954.cfm#SAF
Sponsor: Community Foundation Serving Richmond and Central Virginia
7501 Boulders View Drive, Suite 110
Richmond, VA 23225

Shield-Ayres Foundation Grants   3546
The Shield-Ayres Foundation was established in Texas in 1977 with a primary interest in supporting health, human services, the environment, education, and the arts. The Foundation's focus is to help children and youth, economically disadvantaged, low-income, and students. Types of funding include: advocacy; annual campaigns; capital campaigns; emergency funding; endowments; financial sustainability; fund raising; general operating support; land acquisition; leadership and professional development; outreach; policy and system reform; program development; education; and volunteer development. Interested parties should begin by forwarding a Letter of Inquiry to the Foundation office. It that inquiry results in a favorable response, an invitation to apply will be extended. Online application deadlines are February 15 and August 15.
Requirements: The Foundation funds organizations with programs that focus primarily in Austin, San Antonio, or select other areas where we have special interests. Organizations are eligible to apply once in a twelve-month period, regardless of the approval status of the organization's last application. Organizations are eligible to receive grants from the Shield-Ayres Foundation for up to three consecutive years.
Restrictions: The Foundation does not: lend or grant money to individuals; make grants intended to support candidates for political office; or grant money to private foundations.
Geographic Focus: Texas
Date(s) Application is Due: Feb 15; Aug 15
Amount of Grant: 5,000 - 25,000 USD
Samples: Alamo College Foundation, San Antonio, Texas, $25,000 - support of the Challenge Center (2014); Environmental Defense Fund, New York, New York, $25,000 - support of the Lester Prairie Chicken Habitat Exchange (2014); Greenlights, Austin, Texas, $10,000 - general operating support (2014).
Contact: Cindy Raab; (512) 467-4021; info@shield-ayresfoundation.org
Internet: http://www.shield-ayresfoundation.org/grant-making/funding-priorities/
Sponsor: Shield-Ayres Foundation
3101 Bee Caves Road, Suite 260
Austin, TX 78746-5574

Shopko Foundation Community Charitable Grants   3547
The Shopko Foundation is proud of Shopko's roots as a retail health and optical care provider. To maximize its impact, the Foundation has a narrow focus on areas of giving that support the health of Shopko customers, teammates and communities. The Foundation also recognizes that education is fundamental to an individual's health and functionality in society. To achieve its vision, the Shopko Foundation believes in supporting community projects that may be accessed by, and its contribution made well known to, customers and teammates of Shopko. Funds will support established non-profit organizations with a proven record of success in maintaining solid, critical programs or innovative new organizations and programs supported by established non-profits or successful leadership. Grant funding will be $1,000 or less.
Requirements: Nonprofit 501(c)3 organizations located within 25 miles of a Shopko store are eligible to apply. Grants to accredited publicly/privately funded schools, colleges, and universities will be considered. Grant requests must contain all required information and be submitted at least 45 days prior to the date of the scheduled event to ensure sufficient time for review. Requests should be related to a specific program or project, rather than related to general fundraising.
Restrictions: In general, the Shopko Foundation does not support the following: Programs or events that do not support the Foundation's mission; Programs or events outside of Shopko communities; Sponsorship of cultural exhibits; Events which provide assistance to a specific individual; Advertising in event programs or yearbooks; Religious organizations (however gifts designated for, and restricted to, human services or humanitarian purposes may be eligible); Political or fraternal organizations; Events with multiple or competing business sponsors; Organizations that discriminate on the basis of sex, creed, national origin or religion; Charitable requests in support of raffle, auctions, benefits or similar fundraising events. Applications via postal mail will not be accepted.
Geographic Focus: California, Idaho, Illinois, Indiana, Iowa, Kansas, Kentucky, Michigan, Minnesota, Missouri, Montana, Nebraska, North Dakota, Ohio, Oregon, South Dakota, Utah, Washington, Wisconsin, Wyoming
Amount of Grant: Up to 1,000 USD
Contact: Michelle Hansen, Program Director; (920) 429-4054; fax (920) 496-4225; shopkofoundation@shopko.com or michelle.hansen@shopko.com
Internet: http://www.shopko.com/thumbnail/Company/Community-Giving/Shopko-Foundation/pc/2176/c/2181/2185.uts?&pageSize=
Sponsor: Shopko Foundation
700 Pilgrim Way
Green Bay, WI 54304

Shopko Foundation Green Bay Area Community Grants   3548
The Shopko Foundation is proud of Shopko's roots as a retail health and optical care provider. To maximize its impact, the Foundation has a narrow focus on areas of giving that support the health of Shopko customers, teammates and communities. The Foundation also recognizes that education is fundamental to an individual's health and functionality in society. The Foundation strives to enhance the quality of life in the Green Bay area through charitable causes, events and activities that support healthy lifestyles for residents. Its goal is to make Shopko communities a better place to live by supporting programs and services that improve the health and education of its residents. Grants awarded through this program support larger or long-term community wide projects.
Requirements: Nonprofit 501(c)3 organizations located in the Green Bay area are eligible to apply. Grants to accredited publicly/privately funded schools, colleges, and universities will be considered. Requests should be related to a specific program or project, rather than related to general fundraising.
Restrictions: In general, the Shopko Foundation does not support the following: Programs or events that do not support the Foundation's mission; Programs or events outside of Shopko communities; Sponsorship of cultural exhibits; Events which provide assistance to a specific individual; Advertising in event programs or yearbooks; Religious organizations (however gifts designated for, and restricted to, human services or humanitarian purposes may be eligible); Political or fraternal organizations; Events with multiple or competing business sponsors; Organizations that discriminate on the basis of sex, creed, national origin or religion; Charitable requests in support of raffle, auctions, benefits or similar fundraising events. Applications via postal mail will not be accepted.
Geographic Focus: Wisconsin
Contact: Michelle Hansen, Program Director; (920) 429-4054; fax (920) 496-4225; shopkofoundation@shopko.com or michelle.hansen@shopko.com
Internet: http://www.shopko.com/thumbnail/Company/Community-Giving/Shopko-Foundation/pc/2176/c/2181/2185.uts?&pageSize=
Sponsor: Shopko Foundation
700 Pilgrim Way
Green Bay, WI 54304

Sick Kids Foundation Community Conference Grants   3549
The Sick Kids Foundation Community Conference Grant program supports events which are organized by or for families with children with health challenges, including but not limited to children with acute illness, chronic illness, and disabilities. Support is offered for conferences, workshops or symposia which are relevant to the health of Canada's children. These events focus on information sharing with families and health professionals and/or community organizations. Funding may be awarded for the following items: expenses associated with keynote speakers, i.e. honoraria, economy travel, budget hotel accommodation and meals; audiovisual equipment rental required for presentations; write up of conference for wider distribution after the event (in newsletter, on website etc.); registration fees for parents, children and community groups; on-site babysitting; and conference facilities. Awards are limited to an annual maximum request of $5,000. There are three deadlines for applications each fiscal year: May 31; September 30; and January 31.
Requirements: Eligible events must: address issues that are relevant to child health in Canada; support the parent-child-professional partnership by having a focus on information sharing between families, health professionals, and community organizations; include knowledgeable and credible presenters; take place in Canada; and be sponsored by a registered Canadian charitable organization.
Restrictions: Academic conferences are not eligible for funding. The following items are not eligible for Foundation funding through this program: salaries; computer rentals;

registration fees for speakers or conference planners; planning meetings; or individuals seeking reimbursement for attending or presenting at events. The Foundation will only fund organizations.
Geographic Focus: Canada
Date(s) Application is Due: Jan 31; May 31; Sep 30
Amount of Grant: Up to 5,000 CAD
Contact: Ted Garrard, President; (416) 813-6166 or (800) 661-1083; fax (416) 813-5024; ted.garrard@sickkidsfoundation.com
Internet: http://www.sickkidsfoundation.com/about-us/grants/community-conference-grants
Sponsor: Sick Kids Foundation
525 University Avenue, 14th Floor
Toronto, ON M5G 2L3 Canada

Sick Kids Foundation New Investigator Research Grants 3550
Sick Kids Foundation New Investigator Research Grants are jointly sponsored by SickKids Foundation and the Canadian Institutes of Health Research Institute of Human Development, Child and Youth Health. Grant recipients may obtain up to three years of support for research in the biomedical, clinical, health systems and services, population and public health sectors. The program provides support to child health researchers early in their careers. It is intended that New Investigator Research Grants will enhance the grant recipient's capacity to compete with more senior investigators for research grants from other funders. The Foundation funds research that has the potential to have a significant impact on child health outcomes. Multi-year grants range up to $300,000.
Requirements: The applicant must be a new investigator as of the application deadline. For the purpose of this award, a new investigator is defined as an individual who as a principal or co-principal investigator has not received combined operating grant funding of $500,000 or more (Canadian dollars) and is within five years of their first academic appointment. Academic appointment is defined as an appointment which allows an individual to apply for research grants as an independent investigator.
Geographic Focus: Canada
Amount of Grant: Up to 300,000 CAD
Samples: Dr. Jacob Jaremko, University of Alberta, Edmonton, Alberta, Canada, $257,245 - 3D Ultrasound in Hip Dysplasia (2014); Dr. Alexander Beristain, University of British Columbia, Vancouver, British Columbia, Canada, $299,385 - Effect of Obesity-Associated Inflammation on the Maternal-Fetal Interface in Early Pregnancy (2014); Dr. Sarah Fraser, University of Montreal, Montreal, Quebec, Canada, $283,288 - Collaborative Mental Health Care from Communities, to Professionals, to Policy Change for Inuit of Canada (2014).
Contact: Ted Garrard, President; (416) 813-6166 or (800) 661-1083; fax (416) 813-5024; ted.garrard@sickkidsfoundation.com
Internet: http://www.sickkidsfoundation.com/about-us/grants/new-investigator-research-grants
Sponsor: Sick Kids Foundation
525 University Avenue, 14th Floor
Toronto, ON M5G 2L3 Canada

Sidgmore Family Foundation Grants 3551
The Sidgmore Family Foundation honors the legacy of John W. Sidgmore by taking a proactive approach to helping others succeed. The Foundation desires to use its resources to find creative and innovative solutions so that people may achieve their full potential and become responsible, healthy and productive members of society. In recognition that an impoverished environment limits the possibilities for people to develop and thrive, the Sidgmore Family Foundation is particularly interested in funding organizations that: improve the quality of education and teacher training; further the advancement of knowledge in the field of medicine with a special emphasis on hearing and cardiology; utilize entrepreneurial skills to explore and develop creative, scalable, and sustainable solutions to critical social problems; and, provide support and services to those in need in the Washington D.C. area. Grants are awarded to organizations that have a clear, replicable plan for success, measured sustainable results, and high approval ratings from charity evaluator organizations, such as Charity Navigator. The Foundation also awards multi-year grants that can sustain a program or project.
Requirements: Nonprofit 501(c)3 organizations are eligible. Preference is given to organizations that serve residents in the Washington, D.C. metropolitan area, Maryland, and Virginia. Applicants must begin the process with an initial Letter of Inquiry (LOI). LOIs will receive a response if the Foundation wishes to receive a proposal from your organization.
Restrictions: The Sidgmore Family Foundation does not make grants to individuals, national health organizations, government agencies, or political and public policy advocacy groups.
Geographic Focus: All States
Contact: M. Gelbwaks, Director; (516) 541-2713; SidgmoreFound@aol.com
Internet: http://www.sidgmorefoundation.com/#application_process
Sponsor: Sidgmore Family Foundation
71 Leewater Avenue
Massapequa, NY 11758

Sidney Stern Memorial Trust Grants 3552
The Sidney Stern Memorial Trust provides grants and funding to various non-profit organizations. The board of the trust meets regularly to review grant applications and make funding decisions. Areas of interest for grants include civil rights, children, community development, disabled causes, education, health, science, social services and Native Americans. Recipients of grants include local public media outlets, educational programs, archaeological centers, international aid organizations and performing arts programs for children.
Requirements: The trust prefers all correspondence to be sent by mail. No personal or email requests will be entertained. All organizations applying for funding must be recognized by the IRS as 501(c)3 certified, verifiable via GuideStar Charity Check. Folders containing lengthy brochures, pictorial pamphlets or CDs should not be included. Although the foundation supports projects and organizations across the country, most grants are offered to proposals from California.
Restrictions: The trust does not award grants to individuals, political candidates or campaigns, lobbying projects or programs to directly influence legislation, or for conferences or redistribution.
Geographic Focus: California
Amount of Grant: 750 USD
Contact: Betty Hoffenberg, Chairperson; (800) 352-3705
Internet: http://sidneysternmemorialtrust.org/uploads/Application_and_Guidelines_for_Grant_FINAL.pdf
Sponsor: Sidney Stern Memorial Trust
P.O. Box 457
Pacific Palisades, CA 90272

Sid W. Richardson Foundation Grants 3553
Grants are provided to tax-exempt organizations in Texas in the areas of education (museums, learning centers, day schools, K-12 schools, higher education institutions, and business and economic education), health (medical schools, organ donor registries, hospitals, disease prevention, health science centers, and nursing associations), arts (arts councils, visual and performing arts festivals, museums, ballet, symphony orchestra, and arts education programs), and human services (boys' and girls' clubs, united funds, the elderly, crime prevention, the disabled, housing opportunities, food programs, and drug and alcohol abuse prevention). Types of support include operating budgets, seed grants, building construction funds, equipment acquisition, endowment funds, research, publications, conferences and seminars, matching funds, continuing support, and projects/programs. Award amounts vary depending on proposed projects. Applications must be received no later than January 15.
Requirements: Grant requests will be considered only from tax-exempt organizations described in Section 501(c)3 of the Internal Revenue Code of 1986 and classified as other than a private foundation within the meaning of Section 509(a) of the Code, or from a qualified public entity described in Section 170 of the Code.
Restrictions: Grants are not made to individuals, or for the support of school trips, testimonial dinners, fundraisers, or marketing events. An organization is limited to one application per calendar year.
Geographic Focus: Texas
Date(s) Application is Due: Jan 15
Amount of Grant: 10,000 - 100,000 USD
Samples: Back to School Roundup, Fort Worth, Texas, $10,000 - general support for the 7th annual Tarrant County Back to School Roundup; Baylor College of Medicine Department of Internal Medicine, Houston, Texas, $100,000 - support for Dr. James L. Pool's program in Internal Medicine; Christ's Haven for Children, Keller, Texas, $7,500 - general support for the program assisting foster children.
Contact: Carolyn Johns, (817) 336-0494; fax (817) 332-2176; cjohns@sidrichardson.org or info@sidrichardson.org
Internet: http://www.sidrichardson.org/grants/
Sponsor: Sid W. Richardson Foundation
309 Main Street
Fort Worth, TX 76102

Siebert Lutheran Foundation Grants 3554
Specific areas of interest vary from time to time. At present, the Foundation is supportive of the following: Clergy and lay education and training, community development and outreach, health ministry, evangelism and youth. Grants are occasionally made to provide seed money or start-up costs for a program or project. In such instances, the participation of other donors is desired. Grants are made only to Lutheran organizations exempt under Section 501(c)3 of the Internal Revenue Code and are generally awarded for a one-year period.
Requirements: The Foundation utilizes an on-line grant application. Potential grant applicants, who represent Wisconsin congregations or Wisconsin recognized service organizations in the Lutheran church, may complete the Foundation's formal, online, letter of inquiry. A telephone call is not necessary prior to completing the letter of inquiry for Lutheran organizations in Wisconsin. If your organization does not meet the above criteria and you would like to discuss possible funding, contact the Foundation office (262-754-9160).
Restrictions: The Foundation does not approve grants for the following: endowment funds, fellowships and scholarships, trusts, and other grant-making foundations. Grants are generally not made to churches for capital or operating expenses. No grants are made outside the United States.
Geographic Focus: Wisconsin
Date(s) Application is Due: Mar 1; Jun 1; Sep 1; Dec 1
Contact: Deborah Engel, Administrative Assistant; (262) 754-9160; fax (262) 754-9162; contactus@Siebertfoundation.org
Internet: http://www.siebertfoundation.org/grants.htm
Sponsor: Siebert Lutheran Foundation
300 North Corporate Drive, Suite 200
Brookfield, WI 53045

Sierra Health Foundation Responsive Grants 3555
The Sierra Health Foundation Responsive Grants are designed to promote health and well-being in Northern California communities. Fields of interest include: AIDS; alcoholism; biomedicine; child development, education; child development, services; children/youth, services; community/economic development; crime/violence/ prevention, youth; family services; health care; health organizations, association; human services; leadership

development; medical care, rehabilitation; mental health/crisis services; nutrition; substance abuse, services; and youth development. The Foundation funds employee matching gifts, in-kind gifts, program development, program evaluation; and technical assistance. Examples of Foundation funding include food banks, homeless shelters, senior citizen agencies, youth and family centers, job readiness programs, and faith based organizations.
Requirements: Organizations may contact the Foundation for current application procedures and deadlines.
Restrictions: The Foundation funds the following northern California counties: Alpine, Amador, Butte, Calaveras, Colusa, El Dorado, Glenn, Lassen, Modoc, Mono, Nevada, Placer, Plumas, Sacramento, San Joaquin, Shasta, Sierra, Siskiyou, Solano (eastern), Stanislaus, Sutter, Tehama, Trinity, Tuolumne, Yolo and Yuba. The Foundation does not fund individuals or endowments.
Geographic Focus: California
Amount of Grant: Up to 25,000 USD
Contact: Kathy Mathews, Grants Administrator; (916) 922-4755; fax (916) 922-4024; kmathews@sierrahealth.org or grants@sierrahealth.org
Internet: http://www.sierrahealth.org/doc.aspx?129
Sponsor: Sierra Health Foundation
1321 Garden Highway
Sacramento, CA 95833-9754

Sigma Theta Tau International /American Nurses Foundation Grant   3556
Sigma Theta Tau International (STTI), the only international nursing honor society worldwide, is a global community of nurse leaders dedicated to using knowledge, scholarship, service, and learning to improve the health of the world's people. STTI members belong to 475 chapters in 90 countries. STTI and the American Nurses' Foundation (ANF) have joined forces to provide funding for the STTI/ANF research grant. The purpose of the grant is to encourage the research career development of nurses through support of research conducted by beginning nurse researchers or experienced nurse researchers who are entering a new field of study. Applications must be submitted directly to STTI via their online submission system by the May 1 deadline date. Funding will begin in October. Links to an FAQ, application-review criteria, and help with grant-writing are available at the website.
Requirements: Eligible applicants must meet the following criteria: be a registered nurse with a current license; have a master's or doctoral degree or be enrolled in a doctoral program; have a clinical nursing research topic; submit a completed research application packagbe and signed research agreement; and be ready to implement the research project when funding is received. Preference will be given to Sigma Theta Tau members, other qualifications being equal. Allocation of funds is based on the quality of the proposed research, the future promise of the applicant, and the applicant's research budget.
Restrictions: Funds for this grant do not cover expenses incurred prior to the funding date.
Geographic Focus: All States, All Countries
Date(s) Application is Due: May 1
Amount of Grant: Up to 7,500 USD
Contact: Research Services; (888) 634-7575 or (317) 634-8171; fax (317) 634-8188; research@stti.iupui.edu
Internet: http://www.nursingsociety.org/Research/Grants/Pages/grant_anf.aspx
Sponsor: Sigma Theta Tau International
550 W. North Street
Indianapolis, IN 46202-2156

Sigma Theta Tau International / Council for the Advancement of Nursing Science Grants   3557
Sigma Theta Tau International (STTI), the only international nursing honor society worldwide, is a global community of nurse leaders dedicated to using knowledge, scholarship, service, and learning to improve the health of the world's people. STTI members belong to 475 chapters in 90 countries. STTI and the Council for the Advancement of Nursing Science (CANS) have joined forces to provide funding for the STTI/CANS research grant. The purpose of the grant is to encourage qualified nurses to improve the health of the world's people through research. Proposals for clinical, educational or historical research, including plans for broadly disseminating the research findings, may be submitted for the grant. Applications must be submitted directly to STTI via their online submission system by the July 1 deadline date. Funding will begin in November. Links to an FAQ, application-review criteria, and help with grant-writing are available at the website.
Requirements: Eligible applicants must meet the following criteria: be a member in good standing of both STTI and CANS; hold at minimum a Master's degree or equivalent; submit a completed research application via STTI's on-line submission system; be ready to implement research project when funding is received; submit a final report; submit a completed abstract to Virginia Henderson International Nursing Library; submit the completed project abstract for presentation at the Sigma Theta Tau or Council for Advancement of Nursing Science research meeting; and credit research-grant partners in all publications and presentations of the research.
Restrictions: Funds for this grant do not cover expenses incurred prior to the funding date.
Geographic Focus: All States, All Countries
Date(s) Application is Due: Jul 1
Amount of Grant: Up to 5,000 USD
Contact: Research Services; (888) 634-7575 or (317) 634-8171; research@stti.iupui.edu
Internet: http://www.nursingsociety.org/Research/SmallGrants/Pages/CANS.aspx
Sponsor: Sigma Theta Tau International
550 W. North Street
Indianapolis, IN 46202-2156

Sigma Theta Tau International / Oncology Nursing Society Grant   3558
Sigma Theta Tau International (STTI), the only international nursing honor society worldwide, is a global community of nurse leaders dedicated to using knowledge, scholarship, service, and learning to improve the health of the world's people. STTI members belong to 475 chapters in 90 countries. STTI and the Oncology Nurses Society (ONS) have joined forces to provide funding through ONS's Small Research Grants program for this grant. From the perspectives of both organizations, the purpose of the grant is to stimulate clinically-related oncology-nursing research, increase the knowledge base for oncology-nursing practice, and to prepare future oncology-nurse researchers. Application guidelines are available through ONS; applications must be submitted directly to ONS. Contact information and a link to the ONS website is available at the grant website.
Requirements: Eligible applicants must meet the following criteria: be a registered nurse actively involved in some aspect of cancer-patient care, education or research and have earned a master's degree. Other qualifications being equal, preference will be given to Sigma Theta Tau International members.
Geographic Focus: All States, All Countries
Date(s) Application is Due: Oct 1
Amount of Grant: 10,000 USD
Samples: Kathleen Ruccione, M.P.H., R.N., C.P.O.N.®, F.A.A.N. (USA), $10,000 - "Iron, Anthracyclines and Cardiac Outcomes among Childhood Cancer Survivors"; Shu-Feng Sung, Ph.D., R.N. (USA), $10,000 - " Translaton Cardiovascular Risk and Breast Cancer Outcomes"; Sharon Kozachik, R.N., Ph.D. (USA), $10,000 - "Sleep, HPA Axis Activity and Paclitaxel-Induced Neuropathic Pain".
Contact: Debbie Kubiak, Grants Coordinator; (412) 859-6224 or (866) 357-4667; fax (877) 369-5497; dkubiak@ons.org
Internet: http://www.nursingsociety.org/Research/Grants/Pages/grant_ons.aspx
Sponsor: Sigma Theta Tau International
550 W. North Street
Indianapolis, IN 46202-2156

Sigma Theta Tau International Doris Bloch Research Award   3559
Sigma Theta Tau International (STTI), the only international nursing honor society worldwide, is a global community of nurse leaders dedicated to using knowledge, scholarship, service, and learning to improve the health of the world's people. STTI members belong to 475 chapters in 90 countries. The purpose of STTI's Doris Bloch Research Award is to encourage nurses to contribute to the advancement of nursing through research. Allocation of funds is based on the quality of the proposed research, the future promise of the applicant, and the applicant's research budget. Applications from novice researchers who have received no other national research funds are encouraged and will receive preference for funding, other aspects being equal. Preference will be given to Sigma Theta Tau International members, other qualifications being equal. Applications must be submitted via STTI's online submission system. A link to the submission system will be available in July. Funding will begin in June. Links to an FAQ, application-review criteria, and help with grant-writing are available at the website. Funding is provided by a gift to the honor society's Research Endowment from the Doris Bloch estate.
Requirements: Eligible applicants must meet the following criteria: be a registered nurse with a current license; hold a master's or doctoral degree or be enrolled in a doctoral program; submit a completed research application package and signed research agreement; be ready to implement the research project when funding is received; complete the project within one (1) year of funding; submit to STTI a final report; and submit a completed abstract to STTI's Virginia Henderson International Nursing Research Library.
Geographic Focus: All States, All Countries
Date(s) Application is Due: Dec 1
Amount of Grant: Up to 5,000 USD
Samples: Carina Katigbak, M.S., B.Sc.N. (USA) - "Exploring the role of Community Health Workers upon Hypertension Related Behaviors with Filipino Americans in New York City"; Hsiu-Chin Chen, R.N., Ph.D., Ed.D. (USA) - "The significance of student input to nursing program evaluation: Part II"; Jennifer Hobbs, R.N., Ph.D. (USA) - "Standardizing the Particular in Clinical Information Systems: Nursing's Role in the Development of Clinical Standards during the Late 20th Century".
Contact: Research Services; (888) 634-7575 or (317) 634-8171; fax (317) 634-8188; research@stti.iupui.edu
Internet: http://www.nursingsociety.org/Research/Grants/Pages/grant_bloch.aspx
Sponsor: Sigma Theta Tau International
550 W. North Street
Indianapolis, IN 46202-2156

Sigma Theta Tau Small Grants   3560
Sigma Theta Tau International (STTI), the only international nursing honor society worldwide, is a global community of nurse leaders dedicated to using knowledge, scholarship, service, and learning to improve the health of the world's people. STTI members belong to 475 chapters in 90 countries. The purpose of STTI's Small Grants Program is to encourage nurses to contribute to the advancement of nursing through research. Ten to fifteen grants are awarded annually. Allocation of funds is based on the quality of the proposed research, the future promise of the applicant, and the applicant's research budget. Applications from novice researchers who have received no other national research funds are encouraged and will receive preference for funding, other aspects being equal. Preference will be given to Sigma Theta Tau members, other qualifications being equal. Applications must be submitted via STTI's online submission system. A link to the submission system will be available in July. Funding will begin in June. Links to an FAQ, application-review criteria, and help with grant-writing are available at the website.
Requirements: Eligible applicants must meet the following criteria: be a registered nurse with a current license; have a master's or doctoral degree or be enrolled in a doctoral

program; have a clinical-nursing research topic; submit a completed research application package and signed research agreement; be ready to implement the research project when funding is received; complete the project within one (1) year of funding; submit a final report; and submit a completed abstract to STTI's Virginia Henderson International Nursing Research Library.
Restrictions: Funds for this grant do not cover expenses incurred prior to the funding date.
Geographic Focus: All States, All Countries
Date(s) Application is Due: Dec 1
Amount of Grant: Up to 5,000 USD
Samples: Tara Albrecht, Ph.D., M.S.N., B.S.N. (USA) - "Symptom Management and Psychosocial needs in Patients with Acute Myeloid Leukemia"; Kathleen J. Finlayson, Ph.D., M.S.N. (Australia) - "Identifying Effective New Interventions to Prevent Recurrence of Venous Leg Ulcers in Adults with Chronic Venous Insufficiency"; Orla Marie Smith, M.N. (USA) - "Prevalence and Predictors of Anxiety, Depression and Risk for Post-Traumatic Stress in Family Members of Patients in the Intensive Care Unit: The ICU-Adapts Survey".
Contact: Research Services; (888) 634-7575 or (317) 634-8171; fax (317) 634-8188; research@stti.iupui.edu
Internet: http://www.nursingsociety.org/Research/Grants/Pages/small_grants.aspx
Sponsor: Sigma Theta Tau International
550 W. North Street
Indianapolis, IN 46202-2156

Sigrid Juselius Foundation Grants 3561
The Sigrid Juselius Foundation, offers funding for domestic medical research. To enable established foreign medical researchers to work in a Finnish institution for 1-12 months, the Foundation also offers grants to nationals of other countries. Resources are allocated to senior and post doc researchers. The grants cover all disciplines of medical science. The grants can cover personal expenses (living costs) and travel.
Requirements: The Foundation does not award direct grants to foreign medical research. A foreign scientist interested in coming over to Finland has to get in touch with a Finnish scientist of his particular field, and the application should be made on the part of the Finnish host together with the proposed visiting researcher. A electronic application is available on the Sigrid Juselius Foundation website.
Restrictions: The funds are not available for studies or doctoral theses.
Geographic Focus: All States
Date(s) Application is Due: Apr 15; Sep 15
Contact: Funding Officer; 358 20 710 9082; fax 358 20 710 9089
Internet: http://www.sigridjuselius.fi/index.php?page_id=48
Sponsor: Sigrid Juselius Foundation
Aleksanterinkatu 48 B
Helsinki, FI-00100 Finland

Simeon J. Fortin Charitable Foundation Grants 3562
The Simeon J. Fortin Charitable Foundation was established in 1986 to promote the advancement of cancer research. The Fortin Charitable Foundation supports cancer research organizations dedicated to discovering a cure. To better support the capacity of nonprofit organizations, multi-year funding requests are strongly encouraged. Applicants must apply online at the grant website. Applicants are strongly encouraged to do the following before applying: review the downloadable state application procedures for additional helpful information and clarifications; review the downloadable online-application guidelines at the grant website; review the foundation's funding history (link is available from the grant website); review the online application questions in advance; and review the list of required attachments. These will generally include: a list of board members, financial statements (audited, reviewed, or compiled by independent auditor); an organization summary; a list of other funding sources; an IRS Determination letter; and other required documents. All attachments must be uploaded in the online application as PDF, Word, or Excel files. The application deadline for the Simeon J. Fortin Charitable Foundation is 11:59 p.m. on October 15. Applicants will be notified of grant decisions before December 31.
Geographic Focus: Massachusetts
Date(s) Application is Due: Oct 15
Samples: Childrens Hospital Corporation, Boston, Massachusetts, $90,000; Children's Hospital Corporation, Boston, Massachusetts, $75,000; UMass Memorial Foundation, Worcester, Massachusetts, $50,000.
Contact: Michealle Larkins; (866) 778-6859; michealle.larkins@baml.com
Internet: https://www.bankofamerica.com/philanthropic/fn_search.action
Sponsor: Simeon J. Fortin Charitable Foundation
225 Franklin Street, 4th Floor, MA1-225-04-02
Boston, MA 02110

Simmons Foundation Grants 3563
The foundation awards grants to Maine nonprofits in its areas of interest, including arts, higher education, family services, health care, food services, health organizations, human services, children and youth, services and women. Types of support include general operating support, building construction/renovation, equipment acquisition, and scholarship funds.
Requirements: There are no application forms or deadlines however your initial approach should be in the form of a letter. Applicants should submit the following: detailed description of project and amount of funding requested; brief history of organization and description of its mission; descriptive literature about organization; copy of IRS Determination Letter.
Geographic Focus: Maine
Contact: Suzanne McGuffey, Treasurer; (207) 774-2635
Sponsor: Simmons Foundation
1 Canal Plaza
Portland, ME 04101-4098

Simone and Cino del Duca Grand Prix Awards 3564
The Simone and Cino del Duca Foundation of the Institut de France awards each year a scientific Grand Prix based on the recommendations of a jury of members of the Institut de France. The amount of the Grand Prix will be 364,000 euros: 250,000 euros to finance the work of the award-winning group, 50,000 euros awarded to the director of the research group, and an additional 64,000 euros towards hiring one or more French or foreign postdoctoral researchers to be selected by the group director.
Requirements: The Grand Prix will reward an internationally recognized French or European mathematician proposing an ambitious research project based on the collective work of a group under his/her direction. The proposal should describe the scientific objectives and the means that will be used to implement the project (duration of the project, invitations, meetings, seminars, etc.).
Geographic Focus: All States, Albania, Andorra, Armenia, Austria, Azerbaijan, Belarus, Belgium, Bosnia & Herzegovina, Bulgaria, Croatia, Cyprus, Czech Republic, Denmark, Estonia, Finland, France, Georgia, Germany, Greece, Hungary, Iceland, Ireland, Italy, Kosovo, Latvia, Liechtenstein, Lithuania, Luxembourg, Macedonia, Malta, Moldova, Monaco, Montenegro, Norway, Poland, Portugal, Romania, Russia, San Marino, Serbia, Slovakia, Slovenia, Spain, Sweden, Switzerland, The Netherlands, Turkey, Ukraine, United Kingdom, Vatican City
Date(s) Application is Due: Jan 25
Amount of Grant: Up to 364,000 EUR
Contact: Administrator; 01 47 66 01 21; fondation-del-duca@institut-de-france.fr
Internet: http://www.enseignementsup-recherche.gouv.fr/cid23192/call-for-applications-for-the-grand-prix-scientifique-simone-and-cino-del-duca-2009.html
Sponsor: Institut de France
23, quai Conti
Paris, 75270 France

Simple Advise Education Center Grants 3565
The Simple Advise Education Center supports organizations, centers & services, primarily in North Carolina. Giving to improve the quality of life of citizens with developmental disabilities.
Requirements: There are no specific deadlines with which to adhere. Contact the Foundation for further application information and guidelines.
Geographic Focus: North Carolina
Contact: Barbara J. Spigner, Director; (800) 677-7306
Sponsor: Simple Advise Education Center c/o BJ Hills and Assoc.
P.O. Box 758
Fayetteville, NC 28302-0758

Simpson Lumber Charitable Contributions 3566
Simpson has been in the forest products business since 1890 and currently has two operating subsidiaries: Simpson Door Company and Simpson Tacoma Kraft Company, LLC. The mission of Simpson's contributions program is to improve the quality of life in communities where the company has a significant number of employees living and working; and to serve as a catalyst for employees to become involved and to provide leadership in their communities. The company's broad areas of interest include education, health and human services, and efforts to enhance its operating communities. Contributions are generally made in locations where the company has operations. To the extent possible, contributions will support organizations of interest to, or recommended by, Simpson employees. Simpson prefers to make capital contributions that will benefit the operating communities for the long term as opposed to contributing operating funds. Generally, support is committed for one year at a time and in amounts less than $5,000. Interested applicants should write to Simpson's Public Affairs department to request application materials. Simpson's review committees meet once per year to consider funding applications. The application deadline is May 15. For requests less than $1,000, applicants should contact Simpson's Public Affairs department.
Requirements: 501(c)3 organizations that serve Pierce, Thurston, Lewis, Mason, Grays Harbor, and Cowlitz counties in Washington are eligible to apply. Criteria taken into account in determining the amount of any contributions are as follows: degree of support from company employees; relative size and importance of company operations in the community (balance among Simpson communities); needs of organization or program for which funding is requested; amount of previous company contributions to the organization; amount committed by other companies, foundations, and/or governments (projects should demonstrate broad-based community support); and proximity of the requesting organization to Simpson operations or administrative offices.
Geographic Focus: Washington
Date(s) Application is Due: May 15
Amount of Grant: Up to 5,000 USD
Contact: Raymond P. Tennison, President; (253) 779-6400
Beverly Holland, Public Affairs Manager; (253) 779-6400
Internet: http://www.simpson.com/communitycontribute.cfm
Sponsor: Simpson Lumber Company, LLC
917 East 11th Street
Tacoma, WA 98421

Singing for Change Foundation Grants 3567

The Singing for Change Foundation offers annual competitive grants to progressive U.S. nonprofit organizations that address the root causes of social or environmental problems. SFC is interested in funding projects that improve the quality of life for people and that empower individuals to effect positive change in their communities. Most likely to be considered are organizations that keep their overhead low and collaborate with other groups in their community to find innovative ways of solving common problems. Areas of interest include children and youth--health, education, and protection of children and their families; disenfranchised groups--projects that help people overcome social or economic barriers to education or employment, promote the empowerment of individuals toward self-sufficiency and provide opportunities for personal growth, and demonstrate human equality and encourage people to cross boundary lines to help others and the environment; and the environment--programs that promote environmental awareness and teach people methods of conservation, protection and the responsible use of natural resources. Submit a one-page letter of interest describing the organization and project; full proposals are by invitation.
Requirements: U.S. nonprofit organizations are eligible.
Restrictions: The Singing for Change Foundation does not consider grants to: individuals; government agencies; public or private schools; art, music, or recreational programs, even if offered to disenfranchised groups; political organizations; religious organizations; medical research or disease treatment organizations; basic-needs programs (that exist to supply food or clothing); single service programs such as individual counseling.
Geographic Focus: All States
Amount of Grant: 1,000 - 10,000 USD
Samples: America's Second Harvest of Coastal Georgia, Savannah, Georgia, $5,000 - support for the Kids' Cafe, a unique program dealing with poverty; Bahama Village Music Program, Key West, Florida, $4,009 - support for free music education to the children of Bahama Village residents; Boys Hope Girls Hope of New Orleans, New Orleans, Louisiana, $1,000 - general operations.
Contact: Judith Ranger Smith, Executive Director; (843) 388-7730; judithrangersmith@gmail.com or info@singingforchange.com
Internet: http://www.singingforchange.org/grant_information.html
Sponsor: Singing for Change Foundation
P.O. Box 729
Sullivan's Island, SC 29482

Sioux Falls Area Community Foundation Community Fund Grants 3568

The purpose of the Sioux Falls Area Foundation (SFACF)'s unrestricted grantmaking program is to provide support across a wide spectrum of charitable needs and interests. Grantmaking categories include Arts and Humanities (e.g., theatre, music, arts, dance, cultural development, historic preservation, library programs, and museums); Community Affairs and Development (e.g., citizen participation, public use of parks and recreation, administration of justice, economic development, employment, and training); Education (e.g., lifelong-learning activities in formal educational settings, support of educational facilities and systems, and scholarships); Environment (e.g., protection of natural areas, conservation of energy, prevention and elimination of pollution or hazardous waste, wildlife protection, and water quality); Health (e.g., improvement of healthcare, prevention of substance abuse; support of mental-health needs, and medical research); Human Services (e.g., assistance to families, youth, the elderly, disabled, special groups, social service providers, and those who stand in need); and Religion (e.g., support for churches, religious institutions, and religion programs). SFACF offers two grant programs from its unrestricted funds: Spot Grants for projects up to $3,000 and Community Fund Grants for projects over $3,000. The majority of Community Fund Grants are made in the range of $5,000 - $10,000. Proposals for Community Fund Grants must be submitted using a standard application form (available from the SFACF office or downloadable from the website). Applications are accepted anytime and will be reviewed by the Grants Committee at their next scheduled meeting. Meetings take place six times a year: January, March, May, July, September, and November (a schedule is posted at the SFACF website).
Requirements: Nonprofit organizations serving residents in the Sioux Falls, South Dakota area (Minnehaha, Lincoln, McCook, and Turner counties) are eligible to apply. SFACF considers grant requests for programs that require start-up funds to address important community needs or opportunities, expansion of programs that meet important community needs or opportunities, assistance to organizations weathering unforeseen or unusual financial crises; programs that increase an organization's capacity to advance its mission more efficiently or effectively; and programs or studies that inform the community's understanding of needs or opportunities. Requests are evaluated by the following criteria: comparative benefit to the community; the organization's capacity to achieve the stated objectives; the amount of support requested versus the number of people benefited; a well-planned approach to achieving stated objectives; a reasonable expectation that the program can be sustained over time (where applicable); the organization's history of working collaboratively to address community needs and opportunities; and when applicable, the organization's past SFACF grant performance.
Restrictions: SFACF does not consider grant requests for individuals, national fundraising efforts, political advocacy, and sectarian religious programs. The following types of requests are discouraged: large capital improvements or construction drives; ongoing operational support; reduction or elimination of organizational deficits; reimbursement of expenses undertaken prior to submission of a grant application; computer hardware and software, unless these are the focus of a new or enhanced program; public art for which approval and placement has not yet been secured; and multi-year requests.
Geographic Focus: South Dakota
Date(s) Application is Due: Jan 1; Mar 1; May 1; Jul 1; Sep 1; Nov 1
Samples: Horsepower, Rapid City, South Dakota, $10,000 - to provide therapeutic riding sessions for children who have physical or cognitive challenges and who come from low-income families; Hope Haven International Ministries, Sioux Falls, South Dakota, $7,990 - to provide South Dakota State Penitentiary inmates and volunteers with the proper tools to repair and refurbish wheelchairs to be sent overseas to help those in need; Helpline Center, Sioux Falls, South Dakota, $4,000 - to purchase new volunteer management software for the agency's website.
Contact: Candy Hanson; (605) 336-7055 ext. 12; fax (605) 336-0038; chanson@sfacf.org
Internet: http://www.sfacf.org/AboutGrants.aspx
Sponsor: Sioux Falls Area Community Foundation
300 N Philips Avenue, Suite 102
Sioux Falls, SD 57104-6035

Sioux Falls Area Community Foundation Field-of-Interest and Donor-Advised Grants 3569

The Sioux Falls Area Community Foundation (SFACF) manages over 800 grant-making funds from donors who have specified a particular field of interest (e.g., youth enrichment, the arts, animal welfare, or outdoor recreation) or a specific neighborhood organization or nonprofit. Application guidelines and due dates may vary from donor to donor. Interested applicants are encouraged to call the SFACF's program officer for specific application forms or further information.
Requirements: Requirements may vary from donor to donor. Interested applicants are encouraged to call the SFACF's program officer for further information.
Restrictions: Restrictions may vary from donor to donor. Interested applicants are encouraged to call the SFACF's program officer for further information.
Geographic Focus: South Dakota
Date(s) Application is Due: Aug 15
Samples: Furniture Mission of South Dakota, Sioux Falls, South Dakota - to provide furniture and basic home furnishings for those struggling to make ends meet; Feeding South Dakota, Sioux Falls, South Dakota - to provide weekend meals to school children who rely on free or reduced lunch programs; Almost Home Canine Rescue, Madison, South Dakota - to support the agency's spay and neuter program.
Contact: Patrick Gale; (605) 336-7055 ext. 20; fax (605) 336-0038; pgale@sfacf.org
Internet: http://www.sfacf.org/News.aspx
Sponsor: Sioux Falls Area Community Foundation
300 N Philips Avenue, Suite 102
Sioux Falls, SD 57104-6035

Sioux Falls Area Community Foundation Spot Grants (Unrestricted) 3570

The purpose of the Sioux Falls Area Foundation (SFACF)'s unrestricted grantmaking program is to provide support across a wide spectrum of charitable needs and interests. Grantmaking categories include Arts and Humanities (e.g., theatre, music, arts, dance, cultural development, historic preservation, library programs, and museums); Community Affairs and Development (e.g., citizen participation, public use of parks and recreation, administration of justice, economic development, employment, and training); Education (e.g., lifelong-learning activities in formal educational settings, support of educational facilities and systems, and scholarships); Environment (e.g., protection of natural areas, conservation of energy, prevention and elimination of pollution or hazardous waste, wildlife protection, and water quality); Health (e.g., improvement of healthcare, prevention of substance abuse; support of mental-health needs, and medical research); Human Services (e.g., assistance to families, youth, the elderly, disabled, special groups, social service providers, and those who stand in need); and Religion (e.g., support for churches, religious institutions, and religion programs). SFACF offers two grant programs from its unrestricted funds: Community Fund Grants for projects over $3,000 and Spot Grants for projects up to $3,000. Spot Grant proposals may be submitted at any time and do not require SFACF's standard application form. Applicants should include the following components in their requests: a typed summary of their program in two pages or fewer; signatures of the organization's executive director and board chair; a board of directors roster; a copy of the organization's IRS tax determination letter, and a project budget. In most cases, SFACF will review and respond to Spot Grant requests within two weeks of receipt. SFACF has provided complete guidelines and an informative FAQ at their website.
Requirements: Nonprofit organizations serving residents in the Sioux Falls, South Dakota area (Minnehaha, Lincoln, McCook, and Turner counties) are eligible to apply. SFACF considers grant requests for programs that require start-up funds to address important community needs or opportunities, expansion of programs that meet important community needs or opportunities, assistance to organizations weathering unforeseen or unusual financial crises; programs that increase an organization's capacity to advance its mission more efficiently or effectively; and programs or studies that inform the community's understanding of needs or opportunities. Requests are evaluated by the following criteria: comparative benefit to the community; the organization's capacity to achieve the stated objectives; the amount of support requested versus the number of people benefited; a well-planned approach to achieving stated objectives; a reasonable expectation that the program can be sustained over time (where applicable); the organization's history of working collaboratively to address community needs and opportunities; and when applicable, the organization's past SFACF grant performance.
Restrictions: SFACF does not consider grant requests for individuals, national fundraising efforts, political advocacy, and sectarian religious programs. The following types of requests are discouraged: large capital improvements or construction drives; ongoing operational support; reduction or elimination of organizational deficits; reimbursement of expenses undertaken prior to submission of a grant application; computer hardware and software, unless these are the focus of a new or enhanced program; public art for which approval and placement has not yet been secured; and multi-year requests.
Geographic Focus: South Dakota
Amount of Grant: Up to 3,000 USD
Samples: Hawthorne Elementary School, Sioux Falls, South Dakota, $1,000 - to provide students with milk as part of the school's mid-morning snack program; Here4Youth, Sioux

Falls, South Dakota, $2,481 - to improve efficiency and donor and volunteer relations with the purchase of donor management software; Hillcrest Baptist Church, Sioux Falls, South Dakota, $900 - to provide opportunity for children in the Whittier neighborhood to learn teamwork and good sportsmanship through participation in the Meldrum Soccer League.
Contact: Candy Hanson; (605) 336-7055 ext. 12; fax (605) 336-0038; chanson@sfacf.org
Internet: http://www.sfacf.org/sfacf/aboutgrants.aspx
Sponsor: Sioux Falls Area Community Foundation
300 N Philips Avenue, Suite 102
Sioux Falls, SD 57104-6035

Siragusa Foundation Health Services & Medical Research Grants   3571
The Siragusa Foundation, established in 1950 by Ross D. Siragusa, is a private family foundation that is committed to honoring its founder by sustaining and developing Chicago's extraordinary nonprofit resources. The Foundation's Health Services and Medical Research Program is balanced between medical research organizations, which are concerned with scientific inquiry and the exploration of treatment methods, and health service organizations, which provide direct social and clinical care. The Foundation primarily funds fellowships, single-disease research programs and clinical treatment services. The Foundation believes it is critical to fund research efforts, health support services, patient care services and treatment, and advocacy programs for both adults and youth, all of which are highly complementary to each other.
Requirements: Nonprofit 501(c)3 organization in Chicago are eligible to apply. Before submitting a formal proposal to the Foundation, prospective applicants should submit a two-page letter of inquiry outlining the background of the organization and the proposed project. The letter of inquiry should describe succinctly the project for which funding is requested, including how it relates to the Foundation's interests, the target audience, the estimated budget and request amount. After reviewing the letter of inquiry, the Foundation may or may not request a formal proposal.
Restrictions: The Foundation limits potential grantees from applying for funds to once per calendar year. The Foundation does not accept unsolicited proposals from outside the Chicago area and does not support individuals or political advocacy. The foundation typically does not support capital expenditures or endowments.
Geographic Focus: Illinois
Samples: Arthritis Foundation, Greater Chicago Chapter, Chicago, Illinois, $5,000; Children's Memorial Foundation, Chicago, Illinois, $215,000; Rehabilitation Institute of Chicago, Chicago, Illinois, $16,666.
Contact: Kristen M. Buerster; (312) 755-0064; fax (312) 755-0069
Internet: http://www.siragusa.org/pages/program_areas/13.php
Sponsor: Siragusa Foundation
1 East Wacker Drive, Suite 2910
Chicago, IL 60601

Sir Dorabji Tata Trust Grants for NGOs or Voluntary Organizations   3572
Thoughtful and committed programs initiated by organizations in priority areas are considered for financial support. The Trust draws on the expertise of its staff and specialist consultants who travel widely, study project proposals and keep abreast of developments in each of the sectors (see below).
Requirements: The trust limits its funding to projects within India. This program focuses on specific sectors for its funding: Management of Natural Resources; Livelihood; Education; Health; and Social Development Initiatives. While no specific format is prescribed, applications should preferably include: (a) description of the proposed project with clearly articulated objectives, plans/programs or activities, and mechanisms for impact assessment; (b) profile of the project holder (qualifications, achievements, experience) and a brief background of the core team; (c) Registration Certificates of the organization and Income-tax Exemption Certificates; (d) The financial audited statements of the last three years and Annual Reports.
Geographic Focus: All States, Albania, Andorra, Armenia, Austria, Azerbaijan, Belarus, Belgium, Bosnia & Herzegovina, Bulgaria, Croatia, Cyprus, Czech Republic, Denmark, Estonia, Finland, France, Georgia, Germany, Greece, Hungary, Iceland, Ireland, Italy, Kosovo, Latvia, Liechtenstein, Lithuania, Luxembourg, Macedonia, Malta, Moldova, Monaco, Montenegro, Norway, Poland, Portugal, Romania, Russia, San Marino, Serbia, Slovakia, Slovenia, Spain, Sweden, Switzerland, The Netherlands, Turkey, Ukraine, United Kingdom, Vatican City
Contact: R.M. Lala; 91 22 6665 8282; fax 91 22 2204 5427; sdtt@sdtatatrust.com
Internet: http://www.dorabjitatatrust.org/ngo/ngo_grants.asp
Sponsor: Sir Dorabji Tata Trust
Bombay House, 24 Homi Mody Street
Mumbai, 400 001 India

Sir Dorabji Tata Trust Individual Medical Grants   3573
The Sir Dorabji Tata Trust provides individual medical grants to needy and deserving patients for medical and surgical care. The maximum disbursement for an individual patient is Rs 100,000, covering expenses incurred towards costs of major surgeries like neurosurgery, cardiac surgery, skin graft surgery for burns, cancer, kidney transplant and others. The grants are also available to patients who are physically handicapped. Though the quantum offered is modest and small, the grants bring much-needed financial relief to the patients and their families. The thrust is to help poor patients from government and municipal hospitals. Individual medical grants are one-time grants.
Requirements: The Trust gives priority to cases within Mumbai in view of the difficulties in assessing cases outside Mumbai. To apply for a medical grant from the Sir Dorabji Tata Trust, you are required to visit the Trust Office at Mulla House, 3rd Floor, 51 M.G. Road, Mumbai 400001 between 3:00 pm to 5:00 pm from Monday to Friday, and pick up the application form. Please ensure that you have all the following documents: (a) Doctor's Certificate; (b) Discharge Certificate; (c) Income Certificate; (d) Copies of Hospital/Medical Bills; (e) Proof of Residence i.e. Ration Card or any other document; (e) Recommendation Letter from Medical Social Worker from a general, municipal or government hospital; (f) Employers Reimbursement; (g) Medical Insurance. The completed forms should be submitted in person to the Trust Office along with the required documents within a month of collecting the form.
Restrictions: Limited to India.
Geographic Focus: All States
Amount of Grant: Up to 100,000 INR
Contact: R.M. Lala; 91 22 6665 8282; fax 91 22 2204 5427; sdtt@sdtatatrust.com
Internet: http://www.dorabjitatatrust.org/ind_grants/ind_img.asp
Sponsor: Sir Dorabji Tata Trust
Bombay House, 24 Homi Mody Street
Mumbai, 400 001 India

Sisters of Charity Foundation of Cleveland Reducing Health and Educational   3574
Disparities in the Central Neighborhood Grants
The Sisters of Charity Foundation of Cleveland is committed to helping families and individuals overcome the challenges of poverty. In this grant opportunity the Foundation focuses on health, housing and education as key components to building stronger families and stable neighborhoods. The Foundation seeks to reduce both health and educational disparities in Cuyahoga County. Grant funds may be used for program support, operating support, program-related capital support, planning, or capacity building. Grant periods will vary depending upon the particular grant. Most grants are for one year. Grant periods for planning may be shorter, and multi-year grants are possible.
Requirements: Applicants must be 501(c)3 organizations or governmental units or agencies (such as schools) that primarily serve the Cuyahoga County region.
Restrictions: The Foundation does not make grants to support endowments, fundraising campaigns (including annual appeals), membership drives, debt retirement, and individual scholarships. Further, the Foundation does not make grants to individuals.
Geographic Focus: Ohio
Date(s) Application is Due: Apr 7
Amount of Grant: 40,000 - 150,000 USD
Contact: Ursula Craig; (216) 241-9300; fax (216) 241-9345; ucraig@socfdncleveland.org
Internet: http://www.socfdncleveland.org/sistersofcharity/OurFocusAreas/HealthDisparities/HealthintheCentralNeighborhood/tabid/344/Default.aspx
Sponsor: Sisters of Charity Foundation of Cleveland
The Halle Building, 1228 Euclid Avenue, Suite 330
Cleveland, OH 44115-1834

Sisters of Mercy of North Carolina Foundation Grants   3575
The foundation provides grants to tax-exempt health care, educational, and social service organizations that assist women, children, the elderly, and the poor to improve the quality of their lives. Types of support include start-up grants for new organizations or programs, ongoing operating expenses for individual organizations, program or project expenses, building renovation, and equipment acquisition. Preference will be given to organizations whose efforts are collaborative, ecumenical, and multicultural. Particular attention will be given to organizations that serve the unserved or underserved. Annual deadline dates may vary; contact program staff for exact dates.
Requirements: Tax-exempt health care, education, and social service organizations in North and South Carolina are eligible to apply.
Restrictions: The foundation does not ordinarily support projects, programs, or organizations that serve a limited audience; biomedical or clinical research; units of the federal government; political activities; publication of newsletters, magazines, books and the production of videos; conferences and travel; endowment funds; capital fundraising campaigns; annual giving campaigns; or social events or similar fundraising activities.
Geographic Focus: North Carolina, South Carolina
Date(s) Application is Due: Apr 1; Aug 1; Dec 1
Amount of Grant: 25,000 - 150,000 USD
Samples: Greater Carolinas Chapter of the American Red Cross (Charlotte, NC)--to provide free transportation to and from medical appointments, $50,000.
Contact: Administrator; (704) 366-0087; fax (704) 366-8850; contact@somncfdn.org
Internet: http://www.somncfdn.org/grantseekers.html
Sponsor: Sisters of Mercy of North Carolina Foundation
2115 Rexford Road, Suite 401
Charlotte, NC 28211

Sisters of St. Joseph Healthcare Foundation Grants   3576
The Sisters of St. Joseph Healthcare Foundation is a non-profit, public benefit corporation, which addresses the needs of the working and indigent poor in: Southern California; San Francisco Bay Area; Humboldt County; Fresno County. The Sisters of St. Joseph Healthcare Foundation funds programs which directly serve the needs of the underserved, especially families and children at risk. The Foundation sponsors or supports long-term efforts which are closely identified with the Sisters of St. Joseph of Orange and their mission of bringing unity and healing where divisiveness and oppression exist. The foundation supports the concept of Healthy Communities and desire to fund programs and organizations that: provide direct health-related services; support and transform the individual, social, economic, institutional, and cultural aspects of communities; provide change within larger societal systems to benefit low-income and at-risk populations; and develop the leadership and capacity for self-determination of those served by our funding. The Sisters of St. Joseph Healthcare Foundation is particularly interested in proposals which fall into these funding categories: mental health services; health services; homeless services; violence prevention.

Requirements: Southern California, Humbolt County, Fresno County and San Francisco bay area nonprofits are eligible.
Restrictions: The Foundation does not fund direct support to individuals, annual fund drives, or capital campaigns.
Geographic Focus: California
Amount of Grant: 1,000 - 50,000 USD
Samples: Access OC, Irvine, CA $50,000 - to improve access to specialty care for the safety net population; Boys & Girls Clubs, Garden Grove, CA $30,000 - to fund transportation to improve access to healthcare for low-income families; Casa Teresa, Orange, CA $40,000 - to assist homeless pregnant women with services to help them move from a low-income situation to self-sufficiency;
Contact: Sister Regina Fox; (714) 633-8121, ext. 7109; rfox@csjorange.org
Sponsor: Sisters of St. Joseph Healthcare Foundation
440 South Batavia Street
Orange, CA 92868-3998

Skillman Foundation Good Neighborhoods Grants 3577
The Good Neighborhoods program encourages the creation of safe, healthy, and vibrant neighborhoods where children with the support of caring adults, programs, and experiences can develop fully. Launched in January 2006, the program provides full-scale support to six Detroit neighborhoods where more than 65,000 children live, roughly 30% of the city's child population. Half of the children in these neighborhoods live in poverty. This program encourages the creation of safe, healthy and vibrant neighborhoods where children, with support of caring adults, programs and experiences, can develop fully. The six neighborhoods are: Brightmoor; Cody/Rouge; Northend (also known as Central); Osborn; Chadsey/Condon, in Southwest Detroit; and Vernor, in Southwest Detroit. The goal of the Good Neighborhoods program is to ensure that children experience safe, healthy and high-quality environments where they can thrive: neighborhoods with resources, opportunities and assets for children to develop fully and pursue prosperity.
Requirements: The applicant organizations must be tax exempt and may not be a 509(a) private foundation. The foundation's primary geographic area of focus is the Detroit metropolitan area.
Restrictions: The foundation does not award grants directly to individuals or provide loans of any kind; nor does it support sectarian religious activities, political lobbying, political advocacy, legislative activities, endowments, annual fund drives, basic research, or support of past operating deficits.
Geographic Focus: Michigan
Amount of Grant: 10,000 - 2,000,000 USD
Contact: Tonya Allen; (313) 393-1185; fax (313) 393-1187; info@skillman.org
Internet: http://www.skillman.org/good-neighborhoods/
Sponsor: Skillman Foundation
100 Talon Centre Drive, Suite 100
Detroit, MI 48207

Skin Cancer Foundation Research Grants 3578
Funding is provided annually for basic research, clinical studies, and educational programs related to skin cancer. The foundation's grants program is intended for pilot projects, which, if successful, can be further developed in order to be eligible for larger grants from other sources. Grants include: The Dr. Patricia Wexler Research Grant Awards (one-year, $10,000 grant for a research project); The Theodore S. Tromovitch Memorial Award (one-year, $5,000 grant for a research project); The Melissa K. Bambino Memorial Award (one-year, $8,000 grant for a research project); The Robert N. Cooke Memorial Award (one-year, $3,000 grant for a research project); Pevonia Botanica Research Award (one-year, $10,000 grant for a research project)
Requirements: Candidates must comply with all directions on the application form (available at the website). All projects are selected on the basis of their applicability to the foundation's mission of reducing the incidence, morbidity and mortality of skin cancer. Research must be conducted at institutions within the U.S. Preference is given to projects that address, at the basic science and clinical level, improved methods of prevention, detection, and treatment of skin cancers.
Restrictions: Overhead or indirect costs are not eligible.
Geographic Focus: All States
Date(s) Application is Due: Oct 1
Samples: Galina V. Shurin, PhD, University of Pittsburgh - Phosphatidylserine-based ointment for the topical therapy of skin cancer; Paul Nghiem, MD, PhD, University of Washington - Merkel cell carcinoma: Building a tumor and cell line toolbox; David Polsky, MD, PhD, NYU Medical Center - The use of chloroquine to reduce ultraviolet light-induced DNA damage in the skin
Contact: Grants Administrator; (800) 754-6490; info@skincancer.org
Internet: http://www.skincancer.org/research-grants-2010.html
Sponsor: Skin Cancer Foundation
149 Madison Avenue, Suite 901
New York, NY 10016

Skoll Foundation Awards for Social Entrepreneurship 3579
The Skoll Foundation presents the Skoll Awards for Social Entrepreneurship each year to a select few social entrepreneurs who are solving the world's most pressing problems. The Skoll Award includes a core support grant to the organization to be paid over three years, and a non-cash award to the social entrepreneur presented at the Skoll World Forum on Social Entrepreneurship every spring. The Foundation's focus on the following areas stems from a belief that many of the world's most pressing problems are worsened by inequality between the rich and the poor. Social entrepreneurs provide solutions that address this inequality at a systematic level. This list serves as a guide and is not meant to be comprehensive: economic and social equity; environmental sustainability; health; institutional responsibility; peace and security; and tolerance, justice, and human rights. Qualifying organizations are evaluated against the following criteria for their proposed project: its impact potential, proven approach, innovation, specific issue; leverage; its social entrepreneur; and the project's sustainability. Applications are accepted from January 4 through March 1 of each year.
Requirements: The application process includes several stages. The eligibility quiz helps applicants assess whether they should apply for the award. Organizations that pass the eligibility quiz will be given a URL to the application. Selected applicants (usually ten or fewer each year) will be invited by the program officer to submit a full proposal. This processes includes interviews, a site visit, reference checks, follow-up questions, in-depth financial review, and a discussion of grant objectives. Approximately ten award winners will then be selected.
Restrictions: The Foundation Awards will not support: individuals; programs promoting religious or ideological doctrine, such as those principally sectarian in nature; lobbying (beyond that allowed by law for charitable organizations); film financing; endowments, cash reserves or deficit reductions; government agencies; university-based projects; public schools and school districts; land, site acquisition or facilities construction; institutions that discriminate on the basis of race, creed, age, gender or sexual orientation in policy or practice; grantmaking to other organizations or individuals; event sponsorship; political campaigns; new or early-stage business plans or ideas; organizations whose missions and work focus on a single municipality, province or state; or local offices of parent organizations or specific programs within organizations.
Geographic Focus: All States
Date(s) Application is Due: Mar 1
Contact: Administrator; (650) 331-1031; fax (650) 331-1033; info@skollfoundation.org
Internet: http://www.skollfoundation.org/about/skoll-awards/
Sponsor: Skoll Foundation
250 University Avenue, Suite 200
Palo Alto, CA 94301

Smithsonian Biodiversity Genomics and Bioinformatics Postdoctoral 3580
Fellowships
The Smithsonian Institution (SI) Postdoctoral Fellowships in Biodiversity Genomics and Bioinformatics promote collaborative research in these fields (60%), plus a well-defined outreach component (40%) oriented toward building genomics expertise in the greater SI research community. Research should involve comparative genomic approaches such as phylogenomics, population genomics, metagenomics or transcriptomics, and have a component that involves significant bioinformatics analysis. Your proposal should also detail in at least one page your bioinformatics outreach plan, which can include collaborative work with other SI projects and teams, training workshops or development of software, pipelines or tutorials. We plan to fill three or more Fellowships in the area of Biodiversity Genomics and are especially interested in recruiting a team of Fellows who will work together to advance bioinformatics at the Institution. Coordinated applications dealing with multiple genomic approaches are strongly encouraged. Proposals are due April 1 each year.
Requirements: Applicants must propose to conduct research in-residence for a period of 12 to 24 months. Applicants must have completed or be near completion of the Ph.D. Recipients who have not completed the Ph.D. at the time of application must provide proof of completion of the degree before the fellowship begins.
Geographic Focus: All States
Date(s) Application is Due: Apr 1
Contact: Bryan T. Fair; (202) 275-0655 or (202) 633-7070; siofg@si.edu
Internet: http://www.si.edu/ofg/fell.htm
Sponsor: Smithsonian Institution
470 L'Enfant Plaza SW, Suite 7102, MRC 902 P.O. Box 37012
Washington, D.C. 20013-7012

Smithsonian Museum Conservation Institute Research Post-Doctorate/Post 3581
Graduate Fellowships
Fellowships at the Smithsonian Museum Conservation Institute offers fellowship opportunities involving a variety of artifact analysis, preservation, and conservation treatment specialties. MCI's commitment is to enhance the experience of the fellow or intern, providing specialized technical and scientific training. Such training contributes significantly to the qualifications of the students in their subsequent professional employment. Applications are currently invited for a post-doctoral or a post-graduate fellowship position at the mass spectrom'try and proteomics laboratory at the Smithsonian's Museum Conservation Institute. The aim of the laboratory is to develop mass spectrometry and proteomics technologies relevant to museums' specimens. Projects include but are not limited to: proteomics, biological dating using various MS techniques such as amino acid racemization; analysis of insoluble proteinaceous or polymeric materials; analysis of paints, inks, etc. using surface ionization techniques; and development of portable separation-mass spectrometry devices for onsite chemical/biological analysis. Qualified applicants should send a letter of interest and resume, including lists of publications and references to Dr. Mehdi Moini (moinim@si.edu).
Requirements: The successful applicant will have a Ph.D. or Master's degree in the area of separation, microfabrication, mass spectrometry, or proteomics. Working experience with Thermo LTQ_Orbitrap Veloes, ABI Qstar and 4700 TOF/TOF, as well as with capillary electrophoresis and nano-LC is highly desirable.
Geographic Focus: All States
Contact: Paula T. DePriest, Deputy Director; (301) 238-1206 or (301) 238-1240; fax (301) 238-3709; DepriestP@si.edu or MCIweb@si.edu
Internet: http://si.edu/MCI/english/professional_development/2012ResearchPostDoctorate.html
Sponsor: Smithsonian Institution
4210 Silver Hill Road
Suitland, MD 20746

SNM/ Covidien Seed Grant in Molecular Imaging/Nuclear Medicine Research   3582
The Society of Nuclear Medicine is pleased to offer the SNM/Covidien Seed Grant in Nuclear Medicine Research. This grant is designed to assist researchers in conducting new and innovative pilot projects that have potential for future support from foundations, corporations or government agencies.

Requirements: Applicant Eligibility: basic or clinical scientists employed by academic and research-oriented organizations; applicants from any country may apply; applicants must hold a full-time faculty position in an educational institution when the award period begins; applicants must have completed all advanced training and be no more than five years post training; applicants must not have served as a principal investigator on peer-reviewed grants totaling $50,000 or more in a single calendar year; preference will be given to individuals who have demonstrated great potential for a research career in the field of nuclear medicine/molecular imaging.

Restrictions: No grantee may receive more than one SNM research grant in any one year. The grant will not pay institutional overhead costs or indirect costs. No salary support for principal investigator or co-principal investigators. Grants will be limited to a time span of one year; no extensions are allowed

Geographic Focus: All States
Date(s) Application is Due: Feb 20
Amount of Grant: 25,000 USD
Contact: Nicole Kern, SNM Program Manager; (703) 652-6795, ext. 1255; fax (703) 708-9015; nkern@snm.org
Internet: http://interactive.snm.org/index.cfm?PageID=2255
Sponsor: Society of Nuclear Medicine
1850 Samuel Morse Drive
Reston, VA 20190

SNM Molecular Imaging Research Grant For Junior Medical Faculty   3583
The objective of this program is to provide salary support for one junior faculty member in an academic/research setting to enable them to engage in Molecular Imaging research related to diagnostic or therapeutic applications. The grant will support new, innovative pilot projects in any area of molecular imaging research. The grant may supplement other funds, but not duplicate other support.

Requirements: Applicant Eligibility: applicants must have MD degree (or equivalent) and have completed a residency and be certified in their specialty; applicants must be practicing in an academic/research setting as a faculty member in the United States; applicants must be within 5 years of their initial faculty appointment with an academic rank of instructor or assistant professor (or equivalent); applicants must provide a detailed proposal of the proposed research activity; applicants must not have received grant/contract amounts totaling $50,000.00 or more in a single calendar year as the principle investigator; preference will be given to individuals who have demonstrated great potential for a research career in the field of molecular imaging; multidisciplinary projects are encouraged; only one application per faculty member is accepted per grant; no grantee may receive more than one SNM research grant in any one year. Likewise no other research grant request or continuation will be considered until a satisfactory summary of an earlier grant is received; applicants are required to be a SNM member at the time of application.

Restrictions: The grant will not pay institutional overhead costs or indirect costs
Geographic Focus: All States
Date(s) Application is Due: Jan 20
Amount of Grant: 100,000 USD
Contact: Nicole Kern; (703) 652-6795, ext. 1255; fax (703) 708-9015; nkern@snm.org
Internet: http://interactive.snm.org/index.cfm?PageID=7252
Sponsor: Society of Nuclear Medicine
1850 Samuel Morse Drive
Reston, VA 20190

SNM PDEF Mickey Williams Minority Student Scholarships   3584
This scholarship supports minority students pursuing a two or four year degree in nuclear medicine. It honors the memory of Mickey Williams, a past SNMTS President who immigrated to the United States from Jamaica.

Requirements: To be eligible for the PDEF Mickey Williams Minority Student Scholarship, candidates must: be of African American, Native American, Hispanic American, Asian American or Pacific Islander decent [Eligible Native Americans include American Indian, Eskimo, Hawaiian and Samoan]; hold current U.S. residency status of U.S. citizen, U.S. national, or U.S. permanent resident; be currently enrolled in or accepted into an entry-level nuclear medicine technology program at the time of application; Award is open to students in associate and baccalaureate level programs only; demonstrate financial need; have maintained a cumulative grade point average of 2.5 or better on a four-point scale, with evidence of high academic performance. See SNM website for application.

Restrictions: Individuals with a visitor, student, or G-series visa are not eligible. Individuals with previous certificates or degrees in nuclear medicine sciences are ineligible.
Geographic Focus: All States
Date(s) Application is Due: Jan 15
Amount of Grant: 5,000 USD
Contact: Nicole Kern; (703) 652-6795, ext. 1255; fax (703) 708-9015; nkern@snm.org
Internet: http://interactive.snm.org/index.cfm?PageID=2682&RPID=2259
Sponsor: Society of Nuclear Medicine
1850 Samuel Morse Drive
Reston, VA 20190

SNM Pilot Research Grants   3585
The pilot research grants are designed to help basic or clinical scientists in the early stages of their career conduct research that may lead to further funding. The Mitzi and William Blahd, MD Pilot Research Grant is awarded to the applicant with the highest-ranked proposal. Preference will be given to individuals who have demonstrated great potential for a research career in the field of nuclear medicine/molecular imaging.

Requirements: Applicant Eligibility: basic or clinical scientists employed by academic and research-oriented organizations in North America are eligible; applicants must not be no more than five years post training.
Geographic Focus: All States
Date(s) Application is Due: Feb 20
Amount of Grant: 25,000 USD
Contact: Nicole Mitchell, Program Manager; (703) 652-6795; nkern@snm.org
Internet: http://interactive.snm.org/index.cfm?PageID=2255
Sponsor: Society of Nuclear Medicine
1850 Samuel Morse Drive
Reston, VA 20190

SNM Pilot Research Grants in Nuclear Medicine/Molecular Imaging   3586
The pilot research grants are designed to help basic or clinical scientists in the early stages of their career conduct research that may lead to further funding.

Requirements: Applicant Eligibility: basic or clinical scientists employed by academic and research-oriented organizations in North America are eligible; applicants must be no more than five years post training; applicant must not have served as the principal investigator of a peer-reviewed grant for more than $50,000 in a single calendar year; preference will be given to individuals who have demonstrated great potential for a research career in the field of nuclear medicine/molecular imaging.

Restrictions: The grant will not pay salaries, institutional overhead costs or indirect costs. Funds may supplement other funds, but not duplicate other support. Grants will be limited to a time span of one year. No grantee may receive more than one grant in any one year.
Geographic Focus: All States
Date(s) Application is Due: Feb 20
Amount of Grant: 25,000 USD
Contact: Nicole Kern; (703) 652-6795, ext. 1255; fax (703) 708-9015; nkern@snm.org
Internet: http://interactive.snm.org/index.cfm?PageID=2255
Sponsor: Society of Nuclear Medicine
1850 Samuel Morse Drive
Reston, VA 20190

SNM Postdoctoral Molecular Imaging Scholar Grants   3587
The postdoctoral program will support a two-year research endeavor that promotes integration of molecular imaging into the career of the trainee. The funds maybe used for new, innovative pilot projects in any area of molecular imaging research. The grant may supplement other funds, but not duplicate other support. A summary report and an accounting of funds will be requested at the halfway point and upon completion of the research project. Grantees are expected to submit results of the research as an abstract for presentation at the SNM Annual Meeting and/or as a scientific manuscript to the Journal of Nuclear Medicine. A final report is due within 60 days after completion of the research project.

Requirements: Applicant Eligibility: MD, MD/PhD, and PhD applicants are eligible; applicants from any country may apply; applicants must hold a full-time position in an educational institution when the award starts; applicants must have completed all basic training and be no more than 2 years post training; applicants must not have served as a principal investigator on peer-reviewed grants totaling $50,000 or more in a single calendar year; preference will be given to individuals who have demonstrated great potential for a research career in the field of molecular imaging; only one application per trainee is accepted; no grantee may receive more than one SNM research grant in any one year. Likewise no other research grant request or continuation will be considered until a satisfactory summary of an earlier grant is received.

Restrictions: Grant extensions are not often allowed, but will be reviewed and decided upon by the SNM Grants and Awards Committee if requested. The grant will not pay institutional overhead costs or indirect costs.
Geographic Focus: All States
Date(s) Application is Due: Jan 20
Amount of Grant: 60,000 USD
Contact: Nicole Kern; (703) 652-6795, ext. 1255; fax (703) 708-9015; nkern@snm.org
Internet: http://interactive.snm.org/index.cfm?PageID=7252
Sponsor: Society of Nuclear Medicine
1850 Samuel Morse Drive
Reston, VA 20190

SNM Student Fellowship Awards   3588
The SNM Student Fellowships are designed to stimulate interest in molecular imaging/nuclear medicine by supporting their full-time participation in clinical and basic research activities for three months (or less). These Fellowships are made possible by support of the Education and Research Foundation for SNM.

Requirements: Candidates from any country may apply. Students must be enrolled in a medical, pharmacy or graduate school, or be an undergraduate student demonstrating outstanding competence in nuclear medicine and/or molecular imaging research. Fellowship must be carried out during a three-month that starts no earlier then March and completed no later then December 31st.
Geographic Focus: All States
Date(s) Application is Due: Jan 23

Amount of Grant: 3,000 USD
Contact: Nicole Kern; (703) 652-6795, ext. 1255; fax (703) 708-9015; nkern@snm.org
Internet: http://interactive.snm.org/index.cfm?PageID=2254
Sponsor: Society of Nuclear Medicine
1850 Samuel Morse Drive
Reston, VA 20190

SNM Tetalman Memorial Young Investigator Awards 3589
Established by the family and friends of Dr. Tetalman, this award honors the work of a young investigator who is pursuing a career in nuclear medicine. It is named in memory of a highly respected and productive clinician and researcher, Mark Tetalman, M.D., whose career was cut tragically short. The award is based on a submitted paper supporting current research efforts and accomplishments, teaching, clinical service and administration. Service to the Society and a commitment to academic nuclear medicine will also be considered.
Requirements: Applicants must be 36 years old or younger as of July 1 of the application year and must have obtained certification in Nuclear Medicine or Nuclear Radiology or have completed a PhD program within the last seven years. Complete applications must be must be received, not simply post-marked by November 14th. One original unstapled copy of the completed application must be sent to: SNM Development Office, Attention: SNM Committee on Awards
Geographic Focus: All States
Date(s) Application is Due: Nov 14
Amount of Grant: 2,500 USD
Contact: Nicole Kern; (703) 652-6795, ext. 1255; fax (703) 708-9015; nkern@snm.org
Internet: http://interactive.snm.org/index.cfm?PageID=2253
Sponsor: Society of Nuclear Medicine
1850 Samuel Morse Drive
Reston, VA 20190

SNMTS Clinical Advancement Scholarships 3590
The SNMTS Clinical Advancement Scholarships serve to support technologists who are pursuing clinical advancement through didactic educational programs. Ten (10) $500 scholarships will be awarded. Scholarship application form is available on the SNM website.
Requirements: To be eligible for the SNMTS Clinical Advancement Scholarship, candidates must: be a member of the SNMTS; demonstrate financial need; be currently enrolled in a didactic educational program(s) (ex. CT, DEXA, physics, statistics) which is/are college credit eligible; complete the said class or program. It is not required that class/program completion results in a degree.
Geographic Focus: All States
Date(s) Application is Due: May 1
Amount of Grant: 500 USD
Contact: Nicole Kern; (703) 652-6795, ext. 1255; fax (703) 708-9015; nkern@snm.org
Internet: http://interactive.snm.org/index.cfm?PageID=7270&RPID=2259
Sponsor: Society of Nuclear Medicine Technologist Section
1850 Samuel Morse Drive
Reston, VA 20190

SNMTS Outstanding Educator Awards 3591
The purpose of this award to is to recognize a SNMTS member who has significantly contributed to providing knowledge that advances and promotes the field of nuclear medicine technology through outstanding work in education. The winner will receive $750 along with a plaque to be presented at the SNM Annual Meeting.
Requirements: Eligible candidates include teachers but also educators in the non-traditional sense such as industry and clinical professionals. Evidence of their outstanding work must be shown in the following required supporting documentation: Curriculum Vitae, including listing of publications, presentations, programs, awards and honors; one letter of support from recent student/graduate (within 5 years) or participant in program; one letter of support from current program director or supervisor; one letter of support from a colleague; completed Nomination Form (see website).
Geographic Focus: All States
Date(s) Application is Due: Mar 13
Amount of Grant: 750 USD
Contact: Nicole Kern; (703) 652-6795, ext. 1255; fax (703) 708-9015; nkern@snm.org
Internet: http://interactive.snm.org/index.cfm?PageID=2256
Sponsor: Society of Nuclear Medicine Technologist Section
1850 Samuel Morse Drive
Reston, VA 20190

SNMTS Outstanding Technologist Awards 3592
The purpose of this award to is to recognize an SNMTS (SNM Technologist Section) member who has demonstrated outstanding service and dedication to the field of nuclear medicine technology. The winner will receive $750 along with a plaque to be presented at the SNM Annual Meeting.
Requirements: Eligible candidates include SNMTS members in any area of the field of nuclear medicine technology who have exhibited commitment to advancing the field in their workplace and through their involvement with the Society. Nominees must be involved with the Society at the local, regional and/or national level and have at least five years of experience in nuclear medicine technology. Evidence of their outstanding work must be shown in the following required supporting documentation: curriculum vitae- including listing of publications, presentations, programs, awards and honors; one letter of support from current supervisor; one letter of support from a colleague; one letter of support from a colleague who is a member of the SNMTS; completed Nomination Form (see website for application form).
Geographic Focus: All States
Date(s) Application is Due: Mar 13
Amount of Grant: 750 USD
Contact: Nicole Kern; (703) 652-6795, ext. 1255; fax (703) 708-9015; nkern@snm.org
Internet: http://interactive.snm.org/index.cfm?PageID=2256
Sponsor: Society of Nuclear Medicine
1850 Samuel Morse Drive
Reston, VA 20190

SNMTS Paul Cole Scholarships 3593
The Paul Cole Scholarship is named in memory of Paul Cole, CNMT, who served as President of the SNM Technologist Section (SNMTS) in 1986 and who was known as a champion of education for technologists. This scholarship is funded by the Education and Research Foundation for the SNM.
Requirements: To be eligible for the Paul Cole Scholarship, candidates must: demonstrate financial need; be enrolled in or accepted into an institution accredited through the Joint Review Committee on Educational Programs in Nuclear Medicine Technology (JRCNMT). To see if your institution is accredited with JRCNMT, visit their website at www.jrcnmt.org. Award is open to students in associate, baccalaureate, and certificate level programs; have maintained a minimum cumulative GPA of 2.5 or better (on a 4.0 scale) or B average in a nuclear medicine technology core curriculum.
Geographic Focus: All States
Date(s) Application is Due: Jan 15
Amount of Grant: 1,000 USD
Contact: Nicole Kern; (703) 652-6795, ext. 1255; fax (703) 708-9015; nkern@snm.org
Internet: http://interactive.snm.org/index.cfm?PageID=2681&RPID=2259
Sponsor: Society of Nuclear Medicine Technologist Section
1850 Samuel Morse Drive
Reston, VA 20190

SOBP A.E. Bennett Research Award 3594
The Society of Biological Psychiatry offers an annual award of $5,000 each in basic science and in clinical science for the purpose of stimulating international research in biological psychiatry by young investigators. Candidates must be actively engaged in the research for which the award is sought. Applicants need only write a brief description of their research, submit two to three published or submitted papers and arrange for two letters of support (at least one of the letters needs to be from a member of the Society). Although the research is not to be judged in comparison with the work of the more senior investigators, special consideration will be given to the originality of the approach and independence of thought evident in the submission. Nominations should be submitted on-line on or before September 30.
Requirements: Submissions are welcomed from laboratory researchers in any country who have not passed their 45th birthday or have not been engaged in research for greater than 10 years following award of their terminal degree or the end of formal clinical/fellowship training, whichever is later by deadline date; candidates need not be current members of the society.
Geographic Focus: All States, All Countries
Date(s) Application is Due: Sep 30
Amount of Grant: 5,000 USD
Contact: Maggie Peterson, Executive Director; (904) 953-2842; fax (904) 953-7117; maggie@mayo.edu or sobp@sobp.org
Internet: http://www.sobp.org/i4a/pages/index.cfm?pageID=3379
Sponsor: Society of Biological Psychiatry
4500 San Pablo Road, Birdsall 310
Jacksonville, FL 32224

SOBP Ziskind-Somerfeld Research Awards 3595
The Society of Biological Psychiatry offers an annual award of $5,000 in either basic or clinical research for the purpose of stimulating investigations in biological psychiatry by senior investigators who are members in good standing of the Society of Biological Psychiatry. The Ziskind-Somerfeld Research Foundation makes this award possible. No submission is required to be considered for this award. The research will be judged in comparison with the work of other senior investigators. Special consideration is given to the originality of the approach and independence of thought evident in the submission.
Requirements: Candidates will be at least thirty-five years of age at the time of submission. When more than one investigator is listed as an author on the submitted work, the principal investigators (first and last authors) must be over the age of 45.
Geographic Focus: All States
Amount of Grant: 5,000 USD
Contact: Maggie Peterson, Executive Director; (904) 953-2842; fax (904) 953-7117; maggie@mayo.edu or sobp@sobp.org
Internet: http://www.sobp.org/i4a/pages/index.cfm?pageID=3380#top
Sponsor: Society of Biological Psychiatry
4500 San Pablo Road, Birdsall 310
Jacksonville, FL 32224

SOCFOC Catholic Ministries Grants 3596
The Sisters of Charity Foundation of Cleveland belongs to a community of Catholic health and social service organizations inspired to service by the healing ministry of Jesus Christ. As a supporting foundation to the congregation of the Sisters of Charity of St. Augustine, the Foundation gives priority to the ministries of the Sisters of Charity Health System. A limited amount of remaining funds may be available to other Catholic ministries in health and human services in Cuyahoga County. Funding is limited to Invited Proposals.
Requirements: Each proposal to the Foundation must include a completed application form, proposal narrative, budget, and Budget Narrative. The application form and guidelines

for the proposal narrative may be found on this website under the specific program area. Eligible organizations are 501(c)3 organizations or governmental agencies or units.
Restrictions: The Foundation does not make grants to individuals.
Geographic Focus: Ohio
Amount of Grant: 10,000 - 150,000 USD
Contact: Lynn Berner; (216) 241-9300; fax (216) 241-9345; lberner@socfdncleveland.org
Internet: http://www.socfdncleveland.org/RespondingToOurCommunity/CatholicMinistries/tabid/313/Default.aspx
Sponsor: Sisters of Charity Foundation of Cleveland
The Halle Building, 1228 Euclid Avenue, Suite 330
Cleveland, OH 44115-1834

Society for Imaging Informatics in Medicine (SIIM) Emeritus Mentor Grants 3597
The mission of SIIM, The Society for Imaging Informatics in Medicine, is to advance computer applications and information technology in medical imaging through education and research. In support of this mission, the SIIM Research and Development Committee offers a $20,000 grant for a retired, or semi-retired, expert in imaging informatics to act as a mentor for a trainee (student, resident, fellow) or junior faculty member for hypothesis-driven research. Any area of research involving medical imaging informatics is eligible for support, as well as development of new hardware, software, or processes to support electronic imaging practice, education, or research.
Requirements: A senior, experienced investigator jointly applies with a resident, fellow, graduate student, or junior faculty for a SIIM emeritus mentor grant. Preference will be given to projects at SIIM institutions and to projects that contribute to the development of careers of new or prospective researchers. This grant may not be applied for in conjunction with another SIIM grant covering the same subject material.
Geographic Focus: All States
Date(s) Application is Due: Aug 3
Amount of Grant: 20,000 USD
Contact: Manager; (703) 723-0432; fax (703) 723-0415; grants@siimweb.org
Internet: http://www.scarnet.org/index.cfm?id=63
Sponsor: Society for Imaging Informatics in Medicine
19440 Golf Vista Plaza, Suite 330
Leesburg, VA 20176-8264

Society for Imaging Informatics in Medicine (SIIM) Micro Grants 3598
The mission of SIIM, The Society for Imaging Informatics in Medicine, is to advance computer applications and information technology in medical imaging through education and research. In support of this mission, the SIIM Research and Development Committee offers a $5,000 grant for hypothesis-driven experiments of short duration and high risk. Prior preliminary data is not required. An imaging informatics project that needs resources falls between what a researcher is willing to fund personally or is more than his/her department provides in seed money and what is typically requested in grant applications. Examples include trying unconventional hardware or software in an imaging informatics experiment, trying a particular piece of hardware for a user-interface design, paying for short duration consulting assistance, or paying for an educational course that the researcher needs to enhance a particular study.
Requirements: There are no restrictions on who may apply for the grant. Trainees (residents, fellows, and graduate students) are required to include a letter from a faculty member, acting as an advisor, to show that the applicant's department supports the project. Preference will be given to projects at SIIM institutions and to projects that may yield high value results or provide enough preliminary data to justify additional work and application for traditional grant funding.
Geographic Focus: All States
Date(s) Application is Due: Aug 3
Amount of Grant: 5,000 USD
Contact: Manager; (703) 723-0432; fax (703) 723-0415; grants@siimweb.org
Internet: http://www.scarnet.org/index.cfm?id=63
Sponsor: Society for Imaging Informatics in Medicine
19440 Golf Vista Plaza, Suite 330
Leesburg, VA 20176-8264

Society for Imaging Informatics in Medicine (SIIM) Research Grants 3599
The mission of SIIM, The Society for Imaging Informatics in Medicine, is to advance computer applications and information technology in medical imaging through education and research. In support of this mission, the SIIM Research and Development Committee offers two $50,000 grants for original research. Any area of research involving medical imaging informatics is eligible for support, as well as development of new hardware, software, or processes to support electronic imaging practice, education, or research.
Requirements: Residents, fellows, graduate students, or faculty may apply for SIIM research grants. Preference will be given to projects at SIIM institutions and to projects that contribute to the development of careers of new or prospective researchers. Residents, fellows, and students must conduct their projects under the guidance of experienced investigators. Grant recipients will not be eligible for concurrent support through other SIIM grants.
Geographic Focus: All States
Date(s) Application is Due: Aug 3
Amount of Grant: 50,000 USD
Contact: Manager; (703) 723-0432; fax (703) 723-0415; grants@siimweb.org
Internet: http://www.scarnet.org/index.cfm?id=63
Sponsor: Society for Imaging Informatics in Medicine
19440 Golf Vista Plaza, Suite 330
Leesburg, VA 20176-8264

Society for Imaging Informatics in Medicine (SIIM) Small Grant for Product Support Development 3600
The mission of SIIM, The Society for Imaging Informatics in Medicine, is to advance computer applications and information technology in medical imaging through education and research. In support of this mission, the SIIM Research and Development Committee offers a $20,000 grant to support the development of a product or tool for use in the field of imaging informatics. Any area of product or tool development for use in the field of medical imaging informatics is eligible for support (e.g., open source application development, work flow measurement tools, replacement PACS change management tools).
Requirements: Residents, fellows, graduate students, or faculty may apply for product support development grants. Preference will be given to projects at SIIM institutions and to projects that contribute to the development of careers of new or prospective researchers. Residents, fellows, and students must conduct their projects under the guidance of experienced investigators.
Geographic Focus: All States
Date(s) Application is Due: Aug 3
Amount of Grant: 20,000 USD
Contact: Manager; (703) 723-0432; fax (703) 723-0415; grants@siimweb.org
Internet: http://www.scarnet.org/index.cfm?id=63
Sponsor: Society for Imaging Informatics in Medicine
19440 Golf Vista Plaza, Suite 330
Leesburg, VA 20176-8264

Society for Imaging Informatics in Medicine (SIIM) Small Training Grants 3601
The mission of SIIM, The Society for Imaging Informatics in Medicine, is to advance computer applications and information technology in medical imaging through education and research. In support of this mission, the SIIM Research and Development Committee offers a $20,000 grant targeted at providing training opportunities in the field of imaging informatics for students and imaging informatics professionals. Any area of training within medical imaging informatics is eligible for support. The grant is designed to support hands-on training or experience with a designated mentor or sponsor in the imaging informatics arena. It is not intended to support tuition or other traditional education-related expenses. The period of training must last a minimum of three months and may extend no longer than twelve months.
Requirements: Residents, fellows, graduate students, or faculty members at recognized institutions may apply for SIIM training grants. Preference will be given to training experiences at SIIM institutions and to those that contribute to the careers of new or prospective imaging informatics personnel.
Geographic Focus: All States
Date(s) Application is Due: Aug 3
Amount of Grant: 20,000 USD
Contact: Manager; (703) 723-0432; fax (703) 723-0415; grants@siimweb.org
Internet: http://www.scarnet.org/index.cfm?id=63
Sponsor: Society for Imaging Informatics in Medicine
19440 Golf Vista Plaza, Suite 330
Leesburg, VA 20176-8264

Society for the Arts in Healthcare Consulting Grants 3602
The Society for the Arts in Healthcare has partnered with the National Endowment for the Arts to provide hundreds of consulting grants to arts and/or health care organizations across the US. Through a roster of over 20 trained arts in health care consultants, these grants assist organizations in utilizing a diverse array of arts media including performing arts, visual arts, literature, film, or design to promote the health and well-being of equally diverse populations. Organizations may apply for up to 20 hours of consulting with an experienced leader in the field. This grant pays for all consultant fees and grantees cover any additional costs such as long-distance phone calls or consultant travel and accommodation. Consultants are available to travel to organizations for site visits or to provide information and support via phone or email. The Consulting Service can assist in: program planning/strategic planning; artist training; fundraising and promotion; community collaborations and effective partnerships; evaluation; and improving health care facility design and environments. Projects can range from specific technical assistance such as designing a job description, creating a survey, feedback on a grant application, or developing guidelines for exhibitions to thinking through a strategic plan for a program. Prospective clients can request the number of hours that would be most appropriate for their needs.
Requirements: Applicant organizations must be based in the United States.
Geographic Focus: All States
Date(s) Application is Due: Nov 16
Contact: Anita Boles; (202) 299-9770; fax (202) 299-9887; anita@thesah.org
Internet: http://www.thesah.org/template/page.cfm?page_id=466
Sponsor: Society for the Arts in Healthcare
2437 15th Street NW, B South
Washington, D.C. 20009

Society for the Arts in Healthcare Environmental Arts Research Grant 3603
The American Art Resources and Society for the Arts in Healthcare Environmental Arts Research Grant program offers a cash award sponsored by American Art Resources in Houston, Texas. The program is provided to a Principal Investigator to conduct or supplement Arts in Healthcare research. Selection is made on the merit, innovation and feasibility of the research proposal.
Requirements: Applicant organizations must be based in the United States.
Geographic Focus: All States
Contact: Anita Boles; (202) 299-9770; fax (202) 299-9887; anita@thesah.org
Internet: http://www.thesah.org/template/page.cfm?page_id=570

Sponsor: Society for the Arts in Healthcare
2437 15th Street NW, B South
Washington, D.C. 20009

Society of Cosmetic Chemists Allan B. Black Award Sponsored by Presperse   3604
Dedicated to the advancement of cosmetic science, the Society strives to increase and disseminate scientific information through meetings and publications. By promoting research in cosmetic science and industry, and by setting high ethical, professional and educational standards, it strives to improve the qualifications of cosmetic scientists. Its mission is to further the interests and recognition of cosmetic scientists while maintaining the confidence of the public in the cosmetic and toiletries industry. With the Allan B. Black Award, a scroll and honorarium are awarded annually for the Best Paper on make-up technology either presented at an annual meeting, a seminar, or published in the Journal of Cosmetic Science, the Official Journal of the Society of Cosmetic Chemists.
Geographic Focus: All States
Contact: Theresa Cesario; (212) 668-1500; fax (212) 668-1504; TCesario@SCConline.org
Internet: http://www.scconline.org/website/about_scc/awards_fellowships.shtml
Sponsor: Society of Cosmetic Chemists
120 Wall Street, Suite 2400
New York, NY 10005-4088

Society of Cosmetic Chemists Award Sponsored by McIntyre Group, Ltd.   3605
Dedicated to the advancement of cosmetic science, the Society strives to increase and disseminate scientific information through meetings and publications. By promoting research in cosmetic science and industry, and by setting high ethical, professional and educational standards, it strives to improve the qualifications of cosmetic scientists. Its mission is to further the interests and recognition of cosmetic scientists while maintaining the confidence of the public in the cosmetic and toiletries industry. This award, given in the form of a scroll and honorarium, are awarded for the best paper presented at the Society's Annual Scientific Meeting.
Geographic Focus: All States
Contact: Theresa Cesario; (212) 668-1500; fax (212) 668-1504; TCesario@SCConline.org
Internet: http://www.scconline.org/website/about_scc/awards_fellowships.shtml
Sponsor: Society of Cosmetic Chemists
120 Wall Street, Suite 2400
New York, NY 10005-4088

Society of Cosmetic Chemists Award Sponsored by The HallStar Company   3606
Dedicated to the advancement of cosmetic science, the Society strives to increase and disseminate scientific information through meetings and publications. By promoting research in cosmetic science and industry, and by setting high ethical, professional and educational standards, it strives to improve the qualifications of cosmetic scientists. Its mission is to further the interests and recognition of cosmetic scientists while maintaining the confidence of the public in the cosmetic and toiletries industry. Offered in the form of a scroll and honorarium, this honor is awarded for the Best Paper that makes the greatest scientific contribution to knowledge in the field of protecting against or ameliorating damage to human skin caused by exposure to UV radiation presented at the Society's Annual Scientific Seminar.
Geographic Focus: All States
Contact: Theresa Cesario; (212) 668-1500; fax (212) 668-1504; TCesario@SCConline.org
Internet: http://www.scconline.org/website/about_scc/awards_fellowships.shtml
Sponsor: Society of Cosmetic Chemists
120 Wall Street, Suite 2400
New York, NY 10005-4088

Society of Cosmetic Chemists Chapter Best Speaker Award   3607
Dedicated to the advancement of cosmetic science, the Society strives to increase and disseminate scientific information through meetings and publications. By promoting research in cosmetic science and industry, and by setting high ethical, professional and educational standards, it strives to improve the qualifications of cosmetic scientists. Its mission is to further the interests and recognition of cosmetic scientists while maintaining the confidence of the public in the cosmetic and toiletries industry. The Chapter Best Speaker Award is given for the best scientific paper presented before each Chapter during a given calendar year. Recommendations for this award originates with the Chapters and are submitted to the National Office. A scroll and honorarium ($250 for a manuscript submission and $150 for an abstract submission) is provided by National for presentation by the Chapter.
Geographic Focus: All States
Contact: Theresa Cesario; (212) 668-1500; fax (212) 668-1504; TCesario@SCConline.org
Internet: http://www.scconline.org/website/about_scc/awards_fellowships.shtml
Sponsor: Society of Cosmetic Chemists
120 Wall Street, Suite 2400
New York, NY 10005-4088

Society of Cosmetic Chemists Chapter Merit Award   3608
Dedicated to the advancement of cosmetic science, the Society strives to increase and disseminate scientific information through meetings and publications. By promoting research in cosmetic science and industry, and by setting high ethical, professional and educational standards, it strives to improve the qualifications of cosmetic scientists. Its mission is to further the interests and recognition of cosmetic scientists while maintaining the confidence of the public in the cosmetic and toiletries industry. The Chapter Merit Award is awarded to a Chapter Member for outstanding service, dedication and voluntary services over a period of time. A scroll is provided by National for presentation by the Chapter. Recommendations for this award originates with each Chapter and are submitted to the National Office.
Geographic Focus: All States

Contact: Theresa Cesario; (212) 668-1500; fax (212) 668-1504; TCesario@SCConline.org
Internet: http://www.scconline.org/website/about_scc/awards_fellowships.shtml
Sponsor: Society of Cosmetic Chemists
120 Wall Street, Suite 2400
New York, NY 10005-4088

Society of Cosmetic Chemists Frontier of Science Award   3609
Dedicated to the advancement of cosmetic science, the Society strives to increase and disseminate scientific information through meetings and publications. By promoting research in cosmetic science and industry, and by setting high ethical, professional and educational standards, it strives to improve the qualifications of cosmetic scientists. Its mission is to further the interests and recognition of cosmetic scientists while maintaining the confidence of the public in the cosmetic and toiletries industry. The Frontier of Science Award is an honorarium given to a speaker who has achieved exceptional national or international stature in the scientific community for delivering a lecture at the Annual Scientific Meeting.
Geographic Focus: All States
Contact: Theresa Cesario; (212) 668-1500; fax (212) 668-1504; TCesario@SCConline.org
Internet: http://www.scconline.org/website/about_scc/awards_fellowships.shtml
Sponsor: Society of Cosmetic Chemists
120 Wall Street, Suite 2400
New York, NY 10005-4088

Society of Cosmetic Chemists Hans Schaeffer Award   3610
Dedicated to the advancement of cosmetic science, the Society strives to increase and disseminate scientific information through meetings and publications. By promoting research in cosmetic science and industry, and by setting high ethical, professional and educational standards, it strives to improve the qualifications of cosmetic scientists. Its mission is to further the interests and recognition of cosmetic scientists while maintaining the confidence of the public in the cosmetic and toiletries industry. The Hans Schaeffer Award, offered in the form of a scroll and honorarium, are awarded annually for the Most Innovative Paper presented at either an annual meeting or a seminar.
Geographic Focus: All States
Contact: Theresa Cesario; (212) 668-1500; fax (212) 668-1504; TCesario@SCConline.org
Internet: http://www.scconline.org/website/about_scc/awards_fellowships.shtml
Sponsor: Society of Cosmetic Chemists
120 Wall Street, Suite 2400
New York, NY 10005-4088

Society of Cosmetic Chemists Joseph P. Ciaudelli Award   3611
Dedicated to the advancement of cosmetic science, the Society strives to increase and disseminate scientific information through meetings and publications. By promoting research in cosmetic science and industry, and by setting high ethical, professional and educational standards, it strives to improve the qualifications of cosmetic scientists. Its mission is to further the interests and recognition of cosmetic scientists while maintaining the confidence of the public in the cosmetic and toiletries industry. The Joseph P. Ciaudelli Award, offered in the form of a scroll and honorarium, are bestowed annually for the best article appearing in the Journal of Cosmetic Science on the subject of hair care technology.
Geographic Focus: All States
Contact: Theresa Cesario; (212) 668-1500; fax (212) 668-1504; TCesario@SCConline.org
Internet: http://www.scconline.org/website/about_scc/awards_fellowships.shtml
Sponsor: Society of Cosmetic Chemists
120 Wall Street, Suite 2400
New York, NY 10005-4088

Society of Cosmetic Chemists Keynote Award Lecture Sponsored by Ruger Chemical Corporation   3612
Dedicated to the advancement of cosmetic science, the Society strives to increase and disseminate scientific information through meetings and publications. By promoting research in cosmetic science and industry, and by setting high ethical, professional and educational standards, it strives to improve the qualifications of cosmetic scientists. Its mission is to further the interests and recognition of cosmetic scientists while maintaining the confidence of the public in the cosmetic and toiletries industry. The Keynote Award is an honorarium given to a speaker who has achieved exceptional national or international stature in the scientific community for delivering a lecture at the Annual Scientific Meeting.
Geographic Focus: All States
Contact: Theresa Cesario; (212) 668-1500; fax (212) 668-1504; TCesario@SCConline.org
Internet: http://www.scconline.org/website/about_scc/awards_fellowships.shtml
Sponsor: Society of Cosmetic Chemists
120 Wall Street, Suite 2400
New York, NY 10005-4088

Society of Cosmetic Chemists Literature Award   3613
Dedicated to the advancement of cosmetic science, the Society strives to increase and disseminate scientific information through meetings and publications. By promoting research in cosmetic science and industry, and by setting high ethical, professional and educational standards, it strives to improve the qualifications of cosmetic scientists. Its mission is to further the interests and recognition of cosmetic scientists while maintaining the confidence of the public in the cosmetic and toiletries industry. Through the Literature Award, a scroll and honorarium are presented annually to the author or authors of scientific papers in basic research judged to be an outstanding contribution to cosmetic science and technology.
Geographic Focus: All States
Contact: Theresa Cesario; (212) 668-1500; fax (212) 668-1504; TCesario@SCConline.org

Internet: http://www.scconline.org/website/about_scc/awards_fellowships.shtml
Sponsor: Society of Cosmetic Chemists
120 Wall Street, Suite 2400
New York, NY 10005-4088

Society of Cosmetic Chemists Maison G. de Navarre Medal 3614
Dedicated to the advancement of cosmetic science, the Society strives to increase and disseminate scientific information through meetings and publications. By promoting research in cosmetic science and industry, and by setting high ethical, professional and educational standards, it strives to improve the qualifications of cosmetic scientists. Its mission is to further the interests and recognition of cosmetic scientists while maintaining the confidence of the public in the cosmetic and toiletries industry. The Society's highest honor, the Maison G. de Navarre Medal, recognizes individuals for their technical contributions to cosmetic science. The award is presented annually.
Geographic Focus: All States
Contact: Theresa Cesario; (212) 668-1500; fax (212) 668-1504; TCesario@SCConline.org
Internet: http://www.scconline.org/website/about_scc/awards_fellowships.shtml
Sponsor: Society of Cosmetic Chemists
120 Wall Street, Suite 2400
New York, NY 10005-4088

Society of Cosmetic Chemists Merit Award 3615
Dedicated to the advancement of cosmetic science, the Society strives to increase and disseminate scientific information through meetings and publications. By promoting research in cosmetic science and industry, and by setting high ethical, professional and educational standards, it strives to improve the qualifications of cosmetic scientists. Its mission is to further the interests and recognition of cosmetic scientists while maintaining the confidence of the public in the cosmetic and toiletries industry. The Merit Award is presented annually for outstanding service and distinguished leadership in Society activities.
Geographic Focus: All States
Contact: Theresa Cesario; (212) 668-1500; fax (212) 668-1504; TCesario@SCConline.org
Internet: http://www.scconline.org/website/about_scc/awards_fellowships.shtml
Sponsor: Society of Cosmetic Chemists
120 Wall Street, Suite 2400
New York, NY 10005-4088

Society of Cosmetic Chemists Robert A. Kramer Lifetime Service Award 3616
Dedicated to the advancement of cosmetic science, the Society strives to increase and disseminate scientific information through meetings and publications. By promoting research in cosmetic science and industry, and by setting high ethical, professional and educational standards, it strives to improve the qualifications of cosmetic scientists. Its mission is to further the interests and recognition of cosmetic scientists while maintaining the confidence of the public in the cosmetic and toiletries industry. The Robert A. Kramer Lifetime Service Award, given at the discretion of the Board of Directors, goes to an individual who has given extraordinary service over the course of his or her membership.
Geographic Focus: All States
Contact: Theresa Cesario; (212) 668-1500; fax (212) 668-1504; TCesario@SCConline.org
Internet: http://www.scconline.org/website/about_scc/awards_fellowships.shtml
Sponsor: Society of Cosmetic Chemists
120 Wall Street, Suite 2400
New York, NY 10005-4088

Society of Cosmetic Chemists Shaw Mudge Award 3617
Dedicated to the advancement of cosmetic science, the Society strives to increase and disseminate scientific information through meetings and publications. By promoting research in cosmetic science and industry, and by setting high ethical, professional and educational standards, it strives to improve the qualifications of cosmetic scientists. Its mission is to further the interests and recognition of cosmetic scientists while maintaining the confidence of the public in the cosmetic and toiletries industry. The Shaw Mudge Award, given in the form of a scroll and honorarium, are awarded for the best paper presented at the Society's Annual Scientific Seminar.
Geographic Focus: All States
Contact: Theresa Cesario; (212) 668-1500; fax (212) 668-1504; TCesario@SCConline.org
Internet: http://www.scconline.org/website/about_scc/awards_fellowships.shtml
Sponsor: Society of Cosmetic Chemists
120 Wall Street, Suite 2400
New York, NY 10005-4088

Society of Cosmetic Chemists Student Poster Awards 3618
Dedicated to the advancement of cosmetic science, the Society strives to increase and disseminate scientific information through meetings and publications. By promoting research in cosmetic science and industry, and by setting high ethical, professional and educational standards, it strives to improve the qualifications of cosmetic scientists. Its mission is to further the interests and recognition of cosmetic scientists while maintaining the confidence of the public in the cosmetic and toiletries industry. The Student Poster Award, sponsored by DD-Chemco, Inc., is given annually in conjunction with the Annual Scientific Seminar. Honorariums are given for first, second, third and fourth place Best Student Poster.
Geographic Focus: All States
Contact: Theresa Cesario; (212) 668-1500; fax (212) 668-1504; TCesario@SCConline.org
Internet: http://www.scconline.org/website/about_scc/awards_fellowships.shtml
Sponsor: Society of Cosmetic Chemists
120 Wall Street, Suite 2400
New York, NY 10005-4088

Society of Cosmetic Chemists Young Scientist Awards 3619
The Young Scientist Awards are presented each year to recognize the outstanding achievements of new cosmetic scientists throughout the industry. Each year every manufacturer, supplier or other organization in the personal care industry may select an individual in their employ as a recipient of this award. The criteria for selection is made by the individual company. Information is sent in March of each year and is posted on the SCC Website. The submission deadline is July 1st each year. The awards may be presented at an SCC Chapter Meeting or through an internal corporate presentation. Recipient names are listed in the SCC Winter Newsletter.
Requirements: Recipients be an active member of the Society for at least one year and working in the industry less than 10 years.
Geographic Focus: All States
Date(s) Application is Due: Jul 1
Contact: Theresa Cesario; (212) 668-1500; fax (212) 668-1504; TCesario@SCConline.org
Internet: http://www.scconline.org/website/about_scc/awards_fellowships.shtml
Sponsor: Society of Cosmetic Chemists
120 Wall Street, Suite 2400
New York, NY 10005-4088

Solo Cup Foundation Grants 3620
The foundation awards grants to Illinois nonprofit organizations in its areas of interest, including higher education, health, social services, and Christian organizations and churches. Types of support include capital campaigns, general operating support, and scholarship funds. There are no application deadlines or forms.
Requirements: Illinois nonprofit organizations are eligible.
Geographic Focus: Illinois
Amount of Grant: 5,000 - 85,000 USD
Samples: La Rashida Children's Hospital (Chicago, IL)--for a capital campaign to modernize and upgrade its inpatient facility, $1 million; Immaculate Conception Church (Chicago, IL)--for restoration project, $85,000; Congregation of the Passion (Chicago, IL)--for general support, $50,000.
Contact: Robert M. Korzenski; (847) 831-4800; fax (847) 579-3245; info@solocup.com
Sponsor: Solo Cup Foundation
1700 Old Deerfield Road
Highland Park, IL 60035

Sonoco Foundation Grants 3621
The Sonoco Products Company is the company sponsor of the Sonoco Foundation. The foundation was established in 1983, in the state of South Carolina. The foundation supports organizations involved with education, crime and violence prevention, human services, youth, and community development. Giving primarily in areas of company operations, with emphasis on Hartsville, South Carolina.
Requirements: South Carolina nonprofit organizations are eligible to apply for these grants. There is no application form required. Applicants should submit a detailed description of project and amount of funding requested.
Restrictions: The foundation does not support religious, political, or lobbying organizations. Grants aren't available to individuals, or for educational capital funds programs, endowments, trips, or tours.
Geographic Focus: South Carolina
Amount of Grant: 1,000 - 600,000 USD
Samples: University of South Carolina, Columbia, SC, $600,000; Boy Scouts of America, Florence, SC, $60,000; United Way of Florence County, Florence, SC, $14,000;
Contact: Joyce Beasley, (800) 377-2692 or (843) 383-7851
Internet: http://www.sonoco.com
Sponsor: Sonoco Foundation
1 North Second Street, MS A09
Hartsville, SC 29550-3300

Sonora Area Foundation Competitive Grants 3622
The Foundation's governing board has a pool of discretionary funds, and regularly awards grants to qualified non-profit and public agencies. These grants are competitive and are awarded for worthy projects that help improve the quality of life in Tuolumne County, whether for education, human services, recreation, arts or the environment. The Foundation seeks opportunities to stretch its funding impact by making grants which will work in combination with other funds. This might involve requiring an applicant to provide matching funds from other sources. In addition, the Foundation may choose to make a challenge grant, where additional funds must be secured before Foundation monies are released. The Foundation considers several types of competitive grant requests: Pilot or Demonstration Project (intended to promote creativity, such a project is based on the concept of setting up, testing and evaluating a model approach to dealing with challenges found in the community), New or Expanded Program (based on challenges or needs within Tuolumne County), and Capacity Building (strengthening an applicant organization's ability to achieve its mission or purposes).
Requirements: Nonprofit organizations serving Tuolumne County, CA, may submit applications. Public sector entities or units of government primarily serving Tuolumne County residents are also eligible to apply. Informal associations, community groups or collaboratives may apply through an eligible lead organization acting as the fiscal sponsor. The Sonora Area Foundation Board of Directors reviews grant applications at their scheduled Board meetings (usually the fourth Tuesday) in February, April, June, August, October, December. Applications are due at the Foundation office by the end of the month preceding these Board Meetings.
Restrictions: Grants are not awarded to/for sectarian purposes, private foundations, political purposes, annual or capital campaigns, endowment funds, existing financial obligations or debt retirement, or programs or services for other than Tuolumne County residents.

Geographic Focus: California
Date(s) Application is Due: Jan 31; Mar 31; May 31; Jul 31; Sep 30; Nov 30
Amount of Grant: Up to 200,000 USD
Contact: Lin Freer, Program Manager; (209) 533-2596; fax (209) 533-2412
Internet: http://www.sonora-area.org/grantapplication.html
Sponsor: Sonora Area Foundation
362 South Stewart Street
Sonora, CA 95370

Sony Corporation of America Grants 3623
Sony Corporation of America consists of three operating companies as well as the corporate headquarters, which is based in New York City: Sony Electronics, Inc. (headquarters in San Diego, California); Sony Pictures Entertainment, Inc. (headquarters in Culver City, California); and Sony Music Entertainment, Inc. (headquarters in New York City). Each company, as well as the overall corporation, has its own philanthropic priorities and resources (e.g. grants, product donations, and recordings and screenings) that benefit a multitude of causes. Taken together the corporation's areas of interest cover arts and culture; health and human services; civic and community outreach, education; the environment; disaster response; and volunteerism. The core of Sony's various corporate philanthropy programs are their contributions to the communities in which Sony employees work and live; however, the corporation and subsidiaries contribute to national nonprofits as well. In the past, types of support have included general operating budgets, continuing support, annual campaigns, seed grants, building construction, equipment acquisition, endowment funds, employee matching gifts, internships, and employee-related scholarships. Sony Corporation of America and its subsidiaries welcome requests for support throughout the year. There is no grant application form. Requests must be submitted in writing to the corporation or its operating companies. Contact information is given on this page and at the website. Guidelines for what to include in the application as well as more information on types of programs the corporation and/or its subsidiaries have supported are available at the grant website. Notification of grant-request approval or rejection will be made in writing within one month of receipt of all proposed materials.
Requirements: U.S. nonprofits, including schools and school districts, are eligible.
Restrictions: The corporation does not consider multi-year requests for support. The following types of organizations will not be funded: organizations that discriminate on the basis of race, color, creed, gender, religion, age, national origin, or sexual orientation; partisan political organizations, committees, or candidates and public office holders; religious organizations in support of their sacramental or theological functions; labor unions; endowment or capital campaigns of national origin; organizations whose prime purpose is to influence legislation; testimonial dinners in general; for-profit publications or organizations seeking advertisements or promotional support; individuals seeking self-advancement; foreign or non-U.S.-based organizations; and organizations whose mission is outside of the U.S.
Geographic Focus: All States
Amount of Grant: 1,000 - 100,000 USD
Contact: Janice Pober, Senior Vice President, Corporate Social Responsibility; (310) 244-7737; SPE_CSR@spe.sony.com
Karen E. Kelso, Vice-President; (212) 833-8000; SCA_CSR@sonyusa.com
Julie Wenzel, Senior Manager, Community Relations; (858) 942-2400; SELCommunityAffairs@am.sony.com
Internet: http://www.sony.com/SCA/philanthropy/guidelines.shtml
Sponsor: Sony Corporation of America
550 Madison Avenue, 33rd Floor
New York, NY 10022-3211

Sophia Romero Trust Grants 3624
The Sophia Romero Trust was established in 1948 to support and promote quality human-services and health-care programming for underserved elders living in Bristol County, Massachusetts. Special consideration will be given to organizations that serve older women living in Bristol County, Massachusetts. Grant requests for general operating support or program support are strongly encouraged. The majority of grants from the Romero Trust are one year in duration. Applicants must apply online at the grant website. Applicants are strongly encouraged to do the following before applying: review the downloadable state application procedures for additional helpful information and clarifications; review the downloadable online-application guidelines at the grant website; review the trust's funding history (link is available from the grant website); review the online application questions in advance; and review the list of required attachments. These will generally include: a list of board members, financial statements (audited, reviewed, or compiled by independent auditor); an organization summary; a list of other funding sources; an IRS Determination letter; and other required documents. All attachments must be uploaded in the online application as PDF, Word, or Excel files. The application deadline for the Sophia Romero Trust is 11:59 p.m. on February 1. Applicants will be notified of grant decisions before April 30.
Requirements: Applicants must have 501(c)3 tax-exempt status and serve the residents of Bristol County, Massachusetts.
Restrictions: The fund does not support requests from individuals, organizations attempting to influence policy through direct lobbying, or any political campaigns.
Geographic Focus: Massachusetts
Date(s) Application is Due: Feb 1
Samples: Family Services Association of Greater Fall River, Fall River, Massachusetts, $20,000, continuing support; Southeast Massachusetts Home Health Aides, $4,000, program development; YWCA of Southeastern Massachusetts, New Bedford, Massachusetts, $4,000, Widowed Person Program (WPP).

Contact: Emma Greene, Director; (617) 434-0329; emma.m.greene@baml.com
Internet: https://www.bankofamerica.com/philanthropic/fn_search.action
Sponsor: Sophia Romero Trust
225 Franklin Street, 4th Floor, MA1-225-04-02
Boston, MA 02110

Southbury Community Trust Fund 3625
The Fund awards grants in the areas of arts, community and economic development, education, the environment, and health and human services. High priority is given to program development, program implementation, organizational capacity building, and policy change addressing root causes of problems. Interested applicants should complete a letter of inquiry. Information for submission is available online.
Requirements: Organizations with IRS 501(c)3 status serving Southbury are eligible to submit letters of inquiry.
Restrictions: Funding is not available for: religious purposes; political campaigns; annual appeals; building fund/capital, endowment or memorial campaigns; loans or deficit financing; federal, state, or municipal agencies or departments supported by taxation; individuals.
Geographic Focus: Connecticut
Date(s) Application is Due: Jan 15
Contact: Anne Watkins, Director of Programs; (203) 334-7511; fax (203) 333-4652; awatkins@gbafoundation.org
Internet: http://www.gbafoundation.org/grants/apply.asp
Sponsor: Greater Bridgeport Area Foundation
211 State Street, 3rd Floor
Bridgeport, CT 06604

South Madison Community Foundation Grants 3626
The South Madison Community Foundation (SMCF) Grants serves the residents in Adams, Fall Creek, Green, and Stony Creek Townships of Madison County. The Foundation has developed a procedure for submission and evaluation of grant proposals. The Grants Committee makes its recommendation to the Board of Directors who give final approval for funding. Applicants are notified within one month of the submission deadline of approval or denial of their request for funding. Online applications are available at the Foundation website. Organizations are encouraged to read the Tips for Submitting Effective Proposals on the Foundation website.
Requirements: With limited funds, the Foundation must distribute grant funding that gives the greatest positive impact on the needs and growth of South Madison county. Applicants should focus on projects that: focus on the prevention of problems rather than the symptoms, and make a significant improvement to the community; maintain a proactive focus and an ability to respond to creative ideas; respond to the changing needs of the community; encourage programs that enhance cooperation and collaboration among institutions within the community; and leverage funds through the use of match and challenge grants.
Restrictions: Note that Cycle One of the funding deadlines is funded with a smaller budget than the other funding cycle dates, so that applicants should apply for no more than $1,000. Because of time constraints, all proposals to the Foundation must be submitted in writing, so that no personal presentations are accepted. The following items will not be funded: operational expenses of existing programs; endowments or deficit funding; funds for redistribution by the grantee; conferences, publication, films, television and radio programs unless integral to the project for which the grant is sought; religious or political purposes; travel for individuals or groups such as bands, sports teams, and classes; annual appeals and membership contributions; and major capital improvements.
Geographic Focus: Indiana
Date(s) Application is Due: Feb 1; Jun 1; Oct 1
Amount of Grant: 150 - 10,000 USD
Samples: Indiana State Police Emergency Response Section, purchase of a thermal imaging camera, allowing police to operate during low and no light situations while conducting surveillance and/or pursuing suspects, $1,500 match; Lapel High School, Character Counts Education program, help guide students in their decision-making process in the development of moral values, $5,000; Lapel Optimist Club, help with cost of equipment for participants of Little League Football Program, $1,500.
Contact: Barbara Switzer, Program Director; (765) 778-8444; fax (765) 778-9144; barbara@southmadisonfoundation.org
Internet: http://www.southmadisonfoundation.org/grants.html
Sponsor: South Madison Community Foundation
233 South Main Street
Pendleton, IN 46064

Southwest Florida Community Foundation Competitive Grants 3627
Program/project grants are primarily for new or expanded programs and pilot or demonstration projects, which meet a documented need. Capital grants primarily support the construction, acquisition and renovation of facilities. The foundation offers competitive grants for programs and projects in eight focus areas: animal welfare, arts and culture, community development, conservation and preservation, education, the environment, health, and human services.
Requirements: IRS 501(c)3 tax-exempt organizations undertaking programs to improve the quality of life in Florida's Lee County and contiguous counties, including Charlotte, Collier, Gledes, and Hendry, are eligible. (NOTE: Agencies located in Collier County must also serve one of the other four counties in order to be eligible.) Applicants must employ staff, elect a governing board, and conduct business without discrimination on the basis of race, religion, gender, sexual orientation, age, national origin or disability.
Restrictions: The foundation generally will not fund: Normal operating expenses or existing obligations; Annual campaigns; Endowments; Sectarian purposes (i.e. programs that promote or require a religious doctrine except where designated by a donor); Debt

retirement Monetary awards; Professional conferences, sports team travel, class trips, etc; Individuals; Fraternal organizations, societies, etc; Political organizations or campaigns; Fund-raising or feasibility studies; Research; or, Re-granting.
Geographic Focus: Florida
Date(s) Application is Due: Jun 19
Samples: Calusa Nature Center & Planetarium - $26,000; Fellowship of Christian Athletes - $20,000; Harry Chapin Food Bank - $20,000; Literacy Council of Bonita Springs - $15,000; Take Stock in Children - $13,615; Vineyards Elementary School - $23,000
Contact: Carol McLaughlin, Chief Program Officer; (239) 274-5900, ext. 225; fax (239) 274-5930; cmclaughlin@floridacommunity.org
Internet: http://www.floridacommunity.com/grantseekers/
Sponsor: Southwest Florida Community Foundation
8260 College Parkway, Suite 101
Fort Myers, FL 33919

Southwest Gas Corporation Foundation Grants 3628
The corporate foundation supports nonprofit organizations in its service communities. General support grants and employee matching gifts are made in the areas of education (universities, colleges, and literacy programs) and social services (United Way organizations, youth groups, community service, and volunteer organizations). Additional areas of interest are arts/culture and health. Types of support include general operating grants, projects grants, conferences and seminars, building construction/renovation, capital campaigns, emergency funds, employee matching gifts, research grants, donated equipment, and in-kind services. In southern Arizona, contact Marty Looney, P.O. Box 26500, Tucson, AZ 85726-6500; (520) 794-6416.
Requirements: Nonprofits in Arizona, San Bernardino County, California, and Nevada may apply.
Geographic Focus: Arizona, California, Nevada
Amount of Grant: Up to 620,567 USD
Samples: Phoenix Art Museum (Phoenix, AZ)--for general support, $3000; Boys and Girls Clubs (Tucson, AZ)--for general support, $1000; Community Food Bank (Phoenix, AZ)--for general support, $3500.
Contact: Suzanne Farinas, (702) 876-7247; fax (702) 876-7037
Internet: http://www.swgas.com
Sponsor: Southwest Gas Corporation
P.O. Box 98510
Las Vegas, NV 89193-8510

Special Olympics Health Professions Student Grants 3629
The purpose of this program is to engage health professions students to work with persons with intellectual disabilities as a way of filling in a gap that exists in most health program curricula. The program promotes short-term projects exploring issues that impact the health and well-being of all persons with intellectual disabilities, including, but not limited to, Special Olympics athletes. Projects may include: data collection and analysis on issues impacting persons with intellectual disabilities; measurement of attitudes, opinions and behaviors of health professionals, coaches, family/caregivers and athletes; follow-up assessments of existing programs; or health promotion projects. Projects that involve collaborations with Special Olympics Programs or other CD.C. grant recipients (e.g., state and local health departments) are encouraged. Special Olympics is not currently prioritizing specific research areas for funding, as the program's primary goal is to engage health professions students to work with the population of persons with intellectual disabilities. All awards will be made to the institution on behalf of the student grantee. Funding must include all expenses and no indirect costs may be covered by this grant. Applications are accepted year-round. Application and guidelines are available online.
Requirements: Health professions students at the bachelor, master, doctorate and post-doctorate levels from accredited institutions or programs in the United States or other countries are eligible. Fields or disciplines may include, audiology, dentistry, pre-med, medicine, nursing, optometry, physical therapy, podiatry, public health, social work, law, public administration, or other related health areas. All students??? projects will require a faculty advisor.
Geographic Focus: All States
Amount of Grant: Up to 5,000 USD
Contact: Darcie Mersereau, Manager, Research and Evaluation; (202) 628-3600 or (800) 700-8585; fax (202) 824-0200; 2006hastudentgrant@specialolympics.org
Internet: http://www.specialolympics.org/content.aspx?id=6336&terms=student
Sponsor: Special Olympics
1133 19th Street, NW
Washington, D.C. 20036-3604

Special People in Need Grants 3630
The program awards grants to individuals and organizations operating in the fields of education, health, and assistance to people who are disabled, ill, or economically disadvantaged. There are no application deadlines or forms. Applicants should submit a cover letter, which will be reviewed by the grant committee.
Requirements: Nonprofit organizations are eligible.
Geographic Focus: All States
Amount of Grant: 500 - 5,000 USD
Contact: Gary Kline, (312) 715-5235
Sponsor: Special People in Need Foundation
500 W Madison Street, Suite 3700
Chicago, IL 60661

Special People in Need Scholarships 3631
This scholarship is for undergraduate students with disabilities pursuing a postsecondary education. Applicants must demonstrate a strong financial need and a potential to benefit from an education, which might not otherwise be possible. The institution the student is attending must be willing to serve as a sponsor for disbursement.
Requirements: There are no formal deadlines. To be considered send an application with transcripts, letter of acceptance, and agreement from the institution to act as sponsor, letter of recommendation from an educator, financial statements, biographical data, and other relevant information.
Geographic Focus: All States
Contact: Gary Kline, (312) 715-5235
Sponsor: Special People in Need Foundation
500 W Madison Street, Suite 3700
Chicago, IL 60661

Spencer County Community Foundation Grants 3632
The Spencer County Community Foundation is a nonprofit, public charity created for the people of Spencer County, Indiana. The Foundation helps nonprofits fulfill their missions by strengthening their ability to meet community needs through grants that assist charitable programs, address community issues, support community agencies, launch community initiatives, and support leadership development. Priority funding areas include arts and culture; community development; education; health; human services; environment; recreation; and youth development. Grant proposals are accepted once each year according to the grant cycle. Grants are normally given as one time support of a project but may be considered for additional support for expansions or outgrowths of an initial project. Proposals will be accepted from mid-January through mid-March. Grant awards will be presented in June. The application, supporting materials, and examples of previously funded project are available at the Foundation website.
Requirements: The Foundation welcomes proposals from nonprofit organizations that are tax exempt under sections 501(c)3 and 509(a) and from governmental agencies serving the county. Proposals from nonprofit organizations not classified as a 501(c)3 public charity may be considered if the project is charitable and supports a community need. Proposals submitted by an entity under the auspices of another agency must include a written statement signed by the agency's board president on behalf of the board of directors agreeing to act as the entity's fiscal sponsor, to receive grant monies if awarded, and to oversee the proposed project.
Restrictions: Project areas not considered for funding: religious organizations for religious purposes; political parties or campaigns; endowment creation or debt reduction; operating costs; capital campaigns; annual appeals or membership contributions; travel requests for groups or individuals such as bands, sports teams, or classes.
Geographic Focus: Indiana
Contact: Laura Harmon; (812) 649-5724; laura@spencercommunityfoundation.org
Internet: http://www.spencercommunityfoundation.org/disc-grants
Sponsor: Spencer County Community Foundation
Lincoln Commerce Center
Rockport, IN 47635

Springs Close Foundation Grants 3633
Since it was chartered in 1942, the Springs Close Foundation has contributed over $85 million to a wide variety of charitable and educational causes designed to improve the quality of life and well-being of the people in Chester, Lancaster and York Counties. Support includes, but is not limited to, food, shelter and medical assistance. This temporary change in focus is in response to high levels of unemployment and economic distress in the Foundation's service areas. The Foundation also makes occasional statewide grants in South Carolina. There are three major areas of program interest: Recreation and Environment, Public Education and Early Childhood Development, Community Service and Health
Requirements: Grants are made only to organizations that are tax-exempt under Section 501(c)3 of the Internal Revenue Code. No grants are made to individuals.
Restrictions: The Foundation will only consider grant requests from eligible nonprofit organizations that can effectively deliver emergency and basic support to citizens in Chester and Lancaster counties and a portion of York County.
Geographic Focus: South Carolina
Date(s) Application is Due: Mar 1; Oct 1
Amount of Grant: 5,000 - 50,000 USD
Contact: Angela McCrae, President; (803) 548-2002; fax (803) 548-1797
Internet: http://www.thespringsclosefoundation.org/grants.htm
Sponsor: Springs Close Foundation
1826 Second Baxter Crossing
Fort Mill, SC 29708

Square D Foundation Grants 3634
The foundation makes donations for operating support, capital development needs, and special projects to nonprofit organizations in the areas of education, social welfare, arts and cultural and civic affairs, and health. Each year the foundation supports United Way in communities where the company and its domestic subsidiaries have significant operations. Donations are not normally made to organizations already receiving support through United Way. Support of higher education is achieved through scholarships, endowments for faculty and acquisition or expansion of equipment or facilities, unrestricted operating support, and the matching gift program. Submit letters of application between June and August.
Requirements: Giving primarily to 501(c)3 tax-exempt organizations in areas of company operations, with emphasis on: Illinois; Indiana; Iowa; Kentucky; Missouri; Nebraska; North Carolina; Ohio; South Carolina; Tennessee; and Wisconsin.

Restrictions: Grants are not made to religious organizations for religious purposes, political groups and organizations, labor unions and organizations, organizations making requests by telephone, organizations listed by the U.S. attorney general as subversive, or to individuals.
Geographic Focus: Illinois, Indiana, Iowa, Kentucky, Nebraska, North Carolina, Tennessee, Texas, Wisconsin
Date(s) Application is Due: Aug 31
Amount of Grant: Up to 27,000,000 USD
Contact: Harry Wilson, Secretary; (847) 397-2600
Internet: http://www.squared.com
Sponsor: Square D Foundation
1415 South Roselle Road
Palatine, IL 60067

SRC Medical Research Grants 3635
The Scientific Council for Medicine supports research throughout the discipline of Medical Science, i.e. Medicine, Odontology, Pharmaceutics and Health Sciences. Medical research supported by the Scientific Council ranges from studies at molecular and cell level, and research involving laboratory animal science and patients, to epidemiological studies on groups of human subjects. The Program offers grants for research and junior research positions; grants for research periods for clinical researchers; grants for research stays (sabbaticals) abroad; project research grants (distributed, as a rule, for three years for salary, materials, travel, publication, etc); international postdoctoral fellowships and grants for postdoctoral positions in Sweden; networks; symposium; research equipment; collaboration grant and grant for inviting a visiting scientist.
Geographic Focus: All States
Contact: Paola Norlin; +46 (0)8 546 44 311; fax 46 8 546 44 180; paola.norlin@vr.se
Internet: http://www.vr.se/responsibilities/researchfunding/medicine.4.69f66a93108e85f68d48000837.html
Sponsor: Swedish Research Council
Klarabergsviadukten 82
Stockholm, SE-103 78 Sweden

SSHRC-NSERC College and Community Innovation Enhancement Grants 3636
The objective of the Innovation Enhancement Grants is to increase innovation at the community and/or regional level by enabling Canadian colleges to increase their capacity to work with local small and medium-sized companies. They support applied research and collaborations that facilitate commercialization, as well as technology transfer, adaptation and adoption of new technologies. Grants will stimulate applied research that brings together expertise from diverse fields such as natural sciences and engineering, social sciences and humanities, and/or health sciences to address business-driven challenges and opportunities. Grants provide funding to colleges on a competitive basis to enhance their applied research capacity, and carry out applied research and technology transfer activities in collaboration with, and to the benefit of small to medium size companies in particular. These grants will focus on an area where the college has recognized expertise, that meet local or regional needs, and where there is the potential to increase economic development of the community. Grants are awarded for either a two-year or a five-year period. For the two-year grant, $100,000 per year for each of two years is provided. For the five-year grant, a base grant of up to $500,000 per year is provided for the first three years. During the course of the grants, colleges are expected to increase their effective collaboration with local or regional companies, and other existing community resources. To ensure increased and sustained commitment and involvement from college partners in years four and five of the five-year grants, grants will provide four-fifths of the average annual base funding received in the first three years. It is expected that other sources of support will be available from business and other partners (either cash or in-kind contributions). Additional guidelines, including entry-level and five year grants requirements and descriptions, are available at the NSERC website.
Requirements: Canadian colleges and universities are eligible to apply. These institutions must offer programs in the natural sciences, engineering, social sciences, humanities and/or health sciences. Faculty members involved in the grants must be involved in these disciplines. Institutions must also provide the space, facilities and services to enable its natural sciences, engineering, social sciences, humanities and/or health sciences faculty members to conduct research. Additional requirements and guidelines for businesses are available at the NSERC website.
Geographic Focus: Canada
Contact: Eric Bergeron, Program Officer, Partnerships Portfolio; (613) 996-1422 or (613) 944-5802; fax (613) 947-0223; eric.bergeron@sshrc-crsh.gc.ca
Internet: http://www.nserc-crsng.gc.ca/Professors-Professeurs/RPP-PP/CCI-ICC_eng.asp
Sponsor: Social Sciences and Humanities Research Council of Canada
350 Albert Street, P.O. Box 1610
Ottawa, ON K1P 6G4 Canada

SSHRC Banting Postdoctoral Fellowships 3637
The Banting Postdoctoral Fellowships strive to attract and retain top-tier postdoctoral talent, nationally and internationally; develop their leadership potential; and position them for success as research leaders, contributing to Canada's economic, social, and research based growth through a research-intensive career. Selection criteria is based on research excellence and leadership in the research domain; quality of applicant's proposed research program; and institutional commitment and demonstrated synergy between applicant and institutional strategic priorities. The Fellowships are distributed equally across Canada's three federal granting agencies: the Canadian Institutes of Health Research (CIHR), the Natural Sciences and Engineering Research Council (NSERC) and the Social Sciences and Humanities Research Council (SSHRC). Up to 70 new two-year awards are given annually, with up to 140 awards active at any given time. Candidates may apply through the ResearchNet application process at the Banting website. Additional information and frequently asked questions are also posted on the website.
Requirements: Canadian citizens, permanent residents, and international citizens may apply. Applicants must fulfill or have fulfilled all degree requirements for a Ph.D., its equivalent, or health professional degree (a complete list of eligible health professionals is posted on the website). Applicants must contact the hot research institution whose endorsement they wish to receive, and continue contact to discuss benefits and their rights and responsibilities.
Restrictions: Applicants outside Canada may only hold their Banting Fellowship at a Canadian institution. Canadian or permanent resident applicants who receive their Ph.D. from an international institution may only hold the Banting Fellowship at a Canadian institution. Awards must be taken up no earlier than April 1 and no later than October 1 following the year of application.
Geographic Focus: All States, All Countries, Canada
Date(s) Application is Due: Nov 1
Amount of Grant: 70,000 USD
Contact: Banting Fellowship Coordinator; (613) 943-7777; banting@researchnet-recherchenet.ca or fellowships@sshrc-crsh.gc.ca
Internet: http://www.sshrc-crsh.gc.ca/funding-financement/programs-programmes/fellowships/banting-eng.aspx
Sponsor: Social Sciences and Humanities Research Council of Canada
350 Albert Street, P.O. Box 1610
Ottawa, ON K1P 6G4 Canada

SSRC Collaborative Research Grants on Environment and Health in China 3638
The collaborative grants program was begun to stimulate new research on environment and health in China that is directly linked to policy and practice, to encourage collaboration across disciplines, and to help develop a network of scholars working in this field. Each year seven to eight grants of $10,000 to 15,000 each are given to Chinese institutes for projects that: integrate environment or health perspectives into research currently focused on a single domain by sharing existing data or conducting limited additional data collection; develop a new focus on environment and health within the context of ongoing research on development; broaden the scope of analysis through collaboration with partners who bring additional disciplinary perspectives or international experience; or work with NGO or government partners to make existing research more accessible and useful. Additional information is available at the SSRC website.
Requirements: Each project must address an environmental health problem that affects a substantial number of people and it must be clearly useful in generating or disseminating knowledge that will inform policy or the work of government or civil society actors. In order to ensure feasibility, applicants must demonstrate the necessary access to data, or relationships with local communities or government agencies, to conduct the project successfully. If the project is being carried out at the local level, researchers must either live and work in the same province, or otherwise demonstrate how the project will build local capacity to work on environment and health issues.
Geographic Focus: China
Amount of Grant: 10,000 - 15,000 USD
Contact: Chao Cai; (212) 377-2700; fax (212) 377-2727; chao@ssrc.org
Xiaofang Liu, Program Assistant; (212) 377-2700; fax (212) 377-2727; liu@ssrc.org
Internet: http://www.ssrc.org/fellowships/grants-for-collaborative-research-on-environment-and-health-in-china/
Sponsor: Social Science Research Council
One Pierrepont Plaza, 15th Floor
Brooklyn, NY 11201

St. Joseph Community Health Foundation Burn Care and Prevention Grants 3639
The St. Joseph Community Health Foundation, sponsored by the Poor Handmaids of Jesus Christ, exists to steward resources that strengthen, improve, and sustain long-term community health in Allen County, Indiana. The Foundation mission is to invest, and collaborate with other community agencies to improve the physical, mental, and spiritual health of the poor and under served. The Burn Care and Prevention Grants are restricted funds to promote burn prevention education and to help educate health care professionals and first responders to emergency situations to treat severe burns with best care practices available. Grants typically range in size from $500 to $5,000.
Requirements: Applicants should have a demonstrated history of serving poor and powerless populations residing in Allen County. Applicants should be not-for-profit entities classified as 501(c)3 by the Internal Revenue Service. Some government agencies involved in related service areas may be eligible for support. The Foundation will consider requests for program, program-related equipment, staff continuing education, technical assistance and small capital grants. Applications are available at the Foundation's website.
Restrictions: Grant applications will not be accepted elimination of deficits, support of political activities, individuals, or projects all ready completed.
Geographic Focus: Indiana
Date(s) Application is Due: Mar 1; Jun 1; Sep 1; Dec 1
Samples: Allen County Juvenile Firesetters Task Force, $5,000 - Funding for a seminar pertaining to the identification and treatment of juvenile fire setters; Public Safety Foundation of Northeast Indiana, Inc., $50,000 - Partial funding of construction costs for a new classroom/office building in the Safety Village.
Contact: Meg Distler; (260) 969-2001, ext. 201; fax (260) 969-2004; mdistler@sjchf.org
Internet: http://www.stjosephhealthfdn.org/index.php?option=com_content&view=article&id=72&Itemid=71
Sponsor: St. Joseph Community Health Foundation
2826 South Calhoun Street
Fort Wayne, IN 46807

592 | GRANT PROGRAMS

St. Joseph Community Health Foundation Improving Healthcare Access Grants 3640
The St. Joseph Community Health Foundation was established in 1998 when the Poor Handmaids of Jesus Christ sold the St. Joseph Medical Center and reorganized the St. Joseph Community Health Foundation with a significant share of the proceeds. Proceeds from the Foundation's endowment are redistributed as grants typically ranging from $5,000 to $35,000 to advance programs, projects and partnerships that improve the access to quality health care and the health of the poor and powerless of Allen County, Indiana. Typically, only programs that can demonstrate that greater than 51 percent of their clients are very low income with health issues are considered for these grants. The Foundation also considers these values of the Poor Handmaids as a part of its grant review process: respecting and valuing each person; standing with the poor and powerless; using our talents and resources to respond to the emerging needs of society; nurturing leadership in our efforts to bring peace to the world. The Foundation will consider requests for program support, operations, seed monies, program-related equipment, staff continuing education, technical assistance, and matching funds. The Foundation will commit support on an annual basis only. Additional funding will be contingent upon program performance.
Requirements: Applicants should have a demonstrated history of serving poor and powerless populations with medical, dental, mental, and/or spiritual health care and wellness services. Applicants should be not-for-profit entities classified as 501(c)3 by the Internal Revenue Service. Partnerships between not-for-profits are encouraged. The proposed grant must be operated for the benefit of residents of Allen County, Indiana.
Restrictions: Grant applications will not be accepted for building projects, elimination of deficits, support of political activities, individuals, or projects already completed.
Geographic Focus: Indiana
Date(s) Application is Due: Mar 1; Sep 1
Amount of Grant: 5,000 - 35,000 USD
Samples: AIDS Task Force, Inc., $9,000 - support for emergency medical assistance to people living with HIV and AIDS in Allen County; Boys & Girls Clubs of Fort Wayne, $7,000 - support for the Cavity-Free Zone program promoting dental health and hygiene in low-income children; Cedars H.O.P.E., Inc., $5,000 - matching grant to support the program staff who house mentally ill women suffering from complex psychiatric and medical conditions.
Contact: Meg Distler; (260) 969-2001, ext. 201; fax (260) 969-2004; mdistler@sjchf.org
Internet: http://www.stjosephhealthfdn.org/index.php?option=com_content&view=article&id=79&Itemid=70
Sponsor: St. Joseph Community Health Foundation
2826 South Calhoun Street
Fort Wayne, IN 46807

St. Joseph Community Health Foundation Pfeiffer Fund Grants 3641
The St. Joseph Community Health Foundation, sponsored by the Poor Handmaids of Jesus Christ, exists to steward resources that strengthen, improve, and sustain long-term community health in Allen County, Indiana. The Foundation mission is to invest, and collaborate with other community agencies to improve the physical, mental, and spiritual health of the poor and under served. The Pfeiffer grants are restricted to health serving organizations for capital improvements or acquiring medical & surgical equipment. These funds have primarily been directed to help local clinics acquire medical equipment. All grants will be made to non-profit agencies, not individuals.
Requirements: Applicants should have a demonstrated history of serving poor and powerless populations residing in Allen County, Indiana. Applicants should be not-for-profit entities classified as 501(c)3 by the Internal Revenue Service. Some government agencies involved in related service areas may be eligible for support. Applications are available on the Foundation's website.
Restrictions: Grant applications will not be accepted for elimination of deficits, support of political activities, individuals, or projects all ready completed.
Geographic Focus: Indiana
Date(s) Application is Due: Mar 1; Sep 1
Amount of Grant: 500 - 25,000 USD
Samples: Fort Wayne Sexual Assault Treatment Center, $10,000 - Purchase of sensitive photographic equipment necessary to record evidence of trauma; Matthew 25 Health & Dental Clinic, $51,000 - Medical equipment for the Capital Campaign, with a goal of tripling facility space to serve more patients with expanded services and programs.
Contact: Meg Distler; (260) 969-2001, ext. 201; fax (260) 969-2004; mdistler@sjchf.org
Internet: http://www.stjosephhealthfdn.org/index.php?option=com_content&view=article&id=106&Itemid=129
Sponsor: St. Joseph Community Health Foundation
2826 South Calhoun Street
Fort Wayne, IN 46807

St. Joseph Community Health Foundation Schneider Fellowships 3642
The late Dr. Louis and Mrs. Anne Schneider understood and valued the importance of well-trained health and wellness care providers in the community. As a result, they established these fellowships to assist local not-for-profit agencies build the capacity of their agencies and staffs to provide quality care to the people of Allen County, Indiana. A preference will be to provide Fellowships that broaden the base of community health knowledge with new skills, services, health and wellness care abilities of health professionals in the community. Agencies may apply to use the Fellowships to either send staff or volunteers to continuing education opportunities, or to bring an expert to the community to provide specialized health and wellness care training for a group of staff, volunteers, and/or community members. The Foundation will commit support on an annual basis only.
Requirements: Applicants should have a demonstrated history of primarily serving poor and powerless populations residing in Allen County. Applicants should be not-for-profit entities classified as 501(c)3 by the Internal Revenue Service. The Foundation will consider requests for program, program-related equipment, staff continuing education, and technical assistance. Grant applications are available at the Foundation's website.
Restrictions: The Foundation will not fund anything destructive of human life. Grant applications will not be accepted for building projects, elimination of deficits, support of political activities, individuals, or projects all ready completed.
Geographic Focus: Indiana
Date(s) Application is Due: Mar 1; Jun 1; Sep 1; Dec 1
Amount of Grant: 2,000 USD
Samples: Children's Autism Center, Inc., $2,000 - Funding to present continuing education workshops to staff, volunteers, and parents; IPFW - Personal & Professional Department, $2,000 - Funding for a keynote speaker to speak at a day-long transcultural healthcare conference in Allen County.
Contact: Meg Distler; (260) 969-2001, ext. 201; fax (260) 969-2004; mdistler@sjchf.org
Internet: http://www.stjosephhealthfdn.org/index.php?option=com_content&view=article&id=105&Itemid=123
Sponsor: St. Joseph Community Health Foundation
2826 South Calhoun Street
Fort Wayne, IN 46807

Stackpole-Hall Foundation Grants 3643
The Stackpole-Hall Foundation awards grants to nonprofit organizations and institutions designed to enhance the social welfare of the Pennsylvania area. The Foundation concentrates on the areas of education, health care, cultural, youth development, social welfare, environmental, and community development. Priority is given to Elk County. Grants are awarded in support of matching, seed, partnership, and under certain circumstances, operational grants. The Foundation trustees meet four times a year to award grants. Proposals are due a month prior to the meeting for which they will be considered. Application forms are not required. Detailed instructions about information to include with the letter of inquiry are located at the Foundation website.
Requirements: The Foundation considers requests from organizations which qualify as churches; governmental organizations; or 501(c)3 organizations; or are classified as not being a private foundation under section 509(a) of the federal tax code.
Restrictions: Unless usual circumstances exist, requests for operating or endowment grants are given low priority. Funding is not available for individuals.
Geographic Focus: Pennsylvania
Amount of Grant: Up to 50,000 USD
Contact: William Conrad, Executive Director; (814) 834-1845; fax (814) 834-1869; stackpolehall@windstream.net
Internet: http://www.stackpolehall.org/application.html
Sponsor: Stackpole-Hall Foundation
44 South St. Marys Street
St. Marys, PA 15857

Stan and Sandy Checketts Foundation 3644
Established by Stan and Sandy Checketts in 1998, the Foundation's primary focus is on human services and helping individuals defray medical expenses. Primary fields of interest include: children and youth programs, health care, housing and shelter programs, human services, and recreation. Types of support are general operating funds and grants to individuals. Applicants should submit a detailed description of the project, along with the amount of funding requested. There are no deadlines, and the primary geographic focus is Utah.
Geographic Focus: Utah
Contact: Stan Checketts, President; (435) 752-1987; fax (435) 752-1948
Sponsor: Stan and Sandy Checketts Foundation
350 West 2500 North
Logan, UT 84341-1734

Stark Community Foundation Women's Fund Grants 3645
The Women's Fund is a permanent endowment to benefit future generations and increase vital funding for programs which advance the economic, educational, physical, emotional, social, artistic, and personal growth of women and to educate and inspire women to become leaders in philanthropy. The interest area of this fund includes nonprofit agencies and their programs or projects that meet the needs of women and children in the Stark County area.
Requirements: The Women's Fund would like to fund grassroots nonprofit organizations, churches, and educational components that have unique projects serving women and/or women and children, and their basic needs for daily living.
Geographic Focus: Ohio
Date(s) Application is Due: Mar 5
Amount of Grant: 500 - 3,000 USD
Contact: Cindy Lazor; (330) 454-3426; fax (330) 454-5855; cmlazor@starkcf.org
Internet: http://www.starkcf.org/the_womens_fund.asp
Sponsor: Stark Community Foundation
400 Market Avenue N, Suite 200
Canton, OH 44702

Starke County Community Foundation Grants 3646
The Starke Community Foundation offers grants to schools and teachers with the maximum amount of $500 per school. The following school systems are eligible to apply: Knox, North Judson-San Pierre and Oregon-Davis. Areas of interest include: community development; education; health and human services; youth; environment and recreation; and arts and culture. Applicants may fill out the online application, but are advised to contact the program coordinator to discuss their project before applying.
Requirements: The Foundation favors activities that: reach a broad segment of the community, especially those citizens whose needs are not being met by existing services

that are normally expected to be provided by private rather than government sources; request seed money to realize innovative opportunities to meet needs in the community; stimulate and encourage additional funding; promote cooperation and avoid duplication of effort; help make a charitable organization more effective and efficient and better able to be self-sustaining; and one time projects or needs.
Restrictions: The Foundation will not consider grants for: religious organizations for the sole purpose of furthering that religion (this prohibition does not apply to funds created by donors who have specifically designated religious organizations as beneficiaries of the funds); political activities or those designated to influence legislation; national organizations (unless the monies are to be used solely to benefit citizens of Starke County); grants that directly benefit the donor or the donor's family; fundraising projects; and contributions to endowments.
Geographic Focus: Indiana
Date(s) Application is Due: Sep 30
Amount of Grant: 500 USD
Contact: Grants Coordinator; (574) 223-2227 or (877) 432-6423
Internet: http://www.nicf.org/starke/grants.html
Sponsor: Starke County Community Foundation
1512 South Heaton Street
Knox, IN 46534

Starke County Community Foundation Lilly Endowment Community Scholarships 3647
The Lilly Endowment Scholarship provides funding for full time, four-year tuition for a field of study leading to a baccalaureate degree at any Indiana college or university. The scholarship also includes an $800 per year book stipend.
Requirements: Applicants must be U.S. citizens and residents of Starke County. Candidates can obtain the application from their high school guidance office. All college majors are eligible.
Geographic Focus: Indiana
Contact: Scholarship Contact; (574) 223-2227 or (877) 432-6423; fax (574) 224-3709
Internet: http://www.nicf.org/starke/availablescholarships.html
Sponsor: Starke County Community Foundation
1512 South Heaton Street
Knox, IN 46534

Starke County Community Foundation Scholarships 3648
Starke County offers a wide range of academic scholarships for students of various majors, with specific applications and deadlines. Applicants should refer to the website for application and deadline information or call the scholarship coordinator.
Geographic Focus: Indiana
Contact: Corinne Becknell Lucas; (574) 223-2227 or (877) 432-6423
Internet: http://www.nicf.org/starke/scholarships.html
Sponsor: Starke County Community Foundation
1512 South Heaton Street
Knox, IN 46534

Staunton Farm Foundation Grants 3649
The Staunton Farm Foundation is a family foundation established in 1937 in accordance with the wishes of Matilda Staunton Craig, who wanted her estate to be used to benefit people with mental illness. The Foundation awards grants in the field of mental health in southwestern Pennsylvania. Projects that represent new and different approaches for organizations and ultimately affect patient care are encouraged. Support may be for more than one year, but the project must become self-sustaining following the grant period. Applicants should submit a letter of intent; full proposals are by invitation.
Requirements: Nonprofit organizations in the 10-county area in southwestern Pennsylvania including Washington, Greene, Fayette, Westmoreland, Armstrong, Butler, Lawrence, Beaver, Indiana, and Allegheny are eligible.
Geographic Focus: Pennsylvania
Date(s) Application is Due: Jun 1; Dec 1
Contact: Joni S. Schwager, Executive Director; (412) 281-8020; fax (412) 232-3115; jschwager@stauntonfarm.org
Internet: http://www.stauntonfarm.org
Sponsor: Staunton Farm Foundation
650 Smithfield Street, Suite 210
Pittsburgh, PA 15222

Steele-Reese Foundation Grants 3650
The foundation's available income is divided equally for grants to operating charities in southern Appalachia (particularly Kentucky) and in the Northwest (particularly Idaho). In both areas the funds are devoted to education, health, welfare, and the humanities. Types of support include general operating support, equipment, endowment funds, matching funds, professorships, scholarship funds, and capital campaigns. The foundation gives preference to projects that have, among others, the following characteristics: rural, modest in ambition, narrow in function, unglamorous, based on experience, enjoying community financial support, and essential rather than merely desirable. While the foundation considers southern Kentucky and Idaho to be its territories of primary concern, it makes grants to organizations operating throughout southern Appalachia and in Oregon, Montana, and Wyoming.
Requirements: Applicant should submit a letter requesting guidelines. Only residents of Georgia, Idaho, Kentucky, Montana, North Carolina, Texas, and Wyoming are eligible to apply. Personal and telephone inquiries are not encouraged.
Restrictions: The foundation does not make grants to individuals, to community chest or similar drives, for conferences or workshops, for efforts to influence elections or legislation, for planning purposes or experimental projects, for emergencies, or for permanent support except for occasional endowment grants to organizations where stability is critically important.
Geographic Focus: Georgia, Idaho, Kentucky, Montana, North Carolina, Texas, Wyoming
Date(s) Application is Due: Mar 1
Amount of Grant: 10,000 - 50,000 USD
Samples: Frontier Nursing Service, Wendover, KY, $40,000 - to purchase a new phone system for the Mary Breckenridge Hospital; Mountain Home Montana, Missoula, MT, $28,000 - salary support for the director of a teen-mother residency program; West Central Highlands Resource Conservation and Development Council, Emmett, ID, $10,000 - for upgrades and renovations of the Indian Valley Community Hall.
Contact: Charles U. Buice, (212) 505-2696; charlesbuice@hotmail.com
Internet: http://www.steele-reese.org/what.html
Sponsor: Steele-Reese Foundation
32 Washington Square West
New York, NY 10011

Stella and Charles Guttman Foundation Grants 3651
The majority of grants are made to organizations providing services to people in the New York City metropolitan area. Beginning in 2014, the Stella and Charles Guttman Foundation intends to direct a substantial portion of its grantmaking to programs that serve low income infants, toddlers and preschoolers as they transition to kindergarten. Special emphasis will be placed on programs that improve quality, expand services and create a strong continuum of care for children ages 0 to 3 in high-need neighborhoods. Systemic investments in early childhood programs may include the expansion of evidence-based home visiting programs, infant health and mental health programs and professional development for center-based teachers, as well as home-based caregivers. The Foundation is also committed to fund programs in neighborhoods with high levels of poverty and a large concentration of public housing. In addition to early childhood programs, the Foundation will support programs that work to build a network of education, health and social services for children from birth through college graduation.
Requirements: Charitable 501(c)3 or 170(b)1 organizations are eligible with a strong emphasis on New York City and Israel.
Restrictions: The foundation does not make grants directly to individuals or to organizations not qualified as charitable, for foreign travel or study, to initiate or defend public interest litigation, to support anti-vivisectionist causes, or to religious organizations for religious observances.
Geographic Focus: New York
Amount of Grant: 25,000 - 100,000 USD
Samples: Citizens' Committee for Children of New York, New York, New York, $50,000 - support of educational, policy and advocacy initiatives focused on early childhood education; Good Shepherd Services, Brooklyn, New York, $50,000 - support of the Bronx Opportunity Network, a collaboration of seven community-based organizations; Henry Street Settlement, New York, New York, $100,000 - support of the Henry Street capital campaign and the renovation of the Charles and Stella Guttman Building.
Contact: Elizabeth Olofson, Executive Director; (212) 371-7082; fax (212) 371-8936; eolofson@guttmanfoundation.org
Internet: http://www.guttmanfoundation.org
Sponsor: Stella and Charles Guttman Foundation
122 East 42nd Street, Suite 2010
New York, NY 10168

Sterling-Turner Charitable Foundation Grants 3652
The charitable foundation awards grants to Texas organizations, primarily for higher and secondary education, adult basic education and literacy, social services, youth, the elderly, fine and performing arts groups and other cultural programs, church support and religious programs (Catholic, Jewish, and Protestant), hospitals and health services, hospices, research, conservation, and civic and urban affairs. Grants are awarded for general operating support, annual campaigns, capital campaigns, continuing support, building construction/renovation, equipment acquisition, endowment funds, program and project development, conferences and seminars, curriculum development, fellowships, scholarship funds, research, and matching funds. The board meets in April.
Requirements: Nonprofit Texas organizations in the following counties are eligible to apply: Fort Bend, Harris, Kerr, Tom Green, and Travis. Only those 501(c)3 organizations with offices located within the counties being considered within the State of Texas may submit and all funds must be managed, used and services provided within those counties in the State of Texas. All funds must be used within the requesting county. If the organization is a 509(a)3, there is a template for a required letter of explanation as to why your organization falls under the category that must be submitted for consideration. All documents must be received by 5:00 pm of the deadline date.
Restrictions: Individuals are ineligible.
Geographic Focus: Texas
Date(s) Application is Due: Mar 1
Amount of Grant: 5,000 - 25,000 USD
Contact: Patricia Stilley, Executive Director; (713) 237-1117; fax (713) 223-4638; pstilley@stfdn.org or jarnold@sterlingturnerfoundation.org
Internet: http://sterlingturnerfoundation.org/information_and_instructions.htm
Sponsor: Sterling-Turner Charitable Foundation
5850 San Felipe Street, Suite 125
Houston, TX 77057-3292

Sterling and Shelli Gardner Foundation Grants   3653

The Sterling and Shelli Gardner Foundation was established by co-founders and operators of "Stampin' Up!", a multi-million dollar catalog-based business, in Utah in 2002. The Foundation's major fields of interest include: community and economic development; education; and human services. With a geographic focus throughout the State of Utah, applicants should request a formal application from the Foundation office. The Foundation awards between thirty and forty grants each year, and amounts have recently ranged from $500 to $40,000. Most often, these awards are unrestricted contributions applied toward general operating costs. There are no specified annual deadlines, and applications are taken on a rolling basis.
Requirements: Any 501(c)3 organization serving residents of Utah are eligible to apply.
Geographic Focus: Utah
Amount of Grant: 500 - 40,000 USD
Samples: Ability Found, Salt Lake City, Utah, $5,000 - general operating support for disabled individuals; Friday's Kids Respite, Orem, Utah, $2,500 - general operations; Courage Reins, Highland, Utah, $35,000 - general operations.
Contact: Megan White, Administrator; (801) 717-6789
Sponsor: Sterling and Shelli Gardner Foundation
610 W. Westfield Road
Alpine, UT 84004-1501

Steuben County Community Foundation Grants   3654

The Steuben County Community Foundation works to preserve and enhance the lifestyle and assets of Steuben County, Indiana, for current and future generations by providing ongoing assessment and financial support of identified needs through philanthropic giving and endowment building. The Foundation grants funding for arts and culture, education, health, environment, human services, recreation, and community development. Grants are not limited to these areas, and grant seekers are encouraged to respond to emerging community needs.
Requirements: Prospective applicants are strongly encouraged to discuss their grant requests with Community Foundation staff before beginning the application process. Applications can be obtained on the Foundation's website. The Steuben County Community Foundation encourages projects that: address priority community concerns; encourage more effective use of community resources; test or demonstrate new approaches and techniques in the solution of community problems; are intended to strengthens the management capabilities of agencies, and; promote volunteer participation and citizen involvement in community affairs.
Restrictions: If approved, each grant recipient must sign an agreement that includes the following obligations: public acknowledgment of the Foundation's support; expenditure of the monies as specified in the proposal; return of any unused portion of the grant; completion of an evaluation report; any special conditions as mutually agreed (failure to do so can adversely affect any subsequent requests).
Geographic Focus: Indiana
Date(s) Application is Due: Feb 28; Apr 29; Jun 30; Aug 31; Oct 31; Dec 30
Contact: Bill Stockberger, Program Officer; (260) 665-6656; fax (260) 665-8420
Internet: http://www.steubenfoundation.org/grants/index.html
Sponsor: Steuben County Community Foundation
1701 N Wayne Street
Angola, IN 46703

Stewart Huston Charitable Trust Grants   3655

The purpose of the Trust is to provide funds, technical assistance and collaboration on behalf of non-profit organizations engaged exclusively in religious, charitable or educational work; to extend opportunities to deserving needy persons. Giving primarily in the Savannah, GA, area and Coatesville, PA.
Requirements: 501(c)3 nonprofit organizations are eligible.
Restrictions: Grants are not awarded for scholarship support to individuals, endowment purposes, purchases of tickets or advertising for benefit purposes, coverage of continuing operating deficits, or document publication costs. Support is not provided to intermediate or pass-through organizations (other than United Way) that in turn allocate funds to beneficiaries or to fraternal organizations, political parties or candidates, veterans, labor or local civic groups, volunteer fire companies, or groups engaged in influencing legislation.
Geographic Focus: Georgia, Pennsylvania
Date(s) Application is Due: Jan 15; Mar 1; Sep 1
Amount of Grant: 1,000 - 15,000 USD
Samples: Graystone Society, Coatesville, PA, $325,000 - for building support; Grace United Methodist Church, Savannah, GA, $20,000 - for sponsorships; University of Pennsylvania Health System, Philadelphia, PA, $20,000 - for general support;
Contact: Scott Huston; (610) 384-2666; fax (610) 384-3396; admin@stewarthuston.org
Internet: http://www.stewarthuston.org
Sponsor: Stewart Huston Charitable Trust
50 South First Avenue
Coatesville, PA 19320

Stocker Foundation Grants   3656

The Stocker Foundation aims to lessen the achievement-gap for under-resourced prekindergarten through third grade public school students by investing in programs that strengthen reading literacy. The foundation remains an all-family board, headquartered in Lorain, Ohio. Annual grant distributions focus first on Lorain County, Ohio, the place where assets were generated. Then, in communities where other trustees reside (see below).
Requirements: The Stocker Foundation provides grants to nonprofit organizations qualified under Section 501(c)3 of the Internal Revenue Code, and to selected public sector activities. Grants are considered and decided upon one time annually in the area of improved reading literacy outcomes for under-resourced prekindergarten through fifth grade public school students. Funds are made to organizations in Lorain and Cuyahoga counties, Ohio; Pima County, Arizona; Alameda and San Francisco counties, California; Bernalillo and Dona Ana counties, New Mexico; King County, Washington; and Hartford County, Connecticut. Areas of interest include: supplemental programs that move students toward grade-level reading mastery; comprehensive intervention strategies that increase overall literacy achievement by fourth grade; book distribution programs that increase students' access to print materials, encourage students' reading outside of the classroom, and develops students' life-long love of reading; and, programs that support emerging literacy and reading skills among prekindergarten (children enter kindergarten ready to read). Some limited funding is available for services that can help remove barriers toward reading success. All organizations seeking funding must first submit a Letter of Inquiry (LOI). The specific guidelines for the LOI can be downloaded from the sponsor's website. LOIs are due no later than July 1. For those LOIs moving forward, a full proposal is due by October 1 for a spring decision.
Restrictions: In general, the foundation does not award grants toward: debt reduction, research projects, tickets or advertising for fundraising events, individuals, religious exclusivism, or capital campaigns.
Geographic Focus: Arizona, California, Connecticut, New Mexico, Ohio, Washington
Date(s) Application is Due: Oct 1
Amount of Grant: Up to 100,000 USD
Contact: Patricia O'Brien; (440) 366-4884; pobrien@stockerfoundation.org
Internet: http://www.stockerfoundation.org/grants.aspx
Sponsor: Stocker Foundation
201 Burns Road
Elyria, OH 44035

Straits Area Community Foundation Health Care Grants   3657

Grants from the Straits Area Health Care Fund are for projects and programs that provide general health care services and activities for residents of Cheboygan County and/or Mackinaw City. Applicants may submit requests up to a maximum of $2,000 per application cycle unless otherwise indicated. Mini-grants up to $300 are also available. Applications should be received in the Community Foundation office or postmarked by June 1.
Requirements: IRS 501(c)3 nonprofit medical organizations and other groups who are providing general health care services for the residents of Cheboygan County or Mackinaw City are eligible to apply. An organization may apply each year for a grant.
Restrictions: Grants are not made to individuals.
Geographic Focus: Michigan
Date(s) Application is Due: Jun 1
Amount of Grant: Up to 2,000 USD
Contact: Julie Wiesen, Program Director; (989) 354-6381 or (877) 354-6381; fax (989) 356-3319; wiesenj@cfnem.org
Internet: http://www.cfnem.org/sacf/grants/straits-area-health-care-grant.html
Sponsor: Straits Area Community Foundation
100 N. Ripley, Suite F, P.O. Box 495
Alpena, MI 49707

Strake Foundation Grants   3658

The Foundation supports hospitals, schools, colleges, and Catholic charities, as well as projects focusing in adult basic education and literacy, museums, and arts and culture. Support is considered for operating budgets, capital campaigns, special projects, research, matching funds and general purposes.
Requirements: Awards are made to 501(c)3 organizations located only in the United States, primarily in Texas. Organizations may submit only one request per calendar year. There are no set amounts for requests, however awards generally range between $2,000 and $20,000 with a few exceptions as high as $50,000.
Restrictions: Awards are not made to support individuals, nor for deficit financing, consulting services, technical assistance, publications, or loans.
Geographic Focus: All States
Sample: Corporation for Educational Radio and Television, New York, NY, $2,500–to support the 'Black American Conservatism' companion website; El Centro de Corazon, Houston, TX, $$7,500–operating support; Fund for American Studies, Washington, D.C., $5,000–scholarship support.
Contact: George Strake, Jr., President; (713) 216-2400; foundation@strake.org
Sponsor: Strake Foundation
712 Main Street, Suite 3300
Houston, TX 77002

Stranahan Foundation Grants   3659

The Stranahan Foundation was created in 1944 by brothers Frank D. and Robert A. Stranahan, founders of the Champion Spark Plug Company in Toledo, Ohio. The purpose of the foundation is to assist individuals and groups in their efforts to become more self-sufficient and contribute to the improvement of society and the environment. The foundation supports a multitude of important programs that fit within five priority areas of interest: Human Services, Ecological Well-Being, Arts & Culture, Education, and Mental & Physical Health. Grant funds may be used for start up support for a new program, operating support, expansion or capacity building, or capital support.
Requirements: Nonprofit organizations with 501(c)3 tax-exempt status are eligible to apply. While the foundation awards funds nationwide, its focus is on the Toledo, Ohio area. All applicants must, as a first step, submit a letter of inquiry to the Stranahan Foundation. Full proposals are by invitation only and may only be submitted by organizations that are invited to apply after their letter of inquiry has been accepted and reviewed. The Foundation will contact those organizations invited to submit a full proposal and notify

those that are not eligible to apply. Instructions and forms for letters of inquiry and full grant proposals can be found on the website.
Restrictions: The Stranahan Foundation does not normally consider proposals for funding in the following areas: personal businesses; reduction or elimination of deficits; projects that are located outside of the United States; endowment fund campaigns; government sponsored or controlled projects; or individuals. Additionally, the foundation will not support organizations that discriminate in the leadership, staffing or service provision on the basis of age, gender, race, ethnicity, sexual orientation, disability, national origin, political affiliation or religious beliefs.
Geographic Focus: All States
Date(s) Application is Due: Jul 1
Amount of Grant: 1,000 USD
Contact: Pam Roberts; (419) 882-5575; fax (419) 882-2072; proberts@stranahanfoundation.org
Internet: http://www.stranahanfoundation.org/index.php?src=gendocs&ref=GrantmakingPriorities&category=Main
Sponsor: Stranahan Foundation
4169 Holland-Sylvania Road, #201
Toledo, OH 43623

Streilein Foundation for Ocular Immunology Visiting Professorships 3660
The SFOI Visiting Professorship is intended to stimulate the spread of knowledge concerning the immunological causes and approaches for managing ocular diseases. The SFOI Visiting Professorship should provide a clear scientific and educational benefit for the host institution. The Professorship will support a short visit of a senior scientist or clinician with a track record of ocular immunology research to a host institution. An important feature of the Visiting Professorship is to encourage and mentor young investigators. The visiting professor shall contribute one full day of teaching and interacting with students, residents, postdoctoral fellows, and faculty. The host institution will receive a $5,000 award to pay for coach airfare, and/or standard ground transportation, hotel (up to two nights), meals, honorarium, and expenses associated with this visit (e.g., refreshments, advertising). Transportation by personal car will be reimbursed at $0.51/mile. Meal allowance will be set by the host institution. Also included is an honorarium of $1,500 paid at the time expense reimbursement is submitted.
Requirements: Any institution of higher education or research institute is eligible to apply. The application for a SFOI Visiting Professorship must be submitted by a departmental Chairman of an educational institution, the Director of a research institute, or an official of similar stature from the host organization. The application must be received by the SFOI by July 1st of the calendar year in which the professorship is intended.
Geographic Focus: All States
Date(s) Application is Due: Jul 1
Amount of Grant: 6,500 USD
Contact: Joan Stein-Streilein, Ph.D., Treasurer; (617) 912-7494; fax (617) 912-0105; jstein@vision.eri.harvard.edu
Internet: http://www.streilein-foundation.org/fellowships.html
Sponsor: Streilein Foundation for Ocular Immunology
P.O. Box 6104
Boston, MA 02114

Stroke Association Allied Health Professional Research Bursaries 3661
The Research Bursary Awards are primarily intended for the purpose of training nurses and therapists, but consideration will be given to other health professionals. These bursaries will be awarded to UK departments that can demonstrate a track record and current participation in stroke research. The host department will be required to provide appropriate supervision and a research training program suitable for equipping the candidate for a career in stroke research, and to enable them to obtain a post graduate qualification. Bursaries may be taken up on a part-time, pro-rata basis.
Requirements: Research must take place in the United Kingdom.
Geographic Focus: All States, United Kingdom
Amount of Grant: 30,000 - 60,000 GBP
Samples: Trudi Pelton, University of Birmingham; Holly Robson, University of Manchester, Linda Hammett, St. George's, University of London; Andrea Atzori, University of Southampton; Emma Jones, University of Sheffield.
Contact: Research Department; (020) 7566-0345 or (020) 7490-2686; fax (020) 7251-9096; research@stroke.org.uk
Internet: http://www.stroke.org.uk/research/research_we_fund/allied_health_professional_bursaries/index.html
Sponsor: Stroke Association
Stroke House, 240 City Road
London, EC1V 2PR England

Stroke Association Clinical Fellowships 3662
Fellowships are awarded to individuals for training in stroke medicine, and are intended to enable a specialist registrar to gain the appropriate clinical experience and accreditation required for a career in the prevention, treatment or rehabilitation of stroke.
Requirements: Applicants are required to nominate a supervisor and training program in a host institution that has been approved for stroke medicine training by the Stroke Medicine Sub-specialty Advisory Committee (SSAC). The funding is not to be taken as part of a one year taught course or for those who have completed the SpR training program and have certification.
Geographic Focus: All States, United Kingdom
Date(s) Application is Due: Jan 6
Amount of Grant: 40,000 GBP
Contact: Research Department; (020) 7566-0345 or (020) 7490-2686; fax (020) 7251-9096; research@stroke.org.uk
Internet: http://www.stroke.org.uk/research/research_we_fund/fellowships/index.html
Sponsor: Stroke Association
Stroke House, 240 City Road
London, EC1V 2PR England

Stroke Association Research Project Grants 3663
Research project grants are awarded to doctors, professors, research fellows, therapists, psychologists and nurses at universities and hospitals across the UK. Project grants can last for up to three years. The Stroke Association's priorities for funding include research into the prevention, treatment, rehabilitation and long-term care of stroke patients. The majority of funded projects are patient orientated. The Stroke Association has recently undertaken a review of its activities and has developed a new Research and Development Strategy. As a result, applications for Research Project Grants have been revised and new application forms have been developed. All applications for funding by The Stroke Association will now undergo review by a panel of service users in addition to expert peer review. Therefore all applicants will now be required to complete a Project Grant Lay Summary Form, as well as the main Research Project Grant Application Form. Details of the new procedures and forms can be found on online.
Requirements: Research must take place in the United Kingdom.
Restrictions: Applications that are not submitted on the new forms will not be accepted for this application round.
Geographic Focus: All States, United Kingdom
Date(s) Application is Due: Feb 5
Amount of Grant: 70,000 - 210,000 GBP
Contact: Research Department; (020) 7566-0345 or (020) 7490-2686; fax (020) 7251-9096; research@stroke.org.uk
Internet: http://www.stroke.org.uk/research/apply_for_funding/index.html
Sponsor: Stroke Association
Stroke House, 240 City Road
London, EC1V 2PR England

Strowd Roses Grants 3664
Strowd Roses, Inc. is a private charitable foundation which was established in 2001 under the will of Mrs. Irene Harrison Strowd of Chapel Hill, North Carolina. The Board of Directors of Strowd Roses, Inc. makes grants to qualified tax-exempt charitable, educational, religious and public organizations that are based in Chapel Hill or Carrboro, or are devoted primarily to benefiting the citizens of those communities. Grants may also be made to individuals engaged in projects designed to enhance the Chapel Hill/Carrboro community. Grants may, at the Board's discretion, include support for operating as well as capital expenditures, seed money and matching grants. Grants to individuals may be used for scholarships or fellowships; to produce a report or similar product; or to improve or enhance a literary, artistic, musical, scientific, teaching or similar capacity, skill or talent which will be used to benefit the life of the community. Particular consideration is given to those projects and purposes which further the interests of Mr. and Mrs. Strowd, including the welfare of children and youth, the enhancement of the environment, and the promotion of a sense of civic duty.
Requirements: Applicants should submit the appropriate application form (see foundations website), along with a proposed budget and evidence of their tax-exempt status (IRS determination letter or comparable documentation) to: Board of Directors, Strowd Roses, Inc., P.O. Box 3558, Chapel Hill, NC, 27515-3558.
Geographic Focus: North Carolina
Date(s) Application is Due: Jan 31; Apr 30; Jul 31; Oct 31
Amount of Grant: 10,000 USD
Contact: Jennifer Boger; (919) 929-1984; fax (919) 929-1990; jboger@strowdroses.org
Internet: http://www.strowdroses.org/grantApp.htm
Sponsor: Strowd Roses
P.O. Box 3558
Chapel Hill, NC 27515-3558

STTI/ATI Educational Assessment Nursing Research Grants 3665
The purpose of this grant, co-sponsored by Assessment Technologies Institute (ATI), with the Honor Society of Nursing, Sigma Theta Tau International will support research that encourages the appropriate use of a standardized assessment program in a nursing education curriculum, including, but not limited to: use of formative assessment to drive remediation; use of summative assessment to guide curricular evaluation and revision; timing of assessment administration within a program of study; identification of key markers of at-risk student status.
Requirements: Eligibility Criteria: registered nurse with current license; have a master's or doctoral degree or be enrolled in a doctoral program; submit a completed research application package and signed research agreement via the online submission system; ready to implement research project when funding is received; submit a final report; submit completed abstract to the Virginia Henderson Research Library and credit research grant partners in all publications and presentations of the research.
Geographic Focus: All States
Date(s) Application is Due: Jul 1
Amount of Grant: 4,500 USD
Contact: Research Services; (888) 634-7575 or (317) 634-8171; fax (317) 634-8188; research@stti.iupui.edu
Internet: http://www.nursingsociety.org/Research/SmallGrants/Pages/STTIATI.aspx
Sponsor: Sigma Theta Tau International
550 W. North Street
Indianapolis, IN 46202-2156

STTI Emergency Nurses Association Foundation Grant 3666

The ENA Foundation and Sigma Theta Tau International, Inc. (STTI) have combined resources to offer this annual research grant. The ENA Foundation/STTI research grant provides funding for research that will advance the specialized practice of emergency nursing. All relevant research topics will be considered. Priority will be given to research studies relating to the ENA/ENA Foundation research initiatives.

Requirements: Eligibility Criteria: principal investigator must be a registered nurse; team members may be from other disciplines; principal investigator must have a master's degree; submit a completed research application package and signed research agreement; be ready to start, or have already started, the research project. An electronic Research Grant Application is available on the ENA Foundation website. All applications MUST be returned to ENA Foundation.

Restrictions: Application packets submitted to Sigma Theta Tau International will not be reviewed.

Geographic Focus: All States
Date(s) Application is Due: Mar 1
Amount of Grant: 6,000 USD
Contact: Department of Research; (847) 460-4119; fax (847) 460-4005; res@ena.org
Internet: http://www.nursingsociety.org/Research/SmallGrants/Pages/grant_ena.aspx
Sponsor: Emergency Nurses Association Foundation
915 Lee Street
Des Plaines, IL 60016-6569

STTI Environment of Elder Care Nursing Research Grants 3667

The Environment of Elder Care Nursing Research Grant co-sponsored by Hill-Rom Company will advance the science of nursing through research focused on critical aspects of elder care including clear lungs, no falls, safe skin, patient comfort and ease-of-use. Applications from novice researchers who have received no other national research funds are encouraged. Preference will be given to Sigma Theta Tau International members, other qualifications being equal. Allocation of funds is based on the quality of the proposed research, the future promise of the applicant, and the applicant's research.

Requirements: Eligibility Criteria: registered nurse with a current license; hold at minimum a Master's degree or its equivalent, and/or be enrolled in a Doctoral program; submission of completed research application packet and a signed research agreement via our on-line submission system; ready to implement research project when funding is received; submit to STTI a final report; submit completed abstract to STTI's Virginia Henderson International Nursing Research Library; and submit an abstract for publication to Hill-Rom regarding the research done as a product of the grant, and credit research grant partners in all publications and presentations of the research.

Geographic Focus: All States
Date(s) Application is Due: Jul 1
Amount of Grant: Up to 9,000 USD
Contact: Research Services; (888) 634-7575 or (317) 634-8171; fax (317) 634-8188; research@stti.iupui.edu
Internet: http://www.nursingsociety.org/Research/SmallGrants/Pages/Environmentof Care.aspx
Sponsor: Sigma Theta Tau International
550 W. North Street
Indianapolis, IN 46202-2156

STTI Joan K. Stout RN Research Grants 3668

The purpose of the Sigma Theta Tau International/Joan K. Stout, RN, Research Grant is to advance ongoing evidence-based study by nurse researchers on the impact of the practice of simulation education in schools of nursing and clinical care settings. The allocation of funds is based upon a research project that is ready for implementation. The proposed research project should be designed to ensure the ongoing practice of nurse-led simulation in improving quality of care in clinical and/or academic settings with the potential for further funding and ongoing research. Funds provided by interest from a research endowment established at Sigma Theta Tau International Foundation for Nursing, with a donation by the Hugoton Foundation.

Requirements: Eligibility Criteria: registered nurse with current license; have a master's degree, and/or be enrolled in a Doctoral program; submit a completed research application via online submission system; ready to implement research project when funding is received; complete project within one (1) year of funding; submit to STTI a final report; submit completed abstract to STTI's Virginia Henderson International Nursing Research Library.

Geographic Focus: All States
Date(s) Application is Due: Jul 1
Amount of Grant: 5,000 USD
Contact: Research Services; (888) 634-7575 or (317) 634-8171; fax (317) 634-8188; research@stti.iupui.edu
Internet: http://www.nursingsociety.org/Research/SmallGrants/Pages/JoanKStout.aspx
Sponsor: Sigma Theta Tau International
550 W. North Street
Indianapolis, IN 46202-2156

STTI National League for Nursing Grants 3669

Funds for this grant are made possible from a gift from Dr. Diane Billings to Sigma Theta Tau International Foundation for Nursing's research endowment. The purpose of this grant is to support research that advances the science of nursing education and learning through the use of technology in dissemination of knowledge.

Requirements: Eligibility Criteria: preference given to Sigma Theta Tau International and/or National League for Nursing members; grant proposals must advance the science of nursing education through the use of technology in the dissemination of knowledge; registered nurse with current license; hold a master's or doctoral degree or be enrolled in a doctoral program; submit a completed research application package and signed research agreement; ready to implement research project when funding is received. All applications must be submitted via the online submission system (see STTI website)

Geographic Focus: All States
Date(s) Application is Due: Jun 1
Amount of Grant: 5,000 USD
Contact: Research Services; (888) 634-7575 or (317) 634-8171; fax (317) 634-8188; research@stti.iupui.edu
Internet: http://www.nursingsociety.org/Research/SmallGrants/Pages/grant_nln.aspx
Sponsor: Sigma Theta Tau International
550 W. North Street
Indianapolis, IN 46202-2156

STTI Rosemary Berkel Crisp Research Awards 3670

The purpose of the STTI Rosemary Berkel Crisp Research Awards is to support nursing research in the critical areas of women's health, oncology, and pediatrics. The allocation of funds is based upon a research project that is ready for implementation in the area of women's health, oncology, or pediatrics; the quality of the proposed research, future potential of the application, appropriateness of the research budget, and feasibility of time frame. Funds provided by interest from a gift from The Harry L. Crisp, II and Rosemary Berkel Crisp Foundation to the honor society's Research Endowment.

Requirements: Eligibility Criteria: registered nurse with current license; hold a master's or doctoral degree or be enrolled in a doctoral program; submit a completed research application package and signed research agreement via the online submission system; ready to implement research project when funding is received; member of Sigma Theta Tau International; Note: Some preference will be given to applicants residing in Illinois, Missouri, Arkansas, Kentucky, and Tennessee.

Geographic Focus: All States
Date(s) Application is Due: Dec 1
Amount of Grant: Up to 5,000 USD
Contact: Research Services; (888) 634-7575 or (317) 634-8171; fax (317) 634-8188; research@stti.iupui.edu
Internet: http://www.nursingsociety.org/Research/SmallGrants/Pages/grant_crisp.aspx
Sponsor: Sigma Theta Tau International
550 W. North Street
Indianapolis, IN 46202-2156

STTI Virginia Henderson Clinical Research Grants 3671

The purpose of the Sigma Theta Tau International Virginia Henderson Clinical Research Grant is to encourage the research career development of clinically based nurses through support of clinically oriented research. Allocation of funds is based on a research project ready for implementation, the quality of the proposed research, the future potential of the applicant, appropriateness of the research budget and feasibility of the time frame. Funds for this biennial grant are provided by interest from the Virginia Henderson Clinical Research Endowment Fund.

Requirements: Eligibility Criteria: registered nurse actively involved in some aspect of health care delivery, education or research in a clinical setting; hold a master's or doctoral degree or be enrolled in a doctoral program; submit a completed research application package and signed research agreement via the online submission system; ready to implement research project when funding is received; member of Sigma Theta Tau International.

Geographic Focus: All States
Date(s) Application is Due: Dec 1
Amount of Grant: Up to 5,000 USD
Contact: Research Services; (888) 634-7575 or (317) 634-8171; fax (317) 634-8188; research@stti.iupui.edu
Internet: http://www.nursingsociety.org/Research/SmallGrants/Pages/grant_VHL.aspx
Sponsor: Sigma Theta Tau International
550 W. North Street
Indianapolis, IN 46202-2156

Subaru of Indiana Automotive Foundation Grants 3672

The Subaru of Indiana Automotive Foundation is committed to making gifts to qualifying organizations, institutions, or entities within Indiana that will improve the quality of life and help meet the needs of the residents of the state. The foundation awards cash grants that are used to support the funding of specific capital projects in the areas of: arts and culture, education, and health and welfare. Grants must be used for investments in facilities, equipment, or real estate made by qualifying organizations. Grant requests must be for $1,000 or more, with a maximum requested amount of $10,000. Funding can be used for investments in facilities, equipment, or real estate (non-operation funding). Applications for grants will be accepted from January 1 through March 31 to be considered for a grant to be dispersed by June 15. Applications will be accepted from July 1 through September 30 to be considered for a grant to be dispersed by December 15.

Requirements: Applying organizations must be 501(c)3 tax-exempt, with a chapter or office in Indiana; an education institution located in Indiana; or an Indiana governmental or quasi-governmental entity.

Restrictions: Support will not be provided for operating costs, routine expenses, or deficit reduction; endowments or memorials; fundraising events, conferences, meals, or travel; or annual fund drives. Support will not be provided to individuals; organizations located outside of Indiana or organizations that are not tax-exempt; for-profit businesses; organizations whose primary purpose is to influence legislation or support political candidates; religious institutions for religious purposes or fraternal organizations; or

organizations that discriminate in the provision of services on the basis of race, sex, color, national origin, disability, age, religious affiliation, or any other unlawful basis.
Geographic Focus: Indiana
Date(s) Application is Due: Mar 31; Sep 30
Amount of Grant: 1,000 - 10,000 USD
Contact: Shannon Walker; (765) 449-6565; shannon.walker@subaru-sia.com
Internet: http://www.siafoundation.org
Sponsor: Subaru of Indiana Automotive Foundation
P.O. Box 6479
Lafayette, IN 47903

Summer Medical & Dental Education Program: Interprofessional Pilot Grants   3673
The Summer Medical and Dental Education Program (SMDEP) is a free, summer enrichment program to strengthen the academic proficiency and career development of rising sophomore and junior college students, and community college students from backgrounds that are underrepresented in the health professions and who are interested in pursuing health-related careers. Through this call for proposals, RWJF seeks to identify an increased number of institutions to participate as SMDEP sites—providing this proven model for science, technology, engineering, and math (STEM) enrichment, life-skills development, exposure to health-related professions, and research and leadership training. Up to $415,000 will be awarded to support planning and implementation. A required brief proposal is due by March 18, and invited full proposals are due by May 18.
Requirements: RWJF is soliciting applications from both: institutions that seek to be part of the SMDEP program and receive program-related support from the national program office, but do not need additional financial support (self-funded site); and institutions that need financial support to develop and implement an SMDEP site (RWJF-funded site). Applicants should include: a school of medicine; a school of dentistry; and at least one other health profession program/school from the Interprofessional Education Collaborative. All accredited medical and dental schools are eligible to apply, along with an accredited health professions school.
Geographic Focus: All States
Date(s) Application is Due: May 18
Amount of Grant: Up to 415,000 USD
Contact: Peggy Geigher; (202) 828-0401; smdep_cfp@aamc.org
Internet: http://www.rwjf.org/en/library/funding-opportunities/2016/summer-medical-and-dental-education-program--site-r.html
Sponsor: Robert Wood Johnson Foundation
Route 1 and College Road East, P.O. Box 2316
Princeton, NJ 08543-2316

Summit Foundation Grants   3674
The Summit Foundation supports charitable organizations that enhance the quality of life in Summit County, Colorado, and neighboring communities. Grants are awarded twice each year, in June and in December, to agencies providing programs or services in the areas of health and human service, art and culture, education, environment and sports. The Foundation funds specific projects and programs which have measurable results.
Requirements: Applicants for funding must be tax exempt under the provisions of section 501(c)3 and 170(b)1(a)(i.V.I.) of the Internal Revenue Code.
Restrictions: The Foundation will not fund any political campaign on behalf of any issue or candidate. Additionally, The Foundation does not fund religious programs. Organizations requesting funding support will only be considered for funding once in a calendar year. Requests for programs or projects already completed are not eligible for funding.
Geographic Focus: Colorado
Date(s) Application is Due: Apr 20; Oct 15
Contact: Megan Nuttelman, Program Officer; (970) 453-5970; fax (970) 453-1423; sumfound@colorado.net or megan@summitfoundation.org
Internet: http://www.summitfoundation.org/?page_id=65#landhere
Sponsor: Summit Foundation
103 S Harris Street, Suites 201 and 204, P.O. Box 4000
Breckenridge, CO 80424

Sunderland Foundation Grants   3675
The foundation makes grants annually to publicly supported charitable organizations in its areas of interest, including higher education; churches; youth serving agencies; health facilities; community buildings; museums; civic projects; and and low maintenance, energy efficient housing projects sponsored by qualified tax-exempt organizations. The Foundation generally awards grants to larger, well-established nonprofit organizations. The Sunderland Foundation primarily makes grants in the geographic areas that have connections to the Ash Grove Cement Company. Grants are awarded in Western Missouri, Kansas, Nebraska, Arkansas, and Western Iowa, and to a lesser extent, in Idaho, Oregon, Washington, Utah and Montana.
Requirements: All grant requests should be submitted in writing, and proposals should include the following: a clear description of the project for which funds are being requested, including program goals and objectives, documentation of need and expected outcomes; a brief background on the proposing organization or agency; a detailed expense budget for the project indicating how the funds would be spent and over what time period; an income statement showing other sources of project support, public and/or private, which have been or will be solicited, including a statement of funds that have been received or pledged to date; a financial plan showing how the project will be supported beyond the grant period; the organization's current board of directors and their terms of office; a copy of the organization's most recent 501(c)3 tax-exempt ruling from the IRS; the organization's most recent certified audit or audited financial statement, where applicable. Proposals should be submitted by mail to the address listed in the "Sponsoring Organization" category. Proposals will also be accepted by email. Send a PDF file to the following address: sunderlandfoundation@ashgrove.com.
Restrictions: The Foundation does not award grants: for annual operating expenses; for programs or endowments; provide sponsorship for special events; to primary and secondary schools are generally not considered; to individuals. No scholarships are available from the Foundation.
Geographic Focus: Arkansas, Idaho, Iowa, Kansas, Missouri, Montana, Nebraska, Oregon, Utah, Washington
Samples: Pleasant Valley Baptist Church, Liberty, MO, $50,000 (2006%2Proxy-Connection: keep-alive Cache-Control: max-age=0; Salvation Army of Greater Kansas City, Kansas City, MO, $50,000; MD Anderson Cancer Center, Houston, TX, $100,000; Donnelly College, Kansas City, KS, $25,000;
Contact: Kent Sunderland; (913) 451-8900; sunderlandfoundation@ashgrove.com
Internet: http://www.sunderlandfoundation.org
Sponsor: Sunderland Foundation
P.O. Box 25900
Overland Park, KS 66225

Sunflower Foundation Bridge Grants   3676
Bridge Grant funding is intended to provide core transitional financial support for new or expanded medical, mental, dental and/or public health services that provide primary care or primary prevention services to the uninsured and the underinsured. Funds are available for: salaries, benefits, allowable indirect expenses appropriate to the health care practitioners needed for the program; administrative services (e.g. financial & medical record clerks, clerical staff) resulting from the new or expanded services for which funding is being requested from the Foundation. Applicants must demonstrate a cost share in the project through a cash match of $1 to $1 of the total amount requested. Grants will not exceed $200,000 each and may be for projects up to three years in length.
Requirements: Proposals are invited from private nonprofit or public organizations that have a mission to improve health status or access to quality, affordable health care for Kansans.
Restrictions: The foundation will not fund ongoing general operating expenses or existing deficits; capital, endowment or specific fund-raising campaigns; fund raising events; routine continuing education; travel to conferences not directly related to the project; programs that are not Kansas-based; programs that require additional staff but demonstrate no clear means of sustainability after foundation funding; individual medical care or support; medical equipment; capital equipment (except for allowable technology); political purposes; or support of organizations that practice discrimination. Additional exclusions include: requests for new construction; and facility renovations. Any Safety Net clinic that qualifies for the Dental Hub initiative should contact the Sunflower Foundation program staff prior to application (there may be some restrictions for Dental Bridge Grant funding for organizations eligible for Dental Hub funding).
Geographic Focus: Kansas
Date(s) Application is Due: Oct 23
Amount of Grant: Up to 200,000 USD
Contact: Cheryl Bean; (785) 232-3000, ext. 104; cbean@sunflowerfoundation.org
Internet: http://www.sunflowerfoundation.org/areas_of_interest-health_care_access.php
Sponsor: Sunflower Foundation: Health Care for Kansans
1200 SW Executive Drive, Suite 100
Topeka, KS 66615-3850

Sunflower Foundation Capacity Building Grants   3677
The Foundation funds two types of Capacity Building grants: Assessment (grant limit is based on scope of work) --the first step of any effort to build capacity within an organization is an assessment of present capacity to identify what is working well and what needs to be strengthened (the Foundation will fund technical assistance, consulting, and other expenses related to conducting a formal assessment of current organizational capacity); and Implementation ($20,000 grant limit)--the Foundation will fund the implementation of capacity building strategies. Allowable expenses typically include such items as technical assistance/consulting, technology (hardware and/or software), limited equipment (e.g. telephone systems), training and assessment/evaluation.
Requirements: Proposals are invited from private nonprofit or public organizations that have a mission to improve health status or access to quality, affordable health care for Kansans.
Restrictions: The foundation will not fund ongoing general operating expenses or existing deficits; capital, endowment or specific fund-raising campaigns; fund raising events; routine continuing education; travel to conferences not directly related to the project; programs that are not Kansas-based; programs that require additional staff but demonstrate no clear means of sustainability after foundation funding; individual medical care or support; medical equipment; capital equipment (except for allowable technology); political purposes; or support of organizations that practice discrimination. Additional exclusions include: program expenses; existing staff salaries/benefits; Electronic Medical Records (EMR) technology; medical equipment; capital equipment (except for allowable technology); routine continuing education; and travel to conferences not directly related to the project.
Geographic Focus: Kansas
Date(s) Application is Due: Jun 19
Amount of Grant: Up to 20,000 USD
Contact: Grants Administrator; (785) 232-3000 or (866) 232-3020; info@sunflowerfoundation.org
Internet: http://www.sunflowerfoundation.org/flash/request.html
Sponsor: Sunflower Foundation: Health Care for Kansans
1200 SW Executive Drive, Suite 100
Topeka, KS 66615-3850

Sunflower Foundation Walking Trails Grants 3678
The Foundation is seeking projects that bring together public and private support to construct Walking Trails as a means to increase physical activity among Kansans. Grant funds may be used for Walking Trails building materials only. Walking Trails are generally expected to be at least ???? mile in length. Grants awards will be for one year and will not exceed $15,000. Organizations that are seeking matching funding to secure state or federal grants for larger Trail projects should contact the Foundation before submitting a proposal.
Requirements: Proposals are invited from private nonprofit or public organizations that have a mission to improve health status or access to quality, affordable health care for Kansans. Public access to the Walking Trail is required. Applicants must demonstrate a cost share in the project through a cash match of $1 to $1 of the total amount requested from the Sunflower Foundation.
Restrictions: The foundation will not fund ongoing general operating expenses or existing deficits; capital, endowment or specific fund-raising campaigns; fund raising events; routine continuing education; travel to conferences not directly related to the project; programs that are not Kansas-based; programs that require additional staff but demonstrate no clear means of sustainability after foundation funding; individual medical care or support; medical equipment; capital equipment (except for allowable technology); political purposes; or support of organizations that practice discrimination. Additional exclusions include: requests for physical fitness equipment; and fitness facility dues/membership fees.
Geographic Focus: Kansas
Date(s) Application is Due: Jun 29
Amount of Grant: Up to 15,000 USD
Contact: Administrator; (785) 232-3000 or (866) 232-3020; info@sunflowerfoundation.org
Internet: http://www.sunflowerfoundation.org/flash/request.html
Sponsor: Sunflower Foundation: Health Care for Kansans
1200 SW Executive Drive, Suite 100
Topeka, KS 66615-3850

Sunoco Foundation Grants 3679
The Sunoco Foundation invests in projects that promote local education and workforce development or that make communities great places to live and work. Grants are primarily awarded to nonprofits located where Sunoco has a major presence. The foundation considers all types of efforts, including homelessness, housing, community development, seniors and education. The Sunoco Foundation aligns giving and business strategy with focus on three key areas--Fueling Minds: Educate and Develop Skills for the Workforce; Fueling the Planet: Promote Environmental Stewardship and Responsibility; and, Fueling Communities: Make them great places to live and work.
Requirements: Applicants must successfully complete an eligibility quiz before submitting an online letter of inquiry. If the letter of inquiry matches the foundation's priorities, the applicant will be invited to submit a full proposal. The Foundation prefers to fund specific projects rather than operating budgets.
Restrictions: Requests for deficit funding, individuals, benefit fundraisers, endowments, surveys, studies, religious groups, fraternal organizations, athletic groups, schools, single diseases, and non-tax exempt organizations are generally not considered.
Geographic Focus: Alabama, Connecticut, Delaware, District of Columbia, Florida, Georgia, Indiana, Kentucky, Maine, Maryland, Massachusetts, Michigan, New Hampshire, New Jersey, New York, North Carolina, Ohio, Pennsylvania, Rhode Island, South Carolina, Tennessee, Vermont, Virginia, West Virginia
Contact: Ruth Clauser; (215) 977-3000; fax (215) 977-3409; raclauser@sunocoinc.com
Internet: https://online.foundationsource.com/public/home/sunoco
Sponsor: Sunoco Foundation
1735 Market Street, Suite L1
Philadelphia, PA 19103

SunTrust Bank Trusteed Foundations Florence C. and Harry L. English Memorial Fund Grants 3680
Harry L. English was the son of Capt. James Warren English, a Civil War hero and mayor of Atlanta from 1882-1884. Capt. English gained prominence in Atlanta through the real estate business, and Harry helped found the Chattahoochee Brick Company. Harry and his wife Florence built a house on West Paces Ferry Road across from the Governor's Mansion. The house still stands today. The couple had a daughter, Emily, who married James D. Robinson, Sr., the founder and CEO of 4th National Bank. Harry died in 1937, and in 1964, The Florence C. and Harry L. English Memorial Fund was established to honor Harry and Florence. The geographic focus of the SunTrust Bank Trusteed Foundations is metropolitan Atlanta. The Trustees will consider requests for capital improvements such as buildings, furniture and equipment, and alterations to existing structures. Applications also will be considered for special projects of a community nature, special studies, surveys, research and pilot programs which do not commit the funds to recurring expenditures. Community benefit and return on investment are primary considerations in distribution decisions. Other Distribution Committee considerations include: emphasis on metropolitan Atlanta; organization/community coordination and support; timeliness and precedence; organization management and governance; grant multiplier effect; human value and self-help emphasis; ultimate benefit to the community; financial management; and implementation of a strategic plan. The three annual deadlines for submission have been identified as March 31, August 31, and November 30.
Requirements: Nonprofits in areas where SunTrust Bank has a major presence are eligible.
Restrictions: These foundations do not accept unsolicited grant applications from outside metropolitan Atlanta unless an organization has been specifically named by the donor as an eligible recipient.
Geographic Focus: Georgia
Date(s) Application is Due: Mar 31; Aug 31; Nov 30
Amount of Grant: 1,000 - 20,000 USD
Contact: Kay Miller, Secretary of the Distribution Committee; (404) 588-8250
Internet: https://www.suntrust.com/Microsites/foundation/funds.htm
Sponsor: SunTrust Bank Trusteed Foundations
P.O. Box 4418, Mail Code 041
Atlanta, VA 30302

SunTrust Bank Trusteed Foundations Greene-Sawtell Grants 3681
The Greene-Sawtell Foundation was established by Forest Greene and Alice Greene Sawtell, who served as the first Advisory Committee. Under the terms of the document which established the Foundation, the advisory committee was created to select the charitable organizations which the Foundation would support. SunTrust Bank, as Trustee, now determines which charitable, religious and educational organizations will receive support from the Foundation. The Bank's Foundation Trustees consider organizations which the Foundation has supported in the past, those in which Alice Sawtell had a personal interest, and those which are of a similar nature to the organizations in which she had an interest. The geographic focus of the Trust is metropolitan Atlanta. The Trustees will consider requests for capital improvements such as buildings, furniture and equipment, and alterations to existing structures. Applications also will be considered for special projects of a community nature, special studies, surveys, research and pilot programs which do not commit the funds to recurring expenditures. Community benefit and return on investment are primary considerations in distribution decisions. Other Distribution Committee considerations include: emphasis on metropolitan Atlanta; organization/community coordination and support; timeliness and precedence; organization management and governance; grant multiplier effect; human value and self-help emphasis; ultimate benefit to the community; financial management; and implementation of a strategic plan. The three annual deadlines for submission have been identified as March 31, August 31, and November 30.
Requirements: Nonprofits in areas where SunTrust Bank has a major presence are eligible.
Restrictions: These foundations do not accept unsolicited grant applications from outside metropolitan Atlanta unless an organization has been specifically named by the donor as an eligible recipient.
Geographic Focus: Georgia
Date(s) Application is Due: Mar 31; Aug 31; Nov 30
Amount of Grant: 1,000 - 20,000 USD
Contact: Kay Miller, Secretary of the Distribution Committee; (404) 588-8250
Internet: https://www.suntrust.com/Microsites/foundation/funds.htm
Sponsor: SunTrust Bank Trusteed Foundations
P.O. Box 4418, Mail Code 041
Atlanta, VA 30302

SunTrust Bank Trusteed Foundations Harriet McDaniel Marshall Tust Grants 3682
Harriett McDaniel Marshall created this Trust in 1962 to honor her father, Sanders McDaniel, a Monroe, Georgia, native and an Atlanta attorney. Sanders McDaniel was the son of former Governor Henry D. McDaniel (1883-1888), under whose administration the State Capitol Building in Atlanta was completed. Harriett married Rembert Marshall, one of her father's law partners. The geographic focus of the Trust is metropolitan Atlanta. The Trustees will consider requests for capital improvements such as buildings, furniture and equipment, and alterations to existing structures. Applications also will be considered for special projects of a community nature, special studies, surveys, research and pilot programs which do not commit the funds to recurring expenditures. Community benefit and return on investment are primary considerations in distribution decisions. Other Distribution Committee considerations include: emphasis on metropolitan Atlanta; organization/community coordination and support; timeliness and precedence; organization management and governance; grant multiplier effect; human value and self-help emphasis; ultimate benefit to the community; financial management; and implementation of a strategic plan. The three annual deadlines for submission have been identified as March 31, August 31, and November 30.
Requirements: Nonprofit organizations in areas where SunTrust Bank has a major presence are eligible.
Restrictions: These foundations do not accept unsolicited grant applications from outside metropolitan Atlanta unless an organization has been specifically named by the donor as an eligible recipient.
Geographic Focus: Georgia
Date(s) Application is Due: Mar 31; Aug 31; Nov 30
Amount of Grant: 1,000 - 20,000 USD
Contact: Kay Miller, Secretary of the Distribution Committee; (404) 588-8250
Internet: https://www.suntrust.com/Microsites/foundation/funds.htm
Sponsor: SunTrust Bank Trusteed Foundations
P.O. Box 4418, Mail Code 041
Atlanta, VA 30302

SunTrust Bank Trusteed Foundations Nell Warren Elkin & William Simpson Elkin Grants 3683
The Nell Warren Elkin and William Simpson Elkin Foundation was created as a memorial by Miss Margaret R. Warren, Miss Charlotte L. Warren and Mrs. Josephine Warren Asbury, sisters of Nell Warren Elkin. The following charitable institutions were suggested by the Foundation's founders as worthy and valuable organizations. However, the Foundation's trustee is not required to make grants to these organizations and will entertain proposals from other organizations: Robert Winship Cancer Clinic of Atlanta; A.G. Rhodes of Atlanta; Georgia Heart Association of Atlanta; Bible Study Hour of Philadelphia; American Bible Society of New York City; and Emory University of Atlanta. The Trustees will consider requests for capital improvements such as buildings, furniture and equipment, and alterations to existing structures.

Applications also will be considered for special projects of a community nature, special studies, surveys, research and pilot programs which do not commit the funds to recurring expenditures. Community benefit and return on investment are primary considerations in distribution decisions. Other Distribution Committee considerations include: emphasis on metropolitan Atlanta; organization/community coordination and support; timeliness and precedence; organization management and governance; grant multiplier effect; human value and self-help emphasis; ultimate benefit to the community; financial management; and implementation of a strategic plan. The three annual deadlines for submission have been identified as March 31, August 31, and November 30.
Requirements: Nonprofits in areas where SunTrust Bank has a major presence are eligible.
Restrictions: These foundations do not accept unsolicited grant applications from outside metropolitan Atlanta unless an organization has been specifically named by the donor as an eligible recipient.
Geographic Focus: All States, Georgia, New York, Pennsylvania
Date(s) Application is Due: Mar 31; Aug 31; Nov 30
Amount of Grant: 1,000 - 20,000 USD
Contact: Kay Miller, Secretary of the Distribution Committee; (404) 588-8250
Internet: https://www.suntrust.com/Microsites/foundation/funds.htm
Sponsor: SunTrust Bank Trusteed Foundations
P.O. Box 4418, Mail Code 041
Atlanta, VA 30302

SunTrust Bank Trusteed Foundations Thomas Guy Woolford Charitable Trust Grants 3684

The Woolford Trust was created under the will of Thomas Guy Woolford, a member of the family who founded Retail Credit Company (1899) which ultimately became Equifax. This Trust was established to make gifts to institutions that are organized and operated exclusively for charitable, religious, educational or scientific purposes. Applications also will be considered for special projects of a community nature, special studies, surveys, research and pilot programs which do not commit the funds to recurring expenditures. Because of the family's ties to the Atlanta area, the Distribution Committee will give priority to charitable, religious, educational or scientific organizations in the metropolitan Atlanta area. Community benefit and return on investment are primary considerations in distribution decisions. Other Distribution Committee considerations include: emphasis on metropolitan Atlanta; organization/community coordination and support; timeliness and precedence; organization management and governance; grant multiplier effect; human value and self-help emphasis; ultimate benefit to the community; financial management; and implementation of a strategic plan. The three annual deadlines for submission have been identified as March 31, August 31, and November 30.
Requirements: Nonprofits in areas where SunTrust Bank has a major presence are eligible.
Restrictions: These foundations do not accept unsolicited grant applications from outside metropolitan Atlanta unless an organization has been specifically named by the donor as an eligible recipient.
Geographic Focus: Georgia
Date(s) Application is Due: Mar 31; Aug 31; Nov 30
Amount of Grant: 1,000 - 20,000 USD
Contact: Kay Miller, Secretary of the Distribution Committee; (404) 588-8250
Internet: https://www.suntrust.com/Microsites/foundation/funds.htm
Sponsor: SunTrust Bank Trusteed Foundations
P.O. Box 4418, Mail Code 041
Atlanta, VA 30302

SunTrust Bank Trusteed Foundations Walter H. & Marjory M. Rich Memorial Fund Grants 3685

Walter H. Rich was the second president of Rich's department store (1926-1947), succeeding his uncle, Morris Rich, who founded the company as M. Rich and Brothers in 1867. Morris Rich changed the name to Rich's when a new store was built on the corner of Broad and Alabama Streets in downtown Atlanta in 1924. Both Morris and Walter as well as Daniel Rich, Morris' brother and Walter's father, are buried in Atlanta's historic Oakland Cemetery. The Walter H. and Marjory M. Rich Memorial Fund was established in 1959. The geographic focus of the Memorial Fund is metropolitan Atlanta. The Trustees will consider requests for capital improvements such as buildings, furniture and equipment, and alterations to existing structures. Applications also will be considered for special projects of a community nature, special studies, surveys, research and pilot programs which do not commit the funds to recurring expenditures. Community benefit and return on investment are primary considerations in distribution decisions. Other Distribution Committee considerations include: emphasis on metropolitan Atlanta; organization/community coordination and support; timeliness and precedence; organization management and governance; grant multiplier effect; human value and self-help emphasis; ultimate benefit to the community; financial management; and implementation of a strategic plan. The three annual deadlines for submission have been identified as March 31, August 31, and November 30.
Requirements: Nonprofits in areas where SunTrust Bank has a major presence are eligible.
Restrictions: These foundations do not accept unsolicited grant applications from outside metropolitan Atlanta unless an organization has been specifically named by the donor as an eligible recipient.
Geographic Focus: Georgia
Date(s) Application is Due: Mar 31; Aug 31; Nov 30
Amount of Grant: 1,000 - 20,000 USD
Contact: Kay Miller, Secretary of the Distribution Committee; (404) 588-8250
Internet: https://www.suntrust.com/Microsites/foundation/funds.htm
Sponsor: SunTrust Bank Trusteed Foundations
P.O. Box 4418, Mail Code 041
Atlanta, VA 30302

Support Our Aging Religious (SOAR) Grants 3686

Support Our Aging Religious (SOAR) was established to raise national public awareness of the religious retirement crisis, to network concerned laity, and to provide financial resources to religious communities to assist them in the care of their elderly members. Each year the program awards grants to congregations across the country in its areas of interest, including the fields of education, social services, and health care. Grant categories include: building renovations to allow for handicapped accessibility, physical therapy, and medical equipment; and safety needs--fire alarms, sprinklers, hospital beds, and nurse call systems.
Requirements: SOAR welcomes applications from religious institutes that need assistance to provide critically needed items to care for their retired and infirm members.
Restrictions: SOAR is unable to provide: ongoing living expenses of retired religious in assisted living or skilled nursing facilities; major contributions to retirement funds; funds for construction of new buildings; funds for operational expenses; or cosmetic improvements that do not directly relate to the physical well-being of the retired. Proposals which request partial funding for large projects if additional funds have not been obtained cannot be supported.
Geographic Focus: All States
Date(s) Application is Due: Nov 1
Amount of Grant: 500 - 30,000 USD
Samples: Sisters, Servants of the IHM, Monroe, MI -- handicapped accessible van; Society of the Sacred Heart, Atherton, CA -- backup generator; Conception Abbey, Inc., Conception, Missouri -- elevator upgrade.
Contact: Deborah Hudson Vornbrock, Grants Administrator; (202) 529-7627; fax (202) 529-7633; dvornbrock@soar-usa.org or info@soar-usa.org
Internet: http://www.soar-usa.org/published/index.html
Sponsor: Support Our Aging Religious
900 Varnum Street NE
Washington, D.C. 20017-2145

Susan G. Komen Breast Cancer Foundation Brinker Awards for Scientific Distinction 3687

Established through a partnership between Susan G. Komen for the Cure and Brinker International in 1992, the Brinker Award for Scientific Distinction is a marquee award that recognizes pioneering work in breast cancer research and treatment. The Brinker Award honors leading scientists for their significant achievements and contributions to basic and translational science and clinical practice that have advanced the fight to save lives and realize our vision of a world without breast cancer. Since 1992, the roster of Komen Brinker Award laureates has grown to include names of researchers who have made the most significant advances in breast cancer research and medicine. The Brinker Award for Scientific Distinction is given annually in two categories. The Basic Science Award is presented to a researcher whose scientific discoveries or novel technologies have added substantively to our understanding of breast cancer and the intrinsic molecular processes that drive the disease, and/or whose work has bridged the gap between basic research and patient care. The Clinical Research Award is presented to a clinical or translational researcher who has advanced the identification of new prevention, detection or treatment approaches for breast cancer and promoted their incorporation into clinical care.
Requirements: Breast cancer researchers who are affiliated with nonprofit institutions (academic, industry, government, or other) in the United States and abroad are eligible.
Geographic Focus: All States
Date(s) Application is Due: Jun 30
Amount of Grant: 20,000 USD
Contact: Award Administrator; (972) 855-1600; fax (972) 855-1605
Internet: http://ww5.komen.org/ResearchGrants/BrinkerAward.html
Sponsor: Susan G. Komen Breast Cancer Foundation
5005 LBJ Freeway, Suite 250
Dallas, TX 75244

Susan G. Komen Breast Cancer Foundation Career Catalyst Research Grants 3688

Career Catalyst Research (CCR) Grants are intended to foster promising breast cancer researchers who are in the early stages of their faculty careers by providing support for up to three years of "protected time" for research career development under the guidance of a mentor committee. These Grants provide funding for research projects that have significant potential to advance our understanding of breast cancer and lead to reductions in breast cancer incidence and/or mortality within the next decade. It is expected that awardees will launch independent research careers and successfully compete for subsequent research project funding in breast cancer following the successful completion of a Career Catalyst Research Grant. CCRs provide support for hypothesis-driven research projects that have significant potential to advance our understanding of breast cancer and lead to reductions in breast cancer incidence and/or mortality within the next decade. Studies focusing on quality of life or survivorship issues are not appropriate for this mechanism. Applicants may request funding of up to $150,000 per year (combined direct and indirect costs) for up to three years.
Requirements: Applicants and Institutions must conform to the eligibility criteria to apply for a CCR Grant (eligibility requirements must be met at the time of full application submission). Applicants must: have a doctoral degree, including MD, PhD, DrPH, DO, or equivalent; currently hold a faculty appointment or have a formal offer letter at the time of application, verified by the Letter of Institutional Support; not have held any faculty appointment, including non-tenure and tenure track appointments combined, for more than a total of 6 years at the time of application, verified by the Letter of Institutional Support; not currently hold or simultaneously apply for a Komen Investigator-Initiated Research Grant; not currently be a Principal Investigator on an existing R01 research grant; have adequate space to conduct proposed research and protected time for research, verified by the Letter of Institutional Support; and ensure that all past and current

Komen-funded grants are up to date and in compliance with all Komen requirements; e.g. Applicants are not required to be U.S. citizens or residents. Institutions must be a nonprofit institution or organization anywhere in the world.
Geographic Focus: All States, All Countries
Date(s) Application is Due: Dec 17
Amount of Grant: Up to 450,000 USD
Contact: Director, Grants and Sponsored Programs; (972) 855-1600 or (866) 921-9678; fax (972) 855-1605; helpdesk@komengrantsaccess.org
Internet: http://ww5.komen.org/uploadedFiles/Content/ResearchGrants/GrantPrograms/new-CareerCatalystRFA2012-2.pdf
Sponsor: Susan G. Komen Breast Cancer Foundation
5005 LBJ Freeway, Suite 250
Dallas, TX 75244

Susan G. Komen Breast Cancer Foundation Challege Grants: Breast Cancer and the Environment   3689

Major advances have been made in understanding the biology and diversity of breast cancer, but much more remains to be discovered about the many causes of breast cancer – particularly what contributions a diverse array of environmental factors may be making – and how to prevent it. The challenges are many: The scientific community has been presented with conflicting and inconclusive results from past studies. With increased knowledge of the complexity of breast cancer biology, it has become apparent that future research into environmental influences will need to focus on early-life exposures, associations with specific tumor types, and gene-environment interactions. Adding to the complexity of this task is the fact that for a wide array of exposures the assessment methodologies, tools and resources are limited. Through its Challenge Grants: Breast Cancer and the Environment program, the Foundation seeks to address key research needs in the field of environmental contributions to breast cancer risk. Susan G. Komen for the Cure and its Scientific Advisory Board have requested that the Institute of Medicine (IOM) review the current evidence on environmental risk factors for breast cancer, consider gene-environment interactions in breast cancer, explore evidence-based actions that might reduce the risk of breast cancer, and recommend research needed in these areas. At the conclusion of their report, the IOM issued 13 recommendations for further research, of which three have been chosen by the Komen Scientific Advisory Board as the subject of Challenge Grants: Studies of Occupational Cohorts and Other Highly Exposed Populations; New Exposure Assessment Tools; and Minimizing Exposure to Ionizing Radiation.
Requirements: U.S. nonprofits and international institutions will be eligible to apply.
Geographic Focus: All States
Amount of Grant: Up to 250,000 USD
Contact: Director, Grants and Sponsored Programs; (972) 855-1600; fax (972) 855-1605; grants@komen.org
Internet: http://ww5.komen.org/ResearchGrants/FundingOpportunities.html
Sponsor: Susan G. Komen Breast Cancer Foundation
5005 LBJ Freeway, Suite 250
Dallas, TX 75244

Susan G. Komen Breast Cancer Foundation Challege Grants: Investigator Initiated Research   3690

Susan G. Komen Breast Cancer Foundation Challenge Grants seek to stimulate exploration of new ideas and novel approaches to research on breast cancer and the environment, through a life course approach, that have significant potential to lead to reductions in breast cancer incidence and/or mortality within the next decade. In the area of Studies of Occupational Cohorts and Other Highly-Exposed Populations, the research seeking to study populations exposed to potentially cancer-causing agents or external toxins, chemicals, hormones or radiation, such as occupational cohorts, persons with event-related high exposures, or patient groups receiving high-dose or long-term medical treatments, including medical radiation. In the area of New Exposure Assessment Tools, the research is seeking to improve methodologies for measuring, across the life course, personal exposure to and biologically effective doses of environmental factors that may alter risk for or susceptibility to breast cancer. In the area of Minimizing Exposure to Ionizing Radiation, comparative effectiveness research is seeking to assess the relative benefits and harms of medical radiation, including imaging procedures and diagnostic/follow-up radiologic tests using algorithms in common practice. Applicants may request either three or four years of funding as follows: up to a total of $750,000 over three years (combined direct and indirect costs); or up to a total of $1,000,000 over four years (combined direct and indirect costs). Reasonable compensation of advocates is allowed when advocates perform services that would otherwise be a contracted expense. Compensation may be in the form of salary, per-hour compensation, or honoraria. Additionally, grant funds can be used for advocate participation in scientific conferences that would enhance their knowledge and skills related to the research project.
Requirements: A wide range of nonprofits, including research groups and health and human service agencies, are eligible to apply. Budgets are not required to be equivalent across each year of the Grant, but rather should reflect the costs appropriate to support the research project each year.
Restrictions: Personnel on the project are limited to a base salary at or below $250,000 per year, and PIs must provide a 10% minimal level of effort. Equipment costs are limited to no more than 30% of total direct costs. Indirect costs cannot exceed 25% of total direct costs (including any indirect costs paid through subcontracts or consortia). Costs that are not allowed include: travel costs; publication costs and meeting-related poster printing costs are allowed; graduate and postdoctoral fellow tuition; visa costs; or professional membership dues.
Geographic Focus: All States
Date(s) Application is Due: Dec 20

Amount of Grant: Up to 1,000,000 USD
Contact: Director, Grants and Sponsored Programs; (972) 855-1600 or (866) 921-9678; fax (972) 855-1605; helpdesk@komengrantaccess.org
Internet: http://ww5.komen.org/uploadedFiles/Content/ResearchGrants/GrantPrograms/new-Challenge-InvestInitiatedRFA2012-2.pdf
Sponsor: Susan G. Komen Breast Cancer Foundation
5005 LBJ Freeway, Suite 250
Dallas, TX 75244

Susan G. Komen Breast Cancer Foundation Challenge Grants: Career Catalyst Research   3691

Challenge Grants-Career Catalyst Research are intended to foster promising breast cancer researchers who are in the early stages of their faculty careers by providing support for up to three years of protected time for research career development under the guidance of a mentor committee. It is expected that awardees will launch independent research careers and successfully compete for subsequent research project funding in breast cancer following the successful completion of the Grant. These Grants provide funding for research projects that have significant potential to advance our understanding of breast cancer and the environment, through a life course approach, that have significant potential to lead to reductions in breast cancer incidence and/or mortality within the next decade. Applications will be accepted within the following three focus areas: studies of occupational cohorts and other highly-exposed populations; new exposure assessment tools; and minimizing exposure to ionizing radiation. The pre-application deadline is September 10 annually, and there is a limit of $450,000 combined direct and indirect costs ($150,000 per year for up to 3 years).
Requirements: CCR grants are open to scientists who have held faculty positions for no more than six years at the time of full application to achieve research independence. Applicants and Institutions must conform to the eligibility criteria to apply for a CCR Grant (eligibility requirements must be met at the time of full application submission). Applicants must: have a doctoral degree, including MD, PhD, DrPH, DO, or equivalent; currently hold a faculty appointment or have a formal offer letter at the time of application, verified by the Letter of Institutional Support; not have held any faculty appointment, including non-tenure and tenure track appointments combined, for more than a total of 6 years at the time of application, verified by the Letter of Institutional Support; not currently hold or simultaneously apply for a Komen Investigator-Initiated Research Grant; not currently be a Principal Investigator on an existing R01 research grant; have adequate space to conduct proposed research and protected time for research, verified by the Letter of Institutional Support; and ensure that all past and current Komen-funded grants are up to date and in compliance with all Komen requirements; e.g. Applicants are not required to be U.S. citizens or residents. Institutions must be a nonprofit institution or organization anywhere in the world.
Geographic Focus: All States, All Countries
Date(s) Application is Due: Dec 17
Amount of Grant: Up to 450,000 USD
Contact: Director, Grants and Sponsored Programs; (972) 855-1600 or (866) 921-9678; fax (972) 855-1605; helpdesk@komengrantaccess.org
Internet: http://ww5.komen.org/uploadedFiles/Content/ResearchGrants/GrantPrograms/Challenge-CareerCatalystRFA2012-2.pdf
Sponsor: Susan G. Komen Breast Cancer Foundation
5005 LBJ Freeway, Suite 250
Dallas, TX 75244

Susan G. Komen Breast Cancer Foundation College Scholarships   3692

In 2001, Komen established the college scholarship program to help students who would find attending college to be a significant financial burden due to the loss of a parent or guardian to breast cancer or their own breast cancer diagnosis at age 25 or younger. The program has evolved into a dynamic partnership between Komen and students who serve in their communities, sharing Komen's message of breast cancer awareness and raising funds while pursuing their baccalaureate degrees. The award is a generous scholarship of $10,000 a year for up to four years to attend a state university in pursuit of a baccalaureate degree. Scholarship recipients are selected based on scholastic achievement, community service, financial need and demonstrated leadership potential. The annual deadline is October 15.
Requirements: Applicants must meet all of the following criteria to be eligible for this scholarship: have lost a parent/guardian to breast cancer or be a breast cancer survivor diagnosed at 25 years or younger; be a high school senior, college freshman, sophomore or junior; plan to attend a state-supported college or university in the state where they permanently reside (students in Washington D.C. can attend state-supported schools in Maryland or Virginia); have a high school and/or college GPA of 2.8 on a 4.0 scale; be no older than 25 years old by May of the application year; be a U.S. citizen, or documented permanent resident of the U.S. (or U.S. Territory); never at any time have been subject to any disciplinary action by any institution or entity, including, but not limited to, any educational or law enforcement agency; and be a driven advocate for breast cancer education and research who is able to complete a minimum of 20 hours of community service per semester.
Restrictions: For-profit organizations are not eligible.
Geographic Focus: All States
Date(s) Application is Due: Oct 15
Amount of Grant: 10,000 USD
Contact: Director; (972) 855-1600; fax (972) 855-1605; contactus@applyists.com
Internet: http://ww5.komen.org/ResearchGrants/CollegeScholarshipAward.html
Sponsor: Susan G. Komen Breast Cancer Foundation
5005 LBJ Freeway, Suite 250
Dallas, TX 75244

Susan G. Komen Breast Cancer Foundation Investigator Initiated Research Grants   3693

Susan G. Komen Breast Cancer Foundation Investigator Initiated Research Grants seek to stimulate exploration of new ideas and novel approaches in breast cancer research and clinical practice that have significant potential to lead to reductions in breast cancer incidence and/or mortality within the next decade. Applications addressing topics other than those described below will be administratively withdrawn from consideration, and will not be reviewed or scored: Novel Therapeutics and/or Resistance - Therapeutic Implications of Tumor Genomics; Biology of Breast Cancer - Implications of the Immune System in Breast Cancer Biology; Prevention/Early Detection - New Strategies for Early Detection; and Disparities in Breast Cancer Outcomes - Outcomes of Specific Populations after Diagnosis. The pre-application submission deadline is September 10, with the full application due December 17. $1,000,000 in combined direct and indirect costs are allowable, which equals $250,000 per year for up to 4 years.
Requirements: Grants will be awarded to a single Principal Investigator (PI) or one PI and one Co-Principal Investigator (Co-PI). PIs and Co-PIs: must have a doctoral degree, including MD, PhD, DrPH, DO, or equivalent; must have a full time faculty appointment at the time of application; may only apply in the final year of their funding and may not hold both grants simultaneously; must ensure that all past and current Komen-funded Grants are up to date and in compliance with all Komen requirements; and are not required to be U.S. citizens or residents. Primary Institution must be a nonprofit institution or organization anywhere in the world.
Geographic Focus: All States, All Countries
Date(s) Application is Due: Dec 17
Amount of Grant: Up to 1,000,000 USD
Contact: Director, Grants and Sponsored Programs; (972) 855-1600 or (866) 921-9678; fax (972) 855-1605; helpdesk@komengrantsaccess.org
Internet: http://ww5.komen.org/uploadedFiles/Content/ResearchGrants/GrantPrograms/InvestInitiatedRFA2012-2-new2a.pdf
Sponsor: Susan G. Komen Breast Cancer Foundation
5005 LBJ Freeway, Suite 250
Dallas, TX 75244

Susan Mott Webb Charitable Trust Grants   3694

The trust awards grants to eligible Alabama nonprofit organizations in its areas of interest, including the arts, civic and public affairs, animals/wildlife, community development, education, health care, religion, social services, youth programs and homelessness. Types of support include annual campaigns, building construction/renovation, capital campaigns, continuing support, curriculum development, emergency funds, endowments, equipment, general/operating support, internship funds, program development, publication, and technical assistance. Contact the Foundation for further application information and guidelines.
Requirements: Giving limited to the greater Birmingham, AL, area.
Restrictions: Grants do not support advertising, individuals, international organizations, or political organizations.
Geographic Focus: Alabama
Date(s) Application is Due: Apr 1; Oct 1
Amount of Grant: 2,000 - 100,000 USD
Samples: Alabama Symphonic Association, Birmingham, AL, $40,000; Crisis Resource Center of Southeast Kansas, Pittsburg, KS, $25,000;
Contact: Laura Wainwright, Vice President, c/o Regions Bank; (205) 801-0380; fax (205) 581-7433; laura.wainwright@regions.com
Sponsor: Susan Mott Webb Charitable Trust
P.O. Box 11426
Birmingham, AL 35202-1426

Susan Vaughan Foundation Grants   3695

The Susan Vaughan Foundation established in 1952, supports non-profits involved with: education, particularly to a library; arts; environment, natural resources; higher education; human services. The Foundation gives primarily in Houston and Austin Texas area with support in the form of: annual campaigns; building/renovation; capital campaigns; general/operating support; matching/challenge support.
Requirements: There is no formal application form required, applicant must be a non-profit in the Houston and Austin, Texas region. Initial contact should be made through a Letter of Inquiry.
Geographic Focus: Texas
Amount of Grant: 2,500 - 630,000 USD
Samples: Clayton Library Friends, Houston, TX, $630,000; Houston Downtown Park Conservancy, Houston, TX, $50,000; Trees for Houston, Houston, TX, $30,000;
Contact: Jennifer Grosvenor, Grant Coordinator c/o Legacy Trust Co; (713) 651-8980; jgrosvenor@legacytrust.com
Sponsor: Susan Vaughan Foundation
600 Jefferson Street, Suite 300
Houston, TX 77002-7377

Highmark BCBS Challenge for Healthier Schools in Western Pennsylvania   3696

The program addresses the problem of overweight and inactive children and the subsequent health consequences of being overweight. Schools in Western Pennsylvania will receive support for innovative, evidence-based model program ideas dealing with increased physical activity or nutrition education. The nine most promising plans will win cash grants to be used to implement their programs at the school. Four cash awards will be made to elementary schools, three cash award will be made to middles schools, and two cash awards will be made to high schools. The program encourages collaborative efforts among teachers, students, and parents to create innovative approaches to student wellness.
Requirements: To be eligible, schools must be located within the 29 counties of the Western Pennsylvania region, and become a Caring Team School through the Western Pennsylvania Caring Foundation (http://www.highmarkcaringfoundation.com/chooseRegion.shtml).
Geographic Focus: Pennsylvania
Date(s) Application is Due: Jun 1
Amount of Grant: 10,000 - 20,000 USD
Contact: Aaron Billger, (412) 544-7826; aaron.billger@highmark.com
Internet: https://www.highmark.com/hmk2/community/index.shtml
Sponsor: Highmark
120 5th Avenue, Suite 2112
Pittsburgh, PA 15222-3099

Swindells Charitable Foundation   3697

The Swindells Charitable Foundation was established in 1933 to support and promote quality health and human-services programming for underserved children and adults. The Swindells Charitable Foundation also makes grants to public charitable hospitals. Preference is given to organizations that serve sick or economically disadvantaged children or older adults. Special consideration is given to organizations that provide for the "basic human needs" of individuals. Grants from the Swindells Charitable Foundation are one year in duration. Applicants must apply online at the grant website. Applicants are strongly encouraged to do the following before applying: review the downloadable state application procedures for additional helpful information and clarifications; review the downloadable online-application guidelines at the grant website; review the foundation's funding history (link is available from the grant website); review the online application questions in advance; and review the list of required attachments. These will generally include: a list of board members, financial statements (audited, reviewed, or compiled by independent auditor); an organization summary; a list of other funding sources; an IRS Determination letter; and other required documents. All attachments must be uploaded in the online application as PDF, Word, or Excel files. The Swindells Charitable Foundation has biannual deadlines of February 1 and August 1. Applications are by 11:59 p.m. on the deadline dates. Applicants will be notified of grant decisions by letter within two to three months after each respective proposal deadline.
Requirements: Applicants must have 501(c)3 tax-exempt status.
Restrictions: Applicants will not be awarded a grant for more than three consecutive years. The trust does not support requests from individuals, organizations attempting to influence policy through direct lobbying, or any political campaigns. Capital requests will not be considered.
Geographic Focus: Connecticut
Date(s) Application is Due: Feb 1; Aug 1
Contact: Kate Kerchaert; (860) 657-7016; kate.kerchaert@baml.com
Internet: https://www.bankofamerica.com/philanthropic/fn_search.action
Sponsor: Swindells Charitable Foundation
200 Glastonbury Boulevard, Suite # 200, CT2-545-02-05
Glastonbury, CT 06033-4056

Sybil G. Jacobs Award for Outstanding Use of Tobacco Industry Documents   3698

The award recognizes a person who has made a significant and well-recognized contribution to the health of the public in the recent past through use of tobacco documents. The award also honors innovation in the use and application of tobacco industry documents to improve the public's health and, where applicable, to further the goals of tobacco prevention and control in order to help build a world where young people reject tobacco and anyone can quit. Those nominated should be individuals who have made a notable impact through innovative use of tobacco industry documents as applied to research, policy, or advocacy.
Requirements: At least one of the following two criteria must be met by the nominated individual: Nominees must have made a remarkable research, policy, or advocacy contribution with the use of tobacco industry documents, and/or; Nominees must have employed innovative, creative approaches to the employment of tobacco industry documents that result in an improvement in the health or public awareness of a community or nation. Nominations may be made by individuals such as colleagues, peers, coworkers, instructors/professors, governments, or CBOs that are qualified to make such a nomination because of their familiarity with the nominee's work and contribution.
Restrictions: Nominees must not have any affiliation with the tobacco industry. Members of the awards committee may not nominate potential recipients, although they may provide written support as part of a nomination package. Employees and directors of the American Legacy Foundation and their relatives are not eligible.
Geographic Focus: All States
Date(s) Application is Due: Feb 6
Amount of Grant: 7,500 USD
Contact: Virginia Lockmuller; (202) 454-5555; awards@americanlegacy.org
Internet: http://www.legacyforhealth.org/awards/
Sponsor: American Legacy Foundation
1724 Massachusetts Avenue NW
Washington, D.C. 20036

Symantec Community Relations and Corporate Philanthropy Grants   3699

Symantec's Community Relations and Corporate Philanthropy Program strives to have a positive impact on our local communities around the globe. Symantec supports four main philanthropic areas, all with the objective of creating a sustainable and diverse future for the technology industry; science, technology, engineering, and math education; equal access to education; diversity in engineering; environmental responsibility; and online safety.
Requirements: 501(c)3 entities and/or educational institutions that have philosophy programs that benefit the lives of children are encouraged to apply. Nonprofit organizations in Minneapolis, Minnesota; San Francisco Bay Area, California; Orlando, Florida; and Vienna, Virginia, are eligible.

Restrictions: Grants will not be made for political or religious purposes. Grants will not be made to organizations that discriminate on the basis of age, disability, religion, ethnic origin, sex or sexual orientation.
Geographic Focus: California, Florida, Minnesota, Virginia
Contact: Grants Director; (800) 327-2232 or (650) 527-8000; fax (650) 527-2908; community_relations@Symantec.com
Internet: http://www.symantec.com/about/profile/responsibility/community/index.jsp
Sponsor: Symantec Corporation
350 Ellis Street
Mountain View, CA 94043

T.L.L. Temple Foundation Grants　　3700
The T.L.L. Temple Foundation was established in 1962 by Georgie Temple Munz in honor of her father, Thomas Lewis Latané Temple, an East Texas lumberman and founder of Southern Pine Lumber Company, which later became Temple Industries. It was her wish to create a charitable foundation that would operate primarily to improve the quality of life for the inhabitants of Deep East Texas. The foundation supports organizations devoted to programs in the areas in education, public health, public affairs, human services, arts and culture, and the environment. Since its inception, the T.L.L. Temple Foundation has been committed to supporting environmental initiatives devoted to the conservation of forest lands and river systems, and the preservation of native plant and wildlife species—to protect and ensure the perpetuity of these significant natural resources.
Requirements: The foundation primarily makes grants to projects located and/or to be operated in the area constituting the East Texas pine timber belt and Miller County, Arkansas in which T.L.L. Temple founded and operated his timber production and manufacturing enterprises. Governmental units exempt under the IRS code and 501(c)3 nonprofit organizations (not classified as a private foundation) are eligible to apply. There are no specific deadlines.
Restrictions: Grants do not support private foundations. Grants are not made to individuals for scholarships, research or other purposes.
Geographic Focus: Arkansas, Texas
Contact: Millard F. Zeagler, Executive Director; (936) 634-3900; fax (936) 639-5199
Sponsor: T.L.L. Temple Foundation
204 Champions Drive
Lufkin, TX 75901-7321

T. Spencer Shore Foundation Grants　　3701
The T. Spencer Shore Foundation offers financial support in the New England region, as well as New York and Ohio. Its primary fields of interest is funding for higher education, religion, and heath care. Grants are given for general operating support only. A specific application form is required, though there are no annual deadlines. Applicants should contact the office for further instructions. Amounts range from $500 to $3,000.
Requirements: Any college or university in Massachusetts, New Hampshire, Connecticut, Vermont, Rhode Island, Maine, Ohio, or New York are eligible to apply.
Geographic Focus: Connecticut, Maine, Massachusetts, New Hampshire, New York, Ohio, Rhode Island, Vermont
Amount of Grant: 500 - 3,000 USD
Samples: Carver Center, New York, New York, $1,000 - general operations; Rye Presbyterian Church, Rye, New York, $3,000 - general operations; Wellesley College, Wellesley, Massachusetts, $1,000 - general operations.
Contact: Thomas S. Shore, Jr., Trustee; (207) 967-0129
Sponsor: T. Spencer Shore Foundation
P.O. Box 629
Kennebunkport, ME 04046-0629

Taubman Endowment for the Arts Foundation Grants　　3702
The Taubman Endowment for the Arts was initially established to support the arts in Florida, Maryland, Michigan, and New York, through grants awards. More recently, it has been renamed the Taubman Foundation, with its current primary fields of support including: art museums; elementary and secondary education; higher education; hospital care; human services; and Judaism. Awards typically are given in the form of general operating support, program development, endowments, and equipment purchase. There are no formal application materials required, so interested parties should submit: a detailed description of project and amount of funding requested; a list of additional sources and amount of support; and a brief explanation of what the applicant expects to receive from the foundation and why. Submissions are accepted at any time. Most recent awards have ranged from $5,000 to $80,000.
Restrictions: Grants are not made to individuals.
Geographic Focus: Florida, Maryland, Michigan, New York
Amount of Grant: 5,000 - 80,000 USD
Contact: Glenda Cole, Vice President, Marketing & Sponsorship; (248) 258-6800 or (248) 258-7618; gcole@taubman.com
Sponsor: Taubman Foundation
200 East Long Lake Road, Suite 300
Bloomfield Hills, MI 48304

Tauck Family Foundation Grants　　3703
The Tauck Family Foundation Grants focus on organizations and programs that are committed to Bridgeport's elementary school children, and to strengthening their capacity to help these children build the social and emotional skills they need to succeed throughout their educations. The desire is to help create better outcomes for children from low?income families and to focus the foundation's efforts in a way that would strengthen its nonprofit investees over time.
Requirements: The Tauck Family Foundation has selected its first portfolio of investees and has made a multi-year commitment to those organizations. As a result the foundation is not accepting proposals at this time. However, the Tauck Family Founation continues to be interested in learning about organizations developing the essential life skills of Bridgeport children. Nonprofit 501(c)3 organizations that meet the eligibility guidelines are free to submit an inquiry (via the foundation's website). Eligibility guidelines are as follows: be a non-profit, 501(c)3 organization; work (or have a specific strategy to work) with elementary (kindergarten through fifth grade) students in Bridgeport, Connecticut (organizations may be based in other cities/towns as long as they work with a significant number - more than 100 - of Bridgeport children in grades K-5); have an operating budget of $300,000 or more; consider the development of life skills of youth from low-income families to be core to the organization's mission; be interested in developing organizationally and receiving capacity building support, as needed and including: (a) going through an intensive Theory of Change process with the Tauck Family Foundation's consultants, and (b) implementing performance management systems to measure and monitor the development of children's self-control, persistence, mastery orientation, and academic self-efficacy.
Geographic Focus: Connecticut
Contact: Mirellise Vazquez, Program Officer; (203) 899-6824; fax (203) 286-1340; mirellise@tauckfoundation.org
Internet: http://www.tauckfamilyfoundation.org/how-to-apply
Sponsor: Tauck Family Foundation
P.O. Box 5020
Norwalk, CT 06856

Taylor S. Abernathy and Patti Harding Abernathy Charitable Trust Grants　　3704
The Taylor S. and Patti Harding Abernathy Charitable Trust was established in 1988 to support religious, charitable, scientific, and educational purposes. To that end, the trust provides grants that support and promote quality educational, cultural, human services, and health care programming. Grant requests for general operating support and program support will be considered. The Trust generally supports organizations that serve residents of the Greater Kansas City Metropolitan area. There are no application deadlines for the Abernathy Charitable Trust. Proposals are reviewed on an ongoing basis. Grants are one year in duration. Grant requests for general operating support and program support will be considered.
Requirements: 501(c)3 organizations serving the residents of the Greater Kansas City Metropolitan area are welcome to apply.
Restrictions: Grant requests for capital support will not be considered.
Geographic Focus: Kansas, Missouri
Amount of Grant: Up to 25,000 USD
Contact: Scott Berghaus; (816) 292-4300; scott.berghaus@ustrust.com
Internet: https://www.bankofamerica.com/philanthropic/foundation.go?fnId=131
Sponsor: Taylor S. and Patti Harding Abernathy Charitable Trust
P.O. Box 831041
Dallas, TX 75283-1041

TE Foundation Grants　　3705
The foundation provides grants to nonprofit organizations in geographic areas where Tyco Electronics has a significant employee population and for specific projects or programs in broad categories, including education (with an emphasis on math and science), community impact, and arts and culture. In addition to a matching gifts program for employee contributions to accredited high schools, colleges, and universities, the foundation makes direct grants for programs that address a business or community concern of Tyco Electronics. Organizations that support pre-college math and science education receive special attention. Agencies that promote personal growth, career opportunities, and economic self-sufficiency are encouraged to apply, as are local chapters of health- and civic-related organizations. Special attention is given to community-wide arts organizations that solicit and allocate funds for a number of arts groups and institutions. Local public television and radio stations are encouraged to apply for funding of specific education initiatives. Capital campaigns of significant arts and cultural organizations serving communities in which the corporation has a major presence also will receive consideration. Grants also are awarded to support general operations, program development, and employee matching gifts. Applications are accepted throughout the year but are considered on the listed application deadlines.
Requirements: The TE Foundation limits grants to U.S. organizations that qualify as nonprofit under Section 501(c)3 of the Internal Revenue Code. Requests receive preferential review if the organization is supported by TE employees as volunteers.
Restrictions: The foundation generally will not support organizations in geographic areas where Tyco Electronics has few or no employees; individuals, private foundations, national organizations, or service clubs; fraternal, social, labor or trade organizations; organizations that discriminate on the basis of race, religion, color, national origin, physical or mental conditions, veteran or marital status, age, or sex; churches or religious organizations, with the exception of nondenominational programs sponsored by a church or religious group such as a food bank, youth center or non-sectarian education programs; political campaigns; loans or investments; or programs that pose a potential conflict of interest.
Geographic Focus: California, Massachusetts, Michigan, North Carolina, Pennsylvania, South Carolina, Texas, Virginia
Date(s) Application is Due: Mar 15; Jun 15; Sep 15; Dec 15
Amount of Grant: 250 - 25,000 USD
Contact: Mary Rakoczy, (717) 592-4869; fax (717) 592-4022; TEfoundation@te.com
Internet: http://www.te.com/en/about-te/responsibility/community.html
Sponsor: Tyco Electronics Foundation
c/o TE Corporation
Harrisburg, PA 17105-3608

Tellabs Foundation Grants 3706
The foundation awards grants to eligible nonprofit organizations in the following areas: educational programs with a particular focus on local and national programs and curricula for engineering, science, mathematics and technology; health and human services programs for projects involving health and wellness-related, research, education and treatment in the United States (The primary focus is on projects involving hospitals and health care facilities); and environmental programs that encourage the understanding and protection of the environment. Generally, grants are awarded for specific programs rather than for general operating funds or capital projects. The primary focus of the foundation is to support programs in areas in which Tellabs employees live and work. Tellabs Inc and its affiliates also have direct giving programs, including an employee matching gift program, giving to the United Way, and limited direct corporate grants. The board meets quarterly (January, April, July and October); submit proposals at least four weeks before scheduled meetings.
Requirements: Unless invited by the foundation board to submit a full grant proposal, all applicants or programs must first submit a letter of inquiry. There is no set format, but it should be 1-2 pages in length. All grant recipients must be recognized by the IRS as tax exempt, not-for-profit organizations under Section 501(c)3 of the Internal Revenue Code. Contact program staff for company locations.
Restrictions: The foundation generally will not consider requests from: Organizations which do not have a 501(c)3 status; Political organizations or parties, candidates for political office, and organizations whose primary purpose is to influence legislation; Labor unions or organizations; Local athletic or sports programs; Service organizations raising money for community purposes; Individuals; Travel funds for tours, expeditions, or trips by individuals or groups; Dues or gifts to national or local alumni groups, clubs or fraternities; Institutional memberships or subscription fees for publications; Gifts to individual churches, synagogues or other entities organized exclusively for religious purposes; Donations for benefit events, raffle tickets or fundraising efforts that involve value returned to the donor; Organizations not operating for the benefit of the general public or that have discriminatory practices; Any grantee who receives funds of $50,000 or more shall not be entitled to submit another grant proposal for three years from the date of grant.
Geographic Focus: All States
Amount of Grant: 10,000 USD
Samples: Purdue U (West Lafayette, IN)--to construct a facility for the department of computer sciences, $30,000.
Contact: Meredith Hilt, Executive Director; (630) 798-2506; fax (630) 798-4778; meredith.hilt@tellabs.com
Internet: http://www.tellabs.com/about/foundation.shtml
Sponsor: Tellabs Foundation
1415 W Diehl Road, Mail Stop 10
Naperville, IL 60563

Tension Envelope Foundation Grants 3707
Incorporated in 1954 in Missouri, the Tension Envelope Foundation supports nonprofits in company-operating areas, with an emphasis on Jewish welfare funds, community funds, higher education, health, civic affairs, culture and the arts, and youth. Funding typically is provided for general operating costs. There are no annual deadlines or guidelines, and the foundation does not distribute an annual report. To apply for a grant, send proposal with an overview of the project, budget, and proof of 501(c)3 status, requesting an application form in the process. The board meets several times each year.
Requirements: 501(c)3 nonprofits in California, Iowa, Kansas, Minnesota, Missouri, North Carolina, Tennessee, and Texas are eligible. Primary consideration is in the Kansas City metro area.
Geographic Focus: California, Iowa, Kansas, Minnesota, Missouri, North Carolina, Tennessee, Texas
Amount of Grant: 100 - 70,000 USD
Contact: William L. Berkley, Director; (816) 471-3800
Sponsor: Tension Envelope Foundation
819 E 19th Street, 5th Floor
Kansas City, MO 64108-1781

Texas Commission on the Arts Arts Respond Project Grants 3708
This program provides project assistance grants on a short-term basis and may include administrative costs directly related to the project. Projects must address one of the following priority areas: education, health and human services, economic development, public safety, criminal justice, natural resources, or agriculture. Organizations that are eligible are: Arts Organizations, Established Arts Organizations, Minority Arts Organizations, Rural Arts Providers.
Requirements: To be eligible for TCA grants, an organization must: be a tax-exempt nonprofit organization as designated by the Internal Revenue Service and/or must be an entity of government; be incorporated in Texas; have fulfilled all its outstanding contractual obligations to the State of Texas (i.e. student loans, child support, taxes, etc.); and comply with regulations pertaining to federal grant recipients including Title VI of the Civil Rights Act of 1964, Section 504 of the Rehabilitation Act of 1973, the Age Discrimination Act of 1975, the Education Amendments of 1972, the Americans with Disabilities Act of 1990, and the Drug Free Workplace Act of 1988.
Geographic Focus: Texas
Amount of Grant: Up to 45,000 USD
Contact: Director; (512) 463-5535 or (800) 252-9415; front.desk@arts.state.tx.us
Internet: http://www.arts.state.tx.us/index.php?option=com_wrapper&view=wrapper&Itemid=86
Sponsor: Texas Commission on the Arts
P.O. Box 13406
Austin, TX 78711-3406

Texas Instruments Corporation Health and Human Services Grants 3709
The Corporation builds healthy communities by funding health and human service programs that address the needs of the community. The Health and Human Services grants are primarily program specific, with limited support to United Way agencies. The Corporation board meets quarterly, and funding decisions are made within three weeks of each meeting.
Requirements: Applicants must be 501(c)3 tax-exempt nonprofit organizations. Grant applications are only accepted online. Applicants access the eligibility quiz at the website and if they are eligible, they gain access to the application.
Restrictions: Corporate grant funding is prioritized based on opportunities for TI employee/retiree involvement. Due to the TI Foundation's and TI employees' long-time support of United Way, only limited support of United Way agencies will be considered. The Corporation does not support the following: organizations that are not 501(c)3 charities; grants to individuals, including sponsorships; private foundations or endowment funds; sectarian or denominational religious organizations; political activities, parties or candidates; veteran organizations; hospitals; fraternal or labor organizations; courtesy advertising, including program books and yearbooks; entertainment events, scholarships or conferences; sporting events or teams; golf tournaments; unrestricted gifts to national or international organizations; travel or tours; or table sponsorships.
Geographic Focus: Texas
Contact: Andy Smith; (214) 480-3462; fax (214) 480-2920; wasmith@ti.com
Internet: http://www.ti.com/corp/docs/csr/giving.shtml
Sponsor: Texas Instruments Corporation
12500 TI Boulevard
Dallas, TX 75266-0199

Texas Instruments Foundation Community Services Grants 3710
The Foundation offers funding to enrich health and human services programs that meet the greatest community needs. This includes capital or civic campaigns. The Foundation considers supporting such campaigns based on community needs and economic impact. The Foundation primarily supports services through its annual United Way donations. Beyond United Way, it also provides limited support to organizations meeting the greatest community needs. The Foundation board meets quarterly (March, June, September, and November) and makes funding decisions within three weeks of each meeting.
Requirements: Applicants must be 501(c)3 nonprofit organizations. Grant applications are only accepted online. Applicants access the eligibility quiz at the website and if they are eligible, they gain access to the application. The Foundation primarily supports programs in the Dallas area.
Restrictions: The Foundation does not support the following: organizations that are not 501(c)3 charities; grants to individuals, including sponsorships; private foundations or endowment funds; sectarian or denominational religious organizations; political activities, parties or candidates; veteran organizations; hospitals; fraternal or labor organizations; courtesy advertising, including program books and yearbooks; entertainment events, scholarships or conferences; sporting events or teams; golf tournaments; unrestricted gifts to national or international organizations; travel or tours; or table sponsorships.
Geographic Focus: Texas
Contact: Ann Pomykal; (214) 480-6873; fax (214) 480-6820; giving@ti.com
Internet: http://www.ti.com/corp/docs/csr/giving.shtml
Sponsor: Texas Instruments Foundation
12500 TI Boulevard
Dallas, TX 75266-0199

Texas Instruments Foundation STEM Education Grants 3711
The Foundation funds programs that emphasize math and science education, including STEM (science, technology, engineering and math) teacher effectiveness. Grant decisions are made within three weeks of the Foundation board meetings (March, June, September, and November).
Requirements: Applicants must be 501(c)3 tax-exempt nonprofit organizations. Grant applications are only accepted online. Applicants access the eligibility quiz at the website and if they are eligible, they will then gain access to the application. The Foundation primarily supports programs in the Dallas area.
Restrictions: The Foundation does not support the following: organizations that are not 501(c)3 charities; grants to individuals, including sponsorships; private foundations or endowment funds; sectarian or denominational religious organizations; political activities, parties or candidates; veteran organizations; hospitals; fraternal or labor organizations; courtesy advertising, including program books and yearbooks; entertainment events, scholarships or conferences; sporting events or teams; golf tournaments; unrestricted gifts to national or international organizations; travel or tours; or table sponsorships.
Geographic Focus: Texas
Contact: Ann Pomykal; (214) 480-6873; fax (214) 480-6820; giving@ti.com
Internet: http://www.ti.com/corp/docs/csr/giving.shtml
Sponsor: Texas Instruments Foundation
12500 TI Boulevard
Dallas, TX 75266-0199

Textron Corporate Contributions Grants 3712
Textron Inc. was founded in 1923 as a small textile company and has since become one of the world's best known multi-industry companies. Textron focuses philanthropic giving in the following areas: workforce development and education; healthy families/vibrant communities; and sponsorships. In the area of workforce development and education, the company focuses on job-training and employment development (eg., school-to-work, welfare-to-work, job-training for underserved-audiences, literacy, and English-as a-Second-Language programs). In the area of healthy families/vibrant communities, the company focuses on arts and culture (with emphasis on outreach programs that enhance

learning and target low- and moderate-income individuals), community revitalization (eg., affordable housing and economic development in low-income areas), and health and human-service organizations (eg., food pantries, homeless shelters, and services for low-income residents). In the area of sponsorships, the company encourages volunteerism and sponsors worthwhile events that benefit the communities where employees live and work. Textron's grant history has included funding of general-operating costs, capital campaigns, building construction/renovation, equipment acquisition, program development, conferences and seminars, publication, seed money, fellowships, scholarship funds, research, technical assistance, consulting services, and matching funds. Downloadable guidelines (PDF) and application (Word document) are available online. The completed application and required accompanying documentation must be received via mail by the deadline date.
Requirements: Textron targets its giving to nonprofit agencies located in its headquarters state of Rhode Island and those locations where the company has divisional operations. Organizations outside of Rhode Island should contact the Textron company in their area; a listing of Textron businesses along with their contact information can be accessed by clicking the "Contact Us" link at the Textron website.
Restrictions: Textron will review only one request per organization during a 12-month period. Contributions will not be made to the following types of organizations: organizations without 501(c)3 status as defined by the Internal Revenue Service; individuals; political, fraternal or veterans organizations; religious institutions when the grant would support sectarian activities; and organizations that discriminate by race, creed, gender, ethnicity, sexual orientation, disability, age or any other basis prohibited by law.
Geographic Focus: Georgia, Illinois, Kansas, Louisiana, Maryland, Massachusetts, New York, North Carolina, Pennsylvania, Rhode Island, Texas, Germany, Great Britain
Date(s) Application is Due: Mar 1; Sep 1
Samples: The Providence Center, Providence, Rhode Island, $10,000 - to provide training opportunities in receptionist, production, driver, and computer vocations.
Contact: Karen Warfield; (401) 421-2800; fax (401) 457-2225
Internet: http://www.textron.com/about/commitment/corp-giving/
Sponsor: Textron Charitable Trust
40 Westminster Street
Providence, RI 02903

Thelma Braun and Bocklett Family Foundation Grants 3713
The Thelma Braun and Bocklett Family Foundation was established in 1980 to support and promote quality education, cultural, human-services, and health-care programming for underserved populations. Miss Braun was a gifted musician who spent many years playing the organ and piano for area organizations and churches. She never married and had no children of her own, but she truly loved the young people in the area. Special consideration is given to charitable organizations that serve the people of Grayson County, Texas, especially in the areas of arts and education. Applicants must apply online at the grant website. The Thelma Braun and Bocklett Family Foundation has rolling deadlines and decision dates. Applicants are strongly encouraged to do the following before applying: review the downloadable state application procedures at the grant website; review the downloadable online-application guidelines at the grant website; review the foundation's funding history (link is available from the grant website); review the online application questions in advance; and review the list of required attachments. These will generally include: a list of board members, financial statements (audited, reviewed, or compiled by independent auditor); an organization summary; a list of other funding sources; an IRS Determination letter; and other required documents. All attachments must be uploaded in the online application as PDF, Word, or Excel files. Two annual deadlines have been specified: February 1 and August 1.
Requirements: Applicants must have 501(c)3 tax-exempt status.
Restrictions: The foundation does not support requests from individuals, organizations attempting to influence policy through direct lobbying, or any political campaigns.
Geographic Focus: Texas
Date(s) Application is Due: Feb 1; Aug 1
Contact: Debi Allen, Philanthropic Administrative Team Leader; (214) 209-1965 or (214) 209-1370; tx.philanthropic@ustrust.com
Internet: https://www.bankofamerica.com/philanthropic/fn_search.action
Sponsor: Thelma Braun and Bocklett Family Foundation
901 Main Street, 19th Floor
Dallas, TX 75202-3714

Thelma Doelger Charitable Trust Grants 3714
Established in California in 1995, the Thelma Doelger Charitable Trust awards grants in the San Francisco Bay. Giving is primarily aimed at animal welfare, social services, medical centers, and children and youth services. The Trust's major fields of interest include: aging centers and services; animal welfare; boys and girls clubs; children and youth services; higher education; hospitals; human services; museums; and zoos. Grants typically take the form of general purposes and support of operating budgets. Most recently, awards have ranged from $3,000 to $50,000. Though a formal application is required, there are no annual submission deadlines. The initial approach should be by letter or telephone, requesting the application.
Requirements: Nonprofit organizations in the San Francisco Bay area of northern California may submit grant requests.
Restrictions: Individuals are not eligible.
Geographic Focus: California
Amount of Grant: 3,000 - 50,000 USD
Samples: Seton Medical Center, Daly City, California, $35,000 - general operating costs; Marin Humane Society, Novato, California, $50,000 - general operating costs; Curi Odyssey (formerly Coyote Point Museum), San Mateo, California, $50,000 - general operating costs.
Contact: D. Eugene Richard, Trustee; (650) 755-2333
Sponsor: Thelma Doelger Charitable Trust
950 Daly Boulevard, Suite 300
Daly City, CA 94015-3004

Theodore Edson Parker Foundation Grants 3715
The Theodore Edson Parker Foundation's primary goal is to make effective grants that benefit the city of Lowell, Massachusetts, and its residents. Grants are made for a variety of purposes including social services, cultural programs, community development activities, education, community health needs, and urban environmental projects. The Foundation funds specific needs including special programs and projects, capital improvements and equipment purchases, and technical assistance. In his will, Mr. Parker suggested that his trustees give special consideration to the welfare of children, disadvantaged young women, and the elderly. The Foundation is especially committed to assisting these and other underserved individuals including refugees, immigrants, and people of color. Preference is given to applicants who have formulated creative approaches to societal problems and can provide leverage for Foundation funds. The Foundation favors projects that can demonstrate good prospects for continuation after the conclusion of Foundation funding. Interested applicants are welcome to contact staff prior to applying. All applications are accepted online.
Requirements: Eligible applicants are 501(c)3 organizations in Lowell, Massachusetts.
Restrictions: The Foundation does not usually fund the operating expenses of charitable organizations, endowments, or fund deficits. Applicants are limited to one per year. Grant recipients should expect to wait two years before submitting a new request.
Geographic Focus: Massachusetts
Date(s) Application is Due: Jan 15; May 15; Sep 15
Amount of Grant: 10,000 - 50,000 USD
Samples: 911 Electronic Media Arts, Lowell, Massachusetts, $10,000 - support for paid coordinator for small, volunteer-led gallery; Lowell Wish Project, Inc., Lowell, Massachusetts, $20,000 - expansion of the organizaion's Emergency Support Program; and Girls Incorporated for Greater Lowell, Lowell, Massachusetts, $50,000 - renovations to create space for a program serving older girls.
Contact: Philip Hall; (617) 391-3097 or (617) 426-7087; phall@grantsmanagement.com
Internet: http://parkerfoundation.gmafoundations.com/
Sponsor: Theodore Edson Parker Foundation
77 Summer Street, 8th Floor
Boston, MA 02110-1006

The Ray Charles Foundation Grants 3716
Although musician Ray Charles was blind since the age of seven, he always maintained that blindness was not a handicap, but rather that the inability to hear music constituted the truer loss. Because of this philosophy, he began making contributions to the field of hearing impairment and often anonymously contributed to cochlear implants for individuals who couldn't afford such surgeries. In 1986, Mr. Charles founded "The Robinson Foundation for Hearing Disorders" which was later renamed "The Ray Charles Foundation." Convinced of the tremendous need of education for youth, Mr. Charles also directed the foundation to make donations and support institutions and organizations for educational purposes. Over the past 23 years, the Ray Charles Foundation has provided financial donations to various institutions involved in the areas of hearing disorders as well as education. The foundation prefers to fund programs and projects that show promise of strengthening the community beyond the grant period and that offer maximum community impact to achieve long-term results. Types of funding the foundation provides are for capital, program, and core support. Capital funding is available for land, facility, and equipment purchases and renovations or new construction. Renovation and construction projects will be invited only for organizations that have already raised a substantial amount of their fundraising goal. The project must either be in construction or have a firm construction start date. Program funding is available for both new and expansion projects. While these must evidence a viable fundraising and sustainability plan, requests for program development or enhancement activiies will still be considered. Core support funding is available for established and well-managed organizations and programs. Communication with Foundation Directors is discouraged. Applicants to the Ray Charles Foundation should simply fill out and submit the foundation's Grant Application Packet via a mail carrier to the address given. Guidelines, instructions, and links to form are given at the website. Applications are accepted continually throughout the year.
Requirements: 501(c)3 organizations that have been in existence for at least three years and that are incorporated and delivering services to the United States are eligible to apply. Programs must advance the mission of the Foundation in the following program categories: Education, Hearing Disorders, and Culture and the Arts. Educational programs eligible for funding include, but are not limited to, after-school programs and academic activities aimed at college preparation for underprivileged youth, and academic/therapy services for youth who are blind and/or deaf. Hearing Disorder programs eligible for funding include research and treatment in the area of hearing disorders and educational programs and resources for youth with hearing disorders. Culture and Arts programs eligible for funding include institutions and museums whose mission it is to provide musical and cultural education and access to underprivileged and disadvantaged youth.
Restrictions: The Foundation does not fund organizations and programs outside of the United States. Most grants will be limited to one year in duration. Multi-year funding will be considered on an exceptional case by case basis. The Foundation generally does not approve grants to organizations on a continuing annual basis. If an agency applies for a 'one year grant' more than three years in a row, they will be declined and asked to wait at least two years before applying again. More obligations and responsibilities of grant recipients are listed at the website.
Geographic Focus: All States
Contact: Coordinator; (323) 737-8000; info@theraycharlesfoundation.org

Internet: http://www.theraycharlesfoundation.org/GrantQualifications.html
Sponsor: Ray Charles Foundation
2107 W. Washington Boulevard
Los Angeles, CA 90018

Thomas and Dorothy Leavey Foundation Grants 3717
Thomas Leavey and partner John C. Tyler began the Farmer's Insurance Company in 1928. The foundation was begun in 1952 by Leavey and his wife, Dorothy. Thomas Leavey died in 1980 and Dorothy actively led the foundation until her death in 1998. Giving primarily in southern California, the Foundation offers support for: hospitals, medical research, higher and secondary education, Catholic church groups; and provides scholarships to children of employees of Farmers Group.
Requirements: The foundation gives primarily in southern California.
Geographic Focus: California
Amount of Grant: 10,000 - 2,000,000 USD
Contact: Kathleen Leavey McCarthy, Chair; (310) 551-9936
Sponsor: Thomas and Dorothy Leavey Foundation
10100 Santa Monica Boulevard, Suite 610
Los Angeles, CA 90067-4110

Thomas Austin Finch, Sr. Foundation Grants 3718
The purpose of the Finch Foundation is the improvement of the mental, moral and physical well-being of the inhabitants of Thomasville, North Carolina, with emphasis on improving education, improving health related facilities and attracting new business to the community. However, grants are made for a wide variety of charitable causes throughout the greater Thomasville area. In general, grants are awarded for seed money, matching funds and general purposes. Primary consideration is given to projects of a non-recurring nature or to start up funding of limited duration. The foundation meets twice a year to review grant requests. Requests must be submitted by February 15th for the Spring Meeting and October 15th for the Fall Meeting. Application forms are available online. Applicants will receive notice acknowledging receipt of the grant request, and subsequently be notified of the grant declination or approval.
Requirements: Eligible applicants are IRS 501(c)3 non-profit organizations located within the corporate limits of the City of Thomasville, North Carolina or, organizations where the benefits of the grant will inure primarily to the residents of Thomasville. Proposals should be submitted in the following format: completed Common Grant Application Form; an original Proposal Statement; an audited financial report and a current year operating budget; a copy of your official IRS Letter with your tax determination; a listing of your Board of Directors. Proposal Statements (second item in the above Format) should answer these questions: what are the objectives and expected outcomes of this program/project/request; what strategies will be used to accomplish your objective; what is the timeline for completion; if this is part of an on-going program, how long has it been in operation; what criteria will you use to measure success; if the request is not fully funded, what other sources can you engage; an Itemized budget should be included; please describe any collaborative ventures. Prior to the distribution of funds, all approved grantees must sign and return a Grant Agreement Form, stating that the funds will be used for the purpose intended. Progress reports and Completion reports must also be filed as required for your specific grant. All current grantees must be in good standing with required documentation prior to submitting new proposals to any foundation.
Restrictions: Grants are generally not made for typical operational or maintenance-oriented purposes, political purposes, nor to organizations which discriminate on the basis of race, ethnic origin, sexual or religious preference, age or gender.
Geographic Focus: North Carolina
Date(s) Application is Due: Feb 15; Oct 15
Amount of Grant: 1,500 - 100,000 USD
Samples: Memorial United Methodist Church, $30,000--general operating funds; Hospice of Davidson County, Inc., $60,000--construction of inpatient facility; Thomasville City Schools, $24,400--active classroom project.
Contact: Wachovia Bank, N.A., Trustee; grantinquiries6@wachovia.com
Sponsor: Thomas Austin Finch, Sr. Foundation
Wachovia Bank, NC6732, 100 North Main Street
Winston Salem, NC 27150

Thomas C. Ackerman Foundation Grants 3719
The Thomas C. Ackerman Foundation was founded in 1991. By Spring, 1992, substantial assets had been received from the Thomas C. Ackerman Trust, funded as a result of Thomas Ackerman's death on February 13, 1991; the Board of Directors was fully constituted; and the guidelines and objectives for grant making were adopted. The Foundation's areas of interest include: arts and culture; education; and health and human services. To be considered for a grant from the Foundation, an organization must initially submit a Letter of Intent online. Directions for completing the grant application will be provided to an organization, if it is invited to submit one. Awards generally range from $1,000 to $40,000.
Requirements: California nonprofits, primarily in San Diego County, are eligible.
Restrictions: It is a policy of the Foundation not to provide continuous support for any project to the extent that the project becomes dependent upon the Foundation for its continued existence. The Foundation may make multi-year commitments on occasion. Grants are not made to individuals. Generally, the Foundation does not support conferences or symposia. While occasional grant support is given to religious organizations, those grants are made for direct support of nonsectarian educational or service projects and not for projects which are of primary benefit to members of a particular religion or belief or which primarily promote a particular religion or belief. The Foundation will not consider grants relating to human medical or biomedical research.
Geographic Focus: California
Amount of Grant: 1,000 - 40,000 USD
Samples: San Diego Public Library, San Diego, California, $35,000 - community development; San Diego Food Bank, San Diego, California, $15,000 - in support of the Food 4 Kids program; Elementary Institute of Science, San Diego, California, $5,000 - educational expansion.
Contact: Lynne Newman; (619) 741-0113; info@AckermanFoundation.org
Internet: http://www.ackermanfoundation.org/
Sponsor: Thomas C. Ackerman Foundation
3755 Avocado Boulevard, #518
La Mesa, CA 91941-7301

Thomas C. Burke Foundation Grants 3720
The mission of The Thomas C. Burke Foundation is to support charitable organizations that ease human suffering and promote the health of people afflicted by disease, physical weakness, physical disability, and physical injury. The Foundation has a strong interest in programs focused on cancer prevention and treatment. Grants from The Thomas C. Burke Foundation are primarily one year in duration. On occasion, multi-year support is awarded. Applications should be submitted through the online application system by 11:59 p.m. on July 1. The link to the online application system is available at the grant website. Applicants are strongly encouraged to do the following before applying: review the downloadable state application procedures for additional helpful information and clarifications; review the downloadable online-application guidelines at the grant website; review the foundation's funding history (link is available from the grant website); review the online application questions in advance; and review the list of required attachments. These will generally include: a list of board members, financial statements (audited, reviewed, or compiled by independent auditor); an organization summary; a list of other funding sources; an IRS Determination letter; and other required documents. All attachments must be uploaded in the online application as PDF, Word, or Excel files. Applicants will be notified of grant decisions by letter within three to four months after the deadline.
Requirements: Charitable organizations must have 501(c)3 tax-exempt status and serve Bibb County and its surrounding communities. A breakdown of number/percentage of people served by specific counties is required on the online application.
Restrictions: Applicants will not be awarded a grant for more than three consecutive years. The foundation does not support requests from individuals, organizations attempting to influence policy through direct lobbying, or any political campaigns.
Geographic Focus: Georgia
Date(s) Application is Due: Jul 1
Contact: Mark S. Drake, Vice President; (404) 264.1377; mark.s.drake@ustrust.com
Internet: https://www.bankofamerica.com/philanthropic/fn_search.action
Sponsor: Thomas C. Burke Foundation
3414 Peachtree Road, N.E., Suite 1475 GA7-813-14-04
Atlanta, GA 30326-1113

Thomas J. Atkins Memorial Trust Fund Grants 3721
The Thomas J. Atkins Memorial Trust Fund was established in 1946 to support and promote quality educational, human services, and health care programming for underserved populations. The Trust Fund specifically serves the people of Middlesex County, Connecticut. Preference is given to organizations and programs that provide for the relief of human distress and suffering. The deadline for application is October 1. Applicants will be notified of grant decisions by letter within 2 to 3 months after the proposal deadline. Grants from the Atkins Memorial Trust Fund are 1 year in duration.
Requirements: 501(c)3 organizations serving the residents of Middlesex, Connecticut, are eligible to apply.
Geographic Focus: Connecticut
Date(s) Application is Due: Oct 1
Contact: Kate Kerchaert; (860) 657-7016; kate.kerchaert@baml.com
Internet: https://www.bankofamerica.com/philanthropic/grantmaking.action
Sponsor: Thomas J. Atkins Memorial Trust Fund
200 Glastonbury Boulevard, Suite #200, CT2-545-02-05
Glastonbury, CT O6033-4056

Thomas J. Long Foundation Community Grants 3722
Through its Community Grants Program, the Foundation practices responsive grant making and awards grants to charitable organizations in the San Francisco East Bay, and in five selected fields of interest — Arts & Culture, Conservation, Education, Health, and Human Services. The Foundation gives preference to proposals received from organizations that provide direct benefit to low income children and youth, the elderly and disabled persons. Safety net services, and programs focused on economic self-sufficiency are also favored.
Requirements: The Foundation awards grants to selected tax-exempt charitable organizations that have been recognized by the IRS as being described in Section 501(c)3 and 509(a)1 or 509(a)2 of the Internal Revenue Code. The Foundation primarily funds charitable programs and services which are of particular benefit to residents of the East Bay counties of Alameda and Contra Costa. Organizations that are considering a proposal for grant support should have an established presence in these communities including an office and/or dedicated staff permanently assigned to the East Bay. Proposals which do not focus on the East Bay are unlikely to be funded and it is rarely worthwhile to submit a proposal.
Restrictions: The Foundation does not ordinarily award grants for the following: 509(a)3 supporting organizations; government entities; individuals; advocacy or influencing public policy; endowments; international grants; loan repayments; research or studies; capital campaigns (invitation only); projects or programs that have already taken place; or, national organizations headquartered outside of the San Francisco Bay Area. The Thomas J. Long Foundation uses a formal grant process to administer its grant making program. All

forms used in the grant process are available only through the Foundation's on-line grants management system. Requests for funding or reports submitted in any other manner will not be accepted. Grant proposals are accepted on a continuous basis throughout the year.
Geographic Focus: California
Contact: Nancy Shillis, Program Officer; (925) 944-3800; fax (925) 944-3573
Internet: http://www.thomasjlongfdn.org/?q=grants
Sponsor: Thomas J. Long Foundation
2950 Buskirk Avenue, Suite 160
Walnut Creek, CA 94597

Thomas Jefferson Rosenberg Foundation Grants — 3723
The Thomas Jefferson Rosenberg Foundation was established in New York in 1989, with a primary purpose of supporting higher education, Jewish agencies, and human service organizations. The Foundation's primary fields of interest currently include: arts and culture; diseases; human services; Judaism; and undergraduate education. There are no annual application deadlines, and the Foundation does not accept uninvited applications. Interested parties should begin by contacting the office directly. Recent awards have ranged from $250 to $100,000.
Requirements: 501(c)3 organizations serving the residents of New York, Massachusetts, Connecticut, the District of Columbia, Texas, Florida, Kentucky, and California have been the most recent recipients of grants.
Geographic Focus: California, Connecticut, District of Columbia, Florida, Kentucky, Massachusetts, Texas
Amount of Grant: 250 - 100,000 USD
Samples: Cornell Center for Integrated Medicine, New York, New York, $100,000 - general operating support (2014); Friends of Israel Defense, New York, New York, $45,000 - general operating support (2014); High Adventure Ministries, Louisville, Kentucky, $50,000 - general operating support (2014).
Contact: Henry Rosenberg, President; (212) 869-1490
Sponsor: Thomas Jefferson Rosenberg Foundation
5 Hanover Square, 23rd Floor
New York, NY 10004-2614

Thomas Sill Foundation Grants — 3724
Mr. Sill lived his entire life in Winnipeg and practiced as a chartered accountant for many years. He was an astute investor who built a fortune which became the basis for the Thomas Sill Foundation. The foundation provides encouragement and financial support to qualifying Manitoba organizations that strive to improve the quality of life in the province. The foundation awards grants in the following areas of interest: Responses to Community (agencies addressing poverty, women's shelters, qualifying daycares, mentally and physically challenged people, and community well-being); Health (eye care, palliative care, mental illness); Education (students at risk, including adults); Arts and Culture; Heritage (museums, architecture, projects); and, Environment (water issues). Grants awarded may be capital, operating or project in nature.
Requirements: Registered charities may obtain an application form by phoning the foundation office, at which time a preliminary discussion will determine eligibility. There are no deadlines by which applications must be submitted. Applicants should allow four months, from the submission of a request, to receive a response.
Restrictions: Successful applicants must wait two years before submitting another request.
Geographic Focus: Canada
Amount of Grant: 1,000 - 50,000 CAD
Contact: Hugh Arklie; (204) 947-3782; fax (204) 956-4702; hugha@tomsill.ca
Internet: http://thomassillfoundation.com/guidelines/
Sponsor: Thomas Sill Foundation
206-1661 Portage Avenue
Winnipeg, MB R3J 3T7 Canada

Thomas Thompson Trust Grants — 3725
The Thompson Trust limits its distribution of funds to organizations located in Windham County, Vermont, (primarily the Town of Brattleboro) or in Dutchess County, New York (primarily the Town of Rhinebeck) which predominately serve residents located in those areas. The Will of Thomas Thompson was executed in 1867 and defined his charitable purposes rather narrowly as was the practice with 19th Century philanthropy. By successive court decrees in the 20th Century, the Trustees are now authorized to make grants to charitable organizations whose work and purposes promote health, education or the general social or civic betterment in the stated geographical areas; but the Trustees will continue to place particular emphasis on healthcare and other social services. The Trustees generally meet on the fourth Thursday in January, April, July and October. Applications are due by the first day of the month of the meeting.
Requirements: Organizations must have operated for three consecutive years before applying and be located in Windham County, VT, with preference given to Brattleboro; or Duchess County, New York, with preference given to Rhinebeck.
Restrictions: Grants are not made for general operating support, seed money, endowment purposes, or loans.
Geographic Focus: New York, Vermont
Date(s) Application is Due: Jan 1; Apr 1; Jul 1; Oct 1
Amount of Grant: 1,000 - 20,000 USD
Contact: Susan T. Monahan; (617) 951-1108; fax (617) 542-7437; smonahan@rackemann.com
Internet: http://www.cybergrants.com/thompson/grant.html
Sponsor: Thomas Thompson Trust
160 Federal Street, 13th Floor
Boston, MA 02110-1700

Thomas W. Bradley Foundation Grants — 3726
Established in 1976 in Maryland, the Thomas W. Bradley Foundation offer grants in the greater metropolitan Baltimore, Maryland, area. Funding is limited to organizations which work with or benefit mentally or physically handicapped children. Primary fields of interest include: children, children's services, and the developmentally disabled. Funding comes in the form of general operating support. Application forms are required, and applicants should submit the following: name, address and phone number of organization; copy of IRS Determination Letter; a brief history of the organization and description of its mission; a listing of board of directors, trustees, officers and other key people and their affiliations; a detailed description of project and amount of funding requested; and a copy of the current year's organizational budget and project budget. There are no specific deadlines with which to adhere.
Requirements: Only 501(c)3 organizations that serve the residents of Maryland should apply, and the Foundation gives preference in the Baltimore metropolitan area.
Restrictions: No grants, scholarships, fellowships, prizes, or similar benefits are made to individuals.
Geographic Focus: Maryland
Amount of Grant: 3,000 - 6,000 USD
Contact: Robert L. Pierson, Trustee; (410) 821-3006; fax (410) 821-3007; info
Sponsor: Thomas W. Bradley Foundation
305 W. Chesapeake Avenue, Suite 308
Towson, MD 21204-4440

Thompson Charitable Foundation Grants — 3727
Established in 1987, the Foundation supports organizations involved with education, Christian organizations, Autism research, health, human services, youth, including funding for capital and building improvements for human service organizations and educational institutions. Giving is limited to Bell, Clay, Laurel, and Leslie counties, Kentucky; Anderson, Blount, Knox, and Scott counties, Tennessee; and Buchanan and Tazewell counties, Virginia.
Requirements: Nonprofits operating in Bell, Clay, Laurel, and Leslie counties, Kentucky; Anderson, Blount, Knox, and Scott counties, Tennessee; and Buchanan and Tazewell counties, Virginia are eligible to apply. Initial contact should be a letter, no more then 2 pages long. The letter should include: a statement of the problem, the project will address; a detailed description of project, and amount of funding requested.
Restrictions: No support for religious or political organizations, budget deficits or endowments.
Geographic Focus: Kentucky, Tennessee, Virginia
Date(s) Application is Due: Mar 31; Sep 1
Amount of Grant: 2,000 - 500,000 USD
Contact: Debbie Black; (865) 588-0491; fax (865) 588-4496; debbie@cf.org
Sponsor: Thompson Charitable Foundation
800 South Gay Street, Suite 2021
Knoxville, TN 37929-9710

Thomson Reuters / MLA Doctoral Fellowships — 3728
The purpose of the Thomson Reuters/MLA Doctoral Fellowship is to foster and encourage superior students to conduct doctoral work in health-sciences librarianship or information science by providing support to individuals who have been admitted to candidacy. The award was originally established in 1986 by the Institute of Scientific Information (ISI, which is now a part of Thomson Reuters) and is administered by the Medical Library Association (MLA). The award is granted bi-annually in even-numbered years. The fellowship pays $2,000 in one payment toward the following types of expenses: project expenses; related travel expenses; or augmentation of another larger, separately-funded project relevant to health-science librarianship and which is part of the requirements of a doctoral degree. Guidelines and application materials are available at the website. All materials and letters of support must be received by MLA via email, fax, or mail by December 1. email applications and documents will be accepted as PDF or MS Word files only. Founded in 1898, the Medical Library Association (MLA) is a nonprofit, educational organization of more than 1,100 institutions and 3,600 individual members in the health sciences information field committed to educating health-information professionals, supporting health-information research, promoting access to the world's health-sciences information, and working to ensure that the best health information is available to all.
Requirements: Applicants must satisfy the following eligibility criteria: be graduates of an American Library Association (ALA)-accredited school of library science; be candidates in a Ph.D. program with emphasis on biomedical and health-related information science; and be citizens of or have permanent residence status in the United States or Canada. Preference will be given to applicants who have at least 75% of their course work completed and dissertation prospectus either approved or in the approval process.
Restrictions: Past recipients of the Thomson Reuters / MLA Doctoral Fellowship are ineligible. The following types of expenses are not supported: tuition fees; tuition-related expenses; and living expenses.
Geographic Focus: All States, Canada
Date(s) Application is Due: Dec 1
Amount of Grant: Up to 2,000 USD
Contact: Carla Funk, Program Contact; (312) 419-9094, ext. 14; fax (312) 419-8950; mlapd2@mlahq.org or grants@mlhq.org
Internet: http://www.mlanet.org/awards/grants/index.html
Sponsor: Medical Library Association
65 East Wacker Place, Suite 1900
Chicago, IL 60601-7246

Thoracic Foundation Grants 3729
The major emphasis of the Thoracic Foundation of Boston is to offer support for medical research in thoracic diseases. Contributions are also made to hospitals providing service to thoracic and cardiac cases. Fields of interest include: asthma research; cancer research; children and youth services; heart and circulatory diseases; specialty hospitals; medical school education; and smoking cessation. In addition, funds are awarded to assist with publications, conferences, and seminars in the same field. Grants are awarded thoughout Massachusetts, but primarily in the Boston metropolitan area. A letter of application may be submitted at any time, offering a detailed description of the project and amount of funding requested. Awards range from $5,000 to $75,000.
Geographic Focus: District of Columbia, Massachusetts
Sample: Action on Smoking and Health, Washington, D.C., $5,000 - funding for non-smokers' rights; Boston Biomedical Research Institution, Boston, Massachusetts, $75,000 - research and development related to thoracic disease; Beth Israel Deaconess Medical Center, Boston, Massachusetts, $35,000 - experimental radiology program.
Contact: Bradley R. Cook, Trustee, c/o Taylor, Ganson, and Perrin, LLP; (617) 951-2777; fax (617) 951-0989
Sponsor: Thoracic Foundation
160 Federal Street, 20th Floor
Boston, MA 02110-1700

Thorman Boyle Foundation Grants 3730
The Thorman Boyle Foundation was established in California in 2000, with its primary fields of interested designated as: the arts; human services; and youth development. There are no formal application materials required, and interested parties should forward a letter of application stating their organization's purpose, as well as overall budgetary needs. Most recently, grant awards have ranged from $100 to a maximum of $2,000. No annual deadlines for submission have been identified.
Requirements: Though giving is primarily limited to 501(c)3 organizations either located in, or serving the residents of, California, the Foundation has, on occasion, awarded grants outside of this region.
Geographic Focus: All States
Amount of Grant: 100 - 2,000 USD
Samples: Fresh Lifelines for Youth, San Jose, California, $2,000 - general operations; Arts in Action, Menlo Park, California, $500 - general operations; Heifer International, Little Rock, Arkansas, $1,000 - general operations.
Contact: Mary E. Boyle, President; (650) 856-7445 or (650) 799-4300
Sponsor: Thorman Boyle Foundation
P.O. Box 2757
Cupertino, CA 95015

Thrasher Research Fund Early Career Awards 3731
The Fund recognizes that young investigators may find it difficult to remain in pediatric research because of a lack of funding. Therefore, the purpose of this program is to encourage the development of medical research in child health by awarding small grants to new researchers, helping them gain a foothold in this important area. The Fund is open to a variety of research topics important to children's health. Both incidence and severity are considered when determining the significance of a problem being studied. The program is particularly interested in applicants that show great potential to impact that field of children's health through medical research. Both an applicant's aptitude and inclination toward research are considered. The quality of the mentor and the mentoring relationship are also considered to be important predictors of success. There are three funding cycles per year for Early Career Award Program grants.
Requirements: Eligible applicants are: physicians who are in a residency/fellowship training program, or who completed that program no more than one year before the date of submission of the concept paper; post-doctoral researchers who received the doctoral level degree no more than three years prior to the date of submission of the concept paper. Each project needs to be under the guidance of a mentor. The qualifications and experience of the mentor will be considered in the evaluation of the application. A mentor may have only one Thrasher Research Fund Early Career Awardee at a time. There are no restrictions with regard to citizenship. The Fund is open to applications from institutions both inside and outside the United States.
Restrictions: A new investigator who holds a National Institutes of Health (NIH) K award or a Clinical and Translational Science Award (CTSA) is not eligible to apply for the Thrasher Early Career Award.
Geographic Focus: All States, All Countries
Date(s) Application is Due: Mar 30; Jul 27; Nov 30
Contact: Megan Duncan; (801) 240-4720; DuncanME@thrasherresearch.org
Internet: http://www.thrasherresearch.org/sites/www_thrasherresearch_org/Default.aspx?page=231
Sponsor: Thrasher Research Fund
68 S. Main Street, Suite 400
Salt Lake City, UT 84101

Thrasher Research Fund Grants 3732
The Thrasher Research Fund provides grants for pediatric medical research. Because significant solutions for many children's health problems remain undiscovered, the Fund invites a broad array of applications. The Thrasher Research Fund seeks to foster an environment of creativity and discovery aimed at finding solutions to children's health problems. The Fund awards grants for research that offers substantial promise for meaningful advances in prevention and treatment of children's diseases, particularly research that offers broad-based applications.
Requirements: Principal Investigators must be qualified in terms of education and experience to conduct research. A doctoral-level degree is required. There are no citizenship or residency requirements. The Fund is open to applications from institutions both inside and outside the United States.
Restrictions: The fund does not award grants for educational programs; general operating expenses; general bridge funding for incomplete projects; construction or renovation of buildings or facilities; loans, student aid, scholarships, tuition; or, support of other funding organizations. Grants generally are not awarded for conferences, workshops, or symposia. Proposals in the areas of research on human fetal tissue or behavioral science research will not be considered.
Geographic Focus: All States, All Countries
Amount of Grant: 100,000 - 350,000 USD
Contact: Megan Duncan; (801) 240-4720; DuncanME@thrasherresearch.org
Internet: http://www.thrasherresearch.org/sites/www_thrasherresearch_org/Default.aspx?page=28
Sponsor: Thrasher Research Fund
68 S. Main Street, Suite 400
Salt Lake City, UT 84101

Tifa Foundation Grants 3733
The Tifa Foundation has two grant making mechanism. The first, known as Pro-Active Grants, is through a proactive approach in which Tifa assumes an active role in working together with other organizations that correspond with Tifa's mission. The second, known as Open Grants, is through the endorsement of program proposals that conform to Tifa's strategic direction and priority/supporting programs.
Requirements: Grants are awarded to: organizations with proposed program activities that are in accordance with Tifa's vision and mission; organizations that possess unblemished an track record during their involvement with Tifa as well as with other donor institutions; organizations with legal entities; organizations that cooperate with government agencies; and institutions under the control of a particular university.
Restrictions: Grants are not eligible for: individuals, including proposals for scholarships, trips, conferences and others; all government agencies, legislative and judiciary bodies, political parties, military or those with affiliation/direct relationship with the aforementioned bodies; all organizations with profit-seeking interests; program proposals where the end product or outcome is sold to the public (except upon Tifa's approval); organizations that are still engaged in cooperation with Tifa for the ongoing period except upon prior approval.
Geographic Focus: All States
Date(s) Application is Due: Jan 15; Jul 15
Contact: Yayasan Tifa, (62) 021 829 2776; fax (62) 021 837 83648; public@tifafoundation.org
Internet: http://www.tifafoundation.org/
Sponsor: Tifa Foundation
Jl. Jaya Mandala II No. 14 E
Menteng Dalam, South Jakarta, 12870 Indonesia

Tipton County Foundation Grants 3734
Groups engaged in charitable projects or programs are invited to apply for a Tipton County Foundation Grant. Grant are made only to groups, organizations, or projects: nonprofit and charitable; that document their responsible fiscal management and adequate accounting procedures; proposed in writing by the organization's governing body. Proposals are reviewed quarterly on the first Monday of January, April, August, and October.
Requirements: Applicants must submit a letter of intent before applying for a grant with the Foundation. The LOI must include a detailed description of the organization, what the organization would like to do, who will benefit, and its cost, along with a timeline of the project, and a copy of the IRS tax exempt letter. If the organization submits a proposal, it must include its rationale, collaboration, and organizational strength in completing the project. The following attachments must be included with the proposal: a detailed list of the board of directors; evidence that the proposal has the board's approval; year-end and current financial statements; organizational budget; and a detailed project budget that describes revenue, expenses, in-kind revenue/expenses, and whether the revenue is pending, received, or approved.
Restrictions: The Foundation does not make grants to individuals or for the following: political purposes; programs or equipment that were committed to prior to the grant proposal period; debt reduction; annual appeals or membership contributions; building of endowments, unless located at the Tipton Foundation; projects that benefit only a few persons; travel expenses for individuals or groups.
Geographic Focus: Indiana
Contact: Lori Tragesser; (765) 675-1941; fax (765) 675-8488; tcf@tipton.org
Internet: http://www.tiptoncf.org/grants.htm
Sponsor: Tipton County Foundation
1020 West Jefferson Street
Tipton, IN 46072

TJX Foundation Grants 3735
The primary mission of the foundation is to support programs that provide basic-need services to disadvantaged children, women, and families in communities in which TJX does business. The TJX Foundation is currently focused on supporting 501(c)3 charities that conform to the following giving guidelines: Civic/Community - Emphasis is on programs that teach disadvantaged persons independent living skills and that work to improve race/cultural relations; Domestic Violence Prevention - Support will target immediate emergency services and shelter accommodations for victims and family members affected by abusive situations as well as programming that works to break the cycle of violence; Education - Support will target programs that provide academic and vocational opportunities for the disadvantaged, including early intervention, mentoring, tutoring, GED and college coursework, as well as programs that teach English; Health - Support will target programs that provide early and comprehensive prenatal services and healthy-baby education and, in

some cases, research; Social Services - support for programs for disadvantaged children and families, including those that provide food and other basic needs, counseling and family support, adoption services, and youth development--also programs that provide direct services to those with mental or physical impairments. Emergency assistance programs, such as disaster relief intervention projects, are also supported through the Foundation.
Requirements: IRS 501(c)3 tax-exempt organizations are eligible. Support is focused on organizations in communities where one or more of TJX's divisions operates a home office, a store or distribution facility, and whose programs help to: promote strong families, provide emergency shelter, enhance education/job readiness, and build community ties. A required Eligibility Questionnaire can be found at the sponsor's website.
Restrictions: The foundation will not fund international organizations, other giving organizations, prison populations/offenders and ex-offenders, capital campaign requests, cash reserves, computer purchases, conferences/seminars, consultant fees/salaries, conventions, education loans, endowments, fellowships, films/photography, individual requests, new construction, political organizations, programs in operation for less than 12 months, publications, public policy research/advocacy, renovations, building expansions, salary-only requests, seed money/start-up costs, training money/stipend, travel grants/transportation, and unrestricted grants.
Geographic Focus: All States
Date(s) Application is Due: Mar 3; Jul 7
Contact: Christine Strickland; (774) 308-3199; tjx_foundation@tjx.com
Internet: http://www.tjx.com/corporate_community_foundation.asp
Sponsor: TJX Foundation
770 Cochituate Road, Route 300-1BN
Framingham, MA 01701

Todd Brock Family Foundation Grants 3736
Established in Texas in 2007, the Todd Brock Family Foundation offers support for educational programs, children, athletics, research, and community projects, as well Protestant agencies and churches. A formal application is required, and applicants should forward the entire proposal to the office. The Foundation rarely offers funding outside of Texas. There are no deadlines for submitting a completed proposal, and grants have most recently ranged from $300 to $250,000.
Geographic Focus: Texas
Contact: Todd O. Brock, President; (409) 833-6226; fax (409) 832-3019
Sponsor: Todd Brock Family Foundation
1670 E Cardinal Drive, P.O. Box 306
Beaumont, TX 77704-0306

Tommy Hilfiger Corporate Foundation Grants 3737
The foundation gives to children's programs nationally in areas of education, health, human services, and the arts, with the main focus being on education. Proposals for educational and cultural programs must address the following priorities: target K-12 and college students; expose students to career opportunities; develop skills in new technologies; leverage teacher, administrator, parental, and community involvement; include hands-on program activities; lead to comprehensive, systemic change on a regional and/or national basis; involve collaborative partnerships; demonstrate capacity to gain continuing support; result in dissemination and replication of lessons learned; have broad and positive impact on diverse populations with a special emphasis on women, minorities, and at-risk students; and develop evaluation component with measurable results Proposals that are health-related should be centered on the following: medical centers or hospitals; health education; and innovative therapy programs.
Requirements: 501(c)3 nonprofit organizations are eligible. Grant requests must be made in writing. Proposal seekers are advised to submit a three- to five-page proposal including a detailed budget (all pages must be numbered). A separate cover letter summarizing the proposal request is also required.
Restrictions: Grants are not made to/for individuals; political parties, campaigns, or causes; endowment funds or campaigns; film, video, television, and radio projects; or any organization that discriminates on the basis of race, creed, gender, or national origin.
Geographic Focus: All States
Date(s) Application is Due: Apr 1; Oct 1
Amount of Grant: 10,000 - 30,000 USD
Contact: Program Officer; (212) 840-8888
Internet: http://companyinfo.tommy.com/#/tommy_foundation/grant_guidelines
Sponsor: Tommy Hilfiger Corporate Foundation
601 W 26th Street, 6th Floor
New York, NY 10001-1101

Topeka Community Foundation Kansas Blood Services Scholarships 3738
The Kansas Blood Services Fund with the Topeka Community Foundation provides scholarships for individuals in the medical field of blood banking or a related field. The scholarship can be used to cover expenses related to attending a professional conference or meeting (including conference fees, hotel, meals and transportation).
Requirements: Residents of Nemaha, Riley, Pottawatomie, Jackson, Jefferson, Douglas, Shawnee, Wabaunsee, Osage and Lyon counties in Kansas are eligible to apply.
Geographic Focus: Kansas
Contact: Marsha Pope; (785) 272-4804; pope@topekacommunityfoundation.org
Internet: http://www.topekacommunityfoundation.org/newsarticle.cfm?articleid=1002 3450&ptsidebaroptid=0&returnto=page31525.cfm&returntoname=Grant Informatio n&siteid=1897&pageid=31520&sidepageid=31525
Sponsor: Topeka Community Foundation
5431 SW 29th Street, Suite 300
Topeka, KS 66614

Topfer Family Foundation Grants 3739
The mission of the foundation (TFF) is to fund programs and organizations that connect people to the tools and resources they need to build self-sufficient and fulfilling lives. Programs areas of interest include child abuse prevention and treatment; youth enrichment; job training and support services; children's health; and aging in place. The Topfer family is keenly interested in understanding the unique needs of the communities in which they invest foundation resources. Therefore, contributions are limited to the communities in which the family resides: the greater Austin, TX, and the greater Chicago, IL, metropolitan areas. In Illinois, preference is given to organizations serving Du Page and Cook Counties. Note to Chicago area applicants: The foundation is not accepting unsolicited applications in the Job Training or Children's Health program areas from the Chicago area at this time. Guidelines and application are available online.
Requirements: 501(c)3 nonprofit organizations in Illinois and Texas are eligible.
Restrictions: The foundation does not generally support: grants or loans to individuals; loans to charitable organizations; more than 20 percent of an organization's operating budget; administrative costs; organizations that exclude participants based on race or religion; the purchase of dinner, gala, or raffle tickets; school fundraisers or events (including sports and other extracurricular activities); tax-generating entities (municipalities, school districts, etc.) for services within their normal responsibilities; public or private educational institutions for recurrent operating, administrative, or capital expenses (i.e., acquisition, construction, improvement, and maintenance of buildings and equipment), or scholarships that subsidize existing funding base; academic or scientific research; advertising or marketing efforts; political campaigns or purposes; or any one program more than once a calendar year.
Geographic Focus: Illinois, Texas
Samples: Foundation Communities (Austin, TX)--to develop efficiency apartments for people with very low incomes, most of whom are homeless or in imminent danger of becoming homeless, $100,000; Bridge Communities (Du Page County, IL)--to support after school tutoring and educational enrichment for at-risk children, $20,000; Dress for Success (Travis County, TX)--to provide clothing for interviews for women in search of employment, $5000; Community Care Options (Berwyn, IL)--to support case management services for seniors, $50,000.
Contact: Grants Administrator; (866) 897-0298 or (512) 329-0009
Internet: http://www.topferfamilyfoundation.org
Sponsor: Topfer Family Foundation
5000 Plaza on the Lake, Suite 170
Austin, TX 78746

Tourette Syndrome Association Post-Doctoral Fellowships 3740
The Tourette Syndrome Association is interested in supporting any research that has the potential to contribute to the understanding of Tourette Syndrome, including genetics, pathogenesis, pathophysiology, clinical treatments and animal model development. Post-doctoral research fellowships of up to $40,000 for one year are awarded. Applications are evaluated on the following information: candidate's qualifications and objectives; candidate's experience relevant to the project; methodology; significance and/or relevance to the TS field; percentage of time to be devoted to project; adequacy and availability of research facilities and other project support; and the ability to complete the project in a timely manner. The application and deadline information are available at the website.
Requirements: Candidates must have a M.D., Ph.D. or equivalent. Previous experience in the field of movement disorders is desirable, but not essential. International applicants are welcome.
Geographic Focus: All States, All Countries
Amount of Grant: Up to 40,000 USD
Contact: Heather Cowley, Fellowship Administrator; (718) 224-2999, ext. 247 or (718) 224-2999, ext. 222; fax (718) 279-9596; heather.cowley@tsa-usa.org
Internet: http://tsa-usa.org/research.html
Sponsor: Tourette Syndrome Association
42-40 Bell Boulevard
Bayside, NY 11361-2820

Tourette Syndrome Association Research Grants 3741
The Tourette Syndrome Association (TSA) is interested in receiving applications from any research study that may contribute to the understanding of Tourette Syndrome, including genetics, pathogenesis, pathophysiology, clinical treatments and animal model development. Research grants of up to $75,000 for one year or up to $150,000 for two years are awarded. The TSA will consider reasonable requests for the following: salaries (NIH salary caps apply); laboratory supplies and other research-related expenses; travel expenses for investigators (up to $1,500 per year); research equipment up to $5,000; and indirect costs of up to 10% may be included. Applicants are evaluated on the following information: candidate's qualifications and objectives; their relevant experience to the project; methodology; significance/relevance to the TS field; percentage of time to be devoted to the project; adequacy of research facilities and other project support; and the ability to complete the project in a timely manner. For preliminary screening, a pre-proposal briefly describing the scientific basis and relevance of the proposed project is required. Preproposal applications are available at the website. Applicants for grants involving clinical treatment are encourage to read the posted clinical studies documents at the website. Payment schedules will vary according to one year or two year awards. Applicants invited to submit a full research proposal will be contacted with instructions.
Requirements: Investigators are required to have an advanced degree, such as a Ph.D., M.D., or equivalent. International applicants are welcome to apply. Investigators for nonprofit and for-profit organizations may apply. Previous experience in the field of movement disorders is desirable, but not essential.
Geographic Focus: All States, All Countries
Date(s) Application is Due: Nov 1
Amount of Grant: 75,000 - 150,000 USD

Contact: Heather Cowley, TSA Research Grant Coordinator; (718) 224-2999, ext. 247 or (718) 224-2999, ext. 222; fax (718) 279-9596; heather.cowley@tsa-usa.org
Internet: http://tsa-usa.org/research.html
Sponsor: Tourette Syndrome Association
42-40 Bell Boulevard
Bayside, NY 11361-2820

Toyota Motor Engineering & Manufacturing North America Grants   3742
Toyota Motor Engineering & Manufacturing North America oversees the various North American manufacturing operations of Toyota Motor Corporation. Funding from this facility is centered in the tri-state region of Greater Cincinnati, Northern Kentucky and southeastern Indiana, primarily supporting education, environment and safety, but also funding civic groups, health and human services, arts and culture. TEMA prefers to support programs, rather than events. Grants are provided to support the development and implementation of programs that generally range from $50,000 to $200,000. Submission deadlines for applications are January 1, April 1, July 1, October 1. The review process can take up to six months.
Requirements: Communities served must be in Greater Cincinnati, Northern Kentucky or Southeast Indiana.
Restrictions: EMA does not make grants for publications, lobbying activities, advertising, capital campaigns or endowments. Individuals are ineligible to apply. Toyota will not make grants to the following types of organizations: those not recognized as 501(c)3 by the Internal Revenue Service; those that practice discrimination by race, creed, color, sex, age or national origin; those that serve only their own memberships, such as fraternal organizations, labor organizations or religious groups; or political parties or candidates.
Geographic Focus: Indiana, Kentucky, Ohio
Date(s) Application is Due: Jan 1; Apr 1; Jul 1; Oct 1
Amount of Grant: 50,000 - 200,000 USD
Contact: Grants Administrator; (859) 746-4000; fax (859) 746-4190
Internet: http://www.toyota.com/about/philanthropy/guidelines/index.html#tmmk
Sponsor: Toyota Motor Engineering & Manufacturing North America
25 Atlantic Avenue
Erlanger, KY 41018

Toyota Motor Manufacturing of Alabama Grants   3743
Established in 2001, Toyota Motor Manufacturing of Alabama manufactures V6 and V8 engines for light trucks, with operations include machining and assembly. TMMAL believes in becoming an integral part of the community by improving the quality of life where its team members live and work. TMMAL provides funding to education, health and human services, civic affairs, arts and culture, and environmental organizations. TMMAL prefers to support programs that are sustainable, diverse, and have an educational focus. Grants are provided to support the development and implementation of programs that generally range from $50,000 to $200,000. Applications are reviewed quarterly, in May, August, November, and February. Submission deadlines for applications are April 15, July 15, October 15, and January 15. The review process can take up to six months.
Requirements: Applicant organizations must have 501(c)3 tax-exempt status, be located within or serve population(s) of Madison County, Alabama, and present a proposal that satisfies the mission, guidelines and limitations of the corporation.
Restrictions: The Toyota Motor Manufacturing of Alabama does not make grants for publications, lobbying activities, advertising, capital campaigns or endowments. Individuals are ineligible to apply. Toyota will not make grants to the following types of organizations: those not recognized as 501(c)3 by the Internal Revenue Service; those that practice discrimination by race, creed, color, sex, age or national origin; those that serve only their own memberships, such as fraternal organizations, labor organizations or religious groups; or political parties or candidates.
Geographic Focus: Alabama
Date(s) Application is Due: Jan 15; Apr 15; Jul 15; Oct 15
Amount of Grant: 50,000 - 200,000 USD
Contact: Grants Administrator; (256) 746-5000; fax (256) 746-5906
Internet: http://www.toyota.com/about/our_business/engineering_and_manufacturing/tmmal/
Sponsor: Toyota Motor Manufacturing of Alabama
1 Cottonvalley Drive
Huntsville, AL 35810

Toyota Motor Manufacturing of Indiana Grants   3744
Toyota Motor Manufacturing of Indiana offers funding in support of residents from the Indiana counties of Daviess, Dubois, Gibson, Knox, Pike, Posey, Spencer, Warrick, and Vanderburgh. Counties in Illinois include Wabash and White, as well as the Kentucky counties of Daviess and Henderson. TMMI supports a variety of programs, including youth and education, health and human services, civic and community, the environment, and arts and culture. Grants are provided to support the development and implementation of programs that generally range from $50,000 to $200,000. Submission deadlines for applications are February 15, May 15, August 15, and November 15, with notification of results by the end of the month following the deadline.
Requirements: Applicant organizations must have 501(c)3 tax-exempt status, and be located within or serve population(s) of: Indiana, specifically Daviess, Dubois, Gibson, Knox, Pike, Posey, Spencer, Warrick, and Vanderburgh counties; Illinois, specifically Wabash and White counties; or Kentucky, specifically Daviess and Henderson ountiesdoes not make grants for publications, lobbying activities, advertising, capital campaigns or endowments. Individuals are ineligible to apply. Toyota will not make grants to the following types of organizations: those not recognized as 501(c)3 by the Internal Revenue Service; those that practice discrimination by race, creed, color, sex, age or national origin; those that serve only their own memberships, such as fraternal organizations, labor organizations or religious groups; or political parties or candidates. . Applicants must present a proposal that satisfies the mission, guidelines and limitations of the corporation.
Restrictions: Toyota Motor Manufacturing of Indiana does not make grants for publications, lobbying activities, advertising, capital campaigns or endowments. Individuals are ineligible to apply. Toyota will not make grants to the following types of organizations: those not recognized as 501(c)3 by the Internal Revenue Service; those that practice discrimination by race, creed, color, sex, age or national origin; those that serve only their own memberships, such as fraternal organizations, labor organizations or religious groups; or political parties or candidates.
Geographic Focus: Illinois, Indiana, Kentucky
Date(s) Application is Due: Feb 15; May 15; Aug 15; Nov 15
Amount of Grant: 50,000 - 200,000 USD
Contact: Grants Administrator; (812) 387-2000 or (812) 387-2266; fax (812) 387-2002
Internet: http://www.toyota.com/about/philanthropy/guidelines/index.html#limitations
Sponsor: Toyota Motor Manufacturing of Indiana
4000 Tulip Tree Drive
Princeton, IN 47670

Toyota Motor Manufacturing of Kentucky Grants   3745
Toyota Motor Manufacturing of Kentucky proves its commitment to the community, as well as to the state, through both monetary contributions and personal involvement of TMMK team members. Besides being a major contributor to United Way of the Bluegrass, which serves eight central Kentucky counties, many employees are members of and serve on community organization boards. Its funding priorities include: the committed to the education of people of all ages - in particular, TMMK participates in educational programs that will help ensure the success of Kentucky's reform related programs; health and human services, by supporting the advancement of physical and mental health for people of all ages; arts and culture, by preservation and advancement of the arts and culture, particularly for our children; the environment, by supporting a variety of efforts to provide for the education, sustainability and preservation of our environment; civic and community progress, by helping make a difference by supporting groups that address local and state issues as well as provide leadership programs for developing human capital; and minorities and diversity, by supporting the advancement and growth of opportunity, inclusion, respect, equality and justice for all people. Grants are provided to support the development and implementation of programs that generally range from $50,000 to $200,000. Submission deadlines for applications are February 1, May 1, August 1, and November 1.
Requirements: Applicant organizations must have 501(c)3 tax-exempt status, and be located within or serve population(s) of any county in Kentucky, with the exception of Boone, Kenton and Campbell. Applicants must also present a proposal that satisfies the mission, guidelines and limitations of the corporation.
Restrictions: Toyota Motor Manufacturing of Kentucky does not make grants for publications, lobbying activities, advertising, capital campaigns or endowments. Individuals are ineligible to apply. Toyota will not make grants to the following types of organizations: those not recognized as 501(c)3 by the Internal Revenue Service; those that practice discrimination by race, creed, color, sex, age or national origin; those that serve only their own memberships, such as fraternal organizations, labor organizations or religious groups; or political parties or candidates.
Geographic Focus: Kentucky
Date(s) Application is Due: Feb 1; May 1; Aug 1; Nov 1
Amount of Grant: 50,000 - 200,000 USD
Contact: Grants Administrator; (502) 868-2000; fax (502) 868-3060
Internet: http://www.toyotageorgetown.com/comm2.asp
Sponsor: Toyota Motor Manufacturing of Kentucky
1001 Cherry Blossom Way
Georgetown, KY 40324

Toyota Motor Manufacturing of Mississippi Grants   3746
Toyota Motor Manufacturing of Mississippi supports sustainable and diverse programs and organizations focusing on youth and education, health and human services, civic and community, the environment, and arts and culture. TMMMS prefers to support program specific grants rather than event sponsorships. Grants are provided to support the development and implementation of programs that generally range from $50,000 to $200,000. Submission deadlines for applications are February 1, May 15, August 15, and November 15. Grant request status notification will be provided to applicants within 45 days of the application deadline.
Requirements: The geographic scope of funding is centered around Pontotoc County, Union County, and Lee County, all in West Virginia.
Restrictions: Toyota Motor Manufacturing of Mississippi does not make grants for publications, lobbying activities, advertising, capital campaigns or endowments. Individuals are ineligible to apply. Toyota will not make grants to the following types of organizations: those not recognized as 501(c)3 by the Internal Revenue Service; those that practice discrimination by race, creed, color, sex, age or national origin; those that serve only their own memberships, such as fraternal organizations, labor organizations or religious groups; or political parties or candidates.
Geographic Focus: All States
Date(s) Application is Due: Feb 1; May 15; Aug 15; Nov 15
Amount of Grant: 50,000 - 200,000 USD
Contact: Grants Administrator; (662) 317-3281 or (662) 538-5902
Internet: http://www.toyota.com/about/philanthropy/guidelines/index.html#limitations
Sponsor: Toyota Motor Manufacturing of Mississippi
1200 Magnolia Drive
Blue Springs, MS 38828

Toyota Motor Manufacturing of Texas Grants 3747

As a good corporate citizen, Toyota Motor Manufacturing of Texas contributes to economic and social development in local communities. Giving back to the communities where its team members live and work is a priority. The corporation believes in helping people improve the quality of life in their communities through educational and family literacy programs. It also partners with leading organizations that educate children and their families about creating a cleaner, greener and healthier world. TMMTX makes grants to support programs and events benefiting the following categories: youth and education, health and human services, arts and culture, civic and community, and the environment. Requests are reviewed quarterly. Grants are provided to support the development and implementation of programs that generally range from $50,000 to $200,000. Submission deadlines for applications are January 31, April 30, September 30, and October 31 of every calendar year.
Requirements: Applicant organizations must have 501(c)3 tax-exempt status, and be located within or serve population(s) of Bear County, Texas, and its adjacent counties. Applicants must also present a proposal that satisfies the mission, guidelines and limitations of the corporation.
Restrictions: TMMTX does not donate vehicles. Nor does it make grants for publications, lobbying activities, advertising, capital campaigns or endowments. Individuals are ineligible to apply. Toyota will not make grants to the following types of organizations: those not recognized as 501(c)3 by the Internal Revenue Service; those that practice discrimination by race, creed, color, sex, age or national origin; those that serve only their own memberships, such as fraternal organizations, labor organizations or religious groups; or political parties or candidates.
Geographic Focus: Texas
Date(s) Application is Due: Jan 31; Apr 30; Sep 30; Oct 31
Amount of Grant: 50,000 - 200,000 USD
Contact: Grants Administrator; (210) 263-4000
Internet: http://www.toyotatexas.com/index.php?option=com_content&view=article&id=5&Itemid=4
Sponsor: Toyota Motor Manufacturing of Texas
1 Lone Star Pass
San Antonio, TX 78264

Toyota Motor Manufacturing of West Virginia Grants 3748

Toyota Motor Manufacturing of West Virginia is committed to continuing the worldwide Toyota tradition of community involvement and support. In response to the needs and interests of its team members and the community, Toyota has designed a corporate donations program focusing on specific issues and geographic areas with its first priority of improving the quality of life in the community in which it operates. TMMWV provides funding to education, health and human services, civic and community, arts and culture, and environmental organizations. TMMWV's education donations focus primarily on projects that serve K-12 students in the public setting. Community development activities must improve the economy or the quality of life for the people in the region. Health and human services projects must be aimed at significantly improving health or health care, or providing assistance to families in need. Environmental activities must be designed to preserve, restore, or improve the quality of air, natural or wildlife resources. Grants are provided to support the development and implementation of programs that generally range from $50,000 to $200,000. Grants are reviewed quarterly, in March, June, September, and December. Submission deadlines for applications are March 1, June 1, October 1, and January 1.
Requirements: Top priority is given to Putnam County; second priority is given to Cabell, Jackson, Kanawha, Lincoln and Mason counties; and there is a limited participation in important statewide projects.
Restrictions: Toyota Motor Manufacturing of West Virginia does not make grants for publications, lobbying activities, advertising, capital campaigns or endowments. Individuals are ineligible to apply. Toyota will not make grants to the following types of organizations: those not recognized as 501(c)3 by the Internal Revenue Service; those that practice discrimination by race, creed, color, sex, age or national origin; those that serve only their own memberships, such as fraternal organizations, labor organizations or religious groups; or political parties or candidates.
Geographic Focus: West Virginia
Date(s) Application is Due: Jan 1; Mar 1; Jun 1; Oct 1
Amount of Grant: 50,000 - 200,000 USD
Contact: Grants Administrator; (304) 937-7000
Internet: http://www.toyota.com/about/philanthropy/guidelines/index.html#limitations
Sponsor: Toyota Motor Manufacturing of West Virginia
1 Sugar Maple Lane
Buffalo, WV 25033

Toyota Motor North America of New York Grants 3749

Toyota Motor North America offers grant funding nationally, focusing on three primary areas: the environment, safety, and education. National programs in these areas must have a broad reach by impacting several major U.S. cities, communities or groups. In the local New York City area, Toyota also focuses on those three major areas, and provides other local assistance as well, including arts and culture, civic and community, health and human services and leadership development. Toyota prefers to support programs, rather than sponsor events. Organizations must apply for each new grant requested, and subsequent funding is contingent upon evaluation of previous activities. The geographic scope is the continental U.S. for programs national in scope and the New York City area for community-based programs. Only online applications are accepted.
Requirements: Applicant organizations must have 501(c)3 tax-exempt status, be located within or serve population(s) either on a national scope or specifically in the New York City area, and present a proposal that satisfies the mission, guidelines and limitations of the corporation.
Restrictions: Toyota Motor North America of New York does not make grants for publications, lobbying activities, advertising, capital campaigns or endowments. Individuals are ineligible to apply. Toyota will not make grants to the following types of organizations: those not recognized as 501(c)3 by the Internal Revenue Service; those that practice discrimination by race, creed, color, sex, age or national origin; those that serve only their own memberships, such as fraternal organizations, labor organizations or religious groups; or political parties or candidates.
Geographic Focus: All States
Amount of Grant: 50,000 - 200,000 USD
Contact: Grants Administrator; (212) 223-0303; fax (212) 759-7670
Internet: http://www.toyota.com/about/philanthropy/guidelines/index.html
Sponsor: Toyota Motor North America
601 Lexington Avenue, 49th Floor
New York, NY 10022

Toyota Motor Sales, USA Grants 3750

Toyota Motor Sales, USA, offers grant funding within the Torrance, California for community?based programs. The program's funding scope primarily supports education, environment and safety, but also funds civic groups, arts and culture, health and human services. Toyota Motor Sales prefers to support programs, rather than sponsor events. Organizations must apply for each new grant requested, and subsequent funding is contingent upon evaluation of previous activities. Only online applications are accepted.
Requirements: Applicant organizations must have 501(c)3 tax-exempt status, be located within or serve population(s) in the Torrance, California, region, and present a proposal that satisfies the mission, guidelines and limitations of the corporation.
Restrictions: Toyota Motor Sales, USA does not make grants for publications, lobbying activities, advertising, capital campaigns or endowments. Individuals are ineligible to apply. Toyota will not make grants to the following types of organizations: those not recognized as 501(c)3 by the Internal Revenue Service; those that practice discrimination by race, creed, color, sex, age or national origin; those that serve only their own memberships, such as fraternal organizations, labor organizations or religious groups; or political parties or candidates.
Geographic Focus: California
Amount of Grant: 50,000 - 200,000 JPY
Contact: Grants Administrator; (310) 468-5249 or (310) 468-4216; fax (310) 468-7840
Internet: http://www.toyota.com/about/philanthropy/guidelines/index.html
Sponsor: Toyota Motor Sales, USA
19001 S. Western Avenue
Torrance, 90501

Toys R Us Children's Fund Grants 3751

The Toys R Us Children's Fund is a public charity affiliated with Toys R Us, Inc. Since it was founded in 1992, the Children's Fund has contributed millions of dollars annually to qualified organizations that keep children safe and help them in times of need, including those providing disaster relief to children and families who are victims of large-scale crises. The Fund also provides grants to leading organizations that support children with special needs. Primary fund raising activities include the annual Toys R Us Children's Fund Gala, a private, invitation-only event that showcases the work of the Fund and raises money to support its many beneficiaries.
Requirements: U.S. 501(c)3 tax-exempt organizations are eligible.
Geographic Focus: All States
Contact: Children's Fund Coordinator; (973) 617-3500 or (800) 869-7787; fax (973) 617-4040; TRUCF@toysrus.com
Internet: http://www9.toysrus.com/about/corpPhilanthropy.cfm
Sponsor: Toys R Us
1 Geoffrey Way
Wayne, NJ 07470-2030

Trauma Center Association of America Finance Fellowship 3752

The Trauma Center Association of America Board of Directors and staff offer a once-in-a-lifetime educational opportunity for Trauma Program Leaders seeking to build upon their existing business acumen in trauma finance through a two-week Fellowship funded by the Association. This member opportunity consists of developing business and management skills, building a trauma finance knowledge base, interpreting data and statistics, and much, much more. Enrollment is now open for the Trauma Finance Fellowship Program. All applications must be submitted by February 13.
Geographic Focus: All States
Date(s) Application is Due: Feb 13
Contact: Ann Bellows; (575) 525-9511; fax (575) 647-9600; ann@traumacenters.org
Internet: http://tcaa.site-ym.com/?page=Fellowship
Sponsor: Trauma Center Association of America
1155 South Telshor Boulevard, Suite 201
Las Cruces, NM 88011

TRDRP California Research Awards 3753

This program is focused on research projects that address questions specific to tobacco-related disease or tobacco control issues in California. The awards will provide funding for average annual direct costs of $140,000; $170,000 per year for research involving human subjects. Full indirect costs are paid to eligible institutions. Maximum duration is 3 years.
Requirements: Principal Investigators from California not-for-profit organizations are eligible for funding. Research undertaken with TRDRP funds must be conducted primarily in California. The Principal Investigator must conduct and supervise the research project directly and in person. Underrepresented racial and ethnic investigators are encouraged to

submit applications. To be competitive, proposals must present a compelling justification, including evidence if available, supporting the California-specific criterion. The proposal should be fully developed, scientifically rigorous, and include sound background information, hypothesis, and promising preliminary studies or preliminary data.
Restrictions: All submitted applications will be screened for their direct relevance to tobacco use or tobacco-related disease. Proposed research that is not relevant will not undergo peer review for scientific merit. It is incumbent on the applicant to make a compelling case that the proposed work is directly relevant to the mission of TRDRP.
Geographic Focus: California
Date(s) Application is Due: Jan 15
Amount of Grant: 140,000 - 170,000 USD
Contact: M.F. Bowen, (510) 987-9811; mf.bowen@ucop.edu
Internet: http://www.trdrp.org/fundingopps/callawardmechs.asp#calrchawd
Sponsor: Tobacco-Related Disease Research Program
Office of the President, 300 Lakeside Drive, 6th Floor
Oakland, CA 94612-3550

TRDRP Exploratory/Developmental Research Awards   3754
The purpose of these grants is to gather preliminary data or demonstrate proof-of-principle (i.e., pilot projects), or to conduct a research project within the specified limits of money and time. The ultimate goal of these awards is to provide the foundation for proposals for fully-developed research project awards from other funding programs. Total direct costs of $250,000 for the entire project duration (maximum of 2 years). Indirect costs are paid to eligible institutions.
Requirements: Principal Investigators from California not-for-profit organizations are eligible for funding. Research undertaken with TRDRP funds must be conducted primarily in California. The Principal Investigator must conduct and supervise the research project directly and in person. Underrepresented racial and ethnic investigators are encouraged to submit applications. Although Principal Investigators may submit as many proposals for these awards as they wish for separate projects, no more than one Exploratory/Developmental Research Award will be made to any individual PI in the same year. Only proposals that address one of TRDRP's Primary research area will be eligible and will undergo peer review. Exploratory/Developmental Research Awards are for new projects only; that is, competing renewal applications will not be accepted.
Restrictions: All submitted applications will be screened for their direct relevance to tobacco use or tobacco-related disease. Proposed research that is not relevant will not undergo peer review for scientific merit. It is incumbent on the applicant to make a compelling case that the proposed work is directly relevant to the mission of TRDRP.
Geographic Focus: All States
Date(s) Application is Due: Jan 15
Amount of Grant: 250,000 USD
Contact: M.F. Bowen, (510) 987-9811; mf.bowen@ucop.edu
Internet: http://www.trdrp.org/fundingopps/callawardmechs.asp#explordev
Sponsor: Tobacco-Related Disease Research Program
Office of the President, 300 Lakeside Drive, 6th Floor
Oakland, CA 94612-3550

TRDRP New Investigator Awards   3755
This award mechanism is specifically designed to enable new investigators to initiate an independent research program in a tobacco-related field. These are awards for investigators who have a doctoral degree, have not yet served as a principal investigator of a research project grant (e.g., an NIH R01-type grant), and have had no more than five years experience as an independent investigator, i.e., no more than five years since completing formal postdoctoral training, or since the doctoral degree if no postdoctoral training. Direct costs of $90,000 per year can be averaged over the duration of the award. The indirect cost rate is capped at 8% for eligible institutions.
Requirements: New investigators must commit to a minimum of 50 percent time to the research project for the award duration.
Restrictions: Current or past new investigator awardees from TRDRP, other UC Special Research Programs, or other agencies (for example, current or past recipients of NIH K-series career development awards) are not eligible.
Geographic Focus: California
Date(s) Application is Due: Jan 15
Amount of Grant: 90,000 USD
Contact: M.F. Bowen, (510) 987-9811; mf.bowen@ucop.edu
Internet: http://www.trdrp.org/fundingopps/callawardmechs.asp
Sponsor: Tobacco-Related Disease Research Program
Office of the President, 300 Lakeside Drive, 6th Floor
Oakland, CA 94612-3550

Tri-State Community Twenty-first Century Endowment Fund Grants   3756
The Foundation for the Tri-State Community makes grants in support of educational, cultural, scientific and other charitable purposes in Kentucky, Ohio, and West Virginia. Specifically, the Foundation encourages project proposals with the following goals: improve the quality of life in the Tri-State area; test or demonstrate new approaches and techniques to find solutions to important community problems; promote community volunteerism; strengthen non-profit agencies and institutions by reducing operating costs, increasing public financial support, and/or improving internal management; and use community resources to promote coordination, cooperation, and sharing among organizations to eliminate duplicate services.
Requirements: 501(c)3 tax-exempt organizations that serve Boyd and Greenup Counties, Kentucky; Lawrence County, Ohio; and Cabell and Wayne Counties, West Virginia are eligible. Applicants should contact the Foundation to obtain a copy of the grant application and to discuss the proposed project.
Restrictions: The foundation does not fund general operating support grants, endowment funds, individuals, operational deficits, or sectarian activities of religious organizations. Also, organizations may submit only one application per funding cycle, and past awardees should wait at least 12 months from the approval of their award to submit another request. The 21st Century Endowment grants also do not provide funding for components of projects begun before grant award decisions are made.
Geographic Focus: Kentucky, Ohio, West Virginia
Date(s) Application is Due: Jan 15; Apr 15; Jul 15; Oct 15
Amount of Grant: 250 - 5,000 USD
Sample: Helping Hands of Greenup County, Kentucky, $2,000 - for the purchase of a color printer; Huntington Area Food Bank, Tri-State area, $4,000 - Primarius software program; LED.C. Symmes Creek Restoration Committee, Ohio, $2,121 - funding to develop and publish a Symmes Creek Canoe Trail Map.
Contact: Mary Witten Wiseman, President; (304) 942-0046; fax (304) 942-0048; FTSC_MWWiseman@yahoo.com
Internet: http://www.tristatefoundation.org/html/grants.html
Sponsor: Foundation for the Tri-State Community
P.O. Box 2096
Ashland, KY 41105-2096

Triangle Community Foundation Donor-Advised Grants   3757
Established in 1983, the Triangle Community Foundation is a nonprofit organization that serves Chatham, Durham, Orange and Wake counties of North Carolina. There's no application process for the donor-advised grants, rather, fundholders advise Foundation staff as to where they would like their grants made. Often, fundholders seek guidance from staff about which nonprofits are doing work matching their charitable interests. For this reason, it is important nonprofits keep members of the Philanthropic Services Team informed about their organizations, including special projects and funding needs. For more information, nonprofit organizations should contact the following staff: Arts, Culture, & Humanites - Lori O'Keefe, lori@trianglecf.org; Community Building - Agnes Vishnevkin, agnes@trianglecf.org; Education & Youth Development - Robyn Fehrman, robyn@trianglecf.org; Health & Human Services - Sandra Rodriguez, sandra@trianglecf.org; Law, Justice, & Public Action - Robyn Fehrman, robyn@trianglecf.org; Natural Resources - Tracy Joseph, tracy@trianglecf.org; Volunteerism & Community Support - Libby Long, libby@trianglecf.org; Religious & Spiritual Development - Tracy Joseph, tracy@trianglecf.org.
Requirements: North Carolina 501(c)3 tax-exempt organizations, government units, and religious congregations planning projects that benefit residents of Chatham, Durham, Orange and Wake counties of North Carolina are eligible.
Geographic Focus: North Carolina
Contact: Lori O'Keefe, Director of Philanthropic Services; (919) 474-8370, ext. 144; fax (919) 941-9208; lori@trianglecf.org
Internet: http://www.trianglecf.org/page33699.cfm
Sponsor: Triangle Community Foundation
324 Blackwell Street, Suite 1220
Durham, NC 27701

Triangle Community Foundation Shaver-Hitchings Scholarship   3758
The Shaver-Hitchings Scholarship Fund was created in 1990 by Dr. George H. Hitchings and Dr. Joyce Shaver Hitchings to honor and reward individuals with a commitment to helping others in the area of drug and alcohol addiction. The Scholarship is available to any graduate student, physician's assistant, or medical student in the Triangle area who has worked (preferably as a volunteer) to help others with alcoholism, drug abuse and addictive disorder treatment or with preventive education on the subject of addiction. The fund is designed to recognize the recipient for his or her work or volunteer activity and to provide financial aid. Excellent candidates who have demonstrated a commitment to working with others in the area of addictive disorders, especially those who have done so as volunteers, are sought. One Scholarship of $1,500 will be awarded each year. The Scholarship may be applied toward expenses for tuition, required fees, and required books or materials not covered by other financial aid.
Requirements: Applicant must be a resident of Durham, Orange, Wake, or Chatham County.
Restrictions: The Scholarship does not cover living expenses such as room and board, transportation, or childcare.
Geographic Focus: North Carolina
Date(s) Application is Due: Mar 15
Contact: Libby Long Richards, Program Coordinator; (919) 474-8370, ext. 134; fax (919) 941-9208; libby@trianglecf.org
Internet: http://www.trianglecf.org/grants_support/view_scholarships/shaver-hitchings_scholarship/
Sponsor: Triangle Community Foundation
324 Blackwell Street, Suite 1220
Durham, NC 27701

Trinity Lutheran School of Nursing Alumnae Scholarships   3759
The Trinity Lutheran Hospital School of Nursing Alumnae Scholarship fund was established to assist persons who wish to pursue a career in nursing (such as undergraduate, advanced degree, etc.) and need financial assistance. Applications for the scholarship must be received by May 1st. Renewal of the scholarship is possible each year for those who reapply and meet the criteria.
Requirements: Applicants must: be a student currently enrolled in an accredited RN school of nursing or in an accredited higher education nursing program; have completed at least one semester at an accredited school; be a full or part-time student and the course work must be continuous; and have a cumulative GPA of 2.5.

Geographic Focus: All States
Date(s) Application is Due: May 1
Contact: Becky Schaid; (816) 276-7515 or (816) 276-7555; becky@btllf.org
Internet: http://www.btllf.org/funding.html
Sponsor: Baptist-Trinity Lutheran Legacy Foundation
6601 Rockhill Road
Kansas City, MO 64131

Triological Society Research Career Development Awards 3760
The purpose of this award is to provide support for the research career development of otolaryngologists (head and neck surgeons) who have made a commitment to focus their research endeavors on patient-oriented research, e.g., clinical trials, translational research. Research training supported by this award may be related to any research questions relevant to the specialty of otolaryngology, as long as it is demonstrated that the training will have a direct impact on the applicant's ability to pursue his/her long-range research objectives. Grants can be for one to two years, are non-renewable, and have a $40,000 maximum award limitation. There are approximately six grants awarded annually. Letters of Intent are due by December 15, with full proposals received by January 15 annually.
Requirements: Otolaryngologists-head and neck surgeons who hold full-time, part-time or contributed service medical school faculty appointments and who have made a commitment to focus their research endeavors on patient-oriented research may apply. Applicants must be sponsored by the Chair of his/her division or department and by an official representative of the institution which would administer the award and in whose name the application is formally submitted.
Geographic Focus: All States
Date(s) Application is Due: Jan 15; Dec 15
Amount of Grant: Up to 40,000 USD
Contact: Stephanie L. Jones; (703) 519-1586 or (703) 836-4444; sljones@entnet.org
Internet: http://www.triological.org/researchgrants.htm
Sponsor: Triological Society
555 North 30th Street
Omaha, NE 68131

Trull Foundation Grants 3761
The Trull Foundation was established to share God's bounty in Texas and beyond. The Foundation supports organizations involved with: health care, the coastal Texas environment, seniors, child abuse, neglect, farming, ranching, agriculture, birds, hunger, substance abuse, religion, charity, education, improving the quality of life, especially for those living in poor or oppressed conditions, with a special concern for the needs of Palacios and Matagorda County where the Foundation has its roots.
Requirements: 501(c)3 tax-exempt organizations and departments, agencies, and other services operated within federal, state, or local government agencies and institutions and agencies affiliated with organized religions and religious bodies are eligible. Proposals should be mailed directly to the foundation, faxed or emailed proposals will not be excepted. See foundation's website for additional proposal guidelines and application forms.
Restrictions: The foundation usually will not make long term commitments; make grants for buildings, endowments, or research; repeat grants to the same project longer than three years; or fund operational expenses except during initial years. Any scholarship funds granted are made to institutions and are administered by those institutions. No scholarships or other funds are granted to individuals.
Geographic Focus: All States
Amount of Grant: 250 - 20,000 USD
Samples: El Buen Pastor Early Childhood Development Center, Austin, TX, $8,000 - Student Tuition Scholarships; Palmer Drug Abuse Program of Corpus Christi, Inc., Corpus Christi, TX, $7,000 - Reaching out to the Rural Areas of South Texas; Ducks Unlimited, Inc., Memphis, TN, $10,000 - Texas Prairie Wetlands; Louise Independent School District, Louise, TX, $3,000 - TI-83 calculators for students in Grades 9-12; Blessing Historical Foundation, Blessing, TX, $15,000 - Hotel Restoration;
Contact: E. Gail Purvis; (361) 972-5241; fax (361) 972-1109; gpurvis@trullfoundation.org
Internet: http://www.trullfoundation.org
Sponsor: Trull Foundation
404 Fourth Street
Palacios, TX 77465

Tull Charitable Foundation Grants 3762
Priority interest areas for awarding grants, in Georgia, are education, health and human services, youth development, and the arts. Requests for major capital projects are eligible for consideration. Requests for endowments and for the start-up of new initiatives are sometimes considered.
Requirements: The Foundation's Trustees limit grant awards to 501(c)3 organizations based in the State of Georgia. Requests from organizations located outside of Georgia are not considered. It is preferred that an applicant organization be able to demonstrate a broad base of financial support for a proposed project from its own community and constituency prior to asking for support of that project from the Foundation. Prior to submitting a full proposal, it is recommended (but not required) that an applicant organization contact the Foundation via a concise letter-of-intent in order to determine the potential eligibility of the request.
Restrictions: Grants are not available for operating support, research, conferences and seminars, legislative lobbying or other political purposes, special events, or individuals; nor to churches, sports booster clubs, to sponsor events or to retire accumulated debt. The Foundation does not utilize fiscal agents to handle funds for organizations that do not have an IRS certification letter.
Geographic Focus: Georgia

Date(s) Application is Due: Mar 3; Jun 2; Sep 2; Dec 2
Contact: Carol D. Aiken; (404) 659-7079; carol@tullfoundation.org
Internet: http://www.tullfoundation.org/app_procedures.asp
Sponsor: Tull Charitable Foundation
50 Hurt Plaza, Suite 1245
Atlanta, GA 30303

Turner Foundation Grants 3763
The Turner Foundation believes each organization to be important and that the best person to convey the passion for your organization is you. The Foundation supports initiatives that enhance the quality of life in the greater Springfield/Clark County community through artistic, educational, environmental, recreational, family, healthcare, historic preservation, community beautification and revitalization programs. In most cases, discretionary grants fall in the range of $5,000 to $15,000.
Requirements: To be considered for a discretionary grant, each organization must serve the people of Clark County Ohio, be a tax-exempt non-profit with a 501(c)3 classification from the IRS, and have a governing board.
Restrictions: The Turner Foundation does not fund individuals, churches, legislative action groups, annual fundraising campaigns, scholarships, fraternal groups or political groups and issues.
Geographic Focus: Ohio
Date(s) Application is Due: Sep 14
Amount of Grant: 5,000 - 30,000 USD
Contact: Director; (937) 325-1300; fax (937) 325-0100; email@hmturnerfoundation.org
Internet: http://www.hmturnerfoundation.org/grants.html
Sponsor: Turner Foundation
4 West Main Street, Suite 800
Springfield, OH 45502

Tylenol Future Care Scholarships 3764
The makers of the Tylenol Family of Products will award 40 scholarships in the amounts of $5,000 and $10,000 for higher education to students who demonstrate leadership in community activities and school activities and who intend to major in areas that will lead to health-related fields. Apply online at the sponsor's website.
Requirements: Application is open to students who will be attending an undergraduate or graduate course of study in the fall at an accredited two- or four-year college, university or vocational-technical school. This includes those students currently enrolled in an undergraduate or graduate course of study and have one or more years of school remaining. Applicants will be judged on leadership qualities and academic performance.
Geographic Focus: All States
Date(s) Application is Due: May 15
Amount of Grant: 5,000 - 10,000 USD
Contact: Scholarship Program Administrators; (888) 543-8255
Internet: http://www.tylenol.com/news/scholarship?id=tylenol/news/subptyschol.inc&s_kwcid=ContentNetwork|1522864193
Sponsor: McNeil Consumer Healthcare
7050 Camp Hill Road
Fort Washington, PA 19034

Tyler Aaron Bookman Memorial Foundation Trust Grants 3765
The Tyler Aaron Bookman Memorial Foundation was established by Neil and Jill Bookman in memory of their eleven-year-old son, Tyler, who died of a brain tumor. His parents, based on Tyler's love of learning, started the Foundation with the money they had saved for his college education. To them, education is key to breaking the cycle of poverty. With funding centered in Camden, New Jersey, and adjoining states, the Foundation's primary fields of interest is support of health care and research, Catholic churches and agencies, and children and youth education. There are no specific deadlines with which to adhere, and applicants should initially approach the Foundation by way of letter requesting an application form.
Restrictions: No grants are given to individuals.
Geographic Focus: New Jersey, North Carolina, Pennsylvania
Amount of Grant: Up to 30,000 USD
Samples: La Salle University, Philadelphia, Pennsylvania, $12,699; Pediatric Brain Tumor Foundation, Asheville, North Carolina, $27,000; Children's Hospital of Philadelphia, Philadelphia, Pennsylvania, $15,000.
Contact: Neil S. Bookman; (215) 646-2192 or (267) 216-7718; fax (215) 654-6060
Sponsor: Tyler Aaron Bookman Memorial Foundation
426 Newbold Road
Jenkintown, PA 19046-2851

U.S. Cellular Corporation Grants 3766
The U.S. Cellular Corporation strives to build a connection with communities that extend beyond its business by supporting causes that strengthens every neighborhood where it's customers live, work and play. The Corporation awards grants to projects or programs that have significant relevance within its operating communities and that relate to the following strategic areas of concern: civic and community, education, health and human service, environment, and arts and culture. Furthermore, through its Associate Matching program, it matches associates' charitable donations to nonprofit organizations dollar-for-dollar, up to $2,500 per year, with an annual cap of $250,000.
Requirements: 501(c)3 tax-exempt organizations within the Corporation's areas of operation are eligible to apply. This includes parts of: Iowa; Missouri; Oklahoma; northern Texas; Nebraska; southern Wisconsin; northwestern Illinois; north central

Indiana; Oregon; northern California; southern Washington; eastern Tennessee; North Carlina; western Virginia; western Maryland; and eastern West Virginia.
Restrictions: The corporation will not make charitable contributions to the following entities: individuals; agencies that discriminate on the basis of race, color, creed, or national origin; political causes, candidates, legislative lobbying, or advocacy groups; endowments or memorials; construction or renovation projects (except Habitat for Humanity); religious organizations seeking to further a denominational or sectarian purpose; social, labor, alumni, or fraternal organizations serving a limited constituency; primary, secondary, or charter schools; special occasion, good-will, and single-interest magazines; local athletic or sports programs, walk-a-thons, and little leagues; or travel funds for tours, expeditions, or trips.
Geographic Focus: California, Illinois, Indiana, Iowa, Maryland, Missouri, Nebraska, North Carolina, Oklahoma, Oregon, Tennessee, Texas, Virginia, Washington, West Virginia, Wisconsin
Contact: External Communications Department; (773) 399-8900; fax (773) 399-8937; communityprograms@uscellular.com
Internet: http://www.uscellular.com/about/community-outreach/index.html
Sponsor: U.S. Cellular Corporation
8410 W Bryn Mawr, Suite 700
Chicago, IL 60634

U.S. Department of Education 21st Century Community Learning Centers 3767
This program supports the creation of community learning centers that provide academic enrichment opportunities for children, particularly students who attend high-poverty and low-performing schools. The program helps students meet state and local student standards in core academic subjects, such as reading and math; offers students a broad array of enrichment activities that can complement their regular academic programs; and offers literacy and other educational services to the families of participating children.
Requirements: Awards are made to State Education Agencies (SEAs). Local education agencies (LEAs) and nonprofit organization may apply to states for subgrants. Formula grants are awarded to State educational agencies, which in turn manage statewide competitions and award grants to eligible entities. For this program, eligible entity means a local educational agency, community-based organization, another public or private entity, or a consortium of two or more such agencies, organizations, or entities. States must give priority to applications that are jointly submitted by a local educational agency and a community-based organization or other public or private entity. Consistent with this definition of eligible entities, faith-based organizations are eligible to participate in the program. Each eligible entity that receives an award from the state may use the funds to carry out a broad array of before- and after-school activities (including those held during summer recess periods) to advance student achievement. Many states around the country are conducting competitions to award 21st Century Community Learning Center grants. The State Contact List (http://www.ed.gov/programs/21stcclc/contacts.html#state) now includes links to State websites and application due dates.
Geographic Focus: All States
Contact: Peter Eldridge, Acting Program Contact; (202) 260-2514; fax (202) 260-8969; 21stCCLC@ed.gov or Peter.Eldridge@ed.gov
Internet: http://www.ed.gov/programs/21stcclc/index.html
Sponsor: U.S. Department of Education
400 Maryland Avenue SW, Room 3E246
Washington, D.C. 20202-6200

U.S. Department of Education Erma Byrd Scholarships 3768
The Erma Byrd Scholarship Program provides scholarships to individuals pursuing a course of study that will lead to a career in industrial health and safety occupations, including mine safety. This program is designed to increase the skilled workforce in these fields at both the fundamental skills level and the advanced skills level. The program has a service obligation component, requiring recipients of the scholarship to be employed in a career position directly related to industrial health and safety, including mine safety, for a period of one year upon completion of the degree program. The selected areas of study are: mining and mineral engineering, industrial engineering, occupational safety and health technology/technician, quality control technology/technician, industrial safety technology/technician, hazardous materials information systems technology/technician, mining technology/technician, and occupational health and industrial hygiene. Scholarship recipients must begin such employment no more than six months after the completion of their degree program. Average awards will be as follows: $2,500 associate's degree; $5,000 bachelor's degree; $10,000 graduate degree.
Requirements: Applicants must be: U.S. citizen, national, or permanent resident; pursuing a degree in mining and mineral engineering, industrial engineering, occupational safety and health technology/technician, quality control technology/technician, industrial safety technology/technician, hazardous materials information systems technology/technician, mining technology/technician, or occupational health and industrial hygiene; and, within two years of completing an associate's, bachelor's, or graduate degree at an accredited U.S. institution of higher education. As part of your application package, your Social Security Number must be submitted to the U.S. Department of Education via postal mail using the form on page 34 of the application package. For security reasons, DO NOT submit this form via email with the rest of your application.
Geographic Focus: All States
Date(s) Application is Due: Jul 31
Amount of Grant: 2,500 - 10,000 USD
Contact: Lauren Kennedy, (202) 502-7630; ermabyrdprogram@ed.gov
Internet: http://www.ed.gov/programs/ermabyrd/index.html
Sponsor: U.S. Department of Education
1990 K Street NW, Room 6121
Washington, D.C. 20006-8524

U.S. Department of Education Rehabilitation Engineering Research Centers 3769 Grants
RERCs support activities that: (1) lead to the development of methods, procedures, and devices that will benefit individuals with disabilities, especially those with the most severe disabilities; or (2) involve technology for the purposes of enhancing opportunities for meeting the needs of and addressing the barriers confronted by individuals with disabilities in all aspects of their lives. Types of activities supported include: the development of technological systems for persons with disabilities; stimulation of the production and distribution of equipment in the private sector; and clinical evaluations of equipment. Awards are for five years, except that grants to new recipients or to support new or innovative research may be made for fewer than five years.
Requirements: States, Institutions of Higher Education (IHEs), Nonprofit Organizations, public or private agencies, including for-profit agencies, Indian tribes, and tribal organizations may apply. RERCs must be operated by or in collaboration with one or more institutions of higher education or nonprofit organizations.
Geographic Focus: All States
Date(s) Application is Due: Apr 1; Aug 1
Amount of Grant: 850,000 - 1,000,000 USD
Contact: Donna Nangle, (202) 245-7462; fax (202) 245-7323; donna.nangle@ed.gov
Internet: http://www.ed.gov/programs/rerc/index.html
Sponsor: U.S. Department of Education
400 Maryland Avenue SW, Room 5051, PCP
Washington, D.C. 20202-2700

U.S. Department of Education Rehabilitation Research Training Centers 3770
The RRTCs conduct coordinated and advanced programs of research, training, and information dissemination. Each RRTC has a major program of research in a particular area, such as mental illness, vocational rehabilitation (VR), or independent living, which is specified by the National Institute on Disability and Rehabilitation Research (NIDRR). The RRTCs must serve as centers of national excellence and national or regional resources for providers and individuals with disabilities and their representatives. RRTC awards are for five years, except that grants to new recipients or to support new or innovative research may be made for fewer than five years.
Requirements: Note: Each year, competitions are held in specific areas that determine the types of projects. Eligible applicants include Institutions of Higher Education and Nonprofit Organizations. States, public or private agencies - including for-profit agencies, Indian tribes, and tribal organizations may also apply. Rehabilitation research and training centers must be operated by or in collaboration with: (1) one or more institutions of higher education or (2) one or more providers of rehabilitation or other appropriate services.
Geographic Focus: All States
Date(s) Application is Due: Aug 21; Aug 22
Amount of Grant: Up to 675,000 USD
Contact: Donna Nangle, (202) 245-7462; fax (202) 245-7323; donna.nangle@ed.gov
Internet: http://www.ed.gov/programs/rrtc/index.html
Sponsor: U.S. Department of Education
400 Maryland Avenue SW, Room 5051, PCP
Washington, D.C. 20202-2700

U.S. Department of Education Rehabilitation Training--Rehabilitation 3771 Continuing Education Programs--Institute on Rehabilitation Issues
The purpose of the Technical Assistance and Dissemination To Improve Services and Results For Children With Disabilities program is to promote academic achievement and to improve results for children with disabilities by providing technical assistance (TA), supporting model demonstration projects, disseminating useful information, and implementing activities that are supported by scientifically based research. The purpose of this priority is to support the establishment and operation of State Technical Assistance Projects To Improve Services and Results for Children Who Are Deaf-Blind (projects). Grants are available to support projects in the District of Columbia; Puerto Rico; and the Virgin Islands.
Requirements: Eligible applicants include: SEAs; LEAs, including public charter schools that are considered LEAs under State law; IHEs; other public agencies; private nonprofit organizations; outlying areas; FAS; Indian tribes or tribal organizations; and for-profit organizations. The projects funded under this competition must make positive efforts to employ and advance in employment qualified individuals with disabilities (see section 606 of IDEA). Applicants and grant recipients funded under this competition must involve individuals with disabilities or parents of individuals with disabilities ages birth through 26 in planning, implementing, and evaluating the projects (see section 682(a)(1)(A) of IDEA).
Restrictions: Funds awarded under this priority may not be used to provide direct early intervention services under Part C of IDEA, or direct special education and related services under Part B of IDEA. The Department will reject an application for a State project that proposes a budget exceeding the funding level for any single budget period of 12 months.
Geographic Focus: District of Columbia, Puerto Rico, U.S. Virgin Islands
Date(s) Application is Due: Jul 17
Amount of Grant: 30,000 - 65,000 USD
Contact: Lisa Gorove, (202) 245-7357; fax (202) 245-7617; Lisa.Gorove@ed.gov
Internet: http://www.ed.gov/programs/rsatrain/applicant.html#84264c
Sponsor: U.S. Department of Education
400 Maryland Avenue SW, Room 5038, PCP
Washington, D.C. 20202-2800

U.S. Department of Education Special Education--National Activities--Parent Information Centers 3772

The purpose of this program is to ensure that parents of children with disabilities receive training and information to help improve results for their children. Awards are made for parent information centers, community parent centers, and for technical assistance to such centers. This program includes the following grants competitions: Community Parent Resource Centers (CFDA 84.328C); Parent Training and Information Centers (CFDA 84.328M); and, Technical Assistance for the Parent Centers (CFDA 84.328R).
Requirements: Nonprofit organizations are eligible to apply. For Parent Training and Information (PTI) Centers (# 84.328M), parent organizations may apply. A parent organization is a private nonprofit organization (other than an institution of higher education [IHE]) that: (a) has a board of directors (1) the majority of whom are parents of children with disabilities ages birth through 26; (2) that includes (i) individuals working in the fields of special education, related services, and early intervention, (ii) individuals with disabilities, and (iii) the parent and professional members of which are broadly representative of the population to be served, including low-income parents of limited English proficient (LEP) children; and (b) has as its mission serving families of children with disabilities who are ages birth through 26, and have the full range of disabilities described in Sec. 602(3) of the Individuals with Disabilities Education Act(IDEA). For Community Parent Resource Centers (CPRC; # 84.328C), local parent organizations may apply (same requirements as listed above for PTI Centers). Additionally, *Community to be served* refers to a community whose members experience significant isolation from available sources of information and support as a result of cultural, economic, linguistic, or other circumstances deemed appropriate by the secretary of education. For Technical Assistance for Parent Training and Information Centers (# 84.328R), state education agencies (SEAs); local education agencies (LEAs); public charter schools that are LEAs under state law; IHEs; other public agencies; private nonprofit organizations; outlying areas (American Samoa, Guam, the Northern Mariana Islands, and the U.S. Virgin Islands); freely associated states; Indian tribes or tribal organizations; and for-profit organizations may apply.
Geographic Focus: All States
Date(s) Application is Due: Feb 26; Apr 18
Amount of Grant: 100,000 - 500,000 USD
Contact: Lisa Gorove, (202) 245-7357; fax (202) 245-7617; Lisa.Gorove@ed.gov
Internet: http://www.ed.gov/programs/oseppic/index.html
Sponsor: U.S. Department of Education
400 Maryland Avenue SW, Room 5051, PCP
Washington, D.C. 20202-2700

U.S. Department of Education United States-Russia Program: Improving Research and Educational Activities in Higher Education 3773

The program provides grants that demonstrate partnerships between Russian and American institutions of higher education that contribute to the development and promotion of educational opportunities between the two nations, particularly in the areas of mutual foreign language learning and the cooperative study of mathematics and science. Collaborative projects comprised of partnerships of Russian and American universities and colleges will be supported.
Requirements: Institutions of Higher Education (IHEs) or combinations of IHEs and other public and private nonprofit institutions and agencies are eligible to apply. Applicants must select one of three academic discipline as the subject area for their grant proposal: (A) Environmental Science Studies; (B) Biotechnology; or, (C) Any discipline, other than (A) and (B). Applications are invited from institutions of higher education with the capacity to contribute to a collaborative project in the areas listed with a Russian institution. The consortium partners, through promoting the study of and communication in foreign languages, are expected to increase awareness and understanding of the two cultures, and to strengthen the professional and scholarly ties between the two countries. Russian institutions will apply to The Russian Ministry of Education and Science for funding.
Restrictions: Any application that proposes a budget exceeding $150,000 for a single budget period of 12 months will be rejected. The Assistant Secretary for Postsecondary Education may change the maximum amount through a notice published in the Federal Register.
Geographic Focus: All States, Russia
Date(s) Application is Due: Jul 7
Amount of Grant: 189,000 USD
Samples: 2008: Stanford University, CA - $188,718
Contact: Krish Mathur; (202) 502-7512; fax (202) 502-7877; Krish.Mathur@ed.gov
Internet: http://www.ed.gov/programs/fipserussia/index.html
Sponsor: U.S. Department of Education
1990 K Street NW, Room 6155
Washington, D.C. 20006-8544

U.S. Department of Education Vocational Rehabilitation Services Projects for American Indians with Disabilities Grants 3774

The program provides financial assistance for the establishment and operations of vocational rehabilitation (VR) services programs for American Indians with disabilities living on or near a federal or state reservation. The purpose of this program is to assist tribal governments to develop or to increase their capacity to provide a program of vocational rehabilitation services, in a culturally relevant manner, to American Indians with disabilities residing on or near federal or state reservations. The program's goal is to enable these individuals, consistent with their individual strengths, resources, priorities, concerns, abilities, capabilities, and informed choice, to prepare for and engage in gainful employment. Program services are provided under an individualized plan for employment and may include native healing services.
Requirements: The governing body of an Indian tribe or consortia of such governing bodies located on federal and state reservations may apply.
Geographic Focus: All States
Date(s) Application is Due: May 5; Jun 12
Amount of Grant: 350,000 - 450,000 USD
Contact: Alfreda Reeves, (202) 245-7485; alfreda.reeves@ed.gov
Internet: http://www.ed.gov/programs/vramerind/index.html
Sponsor: U.S. Department of Education
400 Maryland Avenue SW, Room 5051, PCP
Washington, D.C. 20202-2700

U.S. Lacrosse EAD Grants 3775

U.S. Lacrosse is playing a leading role in ongoing research and is committed to educating the national lacrosse community about the life-saving value of having Automatic External Defibrillators (AEDs) available during lacrosse games and practices. In order to achieve that goal, U.S. Lacrosse, the national governing body of the sport, and Cardiac Science, a leading manufacturer of cardiology products, are offering an AED grant program which provides an AED and comprehensive management of AED and CPR training to lacrosse leagues or U.S. Lacrosse chapters.
Requirements: Applicant must be a current member of U.S. Lacrosse. Awarded group must budget $1295 towards the AED Grant Program.
Geographic Focus: All States
Date(s) Application is Due: Nov 30
Contact: Sarah Newman; (410) 235-6882; fax (410) 366-6735; snewman@uslacrosse.org
Internet: http://www.uslacrosse.org/programs/AEDgrantprogram.phtml
Sponsor: U.S. Lacrosse
113 W University Parkway
Baltimore, MD 21210

UCLA / RAND Corporation Post-Doctoral Fellowships 3776

The post-doctoral training program is jointly housed in the RAND Health Sciences program and the UCLA School of Public Health. Post-doctoral fellows have the option of pursuing a Masters degree in the UCLA Department of Health Services. The grant covers all tuition and fees, as well as a stipend, which currently ranges from $35,568 to $51,036, depending on the number of years of post-doctoral experience. Travel money is also available for fellows to attend professional meetings.
Requirements: To be eligible, candidates must have obtained a Ph.D., medical degree, or other professional doctoral degree. Awards will last two years. Fellows will be required to attend at least 8 courses at UCLA or RAND, including three courses in health services research methods, two in statistics, and three in health services issue areas. Those pursuing a Masters degree must complete a total of 12 courses. Upon arrival, fellows will align with an ongoing research project at RAND or UCLA. RAND and UCLA have numerous ongoing projects in the health services research area. These fall into several areas, including: access/utilization among the underserved; assessing quality, effectiveness, outcomes; health behaviors; alternative delivery systems; hospital and physician payment; health care cost containment; technology assessment; private health insurance; international comparisons. By the end of the first year, the fellow will identify a research idea and write a proposal. The proposal may be addressed to the principal investigator of an ongoing research project, or to an outside funding agency. By the end of the program, the fellow will prepare an article on research conducted during the fellowship, and submit it to a peer reviewed journal. This can either be done alone, or in collaboration with the research director or other member of the fellow's research team.
Geographic Focus: All States
Amount of Grant: 35,568 - 51,036 USD
Contact: Communications Office; (310) 393-0411, ext. 7775; RAND_Health@rand.org
Internet: http://www.ph.ucla.edu/hs/post_doctoral.html
Sponsor: RAND Corporation
1776 Main Street, P.O. Box 2138
Santa Monica, CA 90407-2138

UCT Scholarship Program 3777

For more than 50 years, United Commercial Travelers (UCT) has been dedicated to assisting people in the United States and Canada with intellectual disabilities, primarily by providing scholarships through the UCT Scholarship Program to individuals pursuing college degrees or certification to teach persons with intellectual disabilities. The scholarships provide up to $2,500 to any one applicant in any one calendar year. Individuals interested in applying for a UCT scholarship should download the application from the website, complete the application, and mail it to the address given. Applications are reviewed on a monthly basis. Scholarship recipients will receive payment once positive documentation that they have actually registered and paid for applicable college or university courses has been received by UCT. Recipients should allow two months for distribution of scholarship assistance.
Requirements: To be eligible, applicants must meet one or more of the following criteria: be a teacher of people with intellectual disabilities who needs additional course work to be certified or to retain certification; be an experienced teacher who wishes to become certified to teach people with intellectual disabilities; have a bachelor's or master's degree and plan to pursue graduate work in special education with an emphasis on teaching people with intellectual disabilities; be a college junior or senior whose course of undergraduate study is special education specifically focusing on teaching people with intellectual disabilities; be enrolled in courses to become a certified instructor under a structured trade, vocational, or recreation program at a facility for people with intellectual disabilities; or be in the second year of a two-year associate's degree program in an accredited school, with the course of study specifically focusing on teaching people with intellectual disabilities. Applicants must

plan to be of service to people with intellectual disabilities in the United States or Canada. Applicants must demonstrate a justifiable need for financial assistance.
Restrictions: The scholarship assistance is a reimbursement to help cover registration fees, tuition, and textbooks only. Applicants teaching or studying special education focusing on areas other than intellectual disabilities are not eligible.
Geographic Focus: All States, Canada
Amount of Grant: Up to 2,500 USD
Contact: Ann Marshall, Fraternal Department; (800) 848-0123, ext. 126 or (614) 487-9680; fax (614) 487-9688; amarshall@uct.org
Internet: http://www.uct.org/scholarships.html
Sponsor: Order of United Commercial Travelers of America
1801 Watermark Drive
Columbus, OH 43215

UICC American Cancer Society International Fellowships for Beginning Investigators   3778
The international fellowships fund 12-month basic or clinical research projects in the areas of epidemiology, prevention, detection, diagnosis, treatment, and psycho-oncology that foster a bidirectional flow of knowledge, experience, expertise, and innovation to and from the United States. Candidates should be in the early stages of their careers. Applications that are geared to the development of specific cancer control measures in developing and central and east European countries are particularly encouraged. Six to eight fellowships, with tenures beginning in April, will be awarded each year.
Requirements: Eligible candidates should hold assistant professorships or similar positions at their home institutes and have, ordinarily, a minimum of two and a maximum of ten years of postdoctoral experience after obtaining their MD or PhD degrees or equivalents.
Restrictions: Candidates who are physically present at the proposed host institute while their applications are under consideration are not eligible for UICC fellowships.
Geographic Focus: All States
Date(s) Application is Due: Dec 1
Amount of Grant: 45,000 USD
Samples: Paula Ximena Fernandez-Calotti, from Laboratory of Oncohematology Institute of Hematology Research, M.R. Castex, Argentina, to the Department of Biochemistry and Molecular Biology, University of Barcelona, Spain; Nikita A. Makretsov, from the Department of Pathology & Laboratory Medicine, Memorial University of Newfoundland, Canada, to the Department of Oncology Hutchinson / MRC Research Centre, United Kingdom.
Contact: Beate Vought; +41 22 809 1843; fax +41 22 809 1810; fellows@uicc
Internet: http://www.uicc.org/index.php?option=com_content&task=view&id=14259&Itemid=182
Sponsor: International Union Against Cancer
62, route de Frontenex
Geneva, CH-1207 Switzerland

UICC International Cancer Technology Transfer (ICRETT) Fellowships   3779
The program helps to facilitate rapid international transfer of cancer research techiques and technology and clinical management skills. It is hoped that funding will support the exchange of knowledge and enhance skills in basic, clinical, behavioral, and epidemiological areas of cancer research, in cancer control and prevention (including tobacco control). Participants should acquire appropriate clinical management, diagnostic and therapeutic skills for effective application and use in the home organization upon their return. There are 120 to 150 awards available annually, with applications accepted at any time and notification generally within 60 days of registration.
Requirements: Eligible applicants include appropriately qualified investigators, pathologists, epidemiologists, cancer registrars, behavioral scientists, tobacco control activists, laboratory technicians, clinicians and general practitioners, generally in the early stages of their careers.
Geographic Focus: All States
Amount of Grant: 3,400 USD
Contact: Beate Vought; +41 22 809 1843; fax +41 22 809 1810; fellows@uicc.ch
Internet: http://www.uicc.org/index.php?option=com_content&task=view&id=14262&Itemid=185
Sponsor: International Union Against Cancer
62, route de Frontenex
Geneva, CH-1207 Switzerland

UICC Raisa Gorbachev Memorial International Cancer Fellowships   3780
This award is funded by UICC International Cancer Foundation and enables qualified beginning pathologists to carry out specific research or training project relating to the many fields of oncology, and should be centred on an appropriate training period in modern diagnostic tools in the field of pathology. The Fellowship is conditional on Fellows returning to the home institutes/country at the end of the fellowship period. The original Fellowships cannot be prolonged, or run concurrently with other awards, even those funded by other agencies. They can however, be extended subject to the written approval of home and host supervisors and at no additional cost to UICC i.e. up to a total duration of 3 months. Should return of the Fellow to their home country be delayed past these 3 months, 50 percent of the travel award must be reimbursed to UICC.
Requirements: The candidate must possess the appropriate professional qualifications and experience according to the specifications described on the website and must currently be engaged in cancer research, clinical oncology practice, or cancer society work. To permit effective communication at the host institute, the candidate must have adequate fluency in a common language. The candidate must also be on the staff payroll of a university, research laboratory or institute, hospital, oncology unit to where they will return at the end of a fellowship.
Restrictions: Candidates attached to commercial entities are not eligible. Fellowships cannot be granted to candidates who are already physically present at the proposed host institute whilst their applications are under consideration.
Geographic Focus: All States
Contact: Beate Vought; +41 22 809 1843; fax +41 22 809 1810; fellows@uicc.ch
Internet: http://www.uicc.org/index.php?option=com_content&task=view&id=15956&Itemid=301
Sponsor: International Union Against Cancer
62, route de Frontenex
Geneva, CH-1207 Switzerland

UICC Yamagiwa-Yoshida Memorial International Cancer Study Grants   3781
These study grants enable cancer investigators from any country to establish bilateral research projects abroad that exploit complementary materials or skills, including advanced training in experimental methods or special techniques. There are 14 to 16 three-to six-month grants awarded each year. Selection takes place twice annually, with recipients notified in mid-April and mid-October. Grants should be taken up within six months following notification. Travel awards contribute toward the least expensive international two-way airfare or other appropriate form of transport. Financial support is not provided for dependents. Application materials are available on the Web site.
Requirements: Candidates should have appropriate scientific or medical qualifications and be actively engaged in cancer research.
Geographic Focus: All States
Date(s) Application is Due: Jan 15; Jul 1
Amount of Grant: 30,000 - 60,000 USD
Samples: Dr. Hakan Akbulut, from Department of Medical Oncology, Cebeci Hospital, Turkey, to Sydney Kimmel Cancer Center, USA; Dr. Gil Ast, from Department of Human Molecular Genetics & Biochemistry, Sackler School of Medicine, Israel, to Department of Biochemistry, School of Medicine, University of Washington, USA.
Contact: Beate Vought; +41 22 809 1843; fax +41 22 809 1810; fellows@uicc.ch
Internet: http://www.uicc.org/index.php?option=com_content&task=view&id=15958&Itemid=302
Sponsor: International Union Against Cancer
62, route de Frontenex
Geneva, CH-1207 Switzerland

UniHealth Foundation Community Health Improvement Grants   3782
UniHealth Foundation is an independent private health care foundation established in 1998 whose mission is to support and facilitate activities that significantly improve the health and well being of individuals and communities they serve. The objectives of the Community Health Improvement Grants are to: enable informed consumer health decisions, including appropriate use of healthcare services; promote improved individual health behaviors; identify and reduce risk factors for disease and injury; and ameliorate the impact of disease and enhance the quality of life. Grants supports community health improvement activities of hospitals by funding health education, prevention and treatment programs. Health education includes information dissemination on a broad range of topics, including health systems navigation and informed consumerism or the development of culturally sensitive educational materials. Prevention, in addition to clinical preventive services, might include injury and violence prevention or caregiver support services. Funding for treatment services targets activities aimed at improving health status and quality of life, including aspects of chronic condition management, dental care, end-of-life care and rehabilitation services. Grants range from $50,000 to $350,000. Throughout the year the Foundation announces other initiatives that focus on specific goals and objectives to promote foundation-identified priorities. New requests for proposals are listed on the Foundation's website.
Requirements: Most grants are made for the purpose of funding healthcare services and programs provided by or through qualified charitable hospitals in specified service areas in Los Angeles and northern Orange Counties. The service areas are: San Fernando and Santa Clarita Valley; Westside and Downtown Los Angeles; San Gabriel Valley; and Long Beach and Orange County. Applicants are required to submit a preliminary letter of inquiry. There are no deadlines. Those applicants invited to submit a full proposal will receive detailed information.
Restrictions: Individuals are not eligible for grants. Grants will not be given for propagandizing and/or influencing legislation or elections, political campaigns, programs that promote religious doctrine, events, endowments, annual fund drives, non-applied research, general operating expenses, or retirement of debt. Indirect costs for up to a maximum of ten percent of direct costs is allowable.
Geographic Focus: California
Contact: Caroline Chung, Grants Manager; (213) 630-6500; fax (213) 630-6509; Webadmin@unihealthfoundation.org
Internet: http://unihealthfoundation.org/program_areas.html
Sponsor: UniHealth Foundation
800 Wilshire Boulevard, Suite 1300
Los Angeles, CA 90017

Unihealth Foundation Grants for Non-Profit Oranizations   3783
UniHealth Foundation is an independent private health care foundation established in 1998 whose mission is to support and facilitate activities that significantly improve the health and well being of individuals and communities they serve. The objectives of the Grants for Non-Profit Organizations are: to promote positive health behaviors; to overcome systemic barriers to healthy living; and to advance informed consumerism. Funding is provided for traditional health education and disease prevention strategies on a broad range of topics as well as health system navigation, caregiver support services

or the development of culturally sensitive educational materials. Grants range from $10,000 to $350,000. Throughout the year the Foundation announces other initiatives that focus on specific goals and objectives to promote foundation-identified priorities. New requests for proposals are listed on the Foundation's website.
Requirements: Qualified nonprofit organizations providing health-related services consistent with the Foundation's mission are eligible. Applicants are required to submit a preliminary letter of inquiry. There are no deadlines. Those applicants invited to submit a full proposal will receive detailed information.
Restrictions: Individuals are not eligible for grants. Grants will not be given for propagandizing and/or influencing legislation or elections, political campaigns, programs that promote religious doctrine, events, endowments, annual fund drives, non-applied research, general operating expenses, or retirement of debt. Indirect costs for up to a maximum of ten percent of direct costs is allowable.
Geographic Focus: California
Samples: Bet Tzedek - The House of Justice, Los Angeles, California, $352,065, transitions and Children's Bureau, Los Angeles, California, $50,000, health access.
Contact: Caroline Chung, Grants Manager; (213) 630-6500; fax (213) 630-6509; webadmin@unihealthfoundation.org
Internet: http://unihealthfoundation.org/program_areas.html
Sponsor: UniHealth Foundation
800 Wilshire Boulevard, Suite 1300
Los Angeles, CA 90017

UniHealth Foundation Healthcare Systems Enhancements Grants   3784
UniHealth Foundation is an independent private health care foundation established in 1998 whose mission is to support and facilitate activities that significantly improve the health and well being of individuals and communities they serve. The objectives of the Healthcare Systems Enhancements Grants are to: increase access to comprehensive, coordinated care at the appropriate point of service; improve the quality of healthcare services in terms of patient satisfaction and clinical outcomes; and promote the financial stability and viability of not-for-profit hospitals. Support is provided for programs which address access to health care, quality of clinical services, patient satisfaction and organizational capacity building. Such programs might include partnerships with other hospitals and/or community clinics and community based organizations to coordinate care. Projects that focus on administrative systems improvement, organizational development, change management and capital needs associated with program delivery will also be considered. Grants range from $50,000 to $350,000. Throughout the year the Foundation may announce other initiatives that focus on specific goals and objectives to promote Foundation-identified priorities. New requests for proposals are listed on the Foundation's website.
Requirements: Most grants are made for the purpose of funding healthcare services and programs provided by or through qualified charitable hospitals in specified service areas in Los Angeles and northern Orange Counties. The service areas are: San Fernando and Santa Clarita Valley; Westside and Downtown Los Angeles; San Gabriel Valley; and Long Beach and Orange County. Applicants are required to submit a preliminary letter of inquiry. There are no deadlines. Those applicants invited to submit a full proposal will receive detailed information.
Restrictions: Individuals are not eligible for grants. Grants will not be given for propagandizing and/or influencing legislation or elections, political campaigns, programs that promote religious doctrine, events, endowments, annual fund drives, non-applied research, general operating expenses, or retirement of debt. Indirect costs for up to a maximum of ten percent of direct costs is allowable.
Geographic Focus: California
Samples: Gateways Hospital and Mental Health Center, Los Angeles, California, $55,000, child and adolescent evidence based treatment program; Children's Hospital of Orange County, CHOC Children's positive patient identification pilot project, Orange, California, $350,000; and Children's Hospital of Orange County, Orange, California, $96,000, NICU extremely low birth-rate baby unit.
Contact: Caroline Chung, Grants Manager; (213) 630-6500; fax (213) 630-6509; webadmin@unihealthfoundation.org
Internet: http://www.unihealthfoundation.org/program_areas.html
Sponsor: UniHealth Foundation
800 Wilshire Boulevard, Suite 1300
Los Angeles, CA 90017

UniHealth Foundation Innovation Fund Grants   3785
UniHealth Foundation is an independent private health care foundation established in 1998 whose mission is to support and facilitate activities that significantly improve the health and well being of individuals and communities they serve. The objective of Innovation Fund Grants is to encourage new, creative approaches that address complex and enduring health challenges. The Foundation encourages the development of new ways of delivering and financing healthcare programs and services. Grant funding is designed to provide an opportunity for nonprofit organizations to test creative ideas and promising new practices through interventions which address demonstrated need. Grants range from $10,000 to $350,000. Throughout the year the Foundation announces other initiatives that focus on specific goals and objectives to promote foundation-identified priorities. New requests for proposals are listed on the Foundation's website.
Requirements: Qualified nonprofit organizations providing health-related services consistent with the Foundation's mission are eligible. Applicants are required to submit a preliminary letter of inquiry. There are no deadlines. Those applicants invited to submit a full proposal will receive detailed information.
Restrictions: Individuals are not eligible for grants. Grants will not be given for propagandizing and/or influencing legislation or elections, political campaigns, programs that promote religious doctrine, events, endowments, annual fund drives, non-applied research, general operating expenses, or retirement of debt. Indirect costs for up to a maximum of ten percent of direct costs is allowable.
Geographic Focus: California
Contact: Caroline Chung, Grants Manager; (213) 630-6500; fax (213) 630-6509; webadmin@unihealthfoundation.org
Internet: http://unihealthfoundation.org/program_areas.html
Sponsor: UniHealth Foundation
800 Wilshire Boulevard, Suite 1300
Los Angeles, CA 90017

UniHealth Foundation Workforce Development Grants   3786
UniHealth Foundation is an independent private health care foundation established in 1998 whose mission is to support and facilitate activities that significantly improve the health and well being of individuals and communities they serve. The objectives of the Workforce Development Grants are to: increase the size and quality of the healthcare workforce; encourage and enable the practice of medicine in underserved communities; promote the training and retention of competent clinical staff; and foster skills improvement and knowledge enhancement through continuous education. Grants support programs to train and retain qualified clinical staff, foster ongoing professional development, including management and leadership skills, and promote allied health professions training. Throughout the year the Foundation announces other initiatives that focus on specific goals and objectives to promote foundation-identified priorities. New requests for proposals are listed on the Foundation's website.
Requirements: Most grants are made for the purpose of funding healthcare services and programs provided by or through qualified charitable hospitals in specified service areas in Los Angeles and northern Orange Counties. The service areas are: San Fernando and Santa Clarita Valley; Westside and Downtown Los Angeles; San Gabriel Valley; and Long Beach and Orange County. Applicants are required to submit a preliminary letter of inquiry. There are no deadlines. Those applicants invited to submit a full proposal will receive detailed information.
Restrictions: Individuals are not eligible for grants. Grants will not be given for propagandizing and/or influencing legislation or elections, political campaigns, programs that promote religious doctrine, events, endowments, annual fund drives, non-applied research, general operating expenses, or retirement of debt. Indirect costs for up to a maximum of ten percent of direct costs is allowable.
Geographic Focus: California
Contact: Caroline Chung, Grants Manager; (213) 630-6500; fax (213) 630-6509; webadmin@unihealthfoundation.org
Internet: http://unihealthfoundation.org/program_areas.html
Sponsor: UniHealth Foundation
800 Wilshire Boulevard, Suite 1300
Los Angeles, CA 90017

Union Bank, N.A. Corporate Sponsorships and Donations   3787
As part of its ten-year community committment, Union Bank has pledged to annually distribute at least two percent of its annual after-tax net profit to help meet the needs of the communities it serves, a commitment that has resulted in donations exceeding $72 million dollars during the first six years. This two-percent charitable commitment is achieved through contributions and sponsorships made directly by the bank (through the corporate contribution program) and through grants and investments made by the bank's foundation (Union Bank Foundation). The bank's corporate contributions program is intended to enhance the bank's reputation and visibility by supporting the charitable work of its employees and clients. The bank funds donations and sponsorships supporting a broad range of charitable categories, including community economic development, affordable housing, education, health and human services, culture and arts, emergency services, and the environment. The bank is particularly interested in donations and sponsorships that support low-income populations and promote and enhance diversity in all its forms. The bank's local-area contribution committees consider applications at their monthly meetings. Applications are accepted via an online application system accessible from the bank website. Please note that event-sponsorship applications should be submitted at least ninety days in advance of the event date. Prospective applicants should review the bank's application guidelines and instructions, which are available at the website. Questions may be directed to the foundation officers listed on this page.
Requirements: 501(c)3 nonprofits in company-operating areas in California and the Pacific Northwest are eligible (e.g., San Diego, San Francisco, Los Angeles, Anaheim, Berkeley, Del Mar, Fresno, Irvine, Mission Grove, Pasadena, Sacramento, Salinas, San Jose, Santa Ana, and Torrance, California). A branch locator is available at the bank's website.
Restrictions: The bank does not support the following requests from the following entities or for the following items: individuals; veterans, military, fraternal, or professional organizations; political organizations or programs; service club activities; other intermediary foundations (i.e., foundations which, in turn, make grants to other charities); churches or religious groups (except separately incorporated community development corporations); educational institution operating funds; and individual elementary or secondary schools.
Geographic Focus: California, Oregon, Washington
Contact: J.R. Raines, Assistant Vice President; (619) 230-3105; CSRGroup@unionbank.com or charitablegiving@unionbank.com
Internet: https://www.unionbank.com/global/about/corporate-social-responsibility/foundation/foundation-grants.jsp#products-tab-item-2
Sponsor: Union Bank, N.A.
350 California Street
San Francisco, CA 94104

Union Bank, N.A. Foundation Grants 3788
As part of its ten-year community committment, Union Bank has pledged to annually distribute at least two percent of its annual after-tax net profit to help meet the needs of the communities it serves, a commitment that has resulted in donations exceeding $72 million dollars during the first six years. The two-percent charitable commitment is achieved through contributions and sponsorships made directly by the bank and through grants and investments made by the Union Bank Foundation, a nonprofit public-benefit corporation which serves as an agent for the bank. Because of its belief that the long-term success of the Union Bank business-model is dependent upon the existence of healthy communities, Union Bank Foundation focuses its philanthropy on building innovative initiatives and partnerships to cultivate healthy communities, which it identifies as possessing the following characteristics: stable families with high rates of home ownership; availability of affordable housing; livable-wage job opportunities; accessible public transportation; convenient access to professional services (e.g., doctors, lawyers, and accountants); adequate public services (e.g., police, fire, and sanitation); safe public places to relax and recreate (e.g., parks, libraries, theaters); clean air and water supplies; a high level of civic engagement; a community constituency possessing diverse income levels; well-funded public schools; successful small business owners who live in the community; a variety of retail shops and restaurants; and traditional financial institutions providing access to capital in or adjacent to the community. With an eye toward being an agent of positive change in Union Bank communities, the foundation focuses on the following strategic program areas (targeting resources especially to benefit low- to moderate-income populations): Affordable Housing; Community Economic Development; Education; and Environment. In the area of Affordable Housing, the foundation focuses on for-sale housing, rental housing, special-needs housing, senior housing, transitional-living facilities, emergency/homeless shelters, youth housing, self-help housing, farm-worker housing, pre-development funding to nonprofit developers, and capacity building for nonprofit housing organizations. In the area of Community Economic Development area, the foundation focuses on small business development, individual development, and neighborhood development. Small business development includes micro-enterprise development and support, technical assistance/entrepreneurial training, organizations that promote access to capital for business or farms meeting Small Business Administration criteria, and job creation. Individual development includes job training/apprenticeship, welfare-to-work programs, wealth-accumulation/asset-building programs, life-skills training, financial-literacy/credit-counseling programs, mortgage credit counseling, business education, and intervention/prevention programs for at-risk youth. Neighborhood Development includes gang prevention/gang intervention programs, crime prevention, dispute resolution/mediation/violence prevention, reduction of liquor outlets, improved quality of food in local markets, childcare and daycare programs, drug- and alcohol-rehabilitation programs, independent living programs, organizational capacity building and funding for operating/administrative expenses, and community organizing to engage, inform and empower citizenry. In the area of Education, the foundation focuses on scholarship programs, tutoring programs, general education degree (GED) preparation, English as a second language (ESL) programs, computer education, support for the teaching profession, teacher training, literacy programs, parent education, visual- and performing-arts-organizations outreach programs, enrichment programs, and capacity-building. In the area of Environment the foundation focuses on brown-field remediation, science and education relevant to green building, energy upgrade and conservation, rehabilitation and cleanup, coastal/creek- and reserve-cleanup and preservation, urban green-space projects, environmental education, aquariums and museums, ecology and recycling centers, and state parks, nature centers, conservancy centers, botanical gardens, and wildlife centers. The Union Bank Foundation prefers program grants, but will consider requests for core operating support and/or capacity-building grants to support exceptional work within its strategic funding categories. The foundation considers applications at its bimonthly board meetings. Applications are accepted via an online application system accessible from the foundation website. Applicants must choose from three categories when they apply. These are requests for $1,000 or less, requests for $1001 to $25,000, and requests for over $25,000. Prospective applicants should review the foundation's application guidelines and instructions, which are available at the foundation website. Questions may be directed to the foundation officers listed on this page.
Requirements: 501(c)3 nonprofits in company-operating areas in California and the Pacific Northwest are eligible (e.g., San Diego, San Francisco, Los Angeles, Anaheim, Berkeley, Del Mar, Fresno, Irvine, Mission Grove, Pasadena, Sacramento, Salinas, San Jose, Santa Ana, and Torrance, California). A branch locator is available at the sponsor website.
Restrictions: The foundation does not support the following requests from the following entities or for the following items: individuals; veterans, military, fraternal, or professional organizations; political organizations or programs; service club activities; other intermediary foundations (i.e., foundations which, in turn, make grants to other charities); churches or religious groups (except separately incorporated community development corporations); educational institution operating funds; and individual elementary or secondary schools.
Geographic Focus: California, Oregon, Washington
Amount of Grant: 5,000 - 25,000 USD
Samples: Asian Business Center, Los Angeles, California; Catholic Charities of San Diego, San Diego, California; Audubon California, Sacramento, California.
Contact: J.R. Raines; (619) 230-3105; charitablegiving@unionbank.com
Internet: https://www.unionbank.com/global/about/corporate-social-responsibility/foundation/foundation-grants.jsp
Sponsor: Union Bank Foundation
P.O. Box 45174
San Francisco, CA 94145-0174

Union Benevolent Association Grants 3789
The Union Benevolent Association attempts to distribute its limited resources as widely as is consistent with the needs of its recipients. It tends to select projects that aid large groups of individuals among the disadvantaged. Grants have frequently been given to help support summer camping and educational experiences for inner city youths, to aid programs for the elderly and handicapped, to assist those involved in legal, medical, and social difficulties, to improve the quality of life in deteriorating neighborhoods, and to provide deprived young people with employable skills and artistic and cultural experiences. The Board prefers to make grants for specific projects and, as a rule, does not contribute toward general operating revenues or major building programs. It limits its grants to projects serving the needs of residents of Philadelphia. Preference is given to organizations with budgets of 2 million and below, and which leverage the use of volunteers.
Requirements: The Association makes grants only to 501(c)3 non-profit organizations,that work with people in need, within the city of Philadelphia, PA. Submitted proposals must be on the Delaware Valley Grantmakers Common Application Form (available on the Union Benevolent Association website) and be received (not posted) by email AND regular mail on or before 5:00PM on the deadline date.
Restrictions: The Association does not make grants to support capital renovations or to individuals.
Geographic Focus: Pennsylvania
Date(s) Application is Due: Apr 30; Sep 30
Contact: Fernando Chang-Muy; (215) 763-7670; fax (215) 731-9457; info@uba1831.org
Internet: http://www.uba1831.org/grant_guide.html
Sponsor: Union Benevolent Association
1616 Walnut Street, Suite 800
Philadelphia, PA 19103

Union County Community Foundation Grants 3790
Established in 1989, the Union County Community Foundation is a regional affiliate of Foundation for the Carolinas. The Foundation assists donors in making charitable gifts to the community, provides services for nonprofit organizations to create new or manage existing endowments and makes grants for new projects. Preference will be given to programs that meet the following criteria: serve those who are mentally or physically challenged as a result of visual and/or hearing impairments; provide professional development for Union County public school teachers; or address community needs through innovative approaches (funding available only to programs that cannot be covered by normal budgets of existing charitable agencies). Giving limited for the benefit of Union County, NC.
Requirements: Non-profits in or benefiting the community of Union County, North Carolina are eligible to apply. Contact the Foundation, an application form is required to submit a proposal. Applicants should submit the following: signature and title of chief executive officer; copy of most recent annual report/audited financial statement/990; copy of current year's organizational budget and/or project budget; listing of board of directors, trustees, officers and other key people and their affiliations; copy of IRS Determination Letter.
Restrictions: No grants to individuals, or for capital campaigns, ongoing operating budgets beyond the seed level, publication of books, conferences, or endowment funds.
Geographic Focus: North Carolina
Contact: Karen Coppadge, Grants Specialist; (704) 973-4559; kcoppadge@fftc.org
Sponsor: Union County Community Foundation
217 South Tryon Street
Charlotte, NC 28202-3201

Union Labor Health Foundation Angel Fund Grants 3791
The Union Labor Health Foundation Angel Fund provides small grants to meet immediate medical or health related needs of individuals living in Humboldt County. The Fund serves the needs of children and adults. Decisions about grant requests are contingent upon funding criteria and the availability of current funds. Most grants are approved in the $25-$250 range. Grants in excess of $500 are seldom approved. The following items are typically funded: medically related examinations, procedures and equipment; eye exams and eyeglasses ($130 maximum, requires a copy of the perscription if exam has been completed); medically related travel ($200 maximum) (non-emergency requires a form completed by a physician); prescriptions (emergency, short-term); items to improve accessibility/independence for disabled or elderly individuals; orthopedic needs ($180 maximum), compression stockings ($160 maximum); dental procedures for youth through age 19. Other items that are occasionally funded include the following: local bus tickets (for transportation to medical appointments); wheelchair accessibility items; pool therapy; therapy equipment; minor home repairs (if related to a medical, safety, accessibility or independence need); camps for kids with special needs, when other sources of funding (i.e. scholarships) have been researched and are not available; psychological counseling/evaluation, only on an emergency basis and with a long term management plan; and children's bike helmets (distributed through the Eureka Police Department). Applications are accepted at any time and are reviewed every Wednesday. There is generally a two week turn-around time on application review and check processing. The application and additional information are available at the website.
Requirements: Applications must be made through a qualified sponsor, such as a recognized social service agency, school counselor, or medical provider. The sponsor will then help to administer the funds which are granted. Applications should be filled out as specifically as possible, including the medical condition of the applicant and whether they can pay a portion of the cost of the item or service. Applications should also include any comparison shopping for the item, and whether additional sources have been sought. If the applicant is a cancer survivor, the application should indicate if they are in contact with the American Cancer Society, and if they have been approved for financial assistance. For children with medical needs, the application should indicate if the California Children's Services is involved.

Restrictions: The following items are usually not funded: adult dental care; dentures; hearing aids; lift chair; acupuncture/massage; care providers; weight loss programs; CPAP machines; counseling (long term); aerochamber; burial/cremation/funeral expenses; tattoo removal; iPad/iPod; cell phone; vision therapy; rent/deposits/utility bills; car payments/vehicle maintenance; birth certificates; green cards; driver's license or driver education courses; waste removal; smoke alarms; wood for stove; baby items; child care; parenting classes; woodstove barriers/fireplace gates; summer camp; or dog training.
Geographic Focus: California
Amount of Grant: 25 - 250 USD
Contact: Jill Moore, Program Coordinator, Community Strategies; (707) 442-2993, ext. 314; fax (707) 442-9072; jillm@hafoundation.org
Internet: http://www.ulhf.org/content/view/91/82/
Sponsor: Union Labor Health Foundation
363 Indianola Road
Bayside, CA 95524

Union Labor Health Foundation Community Grants 3792
The Union Labor Health Foundation Community Grants support health programs that make a difference in the lives of Humboldt County residents. The Foundation provides funding to projects or institutions that nurture, foster, encourage, support and educate in order to enhance the wellbeing of each individual within the Humboldt County region. The Foundation is interested in projects that concentrate on the following: nurture, foster, encourage, support and educate, in order to enhance the wellbeing of residents of the County of Humboldt; increase access and/or service delivery for underserved populations; and have a lasting impact or a plan for sustainability after the Foundation grant funding is expended. Detailed instructions about application submission are available at the website.
Requirements: Applicants must be nonprofit charitable or federal tax exempt organizations, public schools, government agencies, Indian tribal governments or have a qualified fiscal sponsor. Projects must benefit the communities within Humboldt County. All organizations from outside this service area must demonstrate that they are working with a county based group to develop and implement the proposed project.
Restrictions: Grants cannot be made for the following: expenses outside the service area such as travel expenses of schools or groups for trips, good will ambassadors, or scholarships and fellowships to other countries; infrastructure, deferred maintenance or annual operating costs of public institutions, churches, services of special tax districts, or government agencies; religious activities or projects that exclusively benefit the members of sectarian or religious organizations; or expenses that have already been incurred.
Geographic Focus: California
Date(s) Application is Due: Sep 4
Amount of Grant: 4,500 - 10,000 USD
Contact: Amy Jester, Program Manager, Health and Nonprofit Resources; (707) 442-2993, ext. 374; fax (707) 442-9072; amyj@hafoundation.org
Internet: http://www.ulhf.org/content/view/92/83/
Sponsor: Union Labor Health Foundation
363 Indianola Road
Bayside, CA 95524

Union Labor Health Foundation Dental Angel Fund Grants 3793
The Union Labor Health Foundation (ULHF), a supporting organization of the Humboldt Area Foundation, provides funds for the Dental Angel Fund. The Dental Angel Fund was created to meet dental-related needs of individual youth and youth groups in Humboldt County in situations where emergency funds are required. The Fund is designed to meet the needs of children up to the age of 19 who are residents of Humboldt County. Applications are accepted at any time and reviewed periodically, with a two week turn-around time. The application is available at the website.
Requirements: The application should include a pre-treatment plan from the treating dentist.
Restrictions: A third party must apply for the individual. Funding is not available for orthodontia or cosmetic dental needs. Service providers may not make requests for their own reimbursement and individuals may not apply on their own behalf.
Geographic Focus: California
Contact: Jill Moore, Program Coordinator, Community Strategies; (707) 442-2993, ext. 314; fax (707) 442-9072; jillm@hafoundation.org
Internet: http://www.ulhf.org/content/view/91/82/
Sponsor: Union Labor Health Foundation
363 Indianola Road
Bayside, CA 95524

Union Pacific Foundation Health and Human Services Grants 3794
The Foundation has a strong interest in promoting organizational effectiveness among nonprofits. To that end, this foundation dedicates the majority of their grants to help nonprofit organizations build their capacity, increase their impact, and operate more efficiently and effectively. Grants are made primarily to proposals in the areas of community and civic, and health and human services. The health and human services category assists organizations dedicated to improving the level of health care or providing human services in the community. Types of support include general operating support, continuing support, capital campaigns, building construction and renovation, curriculum development, equipment acquisition, program development, and matching funds.
Requirements: The Foundation will accept only online applications; printed copies of the application are not available and will not be accepted. Grants are made to institutions located in communities served by Union Pacific Corporation and its operating company Union Pacific Railroad Company.
Restrictions: The Foundation generally will not consider a request from or for: individuals; organizations/projects/programs that do not fit within the Foundation's funding priorities; organizations without a Section 501(c)3 public charity determination letter from the Internal Revenue Service; organizations that channel grant funds to third parties; organizations whose dominant purpose is to influence legislation or participate/intervene in political campaigns on behalf of or against any candidate for public office; organization/projects/programs for which the Foundation is asked to serve as the sole funder; organizations that already have an active multi-year Union Pacific Foundation grant; religious organizations for non-secular programs; organizational deficits; local affiliates of national health/disease-specific organizations; non U.S.-based charities; organizations whose program activities are mainly international; elementary or secondary schools; athletic programs or events; donations of railroad equipment; conventions, conferences or seminars; fellowships or research; loans; labor organizations; or organizations whose programs have a national scope.
Geographic Focus: Arizona, Arkansas, California, Colorado, Idaho, Iowa, Kansas, Louisiana, Minnesota, Missouri, Montana, Nebraska, Nevada, New Mexico, Oklahoma, Oregon, Tennessee, Texas, Utah, Washington, Wisconsin, Wyoming
Date(s) Application is Due: Aug 15
Contact: Darlynn Myers, Director; (402) 271-5600; fax (402) 501-2291; upf@up.com
Internet: http://www.up.com/found/grants.shtml
Sponsor: Union Pacific Foundation
1400 Douglas Street, Stop 1560
Omaha, NE 68179

UnitedHealthcare Children's Foundation Grants 3795
UnitedHealthcare Children's Foundation grants provide financial help and assistance for families with children that have medical needs not covered or not fully covered by their commercial health insurance plan. The Foundation aims to fill the gap between what medical services and items a child needs and what their commercial health benefit plan will pay for. The amount awarded to an individual within a 12-month period is limited to either $5,000 or 85% of the fund balance, whichever amount is less. Awards to any one individual are limited to a lifetime maximum of $10,000.
Requirements: The applicant must: be 16 years old or younger and live in the United States and receive and pay for care/items in the United States; and be covered by a commercial health insurance plan and limits for the requested service are either exceeded, or no coverage is available and/or the costs are a serious financial burden on the family. An application must be submitted prior to the child's 17th birthday.
Restrictions: The following set of items are excluded from grant consideration: alternative treatments, including listening, vision, cognitive, neuro-feedback and social skills therapy; dental or orthodontic treatment, unless related to a serious medical condition (such as cleft palate); drugs not licensed by the FDA; educational and tutoring programs; camps; home improvements and modifications; vehicles (cars, vans, trucks, etc.); service dogs or other animals, unless to support the visually impaired; biofeedback; biomedical consultations; chelation therapy, unless for proven medical indication of lead or copper or iron; heavy metal toxicity testing; hyperbaric oxygen treatment; herbal testing; Relationship Development Intervention (RDI); infertility, pregnancy, and birthing; autopsy; burial costs; funeral costs; clinical trials; or tablets (such as iPads,etc.), computers, mobile devices and other recreational electronics not specifically designed for medical or clinical treatment purposes.
Geographic Focus: All States
Amount of Grant: Up to 5,000 USD
Contact: Customer Servive; (855) 698-4223; customerservice@uhccf.org
Internet: http://www.uhccf.org/apply/
Sponsor: UnitedHealthcare Children's Foundation
P.O. Box 41, 9700 Health Care Lane
Minneapolis, MN 55440-0041

United Healthcare Community Grants in Michigan 3796
United Healthcare is dedicated to helping people nationwide live healthier lives by simplifying the health care experience, meeting consumer health and wellness needs, and sustaining trusted relationships with care providers. United Healthcare Community grants offer funds for specific proposals from community organizations to improve nutrition and physical activity resources for Michigan's youth. Initial grants will range from $20,000 to $30,000. Proposals are due May 1, and successful awardees will be notified September 30.
Requirements: 501(c)3 organizations either located in, or serving the residents of, Michigan are eligible to apply.
Geographic Focus: All States
Date(s) Application is Due: May 1
Amount of Grant: 20,000 - 30,000 USD
Contact: Molly McMillen, (952) 931-6029 or (866) 633-2446; molly.mcmillen@uhc.com or communitygrants@uhc.com
Internet: http://www.businesswire.com/news/home/20150420006768/en/UnitedHealthcare-Award-120000-Grants-Michigan-Organizations-Dedicated#.VUKOS2bfit8
Sponsor: United Healthcare Services
5901 Lincoln Drive
Minneapolis, MN 55436

United Hospital Fund of New York Health Care Improvement Grants 3797
United Hospital Fund awards provides funding to not-for-profit and public hospitals, and health care, academic, and public interest organizations, to improve health care in New York City. The Fund supports the development of model projects, sponsors research to analyze systemic problems, and fosters innovative solutions to health care issues. Health Care Improvement Grants are awarded to develop and evaluate innovative health care projects and conduct research and analysis of significant health systems issues that will shape the future of health care in New York City. Grants are intended to increase access

to and use of appropriate, high quality, and efficient health care services, especially for low-income and other vulnerable persons. Applications should address one of the following issue areas: expanding health insurance coverage; strengthening health care finances; improving quality of care; or redesigning health care services. Projects should test innovative or creative responses to an important aspect of one of these issues. They also should yield information and lessons of significance to a broader community of health care providers and policymakers.
Requirements: Applicants must be not-for-profit organizations in New York City. Qualifying organizations may submit only one application per grant cycle. Applicants are available at the Fund's website.
Restrictions: Unallowable cost include: overhead or other indirect costs; general operating support; and capital costs (such as construction, acquisition, or renovation of facilities). Equipment expenses must be integral to the project and justified in the budget narrative and may not exceed five percent of the total request.
Geographic Focus: New York
Date(s) Application is Due: Mar 1; Jul 1; Nov 1
Samples: Fund for Public Health in New York, New York, New York $33,000, improving quality of care; Highbridge Community Life Center, New York, New York, $75,000, redesigning health care services; New York City AIDS Fund, New York, New York, $25,000, improving the quality of care.
Contact: Hollis Holmes; (212) 494-0700; fax (212) 494-0800; hholmes@uhfnyc.org
Internet: http://www.uhfnyc.org/grants
Sponsor: United Hospital Fund of New York
1411 Broadway, 12th Floor
New York, NY 10018

United Methodist Committee on Relief Global AIDS Fund Grants   3798
The United Methodist Global AIDS Fund supports education, prevention and care programs for people living with HIV and AIDS around the world. The fund currently supports over 200 HIV/AIDS church oriented and Christ centered ministries in 37 countries, including the United States. The Fund develops appropriate promotional materials and funding guidelines, advocates for social justice, and encourages partnerships between congregations and organizations globally that are engaged in the struggle against HIV/AIDS. The Fund is guided by an inter-agency committee comprised of representatives from the Council of Bishops, General Board of Church and Society, General Board of Global Ministries, General Commission on Communications, the Division on Ministries with Young People and the Office of Christian Unity and Interreligious Concerns.
Requirements: Overseas programs fitting the mission of the Global AIDS Fund may apply to receive a grant. The application should propose work in one or more of the following types of activities: promoting HIV/AIDS awareness and prevention by education and information; behavior modification; voluntary testing and counseling; treatment of persons with AIDS; home based care; and care of orphans and vulnerable children.
Restrictions: This program is only for grants covering primarily United Methodist or Methodist-related projects for up to a one-year period.
Geographic Focus: All States
Contact: Shannon Trilli, Director of Global Health Initiatives; (212) 870-3870 or (212) 870-3951; strilli@umcor.org or umcor@umcor.org
Internet: http://www.umcor.org/UMCOR/Programs/Global-Health/HIV-AIDS
Sponsor: United Methodist Committee on Relief
475 Riverside Drive, Room 1520
New York, NY 10115

United Methodist Committee on Relief Global Health Grants   3799
UMCOR Global Health programs work internationally with more than 300 United Methodist hospitals and clinics, using education as well as preventative and curative measures to confront major health issues such as malnutrition, maternal and child mortality, HIV and AIDS, and malaria. Our programs also seek to increase access to clean water and better sanitation. In the United States, UMCOR Global Health programs connect people to resources and support networks and advocate for people with disabilities. UMCOR's holistic approach to health empowers people to take charge of their own lives. This strategy emphasizes education and the development of local resources so that improvements in public health are sustained over time. UMCOR's programs reach entire communities, especially the most vulnerable, and value local cultural practices as sources of preventative and curative health care. The goals are to: end poverty and hunger; attain universal education; achieve gender equality; promote child health; promote maternal health; combat HIV/AIDS and other disease of poverty; attain environmental sustainability; and build a global partnership.
Restrictions: This program is only for United Methodist and Methodist-related projects outside the U.S.
Geographic Focus: All States, All Countries
Contact: Shannon Trilli, Director of Global Health Initiatives; (212) 870-3870 or (212) 870-3951; strilli@umcor.org or umcor@umcor.org
Internet: http://new.gbgm-umc.org/umcor/http://www.umcor.org/UMCOR/Programs/Global-Health/Global-Health-and-Developmentwork/health/malaria/
Sponsor: United Methodist Committee on Relief
475 Riverside Drive, Room 1520
New York, NY 10115

United Methodist Health Child Mental Health Grants   3800
Brain imaging research reveals the astonishing impact early childhood experiences have on the physical architecture of the brain. Other research shows the indivisible interrelationship of early growth in cognitive, physical and mental health. The foundational importance of social and emotional development in early childhood to success in learning, regulating behavior, and building positive relationships throughout life is increasingly recognized. The neuroscience of the first five years of a child's life is clear. Public policies and programs intended to support the healthy development of children have not kept pace with the science. Establishing systems for identifying potential social and emotional development issues as early as possible and assuring access to timely and appropriate early intervention is not only more economical and effective than treating a problem later - such as school failure or incarceration - but early support also minimizes the damage to a child's potential and a family's well-being. The Child Mental Health Grants funding opportunity is intended to support the design and implementation of a coordinated system of early screening and intervention for the healthy social and emotional development of young children in three to four Kansas communities. The specific goals of the funding are: to stimulate coordinated community-wide screening of pregnant women, new mothers, and young children; and to improve coordination among relevant child-serving organizations to assure timely and appropriate promotion, prevention, and treatment for young children's social and emotional development. The maximum individual grant award will be $200,000.
Requirements: Grants can be awarded only to 501(c)3 organizations or governmental entities with projects to benefit the health of Kansas residents. Grants have one or more of the following purposes: to develop new or expanded, sustainable program resources to provide quality services; to change the delivery system to meet demands, improve access/quality, or reduce cost; to test innovative ideas for improved service delivery; to offer public education for improvement of individual and community health care; to provide group opportunities for health care providers to improve critical skills; or to develop technical expertise, collaborations, and similar supports for improvement and change in health care service delivery and education. Contact the sponsor before beginning a grant application - the Health Fund will not accept unsolicited and unauthorized applications, and will be automatically rejected them without consideration.
Restrictions: Grants are not awarded to individuals or for individual medical, dental, or other personal care treatment. Generally, capital projects and endowments are not funded. The fund does not provide regular operating expenses of on-going projects. The Health Fund does not fund projects in other U.S. states (outside of Kansas) or foreign nations.
Geographic Focus: Kansas
Contact: Virginia Elliott; (620) 662-8586; fax (620) 662-8597; velliot@healthfund.org
Internet: http://www.healthfund.org/ycsed2012.php
Sponsor: United Methodist Health Ministry Fund
100 East First Avenue, P.O. Box 1384
Hutchinson, KS 67504-1384

United Methodist Health Ministry Fund Grants   3801
Grantmaking is the Health Fund's primary means of achieving its mission, and we see each grant as an opportunity to move toward the goal of healthy Kansans. To maximize impact with available resources, the Health Fund is targeting the following three areas for funding: access; fit kids; ready for life. In addition to these general focus areas, the Health Fund seeks to support Kansas United Methodist churches in health ministry work through the Healthy Congregations Covenant program. Grants are awarded to health care projects proposed by eligible organizations to respond to needs and build on assets of local, regional, and state situations. These grants generally have one or more of the following purposes: develop new or expanded, sustainable program resources to provide quality services; change the delivery system to meet demands, improve access/quality, or reduce cost; test innovative ideas for improved service delivery; offer public education for improvement of individual and community health care; provide group opportunities for health care providers to improve critical skills; and develop technical expertise, collaborations, and similar supports for improvement and change in health care service delivery and education.
Geographic Focus: Kansas
Contact: Virginia Elliott; (620) 662-8586; fax (620) 662-8597; velliott@healthfund.org
Internet: http://www.healthfund.org/funding.php
Sponsor: United Methodist Health Ministry Fund
100 East First Avenue, P.O. Box 1384
Hutchinson, KS 67504-1384

United States Institute of Peace - Jennings Randolph Senior Fellowships   3802
The Jennings Randolph Senior Fellowships provide scholars, policy analysts, policy makers, journalists, and other experts with opportunities to spend time in residence at the Institute, reflecting and writing on pressing international peace and security challenges. Applications are invited from all disciplines and professions. Senior Fellowships usually last for ten months, starting in October, but shorter-term fellowships are also available. Fellowships are open to citizens of any country. The Institute awards between 10 and 12 fellowships per year. The program attempts to match the recipient's earned income during the year preceding the fellowships, up to a maximum of $100,000 for 10 months. The Institute will provide coverage of 80% of health premiums for the Fellow and his/her eligible dependents, with a cap of $500 per month. The Institute will also cover travel to and from Washington, D.C., for Fellows and their dependents. Each Fellow is provided with a part-time research assistant during his/her fellowship. The Institute does not provide housing in Washington D.C., but it provides information on housing, schools and daycare.
Requirements: Citizens of any country may apply. Non-U.S. citizens without permanent resident status must obtain a J-1 exchange visitor visa to participate in the Fellowship Program. This status requires recipients to reside in their home country for two years following the fellowship before applying for the H or L visa, or for permanent residency in the United States. There is no specific educational degree requirement for candidates. Fellows come from a variety of professional backgrounds and from early, middle and late stages of their careers. Priority is given to proposals deemed likely to make timely and significant contributions to the understanding and resolution of ongoing and emerging conflicts and other challenges to international peace and security. Historical topics are

appropriate if they promise to shed light on contemporary issues. Area studies projects and single-case studies will be competitive if they focus on conflict and its resolution, apply to other regions and cases, or both. Fellows are expected to be at the Institute and participate in its daily life. Fellows are expected to devote full attention to their Fellowship work in order to complete their projects within the period of residency. The Institute requires first right of review for manuscripts produced as a result of Fellowship support.
Restrictions: Senior Fellow awards may not be granted for projects that constitute policymaking for a government agency or private organization, focus to any substantial degree on conflicts within U.S. domestic society, or adopt a partisan, advocacy, or activist stance. Joint applications (two or more applicants for a single project) will not be accepted. An Institute Fellowship may not be deferred or combined with any other major award.
Geographic Focus: All States, All Countries
Date(s) Application is Due: Sep 10
Amount of Grant: Up to 100,000 USD
Contact: Elizabeth Cole; (202) 429-3869 or (202) 457-1700; interviews@usip.org
Internet: http://www.usip.org/grants-fellowships/jennings-randolph-senior-fellowship-program
Sponsor: United States Institute of Peace
2301 Constitution Avenue, NW
Washington, D.C. 20037

United Technologies Corporation Grants   3803
United Technologies Corporation (UTC) grant-making is centered on the geographic regions where its employees live and work. Additionally the corporation focuses its grants on the following four areas of interest: supporting vibrant communities, building sustainable cities, advancing STEM education, and investing in emerging markets. A more detailed description of each area follows. UTC supports vibrant communities by supporting community revitalization initiatives, health and human service programs, and arts and culture. UTC defines sustainable cities as those that are safe and energy efficient to protect people, assets, and natural resources. In support of sustainable cities, UTC focuses on sustainable building practices, urban green space, and preservation of natural habitats to offset green-house gas emissions. To advance science, technology, engineering, and mathematics (STEM) education and to develop the next generation of engineers and scientists, UTC targets programs that include employee volunteerism to spark students' interest and inspire innovation, especially in minorities and women. As it invests in emerging markets, UTC seeks to lay a foundation for responsible citizenship from the inception of business expansion by supporting communities through employee engagement in China and India. Grant seekers have the option of applying either to UTC corporate headquarters or to a UTC business unit (Pratt & Whitney, Otis, Carrier, Sikorsky, UTC Fire & Security, and Hamilton Sundstrand). Either way, application is made online at the grant website given. Grant applications are accepted between January 1 and June 30 each year. Awardees will receive notification within one quarter of their application submission (or in the case of a UTC-business-unit application, in the first quarter of the calendar year in which funding will occur).
Requirements: 501(c)3 organizations in the U.S. and equivalent nonprofit organizations in the corporation's emerging markets are eligible.
Restrictions: Non-profit organizations may apply only once a year and to only one UTC business - either Corporate Headquarters or one of the UTC business units (see description section). UTC will not fund individuals, religious activities or organizations, municipalities, alumni groups (unless the award is distributed to the eligible higher education institution), booster clubs, sororities or fraternities, political groups, organizations engaged in or advocating illegal action, or any organization determined by UTC to have a conflict of interest. Additionally the corporation will not support fees for publication or merchandise.
Geographic Focus: All States, All Countries
Date(s) Application is Due: Jun 30
Contact: Andrew Olivastro; (860) 728-7000; contribu@corphq.utc.com
Internet: http://www.utc.com/Corporate+Responsibility/Community/Apply+for+a+grant
Sponsor: United Technologies Corporation
United Technologies Building
Hartford, CT 06101

Unity Foundation Of LaPorte County Grants   3804
The Unity Foundation of LaPorte County makes discretionary and field of interest grants to charitable organizations in the area of the arts, education, health and human services, the environment, and the community for the benefit of the citizens of LaPorte County, Indiana. The Foundation is particularly interested in funding: projects that are not adequately being serviced by existing community resources; start-up costs for new programs; one-time projects or needs; projects that provide leverage for generating other funds and community resources; projects that facilitate cooperation and collaboration between organizations and the communities within LaPorte County.
Requirements: Only charitable organizations with a verifiable 501(c)3 IRS status operating or offering programs in LaPorte County will be considered for funding.
Restrictions: No grants will be made: to churches for sectarian religious programs; for operating budgets or for basic municipal or educational functions and services; for endowment campaigns or for previously incurred debts; to provide long-term funding. No grants will be made for post-event or after-the-fact situations.
Geographic Focus: Indiana
Date(s) Application is Due: Jul 18
Contact: Margaret Spartz, Primary Contact; (219) 879-0327 or (888) 898-6489; mspartz@uflc.net or unity@uflc.net
Internet: http://uflc.net/grants-scholarships/
Sponsor: Unity Foundation of LaPorte County
619 Franklin Street, P.O. Box 527
Michigan City, IN 46361

USAID Accelerating Progress Against Tuberculosis in Kenya Grants   3805
The United States Agency for International Development (USAID) is seeking applications for a Cooperative Agreement to provide funding in support of a program entitled Accelerating Progress Against TB (APA-K) in Kenya. The main objective of this program is to extend access to quality-assured TB services in all districts and for all forms of TB, through the identification and implementation of evidence-based activities that support and/or complement the activities of the Government of Kenya's Division of Leprosy, TB and Lung Diseases. It is expected that a single grant with be funded in the amount of $40,000,000. The deadline for application is April 15.
Requirements: In response to this RFA, only local Kenyan organizations are eligible to apply.
Geographic Focus: Kenya
Date(s) Application is Due: Apr 15
Amount of Grant: 40,000,000 USD
Contact: Jennifer Kiiru, Acquisitions and Assistance Specialist; +254-20-862-2848 or +254-20-862-2000; fax +254-20-862-2680; jkiiru@usaid.gov or usaidke@usaid.gov
Internet: http://www.grants.gov/search/search.do;jsessionid=Ty6nRyTfyLlh9Bp501Fpjp txyhJL02HR2D01sPWcc1s9JVyBsr3x!1962156644?oppId=223813&mode=VIEW
Sponsor: U.S. Agency for International Development
Ronald Reagan Building
Washington, D.C. 20523-1000

USAID Child, Newborn, and Maternal Health Project Grants   3806
USAID expects in this 5-year Award to impact child and maternal health status in Pakistan significantly by increasing demand for, access to, use of, and quality of maternal and child health services – especially for Punjab and Sindh Provinces. By integrating essential child health services with critical elements of maternal and newborn healthcare, USAID expects to create sustainable linkages for ongoing improvements to health outcomes for women, infants, and children. The expected number of awards is one (1), to be funded at $60,000,000. The closing date for electronic applications is May 20, while hard copies must be received by May 23.
Requirements: The United States Agency for International Development (USAID) Mission to Pakistan invites interested non-governmental organizations (U.S. and Non-U.S.), private voluntary organizations (PVOs), faith-based organizations (FBOs), public international organizations (PIOs), and for profit organizations to submit an application for the Child, Newborn, and Maternal Health Project.
Geographic Focus: All States, All Countries
Amount of Grant: 60,000,000 USD
Contact: Sara Saqib; +92-51-208-1277; ssaqib@usaid.gov
Internet: http://www.grants.gov/search/search.do;jsessionid=sc1DRwpLDV4C4tmQM2Q 6mJBf3SHMkwKghf75HcJlrrhtqHvTfTzN!-1898900577?oppId=228973&mode=VIEW
Sponsor: U.S. Agency for International Development
Ronald Reagan Building
Washington, D.C. 20523-1000

USAID Comprehensive District-Based Support for Better HIV/TB Patient Outcomes Grants   3807
The United States Government, as represented by the United States Agency for International Development (USAID) Mission to Southern Africa, is seeking applications from organizations interested in implementing a five-year comprehensive district-based HIV-related services support program for better patient outcomes, as fully described in this Request for Applications (RFA). The purpose of this program is to strengthen SAG systems in order to improve patient outcomes and prevent HIV by supporting comprehensive clinic-based HIV-related services in select districts. All support will be in line with SAG national, provincial and district policies, standards, guidelines, and implementation plans as well as the Partnership Framework between the USG and SAG. The program will build and capitalize on the accomplishments and lessons learned since the national ARV rollout began in 2004. It will focus on institutionalization of routine and consistent use of systems designed to improve efficiencies in district and sub-district management that ultimately improve patient and public health outcomes.
Requirements: Any local South African non-governmental organizations (NGOs), private voluntary organizations (PVOs), Faith-based and community organizations, and for-profit companies willing to forego profit will be eligible for award under this RFA. In support of the Agency's interest in fostering a larger assistance base and expanding the number and sustainability of development partners, USAID encourages applications from potential new partners.
Geographic Focus: South Africa
Amount of Grant: Up to 20,000,000 USD
Contact: Beatrice Lumanade, Acquisitions and Assistance Specialist; +27 12 452 2377 or +27 12 452 2000; blumande@usaid.gov or applications4@usaid.gov
Internet: http://www07.grants.gov/search/search.do;jsessionid=1v8kR3HpjBFsc7Z47pRgj fSFjbcHqr2Jf6p2BXMYZTRoadnTrpWtT4!1321711693?oppId=169154&mode=VIEW
Sponsor: U.S. Agency for International Development
Ronald Reagan Building
Washington, D.C. 20523-1000

USAID Development Innovation Accelerator Grants   3808
This Broad Agency Announcement seeks opportunities to co-create, co-design, co-invest, and collaborate in basic and applied research and development for Science, Technology, Innovation, and Partnership (STIP). The United States Agency for International Development (USAID) invites organizations and companies to participate with USAID in response to a Critical Development Challenge Addendum issued under this BAA, to provide innovations and technologies that further USAID's development goal of dramatically improving or saving the lives of over 200 million people over the next five years. Nothing in the BAA precludes

reasonable cost sharing, matching, leveraging, or other exchange of resource arrangements, and proposers are encouraged to suggest creative approaches to resourcing projects.
Requirements: Public, private, for-profit, and non-profit organizations, as well as institutions of higher education, public international organizations, non-governmental organizations, U.S. and non-U.S. government organizations, and international donor organizations are eligible to apply.
Geographic Focus: All States, All Countries
Contact: Agreements Specialist; DIAglobalscale@usaid.gov
Internet: http://www.grants.gov/web/grants/view-opportunity.html?oppId=255248
Sponsor: U.S. Agency for International Development
Ronald Reagan Building
Washington, D.C. 20523-1000

USAID Ebola Response, Recovery and Resilience in West Africa Grants   3809
USAID Ebola Response, Recovery and Resilience in West Africa Grant program is an addendum to an existing announcement. All interested organizations should carefully review both this addendum and the full announcement, APS-OAA-14-000001 (where a duplicate of this addendum is also posted). Important information contained in the full worldwide announcement is not repeated in this specific addendum. Through this Addendum to the Global Development Alliance, USAID is hereby requesting the submission of concept papers that offer solutions to address challenges faced in the ongoing response, recovery and resilience efforts in the three countries where Ebola Virus Disease (EVD) has had the greatest impact - Liberia, Guinea and Sierra Leone.
Requirements: Eligible applicants include: nonprofits having a 501(c)3 status with the IRS, other than institutions of higher education; public and state controlled institutions of higher education; small businesses; for profit organizations other than small businesses; public housing authorities; Indian housing authorities; nonprofits that do not have a 501(c)3 status with the IRS, other than institutions of higher education; Native American tribal governments (Federally recognized); and private institutions of higher education.
Geographic Focus: All States
Date(s) Application is Due: Sep 15
Contact: Ken Lee, Agreement Specialist; (202) 712-5158; klee@usaid.gov
Internet: http://www.grants.gov/web/grants/view-opportunity.html?oppId=277276
Sponsor: U.S. Agency for International Development
Ronald Reagan Building
Washington, D.C. 20523-1000

USAID Family Health Plus Project Grants   3810
The purpose of this activity is to expand contraceptive choices in family planning and reproductive health by increasing the availability of long acting methods of family planning in Nigeria. This activity will support USAID/Nigeria's goal of increasing the use of high impact interventions, within the Investing in People Foreign Assistance objective. The project will identify suitable training venues to provide experience in implants and IUDs and adopt a practical curriculum to cover these topics. The project will visit the proposed training venues to ensure that there is adequate patient flow and that good quality procedures are being used and taught. Project staff themselves do not need to provide this training but must be able to assess the capacity of other trainers to provide good quality training. If possible, sites within the trainees' own state would be preferred; especially those also used as training sites for nursing or midwife schools. A single funding award is anticipated in the amount of $9,000,000. The annual posted deadline for applications is May 23.
Requirements: Nigerian non-governmental organizations are eligible to submit applications. Applicants must have established financial management, monitoring and evaluation, internal control systems, and policies and procedures that comply with established U.S. Government standards, laws, and regulations. All potential awardees will be subject to a responsibility determination (this may include a pre-award survey) issued by a warranted Agreements Officer in USAID.
Geographic Focus: Nigeria
Contact: Ugo Oguejiofor; +234 803 900 9300; abujafhplus@usaid.gov
Internet: http://www07.grants.gov/search/search.do;jsessionid=3YLtRvGZL1CMpzgyybZ8 WNC1DsyRSGC97kmvZ9BS6c5VQ1SvFSbL!1866286003?oppId=229613&mode=VIEW
Sponsor: U.S. Agency for International Development
Ronald Reagan Building
Washington, D.C. 20523-1000

USAID Family Planning and Reproductive Health Methods Grants   3811
The United States Agency for International Development (USAID) is seeking concept papers from qualified U.S. nonprofit non-Governmental Organizations (NGOs) for a program titled Family Planning and Reproductive Health Methods to Address Unmet Need, for funding of Cooperative Agreements. The purpose of this APS is to support the research, development, and introduction of technologies and approaches that better meet the needs of women and girls as their sexual and reproductive health concerns change over time. The General Objectives of the APS are to: refine existing FP methods to address method-related reasons for non-use; respond to product-related issues about currently available FP methods that arise at purchase and/or from the field, and that may affect provider/user perceptions and/or the supply chain; develop new FP methods that address method-related reasons for non-use, and/or fill gaps in the existing method mix; conduct research to foster the introduction and uptake of new and/or underutilized woman-initiated methods, particularly non-hormonal barriers, contraceptive vaginal rings, and fertility awareness methods based on knowledge and monitoring of the menstrual cycle; and develop multipurpose prevention technologies (MPTs) that address the simultaneous risks of unintended pregnancy, HIV, and other sexually transmitted infections (STIs) – particularly Herpes Simplex Virus (HSV) and Human Papillomavirus (HPV). The funding range of awards is $500,000 to $80,000,000.
Requirements: Eligible organizations include registered U.S. and non-U.S. private non-governmental organizations, non-profit organizations, and for-profit organizations willing to forego profit.
Geographic Focus: All States
Date(s) Application is Due: Jan 14
Amount of Grant: 500,000 - 80,000,000 USD
Contact: Jacquelin Bell, Agreement Specialist; (202) 567-4406; jacbell@usaid.gov
Internet: http://www.grants.gov/web/grants/view-opportunity.html?oppId=215873
Sponsor: U.S. Agency for International Development
Ronald Reagan Building
Washington, D.C. 20523-1000

USAID Fighting Ebola Grants   3812
This Broad Agency Announcement (BAA) seeks opportunities to co-create, co-design, co-invest, and collaborate in the development, testing, and scaling of practical and cost-effective innovations that can help healthcare workers on the front lines provide better care and stop the spread of Ebola. The United States Agency for International Development (USAID) invites organizations and companies to participate with USAID, in cooperation with its partners, in response to Fighting Ebola Challenge Addenda issued under this BAA, to provide innovations and technologies that further the U.S. Government's commitment to addressing the Ebola epidemic. No annual deadline for submission of applications has been identified. Multiple awards are anticipated.
Requirements: Public, private, for-profit, and non-profit organizations, as well as institutions of higher education, public international organizations, non-governmental organizations, U.S. and non-U.S. government organizations, and international donor organizations are eligible under this BAA.
Geographic Focus: All States, All Countries
Contact: Program Specialist; submitebolaEOI@usaid.gov
Internet: http://www.grants.gov/web/grants/view-opportunity.html?oppId=268349
Sponsor: U.S. Agency for International Development
Ronald Reagan Building
Washington, D.C. 20523-1000

USAID Global Health Development Innovation Accelerator Broad Agency Announcement Grants   3813
The Global Health Development Innovation Accelerator Broad Agency Announcement seeks opportunities to co-create, co-design, co-invest, and collaborate in the development, testing, and scaling of innovative approaches that address critical global health challenges. The United States Agency for International Development invites organizations and companies to participate with USAID, in cooperation with its partners, in response to a Global Health Challenge Addenda issued under this BAA to provide innovative interventions and technologies that further the U.S. Government's commitment to prevent and manage critical global health challenges. Multiple awards are anticipated. The amount of resources made available under this BAA will depend on the concepts received and the availability of funds. Funding levels will correspond to the scope and scale of the innovation but will generally be $250,000 per seed project, $250,000 per validation project, and $2,000,000 for transition awards. Seed and validation projects will be funded up to two years and transition to scale projects will be funded up to four years.
Requirements: Public, private, for-profit, and non-profit organizations, as well as institutions of higher education, public international organizations, non-governmental organizations, U.S. and non-U.S. government organizations, and international donor organizations are eligible to apply.
Restrictions: Individuals are not eligible to apply. Applicants from organizations based in, or applications with an operational focus in, the following countries are not eligible: Cuba, Iran, North Korea and Syria.
Geographic Focus: All States, All Countries
Contact: Agreement Officer; savinglivesatbirth@usaid.gov
Internet: http://www.grants.gov/web/grants/view-opportunity.html?oppId=273354
Sponsor: U.S. Agency for International Development
Ronald Reagan Building
Washington, D.C. 20523-1000

USAID Health System Strengthening Project Grants   3814
Pakistan (USAID/Pakistan) seeks applications from international and local non-governmental organizations, non-profit or voluntary organizations for the Health Systems Strengthening (HSS) component of a large Maternal and Child Health (MCH) Program. This APS provides prospective applicants with a fair opportunity to submit applications to USAID/Pakistan for a range of activities to build and strengthen the underlying health systems that are the backbone of equitable and quality MCH services. The expected number of awards is three (3), with each award ranging from $1,000,000 to $21,900,000. The deadline for application submission is October 23.
Requirements: Eligibility to apply is unrestricted.
Geographic Focus: All States, All Countries
Amount of Grant: 1,000,000 - 21,900,000 USD
Contact: Sara Saqib; +92-51-208-1277; ssaqib@usaid.gov
Internet: http://www.grants.gov/search/search.do;jsessionid=sc1DRwpLDV4C4tmQM2Q 6mJBf3SHMkwKghf75HcJlrrhtqHvTfTzN!-1898900577?oppId=204673&mode=VIEW
Sponsor: U.S. Agency for International Development
Ronald Reagan Building
Washington, D.C. 20523-1000

USAID Higher Education Partnerships for Innovation and Impact Grants 3815

The United States Agency for International Development (USAID) is seeking concept papers from qualified U.S. and non-U.S. higher education institutions (HEIs) to work with USAID to advance strategic priorities and objectives and achieve sustainable development outcomes, results, and impact. This Annual Program Statement (APS) has the flexibility to award Cooperative Agreements, Grants, Fixed Amount Awards, and leader with Associate Awards. USAID seeks to optimize its relationship with HEIs by identifying and promoting successful partnerships and collaboration models, and increasing USAID's access to higher education technical resources. The purpose of this APS is to promote opportunities for leveraging HEI capabilities across USAID's portfolio and its program cycle, and strengthen developing country HEI capabilities to respond to and solve critical development challenges. The posted deadline for applications is June 29.
Requirements: The Agency requests Concept Papers from U.S. and non-U.S. higher education institutions. All concept papers and subsequent full applications, if requested, will be expected to feature an higher education institution as the lead implementer.
Restrictions: This APS is not supported by specific funding, and any funding for any USAID-HEI partnership proposed under this APS would have to be requested from the specific USAID Mission, Bureau, or Independent Office with which the prospective applicant seeks to collaborate and to which the Concept Paper will be submitted.
Geographic Focus: All States
Date(s) Application is Due: Jun 29
Contact: Michele Maximilien; (202) 567-5073; mmaximilien@usaid.gov
Internet: http://www.grants.gov/web/grants/view-opportunity.html?oppId=277586
Sponsor: U.S. Agency for International Development
Ronald Reagan Building
Washington, D.C. 20523-1000

USAID HIV Prevention with Key Populations - Mali Grants 3816

The USAID/Mali HIV Prevention Program aims to provide HIV prevention services to one or more of the following key populations (KPs): men who have sex with men (MSM); commercial sex workers (CSWs); and/or HIV positive individuals through Positive Health Dignity and Prevention (PHDP). The primary goal is to reduce HIV prevalence among KPs in Mali and among the general population through behavior change communication (BCC), HIV counseling and testing, sexually transmitted infections (STI) management, and distribution and promotion of condoms and lubricants. Activities to be funded through this RFA should be in line with the U.S. Government (USG) Mali HIV/AIDS Strategic Plan 2010-2015 and will support the United States' President's Emergency Plan for AIDS Relief (PEPFAR) initiative. It is anticipated that three (3) awards will be funded, ranging from $600,000 to $2,000,000. The next upcoming deadline is May 8.
Requirements: All local Mali organizations are eligible to apply.
Geographic Focus: Mali
Date(s) Application is Due: May 8
Amount of Grant: 600,000 - 2,000,000 USD
Contact: Zachary Clarke, Contracting and Agreement Officer; +223 20 70 27 78 or +223 20 70 23 00; fax +223 20 22 39 33; zclarke@usaid.gov
Internet: http://www.grants.gov/search/search.do;jsessionid=k9ywRnXcGzJNZf5yyFK5M kJyHQhBHGTQlcyJyyVN9yBxlqx5pC5V!1866286003?oppId=225213&mode=VIEW
Sponsor: U.S. Agency for International Development
Ronald Reagan Building
Washington, D.C. 20523-1000

USAID Integration of Care and Support within the Health System to Support Better Patient Outcomes Grants 3817

The purpose of the USAID Integration of Care and Support within the Health System to Support Better Patient Outcomes Grants program is to support and strengthen integration of care and support services within the broader health system, and to strengthen community systems and organizations to ensure the provision of a continuum of comprehensive care and support services (palliative care). Two awards are anticipated, with each ranging up to $25,000,000. The posted annual application deadline is May 16.
Requirements: Any local South African non-governmental organization (NGO) and for-profit organization meeting the criteria in section III is eligible to apply.
Geographic Focus: South Africa
Date(s) Application is Due: May 16
Amount of Grant: Up to 25,000,000 USD
Contact: Martha Zhou, Grantor; +012-452-2179; mzhou@usaid.gov
Internet: http://www07.grants.gov/search/search.do;jsessionid=1v8kR3HpjBFsc7Z47pRgj fSFjbcHqr2Jf6p2BXMYZTRoadnTrpWtT4!1321711693?oppId=230173&mode=VIEW
Sponsor: U.S. Agency for International Development
Ronald Reagan Building
Washington, D.C. 20523-1000

USAID Land Use Change and Disease Emergence Grants 3818

The U.S. Agency for International Development's Regional Development Mission for Asia (USAID/RDMA) seeks to advance strategic support for multi-disciplinary solutions to pressing regional development challenges. Emerging infectious disease of pandemic potential and unchecked climate change represent significant sources of social and economic instability and are impediments to sustainable development. USAID/RDMA, in line with the objectives of the United States Government's Global Health Initiative and Global Climate Change Initiative, is soliciting applications for a program targeting enhanced economic justification for the creation and maintenance of healthy, sustainable landscapes. This new Land Use Change and Disease Emergence program (the Program) will generate a model capable of quantifying the infectious disease modulating value of ecosystems across a gradient of land use, providing a more comprehensive understanding of ecosystems services valuation, and enabling actionable, evidence-based tools promoting reduced-impact land utilization. One award is expected, ranging up to $2,000,000. The posted application deadline is May 29.
Requirements: U.S. or non-U.S. non-governmental organizations (NGOs), including universities, academic institutions, and/or a consortia, and Public International Organizations (PIOs) are eligible to apply.
Geographic Focus: All States, All Countries
Date(s) Application is Due: May 29
Amount of Grant: Up to 20,000,000 USD
Contact: Praveena ViraSingh; 66-2-257-3000; pvirasingh@usaid.gov
Internet: http://www07.grants.gov/search/search.do;jsessionid=1v8kR3HpjBFsc7Z47pRgj fSFjbcHqr2Jf6p2BXMYZTRoadnTrpWtT4!1321711693?oppId=229353&mode=VIEW
Sponsor: U.S. Agency for International Development
Ronald Reagan Building
Washington, D.C. 20523-1000

USAID Microbicides Research, Development, and Introduction Grants 3819

Clinical trials of vaginal microbicide have been tested recently in South Africa to determine efficacy of reduction of HIV infection. Microbicide research, development, and introduction activities which could potentially be supported through this APS mechanism are expected to contribute substantially to the Principles of the Global Health Initiative (GHI), particularly in promoting research and innovation; implementing a woman- and girl-centered approach; strengthening and leveraging inputs from multilateral organizations, global health partnerships, and the private sector; and encouraging country ownership and leadership. Potential activities could also contribute significantly to strengthening the health system building blocks related to commodities and procurement, service delivery, and leadership and governance. It is expected that gender analyses and other gender considerations will continue to inform the design and implementation of activities implemented as part of the Agency's microbicide program. All awards will also be subject to USAID environmental requirements, including completion of an initial environmental examination. The award floor is $100,000, with a maximum grant award of $2,250,000.
Requirements: To qualify for funding, applicants must: be either U.S. or non-U.S. organizations registered and working in USAID priority countries; and have substantial experience (5 or more years) working directly with relevant HIV/AIDS-related activities in multiple countries in sub-Saharan Africa. Relevancy of experience will be determined by USAID.
Restrictions: Non-eligible entities include other U.S. Government agencies and departments. Other U.S. Government agencies and departments may not apply for USAID funding under this APS.
Geographic Focus: All States, All Countries
Amount of Grant: 100,000 - 2,250,000 USD
Contact: Kathleen Bumpass; (202) 567-5008; kbumpass@usaid.gov
Internet: http://www.grants.gov/web/grants/view-opportunity.html?oppId=187913
Sponsor: U.S. Agency for International Development
Ronald Reagan Building
Washington, D.C. 20523-1000

USAID Policy, Advocacy, and Communication Enhanced for Population and Reproductive Health Grants 3820

The United States Agency for International Development (USAID) seeks concept papers from qualified applicants. This APS publicizes the intention of the United States Government (USG) to fund up to three (3) awards through the USAID, Global Health Bureau to address the overarching APS Program Purpose of assuring family planning and population issues are included in policies and programs as key to sustainable and equitable economic growth and development. The award(s) under this APS will advance the Agency's goals of ending extreme poverty and ending preventable child and maternal deaths by ensuring that decision makers understand the health and economic benefits of voluntary family planning and in turn, implement policies and programs to help meet unmet demand for family planning. Awards will range up to a maximum of $55,000,000. The annual closing date for applications is April 15.
Requirements: To qualify for funding, organizations must be: a U.S. organization or non-U.S. organization registered and working in USAID-assisted countries; a U.S. or non-U.S. academic institution (university, professional school, or two- or four-year college, including community colleges); or a research institute, professional society or similar organization.
Geographic Focus: All States
Date(s) Application is Due: Mar 12
Amount of Grant: Up to 55,000,000 USD
Contact: Samantha Corey, Agreement Specialist; (202) 567-4517; scorey@usaid.gov
Internet: http://www.grants.gov/web/grants/view-opportunity.html?oppId=275168
Sponsor: U.S. Agency for International Development
Ronald Reagan Building
Washington, D.C. 20523-1000

USAID Research and Innovation for Health Supply Chain Systems and Commodity Security Grants 3821

The purpose of this APS is to publicize the United States Government's (USG) plan to fund a limited number of awards through the U.S. Agency for International Development's (USAID) Washington Global Health Bureau (GH) to support research on a focused set of health supply chain systems and related commodity security issues in low and middle income countries. Funds may support: formative, analytical research on challenges to improving supply chain systems and the enabling environments for commodity security; and/or development and testing of interventions in the field based on formative research

conducted under this APS or from other sources. The awards under this APS will contribute to advancing and supporting health programs worldwide, contributing to ending preventable child and maternal deaths, creating an AIDS-free generation, and protecting communities from infectious diseases. In addition, this APS will address several key focus areas of the Global Health Initiative (GHI), including building sustainability through health systems strengthening and accelerating results through research and innovation. The objectives of the APS are listed below: improve health supply chain systems through transformative changes that use industry best practice; improve the quantity, reliability, and efficiency of financing for health commodities and supply chain systems; and improve governance and accountability for commodity security.
Requirements: To qualify for funding, organizations must be: U.S. organizations or non-U.S. organizations registered and working in USAID-assisted countries; and academic organizations (universities, professional schools, or two- or four-year colleges, including community colleges), or research institutes, professional societies or similar organizations that are directly associated with research activities.
Geographic Focus: All States
Date(s) Application is Due: May 29
Amount of Grant: 500,000 - 2,000,000 USD
Contact: Lesley Stewart; 202-567-4762 or 202-567-4664; lestewart@usaid.gov
Internet: http://www.federalgrants.com/Research-and-Innovation-for-Health-Supply-Chain-Systems-and-Commodity-Security-46262.html
Sponsor: U.S. Agency for International Development
Ronald Reagan Building
Washington, D.C. 20523-1000

USAID Social Science Research in Population & Reproductive Health Grants   3822
The purpose of the Social Science Research in Population and Reproductive Health grant is to publicize the United States Government's (USG) plan to fund one award through USAID/Washington's Office of Population and Reproductive Health (PRH) to support a broad range of social science and behavioral research and technical assistance on a focused set of family planning and reproductive health (FP/RH) issues. Funds may support activities that develop and test scalable interventions to achieve social and normative change for FP/RH outcomes; build on existing activities to test evidence-based practices/interventions in new settings; provide technical assistance for replication and scale up of successful interventions; and monitor and evaluate the impact and sustainability of scale up of evidence-based interventions, including evaluation of current or previous activities funded by USAID or other donors. The award under this APS will contribute to PRH's strategic objective of advancing and supporting FP/RH programs worldwide and health objectives of reducing unintended pregnancies, abortion, and improving maternal and child health. This APS will address two cross-cutting issues for PRH, gender and youth, which are also Agency-wide focal areas as highlighted in USAID's Youth in Development Policy and Gender Equality and Female Empowerment Policy. The estimated award is $30,000,000, with a deadline for application submission specified as November 11.
Requirements: U.S. organizations or non-U.S. organizations registered and working in USAID priority countries are eligible to apply.
Geographic Focus: All States, All Countries
Date(s) Application is Due: Nov 11
Contact: Terry Vann Ellis; (202) 712-0014; fax (202) 216-3132; tellis@usaid.gov
Internet: http://www.grants.gov/web/grants/view-opportunity.html?oppId=269551
Sponsor: U.S. Agency for International Development
Ronald Reagan Building
Washington, D.C. 20523-1000

USAID Support for International Family Planning Organizations Grants   3823
USAID International Family Planning (FP) Organizations Grants are made to increase the use of FP services globally through strengthening selected international family planning organizations which have a global reach and an extensive, multi-country network of FP clinics in order to achieve maximum program impact and synergies. Applicant are expected to address four results to achieve project goals. The results are: srengthened organizational capacity to deliver quality FP services to target groups; internal quality assurance standards and results quantified and disseminated to strengthen FP performance at a global level; increased organizational sustainability of country level programs, including internal South to South support and technical assistance; and gender-sensitive FP services targeting youth strengthened at a global level. If an award is made, implementing partners will be supported to enhance their capacity to manage and scale up FP interventions worldwide and in specific targeted country settings. There is currently no closing date for applications. It is expected that three awards will be given, with a maximum value of $40,000,000.
Requirements: Eligible organizations must be: United States organizations or non-United States organizations registered and working in USAID priority countries with a worldwide network of FP clinics and a recognized track record in delivering FP services; have documented experience in implementing FP programs in developing countries with FP being a central part of the organization's mission; have the capacity to provide south to south training via networks of clinics in multiple countries and regions; have links with community based support systems and non-government organizations and have a strong and a documented track record in training these organizations in FP; and be able to expand USAID's reach by operating in countries in which USAID does not currently have programming, particularly in Sub-Saharan Africa and South Asia.
Geographic Focus: All States, All Countries
Contact: Alisa Dunn, Contract Specialist; (212) 712-0908; adunn@usaid.gov
Internet: http://www.grants.gov/web/grants/view-opportunity.html?oppId=50154
Sponsor: U.S. Agency for International Development
Ronald Reagan Building
Washington, D.C. 20523-1000

USAID Systems Strengthening for Better HIV/TB Patient Outcomes Grants   3824
USAID-South Africa Systems Strengthening for Better HIV/TB Patient Outcomes Grants are intended to strengthen South African Government systems in order to improve patient outcomes and prevent HIV by supporting comprehensive clinic-based (hospitals, community health centers, and primary health care clinics) HIV-related services. Applicants may submit a concept paper, and approved applicants will be invited to submit a full application. Eight awards are anticipated, ranging from $10,000,000 to $150,000,000 each.
Requirements: Any local South African non-governmental organizations, private voluntary organizations, faith-based and community organizations, and for-profit companies willing to forego profit is eligible. Applications from potential new partners are encouraged.
Geographic Focus: South Africa
Date(s) Application is Due: Jan 6
Amount of Grant: 10,000,000 - 100,000,000 USD
Contact: Beatrice Lumanade; +27 12 452 2377; fax Applications4@usaid.gov
Internet: http://www07.grants.gov/search/search.do;jsessionid=1v8kR3HpjBFsc7Z47pRgjfSFjbcHqr2Jf6p2BXMYZTRoadnTrpWtT4!1321711693?oppId=133033&mode=VIEW
Sponsor: U.S. Agency for International Development
Ronald Reagan Building
Washington, D.C. 20523-1000

US Airways Community Contributions   3825
U.S. Airways invests in community organizations and initiatives to enhance quality of life in markets served by the airline. The airline's funding priorities include arts and culture, health and human services, and education and environment in the following hub markets and focus cities: Boston, Charlotte, Las Vegas, New York, Philadelphia, Phoenix, Pittsburgh and Washington, D.C..
Requirements: Tax exempt organizations in the airline's market areas are eligible to apply. All requests must be submitted in writing on the organization's letterhead 6-8 weeks prior to the event and any print deadlines. All completed proposals will receive a written response within 6-8 weeks from receipt. The Contributions Committee meets bi-weekly to review requests.
Restrictions: Funding will not be considered for: Pass through and start-up organizations; Individuals, including support for specific students, researchers or conference fees; Endowment programs; Foundations which are themselves grant-making entities; Professional associations; Debt-reduction campaigns; Fundraising activities related to individual sponsorship (i.e. walk-a-thons); Service clubs (i.e. Lions, Rotary, Kiwanis); Sports teams, political, labor or fraternal organizations, scouting groups and religious organizations; Organizations or programs funded 50% or more by government sources; Individual schools; district and system-wide education programs will be considered; Capital and building campaigns.
Geographic Focus: Arizona, District of Columbia, Massachusetts, Nevada, New York, North Carolina, Pennsylvania
Contact: Community Relations Manager; community.relations@usairways.com
Internet: http://www.usairways.com/awa/content/aboutus/corporategiving/default.aspx
Sponsor: U.S. Airways Community Foundation
4000 E Sky Harbor Boulevard
Phoenix, AZ 85034

US Airways Community Foundation Grants   3826
The community foundation supports multi-year capital and building campaigns by 501(c)3 non-profit organizations operating in U.S. Airways' hub markets of Charlotte, Philadelphia and Phoenix. Grant application deadlines are April 1 and October 1 of each year. Grants will be made in late May and November.
Requirements: Grantees must be 501(c)3 nonprofit organizations that operate in one of U.S. Airways' hub cities (Charlotte, Philadelphia and Phoenix) and improve the quality and availability of charitable health care; artistic and cultural organizations; education; and/or community services. Support includes multi-year capital and building campaigns only
Restrictions: The foundation will not consider funding for endowment programs; professional associations; debt-reduction campaigns; service clubs (i.e. Lions, Rotary and Kiwanis); political, labor/fraternal or sports organizations; individual schools; general operating support; staff/consultant fees; hardware, software; infrastructure; religious activities.
Geographic Focus: Arizona, North Carolina, Pennsylvania
Date(s) Application is Due: Apr 1; Oct 1
Contact: Julie Coleman; (480) 693-3652; community.relations@usairways.com
Internet: http://www.usairways.com/awa/content/aboutus/corporategiving/default.aspx
Sponsor: U.S. Airways Community Foundation
4000 E Sky Harbor Boulevard
Phoenix, AZ 85034

USCM HIV/AIDS Prevention Grants   3827
The USCM HIV Prevention Grants Program, in cooperation with the Centers for Disease Control and Prevention, provides financial and technical assistance to local health departments and community-based organizations to implement HIV prevention programs. The goal of this effort is to strengthen the capacity of local service providers to carry out effective HIV/AIDS prevention activities.
Requirements: Grants are awarded on the recommendation of an independent external review panel composed of HIV/AIDS experts drawn from local and state health departments, national organizations, and local community-based organizations.
Geographic Focus: All States
Contact: Crystal D. Swann; (202) 861-6707; fax (202) 293-2352; cswann@usmayors.org
Internet: http://www.usmayors.org/hivprevention/hiv_prevention_grant.asp
Sponsor: United States Conference of Mayors
1620 I Street NW
Washington, D.C. 20006

US CRDF Leishmaniasis: Collaborative Research Opportunities in North Africa and the Middle East — 3828

The U.S. Civilian Research and Development Foundation (CRDF) is a nonprofit organization authorized by the U.S. Congress and established in 1995 by the National Science Foundation. This unique organization promotes international scientific and technical collaboration through grants, technical resources, and training. CRDF, with funding from the National Institute for Allergy and Infectious Diseases (NIAID), announces a targeted, collaborative grant competition on Leishmaniasis in the North Africa and the Middle East (MENA) Region. This grant competition is a follow-on activity to the June 2009 interdisciplinary research conference in Tunisia sponsored by NIAID that brought together regional and international scientists to discuss research opportunities on sand fly biology, leishmania parasites, immunology, leishmaniasis epidemiology, animal reservoirs, clinical manifestations, genetics, diagnostics, treatment, and prevention of leishmaniasis in North Africa and the Middle East. These follow-on grants will provide up to one- year's support to joint research teams of U.S. and MENA scientists, with the possibility of up to a one year no-cost extension, if justified. CRDF and NIAID anticipate that the research implemented through these awards will lay the foundation for future grant proposals to NIAID and other funding agencies. The average anticipated grant will provide up to $30,000 U.S. dollars for a period of one year. At least 80% of the funds awarded to each project will be used for project-related expenses of the MENA component of the research team, including institutional support. No more than 20% may be used for U.S. team expenses. As part of the grant, the CRDF strongly encourages a member of the MENA team to visit the U.S. team's laboratory, and members of the U.S. team are encouraged to visit the MENA laboratory.
Requirements: Each proposal submitted to the CRDF must have at least one MENA Principal Investigator and at least one U.S. Principal Investigator, who share overall responsibility for the project in their respective countries, coordinating all project participants and institutions. Each proposal must meet each of the following eligibility criteria: 1) Only workshop attendees and their institutions are eligible to apply. 2) Each Principal Investigator must work full-time in a civilian research environment. (U.S. Government Laboratories working on civilian-oriented R&D projects are eligible to apply to this program.) 3) The U.S. Principal Investigator and participants of the U.S. team may be foreign nationals, but must reside in the U.S during the course of the grant award. Graduate students on the U.S. team may be foreign nationals, but they must be enrolled in an accredited degree program at a U.S. institution. A PI from Navy Medical Research Unit 3 (NAMRU3) may serve as either a U.S. or MENA researcher. 4) All projects must be in one of the following subject areas: Sand fly biology; Leishmania Parasites; Immunology; Leishmaniasis Epidemiology; Animal reservoirs; Clinical manifestations, treatment, and prevention of leishmaniasis in North Africa and the Middle East (but not clinical trials involving interventions); Genetics; Diagnostics. 5) Only one proposal will be accepted from a U.S. or MENA PI. 6) All projects must be oriented toward non-military objectives and must be carried out in a civilian research environment. Use of human subjects requires appropriate approvals from all institutions involved including Local Ethics Committee (LEC) and U.S. IBR. Middle East countries: Afghanistan, Behrain, Iran, Iraq, Israel, Jordan, Kyrgyzstan, Kuwait, Lebanon, Oman, Pakistan, Palestine, Qatar, Saudi Arabia, Syria, Tajikistan, Turkey, Turkmenistan, United Arab Emirates, Uzbekistan, Yemen. North African countries: Algeria, Egypt, Libya, Morocco, Sudan, Tunisia, Western Sahara.
Restrictions: No clinical trials involving interventions will be accepted for this competition.
Geographic Focus: All States
Date(s) Application is Due: Sep 30
Amount of Grant: Up to 30,000 USD
Contact: Mr. Stuart Politi; (703) 526-9720; fax (703) 526-9721; cgp@crdf.org
Internet: http://www.crdf.org/funding/funding_show.htm?doc_id=949719
Sponsor: U.S. Civilian Research and Development Foundation
1530 Wilson Boulevard, 3rd Floor
Arlington, VA 22209

USDA Delta Health Care Services Grants — 3829

The Delta Health Care Services Grant Program is designed to provide financial assistance to address the continued unmet health need in the Delta Region through cooperation among health care professionals, institutions of higher education, research institutions, and other entities in the Delta Region. Each fiscal year, applications are solicited through a Notice of Funds Availability (NOFA) published in the Federal Register. Grant funds must be used for projects in rural areas within the Delta Region to develop: a health care cooperative; health care services; health education programs; health care training programs; and expand public health-related facilities in the Delta Region to address longstanding and unmet health needs of the region.
Requirements: Consortiums of regional institutions of higher education, academic health and research institutes, and economic development entities located in the Delta Region that have experience in addressing the health care issues in the region are eligible. The Delta region includes specified counties in: Alabama, Arkansas, Illinois, Kentucky, Louisiana, Mississippi, Missouri, and Tennessee.
Restrictions: Individuals are not eligible for this program.
Geographic Focus: Alabama, Arkansas, Illinois, Kentucky, Louisiana, Mississippi, Missouri, Tennessee
Contact: Claudette Fernandez, Director; (202) 690-4730; fax (202) 690-4737; claudette.fernandez@wdc.usda.gov
Internet: http://www.rurdev.usda.gov/BCP_DeltaHealthCare.html
Sponsor: U.S. Department of Agriculture
1400 Independence Avenue, SW
Washington, D.C. 20250

USDA Organic Agriculture Research and Extension Initiative — 3830

The USDA Organic Agriculture Research and Extension Initiative (OREI) seeks to solve critical organic agriculture issues, priorities, or problems through the integration of research and extension activities. The purpose of this program is to fund projects that will enhance the ability of producers and processors who have already adopted organic standards to grow and market high quality organic agricultural products. Priority concerns include biological, physical, and social sciences, including economics. The OREI is particularly interested in projects that emphasize research and outreach that assist farmers and ranchers with whole farm planning. Projects should plan to deliver applied production information to producers. Fieldwork must be done on certified organic land or on land in transition to organic certification, as appropriate to project goals and objectives. Funding amounts are based on specific project categories.
Requirements: The following are eligible to apply for the initiative: public and state controlled institutions of higher education; nonprofits having a 501(c)3 status with the IRS, other than institutions of higher education; nonprofits that do not have a 501(c)3 status with the IRS, other than institutions of higher education; private institutions of higher education; individuals; for profit organizations other than small businesses; and small businesses. The following entities are also eligible to apply: state agricultural experiment stations; colleges and universities; university research foundations; other research institutions and organizations; federal agencies; national laboratories; private organizations or corporations; individuals who are United States citizens or nationals; or a group consisting of two or more of the above entities.
Restrictions: Recipients are required to provide funds or in-kind support from non-federal sources in an amount at least equal to the amount provided by the funding agency.
Geographic Focus: All States
Date(s) Application is Due: Mar 9
Amount of Grant: 50,000 - 2,000,000 USD
Contact: Steven Smith; (202 401-6134; fax (202) 401-1782; sismith@nifa.usda.gov
Internet: http://www.nifa.usda.gov/funding/rfas/OREI.html
Sponsor: U.S. Department of Agriculture
1400 Independence Avenue, SW, Stop 2201
Washington, D.C. 20250-2201

USDD Breast Cancer Clinical Translational Research Grants — 3831

Applications for the Breast Cancer Research Program (BCRP) are being solicited by the Assistant Secretary of Defense for Health Affairs, Defense Health Program. The BCRP was established in 1992 to support innovative research focused on ending breast cancer. The BCRP challenges the scientific community to design research that will address the urgency of the vision to end breast cancer. Specifically, the BCRP seeks to accelerate high-impact research, encourage innovation and stimulate creativity, bring new investigators into the breast cancer field, and facilitate multidisciplinary collaborations. The BCRP Clinical Translational Research (CTR) Grants are intended to promote significant improvements over current approaches to breast cancer prevention and therapy, such as studies that will prevent primary or recurrent breast cancer and studies that may result in a new treatment to prevent breast cancer progression to metastasis. CTR Grants support research projects that are likely to have a major impact on breast cancer by applying promising research findings to patients with, or populations at risk for, breast cancer. Principal Investigators (PIs) and their collaborators may have originated projects in their laboratories that will form the basis for 1-2 years of advanced translational research leading to a prospective clinical trial to be conducted during this award period. Alternatively, PIs may leverage partnerships with industry. The ability to conduct the required translational research and early phase clinical trial during the award period must be demonstrated. Investigators must demonstrate availability of and access to an appropriate patient population that will support meaningful clinical outcomes. Submission is a multi-step process requiring both pre-application submission through the CDMRP eReceipt System (https://cdmrp.org/) and application submission through Grants.gov (http://www.grants.gov/). The maximum period of performance is 5 years. The maximum allowable direct costs for the entire period is $12M plus indirect costs.
Requirements: PIs must be at or above the level of assistant professor (or equivalent).
Restrictions: Applicants are encouraged to sign up on Grants.gov for "send me change notification emails" by following the link on the Synopsis page for the Program Announcement/Funding Opportunity. If the application package is updated or changed, the original version of the application package may not be accepted by Grants.gov.
Geographic Focus: All States
Date(s) Application is Due: Aug 15
Contact: Grants Officer; (301) 682-5507; help@cdmrp.org
Internet: http://www.grants.gov/search/search.do;jsessionid=G6z2PM6MT4vFzPGvmYl7TPJTc9lpLwG0CHlsTPC1jw2TJXJRC87n!488296553?oppId=146134&mode=VIEW
Sponsor: U.S. Department of Defense
1400 Defense Pentagon
Washington, D.C. 20301-1400

USDD Breast Cancer Era of Hope Scholar Grants — 3832

The Era of Hope Scholar Grants support individuals who are early in their careers and have high potential for innovation in breast cancer research. Since the intent is to recognize creative and innovative individuals rather than projects, the central feature of the award is the innovative contribution that the Principal Investigator (PI) can make toward ending breast cancer. The PI should articulate a vision that challenges current dogma and demonstrates an ability to look beyond tradition and convention. Experience in breast cancer research is not required; however, the application must focus on breast cancer and the PI must commit at least 50% of his/her full-time professional effort during the award period to breast cancer research. Five awards are anticipated. The maximum period of performance is 5 years, and the maximum direct costs for the entire period is $2.5M plus indirect costs.

Requirements: Individuals should be exceptionally talented scientists who have demonstrated that they are the "best and brightest" in their field(s) through extraordinary creativity, vision, and productivity. They should have demonstrated experience in forming effective partnerships and collaborations and should exhibit strong potential for leadership in the breast cancer research community. The PI must be an independent investigator within 6 years of his/her last mentored position as of the deadline.
Restrictions: Postdoctoral fellows, clinical fellows (including residents and interns), and other researchers in training positions are not eligible for this award.
Geographic Focus: All States
Date(s) Application is Due: Apr 19
Contact: Grants Officer; (301) 682-5507; help@cdmrp.org
Internet: http://www07.grants.gov/search/search.do;jsessionid=kvB0PNRhp6q1GBx0m TyPhVY6Lp0lyn05T431P0d9jvdVvysv7tSH!488296553?oppId=146114&mode=VIEW
Sponsor: U.S. Department of Defense
1400 Defense Pentagon
Washington, D.C. 20301-1400

USDD Breast Cancer Idea Grants 3833
The Idea Award is designed to promote new ideas that are still in the early stages of development and have the potential to yield highly impactful data and new avenues of investigation. This mechanism supports conceptually innovative, high-risk/high-reward research that could lead to critical discoveries or major advancements that will accelerate progress toward ending breast cancer. Applications should include a well-formulated, testable hypothesis based on strong scientific rationale. Innovation and impact are the most important aspects of the Idea Award. Applications that demonstrate exceptional scientific merit but lack innovation and high potential impact do not meet the intent of the Idea Award. Presentation of preliminary data is not consistent with the intent of the Idea Award mechanism. While the inclusion of preliminary data is not prohibited, the strength of the application should not rely on preliminary data, but on the innovative idea and the potential impact it will provide. Submission is a multi-step process requiring both pre-application submission through the CDMRP eReceipt System (https://cdmrp.org/) and application submission through Grants.gov (http://www.grants.gov/). Approximately fifty awards are anticipated. The maximum period of performance is 3 years. The maximum allowable direct costs for the entire period is $375,000 plus indirect costs.
Requirements: Investigators at all academic levels (or equivalent) are eligible.
Restrictions: Applicants are encouraged to sign up on Grants.gov for "send me change notification emails" by following the link on the Synopsis page for the Program Announcement/Funding Opportunity. If the application package is updated or changed, the original version of the application package may not be accepted by Grants.gov.
Geographic Focus: All States
Date(s) Application is Due: Aug 15
Contact: Grants Officer; (301) 682-5507; help@cdmrp.org
Internet: http://www07.grants.gov/search/search.do;jsessionid=xrZyPNZPBMTJ6V24yPlrLt PxyWHP8L5d3WzzXHTGqVwCSZdY5VyK!488296553?oppId=146113&mode=VIEW
Sponsor: U.S. Department of Defense
1400 Defense Pentagon
Washington, D.C. 20301-1400

USDD Breast Cancer Impact Grants 3834
The Breast Cancer Research Program (BCRP) Impact Grants support research projects (from small to largescale) that specifically focus on scientific and clinical breast cancer issues, which, if successfully addressed, will revolutionize the understanding, prevention, and/or treatment of breast cancer and make major advances toward the goal of ending the disease. The BCRP particularly encourages applications that focus on poorly understood issues that are of critical significance in breast cancer, such as those related to prevention, susceptibility, recurrence, or metastasis. The proposed work must be based on sound overall research and fully supported by preliminary data and/or published reports. The most important aspect is the potential of the proposed research to have an unprecedented impact on reducing breast cancer incidence, recurrence, and/or mortality. The research project may be from any discipline or combination of disciplines, including basic, translational, clinical (clinical trials are allowed), behavioral, and/or epidemiological research. The potential impact may be near term or long term, but it must be significant and non-incremental. Applications are encouraged that include meaningful and productive collaborations between investigators. The Partnering PI Option is structured to accommodate two PIs, who will each receive a separate award. One partner is identified as the Initiating PI and the other partner is identified as the Partnering PI. The Initiating and Partnering PIs have different submission requirements; however, both PIs should contribute to the preparation of a single application. The collaborative partners may have expertise in similar or disparate scientific disciplines, but each partner is expected to bring different strengths to the application. New collaborations are encouraged but not required. It is the responsibility of the collaborating investigators to describe how their combined expertise in the collaboration will better address the research question and explain why the work should be done together rather than through separate efforts. Submission is a multi-step process requiring both pre-application submission through the CDMRP eReceipt System (https://cdmrp.org/) and application submission through Grants.gov (http://www.grants. gov/). Five awards are anticipated. The maximum period of performance is 5 years. The maximum allowable direct costs for the entire period is $2M plus indirect costs.
Requirements: PIs must be at or above the level of Assistant Professor (or equivalent).
Restrictions: Applicants are encouraged to sign up on Grants.gov for "send me change notification emails" by following the link on the Synopsis page for the Program Announcement/Funding Opportunity. If the application package is updated or changed, the original version of the application package may not be accepted by Grants.gov.
Geographic Focus: All States
Date(s) Application is Due: Aug 15
Contact: Grants Officer; (301) 682-5507; help@cdmrp.org
Internet: http://www07.grants.gov/search/search.do;jsessionid=3JX5PNlGM67F6dpRvnM 4TBpcZtQzdPD1pnJYwTGzv4FVp28xw3xg!488296553?oppId=146133&mode=VIEW
Sponsor: U.S. Department of Defense
1400 Defense Pentagon
Washington, D.C. 20301-1400

USDD Breast Cancer Innovation Grants 3835
The Breast Cancer Research Program (BCRP) Innovator Grants support visionary individuals who have demonstrated creativity, innovative work, and leadership in any field including, but not limited to, breast cancer. The Innovator Grants will provide these individuals with the funding and freedom to pursue their most novel, visionary, high-risk ideas that could ultimately lead to ending breast cancer. Since the intent of the Innovator Grant mechanism is to recognize creative and innovative individuals rather than projects, the central feature of the award is the innovative contribution that the Principal Investigator (PI) can make toward ending breast cancer. Experience in breast cancer research is not required; however, the application must focus on breast cancer, and the PI must commit at least 50% of his/her full-time professional effort during the award period to breast cancer research. Submission is a multi-step process requiring both pre-application submission through the CDMRP eReceipt System (https://cdmrp.org/) and application submission through Grants.gov (http://www.grants.gov/). Three awards are anticipated. The maximum period of performance is 5 years. The maximum allowable direct costs for the entire period is $5M plus indirect costs.
Requirements: PIs must be at or above the level of associate professor (or equivalent). The PI should have a past record of creativity, promise for continued innovation in future work, and a vision that challenges current dogma and demonstrates an ability to look beyond tradition and convention. The PI is also expected to have demonstrated success at forming and leading effective partnerships and collaborations.
Restrictions: Previous recipients of the BCRP Innovator Grants and BCRP Integration Panel members are ineligible. Applicants are encouraged to sign up on Grants.gov for "send me change notification emails" by following the link on the Synopsis page for the Program Announcement/Funding Opportunity. If the application package is updated or changed, the original version of the application package may not be accepted by Grants.gov.
Geographic Focus: All States
Date(s) Application is Due: Aug 2
Contact: Grants Officer; (301) 682-5507; help@cdmrp.org
Internet: http://www07.grants.gov/search/search.do;jsessionid=TR2GPNndhVCGQGnL6n9 z32GJB9hkxn2NDRVlLLqDhVQGNF5Fhc2B!488296553?oppId=146115&mode=VIEW
Sponsor: U.S. Department of Defense
1400 Defense Pentagon
Washington, D.C. 20301-1400

USDD Breast Cancer Transformative Vision Grant 3836
The Breast Cancer Research Program (BCRP) Transformative Vision Grant supports research projects that will fulfill an extraordinary vision for dramatically affecting the prevention or treatment of breast cancer. The Grant requires a plan that will test and achieve the vision as quickly as possible through the translation of the research ideas to individuals with, and/or those at risk for, breast cancer. The scope of the effort may include a broad spectrum of research spanning from basic to clinical studies. There are two critical components. The first is vision and impact. A vision for a new approach that will have a revolutionary impact on the prevention or treatment of breast cancer must be articulated. The time to the final impact may vary, but the success of the vision must be transformative and significantly advance the goal of ending breast cancer. The second is implementation. The vision must be supported by a detailed plan that identifies critical milestones, outlines the innovations and technical solutions that will be implemented to accomplish the milestones, and explains how these solutions will be translated to individuals with, and/or those at risk for, breast cancer. It is expected that the proposed plan will present an exceptional level of innovation and creativity and that the Principal Investigator (PI) will assemble the team necessary to realize the vision. Submission is a multi-step process requiring both pre-application submission through the CDMRP eReceipt System (https://cdmrp.org/) and application submission through Grants.gov (http://www.grants.gov/). One award is anticipated. The maximum period of performance is 5 years. The maximum allowable direct costs for the entire period is $12M plus indirect costs.
Requirements: PIs must be at or above the level of Assistant Professor (or equivalent).
Restrictions: Applicants are encouraged to sign up on Grants.gov for "send me change notification emails" by following the link on the Synopsis page for the Program Announcement/Funding Opportunity. If the application package is updated or changed, the original version of the application package may not be accepted by Grants.gov.
Geographic Focus: All States
Date(s) Application is Due: Aug 15
Contact: Grants Officer; (301) 682-5507; help@cdmrp.org
Internet: http://www.grants.gov/search/search.do;jsessionid=vQJqPN4Q9PjYKC9JVyhJ gwGfQj3dL3Sn40jyCLgQs42pN0X6wdsb!488296553?oppId=146116&mode=VIEW
Sponsor: U.S. Department of Defense
1400 Defense Pentagon
Washington, D.C. 20301-1400

USDD Broad Agency Announcement Grants 3837
The Department of Defense HIV/AIDS Prevention Program (DHAPP) has successfully engaged over 80 countries in efforts to combat HIV/AIDS among their respective military services. DHAPP is a partner collaborating with the United States State Department, Health and Human Services, United States Agency for International Development, and

Centers for Disease Control and Prevention, in the President's Emergency Plan for AIDS Relief (PEPFAR). This BAA is intended to solicit existing partners and establish new partners in order to expand the DHAPP program. Proposals should focus on rapidly extending HIV/AIDS services. Applicants are encouraged to target specific needs with a practical business plan, using small grass-roots organizations to provide community-based services as a way to enhance organic capabilities and sustainability. Concept Papers may be submitted at any time during this period. Early submission is strongly encouraged due to the possible lack of funding for the later evaluation periods. The period of performance is generally one year, with up to two additional phases subject to availability of funding. Applicants are encouraged to propose additional phases if appropriate. This is a multiple year BAA which may be revised as needed. Changes will be posted at Grants.gov (http://grants.gov/applicants/find_grant_opportunities.jsp). Applicants are encouraged to sign up on Grants.gov for "send me change notification emails" by following the link on the synopsis page for the program announcement/funding opportunity.

Requirements: All responsible sources from academia, industry, and non-governmental organizations are eligible. Applicants respondents must demonstrate the active support of the in-country military and the Department of Defense representative in the corresponding United States Embassy in the planning and execution of proposals.

Restrictions: Funding is available for Angola, Botswana, Caribbean Regional, Central America Regional, Cote d' Ivoire, Democratic Republic of the Congo, Dominican Republic, Ethiopia, Ghana, Guyana, India, Indonesia, Lesotho, Malawi, Mozambique, Namibia, Rwanda, South Africa, Swaziland, Uganda, Ukraine, Vietnam and Zambia. A description of the work needed, along with the program areas and estimated funding is available in the BAA found at the website. All questions should be submitted to the points of contact by email.

Geographic Focus: All States
Date(s) Application is Due: Sep 30
Amount of Grant: 300,000 - 700,000 USD
Contact: Latrice Rubenstein; (619) 532-4357; Latrice.Rubenstein@navy.mil
Internet: http://www.grants.gov/search/search.do;jsessionid=1GhLPGQSVtvhVkKQYJgRnyFGWgBT8SQvFXjYv31mW6Ky7klFQ1Vx!-757993493?oppId=67733&mode=VIEW
Sponsor: U.S. Department of Defense
5450 Carlistle Park, P.O. Box 2050
Mechanicsburg, PA 17055-0791

USDD Broad Agency Announcement HIV/AIDS Prevention Grants 3838

The Department of Defense HIV/AIDS Prevention Program (DHAPP) has successfully engaged over 80 countries in efforts to combat HIV/AIDS among their respective military services. DHAPP is a partner collaborating with the United States State Department, Health and Human Services, United States Agency for International Development, and Centers for Disease Control and Prevention, in the President's Emergency Plan for AIDS Relief (PEPFAR). This Broad Agency Announcement (BAA) allows DHAPP to assist countries where militaries are not funded under PEPFAR by soliciting existing partners and establishing new partners in furtherance of DHAPP and partner military program goals. Proposals should focus on rapidly extending HIV/AIDS services. Applicants are encouraged to target specific needs with a practical business plan, using small grass-roots organizations to provide community-based services as a way to enhance organic capabilities and sustainability. Concept papers are required and full proposals are by invitation only. Early submission is strongly encouraged. The period of performance is generally one year, with up to two additional phases subject to availability of funding. Applicants are encouraged to propose additional phases if appropriate. This is a multiple year BAA which may be revised as needed. Based on the availability of funds, up to three evaluation cycles will be conducted. Changes will be posted at Grants.gov (http://grants.gov/applicants/find_grant_opportunities.jsp). Applicants are encouraged to sign up on Grants.gov for "send me change notification emails" by following the link on the synopsis page for the program announcement/funding opportunity.

Requirements: All responsible sources from academia, industry, and non-governmental organizations are eligible. Applicants must demonstrate the active support of the in-country military and the USDD representative in the corresponding United States Embassy in the planning and execution of proposals.

Restrictions: Funding is available for Ghana and El Salvador. A description of the work needed, along with the program areas and estimated funding is available in the BAA found at the website. All questions should be submitted to the points of contact by email.

Geographic Focus: All States
Amount of Grant: 60,000 - 500,000 USD
Contact: Latrice Rubenstein; (619) 556-5276; Latrice.Rubenstein@navy.mil
Internet: http://www07.grants.gov/search/search.do;jsessionid=DBT6PFBSpD4JnP9CyYnwCqZjv7ZnwjK79s2tl1KPTPCTvZRwpQJn!1471941753?oppId=141233&mode=VIEW
xLjJrqQ0vdnQM3Wt08p5Y171y8Gj4qPxb!748281696?oppId=143433&mode=VIEW
Sponsor: U.S. Department of Defense
5450 Carlistle Park, P.O. Box 2050
Mechanicsburg, PA 17055-0791

USDD Care for the Critically Injured Burn Patient Grants 3839

The Assistant Secretary of Defense for Health Affairs, Defense Health Program Medical Research and Development Office is soliciting proposals for the Defense Medical Research and Development Program (DMRDP). The Telemedicine and Advanced Technology Research Center (TATRC) is administering the application process, and the United States Army Medical Research Acquisition Activity (USAMRAA) is issuing this program announcement and will be negotiating all resulting awards. The goal is to advance the state of medical science in those areas of most pressing need and relevance to today's battlefield experience, in this case burn injuries. The objectives of the DMRDP are to discover and explore innovative approaches to protect, support, and advance the health and welfare of military personnel, families, and communities; to accelerate the transition of medical technologies into deployable products; and to accelerate the translation of advances in knowledge into new standards of care for injury prevention, treatment of casualties, rehabilitation, and training systems that can be applied in theater or in the clinical facilities of the Military Health System. The research results are expected to increase the body of knowledge available to professionals and practitioners in health, medical science and related fields. The research impact is expected to benefit the military community and civilian community, particularly as it relates to the following priority areas: (1) multicenter development and assessment of the impact of checklists on standardization of burn care and improvement of outcomes in burn patients; (2) multicenter studies of ICU-based rehabilitation outcomes, i.e., evaluation of physical and rehabilitation technologies/therapies directed at short-term ICU outcomes, understanding the effect of rehabilitation on pathophysiology in the ICU burn patient, and documenting the efficacy of these immediate ICU interventions on long-term quality-of-life outcomes such as the return-to-work/duty rate; (3) device/drug development and validation for inhalation lung injury; and (4) research into the formation of hypertrophic scarring and the development of therapies to manage the condition. Pre-applications are due by December 2, and applications are by invitation only. Three to six awards are anticipated.

Requirements: Principal investigators must be independent investigators at any academic level from academia, research institutions, industry, local, state or federal governments, and private foundations that possess the skills, knowledge, and resources necessary to carry out the proposed research. Applications from investigators within the military services and other federal agencies are encouraged, as are applications involving multidisciplinary collaborations among academia, industry, the military services, the Department of Veterans Affairs, and other federal government agencies.

Geographic Focus: All States
Date(s) Application is Due: Mar 9
Amount of Grant: 1,500,000 USD
Contact: Mary Rico, Grants Officer; (703) 518-4726; mary.rico@amedd.army.mil
Internet: http://www07.grants.gov/search/search.do;jsessionid=YYJpP9pQHVhR7BFjdtNMBm1grFHnpNT4H8rRHtKrSJtHcGVFsQnJ!712472910?oppId=129414&mode=VIEW
Sponsor: U.S. Department of Defense
1400 Defense Pentagon
Washington, D.C. 20301-1400

USDD Clinical Functional Assessment Investigator-Initiated Grants 3840

The United States Army Medical Research and Materiel Command (USAMRMC) is seeking health information technology research that may lead to outcomes and products that may qualify as advanced development activities within the Department of Defense (USDD). The results of this research will support the Military Health System in the adoption and implementation of clinical functional disability assessments and standards, by leveraging the skills and talents of United States based businesses and teaching/research academic institutions for success. The intent is to create enduring initiatives and collaborations that will develop and support sustained business and enterprise solutions for the USDD as well as organizations such as the Veterans Benefits Administration, Centers for Medicare and Medicaid Services, and the general public. Proposals are invited that develop products and systems and generate new knowledge that supports medical protection and treatment for USDD personnel and beneficiaries. Military relevance will be a primary criterion for evaluation in all areas. Pre-proposals are due by November 30, and full proposals are by invitation only. One or two awards are anticipated.

Requirements: Eligible applicants include United States based businesses, teaching research academic institutions, and multi-center collaborations, to include international partners, that develop products and systems and generate new knowledge that supports medical protection and treatment for USDD personnel and beneficiaries. Principal investigators must be independent investigators at any academic level from academia, research institutions, government, industry, and private foundations that possess the skills, knowledge, and resources necessary to carry out the proposed research.

Geographic Focus: All States
Date(s) Application is Due: Mar 15
Amount of Grant: 1,500,000 USD
Contact: Mary Rico, Grants Officer; (703) 518-4726; mary.rico@amedd.army.mil
Internet: http://www07.grants.gov/search/search.do;jsessionid=YYJpP9pQHVhR7BFjdtNMBm1grFHnpNT4H8rRHtKrSJtHcGVFsQnJ!712472910?oppId=127813&mode=VIEW
Sponsor: U.S. Department of Defense
1400 Defense Pentagon
Washington, D.C. 20301-1400

USDD HIV/AIDS Prevention: Military Specific HIV/AIDS Prevention, Care, and Treatment for Non-PEPFAR Funded Countries Grants 3841

The Department of Defense HIV/AIDS Prevention Program (DHAPP) has successfully engaged over 80 countries in efforts to combat HIV/AIDS among their respective military services. DHAPP is a partner collaborating with the United States State Department, Health and Human Services, United States Agency for International Development, and Centers for Disease Control and Prevention, in the President's Emergency Plan for AIDS Relief (PEPFAR). This Broad Agency Announcement (BAA) allows DHAPP to assist countries where militaries are not funded under PEPFAR and to solicit existing partners and establish new partners in furtherance of DHAPP and partner military program goals. Proposals should focus on rapidly extending HIV/AIDS services. Applicants are encouraged to target specific needs with a practical business plan, using small grass-roots organizations to provide community-based services as a way to enhance organic capabilities and sustainability. Concept papers are required and full proposals are by invitation only. Submission within the first cycle, ending February 21, is strongly encouraged. The period of performance is generally one year, with up to two additional phases subject to availability of funding. Applicants are encouraged to propose additional phases if appropriate. This is

a multiple year BAA which may be revised as needed. Based on the availability of funds, up to three evaluation cycles will be conducted. Changes will be posted at Grants.gov (http://grants.gov/applicants/find_grant_opportunities.jsp). Applicants are encouraged to sign up on Grants.gov for "send me change notification emails" by following the link on the synopsis page for the program announcement/funding opportunity.
Requirements: All responsible sources from academia, industry, and non-governmental organizations are eligible. Applicants respondents must demonstrate the active support of the in-country military and the Department of Defense representative in the corresponding United States Embassy in the planning and execution of proposals.
Restrictions: Funding is available for Chad, Niger, Mali, and Burkina Faso. A description of the work needed, along with the program areas and estimated funding is available in the BAA found at the website. All questions should be submitted to the points of contact by email.
Geographic Focus: All States
Date(s) Application is Due: Sep 30
Amount of Grant: 30,000 - 290,000 USD
Contact: Latrice Rubenstein; (619) 556-5276; Latrice.Rubenstein@navy.mil
Internet: http://www.grants.gov/search/search.do;jsessionid=2zGTPFTZnGnR2n0hFVQxLjJrqQ0vdnQM3Wt08p5Y171y8Gj4qPxb!748281696?oppId=141233&mode=VIEW
Sponsor: U.S. Department of Defense
5450 Carlistle Park, P.O. Box 2050
Mechanicsburg, PA 17055-0791

USDD HIV/AIDS Prevention Program Information Systems Development Grants 3842

The Department of Defense HIV/AIDS Prevention Program (DHAPP) has successfully engaged over 80 countries in efforts to combat HIV/AIDS among their respective military services. DHAPP is a partner collaborating with the United States State Department, Health and Human Services, United States Agency for International Development, and Centers for Disease Control and Prevention, in the President's Emergency Plan for AIDS Relief. This BAA is intended to solicit existing partners and establish new partners to support the development of a DHAPP.org web site. This site will serve as a source for program updates, with both a public side and a password protected intranet side. A secure online database is needed and needs to aggregate all individual partners' submissions into a single file for download by the DHAPP operations staff in a common format, such as XL and CSV and TXT. Other site capacities should include a file structure to upload and download files, internet and email messaging, and database functions to support information gathering and analysis. Proposed work should include an infrastructure plan and a plan for support and/or turnover of the site. The period of performance is generally one year, with additional phases at the discretion of the government. Applicants are encouraged to propose additional phases if appropriate. This is a multiple year BAA which may be revised as needed. Based on the availability of funds, up to three evaluation cycles will be conducted. Changes will be posted at Grants.gov (http://grants.gov/applicants/find_grant_opportunities.jsp). Applicants are encouraged to sign up on Grants.gov for "send me change notification emails" by following the link on the synopsis page for the program announcement/funding opportunity.
Requirements: All responsible sources from academia, industry, and non-governmental organizations are eligible. Applicants must demonstrate the active support of the in-country military and the USDD representative in the corresponding United States Embassy in the planning and execution of proposals. Applicants should have prior experience with development and hosting of websites preferably for funding multiple partners, from many countries and cultures and languages, and experience in training staff to use and maintain web based databases. Applicants should also have experience with development and hosting of USDD websites and be familiar with the most recent Information Assurance requirements.
Geographic Focus: All States
Date(s) Application is Due: Sep 30
Amount of Grant: 80,000 - 100,000 USD
Contact: Latrice Rubenstein; (619) 556-5276; Latrice.Rubenstein@navy.mil
Internet: http://www.afosr.af.mil/pdfs/BAA2005-1.pdf
Sponsor: U.S. Department of Defense
5450 Carlistle Park, P.O. Box 2050
Mechanicsburg, PA 17055-0791

USDD Medical Practice Initiative Breadth of Medical Practice and Disease Frequency Exposure Grants 3843

USDD Medical Practice Initiative Breadth of Medical Practice and Disease Frequency Exposure (MPI-BMP) Grants are being solicited by the Assistant Secretary of Defense for Health Affairs, Defense Health Program (DHP). The MPI is primarily focused on the development of medical training systems and competency assessment tools for the sustainment of military medical readiness. The preliminary focus will be on determining the viability of extracting relevant metrics from existing data for use in the determination of skills degradation and using this data to assist in identifying gaps. The MPI-BMP does not only concern itself with degradation from being removed from practice, but also looks at the breadth of clinical exposures and looks across each physician's expected clinical knowledge domain. The data collected in the course of this research will aid both public and private sector organizations in their efforts to improve the quality of care provided in the outpatient setting. The primary, but not exclusive, purpose is research that will: (1) identify when and why degradation of cognitive clinical skills occurs; (2) create validated analytical tools that can predict the probable onset of cognitive skills degradation and determine, with specificity, when skill or knowledge areas have degraded or will be likely to degrade; and (3) propose methods and tools which will enable physicians to preemptively refresh knowledge and maintain familiarity and fluency across the expected competencies for their specialty. Pre-proposals are due December 22, and proposals are by invitation only. Two awards are anticipated. The maximum period of performance is three years.
Requirements: Eligible organizations must be extramural and include nonprofit, public, and private organizations, such as universities, colleges, hospitals, laboratories, and commercial firms. Historically black colleges and universities/minority institutions are encouraged to apply.
Geographic Focus: All States
Date(s) Application is Due: Apr 19
Contact: Mary Rico, Grants Officer; (703) 518-4726; mary.rico@amedd.army.mil
Internet: http://www07.grants.gov/search/search.do;jsessionid=YYJpP9pQHVhR7BFjdtNMBm1grFHnpNT4H8rRHtKrSJtHcGVFsQnJ!712472910?oppId=127853&mode=VIEW
Sponsor: U.S. Department of Defense
1400 Defense Pentagon
Washington, D.C. 20301-1400

USDD Medical Practice Initiative Procedural Skill Decay and Maintenance Grants 3844

USDD Medical Practice Initiative Procedural Skill Decay and Maintenance (MPI-PSD) Grants are awarded by the United States Army Medical Research Acquisition Activity (USAMRAA) and managed by the Army's Telemedicine and Advanced Technology Research Center. The MPI is primarily focused on the development of medical training systems and competency assessment tools for the sustainment of military medical readiness. The preliminary focus will be on determining the viability of extracting relevant metrics from existing data for use in the determination of skills degradation and using this data to assist in identifying gaps. The data collected in the course of this research will aid both public and private sector organizations in their efforts to improve the quality of care provided in the outpatient setting. The primary, but not exclusive, purpose is research that will: (1) identify when and why degradation of psychomotor and procedural clinical skills occurs; (2) create validated analytical tools that can predict the probable onset of skills2 degradation and determine, with specificity, when skills have degraded or will be likely to degrade; and (3) propose methods and tools which will enable physicians or surgeons to preemptively refresh the expected competencies for their specialty. The study must focus on one of more of the following physician practices: (1) surgical; (2) intensive care medicine; and (3) outpatient invasive procedures. Pre-proposals are due December 22, and proposals are by invitation only. Two awards are anticipated. The maximum period of performance is three years. The maximum period of performance is three years.
Requirements: Eligible organizations must be extramural and include nonprofit, public, and private organizations, such as universities, colleges, hospitals, laboratories, and commercial firms. Historically black colleges and universities/minority institutions are encouraged to apply.
Geographic Focus: All States
Date(s) Application is Due: Apr 19
Contact: Mary Rico, Grants Officer; (703) 518-4726; mary.rico@amedd.army.mil
Internet: http://www07.grants.gov/search/search.do;jsessionid=YYJpP9pQHVhR7BFjdtNMBm1grFHnpNT4H8rRHtKrSJtHcGVFsQnJ!712472910?oppId=127873&mode=VIEW
Sponsor: U.S. Department of Defense
1400 Defense Pentagon
Washington, D.C. 20301-1400

USDD President's Emergency Plan for AIDS Relief Military Specific HIV Seroprevalence and Behavioral Epidemiology Risk Suvey Grants 3845

The Department of Defense HIV/AIDS Prevention Program (DHAPP) has successfully engaged over 80 countries in efforts to combat HIV/AIDS among their respective military services. DHAPP is a partner collaborating with the United States State Department, Health and Human Services, United States Agency for International Development, and Centers for Disease Control and Prevention, in the President's Emergency Plan for AIDS Relief (PEPFAR). This Broad Agency Announcement (BAA) is to solicit existing partners and establish new partners to assist select military troops with performing an independent HIV Seroprevalence and Behavioral Epidemiology Risk Survey. Proposals should focus on rapidly extending HIV/AIDS services. Applicants are encouraged to target specific needs with a practical business plan, using small grass-roots organizations to provide community-based services as a way to enhance organic capabilities and sustainability. Concept papers are required and full proposals are by invitation only. Concept papers may be submitted at any time. Early submission is strongly encouraged due to the possible lack of funding for the later evaluation periods. The period of performance is generally one year, with up to two additional phases subject to availability of funding. Applicants are encouraged to propose additional phases if appropriate. This is a multiple year BAA which may be revised as needed. Based on the availability of funds, up to three evaluation cycles will be conducted. Changes will be posted at Grants.gov (http://grants.gov/applicants/find_grant_opportunities.jsp). Applicants are encouraged to sign up on Grants.gov for "send me change notification emails" by following the link on the synopsis page for the program announcement/funding opportunity.
Requirements: All responsible sources from academia, industry, and non-governmental organizations are eligible. Applicants must demonstrate the active support of the in-country military and the USDD representative in the corresponding United States Embassy in the planning and execution of proposals.
Restrictions: Countries covered by this BAA are: Democratic Republic of Congo, Liberia, Malawi, Mali, Sierra Leone, South Africa, and Vietnam.
Geographic Focus: All States
Date(s) Application is Due: Sep 30
Contact: Latrice Rubenstein; (619) 556-5276; Latrice.Rubenstein@navy.mil
Internet: http://www.grants.gov/search/search.do;jsessionid=1GhLPGQSVtvhVkKQYJgRnyFGWgBT8SQyFXjYv31mW6Ky7klFQ1Vx!-757993493?oppId=65295&mode=VIEW
Sponsor: U.S. Department of Defense
5450 Carlistle Park, P.O. Box 2050
Mechanicsburg, PA 17055-0791

USDD U.S. Army Medical Research and Materiel Command Broad Agency Announcement for Extramural Medical Research Grants 3846

The United States Army Medical Research and Materiel Command's mission is to provide solutions to medical problems of importance to the American warfighter at home and abroad. The scope of this effort and the priorities attached to specific projects are influenced by changes in military and civilian medical science and technology, operational requirements, military threat assessments, and national defense strategies. Research areas of interest for the Extramural Medical Research Grants include the following: military infectious diseases; combat casualty care; military operational medicine; clinical and rehabilitative medicine; medical biological defense; medical chemical defense; telemedicine and advanced technology; and special programs (including health care delivery; detection, diagnosis, control or eradication of specified diseases, conditions, or syndromes; and the advancement of military medical interests). This Broad Agency Announcement (BAA) is intended to solicit extramural research and development ideas. Funding has not been set aside specifically for this announcement and the number of awards has not been determined. Applicants are encouraged to submit proposals that span their entire research project up to five years. Pre-proposals and full proposals may be submitted at any time until the deadline.
Requirements: Eligible applicants include the following: private/public/state controlled institutions of higher education; Hispanic-serving institutions; historically black colleges and universities/minority institutions; tribally controlled colleges and universities; Alaska Native and Native Hawaiian serving institutions; nonprofits with 501(c)3 status (other than institutions of higher education; small businesses; for-profit organizations (other than small businesses); Indian/Native American tribal governments; Indian/Native American tribally designated organizations; non-domestic (non-United States) entities (foreign organizations); and federal, state, and local government agencies.
Geographic Focus: All States
Date(s) Application is Due: Sep 30
Contact: Grants Officer; (703) 518-4726; (QA.BAA@amedd.army.mil);
Internet: http://www07.grants.gov/search/search.do;jsessionid=24YPPC4SdRy7mYT61ph b5n46fQty11L2LySnpQGjn4qRpnSwgvG0!712472910?oppId=125873&mode=VIEW
Sponsor: U.S. Department of Defense
1400 Defense Pentagon
Washington, D.C. 20301-1400

USDD Workplace Violence in the Military Grants 3847

The Defense Health Program is seeking applications for the Psychological Health and Traumatic Brain Injury (PH/TBI) Research Program for Workplace Violence in the Military. The PH/TBI Research Program was established to complement ongoing Department of Defense (USDD) efforts towards promoting optimal care for PH (including workplace violence across the USDD service areas) and TBI in the areas of prevention, detection and intervention. United States warfighters endure increasingly demanding and high tempo conditions both in garrison and in the combat field of operations in order to keep pace with ongoing wartime mission requirements. In addition to the loss of lives overseas, warfighters have lost their lives on an Army post stemming from acts of violence in the workplace. The tragic shooting of military personnel at Fort Hood in November 2009 highlighted the need for the USDD to thoroughly review its approach to force health protection and to broaden their historical focus on hostile external threats. The USDD force protection policies, programs, and procedures need to be improved and integrated into a cohesive and comprehensive approach. This Grant program stems directly from recommendations yielded from the Independent Review Related to Fort Hood, which determined that USDD programs, policies, process and procedures that address identification of indicators for violence and radicalization are outdated, incomplete, and fail to include key indicators of potentially violent behaviors. It was also determined that currently the USDD has no risk assessment system available to supervisors and commanders to assist them in the identification, mitigation, and tracking of internal threats. The Independent Review outlined actions needed to identify behavioral indicators of violence. Applications are sought to support a retrospective study and a prospective study across USDD service areas to develop a set of empirical behavioral-based predictors of potential violence. Applications may include a combined, two-tiered approach that would first involve the retrospective study with those results informing the design of the prospective study. Submitters may also choose to propose only a retrospective or a prospective study. Approximately $8M is available, and five awards are anticipated. Pe-applications should be submitted by February 27.
Requirements: Applications from investigators within the military services and other federal agencies are encouraged, as are applications involving multidisciplinary collaborations among academia, industry, the military services, the Department of Veterans Affairs, and other federal government agencies. A principal investigator must be an independent investigator at any academic level from government, academia, research institutions, industry, and private foundations that possess the skills, knowledge, and resources necessary to carry out the proposed research.
Geographic Focus: All States
Date(s) Application is Due: Apr 26
Amount of Grant: 3,000,000 USD
Contact: Grants Manager; (703) 674-2500; programannouncements@tatrc.org.
Internet: http://www.grants.gov/search/search.do;jsessionid=Q2lZP9lV4mvG5TgvVqhZ2Y cTpzV1NQypcLBss0zn38199gQ1WYBM!-2099600874?oppId=140173&mode=VIEW
Sponsor: U.S. Department of Defense
1400 Defense Pentagon
Washington, D.C. 20301-1400

USG Foundation Grants 3848

The Foundation is committed to social responsibility and supports local and national charitable organizations that serve and educate the communities in which USG Corporation operates. The Foundation originated in 1979 to enrich the lives of families, friends, colleagues, customers and citizens where USG employees live and work. It provides financial assistance to non-profit organizations with solutions in mind for social, health and educational issues. For more information contact the Foundation.
Geographic Focus: All States
Contact: Administrator; (312) 606-4297; usgfoundation@usg.com
Internet: http://www.usg.com/company/corporate-responsibility.html
Sponsor: USG Foundation
550 West Adams
Chicago, IL 60606

Vancouver Foundation Grants and Community Initiatives Program 3849

Vancouver Foundation's Grants and Community Initiatives program provides Community Impact Grants in nine fields of interest: Animal Welfare; Arts and Culture; Children, Youth and Families; Education; Environment; Health and Social Development; Health and Medical Research; Youth Homelessness; Youth Philanthropy. The Vancouver Foundation has a two-stage application process. The first stage is a letter of intent. This consists of a brief proposal to determine basic suitability. The second stage is submission of a full grant application by a specific date. The letter of intent may be submitted online. You will be notified, if you are invited to submit a full grant application. The following is a list of deadline dates for the letters of intent and applications. Animal Welfare, Arts and Culture, Children, Youth and Families, Education, Environment, and Health and Social Development grants have a deadline date of: Letter of Intent, June 18, Application, August 6. Youth Homelessness grants: Letter of Intent, August 6, Application, September 24. Youth Philanthropy grants: Letter of Intent, September 17, Application, October 1. Health and Medical Research grants deadline dates are undetermined, at this time.
Requirements: Under our Grants and Community Initiatives program, Vancouver Foundation funds eligible applicants, which include: registered charities and qualified donees under the Income Tax Act; First Nations that may be considered a public body performing a function of government.
Restrictions: The Foundation does not fund the following: 100% of a proposal's costs; an organization's ongoing operational or core expenses; retroactive funding, or for any expenses to be incurred prior to the Foundation's decision date; debt retirement or reserves; mortgage pay-downs; conferences, competitions, symposia, annual events, or travel to/attendance at such events; office equipment and furniture; activities of religious organizations that serve primarily their membership and/or their direct religious purposes, unless the community at large will benefit significantly; sabbatical leaves, student exchanges; medical facilities or equipment; league-based sports and recreation programs; library acquisitions and construction; school construction; publication of books; academic or dissertation research; research with human subjects; endowment matching grants; capital requests, with the exceptions noted above; activities previously supported through government funding; bursaries, scholarships and awards; school trips, annual events or equipment; grants to individuals or businesses; funding requests if reports on any previous grants are overdue.
Geographic Focus: All States, Canada
Contact: Mark Gifford, Director; (604) 629-5362; markg@vancouverfoundation.ca
Internet: http://www.vancouverfoundation.ca/grants/communityinitiatives.htm
Sponsor: Vancouver Foundation
555 West Hastings Street, Suite 1200
Vancouver, BC V6B 4N6 Canada

Vancouver Foundation Public Health Bursary Package 3850

With funding from the BC Associated Boards of Health Dr. Ken Benson Memorial Bursary and the Isabel Loucks Foster Public Health Bursary Funds, bursaries are available to qualified students who are or wish to engage in post-graduate study in a field of Public or Community health practice within any recognized discipline. The funds are managed by the Vancouver Foundation and the Bursary funds will be released to a successful applicant(s) annually on the recommendation of the Health Officers' Council. The Health Officers' Council is comprised of the Medical Health Officers and public health physicians of British Columbia. Bursaries in an amount up to $3,000 are available to eligible applicants on a one- time-only basis. Successful candidates will be notified by October 31st.
Requirements: Eligibility for the Isabel Loucks Foster Public Health Bursary Fund: full-time students engaged in post-graduate study in a field of Public/Community Health Practice within an applicable discipline (e.g. nurses, nutritionists, environmental health officers, physicians, etc.); persons must be ordinarily resident in B.C; persons must have demonstrated excellence in performance and leadership ability; studies may be undertaken at any suitable educational institution either within or outside British Columbia.
Geographic Focus: All States, Canada
Date(s) Application is Due: Sep 4
Amount of Grant: 3,000 USD
Contact: Faye Wightman; (604) 688-2204; info@vancouverfoundation.ca
Internet: http://www.vancouverfoundation.ca/grants/specialprograms.htm
Sponsor: Vancouver Foundation
555 West Hastings Street, Suite 1200
Vancouver, BC V6B 4N6 Canada

Vancouver Sun Children's Fund Grants 3851

In partnership with Vancouver Foundation, the Vancouver Sun Children's Fund provides grants twice yearly to registered charities in British Columbia that directly serve the needs of children and youth (up to18 years) with special challenges. Grants range from $100

to $5,000 and recommendations are made by Vancouver Foundation's Children, Youth and Families Advisory Committee whose members have knowledge and experience in the fields of children's physical, mental and emotional health.
Requirements: Eligible applicants are registered charities and other qualified organizations under the Income Tax Act. They must demonstrate fiscal responsibility and effective management. To apply, submit your Letter of Intent electronically.
Restrictions: Grants are not made to individuals or businesses.
Geographic Focus: All States, Canada
Amount of Grant: 100 - 5,000 USD
Contact: Mark Gifford Director, Grants and Community Initiatives; (604) 688-2204; fax (604) 688-4170; markg@vancouverfoundation.ca
Internet: http://www.vancouverfoundation.ca/grants/specialprograms.htm
Sponsor: Vancouver Sun Children's Fund c/o Vancouver Foundation
Harbour Centre, 555 West Hastings Street, Suite 1200, Box 12132
Vancouver, BC V6B 4N6 Canada

Vanderburgh Community Foundation Grants 3852
The Vanderburgh Community Foundation is a nonprofit, public charity created for the people of Vanderburgh County, Indiana. The Foundation allows nonprofits to fulfill their missions by strengthening their ability to meet community needs through grants that assist charitable programs, address community issues, support community agencies, launch community initiatives, and support leadership development. Funding priorities include arts and culture; community development; education; health; human services; environment; recreation; and youth development. The Foundation's grant cycle is announced at the end of July. The preliminary proposal form is available at the website.
Requirements: The Foundation gives primarily to 501(c)3 tax-exempt organizations in Vanderburgh County, Indiana. Organizations must first submit a preliminary proposal to determine their eligibility. Proposals will be reviewed within 30 days, then select applicants will receive a formal invitation to submit a full proposal. Grant checks are distributed the following January.
Restrictions: The following project areas are not considered for funding: religious organizations for religious purposes; political parties or campaigns; endowment creation or debt reduction; capital campaign; operating costs not directly related to the proposed project; annual appeals or membership contributions; and travel requests for groups or individuals such as bands, sports teams, or classes.
Geographic Focus: Indiana
Date(s) Application is Due: Aug 5
Amount of Grant: 1,000 - 10,000 USD
Contact: Carol Pace, Program Officer; (812) 422-1245; fax (812) 429-0840
Internet: http://www.vanderburghcommunityfoundation.org/disc-grants
Sponsor: Vanderburgh Community Foundation
401 South East 6th Street, Suite 203
Evansville, IN 47708

Vanderburgh Community Foundation Women's Fund 3853
The Vanderburgh Community Foundation Women's Fund pools gifts of all sizes from a broad base of donors to support Vanderburgh grantmaking, education and advocacy work. The Women's Fund endowment is a permanent resource created to address the changing needs and priorities of women and children in the Vanderburgh County. Project areas considered including community development; education; health; human services; capital projects; endowment creation; or other activities that improve the quality of life for women, children, or both in Vanderburgh County. The preliminary proposal form is available at the Foundation website.
Requirements: Consideration will be given to nonprofit organizations that are 501(c)3 tax exempt. Organizations not classified as tax exempt may be considered with a fiscal sponsor. A preliminary proposal is required before submitting a full proposal.
Restrictions: Project areas not considered for funding include the following: projects or programs of organizations that are not committed to gender equity; funding to reduce or retire debt of the organization; projects that focus solely on the spiritual needs and growth of a church congregation or members of other religious organizations; political parties or campaigns; operating costs not directly related to the proposed project (the organization's general operating expenses including equipment, staff salary, rent, and utilities); event sponsorships, annual appeals, and membership contributions; travel expenses for groups or individuals such as bands, sports teams, or classes; scholarships or other grants to individuals.
Geographic Focus: Indiana
Date(s) Application is Due: Jul 13
Amount of Grant: 1,000 - 55,000 USD
Contact: Carol Pace, Program Officer; (812) 422-1245; fax (812) 429-0840
Internet: http://www.womensfundvc.org/
Sponsor: Vanderburgh Community Foundation
401 South East 6th Street, Suite 203
Evansville, IN 47708

VDH Commonwealth of Virginia Nurse Educator Scholarships 3854
Scholarships will be provided to full/part-time graduate nursing students who are accepted to or enrolled in a master's or doctoral level nursing program in the Commonwealth of Virginia. The purpose of this program is to increase the number of nursing faculty by providing master's/doctoral students with financial support. Additionally, this program will assist Virginia nursing schools to recruit and retain new nursing faculty, which enables the schools to increase enrollment.
Requirements: Full/part-time master's or doctoral nursing students enrolled or accepted in nursing program in Virginia will be eligible to apply. Students must complete their degree requirements in two years or less. Priority shall be given to master's degree candidates who will teach in community colleges. Applications will be accepted from June 1 through July 31 each calendar year. Complete and submit a Nursing Scholarship Application along with all required materials. Applications will be available online at: http://www.vdh.virginia.gov/healthpolicy/primarycare/incentives/nursing/index.htm
Restrictions: Recipients will be required to give two years of teaching service as a faculty member for every year they receive a scholarship. Recipients who do not take teaching positions at a Virginia nursing program will be required to pay back the scholarship with a 9% interest penalty. Repayment must be paid in full within two years of degree completion. Recipients must begin their teaching service within three months of completing their educational program.
Geographic Focus: Virginia
Date(s) Application is Due: Jul 31
Amount of Grant: 20,000 USD
Contact: Aileen Harris; (804) 864-7436; Aileen.Harris@vdh.virginia.gov
Internet: http://www.vdh.virginia.gov/healthpolicy/primarycare/incentives/nursing/va_nsp_guidelines.htm
Sponsor: Virginia Department of Health
P.O. Box 2448
Richmond, VA 23218

VDH Emergency Medical Services Training Fund (EMSTF) 3855
The EMS Training Fund program is designed to provide financial assistance for Virginia Certified EMS providers and Virginia Office of EMS approved Emergency Medical Services courses. These funds shall supplement local support for Emergency Medical Services Courses. The Emergency Medical Services Training Funds are monies available for student expenses related to attending EMS Certification programs, tuition reimbursement, auxiliary programs and continuing education programs whose lessons are based upon or resemble the learning objectives in the United States Department of Transportation's Intermediate-99 and Paramedic curricula and the Enhanced curricula as defined in 12VAC5-31. See website for application.
Requirements: These funds are designed for non-profit entities and individuals participating in Virginia's EMS System.
Geographic Focus: Virginia
Contact: Jackie Hunter; (804) 864-7585; Jacqueline.Hunter@vdh.virginia.gov
Internet: http://www.vdh.virginia.gov/OEMS/Training/EMSTF.htm
Sponsor: Virginia Department of Health
P.O. Box 2448
Richmond, VA 23218

VDH Rescue Squad Assistance Fund Grants 3856
The Financial Assistance for Emergency Medical Services Grants Program, known as the Rescue Squad Assistance Fund (RSAF) Grant Program is a multi-million dollar matching grant program for Virginia governmental, non-profit EMS agencies and organizations to provide financial assistance based on a demonstrated need. Items eligible for funding include EMS equipment and vehicles, computers, EMS management programs, courses/classes and projects benefiting the recruitment and retention of EMS members. RSAF is primarily a reimbursement grant that requires the grantee to make the purchase for the awarded item(s) and then submit an invoice for reimbursement.
Requirements: Applicant must be a Virginia non-profit agency/organization or governmental organization involved in emergency medical service (ems); applicant must submit verification of its FIN. Verification can be provided in the following formats: copy of the original letter from the IRS issuing FIN, copy of the latest tax returns (1st page only), statement from the County Administrator or City Manager of the municipality stating that the application is non-profit or a government agency and verifies their FIN; applicant must submit a copy (1st page only) of the most recent Federal Tax Return from the IRS (Form 990); applicant must submit the Virginia Office of EMS Affirmation Page in its entirety including the original signature of the Authorized Agent, the Fiscal Officer (Treasurer) and the Operational Medical Director (OMD). All Grant Applications must be submitted via the Internet. The Affirmation page is due by September 15 with original signatures. Faxed Applications will not be accepted. Submit all grant information to: Office of Emergency Medical Services, Attn: Amanda Davis - Grant Manager, 109 Governor Street, Suite UB-55, Richmond, VA 23219.
Restrictions: Applications submitted with line items less then $500.00 will be disqualified.
Geographic Focus: Virginia
Date(s) Application is Due: Sep 15
Contact: Amanda Davis, Grant Manager; (804) 864-7600; fax (804) 864-7580
Internet: http://www.vdh.virginia.gov/OEMS/Grants/index.htm
Sponsor: Virginia Department of Health
P.O. Box 2448
Richmond, VA 23218

Vectren Foundation Grants 3857
The foundation focuses its efforts in four major areas of giving: Education, Arts & Culture, Civic, and Health & Human Services. Priority will be given to those organizations and activities that serve to improve the quality of life in the communities Vectren serves.
Requirements: The only awards grants to organizations exempt from Federal Tax under Section 501(c)3 of the IRS code. The funding request application can be downloaded from the website. Applications should include a copy of your tax-exempt ruling from the IRS; a list of current board members and annual report, if available; and, a current operating and project budget. Applications for contributions to capital fund drives or to purchase equipment should also provide a clear description of the project, including its goals and objectives; and, a financial plan showing how the project will be funded. The plan should include other possible sources of support and funds that have been pledged or received to date.

Restrictions: The foundation will not: support programs, activities or campaigns that advance a specific religious agenda; make contributions to political, fraternal, labor or veteran's organizations, unless the activity is for direct community involvement; make contributions to issues-oriented organizations; will not award scholarships or grants that benefit individuals (This includes individuals raising funds for charitable purposes, student ambassador programs and sponsorship of individual sports or academic teams.); or, award grants involving travel outside of the Vectren Service Territory.
Geographic Focus: Indiana
Contact: Mark H. Miller; (812) 491-4176; fax (812) 491-4078; mmiller@vectren.com
Internet: http://www.vectren.com/web/holding/discover/foundation/foundation_i.jsp
Sponsor: Vectren Foundation
One Vectren Square
Evansville, IN 47708

Verizon Foundation Health Care and Accessibility Grants 3858
The Verizon Foundation invests in projects that provide technology to help under served populations and people with disabilities access information on critical health issues. It also supports innovative technology that helps health care providers increase their efficiency, effectiveness and reach. The Foundation only accepts electronic proposals through its online process, and reviews unsolicited proposals on a continuous calendar year basis from January 1st through October 31st.
Requirements: Proposals will be considered from eligible tax-exempt organizations in certain 501(c)3 subsections as defined by the Internal Revenue Service (IRS). Proposals will also be considered from elementary and secondary schools (public and private) that are registered with the National Center for Education Statistics (NCES). Proposals may also be considered from eligible tax-exempt organizations in the subsection 170(B)(1)(a)(i)--Church, provided that the proposal will benefit a large portion of a community without regard to religious affiliation and does not duplicate the work of other agencies in the community.
Restrictions: The Verizon Foundation does not provide funding for any of the following: individuals; private charity or foundation; organizations not exempt under Section 501(c)3 of the Internal Revenue Code, and not eligible for tax-deductible support; religious organizations, unless the particular program will benefit a large portion of a community without regard to religious affiliation and does not duplicate the work of other agencies in the community; political causes, candidates, organizations or campaigns; organizations that discriminate on the basis of age, color, citizenship, disability, disabled veteran status, gender, race, religion, national origin, marital status, sexual orientation, military service or status or Vietnam-era veteran status; organizations whose primary purpose is to influence legislation; endowments or capital campaigns; film, music, TV, video and media production projects or broadcast program underwriting; research studies, unless related to projects we are already supporting; sports sponsorships; performing arts tours; association memberships; or organizations that have received a grant from the Verizon Foundation in the last three consecutive years - organization may reapply after a one-year hiatus.
Geographic Focus: All States
Amount of Grant: 5,000 - 10,000 USD
Contact: Patrick Gaston, President; (800) 360-7955; fax (908) 630-2660; patrick.g.gaston@erizon.com
Internet: http://foundation.verizon.com/core/health.shtml
Sponsor: Verizon Foundation
1 Verizon Way
Basking Ridge, NJ 07920

Verizon Foundation Maryland Grants 3859
The foundation invites applications, in Maryland, to receive financial grants that support education, literacy, health care access, workforce development, and the reduction of domestic violence. Consideration will be given to proposals for programs that provide job training for individuals, including welfare-to-work and school-to-work training programs; and assist people with disabilities in acquiring computer skills through adaptive technology that prepares them to enter the workforce. The Foundation only accepts electronic proposals through its online process, and reviews unsolicited proposals on a continuous calendar year basis from January 1st through October 31st.
Requirements: Proposals will be considered from eligible tax-exempt organizations in certain 501(c)3 subsections as defined by the Internal Revenue Service (IRS). Proposals will also be considered from elementary and secondary schools (public and private) that are registered with the National Center for Education Statistics (NCES). Proposals may also be considered from eligible tax-exempt organizations in the subsection 170(B)(1)(a)(i)--Church, provided that the proposal will benefit a large portion of a community without regard to religious affiliation and does not duplicate the work of other agencies in the community.
Restrictions: The Verizon Foundation does not provide funding for any of the following: individuals; private charity or foundation; organizations not exempt under Section 501(c)3 of the Internal Revenue Code, and not eligible for tax-deductible support; religious organizations, unless the particular program will benefit a large portion of a community without regard to religious affiliation and does not duplicate the work of other agencies in the community; political causes, candidates, organizations or campaigns; organizations that discriminate on the basis of age, color, citizenship, disability, disabled veteran status, gender, race, religion, national origin, marital status, sexual orientation, military service or status or Vietnam-era veteran status; organizations whose primary purpose is to influence legislation; endowments or capital campaigns; film, music, TV, video and media production projects or broadcast program underwriting; research studies, unless related to projects we are already supporting; sports sponsorships; performing arts tours; association memberships; or organizations that have received a grant from the Verizon Foundation in the last three consecutive years - organization may reapply after a one-year hiatus.
Geographic Focus: Maryland
Amount of Grant: 5,000 - 10,000 USD
Contact: Diane Miles, Director; (410) 393-7450 or (800) 360-7955; fax (908) 630-2660; diane.f.miles@verizon.com
Internet: http://foundation.verizon.com/cybergrants/plsql/vznadmin.vz_www.iyc?x_zip=21202
Sponsor: Verizon Foundation
1 East Pratt Street, 10E
Baltimore, MD 21202

Verizon Foundation New York Grants 3860
The Verizon Foundation--the philanthropic arm of Verizon Communications--supports a variety of programs and events in communities from Buffalo to Brooklyn. As one of the largest corporate contributors in New York, Verizon is proud to have directed more than 6,000 grants totaling more than $14 million to some 2,700 non-profits across the state. Those non-profits include school districts and universities, multicultural groups and hospitals, children's organizations and those assisting the disabled and victims of domestic violence. Through the Foundation, Verizon New York underscores the company's commitment to key funding priorities--combating literacy and domestic violence, empowering communities with technology, bridging the digital divide, creating a skilled work force, and galvanizing employees to volunteer. The Foundation only accepts electronic proposals through its online process, and reviews unsolicited proposals on a continuous calendar year basis from January 1st through October 31st.
Requirements: Proposals will be considered from eligible tax-exempt organizations in certain 501(c)3 subsections as defined by the Internal Revenue Service (IRS). Proposals will also be considered from elementary and secondary schools (public and private) that are registered with the National Center for Education Statistics (NCES). Proposals may also be considered from eligible tax-exempt organizations in the subsection 170(B)(1)(a)(i)--Church, provided that the proposal will benefit a large portion of a community without regard to religious affiliation and does not duplicate the work of other agencies in the community.
Restrictions: The Verizon Foundation does not provide funding for any of the following: individuals; private charity or foundation; organizations not exempt under Section 501(c)3 of the Internal Revenue Code, and not eligible for tax-deductible support; religious organizations, unless the particular program will benefit a large portion of a community without regard to religious affiliation and does not duplicate the work of other agencies in the community; political causes, candidates, organizations or campaigns; organizations that discriminate on the basis of age, color, citizenship, disability, disabled veteran status, gender, race, religion, national origin, marital status, sexual orientation, military service or status or Vietnam-era veteran status; organizations whose primary purpose is to influence legislation; endowments or capital campaigns; film, music, TV, video and media production projects or broadcast program underwriting; research studies, unless related to projects we are already supporting; sports sponsorships; performing arts tours; association memberships; or organizations that have received a grant from the Verizon Foundation in the last three consecutive years - organization may reapply after a one-year hiatus.
Geographic Focus: New York
Amount of Grant: 5,000 - 10,000 USD
Contact: Sandy Wilson; (800) 360-7955; fax (908) 630-2660; sandra.c.wilson@verizon.com
Internet: http://foundation.verizon.com/cybergrants/plsql/vznadmin.vz_www.iyc?x_zip=10007
Sponsor: Verizon Foundation
140 West Street, 26th Floor, Room 2621
New York, NY 10007

Verizon Foundation Northeast Region Grants 3861
The Verizon Foundation is committed to supporting programs and projects that leverage technology to improve basic computer literacy and domestic violence initiatives, seeking organizations that serve the needs of diverse communities, people with disabilities, and the economically and socially challenged. The Foundation only accepts electronic proposals through its online process, and reviews unsolicited proposals on a continuous calendar year basis from January 1st through October 31st.
Requirements: Proposals will be considered from eligible tax-exempt organizations in certain 501(c)3 subsections as defined by the Internal Revenue Service (IRS). Proposals will also be considered from elementary and secondary schools (public and private) that are registered with the National Center for Education Statistics (NCES). Proposals may also be considered from eligible tax-exempt organizations in the subsection 170(B)(1)(a)(i)--Church, provided that the proposal will benefit a large portion of a community without regard to religious affiliation and does not duplicate the work of other agencies in the community.
Restrictions: The Verizon Foundation does not provide funding for any of the following: individuals; private charity or foundation; organizations not exempt under Section 501(c)3 of the Internal Revenue Code, and not eligible for tax-deductible support; religious organizations, unless the particular program will benefit a large portion of a community without regard to religious affiliation and does not duplicate the work of other agencies in the community; political causes, candidates, organizations or campaigns; organizations that discriminate on the basis of age, color, citizenship, disability, disabled veteran status, gender, race, religion, national origin, marital status, sexual orientation, military service or status or Vietnam-era veteran status; organizations whose primary purpose is to influence legislation; endowments or capital campaigns; film, music, TV, video and media production projects or broadcast program underwriting; research studies, unless related to projects we are already supporting; sports sponsorships; performing arts tours; association memberships; or organizations that have received a grant from the Verizon Foundation in the last three consecutive years - organization may reapply after a one-year hiatus.
Geographic Focus: Massachusetts, Rhode Island
Amount of Grant: 5,000 - 10,000 USD

Contact: Rick Colon; (781) 849-2046; richard.b.colon@verizon.com
Internet: http://foundation.verizon.com/cybergrants/plsql/vznadmin.vz_www.iyc?x_zip=02184
Sponsor: Verizon Foundation
125 Lundquist Drive, Room 1730
Braintree, MA 02184

Verizon Foundation Pennsylvania Grants 3862
Verizon Foundation Pennsylvania is investing in online resources, community-based initiatives and state-wide programs that teach non readers to read and advance the complex skills necessary for educational achievement and job success in the 21st Century. The grants have enabled the recipient organizations to improve basic literacy skills, create new and diverse approaches to technology-based learning, and help domestic violence victims rebuild their lives. The Foundation only accepts electronic proposals through its online process, and reviews unsolicited proposals on a continuous calendar year basis from January 1st through October 31st.
Requirements: Proposals will be considered from eligible tax-exempt organizations in certain 501(c)3 subsections as defined by the Internal Revenue Service (IRS). Proposals will also be considered from elementary and secondary schools (public and private) that are registered with the National Center for Education Statistics (NCES). Proposals may also be considered from eligible tax-exempt organizations in the subsection 170(B)(1)(a)(i)--Church, provided that the proposal will benefit a large portion of a community without regard to religious affiliation and does not duplicate the work of other agencies in the community.
Restrictions: The Verizon Foundation does not provide funding for any of the following: individuals; private charity or foundation; organizations not exempt under Section 501(c)3 of the Internal Revenue Code, and not eligible for tax-deductible support; religious organizations, unless the particular program will benefit a large portion of a community without regard to religious affiliation and does not duplicate the work of other agencies in the community; political causes, candidates, organizations or campaigns; organizations that discriminate on the basis of age, color, citizenship, disability, disabled veteran status, gender, race, religion, national origin, marital status, sexual orientation, military service or status or Vietnam-era veteran status; organizations whose primary purpose is to influence legislation; endowments or capital campaigns; film, music, TV, video and media production projects or broadcast program underwriting; research studies, unless related to projects we are already supporting; sports sponsorships; performing arts tours; association memberships; or organizations that have received a grant from the Verizon Foundation in the last three consecutive years - organization may reapply after a one-year hiatus.
Geographic Focus: Pennsylvania
Amount of Grant: 5,000 - 10,000 USD
Contact: Denise Loughlin, Director; (215) 466-3351; fax (215) 466-3351; denise.g.loughlin@verizon.com
Internet: http://www22.verizon.com/about/community/pa/community/comm_index.html
Sponsor: Verizon Foundation
3478 Kirkwood Road
Philadelphia, PA 19114

Verizon Foundation Vermont Grants 3863
The Foundation is committed to supporting programs and projects that leverage technology to improve basic computer literacy and domestic violence initiatives, seeking organizations that serve the needs of diverse communities, people with disabilities, and the economically and socially challenged. Last year, Verizon Foundation donated close to $250,000 to Vermont nonprofit organizations and our employees clocked more than 16,000 volunteer hours to community organizations. The Foundation only accepts electronic proposals through its online process, and reviews unsolicited proposals on a continuous calendar year basis from January 1st through October 31st.
Requirements: Proposals will be considered from eligible tax-exempt organizations in certain 501(c)3 subsections as defined by the Internal Revenue Service (IRS). Proposals will also be considered from elementary and secondary schools (public and private) that are registered with the National Center for Education Statistics (NCES). Proposals may also be considered from eligible tax-exempt organizations in the subsection 170(B)(1)(a)(i)--Church, provided that the proposal will benefit a large portion of a community without regard to religious affiliation and does not duplicate the work of other agencies in the community.
Restrictions: The Verizon Foundation does not provide funding for any of the following: individuals; private charity or foundation; organizations not exempt under Section 501(c)3 of the Internal Revenue Code, and not eligible for tax-deductible support; religious organizations, unless the particular program will benefit a large portion of a community without regard to religious affiliation and does not duplicate the work of other agencies in the community; political causes, candidates, organizations or campaigns; organizations that discriminate on the basis of age, color, citizenship, disability, disabled veteran status, gender, race, religion, national origin, marital status, sexual orientation, military service or status or Vietnam-era veteran status; organizations whose primary purpose is to influence legislation; endowments or capital campaigns; film, music, TV, video and media production projects or broadcast program underwriting; research studies, unless related to projects we are already supporting; sports sponsorships; performing arts tours; association memberships; or organizations that have received a grant from the Verizon Foundation in the last three consecutive years - organization may reapply after a one-year hiatus.
Geographic Focus: Vermont
Amount of Grant: 5,000 - 10,000 USD
Contact: Stephanie Lee; (617) 743-5440; stephanie.s.lee@verizon.com
Internet: http://foundation.verizon.com/cybergrants/plsql/vznadmin.vz_www.iyc?x_zip=05403
Sponsor: Verizon Foundation
185 Franklin Street, Room 1702
Boston, MA 02110

Verizon Foundation Virginia Grants 3864
Verizon Foundation grants have touched and will continue to affect thousands of lives in Virginia. Verizon is committed to investing in the technological, financial, and human resources necessary to build more literate communities that can prosper in the information era, and to foster domestic violence awareness, prevention, and recovery. The Foundation will continue to form strong partnerships with Virginia's non-profit community to achieve its funding priority goals and improve the quality of life for people across the Commonwealth. The Foundation only accepts electronic proposals through its online process, and reviews unsolicited proposals on a continuous calendar year basis from January 1st through October 31st.
Requirements: Proposals will be considered from eligible tax-exempt organizations in certain 501(c)3 subsections as defined by the Internal Revenue Service (IRS). Proposals will also be considered from elementary and secondary schools (public and private) that are registered with the National Center for Education Statistics (NCES). Proposals may also be considered from eligible tax-exempt organizations in the subsection 170(B)(1)(a)(i)--Church, provided that the proposal will benefit a large portion of a community without regard to religious affiliation and does not duplicate the work of other agencies in the community.
Restrictions: The Verizon Foundation does not provide funding for any of the following: individuals; private charity or foundation; organizations not exempt under Section 501(c)3 of the Internal Revenue Code, and not eligible for tax-deductible support; religious organizations, unless the particular program will benefit a large portion of a community without regard to religious affiliation and does not duplicate the work of other agencies in the community; political causes, candidates, organizations or campaigns; organizations that discriminate on the basis of age, color, citizenship, disability, disabled veteran status, gender, race, religion, national origin, marital status, sexual orientation, military service or status or Vietnam-era veteran status; organizations whose primary purpose is to influence legislation; endowments or capital campaigns; film, music, TV, video and media production projects or broadcast program underwriting; research studies, unless related to projects we are already supporting; sports sponsorships; performing arts tours; association memberships; or organizations that have received a grant from the Verizon Foundation in the last three consecutive years - organization may reapply after a one-year hiatus.
Geographic Focus: Virginia
Amount of Grant: 5,000 - 10,000 USD
Contact: Stephan Clementi, (804) 772-1673 or (800) 360-7955; fax (908) 630-2660; steve.clementi@verizon.com
Internet: http://foundation.verizon.com/cybergrants/plsql/vznadmin.vz_www.iyc?x_zip=24012
Sponsor: Verizon Foundation
703 East Grace Street, 7th Floor
Richmond, VA 23219

Vermont Community Foundation Grants 3865
The foundation has broad program interests and will consider any project that meets a clearly defined community need in Vermont. Categories of support include, but are not limited to: the arts and humanities; education; the environment; historic resources; health; public affairs and community development; and social services.
Requirements: To be eligible, an organization must: have 501(c)3 tax-exempt status and not be classified as a private foundation; be located in or serve the people of Vermont; and employ staff and provide services without discrimination on the basis of race, religion, sex, age, national origin, disability, or sexual orientation. See Foundations website for additional guidelines, including rolling deadlines, and application forms.
Restrictions: The foundation does not make grants for endowments, annual operating or capital campaigns, religious purposes, individuals, debts, or equipment unless it is an integral part of an otherwise eligible project.
Geographic Focus: Vermont
Amount of Grant: Up to 25,000 USD
Contact: Peter Espenshade; (802) 388-3355, ext. 248; pespenshade@vermontcf.org
Internet: http://www.vermontcf.org/apply/
Sponsor: Vermont Community Foundation
3 Court Street, P.O. Box 30
Middlebury, VT 05753

Vernon K. Krieble Foundation Grants 3866
The Foundation was established in 1984 by the family of Professor Vernon K. Krieble, scientist, educator, inventor, and entrepreneur. Recognizing that the Foundation's assets are the product of a free and democratic society, the founders considered it fitting that those assets be used "to further democratic capitalism and the preserve and promote a society of free, educated, healthy and creative individuals." The Foundation offers support to non-profit charitable and educational organizations that demonstrate leadership in furthering the original objectives, so that future generations can aspire to and achieve their full potential in a free society. Awards range from $2,500 to $50,000. There are no deadlines.
Requirements: Nonprofit 501(c)3 organizations are eligible. Funding is provided only for those organizations and projects which involve public policy research and education on issues supporting the preservation, and in some cases the restoration, or freedom and democracy in the United States, according to the principles of the Founding Fathers. Written proposals should include a summary of the project, the project budget, the amount requested, the qualifications of individuals involved, and a copy of the organization's Internal Revenue Service determination letter.
Geographic Focus: All States
Contact: Helen E. Krieble, President; (303) 758-3956; fax (303) 488-0068
Internet: http://www.krieble.org/grants
Sponsor: Vernon K. Krieble Foundation
1777 S Harrison Street, Suite 807
Denver, CO 80210

VHA Health Foundation Grants　　3867

The Foundation has dedicated its resources to create a legacy strategy, which will build on its vision of the last decade by focusing on two critical issues that will significantly improve the delivery of health care: patient safety and disaster relief. Programs submitted for funding consideration must represent a new and innovative approach to hospital emergency preparedness and be beyond the concept stage - ready to test, implement, expand, or diffuse. Innovations should be either transformational (i.e. involve significant change in current operations) and/or disruptive (i.e., have profound impact on services and economics) in nature.
Requirements: Unsolicited inquires may be submitted at any time during the year. The Letter of Inquiry is the required first step in the process. Eligible programs should clearly incorporate four key elements: Innovation, Impact, Replicability and Sustainability. Grants are available to U.S. health care providers, including hospitals, health care systems, clinics, and medical practices with IRS tax-exempt 501(c)3 status. Local partnerships are encouraged to apply, but, since the hospital is the foundation's primary audience, the health care provider must serve as fiduciary agent and play an integral role in the program. An organization may submit multiple inquiries for consideration. Previously funded organizations may apply. Applicants are required to financially invest in the program with cash from any source and/or through in-kind contributions specific to the grant period. The minimum level of the investment is equal to one-half of the funding requested from the foundation.
Restrictions: The foundation will not consider requests for seed money or planning grants; primary research and development of technology; construction or renovation; indirect costs or overhead; endowments; political activities or attempts to influence specific legislation; individual scholarships or tuition assistance; or programs outside the United States.
Geographic Focus: All States
Amount of Grant: 100,000 - 250,000 USD
Samples: MeritCare Health System, Fargo, ND and Medcenter One Health System, Bismarck, ND, $62,500--disaster relief, North Dakota flooding; Arkansas Methodist Medical Center, Paragould, AR, St. Bernard's Regional Medical Center, Jonesboro, AR, and Washington Regional Medical Center, Fayetteville, AR, $170,850--disaster relief, Arkansas winter storm.
Contact: Michael Regier, Senior Vice President; (877) 847-1450 or (972) 830-0422; fax (972) 830-0332; vhahealthfoundation@vha.com
Internet: https://www.vhafoundation.org/portal/server.pt?open=512&objID=1204&PageID=0&cached=true&mode=2&userID=504686
Sponsor: VHA Health Foundation
220 Easr Las Colinas Boulevard
Irving, TX 75039-5500

Victor E. Speas Foundation Grants　　3868

The Victor E. Speas Foundation was established to provide medical care for the needy, to further medical research, and to support and promote quality educational, cultural, human-services, and health-care programming. In the area of arts, culture, and humanities, the foundation supports programming that: fosters the enjoyment and appreciation of the visual and performing arts; strengthens humanities and arts-related education programs; provides affordable access; enhances artistic elements in communities; and nurtures a new generation of artists. In the area of education, the foundation supports programming that: promotes effective teaching; improves the academic achievement of, or expands educational opportunities for, disadvantaged students; improves governance and management; strengthens nonprofit organizations, school leadership, and teaching; and bolsters strategic initiatives of area colleges and universities. In the area of health, the foundation supports programming that improves the delivery of health care to the indigent, uninsured, and other vulnerable populations and addresses health and health-care problems that intersect with social factors. In the area of human services, the foundation funds programming that: strengthens agencies that deliver critical human services and maintains the community's safety net and helps agencies respond to federal, state, and local public policy changes. In the area of community improvement, the foundation funds capacity-building and infrastructure-development projects including: assessments, planning, and implementation of technology for management and programmatic functions within an organization; technical assistance on wide-ranging topics, including grant writing, strategic planning, financial management services, business development, board and volunteer management, and marketing; and mergers, affiliations, or other restructuring efforts. Grant requests for general operating support and program support will be considered. Grants from the foundation are one year in duration. Application materials are available for download at the grant website. There are no application deadlines for the Victor E. Speas Foundation. Proposals are reviewed on an ongoing basis.
Requirements: Applicants must have 501(c)3 tax-exempt status and serve the residents of Kansas City, Missouri. Two copies of the completed application must be mailed.
Restrictions: Grant requests for capital support will not be considered. The trust does not support requests from individuals, organizations attempting to influence policy through direct lobbying, or any political campaigns.
Geographic Focus: Missouri
Samples: Childrens Mercy Hospital Foundation, Kansas City, Missouri, $100,000, to fund research and clinical trials of reformulated adult drugs to benefit pediatric patients with rare and neglected blood and other cancers; Open Options, Kansas City, Missouri, $25,000, United Cerebral Palsy of Greater Kansas City technology upgrade; Metropolitan Community Colleges Foundation, Kansas City, Missouri, $50,000, for medical equipment for the Health Science Institute training facility.
Contact: Spence Heddens; (816) 292-4301; Spence.heddens@baml.com
Internet: https://www.bankofamerica.com/philanthropic/fn_search.action
Sponsor: Victor E. Speas Foundation
1200 Main Street, 14th Floor, P.O. Box 219119
Kansas City, MO 64121-9119

Victor Grifols i Lucas Foundation Ethics and Science Awars for Educational Institutions　　3869

The Victor Grifols i Lucas Foundation awards three prizes annually to educational institutions with the intention of providing recognition and support to projects that encourage ethical awareness and sensitivity among students in relation to scientific issues. All educational institutions in Catalonia that teach primary and secondary education, or pre-university education, are eligible for the prize. The prizes are: First prize - 5,000 euros; Second prize - 2,500 euros; and Third prize - 1,000 euros. The annual deadline for applications is May 31.
Requirements: Any primary or secondary education program in Catalonia is eligible to apply.
Geographic Focus: Spain
Date(s) Application is Due: May 31
Amount of Grant: 1,000 - 5,000 USD
Contact: Grant Director; +34 935 710 410; fundacio.grifols@grifols.com
Internet: http://www.fundaciongrifols.org/portal/en/2/premi-etica-i-ciencia
Sponsor: Victor Grifols i Lucas Foundation
Carrer Jesús i Maria, 6
Barcelona, 08022 Spain

Victor Grifols i Lucas Foundation Prize for Journalistic Work on Bioethics　　3870

The Victor Grifols i Lucas Foundation awards one annual Prize for Journalistic Work on Bioethics. The Prize, worth 3000 euros, is open to pieces of journalism tackling bioethical issues which have been published throughout the year in any medium. The annual deadline for entries is May 31. Judging of the competition will be communicated via the website during the first half of October.
Geographic Focus: All States, Spain
Date(s) Application is Due: May 31
Amount of Grant: 3,000 EUR
Contact: Grant Director; +34 935 710 410; fundacio.grifols@grifols.com
Internet: http://www.fundaciongrifols.org/portal/en/2/premio_periodistico
Sponsor: Victor Grifols i Lucas Foundation
Carrer Jesús i Maria, 6
Barcelona, 08022 Spain

Victor Grifols i Lucas Foundation Research Grants　　3871

The Víctor Grífols i Lucas Foundation awards six grants for research in bioethics. The grants, each worth 5000 euros, are awarded to research projects to be undertaken on any issue related to bioethics and its practical application. Awards will be paid as follows: one-third of the amount when the grant is initially awarded; two-thirds of the award upon completion of the project research. Projects may be proposed individually or by teams. The annual deadline is May 31.
Requirements: If a project is not completed, the grant recipient must return the initial one-third of the award given.
Geographic Focus: All States, Spain
Date(s) Application is Due: May 31
Amount of Grant: 5,000 EUR
Contact: Grant Director; +34 935 710 410; fundacio.grifols@grifols.com
Internet: http://www.fundaciongrifols.org/portal/en/2/becas_investigacion
Sponsor: Victor Grifols i Lucas Foundation
Carrer Jesús i Maria, 6
Barcelona, 08022 Spain

Victor Grifols i Lucas Foundation Secondary School Prizes　　3872

The Victor Grifols i Lucas Foundation awards three prizes annually for research in bioethics conducted by senior high school students. It is hoped that these prizes will promote the inclusion of the ethical dimension in research in professional education from the outset. Prizes will be awarded to the three best research projects in bioethics completed during the academic year in public and private secondary schools in Catalonia. The prizes are as follows: First Prize - University fees for the first two years of university study; Second Prize - an IPad; and Third Prize - a Netbook. The deadline for completed work is May 31.
Geographic Focus: Spain
Date(s) Application is Due: May 31
Contact: Grant Director; +34 935 710 410; fundacio.grifols@grifols.com
Internet: http://www.fundaciongrifols.org/portal/en/2/secondary_school_prize
Sponsor: Victor Grifols i Lucas Foundation
Carrer Jesús i Maria, 6
Barcelona, 08022 Spain

Vigneron Memorial Fund Grants　　3873

The Vigneron Memorial Fund was established in 1959 to support charitable organizations that work to improve the lives of physically disabled children and adults. Preference is given to charitable organizations that serve the people of the city of Providence or the town of Narragansett, Rhode Island. Capital requests that fund handicapped assistive devices (wheelchairs, walkers, etc.) or adaptive equipment (lift installation, ramp installation, etc.) are strongly encouraged. Grant requests for general operating or program support will also be considered. The majority of grants from the Vigneron Memorial Fund are one year in duration. Applicants must apply online at the grant website. Applicants are strongly encouraged to do the following before applying: review the downloadable state application procedures for additional helpful information and clarifications; review the downloadable online-application guidelines at the grant website; review the foundation's funding history (link is available from the grant website); review the online application questions in advance; and review the list of required attachments. These will generally include: a list of board members, financial statements (audited, reviewed, or compiled by independent auditor); an organization summary; a list of other funding sources; an IRS

Determination letter; and other required documents. All attachments must be uploaded in the online application as PDF, Word, or Excel files. The Vigneron Memorial Fund shares a mission and grantmaking focus with the John D. & Katherine A. Johnston Foundation. Both foundations have the same proposal deadline date of 11:59 p.m. on April 1. Applicants will be notified of grant decisions before May 31.
Requirements: Applicants must have 501(c)3 tax-exempt status.
Restrictions: The fund does not support requests from individuals, organizations attempting to influence policy through direct lobbying, or any political campaigns.
Geographic Focus: Rhode Island
Date(s) Application is Due: Apr 1
Samples: Homefront Health Care, Providence, Rhode Island, $10,000, general operations; Nickerson Community Center, Providence, Rhode Island, $5,000, for disability access ramp for the main entrance; WaterFire Providence, Providence, Rhode Island, WaterFire Access Program (a highly successful outreach program for children and adults with disabilities and/or mobility restrictions).
Contact: Emma Greene, Director; (617) 434-0329; emma.m.greene@baml.com
Internet: https://www.bankofamerica.com/philanthropic/fn_search.action
Sponsor: Vigneron Memorial Fund
225 Franklin Street, 4th Floor, MA1-225-04-02
Boston, MA 02110

Virginia Department of Health Mary Marshall Nursing Scholarships for Licensed Practical Nurses 3874

The Mary Marshall Nursing Scholarships for Licensed Practical Nurses are funded by the Virginia General Assembly along with a donation from The Board of Nursing. The scholarship amount is dependent on amount of scholarship source and the number of qualified applicants. It is important that all applicants fully understand the conditions of accepting a Mary Marshall Nursing Scholarship. These awards are not gifts. Scholarship recipients must agree to engage in full time nursing in Virginia for one month for every $100 received.
Requirements: To be considered for a Mary Marshall Nursing Scholarship, an applicant must meet the following criteria: residency in Virginia for at least one year; acceptance or enrollment as a full-time or part-time student in a practical school of nursing in the state of Virginia; and have submitted a completed application form and a recommendation from the Program Director regarding scholastic attainment and financial need prior to June 30. Applications and guidelines are available online from May 1 to June 30 every year. Applications must be typed, printed and mailed (with original signatures) to the: Virginia Department of Health, Office of Minority Health and Public Health Policy, ATTN: Nursing Scholarship, 109 Governor Street, Suite 1016, East Richmond, Virginia 23219.
Restrictions: The award recipient has 60 days from the date of graduation to obtain his/her license. Full time employment must begin within 90 days of the recipient's license date. Voluntary military service, even if stationed in Virginia, cannot be used to repay scholarship awards.
Geographic Focus: Virginia
Date(s) Application is Due: Jun 30
Contact: Aileen Harris, Health Care Workforce Manager; (804) 864-7436; fax (804) 864-7440; Aileen.Harris@vdh.virginia.gov
Internet: http://www.vdh.virginia.gov/healthpolicy/primarycare/incentives/nursing/lpn_guidelines.htm
Sponsor: Virginia Department of Health
P.O. Box 2448
Richmond, VA 23218

Virginia Department of Health Mary Marshall Nursing Scholarships for Registered Nurses 3875

The Mary Marshall Nursing Scholarships are funded by the Virginia General Assembly along with a donation from The Board of Nursing. The scholarship amount is dependent upon amount of scholarship source and the number of qualified applicants. It is important that all applicants fully understand the conditions of accepting a Mary Marshall Nursing Scholarship. These awards are not gifts. Scholarship recipients must agree to engage in full time nursing in Virginia for one month for every $100 received.
Requirements: To be considered for a Mary Marshall Nursing Scholarship, an applicant must meet the following criteria: residency in Virginia for at least one year; acceptance or enrollment as a full time or part time student in a school of nursing in the state of Virginia; demonstration of a cumulative grade point average of at least 3.0 in required courses, not electives; demonstration of financial need, verified by the Financial Aid Office/authorized person at the applicant's nursing school; and have submitted a completed application form and an official grade transcript. Applications and guidelines are available online from May 1 to June 30 every year. Applications must be typed, printed and mailed (with original signatures) to the: Virginia Department of Health, Office of Minority Health and Public Health Policy, ATTN: Nursing Scholarship, 109 Governor Street, Suite 1016, East Richmond, Virginia 23219.
Restrictions: The award recipient has 60 days from the date of graduation to obtain his/her license. Full time employment must begin within 90 days of the recipient's license date. Voluntary military service, even if stationed in Virginia, cannot be used to repay scholarship awards.
Geographic Focus: Virginia
Date(s) Application is Due: Jun 30
Contact: Aileen Harris, Health Care Workforce Manager; (804) 864-7436; fax (804) 864-7440; Aileen.Harris@vdh.virginia.gov
Internet: http://www.vdh.virginia.gov/healthpolicy/primarycare/incentives/nursing/rn_guidelines.htm
Sponsor: Virginia Department of Health
P.O. Box 2448
Richmond, VA 23218

Virginia Department of Health Nurse Practitioner/Nurse Midwife Scholarships 3876

The program awards nurse practitioner/nurse midwife scholarships on a competitive basis. Considerations for award selections include scholastic achievement; character; and stated commitment to postgraduate employment in a medically underserved area of Virginia, in an employment setting that provides services to persons who are unable to pay for the service and participates in all government-sponsored insurance programs designed to assure access to medical care services for covered persons. Student recipients must agree to engage in full-time practice in a designated medically underserved area for a period of years equal to the number of annual scholarships received. Preference will be given to residents of the commonwealth; minority students; students enrolled in family practice, obstetrics and gynecology, pediatric, adult health, and geriatric nurse practitioner programs; and residents of medically underserved areas of Virginia, as determined by the board of health, in accordance with the provisions of its regulations for that purpose. The application period for this scholarship is May 1st - June 30th.
Requirements: The applicant must meet the following criteria: residency in Virginia for at least one year; acceptance or enrollment as a full-time student in a nurse practitioner/nurse midwifery program in Virginia or a nurse midwifery program in a nearby state; demonstration of a cumulative grade-point average of at least 3.0 in graduate and/or undergraduate courses; have submitted a completed application form, official grade transcript of graduate and/or undergraduate courses, and a statement of intent to practice as a nurse practitioner/nurse midwife in an undeserved area of Virginia following graduation; submission of two reference letters; and submission of all materials to the Virginia Department of Health, Office of Minority Health and Public Health Policy, ATTN: Nursing Scholarship, 109 Governor Street, Suite 1016, East Richmond, Virginia 23219
Geographic Focus: Virginia
Date(s) Application is Due: Jun 30
Contact: Aileen Harris, Health Care Workforce Manager; (804) 864-7436; fax (804) 864-7440; Aileen.Harris@vdh.virginia.gov
Internet: http://www.vdh.virginia.gov/healthpolicy/primarycare/incentives/nursing/index.htm
Sponsor: Virginia Department of Health
P.O. Box 2448
Richmond, VA 23218

Virginia L. and William K. Beatty MLA Volunteer Service Award 3877

This award, established in 2007, is named to honor the Beattys' significant contributions to the Medical Library Association (MLA) and to the profession as long-time volunteers to the association. The purpose of the award is to recognize a medical librarian who has demonstrated outstanding, sustained service to MLA and the health-sciences library profession. The recipient receives a certificate of outstanding service at the association's annual meeting and a cash award of $1,000 after the annual meeting. The recipient assumes all costs of attending the meeting and the ceremony at which the presentation is made. Guidelines and a nomination form are available at the website. In order for the nomination to be considered, all materials and letters of support must be received by MLA via email, fax, or mail (in order of preference) by November 1. The recipient will be notified in March before the annual meeting. Founded in 1898, the Medical Library Association (MLA) is a nonprofit, educational organization of more than 1,100 institutions and 3,600 individual members in the health sciences information field committed to educating health-information professionals, supporting health-information research, promoting access to the world's health-sciences information, and working to ensure that the best health information is available to all.
Requirements: The nominee must be a member of MLA for each of the last ten years immediately preceding the nomination. The nomination must be made for exceptional contribution(s) to furthering the mission, goals, and objectives of MLA and the profession as demonstrated by outstanding and significant service in the association's leadership, publications, research, and special projects (or a combination of these four elements).
Restrictions: The nominee must be an "unsung hero" of MLA and have not served in an elected national-leadership position or received a national MLA award prior to or at the time of nomination. Elected national leadership positions include serving on the MLA Board, Nominating Committee, or as President-Elect.
Geographic Focus: All States, All Countries
Date(s) Application is Due: Nov 1
Amount of Grant: 1,000 USD
Contact: Carla Funk, Program Contact; (312) 419-9094, ext. 14; fax (312) 419-8950; mlapd2@mlahq.org or awards@mlahq.org
Internet: http://www.mlanet.org/awards/honors/index.html
Sponsor: Medical Library Association
65 East Wacker Place, Suite 1900
Chicago, IL 60601-7246

Virginia W. Kettering Foundation Grants 3878

The Virginia W. Kettering Foundation, a private family foundation in Dayton, Ohio, was activated in 2003 when Virginia Kettering passed away at the age of 95. To continue her legacy of confident determination and passionate commitment to the Dayton area, Virginia Kettering directed the stewards of her foundation to support charitable organizations for charitable purposes within Montgomery, Greene, Clark, Miami, Darke, Preble, Butler and Warren Counties. The Foundation's Distribution Committee meets biannually (April and October) to make funding decisions. Primary areas of support include: arts, culture and humanities; education; environment; health and medical needs; human services; and public and societal benefit. There are two annual deadlines for invited applicants: March 15 and September 15.
Requirements: 501(c)3 organizations located in, or that serve, the counties of Montgomery, Greene, Clark, Miami, Darke, Preble, Butler and Warren in Ohio.

Restrictions: A Request Summary will not be accepted for any of the following purposes: religious organizations for religious purposes; individual public elementary or secondary schools or public school districts; multi-year grants; grants or loans to individuals; tickets, advertising or sponsorships of fundraising events; efforts to carry on propaganda or otherwise attempt to influence legislation; or activities of 509(a)3 Type III Supporting Organizations.
Geographic Focus: Ohio
Date(s) Application is Due: Mar 15; Sep 15
Amount of Grant: 5,000 - 50,000 USD
Contact: Judith M. Thompson, Executive Director; (973) 228-1021; fax (888) 719-1185; info@ketteringfamilyphilanthropies.org
Internet: https://www.cfketteringfamilies.com/vwk/application-process
Sponsor: Virginia W. Kettering Foundation
40 North Main Street, #1480
Dayton, OH 45423

Visiting Nurse Foundation Grants 3879
The foundation awards grants in support of home- and community-based health care for the medically underserved in Cook and the collar counties of Lake, McHenry, DuPage, Kane and Will, with a focus on Chicago. Priority areas are home health care services; prevention and health promotion; and early intervention. Preference is given to programs using nurses to provide the care. Types of support include program, capital, and general operating grants. Letters of intent should be submitted by deadline dates listed, and full proposals will be by invitation only.
Requirements: Nonprofit organizations in Cook, Lake, McHenry, DuPage, Kane, and Will counties of Illinois are eligible.
Restrictions: No grants are made to individuals.
Geographic Focus: Illinois
Date(s) Application is Due: Jan 19; Apr 19; Oct 20
Amount of Grant: 15,000 - 80,000 USD
Contact: Robert N. DiLeonardi; (312) 214-1521; fax (312) 214-1529; info@vnafoundation.net
Internet: http://www.vnafoundation.net
Sponsor: Visiting Nurse Association Foundation
20 N Wacker Drive, Suite 3118
Chicago, IL 60606

Volkswagen of America Corporate Contributions Grants 3880
The corporation accepts proposals broadly from programs servicing needs in healthcare, wellness, children, medical research, education and diversity.
Requirements: There are no specific deadlines, however the review committee does meet quarterly to go over proposals received. The committee considers the type of organization, its mission and the intended use of the funds.
Restrictions: Grants are available to 501(c)3 organizations only.
Geographic Focus: All States
Amount of Grant: 1,000 - 2,500 USD
Contact: Tara Jones, (248) 754-4693; tara.jones@vw.com
Sponsor: Volkswagen of America
3800 Hamlin Road
Auburn Hills, MI 48326

W.C. Griffith Foundation Grants 3881
The foundation awards grants, primarily in Indiana, to nonprofit organizations in its areas of interest, including arts and culture, cancer and medical research, children and youth services, Christian churches and organizations, community development, education, environmental programs, health organizations and hospitals, higher education, homeless services, human reproduction and fertility, library science and libraries, minorities, museums, music and performing arts, secondary education, and social services delivery. Types of support include building construction/renovation, capital campaigns, and continuing support. There are no application forms or deadlines. The board meets in June and November; proposals should be received in May and October.
Requirements: Nonprofit organizations are eligible. Preference is given to requests from Indianapolis, Indiana.
Geographic Focus: All States
Amount of Grant: 500 - 50,000 USD
Contact: Curt Farran, c/o National City Bank of Indiana; (317) 267-7262
Sponsor: W.C. Griffith Foundation
101 West Washington Street, Suite 600 East
Indianapolis, IN 46255

W. C. Griffith Foundation Grants 3882
The W. C. Griffith Foundation was established in 1959 by a donation of William C. Griffith, and Ruth Perry Griffith. The Foundation gives primarily in Indianapolis-Carmel, Indiana area. The Foundation provides support primarily for hospitals, health associations, medical and cancer research, the arts, including music and museums, community funds and development, higher, secondary, and other education, family planning services, child welfare, the homeless, the environment, libraries, and Christian religious organizations.
Requirements: Applicants should submit the following: detailed description of project and amount of funding requested; 1 Copy of the proposal.
Geographic Focus: Indiana
Amount of Grant: 1,000 - 45,000 USD
Contact: Trust Coordinator; fax (317) 693-2504
Sponsor: W.C. Griffith Foundation
101 West Ohio Street, Suite 1450
Indianapolis, IN 46204-1998

W. Clarke Swanson, Jr. Foundation Grants 3883
The Foundation, established in 1998, is the brain-child of W. Clarke Swanson, Jr., founder of Swanson Vineyards. The Foundation is primarily interested in supporting the arts, education, and health care in California, there is some national giving. The type of funding typically comes in the form of general operating support. Although there is no specific application form required, applicants should submit a detailed description of the project and amount of funding requested.
Geographic Focus: All States
Amount of Grant: 3,000 - 250,000 USD
Samples: Friends of Lincoln Theater, Yountville, California, $172,200; University of California at San Francisco Foundation, San Francisco, California, $50,000.
Contact: W. Clarke Swanson; (707) 944-0905, ext. 20; clarke@swansonvineyards.com
Sponsor: W. Clarke Swanson, Jr. Foundation
1050 Oakville Cross Road, P.O. Box 148
Oakville, CA 94562

W.H. and Mary Ellen Cobb Charitable Trust Grants 3884
The Cobb's were originally from Kentucky, but moved to the Texas Panhandle where they owned and operated several clothing stores. Although they did not have children of their own, they were interested in the well-being of children and therefore established their charitable trust to benefit local charities in the Panhandle whose mission has a strong "emphasis on helping children." This foundation makes approximately five-seven awards each year and grants are typically between $1,000 and $25,000. Applicants must apply online at the grant website. Applicants are strongly encouraged to do the following before applying: review the downloadable state application procedures for additional helpful information and clarifications; review the downloadable online-application guidelines at the grant website; review the trust's funding history (link is available from the grant website); review the online application questions in advance; and review the list of required attachments. These will generally include: a list of board members, financial statements (audited, reviewed, or compiled by independent auditor); an organization summary; a list of other funding sources; an IRS Determination letter; and other required documents. All attachments must be uploaded in the online application as PDF, Word, or Excel files. The application deadline for this trust is 11:59 p.m. on September 30.
Requirements: The W.H. & Mary Ellen Cobb Charitable Trust considers requests from charitable organizations whose primary focus is the provision of services that benefit children. Organizations must: be geographically located within the Texas Panhandle; serve residents of the Panhandle, Amarillo, and vicinity; and have 501(c)3 tax-exempt status.
Restrictions: The trust does not support requests from individuals, organizations attempting to influence policy through direct lobbying, or any political campaigns.
Geographic Focus: Texas
Date(s) Application is Due: Sep 30
Amount of Grant: 1,000 - 25,000 USD
Samples: St. Andrews Episcopal School, Amarillo, Texas, $10,000; High Plains Childrens Home and Family Services, Amarillo, Texas, $7,000; Maverick Boys and Girls Club of Amarillo, Amarillo, Texas, $5,000.
Contact: Mark J. Smith; (817) 390-6028; tx.philanthropic@baml.com
Internet: https://www.bankofamerica.com/philanthropic/fn_search.action
Sponsor: W.H. and Mary Ellen Cobb Charitable Trust
500 West 7th Street, 15th Floor, TX1-497-15-08
Fort Worth, TX 76102-4700

W. James Spicer Scholarship 3885
The W. James Spicer Scholarship was established to recognize the teaching career of Mr. Spicer who taught in the Gary School Corporation for over 40 years. The one-time, $1,000 scholarship is awarded to a graduating senior at Wirt/Emerson School for the Visual and Performing Arts who will be pursuing higher education. All college majors are eligible.
Requirements: Applicants must be pursuing higher education, with a cumulative GPA of at least 2.0. Along with the completed application, candidates must submit a two page essay, the Scholastic Profile Form completed by their guidance counselor, an official high school transcript, a letter of recommendation from a teacher, a recommendation letter from someone other than a family member, and a letter of acceptance from the school they plan to attend.
Geographic Focus: Indiana
Date(s) Application is Due: Apr 15
Amount of Grant: 1,000 USD
Contact: Spicer Scholarship Contact; (219) 736-1880; fax (219) 736-1940; legacy@legacyfoundationlakeco.org
Internet: http://www.legacyfoundationlakeco.org/scholarships/index.htm
Sponsor: Legacy Foundation
1000 East 80th Place, 302 South
Merrillville, IN 46410

W.K. Kellogg Foundation Healthy Kids Grants 3886
The Foundation believes that children need nutrition, stimulation, healthy living conditions and access to quality health care. It helps many of them get all four, by funding organizations that improve birth outcomes and first food experiences, create access to healthy foods, improve health services, and educate families and communities about the inter-related factors that determine well-being. The Foundation especially focuses on children who are disadvantaged by multiple societal factors, a disproportionate percentage of whom are children of color.
Requirements: To be eligible for support, your organization or institution, as well as the purpose of the proposed project, must qualify under regulations of the United States Internal Revenue Service. The Kellogg Foundation does not have any submission deadlines. Grant applications are accepted throughout the year and are reviewed at its headquarters in Battle Creek, Michigan, or in its regional office in Mexico.

Restrictions: In general, the Foundation does not provide funding for: operational phases of established programs; capital requests (which includes the construction, purchase, renovation, and/or furnishing of facilities); equipment; conferences and workshops; films, television and/or radio programs; endowments; development campaigns; or research/studies (unless they are an integral part of a larger program budget being considered for funding).
Geographic Focus: All States
Contact: Deborah A. Rey; (269) 969-2133 or (269) 968-1611; fax (269) 968-0413
Internet: http://www.wkkf.org/what-we-support/healthy-kids.aspx
Sponsor: W.K. Kellogg Foundation
1 Michigan Avenue East
Battle Creek, MI 49017-4012

W.K. Kellogg Foundation Secure Families Grants 3887
The primary needs of the family must be addressed to create pathways out of poverty for children. The Foundation supports programs that increase family stability, foster quality jobs, careers and entrepreneurship, and promote secondary achievement and financial independence. It further supports strategies that increase income, assets, and aspirations of vulnerable children and their families and reduce disparities based on class, gender and race. The Foundation focuses on three areas, each with a direct impact on the ability of families to generate income and accumulate assets: family stability; career ladders; and financial independence.
Requirements: To be eligible for support, your organization or institution, as well as the purpose of the proposed project, must qualify under regulations of the United States Internal Revenue Service. The Kellogg Foundation does not have any submission deadlines. Grant applications are accepted throughout the year and are reviewed at its headquarters in Battle Creek, Michigan, or in its regional office in Mexico.
Restrictions: In general, the Foundation does not provide funding for: operational phases of established programs; capital requests (which includes the construction, purchase, renovation, and/or furnishing of facilities); equipment; conferences and workshops; films, television and/or radio programs; endowments; development campaigns; or research/studies (unless they are an integral part of a larger program budget being considered for funding).
Geographic Focus: All States
Contact: Deborah A. Rey; (269) 969-2133 or (269) 968-1611; fax (269) 968-0413
Internet: http://www.wkkf.org/what-we-support/secure-families.aspx
Sponsor: W.K. Kellogg Foundation
1 Michigan Avenue East
Battle Creek, MI 49017-4012

W.M. Keck Foundation Medical Research Grants 3888
The W.M. Keck Foundation concentrates on strengthening studies and programs in accredited universities and colleges and major, independent medical research institutions. Medical Research grants seek to advance the frontiers of the life sciences by supporting basic research that is high-risk and has the potential to transform its field. Successful projects are distinctive and novel in their approach to problems, push the edge of their field, or question the prevailing paradigm. Past grants have been awarded to major universities, medical schools, and independent research institutions to support pioneering biological research, including the development of new technologies, instrumentation, or methodologies. Historically, grants range from $500,000 to $5 million, and are typically $2 million or less. Applicants are strongly urged to contact Foundation staff well in advance of submitting a Phase I Application. The best times for these contacts are between January 1 and February 15 leading up to a May 1 submittal, or between July 1 and August 15 leading up to a November 1 submittal.
Requirements: Research universities, medical colleges and major private independent scientific and medical research institutes are eligible. Organizations must be exempt from federal taxation as defined by Section 501(c)3 of the Internal Revenue Code and be designated as a public charity (and not a private foundation) as defined by Section 509(a)1 or 509(a)2 or 170(b) of the Internal Revenue Code. If the institution is located in the State of California, the organization must also be exempt from California State Franchise or Income Tax under Section 23701(d) of the Revenue and Taxation Code.
Restrictions: Grants are not considered for: general operating expenses, endowments or deficit reduction; general and federated campaigns, including fundraising events, dinners or mass mailings; individuals; conference or seminar sponsorship; book publication and film or theater productions; public policy research; institutions that are located outside the United States; institutions that do not have at least three consecutive full, certified, audited financial statements; conduit organizations, unified funds or organizations that use grant funds from donors to support other organizations or individuals; or institutions that are subsidiaries or affiliates of larger entities that do not have a separate board of directors and independent audited financial statements.
Geographic Focus: All States
Date(s) Application is Due: May 1; Nov 1
Amount of Grant: 500,000 - 5,000,000 USD
Samples: Dirk Englund, Jonathan Owen, & Rafael Yuste, Columbia University, New York, New York, $1,000,000 - to develop a novel method for direct, real-time imaging of neuronal voltage signals; Laura Ranum & Maurice Swanson, University of Florida, Gainesville, Florida, $1,000,000 - to pursue critical questions: How does RAN translation work? Is RAN translation a key, previously unrecognized cellular process? Are repetitive sequences throughout the genome translated into proteins and if so, what is their function?.
Contact: Margie Antonetti, Grants and Database Specialist; (213) 680-3833; fax (213) 614-0934; mantonetti@wmkeck.org or MedRsch@wmkeck.org
Internet: http://www.wmkeck.org/grant-programs/medicalresearch.html
Sponsor: W.M. Keck Foundation
550 S Hope Street, Suite 2500
Los Angeles, CA 90071

W.M. Keck Foundation Science and Engineering Research Grants 3889
The W.M. Keck Foundation concentrates on strengthening studies and programs in accredited universities and colleges and major, independent medical research institutions. Science and Engineering Research grants seek to benefit humanity by supporting projects that are distinctive and novel in their approach, question the prevailing paradigm, or have the potential to break open new territory in their field. Past grants have been awarded to major universities and independent research institutions to support pioneering science and engineering research, the development of new technologies, and to facilitate the purchase of advanced instruments where such instruments would further specific research ventures. Historically, grants range from $500,000 to $5 million, and are typically $2 million or less. Applicants are strongly urged to contact Foundation staff well in advance of submitting a Phase I Application. The best times for these contacts are between January 1 and February 15 leading up to a May 1 submittal, or between July 1 and August 15 leading up to a November 1 submittal.
Requirements: Research universities, medical colleges and major private independent scientific and medical research institutes are eligible. Organizations must be exempt from federal taxation as defined by Section 501(c)3 of the Internal Revenue Code and be designated as a public charity (and not a private foundation) as defined by Section 509(a)1 or 509(a)2 or 170(b) of the Internal Revenue Code. If the institution is located in the State of California, the organization must also be exempt from California State Franchise or Income Tax under Section 23701(d) of the Revenue and Taxation Code.
Restrictions: Grants are not considered for: general operating expenses, endowments or deficit reduction; general and federated campaigns, including fundraising events, dinners or mass mailings; individuals; conference or seminar sponsorship; book publication and film or theater productions; public policy research; institutions that are located outside the United States; institutions that do not have at least three consecutive full, certified, audited financial statements; conduit organizations, unified funds or organizations that use grant funds from donors to support other organizations or individuals; or institutions that are subsidiaries or affiliates of larger entities that do not have a separate board of directors and independent audited financial statements.
Geographic Focus: All States
Date(s) Application is Due: May 1; Nov 1
Amount of Grant: 500,000 - 5,000,000 USD
Samples: Alexander Fridman, Drexel University, Philadelphia, Pennsylvania, $1,000,000 - to develop a fundamental understanding of bio-molecular interaction pathways between strongly non-equilibrium electrical plasma and living cells; Jules S. Jaffe, University of California, San Diego, California, $1,000,000 - to build the world's first high resolution, 3D, in situ, underwater microscope.
Contact: Margie Antonetti, Grants and Database Specialist; (213) 680-3833; fax (213) 614-0934; mantonetti@wmkeck.org or SciEng@wmkeck.org
Internet: http://www.wmkeck.org/grant-programs/science-engineering.html
Sponsor: W.M. Keck Foundation
550 S Hope Street, Suite 2500
Los Angeles, CA 90071

W.M. Keck Foundation Southern California Grants 3890
The Southern California program seeks to promote the education and healthy development of children and youth, strengthening families and enhancing the lives of people in the Greater Los Angeles area through its support of organizations that provide arts and cultural enrichment, civic and community services, early childhood and pre-college education, and health care. A special emphasis is placed on projects that focus on children and youth from low-income families, special needs populations, and safety-net services. Collaborative initiatives, as well as projects arising from the vision of one organization's strong leadership, are supported. Historically, grants range from $100,000 to $1 million, and typically are under $500,000. Applicants are strongly urged to contact Foundation staff well in advance of submitting a Phase I Application. The best times for these contacts are between January 1 and February 15 leading up to a May 1 submittal, or between July 1 and August 15 leading up to a November 1 submittal.
Requirements: Nonprofit organizations, including colleges and universities, pursuing relevant projects in Los Angeles County are eligible to apply.
Restrictions: Grants are not considered for: general operating expenses, endowments or deficit reduction; general and federated campaigns, including fundraising events, dinners or mass mailings; individuals; conference or seminar sponsorship; book publication and film or theater productions; public policy research; institutions that are located outside the United States; institutions that do not have at least three consecutive full, certified, audited financial statements; conduit organizations, unified funds or organizations that use grant funds from donors to support other organizations or individuals; or institutions that are subsidiaries or affiliates of larger entities that do not have a separate board of directors and independent audited financial statements.
Geographic Focus: California
Date(s) Application is Due: May 1; Nov 1
Amount of Grant: 100,000 - 1,000,000 USD
Samples: Alliance for a Better Community, Los Angeles, California, $200,000 - to build the capacity of parents to participate in various forms of civic learning and engagement in order to support their student's achievement; Alliance for Children's Rights, Los Angeles, California, $250,000 - o ensure that implementation of the California Fostering Connections to Success Act.
Contact: Margie Antonetti, Grants and Database Specialist; (213) 680-3833; fax (213) 614-0934; mantonetti@wmkeck.org or SoCal@wmkeck.org
Internet: http://www.wmkeck.org/grant-programs/southern-california-program.html
Sponsor: W.M. Keck Foundation
550 S Hope Street, Suite 2500
Los Angeles, CA 90071

W.M. Keck Foundation Undergraduate Education Grants 3891

The W.M. Keck Foundation concentrates on strengthening studies and programs in accredited universities and colleges and major, independent medical research institutions. The Foundation's Undergraduate Education program promotes distinctive learning and research experiences in science, engineering, and the liberal arts at four-year undergraduate colleges only in Foundation designated states, or through national organizations that address undergraduate needs. Historically, grants range from $200,000 to $1 million, and are typically under $500,000. On a case-by-case basis, the Foundation may consider the award of smaller seed grants for planning and piloting. Applicants are strongly urged to contact Foundation staff well in advance of submitting a Phase I Application. The best times for these contacts are between January 1 and February 15 leading up to a May 1 submittal, or between July 1 and August 15 leading up to a November 1 submittal.
Requirements: Public institutions and research universities located in the designated states of Alaska, Arizona, California, Colorado, Hawaii, Idaho, Kansas, Louisiana, Mississippi, Montana, Nevada, New Mexico, Oklahoma, Oregon, Texas, Utah, Washington, and Wyoming may apply.
Restrictions: Grants are not considered for: general operating expenses, endowments or deficit reduction; general and federated campaigns, including fundraising events, dinners or mass mailings; individuals; conference or seminar sponsorship; book publication and film or theater productions; public policy research; institutions that are located outside the United States; institutions that do not have at least three consecutive full, certified, audited financial statements; conduit organizations, unified funds or organizations that use grant funds from donors to support other organizations or individuals; or institutions that are subsidiaries or affiliates of larger entities that do not have a separate board of directors and independent audited financial statements.
Geographic Focus: Alaska, Arizona, California, Colorado, Hawaii, Idaho, Kansas, Louisiana, Mississippi, Montana, Nevada, New Mexico, Oklahoma, Oregon, Texas, Utah, Washington, Wyoming
Date(s) Application is Due: May 1; Nov 1
Amount of Grant: 200,000 - 1,000,000 USD
Contact: Margie Antonetti, Grants and Database Specialist; (213) 680-3833; fax (213) 614-0934; mantonetti@wmkeck.org or UGProgram@wmkeck.org
Internet: http://www.wmkeck.org/grant-programs/undergraduate-program.html
Sponsor: W.M. Keck Foundation
550 S Hope Street, Suite 2500
Los Angeles, CA 90071

W.P. and Bulah Luse Foundation Grants 3892

The W.P. and Bulah Luse Foundation was established in 1947. Mr. Luse, a self-made wildcatter in the early Texas oilfields, and his wife Bulah created this foundation to support and promote quality education, human-services, and health-care programming for underserved populations. Special consideration is given to charitable organizations that serve the people of Dallas, Texas, and its surrounding communities. The majority of grants from the Luse Foundation are one year in duration. Applicants must apply online at the grant website. Applicants are strongly encouraged to do the following before applying: review the downloadable state application procedures for additional helpful information and clarifications; review the downloadable online-application guidelines at the grant website; review the foundation's funding history (link is available from the grant website); review the online application questions in advance; and review the list of required attachments. These will generally include: a list of board members, financial statements (audited, reviewed, or compiled by independent auditor); an organization summary; a list of other funding sources; an IRS Determination letter; and other required documents. All attachments must be uploaded in the online application as PDF, Word, or Excel files. The W.P. & Bulah Luse Foundation has biannual deadlines of June 30 and December 31. Applications must be submitted by 11:59 p.m. on the deadline dates. Grant applicants for the June deadline will be notified of grant decisions by October 31. Grant applicants for the December deadline will be notified of grant decisions by April 30 of the following year.
Requirements: Applicants must have 501(c)3 tax-exempt status.
Restrictions: If an organization receives a grant from the Luse Foundation, it must skip a year before submitting a subsequent application. The foundation does not support requests from individuals, organizations attempting to influence policy through direct lobbying, or any political campaigns.
Geographic Focus: Texas
Date(s) Application is Due: Jun 30; Dec 31
Samples: Dallas Lighthouse for the Blind, Dallas, Texas, $10,000; Texas A & M Foundation, College Station, Texas, $40,000; Methodist Health System Foundation, Dallas, Texas, $15,000.
Contact: David Ross, Senior Vice President; tx.philanthropic@baml.com
Internet: https://www.bankofamerica.com/philanthropic/fn_search.action
Sponsor: W.P. and Bulah Luse Foundation
901 Main Street, 19th Floor, TX1-492-19-11
Dallas, TX 75202-3714

Wabash Valley Community Foundation Grants 3893

The Foundation's mission is to promote community investment for a better tomorrow. Giving primarily for arts and culture, education, human services, community development, and religion. Each year, the Foundation provides grants to nonprofit organizations for projects that meet the charitable needs in the Wabash Valley. With limited unrestricted funding available, the Foundation focuses its priorities on the following: the prevention of problems rather than on the symptoms; to maintain a proactive focus and an ability to respond to creative ideas from grant seekers; to assist grant seekers to better respond to the changing needs of the community; to encourage program that enhance cooperation and collaboration among institutions within the community; to leverage funds through the use of seed money, match, and challenge grants; and to fund projects that will make a significant improvement to the community.
Requirements: A letter of intent is required before submitting a grant request, along with a tax-exempt verification. Selected organizations will then be invited tot submit a full proposal.
Restrictions: Giving is restricted to Clay, Sullivan, and Vigo counties in Indiana. Funding is not available for the following: operational expenses of existing programs; endowments or deficit funding; funds for redistribution by the grantee; conferences, publications, films, television and radio programs unless integral to the project for which the grant is sought; religious purposes; travel for individuals or groups such as bands, sports teams and classes; annual appeals and membership contributions; commonly accepted community services, which are tax supported such as fire and police protection, welfare and library service, etc; or political campaigns or lobbying activities
Geographic Focus: Indiana
Date(s) Application is Due: Jun 1; Nov 1
Contact: Beth Tevlin; (812) 232-2234 or (877) 232-2230; beth@wvcf.com
Internet: http://www.wvcf.com/wvcf_index3.htm
Sponsor: Wabash Valley Community Foundation
2901 Ohio Boulevard, Suite 153
Terre Haute, IN 47803

Waitt Family Foundation Grants 3894

The Waitt Family Foundation is looking for programs with real potential to improve the lives of people in at-risk populations throughout the world. These programs, supported by the Waitt Helping Hands Fund, will encourage collaboration across various disciplines or related organizations in order to change the way people see themselves, others, and the opportunities before them. The Foundation is interested in programs that foster independence, self-help, and self-reliance, and want people to obtain the skills necessary to better themselves and their communities. Most importantly, the Foundation is looking for projects that create long-term change to help the greatest number of people. There is a $100,000 minimum for all grant requests. Multi-year proposals will be considered.
Requirements: Organizations must have a U.S. tax identification number and be able to provide a current copy of their IRS 501(c)3 determination letter. Only organizations with federal non-profit agency tax status can receive financial support.
Restrictions: The Foundation does not fund the following categories: individuals; for-profit organizations; capital campaigns; endowments; debt reduction; arts; religious organizations; public education institutions; health care organizations or hospitals; or lobbying prohibited by the Internal Revenue Code.
Geographic Focus: California, Iowa, Nebraska, South Dakota
Date(s) Application is Due: Feb 1; May 1; Aug 1; Oct 1
Amount of Grant: 100,000 USD
Contact: Cherie Jacobson, (858) 551-4400; fax (858) 551-6871; grants@waittfoundation.org
Internet: http://waittfoundation.org/grants/guidelines.html
Sponsor: Waitt Family Foundation
P.O. Box 1948
La Jolla, CA 92038-1948

Walker Area Community Foundation Grants 3895

The Walker Area Community Foundation supports nonprofit organizations in specific fields of interest, which include: arts and humanities; children and youth; education; elder care; the environment; health and medicine; recreation; and social services. In order to make the greatest impact with the funds available, the Foundation prefers requests that: address a critical community need; get at the root causes of a problem; do not duplicate existing services unless they serve a population not already being served; are pilot projects which, if successful, can be expanded to serve a wider population or be duplicated by other organizations; involve collaboration and cooperation with other organizations and agencies; and include an effective mechanism for measuring the impact of The Community Foundation's investment.
Requirements: Requests from public school systems must come from the superintendent's office. If the request is for an individual school, it must be for a pilot project that, if successful, would be duplicated in other schools in the system. Requests from large organizations with many branches or departments (e.g. colleges, universities, the YMCA and public libraries) must come through the development or president's office and have the approval of the head of that office. Ordinarily, the Foundation does not make grants for: operating expenses unless they are for the initial stages of a pilot project; program expenses that occur on a regular basis; or regularly supported activities of fundraising organizations.
Restrictions: No grants are made to or for: individuals; religious organizations for religious purposes; dinners, balls or other ticketed events; political purposes; lobbying activities; replacement of government grants or funding; endowments; or other discretionary pools.
Geographic Focus: Alabama
Contact: Mimi Hudson; (205) 302-0001; fax (205) 302-0424; mhudson@wacf.org
Internet: http://www.wacf.org/how-to-apply-for-a-grant.html
Sponsor: Walker Area Community Foundation
611 8th Avenue
Jasper, AL 35501

Wallace Alexander Gerbode Foundation Grants 3896

The Foundation is interested in programs and projects offering potential for significant impact. The primary focus is on the California San Francisco Bay area and Hawaii. The Foundation's interests generally fall under the following categories: arts and culture, environment, reproductive rights and health, citizen participation/building communities/inclusiveness, strength of the philanthropic process and the nonprofit sector, and special projects. Letter of inquiry are accepted on an ongoing basis. Awards range from $1,600 to $75,000. The Foundation also has a Special Awards program to commission new works of art. Special Awards program announcements, including

guidelines, an application and deadlines, are posted on the Foundation's website. Generally Special Award programs are announced in May, applications are due in late August and awards are announced in January.
Requirements: Eligible organizations must be 501(c)3 nonprofits. Letters of inquiry should be sent to the Administrative Manager and include a brief description of the proposed work, a description of the organization, and a budget.
Restrictions: The Foundation generally does not support direct services, deficit budgets, general operating funds, building or equipment funds, general fundraising campaigns, religious purposes, private schools, publications, scholarships, or grants to individuals.
Geographic Focus: California, Hawaii
Samples: Alonzo Kings LINES Ballet, San Francisco, California, $75,000, support of commission; Aging In Place, San Francisco, California, $20,000, support of San Francisco Village-Northside; and Hastings College of the Law, San Francisco, California, support of the Center for WorkLife Law.
Contact: Olivia Malabuyo Tablante; (415) 391-0911; olivia@gerbode.org
Internet: http://fdncenter.org/grantmaker/gerbode
Sponsor: Wallace Alexander Gerbode Foundation
111 Pine Street, Suite 1515
San Francisco, CA 94111

Walmart Foundation Community Giving Grants 3897
Walmart's founder, Sam Walton, introduced the charitable philosophy, operate globally, give back locally. With that in mind, the Walmart Foundation supports local programs in an effort to impact neighborhoods where its employees live and work. Areas of focus include education, workforce development/economic opportunity, health and wellness, and environmental sustainability. Awards range from $250 to $2,500. The annual grant cycle begins Feb. 1 and the application deadline is December 31. Organizations are encouraged to limit the number of pending applications to 25.
Requirements: Applicants must fit within one of the Foundation's four focus areas and be one of the following: hold a current tax-exempt status under Section 501(c)3, 4, 6 or 19 of the Internal Revenue Code; be a recognized government entity that is requesting funds exclusively for public purposes; be a K-12 public/private school, charter school, community/junior college, state/private college or university; or be a church or other faith based organization with proposed projects that address and benefit the needs of the community at large.
Restrictions: Applicants must offer programs that benefit the local community. Applicants receiving sponsorship at the national level are excluded from applying. Applications are online.
Geographic Focus: All States
Date(s) Application is Due: Dec 31
Amount of Grant: 250 - 2,500 USD
Contact: Julie Gehrki; (479) 273-4000 or (800) 530-9925; fax (479) 273-6850
Internet: http://walmartstores.com/CommunityGiving/238.aspx
Sponsor: Walmart Foundation
702 SW 8th Street, Department 8687, No. 0555
Bentonville, AR 72716-0555

Walmart Foundation National Giving Grants 3898
The Walmart Foundation's mission is to create opportunities so people can live better. The National Giving program allows the Foundation to work strategically with organizations working across one or more states to address social issues strongly aligned with its focus areas. The Foundation often provides funds to organizations that have local affiliates around the country, and the majority of grants from this program include re-grants to implement programs in local communities. These grants support initiatives focused on: hunger relief and healthy eating; sustainable food production and supply; women's economic empowerment; and career opportunities. The Program awards grants of $250,000 and above. Applicants should submit program ideas using the letter of inquiry (LOI) format.
Requirements: Applicants must be 501(c)3 organizations; recognized by the Internal Revenue Service as a public charity within the meaning of either Section 509(a)1 or 509(a)2 of the Internal Revenue Code; and must operate on a national scope through the existence of chapters or affiliates in a large number of states around the country; or possess a regional/local focus, but seek funding to replicate program activities nationally. In the case of proposals seeking funding for replication, organizations must demonstrate the capacity to support national expansion and be ready to begin the replication process.
Restrictions: The Foundation does not accept unsolicited proposals but does accept letters of inquiry providing a general understanding of the problem or issue, the need for the proposed solution and the applicant's capacity to carry out the work. If the applicant is selected to submit a full proposal, guidelines and additional information will be provided. The letter of intent is submitted online. Funding exclusions include: association and chamber memberships; athletic sponsorships (teams/events); capital campaigns and endowments (defined as any plans to raise funds for a significant purchase or expense, such as new construction, major renovations or to help fund normal budgetary items); faith-based organizations when the proposed grant will only benefit the organization or its members; general operating expenses; political causes, candidates, organizations or campaigns; research projects; scholarships (tuition, room and board or any other expense related to college, university, or vocational school attendance); and sponsorship of fundraising events (galas, walks, races, tournaments).
Geographic Focus: All States
Contact: Julie Gehrki; (479) 273-4000 or (800) 530-9925; fax (479) 273-6850
Internet: http://foundation.walmart.com/apply-for-grants/national-giving
Sponsor: Walmart Foundation
702 SW 8th Street, Department 8687, No. 0555
Bentonville, AR 72716-0555

Walmart Foundation Northwest Arkansas Giving 3899
As the largest global retailer, Walmart has an opportunity and responsibility to make a difference on the issues that affect the communities it serves around the world as well as those that impact its friends, neighbors and customers in Northwest Arkansas. Together, Walmart and the Walmart Foundation want to build on the tradition of supporting the needs of the Northwest Arkansas community. Specifically, they have established the following strategic focus areas in this region: hunger relief and healthy eating services; access to health care; and quality of life improvement programs. Annual deadlines for hunger relief and healthy eating services applications are February 1, April 1, June 1, August 1, October 1, and December 1; annual deadlines for access to health care applications are February 1, April 1, June 1, August 1, October 1, and December 1; and annual deadlines for quality of life improvement programs are February 10, May 15, August 4, and November 20. The minimum grant amount is $10,000.
Requirements: Applicants must be 501(c)3 organizations; recognized by the Internal Revenue Service as a public charity within the meaning of either Section 509(a)1 or 509(a)2 of the Internal Revenue Code; and must operate within the northwest Arkansas region. In the case of proposals seeking funding for replication, organizations must demonstrate the capacity to support expansion and be ready to begin the replication process.
Geographic Focus: Arkansas
Date(s) Application is Due: Feb 1; Apr 1; Jun 1; Aug 1; Oct 1; Dec 1
Amount of Grant: 10,000 - 1,000,000 USD
Contact: Julie Gehrki; (479) 273-4000 or (800) 530-9925; fax (479) 273-6850
Internet: http://foundation.walmart.com/apply-for-grants/northwest-arkansas-giving-program
Sponsor: Walmart Foundation
702 SW 8th Street, Department 8687, No. 0555
Bentonville, AR 72716-0555

Walmart Foundation State Giving Grants 3900
The Walmart Foundation's mission is to create opportunities so people can live better. State Giving Grants support initiatives aligned with their mission and having a long-lasting, positive impact within a state or region. Interest areas include: hunger relief and healthy eating; sustainable food production and supply; women's economic empowerment; and career opportunities. The Foundation has a particular interest in supporting the following populations: veterans and military families, traditionally under-served groups, individuals with disabilities, and people impacted by natural disasters. Awards range between $25,000 and $250,000. There are four distinct application cycles (each open for about five days) focused on different giving areas and geographical areas. See the web site for specifics.
Requirements: Applicants must be 501(c)3 organizations. Applications accepted online only.
Restrictions: Funding awarded in a particular state must be fully allocated within that state. Applications accepted online only.
Geographic Focus: All States
Date(s) Application is Due: Jan 30; May 1; Jul 17; Sep 18
Amount of Grant: 25,000 - 250,000 USD
Contact: Julie Gehrki; (479) 273-4000 or (800) 530-9925; fax (479) 273-6850
Internet: http://foundation.walmart.com/apply-for-grants/state-giving
Sponsor: Walmart Foundation
702 SW 8th Street, Department 8687, No. 0555
Bentonville, AR 72716-0555

Walter L. Gross III Family Foundation Grants 3901
The foundation awards grants to eligible Ohio and Kentucky nonprofit organizations in its areas of interest, including animals and wildlife, Christian churches and organizations, community outreach programs, education, environment, health care and medical, and social services. Grants are considered on a case-by-case basis and have been awarded to support building construction/renovation and general operating support. There are no application forms or deadlines with which to adhere, and applicants should begin by contacting the office directly.
Requirements: Ohio and Kentucky 501(c)3 nonprofit organizations are eligible to apply.
Geographic Focus: Kentucky, Ohio
Amount of Grant: 1,000 - 20,000 USD
Samples: Habitat for Humanity, Cincinnati, Ohio--for general support, $10,000; Miami University Foundation, Oxford, Ohio--for general support, $300,000; Our Daily Bread, Cincinnati, Ohio--for general support, $2,000.
Contact: Walter L. Gross III; (513) 785-6060 or (513) 785-6072; fax (513) 683-9467
Sponsor: Walter L. Gross III Family Foundation
9435 Waterstone Boulevard, Suite 390
Cincinnati, OH 45249-8227

Warren County Community Foundation Grants 3902
The Warren County Community Foundation Grants fund projects with focus on: innovative approaches or making significant improvements in solving Warren County problems; maintaining a proactive awareness with the ability to respond to creative ideas; and preventing problems rather than treating symptoms of problems. Grant categories include education, human services, recreation, arts and culture; citizenship, environment, and economic development.
Requirements: When applying for a Warren County Community Grant, organizations should first submit a pre-application letter, which will be reviewed by the Foundation's grant committee. The pre-application letter should be one page, provide a detailed description of the project, including the targeted group it will benefit, the amount of funding requested, and the projected time line, along with the applicant's contact information. Pre-application letters are accepted on the third Thursday in January for April funding, and the third Thursday in May for funding in August. If the project meets the guidelines and funds are likely to be available, the organization is asked to submit eleven copies of the Grant Application form and any attachments. Organizations should contact the Foundation office for a current application.

Restrictions: The Foundation will not fund any of the following: travel or expenses for individuals and groups such as clubs, sports and teams, and classes; projects outside of Warren County (unless it is significant to Warren County residents); organizations fully-funded by government; projects for day-to-day operations, endowments, or excessively long-term projects; religious activities or programs serving one denomination; political organizations or activities; nationwide and/or statewide fund raising efforts.
Geographic Focus: Indiana
Amount of Grant: 300 - 185,000 USD
Contact: Carol Clark, Executive Director; (765) 764-1501; fax (765) 764-1501; warrencountyfoundation@yahoo.com
Internet: http://www.warrencountyfoundation.com/Grants.html
Sponsor: Warren County Community Foundation
31 North Monroe Street
Williamsport, IN 47993

Warren County Community Foundation Mini-Grants 3903
The Warren County Community Foundation funds community projects that have the greatest impact on meeting the needs of Warren County. Projects should focus on: innovative approaches or making significant improvements in solving Warren County problems; maintaining a proactive awareness with the ability to respond to creative ideas; and preventing problems rather than treating symptoms of problems. Grant priorities include: education, human services, recreation, arts and culture, citizenship, environment, and economic development. In order for the Foundation to react in a more timely matter to needs in the community, the Board has appropriated funds to the Mini-Grant Program. The Mini-Grants Program funds requests of $250 or less with a 30 day approval process. Established grant policies and exclusions will still apply.
Requirements: Applicants should submit a pre-application letter: a one page summary of the proposed project, the targeted group it will benefit, the amount of funding requested; and the projected time frame to complete the project. The applicant will be contacted to submit a grant application if the Foundation decides that the project is appropriate for funding.
Restrictions: Grants are made on a first-come, first-served basis, until the funds have been expended for that calendar year.
Geographic Focus: Indiana
Amount of Grant: Up to 250 USD
Contact: Carol Clark, Executive Director; (765) 764-1501; fax (765) 764-1501; warrencountyfoundation@yahoo.com
Internet: http://www.warrencountyfoundation.com/Grants.html
Sponsor: Warren County Community Foundation
31 North Monroe Street
Williamsport, IN 47993

Warrick County Community Foundation Grants 3904
The Warrick County Community Foundation is a nonprofit, public charity created for the people of Warrick County, Indiana. The Foundation helps nonprofits fulfill their missions by strengthening their ability to meet community needs through grants that assist charitable programs, address community issues, support community agencies, launch community initiatives, and support leadership development. Program areas considered for funding including arts and culture; community development; education; health; human services; environment; recreation; and youth development. The application, supplemental forms needed, and examples of recent grants are available at the website.
Requirements: The Foundation welcomes proposals from nonprofit organizations that are deemed tax-exempt under sections 501(c)3 and 509(a) of the IRS code and from governmental agencies serving the Warrick County. Proposals from nonprofit organizations not classified as a 501(c)3 public charity may be considered if the project is charitable and supports a community need.
Restrictions: Project areas not considered for funding include religious organizations for religious purposes; political parties or campaigns; endowment creation or debt reduction; operating costs; capital campaigns; annual appeals or membership contributions; travel requests for groups or individuals such as bands, sports teams, or classes.
Geographic Focus: Indiana
Date(s) Application is Due: Sep 4
Amount of Grant: Up to 2,000 USD
Contact: Karen Embry, Administrative Assistant; (812) 897-2030
Internet: http://www.warrickcommunityfoundation.org/disc-grants
Sponsor: Warrick County Community Foundation
224 West Main Street, P.O. Box 215
Boonville, IN 47601

Warrick County Community Foundation Women's Fund 3905
The Warrick County Community Foundation Women's Fund distributes grants supporting a variety of resources serving women of all ages that will improve self-esteem and self-image among women and girls; promote health and improve accessibility and affordability of health care; help mothers improve parenting skills; provide avenues for career advancement through education and training; and create positive change for Warrick County's women and girls.
Requirements: A letter of inquiry of no more than two typed papers is required to apply for a Women's Fund grant. Funding requests will be considered from nonprofit, charitable organizations serving Warrick County women and girls. The letter should describe the project and funding requested, the need or issue the project will address, what will be accomplished, and the project's timetable. Organizations should also include contact information and an IRS determination letter (or an explanation of why no IRS letter is included).
Geographic Focus: Indiana
Contact: Susan Sublett, Director; (812) 897-2030
Internet: http://www.warrickcommunityfoundation.org/womens-fund
Sponsor: Warrick County Community Foundation
224 West Main Street, P.O. Box 215
Boonville, IN 47601

Washington County Community Foundation Grants 3906
The Washington County Community Foundation offers grants to nonprofit organizations in Washington County. Organizations may sign up to be notified of current grant cycles and criteria. The application and examples of previously funded projects are available at the Foundation website.
Requirements: Funding will be made to nonprofit organizations whose programs benefit Washington County residents. Grants from Washington County Foundation must meet legal and tax requirements as to purpose and may be made only to non-profit organizations and causes. Grant recipients must show that their financial affairs are being properly administered and may be required to submit audited balance sheets and operating statements.
Restrictions: Applicants are required to attend an orientation meeting before beginning the grant writing process. Funding is not available for the following: political parties or campaigns; sectarian religious purposes that do not serve the general public; programs or equipment that was committed to prior to the grant application submission; or endowment creation or debt reduction.
Geographic Focus: Indiana
Contact: Judy Johnson, Executive Director; (812) 883-7334; info@wccf.biz
Internet: http://wccf.biz/grants/grant-criteria.html
Sponsor: Washington County Community Foundation
1707 North Shelby Street, P.O. Box 50
Salem, IN 47167

Washington County Community Foundation Youth Grants 3907
The purpose of the Washington County Youth Foundation is to improve the quality of life in Washington County. The principle consideration in a grant is its effectiveness in meeting demonstrated or perceived health, education, cultural, recreational, environmental, or social needs in the community, or in making studies to determine how best to meet such needs. The participation of other contributions by using matching, challenge, and other grant techniques is strongly encouraged. The application and examples of previously funded grants are available at the Foundation website.
Requirements: Grants from Washington County Youth Foundation must meet legal and tax requirements and may be made only to non-profit organizations and causes. Grant recipients must show that their financial affairs are being properly administered and may be required to submit audited balance sheets and operating statements. To qualify for funding, projects must address a community need, be planned and implemented by a youth group in the county, and associated with a non-profit organization.
Restrictions: Funding is not available for the following: political parties or campaigns; sectarian religious purposes that do not serve the general community; endowment creation and debt reduction; programs or equipment that was committed to prior to the grant application submission; meeting routine budgets.
Geographic Focus: Indiana
Contact: Judy Johnson, Executive Director; (812) 883-7334; info@wccf.biz
Internet: http://wccf.biz/youth_foundation/yf_grant-criteria.html
Sponsor: Washington County Community Foundation
1707 North Shelby Street, P.O. Box 50
Salem, IN 47167

Washington Gas Charitable Giving Contributions 3908
For more than 160 years, Washington Gas has been an integral part of the growing metropolitan region of Washington D.C. Chartered by the 30th Congress and signed into law by President James Polk in 1848, the company has developed a Charitable Giving Program that is designed to make a meaningful and lasting impact on the communities it serves. Washington Gas focuses on three primary areas: education, the environment, and health. Types of support offered are grants, in-kind contributions, and volunteer resources. In the area of education, emphasis is placed on the development of math, science, technology, and business skills in K-12 youth. Consideration is also given to arts-related programs. In the area of the environment, emphasis is placed on programs that promote cleaner air and water programs that protect and preserve the ecological system of the metropolitan area. In the area of health, consideration is given to health organizations that strive to improve the health and well-being of individuals within the community. Emphasis is also placed on energy assistance programs for low-income residents to heat and cool their homes, reducing illness and casualties resulting from exposure to extreme temperatures. The company accepts applications on a rolling basis. Basic guidelines are listed at the website and provided in a downloadable PDF file. Applications must be mailed to the contact information given. Notification of acceptance or rejection generally takes fifteen business days or longer.
Requirements: 501(c)3 organizations are eligible to apply. Support is provided primarily in Washington, D.C., Maryland, and Virginia. The company prefers to support specific programs over general funding.
Restrictions: Support is not provided to religious organizations for sectarian purposes, political associations, organizations with strictly a sports focus, individuals, and requests for capital or endowment campaigns.
Geographic Focus: District of Columbia, Maryland, Virginia
Contact: Tracye Funn, (703) 750-1000; tfunn@washgas.com
Internet: http://www.washgas.com/pages/CharitableGiving
Sponsor: Washington Gas Company
101 Constitution Avenue, North West, 3rd Floor
Washington, D.C. 20080

Wayne County Foundation Grants   3909
The Foundation seeks to serve the charitable, cultural, educational, and community needs of the citizens in Wayne County, Indiana. Areas of interest include animal welfare; agriculture; the performing arts; cancer and glaucoma research/ child abuse prevention; delinquency prevention; the environment; historic preservation; homelessness; human service needs; literacy; mental health; special needs therapy; and senior citizens. Grants will be made primarily to established organizations that serve the county. The programs of such organizations will reflect the concerns of community leadership. Types of support include equipment, program development, conferences and seminars, publication, seed money, and scholarship funds. Normally, grant commitments are for one year. Deadlines vary according to program. The application and samples of previously funded projects available at the website.
Requirements: The following information must be submitted to apply: the completed application cover; a statement of need; project description and anticipated community impact; a brief description of the applicant's history, purpose, and population served; a plan to evaluate the success or effectiveness of the proposed project; the project budget and narrative budget; the 501(c)3 determination letter; a list of Board members with contact information; evidence of the Board's approval for application; statement of financial position and operating statement; and current Form 990.
Restrictions: The Foundation will not support annual fund campaigns; operating or capital debt reduction; religious purposes; grants to individuals or for travel purposes; services commonly regarded as the responsibility of government such as fire and police protection; public school services required by state law; standard instructional or regular operating costs of nonpublic schools; or repeat funding of projects previously supported in recent grant periods.
Geographic Focus: Indiana
Date(s) Application is Due: Apr 1; Jul 30; Oct 1
Contact: Stephen C. Borchers, Executive Director; (765) 962-1638; fax (765) 966-0882; steve@waynecountyfoundation.org
Internet: http://www.waynecountyfoundation.org/index.html
Sponsor: Wayne County Foundation
33 South 7th Street, Suite 1
Richmond, IN 47374

Weaver Popcorn Foundation Grants   3910
Established in Indiana in 1997, the Weaver Popcorn Foundation offers funding throughout the State of Indiana. Its primary fields of interest include: boy scouts; children and youth services; education; family services; domestic violence; health care; higher education; human services; and secondary school programs. The primary type of funding is general operating support. A formal application is not required, and interested parties should submit a brief overview/history of the organization, its mission, and an outline of budgetary needs. There are no annual deadlines specified. Typical awards range from $250 to $20,000, though some grants have reached as much as $125,000.
Requirements: Any Indiana non-profit is eligible to apply.
Geographic Focus: Indiana
Amount of Grant: 250 - 125,000 USD
Contact: Brian Hamilton, (317) 915-4050
Sponsor: Weaver Popcorn Foundation
14470 Bergen Boulevard, Suite 100
Noblesville, IN 46060-3377

Webster Cornwell Memorial Scholarship   3911
The Webster Cornwell Memorial Scholarship is awarded each year to a graduating senior from Hammond High School. The $1,000 scholarship is designated for tuition, fees, and/or books, and is renewable for three years. A student will be selected based upon their cumulative GPA, high school and community activities, and an essay. All college majors are eligible.
Requirements: Applicants are required to have at least a 2.5 GPA and full time enrollment in a program of study leading to an associate or bachelor degree at an accredited college or university. In addition to the application, candidates must submit their high school transcript and the answer to an essay question posted in the application.
Geographic Focus: Indiana
Date(s) Application is Due: Apr 15
Amount of Grant: 1,000 USD
Contact: Cornwell Scholarship Contact; (219) 736-1880; fax (219) 736-1940; legacy@legacyfoundationlakeco.org
Internet: http://www.legacyfoundationlakeco.org/scholarships/index.htm
Sponsor: Legacy Foundation
1000 East 80th Place, 302 South
Merrillville, IN 46410

Wege Foundation Grants   3912
The Wege Foundation awards grants in Michigan in five specific categories: education, environment, arts and culture, health care, and human services. Areas of interest include children and youth services; Christian agencies and churches; community development; elementary and secondary education; environmental resources; higher education; hospitals (general); human services; museums; and performing arts. The Foundation strives to be an inspiration for other communities. According to the Foundation's vision, "Grand Rapids inspires others to create communities that forge a balance in the environment, health, education, and arts to encourage healthier lives in mind, body, and spirit."
Requirements: Grants are awarded to nonprofit organizations in the greater Kent County, MI, area, with emphasis on the Grand Rapids area. The applicant may access the application after an online quiz.
Restrictions: The Foundation only funds organizations classified as tax-exempt under section 501(c)3 of the Internal Revenue Code. Grants are not awarded for operating budgets.
Geographic Focus: Michigan
Date(s) Application is Due: Feb 15; Sep 15
Contact: Jody Price, Corporate Financial Officer; (616) 957-0480; fax (616) 957-0616; jprice@wegefoundation.org
Internet: http://www.wegefoundation.com/seekingagrant/seekingagrant.html
Sponsor: Wege Foundation
P.O. Box 6388
Grand Rapids, MI 49516-6388

Weingart Foundation Grants   3913
Weingart Foundation makes grants to assist organizations that work in the areas of health, human services, and education. The Foundation gives highest priority to activities that provide greater access to people who are economically disadvantaged and underserved. Of particular interest to the Foundation are applications that specifically address the needs of low-income children and youth, older adults, and people affected by disabilities and homelessness. The Foundation also funds activities that benefit the general community and improve the quality of life for all individuals in Southern California. Weingart Foundation offers the following types of grants: core support; capital; capacity building; program; and a small grant program. See website for a description of each grant type: http://www.weingartfnd.org/default.asp?PID=5
Requirements: An organization that is certified as tax exempt under Section 501(c)3 of the U.S. Internal Revenue Code and is not a private foundation as defined in section 509(a) of that Code is eligible for consideration. Preference is given to organizations providing services in the following six Southern California counties: Los Angeles, Orange, Riverside, Santa Barbara, San Bernardino, and Ventura.
Restrictions: Grants are not made: for propagandizing, influencing legislation and/or elections, promoting voter registration; for political candidates, political campaigns; for litigation; to institutions limiting their services to persons of a single religious sect or denomination; for social or political issues outside the United States of America; to individuals; to federated appeals or for the collection of funds for redistribution to other nonprofit groups; for conferences, workshops, temporary exhibits, travel, surveys, films or publishing activities; for endowment funds; for contingencies, deficits or debt reduction; for fundraising dinners or events; for research. Grants generally are not approved for: national organizations that do not have local chapters operating in the geographic area of grant focus; projects or programs normally financed by government sources; refugee or religious programs, consumer interest or environmental advocacy; feasibility studies. The Foundation does not fund Section 509(a)(3) Type III non-functionally integrated supporting organizations.
Geographic Focus: California
Samples: National Urban Fellows, $25,000--program support for the Southern California Capacity Building Initiative; Performing Arts Center of Los Angeles, $150,000--core support for Music Center education programs; JWCH Medical Clinic, $900,000--capacity building support to strengthen administrative infrastructure; University of Southern California, $10,000--support for the Los Angeles Leadership Group; Church of Our Saviour, $150,000--core support.
Contact: Administrator; (213) 688-7799; fax (213) 688-1515; info@weingartfnd.org
Internet: http://www.weingartfnd.org
Sponsor: Weingart Foundation
1055 W Seventh Street, Suite 3050
Los Angeles, CA 90017-2305

Welborn Baptist Foundation General Opportunity Grants   3914
The Welborn Baptist Foundation is seeking to fund lasting change in the communities it serves. Towards that end, the Foundation will assess grant proposals based on: demonstrated long-term results, or exceptional potential for such results based on well-established research; addressing root causes rather than symptoms; fit with the Foundation's target areas of emphasis; deep, established collaborations with (not just referrals from) other like-minded organizations; excellent prospects for long-term sustainability without Foundation resources; and a clear implementation plan that enables successes to be replicated elsewhere. To meet needs that do not naturally fit into the other priority target areas, but which are consistent with the Foundation's mission, the General Opportunity target area has been established. In the past, General Opportunity grants have focused on opportunities that enhance the general health and/or the health educational status of the community. Ideal opportunities for consideration in this category are those which address both health and health education. Proposals which have only a peripheral relationship to health or health education are not likely to be funded.
Requirements: Giving is limited to: Gallatin, Saline, Wabash, Wayne and White counties in Illinous; Dubois, Gibson, Perry, Pike, Posey, Spencer, Vanderburgh, and Warrick counties in Indiana; and Henderson County, Kentucky. Applicants must submit a Letter of Interest (LOI). All secular, church and other faith-based not-for-profit organizations that are tax exempt under section 501(c)3 of the IRS Code are eligible. Participation beyond the Letter of Interest stage is by invitation only.
Restrictions: The following program and project areas will not be considered for funding: scholarships, loans, grants or fellowship support directly to or for the benefit of specific and known individuals; establishment of, or contributions to, a permanent endowment, foundation, trust or permanent interest-bearing account; carrying on of propaganda or attempt to influence legislation or public elections; restricting the services, facilities or employment provided by the grant to individuals based on race, creed, color, sex, or national origin; any governmental agencies reporting to an elected or appointed official (except for schools governed by citizens boards); any requests for funding for deficits or retirement of debt; fundraising events; annual fund drives; venture capital for competitive profit making ventures; or basic scientific research. Additionally, it should be noted that the Foundation does not fund applications seeking replacement dollars (i.e., funding to substitute for dollars lost from another grantor).
Geographic Focus: All States

Contact: Kevin Bain; (812) 437-8260 or (877) 437-8260; info@welbornfdn.org
Internet: http://www.welbornfdn.org/grant-process/funding-targets
Sponsor: Welborn Baptist Foundation
Twenty-One Southeast Third Street, Suite 610
Evansville, IN 47708

Welborn Baptist Foundation Health Ministries Grants **3915**

The Welborn Baptist Foundation is seeking to fund lasting change in the communities it serves. Towards that end, the Foundation will assess grant proposals based on: demonstrated long-term results, or exceptional potential for such results based on well-established research; addressing root causes rather than symptoms; fit with the Foundation's target areas of emphasis; deep, established collaborations with (not just referrals from) other like-minded organizations; excellent prospects for long-term sustainability without Foundation resources; and a clear implementation plan that enables successes to be replicated elsewhere. A subcategory of the Foundation's Community Health Status target is Health Ministries. A health ministry integrates faith and health within a church community for its members and the community it serves. It is an integral part of the overall ministry of the church, promoting physical, emotional, social and spiritual well-being. Health ministers are many – clergy, nurses and laity – coming together to share the compassionate love and grace of Jesus Christ. Furthermore, the Foundation will continue to offer technical assistance, training and seed funding to establish health ministries in congregations throughout the Foundation's Service Area.
Requirements: Giving is limited to: Gallatin, Saline, Wabash, Wayne and White counties in Illinous; Dubois, Gibson, Perry, Pike, Posey, Spencer, Vanderburgh, and Warrick counties in Indiana; and Henderson County, Kentucky. Applicants must submit a Letter of Interest (LOI). All secular, church and other faith-based not-for-profit organizations that are tax exempt under section 501(c)3 of the IRS Code are eligible. Participation beyond the Letter of Interest stage is by invitation only.
Restrictions: The following program and project areas will not be considered for funding: scholarships, loans, grants or fellowship support directly to or for the benefit of specific and known individuals; establishment of, or contributions to, a permanent endowment, foundation, trust or permanent interest-bearing account; carrying on of propaganda or attempt to influence legislation or public elections; restricting the services, facilities or employment provided by the grant to individuals based on race, creed, color, sex, or national origin; any governmental agencies reporting to an elected or appointed official (except for schools governed by citizens boards); any requests for funding for deficits or retirement of debt; fundraising events; annual fund drives; venture capital for competitive profit making ventures; or basic scientific research. Additionally, it should be noted that the Foundation does not fund applications seeking replacement dollars (i.e., funding to substitute for dollars lost from another grantor).
Geographic Focus: Illinois, Indiana, Kentucky
Contact: Kevin Bain, Executive Director; (812) 437-8260 or (877) 437-8260; fax (812) 437-8269; info@welbornfdn.org
Internet: http://www.welbornfdn.org/grant-process/funding-targets
Sponsor: Welborn Baptist Foundation
Twenty-One Southeast Third Street, Suite 610
Evansville, IN 47708

Welborn Baptist Foundation Improvements to Community Health Status Grants **3916**

The Welborn Baptist Foundation is seeking to fund lasting change in the communities it serves. Towards that end, the Foundation will assess grant proposals based on: demonstrated long-term results, or exceptional potential for such results based on well-established research; addressing root causes rather than symptoms; fit with the Foundation's target areas of emphasis; deep, established collaborations with (not just referrals from) other like-minded organizations; excellent prospects for long-term sustainability without Foundation resources; and a clear implementation plan that enables successes to be replicated elsewhere. in the area of Improvements to Community Health Status, the Foundation has initiated a community-wide coalition designed to achieve long-term reduction in the proportion of residents that are either overweight or obese, through increased physical activity and healthier eating. Beyond this, the Foundation's funding objective in this target area is for all community residents to proactively take the appropriate steps that can lead to lower levels of chronic illness. The primary focus is therefore on: programs that increase the access of high-risk populations to health care services; and programs that address the prevention and successful management of chronic disease states.
Requirements: Giving is limited to: Gallatin, Saline, Wabash, Wayne and White counties in Illinous; Dubois, Gibson, Perry, Pike, Posey, Spencer, Vanderburgh, and Warrick counties in Indiana; and Henderson County, Kentucky. Applicants must submit a Letter of Interest (LOI). All secular, church and other faith-based not-for-profit organizations that are tax exempt under section 501(c)3 of the IRS Code are eligible. Participation beyond the Letter of Interest stage is by invitation only.
Restrictions: The following program and project areas will not be considered for funding: scholarships, loans, grants or fellowship support directly to or for the benefit of specific and known individuals; establishment of, or contributions to, a permanent endowment, foundation, trust or permanent interest-bearing account; carrying on of propaganda or attempt to influence legislation or public elections; restricting the services, facilities or employment provided by the grant to individuals based on race, creed, color, sex, or national origin; any governmental agencies reporting to an elected or appointed official (except for schools governed by citizens boards); any requests for funding for deficits or retirement of debt; fundraising events; annual fund drives; venture capital for competitive profit making ventures; or basic scientific research. Additionally, it should be noted that the Foundation does not fund applications seeking replacement dollars (i.e., funding to substitute for dollars lost from another grantor).
Geographic Focus: Illinois, Indiana, Kentucky
Contact: Kevin Bain, Executive Director; (812) 437-8260 or (877) 437-8260; fax (812) 437-8269; info@welbornfdn.org
Internet: http://www.welbornfdn.org/grant-process/funding-targets
Sponsor: Welborn Baptist Foundation
Twenty-One Southeast Third Street, Suite 610
Evansville, IN 47708

Welborn Baptist Foundation School Based Health Grants **3917**

The Welborn Baptist Foundation is seeking to fund lasting change in the communities it serves. Towards that end, the Foundation will assess grant proposals based on: demonstrated long-term results, or exceptional potential for such results based on well-established research; addressing root causes rather than symptoms; fit with the Foundation's target areas of emphasis; deep, established collaborations with (not just referrals from) other like-minded organizations; excellent prospects for long-term sustainability without Foundation resources; and a clear implementation plan that enables successes to be replicated elsewhere. The School Based Health target area, known as HEROES, is a three-year commitment, during which a school will focus on implementing the Centers for Disease Control's Coordinated School Health Model (CSH) into their school environment and begin to change their school's culture. The funding from WBF is geared towards the Physical Education/Physical Activity and Nutrition components of CSH, but the schools are also expected to begin making changes in the other six components as well.
Requirements: Giving is limited to: Gallatin, Saline, Wabash, Wayne and White counties in Illinous; Dubois, Gibson, Perry, Pike, Posey, Spencer, Vanderburgh, and Warrick counties in Indiana; and Henderson County, Kentucky. Applicants must submit a Letter of Interest (LOI). All secular, church and other faith-based not-for-profit organizations that are tax exempt under section 501(c)3 of the IRS Code are eligible. Participation beyond the Letter of Interest stage is by invitation only.
Restrictions: The following program and project areas will not be considered for funding: scholarships, loans, grants or fellowship support directly to or for the benefit of specific and known individuals; establishment of, or contributions to, a permanent endowment, foundation, trust or permanent interest-bearing account; carrying on of propaganda or attempt to influence legislation or public elections; restricting the services, facilities or employment provided by the grant to individuals based on race, creed, color, sex, or national origin; any governmental agencies reporting to an elected or appointed official (except for schools governed by citizens boards); any requests for funding for deficits or retirement of debt; fundraising events; annual fund drives; venture capital for competitive profit making ventures; or basic scientific research. Additionally, it should be noted that the Foundation does not fund applications seeking replacement dollars (i.e., funding to substitute for dollars lost from another grantor).
Geographic Focus: Illinois, Indiana, Kentucky
Contact: Kevin Bain, Executive Director; (812) 437-8260 or (877) 437-8260; fax (812) 437-8269; info@welbornfdn.org
Internet: http://www.welbornfdn.org/grant-process/funding-targets
Sponsor: Welborn Baptist Foundation
Twenty-One Southeast Third Street, Suite 610
Evansville, IN 47708

Wellcome Trust Biomedical Science Grants **3918**

Wellcome Trust supports research into all aspects of science: from molecules and cells vital to life, through the spread of diseases or the vectors of disease across the globe, to clinical and public health research to improve the quality of healthcare. Research can be based in the laboratory, the clinic or the field, and may involve experimental or theoretical approaches. The Trust not only funds scientists but also supports undergraduate and doctoral students, clinicians, dentists and veterinarians. It funds the best researchers with the most innovative and exciting ideas in the form of personal and research support, collaborations and support for symposia, conferences and workshops. A significant proportion of the funding is devoted to support those working overseas.
Requirements: Overseas applicants should fulfill the eligibility criteria for principal applicants. Principal applicants must be established researchers who are expected to apply from an eligible organization, able to sign up to the Wellcome Trust's Grant Conditions, and should normally: hold an academic or research post (or equivalent); have at least five years' postdoctoral (or equivalent) research experience (ndividuals who hold an established lectureship or a well-recognized fellowship, but who have less than five years postdoctoral experience are also eligible; other researchers with less than five years' experience should contact the Grants Information Desk for further advice); be in receipt of salary funding for the duration of the grant requested or, if this is not in place, have a position that is underwritten by their employing institution for the duration of the grant.
Geographic Focus: All States, All Countries
Contact: Kevin Moses, Director of Science; +44 (0)20 7611 5757 or +44 (0)20 7611 8888; fax +44 (0)20 7611 8258; sciencegrants@wellcome.ac.uk
Internet: http://www.wellcome.ac.uk/Funding/Biomedical-science/index.htm
Sponsor: Wellcome Trust
Gibbs Building, 215 Euston Road
London, GLONDON NW1 2BE United Kingdom

Wellcome Trust New Investigator Awards **3919**

The Wellcome Trust is particularly interested in receiving applications from university lecturers within the first five years of their independent research careers. New Investigator Awards are intended to support strong researchers who are in the early stages of their independent research careers and have already shown that they can innovate and drive advances in their field of study. Candidates should be no more than five years from

appointment to their first academic position. These Awards provide flexible support at a level and length appropriate to enable the best researchers to address the most important questions about health and disease. Awards may equally be small or large, but candidates should be able to articulate a compelling vision for their research, while ensuring that their proposal and requested funding is appropriate to their research experience to date. The research should contribute to the Trust's vision to achieve extraordinary improvements in human and animal health, and the proposal should describe how it will address one or more of our five major challenges. Prospective applicants are encouraged to contact the Trust if they wish to discuss their proposal or their suitability for this scheme.
Requirements: To be eligible for a New Investigator Award you should be based in the UK, Republic of Ireland or a low- or middle-income country, and you should: have an established academic post at an eligible higher education or research institution (you are employed on a permanent, open-ended or long-term rolling contract, salaried by your host institution); and be no more than five years from appointment to your first established academic post on the date you submit your application. More experienced researchers who have obtained significant, independent research support prior to their first academic appointment will be advised to apply for a Senior Investigator Award. Applicants are also eligible to apply if you have a written guarantee of an
Geographic Focus: Afghanistan, Albania, Algeria, American Samoa, Angola, Argentina, Armenia, Azerbaijan, Bangladesh, Belarus, Belize, Benin, Bhutan, Bolivia, Bosnia & Herzegovina, Botswana, Brazil, Bulgaria, Burkina Faso, Burundi, Cambodia, Cameroon, Cape Verde, Central African Republic, Chad, China, Colombia, Comoros, Congo, Congo, Democratic Republic of, Costa Rica, Cote d' Ivoire (Ivory Coast), Cuba, Djibouti, Dominica, Dominican Republic, Ecuador, Egypt, El Salvador, Eritrea, Ethiopia, Fiji, Gabon, Gambia, Ghana, Grenada, Guatemala, Guinea, Guinea-Bissau, Guyana, Haiti, Honduras, India, Indonesia, Iran, Iraq, Ireland, Jamaica, Jordan, Kazakhstan, Kenya, Kiribati, Korea, Kosovo, Kyrgyzstan, Laos, Lebanon, Lesotho, Liberia, Libya, Macedonia, Madagascar, Malawi, Malaysia, Maldives, Mali, Mauritania, Mauritius, Mexico, Micronesia, Moldova, Mongolia, Montenegro, Morocco, Mozambique, Myanmar (Burma), Namibia, Nepal, Nicaragua, Niger, Nigeria, Pakistan, Palau, Panama, Papua New Guinea, Paraguay, Peru, Philippines, Romania, Rwanda, Sao Tome & Principe, Senegal, Serbia, Seychelles, Sierra Leone, Solomon Islands, Somalia, South Africa, South Sudan, Sri Lanka, St. Lucia, St. Vincent and the Grenadines, Suriname, Swaziland, Syrian Arab Republic, Tajikistan, Tanzania, Thailand, Timor-Lester, Togo, Tonga, Tunisia, Turkey, Turkmenistan, Tuvalu, Uganda, Ukraine, United Kingdom, Uzbekistan, Vanuatu, Venezuela, Vietnam, Yemen, Zambia, Zimbabwe
Date(s) Application is Due: Mar 14; Jul 14; Nov 14
Contact: Kevin Moses, Director of Science; +44 (0)20 7611 5757 or +44 (0)20 7611 8888; fax +44 (0)20 7611 8258; sciencegrants@wellcome.ac.uk
Internet: http://www.wellcome.ac.uk/Funding/Biomedical-science/Funding-schemes/Investigator-Awards/WTX059284.htm
Sponsor: Wellcome Trust
Gibbs Building, 215 Euston Road
London, GLONDON NW1 2BE United Kingdom

Wells County Foundation Grants 3920
The Wells County Foundation Grants give priority to programs having a positive effect on the Wells County community. Grant making fields of interest include arts and culture, education, civic affairs, youth, environment, community development, animal welfare, recreation, and health and human services. In reviewing grant proposals, the Foundation gives careful consideration to: the potential impact of the request and the number of people benefited; an innovative approach; the degree to which the applicant works with or complements the services of other community organizations; the extent of local involvement and support for the project; the organization's demonstrated fiscal responsibility and management qualification; and the organization's ability to obtain necessary additional and future funding.
Requirements: Applicants are encouraged to contact the Foundation to discuss their project. They should be prepared to give a brief discussion to enable the Foundation to determine whether the request falls within the grant-making guidelines. In addition, the staff will inform the applicant of the deadline for the most current grant cycle. Grant proposals for projects should include the following: a title page with the organization's contact information, the project's title and amount requested; a proposal narrative, summarizing what issues the project will address, its expectations, how many will benefit, the role of volunteers, its planned evaluation, and a signed endorsement by the Board of Directors; financial information, with the project budget, two pricing quotes for equipment requested, a list of other funding sources, and the project's funding for the future; and organizational information, including a brief history with mission and purpose, list of officers, financial statement or audit, and a copy of the tax exemption the IRS.
Restrictions: No grants will be made to any political organization or to support attempts to influence the legislation of any governing body other than through making available to the community at large the results of non-partisan analysis, study and/or research.
Geographic Focus: Indiana
Date(s) Application is Due: Feb 15; Jun 15; Oct 17
Samples: Bluffton Parks & Recreation Department, purchase of an ADA compliant pool lift chair for the Wells Community swimming pool so that patrons can enter and exit the pool independently, $7,000; Community Action of Northeast Indiana (CANI), provide emergency assistance to Wells County families through CANI's Homeles Prevention Rapid Re-housing Program so that families can work toward self-sufficiency, $3,000; Family Centered Services, Inc, to provide administrative funding for the Youth As Resources program so that young people can design and carry out service projects that address specific community needs, $7,000.
Contact: Tammy Slater, Chief Executive Officer; (260) 824-8620; fax (260) 824-3981; wellscountyfound@wellscountyfound.org
Internet: http://www.wellscountyfound.org/Grants.html
Sponsor: Wells County Foundation
360 North Main Street, Suite C
Bluffton, IN 46714

Western Indiana Community Foundation Grants 3921
The Western Indiana Community Foundation focuses its attention on local needs within the geographical boundaries as set by the Board of Directors. Primary fields of interest include: health, charitable service, education, cultural affairs, and community improvement. The Foundation is especially interested in learning of plans for: start-up costs for new programs; one-time projects or needs; and capital needs beyond an applicant's capabilities and needs. Grant applications may be submitted throughout the year. Applicants will be notified immediately following the Board of Directors decision.
Requirements: Organizations must fill out the online application and include the following information: their full contact information, description of the project, and grant request amount; other funding sources for the project, an itemized expenses list and project timeline; the organization's IRS tax exempt status, how they plan to evaluate the project, and a description of public relations plans/foundation funding. The application packet should then be mailed to the Foundation,
Restrictions: The Foundation will not consider: grants for individuals; organizations for political or religious purposes; support for regular operating budgets; contributions to endowments; providing for long term funding; post-event situations; or apparel such as school/sport uniforms.
Geographic Focus: Indiana
Samples: City of Covington - $1,600 for mobile life guard chairs at the city swimming pool; Friendship Circle Center - $3,373 annual payout for operations and parking lot resurfacing; Fountain County Mentoring - $2,605 for youth field trip to Chicago's Shedd Aquarium and Museum of Science and Industry; Covington High School - $500 - for SAFE-TALK training session for faculty and staff to help identify students who are high risk for suicide or other self-destructive behaviors.
Contact: Dale White; (765)-793-0702; fax (765)-793-0703; dwhite@wicf-inc.org
Internet: http://www.wicf-inc.org/grant_guidelines.asp
Sponsor: Western Indiana Community Foundation
135 South Stringtown Road
Covington, IN 47932-0175

Western New York Foundation Grants 3922
The Foundation supports sustainable organizations that improve the quality of life in Western New York. The Foundation makes investments that build on nonprofits' proven strengths in order to improve their effectiveness and their ability to fulfill their missions. Funding is provided in these categories: human services; education; urban and rural development; arts, culture, and humanities; and housing, park and land use. Deadline dates for the Alignment Determination Application (Letter of Inquiry) submissions are June 30 and November 30. If approved, a formal go-ahead will be given to submit an Organizational Assessment Application, with a deadline three weeks after receiving the go-ahead.
Requirements: Western New York State 501(c)3 organizations located within one of the following counties are eligible: Allegany, Cattaraugus, Chautauqua, Erie, Genesee, Niagara, and Wyoming. All applicants must have three years of 990 filings in order to be eligible to apply for a grant.
Restrictions: Only one application may be submitted at a time. Funded organizations may reapply two years following the final payment of an award. The following are ineligible: religious organizations for religious purposes; political organizations, campaigns, and candidates; municipal and government entities; grants or loans to individuals; fund-raising events, i.e. sponsorships, tables, dinners, and telethons; endowments; scholarships; operating expenses; hospital capital campaigns; and general capital campaigns.
Geographic Focus: New York
Contact: Beth Kinsman Gosch; (716) 839-4225; bgosch@wnyfoundation.org
Internet: http://www.wnyfoundation.org
Sponsor: Western New York Foundation
11 Summer Street, Fourth Floor
Buffalo, NY 14209

Western Union Foundation Grants 3923
The Foundation generally awards grants in the United States and internationally in its areas of interest, including education; and health and human services addressing literacy, health care for the uninsured, nutrition, pre- and postnatal care, childhood immunizations, poverty, language barriers, and cultural adjustment. In order to maximize resources and create a greater impact in communities, the Foundation will not accept unsolicited proposals in 2010. The Foundation will work with its current NGO partners and solicit proposals from NGOs that closely align with the Foundation's education and economic development program areas.
Restrictions: In general grants do not support general operating support, individuals, endowments, special events, capital projects, early childhood education, other post-secondary scholarship programs, deficits or retirement of debt, re-granting agencies, and awards.
Geographic Focus: Colorado, Florida, Nebraska, New York, Texas
Date(s) Application is Due: Mar 1; Jun 1; Dec 1
Amount of Grant: 1,000 - 25,000 USD
Contact: Luella Chavez D'Angelo, President; (720) 332-4763 or (720) 332-6606; fax (720) 332-4772; luella.dangelo@westernunion.com
Internet: http://foundation.westernunion.com/ourProgramsDirect.html
Sponsor: Western Union Foundation
12500 East Belford Avenue, Suite 1-I
Englewood, CO 80112

WestWind Foundation Reproductive Health and Rights Grants   3924
Concerned with an exploding population in the Latin American and Caribbean region caused by inadequate access to reproductive health services, the Trustees of the WestWind Foundation decided to create a program that would support NGOs that work to provide services and improve access both in the region and in other parts of the globe. The program operates with the understanding of the international agreement made in Cairo in 1994 at the International Conference on Population and Development (ICPD), when the global community affirmed its commitment to address population growth by working to ensure individual rights and freedoms rather than demographic targets. In lieu of these trends, the goals of the Reproductive Health and Rights program are to support NGOs that seek to: improve access to reproductive health services, particularly in the LAC region; promote reproductive health and rights, both domestically and abroad; and promote adolescent sexual and reproductive health, both domestically and in the LAC region. he foundation supports organizations that seek to advance a range of reproductive health issues, including, but not limited to: supporting emergency contraception; promoting adolescent sexuality education and empowerment; preventing maternal mortality; and providing post-abortion care. Applicants should submit an online Letter of Inquiry (LOI) prior to the annual deadline. Typically, WestWind will respond to LOIs between 4 to 6 weeks after the letter has been received.
Requirements: The RHR Program currently supports non-governmental organizations (NGOs) that work both domestically and abroad to improve women's access to reproductive health services. Grants in the RHRP area are made primarily to U.S. based organizations for: international projects in the Latin America and Caribbean region; and projects that have national significance. There is also a small portion of funds that are available for global, opportunistic projects.
Geographic Focus: All States
Date(s) Application is Due: Mar 1
Amount of Grant: 5,000 - 225,000 USD
Contact: Kristen Miller, Program Consultant; (434) 977-5762, ext. 24; fax (434) 977-3176; bonnell@westwindfoundation.org or info@westwindfoundation.org
Internet: http://www.westwindfoundation.org/program-areas/reproductive-health-rights/
Sponsor: WestWind Foundation
204 East High Street
Charlottesville, VA 22902

Weyerhaeuser Company Foundation Grants   3925
The foundation's mission is to improve the quality of life in communities where Weyerhaeuser has a major presence and to provide leadership that increases public understanding of issues where society's needs intersect with the interests of the forest products industry. Community service grants include awards in the fields of education and youth, health and welfare, civic and community improvement, and culture and the arts. In addition to community grants, industry-related awards are made to educational institutions, environmental groups, and professional organizations that promote further understanding of how the forest products industry responds to a changing society. Types of support include seed money, building construction/renovation, equipment acquisition, employee-related scholarships, publication, conferences and seminars, fellowships, lectureships, operating budgets, research, program development, employee matching gifts, and technical assistance. Applicants should send a short letter that introduces the project and sponsoring organization and provide tax-exempt status evidence. If further consideration is warranted, the foundation may ask for additional information or a formal proposal. Proposals may be submitted at any time.
Requirements: Applying organizations must have nonprofit, tax-exempt status. To be considered for funding, an organization must: serve a community within a 50-mile radius of a major Weyerhaeuser facility; support a state-wide issue of interest to the Foundation and Weyerhaeuser in the key states of Alabama, Arkansas, Louisiana, Mississippi, North Carolina, Oklahoma, Oregon or Washington; or support a selected, high-priority national or international initiative directly related to the sustainability and importance of working forests. A limited number of smaller awards are also made to other locales where fewer employees are based.
Restrictions: Grants are not awarded to individuals or for political campaigns, activities that influence legislation, religious organizations seeking funds for theological purposes, or funds to purchase tickets or tables at fundraising benefits.
Geographic Focus: Alabama, Arizona, Arkansas, California, Colorado, District of Columbia, Georgia, Idaho, Illinois, Kentucky, Louisiana, Maryland, Michigan, Minnesota, Mississippi, Missouri, New Hampshire, New Jersey, New Mexico, North Carolina, Ohio, Oklahoma, Oregon, Pennsylvania, South Carolina, Texas, Utah, Virginia, Washington, West Virginia, Wisconsin
Date(s) Application is Due: Aug 31
Contact: Anne Levya, Team Coordinator; (253) 924-3159; fax (253) 924-3658; foundation@weyerhaeuser.com or anne.leyva@weyerhaeuser.com
Internet: http://www.weyerhaeuser.com/Sustainability/Foundation
Sponsor: Weyerhaeuser Company Foundation
P.O. Box 9777
Federal Way, WA 98063-9777

Weyerhaeuser Family Foundation Health Grants   3926
The Weyerhaeuser Family Foundation supports programs of national and international significance that promote the welfare of human and natural resources. These efforts will enhance the creativity, strengths and skills already possessed by those in need and reinforce the sustaining processes inherent in nature. The Foundation supports multi-site, national or international projects dealing with mental health, chemical dependency and population and family planning. Multi-site, national and international educational projects will also be considered. The Letter of Intent is the first step in the application process and should be no more than two pages, to which you must attach a one-page budget summary and an Application Cover Sheet-General program. The average grant can range up to $25,000.
Requirements: U.S. nonprofit, tax-exempt organizations are eligible.
Restrictions: The General Program does not fund projects serving only local or regional domestic areas; to be eligible, projects must be multi-site, national or international. In addition, the Foundation will not consider proposals in the following areas: books or media projects, unless the project is connected to other areas of Foundation interest; capital projects; individuals, scholarships or fellowships; land acquisitions or trades; lobbying activity; ongoing projects or general operating support for an organization; organizations located outside the United States; or research projects.
Geographic Focus: All States
Date(s) Application is Due: Apr 1; Aug 1
Amount of Grant: Up to 25,000 USD
Contact: Peter A. Konrad; (303) 993-5385; pkonrad@konradconsulting.com
Internet: http://www.wfamilyfoundation.org/general_program.html
Sponsor: Weyerhaeuser Family Foundation
2000 Wells Fargo Place, 30 East Seventh Street
St. Paul, MN 55101-4930

WHAS Crusade for Children Grants   3927
The organization, established in Kentucky in 1954, is dedicated to helping the handicapped children of today and to expanding the ability to care for and prevent the handicaps of tomorrow. The crusade supports organizations which serve children suffering from a variety of mental and physical disabilities. WHAS Crusade for Children grants are made to non-profit agencies, schools and hospitals to help children (up to age 18) overcome physical, mental, emotional or medical challenges. Grants are for direct services only, and are made for specific programs or equipment that provide direct benefit to special needs children. Applications are mailed out each year in November and December. All agencies that received a grant this year receive an application for next year. To request an application, a non-profit agency, school or hospital should send a letter (on letterhead signed by the agency director or school superintendent) to the WHAS Crusade for Children at the address listed. The deadline for applications is late January each year.
Requirements: Agencies serving special needs children from Kentucky and southern Indiana are eligible to receive grants. The Crusade must have support from a county before grants can be made to a school, hospital or agency located in that county.
Restrictions: Grants are not made to individuals, and are not considered operating grants.
Geographic Focus: Indiana, Kentucky
Contact: Dawn Lee, President; (502) 582-7706; fax (502) 582-7712; dlee@whas11.com or contact-crusade@whascrusade.org
Internet: http://www.whascrusade.org/grants/
Sponsor: WHAS Crusade for Children
P.O. Box 1100
Louisville, KY 40201-1100

White County Community Foundation - Adam Krintz Memorial Scholarship   3928
The White County Community Foundation - Adam Krintz Memorial Scholarship is available to students entering any field of study at a college, university or vocational school. Applicants must be graduates of Twin Lakes High School, with well-rounded activities in school, community, and church. They should also have a GPA of at least 3.0.
Requirements: Along with the application, students are required to submit a personal insight essay; an official high school transcript; a copy of a letter of acceptance from their college or university of choice; and two recommendation letters. The application packet can be returned to their guidance office or the Foundation.
Geographic Focus: Indiana
Date(s) Application is Due: Feb 27
Amount of Grant: 500 USD
Contact: Lesley Wineland Goss; (574) 583-6911; fax (574) 583-8757; director@whitecf.org
Internet: http://www.whitecf.org/GrantScol.html
Sponsor: White County Community Foundation
1001 South Main Street
Monticello, IN 47960

White County Community Foundation - Annie Horton Scholarship   3929
The White County Community Foundation - Annie Horton Scholarship is available for a White County high school freshman, sophomore, or junior accepted to a summer program in the field of academics or leadership. The award will provide up to $500 for the cost of a summer program, which may include tuition, room and board. Preference will be given (but not limited) to a student accepted at a summer program in the field of journalism, art, music, science or leadership. Enthusiasm and interest in the summer program will be the most important criterion for acceptance, followed by academic standing and financial need.
Requirements: In addition to the application, students submit a personal essay and two letters of recommendation.
Geographic Focus: Indiana
Date(s) Application is Due: Apr 18
Amount of Grant: Up to 500 USD
Contact: Lesley Wineland Goss; (574) 583-6911; fax (574) 583-8757; director@whitecf.org
Internet: http://www.whitecf.org/GrantScol.html
Sponsor: White County Community Foundation
1001 South Main Street
Monticello, IN 47960

White County Community Foundation - Landis Memorial Scholarship 3930
The White County Community Foundation - Landis Memorial Scholarship is available to Twin Lakes graduating seniors intending to pursue higher education at a college, university, or school of training. Applicants must be well-rounded students who are active in their school, community, or church, with a B average or better. Financial need will be a strong consideration. Any college major is eligible.
Requirements: In addition to the online application, student submit a personal insight essay; an official high school transcript; a copy of their college acceptance letter; and a written recommendation from a teacher. The application packet may be sent to the student's guidance office by February 7 or to the Foundation by February 29.
Geographic Focus: Indiana
Date(s) Application is Due: Feb 7; Feb 29
Amount of Grant: 10,000 USD
Contact: Lesley Wineland Goss; (574) 583-6911; fax (574) 583-8757; director@whitecf.org
Internet: http://www.whitecf.org/GrantScol.html
Sponsor: White County Community Foundation
1001 South Main Street
Monticello, IN 47960

White County Community Foundation - Lilly Endowment Scholarships 3931
The Lilly Endowment Community Scholarship Program is designed to raise the level of educational attainment in Indiana and increase awareness of the potential of Indiana's community foundations to improve the quality of life of the state's residents. Scholarships pay for full academic tuition and required fees plus an annual $800 allocation for books and equipment for four years at any accredited Indiana college or university selected by the recipient. The scholarships are sponsored by Lilly Endowment Inc., and administered by community foundations throughout Indiana. Applicants must: be a U.S. citizen and have graduated from an accredited White County high school; have at least a 2.75 GPA; and have a combined SAT score of 800 or above (critical reading & math). Selection considerations include: financial need; community and school activities and evidence of leadership; the essay; grade point average and difficulty of high school courses taken; and the interview of finalists. Other scholarships, such as 21st Century Scholars, will be taken into consideration when calculating the applicant's financial need. Any college major may apply.
Requirements: In addition to the online application, students must submit a completed financial aid form, high school transcript, spring class schedule, an essay, and three letters of recommendation. The application packet should be submitted to the Foundation in early January.
Geographic Focus: Indiana
Date(s) Application is Due: Jan 17
Contact: Lesley Wineland Goss; (574) 583-6911; fax (574) 583-8757; director@whitecf.org
Internet: http://www.whitecf.org/GrantScol.html
Sponsor: White County Community Foundation
1001 South Main Street
Monticello, IN 47960

White County Community Foundation - Tri-County Educational Scholarships 3932
The White County Community Foundation - Tri-County Educational Scholarships offer funding for Tri-County seniors pursuing higher education at a college, university, or school of training. Applications will be reviewed using the following criteria: academic record; school-sponsored activities; community service and personal character; and a personal essay. Several memorial scholarships are available from $500 to $1,000. Any college major is eligible.
Requirements: Along with the online application, students must submit a personal insight essay, and an official high school transcript. The application packet is submitted to the student's guidance office by March 1.
Restrictions: Applicants must be graduating seniors from Tri-County High School.
Geographic Focus: Indiana
Date(s) Application is Due: Mar 1
Amount of Grant: 500 - 1,000 USD
Contact: Lesley Wineland Goss; (574) 583-6911; fax (574) 583-8757; director@whitecf.org
Internet: http://www.whitecf.org/GrantScol.html
Sponsor: White County Community Foundation
1001 South Main Street
Monticello, IN 47960

White County Community Foundation - Women Giving Together Grants 3933
The White County Community Foundation - Women Giving Together Grants address White County's women and families, and inspire women to strengthen White County through charitable giving. Members of Women Giving Together believe that high priority issues should be addressed thoughtfully to inspire social change and improve the quality of life in White County. Grant funding focuses on community enhancement, education, and social/human services. Program priorities: must serve a charitable purpose; serve the needs of the women and families of the greater White County area; demonstrate an innovative and unique approach; and indicate other funding sources.
Requirements: Applicants are encouraged to contact the Foundation to be certain their project meets the Foundation's guidelines. Organizations submit the online application to include the following information: their organization's contact information and organization's mission statement; amount requested for the project with its timeline; a narrative that describes the organization, its need, project description, and how it will benefit the White County area; a list of those responsible, with timeline and funding sources other than the WGT; how the project will be financed in the future and how it will affect the community. Organizations should also submit a detailed budget; financial statements; the IRS determination letter; and a list of board members and their authorization for the project.
Restrictions: Projects not funded include: non-charitable projects and organizations; projects fully funded by local government; projects to fund an endowment, ongoing operating budgets, existing deficit, debt reduction, or multi-year, long-term funding; religious activities that do not serve the community as a whole; special events such as parades, festivals, or sporting activities; political organizations or campaigns; national and state fundraising efforts; project that indicate discrimination.
Geographic Focus: Indiana
Date(s) Application is Due: Oct 1
Contact: Lesley Wineland Goss; (574) 583-6911; fax (574) 583-8757; director@whitecf.org
Internet: http://www.whitecf.org/WGTGrant.html
Sponsor: White County Community Foundation
1001 South Main Street
Monticello, IN 47960

White County Community Foundation Grants 3934
The White County Community Foundation Grants fund projects in the following categories: education - to develop the untapped capacities of those of all ages, families, or communities; human services - to achieve a positive change in the conditions that adversely affect the elderly, disabled, or economically disadvantaged; recreation - to create or expand family-oriented leisure time opportunities; arts and culture - to encourage and stimulate the arts and historical preservation; citizenship - to generate an increased level of volunteerism, increased county-wide cooperation, and new leadership; environment - to help protect or enhance the environment that must support our life systems beyond the next century; and economic - to generate economic development activity within the area of direct job creation opportunity in White County. Priority is given to projects that: reach as many people as possible; identify and address an immediate community need; improve the ability of the organization to serve the community over the long-term; are run by non-profit organization; serve White County; and demonstrate fundraising from sources other that the Foundation as well as support from within the requesting organization.
Requirements: Before applying, organizations should contact the Foundation to verify deadlines and funding priorities, and to ensure they have the current application form. Once applicants reach this stage, they should include the following information with the online or hard copy application: a brief description of the organization; a narrative on the nature, purpose, and benefits of the project; a description of the project coordination; a project timeline with anticipated start and completion dates; a detailed budget; funding sources and future plan for funding; the impact of the project on the community; organization's most recent year-end financial statements and current operating budget; IRS determination letter; organization's board members; board member authorization for project funding.
Restrictions: The following projects are not eligible for funding: individuals; projects outside the White County area; programs normally funded by local government; projects to fund an endowment, ongoing operating budget, existing deficit, debt reduction; or multi-year, long-term funding; religious activities that predominantly serve one denomination; special events such as parades, festivals, and sporting activities; political organizations or campaigns; national and state fundraising efforts; and projects that indicate discrimination.
Geographic Focus: Indiana
Contact: Lesley Wineland Goss; (574) 583-6911; fax (574) 583-8757; director@whitecf.org
Internet: http://www.whitecf.org/CommGrant.html
Sponsor: White County Community Foundation
1001 South Main Street
Monticello, IN 47960

Whitley County Community Foundation - Lilly Endowment Scholarship 3935
The Whitley County Lilly Endowment Scholarship covers full tuition and $800 per year for books and equipment for four years. Applicants must be Whitley County residents, high school graduates, pursuing a full-time baccalaurate course of study at a public or private college or university in Indiana. They must also have financial need and demonstrate pride in their community, responsibility, perseverance, self-esteem, and the desire to contribute through work or volunteerism. Any four year field of study is eligible.
Requirements: In addition to the application, students should include the following information: their high school transcript and SAT scores; two letters of recommendation; their parent or guardian's current tax forms; information regarding all financial resources available for funding their education, including other scholarships received or pending; and two essay discussions.
Geographic Focus: Indiana
Date(s) Application is Due: Jan 1
Contact: September McConnell; (260) 244-5224; fax (260) 244-5724; sepwccf@gmail.com
Internet: http://whitleycountycommunityfoundation.org/lilly.html
Sponsor: Whitley County Community Foundation
400 North Whitley Street
Columbia City, IN 46725

Whitley County Community Foundation Grants 3936
The Whitley County Community Foundation directs grants to charitable projects that will make a positive impact on Whitley County and its people, with particular interest in projects that shed new light on local needs and provide innovative, long-term solutions. Categories of support include: arts & culture, health, civic affairs, recreation, community development, welfare, and education. Grant reviews are scheduled on May 1 and December 1, but applicants are encouraged to submit proposals well in advance.
Requirements: The grant application must include a copy of the organization's IRS tax exempt letter, its annual budget and its projected budget. The cover letter should include: an introduction that establishes the organization's purpose and credibility, with background and accomplishments; a statement of need addressing why the project is necessary, documenting the need with statistics; what the organization hopes to accomplish with measurable objectives; what methods will be used to analyze results and refine the program; and how the project will be funded or maintained in the future.

The packet may also include endorsement letters, a list of the board of directors, and a list of past support from other funders.
Restrictions: The Foundation is unlikely to support: annual campaigns, political activities, private schools, advertising, religious/sectarian causes, organizations outside the Whitley service area, and debt retirement.
Geographic Focus: Indiana
Date(s) Application is Due: May 1; Dec 1
Amount of Grant: 500 - 15,000 USD
Contact: September McConnell; (260) 244-5224; fax (260) 244-5724; sepwccf@gmail.com
Internet: http://whitleycountycommunityfoundation.org/grantseekers.html
Sponsor: Whitley County Community Foundation
400 North Whitley Street
Columbia City, IN 46725

Whitley County Community Foundation Scholarships 3937
The Whitley County Community Foundation offers a variety of scholarships with criteria specific to the particular scholarships. The Foundation posts a list of scholarships categorized by high school, county, college major, and non-traditional student. Applicants should carefully read the information to see if they should submit a hardcopy of the application materials, submit the information as an email attachment to the community foundation, or if there is a specific business address to send the application materials.
Restrictions: Scholarships are available only to high school seniors from Whitko, Columbia City, and Churubusco.
Geographic Focus: Indiana
Contact: September McConnell; (260) 244-5224; fax (260) 244-5724; sepwccf@gmail.com
Internet: http://whitleycountycommunityfoundation.org/whitley.html
Sponsor: Whitley County Community Foundation
400 North Whitley Street
Columbia City, IN 46725

Whitney Foundation Grants 3938
The Foundation awards grants to nonprofit organizations primarily in Minnesota in its areas of interest, including AIDS prevention, arts, children and youth services, education, and human services. Types of support include program development, annual campaigns, and continuing support. Application forms are not required. Awards range from $100 to $5,000.
Requirements: Eligible applicants must be 501(c)3 organizations.
Geographic Focus: Minnesota
Contact: Carol VanOrnum, (952) 835-2577
Sponsor: Whitney Foundation
601 Carlson Parkway
Minnetonka, MN 55305

WHO Foundation General Grants 3939
The WHO Foundation nationally supports grass-roots charities serving the overlooked needs of women and children. Grants are provided to organizations serving women and/or children in the United States and Puerto Rico. Specific projects and programs addressing health, education, and social service needs are the priority. The foundation recognizes the value of new programs created to respond to changing needs and will consider funding projects of an original or pioneering nature within an existing organization. Application and guidelines are available for download at the sponsor's website.
Requirements: 501(c)3 nonprofit organizations in the United States and Puerto Rico are eligible. Organizations must have been incorporated for a minimum of three years prior to application. Preference will be given to organizations with an operating budget of $3 million or less, those not dependent upon government grants, and those with greater organizational program costs than personnel costs. Funding requests must be made using the WHO Foundation application. Electronic and faxed submissions will not be accepted.
Restrictions: The following types of organizations, activities or purposes will not be considered: personal requests, loans or scholarships to individuals; educational institutions; endowment campaigns; international programs or projects; government agencies; fiscal agents; religious organizations (including young Men's and Women's Christian Association); political causes, candidates, organizations or campaigns; foundations that are grant making institutions; advertising in charitable publications; sports organizations; labor groups; research projects; travel for individuals or groups; conferences, galas, charity balls, sponsorships, seminars, or reunions; capital campaigns; salaries; or building campaigns (i.e. Habitat for Humanity).
Geographic Focus: All States, Puerto Rico
Amount of Grant: 5,000 - 30,000 USD
Contact: Cindy Turek; (800) 946-4663 or (972) 341-3019; who@beauticontrol.com
Internet: http://www.whofoundation.org/WHO_FundingCriteria.htm
Sponsor: Women Helping Others Foundation
P.O. Box 816029
Dallas, TX 75381-6029

WHO Foundation Volunteer Service Grants 3940
The WHO Foundation nationally supports grass-roots charities serving the overlooked needs of women and children. Grants are provided to organizations serving women and/or children in the United States and Puerto Rico. Specific projects and programs addressing health and social service needs are its priority. The Foundation recognizes the value of new programs created to respond to changing needs and will consider funding projects of an original or pioneering nature within an existing organization. Funding will be considered for: human services, including homelessness, abuse and neglect, hunger, and domestic violence; health services, including the medically uninsured, therapeutic programs, and physical or mental disabilities; and education, including free afterschool programs, adult education, job training, GED programs, and literacy. Applications must be received no later than June 2. Grant requests from $1,000 up to $40,000 will be considered.
Requirements: In order to qualify for funding, an organization must have a 501(c)3 rating in their name (no affiliates or fiscal agents accepted) for a minimum of three (3) years prior to application. All funds must be used in the calendar year in which they are received.
Restrictions: Funding will not be considered for: personal requests, loans or scholarships to individuals; international programs or projects; conferences or seminars; travel for individuals or groups; educational institutions and their foundations; political causes, candidates, organizations or campaigns; religious purposes, including church groups and activities; foundations that are grant making institutions; research projects; annual or capital fund campaigns, underwriting or sponsor events; salaries; endowment funds; sports organizations or athletic activities; or animal welfare.
Geographic Focus: All States, Puerto Rico
Date(s) Application is Due: Jun 2
Amount of Grant: 1,000 - 40,000 USD
Contact: Cindy Turek; (800) 946-4663 or (972) 341-3019; who@beauticontrol.com
Internet: http://www.whofoundation.org/Funding/index.asp
Sponsor: Women Helping Others Foundation
P.O. Box 816029
Dallas, TX 75381-6029

Wilhelmina W. Jackson Trust Scholarships 3941
The Wilhelmina W. Jackson Trust offers scholarships to residents of Swampscott and Marblehead, Massachusetts, for the study of medicine, and for the study of any form of the creative arts at a college, university or art school of the student's choice. Selection is based on financial need, artistic ability, character and work habits, and GPA. The scholarship is awarded for up to four years. Forms of the creative arts included are painting, sculpture, graphics, printmaking, industrial design, illustration, photography, ceramics, and art history. Recent awards have ranged from $1,000 to $10,000. A formal application is required, and the annual application deadline is February 15.
Requirements: Giving limited to residents of Swampscott and Marblehead, Massachusetts.
Geographic Focus: Massachusetts
Date(s) Application is Due: Feb 15
Amount of Grant: 1,000 - 10,000 USD
Contact: Juanita Lamby; (781) 639-3100, ext. 2124; lamby.juanita@marbleheadschools.org
Sponsor: Wilhelmina W. Jackson Trust
P.O. Box 1802
Providence, RI 02901-1802

Wilhelm Sander-Stiftung Foundation Grants 3942
The Wilhelm Sanders Foundation was established in Germany in 1973. The purpose of the Foundation grants are to support research in the field of human medicine, with a strong focus on cancer research, both clinical and clinical-experimental. Applications are accepted electronically (online) or via hard copy. There are no annual deadlines specified, and interested applicants should begin by contacting the Foundation directly.
Restrictions: The support by the Wilhelm Sander-Stiftung is limited to Germany and Switzerland.
Geographic Focus: Germany, Sweden
Contact: Bernhard Knappe; +49 (089) 544-1870; fax +49 (089) 544-18720; info@sanst.de
Internet: http://www.wilhelm-sander-stiftung.de
Sponsor: Wilhelm Sander-Stiftung Foundation
Goethe route 74
Munich, 80336 Germany

Willard and Pat Walker Charitable Foundation Grants 3943
The Willard and Pat Walker Charitable Foundation awards grants to Arkansas nonprofit organizations in its areas of interest, including: the arts; children and youth services; health organizations; higher education; residential and custodial hospice care; and social services. Types of support include: building construction and renovation; capital campaigns; continuing support; endowments; equipment; general operating grants; matching grants; program development; and scholarship endowment funds. There are no application forms, so applicants should provide a detailed description of their project and the amount of funding requested. There are two annual deadlines of March 1 and October 1. The Board meets twice per year, in April and November
Requirements: Arkansas nonprofit organizations are eligible to apply.
Restrictions: Individuals are not eligible.
Geographic Focus: Arkansas
Date(s) Application is Due: Mar 1; Oct 1
Amount of Grant: Up to 1,000,000 USD
Contact: John M. Walker; (479) 582-2310; fax (479) 582-2292; walkerfamily1@sbcglobal.net
Sponsor: Willard and Pat Walker Charitable Foundation
P.O. Box 10500
Fayetteville, AR 72703-2857

Willary Foundation Grants 3944
The Willary Foundation supports projects that are interesting, creative, and imaginative and that benefit communities in northeastern Pennsylvania. The Foundation is particularly interested in projects that support leadership and the development of leadership in business, the economy, education, human services, government, the arts, media, and research. The Foundation gives preference to efforts that have a ripple effect in the community and those that are conducted in conjunction with other sources of funding. Willary seeks to foster both individuals and groups with unique, innovative, or unusual ideas and efforts. The application and samples of previously funded projects are available at the Foundation website.

Requirements: Individuals and nonprofits in Lackawanna and Luzerne counties, Pennsylvania, are eligible to apply.
Restrictions: Willary will not consider applications for capital campaigns and annual drives.
Geographic Focus: Pennsylvania
Date(s) Application is Due: Mar 10; Aug 27
Amount of Grant: 1,200 - 50,000 USD
Contact: Linda Donovan; (570) 961-6952; fax (570) 961-7269; info@willary.org
Internet: http://www.willary.org
Sponsor: Willary Foundation
P.O. Box 283
Scranton, PA 18501-0937

William A. Badger Foundation Grants 3945
The Nabors to Neighbors Foundation was created in 2007 to assist charitable organizations focusing on need based projects for direct programming, capital and operating initiatives. This is a family foundation dedicated to organizations that deliver measurable results, seek partnerships and collaborations, utilize their resources within their respective communities while working to increase equity for those most in need. The Foundation encourages nonprofit organizations to apply who specialize in, though not limited to, improving the lives of children, education and medical initiatives. The Foundation makes no geographic restrictions on distributions. However, it has been the practice of the Trustees to make grants within Whitfield County, specifically Dalton, located in north Georgia. Requests must be postmarked by February 1, July 9 or October 1 in order to be considered.
Requirements: The Nabors to Neighbors Foundation makes grants to qualified 501(c)3 organizations. All requests must include: background information on the organization, including a brief history, the organization's current address and phone number and the name and title or the primary contact; the goals, objectives, and budget for the one project or program for which funds are being requested; the amount of the grant requested; summary of how the funds will be used; supporting financial information on the organization, to include current financial status and listing of Board of Trustees; copy of organization's tax exemption letter from the Internal Revenue Service. Application form is available online and all requests must be postmarked by February 1, July 9 or October 1 in order to be considered.
Restrictions: Grants are not made to: individuals; an organization to be used as pass-through funds for an ineligible organization Faith-based organizations without a 501(c)3 exemption; organizations with political purposes, nor to organizations which discriminate on the basis of race, ethnic origin, sexual or religious preference, age or gender.
Geographic Focus: All States
Date(s) Application is Due: Feb 1; Jul 9; Oct 1
Contact: Trustee, c/o Wachovia Bank; grantinquiries8@wachovia.com
Internet: https://www.wachovia.com/foundation/v/index.jsp?vgnextoid=108bf296ac212210VgnVCM100000617d6fa2RCRD&vgnextfmt=default
Sponsor: Nabors to Neighbors Foundation
3280 Peachtree Road NE, Suite 400, MC G0141-041
Atlanta, GA 30305

William B. Dietrich Foundation Grants 3946
The Foundation awards funding to nonprofits preferably for local needs. There are no submission deadlines. Areas of interest include children, the elderly, AIDS, museums, and libraries.
Requirements: Applicants may apply in writing, outlining the nature of the organization, the intended use of funds requested. A copy of the Internal Revenue Service determination letter should be submitted.
Restrictions: No grants are provided for individuals.
Geographic Focus: All States
Contact: Frank G. Cooper, President; (215) 979-1919
Sponsor: William B. Dietrich Foundation
P.O. Box 58177
Philadelphia, PA 19102-8177

William B. Stokely Jr. Foundation Grants 3947
The Foundation participates in scholarship funding at various colleges and universities mainly in eastern Tennessee. Grants are made to the educational institutions, which then distribute funds through their scholarship programs. Consideration also is given to requests from the arts, health service, civic organizations, and youth services. The Foundation does not require completion of a formal application, nor are there established deadline dates. All proposals must be submitted in writing.
Geographic Focus: Tennessee
Amount of Grant: 250 - 100,000 USD
Contact: William Stokely III, President; (865) 966-4878
Sponsor: William B. Stokely Jr. Foundation
620 Campbell Station Road, Suite 27
Knoxville, TN 37922-1636

William Blair and Company Foundation Grants 3948
Contributing to the community is an important part of the culture of William Blair and Company. The Foundation was officially established in Illinois in 1980, with giving primarily centered around metropolitan Chicago. All partners of the firm contribute part of their individual share of profits to the Foundation. Donation requests are made to the Foundation by partners and employees. The Foundation supports a broad range of causes including: civic affairs; public safety; arts and culture; higher education; youth-oriented activities; healthcare research; cultural affairs; and civic charities. Types of support include: annual campaigns; building and renovation; capital campaigns; general operating support; endowments; fellowships; internship programs; and scholarship funding. There are no specific deadlines or applications forms required. Funding typically ranges from $500 to $25,000.

Requirements: Requests can be made by sending a letter with a general description of the organization and its special purpose. Activities should have a significant impact on the Chicago metropolitan area.
Geographic Focus: Illinois
Contact: E. David Coolidge III, Vice President; (312) 236-1600
Sponsor: William Blair and Company Foundation
222 W Adams Street, 28th Floor
Chicago, IL 60606

William D. Laurie, Jr. Charitable Foundation Grants 3949
The William D. Laurie, Jr. Charitable Foundation, named after the Vice President and Detroit Manager of J. Walter Thompson Company, was established in Rhode Island in 2002. The Foundation's primary fields of interest include the support of: animal and wildlife programs; education; and health organizations. There are no formal application requirements, and interested parties should begin by forwarding a proposal letter to the Foundation office. No annual deadlines have been identified. Typical awards have recently ranged from $250 to $2,000.
Requirements: Any 501(c)3 organization located in, or supporting residents of, Rhode Island are welcome to apply.
Geographic Focus: Rhode Island
Amount of Grant: 250 - 2,000 USD
Contact: David H. Laurie, Trustee; (401) 423-1811 or (401) 423-0403
Sponsor: William D. Laurie, Jr. Charitable Foundation
15 Dumplings Drive
Jamestown, RI 02835-2904

William G. and Helen C. Hoffman Foundation Grants 3950
Helen C. Hoffman resided in the Village of South Orange, New Jersey. Her foundation was established in 1998 in memory of herself and her husband after the death of their daughter Corinne Blair. Her testamentary wish was to establish this foundation for charitable, religious, scientific, literary, and educational purposes. Her preference was to support blindness and its cure. Approximately 90% of the grants will provide support to the blind and, to medical research for the prevention of blindness. The remaining 10% will fund annual grants in the following areas of interest: education, the arts, environment and, social/civic causes. Requests must be received by January 15 for the March meeting, and August 22 for the October meeting. Application forms are available online. Applicants will receive notice acknowledging receipt of the grant request, and subsequently be notified of the grant declination or approval.
Requirements: Any U.S. 501(c)3 non-profit organizations may apply. Proposals should be submitted in the following format: completed Common Grant Application Form; an original Proposal Statement; an audited financial report and a current year operating budget; a copy of your official IRS Letter with your tax determination; a listing of your Board of Directors. Proposal Statements (second item in the above Format) should answer these questions: what are the objectives and expected outcomes of this program/project/request; what strategies will be used to accomplish your objective; what is the timeline for completion; if this is part of an on-going program, how long has it been in operation; what criteria will you use to measure success; if the request is not fully funded, what other sources can you engage; an Itemized budget should be included; please describe any collaborative ventures. Prior to the distribution of funds, all approved grantees must sign and return a Grant Agreement Form, stating that the funds will be used for the purpose intended. Progress reports and Completion reports must also be filed as required for your specific grant. All current grantees must be in good standing with required documentation prior to submitting new proposals to any foundation.
Restrictions: Grants are not made for political purposes, nor to organizations which discriminate on the basis of race, ethnic origin, sexual or religious preference, age or gender.
Geographic Focus: All States
Date(s) Application is Due: Jan 15; Aug 22
Amount of Grant: 5,000 - 15,000 USD
Contact: Wachovia Bank, N.A., Trustee; grantinquiries2@wachovia.com
Internet: https://www.wachovia.com/foundation/v/index.jsp?vgnextoid=522852199c0aa110VgnVCM1000004b0d1872RCRD&vgnextfmt=default
Sponsor: William G. and Helen C. Hoffman Foundation
190 River Road, NJ3132
Summit, NJ 07901

William G. Gilmore Foundation Grants 3951
The foundation supports nonprofits primarily in the San Francisco Bay Area, CA; some funding also in Pueblo, CO and Portland, OR. Giving primarily for the arts, health, and children, youth, and social services.
Requirements: California, Colorado, and Oregon nonprofits are eligible to apply. Contact the Foundation for additional application information.
Restrictions: No grants to individuals.
Geographic Focus: California, Colorado, Oregon
Amount of Grant: 1,000 - 50,000 USD
Contact: Faye Wilson, Executive Director; (415) 546-1400; fax (415) 391-8732
Sponsor: William G. Gilmore Foundation
120 Montgomery Street, Suite 1880
San Francisco, CA 94104-4317

William H. Adams Foundation for ALS Grants 3952
Established in 1999, the William H. Adams Foundation researches and funds programs specializing in the treatment of Amyotrophic Lateral Sclerosis (ALS) or Lou Gehrig's disease. Since there are no formal applications required, interested parties should

contact the Foundation directly. Most recent grant amounts have ranged from $10,000 to $250,000, and are either in the form of general donations or research specific. There are no specific annual deadlines for making application.
Geographic Focus: All States
Amount of Grant: 10,000 - 250,000 USD
Contact: Ellen Adams, President; (415) 592-8151
Sponsor: William H. Adams Foundation for ALS
1918 43rd Avenue
San Francisco, CA 94116-1025

William H. Hannon Foundation Grants 3953
The Hannon Foundation primarily funds non-profit organizations whose works address the goals of its founder as described in the Foundation's Mission Statement. The Foundation's mission is: to enhance the welfare and education of students in both public and private elementary schools, high schools, and universities; to aid in the advancement of health and human services; to address the needs of the disadvantaged, aged, sick, and homeless; and to support and promote the values of William Hannon's faith through support of the Roman Catholic Church. Large grants only are made to institutions and organizations that played an important role in Hannon's life, and the Foundation always keeps his mission and legacy in mind when awarding grants. The Foundation limits its grants to programs primarily in the greater Los Angeles area. Grant proposals are considered at each quarterly Board of Directors' Meeting. Request letters should be received by August 1 for the September meeting; November 1 for the December meeting; February 1 for the March meeting; or May 1 for the June meeting. Applicants will be notified of the decision in writing within thirty days of the quarterly Board Meeting. Site visits may be a part of our evaluation process.
Requirements: Only Internal Revenue Service certified, non-profit public charities are eligible for grants.
Restrictions: Grant requests are not considered for individuals, underwriting parties, travel funds, advertisements, the advancement of political agendas, and radio or TV programming.
Geographic Focus: California
Date(s) Application is Due: May 1; Aug 1; Nov 1
Contact: Kathleen Hannon Aikenhead, President; (310) 260-2470; fax (310) 260-9740; williamhannonfdn@yahoo.com
Internet: http://www.hannonfoundation.org/grantmaking.html
Sponsor: William H. Hannon Foundation
729 Montana Avenue, Suite 5
Santa Monica, CA 90403

William J. and Tina Rosenberg Foundation Grants 3954
The William J. and Tina Rosenberg Foundation was established in Florida in 1970, with giving centered in the Dade County area. Currently, the Foundation's primary fields of interest include: public education; the environment; disadvantaged; social services; cultural programs; museums; and health care. Types of support are: emergency funds; general operating support; and seed money. There are no specific annual deadlines, and interested parties should send a letter of application directly to the Foundation. Most recent awards have ranged from $5,000 to $50,000.
Restrictions: No grant support is given to private schools or individuals.
Geographic Focus: Florida
Amount of Grant: 5,000 - 50,000 USD
Contact: Ruth Admire, Administrator; (305) 444-6121; fax (305) 444-5508; info@sullivanadmire.com or ruth.admire@sullivanadmire.com
Internet: http://www.sullivanadmire.com/charitable.html
Sponsor: William J. and Tina Rosenberg Foundation
255 Ponce de Leon Boulevard, Suite 320
Coral Gables, FL 33134

William J. Brace Charitable Trust 3955
The William J. Brace Charitable Trust was established in 1958 to support and promote quality educational, cultural, human-services, and health-care programming, with a preference for the following three areas: the education and health of children; the health and care of older adults; and hospitals in Kansas City, Missouri. Grant requests for general operating support and program support will be considered. Grants from the Trust are one year in duration. There are no application deadlines for the Brace Charitable Trust. Proposals are reviewed on an ongoing basis. Applicants may download the application and Missouri state guidelines at the grant website. Applicants are strongly encouraged to review the state guidelines for additional helpful information and clarifications on the three areas of interest. The annual deadline for submissions is October 31. Most recent awards have ranged from $8,000 to as high as $50,000.
Restrictions: The Trust generally supports organizations that serve the residents of Kansas City, Missouri. Grant requests for capital support will not be considered.
Geographic Focus: Missouri
Date(s) Application is Due: Oct 31
Amount of Grant: 8,000 - 50,000 USD
Contact: Scott Berghaus; (816) 292-4300 or (816) 292-4301; scott.berghaus@ustrust.com
Internet: https://www.bankofamerica.com/philanthropic/foundation.go?fnId=127
Sponsor: William J. Brace Charitable Trust
1200 Main Street, 14th Floor, P.O. Box 219119
Kansas City, MO 64121-9119

William Ray and Ruth E. Collins Foundation Grants 3956
The Collins Foundation was established by William Ray and Ruth E. Collins for charitable, religious and educational purposes of organizations within Boulder County, Colorado. The Collins' were residents of Boulder, and they were co-owners of a clothing and shoe store. The Foundation's primary fields of interest include: the arts; culture; humanities; education; the environment; animals; health; human services; and religion. The average funding range is $1,000 to $5,000, with approximately thirty awards given each year. Applications are accepted year-round, though they must be submitted by November 30 to be reviewed at the annual grant meeting that occurs in January.
Requirements: Giving is limited to charitable organizations operating in or supporting Boulder County, Colorado. Grantees must be qualified as public charities under section 501(c)3. Applications must be submitted through the online grant application form or alternative accessible application designed for assistive technology users.
Geographic Focus: Colorado
Date(s) Application is Due: Nov 30
Amount of Grant: 1,000 - 5,000 USD
Contact: George Weaver, Special Trustee; (888) 234-1999; fax (877) 746-5889; grantadministration@wellsfargo.com
Internet: https://www.wellsfargo.com/privatefoundationgrants/collins
Sponsor: William Ray and Ruth E. Collins Foundation
1740 Broadway
Denver, CO 80274

Williams Companies Foundation Homegrown Giving Grants 3957
The Williams Companies Foundation Homegrown Giving program is designed to help employees meet the unique needs of their local communities, funding is available to support eligible non-profit organizations where Williams' current employees are involved on a regular basis. The Public Outreach Business Partner determines funding based on program guidelines and within established parameters. Donations range from $100 to $5,000.
Requirements: Nonprofits in Williams operating communities are eligible.
Restrictions: No grants are made to individuals.
Geographic Focus: Alabama, Louisiana, New Jersey, North Carolina, Pennsylvania, South Carolina, Texas, Virginia
Amount of Grant: 100 - 5,000 USD
Contact: Beth Stewart; (918) 573-1190 or (800) 945-5426; communityrelationstulsa@williams.com or williamscommunityoutreach@williams.com
Internet: http://www.williams.com/community/foundation.asp
Sponsor: Williams Companies Foundation
One Williams Center, MD 50-5
Tulsa, OK 74172

William T. Grant Foundation Research Grants 3958
The foundation supports research to improve the lives of youth ages 8 to 25 in the United States. This is done primarily by investing in high quality empirical studies. The foundation's Current Research Interests are understanding and improving social settings (i.e., families, schools, peer groups, organizations, programs, etc.), their effects on youth, and the use and influence of scientific evidence. Within these interests, the foundation is interested in descriptive and intervention studies. There are no geographical boundaries for its support of research projects.
Requirements: The grants are limited, without exception, to established tax-exempt, private organizations and institutions. To apply for a major grant, the Principal Investigator must submit a Letter of Inquiry to the grants coordinator briefly describing the project or program and the financial needs. Letters of Inquiry are due on the deadline dates listed. If it is determined that the project falls within the current program interests and priorities, a full proposal will be requested. Full proposal deadlines are four to six months prior to the quarterly board meetings in March and October.
Restrictions: The foundation does not support or make contributions to building funds, fundraising drives, endowment funds, general operating budgets, or scholarships. Grants are made to organizations or institutions, not individuals.
Geographic Focus: All States
Date(s) Application is Due: Jan 3; Apr 1; Nov 1
Contact: Nancy Rivera-Torres; (212) 752-0071; fax (212) 752-1398; nrivera@wtgrantfdn.org
Internet: http://www.wtgrantfoundation.org/funding_opportunities/research_grants
Sponsor: William T. Grant Foundation
570 Lexington Avenue, 18th Floor
New York, NY 10022-6837

Willis C. Helm Charitable Trust Grants 3959
The Willis C. Helm Charitable Trust was created in 1954 by Katherine B. Helm in memory of her husband, Willis C. Helm. Mrs. Helm's desire was to support charities which do constructive work in the field of juvenile delinquency and development of boys from broken homes, as well as specifically named charities. Those charities include: Berea College, Berea, Kentucky; Piney Woods Country Life School, Piney Woods, Mississippi; Father Flanagans Boys Home, Boys Town, Nebraska; and Vision Loss Resources, Minneapolis, Minnesota. For all other applicants, the average grant will range from $5,000 to $15,000, with the annual number of awards being approximately five. Applications must be submitted by August 31 to be reviewed at the annual grant meeting.
Requirements: Grantees must be qualified as public charities under Internal Revenue section 501(c)3. Applications must be submitted through the online grant application form or alternative accessible application designed for assistive technology users.
Geographic Focus: All States
Date(s) Application is Due: Aug 31
Contact: Jason Craig; (888) 234-1999; grantadministration@wellsfargo.com
Internet: https://www.wellsfargo.com/privatefoundationgrants/helm
Sponsor: Willis C. Helm Charitable Trust
1740 Broadway
Denver, CO 80274-0001

Wilson-Wood Foundation Grants 3960
The Wilson-Wood Foundation established in 1983, supports organizations involved with: adult education; aging, centers/services; children/youth, services; education; health care; housing/shelter, development; human services; nutrition; women, centers/services. Giving is limited to the Manatee-Sarasota, Florida area, for for the underprivileged and the less fortunate in the community.
Requirements: Non-profits in the Manatee-Sarasota, Florida area are eligible to apply. Initial approach should be a phone call to the Foundation prior to submitting a the letter of inquiry (must be received by June 1). There is no application form required. Applicants must submit two copies of the proposal. Application must include the following information: timetable for implementation and evaluation of project; qualifications of key personnel; name, address and phone number of organization; copy of IRS Determination Letter, must be dated within the past 10 years; brief history of organization and description of its mission; copy of most recent annual report/audited financial statement/990; descriptive literature about organization; listing of board of directors, trustees, officers and other key people and their affiliations; detailed description of project and amount of funding requested; copy of current year's organizational budget and/or project budget; listing of additional sources and amount of support.
Restrictions: No support for foreign organizations, supporting organizations, or private foundations. No grants to individuals, or for endowment funds, deficit financing, travel projects, research, fundraising costs, multi-year projects, conferences, emergency funding or start up costs.
Geographic Focus: Florida
Date(s) Application is Due: Jun 1
Amount of Grant: 8,000 - 30,000 USD
Contact: Susan Wood, Executive Director; (941) 966-3635
Sponsor: Wilson-Wood Foundation
930 Scherer Way
Osprey, FL 34229-6867

Windham Foundation Grants 3961
Since its founding, the Windham Foundation has provided more than $10 million in grants to non-profit organizations to serve its mission of promoting Vermont's rural communities. Particular emphasis is given to projects which enhance the unique qualities of Vermont's small town life, support its natural and working landscape, sustain Vermont's social, cultural and natural resources or preserve its history and traditions while enhancing day-to-day community life. Grants are made to nonprofit organizations (501(c)3 with programs active in Vermont. The Foundation assists organizations in the following areas: agriculture and the food systems; disadvantaged youth; environmental enhancement; education (pre-K through college); public policy issues; promotion of the arts, crafts, and Vermont traditions; human services; and historical preservation. Proposals are evaluated on their fit for the Foundation; likelihood of success; fiscal strength; evidence of community support; capital and historical preservation; land conservation and farm viability. There is no fixed limit on the amount of grant requests although most will be within the $5,000 to $10,000 range. Grants are reviewed on a quarterly basis by the Board of Trustees. The committee recommendations are submitted to the Board for final approval at quarterly meetings. Although applications are reviewed quarterly, grant requests may be submitted at any time of the year.
Requirements: The Foundation does not pre-screen applications although it will respond to letters of inquiry or questions directed to the grants administrator. Personal interviews or site visits are not required but may be requested by the Foundation as part of the process. All applications must be submitted electronically. Applicants will be notified by mail of the Foundation's decision 10 to 12 weeks after the deadline.
Restrictions: Funding is not available for endowment campaigns; sporting activities, outings or events; fraternal or religious organizations, including schools with religious affiliation; individual fellowships or scholarships; summer camps, playgrounds or day care facilities, and skate parks unless part of a comprehensive after school program; specific cultural performances; publications or surveys; or affiliates of national organizations focused on particular diseases or those that provide emergency relief efforts.
Geographic Focus: Vermont
Date(s) Application is Due: Feb 15; May 3; Aug 2; Nov 2
Amount of Grant: 5,000 - 10,000 USD
Contact: Becky Nystrom, Executive Assistant; (802) 843-2211, ext. 10; fax (802) 843-2205; info@windham-foundation.net
Internet: http://www.windham-foundation.org/programs/grants.html
Sponsor: Windham Foundation
225 Townshend Road
Grafton, VT 05146

Winifred & Harry B. Allen Foundation Grants 3962
Established in 1963 in honor of Harry B. Allen, president of the Belvedere Land Company, and his wife, the Foundation has as its primary fields of interest: animal and wildlife preservation; education; the environmental programs; environmental education; protection of natural resources; health care; health organizations; human services; immigrants/refugees; performing arts; and visual arts. Though giving is centered around Marin County, California, the Foundation has also supported 501(c)3 organizations in Massachusetts, Georgia, New York, and the District of Columbia. Applicants should contact the Foundation in writing, stating the purpose and offering proof of tax exempt status. Though there are no specific deadlines, the Board meets four times annually, including April 15, June 15, September 15, and December 15.
Restrictions: No grants are given to individuals.
Geographic Focus: California, District of Columbia, Georgia, Massachusetts, New York
Amount of Grant: 500 - 5,000 USD
Contact: Howard B. Allen, (415) 435-2439; fax (415) 435-3166
Sponsor: Winifred and Harry B. Allen Foundation
83 Beach Road, P.O. Box 380
Belvedere, CA 94920-0380

Winn Feline Foundation/AVMF Excellence In Feline Research Award 3963
The Winn Foundation has partnered with the American Veterinary Medical Foundation (AVMF) to recognize the veterinary research scientist whose work mirrors Winn's efforts to improve the lives of "Every Cat, Every Day". The Winn/AVMF Excellence in Feline Research Award is accompanied by a $2,500 cash award, a crystal cat sculpture, plus up to $1,000 to offset round trip airfare and other travel expenses. The Research Award is paired with a matching scholarship award by the AVMF for a veterinary student interested in feline medicine. For nomination forms and additional information related to supporting materials, award criteria, and the selection process, visit www.avma.org/awards.
Requirements: Applicants should include a cover letter, a list of significant research achievements, and a concise curriculum vita (CV) including relevant publications. The cover letter must include the name of the award; the name, mailing address, telephone number(s), email address, and college and year of graduation of the nominee; the nature of the nominee's professional activity (type of practice or type of salaried work); organizational memberships (professional and scientific) of the nominee; a narrative sketch of the nominee's professional background; and a statement pertaining to the nominee's qualifications for the award. To be considered, nominees must have demonstrated Winn's mission to advance feline health and welfare through research; have a pattern of following good research study protocols; complete progress reports in a timely manner; and have their research published and acknowledge Winn as a donor of grants when appropriate.
Restrictions: No self-nominations will be considered.
Geographic Focus: All States
Date(s) Application is Due: Feb 1
Contact: Janet Wolf, Executive Director; (856) 447-9798; winnfeline@aol.com
Eileen Hoblit, Award Contact; (800) 248-2862, ext. 6778; fax (847) 925-1329
Internet: http://www.winnfelinehealth.org/Pages/Researchers.html
Sponsor: Winn Feline Foundation
390 Amwell Road, Suite 402
Hillsborough, NJ 08844

Winn Feline Foundation Grants 3964
The foundation supports research studies into medical problems affecting cats with the Winn Feline Foundation Grant and the Miller Trust Grant. The Foundation supports research on feline asthma, anemia, inflammatory bowel disease, hyperthyroidism, FIP, megacolon, polycystic kidney disease, transdermal drug delivery, liver disease, and vaccine-associated fibrosarcoma. Studies general in scope are primarily encouraged; however, the foundation is also interested in projects that address problems unique to individual breeds. Continuation of grants awarded will be considered. Application guidelines and current deadlines are available online.
Requirements: Applicants may be faculty veterinarians, postdoctoral fellows, practicing veterinarians, or veterinary students.
Restrictions: The foundation does not fund salaries of principal investigators, major equipment expenditures, travel or indirect costs.
Geographic Focus: All States
Contact: Janet Wolf; (908) 359-1184; fax (908) 359-7619; winn@winnfelinehealth.org
Internet: http://www.winnfelinehealth.org
Sponsor: Winn Feline Foundation
390 Amwell Road, Suite 402
Hillsborough, NJ 08844

Winston-Salem Foundation Competitive Grants 3965
The foundation makes competitive grant awards to tax-exempt, nonprofit agencies in the greater Forsyth County, North Carolina, area. Its charitable purposes include: public interest, education and recreation, health, human services, arts and culture, youth, and older adults. Grants ordinarily are made for proposals that initiate, expand, or improve direct services to people and for endowment purposes under certain conditions. Additional types of support include program development, seed grants, scholarship funds, technical assistance, employee matching gifts, scholarships to individuals, and matching funds. Except in unusual cases, grants will be approved for one year at a time. Potential applicants should contact the office for an appointment to introduce the proposed project.
Requirements: Organizations in Davidson, Davie, Forsyth, Surry, Stokes, Wilkes, and Yadkin counties in North Carolina are eligible to apply.
Restrictions: Grants for major equipment purposes and for capital campaigns in support of local community facilities are of low priority.
Geographic Focus: North Carolina
Contact: Brittney Gaspari, Grants Director; (336) 725-2382; fax (336) 727-0581; bgaspari@wsfoundation.org or info@wsfoundation.org
Internet: http://www.wsfoundation.org/grant-seekers/types-of-grants/competitive-grants/
Sponsor: Winston-Salem Foundation
860 W Fifth Street
Winston-Salem, NC 27101-2506

Winston-Salem Foundation Elkin/Tri-County Grants 3966
The Funds consist of three component trusts established by Richard T. Chatham, Lucy Hanes Chatham, and citizens of Elkin to benefit the community. Ordinarily, grants are made for projects that initiate, expand or improve direct service to people. Applicants should send a two or three page letter describing the need for the project, its goals and objectives, its cost and the portion of the cost requested.

Requirements: A list of board members is required, as well as organizational and project budgets, and evidence of non-profit 501(c)3 tax-exempt status. The chief board officer must sign the proposal or write an endorsement cover letter.
Restrictions: Grants will be made only to legally recognized non-profit organizations and educational institutions in Wilkes, Surry, and Yadkin Counties. Those that benefit residents of the Elkin-Jonesville-Dobson-Roaring River radius will be of priority interest to the advisory committee. Grants will not be made for on-going operating expenses. Foundation funds will not be granted to supplant tax or government funds for projects or institutions that would ordinarily receive public funding support.
Geographic Focus: North Carolina
Date(s) Application is Due: Jun 15
Amount of Grant: 500 - 7,000 USD
Samples: Communities in Schools of Wilkes, $3,500; Hugh Chatham Memorial Hospital Foundation, $7,000; Mountain Valley Hospice and Palliative Care, $2,500.
Contact: Brittney Gaspari, Grants Director; (336) 725-2382; fax (336) 727-0581; bgaspari@wsfoundation.org or info@wsfoundation.org
Internet: http://www.wsfoundation.org/grant-seekers/types-of-grants/elkintri-county-grants/
Sponsor: Winston-Salem Foundation
860 W Fifth Street
Winston-Salem, NC 27101-2506

Winston-Salem Foundation Stokes County Grants   3967
The Trust provides grants to non-profit organizations or informal groups in Stokes County for worthy public and charitable purposes with an emphasis on developing leaders and inspiring others. Leadership can be either formal and traditional, or demonstrated in less traditional ways that are creative and innovative, including leading by example. The trust is designed both to work with established organizations and to encourage new initiatives. Special attention will be given to encouraging leadership in certain focus areas: education; arts; environmental protection and recreational use; historical preservation; local government and community services; health; and organizations and issues affecting minorities and low-resource communities.
Requirements: Applicants can be informal groups of people with innovative ideas, including those that might grow into nonprofit organizations; or established organizations with 501(c)3 tax-exempt status. The Advisory Board will also consider letters from small groups of people with good ideas who need guidance about turning them into formal grant proposals in the future.
Restrictions: Grants will not be made for projects or institutions that would ordinarily receive public funding. However, public institutions such as schools or parks may submit proposals for innovative projects for which funds are not ordinarily provided.
Geographic Focus: North Carolina
Date(s) Application is Due: Sep 11
Amount of Grant: Up to 3,000 USD
Contact: Brittney Gaspari, Grants Director; (336) 725-2382; fax (336) 727-0581; bgaspari@wsfoundation.org or info@wsfoundation.org
Internet: http://www.wsfoundation.org/grant-seekers/types-of-grants/stokes-county-grants/
Sponsor: Winston-Salem Foundation
860 W Fifth Street
Winston-Salem, NC 27101-2506

Winston-Salem Foundation Victim Assistance Grants   3968
The fund provides financial assistance to victims of violent crime to help minimize complications from the event. The fund also provides grants for statewide organizations that provide information in a supportive manner to victims of violent crime. Victims of violent crime such as rape or assault are eligible for assistance from the fund. The fund is also available to assist surviving family members of homicide victims and victims of violent crime resulting from domestic violence. The victim can receive monetary support for expenses such as the following: relocation expenses; locks, alarms, or repair of damages from break-ins; clothes, especially for rape victims when clothes are retained as evidence; transportation to counseling visits; lost wages that may be reimbursed to the Fund at a later date by North Carolina Crime Victims Compensation; and help with payment of bills that are late because of the consequences of the crime.
Requirements: Victims should be residents of Forsyth, Davie, or Stokes counties (or the area being served by Family Services, Inc.).
Restrictions: The fund is not for assistance to organizations serving victims (except, as noted in the purpose, for nonprofit statewide organizations that provide information in a supportive manner to victims of violent crime).
Geographic Focus: North Carolina
Contact: Betty Gray Davis, (336) 725-2382; fax (336) 727-0581; bgdavis@wsfoundation.org
Internet: http://www.wsfoundation.org/grant-seekers/types-of-grants/victim-assistance-grants/
Sponsor: Winston-Salem Foundation
860 W Fifth Street
Winston-Salem, NC 27101-2506

Wolfe Associates Grants   3969
Wolfe Associates Grants are made in the following fields: health and medicine; religion; education; culture, community service, youth skills development, and business. Awards range from $500 to $25,000.
Requirements: Applicants must be 501(c)3 organizations. Funding is primarily made to, but is not limited to, Ohio organizations. Application is made by sending a cover with a brief summary of the request, the amount requested, a copy of the organization's Internal Revenue Service determination letter and most recent form 990, and financial statements. There are no submission deadlines.
Geographic Focus: Ohio

Contact: Rita J. Wolfe, Vice President; (614) 460-3782
Sponsor: Wolfe Associates
34 S Third Street
Columbus, OH 43215

Women of the ELCA Opportunity Scholarships for Lutheran Laywomen   3970
The purpose of these scholarships is to provide assistance to women studying for a career other than the ordained ministry. The following scholarships are offered: Amelia Kemp Scholarship-- for ELCA women of color in undergraduate, graduate, professional, or vocational courses of study; Belmer, Flora Prince Scholarships--for ELCA women studying for ELCA service abroad; Kahler, Vickers/Raup, Emma Wettstein Scholarships--for ELCA women studying for service in health professions associated with ELCA projects abroad; Irene Drinkall Franke, Mary Seeley Knudstrup Scholarships--for ELCA women in graduate courses of study preparing for occupations in Christian service; and Cronk Memorial, First Triennium Board, General, Mehring, Paepke, Piero/Wade/Wade, Edwin, and Edna Robeck Scholarships--for ELCA women in undergraduate, graduate, professional, or vocational courses of study.
Requirements: The applicant must be at least 21 years old, be a U.S. citizen, hold membership in the Evangelical Lutheran Church in America, and have experienced an interruption in education of two or more years since completion of high school. Scholarship funds must be used within the year following the award.
Geographic Focus: All States
Amount of Grant: 800 - 1,000 USD
Contact: Valora Starr; (800) 638-3522, ext. 2741; Valora.Starr@elca.org
Internet: http://www.elca.org/Growing-In-Faith/Ministry/Women-of-the-ELCA/Engage-in-action-and-support-one-another-in-our-callings/Scholarships/For-Lutheran-Laywomen.aspx
Sponsor: Evangelical Lutheran Church in America
8765 W Higgins Road
Chicago, IL 60631-4189

Wood-Claeyssens Foundation Grants   3971
The Foundation awards grants to eligible California nonprofit organizations. Types of support include annual campaigns, capital campaigns, continuing support, and general operating support. An application is available on the website.
Requirements: California 501(c)3 organizations serving Santa Barbara and Ventura Counties are eligible.
Restrictions: Funding is not available to individuals or to organizations that discriminate on the basis of age, gender, race, ethnicity, sexual orientation, disability, national origin, political affiliation or religious belief.
Geographic Focus: California
Date(s) Application is Due: Jun 30
Contact: Noelle Claeyssens Burkey; (805) 966-0543; wcf0543@gmail.com
Internet: http://www.woodclaeyssensfoundation.com/Funding.htm
Sponsor: Wood-Claeyssens Foundation
P.O. Box 30586
Santa Barbara, CA 93130-0586

Wood Family Charitable Trust Grants   3972
The Wood Family Charitable Trust was established in 2007 to support and promote quality educational, human-services, and health-care programming for underserved populations. The Wood Charitable Trust specifically serves the people of Vale, Oregon and its surrounding communities. Grant requests for general operating support are strongly encouraged. Program support will also be considered. Small, program-related capital expenses may be included in general operating or program requests. The majority of grants from the Wood Charitable Trust are one year in duration; on occasion, multi-year support is awarded. Application materials are available for download at the grant website. Applicants are strongly encouraged to review the state application guidelines for additional helpful information and clarifications before applying. Applicants are also encouraged to review the trust's funding history (link is available from the grant website). The application deadline for the Wood Family Charitable Trust is March 15. Applicants will be notified of grant decisions by May 15.
Requirements: The Wood Charitable Trust specifically serves the people of Vale, Oregon and its surrounding communities. Applicants must have 501(c)3 tax-exempt status.
Restrictions: The trust does not support requests from individuals, organizations attempting to influence policy through direct lobbying, or any political campaigns.
Geographic Focus: Oregon
Date(s) Application is Due: Mar 15
Contact: Cindy Keyser, Vice President; (800) 848-7177; cindy.s.keyser@baml.com
Internet: https://www.bankofamerica.com/philanthropic/fn_search.action
Sponsor: Wood Family Charitable Trust
800 5th Avenue, WA1-501-33-23
Seattle, WA 98104

Woodward Fund Grants   3973
Grants support nonprofit institutions, corporations, and associations that are located in Georgia or one of its neighboring states and are organized and operated exclusively for religious, educational, and charitable and scientific purposes. Grantmaking focuses on capital projects. The distribution committee makes its decisions based on materials provided in the grant request. The trust does not require a formal application form, but requests a proposal outlining the project or program and containing the following information: project/program description; complete itemized project budget, including project schedule; need for the project; other funding sources, including the amount received from each source; names and qualifications of those conducting the project; objectives and how they will be achieved;

a brief description of the applicant organization; method and criteria for project evaluation; list of officers, board of directors, or trustees; copy of the 501(c)3 designation letter; and plans for recognition of the trust in the project. Applicants should not contact committee members personally concerning a proposal or anticipated approach to the fund.
Requirements: 501(c)3 nonprofit organizations in Georgia and its neighboring states are eligible. Government agencies to which contributions by individuals are made deductible from income by IRS laws also are eligible.
Restrictions: Grants are not awarded to individuals or for scholarships or student loans.
Geographic Focus: Alabama, Florida, Georgia, North Carolina, South Carolina, Tennessee
Date(s) Application is Due: May 1; Nov 1
Amount of Grant: 5,000 - 300,000 USD
Contact: Alice Sheets; grantinquiries8@wachovia.com
Internet: https://www.wachovia.com/charitable_services/woodward_overview.asp
Sponsor: David, Helen, and Marian Woodward Fund
3280 Peachtree Road NE, Suite 400, MC G0141-041
Atlanta, GA 30305

World of Children Health Award 3974
The World of Children Awards program was created to recognize and elevate those selfless individuals who make a difference in the lives of children here in the USA and across the globe, regardless of political, religious or geographical boundaries. These courageous leaders recognize that our children are the world's most important asset. The Health Award, in the amount of up to $50,000, recognizes individuals making extraordinary contributions to children through the fields of health, medicine, or the sciences. The Award honors this courageous leader at an annual Awards Ceremony and grants them funds to elevate their work. The annual deadline for nominations is April 1.
Requirements: Nominations must be submitted in English. Nominees for the Award must have; created, managed or otherwise supported a sustainable program which has significantly contributed to the improved health of children; do this work over and above their normal employment, or work for little or no pay; have been doing this for a minimum of 10 years; and have an existing non-profit organization in good standing, which can receive grant funds if awarded.
Restrictions: Organizations are not eligible; however a nominee or group of nominees may be part of an organization.
Geographic Focus: All States
Date(s) Application is Due: Apr 1
Amount of Grant: 50,000 USD
Contact: Lynn Wallace Naylor, Executive Director; (925) 452-8272; fax (925) 452-8229; lynn@worldofchildren.org or contact@worldofchildren.org
Internet: http://www.worldofchildren.org/theaward/awards-we-give/health-award/
Sponsor: World of Children
11501 Dublin Boulevard, Suite 200
Dublin, CA 94568

World of Children Humanitarian Award 3975
The World of Children Awards program was created to recognize and elevate those selfless individuals who make a difference in the lives of children here in the USA and across the globe, regardless of political, religious or geographical boundaries. These courageous leaders recognize that our children are the world's most important asset. The Humanitarian Award, with a maximum of $50,000, recognizes an individual who has made a significant contribution to children in the areas of social services, education or humanitarian services. The Award honors this humanitarian leader at an annual Awards Ceremony and grants them funds to elevate their work. The annual deadline for nominations is April 1.
Requirements: Nominations must be submitted in English. The nominee must: have created, managed or otherwise supported a sustainable program which has significantly contributed to children's opportunities to be safe, to learn, and to grow; do this work over and above their normal employment, OR work for little or no pay; have been doing this for a minimum of 10 years; and have an existing non-profit organization in good standing, which can receive grant funds if awarded.
Restrictions: Organizations are not eligible; however a nominee or group of nominees may be part of an organization.
Geographic Focus: All States
Date(s) Application is Due: Apr 1
Amount of Grant: 50,000 USD
Contact: Lynn Wallace Naylor, Executive Director; (925) 452-8272; fax (925) 452-8229; lynn@worldofchildren.org or contact@worldofchildren.org
Internet: http://www.worldofchildren.org/theaward/awards-we-give/humanitarian-award/
Sponsor: World of Children
11501 Dublin Boulevard, Suite 200
Dublin, CA 94568

Xoran Technologies Resident Research Grant 3976
The purpose of this grant is to stimulate original resident research in otolaryngology projects that are well-conceived and scientifically valid, with potential to clarify the role of head and neck computed tomography in diagnosis or treatment. Awards are for one year, are non renewable, and have a $10,000 maximum limit. Generally, a single award is made available annually. Letters of Intent are due on December 15, with full proposals due by January 15.
Requirements: Application is open to residents of an accredited otolaryngolgoy-head and neck surgery training program in the U.S. and Canada for projects that are well-conceived and scientifically valid, with potential to clarify the role of the head and neck computed tomography in diagnosis or treatment.
Geographic Focus: All States
Date(s) Application is Due: Jan 15; Dec 15
Amount of Grant: Up to 10,000 USD
Contact: Stephanie L. Jones; (703) 519-1586 or (703) 836-4444; sljones@entnet.org
Internet: http://www.entnet.org/EducationAndResearch/coreGrants.cfm
Sponsor: Xoran Technologies
One Prince Street
Alexandria, VA 22314-3357

Yampa Valley Community Foundation Cody St. John Scholarships 3977
The Foundation supports programs benefiting the Yampa Valley community. The Cody St. John Scholarships are administered by the Foundation to assist professional ski patrollers in furthering their medical education. This is a new Scholarship and details are still being worked out. Applicants may call the Foundation for further information.
Geographic Focus: All States
Contact: Jennifer Shea, Program Manager; (907) 879-8632; jennifer@yvcf.org
Internet: http://www.yvcf.org/scholarship-apply.php
Sponsor: Yampa Valley Community Foundation
465 Anglers Drive, Suite 2-G
Steamboat Springs, CO 80488

Yampa Valley Community Foundation Grants 3978
The Foundation supports innovative programs benefiting the Yampa Valley community. Funding is provided for proposals that: promote the mission of the Foundation; serve either a broad or underserved population in the Yampa Valley; demonstrate the anticipated impact in the Yampa Valley; define the measurement and evaluation process to be used; effectively leverage the Foundation's resources; and exhibit sound business and financial practices. Grants may be awarded for innovative programs demonstrating progress toward achieving community goals in five focus areas: arts and culture, education, environment, health and human services, and recreation. Initially a letter of intent, providing an overview of the proposed project or program and the funding requested, should be submitted. Selected applicants will be invited to submit a full proposal.
Requirements: Eligible organizations must be a nonprofit 501(c)3 organization or fiscally sponsored by a qualifying organization in Routt or Moffat Counties that benefit the Yampa Valley. Only grant requests for charitable purposes that have a public benefit are considered.
Restrictions: The Foundation does not grant for debt reduction, endowments, political purposes or religious purposes. Proposals are typically declined for individual or professional development, team or travel expenses, retro-active grants for projects already completed or in process, or 100% of funding for a project. Funded programs and services may not include any political or religious intentions.
Geographic Focus: Colorado
Date(s) Application is Due: Jun 1
Samples: Advocates Building Peaceful Communities, Steamboat Springs, Colorado, $6,550 - general support; Healthcare Foundation for the Yampa Valley, Steamboat Springs, Colorado, $11,250 - CT scanner challenge match; and South Routt School District Oak Creek, Colorado, $10,000 - community greenhouse.
Contact: Jennifer Shea, Program Manager; (970) 879-8632; fax (970) 871-0431; jennifer@yvcf.org or nfo@yvcf.org
Internet: http://www.yvcf.org/grants.php
Sponsor: Yampa Valley Community Foundation
465 Anglers Drive, Suite 2-G
Steamboat Springs, CO 80488

Yampa Valley Community Foundation Volunteer Firemen Scholarships 3979
The Foundation supports programs benefiting the Yampa Valley community. The Steamboat Springs Volunteer Firemen sponsor two scholarships: the Rusty Chandler Memorial Scholarship and the Dave Linner Memorial Scholarship. The Rusty Chandler Memorial Scholarship is awarded to a high school senior and emphasizes community service. The Dave Linner Memorial Scholarship is awarded to a high school senior entering the EMS or medical related fields. Applications are available on the website.
Requirements: Applicants must be seniors graduating from Steamboat Springs High and have attended Steamboat Springs High School for at least two full semesters
Geographic Focus: Colorado
Date(s) Application is Due: Apr 15
Contact: Jennifer Shea, Program Manager; (907) 879-8632; jennifer@yvcf.org
Internet: http://www.yvcf.org/scholarship-apply.php
Sponsor: Yampa Valley Community Foundation
465 Anglers Drive, Suite 2-G
Steamboat Springs, CO 80488

Yawkey Foundation Grants 3980
The Yawkey Foundation is committed to continuing the legacy of Tom and Jean Yawkey by making significant and positive impacts on the quality of life of children, families, and the underserved in the areas of New England and Georgetown County, South Carolina. Funding supports the areas of health care, education, human services, youth and amateur athletics, arts and culture, and conservation and wildlife. Request should be limited to $25,000 unless otherwise directed. Applications are currently accepted only from organizations previously funded by the Foundation. Deadlines are: arts and culture, conservation, and health care March 1; human services June 15; education September 1; and youth and amateur athletics November 15.
Requirements: Eligible applicants must be tax-exempt 501(c)3 organizations. Proposals must provide significant benefits to a broad constituency either in New England or Georgetown County, South Carolina. The Foundation has a particular concern for organizations that serve disadvantaged children and families and also considers the following: relevance of the

proposed project or program to the Foundation's areas of interest; need outlined in the proposal and how the organization has and will continue to address that need; the organization's fiscal health and ability to manage its resources effectively; ability of the project or program to leverage funding and support from other sources; ability of the organization and its staff to achieve the desired results; adequacy of proposed activities, budget, and timetable to achieve the desired results; and evidence of appropriate cooperation with other organizations.
Restrictions: All final reports for prior funding must be submitted before applying for additional funding. Only one request may be submitted during a twelve-month period. Organizations that have received three or more years of consecutive funding will not be eligible to reapply for funding for a one-year period. The Foundation does not make grants to: organizations that are not tax-exempt under section 501(c)3 of the Internal Revenue Service; individuals; private foundations; 509(a)3 Type III non-functionally integrated organizations; organizations or programs that provide benefits outside of the United States; legislative lobbying; foundations created by political or governmental or for-profit organizations; political campaigns and causes; government agencies, or agencies directly benefiting public entities; public school districts and public schools (including charter schools); community and economic development corporations or programs; advocacy groups; operating deficits or retirement of debt; general endowments; general capital campaigns; events, conferences, seminars, and group travel; awards, prizes, and monuments; fraternal, trade, civic, or labor organizations; music, video, or film production; feasibility or research studies; pass-through, intermediary organizations or foundations; religious organizations for sectarian purposes; and workforce development programs.
Geographic Focus: Connecticut, Maine, Massachusetts, New Hampshire, Rhode Island, South Carolina, Vermont
Contact: Nancy Brodnicki, Grants Program Administrator; (781) 329-7470
Internet: http://www.yawkeyfoundation.org/grant_guidelines.html
Sponsor: Yawkey Foundation
990 Washington Street
Dedham, MA 02026-6716

Young Ambassador Scholarship In Memory of Christopher Nordquist 3981
The annual $1,000 award to an eligible student is made possible through private contributions given to The Eye-Bank in memory of Christopher, who was two when he died and left the gift of sight. Through the years, his mother Andrea Nordquist has shared the story of their family's loss and subsequent decision to donate Christopher's corneas in hopes of increasing awareness of the good that eye donation can do. In this way, Christopher became one of The Eye-Bank's first Young Ambassadors. The Scholarship in his memory is intended to encourage young people to pursue learning and to help spread the message about the priceless gift almost anyone can leave after death – the gift of sight through eye donation. The scholarship award can be used for continuing education at an accredited university, college, trade or technical school certificate program.
Requirements: To apply, students must meet all of the following *Requirements:* Reside within The Eye-Bank's service area (New York City, Nassau, Suffolk, Westchester, Rockland, Duchess or Putnam Counties of New York State); and, be a student under 25 years of age who will be, or is presently enrolled, in a 2 or 4-year college, university, trade or technical school for the current academic year at the time of application deadline.
Geographic Focus: New York
Date(s) Application is Due: Jun 1
Amount of Grant: 1,000 - 1,000 USD
Samples: Jessi-Ann Bettcher-2009; Ian Phillips-2007; Andrew Perez-2006
Contact: Andrea Nordquist
Internet: http://www.eyedonation.org/scholarship_program.html
Sponsor: Eye-Bank for Sight Restoration
120 Wall Street
New York, NY 10005-3902

Z. Smith Reynolds Foundation Small Grants 3982
The Z. Smith Reynolds Foundation now offers a Small Grants Process for grant requests of up to $35,000 per year for up to two years. Small grants are made in the areas of community building and economic development, the environment, governance, public policy and civic engagement, pre-collegiate education, and social justice or equity. In addition to funding projects that achieve the goals of each focus area, the foundation has an interest in building the capacity of organizations and in promoting organizational development. New programs, rather than those that are well-established and well-funded, receive priority consideration. Types of support include operating budgets, continuing support, annual campaigns, seed grants, matching funds, projects/programs, conferences and seminars, and technical assistance.
Requirements: The foundation makes grants only to nonprofit, tax-exempt, charitable organizations and institutions in North Carolina.
Restrictions: The foundation does not give priority to: the arts; capital campaigns; computer hardware or software purchases; conferences, seminars, or symposiums; crisis intervention programs; fund raising events; historic preservation; local food banks; or substance abuse treatment programs.
Geographic Focus: North Carolina
Date(s) Application is Due: Aug 3
Amount of Grant: Up to 35,000 USD
Contact: Leslie Winner, Executive Director; (800) 443-8319, ext. 105 or (336) 725-7541, ext. 105; fax (336) 725-6069; lwinner@zsr.org or info@zsr.org
Internet: http://www.zsr.org/small_grants.htm
Sponsor: Z. Smith Reynolds Foundation
147 South Cherry Street, Suite 200
Winston-Salem, NC 27101-5287

Zane's Foundation Grants 3983
The Zane's Foundation, situated in northeast Ohio, was established in memory of the Youssef family's youngest son, who had severe disabilities. Funding is aimed at assisting families who have children with special needs. Though there are no specified annual deadlines or amount limits, an online application process is utilized. All types of programs and support will be considered, including equipment, summer camp, medical needs, and family assistance.
Requirements: Applicants must reside in the northeast Ohio region.
Geographic Focus: Ohio
Amount of Grant: Up to 2,500 USD
Contact: Stacy Youssef, President; (330) 677-9263
Internet: http://zanesfoundation.org/site/
Sponsor: Zane's Foundation
P.O. Box 1642
Stow, OH 44224

Zellweger Baby Support Network Grants 3984
This organization was started by a small group of parents whose lives have been affected in some way by a rare disorder. The Network provides support to families dealing with Zellweger syndrome. Its mission is to promote, advance, and improve awareness of Zellweger syndrome and other peroxisomal disorders, to assist, support, and aid, financially or otherwise, individuals and families affected by Zellweger syndrome. Grants are given to individuals. Grant amounts range from $100 to $1,000, and applications are available at the Foundation website.
Geographic Focus: All States
Amount of Grant: 100 - 1,000 USD
Contact: Pam Freeth, President; (605) 645-2983
Internet: http://www.zbsn.org/
Sponsor: Zellweger Baby Support Network
530 W. Jackson Boulevard
Spearfish, TX 57783

ZYTL Foundation Grants 3985
Established in Colorado in 1998, the ZYTL Foundation offers funding primarily in its home state. The Foundation's major fields of interest include: health services; human services; and religious programs. The type of funding given is always general operating support, and typically ranges from $100 up to a maximum of $5,000. A formal application is required, and can be secured by contacting the Foundation directly. There are no specified annual deadlines for the submission of application materials.
Restrictions: Grant awards are nor given directly to individuals.
Geographic Focus: Colorado
Amount of Grant: 100 - 5,000 USD
Samples: Lighthouse Pregnancy Center, Denver, Colorado, $1,500 - general operating support; Marriage Missionaries, Parker, Colorado, $2,000 - general operating support; Focus, Golden, Colorado, $5,000 - general operating support.
Contact: Janice K. Zapapas, President; (303) 770-1974
Sponsor: ZYTL Foundation
7975 S. Eudora Circle
Centennial, CO 80122

Subject Index

AIDS
25th Anniversary Foundation Grants, 8
Abbott Fund Global AIDS Care Grants, 202
Abbott Fund Science Education Grants, 203
Actors Fund HIV/AIDS Initiative Grants, 283
AEC Trust Grants, 315
Aid for Starving Children International Grants, 368
AIDS Vaccine Advocacy Coalition (AVAC) Fund Grants, 369
Alexander Fndn Insurance Continuation Grants, 422
Alexis Gregory Foundation Grants, 426
AMA Foundation Seed Grants for Research, 494
amfAR Fellowships, 567
amfAR Global Initiatives Grants, 568
amfAR Mathilde Krim Fellowships in Basic Biomedical Research, 569
amfAR Public Policy Grants, 570
amfAR Research Grants, 571
Ann and Robert H. Lurie Family Fndn Grants, 593
ASM/CDC Fellowships in Infectious Disease and Public Health Microbiology, 716
ASM Merck Irving S. Sigal Memorial Awards, 739
ASPH/CDC Allan Rosenfield Global Health Fellowships, 780
Assurant Foundation Grants, 788
Blowitz-Ridgeway Foundation Grants, 913
Bodenwein Public Benevolent Foundation Grants, 928
Boston Foundation Grants, 936
Bristol-Myers Squibb Foundation Fellowships, 956
Bristol-Myers Squibb Foundation Global HIV/AIDS Initiative Grants, 957
Bristol-Myers Squibb Foundation Health Disparities Grants, 958
Cable Positive's Tony Cox Community Grants, 992
CDC Epidemic Intell Service Training Grants, 1055
CDC Epidemiology Elective Rotation, 1056
Charles Delmar Foundation Grants, 1115
Children Affected by AIDS Foundation Camp Network Grants, 1144
Children Affected by AIDS Foundation Domestic Grants, 1145
Children Affected by AIDS Foundation Family Assistance Emergency Fund Grants, 1146
Community Fndn AIDS Endowment Awards, 1237
Community Foundation for Greater Atlanta AIDS Fund Grants, 1239
Community Fndn for the Capital Region Grants, 1259
Community Foundation of Louisville AIDS Project Fund Grants, 1289
Cooper Industries Foundation Grants, 1345
Coors Brewing Corporate Contributions Grants, 1346
Cornerstone Foundation of Northeastern Wisconsin Grants, 1347
Cowles Charitable Trust Grants, 1354
Dade Community Foundation Community AIDS Partnership Grants, 1409
Dade Community Foundation Grants, 1410
Dallas Women's Foundation Grants, 1417
Danellie Foundation Grants, 1421
David Geffen Foundation Grants, 1431
DIFFA/Chicago Grants, 1477
Doris Duke Charitable Foundation Clinical Scientist Development Award, 1494
Doris Duke Charitable Foundation Operations Research on AIDS Care and Treatment in Africa (ORACTA) Grants, 1497
Duchossois Family Foundation Grants, 1518
Dyson Foundation Mid-Hudson Valley Project Support Grants, 1534
Eddie C. & Sylvia Brown Family Fndn Grants, 1548
Elizabeth Glaser Int'l Leadership Awards, 1566
Elizabeth Glaser Scientist Award, 1567
Elton John AIDS Foundation Grants, 1585
Emily Davie and Joseph S. Kornfeld Foundation Grants, 1591
Ensworth Charitable Foundation Grants, 1595
Firelight Foundation Grants, 1667
Frances L. & Edwin L. Cummings Fund Grants, 1717

Frank E. and Seba B. Payne Foundation Grants, 1722
Gates Award for Global Health, 1775
Gill Foundation - Gay and Lesbian Fund Grants, 1821
Green Bay Packers Foundation Grants, 1890
Hagedorn Fund Grants, 1934
Hasbro Children's Fund Grants, 1973
Horizon Foundation for New Jersey Grants, 2048
HRSA Resource and Technical Assistance Center for HIV Prevention and Care for Black Men who Have Sex with Men Cooperative Agreement, 2069
HRSA Ryan White HIV AIDS Drug Assistance Grants, 2070
Indiana AIDS Fund Grants, 2155
Ittleson Foundation AIDS Grants, 2166
Jerome Robbins Foundation Grants, 2223
John M. Lloyd Foundation Grants, 2244
Johnson & Johnson Corporate Contributions, 2253
Johnson & Johnson/SAH Arts & Healing Grants, 2254
M-A-C AIDS Fund Grants, 2400
M. Bastian Family Foundation Grants, 2402
McCarthy Family Foundation Grants, 2490
Meta and George Rosenberg Foundation Grants, 2525
NIA AIDS and Aging: Behavioral Sciences Prevention Research Grants, 2747
NIA Medical Management of Older Patients with HIV/AIDS Grants, 2770
NIDDK Transmission of Human Immunodeficiency Virus (HIV) In Semen, 2851
NIMH AIDS and Aging: Behavioral Sciences Prevention Research Grants, 2915
NINR AIDS and Aging: Behavioral Sciences Prevention Research Grants, 2927
NYCT AIDS/HIV Grants, 3048
NYCT Girls and Young Women Grants, 3053
Oppenstein Brothers Foundation Grants, 3078
OSF International Harm Reduction Development Program Grants, 3090
OSF International Palliative Care Grants, 3091
Paul Rapoport Foundation Grants, 3125
Pfizer Healthcare Charitable Contributions, 3180
Pittsburgh Foundation Medical Research Grants, 3231
Playboy Foundation Grants, 3237
Puerto Rico Community Foundation Grants, 3278
Quantum Corporation Snap Server Grants, 3289
Questar Corporate Contributions Grants, 3291
Rhode Island Foundation Grants, 3362
Rockefeller Brothers Fund Pivotal Places Grants: South Africa, 3395
Ruth Anderson Foundation Grants, 3425
S. Mark Taper Foundation Grants, 3449
San Diego HIV Funding Collaborative Grants, 3473
San Francisco Fndn Community Health Grants, 3478
United Hospital Fund of New York Health Care Improvement Grants, 3797
United Methodist Committee on Relief Global AIDS Fund Grants, 3798
USAID Comprehensive District-Based Support for Better HIV/TB Patient Outcomes Grants, 3807
USAID HIV Prevention with Key Populations - Mali Grants, 3816
USAID Microbicides Research, Development, and Introduction Grants, 3819
USAID Systems Strengthening for Better HIV/TB Patient Outcomes Grants, 3824
USCM HIV/AIDS Prevention Grants, 3827
USDD Broad Agency Announcement Grants, 3837
USDD Broad Agency Announcement HIV/AIDS Prevention Grants, 3838
USDD HIV/AIDS Prevention: Military Specific HIV/AIDS Prevention, Care, and Treatment for Non-PEPFAR Funded Countries Grants, 3841
USDD HIV/AIDS Prevention Program Information Systems Development Grants, 3842
USDD President's Emergency Plan for AIDS Relief Military Specific HIV Seroprevalence and Behavioral Epidemiology Risk Suvey Grants, 3845
Victor E. Speas Foundation Grants, 3868

AIDS Counseling
Actors Fund HIV/AIDS Initiative Grants, 283
AIDS Vaccine Advocacy Coalition Fund Grants, 369
Ann and Robert H. Lurie Family Fndn Grants, 593
Children Affected by AIDS Foundation Domestic Grants, 1145
Community Fndn AIDS Endowment Awards, 1237
Community Foundation for Greater Atlanta AIDS Fund Grants, 1239
Community Fndn for the Capital Region Grants, 1259
Community Foundation of Louisville AIDS Project Fund Grants, 1289
Dade Community Foundation Community AIDS Partnership Grants, 1409
DHHS Adolescent Family Life Demonstration Projects, 1472
DIFFA/Chicago Grants, 1477
Dyson Foundation Mid-Hudson Valley Project Support Grants, 1534
Eddie C. & Sylvia Brown Family Fndn Grants, 1548
Elton John AIDS Foundation Grants, 1585
HRSA Resource and Technical Assistance Center for HIV Prevention and Care for Black Men who Have Sex with Men Cooperative Agreement, 2069
HRSA Ryan White HIV AIDS Drug Assistance Grants, 2070
M-A-C AIDS Fund Grants, 2400
Ms. Foundation for Women Health Grants, 2597
NIA Medical Management of Older Patients with HIV/AIDS Grants, 2770
NYCT AIDS/HIV Grants, 3048
OSF International Palliative Care Grants, 3091
Rockefeller Brothers Fund Pivotal Places Grants: South Africa, 3395
San Diego HIV Funding Collaborative Grants, 3473
United Methodist Committee on Relief Global AIDS Fund Grants, 3798
USAID HIV Prevention with Key Populations - Mali Grants, 3816
USAID Microbicides Research, Development, and Introduction Grants, 3819

AIDS Education
Abbott Fund Science Education Grants, 203
Actors Fund HIV/AIDS Initiative Grants, 283
Aid for Starving Children International Grants, 368
AIDS Vaccine Advocacy Coalition (AVAC) Fund Grants, 369
Alexis Gregory Foundation Grants, 426
Alfred E. Chase Charitable Foundation Grants, 430
amfAR Fellowships, 567
amfAR Global Initiatives Grants, 568
amfAR Mathilde Krim Fellowships in Basic Biomedical Research, 569
amfAR Public Policy Grants, 570
amfAR Research Grants, 571
Cable Positive's Tony Cox Community Grants, 992
Community Fndn AIDS Endowment Awards, 1237
Community Foundation for Greater Atlanta AIDS Fund Grants, 1239
Community Fndn for the Capital Region Grants, 1259
Community Foundation of Louisville AIDS Project Fund Grants, 1289
Dade Community Foundation Community AIDS Partnership Grants, 1409
DIFFA/Chicago Grants, 1477
Dyson Foundation Mid-Hudson Valley Project Support Grants, 1534
Eddie C. & Sylvia Brown Family Fndn Grants, 1548
Elton John AIDS Foundation Grants, 1585
Fairlawn Foundation Grants, 1626
Frances L. and Edwin L. Cummings Grants, 1717
Gill Foundation - Gay and Lesbian Fund Grants, 1821
GNOF IMPACT Kahn-Oppenheim Grants, 1842
Health Fndn of Greater Indianapolis Grants, 1984
HRSA Resource and Technical Assistance Center for HIV Prevention and Care for Black Men who Have Sex with Men Cooperative Agreement, 2069

652 / AIDS Education

HRSA Ryan White HIV AIDS Drug Assistance Grants, 2070
Ittleson Foundation AIDS Grants, 2166
Jerome Robbins Foundation Grants, 2223
M-A-C AIDS Fund Grants, 2400
McCarthy Family Foundation Grants, 2490
Meta and George Rosenberg Foundation Grants, 2525
Morris & Gwendolyn Cafritz Fndn Grants, 2594
Ms. Foundation for Women Health Grants, 2597
NYCT AIDS/HIV Grants, 3048
OSF International Harm Reduction Development Program Grants, 3090
OSF International Palliative Care Grants, 3091
Paul Rapoport Foundation Grants, 3125
PC Fred H. Bixby Fellowships, 3126
Portland Fndn Women's Giving Circle Grant, 3249
Pride Foundation Grants, 3268
Rockefeller Brothers Fund Pivotal Places Grants: South Africa, 3395
San Diego HIV Funding Collaborative Grants, 3473
Sierra Health Foundation Responsive Grants, 3555
United Methodist Committee on Relief Global AIDS Fund Grants, 3798
USAID HIV Prevention with Key Populations - Mali Grants, 3816
USAID Microbicides Research, Development, and Introduction Grants, 3819
USAID Systems Strengthening for Better HIV/TB Patient Outcomes Grants, 3824
USCM HIV/AIDS Prevention Grants, 3827

AIDS Prevention

Abbott Fund Global AIDS Care Grants, 202
Abbott Fund Science Education Grants, 203
Actors Fund HIV/AIDS Initiative Grants, 283
Aid for Starving Children International Grants, 368
AIDS Vaccine Advocacy Coalition (AVAC) Fund Grants, 369
Alexander Fndn Insurance Continuation Grants, 422
Alexis Gregory Foundation Grants, 426
Cable Positive's Tony Cox Community Grants, 992
CDC Public Health Prev Service Fellowships, 1069
CDC Public Health Prevention Service Fellowship Sponsorships, 1070
Community Fndn AIDS Endowment Awards, 1237
Community Foundation for Greater Atlanta AIDS Fund Grants, 1239
Community Fndn for the Capital Region Grants, 1259
Community Foundation of Louisville AIDS Project Fund Grants, 1289
Dade Community Foundation Community AIDS Partnership Grants, 1409
DIFFA/Chicago Grants, 1477
Dyson Foundation Mid-Hudson Valley Project Support Grants, 1534
Eddie C. & Sylvia Brown Family Fndn Grants, 1548
Elizabeth Glaser Int'l Leadership Awards, 1566
Elton John AIDS Foundation Grants, 1585
Frances L. and Edwin L. Cummings Memorial Fund Grants, 1717
GNOF IMPACT Kahn-Oppenheim Grants, 1842
HRSA Resource and Technical Assistance Center for HIV Prevention and Care for Black Men who Have Sex with Men Cooperative Agreement, 2069
HRSA Ryan White HIV AIDS Drug Assistance Grants, 2070
Ittleson Foundation AIDS Grants, 2166
Jerome Robbins Foundation Grants, 2223
M-A-C AIDS Fund Grants, 2400
Meta and George Rosenberg Foundation Grants, 2525
Morris & Gwendolyn Cafritz Fndn Grants, 2594
Ms. Foundation for Women Health Grants, 2597
NIA AIDS and Aging: Behavioral Sciences Prevention Research Grants, 2747
NIDDK Transmission of Human Immunodeficiency Virus (HIV) In Semen, 2851
NIMH AIDS and Aging: Behavioral Sciences Prevention Research Grants, 2915
NINR AIDS and Aging: Behavioral Sciences Prevention Research Grants, 2927

NYCT AIDS/HIV Grants, 3048
OSF International Harm Reduction Development Program Grants, 3090
Paul Rapoport Foundation Grants, 3125
Questar Corporate Contributions Grants, 3291
Rainbow Endowment Grants, 3300
RCF General Community Grants, 3312
Rockefeller Brothers Fund Pivotal Places Grants: South Africa, 3395
San Diego HIV Funding Collaborative Grants, 3473
United Methodist Committee on Relief Global AIDS Fund Grants, 3798
USAID HIV Prevention with Key Populations - Mali Grants, 3816
USAID Microbicides Research, Development, and Introduction Grants, 3819
USAID Systems Strengthening for Better HIV/TB Patient Outcomes Grants, 3824
USCM HIV/AIDS Prevention Grants, 3827
USDD Broad Agency Announcement Grants, 3837
USDD Broad Agency Announcement HIV/AIDS Prevention Grants, 3838
USDD HIV/AIDS Prevention: Military Specific HIV/AIDS Prevention, Care, and Treatment for Non-PEPFAR Funded Countries Grants, 3841
USDD HIV/AIDS Prevention Program Information Systems Development Grants, 3842
USDD President's Emergency Plan for AIDS Relief Military Specific HIV Seroprevalence and Behavioral Epidemiology Risk Suvey Grants, 3845

Aboriginal Studies

AOCS Analytical Division Student Award, 612
Australasian Institute of Judicial Administration Seed Funding Grants, 806
Vancouver Foundation Grants and Community Initiatives Program, 3849
Vanderburgh Community Fndn Women's Fund, 3853

Abortion

General Service Reproductive Justice Grants, 1785
Huber Foundation Grants, 2071
Lalor Foundation Anna Lalor Burdick Grants, 2335
Lalor Foundation Postdoctoral Fellowships, 2336
Playboy Foundation Grants, 3237

Academic Achievement

3M Company Fndn Community Giving Grants, 6
Albertson's Charitable Giving Grants, 410
Albert W. Rice Charitable Foundation Grants, 411
Alfred E. Chase Charitable Foundation Grants, 430
Alice Tweed Tuohy Foundation Grants Program, 437
AON Foundation Grants, 623
ARCO Foundation Education Grants, 652
Cargill Citizenship Fund Corporate Giving, 1018
Carnegie Corporation of New York Grants, 1033
Charles H. Pearson Foundation Grants, 1121
Clarence E. Heller Charitable Foundation Grants, 1166
Community Foundation Serving Riverside and San Bernardino Counties Impact Grants, 1319
Fluor Foundation Grants, 1682
Foundation for the Mid South Community Development Grants, 1699
Frank Reed and Margaret Jane Peters Memorial Fund II Grants, 1728
Gamble Foundation Grants, 1769
General Mills Foundation Grants, 1782
George W. Wells Foundation Grants, 1804
Greater Milwaukee Foundation Grants, 1883
Guy I. Bromley Trust Grants, 1919
Harold Brooks Foundation Grants, 1952
Harvest Foundation Grants, 1971
Helena Rubinstein Foundation Grants, 1991
Jane's Trust Grants, 2206
Jessie B. Cox Charitable Trust Grants, 2225
John W. Speas and Effie E. Speas Grants, 2264
Lewis H. Humphreys Charitable Trust Grants, 2361
Louetta M. Cowden Foundation Grants, 2377
Louis and Elizabeth Nave Flarsheim Charitable Foundation Grants, 2379

Lynde & Harry Bradley Foundation Fellowships, 2396
Lynde and Harry Bradley Foundation Grants, 2397
Lynde and Harry Bradley Foundation Prizes: Bradley Prizes, 2398
Marie C. and Joseph C. Wilson Foundation Rochester Small Grants, 2431
Mattel Children's Foundation Grants, 2454
Mattel International Grants Program, 2455
NMF Franklin C. McLean Award, 2955
NMF Hugh J. Andersen Memorial Scholarship, 2958
NMF Metropolitan Life Foundation Awards Program for Academic Excellence in Medicine, 2961
NMF National Medical Association Awards for Medical Journalism, 2962
NMF National Medical Association Emerging Scholar Awards, 2963
NMF National Medical Association Patti LaBelle Award, 2964
NMF Ralph W. Ellison Prize, 2966
NMF William and Charlotte Cadbury Award, 2967
Quantum Foundation Grants, 3290
RGk Foundation Grants, 3361
S. Mark Taper Foundation Grants, 3449
SfN Janett Rosenberg Trubatch Career Development Award, 3524
SfN Louise Hanson Marshall Specific Recognition Award, 3527
SfN Nemko Prize in Cellular or Molecular Neuroscience, 3529
Victor E. Speas Foundation Grants, 3868
Z. Smith Reynolds Foundation Small Grants, 3982

Accidents

Jacob and Valeria Langeloth Foundation Grants, 2182

Acoustics

ASHFoundation Research Grant in Speech, 709

Addictions

AACAP-NIDA K12 Career Development Awards, 49
AACAP Educational Outreach Program for Residents in Alcohol Research, 54
Abell Foundation Criminal Justice and Addictions Grants, 208
Actors Addiction & Recovery Services Grants, 282
American Psychiatric Association Minority Medical Student Summer Externship in Addiction Psychiatry, 545
Boyd Gaming Corporation Contributions Program, 941
Bush Fndn Health & Human Services Grants, 979
Community Foundation for Northeast Michigan Tobacco Settlement Grants, 1256
CTCRI Idea Grants, 1369
Griffin Foundation Grants, 1903
Johnson & Johnson Corporate Contributions, 2253
Lisa Higgins-Hussman Foundation Grants, 2370
Lydia deForest Charitable Trust Grants, 2393
MGM Resorts Foundation Community Grants, 2541
National Center for Responsible Gaming Grants, 2629
National Center for Responsible Gaming Postdoctoral Fellowships, 2630
National Center for Resp Gaming Seed Grants, 2631
NIDA Pilot and Feasibility Studies in Preparation for Drug Abuse Prevention Trials, 2794
OSF Tackling Addiction Grants in Baltimore, 3096
Pfizer Global Research Awards for Nicotine Independence, 3179
Rainbow Fund Grants, 3301
TRDRP California Research Awards, 3753
TRDRP New Investigator Awards, 3755
Triangle Community Foundation Shaver-Hitchings Scholarship, 3758
Weyerhaeuser Family Foundation Health Grants, 3926

Adolescent Health

Agnes M. Lindsay Trust Grants, 346
Appalachian Regional Commission Health Grants, 637
Benton Community Fndn Cookie Jar Grant, 877
Blowitz-Ridgeway Foundation Grants, 913
CFFVR Basic Needs Giving Partnership Grants, 1094

SUBJECT INDEX

Cigna Civic Affairs Sponsorships, 1159
Community Fndn of Bloomington & Monroe County - Precision Health Network Cycle Grants, 1264
Community Fndn of Louisville Health Grants, 1295
Cone Health Foundation Grants, 1323
Crail-Johnson Foundation Grants, 1355
Delta Air Lines Foundation Health and Wellness Grants, 1455
DHHS Oral Health Promotion Research Across the Lifespan, 1474
DTE Energy Foundation Health and Human Services Grants, 1516
E.W. "Al" Thrasher Awards, 1540
Fairlawn Foundation Grants, 1626
Foundation for Seacoast Health Grants, 1698
Foundations of East Chicago Health Grants, 1709
General Service Foundation Human Rights and Economic Justice Grants, 1784
Gibson County Community Foundation Women's Fund Grants, 1817
Grand Rapids Community Foundation Ionia County Youth Fund Grants, 1871
Grand Rapids Community Foundation Southeast Ottawa Youth Fund Grants, 1874
Grand Rapids Community Foundation Sparta Youth Fund Grants, 1876
Greater Tacoma Community Foundation Ryan Alan Hade Endowment Fund, 1887
Humana Foundation Grants, 2077
John Edward Fowler Memorial Fndn Grants, 2235
Johnson & Johnson Community Health Grants, 2252
Kansas Health Fndn Major Initiatives Grants, 2288
Lloyd A. Fry Foundation Health Grants, 2373
Marie C. and Joseph C. Wilson Foundation Rochester Small Grants, 2431
Mary Black Fndn Community Health Grants, 2441
Medtronic Foundation Strengthening Health Systems Grants, 2515
MetroWest Health Foundation Grants to Reduce the Incidence of High Risk Behaviors Among Adolescents, 2528
MMS and Alliance Charitable Foundation Grants for Community Action and Care for the Medically Uninsured, 2587
NAPNAP Foundation Shourd Parks Immunization Project Small Grants, 2617
NINR Chronic Illness Self-Management in Children and Adolescents Grants, 2928
Partnership for Cures Two Years To Cures Grants, 3118
Philip L. Graham Fund Health and Human Services Grants, 3190
Porter County Health and Wellness Grant, 3246
RCF The Women's Fund Grants, 3315
Robert and Joan Dircks Foundation Grants, 3379
Seattle Foundation Health and Wellness Grants, 3507
United Healthcare Comm Grants in Michigan, 3796
United Hospital Fund of New York Health Care Improvement Grants, 3797
United Methodist Health Ministry Fund Grants, 3801

Adolescent Psychiatry
AACAP Child and Adolescent Psychiatry (CAP) Teaching Scholarships, 51
AACAP Educational Outreach Program for Child and Adolescent Psychiatry Residents, 52
AACAP Educational Outreach Program for General Psychiatry Residents, 53
AACAP Elaine Schlosser Lewis Award for Research on Attention-Deficit Disorder, 55
AACAP George Tarjan Award for Contributions in Developmental Disabilities, 56
AACAP Irving Philips Award for Prevention, 57
AACAP Jeanne Spurlock Lecture and Award on Diversity and Culture, 58
AACAP Jeanne Spurlock Minority Med Student Clinical Fellow in Child & Adolescent Psych, 59
AACAP Klingenstein Third Generation Foundation Award for Research in Depression or Suicide, 62
AACAP Mary Crosby Congressional Fellowships, 64

AACAP Norbert and Charlotte Rieger Award for Scientific Achievement, 65
AACAP Pilot Research Awards, Supported by Eli Lilly and Company, 68
AACAP Rieger Psychody Psychotherapy Award, 70
AACAP Rieger Service Award for Excellence, 71
AACAP Rob Cancro Academic Leadership Award, 72
AACAP Sidney Berman Award for the School-Based Study and Intervention for Learning Disorders and Mental Illness, 73
AACAP Simon Wile Leader in Consultation Award, 74
AACAP Systems of Care Special Scholarships, 76
American Psychiatric Foundation Grantss, 552
American Psychiatric Foundation James H. Scully Jr., M.D., Educational Fund Grants, 555
American Psychiatric Foundation Typical or Troubled School Mental Health Education Grants, 558
Blowitz-Ridgeway Foundation Grants, 913
NARSAD Ruane Prize for Child and Adolescent Psychiatric Research, 2624
SOBP A.E. Bennett Research Award, 3594

Adolescent Psychology
Blowitz-Ridgeway Foundation Grants, 913
Greater Tacoma Community Foundation Ryan Alan Hade Endowment Fund, 1887
NARSAD Ruane Prize for Child and Adolescent Psychiatric Research, 2624
NIMH Early Identification and Treatment of Mental Disorders in Children and Adolescents Grants, 2918
William T. Grant Foundation Research Grants, 3958

Adolescents
AON Foundation Grants, 623
AT&T Foundation Civic and Community Service Program Grants, 793
Connecticut Community Foundation Grants, 1324
FCD New American Children Grants, 1637
Greater Worcester Community Foundation Discretionary Grants, 1888
HRAMF Deborah Munroe Noonan Memorial Research Grants, 2058
Laura B. Vogler Foundation Grants, 2345
MetroWest Health Foundation Grants to Reduce the Incidence of High Risk Behaviors Among Adolescents, 2528
USAID Comprehensive District-Based Support for Better HIV/TB Patient Outcomes Grants, 3807

Adoption
Abell-Hanger Foundation Grants, 207
Allan C. and Lelia J. Garden Foundation Grants, 438
DHHS Adolescent Family Life Demonstration Projects, 1472

Adult Basic Education
Charles Nelson Robinson Fund Grants, 1125
Community Fndn for the Capital Region Grants, 1259
IIE Western Union Family Scholarships, 2143
McCarthy Family Foundation Grants, 2490
Nationwide Insurance Foundation Grants, 2637
PacifiCare Foundation Grants, 3102
PepsiCo Foundation Grants, 3137
Reinberger Foundation Grants, 3357
WHO Foundation Volunteer Service Grants, 3940

Adult Development
George Family Foundation Grants, 1794
IIE Western Union Family Scholarships, 2143
NSF Social Psychology Research Grants, 3023

Adult and Continuing Education
Ahmanson Foundation Grants, 355
Allen P. & Josephine B. Green Fndn Grants, 443
Allyn Foundation Grants, 448
Arkell Hall Foundation Grants, 661
Arlington Community Foundation Grants, 662
Atlanta Foundation Grants, 796
Auburn Foundation Grants, 798
Ball Brothers Foundation General Grants, 820

Battle Creek Community Foundation Grants, 839
Bayer Hemophilia Caregivers Education Award, 847
Benton Community Foundation Grants, 878
Berrien Community Foundation Grants, 884
Besser Foundation Grants, 887
Blue Mountain Community Foundation Grants, 921
Blue River Community Foundation Grants, 922
Bodenwein Public Benevolent Foundation Grants, 928
Boettcher Foundation Grants, 931
Booth-Bricker Fund Grants, 933
Boston Foundation Grants, 936
Brown County Community Foundation Grants, 970
Carl B. and Florence E. King Foundation Grants, 1022
Cemala Foundation Grants, 1077
CFFVR Basic Needs Giving Partnership Grants, 1094
CFFVR Schmidt Family G4 Grants, 1103
Charles Nelson Robinson Fund Grants, 1125
CIGNA Foundation Grants, 1160
Community Foundation of Bartholomew County Heritage Fund Grants, 1262
Community Foundation of Bartholomew County James A. Henderson Award for Fundraising, 1263
Community Fndn of Central Illinois Grants, 1267
Community Foundation of Greater Fort Wayne - Community Endowment and Clarke Endowment Grants, 1274
Connelly Foundation Grants, 1326
Constantin Foundation Grants, 1341
Coors Brewing Corporate Contributions Grants, 1346
Cornerstone Foundation of Northeastern Wisconsin Grants, 1347
Cowles Charitable Trust Grants, 1354
Cruise Industry Charitable Foundation Grants, 1363
Dayton Power and Light Foundation Grants, 1437
eBay Foundation Community Grants, 1543
Emma B. Howe Memorial Foundation Grants, 1593
Essex County Community Foundation Merrimack Valley General Fund Grants, 1603
Evjue Foundation Grants, 1617
Field Foundation of Illinois Grants, 1660
Fisher Foundation Grants, 1670
Frances L. and Edwin L. Cummings Memorial Fund Grants, 1717
G.N. Wilcox Trust Grants, 1768
George Foundation Grants, 1795
Gibson County Community Foundation Women's Fund Grants, 1817
Goodrich Corporation Foundation Grants, 1854
Guido A. & Elizabeth H. Binda Fndn Grants, 1913
Hallmark Corporate Foundation Grants, 1936
Harold Simmons Foundation Grants, 1955
Helen Steiner Rice Foundation Grants, 1998
Howard and Bush Foundation Grants, 2054
HRSA Nurse Education, Practice, Quality and Retention (NEPQR) Grants, 2068
IIE Western Union Family Scholarships, 2143
James A. and Faith Knight Foundation Grants, 2186
James Ford Bell Foundation Grants, 2187
John H. and Wilhelmina D. Harland Charitable Foundation Children and Youth Grants, 2238
Joseph H. & Florence A. Roblee Fndn Grants, 2269
LaGrange County Community Foundation Scholarships, 2330
Liberty Bank Foundation Grants, 2362
Lubrizol Foundation Grants, 2384
Mardag Foundation Grants, 2428
Marie C. and Joseph C. Wilson Foundation Rochester Small Grants, 2431
MLA Continuing Education Grants, 2559
Nationwide Insurance Foundation Grants, 2637
NBME Stemmler Med Ed Research Grants, 2638
Norcliffe Foundation Grants, 2979
Oppenstein Brothers Foundation Grants, 3078
PacifiCare Foundation Grants, 3102
Parkersburg Area Community Foundation Action Grants, 3116
Perry County Community Foundation Grants, 3142
Peyton Anderson Foundation Grants, 3160
Pfizer Medical Education Track One Grants, 3182
Pfizer Medical Education Track Two Grants, 3183

654 / Adult and Continuing Education

Phi Upsilon Lillian P. Schoephoerster Award, 3196
Pike County Community Foundation Grants, 3222
Piper Jaffray Fndn Communities Giving Grants, 3227
Principal Financial Group Foundation Grants, 3273
Pulaski County Community Foundation Grants, 3279
Samuel S. Johnson Foundation Grants, 3468
San Antonio Area Foundation Grants, 3469
Schlessman Family Foundation Grants, 3493
Sioux Falls Area Community Foundation Community Fund Grants (Unrestricted), 3568
Sioux Falls Area Community Foundation Spot Grants (Unrestricted), 3570
Sisters of Charity Foundation of Cleveland Reducing Health and Educational Disparities in the Central Neighborhood Grants, 3574
Sonora Area Foundation Competitive Grants, 3622
Strake Foundation Grants, 3658
Thomas Sill Foundation Grants, 3724
Vanderburgh Community Foundation Grants, 3852
Warrick County Community Foundation Grants, 3904
WHO Foundation Volunteer Service Grants, 3940
Wilson-Wood Foundation Grants, 3960

Adults
Carrie E. and Lena V. Glenn Foundation Grants, 1036
Charles Delmar Foundation Grants, 1115
Charles Nelson Robinson Fund Grants, 1125
Del Mar Healthcare Fund Grants, 1453
Mabel H. Flory Charitable Trust Grants, 2408
NHLBI Targeted Approaches to Weight Control for Young Adults Grants, 2736
Piper Trust Older Adults Grants, 3229
Union Labor Health Fndn Angel Fund Grants, 3791

Advanced Imaging
Grass Foundation Marine Biological Laboratory Advanced Imaging Fellowships, 1877

Advertising
Google Grants Beta, 1857

Aerospace
ACSM NASA Space Physiology Research Grants, 271

Africa
Aid for Starving Children International Grants, 368
Annenberg Foundation Grants, 596
ASM-UNESCO Leadership Grant for International Educators, 715
Besser Foundation Grants, 887
Doris Duke Charitable Foundation Operations Research on AIDS Care and Treatment in Africa (ORACTA) Grants, 1497
Elizabeth Glaser Int'l Leadership Awards, 1566
ExxonMobil Foundation Malaria Grants, 1621
Fulbright Traditional Scholar Program in Sub-Saharan Africa, 1756
IIE African Center of Excellence for Women's Leadership Grants, 2124
USAID Ebola Response, Recovery and Resilience in West Africa Grants, 3809

Africa, Eastern
Acumen East Africa Fellowship, 284

Africa, Northern
Fulbright Traditional Scholar Program in the Near East and North Africa Region, 1758

Africa, Sub-Saharan
Firelight Foundation Grants, 1667
Fulbright Traditional Scholar Program in Sub-Saharan Africa, 1756
IIE Hewlett Fnd/IIE Dissertation Fellowship, 2132
USAID Support for International Family Planning Organizations Grants, 3823

Africa, Western
NMF General Electric Med Fellowships, 2956

African American Students
AMA Foundation Minority Scholars Awards, 491
MLA Grad Scholarship for Minority Students, 2565
NMF National Medical Association Patti LaBelle Award, 2964

African Americans
ADA Found Minority Dental Scholarships, 292
American Sociological Association Minority Fellowships, 563
Arkansas Community Foundation Arkansas Black Hall of Fame Grants, 659
ASM Robert D. Watkins Graduate Research Fellowship, 747
Boston Foundation Grants, 936
Charles Delmar Foundation Grants, 1115
CMS Historically Black Colleges and Universities (HBCU) Health Services Research Grants, 1187
Coca-Cola Foundation Grants, 1203
Connecticut Community Foundation Grants, 1324
Effie and Wofford Cain Foundation Grants, 1563
FDHN Student Research Fellowships, 1654
GNOF IMPACT Grants for Health and Human Services, 1839
NMF Franklin C. McLean Award, 2955
Philadelphia Foundation Organizational Effectiveness Grants, 3189
RWJF Harold Amos Medical Faculty Development Research Grants, 3434
Saint Luke's Health Initiatives Grants, 3455
SNM PDEF Mickey Williams Minority Student Scholarships, 3584
Strowd Roses Grants, 3664

African Art
Christensen Fund Regional Grants, 1149

After-School Programs
Boston Globe Foundation Grants, 937
Carnegie Corporation of New York Grants, 1033
CFFVR Basic Needs Giving Partnership Grants, 1094
Community Foundation of Greenville Hollingsworth Funds Program/Project Grants, 1285
CVS Community Grants, 1379
Edward W. and Stella C. Van Houten Memorial Fund Grants, 1559
Emma B. Howe Memorial Foundation Grants, 1593
Fargo-Moorhead Area Fndn Woman's Grants, 1633
Foellinger Foundation Grants, 1684
Frank B. Hazard General Charity Fund Grants, 1721
Frank Loomis Palmer Fund Grants, 1726
Fremont Area Community Fndn General Grants, 1746
Ginn Foundation Grants, 1822
Global Fund for Children Grants, 1832
Helen Bader Foundation Grants, 1992
J.C. Penney Company Grants, 2169
Jeffris Wood Foundation Grants, 2217
Jim Moran Foundation Grants, 2228
John Clarke Trust Grants, 2232
Mattel Children's Foundation Grants, 2454
Mattel International Grants Program, 2455
Milagro Foundation Grants, 2552
Packard Foundation Children, Families, and Communities Grants, 3104
Pinkerton Foundation Grants, 3224
Priddy Foundation Program Grants, 3266
RCF Summertime Kids Grants, 3314
Rohm and Haas Company Grants, 3403
Samuel S. Johnson Foundation Grants, 3468
Southbury Community Trust Fund, 3625
Tauck Family Foundation Grants, 3703
Tri-State Community Twenty-first Century Endowment Fund Grants, 3756
U.S. Department of Education 21st Century Community Learning Centers, 3767
W.H. & Mary Ellen Cobb Trust Grants, 3884
WHO Foundation Volunteer Service Grants, 3940
William Blair and Company Foundation Grants, 3948
William G. & Helen C. Hoffman Fndn Grants, 3950

Age Discrimination
Allstate Corporate Giving Grants, 446
Allstate Corp Hometown Commitment Grants, 447

Aging/Gerontology
AAFCS International Graduate Fellowships, 129
AAFCS National Graduate Fellowships, 130
AAFCS National Undergraduate Scholarships, 131
Abington Foundation Grants, 211
ACL Learning Collaboratives for Advanced Business Acumen Skills Grants, 244
Administration on Aging Senior Medicare Patrol Project Grants, 309
AFAR CART Fund Grants, 334
AFAR Medical Student Training in Aging Research Program, 335
AFAR Paul Beeson Career Development Awards in Aging Research for the Island of Ireland, 336
AFAR Research Grants, 337
AGHE Graduate Scholarships and Fellowships, 343
Ahn Family Foundation Grants, 356
Alpha Natural Resources Corporate Giving, 450
American Academy of Nursing Building Academic Geriatric Nursing Capacity Scholarships, 505
American Academy of Nursing Claire M. Fagin Fellowships, 506
American Academy of Nursing Mayday Grants, 507
American Academy of Nursing MBA Scholarships, 508
American Society on Aging Graduate Student Research Award, 560
American Society on Aging Mental Health and Aging Awards, 561
American Society on Aging NOMA Award for Excellence in Multicultural Aging, 562
APSA Congressional Health and Aging Policy Fellowships, 644
Arlington Community Foundation Grants, 662
ASA Metlife Foundation MindAlert Awards, 676
Austin-Bailey Health and Wellness Fndn Grants, 803
Bender Foundation Grants, 876
Brookdale Foundation Leadership in Aging Fellowships, 965
Brookdale Fnd National Group Respite Grants, 966
Bupa Foundation Medical Research Grants, 974
Caesars Foundation Grants, 995
Carl B. and Florence E. King Foundation Grants, 1022
Carl W. and Carrie Mae Joslyn Trust Grants, 1031
Catherine Kennedy Home Foundation Grants, 1043
Charles Delmar Foundation Grants, 1115
Claude Pepper Foundation Grants, 1177
CNCS Senior Companion Program Grants, 1191
Community Foundation of Riverside and San Bernardino County James Bernard and Mildred Jordan Tucker Grants, 1305
Cralle Foundation Grants, 1356
Daniels Fund Grants-Aging, 1424
Daphne Seybolt Culpeper Fndn Grants, 1425
Donald W. Reynolds Foundation Aging and Quality of Life Grants, 1487
Duke University Postdoctoral Research Fellowships in Aging, 1528
Edward N. and Della L. Thome Memorial Foundation Grants, 1557
Effie and Wofford Cain Foundation Grants, 1563
Ellison Medical Foundation/AFAR Julie Martin Mid-Career Award in Aging Research, 1576
Ellison Medical Foundation/AFAR Postdoctoral Fellows in Aging Research Program, 1577
Fallon OrNda Community Health Fund Grants, 1627
Glenn/AFAR Gerontology Awards, 1831
Greater Worcester Community Foundation Jeppson Memorial Fund for Brookfield Grants, 1889
Harry Kramer Memorial Fund Grants, 1965
Hawaii Community Foundation Health Education and Research Grants, 1976
Helen Bader Foundation Grants, 1992
Helen Steiner Rice Foundation Grants, 1998
Henrietta Tower Wurts Memorial Fndn Grants, 2002
Horace A. Kimball and S. Ella Kimball Foundation Grants, 2046

HRAMF Thorne Foundation Awards in Age-Related Macular Degeneration Research, 2065
HRAMF Thorne Foundation Awards in Alzheimer's Disease Drug Discovery Research, 2066
James H. Cummings Foundation Grants, 2190
Jerome and Mildred Paddock Foundation Grants, 2222
John G. Martin Foundation Grants, 2237
Kavli Foundation Research Grants, 2296
Leonard and Helen R. Stulman Charitable Foundation Grants, 2356
M.B. and Edna Zale Foundation Grants, 2401
M.D. Anderson Foundation Grants, 2403
Marjorie Moore Charitable Foundation Grants, 2438
Maurice J. Masserini Charitable Trust Grants, 2456
McLean Contributionship Grants, 2497
Mericos Foundation Grants, 2518
MetroWest Health Foundation Grants--Healthy Aging, 2527
Metzger-Price Fund Grants, 2530
MMAAP Foundation Geriatric Project Awards, 2582
NHLBI Independent Scientist Award, 2701
NIA Aging, Oxidative & Cell Death Grants, 2745
NIA Aging Research Dissertation Awards to Increase Diversity, 2746
NIA AIDS and Aging: Behavioral Sciences Prevention Research Grants, 2747
NIA Alzheimer's Disease Core Centers Grants, 2748
NIA Archiving and Development of Socialbehavioral Datasets in Aging Related Studies Grants, 2752
NIA Awards to Support Research on the Biology of Aging in Invertebrates Grants, 2753
NIA Bioenergetics, Fatigability, and Activity Limitations in Aging, 2755
NIA Diversity in Medication Use and Outcomes in Aging Populations Grants, 2756
NIA Effects of Gene-Social Environment Interplay on Health and Behavior in Later Life Grants, 2757
NIA Harmonization of Longitudinal Cross-National Surveys of Aging Grants, 2760
NIA Health Behaviors and Aging Grants, 2761
NIA Healthy Aging through Behavioral Economic Analyses of Situations Grants, 2762
NIA Higher-Order Cognitive Functioning and Aging Grants, 2763
NIA Human Biospecimen Resources for Aging Research Grants, 2764
NIA Independent Scientist Award, 2766
NIA Malnutrition in Older Persons Research, 2767
NIA Mechanisms, Measurement, and Management of Pain in Aging: from Molecular to Clinical, 2768
NIA Medical Management of Older Patients with HIV/AIDS Grants, 2770
NIA Mentored Research Scientist Devel Award, 2772
NIA Minority Dissert Aging Research Grants, 2773
NIA Multicenter Study on Exceptional Survival in Families: The Long Life Family Study (LLFS) Grants, 2775
NIA Nathan Shock Centers of Excellence in Basic Biology of Aging Grants, 2776
NIA Network Infrastructure Support for Emerging Behavioral & Social Research in Aging, 2777
NIA Pilot Research Grant Program, 2778
NIA Postdoctoral Research on Aging in Canada, 2779
NIA Promoting Careers in Aging and Health Disparities Research Grants, 2780
NIA Renal Function and Chronic Kidney Disease in Aging Grants, 2781
NIA Role of Apolipoprotein E, Lipoprotein Receptors and CNS Lipid Homeostasis in Brain Aging and Alzheimer's Disease, 2782
NIA Role of Nuclear Receptors in Tissue and Organismal Aging Grants, 2783
NIA Support of Scientific Meetings as Cooperative Agreements, 2784
NIA Thyroid in Aging Grants, 2785
NIA Transdisciplinary Research on Fatigue and Fatigability in Aging Grants, 2786
NIA Translational Research at the Aging/Cancer Interface (TRACI) Grants, 2787

NIA Vulnerable Dendrites and Synapses in Aging and Alzheimer's Disease Grants, 2788
NIH Research on Sleep and Sleep Disorders, 2898
NIMH AIDS and Aging: Behavioral Sciences Prevention Research Grants, 2915
NINR AIDS and Aging: Behavioral Sciences Prevention Research Grants, 2927
NSF Biomedical Engineering and Engineering Healthcare Grants, 3009
Ordean Foundation Grants, 3080
OSF International Palliative Care Grants, 3091
Paul Balint Charitable Trust Grants, 3122
Piper Trust Healthcare & Med Research Grants, 3228
Piper Trust Older Adults Grants, 3229
Posey Community Fndn Women's Fund Grants, 3251
Posey County Community Foundation Grants, 3252
Powell Foundation Grants, 3255
Quantum Foundation Grants, 3290
Retirement Research Foundation General Program Grants, 3359
Rhode Island Foundation Grants, 3362
Rosalinde and Arthur Gilbert Foundation/AFAR New Investigator Awards in Alzheimer's Disease, 3406
Rose Community Foundation Aging Grants, 3410
S. Mark Taper Foundation Grants, 3449
Strowd Roses Grants, 3664
STTI Environment of Elder Care Nursing Research Grants, 3667
Thelma Doelger Charitable Trust Grants, 3714
Vigneron Memorial Fund Grants, 3873
Walker Area Community Foundation Grants, 3895
Wilson-Wood Foundation Grants, 3960

Agribusiness
CLIF Bar Family Foundation Grants, 1183

Agricultural Commodities
Aid for Starving Children International Grants, 368

Agricultural Extension
Phi Upsilon Omicron Florence Fallgatter Distinguished Service Award, 3192

Agricultural Planning/Policy
American Jewish World Service Grants, 533
Blue Cross Blue Shield of Minnesota Foundation - Healthy Equity: Health Impact Assessment Demonstration Project Grants, 916
Blue Cross Blue Shield of Minnesota Foundation - Healthy Equity: Health Impact Assessment Program Grants, 917
Browning-Kimball Foundation Grants, 971
Clarence E. Heller Charitable Foundation Grants, 1166
Institute for Agriculture and Trade Policy Food and Society Fellowships, 2159

Agriculture
Agway Foundation Grants, 348
ALSAM Foundation Grants, 454
Canada-U.S. Fulbright New Century Scholars Program Grants, 1008
Canada-U.S. Fulbright Senior Specialists Grants, 1009
Canada Graduate Scholarships (CGS) and NSERC Postgraduate Scholarships (PGS), 1010
CLIF Bar Family Foundation Grants, 1183
Community Foundation of Greater Fort Wayne - Lilly Endowment Scholarships, 1275
Donna K. Yundt Memorial Scholarship, 1488
Dr. John Maniotes Scholarship, 1506
Dr. R.T. White Scholarship, 1510
Elaine Feld Stern Charitable Trust Grants, 1565
Flinn Foundation Scholarships, 1675
Fred and Louise Latshaw Scholarship, 1735
Fulbright Alumni Initiatives Awards, 1749
Fulbright Distinguished Chairs Awards, 1750
Fulbright New Century Scholars Grants, 1753
Fulbright Specialists Program Grants, 1754
Fulbright Scholars in Europe and Eurasia, 1755
Fulbright Traditional Scholar Program in Sub-Saharan Africa, 1756

Fulbright Traditional Scholar Program in the East Asia/Pacific Region, 1757
Fulbright Traditional Scholar Program in the Near East and North Africa Region, 1758
Fulbright Traditional Scholar Program in the South and Central Asia Region, 1759
Fulbright Traditional Scholar Program in the Western Hemisphere, 1760
Fulton County Community Foundation Paul and Dorothy Arven Memorial Scholarship, 1765
Gamble Foundation Grants, 1769
Gruber Foundation Weizmann Institute Awards, 1911
Hammond Common Council Scholarships, 1938
IIE Lotus Scholarships, 2138
IIE Mattel Global Scholarship, 2139
IIE New Leaders Group Award for Mutual Understanding, 2141
Institute for Agriculture and Trade Policy Food and Society Fellowships, 2159
John Deere Foundation Grants, 2234
LaGrange County Lilly Endowment Community Scholarship, 2331
Lake County Athletic Officials Association Scholarships, 2332
PepsiCo Foundation Grants, 3137
Perry County Community Foundation Grants, 3142
Posey County Community Foundation Grants, 3252
Prospect Burma Scholarships, 3275
Puerto Rico Community Foundation Grants, 3278
Richland County Bank Grants, 3371
RSC Thomas W. Eadie Medal, 3423
Samuel S. Johnson Foundation Grants, 3468
Schrage Family Foundation Scholarships, 3494
Seneca Foods Foundation Grants, 3511
Sengupta Family Scholarship, 3512
Trull Foundation Grants, 3761
Whitley County Community Foundation - Lilly Endowment Scholarship, 3935
Whitley County Community Fndn Scholarships, 3937
Windham Foundation Grants, 3961

Agriculture Education
Charles Nelson Robinson Fund Grants, 1125
Conservation, Food, and Health Foundation Grants for Developing Countries, 1338
Dean Foods Community Involvement Grants, 1440
Fulton County Community Foundation Paul and Dorothy Arven Memorial Scholarship, 1765
Harden Foundation Grants, 1945
Posey County Community Foundation Grants, 3252
Richland County Bank Grants, 3371
USDA Organic Agriculture Research Grants, 3830

Airplanes
Northwest Airlines KidCares Medical Travel Assistance, 2998

Alaska
Kenai Peninsula Foundation Grants, 2298

Alaskan Natives
ACF Native American Social and Economic Development Strategies Grants, 231
AMA Foundation Minority Scholars Awards, 491
ASM Robert D. Watkins Graduate Research Fellowship, 747
CDC Increasing Breast and Cervical Cancer Screening Services for Urban American Indian/Alaska Native Women, 1063
FDHN Student Research Fellowships, 1654
MLA Grad Scholarship for Minority Students, 2565
NHLBI Training Program Grants for Institutions that Promote Diversity, 2738

Alcohol Education
AACAP Educational Outreach Program for Residents in Alcohol Research, 54
CFFVR Alcoholism and Drug Abuse Grants, 1093
Coors Brewing Corporate Contributions Grants, 1346
Fairlawn Foundation Grants, 1626

656 / Alcohol Education

George P. Davenport Trust Fund Grants, 1800
NYCT Girls and Young Women Grants, 3053
NYCT Substance Abuse Grants, 3057
Sierra Health Foundation Responsive Grants, 3555
Weyerhaeuser Family Foundation Health Grants, 3926

Alcohol/Alcoholism
AACAP Educational Outreach Program for Residents in Alcohol Research, 54
ACF Native American Social and Economic Development Strategies Grants, 231
Achelis Foundation Grants, 234
Actors Addiction & Recovery Services Grants, 282
Adaptec Foundation Grants, 304
Alcohol Misuse and Alcoholism Research Grants, 415
Alliance Healthcare Foundation Grants, 444
AMA Foundation Fund for Better Health Grants, 484
amfAR Fellowships, 567
amfAR Global Initiatives Grants, 568
amfAR Mathilde Krim Fellowships in Basic Biomedical Research, 569
amfAR Public Policy Grants, 570
amfAR Research Grants, 571
Barberton Community Foundation Grants, 829
Berks County Community Foundation Grants, 881
Bodman Foundation Grants, 929
Bupa Foundation Multi-Country Grant, 975
Bush Fndn Health & Human Services Grants, 979
Cambridge Community Foundation Grants, 1004
Carlisle Foundation Grants, 1025
Carpenter Foundation Grants, 1035
CFFVR Alcoholism and Drug Abuse Grants, 1093
Constantin Foundation Grants, 1341
Coors Brewing Corporate Contributions Grants, 1346
D.F. Halton Foundation Grants, 1405
Dallas Women's Foundation Grants, 1417
Dennis & Phyllis Washington Fndn Grants, 1458
Farmers Insurance Corporate Giving Grants, 1634
Foundation for a Healthy Kentucky Grants, 1691
Frances and John L. Loeb Family Fund Grants, 1716
George P. Davenport Trust Fund Grants, 1800
Greater Worcester Community Foundation Discretionary Grants, 1888
Hasbro Children's Fund Grants, 1973
Health Foundation of Greater Cincinnati Grants, 1983
Huffy Foundation Grants, 2073
Johnson & Johnson Corporate Contributions, 2253
Joseph Drown Foundation Grants, 2268
Joseph H. & Florence A. Roblee Fndn Grants, 2269
L. W. Pierce Family Foundation Grants, 2328
Lester Ray Fleming Scholarships, 2360
Mary Owen Borden Foundation Grants, 2446
Memorial Foundation Grants, 2516
Mid-Iowa Health Foundation Community Response Grants, 2549
NIAAA Independent Scientist Award, 2742
NIAAA Mechanisms of Alcohol and Drug-Induced Pancreatitis Grants, 2743
NIH Research on Sleep and Sleep Disorders, 2898
NIMH Early Identification and Treatment of Mental Disorders in Children and Adolescents Grants, 2918
NYCT Substance Abuse Grants, 3057
Ordean Foundation Grants, 3080
Pajaro Valley Community Health Health Trust Insurance/Coverage & Education on Using the System Grants, 3108
Peter and Elizabeth C. Tower Foundation Mental Health Reference and Resource Materials Mini-Grants, 3147
Peter and Elizabeth C. Tower Foundation Substance Abuse Grants, 3152
Peter F. McManus Charitable Trust Grants, 3153
Puerto Rico Community Foundation Grants, 3278
Quantum Corporation Snap Server Grants, 3289
Questar Corporate Contributions Grants, 3291
Rayonier Foundation Grants, 3311
Ruth Anderson Foundation Grants, 3425
RWJF New Jersey Health Initiatives Grants, 3440
S.H. Cowell Foundation Grants, 3447
SAMHSA Conference Grants, 3462
SAMHSA Drug Free Communities Support Program Grants, 3463
SAMHSA Strategic Prevention Framework State Incentive Grants, 3464
SfN Jacob P. Waletzky Award, 3523
Sid W. Richardson Foundation Grants, 3553
Sioux Falls Area Community Foundation Spot Grants (Unrestricted), 3570
Southbury Community Trust Fund, 3625
Stewart Huston Charitable Trust Grants, 3655
Triangle Community Foundation Shaver-Hitchings Scholarship, 3758
Union Bank, N.A. Foundation Grants, 3788
Victor E. Speas Foundation Grants, 3868
Weyerhaeuser Family Foundation Health Grants, 3926
Whitney Foundation Grants, 3938

Allergy
AAAAI Allied Health Prof Recognition Award, 19
AAAAI ARTrust Mini Grants for Allied Health, 20
AAAAI ARTrust Grants for Allied Health Travel, 21
AAAAI ARTrust Grants for Clinical Research, 22
AAAAI ARTrust Grants in Faculty Development, 23
AAAAI Distinguished Clinician Award, 24
AAAAI Distinguished Layperson Award, 25
AAAAI Distinguished Scientist Award, 26
AAAAI Distinguished Service Award, 27
AAAAI Fellows-in-Training Abstract Award, 28
AAAAI Fellows-in-Training Grants, 29
AAAAI Fellows-in-Training Travel Grants, 30
AAAAI Mentorship Award, 31
AAAAI Outstanding Vol Clinical Faculty Award, 32
AAAAI RSLAAIS Leadership Award, 33
AAAAI Special Recognition Award, 34
Coca-Cola Foundation Grants, 1203
Everyone Breathe Asthma Education Grants, 1616
NIAID Independent Scientist Award, 2765

Allied Health
Carla J. Funk Governmental Relations Award, 1020
EBSCO / MLA Annual Meeting Grants, 1545
June Pangburn Memorial Scholarship, 2283
Mary Black Fndn Community Health Grants, 2441
Medical Informatics Section/MLA Career Development Grant, 2508
MLA David A. Kronick Traveling Fellowship, 2561
MLA Janet Doe Lectureship Award, 2567
MLA Murray Gottlieb Prize Essay Award, 2571
MLA Section Project of the Year Award, 2574
MLA T. Mark Hodges Int'l Service Award, 2575
NIH Mentored Clinical Scientist Research Career Development Award, 2887

Allied Health Education
AAHPERD-AAHE Barb A. Cooley Scholarships, 148
AAHPERD-AAHE Del Oberteuffer Scholarships, 149
ACL Self-Advocacy Resource Center Grants, 247
APSA Robert Wood Johnson Foundation Health Policy Fellowships, 645
Benton County Foundation - Fitzgerald Family Scholarships, 879
Biogen Foundation Patient Educational Grants, 901
CDC Public Health Informatics Fellowships, 1068
CVS All Kids Can Grants, 1377
CVS Community Grants, 1379
DHHS Oral Health Promotion Research Across the Lifespan, 1474
Health Fndn of Greater Indianapolis Grants, 1984
Hospital Libraries Section / MLA Professional Development Grants, 2052
J.M. Long Foundation Grants, 2173
James G.K. McClure Educational and Development Fund Grants, 2188
Josiah Macy Jr. Foundation Grants, 2277
MLA Continuing Education Grants, 2559
MLA David A. Kronick Traveling Fellowship, 2561
MLA Donald A.B. Lindberg Fellowship, 2562
MLA Janet Doe Lectureship Award, 2567
MLA Lucretia W. McClure Excellence in Education Award, 2570
MLA T. Mark Hodges Int'l Service Award, 2575
NIH Mentored Clinical Scientist Research Career Development Award, 2887
Pajaro Valley Community Health Trust Diabetes and Contributing Factors Grants, 3109
Porter County Health Occupations Scholarship, 3247
Schering-Plough Foundation Health Grants, 3492
UniHealth Foundation Community Health Improvement Grants, 3782

Alternative Fuels
Alfred P. Sloan Foundation Research Fellowships, 432
BMW of North America Charitable Contributions, 926

Alternative Medicine
AIHS Alberta/Pfizer Translat Research Grants, 374
Blowitz-Ridgeway Foundation Grants, 913
Bravewell Leadership Award, 946
Fannie E. Rippel Foundation Grants, 1630
James Hervey Johnson Charitable Educational Trust Grants, 2191
NCCAM Exploratory Developmental Grants for Complementary and Alternative Medicine (CAM) Studies of Humans, 2639
NCCAM Ruth L. Kirschstein National Research Service Awards for Postdoctoral Training in Complementary and Alternative Medicine, 2640
NINR Quality of Life for Individuals at the End of Life Grants, 2933
RACGP Vicki Kotsirilos Integrative Med Grants, 3297
Susan G. Komen Breast Cancer Foundation Challege Grants: Investigator Initiated Research, 3690
U.S. Department of Education Vocational Rehabilitation Services Projects for American Indians with Disabilities Grants, 3774

Alternative Modes of Education
James R. Thorpe Foundation Grants, 2199
U.S. Department of Education Vocational Rehabilitation Services Projects for American Indians with Disabilities Grants, 3774

Alzheimer's Disease
ACL Alzheimer's Disease Initiative Grants, 237
AHAF Alzheimer's Disease Research Grants, 349
AIHS Alberta/Pfizer Translat Research Grants, 374
Alzheimer's Association Conference Grants, 463
Alzheimer's Association Development of New Cognitive and Functional Instruments Grants, 464
Alzheimer's Association Everyday Technologies for Alzheimer Care Grants, 465
Alzheimer's Association Investigator-Initiated Research Grants, 466
Alzheimer's Association Mentored New Investigator Research Grants to Promote Diversity, 467
Alzheimer's Association Neuronal Hyper Excitability and Seizures in Alzheimer's Disease Grants, 468
Alzheimer's Association New Investigator Research Grants, 469
Alzheimer's Association New Investigator Research Grants to Promote Diversity, 470
Alzheimer's Association U.S.-U.K. Young Investigator Exchange Fellowships, 471
Alzheimer's Association Zenith Fellows Awards, 472
Austin S. Nelson Foundation Grants, 805
Balfe Family Foundation Grants, 819
Brookdale Foundation Leadership in Aging Fellowships, 965
Brookdale Fnd National Group Respite Grants, 966
Caesars Foundation Grants, 995
CFFVR Robert and Patricia Endries Family Foundation Grants, 1102
CNCS Senior Companion Program Grants, 1191
Dyson Foundation Mid-Hudson Valley Project Support Grants, 1534
Edward N. and Della L. Thome Memorial Foundation Grants, 1557
F.M. Kirby Foundation Grants, 1624
Fairlawn Foundation Grants, 1626
Fremont Area Community Fndn General Grants, 1746

SUBJECT INDEX

G.N. Wilcox Trust Grants, 1768
GNOF IMPACT Harold W. Newman, Jr. Charitable Trust Grants, 1841
Grand Rapids Area Community Fndn Grants, 1865
Hawaii Community Foundation Health Education and Research Grants, 1976
Helen Bader Foundation Grants, 1992
Henrietta Lange Burk Fund Grants, 2001
HRAMF Thome Foundation Awards in Alzheimer's Disease Drug Discovery Research, 2066
NIA Alzheimer's Disease Core Centers Grants, 2748
NIA Alzheimer's Disease Drug Development Program Grants, 2749
NIA Alzheimer's Disease Pilot Clinical Trials, 2750
NIA Alzheimer's Disease Centers Grants, 2751
NIA Grants for Alzheimer's Disease Drugs, 2759
NIA Role of Apolipoprotein E, Lipoprotein Receptors and CNS Lipid Homeostasis in Brain Aging and Alzheimer's Disease, 2782
NIA Vulnerable Dendrites and Synapses in Aging and Alzheimer's Disease Grants, 2788
Northwestern Mutual Foundation Grants, 2999
Peyton Anderson Foundation Grants, 3160
Pfizer Australia Neuroscience Research Grants, 3175
Pittsburgh Foundation Medical Research Grants, 3231
Rosalinde and Arthur Gilbert Foundation/AFAR New Investigator Awards in Alzheimer's Disease, 3406
Rose Community Foundation Aging Grants, 3410
SfN Julius Axelrod Prize, 3526

American History
Radcliffe Institute Carol K. Pforzheimer Student Fellowships, 3298

American Recovery and Reinvestment Act
NIH Recovery Act Limited Competition: Academic Research Enhancement Awards, 2892
NIH Recovery Act Limited Competition: Biomedical Research, Development, and Growth to Spur the Acceleration of New Technologies Grants, 2893
NIH Recovery Act Limited Competition: Small Business Catalyst Awards for Accelerating Innovative Research Grants, 2895
NIH Recovery Act Limited Competition: Supporting New Faculty Recruitment to Enhance Research Resources through Biomedical Research Core Centers Grants, 2896

American Studies
Canada-U.S. Fulbright New Century Scholars Program Grants, 1008
Canada-U.S. Fulbright Senior Specialists Grants, 1009
Fulbright Alumni Initiatives Awards, 1749
Fulbright Distinguished Chairs Awards, 1750
Fulbright New Century Scholars Grants, 1753
Fulbright Specialists Program Grants, 1754
Fulbright Scholars in Europe and Eurasia, 1755
Fulbright Traditional Scholar Program in Sub-Saharan Africa, 1756
Fulbright Traditional Scholar Program in the East Asia/Pacific Region, 1757
Fulbright Traditional Scholar Program in the Near East and North Africa Region, 1758
Fulbright Traditional Scholar Program in the South and Central Asia Region, 1759
Fulbright Traditional Scholar Program in the Western Hemisphere, 1760

Amyotrophic Lateral Sclerosis
Judith and Jean Pape Adams Charitable Foundation ALS Grants, 2279
William H. Adams Foundation for ALS Grants, 3952

Analytical Techniques
AOCS George Schroepfer Medal, 613

Anatomy
PhRMA Foundation Pharmacology/Toxicology Post Doctoral Fellowships, 3215

Anemia
HRAMF Taub Fnd Grants for MDS Research, 2064

Anesthesiology
Duke University Adult Cardiothoracic Anesthesia and Critical Care Medicine Fellowships, 1520
Duke University Ambulatory and Regional Anesthesia Fellowships, 1521
Duke Univ Obstetric Anesthesia Fellowships, 1525
Duke University Ped Anesthesiology Fellowship, 1527

Animal Care
Adam Richter Charitable Trust Grants, 298
Alberto Culver Corporate Contributions Grants, 408
Batchelor Foundation Grants, 837
Bearemy's Kennel Pals Grants, 863
Benton County Foundation Grants, 880
Champlin Foundations Grants, 1110
Charles H. Hall Foundation, 1120
Charlotte County (FL) Community Foundation Grants, 1126
Community Foundation of Tampa Bay Grants, 1312
Dade Community Foundation Grants, 1410
Dean Foods Community Involvement Grants, 1440
DeRoy Testamentary Foundation Grants, 1469
Earl and Maxine Claussen Trust Grants, 1541
Fargo-Moorhead Area Foundation Grants, 1632
Gamble Foundation Grants, 1769
Green Bay Packers Foundation Grants, 1890
Greenfield Foundation of Maine Grants, 1893
Huisking Foundation Grants, 2076
Jane's Trust Grants, 2206
Kinsman Foundation Grants, 2309
Lubbock Area Foundation Grants, 2383
Lucy Downing Nisbet Charitable Fund Grants, 2387
Maddie's Fund Medical Equipment Grants, 2411
Margaret T. Morris Foundation Grants, 2429
Matilda R. Wilson Fund Grants, 2453
Natalie W. Furniss Charitable Trust Grants, 2626
Owen County Community Foundation Grants, 3100
Perkins Charitable Foundation Grants, 3140
Perry County Community Foundation Grants, 3142
Pet Care Trust Sue Busch Memorial Award, 3143
PMI Foundation Grants, 3239
Puerto Rico Community Foundation Grants, 3278
San Antonio Area Foundation Grants, 3469
Santa Fe Community Foundation Seasonal Grants-Fall Cycle, 3482
Seattle Foundation Benjamin N. Phillips Grants, 3504
Sonora Area Foundation Competitive Grants, 3622
Southwest Florida Community Foundation Competitive Grants, 3627
Vancouver Foundation Grants and Community Initiatives Program, 3849
Walter L. Gross III Family Foundation Grants, 3901
William D. Laurie, Jr. Charitable Fndn Grants, 3949
Winn Feline Foundation/AVMF Excellence In Feline Research Award, 3963
Winn Feline Foundation Grants, 3964

Animal Development
NEAVS Fellowsips in Women's Health and Sex Differences, 2659

Animal Diseases/Pathology
ACVIM Foundation Clinical Investigation Grants, 286
AMHPS Dr. James A. Ferguson Emerging Infectious Diseases Fellowships, 573
CDC Epidemic Intell Service Training Grants, 1055
CDC Epidemiology Elective Rotation, 1056
Emma Barnsley Foundation Grants, 1594
Maddie's Fund Medical Equipment Grants, 2411
Winn Feline Foundation/AVMF Excellence In Feline Research Award, 3963

Animal Ecology
Greenfield Foundation of Maine Grants, 1893

Animal Genetics/Breeding
Emma Barnsley Foundation Grants, 1594

Animal Rescue
Boyle Foundation Grants, 942
Emma Barnsley Foundation Grants, 1594
Lucy Downing Nisbet Charitable Fund Grants, 2387
Sioux Falls Area Community Foundation Field-of-Interest and Donor-Advised Grants, 3569
William Ray & Ruth E. Collins Fndn Grants, 3956

Animal Research Policy
Alternatives Research and Development Foundation Alternatives in Education Grants, 455
Alternatives Research and Development Foundation Grants, 456
NCRR Novel Approaches to Enhance Animal Stem Cell Research, 2658
NEAVS Fellowsips in Women's Health and Sex Differences, 2659
Vancouver Foundation Grants and Community Initiatives Program, 3849

Animal Rights
Adam Richter Charitable Trust Grants, 298
Ann Jackson Family Foundation Grants, 598
Aragona Family Foundation Grants, 648
Blue Mountain Community Foundation Grants, 921
Bodenwein Public Benevolent Foundation Grants, 928
Chamberlain Foundation Grants, 1108
Charles H. Hall Foundation, 1120
Cleveland-Cliffs Foundation Grants, 1181
Emma Barnsley Foundation Grants, 1594
Frank Stanley Beveridge Foundation Grants, 1730
Harrison County Community Foundation Grants, 1957
Harrison County Community Foundation Signature Grants, 1958
Horace A. Kimball and S. Ella Kimball Foundation Grants, 2046
Natalie W. Furniss Charitable Trust Grants, 2626
Perkins Charitable Foundation Grants, 3140
Rhode Island Foundation Grants, 3362
Thelma Doelger Charitable Trust Grants, 3714
Vancouver Foundation Grants and Community Initiatives Program, 3849
Walter L. Gross III Family Foundation Grants, 3901

Animal Science
ACVIM Foundation Clinical Investigation Grants, 286
NSF Animal Developmental Mechanisms Grants, 3007
NSF Cell Systems Cluster Grants, 3011

Animal Welfare
Annenberg Foundation Grants, 596
Arkansas Community Foundation Grants, 660
Austin S. Nelson Foundation Grants, 805
Boyle Foundation Grants, 942
Central Okanagan Foundation Grants, 1080
Community Foundation of Muncie and Delaware County Maxon Grants, 1301
Cresap Family Foundation Grants, 1359
Earl and Maxine Claussen Trust Grants, 1541
Emma Barnsley Foundation Grants, 1594
Greygates Foundation Grants, 1901
Guy I. Bromley Trust Grants, 1919
Harden Foundation Grants, 1945
Harrison County Community Foundation Signature Grants, 1958
Herbert A. & Adrian W. Woods Fndn Grants, 2007
Katharine Matthies Foundation Grants, 2291
Lil and Julie Rosenberg Foundation Grants, 2364
Lucy Downing Nisbet Charitable Fund Grants, 2387
Maddie's Fund Medical Equipment Grants, 2411
Matilda R. Wilson Fund Grants, 2453
NEAVS Fellowsips in Women's Health and Sex Differences, 2659
PMI Foundation Grants, 3239
Robert R. Meyer Foundation Grants, 3387
Santa Fe Community Foundation Seasonal Grants-Fall Cycle, 3482
Sarkeys Foundation Grants, 3485
Seattle Foundation Benjamin N. Phillips Memorial Fund Grants, 3504

658 / Animal Welfare

Sioux Falls Area Community Foundation Field-of-Interest and Donor-Advised Grants, 3569
Thelma Doelger Charitable Trust Grants, 3714
Thorman Boyle Foundation Grants, 3730
Wells County Foundation Grants, 3920
William D. Laurie, Jr. Charitable Fndn Grants, 3949
William Ray & Ruth E. Collins Fndn Grants, 3956

Animals as Pets
Bearemy's Kennel Pals Grants, 863
Charles H. Hall Foundation, 1120
Emma Barnsley Foundation Grants, 1594
Mericos Foundation Grants, 2518
Santa Fe Community Foundation Seasonal Grants-Fall Cycle, 3482

Animals for Assistance/Therapy
Bearemy's Kennel Pals Grants, 863
Blum-Kovler Foundation Grants, 924
Lumpkin Family Fnd Healthy People Grants, 2391
May and Stanley Smith Charitable Trust Grants, 2459
NEAVS Fellowsips in Women's Health and Sex Differences, 2659
Reader's Digest Partners for Sight Fndn Grants, 3345
Robert R. Meyer Foundation Grants, 3387

Anorexia Nervosa
Hilda and Preston Davis Foundation Postdoctoral Fellowships in Eating Disorders Research, 2029
Klarman Family Foundation Grants in Eating Disorders Research Grants, 2310
NIMH Early Identification and Treatment of Mental Disorders in Children and Adolescents Grants, 2918

Anthropology
Australasian Institute of Judicial Administration Seed Funding Grants, 806
Canada-U.S. Fulbright New Century Scholars Program Grants, 1008
Canada-U.S. Fulbright Senior Specialists Grants, 1009
Fulbright Alumni Initiatives Awards, 1749
Fulbright Distinguished Chairs Awards, 1750
Fulbright New Century Scholars Grants, 1753
Fulbright Specialists Program Grants, 1754
Fulbright Scholars in Europe and Eurasia, 1755
Fulbright Traditional Scholar Program in Sub-Saharan Africa, 1756
Fulbright Traditional Scholar Program in the East Asia/Pacific Region, 1757
Fulbright Traditional Scholar Program in the Near East and North Africa Region, 1758
Fulbright Traditional Scholar Program in the South and Central Asia Region, 1759
Fulbright Traditional Scholar Program in the Western Hemisphere, 1760
Harry Frank Guggenheim Foundation Dissertation Fellowships, 1963

Anthropology, Cultural
Australasian Institute of Judicial Administration Seed Funding Grants, 806

Anthropology, Social
Australasian Institute of Judicial Administration Seed Funding Grants, 806

Antibiotics
Pfizer Anti Infective Research EU Grants, 3164
Pfizer ASPIRE EU Emerging Mechanisms of Resistance Antibacterial Research Awards, 3167
Pfizer ASPIRE EU MRSA Nosocomial Pneumonia & MRSA Complicated Skin & Soft Tissue Infections Antibacterial Research Awards, 3168
Pfizer ASPIRE North America Broad Spectrum Antibiotics for the Treatment of Gram-Negative or Polymicrobial Infections Research Awards, 3169
Pfizer ASPIRE North America Narrow Spectrum Antibiotics for the Treatment of MRSA Research Awards, 3171
Pfizer Healthcare Charitable Contributions, 3180

Antivirals
Doris Duke Charitable Fndn Operations Research on AIDS Care and Treatment in Africa Grants, 1497
Pfizer Anti Infective Research EU Grants, 3164

Appalachia
Foundation for Appalachian Ohio Susan K. Ipacs Nursing Legacy Scholarships, 1693
Foundation for Appalachian Ohio Zelma Gray Medical School Scholarship, 1694

Aquariums
Union Bank, N.A. Foundation Grants, 3788

Archaeology
Australasian Institute of Judicial Administration Seed Funding Grants, 806
Canada-U.S. Fulbright New Century Scholars Program Grants, 1008
Canada-U.S. Fulbright Senior Specialists Grants, 1009
Chamberlain Foundation Grants, 1108
Colonel Stanley R. McNeil Foundation Grants, 1211
Fulbright Alumni Initiatives Awards, 1749
Fulbright Distinguished Chairs Awards, 1750
Fulbright New Century Scholars Grants, 1753
Fulbright Specialists Program Grants, 1754
Fulbright Scholars in Europe and Eurasia, 1755
Fulbright Traditional Scholar Program in Sub-Saharan Africa, 1756
Fulbright Traditional Scholar Program in the East Asia/Pacific Region, 1757
Fulbright Traditional Scholar Program in the Near East and North Africa Region, 1758
Fulbright Traditional Scholar Program in the South and Central Asia Region, 1759
Fulbright Traditional Scholar Program in the Western Hemisphere, 1760
Stewart Huston Charitable Trust Grants, 3655

Architecture
Denver Foundation Community Grants, 1461

Archives
Grammy Foundation Grants, 1863
Lucy Downing Nisbet Charitable Fund Grants, 2387

Arrhythmia
NHLBI New Approaches to Arrhythmia Detection and Treatment Grants for SBIR, 2715
NHLBI New Approaches to Arrhythmia Detection and Treatment Grants for STTR, 2716
NHLBI Targeting Calcium Regulatory Molecules for Arrhythmia Prevention Grants, 2737

Art Appreciation
Benton County Foundation Grants, 880
Boyd Gaming Corporation Contributions Program, 941
Carl B. and Florence E. King Foundation Grants, 1022
Carrier Corporation Contributions Grants, 1038
Florida Division of Cultural Affairs Arts In Education Arts Partnership Grants, 1679
Guy I. Bromley Trust Grants, 1919
Helen Bader Foundation Grants, 1992
Helen Gertrude Sparks Charitable Trust Grants, 1993
Honda of America Manufacturing Fndn Grants, 2045
John W. Speas and Effie E. Speas Memorial Trust Grants, 2264
Lewis H. Humphreys Charitable Trust Grants, 2361
Louetta M. Cowden Foundation Grants, 2377
Louis and Elizabeth Nave Flarsheim Charitable Foundation Grants, 2379
Perry County Community Foundation Grants, 3142
Pike County Community Foundation Grants, 3222
PMI Foundation Grants, 3239
San Diego Women's Foundation Grants, 3474
Warrick County Community Foundation Grants, 3904

Art Conservation
Christensen Fund Regional Grants, 1149
Pike County Community Foundation Grants, 3222

SUBJECT INDEX

Art Criticism
Florida Division of Cultural Affairs Arts In Education Arts Partnership Grants, 1679

Art Education
2 Depot Square Ipswich Charitable Fndn Grants, 4
3M Company Fndn Community Giving Grants, 6
Abbott Fund Community Grants, 201
Abell-Hanger Foundation Grants, 207
Ahn Family Foundation Grants, 356
Alaska Airlines Foundation Grants, 403
Angels Baseball Foundation Grants, 590
Anna Fitch Ardenghi Trust Grants, 592
Annenberg Foundation Grants, 596
APS Foundation Grants, 646
Ashland Corporate Contributions Grants, 713
Assurant Health Foundation Grants, 789
Bayer Foundation Grants, 846
Benton County Foundation Grants, 880
Boyd Gaming Corporation Contributions Program, 941
Christensen Fund Regional Grants, 1149
Connecticut Community Foundation Grants, 1324
Cooper Industries Foundation Grants, 1345
DAAD Research Stays for University Academics and Scientists, 1408
Eisner Foundation Grants, 1564
Erie Community Foundation Grants, 1601
Ethel S. Abbott Charitable Foundation Grants, 1607
Florida Division of Cultural Affairs Arts In Education Arts Partnership Grants, 1679
Ford Motor Company Fund Grants Program, 1687
George A Ohl Jr. Foundation Grants, 1791
George Gund Foundation Grants, 1796
Grand Rapids Area Community Fndn Grants, 1865
Grand Rapids Area Community Foundation Nashwauk Area Endowment Fund Grants, 1866
Greater Worcester Community Foundation Discretionary Grants, 1888
Guy I. Bromley Trust Grants, 1919
H & R Foundation Grants, 1920
Helen Bader Foundation Grants, 1992
Helen Gertrude Sparks Charitable Trust Grants, 1993
Herman Goldman Foundation Grants, 2012
Honda of America Manufacturing Fndn Grants, 2045
Horace Moses Charitable Foundation Grants, 2047
Horizon Foundation for New Jersey Grants, 2048
Jane's Trust Grants, 2206
Jayne and Leonard Abess Foundation Grants, 2214
John W. Speas and Effie E. Speas Memorial Trust Grants, 2264
Judith and Jean Pape Adams Charitable Foundation Tulsa Area Grants, 2280
Kathryne Beynon Foundation Grants, 2294
Leon and Thea Koerner Foundation Grants, 2355
Lincoln Financial Foundation Grants, 2367
Louetta M. Cowden Foundation Grants, 2377
Louis and Elizabeth Nave Flarsheim Charitable Foundation Grants, 2379
Lubrizol Foundation Grants, 2384
Marie H. Bechtel Charitable Remainder Uni-Trust Grants, 2432
Marion I. and Henry J. Knott Foundation Discretionary Grants, 2435
Marion I. and Henry J. Knott Foundation Standard Grants, 2436
Mead Johnson Nutritionals Evansville-Area Organizations Grants, 2503
Milagro Foundation Grants, 2552
Mimi and Peter Haas Fund Grants, 2557
Morris & Gwendolyn Cafritz Fndn Grants, 2594
NHSCA Arts in Health Care Project Grants, 2741
North Carolina GlaxoSmithKline Fndn Grants, 2991
Peacock Foundation Grants, 3133
Pike County Community Foundation Grants, 3222
PMI Foundation Grants, 3239
Rasmuson Foundation Tier Two Grants, 3308
Robert W. Woodruff Foundation Grants, 3389
Rockwell International Corporate Trust Grants Program, 3399
San Diego Women's Foundation Grants, 3474

SUBJECT INDEX

Sid W. Richardson Foundation Grants, 3553
Taylor S. Abernathy and Patti Harding Abernathy Charitable Trust Grants, 3704
Vancouver Foundation Grants and Community Initiatives Program, 3849
Washington Gas Charitable Giving, 3908
Wood Family Charitable Trust Grants, 3972

Art History
Florida Division of Cultural Affairs Arts In Education Arts Partnership Grants, 1679
Pike County Community Foundation Grants, 3222

Art Museums
Blum-Kovler Foundation Grants, 924
Helen Pumphrey Denit Charitable Trust Grants, 1996
Mericos Foundation Grants, 2518
Meyer and Stephanie Eglin Foundation Grants, 2531
William J. & Tina Rosenberg Fndn Grants, 3954

Art Therapy
Community Fndn for the Capital Region Grants, 1259
Deborah Munroe Noonan Memorial Grants, 1443

Art in Public Places
Abbott Fund Community Grants, 201
Amador Community Foundation Grants, 481
Boyd Gaming Corporation Contributions Program, 941
Golden Heart Community Foundation Grants, 1850
Grand Rapids Area Community Fndn Grants, 1865
Grand Rapids Area Community Foundation Nashwauk Area Endowment Fund Grants, 1866
Helen S. Boylan Foundation Grants, 1997

Art, Experimental
Connecticut Community Foundation Grants, 1324

Arthritis
Adams Foundation Grants, 301
Balfe Family Foundation Grants, 819
Dorr Institute for Arthritis Research & Educ, 1502
Fairlawn Foundation Grants, 1626
Mabel H. Flory Charitable Trust Grants, 2408
NIAMS Pilot and Feasibility Clinical Research Grants in Arthritis, Musculoskeletal & Skin Diseases, 2774
Pfizer ASPIRE North America Rheumatology Research Awards, 3172
Pfizer Healthcare Charitable Contributions, 3180
Pfizer Inflammation Competitive Research Awards (UK), 3181
Pittsburgh Foundation Medical Research Grants, 3231

Artificial Reproduction
USAID Family Planning and Reproductive Health Methods Grants, 3811

Artists in Residence
Florida Division of Cultural Affairs Arts In Education Arts Partnership Grants, 1679
Marjorie Moore Charitable Foundation Grants, 2438
NHSCA Arts in Health Care Project Grants, 2741

Arts Administration
Alcatel-Lucent Technologies Foundation Grants, 413
Florida Division of Cultural Affairs Arts In Education Arts Partnership Grants, 1679
J.L. Bedsole Foundation Grants, 2172
Procter and Gamble Fund Grants, 3274
Sid W. Richardson Foundation Grants, 3553
SunTrust Bank Trusteed Foundations Florence C. and Harry L. English Memorial Fund Grants, 3680
SunTrust Bank Trusteed Foundations Harriet McDaniel Marshall Tust Grants, 3682
SunTrust Bank Trusteed Foundations Walter H. and Marjory M. Rich Memorial Fund Grants, 3685

Arts Festivals
Angels Baseball Foundation Grants, 590
CONSOL Energy Community Dev Grants, 1339
Honda of America Manufacturing Fndn Grants, 2045

Leon and Thea Koerner Foundation Grants, 2355
Macquarie Bank Foundation Grants, 2410
OceanFirst Foundation Major Grants, 3062

Arts and Culture
1st Source Foundation Grants, 2
3M Company Fndn Community Giving Grants, 6
ACF Native American Social and Economic Development Strategies Grants, 231
Ackerman Foundation Grants, 236
Adams Foundation Grants, 301
Adelaide Breed Bayrd Foundation Grants, 306
Amica Companies Foundation Grants, 574
Annenberg Foundation Grants, 596
ARS Foundation Grants, 667
Arthur F. & Arnold M. Frankel Fndn Grants, 672
Assisi Fndn of Memphis Capital Project Grants, 786
Assisi Foundation of Memphis General Grants, 787
Atherton Family Foundation Grants, 794
Avista Foundation Economic and Cultural Vitality Grants, 812
Bank of America Fndn Matching Gifts, 825
BBVA Compass Foundation Charitable Grants, 848
Beckman Coulter Foundation Grants, 868
Belvedere Community Foundation Grants, 874
Blackford County Community Foundation Grants, 906
Blumenthal Foundation Grants, 925
Boeing Company Contributions Grants, 930
Bonfils-Stanton Foundation Grants, 932
Boston Globe Foundation Grants, 937
Bradley C. Higgins Foundation Grants, 945
Browning-Kimball Foundation Grants, 971
Burlington Industries Foundation Grants, 977
Business Bank of Nevada Community Grants, 981
Cardinal Health Foundation Grants, 1017
Centerville-Washington Foundation Grants, 1078
CFFVR Chilton Area Community Fndn Grants, 1096
Charles H. Hall Foundation, 1120
Chilkat Valley Community Foundation Grants, 1148
CICF City of Noblesville Community Grant, 1155
CICF James Proctor Grant for Aged, 1157
Clayton Fund Grants, 1180
CNO Financial Group Community Grants, 1201
Community Foundation for Greater Atlanta Clayton County Fund Grants, 1240
Community Foundation for Greater Atlanta Common Good Funds Grants, 1241
Community Foundation for Greater Atlanta Morgan County Fund Grants, 1244
Community Foundation for Greater Atlanta Newton County Fund Grants, 1245
Community Foundation for Greater Atlanta Strategic Restructuring Fund Grants, 1246
Community Foundation for Northeast Michigan Mini-Grants, 1255
Community Fndn for San Benito County Grants, 1257
Community Foundation for Southeast Michigan Grants, 1258
Community Foundation of Bloomington and Monroe County Grants, 1265
Community Foundation of Boone County Grants, 1266
Community Foundation of Greater Greensboro Women to Women Fund Grants, 1278
Community Fndn of Greater Lafayette Grants, 1280
Community Fndn of Jackson County Grants, 1287
Community Foundation of Louisville Anna Marble Memorial Fund for Princeton Grants, 1290
Community Foundation of Muncie and Delaware County Grants, 1300
Community Foundation of Muncie and Delaware County Maxon Grants, 1301
Community Foundation of Riverside & San Bernardino County Impact Grants, 1303
Community Fndn of Riverside & San Bernardino County Irene S. Rockwell Grants, 1304
Community Foundation Partnerships - Lawrence County Grants, 1317
Community Foundation Partnerships - Martin County Grants, 1318

Community Foundation Serving Riverside and San Bernardino Counties Impact Grants, 1319
Covenant Educational Foundation Grants, 1348
Crane Fund Grants, 1358
Cresap Family Foundation Grants, 1359
Crown Point Community Foundation Grants, 1361
CSRA Community Foundation Grants, 1366
D.V. and Ida J. McEachern Trust Grants, 1406
Dana Brown Charitable Trust Grants, 1419
Daniel & Nanna Stern Family Fndn Grants, 1422
Davis Family Foundation Grants, 1435
Dayton Power and Light Foundation Grants, 1437
Decatur County Community Foundation Large Project Grants, 1444
Decatur County Community Foundation Small Project Grants, 1445
DeKalb County Community Foundation Grants, 1447
Dyson Foundation Mid-Hudson Valley Project Support Grants, 1534
E. Clayton and Edith P. Gengras, Jr. Foundation, 1536
E.J. Grassmann Trust Grants, 1537
Earl and Maxine Claussen Trust Grants, 1541
Elaine Feld Stern Charitable Trust Grants, 1565
Elkhart County Community Foundation Fund for Elkhart County, 1572
El Pomar Foundation Grants, 1582
Ensworth Charitable Foundation Grants, 1595
Fifth Third Foundation Grants, 1661
FIU Global Civic Engagement Mini Grants, 1673
Four County Community Fndn General Grants, 1710
Four J Foundation Grants, 1713
Franklin H. Wells and Ruth L. Wells Foundation Grants, 1725
Frank Loomis Palmer Fund Grants, 1726
Gardner Foundation Grants, 1771
Garland D. Rhoads Foundation, 1774
George A. and Grace L. Long Foundation Grants, 1788
George H.C. Ensworth Memorial Fund Grants, 1797
George Kress Foundation Grants, 1799
GNOF IMPACT Grants for Arts and Culture, 1838
Golden Heart Community Foundation Grants, 1850
Graham Foundation Grants, 1862
Greater Sitka Legacy Fund Grants, 1885
Greene County Community Foundation Grants, 1892
Green River Area Community Fndn Grants, 1895
Grundy Foundation Grants, 1912
Hancock County Community Foundation - Field of Interest Grants, 1943
Harden Foundation Grants, 1945
Hardin County Community Foundation Grants, 1946
Harmony Project Grants, 1948
Harold and Arlene Schnitzer CARE Foundation Grants, 1950
Harrison County Community Foundation Grants, 1957
Harrison County Community Foundation Signature Grants, 1958
Harvey Randall Wickes Foundation Grants, 1972
Hawaiian Electric Industries Charitable Foundation Grants, 1975
Helen Bader Foundation Grants, 1992
Hendricks County Community Fndn Grants, 2000
Herbert A. & Adrian W. Woods Fndn Grants, 2007
Hutton Foundation Grants, 2087
Ike and Roz Friedman Foundation Grants, 2145
Inasmuch Foundation Grants, 2149
Intergrys Corporation Grants, 2160
James L. and Mary Jane Bowman Charitable Trust Grants, 2194
Jane's Trust Grants, 2206
Jayne and Leonard Abess Foundation Grants, 2214
Jennings County Community Foundation Women's Giving Circle Grant, 2221
Jerome Robbins Foundation Grants, 2223
Jessica Stevens Community Foundation Grants, 2224
John P. Murphy Foundation Grants, 2248
John W. and Anna H. Hanes Foundation Grants, 2261
Joseph Henry Edmondson Foundation Grants, 2270
Kenai Peninsula Foundation Grants, 2298
Ketchikan Community Foundation Grants, 2303
Kettering Family Foundation Grants, 2304

660 / Arts and Culture

Kettering Fund Grants, 2305
Kinsman Foundation Grants, 2309
Kodiak Community Foundation Grants, 2314
Kuntz Foundation Grants, 2326
Lake County Community Fund Grants, 2333
Legler Benbough Foundation Grants, 2352
Leo Goodwin Foundation Grants, 2354
Leon and Thea Koerner Foundation Grants, 2355
Libra Foundation Grants, 2363
Lucy Downing Nisbet Charitable Fund Grants, 2387
Madison County Community Foundation - City of Anderson Quality of Life Grant, 2412
Marion I. and Henry J. Knott Foundation Discretionary Grants, 2435
Marion I. and Henry J. Knott Foundation Standard Grants, 2436
Marjorie Moore Charitable Foundation Grants, 2438
Maurice J. Masserini Charitable Trust Grants, 2456
McGraw-Hill Companies Community Grants, 2494
MeadWestvaco Foundation Sustainable Communities Grants, 2506
Merrick Foundation Grants, 2521
Merrick Foundation Grants, 2522
Meta and George Rosenberg Foundation Grants, 2525
Meyer Memorial Trust Grassroots Grants, 2536
Meyer Memorial Trust Responsive Grants, 2537
Middlesex Savings Charitable Foundation Capacity Building Grants, 2551
Miller Foundation Grants, 2556
Moline Foundation Community Grants, 2590
Norcliffe Foundation Grants, 2979
Nord Family Foundation Grants, 2980
Norfolk Southern Foundation Grants, 2982
North Dakota Community Foundation Grants, 2994
OceanFirst Foundation Major Grants, 3062
Olive Higgins Prouty Foundation Grants, 3072
Orange County Community Foundation Grants, 3079
Oregon Community Fndn Community Grants, 3082
Owen County Community Foundation Grants, 3100
Pacific Life Foundation Grants, 3103
Packard Foundation Local Grants, 3105
Palmer Foundation Grants, 3112
Parke County Community Foundation Grants, 3115
Perkin Fund Grants, 3139
Petersburg Community Foundation Grants, 3156
Peyton Anderson Foundation Grants, 3160
Piedmont Natural Gas Corporate and Charitable Contributions, 3220
PMI Foundation Grants, 3239
Pokagon Fund Grants, 3242
Portland General Electric Foundation Grants, 3250
PPG Industries Foundation Grants, 3257
Price Chopper's Golub Foundation Grants, 3261
Priddy Foundation Program Grants, 3266
Pulido Walker Foundation, 3282
Quaker Chemical Foundation Grants, 3286
Rasmuson Foundation Tier Two Grants, 3308
Robbins Charitable Foundation Grants, 3378
Robert F. Stoico/FIRSTFED Charitable Foundation Grants, 3382
Robert W. Woodruff Foundation Grants, 3389
Rohm and Haas Company Grants, 3403
Rose Hills Foundation Grants, 3413
Sain-Orr and Royak-DeForest Steadman Foundation Grants, 3452
Samuel N. and Mary Castle Foundation Grants, 3467
Sandy Hill Foundation Grants, 3477
Sarkeys Foundation Grants, 3485
Seward Community Foundation Grants, 3514
Seward Community Foundation Mini-Grants, 3515
Shell Oil Company Foundation Community Development Grants, 3544
Shield-Ayres Foundation Grants, 3546
Sioux Falls Area Community Foundation Community Fund Grants (Unrestricted), 3568
Sioux Falls Area Community Foundation Field-of-Interest and Donor-Advised Grants, 3569
Sioux Falls Area Community Foundation Spot Grants (Unrestricted), 3570
Sony Corporation of America Grants, 3623

Starke County Community Foundation Grants, 3646
Sterling and Shelli Gardner Foundation Grants, 3653
Stewart Huston Charitable Trust Grants, 3655
SunTrust Bank Trusteed Foundations Greene-Sawtell Grants, 3681
SunTrust Bank Trusteed Foundations Nell Warren Elkin and William Simpson Elkin Grants, 3683
SunTrust Bank Trusteed Foundations Thomas Guy Woolford Charitable Trust Grants, 3684
Textron Corporate Contributions Grants, 3712
The Ray Charles Foundation Grants, 3716
Thomas J. Long Foundation Community Grants, 3722
Thomas Jefferson Rosenberg Foundation Grants, 3723
Toyota Motor Engineering & Manufacturing North America Grants, 3742
Toyota Motor Manufacturing of Alabama Grants, 3743
Toyota Motor Manufacturing of Indiana Grants, 3744
Toyota Motor Manuf of Mississippi Grants, 3746
Toyota Motor Manufacturing of Texas Grants, 3747
Toyota Motor Manuf of West Virginia Grants, 3748
Toyota Motor N America of New York Grants, 3749
Toyota Motor Sales, USA Grants, 3750
Union Bank, N.A. Corporate Sponsorships and Donations, 3787
United Technologies Corporation Grants, 3803
Vanderburgh Community Foundation Grants, 3852
Victor E. Speas Foundation Grants, 3868
Virginia W. Kettering Foundation Grants, 3878
W.M. Keck Fndn So California Grants, 3890
Wabash Valley Community Foundation Grants, 3893
Walker Area Community Foundation Grants, 3895
Wallace Alexander Gerbode Foundation Grants, 3896
Warren County Community Foundation Grants, 3902
Warren County Community Fndn Mini-Grants, 3903
Washington County Community Fndn Grants, 3906
Washington County Community Foundation Youth Grants, 3907
Wells County Foundation Grants, 3920
Western New York Foundation Grants, 3922
White County Community Foundation Grants, 3934
Whitley County Community Foundation Grants, 3936
Willary Foundation Grants, 3944
William B. Stokely Jr. Foundation Grants, 3947
William J. & Tina Rosenberg Fndn Grants, 3954
William Ray & Ruth E. Collins Fndn Grants, 3956
Williams Companies Foundation Homegrown Giving Grants, 3957
Winston-Salem Foundation Competitive Grants, 3965
Yawkey Foundation Grants, 3980

Arts, Fine
Charles H. Hall Foundation, 1120
Crystelle Waggoner Charitable Trust Grants, 1364
GNOF IMPACT Grants for Arts and Culture, 1838
Helen Bader Foundation Grants, 1992
Helen Gertrude Sparks Charitable Trust Grants, 1993
Helen Pumphrey Denit Charitable Trust Grants, 1996
Katrine Menzing Deakins Trust Grants, 2295
Lewis H. Humphreys Charitable Trust Grants, 2361
Thelma Braun & Bocklett Family Fndn Grants, 3713

Arts, General
1st Source Foundation Grants, 2
2 Depot Square Ipswich Charitable Fndn Grants, 4
A. Gary Anderson Family Foundation Grants, 16
Aaron Foundation Grants, 193
AAUW American Dissertation Fellowships, 195
AAUW American Postdoctoral Research Leave Fellowships, 196
Abbot and Dorothy H. Stevens Foundation Grants, 197
Abbott Fund Community Grants, 201
Abramson Family Foundation Grants, 217
ACF Foundation Grants, 230
Achelis Foundation Grants, 234
Adam Richter Charitable Trust Grants, 298
Adams County Community Foundation of Pennsylvania Grants, 300
Adobe Community Investment Grants, 310
AEC Trust Grants, 315
AFG Industries Grants, 340

SUBJECT INDEX

A Friends' Foundation Trust Grants, 341
Agnes Gund Foundation Grants, 345
Ahmanson Foundation Grants, 355
Ahn Family Foundation Grants, 356
Air Products and Chemicals Grants, 399
Alabama Power Foundation Grants, 400
Alaska Airlines Foundation Grants, 403
Alberto Culver Corporate Contributions Grants, 408
Albuquerque Community Foundation Grants, 412
Alcatel-Lucent Technologies Foundation Grants, 413
Alcoa Foundation Grants, 414
Alexander & Baldwin Fnd Mainland Grants, 417
Alexander and Baldwin Foundation Hawaiian and Pacific Island Grants, 418
Alexis Gregory Foundation Grants, 426
Alice Tweed Tuohy Foundation Grants Program, 437
Allegheny Technologies Charitable Trust, 440
Alpha Natural Resources Corporate Giving, 450
Altman Foundation Health Care Grants, 457
Alvin and Fanny Blaustein Thalheimer Foundation Baltimore Communal Grants, 459
Amador Community Foundation Grants, 481
American Schlafhorst Foundation Grants, 559
Amerigroup Foundation Grants, 565
Andrew Family Foundation Grants, 588
Angels Baseball Foundation Grants, 590
Anna Fitch Ardenghi Trust Grants, 592
Annenberg Foundation Grants, 596
Annie Sinclair Knudsen Memorial Fund/Kaua'i Community Grants Program, 597
AON Foundation Grants, 623
APS Foundation Grants, 646
AptarGroup Foundation Grants, 647
Aratani Foundation Grants, 649
Arizona Cardinals Grants, 655
Arizona Public Service Corporate Giving Program Grants, 658
Arkansas Community Foundation Grants, 660
Arlington Community Foundation Grants, 662
Arthur Ashley Williams Foundation Grants, 671
Arthur F. & Arnold M. Frankel Fndn Grants, 672
Assurant Health Foundation Grants, 789
Atherton Family Foundation Grants, 794
Athwin Foundation Grants, 795
Atlanta Foundation Grants, 796
Audrey & Sydney Irmas Foundation Grants, 799
Aurora Foundation Grants, 802
Autodesk Community Relations Grants, 809
Autzen Foundation Grants, 810
Avista Foundation Economic and Cultural Vitality Grants, 812
Bacon Family Foundation Grants, 817
Bailey Foundation Grants, 818
BancorpSouth Foundation Grants, 823
Banfi Vintners Foundation Grants, 824
Bank of America Fndn Matching Gifts, 825
Barberton Community Foundation Grants, 829
Barker Welfare Foundation Grants, 831
Barnes Group Foundation Grants, 832
Barra Foundation Community Fund Grants, 833
Barra Foundation Project Grants, 834
Batchelor Foundation Grants, 837
Baton Rouge Area Foundation Grants, 838
Battle Creek Community Foundation Grants, 839
Batts Foundation Grants, 840
Bayer Foundation Grants, 846
Beazley Foundation Grants, 865
Beckley Area Foundation Grants, 867
Belk Foundation Grants, 872
Ben B. Cheney Foundation Grants, 875
Benton Community Foundation Grants, 878
Benton County Foundation Grants, 880
Berks County Community Foundation Grants, 881
Berrien Community Foundation Grants, 884
Besser Foundation Grants, 887
Bildner Family Foundation Grants, 891
Blanche and Irving Laurie Foundation Grants, 910
Blanche and Julian Robertson Family Foundation Grants, 911
Blue Grass Community Fndn Harrison Grants, 919

SUBJECT INDEX

Arts, General /661

Blue Grass Community Foundation Hudson-Ellis Fund Grants, 920
Blue Mountain Community Foundation Grants, 921
Blue River Community Foundation Grants, 922
Bodenwein Public Benevolent Foundation Grants, 928
Borkee-Hagley Foundation Grants, 934
Bosque Foundation Grants, 935
Boston Foundation Grants, 936
Bradley-Turner Foundation Grants, 944
Bradley C. Higgins Foundation Grants, 945
Bridgestone/Firestone Trust Fund Grants, 949
Bright Family Foundation Grants, 951
Brook J. Lenfest Foundation Grants, 967
Brown Advisory Charitable Foundation Grants, 969
Brown County Community Foundation Grants, 970
Browning-Kimball Foundation Grants, 971
Bruce and Adele Greenfield Foundation Grants, 972
Burlington Northern Santa Fe Foundation Grants, 978
Byron W. & Alice L. Lockwood Fnd Grants, 991
Cabot Corporation Foundation Grants, 993
Caleb C. and Julia W. Dula Educational and Charitable Foundation Grants, 997
Callaway Foundation Grants, 1002
Cambridge Community Foundation Grants, 1004
Campbell Soup Foundation Grants, 1007
Canada-U.S. Fulbright New Century Scholars Program Grants, 1008
Canada-U.S. Fulbright Senior Specialists Grants, 1009
Capital Region Community Foundation Grants, 1014
Carl & Eloise Pohlad Family Fndn Grants, 1021
Carl B. and Florence E. King Foundation Grants, 1022
Carl C. Icahn Foundation Grants, 1023
Carl M. Freeman Foundation FACES Grants, 1026
Carolyn Foundation Grants, 1034
Carpenter Foundation Grants, 1035
Carrie E. and Lena V. Glenn Foundation Grants, 1036
Cemala Foundation Grants, 1077
Centerville-Washington Foundation Grants, 1078
Central Okanagan Foundation Grants, 1080
Cessna Foundation Grants Program, 1081
CFFVR Clintonville Area Foundation Grants, 1097
CFFVR Frank C. Shattuck Community Grants, 1098
CFFVR Project Grants, 1101
CFFVR Robert and Patricia Endries Family Foundation Grants, 1102
CFFVR Shawano Area Community Foundation Grants, 1104
CFFVR Waupaca Area Community Foundation Grants, 1105
CFFVR Women's Fund for the Fox Valley Region Grants, 1107
Chamberlain Foundation Grants, 1108
Champlin Foundations Grants, 1110
Charles H. Dater Foundation Grants, 1118
Charles H. Hall Foundation, 1120
Charles Lafitte Foundation Grants, 1123
Charlotte County (FL) Community Foundation Grants, 1126
Chazen Foundation Grants, 1129
Chiles Foundation Grants, 1147
Christy-Houston Foundation Grants, 1153
Chula Vista Charitable Foundation Grants, 1154
CICF Legacy Fund Grants, 1158
CIGNA Foundation Grants, 1160
CIT Corporate Giving Grants, 1161
Citizens Bank Mid-Atlantic Charitable Foundation Grants, 1162
Clark-Winchcole Foundation Grants, 1173
Clark County Community Foundation Grants, 1175
Claude Worthington Benedum Fndn Grants, 1178
Clayton Fund Grants, 1180
Cleveland-Cliffs Foundation Grants, 1181
Cleveland Browns Foundation Grants, 1182
CNA Foundation Grants, 1189
Coastal Community Foundation of South Carolina Grants, 1202
Coca-Cola Foundation Grants, 1203
Cockrell Foundation Grants, 1204
Coeta and Donald Barker Foundation Grants, 1205
Collins Foundation Grants, 1210

Columbus Foundation Competitive Grants, 1217
Columbus Foundation Mary Eleanor Morris Fund Grants, 1220
Columbus Foundation Paul G. Duke Grants, 1221
Communities Foundation of Texas Grants, 1236
Community Foundation Alliance City of Evansville Endowment Fund Grants, 1238
Community Foundation for Greater Atlanta Common Good Funds Grants, 1241
Community Foundation for Greater Atlanta Strategic Restructuring Fund Grants, 1246
Community Fndn for Greater Buffalo Grants, 1247
Community Fndn for Monterey County Grants, 1253
Community Fndn for Muskegon County Grants, 1254
Community Foundation for Northeast Michigan Mini-Grants, 1255
Community Foundation for the National Capital Region Community Leadership Grants, 1260
Community Foundation of Bartholomew County Heritage Fund Grants, 1262
Community Foundation of Bartholomew County James A. Henderson Award for Fundraising, 1263
Community Fndn of Central Illinois Grants, 1267
Community Fndn of East Central Illinois Grants, 1268
Community Foundation of Eastern Connecticut General Southeast Grants, 1269
Community Foundation of Greater Birmingham Grants, 1272
Community Foundation of Greater Flint Grants, 1273
Community Foundation of Greater Fort Wayne - Community Endowment and Clarke Endowment Grants, 1274
Community Foundation of Greater Greensboro Community Grants, 1277
Community Foundation of Greater Greensboro Women to Women Fund Grants, 1278
Community Foundation of Greater New Britain Grants, 1281
Community Fndn of Greater Tampa Grants, 1282
Community Foundation of Greenville-Greenville Women Giving Grants, 1283
Community Foundation of Greenville Community Enrichment Grants, 1284
Community Foundation of Greenville Hollingsworth Funds Program/Project Grants, 1285
Community Foundation of Louisville Anna Marble Memorial Fund for Princeton Grants, 1290
Community Foundation of Mount Vernon and Knox County Grants, 1299
Community Fndn of Randolph County Grants, 1302
Community Fndn of Riverside & San Bernardino County Irene S. Rockwell Grants, 1304
Community Fndn of So Alabama Grants, 1306
Community Fndn of South Puget Sound Grants, 1307
Community Fndn of Switzerland County Grants, 1311
Community Foundation of Tampa Bay Grants, 1312
Community Foundation of the Eastern Shore Community Needs Grants, 1313
Community Foundation of the Verdugos Educational Endowment Fund Grants, 1314
Community Foundation of the Verdugos Grants, 1315
Community Fndn of Wabash County Grants, 1316
Connecticut Community Foundation Grants, 1324
Connelly Foundation Grants, 1326
ConocoPhillips Foundation Grants, 1327
Constantin Foundation Grants, 1341
Consumers Energy Foundation, 1343
Cooke Foundation Grants, 1344
Cooper Industries Foundation Grants, 1345
Cornerstone Foundation of Northeastern Wisconsin Grants, 1347
Cowles Charitable Trust Grants, 1354
Cralle Foundation Grants, 1356
Crane Fund Grants, 1358
CSX Corporate Contributions Grants, 1368
Cudd Foundation Grants, 1370
Cumberland Community Foundation Grants, 1374
CUNA Mutual Group Fndn Community Grants, 1375
Curtis Foundation Grants, 1376
Cyrus Eaton Foundation Grants, 1380

Dade Community Foundation Grants, 1410
DaimlerChrysler Corporation Fund Grants, 1412
Dairy Queen Corporate Contributions Grants, 1413
Dale and Edna Walsh Foundation Grants, 1415
Dana Brown Charitable Trust Grants, 1419
Daniel & Nanna Stern Family Fndn Grants, 1422
David Geffen Foundation Grants, 1431
Dayton Power and Light Foundation Grants, 1437
Daywood Foundation Grants, 1438
Deaconess Community Foundation Grants, 1439
Dearborn Community Foundation City of Lawrenceburg Community Grants, 1441
Dearborn Community Foundation County Progress Grants, 1442
Delaware Community Foundation Grants, 1449
DeMatteis Family Foundation Grants, 1457
Denver Foundation Community Grants, 1461
DeRoy Testamentary Foundation Grants, 1469
Deutsche Banc Alex Brown and Sons Charitable Foundation Grants, 1471
Donald and Sylvia Robinson Family Foundation Grants, 1486
Dorothy Rider Pool Health Care Grants, 1500
Dr. Leon Bromberg Charitable Trust Grants, 1509
Drs. Bruce and Lee Foundation Grants, 1513
Dubois County Community Foundation Grants, 1517
DuPage Community Foundation Grants, 1531
Dyson Foundation Mid-Hudson Valley Project Support Grants, 1534
E. Clayton and Edith P. Gengras, Jr. Foundation, 1536
E.J. Grassmann Trust Grants, 1537
E.L. Wiegand Foundation Grants, 1538
Earl and Maxine Claussen Trust Grants, 1541
Eastman Chemical Company Foundation Grants, 1542
Eberly Foundation Grants, 1544
Eddie C. & Sylvia Brown Family Fndn Grants, 1548
EDS Foundation Grants, 1552
Educational Foundation of America Grants, 1553
Edward W. and Stella C. Van Houten Memorial Fund Grants, 1559
Edwin W. and Catherine M. Davis Fndn Grants, 1561
Eisner Foundation Grants, 1564
Elaine Feld Stern Charitable Trust Grants, 1565
Elizabeth Morse Genius Charitable Trust Grants, 1569
Elkhart County Community Foundation Grants, 1573
El Paso Community Foundation Grants, 1579
El Paso Corporate Foundation Grants, 1580
El Pomar Foundation Anna Keesling Ackerman Fund Grants, 1581
El Pomar Foundation Grants, 1582
Elsie H. Wilcox Foundation Grants, 1583
Elsie Lee Garthwaite Memorial Fndn Grants, 1584
Emerson Electric Company Contributions Grants, 1590
Ensworth Charitable Foundation Grants, 1595
Entergy Corporation Micro Grants, 1596
Erie Chapman Foundation Grants, 1600
Erie Community Foundation Grants, 1601
Essex County Community Foundation Discretionary Fund Grants, 1602
Essex County Community Foundation Merrimack Valley General Fund Grants, 1603
Essex County Community Foundation Webster Family Fund Grants, 1604
Estee Lauder Grants, 1606
Ethel S. Abbott Charitable Foundation Grants, 1607
Ethel Sergeant Clark Smith Foundation Grants, 1608
Eugene M. Lang Foundation Grants, 1611
Evanston Community Foundation Grants, 1615
F.M. Kirby Foundation Grants, 1624
Fairfield County Community Foundation Grants, 1625
Fargo-Moorhead Area Foundation Grants, 1632
Farmers Insurance Corporate Giving Grants, 1634
Faye McBeath Foundation Grants, 1635
Fayette County Community Foundation Grants, 1636
Ferree Foundation Grants, 1658
Fidelity Foundation Grants, 1659
Fishman Family Foundation Grants, 1672
FIU Global Civic Engagement Mini Grants, 1673
Foellinger Foundation Grants, 1684
Fondren Foundation Grants, 1685

662 / Arts, General

Ford Motor Company Fund Grants Program, 1687
Forrest C. Lattner Foundation Grants, 1688
Four County Community Fndn General Grants, 1710
Frances and John L. Loeb Family Fund Grants, 1716
Francis L. Abreu Charitable Trust Grants, 1719
Franklin County Community Foundation Grants, 1724
Frank Stanley Beveridge Foundation Grants, 1730
Fred Baldwin Memorial Foundation Grants, 1736
Fred C. & Katherine B. Andersen Grants, 1737
Fred L. Emerson Foundation Grants, 1743
Fremont Area Community Foundation Amazing X Grants, 1744
Fremont Area Community Fndn General Grants, 1746
Fuji Film Grants, 1748
Fulbright Alumni Initiatives Awards, 1749
Fulbright Distinguished Chairs Awards, 1750
Fulbright New Century Scholars Grants, 1753
Fulbright Specialists Program Grants, 1754
Fulbright Scholars in Europe and Eurasia, 1755
Fulbright Traditional Scholar Program in Sub-Saharan Africa, 1756
Fulbright Traditional Scholar Program in the East Asia/Pacific Region, 1757
Fulbright Traditional Scholar Program in the Near East and North Africa Region, 1758
Fulbright Traditional Scholar Program in the South and Central Asia Region, 1759
Fulbright Traditional Scholar Program in the Western Hemisphere, 1760
Fulton County Community Foundation Grants, 1763
Gardner Foundation Grants, 1771
Garland D. Rhoads Foundation, 1774
Gebbie Foundation Grants, 1777
GenCorp Foundation Grants, 1779
General Mills Foundation Grants, 1782
General Motors Foundation Grants, 1783
George A. and Grace L. Long Foundation Grants, 1788
George and Ruth Bradford Foundation Grants, 1789
George A Ohl Jr. Foundation Grants, 1791
George Gund Foundation Grants, 1796
George H.C. Ensworth Memorial Fund Grants, 1797
George S. & Dolores Dore Eccles Fndn Grants, 1801
George W. Brackenridge Foundation Grants, 1802
George W. Codrington Charitable Fndn Grants, 1803
Gheens Foundation Grants, 1814
Gibson Foundation Grants, 1818
Gill Foundation - Gay and Lesbian Fund Grants, 1821
Ginn Foundation Grants, 1822
Giving Sum Annual Grant, 1825
Gladys Brooks Foundation Grants, 1826
GlaxoSmithKline Corporate Grants, 1829
GNOF Exxon-Mobil Grants, 1836
GNOF Freeman Challenge Grants, 1837
GNOF IMPACT Grants for Arts and Culture, 1838
GNOF Norco Community Grants, 1845
Golden Heart Community Foundation Grants, 1850
Goodrich Corporation Foundation Grants, 1854
Goodyear Tire Grants, 1856
Grace and Franklin Bernsen Foundation Grants, 1858
Graham Foundation Grants, 1862
Grand Haven Area Community Fndn Grants, 1864
Grand Rapids Area Community Fndn Grants, 1865
Grand Rapids Area Community Foundation Nashwauk Area Endowment Fund Grants, 1866
Grand Rapids Area Community Foundation Wyoming Grants, 1867
Grand Rapids Area Community Foundation Wyoming Youth Fund Grants, 1868
Grand Rapids Community Foundation Ionia County Grants, 1870
Grand Rapids Community Foundation Lowell Area Fund Grants, 1872
Grand Rapids Community Foundation Southeast Ottawa Grants, 1873
Grand Rapids Community Fndn Sparta Grants, 1875
Greater Cincinnati Foundation Priority and Small Projects/Capacity-Building Grants, 1880
Greater Green Bay Community Fndn Grants, 1881
Greater Kanawha Valley Foundation Grants, 1882
Greater Milwaukee Foundation Grants, 1883
Greater Saint Louis Community Fndn Grants, 1884
Greater Sitka Legacy Fund Grants, 1885
Greater Tacoma Community Foundation Grants, 1886
Greater Worcester Community Foundation Jeppson Memorial Fund for Brookfield Grants, 1889
Green Diamond Charitable Contributions, 1891
Green River Area Community Fndn Grants, 1895
Griffin Foundation Grants, 1903
Guido A. & Elizabeth H. Binda Fndn Grants, 1913
Gulf Coast Community Foundation Grants, 1915
H & R Foundation Grants, 1920
H.A. & Mary K. Chapman Trust Grants, 1921
H.B. Fuller Foundation Grants, 1922
H.J. Heinz Company Foundation Grants, 1923
H. Leslie Hoffman and Elaine S. Hoffman Foundation Grants, 1924
Hallmark Corporate Foundation Grants, 1936
Hardin County Community Foundation Grants, 1946
Harley Davidson Foundation Grants, 1947
Harold Alfond Foundation Grants, 1949
Harold Simmons Foundation Grants, 1955
Harris and Eliza Kempner Fund Grants, 1956
Harrison County Community Foundation Signature Grants, 1958
Harry Bramhall Gilbert Charitable Trust Grants, 1961
Harry Kramer Memorial Fund Grants, 1965
Harry W. Bass, Jr. Foundation Grants, 1968
Hartford Courant Foundation Grants, 1969
Hartford Foundation Regular Grants, 1970
Hawaiian Electric Industries Charitable Foundation Grants, 1975
Hawaii Community Foundation West Hawaii Fund Grants, 1978
Hawn Foundation Grants, 1979
HCA Foundation Grants, 1980
Heckscher Foundation for Children Grants, 1988
Helena Rubinstein Foundation Grants, 1991
Helen Bader Foundation Grants, 1992
Helen Gertrude Sparks Charitable Trust Grants, 1993
Helen S. Boylan Foundation Grants, 1997
Henrietta Tower Wurts Memorial Fndn Grants, 2002
Herbert H. & Grace A. Dow Fndn Grants, 2008
Herman Goldman Foundation Grants, 2012
Hershey Company Grants, 2013
High Meadow Foundation Grants, 2027
Hill Crest Foundation Grants, 2030
Hillman Foundation Grants, 2032
Hillsdale County Community General Adult Foundation Grants, 2034
Hillsdale Fund Grants, 2035
Hilton Head Island Foundation Grants, 2036
Hoglund Foundation Grants, 2040
Holland/Zeeland Community Fndn Grants, 2041
Honda of America Manufacturing Fndn Grants, 2045
Horace A. Kimball and S. Ella Kimball Foundation Grants, 2046
Horizons Community Issues Grants, 2049
Houston Endowment Grants, 2053
Howard and Bush Foundation Grants, 2054
Howe Foundation of North Carolina Grants, 2055
Hudson Webber Foundation Grants, 2072
Huffy Foundation Grants, 2073
Hugh J. Andersen Foundation Grants, 2074
Humana Foundation Grants, 2077
Huntington County Community Foundation Make a Difference Grants, 2083
Huntington National Bank Community Grants, 2084
Hutchinson Community Foundation Grants, 2085
Hut Foundation Grants, 2086
I.A. O'Shaughnessy Foundation Grants, 2088
Idaho Community Foundation Eastern Region Competitive Grants, 2114
Idaho Power Company Corporate Contributions, 2115
IIE 911 Armed Forces Scholarships, 2122
IIE Freeman Foundation Indonesia Internships, 2131
IIE Rockefeller Foundation Bellagio Center Residencies, 2142
Ike and Roz Friedman Foundation Grants, 2145
Illinois Tool Works Foundation Grants, 2147
Irving S. Gilmore Foundation Grants, 2163
Irvin Stern Foundation Grants, 2164
J.B. Reynolds Foundation Grants, 2168
J.M. Long Foundation Grants, 2173
J. Walton Bissell Foundation Grants, 2177
Jackson County Community Foundation Unrestricted Grants, 2180
Jacobs Family Village Neighborhoods Grants, 2184
James & Abigail Campbell Family Fndn Grants, 2185
James A. and Faith Knight Foundation Grants, 2186
James Ford Bell Foundation Grants, 2187
James Graham Brown Foundation Grants, 2189
James M. Collins Foundation Grants, 2195
James R. Thorpe Foundation Grants, 2199
James S. Copley Foundation Grants, 2200
Janus Foundation Grants, 2211
Jasper Foundation Grants, 2212
Jay and Rose Phillips Family Foundation Grants, 2213
Jayne and Leonard Abess Foundation Grants, 2214
Jean and Louis Dreyfus Foundation Grants, 2215
JELD-WEN Foundation Grants, 2218
Jennings County Community Foundation Grants, 2220
Jessie Ball Dupont Fund Grants, 2226
John Ben Snow Memorial Trust Grants, 2231
John Deere Foundation Grants, 2234
John G. Duncan Charitable Trust Grants, 2236
John I. Smith Charities Grants, 2240
John J. Leidy Foundation Grants, 2241
John Jewett and Helen Chandler Garland Foundation Grants, 2242
John Lord Knight Foundation Grants, 2243
Johns Manville Fund Grants, 2251
Johnson & Johnson/SAH Arts & Healing Grants, 2254
John W. Alden Trust Grants, 2260
John W. Anderson Foundation Grants, 2262
John W. Speas and Effie E. Speas Memorial Trust Grants, 2264
Joseph Alexander Foundation Grants, 2265
Joseph Drown Foundation Grants, 2268
Josephine S. Gumbiner Foundation Grants, 2273
Josephine Schell Russell Charitable Trust Grants, 2274
Josiah W. and Bessie H. Kline Foundation Grants, 2278
Judith and Jean Pape Adams Charitable Foundation Tulsa Area Grants, 2280
Judith Clark-Morrill Foundation Grants, 2281
Julius N. Frankel Foundation Grants, 2282
Kahuku Community Fund, 2286
K and F Baxter Family Foundation Grants, 2287
Katharine Matthies Foundation Grants, 2291
Katherine John Murphy Foundation Grants, 2293
Kathryne Beynon Foundation Grants, 2294
Kenai Peninsula Foundation Grants, 2298
Kenneth T. & Eileen L. Norris Fndn Grants, 2300
Ketchikan Community Foundation Grants, 2303
Kettering Fund Grants, 2305
Kinsman Foundation Grants, 2309
Kodiak Community Foundation Grants, 2314
Kosciusko County Community Fndn Grants, 2319
Kovler Family Foundation Grants, 2323
Kuntz Foundation Grants, 2326
Land O'Lakes Foundation Mid-Atlantic Grants, 2341
Leo Niessen Jr., Charitable Trust Grants, 2358
Lester E. Yeager Charitable Trust B Grants, 2359
Lewis H. Humphreys Charitable Trust Grants, 2361
Liberty Bank Foundation Grants, 2362
Lillian S. Wells Foundation Grants, 2365
Lincoln Financial Foundation Grants, 2367
Lotus 88 Fnd for Women & Children Grants, 2376
Louie M. & Betty M. Phillips Fndn Grants, 2378
Louis and Elizabeth Nave Flarsheim Charitable Foundation Grants, 2379
Lowe Foundation Grants, 2381
Lowell Berry Foundation Grants, 2382
Lubbock Area Foundation Grants, 2383
Lubrizol Foundation Grants, 2384
Lucile Horton Howe and Mitchell B. Howe Foundation Grants, 2385
Lucy Downing Nisbet Charitable Fund Grants, 2387
Lucy Gooding Charitable Fndn Grants, 2388
Ludwick Family Foundation Grants, 2389
M. Bastian Family Foundation Grants, 2402

SUBJECT INDEX

M.J. Murdock Charitable Trust General Grants, 2405
Macquarie Bank Foundation Grants, 2410
Madison County Community Foundation General Grants, 2413
Manuel D. & Rhoda Mayerson Fndn Grants, 2420
Marathon Petroleum Corporation Grants, 2422
Marcia and Otto Koehler Foundation Grants, 2427
Mardag Foundation Grants, 2428
Margaret T. Morris Foundation Grants, 2429
Marie H. Bechtel Charitable Remainder Uni-Trust Grants, 2432
Mary K. Chapman Foundation Grants, 2443
Mary Owen Borden Foundation Grants, 2446
Mary S. and David C. Corbin Foundation Grants, 2447
Matilda R. Wilson Fund Grants, 2453
Max and Victoria Dreyfus Foundation Grants, 2457
McCombs Foundation Grants, 2491
McConnell Foundation Grants, 2492
McInerny Foundation Grants, 2495
McKesson Foundation Grants, 2496
Mead Johnson Nutritionals Evansville-Area Organizations Grants, 2503
Mead Witter Foundation Grants, 2507
Mericos Foundation Grants, 2518
Meriden Foundation Grants, 2519
Merrick Foundation Grants, 2521
Merrick Foundation Grants, 2522
Mervin Bovaird Foundation Grants, 2523
Meyer Memorial Trust Special Grants, 2538
MGN Family Foundation Grants, 2542
Middlesex Savings Charitable Foundation Capacity Building Grants, 2551
Mimi and Peter Haas Fund Grants, 2557
Moline Foundation Community Grants, 2590
Morris & Gwendolyn Cafritz Fndn Grants, 2594
Nathan Cummings Foundation Grants, 2627
Nelda C. and H.J. Lutcher Stark Fndn Grants, 2672
Newton County Community Foundation Grants, 2681
New York University Steinhardt School of Education Fellowships, 2684
NHSCA Arts in Health Care Project Grants, 2741
Nicholas H. Noyes Jr. Memorial Fndn Grants, 2791
Nina Mason Pulliam Charitable Trust Grants, 2921
Noble County Community Foundation Grants, 2975
Nordson Corporation Foundation Grants, 2981
North Carolina Community Foundation Grants, 2990
Northwestern Mutual Foundation Grants, 2999
Norwin S. & Elizabeth N. Bean Fndn Grants, 3001
OceanFirst Foundation Major Grants, 3062
Oceanside Charitable Foundation Grants, 3063
Oleonda Jameson Trust Grants, 3071
Olive Higgins Prouty Foundation Grants, 3072
Oppenstein Brothers Foundation Grants, 3078
Oregon Community Fndn Community Grants, 3082
Oscar Rennebohm Foundation Grants, 3085
Parkersburg Area Community Foundation Action Grants, 3116
Patrick and Anna M. Cudahy Fund Grants, 3120
Peacock Foundation Grants, 3133
Perpetual Trust for Charitable Giving Grants, 3141
Perry County Community Foundation Grants, 3142
Peter Kiewit Foundation General Grants, 3154
Peter Kiewit Foundation Small Grants, 3155
Petersburg Community Foundation Grants, 3156
Peyton Anderson Foundation Grants, 3160
Phoenix Suns Charities Grants, 3201
Pike County Community Foundation Grants, 3222
Pinellas County Grants, 3223
Pinkerton Foundation Grants, 3224
Piper Jaffray Fndn Communities Giving Grants, 3227
Pittsburgh Foundation Community Fund Grants, 3230
Plough Foundation Grants, 3238
PMI Foundation Grants, 3239
Pokagon Fund Grants, 3242
Polk Bros. Foundation Grants, 3243
Posey Community Fndn Women's Fund Grants, 3251
Posey County Community Foundation Grants, 3252
Powell Foundation Grants, 3255
Price Chopper's Golub Foundation Grants, 3261
Price Family Charitable Fund Grants, 3262

Pride Foundation Grants, 3268
Prince Charitable Trusts Chicago Grants, 3270
Prince Charitable Trusts DC Grants, 3271
Princeton Area Community Foundation Greater Mercer Grants, 3272
Principal Financial Group Foundation Grants, 3273
Puerto Rico Community Foundation Grants, 3278
Pulaski County Community Foundation Grants, 3279
Putnam County Community Foundation Grants, 3283
Quaker Chemical Foundation Grants, 3286
Rajiv Gandhi Foundation Grants, 3302
Ralphs Food 4 Less Foundation Grants, 3305
Rasmuson Foundation Tier One Grants, 3307
Rasmuson Foundation Tier Two Grants, 3308
Rathmann Family Foundation Grants, 3309
Rhode Island Foundation Grants, 3362
Richard and Helen DeVos Foundation Grants, 3364
Richard & Susan Smith Family Fndn Grants, 3365
Richard D. Bass Foundation Grants, 3366
Ripley County Community Foundation Grants, 3376
Ripley County Community Foundation Small Project Grants, 3377
Robbins Charitable Foundation Grants, 3378
Robert B McMillen Foundation Grants, 3381
Robert R. Meyer Foundation Grants, 3387
Rochester Area Community Foundation Grants, 3391
Rochester Area Foundation Grants, 3392
Rockwell International Corporate Trust Grants Program, 3399
Rogers Family Foundation Grants, 3401
Ronald McDonald House Charities Grants, 3405
Roy & Christine Sturgis Charitable Grants, 3415
Rush County Community Foundation Grants, 3424
Ruth and Vernon Taylor Foundation Grants, 3426
Ruth H. & Warren A. Ellsworth Fndn Grants, 3429
S. Livingston Mather Charitable Trust Grants, 3448
S. Mark Taper Foundation Grants, 3449
Saint Louis Rams Fndn Community Donations, 3453
Salem Foundation Charitable Trust Grants, 3456
Salem Foundation Grants, 3457
Samuel N. and Mary Castle Foundation Grants, 3467
Samuel S. Johnson Foundation Grants, 3468
San Antonio Area Foundation Grants, 3469
San Juan Island Community Foundation Grants, 3480
Santa Fe Community Foundation Seasonal Grants-Fall Cycle, 3482
Sartain Lanier Family Foundation Grants, 3486
Sasco Foundation Grants, 3487
Schering-Plough Foundation Community Initiatives Grants, 3491
Schramm Foundation Grants, 3495
Schurz Communications Foundation Grants, 3496
Scott County Community Foundation Grants, 3500
Seabury Foundation Grants, 3501
Seagate Tech Corp Capacity to Care Grants, 3502
Seaver Institute Grants, 3509
Self Foundation Grants, 3510
Sensient Technologies Foundation Grants, 3513
Shield-Ayres Foundation Grants, 3546
Shopko Fndn Community Charitable Grants, 3547
Simmons Foundation Grants, 3563
Sioux Falls Area Community Foundation Community Fund Grants (Unrestricted), 3568
Sioux Falls Area Community Foundation Spot Grants (Unrestricted), 3570
Society for the Arts in Healthcare Grants, 3602
Society for the Arts in Healthcare Environmental Arts Research Grant, 3603
Sonoco Foundation Grants, 3621
Sonora Area Foundation Competitive Grants, 3622
Sony Corporation of America Grants, 3623
Southbury Community Trust Fund, 3625
South Madison Community Foundation Grants, 3626
Southwest Florida Community Foundation Competitive Grants, 3627
Southwest Gas Corporation Foundation Grants, 3628
Spencer County Community Foundation Grants, 3632
Square D Foundation Grants, 3634
Stackpole-Hall Foundation Grants, 3643
Stark Community Fndn Women's Grants, 3645

Stella and Charles Guttman Foundation Grants, 3651
Sterling and Shelli Gardner Foundation Grants, 3653
Steuben County Community Foundation Grants, 3654
Strake Foundation Grants, 3658
Stranahan Foundation Grants, 3659
Strowd Roses Grants, 3664
Subaru of Indiana Automotive Foundation Grants, 3672
Summit Foundation Grants, 3674
Sunderland Foundation Grants, 3675
Sunoco Foundation Grants, 3679
SunTrust Bank Trusteed Foundations Florence C. and Harry L. English Memorial Fund Grants, 3680
SunTrust Bank Trusteed Foundations Greene-Sawtell Grants, 3681
SunTrust Bank Trusteed Foundations Harriet McDaniel Marshall Tust Grants, 3682
SunTrust Bank Trusteed Foundations Nell Warren Elkin and William Simpson Elkin Grants, 3683
SunTrust Bank Trusteed Foundations Thomas Guy Woolford Charitable Trust Grants, 3684
SunTrust Bank Trusteed Foundations Walter H. and Marjory M. Rich Memorial Fund Grants, 3685
Susan Mott Webb Charitable Trust Grants, 3694
Susan Vaughan Foundation Grants, 3695
Taubman Endowment for the Arts Fndn Grants, 3702
Tauck Family Foundation Grants, 3703
Taylor S. Abernathy and Patti Harding Abernathy Charitable Trust Grants, 3704
TE Foundation Grants, 3705
Tension Envelope Foundation Grants, 3707
Thelma Braun & Bocklett Family Fndn Grants, 3713
Thomas C. Ackerman Foundation Grants, 3719
Tommy Hilfiger Corporate Foundation Grants, 3737
Toyota Motor Manufacturing of Alabama Grants, 3743
Toyota Motor Manufacturing of Indiana Grants, 3744
Toyota Motor Manuf of Mississippi Grants, 3746
Toyota Motor Manufacturing of Texas Grants, 3747
Toyota Motor Manuf of West Virginia Grants, 3748
Toyota Motor Sales, USA Grants, 3750
Triangle Community Foundation Donor-Advised Grants, 3757
Trull Foundation Grants, 3761
Tull Charitable Foundation Grants, 3762
Turner Foundation Grants, 3763
U.S. Cellular Corporation Grants, 3766
Union Benevolent Association Grants, 3789
United Technologies Corporation Grants, 3803
Unity Foundation Of LaPorte County Grants, 3804
US Airways Community Contributions, 3825
US Airways Community Foundation Grants, 3826
Vancouver Foundation Grants and Community Initiatives Program, 3849
Vectren Foundation Grants, 3857
Vermont Community Foundation Grants, 3865
W.C. Griffith Foundation Grants, 3881
W. C. Griffith Foundation Grants, 3882
W. Clarke Swanson, Jr. Foundation Grants, 3883
Walker Area Community Foundation Grants, 3895
Wallace Alexander Gerbode Foundation Grants, 3896
Walter L. Cross III Family Foundation Grants, 3901
Warrick County Community Foundation Grants, 3904
Wayne County Foundation Grants, 3909
Weingart Foundation Grants, 3913
Western New York Foundation Grants, 3922
Weyerhaeuser Company Foundation Grants, 3925
White County Community Foundation - Annie Horton Scholarship, 3929
Willard & Pat Walker Charitable Fndn Grants, 3943
William G. & Helen C. Hoffman Fndn Grants, 3950
William G. Gilmore Foundation Grants, 3951
William Ray & Ruth E. Collins Fndn Grants, 3956
Williams Companies Foundation Homegrown Giving Grants, 3957
Winston-Salem Fndn Stokes County Grants, 3967

Asia

ASM-UNESCO Leadership Grant for International Educators, 715
Fulbright Traditional Scholar Program in the East Asia/Pacific Region, 1757

664 / Asia, East (Far East)

Asia, East (Far East)
Fulbright Traditional Scholar Program in the East Asia/Pacific Region, 1757

Asia, Southeast
Fulbright Traditional Scholar Program in the East Asia/Pacific Region, 1757

Asian Americans
American Sociological Association Minority Fellowships, 563
Ben B. Cheney Foundation Grants, 875
John R. Oishei Foundation Grants, 2249
MLA Grad Scholarship for Minority Students, 2565
Philadelphia Foundation Organizational Effectiveness Grants, 3189
SNM PDEF Mickey Williams Minority Student Scholarships, 3584

Asian Arts
Christensen Fund Regional Grants, 1149
E. Rhodes & Leona B. Carpenter Grants, 1539

Assisted-Living Programs
ACL Centers for Independent Living Competition Grants, 239
Adelaide Breed Bayrd Foundation Grants, 306
California Endowment Innovative Ideas Challenge Grants, 1000
CFFVR Basic Needs Giving Partnership Grants, 1094
Christine & Katharina Pauly Trust Grants, 1150
Clark and Ruby Baker Foundation Grants, 1174
David N. Lane Trust Grants for Aged and Indigent Women, 1433
Edward N. and Della L. Thome Memorial Foundation Grants, 1557
Frank B. Hazard General Charity Fund Grants, 1721
James R. Thorpe Foundation Grants, 2199
Jenkins Foundation: Improving the Health of Greater Richmond Grants, 2219
Marjorie Moore Charitable Foundation Grants, 2438
Mary Black Foundation Active Living Grants, 2440
May and Stanley Smith Charitable Trust Grants, 2459
McLean Contributionship Grants, 2497
MetroWest Health Foundation Grants--Healthy Aging, 2527
Priddy Foundation Program Grants, 3266
Reinberger Foundation Grants, 3357
Sheltering Arms Fund Grants, 3545
Sophia Romero Trust Grants, 3624
Union Bank, N.A. Corporate Sponsorships and Donations, 3787
Union Bank, N.A. Foundation Grants, 3788

Assistive Technology
Christopher & Dana Reeve Foundation Quality of Life Grants, 1152
HRAMF Deborah Munroe Noonan Memorial Research Grants, 2058
John D. & Katherine A. Johnston Fndn Grants, 2233
May and Stanley Smith Charitable Trust Grants, 2459
Reader's Digest Partners for Sight Fndn Grants, 3345
U.S. Department of Education Rehabilitation Engineering Research Centers Grants, 3769
Vigneron Memorial Fund Grants, 3873

Asthma
AAAAI ARTrust Mini Grants for Allied Health, 20
AAAAI ARTrust Grants for Clinical Research, 22
AAAAI Distinguished Service Award, 27
AAAAI Fellows-in-Training Abstract Award, 28
AAAAI Fellows-in-Training Grants, 29
AAAAI Mentorship Award, 31
AAAAI RSLAAIS Leadership Award, 33
AAAAI Special Recognition Award, 34
AAFA Investigator Research Grants, 128
ACAAI Foundation Research Grants, 219
Everyone Breathe Asthma Education Grants, 1616
Fairlawn Foundation Grants, 1626
GNOF IMPACT Kahn-Oppenheim Grants, 1842

Kathryne Beynon Foundation Grants, 2294
Medtronic Foundation Strengthening Health Systems Grants, 2515
NHLBI Airway Smooth Muscle Function and Targeted Therapeutics in Human Asthma Grants, 2687
NHLBI Clinical Centers for the NHLBI Asthma Network (AsthmaNet) Grants, 2697
Quantum Foundation Grants, 3290
RACGP Nat Asthma Council Research Award, 3296
Sandler Program for Asthma Research Grants, 3475

Astronomy
Alfred P. Sloan Foundation International Science Engagement Grants, 431
Canada Graduate Scholarships (CGS) and NSERC Postgraduate Scholarships (PGS), 1010
Gruber Foundation Cosmology Prize, 1907
Perkin Fund Grants, 3139
Roy J. Carver Charitable Trust Medical and Science Research Grants, 3416

Astrophysics
Brinson Foundation Grants, 953
Royal Norwegian Embassy Kavli Prizes, 3414

At-Risk Students
Advance Auto Parts Corporate Giving Grants, 312
Alfred E. Chase Charitable Foundation Grants, 430
Charles H. Pearson Foundation Grants, 1121
CUNA Mutual Group Fndn Community Grants, 1375
Elizabeth Morse Genius Charitable Trust Grants, 1569
Golden Heart Community Foundation Grants, 1850
Greater Sitka Legacy Fund Grants, 1885
Guy I. Bromley Trust Grants, 1919
Helen Irwin Littauer Educational Trust Grants, 1994
Hilda and Preston Davis Foundation Grants, 2028
John W. Speas and Effie E. Speas Memorial Trust Grants, 2264
Kenai Peninsula Foundation Grants, 2298
Ketchikan Community Foundation Grants, 2303
Kodiak Community Foundation Grants, 2314
Lewis H. Humphreys Charitable Trust Grants, 2361
Louetta M. Cowden Foundation Grants, 2377
Louis and Elizabeth Nave Flarsheim Charitable Foundation Grants, 2379
Mardag Foundation Grants, 2428
MetroWest Health Foundation Grants to Reduce the Incidence of High Risk Behaviors Among Adolescents, 2528
Michael Reese Health Trust Responsive Grants, 2545
Petersburg Community Foundation Grants, 3156
Portland General Electric Foundation Grants, 3250
Sidgmore Family Foundation Grants, 3551
Sioux Falls Area Community Foundation Field-of-Interest and Donor-Advised Grants, 3569
Stocker Foundation Grants, 3656
Tauck Family Foundation Grants, 3703
The Ray Charles Foundation Grants, 3716
Thomas Sill Foundation Grants, 3724
Union Bank, N.A. Foundation Grants, 3788
Victor E. Speas Foundation Grants, 3868

At-Risk Youth
Advance Auto Parts Corporate Giving Grants, 312
Aid for Starving Children Emergency Assistance Fund Grants, 367
Alfred E. Chase Charitable Foundation Grants, 430
Alliance Healthcare Foundation Grants, 444
Amelia Sillman Rockwell and Carlos Perry Rockwell Charities Fund Grants, 501
Andre Agassi Charitable Foundation Grants, 587
Bernard and Audre Rapoport Foundation Health Grants, 882
Boston Foundation Grants, 936
Bright Promises Foundation Grants, 952
CFFVR Basic Needs Giving Partnership Grants, 1094
CFFVR Robert and Patricia Endries Family Foundation Grants, 1102
CFFVR Schmidt Family G4 Grants, 1103
Charles H. Pearson Foundation Grants, 1121

SUBJECT INDEX

Charles Lafitte Foundation Grants, 1123
Christine & Katharina Pauly Trust Grants, 1150
CIT Corporate Giving Grants, 1161
Citizens Bank Mid-Atlantic Charitable Foundation Grants, 1162
CNO Financial Group Community Grants, 1201
Columbus Foundation Traditional Grants, 1223
Community Foundation for Northeast Michigan Tobacco Settlement Grants, 1256
Community Foundation of Greater Greensboro Community Grants, 1277
Constellation Energy Corporate Grants, 1342
Cruise Industry Charitable Foundation Grants, 1363
CUNA Mutual Group Fndn Community Grants, 1375
Dallas Women's Foundation Grants, 1417
Deborah Munroe Noonan Memorial Grants, 1443
Denver Broncos Charities Fund Grants, 1460
Educational Foundation of America Grants, 1553
Edward and Romell Ackley Foundation Grants, 1555
Elizabeth Morse Genius Charitable Trust Grants, 1569
EPA Children's Health Protection Grants, 1598
Express Scripts Foundation Grants, 1620
Four County Community Fndn General Grants, 1710
Four County Community Foundation Healthy Senior/Healthy Youth Fund Grants, 1711
Frank Reed and Margaret Jane Peters Memorial Fund II Grants, 1728
Frederick W. Marzahl Memorial Fund Grants, 1741
Global Fund for Children Grants, 1832
Golden Heart Community Foundation Grants, 1850
Grace and Franklin Bernsen Foundation Grants, 1858
Greater Sitka Legacy Fund Grants, 1885
Greater Tacoma Community Foundation Ryan Alan Hade Endowment Fund, 1887
Guy I. Bromley Trust Grants, 1919
Hasbro Children's Fund Grants, 1973
Health Foundation of Greater Cincinnati Grants, 1983
Helen Irwin Littauer Educational Trust Grants, 1994
Herbert A. & Adrian W. Woods Fndn Grants, 2007
Hilda and Preston Davis Foundation Grants, 2028
Horace Moses Charitable Foundation Grants, 2047
Jack H. & William M. Light Trust Grants, 2179
Janus Foundation Grants, 2211
Jim Moran Foundation Grants, 2228
John Edward Fowler Memorial Fndn Grants, 2235
Josephine Schell Russell Charitable Trust Grants, 2274
Kenai Peninsula Foundation Grants, 2298
Ketchikan Community Foundation Grants, 2303
Kodiak Community Foundation Grants, 2314
Leo Goodwin Foundation Grants, 2354
Leon and Thea Koerner Foundation Grants, 2355
Lillian S. Wells Foundation Grants, 2365
Louis H. Aborn Foundation Grants, 2380
Mabel A. Horne Trust Grants, 2406
Mardag Foundation Grants, 2428
Mary Black Foundation Early Childhood Development Grants, 2442
Mary Owen Borden Foundation Grants, 2446
Mattel Children's Foundation Grants, 2454
Mattel International Grants Program, 2455
McKesson Foundation Grants, 2496
MetroWest Health Foundation Grants to Reduce the Incidence of High Risk Behaviors Among Adolescents, 2528
MGM Resorts Foundation Community Grants, 2541
Michael Reese Health Trust Responsive Grants, 2545
Mid-Iowa Health Foundation Community Response Grants, 2549
Milagro Foundation Grants, 2552
Nationwide Insurance Foundation Grants, 2637
OneFamily Foundation Grants, 3075
Peter and Elizabeth C. Tower Foundation Annual Intellectual Disabilities Grants, 3144
Peter and Elizabeth C. Tower Foundation Annual Mental Health Grants, 3145
Peter and Elizabeth C. Tower Foundation Mental Health Reference and Resource Materials Mini-Grants, 3147
Peter and Elizabeth C. Tower Foundation Social and Emotional Preschool Curriculum Grants, 3151

SUBJECT INDEX

Petersburg Community Foundation Grants, 3156
PeyBack Foundation Grants, 3159
Pinkerton Foundation Grants, 3224
Porter County Health and Wellness Grant, 3246
Portland General Electric Foundation Grants, 3250
Raymond John Wean Foundation Grants, 3310
Richard Davoud Donchian Foundation Grants, 3367
Robins Foundation Grants, 3390
Ronald McDonald House Charities Grants, 3405
Santa Barbara Foundation Strategy Grants - Core Support, 3481
Schlessman Family Foundation Grants, 3493
Sidgmore Family Foundation Grants, 3551
Sioux Falls Area Community Foundation Field-of-Interest and Donor-Advised Grants, 3569
Stella and Charles Guttman Foundation Grants, 3651
Stewart Huston Charitable Trust Grants, 3655
Stocker Foundation Grants, 3656
Tauck Family Foundation Grants, 3703
The Ray Charles Foundation Grants, 3716
Thomas Sill Foundation Grants, 3724
Topfer Family Foundation Grants, 3739
Tri-State Community Twenty-first Century Endowment Fund Grants, 3756
Union Bank, N.A. Foundation Grants, 3788
Victor E. Speas Foundation Grants, 3868
W.K. Kellogg Foundation Healthy Kids Grants, 3886
Wood-Claeyssens Foundation Grants, 3971
Z. Smith Reynolds Foundation Small Grants, 3982

Athletics
Abby's Legendary Pizza Foundation Grants, 204
ACSM Dr. Raymond A. Weiss Research Endowment Grant, 262
Adler-Clark Electric Community Commitment Foundation Grants, 307
El Pomar Foundation Grants, 1582
Four County Community Fndn General Grants, 1710
Textron Corporate Contributions Grants, 3712
Todd Brock Family Foundation Grants, 3736
Yawkey Foundation Grants, 3980

Atmospheric Sciences
W.M. Keck Foundation Science and Engineering Research Grants, 3889
Wellcome Trust Biomedical Science Grants, 3918

Attention Deficit Hyperactivity Disorder
AACAP Elaine Schlosser Lewis Award for Research on Attention-Deficit Disorder, 55
AACAP Pilot Research Award for Attention-Deficit Disorder, 66
Abracadabra Foundation Grants, 216

Audience Development
Air Products and Chemicals Grants, 399
Alvin and Fanny Blaustein Thalheimer Foundation Baltimore Communal Grants, 459
Connecticut Community Foundation Grants, 1324
McKesson Foundation Grants, 2496
SfN Science Educator Award, 3535

Audiology
ASHA Multicultural Activities Projects Grants, 697
ASHA Students Preparing for Academic & Research Careers (SPARC) Award, 700
ASHFoundation Graduate Student International Scholarship, 702
ASHFoundation Graduate Student Scholarships for Minority Students, 704
ASHFoundation Graduate Student with a Disability Scholarship, 705
ASHFoundation New Century Scholars Program Doctoral Scholarships, 706
ASHFoundation New Century Scholars Research Grant, 707
ASHFoundation Grant for New Investigators, 708
ASHFoundation Student Research Grants in Audiology, 710
Special Olympics Health Profess Student Grants, 3629

Audiovisual Materials
Vermont Community Foundation Grants, 3865

Australia
American College of Surgeons Australia and New Zealand Chapter Travelling Fellowships, 519
Australasian Institute of Judicial Administration Seed Funding Grants, 806

Autism
CFFVR Wisconsin King's Daus & Sons Grants, 1106
Horizon Foundation for New Jersey Grants, 2048
HRAMF Deborah Munroe Noonan Memorial Research Grants, 2058
Ireland Family Foundation Grants, 2162
Lisa Higgins-Hussman Foundation Grants, 2370
NWHF Mark O. Hatfield Research Fellowship, 3035
Partnership for Cures Two Years To Cures Grants, 3118
Thompson Charitable Foundation Grants, 3727

Autoimmunity
AIHS Alberta/Pfizer Translat Research Grants, 374
Glaucoma Research Pilot Project Grants, 1828
Scleroderma Foundation Established Investigator Grants, 3497

Automotive Engineering
ArvinMeritor Grants, 675
BMW of North America Charitable Contributions, 926

Bacteriology
ASM/CDC Fellowships in Infectious Disease and Public Health Microbiology, 716
Pfizer ASPIRE EU Emerging Mechanisms of Resistance Antibacterial Research Awards, 3167
Pfizer ASPIRE EU MRSA Nosocomial Pneumonia & MRSA Complicated Skin & Soft Tissue Infections Antibacterial Research Awards, 3168
Pfizer ASPIRE North America Narrow Spectrum Antibiotics for the Treatment of MRSA Research Awards, 3171

Bangladesh
Fulbright Traditional Scholar Program in the South and Central Asia Region, 1759

Baptist Church
Booth-Bricker Fund Grants, 933
Bosque Foundation Grants, 935
Bradley-Turner Foundation Grants, 944
Effie and Wofford Cain Foundation Grants, 1563

Basic Living Expenses
ADA Foundation Relief Grants, 293
Amica Companies Foundation Grants, 574
Ben B. Cheney Foundation Grants, 875
Carl R. Hendrickson Family Foundation Grants, 1028
CICF City of Noblesville Community Grant, 1155
IBRO-PERC InEurope Short Stay Grants, 2097
IBRO Asia Reg APRC Exchange Fellowships, 2101
Mary L. Peyton Foundation Grants, 2445
MBL Burr & Susie Steinbach Fellowship, 2467
Swindells Charitable Foundation, 3697

Basic Skills Education
Ahmanson Foundation Grants, 355
Allen P. & Josephine B. Green Fndn Grants, 443
Allyn Foundation Grants, 448
Alpha Natural Resources Corporate Giving, 450
Arkell Hall Foundation Grants, 661
Atlanta Foundation Grants, 796
Auburn Foundation Grants, 798
Ball Brothers Foundation General Grants, 820
Baptist Community Ministries Grants, 828
Battle Creek Community Foundation Grants, 839
Benton Community Foundation Grants, 878
Berrien Community Foundation Grants, 884
Blue Mountain Community Foundation Grants, 921
Blue River Community Foundation Grants, 922
Bodenwein Public Benevolent Foundation Grants, 928

Boettcher Foundation Grants, 931
Booth-Bricker Fund Grants, 933
Boston Foundation Grants, 936
Brown County Community Foundation Grants, 970
Carl B. and Florence E. King Foundation Grants, 1022
Cemala Foundation Grants, 1077
CICF City of Noblesville Community Grant, 1155
CIGNA Foundation Grants, 1160
Community Foundation of Bartholomew County Heritage Fund Grants, 1262
Community Foundation of Bartholomew County James A. Henderson Award for Fundraising, 1263
Community Fndn of Central Illinois Grants, 1267
Community Foundation of Greater Fort Wayne - Community Endowment and Clarke Endowment Grants, 1274
Cornerstone Foundation of Northeastern Wisconsin Grants, 1347
Cowles Charitable Trust Grants, 1354
Cruise Industry Charitable Foundation Grants, 1363
Dayton Power and Light Foundation Grants, 1437
Evjue Foundation Grants, 1617
Field Foundation of Illinois Grants, 1660
Frances L. and Edwin L. Cummings Memorial Fund Grants, 1717
G.N. Wilcox Trust Grants, 1768
George Foundation Grants, 1795
George W. Wells Foundation Grants, 1804
Guido A. & Elizabeth H. Binda Fndn Grants, 1913
Hallmark Corporate Foundation Grants, 1936
Harold Simmons Foundation Grants, 1955
Helen Steiner Rice Foundation Grants, 1998
Howard and Bush Foundation Grants, 2054
John H. and Wilhelmina D. Harland Charitable Foundation Children and Youth Grants, 2238
John I. Smith Charities Grants, 2240
Joseph H. & Florence A. Roblee Fndn Grants, 2269
Mardag Foundation Grants, 2428
Mary Wilmer Covey Charitable Trust Grants, 2448
May and Stanley Smith Charitable Trust Grants, 2459
Norcliffe Foundation Grants, 2979
Oppenstein Brothers Foundation Grants, 3078
PacifiCare Foundation Grants, 3102
Parkersburg Area Community Foundation Action Grants, 3116
PepsiCo Foundation Grants, 3137
Peyton Anderson Foundation Grants, 3160
Principal Financial Group Foundation Grants, 3273
San Antonio Area Foundation Grants, 3469
Sony Corporation of America Grants, 3623
Sterling-Turner Charitable Foundation Grants, 3652
Strake Foundation Grants, 3658
U.S. Department of Education Rehabilitation Training - Rehabilitation Continuing Education - Institute on Rehabilitation Issues, 3771
Wilson-Wood Foundation Grants, 3960

Beautification
Hardin County Community Foundation Grants, 1946

Behavioral Medicine
ACSM Coca-Cola Company Doctoral Student Grant on Behavior Research, 261
Allen P. & Josephine B. Green Fndn Grants, 443
Bravewell Leadership Award, 946
Cystic Fibrosis Canada Fellowships, 1384
James S. McDonnell Fnd Research Grants, 2203
James S. McDonnell Foundation Scholar Awards, 2204
NCI Stages of Breast Development: Normal to Metastatic Disease Grants, 2655
NEI Ruth L. Kirschstein National Research Service Award Short-Term Institutional Research Training Grants, 2668
NHLBI Ruth L. Kirschstein National Research Service Awards for Individual Postdoctoral Fellows, 2729
NHLBI Ruth L. Kirschstein National Research Service Awards for Individual Predoctoral Fellowships to Promote Diversity in Health-Related Research, 2730
NIA AIDS and Aging: Behavioral Sciences Prevention Research Grants, 2747

NIH Research on Sleep and Sleep Disorders, 2898
NIH Ruth L. Kirschstein National Research Service Awards for Individual Postdoc Fellowships, 2902
NIMH AIDS and Aging: Behavioral Sciences Prevention Research Grants, 2915
NINR AIDS and Aging: Behavioral Sciences Prevention Research Grants, 2927
Premera Blue Cross Grants, 3258

Behavioral Sciences
ACS Research Scholar Grants in Psychosocial and Behavioral and Cancer Control Research, 281
Alzheimer's Association Investigator-Initiated Research Grants, 466
APA Congressional Fellowships, 624
Arizona Diamondbacks Charities Grants, 657
Cystic Fibrosis Canada Fellowships, 1384
Cystic Fibrosis Canada Studentships, 1390
Cystic Fibrosis Canada Summer Studentships, 1391
Duke University Postdoctoral Research Fellowships in Aging, 1528
Health Foundation of Greater Cincinnati Grants, 1983
James S. McDonnell Foundation Scholar Awards, 2204
James S. McDonnell Foundation Understanding Human Cognition Awards, 2205
National Center for Responsible Gaming Grants, 2629
National Center for Responsible Gaming Postdoctoral Fellowships, 2630
National Center for Resp Gaming Seed Grants, 2631
NCI Cancer Education and Career Development Program, 2648
NCI Exploratory Grants for Behavioral Research in Cancer Control, 2651
NCI Ruth L. Kirschstein National Research Service Award Institutional Training Grants, 2654
NHLBI Mentored Quantitative Research Career Development Awards, 2711
NHLBI Pathway to Independence Awards, 2718
NHLBI Short-Term Research Education Program to Increase Diversity in Health-Related Research, 2733
NHLBI Translating Basic Behavioral and Social Science Discoveries into Interventions to Reduce Obesity: Centers for Behavioral Intervention Development Grants, 2739
NIA Alzheimer's Disease Core Centers Grants, 2748
NIA Archiving and Development of Socialbehavioral Datasets in Aging Related Studies Grants, 2752
NIA Network Infrastructure Support for Emerging Behavioral & Social Research in Aging, 2777
NIDDK Planning Grants For Translational Research For The Prevention And Control Of Diabetes And Obesity, 2837
NIGMS Ruth L. Kirschstein National Research Service Awards for Individual Predoctoral Fellows in PharmD/PhD Grants, 2863
NIH Biobehavioral Research for Effective Sleep, 2865
NIH Enhancing Adherence to Diabetes Self-Management Behaviors, 2875
NIH Exploratory/Devel Research Grant, 2877
NIH Mentored Research Scientist Dev Awards, 2889
NIMH Short Courses in Neuroinformatics, 2920
NSF Alan T. Waterman Award, 3006
NSF Doctoral Dissertation Improvement Grants in the Directorate for Biological Sciences (DDIG), 3013
OBSSR Behavioral and Social Science Research on Understanding and Reducing Health Disparities Grants, 3061
SfN Donald B. Lindsley Prize in Behavioral Neuroscience, 3521
SfN Neuroscience Fellowships, 3530
Simone and Cino del Duca Grand Prix Awards, 3564

Behavioral/Experimental Psychology
NARSAD Goldman-Rakic Prize for Cognitive Neuroscience, 2621
NIA Behavioral and Social Research Grants on Disasters and Health, 2754
SfN Swartz Prize for Theoretical and Computational Neuroscience, 3537

Behavioral/Social Sciences
ACSM Coca-Cola Company Doctoral Student Grant on Behavior Research, 261
National Center for Responsible Gaming Grants, 2629
National Center for Responsible Gaming Postdoctoral Fellowships, 2630
National Center for Resp Gaming Seed Grants, 2631
NCHS National Center for Health Statistics Postdoctoral Research Appointments, 2642

Beverages, Alcoholic
Alcohol Misuse and Alcoholism Research Grants, 415

Biochemistry
AHAF National Glaucoma Research Grants, 351
AOCS A. Richard Baldwin Award, 610
AOCS George Schroepfer Medal, 613
AOCS Health & Nutrition Div Student Award, 614
AOCS Health and Nutrition Poster Competition, 615
AOCS Processing Division Student Award, 618
AOCS Protein and Co-Products Division Student Poster Competition, 619
AOCS Holman Lifetime Achievement Award, 620
AOCS Supelco/Nicholas Pelick-AOCS Research Award, 621
Arnold & Mabel Beckman Fndn Scholars Grants, 665
Barth Syndrome Foundation Research Grants, 836
Greater Milwaukee Foundation Grants, 1883
HRAMF Jeffress Trust Awards in Interdisciplinary Research, 2060
HRAMF Smith Family Awards for Excellence in Biomedical Research, 2063
Lalor Foundation Postdoctoral Fellowships, 2336
NIH Ruth L. Kirschstein National Research Service Awards for Individual Postdoc Fellowships, 2902
NSF-NIST Interaction in Chemistry, Materials Research, Molecular Biosciences, Bioengineering, and Chemical Engineering, 3004
NSF Biomolecular Systems Cluster Grants, 3010
NSF Chemistry Research Experiences for Undergraduates (REU), 3012
NSF Genes and Genome Systems Cluster Grants, 3015
Pittsburgh Foundation Medical Research Grants, 3231
Ralph F. Hirschmann Award in Peptide Chem, 3303
Searle Scholars Program Grants, 3503
Sigrid Juselius Foundation Grants, 3561

Biodiversity
James S. McDonnell Foundation Complex Systems Collaborative Activity Awards, 2202
Rohm and Haas Company Grants, 3403

Bioenergetics
NIA Bioenergetics, Fatigability, and Activity Limitations in Aging, 2755

Bioengineering
ASME H.R. Lissner Award, 725
NHGRI Mentored Research Scientist Development Award, 2686
NHLBI Bioengineering and Obesity Grants, 2689
NHLBI Bioengineering Approaches to Energy Balance and Obesity Grants for SBIR, 2690
NHLBI Bioengineering Approaches to Energy Balance and Obesity Grants for STTR, 2691
NHLBI Mentored Quantitative Research Career Development Awards, 2711
NIA Human Biospecimen Resources for Aging Research Grants, 2764
NSF Emerging Frontiers in Research and Innovation (EFRI) Grants, 3014

Bioethics
Victor Grifols i Lucas Foundation Ethics and Science Awars for Educational Institutions, 3869
Victor Grifols i Lucas Foundation Prize for Journalistic Work on Bioethics, 3870
Victor Grifols I Lucas Fndn Research Grants, 3871
Victor Grifols i Lucas Foundation Secondary School Prizes, 3872

Bioinformatics
NLM Express Research Grants in Biomedical Informatics, 2943
NLM Limited Competition for Continuation of Biomedical Informatics/Bioinformatics Resource Grants, 2947
Smithsonian Biodiversity Genomics and Bioinformatics Postdoctoral Fellowships, 3580

Biological Oceanography
MBL Frederik B. and Betsy G. Bang Summer Fellowships, 2473

Biological Sciences
AAAS Eppendorf Science Prize for Neurobiology, 38
AAAS Science and Technology Policy Fellowships: Global Health and Development, 41
Arnold and Mabel Beckman Foundation Scholars Grants, 665
Brookdale Foundation Leadership in Aging Fellowships, 965
CDC Collegiate Leaders in Environmental Health Internships, 1048
Ellison Medical Foundation/AFAR Postdoctoral Fellows in Aging Research Program, 1577
EMBO Installation Grants, 1586
HHMI Grants and Fellowships Programs, 2020
James Ford Bell Foundation Grants, 2187
John Stauffer Charitable Trust Grants, 2258
MBL Baxter Postdoctoral Summer Fellowship, 2466
MBL Scholarships and Awards, 2486
NCI Application of Metabolomics for Translational and Biological Research Grants, 2646
NHLBI Ruth L. Kirschstein National Research Service Awards for Individual Senior Fellows, 2732
NLM Manufacturing Processes of Medical, Dental, and Biological Technologies (SBIR) Grants, 2948
NSF Alan T. Waterman Award, 3006
NSF Doctoral Dissertation Improvement Grants in the Directorate for Biological Sciences (DDIG), 3013
NSF Instrument Development for Bio Research, 3017
NSF Postdoc Research Fellowships in Biology, 3020
NSF Undergraduate Research and Mentoring in the Biological Sciences (URM), 3025
PhRMA Foundation Informatics Research Starter Grants, 3208
PhRMA Fndn Info Sabbatical Fellowships, 3209
Richard King Mellon Foundation Grants, 3369
Scleroderma Foundation Established Investigator Grants, 3497
Searle Scholars Program Grants, 3503
Smithsonian Museum Conservation Institute Research Post-Doctorate/Post-Graduate Fellowships, 3581
SOBP Ziskind-Somerfeld Research Awards, 3595
Society of Cosmetic Chemists Allan B. Black Award Sponsored by Prespere, 3604
Society of Cosmetic Chemists Award Sponsored by McIntyre Group, Ltd., 3605
Society of Cosmetic Chemists Award Sponsored by The HallStar Company, 3606
Society of Cosmetic Chemists Chapter Best Speaker Award, 3607
Society of Cosmetic Chem Chapter Merit Award, 3608
Society of Cosmetic Chemists Frontier of Science Award, 3609
Society of Cosmetic Chem Hans Schaeffer Award, 3610
Society of Cosmetic Chem Jos Ciaudelli Award, 3611
Society of Cosmetic Chemists Keynote Award Lecture Sponsored by Ruger Chemical Corporation, 3612
Society of Cosmetic Chemists Literature Award, 3613
Society of Cosmetic Chemists Maison G. de Navarre Medal, 3614
Society of Cosmetic Chemists Merit Award, 3615
Society of Cosmetic Chemists Robert A. Kramer Lifetime Service Award, 3616
Society of Cosmetic Chem Shaw Mudge Award, 3617
Society of Cosmetic Chem Stud Poster Awards, 3618
Society of Cosmetic Chem You Scientist Awards, 3619
Wellcome Trust Biomedical Science Grants, 3918

SUBJECT INDEX

Biological/Chemical Warfare
APHL Emerging Infectious Diseases Fellowships, 636

Biology
AAAS Eppendorf Science Prize for Neurobiology, 38
Alfred P. Sloan Foundation Research Fellowships, 432
American Chemical Society Award in Separations Science and Technology, 514
APHL Emerging Infectious Diseases Fellowships, 636
Australasian Institute of Judicial Administration Seed Funding Grants, 806
BWF Ad Hoc Grants, 982
BWF Investigators in the Pathogenesis of Infectious Disease Awards, 988
Canada-U.S. Fulbright New Century Scholars Program Grants, 1008
Canada-U.S. Fulbright Senior Specialists Grants, 1009
Canada Graduate Scholarships (CGS) and NSERC Postgraduate Scholarships (PGS), 1010
Cedar Tree Foundation David H. Smith Conservation Research Fellowship, 1076
CFF Postdoctoral Research Fellowships, 1089
Collins Foundation Grants, 1210
FDHN Funderburg Research Scholar Award in Gastric Biology Related to Cancer, 1645
Fulbright Alumni Initiatives Awards, 1749
Fulbright Distinguished Chairs Awards, 1750
Fulbright New Century Scholars Grants, 1753
Fulbright Specialists Program Grants, 1754
Fulbright Scholars in Europe and Eurasia, 1755
Fulbright Traditional Scholar Program in Sub-Saharan Africa, 1756
Fulbright Traditional Scholar Program in the East Asia/Pacific Region, 1757
Fulbright Traditional Scholar Program in the Near East and North Africa Region, 1758
Fulbright Traditional Scholar Program in the South and Central Asia Region, 1759
Fulbright Traditional Scholar Program in the Western Hemisphere, 1760
Harry Frank Guggenheim Foundation Dissertation Fellowships, 1963
HHMI Grants and Fellowships Programs, 2020
HRAMF Jeffress Trust Awards in Interdisciplinary Research, 2060
IIE Central Europe Summer Research Institute Summer Research Fellowship, 2127
James S. McDonnell Foundation Complex Systems Collaborative Activity Awards, 2202
Marion I. and Henry J. Knott Foundation Standard Grants, 2436
NCI Stages of Breast Development: Normal to Metastatic Disease Grants, 2655
NHGRI Mentored Research Scientist Development Award, 2686
NHLBI Mentored Quantitative Research Career Development Awards, 2711
NHLBI Progenitor Cell Biology Consortium Administrative Coordinating Center Grants, 2722
NIA Awards to Support Research on the Biology of Aging in Invertebrates Grants, 2753
NIAMS Pilot and Feasibility Clinical Research Grants in Arthritis, Musculoskeletal & Skin Diseases, 2774
NIA Nathan Shock Centers of Excellence in Basic Biology of Aging Grants, 2776
NIDDK Developmental Biology and Regeneration of the Liver Grants, 2807
NIH Enhancing Adherence to Diabetes Self-Management Behaviors, 2875
North Carolina GlaxoSmithKline Fndn Grants, 2991
NSF Biological Physics (BP), 3008
NSF Postdoc Research Fellowships in Biology, 3020
SOBP Ziskind-Somerfeld Research Awards, 3595
US CRDF Leishmaniasis: Collaborative Research Opportunities in N Africa & Middle East, 3828

Biology Education
Amgen Foundation Grants, 572
ANA Distinguished Neurology Teacher Award, 580
CCFF Life Sciences Student Awards, 1045
Mt. Sinai Health Care Foundation Academic Medicine and Bioscience Grants, 2598
NSF Undergraduate Research and Mentoring in the Biological Sciences (URM), 3025

Biology, Behavioral
NIH Ruth L. Kirschstein National Research Service Awards for Individual Predoctoral Fellows, 2903

Biology, Cellular
AAAS Eppendorf Science Prize for Neurobiology, 38
AABB Dale A. Smith Memorial Award, 43
AABB Hemphill-Jordan Leadership Award, 44
AABB John Elliott Memorial Award, 45
AABB Karl Landsteiner Memorial Award and Lectureship, 46
AABB Sally Frank Award & Lectureship, 47
AABB Tibor Greenwalt Memorial Award and Lectureship, 48
AFAR Research Grants, 337
AHAF National Glaucoma Research Grants, 351
APHL Emerging Infectious Diseases Fellowships, 636
CFF Postdoctoral Research Fellowships, 1089
CFF Research Grants, 1090
FDHN Fellowship to Faculty Transition Awards, 1644
Fndn for Appalachian Ohio Bachtel Scholarships, 1692
Glaucoma Research Pilot Project Grants, 1828
Grass Foundation Marine Biological Laboratory Advanced Imaging Fellowships, 1877
Hereditary Disease Foundation John J. Wasmuth Postdoctoral Fellowships, 2009
Hereditary Disease Fnd Lieberman Award, 2010
Hereditary Disease Foundation Research Grants, 2011
HRAMF Taub Fnd Grants for MDS Research, 2064
IARC Expertise Transfer Fellowship, 2090
IARC Postdoctoral Fellowships for Training in Cancer Research, 2091
IARC Visiting Award for Senior Scientists, 2092
LAM Fnd Established Investigator Awards, 2337
LAM Foundation Pilot Project Grants, 2338
LAM Foundation Postoctoral Fellowships, 2339
MBL Albert & Ellen Grass Fellowships, 2463
MBL Associates Summer Fellowships, 2465
MBL Burr & Susie Steinbach Fellowship, 2467
MBL E.E. Just Summer Fellowship for Minority Scientists, 2468
MBL Erik B. Fries Summer Fellowships, 2469
MBL Evelyn and Melvin Spiegal Fellowship, 2471
MBL Frank R. Lillie Summer Fellowship, 2472
MBL Fred Karush Library Readership, 2474
MBL Gruss Lipper Family Foundation Summer Fellowship, 2475
MBL H. Keffer Hartline and Edward F. MacNichol, Jr. Fellowships, 2476
MBL Herbert W. Rand Summer Fellowship, 2477
MBL James E. and Faith Miller Memorial Summer Fellowship, 2478
MBL John M. Arnold Award, 2479
MBL Laura and Arthur Colwin Fellowships, 2480
MBL Lucy B. Lemann Summer Fellowship, 2481
MBL M.G.F. Fuortes Summer Fellowships, 2482
MBL Nikon Summer Fellowship, 2483
MBL Robert Day Allen Summer Fellowship, 2485
MBL Stephen W. Kuffler Summer Fellowships, 2487
MBL William Townsend Porter Summer Fellowships for Minority Investigators, 2488
NARSAD Goldman-Rakic Prize for Cognitive Neuroscience, 2621
NHLBI Progenitor Cell Biology Consortium Administrative Coordinating Center Grants, 2722
NHLBI Research on the Role of Cardiomyocyte Mitochondria in Heart Disease: An Integrated Approach, 2726
NIA Nathan Shock Centers of Excellence in Basic Biology of Aging Grants, 2776
NIDDK Cell-Specific Delineation of Prostate and Genitourinary Development, 2804
NIH Research on Sleep and Sleep Disorders, 2898
NSF Biomedical Engineering and Engineering Healthcare Grants, 3009
NSF Cell Systems Cluster Grants, 3011
PhRMA Foundation Pharmacology/Toxicology Post Doctoral Fellowships, 3215
PhRMA Foundation Pharmacology/Toxicology Pre Doctoral Fellowships, 3216
PhRMA Foundation Pharmacology/Toxicology Research Starter Grants, 3217
PhRMA Foundation Pharmacology/Toxicology Sabbatical Fellowships, 3218
Searle Scholars Program Grants, 3503
SfN Ralph W. Gerard Prize in Neuroscience, 3534
Simone and Cino del Duca Grand Prix Awards, 3564

Biology, Conservation
Christensen Fund Regional Grants, 1149

Biology, Developmental/Evolutionary
AHAF Alzheimer's Disease Research Grants, 349
Alfred P. Sloan Foundation International Science Engagement Grants, 431
MBL Albert & Ellen Grass Fellowships, 2463
MBL Associates Summer Fellowships, 2465
MBL Burr & Susie Steinbach Fellowship, 2467
MBL E.E. Just Summer Fellowship for Minority Scientists, 2468
MBL Erik B. Fries Summer Fellowships, 2469
MBL Evelyn and Melvin Spiegal Fellowship, 2471
MBL Frank R. Lillie Summer Fellowship, 2472
MBL Fred Karush Library Readership, 2474
MBL Gruss Lipper Family Foundation Summer Fellowship, 2475
MBL H. Keffer Hartline and Edward F. MacNichol, Jr. Fellowships, 2476
MBL Herbert W. Rand Summer Fellowship, 2477
MBL James E. and Faith Miller Memorial Summer Fellowship, 2478
MBL John M. Arnold Award, 2479
MBL Laura and Arthur Colwin Fellowships, 2480
MBL Lucy B. Lemann Summer Fellowship, 2481
MBL M.G.F. Fuortes Summer Fellowships, 2482
MBL Nikon Summer Fellowship, 2483
MBL Stephen W. Kuffler Summer Fellowships, 2487
MBL William Townsend Porter Summer Fellowships for Minority Investigators, 2488
NICHD Ruth L. Kirschstein National Research Service Award (NRSA) Institutional Predoctoral Training Program in Systems Biology of Developmental Biology & Birth Defects, 2790
NSF Animal Developmental Mechanisms Grants, 3007

Biology, Molecular
AAAS Eppendorf Science Prize for Neurobiology, 38
AAAS GE & Science Prize for Yng Life Scientists, 39
AHAF Alzheimer's Disease Research Grants, 349
AHAF National Glaucoma Research Grants, 351
Alfred P. Sloan Foundation International Science Engagement Grants, 431
EMBO Installation Grants, 1586
EMBO Long-Term Fellowships, 1587
EMBO Short Term Fellowships, 1588
FDHN Fellowship to Faculty Transition Awards, 1644
Glaucoma Research Pilot Project Grants, 1828
Grass Foundation Marine Biological Laboratory Advanced Imaging Fellowships, 1877
Hilda and Preston Davis Foundation Postdoctoral Fellowships in Eating Disorders Research, 2029
IARC Expertise Transfer Fellowship, 2090
IARC Postdoctoral Fellowships for Training in Cancer Research, 2091
IARC Visiting Award for Senior Scientists, 2092
LAM Fnd Established Investigator Awards, 2337
LAM Foundation Pilot Project Grants, 2338
LAM Foundation Postoctoral Fellowships, 2339
MBL Frederik B. and Betsy G. Bang Summer Fellowships, 2473
NCI Academic-Industrial Partnerships for Devel and Validation of In Vivo Imaging Systems and Methods for Cancer Investigations, 2644
NCI Application and Use of Transformative Emerging Technologies in Cancer Research, 2645

Biology, Molecular

NCI Cancer Education and Career Development Program, 2648
NIA Nathan Shock Centers of Excellence in Basic Biology of Aging Grants, 2776
NIDCD Mentored Research Scientist Development Award, 2796
NIH Research on Sleep and Sleep Disorders, 2898
NSF Biomolecular Systems Cluster Grants, 3010
NSF Systematic Biology and Biodiversity Inventories Grants, 3024
PhRMA Foundation Pharmacology/Toxicology Post Doctoral Fellowships, 3215
PhRMA Foundation Pharmacology/Toxicology Pre Doctoral Fellowships, 3216
PhRMA Foundation Pharmacology/Toxicology Research Starter Grants, 3217
PhRMA Foundation Pharmacology/Toxicology Sabbatical Fellowships, 3218
SfN Ralph W. Gerard Prize in Neuroscience, 3534
Simone and Cino del Duca Grand Prix Awards, 3564

Biology, Reproductive

Lalor Foundation Postdoctoral Fellowships, 2336
NSF Animal Developmental Mechanisms Grants, 3007
USAID Family Planning and Reproductive Health Methods Grants, 3811
USAID Social Science Research in Population and Reproductive Health Grants, 3822

Biology, Structural

NIH Structural Bio of Membrane Proteins Grant, 2911
NSF Emerging Frontiers in Research and Innovation (EFRI) Grants, 3014

Biology, Systematic

AACR Team Science Award, 126
NICHD Ruth L. Kirschstein National Research Service Award (NRSA) Institutional Predoctoral Training Program in Systems Biology of Developmental Biology & Birth Defects, 2790
NIH Biology, Development, and Progression of Malignant Prostate Disease Research Grants, 2866
NSF Emerging Frontiers in Research and Innovation (EFRI) Grants, 3014
NSF Systematic Biology and Biodiversity Inventories Grants, 3024

Biomarkers

NIDCD Mentored Research Scientist Development Award, 2796
NIDDK Devel of Disease Biomarkers Grants, 2809
NIH Nonhuman Primate Immune Tolerance Cooperative Study Group Grants, 2890
Pfizer ASPIRE EU Antifungal Research Awards, 3165
Pfizer ASPIRE Worldwide Endocrine Young Investigator Grants, 3173

Biomechanics

ASME H.R. Lissner Award, 725

Biomedical Education

AIHS Full-Time M.D./Ph.D. Studentships, 385
AIHS Knowledge Exchange Visiting Professors, 390
AIHS Knowledge Exchange - Visiting Scientists, 391
AIHS Media Summer Fellowship, 392
AIHS Part-Time Studentships, 394
AlohaCare Believes in Me Scholarship, 449
ANA Distinguished Neurology Teacher Award, 580
Biogen Foundation Patient Educational Grants, 901
Bristol-Myers Squibb Clinical Outcomes and Research Grants, 955
CDC Epidemiology Elective Rotation, 1056
HHMI Biomedical Research Grants for International Scientists: Infectious Diseases & Parasitology, 2017
HHMI Biomedical Research Grants for International Scientists in Canada and Latin America, 2018
HHMI Biomedical Research Grants for International Scientists in the Baltics, Central and Eastern Europe, Russia, and Ukraine, 2019
HHMI Physician-Scientist Early Career Award, 2023

Komen Greater NYC Small Grants, 2317
Kosciuszko Foundation Dr. Marie E. Zakrzewski Medical Scholarship, 2321
Lynn and Rovena Alexander Family Foundation Grants, 2399
Mt. Sinai Health Care Foundation Academic Medicine and Bioscience Grants, 2598
NCI Ruth L. Kirschstein National Research Service Award Institutional Training Grants, 2654
NHLBI Biomedical Research Training Program for Individuals from Underrepresented Groups, 2692
NHLBI Summer Internship Program in Biomedical Research, 2735
NIH Summer Internship Program in Biomedical Research, 2912
NYCT Biomedical Research Grants, 3049
PC Fred H. Bixby Fellowships, 3126
Piper Trust Healthcare & Med Research Grants, 3228
RCPSC Tom Dignan Indigenous Health Award, 3321
RCPSC International Residency Educator of the Year Award, 3325
RCPSC Int'l Resident Leadership Award, 3326
RCPSC Kristin Sivertz Res Leadership Award, 3332
RCPSC Peter Warren Traveling Scholarship, 3336
RCPSC Prix D'excellence (Specialist of the Year) Award, 3337
RCPSC Program Administrator Award for Innovation and Excellence, 3339
Sierra Health Foundation Responsive Grants, 3555
Topeka Community Foundation Kansas Blood Services Scholarships, 3738

Biomedical Engineering

HRAMF Smith Family Awards for Excellence in Biomedical Research, 2063
IIE Whitaker Int'l Fellowships & Scholarships, 2144
MBL Albert & Ellen Grass Fellowships, 2463
MBL Associates Summer Fellowships, 2465
MBL Burr & Susie Steinbach Fellowship, 2467
MBL E.E. Just Summer Fellowship for Minority Scientists, 2468
MBL Erik B. Fries Summer Fellowships, 2469
MBL Eugene and Millicent Bell Fellowships, 2470
MBL Evelyn and Melvin Spiegal Fellowship, 2471
MBL Frank R. Lillie Summer Fellowship, 2472
MBL Fred Karush Library Readership, 2474
MBL Gruss Lipper Family Foundation Summer Fellowship, 2475
MBL H. Keffer Hartline and Edward F. MacNichol, Jr. Fellowships, 2476
MBL Herbert W. Rand Summer Fellowship, 2477
MBL James E. and Faith Miller Memorial Summer Fellowship, 2478
MBL John M. Arnold Award, 2479
MBL Lucy B. Lemann Summer Fellowship, 2481
MBL M.G.F. Fuortes Summer Fellowship, 2482
MBL Nikon Summer Fellowship, 2483
MBL Scholarships and Awards, 2486
MBL Stephen W. Kuffler Summer Fellowships, 2487
MBL William Townsend Porter Summer Fellowships for Minority Investigators, 2488
NHLBI Intramural Research Training Awards, 2702
NSF Biomedical Engineering and Engineering Healthcare Grants, 3009
NSF Emerging Frontiers in Research and Innovation (EFRI) Grants, 3014
SfN Neuroscience Fellowships, 3530
W.M. Keck Foundation Science and Engineering Research Grants, 3889

Biomedical Research

25th Anniversary Foundation Grants, 8
100 Mile Man Foundation Grants, 10
A.O. Smith Foundation Community Grants, 17
AACR Award for Outstanding Achievement in Cancer Research, 103
AACR Team Science Award, 126
AAMC Award for Distinguished Research, 155
Achelis Foundation Grants, 234

Acid Maltase Deficiency Association Helen Walker Research Grant, 235
Adaptec Foundation Grants, 304
AFAR CART Fund Grants, 334
Affymetrix Corporate Contributions Grants, 339
AHRQ Individual Awards for Postdoc Fellows Ruth L. Kirschstein National Research Service Awards, 366
AIHS Alberta/Pfizer Translat Research Grants, 374
AIHS Clinical Fellowships, 375
AIHS Fast-Track Fellowships, 379
AIHS ForeFront Internships, 380
AIHS Full-Time Fellowships, 383
AIHS Full-Time Health Research Studentships, 384
AIHS Full-Time M.D./Ph.D. Studentships, 385
AIHS Full-Time Studentships, 386
AIHS Heritage Youth Researcher Summer Science Program, 387
AIHS Interdisciplinary Team Grants, 388
AIHS Knowledge Exchange Visiting Professors, 390
AIHS Knowledge Exchange - Visiting Scientists, 391
AIHS Part-Time Fellowships, 393
AIHS Part-Time Studentships, 394
AIHS Proposals for Special Initiatives, 395
AIHS Research Prize, 396
AIHS Summer Studentships, 397
Alaska Airlines Corporate Giving Medical Emergency and Research Grants, 402
Albany Medical Center Prize in Medicine and Biomedical Research, 404
Albert and Mary Lasker Foundation Awards, 406
Albert and Mary Lasker Foundation Clinical Research Scholars, 407
Alcoa Foundation Grants, 414
Allyn Foundation Grants, 448
ALSAM Foundation Grants, 454
AMDA Foundation Quality Improvement Award, 499
American Chemical Society Award in Separations Science and Technology, 514
American Foodservice Charitable Trust Grants, 532
American Psychiatric Foundation Disaster Recovery Fund for Psychiatrists, 553
Amgen Foundation Grants, 572
AMHPS Dr. James A. Ferguson Emerging Infectious Diseases Fellowships, 573
ANA/Grass Foundation Award in Neuroscience, 578
ANA Derek Denny-Brown Neurological Scholar Award, 579
ANA Wolfe Neuropathy Research Prize, 585
Annenberg Foundation Grants, 596
Anthony R. Abraham Foundation Grants, 608
AOCS George Schroepfer Medal, 613
AOCS Industrial Oil Products Division Student Award, 616
APHL Emerging Infectious Diseases Fellowships, 636
Arizona Cardinals Grants, 655
Arnold and Mabel Beckman Foundation Beckman-Argyros Award in Vision Research, 664
ASCO/UICC International Cancer Technology Transfer (ICRETT) Fellowships, 677
ASCO Advanced Clinical Research Award in Colorectal Cancer, 678
ASCO Advanced Clinical Research Awards in Breast Cancer, 679
ASCO Advanced Clinical Research Awards in Sarcoma, 680
ASCO Long-Term International Fellowships, 681
ASM Abbott Laboratories Award in Clinical and Diagnostic Immunology, 717
Australian Academy of Science Grants, 807
Barth Syndrome Foundation Research Grants, 836
Batchelor Foundation Grants, 837
BCBSNC Foundation Grants, 860
Beckman Coulter Foundation Grants, 868
Biogen Foundation Scientific Research Grants, 902
Blue Shield of California Grants, 923
Blum-Kovler Foundation Grants, 924
Bodman Foundation Grants, 929
Booth-Bricker Fund Grants, 933
Bosque Foundation Grants, 935
Bright Family Foundation Grants, 951

SUBJECT INDEX

Brinson Foundation Grants, 953
Bristol-Myers Squibb Clinical Outcomes and Research Grants, 955
Bristol-Myers Squibb Foundation Fellowships, 956
Bristol-Myers Squibb Foundation Global HIV/AIDS Initiative Grants, 957
Bristol-Myers Squibb Foundation Health Disparities Grants, 958
Brookdale Foundation Leadership in Aging Fellowships, 965
Brown Advisory Charitable Foundation Grants, 969
Burden Trust Grants, 976
BWF Institutional Program Unifying Population and Laboratory Based Sciences Grants, 987
BWF Preterm Birth Initiative Grants, 990
Byron W. & Alice L. Lockwood Fnd Grants, 991
Callaway Golf Company Foundation Grants, 1003
Campbell Soup Foundation Grants, 1007
Carl C. Icahn Foundation Grants, 1023
Carrie Estelle Doheny Foundation Grants, 1037
Catherine Holmes Wilkins Foundation Grants, 1042
CCFF Chairman's Dist Life Sciences Award, 1044
Charles Edison Fund Grants, 1116
Charles H. Revson Foundation Grants, 1122
Charles Lafitte Foundation Grants, 1123
Chest Foundation Geriatric Development Research Awards, 1133
Children's Leukemia Research Association Research Grants, 1138
Children's Tumor Fndn Clin Research Awards, 1140
Children's Tumor Foundation Drug Discovery Initiative Awards, 1141
Children's Tumor Foundation Schwannomatosis Awards, 1142
Children's Tumor Foundation Young Investigator Awards, 1143
Chiles Foundation Grants, 1147
CJ Foundation for SIDS Research Grants, 1164
Claude Pepper Foundation Grants, 1177
CMS Research and Demonstration Grants, 1188
Columbus Foundation Competitive Grants, 1217
Community Foundation of Louisville Bobbye M. Robinson Fund Grants, 1291
Community Foundation of Louisville Dr. W. Barnett Owen Memorial Fund for the Children of Louisville and Jefferson County Grants, 1294
Community Foundation of Louisville Irving B. Klempner Fund Grants, 1296
Community Foundation of Louisville Lee Look Fund for Spinal Injury Grants, 1297
Connecticut Health Foundation Health Initiative Grants, 1325
Conquer Cancer Foundation of ASCO Career Development Award, 1328
Conquer Cancer Foundation of ASCO Comparative Effectiveness Research Professorship in Breast Cancer, 1329
Conquer Cancer Foundation of ASCO Improving Cancer Care Grants, 1331
Conquer Cancer Foundation of ASCO International Innovation Grant, 1334
Conquer Cancer Foundation of ASCO Medical Student Rotation Grants, 1335
Conquer Cancer Foundation of ASCO Translational Research Professorships, 1337
CSTE CDC/CSTE Applied Epidemiology Fellowships, 1367
Cystic Fibrosis Canada Clinical Fellowships, 1381
Cystic Fibrosis Canada Fellowships, 1384
Cystic Fibrosis Canada Research Grants, 1385
Cystic Fibrosis Canada Senior Scientist Research Training Awards, 1387
Cystic Fibrosis Canada Studentships, 1390
Cystic Fibrosis Canada Summer Studentships, 1391
Cystic Fibrosis Canada Visiting Allied Health Professional Awards, 1394
Cystic Fibrosis Canada Visiting Scientist Awards, 1396
Cystic Fibrosis Research Eliz Nash Fellowships, 1401
Cystic Fibrosis Research New Horizons Campaign Grants, 1402

Cystic Fibrosis Trust Research Grants, 1404
DAAD Research Stays for University Academics and Scientists, 1408
Dana Foundation Science and Health Grants, 1420
Del E. Webb Foundation Grants, 1450
DeMatteis Family Foundation Grants, 1457
Denton A. Cooley Foundation Grants, 1459
Doris Duke Charitable Foundation Clinical Interfaces Award Program, 1492
Doris Duke Charitable Foundation Clinical Research Fellowships for Medical Students, 1493
Doris Duke Charitable Foundation Clinical Scientist Development Award (CSDA) Bridge Grants, 1495
Doris Duke Charitable Foundation Distinguished Clinical Scientist Award Program, 1496
Dorothea Haus Ross Foundation Grants, 1498
Dorr Institute for Arthritis Research & Educ, 1502
Dr. Scholl Foundation Grants Program, 1511
Duke University Postdoctoral Research Fellowships in Aging, 1528
Edwin S. Webster Foundation Grants, 1560
Effie and Wofford Cain Foundation Grants, 1563
Ellison Medical Foundation/AFAR Postdoctoral Fellows in Aging Research Program, 1577
Emerson Charitable Trust Grants, 1589
Emma B. Howe Memorial Foundation Grants, 1593
Eugene M. Lang Foundation Grants, 1611
ExxonMobil Foundation Malaria Grants, 1621
Eye-Bank for Sight Restoration and Fight for Sight Summer Student Research Fellowship, 1622
Fairlawn Foundation Grants, 1626
FAMRI Clinical Innovator Awards, 1628
FAMRI Young Clinical Scientist Awards, 1629
Fidelity Foundation Grants, 1659
Fight for Sight-Streilein Foundation for Ocular Immunology Research Award, 1662
Fight for Sight Grants-in-Aid, 1663
Fight for Sight Post-Doctoral Awards, 1664
Fight for Sight Summer Student Fellowships, 1665
Fishman Family Foundation Grants, 1672
Forrest C. Lattner Foundation Grants, 1688
Foundation for Pharmaceutical Sciences Herb and Nina Demuth Grant, 1697
Foundation of Orthopedic Trauma Grants, 1707
Foundation of the American Thoracic Society Research Grants, 1708
Frank Stanley Beveridge Foundation Grants, 1730
Frederick Gardner Cottrell Foundation Grants, 1739
Fritz B. Burns Foundation Grants, 1747
Gebbie Foundation Grants, 1777
George Foundation Grants, 1795
George W. Brackenridge Foundation Grants, 1802
Gerber Foundation Grants, 1807
Gheens Foundation Grants, 1814
Gil and Dody Weaver Foundation Grants, 1819
Gilbert Memorial Fund Grants, 1820
Good Samaritan Inc Grants, 1855
Greenspun Family Foundation Grants, 1896
Grifols Community Outreach Grants, 1904
H.A. & Mary K. Chapman Trust Grants, 1921
Hampton Roads Community Foundation Mental Health Research Grants, 1942
Harold R. Bechtel Testamentary Charitable Trust Grants, 1954
Harry A. & Margaret D. Towsley Fndn Grants, 1959
Harry Edison Foundation, 1962
Hawn Foundation Grants, 1979
Healthcare Foundation of New Jersey Grants, 1982
Hearst Foundations Health Grants, 1987
HHMI-NIBIB Interfaces Initiative Grants, 2014
HHMI-NIH Cloister Research Scholars, 2015
HHMI Biomedical Research Grants for International Scientists: Infectious Diseases & Parasitology, 2017
HHMI Biomedical Research Grants for International Scientists in Canada and Latin America, 2018
HHMI Biomedical Research Grants for International Scientists in the Baltics, Central and Eastern Europe, Russia, and Ukraine, 2019
HHMI Grants and Fellowships Programs, 2020
HHMI International Research Scholars Program, 2021

HHMI Physician-Scientist Early Career Award, 2023
HHMI Research Training Fellowships, 2024
Hilda and Preston Davis Foundation Grants, 2028
HomeBanc Foundation Grants, 2042
HRAMF Charles A. King Trust Postdoctoral Research Fellowships, 2056
HRAMF Jeffress Trust Awards in Interdisciplinary Research, 2060
HRAMF Ralph and Marian Falk Medical Research Trust Catalyst Awards, 2061
HRAMF Ralph and Marian Falk Medical Research Trust Transformational Awards, 2062
HRAMF Smith Family Awards for Excellence in Biomedical Research, 2063
IBRO-PERC Support for European Workshops, Symposia and Meetings, 2098
IBRO Latin America Regional Funding for Short Research Stays, 2108
J.B. Reynolds Foundation Grants, 2168
J.W. Kieckhefer Foundation Grants, 2176
Jack H. & William M. Light Trust Grants, 2179
James H. Cummings Foundation Grants, 2190
Joe W. and Dorothy Dorsett Brown Fndn Grants, 2230
John R. Oishei Foundation Grants, 2249
John S. Dunn Research Foundation Grants, 2250
Johnson & Johnson Corporate Contributions, 2253
Joseph Alexander Foundation Grants, 2265
Joseph Drown Foundation Grants, 2268
Josiah W. and Bessie H. Kline Foundation Grants, 2278
Kavli Foundation Research Grants, 2296
Kenneth T. & Eileen L. Norris Fndn Grants, 2300
Kettering Fund Grants, 2305
Klarman Family Foundation Grants in Eating Disorders Research Grants, 2310
Komen Greater NYC Clinical Research Enrollment Grants, 2315
Komen Greater NYC Community Breast Health Grants, 2316
Lillian S. Wells Foundation Grants, 2365
Lisa Higgins-Hussman Foundation Grants, 2370
Lucile Horton Howe and Mitchell B. Howe Foundation Grants, 2385
Lumpkin Family Fnd Healthy People Grants, 2391
Lymphatic Education and Research Network Additional Support Grants for NIH-funded F32 Postdoctoral Fellows, 2394
Lymphatic Education and Research Network Postdoctoral Fellowships, 2395
M.J. Murdock Charitable Trust General Grants, 2405
March of Dimes Newborn Screening Awards, 2425
Margaret T. Morris Foundation Grants, 2429
Mary K. Chapman Foundation Grants, 2443
Mary Kay Foundation Cancer Research Grants, 2444
Mary S. and David C. Corbin Foundation Grants, 2447
MBL Associates Summer Fellowships, 2465
MBL Baxter Postdoctoral Summer Fellowship, 2466
MBL Burr & Susie Steinbach Fellowship, 2467
MBL Eugene and Millicent Bell Fellowships, 2470
MBL Evelyn and Melvin Spiegal Fellowship, 2471
MBL Frank R. Lillie Summer Fellowship, 2472
MBL Frederik B. and Betsy G. Bang Summer Fellowships, 2473
MBL Gruss Lipper Family Foundation Summer Fellowship, 2475
MBL Herbert W. Rand Summer Fellowship, 2477
MBL James E. and Faith Miller Memorial Summer Fellowship, 2478
MBL John M. Arnold Award, 2479
MBL Lucy B. Lemann Summer Fellowship, 2481
MBL William Townsend Porter Summer Fellowships for Minority Investigators, 2488
McCarthy Family Foundation Grants, 2490
McCombs Foundation Grants, 2491
McLean Contributionship Grants, 2497
Mericos Foundation Grants, 2518
Merrick Foundation Grants, 2521
MGN Family Foundation Grants, 2542
MLA Donald A.B. Lindberg Fellowship, 2562
MMAAP Foundation Geriatric Project Awards, 2582
MMAAP Fndn Hematology Project Awards, 2583

MMAAP Foundation Research Project Award in Reproductive Medicine, 2584
Morehouse PHSI Project Imhotep Internships, 2592
Mt. Sinai Health Care Foundation Academic Medicine and Bioscience Grants, 2598
National MPS Society Grants, 2634
NCI Ruth L. Kirschstein National Research Service Award Institutional Training Grants, 2654
NEAVS Fellowsips in Women's Health and Sex Differences, 2659
NEI Ruth L. Kirschstein National Service Award Short-Term Institutional Training Grants, 2668
Nell J. Redfield Foundation Grants, 2673
NFL Charities Medical Grants, 2685
NHLBI Ancillary Studies in Clinical Trials, 2688
NHLBI Biomedical Research Training Program for Individuals from Underrepresented Groups, 2692
NHLBI Independent Scientist Award, 2701
NHLBI Intramural Research Training Awards, 2702
NHLBI Investigator Initiated Multi-Site Clinical Trials, 2703
NHLBI Lymphatics in Health and Disease in the Digestive, Urinary, Cardiovascular and Pulmonary Systems, 2705
NHLBI Mentored Career Dev Award to Promote Faculty Diversity in Biomed Research, 2708
NHLBI Mentored Clinical Scientist Research Career Development Awards, 2709
NHLBI Mentored Patient-Oriented Research Career Development Awards, 2710
NHLBI Mentored Quantitative Research Career Development Awards, 2711
NHLBI Midcareer Investigator Award in Patient-Oriented Research, 2713
NHLBI Pathway to Independence Awards, 2718
NHLBI Research Demonstration and Dissemination Grants, 2724
NHLBI Ruth L. Kirschstein National Research Service Awards for Individual Postdoctoral Fellows, 2729
NHLBI Ruth L. Kirschstein National Research Service Awards for Individual Predoctoral Fellowships to Promote Diversity in Health-Related Research, 2730
NHLBI Ruth L. Kirschstein National Research Service Awards for Individual Predoctoral MD/PhD Fellows and Other Dual Degree Fellows, 2731
NHLBI Ruth L. Kirschstein National Research Service Awards for Individual Senior Fellows, 2732
NHLBI Short-Term Research Education Program to Increase Diversity in Health-Related Research, 2733
NHLBI Summer Internship Program in Biomedical Research, 2735
NIAAA Independent Scientist Award, 2742
NIAAA Mentored Clinical Scientist Research Career Development Awards, 2744
NIA Alzheimer's Disease Core Centers Grants, 2748
NIAID Independent Scientist Award, 2765
NIA Independent Scientist Award, 2766
NIA Mentored Clinical Scientist Research Career Development Awards, 2771
NIDCD Mentored Clinical Scientist Research Career Development Award, 2795
NIDDK Ancillary Studies to Major Ongoing NIDDK and NHLBI Clinical Research Studies, 2800
NIDDK Erythropoiesis: Components and Mechanisms Grants, 2817
NIEHS Hazardous Materials Worker Health and Safety Training Grants, 2854
NIEHS Hazmat Training at Doe Nuclear Weapons Complex Grants, 2855
NIEHS Independent Scientist Award, 2856
NIEHS SBIR E-Learning for HAZMAT and Emergency Response Grants, 2859
NIGMS Ruth L. Kirschstein National Research Service Awards for Individual Predoctoral Fellows in PharmD/PhD Grants, 2863
NIH Academic Research Enhancement Awards, 2864
NIH Biomedical Research on the International Space Station, 2867
NIH Earth-Based Research Relevant to the Space Environment, 2874

NIH Exceptional, Unconventional Research Enabling Knowledge Acceleration (EUREKA) Grants, 2876
NIH Exploratory/Devel Research Grant, 2877
NIH Exploratory Innovations in Biomedical Computational Science & Technology Grants, 2878
NIH Independent Scientist Award, 2882
NIH Innovations in Biomedical Computational Science and Technology Grants, 2883
NIH Innovations in Biomedical Computational Science and Technology Initiative Grants for SBIR, 2884
NIH Innovations in Biomedical Computational Science and Technology Initiative Grants for STTR, 2885
NIH Mentored Clinical Scientist Devel Award, 2886
NIH Mentored Research Scientist Dev Awards, 2889
NIH Recovery Act Limited Competition: Academic Research Enhancement Awards, 2892
NIH Recovery Act Limited Competition: Biomedical Research, Development, and Growth to Spur the Acceleration of New Technologies Grants, 2893
NIH Recovery Act Limited Competition: Small Business Catalyst Awards for Accelerating Innovative Research Grants, 2895
NIH Recovery Act Limited Competition: Supporting New Faculty Recruitment to Enhance Research Resources through Biomedical Research Core Centers Grants, 2896
NIH Ruth L. Kirschstein National Research Service Award Institutional Research Training Grants, 2900
NIH Ruth L. Kirschstein National Research Service Awards (NRSA) for Individual Senior Fellows, 2901
NIH Ruth L. Kirschstein National Research Service Awards for Individual Predoctoral Fellows, 2903
NIH Ruth L. Kirschstein National Research Service Awards for Individual Predoctoral Fellowships to Promote Diversity in Research, 2904
NIH Ruth L. Kirschstein National Research Service Awards for Individual Predoctoral MD/PhD and Other Dual Doctoral Degree Fellows, 2905
NIH Ruth L. Kirschstein Nat Research Service Award Short-Term Institutional Training Grants, 2906
NIH Summer Internship Program in Biomedical Research, 2912
NIMH Basic and Translational Research in Emotion Grants, 2916
NLM Academic Research Enhancement Awards, 2940
NLM Exceptional, Unconventional Research Grants Enabling Knowledge Acceleration, 2941
NLM Express Research Grants in Biomedical Informatics, 2943
NLM Grants for Scholarly Works in Biomedicine and Health, 2944
NLM Internet Connection for Medical Institutions Grants, 2946
NLM Limited Competition for Continuation of Biomedical Informatics/Bioinformatics Resource Grants, 2947
NLM Predictive Multiscale Models of the Physiome in Health and Disease Grants, 2950
NLM Research Supplements to Promote Reentry in Health-Related Research, 2952
Norcliffe Foundation Grants, 2979
Notsew Orm Sands Foundation Grants, 3002
NSF Accelerating Innovation Research, 3005
NSF Alan T. Waterman Award, 3006
NSF Emerging Frontiers in Research and Innovation (EFRI) Grants, 3014
NSF Instrument Development for Bio Research, 3017
NSF Postdoc Research Fellowships in Biology, 3020
Nuffield Foundation Africa Program Grants, 3026
Nuffield Foundation Law and Society Grants, 3028
Nuffield Foundation Small Grants, 3031
NYCT Biomedical Research Grants, 3049
Olympus Corporation of Americas Fellowships, 3074
Orthopaedic Trauma Association Kathy Cramer Young Clinician Memorial Fellowships, 3083
Orthopaedic Trauma Assoc Research Grants, 3084
Partnership for Cures Two Years To Cures Grants, 3118
Patron Saints Foundation Grants, 3121
PC Fred H. Bixby Fellowships, 3126
Peabody Foundation Grants, 3132

Pediatric Brain Tumor Fndn Research Grants, 3134
Pediatric Cancer Research Foundation Grants, 3135
Perkin Fund Grants, 3139
Perpetual Trust for Charitable Giving Grants, 3141
Pew Charitable Trusts Biomed Research Grants, 3157
Pfizer Australia Paediatric Endocrine Care Research Grants, 3176
Pfizer Research Initiative Dermatology Grants (Germany), 3184
Piper Trust Healthcare & Med Research Grants, 3228
PKD Foundation Lillian Kaplan International Prize for the Advancement in the Understanding of Polycystic Kidney Disease, 3232
Presbyterian Health Foundation Bridge, Seed and Equipment Grants, 3259
Price Family Charitable Fund Grants, 3262
Rajiv Gandhi Foundation Grants, 3302
RCPSC Detweiler Traveling Fellowships, 3320
RCPSC Medical Education Research Grants, 3334
RCPSC Peter Warren Traveling Scholarship, 3336
RCPSC Robert Maudsley Fellowship for Studies in Medical Education, 3341
Reinberger Foundation Grants, 3357
RGk Foundation Grants, 3361
Richard & Susan Smith Family Fndn Grants, 3365
Robert Leet & Clara Guthrie Patterson Grants, 3383
Rockefeller Archive Center Research Grants, 3393
Ronald McDonald House Charities Grants, 3405
Roy J. Carver Charitable Trust Medical and Science Research Grants, 3416
RSC McLaughlin Medal, 3421
RSC Thomas W. Eadie Medal, 3423
RWJF Harold Amos Medical Faculty Development Research Grants, 3434
S.D. Bechtel, Jr. Fndn/Stephen Bechtel Preventive Healthcare and Selected Research Grants, 3446
Saint Luke's Foundation Grants, 3454
San Antonio Area Foundation Grants, 3469
Sarkeys Foundation Grants, 3485
Savoy Foundation Studentships, 3490
Schering-Plough Foundation Health Grants, 3492
Searle Scholars Program Grants, 3503
Sensient Technologies Foundation Grants, 3513
SfN Janett Rosenberg Trubatch Career Development Award, 3524
SfN Julius Axelrod Prize, 3526
SfN Neuroscience Fellowships, 3530
SfN Peter and Patricia Gruber International Research Award, 3533
SfN Young Investigator Award, 3541
Sick Kids Fndn Community Conference Grants, 3549
Sick Kids Foundation New Investigator Research Grants, 3550
Sidgmore Family Foundation Grants, 3551
Sigma Theta Tau International / Oncology Nursing Society Grant, 3558
Sigrid Juselius Foundation Grants, 3561
Simeon J. Fortin Charitable Foundation Grants, 3562
Simone and Cino del Duca Grand Prix Awards, 3564
Skin Cancer Foundation Research Grants, 3578
Society of Cosmetic Chemists Allan B. Black Award Sponsored by Presperse, 3604
Society of Cosmetic Chemists Award Sponsored by McIntyre Group, Ltd., 3605
Society of Cosmetic Chemists Award Sponsored by The HallStar Company, 3606
Society of Cosmetic Chemists Chapter Best Speaker Award, 3607
Society of Cosmetic Chem Chapter Merit Award, 3608
Society of Cosmetic Chemists Frontier of Science Award, 3609
Society of Cosmetic Chem Hans Schaeffer Award, 3610
Society of Cosmetic Chem Jos Ciaudelli Award, 3611
Society of Cosmetic Chemists Keynote Award Lecture Sponsored by Ruger Chemical Corporation, 3612
Society of Cosmetic Chemists Literature Award, 3613
Society of Cosmetic Chemists Maison G. de Navarre Medal, 3614
Society of Cosmetic Chemists Merit Award, 3615

SUBJECT INDEX

Biotechnology /671

Society of Cosmetic Chemists Robert A. Kramer Lifetime Service Award, 3616
Society of Cosmetic Chem Shaw Mudge Award, 3617
Society of Cosmetic Chem Stud Poster Awards, 3618
Society of Cosmetic Chem You Scientist Awards, 3619
Susan G. Komen Breast Cancer Foundation Brinker Awards for Scientific Distinction, 3687
Susan G. Komen Breast Cancer Foundation Career Catalyst Research Grants, 3688
Susan G. Komen Breast Cancer Foundation Investigator Initiated Research Grants, 3693
Tellabs Foundation Grants, 3706
Thomas and Dorothy Leavey Foundation Grants, 3717
Thomas Austin Finch, Sr. Foundation Grants, 3718
Tourette Syndrome Assoc Post-Doc Fellowships, 3740
Tourette Syndrome Association Research Grants, 3741
USAID Fighting Ebola Grants, 3812
Victor E. Speas Foundation Grants, 3868
Victor Grifols I Lucas Fndn Research Grants, 3871
W.C. Griffith Foundation Grants, 3881
Wellcome Trust New Investigator Awards, 3919
William G. & Helen C. Hoffman Fndn Grants, 3950
William H. Adams Foundation for ALS Grants, 3952
William H. Hannon Foundation Grants, 3953

Biomedical Research Resources
AES Robert S. Morison Fellowship, 323
AIHS ForeFront Internships, 380
AIHS Proposals for Special Initiatives, 395
Cystic Fibrosis Research New Horizons Campaign Grants, 1402
Gilbert Memorial Fund Grants, 1820
MBL Burr & Susie Steinbach Fellowship, 2467
MBL Erik B. Fries Summer Fellowships, 2469
MBL Frank R. Lillie Summer Fellowship, 2472
MBL Fred Karush Library Readership, 2474
MBL Herbert W. Rand Summer Fellowship, 2477
MBL John M. Arnold Award, 2479
Nestle Foundation Training Grant, 2679
NHLBI Biomedical Research Training Program for Individuals from Underrepresented Groups, 2692
NHLBI Intramural Research Training Awards, 2702
NHLBI Mentored Clinical Scientist Research Career Development Awards, 2709
NHLBI Mentored Patient-Oriented Research Career Development Awards, 2710
NHLBI Midcareer Investigator Award in Patient-Oriented Research, 2713
NHLBI Research Demonstration and Dissemination Grants, 2724
NIAAA Mentored Clinical Scientist Research Career Development Awards, 2744
NIDCD Mentored Clinical Scientist Research Career Development Award, 2795
NIH Exploratory Innovations in Biomedical Computational Science & Technology Grants, 2878
NIH Innovations in Biomedical Computational Science and Technology Grants, 2883
NIH Innovations in Biomedical Computational Science and Technology Initiative Grants for SBIR, 2884
NIH Innovations in Biomedical Computational Science and Technology Initiative Grants for STTR, 2885
NIH Recovery Act Limited Competition: Supporting New Faculty Recruitment to Enhance Research Resources through Biomedical Research Core Centers Grants, 2896
NIH Ruth L. Kirschstein National Research Service Award Institutional Research Training Grants, 2900
NIH Ruth L. Kirschstein Nat Research Service Award Short-Term Institutional Training Grants, 2906
NLM Academic Research Enhancement Awards, 2940
NLM Limited Competition for Continuation of Biomedical Informatics/Bioinformatics Resource Grants, 2947
NSF Accelerating Innovation Research, 3005
NSF Emerging Frontiers in Research and Innovation (EFRI) Grants, 3014
Partnership for Cures Two Years To Cures Grants, 3118
Pew Charitable Trusts Biomed Research Grants, 3157
USAID Fighting Ebola Grants, 3812

Biomedical Research Training
AES Epilepsy Research Recognition Awards, 318
AIHS Heritage Youth Researcher Summer Science Program, 387
AIHS Proposals for Special Initiatives, 395
Alaska Airlines Corporate Giving Medical Emergency and Research Grants, 402
Albert and Mary Lasker Foundation Clinical Research Scholars, 407
APHL Emerging Infectious Diseases Fellowships, 636
CDC Epidemic Intell Service Training Grants, 1055
CFF Shwachman Clinical Investigator Award, 1086
Cystic Fibrosis Canada Visiting Allied Health Professional Awards, 1394
Harmony Project Grants, 1948
IBRO Latin America Regional Funding for Short Research Stays, 2108
IBRO Regional Grants for Int'l Fellowships to U.S. Laboratory Summer Neuroscience Courses, 2110
Morehouse PHSI Project Imhotep Internships, 2592
Mt. Sinai Health Care Foundation Academic Medicine and Bioscience Grants, 2598
NEI Ruth L. Kirschstein National Research Service Award Short-Term Institutional Research Training Grants, 2668
NHLBI Biomedical Research Training Program for Individuals from Underrepresented Groups, 2692
NHLBI Intramural Research Training Awards, 2702
NHLBI Ruth L. Kirschstein National Research Service Awards for Individual Postdoctoral Fellows, 2729
NHLBI Ruth L. Kirschstein National Research Service Awards for Individual Predoctoral Fellowships to Promote Diversity in Health-Related Research, 2730
NHLBI Summer Internship Program in Biomedical Research, 2735
NHLBI Training Program Grants for Institutions that Promote Diversity, 2738
NIH Ruth L. Kirschstein National Research Service Award Institutional Research Training Grants, 2900
NIH Ruth L. Kirschstein Nat Research Service Award Short-Term Institutional Training Grants, 2906
NSF Postdoc Research Fellowships in Biology, 3020
Osteosynthesis and Trauma Care Foundation European Visiting Fellowships, 3097
PhRMA Fndn Info Sabbatical Fellowships, 3209
PhRMA Foundation Pharmaceutics Postdoctoral Fellowships, 3211
PhRMA Foundation Pharmacology/Toxicology Pre Doctoral Fellowships, 3216
Piper Trust Healthcare & Med Research Grants, 3228
Topeka Community Foundation Kansas Blood Services Scholarships, 3738

Biomedicine
Alpha Research Foundation Grants, 452
ASPEN Rhoads Research Foundation Abbott Nutrition Research Grants, 761
ASPEN Rhoads Research Foundation Baxter Parenteral Nutrition Research Grant, 762
ASPEN Rhoads Research Foundation C. Richard Fleming Grant, 763
ASPEN Rhoads Research Foundation Maurice Shils Grant, 764
ASPEN Rhoads Research Foundation Norman Yoshimura Grant, 765
BWF Postdoctoral Enrichment Grants, 989
HHMI-NIBIB Interfaces Initiative Grants, 2014
HRAMF Smith Family Awards for Excellence in Biomedical Research, 2063
Komen Greater NYC Small Grants, 2317
Lumpkin Family Fnd Healthy People Grants, 2391
Mary K. Chapman Foundation Grants, 2443
MBL Frederik B. and Betsy G. Bang Summer Fellowships, 2473
MMAAP Fndn Dermatology Project Awards, 2578
MMAAP Foundation Fellowship Award in Translational Medicine, 2581
NIH Earth-Based Research Relevant to the Space Environment, 2874
NIH Exploratory Innovations in Biomedical Computational Science & Technology Grants, 2878
NIH Innovations in Biomedical Computational Science and Technology Grants, 2883
NIH Innovations in Biomedical Computational Science and Technology Initiative Grants for SBIR, 2884
NIH Innovations in Biomedical Computational Science and Technology Initiative Grants for STTR, 2885
NIH Ruth L. Kirschstein National Research Service Awards for Individual Predoctoral Fellows, 2903
NLM Academic Research Enhancement Awards, 2940
NLM Exceptional, Unconventional Research Grants Enabling Knowledge Acceleration, 2941
NLM Limited Competition for Continuation of Biomedical Informatics/Bioinformatics Resource Grants, 2947
NLM Predictive Multiscale Models of the Physiome in Health and Disease Grants, 2950
NYCT Biomedical Research Grants, 3049
Partnership for Cures Two Years To Cures Grants, 3118
PC Fred H. Bixby Fellowships, 3126
Prudential Foundation Education Grants, 3276
RSC Abbyann D. Lynch Medal in Bioethics, 3419
Thrasher Research Fund Early Career Awards, 3731
Topeka Community Foundation Kansas Blood Services Scholarships, 3738
USAID Fighting Ebola Grants, 3812

Biometry
NCI Basic Cancer Research in Cancer Health Disparities Grants, 2647
NEI Clinical Vision Research Devel Award, 2664
NHLBI National Research Service Award Programs in Cardiovascular Epidemiology and Biostatistics, 2714
NIA Nathan Shock Centers of Excellence in Basic Biology of Aging Grants, 2776

Biophysics
Grass Foundation Marine Biological Laboratory Fellowships, 1878
HRAMF Smith Family Awards for Excellence in Biomedical Research, 2063
NHLBI Career Transition Awards, 2694
NIH Structural Bio of Membrane Proteins Grant, 2911
NSF Biological Physics (BP), 3008
NSF Biomolecular Systems Cluster Grants, 3010
PhRMA Foundation Pharmaceutics Postdoctoral Fellowships, 3211
PhRMA Foundation Pharmaceutics Pre Doctoral Fellowships, 3212

Biostatistics
Morehouse PHSI Project Imhotep Internships, 2592
NHLBI Summer Inst for Training in Biostatistics, 2734
NLM Express Research Grants in Biomedical Informatics, 2943

Biosynthesis
NSF Biomolecular Systems Cluster Grants, 3010

Biotechnology
AIHS ForeFront MBT Studentship Awards, 382
ASM Promega Biotechnology Research Award, 743
Canada Graduate Scholarships (CGS) and NSERC Postgraduate Scholarships (PGS), 1010
CCFF Chairman's Dist Life Sciences Award, 1044
Kavli Foundation Research Grants, 2296
Lumpkin Family Fnd Healthy People Grants, 2391
NHLBI Summer Inst for Training in Biostatistics, 2734
NINR Acute and Chronic Care During Mechanical Ventilation Research Grants, 2926
North Carolina Biotechnology Center Event Sponsorship Grants, 2984
North Carolina Biotechnology Center Grantsmanship Training Grants, 2985
North Carolina Biotech Center Meeting Grants, 2986
North Carolina Biotechnology Center Regional Development Grants, 2987
North Carolina Biotechnology Centers of Innovation Grants, 2988

672 / Biotechnology

North Carolina Biotechnology Center Technology Enhancement Grants, 2989
NSF Biomedical Engineering and Engineering Healthcare Grants, 3009
NSF Chemistry Research Experiences for Undergraduates (REU), 3012
NSF Emerging Frontiers in Research and Innovation (EFRI) Grants, 3014
PhRMA Foundation Pharmaceutics Research Starter Grants, 3213
PhRMA Foundation Pharmaceutics Sabbatical Fellowships, 3214

Bipolar Disorder
AACAP Quest for the Test Bipolar Disorder Pilot Research Award, 69
American Psychiatric Association/AstraZeneca Young Minds in Psychiatry International Awards, 538
NARSAD Colvin Prize for Outstanding Achievement in Mood Disorders Research, 2619
NARSAD Distinguished Investigator Grants, 2620
NARSAD Independent Investigator Grants, 2622
NARSAD Young Investigator Grants, 2625
Peter and Elizabeth C. Tower Foundation Annual Mental Health Grants, 3145

Birth/Congenital Defects
BWF Preterm Birth Initiative Grants, 990
CDC Epidemic Intell Service Training Grants, 1055
CDC Epidemiology Elective Rotation, 1056
CDI Interdisciplinary Research Initiatives Grants, 1074
CDI Postdoctoral Fellowships, 1075
HRAMF Deborah Munroe Noonan Memorial Research Grants, 2058
Little Life Foundation Grants, 2372
March of Dimes Program Grants, 2426
NICHD Ruth L. Kirschstein National Research Service Award (NRSA) Institutional Predoctoral Training Program in Systems Biology of Developmental Biology & Birth Defects, 2790
TJX Foundation Grants, 3735
USAID Global Health Development Innovation Accelerator Broad Agency Grants, 3813

Bisexuals
Ms. Fndn for Women Ending Violence Grants, 2596
Ms. Foundation for Women Health Grants, 2597

Blacks
AUPHA Corris Boyd Scholarships, 800
HRSA Resource and Technical Assistance Center for HIV Prevention and Care for Black Men who Have Sex with Men Cooperative Agreement, 2069
Theodore Edson Parker Foundation Grants, 3715

Bladder
NIDDK Basic Research Grants in the Bladder and Lower Urinary Tract, 2802
Pfizer Healthcare Charitable Contributions, 3180
Pfizer Medical Education Track Two Grants, 3183

Blood Coagulation
AABB Sally Frank Award & Lectureship, 47
ACCP Anticoagulation Training Program Grants, 227
HRAMF Taub Fnd Grants for MDS Research, 2064
MMAAP Foundation Fellowship Award in Hematology, 2579
NHLBI Immunomodulatory, Inflammatory, & Vasoregulatory Properties of Transfused Red Blood Cell Units as a Function of Preparation & Storage Grants, 2700

Blood Pressure
Fairlawn Foundation Grants, 1626
NIDDK Grants for Basic Research in Glomerular Diseases, 2819

Bolivia
Fulbright Traditional Scholar Program in the Western Hemisphere, 1760

Bone Disorders
NIH Bone and the Hematopoietic and Immune Systems Research Grants, 2868
NIH Receptors and Signaling in Bone in Health and Disease, 2891

Bone Marrow
HAF Barry F. Phelps Fund Grants, 1927
HRAMF Taub Fnd Grants for MDS Research, 2064
NIH Bone and the Hematopoietic and Immune Systems Research Grants, 2868

Book Awards
AAAS/Subaru SB&F Prize for Excellence in Science Books, 35

Books
AAAS/Subaru SB&F Prize for Excellence in Science Books, 35
ALA Donald G. Davis Article Award, 401
Greater Sitka Legacy Fund Grants, 1885
Kenai Peninsula Foundation Grants, 2298
Petersburg Community Foundation Grants, 3156

Botanical Gardens
Auburn Foundation Grants, 798
Bruce and Adele Greenfield Foundation Grants, 972
Ethel S. Abbott Charitable Foundation Grants, 1607
Herbert A. & Adrian W. Woods Fndn Grants, 2007
Kelvin and Eleanor Smith Foundation Grants, 2297
Perry County Community Foundation Grants, 3142
PMI Foundation Grants, 3239
Posey County Community Foundation Grants, 3252
Spencer County Community Foundation Grants, 3632
Union Bank, N.A. Foundation Grants, 3788
Warrick County Community Foundation Grants, 3904

Botanicals
NIH Dietary Supplement Research Centers: Botanicals Grants, 2873

Botany
Australasian Institute of Judicial Administration Seed Funding Grants, 806

Brain
A-T Children's Project Grants, 13
A-T Children's Project Post Doctoral Fellowships, 14
Alzheimer's Association Investigator-Initiated Research Grants, 466
CDC Cooperative Agreement for Continuing Enhanced National Surveillance for Prion Diseases in the United States, 1049
Dana Foundation Science and Health Grants, 1420
Gruber Foundation Neuroscience Prize, 1909
James S. McDonnell Foundation Brain Cancer Research Collaborative Activity Awards, 2201
James S. McDonnell Fnd Research Grants, 2203
James S. McDonnell Foundation Scholar Awards, 2204
James S. McDonnell Foundation Understanding Human Cognition Awards, 2205
Lumosity Human Cognition Grant, 2390
NIMH Short Courses in Neuroinformatics, 2920
Robert Leet & Clara Guthrie Patterson Grants, 3383
SfN Jacob P. Waletzky Award, 3523
SfN Julius Axelrod Prize, 3526
SfN Nemko Prize in Cellular or Molecular Neuroscience, 3529
SfN Ralph W. Gerard Prize in Neuroscience, 3534
SfN Science Educator Award, 3535
SfN Swartz Prize for Theoretical and Computational Neuroscience, 3537
USDD Workplace Violence in Military Grants, 3847

Brain Tumors
Children's Brain Tumor Fnd Research Grants, 1136
Children's Tumor Fndn Clin Research Awards, 1140
Children's Tumor Foundation Schwannomatosis Awards, 1142

SUBJECT INDEX

Children's Tumor Foundation Young Investigator Awards, 1143
Goldhirsh Fnd Brain Tumor Research Grants, 1852
James S. McDonnell Foundation Brain Cancer Research Collaborative Activity Awards, 2201
Pediatric Brain Tumor Fndn Research Grants, 3134

Brazil
Fulbright Traditional Scholar Program in the Western Hemisphere, 1760

Breast Cancer
1 in 9: Long Island Breast Cancer Action Coalition Grants, 1
AACR Outstanding Investigator Award for Breast Cancer Research, 121
ASCO/UICC International Cancer Technology Transfer (ICRETT) Fellowships, 677
ASCO Advanced Clinical Research Awards in Breast Cancer, 679
ASCO Young Investigator Award, 683
Avon Foundation Breast Care Fund Grants, 813
Avon Products Foundation Grants, 814
BCRF-AACR Grants for Translational Breast Cancer Research, 861
BCRF Research Grants, 862
Breast Cancer Fund Grants, 947
Campbell Soup Foundation Grants, 1007
CDC Increasing Breast and Cervical Cancer Screening Services for Urban American Indian/Alaska Native Women, 1063
CDC Public Health Prev Service Fellowships, 1069
CDC Public Health Prevention Service Fellowship Sponsorships, 1070
CFFVR Women's Fund for the Fox Valley Region Grants, 1107
Conquer Cancer Foundation of ASCO Career Development Award, 1328
Conquer Cancer Foundation of ASCO Comparative Effectiveness Research Professorship in Breast Cancer, 1329
Conquer Cancer Foundation of ASCO Drug Development Research Professorship, 1330
Conquer Cancer Foundation of ASCO Improving Cancer Care Grants, 1331
Conquer Cancer Foundation of ASCO International Development and Education Awards, 1333
DOD HBCU/MI Partnership Training Award, 1480
GNOF IMPACT Harold W. Newman, Jr. Charitable Trust Grants, 1841
IBCAT Screening Mammography Grants, 2095
Komen Greater NYC Clinical Research Enrollment Grants, 2315
Komen Greater NYC Community Breast Health Grants, 2316
Komen Greater NYC Small Grants, 2317
Kroger Foundation Women's Health Grants, 2325
Mary Kay Foundation Cancer Research Grants, 2444
Meyer and Stephanie Eglin Foundation Grants, 2531
Miles of Hope Breast Cancer Foundation Grants, 2553
Nancy R. Gelman Foundation Breast Cancer Seed Grants, 2608
NCI Stages of Breast Development: Normal to Metastatic Disease Grants, 2655
Penny Severns Breast, Cervical and Ovarian Cancer Research Grants, 3136
Pezcoller Foundation-AACR International Award for Cancer Research, 3161
Pfizer Healthcare Charitable Contributions, 3180
Pfizer Medical Education Track Two Grants, 3183
Premera Blue Cross Grants, 3258
Seattle Foundation Medical Funds Grants, 3508
Susan G. Komen Breast Cancer Foundation Brinker Awards for Scientific Distinction, 3687
Susan G. Komen Breast Cancer Foundation Career Catalyst Research Grants, 3688
Susan G. Komen Breast Cancer Foundation Challenge Grants: Breast Cancer and the Environment, 3689
Susan G. Komen Breast Cancer Foundation Challenge Grants: Investigator Initiated Research, 3690

SUBJECT INDEX

Susan G. Komen Breast Cancer Foundation Challenge Grants: Career Catalyst Research, 3691
Susan G. Komen Breast Cancer Foundation College Scholarships, 3692
Susan G. Komen Breast Cancer Foundation Investigator Initiated Research Grants, 3693
USDD Breast Cancer Clinical Translational Research Grants, 3831
USDD Breast Cancer Era of Hope Grants, 3832
USDD Breast Cancer Idea Grants, 3833
USDD Breast Cancer Impact Grants, 3834
USDD Breast Cancer Innovation Grants, 3835
USDD Breast Cancer Transformative Grant, 3836

Broadcast Media
AMA-RFS and AMA Foundation Medical Student & Resident/Fellow Elective-Medicine & Media, 478
American Academy of Nursing Media Awards, 509
OSF Health Media Initiative Grants, 3089
Reinberger Foundation Grants, 3357
SfN Science Educator Award, 3535
SfN Science Journalism Student Awards, 3536

Building/Construction
Charles H. Farnsworth Trust Grants, 1119
Clark and Ruby Baker Foundation Grants, 1174
Eugene Straus Charitable Trust, 1612
Gladys Brooks Foundation Grants, 1826
GNOF Albert N. & Hattie M. McClure Grants, 1834
GNOF Exxon-Mobil Grants, 1836
GNOF Norco Community Grants, 1845
Hillcrest Foundation Grants, 2031
Home Building Industry Disaster Relief Fund, 2043
IDPH Hosptial Capital Investment Grants, 2119
Janson Foundation Grants, 2210
Jennings County Community Foundation Grants, 2220
Josephine Schell Russell Charitable Trust Grants, 2274
Katharine Matthies Foundation Grants, 2291
Kosciusko County Community Foundation REMC Operation Round Up Grants, 2320
Lotus 88 Fnd for Women & Children Grants, 2376
MetroWest Health Foundation Capital Grants for Health-Related Facilities, 2526
Norcliffe Foundation Grants, 2979
Paul Ogle Foundation Grants, 3124
Potts Memorial Foundation Grants, 3254
Priddy Foundation Capital Grants, 3263
Robert R. Meyer Foundation Grants, 3387
SunTrust Bank Trusteed Foundations Walter H. and Marjory M. Rich Memorial Fund Grants, 3685
Vigneron Memorial Fund Grants, 3873

Burma
Prospect Burma Scholarships, 3275

Burns
Austin S. Nelson Foundation Grants, 805
IAFF Burn Foundation Research Grants, 2089
John S. Dunn Research Foundation Grants, 2250
Pfizer ASPIRE North America Broad Spectrum Antibiotics for the Treatment of Gram-Negative or Polymicrobial Infections Research Awards, 3169
St. Joseph Community Health Foundation Burn Care and Prevention Grants, 3639
USDD Care for the Critically Injured Burn Patient Grants, 3839

Business
Archer Daniels Midland Foundation Grants, 651
ARCO Foundation Education Grants, 652
AT&T Foundation Civic and Community Service Program Grants, 793
Beldon Fund Grants, 871
Boyd Gaming Corporation Contributions Program, 941
Collins C. Diboll Private Foundation Grants, 1209
Community Foundation of Greater Fort Wayne - Lilly Endowment Scholarships, 1275
Constellation Energy Corporate Grants, 1342
E.L. Wiegand Foundation Grants, 1538
Ford Motor Company Fund Grants Program, 1687

Fred and Louise Latshaw Scholarship, 1735
Graco Foundation Grants, 1860
IIE Iraq Scholars and Leaders Scholarships, 2133
John Ben Snow Memorial Trust Grants, 2231
LaGrange County Lilly Endowment Community Scholarship, 2331
NHSCA Arts in Health Care Project Grants, 2741
Noble County Community Foundation - Kathleen June Earley Memorial Scholarship, 2973
Phoenix Suns Charities Grants, 3201
Procter and Gamble Fund Grants, 3274
Prudential Foundation Education Grants, 3276
Schrage Family Foundation Scholarships, 3494
Sengupta Family Scholarship, 3512
United Technologies Corporation Grants, 3803
W. James Spicer Scholarship, 3885
Whitley County Community Foundation - Lilly Endowment Scholarship, 3935
Whitley County Community Fndn Scholarships, 3937
Willary Foundation Grants, 3944
Wolfe Associates Grants, 3969

Business Administration
Advocate HealthCare Post Graduate Administrative Fellowship, 314
American Academy of Nursing Claire M. Fagin Fellowships, 506
American Academy of Nursing Mayday Grants, 507
American Academy of Nursing MBA Scholarships, 508
Canada-U.S. Fulbright New Century Scholars Program Grants, 1008
Canada-U.S. Fulbright Senior Specialists Grants, 1009
Cargill Citizenship Fund Corporate Giving, 1018
Fulbright Alumni Initiatives Awards, 1749
Fulbright Distinguished Chairs Awards, 1750
Fulbright New Century Scholars Grants, 1753
Fulbright Specialists Program Grants, 1754
Fulbright Scholars in Europe and Eurasia, 1755
Fulbright Traditional Scholar Program in Sub-Saharan Africa, 1756
Fulbright Traditional Scholar Program in the East Asia/Pacific Region, 1757
Fulbright Traditional Scholar Program in the Near East and North Africa Region, 1758
Fulbright Traditional Scholar Program in the South and Central Asia Region, 1759
Fulbright Traditional Scholar Program in the Western Hemisphere, 1760
Mayo Clinic Administrative Fellowship, 2460
Mayo Clinic Admin Fellowship - Eau Claire, 2461
Mayo Clinic Business Consulting Fellowship, 2462
Richard Davoud Donchian Foundation Grants, 3367
SfN Louise Hanson Marshall Specific Recognition Award, 3527

Business Development
ACF Native American Social and Economic Development Strategies Grants, 231
Arkansas Community Foundation Arkansas Black Hall of Fame Grants, 659
Bailey Foundation Grants, 818
Boyd Gaming Corporation Contributions Program, 941
General Motors Foundation Grants, 1783
Grand Haven Area Community Fndn Grants, 1864
HAF Technical Assistance Program Grants, 1933
Harold Alfond Foundation Grants, 1949
Helen Bader Foundation Grants, 1992
Indiana 21st Century Research and Technology Fund Awards, 2154
Jacobs Family Village Neighborhoods Grants, 2184
Lewis H. Humphreys Charitable Trust Grants, 2361
Louis and Elizabeth Nave Flarsheim Charitable Foundation Grants, 2379
Priddy Foundation Organizational Development Grants, 3265
Pulaski County Community Foundation Grants, 3279
Southbury Community Trust Fund, 3625
SunTrust Bank Trusteed Foundations Florence C. and Harry L. English Memorial Fund Grants, 3680

SunTrust Bank Trusteed Foundations Greene-Sawtell Grants, 3681
SunTrust Bank Trusteed Foundations Harriet McDaniel Marshall Tust Grants, 3682
SunTrust Bank Trusteed Foundations Nell Warren Elkin and William Simpson Elkin Grants, 3683
SunTrust Bank Trusteed Foundations Walter H. and Marjory M. Rich Memorial Fund Grants, 3685
Thomas Austin Finch, Sr. Foundation Grants, 3718
U.S. Department of Education Vocational Rehabilitation Services Projects for American Indians with Disabilities Grants, 3774

Business Education
3M Company Fndn Community Giving Grants, 6
Abell-Hanger Foundation Grants, 207
Alcoa Foundation Grants, 414
Allstate Corporate Giving Grants, 446
Allstate Corp Hometown Commitment Grants, 447
American Academy of Nursing Claire M. Fagin Fellowships, 506
American Academy of Nursing Mayday Grants, 507
American Academy of Nursing MBA Scholarships, 508
AMI Semiconductors Corporate Grants, 576
Benton Community Foundation Grants, 878
Blue River Community Foundation Grants, 922
Boettcher Foundation Grants, 931
Bright Family Foundation Grants, 951
Brown County Community Foundation Grants, 970
Chazen Foundation Grants, 1129
Coca-Cola Foundation Grants, 1203
Community Foundation of Bartholomew County Heritage Fund Grants, 1262
Community Foundation of Bartholomew County James A. Henderson Award for Fundraising, 1263
Community Foundation of Greater Fort Wayne - Community Endowment and Clarke Endowment Grants, 1274
Consumers Energy Foundation, 1343
D.F. Halton Foundation Grants, 1405
DaimlerChrysler Corporation Fund Grants, 1412
Essex County Community Foundation Women's Fund Grants, 1605
F.M. Kirby Foundation Grants, 1624
FMC Foundation Grants, 1683
GenCorp Foundation Grants, 1779
General Motors Foundation Grants, 1783
James M. Collins Foundation Grants, 2195
Richard and Helen DeVos Foundation Grants, 3364
Robert W. Woodruff Foundation Grants, 3389
Sid W. Richardson Foundation Grants, 3553
Union Bank, N.A. Foundation Grants, 3788
Washington Gas Charitable Giving, 3908

Business Ethics
Richard Davoud Donchian Foundation Grants, 3367

Business and Commerce
CONSOL Energy Community Dev Grants, 1339
GNOF New Orleans Works Grants, 1844
SSHRC-NSERC College and Community Innovation Enhancement Grants, 3636

Canada
Cystic Fibrosis Canada Senior Scientist Research Training Awards, 1387
Cystic Fibrosis Canada Small Conference Grants, 1388
Cystic Fibrosis Canada Summer Studentships, 1391
Cystic Fibrosis Canada Visiting Clinician Awards, 1395
Cystic Fibrosis Canada Visiting Scientist Awards, 1396
Fulbright Traditional Scholar Program in the Western Hemisphere, 1760
NIA Postdoctoral Research on Aging in Canada, 2779
RCPSC Balfour M. Mount Visiting Professorship in Palliative Medicine, 3318
RCPSC Continuing Proffesional Dev Grants, 3319
RCPSC Detweiler Traveling Fellowships, 3320
RCPSC Tom Dignan Indigenous Health Award, 3321
RCPSC Harry S. Morton Traveling Fellowship in Surgery, 3324

674 / Canada

RCPSC International Residency Educator of the Year Award, 3325
RCPSC Int'l Resident Leadership Award, 3326
RCPSC James H. Graham Award of Merit, 3327
RCPSC Janes Visiting Professorship in Surgery, 3328
RCPSC John G. Wade Visiting Professorship in Patient Safety & Simulation-Based Med Education, 3329
RCPSC KJR Wightman Award for Scholarship in Ethics, 3330
RCPSC KJR Wightman Visiting Professorship in Medicine, 3331
RCPSC Kristin Sivertz Res Leadership Award, 3332
RCPSC Medical Education Research Grants, 3334
RCPSC Mentor of the Year Award, 3335
RCPSC Peter Warren Traveling Scholarship, 3336
RCPSC Prix D'excellence (Specialist of the Year) Award, 3337
RCPSC Professional Development Grants, 3338
RCPSC Program Administrator Award for Innovation and Excellence, 3339
RCPSC Program Director of the Year Award, 3340
RCPSC Robert Maudsley Fellowship for Studies in Medical Education, 3341
RCPSC Royal College Medal Award in Medicine, 3342
RCPSC Royal College Medal Award in Surgery, 3343
RCPSC Teasdale-Corti Humanitarian Award, 3344
RSC Jason A. Hannah Medal, 3420
Sick Kids Foundation New Investigator Research Grants, 3550

Canadian History
ALA Donald G. Davis Article Award, 401

Canadian Studies
Canada-U.S. Fulbright New Century Scholars Program Grants, 1008
Canada-U.S. Fulbright Senior Specialists Grants, 1009

Cancer Detection
1 in 9: Long Island Breast Cancer Action Coalition Grants, 1
2COBS Private Charitable Foundation Grants, 5
A/H Foundation Grants, 18
AAAS Martin and Rose Wachtel Cancer Research Award, 40
AACR-American Cancer Society Award for Research Excellence in Epidemiology and Prevention, 87
AACR-Colorectal Cancer Coalition Fellows Grant, 88
AACR-FNAB Fellows Grant for Translational Pancreatic Cancer Research, 89
AACR-FNAB Foundation Career Development Award for Translational Cancer Research, 90
AACR-Genentech BioOncology Career Development Award for Cancer Research on the HER Family Pathway, 91
AACR-Genentech BioOncology Fellowship for Cancer Research in Angiogenesis, 92
AACR-GlaxoSmithKline Clinical Cancer Research Scholar Awards, 93
AACR-National Brain Tumor Fnd Fellows Grant, 95
AACR-NCI International Investigator Opportunity Grants, 96
AACR-Pancreatic Cancer Action Network Career Development Award for Research, 97
AACR-Pancreatic Cancer Action Network Innovative Grants, 98
AACR-Prevent Cancer Foundation Award for Excellence in Cancer Prevention Research, 99
AACR-WICR Scholar Awards, 100
AACR Award for Lifetime Achievement in Cancer Research, 102
AACR Award for Outstanding Achievement in Chemistry in Cancer Research, 104
AACR Brigid G. Leventhal Scholar in Cancer Research Awards, 106
AACR Centennial Postdoctoral Fellowships in Cancer Research, 108
AACR Centennial Pre-doctoral Fellowships in Cancer Research, 109
AACR Fellows Grants, 111

AACR G.H.A. Clowes Memorial Award, 112
AACR Gertrude Elion Cancer Research Award, 113
AACR Shepard Bladder Cancer Research Grants, 114
AACR Joseph H. Burchenal Memorial Award, 115
AACR Kirk A. Landon and Dorothy P. Landon Foundation Prizes, 117
AACR Foti Award for Leadership & Extraordinary Achievements in Cancer Research, 118
AACR Minority-Serving Institution Faculty Scholar in Cancer Research Awards, 119
AACR Min Scholar in Cancer Research Award, 120
AACR Outstanding Investigator Award for Breast Cancer Research, 121
AACR Princess Takamatsu Memorial Lectureship, 122
AACR Team Science Award, 126
ACGT Young Investigator Grants, 233
Angel Kiss Foundation Grants, 589
Angels Baseball Foundation Grants, 590
Ann and Robert H. Lurie Family Fndn Grants, 593
Austin S. Nelson Foundation Grants, 805
Avon Foundation Breast Care Fund Grants, 813
Avon Products Foundation Grants, 814
Balfe Family Foundation Grants, 819
Breast Cancer Fund Grants, 947
CDC Increasing Breast and Cervical Cancer Screening Services for Urban American Indian/Alaska Native Women, 1063
CDC Public Health Prev Service Fellowships, 1069
CDI Interdisciplinary Research Initiatives Grants, 1074
CDI Postdoctoral Fellowships, 1075
CFFVR Jewelers Mutual Charitable Giving, 1099
Coleman Foundation Cancer Care Grants, 1206
Community Foundation of Louisville Bobbye M. Robinson Fund Grants, 1291
Community Foundation of Louisville Irving B. Klempner Fund Grants, 1296
Conquer Cancer Foundation of ASCO International Development and Education Awards, 1333
Expect Miracles Foundation Grants, 1619
FDHN Funderburg Research Scholar Award in Gastric Biology Related to Cancer, 1645
Foundation of CVPH Chelsea's Rainbow Grants, 1702
General Motors Foundation Grants, 1783
Gil and Dody Weaver Foundation Grants, 1819
GNOF IMPACT Harold W. Newman, Jr. Charitable Trust Grants, 1841
Gruber Foundation Weizmann Institute Awards, 1911
Hearst Foundations Health Grants, 1987
IARC Expertise Transfer Fellowship, 2090
IARC Postdoctoral Fellowships for Training in Cancer Research, 2091
IARC Visiting Award for Senior Scientists, 2092
James S. McDonnell Foundation Brain Cancer Research Collaborative Activity Awards, 2201
Kroger Foundation Women's Health Grants, 2325
Littlefield-AACR Grants in Metastatic Colon Cancer Research, 2371
Lustgarten Foundation for Pancreatic Cancer Research Grants, 2392
Mary Kay Foundation Cancer Research Grants, 2444
Morgan Adams Foundation Grants, 2593
NCI Centers of Excellence in Cancer Communications Research, 2649
NCI Diet, Epigenetic Events, and Cancer Prevention Grants, 2650
NIH Biology, Development, and Progression of Malignant Prostate Disease Research Grants, 2866
NYCT Biomedical Research Grants, 3049
Paul Marks Prizes for Cancer Research, 3123
Pediatric Cancer Research Foundation Grants, 3135
Penny Severns Breast, Cervical and Ovarian Cancer Research Grants, 3136
Pezcoller Foundation-AACR International Award for Cancer Research, 3161
Ruth Estrin Goldberg Memorial for Cancer Research Grants, 3428
Ryan Gibson Foundation Grants, 3445
Seattle Foundation Medical Funds Grants, 3508
Susan G. Komen Breast Cancer Foundation Challege Grants: Breast Cancer and the Environment, 3689

SUBJECT INDEX

Susan G. Komen Breast Cancer Foundation Investigator Initiated Research Grants, 3693
T. Spencer Shore Foundation Grants, 3701
Thomas C. Burke Foundation Grants, 3720
UICC American Cancer Society International Fellowships for Beginning Investigators, 3778
USDD Breast Cancer Clinical Translational Research Grants, 3831
USDD Breast Cancer Impact Grants, 3834
USDD Breast Cancer Innovation Grants, 3835
USDD Breast Cancer Transformative Grant, 3836
W. C. Griffith Foundation Grants, 3882

Cancer Prevention
1 in 9: Long Island Breast Cancer Action Coalition Grants, 1
2COBS Private Charitable Foundation Grants, 5
A/H Foundation Grants, 18
AAAS Martin and Rose Wachtel Cancer Research Award, 40
AACR-American Cancer Society Award for Research Excellence in Epidemiology and Prevention, 87
AACR-Colorectal Cancer Coalition Fellows Grant, 88
AACR-FNAB Fellows Grant for Translational Pancreatic Cancer Research, 89
AACR-FNAB Foundation Career Development Award for Translational Cancer Research, 90
AACR-Genentech BioOncology Career Development Award for Cancer Research on the HER Family Pathway, 91
AACR-Genentech BioOncology Fellowship for Cancer Research in Angiogenesis, 92
AACR-GlaxoSmithKline Clinical Cancer Research Scholar Awards, 93
AACR-National Brain Tumor Fnd Fellows Grant, 95
AACR-NCI International Investigator Opportunity Grants, 96
AACR-Pancreatic Cancer Action Network Career Development Award for Research, 97
AACR-Pancreatic Cancer Action Network Innovative Grants, 98
AACR-Prevent Cancer Foundation Award for Excellence in Cancer Prevention Research, 99
AACR-WICR Scholar Awards, 100
AACR Award for Lifetime Achievement in Cancer Research, 102
AACR Award for Outstanding Achievement in Chemistry in Cancer Research, 104
AACR Basic Cancer Research Fellowships, 105
AACR Brigid G. Leventhal Scholar in Cancer Research Awards, 106
AACR Career Development Awards for Pediatric Cancer Research, 107
AACR Centennial Postdoctoral Fellowships in Cancer Research, 108
AACR Centennial Pre-doctoral Fellowships in Cancer Research, 109
AACR Clinical & Translat Research Fellowships, 110
AACR Fellows Grants, 111
AACR G.H.A. Clowes Memorial Award, 112
AACR Gertrude Elion Cancer Research Award, 113
AACR Shepard Bladder Cancer Research Grants, 114
AACR Joseph H. Burchenal Memorial Award, 115
AACR Judah Folkman Career Development Award for Anti-Angiogenesis Research, 116
AACR Kirk A. Landon and Dorothy P. Landon Foundation Prizes, 117
AACR Foti Award for Leadership & Extraordinary Achievements in Cancer Research, 118
AACR Minority-Serving Institution Faculty Scholar in Cancer Research Awards, 119
AACR Min Scholar in Cancer Research Award, 120
AACR Outstanding Investigator Award for Breast Cancer Research, 121
AACR Princess Takamatsu Memorial Lectureship, 122
AACR Team Science Award, 126
ACGT Young Investigator Grants, 233
Angel Kiss Foundation Grants, 589
Angels Baseball Foundation Grants, 590
Ann and Robert H. Lurie Family Fndn Grants, 593

SUBJECT INDEX

ASCO Advanced Clinical Research Award in Colorectal Cancer, 678
ASCO Advanced Clinical Research Awards in Breast Cancer, 679
ASCO Advanced Clinical Research Awards in Sarcoma, 680
ASCO Long-Term International Fellowships, 681
Austin S. Nelson Foundation Grants, 805
Balfe Family Foundation Grants, 819
Breast Cancer Fund Grants, 947
CDC Increasing Breast and Cervical Cancer Screening Services for Urban American Indian/Alaska Native Women, 1063
CDC Public Health Prev Service Fellowships, 1069
CDC Public Health Prevention Service Fellowship Sponsorships, 1070
CFFVR Jewelers Mutual Charitable Giving, 1099
Coleman Foundation Cancer Care Grants, 1206
Community Foundation of Louisville Bobbye M. Robinson Fund Grants, 1291
Conquer Cancer Foundation of ASCO Comparative Effectiveness Research Professorship in Breast Cancer, 1329
Conquer Cancer Foundation of ASCO Drug Development Research Professorship, 1330
Conquer Cancer Foundation of ASCO Improving Cancer Care Grants, 1331
Conquer Cancer Foundation of ASCO International Innovation Grant, 1334
Conquer Cancer Foundation of ASCO Medical Student Rotation Grants, 1335
Conquer Cancer Foundation of ASCO Translational Research Professorships, 1337
Cowles Charitable Trust Grants, 1354
Expect Miracles Foundation Grants, 1619
Fairlawn Foundation Grants, 1626
FDHN Funderburg Research Scholar Award in Gastric Biology Related to Cancer, 1645
Foundation of CVPH Chelsea's Rainbow Grants, 1702
Georgia Power Foundation Grants, 1806
Gerber Foundation Grants, 1807
Gil and Dody Weaver Foundation Grants, 1819
GNOF IMPACT Harold W. Newman, Jr. Charitable Trust Grants, 1841
GNOF IMPACT Kahn-Oppenheim Grants, 1842
Greenfield Foundation of Maine Grants, 1893
Greenwall Foundation Bioethics Grants, 1897
Gregory Family Foundation Grants (Florida), 1899
Gruber Foundation Weizmann Institute Awards, 1911
Hearst Foundations Health Grants, 1987
IARC Expertise Transfer Fellowship, 2090
IARC Postdoctoral Fellowships for Training in Cancer Research, 2091
IARC Visiting Award for Senior Scientists, 2092
IDPH Carolyn Adams Ticket for the Cure Community Grants, 2117
James S. McDonnell Foundation Brain Cancer Research Collaborative Activity Awards, 2201
Kroger Foundation Women's Health Grants, 2325
Landon Foundation-AACR Innovator Award for Cancer Prevention Research, 2342
Landon Foundation-AACR Innovator Award for International Collaboration, 2343
Lisa Higgins-Hussman Foundation Grants, 2370
Littlefield-AACR Grants in Metastatic Colon Cancer Research, 2371
Mary Kay Foundation Cancer Research Grants, 2444
Medtronic Foundation Strengthening Health Systems Grants, 2515
Morgan Adams Foundation Grants, 2593
NCI Basic Cancer Research in Cancer Health Disparities Grants, 2647
NCI Cancer Education and Career Development Program, 2648
NCI Centers of Excellence in Cancer Communications Research, 2649
NCI Diet, Epigenetic Events, and Cancer Prevention Grants, 2650
NCI Exploratory Grants for Behavioral Research in Cancer Control, 2651

NYCT Biomedical Research Grants, 3049
Paul Marks Prizes for Cancer Research, 3123
Pediatric Cancer Research Foundation Grants, 3135
Penny Severns Breast, Cervical and Ovarian Cancer Research Grants, 3136
Pezcoller Foundation-AACR International Award for Cancer Research, 3161
RCF General Community Grants, 3312
Ryan Gibson Foundation Grants, 3445
Susan G. Komen Breast Cancer Foundation Brinker Awards for Scientific Distinction, 3687
Susan G. Komen Breast Cancer Foundation Career Catalyst Research Grants, 3688
Susan G. Komen Breast Cancer Foundation Challege Grants: Breast Cancer and the Environment, 3689
T. Spencer Shore Foundation Grants, 3701
Thomas C. Burke Foundation Grants, 3720
UICC American Cancer Society International Fellowships for Beginning Investigators, 3778
USDD Breast Cancer Clinical Translational Research Grants, 3831
USDD Breast Cancer Idea Grants, 3833
USDD Breast Cancer Impact Grants, 3834
USDD Breast Cancer Innovation Grants, 3835
USDD Breast Cancer Transformative Grant, 3836
W. C. Griffith Foundation Grants, 3882

Cancer/Carcinogenesis

2COBS Private Charitable Foundation Grants, 5
AAAS Martin and Rose Wachtel Cancer Research Award, 40
AACR-American Cancer Society Award for Research Excellence in Epidemiology and Prevention, 87
AACR-Colorectal Cancer Coalition Fellows Grant, 88
AACR-FNAB Fellows Grant for Translational Pancreatic Cancer Research, 89
AACR-FNAB Foundation Career Development Award for Translational Cancer Research, 90
AACR-Genentech BioOncology Career Development Award for Cancer Research on the HER Family Pathway, 91
AACR-GlaxoSmithKline Clinical Cancer Research Scholar Awards, 93
AACR-Minorities in Cancer Research Jane Cooke Wright Lectureship Awards, 94
AACR-National Brain Tumor Fnd Fellows Grant, 95
AACR-NCI International Investigator Opportunity Grants, 96
AACR-Pancreatic Cancer Action Network Career Development Award for Research, 97
AACR-Pancreatic Cancer Action Network Innovative Grants, 98
AACR-Prevent Cancer Foundation Award for Excellence in Cancer Prevention Research, 99
AACR-WICR Scholar Awards, 100
AACR-Women in Cancer Research Charlotte Friend Memorial Lectureship Awards, 101
AACR Award for Lifetime Achievement in Cancer Research, 102
AACR Award for Outstanding Achievement in Cancer Research, 103
AACR Award for Outstanding Achievement in Chemistry in Cancer Research, 104
AACR Basic Cancer Research Fellowships, 105
AACR Career Development Awards for Pediatric Cancer Research, 107
AACR Centennial Postdoctoral Fellowships in Cancer Research, 108
AACR Centennial Pre-doctoral Fellowships in Cancer Research, 109
AACR Clinical & Translat Research Fellowships, 110
AACR Fellows Grants, 111
AACR G.H.A. Clowes Memorial Award, 112
AACR Gertrude Elion Cancer Research Award, 113
AACR Shepard Bladder Cancer Research Grants, 114
AACR Joseph H. Burchenal Memorial Award, 115
AACR Judah Folkman Career Development Award for Anti-Angiogenesis Research, 116
AACR Kirk A. Landon and Dorothy P. Landon Foundation Prizes, 117

Cancer/Carcinogenesis /675

AACR Foti Award for Leadership & Extraordinary Achievements in Cancer Research, 118
AACR Minority-Serving Institution Faculty Scholar in Cancer Research Awards, 119
AACR Min Scholar in Cancer Research Award, 120
AACR Outstanding Investigator Award for Breast Cancer Research, 121
AACR Princess Takamatsu Memorial Lectureship, 122
AACR Richard & Hinda Rosenthal Mem Awards, 123
AACR Team Science Award, 126
ACGT Investigators Grants, 232
ACGT Young Investigator Grants, 233
ACS Cancer Control Career Development Awards for Primary Care Physicians, 250
ACS Clinical Research Professor Grants, 251
ACS Doc Degree Scholarships in Cancer Nursing, 252
ACS Doctoral Scholarships in Cancer Nursing, 253
ACS Doctoral Grants in Oncology Social Work, 254
ACS Scholarships in Cancer Nursing Practice, 255
ACS Institutional Research Grants, 256
ACS Master's Training Grants in Clinical Oncology Social Work, 258
ACS MEN2 Thyroid Cancer Professorship Grants, 263
ACS Mentored Research Scholar Grant in Applied and Clinical Research, 264
ACS Physician Training Awards in Prevent Med, 275
ACS Pilot and Exploratory Projects in Palliative Care of Cancer Patients and Their Families Grants, 276
ACS Postdoctoral Fellowships, 277
ACS Research Professor Grants, 278
ACS Research Scholar Grants for Health Services and Health Policy Research, 279
ACS Research Scholar Grants in Basic, Preclinical, Clinical and Epidemiology Research, 280
ACS Research Scholar Grants in Psychosocial and Behavioral and Cancer Control Research, 281
Affymetrix Corporate Contributions Grants, 339
Alexander and Margaret Stewart Trust Grants, 419
AMA Foundation Seed Grants for Research, 494
Angels Baseball Foundation Grants, 590
Annunziata Sanguinetti Foundation Grants, 600
Aragona Family Foundation Grants, 648
ArvinMeritor Foundation Health Grants, 674
ASCO/UICC International Cancer Technology Transfer (ICRETT) Fellowships, 677
ASCO Advanced Clinical Research Award in Colorectal Cancer, 678
ASCO Advanced Clinical Research Awards in Breast Cancer, 679
ASCO Advanced Clinical Research Awards in Sarcoma, 680
ASCO Long-Term International Fellowships, 681
ASCO Merit Awards, 682
ASCO Young Investigator Award, 683
Bodenwein Public Benevolent Foundation Grants, 928
Booth-Bricker Fund Grants, 933
Bristol-Myers Squibb Foundation Health Disparities Grants, 958
Callaway Golf Company Foundation Grants, 1003
Caplow Applied Science Carcinogen Prize, 1015
CFFVR Robert and Patricia Endries Family Foundation Grants, 1102
Charles Lafitte Foundation Grants, 1123
Coleman Foundation Cancer Care Grants, 1206
Community Foundation of Louisville Bobbye M. Robinson Fund Grants, 1291
Community Foundation of Louisville Irving B. Klempner Fund Grants, 1296
Conquer Cancer Foundation of ASCO Career Development Award, 1328
Conquer Cancer Foundation of ASCO Comparative Effectiveness Research Professorship in Breast Cancer, 1329
Conquer Cancer Foundation of ASCO Drug Development Research Professorship, 1330
Conquer Cancer Foundation of ASCO Improving Cancer Care Grants, 1331
Conquer Cancer Foundation of ASCO International Development and Education Award in Palliative Care, 1332

676 / Cancer/Carcinogenesis

Conquer Cancer Foundation of ASCO International Innovation Grant, 1334
Conquer Cancer Foundation of ASCO Medical Student Rotation Grants, 1335
Conquer Cancer Foundation of ASCO Resident Travel Award for Underrepresented Populations, 1336
Conquer Cancer Foundation of ASCO Translational Research Professorships, 1337
Cooper Industries Foundation Grants, 1345
Coors Brewing Corporate Contributions Grants, 1346
CRH Foundation Grants, 1360
D.F. Halton Foundation Grants, 1405
DeRoy Testamentary Foundation Grants, 1469
DOD HBCU/MI Partnership Training Award, 1480
Doris Duke Charitable Foundation Clinical Scientist Development Award, 1494
Duchossois Family Foundation Grants, 1518
E.L. Wiegand Foundation Grants, 1538
Educational Foundation of America Grants, 1553
Elizabeth McGraw Foundation Grants, 1568
Emerson Charitable Trust Grants, 1589
Emma B. Howe Memorial Foundation Grants, 1593
Fannie E. Rippel Foundation Grants, 1630
Farmers Insurance Corporate Giving Grants, 1634
Fondren Foundation Grants, 1685
Fndn for Appalachian Ohio Bachtel Scholarships, 1692
Fred C. & Katherine B. Andersen Grants, 1737
General Motors Foundation Grants, 1783
Georgia Power Foundation Grants, 1806
GNOF IMPACT Harold W. Newman, Jr. Charitable Trust Grants, 1841
Greenspun Family Foundation Grants, 1896
Greenwall Foundation Bioethics Grants, 1897
Grifols Community Outreach Grants, 1904
Hagedorn Fund Grants, 1934
Hawaii Community Foundation Health Education and Research Grants, 1976
Hearst Foundations Health Grants, 1987
HHMI Biomedical Research Grants for International Scientists: Infectious Diseases & Parasitology, 2017
HHMI Biomedical Research Grants for International Scientists in Canada and Latin America, 2018
HHMI Biomedical Research Grants for International Scientists in the Baltics, Central and Eastern Europe, Russia, and Ukraine, 2019
Hillman Foundation Grants, 2032
HomeBanc Foundation Grants, 2042
Horizon Foundation for New Jersey Grants, 2048
IARC Expertise Transfer Fellowship, 2090
IARC Postdoctoral Fellowships for Training in Cancer Research, 2091
IARC Visiting Award for Senior Scientists, 2092
James S. McDonnell Foundation Brain Cancer Research Collaborative Activity Awards, 2201
James S. McDonnell Fnd Research Grants, 2203
Janus Foundation Grants, 2211
John S. Dunn Research Foundation Grants, 2250
K21 Health Foundation Cancer Care Grants, 2284
Komen Greater NYC Clinical Research Enrollment Grants, 2315
Komen Greater NYC Community Breast Health Grants, 2316
Landon Foundation-AACR Innovator Award for Cancer Prevention Research, 2342
Landon Foundation-AACR Innovator Award for International Collaboration, 2343
Lillian S. Wells Foundation Grants, 2365
Littlefield-AACR Grants in Metastatic Colon Cancer Research, 2371
Lustgarten Foundation for Pancreatic Cancer Research Grants, 2392
Mary Kay Foundation Cancer Research Grants, 2444
Mesothelioma Applied Research Fndn Grants, 2524
NCI/NCCAM Quick-Trials for Novel Cancer Therapies, 2643
NCI Academic-Industrial Partnerships for Devel and Validation of In Vivo Imaging Systems and Methods for Cancer Investigations, 2644
NCI Application and Use of Transformative Emerging Technologies in Cancer Research, 2645

NCI Basic Cancer Research in Cancer Health Disparities Grants, 2647
NCI Cancer Education and Career Development Program, 2648
NCI Centers of Excellence in Cancer Communications Research, 2649
NCI Exploratory Grants for Behavioral Research in Cancer Control, 2651
NCI Stages of Breast Development: Normal to Metastatic Disease Grants, 2655
NCI Technologies and Software to Support Integrative Cancer Biology Research (SBIR) Grants, 2656
NIEHS Small Business Innovation Research (SBIR) Program Grants, 2860
NYCT Biomedical Research Grants, 3049
OSF International Palliative Care Grants, 3091
Pancreatic Cancer Action Network Fellowship, 3113
Pancreatic Cancer Action Network-AACR Pathway to Leadership Grants, 3114
Partnership for Cures Two Years To Cures Grants, 3118
Paul Marks Prizes for Cancer Research, 3123
Pediatric Cancer Research Foundation Grants, 3135
Pezcoller Foundation-AACR International Award for Cancer Research, 3161
Pfizer Australia Cancer Research Grants, 3174
Pittsburgh Foundation Medical Research Grants, 3231
Premera Blue Cross Grants, 3258
Robert R. Meyer Foundation Grants, 3387
Ruth Estrin Goldberg Memorial for Cancer Research Grants, 3428
Sara Elizabeth O'Brien Trust Grants, 3483
Seattle Foundation Medical Funds Grants, 3508
Simeon J. Fortin Charitable Foundation Grants, 3562
Skin Cancer Foundation Research Grants, 3578
Stella and Charles Guttman Foundation Grants, 3651
Susan G. Komen Breast Cancer Foundation Challege Grants: Breast Cancer and the Environment, 3689
Susan G. Komen Breast Cancer Foundation Challege Grants: Investigator Initiated Research, 3690
Susan G. Komen Breast Cancer Foundation Challege Grants: Career Catalyst Research, 3691
Susan G. Komen Breast Cancer Foundation College Scholarships, 3692
Susan G. Komen Breast Cancer Foundation Investigator Initiated Research Grants, 3693
Thompson Charitable Foundation Grants, 3727
UICC American Cancer Society International Fellowships for Beginning Investigators, 3778
UICC International Cancer Technology Transfer (ICRETT) Fellowships, 3779
UICC Raisa Gorbachev Memorial International Cancer Fellowships, 3780
UICC Yamagiwa-Yoshida Memorial International Cancer Study Grants, 3781
Union Bank, N.A. Foundation Grants, 3788
United Hospital Fund of New York Health Care Improvement Grants, 3797
USDD Breast Cancer Transformative Grant, 3836
Victor E. Speas Foundation Grants, 3868
Visiting Nurse Foundation Grants, 3879
W.C. Griffith Foundation Grants, 3881
W. C. Griffith Foundation Grants, 3882
Wilhelm Sander-Stiftung Foundation Grants, 3942

Carbohydrates
American Chem Society Claude Hudson Awards, 515

Cardiology
Abbott Fund Science Education Grants, 203
ACCP Heart Failure Training Program Grants, 228
AFAR Research Grants, 337
Duke University Clinical Cardiac Electrophysiology Fellowships, 1522
Duke Univ Interventional Cardio Fellowships, 1524
Edna G. Kynett Memorial Foundation Grants, 1551
Genentech Corp Charitable Contributions, 1780
Harold R. Bechtel Charitable Remainder Uni-Trust Grants, 1953
Horizon Foundation for New Jersey Grants, 2048
Joseph Drown Foundation Grants, 2268

NHLBI Ancillary Studies in Clinical Trials, 2688
NHLBI Cardiac Devel Consortium Grants, 2693
NHLBI Pediatric Cardiac Genomics Consortium Grants, 2719
NHLBI Research on the Role of Cardiomyocyte Mitochondria in Heart Disease: An Integrated Approach, 2726
NIH Sarcoidosis: Research into the Cause of Multi-Organ Disease and Clinical Strategies for Therapy Grants, 2907
Thoracic Foundation Grants, 3729

Cardiomyocyte Mitochondria
NHLBI Research on the Role of Cardiomyocyte Mitochondria in Heart Disease: An Integrated Approach, 2726

Cardiovascular Diseases
Abbott Fund Science Education Grants, 203
AIHS Alberta/Pfizer Translat Research Grants, 374
Alcohol Misuse and Alcoholism Research Grants, 415
AMA Foundation Seed Grants for Research, 494
Balfe Family Foundation Grants, 819
Bristol-Myers Squibb Clinical Outcomes and Research Grants, 955
Chest Foundation/LUNGevity Foundation Clinical Research in Lung Cancer Grants, 1131
Children's Cardiomyopathy Foundation Research Grants, 1137
CVS All Kids Can Grants, 1377
D.F. Halton Foundation Grants, 1405
Doris Duke Charitable Foundation Clinical Scientist Development Award, 1494
Duke Univ Interventional Cardio Fellowships, 1524
E.L. Wiegand Foundation Grants, 1538
Educational Foundation of America Grants, 1553
Emma B. Howe Memorial Foundation Grants, 1593
Fannie E. Rippel Foundation Grants, 1630
Fndn for Appalachian Ohio Bachtel Scholarships, 1692
Gerber Foundation Grants, 1807
GNOF IMPACT Harold W. Newman, Jr. Charitable Trust Grants, 1841
GNOF IMPACT Kahn-Oppenheim Grants, 1842
Harold R. Bechtel Charitable Remainder Uni-Trust Grants, 1953
Hawaii Community Foundation Health Education and Research Grants, 1976
Henrietta Lange Burk Fund Grants, 2001
HRAMF Harold S. Geneen Charitable Trust Awards for Coronary Heart Disease Research, 2059
Jacob and Valeria Langeloth Foundation Grants, 2182
Johnson & Johnson Community Health Grants, 2252
Lucy Downing Nisbet Charitable Fund Grants, 2387
Medtronic Foundation HeartRescue Grants, 2513
Medtronic Foundation Strengthening Health Systems Grants, 2515
NHLBI Ancillary Studies in Clinical Trials, 2688
NHLBI Cardiac Devel Consortium Grants, 2693
NHLBI Career Transition Awards, 2694
NHLBI Lymphatics in Health and Disease in the Digestive, Urinary, Cardiovascular and Pulmonary Systems, 2705
NHLBI Mentored Career Award for Faculty at Institutions that Promote Diversity (K01), 2706
NHLBI Mentored Career Award For Faculty At Minority Institutions, 2707
NHLBI National Research Service Award Programs in Cardiovascular Epidemiology and Biostatistics, 2714
NHLBI Research on the Role of Cardiomyocyte Mitochondria in Heart Disease: An Integrated Approach, 2726
NHLBI Short-Term Research Education Program to Increase Diversity in Health-Related Research, 2733
NHLBI Training Program Grants for Institutions that Promote Diversity, 2738
NIH Research on Sleep and Sleep Disorders, 2898
Pfizer ASPIRE North America Broad Spectrum Antibiotics for the Treatment of Gram-Negative or Polymicrobial Infections Research Awards, 3169

SUBJECT INDEX

Pfizer ASPIRE North America Narrow Spectrum Antibiotics for the Treatment of MRSA Research Awards, 3171
Pfizer Healthcare Charitable Contributions, 3180
Pfizer Medical Education Track Two Grants, 3183
Premera Blue Cross Grants, 3258
RACGP Cardiovascular Research Grants in General Practice, 3295
Saint Luke's Health Initiatives Grants, 3455
Seattle Foundation Medical Funds Grants, 3508

Cardiovascular Health
CDC Fnd Tobacco Network Lab Fellowship, 1062
CDI Interdisciplinary Research Initiatives Grants, 1074
CDI Postdoctoral Fellowships, 1075
Delta Air Lines Foundation Health and Wellness Grants, 1455
DTE Energy Foundation Health and Human Services Grants, 1516
GNOF IMPACT Kahn-Oppenheim Grants, 1842
Mary Black Fndn Community Health Grants, 2441
Medtronic Foundation Strengthening Health Systems Grants, 2515
NFL Charities Medical Grants, 2685
Premera Blue Cross Grants, 3258

Cardiovascular System
Chest Foundation/LUNGevity Foundation Clinical Research in Lung Cancer Grants, 1131
Children's Cardiomyopathy Foundation Research Grants, 1137
Duke Univ Interventional Cardio Fellowships, 1524
NHLBI Lymphatics in Health and Disease in the Digestive, Cardio & Pulmonary Systems, 2704
NHLBI Mentored Career Dev Award to Promote Faculty Diversity in Biomed Research, 2708
NHLBI Research on the Role of Cardiomyocyte Mitochondria in Heart Disease: An Integrated Approach, 2726
NHLBI Ruth L. Kirschstein National Research Service Awards for Individual Postdoctoral Fellows, 2729
Simone and Cino del Duca Grand Prix Awards, 3564

Care Givers
ACL Learning Collaboratives for Advanced Business Acumen Skills Grants, 244
Brookdale Fnd National Group Respite Grants, 966
Rosalynn Carter Institute Georgia Caregiver of the Year Awards, 3407
Unihealth Foundation Grants for Non-Profit Oranizations, 3783

Career Education and Planning
Adaptec Foundation Grants, 304
AFAR Paul Beeson Career Development Awards in Aging Research for the Island of Ireland, 336
AHNS/AAO-HNSF Surgeon Scientist Combined Award, 357
AHNS/AAO-HNSF Young Investigator Combined Award, 358
AIHS Heritage Youth Researcher Summer Science Program, 387
AMD Corporate Contributions Grants, 500
Cleveland Browns Foundation Grants, 1182
Community Foundation of Greater Flint Grants, 1273
DaimlerChrysler Corporation Fund Grants, 1412
Essex County Community Foundation Women's Fund Grants, 1605
Ford Motor Company Fund Grants Program, 1687
Grifols Community Outreach Grants, 1904
Highmark Corporate Giving Grants, 2025
KeyBank Foundation Grants, 2307
Lincoln Financial Foundation Grants, 2367
McCarthy Family Foundation Grants, 2490
NCI Cancer Education and Career Development Program, 2648
NCI Centers of Excellence in Cancer Communications Research, 2649
NCI Mentored Career Development Award to Promote Diversity, 2653

NEI Scholars Program, 2669
NHGRI Mentored Research Scientist Development Award, 2686
NIA Mentored Research Scientist Devel Award, 2772
NIEHS Mentored Research Scientist Development Award, 2858
NINR Mentored Research Scientist Development Award, 2932
NSF Presidential Early Career Awards for Scientists and Engineers (PECASE) Grants, 3021
Piper Jaffray Fndn Communities Giving Grants, 3227
TE Foundation Grants, 3705
Tommy Hilfiger Corporate Foundation Grants, 3737
U.S. Department of Education 21st Century Community Learning Centers, 3767

Caries
DHHS Oral Health Promotion Research Across the Lifespan, 1474
NIH Ruth L. Kirschstein National Research Service Awards for Individual Postdoc Fellowships, 2902

Cataloging and Classification
ASM USFCC/J. Roger Porter Award, 756

Catholic Church
Anthony R. Abraham Foundation Grants, 608
Archer Daniels Midland Foundation Grants, 651
Booth-Bricker Fund Grants, 933
Carrie Estelle Doheny Foundation Grants, 1037
Charles Delmar Foundation Grants, 1115
Collins C. Diboll Private Foundation Grants, 1209
Connelly Foundation Grants, 1326
Dolan Children's Foundation Grants, 1482
Dorothea Haus Ross Foundation Grants, 1498
Eugene B. Casey Foundation Grants, 1609
G.A. Ackermann Memorial Fund Grants, 1767
Hackett Foundation Grants, 1926
Huiskjng Foundation Grants, 2076
I.A. O'Shaughnessy Foundation Grants, 2088
Ida Alice Ryan Charitable Trust Grants, 2113
James J. & Joan A. Gardner Family Fndn Grants, 2193
Kevin P. and Sydney B. Knight Family Foundation Grants, 2306
Kuntz Foundation Grants, 2326
Marion I. and Henry J. Knott Foundation Discretionary Grants, 2435
Marion I. and Henry J. Knott Foundation Standard Grants, 2436
Raskob Foundation for Catholic Activities Grants, 3306
Richard D. Bass Foundation Grants, 3366
Strake Foundation Grants, 3658
Thomas and Dorothy Leavey Foundation Grants, 3717
Tyler Aaron Bookman Memorial Foundation Trust Grants, 3765
William H. Hannon Foundation Grants, 3953

Cell Death
NIA Aging, Oxidative & Cell Death Grants, 2745

Ceramics (Visual Arts)
George W. Wells Foundation Grants, 1804

Cerebral Palsy
Austin S. Nelson Foundation Grants, 805
CFFVR Robert and Patricia Endries Family Foundation Grants, 1102
F.M. Kirby Foundation Grants, 1624
Hampton Roads Community Foundation Developmental Disabilities Grants, 1939
HRAMF Deborah Munroe Noonan Memorial Research Grants, 2058
Robert R. Meyer Foundation Grants, 3387

Cervical Cancer
Conquer Cancer Foundation of ASCO International Development and Education Awards, 1333
GNOF IMPACT Harold W. Newman, Jr. Charitable Trust Grants, 1841

Penny Severns Breast, Cervical and Ovarian Cancer Research Grants, 3136
Seattle Foundation Medical Funds Grants, 3508

Chamber Music
Ann Arbor Area Community Foundation Grants, 594
Clarence E. Heller Charitable Foundation Grants, 1166

Charter Schools
Achelis Foundation Grants, 234
Bodman Foundation Grants, 929

Chemical Dynamics
NSF Chemistry Research Experiences for Undergraduates (REU), 3012

Chemical Effects
Weyerhaeuser Family Foundation Health Grants, 3926

Chemical Engineering
IIE KAUST Graduate Fellowships, 2134
Lubrizol Foundation Grants, 2384
NSF-NIST Interaction in Chemistry, Materials Research, Molecular Biosciences, Bioengineering, and Chemical Engineering, 3004

Chemical Sciences
CDC Collegiate Leaders in Environmental Health Internships, 1048
IIE KAUST Graduate Fellowships, 2134
PhRMA Foundation Informatics Research Starter Grants, 3208
PhRMA Fndn Info Sabbatical Fellowships, 3209
W.M. Keck Foundation Science and Engineering Research Grants, 3889
Wellcome Trust Biomedical Science Grants, 3918

Chemistry
AACR Award for Outstanding Achievement in Chemistry in Cancer Research, 104
AACR Team Science Award, 126
AHAF Alzheimer's Disease Research Grants, 349
Alfred P. Sloan Foundation Research Fellowships, 432
American Chemical Society Alfred Burger Award in Medicinal Chemistry, 510
American Chemical Society ANYL Award for Distinguished Service in the Advancement of Analytical Chemistry, 512
American Chemical Society Award in Separations Science and Technology, 514
American Chem Society Claude Hudson Awards, 515
American Chemical Society GCI Pharmaceutical Roundtable Research Grants, 516
AOCS A. Richard Baldwin Award, 610
AOCS Alton E. Bailey Award, 611
AOCS Supelco/Nicholas Pelick-AOCS Research Award, 621
APHL Emerging Infectious Diseases Fellowships, 636
Arnold and Mabel Beckman Foundation Scholars Grants, 665
Arnold and Mabel Beckman Foundation Young Investigators Grants, 666
BWF Career Awards at the Scientific Interface, 983
Canada-U.S. Fulbright New Century Scholars Program Grants, 1008
Canada-U.S. Fulbright Senior Specialists Grants, 1009
Canada Graduate Scholarships (CGS) and NSERC Postgraduate Scholarships (PGS), 1010
Frederic Stanley Kipping Award in Silicon Chem, 1742
Fulbright Alumni Initiatives Awards, 1749
Fulbright Distinguished Chairs Awards, 1750
Fulbright New Century Scholars Grants, 1753
Fulbright Specialists Program Grants, 1754
Fulbright Scholars in Europe and Eurasia, 1755
Fulbright Traditional Scholar Program in Sub-Saharan Africa, 1756
Fulbright Traditional Scholar Program in the East Asia/Pacific Region, 1757
Fulbright Traditional Scholar Program in the Near East and North Africa Region, 1758

678 / Chemistry

Fulbright Traditional Scholar Program in the South and Central Asia Region, 1759
Fulbright Traditional Scholar Program in the Western Hemisphere, 1760
HHMI-NIBIB Interfaces Initiative Grants, 2014
HRAMF Jeffress Trust Awards in Interdisciplinary Research, 2060
HRAMF Smith Family Awards for Excellence in Biomedical Research, 2063
IIE Central Europe Summer Research Institute Summer Research Fellowship, 2127
John Stauffer Charitable Trust Grants, 2258
Lubrizol Foundation Grants, 2384
Marion I. and Henry J. Knott Standard Grants, 2436
MBL Eugene and Millicent Bell Fellowships, 2470
NHGRI Mentored Research Scientist Development Award, 2686
NHLBI Mentored Quantitative Research Career Development Awards, 2711
North Carolina GlaxoSmithKline Fndn Grants, 2991
NSF Chemistry Research Experiences for Undergraduates (REU), 3012
Ralph F. Hirschmann Award in Peptide Chem, 3303
Searle Scholars Program Grants, 3503
Society of Cosmetic Chemists Allan B. Black Award Sponsored by Presperse, 3604
Society of Cosmetic Chemists Award Sponsored by McIntyre Group, Ltd., 3605
Society of Cosmetic Chemists Award Sponsored by The HallStar Company, 3606
Society of Cosmetic Chemists Chapter Best Speaker Award, 3607
Society of Cosmetic Chem Chapter Merit Award, 3608
Society of Cosmetic Chemists Frontier of Science Award, 3609
Society of Cosmetic Chem Hans Schaeffer Award, 3610
Society of Cosmetic Chem Jos Ciaudelli Award, 3611
Society of Cosmetic Chemists Keynote Award Lecture Sponsored by Ruger Chemical Corporation, 3612
Society of Cosmetic Chemists Literature Award, 3613
Society of Cosmetic Chemists Maison G. de Navarre Medal, 3614
Society of Cosmetic Chemists Merit Award, 3615
Society of Cosmetic Chemists Robert A. Kramer Lifetime Service Award, 3616
Society of Cosmetic Chem Shaw Mudge Award, 3617
Society of Cosmetic Chem Stud Poster Awards, 3618
Society of Cosmetic Chem You Scientist Awards, 3619

Chemistry Education
CCFF Life Sciences Student Awards, 1045
FMC Foundation Grants, 1683
North Carolina GlaxoSmithKline Fndn Grants, 2991

Chemistry, Analytical
American Chemical Society ANYL Arthur F. Findeis Award for Achievements by a Young Analytical Scientist, 511
American Chemical Society ANYL Award for Distinguished Service in the Advancement of Analytical Chemistry, 512
AOCS Analytical Division Student Award, 612
AOCS George Schroepfer Medal, 613
NSF Chemistry Research Experiences for Undergraduates (REU), 3012

Chemistry, Environmental
American Chemical Society Award for Creative Advances in Environmental Science and Technology, 513
NSF Chemistry Research Experiences for Undergraduates (REU), 3012

Chemistry, Inorganic
NSF Chemistry Research Experiences for Undergraduates (REU), 3012

Chemistry, Materials
NSF Chemistry Research Experiences for Undergraduates (REU), 3012

Chemistry, Organic
NSF Chemistry Research Experiences for Undergraduates (REU), 3012
PhRMA Foundation Pharmaceutics Postdoctoral Fellowships, 3211
PhRMA Foundation Pharmaceutics Pre Doctoral Fellowships, 3212

Chemistry, Physical
NIH Ruth L. Kirschstein National Research Service Awards for Individual Postdoc Fellowships, 2902
NSF Chemistry Research Experiences for Undergraduates (REU), 3012
PhRMA Foundation Pharmaceutics Postdoctoral Fellowships, 3211
PhRMA Foundation Pharmaceutics Pre Doctoral Fellowships, 3212

Chemistry, Separations
American Chemical Society Award in Separations Science and Technology, 514

Chemistry, Surface
NSF Chemistry Research Experiences for Undergraduates (REU), 3012

Chemotherapy
ASM sanofi-aventis ICAAC Award, 749

Child Abuse
ACF Native American Social and Economic Development Strategies Grants, 231
Austin S. Nelson Foundation Grants, 805
Baxter International Foundation Grants, 843
Carl B. and Florence E. King Foundation Grants, 1022
Carl C. Icahn Foundation Grants, 1023
CFFVR Schmidt Family G4 Grants, 1103
Charles H. Hall Foundation, 1120
Charles Lafitte Foundation Grants, 1123
Children's Trust Fund of Oregon Fndn Grants, 1139
Cleveland-Cliffs Foundation Grants, 1181
Community Foundation for Greater New Haven Women & Girls Grants, 1252
ConocoPhillips Foundation Grants, 1327
Dade Community Foundation Grants, 1410
Delaware Community Foundation-Youth Philanthropy Board for Kent County, 1448
Doris and Victor Day Foundation Grants, 1491
Elkhart County Community Foundation Fund for Elkhart County, 1572
Elliot Foundation Inc Grants, 1575
Faye McBeath Foundation Grants, 1635
Global Fund for Children Grants, 1832
Greater Tacoma Community Foundation Ryan Alan Hade Endowment Fund, 1887
Green Bay Packers Foundation Grants, 1890
Hasbro Children's Fund Grants, 1973
Herbert A. & Adrian W. Woods Fndn Grants, 2007
Hoglund Foundation Grants, 2040
Huie-Dellmon Trust Grants, 2075
James R. Thorpe Foundation Grants, 2199
Jim Moran Foundation Grants, 2228
John W. Speas and Effie E. Speas Grants, 2264
Katharine Matthies Foundation Grants, 2291
Linford and Mildred White Charitable Grants, 2368
McCarthy Family Foundation Grants, 2490
Meyer Foundation Healthy Communities Grants, 2533
Oak Foundation Child Abuse Grants, 3059
Peacock Foundation Grants, 3133
Questar Corporate Contributions Grants, 3291
Reynolds & Reynolds Associate Fndn Grants, 3360
Robert R. McCormick Trib Community Grants, 3385
Robert R. Meyer Foundation Grants, 3387
Southbury Community Trust Fund, 3625
Topfer Family Foundation Grants, 3739
WHO Foundation Volunteer Service Grants, 3940

Child Care
Alexander and Margaret Stewart Trust Grants, 419
Allan C. and Lelia J. Garden Foundation Grants, 438
Frank Loomis Palmer Fund Grants, 1726
GNOF Stand Up For Our Children Grants, 1847
Leo Goodwin Foundation Grants, 2354
Medtronic Foundation Community Link Human Services Grants, 2511
Phi Upsilon Omicron Frances Morton Holbrook Alumni Award, 3193
Union Bank, N.A. Foundation Grants, 3788

Child Development
Abby's Legendary Pizza Foundation Grants, 204
Bernard and Audre Rapoport Foundation Health Grants, 882
Children's Trust Fund of Oregon Fndn Grants, 1139
Mericos Foundation Grants, 2518
Peter and Elizabeth C. Tower Foundation Social and Emotional Preschool Curriculum Grants, 3151
Piper Trust Healthcare & Med Research Grants, 3228
Vanderburgh Community Foundation Grants, 3852

Child Psychiatry
AACAP Beatrix A. Hamburg Award for the Best New Research Poster by a Child and Adolescent Psychiatry Resident, 50
AACAP Child and Adolescent Psychiatry (CAP) Teaching Scholarships, 51
AACAP Educational Outreach Program for Child and Adolescent Psychiatry Residents, 52
AACAP Educational Outreach Program for General Psychiatry Residents, 53
AACAP Elaine Schlosser Lewis Award for Research on Attention-Deficit Disorder, 55
AACAP George Tarjan Award for Contributions in Developmental Disabilities, 56
AACAP Irving Philips Award for Prevention, 57
AACAP Jeanne Spurlock Lecture and Award on Diversity and Culture, 58
AACAP Jeanne Spurlock Minority Med Student Clinical Fellow in Child & Adolescent Psych, 59
AACAP Junior Investigator Awards, 61
AACAP Klingenstein Third Generation Foundation Award for Research in Depression or Suicide, 62
AACAP Mary Crosby Congressional Fellowships, 64
AACAP Norbert and Charlotte Rieger Award for Scientific Achievement, 65
AACAP Pilot Research Award for Attention-Deficit Disorder, 66
AACAP Pilot Research Awards, Supported by Eli Lilly and Company, 68
AACAP Quest for the Test Bipolar Disorder Pilot Research Award, 69
AACAP Rieger Psychody Psychotherapy Award, 70
AACAP Rieger Service Award for Excellence, 71
AACAP Rob Cancro Academic Leadership Award, 72
AACAP Sidney Berman Award for the School-Based Study and Intervention for Learning Disorders and Mental Ilness, 73
AACAP Simon Wile Leader in Consultation Award, 74
AACAP Systems of Care Special Scholarships, 76
American Psychiatric Foundation Grantss, 552
American Psychiatric Foundation James H. Scully Jr., M.D., Educational Fund Grants, 555
American Psychiatric Foundation Minority Fellowships, 557
American Psychiatric Foundation Typical or Troubled School Mental Health Education Grants, 558
Community Foundation for Greater New Haven Women & Girls Grants, 1252
Greater Tacoma Community Foundation Ryan Alan Hade Endowment Fund, 1887
NARSAD Ruane Prize for Child and Adolescent Psychiatric Research, 2624
SOBP A.E. Bennett Research Award, 3594
WHAS Crusade for Children Grants, 3927

Child Psychology/Development
Allen Foundation Educational Nutrition Grants, 442
AON Foundation Grants, 623
Assurant Foundation Grants, 788
Carnegie Corporation of New York Grants, 1033

SUBJECT INDEX

Community Foundation of Greater New Britain
    Grants, 1281
Connecticut Community Foundation Grants, 1324
CVS All Kids Can Grants, 1377
Effie and Wofford Cain Foundation Grants, 1563
FCD New American Children Grants, 1637
Foundation for Seacoast Health Grants, 1698
George Family Foundation Grants, 1794
Gerber Foundation Grants, 1807
Harold Simmons Foundation Grants, 1955
Helen Bader Foundation Grants, 1992
Henrietta Tower Wurts Memorial Fndn Grants, 2002
Hoglund Foundation Grants, 2040
John Merck Scholars Awards, 2246
Mericos Foundation Grants, 2518
Mimi and Peter Haas Fund Grants, 2557
NARSAD Ruane Prize for Child and Adolescent
    Psychiatric Research, 2624
NIEHS Small Business Innovation Research (SBIR)
    Program Grants, 2860
NIMH Early Identification and Treatment of Mental
    Disorders in Children and Adolescents Grants, 2918
NSF Perception, Action & Cognition Grants, 3019
NSF Social Psychology Research Grants, 3023
Piper Jaffray Fndn Communities Giving Grants, 3227
Salem Foundation Charitable Trust Grants, 3456
Seabury Foundation Grants, 3501
WHAS Crusade for Children Grants, 3927
William T. Grant Foundation Research Grants, 3958

Child Sexual Abuse
Austin S. Nelson Foundation Grants, 805
CFFVR Schmidt Family G4 Grants, 1103
Greater Tacoma Community Foundation Ryan Alan
    Hade Endowment Fund, 1887
Ms. Fndn for Women Ending Violence Grants, 2596
Oak Foundation Child Abuse Grants, 3059
Reynolds & Reynolds Associate Fndn Grants, 3360
Robert R. Meyer Foundation Grants, 3387

Child Welfare
Agnes M. Lindsay Trust Grants, 346
A Good Neighbor Foundation Grants, 347
Allen P. & Josephine B. Green Fndn Grants, 443
Alliance Healthcare Foundation Grants, 444
Ann Jackson Family Foundation Grants, 598
Bernard and Audre Rapoport Foundation Health
    Grants, 882
Blum-Kovler Foundation Grants, 924
Bodenwein Public Benevolent Foundation Grants, 928
Bodman Foundation Grants, 929
Caleb C. and Julia W. Dula Educational and Charitable
    Foundation Grants, 997
Caplow Applied Science Children's Prize, 1016
Carl B. and Florence E. King Foundation Grants, 1022
Carl C. Icahn Foundation Grants, 1023
Carls Foundation Grants, 1030
Carolyn Foundation Grants, 1034
Carrie Estelle Doheny Foundation Grants, 1037
Charles Lafitte Foundation Grants, 1123
Clayton Fund Grants, 1180
Colonel Stanley R. McNeil Foundation Grants, 1211
Community Fndn for Muskegon County Grants, 1254
Community Fndn for the Capital Region Grants, 1259
Community Foundation of Louisville Dr. W. Barnett
    Owen Memorial Fund for the Children of Louisville
    and Jefferson County Grants, 1294
Cumberland Community Foundation Grants, 1374
D. W. McMillan Foundation Grants, 1407
David M. and Marjorie D. Rosenberg Foundation
    Grants, 1432
Detlef Schrempf Foundation Grants, 1470
DHHS Oral Health Promotion Research Across the
    Lifespan, 1474
Dorothea Haus Ross Foundation Grants, 1498
Gardner Foundation Grants, 1772
Greygates Foundation Grants, 1901
Hawaii Community Foundation Reverend Takie
    Okumura Family Grants, 1977
Hoblitzelle Foundation Grants, 2038

Huffy Foundation Grants, 2073
Ike and Roz Friedman Foundation Grants, 2145
J.W. Kieckhefer Foundation Grants, 2176
Janirve Foundation Grants, 2209
John H. and Wilhelmina D. Harland Charitable
    Foundation Children and Youth Grants, 2238
John I. Smith Charities Grants, 2240
Joseph Henry Edmondson Foundation Grants, 2270
Josephine Goodyear Foundation Grants, 2272
Katherine John Murphy Foundation Grants, 2293
Kathryne Beynon Foundation Grants, 2294
L. W. Pierce Family Foundation Grants, 2328
Lil and Julie Rosenberg Foundation Grants, 2364
Lisa Higgins-Hussman Foundation Grants, 2370
Lucile Horton Howe and Mitchell B. Howe Foundation
    Grants, 2385
Nuffield Foundation Children & Families Grants, 3027
Oak Foundation Child Abuse Grants, 3059
Olive Higgins Prouty Foundation Grants, 3072
OneFamily Foundation Grants, 3075
Ricks Family Charitable Trust Grants, 3373
Robert R. McCormick Trib Community Grants, 3385
Ruth Eleanor Bamberger and John Ernest Bamberger
    Memorial Foundation Grants, 3427
Scott B. & Annie P. Appleby Charitable Grants, 3499
Stackpole-Hall Foundation Grants, 3643
United Methodist Committee on Relief Global Health
    Grants, 3799
W. C. Griffith Foundation Grants, 3882
W.H. & Mary Ellen Cobb Trust Grants, 3884
W.K. Kellogg Foundation Secure Families Grants, 3887
William J. & Tina Rosenberg Fndn Grants, 3954
Zane's Foundation Grants, 3983

Child/Maternal Health
AAP Community Access To Child Health (CATCH)
    Advocacy Training Grants, 177
AAP Community Access to Child Health (CATCH)
    Planning Grants, 179
AAP Community Access To Child Health (CATCH)
    Residency Training Grants, 180
AAP Community Access To Child Health (CATCH)
    Resident Grants, 181
AAP Leonard P. Rome Community Access to Child
    Health (CATCH) Visiting Professorships, 186
Abbott Fund Access to Health Care Grants, 199
ADA Foundation Samuel Harris Children's Dental
    Health Grants, 294
AEGON Transamerica Foundation Health and
    Welfare Grants, 316
AHRQ Independent Scientist Award, 365
Alcatel-Lucent Technologies Foundation Grants, 413
American Psychiatric Foundation Jeanne Spurlock
    Congressional Fellowship, 556
American Psychiatric Foundation Minority
    Fellowships, 557
Batchelor Foundation Grants, 837
BCBSNC Foundation Grants, 860
Bernard and Audre Rapoport Foundation Health
    Grants, 882
Blackford County Community Foundation - WOW
    Grants, 905
Blowitz-Ridgeway Foundation Grants, 913
California Endowment Innovative Ideas Challenge
    Grants, 1000
California Wellness Foundation Work and Health
    Program Grants, 1001
CFF Research Grants, 1090
CFFVR Schmidt Family G4 Grants, 1103
Charles H. Dater Foundation Grants, 1118
Children's Brain Tumor Fnd Research Grants, 1136
Children Affected by AIDS Foundation Domestic
    Grants, 1145
Cleveland Browns Foundation Grants, 1182
Community Foundation for Greater New Haven
    Women & Girls Grants, 1252
Community Fndn of Bloomington & Monroe County
    - Precision Health Network Cycle Grants, 1264
Connecticut Health Foundation Health Initiative
    Grants, 1325

Child/Maternal Health /679

CSTE CDC/CSTE Applied Epidemiology
    Fellowships, 1367
Dallas Women's Foundation Grants, 1417
Delta Air Lines Foundation Health and Wellness
    Grants, 1455
DTE Energy Foundation Health and Human Services
    Grants, 1516
E.W. "Al" Thrasher Awards, 1540
Edwards Memorial Trust Grants, 1558
Edward W. and Stella C. Van Houten Memorial Fund
    Grants, 1559
El Paso Community Foundation Grants, 1579
EPA Children's Health Protection Grants, 1598
Express Scripts Foundation Grants, 1620
F.M. Kirby Foundation Grants, 1624
Fairlawn Foundation Grants, 1626
Foundation for Seacoast Health Grants, 1698
Frances and John L. Loeb Family Fund Grants, 1716
Gates Award for Global Health, 1775
Gerber Foundation Grants, 1807
Gibson County Community Foundation Women's Fund
    Grants, 1817
GlaxoSmithKline Corporate Grants, 1829
GNOF Stand Up For Our Children Grants, 1847
Healthcare Foundation of New Jersey Grants, 1982
Health Foundation of Greater Cincinnati Grants, 1983
Health Fndn of Greater Indianapolis Grants, 1984
Hoglund Foundation Grants, 2040
HRAMF Charles H. Hood Foundation Child Health
    Research Awards, 2057
Illinois Children's Healthcare Foundation Grants, 2146
Jane Bradley Pettit Foundation Health Grants, 2208
Jewish Fund Grants, 2227
Johnson & Johnson Community Health Grants, 2252
Johnson & Johnson Corporate Contributions, 2253
Josiah Macy Jr. Foundation Grants, 2277
Kansas Health Fndn Major Initiatives Grants, 2288
Kansas Health Foundation Recognition Grants, 2289
Lloyd A. Fry Foundation Health Grants, 2373
Long Island Community Foundation Grants, 2375
Lowe Foundation Grants, 2381
March of Dimes Agnes Higgins Award, 2423
March of Dimes Program Grants, 2426
Mary Black Foundation Active Living Grants, 2440
Medtronic Foundation Patient Link Grants, 2514
Michael Reese Health Trust Responsive Grants, 2545
Mid-Iowa Health Foundation Community Response
    Grants, 2549
NAPNAP Foundation Innovative Health Care Small
    Grant, 2611
NAPNAP Foundation McNeil Grant-in-Aid, 2612
NAPNAP Foundation Nursing Research Grants, 2615
Nestle Foundation Large Research Grants, 2675
Nestle Foundation Pilot Grants, 2676
Nestle Foundation Re-entry Grants, 2677
Nestle Foundation Small Research Grants, 2678
Nestle Foundation Training Grant, 2679
NIH Research on Sleep and Sleep Disorders, 2898
PACCAR Foundation Grants, 3101
Pajaro Valley Community Health Health Trust
    Insurance/Coverage & Education on Using the
    System Grants, 3108
Pediatric Brain Tumor Fndn Research Grants, 3134
Pediatric Cancer Research Foundation Grants, 3135
Porter County Health and Wellness Grant, 3246
Porter County Women's Grant, 3248
Premera Blue Cross Grants, 3258
Prince Charitable Trusts Chicago Grants, 3270
Quantum Corporation Snap Server Grants, 3289
Quantum Foundation Grants, 3290
RGk Foundation Grants, 3361
Robert R. Meyer Foundation Grants, 3387
Ronald McDonald House Charities Grants, 3405
Saigh Foundation Grants, 3451
Seattle Foundation Health and Wellness Grants, 3507
Sidgmore Family Foundation Grants, 3551
Stewart Huston Charitable Trust Grants, 3655
Thrasher Research Fund Early Career Awards, 3731
TJX Foundation Grants, 3735
Topfer Family Foundation Grants, 3739

680 / Child/Maternal Health

United Healthcare Comm Grants in Michigan, 3796
United Hospital Fund of New York Health Care Improvement Grants, 3797
United Methodist Committee on Relief Global Health Grants, 3799
USAID Child, Newborn, and Maternal Health Project Grants, 3806
USAID Family Health Plus Project Grants, 3810
USAID Family Planning and Reproductive Health Methods Grants, 3811
USAID Health System Strengthening Grants, 3814
WHO Foundation General Grants, 3939

Children (Patients)
1st Touch Foundation Grants, 3
A-T Children's Project Post Doctoral Fellowships, 14
Bernard and Audre Rapoport Foundation Health Grants, 882
Children's Tumor Fndn Clin Research Awards, 1140
Children's Tumor Foundation Young Investigator Awards, 1143
Dyson Foundation Mid-Hudson Valley Project Support Grants, 1534
E.W. "Al" Thrasher Awards, 1540
Elizabeth Glaser Int'l Leadership Awards, 1566
Ford Family Foundation Grants - Access to Health and Dental Services, 1686
Giant Food Charitable Grants, 1816
HAF Barry F. Phelps Fund Grants, 1927
HAF JoAllen K. Twiddy-Wood Memorial Fund Grants, 1929
HAF Phyllis Nilsen Leal Memorial Fund Gifts, 1930
Maggie Welby Foundation Grants, 2414
Medtronic Foundation Patient Link Grants, 2514
Northwest Airlines KidCares Medical Travel Assistance, 2998
NYAM Brock Lecture, Award and Visiting Professorship in Pediatrics, 3038
Portland Fndn Women's Giving Circle Grant, 3249
Richard Davoud Donchian Foundation Grants, 3367
Salt River Health & Human Services Grants, 3459
Sara Elizabeth O'Brien Trust Grants, 3483
Thrasher Research Fund Early Career Awards, 3731
Union Labor Health Fndn Angel Fund Grants, 3791
UnitedHealthcare Children's Foundation Grants, 3795
Zane's Foundation Grants, 3983

Children and Youth
1st Touch Foundation Grants, 3
2 Depot Square Ipswich Charitable Fndn Grants, 4
3M Company Fndn Community Giving Grants, 6
3M Fndn Health & Human Services Grants, 7
1976 Foundation Grants, 12
A-T Children's Project Grants, 13
A-T Children's Project Post Doctoral Fellowships, 14
AACAP-NIDA K12 Career Development Awards, 49
AACAP Junior Investigator Awards, 61
AACAP Pilot Research Award for Attention-Deficit Disorder, 66
AACAP Quest for the Test Bipolar Disorder Pilot Research Award, 69
AAP Anne E. Dyson Child Advocacy Awards, 176
Aaron & Cecile Goldman Family Fndn Grants, 191
Aaron Foundation Grants, 194
Abbot and Dorothy H. Stevens Foundation Grants, 197
Abbott Fund Community Grants, 201
Abby's Legendary Pizza Foundation Grants, 204
AcademyHealth Nemours Child Health Services Research Awards, 224
ACE Charitable Foundation Grants, 229
ACF Foundation Grants, 230
ACF Native American Social and Economic Development Strategies Grants, 231
Achelis Foundation Grants, 234
Adams and Reese Corporate Giving Grants, 299
Adams Rotary Memorial Fund A Grants, 302
Adaptec Foundation Grants, 304
Adelaide Breed Bayrd Foundation Grants, 306
Administaff Community Affairs Grants, 308
Advance Auto Parts Corporate Giving Grants, 312

AEGON Transamerica Foundation Health and Welfare Grants, 316
Aetna Foundation Obesity Grants, 331
Aetna Foundation Regional Health Grants, 333
A Friends' Foundation Trust Grants, 341
Agnes M. Lindsay Trust Grants, 346
A Good Neighbor Foundation Grants, 347
Agway Foundation Grants, 348
Albert W. Rice Charitable Foundation Grants, 411
Alcatel-Lucent Technologies Foundation Grants, 413
Alcohol Misuse and Alcoholism Research Grants, 415
Alfred and Tillie Shemanski Testamentary Trust Grants, 428
Alfred E. Chase Charitable Foundation Grants, 430
Alice Tweed Tuohy Foundation Grants Program, 437
Allan C. and Lelia J. Garden Foundation Grants, 438
Allegis Group Foundation Grants, 441
Allen Foundation Educational Nutrition Grants, 442
Alliance Healthcare Foundation Grants, 444
Alpha Natural Resources Corporate Giving, 450
Amelia Sillman Rockwell and Carlos Perry Rockwell Charities Fund Grants, 501
American Academy of Dermatology Camp Discovery Scholarships, 503
American Foodservice Charitable Trust Grants, 532
American Psychiatric Foundation Jeanne Spurlock Congressional Fellowship, 556
American Psychiatric Foundation Minority Fellowships, 557
American Schlafhorst Foundation Grants, 559
Amerigroup Foundation Grants, 565
Andre Agassi Charitable Foundation Grants, 587
Andrew Family Foundation Grants, 588
Angels Baseball Foundation Grants, 590
Anheuser-Busch Foundation Grants, 591
Anna Fitch Ardenghi Trust Grants, 592
Ann and Robert H. Lurie Family Fndn Grants, 593
Anne J. Caudal Foundation Grants, 595
Annenberg Foundation Grants, 596
Annunziata Sanguinetti Foundation Grants, 600
Anthony R. Abraham Foundation Grants, 608
Antone & Edene Vidinha Charitable Trust Grants, 609
Appalachian Regional Commission Health Grants, 637
APS Foundation Grants, 646
Aragona Family Foundation Grants, 648
Aratani Foundation Grants, 649
Arcadia Foundation Grants, 650
Arizona Cardinals Grants, 655
Arizona Diamondbacks Charities Grants, 657
Arizona Public Service Corporate Giving Program Grants, 658
Arlington Community Foundation Grants, 662
Arthur M. Blank Family Foundation Inspiring Spaces Grants, 673
Assurant Foundation Grants, 788
AT&T Foundation Civic and Community Service Program Grants, 793
Athwin Foundation Grants, 795
Austin-Bailey Health and Wellness Fndn Grants, 803
Austin College Leadership Award, 804
Austin S. Nelson Foundation Grants, 805
Avon Products Foundation Grants, 814
Babcock Charitable Trust Grants, 816
Bailey Foundation Grants, 818
Baltimore Washington Center Child Psychotherapy Fellowships, 822
BancorpSouth Foundation Grants, 823
Bank of America Fndn Matching Gifts, 825
Baptist Community Ministries Grants, 828
Barberton Community Foundation Grants, 829
Barker Foundation Grants, 830
Barr Fund Grants, 835
Barth Syndrome Foundation Research Grants, 836
Batchelor Foundation Grants, 837
Batts Foundation Grants, 840
Baxter International Foundation Grants, 843
BCBSM Building Healthy Communities Engaging Elementary Schools and Community Partners Grants, 849
BCBSM Corporate Contributions Grants, 851

Bella Vista Foundation Grants, 873
Bender Foundation Grants, 876
Benton County Foundation Grants, 880
Bernard and Audre Rapoport Foundation Health Grants, 882
Berrien Community Foundation Grants, 884
Bikes Belong Foundation Paul David Clark Bicycling Safety Grants, 890
Bill and Melinda Gates Foundation Water, Sanitation and Hygiene Grants, 894
Bindley Family Foundation Grants, 896
Birmingham Foundation Grants, 903
BJ's Charitable Foundation Grants, 904
Blackford County Community Foundation - WOW Grants, 905
Blackford County Community Foundation Grants, 906
Blanche and Irving Laurie Foundation Grants, 910
Blowitz-Ridgeway Foundation Grants, 913
Blue Cross Blue Shield of Minnesota Foundation - Healthy Children: Growing Up Healthy Grants, 915
BMW of North America Charitable Contributions, 926
Bodenwein Public Benevolent Foundation Grants, 928
Bodman Foundation Grants, 929
Booth-Bricker Fund Grants, 933
Borkee-Hagley Foundation Grants, 934
Boston Globe Foundation Grants, 937
Boston Psychoanalytic Society and Institute Fellowship in Child Psychoanalytic Psychotherapy, 939
Bradley-Turner Foundation Grants, 944
Brian G. Dyson Foundation Grants, 948
Bright Family Foundation Grants, 951
Bristol-Myers Squibb Foundation Health Education Grants, 959
Brooklyn Community Foundation Caring Neighbors Grants, 968
Browning-Kimball Foundation Grants, 971
Burden Trust Grants, 976
Burlington Industries Foundation Grants, 977
Caesars Foundation Grants, 995
Cailloux Foundation Grants, 996
Callaway Golf Company Foundation Grants, 1003
Cambridge Community Foundation Grants, 1004
Campbell Soup Foundation Grants, 1007
Caplow Applied Science Children's Prize, 1016
Caring Foundation Grants, 1019
Carl B. and Florence E. King Foundation Grants, 1022
Carlisle Foundation Grants, 1025
Carls Foundation Grants, 1030
Carl W. and Carrie Mae Joslyn Trust Grants, 1031
Carolyn Foundation Grants, 1034
Carpenter Foundation Grants, 1035
Carrie E. and Lena V. Glenn Foundation Grants, 1036
Carrie Estelle Doheny Foundation Grants, 1037
Carroll County Community Foundation Grants, 1039
Catherine Holmes Wilkins Foundation Grants, 1042
Cemala Foundation Grants, 1077
Central Okanagan Foundation Grants, 1080
Cessna Foundation Grants Program, 1081
CFFVR Basic Needs Giving Partnership Grants, 1094
CFFVR Frank C. Shattuck Community Grants, 1098
CFFVR Myra M. and Robert L. Vandehey Foundation Grants, 1100
CFFVR Project Grants, 1101
CFFVR Robert and Patricia Endries Family Foundation Grants, 1102
CFFVR Schmidt Family G4 Grants, 1103
CFFVR Shawano Area Community Foundation Grants, 1104
CFFVR Women's Fund for the Fox Valley Region Grants, 1107
Chamberlain Foundation Grants, 1108
Champ-A Champion Fur Kids Grants, 1109
Chapman Charitable Foundation Grants, 1112
CharityWorks Grants, 1113
Charles Delmar Foundation Grants, 1115
Charles H. Dater Foundation Grants, 1118
Charles H. Hall Foundation, 1120
Charles H. Pearson Foundation Grants, 1121
Charles Lafitte Foundation Grants, 1123

SUBJECT INDEX

Children and Youth /681

Children's Cardiomyopathy Foundation Research Grants, 1137
Children's Trust Fund of Oregon Fndn Grants, 1139
Children's Tumor Fndn Clin Research Awards, 1140
Children's Tumor Foundation Young Investigator Awards, 1143
Children Affected by AIDS Foundation Camp Network Grants, 1144
Children Affected by AIDS Foundation Domestic Grants, 1145
Children Affected by AIDS Foundation Family Assistance Emergency Fund Grants, 1146
Christine & Katharina Pauly Trust Grants, 1150
Chula Vista Charitable Foundation Grants, 1154
CICF Indianapolis Fndn Community Grants, 1156
CIT Corporate Giving Grants, 1161
Citizens Bank Mid-Atlantic Charitable Foundation Grants, 1162
Clark-Winchcole Foundation Grants, 1173
Clayton Baker Trust Grants, 1179
Cleveland Browns Foundation Grants, 1182
CLIF Bar Family Foundation Grants, 1183
CNA Foundation Grants, 1189
CNCS Social Innovation Grants, 1192
CNO Financial Group Community Grants, 1201
Coeta and Donald Barker Foundation Grants, 1205
Collective Brands Foundation Grants, 1208
Colonel Stanley R. McNeil Foundation Grants, 1211
Columbus Foundation Central Benefits Health Care Foundation Grants, 1216
Columbus Foundation Paul G. Duke Grants, 1221
Columbus Foundation Traditional Grants, 1223
Commonwealth Fund Harkness Fellowships in Health Care Policy and Practice, 1229
Communities Foundation of Texas Grants, 1236
Community Fndn for Greater Buffalo Grants, 1247
Community Foundation for Greater New Haven Women & Girls Grants, 1252
Community Fndn for Muskegon County Grants, 1254
Community Foundation for Northeast Michigan Tobacco Settlement Grants, 1256
Community Fndn for San Benito County Grants, 1257
Community Foundation for the National Capital Region Community Leadership Grants, 1260
Community Foundation of Bloomington and Monroe County Grants, 1265
Community Foundation of Eastern Connecticut General Southeast Grants, 1269
Community Foundation of Grant County Grants, 1271
Community Foundation of Greater Flint Grants, 1273
Community Fndn of Greater Lafayette Grants, 1280
Community Foundation of Greater New Britain Grants, 1281
Community Fndn of Greater Tampa Grants, 1282
Community Foundation of Jackson County Seymour Noon Lions Club Grant, 1288
Community Foundation of Louisville Diller B. and Katherine P. Groff Fund for Pediatric Surgery Grants, 1293
Community Foundation of Louisville Dr. W. Barnett Owen Memorial Fund for the Children of Louisville and Jefferson County Grants, 1294
Community Foundation of Mount Vernon and Knox County Grants, 1299
Community Foundation of Muncie and Delaware County Maxon Grants, 1301
Community Fndn of Randolph County Grants, 1302
Community Foundation of Riverside & San Bernardino County Impact Grants, 1303
Community Foundation of the Verdugos Educational Endowment Fund Grants, 1314
Community Foundation Partnerships - Lawrence County Grants, 1317
Community Foundation Serving Riverside and San Bernardino Counties Impact Grants, 1319
Comprehensive Health Education Fndn Grants, 1320
ConAgra Foods Fndn Community Impact Grants, 1321
ConAgra Foods Nourish our Community Grants, 1322
Constellation Energy Corporate Grants, 1342
Crail-Johnson Foundation Grants, 1355

Cralle Foundation Grants, 1356
Crane Fund Grants, 1358
Cresap Family Foundation Grants, 1359
CSRA Community Foundation Grants, 1366
CSX Corporate Contributions Grants, 1368
Cudd Foundation Grants, 1370
Cumberland Community Foundation Grants, 1374
CUNA Mutual Group Fndn Community Grants, 1375
CVS Community Grants, 1379
Cyrus Eaton Foundation Grants, 1380
D.V. and Ida J. McEachern Trust Grants, 1406
D. W. McMillan Foundation Grants, 1407
Dade Community Foundation Grants, 1410
Dairy Queen Corporate Contributions Grants, 1413
Daisy Marquis Jones Foundation Grants, 1414
Dallas Mavericks Foundation Grants, 1416
Dammann Fund Grants, 1418
Dana Brown Charitable Trust Grants, 1419
Danellie Foundation Grants, 1421
Daphne Seybolt Culpeper Fndn Grants, 1425
Dayton Power and Light Company Foundation Signature Grants, 1436
Dayton Power and Light Foundation Grants, 1437
Daywood Foundation Grants, 1438
Deaconess Community Foundation Grants, 1439
Deborah Munroe Noonan Memorial Grants, 1443
Decatur County Community Foundation Small Project Grants, 1445
Delaware Community Foundation-Youth Philanthropy Board for Kent County, 1448
Delmarva Power and Light Contributions, 1454
Dennis & Phyllis Washington Fndn Grants, 1458
Denver Foundation Community Grants, 1461
Dept of Ed Safe and Drug-Free Schools and Communities State Grants, 1464
Dept of Ed Special Education--Personnel Development to Improve Services and Results for Children with Disabilities, 1465
Dept of Ed Special Education-National Activities-Technology and Media Services for Individuals with Disabilities, 1468
Detlef Schrempf Foundation Grants, 1470
DHHS Oral Health Promotion Research Across the Lifespan, 1474
Dickson Foundation Grants, 1476
Dole Food Company Charitable Contributions, 1483
Dora Roberts Foundation Grants, 1489
Dorothea Haus Ross Foundation Grants, 1498
Dr. John T. Macdonald Foundation Grants, 1507
Dr. Scholl Foundation Grants Program, 1511
Duke University Ped Anesthesiology Fellowship, 1527
Dyson Foundation Mid-Hudson Valley Project Support Grants, 1534
E.L. Wiegand Foundation Grants, 1538
eBay Foundation Community Grants, 1543
Eckerd Corporation Foundation Grants, 1547
Edina Realty Foundation Grants, 1550
Edward and Helen Bartlett Foundation Grants, 1554
Edward and Romell Ackley Foundation Grants, 1555
Edwards Memorial Trust Grants, 1558
Edward W. and Stella C. Van Houten Memorial Fund Grants, 1559
Edyth Bush Charitable Foundation Grants, 1562
Eisner Foundation Grants, 1564
Elizabeth Morse Genius Charitable Trust Grants, 1569
Elliot Foundation Inc Grants, 1575
El Paso Community Foundation Grants, 1579
Elsie H. Wilcox Foundation Grants, 1583
Elsie Lee Garthwaite Memorial Fndn Grants, 1584
Emma B. Howe Memorial Foundation Grants, 1593
Entergy Corporation Micro Grants, 1596
Entergy Corporation Open Grants for Healthy Families, 1597
Essex County Community Foundation Merrimack Valley General Fund Grants, 1603
Ethel Sergeant Clark Smith Foundation Grants, 1608
Evanston Community Foundation Grants, 1615
Express Scripts Foundation Grants, 1620
Ezra M. Cutting Trust Grants, 1623
Fairfield County Community Foundation Grants, 1625

Fallon OrNda Community Health Fund Grants, 1627
FAR Fund Grants, 1631
Fargo-Moorhead Area Foundation Grants, 1632
Fargo-Moorhead Area Fndn Woman's Grants, 1633
Faye McBeath Foundation Grants, 1635
Fayette County Foundation Grants, 1636
FCD New American Children Grants, 1637
Federal Express Corporate Contributions, 1657
Ferree Foundation Grants, 1658
Firelight Foundation Grants, 1667
Florence Hunt Maxwell Foundation Grants, 1676
Florida BRAIVE Fund of Dade Community Foundation, 1678
Floyd A. & Kathleen C. Cailloux Fnd Grants, 1681
Foellinger Foundation Grants, 1684
Ford Family Foundation Grants - Access to Health and Dental Services, 1686
Foster Foundation Grants, 1689
Foundation of CVPH Chelsea's Rainbow Grants, 1702
Foundation of CVPH Roger Senecal Endowment Fund Grants, 1705
Four County Community Fndn General Grants, 1710
Four County Community Foundation Healthy Senior/Healthy Youth Fund Grants, 1711
Fourjay Foundation Grants, 1712
Frances L. and Edwin L. Cummings Memorial Fund Grants, 1717
Francis L. Abreu Charitable Trust Grants, 1719
Francis T. & Louise T. Nichols Fndn Grants, 1720
Frank B. Hazard General Charity Fund Grants, 1721
Frank E. and Seba B. Payne Foundation Grants, 1722
Frank Reed and Margaret Jane Peters Memorial Fund I Grants, 1727
Frank Reed and Margaret Jane Peters Memorial Fund II Grants, 1728
Frank S. Flowers Foundation Grants, 1729
Frank Stanley Beveridge Foundation Grants, 1730
Fred Baldwin Memorial Foundation Grants, 1736
Fred C. & Katherine B. Andersen Grants, 1737
Freddie Mac Foundation Grants, 1738
Frederick McDonald Trust Grants, 1740
Frederick W. Marzahl Memorial Fund Grants, 1741
Fred L. Emerson Foundation Grants, 1743
Fremont Area Community Foundation Amazing X Grants, 1744
Fremont Area Community Fndn General Grants, 1746
G.A. Ackermann Memorial Fund Grants, 1767
Gamble Foundation Grants, 1769
Gebbie Foundation Grants, 1777
General Mills Champs for Healthy Kids Grants, 1781
General Mills Foundation Grants, 1782
George and Ruth Bradford Foundation Grants, 1789
George A Ohl Jr. Foundation Grants, 1791
George Foundation Grants, 1795
George H.C. Ensworth Memorial Fund Grants, 1797
George Kress Foundation Grants, 1799
George P. Davenport Trust Fund Grants, 1800
George S. & Dolores Dore Eccles Fndn Grants, 1801
George W. Brackenridge Foundation Grants, 1802
George W. Codrington Charitable Fndn Grants, 1803
George W. Wells Foundation Grants, 1804
Gerber Foundation Grants, 1807
Giant Food Charitable Grants, 1816
Gil and Dody Weaver Foundation Grants, 1819
Ginn Foundation Grants, 1822
Giving in Action Society Children & Youth with Special Needs Grants, 1823
Global Fund for Children Grants, 1832
GNOF Exxon-Mobil Grants, 1836
GNOF Norco Community Grants, 1845
Godfrey Foundation Grants, 1849
Golden Heart Community Foundation Grants, 1850
Golden State Warriors Foundation Grants, 1851
Grace and Franklin Bernsen Foundation Grants, 1858
Grace Bersted Foundation Grants, 1859
Graco Foundation Grants, 1860
Graham and Carolyn Holloway Family Foundation Grants, 1861
Grammy Foundation Grants, 1863
Grand Haven Area Community Fndn Grants, 1864

682 / Children and Youth

Grand Rapids Area Community Fndn Grants, 1865
Grand Rapids Area Community Foundation Nashwauk Area Endowment Fund Grants, 1866
Grand Rapids Area Community Foundation Wyoming Grants, 1867
Grand Rapids Area Community Foundation Wyoming Youth Fund Grants, 1868
Grand Rapids Community Foundation Grants, 1869
Grand Rapids Community Foundation Ionia County Grants, 1870
Grand Rapids Community Foundation Southeast Ottawa Grants, 1873
Grand Rapids Community Foundation Southeast Ottawa Youth Fund Grants, 1874
Grand Rapids Community Fndn Sparta Grants, 1875
Grand Rapids Community Foundation Sparta Youth Fund Grants, 1876
Greater Sitka Legacy Fund Grants, 1885
Greater Tacoma Community Foundation Ryan Alan Hade Endowment Fund, 1887
Greater Worcester Community Foundation Discretionary Grants, 1888
Greater Worcester Community Foundation Jeppson Memorial Fund for Brookfield Grants, 1889
Green Bay Packers Foundation Grants, 1890
Green River Area Community Fndn Grants, 1895
Greenspun Family Foundation Grants, 1896
Gregory B. Davis Foundation Grants, 1898
Gregory L. Gibson Charitable Fndn Grants, 1900
Greygates Foundation Grants, 1901
Grotto Foundation Project Grants, 1905
Gulf Coast Foundation of Community Operating Grants, 1916
Gulf Coast Foundation of Community Program Grants, 1917
Guy I. Bromley Trust Grants, 1919
H.B. Fuller Foundation Grants, 1922
H.J. Heinz Company Foundation Grants, 1923
H. Leslie Hoffman and Elaine S. Hoffman Foundation Grants, 1924
Hagedorn Fund Grants, 1934
Hall-Perrine Foundation Grants, 1935
Hancock County Community Foundation - Field of Interest Grants, 1943
Harden Foundation Grants, 1945
Harold and Arlene Schnitzer CARE Foundation Grants, 1950
Harold Brooks Foundation Grants, 1952
Harold R. Bechtel Charitable Remainder Uni-Trust Grants, 1953
Harold R. Bechtel Testamentary Charitable Trust Grants, 1954
Harris and Eliza Kempner Fund Grants, 1956
Harry B. and Jane H. Brock Foundation Grants, 1960
Harry Frank Guggenheim Fnd Research Grants, 1964
Harry W. Bass, Jr. Foundation Grants, 1968
Hartford Courant Foundation Grants, 1969
Hartford Foundation Regular Grants, 1970
Harvey Randall Wickes Foundation Grants, 1972
Hasbro Children's Fund Grants, 1973
Hasbro Corporation Gift of Play Hospital and Pediatric Health Giving, 1974
Hawaii Community Foundation Reverend Takie Okumura Family Grants, 1977
HCA Foundation Grants, 1980
Heckscher Foundation for Children Grants, 1988
Hedco Foundation Grants, 1989
Heineman Foundation for Research, Education, Charitable and Scientific Purposes, 1990
Helena Rubinstein Foundation Grants, 1991
Helen Bader Foundation Grants, 1992
Helen Gertrude Sparks Charitable Trust Grants, 1993
Helen S. Boylan Foundation Grants, 1997
Hendrick Foundation for Children Grants, 1999
Henrietta Tower Wurts Memorial Fndn Grants, 2002
Henry County Community Foundation Grants, 2005
Henry L. Guenther Foundation Grants, 2006
Herbert A. & Adrian W. Woods Fndn Grants, 2007
Herbert H. & Grace A. Dow Fndn Grants, 2008
Herman Goldman Foundation Grants, 2012
Hershey Company Grants, 2013
Highmark BCBS Challenge for Healthier Schools in Western Pennsylvania, 3696
Hilda and Preston Davis Foundation Grants, 2028
Hillman Foundation Grants, 2032
Hillsdale County Community Foundation Healthy Senior/Healthy Youth Fund Grants, 2033
Hilton Head Island Foundation Grants, 2036
Hilton Hotels Corporate Giving Program Grants, 2037
Hoglund Foundation Grants, 2040
Holland/Zeeland Community Fndn Grants, 2041
Horace A. Kimball and S. Ella Kimball Foundation Grants, 2046
Horizons Community Issues Grants, 2049
Howard and Bush Foundation Grants, 2054
HRAMF Deborah Munroe Noonan Memorial Research Grants, 2058
Hugh J. Andersen Foundation Grants, 2074
Humana Foundation Grants, 2077
Huntington Beach Police Officers Fndn Grants, 2081
Huntington Clinical Foundation Grants, 2082
Huntington National Bank Community Grants, 2084
Hutchinson Community Foundation Grants, 2085
Hut Foundation Grants, 2086
Hutton Foundation Grants, 2087
Illinois Children's Healthcare Foundation Grants, 2146
Illinois Tool Works Foundation Grants, 2147
Impact 100 Grants, 2148
Institute for Agriculture and Trade Policy Food and Society Fellowships, 2159
Irving S. Gilmore Foundation Grants, 2163
Isabel Allende Foundation Esperanza Grants, 2165
J.C. Penney Company Grants, 2169
Jack H. & William M. Light Trust Grants, 2179
Jacob G. Schmidlapp Trust Grants, 2183
Jacobs Family Village Neighborhoods Grants, 2184
James & Abigail Campbell Family Fndn Grants, 2185
James Ford Bell Foundation Grants, 2187
James H. Cummings Foundation Grants, 2190
James J. & Angelia M. Harris Fndn Grants, 2192
James M. Collins Foundation Grants, 2195
James M. Cox Foundation of Georgia Grants, 2196
James R. Thorpe Foundation Grants, 2199
James S. Copley Foundation Grants, 2200
Janirve Foundation Grants, 2209
Janus Foundation Grants, 2211
Jay and Rose Phillips Family Foundation Grants, 2213
JELD-WEN Foundation Grants, 2218
Jerome and Mildred Paddock Foundation Grants, 2222
Jim Moran Foundation Grants, 2228
JM Foundation Grants, 2229
John Clarke Trust Grants, 2232
John Edward Fowler Memorial Fndn Grants, 2235
John H. and Wilhelmina D. Harland Charitable Foundation Children and Youth Grants, 2238
John Jewett and Helen Chandler Garland Foundation Grants, 2242
John Merck Scholars Awards, 2246
John P. McGovern Foundation Grants, 2247
John W. Alden Trust Grants, 2260
John W. and Anna H. Hanes Foundation Grants, 2261
John W. Anderson Foundation Grants, 2262
John W. Speas and Effie E. Speas Memorial Trust Grants, 2264
Joseph H. & Florence A. Roblee Fndn Grants, 2269
Joseph Henry Edmondson Foundation Grants, 2270
Josephine G. Russell Trust Grants, 2271
Josephine Goodyear Foundation Grants, 2272
Josephine S. Gumbiner Foundation Grants, 2273
Josephine Schell Russell Charitable Trust Grants, 2274
Josiah W. and Bessie H. Kline Foundation Grants, 2278
Judith Clark-Morrill Foundation Grants, 2281
Julius N. Frankel Foundation Grants, 2282
K and F Baxter Family Foundation Grants, 2287
Katharine Matthies Foundation Grants, 2291
Kathryne Beynon Foundation Grants, 2294
Katrine Menzing Deakins Trust Grants, 2295
Kenai Peninsula Foundation Grants, 2298
Ketchikan Community Foundation Grants, 2303
Kevin P. and Sydney B. Knight Family Foundation Grants, 2306
Kimball International-Habig Foundation Health and Human Services Grants, 2308
Kosair Charities Grants, 2318
Kosciusko County Community Fndn Grants, 2319
Kovler Family Foundation Grants, 2323
L. W. Pierce Family Foundation Grants, 2328
Land O'Lakes Foundation Mid-Atlantic Grants, 2341
Lands' End Corporate Giving Program, 2344
Laura B. Vogler Foundation Grants, 2345
Laura Moore Cunningham Foundation Grants, 2346
Layne Beachley Aim for the Stars Fnd Grants, 2349
Leo Goodwin Foundation Grants, 2354
Leon and Thea Koerner Foundation Grants, 2355
Leo Niessen Jr., Charitable Trust Grants, 2358
Lisa and Douglas Goldman Fund Grants, 2369
Long Island Community Foundation Grants, 2375
Lotus 88 Fnd for Women & Children Grants, 2376
Louetta M. Cowden Foundation Grants, 2377
Louis H. Aborn Foundation Grants, 2380
Lowell Berry Foundation Grants, 2382
Lucile Horton Howe and Mitchell B. Howe Foundation Grants, 2385
Lucile Packard Foundation for Children's Health Grants, 2386
Lucy Gooding Charitable Fndn Grants, 2388
Ludwick Family Foundation Grants, 2389
M.E. Raker Foundation Grants, 2404
Mabel A. Horne Trust Grants, 2406
Mabel H. Flory Charitable Trust Grants, 2408
Maggie Welby Foundation Grants, 2414
Manuel D. & Rhoda Mayerson Fndn Grants, 2420
Marathon Petroleum Corporation Grants, 2422
Mardag Foundation Grants, 2428
Margaret T. Morris Foundation Grants, 2429
Marie H. Bechtel Charitable Remainder Uni-Trust Grants, 2432
Mary Black Foundation Early Childhood Development Grants, 2442
Mary Wilmer Covey Charitable Trust Grants, 2448
Mattel Children's Foundation Grants, 2454
Mattel International Grants Program, 2455
Maurice J. Masserini Charitable Trust Grants, 2456
Maxon Charitable Foundation Grants, 2458
May and Stanley Smith Charitable Trust Grants, 2459
McCarthy Family Foundation Grants, 2490
McCombs Foundation Grants, 2491
McConnell Foundation Grants, 2492
McKesson Foundation Grants, 2496
McLean Contributionship Grants, 2497
McLean Foundation Grants, 2498
McMillen Foundation Grants, 2499
Mead Johnson Nutritionals Evansville-Area Organizations Grants, 2503
Medtronic Foundation Community Link Human Services Grants, 2511
Medtronic Foundation Patient Link Grants, 2514
Memorial Foundation Grants, 2516
Mercedes-Benz USA Corporate Contributions, 2517
Mericos Foundation Grants, 2518
Meriden Foundation Grants, 2519
Merrick Foundation Grants, 2521
Meta and George Rosenberg Foundation Grants, 2525
Metzger-Price Fund Grants, 2530
Meyer Memorial Trust Special Grants, 2538
MGM Resorts Foundation Community Grants, 2541
MGN Family Foundation Grants, 2542
Mid-Iowa Health Foundation Community Response Grants, 2549
Milagro Foundation Grants, 2552
Milken Family Foundation Grants, 2555
Minneapolis Foundation Community Grants, 2558
Mockingbird Foundation Grants, 2589
Morgan Adams Foundation Grants, 2593
Morris & Gwendolyn Cafritz Fndn Grants, 2594
NAPNAP Fndn Elaine Gelman Scholarship, 2609
NAPNAP Fndn Grad Research Grant, 2610
NAPNAP Foundation Innovative Health Care Small Grant, 2611

SUBJECT INDEX

Children and Youth /683

NAPNAP Foundation McNeil Grant-in-Aid, 2612
NAPNAP Foundation McNeil PNP Scholarships, 2613
NAPNAP Foundation McNeil Rural and Underserved Scholarships, 2614
NAPNAP Foundation Nursing Research Grants, 2615
NAPNAP Foundation Reckitt Benckiser Student Scholarship, 2616
Nathan Cummings Foundation Grants, 2627
Nelda C. and H.J. Lutcher Stark Fndn Grants, 2672
Nicholas H. Noyes Jr. Memorial Fndn Grants, 2791
Nina Mason Pulliam Charitable Trust Grants, 2921
NINR Chronic Illness Self-Management in Children and Adolescents Grants, 2928
Nordson Corporation Foundation Grants, 2981
North Dakota Community Foundation Grants, 2994
Northwest Airlines KidCares Medical Travel Assistance, 2998
Northwestern Mutual Foundation Grants, 2999
Nuffield Foundation Children & Families Grants, 3027
NYCT Children/Youth with Disabilities Grants, 3052
NYCT Girls and Young Women Grants, 3053
NYCT Mental Health and Retardation Grants, 3056
Oak Foundation Child Abuse Grants, 3059
Oceanside Charitable Foundation Grants, 3063
Office Depot Corporation Community Relations Grants, 3065
Ohio County Community Foundation Grants, 3068
Ohio County Community Fndn Mini-Grants, 3069
Oleonda Jameson Trust Grants, 3071
OneFamily Foundation Grants, 3075
Ordean Foundation Grants, 3080
Packard Foundation Children, Families, and Communities Grants, 3104
Packard Foundation Local Grants, 3105
Pajaro Valley Community Health Health Trust Insurance/Coverage & Education on Using the System Grants, 3108
Palmer Foundation Grants, 3112
Patrick and Anna M. Cudahy Fund Grants, 3120
PC Fred H. Bixby Fellowships, 3126
Peabody Foundation Grants, 3132
Pediatric Brain Tumor Fndn Research Grants, 3134
Pediatric Cancer Research Foundation Grants, 3135
PepsiCo Foundation Grants, 3137
Perkins Charitable Foundation Grants, 3140
Perpetual Trust for Charitable Giving Grants, 3141
Peter and Elizabeth C. Tower Foundation Annual Mental Health Grants, 3145
Peter and Elizabeth C. Tower Foundation Learning Disability Grants, 3146
Peter and Elizabeth C. Tower Foundation Mental Health Reference and Resource Materials Mini-Grants, 3147
Peter and Elizabeth C. Tower Foundation Phase II Technology Initiative Grants, 3149
Peter and Elizabeth C. Tower Foundation Phase I Technology Initiative Grants, 3150
Peter and Elizabeth C. Tower Foundation Social and Emotional Preschool Curriculum Grants, 3151
Peter and Elizabeth C. Tower Foundation Substance Abuse Grants, 3152
Petersburg Community Foundation Grants, 3156
Pew Charitable Trusts Children & Youth Grants, 3158
PeyBack Foundation Grants, 3159
Peyton Anderson Foundation Grants, 3160
Philadelphia Foundation Organizational Effectiveness Grants, 3189
Phoenix Suns Charities Grants, 3201
Piedmont Health Foundation Grants, 3219
Pike County Community Foundation Grants, 3222
Pinellas County Grants, 3223
Pinkerton Foundation Grants, 3224
Pinnacle Foundation Grants, 3225
Pittsburgh Foundation Community Fund Grants, 3230
Plough Foundation Grants, 3238
Pohlad Family Foundation, 3241
Polk Bros. Foundation Grants, 3243
Pollock Foundation Grants, 3244
Porter County Health and Wellness Grant, 3246
Portland Fndn Women's Giving Circle Grant, 3249

Posey Community Fndn Women's Fund Grants, 3251
Posey County Community Foundation Grants, 3252
Pott Foundation Grants, 3253
Powell Foundation Grants, 3255
Price Chopper's Golub Foundation Grants, 3261
Price Family Charitable Fund Grants, 3262
Prince Charitable Trusts Chicago Grants, 3270
Prudential Foundation Education Grants, 3276
Pulaski County Community Foundation Grants, 3279
Putnam County Community Foundation Grants, 3283
Qualcomm Grants, 3287
Quantum Foundation Grants, 3290
Rainbow Endowment Grants, 3300
Rajiv Gandhi Foundation Grants, 3302
Ralph M. Parsons Foundation Grants, 3304
Rathmann Family Foundation Grants, 3309
Raymond John Wean Foundation Grants, 3310
Rayonier Foundation Grants, 3311
RCF General Community Grants, 3312
RCF Summertime Kids Grants, 3314
Rehab Therapy Foundation Grants, 3356
Reinberger Foundation Grants, 3357
Reynolds & Reynolds Associate Fndn Grants, 3360
RGk Foundation Grants, 3361
Rhode Island Foundation Grants, 3362
Richard & Susan Smith Family Fndn Grants, 3365
Richard Davoud Donchian Foundation Grants, 3367
Richard King Mellon Foundation Grants, 3369
Richland County Bank Grants, 3371
Ricks Family Charitable Trust Grants, 3373
Robert and Joan Dircks Foundation Grants, 3379
Robert B McMillen Foundation Grants, 3381
Robert R. McCormick Trib Community Grants, 3385
Robert R. Meyer Foundation Grants, 3387
Robert W. Woodruff Foundation Grants, 3389
Robins Foundation Grants, 3390
Rochester Area Community Foundation Grants, 3391
Rochester Area Foundation Grants, 3392
Rockwell Collins Charitable Corporation Grants, 3397
Ronald McDonald House Charities Grants, 3405
Roy & Christine Sturgis Charitable Grants, 3415
Rush County Community Foundation Grants, 3424
Ruth Anderson Foundation Grants, 3425
Ruth H. & Warren A. Ellsworth Fndn Grants, 3429
RWJF Childhood Obesity Grants, 3431
RWJF Healthy Eating Research Grants, 3437
S.H. Cowell Foundation Grants, 3447
S. Livingston Mather Charitable Trust Grants, 3448
S. Mark Taper Foundation Grants, 3449
Saigh Foundation Grants, 3451
Salmon Foundation Grants, 3458
Salt River Health & Human Services Grants, 3459
SAMHSA Child Mental Health Initiative (CMHI) Grants, 3461
SAMHSA Strategic Prevention Framework State Incentive Grants, 3464
Samuel N. and Mary Castle Foundation Grants, 3467
Samuel S. Johnson Foundation Grants, 3468
San Diego Foundation for Change Grants, 3470
Sands Foundation Grants, 3476
San Juan Island Community Foundation Grants, 3480
Sara Elizabeth O'Brien Trust Grants, 3483
Sarkeys Foundation Grants, 3485
Schramm Foundation Grants, 3495
Schurz Communications Foundation Grants, 3496
Scott County Community Foundation Grants, 3500
Seabury Foundation Grants, 3501
Seattle Foundation Benjamin N. Phillips Memorial Fund Grants, 3504
Self Foundation Grants, 3510
Seneca Foods Foundation Grants, 3511
Shell Oil Company Foundation Community Development Grants, 3544
Shield-Ayres Foundation Grants, 3546
Shopko Fndn Community Charitable Grants, 3547
Sick Kids Fndn Community Conference Grants, 3549
Sick Kids Foundation New Investigator Research Grants, 3550
Sidgmore Family Foundation Grants, 3551
Sidney Stern Memorial Trust Grants, 3552

Sid W. Richardson Foundation Grants, 3553
Siebert Lutheran Foundation Grants, 3554
Sierra Health Foundation Responsive Grants, 3555
Simmons Foundation Grants, 3563
Singing for Change Foundation Grants, 3567
Sioux Falls Area Community Foundation Community Fund Grants (Unrestricted), 3568
Sioux Falls Area Community Foundation Field-of-Interest and Donor-Advised Grants, 3569
Sioux Falls Area Community Foundation Spot Grants (Unrestricted), 3570
Sisters of Mercy of North Carolina Fndn Grants, 3575
Sisters of St. Joseph Healthcare Fndn Grants, 3576
Sonoco Foundation Grants, 3621
Southbury Community Trust Fund, 3625
Spencer County Community Foundation Grants, 3632
Stark Community Fndn Women's Grants, 3645
Stella and Charles Guttman Foundation Grants, 3651
Sterling and Shelli Gardner Foundation Grants, 3653
Strowd Roses Grants, 3664
Sunderland Foundation Grants, 3675
Susan Mott Webb Charitable Trust Grants, 3694
Swindells Charitable Foundation, 3697
Symantec Community Relations and Corporate Philanthropy Grants, 3699
Tauck Family Foundation Grants, 3703
Textron Corporate Contributions Grants, 3712
Thelma Braun & Bocklett Family Fndn Grants, 3713
Thelma Doelger Charitable Trust Grants, 3714
Theodore Edson Parker Foundation Grants, 3715
The Ray Charles Foundation Grants, 3716
Thomas Austin Finch, Sr. Foundation Grants, 3718
Thomas Sill Foundation Grants, 3724
Thomas W. Bradley Foundation Grants, 3726
Thompson Charitable Foundation Grants, 3727
Thoracic Foundation Grants, 3729
TJX Foundation Grants, 3735
Todd Brock Family Foundation Grants, 3736
Tommy Hilfiger Corporate Foundation Grants, 3737
Topfer Family Foundation Grants, 3739
Toyota Motor Manufacturing of Indiana Grants, 3744
Toyota Motor Manufacturing of Kentucky Grants, 3745
Toyota Motor Manuf of Mississippi Grants, 3746
Toyota Motor Manufacturing of Texas Grants, 3747
Toys R Us Children's Fund Grants, 3751
Triangle Community Foundation Donor-Advised Grants, 3757
Trull Foundation Grants, 3761
Tyler Aaron Bookman Memorial Foundation Trust Grants, 3765
U.S. Lacrosse EAD Grants, 3775
Union Bank, N.A. Corporate Sponsorships and Donations, 3787
Union Bank, N.A. Foundation Grants, 3788
Union Benevolent Association Grants, 3789
Union Labor Health Fndn Angel Fund Grants, 3791
Union Labor Health Foundation Dental Angel Fund Grants, 3793
UnitedHealthcare Children's Foundation Grants, 3795
United Healthcare Comm Grants in Michigan, 3796
United Methodist Child Mental Health Grants, 3800
United Methodist Health Ministry Fund Grants, 3801
Vancouver Foundation Grants and Community Initiatives Program, 3849
Vancouver Sun Children's Fund Grants, 3851
Vanderburgh Community Foundation Grants, 3852
Victor E. Speas Foundation Grants, 3868
Vigneron Memorial Fund Grants, 3873
Volkswagen of America Corporate Contributions, 3880
W.C. Griffith Foundation Grants, 3881
W. C. Griffith Foundation Grants, 3882
W.H. & Mary Ellen Cobb Trust Grants, 3884
W.K. Kellogg Foundation Secure Families Grants, 3887
W.M. Keck Fndn So California Grants, 3890
Waitt Family Foundation Grants, 3894
Walker Area Community Foundation Grants, 3895
Walmart Foundation National Giving Grants, 3898
Warrick County Community Foundation Grants, 3904
Warrick County Community Foundation Women's Fund, 3905

684 / Children and Youth

Weaver Popcorn Foundation Grants, 3910
Weingart Foundation Grants, 3913
Wells County Foundation Grants, 3920
Western Union Foundation Grants, 3923
Weyerhaeuser Family Foundation Health Grants, 3926
WHAS Crusade for Children Grants, 3927
Whitney Foundation Grants, 3938
William A. Badger Foundation Grants, 3945
William B. Stokely Jr. Foundation Grants, 3947
William G. & Helen C. Hoffman Fndn Grants, 3950
William J. Brace Charitable Trust, 3955
Williams Companies Foundation Homegrown Giving Grants, 3957
William T. Grant Foundation Research Grants, 3958
Wilson-Wood Foundation Grants, 3960
Winston-Salem Foundation Competitive Grants, 3965
Winston-Salem Fndn Elkin/Tri-County Grants, 3966
Wolfe Associates Grants, 3969
Wood-Claeyssens Foundation Grants, 3971
Yawkey Foundation Grants, 3980
Z. Smith Reynolds Foundation Small Grants, 3982
Zane's Foundation Grants, 3983
Zellweger Baby Support Network Grants, 3984

Children's Museums
Angels Baseball Foundation Grants, 590
Anschutz Family Foundation Grants, 602
Ben B. Cheney Foundation Grants, 875
Cabot Corporation Foundation Grants, 993
Ike and Roz Friedman Foundation Grants, 2145
Lewis H. Humphreys Charitable Trust Grants, 2361
Phoenix Suns Charities Grants, 3201

China
SSRC Collaborative Research Grants on Environment and Health in China, 3638

Chronic Illness
Aetna Foundation Health Grants in Connecticut, 327
Bupa Foundation Medical Research Grants, 974
CDC Experience Epidemiology Fellowships, 1059
CFF Research Grants, 1090
CSTE CDC/CSTE Applied Epidemiology Fellowships, 1367
Emma B. Howe Memorial Foundation Grants, 1593
Gerber Foundation Grants, 1807
Graham and Carolyn Holloway Family Foundation Grants, 1861
Health Foundation of Greater Cincinnati Grants, 1983
Herbert A. & Adrian W. Woods Fndn Grants, 2007
HRAMF Deborah Munroe Noonan Memorial Research Grants, 2058
Josiah Macy Jr. Foundation Grants, 2277
Leonard and Helen R. Stulman Charitable Foundation Grants, 2356
Mary L. Peyton Foundation Grants, 2445
Mary Wilmer Covey Charitable Trust Grants, 2448
Medtronic Foundation Patient Link Grants, 2514
Medtronic Foundation Strengthening Health Systems Grants, 2515
Mid-Iowa Health Foundation Community Response Grants, 2549
NAPNAP Foundation Nursing Research Grants, 2615
NIGMS Advancing Basic Behavioral and Social Research on Resilience: an Integrative Science Approach, 2861
NIH Biobehavioral Research for Effective Sleep, 2865
NIH Research on Sleep and Sleep Disorders, 2898
NINR Acute and Chronic Care During Mechanical Ventilation Research Grants, 2926
NINR Chronic Illness Self-Management in Children and Adolescents Grants, 2928
NSF Biomedical Engineering and Engineering Healthcare Grants, 3009
RWJF New Jersey Health Initiatives Grants, 3440
Special People in Need Grants, 3630
Thomas C. Burke Foundation Grants, 3720
TJX Foundation Grants, 3735
Welborn Baptist Foundation Improvements to Community Health Status Grants, 3916

Churches
Abel Foundation Grants, 206
Adam Richter Charitable Trust Grants, 298
Adams County Community Foundation of Pennsylvania Grants, 300
ALSAM Foundation Grants, 454
American Foodservice Charitable Trust Grants, 532
Anthony R. Abraham Foundation Grants, 608
Antone & Edene Vidinha Charitable Trust Grants, 609
Aragona Family Foundation Grants, 648
Archer Daniels Midland Foundation Grants, 651
Austin S. Nelson Foundation Grants, 805
Booth-Bricker Fund Grants, 933
Bosque Foundation Grants, 935
Boston Foundation Grants, 936
Caddock Foundation Grants, 994
Callaway Foundation Grants, 1002
Carnahan-Jackson Foundation Grants, 1032
Catherine Kennedy Home Foundation Grants, 1043
Champlin Foundations Grants, 1110
Chiles Foundation Grants, 1147
Clark-Winchcole Foundation Grants, 1173
Community Fndn of East Central Illinois Grants, 1268
Community Foundation of Greenville Community Enrichment Grants, 1284
Community Foundation of Greenville Hollingsworth Funds Program/Project Grants, 1285
Danellie Foundation Grants, 1421
Gardner Foundation Grants, 1771
General Mills Foundation Grants, 1782
George Kress Foundation Grants, 1799
George W. Brackenridge Foundation Grants, 1802
Gregory Family Foundation Grants (Florida), 1899
Griffin Family Foundation Grants, 1902
Helen Steiner Rice Foundation Grants, 1998
High Meadow Foundation Grants, 2027
Howe Foundation of North Carolina Grants, 2055
I.A. O'Shaughnessy Foundation Grants, 2088
James M. Collins Foundation Grants, 2195
Jessie Ball Dupont Fund Grants, 2226
John H. and Wilhelmina D. Harland Charitable Foundation Children and Youth Grants, 2238
John I. Smith Charities Grants, 2240
Kathryne Beynon Foundation Grants, 2294
Maxon Charitable Foundation Grants, 2458
Mervin Bovaird Foundation Grants, 2523
Perkins Charitable Foundation Grants, 3140
Richard and Helen DeVos Foundation Grants, 3364
Sioux Falls Area Community Foundation Community Fund Grants (Unrestricted), 3568
Sioux Falls Area Community Foundation Field-of-Interest and Donor-Advised Grants, 3569
Sioux Falls Area Community Foundation Spot Grants (Unrestricted), 3570
Sunderland Foundation Grants, 3675
Support Our Aging Religious (SOAR) Grants, 3686
Susan Mott Webb Charitable Trust Grants, 3694
W.C. Griffith Foundation Grants, 3881
W. C. Griffith Foundation Grants, 3882
Walter L. Gross III Family Foundation Grants, 3901
Wege Foundation Grants, 3912
William H. Hannon Foundation Grants, 3953

Circadian Rhythms
NHLBI Circadian-Coupled Cellular Function in Heart, Lung, and Blood Tissue Grants, 2696
NIH Research on Sleep and Sleep Disorders, 2898

Circulatory Disease
Delmarva Power and Light Contributions, 1454
Edna G. Kynett Memorial Foundation Grants, 1551
HRAMF Harold S. Geneen Charitable Trust Awards for Coronary Heart Disease Research, 2059

Citizenship
Battle Creek Community Foundation Grants, 839
Carnegie Corporation of New York Grants, 1033
Citizens Bank Mid-Atlantic Charitable Foundation Grants, 1162
Harvest Foundation Grants, 1971

John Edward Fowler Memorial Fndn Grants, 2235
Kenneth T. & Eileen L. Norris Fndn Grants, 2300
Lubrizol Foundation Grants, 2384
Oregon Community Fndn Community Grants, 3082
Piper Jaffray Fndn Communities Giving Grants, 3227
Richard King Mellon Foundation Grants, 3369
Wallace Alexander Gerbode Foundation Grants, 3896
Warren County Community Foundation Grants, 3902
Warren County Community Fndn Mini-Grants, 3903
White County Community Foundation Grants, 3934

Civic Affairs
A.O. Smith Foundation Community Grants, 17
Abbott Fund Global AIDS Care Grants, 202
Adams Foundation Grants, 301
Advanced Micro Devices Comm Affairs Grants, 313
AFG Industries Grants, 340
Alabama Power Foundation Grants, 400
Alaska Airlines Foundation Grants, 403
Alberto Culver Corporate Contributions Grants, 408
Albert Pick Jr. Fund Grants, 409
Alcatel-Lucent Technologies Foundation Grants, 413
Alcoa Foundation Grants, 414
Alice Tweed Tuohy Foundation Grants Program, 437
Allegheny Technologies Charitable Trust, 440
Annenberg Foundation Grants, 596
Arie and Ida Crown Memorial Grants, 654
Arizona Cardinals Grants, 655
Arkansas Community Foundation Grants, 660
ArvinMeritor Grants, 675
Ashland Corporate Contributions Grants, 713
Autodesk Community Relations Grants, 809
Banfi Vintners Foundation Grants, 824
Barker Welfare Foundation Grants, 831
Bayer Foundation Grants, 846
Bechtel Group Foundation Building Positive Community Relationships Grants, 866
Bertha Russ Lytel Foundation Grants, 885
Besser Foundation Grants, 887
Blum-Kovler Foundation Grants, 924
Blumenthal Foundation Grants, 925
Boeing Company Contributions Grants, 930
Boston Globe Foundation Grants, 937
Bridgestone/Firestone Trust Fund Grants, 949
Burlington Industries Foundation Grants, 977
Burlington Northern Santa Fe Foundation Grants, 978
Cailloux Foundation Grants, 996
Carla J. Funk Governmental Relations Award, 1020
Centerville-Washington Foundation Grants, 1078
Central Carolina Community Foundation Community Impact Grants, 1079
CFFVR Basic Needs Giving Partnership Grants, 1094
Chatlos Foundation Grants Program, 1128
CICF Indianapolis Fndn Community Grants, 1156
CICF Legacy Fund Grants, 1158
CIGNA Foundation Grants, 1160
Clayton Baker Trust Grants, 1179
Cleveland-Cliffs Foundation Grants, 1181
Clinton County Community Foundation Grants, 1184
CNA Foundation Grants, 1189
Cockrell Foundation Grants, 1204
Columbus Foundation Mary Eleanor Morris Fund Grants, 1220
Comerica Charitable Foundation Grants, 1224
Commonwealth Edison Grants, 1225
Community Fndn for Greater Buffalo Grants, 1247
Community Foundation for Greater New Haven Valley Neighborhood Grants, 1251
Community Foundation for Northeast Michigan Mini-Grants, 1255
Community Foundation for Southeast Michigan Grants, 1258
Community Foundation for the National Capital Region Community Leadership Grants, 1260
Community Foundation of Eastern Connecticut General Southeast Grants, 1269
Community Foundation of Greater Greensboro Community Grants, 1277
Community Fndn of Howard County Grants, 1286

SUBJECT INDEX

Civil Society /685

Community Foundation of Muncie and Delaware County Maxon Grants, 1301
Community Foundation of Riverside & San Bernardino County Impact Grants, 1303
Community Fndn of So Alabama Grants, 1306
Community Foundation of St. Joseph County Special Project Challenge Grants, 1310
Community Foundation of the Verdugos Grants, 1315
Community Foundation Partnerships - Lawrence County Grants, 1317
Community Foundation Partnerships - Martin County Grants, 1318
Community Foundation Serving Riverside and San Bernardino Counties Impact Grants, 1319
Connelly Foundation Grants, 1326
Consumers Energy Foundation, 1343
Cooper Industries Foundation Grants, 1345
Crown Point Community Foundation Grants, 1361
Cruise Industry Charitable Foundation Grants, 1363
CSRA Community Foundation Grants, 1366
CSX Corporate Contributions Grants, 1368
Cumberland Community Foundation Grants, 1374
DaimlerChrysler Corporation Fund Grants, 1412
Dayton Power and Light Foundation Grants, 1437
Decatur County Community Foundation Large Project Grants, 1444
Denver Foundation Community Grants, 1461
Deutsche Banc Alex Brown and Sons Charitable Foundation Grants, 1471
Dr. Scholl Foundation Grants Program, 1511
Dunspaugh-Dalton Foundation Grants, 1530
Eastman Chemical Company Foundation Grants, 1542
eBay Foundation Community Grants, 1543
Elmer L. & Eleanor J. Andersen Fndn Grants, 1578
El Paso Community Foundation Grants, 1579
El Pomar Foundation Anna Keesling Ackerman Fund Grants, 1581
El Pomar Foundation Grants, 1582
Ethel Sergeant Clark Smith Foundation Grants, 1608
Evan and Susan Bayh Foundation Grants, 1614
F.M. Kirby Foundation Grants, 1624
Fargo-Moorhead Area Fndn Woman's Grants, 1633
Farmers Insurance Corporate Giving Grants, 1634
Faye McBeath Foundation Grants, 1635
Ferree Foundation Grants, 1658
Fifth Third Foundation Grants, 1661
Finance Factors Foundation Grants, 1666
FirstEnergy Foundation Community Grants, 1668
Floyd A. & Kathleen C. Cailloux Fnd Grants, 1681
Fluor Foundation Grants, 1682
Ford Motor Company Fund Grants Program, 1687
Frank S. Flowers Foundation Grants, 1729
Fulton County Community Foundation Grants, 1763
GenCorp Foundation Grants, 1779
George Gund Foundation Grants, 1796
Gibson County Community Foundation Women's Fund Grants, 1817
Global Fund for Women Grants, 1833
Goodrich Corporation Foundation Grants, 1854
Goodyear Tire Grants, 1856
Grace and Franklin Bernsen Foundation Grants, 1858
Grand Haven Area Community Fndn Grants, 1864
Greater Tacoma Community Foundation Grants, 1886
Greater Worcester Community Foundation Discretionary Grants, 1888
Green Bay Packers Foundation Grants, 1890
Gulf Coast Community Foundation Grants, 1915
H.A. & Mary K. Chapman Trust Grants, 1921
Hallmark Corporate Foundation Grants, 1936
Hancock County Community Foundation - Field of Interest Grants, 1943
Helen K. & Arthur E. Johnson Fndn Grants, 1995
Helen S. Boylan Foundation Grants, 1997
Hillman Foundation Grants, 2032
Hilton Hotels Corporate Giving Program Grants, 2037
Hoblitzelle Foundation Grants, 2038
Howard and Bush Foundation Grants, 2054
Hugh J. Andersen Foundation Grants, 2074
Huntington County Community Foundation Make a Difference Grants, 2083

Hutton Foundation Grants, 2087
Intergrys Corporation Grants, 2160
J.L. Bedsole Foundation Grants, 2172
James L. and Mary Jane Bowman Charitable Trust Grants, 2194
John Deere Foundation Grants, 2234
John J. Leidy Foundation Grants, 2241
John Merck Fund Grants, 2245
Katharine Matthies Foundation Grants, 2291
Kosciusko County Community Fndn Grants, 2319
Lake County Community Fund Grants, 2333
Land O'Lakes Foundation Mid-Atlantic Grants, 2341
Lisa and Douglas Goldman Fund Grants, 2369
Lockheed Martin Corp Foundation Grants, 2374
Louie M. & Betty M. Phillips Fndn Grants, 2378
Lubbock Area Foundation Grants, 2383
Lubrizol Foundation Grants, 2384
Lynde & Harry Bradley Foundation Fellowships, 2396
Lynde and Harry Bradley Foundation Grants, 2397
Lynde and Harry Bradley Foundation Prizes: Bradley Prizes, 2398
Madison County Community Foundation - City of Anderson Quality of Life Grant, 2412
Madison County Community Foundation General Grants, 2413
Mary K. Chapman Foundation Grants, 2443
McCombs Foundation Grants, 2491
MeadWestvaco Foundation Sustainable Communities Grants, 2506
Meriden Foundation Grants, 2519
Merrick Foundation Grants, 2522
MGM Resorts Foundation Community Grants, 2541
Morris & Gwendolyn Cafritz Fndn Grants, 2594
NHSCA Arts in Health Care Project Grants, 2741
Nina Mason Pulliam Charitable Trust Grants, 2921
Noble County Community Foundation Grants, 2975
Norcliffe Foundation Grants, 2979
Nord Family Foundation Grants, 2980
Nordson Corporation Foundation Grants, 2981
PACCAR Foundation Grants, 3101
Pacific Life Foundation Grants, 3103
Paul Ogle Foundation Grants, 3124
Percy B. Ferebee Endowment Grants, 3138
Piedmont Natural Gas Corporate and Charitable Contributions, 3220
Piper Jaffray Fndn Communities Giving Grants, 3227
PMI Foundation Grants, 3239
PPCF Community Grants, 3256
Princeton Area Community Foundation Greater Mercer Grants, 3272
Procter and Gamble Fund Grants, 3274
Puerto Rico Community Foundation Grants, 3278
Quaker Chemical Foundation Grants, 3286
Rainbow Endowment Grants, 3300
Ralph M. Parsons Foundation Grants, 3304
RCF General Community Grants, 3312
RGk Foundation Grants, 3361
Rice Foundation Grants, 3363
Richard King Mellon Foundation Grants, 3369
Robert W. Woodruff Foundation Grants, 3389
Rochester Area Community Foundation Grants, 3391
Rockefeller Brothers Peace & Security Grants, 3394
Rockwell Collins Charitable Corporation Grants, 3397
Rohm and Haas Company Grants, 3403
Ronald McDonald House Charities Grants, 3405
Rose Hills Foundation Grants, 3413
Ruth and Vernon Taylor Foundation Grants, 3426
San Diego Women's Foundation Grants, 3474
Schramm Foundation Grants, 3495
Seagate Tech Corp Capacity to Care Grants, 3502
Shell Oil Company Foundation Community Development Grants, 3544
Sony Corporation of America Grants, 3623
Spencer County Community Foundation Grants, 3632
Square D Foundation Grants, 3634
Stackpole-Hall Foundation Grants, 3643
Stewart Huston Charitable Trust Grants, 3655
Sunderland Foundation Grants, 3675
Sunoco Foundation Grants, 3679

SunTrust Bank Trusteed Foundations Florence C. and Harry L. English Memorial Fund Grants, 3680
SunTrust Bank Trusteed Foundations Greene-Sawtell Grants, 3681
SunTrust Bank Trusteed Foundations Harriet McDaniel Marshall Tust Grants, 3682
SunTrust Bank Trusteed Foundations Nell Warren Elkin and William Simpson Elkin Grants, 3683
SunTrust Bank Trusteed Foundations Thomas Guy Woolford Charitable Trust Grants, 3684
SunTrust Bank Trusteed Foundations Walter H. and Marjory M. Rich Memorial Fund Grants, 3685
TE Foundation Grants, 3705
Tension Envelope Foundation Grants, 3707
TJX Foundation Grants, 3735
Toyota Motor Engineering & Manufacturing North America Grants, 3742
Toyota Motor Manuf of Mississippi Grants, 3746
Toyota Motor Manuf of West Virginia Grants, 3748
Toyota Motor N America of New York Grants, 3749
Toyota Motor Sales, USA Grants, 3750
U.S. Cellular Corporation Grants, 3766
United Technologies Corporation Grants, 3803
Vectren Foundation Grants, 3857
Vernon K. Krieble Foundation Grants, 3866
Wells County Foundation Grants, 3920
Whitley County Community Foundation Grants, 3936
William Blair and Company Foundation Grants, 3948
William G. & Helen C. Hoffman Fndn Grants, 3950
Z. Smith Reynolds Foundation Small Grants, 3982

Civic Engagement
Ben B. Cheney Foundation Grants, 875
Carla J. Funk Governmental Relations Award, 1020
Decatur County Community Foundation Small Project Grants, 1445
FIU Global Civic Engagement Mini Grants, 1673
GNOF Stand Up For Our Children Grants, 1847
Manuel D. & Rhoda Mayerson Fndn Grants, 2420
Piper Trust Older Adults Grants, 3229
PMI Foundation Grants, 3239
Rockefeller Brothers Peace & Security Grants, 3394
Sioux Falls Area Community Foundation Community Fund Grants (Unrestricted), 3568
Sioux Falls Area Community Foundation Spot Grants (Unrestricted), 3570
Union Bank, N.A. Foundation Grants, 3788
Walmart Foundation State Giving Grants, 3900

Civics
Ben B. Cheney Foundation Grants, 875
Gill Foundation - Gay and Lesbian Fund Grants, 1821
Hershey Company Grants, 2013
Toyota Motor Engineering & Manufacturing North America Grants, 3742
Toyota Motor Manufacturing of Texas Grants, 3747
Toyota Motor Manuf of West Virginia Grants, 3748
Toyota Motor N America of New York Grants, 3749
Toyota Motor Sales, USA Grants, 3750

Civics Education
Ben B. Cheney Foundation Grants, 875
Richard King Mellon Foundation Grants, 3369

Civil Engineering
NSF-NIST Interaction in Chemistry, Materials Research, Molecular Biosciences, Bioengineering, and Chemical Engineering, 3004

Civil Law
John Merck Fund Grants, 2245
OSF International Harm Reduction Development Program Grants, 3090

Civil Service
John Merck Fund Grants, 2245

Civil Society
Cowles Charitable Trust Grants, 1354
Prospect Burma Scholarships, 3275

686 / Civil/Human Rights

Civil/Human Rights
Allstate Corporate Giving Grants, 446
Allstate Corp Hometown Commitment Grants, 447
American Jewish World Service Grants, 533
Assisi Fndn of Memphis Capital Project Grants, 786
Assisi Foundation of Memphis General Grants, 787
Blumenthal Foundation Grants, 925
Carrier Corporation Contributions Grants, 1038
Clayton Baker Trust Grants, 1179
Coastal Community Foundation of South Carolina Grants, 1202
Community Foundation of Greater Greensboro Community Grants, 1277
Cowles Charitable Trust Grants, 1354
David Geffen Foundation Grants, 1431
Fargo-Moorhead Area Foundation Grants, 1632
Ford Motor Company Fund Grants Program, 1687
Frank Stanley Beveridge Foundation Grants, 1730
Gill Foundation - Gay and Lesbian Fund Grants, 1821
Global Fund for Children Grants, 1832
Global Fund for Women Grants, 1833
H.J. Heinz Company Foundation Grants, 1923
Harold Simmons Foundation Grants, 1955
Herman Goldman Foundation Grants, 2012
Horizons Community Issues Grants, 2049
James J. & Joan A. Gardner Family Fndn Grants, 2193
Jay and Rose Phillips Family Foundation Grants, 2213
John M. Lloyd Foundation Grants, 2244
John Merck Fund Grants, 2245
Knox County Community Foundation Grants, 2313
Lisa and Douglas Goldman Fund Grants, 2369
Minneapolis Foundation Community Grants, 2558
MLA David A. Kronick Traveling Fellowship, 2561
Norman Foundation Grants, 2983
OSF Accountability and Monitoring in Health Initiative Grants, 3087
OSF Health Media Initiative Grants, 3089
OSF International Harm Reduction Development Program Grants, 3090
OSF Law and Health Initiative Grants, 3092
OSF Roma Health Project Grants, 3095
Otto Bremer Foundation Grants, 3099
Philadelphia Foundation Organizational Effectiveness Grants, 3189
Playboy Foundation Grants, 3237
Prospect Burma Scholarships, 3275
Rainbow Endowment Grants, 3300
Rajiv Gandhi Foundation Grants, 3302
Samuel S. Johnson Foundation Grants, 3468
San Diego Foundation for Change Grants, 3470
Sidney Stern Memorial Trust Grants, 3552
Sir Dorabji Tata Trust Grants for NGOs or Voluntary Organizations, 3572
Skoll Fndn Awards for Social Entrepreneurship, 3579
Tifa Foundation Grants, 3733
Vernon K. Krieble Foundation Grants, 3866
Warrick County Community Foundation Grants, 3904
Z. Smith Reynolds Foundation Small Grants, 3982

Classroom Instruction
Coca-Cola Foundation Grants, 1203
IBRO-PERC Support for European Workshops, Symposia and Meetings, 2098
Kettering Fund Grants, 2305
Long Island Community Foundation Grants, 2375
NSF Presidential Early Career Awards for Scientists and Engineers (PECASE) Grants, 3021
SfN Science Educator Award, 3535
Vectren Foundation Grants, 3857

Climatology
Bullitt Foundation Grants, 973
IIE Rockefeller Foundation Bellagio Center Residencies, 2142
James S. McDonnell Foundation Complex Systems Collaborative Activity Awards, 2202

Clinical Medicine, General
ACSM Foundation Clinical Sports Medicine Endowment Grants, 265
AIHS Clinical Fellowships, 375
AIHS Fast-Track Fellowships, 379
American College of Surgeons and The Triological Society Clinical Scientist Development Awards, 518
American Philosophical Society Daland Fellowships in Clinical Investigation, 537
Biogen Foundation General Donations, 899
Bristol-Myers Squibb Clinical Outcomes and Research Grants, 955
CDC Evidence-Based Laboratory Medicine: Quality/Performance Measure Evaluation, 1058
CFF Shwachman Clinical Investigator Award, 1086
CVS All Kids Can Grants, 1377
Doris Duke Charitable Foundation Clinical Interfaces Award Program, 1492
Doris Duke Charitable Foundation Clinical Research Fellowships for Medical Students, 1493
Doris Duke Charitable Foundation Clinical Scientist Development Award (CSDA) Bridge Grants, 1495
Doris Duke Charitable Foundation Distinguished Clinical Scientist Award Program, 1496
E.W. "Al" Thrasher Awards, 1540
Mayo Clinic Administrative Fellowship, 2460
MMS and Alliance Charitable Foundation International Health Studies Grants, 2588
NCI Ruth L. Kirschstein National Research Service Award Institutional Training Grants, 2654
NCI Stages of Breast Development: Normal to Metastatic Disease Grants, 2655
NEI Clinical Vision Research Devel Award, 2664
NHLBI Mentored Patient-Oriented Research Career Development Awards, 2710
NIA Alzheimer's Disease Core Centers Grants, 2748
NIDDK Pilot and Feasibility Clinical Research Grants in Kidney or Urologic Diseases, 2835
NIH Mentored Research Scientist Dev Awards, 2889
NLM Internet Connection for Medical Institutions Grants, 2946
NYAM Mary and David Hoar Fellowship, 3045
OSF Access to Essential Medicines Initiative, 3086
Peter F. McManus Charitable Trust Grants, 3153
PhRMA Foundation Health Outcomes Pre-Doctoral Fellowships, 3203
PhRMA Foundation Paul Calabresi Medical Student Research Fellowship, 3210
Piper Trust Healthcare & Med Research Grants, 3228
Regenstrief Master Of Science In Clinical Research/CITE Program, 3353
Robert B. Adams Foundation Scholarships, 3380
Robert Leet & Clara Guthrie Patterson Grants, 3383
SfN Neuroscience Fellowships, 3530

Clinical Psychology
NHLBI Mentored Patient-Oriented Research Career Development Awards, 2710

Clinical Research
AAAAI ARTrust Mini Grants for Allied Health, 20
AAAAI ARTrust Grants for Clinical Research, 22
AAAAI ARTrust Grants in Faculty Development, 23
ACSM Foundation Clinical Sports Medicine Endowment Grants, 265
AES-Grass Young Investigator Travel Awards, 317
AES Epilepsy Research Recognition Awards, 318
AES Research and Training Workshop Awards, 320
AES Research Infrastructure Awards, 321
AES Research Initiative Awards, 322
AES Robert S. Morison Fellowship, 323
AES Susan S. Spencer Clinical Research Epilepsy Training Fellowship, 325
AMSSM Foundation Research Grants, 577
ANA/Grass Foundation Award in Neuroscience, 578
ANA Derek Denny-Brown Neurological Scholar Award, 579
ANA Distinguished Neurology Teacher Award, 580
ANA Junior Academic Neurologist Scholarships, 582
ANA Wolfe Neuropathy Research Prize, 585
BCBSM Foundation Physician Investigator Research Awards, 855
Children's Tumor Fndn Clin Research Awards, 1140

SUBJECT INDEX

Children's Tumor Foundation Schwannomatosis Awards, 1142
Conquer Cancer Foundation of ASCO Career Development Award, 1328
Conquer Cancer Foundation of ASCO International Innovation Grant, 1334
FAMRI Clinical Innovator Awards, 1628
FAMRI Young Clinical Scientist Awards, 1629
Fight for Sight-Streilein Foundation for Ocular Immunology Research Award, 1662
Fight for Sight Grants-in-Aid, 1663
Fight for Sight Summer Student Fellowships, 1665
Foundation of the American Thoracic Society Research Grants, 1708
Gilbert Memorial Fund Grants, 1820
HRAMF Charles A. King Trust Postdoctoral Research Fellowships, 2056
HRAMF Deborah Munroe Noonan Memorial Research Grants, 2058
HRAMF Ralph and Marian Falk Medical Research Trust Catalyst Awards, 2061
HRAMF Ralph and Marian Falk Medical Research Trust Transformational Awards, 2062
HRAMF Thome Foundation Awards in Age-Related Macular Degeneration Research, 2065
IBRO Return Home Fellowships, 2112
Klarman Family Foundation Grants in Eating Disorders Research Grants, 2310
Lymphatic Education and Research Network Additional Support Grants for NIH-funded F32 Postdoctoral Fellows, 2394
Lymphatic Education and Research Network Postdoctoral Fellowships, 2395
Mayo Clinic Administrative Fellowship, 2460
MMS and Alliance Charitable Foundation International Health Studies Grants, 2588
NARSAD Ruane Prize for Child and Adolescent Psychiatric Research, 2624
National Parkinson Foundation Clinical Research Fund Grants, 2635
NFL Charities Medical Grants, 2685
NHLBI Ancillary Studies in Clinical Trials, 2688
NHLBI Ruth L. Kirschstein National Research Service Awards for Individual Postdoctoral Fellows, 2729
NHLBI Ruth L. Kirschstein National Research Service Awards for Individual Predoctoral MD/PhD Fellows and Other Dual Degree Fellows, 2731
NSF Accelerating Innovation Research, 3005
Perpetual Trust for Charitable Giving Grants, 3141
Pfizer Advancing Research in Transplantation Science (ARTS) Research Awards, 3163
Pfizer Anti Infective Research EU Grants, 3164
Pfizer ASPIRE EU Antifungal Research Awards, 3165
Pfizer ASPIRE EU Dupuytren's Contracture Research Awards, 3166
Pfizer ASPIRE EU Emerging Mechanisms of Resistance Antibacterial Research Awards, 3167
Pfizer ASPIRE EU MRSA Nosocomial Pneumonia & MRSA Complicated Skin & Soft Tissue Infections Antibacterial Research Awards, 3168
Pfizer ASPIRE North America Broad Spectrum Antibiotics for the Treatment of Gram-Negative or Polymicrobial Infections Research Awards, 3169
Pfizer ASPIRE North America Hemophilia Research Awards, 3170
Pfizer ASPIRE North America Narrow Spectrum Antibiotics for the Treatment of MRSA Research Awards, 3171
Pfizer ASPIRE Worldwide Endocrine Young Investigator Grants, 3173
Pfizer Australia Cancer Research Grants, 3174
Pfizer Australia Neuroscience Research Grants, 3175
Pfizer Australia Paediatric Endocrine Care Research Grants, 3176
PFizer Compound Transfer Agreements, 3177
Pfizer Global Investigator Research Grants, 3178
Pfizer Medical Education Track Two Grants, 3183
SfN Jacob P. Waletzky Award, 3523
SfN Janett Rosenberg Trubatch Career Development Award, 3524

SUBJECT INDEX

SfN Julius Axelrod Prize, 3526
SfN Peter and Patricia Gruber International Research Award, 3533
SfN Young Investigator Award, 3541
Sick Kids New Investigator Research Grants, 3550
Sigma Theta Tau International /American Nurses Foundation Grant, 3556
Sigma Theta Tau International / Council for the Advancement of Nursing Science Grants, 3557
Sigma Theta Tau International / Oncology Nursing Society Grant, 3558
Sigma Theta Tau International Doris Bloch Research Award, 3559
Sigma Theta Tau Small Grants, 3560
Victor E. Speas Foundation Grants, 3868
Wellcome Trust New Investigator Awards, 3919

Clinical Sociology
Peter F. McManus Charitable Trust Grants, 3153

Clinical Trials
Alton Ochsner Award, 458
ASCO Advanced Clinical Research Award in Colorectal Cancer, 678
ASCO Advanced Clinical Research Awards in Breast Cancer, 679
ASCO Advanced Clinical Research Awards in Sarcoma, 680
ASCO Long-Term International Fellowships, 681
Bristol-Myers Squibb Clinical Outcomes and Research Grants, 955
CDC Evidence-Based Laboratory Medicine: Quality/Performance Measure Evaluation, 1058
Conquer Cancer Foundation of ASCO International Innovation Grant, 1334
Conquer Cancer Foundation of ASCO Medical Student Rotation Grants, 1335
NCI/NCCAM Quick-Trials for Novel Cancer Therapies, 2643
NEI Clinical Study Planning Grant, 2663
NEI Clinical Vision Research Devel Award, 2664
NHLBI Investigator Initiated Multi-Site Clinical Trials, 2703
NHLBI Mentored Clinical Scientist Research Career Development Awards, 2709
NIAAA Mentored Clinical Scientist Research Career Development Awards, 2744
NICHD Innovative Therapies and Clinical Studies for Screenable Disorders Grants, 2789
NIDCD Mentored Clinical Scientist Research Career Development Award, 2795
NIDDK Ancillary Studies of Kidney Disease Accessing Information from Clinical Trials, Epidemiological Studies, and Databases Grants, 2799
NIDDK Ancillary Studies to Major Ongoing NIDDK and NHLBI Clinical Research Studies, 2800
NIDDK Basic and Clinical Studies of Congenital Urinary Tract Obstruction Grants, 2801
NIDDK Exploratory/Developmental Clinical Research Grants in Obesity, 2818
NIDDK Multi-Center Clinical Study Cooperative Agreements, 2826
NIH Mentored Clinical Scientist Devel Award, 2886
NIH Sarcoidosis: Research into the Cause of Multi-Organ Disease and Clinical Strategies for Therapy Grants, 2907
NKF Franklin McDonald/Fresenius Medical Care Clinical Research Grants, 2936
PDF Postdoctoral Fellowships for Neurologists, 3130
Pfizer ASPIRE North America Broad Spectrum Antibiotics for the Treatment of Gram-Negative or Polymicrobial Infections Research Awards, 3169
Pfizer ASPIRE North America Narrow Spectrum Antibiotics for the Treatment of MRSA Research Awards, 3171
PFizer Compound Transfer Agreements, 3177
Pfizer Global Investigator Research Grants, 3178
Tourette Syndrome Association Post-Doctoral Fellowships, 3740
Tourette Syndrome Association Research Grants, 3741

Clinics
Ann Arbor Area Community Foundation Grants, 594
California Community Foundation Health Care Grants, 998
CDC Evaluation of the Use of Rapid Testing For Influenza in Outpatient Medical Settings, 1057
Clark and Ruby Baker Foundation Grants, 1174
Colorado Resource for Emergency Education and Trauma Grants, 1212
Covidien Medical Product Donations, 1352
Crystelle Waggoner Charitable Trust Grants, 1364
Cystic Fibrosis Canada Clinic Incentive Grants, 1383
Cystic Fibrosis Canada Small Conference Grants, 1388
Deborah Munroe Noonan Memorial Grants, 1443
E. Rhodes & Leona B. Carpenter Grants, 1539
Fondren Foundation Grants, 1685
Foundation for Health Enhancement Grants, 1696
George E. Hatcher, Jr. and Ann Williams Hatcher Foundation Grants, 1792
George W. Wells Foundation Grants, 1804
Gladys Brooks Foundation Grants, 1826
GNOF IMPACT Gulf States Eye Surgery Fund, 1840
Helen Irwin Littauer Educational Trust Grants, 1994
Hospital Libraries Section / MLA Professional Development Grants, 2052
Lumpkin Family Fnd Healthy People Grants, 2391
Lydia deForest Charitable Trust Grants, 2393
Marin Community Foundation Improving Community Health Grants, 2433
Meyer Foundation Healthy Communities Grants, 2533
Mt. Sinai Health Care Foundation Health of the Urban Community Grants, 2600
Norfolk Southern Foundation Grants, 2982
Piper Trust Healthcare & Med Research Grants, 3228
RCF General Community Grants, 3312
Reinberger Foundation Grants, 3357
Shell Deer Park Grants, 3543
Sheltering Arms Fund Grants, 3545

Coastal Processes
GNOF Bayou Communities Grants, 1835

Cognitive Development/Processes
CVS Community Grants, 1379
FRAXA Research Foundation Program Grants, 1733
Gerber Foundation Grants, 1807
James S. McDonnell Fnd Research Grants, 2203
James S. McDonnell Foundation Scholar Awards, 2204
James S. McDonnell Foundation Understanding Human Cognition Awards, 2205
NARSAD Goldman-Rakic Prize for Cognitive Neuroscience, 2621
National Center for Responsible Gaming Grants, 2629
National Center for Responsible Gaming Postdoctoral Fellowships, 2630
National Center for Resp Gaming Seed Grants, 2631
NCHS National Center for Health Statistics Postdoctoral Research Appointments, 2642
NIA Factors Affecting Cognitive Function in Adults with Down Syndrome Grants, 2758
NIA Higher-Order Cognitive Functioning and Aging Grants, 2763
NLM Research Project Grants, 2951
NSF Perception, Action & Cognition Grants, 3019
Peter and Elizabeth C. Tower Foundation Phase II Technology Initiative Grants, 3149
Peter and Elizabeth C. Tower Foundation Phase I Technology Initiative Grants, 3150
SfN Nemko Prize in Cellular or Molecular Neuroscience, 3529

Colitis
Broad Foundation IBD Research Grants, 964
Partnership for Cures Two Years To Cures Grants, 3118

Collaboration
AES Research Infrastructure Awards, 321
AES Research Initiative Awards, 322
AIHS Collaborative Research and Innovation Grants - Collaborative Program, 376
AIHS Collaborative Research and Innovation Grants - Collaborative Project, 377
AIHS Collaborative Research and Innovation Grants - Collaborative Team, 378
California Endowment Innovative Ideas Challenge Grants, 1000
CNCS Social Innovation Grants, 1192
GNOF IMPACT Grants for Health and Human Services, 1839
GNOF Norco Community Grants, 1845
GNOF Organizational Effectiveness Grants and Workshops, 1846
GNOF Stand Up For Our Children Grants, 1847
Helen Bader Foundation Grants, 1992
IBRO-PERC InEurope Short Stay Grants, 2097
IBRO Latin America Regional Funding for PROLAB Collaborations, 2106
Maine Community Foundation Hospice Grants, 2416
MBL Albert & Ellen Grass Fellowships, 2463
MBL Gruss Lipper Family Foundation Summer Fellowship, 2475
MLA Section Project of the Year Award, 2574
SSHRC-NSERC College and Community Innovation Enhancement Grants, 3636

College Students
Alcohol Misuse and Alcoholism Research Grants, 415
Community Foundation of Greater Fort Wayne Scholarships, 1276
IIE Adell and Hancock Scholarships, 2123
IIE Freeman Foundation Indonesia Internships, 2131
Tommy Hilfiger Corporate Foundation Grants, 3737

College-Preparatory Education
3M Company Fndn Community Giving Grants, 6
AIHS Heritage Youth Researcher Summer Science Program, 387
Florida High School/High Tech Project Grants, 1680
Mary Wilmer Covey Charitable Trust Grants, 2448

Colombia
Fulbright Traditional Scholar Program in the Western Hemisphere, 1760

Communications
AAFCS International Graduate Fellowships, 129
AAFCS National Graduate Fellowships, 130
AAFCS National Undergraduate Scholarships, 131
Abell-Hanger Foundation Grants, 207
AMA-RFS and AMA Foundation Medical Student & Resident/Fellow Elective-Medicine & Media, 478
ASHFoundation New Century Scholars Research Grant, 707
ASM Public Communications Award, 744
AVDF Health Care Grants, 811
Canada-U.S. Fulbright New Century Scholars Program Grants, 1008
Canada-U.S. Fulbright Senior Specialists Grants, 1009
Canada Graduate Scholarships (CGS) and NSERC Postgraduate Scholarships (PGS), 1010
Community Foundation for the National Capital Region Community Leadership Grants, 1260
DOL Occupational Safety and Health--Susan Harwood Training Grants, 1484
EPA Children's Health Protection Grants, 1598
Fulbright Alumni Initiatives Awards, 1749
Fulbright Distinguished Chairs Awards, 1750
Fulbright New Century Scholars Grants, 1753
Fulbright Specialists Program Grants, 1754
Fulbright Scholars in Europe and Eurasia, 1755
Fulbright Traditional Scholar Program in Sub-Saharan Africa, 1756
Fulbright Traditional Scholar Program in the East Asia/Pacific Region, 1757
Fulbright Traditional Scholar Program in the Near East and North Africa Region, 1758
Fulbright Traditional Scholar Program in the South and Central Asia Region, 1759
Fulbright Traditional Scholar Program in the Western Hemisphere, 1760

688 / Communications

Harry Frank Guggenheim Foundation Dissertation Fellowships, 1963
Heineman Foundation for Research, Education, Charitable and Scientific Purposes, 1990
Josiah Macy Jr. Foundation Grants, 2277
NCI Centers of Excellence in Cancer Communications Research, 2649
New York University Steinhardt School of Education Fellowships, 2684
NLM Internet Connection for Medical Institutions Grants, 2946
NSF Chemistry Research Experiences for Undergraduates (REU), 3012
RSC Thomas W. Eadie Medal, 3423
Tellabs Foundation Grants, 3706

Communicative Disorders, Hearing
ASHA Advancing Academic-Research Award, 695
ASHA Minority Student Leadership Awards, 696
ASHA Research Mentoring-Pair Travel Award, 698
ASHA Student Research Travel Award, 699
ASHA Students Preparing for Academic & Research Careers (SPARC) Award, 700
ASHFoundation Clinical Research Grants, 701
ASHFoundation Graduate Int'l Scholarship, 702
ASHFoundation Graduate Student Scholarships, 703
ASHFoundation Graduate Student Scholarships for Minority Students, 704
ASHFoundation Graduate Student with a Disability Scholarship, 705
ASHFoundation New Century Scholars Program Doctoral Scholarships, 706
ASHFoundation New Century Scholars Research Grant, 707
ASHFoundation Grant for New Investigators, 708
ASHFoundation Student Research Grants in Audiology, 710
Community Foundation of Jackson County Seymour Noon Lions Club Grant, 1288
NIDCD Mentored Clinical Scientist Research Career Development Award, 2795
Oticon Focus on People Awards, 3098
Reinberger Foundation Grants, 3357
The Ray Charles Foundation Grants, 3716

Communicative Disorders, Speech
ASHA Advancing Academic-Research Award, 695
ASHA Minority Student Leadership Awards, 696
ASHA Research Mentoring-Pair Travel Award, 698
ASHA Student Research Travel Award, 699
ASHA Students Preparing for Academic & Research Careers (SPARC) Award, 700
ASHFoundation Clinical Research Grants, 701
ASHFoundation Graduate Student International Scholarship, 702
ASHFoundation Graduate Student Scholarships, 703
ASHFoundation Graduate Student Scholarships for Minority Students, 704
ASHFoundation Graduate Student with a Disability Scholarship, 705
ASHFoundation New Century Scholars Program Doctoral Scholarships, 706
ASHFoundation New Century Scholars Research Grant, 707
ASHFoundation Grant for New Investigators, 708
ASHFoundation Research Grant in Speech, 709
Community Foundation of Jackson County Seymour Noon Lions Club Grant, 1288
GFWC of Massachusetts Communication Disorder/ Speech Therapy Scholarship, 1812
NIDCD Mentored Clinical Scientist Research Career Development Award, 2795

Community Colleges
GNOF New Orleans Works Grants, 1844
Guy I. Bromley Trust Grants, 1919
Louetta M. Cowden Foundation Grants, 2377
Marion Gardner Jackson Charitable Trust Grants, 2434
Norfolk Southern Foundation Grants, 2982
Reinberger Foundation Grants, 3357

Community Development
2 Depot Square Ipswich Charitable Fndn Grants, 4
3M Fndn Health & Human Services Grants, 7
A/H Foundation Grants, 18
AAP Resident Initiative Fund Grants, 189
Abbot and Dorothy H. Stevens Foundation Grants, 197
Abbott Fund Access to Health Care Grants, 199
Abbott Fund Community Grants, 201
Abernethy Family Foundation Grants, 210
Able To Serve Grants, 212
ACF Foundation Grants, 230
ACF Native American Social and Economic Development Strategies Grants, 231
Ackerman Foundation Grants, 236
ACL Centers for Independent Living Competition Grants, 239
ACL Diversity Community of Practice Grants, 240
Adams Foundation Grants, 301
Adobe Community Investment Grants, 310
AEC Trust Grants, 315
Agnes M. Lindsay Trust Grants, 346
Air Products and Chemicals Grants, 399
Alberto Culver Corporate Contributions Grants, 408
Albert Pick Jr. Fund Grants, 409
Alexander & Baldwin Fnd Mainland Grants, 417
Alexander and Baldwin Foundation Hawaiian and Pacific Island Grants, 418
Alexander Eastman Foundation Grants, 420
Alice Tweed Tuohy Foundation Grants Program, 437
Allyn Foundation Grants, 448
Altman Foundation Health Care Grants, 457
AMA-MSS Chapter Involvement Grants, 473
AMA-MSS Chapter of the Year (COTY) Award, 474
Amador Community Foundation Grants, 481
AMA Foundation Fund for Better Health Grants, 484
AMA Foundation Healthy Communities/Healthy America Grants, 486
AMA Foundation Jack B. McConnell, MD Awards for Excellence in Volunteerism, 487
AMA Foundation Leadership Awards, 490
AMD Corporate Contributions Grants, 500
American Express Community Service Grants, 531
American Foodservice Charitable Trust Grants, 532
American Jewish World Service Grants, 533
Amerigroup Foundation Grants, 565
Amica Companies Foundation Grants, 574
Angels Baseball Foundation Grants, 590
Anheuser-Busch Foundation Grants, 591
Ann and Robert H. Lurie Family Fndn Grants, 593
Ann Arbor Area Community Foundation Grants, 594
Annie Sinclair Knudsen Memorial Fund/Kaua'i Community Grants Program, 597
APS Foundation Grants, 646
Aragona Family Foundation Grants, 648
Aratani Foundation Grants, 649
Arizona Public Service Corporate Giving Program Grants, 658
Arkansas Community Foundation Grants, 660
Arlington Community Foundation Grants, 662
Autauga Area Community Foundation Grants, 808
Autodesk Community Relations Grants, 809
Avista Foundation Economic and Cultural Vitality Grants, 812
Babcock Charitable Trust Grants, 816
Bacon Family Foundation Grants, 817
Bailey Foundation Grants, 818
Ball Brothers Foundation General Grants, 820
BancorpSouth Foundation Grants, 823
Bank of America Fndn Matching Gifts, 825
Barberton Community Foundation Grants, 829
Barnes Group Foundation Grants, 832
Baton Rouge Area Foundation Grants, 838
Battle Creek Community Foundation Grants, 839
Batts Foundation Grants, 840
BCBSM Building Healthy Communities Engaging Elementary Schools and Community Partners Grants, 849
BCBSM Foundation Community Health Matching Grants, 852
BCBSM Fndn Proposal Development Awards, 857

Beazley Foundation Grants, 865
Beckley Area Foundation Grants, 867
Beckman Coulter Foundation Grants, 868
Ben B. Cheney Foundation Grants, 875
Bernard and Audre Rapoport Foundation Health Grants, 882
Bikes Belong Foundation Paul David Clark Bicycling Safety Grants, 890
Biogen Foundation General Donations, 899
Birmingham Foundation Grants, 903
Blackford County Community Foundation Grants, 906
Black River Falls Area Foundation Grants, 909
Blanche and Julian Robertson Family Foundation Grants, 911
Blue Cross Blue Shield of Minnesota Foundation - Healthy Children: Growing Up Healthy Grants, 915
Blue Grass Community Fndn Harrison Grants, 919
Blue Grass Community Foundation Hudson-Ellis Fund Grants, 920
Blumenthal Foundation Grants, 925
Bodenwein Public Benevolent Foundation Grants, 928
Boeing Company Contributions Grants, 930
Bonfils-Stanton Foundation Grants, 932
Bosque Foundation Grants, 935
Boyd Gaming Corporation Contributions Program, 941
BP Foundation Grants, 943
Bradley C. Higgins Foundation Grants, 945
Brian G. Dyson Foundation Grants, 948
Bridgestone/Firestone Trust Fund Grants, 949
Brooklyn Community Foundation Caring Neighbors Grants, 968
Brown Advisory Charitable Foundation Grants, 969
Burlington Industries Foundation Grants, 977
Business Bank of Nevada Community Grants, 981
Caleb C. and Julia W. Dula Educational and Charitable Foundation Grants, 997
California Community Foundation Health Care Grants, 998
California Community Foundation Human Development Grants, 999
California Endowment Innovative Ideas Challenge Grants, 1000
Callaway Foundation Grants, 1002
Cambridge Community Foundation Grants, 1004
Campbell Soup Foundation Grants, 1007
Caring Foundation Grants, 1019
Carl B. and Florence E. King Foundation Grants, 1022
Carl M. Freeman Foundation FACES Grants, 1026
Carl M. Freeman Foundation Grants, 1027
Carnahan-Jackson Foundation Grants, 1032
Carrie E. and Lena V. Glenn Foundation Grants, 1036
Carrie Estelle Doheny Foundation Grants, 1037
Carrier Corporation Contributions Grants, 1038
Carroll County Community Foundation Grants, 1039
Cass County Community Foundation Grants, 1041
Catherine Kennedy Home Foundation Grants, 1043
CCHD Community Development Grants, 1046
Central Okanagan Foundation Grants, 1080
Cessna Foundation Grants Program, 1081
CFFVR Basic Needs Giving Partnership Grants, 1094
CFFVR Capital Credit Union Charitable Giving Grants, 1095
CFFVR Chilton Area Community Fndn Grants, 1096
CFFVR Clintonville Area Foundation Grants, 1097
CFFVR Frank C. Shattuck Community Grants, 1098
CFFVR Myra M. and Robert L. Vandehey Foundation Grants, 1100
CFFVR Project Grants, 1101
CFFVR Robert and Patricia Endries Family Foundation Grants, 1102
CFFVR Schmidt Family G4 Grants, 1103
CFFVR Shawano Area Community Foundation Grants, 1104
CFFVR Waupaca Area Community Foundation Grants, 1105
CFFVR Wisconsin King's Daus & Sons Grants, 1106
CFFVR Women's Fund for the Fox Valley Region Grants, 1107
Champlin Foundations Grants, 1110
Charles Delmar Foundation Grants, 1115

SUBJECT INDEX
Community Development /689

Charles H. Farnsworth Trust Grants, 1119
Charles M. & Mary D. Grant Fndn Grants, 1124
Charlotte County (FL) Community Foundation Grants, 1126
Chemtura Corporation Contributions Grants, 1130
Chilkat Valley Community Foundation Grants, 1148
Christy-Houston Foundation Grants, 1153
Chula Vista Charitable Foundation Grants, 1154
CICF City of Noblesville Community Grant, 1155
Clarence T.C. Ching Foundation Grants, 1167
Clark County Community Foundation Grants, 1175
Claude Worthington Benedum Fndn Grants, 1178
Clinton County Community Foundation Grants, 1184
CNA Foundation Grants, 1189
CNCS AmeriCorps VISTA Project Grants, 1190
CNCS Social Innovation Grants, 1192
Coeta and Donald Barker Foundation Grants, 1205
Coleman Foundation Cancer Care Grants, 1206
Comerica Charitable Foundation Grants, 1224
Communities Foundation of Texas Grants, 1236
Community Foundation Alliance City of Evansville Endowment Fund Grants, 1238
Community Foundation for Greater Atlanta Clayton County Fund Grants, 1240
Community Foundation for Greater Atlanta Common Good Funds Grants, 1241
Community Foundation for Greater Atlanta Managing For Excellence Award, 1242
Community Foundation for Greater Atlanta Metropolitan Atlanta An Extra Wish Grants, 1243
Community Foundation for Greater Atlanta Morgan County Fund Grants, 1244
Community Foundation for Greater Atlanta Newton County Fund Grants, 1245
Community Foundation for Greater Atlanta Strategic Restructuring Fund Grants, 1246
Community Fndn for Greater Buffalo Grants, 1247
Community Foundation for Greater New Haven $5,000 and Under Grants, 1248
Community Foundation for Greater New Haven Responsive New Grants, 1249
Community Foundation for Greater New Haven Sponsorship Grants, 1250
Community Foundation for Greater New Haven Valley Neighborhood Grants, 1251
Community Foundation for Greater New Haven Women & Girls Grants, 1252
Community Fndn for Muskegon County Grants, 1254
Community Fndn for San Benito County Grants, 1257
Community Foundation for the National Capital Region Community Leadership Grants, 1260
Community Fndn of Bloomington & Monroe County - Precision Health Network Cycle Grants, 1264
Community Foundation of Bloomington and Monroe County Grants, 1265
Community Foundation of Boone County Grants, 1266
Community Fndn of East Central Illinois Grants, 1268
Community Foundation of Grant County Grants, 1271
Community Foundation of Greater Birmingham Grants, 1272
Community Foundation of Greater Greensboro Community Grants, 1277
Community Foundation of Greater Greensboro Women to Women Fund Grants, 1278
Community Foundation of Greater New Britain Grants, 1281
Community Fndn of Greater Tampa Grants, 1282
Community Foundation of Greenville-Greenville Women Giving Grants, 1283
Community Foundation of Greenville Community Enrichment Grants, 1284
Community Foundation of Greenville Hollingsworth Funds Program/Project Grants, 1285
Community Fndn of Howard County Grants, 1286
Community Foundation of Louisville Anna Marble Memorial Fund for Princeton Grants, 1290
Community Foundation of Mount Vernon and Knox County Grants, 1299
Community Foundation of Muncie and Delaware County Grants, 1300

Community Foundation of Muncie and Delaware County Maxon Grants, 1301
Community Fndn of Randolph County Grants, 1302
Community Fndn of Riverside & San Bernardino County Irene S. Rockwell Grants, 1304
Community Foundation of St. Joseph County Special Project Challenge Grants, 1310
Community Fndn of Switzerland County Grants, 1311
Community Foundation of Tampa Bay Grants, 1312
Community Foundation of the Eastern Shore Community Needs Grants, 1313
Community Foundation of the Verdugos Educational Endowment Fund Grants, 1314
Community Foundation of the Verdugos Grants, 1315
Community Fndn of Wabash County Grants, 1316
Community Foundation Partnerships - Martin County Grants, 1318
ConAgra Foods Fndn Community Impact Grants, 1321
ConAgra Foods Nourish our Community Grants, 1322
Connelly Foundation Grants, 1326
ConocoPhillips Foundation Grants, 1327
CONSOL Energy Community Dev Grants, 1339
Constantin Foundation Grants, 1341
Constellation Energy Corporate Grants, 1342
Cornerstone Foundation of Northeastern Wisconsin Grants, 1347
Covidien Partnership for Neighborhood Wellness Grants, 1353
Cralle Foundation Grants, 1356
Crane Fund Grants, 1358
Cruise Industry Charitable Foundation Grants, 1363
CSL Behring Local Empowerment for Advocacy Development (LEAD) Grants, 1365
CSRA Community Foundation Grants, 1366
CSX Corporate Contributions Grants, 1368
Cuesta Foundation Grants, 1371
Cumberland Community Foundation Grants, 1374
Curtis Foundation Grants, 1376
CVS All Kids Can Grants, 1377
D.F. Halton Foundation Grants, 1405
D. W. McMillan Foundation Grants, 1407
Dade Community Foundation Community AIDS Partnership Grants, 1409
Dade Community Foundation Grants, 1410
Daisy Marquis Jones Foundation Grants, 1414
Danellie Foundation Grants, 1421
David M. and Marjorie D. Rosenberg Foundation Grants, 1432
Daviess County Community Foundation Health Grants, 1434
Dayton Power and Light Company Foundation Signature Grants, 1436
Daywood Foundation Grants, 1438
Dearborn Community Foundation City of Lawrenceburg Community Grants, 1441
Decatur County Community Foundation Large Project Grants, 1444
Decatur County Community Foundation Small Project Grants, 1445
DeKalb County Community Foundation Grants, 1447
Delaware Community Foundation Grants, 1449
Dept of Ed Safe and Drug-Free Schools and Communities State Grants, 1464
DeRoy Testamentary Foundation Grants, 1469
Dole Food Company Charitable Contributions, 1483
Duneland Health Council Incorporated Grants, 1529
Dyson Foundation Mid-Hudson Valley General Operating Support Grants, 1533
Dyson Foundation Mid-Hudson Valley Project Support Grants, 1534
E. Clayton and Edith P. Gengras, Jr. Foundation, 1536
Eberly Foundation Grants, 1544
Eddie C. & Sylvia Brown Family Fndn Grants, 1548
Edina Realty Foundation Grants, 1550
Edward W. and Stella C. Van Houten Memorial Fund Grants, 1559
Edwin S. Webster Foundation Grants, 1560
Edyth Bush Charitable Foundation Grants, 1562
Elkhart County Community Foundation Fund for Elkhart County, 1572

Elkhart County Community Foundation Grants, 1573
Elliot Foundation Inc Grants, 1575
El Paso Community Foundation Grants, 1579
El Paso Corporate Foundation Grants, 1580
El Pomar Foundation Anna Keesling Ackerman Fund Grants, 1581
El Pomar Foundation Grants, 1582
Elsie H. Wilcox Foundation Grants, 1583
Ensworth Charitable Foundation Grants, 1595
Entergy Corporation Micro Grants, 1596
Entergy Corporation Open Grants for Healthy Families, 1597
Essex County Community Foundation Merrimack Valley General Fund Grants, 1603
Ethel S. Abbott Charitable Foundation Grants, 1607
Ethel Sergeant Clark Smith Foundation Grants, 1608
Eugene Straus Charitable Trust, 1612
Evanston Community Foundation Grants, 1615
Evjue Foundation Grants, 1617
Expect Miracles Foundation Grants, 1619
Fairfield County Community Foundation Grants, 1625
Fargo-Moorhead Area Foundation Grants, 1632
Fayette County Foundation Grants, 1636
Ferree Foundation Grants, 1658
Fidelity Foundation Grants, 1659
Fifth Third Foundation Grants, 1661
Finance Factors Foundation Grants, 1666
FirstEnergy Foundation Community Grants, 1668
Fisher Foundation Grants, 1670
Fishman Family Foundation Grants, 1672
FIU Global Civic Engagement Mini Grants, 1673
Flextronics Foundation Disaster Relief Grants, 1674
Florida BRAIVE Fund of Dade Community Foundation, 1678
FMC Foundation Grants, 1683
Foellinger Foundation Grants, 1684
Fondren Foundation Grants, 1685
Ford Motor Company Fund Grants Program, 1687
Four County Community Fndn General Grants, 1710
Four County Community Foundation Healthy Senior/ Healthy Youth Fund Grants, 1711
Fourjay Foundation Grants, 1712
France-Merrick Foundation Health and Human Services Grants, 1715
Frances L. and Edwin L. Cummings Memorial Fund Grants, 1717
Frances W. Emerson Foundation Grants, 1718
Franklin County Community Foundation Grants, 1724
Franklin H. Wells and Ruth L. Wells Foundation Grants, 1725
Frank Stanley Beveridge Foundation Grants, 1730
Fred Baldwin Memorial Foundation Grants, 1736
Freddie Mac Foundation Grants, 1738
Frederick McDonald Trust Grants, 1740
Fremont Area Community Foundation Amazing X Grants, 1744
Fremont Area Community Foundation Elderly Needs Grants, 1745
Fremont Area Community Fndn General Grants, 1746
Gamble Foundation Grants, 1769
Gardner Foundation Grants, 1772
Gebbie Foundation Grants, 1777
Genentech Corp Charitable Contributions, 1780
George A. and Grace L. Long Foundation Grants, 1788
George and Sarah Buchanan Foundation Grants, 1790
George A Ohl Jr. Foundation Grants, 1791
George Gund Foundation Grants, 1796
George Kress Foundation Grants, 1799
Georgiana Goddard Eaton Memorial Grants, 1805
Gertrude and William C. Wardlaw Fund Grants, 1808
Gertrude E. Skelly Charitable Foundation Grants, 1810
Gheens Foundation Grants, 1814
Giant Eagle Foundation Grants, 1815
Gibson County Community Foundation Women's Fund Grants, 1817
Ginn Foundation Grants, 1822
Giving Sum Annual Grant, 1825
Gladys Brooks Foundation Grants, 1826
GlaxoSmithKline Corporate Grants, 1829
GNOF Bayou Communities Grants, 1835

690 / Community Development

GNOF Exxon-Mobil Grants, 1836
GNOF IMPACT Grants for Arts and Culture, 1838
GNOF IMPACT Grants for Health and Human Services, 1839
GNOF New Orleans Works Grants, 1844
GNOF Norco Community Grants, 1845
Godfrey Foundation Grants, 1849
Grace and Franklin Bernsen Foundation Grants, 1858
Graco Foundation Grants, 1860
Grand Rapids Area Community Foundation Wyoming Grants, 1867
Grand Rapids Area Community Foundation Wyoming Youth Fund Grants, 1868
Grand Rapids Community Foundation Grants, 1869
Grand Rapids Community Foundation Ionia County Grants, 1870
Grand Rapids Community Foundation Ionia County Youth Fund Grants, 1871
Grand Rapids Community Foundation Lowell Area Fund Grants, 1872
Grand Rapids Community Foundation Southeast Ottawa Grants, 1873
Grand Rapids Community Foundation Southeast Ottawa Youth Fund Grants, 1874
Grand Rapids Community Fndn Sparta Grants, 1875
Grand Rapids Community Foundation Sparta Youth Fund Grants, 1876
Greater Cincinnati Foundation Priority and Small Projects/Capacity-Building Grants, 1880
Greater Green Bay Community Fndn Grants, 1881
Greater Milwaukee Foundation Grants, 1883
Greater Saint Louis Community Fndn Grants, 1884
Greater Sitka Legacy Fund Grants, 1885
Greater Worcester Community Foundation Jeppson Memorial Fund for Brookfield Grants, 1889
Green Diamond Charitable Contributions, 1891
Greene County Foundation Grants, 1892
Green River Area Community Fndn Grants, 1895
Gregory L. Gibson Charitable Fndn Grants, 1900
Grover Hermann Foundation Grants, 1906
Grundy Foundation Grants, 1912
Guido A. & Elizabeth H. Binda Fndn Grants, 1913
Gulf Coast Community Foundation Grants, 1915
Gulf Coast Foundation of Community Operating Grants, 1916
Gulf Coast Foundation of Community Program Grants, 1917
Guy I. Bromley Trust Grants, 1919
H & R Foundation Grants, 1920
H.B. Fuller Foundation Grants, 1922
H. Leslie Hoffman and Elaine S. Hoffman Foundation Grants, 1924
HAF Technical Assistance Program Grants, 1933
Hagedorn Fund Grants, 1934
Hannaford Charitable Foundation Grants, 1944
Hardin County Community Foundation Grants, 1946
Harley Davidson Foundation Grants, 1947
Harold Alfond Foundation Grants, 1949
Harold and Arlene Schnitzer CARE Foundation Grants, 1950
Harold R. Bechtel Charitable Remainder Uni-Trust Grants, 1953
Harold R. Bechtel Testamentary Charitable Trust Grants, 1954
Harris and Eliza Kempner Fund Grants, 1956
Harrison County Community Foundation Signature Grants, 1958
Harry B. and Jane H. Brock Foundation Grants, 1960
Harry Bramhall Gilbert Charitable Trust Grants, 1961
Harry Kramer Memorial Fund Grants, 1965
Harry S. Black and Allon Fuller Fund Grants, 1966
Harry W. Bass, Jr. Foundation Grants, 1968
Hartford Courant Foundation Grants, 1969
Hartford Foundation Regular Grants, 1970
Harvest Foundation Grants, 1971
Hawaiian Electric Industries Charitable Foundation Grants, 1975
Hawaii Community Foundation Reverend Takie Okumura Family Grants, 1977
Hawaii Community Foundation West Grants, 1978

Health Foundation of Greater Cincinnati Grants, 1983
Hedco Foundation Grants, 1989
Helen Bader Foundation Grants, 1992
Helen S. Boylan Foundation Grants, 1997
Hendricks County Community Fndn Grants, 2000
Henrietta Tower Wurts Memorial Fndn Grants, 2002
Henry A. and Mary J. MacDonald Foundation, 2003
Henry County Community Foundation Grants, 2005
Herbert H. & Grace A. Dow Fndn Grants, 2008
Hershey Company Grants, 2013
Highmark Corporate Giving Grants, 2025
High Meadow Foundation Grants, 2027
Hill Crest Foundation Grants, 2030
Hillman Foundation Grants, 2032
Hillsdale County Community General Adult Foundation Grants, 2034
Hilton Head Island Foundation Grants, 2036
Hoglund Foundation Grants, 2040
Holland/Zeeland Community Fndn Grants, 2041
Homer Foundation Grants, 2044
Honda of America Manufacturing Fndn Grants, 2045
Horace A. Kimball and S. Ella Kimball Foundation Grants, 2046
Horizon Foundation for New Jersey Grants, 2048
Hormel Foundation Grants, 2051
Houston Endowment Grants, 2053
Howard and Bush Foundation Grants, 2054
Howe Foundation of North Carolina Grants, 2055
Hudson Webber Foundation Grants, 2072
Hugh J. Andersen Foundation Grants, 2074
Humana Foundation Grants, 2077
Huntington Clinical Foundation Grants, 2082
Huntington County Community Foundation Make a Difference Grants, 2083
Hutchinson Community Foundation Grants, 2085
Hut Foundation Grants, 2086
Hutton Foundation Grants, 2087
I.A. O'Shaughnessy Foundation Grants, 2088
Idaho Community Foundation Eastern Region Competitive Grants, 2114
Idaho Power Company Corporate Contributions, 2115
IIE Eurobank EFG Scholarships, 2130
Ike and Roz Friedman Foundation Grants, 2145
Illinois Tool Works Foundation Grants, 2147
Impact 100 Grants, 2148
Inasmuch Foundation Grants, 2149
Independence Blue Cross Charitable Medical Care Grants, 2150
Independence Community Foundation Community Quality of Life Grant, 2153
Intergrys Corporation Grants, 2160
Ireland Family Foundation Grants, 2162
Irving S. Gilmore Foundation Grants, 2163
Irvin Stern Foundation Grants, 2164
Isabel Allende Foundation Esperanza Grants, 2165
J.B. Reynolds Foundation Grants, 2168
J.C. Penney Company Grants, 2169
J.E. and L.E. Mabee Foundation Grants, 2170
J.L. Bedsole Foundation Grants, 2172
J.W. Kieckhefer Foundation Grants, 2176
Jackson County Community Foundation Unrestricted Grants, 2180
Jacob G. Schmidlapp Trust Grants, 2183
Jacobs Family Village Neighborhoods Grants, 2184
James & Abigail Campbell Family Fndn Grants, 2185
James A. and Faith Knight Foundation Grants, 2186
James Ford Bell Foundation Grants, 2187
James Graham Brown Foundation Grants, 2189
James J. & Angelia M. Harris Fndn Grants, 2192
James R. Dougherty Jr. Foundation Grants, 2198
James S. Copley Foundation Grants, 2200
Janirve Foundation Grants, 2209
Janus Foundation Grants, 2211
Jasper Foundation Grants, 2212
JELD-WEN Foundation Grants, 2218
Jennings County Community Foundation Grants, 2220
Jessica Stevens Community Foundation Grants, 2224
Joe W. and Dorothy Dorsett Brown Fndn Grants, 2230
John Ben Snow Memorial Trust Grants, 2231
John Deere Foundation Grants, 2234

John G. Duncan Charitable Trust Grants, 2236
John H. Wellons Foundation Grants, 2239
John Jewett and Helen Chandler Garland Foundation Grants, 2242
John P. McGovern Foundation Grants, 2247
John S. Dunn Research Foundation Grants, 2250
Johnson County Community Foundation Grants, 2256
John W. Alden Trust Grants, 2260
John W. Anderson Foundation Grants, 2262
John W. Speas and Effie E. Speas Memorial Trust Grants, 2264
Joseph Henry Edmondson Foundation Grants, 2270
Josephine G. Russell Trust Grants, 2271
Josephine Goodyear Foundation Grants, 2272
Josephine Schell Russell Charitable Trust Grants, 2274
Judith Clark-Morrill Foundation Grants, 2281
K21 Health Foundation Cancer Care Grants, 2284
K21 Health Foundation Grants, 2285
Kahuku Community Fund, 2286
Katharine Matthies Foundation Grants, 2291
Kenai Peninsula Foundation Grants, 2298
Kent D. Steadley and Mary L. Steadley Memorial Trust Grants, 2301
Ketchikan Community Foundation Grants, 2303
Knox County Community Foundation Grants, 2313
Kodiak Community Foundation Grants, 2314
Kosciusko County Community Fndn Grants, 2319
Kosciusko County Community Foundation REMC Operation Round Up Grants, 2320
Kuntz Foundation Grants, 2326
L. W. Pierce Family Foundation Grants, 2328
Lake County Community Fund Grants, 2333
Lands' End Corporate Giving Program, 2344
Legler Benbough Foundation Grants, 2352
Leo Niessen Jr., Charitable Trust Grants, 2358
Lewis H. Humphreys Charitable Trust Grants, 2361
Liberty Bank Foundation Grants, 2362
Lillian S. Wells Foundation Grants, 2365
Long Island Community Foundation Grants, 2375
Lotus 88 Fnd for Women & Children Grants, 2376
Louetta M. Cowden Foundation Grants, 2377
Louie M. & Betty M. Phillips Fndn Grants, 2378
Louis and Elizabeth Nave Flarsheim Charitable Foundation Grants, 2379
Lowell Berry Foundation Grants, 2382
Lubbock Area Foundation Grants, 2383
Lucy Downing Nisbet Charitable Fund Grants, 2387
Lucy Gooding Charitable Fndn Grants, 2388
M.B. and Edna Zale Foundation Grants, 2401
Mabel F. Hoffman Charitable Trust Grants, 2407
Mabel Y. Hughes Charitable Trust Grants, 2409
Madison County Community Foundation General Grants, 2413
Marathon Petroleum Corporation Grants, 2422
Marie C. and Joseph C. Wilson Foundation Rochester Small Grants, 2431
Marie H. Bechtel Charitable Remainder Uni-Trust Grants, 2432
Marion Gardner Jackson Charitable Trust Grants, 2434
Marshall County Community Foundation Grants, 2439
Massage Therapy Foundation Community Service Grants, 2449
Maxon Charitable Foundation Grants, 2458
McCallum Family Foundation Grants, 2489
McConnell Foundation Grants, 2492
McMillen Foundation Grants, 2499
Mead Johnson Nutritionals Evansville-Area Organizations Grants, 2503
MeadWestvaco Foundation Sustainable Communities Grants, 2506
Merrick Foundation Grants, 2522
Merrick Foundation Grants, 2521
Mervin Bovaird Foundation Grants, 2523
Metzger-Price Fund Grants, 2530
Meyer Fndn Management Assistance Grants, 2534
Meyer Memorial Trust Responsive Grants, 2537
Meyer Memorial Trust Special Grants, 2538
MGM Resorts Foundation Community Grants, 2541
Miami County Community Foundation - Operation Round Up Grants, 2543

SUBJECT INDEX

Community Development /691

Mid-Iowa Health Foundation Community Response Grants, 2549
Middlesex Savings Charitable Foundation Basic Human Needs Grants, 2550
Montgomery County Community Fndn Grants, 2591
Morris & Gwendolyn Cafritz Fndn Grants, 2594
Natalie W. Furniss Charitable Trust Grants, 2626
Nelda C. and H.J. Lutcher Stark Fndn Grants, 2672
Newton County Community Foundation Grants, 2681
Noble County Community Foundation - Delta Theta Tau Sorority IOTA IOTA Chapter Riecke Scholarship, 2971
Noble County Community Foundation Grants, 2975
Norfolk Southern Foundation Grants, 2982
North Carolina Biotech Center Meeting Grants, 2986
North Carolina Biotechnology Center Regional Development Grants, 2987
North Carolina Community Foundation Grants, 2990
North Central Health Services Grants, 2992
North Central Health Services Grants, 2993
Northern Chautauqua Community Foundation Community Grants, 2995
Northern New York Community Fndn Grants, 2996
Northwestern Mutual Foundation Grants, 2999
Northwest Minnesota Foundation Women's Fund Grants, 3000
Norwin S. & Elizabeth N. Bean Fndn Grants, 3001
NWHF Community-Based Participatory Research Grants, 3032
NWHF Kaiser Permanente Community Grants, 3034
NYCT Girls and Young Women Grants, 3053
NYCT Grants for the Elderly, 3054
NYCT Technical Assistance Grants, 3058
OceanFirst Foundation Major Grants, 3062
Oceanside Charitable Foundation Grants, 3063
Office Depot Corporation Community Relations Grants, 3065
Ohio County Community Foundation Board of Directors Grants, 3067
Ohio County Community Foundation Grants, 3068
Ohio County Community Fndn Mini-Grants, 3069
Oleonda Jameson Trust Grants, 3071
OneStar Foundation AmeriCorps Grants, 3077
Oregon Community Fndn Community Grants, 3082
Oscar Rennebohm Foundation Grants, 3085
Otto Bremer Foundation Grants, 3099
Owen County Community Foundation Grants, 3100
Packard Foundation Local Grants, 3105
Parke County Community Foundation Grants, 3115
Parkersburg Area Community Foundation Action Grants, 3116
Patrick and Anna M. Cudahy Fund Grants, 3120
Paul Ogle Foundation Grants, 3124
Peacock Foundation Grants, 3133
Percy B. Ferebee Endowment Grants, 3138
Perkins Charitable Foundation Grants, 3140
Perpetual Trust for Charitable Giving Grants, 3141
Perry County Community Foundation Grants, 3142
Pet Care Trust Sue Busch Memorial Award, 3143
Peter and Elizabeth C. Tower Foundation Organizational Scholarships, 3148
Peter and Elizabeth C. Tower Foundation Social and Emotional Preschool Curriculum Grants, 3151
Peter Kiewit Foundation General Grants, 3154
Peter Kiewit Foundation Small Grants, 3155
Petersburg Community Foundation Grants, 3156
Phelps County Community Foundation Grants, 3188
Philadelphia Foundation Organizational Effectiveness Grants, 3189
Piedmont Health Foundation Grants, 3219
Piedmont Natural Gas Corporate and Charitable Contributions, 3220
Pike County Community Foundation Grants, 3222
Pinellas County Grants, 3223
Piper Trust Older Adults Grants, 3229
Pittsburgh Foundation Community Fund Grants, 3230
Plough Foundation Grants, 3238
PNC Foundation Community Services Grants, 3240
Polk Bros. Foundation Grants, 3243
Portland General Electric Foundation Grants, 3250

Posey Community Fndn Women's Fund Grants, 3251
Posey County Community Foundation Grants, 3252
Pott Foundation Grants, 3253
PPCF Community Grants, 3256
PPG Industries Foundation Grants, 3257
Price Family Charitable Fund Grants, 3262
Priddy Foundation Operating Grants, 3264
Priddy Foundation Organizational Development Grants, 3265
Priddy Foundation Program Grants, 3266
Prince Charitable Trusts DC Grants, 3271
Princeton Area Community Foundation Greater Mercer Grants, 3272
Puerto Rico Community Foundation Grants, 3278
Pulaski County Community Foundation Grants, 3279
Putnam County Community Foundation Grants, 3283
Quaker Chemical Foundation Grants, 3286
Quantum Foundation Grants, 3290
Raskob Foundation for Catholic Activities Grants, 3306
Rasmuson Foundation Tier One Grants, 3307
Rasmuson Foundation Tier Two Grants, 3308
Rayonier Foundation Grants, 3311
RCF General Community Grants, 3312
RCF Individual Assistance Grants, 3313
Regence Fndn Access to Health Care Grants, 3346
Regence Foundation Health Care Community Awareness and Engagement Grants, 3347
Regence Fndn Health Care Connections Grants, 3348
Regence Fndn Improving End-of-Life Grants, 3349
Regence Fndn Tools & Technology Grants, 3350
Retirement Research Foundation General Program Grants, 3359
Reynolds & Reynolds Associate Fndn Grants, 3360
Rhode Island Foundation Grants, 3362
Richard D. Bass Foundation Grants, 3366
Richard M. Fairbanks Foundation Grants, 3370
Richland County Bank Grants, 3371
Ripley County Community Foundation Grants, 3376
Ripley County Community Foundation Small Project Grants, 3377
Robbins Charitable Foundation Grants, 3378
Robert and Joan Dircks Foundation Grants, 3379
Robert B McMillen Foundation Grants, 3381
Robert R. Meyer Foundation Grants, 3387
Rochester Area Community Foundation Grants, 3391
Rochester Area Foundation Grants, 3392
Rockefeller Foundation Grants, 3396
Rockwell Fund, Inc. Grants, 3398
Rockwell International Corporate Trust Grants Program, 3399
Rush County Community Foundation Grants, 3424
Ruth Anderson Foundation Grants, 3425
Ruth H. & Warren A. Ellsworth Fndn Grants, 3429
RWJF Vulnerable Populations Portfolio Grants, 3444
S. Livingston Mather Charitable Trust Grants, 3448
Saginaw Community Foundation Senior Citizen Enrichment Fund, 3450
Samuel S. Johnson Foundation Grants, 3468
San Diego Foundation Paradise Valley Hospital Community Fund Grants, 3472
San Diego Women's Foundation Grants, 3474
Sarkeys Foundation Grants, 3485
Sartain Lanier Family Foundation Grants, 3486
Schering-Plough Foundation Community Initiatives Grants, 3491
Schramm Foundation Grants, 3495
Schurz Communications Foundation Grants, 3496
Scott County Community Foundation Grants, 3500
Seattle Foundation Benjamin N. Phillips Memorial Fund Grants, 3504
Seattle Foundation Health and Wellness Grants, 3507
Self Foundation Grants, 3510
Sensient Technologies Foundation Grants, 3513
Seward Community Foundation Grants, 3514
Seward Community Foundation Mini-Grants, 3515
Shaw's Supermarkets Donations, 3542
Shell Deer Park Grants, 3543
Shell Oil Company Foundation Community Development Grants, 3544
Shield-Ayres Foundation Grants, 3546

Shopko Foundation Green Bay Area Community Grants, 3548
Sidgmore Family Foundation Grants, 3551
Sidney Stern Memorial Trust Grants, 3552
Siebert Lutheran Foundation Grants, 3554
Sierra Health Foundation Responsive Grants, 3555
Simple Advise Education Center Grants, 3565
Simpson Lumber Charitable Contributions, 3566
Sioux Falls Area Community Foundation Community Fund Grants (Unrestricted), 3568
Sioux Falls Area Community Foundation Spot Grants (Unrestricted), 3570
Sir Dorabji Tata Trust Grants for NGOs or Voluntary Organizations, 3572
Sisters of Charity Foundation of Cleveland Reducing Health and Educational Disparities in the Central Neighborhood Grants, 3574
Skillman Fndn Good Neighborhoods Grants, 3577
SOCFOC Catholic Ministries Grants, 3596
Sonoco Foundation Grants, 3621
Southbury Community Trust Fund, 3625
South Madison Community Foundation Grants, 3626
Southwest Florida Community Foundation Competitive Grants, 3627
Southwest Gas Corporation Foundation Grants, 3628
Spencer County Community Foundation Grants, 3632
Springs Close Foundation Grants, 3633
St. Joseph Community Health Foundation Improving Healthcare Access Grants, 3640
Stackpole-Hall Foundation Grants, 3643
Starke County Community Foundation Grants, 3646
Sterling and Shelli Gardner Foundation Grants, 3653
Steuben County Community Foundation Grants, 3654
Stocker Foundation Grants, 3656
Stranahan Foundation Grants, 3659
Strowd Roses Grants, 3664
Summit Foundation Grants, 3674
Sunderland Foundation Grants, 3675
SunTrust Bank Trusteed Foundations Walter H. and Marjory M. Rich Memorial Fund Grants, 3685
Susan Mott Webb Charitable Trust Grants, 3694
Susan Vaughan Foundation Grants, 3695
Taylor S. Abernathy and Patti Harding Abernathy Charitable Trust Grants, 3704
TE Foundation Grants, 3705
Tension Envelope Foundation Grants, 3707
Thelma Braun & Bocklett Family Fndn Grants, 3713
Theodore Edson Parker Foundation Grants, 3715
Thomas and Dorothy Leavey Foundation Grants, 3717
Thomas Austin Finch, Sr. Foundation Grants, 3718
Thomas C. Ackerman Foundation Grants, 3719
Thompson Charitable Foundation Grants, 3727
Toyota Motor Engineering & Manufacturing North America Grants, 3742
Toyota Motor Manufacturing of Alabama Grants, 3743
Toyota Motor Manufacturing of Indiana Grants, 3744
Toyota Motor Manufacturing of Kentucky Grants, 3745
Toyota Motor Manuf of Mississippi Grants, 3746
Toyota Motor Manuf of West Virginia Grants, 3748
Toyota Motor N America of New York Grants, 3749
Toyota Motor Sales, USA Grants, 3750
Triangle Community Foundation Donor-Advised Grants, 3757
Trull Foundation Grants, 3761
Turner Foundation Grants, 3763
U.S. Department of Education 21st Century Community Learning Centers, 3767
Union Bank, N.A. Corporate Sponsorships and Donations, 3787
Union Bank, N.A. Foundation Grants, 3788
Union Benevolent Association Grants, 3789
Union County Community Foundation Grants, 3790
Unity Foundation Of LaPorte County Grants, 3804
USCM HIV/AIDS Prevention Grants, 3827
Vancouver Foundation Grants and Community Initiatives Program, 3849
Vancouver Sun Children's Fund Grants, 3851
Vanderburgh Community Foundation Grants, 3852
Vermont Community Foundation Grants, 3865
W. C. Griffith Foundation Grants, 3882

692 / Community Development

W.C. Griffith Foundation Grants, 3881
W.K. Kellogg Foundation Secure Families Grants, 3887
Wabash Valley Community Foundation Grants, 3893
Walker Area Community Foundation Grants, 3895
Wallace Alexander Gerbode Foundation Grants, 3896
Walmart Foundation Community Giving Grants, 3897
Walmart Foundation National Giving Grants, 3898
Walmart Foundation Northwest Arkansas Giving, 3899
Walmart Foundation State Giving Grants, 3900
Walter L. Gross III Family Foundation Grants, 3901
Warrick County Community Foundation Grants, 3904
Wayne County Foundation Grants, 3909
Wells County Foundation Grants, 3920
Western Indiana Community Foundation Grants, 3921
Western New York Foundation Grants, 3922
Weyerhaeuser Company Foundation Grants, 3925
White County Community Foundation - Women Giving Together Grants, 3933
Whitley County Community Foundation Grants, 3936
Willary Foundation Grants, 3944
William A. Badger Foundation Grants, 3945
William Blair and Company Foundation Grants, 3948
William G. & Helen C. Hoffman Fndn Grants, 3950
William G. Gilmore Foundation Grants, 3951
Wilson-Wood Foundation Grants, 3960
Winston-Salem Fndn Elkin/Tri-County Grants, 3966
Yampa Valley Community Foundation Grants, 3978

Community Education
ACF Native American Social and Economic Development Strategies Grants, 231
Autauga Area Community Foundation Grants, 808
Avista Foundation Economic and Cultural Vitality Grants, 812
Beckman Coulter Foundation Grants, 868
Ben B. Cheney Foundation Grants, 875
Blackford County Community Foundation Grants, 906
Blue Cross Blue Shield of Minnesota Foundation - Health Equity: Building Health Equity Together Grants, 914
Blue Cross Blue Shield of Minnesota Foundation - Healthy Equity: Health Impact Assessment Demonstration Project Grants, 916
BP Foundation Grants, 943
Bright Promises Foundation Grants, 952
Brown Advisory Charitable Foundation Grants, 969
California Endowment Innovative Ideas Challenge Grants, 1000
Charles H. Farnsworth Trust Grants, 1119
CJ Foundation for SIDS Program Services Grants, 1163
Community Foundation for Greater Atlanta Clayton County Fund Grants, 1240
Community Foundation for Greater Atlanta Morgan County Fund Grants, 1244
Community Foundation for Greater Atlanta Newton County Fund Grants, 1245
Community Fndn of Bloomington & Monroe County - Precision Health Network Cycle Grants, 1264
Community Fndn of Riverside & San Bernardino County Irene S. Rockwell Grants, 1304
Crane Fund Grants, 1358
E. Clayton and Edith P. Gengras, Jr. Foundation, 1536
Four County Community Fndn General Grants, 1710
Four County Community Foundation Healthy Senior/Healthy Youth Fund Grants, 1711
Guy I. Bromley Trust Grants, 1919
Helen Bader Foundation Grants, 1992
Kansas Health Fndn Major Initiatives Grants, 2288
Kenai Peninsula Foundation Grants, 2298
Ketchikan Community Foundation Grants, 2303
Louetta M. Cowden Foundation Grants, 2377
Lucy Downing Nisbet Charitable Fund Grants, 2387
Marion Gardner Jackson Charitable Trust Grants, 2434
Michael Reese Health Trust Core Grants, 2544
Michael Reese Health Trust Responsive Grants, 2545
MMS and Alliance Charitable Foundation Grants for Community Action and Care for the Medically Uninsured, 2587
Mt. Sinai Health Care Foundation Health of the Urban Community Grants, 2600

Norfolk Southern Foundation Grants, 2982
North Central Health Services Grants, 2992
OceanFirst Foundation Major Grants, 3062
Ohio County Community Foundation Board of Directors Grants, 3067
Peter and Elizabeth C. Tower Foundation Social and Emotional Preschool Curriculum Grants, 3151
Peter Kiewit Foundation Small Grants, 3155
Portland Fndn Women's Giving Circle Grant, 3249
PPG Industries Foundation Grants, 3257
Rasmuson Foundation Tier Two Grants, 3308
RCF General Community Grants, 3312
RCF Individual Assistance Grants, 3313
Richland County Bank Grants, 3371
Seattle Foundation Benjamin N. Phillips Memorial Fund Grants, 3504
Seward Community Foundation Grants, 3514
Seward Community Foundation Mini-Grants, 3515
SfN Science Journalism Student Awards, 3536
Shield-Ayres Foundation Grants, 3546
Walker Area Community Foundation Grants, 3895
Walmart Foundation Community Giving Grants, 3897
Walmart Foundation National Giving Grants, 3898
Walmart Foundation Northwest Arkansas Giving, 3899
Walmart Foundation State Giving Grants, 3900

Community Outreach
3M Company Fndn Community Giving Grants, 6
AAAAI RSLAAIS Leadership Award, 33
Able To Serve Grants, 212
ACL Medicare Improvements for Patients & Providers Funding for Beneficiary Outreach & Assistance for Title VI Native Americans, 245
ADA Foundation Thomas J. Zwemer Award, 297
Alliant Energy Corporation Contributions, 445
Allstate Corporate Giving Grants, 446
Allstate Corp Hometown Commitment Grants, 447
American Express Community Service Grants, 531
American Foodservice Charitable Trust Grants, 532
Angels Baseball Foundation Grants, 590
Antone & Edene Vidinha Charitable Trust Grants, 609
Aragona Family Foundation Grants, 648
ASU Graduate College Science Foundation Arizona Bisgrove Postdoctoral Scholars, 792
Autauga Area Community Foundation Grants, 808
Avista Foundation Economic and Cultural Vitality Grants, 812
BancorpSouth Foundation Grants, 823
Batchelor Foundation Grants, 837
Beckman Coulter Foundation Grants, 868
Ben B. Cheney Foundation Grants, 875
Bernard and Audre Rapoport Foundation Health Grants, 882
Blue Cross Blue Shield of Minnesota Fndn Healthy Equity: Public Libraries for Health Grants, 918
Boston Foundation Grants, 936
Bright Promises Foundation Grants, 952
Brookdale Fnd National Group Respite Grants, 966
Cable Positive's Tony Cox Community Grants, 992
California Community Foundation Human Development Grants, 999
California Endowment Innovative Ideas Challenge Grants, 1000
Cambridge Community Foundation Grants, 1004
Campbell Soup Foundation Grants, 1007
Caring Foundation Grants, 1019
Carl B. and Florence E. King Foundation Grants, 1022
Carrie E. and Lena V. Glenn Foundation Grants, 1036
Carrier Corporation Contributions Grants, 1038
Catherine Kennedy Home Foundation Grants, 1043
CCHD Community Development Grants, 1046
Cessna Foundation Grants Program, 1081
CFFVR Alcoholism and Drug Abuse Grants, 1093
CFFVR Basic Needs Giving Partnership Grants, 1094
CFFVR Capital Credit Union Charitable Giving Grants, 1095
CFFVR Chilton Area Community Fndn Grants, 1096
CFFVR Clintonville Area Foundation Grants, 1097
CFFVR Frank C. Shattuck Community Grants, 1098
CFFVR Schmidt Family G4 Grants, 1103

SUBJECT INDEX

CFFVR Shawano Area Community Foundation Grants, 1104
CFFVR Waupaca Area Community Foundation Grants, 1105
CFFVR Wisconsin King's Daus & Sons Grants, 1106
CFFVR Women's Fund for the Fox Valley Region Grants, 1107
Charles Delmar Foundation Grants, 1115
CICF City of Noblesville Community Grant, 1155
Clark-Winchcole Foundation Grants, 1173
Colonel Stanley R. McNeil Foundation Grants, 1211
Columbus Foundation Traditional Grants, 1223
Community Foundation for Greater New Haven Women & Girls Grants, 1252
Community Fndn for San Benito County Grants, 1257
Community Fndn of Bloomington & Monroe County - Precision Health Network Cycle Grants, 1264
Community Foundation of Greater New Britain Grants, 1281
Community Foundation of Greenville Community Enrichment Grants, 1284
Community Foundation of Greenville Hollingsworth Funds Program/Project Grants, 1285
Community Fndn of Randolph County Grants, 1302
Community Fndn of Switzerland County Grants, 1311
ConAgra Foods Nourish our Community Grants, 1322
CONSOL Energy Community Dev Grants, 1339
Cowles Charitable Trust Grants, 1354
Cralle Foundation Grants, 1356
Crane Fund Grants, 1358
Curtis Foundation Grants, 1376
D. W. McMillan Foundation Grants, 1407
Dade Community Foundation Community AIDS Partnership Grants, 1409
Danellie Foundation Grants, 1421
Delaware Community Foundation-Youth Philanthropy Board for Kent County, 1448
E. Clayton and Edith P. Gengras, Jr. Foundation, 1536
Edina Realty Foundation Grants, 1550
Edward W. and Stella C. Van Houten Memorial Fund Grants, 1559
Eisner Foundation Grants, 1564
Elkhart County Community Foundation Fund for Elkhart County, 1572
El Pomar Foundation Grants, 1582
Elsie H. Wilcox Foundation Grants, 1583
EPA Children's Health Protection Grants, 1598
Ethel Sergeant Clark Smith Foundation Grants, 1608
Evan and Susan Bayh Foundation Grants, 1614
FAR Fund Grants, 1631
Fargo-Moorhead Area Fndn Woman's Grants, 1633
Florida BRAIVE Fund of Dade Community Foundation, 1678
Foundation for the Mid South Community Development Grants, 1699
Four County Community Fndn General Grants, 1710
Four County Community Foundation Healthy Senior/Healthy Youth Fund Grants, 1711
Frederick W. Marzahl Memorial Fund Grants, 1741
Fremont Area Community Foundation Amazing X Grants, 1744
Fremont Area Community Foundation Elderly Needs Grants, 1745
George and Sarah Buchanan Foundation Grants, 1790
George A Ohl Jr. Foundation Grants, 1791
George Family Foundation Grants, 1794
Georgiana Goddard Eaton Memorial Grants, 1805
Giant Eagle Foundation Grants, 1815
Giving Sum Annual Grant, 1825
GNOF IMPACT Grants for Health and Human Services, 1839
Grand Rapids Area Community Foundation Wyoming Youth Fund Grants, 1868
Grand Rapids Community Foundation Ionia County Grants, 1870
Grand Rapids Community Foundation Ionia County Youth Fund Grants, 1871
Grand Rapids Community Foundation Southeast Ottawa Youth Fund Grants, 1874

SUBJECT INDEX

Grand Rapids Community Foundation Sparta Youth Fund Grants, 1876
Greater Sitka Legacy Fund Grants, 1885
Gulf Coast Foundation of Community Operating Grants, 1916
Gulf Coast Foundation of Community Program Grants, 1917
Guy I. Bromley Trust Grants, 1919
Hardin County Community Foundation Grants, 1946
Harley Davidson Foundation Grants, 1947
Harry Kramer Memorial Fund Grants, 1965
Hartford Courant Foundation Grants, 1969
HCA Foundation Grants, 1980
Health Foundation of Greater Cincinnati Grants, 1983
Helen Bader Foundation Grants, 1992
Henrietta Tower Wurts Memorial Fndn Grants, 2002
High Meadow Foundation Grants, 2027
Hilton Head Island Foundation Grants, 2036
Horace A. Kimball and S. Ella Kimball Foundation Grants, 2046
Ireland Family Foundation Grants, 2162
James & Abigail Campbell Family Fndn Grants, 2185
Janirve Foundation Grants, 2209
Jennings County Community Foundation Grants, 2220
John H. Wellons Foundation Grants, 2239
Johnson County Community Foundation Grants, 2256
Johnson Foundation Wingspread Conference Support Program, 2257
John W. Anderson Foundation Grants, 2262
John W. Speas and Effie E. Speas Memorial Trust Grants, 2264
Joseph Henry Edmondson Foundation Grants, 2270
Judith Clark-Morrill Foundation Grants, 2281
K21 Health Foundation Cancer Care Grants, 2284
K21 Health Foundation Grants, 2285
Ketchikan Community Foundation Grants, 2303
Kosciusko County Community Fndn Grants, 2319
Kuntz Foundation Grants, 2326
L. W. Pierce Family Foundation Grants, 2328
Leo Niessen Jr., Charitable Trust Grants, 2358
Lewis H. Humphreys Charitable Trust Grants, 2361
Lotus 88 Fnd for Women & Children Grants, 2376
Louetta M. Cowden Foundation Grants, 2377
Louis and Elizabeth Nave Flarsheim Charitable Foundation Grants, 2379
Lucy Downing Nisbet Charitable Fund Grants, 2387
Lucy Gooding Charitable Fndn Grants, 2388
M.B. and Edna Zale Foundation Grants, 2401
Mabel F. Hoffman Charitable Trust Grants, 2407
Madison County Community Foundation General Grants, 2413
Maine Community Foundation Penobscot Valley Health Association Grants, 2417
Marion Gardner Jackson Charitable Trust Grants, 2434
Maxon Charitable Foundation Grants, 2458
Mead Johnson Nutritionals Evansville-Area Organizations Grants, 2503
MedImmune Charitable Grants, 2509
Merrick Foundation Grants, 2522
Michael Reese Health Trust Responsive Grants, 2545
Mid-Iowa Health Foundation Community Response Grants, 2549
Military Ex-Prisoners of War Foundation Grants, 2554
Mt. Sinai Health Care Foundation Health of the Urban Community Grants, 2600
NAPNAP Foundation Wyeth Pediatric Immunization Grant, 2618
NEI Innovative Patient Outreach Programs And Ocular Screening Technologies To Improve Detection Of Diabetic Retinopathy Grants, 2665
Newton County Community Foundation Grants, 2681
NIH Recovery Act Limited Competition: Building Sustainable Community-Linked Infrastructure to Enable Health Science Research, 2894
Norfolk Southern Foundation Grants, 2982
North Central Health Services Grants, 2993
Northwestern Mutual Foundation Grants, 2999
NWHF Community-Based Participatory Research Grants, 3032

Ohio County Community Foundation Board of Directors Grants, 3067
Parke County Community Foundation Grants, 3115
Patrick and Anna M. Cudahy Fund Grants, 3120
Peacock Foundation Grants, 3133
PepsiCo Foundation Grants, 3137
Percy B. Ferebee Endowment Grants, 3138
Perpetual Trust for Charitable Giving Grants, 3141
Perry County Community Foundation Grants, 3142
Petersburg Community Foundation Grants, 3156
Pike County Community Foundation Grants, 3222
Pinellas County Grants, 3223
Piper Trust Older Adults Grants, 3229
Plough Foundation Grants, 3238
Polk Bros. Foundation Grants, 3243
Posey Community Fndn Women's Fund Grants, 3251
Posey County Community Foundation Grants, 3252
Pott Foundation Grants, 3253
Powell Foundation Grants, 3255
PPG Industries Foundation Grants, 3257
Price Family Charitable Fund Grants, 3262
Prince Charitable Trusts DC Grants, 3271
Princeton Area Community Foundation Greater Mercer Grants, 3272
Pulaski County Community Foundation Grants, 3279
Qualcomm Grants, 3287
Rasmuson Foundation Tier One Grants, 3307
Rasmuson Foundation Tier Two Grants, 3308
RCF General Community Grants, 3312
RCF Individual Assistance Grants, 3313
Rhode Island Foundation Grants, 3362
Richard and Helen DeVos Foundation Grants, 3364
Richard D. Bass Foundation Grants, 3366
Rush County Community Foundation Grants, 3424
Sain-Orr and Royak-DeForest Steadman Foundation Grants, 3452
Saint Luke's Foundation Grants, 3454
Saint Luke's Health Initiatives Grants, 3455
San Diego Women's Foundation Grants, 3474
Sarasota Memorial Healthcare Fndn Grants, 3484
Schurz Communications Foundation Grants, 3496
Scott County Community Foundation Grants, 3500
Seward Community Foundation Grants, 3514
Seward Community Foundation Mini-Grants, 3515
SfN Science Educator Award, 3535
SfN Science Journalism Student Awards, 3536
Shield-Ayres Foundation Grants, 3546
Shopko Foundation Green Bay Area Community Grants, 3548
Siebert Lutheran Foundation Grants, 3554
Simple Advise Education Center Grants, 3565
SOCFOC Catholic Ministries Grants, 3596
Sonoco Foundation Grants, 3621
Sony Corporation of America Grants, 3623
Spencer County Community Foundation Grants, 3632
St. Joseph Community Health Foundation Improving Healthcare Access Grants, 3640
Stocker Foundation Grants, 3656
Strowd Roses Grants, 3664
Susan Mott Webb Charitable Trust Grants, 3694
Susan Vaughan Foundation Grants, 3695
Thomas Austin Finch, Sr. Foundation Grants, 3718
Thompson Charitable Foundation Grants, 3727
Toyota Motor N America of New York Grants, 3749
Union Benevolent Association Grants, 3789
Unity Foundation Of LaPorte County Grants, 3804
USCM HIV/AIDS Prevention Grants, 3827
Vancouver Foundation Grants and Community Initiatives Program, 3849
Vancouver Sun Children's Fund Grants, 3851
VHA Health Foundation Grants, 3867
W. C. Griffith Foundation Grants, 3882
Walker Area Community Foundation Grants, 3895
Walmart Foundation Northwest Arkansas Giving, 3899
Walmart Foundation State Giving Grants, 3900
Walter L. Gross III Family Foundation Grants, 3901
Warrick County Community Foundation Grants, 3904
William G. & Helen C. Hoffman Fndn Grants, 3950
William G. Gilmore Foundation Grants, 3951
Wilson-Wood Foundation Grants, 3960

Community Services
2COBS Private Charitable Foundation Grants, 5
3M Fndn Health & Human Services Grants, 7
49ers Foundation Grants, 9
A/H Foundation Grants, 18
AAMC Caring for Community Grants, 156
Abbot and Dorothy H. Stevens Foundation Grants, 197
Abbott Fund Community Grants, 201
Abernethy Family Foundation Grants, 210
Able To Serve Grants, 212
ACF Native American Social and Economic Development Strategies Grants, 231
Ackerman Foundation Grants, 236
ACL Alzheimer's Disease Initiative Grants, 237
ACL Centers for Independent Living Competition Grants, 239
Adelaide Breed Bayrd Foundation Grants, 306
Agnes M. Lindsay Trust Grants, 346
Agway Foundation Grants, 348
Aid for Starving Children Emergency Assistance Fund Grants, 367
AIHS Full-Time Health Research Studentships, 384
Alliant Energy Corporation Contributions, 445
Allyn Foundation Grants, 448
AMA Foundation Fund for Better Health Grants, 484
American Express Community Service Grants, 531
American Psychiatric Foundation Helping Hands Grants, 554
Amerisure Insurance Community Service Grants, 566
Amica Insurance Company Community Grants, 575
Andersen Corporate Foundation, 586
Angels Baseball Foundation Grants, 590
Anheuser-Busch Foundation Grants, 591
Ann and Robert H. Lurie Family Fndn Grants, 593
Anthem Blue Cross and Blue Shield Grants, 607
Aragona Family Foundation Grants, 648
Ashland Corporate Contributions Grants, 713
AT&T Foundation Civic and Community Service Program Grants, 793
Austin College Leadership Award, 804
Autauga Area Community Foundation Grants, 808
Avista Foundation Economic and Cultural Vitality Grants, 812
Bailey Foundation Grants, 818
BancorpSouth Foundation Grants, 823
Bank of America Fndn Matching Gifts, 825
Beckman Coulter Foundation Grants, 868
Ben B. Cheney Foundation Grants, 875
Bernard and Audre Rapoport Foundation Health Grants, 882
Biogen Foundation General Donations, 899
Birmingham Foundation Grants, 903
BJ's Charitable Foundation Grants, 904
Blackford County Community Foundation Grants, 906
Black River Falls Area Foundation Grants, 909
Blue Cross Blue Shield of Minnesota Foundation - Health Equity: Building Health Equity Together Grants, 914
Blue Cross Blue Shield of Minnesota Foundation - Healthy Children: Growing Up Healthy Grants, 915
Blue Cross Blue Shield of Minnesota Foundation - Healthy Equity: Health Impact Assessment Demonstration Project Grants, 916
Blue Grass Community Fndn Harrison Grants, 919
Blue Grass Community Foundation Hudson-Ellis Fund Grants, 920
Blue Mountain Community Foundation Grants, 921
Blumenthal Foundation Grants, 925
Bonfils-Stanton Foundation Grants, 932
Boston Globe Foundation Grants, 937
BP Foundation Grants, 943
Bradley C. Higgins Foundation Grants, 945
Bright Promises Foundation Grants, 952
Caddock Foundation Grants, 994
Cailloux Foundation Grants, 996
California Community Foundation Health Care Grants, 998
California Community Foundation Human Development Grants, 999

Community Service

California Endowment Innovative Ideas Challenge Grants, 1000
Callaway Foundation Grants, 1002
Caring Foundation Grants, 1019
Carl R. Hendrickson Family Foundation Grants, 1028
Carrie E. and Lena V. Glenn Foundation Grants, 1036
CCHD Community Development Grants, 1046
Central Okanagan Foundation Grants, 1080
CFFVR Chilton Area Community Fndn Grants, 1096
CFFVR Waupaca Area Community Foundation Grants, 1105
Charles Delmar Foundation Grants, 1115
Charles H. Dater Foundation Grants, 1118
Charles H. Farnsworth Trust Grants, 1119
Charlotte County (FL) Community Foundation Grants, 1126
Chilkat Valley Community Foundation Grants, 1148
Clark County Community Foundation Grants, 1175
Coleman Foundation Cancer Care Grants, 1206
Colonel Stanley R. McNeil Foundation Grants, 1211
Community Foundation for Greater Atlanta Clayton County Fund Grants, 1240
Community Foundation for Greater Atlanta Common Good Funds Grants, 1241
Community Foundation for Greater Atlanta Metropolitan Atlanta An Extra Wish Grants, 1243
Community Foundation for Greater Atlanta Morgan County Fund Grants, 1244
Community Foundation for Greater Atlanta Newton County Fund Grants, 1245
Community Foundation for Greater Atlanta Strategic Restructuring Fund Grants, 1246
Community Foundation for Greater New Haven Women & Girls Grants, 1252
Community Fndn for San Benito County Grants, 1257
Community Fndn of Central Illinois Grants, 1267
Community Foundation of Greater Flint Grants, 1273
Community Foundation of Louisville Anna Marble Memorial Fund for Princeton Grants, 1290
Community Foundation of Louisville Morris and Esther Lee Fund Grants, 1298
Community Foundation of Muncie and Delaware County Maxon Grants, 1301
Community Fndn of Riverside & San Bernardino County Irene S. Rockwell Grants, 1304
Community Foundation of St. Joseph County Special Project Challenge Grants, 1310
Community Fndn of Switzerland County Grants, 1311
ConAgra Foods Nourish our Community Grants, 1322
ConocoPhillips Foundation Grants, 1327
CONSOL Energy Community Dev Grants, 1339
Cralle Foundation Grants, 1356
Crane Fund Grants, 1358
Cruise Industry Charitable Foundation Grants, 1363
Cuesta Foundation Grants, 1371
Cullen Foundation Grants, 1372
CVS Caremark Charitable Trust Grants, 1378
CVS Community Grants, 1379
D. W. McMillan Foundation Grants, 1407
Dade Community Foundation Community AIDS Partnership Grants, 1409
Dairy Queen Corporate Contributions Grants, 1413
Dale and Edna Walsh Foundation Grants, 1415
Dallas Mavericks Foundation Grants, 1416
Daviess County Community Foundation Health Grants, 1434
Dearborn Community Foundation County Progress Grants, 1442
Dennis & Phyllis Washington Fndn Grants, 1458
Dominion Foundation Grants, 1485
Do Something Awards, 1503
E. Clayton and Edith P. Gengras, Jr. Foundation, 1536
Eastman Chemical Company Foundation Grants, 1542
Edina Realty Foundation Grants, 1550
Effie and Wofford Cain Foundation Grants, 1563
Elkhart County Community Foundation Fund for Elkhart County, 1572
El Pomar Foundation Grants, 1582
Emerson Charitable Trust Grants, 1589
Ensworth Charitable Foundation Grants, 1595
Entergy Corporation Micro Grants, 1596
Essex County Community Foundation Discretionary Fund Grants, 1602
Ethel S. Abbott Charitable Foundation Grants, 1607
Ethel Sergeant Clark Smith Foundation Grants, 1608
Eugene B. Casey Foundation Grants, 1609
Evan and Susan Bayh Foundation Grants, 1614
Faye McBeath Foundation Grants, 1635
Finance Factors Foundation Grants, 1666
Florida BRAIVE Fund of Dade Community Foundation, 1678
Floyd A. & Kathleen C. Cailloux Fnd Grants, 1681
Foster Foundation Grants, 1689
Foundation for the Mid South Community Development Grants, 1699
Four County Community Fndn General Grants, 1710
Four County Community Foundation Healthy Senior/Healthy Youth Fund Grants, 1711
France-Merrick Foundation Health and Human Services Grants, 1715
Frank G. and Freida K. Brotz Family Foundation Grants, 1723
George A. and Grace L. Long Foundation Grants, 1788
George A Ohl Jr. Foundation Grants, 1791
Georgiana Goddard Eaton Memorial Grants, 1805
Gil and Dody Weaver Foundation Grants, 1819
Giving Sum Annual Grant, 1825
GNOF IMPACT Grants for Health and Human Services, 1839
GNOF Maison Hospitaliere Grants, 1843
Godfrey Foundation Grants, 1849
Goodrich Corporation Foundation Grants, 1854
Goodyear Tire Grants, 1856
Grand Rapids Community Foundation Ionia County Grants, 1870
Greater Sitka Legacy Fund Grants, 1885
Green Bay Packers Foundation Grants, 1890
Green Diamond Charitable Contributions, 1891
Green River Area Community Fndn Grants, 1895
Gulf Coast Foundation of Community Operating Grants, 1916
Gulf Coast Foundation of Community Program Grants, 1917
Hagedorn Fund Grants, 1934
Hannaford Charitable Foundation Grants, 1944
Hardin County Community Foundation Grants, 1946
Harold R. Bechtel Charitable Remainder Uni-Trust Grants, 1953
Harold Simmons Foundation Grants, 1955
Harrison County Community Foundation Grants, 1957
Harrison County Community Foundation Signature Grants, 1958
Harvest Foundation Grants, 1971
Hasbro Corporation Gift of Play Hospital and Pediatric Health Giving, 1974
Healthcare Fndn for Orange County Grants, 1981
Health Foundation of Greater Cincinnati Grants, 1983
Helena Rubinstein Foundation Grants, 1991
Helen Bader Foundation Grants, 1992
Helen K. & Arthur E. Johnson Fndn Grants, 1995
Henrietta Tower Wurts Memorial Fndn Grants, 2002
Henry A. and Mary J. MacDonald Foundation, 2003
Henry L. Guenther Foundation Grants, 2006
Herbert A. & Adrian W. Woods Fndn Grants, 2007
Horace A. Kimball and S. Ella Kimball Foundation Grants, 2046
Horizons Community Issues Grants, 2049
Ida Alice Ryan Charitable Trust Grants, 2113
Ike and Roz Friedman Foundation Grants, 2145
Ireland Family Foundation Grants, 2162
James A. and Faith Knight Foundation Grants, 2186
James L. and Mary Jane Bowman Charitable Trust Grants, 2194
James M. Cox Foundation of Georgia Grants, 2196
Janus Foundation Grants, 2211
Jennings County Community Foundation Grants, 2220
Jennings County Community Foundation Women's Giving Circle Grant, 2221
Jessica Stevens Community Foundation Grants, 2224
John Clarke Trust Grants, 2232
Johnson County Community Foundation Grants, 2256
John W. Anderson Foundation Grants, 2262
John W. Speas and Effie E. Speas Memorial Trust Grants, 2264
Joseph Drown Foundation Grants, 2268
Joseph Henry Edmondson Foundation Grants, 2270
Josephine Schell Russell Charitable Trust Grants, 2274
Judith Clark-Morrill Foundation Grants, 2281
K21 Health Foundation Grants, 2285
Kenai Peninsula Foundation Grants, 2298
Ketchikan Community Foundation Grants, 2303
Klarman Family Foundation Grants in Eating Disorders Research Grants, 2310
Knox County Community Foundation Grants, 2313
Kodiak Community Foundation Grants, 2314
Kuntz Foundation Grants, 2326
L. W. Pierce Family Foundation Grants, 2328
Leo Niessen Jr., Charitable Trust Grants, 2358
Lewis H. Humphreys Charitable Trust Grants, 2361
Liberty Bank Foundation Grants, 2362
Louie M. & Betty M. Phillips Fndn Grants, 2378
Louis and Elizabeth Nave Flarsheim Charitable Foundation Grants, 2379
Lucy Gooding Charitable Fndn Grants, 2388
Madison County Community Foundation General Grants, 2413
Maine Community Foundation Penobscot Valley Health Association Grants, 2417
Marathon Petroleum Corporation Grants, 2422
Marion Gardner Jackson Charitable Trust Grants, 2434
McCallum Family Foundation Grants, 2489
McCune Foundation Human Services Grants, 2493
Mead Johnson Nutritionals Evansville-Area Organizations Grants, 2503
MedImmune Charitable Grants, 2509
Merrick Foundation Grants, 2521
Metzger-Price Fund Grants, 2530
Meyer and Stephanie Eglin Foundation Grants, 2531
Michael Reese Health Trust Core Grants, 2544
Michael Reese Health Trust Responsive Grants, 2545
Middlesex Savings Charitable Foundation Basic Human Needs Grants, 2550
Middlesex Savings Charitable Foundation Capacity Building Grants, 2551
MMS and Alliance Charitable Foundation Grants for Community Action and Care for the Medically Uninsured, 2587
Newton County Community Foundation Grants, 2681
NMF Franklin C. McLean Award, 2955
NMF William and Charlotte Cadbury Award, 2967
Norfolk Southern Foundation Grants, 2982
North Central Health Services Grants, 2992
North Central Health Services Grants, 2993
Northwestern Mutual Foundation Grants, 2999
NYC Managed Care Consumer Assistance Workshop Re-Design Grants, 3047
NYCT Grants for the Elderly, 3054
NYCT Technical Assistance Grants, 3058
Office Depot Corporation Community Relations Grants, 3065
Ohio County Community Foundation Board of Directors Grants, 3067
Ohio County Community Foundation Grants, 3068
Ohio County Community Fndn Mini-Grants, 3069
Oregon Community Fndn Community Grants, 3082
Patrick and Anna M. Cudahy Fund Grants, 3120
Peacock Foundation Grants, 3133
Percy B. Ferebee Endowment Grants, 3138
Perpetual Trust for Charitable Giving Grants, 3141
Petersburg Community Foundation Grants, 3156
Pfizer Healthcare Charitable Contributions, 3180
Pike County Community Foundation Grants, 3222
Pinnacle Foundation Grants, 3225
Piper Trust Healthcare & Med Research Grants, 3228
Piper Trust Older Adults Grants, 3229
PNC Foundation Community Services Grants, 3240
Polk Bros. Foundation Grants, 3243
Pott Foundation Grants, 3253
Powell Foundation Grants, 3255
PPG Industries Foundation Grants, 3257

SUBJECT INDEX

Price Family Charitable Fund Grants, 3262
Princeton Area Community Foundation Greater Mercer Grants, 3272
Putnam County Community Foundation Grants, 3283
Quality Health Foundation Grants, 3288
Radcliffe Institute Carol K. Pforzheimer Student Fellowships, 3298
RCF General Community Grants, 3312
RCF Individual Assistance Grants, 3313
Regence Fndn Access to Health Care Grants, 3346
Regence Foundation Health Care Community Awareness and Engagement Grants, 3347
Regence Fndn Health Care Connections Grants, 3348
Regence Fndn Improving End-of-Life Grants, 3349
Regence Fndn Tools & Technology Grants, 3350
Richland County Bank Grants, 3371
Robert R. McCormick Tribune Veterans Initiative Grants, 3386
Rochester Area Foundation Grants, 3392
Rose Hills Foundation Grants, 3413
RWJF Community Health Leaders Awards, 3432
RWJF Vulnerable Populations Portfolio Grants, 3444
Sain-Orr and Royak-DeForest Steadman Foundation Grants, 3452
Salem Foundation Charitable Trust Grants, 3456
Salem Foundation Grants, 3457
SAMHSA Strategic Prevention Framework State Incentive Grants, 3464
San Antonio Area Foundation Grants, 3469
San Diego Women's Foundation Grants, 3474
Sandy Hill Foundation Grants, 3477
San Juan Island Community Foundation Grants, 3480
Sarkeys Foundation Grants, 3485
Sasco Foundation Grants, 3487
Schering-Plough Foundation Community Initiatives Grants, 3491
Seattle Foundation Benjamin N. Phillips Memorial Fund Grants, 3504
Seward Community Foundation Grants, 3514
Seward Community Foundation Mini-Grants, 3515
Shaw's Supermarkets Donations, 3542
Shield-Ayres Foundation Grants, 3546
Shopko Foundation Green Bay Area Community Grants, 3548
Sidgmore Family Foundation Grants, 3551
Sierra Health Foundation Responsive Grants, 3555
Sir Dorabji Tata Trust Grants for NGOs or Voluntary Organizations, 3572
SOCFOC Catholic Ministries Grants, 3596
Southwest Gas Corporation Foundation Grants, 3628
St. Joseph Community Health Foundation Improving Healthcare Access Grants, 3640
Stan and Sandy Checketts Foundation, 3644
Sterling and Shelli Gardner Foundation Grants, 3653
Stocker Foundation Grants, 3656
Tauck Family Foundation Grants, 3703
Thelma Braun & Bocklett Family Fndn Grants, 3713
Thomas Austin Finch, Sr. Foundation Grants, 3718
Thompson Charitable Foundation Grants, 3727
Toyota Motor Engineering & Manufacturing North America Grants, 3742
Toyota Motor Manufacturing of Alabama Grants, 3743
Toyota Motor Manufacturing of Indiana Grants, 3744
Toyota Motor Manufacturing of Kentucky Grants, 3745
Toyota Motor Manuf of Mississippi Grants, 3746
Toyota Motor Manuf of West Virginia Grants, 3748
Toyota Motor N America of New York Grants, 3749
Toyota Motor Sales, USA Grants, 3750
Triangle Community Foundation Donor-Advised Grants, 3757
Tull Charitable Foundation Grants, 3762
U.S. Cellular Corporation Grants, 3766
U.S. Dept of Education Special Ed--National Activities--Parent Information Centers, 3772
Union Pacific Foundation Health and Human Services Grants, 3794
US Airways Community Foundation Grants, 3826
Vigneron Memorial Fund Grants, 3873
Visiting Nurse Foundation Grants, 3879
W.K. Kellogg Foundation Healthy Kids Grants, 3886

Walker Area Community Foundation Grants, 3895
Walmart Foundation Community Giving Grants, 3897
Walmart Foundation National Giving Grants, 3898
Walmart Foundation Northwest Arkansas Giving, 3899
Walmart Foundation State Giving Grants, 3900
Warrick County Community Foundation Grants, 3904
Weaver Popcorn Foundation Grants, 3910
Western New York Foundation Grants, 3922
William G. & Helen C. Hoffman Fndn Grants, 3950
Wilson-Wood Foundation Grants, 3960
Winston-Salem Foundation Competitive Grants, 3965
Winston-Salem Fndn Stokes County Grants, 3967
Wolfe Associates Grants, 3969

Community and School Relations
American Express Community Service Grants, 531
DOJ Gang-Free Schools and Communities Intervention Grants, 1481
Evanston Community Foundation Grants, 1615
Farmers Insurance Corporate Giving Grants, 1634
Joe W. and Dorothy Dorsett Brown Fndn Grants, 2230
Mead Johnson Nutritionals Evansville-Area Organizations Grants, 2503
Mt. Sinai Health Care Foundation Health of the Urban Community Grants, 2600
OneFamily Foundation Grants, 3075
Princeton Area Community Foundation Greater Mercer Grants, 3272
Self Foundation Grants, 3510
Vancouver Foundation Grants and Community Initiatives Program, 3849

Compensatory Education
William T. Grant Foundation Research Grants, 3958

Complementary Medicine
Bernard and Audre Rapoport Foundation Health Grants, 882
NCCAM Ruth L. Kirschstein National Research Service Awards for Postdoctoral Training in Complementary and Alternative Medicine, 2640

Complex Systems
James S. McDonnell Foundation Complex Systems Collaborative Activity Awards, 2202

Compulsive Behavior
Peter and Elizabeth C. Tower Foundation Annual Mental Health Grants, 3145

Computational Analysis
James S. McDonnell Foundation Complex Systems Collaborative Activity Awards, 2202
SfN Swartz Prize for Theoretical and Computational Neuroscience, 3537

Computer Arts
AAAS Science Prize for Online Resources in Educ, 42

Computer Education/Literacy
Achelis Foundation Grants, 234
Benton Community Fndn Cookie Jar Grant, 877
Blue Cross Blue Shield of Minnesota Fndn Healthy Equity: Public Libraries for Health Grants, 918
Bodman Foundation Grants, 929
Farmers Insurance Corporate Giving Grants, 1634
IIE Western Union Family Scholarships, 2143
Robert R. McCormick Trib Community Grants, 3385
Union Bank, N.A. Foundation Grants, 3788
Verizon Foundation Maryland Grants, 3859
Verizon Foundation New York Grants, 3860
Verizon Foundation Northeast Region Grants, 3861
Verizon Foundation Pennsylvania Grants, 3862
Verizon Foundation Vermont Grants, 3863
Verizon Foundation Virginia Grants, 3864

Computer Engineering
Claude Worthington Benedum Fndn Grants, 1178

Computer Science
AACR Team Science Award, 126
Alfred P. Sloan Foundation Research Fellowships, 432
BWF Career Awards at the Scientific Interface, 983
Canada Graduate Scholarships (CGS) and NSERC Postgraduate Scholarships (PGS), 1010
CDC Public Health Informatics Fellowships, 1068
Gates Millennium Scholars Program, 1776
IIE Central Europe Summer Research Institute Summer Research Fellowship, 2127
IIE KAUST Graduate Fellowships, 2134
IIE Mattel Global Scholarship, 2139
NHGRI Mentored Research Scientist Development Award, 2686
NHLBI Mentored Quantitative Research Career Development Awards, 2711
NIH Exploratory Innovations in Biomedical Computational Science & Technology Grants, 2878
NIH Innovations in Biomedical Computational Science and Technology Grants, 2883
NIH Innovations in Biomedical Computational Science and Technology Initiative Grants for SBIR, 2884
NIH Innovations in Biomedical Computational Science and Technology Initiative Grants for STTR, 2885
NIMH Curriculum Development Award in Neuroinformatics Research and Analysis, 2917
NLM Academic Research Enhancement Awards, 2940
NLM Research Project Grants, 2951
North Carolina GlaxoSmithKline Fndn Grants, 2991
NSF Chemistry Research Experiences for Undergraduates (REU), 3012
Rockwell International Corporate Trust Grants Program, 3399

Computer Software
Alice Tweed Tuohy Foundation Grants Program, 437
McLean Contributionship Grants, 2497

Computer Technology
AAAS Science Prize for Online Resources in Educ, 42
IIE Mattel Global Scholarship, 2139
John R. Oishei Foundation Grants, 2249
Microsoft Research Cell Phone as a Platform for Healthcare Grants, 2548
NIH Exploratory Innovations in Biomedical Computational Science & Technology Grants, 2878
NIH Innovations in Biomedical Computational Science and Technology Grants, 2883
NIH Innovations in Biomedical Computational Science and Technology Initiative Grants for SBIR, 2884
NIH Innovations in Biomedical Computational Science and Technology Initiative Grants for STTR, 2885
NLM Academic Research Enhancement Awards, 2940
Peter and Elizabeth C. Tower Foundation Phase II Technology Initiative Grants, 3149
Peter and Elizabeth C. Tower Foundation Phase I Technology Initiative Grants, 3150
Society for Imaging Informatics in Medicine (SIIM) Emeritus Mentor Grants, 3597
Society for Imaging Informatics in Medicine (SIIM) Micro Grants, 3598
Society for Imaging Informatics in Medicine (SIIM) Research Grants, 3599
Society for Imaging Informatics in Med Small Grant for Product Support Development, 3600
Society for Imaging Informatics in Medicine (SIIM) Small Training Grants, 3601
Verizon Foundation Maryland Grants, 3859
Verizon Foundation New York Grants, 3860
Verizon Foundation Northeast Region Grants, 3861
Verizon Foundation Pennsylvania Grants, 3862
Verizon Foundation Vermont Grants, 3863
Verizon Foundation Virginia Grants, 3864
Wood Family Charitable Trust Grants, 3972

Computer-Aided Instruction
Blue Cross Blue Shield of Minnesota Fndn Healthy Equity: Public Libraries for Health Grants, 918

Conferences
AIHS Knowledge Exchange - Conference Grants, 389
Alzheimer's Association Conference Grants, 463
ASHA Minority Student Leadership Awards, 696
ASHA Research Mentoring-Pair Travel Award, 698
ASHA Student Research Travel Award, 699
ASHA Students Preparing for Academic & Research Careers (SPARC) Award, 700
Cystic Fibrosis Canada Small Conference Grants, 1388
Cystic Fibrosis Canada Travel Supplement Grants, 1393
IBRO-PERC Support for Site Lectures, 2099
IBRO Latin America Regional Travel Grants, 2109
MLA Lucretia W. McClure Excellence in Education Award, 2570
NIDDK Traditional Conference Grants, 2848
Oppenstein Brothers Foundation Grants, 3078
Orthopaedic Trauma Association Kathy Cramer Young Clinician Memorial Fellowships, 3083
Pfizer Medical Education Track Two Grants, 3183
Potts Memorial Foundation Grants, 3254
SfN Science Journalism Student Awards, 3536
Sick Kids Fndn Community Conference Grants, 3549
South Madison Community Foundation Grants, 3626
Topeka Community Foundation Kansas Blood Services Scholarships, 3738

Conferences, Travel to
AACAP Systems of Care Special Scholarships, 76
ACSM International Student Awards, 269
Acumen East Africa Fellowship, 284
AES Research and Training Workshop Awards, 320
AES William G. Lennox Award, 326
AIHS Knowledge Exchange - Conference Grants, 389
ALFJ Astraea U.S. and Int'l Movement Fund, 427
ANA Junior Academic Neurologist Scholarships, 582
ANA Soriano Lectureship, 584
AOCS A. Richard Baldwin Award, 610
AOCS Industrial Oil Products Division Student Award, 616
AOCS Lipid Oxidation and Quality Division Poster Competition, 617
AOCS Surfactants and Detergents Division Student Award, 622
ASHA Minority Student Leadership Awards, 696
ASHA Research Mentoring-Pair Travel Award, 698
ASHA Student Research Travel Award, 699
ASHA Students Preparing for Academic & Research Careers (SPARC) Award, 700
ASM-UNESCO Leadership Grant for International Educators, 715
ASM Millis-Colwell Postgraduate Travel Grant, 741
BWF Collaborative Research Travel Grants, 985
Conquer Cancer Foundation of ASCO Resident Travel Award for Underrepresented Populations, 1336
Cystic Fibrosis Canada Special Travel Grants For Fellows and Students, 1389
Cystic Fibrosis Canada Travel Supplement Grants, 1393
EBSCO / MLA Annual Meeting Grants, 1545
GNOF Organizational Effectiveness Grants, 1846
IBRO/SfN International Travel Grants, 2100
IBRO International Travel Grants, 2104
International Positive Psychology Association Student Scholarships, 2161
MLA Janet Doe Lectureship Award, 2567
NIDDK Traditional Conference Grants, 2848
Orthopaedic Trauma Association Kathy Cramer Young Clinician Memorial Fellowships, 3083
Phi Upsilon Omicron Florence Fallgatter Distinguished Service Award, 3192
Phi Upsilon Omicron Frances Morton Holbrook Alumni Award, 3193
SfN Bernice Grafstein Award for Outstanding Accomplishments in Mentoring, 3518
SfN Chapter Grants, 3520
SfN Donald B. Lindsley Prize in Behavioral Neuroscience, 3521
SfN Federation of European Neuroscience Societies Forum Travel Awards, 3522
SfN Japan Neuroscience Society Meeting Travel Awards, 3525
SfN Louise Hanson Marshall Specific Recognition Award, 3527
SfN Next Generation Award, 3531
SfN Science Educator Award, 3535
SfN Science Journalism Student Awards, 3536
SfN Trainee Professional Development Awards, 3538
SfN Travel Awards for the International Brain Research Organization World Congress, 3539
SfN Undergrad Brain Awareness Travel Award, 3540
St. Joseph Community Health Foundation Schneider Fellowships, 3642
Topeka Community Foundation Kansas Blood Services Scholarships, 3738

Conflict/Dispute Resolution
Allstate Corporate Giving Grants, 446
Allstate Corp Hometown Commitment Grants, 447
American-Scandinavian Foundation Visiting Lectureship Grants, 502
Harry Frank Guggenheim Fnd Research Grants, 1964
Mary Owen Borden Foundation Grants, 2446
Sioux Falls Area Community Foundation Spot Grants (Unrestricted), 3570

Conservation
Centerville-Washington Foundation Grants, 1078
Columbus Foundation Competitive Grants, 1217
ConocoPhillips Foundation Grants, 1327
Packard Foundation Local Grants, 3105
Prospect Burma Scholarships, 3275
William B. Stokely Jr. Foundation Grants, 3947

Conservation, Agriculture
Bullitt Foundation Grants, 973
Clarence E. Heller Charitable Foundation Grants, 1166
CLIF Bar Family Foundation Grants, 1183
Conservation, Food, and Health Foundation Grants for Developing Countries, 1338
Dean Foods Community Involvement Grants, 1440
Norman Foundation Grants, 2983
PepsiCo Foundation Grants, 3137

Conservation, Natural Resources
Abel Foundation Grants, 206
ACF Native American Social and Economic Development Strategies Grants, 231
Alberto Culver Corporate Contributions Grants, 408
Allen P. & Josephine B. Green Fndn Grants, 443
Amgen Foundation Grants, 572
Archer Daniels Midland Foundation Grants, 651
Ashland Corporate Contributions Grants, 713
Australian Academy of Science Grants, 807
Batchelor Foundation Grants, 837
Bella Vista Foundation Grants, 873
Blue Cross Blue Shield of Minnesota Foundation - Healthy Equity: Health Impact Assessment Demonstration Project Grants, 916
BMW of North America Charitable Contributions, 926
Boeing Company Contributions Grants, 930
Bullitt Foundation Grants, 973
Carls Foundation Grants, 1030
Carpenter Foundation Grants, 1035
Central Okanagan Foundation Grants, 1080
Champlin Foundations Grants, 1110
Clarence E. Heller Charitable Foundation Grants, 1166
CLIF Bar Family Foundation Grants, 1183
Coeta and Donald Barker Foundation Grants, 1205
Collins Foundation Grants, 1210
Columbus Foundation Mary Eleanor Morris Fund Grants, 1220
Conservation, Food, and Health Foundation Grants for Developing Countries, 1338
Constellation Energy Corporate Grants, 1342
Cumberland Community Foundation Grants, 1374
Cyrus Eaton Foundation Grants, 1380
Donald and Sylvia Robinson Family Foundation Grants, 1486
Dorrance Family Foundation Grants, 1501
Evan and Susan Bayh Foundation Grants, 1614
Field Foundation of Illinois Grants, 1660
Ford Motor Company Fund Grants Program, 1687
Frank Stanley Beveridge Foundation Grants, 1730
George Gund Foundation Grants, 1796
Giving Sum Annual Grant, 1825
Greater Milwaukee Foundation Grants, 1883
Green Diamond Charitable Contributions, 1891
Harry A. & Margaret D. Towsley Fndn Grants, 1959
J.W. Kieckhefer Foundation Grants, 2176
Jane's Trust Grants, 2206
Jessie B. Cox Charitable Trust Grants, 2225
Jessie Ball Dupont Fund Grants, 2226
M.E. Raker Foundation Grants, 2404
Mary Owen Borden Foundation Grants, 2446
McLean Contributionship Grants, 2497
Nelda C. and H.J. Lutcher Stark Fndn Grants, 2672
Nina Mason Pulliam Charitable Trust Grants, 2921
Norcliffe Foundation Grants, 2979
Norman Foundation Grants, 2983
Ogden Codman Trust Grants, 3066
Oscar Rennebohm Foundation Grants, 3085
Posey County Community Foundation Grants, 3252
R.T. Vanderbilt Trust Grants, 3294
Rajiv Gandhi Foundation Grants, 3302
Robert W. Woodruff Foundation Grants, 3389
Rohm and Haas Company Grants, 3403
San Juan Island Community Foundation Grants, 3480
Shield-Ayres Foundation Grants, 3546
Singing for Change Foundation Grants, 3567
Susan Vaughan Foundation Grants, 3695
USDA Organic Agriculture Research Grants, 3830
Wege Foundation Grants, 3912

Consumer Behavior
Bodman Foundation Grants, 929
MLA Donald A.B. Lindberg Fellowship, 2562

Consumer Education/Information
Achelis Foundation Grants, 234
BMW of North America Charitable Contributions, 926
Bodman Foundation Grants, 929
Covidien Partnership for Neighborhood Wellness Grants, 1353
MLA Donald A.B. Lindberg Fellowship, 2562
NYC Managed Care Consumer Assistance Workshop Re-Design Grants, 3047
Phi Upsilon Omicron Florence Fallgatter Distinguished Service Award, 3192

Consumer Sciences
AAFCS International Graduate Fellowships, 129
AAFCS National Graduate Fellowships, 130
AAFCS National Undergraduate Scholarships, 131
Phi Upsilon Omicron Alumni Research Grant, 3191
Phi Upsilon Omicron Florence Fallgatter Distinguished Service Award, 3192
Phi Upsilon Omicron Frances Morton Holbrook Alumni Award, 3193
Phi Upsilon Omicron Geraldine Clewell Senior Awards, 3194
Phi Upsilon Omicron Janice Cory Bullock Collegiate Award, 3195
Phi Upsilon Lillian P. Schoephoerster Award, 3196
Phi Upsilon Omicron Margaret Jerome Sampson Scholarships, 3197
Phi Upsilon Omicron Orinne Johnson Writing Award, 3198
Phi Upsilon Omicron Sarah Thorniley Phillips Leader Awards, 3199
Phi Upsilon Omicron Undergraduate Karen P. Goebel Conclave Award, 3200
W.M. Keck Foundation Science and Engineering Research Grants, 3889

Consumer Services
California Wellness Foundation Work and Health Program Grants, 1001
EPA Children's Health Protection Grants, 1598
IIE New Leaders Group Award for Mutual Understanding, 2141
Jacob and Valeria Langeloth Foundation Grants, 2182

SUBJECT INDEX

Contraceptives
Lalor Foundation Anna Lalor Burdick Grants, 2335
Lalor Foundation Postdoctoral Fellowships, 2336
PC Fred H. Bixby Fellowships, 3126

Cooperative Education
IBRO Latin America Regional Funding for Short Courses, Workshops, and Symposia, 2107

Corporate/Strategic Planning
Blue Cross Blue Shield of Minnesota Foundation - Healthy Equity: Health Impact Assessment Program Grants, 917
GNOF Organizational Effectiveness Grants and Workshops, 1846
Middlesex Savings Charitable Foundation Capacity Building Grants, 2551
Peter and Elizabeth C. Tower Foundation Phase II Technology Initiative Grants, 3149
Peter and Elizabeth C. Tower Foundation Phase I Technology Initiative Grants, 3150

Corporations
GNOF New Orleans Works Grants, 1844

Cosmetic Science
Society of Cosmetic Chemists Allan B. Black Award Sponsored by Presperse, 3604
Society of Cosmetic Chemists Award Sponsored by McIntyre Group, Ltd., 3605
Society of Cosmetic Chemists Award Sponsored by The HallStar Company, 3606
Society of Cosmetic Chemists Chapter Best Speaker Award, 3607
Society of Cosmetic Chem Chapter Merit Award, 3608
Society of Cosmetic Chemists Frontier of Science Award, 3609
Society of Cosmetic Chem Hans Schaeffer Award, 3610
Society of Cosmetic Chem Jos Ciaudelli Award, 3611
Society of Cosmetic Chemists Keynote Award Lecture Sponsored by Ruger Chemical Corporation, 3612
Society of Cosmetic Chemists Literature Award, 3613
Society of Cosmetic Chemists Maison G. de Navarre Medal, 3614
Society of Cosmetic Chemists Merit Award, 3615
Society of Cosmetic Chemists Robert A. Kramer Lifetime Service Award, 3616
Society of Cosmetic Chem Shaw Mudge Award, 3617
Society of Cosmetic Chem Stud Poster Awards, 3618
Society of Cosmetic Chem You Scientist Awards, 3619

Cosmology
Brinson Foundation Grants, 953
Gruber Foundation Cosmology Prize, 1907

Counseling/Guidance
CJ Foundation for SIDS Program Services Grants, 1163
Edina Realty Foundation Grants, 1550
Greater Worcester Community Foundation Discretionary Grants, 1888
Gulf Coast Foundation of Community Operating Grants, 1916
Gulf Coast Foundation of Community Program Grants, 1917
HRSA Ryan White HIV AIDS Drug Assistance Grants, 2070
Hudson Webber Foundation Grants, 2072
Independence Community Foundation Community Quality of Life Grant, 2153
James R. Thorpe Foundation Grants, 2199
Kimball International-Habig Foundation Health and Human Services Grants, 2308
Leo Niessen Jr., Charitable Trust Grants, 2358
Mary Owen Borden Foundation Grants, 2446
MGM Resorts Foundation Community Grants, 2541
MGN Family Foundation Grants, 2542
OneFamily Foundation Grants, 3075
Robert R. Meyer Foundation Grants, 3387
Ruth H. & Warren A. Ellsworth Fndn Grants, 3429
Seabury Foundation Grants, 3501
TE Foundation Grants, 3705
Textron Corporate Contributions Grants, 3712
Vermont Community Foundation Grants, 3865
Z. Smith Reynolds Foundation Small Grants, 3982

Counseling/Guidance Education
Medtronic Foundation Community Link Human Services Grants, 2511
Robert R. McCormick Trib Community Grants, 3385

Craniofacial Anomalies
DHHS Oral Health Promotion Research Across the Lifespan, 1474
Ronald McDonald House Charities Grants, 3405

Creative Writing
AAAS/Subaru SB&F Prize for Excellence in Science Books, 35
IIE Rockefeller Foundation Bellagio Center Residencies, 2142
Leon and Thea Koerner Foundation Grants, 2355
McKesson Foundation Grants, 2496
Olive Higgins Prouty Foundation Grants, 3072

Creativity
MLA Section Project of the Year Award, 2574
United Hospital Fund of New York Health Care Improvement Grants, 3797

Crime Control
Cemala Foundation Grants, 1077
DOJ Gang-Free Schools and Communities Intervention Grants, 1481
Sonora Area Foundation Competitive Grants, 3622

Crime Prevention
Abbot and Dorothy H. Stevens Foundation Grants, 197
Abell Foundation Criminal Justice and Addictions Grants, 208
ACL Empowering Seniors to Prevent Health Care Fraud Grants, 241
Austin S. Nelson Foundation Grants, 805
Baptist Community Ministries Grants, 828
Beazley Foundation Grants, 865
BJ's Charitable Foundation Grants, 904
ConocoPhillips Foundation Grants, 1327
Constantin Foundation Grants, 1341
Cornerstone Foundation of Northeastern Wisconsin Grants, 1347
Daphne Seybolt Culpeper Fndn Grants, 1425
Elliot Foundation Inc Grants, 1575
Farmers Insurance Corporate Giving Grants, 1634
Frank Stanley Beveridge Foundation Grants, 1730
G.N. Wilcox Trust Grants, 1768
Herman Goldman Foundation Grants, 2012
Hudson Webber Foundation Grants, 2072
Joseph H. & Florence A. Roblee Fndn Grants, 2269
Kahuku Community Fund, 2286
Knox County Community Foundation Grants, 2313
Ordean Foundation Grants, 3080
Perry County Community Foundation Grants, 3142
Pike County Community Foundation Grants, 3222
Plough Foundation Grants, 3238
Posey County Community Foundation Grants, 3252
R.C. Baker Foundation Grants, 3292
RCF General Community Grants, 3312
Sid W. Richardson Foundation Grants, 3553
Sonoco Foundation Grants, 3621
Union Bank, N.A. Foundation Grants, 3788

Crime Victims
ACL Empowering Seniors to Prevent Health Care Fraud Grants, 241
Austin S. Nelson Foundation Grants, 805
Carrie E. and Lena V. Glenn Foundation Grants, 1036
Cralle Foundation Grants, 1356
John W. Speas and Effie E. Speas Memorial Trust Grants, 2264
Linford and Mildred White Charitable Grants, 2368
Winston-Salem Fndn Victim Assistance Grants, 3968

Criminal Behavior
Abell Foundation Criminal Justice and Addictions Grants, 208
John R. Oishei Foundation Grants, 2249
Ordean Foundation Grants, 3080

Criminal Justice
Abell Foundation Criminal Justice and Addictions Grants, 208
Baptist Community Ministries Grants, 828
Carroll County Community Foundation Grants, 1039
Coastal Community Foundation of South Carolina Grants, 1202
Daisy Marquis Jones Foundation Grants, 1414
Good Samaritan Inc Grants, 1855
Henry County Community Foundation Grants, 2005
Marion I. and Henry J. Knott Foundation Standard Grants, 2436
Puerto Rico Community Foundation Grants, 3278
Sioux Falls Area Community Foundation Community Fund Grants (Unrestricted), 3568
Sioux Falls Area Community Foundation Spot Grants (Unrestricted), 3570
Texas Commission on the Arts Arts Respond Project Grants, 3708

Criminology
Harry Frank Guggenheim Foundation Dissertation Fellowships, 1963

Crisis Counseling
Adams Foundation Grants, 301
Austin S. Nelson Foundation Grants, 805
Barr Fund Grants, 835
Charles H. Pearson Foundation Grants, 1121
D. W. McMillan Foundation Grants, 1407
Duchossois Family Foundation Grants, 1518
Edwards Memorial Trust Grants, 1558
Kimball International-Habig Foundation Health and Human Services Grants, 2308
Medtronic Foundation Community Link Human Services Grants, 2511
Nationwide Insurance Foundation Grants, 2637
Ordean Foundation Grants, 3080
Perry County Community Foundation Grants, 3142
Pike County Community Foundation Grants, 3222
Sensient Technologies Foundation Grants, 3513
Sierra Health Foundation Responsive Grants, 3555
Spencer County Community Foundation Grants, 3632
Warrick County Community Foundation Grants, 3904

Critical Care Medicine
AACN-Edwards Lifesciences Nurse-Driven Clinical Practice Outcomes Grants, 77
AACN-Philips Medical Systems Clinical Outcomes Grants, 78
AACN-Sigma Theta Tau Critical Care Grant, 79
AACN Clinical Inquiry Fund Grants, 80
AACN Clinical Practice Grants, 81
AACN Critical Care Grants, 82
AACN End of Life/Palliative Care Small Projects Grants, 83
AACN Evidence-Based Clinical Practice Grants, 84
AACN Mentorship Grant, 85
AACN Physio-Control Small Projects Grants, 86
Baptist Community Ministries Grants, 828
Chest Fndn Eli Lilly & Company Distinguished Scholar in Critical Care Med Award, 1132
Clarian Health Critical/Progressive Care Interns, 1168
Duke University Adult Cardiothoracic Anesthesia and Critical Care Medicine Fellowships, 1520
Duke University Ped Anesthesiology Fellowship, 1527
E.L. Wiegand Foundation Grants, 1538
Herman Goldman Foundation Grants, 2012
Lowe Foundation Grants, 2381
OSF Access to Essential Medicines Initiative, 3086
Piper Trust Healthcare & Med Research Grants, 3228
Questar Corporate Contributions Grants, 3291
Stewart Huston Charitable Trust Grants, 3655

698 / Critical Care Medicine

STTI Emergency Nurses Association Foundation Grant, 3666
STTI Environment of Elder Care Nursing Research Grants, 3667

Crohns Disease
Partnership for Cures Two Years To Cures Grants, 3118

Crop Science
BWF Innovation in Regulatory Science Grants, 986

Cross-Cultural Studies
eBay Foundation Community Grants, 1543

Crystallography
NSF Small Business Innovation Research (SBIR) Grants, 3022

Cuba
American Sociological Association Minority Fellowships, 563

Culinary Arts
Adolph Coors Foundation Grants, 311
Carl B. and Florence E. King Foundation Grants, 1022
Radcliffe Institute Carol K. Pforzheimer Student Fellowships, 3298

Cultural Activities/Programs
2 Depot Square Ipswich Charitable Fndn Grants, 4
49ers Foundation Grants, 9
A. Gary Anderson Family Foundation Grants, 16
A.O. Smith Foundation Community Grants, 17
Aaron Foundation Grants, 193
Abell-Hanger Foundation Grants, 207
Abington Foundation Grants, 211
Achelis Foundation Grants, 234
Adelaide Breed Bayrd Foundation Grants, 306
Adobe Community Investment Grants, 310
AEC Trust Grants, 315
Agnes Gund Foundation Grants, 345
Ahmanson Foundation Grants, 355
Air Products and Chemicals Grants, 399
Alabama Power Foundation Grants, 400
Alaska Airlines Foundation Grants, 403
Alberto Culver Corporate Contributions Grants, 408
Albert Pick Jr. Fund Grants, 409
Albuquerque Community Foundation Grants, 412
Alcatel-Lucent Technologies Foundation Grants, 413
Alcoa Foundation Grants, 414
Allegheny Technologies Charitable Trust, 440
Allen P. & Josephine B. Green Fndn Grants, 443
Alvin and Fanny Blaustein Thalheimer Foundation Baltimore Communal Grants, 459
Anheuser-Busch Foundation Grants, 591
Anna Fitch Ardenghi Trust Grants, 592
Ann Arbor Area Community Foundation Grants, 594
Ann Peppers Foundation Grants, 599
AON Foundation Grants, 623
AptarGroup Foundation Grants, 647
Aratani Foundation Grants, 649
Archer Daniels Midland Foundation Grants, 651
Argyros Foundation Grants Program, 653
Arie and Ida Crown Memorial Grants, 654
Arizona Cardinals Grants, 655
Arizona Public Service Corporate Giving Program Grants, 658
Arkansas Community Foundation Grants, 660
ArvinMeritor Grants, 675
ASHA Multicultural Activities Projects Grants, 697
Ashland Corporate Contributions Grants, 713
Atlanta Foundation Grants, 796
Audrey & Sydney Irmas Foundation Grants, 799
Bacon Family Foundation Grants, 817
Ball Brothers Foundation General Grants, 820
Barberton Community Foundation Grants, 829
Barker Welfare Foundation Grants, 831
Barra Foundation Project Grants, 834
Batts Foundation Grants, 840
BCBSNC Foundation Grants, 860

Bechtel Group Foundation Building Positive Community Relationships Grants, 866
Beckley Area Foundation Grants, 867
Belk Foundation Grants, 872
Benton Community Foundation Grants, 878
Benton County Foundation Grants, 880
Berks County Community Foundation Grants, 881
Berrien Community Foundation Grants, 884
Bertha Russ Lytel Foundation Grants, 885
Besser Foundation Grants, 887
Blue Grass Community Fndn Harrison Grants, 919
Blue Grass Community Foundation Hudson-Ellis Fund Grants, 920
Blue River Community Foundation Grants, 922
Blum-Kovler Foundation Grants, 924
Bodenwein Public Benevolent Foundation Grants, 928
Bodman Foundation Grants, 929
Boeing Company Contributions Grants, 930
Boettcher Foundation Grants, 931
Boston Foundation Grants, 936
Bradley-Turner Foundation Grants, 944
Brown County Community Foundation Grants, 970
Browning-Kimball Foundation Grants, 971
Burlington Northern Santa Fe Foundation Grants, 978
Cailloux Foundation Grants, 996
Cambridge Community Foundation Grants, 1004
Carl & Eloise Pohlad Family Fndn Grants, 1021
Carl B. and Florence E. King Foundation Grants, 1022
Carl C. Icahn Foundation Grants, 1023
Carolyn Foundation Grants, 1034
Carroll County Community Foundation Grants, 1039
Cemala Foundation Grants, 1077
Central Carolina Community Foundation Community Impact Grants, 1079
Central Okanagan Foundation Grants, 1080
Cessna Foundation Grants Program, 1081
Chamberlain Foundation Grants, 1108
Charles H. Dater Foundation Grants, 1118
Charlotte County (FL) Community Foundation Grants, 1126
Chiles Foundation Grants, 1147
Christy-Houston Foundation Grants, 1153
CICF Indianapolis Fndn Community Grants, 1156
CICF Legacy Fund Grants, 1158
CIGNA Foundation Grants, 1160
Citizens Bank Mid-Atlantic Charitable Foundation Grants, 1162
Clark County Community Foundation Grants, 1175
Cleveland-Cliffs Foundation Grants, 1181
Cleveland Browns Foundation Grants, 1182
Clinton County Community Foundation Grants, 1184
CNA Foundation Grants, 1189
Coastal Community Foundation of South Carolina Grants, 1202
Cockrell Foundation Grants, 1204
Collins Foundation Grants, 1210
Columbus Foundation Robert E. and Genevieve B. Schaefer Fund Grants, 1222
Comerica Charitable Foundation Grants, 1224
Commonwealth Edison Grants, 1225
Communities Foundation of Texas Grants, 1236
Community Foundation Alliance City of Evansville Endowment Fund Grants, 1238
Community Foundation for Northeast Michigan Mini-Grants, 1255
Community Foundation for the National Capital Region Community Leadership Grants, 1260
Community Foundation of Bartholomew County Heritage Fund Grants, 1262
Community Foundation of Bartholomew County James A. Henderson Award for Fundraising, 1263
Community Foundation of Bloomington and Monroe County Grants, 1265
Community Foundation of Eastern Connecticut General Southeast Grants, 1269
Community Foundation of Greater Birmingham Grants, 1272
Community Foundation of Greater Fort Wayne - Community Endowment and Clarke Endowment Grants, 1274

Community Foundation of Greater Greensboro Community Grants, 1277
Community Fndn of Greater Tampa Grants, 1282
Community Fndn of Howard County Grants, 1286
Community Foundation of Mount Vernon and Knox County Grants, 1299
Community Fndn of Randolph County Grants, 1302
Community Fndn of South Puget Sound Grants, 1307
Community Foundation of the Verdugos Grants, 1315
Community Foundation Serving Riverside and San Bernardino Counties Impact Grants, 1319
Connecticut Community Foundation Grants, 1324
Connelly Foundation Grants, 1326
Constantin Foundation Grants, 1341
Consumers Energy Foundation, 1343
Cooke Foundation Grants, 1344
Cooper Industries Foundation Grants, 1345
Cornerstone Foundation of Northeastern Wisconsin Grants, 1347
Cowles Charitable Trust Grants, 1354
CSX Corporate Contributions Grants, 1368
Cudd Foundation Grants, 1370
Cullen Foundation Grants, 1372
Cultural Society of Filipino Americans Grants, 1373
Cumberland Community Foundation Grants, 1374
Cyrus Eaton Foundation Grants, 1380
DaimlerChrysler Corporation Fund Grants, 1412
Dairy Queen Corporate Contributions Grants, 1413
Dana Brown Charitable Trust Grants, 1419
Dayton Power and Light Foundation Grants, 1437
Denver Foundation Community Grants, 1461
Dr. Scholl Foundation Grants Program, 1511
Dunspaugh-Dalton Foundation Grants, 1530
DuPage Community Foundation Grants, 1531
Eastman Chemical Company Foundation Grants, 1542
Eberly Foundation Grants, 1544
Edwin S. Webster Foundation Grants, 1560
Elkhart County Community Foundation Grants, 1573
Elmer L. & Eleanor J. Andersen Fndn Grants, 1578
Emerson Charitable Trust Grants, 1589
Entergy Corporation Micro Grants, 1596
Erie Community Foundation Grants, 1601
Essex County Community Foundation Discretionary Fund Grants, 1602
Ethel Sergeant Clark Smith Foundation Grants, 1608
Evjue Foundation Grants, 1617
Fargo-Moorhead Area Foundation Grants, 1632
Farmers Insurance Corporate Giving Grants, 1634
Ferree Foundation Grants, 1658
Fidelity Foundation Grants, 1659
Field Foundation of Illinois Grants, 1660
FirstEnergy Foundation Community Grants, 1668
Fishman Family Foundation Grants, 1672
Floyd A. & Kathleen C. Cailloux Fnd Grants, 1681
Fluor Foundation Grants, 1682
Fondren Foundation Grants, 1685
Francis L. Abreu Charitable Trust Grants, 1719
Frank E. and Seba B. Payne Foundation Grants, 1722
Fred C. & Katherine B. Andersen Grants, 1737
Fred L. Emerson Foundation Grants, 1743
Fremont Area Community Fndn General Grants, 1746
Fuji Film Grants, 1748
Fulton County Community Foundation Grants, 1763
G.N. Wilcox Trust Grants, 1768
Gebbie Foundation Grants, 1777
General Mills Foundation Grants, 1782
General Motors Foundation Grants, 1783
Genuardi Family Foundation Grants, 1787
George W. Codrington Charitable Fndn Grants, 1803
Georgia Power Foundation Grants, 1806
Gertrude and William C. Wardlaw Fund Grants, 1808
Giving Sum Annual Grant, 1825
GlaxoSmithKline Corporate Grants, 1829
GNOF Exxon-Mobil Grants, 1836
Golden State Warriors Foundation Grants, 1851
Goodrich Corporation Foundation Grants, 1854
Goodyear Tire Grants, 1856
Grand Haven Area Community Fndn Grants, 1864
Grand Rapids Community Foundation Grants, 1869

SUBJECT INDEX

Greater Cincinnati Foundation Priority and Small Projects/Capacity-Building Grants, 1880
Greater Green Bay Community Fndn Grants, 1881
Greater Milwaukee Foundation Grants, 1883
Greater Tacoma Community Foundation Grants, 1886
Greater Worcester Community Foundation Discretionary Grants, 1888
Guido A. & Elizabeth H. Binda Fndn Grants, 1913
Gulf Coast Community Foundation Grants, 1915
Guy I. Bromley Trust Grants, 1919
H.B. Fuller Foundation Grants, 1922
H.J. Heinz Company Foundation Grants, 1923
Harley Davidson Foundation Grants, 1947
Harry W. Bass, Jr. Foundation Grants, 1968
Helen K. & Arthur E. Johnson Fndn Grants, 1995
High Meadow Foundation Grants, 2027
Hill Crest Foundation Grants, 2030
Hillman Foundation Grants, 2032
Hillsdale Fund Grants, 2035
Hoblitzelle Foundation Grants, 2038
Holland/Zeeland Community Fndn Grants, 2041
Homer Foundation Grants, 2044
Houston Endowment Grants, 2053
Howard and Bush Foundation Grants, 2054
Hudson Webber Foundation Grants, 2072
Huffy Foundation Grants, 2073
Hugh J. Andersen Foundation Grants, 2074
Humana Foundation Grants, 2077
Huntington County Community Foundation Make a Difference Grants, 2083
I.A. O'Shaughnessy Foundation Grants, 2088
Idaho Community Foundation Eastern Region Competitive Grants, 2114
Idaho Power Company Corporate Contributions, 2115
Impact 100 Grants, 2148
Intergrys Corporation Grants, 2160
J.M. Long Foundation Grants, 2173
J.W. Kieckhefer Foundation Grants, 2176
James L. and Mary Jane Bowman Charitable Trust Grants, 2194
Jane's Trust Grants, 2206
Janus Foundation Grants, 2211
Jessie Ball Dupont Fund Grants, 2226
John Ben Snow Memorial Trust Grants, 2231
John Deere Foundation Grants, 2234
John J. Leidy Foundation Grants, 2241
John R. Oishei Foundation Grants, 2249
Judith and Jean Pape Adams Charitable Foundation Tulsa Area Grants, 2280
Katharine Matthies Foundation Grants, 2291
Katherine John Murphy Foundation Grants, 2293
Kenneth T. & Eileen L. Norris Fndn Grants, 2300
Kodiak Community Foundation Grants, 2314
Land O'Lakes Foundation Mid-Atlantic Grants, 2341
Liberty Bank Foundation Grants, 2362
Lincoln Financial Foundation Grants, 2367
Lockheed Martin Corp Foundation Grants, 2374
Lotus 88 Fnd for Women & Children Grants, 2376
Lowell Berry Foundation Grants, 2382
Lubbock Area Foundation Grants, 2383
Lynde & Harry Bradley Foundation Fellowships, 2396
Lynde and Harry Bradley Foundation Grants, 2397
Lynde and Harry Bradley Foundation Prizes: Bradley Prizes, 2398
Mabel Y. Hughes Charitable Trust Grants, 2409
Madison County Community Foundation General Grants, 2413
Marcia and Otto Koehler Foundation Grants, 2427
Margaret T. Morris Foundation Grants, 2429
Marion I. and Henry J. Knott Foundation Discretionary Grants, 2435
Marion I. and Henry J. Knott Foundation Standard Grants, 2436
Mary K. Chapman Foundation Grants, 2443
Mary Owen Borden Foundation Grants, 2446
Mary S. and David C. Corbin Foundation Grants, 2447
Maurice J. Masserini Charitable Trust Grants, 2456
Max and Victoria Dreyfus Foundation Grants, 2457
Maxon Charitable Foundation Grants, 2458
McConnell Foundation Grants, 2492

McInerny Foundation Grants, 2495
McKesson Foundation Grants, 2496
Miami County Community Foundation - Operation Round Up Grants, 2543
Nicholas H. Noyes Jr. Memorial Fndn Grants, 2791
Nina Mason Pulliam Charitable Trust Grants, 2921
Noble County Community Foundation Grants, 2975
Norcliffe Foundation Grants, 2979
Nordson Corporation Foundation Grants, 2981
Norfolk Southern Foundation Grants, 2982
Northern New York Community Fndn Grants, 2996
Ogden Codman Trust Grants, 3066
Oleonda Jameson Trust Grants, 3071
Oppenstein Brothers Foundation Grants, 3078
PACCAR Foundation Grants, 3101
Parkersburg Area Community Foundation Action Grants, 3116
Percy B. Ferebee Endowment Grants, 3138
Perkins Charitable Foundation Grants, 3140
Peter Kiewit Foundation General Grants, 3154
Peter Kiewit Foundation Small Grants, 3155
Peyton Anderson Foundation Grants, 3160
Phelps County Community Foundation Grants, 3188
Philadelphia Foundation Organizational Effectiveness Grants, 3189
Phoenix Suns Charities Grants, 3201
Piper Jaffray Fndn Communities Giving Grants, 3227
Pittsburgh Foundation Community Fund Grants, 3230
Polk Bros. Foundation Grants, 3243
Pollock Foundation Grants, 3244
Prince Charitable Trusts Chicago Grants, 3270
Prince Charitable Trusts DC Grants, 3271
Principal Financial Group Foundation Grants, 3273
Procter and Gamble Fund Grants, 3274
Puerto Rico Community Foundation Grants, 3278
Quaker Chemical Foundation Grants, 3286
Qualcomm Grants, 3287
R.C. Baker Foundation Grants, 3292
R.T. Vanderbilt Trust Grants, 3294
Rainbow Endowment Grants, 3300
Ralph M. Parsons Foundation Grants, 3304
Ralphs Food 4 Less Foundation Grants, 3305
Rasmuson Foundation Tier One Grants, 3307
RCF General Community Grants, 3312
Reinberger Foundation Grants, 3357
Richard & Susan Smith Family Fndn Grants, 3365
Richard King Mellon Foundation Grants, 3369
Riley Foundation Grants, 3374
Robert B McMillen Foundation Grants, 3381
Robert W. Woodruff Foundation Grants, 3389
Robins Foundation Grants, 3390
Rochester Area Community Foundation Grants, 3391
Rochester Area Foundation Grants, 3392
Rockefeller Brothers Peace & Security Grants, 3394
Rockefeller Foundation Grants, 3396
Rockwell Collins Charitable Corporation Grants, 3397
Rockwell International Corporate Trust Grants Program, 3399
Roy & Christine Sturgis Charitable Grants, 3415
Ruth Anderson Foundation Grants, 3425
S. Livingston Mather Charitable Trust Grants, 3448
Salem Foundation Grants, 3457
San Antonio Area Foundation Grants, 3469
Sarkeys Foundation Grants, 3485
Schering-Plough Foundation Community Initiatives Grants, 3491
Schlessman Family Foundation Grants, 3493
Schurz Communications Foundation Grants, 3496
Scott B. & Annie P. Appleby Charitable Grants, 3499
Scott County Community Foundation Grants, 3500
Seabury Foundation Grants, 3501
Seagate Tech Corp Capacity to Care Grants, 3502
Seaver Institute Grants, 3509
Self Foundation Grants, 3510
Shell Deer Park Grants, 3543
Shell Oil Company Foundation Community Development Grants, 3544
Sonoco Foundation Grants, 3621
Sonora Area Foundation Competitive Grants, 3622
Sony Corporation of America Grants, 3623

Cultural Diversity /699

Southwest Florida Community Foundation Competitive Grants, 3627
Southwest Gas Corporation Foundation Grants, 3628
Spencer County Community Foundation Grants, 3632
Square D Foundation Grants, 3634
Stackpole-Hall Foundation Grants, 3643
Stella and Charles Guttman Foundation Grants, 3651
Stewart Huston Charitable Trust Grants, 3655
Strake Foundation Grants, 3658
Stranahan Foundation Grants, 3659
Summit Foundation Grants, 3674
Sunderland Foundation Grants, 3675
Sunoco Foundation Grants, 3679
SunTrust Bank Trusteed Foundations Florence C. and Harry L. English Memorial Fund Grants, 3680
SunTrust Bank Trusteed Foundations Greene-Sawtell Grants, 3681
SunTrust Bank Trusteed Foundations Harriet McDaniel Marshall Tust Grants, 3682
SunTrust Bank Trusteed Foundations Nell Warren Elkin and William Simpson Elkin Grants, 3683
SunTrust Bank Trusteed Foundations Thomas Guy Woolford Charitable Trust Grants, 3684
SunTrust Bank Trusteed Foundations Walter H. and Marjory M. Rich Memorial Fund Grants, 3685
Tauck Family Foundation Grants, 3703
Taylor S. Abernathy and Patti Harding Abernathy Charitable Trust Grants, 3704
TE Foundation Grants, 3705
Tension Envelope Foundation Grants, 3707
Theodore Edson Parker Foundation Grants, 3715
Thomas C. Ackerman Foundation Grants, 3719
Thomas J. Long Foundation Community Grants, 3722
TJX Foundation Grants, 3735
Tri-State Community Twenty-first Century Endowment Fund Grants, 3756
U.S. Cellular Corporation Grants, 3766
U.S. Department of Education 21st Century Community Learning Centers, 3767
Union Bank, N.A. Corporation Grants, 3788
United Technologies Corporation Grants, 3803
US Airways Community Contributions, 3825
US Airways Community Foundation Grants, 3826
Vancouver Foundation Grants and Community Initiatives Program, 3849
Vectren Foundation Grants, 3857
Wallace Alexander Gerbode Foundation Grants, 3896
Washington County Community Foundation Youth Grants, 3907
Wayne County Foundation Grants, 3909
Western Indiana Community Foundation Grants, 3921
Weyerhaeuser Company Foundation Grants, 3925
William B. Stokely Jr. Foundation Grants, 3947
William Blair and Company Foundation Grants, 3948
William Ray & Ruth E. Collins Fndn Grants, 3956
Winston-Salem Foundation Competitive Grants, 3965
Z. Smith Reynolds Foundation Small Grants, 3982

Cultural Diversity

Ackerman Foundation Grants, 236
ACL Diversity Community of Practice Grants, 240
ACL University Centers for Excellence in Developmental Network Diversity and Inclusion Training Action Planning Grants, 249
Alcoa Foundation Grants, 414
American Society on Aging NOMA Award for Excellence in Multicultural Aging, 562
ASHA Multicultural Activities Projects Grants, 697
AT&T Foundation Civic and Community Service Program Grants, 793
AVDF Health Care Grants, 811
Breast Cancer Fund Grants, 947
California Endowment Innovative Ideas Challenge Grants, 1000
Cargill Citizenship Fund Corporate Giving, 1018
Community Foundation of Greater Greensboro Community Grants, 1277
Cultural Society of Filipino Americans Grants, 1373
Delta Air Lines Foundation Prize for Global Understanding, 1456

700 / Cultural Diversity

George Foundation Grants, 1795
Henry County Community Foundation Grants, 2005
Lillian S. Wells Foundation Grants, 2365
Nathan Cummings Foundation Grants, 2627
Noble County Community Foundation Grants, 2975
Philadelphia Foundation Organizational Effectiveness Grants, 3189
Pittsburgh Foundation Community Fund Grants, 3230
Portland General Electric Foundation Grants, 3250
Seabury Foundation Grants, 3501
Sioux Falls Area Community Foundation Community Fund Grants (Unrestricted), 3568
Sioux Falls Area Community Foundation Spot Grants (Unrestricted), 3570
Vancouver Foundation Grants and Community Initiatives Program, 3849
Washington County Community Foundation Youth Grants, 3907

Cultural Heritage
Adolph Coors Foundation Grants, 311
Central Okanagan Foundation Grants, 1080
Charlotte County (FL) Community Foundation Grants, 1126
Cultural Society of Filipino Americans Grants, 1373
Delta Air Lines Foundation Prize for Global Understanding, 1456
Giving Sum Annual Grant, 1825
GNOF IMPACT Grants for Arts and Culture, 1838
Greater Sitka Legacy Fund Grants, 1885
Hawaii Community Foundation Reverend Takie Okumura Family Grants, 1977
Homer Foundation Grants, 2044
Infinity Foundation Grants, 2158
Judith and Jean Pape Adams Charitable Foundation Tulsa Area Grants, 2280
Kenai Peninsula Foundation Grants, 2298
Ketchikan Community Foundation Grants, 2303
Kodiak Community Foundation Grants, 2314
Leon and Thea Koerner Foundation Grants, 2355
Lotus 88 Fnd for Women & Children Grants, 2376
Mary K. Chapman Foundation Grants, 2443
Petersburg Community Foundation Grants, 3156
Schering-Plough Foundation Community Initiatives Grants, 3491
W.M. Keck Fndn So California Grants, 3890
Winston-Salem Fndn Stokes County Grants, 3967

Cultural Identity
Cultural Society of Filipino Americans Grants, 1373
Delta Air Lines Foundation Prize for Global Understanding, 1456
Homer Foundation Grants, 2044
James & Abigail Campbell Family Fndn Grants, 2185
W.M. Keck Fndn So California Grants, 3890

Cultural Outreach
2 Depot Square Ipswich Charitable Fndn Grants, 4
3M Company Fndn Community Giving Grants, 6
Abney Foundation Grants, 214
Agnes Gund Foundation Grants, 345
ALA Donald G. Davis Article Award, 401
Alexander & Baldwin Fnd Mainland Grants, 417
Alexander and Baldwin Foundation Hawaiian and Pacific Island Grants, 418
Anheuser-Busch Foundation Grants, 591
Anna Fitch Ardenghi Trust Grants, 592
Annie Sinclair Knudsen Memorial Fund/Kaua'i Community Grants Program, 597
Avon Products Foundation Grants, 814
Ball Brothers Foundation General Grants, 820
Barnes Group Foundation Grants, 832
Baxter International Foundation Grants, 843
Bayer Foundation Grants, 846
Benton County Foundation Grants, 880
Bridgestone/Firestone Trust Fund Grants, 949
Browning-Kimball Foundation Grants, 971
Carl B. and Florence E. King Foundation Grants, 1022
CFFVR Clintonville Area Foundation Grants, 1097
Clinton County Community Foundation Grants, 1184

Columbus Foundation Robert E. and Genevieve B. Schaefer Fund Grants, 1222
Community Foundation of Greater New Britain Grants, 1281
Community Foundation of Greenville-Greenville Women Giving Grants, 1283
Community Fndn of Howard County Grants, 1286
Community Foundation of the Eastern Shore Community Needs Grants, 1313
Cullen Foundation Grants, 1372
Cultural Society of Filipino Americans Grants, 1373
Dade Community Foundation Grants, 1410
Edward Bangs Kelley and Elza Kelley Foundation Grants, 1556
Eisner Foundation Grants, 1564
El Paso Corporate Foundation Grants, 1580
Essex County Community Foundation Merrimack Valley General Fund Grants, 1603
Essex County Community Foundation Webster Family Fund Grants, 1604
Evjue Foundation Grants, 1617
Fairfield County Community Foundation Grants, 1625
Fallon OrNda Community Health Fund Grants, 1627
Federal Express Corporate Contributions, 1657
Fishman Family Foundation Grants, 1672
Fred Baldwin Memorial Foundation Grants, 1736
Fulbright International Education Administrators (IEA) Seminar Program Grants, 1752
Greater Sitka Legacy Fund Grants, 1885
Greater Worcester Community Foundation Jeppson Memorial Fund for Brookfield Grants, 1889
Guy I. Bromley Trust Grants, 1919
H.A. & Mary K. Chapman Trust Grants, 1921
H.J. Heinz Company Foundation Grants, 1923
Harry Bramhall Gilbert Charitable Trust Grants, 1961
Hartford Foundation Regular Grants, 1970
Hawaii Community Foundation West Hawaii Fund Grants, 1978
Heckscher Foundation for Children Grants, 1988
Helen Bader Foundation Grants, 1992
Hershey Company Grants, 2013
Highmark Corporate Giving Grants, 2025
Hilton Head Island Foundation Grants, 2036
Homer Foundation Grants, 2044
Horizons Community Issues Grants, 2049
Humana Foundation Grants, 2077
Hutchinson Community Foundation Grants, 2085
Hut Foundation Grants, 2086
Infinity Foundation Grants, 2158
Irving S. Gilmore Foundation Grants, 2163
Irvin Stern Foundation Grants, 2164
Jackson County Community Foundation Unrestricted Grants, 2180
Jacob G. Schmidlapp Trust Grants, 2183
Jacobs Family Village Neighborhoods Grants, 2184
James Ford Bell Foundation Grants, 2187
James S. Copley Foundation Grants, 2200
JELD-WEN Foundation Grants, 2218
Joseph H. & Florence A. Roblee Fndn Grants, 2269
Judith and Jean Pape Adams Charitable Foundation Tulsa Area Grants, 2280
Kahuku Community Fund, 2286
Katrine Menzing Deakins Trust Grants, 2295
Kenai Peninsula Foundation Grants, 2298
Ketchikan Community Foundation Grants, 2303
Kodiak Community Foundation Grants, 2314
Ludwick Family Foundation Grants, 2389
M.B. and Edna Zale Foundation Grants, 2401
M.J. Murdock Charitable Trust General Grants, 2405
Maine Community Foundation Charity Grants, 2415
Mary K. Chapman Foundation Grants, 2443
Nathan Cummings Foundation Grants, 2627
Pinellas County Grants, 3223
Qualcomm Grants, 3287
Rasmuson Foundation Tier Two Grants, 3308
Richard & Susan Smith Family Fndn Grants, 3365
Salem Foundation Charitable Trust Grants, 3456
Samuel N. and Mary Castle Foundation Grants, 3467
Southbury Community Trust Fund, 3625
Textron Corporate Contributions Grants, 3712

SUBJECT INDEX

Thomas J. Long Foundation Community Grants, 3722
Trull Foundation Grants, 3761
Union Benevolent Association Grants, 3789
W.C. Griffith Foundation Grants, 3881
Wallace Alexander Gerbode Foundation Grants, 3896
Washington County Community Foundation Youth Grants, 3907
William Ray & Ruth E. Collins Fndn Grants, 3956

Curriculum Development
A.O. Smith Foundation Community Grants, 17
Adobe Community Investment Grants, 310
Amgen Foundation Grants, 572
Bayer Foundation Grants, 846
Bristol-Myers Squibb Foundation Science Education Grants, 961
Cambridge Community Foundation Grants, 1004
Carnahan-Jackson Foundation Grants, 1032
Christensen Fund Regional Grants, 1149
Crail-Johnson Foundation Grants, 1355
Florida Division of Cultural Affairs Arts In Education Arts Partnership Grants, 1679
Ford Motor Company Fund Grants Program, 1687
GlaxoSmithKline Corporate Grants, 1829
Huntington Beach Police Officers Fndn Grants, 2081
Jack H. & William M. Light Trust Grants, 2179
Jennings County Community Foundation Grants, 2220
Josiah Macy Jr. Foundation Grants, 2277
Lynde & Harry Bradley Foundation Fellowships, 2396
Lynde and Harry Bradley Foundation Grants, 2397
Lynde and Harry Bradley Foundation Prizes: Bradley Prizes, 2398
NCI Cancer Education and Career Development Program, 2648
NIH School Interventions to Prevent Obesity, 2908
NIMH Curriculum Development Award in Neuroinformatics Research and Analysis, 2917
Polk Bros. Foundation Grants, 3243
Procter and Gamble Fund Grants, 3274
Rathmann Family Foundation Grants, 3309
SfN Science Educator Award, 3535
South Madison Community Foundation Grants, 3626
STTI/ATI Educational Assessment Nursing Research Grants, 3665
Tellabs Foundation Grants, 3706
Vancouver Foundation Grants and Community Initiatives Program, 3849
Vectren Foundation Grants, 3857
Wege Foundation Grants, 3912
Western Indiana Community Foundation Grants, 3921
Weyerhaeuser Company Foundation Grants, 3925

Cystic Fibrosis
Abbott Fund CFCareForward Scholarships, 200
CFF-NIH Funding Grants, 1083
CFF Clinical Research Grants, 1084
CFF First- and Second-Year Clinical Fellowships, 1085
CFF Shwachman Clinical Investigator Award, 1086
CFF Leroy Matthews Physician-Scientist Awards, 1087
CFF Pilot and Feasibility Awards, 1088
CFF Postdoctoral Research Fellowships, 1089
CFF Research Grants, 1090
CFF Student Traineeships, 1091
CFF 3rd through 5th Year Clinical Fellowships, 1092
Cystic Fibrosis Canada Clinical Fellowships, 1381
Cystic Fibrosis Canada Clinical Project Grants, 1382
Cystic Fibrosis Canada Clinic Incentive Grants, 1383
Cystic Fibrosis Canada Fellowships, 1384
Cystic Fibrosis Canada Research Grants, 1385
Cystic Fibrosis Canada Scholarships, 1386
Cystic Fibrosis Canada Senior Scientist Research Training Awards, 1387
Cystic Fibrosis Canada Small Conference Grants, 1388
Cystic Fibrosis Canada Special Travel Grants For Fellows and Students, 1389
Cystic Fibrosis Canada Studentships, 1390
Cystic Fibrosis Canada Summer Studentships, 1391
Cystic Fibrosis Canada Transplant Center Incentive Grants, 1392
Cystic Fibrosis Canada Travel Supplement Grants, 1393

SUBJECT INDEX

Cystic Fibrosis Canada Visiting Allied Health
    Professional Awards, 1394
Cystic Fibrosis Canada Visiting Clinician Awards, 1395
Cystic Fibrosis Canada Visiting Scientist Awards, 1396
Cystic Fibrosis Lifestyle Foundation Individual
    Recreation Grants, 1397
Cystic Fibrosis Lifestyle Foundation Loretta Morris
    Memorial Fund Grants, 1398
Cystic Fibrosis Lifestyle Foundation Mentored
    Recreation Grants, 1399
Cystic Fibrosis Lifestyle Foundation Peer Support
    Grants, 1400
Cystic Fibrosis Research Eliz Nash Fellowships, 1401
Cystic Fibrosis Research New Horizons Campaign
    Grants, 1402
Cystic Fibrosis Scholarships, 1403
Cystic Fibrosis Trust Research Grants, 1404
Elizabeth Nash Foundation Scholarships, 1570
Elizabeth Nash Fnd Summer Research Awards, 1571
Foundation of CVPH April LaValley Grants, 1701
Robert R. Meyer Foundation Grants, 3387

Cytoskeleton
NSF Cell Systems Cluster Grants, 3011

DNA
APHL Emerging Infectious Diseases Fellowships, 636

Dance
AAHPERD Abernathy Presidential Scholarships, 151
Achelis Foundation Grants, 234
Adaptec Foundation Grants, 304
Agnes Gund Foundation Grants, 345
Bodman Foundation Grants, 929
Carnahan-Jackson Foundation Grants, 1032
Cooke Foundation Grants, 1344
Crystelle Waggoner Charitable Trust Grants, 1364
Jerome Robbins Foundation Grants, 2223
Kenneth T. & Eileen L. Norris Fndn Grants, 2300
Leon and Thea Koerner Foundation Grants, 2355
Procter and Gamble Fund Grants, 3274
Puerto Rico Community Foundation Grants, 3278
Richard D. Bass Foundation Grants, 3366
Robert R. Meyer Foundation Grants, 3387
Sid W. Richardson Foundation Grants, 3553
Sioux Falls Area Community Foundation Community
    Fund Grants (Unrestricted), 3568
Sioux Falls Area Community Foundation Spot Grants
    (Unrestricted), 3570
Southbury Community Trust Fund, 3625

Dance Education
AAHPERD Abernathy Presidential Scholarships, 151
Jerome Robbins Foundation Grants, 2223
Leon and Thea Koerner Foundation Grants, 2355

Data Analysis
AES Research Infrastructure Awards, 321
NCI Academic Industrial Partnerships for Devel and
    Validation of In Vivo Imaging Systems and Methods
    for Cancer Investigations, 2644
NCI Application and Use of Transformative Emerging
    Technologies in Cancer Research, 2645
NEI Research Grant For Secondary Data Analysis
    Grants, 2667
NIMH Short Courses in Neuroinformatics, 2920
NLM Express Research Grants in Biomedical
    Informatics, 2943
NSF Doctoral Dissertation Improvement Grants in the
    Directorate for Biological Sciences (DDIG), 3013

Data Processing
NLM Express Research Grants in Biomedical
    Informatics, 2943

Databases
AES Research Infrastructure Awards, 321
NEI Research Grant For Secondary Data Analysis
    Grants, 2667

NIDDK Ancillary Studies of Kidney Disease Accessing
    Information from Clinical Trials, Epidemiological
    Studies, and Databases Grants, 2799
NLM Internet Connection for Medical Institutions
    Grants, 2946
NLM Research Project Grants, 2951

Day Care
Carnegie Corporation of New York Grants, 1033
CFFVR Basic Needs Giving Partnership Grants, 1094
Community Foundation of the Verdugos Educational
    Endowment Fund Grants, 1314
Fargo-Moorhead Area Fndn Woman's Grants, 1633
FCD New American Children Grants, 1637
Florida BRAIVE Fund of Dade Community
    Foundation, 1678
Fred & Gretel Biel Charitable Trust Grants, 1734
Gulf Coast Foundation of Community Operating
    Grants, 1916
Gulf Coast Foundation of Community Program
    Grants, 1917
Hasbro Children's Fund Grants, 1973
Josephine S. Gumbiner Foundation Grants, 2273
Kathryne Beynon Foundation Grants, 2294
Knox County Community Foundation Grants, 2313
Kosciusko County Community Foundation REMC
    Operation Round Up Grants, 2320
Mary Owen Borden Foundation Grants, 2446
OneFamily Foundation Grants, 3075
Perry County Community Foundation Grants, 3142
Pike County Community Foundation Grants, 3222
PMI Foundation Grants, 3239
Porter County Women's Grant, 3248
Posey County Community Foundation Grants, 3252
RCF Summertime Kids Grants, 3314
Spencer County Community Foundation Grants, 3632
Thomas Sill Foundation Grants, 3724
U.S. Department of Education 21st Century
    Community Learning Centers, 3767
Union Bank, N.A. Foundation Grants, 3788
W.H. & Mary Ellen Cobb Trust Grants, 3884

Death/Mortality
AAHPM Hospice & Palliative Med Fellowships, 152
Caplow Applied Science Children's Prize, 1016
CIGNA Foundation Grants, 1160
George E. Hatcher, Jr. and Ann Williams Hatcher
    Foundation Grants, 1792
Greenwall Foundation Bioethics Grants, 1897
NCI Centers of Excellence in Cancer Communications
    Research, 2649
NIA Health Behaviors and Aging Grants, 2761
NINR Quality of Life for Individuals at the End of Life
    Grants, 2933
Quantum Corporation Snap Server Grants, 3289

Decision Sciences
CDC Steven M. Teutsch Prevention Effectiveness
    Fellowships, 1071
Jane Beattie Memorial Scholarship, 2207
NCI Centers of Excellence in Cancer Communications
    Research, 2649
NLM Research Project Grants, 2951

Defense Technology
AAAS Science and Technology Policy Fellowships:
    Global Health and Development, 41

Dementia
Brookdale Fnd National Group Respite Grants, 966
Helen Bader Foundation Grants, 1992
Henrietta Lange Burk Fund Grants, 2001
Pfizer Australia Neuroscience Research Grants, 3175
Pfizer Healthcare Charitable Contributions, 3180

Democracy
Vernon K. Krieble Foundation Grants, 3866

Dental Health and Hygiene /701

Demography
James S. McDonnell Foundation Complex Systems
    Collaborative Activity Awards, 2202
NCHS National Center for Health Statistics
    Postdoctoral Research Appointments, 2642

Dental Education
ADA Foundation Bud Tarrson Dental School Student
    Community Leadership Awards, 287
ADA Foundation Dental Student Scholarships, 288
ADA Foundation Dentsply International Research
    Fellowships, 289
ADA Found Minority Dental Scholarships, 292
ADA Foundation Scientist in Training Fellowship, 295
ADA Foundation Summer Scholars Fellowships, 296
ADA Foundation Thomas J. Zwemer Award, 297
Alpha Omega Foundation Grants, 451
Baxter International Foundation Grants, 843
Community Foundation of Louisville Delta Dental of
    Kentucky Fund Grants, 1292
DeKalb County Community Foundation - Garrett
    Hospital Aid Foundation Grants, 1446
Ford Family Foundation Grants - Access to Health and
    Dental Services, 1686
HHMI-NIH Cloister Research Scholars, 2015
HHMI Physician-Scientist Early Career Award, 2023
MLA Donald A.B. Lindberg Fellowship, 2562
NHLBI Biomedical Research Training Program for
    Individuals from Underrepresented Groups, 2692
NIH Ruth L. Kirschstein National Research Service
    Award Institutional Research Training Grants, 2900
NIH Ruth L. Kirschstein Nat Research Service Award
    Short-Term Institutional Training Grants, 2906
NYAM David E. Rogers Fellowships, 3040
Pajaro Valley Community Health Health Trust
    Insurance/Coverage & Education on Using the
    System Grants, 3108
Pajaro Valley Community Health Trust Oral Health:
    Prevention & Access Grants, 3110
Pollock Foundation Grants, 3244
Summer Medical and Dental Education Program:
    Interprofessional Pilot Grants, 3673

Dental Health and Hygiene
ADA Foundation Disaster Assistance Grants, 290
ADA Foundation Samuel Harris Children's Dental
    Health Grants, 294
Aetna Foundation Regional Health Grants, 333
Alexander Eastman Foundation Grants, 420
Alpha Omega Foundation Grants, 451
Campbell Hoffman Foundation Grants, 1006
Community Foundation of Louisville Delta Dental of
    Kentucky Fund Grants, 1292
Delta Air Lines Foundation Health and Wellness
    Grants, 1455
DHHS Oral Health Promotion Research Across the
    Lifespan, 1474
Elkhart County Community Foundation Fund for
    Elkhart County, 1572
HAF JoAllen K. Twiddy Wood Memorial Fund
    Grants, 1929
Health Fndn of Greater Indianapolis Grants, 1984
Henrietta Tower Wurts Memorial Fndn Grants, 2002
Jenkins Foundation: Improving the Health of Greater
    Richmond Grants, 2219
Kansas Health Fndn Major Initiatives Grants, 2288
Lloyd A. Fry Foundation Health Grants, 2373
Nelda C. and H.J. Lutcher Stark Fndn Grants, 2672
NIH Ruth L. Kirschstein National Research Service
    Awards for Individual Postdoc Fellowships, 2902
NLM Manufacturing Processes of Medical, Dental,
    and Biological Technologies (SBIR) Grants, 2948
NYAM David E. Rogers Fellowships, 3040
Pajaro Valley Community Health Health Trust
    Insurance/Coverage & Education on Using the
    System Grants, 3108
Pajaro Valley Community Health Trust Oral Health:
    Prevention & Access Grants, 3110
Pew Charitable Trusts Children & Youth Grants, 3158

Dental Health and Hygiene

Ruth Eleanor Bamberger and John Ernest Bamberger Memorial Foundation Grants, 3427
Saint Luke's Health Initiatives Grants, 3455
St. Joseph Community Health Foundation Improving Healthcare Access Grants, 3640
Union Labor Health Foundation Dental Angel Fund Grants, 3793

Dentistry

ADA Foundation George C. Paffenbarger Student Research Awards, 291
AIHS Clinical Fellowships, 375
Alpha Omega Foundation Grants, 451
CDC Public Health Informatics Fellowships, 1068
DHHS Oral Health Promotion Research Across the Lifespan, 1474
Foundation for Health Enhancement Grants, 1696
NIA Mentored Clinical Scientist Research Career Development Awards, 2771
NYAM David E. Rogers Fellowships, 3040
PhRMA Foundation Health Outcomes Pre-Doctoral Fellowships, 3203
PhRMA Foundation Paul Calabresi Medical Student Research Fellowship, 3210
Special Olympics Health Profess Student Grants, 3629

Dentistry, Preventive

Ford Family Foundation Grants - Access to Health and Dental Services, 1686
Foundation for Health Enhancement Grants, 1696

Depression

AACAP Klingenstein Third Generation Foundation Award for Research in Depression or Suicide, 62
Aetna Foundation Regional Health Grants, 333
George Foundation Grants, 1795
Grand Rapids Community Foundation Southeast Ottawa Youth Fund Grants, 1874
Horizon Foundation for New Jersey Grants, 2048
NARSAD Colvin Prize for Outstanding Achievement in Mood Disorders Research, 2619
NARSAD Distinguished Investigator Grants, 2620
NARSAD Independent Investigator Grants, 2622
NARSAD Young Investigator Grants, 2625
Nick Traina Foundation Grants, 2792
NIMH Early Identification and Treatment of Mental Disorders in Children and Adolescents Grants, 2918
NINR Quality of Life for Individuals at the End of Life Grants, 2933
Peter and Elizabeth C. Tower Foundation Annual Mental Health Grants, 3145
Pfizer Australia Neuroscience Research Grants, 3175
Pfizer Healthcare Charitable Contributions, 3180

Dermatology

American Academy of Dermatology Camp Discovery Scholarships, 503
MMAAP Foundation Dermatology Fellowships, 2577
MMAAP Fndn Dermatology Project Awards, 2578
National Psoriasis Foundation Research Grants, 2636
Pfizer Research Initiative Dermatology Grants (Germany), 3184
Skin Cancer Foundation Research Grants, 3578

Design Arts

Jayne and Leonard Abess Foundation Grants, 2214

Developing/Underdeveloped Nations

ADA Foundation Thomas J. Zwemer Award, 297
American Jewish World Service Grants, 533
Bill and Melinda Gates Foundation Emergency Response Grants, 892
Bill and Melinda Gates Foundation Water, Sanitation and Hygiene Grants, 894
Conservation, Food, and Health Foundation Grants for Developing Countries, 1338
Covidien Medical Product Donations, 1352
Elizabeth Glaser Int'l Leadership Awards, 1566
Gates Award for Global Health, 1775
Harold Simmons Foundation Grants, 1955
IBRO Asia Reg APRC Exchange Fellowships, 2101
IBRO Asia Regional APRC Travel Grants, 2103
IBRO Return Home Fellowships, 2112
John Deere Foundation Grants, 2234
MAP International Medical Fellowships, 2421
MLA T. Mark Hodges Int'l Service Award, 2575
Nestle Foundation Large Research Grants, 2675
Nestle Foundation Pilot Grants, 2676
Nestle Foundation Re-entry Grants, 2677
Nestle Foundation Training Grant, 2679
Packard Foundation Population and Reproductive Health Grants, 3106
PC Fred H. Bixby Fellowships, 3126
Sir Dorabji Tata Trust Grants for NGOs or Voluntary Organizations, 3572
UICC American Cancer Society International Fellowships for Beginning Investigators, 3778

Developmental Psychology

NSF Social Psychology Research Grants, 3023

Developmentally Disabled

ACL Partnerships in Employment Systems Change Grants, 246
Brighter Tomorrow Foundation Grants, 950
Caesars Foundation Grants, 995
Catherine Kennedy Home Foundation Grants, 1043
CDC Epidemic Intell Service Training Grants, 1055
CDC Epidemiology Elective Rotation, 1056
Coleman Foundation Developmental Disabilities Grants, 1207
Columbus Foundation Traditional Grants, 1223
CVS Community Grants, 1379
Dr. Scholl Foundation Grants Program, 1511
George W. Wells Foundation Grants, 1804
Grace Bersted Foundation Grants, 1859
Graham and Carolyn Holloway Family Foundation Grants, 1861
H & R Foundation Grants, 1920
Hampton Roads Community Foundation Developmental Disabilities Grants, 1939
HRAMF Deborah Munroe Noonan Memorial Research Grants, 2058
John D. & Katherine A. Johnston Fndn Grants, 2233
John Merck Fund Grants, 2245
John Merck Scholars Awards, 2246
John W. Anderson Foundation Grants, 2262
Leon and Thea Koerner Foundation Grants, 2355
Lewis H. Humphreys Charitable Trust Grants, 2361
Lucy Gooding Charitable Fndn Grants, 2388
Marjorie Moore Charitable Foundation Grants, 2438
North Carolina GlaxoSmithKline Fndn Grants, 2991
Peter and Elizabeth C. Tower Foundation Annual Intellectual Disabilities Grants, 3144
Peter and Elizabeth C. Tower Foundation Mental Health Reference and Resource Materials Mini-Grants, 3147
Peter and Elizabeth C. Tower Foundation Phase II Technology Initiative Grants, 3149
Peter and Elizabeth C. Tower Foundation Phase I Technology Initiative Grants, 3150
Peter and Elizabeth C. Tower Foundation Social and Emotional Preschool Curriculum Grants, 3151
Robert and Joan Dircks Foundation Grants, 3379
Simple Advise Education Center Grants, 3565
Spencer County Community Foundation Grants, 3632
Strowd Roses Grants, 3664
Thomas W. Bradley Foundation Grants, 3726

Diabetes

Abbott Fund Access to Health Care Grants, 199
Abbott Fund Science Education Grants, 203
Aetna Foundation Regional Health Grants, 333
AIHS Alberta/Pfizer Translat Research Grants, 374
Austin S. Nelson Foundation Grants, 805
Balfe Family Foundation Grants, 819
Campbell Soup Foundation Grants, 1007
CFFVR Basic Needs Giving Partnership Grants, 1094
CFFVR Jewelers Mutual Charitable Giving, 1099
CFFVR Robert and Patricia Endries Family Foundation Grants, 1102
CNCS Senior Companion Program Grants, 1191
Community Foundation of Jackson County Seymour Noon Lions Club Grant, 1288
Dorothy Rider Pool Health Care Grants, 1500
Fairlawn Foundation Grants, 1626
Florence Hunt Maxwell Foundation Grants, 1676
Foundation for the Mid South Health and Wellness Grants, 1700
Gerber Foundation Grants, 1807
Gheens Foundation Grants, 1814
GNOF IMPACT Kahn-Oppenheim Grants, 1842
Greenwall Foundation Bioethics Grants, 1897
Hawaii Community Foundation Health Education and Research Grants, 1976
Horizon Foundation for New Jersey Grants, 2048
J. Spencer Barnes Memorial Foundation Grants, 2175
Johnson & Johnson Community Health Grants, 2252
Kovler Family Foundation Grants, 2323
M. Bastian Family Foundation Grants, 2402
Medtronic Foundation Strengthening Health Systems Grants, 2515
NEI Innovative Patient Outreach Programs And Ocular Screening Technologies To Improve Detection Of Diabetic Retinopathy Grants, 2665
NIDDK Adverse Metabolic Side Effects of Second Generation Psychotropic Medications Leading to Obesity and Increased Diabetes Risk Grants, 2798
NIDDK Collaborative Interdisciplinary Team Science in Diabetes, Endocrinology and Metabolic Diseases Grants, 2806
NIDDK Devel of Disease Biomarkers Grants, 2809
NIDDK Diabetes, Endocrinology, and Metabolic Diseases NRSAs--Individual, 2810
NIDDK Diabetes Research Centers, 2811
NIDDK Education Program Grants, 2813
NIDDK Enhancing Zebrafish Research with Research Tools and Techniques, 2815
NIDDK Health Disparities in NIDDK Diseases, 2820
NIDDK Identifying & Reducing Diabetes & Obesity Related Disparities in Healthcare Systems, 2821
NIDDK Insulin Signaling and Receptor Cross Talk Grants, 2822
NIDDK Mentored Research Scientist Development Award, 2825
NIDDK Multi-Center Clinical Study Cooperative Agreements, 2826
NIDDK Multi-Center Clinical Study Implementation Planning Grants, 2827
NIDDK Neurobiology of Diabetic Complications Grants, 2828
NIDDK Non-Invasive Methods for Diagnosis and Progression of Diabetes, Kidney, Urological, Hematological and Digestive Diseases, 2832
NIDDK Pancreatic Development and Regeneration Grants: Cellular Therapies for Diabetes, 2833
NIDDK Feasibility Clinical Research Grants in Diabetes, Endocrine and Metabolic Diseases, 2834
NIDDK Planning Grants For Translational Research For The Prevention And Control Of Diabetes And Obesity, 2837
NIDDK Proteomics: Diabetes, Obesity, And Endocrine, Digestive, Kidney, Urologic, And Hematologic Diseases, 2838
NIDDK Role of Gastrointestinal Surgical Procedures in Amelioration of Obesity-Related Insulin Resistance & Diabetes Weight Loss, 2841
NIDDK Secondary Analyses in Obesity, Diabetes and Digestive and Kidney Diseases Grants, 2842
NIDDK Seeding Collaborative Interdisciplinary Team Science in Diabetes, Endocrinology and Metabolic Diseases Grants, 2843
NIDDK Small Business Innov Research to Develop New Therapeutics & Monitoring Tech for Type 1 Diabetes Towards an Artificial Pancreas, 2845
NIDDK Small Grants for K08/K23 Recipients, 2846
NIDDK Small Grants for Underrepresented Minority Scientists in Diabetes and Digestive and Kidney Diseases, 2847

SUBJECT INDEX

NIDDK Translational Research for the Prevention and Control of Diabetes and Obesity, 2850
NIDDK Type 2 Diabetes in the Pediatric Population Research Grants, 2852
NIH Enhancing Adherence to Diabetes Self-Management Behaviors, 2875
NIH Fine Mapping and Function of Genes for Type 1 Diabetes Grants, 2879
NIH Nonhuman Primate Immune Tolerance Cooperative Study Group Grants, 2890
NINR Diabetes Self-Management in Minority Populations Grants, 2930
OneSight Research Foundation Block Grants, 3076
Pajaro Valley Community Health Trust Diabetes and Contributing Factors Grants, 3109
Partnership for Cures Two Years To Cures Grants, 3118
Pfizer ASPIRE North America Broad Spectrum Antibiotics for the Treatment of Gram-Negative or Polymicrobial Infections Research Awards, 3169
Pfizer ASPIRE North America Narrow Spectrum Antibiotics for the Treatment of MRSA Research Awards, 3171
Pfizer Australia Paediatric Endocrine Care Research Grants, 3176
Pfizer Healthcare Charitable Contributions, 3180
Pfizer Medical Education Track Two Grants, 3183
Pittsburgh Foundation Medical Research Grants, 3231
Premera Blue Cross Grants, 3258
Richard & Susan Smith Family Fndn Grants, 3365
Seattle Foundation Medical Funds Grants, 3508
William G. & Helen C. Hoffman Fndn Grants, 3950

Diabetic Retinopathy
AIHS Alberta/Pfizer Translat Research Grants, 374
NEI Innovative Patient Outreach Programs And Ocular Screening Technologies To Improve Detection Of Diabetic Retinopathy Grants, 2665
Sara Elizabeth O'Brien Trust Grants, 3483

Diagnosis, Medical
ADA Foundation Thomas J. Zwemer Award, 297
Alexander and Margaret Stewart Trust Grants, 419
ALVRE Casselberry Award, 460
ALVRE Grant, 461
ALVRE Seymour R. Cohen Award, 462
ASM Scherago-Rubin Award, 750
Children's Cardiomyopathy Foundation Research Grants, 1137
Covidien Partnership for Neighborhood Wellness Grants, 1353
Emma B. Howe Memorial Foundation Grants, 1593
HRSA Ryan White HIV AIDS Drug Assistance Grants, 2070
Lymphatic Education and Research Network Postdoctoral Fellowships, 2395
Maddie's Fund Medical Equipment Grants, 2411
NCI Exploratory Grants for Behavioral Research in Cancer Control, 2651
NEI Innovative Patient Outreach Programs And Ocular Screening Technologies To Improve Detection Of Diabetic Retinopathy Grants, 2665
NHLBI Investigator Initiated Multi-Site Clinical Trials, 2703
NIA Alzheimer's Disease Core Centers Grants, 2748
OSF Access to Essential Medicines Initiative, 3086
Partnership for Cures Two Years To Cures Grants, 3118
Pfizer ASPIRE EU Antifungal Research Awards, 3165
Pfizer Inflammation Competitive Research Awards (UK), 3181
Piper Trust Healthcare & Med Research Grants, 3228
Susan G. Komen Breast Cancer Foundation Challenge Grants: Investigator Initiated Research, 3690
UICC American Cancer Society International Fellowships for Beginning Investigators, 3778
US CRDF Leishmaniasis: Collaborative Research Opportunities in N Africa & Middle East, 3828

Dietary Foods
Institute for Agriculture and Trade Policy Food and Society Fellowships, 2159

NCI Diet, Epigenetic Events, and Cancer Prevention Grants, 2650
NCI Improving Diet and Physical Activity Assessment Grants, 2652
NHLBI Targeted Approaches to Weight Control for Young Adults Grants, 2736
NIH Dietary Supplement Research Centers: Botanicals Grants, 2873

Dietary Supplements
NIH Dietary Supplement Research Centers: Botanicals Grants, 2873
Pfizer Medical Education Track Two Grants, 3183

Dietetics/Nutrition
AOCS Health & Nutrition Div Student Award, 614
AOCS Health and Nutrition Poster Competition, 615
AOCS Holman Lifetime Achievement Award, 620
Bupa Foundation Multi-Country Grant, 975
GNOF IMPACT Kahn-Oppenheim Grants, 1842
PepsiCo Foundation Grants, 3137
Phi Upsilon Omicron Frances Morton Holbrook Alumni Award, 3193
Phi Upsilon Omicron Geraldine Clewell Senior Awards, 3194
Phi Upsilon Omicron Janice Cory Bullock Collegiate Award, 3195
Phi Upsilon Lillian P. Schoephoerster Award, 3196
Phi Upsilon Omicron Margaret Jerome Sampson Scholarships, 3197
Phi Upsilon Orinne Johnson Writing Award, 3198
Phi Upsilon Omicron Sarah Thorniley Phillips Leader Awards, 3199
Phi Upsilon Omicron Undergraduate Karen P. Goebel Conclave Award, 3200
RWJF Healthy Eating Research Grants, 3437
Union Bank, N.A. Foundation Grants, 3788

Digestive Diseases and Disorders
ASGE Don Wilson Award, 685
ASGE Endoscopic Research Awards, 686
ASGE Endoscopic Research Career Development Awards, 687
ASGE Given Capsule Endoscopy Research Award, 688
CFF First- and Second-Year Clinical Fellowships, 1085
CFF Postdoctoral Research Fellowships, 1089
CFF Research Grants, 1090
CFF 3rd through 5th Year Clinical Fellowships, 1092
FDHN Bridging Grants, 1638
FDHN Centocor International Research Fellowship in Gastrointestinal Inflammation & Immunology, 1639
FDHN Designated Research Award in Geriatric Gastroenterology, 1641
FDHN Designated Research Award in Research Related to Pancreatitis, 1642
FDHN Fellow Abstract Prizes, 1643
FDHN Fellowship to Faculty Transition Awards, 1644
FDHN Graduate Student Awards, 1646
FDHN Isenberg Int'l Research Scholar Award, 1647
FDHN June & Donald O. Castell MD, Esophageal Clinical Research Award, 1648
FDHN Moti L. & Kamla Rustgi International Travel Awards, 1649
FDHN Non-Career Research Awards, 1650
FDHN Student Abstract Prizes, 1653
FDHN Student Research Fellowships, 1654
FDHN TAP Endowed Designated Research Award in Acid-Related Diseases, 1655
NHLBI Lymphatics in Health and Disease in the Digestive, Urinary, Cardiovascular and Pulmonary Systems, 2705
NIDDK Education Program Grants, 2813
NIDDK Enhancing Zebrafish Research with Research Tools and Techniques, 2815
NIDDK Health Disparities in NIDDK Diseases, 2820
NIDDK Intestinal Failure, Short Gut Syndrome and Small Bowel Transplantation Grants, 2823
NIDDK Intestinal Stem Cell Consortium Grants, 2824
NIDDK Mentored Research Scientist Development Award, 2825

Disabled /703

NIDDK Multi-Center Clinical Study Cooperative Agreements, 2826
NIDDK Multi-Center Clinical Study Implementation Planning Grants, 2827
NIDDK Non-Invasive Methods for Diagnosis and Progression of Diabetes, Kidney, Urological, Hematological and Digestive Diseases, 2832
NIDDK Pilot and Feasibility Clinical Research Studies in Digestive Diseases and Nutrition, 2836
NIDDK Proteomics: Diabetes, Obesity, And Endocrine, Digestive, Kidney, Urologic, And Hematologic Diseases, 2838
NIDDK Secondary Analyses in Obesity, Diabetes and Digestive and Kidney Diseases Grants, 2842
NIDDK Silvio O. Conte Digestive Diseases Research Core Centers Grants, 2844
NIDDK Small Grants for K08/K23 Recipients, 2846
NIDDK Small Grants for Underrepresented Minority Scientists in Diabetes and Digestive and Kidney Diseases, 2847

Digestive System
ASGE Endoscopic Research Awards, 686
ASGE Endoscopic Research Career Development Awards, 687
ASGE Given Capsule Endoscopy Research Award, 688
FDHN Centocor International Research Fellowship in Gastrointestinal Inflammation & Immunology, 1639
FDHN Designated Research Award in Geriatric Gastroenterology, 1641
FDHN Fellow Abstract Prizes, 1643
FDHN June & Donald O. Castell MD, Esophageal Clinical Research Award, 1648
FDHN Moti L. & Kamla Rustgi International Travel Awards, 1649
FDHN Non-Career Research Awards, 1650
FDHN TAP Endowed Designated Research Award in Acid-Related Diseases, 1655
NHLBI Lymphatics in Health and Disease in the Digestive, Urinary, Cardiovascular and Pulmonary Systems, 2705
NIDDK Devel of Disease Biomarkers Grants, 2809
NIDDK Intestinal Failure, Short Gut Syndrome and Small Bowel Transplantation Grants, 2823
NIDDK Pilot and Feasibility Clinical Research Studies in Digestive Diseases and Nutrition, 2836
NIDDK Secondary Analyses in Obesity, Diabetes and Digestive and Kidney Diseases Grants, 2842
NIDDK Silvio O. Conte Digestive Diseases Research Core Centers Grants, 2844

Disabled
AAPD Henry B. Betts Award, 182
AAPD Paul G. Hearne Leadership Award, 183
Abbot and Dorothy H. Stevens Foundation Grants, 197
Abell-Hanger Foundation Grants, 207
Able To Serve Grants, 212
Able Trust Vocational Rehabilitation Grants for Individuals, 213
ACL Learning Collaboratives for Advanced Business Acumen Skills Grants, 244
Adams Rotary Memorial Fund A Grants, 302
Aetna Foundation Health Grants in Connecticut, 327
Alberto Culver Corporate Contributions Grants, 408
Albert W. Rice Charitable Foundation Grants, 411
Albuquerque Community Foundation Grants, 412
Alexander and Margaret Stewart Trust Grants, 419
Alfred E. Chase Charitable Foundation Grants, 430
Amelia Sillman Rockwell and Carlos Perry Rockwell Charities Fund Grants, 501
Amerigroup Foundation Grants, 565
Anne J. Caudal Foundation Grants, 595
Ann Jackson Family Foundation Grants, 598
Ann Peppers Foundation Grants, 599
Annunziata Sanguinetti Foundation Grants, 600
Arthur Ashley Williams Foundation Grants, 671
ASHFoundation Graduate Student Scholarships for Minority Students, 704
ASHFoundation Graduate Student with a Disability Scholarship, 705

704 / Disabled

Austin S. Nelson Foundation Grants, 805
BancorpSouth Foundation Grants, 823
Baxter International Foundation Grants, 843
Berrien Community Foundation Grants, 884
Bodenwein Public Benevolent Foundation Grants, 928
Boston Foundation Grants, 936
Boston Globe Foundation Grants, 937
Callaway Golf Company Foundation Grants, 1003
Carl R. Hendrickson Family Foundation Grants, 1028
Carls Foundation Grants, 1030
Carl W. and Carrie Mae Joslyn Trust Grants, 1031
Carnahan-Jackson Foundation Grants, 1032
Carrie E. and Lena V. Glenn Foundation Grants, 1036
Carrie Estelle Doheny Foundation Grants, 1037
Catherine Kennedy Home Foundation Grants, 1043
Cessna Foundation Grants Program, 1081
CFFVR Robert and Patricia Endries Family Foundation Grants, 1102
Champ-A Champion Fur Kids Grants, 1109
Charles Delmar Foundation Grants, 1115
Charles H. Hall Foundation, 1120
Charles Nelson Robinson Fund Grants, 1125
Chatlos Foundation Grants Program, 1128
Children's Trust Fund of Oregon Fndn Grants, 1139
Christopher & Dana Reeve Foundation Quality of Life Grants, 1152
CICF Indianapolis Fndn Community Grants, 1156
Clara Blackford Smith and W. Aubrey Smith Charitable Foundation Grants, 1165
Clark-Winchcole Foundation Grants, 1173
CNA Foundation Grants, 1189
CNCS Senior Companion Program Grants, 1191
Coleman Foundation Developmental Disabilities Grants, 1207
Collins Foundation Grants, 1210
Columbus Foundation Traditional Grants, 1223
Community Foundation of Greater Birmingham Grants, 1272
Community Foundation of the Verdugos Educational Endowment Fund Grants, 1314
Community Foundation of the Verdugos Grants, 1315
Cooper Industries Foundation Grants, 1345
Cornerstone Foundation of Northeastern Wisconsin Grants, 1347
Cralle Foundation Grants, 1356
D. W. McMillan Foundation Grants, 1407
Dammann Fund Grants, 1418
Daphne Seybolt Culpeper Fndn Grants, 1425
Denver Broncos Charities Fund Grants, 1460
Dept of Ed Rehabilitation Training Grants, 1463
Dept of Ed Special Education--Personnel Development to Improve Services and Results for Children with Disabilities, 1465
Dept of Ed Special Education--Studies and Evaluations, 1466
Dept of Ed Special Education--Technical Assistance and Dissemination to Improve Services and Results for Children with Disabilities, 1467
Dept of Ed Special Education-National Activities-Technology and Media Services for Individuals with Disabilities, 1468
Different Needz Foundation Grants, 1478
Disable American Veterans Charitable Grants, 1479
Dolan Children's Foundation Grants, 1482
Doree Taylor Charitable Foundation, 1490
Dorothea Haus Ross Foundation Grants, 1498
Edwards Memorial Trust Grants, 1558
Effie and Wofford Cain Foundation Grants, 1563
Eisner Foundation Grants, 1564
El Paso Community Foundation Grants, 1579
Elsie H. Wilcox Foundation Grants, 1583
Emma B. Howe Memorial Foundation Grants, 1593
Eva L. & Joseph M. Bruening Fndn Grants, 1613
Florence Hunt Maxwell Foundation Grants, 1676
Foellinger Foundation Grants, 1684
Fourjay Foundation Grants, 1712
Fremont Area Community Foundation Amazing X Grants, 1744
George A Ohl Jr. Foundation Grants, 1791
Gheens Foundation Grants, 1814

Giving in Action Society Children & Youth with Special Needs Grants, 1823
Giving in Action Society Family Indep Grants, 1824
GNOF IMPACT Grants for Health and Human Services, 1839
Grace and Franklin Bernsen Foundation Grants, 1858
Grace Bersted Foundation Grants, 1859
Graham and Carolyn Holloway Family Foundation Grants, 1861
Greater Milwaukee Foundation Grants, 1883
Green Bay Packers Foundation Grants, 1890
Greygates Foundation Grants, 1901
Gulf Coast Community Foundation Grants, 1915
Hackett Foundation Grants, 1926
Harold and Rebecca H. Gross Foundation Grants, 1951
Harold Brooks Foundation Grants, 1952
Harry Kramer Memorial Fund Grants, 1965
Harry S. Black and Allon Fuller Fund Grants, 1966
Hasbro Children's Fund Grants, 1973
Helen Gertrude Sparks Charitable Trust Grants, 1993
Helen Irwin Littauer Educational Trust Grants, 1994
Helen Steiner Rice Foundation Grants, 1998
Henrietta Lange Burk Fund Grants, 2001
Herbert A. & Adrian W. Woods Fndn Grants, 2007
Hilda and Preston Davis Foundation Grants, 2028
Hoblitzelle Foundation Grants, 2038
Horace A. Kimball and S. Ella Kimball Foundation Grants, 2046
Howe Foundation of North Carolina Grants, 2055
HRAMF Deborah Munroe Noonan Memorial Research Grants, 2058
Ireland Family Foundation Grants, 2162
J.W. Kieckhefer Foundation Grants, 2176
J. Walton Bissell Foundation Grants, 2177
Jacob and Valeria Langeloth Foundation Grants, 2182
James Ford Bell Foundation Grants, 2187
Janirve Foundation Grants, 2209
Jay and Rose Phillips Family Foundation Grants, 2213
John D. & Katherine A. Johnston Fndn Grants, 2233
John H. and Wilhelmina D. Harland Charitable Foundation Children and Youth Grants, 2238
John H. Wellons Foundation Grants, 2239
John I. Smith Charities Grants, 2240
John J. Leidy Foundation Grants, 2241
John W. Alden Trust Grants, 2260
Joseph P. Kennedy Jr. Foundation Grants, 2275
Josiah W. and Bessie H. Kline Foundation Grants, 2278
Kenneth T. & Eileen L. Norris Fndn Grants, 2300
Kessler Fnd Signature Employment Grants, 2302
Lands' End Corporate Giving Program, 2344
Ludwick Family Foundation Grants, 2389
Lydia deForest Charitable Trust Grants, 2393
M.E. Raker Foundation Grants, 2404
M.J. Murdock Charitable Trust General Grants, 2405
Mabel H. Flory Charitable Trust Grants, 2408
Maine Community Foundation Charity Grants, 2415
Manuel D. & Rhoda Mayerson Fndn Grants, 2420
Margaret Wiegand Trust Grants, 2430
Marie C. and Joseph C. Wilson Foundation Rochester Small Grants, 2431
Marjorie Moore Charitable Foundation Grants, 2438
Mary Wilmer Covey Charitable Trust Grants, 2448
May and Stanley Smith Charitable Trust Grants, 2459
Medtronic Foundation Patient Link Grants, 2514
MGM Resorts Foundation Community Grants, 2541
Mid-Iowa Health Foundation Community Response Grants, 2549
Military Ex-Prisoners of War Foundation Grants, 2554
Mockingbird Foundation Grants, 2589
Mt. Sinai Health Care Foundation Health of the Jewish Community Grants, 2599
Nell J. Redfield Foundation Grants, 2673
NIAMS Pilot and Feasibility Clinical Research Grants in Arthritis, Musculoskeletal & Skin Diseases, 2774
Nina Mason Pulliam Charitable Trust Grants, 2921
NSF Biomedical Engineering and Engineering Healthcare Grants, 3009
NYCT Children/Youth with Disabilities Grants, 3052
Oppenstein Brothers Foundation Grants, 3078
Ordean Foundation Grants, 3080

Oticon Focus on People Awards, 3098
Paul Balint Charitable Trust Grants, 3122
Peabody Foundation Grants, 3132
Peacock Foundation Grants, 3133
Pittsburgh Foundation Community Fund Grants, 3230
R.C. Baker Foundation Grants, 3292
R.S. Gernon Trust Grants, 3293
Rajiv Gandhi Foundation Grants, 3302
Rayonier Foundation Grants, 3311
Reader's Digest Partners for Sight Fndn Grants, 3345
Rehab Therapy Foundation Grants, 3356
Robert R. Meyer Foundation Grants, 3387
Rockwell International Corporate Trust Grants Program, 3399
Roeher Institute Research Grants, 3400
Roy & Christine Sturgis Charitable Grants, 3415
S. Mark Taper Foundation Grants, 3449
Salem Foundation Grants, 3457
San Francisco Foundation Disability Rights Advocate Fund Emergency Grants, 3479
Schlessman Family Foundation Grants, 3493
Schurz Communications Foundation Grants, 3496
Seabury Foundation Grants, 3501
Shell Deer Park Grants, 3543
Sheltering Arms Fund Grants, 3545
Shopko Fndn Community Charitable Grants, 3547
Sidney Stern Memorial Trust Grants, 3552
Sid W. Richardson Foundation Grants, 3553
Simple Advise Education Center Grants, 3565
Sioux Falls Area Community Foundation Community Fund Grants (Unrestricted), 3568
Sioux Falls Area Community Foundation Spot Grants (Unrestricted), 3570
Sophia Romero Trust Grants, 3624
Special Olympics Health Profess Student Grants, 3629
Special People in Need Grants, 3630
Special People in Need Scholarships, 3631
Stella and Charles Guttman Foundation Grants, 3651
Strowd Roses Grants, 3664
Thomas Austin Finch, Sr. Foundation Grants, 3718
Thomas Sill Foundation Grants, 3724
TJX Foundation Grants, 3735
U.S. Department of Education 21st Century Community Learning Centers, 3767
U.S. Department of Education Rehabilitation Engineering Research Centers Grants, 3769
U.S. Department of Education Rehabilitation Research Training Centers (RRTCs), 3770
U.S. Department of Education Rehabilitation Training - Rehabilitation Continuing Education - Institute on Rehabilitation Issues, 3771
U.S. Dept of Education Special Ed--National Activities--Parent Information Centers, 3772
U.S. Department of Education Vocational Rehabilitation Services Projects for American Indians with Disabilities Grants, 3774
Union Benevolent Association Grants, 3789
Union County Community Foundation Grants, 3790
Union Labor Health Fndn Angel Fund Grants, 3791
Vancouver Sun Children's Fund Grants, 3851
Victor E. Speas Foundation Grants, 3868
Vigneron Memorial Fund Grants, 3873
W.P. and Bulah Luse Foundation Grants, 3892
Weingart Foundation Grants, 3913
Wilson-Wood Foundation Grants, 3960

Disabled (Target Groups)

Anne J. Caudal Foundation Grants, 595
Autodesk Community Relations Grants, 809
Catherine Holmes Wilkins Foundation Grants, 1042
Grifols Community Outreach Grants, 1904
Kessler Fnd Signature Employment Grants, 2302
Rosalynn Carter Institute John and Betty Pope Fellowships, 3408
SfN Science Educator Award, 3535
Singing for Change Foundation Grants, 3567
Verizon Foundation Maryland Grants, 3859
Verizon Foundation New York Grants, 3860
Verizon Foundation Northeast Region Grants, 3861
Verizon Foundation Pennsylvania Grants, 3862

SUBJECT INDEX

Verizon Foundation Vermont Grants, 3863
Verizon Foundation Virginia Grants, 3864

Disabled Student Support
Able Trust Vocational Rehabilitation Grants for Individuals, 213
AFB Rudolph Dillman Memorial Scholarship, 338
Allan C. and Lelia J. Garden Foundation Grants, 438
ASHFoundation Graduate Student Scholarships, 703
Champ-A Champion Fur Kids Grants, 1109
CVS All Kids Can Grants, 1377
CVS Community Grants, 1379
Florida High School/High Tech Project Grants, 1680
Guy I. Bromley Trust Grants, 1919
Lewis H. Humphreys Charitable Trust Grants, 2361
Louetta M. Cowden Foundation Grants, 2377
Peacock Foundation Grants, 3133
Peter and Elizabeth C. Tower Foundation Annual Intellectual Disabilities Grants, 3144
Roy & Christine Sturgis Charitable Grants, 3415
Special People in Need Scholarships, 3631
William G. & Helen C. Hoffman Fndn Grants, 3950

Disabled, Accessibility for
AAPD Henry B. Betts Award, 182
AAPD Paul G. Hearne Leadership Award, 183
ACL Business Acumen for Disability Organizations Grants, 238
ACL Training and Technical Assistance Center for State Intellectual and Developmental Disabilities Delivery Systems Grants, 248
Adolph Coors Foundation Grants, 311
Anne J. Caudal Foundation Grants, 595
Christopher & Dana Reeve Foundation Quality of Life Grants, 1152
Coleman Foundation Developmental Disabilities Grants, 1207
CVS Community Grants, 1379
Giving in Action Society Children & Youth with Special Needs Grants, 1823
Giving in Action Society Family Indep Grants, 1824
Harry S. Black and Allon Fuller Fund Grants, 1966
Hugh J. Andersen Foundation Grants, 2074
Jay and Rose Phillips Family Foundation Grants, 2213
John W. Speas and Effie E. Speas Memorial Trust Grants, 2264
Kessler Fnd Signature Employment Grants, 2302
Lewis H. Humphreys Charitable Trust Grants, 2361
Louis and Elizabeth Nave Flarsheim Charitable Foundation Grants, 2379
Mary Wilmer Covey Charitable Trust Grants, 2448
Multiple Sclerosis Foundation Brighter Tomorrow Grants, 2602
NIDRR Field-Initiated Projects, 2853
NYCT Children/Youth with Disabilities Grants, 3052
Robins Foundation Grants, 3390
SfN Science Educator Award, 3535
Sheltering Arms Fund Grants, 3545
Sioux Falls Area Community Foundation Field-of-Interest and Donor-Advised Grants, 3569
The Ray Charles Foundation Grants, 3716
U.S. Department of Education Rehabilitation Engineering Research Centers Grants, 3769
Vigneron Memorial Fund Grants, 3873
William G. & Helen C. Hoffman Fndn Grants, 3950

Disabled, Education
Charles Lafitte Foundation Grants, 1123
Christopher & Dana Reeve Foundation Quality of Life Grants, 1152
DeKalb County Community Foundation - Garrett Hospital Aid Foundation Grants, 1446
Emily Hall Tremaine Foundation Learning Disabilities Grants, 1592
Jay and Rose Phillips Family Foundation Grants, 2213
Marion I. and Henry J. Knott Foundation Discretionary Grants, 2435
Marion I. and Henry J. Knott Foundation Standard Grants, 2436
Marjorie Moore Charitable Foundation Grants, 2438

Mary Wilmer Covey Charitable Trust Grants, 2448
Roy & Christine Sturgis Charitable Grants, 3415
Singing for Change Foundation Grants, 3567

Disabled, Higher Education
Greenspun Family Foundation Grants, 1896

Disadvantaged, Economically
Abbot and Dorothy H. Stevens Foundation Grants, 197
Achelis Foundation Grants, 234
ADA Foundation Bud Tarrson Dental School Student Community Leadership Awards, 287
ADA Foundation Thomas J. Zwemer Award, 297
Ahmanson Foundation Grants, 355
Aid for Starving Children Emergency Assistance Fund Grants, 367
Alberto Culver Corporate Contributions Grants, 408
Albert W. Rice Charitable Foundation Grants, 411
Albuquerque Community Foundation Grants, 412
Alfred and Tillie Shemanski Testamentary Trust Grants, 428
Alfred E. Chase Charitable Foundation Grants, 430
Allan C. and Lelia J. Garden Foundation Grants, 438
Allegis Group Foundation Grants, 441
AlohaCare Believes in Me Scholarship, 449
Amelia Sillman Rockwell and Carlos Perry Rockwell Charities Fund Grants, 501
Amerigroup Foundation Grants, 565
Andrew Family Foundation Grants, 588
Anschutz Family Foundation Grants, 602
Ansell, Zaro, Grimm & Aaron Foundation Grants, 603
ARCO Foundation Education Grants, 652
Austin-Bailey Health and Wellness Fndn Grants, 803
Batchelor Foundation Grants, 837
BBVA Compass Foundation Charitable Grants, 848
BCBSM Foundation Community Health Matching Grants, 852
Bill and Melinda Gates Foundation Emergency Response Grants, 892
Blue Cross Blue Shield of Minnesota Foundation - Health Equity: Building Health Equity Together Grants, 914
Blue Cross Blue Shield of Minnesota Fndn Healthy Equity: Public Libraries for Health Grants, 918
Bodenwein Public Benevolent Foundation Grants, 928
Bodman Foundation Grants, 929
Boston Globe Foundation Grants, 937
Bright Promises Foundation Grants, 952
Brinson Foundation Grants, 953
Callaway Golf Company Foundation Grants, 1003
Carl B. and Florence E. King Foundation Grants, 1022
Carl R. Hendrickson Family Foundation Grants, 1028
Carls Foundation Grants, 1030
Carolyn Foundation Grants, 1034
Carrie E. and Lena V. Glenn Foundation Grants, 1036
Catherine Holmes Wilkins Foundation Grants, 1042
CCHD Community Development Grants, 1046
Cessna Foundation Grants Program, 1081
CFFVR Basic Needs Giving Partnership Grants, 1094
CFFVR Jewelers Mutual Charitable Giving, 1099
CFFVR Robert and Patricia Endries Family Foundation Grants, 1102
CFFVR Schmidt Family G4 Grants, 1103
CFFVR Waupaca Area Community Foundation Grants, 1105
Charles Delmar Foundation Grants, 1115
Charles H. Dater Foundation Grants, 1118
Charles H. Hall Foundation, 1120
Charles H. Pearson Foundation Grants, 1121
Charles Nelson Robinson Fund Grants, 1125
CIT Corporate Giving Grants, 1161
Clara Blackford Smith and W. Aubrey Smith Charitable Foundation Grants, 1165
Clark and Ruby Baker Foundation Grants, 1174
Claude Pepper Foundation Grants, 1177
Clayton Baker Trust Grants, 1179
CNA Foundation Grants, 1189
CNCS AmeriCorps VISTA Project Grants, 1190
CNCS Senior Companion Program Grants, 1191
Colonel Stanley R. McNeil Foundation Grants, 1211

Columbus Foundation Central Benefits Health Care Foundation Grants, 1216
Community Fndn for Greater Buffalo Grants, 1247
Community Foundation of Eastern Connecticut General Southeast Grants, 1269
Community Foundation of Greater Birmingham Grants, 1272
Community Foundation of Greenville Hollingsworth Funds Program/Project Grants, 1285
Cone Health Foundation Grants, 1323
Constantin Foundation Grants, 1341
Constellation Energy Corporate Grants, 1342
Cooper Industries Foundation Grants, 1345
Cornerstone Foundation of Northeastern Wisconsin Grants, 1347
Covidien Medical Product Donations, 1352
Crail-Johnson Foundation Grants, 1355
Cralle Foundation Grants, 1356
CVS All Kids Can Grants, 1377
CVS Community Grants, 1379
D. W. McMillan Foundation Grants, 1407
Danellie Foundation Grants, 1421
Daniels Fund Grants-Aging, 1424
Daphne Seybolt Culpeper Fndn Grants, 1425
David N. Lane Trust Grants for Aged and Indigent Women, 1433
Decatur County Community Foundation Small Project Grants, 1445
Dennis & Phyllis Washington Fndn Grants, 1458
Doree Taylor Charitable Foundation, 1490
Dorothea Haus Ross Foundation Grants, 1498
Dr. & Mrs. Paul Pierce Memorial Fndn Grants, 1505
Edina Realty Foundation Grants, 1550
Edward W. and Stella C. Van Houten Memorial Fund Grants, 1559
Eisner Foundation Grants, 1564
El Paso Community Foundation Grants, 1579
Elsie Lee Garthwaite Memorial Fndn Grants, 1584
Ensworth Charitable Foundation Grants, 1595
Faye McBeath Foundation Grants, 1635
Florence Hunt Maxwell Foundation Grants, 1676
Foundation for the Mid South Community Development Grants, 1699
Fourjay Foundation Grants, 1712
Four J Foundation Grants, 1713
Frank Loomis Palmer Fund Grants, 1726
Frank Reed and Margaret Jane Peters Memorial Fund I Grants, 1727
Frank Reed and Margaret Jane Peters Memorial Fund II Grants, 1728
Fred & Gretel Biel Charitable Trust Grants, 1734
Frederick McDonald Trust Grants, 1740
Frederick W. Marzahl Memorial Fund Grants, 1741
Gamble Foundation Grants, 1769
Gates Millennium Scholars Program, 1776
George A. and Grace L. Long Foundation Grants, 1788
George A Ohl Jr. Foundation Grants, 1791
George W. Wells Foundation Grants, 1804
Georgiana Goddard Eaton Memorial Grants, 1805
Global Fund for Children Grants, 1832
GNOF IMPACT Grants for Health and Human Services, 1839
GNOF IMPACT Gulf States Eye Surgery Fund, 1840
GNOF Maison Hospitaliere Grants, 1843
Greygates Foundation Grants, 1901
Grover Hermann Foundation Grants, 1906
Guy I. Bromley Trust Grants, 1919
H.A. & Mary K. Chapman Trust Grants, 1921
H.J. Heinz Company Foundation Grants, 1923
Harold and Arlene Schnitzer CARE Grants, 1950
Harold Brooks Foundation Grants, 1952
Harry S. Black and Allon Fuller Fund Grants, 1966
Hasbro Children's Fund Grants, 1973
Hearst Foundations Health Grants, 1987
Helen Gertrude Sparks Charitable Trust Grants, 1993
Helen Irwin Littauer Educational Trust Grants, 1994
Henrietta Lange Burk Fund Grants, 2001
Henrietta Tower Wurts Memorial Fndn Grants, 2002
Herbert A. & Adrian W. Woods Fndn Grants, 2007
Herman Goldman Foundation Grants, 2012

706 / Disadvantaged, Economically

Hillcrest Foundation Grants, 2031
Horace A. Kimball and S. Ella Kimball Foundation Grants, 2046
Horace Moses Charitable Foundation Grants, 2047
Howard and Bush Foundation Grants, 2054
IBCAT Screening Mammography Grants, 2095
IIE David L. Boren Fellowships, 2129
Indiana Minority Teacher/Special Services Scholarships, 2156
Ittleson Foundation AIDS Grants, 2166
James Ford Bell Foundation Grants, 2187
James H. Cummings Foundation Grants, 2190
James R. Thorpe Foundation Grants, 2199
Jane's Trust Grants, 2206
Janirve Foundation Grants, 2209
Jeffris Wood Foundation Grants, 2217
Jerome and Mildred Paddock Foundation Grants, 2222
Jessie B. Cox Charitable Trust Grants, 2225
Jim Moran Foundation Grants, 2228
John Clarke Trust Grants, 2232
John D. & Katherine A. Johnston Fndn Grants, 2233
John P. Murphy Foundation Grants, 2248
John W. Boynton Fund Grants, 2263
John W. Speas and Effie E. Speas Memorial Trust Grants, 2264
Joseph H. & Florence A. Roblee Fndn Grants, 2269
Josephine G. Russell Trust Grants, 2271
Josephine Goodyear Foundation Grants, 2272
Kahuku Community Fund, 2286
KeyBank Foundation Grants, 2307
Lands' End Corporate Giving Program, 2344
Legler Benbough Foundation Grants, 2352
Leo Niessen Jr., Charitable Trust Grants, 2358
Lewis H. Humphreys Charitable Trust Grants, 2361
Liberty Bank Foundation Grants, 2362
Linford and Mildred White Charitable Grants, 2368
Long Island Community Foundation Grants, 2375
Louetta M. Cowden Foundation Grants, 2377
Louis and Elizabeth Nave Flarsheim Charitable Foundation Grants, 2379
Lucy Downing Nisbet Charitable Fund Grants, 2387
Lucy Gooding Charitable Fndn Grants, 2388
Lydia deForest Charitable Trust Grants, 2393
M.D. Anderson Foundation Grants, 2403
Mabel A. Horne Trust Grants, 2406
Mabel F. Hoffman Charitable Trust Grants, 2407
Maine Community Foundation Charity Grants, 2415
Marathon Petroleum Corporation Grants, 2422
Marin Community Foundation Improving Community Health Grants, 2433
Mary K. Chapman Foundation Grants, 2443
Mary Owen Borden Foundation Grants, 2446
Mattel Children's Foundation Grants, 2454
Mattel International Grants Program, 2455
May and Stanley Smith Charitable Trust Grants, 2459
MBL William Townsend Porter Summer Fellowships for Minority Investigators, 2488
Medtronic Foundation CommunityLink Health Grants, 2510
Medtronic Foundation Community Link Human Services Grants, 2511
Mercedes-Benz USA Corporate Contributions, 2517
Meyer Foundation Healthy Communities Grants, 2533
MGM Resorts Foundation Community Grants, 2541
MGN Family Foundation Grants, 2542
Middlesex Savings Charitable Foundation Basic Human Needs Grants, 2550
Milagro Foundation Grants, 2552
Minneapolis Foundation Community Grants, 2558
MMS and Alliance Charitable Foundation Grants for Community Action and Care for the Medically Uninsured, 2587
Nationwide Insurance Foundation Grants, 2637
Nelda C. and H.J. Lutcher Stark Fndn Grants, 2672
Nicholas H. Noyes Jr. Memorial Fndn Grants, 2791
Norman Foundation Grants, 2983
Northwestern Mutual Foundation Grants, 2999
NYC Managed Care Consumer Assistance Workshop Re-Design Grants, 3047
Oleonda Jameson Trust Grants, 3071

Oppenstein Brothers Foundation Grants, 3078
Ordean Foundation Grants, 3080
Peacock Foundation Grants, 3133
PepsiCo Foundation Grants, 3137
Perpetual Trust for Charitable Giving Grants, 3141
PeyBack Foundation Grants, 3159
Pfizer Healthcare Charitable Contributions, 3180
Piper Trust Healthcare & Med Research Grants, 3228
PMI Foundation Grants, 3239
Pohlad Family Foundation, 3241
Porter County Women's Grant, 3248
Powell Foundation Grants, 3255
Premera Blue Cross Grants, 3258
Prince Charitable Trusts DC Grants, 3271
Prudential Foundation Education Grants, 3276
R.S. Gernon Trust Grants, 3293
Ralph M. Parsons Foundation Grants, 3304
Raskob Foundation for Catholic Activities Grants, 3306
Rayonier Foundation Grants, 3311
Rhode Island Foundation Grants, 3362
Richard & Susan Smith Family Fndn Grants, 3365
Robert and Joan Dircks Foundation Grants, 3379
Robert B McMillen Foundation Grants, 3381
Robert R. McCormick Tribune Veterans Initiative Grants, 3386
Robert R. Meyer Foundation Grants, 3387
Rockefeller Foundation Grants, 3396
RWJF Vulnerable Populations Portfolio Grants, 3444
S. Mark Taper Foundation Grants, 3449
Saigh Foundation Grants, 3451
Salt River Health & Human Services Grants, 3459
Sands Foundation Grants, 3476
Schlessman Family Foundation Grants, 3493
SfN Neuroscience Fellowships, 3530
SfN Science Educator Award, 3535
Sheltering Arms Fund Grants, 3545
Shopko Fndn Community Charitable Grants, 3547
Singing for Change Foundation Grants, 3567
Sioux Falls Area Community Foundation Community Fund Grants (Unrestricted), 3568
Sioux Falls Area Community Foundation Field-of-Interest and Donor-Advised Grants, 3569
Sioux Falls Area Community Foundation Spot Grants (Unrestricted), 3570
Sir Dorabji Tata Trust Grants for NGOs or Voluntary Organizations, 3572
Sir Dorabji Tata Trust Individual Medical Grants, 3573
Sisters of Mercy of North Carolina Fndn Grants, 3575
Sisters of St. Joseph Healthcare Fndn Grants, 3576
Sophia Romero Trust Grants, 3624
Special People in Need Scholarships, 3631
Stark Community Fndn Women's Grants, 3645
Stewart Huston Charitable Trust Grants, 3655
Strowd Roses Grants, 3664
Swindells Charitable Foundation, 3697
Tauck Family Foundation Grants, 3703
Textron Corporate Contributions Grants, 3712
Thelma Braun & Bocklett Family Fndn Grants, 3713
Theodore Edson Parker Foundation Grants, 3715
The Ray Charles Foundation Grants, 3716
Thompson Charitable Foundation Grants, 3727
Trull Foundation Grants, 3761
Union Bank, N.A. Foundation Grants, 3788
Union Benevolent Association Grants, 3789
Union Labor Health Fndn Angel Fund Grants, 3791
Union Labor Health Fndn Community Grants, 3792
Union Labor Health Foundation Dental Angel Fund Grants, 3793
Volkswagen of America Corporate Contributions, 3880
Washington Gas Charitable Giving, 3908
Western Union Foundation Grants, 3923
William G. & Helen C. Hoffman Fndn Grants, 3950
William H. Hannon Foundation Grants, 3953
William J. & Tina Rosenberg Fndn Grants, 3954
William J. Brace Charitable Trust, 3955
Wilson-Wood Foundation Grants, 3960

Disaster Preparedness

3M Fndn Health & Human Services Grants, 7
Advance Auto Parts Corporate Giving Grants, 312

Baxter International Corporate Giving Grants, 841
Bill and Melinda Gates Foundation Emergency Response Grants, 892
CDC Foundation Emergency Response Grants, 1061
Delmarva Power and Light Contributions, 1454
Elizabeth Morse Genius Charitable Trust Grants, 1569
GNOF Bayou Communities Grants, 1835
J.H. Robbins Foundation Grants, 2171
Nationwide Insurance Foundation Grants, 2637

Disaster Relief

3M Fndn Health & Human Services Grants, 7
ADA Foundation Disaster Assistance Grants, 290
Adler-Clark Electric Community Commitment Foundation Grants, 307
Advance Auto Parts Corporate Giving Grants, 312
Aid for Starving Children International Grants, 368
AIG Disaster Relief Fund Grants, 370
Albertson's Charitable Giving Grants, 410
Alice Tweed Tuohy Foundation Grants Program, 437
Allstate Corporate Giving Grants, 446
Allstate Corp Hometown Commitment Grants, 447
American Psychiatric Foundation Disaster Recovery Fund for Psychiatrists, 553
Anheuser-Busch Foundation Grants, 591
AT&T Foundation Civic and Community Service Program Grants, 793
Avon Products Foundation Grants, 814
Baxter International Corporate Giving Grants, 841
Beazley Foundation Grants, 865
Becton Dickinson and Company Grants, 869
Bill and Melinda Gates Foundation Emergency Response Grants, 892
BJ's Charitable Foundation Grants, 904
Boeing Company Contributions Grants, 930
BP Foundation Grants, 943
California Endowment Innovative Ideas Challenge Grants, 1000
Callaway Golf Company Foundation Grants, 1003
Campbell Soup Foundation Grants, 1007
Cargill Citizenship Fund Corporate Giving, 1018
Carnegie Corporation of New York Grants, 1033
CDC Foundation Emergency Response Grants, 1061
CNCS AmeriCorps VISTA Project Grants, 1190
Coca-Cola Foundation Grants, 1203
Community Fndn for the Capital Region Grants, 1259
Covidien Medical Product Donations, 1352
Cultural Society of Filipino Americans Grants, 1373
Delmarva Power and Light Contributions, 1454
Elizabeth Morse Genius Charitable Trust Grants, 1569
Farmers Insurance Corporate Giving Grants, 1634
Federal Express Corporate Contributions, 1657
Flextronics Foundation Disaster Relief Grants, 1674
GNOF Bayou Communities Grants, 1835
Gulf Coast Community Foundation Grants, 1915
H. Schaffer Foundation Grants, 1925
Harry Kramer Memorial Fund Grants, 1965
HHMI Grants and Fellowships Programs, 2020
Home Building Industry Disaster Relief Fund, 2043
Humana Foundation Grants, 2077
J.H. Robbins Foundation Grants, 2171
Jessie Ball Dupont Fund Grants, 2226
John Deere Foundation Grants, 2234
Lawrence Foundation Grants, 2347
M.J. Murdock Charitable Trust General Grants, 2405
Merkel Foundation Grants, 2520
Nationwide Insurance Foundation Grants, 2637
Noble County Community Foundation Grants, 2975
Perry County Community Foundation Grants, 3142
Pike County Community Foundation Grants, 3222
Posey County Community Foundation Grants, 3252
Procter and Gamble Fund Grants, 3274
Ralphs Food 4 Less Foundation Grants, 3305
Robert R. Meyer Foundation Grants, 3387
Rockefeller Foundation Grants, 3396
Schering-Plough Foundation Health Grants, 3492
Seneca Foods Foundation Grants, 3511
Shell Deer Park Grants, 3543
Sony Corporation of America Grants, 3623
Spencer County Community Foundation Grants, 3632

SUBJECT INDEX

Sunoco Foundation Grants, 3679
Thompson Charitable Foundation Grants, 3727
Union Bank, N.A. Foundation Grants, 3788
VHA Health Foundation Grants, 3867
Warrick County Community Foundation Grants, 3904

Disasters
3M Fndn Health & Human Services Grants, 7
ADA Foundation Disaster Assistance Grants, 290
Adler-Clark Electric Community Commitment Foundation Grants, 307
AIG Disaster Relief Fund Grants, 370
American Foodservice Charitable Trust Grants, 532
American Psychiatric Foundation Disaster Recovery Fund for Psychiatrists, 553
Baxter International Corporate Giving Grants, 841
Bill and Melinda Gates Foundation Emergency Response Grants, 892
CDC Foundation Emergency Response Grants, 1061
Community Fndn for the Capital Region Grants, 1259
Delmarva Power and Light Contributions, 1454
Elizabeth Morse Genius Charitable Trust Grants, 1569
FAR Fund Grants, 1631
Flextronics Foundation Disaster Relief Grants, 1674
Francis T. & Louise T. Nichols Fndn Grants, 1720
GNOF Bayou Communities Grants, 1835
Harry Kramer Memorial Fund Grants, 1965
Hormel Foods Charitable Trust Grants, 2050
J.H. Robbins Foundation Grants, 2171
NIA Behavioral and Social Research Grants on Disasters and Health, 2754
NIGMS Advancing Basic Behavioral and Social Research on Resilience: an Integrative Science Approach, 2861
Rohm and Haas Company Grants, 3403

Discrimination
Allstate Corporate Giving Grants, 446
Allstate Corp Hometown Commitment Grants, 447
Georgia Power Foundation Grants, 1806
Otto Bremer Foundation Grants, 3099
San Diego Foundation for Change Grants, 3470
Sioux Falls Area Community Foundation Spot Grants (Unrestricted), 3570
Waitt Family Foundation Grants, 3894

Disease, Chronic
Bupa Foundation Medical Research Grants, 974
CDC Epidemic Intell Service Training Grants, 1055
CDC Epidemiology Elective Rotation, 1056
CDC Experience Epidemiology Fellowships, 1059
CDC Public Health Associates, 1065
CDC Public Health Associates Hosts, 1066
CSTE CDC/CSTE Applied Epidemiology Fellowships, 1367
George E. Hatcher, Jr. and Ann Williams Hatcher Foundation Grants, 1792
HRAMF Deborah Munroe Noonan Memorial Research Grants, 2058
Mary Wilmer Covey Charitable Trust Grants, 2448
Medtronic Foundation Strengthening Health Systems Grants, 2515
NIGMS Advancing Basic Behavioral and Social Research on Resilience: an Integrative Science Approach, 2861
NIH Self-Management Strategies Across Chronic Diseases Grants, 2909
Thomas C. Burke Foundation Grants, 3720
TJX Foundation Grants, 3735

Diseases
A-T Children's Project Grants, 13
A-T Children's Project Post Doctoral Fellowships, 14
Acid Maltase Deficiency Association Helen Walker Research Grant, 235
Affymetrix Corporate Contributions Grants, 339
Albertson's Charitable Giving Grants, 410
Batchelor Foundation Grants, 837
Bill and Melinda Gates Foundation Water, Sanitation and Hygiene Grants, 894
CDC-Hubert Global Health Fellowship, 1047
CDC Epidemic Intell Service Training Grants, 1055
Clayton Fund Grants, 1180
David M. and Marjorie D. Rosenberg Foundation Grants, 1432
Doris Duke Charitable Foundation Clinical Scientist Development Award, 1494
Garland D. Rhoads Foundation, 1774
Gates Award for Global Health, 1775
George E. Hatcher, Jr. and Ann Williams Hatcher Foundation Grants, 1792
Hearst Foundations Health Grants, 1987
HHMI Med into Grad Initiative Grants, 2022
HRAMF Charles A. King Trust Postdoctoral Research Fellowships, 2056
HRAMF Harold S. Geneen Charitable Trust Awards for Coronary Heart Disease Research, 2059
HRAMF Taub Fnd Grants for MDS Research, 2064
Ike and Roz Friedman Foundation Grants, 2145
J.N. and Macie Edens Foundation Grants, 2174
James J. & Joan A. Gardner Family Fndn Grants, 2193
Lisa Higgins-Hussman Foundation Grants, 2370
Mary Wilmer Covey Charitable Trust Grants, 2448
NIGMS Advancing Basic Behavioral and Social Research on Resilience: an Integrative Science Approach, 2861
PFizer Compound Transfer Agreements, 3177
Pfizer Global Investigator Research Grants, 3178
Premera Blue Cross Grants, 3258
Robert R. Meyer Foundation Grants, 3387
Samueli Institute Scientific Research Grants, 3466
Thomas C. Burke Foundation Grants, 3720
Thomas Jefferson Rosenberg Foundation Grants, 3723
USAID Ebola Response, Recovery and Resilience in West Africa Grants, 3809
USAID Fighting Ebola Grants, 3812
USAID Land Use Change and Disease Emergence Grants, 3818
USDD Medical Practice Breadth of Medical Practice and Disease Frequency Exposure Grants, 3843
Wellcome Trust New Investigator Awards, 3919
Zellweger Baby Support Network Grants, 3984

Distance Learning
Alfred P. Sloan Foundation Research Fellowships, 432
Chatlos Foundation Grants Program, 1128
Hillsdale Fund Grants, 2035

Diversity
ACL Diversity Community of Practice Grants, 240
ACL University Centers for Excellence in Developmental Network Diversity and Inclusion Training Action Planning Grants, 249
BBVA Compass Foundation Charitable Grants, 848
Benton Community Fndn Cookie Jar Grant, 877
Boeing Company Contributions Grants, 930
Community Fndn of Greater Lafayette Grants, 1280
KeyBank Foundation Grants, 2307
NHLBI Mentored Career Dev Award to Promote Faculty Diversity in Biomed Research, 2708
NHLBI Ruth L. Kirschstein National Research Service Awards for Individual Senior Fellows, 2732
PepsiCo Foundation Grants, 3137
Seagate Tech Corp Capacity to Care Grants, 3502
SfN Louise Hanson Marshall Specific Recognition Award, 3527
SfN Mika Salpeter Lifetime Achievement Award, 3528
Sisters of Mercy of North Carolina Fndn Grants, 3575
Stewart Huston Charitable Trust Grants, 3655
Volkswagen of America Corporate Contributions, 3880

Documentaries
Connecticut Community Foundation Grants, 1324

Domestic Violence
Abbot and Dorothy H. Stevens Foundation Grants, 197
Adaptec Foundation Grants, 304
Alberto Culver Corporate Contributions Grants, 408
Amelia Sillman Rockwell and Carlos Perry Rockwell Charities Fund Grants, 501

Dramatic/Theater Arts /707

Austin-Bailey Health and Wellness Fndn Grants, 803
Austin S. Nelson Foundation Grants, 805
Baxter International Foundation Grants, 843
Blue Shield of California Grants, 923
Boston Jewish Community Women's Fund Grants, 938
Bush Fndn Health & Human Services Grants, 979
Cambridge Community Foundation Grants, 1004
Carlisle Foundation Grants, 1025
Carl M. Freeman Foundation FACES Grants, 1026
Carrie E. and Lena V. Glenn Foundation Grants, 1036
Catherine Kennedy Home Foundation Grants, 1043
Charles Nelson Robinson Fund Grants, 1125
Children's Trust Fund of Oregon Fndn Grants, 1139
Community Foundation of Eastern Connecticut Northeast Women and Girls Grants, 1270
Crail-Johnson Foundation Grants, 1355
Dallas Women's Foundation Grants, 1417
Dammann Fund Grants, 1418
Dennis & Phyllis Washington Fndn Grants, 1458
Edina Realty Foundation Grants, 1550
Faye McBeath Foundation Grants, 1635
Frank B. Hazard General Charity Fund Grants, 1721
Fremont Area Community Fndn General Grants, 1746
Global Fund for Women Grants, 1833
Greater Tacoma Community Foundation Ryan Alan Hade Endowment Fund, 1887
Greater Worcester Community Foundation Discretionary Grants, 1888
Harry Frank Guggenheim Fnd Research Grants, 1964
Health Fndn of Greater Indianapolis Grants, 1984
Herbert A. & Adrian W. Woods Fndn Grants, 2007
Hilton Head Island Foundation Grants, 2036
Hugh J. Andersen Foundation Grants, 2074
Jacob and Valeria Langeloth Foundation Grants, 2182
James R. Dougherty Jr. Foundation Grants, 2198
Jeffris Wood Foundation Grants, 2217
Jim Moran Foundation Grants, 2228
John W. Speas and Effie E. Speas Memorial Trust Grants, 2264
Kahuku Community Fund, 2286
Lucy Downing Nisbet Charitable Fund Grants, 2387
Mabel A. Horne Trust Grants, 2406
Mardag Foundation Grants, 2428
Meyer Foundation Healthy Communities Grants, 2533
Morris & Gwendolyn Cafritz Fndn Grants, 2594
OneFamily Foundation Grants, 3075
Piper Jaffray Fndn Communities Giving Grants, 3227
Portland General Electric Foundation Grants, 3250
Posey Community Fndn Women's Fund Grants, 3251
Quantum Foundation Grants, 3290
R.S. Gernon Trust Grants, 3293
Reinberger Foundation Grants, 3357
Reynolds & Reynolds Associate Fndn Grants, 3360
ROSE Fund Grants, 3412
S. Mark Taper Foundation Grants, 3449
TJX Foundation Grants, 3735
Verizon Foundation Maryland Grants, 3859
Verizon Foundation New York Grants, 3860
Verizon Foundation Northeast Region Grants, 3861
Verizon Foundation Pennsylvania Grants, 3862
Verizon Foundation Vermont Grants, 3863
Verizon Foundation Virginia Grants, 3864
Warrick County Community Foundation Women's Fund, 3905
WHO Foundation Volunteer Service Grants, 3940

Drama
Herman Goldman Foundation Grants, 2012

Dramatic/Theater Arts
Adaptec Foundation Grants, 304
Alcatel-Lucent Technologies Foundation Grants, 413
Axe-Houghton Foundation Grants, 815
Blanche and Irving Laurie Foundation Grants, 910
F.M. Kirby Foundation Grants, 1624
High Meadow Foundation Grants, 2027
Huffy Foundation Grants, 2073
Jerome Robbins Foundation Grants, 2223
Kenneth T. & Eileen L. Norris Fndn Grants, 2300
Leon and Thea Koerner Foundation Grants, 2355

Meyer Memorial Trust Special Grants, 2538
Norcliffe Foundation Grants, 2979
Peyton Anderson Foundation Grants, 3160
Procter and Gamble Fund Grants, 3274
Robert W. Woodruff Foundation Grants, 3389

Driver Education
American Trauma Society, Pennsylvania Division Mini-Grants, 564
BMW of North America Charitable Contributions, 926

Dropouts
Adaptec Foundation Grants, 304
Coca-Cola Foundation Grants, 1203
Joseph Drown Foundation Grants, 2268
PepsiCo Foundation Grants, 3137
Rockwell Fund, Inc. Grants, 3398

Drug Design
Conquer Cancer Foundation of ASCO Drug Development Research Professorship, 1330
MMAAP Foundation Fellowship Award in Translational Medicine, 2581
MMAAP Foundation Research Project Award in Translational Medicine, 2585
NIA Alzheimer's Disease Drug Development Program Grants, 2749

Drug Education
Alcatel-Lucent Technologies Foundation Grants, 413
CFFVR Alcoholism and Drug Abuse Grants, 1093
Dept of Ed Safe and Drug-Free Schools and Communities State Grants, 1464
GNOF IMPACT Kahn-Oppenheim Grants, 1842
Grand Rapids Area Community Foundation Wyoming Youth Fund Grants, 1868
Grand Rapids Community Foundation Ionia County Youth Fund Grants, 1871
Grand Rapids Community Foundation Southeast Ottawa Youth Fund Grants, 1874
Grand Rapids Community Foundation Sparta Youth Fund Grants, 1876
MetroWest Health Foundation Grants to Reduce the Incidence of High Risk Behaviors Among Adolescents, 2528
NIDA Pilot and Feasibility Studies in Preparation for Drug Abuse Prevention Trials, 2794
NYCT Girls and Young Women Grants, 3053
NYCT Substance Abuse Grants, 3057
OSF International Harm Reduction Development Program Grants, 3090
OSF Tackling Addiction Grants in Baltimore, 3096
Weyerhaeuser Family Foundation Health Grants, 3926

Drug Metabolism
ASPET Brodie Award in Drug Metabolism, 770

Drug Testing
BWF Innovation in Regulatory Science Grants, 986
MMAAP Foundation Fellowship Award in Translational Medicine, 2581
NIA Alzheimer's Disease Drug Development Program Grants, 2749
NIA Grants for Alzheimer's Disease Drugs, 2759
NIDDK Development of Assays for High-Throughput Drug Screening Grants, 2808

Drugs/Drug Abuse
AACAP Jeanne Spurlock Research Fellowship in Substance Abuse and Addiction for Minority Medical Students, 60
Achelis Foundation Grants, 234
Actors Addiction & Recovery Services Grants, 282
Adaptec Foundation Grants, 304
Alliance Healthcare Foundation Grants, 444
amfAR Fellowships, 567
amfAR Global Initiatives Grants, 568
amfAR Mathilde Krim Fellowships in Basic Biomedical Research, 569
amfAR Public Policy Grants, 570

amfAR Research Grants, 571
Assurant Foundation Grants, 788
Audrey & Sydney Irmas Foundation Grants, 799
Austin S. Nelson Foundation Grants, 805
Barberton Community Foundation Grants, 829
Battle Creek Community Foundation Grants, 839
Benton Community Foundation Grants, 878
Berks County Community Foundation Grants, 881
Blue River Community Foundation Grants, 922
Bodman Foundation Grants, 929
Brown County Community Foundation Grants, 970
Cambridge Community Foundation Grants, 1004
Carnahan-Jackson Foundation Grants, 1032
Carpenter Foundation Grants, 1035
Carrie E. and Lena V. Glenn Foundation Grants, 1036
CFFVR Alcoholism and Drug Abuse Grants, 1093
Community Foundation of Bartholomew County Heritage Fund Grants, 1262
Community Foundation of Bartholomew County James A. Henderson Award for Fundraising, 1263
Community Foundation of Greater Birmingham Grants, 1272
Community Foundation of Greater Fort Wayne - Community Endowment and Clarke Endowment Grants, 1274
ConocoPhillips Foundation Grants, 1327
Constantin Foundation Grants, 1341
Cornerstone Foundation of Northeastern Wisconsin Grants, 1347
D.F. Halton Foundation Grants, 1405
Danellie Foundation Grants, 1421
Delaware Community Foundation-Youth Philanthropy Board for Kent County, 1448
Dennis & Phyllis Washington Fndn Grants, 1458
Dept of Ed Safe and Drug-Free Schools and Communities State Grants, 1464
DeRoy Testamentary Foundation Grants, 1469
DPA Promoting Policy Change Advocacy Grants, 1504
eBay Foundation Community Grants, 1543
Eva L. & Joseph M. Bruening Fndn Grants, 1613
Farmers Insurance Corporate Giving Grants, 1634
Florida BRAIVE Fund of Dade Community Foundation, 1678
Foundation for a Healthy Kentucky Grants, 1691
Fourjay Foundation Grants, 1712
Frances and John L. Loeb Family Fund Grants, 1716
GEICO Public Service Awards, 1778
GNOF IMPACT Kahn-Oppenheim Grants, 1842
Grand Rapids Community Foundation Southeast Ottawa Youth Fund Grants, 1874
Greater Worcester Community Foundation Discretionary Grants, 1888
Gulf Coast Foundation of Community Operating Grants, 1916
Gulf Coast Foundation of Community Program Grants, 1917
Harry Frank Guggenheim Fnd Research Grants, 1964
Hasbro Children's Fund Grants, 1973
Health Foundation of Greater Cincinnati Grants, 1983
Huffy Foundation Grants, 2073
Johnson & Johnson Corporate Contributions, 2253
Joseph Drown Foundation Grants, 2268
Joseph H. & Florence A. Roblee Fndn Grants, 2269
Kahuku Community Fund, 2286
L. W. Pierce Family Foundation Grants, 2328
Lester Ray Fleming Scholarships, 2360
Lillian S. Wells Foundation Grants, 2365
Lucile Horton Howe & Mitchell B. Howe Grants, 2385
March of Dimes Program Grants, 2426
Mary Owen Borden Foundation Grants, 2446
Memorial Foundation Grants, 2516
MetroWest Health Foundation Grants to Reduce the Incidence of High Risk Behaviors Among Adolescents, 2528
Mid-Iowa Health Foundation Community Response Grants, 2549
NIAAA Mechanisms of Alcohol and Drug-Induced Pancreatitis Grants, 2743
NIDA Mentored Clinical Scientist Research Career Development Awards, 2793

NIDA Pilot and Feasibility Studies in Preparation for Drug Abuse Prevention Trials, 2794
NIH Research on Sleep and Sleep Disorders, 2898
NIMH Early Identification and Treatment of Mental Disorders in Children and Adolescents Grants, 2918
NYCT Substance Abuse Grants, 3057
Ordean Foundation Grants, 3080
OSF International Harm Reduction Development Program Grants, 3090
Patrick and Anna M. Cudahy Fund Grants, 3120
Peter and Elizabeth C. Tower Foundation Mental Health Reference and Resource Materials Mini-Grants, 3147
Peter and Elizabeth C. Tower Foundation Substance Abuse Grants, 3152
Peter F. McManus Charitable Trust Grants, 3153
Phoenix Suns Charities Grants, 3201
Piedmont Natural Gas Foundation Health and Human Services Grants, 3221
Puerto Rico Community Foundation Grants, 3278
Quantum Corporation Snap Server Grants, 3289
Questar Corporate Contributions Grants, 3291
Rayonier Foundation Grants, 3311
Ruth Anderson Foundation Grants, 3425
RWJF New Jersey Health Initiatives Grants, 3440
SAMHSA Campus Suicide Prevention Grants, 3460
SAMHSA Conference Grants, 3462
SAMHSA Drug Free Communities Support Program Grants, 3463
SAMHSA Strategic Prevention Framework State Incentive Grants, 3464
Seabury Foundation Grants, 3501
Sid W. Richardson Foundation Grants, 3553
Sioux Falls Area Community Foundation Community Fund Grants (Unrestricted), 3568
Sioux Falls Area Community Foundation Spot Grants (Unrestricted), 3570
Southbury Community Trust Fund, 3625
Stackpole-Hall Foundation Grants, 3643
Stella and Charles Guttman Foundation Grants, 3651
Stewart Huston Charitable Trust Grants, 3655
Triangle Community Foundation Shaver-Hitchings Scholarship, 3758
Trull Foundation Grants, 3761
Union Bank, N.A. Foundation Grants, 3788
Victor E. Speas Foundation Grants, 3868
Whitney Foundation Grants, 3938

Dyslexia
Good Samaritan Inc Grants, 1855

Early Childhood Development
ACF Native American Social and Economic Development Strategies Grants, 231
Bernard and Audre Rapoport Foundation Health Grants, 882
Blue Cross Blue Shield of Minnesota Foundation - Healthy Children: Growing Up Healthy Grants, 915
Piper Trust Healthcare & Med Research Grants, 3228
TJX Foundation Grants, 3735
Vanderburgh Community Foundation Grants, 3852

Early Childhood Education
ACF Native American Social and Economic Development Strategies Grants, 231
BMW of North America Charitable Contributions, 926
Community Foundation of Louisville Dr. W. Barnett Owen Memorial Fund for the Children of Louisville and Jefferson County Grants, 1294
Express Scripts Foundation Grants, 1620
GNOF Stand Up For Our Children Grants, 1847
Harry A. & Margaret D. Towsley Fndn Grants, 1959
Leon and Thea Koerner Foundation Grants, 2355
Louis H. Aborn Foundation Grants, 2380
Packard Foundation Local Grants, 3105
Phi Upsilon Omicron Alumni Research Grant, 3191
Phi Upsilon Omicron Florence Fallgatter Distinguished Service Award, 3192
Phi Upsilon Omicron Geraldine Clewell Senior Awards, 3194

SUBJECT INDEX

Phi Upsilon Lillian P. Schoephoerster Award, 3196
Phi Upsilon Orinne Johnson Writing Award, 3198
Phi Upsilon Omicron Undergraduate Karen P. Goebel Conclave Award, 3200
Reinberger Foundation Grants, 3357
Robert R. McCormick Trib Community Grants, 3385
Robert R. Meyer Foundation Grants, 3387
Rollins-Luetkemeyer Foundation Grants, 3404
Samuel N. and Mary Castle Foundation Grants, 3467
SfN Science Educator Award, 3535
Stocker Foundation Grants, 3656
TJX Foundation Grants, 3735
Toys R Us Children's Fund Grants, 3751

Earth Science Education
W.M. Keck Foundation Science and Engineering Research Grants, 3889

Earth Sciences
IIE KAUST Graduate Fellowships, 2134

Eating Disorders
Hilda and Preston Davis Foundation Grants, 2028
Hilda and Preston Davis Foundation Postdoctoral Fellowships in Eating Disorders Research, 2029
Klarman Family Foundation Grants in Eating Disorders Research Grants, 2310
NEDA/AED Charron Family Research Grant, 2660
NEDA/AED Joan Wismer Research Grant, 2661
NEDA/AED Tampa Bay Eating Disorders Task Force Award, 2662

Ebola
USAID Ebola Response, Recovery and Resilience in West Africa Grants, 3809
USAID Fighting Ebola Grants, 3812

Ecology
Beirne Carter Foundation Grants, 870
Carnahan-Jackson Foundation Grants, 1032
Clarence E. Heller Charitable Foundation Grants, 1166
Coastal Community Foundation of South Carolina Grants, 1202
Conservation, Food, and Health Foundation Grants for Developing Countries, 1338
MBL Albert & Ellen Grass Fellowships, 2463
MBL Associates Summer Fellowships, 2465
MBL Burr & Susie Steinbach Fellowship, 2467
MBL E.E. Just Summer Fellowship for Minority Scientists, 2468
MBL Evelyn and Melvin Spiegal Fellowship, 2471
MBL Frank R. Lillie Summer Fellowship, 2472
MBL Gruss Lipper Family Foundation Summer Fellowship, 2475
MBL H. Keffer Hartline and Edward F. MacNichol, Jr. Fellowships, 2476
MBL Herbert W. Rand Summer Fellowship, 2477
MBL James E. and Faith Miller Memorial Summer Fellowship, 2478
MBL Lucy B. Lemann Summer Fellowship, 2481
MBL M.G.F. Fuortes Summer Fellowships, 2482
MBL William Townsend Porter Summer Fellowships for Minority Investigators, 2488
NSF Doctoral Dissertation Improvement Grants in the Directorate for Biological Sciences (DDIG), 3013
Piedmont Natural Gas Corporate and Charitable Contributions, 3220
Prospect Burma Scholarships, 3275
Union Bank, N.A. Foundation Grants, 3788

Ecology, Environmental Education
Norfolk Southern Foundation Grants, 2982
Tri-State Community Twenty-first Century Endowment Fund Grants, 3756

Economic Development
A/H Foundation Grants, 18
ACF Native American Social and Economic Development Strategies Grants, 231
Achelis Foundation Grants, 234

Air Products and Chemicals Grants, 399
Alabama Power Foundation Grants, 400
Alcatel-Lucent Technologies Foundation Grants, 413
Allstate Corporate Giving Grants, 446
Allstate Corp Hometown Commitment Grants, 447
Alvin and Fanny Blaustein Thalheimer Foundation Baltimore Communal Grants, 459
Amador Community Foundation Grants, 481
American Jewish World Service Grants, 533
AMI Semiconductors Corporate Grants, 576
Aragona Family Foundation Grants, 648
Arkansas Community Foundation Arkansas Black Hall of Fame Grants, 659
Bacon Family Foundation Grants, 817
Barberton Community Foundation Grants, 829
Bayer Foundation Grants, 846
Berks County Community Foundation Grants, 881
Bodman Foundation Grants, 929
Boeing Company Contributions Grants, 930
BP Foundation Grants, 943
California Endowment Innovative Ideas Challenge Grants, 1000
Campbell Soup Foundation Grants, 1007
Carl & Eloise Pohlad Family Fndn Grants, 1021
Carlisle Foundation Grants, 1025
Carnegie Corporation of New York Grants, 1033
CharityWorks Grants, 1113
Charles M. & Mary D. Grant Fndn Grants, 1124
Chemtura Corporation Contributions Grants, 1130
CIT Corporate Giving Grants, 1161
Claude Worthington Benedum Fndn Grants, 1178
CNCS Social Innovation Grants, 1192
Columbus Foundation Robert E. and Genevieve B. Schaefer Fund Grants, 1222
Comerica Charitable Foundation Grants, 1224
Commonwealth Edison Grants, 1225
Community Foundation for Greater Atlanta Clayton County Fund Grants, 1240
Community Foundation for Greater Atlanta Common Good Funds Grants, 1241
Community Foundation for Greater Atlanta Morgan County Fund Grants, 1244
Community Foundation for Greater Atlanta Newton County Fund Grants, 1245
Community Fndn for San Benito County Grants, 1257
Community Foundation for Southeast Michigan Grants, 1258
Community Foundation of Greater New Britain Grants, 1281
Community Fndn of Jackson County Grants, 1287
Community Foundation of Muncie and Delaware County Grants, 1300
Community Fndn of Wabash County Grants, 1316
Constellation Energy Corporate Grants, 1342
CSRA Community Foundation Grants, 1366
Curtis Foundation Grants, 1376
Dade Community Foundation Grants, 1410
Dayton Power and Light Foundation Grants, 1437
Daywood Foundation Grants, 1438
Denver Foundation Community Grants, 1461
eBay Foundation Community Grants, 1543
El Paso Community Foundation Grants, 1579
Erie Community Foundation Grants, 1601
Ezra M. Cutting Trust Grants, 1623
Fayette County Foundation Grants, 1636
Ferree Foundation Grants, 1658
FIU Global Civic Engagement Mini Grants, 1673
Foundation for the Mid South Community Development Grants, 1699
Four County Community Fndn General Grants, 1710
Frances W. Emerson Foundation Grants, 1718
Franklin H. Wells and Ruth L. Wells Foundation Grants, 1725
Frederick McDonald Trust Grants, 1740
Gardner Foundation Grants, 1772
Gebbie Foundation Grants, 1777
George Gund Foundation Grants, 1796
Gheens Foundation Grants, 1814
GNOF Bayou Communities Grants, 1835
GNOF New Orleans Works Grants, 1844

Greater Sitka Legacy Fund Grants, 1885
Greater Tacoma Community Foundation Grants, 1886
Green Diamond Charitable Contributions, 1891
Grotto Foundation Project Grants, 1905
Grover Hermann Foundation Grants, 1906
HAF Technical Assistance Program Grants, 1933
Harold R. Bechtel Testamentary Charitable Trust Grants, 1954
Harry B. and Jane H. Brock Foundation Grants, 1960
Hartford Foundation Regular Grants, 1970
Harvest Foundation Grants, 1971
Hawaii Community Foundation West Hawaii Fund Grants, 1978
Helen Bader Foundation Grants, 1992
Helen S. Boylan Foundation Grants, 1997
Holland/Zeeland Community Fndn Grants, 2041
Hormel Foundation Grants, 2051
Howard and Bush Foundation Grants, 2054
Hudson Webber Foundation Grants, 2072
Huntington National Bank Community Grants, 2084
Hutchinson Community Foundation Grants, 2085
Hut Foundation Grants, 2086
IIE David L. Boren Fellowships, 2129
IIE Freeman Foundation Indonesia Internships, 2131
IIE Hewlett Fnd/IIE Dissertation Fellowship, 2132
J.L. Bedsole Foundation Grants, 2172
Jackson County Community Foundation Unrestricted Grants, 2180
Jacobs Family Village Neighborhoods Grants, 2184
Janirve Foundation Grants, 2209
Jennings County Community Foundation Grants, 2220
John Deere Foundation Grants, 2234
Judith Clark-Morrill Foundation Grants, 2281
Kenai Peninsula Foundation Grants, 2298
Kent D. Steadley and Mary L. Steadley Memorial Trust Grants, 2301
Ketchikan Community Foundation Grants, 2303
Liberty Bank Foundation Grants, 2362
Lincoln Financial Foundation Grants, 2367
Long Island Community Foundation Grants, 2375
Lynde & Harry Bradley Foundation Fellowships, 2396
Lynde and Harry Bradley Foundation Grants, 2397
Lynde and Harry Bradley Foundation Prizes: Bradley Prizes, 2398
M.J. Murdock Charitable Trust General Grants, 2405
Madison County Community Foundation - City of Anderson Quality of Life Grant, 2412
Madison County Community Foundation General Grants, 2413
Marathon Petroleum Corporation Grants, 2422
Marie H. Bechtel Charitable Remainder Uni-Trust Grants, 2432
Maxon Charitable Foundation Grants, 2458
Metzger-Price Fund Grants, 2530
MGM Resorts Foundation Community Grants, 2541
Miller Foundation Grants, 2556
Minneapolis Foundation Community Grants, 2558
Morris & Gwendolyn Cafritz Fndn Grants, 2594
Norman Foundation Grants, 2983
Northwest Minnesota Foundation Women's Fund Grants, 3000
NYCT Girls and Young Women Grants, 3053
Owen County Community Foundation Grants, 3100
Paul Ogle Foundation Grants, 3124
Percy B. Ferebee Endowment Grants, 3138
Petersburg Community Foundation Grants, 3156
Philadelphia Foundation Organizational Effectiveness Grants, 3189
Piedmont Natural Gas Corporate and Charitable Contributions, 3220
Pittsburgh Foundation Community Fund Grants, 3230
Plough Foundation Grants, 3238
PMI Foundation Grants, 3239
Priddy Foundation Organizational Development Grants, 3265
Prince Charitable Trusts DC Grants, 3271
Princeton Area Community Foundation Greater Mercer Grants, 3272
Procter and Gamble Fund Grants, 3274
Puerto Rico Community Foundation Grants, 3278

710 / Economic Development

Pulaski County Community Foundation Grants, 3279
Putnam County Community Foundation Grants, 3283
RCF General Community Grants, 3312
Rhode Island Foundation Grants, 3362
Richard D. Bass Foundation Grants, 3366
Richard King Mellon Foundation Grants, 3369
Riley Foundation Grants, 3374
Robert F. Stoico/FIRSTFED Charitable Foundation Grants, 3382
Robert W. Woodruff Foundation Grants, 3389
Rockefeller Foundation Grants, 3396
S. Mark Taper Foundation Grants, 3449
San Diego Foundation for Change Grants, 3470
Sandy Hill Foundation Grants, 3477
Sioux Falls Area Community Foundation Community Fund Grants (Unrestricted), 3568
Sioux Falls Area Community Foundation Spot Grants (Unrestricted), 3570
Sonoco Foundation Grants, 3621
Southbury Community Trust Fund, 3625
Sterling and Shelli Gardner Foundation Grants, 3653
Sunoco Foundation Grants, 3679
SunTrust Bank Trusteed Foundations Florence C. and Harry L. English Memorial Fund Grants, 3680
SunTrust Bank Trusteed Foundations Greene-Sawtell Grants, 3681
SunTrust Bank Trusteed Foundations Harriet McDaniel Marshall Tust Grants, 3682
SunTrust Bank Trusteed Foundations Nell Warren Elkin and William Simpson Elkin Grants, 3683
SunTrust Bank Trusteed Foundations Thomas Guy Woolford Charitable Trust Grants, 3684
SunTrust Bank Trusteed Foundations Walter H. and Marjory M. Rich Memorial Fund Grants, 3685
Texas Commission on the Arts Arts Respond Project Grants, 3708
Textron Corporate Contributions Grants, 3712
Thomas and Dorothy Leavey Foundation Grants, 3717
Union Bank, N.A. Corporate Sponsorships and Donations, 3787
Union Bank, N.A. Foundation Grants, 3788
Union County Community Foundation Grants, 3790
W.K. Kellogg Foundation Healthy Kids Grants, 3886
Waitt Family Foundation Grants, 3894
Walker Area Community Foundation Grants, 3895
Walmart Foundation National Giving Grants, 3898
Warren County Community Foundation Grants, 3902
Warren County Community Fndn Mini-Grants, 3903
Western New York Foundation Grants, 3922
White County Community Foundation Grants, 3934
Z. Smith Reynolds Foundation Small Grants, 3982

Economic Justice
GNOF New Orleans Works Grants, 1844
Otto Bremer Foundation Grants, 3099
Skoll Fndn Awards for Social Entrepreneurship, 3579

Economic Opportunities
ACF Native American Social and Economic Development Strategies Grants, 231
CNCS AmeriCorps VISTA Project Grants, 1190
CNCS Social Innovation Grants, 1192
Ezra M. Cutting Trust Grants, 1623
GNOF IMPACT Grants for Arts and Culture, 1838
GNOF New Orleans Works Grants, 1844
Legler Benbough Foundation Grants, 2352
Walmart Foundation Community Giving Grants, 3897
Walmart Foundation National Giving Grants, 3898
Walmart Foundation State Giving Grants, 3900

Economic Science
W.M. Keck Foundation Science and Engineering Research Grants, 3889

Economic Self-Sufficiency
Abington Foundation Grants, 211
Adolph Coors Foundation Grants, 311
Battle Creek Community Foundation Grants, 839
BBVA Compass Foundation Charitable Grants, 848
Boeing Company Contributions Grants, 930

CFFVR Waupaca Area Community Foundation Grants, 1105
Citizens Bank Mid-Atlantic Charitable Foundation Grants, 1162
Community Foundation of Eastern Connecticut Northeast Women and Girls Grants, 1270
Emma B. Howe Memorial Foundation Grants, 1593
Ezra M. Cutting Trust Grants, 1623
FCD New American Children Grants, 1637
Foundation for the Mid South Community Development Grants, 1699
Graco Foundation Grants, 1860
Kahuku Community Fund, 2286
KeyBank Foundation Grants, 2307
Legler Benbough Foundation Grants, 2352
May and Stanley Smith Charitable Trust Grants, 2459
Medtronic Foundation Community Link Human Services Grants, 2511
Pinellas County Grants, 3223
Porter County Women's Grant, 3248
Retirement Research Foundation General Program Grants, 3359
Robins Foundation Grants, 3390
Rockwell Fund, Inc. Grants, 3398
Salt River Health & Human Services Grants, 3459
Samuel S. Johnson Foundation Grants, 3468
Seabury Foundation Grants, 3501
TE Foundation Grants, 3705
U.S. Department of Education Rehabilitation Research Training Centers (RRTCs), 3770
Union Bank, N.A. Foundation Grants, 3788

Economic Stimulus
Ezra M. Cutting Trust Grants, 1623
NIH Recovery Act Limited Competition: Academic Research Enhancement Awards, 2892
NIH Recovery Act Limited Competition: Biomedical Research, Development, and Growth to Spur the Acceleration of New Technologies Grants, 2893
NIH Recovery Act Limited Competition: Small Business Catalyst Awards for Accelerating Innovative Research Grants, 2895
NIH Recovery Act Limited Competition: Supporting New Faculty Recruitment to Enhance Research Resources through Biomedical Research Core Centers Grants, 2896

Economics
Alfred P. Sloan Foundation Research Fellowships, 432
AT&T Foundation Civic and Community Service Program Grants, 793
BP Foundation Grants, 943
Canada-U.S. Fulbright New Century Scholars Program Grants, 1008
Canada-U.S. Fulbright Senior Specialists Grants, 1009
Carnegie Corporation of New York Grants, 1033
CDC Steven M. Teutsch Prevention Effectiveness Fellowships, 1071
Community Foundation of Greater Fort Wayne - Lilly Endowment Community Scholarships, 1275
Community Foundation of St. Joseph County Lilly Endowment Community Scholarship, 1308
Donna K. Yundt Memorial Scholarship, 1488
Dr. John Maniotes Scholarship, 1506
Dr. R.T. White Scholarship, 1510
Dr. Scholl Foundation Grants Program, 1511
Elkhart County Foundation Lilly Endowment Community Scholarships, 1574
Fairfield County Community Foundation Grants, 1625
Flinn Foundation Scholarships, 1675
Ford Motor Company Fund Grants Program, 1687
Fred and Louise Latshaw Scholarship, 1735
Fulbright Alumni Initiatives Awards, 1749
Fulbright Distinguished Chairs Awards, 1750
Fulbright New Century Scholars Grants, 1753
Fulbright Specialists Program Grants, 1754
Fulbright Scholars in Europe and Eurasia, 1755
Fulbright Traditional Scholar Program in Sub-Saharan Africa, 1756

SUBJECT INDEX

Fulbright Traditional Scholar Program in the East Asia/Pacific Region, 1757
Fulbright Traditional Scholar Program in the Near East and North Africa Region, 1758
Fulbright Traditional Scholar Program in the South and Central Asia Region, 1759
Fulbright Traditional Scholar Program in the Western Hemisphere, 1760
Fulton County Community Foundation 4Community Higher Education Scholarship, 1762
George S. & Dolores Dore Eccles Fndn Grants, 1801
Hammond Common Council Scholarships, 1938
Harry Frank Guggenheim Foundation Dissertation Fellowships, 1963
IBEW Local Union #697 Memorial Scholarships, 2096
IIE Iraq Scholars and Leaders Scholarships, 2133
IIE Klein Family Scholarship, 2135
IIE Leonora Lindsley Memorial Fellowships, 2136
IIE Lingnan Foundation W.T. Chan Fellowship, 2137
IIE Lotus Scholarships, 2138
IIE Mattel Global Scholarship, 2139
IIE Nancy Petry Scholarship, 2140
IIE New Leaders Group Award for Mutual Understanding, 2141
IIE Western Union Family Scholarships, 2143
James S. McDonnell Foundation Complex Systems Collaborative Activity Awards, 2202
John V. and George Primich Family Scholarship, 2259
Joshua Benjamin Cohen Memorial Scholarship, 2276
LaGrange County Lilly Endowment Community Scholarship, 2331
Lake County Athletic Officials Association Scholarships, 2332
Lake County Lilly Endowment Community Scholarships, 2334
LEGENDS Scholarship, 2351
Nancy J. Pinnick Memorial Scholarship, 2607
NCHS National Center for Health Statistics Postdoctoral Research Appointments, 2642
NIA Healthy Aging through Behavioral Economic Analyses of Situations Grants, 2762
Noble County Community Foundation - Kathleen June Earley Memorial Scholarship, 2973
PhRMA Foundation Health Outcomes Post Doctoral Fellowships, 3202
PhRMA Foundation Informatics Post Doctoral Fellowships, 3206
Porter County Foundation Lilly Endowment Community Scholarships, 3245
Prospect Burma Scholarships, 3275
Pulaski County Community Fndn Scholarships, 3281
Rajiv Gandhi Foundation Grants, 3302
Rayonier Foundation Grants, 3311
RSC Thomas W. Eadie Medal, 3423
Schrage Family Foundation Scholarships, 3494
Sengupta Family Scholarship, 3512
W. James Spicer Scholarship, 3885
Webster Cornwell Memorial Scholarship, 3911
Whitley County Community Fndn Scholarships, 3937

Economics Education
3M Company Fndn Community Giving Grants, 6
Air Products and Chemicals Grants, 399
Alcoa Foundation Grants, 414
Allstate Corporate Giving Grants, 446
Allstate Corp Hometown Commitment Grants, 447
CNO Financial Group Community Grants, 1201
Coca-Cola Foundation Grants, 1203
FMC Foundation Grants, 1683
George W. Codrington Charitable Fndn Grants, 1803
J.L. Bedsole Foundation Grants, 2172
PACCAR Foundation Grants, 3101
Procter and Gamble Fund Grants, 3274
Sid W. Richardson Foundation Grants, 3553

Ecosystems
Blue Cross Blue Shield of Minnesota Foundation - Healthy Equity: Health Impact Assessment Demonstration Project Grants, 916

SUBJECT INDEX

Conservation, Food, and Health Foundation Grants for Developing Countries, 1338
Washington Gas Charitable Giving, 3908

Editing
AMA Virtual Mentor Theme Issue Editor Grants, 496

Education
1st Source Foundation Grants, 2
2 Depot Square Ipswich Charitable Fndn Grants, 4
49ers Foundation Grants, 9
1675 Foundation Grants, 11
1976 Foundation Grants, 12
AAP Resident Initiative Fund Grants, 189
Aaron & Cecile Goldman Family Fndn Grants, 191
Abbot and Dorothy H. Stevens Foundation Grants, 197
Abbott Fund Community Grants, 201
Abbott Fund Global AIDS Care Grants, 202
Abernethy Family Foundation Grants, 210
Abington Foundation Grants, 211
Able To Serve Grants, 212
Abney Foundation Grants, 214
Aboudane Family Foundation Grants, 215
Abramson Family Foundation Grants, 217
Abundance Foundation International Grants, 218
ACE Charitable Foundation Grants, 229
ACF Foundation Grants, 230
Achelis Foundation Grants, 234
Ackerman Foundation Grants, 236
ACL University Centers for Excellence in Developmental Network Diversity and Inclusion Training Action Planning Grants, 249
Adams County Community Foundation of Pennsylvania Grants, 300
Adams Foundation Grants, 301
Adelaide Breed Bayrd Foundation Grants, 306
Adler-Clark Electric Community Commitment Foundation Grants, 307
Administaff Community Affairs Grants, 308
Administration on Aging Senior Medicare Patrol Project Grants, 309
Adobe Community Investment Grants, 310
Advance Auto Parts Corporate Giving Grants, 312
Advanced Micro Devices Comm Affairs Grants, 313
AFG Industries Grants, 340
A Friends' Foundation Trust Grants, 341
A Fund for Women Grants, 342
Agnes B. Hunt Trust Grants, 344
Agnes Gund Foundation Grants, 345
Agnes M. Lindsay Trust Grants, 346
Ahn Family Foundation Grants, 356
Air Products and Chemicals Grants, 399
Alaska Airlines Foundation Grants, 403
Albert and Margaret Alkek Foundation Grants, 405
Albert Pick Jr. Fund Grants, 409
Albertson's Charitable Giving Grants, 410
Albert W. Rice Charitable Foundation Grants, 411
Albuquerque Community Foundation Grants, 412
Alcoa Foundation Grants, 414
Alcon Foundation Grants Program, 416
Alexander & Baldwin Fnd Mainland Grants, 417
Alexander and Baldwin Foundation Hawaiian and Pacific Island Grants, 418
Alfred and Tillie Shemanski Testamentary Trust Grants, 428
Alfred Bersted Foundation Grants, 429
Alfred E. Chase Charitable Foundation Grants, 430
Alice Tweed Tuohy Foundation Grants Program, 437
Allan C. and Lelia J. Garden Foundation Grants, 438
Allegan County Community Foundation Grants, 439
Allegheny Technologies Charitable Trust, 440
Allegis Group Foundation Grants, 441
Allen P. & Josephine B. Green Fndn Grants, 443
Allstate Corporate Giving Grants, 446
Allstate Corp Hometown Commitment Grants, 447
Alpha Natural Resources Corporate Giving, 450
Altman Foundation Health Care Grants, 457
Alvin and Fanny Blaustein Thalheimer Foundation Baltimore Communal Grants, 459
AMA-MSS Chapter Involvement Grants, 473

AMA-MSS Government Relations Advocacy Fellowship, 475
AMA-MSS Government Relations Internships, 476
AMA-RFS and AMA Foundation Medical Student & Resident/Fellow Elective-Medicine & Media, 478
AMA-RFS Legislative Awareness Internships, 479
Amador Community Foundation Grants, 481
AMD Corporate Contributions Grants, 500
American Electric Power Foundation Grants, 530
American Foodservice Charitable Trust Grants, 532
American Jewish World Service Grants, 533
American Psychiatric Foundation Grantss, 552
American Psychiatric Foundation Typical or Troubled School Mental Health Education Grants, 558
American Schlafhorst Foundation Grants, 559
American Trauma Society, Pennsylvania Division Mini-Grants, 564
Amerigroup Foundation Grants, 565
Amerisure Insurance Community Service Grants, 566
Amgen Foundation Grants, 572
Amica Insurance Company Community Grants, 575
Andersen Corporate Foundation, 586
Andre Agassi Charitable Foundation Grants, 587
Andrew Family Foundation Grants, 588
Angels Baseball Foundation Grants, 590
Anheuser-Busch Foundation Grants, 591
Anna Fitch Ardenghi Trust Grants, 592
Ann Arbor Area Community Foundation Grants, 594
Annenberg Foundation Grants, 596
Annie Sinclair Knudsen Memorial Fund/Kaua'i Community Grants Program, 597
Ann Peppers Foundation Grants, 599
Annunziata Sanguinetti Foundation Grants, 600
Anthony R. Abraham Foundation Grants, 608
Antone & Edene Vidinha Charitable Trust Grants, 609
AON Foundation Grants, 623
APS Foundation Grants, 646
AptarGroup Foundation Grants, 647
Aragona Family Foundation Grants, 648
Aratani Foundation Grants, 649
Arcadia Foundation Grants, 650
ARCO Foundation Education Grants, 652
Arie and Ida Crown Memorial Grants, 654
Arizona Cardinals Grants, 655
Arizona Diamondbacks Charities Grants, 657
Arizona Public Service Corporate Giving Program Grants, 658
Arkansas Community Foundation Arkansas Black Hall of Fame Grants, 659
Arkansas Community Foundation Grants, 660
ARS Foundation Grants, 667
Arthur and Rochelle Belfer Foundation Grants, 670
Arthur Ashley Williams Foundation Grants, 671
ArvinMeritor Grants, 675
ASGH Award for Excellence in Human Genetics Education, 689
ASHA Multicultural Activities Projects Grants, 697
Ashland Corporate Contributions Grants, 713
ASM Undergraduate Research Fellowship, 755
Assisi Fndn of Memphis Capital Project Grants, 786
Assisi Fndn of Memphis General Grants, 787
Assurant Health Foundation Grants, 789
AT&T Foundation Civic and Community Service Program Grants, 793
Atherton Family Foundation Grants, 794
Athwin Foundation Grants, 795
Atlanta Foundation Grants, 796
Aurora Foundation Grants, 802
Austin College Leadership Award, 804
Australasian Institute of Judicial Administration Seed Funding Grants, 806
Australian Academy of Science Grants, 807
Autodesk Community Relations Grants, 809
Axe-Houghton Foundation Grants, 815
Babcock Charitable Trust Grants, 816
Bacon Family Foundation Grants, 817
Balfe Family Foundation Grants, 819
BancorpSouth Foundation Grants, 823
Bank of America Fndn Matching Gifts, 825
Baptist Community Ministries Grants, 828

Barberton Community Foundation Grants, 829
Barker Foundation Grants, 830
Barker Welfare Foundation Grants, 831
Barra Foundation Community Fund Grants, 833
Barra Foundation Project Grants, 834
Batchelor Foundation Grants, 837
Baton Rouge Area Foundation Grants, 838
Batts Foundation Grants, 840
Baxter International Corporate Giving Grants, 841
Bayer Foundation Grants, 846
BBVA Compass Foundation Charitable Grants, 848
BCBSM Foundation Student Award Program, 858
Beazley Foundation Grants, 865
Bechtel Group Foundation Building Positive Community Relationships Grants, 866
Beckley Area Foundation Grants, 867
Beckman Coulter Foundation Grants, 868
Beirne Carter Foundation Grants, 870
Belvedere Community Foundation Grants, 874
Ben B. Cheney Foundation Grants, 875
Benton County Foundation - Fitzgerald Family Scholarships, 879
Benton County Foundation Grants, 880
Berks County Community Foundation Grants, 881
Berrien Community Foundation Grants, 884
Besser Foundation Grants, 887
BHHS Legacy Foundation Grants, 888
BibleLands Grants, 889
Bill Hannon Foundation Grants, 895
Biogen Foundation Healthcare Professional Education Grants, 900
Biogen Foundation Patient Educational Grants, 901
BJ's Charitable Foundation Grants, 904
Blackford County Community Foundation Grants, 906
Blanche and Irving Laurie Foundation Grants, 910
Blanche and Julian Robertson Family Foundation Grants, 911
Blowitz-Ridgeway Foundation Early Childhood Development Research Award, 912
Blowitz-Ridgeway Foundation Grants, 913
Blue Cross Blue Shield of Minnesota Foundation - Health Equity: Building Health Equity Together Grants, 914
Blue Grass Community Fndn Harrison Grants, 919
Blue Grass Community Foundation Hudson-Ellis Fund Grants, 920
Blumenthal Foundation Grants, 925
BMW of North America Charitable Contributions, 926
Bob & Delores Hope Foundation Grants, 927
Boettcher Foundation Grants, 931
Booth-Bricker Fund Grants, 933
Boston Globe Foundation Grants, 937
Boston Jewish Community Women's Fund Grants, 938
Boyle Foundation Grants, 942
Bradley-Turner Foundation Grants, 944
Bradley C. Higgins Foundation Grants, 945
Bridgestone/Firestone Trust Fund Grants, 949
Bright Family Foundation Grants, 951
Brinson Foundation Grants, 953
Bristol-Myers Squibb Foundation Health Education Grants, 959
Bristol-Myers Squibb Foundation Science Education Grants, 961
Brook J. Lenfest Foundation Grants, 967
Brown Advisory Charitable Foundation Grants, 969
Burden Trust Grants, 976
Business Bank of Nevada Community Grants, 981
BWF Postdoctoral Enrichment Grants, 989
Byron W. & Alice L. Lockwood Fnd Grants, 991
Cabot Corporation Foundation Grants, 993
Cailloux Foundation Grants, 996
Callaway Foundation Grants, 1002
Callaway Golf Company Foundation Grants, 1003
Cambridge Community Foundation Grants, 1004
Campbell Soup Foundation Grants, 1007
Canada-U.S. Fulbright New Century Scholars Program Grants, 1008
Canada-U.S. Fulbright Senior Specialists Grants, 1009
Capital Region Community Foundation Grants, 1014
Cardinal Health Foundation Grants, 1017

712 / Education

Cargill Citizenship Fund Corporate Giving, 1018
Caring Foundation Grants, 1019
Carl & Eloise Pohlad Family Fndn Grants, 1021
Carl B. and Florence E. King Foundation Grants, 1022
Carl C. Icahn Foundation Grants, 1023
Carl Gellert and Celia Berta Gellert Foundation Grants, 1024
Carl M. Freeman Foundation FACES Grants, 1026
Carl R. Hendrickson Family Foundation Grants, 1028
Carl W. and Carrie Mae Joslyn Trust Grants, 1031
Carnahan-Jackson Foundation Grants, 1032
Carolyn Foundation Grants, 1034
Carpenter Foundation Grants, 1035
Carrie E. and Lena V. Glenn Foundation Grants, 1036
Carrier Corporation Contributions Grants, 1038
Carroll County Community Foundation Grants, 1039
Cass County Community Foundation Grants, 1041
Centerville-Washington Foundation Grants, 1078
Central Carolina Community Foundation Community Impact Grants, 1079
Central Okanagan Foundation Grants, 1080
Cessna Foundation Grants Program, 1081
CFFVR Basic Needs Giving Partnership Grants, 1094
CFFVR Chilton Area Community Fndn Grants, 1096
CFFVR Clintonville Area Foundation Grants, 1097
CFFVR Frank C. Shattuck Community Grants, 1098
CFFVR Jewelers Mutual Charitable Giving, 1099
CFFVR Myra M. and Robert L. Vandehey Foundation Grants, 1100
CFFVR Project Grants, 1101
CFFVR Shawano Area Community Foundation Grants, 1104
CFFVR Waupaca Area Community Foundation Grants, 1105
CFFVR Wisconsin King's Daus & Sons Grants, 1106
Chamberlain Foundation Grants, 1108
Champlin Foundations Grants, 1110
Chapman Charitable Foundation Grants, 1112
CharityWorks Grants, 1113
Charles Delmar Foundation Grants, 1115
Charles F. Bacon Trust Grants, 1117
Charles H. Dater Foundation Grants, 1118
Charles H. Hall Foundation, 1120
Charles H. Pearson Foundation Grants, 1121
Charles H. Revson Foundation Grants, 1122
Charles Lafitte Foundation Grants, 1123
Charlotte County (FL) Community Foundation Grants, 1126
Chase Paymentech Corporate Giving Grants, 1127
Chazen Foundation Grants, 1129
Chemtura Corporation Contributions Grants, 1130
Chilkat Valley Community Foundation Grants, 1148
Christensen Fund Regional Grants, 1149
Christy-Houston Foundation Grants, 1153
Chula Vista Charitable Foundation Grants, 1154
CICF Indianapolis Fndn Community Grants, 1156
CICF Legacy Fund Grants, 1158
CIT Corporate Giving Grants, 1161
Citizens Bank Mid-Atlantic Charitable Foundation Grants, 1162
Clara Blackford Smith and W. Aubrey Smith Charitable Foundation Grants, 1165
Clarence E. Heller Charitable Foundation Grants, 1166
Clarence T.C. Ching Foundation Grants, 1167
Clarian Health Scholarships for LPNs, 1171
Clark County Community Foundation Grants, 1175
Claude Worthington Benedum Fndn Grants, 1178
Clayton Baker Trust Grants, 1179
Cleveland-Cliffs Foundation Grants, 1181
Clinton County Community Foundation Grants, 1184
CNA Foundation Grants, 1189
CNCS AmeriCorps VISTA Project Grants, 1190
Coastal Community Foundation of South Carolina Grants, 1202
Coca-Cola Foundation Grants, 1203
Collins C. Diboll Private Foundation Grants, 1209
Collins Foundation Grants, 1210
Colonel Stanley R. McNeil Foundation Grants, 1211
Columbus Foundation Competitive Grants, 1217

Columbus Foundation Mary Eleanor Morris Fund Grants, 1220
Columbus Foundation Paul G. Duke Grants, 1221
Columbus Foundation Traditional Grants, 1223
Comerica Charitable Foundation Grants, 1224
Commonwealth Edison Grants, 1225
Communities Foundation of Texas Grants, 1236
Community Foundation Alliance City of Evansville Endowment Fund Grants, 1238
Community Fndn for Greater Buffalo Grants, 1247
Community Fndn for Monterey County Grants, 1253
Community Fndn for Muskegon County Grants, 1254
Community Fndn for San Benito County Grants, 1257
Community Foundation for the National Capital Region Community Leadership Grants, 1260
Community Foundation of Boone County Grants, 1266
Community Fndn of East Central Illinois Grants, 1268
Community Foundation of Eastern Connecticut General Southeast Grants, 1269
Community Foundation of Grant County Grants, 1271
Community Foundation of Greater Birmingham Grants, 1272
Community Foundation of Greater Flint Grants, 1273
Community Foundation of Greater Fort Wayne Scholarships, 1276
Community Fndn of Greater Lafayette Grants, 1280
Community Foundation of Greater New Britain Grants, 1281
Community Fndn of Greater Tampa Grants, 1282
Community Foundation of Greenville-Greenville Women Giving Grants, 1283
Community Foundation of Greenville Community Enrichment Grants, 1284
Community Foundation of Greenville Hollingsworth Funds Program/Project Grants, 1285
Community Fndn of Howard County Grants, 1286
Community Fndn of Jackson County Grants, 1287
Community Foundation of Mount Vernon and Knox County Grants, 1299
Community Foundation of Muncie and Delaware County Grants, 1300
Community Foundation of Muncie and Delaware County Maxon Grants, 1301
Community Fndn of Randolph County Grants, 1302
Community Fndn of So Alabama Grants, 1306
Community Fndn of South Puget Sound Grants, 1307
Community Foundation of St. Joseph County Special Project Challenge Grants, 1310
Community Fndn of Switzerland County Grants, 1311
Community Foundation of Tampa Bay Grants, 1312
Community Foundation of the Verdugos Educational Endowment Fund Grants, 1314
Community Foundation of the Verdugos Grants, 1315
Community Fndn of Wabash County Grants, 1316
Community Foundation Partnerships - Lawrence County Grants, 1317
Community Foundation Partnerships - Martin County Grants, 1318
Connecticut Community Foundation Grants, 1324
Conquer Cancer Foundation of ASCO Comparative Effectiveness Research Professorship in Breast Cancer, 1329
Conquer Cancer Foundation of ASCO Translational Research Professorships, 1337
Consumers Energy Foundation, 1343
Cooke Foundation Grants, 1344
Cooper Industries Foundation Grants, 1345
Cornerstone Foundation of Northeastern Wisconsin Grants, 1347
Cowles Charitable Trust Grants, 1354
Cralle Foundation Grants, 1356
Crane Foundation Grants, 1357
Cresap Family Foundation Grants, 1359
Crown Point Community Foundation Grants, 1361
Cruise Industry Charitable Foundation Grants, 1363
CSRA Community Foundation Grants, 1366
CSX Corporate Contributions Grants, 1368
Cudd Foundation Grants, 1370
Cullen Foundation Grants, 1372
Cultural Society of Filipino Americans Grants, 1373

SUBJECT INDEX

Cumberland Community Foundation Grants, 1374
CUNA Mutual Group Fndn Community Grants, 1375
CVS Caremark Charitable Trust Grants, 1378
CVS Community Grants, 1379
Cyrus Eaton Foundation Grants, 1380
D.F. Halton Foundation Grants, 1405
D.V. and Ida J. McEachern Trust Grants, 1406
DAAD Research Stays for University Academics and Scientists, 1408
Dade Community Foundation Grants, 1410
DaimlerChrysler Corporation Fund Grants, 1412
Dairy Queen Corporate Contributions Grants, 1413
Dallas Mavericks Foundation Grants, 1416
Dammann Fund Grants, 1418
Dana Brown Charitable Trust Grants, 1419
Dana Foundation Science and Health Grants, 1420
Danellie Foundation Grants, 1421
Daniel & Nanna Stern Family Fndn Grants, 1422
Daphne Seybolt Culpeper Fndn Grants, 1425
Davis Family Foundation Grants, 1435
Dayton Power and Light Company Foundation Signature Grants, 1436
Dayton Power and Light Foundation Grants, 1437
Deaconess Community Foundation Grants, 1439
Dean Foods Community Involvement Grants, 1440
Dearborn Community Foundation City of Lawrenceburg Community Grants, 1441
Dearborn Community Foundation County Progress Grants, 1442
Deborah Munroe Noonan Memorial Grants, 1443
Decatur County Community Foundation Large Project Grants, 1444
DeKalb County Community Foundation Grants, 1447
Delaware Community Foundation-Youth Philanthropy Board for Kent County, 1448
Delaware Community Foundation Grants, 1449
Del E. Webb Foundation Grants, 1450
Delmarva Power and Light Contributions, 1454
DeMatteis Family Foundation Grants, 1457
Denver Broncos Charities Fund Grants, 1460
Denver Foundation Community Grants, 1461
Dept of Ed Safe and Drug-Free Schools and Communities State Grants, 1464
Dept of Ed Special Education--Personnel Development to Improve Services and Results for Children with Disabilities, 1465
DeRoy Testamentary Foundation Grants, 1469
Deutsche Banc Alex Brown and Sons Charitable Foundation Grants, 1471
Dickson Foundation Grants, 1476
DOJ Gang-Free Schools and Communities Intervention Grants, 1481
Dolan Children's Foundation Grants, 1482
Dole Food Company Charitable Contributions, 1483
Dora Roberts Foundation Grants, 1489
Doris and Victor Day Foundation Grants, 1491
Dorothy Hooper Beattie Foundation Grants, 1499
Dorrance Family Foundation Grants, 1501
Dr. & Mrs. Paul Pierce Memorial Fndn Grants, 1505
Dr. Leon Bromberg Charitable Trust Grants, 1509
Drs. Bruce and Lee Foundation Grants, 1513
Dubois County Community Foundation Grants, 1517
Duneland Health Council Incorporated Grants, 1529
DuPage Community Foundation Grants, 1531
DuPont Pioneer Community Giving Grants, 1532
Dyson Foundation Mid-Hudson Valley General Operating Support Grants, 1533
Dyson Foundation Mid-Hudson Valley Project Support Grants, 1534
E.L. Wiegand Foundation Grants, 1538
Earl and Maxine Claussen Trust Grants, 1541
Eastman Chemical Company Foundation Grants, 1542
eBay Foundation Community Grants, 1543
Eberly Foundation Grants, 1544
Eckerd Corporation Foundation Grants, 1547
Eddie C. & Sylvia Brown Family Fndn Grants, 1548
Eden Hall Foundation Grants, 1549
Edna G. Kynett Memorial Foundation Grants, 1551
EDS Foundation Grants, 1552
Educational Foundation of America Grants, 1553

SUBJECT INDEX

Edward and Helen Bartlett Foundation Grants, 1554
Edward Bangs Kelley and Elza Kelley Foundation Grants, 1556
Edward W. and Stella C. Van Houten Memorial Fund Grants, 1559
Edwin S. Webster Foundation Grants, 1560
Edyth Bush Charitable Foundation Grants, 1562
Elizabeth Morse Genius Charitable Trust Grants, 1569
Elkhart County Community Foundation Grants, 1573
Elmer L. & Eleanor J. Andersen Fndn Grants, 1578
El Paso Community Foundation Grants, 1579
El Paso Corporate Foundation Grants, 1580
El Pomar Foundation Anna Keesling Ackerman Fund Grants, 1581
El Pomar Foundation Grants, 1582
Elsie H. Wilcox Foundation Grants, 1583
Elsie Lee Garthwaite Memorial Fndn Grants, 1584
Emerson Electric Company Contributions Grants, 1590
Emily Hall Tremaine Foundation Learning Disabilities Grants, 1592
Emma B. Howe Memorial Foundation Grants, 1593
Ensworth Charitable Foundation Grants, 1595
Entergy Corporation Micro Grants, 1596
Erie Community Foundation Grants, 1601
Essex County Community Foundation Discretionary Fund Grants, 1602
Essex County Community Foundation Merrimack Valley General Fund Grants, 1603
Estee Lauder Grants, 1606
Ethel Sergeant Clark Smith Foundation Grants, 1608
Eugene G. and Margaret M. Blackford Memorial Fund Grants, 1610
Eugene M. Lang Foundation Grants, 1611
Evan and Susan Bayh Foundation Grants, 1614
Evanston Community Foundation Grants, 1615
Ewing Halsell Foundation Grants, 1618
Express Scripts Foundation Grants, 1620
F.M. Kirby Foundation Grants, 1624
Fairfield County Community Foundation Grants, 1625
Fargo-Moorhead Area Foundation Grants, 1632
Fargo-Moorhead Area Fndn Woman's Grants, 1633
Farmers Insurance Corporate Giving Grants, 1634
Faye McBeath Foundation Grants, 1635
Fayette County Foundation Grants, 1636
FCD New American Children Grants, 1637
Federal Express Corporate Contributions, 1657
Ferree Foundation Grants, 1658
Fidelity Foundation Grants, 1659
Fifth Third Foundation Grants, 1661
Finance Factors Foundation Grants, 1666
FirstEnergy Foundation Community Grants, 1668
Fishman Family Foundation Grants, 1672
FIU Global Civic Engagement Mini Grants, 1673
Florian O. Bartlett Trust Grants, 1677
Florida BRAIVE Fund of Dade Community Foundation, 1678
Floyd A. & Kathleen C. Cailloux Fnd Grants, 1681
Fluor Foundation Grants, 1682
FMC Foundation Grants, 1683
Foellinger Foundation Grants, 1684
Fondren Foundation Grants, 1685
Ford Motor Company Fund Grants Program, 1687
Forrest C. Lattner Foundation Grants, 1688
Foster Foundation Grants, 1689
Fourjay Foundation Grants, 1712
Frances and John L. Loeb Family Fund Grants, 1716
Frances W. Emerson Foundation Grants, 1718
Francis T. & Louise T. Nichols Fndn Grants, 1720
Frank B. Hazard General Charity Fund Grants, 1721
Frank E. and Seba B. Payne Foundation Grants, 1722
Franklin County Community Foundation Grants, 1724
Franklin H. Wells and Ruth L. Wells Foundation Grants, 1725
Frank Loomis Palmer Fund Grants, 1726
Frank Reed and Margaret Jane Peters Memorial Fund I Grants, 1727
Frank Reed and Margaret Jane Peters Memorial Fund II Grants, 1728
Frank Stanley Beveridge Foundation Grants, 1730
Frank W. & Carl S. Adams Memorial Grants, 1731

Fraser-Parker Foundation Grants, 1732
Fred Baldwin Memorial Foundation Grants, 1736
Freddie Mac Foundation Grants, 1738
Frederick W. Marzahl Memorial Fund Grants, 1741
Fremont Area Community Foundation Amazing X Grants, 1744
Fremont Area Community Fndn General Grants, 1746
Fritz B. Burns Foundation Grants, 1747
Fuji Film Grants, 1748
Fulbright Alumni Initiatives Awards, 1749
Fulbright Distinguished Chairs Awards, 1750
Fulbright International Education Administrators (IEA) Seminar Program Grants, 1752
Fulbright New Century Scholars Grants, 1753
Fulbright Specialists Program Grants, 1754
Fulbright Scholars in Europe and Eurasia, 1755
Fulbright Traditional Scholar Program in Sub-Saharan Africa, 1756
Fulbright Traditional Scholar Program in the East Asia/Pacific Region, 1757
Fulbright Traditional Scholar Program in the Near East and North Africa Region, 1758
Fulbright Traditional Scholar Program in the South and Central Asia Region, 1759
Fulbright Traditional Scholar Program in the Western Hemisphere, 1760
Fuller E. Callaway Foundation Grants, 1761
G.N. Wilcox Trust Grants, 1768
Gardner Foundation Grants, 1770
Garland D. Rhoads Foundation, 1774
Gates Millennium Scholars Program, 1776
Gebbie Foundation Grants, 1777
GEICO Public Service Awards, 1778
Genentech Corp Charitable Contributions, 1780
General Mills Foundation Grants, 1782
General Motors Foundation Grants, 1783
Genesis Foundation Grants, 1786
Genuardi Family Foundation Grants, 1787
George A. and Grace L. Long Foundation Grants, 1788
George and Ruth Bradford Foundation Grants, 1789
George A Ohl Jr. Foundation Grants, 1791
George Family Foundation Grants, 1794
George Gund Foundation Grants, 1796
George H.C. Ensworth Memorial Fund Grants, 1797
George Kress Foundation Grants, 1799
George P. Davenport Trust Fund Grants, 1800
George W. Brackenridge Foundation Grants, 1802
George W. Codrington Charitable Fndn Grants, 1803
George W. Wells Foundation Grants, 1804
Georgiana Goddard Eaton Memorial Grants, 1805
Georgia Power Foundation Grants, 1806
Gerber Foundation Grants, 1807
Gertrude and William C. Wardlaw Fund Grants, 1808
GFWC of Massachusetts Memorial Education Scholarship, 1813
Gheens Foundation Grants, 1814
Gibson Foundation Grants, 1818
Gil and Dody Weaver Foundation Grants, 1819
Ginn Foundation Grants, 1822
Giving Sum Annual Grant, 1825
GlaxoSmithKline Corporate Grants, 1829
Global Fund for Children Grants, 1832
Global Fund for Women Grants, 1833
GNOF Bayou Communities Grants, 1835
GNOF Exxon-Mobil Grants, 1836
GNOF Freeman Challenge Grants, 1837
GNOF IMPACT Kahn-Oppenheim Grants, 1842
Godfrey Foundation Grants, 1849
Golden State Warriors Foundation Grants, 1851
Goodrich Corporation Foundation Grants, 1854
Goodyear Tire Grants, 1856
Grace and Franklin Bernsen Foundation Grants, 1858
Grace Bersted Foundation Grants, 1859
Graco Foundation Grants, 1860
Graham Foundation Grants, 1862
Grand Haven Area Community Fndn Grants, 1864
Grand Rapids Area Community Foundation Wyoming Grants, 1867
Grand Rapids Area Community Foundation Wyoming Youth Fund Grants, 1868

Grand Rapids Community Foundation Grants, 1869
Grand Rapids Community Foundation Ionia County Grants, 1870
Grand Rapids Community Foundation Ionia County Youth Fund Grants, 1871
Grand Rapids Community Foundation Lowell Area Fund Grants, 1872
Grand Rapids Community Foundation Southeast Ottawa Grants, 1873
Grand Rapids Community Foundation Southeast Ottawa Youth Fund Grants, 1874
Grand Rapids Community Fndn Sparta Grants, 1875
Grand Rapids Community Foundation Sparta Youth Fund Grants, 1876
Great-West Life Grants, 1879
Greater Cincinnati Foundation Priority and Small Projects/Capacity-Building Grants, 1880
Greater Green Bay Community Fndn Grants, 1881
Greater Kanawha Valley Foundation Grants, 1882
Greater Milwaukee Foundation Grants, 1883
Greater Saint Louis Community Fndn Grants, 1884
Greater Sitka Legacy Fund Grants, 1885
Greater Tacoma Community Foundation Grants, 1886
Greater Worcester Community Foundation Discretionary Grants, 1888
Greater Worcester Community Foundation Jeppson Memorial Fund for Brookfield Grants, 1889
Green Bay Packers Foundation Grants, 1890
Green Diamond Charitable Contributions, 1891
Greene County Foundation Grants, 1892
Gregory L. Gibson Charitable Fndn Grants, 1900
Griffin Family Foundation Grants, 1902
Grifols Community Outreach Grants, 1904
Guido A. & Elizabeth H. Binda Fndn Grants, 1913
Guitar Center Music Foundation Grants, 1914
Gulf Coast Community Foundation Grants, 1915
Gulf Coast Foundation of Community Operating Grants, 1916
Gulf Coast Foundation of Community Program Grants, 1917
Guy I. Bromley Trust Grants, 1919
H & R Foundation Grants, 1920
H.A. & Mary K. Chapman Trust Grants, 1921
H.B. Fuller Foundation Grants, 1922
H.J. Heinz Company Foundation Grants, 1923
H. Leslie Hoffman and Elaine S. Hoffman Foundation Grants, 1924
Hagedorn Fund Grants, 1934
Hall-Perrine Foundation Grants, 1935
Hallmark Corporate Foundation Grants, 1936
Hammond Common Council Scholarships, 1938
Hancock County Community Foundation - Field of Interest Grants, 1943
Hardin County Community Foundation Grants, 1946
Harmony Project Grants, 1948
Harold Alfond Foundation Grants, 1949
Harold and Arlene Schnitzer CARE Foundation Grants, 1950
Harold R. Bechtel Charitable Remainder Uni-Trust Grants, 1953
Harris and Eliza Kempner Fund Grants, 1956
Harry A. & Margaret D. Towsley Fndn Grants, 1959
Harry B. and Jane H. Brock Foundation Grants, 1960
Harry Bramhall Gilbert Charitable Trust Grants, 1961
Harry Frank Guggenheim Foundation Dissertation Fellowships, 1963
Harry W. Bass, Jr. Foundation Grants, 1968
Hartford Courant Foundation Grants, 1969
Hartford Foundation Regular Grants, 1970
Harvest Foundation Grants, 1971
Hawaiian Electric Industries Charitable Foundation Grants, 1975
Hawaii Community Foundation West Hawaii Fund Grants, 1978
Hawn Foundation Grants, 1979
Heckscher Foundation for Children Grants, 1988
Hedco Foundation Grants, 1989
Helena Rubinstein Foundation Grants, 1991
Helen Bader Foundation Grants, 1992
Helen Gertrude Sparks Charitable Trust Grants, 1993

Education

Helen K. & Arthur E. Johnson Fndn Grants, 1995
Helen Pumphrey Denit Charitable Trust Grants, 1996
Helen S. Boylan Foundation Grants, 1997
Hendricks County Community Fndn Grants, 2000
Henrietta Tower Wurts Memorial Fndn Grants, 2002
Henry County Community Foundation Grants, 2005
Herbert H. & Grace A. Dow Fndn Grants, 2008
HHMI Biomedical Research Grants for International Scientists: Infectious Diseases & Parasitology, 2017
HHMI Biomedical Research Grants for International Scientists in Canada and Latin America, 2018
HHMI Biomedical Research Grants for International Scientists in the Baltics, Central and Eastern Europe, Russia, and Ukraine, 2019
Highmark Corporate Giving Grants, 2025
Hilda and Preston Davis Foundation Grants, 2028
Hillcrest Foundation Grants, 2031
Hillman Foundation Grants, 2032
Hillsdale County Community General Adult Foundation Grants, 2034
Hillsdale Fund Grants, 2035
Hilton Head Island Foundation Grants, 2036
Hilton Hotels Corporate Giving Program Grants, 2037
Hoblitzelle Foundation Grants, 2038
Hoglund Foundation Grants, 2040
Holland/Zeeland Community Fndn Grants, 2041
HomeBanc Foundation Grants, 2042
Homer Foundation Grants, 2044
Honda of America Manufacturing Fndn Grants, 2045
Horace A. Kimball and S. Ella Kimball Foundation Grants, 2046
Horace Moses Charitable Foundation Grants, 2047
Hormel Foods Charitable Trust Grants, 2050
Hormel Foundation Grants, 2051
Howard and Bush Foundation Grants, 2054
Howe Foundation of North Carolina Grants, 2055
Hudson Webber Foundation Grants, 2072
Huffy Foundation Grants, 2073
Human Source Foundation Grants, 2078
Huntington Beach Police Officers Fndn Grants, 2081
Huntington Clinical Foundation Grants, 2082
Huntington County Community Foundation Make a Difference Grants, 2083
Huntington National Bank Community Grants, 2084
Hutchinson Community Foundation Grants, 2085
Hut Foundation Grants, 2086
I.A. O'Shaughnessy Foundation Grants, 2088
IBEW Local Union #697 Memorial Scholarships, 2096
Idaho Community Foundation Eastern Region Competitive Grants, 2114
Idaho Power Company Corporate Contributions, 2115
Ida S. Barter Trust Grants, 2116
IIE Freeman Foundation Indonesia Internships, 2131
Illinois Tool Works Foundation Grants, 2147
Impact 100 Grants, 2148
Inasmuch Foundation Grants, 2149
Intergrys Corporation Grants, 2160
Ireland Family Foundation Grants, 2162
Irving S. Gilmore Foundation Grants, 2163
Irvin Stern Foundation Grants, 2164
Isabel Allende Foundation Esperanza Grants, 2165
Ittleson Foundation Mental Health Grants, 2167
J.E. and L.E. Mabee Foundation Grants, 2170
J.W. Kieckhefer Foundation Grants, 2176
J. Willard Marriott, Jr. Foundation Grants, 2178
Jackson County Community Foundation Unrestricted Grants, 2180
Jacob and Valeria Langeloth Foundation Grants, 2182
Jacob G. Schmidlapp Trust Grants, 2183
Jacobs Family Village Neighborhoods Grants, 2184
James & Abigail Campbell Family Fndn Grants, 2185
James A. and Faith Knight Foundation Grants, 2186
James Ford Bell Foundation Grants, 2187
James G.K. McClure Educational and Development Fund Grants, 2188
James J. & Angelia M. Harris Fndn Grants, 2192
James J. & Joan A. Gardner Family Fndn Grants, 2193
James L. and Mary Jane Bowman Charitable Trust Grants, 2194
James M. Cox Foundation of Georgia Grants, 2196
James R. Thorpe Foundation Grants, 2199
James S. Copley Foundation Grants, 2200
Jane's Trust Grants, 2206
Janus Foundation Grants, 2211
Jasper Foundation Grants, 2212
Jay and Rose Phillips Family Foundation Grants, 2213
Jean and Louis Dreyfus Foundation Grants, 2215
JELD-WEN Foundation Grants, 2218
Jennings County Community Foundation Grants, 2220
Jennings County Community Foundation Women's Giving Circle Grant, 2221
Jessica Stevens Community Foundation Grants, 2224
Jessie B. Cox Charitable Trust Grants, 2225
Jessie Ball Dupont Fund Grants, 2226
Jim Moran Foundation Grants, 2228
JM Foundation Grants, 2229
Joe W. and Dorothy Dorsett Brown Fndn Grants, 2230
John Clarke Trust Grants, 2232
John Deere Foundation Grants, 2234
John G. Duncan Charitable Trust Grants, 2236
John G. Martin Foundation Grants, 2237
John Jewett and Helen Chandler Garland Foundation Grants, 2242
John Lord Knight Foundation Grants, 2243
John P. McGovern Foundation Grants, 2247
John R. Oishei Foundation Grants, 2249
Johns Manville Fund Grants, 2251
Johnson & Johnson Corporate Contributions, 2253
John Stauffer Charitable Trust Grants, 2258
John V. and George Primich Family Scholarship, 2259
John W. Alden Trust Grants, 2260
John W. and Anna H. Hanes Foundation Grants, 2261
Joseph Drown Foundation Grants, 2268
Joseph Henry Edmondson Foundation Grants, 2270
Josephine G. Russell Trust Grants, 2271
Josephine S. Gumbiner Foundation Grants, 2273
Josiah W. and Bessie H. Kline Foundation Grants, 2278
Judith and Jean Pape Adams Charitable Foundation Tulsa Area Grants, 2280
Judith Clark-Morrill Foundation Grants, 2281
Kahuku Community Fund, 2286
K and F Baxter Family Foundation Grants, 2287
Kathryne Beynon Foundation Grants, 2294
Katrine Menzing Deakins Trust Grants, 2295
Kavli Foundation Research Grants, 2296
Kelvin and Eleanor Smith Foundation Grants, 2297
Kenai Peninsula Foundation Grants, 2298
Ketchikan Community Foundation Grants, 2303
Kettering Family Foundation Grants, 2304
Kettering Fund Grants, 2305
Kevin P. and Sydney B. Knight Family Foundation Grants, 2306
Knight Family Charitable and Educational Foundation Grants, 2312
Knox County Community Foundation Grants, 2313
Kodiak Community Foundation Grants, 2314
Kosciusko County Community Fndn Grants, 2319
Kovler Family Foundation Grants, 2323
Kuntz Foundation Grants, 2326
Lake County Community Fund Grants, 2333
Land O'Lakes Foundation Mid-Atlantic Grants, 2341
Lands' End Corporate Giving Program, 2344
Laura B. Vogler Foundation Grants, 2345
Laura Moore Cunningham Foundation Grants, 2346
Lawrence Foundation Grants, 2347
Leahi Fund, 2350
Legler Benbough Foundation Grants, 2352
Leonard L. & Bertha U. Abess Fndn Grants, 2357
Leo Niessen Jr., Charitable Trust Grants, 2358
Liberty Bank Foundation Grants, 2362
Lil and Julie Rosenberg Foundation Grants, 2364
Lincoln Financial Foundation Grants, 2367
Lisa and Douglas Goldman Fund Grants, 2369
Long Island Community Foundation Grants, 2375
Lotus 88 Fnd for Women & Children Grants, 2377
Louie M. & Betty M. Phillips Fndn Grants, 2378
Louis H. Aborn Foundation Grants, 2380
Lowell Berry Foundation Grants, 2382
Lucile Horton Howe and Mitchell B. Howe Foundation Grants, 2385
Lucy Gooding Charitable Fndn Grants, 2388
Ludwick Family Foundation Grants, 2389
Lydia deForest Charitable Trust Grants, 2393
Lynde & Harry Bradley Foundation Fellowships, 2396
Lynde and Harry Bradley Foundation Grants, 2397
Lynde and Harry Bradley Foundation Prizes: Bradley Prizes, 2398
M.B. and Edna Zale Foundation Grants, 2401
M.E. Raker Foundation Grants, 2404
M.J. Murdock Charitable Trust General Grants, 2405
Mabel A. Horne Trust Grants, 2406
Mabel F. Hoffman Charitable Trust Grants, 2407
Mabel Y. Hughes Charitable Trust Grants, 2409
Macquarie Bank Foundation Grants, 2410
Madison County Community Foundation - City of Anderson Quality of Life Grant, 2412
Mann T. Lowry Foundation Grants, 2419
Manuel D. & Rhoda Mayerson Fndn Grants, 2420
Marathon Petroleum Corporation Grants, 2422
Marcia and Otto Koehler Foundation Grants, 2427
Mardag Foundation Grants, 2428
Marie H. Bechtel Charitable Remainder Uni-Trust Grants, 2432
Marion Gardner Jackson Charitable Trust Grants, 2434
Marshall County Community Foundation Grants, 2439
Mary K. Chapman Foundation Grants, 2443
Mary Owen Borden Foundation Grants, 2446
Mary Wilmer Covey Charitable Trust Grants, 2448
Mattel Children's Foundation Grants, 2454
Mattel International Grants Program, 2455
Max and Victoria Dreyfus Foundation Grants, 2457
McCarthy Family Foundation Grants, 2490
McCombs Foundation Grants, 2491
McInerny Foundation Grants, 2495
McKesson Foundation Grants, 2496
McLean Contributionship Grants, 2497
McMillen Foundation Grants, 2499
Mead Johnson Nutritionals Evansville-Area Organizations Grants, 2503
Mead Johnson Nutritionals Med Educ Grants, 2504
Mead Witter Foundation Grants, 2507
Medtronic Foundation HeartRescue Grants, 2513
Memorial Foundation Grants, 2516
Mercedes-Benz USA Corporate Contributions, 2517
Mericos Foundation Grants, 2518
Merkel Foundation Grants, 2520
Merrick Foundation Grants, 2522
Mervin Bovaird Foundation Grants, 2523
Metzger-Price Fund Grants, 2530
Meyer Memorial Trust Special Grants, 2538
MGM Resorts Foundation Community Grants, 2541
MGN Family Foundation Grants, 2542
Miami County Community Foundation - Operation Round Up Grants, 2543
Milagro Foundation Grants, 2552
Milken Family Foundation Grants, 2555
Mimi and Peter Haas Fund Grants, 2557
Moline Foundation Community Grants, 2590
Morris & Gwendolyn Cafritz Fndn Grants, 2594
Natalie W. Furniss Charitable Trust Grants, 2626
Nelda C. and H.J. Lutcher Stark Fndn Grants, 2672
Nell J. Redfield Foundation Grants, 2673
Newton County Community Foundation Grants, 2681
New York University Steinhardt School of Education Fellowships, 2684
NFL Charities Medical Grants, 2685
NIH School Interventions to Prevent Obesity, 2908
NIMH Short Courses in Neuroinformatics, 2920
Nina Mason Pulliam Charitable Trust Grants, 2921
Noble County Community Foundation Grants, 2975
Norcliffe Foundation Grants, 2979
Nordson Corporation Foundation Grants, 2981
Norfolk Southern Foundation Grants, 2982
Norman Foundation Grants, 2983
North Carolina Community Foundation Grants, 2990
North Carolina GlaxoSmithKline Fndn Grants, 2991
North Central Health Services Grants, 2993
Northern New York Community Fndn Grants, 2996
Northwestern Mutual Foundation Grants, 2999

SUBJECT INDEX

Education /715

Northwest Minnesota Foundation Women's Fund Grants, 3000
Norwin S. & Elizabeth N. Bean Fndn Grants, 3001
NYCT AIDS/HIV Grants, 3048
NYCT Girls and Young Women Grants, 3053
NYCT Substance Abuse Grants, 3057
Oceanside Charitable Foundation Grants, 3063
Office Depot Corporation Community Relations Grants, 3065
Ohio County Community Foundation Grants, 3068
Ohio County Community Fndn Mini-Grants, 3069
Ohio Learning Network Grants, 3070
Oleonda Jameson Trust Grants, 3071
OneFamily Foundation Grants, 3075
OneStar Foundation AmeriCorps Grants, 3077
Orange County Community Foundation Grants, 3079
Ordean Foundation Grants, 3080
Oregon Community Fndn Community Grants, 3082
OSF Tackling Addiction Grants in Baltimore, 3096
Owen County Community Foundation Grants, 3100
PacifiCare Foundation Grants, 3102
Pacific Life Foundation Grants, 3103
Palmer Foundation Grants, 3112
Parkersburg Area Community Foundation Action Grants, 3116
Patrick and Anna M. Cudahy Fund Grants, 3120
Paul Ogle Foundation Grants, 3124
Peabody Foundation Grants, 3132
Peacock Foundation Grants, 3133
Percy B. Ferebee Endowment Grants, 3138
Perkin Fund Grants, 3139
Perkins Charitable Foundation Grants, 3140
Perpetual Trust for Charitable Giving Grants, 3141
Perry County Community Foundation Grants, 3142
Peter Kiewit Foundation General Grants, 3154
Peter Kiewit Foundation Small Grants, 3155
Petersburg Community Foundation Grants, 3156
Pew Charitable Trusts Children & Youth Grants, 3158
Pfizer Healthcare Charitable Contributions, 3180
Phelps County Community Foundation Grants, 3188
Philadelphia Foundation Organizational Effectiveness Grants, 3189
Piedmont Health Foundation Grants, 3219
Piedmont Natural Gas Corporate and Charitable Contributions, 3220
Pike County Community Foundation Grants, 3222
Pinellas County Grants, 3223
Pinkerton Foundation Grants, 3224
Pinnacle Foundation Grants, 3225
Pittsburgh Foundation Community Fund Grants, 3230
Playboy Foundation Grants, 3237
Plough Foundation Grants, 3238
PMI Foundation Grants, 3239
Pohlad Family Foundation, 3241
Pokagon Fund Grants, 3242
Polk Bros. Foundation Grants, 3243
Portland General Electric Foundation Grants, 3250
Posey County Community Foundation Grants, 3252
Powell Foundation Grants, 3255
PPG Industries Foundation Grants, 3257
Presbyterian Health Foundation Bridge, Seed and Equipment Grants, 3259
Price Chopper's Golub Foundation Grants, 3261
Price Family Charitable Fund Grants, 3262
Pride Foundation Grants, 3268
Prince Charitable Trusts Chicago Grants, 3270
Princeton Area Community Foundation Greater Mercer Grants, 3272
Principal Financial Group Foundation Grants, 3273
Prospect Burma Scholarships, 3275
Prudential Foundation Education Grants, 3276
Puerto Rico Community Foundation Grants, 3278
Pulaski County Community Foundation Grants, 3279
Putnam County Community Foundation Grants, 3283
PVA Education Foundation Grants, 3284
Quaker Chemical Foundation Grants, 3286
Qualcomm Grants, 3287
Quantum Foundation Grants, 3290
Questar Corporate Contributions Grants, 3291
R.C. Baker Foundation Grants, 3292

R.S. Gernon Trust Grants, 3293
R.T. Vanderbilt Trust Grants, 3294
Radcliffe Institute Carol K. Pforzheimer Student Fellowships, 3298
Rajiv Gandhi Foundation Grants, 3302
Ralphs Food 4 Less Foundation Grants, 3305
Rasmuson Foundation Tier One Grants, 3307
Rasmuson Foundation Tier Two Grants, 3308
Rathmann Family Foundation Grants, 3309
Raymond John Wean Foundation Grants, 3310
Rayonier Foundation Grants, 3311
RCF General Community Grants, 3312
RCF Summertime Kids Grants, 3314
Regenstrief General Internal Medicine - General Pediatrics Research Fellowship Program, 3351
Rehabilitation Nursing Fndn Research Grants, 3354
Retirement Research Foundation General Program Grants, 3359
RGk Foundation Grants, 3361
Rhode Island Foundation Grants, 3362
Richard and Helen DeVos Foundation Grants, 3364
Richard & Susan Smith Family Fndn Grants, 3365
Richard D. Bass Foundation Grants, 3366
Richard Davoud Donchian Foundation Grants, 3367
Richard King Mellon Foundation Grants, 3369
Richland County Bank Grants, 3371
Ricks Family Charitable Trust Grants, 3373
Riley Foundation Grants, 3374
Ripley County Community Foundation Grants, 3376
Ripley County Community Foundation Small Project Grants, 3377
Robert and Joan Dircks Foundation Grants, 3379
Robert F. Stoico/FIRSTFED Charitable Foundation Grants, 3382
Robert W. Woodruff Foundation Grants, 3389
Robins Foundation Grants, 3390
Rochester Area Community Foundation Grants, 3391
Rochester Area Foundation Grants, 3392
Rockefeller Brothers Peace & Security Grants, 3394
Rockefeller Brothers Fund Pivotal Places Grants: South Africa, 3395
Rockwell Collins Charitable Corporation Grants, 3397
Rockwell International Corporate Trust Grants Program, 3399
Rogers Family Foundation Grants, 3401
Roget Begnoche Scholarship, 3402
Rohm and Haas Company Grants, 3403
Ronald McDonald House Charities Grants, 3405
Rosalynn Carter Institute John and Betty Pope Fellowships, 3408
Rosalynn Carter Institute Mattie J. T. Stepanek Caregiving Scholarships, 3409
Rose Hills Foundation Grants, 3413
Roy & Christine Sturgis Charitable Grants, 3415
Roy J. Carver Charitable Trust Medical and Science Research Grants, 3416
RRF General Program Grants, 3417
RSC McNeil Medal for the Public Awareness of Science, 3422
Rush County Community Foundation Grants, 3424
Ruth Anderson Foundation Grants, 3425
Ruth and Vernon Taylor Foundation Grants, 3426
Ruth H. & Warren A. Ellsworth Fndn Grants, 3429
S.H. Cowell Foundation Grants, 3447
S. Livingston Mather Charitable Trust Grants, 3448
S. Mark Taper Foundation Grants, 3449
Saigh Foundation Grants, 3451
Saint Louis Rams Fndn Community Donations, 3453
Salem Foundation Charitable Trust Grants, 3456
Salmon Foundation Grants, 3458
Samuel N. and Mary Castle Foundation Grants, 3467
Samuel S. Johnson Foundation Grants, 3468
San Diego Foundation for Change Grants, 3470
San Diego Women's Foundation Grants, 3474
San Francisco Fndn Community Health Grants, 3478
San Juan Island Community Foundation Grants, 3480
Sarkeys Foundation Grants, 3485
Sasco Foundation Grants, 3487
Schering-Plough Foundation Health Grants, 3492
Schlessman Family Foundation Grants, 3493

Schurz Communications Foundation Grants, 3496
Scott County Community Foundation Grants, 3500
Seabury Foundation Grants, 3501
Seagate Tech Corp Capacity to Care Grants, 3502
Seattle Foundation Benjamin N. Phillips Memorial Fund Grants, 3504
Self Foundation Grants, 3510
Seneca Foods Foundation Grants, 3511
Sensient Technologies Foundation Grants, 3513
Seward Community Foundation Grants, 3514
Seward Community Foundation Mini-Grants, 3515
Shell Deer Park Grants, 3543
Shield-Ayres Foundation Grants, 3546
Shopko Fndn Community Charitable Grants, 3547
Sidgmore Family Foundation Grants, 3551
Sidney Stern Memorial Trust Grants, 3552
Siebert Lutheran Foundation Grants, 3554
Simmons Foundation Grants, 3563
Simpson Lumber Charitable Contributions, 3566
Sir Dorabji Tata Trust Grants for NGOs or Voluntary Organizations, 3572
Sisters of Charity Foundation of Cleveland Reducing Health and Educational Disparities in the Central Neighborhood Grants, 3574
Sisters of Mercy of North Carolina Fndn Grants, 3575
SNM PDEF Mickey Williams Minority Student Scholarships, 3584
SNMTS Clinical Advancement Scholarships, 3590
SNMTS Paul Cole Scholarships, 3593
Southbury Community Trust Fund, 3625
South Madison Community Foundation Grants, 3626
Southwest Florida Community Foundation Competitive Grants, 3627
Southwest Gas Corporation Foundation Grants, 3628
Special People in Need Grants, 3630
Spencer County Community Foundation Grants, 3632
Springs Close Foundation Grants, 3633
Square D Foundation Grants, 3634
St. Joseph Community Health Foundation Burn Care and Prevention Grants, 3639
St. Joseph Community Health Foundation Improving Healthcare Access Grants, 3640
Stark Community Fndn Women's Grants, 3645
Steele-Reese Foundation Grants, 3650
Stella and Charles Guttman Foundation Grants, 3651
Sterling and Shelli Gardner Foundation Grants, 3653
Steuben County Community Foundation Grants, 3654
Stewart Huston Charitable Trust Grants, 3655
Stocker Foundation Grants, 3656
Stranahan Foundation Grants, 3659
Strowd Roses Grants, 3664
Subaru of Indiana Automotive Foundation Grants, 3672
Sunoco Foundation Grants, 3679
SunTrust Bank Trusteed Foundations Florence C. and Harry L. English Memorial Fund Grants, 3680
SunTrust Bank Trusteed Foundations Greene-Sawtell Grants, 3681
SunTrust Bank Trusteed Foundations Harriet McDaniel Marshall Tust Grants, 3682
SunTrust Bank Trusteed Foundations Nell Warren Elkin and William Simpson Elkin Grants, 3683
SunTrust Bank Trusteed Foundations Thomas Guy Woolford Charitable Trust Grants, 3684
SunTrust Bank Trusteed Foundations Walter H. and Marjory M. Rich Memorial Fund Grants, 3685
Support Our Aging Religious (SOAR) Grants, 3686
Susan Mott Webb Charitable Trust Grants, 3694
Susan Vaughan Foundation Grants, 3695
Symantec Community Relations and Corporate Philanthropy Grants, 3699
Tauck Family Foundation Grants, 3703
Taylor S. Abernathy and Patti Harding Abernathy Charitable Trust Grants, 3704
TE Foundation Grants, 3705
Tellabs Foundation Grants, 3706
Texas Commission on the Arts Arts Respond Project Grants, 3708
Thelma Braun & Bocklett Family Fndn Grants, 3713
Theodore Edson Parker Foundation Grants, 3715
Thomas Austin Finch, Sr. Foundation Grants, 3718

716 / Education

Thomas C. Ackerman Foundation Grants, 3719
Thomas J. Atkins Memorial Trust Fund Grants, 3721
Thomas J. Long Foundation Community Grants, 3722
Thomas Thompson Trust Grants, 3725
Thompson Charitable Foundation Grants, 3727
Tifa Foundation Grants, 3733
TJX Foundation Grants, 3735
Tommy Hilfiger Corporate Foundation Grants, 3737
Toyota Motor Engineering & Manufacturing North America Grants, 3742
Toyota Motor Manufacturing of Alabama Grants, 3743
Toyota Motor Manufacturing of Indiana Grants, 3744
Toyota Motor Manufacturing of Kentucky Grants, 3745
Toyota Motor Manuf of Mississippi Grants, 3746
Toyota Motor Manufacturing of Texas Grants, 3747
Toyota Motor Manuf of West Virginia Grants, 3748
Toyota Motor N America of New York Grants, 3749
Toyota Motor Sales, USA Grants, 3750
Tri-State Community Twenty-first Century Endowment Fund Grants, 3756
Triangle Community Foundation Donor-Advised Grants, 3757
Triangle Community Foundation Shaver-Hitchings Scholarship, 3758
Trull Foundation Grants, 3761
Tull Charitable Foundation Grants, 3762
Turner Foundation Grants, 3763
U.S. Cellular Corporation Grants, 3766
U.S. Department of Education Erma Byrd Scholarships, 3768
U.S. Department of Education Rehabilitation Training - Rehabilitation Continuing Education - Institute on Rehabilitation Issues, 3771
Union Bank, N.A. Corporate Sponsorships and Donations, 3787
Union Benevolent Association Grants, 3789
United Technologies Corporation Grants, 3803
Unity Foundation Of LaPorte County Grants, 3804
USAID HIV Prevention with Key Populations - Mali Grants, 3816
US Airways Community Foundation Grants, 3826
USG Foundation Grants, 3848
Vancouver Foundation Grants and Community Initiatives Program, 3849
Vanderburgh Community Foundation Grants, 3852
VDH Emergency Med Services Training Grants, 3855
Verizon Foundation Maryland Grants, 3859
Verizon Foundation New York Grants, 3860
Verizon Foundation Northeast Region Grants, 3861
Verizon Foundation Pennsylvania Grants, 3862
Verizon Foundation Vermont Grants, 3863
Verizon Foundation Virginia Grants, 3864
Vermont Community Foundation Grants, 3865
Virginia W. Kettering Foundation Grants, 3878
W.C. Griffith Foundation Grants, 3881
W. C. Griffith Foundation Grants, 3882
W. Clarke Swanson, Jr. Foundation Grants, 3883
W.H. & Mary Ellen Cobb Trust Grants, 3884
W.K. Kellogg Foundation Secure Families Grants, 3887
W.P. and Bulah Luse Foundation Grants, 3892
Wabash Valley Community Foundation Grants, 3893
Walker Area Community Foundation Grants, 3895
Walmart Foundation Community Giving Grants, 3897
Walmart Foundation National Giving Grants, 3898
Walter L. Gross III Family Foundation Grants, 3901
Warren County Community Foundation Grants, 3902
Warren County Community Fndn Mini-Grants, 3903
Warrick County Community Foundation Grants, 3904
Washington County Community Fndn Grants, 3906
Washington County Community Foundation Youth Grants, 3907
Washington Gas Charitable Giving, 3908
Wayne County Foundation Grants, 3909
Weaver Popcorn Foundation Grants, 3910
Wege Foundation Grants, 3912
Welborn Baptist Fndn General Op Grants, 3914
Wells County Foundation Grants, 3920
Western Indiana Community Foundation Grants, 3921
Western Union Foundation Grants, 3923
Weyerhaeuser Company Foundation Grants, 3925

White County Community Foundation - Women Giving Together Grants, 3933
White County Community Foundation Grants, 3934
Whitley County Community Foundation Grants, 3936
Whitney Foundation Grants, 3938
WHO Foundation General Grants, 3939
Willary Foundation Grants, 3944
William A. Badger Foundation Grants, 3945
William D. Laurie, Jr. Charitable Fndn Grants, 3949
William G. & Helen C. Hoffman Fndn Grants, 3950
William H. Hannon Foundation Grants, 3953
William Ray & Ruth E. Collins Fndn Grants, 3956
Williams Companies Foundation Homegrown Giving Grants, 3957
Willis C. Helm Charitable Trust Grants, 3959
Wilson-Wood Foundation Grants, 3960
Winifred & Harry B. Allen Foundation Grants, 3962
Winston-Salem Foundation Competitive Grants, 3965
Winston-Salem Fndn Elkin/Tri-County Grants, 3966
Winston-Salem Fndn Stokes County Grants, 3967
Wood-Claeyssens Foundation Grants, 3971
Woodward Fund Grants, 3973
Yampa Valley Community Foundation Cody St. John Scholarships, 3977
Yampa Valley Community Foundation Volunteer Firemen Scholarships, 3979
Yawkey Foundation Grants, 3980
Z. Smith Reynolds Foundation Small Grants, 3982

Education Reform
Achelis Foundation Grants, 234
Anheuser-Busch Foundation Grants, 591
Bill and Melinda Gates Foundation Policy and Advocacy Grants, 893
Bodman Foundation Grants, 929
Boeing Company Contributions Grants, 930
Carnegie Corporation of New York Grants, 1033
Frances and John L. Loeb Family Fund Grants, 1716
J.C. Penney Company Grants, 2169
John W. Speas and Effie E. Speas Memorial Trust Grants, 2264
Joseph Drown Foundation Grants, 2268
Joseph H. & Florence A. Roblee Fndn Grants, 2269
K and F Baxter Family Foundation Grants, 2287
Lewis H. Humphreys Charitable Trust Grants, 2361
Louis and Elizabeth Nave Flarsheim Charitable Foundation Grants, 2379
Minneapolis Foundation Community Grants, 2558
Portland General Electric Foundation Grants, 3250
Prince Charitable Trusts Chicago Grants, 3270
Richard Davoud Donchian Foundation Grants, 3367
Tommy Hilfiger Corporate Foundation Grants, 3737
William J. Brace Charitable Trust, 3955

Education and Work
Brook J. Lenfest Foundation Grants, 967
PepsiCo Foundation Grants, 3137
Radcliffe Institute Carol K. Pforzheimer Student Fellowships, 3298
SNMTS Outstanding Educator Awards, 3591
Toyota Motor Engineering & Manufacturing North America Grants, 3742
Toyota Motor Manuf of West Virginia Grants, 3748
Vectren Foundation Grants, 3857

Educational Administration
Guy I. Bromley Trust Grants, 1919
IIE Iraq Scholars and Leaders Scholarships, 2133
John W. Speas and Effie E. Speas Memorial Trust Grants, 2264
Lewis H. Humphreys Charitable Trust Grants, 2361
Louetta M. Cowden Foundation Grants, 2377
Louis and Elizabeth Nave Flarsheim Charitable Foundation Grants, 2379

Educational Evaluation/Assessment
A.O. Smith Foundation Community Grants, 17
Blue Cross Blue Shield of Minnesota Foundation - Healthy Equity: Health Impact Assessment Demonstration Project Grants, 916

IBRO Latin America Regional Funding for Neuroscience Schools, 2105
PacifiCare Foundation Grants, 3102
William J. Brace Charitable Trust, 3955

Educational Finance
Allstate Corporate Giving Grants, 446
Allstate Corp Hometown Commitment Grants, 447

Educational Instruction
ASM Int'l Fellowship for Latin America and the Caribbean, 734
ASM Int'l Fellowships for Asia and Africa, 735
Bayer Clinical Scholarship Award, 845
BCBSM Foundation Student Award Program, 858
Bill and Melinda Gates Foundation Policy and Advocacy Grants, 893
Blue Cross Blue Shield of Minnesota Fndn Healthy Equity: Public Libraries for Health Grants, 918
Carroll County Community Foundation Grants, 1039
Charles H. Dater Foundation Grants, 1118
Independence Blue Cross Nursing Internships, 2152
Leonard L. & Bertha U. Abess Fndn Grants, 2357
Mary Wilmer Covey Charitable Trust Grants, 2448
San Diego Women's Foundation Grants, 3474
SfN Science Educator Award, 3535
VDH Rescue Squad Assistance Fund Grants, 3856
Warrick County Community Foundation Grants, 3904

Educational Planning/Policy
Achelis Foundation Grants, 234
Bill and Melinda Gates Foundation Policy and Advocacy Grants, 893
Blue Cross Blue Shield of Minnesota Foundation - Healthy Equity: Health Impact Assessment Demonstration Project Grants, 916
Blue Cross Blue Shield of Minnesota Foundation - Healthy Equity: Health Impact Assessment Program Grants, 917
Bodman Foundation Grants, 929
Bristol-Myers Squibb Foundation Health Education Grants, 959
Everyone Breathe Asthma Education Grants, 1616
Guy I. Bromley Trust Grants, 1919
IBRO Latin America Regional Funding for Neuroscience Schools, 2105
John W. Speas and Effie E. Speas Memorial Trust Grants, 2264
K and F Baxter Family Foundation Grants, 2287
Leonard L. & Bertha U. Abess Fndn Grants, 2357
Lewis H. Humphreys Charitable Trust Grants, 2361
Louetta M. Cowden Foundation Grants, 2377
Louis and Elizabeth Nave Flarsheim Charitable Foundation Grants, 2379

Educational Psychology
Ittleson Foundation Mental Health Grants, 2167

Educational Technology
Alcatel-Lucent Technologies Foundation Grants, 413
Brook J. Lenfest Foundation Grants, 967
Dept of Ed Special Education-National Activities-Technology and Media Services for Individuals with Disabilities, 1468
Hilda and Preston Davis Foundation Grants, 2028
Jessie Ball Dupont Fund Grants, 2226
Thomas J. Long Foundation Community Grants, 3722

Educational Testing/Measurement
NBME Stemmler Med Ed Research Grants, 2638

Educational Theory
Guy I. Bromley Trust Grants, 1919
John W. Speas and Effie E. Speas Memorial Trust Grants, 2264
Lewis H. Humphreys Charitable Trust Grants, 2361
Louetta M. Cowden Foundation Grants, 2377
Louis and Elizabeth Nave Flarsheim Charitable Foundation Grants, 2379

SUBJECT INDEX

Egypt
Fulbright Traditional Scholar Program in the Near East and North Africa Region, 1758

Elder Abuse
Administration on Aging Senior Medicare Patrol Project Grants, 309
Austin S. Nelson Foundation Grants, 805
Community Fndn for the Capital Region Grants, 1259
Peacock Foundation Grants, 3133
Reynolds & Reynolds Associate Fndn Grants, 3360

Elderly
Abbot and Dorothy H. Stevens Foundation Grants, 197
Administration on Aging Senior Medicare Patrol Project Grants, 309
Agnes M. Lindsay Trust Grants, 346
Albert W. Rice Charitable Foundation Grants, 411
Alfred E. Chase Charitable Foundation Grants, 430
Allen P. & Josephine B. Green Fndn Grants, 443
Amelia Sillman Rockwell and Carlos Perry Rockwell Charities Fund Grants, 501
American Schlafhorst Foundation Grants, 559
American Trauma Society, Pennsylvania Division Mini-Grants, 564
AMI Semiconductors Corporate Grants, 576
Ann Arbor Area Community Foundation Grants, 594
Ann Peppers Foundation Grants, 599
Anschutz Family Foundation Grants, 602
Arkell Hall Foundation Grants, 661
Arthur and Rochelle Belfer Foundation Grants, 670
Avon Products Foundation Grants, 814
Bailey Foundation Grants, 818
Barberton Community Foundation Grants, 829
BCBSM Claude Pepper Award, 850
BCBSM Corporate Contributions Grants, 851
BCBSNC Foundation Grants, 860
Ben B. Cheney Foundation Grants, 875
Bender Foundation Grants, 876
Berrien Community Foundation Grants, 884
Bertha Russ Lytel Foundation Grants, 885
Birmingham Foundation Grants, 903
Blackford County Community Foundation Grants, 906
Blanche and Irving Laurie Foundation Grants, 910
Bodenwein Public Benevolent Foundation Grants, 928
Boston Foundation Grants, 936
Brookdale Fnd National Group Respite Grants, 966
Burden Trust Grants, 976
Caleb C. and Julia W. Dula Educational and Charitable Foundation Grants, 997
California Endowment Innovative Ideas Challenge Grants, 1000
Callaway Foundation Grants, 1002
Callaway Golf Company Foundation Grants, 1003
Cambridge Community Foundation Grants, 1004
Carl R. Hendrickson Family Foundation Grants, 1028
Carl W. and Carrie Mae Joslyn Trust Grants, 1031
Carrie E. and Lena V. Glenn Foundation Grants, 1036
Carrie Estelle Doheny Foundation Grants, 1037
Carroll County Community Foundation Grants, 1039
Catherine Kennedy Home Foundation Grants, 1043
CFFVR Basic Needs Giving Partnership Grants, 1094
CFFVR Frank C. Shattuck Community Grants, 1098
Charles F. Bacon Trust Grants, 1117
Charles H. Farnsworth Trust Grants, 1119
Charles Nelson Robinson Fund Grants, 1125
Chatlos Foundation Grants Program, 1128
Christine & Katharina Pauly Trust Grants, 1150
CICF Indianapolis Fndn Community Grants, 1156
CNA Foundation Grants, 1189
CNCS Senior Companion Program Grants, 1191
Columbus Foundation Allen Eiry Fund Grants, 1214
Columbus Foundation J. Floyd Dixon Memorial Fund Grants, 1219
Commonwealth Fund Harkness Fellowships in Health Care Policy and Practice, 1229
Commonwealth Fund Quality of Care for Frail Elders Grants, 1233
Community Foundation of Eastern Connecticut Northeast Women and Girls Grants, 1270

Comprehensive Health Education Fndn Grants, 1320
Cooke Foundation Grants, 1344
Daisy Marquis Jones Foundation Grants, 1414
Danellie Foundation Grants, 1421
Daniels Fund Grants-Aging, 1424
David N. Lane Trust Grants for Aged and Indigent Women, 1433
Della B. Gardner Fund Grants, 1451
Dennis & Phyllis Washington Fndn Grants, 1458
Donald W. Reynolds Foundation Aging and Quality of Life Grants, 1487
Doree Taylor Charitable Foundation, 1490
Dr. Scholl Foundation Grants Program, 1511
Edward N. and Della L. Thome Memorial Foundation Grants, 1557
Edward W. and Stella C. Van Houten Memorial Fund Grants, 1559
Edwin W. and Catherine M. Davis Fndn Grants, 1561
Eisner Foundation Grants, 1564
Elizabeth Morse Genius Charitable Trust Grants, 1569
Essex County Community Foundation Merrimack Valley General Fund Grants, 1603
Eva L. & Joseph M. Bruening Fndn Grants, 1613
Fannie E. Rippel Foundation Grants, 1630
Faye McBeath Foundation Grants, 1635
Florence Hunt Maxwell Foundation Grants, 1676
Foellinger Foundation Grants, 1684
Fourjay Foundation Grants, 1712
Frances and John L. Loeb Family Fund Grants, 1716
Frank Reed and Margaret Jane Peters Memorial Fund I Grants, 1727
Frederick McDonald Trust Grants, 1740
Fremont Area Community Foundation Elderly Needs Grants, 1745
Fremont Area Community Fndn General Grants, 1746
G.N. Wilcox Trust Grants, 1768
George A Ohl Jr. Foundation Grants, 1791
George Foundation Grants, 1795
George P. Davenport Trust Fund Grants, 1800
George W. Wells Foundation Grants, 1804
GNOF IMPACT Grants for Health and Human Services, 1839
Golden Heart Community Foundation Grants, 1850
Graham and Carolyn Holloway Family Foundation Grants, 1861
Greygates Foundation Grants, 1901
Grover Hermann Foundation Grants, 1906
H. Leslie Hoffman and Elaine S. Hoffman Foundation Grants, 1924
Hackett Foundation Grants, 1926
HAF Senior Opportunities Grants, 1932
Hagedorn Fund Grants, 1934
Harold Brooks Foundation Grants, 1952
Harry Kramer Memorial Fund Grants, 1965
Health Fndn of Greater Indianapolis Grants, 1984
Hearst Foundations Health Grants, 1987
Helen Gertrude Sparks Charitable Trust Grants, 1993
Helen Steiner Rice Foundation Grants, 1998
Henrietta Lange Burk Fund Grants, 2001
Henry County Community Foundation Grants, 2005
Herbert H. & Grace A. Dow Fndn Grants, 2008
Hilda and Preston Davis Foundation Grants, 2028
Hoglund Foundation Grants, 2040
Holland/Zeeland Community Fndn Grants, 2041
Hugh J. Andersen Foundation Grants, 2074
J. Walton Bissell Foundation Grants, 2177
Jacob G. Schmidlapp Trust Grants, 2183
James & Abigail Campbell Family Fndn Grants, 2185
James R. Thorpe Foundation Grants, 2199
Janirve Foundation Grants, 2209
Jay and Rose Phillips Family Foundation Grants, 2213
Jean and Louis Dreyfus Foundation Grants, 2215
Jewish Fund Grants, 2227
Jim Moran Foundation Grants, 2228
John Edward Fowler Memorial Fndn Grants, 2235
John G. Martin Foundation Grants, 2237
John H. Wellons Foundation Grants, 2239
John W. Anderson Foundation Grants, 2262
John W. Boynton Fund Grants, 2263
Joseph Henry Edmondson Foundation Grants, 2270

Katharine Matthies Foundation Grants, 2291
Kimball International-Habig Foundation Health and Human Services Grants, 2308
Laura B. Vogler Foundation Grants, 2345
Leon and Thea Koerner Foundation Grants, 2355
Leo Niessen Jr., Charitable Trust Grants, 2358
Long Island Community Foundation Grants, 2375
Lowell Berry Foundation Grants, 2382
Mardag Foundation Grants, 2428
Marie C. and Joseph C. Wilson Foundation Rochester Small Grants, 2431
Marjorie Moore Charitable Foundation Grants, 2438
Mary Black Foundation Active Living Grants, 2440
May and Stanley Smith Charitable Trust Grants, 2459
McLean Foundation Grants, 2498
Medtronic Foundation Patient Link Grants, 2514
Memorial Foundation Grants, 2516
Mericos Foundation Grants, 2518
MetroWest Health Foundation Grants--Healthy Aging, 2527
MGM Resorts Foundation Community Grants, 2541
Michael Reese Health Trust Responsive Grants, 2545
Mid-Iowa Health Foundation Community Response Grants, 2549
Military Ex-Prisoners of War Foundation Grants, 2554
Mt. Sinai Health Care Foundation Health of the Jewish Community Grants, 2599
NCI Stages of Breast Development: Normal to Metastatic Disease Grants, 2655
Nell J. Redfield Foundation Grants, 2673
Nina Mason Pulliam Charitable Trust Grants, 2921
Norcliffe Foundation Grants, 2979
Nuffield Foundation Africa Program Grants, 3026
Nuffield Foundation Law and Society Grants, 3028
Nuffield Foundation Open Door Grants, 3030
Nuffield Foundation Small Grants, 3031
Oppenstein Brothers Foundation Grants, 3078
Paul Balint Charitable Trust Grants, 3122
Phelps County Community Foundation Grants, 3188
Pinellas County Grants, 3223
Piper Trust Older Adults Grants, 3229
Portland Fndn Women's Giving Circle Grant, 3249
Powell Foundation Grants, 3255
Puerto Rico Community Foundation Grants, 3278
Quantum Foundation Grants, 3290
R.C. Baker Foundation Grants, 3292
R.S. Gernon Trust Grants, 3293
Ralph M. Parsons Foundation Grants, 3304
Retirement Research Foundation General Program Grants, 3359
Reynolds & Reynolds Associate Fndn Grants, 3360
Robert & Clara Milton Senior Housing Grants, 3384
Rochester Area Community Foundation Grants, 3391
Rosalynn Carter Institute John and Betty Pope Fellowships, 3408
Rose Community Foundation Aging Grants, 3410
RRF General Program Grants, 3417
RRF Organizational Capacity Building Grants, 3418
Ruth Anderson Foundation Grants, 3425
Saginaw Community Foundation Senior Citizen Enrichment Fund, 3450
Schramm Foundation Grants, 3495
Sid W. Richardson Foundation Grants, 3553
Sisters of Mercy of North Carolina Fndn Grants, 3575
Sophia Romero Trust Grants, 3624
Southwest Gas Corporation Foundation Grants, 3628
Stella and Charles Guttman Foundation Grants, 3651
Stewart Huston Charitable Trust Grants, 3655
STTI Environment of Elder Care Nursing Research Grants, 3667
Support Our Aging Religious (SOAR) Grants, 3686
Swindells Charitable Foundation, 3697
Thelma Braun & Bocklett Family Fndn Grants, 3713
Thelma Doelger Charitable Trust Grants, 3714
Theodore Edson Parker Foundation Grants, 3715
Thomas Austin Finch, Sr. Foundation Grants, 3718
Topfer Family Foundation Grants, 3739
Trull Foundation Grants, 3761
Union Benevolent Association Grants, 3789
Union Labor Health Fndn Angel Fund Grants, 3791

718 / Elderly

Victor E. Speas Foundation Grants, 3868
William H. Hannon Foundation Grants, 3953
William J. Brace Charitable Trust, 3955
Winston-Salem Foundation Competitive Grants, 3965
Wood-Claeyssens Foundation Grants, 3971

Electoral Systems
Baptist Community Ministries Grants, 828
Carnegie Corporation of New York Grants, 1033

Electric Power
DOL Occupational Safety and Health--Susan Harwood Training Grants, 1484

Electronic Media
Reinberger Foundation Grants, 3357
SfN Science Educator Award, 3535
SfN Science Journalism Student Awards, 3536

Electronics/Electrical Engineering
IIE KAUST Graduate Fellowships, 2134
NSF-NIST Interaction in Chemistry, Materials Research, Molecular Biosciences, Bioengineering, and Chemical Engineering, 3004

Electrophysiology
Duke University Clinical Cardiac Electrophysiology Fellowships, 1522

Elementary Education
3M Company Fndn Community Giving Grants, 6
A.O. Smith Foundation Community Grants, 17
Abby's Legendary Pizza Foundation Grants, 204
Abernethy Family Foundation Grants, 210
Advance Auto Parts Corporate Giving Grants, 312
AEC Trust Grants, 315
Affymetrix Corporate Contributions Grants, 339
Ahmanson Foundation Grants, 355
Alabama Power Foundation Grants, 400
Alcatel-Lucent Technologies Foundation Grants, 413
Alpha Natural Resources Corporate Giving, 450
AMD Corporate Contributions Grants, 500
American Electric Power Foundation Grants, 530
American Schlafhorst Foundation Grants, 559
Amerigroup Foundation Grants, 565
Amica Companies Foundation Grants, 574
Amica Insurance Company Community Grants, 575
Archer Daniels Midland Foundation Grants, 651
Atherton Family Foundation Grants, 794
Ball Brothers Foundation General Grants, 820
Baxter International Corporate Giving Grants, 841
Belvedere Community Foundation Grants, 874
Benton Community Fndn Cookie Jar Grant, 877
Benton County Foundation Grants, 880
Berrien Community Foundation Grants, 884
Bertha Russ Lytel Foundation Grants, 885
Blackford County Community Foundation Grants, 906
BMW of North America Charitable Contributions, 926
Boeing Company Contributions Grants, 930
Booth-Bricker Fund Grants, 933
Boston Foundation Grants, 936
BP Foundation Grants, 943
Bristol-Myers Squibb Foundation Health Education Grants, 959
Bristol-Myers Squibb Foundation Science Education Grants, 961
Brown Advisory Charitable Foundation Grants, 969
Burlington Industries Foundation Grants, 977
Callaway Foundation Grants, 1002
Carnegie Corporation of New York Grants, 1033
Carrie E. and Lena V. Glenn Foundation Grants, 1036
Carrie Estelle Doheny Foundation Grants, 1037
Carroll County Community Foundation Grants, 1039
CFFVR Shawano Area Community Foundation Grants, 1104
Chapman Charitable Foundation Grants, 1112
Charles Lafitte Foundation Grants, 1123
Christensen Fund Regional Grants, 1149
Clara Blackford Smith and W. Aubrey Smith Charitable Foundation Grants, 1165

Clarence E. Heller Charitable Foundation Grants, 1166
Clayton Fund Grants, 1180
Clinton County Community Foundation Grants, 1184
Coca-Cola Foundation Grants, 1203
Colonel Stanley R. McNeil Foundation Grants, 1211
Columbus Foundation J. Floyd Dixon Memorial Fund Grants, 1219
Community Fndn of Central Illinois Grants, 1267
Community Foundation of Greater Fort Wayne Scholarships, 1276
Connelly Foundation Grants, 1326
Crane Foundation Grants, 1357
Cresap Family Foundation Grants, 1359
Cullen Foundation Grants, 1372
CVS All Kids Can Grants, 1377
CVS Community Grants, 1379
DaimlerChrysler Corporation Fund Grants, 1412
Dana Brown Charitable Trust Grants, 1419
David M. and Marjorie D. Rosenberg Foundation Grants, 1432
Dean Foods Community Involvement Grants, 1440
Decatur County Community Foundation Large Project Grants, 1444
Dennis & Phyllis Washington Fndn Grants, 1458
Dr. & Mrs. Paul Pierce Memorial Fndn Grants, 1505
Dr. Scholl Foundation Grants Program, 1511
Dunspaugh-Dalton Foundation Grants, 1530
DuPont Pioneer Community Giving Grants, 1532
E.J. Grassmann Trust Grants, 1537
Earl and Maxine Claussen Trust Grants, 1541
Effie and Wofford Cain Foundation Grants, 1563
Eisner Foundation Grants, 1564
Elizabeth McGraw Foundation Grants, 1568
El Pomar Foundation Grants, 1582
Entergy Corporation Micro Grants, 1596
Ferree Foundation Grants, 1658
Field Foundation of Illinois Grants, 1660
Firelight Foundation Grants, 1667
Fisher Foundation Grants, 1670
Florida Division of Cultural Affairs Arts In Education Arts Partnership Grants, 1679
Floyd A. & Kathleen C. Cailloux Fnd Grants, 1681
Four County Community Fndn General Grants, 1710
Four County Community Foundation Healthy Senior/ Healthy Youth Fund Grants, 1711
Four J Foundation Grants, 1713
Frances and John L. Loeb Family Fund Grants, 1716
Frances L. and Edwin L. Cummings Memorial Fund Grants, 1717
Francis T. & Louise T. Nichols Fndn Grants, 1720
Frank B. Hazard General Charity Fund Grants, 1721
Frank Loomis Palmer Fund Grants, 1726
Frank Reed and Margaret Jane Peters Memorial Fund II Grants, 1728
Fred & Gretel Biel Charitable Trust Grants, 1734
Gardner Foundation Grants, 1771
Gardner Foundation Grants, 1772
Gardner Foundation Grants, 1770
GenCorp Foundation Grants, 1779
General Mills Foundation Grants, 1782
George F. Baker Trust Grants, 1793
George Foundation Grants, 1795
GNOF Exxon-Mobil Grants, 1836
GNOF Norco Community Grants, 1845
Goodrich Corporation Foundation Grants, 1854
Grace and Franklin Bersen Foundation Grants, 1858
Green River Area Community Fndn Grants, 1895
Greenspun Family Foundation Grants, 1896
Grundy Foundation Grants, 1912
Harold Alfond Foundation Grants, 1949
Harrison County Community Foundation Grants, 1957
Harrison County Community Foundation Signature Grants, 1958
Helen Bader Foundation Grants, 1992
Herman Goldman Foundation Grants, 2012
Houston Endowment Grants, 2053
Huisking Foundation Grants, 2076
Hutchinson Community Foundation Grants, 2085
Hutton Foundation Grants, 2087
J.C. Penney Company Grants, 2169

SUBJECT INDEX

J.L. Bedsole Foundation Grants, 2172
Janus Foundation Grants, 2211
Jessica Stevens Community Foundation Grants, 2224
John Clarke Trust Grants, 2232
John H. and Wilhelmina D. Harland Charitable Foundation Children and Youth Grants, 2238
John P. Murphy Foundation Grants, 2248
Joseph Henry Edmondson Foundation Grants, 2270
K and F Baxter Family Foundation Grants, 2287
Katharine Matthies Foundation Grants, 2291
Katrine Menzing Deakins Trust Grants, 2295
Kent D. Steadley and Mary L. Steadley Memorial Trust Grants, 2301
Linford and Mildred White Charitable Grants, 2368
Lockheed Martin Corp Foundation Grants, 2374
Lubbock Area Foundation Grants, 2383
Lynn and Rovena Alexander Family Foundation Grants, 2399
Mardag Foundation Grants, 2428
Marion I. and Henry J. Knott Foundation Discretionary Grants, 2435
Marion I. and Henry J. Knott Foundation Standard Grants, 2436
Marjorie Moore Charitable Foundation Grants, 2438
McConnell Foundation Grants, 2492
McGraw-Hill Companies Community Grants, 2494
Mead Johnson Nutritionals Evansville-Area Organizations Grants, 2503
Medtronic Foundation Patient Link Grants, 2514
Mericos Foundation Grants, 2518
Meta and George Rosenberg Foundation Grants, 2525
Meyer Memorial Trust Grassroots Grants, 2536
Meyer Memorial Trust Responsive Grants, 2537
MGM Resorts Foundation Community Grants, 2541
Miller Foundation Grants, 2556
Nicholas H. Noyes Jr. Memorial Fndn Grants, 2791
Nordson Corporation Foundation Grants, 2981
Oppenstein Brothers Foundation Grants, 3078
Pew Charitable Trusts Children & Youth Grants, 3158
Peyton Anderson Foundation Grants, 3160
Phoenix Suns Charities Grants, 3201
Piper Jaffray Fndn Communities Giving Grants, 3227
Plough Foundation Grants, 3238
PMI Foundation Grants, 3239
Powell Foundation Grants, 3255
PPCF Community Grants, 3256
Procter and Gamble Fund Grants, 3274
Quantum Foundation Grants, 3290
R.S. Gernon Trust Grants, 3293
Rajiv Gandhi Foundation Grants, 3302
Raskob Foundation for Catholic Activities Grants, 3306
Reinberger Foundation Grants, 3357
Richard & Susan Smith Family Fndn Grants, 3365
Richland County Bank Grants, 3371
Robert R. McCormick Trib Community Grants, 3385
Robert R. Meyer Foundation Grants, 3387
Robert W. Woodruff Foundation Grants, 3389
Rockefeller Brothers Fund Pivotal Places Grants: South Africa, 3395
Rollins-Luetkemeyer Foundation Grants, 3404
Ronald McDonald House Charities Grants, 3405
Ruth Eleanor Bamberger and John Ernest Bamberger Memorial Foundation Grants, 3427
Saigh Foundation Grants, 3451
Sain-Orr and Royak-DeForest Steadman Foundation Grants, 3452
San Francisco Fndn Community Health Grants, 3478
San Juan Island Community Foundation Grants, 3480
Sartain Lanier Family Foundation Grants, 3486
Schramm Foundation Grants, 3495
Seagate Tech Corp Capacity to Care Grants, 3502
Seneca Foods Foundation Grants, 3511
Seward Community Foundation Grants, 3514
Seward Community Foundation Mini-Grants, 3515
SfN Science Educator Award, 3535
Shopko Fndn Community Charitable Grants, 3547
Sid W. Richardson Foundation Grants, 3553
Sioux Falls Area Community Foundation Community Fund Grants (Unrestricted), 3568

SUBJECT INDEX

Emotional/Mental Health /719

Sioux Falls Area Community Foundation Spot Grants (Unrestricted), 3570
Sonora Area Foundation Competitive Grants, 3622
Sony Corporation of America Grants, 3623
Southwest Gas Corporation Foundation Grants, 3628
Stocker Foundation Grants, 3656
Strowd Roses Grants, 3664
Sunoco Foundation Grants, 3679
Taubman Endowment for the Arts Fndn Grants, 3702
Thelma Braun & Bocklett Family Fndn Grants, 3713
The Ray Charles Foundation Grants, 3716
Trull Foundation Grants, 3761
U.S. Department of Education 21st Century Community Learning Centers, 3767
Union Bank, N.A. Corporate Sponsorships and Donations, 3787
Union Bank, N.A. Foundation Grants, 3788
Union County Community Foundation Grants, 3790
Victor E. Speas Foundation Grants, 3868
Victor Grifols i Lucas Foundation Ethics and Science Awars for Educational Institutions, 3869
W.H. & Mary Ellen Cobb Trust Grants, 3884
Walker Area Community Foundation Grants, 3895
Washington Gas Charitable Giving, 3908
Wege Foundation Grants, 3912
Welborn Baptist Fndn School Health Grants, 3917
WHO Foundation General Grants, 3939
William Blair and Company Foundation Grants, 3948
William D. Laurie, Jr. Charitable Fndn Grants, 3949
William J. Brace Charitable Trust, 3955
Windham Foundation Grants, 3961
Wood Family Charitable Trust Grants, 3972

Embryology
MBL E.E. Just Summer Fellowship for Minority Scientists, 2468

Emergency Preparedness
AIG Disaster Relief Fund Grants, 370
Bill and Melinda Gates Foundation Emergency Response Grants, 892
CDC Experience Epidemiology Fellowships, 1059
CDC Public Health Associates, 1065
CDC Public Health Associates Hosts, 1066
CSTE CDC/CSTE Applied Epidemiology Fellowships, 1367
Delmarva Power and Light Contributions, 1454
Elizabeth Morse Genius Charitable Trust Grants, 1569
IDPH Emergency Medical Services Assistance Fund Grants, 2118
Nationwide Insurance Foundation Grants, 2637
Piedmont Natural Gas Foundation Health and Human Services Grants, 3221
Seattle Foundation Benjamin N. Phillips Memorial Fund Grants, 3504
South Madison Community Foundation Grants, 3626
U.S. Lacrosse EAD Grants, 3775

Emergency Programs
Aid for Starving Children Emergency Assistance Fund Grants, 367
AIG Disaster Relief Fund Grants, 370
Alcatel-Lucent Technologies Foundation Grants, 413
Alexander Foundation Cancer, Catastrophic Illness and Injury Grants, 421
Bill and Melinda Gates Foundation Emergency Response Grants, 892
BJ's Charitable Foundation Grants, 904
Blowitz-Ridgeway Foundation Grants, 913
BP Foundation Grants, 943
Carl R. Hendrickson Family Foundation Grants, 1028
Charles H. Hall Foundation, 1120
Children Affected by AIDS Foundation Family Assistance Emergency Fund Grants, 1146
Colorado Resource for Emergency Education and Trauma Grants, 1212
Community Fndn for the Capital Region Grants, 1259
Community Foundation of Mount Vernon and Knox County Grants, 1299
Cooper Industries Foundation Grants, 1345
Covidien Medical Product Donations, 1352
Delmarva Power and Light Contributions, 1454
Doree Taylor Charitable Foundation, 1490
Elizabeth Morse Genius Charitable Trust Grants, 1569
Flextronics Foundation Disaster Relief Grants, 1674
Florence Hunt Maxwell Foundation Grants, 1676
Frank B. Hazard General Charity Fund Grants, 1721
Frank Loomis Palmer Fund Grants, 1726
Global Fund for Children Grants, 1832
GNOF Albert N. & Hattie M. McClure Grants, 1834
Harold Brooks Foundation Grants, 1952
Helen Irwin Littauer Educational Trust Grants, 1994
Henrietta Lange Burk Fund Grants, 2001
Idaho Community Foundation Eastern Region Competitive Grants, 2114
Jack H. & William M. Light Trust Grants, 2179
James R. Thorpe Foundation Grants, 2199
Jane's Trust Grants, 2206
Kosciusko County Community Foundation REMC Operation Round Up Grants, 2320
Mary L. Peyton Foundation Grants, 2445
McKesson Foundation Grants, 2496
Meyer Memorial Trust Emergency Grants, 2535
Middlesex Savings Charitable Foundation Basic Human Needs Grants, 2550
Middlesex Savings Charitable Foundation Capacity Building Grants, 2551
Nationwide Insurance Foundation Grants, 2637
Packard Foundation Local Grants, 3105
Piedmont Natural Gas Foundation Health and Human Services Grants, 3221
Porter County Women's Grant, 3248
Prince Charitable Trusts DC Grants, 3271
RCF Individual Assistance Grants, 3313
Reinberger Foundation Grants, 3357
Salt River Health & Human Services Grants, 3459
Seattle Foundation Benjamin N. Phillips Memorial Fund Grants, 3504
Union Bank, N.A. Foundation Grants, 3788
Union Labor Health Fndn Angel Fund Grants, 3791
W.H. & Mary Ellen Cobb Trust Grants, 3884

Emergency Services
ADA Foundation Disaster Assistance Grants, 290
ADA Foundation Relief Grants, 293
Aid for Starving Children Emergency Assistance Fund Grants, 367
AIG Disaster Relief Fund Grants, 370
Alexander Foundation Cancer, Catastrophic Illness and Injury Grants, 421
Austin S. Nelson Foundation Grants, 805
Barberton Community Foundation Grants, 829
Bill and Melinda Gates Foundation Emergency Response Grants, 892
BP Foundation Grants, 943
Burlington Industries Foundation Grants, 977
Campbell Soup Foundation Grants, 1007
Carpenter Foundation Grants, 1035
Carrier Corporation Contributions Grants, 1038
Charles H. Hall Foundation, 1120
Colorado Resource for Emergency Education and Trauma Grants, 1212
Community Fndn for the Capital Region Grants, 1259
Community Foundation of Greater Birmingham Grants, 1272
Community Foundation of Louisville Anna Marble Memorial Fund for Princeton Grants, 1290
Community Fndn of Riverside & San Bernardino County Irene S. Rockwell Grants, 1304
Delmarva Power and Light Contributions, 1454
Elizabeth Morse Genius Charitable Trust Grants, 1569
Fayette County Foundation Grants, 1636
Florida BRAIVE Fund of Dade Community Foundation, 1678
Frank B. Hazard General Charity Fund Grants, 1721
Fred & Gretel Biel Charitable Trust Grants, 1734
General Mills Foundation Grants, 1782
Gulf Coast Foundation of Community Operating Grants, 1916
Gulf Coast Foundation of Community Program Grants, 1917
Harold Brooks Foundation Grants, 1952
Huntington Beach Police Officers Fndn Grants, 2081
IDPH Emergency Medical Services Assistance Fund Grants, 2118
IDPH Local Health Department Public Health Emergency Response Grants, 2120
Indiana AIDS Fund Grants, 2155
John W. Boynton Fund Grants, 2263
John W. Speas and Effie E. Speas Memorial Trust Grants, 2264
Lewis H. Humphreys Charitable Trust Grants, 2361
Louis and Elizabeth Nave Flarsheim Charitable Foundation Grants, 2379
McCune Foundation Human Services Grants, 2493
Meyer Memorial Trust Emergency Grants, 2535
Middlesex Savings Charitable Foundation Basic Human Needs Grants, 2550
Morris & Gwendolyn Cafritz Fndn Grants, 2594
Nationwide Insurance Foundation Grants, 2637
Noble County Community Foundation - Art Hutsell Scholarship, 2970
Piedmont Natural Gas Foundation Health and Human Services Grants, 3221
Piper Trust Healthcare & Med Research Grants, 3228
Prince Charitable Trusts Chicago Grants, 3270
Raskob Foundation for Catholic Activities Grants, 3306
RCF Individual Assistance Grants, 3313
Reinberger Foundation Grants, 3357
Rhode Island Foundation Grants, 3362
Robert R. Meyer Foundation Grants, 3387
Salt River Health & Human Services Grants, 3459
San Francisco Foundation Disability Rights Advocate Fund Emergency Grants, 3479
Sioux Falls Area Community Foundation Field-of-Interest and Donor-Advised Grants, 3569
U.S. Lacrosse EAD Grants, 3775
Union Bank, N.A. Corporate Sponsorships and Donations, 3787
Union Bank, N.A. Foundation Grants, 3788
Union Pacific Foundation Health and Human Services Grants, 3794
VDH Emergency Med Services Training Grants, 3855

Emission Control
United Technologies Corporation Grants, 3803

Emotion
NIMH Basic and Translational Research in Emotion Grants, 2916

Emotional/Mental Health
Adolph Coors Foundation Grants, 311
Bella Vista Foundation Grants, 873
Bravewell Leadership Award, 946
Charles H. Pearson Foundation Grants, 1121
Cigna Civic Affairs Sponsorships, 1159
Cone Health Foundation Grants, 1323
Delta Air Lines Foundation Health and Wellness Grants, 1455
DTE Energy Foundation Health and Human Services Grants, 1516
Elizabeth Morse Genius Charitable Trust Grants, 1569
Foundation for the Mid South Health and Wellness Grants, 1700
George W. Wells Foundation Grants, 1804
Gerber Foundation Grants, 1807
Gibson County Community Foundation Women's Fund Grants, 1817
HAF Phyllis Nilsen Leal Memorial Fund Gifts, 1930
Hasbro Children's Fund Grants, 1973
Herbert A. & Adrian W. Woods Fndn Grants, 2007
Illinois Children's Healthcare Foundation Grants, 2146
Kansas Health Fndn Major Initiatives Grants, 2288
Long Island Community Foundation Grants, 2375
Lydia deForest Charitable Trust Grants, 2393
Mary Black Fndn Community Health Grants, 2441
Nick Traina Foundation Grants, 2792

720 / Emotional/Mental Health

Peter and Elizabeth C. Tower Foundation Phase II Technology Initiative Grants, 3149
Peter and Elizabeth C. Tower Foundation Phase I Technology Initiative Grants, 3150
Peter and Elizabeth C. Tower Foundation Social and Emotional Preschool Curriculum Grants, 3151
Peter F. McManus Charitable Trust Grants, 3153
Philip L. Graham Fund Health and Human Services Grants, 3190
Piedmont Natural Gas Foundation Health and Human Services Grants, 3221
Porter County Women's Grant, 3248
Premera Blue Cross Grants, 3258
Robert R. McCormick Tribune Veterans Initiative Grants, 3386
Schlessman Family Foundation Grants, 3493
SOBP A.E. Bennett Research Award, 3594
Sophia Romero Trust Grants, 3624
USDD Workplace Violence in Military Grants, 3847

Emotionally Disturbed
Greater Tacoma Community Foundation Ryan Alan Hade Endowment Fund, 1887
Herbert A. & Adrian W. Woods Fndn Grants, 2007

Employment Opportunity Programs
Able Trust Vocational Rehabilitation Grants for Individuals, 213
Achelis Foundation Grants, 234
ACL Partnerships in Employment Systems Change Grants, 246
A Fund for Women Grants, 342
Albert W. Rice Charitable Foundation Grants, 411
Alfred E. Chase Charitable Foundation Grants, 430
AT&T Foundation Civic and Community Service Program Grants, 793
Baxter International Corporate Giving Grants, 841
Blue Cross Blue Shield of Minnesota Foundation - Health Equity: Building Health Equity Together Grants, 914
Bodman Foundation Grants, 929
Boston Foundation Grants, 936
Bush Fndn Health & Human Services Grants, 979
Charles H. Pearson Foundation Grants, 1121
Coleman Foundation Developmental Disabilities Grants, 1207
Community Fndn for the Capital Region Grants, 1259
Connecticut Community Foundation Grants, 1324
Crail-Johnson Foundation Grants, 1355
Cruise Industry Charitable Foundation Grants, 1363
Edward N. and Della L. Thome Memorial Foundation Grants, 1557
Essex County Community Foundation Merrimack Valley General Fund Grants, 1603
Essex County Community Foundation Women's Fund Grants, 1605
Frank B. Hazard General Charity Fund Grants, 1721
Frank Reed and Margaret Jane Peters Memorial Fund II Grants, 1728
Frank Stanley Beveridge Foundation Grants, 1730
Frederick W. Marzahl Memorial Fund Grants, 1741
Gardner Foundation Grants, 1772
George W. Wells Foundation Grants, 1804
Georgiana Goddard Eaton Memorial Grants, 1805
GNOF New Orleans Works Grants, 1844
Graco Foundation Grants, 1860
Grifols Community Outreach Grants, 1904
Hampton Roads Community Foundation Health and Human Service Grants, 1941
Harold Brooks Foundation Grants, 1952
Helen Bader Foundation Grants, 1992
Henry County Community Foundation Grants, 2005
Johnson & Johnson Corporate Contributions, 2253
Kessler Fnd Signature Employment Grants, 2302
M.D. Anderson Foundation Grants, 2403
NIDRR Field-Initiated Projects, 2853
Perry County Community Foundation Grants, 3142
Pinkerton Foundation Grants, 3224
Priddy Foundation Program Grants, 3266
Richard King Mellon Foundation Grants, 3369

Richard M. Fairbanks Foundation Grants, 3370
Robert F. Stoico/FIRSTFED Charitable Foundation Grants, 3382
Robert R. McCormick Tribune Veterans Initiative Grants, 3386
Rockwell Fund, Inc. Grants, 3398
S. Mark Taper Foundation Grants, 3449
Sioux Falls Area Community Foundation Community Fund Grants (Unrestricted), 3568
Sioux Falls Area Community Foundation Spot Grants (Unrestricted), 3570
Stella and Charles Guttman Foundation Grants, 3651
Textron Corporate Contributions Grants, 3712
U.S. Department of Education Vocational Rehabilitation Services Projects for American Indians with Disabilities Grants, 3774
USDA Organic Agriculture Research Grants, 3830
Verizon Foundation Maryland Grants, 3859
Verizon Foundation New York Grants, 3860
Verizon Foundation Northeast Region Grants, 3861
Verizon Foundation Pennsylvania Grants, 3862
Verizon Foundation Vermont Grants, 3863
Verizon Foundation Virginia Grants, 3864
Warrick County Community Foundation Grants, 3904

Employment/Unemployment Studies
NIGMS Advancing Basic Behavioral and Social Research on Resilience: an Integrative Science Approach, 2861

Endocrine Disorders
NIDDK Devel of Disease Biomarkers Grants, 2809
Pfizer ASPIRE Worldwide Endocrine Young Investigator Grants, 3173
Pfizer Australia Paediatric Endocrine Care Research Grants, 3176

Endocrinology
AHAF National Glaucoma Research Grants, 351
AMA Foundation Seed Grants for Research, 494
Genentech Corp Charitable Contributions, 1780
NIDDK Co-Activators and Co-Repressors in Gene Expression Grants, 2805
NIDDK Collaborative Interdisciplinary Team Science in Diabetes, Endocrinology and Metabolic Diseases Grants, 2806
NIDDK Diabetes, Endocrinology, and Metabolic Diseases NRSAs--Individual, 2810
NIDDK Endoscopic Clinical Research Grants In Pancreatic And Biliary Diseases, 2814
NIDDK Feasibility Clinical Research Grants in Diabetes, Endocrine and Metabolic Diseases, 2834
NIDDK Proteomics: Diabetes, Obesity, And Endocrine, Digestive, Kidney, Urologic, And Hematologic Diseases, 2838
NIDDK Seeding Collaborative Interdisciplinary Team Science in Diabetes, Endocrinology and Metabolic Diseases Grants, 2843
NIDDK Small Grants for Underrepresented Minority Scientists in Diabetes and Digestive and Kidney Diseases, 2847
NIDDK Translational Research for the Prevention and Control of Diabetes and Obesity, 2850
Pfizer ASPIRE Worldwide Endocrine Young Investigator Grants, 3173

Endowments
Clayton Fund Grants, 1180
GNOF Freeman Challenge Grants, 1837
Hearst Foundations Health Grants, 1987
Jack H. & William M. Light Trust Grants, 2179
Margaret T. Morris Foundation Grants, 2429
Roy & Christine Sturgis Charitable Grants, 3415
W.P. and Bulah Luse Foundation Grants, 3892
Willard & Pat Walker Charitable Fndn Grants, 3943

Energy
Canada Graduate Scholarships (CGS) and NSERC Postgraduate Scholarships (PGS), 1010
Clarence E. Heller Charitable Foundation Grants, 1166

SUBJECT INDEX

Conservation, Food, and Health Foundation Grants for Developing Countries, 1338
General Motors Foundation Grants, 1783
Gruber Foundation Weizmann Institute Awards, 1911
NSF-NIST Interaction in Chemistry, Materials Research, Molecular Biosciences, Bioengineering, and Chemical Engineering, 3004

Energy Assistance
Washington Gas Charitable Giving, 3908

Energy Conservation
Bullitt Foundation Grants, 973
Constellation Energy Corporate Grants, 1342
Rohm and Haas Company Grants, 3403
Sioux Falls Area Community Foundation Community Fund Grants (Unrestricted), 3568
Sioux Falls Area Community Foundation Spot Grants (Unrestricted), 3570
Union Bank, N.A. Foundation Grants, 3788

Energy Planning/Policy
Blue Cross Blue Shield of Minnesota Foundation - Healthy Equity: Health Impact Assessment Demonstration Project Grants, 916
Blue Cross Blue Shield of Minnesota Foundation - Healthy Equity: Health Impact Assessment Program Grants, 917

Engineering
AAAS Award for Scie Freedom & Responsibility, 36
AAAS Science and Technology Policy Fellowships: Global Health and Development, 41
AACR Team Science Award, 126
AAUW American Dissertation Fellowships, 195
Alfred P. Sloan Foundation Research Fellowships, 432
American Chemical Society Award in Separations Science and Technology, 514
ARCO Foundation Education Grants, 652
Arnold and Mabel Beckman Foundation Young Investigators Grants, 666
ArvinMeritor Grants, 675
ASU Graduate College Science Foundation Arizona Bisgrove Postdoctoral Scholars, 792
BWF Career Awards at the Scientific Interface, 983
Canada-U.S. Fulbright New Century Scholars Program Grants, 1008
Canada-U.S. Fulbright Senior Specialists Grants, 1009
Canada Graduate Scholarships (CGS) and NSERC Postgraduate Scholarships (PGS), 1010
Community Foundation of Greater Flint Grants, 1273
Community Foundation of Greater Fort Wayne - Lilly Endowment Scholarships, 1275
Community Foundation of St. Joseph County Lilly Endowment Community Scholarship, 1308
DAAD Research Stays for University Academics and Scientists, 1408
Donna K. Yundt Memorial Scholarship, 1488
Dr. John Maniotes Scholarship, 1506
Dr. R.T. White Scholarship, 1510
Elkhart County Foundation Lilly Endowment Community Scholarships, 1574
Flinn Foundation Scholarships, 1675
Fluor Foundation Grants, 1682
Fred and Louise Latshaw Scholarship, 1735
Fulbright Alumni Initiatives Awards, 1749
Fulbright Distinguished Chairs Awards, 1750
Fulbright New Century Scholars Grants, 1753
Fulbright Specialists Program Grants, 1754
Fulbright Scholars in Europe and Eurasia, 1755
Fulbright Traditional Scholar Program in Sub-Saharan Africa, 1756
Fulbright Traditional Scholar Program in the East Asia/Pacific Region, 1757
Fulbright Traditional Scholar Program in the Near East and North Africa Region, 1758
Fulbright Traditional Scholar Program in the South and Central Asia Region, 1759
Fulbright Traditional Scholar Program in the Western Hemisphere, 1760

SUBJECT INDEX

Hammond Common Council Scholarships, 1938
HHMI-NIBIB Interfaces Initiative Grants, 2014
HRAMF Smith Family Awards for Excellence in Biomedical Research, 2063
IIE 911 Armed Forces Scholarships, 2122
IIE AmCham Charitable Foundation U.S. Studies Scholarship, 2125
IIE Brazil Science Without Borders Undergraduate Scholarships, 2126
IIE Central Europe Summer Research Institute Summer Research Fellowship, 2127
IIE Chevron International REACH Scholarships, 2128
IIE David L. Boren Fellowships, 2129
IIE Iraq Scholars and Leaders Scholarships, 2133
IIE Klein Family Scholarship, 2135
IIE Leonora Lindsley Memorial Fellowships, 2136
IIE Lingnan Foundation W.T. Chan Fellowship, 2137
IIE Lotus Scholarships, 2138
IIE Mattel Global Scholarship, 2139
IIE New Leaders Group Award for Mutual Understanding, 2141
IIE Western Union Family Scholarships, 2143
John V. and George Primich Family Scholarship, 2259
Joshua Benjamin Cohen Memorial Scholarship, 2276
LaGrange County Lilly Endowment Community Scholarship, 2331
Lake County Athletic Officials Association Scholarships, 2332
Lake County Lilly Endowment Community Scholarships, 2334
LEGENDS Scholarship, 2351
Medtronic Foundation Fellowships, 2512
Nancy J. Pinnick Memorial Scholarship, 2607
NHGRI Mentored Research Scientist Development Award, 2686
NHLBI Mentored Quantitative Research Career Development Awards, 2711
NIMH Curriculum Development Award in Neuroinformatics Research and Analysis, 2917
NSERC Brockhouse Canada Prize for Interdisciplinary Research in Science and Engineering Grant, 3003
NSF Alan T. Waterman Award, 3006
NSF Emerging Frontiers in Research and Innovation (EFRI) Grants, 3014
NSF Grant Opportunities for Academic Liaison with Industry (GOALI), 3016
NSF Presidential Early Career Awards for Scientists and Engineers (PECASE) Grants, 3021
NSF Small Business Innovation Research (SBIR) Grants, 3022
Porter County Foundation Lilly Endowment Community Scholarships, 3245
Prospect Burma Scholarships, 3275
Pulaski County Community Fndn Scholarships, 3281
Rayonier Foundation Grants, 3311
Rockwell International Corporate Trust Grants, 3399
RSC Thomas W. Eadie Medal, 3423
Schrage Family Foundation Scholarships, 3494
Sengupta Family Scholarship, 3512
Society of Cosmetic Chemists Allan B. Black Award Sponsored by Presperse, 3604
Society of Cosmetic Chemists Award Sponsored by McIntyre Group, Ltd., 3605
Society of Cosmetic Chemists Award Sponsored by The HallStar Company, 3606
Society of Cosmetic Chemists Chapter Best Speaker Award, 3607
Society of Cosmetic Chem Chapter Merit Award, 3608
Society of Cosmetic Chemists Frontier of Science Award, 3609
Society of Cosmetic Chem Hans Schaeffer Award, 3610
Society of Cosmetic Chem Jos Ciaudelli Award, 3611
Society of Cosmetic Chemists Keynote Award Lecture Sponsored by Ruger Chemical Corporation, 3612
Society of Cosmetic Chemists Literature Award, 3613
Society of Cosmetic Chemists Maison G. de Navarre Medal, 3614
Society of Cosmetic Chemists Merit Award, 3615
Society of Cosmetic Chemists Robert A. Kramer Lifetime Service Award, 3616

Society of Cosmetic Chem Shaw Mudge Award, 3617
Society of Cosmetic Chem Stud Poster Awards, 3618
Society of Cosmetic Chem You Scientist Awards, 3619
Texas Instruments Foundation STEM Education Grants, 3711
W. James Spicer Scholarship, 3885
W.M. Keck Foundation Science and Engineering Research Grants, 3889
W.M. Keck Fndn Undergrad Ed Grants, 3891
Webster Cornwell Memorial Scholarship, 3911
Whitley County Community Foundation - Lilly Endowment Scholarship, 3935
Whitley County Community Fndn Scholarships, 3937

Engineering Education
3M Company Fndn Community Giving Grants, 6
Advanced Micro Devices Comm Affairs Grants, 313
Alcatel-Lucent Technologies Foundation Grants, 413
Alcoa Foundation Grants, 414
AMD Corporate Contributions Grants, 500
ArvinMeritor Grants, 675
DaimlerChrysler Corporation Fund Grants, 1412
Dayton Power and Light Foundation Grants, 1437
Emerson Charitable Trust Grants, 1589
FMC Foundation Grants, 1683
Ford Motor Company Fund Grants Program, 1687
Gates Millennium Scholars Program, 1776
Goodrich Corporation Foundation Grants, 1854
HHMI Med into Grad Initiative Grants, 2022
Qualcomm Grants, 3287
Ralph M. Parsons Foundation Grants, 3304
Rockwell Collins Charitable Corporation Grants, 3397
Roy & Christine Sturgis Charitable Grants, 3415
Tellabs Foundation Grants, 3706
United Technologies Corporation Grants, 3803
W.M. Keck Fndn Undergrad Ed Grants, 3891
Washington Gas Charitable Giving, 3908

English as a Second Language
Blue Cross Blue Shield of Minnesota Fndn Healthy Equity: Public Libraries for Health Grants, 918
IIE Western Union Family Scholarships, 2143
Robert R. McCormick Trib Community Grants, 3385
Sophia Romero Trust Grants, 3624
Textron Corporate Contributions Grants, 3712

Enrichment, Student
BWF Postdoctoral Enrichment Grants, 989
Foundation for the Mid South Community Development Grants, 1699

Enteral Nutrition
ASPEN Dudrick Research Scholar Award, 759
ASPEN Harry M. Vars Award, 760
ASPEN Rhoads Research Foundation Abbott Nutrition Research Grants, 761
ASPEN Rhoads Research Foundation Baxter Parenteral Nutrition Research Grant, 762
ASPEN Rhoads Research Foundation C. Richard Fleming Grant, 763
ASPEN Rhoads Research Foundation Maurice Shils Grant, 764
ASPEN Rhoads Research Foundation Norman Yoshimura Grant, 765
ASPEN Scientific Abstracts Awards for Papers or Posters, 766
ASPEN Scientific Abstracts Promising Investigator Awards, 767

Entertainment Industry
Actors Addiction & Recovery Services Grants, 282
Actors Fund HIV/AIDS Initiative Grants, 283

Entrepreneurship
Achelis Foundation Grants, 234
Adolph Coors Foundation Grants, 311
Allstate Corporate Giving Grants, 446
Allstate Corp Hometown Commitment Grants, 447
Alvin and Fanny Blaustein Thalheimer Foundation Baltimore Communal Grants, 459

Bodman Foundation Grants, 929
California Endowment Innovative Ideas Challenge Grants, 1000
Carl R. Hendrickson Family Foundation Grants, 1028
Echoing Green Fellowships, 1546
GNOF New Orleans Works Grants, 1844
JM Foundation Grants, 2229
Minneapolis Foundation Community Grants, 2558
PepsiCo Foundation Grants, 3137
Skoll Fndn Awards for Social Entrepreneurship, 3579
Union Bank, N.A. Foundation Grants, 3788

Environment
Alfred P. Sloan Foundation Science of Learning STEM Grants, 433
Alpha Natural Resources Corporate Giving, 450
Baxter International Corporate Giving Grants, 841
BBVA Compass Foundation Charitable Grants, 848
Belvedere Community Foundation Grants, 874
Bridgestone/Firestone Trust Fund Grants, 949
Caesars Foundation Grants, 995
CDC Collegiate Leaders in Environmental Health Internships, 1048
Centerville-Washington Foundation Grants, 1078
Charles M. & Mary D. Grant Fndn Grants, 1124
Chilkat Valley Community Foundation Grants, 1148
CICF City of Noblesville Community Grant, 1155
Collective Brands Foundation Grants, 1208
Community Foundation for Southeast Michigan Grants, 1258
Community Fndn for the Capital Region Grants, 1259
Community Foundation of Boone County Grants, 1266
Community Fndn of Greater Lafayette Grants, 1280
Community Foundation of Louisville Anna Marble Memorial Fund for Princeton Grants, 1290
Crystelle Waggoner Charitable Trust Grants, 1364
DeKalb County Community Foundation Grants, 1447
DuPage Community Foundation Grants, 1531
Earl and Maxine Claussen Trust Grants, 1541
El Paso Corporate Foundation Grants, 1580
El Pomar Foundation Anna Keesling Ackerman Fund Grants, 1581
El Pomar Foundation Grants, 1582
Erie Community Foundation Grants, 1601
Fairfield County Community Foundation Grants, 1625
FIU Global Civic Engagement Mini Grants, 1673
Four County Community Fndn General Grants, 1710
Fuji Film Grants, 1748
Gardner Foundation Grants, 1771
GNOF Norco Community Grants, 1845
Grand Haven Area Community Fndn Grants, 1864
Greater Cincinnati Foundation Priority and Small Projects/Capacity-Building Grants, 1880
Gruber Foundation Weizmann Institute Awards, 1911
Grundy Foundation Grants, 1912
Harry A. & Margaret D. Towsley Fndn Grants, 1959
Hendricks County Community Fndn Grants, 2000
Intergrys Corporation Grants, 2160
Jasper Foundation Grants, 2212
Jennings County Community Foundation Women's Giving Circle Grant, 2221
Jessica Stevens Community Foundation Grants, 2224
John W. and Anna H. Hanes Foundation Grants, 2261
Lake County Community Fund Grants, 2333
Marathon Petroleum Corporation Grants, 2422
Mericos Foundation Grants, 2518
Norcliffe Foundation Grants, 2979
North Dakota Community Foundation Grants, 2994
Northrop Grumman Corporation Grants, 2997
Orange County Community Foundation Grants, 3079
Oregon Community Fndn Community Grants, 3082
Pacific Life Foundation Grants, 3103
Parke County Community Foundation Grants, 3115
PepsiCo Foundation Grants, 3137
PMI Foundation Grants, 3239
Pokagon Fund Grants, 3242
PPCF Community Grants, 3256
Putnam County Community Foundation Grants, 3283
RCF General Community Grants, 3312
Robbins Charitable Foundation Grants, 3378

722 / Environment

Rohm and Haas Company Grants, 3403
Sartain Lanier Family Foundation Grants, 3486
Seward Community Foundation Grants, 3514
Seward Community Foundation Mini-Grants, 3515
Shield-Ayres Foundation Grants, 3546
Sony Corporation of America Grants, 3623
SSRC Collaborative Research Grants on Environment and Health in China, 3638
Starke County Community Foundation Grants, 3646
Susan G. Komen Breast Cancer Foundation Challenge Grants: Breast Cancer and the Environment, 3689
Toyota Motor Engineering & Manufacturing North America Grants, 3742
Toyota Motor Manufacturing of Alabama Grants, 3743
Toyota Motor Manufacturing of Indiana Grants, 3744
Toyota Motor Manufacturing of Kentucky Grants, 3745
Toyota Motor Manuf of Mississippi Grants, 3746
Toyota Motor Manuf of West Virginia Grants, 3748
Toyota Motor N America of New York Grants, 3749
Toyota Motor Sales, USA Grants, 3750
Tri-State Community Twenty-first Century Endowment Fund Grants, 3756
Union Bank, N.A. Corporate Sponsorships and Donations, 3787
USAID Land Use Change and Disease Emergence Grants, 3818
Vanderburgh Community Foundation Grants, 3852
Walker Area Community Foundation Grants, 3895
Walmart Foundation Community Giving Grants, 3897
Warren County Community Foundation Grants, 3902
Warren County Community Fndn Mini-Grants, 3903
Wells County Foundation Grants, 3920
White County Community Foundation Grants, 3934
William B. Stokely Jr. Foundation Grants, 3947
William J. & Tina Rosenberg Fndn Grants, 3954
William Ray & Ruth E. Collins Fndn Grants, 3956

Environmental Biology
CDC Collegiate Leaders in Environmental Health Internships, 1048
IIE Central Europe Summer Research Institute Summer Research Fellowship, 2127
Northrop Grumman Corporation Grants, 2997
NSF Doctoral Dissertation Improvement Grants in the Directorate for Biological Sciences (DDIG), 3013
Susan G. Komen Breast Cancer Foundation Challenge Grants: Breast Cancer and the Environment, 3689

Environmental Conservation
CNCS AmeriCorps VISTA Project Grants, 1190
Collective Brands Foundation Grants, 1208
Crown Point Community Foundation Grants, 1361
Dean Foods Community Involvement Grants, 1440
George H.C. Ensworth Memorial Fund Grants, 1797
IIE Rockefeller Foundation Bellagio Center Residencies, 2142
Lucy Downing Nisbet Charitable Fund Grants, 2387
Meyer Memorial Trust Grassroots Grants, 2536
Meyer Memorial Trust Responsive Grants, 2537
Northrop Grumman Corporation Grants, 2997
Packard Foundation Local Grants, 3105
Piedmont Natural Gas Corporate and Charitable Contributions, 3220
Rohm and Haas Company Grants, 3403
Sony Corporation of America Grants, 3623
Union Bank, N.A. Foundation Grants, 3788

Environmental Design
Northrop Grumman Corporation Grants, 2997
Vancouver Foundation Grants and Community Initiatives Program, 3849
Windham Foundation Grants, 3961

Environmental Economics
GNOF New Orleans Works Grants, 1844

Environmental Education
Abbott Fund Community Grants, 201
Administaff Community Affairs Grants, 308
AEC Trust Grants, 315

A Friends' Foundation Trust Grants, 341
Ahn Family Foundation Grants, 356
Alaska Airlines Foundation Grants, 403
Alexander & Baldwin Fnd Mainland Grants, 417
Alexander and Baldwin Foundation Hawaiian and Pacific Island Grants, 418
Allen Foundation Educational Nutrition Grants, 442
Alpine Winter Foundation Grants, 453
Amador Community Foundation Grants, 481
Barker Welfare Foundation Grants, 831
Beldon Fund Grants, 871
Blumenthal Foundation Grants, 925
BMW of North America Charitable Contributions, 926
Boeing Company Contributions Grants, 930
Bullitt Foundation Grants, 973
Cabot Corporation Foundation Grants, 993
Cargill Citizenship Fund Corporate Giving, 1018
Carls Foundation Grants, 1030
Carrier Corporation Contributions Grants, 1038
Clarence E. Heller Charitable Foundation Grants, 1166
CSX Corporate Contributions Grants, 1368
Dean Foods Community Involvement Grants, 1440
Del E. Webb Foundation Grants, 1450
Dyson Foundation Mid-Hudson Valley Project Support Grants, 1534
Fargo-Moorhead Area Foundation Grants, 1632
Frederick W. Marzahl Memorial Fund Grants, 1741
Gamble Foundation Grants, 1769
George and Ruth Bradford Foundation Grants, 1789
Hilton Head Island Foundation Grants, 2036
Honda of America Manufacturing Fndn Grants, 2045
Jessica Stevens Community Foundation Grants, 2224
Jessie Ball Dupont Fund Grants, 2226
Johnson Foundation Wingspread Conference Support Program, 2257
Knox County Community Foundation Grants, 2313
Lawrence S. Huntington Fund Grants, 2348
Lester E. Yeager Charitable Trust B Grants, 2359
Northrop Grumman Corporation Grants, 2997
Perry County Community Foundation Grants, 3142
Piedmont Natural Gas Corporate and Charitable Contributions, 3220
Pike County Community Foundation Grants, 3222
Rohm and Haas Company Grants, 3403
Schering-Plough Foundation Community Initiatives Grants, 3491
Scott County Community Foundation Grants, 3500
Seabury Foundation Grants, 3501
Seward Community Foundation Grants, 3514
Seward Community Foundation Mini-Grants, 3515
Spencer County Community Foundation Grants, 3632
TE Foundation Grants, 3705
Union Bank, N.A. Foundation Grants, 3788
Vancouver Foundation Grants and Community Initiatives Program, 3849
Winifred & Harry B. Allen Foundation Grants, 3962

Environmental Effects
Alfred P. Sloan Foundation Research Fellowships, 432
American Chemical Society Award for Creative Advances in Environmental Science and Technology, 513
Carrier Corporation Contributions Grants, 1038
Clarence E. Heller Charitable Foundation Grants, 1166
EPA Children's Health Protection Grants, 1598
Gerber Foundation Grants, 1807
IIE Rockefeller Foundation Bellagio Center Residencies, 2142
March of Dimes Program Grants, 2426
Northrop Grumman Corporation Grants, 2997
NSF-NIST Interaction in Chemistry, Materials Research, Molecular Biosciences, Bioengineering, and Chemical Engineering, 3004
NSF Chemistry Research Experiences for Undergraduates (REU), 3012
Piedmont Natural Gas Corporate and Charitable Contributions, 3220
USAID Land Use Change and Disease Emergence Grants, 3818

SUBJECT INDEX

Environmental Engineering
IIE KAUST Graduate Fellowships, 2134

Environmental Health
AAOHN Found Experienced Researcher Grants, 173
AAOHN Foundation New Investigator Researcher Grants, 174
AAOHN Foundation Professional Development Scholarships, 175
Acumen Global Fellowships, 285
Administaff Community Affairs Grants, 308
Adolph Coors Foundation Grants, 311
AEC Trust Grants, 315
Ahn Family Foundation Grants, 356
Alaska Airlines Foundation Grants, 403
Alexander & Baldwin Fnd Mainland Grants, 417
Alexander and Baldwin Foundation Hawaiian and Pacific Island Grants, 418
Allen Foundation Educational Nutrition Grants, 442
Alliance Healthcare Foundation Grants, 444
Ashland Corporate Contributions Grants, 713
Beldon Fund Grants, 871
Blue Cross Blue Shield of Minnesota Foundation - Healthy Children: Growing Up Healthy Grants, 915
Blue Cross Blue Shield of Minnesota Foundation - Healthy Equity: Health Impact Assessment Demonstration Project Grants, 916
BMW of North America Charitable Contributions, 926
Bravewell Leadership Award, 946
California Wellness Foundation Work and Health Program Grants, 1001
Canada Graduate Scholarships (CGS) and NSERC Postgraduate Scholarships (PGS), 1010
CDC Collegiate Leaders in Environmental Health Internships, 1048
CDC Epidemic Intell Service Training Grants, 1055
CDC Experience Epidemiology Fellowships, 1059
CDC Fnd Tobacco Network Lab Fellowship, 1062
CDC Public Health Associates, 1065
CDC Public Health Associates Hosts, 1066
CDC Summer Graduate Environmental Health Internships, 1072
CDC Summer Program In Environmental Health Internships, 1073
Community Fndn of Switzerland County Grants, 1311
CSTE CDC/CSTE Applied Epidemiology Fellowships, 1367
Donald and Sylvia Robinson Family Foundation Grants, 1486
DuPont Pioneer Community Giving Grants, 1532
EPA Children's Health Protection Grants, 1598
Fremont Area Community Fndn General Grants, 1746
Health Foundation of Greater Cincinnati Grants, 1983
Heineman Foundation for Research, Education, Charitable and Scientific Purposes, 1990
Houston Endowment Grants, 2053
IIE David L. Boren Fellowships, 2129
IIE Rockefeller Foundation Bellagio Center Residencies, 2142
Jessie B. Cox Charitable Trust Grants, 2225
Lawrence S. Huntington Fund Grants, 2348
NIDCD Mentored Research Scientist Development Award, 2796
NIEHS Independent Scientist Award, 2856
NIEHS Mentored Clinical Scientist Research Career Development Awards, 2857
NIEHS Mentored Research Scientist Development Award, 2858
NIEHS Small Business Innovation Research (SBIR) Program Grants, 2860
Peter F. McManus Charitable Trust Grants, 3153
Powell Foundation Grants, 3255
Prince Charitable Trusts DC Grants, 3271
Pulido Walker Foundation, 3282
Rayonier Foundation Grants, 3311
San Diego Foundation for Change Grants, 3470
San Francisco Fndn Community Health Grants, 3478
Schering-Plough Foundation Community Initiatives Grants, 3491

SUBJECT INDEX

SSRC Collaborative Research Grants on Environment and Health in China, 3638
Sunflower Foundation Walking Trails Grants, 3678
Toyota Motor Manufacturing of Texas Grants, 3747
USDA Organic Agriculture Research Grants, 3830
Vancouver Foundation Grants and Community Initiatives Program, 3849
Washington Gas Charitable Giving, 3908
Winifred & Harry B. Allen Foundation Grants, 3962

Environmental Issues
Acumen Global Fellowships, 285
Collective Brands Foundation Grants, 1208
Hillsdale County Community General Adult Foundation Grants, 2034
White County Community Foundation Grants, 3934

Environmental Law
Bullitt Foundation Grants, 973
Norman Foundation Grants, 2983

Environmental Planning/Policy
Achelis Foundation Grants, 234
A Friends' Foundation Trust Grants, 341
Beldon Fund Grants, 871
Blue Cross Blue Shield of Minnesota Foundation - Healthy Equity: Health Impact Assessment Demonstration Project Grants, 916
Blue Cross Blue Shield of Minnesota Foundation - Healthy Equity: Health Impact Assessment Program Grants, 917
Bodman Foundation Grants, 929
Clarence E. Heller Charitable Foundation Grants, 1166
CLIF Bar Family Foundation Grants, 1183
Dyson Foundation Mid-Hudson Valley Project Support Grants, 1534
El Pomar Foundation Grants, 1582
Harrison County Community Foundation Signature Grants, 1958
Honda of America Manufacturing Fndn Grants, 2045
IIE Rockefeller Foundation Bellagio Center Residencies, 2142
Prince Charitable Trusts Chicago Grants, 3270
Windham Foundation Grants, 3961

Environmental Programs
3M Company Fndn Community Giving Grants, 6
1675 Foundation Grants, 11
Abbot and Dorothy H. Stevens Foundation Grants, 197
Abbott Fund Community Grants, 201
Abel Foundation Grants, 206
Abell-Hanger Foundation Grants, 207
ACE Charitable Foundation Grants, 229
Adam Richter Charitable Trust Grants, 298
Administaff Community Affairs Grants, 308
AEC Trust Grants, 315
Ahn Family Foundation Grants, 356
Air Products and Chemicals Grants, 399
Alaska Airlines Foundation Grants, 403
Alberto Culver Corporate Contributions Grants, 408
Albuquerque Community Foundation Grants, 412
Alcoa Foundation Grants, 414
Alexander & Baldwin Fnd Mainland Grants, 417
Alexander and Baldwin Foundation Hawaiian and Pacific Island Grants, 418
Allegan County Community Foundation Grants, 439
Allen Foundation Educational Nutrition Grants, 442
Allen P. & Josephine B. Green Fndn Grants, 443
Alpha Natural Resources Corporate Giving, 450
Alpine Winter Foundation Grants, 453
Amador Community Foundation Grants, 481
American Electric Power Foundation Grants, 530
Amgen Foundation Grants, 572
Andrew Family Foundation Grants, 588
Anheuser-Busch Foundation Grants, 591
Ann Arbor Area Community Foundation Grants, 594
APS Foundation Grants, 646
Archer Daniels Midland Foundation Grants, 651
Arizona Public Service Corporate Giving Program Grants, 658

Arkansas Community Foundation Grants, 660
Arthur Ashley Williams Foundation Grants, 671
Ashland Corporate Contributions Grants, 713
AT&T Foundation Civic and Community Service Program Grants, 793
Athwin Foundation Grants, 795
Autodesk Community Relations Grants, 809
Autzen Foundation Grants, 810
Bacon Family Foundation Grants, 817
Batchelor Foundation Grants, 837
Baton Rouge Area Foundation Grants, 838
Baxter International Corporate Giving Grants, 841
Beazley Foundation Grants, 865
Beldon Fund Grants, 871
Bella Vista Foundation Grants, 873
Bender Foundation Grants, 876
Berks County Community Foundation Grants, 881
Berrien Community Foundation Grants, 884
Bikes Belong Foundation Paul David Clark Bicycling Safety Grants, 890
Blanche and Julian Robertson Family Foundation Grants, 911
BMW of North America Charitable Contributions, 926
Bodenwein Public Benevolent Foundation Grants, 928
Boeing Company Contributions Grants, 930
Borkee-Hagley Foundation Grants, 934
Boston Foundation Grants, 936
Boston Globe Foundation Grants, 937
Brian G. Dyson Foundation Grants, 948
Bullitt Foundation Grants, 973
Caesars Foundation Grants, 995
Callaway Foundation Grants, 1002
Callaway Golf Company Foundation Grants, 1003
Cambridge Community Foundation Grants, 1004
Capital Region Community Foundation Grants, 1014
Carl & Eloise Pohlad Family Fndn Grants, 1021
Carl M. Freeman Foundation FACES Grants, 1026
Carolyn Foundation Grants, 1034
Carrie E. and Lena V. Glenn Foundation Grants, 1036
Carrier Corporation Contributions Grants, 1038
Cemala Foundation Grants, 1077
Cessna Foundation Grants Program, 1081
CFFVR Project Grants, 1101
CFFVR Shawano Area Community Foundation Grants, 1104
Charles Delmar Foundation Grants, 1115
Charlotte County (FL) Community Foundation Grants, 1126
Chilkat Valley Community Foundation Grants, 1148
CICF Indianapolis Fndn Community Grants, 1156
Citizens Bank Mid-Atlantic Charitable Foundation Grants, 1162
Clarence E. Heller Charitable Foundation Grants, 1166
Claude Worthington Benedum Fndn Grants, 1178
Clayton Baker Trust Grants, 1179
Cleveland-Cliffs Foundation Grants, 1181
CLIF Bar Family Foundation Grants, 1183
Coastal Community Foundation of South Carolina Grants, 1202
Collective Brands Foundation Grants, 1208
Community Fndn for Greater Buffalo Grants, 1247
Community Fndn for Monterey County Grants, 1253
Community Fndn for Muskegon County Grants, 1254
Community Foundation for Northeast Michigan Mini-Grants, 1255
Community Fndn of East Central Illinois Grants, 1268
Community Foundation of Eastern Connecticut General Southeast Grants, 1269
Community Foundation of Greater Birmingham Grants, 1272
Community Foundation of Greater Flint Grants, 1273
Community Foundation of Greater Greensboro Community Grants, 1277
Community Fndn of Greater Tampa Grants, 1282
Community Foundation of Greenville-Greenville Women Giving Grants, 1283
Community Foundation of Greenville Community Enrichment Grants, 1284
Community Foundation of Mount Vernon and Knox County Grants, 1299

Environmental Programs /723

Community Fndn of Randolph County Grants, 1302
Community Fndn of South Puget Sound Grants, 1307
Community Foundation of St. Joseph County Special Project Challenge Grants, 1310
Community Foundation of Tampa Bay Grants, 1312
Community Foundation of the Eastern Shore Community Needs Grants, 1313
Community Foundation of the Verdugos Grants, 1315
Community Fndn of Wabash County Grants, 1316
Connecticut Community Foundation Grants, 1324
Constellation Energy Corporate Grants, 1342
Consumers Energy Foundation, 1343
Cooke Foundation Grants, 1344
Cooper Industries Foundation Grants, 1345
Coors Brewing Corporate Contributions Grants, 1346
Cowles Charitable Trust Grants, 1354
Cruise Industry Charitable Foundation Grants, 1363
Crystelle Waggoner Charitable Trust Grants, 1364
CSRA Community Foundation Grants, 1366
Cumberland Community Foundation Grants, 1374
Cyrus Eaton Foundation Grants, 1380
Dade Community Foundation Grants, 1410
DaimlerChrysler Corporation Fund Grants, 1412
Dearborn Community Foundation County Progress Grants, 1442
Delaware Community Foundation Grants, 1449
Dole Food Company Charitable Contributions, 1483
Donald and Sylvia Robinson Family Foundation Grants, 1486
Dorothy Hooper Beattie Foundation Grants, 1499
Dorrance Family Foundation Grants, 1501
Do Something Awards, 1503
Dr. Scholl Foundation Grants Program, 1511
Drs. Bruce and Lee Foundation Grants, 1513
Dubois County Community Foundation Grants, 1517
DuPont Pioneer Community Giving Grants, 1532
Dyson Foundation Mid-Hudson Valley Project Support Grants, 1534
E.J. Grassmann Trust Grants, 1537
eBay Foundation Community Grants, 1543
Edward and Helen Bartlett Foundation Grants, 1554
Edward Bangs Kelley and Elza Kelley Foundation Grants, 1556
Edwin W. and Catherine M. Davis Fndn Grants, 1561
Elliot Foundation Inc Grants, 1575
Elmer L. & Eleanor J. Andersen Fndn Grants, 1578
El Paso Community Foundation Grants, 1579
Ensworth Charitable Foundation Grants, 1595
EPA Children's Health Protection Grants, 1598
Essex County Community Foundation Discretionary Fund Grants, 1602
Estee Lauder Grants, 1606
Evjue Foundation Grants, 1617
Ewing Halsell Foundation Grants, 1618
Ferree Foundation Grants, 1658
Field Foundation of Illinois Grants, 1660
FirstEnergy Foundation Community Grants, 1668
Forrest C. Lattner Foundation Grants, 1688
Foster Foundation Grants, 1689
Four County Community Fndn General Grants, 1710
Frances and John L. Loeb Family Fund Grants, 1716
Franklin County Community Foundation Grants, 1724
Fremont Area Community Fndn General Grants, 1746
Fulton County Community Foundation Grants, 1763
G.N. Wilcox Trust Grants, 1768
Gamble Foundation Grants, 1769
Gebbie Foundation Grants, 1777
GenCorp Foundation Grants, 1779
General Motors Foundation Grants, 1783
George and Ruth Bradford Foundation Grants, 1789
George A Ohl Jr. Foundation Grants, 1791
George Gund Foundation Grants, 1796
George H.C. Ensworth Memorial Fund Grants, 1797
Georgia Power Foundation Grants, 1806
Gibson Foundation Grants, 1818
Giving Sum Annual Grant, 1825
GNOF Bayou Communities Grants, 1835
GNOF Exxon-Mobil Grants, 1836
Good Samaritan Inc Grants, 1855
Grace and Franklin Bernsen Foundation Grants, 1858

724 / Environmental Programs SUBJECT INDEX

Grand Rapids Area Community Foundation Wyoming Grants, 1867
Grand Rapids Community Foundation Grants, 1869
Grand Rapids Community Foundation Ionia County Grants, 1870
Grand Rapids Community Foundation Lowell Area Fund Grants, 1872
Grand Rapids Community Foundation Southeast Ottawa Grants, 1873
Grand Rapids Community Fndn Sparta Grants, 1875
Greater Green Bay Community Fndn Grants, 1881
Greater Saint Louis Community Fndn Grants, 1884
Greater Tacoma Community Foundation Grants, 1886
Greater Worcester Community Foundation Discretionary Grants, 1888
Green Diamond Charitable Contributions, 1891
Greenfield Foundation of Maine Grants, 1893
Guido A. & Elizabeth K. H. Binda Fndn Grants, 1913
H.A. & Mary K. Chapman Trust Grants, 1921
Hagedorn Fund Grants, 1934
Harley Davidson Foundation Grants, 1947
Harrison County Community Foundation Grants, 1957
Harrison County Community Foundation Signature Grants, 1958
Hawaiian Electric Industries Charitable Foundation Grants, 1975
Hawaii Community Foundation West Hawaii Fund Grants, 1978
Heineman Foundation for Research, Education, Charitable and Scientific Purposes, 1990
Helen S. Boylan Foundation Grants, 1997
Hershey Company Grants, 2013
High Meadow Foundation Grants, 2027
Hillman Foundation Grants, 2032
Hoblitzelle Foundation Grants, 2038
Holland/Zeeland Community Fndn Grants, 2041
Homer Foundation Grants, 2044
Honda of America Manufacturing Fndn Grants, 2045
Horace A. Kimball and S. Ella Kimball Foundation Grants, 2046
Houston Endowment Grants, 2053
Hudson Webber Foundation Grants, 2072
Huntington County Community Foundation Make a Difference Grants, 2083
Idaho Power Company Corporate Contributions, 2115
Illinois Tool Works Foundation Grants, 2147
Impact 100 Grants, 2148
Inasmuch Foundation Grants, 2149
Jackson County Community Foundation Unrestricted Grants, 2180
Jacobs Family Village Neighborhoods Grants, 2184
James & Abigail Campbell Family Fndn Grants, 2185
James A. and Faith Knight Foundation Grants, 2186
James Ford Bell Foundation Grants, 2187
James M. Cox Foundation of Georgia Grants, 2196
Jane's Trust Grants, 2206
Janirve Foundation Grants, 2209
Jennings County Community Foundation Grants, 2220
Jessica Stevens Community Foundation Grants, 2224
Jessie B. Cox Charitable Trust Grants, 2225
John Merck Fund Grants, 2245
John W. and Anna H. Hanes Foundation Grants, 2261
Katharine Matthies Foundation Grants, 2291
Kavli Foundation Research Grants, 2296
Kelvin and Eleanor Smith Foundation Grants, 2297
Kenneth T. & Eileen L. Norris Fndn Grants, 2300
Kettering Family Foundation Grants, 2304
Kettering Fund Grants, 2305
Knox County Community Foundation Grants, 2313
Kosciusko County Community Fndn Grants, 2319
Lands' End Corporate Giving Program, 2344
Lawrence Foundation Grants, 2347
Lawrence S. Huntington Fund Grants, 2348
Libra Foundation Grants, 2363
Lisa and Douglas Goldman Fund Grants, 2369
Long Island Community Foundation Grants, 2375
Lotus 88 Fnd for Women & Children Grants, 2376
Lubrizol Foundation Grants, 2384
M.E. Raker Foundation Grants, 2404
Macquarie Bank Foundation Grants, 2410
Margaret T. Morris Foundation Grants, 2429
Marie C. and Joseph C. Wilson Foundation Rochester Small Grants, 2431
Marie H. Bechtel Charitable Remainder Uni-Trust Grants, 2432
Marshall County Community Foundation Grants, 2439
Mary K. Chapman Foundation Grants, 2443
Mary Owen Borden Foundation Grants, 2446
Maxon Charitable Foundation Grants, 2458
McConnell Foundation Grants, 2492
McGraw-Hill Companies Community Grants, 2494
McInerny Foundation Grants, 2495
McKesson Foundation Grants, 2496
McLean Contributionship Grants, 2497
Mead Witter Foundation Grants, 2507
Mervin Bovaird Foundation Grants, 2523
Meyer Memorial Trust Special Grants, 2538
Miami County Community Foundation - Operation Round Up Grants, 2543
Mimi and Peter Haas Fund Grants, 2557
Morris & Gwendolyn Cafritz Fndn Grants, 2594
Natalie W. Furniss Charitable Trust Grants, 2626
Nathan Cummings Foundation Grants, 2627
Nelda C. and H.J. Lutcher Stark Fndn Grants, 2672
Nina Mason Pulliam Charitable Trust Grants, 2921
Norfolk Southern Foundation Grants, 2982
Norman Foundation Grants, 2983
North Carolina Community Foundation Grants, 2990
Northrop Grumman Corporation Grants, 2997
Ogden Codman Trust Grants, 3066
Oregon Community Fndn Community Grants, 3082
Oscar Rennebohm Foundation Grants, 3085
Owen County Community Foundation Grants, 3100
Patrick and Anna M. Cudahy Fund Grants, 3120
Peacock Foundation Grants, 3133
Perkins Charitable Foundation Grants, 3140
Perry County Community Foundation Grants, 3142
Philadelphia Foundation Organizational Effectiveness Grants, 3189
Piedmont Natural Gas Corporate and Charitable Contributions, 3220
Pinellas County Grants, 3223
Pittsburgh Foundation Community Fund Grants, 3230
Posey County Community Foundation Grants, 3252
Powell Foundation Grants, 3255
Prince Charitable Trusts Chicago Grants, 3270
Prince Charitable Trusts DC Grants, 3271
Princeton Area Community Foundation Greater Mercer Grants, 3272
Principal Financial Group Foundation Grants, 3273
Procter and Gamble Fund Grants, 3274
Pulaski County Community Foundation Grants, 3279
Rajiv Gandhi Foundation Grants, 3302
Rathmann Family Foundation Grants, 3309
Rayonier Foundation Grants, 3311
RCF General Community Grants, 3312
Reinberger Foundation Grants, 3357
Rhode Island Foundation Grants, 3362
Rice Foundation Grants, 3363
Richard King Mellon Foundation Grants, 3369
Riley Foundation Grants, 3374
Ripley County Community Foundation Grants, 3376
Ripley County Community Foundation Small Project Grants, 3377
Robbins Charitable Foundation Grants, 3378
Robert R. Meyer Foundation Grants, 3387
Robert W. Woodruff Foundation Grants, 3389
Robins Foundation Grants, 3390
Rochester Area Community Foundation Grants, 3391
Ruth Anderson Foundation Grants, 3425
Ruth and Vernon Taylor Foundation Grants, 3426
S.H. Cowell Foundation Grants, 3447
S. Livingston Mather Charitable Trust Grants, 3448
S. Mark Taper Foundation Grants, 3449
Salem Foundation Grants, 3457
Samuel S. Johnson Foundation Grants, 3468
San Diego Foundation for Change Grants, 3470
San Diego Women's Foundation Grants, 3474
San Francisco Fndn Community Health Grants, 3478
San Juan Island Community Foundation Grants, 3480
Sasco Foundation Grants, 3487
Schering-Plough Foundation Community Initiatives Grants, 3491
Scott County Community Foundation Grants, 3500
Seabury Foundation Grants, 3501
Seagate Tech Corp Capacity to Care Grants, 3502
Seward Community Foundation Grants, 3514
Seward Community Foundation Mini-Grants, 3515
Shaw's Supermarkets Donations, 3542
Singing for Change Foundation Grants, 3567
Sioux Falls Area Community Foundation Spot Grants (Unrestricted), 3570
Sonora Area Foundation Competitive Grants, 3622
Sony Corporation of America Grants, 3623
Southwest Florida Community Foundation Competitive Grants, 3627
Southwest Gas Corporation Foundation Grants, 3628
Spencer County Community Foundation Grants, 3632
Stackpole-Hall Foundation Grants, 3643
Steuben County Community Foundation Grants, 3654
Stewart Huston Charitable Trust Grants, 3655
Strowd Roses Grants, 3664
Summit Foundation Grants, 3674
Sunflower Foundation Walking Trails Grants, 3678
Susan Vaughan Foundation Grants, 3695
Tellabs Foundation Grants, 3706
Theodore Edson Parker Foundation Grants, 3715
Thomas Sill Foundation Grants, 3724
Thompson Charitable Foundation Grants, 3727
Toyota Motor Manufacturing of Alabama Grants, 3743
Toyota Motor Manuf of Mississippi Grants, 3746
Toyota Motor Manuf of West Virginia Grants, 3748
Triangle Community Foundation Donor-Advised Grants, 3757
Trull Foundation Grants, 3761
Turner Foundation Grants, 3763
U.S. Cellular Corporation Grants, 3766
Union Bank, N.A. Corporate Sponsorships and Donations, 3787
Union Benevolent Association Grants, 3789
United Technologies Corporation Grants, 3803
Unity Foundation Of LaPorte County Grants, 3804
USAID Land Use Change and Disease Emergence Grants, 3818
USDA Organic Agriculture Research Grants, 3830
Vancouver Foundation Grants and Community Initiatives Program, 3849
Vermont Community Foundation Grants, 3865
Virginia W. Kettering Foundation Grants, 3878
W. C. Griffith Foundation Grants, 3882
Wallace Alexander Gerbode Foundation Grants, 3896
Walmart Foundation National Giving Grants, 3898
Walter L. Gross III Family Foundation Grants, 3901
Warrick County Community Foundation Grants, 3904
Washington County Community Fndn Grants, 3906
Washington County Community Foundation Youth Grants, 3907
Washington Gas Charitable Giving, 3908
Wege Foundation Grants, 3912
Weyerhaeuser Company Foundation Grants, 3925
William G. & Helen C. Hoffman Fndn Grants, 3950
William Ray & Ruth E. Collins Fndn Grants, 3956
Winifred & Harry B. Allen Foundation Grants, 3962
Winston-Salem Fndn Stokes County Grants, 3967
Z. Smith Reynolds Foundation Small Grants, 3982

Environmental Protection
Alpha Natural Resources Corporate Giving, 450
Caesars Foundation Grants, 995
Collective Brands Foundation Grants, 1208
IIE Freeman Foundation Indonesia Internships, 2131
Lucy Downing Nisbet Charitable Fund Grants, 2387
Piedmont Natural Gas Corporate and Charitable Contributions, 3220
RCF General Community Grants, 3312
Sioux Falls Area Community Foundation Community Fund Grants (Unrestricted), 3568
Sioux Falls Area Community Foundation Spot Grants (Unrestricted), 3570
USDA Organic Agriculture Research Grants, 3830

SUBJECT INDEX

Environmental Research
Acumen Global Fellowships, 285

Environmental Services
Baxter International Corporate Giving Grants, 841
Bill and Melinda Gates Foundation Emergency Response Grants, 892
Carl C. Icahn Foundation Grants, 1023
Elkhart County Community Foundation Fund for Elkhart County, 1572

Environmental Studies
American-Scandinavian Foundation Visiting Lectureship Grants, 502
Annie Sinclair Knudsen Memorial Fund/Kaua'i Community Grants Program, 597
Canada-U.S. Fulbright New Century Scholars, 1008
Canada-U.S. Fulbright Senior Specialists Grants, 1009
CDC Collegiate Leaders in Environmental Health Internships, 1048
Collins Foundation Grants, 1210
Fred Baldwin Memorial Foundation Grants, 1736
Fulbright Alumni Initiatives Awards, 1749
Fulbright Distinguished Chairs Awards, 1750
Fulbright New Century Scholars Grants, 1753
Fulbright Specialists Program Grants, 1754
Fulbright Scholars in Europe and Eurasia, 1755
Fulbright Traditional Scholar Program in Sub-Saharan Africa, 1756
Fulbright Traditional Scholar Program in the East Asia/Pacific Region, 1757
Fulbright Traditional Scholar Program in the Near East and North Africa Region, 1758
Fulbright Traditional Scholar Program in the South and Central Asia Region, 1759
Fulbright Traditional Scholar Program in the Western Hemisphere, 1760
IIE Nancy Petry Scholarship, 2140
Lawrence S. Huntington Fund Grants, 2348
Norwin S. & Elizabeth N. Bean Fndn Grants, 3001
NSF-NIST Interaction in Chemistry, Materials Research, Molecular Biosciences, Bioengineering, and Chemical Engineering, 3004
Society for the Arts in Healthcare Environmental Arts Research Grant, 3603
Southbury Community Trust Fund, 3625
Trull Foundation Grants, 3761
U.S. Department of Education United States-Russia Program: Improving Research and Educational Activities in Higher Education, 3773
Weingart Foundation Grants, 3913

Enzymes/Enzymology
NHLBI Career Transition Awards, 2694
NIDDK Development of Assays for High-Throughput Drug Screening Grants, 2808
NSF Biomolecular Systems Cluster Grants, 3010

Epidemiology
AACR-American Cancer Society Award for Research Excellence in Epidemiology and Prevention, 87
AACR Basic Cancer Research Fellowships, 105
AACR Career Development Awards for Pediatric Cancer Research, 107
AACR Judah Folkman Career Development Award for Anti-Angiogenesis Research, 116
AACR Outstanding Investigator Award for Breast Cancer Research, 121
ACS Mentored Research Scholar Grant in Applied and Clinical Research, 264
ACSM Paffenbarger-Blair Fund for Epidemiological Research on Physical Activity Grants, 273
AHAF Alzheimer's Disease Research Grants, 349
AMSSM Foundation Research Grants, 577
APHL Emerging Infectious Diseases Fellowships, 636
ASM-PAHO Infectious Diseases Epidemiology and Surveillance Fellowships, 714
Bristol-Myers Squibb Foundation Fellowships, 956
Bristol-Myers Squibb Foundation Global HIV/AIDS Initiative Grants, 957
CDC David J. Sencer Museum Teacher Professional Development Workshops, 1053
CDC Disease Detective Camp, 1054
CDC Epidemic Intell Service Training Grants, 1055
CDC Epidemiology Elective Rotation, 1056
CDC Experience Epidemiology Fellowships, 1059
CDC Foundation Atlanta International Health Fellowships, 1060
CDC Preventive Med Residency & Fellowship, 1064
CDC Public Health Informatics Fellowships, 1068
CSTE CDC/CSTE Applied Epidemiology Fellowships, 1367
IARC Expertise Transfer Fellowship, 2090
IARC Postdoctoral Fellowships for Training in Cancer Research, 2091
IARC Visiting Award for Senior Scientists, 2092
IIE Hewlett Fnd/IIE Dissertation Fellowship, 2132
James S. McDonnell Foundation Complex Systems Collaborative Activity Awards, 2202
Landon Foundation-AACR Innovator Award for Cancer Prevention Research, 2342
Landon Foundation-AACR Innovator Award for International Collaboration, 2343
Morehouse PHSI Project Imhotep Internships, 2592
NCI Basic Cancer Research in Cancer Health Disparities Grants, 2647
NCI Cancer Education and Career Development Program, 2648
NCI Stages of Breast Development: Normal to Metastatic Disease Grants, 2655
NEI Clinical Study Planning Grant, 2663
NEI Clinical Vision Research Devel Award, 2664
NHLBI National Research Service Award Programs in Cardiovascular Epidemiology and Biostatistics, 2714
NIAMS Pilot and Feasibility Clinical Research Grants in Arthritis, Musculoskeletal & Skin Diseases, 2774
NIDA Mentored Clinical Scientist Research Career Development Awards, 2793
NIDDK Type 2 Diabetes in the Pediatric Population Research Grants, 2852
NYAM Mary and David Hoar Fellowship, 3045
Pancreatic Cancer Action Network Fellowship, 3113
Pancreatic Cancer Action Network-AACR Pathway to Leadership Grants, 3114
Pfizer Anti Infective Research EU Grants, 3164
Pfizer ASPIRE EU Antifungal Research Awards, 3165
Pfizer ASPIRE EU Emerging Mechanisms of Resistance Antibacterial Research Awards, 3167
Pfizer ASPIRE North America Rheumatology Research Awards, 3172
Pfizer Global Investigator Research Grants, 3178
Pfizer Inflammation Competitive Research Awards (UK), 3181
Simone and Cino del Duca Grand Prix Awards, 3564
Susan G. Komen Breast Cancer Foundation Challege Grants: Investigator Initiated Research, 3690
Susan G. Komen Breast Cancer Foundation Challege Grants: Career Catalyst Research, 3691
UICC American Cancer Society International Fellowships for Beginning Investigators, 3778
US CRDF Leishmaniasis: Collaborative Research Opportunities in N Africa & Middle East, 3828

Epilepsy
AES-Grass Young Investigator Travel Awards, 317
AES Epilepsy Research Recognition Awards, 318
AES JK Penry Excellence in Epilepsy Care Award, 319
AES Research Infrastructure Awards, 321
AES Research Initiative Awards, 322
AES Robert S. Morison Fellowship, 323
AES Service Award, 324
AES Susan S. Spencer Clinical Research Epilepsy Training Fellowship, 325
AES William G. Lennox Award, 326
ASPET Epilepsy Research Award for Outstanding Contributions to the Pharmacology of Antiepileptic Drugs, 772
F.M. Kirby Foundation Grants, 1624
Grass Foundation Marine Biological Laboratory Fellowships, 1878
NIH Research on Sleep and Sleep Disorders, 2898
Savoy Foundation Post-Doctoral and Clinical Research Fellowships, 3488
Savoy Foundation Research Grants, 3489
Savoy Foundation Studentships, 3490

Episcopal Church
Barra Foundation Project Grants, 834
Booth-Bricker Fund Grants, 933
Effie and Wofford Cain Foundation Grants, 1563
Herbert A. & Adrian W. Woods Fndn Grants, 2007

Equal Opportunity
Air Products and Chemicals Grants, 399
Allstate Corporate Giving Grants, 446
Allstate Corp Hometown Commitment Grants, 447
Benton Community Fndn Cookie Jar Grant, 877
DaimlerChrysler Corporation Fund Grants, 1412
Frances and John L. Loeb Family Fund Grants, 1716
Singing for Change Foundation Grants, 3567

Equipment/Instrumentation
AMA Foundation Worldscopes Program, 495
Ben B. Cheney Foundation Grants, 875
Biogen Foundation General Donations, 899
Boeing Company Contributions Grants, 930
Campbell Soup Foundation Grants, 1007
Chatlos Foundation Grants Program, 1128
Collins Foundation Grants, 1210
Community Foundation for Greater Atlanta Metropolitan Atlanta An Extra Wish Grants, 1243
Community Foundation of Riverside and San Bernardino County James Bernard and Mildred Jordan Tucker Grants, 1305
Community Foundation of the Verdugos Educational Endowment Fund Grants, 1314
Different Needz Foundation Grants, 1478
Dr. Scholl Foundation Grants Program, 1511
E.L. Wiegand Foundation Grants, 1538
Fritz B. Burns Foundation Grants, 1747
Gladys Brooks Foundation Grants, 1826
GNOF Albert N. & Hattie M. McClure Grants, 1834
Hamilton Company Syringe Product Grant, 1937
Hampton Roads Community Foundation Faith Community Nursing Grants, 1940
Hillcrest Foundation Grants, 2031
Homer Foundation Grants, 2044
IDPH Hosptial Capital Investment Grants, 2119
Jack H. & William M. Light Trust Grants, 2179
Janson Foundation Grants, 2210
Jennings County Community Foundation Grants, 2220
Katharine Matthies Foundation Grants, 2291
Ludwick Family Foundation Grants, 2389
Marion I. and Henry J. Knott Foundation Discretionary Grants, 2435
Marion I. and Henry J. Knott Foundation Standard Grants, 2436
Norcliffe Foundation Grants, 2979
Norwin S. & Elizabeth N. Bean Fndn Grants, 3001
NSF Doctoral Dissertation Improvement Grants in the Directorate for Biological Sciences (DDIG), 3013
NSF Instrument Development for Bio Research, 3017
Potts Memorial Foundation Grants, 3254
Priddy Foundation Capital Grants, 3263
Ralph M. Parsons Foundation Grants, 3304
Robert Leet & Clara Guthrie Patterson Grants, 3383
RSC Thomas W. Eadie Medal, 3423
Saint Luke's Health Initiatives Grants, 3455
Schering-Plough Foundation Health Grants, 3492
Seattle Foundation Medical Funds Grants, 3508
South Madison Community Foundation Grants, 3626
Square D Foundation Grants, 3634
St. Joseph Community Health Foundation Burn Care and Prevention Grants, 3639
St. Joseph Community Health Foundation Improving Healthcare Access Grants, 3640
St. Joseph Community Health Foundation Pfeiffer Fund Grants, 3641
Tri-State Community Twenty-first Century Endowment Fund Grants, 3756

726 / Equipment/Instrumentation

VDH Rescue Squad Assistance Fund Grants, 3856
Willard & Pat Walker Charitable Fndn Grants, 3943

Ergonomics
BMW of North America Charitable Contributions, 926

Erythropoiesis
NIDDK Erythropoiesis: Components and Mechanisms Grants, 2817

Ethics
AAAS Award for Scie Freedom & Responsibility, 36
Affymetrix Corporate Contributions Grants, 339
Alice Tweed Tuohy Foundation Grants Program, 437
Cargill Citizenship Fund Corporate Giving, 1018
Community Foundation of Greater Flint Grants, 1273
Elizabeth Morse Genius Charitable Trust Grants, 1569
Graco Foundation Grants, 1860
HRAMF Charles A. King Trust Postdoctoral Research Fellowships, 2056
James Hervey Johnson Charitable Educational Trust Grants, 2191
Marion Gardner Jackson Charitable Trust Grants, 2434
NIH Research On Ethical Issues In Human Subjects Research, 2897
Richard Davoud Donchian Foundation Grants, 3367
RSC Abbyann D. Lynch Medal in Bioethics, 3419

Ethnology
Australasian Institute of Judicial Administration Seed Funding Grants, 806

Etiology
AACR Gertrude Elion Cancer Research Award, 113
Alzheimer's Association Conference Grants, 463
Alzheimer's Association Development of New Cognitive and Functional Instruments Grants, 464
Alzheimer's Association Everyday Technologies for Alzheimer Care Grants, 465
Alzheimer's Association Mentored New Investigator Research Grants to Promote Diversity, 467
Alzheimer's Association Neuronal Hyper Excitability and Seizures in Alzheimer's Disease Grants, 468
Alzheimer's Association New Investigator Research Grants to Promote Diversity, 470
Alzheimer's Association U.S.-U.K. Young Investigator Exchange Fellowships, 471
Alzheimer's Association Zenith Fellows Awards, 472
IARC Expertise Transfer Fellowship, 2090
IARC Postdoctoral Fellowships for Training in Cancer Research, 2091
IARC Visiting Award for Senior Scientists, 2092
NIA Alzheimer's Disease Core Centers Grants, 2748
NYAM Edward N. Gibbs Memorial Lecture and Award in Nephrology, 3041
Penny Severns Breast, Cervical and Ovarian Cancer Research Grants, 3136

Europe
Charles Delmar Foundation Grants, 1115
Fulbright Scholars in Europe and Eurasia, 1755

Europe, Central
ASM-UNESCO Leadership Grant for International Educators, 715
UICC American Cancer Society International Fellowships for Beginning Investigators, 3778

Europe, Eastern
ASM-UNESCO Leadership Grant for International Educators, 715
Fulbright Scholars in Europe and Eurasia, 1755
UICC American Cancer Society International Fellowships for Beginning Investigators, 3778

Evolutionary Biology
Alfred P. Sloan Foundation Research Fellowships, 432
MBL Albert & Ellen Grass Fellowships, 2463
MBL Stephen W. Kuffler Summer Fellowships, 2487

NSF Doctoral Dissertation Improvement Grants in the Directorate for Biological Sciences (DDIG), 3013
NSF Genes and Genome Systems Cluster Grants, 3015
NSF Systematic Biology and Biodiversity Inventories Grants, 3024

Exchange Programs, Student
IBRO-PERC InEurope Short Stay Grants, 2097

Exercise
ACSM-GSSI Young Scholar Travel Award, 257
ACSM Carl V. Gisolfi Memorial Fund Grant, 259
ACSM Chas M. Tipton Student Research Award, 260
ACSM Coca-Cola Company Doctoral Student Grant on Behavior Research, 261
ACSM Foundation Research Endowment Grants, 267
ACSM International Student Award, 268
ACSM International Student Awards, 269
ACSM NASA Space Physiology Research Grants, 271
ACSM Oded Bar-Or Int'l Scholar Awards, 272
ACSM Steven M. Horvath Travel Award, 274
Air Products and Chemicals Grants, 399
BCBSNC Foundation Grants, 860
Bikes Belong Foundation Paul David Clark Bicycling Safety Grants, 890
Bright Promises Foundation Grants, 952
Elizabeth Morse Genius Charitable Trust Grants, 1569
Gebbie Foundation Grants, 1777
General Mills Champs for Healthy Kids Grants, 1781
GNOF IMPACT Kahn-Oppenheim Grants, 1842
Healthcare Fndn for Orange County Grants, 1981
Linford and Mildred White Charitable Grants, 2368
NCI Improving Diet and Physical Activity Assessment Grants, 2652
Paso del Norte Health Foundation Grants, 3119
PepsiCo Foundation Grants, 3137
United Methodist Child Mental Health Grants, 3800

Exhibitions, Collections, Performances
Atherton Family Foundation Grants, 794
Christensen Fund Regional Grants, 1149
Crystelle Waggoner Charitable Trust Grants, 1364
Daniel & Nanna Stern Family Fndn Grants, 1422
Greater Worcester Community Foundation Jeppson Memorial Fund for Brookfield Grants, 1889
MLA Louise Darling Medal for Distinguished Achievement in Collection Development in the Health Sciences, 2569
Principal Financial Group Foundation Grants, 3273
Seaver Institute Grants, 3509

Eye Diseases
1st Touch Foundation Grants, 3
1976 Foundation Grants, 12
AHAF Macular Degeneration Research Grants, 350
Alice C. A. Sibley Fund Grants, 435
Austin S. Nelson Foundation Grants, 805
CNIB Baker Applied Research Fund Grants, 1193
CNIB Baker Fellowships, 1194
CNIB Baker New Researcher Fund Grants, 1195
CNIB Barbara Tuck MacPhee Award, 1196
CNIB Canada Glaucoma Clinical Research Council Grants, 1197
CNIB Chanchlani Global Vision Research Award, 1198
CNIB E. (Ben) & Mary Hochhausen Access Technology Research Grants, 1199
CNIB Ross Purse Doctoral Fellowships, 1200
Columbus Foundation Competitive Grants, 1217
Donald and Sylvia Robinson Family Foundation Grants, 1486
E.L. Wiegand Foundation Grants, 1538
Eye-Bank for Sight Restoration and Fight for Sight Summer Student Research Fellowship, 1622
Fairlawn Foundation Grants, 1626
Fight for Sight-Streilein Foundation for Ocular Immunology Research Award, 1662
Fight for Sight Grants-in-Aid, 1663
Fight for Sight Post-Doctoral Awards, 1664
Fight for Sight Summer Student Fellowships, 1665
George Gund Foundation Grants, 1796

SUBJECT INDEX

Glaucoma Foundation Grants, 1827
Glaucoma Research Pilot Project Grants, 1828
GNOF IMPACT Gulf States Eye Surgery Fund, 1840
HRAMF Thome Foundation Awards in Age-Related Macular Degeneration Research, 2065
Margaret Wiegand Trust Grants, 2430
MBL Plum Fndn John E. Dowling Fellowships, 2484
NEI Innovative Patient Outreach Programs And Ocular Screening Technologies To Improve Detection Of Diabetic Retinopathy Grants, 2665
NEI Scholars Program, 2669
NEI Translational Research Program On Therapy For Visual Disorders, 2670
OneSight Research Foundation Block Grants, 3076
Sara Elizabeth O'Brien Trust Grants, 3483
Streilein Foundation for Ocular Immunology Visiting Professorships, 3660
Young Ambassador Scholarship In Memory of Christopher Nordquist, 3981

Facility Support
GNOF Albert N. & Hattie M. McClure Grants, 1834
Michael Reese Health Trust Core Grants, 2544
Western New York Foundation Grants, 3922

Faculty Development
AMDA Foundation Quality Improvement Award, 499
ANA Distinguished Neurology Teacher Award, 580
Arnold and Mabel Beckman Foundation Young Investigators Grants, 666
BWF Institutional Program Unifying Population and Laboratory Based Sciences Grants, 987
Carpenter Foundation Grants, 1035
ERC Starting Grants, 1599
FAMRI Young Clinical Scientist Awards, 1629
HAF Technical Assistance Program Grants, 1933
HHMI/EMBO Start-up Grants for C Europe, 2016
MBL Ann E. Kammer Summer Fellowship, 2464
MBL Burr & Susie Steinbach Fellowship, 2467
MBL E.E. Just Summer Fellowship for Minority Scientists, 2468
MBL Eugene and Millicent Bell Fellowships, 2470
MBL Evelyn and Melvin Spiegal Fellowship, 2471
MBL Frank R. Lillie Summer Fellowship, 2472
MBL Gruss Lipper Family Foundation Summer Fellowship, 2475
MBL Herbert W. Rand Summer Fellowship, 2477
MBL James E. and Faith Miller Memorial Summer Fellowship, 2478
MBL Lucy B. Lemann Summer Fellowship, 2481
MLA T. Mark Hodges Int'l Service Award, 2575
NHLBI Mentored Career Award For Faculty At Minority Institutions, 2707
PhRMA Foundation Health Outcomes Post Doctoral Fellowships, 3202
PhRMA Foundation Pharmaceutics Research Starter Grants, 3213
RCPSC Continuing Proffesional Dev Grants, 3319
Richard M. Fairbanks Foundation Grants, 3370
RWJF Harold Amos Medical Faculty Development Research Grants, 3434
RWJF Harold Amos Medical Faculty Development Scholars, 3435
RWJF Nurse Faculty Scholars, 3441
SfN Next Generation Award, 3531
SNM Molecular Imaging Research Grant For Junior Medical Faculty, 3583
SNM Postdoc Molecular Imaging Scholar Grants, 3587
Western New York Foundation Grants, 3922

Faculty Support
FAMRI Young Clinical Scientist Awards, 1629
NHLBI Ruth L. Kirschstein National Research Service Awards for Individual Predoctoral Fellowships to Promote Diversity in Health-Related Research, 2730

Familial Abuse
Adaptec Foundation Grants, 304
Ahmanson Foundation Grants, 355
Austin S. Nelson Foundation Grants, 805

SUBJECT INDEX

California Endowment Innovative Ideas Challenge Grants, 1000
Elkhart County Community Foundation Fund for Elkhart County, 1572
Herbert A. & Adrian W. Woods Fndn Grants, 2007
John W. Speas and Effie E. Speas Memorial Trust Grants, 2264
Ms. Fndn for Women Ending Violence Grants, 2596
Quantum Foundation Grants, 3290
WHO Foundation Volunteer Service Grants, 3940

Family
3M Company Fndn Community Giving Grants, 6
ACF Native American Social and Economic Development Strategies Grants, 231
Achelis Foundation Grants, 234
ACS Pilot and Exploratory Projects in Palliative Care of Cancer Patients and Their Families Grants, 276
Adaptec Foundation Grants, 304
AEGON Transamerica Foundation Health and Welfare Grants, 316
Agnes M. Lindsay Trust Grants, 346
Alcatel-Lucent Technologies Foundation Grants, 413
Alexander Eastman Foundation Grants, 420
Alfred and Tillie Shemanski Testamentary Trust Grants, 428
Alpha Natural Resources Corporate Giving, 450
Alvin and Fanny Blaustein Thalheimer Foundation Baltimore Communal Grants, 459
Amelia Sillman Rockwell and Carlos Perry Rockwell Charities Fund Grants, 501
Amerigroup Foundation Grants, 565
Andrew Family Foundation Grants, 588
Anne J. Caudal Foundation Grants, 595
Anschutz Family Foundation Grants, 602
AptarGroup Foundation Grants, 647
Arizona Cardinals Grants, 655
Arizona Diamondbacks Charities Grants, 657
AT&T Foundation Civic and Community Service Program Grants, 793
Atlanta Foundation Grants, 796
Auburn Foundation Grants, 798
Barberton Community Foundation Grants, 829
Barker Welfare Foundation Grants, 831
Battle Creek Community Foundation Grants, 839
Benton Community Foundation Grants, 878
Berrien Community Foundation Grants, 884
Bindley Family Foundation Grants, 896
BJ's Charitable Foundation Grants, 904
Blanche and Julian Robertson Family Foundation Grants, 911
Blue Cross Blue Shield of Minnesota Foundation - Health Equity: Building Health Equity Together Grants, 914
Blue River Community Foundation Grants, 922
Blumenthal Foundation Grants, 925
Bodenwein Public Benevolent Foundation Grants, 928
Bodman Foundation Grants, 929
Booth-Bricker Fund Grants, 933
Borkee-Hagley Foundation Grants, 934
Boston Foundation Grants, 936
Boston Globe Foundation Grants, 937
Bradley-Turner Foundation Grants, 944
Brookdale Foundation Leadership in Aging Fellowships, 965
Brookdale Fnd National Group Respite Grants, 966
Brown County Community Foundation Grants, 970
Browning-Kimball Foundation Grants, 971
Burlington Industries Foundation Grants, 977
Bush Fndn Health & Human Services Grants, 979
Cailloux Foundation Grants, 996
California Endowment Innovative Ideas Challenge Grants, 1000
Campbell Soup Foundation Grants, 1007
Cargill Citizenship Fund Corporate Giving, 1018
Carlisle Foundation Grants, 1025
Carolyn Foundation Grants, 1034
Carpenter Foundation Grants, 1035
Catherine Kennedy Home Foundation Grants, 1043
Cemala Foundation Grants, 1077

Central Carolina Community Foundation Community Impact Grants, 1079
Central Okanagan Foundation Grants, 1080
CFFVR Basic Needs Giving Partnership Grants, 1094
CFFVR Myra M. and Robert L. Vandehey Foundation Grants, 1100
CFFVR Waupaca Area Community Foundation Grants, 1105
CharityWorks Grants, 1113
Children's Trust Fund of Oregon Fndn Grants, 1139
CICF Indianapolis Fndn Community Grants, 1156
Citizens Bank Mid-Atlantic Charitable Foundation Grants, 1162
Claude Worthington Benedum Fndn Grants, 1178
Columbus Foundation Paul G. Duke Grants, 1221
Columbus Foundation Traditional Grants, 1223
Community Fndn for Greater Buffalo Grants, 1247
Community Foundation for the National Capital Region Community Leadership Grants, 1260
Community Foundation of Bartholomew County Heritage Fund Grants, 1262
Community Foundation of Bartholomew County James A. Henderson Award for Fundraising, 1263
Community Fndn of Central Illinois Grants, 1267
Community Foundation of Eastern Connecticut General Southeast Grants, 1269
Community Foundation of Greater Fort Wayne - Community Endowment and Clarke Endowment Grants, 1274
Community Foundation of Greenville Community Enrichment Grants, 1284
Community Foundation of Greenville Hollingsworth Funds Program/Project Grants, 1285
Community Foundation of Riverside & San Bernardino County Impact Grants, 1303
Community Foundation Serving Riverside and San Bernardino Counties Impact Grants, 1319
Connecticut Community Foundation Grants, 1324
Constantin Foundation Grants, 1341
Constellation Energy Corporate Grants, 1342
Crail-Johnson Foundation Grants, 1355
Crane Fund Grants, 1358
Cresap Family Foundation Grants, 1359
CSRA Community Foundation Grants, 1366
CSX Corporate Contributions Grants, 1368
D.F. Halton Foundation Grants, 1405
Dairy Queen Corporate Contributions Grants, 1413
Dammann Fund Grants, 1418
Daywood Foundation Grants, 1438
Dennis & Phyllis Washington Fndn Grants, 1458
Dept of Ed Special Education--Technical Assistance and Dissemination to Improve Services and Results for Children with Disabilities, 1467
DHHS Adolescent Family Life Demonstration Projects, 1472
Dominion Foundation Grants, 1485
Donald and Sylvia Robinson Family Foundation Grants, 1486
Edina Realty Foundation Grants, 1550
Educational Foundation of America Grants, 1553
Eisner Foundation Grants, 1564
El Paso Community Foundation Grants, 1579
El Pomar Foundation Grants, 1582
Elsie Lee Garthwaite Memorial Fndn Grants, 1584
Emma B. Howe Memorial Foundation Grants, 1593
Entergy Corporation Micro Grants, 1596
Entergy Corporation Open Grants for Healthy Families, 1597
EPA Children's Health Protection Grants, 1598
Evanston Community Foundation Grants, 1615
Express Scripts Foundation Grants, 1620
Fairfield County Community Foundation Grants, 1625
FAR Fund Grants, 1631
Farmers Insurance Corporate Giving Grants, 1634
Faye McBeath Foundation Grants, 1635
FCD New American Children Grants, 1637
Fidelity Foundation Grants, 1659
Fifth Third Foundation Grants, 1661
Fisher House Fnd Newman's Own Awards, 1671

Florida BRAIVE Fund of Dade Community Foundation, 1678
Floyd A. & Kathleen C. Cailloux Fnd Grants, 1681
Foellinger Foundation Grants, 1684
Foundation for Seacoast Health Grants, 1698
Foundation of CVPH Chelsea's Rainbow Grants, 1702
Fred C. & Katherine B. Andersen Grants, 1737
Fremont Area Community Fndn General Grants, 1746
G.N. Wilcox Trust Grants, 1768
Gebbie Foundation Grants, 1777
General Mills Foundation Grants, 1782
George and Ruth Bradford Foundation Grants, 1789
George Foundation Grants, 1795
George Gund Foundation Grants, 1796
George Kress Foundation Grants, 1799
George W. Wells Foundation Grants, 1804
Georgiana Goddard Eaton Memorial Grants, 1805
GNOF Stand Up For Our Children Grants, 1847
Golden Heart Community Foundation Grants, 1850
Grand Rapids Community Foundation Grants, 1869
Grand Rapids Community Foundation Southeast Ottawa Youth Fund Grants, 1874
Grand Rapids Community Foundation Sparta Youth Fund Grants, 1876
Greater Sitka Legacy Fund Grants, 1885
Gulf Coast Foundation of Community Operating Grants, 1916
Gulf Coast Foundation of Community Program Grants, 1917
H.J. Heinz Company Foundation Grants, 1923
Hagedorn Fund Grants, 1934
Harden Foundation Grants, 1945
Harold and Arlene Schnitzer CARE Foundation Grants, 1950
Harold R. Bechtel Charitable Remainder Uni-Trust Grants, 1953
Harry Frank Guggenheim Fnd Research Grants, 1964
Hartford Courant Foundation Grants, 1969
Hasbro Children's Fund Grants, 1973
Hawaiian Electric Industries Charitable Foundation Grants, 1975
Health Fndn of Greater Indianapolis Grants, 1984
Heckscher Foundation for Children Grants, 1988
Helen Bader Foundation Grants, 1992
Helen Steiner Rice Foundation Grants, 1998
Henrietta Tower Wurts Memorial Fndn Grants, 2002
Horizons Community Issues Grants, 2049
Howard and Bush Foundation Grants, 2054
Hugh J. Andersen Foundation Grants, 2074
Hutchinson Community Foundation Grants, 2085
Hut Foundation Grants, 2086
Hutton Foundation Grants, 2087
Illinois Tool Works Foundation Grants, 2147
Impact 100 Grants, 2148
J.C. Penney Company Grants, 2169
Jacob and Valeria Langeloth Foundation Grants, 2182
Jacob G. Schmidlapp Trust Grants, 2183
Jacobs Family Village Neighborhoods Grants, 2184
James & Abigail Campbell Family Fndn Grants, 2185
James A. and Faith Knight Foundation Grants, 2186
James M. Cox Foundation of Georgia Grants, 2196
James R. Dougherty Jr. Foundation Grants, 2198
Janirve Foundation Grants, 2209
Jay and Rose Phillips Family Foundation Grants, 2213
Jim Moran Foundation Grants, 2228
John P. McGovern Foundation Grants, 2247
Johnson & Johnson Corporate Contributions, 2253
Joseph Henry Edmondson Foundation Grants, 2270
Josiah Macy Jr. Foundation Grants, 2277
Kahuku Community Fund, 2286
Kansas Health Foundation Recognition Grants, 2289
Kenai Peninsula Foundation Grants, 2298
Ketchikan Community Foundation Grants, 2303
Kimball International-Habig Foundation Health and Human Services Grants, 2308
Lands' End Corporate Giving Program, 2344
Leon and Thea Koerner Foundation Grants, 2355
Leo Niessen Jr., Charitable Trust Grants, 2358
Lotus 88 Fnd for Women & Children Grants, 2376

728 / Family

Lucile Horton Howe and Mitchell B. Howe Foundation Grants, 2385
M.B. and Edna Zale Foundation Grants, 2401
Maggie Welby Foundation Grants, 2414
Margaret T. Morris Foundation Grants, 2429
Marie C. and Joseph C. Wilson Foundation Rochester Small Grants, 2431
Mary Owen Borden Foundation Grants, 2446
McCarthy Family Foundation Grants, 2490
McKesson Foundation Grants, 2496
Mervin Bovaird Foundation Grants, 2523
Mimi and Peter Haas Fund Grants, 2557
Minneapolis Foundation Community Grants, 2558
Morris & Gwendolyn Cafritz Fndn Grants, 2594
NAPNAP Fndn Grad Research Grant, 2610
Nationwide Insurance Foundation Grants, 2637
NIA Multicenter Study on Exceptional Survival in Families: The Long Life Family Study (LLFS) Grants, 2775
Nicholas H. Noyes Jr. Memorial Fndn Grants, 2791
NIDRR Field-Initiated Projects, 2853
Nina Mason Pulliam Charitable Trust Grants, 2921
Northwestern Mutual Foundation Grants, 2999
Nuffield Foundation Children & Families Grants, 3027
NYCT Girls and Young Women Grants, 3053
NYCT Substance Abuse Grants, 3057
Ohio County Community Foundation Grants, 3068
Ohio County Community Foundation Mini-Grants, 3069
Oleonda Jameson Trust Grants, 3071
Ordean Foundation Grants, 3080
Parkersburg Area Community Foundation Action Grants, 3116
Perpetual Trust for Charitable Giving Grants, 3141
Peter and Elizabeth C. Tower Foundation Mental Health Reference and Resource Materials Mini-Grants, 3147
Petersburg Community Foundation Grants, 3156
Philadelphia Foundation Organizational Effectiveness Grants, 3189
Phoenix Suns Charities Grants, 3201
Pinkerton Foundation Grants, 3224
Pinnacle Foundation Grants, 3225
Piper Jaffray Fndn Communities Giving Grants, 3227
Pittsburgh Foundation Community Fund Grants, 3230
Plough Foundation Grants, 3238
Polk Bros. Foundation Grants, 3243
Portland General Electric Foundation Grants, 3250
Posey Community Fndn Women's Fund Grants, 3251
Pride Foundation Grants, 3268
Prince Charitable Trusts DC Grants, 3271
Radcliffe Institute Carol K. Pforzheimer Student Fellowships, 3298
Rainbow Endowment Grants, 3300
Ralph M. Parsons Foundation Grants, 3304
Raymond John Wean Foundation Grants, 3310
Rayonier Foundation Grants, 3311
Rhode Island Foundation Grants, 3362
Richard and Helen DeVos Foundation Grants, 3364
Richard King Mellon Foundation Grants, 3369
Robbins Charitable Foundation Grants, 3378
Robins Foundation Grants, 3390
Rochester Area Community Foundation Grants, 3391
Rosalynn Carter Institute Georgia Caregiver of the Year Awards, 3407
Ruth H. & Warren A. Ellsworth Fndn Grants, 3429
S.H. Cowell Foundation Grants, 3447
S. Livingston Mather Charitable Trust Grants, 3448
S. Mark Taper Foundation Grants, 3449
Sarkeys Foundation Grants, 3485
Scott County Community Foundation Grants, 3500
Seabury Foundation Grants, 3501
Seattle Foundation Benjamin N. Phillips Memorial Fund Grants, 3504
Sensient Technologies Foundation Grants, 3513
Shopko Fndn Community Charitable Grants, 3547
Simmons Foundation Grants, 3563
Singing for Change Foundation Grants, 3567
Sophia Romero Trust Grants, 3624
Sunderland Foundation Grants, 3675
Thompson Charitable Foundation Grants, 3727

TJX Foundation Grants, 3735
Toyota Motor Manufacturing of Indiana Grants, 3744
Toyota Motor Manufacturing of Kentucky Grants, 3745
Turner Foundation Grants, 3763
U.S. Dept of Education Special Ed--National Activities--Parent Information Centers, 3772
USAID Family Health Plus Project Grants, 3810
USAID Support for International Family Planning Organizations Grants, 3823
Vancouver Foundation Grants and Community Initiatives Program, 3849
Volkswagen of America Corporate Contributions, 3880
W.K. Kellogg Foundation Secure Families Grants, 3887
Waitt Family Foundation Grants, 3894
Weyerhaeuser Family Foundation Health Grants, 3926
William G. Gilmore Foundation Grants, 3951
Zane's Foundation Grants, 3983

Family Planning
Allyn Foundation Grants, 448
Clayton Fund Grants, 1180
Conservation, Food, and Health Foundation Grants for Developing Countries, 1338
Cowles Charitable Trust Grants, 1354
DHHS Adolescent Family Life Demonstration Projects, 1472
Donald and Sylvia Robinson Family Foundation Grants, 1486
Frances and John L. Loeb Family Fund Grants, 1716
General Service Reproductive Justice Grants, 1785
Health Fndn of Greater Indianapolis Grants, 1984
Huber Foundation Grants, 2071
IIE Hewlett Fnd/IIE Dissertation Fellowship, 2132
J.W. Kieckhefer Foundation Grants, 2176
James Ford Bell Foundation Grants, 2187
Mary Owen Borden Foundation Grants, 2446
Oppenstein Brothers Foundation Grants, 3078
Packard Foundation Population and Reproductive Health Grants, 3106
Perkins Charitable Foundation Grants, 3140
Portland Fndn Women's Giving Circle Grant, 3249
Radcliffe Institute Residential Fellowships, 3299
Robert Sterling Clark Foundation Reproductive Rights and Health Grants, 3388
S.H. Cowell Foundation Grants, 3447
S. Livingston Mather Charitable Trust Grants, 3448
Saint Luke's Health Initiatives Grants, 3455
Samuel S. Johnson Foundation Grants, 3468
Seabury Foundation Grants, 3501
Stella and Charles Guttman Foundation Grants, 3651
Susan Vaughan Foundation Grants, 3695
Union Benevolent Association Grants, 3789
USAID Policy, Advocacy, and Communication Enhanced for Population and Reproductive Health Grants, 3820
USAID Support for International Family Planning Organizations Grants, 3823
W. C. Griffith Foundation Grants, 3882

Family Practice
AAFP Foundation Health Literacy State Grants, 132
AAFP Foundation Joint Grants, 133
AAFP Foundation Pfizer Teacher Devel Awards, 134
AAFP Foundation Wyeth Immunization Awards, 135
AAFP Research Stimulation Grants, 136
ACS Cancer Control Career Development Awards for Primary Care Physicians, 250
Dorothy Rider Pool Health Care Grants, 1500
Josiah Macy Jr. Foundation Grants, 2277
Pfizer/AAFP Foundation Visiting Professorship Program in Family Medicine, 3162
Regenstrief Geriatrics Fellowship, 3352
USAID Support for International Family Planning Organizations Grants, 3823

Family/Marriage Counseling
Amelia Sillman Rockwell and Carlos Perry Rockwell Charities Fund Grants, 501
Charles H. Pearson Foundation Grants, 1121
CJ Foundation for SIDS Program Services Grants, 1163

Daniels Fund Grants-Aging, 1424
Decatur County Community Foundation Small Project Grants, 1445
Florida BRAIVE Fund of Dade Community Foundation, 1678
George W. Wells Foundation Grants, 1804
Gulf Coast Foundation of Community Operating Grants, 1916
Gulf Coast Foundation of Community Program Grants, 1917
Kimball International-Habig Foundation Health and Human Services Grants, 2308
Robert R. McCormick Trib Community Grants, 3385
Sophia Romero Trust Grants, 3624

Farm and Ranch Management
DuPont Pioneer Community Giving Grants, 1532

Farming
Union Bank, N.A. Foundation Grants, 3788

Fashion/Textiles Design
Phi Upsilon Omicron Alumni Research Grant, 3191
Phi Upsilon Omicron Florence Fallgatter Distinguished Service Award, 3192
Phi Upsilon Omicron Geraldine Clewell Senior Awards, 3194
Phi Upsilon Lillian P. Schoephoerster Award, 3196
Phi Upsilon Orinne Johnson Writing Award, 3198
Phi Upsilon Omicron Undergraduate Karen P. Goebel Conclave Award, 3200

Fatigue/Fracture
NIA Bioenergetics, Fatigability, and Activity Limitations in Aging, 2755
NIA Transdisciplinary Research on Fatigue and Fatigability in Aging Grants, 2786

Fellowship Programs, General
A-T Children's Project Post Doctoral Fellowships, 14
AACAP-NIDA K12 Career Development Awards, 49
AACAP Educational Outreach Program for Residents in Alcohol Research, 54
AACAP Jeanne Spurlock Research Fellowship in Substance Abuse and Addiction for Minority Medical Students, 60
AACAP Life Members Mentorship Grants for Medical Students, 63
AACAP Summer Medical Student Fellowships, 75
AACR-Colorectal Cancer Coalition Fellows Grant, 88
AACR Centennial Pre-doctoral Fellowships in Cancer Research, 109
Acumen Global Fellowships, 285
ADA Foundation Dentsply International Research Fellowships, 289
ADA Foundation Scientist in Training Fellowship, 295
ADA Foundation Summer Scholars Fellowships, 296
AES-Grass Young Investigator Travel Awards, 317
AES Robert S. Morison Fellowship, 323
AES Susan S. Spencer Clinical Research Epilepsy Training Fellowship, 325
Aetna Foundation Minority Scholars Grants, 329
Aetna Foundation National Medical Fellowship in Healthcare Leadership, 330
AFAR Medical Student Training in Aging Research Program, 335
AGHE Graduate Scholarships and Fellowships, 343
AHRQ Individual Awards for Postdoc Fellows Ruth L. Kirschstein National Research Service Awards, 366
AIHS Media Summer Fellowship, 392
Alexander von Humboldt Foundation Georg Forster Fellowships for Experienced Researchers, 423
Alexander von Humboldt Foundation Georg Forster Fellowships for Postdoctoral Researchers, 424
AMA-MSS Government Relations Advocacy Fellowship, 475
American Academy of Nursing Claire M. Fagin Fellowships, 506
American College of Surgeons Australia and New Zealand Chapter Travelling Fellowships, 519

SUBJECT INDEX

American College of Surgeons Nizar N. Oweida, MD, FACS, Scholarships, 526
American College of Surgeons Wound Care Management Award, 529
AMHPS Dr. James A. Ferguson Emerging Infectious Diseases Fellowships, 573
APHL Emerging Infectious Diseases Fellowships, 636
APSA Congressional Health and Aging Policy Fellowships, 644
APSA Robert Wood Johnson Foundation Health Policy Fellowships, 645
ASCO Long-Term International Fellowships, 681
ASM-PAHO Infectious Diseases Epidemiology and Surveillance Fellowships, 714
ASM/CDC Fellowships in Infectious Disease and Public Health Microbiology, 716
ASM Congressional Science Fellowships, 720
ASM Int'l Fellowship for Latin America and the Caribbean, 734
ASM Int'l Fellowships for Asia and Africa, 735
ASM Microbiology Undergraduate Research Fellowship, 740
ASM Robert D. Watkins Graduate Research Fellowship, 747
ASM Undergraduate Research Fellowship, 755
ASPH/CDC Allan Rosenfield Global Health Fellowships, 780
ASPH/CDC Public Health Fellowships, 781
ASPO Fellowships, 784
ASU Grad College Reach for the Stars Fellowship, 791
ASU Graduate College Science Foundation Arizona Bisgrove Postdoctoral Scholars, 792
Biogen Foundation Fellowships, 898
Brookdale Foundation Leadership in Aging Fellowships, 965
Carrier Corporation Contributions Grants, 1038
CDC-Hubert Global Health Fellowship, 1047
CDC Epidemic Intell Service Training Grants, 1055
CDC Fnd Tobacco Network Lab Fellowship, 1062
CDC Preventive Med Residency & Fellowship, 1064
CDC Public Health Informatics Fellowships, 1068
CDC Public Health Prev Service Fellowships, 1069
CDC Public Health Prevention Service Fellowship Sponsorships, 1070
Cedar Tree Foundation David H. Smith Conservation Research Fellowship, 1076
Charles H. Revson Foundation Grants, 1122
CSTE CDC/CSTE Applied Epidemiology Fellowships, 1367
Cystic Fibrosis Canada Clinical Fellowships, 1381
Cystic Fibrosis Canada Fellowships, 1384
Daimler and Benz Foundation Fellowships, 1411
Duke University Adult Cardiothoracic Anesthesia and Critical Care Medicine Fellowships, 1520
Duke University Ambulatory and Regional Anesthesia Fellowships, 1521
Duke University Clinical Cardiac Electrophysiology Fellowships, 1522
Duke University Hyperbaric Center Fellowships, 1523
Duke Univ Interventional Cardio Fellowships, 1524
Duke Univ Obstetric Anesthesia Fellowships, 1525
Duke University Pain Management Fellowships, 1526
Duke University Ped Anesthesiology Fellowship, 1527
Duke University Postdoctoral Research Fellowships in Aging, 1528
Edward W. and Stella C. Van Houten Memorial Fund Grants, 1559
EMBO Long-Term Fellowships, 1587
EMBO Short-Term Fellowships, 1588
FAMRI Young Clinical Scientist Awards, 1629
Fragile X Syndrome Postdoc Research Fellows, 1714
Fritz B. Burns Foundation Grants, 1747
Grass Foundation Marine Biological Laboratory Advanced Imaging Fellowships, 1877
Grass Foundation Marine Biological Laboratory Fellowships, 1878
Grover Hermann Foundation Grants, 1906
IBRO Asia Reg APRC Exchange Fellowships, 2101
IBRO Regional Grants for Int'l Fellowships to U.S. Laboratory Summer Neuroscience Courses, 2110

IBRO Research Fellowships, 2111
IBRO Return Home Fellowships, 2112
IHI Quality Improvement Fellowships, 2121
IIE Whitaker Int'l Fellowships & Scholarships, 2144
Lubrizol Foundation Grants, 2384
Lynde & Harry Bradley Foundation Fellowships, 2396
Lynde and Harry Bradley Foundation Grants, 2397
Lynde and Harry Bradley Foundation Prizes: Bradley Prizes, 2398
MAP International Medical Fellowships, 2421
Mayo Clinic Administrative Fellowship, 2460
Mayo Clinic Admin Fellowship - Eau Claire, 2461
Mayo Clinic Business Consulting Fellowship, 2462
MBL Albert & Ellen Grass Fellowships, 2463
MBL Ann E. Kammer Summer Fellowship, 2464
MBL Associates Summer Fellowships, 2465
MBL Baxter Postdoctoral Summer Fellowship, 2466
MBL Burr & Susie Steinbach Fellowship, 2467
MBL E.E. Just Summer Fellowship for Minority Scientists, 2468
MBL Erik B. Fries Summer Fellowships, 2469
MBL Eugene and Millicent Bell Fellowships, 2470
MBL Evelyn and Melvin Spiegal Fellowship, 2471
MBL Frank R. Lillie Summer Fellowship, 2472
MBL Frederik B. and Betsy G. Bang Summer Fellowships, 2473
MBL Fred Karush Library Readership, 2474
MBL Gruss Lipper Family Foundation Summer Fellowship, 2475
MBL H. Keffer Hartline and Edward F. MacNichol, Jr. Fellowships, 2476
MBL Herbert W. Rand Summer Fellowship, 2477
MBL James E. and Faith Miller Memorial Summer Fellowship, 2478
MBL John M. Arnold Award, 2479
MBL Laura and Arthur Colwin Fellowships, 2480
MBL Lucy B. Lemann Summer Fellowship, 2481
MBL M.G.F. Fuortes Summer Fellowships, 2482
MBL Nikon Summer Fellowship, 2483
MBL Plum Fndn John E. Dowling Fellowships, 2484
MBL Robert Day Allen Summer Fellowship, 2485
MBL Stephen W. Kuffler Summer Fellowships, 2487
MBL William Townsend Porter Summer Fellowships for Minority Investigators, 2488
Mead Johnson Nutritionals Scholarships, 2505
MedImmune Charitable Grants, 2509
Medtronic Foundation Fellowships, 2512
Mericos Foundation Grants, 2518
MGFA Post-Doctoral Research Fellowships, 2539
MGFA Student Fellowships, 2540
MLA Cunningham Memorial International Fellowship, 2560
MLA David A. Kronick Traveling Fellowship, 2561
MMAAP Foundation Dermatology Fellowships, 2577
MMAAP Foundation Fellowship Award in Hematology, 2579
MMAAP Foundation Fellowship Award in Reproductive Medicine, 2580
MMAAP Foundation Fellowship Award in Translational Medicine, 2581
MMAAP Foundation Senior Health Fellowship Award in Geriatrics Medicine and Aging Research, 2586
National Center for Responsible Gaming Postdoctoral Fellowships, 2630
NCCAM Ruth L. Kirschstein National Research Service Awards for Postdoctoral Training in Complementary and Alternative Medicine, 2640
NHLBI Ruth L. Kirschstein National Research Service Awards for Individual Predoctoral Fellowships to Promote Diversity in Health-Related Research, 2730
NHLBI Ruth L. Kirschstein National Research Service Awards for Individual Predoctoral MD/PhD Fellows and Other Dual Degree Fellows, 2731
NHLBI Ruth L. Kirschstein National Research Service Awards for Individual Senior Fellows, 2732
NIH Ruth L. Kirschstein National Research Service Awards (NRSA) for Individual Senior Fellows, 2901
NIH Ruth L. Kirschstein National Research Service Awards for Individual Predoctoral Fellowships to Promote Diversity in Research, 2904

NIH Ruth L. Kirschstein National Research Service Awards for Individual Predoctoral MD/PhD and Other Dual Doctoral Degree Fellows, 2905
NINR Ruth L. Kirschstein National Research Service Award for Predoc Fellows in Nursing, 2934
NMF General Electric Med Fellowships, 2956
NSF Postdoc Research Fellowships in Biology, 3020
Olympus Corporation of Americas Fellowships, 3074
Pancreatic Cancer Action Network Fellowship, 3113
PC Fred H. Bixby Fellowships, 3126
PDF-AANF Clinician-Scientist Dev Awards, 3127
PDF Postdoctoral Fellowships for Basic Scientists, 3129
PDF Postdoctoral Fellowships for Neurologists, 3130
PDF Summer Fellowships, 3131
PhRMA Foundation Health Outcomes Pre-Doctoral Fellowships, 3203
PhRMA Foundation Health Outcomes Sabbatical Fellowships, 3205
PhRMA Fndn Info Sabbatical Fellowships, 3209
PhRMA Foundation Paul Calabresi Medical Student Research Fellowship, 3210
Potts Memorial Foundation Grants, 3254
Price Family Charitable Fund Grants, 3262
Pride Foundation Fellowships, 3267
PVA Research Foundation Grants, 3285
Radcliffe Institute Carol K. Pforzheimer Student Fellowships, 3298
Radcliffe Institute Residential Fellowships, 3299
Regenstrief General Internal Medicine - General Pediatrics Research Fellowship Program, 3351
Regenstrief Geriatrics Fellowship, 3352
Rhode Island Foundation Grants, 3362
RWJF Health and Society Scholars, 3436
RWJF Nurse Faculty Scholars, 3441
SfN Neuroscience Fellowships, 3530
Siragusa Foundation Health Services & Medical Research Grants, 3571
Smithsonian Biodiversity Genomics and Bioinformatics Postdoctoral Fellowships, 3580
Smithsonian Museum Conservation Institute Research Post-Doctorate/Post-Graduate Fellowships, 3581
SRC Medical Research Grants, 3635
St. Joseph Community Health Foundation Schneider Fellowships, 3642
Stroke Association Clinical Fellowships, 3662
Thomson Reuters / MLA Doctoral Fellowships, 3728
Trauma Center of America Finance Fellowship, 3752
UCLA / RAND Corporation Post-Doctoral Fellowships, 3776
United States Institute of Peace - Jennings Randolph Senior Fellowships, 3802

Feminism
Radcliffe Institute Residential Fellowships, 3299

Fetus Research
BWF Preterm Birth Initiative Grants, 990
NIH Research on Sleep and Sleep Disorders, 2898

Film Production
Daniel & Nanna Stern Family Fndn Grants, 1422
NLM Grants for Scholarly Works in Biomedicine and Health, 2944
Playboy Foundation Grants, 3237

Films
Canada-U.S. Fulbright New Century Scholars Program Grants, 1008
Canada-U.S. Fulbright Senior Specialists Grants, 1009
Community Foundation of Greater Greensboro Community Grants, 1277
Daniel & Nanna Stern Family Fndn Grants, 1422
Fulbright Alumni Initiatives Awards, 1749
Fulbright Distinguished Chairs Awards, 1750
Fulbright New Century Scholars Grants, 1753
Fulbright Specialists Program Grants, 1754
Fulbright Scholars in Europe and Eurasia, 1755
Fulbright Traditional Scholar Program in Sub-Saharan Africa, 1756

730 / Films

Fulbright Traditional Scholar Program in the East Asia/Pacific Region, 1757
Fulbright Traditional Scholar Program in the Near East and North Africa Region, 1758
Fulbright Traditional Scholar Program in the South and Central Asia Region, 1759
Fulbright Traditional Scholar Program in the Western Hemisphere, 1760

Finance
Community Foundation of St. Joseph County Lilly Endowment Community Scholarship, 1308
John R. Oishei Foundation Grants, 2249
Lincoln Financial Foundation Grants, 2367
Mayo Clinic Administrative Fellowship, 2460
Mayo Clinic Admin Fellowship - Eau Claire, 2461
Trauma Center of America Finance Fellowship, 3752

Financial Aid (Scholarships and Loans)
Abbott Fund CFCareForward Scholarships, 200
Adams Rotary Memorial Fund B Scholarships, 303
Alfred and Tillie Shemanski Testamentary Trust Grants, 428
AlohaCare Believes in Me Scholarship, 449
ASM Robert D. Watkins Graduate Research Fellowship, 747
AUPHA Corris Boyd Scholarships, 800
AUPHA Foster G. McGaw Scholarships, 801
Blackford County Community Foundation Hartford City Kiwanis Scholarship in Memory of Mike McDougall, 907
Blackford County Community Foundation Noble Memorial Scholarship, 908
Crown Point Community Fndn Scholarships, 1362
Dr. John T. Macdonald Foundation Scholarships, 1508
Edward W. and Stella C. Van Houten Memorial Fund Grants, 1559
Elizabeth Nash Foundation Scholarships, 1570
Fndn for Appalachian Ohio Bachtel Scholarships, 1692
Foundation for Appalachian Ohio Susan K. Ipacs Nursing Legacy Scholarships, 1693
Foundation for Appalachian Ohio Zelma Gray Medical School Scholarship, 1694
Greygates Foundation Grants, 1901
Indiana Minority Teacher/Special Services Scholarships, 2156
Indiana Nursing Scholarships, 2157
J. Willard Marriott, Jr. Foundation Grants, 2178
John H. Wellons Foundation Grants, 2239
Joseph and Luella Abell Trust Scholarships, 2266
Joseph Collins Foundation Scholarships, 2267
K21 Health Foundation Cancer Care Grants, 2284
Marathon Petroleum Corporation Grants, 2422
March of Dimes Graduate Nursing Scholarships, 2424
Mary L. Peyton Foundation Grants, 2445
MLA Graduate Scholarship, 2564
MLA Grad Scholarship for Minority Students, 2565
Nesbitt Medical Student Foundation Scholarship, 2674
Pajaro Valley Community Health Trust Promoting Entry & Advancement in the Health Professions Grants, 3111
Phi Upsilon Omicron Margaret Jerome Sampson Scholarships, 3197
Robert B. Adams Foundation Scholarships, 3380
Seattle Fndn Doyne M. Green Scholarships, 3505
Seattle Fndn Hawkins Memorial Scholarship, 3506
Sioux Falls Area Community Foundation Field-of-Interest and Donor-Advised Grants, 3569
Susan G. Komen Breast Cancer Foundation College Scholarships, 3692
W.P. and Bulah Luse Foundation Grants, 3892

Financial Education
Essex County Community Foundation Women's Fund Grants, 1605
Hampton Roads Community Foundation Health and Human Service Grants, 1941
KeyBank Foundation Grants, 2307
McGraw-Hill Companies Community Grants, 2494
Trauma Center of America Finance Fellowship, 3752

Financial Literacy
Alvin and Fanny Blaustein Thalheimer Foundation Baltimore Communal Grants, 459
BBVA Compass Foundation Charitable Grants, 848
Benton Community Fndn Cookie Jar Grant, 877
Blue Cross Blue Shield of Minnesota Fndn Healthy Equity: Public Libraries for Health Grants, 918
CNCS AmeriCorps VISTA Project Grants, 1190
Hampton Roads Community Foundation Health and Human Service Grants, 1941
IIE Western Union Family Scholarships, 2143
KeyBank Foundation Grants, 2307
McGraw-Hill Companies Community Grants, 2494
Medtronic Foundation Community Link Human Services Grants, 2511
Union Bank, N.A. Foundation Grants, 3788

Fine Arts
Abell-Hanger Foundation Grants, 207
Achelis Foundation Grants, 234
Berks County Community Foundation Grants, 881
Bodenwein Public Benevolent Foundation Grants, 928
Bodman Foundation Grants, 929
Charles H. Hall Foundation, 1120
Crystelle Waggoner Charitable Trust Grants, 1364
E.L. Wiegand Foundation Grants, 1538
Elizabeth Morse Genius Charitable Trust Grants, 1569
George A. and Grace L. Long Foundation Grants, 1788
George H.C. Ensworth Memorial Fund Grants, 1797
Helen Gertrude Sparks Charitable Trust Grants, 1993
Helen Pumphrey Denit Charitable Trust Grants, 1996
Henrietta Lange Burk Fund Grants, 2001
John W. Speas and Effie E. Speas Memorial Trust Grants, 2264
Katrine Menzing Deakins Trust Grants, 2295
Leon and Thea Koerner Foundation Grants, 2355
Lewis H. Humphreys Charitable Trust Grants, 2361
Louis and Elizabeth Nave Flarsheim Charitable Foundation Grants, 2379
Lumpkin Family Fnd Healthy People Grants, 2391
Marion Gardner Jackson Charitable Trust Grants, 2434
Puerto Rico Community Foundation Grants, 3278
R.C. Baker Foundation Grants, 3292
Robert R. Meyer Foundation Grants, 3387

Fire
ADA Foundation Disaster Assistance Grants, 290

Fire Prevention
Adler-Clark Electric Community Commitment Foundation Grants, 307
Burlington Northern Santa Fe Foundation Grants, 978
Colorado Resource for Emergency Education and Trauma Grants, 1212
DOL Occupational Safety and Health--Susan Harwood Training Grants, 1484
Francis T. & Louise T. Nichols Fndn Grants, 1720
GEICO Public Service Awards, 1778
RCF General Community Grants, 3312
Seneca Foods Foundation Grants, 3511
Thompson Charitable Foundation Grants, 3727

Fish and Fisheries
GNOF Bayou Communities Grants, 1835
RSC Thomas W. Eadie Medal, 3423

Fishing
GNOF Bayou Communities Grants, 1835

Fitness
ACF Native American Social and Economic Development Strategies Grants, 231
ACSM Michael L. Pollock Student Scholarship, 270
BCBSNC Foundation Grants, 860
Boeing Company Contributions Grants, 930
Collective Brands Foundation Grants, 1208
Foundation for a Healthy Kentucky Grants, 1691
GNOF IMPACT Kahn-Oppenheim Grants, 1842
Highmark BCBS Challenge for Healthier Schools in Western Pennsylvania, 3696

SUBJECT INDEX

Mary Black Foundation Active Living Grants, 2440
Medtronic Foundation CommunityLink Health Grants, 2510
PepsiCo Foundation Grants, 3137
RCF Summertime Kids Grants, 3314
Robert R. Meyer Foundation Grants, 3387

Folklore and Mythology
Canada-U.S. Fulbright New Century Scholars Program Grants, 1008
Canada-U.S. Fulbright Senior Specialists Grants, 1009
Fulbright Alumni Initiatives Awards, 1749
Fulbright Distinguished Chairs Awards, 1750
Fulbright New Century Scholars Grants, 1753
Fulbright Specialists Program Grants, 1754
Fulbright Scholars in Europe and Eurasia, 1755
Fulbright Traditional Scholar Program in Sub-Saharan Africa, 1756
Fulbright Traditional Scholar Program in the East Asia/Pacific Region, 1757
Fulbright Traditional Scholar Program in the Near East and North Africa Region, 1758
Fulbright Traditional Scholar Program in the South and Central Asia Region, 1759
Fulbright Traditional Scholar Program in the Western Hemisphere, 1760

Food Banks
Adams Foundation Grants, 301
Adler-Clark Electric Community Commitment Foundation Grants, 307
Albert W. Rice Charitable Foundation Grants, 411
Alfred E. Chase Charitable Foundation Grants, 430
Alpha Natural Resources Corporate Giving, 450
Anthony R. Abraham Foundation Grants, 608
Aragona Family Foundation Grants, 648
Arthur Ashley Williams Foundation Grants, 671
Batchelor Foundation Grants, 837
Bildner Family Foundation Grants, 891
Caesars Foundation Grants, 995
Callaway Golf Company Foundation Grants, 1003
Campbell Soup Foundation Grants, 1007
Carl M. Freeman Foundation FACES Grants, 1026
Carrie Estelle Doheny Foundation Grants, 1037
Catherine Kennedy Home Foundation Grants, 1043
Cessna Foundation Grants Program, 1081
CFFVR Capital Credit Union Charitable Giving Grants, 1095
CFFVR Jewelers Mutual Charitable Giving, 1099
Charles H. Farnsworth Trust Grants, 1119
Charles H. Pearson Foundation Grants, 1121
Charles Nelson Robinson Fund Grants, 1125
Community Foundation for Greater Atlanta Clayton County Fund Grants, 1240
Community Foundation for Greater Atlanta Morgan County Fund Grants, 1244
Community Foundation for Greater Atlanta Newton County Fund Grants, 1245
CONSOL Energy Community Dev Grants, 1339
Daphne Seybolt Culpeper Fndn Grants, 1425
Denver Foundation Community Grants, 1461
Doree Taylor Charitable Foundation, 1490
Dr. & Mrs. Paul Pierce Memorial Fndn Grants, 1505
Edina Realty Foundation Grants, 1550
Frank B. Hazard General Charity Fund Grants, 1721
Frank Reed and Margaret Jane Peters Memorial Fund II Grants, 1728
Frank Stanley Beveridge Foundation Grants, 1730
George W. Wells Foundation Grants, 1804
Green Bay Packers Foundation Grants, 1890
H.J. Heinz Company Foundation Grants, 1923
Harold Brooks Foundation Grants, 1952
Helen Gertrude Sparks Charitable Trust Grants, 1993
Independence Community Foundation Community Quality of Life Grant, 2153
Kathryne Beynon Foundation Grants, 2294
Kenneth T. & Eileen L. Norris Fndn Grants, 2300
Lincoln Financial Foundation Grants, 2367
Lydia deForest Charitable Trust Grants, 2393

Maurice J. Masserini Charitable Trust Grants, 2456
McCarthy Family Foundation Grants, 2490
Middlesex Savings Charitable Foundation Basic Human Needs Grants, 2550
Norfolk Southern Foundation Grants, 2982
Northwestern Mutual Foundation Grants, 2999
Ordean Foundation Grants, 3080
Packard Foundation Local Grants, 3105
Pott Foundation Grants, 3253
Procter and Gamble Fund Grants, 3274
Reinberger Foundation Grants, 3357
Samuel S. Johnson Foundation Grants, 3468
Shaw's Supermarkets Donations, 3542
Shield-Ayres Foundation Grants, 3546
Sierra Health Foundation Responsive Grants, 3555
Stewart Huston Charitable Trust Grants, 3655
Swindells Charitable Foundation, 3697
Textron Corporate Contributions Grants, 3712
Vancouver Foundation Grants and Community Initiatives Program, 3849
William Blair and Company Foundation Grants, 3948
Wood-Claeyssens Foundation Grants, 3971

Food Consumption
Aetna Foundation Regional Health Grants, 333
Charles H. Hall Foundation, 1120
Coca-Cola Foundation Grants, 1203
General Mills Champs for Healthy Kids Grants, 1781
Johnson Controls Foundation Health and Social Services Grants, 2255
Windham Foundation Grants, 3961

Food Distribution
Albertson's Charitable Giving Grants, 410
Albert W. Rice Charitable Foundation Grants, 411
Bacon Family Foundation Grants, 817
Benton Community Foundation Grants, 878
Blue River Community Foundation Grants, 922
Brown County Community Foundation Grants, 970
Caesars Foundation Grants, 995
Catherine Kennedy Home Foundation Grants, 1043
Charles H. Pearson Foundation Grants, 1121
Christine & Katharina Pauly Trust Grants, 1150
Community Foundation of Bartholomew County Heritage Fund Grants, 1262
Community Foundation of Bartholomew County James A. Henderson Award for Fundraising, 1263
Community Foundation of Greater Fort Wayne - Community Endowment and Clarke Endowment Grants, 1274
Conservation, Food, and Health Foundation Grants for Developing Countries, 1338
David N. Lane Trust Grants for Aged and Indigent Women, 1433
DeRoy Testamentary Foundation Grants, 1469
Donald and Sylvia Robinson Family Foundation Grants, 1486
Doris and Victor Day Foundation Grants, 1491
Florence Hunt Maxwell Foundation Grants, 1676
Frank Reed and Margaret Jane Peters Memorial Fund II Grants, 1728
General Mills Foundation Grants, 1782
George W. Wells Foundation Grants, 1804
Giant Food Charitable Grants, 1816
Helen Gertrude Sparks Charitable Trust Grants, 1993
Horace Moses Charitable Foundation Grants, 2047
John Edward Fowler Memorial Fndn Grants, 2235
John G. Martin Foundation Grants, 2237
Johnson Controls Foundation Health and Social Services Grants, 2255
Lucy Downing Nisbet Charitable Fund Grants, 2387
NYCT Grants for the Elderly, 3054
OneFamily Foundation Grants, 3075
PepsiCo Foundation Grants, 3137
Prince Charitable Trusts DC Grants, 3271
Robert R. Meyer Foundation Grants, 3387
Rockefeller Foundation Grants, 3396
Seagate Tech Corp Capacity to Care Grants, 3502
Sensient Technologies Foundation Grants, 3513
Sid W. Richardson Foundation Grants, 3553

Sioux Falls Area Community Foundation Field-of-Interest and Donor-Advised Grants, 3569
Thomas C. Ackerman Foundation Grants, 3719
United Hospital Fund of New York Health Care Improvement Grants, 3797
W.K. Kellogg Foundation Healthy Kids Grants, 3886

Food Engineering
BWF Innovation in Regulatory Science Grants, 986
NSF-NIST Interaction in Chemistry, Materials Research, Molecular Biosciences, Bioengineering, and Chemical Engineering, 3004

Food Management
Adolph Coors Foundation Grants, 311
Windham Foundation Grants, 3961

Food Preparation
Daniels Fund Grants-Aging, 1424

Food Processing
Institute for Agriculture and Trade Policy Food and Society Fellowships, 2159

Food Production
BWF Innovation in Regulatory Science Grants, 986

Food Safety
Aetna Foundation Regional Health Grants, 333
AOCS Processing Division Student Award, 618
Dean Foods Community Involvement Grants, 1440
IIE Rockefeller Foundation Bellagio Center Residencies, 2142
PepsiCo Foundation Grants, 3137
USDA Organic Agricultural Research Grants, 3830
Windham Foundation Grants, 3961

Food Sciences
AOCS Lipid Oxidation and Quality Division Poster Competition, 617
BWF Innovation in Regulatory Science Grants, 986
Canada Graduate Scholarships (CGS) and NSERC Postgraduate Scholarships (PGS), 1010
Coca-Cola Foundation Grants, 1203
Phi Upsilon Omicron Alumni Research Grant, 3191
Phi Upsilon Omicron Florence Fallgatter Distinguished Service Award, 3192
Phi Upsilon Omicron Frances Morton Holbrook Alumni Award, 3193
Phi Upsilon Omicron Geraldine Clewell Senior Awards, 3194
Phi Upsilon Omicron Janice Cory Bullock Collegiate Award, 3195
Phi Upsilon Lillian P. Schoephoerster Award, 3196
Phi Upsilon Omicron Margaret Jerome Sampson Scholarships, 3197
Phi Upsilon Orinne Johnson Writing Award, 3198
Phi Upsilon Omicron Sarah Thornley Phillips Leader Awards, 3199
Phi Upsilon Omicron Undergraduate Karen P. Goebel Conclave Award, 3200

Food Service Industry
Adler-Clark Electric Community Commitment Foundation Grants, 307

Food Technology
CLIF Bar Family Foundation Grants, 1183
Windham Foundation Grants, 3961

Foods
AAAAI Distinguished Clinician Award, 24
CLIF Bar Family Foundation Grants, 1183
Columbus Foundation Traditional Grants, 1223
ConAgra Foods Fndn Community Impact Grants, 1321
ConAgra Foods Nourish our Community Grants, 1322
Dean Foods Community Involvement Grants, 1440
Elaine Feld Stern Charitable Trust Grants, 1565
Essex County Community Foundation Merrimack Valley General Fund Grants, 1603

Institute for Agriculture and Trade Policy Food and Society Fellowships, 2159
Seneca Foods Foundation Grants, 3511
Simmons Foundation Grants, 3563

Football
Arizona Cardinals Grants, 655
Green Bay Packers Foundation Grants, 1890

Foreign Languages
IIE Eurobank EFG Scholarships, 2130
IIE Leonora Lindsley Memorial Fellowships, 2136

Foreign Languages Education
IIE David L. Boren Fellowships, 2129
IIE Leonora Lindsley Memorial Fellowships, 2136

Forest Products Industry
Green Diamond Charitable Contributions, 1891
Weyerhaeuser Company Foundation Grants, 3925

Forestry Management
RSC Thomas W. Eadie Medal, 3423

Forests and Woodlands
Callaway Foundation Grants, 1002
Canada Graduate Scholarships (CGS) and NSERC Postgraduate Scholarships (PGS), 1010
Eugene B. Casey Foundation Grants, 1609
Green Diamond Charitable Contributions, 1891
Weyerhaeuser Company Foundation Grants, 3925

Foster Care
Alfred and Tillie Shemanski Testamentary Trust Grants, 428
Allan C. and Lelia J. Garden Foundation Grants, 438
Dyson Foundation Mid-Hudson Valley Project Support Grants, 1534
Educational Foundation of America Grants, 1553
Eisner Foundation Grants, 1564
Mockingbird Foundation Grants, 2589
Pew Charitable Trusts Children & Youth Grants, 3158
Peyton Anderson Foundation Grants, 3160

Fund-Raising
AAAAI Distinguished Layperson Award, 25
Alliant Energy Corporation Contributions, 445
GNOF Organizational Effectiveness Grants and Workshops, 1846
Meyer Foundation Benevon Grants, 2532
Middlesex Savings Charitable Foundation Capacity Building Grants, 2551
North Carolina Biotechnology Center Grantsmanship Training Grants, 2985
Pfizer Special Events Grants, 3186
Saint Louis Rams Fndn Community Donations, 3453
Victor E. Speas Foundation Grants, 3868

Gambling
Boyd Gaming Corporation Contributions Program, 941
National Center for Responsible Gaming Grants, 2629
National Center for Responsible Gaming Postdoctoral Fellowships, 2630
National Center for Resp Gaming Seed Grants, 2631

Gangs
DOJ Gang-Free Schools and Communities Intervention Grants, 1481
eBay Foundation Community Grants, 1543
Piedmont Natural Gas Foundation Health and Human Services Grants, 3221
Union Bank, N.A. Foundation Grants, 3788

Gardening
Robert R. Meyer Foundation Grants, 3387

Gastroenterology
ASGE / ConMed Award for Outstanding Manuscript by a Fellow/Resident, 684
ASGE Don Wilson Award, 685

732 / Gastroenterology

ASGE Endoscopic Research Awards, 686
ASGE Endoscopic Research Career Development Awards, 687
ASGE Given Capsule Endoscopy Research Award, 688
CFF First- and Second-Year Clinical Fellowships, 1085
FDHN Bridging Grants, 1638
FDHN Centocor International Research Fellowship in Gastrointestinal Inflammation & Immunology, 1639
FDHN Designated Outcomes Award in Geriatric Gastroenterology, 1640
FDHN Designated Research Award in Geriatric Gastroenterology, 1641
FDHN Designated Research Award in Research Related to Pancreatitis, 1642
FDHN Fellow Abstract Prizes, 1643
FDHN Fellowship to Faculty Transition Awards, 1644
FDHN Funderburg Research Scholar Award in Gastric Biology Related to Cancer, 1645
FDHN Graduate Student Awards, 1646
FDHN Isenberg Int'l Research Scholar Award, 1647
FDHN June & Donald O. Castell MD, Esophageal Clinical Research Award, 1648
FDHN Moti L. & Kamla Rustgi International Travel Awards, 1649
FDHN Non-Career Research Awards, 1650
FDHN Non-Career Research Grants, 1651
FDHN Research Scholar Awards, 1652
FDHN Student Research Fellowships, 1654
FDHN TAP Endowed Designated Research Award in Acid-Related Diseases, 1655
FDHN Translational Research Awards, 1656

Gender
OECD Sexual Health and Rights Project Grants, 3064

Gender Equity
Cambridge Community Foundation Grants, 1004
Dallas Women's Foundation Grants, 1417
Essex County Community Foundation Women's Fund Grants, 1605
Ms. Fndn for Women Ending Violence Grants, 2596
OECD Sexual Health and Rights Project Grants, 3064
PC Fred H. Bixby Fellowships, 3126
SfN Louise Hanson Marshall Specific Recognition Award, 3527
SfN Mika Salpeter Lifetime Achievement Award, 3528
United Methodist Committee on Relief Global Health Grants, 3799

Gender Studies
AAUW American Dissertation Fellowships, 195
Harry Frank Guggenheim Foundation Dissertation Fellowships, 1963
OECD Sexual Health and Rights Project Grants, 3064

Gene Regulation
NHLBI Career Transition Awards, 2694
NIDDK Co-Activators and Co-Repressors in Gene Expression Grants, 2805

Gene Therapy
ACGT Investigators Grants, 232
NHLBI Career Transition Awards, 2694

Genetic Manipulation
Gruber Foundation Rosalind Franklin Young Investigator Award, 1910

Genetics
AFAR CART Fund Grants, 334
AFAR Research Grants, 337
Affymetrix Corporate Contributions Grants, 339
AIHS Alberta/Pfizer Translat Research Grants, 374
ASGH Award for Excellence in Human Genetics Education, 689
ASGH C.W. Cotterman Award, 690
ASGH Charles J. Epstein Trainee Awards for Excellence in Human Genetics Research, 691
ASGH McKusick Leadership Award, 692
ASGH William Allan Award, 693

ASGH William Curt Stern Award, 694
ASHG Public Health Genetics Fellowship, 712
Barth Syndrome Foundation Research Grants, 836
Dana Foundation Science and Health Grants, 1420
Del E. Webb Foundation Grants, 1450
Dr. John T. Macdonald Foundation Grants, 1507
Dr. Scholl Foundation Grants Program, 1511
Fannie E. Rippel Foundation Grants, 1630
Fragile X Syndrome Postdoc Research Fellows, 1714
FRAXA Research Foundation Program Grants, 1733
Glaucoma Research Pilot Project Grants, 1828
Gruber Foundation Gentics Prize, 1908
Gruber Foundation Rosalind Franklin Young Investigator Award, 1910
Hawaii Community Foundation Health Education and Research Grants, 1976
Hereditary Disease Foundation John J. Wasmuth Postdoctoral Fellowships, 2009
Hereditary Disease Fnd Lieberman Award, 2010
Hereditary Disease Foundation Research Grants, 2011
HRAMF Charles H. Hood Foundation Child Health Research Awards, 2057
Libra Foundation Grants, 2363
National MPS Society Grants, 2634
NCI Centers of Excellence in Cancer Communications Research, 2649
NEI Clinical Vision Research Devel Award, 2664
NHGRI Mentored Research Scientist Development Award, 2686
NHLBI Career Transition Awards, 2694
NHLBI Mentored Patient-Oriented Research Career Development Awards, 2710
NHLBI Pediatric Cardiac Genomics Consortium Grants, 2719
NIA Effects of Gene-Social Environment Interplay on Health and Behavior in Later Life Grants, 2757
NIA Nathan Shock Centers of Excellence in Basic Biology of Aging Grants, 2776
NICHD Innovative Therapies and Clinical Studies for Screenable Disorders Grants, 2789
NIDCD Mentored Research Scientist Development Award, 2796
NIDDK Cell-Specific Delineation of Prostate and Genitourinary Development, 2804
NIDDK Co-Activators and Co-Repressors in Gene Expression Grants, 2805
NIDDK Erythroid Lineage Molecular Grants, 2816
NIGMS Enabling Resources for Pharmacogenomics Grants, 2862
NIH Fine Mapping and Function of Genes for Type 1 Diabetes Grants, 2879
NSF Genes and Genome Systems Cluster Grants, 3015
Pediatric Cancer Research Foundation Grants, 3135
PhRMA Fndn Info Sabbatical Fellowships, 3209
Searle Scholars Program Grants, 3503
Sigrid Juselius Foundation Grants, 3561
Susan G. Komen Breast Cancer Foundation Challge Grants: Investigator Initiated Research, 3690
Tourette Syndrome Association Post-Doctoral Fellowships, 3740
Tourette Syndrome Association Research Grants, 3741
US CRDF Leishmaniasis: Collaborative Research Opportunities in N Africa & Middle East, 3828

Genetics, Molecular
APHL Emerging Infectious Diseases Fellowships, 636
ASGH Award for Excellence in Human Genetics Education, 689
ASGH C.W. Cotterman Award, 690
ASGH Charles J. Epstein Trainee Awards for Excellence in Human Genetics Research, 691
ASGH McKusick Leadership Award, 692
ASGH William Allan Award, 693
ASGH William Curt Stern Award, 694
Gruber Foundation Gentics Prize, 1908
Gruber Foundation Rosalind Franklin Young Investigator Award, 1910
Hilda and Preston Davis Foundation Postdoctoral Fellowships in Eating Disorders Research, 2029
IARC Expertise Transfer Fellowship, 2090

IARC Postdoctoral Fellowships for Training in Cancer Research, 2091
IARC Visiting Award for Senior Scientists, 2092
NIDDK Co-Activators and Co-Repressors in Gene Expression Grants, 2805
NIDDK Erythroid Lineage Molecular Grants, 2816
NIH Fine Mapping and Function of Genes for Type 1 Diabetes Grants, 2879
NSF Animal Developmental Mechanisms Grants, 3007
NSF Genes and Genome Systems Cluster Grants, 3015
Pediatric Cancer Research Foundation Grants, 3135
SfN Ralph W. Gerard Prize in Neuroscience, 3534

Genomics
ASHG Public Health Genetics Fellowship, 712
Gruber Foundation Gentics Prize, 1908
Pfizer ASPIRE North America Rheumatology Research Awards, 3172
PhRMA Fndn Info Sabbatical Fellowships, 3209
Smithsonian Biodiversity Genomics and Bioinformatics Postdoctoral Fellowships, 3580

Geography
Canada-U.S. Fulbright New Century Scholars Program Grants, 1008
Canada-U.S. Fulbright Senior Specialists Grants, 1009
Canada Graduate Scholarships (CGS) and NSERC Postgraduate Scholarships (PGS), 1010
Fulbright Alumni Initiatives Awards, 1749
Fulbright Distinguished Chairs Awards, 1750
Fulbright New Century Scholars Grants, 1753
Fulbright Specialists Program Grants, 1754
Fulbright Scholars in Europe and Eurasia, 1755
Fulbright Traditional Scholar Program in Sub-Saharan Africa, 1756
Fulbright Traditional Scholar Program in the East Asia/Pacific Region, 1757
Fulbright Traditional Scholar Program in the Near East and North Africa Region, 1758
Fulbright Traditional Scholar Program in the South and Central Asia Region, 1759
Fulbright Traditional Scholar Program in the Western Hemisphere, 1760
IIE Hewlett Fnd/IIE Dissertation Fellowship, 2132

Geology
American Chemical Society Award in Separations Science and Technology, 514
Canada Graduate Scholarships (CGS) and NSERC Postgraduate Scholarships (PGS), 1010
IIE Iraq Scholars and Leaders Scholarships, 2133
Weingart Foundation Grants, 3913

Geophysics
Brinson Foundation Grants, 953
IIE Iraq Scholars and Leaders Scholarships, 2133

Georgia
Peyton Anderson Foundation Grants, 3160

Geriatrics
AFAR Medical Student Training in Aging Research Program, 335
AHRQ Independent Scientist Award, 365
American Academy of Nursing Building Academic Geriatric Nursing Capacity Scholarships, 505
American Academy of Nursing Claire M. Fagin Fellowships, 506
American Academy of Nursing Mayday Grants, 507
American Academy of Nursing MBA Scholarships, 508
Chest Foundation Geriatric Development Research Awards, 1133
Donald W. Reynolds Foundation Aging and Quality of Life Grants, 1487
Ellison Medical Foundation/AFAR Julie Martin Mid-Career Award in Aging Research, 1576
Ellison Medical Foundation/AFAR Postdoctoral Fellows in Aging Research Program, 1577
FDHN Designated Outcomes Award in Geriatric Gastroenterology, 1640

SUBJECT INDEX

FDHN Designated Research Award in Geriatric Gastroenterology, 1641
Greenwall Foundation Bioethics Grants, 1897
Healthcare Foundation of New Jersey Grants, 1982
Jenkins Foundation: Improving the Health of Greater Richmond Grants, 2219
Leonard and Helen R. Stulman Charitable Foundation Grants, 2356
MMAAP Foundation Geriatric Project Awards, 2582
MMAAP Foundation Senior Health Fellowship Award in Geriatrics Medicine and Aging Research, 2586
Mt. Sinai Health Care Foundation Health of the Jewish Community Grants, 2599
Nell J. Redfield Foundation Grants, 2673
NIA Health Behaviors and Aging Grants, 2761
Piper Trust Older Adults Grants, 3229
RCPSC Balfour M. Mount Visiting Professorship in Palliative Medicine, 3318
Regenstrief General Internal Medicine - General Pediatrics Research Fellowship Program, 3351
United Hospital Fund of New York Health Care Improvement Grants, 3797

German Studies
Fulbright German Studies Seminar Grants, 1751
Fulbright International Education Administrators (IEA) Seminar Program Grants, 1752

Germany
DAAD Research Stays for University Academics and Scientists, 1408
Fulbright German Studies Seminar Grants, 1751
Fulbright International Education Administrators (IEA) Seminar Program Grants, 1752

Gifted/Talented Education
Ford Motor Company Fund Grants Program, 1687
Lynde & Harry Bradley Foundation Fellowships, 2396
Lynde and Harry Bradley Foundation Grants, 2397
Lynde and Harry Bradley Foundation Prizes: Bradley Prizes, 2398
Mary Wilmer Covey Charitable Trust Grants, 2448

Girls
Benton Community Fndn Cookie Jar Grant, 877
Community Foundation of Eastern Connecticut Northeast Women and Girls Grants, 1270
Dallas Women's Foundation Grants, 1417
Frederick McDonald Trust Grants, 1740
Global Fund for Women Grants, 1833
Harmony Project Grants, 1948
Mattel International Grants Program, 2455
RCF The Women's Fund Grants, 3315
USAID Family Planning and Reproductive Health Methods Grants, 3811
USAID Microbicides Research, Development, and Introduction Grants, 3819
Warrick County Community Foundation Women's Fund, 3905

Glaucoma
AHAF National Glaucoma Research Grants, 351
Columbus Foundation Ann Ellis Fund Grants, 1215
Columbus Foundation Competitive Grants, 1217
Glaucoma Foundation Grants, 1827
Glaucoma Research Pilot Project Grants, 1828
GNOF IMPACT Gulf States Eye Surgery Fund, 1840
NYAM Lewis Rudin Glaucoma Grants, 3044

Global Change
Acumen East Africa Fellowship, 284
Bill and Melinda Gates Foundation Emergency Response Grants, 892
Bill and Melinda Gates Foundation Policy and Advocacy Grants, 893
Bill and Melinda Gates Foundation Water, Sanitation and Hygiene Grants, 894
Gates Award for Global Health, 1775
USAID Devel Innovation Accelerator Grants, 3808

Global Issues
Acumen Global Fellowships, 285
Bill and Melinda Gates Foundation Emergency Response Grants, 892
Bill and Melinda Gates Foundation Policy and Advocacy Grants, 893
Bill and Melinda Gates Foundation Water, Sanitation and Hygiene Grants, 894
Gates Award for Global Health, 1775
IIE David L. Boren Fellowships, 2129
IIE Rockefeller Foundation Bellagio Center Residencies, 2142
USAID Devel Innovation Accelerator Grants, 3808

Global Warming
PepsiCo Foundation Grants, 3137

Globalization
Boeing Company Contributions Grants, 930

Glomerular Disease
NIDDK Grants for Basic Research in Glomerular Diseases, 2819

Government
ACF Native American Social and Economic Development Strategies Grants, 231
Arlington Community Foundation Grants, 662
ASM Congressional Science Fellowships, 720
AT&T Foundation Civic and Community Service Program Grants, 793
Baptist Community Ministries Grants, 828
Carnegie Corporation of New York Grants, 1033
Effie and Wofford Cain Foundation Grants, 1563
Eugene B. Casey Foundation Grants, 1609
James S. McDonnell Foundation Complex Systems Collaborative Activity Awards, 2202
Nathan Cummings Foundation Grants, 2627
Playboy Foundation Grants, 3237
Rajiv Gandhi Foundation Grants, 3302
Rockefeller Brothers Peace & Security Grants, 3394
RWJF Childhood Obesity Grants, 3431
Willary Foundation Grants, 3944

Government Regulations
Carla J. Funk Governmental Relations Award, 1020
DPA Promoting Policy Change Advocacy Grants, 1504

Government Studies
APA Congressional Fellowships, 624
APSA Congressional Health and Aging Policy Fellowships, 644
APSA Robert Wood Johnson Foundation Health Policy Fellowships, 645

Government, Comparative
Canada-U.S. Fulbright New Century Scholars Program Grants, 1008
Canada-U.S. Fulbright Senior Specialists Grants, 1009
Fulbright Alumni Initiatives Awards, 1749
Fulbright Distinguished Chairs Awards, 1750
Fulbright New Century Scholars Grants, 1753
Fulbright Specialists Program Grants, 1754
Fulbright Scholars in Europe and Eurasia, 1755
Fulbright Traditional Scholar Program in Sub-Saharan Africa, 1756
Fulbright Traditional Scholar Program in the East Asia/Pacific Region, 1757
Fulbright Traditional Scholar Program in the Near East and North Africa Region, 1758
Fulbright Traditional Scholar Program in the South and Central Asia Region, 1759
Fulbright Traditional Scholar Program in the Western Hemisphere, 1760

Government, Federal
AACAP Mary Crosby Congressional Fellowships, 64
APA Congressional Fellowships, 624
APSA Congressional Health and Aging Policy Fellowships, 644
APSA Robert Wood Johnson Foundation Health Policy Fellowships, 645
Carla J. Funk Governmental Relations Award, 1020
DHHS Emerging Leaders Program Internships, 1473
GEICO Public Service Awards, 1778
IIE David L. Boren Fellowships, 2129
Mt. Sinai Health Care Foundation Health Policy Grants, 2601
Pittsburgh Foundation Community Fund Grants, 3230

Government, Local
Blue Cross Blue Shield of Minnesota Foundation - Health Equity: Building Health Equity Together Grants, 914
Carla J. Funk Governmental Relations Award, 1020
Cone Health Foundation Grants, 1323
DPA Promoting Policy Change Advocacy Grants, 1504
Katharine Matthies Foundation Grants, 2291
Lynde & Harry Bradley Foundation Fellowships, 2396
Lynde and Harry Bradley Foundation Grants, 2397
Lynde and Harry Bradley Foundation Prizes: Bradley Prizes, 2398
Marjorie Moore Charitable Foundation Grants, 2438
NHSCA Arts in Health Care Project Grants, 2741
Parke County Community Foundation Grants, 3115

Government, Municipal
Community Foundation for the National Capital Region Community Leadership Grants, 1260
Four County Community Fndn General Grants, 1710
Four County Community Foundation Healthy Senior/Healthy Youth Fund Grants, 1711
Katharine Matthies Foundation Grants, 2291

Government, State
Carla J. Funk Governmental Relations Award, 1020
Cone Health Foundation Grants, 1323
DPA Promoting Policy Change Advocacy Grants, 1504
Lynde & Harry Bradley Foundation Fellowships, 2396
Lynde and Harry Bradley Foundation Grants, 2397
Lynde and Harry Bradley Foundation Prizes: Bradley Prizes, 2398
Mt. Sinai Health Care Foundation Health Policy Grants, 2601

Governmental Functions
ConocoPhillips Foundation Grants, 1327

Graduate Education
AFB Rudolph Dillman Memorial Scholarship, 338
AIHS Media Summer Fellowship, 392
AOCS Analytical Division Student Award, 612
AOCS Health & Nutrition Div Student Award, 614
AOCS Health and Nutrition Poster Competition, 615
AOCS Processing Division Student Award, 618
AOCS Surfactants and Detergents Division Student Award, 622
BWF Postdoctoral Enrichment Grants, 989
Community Fndn for the Capital Region Grants, 1259
DAR Irene and Daisy MacGregor Memorial Scholarship, 1428
Gates Millennium Scholars Program, 1776
GFWC of Massachusetts Catherine E. Philbin Scholarship, 1811
GFWC of Massachusetts Communication Disorder/Speech Therapy Scholarship, 1812
GFWC of Massachusetts Memorial Education Scholarship, 1813
Grass Foundation Marine Biological Laboratory Advanced Imaging Fellowships, 1877
MBL Gruss Lipper Family Foundation Summer Fellowship, 2475
MBL William Townsend Porter Summer Fellowships for Minority Investigators, 2488
MLA Rittenhouse Award, 2573
SfN Federation of European Neuroscience Societies Forum Travel Awards, 3522
SfN Neuroscience Fellowships, 3530
SfN Science Journalism Student Awards, 3536

Grassroots Leadership
Bank of America Fndn Volunteer Grants, 826
GNOF Stand Up For Our Children Grants, 1847
John Edward Fowler Memorial Fndn Grants, 2235
Miller Foundation Grants, 2556
Norman Foundation Grants, 2983

Great Lakes
George Gund Foundation Grants, 1796

Grief
CJ Foundation for SIDS Program Services Grants, 1163
Robert R. Meyer Foundation Grants, 3387

Growth Factors
NIH Receptors and Signaling in Bone in Health and Disease, 2891

Gun Control
Clayton Baker Trust Grants, 1179

HIV
25th Anniversary Foundation Grants, 8
Abbott Fund Science Education Grants, 203
Actors Fund HIV/AIDS Initiative Grants, 283
AIDS Vaccine Advocacy Coalition (AVAC) Fund Grants, 369
Alexander Fndn Insurance Continuation Grants, 422
Alexis Gregory Foundation Grants, 426
Alfred E. Chase Charitable Foundation Grants, 430
Alliance Healthcare Foundation Grants, 444
Alternatives Research and Development Foundation Alternatives in Education Grants, 455
Alternatives Research and Development Foundation Grants, 456
AMA Foundation Seed Grants for Research, 494
American Psychiatric Association Minority Medical Student Fellowship in HIV Psychiatry, 544
ASM/CDC Fellowships in Infectious Disease and Public Health Microbiology, 716
ASM ICAAC Young Investigator Awards, 732
ASM Merck Irving S. Sigal Memorial Awards, 739
ASPH/CDC Allan Rosenfield Global Health Fellowships, 780
Blowitz-Ridgeway Foundation Grants, 913
Bodenwein Public Benevolent Foundation Grants, 928
Bristol-Myers Squibb Foundation Fellowships, 956
Bristol-Myers Squibb Foundation Global HIV/AIDS Initiative Grants, 957
Bristol-Myers Squibb Foundation Health Disparities Grants, 958
Cable Positive's Tony Cox Community Grants, 992
CDC Epidemic Intell Service Training Grants, 1055
CDC Epidemiology Elective Rotation, 1056
CDC Foundation Atlanta International Health Fellowships, 1060
CDC Public Health Associates, 1065
CDC Public Health Associates Hosts, 1066
CDC Public Health Prev Service Fellowships, 1069
CDC Public Health Prevention Service Fellowship Sponsorships, 1070
Children Affected by AIDS Foundation Camp Network Grants, 1144
Children Affected by AIDS Foundation Domestic Grants, 1145
Children Affected by AIDS Foundation Family Assistance Emergency Fund Grants, 1146
Community Foundation for Greater Atlanta AIDS Fund Grants, 1239
Community Fndn for the Capital Region Grants, 1259
Community Foundation of Louisville AIDS Project Fund Grants, 1289
Dallas Women's Foundation Grants, 1417
DIFFA/Chicago Grants, 1477
Elizabeth Glaser Int'l Leadership Awards, 1566
Elizabeth Glaser Scientist Award, 1567
El Paso Community Foundation Grants, 1579
Elton John AIDS Foundation Grants, 1585
Fairlawn Foundation Grants, 1626
Firelight Foundation Grants, 1667
Frances L. and Edwin L. Cummings Memorial Fund Grants, 1717
Gates Award for Global Health, 1775
Gill Foundation - Gay and Lesbian Fund Grants, 1821
Hagedorn Fund Grants, 1934
Health Fndn of Greater Indianapolis Grants, 1984
Horizon Foundation for New Jersey Grants, 2048
HRSA Resource and Technical Assistance Center for HIV Prevention and Care for Black Men who Have Sex with Men Cooperative Agreement, 2069
HRSA Ryan White HIV AIDS Drug Assistance Grants, 2070
Indiana AIDS Fund Grants, 2155
Jerome Robbins Foundation Grants, 2223
John M. Lloyd Foundation Grants, 2244
Johnson & Johnson Corporate Contributions, 2253
Johnson & Johnson/SAH Arts & Healing Grants, 2254
M-A-C AIDS Fund Grants, 2400
McCarthy Family Foundation Grants, 2490
NHLBI Microbiome of Lung & Respiratory in HIV Infected Individuals & Uninfected Controls, 2712
NIA AIDS and Aging: Behavioral Sciences Prevention Research Grants, 2747
NIA Medical Management of Older Patients with HIV/AIDS Grants, 2770
NIDA Pilot and Feasibility Studies in Preparation for Drug Abuse Prevention Trials, 2794
NIDDK Transmission of Human Immunodeficiency Virus (HIV) In Semen, 2851
NIMH AIDS and Aging: Behavioral Sciences Prevention Research Grants, 2915
NINR AIDS and Aging: Behavioral Sciences Prevention Research Grants, 2927
NYCT AIDS/HIV Grants, 3048
NYCT Girls and Young Women Grants, 3053
OSF International Harm Reduction Development Program Grants, 3090
Paul Rapoport Foundation Grants, 3125
Playboy Foundation Grants, 3237
Questar Corporate Contributions Grants, 3291
Rainbow Endowment Grants, 3300
Rockefeller Brothers Fund Pivotal Places Grants: South Africa, 3395
San Diego HIV Funding Collaborative Grants, 3473
Sexually Transmitted Diseases Postdoctoral Fellowships, 3516
United Methodist Committee on Relief Global AIDS Fund Grants, 3798
USAID Comprehensive District-Based Support for Better HIV/TB Patient Outcomes Grants, 3807
USAID HIV Prevention with Key Populations - Mali Grants, 3816
USAID Microbicides Research, Development, and Introduction Grants, 3819
USAID Systems Strengthening for Better HIV/TB Patient Outcomes Grants, 3824
USCM HIV/AIDS Prevention Grants, 3827
USDD Broad Agency Announcement Grants, 3837
USDD Broad Agency Announcement HIV/AIDS Prevention Grants, 3838
USDD HIV/AIDS Prevention: Military Specific HIV/AIDS Prevention, Care, and Treatment for Non-PEPFAR Funded Countries Grants, 3841
USDD HIV/AIDS Prevention Program Information Systems Development Grants, 3842
USDD President's Emergency Plan for AIDS Relief Military Specific HIV Seroprevalence and Behavioral Epidemiology Risk Suvey Grants, 3845

Habitat Preservation
George H.C. Ensworth Memorial Fund Grants, 1797
Green Diamond Charitable Contributions, 1891
Greygates Foundation Grants, 1901
Sioux Falls Area Community Foundation Community Fund Grants (Unrestricted), 3568
Sioux Falls Area Community Foundation Spot Grants (Unrestricted), 3570
Union Bank, N.A. Foundation Grants, 3788
United Technologies Corporation Grants, 3803

Hawaii
Antone & Edene Vidinha Charitable Trust Grants, 609
Clarence T.C. Ching Foundation Grants, 1167
Elsie H. Wilcox Foundation Grants, 1583

Hawaiian Natives
AMA Foundation Minority Scholars Awards, 491
Antone & Edene Vidinha Charitable Trust Grants, 609
MLA Grad Scholarship for Minority Students, 2565

Hazardous Waste
NIEHS Hazardous Materials Worker Health and Safety Training Grants, 2854
NIEHS Hazmat Training at Doe Nuclear Weapons Complex Grants, 2855
NIEHS SBIR E-Learning for HAZMAT and Emergency Response Grants, 2859
San Diego Foundation for Change Grants, 3470
Sioux Falls Area Community Foundation Community Fund Grants (Unrestricted), 3568
Sioux Falls Area Community Foundation Spot Grants (Unrestricted), 3570
U.S. Department of Education Erma Byrd Scholarships, 3768

Hazardous Waste, Disposal/Clean-Up
NIEHS Hazardous Materials Worker Health and Safety Training Grants, 2854
NIEHS Hazmat Training at Doe Nuclear Weapons Complex Grants, 2855
NIEHS SBIR E-Learning for HAZMAT and Emergency Response Grants, 2859

Headaches
National Headache Foundation Research Grants, 2632
National Headache Foundation Seymour Diamond Clinical Fellowship in Headache Education, 2633

Healing
Ringing Rocks Foundation Discretionary Grants, 3375

Health
3M Company Fndn Community Giving Grants, 6
1675 Foundation Grants, 11
Abundance Foundation International Grants, 218
Ackerman Foundation Grants, 236
ACL Business Acumen for Disability Organizations Grants, 238
ACL Field Initiated Projects Program: Minority-Serving Institution Development Grants, 242
ACL Field Initiated Projects Program: Minority-Serving Institution Research Grants, 243
ACL Self-Advocacy Resource Center Grants, 247
ACSM Michael L. Pollock Student Scholarship, 270
Adams and Reese Corporate Giving Grants, 299
Advance Auto Parts Corporate Giving Grants, 312
AEGON Transamerica Foundation Health and Welfare Grants, 316
Aetna Foundation Obesity Grants, 331
Albert and Margaret Alkek Foundation Grants, 405
Allegis Group Foundation Grants, 441
AMDA Foundation Quality Improvement Award, 499
American Electric Power Foundation Grants, 530
Amerisure Insurance Community Service Grants, 566
Appalachian Regional Commission Health Grants, 637
APSA Robert Wood Johnson Foundation Health Policy Fellowships, 645
Arkansas Community Foundation Arkansas Black Hall of Fame Grants, 659
Armstrong McDonald Foundation Health Grants, 663
Arthur M. Blank Family Foundation Inspiring Spaces Grants, 673
ASHFoundation New Century Scholars Program Doctoral Scholarships, 706
ASHFoundation New Century Scholars Research Grant, 707
ASHFoundation Grant for New Investigators, 708
Austin College Leadership Award, 804
Autodesk Community Relations Grants, 809

SUBJECT INDEX

Health

Bernard and Audre Rapoport Foundation Health Grants, 882
Biogen Foundation Scientific Research Grants, 902
Boyle Foundation Grants, 942
Bradley C. Higgins Foundation Grants, 945
Bright Promises Foundation Grants, 952
Caesars Foundation Grants, 995
Cailloux Foundation Grants, 996
CDC-Hubert Global Health Fellowship, 1047
Centerville-Washington Foundation Grants, 1078
CFFVR Chilton Area Community Fndn Grants, 1096
CFFVR Jewelers Mutual Charitable Giving, 1099
CFFVR Project Grants, 1101
Chapman Charitable Foundation Grants, 1112
Charles M. & Mary D. Grant Fndn Grants, 1124
Chilkat Valley Community Foundation Grants, 1148
CICF James Proctor Grant for Aged, 1157
Cigna Civic Affairs Sponsorships, 1159
CNO Financial Group Community Grants, 1201
Columbus Foundation Competitive Grants, 1217
Community Fndn for San Benito County Grants, 1257
Community Foundation for Southeast Michigan Grants, 1258
Community Fndn for the Capital Region Grants, 1259
Community Fndn of Greater Lafayette Grants, 1280
Community Fndn of Louisville Health Grants, 1295
Community Foundation of Riverside & San Bernardino County Impact Grants, 1303
Community Fndn of South Puget Sound Grants, 1307
Cresap Family Foundation Grants, 1359
CSX Corporate Contributions Grants, 1368
Cuesta Foundation Grants, 1371
D.V. and Ida J. McEachern Trust Grants, 1406
Daisy Marquis Jones Foundation Grants, 1414
Dale and Edna Walsh Foundation Grants, 1415
Dayton Power and Light Company Foundation Signature Grants, 1436
Dayton Power and Light Foundation Grants, 1437
Decatur County Community Foundation Small Project Grants, 1445
Delmarva Power and Light Contributions, 1454
Delta Air Lines Foundation Health and Wellness Grants, 1455
DTE Energy Foundation Health and Human Services Grants, 1516
Dyson Foundation Mid-Hudson Valley Project Support Grants, 1534
Earl and Maxine Claussen Trust Grants, 1541
Edward and Helen Bartlett Foundation Grants, 1554
Edyth Bush Charitable Foundation Grants, 1562
El Paso Corporate Foundation Grants, 1580
El Pomar Foundation Anna Keesling Ackerman Fund Grants, 1581
El Pomar Foundation Grants, 1582
Federal Express Corporate Contributions, 1657
Fidelity Foundation Grants, 1659
Fifth Third Foundation Grants, 1661
FIU Global Civic Engagement Mini Grants, 1673
Foundation for Balance and Harmony Grants, 1695
Foundations of East Chicago Health Grants, 1709
Four County Community Fndn General Grants, 1710
Four County Community Foundation Healthy Senior/ Healthy Youth Fund Grants, 1711
Four J Foundation Grants, 1713
France-Merrick Foundation Health and Human Services Grants, 1715
Freddie Mac Foundation Grants, 1738
Fremont Area Community Foundation Amazing X Grants, 1744
Genentech Corp Charitable Contributions, 1780
George H.C. Ensworth Memorial Fund Grants, 1797
Gil and Dody Weaver Foundation Grants, 1819
GNOF Bayou Communities Grants, 1835
GNOF IMPACT Grants for Health and Human Services, 1839
Grand Haven Area Community Fndn Grants, 1864
Greater Sitka Legacy Fund Grants, 1885
Green River Area Community Fndn Grants, 1895
Gruber Foundation Weizmann Institute Awards, 1911
Hamilton Company Syringe Product Grant, 1937

Hancock County Community Foundation - Field of Interest Grants, 1943
Hannaford Charitable Foundation Grants, 1944
Harden Foundation Grants, 1945
Hardin County Community Foundation Grants, 1946
Harold and Arlene Schnitzer CARE Foundation Grants, 1950
Harrison County Community Foundation Grants, 1957
Harrison County Community Foundation Signature Grants, 1958
Harry Edison Foundation, 1962
Harry S. Black and Allon Fuller Fund Grants, 1966
Harry W. Bass, Jr. Foundation Grants, 1968
Henry and Ruth Blaustein Rosenberg Foundation Health Grants, 2004
Herbert A. & Adrian W. Woods Fndn Grants, 2007
Hilda and Preston Davis Foundation Grants, 2028
Hill Crest Foundation Grants, 2030
Hillsdale County Community General Adult Foundation Grants, 2034
HRK Foundation Health Grants, 2067
Intergrys Corporation Grants, 2160
J.E. and L.E. Mabee Foundation Grants, 2170
J. Willard Marriott, Jr. Foundation Grants, 2178
Jacob and Hilda Blaustein Foundation Health and Mental Health Grants, 2181
Jane Bradley Pettit Foundation Health Grants, 2208
Jasper Foundation Grants, 2212
Jennings County Community Foundation Women's Giving Circle Grant, 2221
Jessica Stevens Community Foundation Grants, 2224
Johnson Controls Foundation Health and Social Services Grants, 2255
Joseph Henry Edmondson Foundation Grants, 2270
Kenai Peninsula Foundation Grants, 2298
Ketchikan Community Foundation Grants, 2303
Kettering Fund Grants, 2305
Kluge Center David B. Larson Fellowship in Health and Spirituality, 2311
Kodiak Community Foundation Grants, 2314
Lake County Community Fund Grants, 2333
Legler Benbough Foundation Grants, 2352
Libra Foundation Grants, 2363
Lisa and Douglas Goldman Fund Grants, 2369
Lloyd A. Fry Foundation Health Grants, 2373
Lucile Packard Foundation for Children's Health Grants, 2386
Lucy Downing Nisbet Charitable Fund Grants, 2387
Lumpkin Family Fnd Healthy People Grants, 2391
Lydia deForest Charitable Trust Grants, 2393
Maine Community Foundation Penobscot Valley Health Association Grants, 2417
Mann T. Lowry Foundation Grants, 2419
Mary Black Fndn Community Health Grants, 2441
Mary Wilmer Covey Charitable Trust Grants, 2448
Mattel International Grants Program, 2455
Medtronic Foundation CommunityLink Health Grants, 2510
Mercedes-Benz USA Corporate Contributions, 2517
Meyer Memorial Trust Grassroots Grants, 2536
Meyer Memorial Trust Responsive Grants, 2537
MGM Resorts Foundation Community Grants, 2541
Morton K. and Jane Blaustein Foundation Health Grants, 2595
Nelda C. and H.J. Lutcher Stark Fndn Grants, 2672
NIGMS Advancing Basic Behavioral and Social Research on Resilience: an Integrative Science Approach, 2861
NIH Building Interdisciplinary Research Careers in Women's Health Grants, 2869
NIH Mentored Clinical Scientist Research Career Development Award, 2887
NIH Ruth L. Kirschstein National Research Service Awards for Individual Predoctoral Fellows, 2903
Norcliffe Foundation Grants, 2979
Northrop Grumman Corporation Grants, 2997
OceanFirst Foundation Major Grants, 3062
Olympus Corporation of Americas Corporate Giving Grants, 3073
OneStar Foundation AmeriCorps Grants, 3077

Orange County Community Foundation Grants, 3079
Oregon Community Fndn Community Grants, 3082
OSF Accountability and Monitoring in Health Initiative Grants, 3087
OSF Global Health Financing Initiative Grants, 3088
OSF Health Media Initiative Grants, 3089
OSF Law and Health Initiative Grants, 3092
OSF Public Health Grants in Kyrgyzstan, 3094
OSF Roma Health Project Grants, 3095
Pacific Life Foundation Grants, 3103
Pajaro Valley Community Health Trust Diabetes and Contributing Factors Grants, 3109
Palmer Foundation Grants, 3112
Petersburg Community Foundation Grants, 3156
Peyton Anderson Foundation Grants, 3160
Philip L. Graham Fund Health and Human Services Grants, 3190
PMI Foundation Grants, 3239
Pokagon Fund Grants, 3242
Porter County Health and Wellness Grant, 3246
PPCF Community Grants, 3256
PPG Industries Foundation Grants, 3257
Pulido Walker Foundation, 3282
Qualcomm Grants, 3287
Ripley County Community Foundation Small Project Grants, 3377
Rockwell Fund, Inc. Grants, 3398
Rogers Family Foundation Grants, 3401
Rohm and Haas Company Grants, 3403
RWJF Policies for Action: Policy and Law Research to Build a Culture of Health Grants, 3443
Salt River Health & Human Services Grants, 3459
Samuel N. and Mary Castle Foundation Grants, 3467
Seattle Foundation Benjamin N. Phillips Memorial Fund Grants, 3504
Seattle Foundation Health and Wellness Grants, 3507
Seward Community Foundation Grants, 3514
Seward Community Foundation Mini-Grants, 3515
Shield-Ayres Foundation Grants, 3546
Sigma Theta Tau International /American Nurses Foundation Grant, 3556
Sigma Theta Tau Small Grants, 3560
Simpson Lumber Charitable Contributions, 3566
Skoll Fndn Awards for Social Entrepreneurship, 3579
Sony Corporation of America Grants, 3623
SSHRC-NSERC College and Community Innovation Enhancement Grants, 3636
SSRC Collaborative Research Grants on Environment and Health in China, 3638
Straits Area Community Foundation Health Care Grants, 3657
Textron Corporate Contributions Grants, 3712
Tifa Foundation Grants, 3733
Toyota Motor Manufacturing of Alabama Grants, 3743
Toyota Motor Manuf of Mississippi Grants, 3746
Toyota Motor Manufacturing of Texas Grants, 3747
Toyota Motor Manuf of West Virginia Grants, 3748
Unihealth Foundation Grants for Non-Profit Oranizations, 3783
UniHealth Foundation Innovation Fund Grants, 3785
UniHealth Foundation Workforce Development Grants, 3786
United Healthcare Comm Grants in Michigan, 3796
United Methodist Health Ministry Fund Grants, 3801
USAID Accelerating Progress Against Tuberculosis in Kenya Grants, 3805
USAID Devel Innovation Accelerator Grants, 3808
USAID Family Health Plus Project Grants, 3810
USAID Global Health Development Innovation Accelerator Broad Agency Grants, 3813
USAID HIV Prevention with Key Populations - Mali Grants, 3816
USAID Integration of Care and Support within the Health System to Support Better Patient Outcomes Grants, 3817
USAID Research and Innovation for Health Supply Chain Systems & Commodity Security Grants, 3821
USDD Clinical Functional Assessment Investigator- Initiated Grants, 3840
USG Foundation Grants, 3848

736 / Health

Walker Area Community Foundation Grants, 3895
Walmart Foundation Community Giving Grants, 3897
Walmart Foundation National Giving Grants, 3898
Walmart Foundation Northwest Arkansas Giving, 3899
Wege Foundation Grants, 3912
Welborn Baptist Fndn General Op Grants, 3914
Welborn Baptist Foundation Improvements to Community Health Status Grants, 3916
Welborn Baptist Fndn School Health Grants, 3917
Wellcome Trust New Investigator Awards, 3919
Whitley County Community Foundation Grants, 3936
William D. Laurie, Jr. Charitable Fndn Grants, 3949
William Ray & Ruth E. Collins Fndn Grants, 3956
Williams Companies Foundation Homegrown Giving Grants, 3957
Willis C. Helm Charitable Trust Grants, 3959
Wolfe Associates Grants, 3969

Health Care
1st Touch Foundation Grants, 3
2 Depot Square Ipswich Charitable Fndn Grants, 4
3M Fndn Health & Human Services Grants, 7
49ers Foundation Grants, 9
100 Mile Man Foundation Grants, 10
A/H Foundation Grants, 18
AACAP Jeanne Spurlock Research Fellowship in Substance Abuse and Addiction for Minority Medical Students, 60
AACAP Summer Medical Student Fellowships, 75
AACR Scholar-in-Training Awards, 124
AACR Scholar-in-Training Awards - Special Conferences, 125
AAFA Investigator Research Grants, 128
AAMC Award for Distinguished Research, 155
AAMC David E. Rogers Award, 157
AAMC Herbert W. Nickens Award, 158
AAMC Herbert W. Nickens Medical Student Scholarships, 160
AAMC Humanism in Medicine Award, 161
AAP Community Access To Child Health (CATCH) Implementation Grants, 178
AAP Resident Initiative Fund Grants, 189
Aaron & Freda Glickman Foundation Grants, 192
Aaron Foundation Grants, 194
Aaron Foundation Grants, 193
Abbott Fund Access to Health Care Grants, 199
Abbott Fund Community Grants, 201
Abbott Fund Global AIDS Care Grants, 202
Abbott Fund Science Education Grants, 203
Abell Fndn Health & Human Services Grants, 209
Abernethy Family Foundation Grants, 210
Abington Foundation Grants, 211
Able To Serve Grants, 212
Abney Foundation Grants, 214
Aboudane Family Foundation Grants, 215
Abramson Family Foundation Grants, 217
ACAAI Foundation Research Grants, 219
AcademyHealth Alice S. Hersh New Investigator Awards, 220
AcademyHealth Alice Hersh Student Scholarships, 221
AcademyHealth Awards, 222
AcademyHealth Nemours Child Health Services Research Awards, 224
AcademyHealth PHSR Interest Group Student Scholarships, 225
AcademyHealth PHSR Research Article of the Year Awards, 226
ACF Foundation Grants, 230
ACGT Young Investigator Grants, 233
Achelis Foundation Grants, 234
ACL Alzheimer's Disease Initiative Grants, 237
ACL Business Acumen for Disability Organizations Grants, 238
ACL Diversity Community of Practice Grants, 240
ACL Field Initiated Projects Program: Minority-Serving Institution Development Grants, 242
ACL Field Initiated Projects Program: Minority-Serving Institution Research Grants, 243
ACL Learning Collaboratives for Advanced Business Acumen Skills Grants, 244

ACS Pilot and Exploratory Projects in Palliative Care of Cancer Patients and Their Families Grants, 276
Adam Richter Charitable Trust Grants, 298
Adams and Reese Corporate Giving Grants, 299
Adams County Community Foundation of Pennsylvania Grants, 300
Addison H. Gibson Foundation Medical Grants, 305
Adelaide Breed Bayrd Foundation Grants, 306
Adler-Clark Electric Community Commitment Foundation Grants, 307
Administaff Community Affairs Grants, 308
Adolph Coors Foundation Grants, 311
Advance Auto Parts Corporate Giving Grants, 312
Advanced Micro Devices Comm Affairs Grants, 313
AEC Trust Grants, 315
AEGON Transamerica Foundation Health and Welfare Grants, 316
Aetna Foundation Health Grants in Connecticut, 327
Aetna Foundation Integrated Health Care Grants, 328
Aetna Foundation Obesity Grants, 331
Aetna Foundation Racial and Ethnic Health Care Equity Grants, 332
AFG Industries Grants, 340
A Friends' Foundation Trust Grants, 341
Agnes Gund Foundation Grants, 345
Agnes M. Lindsay Trust Grants, 346
A Good Neighbor Foundation Grants, 347
Ahmanson Foundation Grants, 355
AHRF Regular Research Grants, 363
Aid for Starving Children International Grants, 368
Air Products and Chemicals Grants, 399
Alabama Power Foundation Grants, 400
Albany Medical Center Prize in Medicine and Biomedical Research, 404
Alberto Culver Corporate Contributions Grants, 408
Albert Pick Jr. Fund Grants, 409
Albertson's Charitable Giving Grants, 410
Albert W. Rice Charitable Foundation Grants, 411
Albuquerque Community Foundation Grants, 412
Alcatel-Lucent Technologies Foundation Grants, 413
Alcoa Foundation Grants, 414
Alcohol Misuse and Alcoholism Research Grants, 415
Alexander Foundation Cancer, Catastrophic Illness and Injury Grants, 421
Alexander Fndn Insurance Continuation Grants, 422
Alfred and Tillie Shemanski Testamentary Trust Grants, 428
Alfred Bersted Foundation Grants, 429
Alfred E. Chase Charitable Foundation Grants, 430
Alice C. A. Sibley Fund Grants, 435
Alice Tweed Tuohy Foundation Grants Program, 437
Allegan County Community Foundation Grants, 439
Allegheny Technologies Charitable Trust, 440
AlohaCare Believes in Me Scholarship, 449
Alpha Natural Resources Corporate Giving, 450
ALSAM Foundation Grants, 454
Altman Foundation Health Care Grants, 457
AMA-MSS Chapter Involvement Grants, 473
AMA-MSS Chapter of the Year (COTY) Award, 474
AMA-MSS Research Poster Award, 477
AMA-RFS and AMA Foundation Medical Student & Resident/Fellow Elective-Medicine & Media, 478
AMA-RFS Legislative Awareness Internships, 479
AMA-WPC Joan F. Giambalvo Memorial Scholarships, 480
Amador Community Foundation Grants, 481
AMA Foundation Dr. Nathan Davis International Awards in Medicine, 483
AMA Foundation Health Literacy Grants, 485
AMA Foundation Healthy Communities/Healthy America Grants, 486
AMA Foundation Jack B. McConnell, MD Awards for Excellence in Volunteerism, 487
AMA Foundation Jordan Fieldman, MD, Awards, 489
AMA Foundation Leadership Awards, 490
AMA Foundation Pride in the Profession Grants, 493
AMA Foundation Worldscopes Program, 495
AMA Virtual Mentor Theme Issue Editor Grants, 496
AMDA Foundation Quality Improvement and Health Outcome Awards, 498

Amelia Sillman Rockwell and Carlos Perry Rockwell Charities Fund Grants, 501
American-Scandinavian Foundation Visiting Lectureship Grants, 502
American Express Community Service Grants, 531
American Foodservice Charitable Trust Grants, 532
American Jewish World Service Grants, 533
American Psychiatric Foundation Disaster Recovery Fund for Psychiatrists, 553
American Schlafhorst Foundation Grants, 559
Amica Companies Foundation Grants, 574
Andersen Corporate Foundation, 586
Angels Baseball Foundation Grants, 590
Anheuser-Busch Foundation Grants, 591
Anna Fitch Ardenghi Trust Grants, 592
Ann and Robert H. Lurie Family Fndn Grants, 593
Annie Sinclair Knudsen Memorial Fund/Kaua'i Community Grants Program, 597
Ann Jackson Family Foundation Grants, 598
Ann Peppers Foundation Grants, 599
Annunziata Sanguinetti Foundation Grants, 600
Anschutz Family Foundation Grants, 602
Anthem Blue Cross and Blue Shield Grants, 607
Antone & Edene Vidinha Charitable Trust Grants, 609
Appalachian Regional Commission Health Grants, 637
APS Foundation Grants, 646
AptarGroup Foundation Grants, 647
Aragona Family Foundation Grants, 648
Aratani Foundation Grants, 649
Arcadia Foundation Grants, 650
Argyros Foundation Grants Program, 653
Arie and Ida Crown Memorial Grants, 654
Arizona Cardinals Grants, 655
Arizona Diamondbacks Charities Grants, 657
Arizona Public Service Corporate Giving Program Grants, 658
Arkell Hall Foundation Grants, 661
Arlington Community Foundation Grants, 662
Armstrong McDonald Foundation Health Grants, 663
ARS Foundation Grants, 667
ArvinMeritor Foundation Health Grants, 674
ArvinMeritor Grants, 675
Assisi Fndn of Memphis Capital Project Grants, 786
Assisi Foundation of Memphis General Grants, 787
Assurant Health Foundation Grants, 789
AT&T Foundation Civic and Community Service Program Grants, 793
Athwin Foundation Grants, 795
Atlanta Foundation Grants, 796
Aurora Foundation Grants, 802
Austin-Bailey Health and Wellness Fndn Grants, 803
Australasian Institute of Judicial Administration Seed Funding Grants, 806
Autzen Foundation Grants, 810
Babcock Charitable Trust Grants, 816
Bacon Family Foundation Grants, 817
Bailey Foundation Grants, 818
Balfe Family Foundation Grants, 819
Ball Brothers Foundation General Grants, 820
Banfi Vintners Foundation Grants, 824
Baptist-Trinity Lutheran Legacy Fndn Grants, 827
Barberton Community Foundation Grants, 829
Barker Foundation Grants, 830
Barker Welfare Foundation Grants, 831
Barra Foundation Community Fund Grants, 833
Barra Foundation Project Grants, 834
Barth Syndrome Foundation Research Grants, 836
Batchelor Foundation Grants, 837
Baton Rouge Area Foundation Grants, 838
Battle Creek Community Foundation Grants, 839
Baxter International Corporate Giving Grants, 841
Baxter International Foundation Foster G. McGaw Prize, 842
Baxter International Foundation Grants, 843
Baxter International Foundation William B. Graham Prize for Health Services Research, 844
BCBSM Claude Pepper Award, 850
BCBSM Corporate Contributions Grants, 851
BCBSM Foundation Community Health Matching Grants, 852

SUBJECT INDEX

Health Care

BCBSM Foundation Excellence in Research Awards for Students, 853
BCBSM Foundation Investigator Initiated Research Grants, 854
BCBSM Foundation Physician Investigator Research Awards, 855
BCBSM Foundation Primary and Clinical Prevention Grants, 856
BCBSM Fndn Proposal Development Awards, 857
BCBSM Foundation Student Award Program, 858
BCBSM Frank J. McDevitt, DO, Excellence in Research Awards, 859
Beazley Foundation Grants, 865
Bechtel Group Foundation Building Positive Community Relationships Grants, 866
Beckley Area Foundation Grants, 867
Becton Dickinson and Company Grants, 869
Beirne Carter Foundation Grants, 870
Ben B. Cheney Foundation Grants, 875
Bender Foundation Grants, 876
Benton County Foundation Grants, 880
Berks County Community Foundation Grants, 881
Berrien Community Foundation Grants, 884
Bert W. Martin Foundation Grants, 886
Besser Foundation Grants, 887
BHHS Legacy Foundation Grants, 888
Bildner Family Foundation Grants, 891
Bindley Family Foundation Grants, 896
Biogen Foundation Healthcare Professional Education Grants, 900
Birmingham Foundation Grants, 903
Blanche and Irving Laurie Foundation Grants, 910
Blanche and Julian Robertson Family Foundation Grants, 911
Blowitz-Ridgeway Foundation Early Childhood Development Research Award, 912
Blowitz-Ridgeway Foundation Grants, 913
Blue Grass Community Fndn Harrison Grants, 919
Blue Grass Community Foundation Hudson-Ellis Fund Grants, 920
Blue Mountain Community Foundation Grants, 921
Blue Shield of California Grants, 923
Blum-Kovler Foundation Grants, 924
Bob & Delores Hope Foundation Grants, 927
Bodenwein Public Benevolent Foundation Grants, 928
Bodman Foundation Grants, 929
Boeing Company Contributions Grants, 930
Boettcher Foundation Grants, 931
Booth-Bricker Fund Grants, 933
Bosque Foundation Grants, 935
Boston Foundation Grants, 936
Boston Globe Foundation Grants, 937
Boston Jewish Community Women's Fund Grants, 938
Boyd Gaming Corporation Contributions Program, 941
Boyle Foundation Grants, 942
Bradley-Turner Foundation Grants, 944
Bradley C. Higgins Foundation Grants, 945
Bravewell Leadership Award, 946
Brian G. Dyson Foundation Grants, 948
Bridgestone/Firestone Trust Fund Grants, 949
Bright Family Foundation Grants, 951
Bristol-Myers Squibb Foundation Health Disparities Grants, 958
Bristol-Myers Squibb Foundation Health Education Grants, 959
Bristol-Myers Squibb Foundation Product Donations Grants, 960
Bristol-Myers Squibb Foundation Women's Health Grants, 962
Bristol-Myers Squibb Patient Assistance Grants, 963
Broad Foundation IBD Research Grants, 964
Brook J. Lenfest Foundation Grants, 967
Brooklyn Community Foundation Caring Neighbors Grants, 968
Bupa Foundation Medical Research Grants, 974
Burlington Industries Foundation Grants, 977
Burlington Northern Santa Fe Foundation Grants, 978
Bush Fndn Health & Human Services Grants, 979
Bush Foundation Medical Fellowships, 980
Byron W. & Alice L. Lockwood Fnd Grants, 991

Caesars Foundation Grants, 995
Caleb C. and Julia W. Dula Educational and Charitable Foundation Grants, 997
California Community Foundation Health Care Grants, 998
Callaway Foundation Grants, 1002
Callaway Golf Company Foundation Grants, 1003
Cambridge Community Foundation Grants, 1004
Camp-Younts Foundation Grants, 1005
Campbell Hoffman Foundation Grants, 1006
Campbell Soup Foundation Grants, 1007
Capital Region Community Foundation Grants, 1014
Cardinal Health Foundation Grants, 1017
Caring Foundation Grants, 1019
Carl & Eloise Pohlad Family Fndn Grants, 1021
Carl C. Icahn Foundation Grants, 1023
Carl Gellert and Celia Berta Gellert Foundation Grants, 1024
Carl M. Freeman Foundation FACES Grants, 1026
Carl R. Hendrickson Family Foundation Grants, 1028
Carlsbad Charitable Foundation Grants, 1029
Carls Foundation Grants, 1030
Carl W. and Carrie Mae Joslyn Trust Grants, 1031
Carolyn Foundation Grants, 1034
Carpenter Foundation Grants, 1035
Carrie Estelle Doheny Foundation Grants, 1037
Carrier Corporation Contributions Grants, 1038
Carroll County Community Foundation Grants, 1039
Carylon Foundation Grants, 1040
Catherine Kennedy Home Foundation Grants, 1043
CDC Cooperative Agreement for Partnership to Enhance Public Health Informatics, 1050
CDC Cooperative Agreement for the Development, Operation, and Evaluation of an Entertainment Education Program, 1051
Cemala Foundation Grants, 1077
Centerville-Washington Foundation Grants, 1078
Central Carolina Community Foundation Community Impact Grants, 1079
Central Okanagan Foundation Grants, 1080
CFFVR Capital Credit Union Charitable Giving Grants, 1095
CFFVR Clintonville Area Foundation Grants, 1097
CFFVR Frank C. Shattuck Community Grants, 1098
CFFVR Jewelers Mutual Charitable Giving, 1099
CFFVR Myra M. and Robert L. Vandehey Foundation Grants, 1100
CFFVR Robert and Patricia Endries Family Foundation Grants, 1102
CFFVR Shawano Area Community Foundation Grants, 1104
CFFVR Women's Fund for the Fox Valley Region Grants, 1107
Chamberlain Foundation Grants, 1108
Champlin Foundations Grants, 1110
Chapman Charitable Foundation Grants, 1112
CharityWorks Grants, 1113
Charles A. Frueauff Foundation Grants, 1114
Charles Delmar Foundation Grants, 1115
Charles F. Bacon Trust Grants, 1117
Charles H. Dater Foundation Grants, 1118
Charles H. Farnsworth Trust Grants, 1119
Charles H. Hall Foundation, 1120
Charles H. Pearson Foundation Grants, 1121
Charles Lafitte Foundation Grants, 1123
Charles Nelson Robinson Fund Grants, 1125
Charlotte County (FL) Community Foundation Grants, 1126
Chase Paymentech Corporate Giving Grants, 1127
Chatlos Foundation Grants Program, 1128
Chemtura Corporation Contributions Grants, 1130
Children's Brain Tumor Fnd Research Grants, 1136
Children Affected by AIDS Foundation Domestic Grants, 1145
Children Affected by AIDS Foundation Family Assistance Emergency Fund Grants, 1146
Chilkat Valley Community Foundation Grants, 1148
Christine & Katharina Pauly Trust Grants, 1150
Christopher & Dana Reeve Foundation Quality of Life Grants, 1152

Christy-Houston Foundation Grants, 1153
Chula Vista Charitable Foundation Grants, 1154
CICF Indianapolis Fndn Community Grants, 1156
CICF Legacy Fund Grants, 1158
Cigna Civic Affairs Sponsorships, 1159
CIGNA Foundation Grants, 1160
CIT Corporate Giving Grants, 1161
Citizens Bank Mid-Atlantic Charitable Foundation Grants, 1162
Clara Blackford Smith and W. Aubrey Smith Charitable Foundation Grants, 1165
Clarence E. Heller Charitable Foundation Grants, 1166
Clarence T.C. Ching Foundation Grants, 1167
Clark-Winchcole Foundation Grants, 1173
Clark and Ruby Baker Foundation Grants, 1174
Clark County Community Foundation Grants, 1175
Claude Bennett Family Foundation Grants, 1176
Claude Worthington Benedum Fndn Grants, 1178
Cleveland-Cliffs Foundation Grants, 1181
CLIF Bar Family Foundation Grants, 1183
CMS Hispanic Health Services Research Grants, 1186
CNA Foundation Grants, 1189
CNO Financial Group Community Grants, 1201
Coastal Community Foundation of South Carolina Grants, 1202
Coca-Cola Foundation Grants, 1203
Cockrell Foundation Grants, 1204
Coeta and Donald Barker Foundation Grants, 1205
Coleman Foundation Cancer Care Grants, 1206
Collins C. Diboll Private Foundation Grants, 1209
Collins Foundation Grants, 1210
Colonel Stanley R. McNeil Foundation Grants, 1211
Colorado Trust Grants, 1213
Columbus Foundation Allen Eiry Fund Grants, 1214
Columbus Foundation Ann Ellis Fund Grants, 1215
Columbus Foundation Central Benefits Health Care Foundation Grants, 1216
Columbus Foundation Estrich Fund Grants, 1218
Columbus Foundation J. Floyd Dixon Memorial Fund Grants, 1219
Columbus Foundation Mary Eleanor Morris Fund Grants, 1220
Columbus Foundation Robert E. and Genevieve B. Schaefer Fund Grants, 1222
Columbus Foundation Traditional Grants, 1223
Comerica Charitable Foundation Grants, 1224
Commonwealth Fund Affordable Health Insurance Grants, 1227
Commonwealth Fund Australian-American Health Policy Fellowships, 1228
Commonwealth Fund Health Care Quality Improvement and Efficiency Grants, 1230
Commonwealth Fund Patient-Centered Coordinated Care Program Grants, 1231
Commonwealth Payment System Reform Grants, 1232
Commonwealth Fund Quality of Care for Frail Elders Grants, 1233
Commonwealth Fund Small Grants, 1234
Commonwealth Fund State High Performance Health Systems Grants, 1235
Communities Foundation of Texas Grants, 1236
Community Foundation Alliance City of Evansville Endowment Fund Grants, 1238
Community Foundation for Greater Atlanta Common Good Funds Grants, 1241
Community Fndn for Greater Buffalo Grants, 1247
Community Fndn for Monterey County Grants, 1253
Community Fndn for Muskegon County Grants, 1254
Community Fndn for the Capital Region Grants, 1259
Community Foundation of Bloomington and Monroe County Grants, 1265
Community Fndn of Central Illinois Grants, 1267
Community Fndn of East Central Illinois Grants, 1268
Community Foundation of Eastern Connecticut General Southeast Grants, 1269
Community Foundation of Greater Birmingham Grants, 1272
Community Foundation of Greater Flint Grants, 1273
Community Foundation of Greater Greensboro Community Grants, 1277

738 / Health Care

Community Foundation of Greater New Britain Grants, 1281
Community Fndn of Greater Tampa Grants, 1282
Community Foundation of Greenville-Greenville Women Giving Grants, 1283
Community Foundation of Greenville Community Enrichment Grants, 1284
Community Foundation of Greenville Hollingsworth Funds Program/Project Grants, 1285
Community Foundation of Louisville AIDS Project Fund Grants, 1289
Community Fndn of Louisville Health Grants, 1295
Community Foundation of Mount Vernon and Knox County Grants, 1299
Community Fndn of Randolph County Grants, 1302
Community Fndn of So Alabama Grants, 1306
Community Foundation of Tampa Bay Grants, 1312
Community Foundation of the Eastern Shore Community Needs Grants, 1313
Community Foundation of the Verdugos Grants, 1315
Community Fndn of Wabash County Grants, 1316
Community Foundation Partnerships - Lawrence County Grants, 1317
Comprehensive Health Education Fndn Grants, 1320
ConAgra Foods Fndn Community Impact Grants, 1321
ConAgra Foods Nourish our Community Grants, 1322
Connecticut Health Foundation Health Initiative Grants, 1325
Connelly Foundation Grants, 1326
Conquer Cancer Foundation of ASCO Improving Cancer Care Grants, 1331
Constantin Foundation Grants, 1341
Cooke Foundation Grants, 1344
Cooper Industries Foundation Grants, 1345
Cornerstone Foundation of Northeastern Wisconsin Grants, 1347
Covenant Foundation of Waterloo Auxiliary Scholarships, 1349
Covenant Matters Foundation Grants, 1350
Crail-Johnson Foundation Grants, 1355
Cresap Family Foundation Grants, 1359
CRH Foundation Grants, 1360
Crystelle Waggoner Charitable Trust Grants, 1364
CSRA Community Foundation Grants, 1366
Cudd Foundation Grants, 1370
Cuesta Foundation Grants, 1371
Cullen Foundation Grants, 1372
Cumberland Community Foundation Grants, 1374
Curtis Foundation Grants, 1376
CVS Caremark Charitable Trust Grants, 1378
Cyrus Eaton Foundation Grants, 1380
D. W. McMillan Foundation Grants, 1407
Dade Community Foundation Grants, 1410
DaimlerChrysler Corporation Fund Grants, 1412
Dairy Queen Corporate Contributions Grants, 1413
Dallas Mavericks Foundation Grants, 1416
Dana Brown Charitable Trust Grants, 1419
Dana Foundation Science and Health Grants, 1420
Danellie Foundation Grants, 1421
Daphne Seybolt Culpeper Fndn Grants, 1425
David Geffen Foundation Grants, 1431
David N. Lane Trust Grants for Aged and Indigent Women, 1433
Daviess County Community Foundation Health Grants, 1434
Davis Family Foundation Grants, 1435
Dayton Power and Light Company Foundation Signature Grants, 1436
Dayton Power and Light Foundation Grants, 1437
Daywood Foundation Grants, 1438
Deaconess Community Foundation Grants, 1439
Dean Foods Community Involvement Grants, 1440
Dearborn Community Foundation County Progress Grants, 1442
Deborah Munroe Noonan Memorial Grants, 1443
Delaware Community Foundation-Youth Philanthropy Board for Kent County, 1448
Delaware Community Foundation Grants, 1449
Della B. Gardner Fund Grants, 1451
Del Mar Healthcare Fund Grants, 1453

Delmarva Power and Light Contributions, 1454
DeMatteis Family Foundation Grants, 1457
Dennis & Phyllis Washington Fndn Grants, 1458
Denton A. Cooley Foundation Grants, 1459
Denver Foundation Community Grants, 1461
DeRoy Testamentary Foundation Grants, 1469
DHHS Adolescent Family Life Demonstration Projects, 1472
DHHS Oral Health Promotion Research Across the Lifespan, 1474
Dickson Foundation Grants, 1476
Dolan Children's Foundation Grants, 1482
Dole Food Company Charitable Contributions, 1483
Dominion Foundation Grants, 1485
Dora Roberts Foundation Grants, 1489
Doree Taylor Charitable Foundation, 1490
Doris and Victor Day Foundation Grants, 1491
Doris Duke Charitable Foundation Clinical Interfaces Award Program, 1492
Dorothea Haus Ross Foundation Grants, 1498
Dorothy Hooper Beattie Foundation Grants, 1499
Dorothy Rider Pool Health Care Grants, 1500
Do Something Awards, 1503
Dr. & Mrs. Paul Pierce Memorial Fndn Grants, 1505
Dr. John T. Macdonald Foundation Grants, 1507
Dr. Leon Bromberg Charitable Trust Grants, 1509
DTE Energy Foundation Health and Human Services Grants, 1516
Dubois County Community Foundation Grants, 1517
Duke Endowment Health Care Grants, 1519
Duke University Pain Management Fellowships, 1526
Duneland Health Council Incorporated Grants, 1529
Dunspaugh-Dalton Foundation Grants, 1530
DuPage Community Foundation Grants, 1531
Dyson Foundation Mid-Hudson Valley General Operating Support Grants, 1533
Dyson Foundation Mid-Hudson Valley Project Support Grants, 1534
E. Clayton and Edith P. Gengras, Jr. Foundation, 1536
E.J. Grassmann Trust Grants, 1537
Earl and Maxine Claussen Trust Grants, 1541
Eckerd Corporation Foundation Grants, 1547
Eddie C. & Sylvia Brown Family Fndn Grants, 1548
Eden Hall Foundation Grants, 1549
EDS Foundation Grants, 1552
Educational Foundation of America Grants, 1553
Edward Bangs Kelley and Elza Kelley Foundation Grants, 1556
Edward N. and Della L. Thome Memorial Foundation Grants, 1557
Edwards Memorial Trust Grants, 1558
Edward W. and Stella C. Van Houten Memorial Fund Grants, 1559
Effie and Wofford Cain Foundation Grants, 1563
Elaine Feld Stern Charitable Trust Grants, 1565
Elizabeth Morse Genius Charitable Trust Grants, 1569
Elkhart County Community Foundation Grants, 1573
El Paso Community Foundation Grants, 1579
El Pomar Foundation Grants, 1582
Elsie H. Wilcox Foundation Grants, 1583
Elsie Lee Garthwaite Memorial Fndn Grants, 1584
Emerson Electric Company Contributions Grants, 1590
Ensworth Charitable Foundation Grants, 1595
Entergy Corporation Open Grants for Healthy Families, 1597
Erie Chapman Foundation Grants, 1600
Erie Community Foundation Grants, 1601
Essex County Community Foundation Webster Family Fund Grants, 1604
Essex County Community Foundation Women's Fund Grants, 1605
Estee Lauder Grants, 1606
Ethel S. Abbott Charitable Foundation Grants, 1607
Ethel Sergeant Clark Smith Foundation Grants, 1608
Eugene M. Lang Foundation Grants, 1611
Evanston Community Foundation Grants, 1615
Ewing Halsell Foundation Grants, 1618
Express Scripts Foundation Grants, 1620
F.M. Kirby Foundation Grants, 1624
Fairfield County Community Foundation Grants, 1625

Fairlawn Foundation Grants, 1626
FAR Fund Grants, 1631
Fargo-Moorhead Area Foundation Grants, 1632
Fargo-Moorhead Area Fndn Woman's Grants, 1633
Farmers Insurance Corporate Giving Grants, 1634
Faye McBeath Foundation Grants, 1635
Fayette County Foundation Grants, 1636
Ferree Foundation Grants, 1658
Fidelity Foundation Grants, 1659
Field Foundation of Illinois Grants, 1660
FirstEnergy Foundation Community Grants, 1668
Fisher Foundation Grants, 1670
Fisher House Fnd Newman's Own Awards, 1671
Flextronics Foundation Disaster Relief Grants, 1674
Florence Hunt Maxwell Foundation Grants, 1676
Florian O. Bartlett Trust Grants, 1677
Floyd A. & Kathleen C. Cailloux Fnd Grants, 1681
Fluor Foundation Grants, 1682
FMC Foundation Grants, 1683
Foellinger Foundation Grants, 1684
Fondren Foundation Grants, 1685
Ford Motor Company Fund Grants Program, 1687
Forrest C. Lattner Foundation Grants, 1688
Foster Foundation Grants, 1689
Fndn for Appalachian Ohio Bachtel Scholarships, 1692
Foundation for Health Enhancement Grants, 1696
Foundation for Seacoast Health Grants, 1698
Foundation of CVPH April LaValley Grants, 1701
Foundation of CVPH Melissa Lahtinen-Penfield Organ Donor Fund Grants, 1703
Foundation of CVPH Rabin Fund Grants, 1704
Foundation of CVPH Roger Senecal Endowment Fund Grants, 1705
Foundation of CVPH Travel Fund Grants, 1706
Foundations of East Chicago Health Grants, 1709
Four County Community Fndn General Grants, 1710
Four County Community Foundation Healthy Senior/Healthy Youth Fund Grants, 1711
Fourjay Foundation Grants, 1712
Four J Foundation Grants, 1713
France-Merrick Foundation Health and Human Services Grants, 1715
Frances and John L. Loeb Family Fund Grants, 1716
Frances L. and Edwin L. Cummings Memorial Fund Grants, 1717
Frances W. Emerson Foundation Grants, 1718
Francis L. Abreu Charitable Trust Grants, 1719
Francis T. & Louise T. Nichols Fndn Grants, 1720
Frank B. Hazard General Charity Fund Grants, 1721
Franklin County Community Foundation Grants, 1724
Franklin H. Wells and Ruth L. Wells Foundation Grants, 1725
Frank Reed and Margaret Jane Peters Memorial Fund I Grants, 1727
Frank Reed and Margaret Jane Peters Memorial Fund II Grants, 1728
Frank S. Flowers Foundation Grants, 1729
Frank Stanley Beveridge Foundation Grants, 1730
Frank W. & Carl S. Adams Memorial Grants, 1731
Fred & Gretel Biel Charitable Trust Grants, 1734
Fred C. & Katherine B. Andersen Grants, 1737
Frederick McDonald Trust Grants, 1740
Frederick W. Marzahl Memorial Fund Grants, 1741
Fremont Area Community Foundation Elderly Needs Grants, 1745
Fremont Area Community Fndn General Grants, 1746
Fuji Film Grants, 1748
Fuller E. Callaway Foundation Grants, 1761
Fulton County Community Foundation Grants, 1763
G.N. Wilcox Trust Grants, 1768
Gardner Foundation Grants, 1771
Gardner Foundation Grants, 1770
Gebbie Foundation Grants, 1777
GenCorp Foundation Grants, 1779
General Mills Foundation Grants, 1782
General Motors Foundation Grants, 1783
General Service Foundation Human Rights and Economic Justice Grants, 1784
Genesis Foundation Grants, 1786
Genuardi Family Foundation Grants, 1787

SUBJECT INDEX

Health Care /739

George A. and Grace L. Long Foundation Grants, 1788
George and Ruth Bradford Foundation Grants, 1789
George A Ohl Jr. Foundation Grants, 1791
George E. Hatcher, Jr. and Ann Williams Hatcher Foundation Grants, 1792
George Family Foundation Grants, 1794
George Foundation Grants, 1795
George H. Hitchings New Investigator Award in Health Research, 1798
George Kress Foundation Grants, 1799
George P. Davenport Trust Fund Grants, 1800
George W. Wells Foundation Grants, 1804
Georgia Power Foundation Grants, 1806
Gertrude and William C. Wardlaw Fund Grants, 1808
Gertrude B. Elion Mentored Medical Student Research Awards, 1809
Gertrude E. Skelly Charitable Foundation Grants, 1810
Gheens Foundation Grants, 1814
Giant Eagle Foundation Grants, 1815
Gibson Foundation Grants, 1818
Gill Foundation - Gay and Lesbian Fund Grants, 1821
Ginn Foundation Grants, 1822
Giving in Action Society Children & Youth with Special Needs Grants, 1823
Giving in Action Society Family Indep Grants, 1824
GlaxoSmithKline Corporate Grants, 1829
GNOF IMPACT Grants for Health and Human Services, 1839
GNOF IMPACT Harold W. Newman, Jr. Charitable Trust Grants, 1841
GNOF Norco Community Grants, 1845
Goddess Scholars Grants, 1848
Godfrey Foundation Grants, 1849
Golden State Warriors Foundation Grants, 1851
Goodrich Corporation Foundation Grants, 1854
Grace and Franklin Bernsen Foundation Grants, 1858
Grace Bersted Foundation Grants, 1859
Grand Rapids Area Community Foundation Wyoming Grants, 1867
Grand Rapids Community Foundation Grants, 1869
Grand Rapids Community Foundation Ionia County Grants, 1870
Grand Rapids Community Foundation Lowell Area Fund Grants, 1872
Grand Rapids Community Foundation Southeast Ottawa Grants, 1873
Grand Rapids Community Fndn Sparta Grants, 1875
Greater Cincinnati Foundation Priority and Small Projects/Capacity-Building Grants, 1880
Greater Green Bay Community Fndn Grants, 1881
Greater Kanawha Valley Foundation Grants, 1882
Greater Milwaukee Foundation Grants, 1883
Greater Saint Louis Community Fndn Grants, 1884
Greater Sitka Legacy Fund Grants, 1885
Greater Tacoma Community Foundation Grants, 1886
Greater Tacoma Community Foundation Ryan Alan Hade Endowment Fund, 1887
Greater Worcester Community Foundation Discretionary Grants, 1888
Green Bay Packers Foundation Grants, 1890
Green Diamond Charitable Contributions, 1891
Greene County Foundation Grants, 1892
Greenfield Foundation of Maine Grants, 1893
Green Foundation Human Services Grants, 1894
Green River Area Community Fndn Grants, 1895
Gregory B. Davis Foundation Grants, 1898
Gregory L. Gibson Charitable Fndn Grants, 1900
Griffin Family Foundation Grants, 1902
Griffin Foundation Grants, 1903
Grotto Foundation Project Grants, 1905
Grover Hermann Foundation Grants, 1906
Guido A. & Elizabeth H. Binda Fndn Grants, 1913
Gulf Coast Community Foundation Grants, 1915
Guy's and St. Thomas' Charity Grants, 1918
Guy I. Bromley Trust Grants, 1919
H & R Foundation Grants, 1920
H.A. & Mary K. Chapman Trust Grants, 1921
H. Leslie Hoffman and Elaine S. Hoffman Foundation Grants, 1924
H. Schaffer Foundation Grants, 1925

Hagedorn Fund Grants, 1934
Hall-Perrine Foundation Grants, 1935
Hallmark Corporate Foundation Grants, 1936
Hampton Roads Community Foundation Faith Community Nursing Grants, 1940
Hannaford Charitable Foundation Grants, 1944
Hardin County Community Foundation Grants, 1946
Harley Davidson Foundation Grants, 1947
Harold Alfond Foundation Grants, 1949
Harold R. Bechtel Charitable Remainder Uni-Trust Grants, 1953
Harold Simmons Foundation Grants, 1955
Harris and Eliza Kempner Fund Grants, 1956
Harrison County Community Foundation Signature Grants, 1958
Harry S. Black and Allon Fuller Fund Grants, 1966
Harry Sudakoff Foundation Grants, 1967
Hartford Courant Foundation Grants, 1969
Hartford Foundation Regular Grants, 1970
Harvest Foundation Grants, 1971
Hawaii Community Foundation Reverend Takie Okumura Family Grants, 1977
Hawaii Community Foundation West Hawaii Fund Grants, 1978
Hawn Foundation Grants, 1979
HCA Foundation Grants, 1980
Healthcare Foundation of New Jersey Grants, 1982
Health Fndn of Greater Indianapolis Grants, 1984
Health Foundation of Southern Florida Responsive Grants, 1985
Hearst Foundations Health Grants, 1987
Heckscher Foundation for Children Grants, 1988
Hedco Foundation Grants, 1989
Heineman Foundation for Research, Education, Charitable and Scientific Purposes, 1990
Helena Rubinstein Foundation Grants, 1991
Helen Gertrude Sparks Charitable Trust Grants, 1993
Helen Irwin Littauer Educational Trust Grants, 1994
Helen K. & Arthur E. Johnson Fndn Grants, 1995
Helen Pumphrey Denit Charitable Trust Grants, 1996
Helen S. Boylan Foundation Grants, 1997
Helen Steiner Rice Foundation Grants, 1998
Hendrick Foundation for Children Grants, 1999
Henrietta Lange Burk Fund Grants, 2001
Henry A. and Mary J. MacDonald Foundation, 2003
Henry and Ruth Blaustein Rosenberg Foundation Health Grants, 2004
Henry County Community Foundation Grants, 2005
Herman Goldman Foundation Grants, 2012
Hershey Company Grants, 2013
HHMI Biomedical Research Grants for International Scientists: Infectious Diseases & Parasitology, 2017
HHMI Biomedical Research Grants for International Scientists in Canada and Latin America, 2018
HHMI Biomedical Research Grants for International Scientists in the Baltics, Central and Eastern Europe, Russia, and Ukraine, 2019
HHMI Med into Grad Initiative Grants, 2022
Highmark Corporate Giving Grants, 2025
High Meadow Foundation Grants, 2027
Hillman Foundation Grants, 2032
Hillsdale Fund Grants, 2035
Hilton Head Island Foundation Grants, 2036
Hilton Hotels Corporate Giving Program Grants, 2037
Hoblitzelle Foundation Grants, 2038
Holland/Zeeland Community Fndn Grants, 2041
Homer Foundation Grants, 2044
Honda of America Manufacturing Fndn Grants, 2045
Horace A. Kimball and S. Ella Kimball Foundation Grants, 2046
Horace Moses Charitable Foundation Grants, 2047
Horizon Foundation for New Jersey Grants, 2048
Horizons Community Issues Grants, 2049
Hormel Foods Charitable Trust Grants, 2050
Hormel Foundation Grants, 2051
Houston Endowment Grants, 2053
Howard and Bush Foundation Grants, 2054
HRK Foundation Health Grants, 2067
Huffy Foundation Grants, 2073
Hugh J. Andersen Foundation Grants, 2074

Huntington Beach Police Officers Fndn Grants, 2081
Huntington Clinical Foundation Grants, 2082
Huntington County Community Foundation Make a Difference Grants, 2083
Huntington National Bank Community Grants, 2084
Hutchinson Community Foundation Grants, 2085
Hut Foundation Grants, 2086
Hutton Foundation Grants, 2087
I.A. O'Shaughnessy Foundation Grants, 2088
Idaho Community Foundation Eastern Region Competitive Grants, 2114
Idaho Power Company Corporate Contributions, 2115
Ida S. Barter Trust Grants, 2116
IDPH Emergency Medical Services Assistance Fund Grants, 2118
IHI Quality Improvement Fellowships, 2121
Illinois Children's Healthcare Foundation Grants, 2146
Illinois Tool Works Foundation Grants, 2147
Inasmuch Foundation Grants, 2149
Independence Blue Cross Charitable Medical Care Grants, 2150
Intergrys Corporation Grants, 2160
Irving S. Gilmore Foundation Grants, 2163
Irvin Stern Foundation Grants, 2164
Isabel Allende Foundation Esperanza Grants, 2165
J.C. Penney Company Grants, 2169
J.H. Robbins Foundation Grants, 2171
J.L. Bedsole Foundation Grants, 2172
J.W. Kieckhefer Foundation Grants, 2176
J. Willard Marriott, Jr. Foundation Grants, 2178
Jackson County Community Foundation Unrestricted Grants, 2180
Jacob and Hilda Blaustein Foundation Health and Mental Health Grants, 2181
Jacob and Valeria Langeloth Foundation Grants, 2182
Jacob G. Schmidlapp Trust Grants, 2183
Jacobs Family Village Neighborhoods Grants, 2184
James & Abigail Campbell Family Fndn Grants, 2185
James Ford Bell Foundation Grants, 2187
James Graham Brown Foundation Grants, 2189
James J. & Angelia M. Harris Fndn Grants, 2192
James L. and Mary Jane Bowman Charitable Trust Grants, 2194
James M. Collins Foundation Grants, 2195
James M. Cox Foundation of Georgia Grants, 2196
James R. Dougherty Jr. Foundation Grants, 2198
James S. Copley Foundation Grants, 2200
Jane Bradley Pettit Foundation Health Grants, 2208
Janirve Foundation Grants, 2209
Janus Foundation Grants, 2211
Jay and Rose Phillips Family Foundation Grants, 2213
Jean and Louis Dreyfus Foundation Grants, 2215
JELD-WEN Foundation Grants, 2218
Jenkins Foundation: Improving the Health of Greater Richmond Grants, 2219
Jennings County Community Foundation Grants, 2220
Jerome and Mildred Paddock Foundation Grants, 2222
Jessica Stevens Community Foundation Grants, 2224
Jessie B. Cox Charitable Trust Grants, 2225
Jessie Ball Dupont Fund Grants, 2226
Jewish Fund Grants, 2227
John Clarke Trust Grants, 2232
John Deere Foundation Grants, 2234
John Edward Fowler Memorial Fndn Grants, 2235
John G. Duncan Charitable Trust Grants, 2236
John G. Martin Foundation Grants, 2237
John H. Wellons Foundation Grants, 2239
John J. Leidy Foundation Grants, 2241
John Jewett and Helen Chandler Garland Foundation Grants, 2242
John Lord Knight Foundation Grants, 2243
John P. McGovern Foundation Grants, 2247
John P. Murphy Foundation Grants, 2248
John R. Oishei Foundation Grants, 2249
John S. Dunn Research Foundation Grants, 2250
Johns Manville Fund Grants, 2251
Johnson & Johnson Corporate Contributions, 2253
Johnson & Johnson/SAH Arts & Healing Grants, 2254
Johnson Controls Foundation Health and Social Services Grants, 2255

740 / Health Care

Johnson County Community Foundation Grants, 2256
John W. Alden Trust Grants, 2260
John W. and Anna H. Hanes Foundation Grants, 2261
John W. Anderson Foundation Grants, 2262
Joseph Drown Foundation Grants, 2268
Joseph Henry Edmondson Foundation Grants, 2270
Josephine G. Russell Trust Grants, 2271
Josephine S. Gumbiner Foundation Grants, 2273
Josephine Schell Russell Charitable Trust Grants, 2274
K21 Health Foundation Grants, 2285
Kahuku Community Fund, 2286
Kate B. Reynolds Trust Health Care Grants, 2290
Katharine Matthies Foundation Grants, 2291
Katherine John Murphy Foundation Grants, 2293
Katrine Menzing Deakins Trust Grants, 2295
Kelvin and Eleanor Smith Foundation Grants, 2297
Kenai Peninsula Foundation Grants, 2298
Kendrick Foundation Grants, 2299
Ketchikan Community Foundation Grants, 2303
Kettering Family Foundation Grants, 2304
Kettering Fund Grants, 2305
Kimball International-Habig Foundation Health and Human Services Grants, 2308
Knight Family Charitable and Educational Foundation Grants, 2312
Kodiak Community Foundation Grants, 2314
Kosair Charities Grants, 2318
L. W. Pierce Family Foundation Grants, 2328
Land O'Lakes Foundation Mid-Atlantic Grants, 2341
Lands' End Corporate Giving Program, 2344
Laura B. Vogler Foundation Grants, 2345
Laura Moore Cunningham Foundation Grants, 2346
Lawrence Foundation Grants, 2347
Lena Benas Memorial Fund Grants, 2353
Leo Niessen Jr., Charitable Trust Grants, 2358
Lester E. Yeager Charitable Trust B Grants, 2359
Liberty Bank Foundation Grants, 2362
Lisa and Douglas Goldman Fund Grants, 2369
Lisa Higgins-Hussman Foundation Grants, 2370
Little Life Foundation Grants, 2372
Lloyd A. Fry Foundation Health Grants, 2373
Lockheed Martin Corp Foundation Grants, 2374
Louetta M. Cowden Foundation Grants, 2377
Louie M. & Betty M. Phillips Fndn Grants, 2378
Louis H. Aborn Foundation Grants, 2380
Lowe Foundation Grants, 2381
Lowell Berry Foundation Grants, 2382
Lubrizol Foundation Grants, 2384
Lucile Packard Foundation for Children's Health Grants, 2386
Ludwick Family Foundation Grants, 2389
Lumpkin Family Fnd Healthy People Grants, 2391
M-A-C AIDS Fund Grants, 2400
M.B. and Edna Zale Foundation Grants, 2401
M. Bastian Family Foundation Grants, 2402
Mabel A. Horne Trust Grants, 2406
Mabel F. Hoffman Charitable Trust Grants, 2407
Mabel H. Flory Charitable Trust Grants, 2408
Mabel Y. Hughes Charitable Trust Grants, 2409
Macquarie Bank Foundation Grants, 2410
MAP International Medical Fellowships, 2421
Marathon Petroleum Corporation Grants, 2422
Marie C. and Joseph C. Wilson Foundation Rochester Small Grants, 2431
Marion I. and Henry J. Knott Foundation Discretionary Grants, 2435
Marion I. and Henry J. Knott Foundation Standard Grants, 2436
Marjorie Moore Charitable Foundation Grants, 2438
Mary Black Fndn Community Health Grants, 2441
Mary K. Chapman Foundation Grants, 2443
Mary L. Peyton Foundation Grants, 2445
Mary Owen Borden Foundation Grants, 2446
Mary S. and David C. Corbin Foundation Grants, 2447
Mary Wilmer Covey Charitable Trust Grants, 2448
Mattel Children's Foundation Grants, 2454
Max and Victoria Dreyfus Foundation Grants, 2457
Maxon Charitable Foundation Grants, 2458
McConnell Foundation Grants, 2492
McInerny Foundation Grants, 2495

McLean Contributionship Grants, 2497
McMillen Foundation Grants, 2499
Mead Johnson Nutritionals Charitable Giving, 2502
MeadWestvaco Foundation Sustainable Communities Grants, 2506
Mead Witter Foundation Grants, 2507
Mericos Foundation Grants, 2518
Meriden Foundation Grants, 2519
Mervin Bovaird Foundation Grants, 2523
Mesothelioma Applied Research Fndn Grants, 2524
MetroWest Health Foundation Capital Grants for Health-Related Facilities, 2526
MetroWest Health Foundation Grants--Healthy Aging, 2527
Metzger-Price Fund Grants, 2530
Meyer Foundation Healthy Communities Grants, 2533
Meyer Memorial Trust Special Grants, 2538
MGN Family Foundation Grants, 2542
Michael Reese Health Trust Core Grants, 2544
Milagro Foundation Grants, 2552
Miles of Hope Breast Cancer Foundation Grants, 2553
Military Ex-Prisoners of War Foundation Grants, 2554
Mimi and Peter Haas Fund Grants, 2557
MMS and Alliance Charitable Foundation Grants for Community Action and Care for the Medically Uninsured, 2587
Moline Foundation Community Grants, 2590
Morton K. and Jane Blaustein Foundation Health Grants, 2595
Mt. Sinai Health Care Foundation Health of the Jewish Community Grants, 2599
NAPNAP Fndn Elaine Gelman Scholarship, 2609
NAPNAP Foundation Innovative Health Care Small Grant, 2611
NAPNAP Foundation McNeil Grant-in-Aid, 2612
NAPNAP Foundation McNeil PNP Scholarships, 2613
NAPNAP Foundation McNeil Rural and Underserved Scholarships, 2614
NAPNAP Foundation Nursing Research Grants, 2615
NAPNAP Foundation Reckitt Benckiser Student Scholarship, 2616
NAPNAP Foundation Shourd Parks Immunization Project Small Grants, 2617
NAPNAP Foundation Wyeth Pediatric Immunization Grant, 2618
Nathan Cummings Foundation Grants, 2627
Nell J. Redfield Foundation Grants, 2673
NHSCA Arts in Health Care Project Grants, 2741
Nicholas H. Noyes Jr. Memorial Fndn Grants, 2791
NIDDK Research Grants on Improving Health Care for Obese Patients, 2840
NIHCM Foundation Health Care Print Journalism Awards, 2870
NIHCM Fndn Health Care Research Awards, 2871
NIHCM Foundation Health Care Television and Radio Journalism Awards, 2872
NIH Exceptional, Unconventional Research Enabling Knowledge Acceleration (EUREKA) Grants, 2876
NIH Research Project Grants, 2899
NIH School Interventions to Prevent Obesity, 2908
Nina Mason Pulliam Charitable Trust Grants, 2921
NINR Mentored Research Scientist Development Award, 2932
Noble County Community Foundation Grants, 2975
Nordson Corporation Foundation Grants, 2981
North Carolina Community Foundation Grants, 2990
North Carolina GlaxoSmithKline Fndn Grants, 2991
North Central Health Services Grants, 2993
Northern New York Community Fndn Grants, 2996
Northrop Grumman Corporation Grants, 2997
Northwestern Mutual Foundation Grants, 2999
Norwin S. & Elizabeth N. Bean Fndn Grants, 3001
Nuffield Foundation Africa Program Grants, 3026
Nuffield Foundation Law and Society Grants, 3028
Nuffield Foundation Small Grants, 3031
NWHF Health Advocacy Small Grants, 3033
NWHF Kaiser Permanente Community Grants, 3034
NWHF Mark O. Hatfield Research Fellowship, 3035
NYAM Brock Lecture, Award and Visiting Professorship in Pediatrics, 3038

SUBJECT INDEX

NYC Managed Care Consumer Assistance Workshop Re-Design Grants, 3047
NYCT AIDS/HIV Grants, 3048
NYCT Blindness and Visual Disabilities Grants, 3050
NYCT Children/Youth with Disabilities Grants, 3052
NYCT Girls and Young Women Grants, 3053
NYCT Grants for the Elderly, 3054
NYCT Health Care Services, Systems, and Policies Grants, 3055
NYCT Mental Health and Retardation Grants, 3056
OceanFirst Foundation Major Grants, 3062
Oceanside Charitable Foundation Grants, 3063
Ogden Codman Trust Grants, 3066
Ohio County Community Foundation Grants, 3068
Ohio County Community Fndn Mini-Grants, 3069
Oleonda Jameson Trust Grants, 3071
Olive Higgins Prouty Foundation Grants, 3072
Olympus Corporation of Americas Corporate Giving Grants, 3073
Oppenstein Brothers Foundation Grants, 3078
Ordean Foundation Grants, 3080
Oregon Community Fndn Community Grants, 3082
Oscar Rennebohm Foundation Grants, 3085
OSF Accountability and Monitoring in Health Initiative Grants, 3087
OSF Health Media Initiative Grants, 3089
OSF International Harm Reduction Development Program Grants, 3090
OSF Mental Health Initiative Grants, 3093
OSF Public Health Grants in Kyrgyzstan, 3094
Owen County Community Foundation Grants, 3100
Pajaro Valley Community Health Health Trust Insurance/Coverage & Education on Using the System Grants, 3108
Pajaro Valley Community Health Trust Oral Health: Prevention & Access Grants, 3110
Pajaro Valley Community Health Trust Promoting Entry & Advancement in the Health Professions Grants, 3111
Parkersburg Area Community Foundation Action Grants, 3116
Patrick and Anna M. Cudahy Fund Grants, 3120
Patron Saints Foundation Grants, 3121
Paul Ogle Foundation Grants, 3124
Peabody Foundation Grants, 3132
Peacock Foundation Grants, 3133
Pediatric Brain Tumor Fndn Research Grants, 3134
Percy B. Ferebee Endowment Grants, 3138
Perkins Charitable Foundation Grants, 3140
Perpetual Trust for Charitable Giving Grants, 3141
Perry County Community Foundation Grants, 3142
Peter Kiewit Foundation General Grants, 3154
Peter Kiewit Foundation Small Grants, 3155
Petersburg Community Foundation Grants, 3156
Peyton Anderson Foundation Grants, 3160
Pfizer Healthcare Charitable Contributions, 3180
Pfizer Medical Education Track Two Grants, 3183
Pfizer Special Events Grants, 3186
Phelps County Community Foundation Grants, 3188
Philadelphia Foundation Organizational Effectiveness Grants, 3189
Philip L. Graham Fund Health and Human Services Grants, 3190
PhRMA Foundation Health Outcomes Research Starter Grants, 3204
PhRMA Foundation Health Outcomes Sabbatical Fellowships, 3205
Piedmont Health Foundation Grants, 3219
Piedmont Natural Gas Foundation Health and Human Services Grants, 3221
Pike County Community Foundation Grants, 3222
Pinnacle Foundation Grants, 3225
Pittsburgh Foundation Community Fund Grants, 3230
Plough Foundation Grants, 3238
Polk Bros. Foundation Grants, 3243
Pollock Foundation Grants, 3244
Posey Community Fndn Women's Fund Grants, 3251
Posey County Community Foundation Grants, 3252
Potts Memorial Foundation Grants, 3254
Powell Foundation Grants, 3255

SUBJECT INDEX

Health Care /741

PPCF Community Grants, 3256
Presbyterian Health Foundation Bridge, Seed and Equipment Grants, 3259
Price Chopper's Golub Foundation Grants, 3261
Price Family Charitable Fund Grants, 3262
Priddy Foundation Program Grants, 3266
Pride Foundation Grants, 3268
Prince Charitable Trusts Chicago Grants, 3270
Principal Financial Group Foundation Grants, 3273
Procter and Gamble Fund Grants, 3274
Puerto Rico Community Foundation Grants, 3278
Pulaski County Community Foundation Grants, 3279
Pulido Walker Foundation, 3282
Putnam County Community Foundation Grants, 3283
Quaker Chemical Foundation Grants, 3286
Qualcomm Grants, 3287
Quality Health Foundation Grants, 3288
Quantum Foundation Grants, 3290
Questar Corporate Contributions Grants, 3291
R.C. Baker Foundation Grants, 3292
R.S. Gernon Trust Grants, 3293
RACGP Cardiovascular Research Grants in General Practice, 3295
Rainbow Endowment Grants, 3300
Rajiv Gandhi Foundation Grants, 3302
Ralph M. Parsons Foundation Grants, 3304
Ralphs Food 4 Less Foundation Grants, 3305
Raskob Foundation for Catholic Activities Grants, 3306
Rasmuson Foundation Tier One Grants, 3307
Rasmuson Foundation Tier Two Grants, 3308
Rathmann Family Foundation Grants, 3309
Rayonier Foundation Grants, 3311
RCPSC Detweiler Traveling Fellowships, 3320
RCPSC James H. Graham Award of Merit, 3327
RCPSC Teasdale-Corti Humanitarian Award, 3344
Regence Fndn Access to Health Care Grants, 3346
Regence Foundation Health Care Community Awareness and Engagement Grants, 3347
Regence Fndn Health Care Connections Grants, 3348
Regence Fndn Improving End-of-Life Grants, 3349
Regence Fndn Tools & Technology Grants, 3350
Regenstrief General Internal Medicine - General Pediatrics Research Fellowship Program, 3351
Rehabilitation Nursing Fndn Research Grants, 3354
Retirement Research Foundation General Program Grants, 3359
Reynolds & Reynolds Associate Fndn Grants, 3360
Rhode Island Foundation Grants, 3362
Richard and Helen DeVos Foundation Grants, 3364
Richard & Susan Smith Family Fndn Grants, 3365
Richard D. Bass Foundation Grants, 3366
Richard E. Griffin Family Foundation Grants, 3368
Richard King Mellon Foundation Grants, 3369
Richard M. Fairbanks Foundation Grants, 3370
Richmond Eye and Ear Fund Grants, 3372
Ricks Family Charitable Trust Grants, 3373
Riley Foundation Grants, 3374
Ripley County Community Foundation Grants, 3376
Ripley County Community Foundation Small Project Grants, 3377
Robert and Joan Dircks Foundation Grants, 3379
Robert W. Woodruff Foundation Grants, 3389
Rochester Area Community Foundation Grants, 3391
Rochester Area Foundation Grants, 3392
Rockefeller Brothers Peace & Security Grants, 3394
Rockefeller Foundation Grants, 3396
Rockwell Collins Charitable Corporation Grants, 3397
Rockwell International Corporate Trust Grants Program, 3399
Rollins-Luetkemeyer Foundation Grants, 3404
Ronald McDonald House Charities Grants, 3405
Rosalynn Carter Institute Mattie J. T. Stepanek Caregiving Scholarships, 3409
Rose Community Foundation Health Grants, 3411
Rose Hills Foundation Grants, 3413
Roy & Christine Sturgis Charitable Grants, 3415
RSC Thomas W. Eadie Medal, 3423
Rush County Community Foundation Grants, 3424
Ruth Anderson Foundation Grants, 3425
Ruth and Vernon Taylor Foundation Grants, 3426

Ruth Eleanor Bamberger and John Ernest Bamberger Memorial Foundation Grants, 3427
Ruth H. & Warren A. Ellsworth Fndn Grants, 3429
RWJF Changes in Health Care Financing and Organization Grants, 3430
RWJF Community Health Leaders Awards, 3432
RWJF Health and Society Scholars, 3436
RWJF Pioneer Portfolio Grants, 3442
RWJF Vulnerable Populations Portfolio Grants, 3444
S.D. Bechtel, Jr. Fndn/Stephen Bechtel Preventive Healthcare and Selected Research Grants, 3446
S. Livingston Mather Charitable Trust Grants, 3448
S. Mark Taper Foundation Grants, 3449
Saigh Foundation Grants, 3451
Sain-Orr and Royak-DeForest Steadman Foundation Grants, 3452
Saint Louis Rams Fndn Community Donations, 3453
Saint Luke's Foundation Grants, 3454
Salem Foundation Grants, 3457
Samueli Institute Scientific Research Grants, 3466
Samuel S. Johnson Foundation Grants, 3468
San Antonio Area Foundation Grants, 3469
San Diego Foundation Health & Human Services Grants, 3471
San Diego Foundation Paradise Valley Hospital Community Fund Grants, 3472
Sandler Program for Asthma Research Grants, 3475
Sands Foundation Grants, 3476
San Francisco Fndn Community Health Grants, 3478
San Juan Island Community Foundation Grants, 3480
Santa Barbara Foundation Strategy Grants - Core Support, 3481
Sarasota Memorial Healthcare Fndn Grants, 3484
Sarkeys Foundation Grants, 3485
Sasco Foundation Grants, 3487
Savoy Foundation Research Grants, 3489
Schering-Plough Foundation Health Grants, 3492
Schramm Foundation Grants, 3495
Schurz Communications Foundation Grants, 3496
Scott County Community Foundation Grants, 3500
Seagate Tech Corp Capacity to Care Grants, 3502
Seattle Foundation Benjamin N. Phillips Memorial Fund Grants, 3504
Seattle Foundation Health and Wellness Grants, 3507
Self Foundation Grants, 3510
Seneca Foods Foundation Grants, 3511
Seward Community Foundation Grants, 3514
Seward Community Foundation Mini-Grants, 3515
Shell Deer Park Grants, 3543
Shell Oil Company Foundation Community Development Grants, 3544
Sheltering Arms Fund Grants, 3545
Shield-Ayres Foundation Grants, 3546
Sidney Stern Memorial Trust Grants, 3552
Siebert Lutheran Foundation Grants, 3554
Sigma Theta Tau International / Council for the Advancement of Nursing Science Grants, 3557
Simmons Foundation Grants, 3563
Simple Advise Education Center Grants, 3565
Singing for Change Foundation Grants, 3567
Sioux Falls Area Community Foundation Community Fund Grants (Unrestricted), 3568
Sioux Falls Area Community Foundation Field-of-Interest and Donor-Advised Grants, 3569
Sioux Falls Area Community Foundation Spot Grants (Unrestricted), 3570
Siragusa Foundation Health Services & Medical Research Grants, 3571
Sir Dorabji Tata Trust Grants for NGOs or Voluntary Organizations, 3572
Sir Dorabji Tata Trust Individual Medical Grants, 3573
Sisters of Mercy of North Carolina Fndn Grants, 3575
SOCFOC Catholic Ministries Grants, 3596
Society for the Arts in Healthcare Grants, 3602
Solo Cup Foundation Grants, 3620
Sonoco Foundation Grants, 3621
Sonora Area Foundation Competitive Grants, 3622
Sophia Romero Trust Grants, 3624
Southbury Community Trust Fund, 3625
South Madison Community Foundation Grants, 3626

Southwest Florida Community Foundation Competitive Grants, 3627
Southwest Gas Corporation Foundation Grants, 3628
Special People in Need Grants, 3630
Spencer County Community Foundation Grants, 3632
Springs Close Foundation Grants, 3633
St. Joseph Community Health Foundation Burn Care and Prevention Grants, 3639
St. Joseph Community Health Foundation Pfeiffer Fund Grants, 3641
St. Joseph Community Health Foundation Schneider Fellowships, 3642
Stackpole-Hall Foundation Grants, 3643
Stan and Sandy Checketts Foundation, 3644
Steele-Reese Foundation Grants, 3650
Steuben County Community Foundation Grants, 3654
Stewart Huston Charitable Trust Grants, 3655
Straits Area Community Foundation Health Care Grants, 3657
Strowd Roses Grants, 3664
Subaru of Indiana Automotive Foundation Grants, 3672
Summit Foundation Grants, 3674
Sunderland Foundation Grants, 3675
Sunflower Foundation Bridge Grants, 3676
Sunflower Foundation Capacity Building Grants, 3677
Sunflower Foundation Walking Trails Grants, 3678
Sunoco Foundation Grants, 3679
SunTrust Bank Trusteed Foundations Florence C. and Harry L. English Memorial Fund Grants, 3680
SunTrust Bank Trusteed Foundations Greene-Sawtell Grants, 3681
SunTrust Bank Trusteed Foundations Harriet McDaniel Marshall Tust Grants, 3682
SunTrust Bank Trusteed Foundations Nell Warren Elkin and William Simpson Elkin Grants, 3683
SunTrust Bank Trusteed Foundations Thomas Guy Woolford Charitable Trust Grants, 3684
SunTrust Bank Trusteed Foundations Walter H. and Marjory M. Rich Memorial Fund Grants, 3685
Support Our Aging Religious (SOAR) Grants, 3686
Susan Mott Webb Charitable Trust Grants, 3694
Susan Vaughan Foundation Grants, 3695
Swindells Charitable Foundation, 3697
Taylor S. Abernathy and Patti Harding Abernathy Charitable Trust Grants, 3704
Tellabs Foundation Grants, 3706
Tension Envelope Foundation Grants, 3707
Texas Instruments Corporation Health and Human Services Grants, 3709
Textron Corporate Contributions Grants, 3712
Thelma Braun & Bocklett Family Fndn Grants, 3713
Theodore Edson Parker Foundation Grants, 3715
Thomas and Dorothy Leavey Foundation Grants, 3717
Thomas Austin Finch, Sr. Foundation Grants, 3718
Thomas C. Ackerman Foundation Grants, 3719
Thomas C. Burke Foundation Grants, 3720
Thomas J. Atkins Memorial Trust Fund Grants, 3721
Thomas J. Long Foundation Community Grants, 3722
Thomas Sill Foundation Grants, 3724
Thomas Thompson Trust Grants, 3725
Thompson Charitable Foundation Grants, 3727
TJX Foundation Grants, 3735
Todd Brock Family Foundation Grants, 3736
Tommy Hilfiger Corporate Foundation Grants, 3737
Toyota Motor Engineering & Manufacturing North America Grants, 3742
Toyota Motor Manufacturing of Alabama Grants, 3743
Toyota Motor Manuf of Mississippi Grants, 3746
Toyota Motor Manufacturing of Texas Grants, 3747
Toyota Motor Manuf of West Virginia Grants, 3748
Toys R Us Children's Fund Grants, 3751
Triangle Community Foundation Donor-Advised Grants, 3757
Tull Charitable Foundation Grants, 3762
Turner Foundation Grants, 3763
Tylenol Future Care Scholarships, 3764
U.S. Department of Education 21st Century Community Learning Centers, 3767
UCLA / RAND Corporation Post-Doctoral Fellowships, 3776

742 / Health Care

UniHealth Foundation Community Health Improvement Grants, 3782
UniHealth Foundation Healthcare Systems Enhancements Grants, 3784
UniHealth Foundation Innovation Fund Grants, 3785
Union Bank, N.A. Corporate Sponsorships and Donations, 3787
Union Bank, N.A. Foundation Grants, 3788
Union Benevolent Association Grants, 3789
Union County Community Foundation Grants, 3790
United Healthcare Comm Grants in Michigan, 3796
United Methodist Committee on Relief Global AIDS Fund Grants, 3798
United Methodist Child Mental Health Grants, 3800
United Technologies Corporation Grants, 3803
USAID Ebola Response, Recovery and Resilience in West Africa Grants, 3809
USAID Family Health Plus Project Grants, 3810
USAID HIV Prevention with Key Populations - Mali Grants, 3816
USAID Integration of Care and Support within the Health System to Support Better Patient Outcomes Grants, 3817
USAID Research and Innovation for Health Supply Chain Systems & Commodity Security Grants, 3821
US Airways Community Foundation Grants, 3826
USDA Delta Health Care Services Grants, 3829
USDD Care for the Critically Injured Burn Patient Grants, 3839
USDD Clinical Functional Assessment Investigator-Initiated Grants, 3840
Vancouver Foundation Grants and Community Initiatives Program, 3849
Vancouver Fndn Public Health Bursary Package, 3850
Vancouver Sun Children's Fund Grants, 3851
Verizon Foundation Health Care and Accessibility Grants, 3858
Vermont Community Foundation Grants, 3865
VHA Health Foundation Grants, 3867
Victor E. Speas Foundation Grants, 3868
Vigneron Memorial Fund Grants, 3873
Virginia W. Kettering Foundation Grants, 3878
Volkswagen of America Corporate Contributions, 3880
W.C. Griffith Foundation Grants, 3881
W. C. Griffith Foundation Grants, 3882
W. Clarke Swanson, Jr. Foundation Grants, 3883
W.H. & Mary Ellen Cobb Trust Grants, 3884
W.K. Kellogg Foundation Secure Families Grants, 3887
W.M. Keck Fndn So California Grants, 3890
W.P. and Bulah Luse Foundation Grants, 3892
Wabash Valley Community Foundation Grants, 3893
Waitt Family Foundation Grants, 3894
Walker Area Community Foundation Grants, 3895
Walmart Foundation Northwest Arkansas Giving, 3899
Walter L. Gross III Family Foundation Grants, 3901
Warrick County Community Foundation Grants, 3904
Washington Gas Charitable Giving, 3908
Wayne County Foundation Grants, 3909
Weaver Popcorn Foundation Grants, 3910
Weingart Foundation Grants, 3913
Welborn Baptist Fndn General Op Grants, 3914
Welborn Baptist Foundation Improvements to Community Health Status Grants, 3916
Western Indiana Community Foundation Grants, 3921
Western Union Foundation Grants, 3923
Weyerhaeuser Company Foundation Grants, 3925
Weyerhaeuser Family Foundation Health Grants, 3926
WHO Foundation General Grants, 3939
WHO Foundation Volunteer Service Grants, 3940
Wilhelm Sander-Stiftung Foundation Grants, 3942
Willard & Pat Walker Charitable Fndn Grants, 3943
William A. Badger Foundation Grants, 3945
William B. Stokely Jr. Foundation Grants, 3947
William Blair and Company Foundation Grants, 3948
William D. Laurie, Jr. Charitable Fndn Grants, 3949
William G. & Helen C. Hoffman Fndn Grants, 3950
William G. Gilmore Foundation Grants, 3951
William H. Hannon Foundation Grants, 3953
William J. & Tina Rosenberg Fndn Grants, 3954
William J. Brace Charitable Trust, 3955
William Ray & Ruth E. Collins Fndn Grants, 3956
William T. Grant Foundation Research Grants, 3958
Willis C. Helm Charitable Trust Grants, 3959
Wilson-Wood Foundation Grants, 3960
Winifred & Harry B. Allen Foundation Grants, 3962
Winston-Salem Foundation Competitive Grants, 3965
Winston-Salem Fndn Elkin/Tri-County Grants, 3966
Winston-Salem Fndn Stokes County Grants, 3967
Wolfe Associates Grants, 3969
Wood-Claeyssens Foundation Grants, 3971
World of Children Health Award, 3974
Yawkey Foundation Grants, 3980
ZYTL Foundation Grants, 3985

Health Care Access

2 Depot Square Ipswich Charitable Fndn Grants, 4
3M Fndn Health & Human Services Grants, 7
1976 Foundation Grants, 12
AAP Community Access To Child Health (CATCH) Advocacy Training Grants, 177
AAP Community Access To Child Health (CATCH) Implementation Grants, 178
AAP Community Access to Child Health (CATCH) Planning Grants, 179
AAP Community Access To Child Health (CATCH) Residency Training Grants, 180
AAP Community Access To Child Health (CATCH) Resident Grants, 181
AAP Leonard P. Rome Community Access to Child Health (CATCH) Visiting Professorships, 186
AAP Resident Initiative Fund Grants, 189
Abbott Fund Community Grants, 201
Abell Fndn Health & Human Services Grants, 209
Abernethy Family Foundation Grants, 210
Abundance Foundation International Grants, 218
Achelis Foundation Grants, 234
ACL Alzheimer's Disease Initiative Grants, 237
ACL Field Initiated Projects Program: Minority-Serving Institution Development Grants, 242
ACL Field Initiated Projects Program: Minority-Serving Institution Research Grants, 243
ADA Foundation Thomas J. Zwemer Award, 297
Adam Richter Charitable Trust Grants, 298
Adams and Reese Corporate Giving Grants, 299
Addison H. Gibson Foundation Medical Grants, 305
Adler-Clark Electric Community Commitment Foundation Grants, 307
Administration on Aging Senior Medicare Patrol Project Grants, 309
Advance Auto Parts Corporate Giving Grants, 312
AEGON Transamerica Foundation Health and Welfare Grants, 316
Aetna Foundation Health Grants in Connecticut, 327
Aetna Foundation Obesity Grants, 331
Aetna Foundation Racial and Ethnic Health Care Equity Grants, 332
A Friends' Foundation Trust Grants, 341
Agnes M. Lindsay Trust Grants, 346
Aid for Starving Children Emergency Assistance Fund Grants, 367
Albert W. Rice Charitable Foundation Grants, 411
Alexander Eastman Foundation Grants, 420
Alfred and Tillie Shemanski Testamentary Trust Grants, 428
Alfred Bersted Foundation Grants, 429
Alfred E. Chase Charitable Foundation Grants, 430
Alice C. A. Sibley Fund Grants, 435
Allan C. and Lelia J. Garden Foundation Grants, 438
Alliance Healthcare Foundation Grants, 444
Alpha Natural Resources Corporate Giving, 450
AMA Foundation Fund for Better Health Grants, 484
AMA Foundation Healthy Communities/Healthy America Grants, 486
Amelia Sillman Rockwell and Carlos Perry Rockwell Charities Fund Grants, 501
American Express Community Service Grants, 531
American Psychiatric Foundation Awards for Advancing Minority Mental Health, 551
American Psychiatric Foundation Disaster Recovery Fund for Psychiatrists, 553
Amica Companies Foundation Grants, 574
Angels Baseball Foundation Grants, 590
Anheuser-Busch Foundation Grants, 591
Appalachian Regional Commission Health Grants, 637
Armstrong McDonald Foundation Health Grants, 663
Assurant Health Foundation Grants, 789
AT&T Foundation Civic and Community Service Program Grants, 793
Autzen Foundation Grants, 810
Babcock Charitable Trust Grants, 816
Balfe Family Foundation Grants, 819
Baptist-Trinity Lutheran Legacy Fndn Grants, 827
Baptist Community Ministries Grants, 828
Barra Foundation Community Fund Grants, 833
Baxter International Corporate Giving Grants, 841
Baxter International Foundation Foster G. McGaw Prize, 842
Baxter International Foundation Grants, 843
BBVA Compass Foundation Charitable Grants, 848
BCBSM Foundation Community Health Matching Grants, 852
BCBSM Foundation Investigator Initiated Research Grants, 854
BCBSM Foundation Physician Investigator Research Awards, 855
BCBSNC Foundation Grants, 860
Benton Community Fndn Cookie Jar Grant, 877
Benton County Foundation Grants, 880
Bernard and Audre Rapoport Foundation Health Grants, 882
BHHS Legacy Foundation Grants, 888
Blue Cross Blue Shield of Minnesota Foundation - Healthy Children: Growing Up Healthy Grants, 915
Blue Grass Community Fndn Harrison Grants, 919
Blue Shield of California Grants, 923
Bodman Foundation Grants, 929
Boyle Foundation Grants, 942
Breast Cancer Fund Grants, 947
Brian G. Dyson Foundation Grants, 948
Bristol-Myers Squibb Foundation Product Donations Grants, 960
Bristol-Myers Squibb Patient Assistance Grants, 963
Bush Fndn Health & Human Services Grants, 979
Caesars Foundation Grants, 995
California Community Foundation Health Care Grants, 998
California Endowment Innovative Ideas Challenge Grants, 1000
California Wellness Foundation Work and Health Program Grants, 1001
Cardinal Health Foundation Grants, 1017
Cargill Citizenship Fund Corporate Giving, 1018
Carl B. and Florence E. King Foundation Grants, 1022
Carl R. Hendrickson Family Foundation Grants, 1028
CFFVR Jewelers Mutual Charitable Giving, 1099
Chapman Charitable Foundation Grants, 1112
Charles H. Hall Foundation, 1120
Charles H. Pearson Foundation Grants, 1121
Charles Nelson Robinson Fund Grants, 1125
Charlotte County (FL) Community Foundation Grants, 1126
Chemtura Corporation Contributions Grants, 1130
Chilkat Valley Community Foundation Grants, 1148
CICF City of Noblesville Community Grant, 1155
Cigna Civic Affairs Sponsorships, 1159
Clark and Ruby Baker Foundation Grants, 1174
Claude Worthington Benedum Fndn Grants, 1178
CMS Hispanic Health Services Research Grants, 1186
CNCS Social Innovation Grants, 1192
Colorado Trust Grants, 1213
Columbus Foundation Allen Eiry Fund Grants, 1214
Columbus Foundation Estrich Fund Grants, 1218
Columbus Foundation J. Floyd Dixon Memorial Fund Grants, 1219
Columbus Foundation Mary Eleanor Morris Fund Grants, 1220
Comerica Charitable Foundation Grants, 1224
Community Foundation for Greater Atlanta Common Good Funds Grants, 1241
Community Fndn for the Capital Region Grants, 1259

SUBJECT INDEX

Health Care Access /743

Community Foundation of Greater Birmingham Grants, 1272
Community Fndn of Howard County Grants, 1286
Community Foundation of Louisville AIDS Project Fund Grants, 1289
Community Fndn of Louisville Health Grants, 1295
Community Fndn of Switzerland County Grants, 1311
Community Foundation Partnerships - Lawrence County Grants, 1317
Community Foundation Serving Riverside and San Bernardino Counties Impact Grants, 1319
Connecticut Health Foundation Health Initiative Grants, 1325
Covidien Medical Product Donations, 1352
Covidien Partnership for Neighborhood Wellness Grants, 1353
Cresap Family Foundation Grants, 1359
Cuesta Foundation Grants, 1371
CVS Community Grants, 1379
D. W. McMillan Foundation Grants, 1407
Dana Brown Charitable Trust Grants, 1419
Daniels Fund Grants-Aging, 1424
David N. Lane Trust Grants for Aged and Indigent Women, 1433
Daviess County Community Foundation Health Grants, 1434
Dayton Power and Light Foundation Grants, 1437
Deborah Munroe Noonan Memorial Grants, 1443
Delmarva Power and Light Contributions, 1454
DIFFA/Chicago Grants, 1477
Dominion Foundation Grants, 1485
Doree Taylor Charitable Foundation, 1490
Dorothy Hooper Beattie Foundation Grants, 1499
Dr. & Mrs. Paul Pierce Memorial Fndn Grants, 1505
DTE Energy Foundation Health and Human Services Grants, 1516
Duke Endowment Health Care Grants, 1519
Dyson Foundation Mid-Hudson Valley Project Support Grants, 1534
E. Clayton and Edith P. Gengras, Jr. Foundation, 1536
E.W. "Al" Thrasher Awards, 1540
Earl and Maxine Claussen Trust Grants, 1541
Edina Realty Foundation Grants, 1550
Edward N. and Della L. Thome Memorial Foundation Grants, 1557
Eisner Foundation Grants, 1564
Elkhart County Community Foundation Fund for Elkhart County, 1572
El Pomar Foundation Grants, 1582
Emerson Electric Company Contributions Grants, 1590
Ensworth Charitable Foundation Grants, 1595
Erie Chapman Foundation Grants, 1600
Essex County Community Foundation Merrimack Valley General Fund Grants, 1603
Eugene G. and Margaret M. Blackford Memorial Fund Grants, 1610
Faye McBeath Foundation Grants, 1635
FCD New American Children Grants, 1637
Finance Factors Foundation Grants, 1666
Florian O. Bartlett Trust Grants, 1677
Ford Family Foundation Grants - Access to Health and Dental Services, 1686
Foundation for a Healthy Kentucky Grants, 1691
Foundation for the Mid South Community Development Grants, 1699
Foundation for the Mid South Health and Wellness Grants, 1700
Foundation of CVPH Melissa Lahtinen-Penfield Organ Donor Fund Grants, 1703
Foundation of CVPH Rabin Fund Grants, 1704
Foundation of CVPH Roger Senecal Endowment Fund Grants, 1705
Foundation of CVPH Travel Fund Grants, 1706
Foundations of East Chicago Health Grants, 1709
Four County Community Fndn General Grants, 1710
Four County Community Foundation Healthy Senior/Healthy Youth Fund Grants, 1711
France-Merrick Foundation Health and Human Services Grants, 1715
Frances W. Emerson Foundation Grants, 1718

Frank B. Hazard General Charity Fund Grants, 1721
Frank Reed and Margaret Jane Peters Memorial Fund II Grants, 1728
Frederick McDonald Trust Grants, 1740
Frederick W. Marzahl Memorial Fund Grants, 1741
Fulton County Community Foundation Women's Giving Circle Grants, 1766
General Service Foundation Human Rights and Economic Justice Grants, 1784
George A. and Grace L. Long Foundation Grants, 1788
George E. Hatcher, Jr. and Ann Williams Hatcher Foundation Grants, 1792
George W. Wells Foundation Grants, 1804
Gerber Foundation Grants, 1807
Gibson County Community Foundation Women's Fund Grants, 1817
Gibson Foundation Grants, 1818
GlaxoSmithKline Corporate Grants, 1829
GNOF Bayou Communities Grants, 1835
GNOF IMPACT Grants for Health and Human Services, 1839
GNOF IMPACT Gulf States Eye Surgery Fund, 1840
GNOF IMPACT Harold W. Newman, Jr. Charitable Trust Grants, 1841
GNOF Norco Community Grants, 1845
Grace Bersted Foundation Grants, 1859
Great-West Life Grants, 1879
Greater Sitka Legacy Fund Grants, 1885
Green Foundation Human Services Grants, 1894
Green River Area Community Fndn Grants, 1895
Gregory B. Davis Foundation Grants, 1898
Gregory L. Gibson Charitable Fndn Grants, 1900
Guy's and St. Thomas' Charity Grants, 1918
Guy I. Bromley Trust Grants, 1919
HAF Barry F. Phelps Fund Grants, 1927
HAF David (Davey) H. Somerville Medical Travel Fund Grants, 1928
HAF JoAllen K. Twiddy-Wood Memorial Fund Grants, 1929
Hampton Roads Community Foundation Faith Community Nursing Grants, 1940
Hannaford Charitable Foundation Grants, 1944
Hardin County Community Foundation Grants, 1946
Harold and Arlene Schnitzer CARE Foundation Grants, 1950
Harold Brooks Foundation Grants, 1952
Harrison County Community Foundation Signature Grants, 1958
Harry Edison Foundation, 1962
Harry S. Black and Allon Fuller Fund Grants, 1966
Harry Sudakoff Foundation Grants, 1967
Harvest Foundation Grants, 1971
Healthcare Fndn for Orange County Grants, 1981
Healthcare Foundation of New Jersey Grants, 1982
Health Foundation of Southern Florida Responsive Grants, 1985
Hearst Foundations Health Grants, 1987
Helen Irwin Littauer Educational Trust Grants, 1994
Henrietta Lange Burk Fund Grants, 2001
Henry A. and Mary J. MacDonald Foundation, 2003
Henry and Ruth Blaustein Rosenberg Foundation Health Grants, 2004
Hill Crest Foundation Grants, 2030
Horace Moses Charitable Foundation Grants, 2047
HRK Foundation Health Grants, 2067
HRSA Nurse Education, Practice, Quality and Retention (NEPQR) Grants, 2068
Huntington Clinical Foundation Grants, 2082
Huntington County Community Foundation Make a Difference Grants, 2083
IBCAT Screening Mammography Grants, 2095
Ida S. Barter Trust Grants, 2116
IDPH Hosptial Capital Investment Grants, 2119
Independence Blue Cross Charitable Medical Care Grants, 2150
J.H. Robbins Foundation Grants, 2171
Jacob and Hilda Blaustein Foundation Health and Mental Health Grants, 2181
James M. Cox Foundation of Georgia Grants, 2196
Jane Bradley Pettit Foundation Health Grants, 2208

Jerome and Mildred Paddock Foundation Grants, 2222
Jessica Stevens Community Foundation Grants, 2224
Jewish Fund Grants, 2227
John Clarke Trust Grants, 2232
John M. Lloyd Foundation Grants, 2244
Johnson & Johnson Community Health Grants, 2252
Johnson & Johnson Corporate Contributions, 2253
Johnson Controls Foundation Health and Social Services Grants, 2255
Joseph Drown Foundation Grants, 2268
Joseph Henry Edmondson Foundation Grants, 2270
Josephine Schell Russell Charitable Trust Grants, 2274
Kansas Health Fndn Major Initiatives Grants, 2288
Kenai Peninsula Foundation Grants, 2298
Kendrick Foundation Grants, 2299
Kenneth T. & Eileen L. Norris Fndn Grants, 2300
Ketchikan Community Foundation Grants, 2303
Kimball International-Habig Foundation Health and Human Services Grants, 2308
Kodiak Community Foundation Grants, 2314
Lena Benas Memorial Fund Grants, 2353
Lisa Higgins-Hussman Foundation Grants, 2370
Lloyd A. Fry Foundation Health Grants, 2373
Louetta M. Cowden Foundation Grants, 2377
Louie M. & Betty M. Phillips Fndn Grants, 2378
Lumpkin Family Fnd Healthy People Grants, 2391
Macquarie Bank Foundation Grants, 2410
Marathon Petroleum Corporation Grants, 2422
Marion Gardner Jackson Charitable Trust Grants, 2434
Marshall County Community Foundation Grants, 2439
Mary Black Fndn Community Health Grants, 2441
Mary K. Chapman Foundation Grants, 2443
Mary Wilmer Covey Charitable Trust Grants, 2448
Mattel Children's Foundation Grants, 2454
Mattel International Grants Program, 2455
Mead Johnson Nutritionals Charitable Giving, 2502
Medtronic Foundation CommunityLink Health Grants, 2510
Medtronic Foundation Strengthening Health Systems Grants, 2515
MetroWest Health Foundation Capital Grants for Health-Related Facilities, 2526
MetroWest Health Foundation Grants--Healthy Aging, 2527
Metzger-Price Fund Grants, 2530
Meyer Foundation Healthy Communities Grants, 2533
MGM Resorts Foundation Community Grants, 2541
Michael Reese Health Trust Responsive Grants, 2545
Mid-Iowa Health Foundation Community Response Grants, 2549
MLA Donald A.B. Lindberg Fellowship, 2562
MMS and Alliance Charitable Foundation Grants for Community Action and Care for the Medically Uninsured, 2587
Moline Foundation Community Grants, 2590
Morris & Gwendolyn Cafritz Fndn Grants, 2594
Morton K. and Jane Blaustein Foundation Health Grants, 2595
Ms. Foundation for Women Health Grants, 2597
Mt. Sinai Health Care Foundation Health of the Jewish Community Grants, 2599
Mt. Sinai Health Care Foundation Health of the Urban Community Grants, 2600
Mt. Sinai Health Care Foundation Health Policy Grants, 2601
NAPNAP Foundation Innovative Health Care Small Grant, 2611
NAPNAP Foundation McNeil Grant-in-Aid, 2612
Northwest Airlines KidCares Medical Travel Assistance, 2998
NYCT AIDS/HIV Grants, 3048
NYCT Blindness and Visual Disabilities Grants, 3050
NYCT Children/Youth with Disabilities Grants, 3052
NYCT Grants for the Elderly, 3054
NYCT Health Care Services, Systems, and Policies Grants, 3055
NYCT Mental Health and Retardation Grants, 3056
OceanFirst Foundation Major Grants, 3062
Ohio County Community Foundation Grants, 3068
Ohio County Community Fndn Mini-Grants, 3069

744 / Health Care Access

Olympus Corporation of Americas Corporate Giving Grants, 3073
Oregon Community Fndn Community Grants, 3082
OSF Accountability and Monitoring in Health Initiative Grants, 3087
OSF Health Media Initiative Grants, 3089
OSF International Harm Reduction Development Program Grants, 3090
OSF Mental Health Initiative Grants, 3093
OSF Public Health Grants in Kyrgyzstan, 3094
OSF Tackling Addiction Grants in Baltimore, 3096
Otto Bremer Foundation Grants, 3099
Pajaro Valley Community Health Health Trust Insurance/Coverage & Education on Using the System Grants, 3108
Pajaro Valley Community Health Trust Oral Health: Prevention & Access Grants, 3110
Petersburg Community Foundation Grants, 3156
Peyton Anderson Foundation Grants, 3160
Pfizer Healthcare Charitable Contributions, 3180
Pfizer Special Events Grants, 3186
Philip L. Graham Fund Health and Human Services Grants, 3190
Piedmont Natural Gas Foundation Health and Human Services Grants, 3221
Piper Trust Healthcare & Med Research Grants, 3228
Piper Trust Older Adults Grants, 3229
Porter County Health and Wellness Grant, 3246
Portland Fndn Women's Giving Circle Grant, 3249
PPCF Community Grants, 3256
Price Chopper's Golub Foundation Grants, 3261
Pulido Walker Foundation, 3282
Qualcomm Grants, 3287
Quality Health Foundation Grants, 3288
Quantum Corporation Snap Server Grants, 3289
Rainbow Endowment Grants, 3300
Ralph M. Parsons Foundation Grants, 3304
Rasmuson Foundation Tier Two Grants, 3308
Raymond John Wean Foundation Grants, 3310
Regence Fndn Access to Health Care Grants, 3346
Regence Foundation Health Care Community Awareness and Engagement Grants, 3347
Regence Fndn Health Care Connections Grants, 3348
Regence Fndn Improving End-of-Life Grants, 3349
Regence Fndn Tools & Technology Grants, 3350
Reynolds & Reynolds Associate Fndn Grants, 3360
Ricks Family Charitable Trust Grants, 3373
Robert F. Stoico/FIRSTFED Charitable Foundation Grants, 3382
Robert R. Meyer Foundation Grants, 3387
Robert Sterling Clark Foundation Reproductive Rights and Health Grants, 3388
Rockefeller Foundation Grants, 3396
Rose Community Foundation Health Grants, 3411
RWJF Changes in Health Care Financing and Organization Grants, 3430
RWJF Disparities Research for Change Grants, 3433
RWJF New Jersey Health Initiatives Grants, 3440
RWJF Pioneer Portfolio Grants, 3442
RWJF Vulnerable Populations Portfolio Grants, 3444
S.D. Bechtel, Jr. Fndn/Stephen Bechtel Preventive Healthcare and Selected Research Grants, 3446
Sain-Orr and Royak-DeForest Steadman Foundation Grants, 3452
Saint Luke's Foundation Grants, 3454
San Diego Foundation Paradise Valley Hospital Community Fund Grants, 3472
Sands Foundation Grants, 3476
San Francisco Fndn Community Health Grants, 3478
Santa Barbara Foundation Strategy Grants - Core Support, 3481
Santa Fe Community Foundation Seasonal Grants-Fall Cycle, 3482
Seattle Foundation Benjamin N. Phillips Memorial Fund Grants, 3504
Seattle Foundation Health and Wellness Grants, 3507
Seward Community Foundation Grants, 3514
Seward Community Foundation Mini-Grants, 3515
Sheltering Arms Fund Grants, 3545
Shield-Ayres Foundation Grants, 3546

Sierra Health Foundation Responsive Grants, 3555
Siragusa Foundation Health Services & Medical Research Grants, 3571
Sir Dorabji Tata Trust Grants for NGOs or Voluntary Organizations, 3572
Sisters of St. Joseph Healthcare Fndn Grants, 3576
Solo Cup Foundation Grants, 3620
Sophia Romero Trust Grants, 3624
Straits Area Community Foundation Health Care Grants, 3657
SunTrust Bank Trusteed Foundations Greene-Sawtell Grants, 3681
SunTrust Bank Trusteed Foundations Harriet McDaniel Marshall Tust Grants, 3682
SunTrust Bank Trusteed Foundations Nell Warren Elkin and William Simpson Elkin Grants, 3683
SunTrust Bank Trusteed Foundations Thomas Guy Woolford Charitable Trust Grants, 3684
SunTrust Bank Trusteed Foundations Walter H. and Marjory M. Rich Memorial Fund Grants, 3685
Texas Instruments Corporation Health and Human Services Grants, 3709
Thelma Braun & Bocklett Family Fndn Grants, 3713
Thomas C. Burke Foundation Grants, 3720
Thomas J. Long Foundation Community Grants, 3722
Topfer Family Foundation Grants, 3739
Toyota Motor Manuf of West Virginia Grants, 3748
Toys R Us Children's Fund Grants, 3751
UCLA / RAND Corporation Post-Doctoral Fellowships, 3776
UniHealth Foundation Healthcare Systems Enhancements Grants, 3784
Union Labor Health Fndn Angel Fund Grants, 3791
Union Labor Health Fndn Community Grants, 3792
United Hospital Fund of New York Health Care Improvement Grants, 3797
United Methodist Child Mental Health Grants, 3800
USAID Accelerating Progress Against Tuberculosis in Kenya Grants, 3805
USAID Devel Innovation Accelerator Grants, 3808
USAID Ebola Response, Recovery and Resilience in West Africa Grants, 3809
USAID Family Health Plus Project Grants, 3810
USAID HIV Prevention with Key Populations - Mali Grants, 3816
USAID Integration of Care and Support within the Health System to Support Better Patient Outcomes Grants, 3817
USAID Research and Innovation for Health Supply Chain Systems & Commodity Security Grants, 3821
USDA Delta Health Care Services Grants, 3829
Vanderburgh Community Foundation Grants, 3852
Verizon Foundation Health Care and Accessibility Grants, 3858
Visiting Nurse Foundation Grants, 3879
W.H. & Mary Ellen Cobb Trust Grants, 3884
W.K. Kellogg Foundation Secure Families Grants, 3887
W.M. Keck Fndn So California Grants, 3890
W.P. and Bulah Luse Foundation Grants, 3892
Walker Area Community Foundation Grants, 3895
Walmart Foundation Northwest Arkansas Giving, 3899
Warrick County Community Foundation Women's Fund, 3905
Washington County Community Foundation Youth Grants, 3907
Weaver Popcorn Foundation Grants, 3910
Welborn Baptist Foundation Improvements to Community Health Status Grants, 3916
Weyerhaeuser Family Foundation Health Grants, 3926
WHO Foundation General Grants, 3939
William J. & Tina Rosenberg Fndn Grants, 3954
William J. Brace Charitable Trust, 3955
William Ray & Ruth E. Collins Fndn Grants, 3956
Willis C. Helm Charitable Trust Grants, 3959
Wolfe Associates Grants, 3969
ZYTL Foundation Grants, 3985

Health Care Administration

AAAAI Outstanding Vol Clinical Faculty Award, 32
Aaron & Freda Glickman Foundation Grants, 192

AMDA Foundation Medical Director of the Year Award, 497
AMDA Foundation Quality Improvement and Health Outcome Awards, 498
American Academy of Nursing Building Academic Geriatric Nursing Capacity Scholarships, 505
AUPHA Corris Boyd Scholarships, 800
AUPHA Foster G. McGaw Scholarships, 801
Beckman Coulter Foundation Grants, 868
Bush Foundation Medical Fellowships, 980
Byron W. & Alice L. Lockwood Fnd Grants, 991
Canada-U.S. Fulbright New Century Scholars Program Grants, 1008
Canada-U.S. Fulbright Senior Specialists Grants, 1009
CDC Public Health Prev Service Fellowships, 1069
CDC Public Health Prevention Service Fellowship Sponsorships, 1070
CMS Historically Black Colleges and Universities (HBCU) Health Services Research Grants, 1187
Commonwealth Payment System Reform Grants, 1232
Covenant Educational Foundation Grants, 1348
Covenant Matters Foundation Grants, 1350
Deborah Munroe Noonan Memorial Grants, 1443
Duke Endowment Health Care Grants, 1519
E. Rhodes & Leona B. Carpenter Grants, 1539
Fulbright Alumni Initiatives Awards, 1749
Fulbright Distinguished Chairs Awards, 1750
Fulbright New Century Scholars Grants, 1753
Fulbright Specialists Program Grants, 1754
Fulbright Scholars in Europe and Eurasia, 1755
Fulbright Traditional Scholar Program in Sub-Saharan Africa, 1756
Fulbright Traditional Scholar Program in the East Asia/Pacific Region, 1757
Fulbright Traditional Scholar Program in the Near East and North Africa Region, 1758
Fulbright Traditional Scholar Program in the South and Central Asia Region, 1759
Fulbright Traditional Scholar Program in the Western Hemisphere, 1760
Guy's and St. Thomas' Charity Grants, 1918
Health Management Scholarships and Grants for Minorities, 1986
IHI Quality Improvement Fellowships, 2121
Mayo Clinic Administrative Fellowship, 2460
Mayo Clinic Admin Fellowship - Eau Claire, 2461
Mayo Clinic Business Consulting Fellowship, 2462
Michael Reese Health Trust Core Grants, 2544
Mt. Sinai Health Care Foundation Health Policy Grants, 2601
NAPNAP Foundation Nursing Research Grants, 2615
National Blood Foundation Research Grants, 2628
NYCT Health Care Services, Systems, and Policies Grants, 3055
OSF Accountability and Monitoring in Health Initiative Grants, 3087
OSF Law and Health Initiative Grants, 3092
OSF Tackling Addiction Grants in Baltimore, 3096
SfN Louise Hanson Marshall Specific Recognition Award, 3527
UCLA / RAND Corporation Post-Doctoral Fellowships, 3776
UniHealth Foundation Healthcare Systems Enhancements Grants, 3784
United Hospital Fund of New York Health Care Improvement Grants, 3797
USAID Devel Innovation Accelerator Grants, 3808

Health Care Assessment

AACAP Jeanne Spurlock Research Fellowship in Substance Abuse and Addiction for Minority Medical Students, 60
AACAP Summer Medical Student Fellowships, 75
Alexander Eastman Foundation Grants, 420
American Psychiatric Foundation Disaster Recovery Fund for Psychiatrists, 553
Blue Cross Blue Shield of Minnesota Foundation - Healthy Equity: Health Impact Assessment Demonstration Project Grants, 916

SUBJECT INDEX

Blue Cross Blue Shield of Minnesota Foundation - Healthy Equity: Health Impact Assessment Program Grants, 917
Commonwealth Fund Harkness Fellowships in Health Care Policy and Practice, 1229
Commonwealth Fund Health Care Quality Improvement and Efficiency Grants, 1230
Community Foundation for the National Capital Region Community Leadership Grants, 1260
Covidien Partnership for Neighborhood Wellness Grants, 1353
Duke Endowment Health Care Grants, 1519
Guy's and St. Thomas' Charity Grants, 1918
Illinois Children's Healthcare Foundation Grants, 2146
Johnson & Johnson Community Health Grants, 2252
Kansas Health Fndn Major Initiatives Grants, 2288
Kansas Health Foundation Recognition Grants, 2289
NYCT Children/Youth with Disabilities Grants, 3052
NYCT Health Care Services, Systems, and Policies Grants, 3055
Piedmont Natural Gas Foundation Health and Human Services Grants, 3221
UCLA / RAND Corporation Post-Doctoral Fellowships, 3776

Health Care Economics
AMI Semiconductors Corporate Grants, 576
CMS Hispanic Health Services Research Grants, 1186
Duke Endowment Health Care Grants, 1519
Foundation for Seacoast Health Grants, 1698
GNOF IMPACT Grants for Health and Human Services, 1839
Guy's and St. Thomas' Charity Grants, 1918
Kendrick Foundation Grants, 2299
Mt. Sinai Health Care Foundation Health Policy Grants, 2601
Percy B. Ferebee Endowment Grants, 3138
PhRMA Foundation Health Outcomes Research Starter Grants, 3204
Robert Leet & Clara Guthrie Patterson Grants, 3383
RWJF Changes in Health Care Financing and Organization Grants, 3430
Sisters of Charity Foundation of Cleveland Reducing Health and Educational Disparities in the Central Neighborhood Grants, 3574
UCLA / RAND Corporation Post-Doctoral Fellowships, 3776
United Hospital Fund of New York Health Care Improvement Grants, 3797

Health Care Financing
Alliance Healthcare Foundation Grants, 444
AUPHA Corris Boyd Scholarships, 800
AUPHA Foster G. McGaw Scholarships, 801
Baptist-Trinity Lutheran Legacy Fndn Grants, 827
BCBSM Foundation Investigator Initiated Research Grants, 854
California Community Foundation Health Care Grants, 998
Covidien Partnership for Neighborhood Wellness Grants, 1353
Duke Endowment Health Care Grants, 1519
Guy's and St. Thomas' Charity Grants, 1918
Mt. Sinai Health Care Foundation Health Policy Grants, 2601
NYCT AIDS/HIV Grants, 3048
OSF Global Health Financing Initiative Grants, 3088
Sarasota Memorial Healthcare Fndn Grants, 3484
UCLA / RAND Corporation Post-Doctoral Fellowships, 3776

Health Care Personnel
AlohaCare Believes in Me Scholarship, 449
AVDF Health Care Grants, 811
Biogen Foundation Healthcare Professional Education Grants, 900
Connecticut Health Foundation Health Initiative Grants, 1325
Covidien Partnership for Neighborhood Wellness Grants, 1353

Pfizer Medical Education Track One Grants, 3182
UniHealth Foundation Workforce Development Grants, 3786

Health Care Promotion
ACL Self-Advocacy Resource Center Grants, 247
Albert W. Rice Charitable Foundation Grants, 411
Armstrong McDonald Foundation Health Grants, 663
Baxter International Foundation Foster G. McGaw Prize, 842
Blue Cross Blue Shield of Minnesota Fndn Healthy Equity: Public Libraries for Health Grants, 918
CDC Steven M. Teutsch Prevention Effectiveness Fellowships, 1071
Cigna Civic Affairs Sponsorships, 1159
Community Fndn of Bloomington & Monroe County - Precision Health Network Cycle Grants, 1264
Covidien Partnership for Neighborhood Wellness Grants, 1353
Foundation for the Mid South Community Development Grants, 1699
France-Merrick Foundation Health and Human Services Grants, 1715
George E. Hatcher, Jr. and Ann Williams Hatcher Foundation Grants, 1792
George W. Wells Foundation Grants, 1804
Gibson County Community Foundation Women's Fund Grants, 1817
GNOF IMPACT Grants for Health and Human Services, 1839
GNOF IMPACT Kahn-Oppenheim Grants, 1842
Hill Crest Foundation Grants, 2030
HRK Foundation Health Grants, 2067
Kansas Health Fndn Major Initiatives Grants, 2288
Medtronic Foundation Strengthening Health Systems Grants, 2515
Meyer Foundation Healthy Communities Grants, 2533
Michael Reese Health Trust Responsive Grants, 2545
MLA T. Mark Hodges Int'l Service Award, 2575
Moline Foundation Community Grants, 2590
Morton K. and Jane Blaustein Foundation Health Grants, 2595
Mt. Sinai Health Care Foundation Health of the Urban Community Grants, 2600
Mt. Sinai Health Care Foundation Health Policy Grants, 2601
Nestle Foundation Training Grant, 2679
Olympus Corporation of Americas Corporate Giving Grants, 3073
OSF Public Health Grants in Kyrgyzstan, 3094
Philip L. Graham Fund Health and Human Services Grants, 3190
Porter County Health and Wellness Grant, 3246
Saint Luke's Foundation Grants, 3454
Sigma Theta Tau International /American Nurses Foundation Grant, 3556
Sigma Theta Tau Small Grants, 3560
Union Labor Health Fndn Community Grants, 3792
United Healthcare Comm Grants in Michigan, 3796
USAID Family Health Plus Project Grants, 3810
USAID Health System Strengthening Grants, 3814
USDA Delta Health Care Services Grants, 3829

Health Disparities
Aetna Foundation Racial and Ethnic Health Care Equity Grants, 332
Appalachian Regional Commission Health Grants, 637
California Endowment Innovative Ideas Challenge Grants, 1000
Community Fndn of Louisville Health Grants, 1295
Delta Air Lines Foundation Health and Wellness Grants, 1455
GNOF IMPACT Grants for Health and Human Services, 1839
Guy I. Bromley Trust Grants, 1919
Louetta M. Cowden Foundation Grants, 2377
Maine Community Foundation Penobscot Valley Health Association Grants, 2417
Medtronic Foundation CommunityLink Health Grants, 2510

Health Personnel/Professions /745

NHLBI Mentored Career Dev Award to Promote Faculty Diversity in Biomed Research, 2708
OBSSR Behavioral and Social Science Research on Understanding and Reducing Health Disparities Grants, 3061
PC Fred H. Bixby Fellowships, 3126
Pfizer ASPIRE North America Rheumatology Research Awards, 3172
Porter County Health and Wellness Grant, 3246
Union Labor Health Fndn Community Grants, 3792
USDD Clinical Functional Assessment Investigator-Initiated Grants, 3840

Health Facilities Studies
NYCT Health Care Services, Systems, and Policies Grants, 3055
UCLA / RAND Corporation Post-Doctoral Fellowships, 3776

Health Insurance
AAP Community Access To Child Health (CATCH) Advocacy Training Grants, 177
AAP Community Access to Child Health (CATCH) Planning Grants, 179
AAP Community Access To Child Health (CATCH) Residency Training Grants, 180
AAP Community Access To Child Health (CATCH) Resident Grants, 181
AAP Leonard P. Rome Community Access to Child Health (CATCH) Visiting Professorships, 186
ACL Empowering Seniors to Prevent Health Care Fraud Grants, 241
ACL Medicare Improvements for Patients & Providers Funding for Beneficiary Outreach & Assistance for Title VI Native Americans, 245
Alliance Healthcare Foundation Grants, 444
AMA Foundation Fund for Better Health Grants, 484
Austin-Bailey Health and Wellness Fndn Grants, 803
BCBSNC Foundation Grants, 860
Blue Shield of California Grants, 923
California Wellness Foundation Work and Health Program Grants, 1001
CMS Hispanic Health Services Research Grants, 1186
CMS Research and Demonstration Grants, 1188
Commonwealth Payment System Reform Grants, 1232
Commonwealth Fund Small Grants, 1234
Edwards Memorial Trust Grants, 1558
Foundation for a Healthy Kentucky Grants, 1691
Guy I. Bromley Trust Grants, 1919
Health Foundation of Greater Cincinnati Grants, 1983
Louetta M. Cowden Foundation Grants, 2377
Michael Reese Health Trust Responsive Grants, 2545
Mid-Iowa Health Foundation Community Response Grants, 2549
MMS and Alliance Charitable Foundation Grants for Community Action and Care for the Medically Uninsured, 2587
Mt. Sinai Health Care Foundation Health Policy Grants, 2601
NYC Managed Care Consumer Assistance Workshop Re-Design Grants, 3047
OSF Global Health Financing Initiative Grants, 3088
Packard Foundation Children, Families, and Communities Grants, 3104
Prince Charitable Trusts Chicago Grants, 3270
Quantum Corporation Snap Server Grants, 3289
Quantum Foundation Grants, 3290
Rehab Therapy Foundation Grants, 3356
RWJF Changes in Health Care Financing and Organization Grants, 3430
UCLA / RAND Corporation Post-Doctoral Fellowships, 3776

Health Personnel/Professions
BCBSM Foundation Student Award Program, 858
MMS and Alliance Charitable Foundation International Health Studies Grants, 2588
NHLBI Short-Term Research Education Program to Increase Diversity in Health-Related Research, 2733
Prince Charitable Trusts Chicago Grants, 3270

746 / Health Personnel/Professions

RWJF Harold Amos Medical Faculty Development Scholars, 3435
RWJF Nurse Faculty Scholars, 3441
Special Olympics Health Profess Student Grants, 3629
Topeka Community Foundation Kansas Blood Services Scholarships, 3738

Health Planning/Policy
3M Fndn Health & Human Services Grants, 7
AAP Community Access To Child Health (CATCH) Advocacy Training Grants, 177
AAP Community Access to Child Health (CATCH) Planning Grants, 179
AAP Community Access To Child Health (CATCH) Residency Training Grants, 180
AAP Community Access To Child Health (CATCH) Resident Grants, 181
AAP Leonard P. Rome Community Access to Child Health (CATCH) Visiting Professorships, 186
AcademyHealth Alice S. Hersh New Investigator Awards, 220
AcademyHealth Alice Hersh Student Scholarships, 221
AcademyHealth HSR Impact Awards, 223
AcademyHealth PHSR Interest Group Student Scholarships, 225
ACS Research Scholar Grants for Health Services and Health Policy Research, 279
Acumen Global Fellowships, 285
Alvin and Fanny Blaustein Thalheimer Foundation Baltimore Communal Grants, 459
American Psychiatric Foundation Jeanne Spurlock Congressional Fellowship, 556
American Psychiatric Foundation Minority Fellowships, 557
Appalachian Regional Commission Health Grants, 637
ASHG Public Health Genetics Fellowship, 712
AUPHA Corris Boyd Scholarships, 800
AUPHA Foster G. McGaw Scholarships, 801
Baxter International Corporate Giving Grants, 841
BCBSM Foundation Excellence in Research Awards for Students, 853
BCBSM Foundation Student Award Program, 858
BCBSM Frank J. McDevitt, DO, Excellence in Research Awards, 859
Becton Dickinson and Company Grants, 869
Blue Cross Blue Shield of Minnesota Foundation - Healthy Equity: Health Impact Assessment Demonstration Project Grants, 916
Blue Cross Blue Shield of Minnesota Foundation - Healthy Equity: Health Impact Assessment Program Grants, 917
Blue Shield of California Grants, 923
California Wellness Foundation Work and Health Program Grants, 1001
Carla J. Funk Governmental Relations Award, 1020
CDC Preventive Med Residency & Fellowship, 1064
CDC Public Health Prev Service Fellowships, 1069
CDC Public Health Prevention Service Fellowship Sponsorships, 1070
CDC Steven M. Teutsch Prevention Effectiveness Fellowships, 1071
Cigna Civic Affairs Sponsorships, 1159
Colorado Trust Grants, 1213
Commonwealth Fund/Harvard University Fellowship in Minority Health Policy, 1226
Commonwealth Fund Affordable Health Insurance Grants, 1227
Commonwealth Fund Australian-American Health Policy Fellowships, 1228
Commonwealth Fund Harkness Fellowships in Health Care Policy and Practice, 1229
Commonwealth Fund Health Care Quality Improvement and Efficiency Grants, 1230
Commonwealth Fund Patient-Centered Coordinated Care Program Grants, 1231
Commonwealth Payment System Reform Grants, 1232
Commonwealth Fund Small Grants, 1234
Commonwealth Fund State High Performance Health Systems Grants, 1235
Community Fndn of Louisville Health Grants, 1295

Covenant Matters Foundation Grants, 1350
Doris Duke Charitable Foundation Operations Research on AIDS Care and Treatment in Africa (ORACTA) Grants, 1497
Entergy Corporation Open Grants for Healthy Families, 1597
Foundation for a Healthy Kentucky Grants, 1691
Fndn for Appalachian Ohio Bachtel Scholarships, 1692
General Service Foundation Human Rights and Economic Justice Grants, 1784
GNOF IMPACT Grants for Health and Human Services, 1839
Greenwall Foundation Bioethics Grants, 1897
Grifols Community Outreach Grants, 1904
Harry S. Black and Allon Fuller Fund Grants, 1966
HHMI Grants and Fellowships Programs, 2020
Hogg Foundation for Mental Health Grants, 2039
Jane Bradley Pettit Foundation Health Grants, 2208
John M. Lloyd Foundation Grants, 2244
Kansas Health Foundation Recognition Grants, 2289
Kinsman Foundation Grants, 2309
Lester Ray Fleming Scholarships, 2360
Mary Black Fndn Community Health Grants, 2441
Meyer Foundation Healthy Communities Grants, 2533
Mt. Sinai Health Care Foundation Health Policy Grants, 2601
NYCT Girls and Young Women Grants, 3053
NYCT Health Care Services, Systems, and Policies Grants, 3055
NYCT Mental Health and Retardation Grants, 3056
OECD Sexual Health and Rights Project Grants, 3064
OSF Accountability and Monitoring in Health Initiative Grants, 3087
OSF Law and Health Initiative Grants, 3092
OSF Roma Health Project Grants, 3095
OSF Tackling Addiction Grants in Baltimore, 3096
Philip L. Graham Fund Health and Human Services Grants, 3190
Rainbow Endowment Grants, 3300
Robert Sterling Clark Foundation Reproductive Rights and Health Grants, 3388
Rose Community Foundation Health Grants, 3411
RWJF Policies for Action: Policy and Law Research to Build a Culture of Health Grants, 3443
Seattle Foundation Health and Wellness Grants, 3507
UCLA / RAND Corporation Post-Doctoral Fellowships, 3776
United Hospital Fund of New York Health Care Improvement Grants, 3797
USAID Child, Newborn, and Maternal Health Project Grants, 3806
USAID Devel Innovation Accelerator Grants, 3808
USAID Global Health Development Innovation Accelerator Broad Agency Grants, 3813
USAID Health System Strengthening Grants, 3814
USAID Research and Innovation for Health Supply Chain Systems & Commodity Security Grants, 3821
William T. Grant Foundation Research Grants, 3958

Health Promotion
49ers Foundation Grants, 9
A-T Medical Research Foundation Grants, 15
AAFP Foundation Health Literacy State Grants, 132
AAFP Research Stimulation Grants, 136
AAMC Abraham Flexner Award, 153
AAMC David E. Rogers Award, 157
Abington Foundation Grants, 211
ACGT Young Investigator Grants, 233
Actors Addiction & Recovery Services Grants, 282
Adolph Coors Foundation Grants, 311
AEGON Transamerica Foundation Health and Welfare Grants, 316
Air Products and Chemicals Grants, 399
Albertson's Charitable Giving Grants, 410
Alexander Eastman Foundation Grants, 420
Allen Foundation Educational Nutrition Grants, 442
Allen P. & Josephine B. Green Fndn Grants, 443
Alvin and Fanny Blaustein Thalheimer Foundation Baltimore Communal Grants, 459
AMA Foundation Health Literacy Grants, 485

American Academy of Dermatology Shade Structure Grants, 504
American Academy of Nursing Media Awards, 509
American Psychiatric Foundation Disaster Recovery Fund for Psychiatrists, 553
Anheuser-Busch Foundation Grants, 591
Ann Arbor Area Community Foundation Grants, 594
Ann Peppers Foundation Grants, 599
AON Foundation Grants, 623
Appalachian Regional Commission Health Grants, 637
AptarGroup Foundation Grants, 647
Armstrong McDonald Foundation Health Grants, 663
ASHG Public Health Genetics Fellowship, 712
Assurant Health Foundation Grants, 789
Autzen Foundation Grants, 810
BCBSM Foundation Investigator Initiated Research Grants, 854
BCBSNC Foundation Grants, 860
Beldon Fund Grants, 871
Bikes Belong Foundation Paul David Clark Bicycling Safety Grants, 890
Bildner Family Foundation Grants, 891
Bright Promises Foundation Grants, 952
Bullitt Foundation Grants, 973
California Endowment Innovative Ideas Challenge Grants, 1000
California Wellness Foundation Work and Health Program Grants, 1001
CDC Public Health Associates, 1065
CDC Public Health Associates Hosts, 1066
CDC Public Health Conference Support Grant, 1067
CDC Public Health Prev Service Fellowships, 1069
CDC Public Health Prevention Service Fellowship Sponsorships, 1070
CDC Steven M. Teutsch Prevention Effectiveness Fellowships, 1071
Champ-A Champion Fur Kids Grants, 1109
Charles H. Pearson Foundation Grants, 1121
Chase Paymentech Corporate Giving Grants, 1127
Christopher & Dana Reeve Foundation Quality of Life Grants, 1152
Cigna Civic Affairs Sponsorships, 1159
CJ Foundation for SIDS Program Services Grants, 1163
CNCS AmeriCorps VISTA Project Grants, 1190
CNCS Social Innovation Grants, 1192
Colonel Stanley R. McNeil Foundation Grants, 1211
Colorado Trust Grants, 1213
Community Fndn of Bloomington & Monroe County - Precision Health Network Cycle Grants, 1264
Comprehensive Health Education Fndn Grants, 1320
Connecticut Community Foundation Grants, 1324
Connecticut Health Foundation Health Initiative Grants, 1325
Conservation, Food, and Health Foundation Grants for Developing Countries, 1338
Daniel Mendelsohn New Investigator Award, 1423
DHHS Oral Health Promotion Research Across the Lifespan, 1474
Dickson Foundation Grants, 1476
Dorothy Rider Pool Health Care Grants, 1500
Edward N. and Della L. Thome Memorial Foundation Grants, 1557
Essex County Community Foundation Discretionary Fund Grants, 1602
Ewing Halsell Foundation Grants, 1618
Express Scripts Foundation Grants, 1620
Floyd A. & Kathleen C. Cailloux Fnd Grants, 1681
Foundation for a Healthy Kentucky Grants, 1691
Frank Reed and Margaret Jane Peters Memorial Fund II Grants, 1728
Gates Award for Global Health, 1775
General Mills Champs for Healthy Kids Grants, 1781
General Mills Foundation Grants, 1782
George E. Hatcher, Jr. and Ann Williams Hatcher Foundation Grants, 1792
Gerber Foundation Grants, 1807
Global Fund for Women Grants, 1833
GNOF IMPACT Kahn-Oppenheim Grants, 1842
GNOF Norco Community Grants, 1845
GNOF Stand Up For Our Children Grants, 1847

SUBJECT INDEX

Health Science /747

Great-West Life Grants, 1879
Harry S. Black and Allon Fuller Fund Grants, 1966
Harvest Foundation Grants, 1971
Hasbro Children's Fund Grants, 1973
Hawn Foundation Grants, 1979
HCA Foundation Grants, 1980
Healthcare Fndn for Orange County Grants, 1981
Health Foundation of Greater Cincinnati Grants, 1983
Health Fndn of Greater Indianapolis Grants, 1984
Health Foundation of Southern Florida Responsive Grants, 1985
Highmark BCBS Challenge for Healthier Schools in Western Pennsylvania, 3696
Hillcrest Foundation Grants, 2031
Hogg Foundation for Mental Health Grants, 2039
HomeBanc Foundation Grants, 2042
Homer Foundation Grants, 2044
Horizon Foundation for New Jersey Grants, 2048
HRAMF Charles H. Hood Foundation Child Health Research Awards, 2057
Institute for Agriculture and Trade Policy Food and Society Fellowships, 2159
James Hervey Johnson Charitable Educational Trust Grants, 2191
Jane's Trust Grants, 2206
Jessie B. Cox Charitable Trust Grants, 2225
Jewish Fund Grants, 2227
Joe W. and Dorothy Dorsett Brown Fndn Grants, 2230
John P. McGovern Foundation Grants, 2247
Johnson & Johnson/SAH Arts & Healing Grants, 2254
Josephine S. Gumbiner Foundation Grants, 2273
Kansas Health Foundation Recognition Grants, 2289
Linford and Mildred White Charitable Grants, 2368
Louie M. & Betty M. Phillips Fndn Grants, 2378
M-A-C AIDS Fund Grants, 2400
Mary Black Foundation Active Living Grants, 2440
Mary Wilmer Covey Charitable Trust Grants, 2448
McCarthy Family Foundation Grants, 2490
Mead Johnson Nutritionals Charitable Giving, 2502
MedImmune Charitable Grants, 2509
Medtronic Foundation CommunityLink Health Grants, 2510
Medtronic Foundation Patient Link Grants, 2514
MetroWest Health Foundation Grants--Healthy Aging, 2527
Michael Reese Health Trust Responsive Grants, 2545
MMS and Alliance Charitable Foundation Grants for Community Action and Care for the Medically Uninsured, 2587
NAPNAP Foundation Innovative Health Care Small Grant, 2611
NAPNAP Foundation McNeil Grant-in-Aid, 2612
NAPNAP Foundation Nursing Research Grants, 2615
Nell J. Redfield Foundation Grants, 2673
Nestle Foundation Pilot Grants, 2676
NIA Health Behaviors and Aging Grants, 2761
NIH Health Promotion Among Racial and Ethnic Minority Males, 2880
NLM Understanding and Promoting Health Literacy Research Grants, 2953
Nuffield Foundation Africa Program Grants, 3026
Nuffield Foundation Law and Society Grants, 3028
Nuffield Foundation Small Grants, 3031
OSF Roma Health Project Grants, 3095
Paso del Norte Health Foundation Grants, 3119
PepsiCo Foundation Grants, 3137
Pfizer Healthcare Charitable Contributions, 3180
Piedmont Health Foundation Grants, 3219
Pinnacle Foundation Grants, 3225
Premera Blue Cross Grants, 3258
Prince Charitable Trusts DC Grants, 3271
Quantum Foundation Grants, 3290
Rainbow Endowment Grants, 3300
RGk Foundation Grants, 3361
Robert R. McCormick Trib Community Grants, 3385
Robert R. Meyer Foundation Grants, 3387
RWJF Health and Society Scholars, 3436
RWJF Healthy Eating Research Grants, 3437
Saigh Foundation Grants, 3451
San Juan Island Community Foundation Grants, 3480

Scott County Community Foundation Grants, 3500
Seagate Tech Corp Capacity to Care Grants, 3502
Sigma Theta Tau International /American Nurses Foundation Grant, 3556
Sigma Theta Tau Small Grants, 3560
Sisters of St. Joseph Healthcare Fndn Grants, 3576
Special Olympics Health Profess Student Grants, 3629
Sunflower Foundation Bridge Grants, 3676
Sunflower Foundation Capacity Building Grants, 3677
Sunflower Foundation Walking Trails Grants, 3678
Sunoco Foundation Grants, 3679
Susan G. Komen Breast Cancer Foundation Career Catalyst Research Grants, 3688
Tellabs Foundation Grants, 3706
Texas Instruments Corporation Health and Human Services Grants, 3709
Textron Corporate Contributions Grants, 3712
Theodore Edson Parker Foundation Grants, 3715
Thomas C. Burke Foundation Grants, 3720
TJX Foundation Grants, 3735
Tri-State Community Twenty-first Century Endowment Fund Grants, 3756
Unihealth Foundation Grants for Non-Profit Oranizations, 3783
UniHealth Foundation Innovation Fund Grants, 3785
United Methodist Child Mental Health Grants, 3800
USAID Child, Newborn, and Maternal Health Project Grants, 3806
USAID Devel Innovation Accelerator Grants, 3808
USAID Global Health Development Innovation Accelerator Broad Agency Grants, 3813
USAID Health System Strengthening Grants, 3814
USAID Research and Innovation for Health Supply Chain Systems & Commodity Security Grants, 3821
VHA Health Foundation Grants, 3867
Visiting Nurse Foundation Grants, 3879
Washington Gas Charitable Giving, 3908
Western Union Foundation Grants, 3923
Whitney Foundation Grants, 3938
World of Children Humanitarian Award, 3975

Health Records
Regenstrief General Internal Medicine - General Pediatrics Research Fellowship Program, 3351
USDD Medical Practice Breadth of Medical Practice and Disease Frequency Exposure Grants, 3843

Health Research
AAAAI Distinguished Layperson Award, 25
AIHS/Mitacs Health Pilot Partnership Interns, 373
AIHS Alberta/Pfizer Translat Research Grants, 374
AIHS Collaborative Research and Innovation Grants - Collaborative Program, 376
AIHS Collaborative Research and Innovation Grants - Collaborative Project, 377
AIHS Collaborative Research and Innovation Grants - Collaborative Team, 378
AIHS Heritage Youth Researcher Summer Science Program, 387
AIHS Sustainability Grants, 398
AMDA Foundation Quality Improvement Award, 499
AMHPS Dr. James A. Ferguson Emerging Infectious Diseases Fellowships, 573
APHL Emerging Infectious Diseases Fellowships, 636
Armstrong McDonald Foundation Health Grants, 663
Bupa Foundation Medical Research Grants, 974
BWF Preterm Birth Initiative Grants, 990
CDC Fnd Tobacco Network Lab Fellowship, 1062
CDC Steven M. Teutsch Prevention Effectiveness Fellowships, 1071
Delta Air Lines Foundation Health and Wellness Grants, 1455
E.W. "Al" Thrasher Awards, 1540
Gilbert Memorial Fund Grants, 1820
Hampton Roads Community Foundation Mental Health Research Grants, 1942
Harry Edison Foundation, 1962
HRAMF Deborah Munroe Noonan Memorial Research Grants, 2058
Kroger Foundation Women's Health Grants, 2325

Lymphatic Education and Research Network Additional Support Grants for NIH-funded F32 Postdoctoral Fellows, 2394
Lymphatic Education and Research Network Postdoctoral Fellowships, 2395
Morehouse PHSI Project Imhotep Internships, 2592
NFL Charities Medical Grants, 2685
NHLBI Ruth L. Kirschstein National Research Service Awards for Individual Predoctoral MD/PhD Fellows and Other Dual Degree Fellows, 2731
Perpetual Trust for Charitable Giving Grants, 3141
Pfizer ASPIRE Worldwide Endocrine Young Investigator Grants, 3173
Philip L. Graham Fund Health and Human Services Grants, 3190
Seattle Foundation Medical Funds Grants, 3508
Sigma Theta Tau International Doris Bloch Research Award, 3559
USAID Family Health Plus Project Grants, 3810
USDD Clinical Functional Assessment Investigator-Initiated Grants, 3840
Wellcome Trust New Investigator Awards, 3919
William D. Laurie, Jr. Charitable Fndn Grants, 3949

Health Sciences
AAHPERD-AAHE Undergraduate Scholarships, 150
AAHPERD Abernathy Presidential Scholarships, 151
AIHS ForeFront MBA Studentship Award, 381
AIHS ForeFront MBT Studentship Awards, 382
Benton County Foundation - Fitzgerald Family Scholarships, 879
Carla J. Funk Governmental Relations Award, 1020
CFFVR Jewelers Mutual Charitable Giving, 1099
Community Foundation of Greater Fort Wayne - Lilly Endowment Scholarships, 1275
Dr. John Maniotes Scholarship, 1506
Dr. R.T. White Scholarship, 1510
Duke University Adult Cardiothoracic Anesthesia and Critical Care Medicine Fellowships, 1520
Duke University Ambulatory and Regional Anesthesia Fellowships, 1521
Duke University Clinical Cardiac Electrophysiology Fellowships, 1522
Duke University Hyperbaric Center Fellowships, 1523
Duke Univ Interventional Cardio Fellowships, 1524
EBSCO / MLA Annual Meeting Grants, 1545
Elkhart County Foundation Lilly Endowment Community Scholarships, 1574
Fulton County Community Foundation 4Community Higher Education Scholarship, 1762
Hammond Common Council Scholarships, 1938
Harry A. & Margaret D. Towsley Fndn Grants, 1959
Hoglund Foundation Grants, 2040
Hospital Libraries Section / MLA Professional Development Grants, 2052
HRAMF Charles A. King Trust Postdoctoral Research Fellowships, 2056
IBEW Local Union #697 Memorial Scholarships, 2096
IIE AmCham Charitable Foundation U.S. Studies Scholarship, 2125
IIE Klein Family Scholarship, 2135
IIE Lingnan Foundation W.T. Chan Fellowship, 2137
IIE Mattel Global Scholarship, 2139
IIE New Leaders Group Award for Mutual Understanding, 2141
John V. and George Primich Family Scholarship, 2259
Joshua Benjamin Cohen Memorial Scholarship, 2276
Lake County Athletic Officials Association Scholarships, 2332
LEGENDS Scholarship, 2351
Majors MLA Chapter Project of the Year, 2418
Medical Informatics Section/MLA Career Development Grant, 2508
MLA Continuing Education Grants, 2559
MLA Cunningham Memorial International Fellowship, 2560
MLA David A. Kronick Traveling Fellowship, 2561
MLA Estelle Brodman Academic Medical Librarian of the Year Award, 2563
MLA Graduate Scholarship, 2564

748 / Health Science

MLA Grad Scholarship for Minority Students, 2565
MLA Ida and George Eliot Prize, 2566
MLA Janet Doe Lectureship Award, 2567
MLA Lois Ann Colaianni Award for Excellence and Achievement in Hospital Librarianship, 2568
MLA Louise Darling Medal for Distinguished Achievement in Collection Development in the Health Sciences, 2569
MLA Lucretia W. McClure Excellence in Education Award, 2570
MLA Murray Gottlieb Prize Essay Award, 2571
MLA Research, Development, and Demonstration Project Grant, 2572
MLA Rittenhouse Award, 2573
MLA Section Project of the Year Award, 2574
MLA T. Mark Hodges Int'l Service Award, 2575
MLA Thomas Reuters / Frank Bradway Rogers Information Advancement Award, 2576
Nancy J. Pinnick Memorial Scholarship, 2607
NHLBI Lymphatics in Health and Disease in the Digestive, Cardio & Pulmonary Systems, 2704
NHLBI Pathway to Independence Awards, 2718
NHLBI Research Supplements to Promote Diversity in Health-Related Research, 2727
NIA Promoting Careers in Aging and Health Disparities Research Grants, 2780
NIH Recovery Act Limited Competition: Building Sustainable Community-Linked Infrastructure to Enable Health Science Research, 2894
NIH Research Project Grants, 2899
NLM Grants for Scholarly Works in Biomedicine and Health, 2944
NLM Research Project Grants, 2951
Noble County Community Foundation - Kathleen June Earley Memorial Scholarship, 2973
North Carolina GlaxoSmithKline Fndn Grants, 2991
NWHF Mark O. Hatfield Research Fellowship, 3035
NYCT Blindness and Visual Disabilities Grants, 3050
Porter County Foundation Lilly Endowment Community Scholarships, 3245
Porter County Health Occupations Scholarship, 3247
Pulaski County Community Fndn Scholarships, 3281
Regenstrief General Internal Medicine - General Pediatrics Research Fellowship Program, 3351
Religion and Health: Effects, Mechanisms, and Interpretation RFP, 3358
San Diego Women's Foundation Grants, 3474
Savoy Foundation Post-Doctoral and Clinical Research Fellowships, 3488
Savoy Foundation Studentships, 3490
Schrage Family Foundation Scholarships, 3494
Scleroderma Foundation Established Investigator Grants, 3497
Sengupta Family Scholarship, 3512
Sigma Theta Tau International / Council for the Advancement of Nursing Science Grants, 3557
Siragusa Foundation Health Services & Medical Research Grants, 3571
SRC Medical Research Grants, 3635
STTI Emergency Nurses Association Foundation Grant, 3666
Texas Instruments Foundation Community Services Grants, 3710
Thomson Reuters / MLA Doctoral Fellowships, 3728
US CRDF Leishmaniasis: Collaborative Research Opportunities in N Africa & Middle East, 3828
Vanderburgh Community Foundation Grants, 3852
Virginia L. and William K. Beatty MLA Volunteer Service Award, 3877

Health Services Delivery
3M Fndn Health & Human Services Grants, 7
A.O. Smith Foundation Community Grants, 17
AAHPM Hospice & Palliative Med Fellowships, 152
Abbott Fund Access to Health Care Grants, 199
Abell-Hanger Foundation Grants, 207
Abell Fndn Health & Human Services Grants, 209
Abington Foundation Grants, 211
Able To Serve Grants, 212

AcademyHealth Alice S. Hersh New Investigator Awards, 220
AcademyHealth HSR Impact Awards, 223
Ackerman Foundation Grants, 236
ACSM Paffenbarger-Blair Fund for Epidemiological Research on Physical Activity Grants, 273
Administration on Aging Senior Medicare Patrol Project Grants, 309
Aetna Foundation Racial and Ethnic Health Care Equity Grants, 332
Agnes B. Hunt Trust Grants, 344
Albertson's Charitable Giving Grants, 410
Allan C. and Lelia J. Garden Foundation Grants, 438
Allen P. & Josephine B. Green Fndn Grants, 443
Alvin and Fanny Blaustein Thalheimer Foundation Baltimore Communal Grants, 459
AMA Foundation Fund for Better Health Grants, 484
American Psychiatric Foundation Awards for Advancing Minority Mental Health, 551
American Psychiatric Foundation Disaster Recovery Fund for Psychiatrists, 553
American Psychiatric Foundation Helping Hands Grants, 554
Amgen Foundation Grants, 572
Amica Insurance Company Community Grants, 575
Anthem Blue Cross and Blue Shield Grants, 607
APA Young Investigator Grant Program, 634
Appalachian Regional Commission Health Grants, 637
ArvinMeritor Grants, 675
Assurant Foundation Grants, 788
Auburn Foundation Grants, 798
AUPHA Corris Boyd Scholarships, 800
AUPHA Foster G. McGaw Scholarships, 801
Barth Syndrome Foundation Research Grants, 836
Baxter International Foundation William B. Graham Prize for Health Services Research, 844
BCBSM Claude Pepper Award, 850
BCBSM Foundation Investigator Initiated Research Grants, 854
BCBSM Fndn Proposal Development Awards, 857
BCBSM Foundation Student Award Program, 858
BCBSM Frank J. McDevitt, DO, Excellence in Research Awards, 859
BCBSNC Foundation Grants, 860
BJ's Charitable Foundation Grants, 904
Blackford County Community Foundation Grants, 906
Blue Shield of California Grants, 923
Burlington Industries Foundation Grants, 977
Business Bank of Nevada Community Grants, 981
California Endowment Innovative Ideas Challenge Grants, 1000
California Wellness Foundation Work and Health Program Grants, 1001
Cardinal Health Foundation Grants, 1017
Carl B. and Florence E. King Foundation Grants, 1022
CDC Public Health Associates Hosts, 1066
Champ-A Champion Fur Kids Grants, 1109
Charles H. Pearson Foundation Grants, 1121
Charlotte County (FL) Community Foundation Grants, 1126
Chase Paymentech Corporate Giving Grants, 1127
Christine & Katharina Pauly Trust Grants, 1150
CICF City of Noblesville Community Grant, 1155
CICF James Proctor Grant for Aged, 1157
Claude Worthington Benedum Fndn Grants, 1178
Clinton County Community Foundation Grants, 1184
CMS Hispanic Health Services Research Grants, 1186
CMS Historically Black Colleges and Universities (HBCU) Health Services Research Grants, 1187
Colorado Trust Grants, 1213
Columbus Foundation Paul G. Duke Grants, 1221
Commonwealth Fund Harkness Fellowships in Health Care Policy and Practice, 1229
Commonwealth Fund Quality of Care for Frail Elders Grants, 1233
Community Foundation for Greater Atlanta Common Good Funds Grants, 1241
Community Foundation of Abilene Celebration of Life Grants, 1261
Community Foundation of Boone County Grants, 1266

SUBJECT INDEX

Community Foundation of Grant County Grants, 1271
Community Foundation of Greenville Hollingsworth Funds Program/Project Grants, 1285
Community Foundation of Louisville Anna Marble Memorial Fund for Princeton Grants, 1290
Community Foundation of Riverside & San Bernardino County Impact Grants, 1303
Community Foundation of St. Joseph County Special Project Challenge Grants, 1310
Community Foundation Serving Riverside and San Bernardino Counties Impact Grants, 1319
Cone Health Foundation Grants, 1323
Connecticut Health Foundation Health Initiative Grants, 1325
Consumers Energy Foundation, 1343
Crown Point Community Foundation Grants, 1361
CVS Community Grants, 1379
Daniels Fund Grants-Aging, 1424
Daviess County Community Foundation Health Grants, 1434
Dearborn Community Foundation City of Lawrenceburg Community Grants, 1441
Decatur County Community Foundation Large Project Grants, 1444
DeKalb County Community Foundation Grants, 1447
Del E. Webb Foundation Grants, 1450
DeMatteis Family Foundation Grants, 1457
Dennis & Phyllis Washington Fndn Grants, 1458
DHHS Oral Health Promotion Research Across the Lifespan, 1474
DIFFA/Chicago Grants, 1477
Doris Duke Charitable Foundation Operations Research on AIDS Care and Treatment in Africa (ORACTA) Grants, 1497
Dorothy Rider Pool Health Care Grants, 1500
DTE Energy Foundation Health and Human Services Grants, 1516
Duke Endowment Health Care Grants, 1519
Eastman Chemical Company Foundation Grants, 1542
Edina Realty Foundation Grants, 1550
Emma B. Howe Memorial Foundation Grants, 1593
Entergy Corporation Open Grants for Healthy Families, 1597
Ewing Halsell Foundation Grants, 1618
Fairlawn Foundation Grants, 1626
Fallon OrNda Community Health Fund Grants, 1627
Finance Factors Foundation Grants, 1666
Foundation for a Healthy Kentucky Grants, 1691
Foundation for Health Enhancement Grants, 1696
Foundation for the Mid South Health and Wellness Grants, 1700
Foundations of East Chicago Health Grants, 1709
Frances and John L. Loeb Family Fund Grants, 1716
Fulton County Community Foundation Women's Giving Circle Grants, 1766
Genesis Foundation Grants, 1786
George E. Hatcher, Jr. and Ann Williams Hatcher Foundation Grants, 1792
Gil and Dody Weaver Foundation Grants, 1819
Giving Sum Annual Grant, 1825
GlaxoSmithKline Foundation IMPACT Awards, 1830
GNOF IMPACT Grants for Health and Human Services, 1839
Goodyear Tire Grants, 1856
Greater Cincinnati Foundation Priority and Small Projects/Capacity-Building Grants, 1880
Green Bay Packers Foundation Grants, 1890
Green Diamond Charitable Contributions, 1891
Grundy Foundation Grants, 1912
HAF Phyllis Nilsen Leal Memorial Fund Gifts, 1930
Hall-Perrine Foundation Grants, 1935
Hampton Roads Community Foundation Faith Community Nursing Grants, 1940
Hasbro Children's Fund Grants, 1973
Hawaii Community Foundation Health Education and Research Grants, 1976
Health Foundation of Greater Cincinnati Grants, 1983
Hearst Foundations Health Grants, 1987
Hedco Foundation Grants, 1989
Helen Pumphrey Denit Charitable Trust Grants, 1996

Helen Steiner Rice Foundation Grants, 1998
Hendricks County Community Fndn Grants, 2000
Highmark Physician eHealth Grants, 2026
Hill Crest Foundation Grants, 2030
Hoglund Foundation Grants, 2040
HomeBanc Foundation Grants, 2042
Honda of America Manufacturing Fndn Grants, 2045
Horace A. Kimball and S. Ella Kimball Foundation Grants, 2046
HRSA Ryan White HIV AIDS Drug Assistance Grants, 2070
IBCAT Screening Mammography Grants, 2095
Idaho Community Foundation Eastern Region Competitive Grants, 2114
Illinois Children's Healthcare Foundation Grants, 2146
James L. and Mary Jane Bowman Charitable Trust Grants, 2194
Jane Bradley Pettit Foundation Health Grants, 2208
Jay and Rose Phillips Family Foundation Grants, 2213
Jessie B. Cox Charitable Trust Grants, 2225
Jewish Fund Grants, 2227
Joe W. and Dorothy Dorsett Brown Fndn Grants, 2230
Johnson & Johnson Corporate Contributions, 2253
Johnson & Johnson/SAH Arts & Healing Grants, 2254
Johnson County Community Foundation Grants, 2256
Josephine S. Gumbiner Foundation Grants, 2273
K21 Health Foundation Grants, 2285
Kendrick Foundation Grants, 2299
Kroger Foundation Women's Health Grants, 2325
LaGrange County Community Fndn Grants, 2329
Liberty Bank Foundation Grants, 2362
Lynde & Harry Bradley Foundation Fellowships, 2396
Lynde and Harry Bradley Foundation Grants, 2397
Lynde and Harry Bradley Foundation Prizes: Bradley Prizes, 2398
M-A-C AIDS Fund Grants, 2400
M.D. Anderson Foundation Grants, 2403
M.J. Murdock Charitable Trust General Grants, 2405
Madison County Community Foundation - City of Anderson Quality of Life Grant, 2412
Marie C. and Joseph C. Wilson Foundation Rochester Small Grants, 2431
Marin Community Foundation Improving Community Health Grants, 2433
McCallum Family Foundation Grants, 2489
McCune Foundation Human Services Grants, 2493
McGraw-Hill Companies Community Grants, 2494
McKesson Foundation Grants, 2496
McLean Foundation Grants, 2498
Mead Johnson Nutritionals Charitable Giving, 2502
Medtronic Foundation CommunityLink Health Grants, 2510
Medtronic Foundation Patient Link Grants, 2514
Medtronic Foundation Strengthening Health Systems Grants, 2515
Memorial Foundation Grants, 2516
MetroWest Health Foundation Capital Grants for Health-Related Facilities, 2526
MetroWest Health Foundation Grants--Healthy Aging, 2527
Michael Reese Health Trust Core Grants, 2544
Michael Reese Health Trust Responsive Grants, 2545
Mid-Iowa Health Foundation Community Response Grants, 2549
Miles of Hope Breast Cancer Foundation Grants, 2553
Miller Foundation Grants, 2556
Minneapolis Foundation Community Grants, 2558
MLA Donald A.B. Lindberg Fellowship, 2562
MMS and Alliance Charitable Foundation Grants for Community Action and Care for the Medically Uninsured, 2587
MMS and Alliance Charitable Foundation International Health Studies Grants, 2588
Ms. Foundation for Women Health Grants, 2597
NAPNAP Fndn Elaine Gelman Scholarship, 2609
NAPNAP Foundation Innovative Health Care Small Grant, 2611
NAPNAP Foundation McNeil Grant-in-Aid, 2612
NAPNAP Foundation McNeil PNP Scholarships, 2613
NAPNAP Foundation McNeil Rural and Underserved Scholarships, 2614
NAPNAP Foundation Nursing Research Grants, 2615
NAPNAP Foundation Reckitt Benckiser Student Scholarship, 2616
NAPNAP Foundation Wyeth Pediatric Immunization Grant, 2618
NCI Exploratory Grants for Behavioral Research in Cancer Control, 2651
NIA Alzheimer's Disease Core Centers Grants, 2748
NMF General Electric Med Fellowships, 2956
North Central Health Services Grants, 2993
Northrop Grumman Corporation Grants, 2997
NYAM Mary and David Hoar Fellowship, 3045
NYCT Children/Youth with Disabilities Grants, 3052
NYCT Girls and Young Women Grants, 3053
OECD Sexual Health and Rights Project Grants, 3064
Office Depot Corporation Community Relations Grants, 3065
PACCAR Foundation Grants, 3101
Perry County Community Foundation Grants, 3142
Pfizer Medical Education Track One Grants, 3182
Pfizer Medical Education Track Two Grants, 3183
Pfizer Special Events Grants, 3186
Piedmont Natural Gas Foundation Health and Human Services Grants, 3221
Piper Trust Older Adults Grants, 3229
Pohlad Family Foundation, 3241
Porter County Health and Wellness Grant, 3246
Premera Blue Cross Grants, 3258
Priddy Foundation Program Grants, 3266
Prince Charitable Trusts DC Grants, 3271
Qualcomm Grants, 3287
Quality Health Foundation Grants, 3288
Quantum Corporation Snap Server Grants, 3289
Quantum Foundation Grants, 3290
RCF General Community Grants, 3312
Ringing Rocks Foundation Discretionary Grants, 3375
Robert R. McCormick Trib Community Grants, 3385
Robert R. Meyer Foundation Grants, 3387
Robert W. Woodruff Foundation Grants, 3389
Rochester Area Community Foundation Grants, 3391
Rockwell International Corporate Trust Grants Program, 3399
Rosalynn Carter Institute John and Betty Pope Fellowships, 3408
Ruth Eleanor Bamberger and John Ernest Bamberger Memorial Foundation Grants, 3427
RWJF Changes in Health Care Financing and Organization Grants, 3430
RWJF Disparities Research for Change Grants, 3433
RWJF Harold Amos Medical Faculty Development Scholars, 3435
RWJF New Jersey Health Initiatives Grants, 3440
Saigh Foundation Grants, 3451
Salem Foundation Charitable Trust Grants, 3456
Salem Foundation Grants, 3457
Samuel S. Johnson Foundation Grants, 3468
San Diego Foundation Health & Human Services Grants, 3471
San Diego Foundation Paradise Valley Hospital Community Fund Grants, 3472
San Diego Women's Foundation Grants, 3474
San Francisco Fndn Community Health Grants, 3478
Santa Fe Community Foundation Seasonal Grants-Fall Cycle, 3482
Sartain Lanier Family Foundation Grants, 3486
Schering-Plough Foundation Health Grants, 3492
Seattle Foundation Benjamin N. Phillips Memorial Fund Grants, 3504
Shell Oil Company Foundation Community Development Grants, 3544
Sierra Health Foundation Responsive Grants, 3555
Sir Dorabji Tata Trust Individual Medical Grants, 3573
Sisters of St. Joseph Healthcare Fndn Grants, 3576
SOCFOC Catholic Ministries Grants, 3596
Sonoco Foundation Grants, 3621
Sony Corporation of America Grants, 3623
Square D Foundation Grants, 3634
Starke County Community Foundation Grants, 3646
Stewart Huston Charitable Trust Grants, 3655
Straits Area Community Foundation Health Care Grants, 3657
Sunflower Foundation Bridge Grants, 3676
Sunflower Foundation Capacity Building Grants, 3677
Sunflower Foundation Walking Trails Grants, 3678
Sunoco Foundation Grants, 3679
TE Foundation Grants, 3705
Textron Corporate Contributions Grants, 3712
Thomas J. Long Foundation Community Grants, 3722
Tipton County Foundation Grants, 3734
U.S. Cellular Corporation Grants, 3766
UCLA / RAND Corporation Post-Doctoral Fellowships, 3776
Unihealth Foundation Grants for Non-Profit Oranizations, 3783
UniHealth Foundation Healthcare Systems Enhancements Grants, 3784
Union Bank, N.A. Corporate Sponsorships and Donations, 3787
Union Labor Health Fndn Community Grants, 3792
Union Pacific Foundation Health and Human Services Grants, 3794
United Methodist Committee on Relief Global AIDS Fund Grants, 3798
United Methodist Child Mental Health Grants, 3800
Unity Foundation Of LaPorte County Grants, 3804
USAID Devel Innovation Accelerator Grants, 3808
USAID Integration of Care and Support within the Health System to Support Better Patient Outcomes Grants, 3817
US Airways Community Contributions, 3825
Vancouver Foundation Grants and Community Initiatives Program, 3849
VHA Health Foundation Grants, 3867
Visiting Nurse Foundation Grants, 3879
W.K. Kellogg Foundation Healthy Kids Grants, 3886
Warrick County Community Foundation Grants, 3904
Washington County Community Fndn Grants, 3906
Wells County Foundation Grants, 3920
Western Union Foundation Grants, 3923
William B. Stokely Jr. Foundation Grants, 3947
Women of the ELCA Opportunity Scholarships for Lutheran Laywomen, 3970
ZYTL Foundation Grants, 3985

Health and Safety Education

AAHPERD Abernathy Presidential Scholarships, 151
ACE Charitable Foundation Grants, 229
Aetna Foundation Regional Health Grants, 333
Alexander and Margaret Stewart Trust Grants, 419
Alexis Gregory Foundation Grants, 426
American Psychiatric Foundation Disaster Recovery Fund for Psychiatrists, 553
American Trauma Society, Pennsylvania Division Mini-Grants, 564
Anthem Blue Cross and Blue Shield Grants, 607
Armstrong McDonald Foundation Health Grants, 663
Austin-Bailey Health and Wellness Grants, 803
Avon Foundation Breast Care Fund Grants, 813
Baxter International Foundation Grants, 843
BCBSNC Foundation Grants, 860
Becton Dickinson and Company Grants, 869
BJ's Charitable Foundation Grants, 904
Blowitz-Ridgeway Foundation Early Childhood Development Research Award, 912
Blue Cross Blue Shield of Minnesota Foundation - Health Equity: Building Health Equity Together Grants, 914
Blue Cross Blue Shield of Minnesota Foundation - Healthy Equity: Health Impact Assessment Demonstration Project Grants, 916
Breast Cancer Fund Grants, 947
CDC Public Health Conference Support Grant, 1067
Champ-A Champion Fur Kids Grants, 1109
Charles Lafitte Foundation Grants, 1123
Cigna Civic Affairs Sponsorships, 1159
CJ Foundation for SIDS Program Services Grants, 1163
CLIF Bar Family Foundation Grants, 1183
Community Fndn of Louisville Health Grants, 1295

750 / Health and Safety Education

Comprehensive Health Education Fndn Grants, 1320
Daviess County Community Foundation Health Grants, 1434
Dearborn Community Foundation City of Lawrenceburg Community Grants, 1441
DIFFA/Chicago Grants, 1477
Dorothy Rider Pool Health Care Grants, 1500
Dr. John T. Macdonald Foundation Grants, 1507
Edina Realty Foundation Grants, 1550
Everyone Breathe Asthma Education Grants, 1616
Express Scripts Foundation Grants, 1620
Fairlawn Foundation Grants, 1626
Foundation for a Healthy Kentucky Grants, 1691
Foundation for Seacoast Health Grants, 1698
Foundations of East Chicago Health Grants, 1709
Frances and John L. Loeb Family Fund Grants, 1716
Global Fund for Children Grants, 1832
GNOF IMPACT Kahn-Oppenheim Grants, 1842
Golden Heart Community Foundation Grants, 1850
Great-West Life Grants, 1879
Greater Sitka Legacy Fund Grants, 1885
Grifols Community Outreach Grants, 1904
Hannaford Charitable Foundation Grants, 1944
Healthcare Foundation of New Jersey Grants, 1982
Health Foundation of Greater Cincinnati Grants, 1983
Health Foundation of Southern Florida Responsive Grants, 1985
Henry and Ruth Blaustein Rosenberg Foundation Health Grants, 2004
Hogg Foundation for Mental Health Grants, 2039
HomeBanc Foundation Grants, 2042
Horizon Foundation for New Jersey Grants, 2048
Indiana AIDS Fund Grants, 2155
John M. Lloyd Foundation Grants, 2244
John P. McGovern Foundation Grants, 2247
John S. Dunn Research Foundation Grants, 2250
Johnson & Johnson Corporate Contributions, 2253
Johnson County Community Foundation Grants, 2256
Kansas Health Foundation Recognition Grants, 2289
Kenai Peninsula Foundation Grants, 2298
Ketchikan Community Foundation Grants, 2303
Knight Family Charitable and Educational Foundation Grants, 2312
Lloyd A. Fry Foundation Health Grants, 2373
M-A-C AIDS Fund Grants, 2400
Mary Black Foundation Active Living Grants, 2440
McCarthy Family Foundation Grants, 2490
MedImmune Charitable Grants, 2509
Medtronic Foundation HeartRescue Grants, 2513
Medtronic Foundation Patient Link Grants, 2514
MMS and Alliance Charitable Foundation Grants for Community Action and Care for the Medically Uninsured, 2587
NAPNAP Foundation Innovative Health Care Small Grant, 2611
NAPNAP Foundation McNeil Grant-in-Aid, 2612
NAPNAP Foundation Nursing Research Grants, 2615
NAPNAP Foundation Shourd Parks Immunization Project Small Grants, 2617
NAPNAP Foundation Wyeth Pediatric Immunization Grant, 2618
National Blood Foundation Research Grants, 2628
NLM Understanding and Promoting Health Literacy Research Grants, 2953
Northrop Grumman Corporation Grants, 2997
Oregon Community Fndn Community Grants, 3082
Pajaro Valley Community Health Trust Diabetes and Contributing Factors Grants, 3109
Petersburg Community Foundation Grants, 3156
Phoenix Suns Charities Grants, 3201
PhRMA Foundation Health Outcomes Sabbatical Fellowships, 3205
Piper Trust Older Adults Grants, 3229
PMI Foundation Grants, 3239
Posey Community Fndn Women's Fund Grants, 3251
PPG Industries Foundation Grants, 3257
Richmond Eye and Ear Fund Grants, 3372
Robert W. Woodruff Foundation Grants, 3389
Rosalynn Carter Institute John and Betty Pope Fellowships, 3408
Saint Luke's Health Initiatives Grants, 3455
Salt River Health & Human Services Grants, 3459
Shaw's Supermarkets Donations, 3542
Sisters of Charity Foundation of Cleveland Reducing Health and Educational Disparities in the Central Neighborhood Grants, 3574
Sisters of St. Joseph Healthcare Fndn Grants, 3576
Special Olympics Health Profess Student Grants, 3629
Susan G. Komen Breast Cancer Foundation Career Catalyst Research Grants, 3688
Tellabs Foundation Grants, 3706
Tommy Hilfiger Corporate Foundation Grants, 3737
Tylenol Future Care Scholarships, 3764
U.S. Department of Education 21st Century Community Learning Centers, 3767
United Healthcare Comm Grants in Michigan, 3796
USAID Devel Innovation Accelerator Grants, 3808
USDA Organic Agriculture Research Grants, 3830
Warrick County Community Foundation Grants, 3904
Welborn Baptist Fndn General Op Grants, 3914
Welborn Baptist Fndn School Health Grants, 3917
Women of the ELCA Opportunity Scholarships for Lutheran Laywomen, 3970

Hearing
AHRF Eugene L. Derlacki Research Grants, 361
AHRF Georgia Birtman Grant, 362
AHRF Wiley H. Harrison Research Award, 364
Marjorie C. Adams Charitable Trust Grants, 2437
Oticon Focus on People Awards, 3098
Reinberger Foundation Grants, 3357

Hearing Impairments
AHRF Eugene L. Derlacki Research Grants, 361
AHRF Georgia Birtman Grant, 362
AHRF Wiley H. Harrison Research Award, 364
ASHFoundation Graduate Student International Scholarship, 702
ASHFoundation Graduate Student Scholarships, 703
ASHFoundation Graduate Student Scholarships for Minority Students, 704
ASHFoundation Graduate Student with a Disability Scholarship, 705
Batchelor Foundation Grants, 837
Carls Foundation Grants, 1030
Carl W. and Carrie Mae Joslyn Trust Grants, 1031
Charles Delmar Foundation Grants, 1115
Columbus Foundation Competitive Grants, 1217
Community Foundation of Jackson County Seymour Noon Lions Club Grant, 1288
Cralle Foundation Grants, 1356
Danellie Foundation Grants, 1421
Herbert A. & Adrian W. Woods Fndn Grants, 2007
Mabel H. Flory Charitable Trust Grants, 2408
Marjorie C. Adams Charitable Trust Grants, 2437
Oticon Focus on People Awards, 3098
The Ray Charles Foundation Grants, 3716
U.S. Department of Education Rehabilitation Training - Rehabilitation Continuing Education - Institute on Rehabilitation Issues, 3771
Union County Community Foundation Grants, 3790

Hematology
Bayer Clinical Scholarship Award, 845
Bayer Hemophilia Caregivers Education Award, 847
Biogen Foundation Healthcare Professional Education Grants, 900
Biogen Foundation Scientific Research Grants, 902
Chest Fnd Grant in Venous Thromboembolism, 1134
CSL Behring Local Empowerment for Advocacy Development (LEAD) Grants, 1365
Doris Duke Charitable Foundation Clinical Scientist Development Award, 1494
Fairlawn Foundation Grants, 1626
Grifols Community Outreach Grants, 1904
MMAAP Foundation Fellowship Award in Hematology, 2579
MMAAP Fndn Hematology Project Awards, 2583
National Blood Foundation Research Grants, 2628
NHLBI Ancillary Studies in Clinical Trials, 2688

NHLBI Characterizing the Blood Stem Cell Niche Grants, 2695
NHLBI Circadian-Coupled Cellular Function in Heart, Lung, and Blood Tissue Grants, 2696
NHLBI Immunomodulatory, Inflammatory, & Vasoregulatory Properties of Transfused Red Blood Cell Units as a Function of Preparation & Storage Grants, 2700
NHLBI Mentored Career Award for Faculty at Institutions that Promote Diversity (K01), 2706
NHLBI Mentored Career Award For Faculty At Minority Institutions, 2707
NHLBI Mentored Career Dev Award to Promote Faculty Diversity in Biomed Research, 2708
NHLBI Ruth L. Kirschstein National Research Service Awards for Individual Postdoctoral Fellows, 2729
NHLBI Short-Term Research Education Program to Increase Diversity in Health-Related Research, 2733
NHLBI Training Program Grants for Institutions that Promote Diversity, 2738
NIDDK Devel of Disease Biomarkers Grants, 2809
NIDDK Education Program Grants, 2813
NIDDK Grants for Basic Research in Glomerular Diseases, 2819
NIDDK Non-Invasive Methods for Diagnosis and Progression of Diabetes, Kidney, Urological, Hematological and Digestive Diseases, 2832
NIDDK Proteomics: Diabetes, Obesity, And Endocrine, Digestive, Kidney, Urologic, And Hematologic Diseases, 2838
NIDDK Small Grants for Underrepresented Minority Scientists in Diabetes and Digestive and Kidney Diseases, 2847
NIH Bone and the Hematopoietic and Immune Systems Research Grants, 2868
NYCT Biomedical Research Grants, 3049
NYCT Blood Disease Research Grants, 3051
Topeka Community Foundation Kansas Blood Services Scholarships, 3738
Victor E. Speas Foundation Grants, 3868

Hemophilia
Baxter International Corporate Giving Grants, 841
Bayer Clinical Scholarship Award, 845
Bayer Hemophilia Caregivers Education Award, 847
CSL Behring Local Empowerment for Advocacy Development (LEAD) Grants, 1365
Pfizer ASPIRE North America Hemophilia Research Awards, 3170
Pfizer Healthcare Charitable Contributions, 3180
Pfizer Medical Education Track Two Grants, 3183

Hepatitis
ASM/CDC Fellowships in Infectious Disease and Public Health Microbiology, 716
Bristol-Myers Squibb Foundation Fellowships, 956
Bristol-Myers Squibb Foundation Health Disparities Grants, 958
CDC Epidemic Intell Service Training Grants, 1055
CDC Epidemiology Elective Rotation, 1056
NIDDK Health Disparities in NIDDK Diseases, 2820
NIDDK Research Grants for Studies of Hepatitis C in the Setting of Renal Disease, 2839

Hepatology
FDHN Non-Career Research Grants, 1651
FDHN Research Scholar Awards, 1652
FDHN Translational Research Awards, 1656

Hereditary Diseases and Disorders
AIHS Alberta/Pfizer Translat Research Grants, 374
Barth Syndrome Foundation Research Grants, 836

Heroism
OceanFirst Foundation Major Grants, 3062

Herpesviruses
Sexually Transmitted Diseases Postdoctoral Fellowships, 3516

SUBJECT INDEX

Higher Education

3M Company Fndn Community Giving Grants, 6
A. Gary Anderson Family Foundation Grants, 16
A/H Foundation Grants, 18
AAHPERD Abernathy Presidential Scholarships, 151
Aaron Foundation Grants, 193
Abbott Fund Global AIDS Care Grants, 202
Abel Foundation Grants, 206
Abell-Hanger Foundation Grants, 207
Abney Foundation Grants, 214
Abramson Family Foundation Grants, 217
Achelis Foundation Grants, 234
ACL Partnerships in Employment Systems Change Grants, 246
Adam Richter Charitable Trust Grants, 298
Adams Foundation Grants, 301
Adaptec Foundation Grants, 304
Adolph Coors Foundation Grants, 311
Advanced Micro Devices Comm Affairs Grants, 313
AEC Trust Grants, 315
AGHE Graduate Scholarships and Fellowships, 343
Agnes Gund Foundation Grants, 345
Agnes M. Lindsay Trust Grants, 346
Ahmanson Foundation Grants, 355
Air Products and Chemicals Grants, 399
Alabama Power Foundation Grants, 400
Alberto Culver Corporate Contributions Grants, 408
Albertson's Charitable Giving Grants, 410
Alexander Eastman Foundation Grants, 420
Alfred and Tillie Shemanski Testamentary Trust Grants, 428
Alfred Bersted Foundation Grants, 429
Allyn Foundation Grants, 448
Alpha Natural Resources Corporate Giving, 450
ALSAM Foundation Grants, 454
AMD Corporate Contributions Grants, 500
American Electric Power Foundation Grants, 530
American Foodservice Charitable Trust Grants, 532
American Schlafhorst Foundation Grants, 559
Amerigroup Foundation Grants, 565
Amica Insurance Company Community Grants, 575
Andrew Family Foundation Grants, 588
Annenberg Foundation Grants, 596
Ann Peppers Foundation Grants, 599
Antone & Edene Vidinha Charitable Trust Grants, 609
AptarGroup Foundation Grants, 647
Archer Daniels Midland Foundation Grants, 651
ARCO Foundation Education Grants, 652
Argyros Foundation Grants Program, 653
Arkell Hall Foundation Grants, 661
ARS Foundation Grants, 667
Arthur and Rochelle Belfer Foundation Grants, 670
ASHA Students Preparing for Academic & Research Careers (SPARC) Award, 700
Ashland Corporate Contributions Grants, 713
ASU Grad College Reach for the Stars Fellowship, 791
Athwin Foundation Grants, 795
Atlanta Foundation Grants, 796
Atran Foundation Grants, 797
Auburn Foundation Grants, 798
Autzen Foundation Grants, 810
Babcock Charitable Trust Grants, 816
Bailey Foundation Grants, 818
Balfe Family Foundation Grants, 819
Ball Brothers Foundation General Grants, 820
BancorpSouth Foundation Grants, 823
Banfi Vintners Foundation Grants, 824
Barnes Group Foundation Grants, 832
Barr Fund Grants, 835
Batchelor Foundation Grants, 837
Batts Foundation Grants, 840
Beazley Foundation Grants, 865
Becton Dickinson and Company Grants, 869
Belk Foundation Grants, 872
Bender Foundation Grants, 876
Bertha Russ Lytel Foundation Grants, 885
Bert W. Martin Foundation Grants, 886
Bill Hannon Foundation Grants, 895
Biogen Foundation Healthcare Professional Education Grants, 900
Blue Mountain Community Foundation Grants, 921
Blum-Kovler Foundation Grants, 924
Blumenthal Foundation Grants, 925
BMW of North America Charitable Contributions, 926
Bodman Foundation Grants, 929
Boeing Company Contributions Grants, 930
Boettcher Foundation Grants, 931
Booth-Bricker Fund Grants, 933
Bosque Foundation Grants, 935
Boston Globe Foundation Grants, 937
BP Foundation Grants, 943
Bradley-Turner Foundation Grants, 944
Brian G. Dyson Foundation Grants, 948
Bright Family Foundation Grants, 951
Brown Advisory Charitable Foundation Grants, 969
Bruce and Adele Greenfield Foundation Grants, 972
Burlington Industries Foundation Grants, 977
Burlington Northern Santa Fe Foundation Grants, 978
BWF Postdoctoral Enrichment Grants, 989
Byron W. & Alice L. Lockwood Fnd Grants, 991
Caesars Foundation Grants, 995
Callaway Foundation Grants, 1002
Callaway Golf Company Foundation Grants, 1003
Camp-Younts Foundation Grants, 1005
Campbell Soup Foundation Grants, 1007
Cargill Citizenship Fund Corporate Giving, 1018
Caring Foundation Grants, 1019
Carl B. and Florence E. King Foundation Grants, 1022
Carnahan-Jackson Foundation Grants, 1032
Carnegie Corporation of New York Grants, 1033
Carylon Foundation Grants, 1040
Cemala Foundation Grants, 1077
Cessna Foundation Grants Program, 1081
CFFVR Clintonville Area Foundation Grants, 1097
CFFVR Shawano Area Community Foundation Grants, 1104
Champlin Foundations Grants, 1110
Chapman Charitable Foundation Grants, 1112
Charles Delmar Foundation Grants, 1115
Charles H. Revson Foundation Grants, 1122
Chazen Foundation Grants, 1129
Chiles Foundation Grants, 1147
Christensen Fund Regional Grants, 1149
CIGNA Foundation Grants, 1160
Clark-Winchcole Foundation Grants, 1173
Clark and Ruby Baker Foundation Grants, 1174
Claude Bennett Family Foundation Grants, 1176
Claude Worthington Benedum Fndn Grants, 1178
Clayton Fund Grants, 1180
Cleveland-Cliffs Foundation Grants, 1181
Coca-Cola Foundation Grants, 1203
Cockrell Foundation Grants, 1204
Coeta and Donald Barker Foundation Grants, 1205
Collins C. Diboll Private Foundation Grants, 1209
Collins Foundation Grants, 1210
Colonel Stanley R. McNeil Foundation Grants, 1211
Community Foundation of Greater Birmingham Grants, 1272
Community Foundation of Greater New Britain Grants, 1281
Community Foundation of St. Joseph County Scholarships, 1309
Community Foundation of the Verdugos Educational Endowment Fund Grants, 1314
Community Fndn of Wabash County Grants, 1316
Connelly Foundation Grants, 1326
ConocoPhillips Foundation Grants, 1327
Conquer Cancer Foundation of ASCO Drug Development Research Professorship, 1330
Constantin Foundation Grants, 1341
Constellation Energy Corporate Grants, 1342
Consumers Energy Foundation, 1343
Cooper Industries Foundation Grants, 1345
Coors Brewing Corporate Contributions Grants, 1346
Cowles Charitable Trust Grants, 1354
Cralle Foundation Grants, 1356
Crane Foundation Grants, 1357
CSRA Community Foundation Grants, 1366
Cullen Foundation Grants, 1372
Cystic Fibrosis Canada Fellowships, 1384
DaimlerChrysler Corporation Fund Grants, 1412
Daphne Seybolt Culpeper Fndn Grants, 1425
David M. and Marjorie D. Rosenberg Foundation Grants, 1432
Davis Family Foundation Grants, 1435
Daywood Foundation Grants, 1438
Decatur County Community Foundation Large Project Grants, 1444
Dennis & Phyllis Washington Fndn Grants, 1458
Dept of Ed College Assistance Migrant Grants, 1462
DeRoy Testamentary Foundation Grants, 1469
Dickson Foundation Grants, 1476
Dolan Children's Foundation Grants, 1482
Dora Roberts Foundation Grants, 1489
Doree Taylor Charitable Foundation, 1490
Duke Univ Interventional Cardio Fellowships, 1524
Dunspaugh-Dalton Foundation Grants, 1530
DuPont Pioneer Community Giving Grants, 1532
Dyson Foundation Mid-Hudson Valley General Operating Support Grants, 1533
E.J. Grassmann Trust Grants, 1537
E.L. Wiegand Foundation Grants, 1538
E. Rhodes & Leona B. Carpenter Grants, 1539
Earl and Maxine Claussen Trust Grants, 1541
Eberly Foundation Grants, 1544
Eckerd Corporation Foundation Grants, 1547
Eden Hall Foundation Grants, 1549
Edna G. Kynett Memorial Foundation Grants, 1551
EDS Foundation Grants, 1552
Edward and Helen Bartlett Foundation Grants, 1554
Edwin W. and Catherine M. Davis Fndn Grants, 1561
El Pomar Foundation Grants, 1582
Elsie H. Wilcox Foundation Grants, 1583
Elsie Lee Garthwaite Memorial Fndn Grants, 1584
Emerson Charitable Trust Grants, 1589
Ethel S. Abbott Charitable Foundation Grants, 1607
Eugene B. Casey Foundation Grants, 1609
Eugene M. Lang Foundation Grants, 1611
Eva L. & Joseph M. Bruening Fndn Grants, 1613
Evjue Foundation Grants, 1617
Ferree Foundation Grants, 1658
FirstEnergy Foundation Community Grants, 1668
Fluor Foundation Grants, 1682
FMC Foundation Grants, 1683
Foellinger Foundation Grants, 1684
Fondren Foundation Grants, 1685
Foster Foundation Grants, 1689
Four County Community Fndn General Grants, 1710
Four County Community Foundation Healthy Senior/Healthy Youth Fund Grants, 1711
Four J Foundation Grants, 1713
Frances L. and Edwin L. Cummings Memorial Fund Grants, 1717
Francis L. Abreu Charitable Trust Grants, 1719
Frank G. and Freida K. Brotz Family Foundation Grants, 1723
Fraser-Parker Foundation Grants, 1732
Fred C. & Katherine B. Andersen Grants, 1737
Frederick Gardner Cottrell Foundation Grants, 1739
Fred L. Emerson Foundation Grants, 1743
Fritz B. Burns Foundation Grants, 1747
Fuller E. Callaway Foundation Grants, 1761
Gardner Foundation Grants, 1770
Gardner W. & Joan G. Heidrick, Jr. Fndn Grants, 1773
Gebbie Foundation Grants, 1777
GenCorp Foundation Grants, 1779
George and Ruth Bradford Foundation Grants, 1789
George F. Baker Trust Grants, 1793
George Family Foundation Grants, 1794
George Kress Foundation Grants, 1799
George P. Davenport Trust Fund Grants, 1800
George S. & Dolores Dore Eccles Fndn Grants, 1801
George W. Brackenridge Foundation Grants, 1802
George W. Codrington Charitable Fndn Grants, 1803
Georgia Power Foundation Grants, 1806
Gertrude and William C. Wardlaw Fund Grants, 1808
Gertrude E. Skelly Charitable Foundation Grants, 1810
Gheens Foundation Grants, 1814
GNOF Exxon-Mobil Grants, 1836
GNOF New Orleans Works Grants, 1844

752 / Higher Education

GNOF Norco Community Grants, 1845
Goodrich Corporation Foundation Grants, 1854
Good Samaritan Inc Grants, 1855
Goodyear Tire Grants, 1856
Grace and Franklin Bernsen Foundation Grants, 1858
Grand Rapids Community Foundation Ionia County Youth Fund Grants, 1871
Grand Rapids Community Foundation Sparta Youth Fund Grants, 1876
Greater Cincinnati Foundation Priority and Small Projects/Capacity-Building Grants, 1880
Greater Milwaukee Foundation Grants, 1883
Green River Area Community Fndn Grants, 1895
Greenspun Family Foundation Grants, 1896
Griffin Foundation Grants, 1903
Grifols Community Outreach Grants, 1904
Grover Hermann Foundation Grants, 1906
Guy I. Bromley Trust Grants, 1919
H & R Foundation Grants, 1920
H. Leslie Hoffman and Elaine S. Hoffman Foundation Grants, 1924
H. Schaffer Foundation Grants, 1925
Hall-Perrine Foundation Grants, 1935
Harold Alfond Foundation Grants, 1949
Harold R. Bechtel Charitable Remainder Uni-Trust Grants, 1953
Harold R. Bechtel Testamentary Charitable Trust Grants, 1954
Harold Simmons Foundation Grants, 1955
Harry B. and Jane H. Brock Foundation Grants, 1960
Harry Kramer Memorial Fund Grants, 1965
Harry Sudakoff Foundation Grants, 1967
Harvey Randall Wickes Foundation Grants, 1972
Hedco Foundation Grants, 1989
Helen Irwin Littauer Educational Trust Grants, 1994
Helen Pumphrey Denit Charitable Trust Grants, 1996
Helen S. Boylan Foundation Grants, 1997
HHMI Grants and Fellowships Programs, 2020
High Meadow Foundation Grants, 2027
Hillcrest Foundation Grants, 2031
Hill Crest Foundation Grants, 2030
Homer Foundation Grants, 2044
Hormel Foundation Grants, 2051
Houston Endowment Grants, 2053
Howe Foundation of North Carolina Grants, 2055
Huffy Foundation Grants, 2073
Huie-Dellmon Trust Grants, 2075
Humana Foundation Grants, 2077
Hutton Foundation Grants, 2087
I.A. O'Shaughnessy Foundation Grants, 2088
Idaho Power Company Corporate Contributions, 2115
Ike and Roz Friedman Foundation Grants, 2145
Illinois Tool Works Foundation Grants, 2147
Intergrys Corporation Grants, 2160
J.B. Reynolds Foundation Grants, 2168
J.C. Penney Company Grants, 2169
J.M. Long Foundation Grants, 2173
J.W. Kieckhefer Foundation Grants, 2176
J. Walton Bissell Foundation Grants, 2177
James Ford Bell Foundation Grants, 2187
James G.K. McClure Educational and Development Fund Grants, 2188
James Hervey Johnson Charitable Educational Trust Grants, 2191
James J. & Angelia M. Harris Fndn Grants, 2192
James J. & Joan A. Gardner Family Fndn Grants, 2193
James M. Collins Foundation Grants, 2195
James R. Dougherty Jr. Foundation Grants, 2198
Janirve Foundation Grants, 2209
JELD-WEN Foundation Grants, 2218
Jennings County Community Foundation Grants, 2220
Joe W. and Dorothy Dorsett Brown Fndn Grants, 2230
John Ben Snow Memorial Trust Grants, 2231
John Deere Foundation Grants, 2234
John H. and Wilhelmina D. Harland Charitable Foundation Children and Youth Grants, 2238
John I. Smith Charities Grants, 2240
John J. Leidy Foundation Grants, 2241
John Lord Knight Foundation Grants, 2243
John P. Murphy Foundation Grants, 2248

John R. Oishei Foundation Grants, 2249
Johnson & Johnson Corporate Contributions, 2253
John Stauffer Charitable Trust Grants, 2258
John W. Anderson Foundation Grants, 2262
John W. Speas and Effie E. Speas Memorial Trust Grants, 2264
Joseph Alexander Foundation Grants, 2265
Joseph Henry Edmondson Foundation Grants, 2270
Josiah W. and Bessie H. Kline Foundation Grants, 2278
Julius N. Frankel Foundation Grants, 2282
Katherine Baxter Memorial Foundation Grants, 2292
Katherine John Murphy Foundation Grants, 2293
Kathryne Beynon Foundation Grants, 2294
Kavli Foundation Research Grants, 2296
Kovler Family Foundation Grants, 2323
L. W. Pierce Family Foundation Grants, 2328
LaGrange County Community Foundation Scholarships, 2330
Leonard and Helen R. Stulman Charitable Foundation Grants, 2356
Leo Niessen Jr., Charitable Trust Grants, 2358
Lester E. Yeager Charitable Trust B Grants, 2359
Lewis H. Humphreys Charitable Trust Grants, 2361
Libra Foundation Grants, 2363
Lillian S. Wells Foundation Grants, 2365
Lockheed Martin Corp Foundation Grants, 2374
Long Island Community Foundation Grants, 2375
Louetta M. Cowden Foundation Grants, 2377
Louis and Elizabeth Nave Flarsheim Charitable Foundation Grants, 2379
Lowe Foundation Grants, 2381
Lubbock Area Foundation Grants, 2383
Lubrizol Foundation Grants, 2384
Lumpkin Family Fnd Healthy People Grants, 2391
Lynde & Harry Bradley Foundation Fellowships, 2396
Lynde and Harry Bradley Foundation Grants, 2397
Lynde and Harry Bradley Foundation Prizes: Bradley Prizes, 2398
Lynn and Rovena Alexander Family Foundation Grants, 2399
M. Bastian Family Foundation Grants, 2402
M.E. Raker Foundation Grants, 2404
M.J. Murdock Charitable Trust General Grants, 2405
Madison County Community Foundation General Grants, 2413
Manuel D. & Rhoda Mayerson Fndn Grants, 2420
Margaret T. Morris Foundation Grants, 2429
Marie H. Bechtel Charitable Remainder Uni-Trust Grants, 2432
Marion Gardner Jackson Charitable Trust Grants, 2434
Mary S. and David C. Corbin Foundation Grants, 2447
Matilda R. Wilson Fund Grants, 2453
Maurice J. Masserini Charitable Trust Grants, 2456
Maxon Charitable Foundation Grants, 2458
McCallum Family Foundation Grants, 2489
McCombs Foundation Grants, 2491
McGraw-Hill Companies Community Grants, 2494
Mead Witter Foundation Grants, 2507
Mericos Foundation Grants, 2518
Meriden Foundation Grants, 2519
Merkel Foundation Grants, 2520
Merrick Foundation Grants, 2521
Meyer and Stephanie Eglin Foundation Grants, 2531
Meyer Memorial Trust Grassroots Grants, 2536
Meyer Memorial Trust Responsive Grants, 2537
Meyer Memorial Trust Special Grants, 2538
MGM Resorts Foundation Community Grants, 2541
MGN Family Foundation Grants, 2542
Military Ex-Prisoners of War Foundation Grants, 2554
Nell J. Redfield Foundation Grants, 2673
NIA Mentored Clinical Scientist Research Career Development Awards, 2771
Nicholas H. Noyes Jr. Memorial Fndn Grants, 2791
Nina Mason Pulliam Charitable Trust Grants, 2921
Noble County Community Fndn Scholarships, 2977
Norcliffe Foundation Grants, 2979
Nordson Corporation Foundation Grants, 2981
Northwestern Mutual Foundation Grants, 2999
Notsew Orm Sands Foundation Grants, 3002

NSF Grant Opportunities for Academic Liaison with Industry (GOALI), 3016
Ohio Learning Network Grants, 3070
Oleonda Jameson Trust Grants, 3071
Olive Higgins Prouty Foundation Grants, 3072
Oppenstein Brothers Foundation Grants, 3078
Oscar Rennebohm Foundation Grants, 3085
Otto Bremer Foundation Grants, 3099
PACCAR Foundation Grants, 3101
Parkersburg Area Community Foundation Action Grants, 3116
Patrick and Anna M. Cudahy Fund Grants, 3120
Paul Ogle Foundation Grants, 3124
PepsiCo Foundation Grants, 3137
Percy B. Ferebee Endowment Grants, 3138
Perkins Charitable Foundation Grants, 3140
Perpetual Trust for Charitable Giving Grants, 3141
Perry County Community Foundation Grants, 3142
Peter F. McManus Charitable Trust Grants, 3153
Peter Kiewit Foundation General Grants, 3154
Peter Kiewit Foundation Small Grants, 3155
Peyton Anderson Foundation Grants, 3160
Phi Upsilon Lillian P. Schoephoerster Award, 3196
Pike County Community Foundation Grants, 3222
Piper Jaffray Fndn Communities Giving Grants, 3227
PMI Foundation Grants, 3239
Posey County Community Foundation Grants, 3252
Pott Foundation Grants, 3253
Powell Foundation Grants, 3255
Price Chopper's Golub Foundation Grants, 3261
Price Family Charitable Fund Grants, 3262
Princeton Area Community Foundation Greater Mercer Grants, 3272
Pulaski County Community Foundation Grants, 3279
Pulido Walker Foundation, 3282
R.C. Baker Foundation Grants, 3292
Ralph M. Parsons Foundation Grants, 3304
Rasmuson Foundation Tier Two Grants, 3308
RCF General Community Grants, 3312
Reinberger Foundation Grants, 3357
RGk Foundation Grants, 3361
Rice Foundation Grants, 3363
Richard & Susan Smith Family Fndn Grants, 3365
Richard D. Bass Foundation Grants, 3366
Richard E. Griffin Family Foundation Grants, 3368
Richard King Mellon Foundation Grants, 3369
Richland County Bank Grants, 3371
Ripley County Community Foundation Grants, 3376
Ripley County Community Foundation Small Project Grants, 3377
Robert B. Adams Foundation Scholarships, 3380
Robert W. Woodruff Foundation Grants, 3389
Robins Foundation Grants, 3390
Rockwell International Corporate Trust Grants Program, 3399
Rollins-Luetkemeyer Foundation Grants, 3404
Roy & Christine Sturgis Charitable Grants, 3415
Rush County Community Foundation Grants, 3424
Ruth H. & Warren A. Ellsworth Fndn Grants, 3429
Sain-Orr and Royak-DeForest Steadman Foundation Grants, 3452
Salem Foundation Grants, 3457
SAMHSA Campus Suicide Prevention Grants, 3460
San Antonio Area Foundation Grants, 3469
Sandy Hill Foundation Grants, 3477
Sarkeys Foundation Grants, 3485
Sartain Lanier Family Foundation Grants, 3486
Schering-Plough Foundation Health Grants, 3492
Schramm Foundation Grants, 3495
Schurz Communications Foundation Grants, 3496
Scott B. & Annie P. Appleby Charitable Grants, 3499
Scott County Community Foundation Grants, 3500
Sensient Technologies Foundation Grants, 3513
Shell Deer Park Grants, 3543
Shield-Ayres Foundation Grants, 3546
Sid W. Richardson Foundation Grants, 3553
Simmons Foundation Grants, 3563
Sioux Falls Area Community Foundation Community Fund Grants (Unrestricted), 3568

SUBJECT INDEX

Sioux Falls Area Community Foundation Spot Grants (Unrestricted), 3570
Sisters of Charity Foundation of Cleveland Reducing Health and Educational Disparities in the Central Neighborhood Grants, 3574
Solo Cup Foundation Grants, 3620
Sonoco Foundation Grants, 3621
Sonora Area Foundation Competitive Grants, 3622
Sony Corporation of America Grants, 3623
Southwest Gas Corporation Foundation Grants, 3628
Special People in Need Scholarships, 3631
Square D Foundation Grants, 3634
St. Joseph Community Health Foundation Schneider Fellowships, 3642
Stackpole-Hall Foundation Grants, 3643
Steele-Reese Foundation Grants, 3650
Stranahan Foundation Grants, 3659
Strowd Roses Grants, 3664
Sunderland Foundation Grants, 3675
Sunoco Foundation Grants, 3679
SunTrust Bank Trusteed Foundations Greene-Sawtell Grants, 3681
SunTrust Bank Trusteed Foundations Harriet McDaniel Marshall Tust Grants, 3682
SunTrust Bank Trusteed Foundations Nell Warren Elkin and William Simpson Elkin Grants, 3683
SunTrust Bank Trusteed Foundations Thomas Guy Woolford Charitable Trust Grants, 3684
SunTrust Bank Trusteed Foundations Walter H. and Marjory M. Rich Memorial Fund Grants, 3685
Susan Mott Webb Charitable Trust Grants, 3694
Susan Vaughan Foundation Grants, 3695
T. Spencer Shore Foundation Grants, 3701
Taubman Endowment for the Arts Fndn Grants, 3702
TE Foundation Grants, 3705
Tension Envelope Foundation Grants, 3707
Textron Corporate Contributions Grants, 3712
Thelma Doelger Charitable Trust Grants, 3714
The Ray Charles Foundation Grants, 3716
Thomas and Dorothy Leavey Foundation Grants, 3717
Thomas Austin Finch, Sr. Foundation Grants, 3718
Thomas J. Long Foundation Community Grants, 3722
Thomas Sill Foundation Grants, 3724
Thompson Charitable Foundation Grants, 3727
U.S. Department of Education United States-Russia Program: Improving Research and Educational Activities in Higher Education, 3773
Union Bank, N.A. Corporate Sponsorships and Donations, 3787
Union Bank, N.A. Foundation Grants, 3788
United Technologies Corporation Grants, 3803
Vancouver Fndn Public Health Bursary Package, 3850
Victor E. Speas Foundation Grants, 3868
W. C. Griffith Foundation Grants, 3882
W.C. Griffith Foundation Grants, 3881
W.K. Kellogg Foundation Healthy Kids Grants, 3886
W.P. and Bulah Luse Foundation Grants, 3892
Walker Area Community Foundation Grants, 3895
Walter L. Gross III Family Foundation Grants, 3901
Weaver Popcorn Foundation Grants, 3910
Wege Foundation Grants, 3912
Weingart Foundation Grants, 3913
Whitney Foundation Grants, 3938
Willard & Pat Walker Charitable Fndn Grants, 3943
William B. Stokely Jr. Foundation Grants, 3947
William Blair and Company Foundation Grants, 3948
William D. Laurie, Jr. Charitable Fndn Grants, 3949
William G. & Helen C. Hoffman Fndn Grants, 3950
William G. Gilmore Foundation Grants, 3951
Windham Foundation Grants, 3961
Yawkey Foundation Grants, 3980

Higher Education Administration
John W. Speas and Effie E. Speas Memorial Trust Grants, 2264
Lewis H. Humphreys Charitable Trust Grants, 2361
Louis and Elizabeth Nave Flarsheim Charitable Foundation Grants, 2379
SfN Louise Hanson Marshall Specific Recognition Award, 3527

Higher Education Studies
AAUW American Postdoctoral Research Leave Fellowships, 196
ADA Foundation Thomas J. Zwemer Award, 297
DAR Irene and Daisy MacGregor Memorial Scholarship, 1428
HAF Riley Frazel Memorial Scholarship, 1931
IBRO Regional Grants for Int'l Fellowships to U.S. Laboratory Summer Neuroscience Courses, 2110
March of Dimes Agnes Higgins Award, 2423
Mayo Clinic Business Consulting Fellowship, 2462
MBL William Townsend Porter Summer Fellowships for Minority Investigators, 2488
Medtronic Foundation Fellowships, 2512
Pride Foundation Fellowships, 3267
Pride Foundation Scholarships, 3269
Scleroderma Foundation Established Investigator Grants, 3497
SfN Neuroscience Fellowships, 3530
Vancouver Fndn Public Health Bursary Package, 3850
White County Community Foundation - Landis Memorial Scholarship, 3930
White County Community Foundation - Tri-County Educational Scholarships, 3932
Whitley County Community Foundation - Lilly Endowment Scholarship, 3935
Whitley County Community Fndn Scholarships, 3937

Higher Education, Private
A/H Foundation Grants, 18
Abell-Hanger Foundation Grants, 207
Burlington Northern Santa Fe Foundation Grants, 978
Charles A. Frueauff Foundation Grants, 1114
Coca-Cola Foundation Grants, 1203
Dr. Scholl Foundation Grants Program, 1511
Dyson Foundation Mid-Hudson Valley General Operating Support Grants, 1533
Goodyear Tire Grants, 1856
Hall-Perrine Foundation Grants, 1935
Huisking Foundation Grants, 2076
Katherine Baxter Memorial Foundation Grants, 2292
Kenneth T. & Eileen L. Norris Fndn Grants, 2300
Marion I. and Henry J. Knott Foundation Discretionary Grants, 2435
Marion I. and Henry J. Knott Foundation Standard Grants, 2436
Nordson Corporation Foundation Grants, 2981
Sandy Hill Foundation Grants, 3477
Starke County Community Fndn Scholarships, 3648
Strake Foundation Grants, 3658
SunTrust Bank Trusteed Foundations Florence C. and Harry L. English Memorial Fund Grants, 3680
T. Spencer Shore Foundation Grants, 3701

Higher Education, Public
A/H Foundation Grants, 18
Dyson Foundation Mid-Hudson Valley General Operating Support Grants, 1533
Goodyear Tire Grants, 1856
Katherine Baxter Memorial Foundation Courses, 2292
Sandy Hill Foundation Grants, 3477
Starke County Community Fndn Scholarships, 3648
T. Spencer Shore Foundation Grants, 3701
William J. & Tina Rosenberg Fndn Grants, 3954

Hispanic Education
General Motors Foundation Grants, 1783
John R. Oishei Foundation Grants, 2249
MLA Grad Scholarship for Minority Students, 2565
NHLBI Training Program Grants for Institutions that Promote Diversity, 2738
NMF Need-Based Scholarships, 2965
S.H. Cowell Foundation Grants, 3447

Hispanic Studies
CMS Hispanic Health Services Research Grants, 1186

Hispanics
ADA Found Minority Dental Scholarships, 292
AMA Foundation Minority Scholars Awards, 491
American Sociological Association Minority Fellowships, 563
ASM Robert D. Watkins Graduate Research Fellowship, 747
AT&T Foundation Civic and Community Service Program Grants, 793
Charles Delmar Foundation Grants, 1115
CMS Hispanic Health Services Research Grants, 1186
Effie and Wofford Cain Foundation Grants, 1563
FDHN Student Research Fellowships, 1654
Healthcare Fndn for Orange County Grants, 1981
NMF Franklin C. McLean Award, 2955
Pfizer ASPIRE North America Rheumatology Research Awards, 3172
Philadelphia Foundation Organizational Effectiveness Grants, 3189
Robert R. Meyer Foundation Grants, 3387
RWJF Harold Amos Medical Faculty Development Research Grants, 3434
Strowd Roses Grants, 3664

Historical Preservation
Blum-Kovler Foundation Grants, 924
Carls Foundation Grants, 1030
Collins C. Diboll Private Foundation Grants, 1209
Community Foundation Partnerships - Lawrence County Grants, 1317
Community Foundation Partnerships - Martin County Grants, 1318
ConocoPhillips Foundation Grants, 1327
Crown Point Community Foundation Grants, 1361
Decatur County Community Foundation Large Project Grants, 1444
Elkhart County Community Foundation Fund for Elkhart County, 1572
Ferree Foundation Grants, 1658
Forrest C. Lattner Foundation Grants, 1688
Foundation for the Mid South Community Development Grants, 1699
George Kress Foundation Grants, 1799
Inasmuch Foundation Grants, 2149
Jane's Trust Grants, 2206
Jasper Foundation Grants, 2212
John W. and Anna H. Hanes Foundation Grants, 2261
Kinsman Foundation Grants, 2309
McCombs Foundation Grants, 2491
McLean Contributionship Grants, 2497
Norcliffe Foundation Grants, 2979
North Carolina Community Foundation Grants, 2990
Parke County Community Foundation Grants, 3115
Rollins-Luetkemeyer Foundation Grants, 3404
Samuel N. and Mary Castle Foundation Grants, 3467
Sioux Falls Area Community Foundation Community Fund Grants (Unrestricted), 3568
Sioux Falls Area Community Foundation Spot Grants (Unrestricted), 3570
Turner Foundation Grants, 3763
Windham Foundation Grants, 3961

History
1675 Foundation Grants, 11
AAHN Grant for Historical Research, 146
AAHN Pre-Doctoral Research Grant, 147
Alice Fisher Society Fellowships, 436
Australasian Institute of Judicial Administration Seed Funding Grants, 806
Blanche and Irving Laurie Foundation Grants, 910
Canada-U.S. Fulbright New Century Scholars Program Grants, 1008
Canada-U.S. Fulbright Senior Specialists Grants, 1009
Charles Edison Fund Grants, 1116
Colonel Stanley R. McNeil Foundation Grants, 1211
Community Foundation of Bloomington and Monroe County Grants, 1265
Cooke Foundation Grants, 1344
Doris and Victor Day Foundation Grants, 1491
E.J. Grassmann Trust Grants, 1537
Fischelis Grants for Research in the History of American Pharmacy, 1669
Fulbright Alumni Initiatives Awards, 1749

754 / History

Fulbright Distinguished Chairs Awards, 1750
Fulbright New Century Scholars Grants, 1753
Fulbright Specialists Program Grants, 1754
Fulbright Scholars in Europe and Eurasia, 1755
Fulbright Traditional Scholar Program in Sub-Saharan Africa, 1756
Fulbright Traditional Scholar Program in the East Asia/Pacific Region, 1757
Fulbright Traditional Scholar Program in the Near East and North Africa Region, 1758
Fulbright Traditional Scholar Program in the South and Central Asia Region, 1759
Fulbright Traditional Scholar Program in the Western Hemisphere, 1760
Harry Frank Guggenheim Foundation Dissertation Fellowships, 1963
Lawrence S. Huntington Fund Grants, 2348
NHSCA Arts in Health Care Project Grants, 2741
Rockefeller Archive Center Research Grants, 3393
Self Foundation Grants, 3510
Sid W. Richardson Foundation Grants, 3553
Turner Foundation Grants, 3763
Vermont Community Foundation Grants, 3865
Warrick County Community Foundation Grants, 3904

Holistic Medicine
AIHS Alberta/Pfizer Translat Research Grants, 374
Bravewell Leadership Award, 946
Samueli Institute Scientific Research Grants, 3466

Home Economics
AAFCS International Graduate Fellowships, 129
AAFCS National Graduate Fellowships, 130
AAFCS National Undergraduate Scholarships, 131
Phi Upsilon Omicron Alumni Research Grant, 3191
Phi Upsilon Omicron Florence Fallgatter Distinguished Service Award, 3192
Phi Upsilon Omicron Frances Morton Holbrook Alumni Award, 3193
Phi Upsilon Omicron Geraldine Clewell Senior Awards, 3194
Phi Upsilon Omicron Janice Cory Bullock Collegiate Award, 3195
Phi Upsilon Lillian P. Schoephoerster Award, 3196
Phi Upsilon Omicron Margaret Jerome Sampson Scholarships, 3197
Phi Upsilon Orinne Johnson Writing Award, 3198
Phi Upsilon Omicron Sarah Thorniley Phillips Leader Awards, 3199
Phi Upsilon Omicron Undergraduate Karen P. Goebel Conclave Award, 3200

Home Economics Education
Phi Upsilon Omicron Florence Fallgatter Distinguished Service Award, 3192

Homeless Shelters
Adaptec Foundation Grants, 304
AEGON Transamerica Foundation Health and Welfare Grants, 316
Alpha Natural Resources Corporate Giving, 450
BJ's Charitable Foundation Grants, 904
Carl B. and Florence E. King Foundation Grants, 1022
Carl M. Freeman Foundation FACES Grants, 1026
Charles H. Hall Foundation, 1120
Clara Blackford Smith and W. Aubrey Smith Charitable Foundation Grants, 1165
Community Foundation of Bloomington and Monroe County Grants, 1265
Community Foundation of Muncie and Delaware County Maxon Grants, 1301
Community Fndn of Switzerland County Grants, 1311
Danellie Foundation Grants, 1421
Dominion Foundation Grants, 1485
Edina Realty Foundation Grants, 1550
Elsie Lee Garthwaite Memorial Fndn Grants, 1584
Frank Loomis Palmer Fund Grants, 1726
Frank Reed and Margaret Jane Peters Memorial Fund II Grants, 1728
Gardner Foundation Grants, 1772

George E. Hatcher, Jr. and Ann Williams Hatcher Foundation Grants, 1792
Georgiana Goddard Eaton Memorial Grants, 1805
GNOF Maison Hospitaliere Grants, 1843
Green Foundation Human Services Grants, 1894
Harold Brooks Foundation Grants, 1952
Helen Irwin Littauer Educational Trust Grants, 1994
Henrietta Lange Burk Fund Grants, 2001
Henrietta Tower Wurts Memorial Fndn Grants, 2002
HomeBanc Foundation Grants, 2042
Horace A. Kimball and S. Ella Kimball Foundation Grants, 2046
Jeffris Wood Foundation Grants, 2217
John H. Wellons Foundation Grants, 2239
Joseph Henry Edmondson Foundation Grants, 2270
L. W. Pierce Family Foundation Grants, 2328
Leo Niessen Jr., Charitable Trust Grants, 2358
Lydia deForest Charitable Trust Grants, 2393
McCarthy Family Foundation Grants, 2490
Medtronic Foundation Community Link Human Services Grants, 2511
Meyer Foundation Healthy Communities Grants, 2533
Norfolk Southern Foundation Grants, 2982
Northwestern Mutual Foundation Grants, 2999
NYCT Grants for the Elderly, 3054
Ordean Foundation Grants, 3080
Packard Foundation Local Grants, 3105
Perpetual Trust for Charitable Giving Grants, 3141
Pinnacle Foundation Grants, 3225
Plough Foundation Grants, 3238
Reinberger Foundation Grants, 3357
Robert R. Meyer Foundation Grants, 3387
Santa Barbara Foundation Strategy Grants - Core Support, 3481
Sierra Health Foundation Responsive Grants, 3555
Sisters of St. Joseph Healthcare Fndn Grants, 3576
Stewart Huston Charitable Trust Grants, 3655
Strowd Roses Grants, 3664
Swindells Charitable Foundation, 3697
Textron Corporate Contributions Grants, 3712
Union Bank, N.A. Foundation Grants, 3788
WHO Foundation Volunteer Service Grants, 3940
Wilson-Wood Foundation Grants, 3960

Homelessness
ACE Charitable Foundation Grants, 229
Achelis Foundation Grants, 234
AEGON Transamerica Foundation Health and Welfare Grants, 316
Anschutz Family Foundation Grants, 602
Arizona Diamondbacks Charities Grants, 657
Assurant Foundation Grants, 788
Audrey & Sydney Irmas Foundation Grants, 799
Barker Welfare Foundation Grants, 831
Baxter International Foundation Grants, 843
Berrien Community Foundation Grants, 884
BJ's Charitable Foundation Grants, 904
Bodman Foundation Grants, 929
Boston Foundation Grants, 936
Brooklyn Community Foundation Caring Neighbors Grants, 968
Carl B. and Florence E. King Foundation Grants, 1022
Carlisle Foundation Grants, 1025
Cemala Foundation Grants, 1077
CFFVR Robert and Patricia Endries Family Foundation Grants, 1102
Charles Delmar Foundation Grants, 1115
Community Fndn for the Capital Region Grants, 1259
Community Foundation of Greenville Hollingsworth Funds Program/Project Grants, 1285
Community Foundation of the Verdugos Grants, 1315
Cralle Foundation Grants, 1356
D. W. McMillan Foundation Grants, 1407
Dade Community Foundation Grants, 1410
Danellie Foundation Grants, 1421
Daphne Seybolt Culpeper Fndn Grants, 1425
David Geffen Foundation Grants, 1431
Delaware Community Foundation Grants, 1449
Denver Broncos Charities Fund Grants, 1460
Denver Foundation Community Grants, 1461

Dominion Foundation Grants, 1485
Edina Realty Foundation Grants, 1550
Effie and Wofford Cain Foundation Grants, 1563
El Pomar Foundation Anna Keesling Ackerman Fund Grants, 1581
El Pomar Foundation Grants, 1582
Elsie Lee Garthwaite Memorial Fndn Grants, 1584
Ensworth Charitable Foundation Grants, 1595
Essex County Community Foundation Merrimack Valley General Fund Grants, 1603
Eugene M. Lang Foundation Grants, 1611
Faye McBeath Foundation Grants, 1635
Fourjay Foundation Grants, 1712
Frances and John L. Loeb Family Fund Grants, 1716
Gardner Foundation Grants, 1772
George A Ohl Jr. Foundation Grants, 1791
Georgiana Goddard Eaton Memorial Grants, 1805
Giant Food Charitable Grants, 1816
Great-West Life Grants, 1879
Green Foundation Human Services Grants, 1894
Greenspun Family Foundation Grants, 1896
H.B. Fuller Foundation Grants, 1922
Hampton Roads Community Foundation Health and Human Service Grants, 1941
Hasbro Children's Fund Grants, 1973
Helen Steiner Rice Foundation Grants, 1998
Henrietta Tower Wurts Memorial Fndn Grants, 2002
Hilton Hotels Corporate Giving Program Grants, 2037
Horace A. Kimball and S. Ella Kimball Foundation Grants, 2046
Horace Moses Charitable Foundation Grants, 2047
Howard and Bush Foundation Grants, 2054
Jacob G. Schmidlapp Trust Grants, 2183
Joe W. and Dorothy Dorsett Brown Fndn Grants, 2230
John Edward Fowler Memorial Fndn Grants, 2235
Joseph H. & Florence A. Roblee Fndn Grants, 2269
Joseph Henry Edmondson Foundation Grants, 2270
L. W. Pierce Family Foundation Grants, 2328
Lands' End Corporate Giving Program, 2344
Laura B. Vogler Foundation Grants, 2345
Leo Niessen Jr., Charitable Trust Grants, 2358
Lucy Gooding Charitable Fndn Grants, 2388
M.J. Murdock Charitable Trust General Grants, 2405
Manuel D. & Rhoda Mayerson Fndn Grants, 2420
McCarthy Family Foundation Grants, 2490
Mervin Bovaird Foundation Grants, 2523
Meyer Foundation Healthy Communities Grants, 2533
Morris & Gwendolyn Cafritz Fndn Grants, 2594
Norfolk Southern Foundation Grants, 2982
NYCT Girls and Young Women Grants, 3053
NYCT Grants for the Elderly, 3054
Oppenstein Brothers Foundation Grants, 3078
Ordean Foundation Grants, 3080
Patrick and Anna M. Cudahy Fund Grants, 3120
Peacock Foundation Grants, 3133
Perpetual Trust for Charitable Giving Grants, 3141
Pinellas County Grants, 3223
Pinnacle Foundation Grants, 3225
Plough Foundation Grants, 3238
Portland General Electric Foundation Grants, 3250
Princeton Area Community Foundation Greater Mercer Grants, 3272
Puerto Rico Community Foundation Grants, 3278
Quantum Foundation Grants, 3290
Reinberger Foundation Grants, 3357
Reynolds & Reynolds Associate Fndn Grants, 3360
Rhode Island Foundation Grants, 3362
Richard & Susan Smith Family Fndn Grants, 3365
Robert R. McCormick Tribune Veterans Initiative Grants, 3386
Robert R. Meyer Foundation Grants, 3387
Ruth Anderson Foundation Grants, 3425
S. Mark Taper Foundation Grants, 3449
Saint Luke's Health Initiatives Grants, 3455
Santa Barbara Foundation Strategy Grants - Core Support, 3481
Schlessman Family Foundation Grants, 3493
Sensient Technologies Foundation Grants, 3513
Singing for Change Foundation Grants, 3567
Sisters of St. Joseph Healthcare Fndn Grants, 3576

SUBJECT INDEX

Strowd Roses Grants, 3664
Susan Mott Webb Charitable Trust Grants, 3694
Swindells Charitable Foundation, 3697
Thomas C. Ackerman Foundation Grants, 3719
Union Bank, N.A. Foundation Grants, 3788
Vancouver Foundation Grants and Community Initiatives Program, 3849
W. C. Griffith Foundation Grants, 3882
W.C. Griffith Foundation Grants, 3881
WHO Foundation Volunteer Service Grants, 3940
William G. & Helen C. Hoffman Fndn Grants, 3950
Wilson-Wood Foundation Grants, 3960

Homeownership
Faye McBeath Foundation Grants, 1635
PMI Foundation Grants, 3239
Union Bank, N.A. Foundation Grants, 3788

Homosexuals, Female
Alexander Foundation Cancer, Catastrophic Illness and Injury Grants, 421
Alexander Fndn Insurance Continuation Grants, 422
Boston Globe Foundation Grants, 937
California Endowment Innovative Ideas Challenge Grants, 1000
Community Foundation for the National Capital Region Community Leadership Grants, 1260
E. Rhodes & Leona B. Carpenter Grants, 1539
Gill Foundation - Gay and Lesbian Fund Grants, 1821
Horizons Community Issues Grants, 2049
Ms. Fndn for Women Ending Violence Grants, 2596
Ms. Foundation for Women Health Grants, 2597
Paul Rapoport Foundation Grants, 3125
Playboy Foundation Grants, 3237
Pride Foundation Grants, 3268
Pride Foundation Scholarships, 3269
Rainbow Endowment Grants, 3300

Homosexuals, Male
Alexander Foundation Cancer, Catastrophic Illness and Injury Grants, 421
Alexander Fndn Insurance Continuation Grants, 422
Boston Globe Foundation Grants, 937
California Endowment Innovative Ideas Challenge Grants, 1000
Community Foundation for the National Capital Region Community Leadership Grants, 1260
David Geffen Foundation Grants, 1431
E. Rhodes & Leona B. Carpenter Grants, 1539
Gill Foundation - Gay and Lesbian Fund Grants, 1821
Horizons Community Issues Grants, 2049
Ms. Fndn for Women Ending Violence Grants, 2596
Ms. Foundation for Women Health Grants, 2597
Paul Rapoport Foundation Grants, 3125
Playboy Foundation Grants, 3237
Pride Foundation Grants, 3268
Pride Foundation Scholarships, 3269
Rainbow Endowment Grants, 3300
San Diego Foundation for Change Grants, 3470
USAID Comprehensive District-Based Support for Better HIV/TB Patient Outcomes Grants, 3807

Hormones
Fndn for Appalachian Ohio Bachtel Scholarships, 1692
NHLBI Lymphatics in Health and Disease in the Digestive, Urinary, Cardiovascular and Pulmonary Systems, 2705
NIDDK Co-Activators and Co-Repressors in Gene Expression Grants, 2805
NIH Receptors and Signaling in Bone in Health and Disease, 2891

Horticulture
Bruce and Adele Greenfield Foundation Grants, 972
Frederick W. Marzahl Memorial Fund Grants, 1741
Hagedorn Fund Grants, 1934
Nina Mason Pulliam Charitable Trust Grants, 2921
Posey County Community Foundation Grants, 3252

Hospice Care
AAHPM Hospice & Palliative Med Fellowships, 152
Adelaide Breed Bayrd Foundation Grants, 306
Arizona Public Service Corporate Giving Program Grants, 658
Bailey Foundation Grants, 818
Bertha Russ Lytel Foundation Grants, 885
Borkee-Hagley Foundation Grants, 934
Burlington Industries Foundation Grants, 977
Carl B. and Florence E. King Foundation Grants, 1022
Carl W. and Carrie Mae Joslyn Trust Grants, 1031
Cemala Foundation Grants, 1077
Christy-Houston Foundation Grants, 1153
Conquer Cancer Foundation of ASCO International Development and Education Award in Palliative Care, 1332
D. W. McMillan Foundation Grants, 1407
Danellie Foundation Grants, 1421
Daniels Fund Grants-Aging, 1424
Del Mar Healthcare Fund Grants, 1453
Dunspaugh-Dalton Foundation Grants, 1530
E. Rhodes & Leona B. Carpenter Grants, 1539
Edward W. and Stella C. Van Houten Memorial Fund Grants, 1559
Gardner Foundation Grants, 1772
George E. Hatcher, Jr. and Ann Williams Hatcher Foundation Grants, 1792
Gregory L. Gibson Charitable Fndn Grants, 1900
Hasbro Corporation Gift of Play Hospital and Pediatric Health Giving, 1974
J.W. Kieckhefer Foundation Grants, 2176
L. W. Pierce Family Foundation Grants, 2328
Lisa Higgins-Hussman Foundation Grants, 2370
Lucy Gooding Charitable Fndn Grants, 2388
Maine Community Foundation Charity Grants, 2415
Maine Community Foundation Hospice Grants, 2416
Margaret T. Morris Foundation Grants, 2429
MGN Family Foundation Grants, 2542
RCPSC Balfour M. Mount Visiting Professorship in Palliative Medicine, 3318
Samuel S. Johnson Foundation Grants, 3468
Sensient Technologies Foundation Grants, 3513
Stewart Huston Charitable Trust Grants, 3655
Thompson Charitable Foundation Grants, 3727
Visiting Nurse Foundation Grants, 3879

Hospital Administration
Advocate HealthCare Post Graduate Administrative Fellowship, 314
Mayo Clinic Administrative Fellowship, 2460
Mayo Clinic Admin Fellowship - Eau Claire, 2461
Mayo Clinic Business Consulting Fellowship, 2462
SfN Louise Hanson Marshall Specific Recognition Award, 3527
UniHealth Foundation Healthcare Systems Enhancements Grants, 3784

Hospitals
1st Source Foundation Grants, 2
Abramson Family Foundation Grants, 217
ACL Business Acumen for Disability Organizations Grants, 238
Adam Richter Charitable Trust Grants, 298
Adelaide Breed Bayrd Foundation Grants, 306
Advocate HealthCare Post Graduate Administrative Fellowship, 314
Alcoa Foundation Grants, 414
Allyn Foundation Grants, 448
American Foodservice Charitable Trust Grants, 532
Andrew Family Foundation Grants, 588
Ann Jackson Family Foundation Grants, 598
Annunziata Sanguinetti Foundation Grants, 600
Anthony R. Abraham Foundation Grants, 608
Antone & Edene Vidinha Charitable Trust Grants, 609
Aragona Family Foundation Grants, 648
Archer Daniels Midland Foundation Grants, 651
Arthur and Rochelle Belfer Foundation Grants, 670
Atran Foundation Grants, 797
Auburn Foundation Grants, 798
Audrey & Sydney Irmas Foundation Grants, 799

Bacon Family Foundation Grants, 817
Banfi Vintners Foundation Grants, 824
Barker Foundation Grants, 830
Batchelor Foundation Grants, 837
Beckman Coulter Foundation Grants, 868
Belk Foundation Grants, 872
Bertha Russ Lytel Foundation Grants, 885
Bert W. Martin Foundation Grants, 886
BibleLands Grants, 889
Bildner Family Foundation Grants, 891
Blum-Kovler Foundation Grants, 924
Blumenthal Foundation Grants, 925
Bob & Delores Hope Foundation Grants, 927
Boettcher Foundation Grants, 931
Burden Trust Grants, 976
Burlington Industries Foundation Grants, 977
Burlington Northern Santa Fe Foundation Grants, 978
Byron W. & Alice L. Lockwood Fnd Grants, 991
Caddock Foundation Grants, 994
Caesars Foundation Grants, 995
Callaway Foundation Grants, 1002
Camp-Younts Foundation Grants, 1005
Caring Foundation Grants, 1019
Carl B. and Florence E. King Foundation Grants, 1022
Carl Gellert and Celia Berta Gellert Foundation Grants, 1024
Carls Foundation Grants, 1030
Carnahan-Jackson Foundation Grants, 1032
Carrie Estelle Doheny Foundation Grants, 1037
Carylon Foundation Grants, 1040
Cessna Foundation Grants Program, 1081
Champlin Foundations Grants, 1110
Charles Delmar Foundation Grants, 1115
Chazen Foundation Grants, 1129
Christy-Houston Foundation Grants, 1153
Clark-Winchcole Foundation Grants, 1173
Clark and Ruby Baker Foundation Grants, 1174
Cockrell Foundation Grants, 1204
Coeta and Donald Barker Foundation Grants, 1205
Colorado Resource for Emergency Education and Trauma Grants, 1212
Columbus Foundation Central Benefits Health Care Foundation Grants, 1216
Commonwealth Edison Grants, 1225
Commonwealth Fund Small Grants, 1234
Communities Foundation of Texas Grants, 1236
Community Foundation of Mount Vernon and Knox County Grants, 1299
Community Foundation of the Verdugos Educational Endowment Fund Grants, 1314
Connelly Foundation Grants, 1326
ConocoPhillips Foundation Grants, 1327
Constantin Foundation Grants, 1341
Cowles Charitable Trust Grants, 1354
Crystelle Waggoner Charitable Trust Grants, 1364
Cystic Fibrosis Canada Fellowships, 1384
Dairy Queen Corporate Contributions Grants, 1413
Dana Brown Charitable Trust Grants, 1419
Daniel & Nanna Stern Family Fndn Grants, 1422
Daphne Seybolt Culpeper Fndn Grants, 1425
David M. and Marjorie D. Rosenberg Foundation Grants, 1432
Davis Family Foundation Grants, 1435
Dennis & Phyllis Washington Fndn Grants, 1458
Denton A. Cooley Foundation Grants, 1459
DeRoy Testamentary Foundation Grants, 1469
Dolan Children's Foundation Grants, 1482
Dora Roberts Foundation Grants, 1489
Dr. Leon Bromberg Charitable Trust Grants, 1509
Dr. Scholl Foundation Grants Program, 1511
Dunspaugh-Dalton Foundation Grants, 1530
E.J. Grassmann Trust Grants, 1537
Eberly Foundation Grants, 1544
Eckerd Corporation Foundation Grants, 1547
Eden Hall Foundation Grants, 1549
Edwards Memorial Trust Grants, 1558
Edwin S. Webster Foundation Grants, 1560
Effie and Wofford Cain Foundation Grants, 1563
Elsie H. Wilcox Foundation Grants, 1583
Eugene B. Casey Foundation Grants, 1609

756 / Hospitals

F.M. Kirby Foundation Grants, 1624
Fannie E. Rippel Foundation Grants, 1630
FirstEnergy Foundation Community Grants, 1668
FMC Foundation Grants, 1683
Ford Motor Company Fund Grants Program, 1687
Foster Foundation Grants, 1689
Fndn for Appalachian Ohio Bachtel Scholarships, 1692
Foundation for Health Enhancement Grants, 1696
Four J Foundation Grants, 1713
Frances L. and Edwin L. Cummings Memorial Fund Grants, 1717
Francis T. & Louise T. Nichols Fndn Grants, 1720
Frank E. and Seba B. Payne Foundation Grants, 1722
Frank G. and Freida K. Brotz Family Foundation Grants, 1723
Frank S. Flowers Foundation Grants, 1729
Fraser-Parker Foundation Grants, 1732
Fritz B. Burns Foundation Grants, 1747
Fuji Film Grants, 1748
G.A. Ackermann Memorial Fund Grants, 1767
G.N. Wilcox Trust Grants, 1768
Gardner Foundation Grants, 1770
Gebbie Foundation Grants, 1777
George E. Hatcher, Jr. and Ann Williams Hatcher Foundation Grants, 1792
George F. Baker Trust Grants, 1793
George Kress Foundation Grants, 1799
George S. & Dolores Dore Eccles Fndn Grants, 1801
George W. Codrington Charitable Fndn Grants, 1803
Gertrude and William C. Wardlaw Fund Grants, 1808
Gladys Brooks Foundation Grants, 1826
GNOF IMPACT Gulf States Eye Surgery Fund, 1840
Goodrich Corporation Foundation Grants, 1854
Greenwall Foundation Bioethics Grants, 1897
Gregory Family Foundation Grants (Florida), 1899
Griffin Family Foundation Grants, 1902
H. Leslie Hoffman and Elaine S. Hoffman Foundation Grants, 1924
Harold Alfond Foundation Grants, 1949
Harry Bramhall Gilbert Charitable Trust Grants, 1961
Harry Edison Foundation, 1962
Harvey Randall Wickes Foundation Grants, 1972
Hawn Foundation Grants, 1979
Healthcare Fndn for Orange County Grants, 1981
Hearst Foundations Health Grants, 1987
Hedco Foundation Grants, 1989
Helen Irwin Littauer Educational Trust Grants, 1994
Helen Pumphrey Denit Charitable Trust Grants, 1996
Henry County Community Foundation Grants, 2005
Henry L. Guenther Foundation Grants, 2006
Hilda and Preston Davis Foundation Grants, 2028
Hoblitzelle Foundation Grants, 2038
Hospital Libraries Section / MLA Professional Development Grants, 2052
Hudson Webber Foundation Grants, 2072
Huffy Foundation Grants, 2073
Huie-Dellmon Trust Grants, 2075
Huisking Foundation Grants, 2076
J.L. Bedsole Foundation Grants, 2172
J. Walton Bissell Foundation Grants, 2177
Jacob and Valeria Langeloth Foundation Grants, 2182
James G.K. McClure Educational and Development Fund Grants, 2188
James H. Cummings Foundation Grants, 2190
James J. & Angelia M. Harris Fndn Grants, 2192
Janirve Foundation Grants, 2209
Joe W. and Dorothy Dorsett Brown Fndn Grants, 2230
John Clarke Trust Grants, 2232
John Jewett and Helen Chandler Garland Foundation Grants, 2242
John Lord Knight Foundation Grants, 2243
John S. Dunn Research Foundation Grants, 2250
Joseph Alexander Foundation Grants, 2265
Josephine Goodyear Foundation Grants, 2272
Josiah W. and Bessie H. Kline Foundation Grants, 2278
Katharine Matthies Foundation Grants, 2291
Katherine John Murphy Foundation Grants, 2293
Kathryne Beynon Foundation Grants, 2294
Kent D. Steadley and Mary L. Steadley Memorial Trust Grants, 2301

Lawrence S. Huntington Fund Grants, 2348
Lillian S. Wells Foundation Grants, 2365
Lubbock Area Foundation Grants, 2383
Lucile Horton Howe and Mitchell B. Howe Foundation Grants, 2385
Lumpkin Family Fnd Healthy People Grants, 2391
Lydia deForest Charitable Trust Grants, 2393
Mabel Y. Hughes Charitable Trust Grants, 2409
Mary S. and David C. Corbin Foundation Grants, 2447
Matilda R. Wilson Fund Grants, 2453
Max and Victoria Dreyfus Foundation Grants, 2457
Maxon Charitable Foundation Grants, 2458
McLean Contributionship Grants, 2497
Mericos Foundation Grants, 2518
Meriden Foundation Grants, 2519
Mervin Bovaird Foundation Grants, 2523
Meyer and Stephanie Eglin Foundation Grants, 2531
Mockingbird Foundation Grants, 2589
Nicholas H. Noyes Jr. Memorial Fndn Grants, 2791
Norcliffe Foundation Grants, 2979
North Central Health Services Grants, 2993
Notsew Orm Sands Foundation Grants, 3002
Olive Higgins Prouty Foundation Grants, 3072
Oppenstein Brothers Foundation Grants, 3078
Peacock Foundation Grants, 3133
Piper Jaffray Fndn Communities Giving Grants, 3227
Pott Foundation Grants, 3253
Procter and Gamble Fund Grants, 3274
Questar Corporate Contributions Grants, 3291
R.C. Baker Foundation Grants, 3292
R.T. Vanderbilt Trust Grants, 3294
Ralphs Food 4 Less Foundation Grants, 3305
Raskob Foundation for Catholic Activities Grants, 3306
Rayonier Foundation Grants, 3311
Reinberger Foundation Grants, 3357
Rice Foundation Grants, 3363
Richard & Susan Smith Family Fndn Grants, 3365
Robert R. Meyer Foundation Grants, 3387
Robins Foundation Grants, 3390
Rockwell International Corporate Trust Grants Program, 3399
Ruth Eleanor Bamberger and John Ernest Bamberger Memorial Foundation Grants, 3427
Ruth H. & Warren A. Ellsworth Fndn Grants, 3429
RWJF New Jersey Health Initiatives Grants, 3440
Saint Luke's Health Initiatives Grants, 3455
Salem Foundation Grants, 3457
Sara Elizabeth O'Brien Trust Grants, 3483
Schering-Plough Foundation Health Grants, 3492
Scott County Community Foundation Grants, 3500
Seabury Foundation Grants, 3501
Sensient Technologies Foundation Grants, 3513
Shell Deer Park Grants, 3543
Shell Oil Company Foundation Community Development Grants, 3544
Sid W. Richardson Foundation Grants, 3553
Solo Cup Foundation Grants, 3620
Sonora Area Foundation Competitive Grants, 3622
Strake Foundation Grants, 3658
Sunderland Foundation Grants, 3675
SunTrust Bank Trusteed Foundations Florence C. and Harry L. English Memorial Fund Grants, 3680
SunTrust Bank Trusteed Foundations Greene-Sawtell Grants, 3681
SunTrust Bank Trusteed Foundations Harriet McDaniel Marshall Tust Grants, 3682
SunTrust Bank Trusteed Foundations Nell Warren Elkin and William Simpson Elkin Grants, 3683
SunTrust Bank Trusteed Foundations Thomas Guy Woolford Charitable Trust Grants, 3684
SunTrust Bank Trusteed Foundations Walter H. and Marjory M. Rich Memorial Fund Grants, 3685
Swindells Charitable Foundation, 3697
Taubman Endowment for the Arts Fndn Grants, 3702
Tellabs Foundation Grants, 3706
Thelma Doelger Charitable Trust Grants, 3714
Thomas and Dorothy Leavey Foundation Grants, 3717
Thomas Austin Finch, Sr. Foundation Grants, 3718
Thompson Charitable Foundation Grants, 3727
Thoracic Foundation Grants, 3729

SUBJECT INDEX

Tommy Hilfiger Corporate Foundation Grants, 3737
Tri-State Community Twenty-first Century Endowment Fund Grants, 3756
United Hospital Fund of New York Health Care Improvement Grants, 3797
Victor E. Speas Foundation Grants, 3868
W.C. Griffith Foundation Grants, 3881
W. C. Griffith Foundation Grants, 3882
Wege Foundation Grants, 3912
Weingart Foundation Grants, 3913
Willard & Pat Walker Charitable Fndn Grants, 3943
William B. Stokely Jr. Foundation Grants, 3947
William H. Hannon Foundation Grants, 3953
Winifred & Harry B. Allen Foundation Grants, 3962

Hotel and Restaurant Management
Coors Brewing Corporate Contributions Grants, 1346
Phi Upsilon Omicron Florence Fallgatter Distinguished Service Award, 3192
Phi Upsilon Omicron Geraldine Clewell Senior Awards, 3194
Phi Upsilon Lillian P. Schoephoerster Award, 3196
Phi Upsilon Orinne Johnson Writing Award, 3198

Housing
Adaptec Foundation Grants, 304
Agnes M. Lindsay Trust Grants, 346
Alexander Foundation Cancer, Catastrophic Illness and Injury Grants, 421
Alpha Natural Resources Corporate Giving, 450
Alvin and Fanny Blaustein Thalheimer Foundation Baltimore Communal Grants, 459
AMD Corporate Contributions Grants, 500
American Electric Power Foundation Grants, 530
Ann Arbor Area Community Foundation Grants, 594
Ann Peppers Foundation Grants, 599
Arizona Cardinals Grants, 655
Arizona Diamondbacks Charities Grants, 657
Arthur Ashley Williams Foundation Grants, 671
Atlanta Foundation Grants, 796
Audrey & Sydney Irmas Foundation Grants, 799
Bacon Family Foundation Grants, 817
BancorpSouth Foundation Grants, 823
Batchelor Foundation Grants, 837
Beazley Foundation Grants, 865
Blue Cross Blue Shield of Minnesota Foundation - Healthy Equity: Health Impact Assessment Demonstration Project Grants, 916
Boston Foundation Grants, 936
Brooklyn Community Foundation Caring Neighbors Grants, 968
Byron W. & Alice L. Lockwood Fnd Grants, 991
Cambridge Community Foundation Grants, 1004
Campbell Soup Foundation Grants, 1007
Carl & Eloise Pohlad Family Fndn Grants, 1021
Carl B. and Florence E. King Foundation Grants, 1022
Carlisle Foundation Grants, 1025
Carl M. Freeman Foundation FACES Grants, 1026
Carnahan-Jackson Foundation Grants, 1032
Carpenter Foundation Grants, 1035
Cemala Foundation Grants, 1077
Cessna Foundation Grants Program, 1081
CFFVR Robert and Patricia Endries Family Foundation Grants, 1102
Children Affected by AIDS Foundation Family Assistance Emergency Fund Grants, 1146
Citizens Bank Mid-Atlantic Charitable Foundation Grants, 1162
Claude Worthington Benedum Fndn Grants, 1178
CNCS AmeriCorps VISTA Project Grants, 1190
Comerica Charitable Foundation Grants, 1224
Community Fndn AIDS Endowment Awards, 1237
Community Foundation Alliance City of Evansville Endowment Fund Grants, 1238
Community Foundation of Greater Birmingham Grants, 1272
Constantin Foundation Grants, 1341
Danellie Foundation Grants, 1421
Delaware Community Foundation Grants, 1449
Delmarva Power and Light Contributions, 1454

SUBJECT INDEX

Dennis & Phyllis Washington Fndn Grants, 1458
Denver Foundation Community Grants, 1461
Dominion Foundation Grants, 1485
Doree Taylor Charitable Foundation, 1490
Edward N. and Della L. Thome Memorial Foundation Grants, 1557
Edwin W. and Catherine M. Davis Fndn Grants, 1561
Ensworth Charitable Foundation Grants, 1595
Evanston Community Foundation Grants, 1615
Fairfield County Community Foundation Grants, 1625
Fisher Foundation Grants, 1670
Fourjay Foundation Grants, 1712
Frank Reed and Margaret Jane Peters Memorial Fund II Grants, 1728
Frank Stanley Beveridge Foundation Grants, 1730
George and Ruth Bradford Foundation Grants, 1789
George A Ohl Jr. Foundation Grants, 1791
George Gund Foundation Grants, 1796
Georgiana Goddard Eaton Memorial Grants, 1805
Ginn Foundation Grants, 1822
Giving in Action Society Children & Youth with Special Needs Grants, 1823
Giving in Action Society Family Indep Grants, 1824
Greater Worcester Community Foundation Discretionary Grants, 1888
Gulf Coast Community Foundation Grants, 1915
Hagedorn Fund Grants, 1934
Hartford Foundation Regular Grants, 1970
HCA Foundation Grants, 1980
Holland/Zeeland Community Fndn Grants, 2041
HomeBanc Foundation Grants, 2042
Home Building Industry Disaster Relief Fund, 2043
Huntington National Bank Community Grants, 2084
Hutchinson Community Foundation Grants, 2085
Idaho Power Company Corporate Contributions, 2115
Illinois Tool Works Foundation Grants, 2147
Jacob G. Schmidlapp Trust Grants, 2183
James R. Thorpe Foundation Grants, 2199
Janirve Foundation Grants, 2209
Joe W. and Dorothy Dorsett Brown Fndn Grants, 2230
John H. Wellons Foundation Grants, 2239
John W. Speas and Effie E. Speas Memorial Trust Grants, 2264
Joseph H. & Florence A. Roblee Fndn Grants, 2269
Josephine S. Gumbiner Foundation Grants, 2273
Kahuku Community Fund, 2286
Katharine Matthies Foundation Grants, 2291
Kathryne Beynon Foundation Grants, 2294
Knox County Community Foundation Grants, 2313
Lena Benas Memorial Fund Grants, 2353
Liberty Bank Foundation Grants, 2362
Lydia deForest Charitable Trust Grants, 2393
Manuel D. & Rhoda Mayerson Fndn Grants, 2420
Mary Owen Borden Foundation Grants, 2446
Mary S. and David C. Corbin Foundation Grants, 2447
Mericos Foundation Grants, 2518
Meyer Foundation Healthy Communities Grants, 2533
Minneapolis Foundation Community Grants, 2558
Morris & Gwendolyn Cafritz Fndn Grants, 2594
Northwestern Mutual Foundation Grants, 2999
Northwest Minnesota Foundation Women's Fund Grants, 3000
NYCT Girls and Young Women Grants, 3053
Oleonda Jameson Trust Grants, 3071
Ordean Foundation Grants, 3080
Patrick and Anna M. Cudahy Fund Grants, 3120
Peacock Foundation Grants, 3133
Perpetual Trust for Charitable Giving Grants, 3141
Perry County Community Foundation Grants, 3142
Peyton Anderson Foundation Grants, 3160
Philadelphia Foundation Organizational Effectiveness Grants, 3189
Pike County Community Foundation Grants, 3222
Pinellas County Grants, 3223
Pinkerton Foundation Grants, 3224
Piper Jaffray Fndn Communities Giving Grants, 3227
Pittsburgh Foundation Community Fund Grants, 3230
Plough Foundation Grants, 3238
PMI Foundation Grants, 3239
Polk Bros. Foundation Grants, 3243

Posey County Community Foundation Grants, 3252
Presbyterian Patient Assistance Program, 3260
Prince Charitable Trusts DC Grants, 3271
Princeton Area Community Foundation Greater Mercer Grants, 3272
Puerto Rico Community Foundation Grants, 3278
Ralphs Food 4 Less Foundation Grants, 3305
Reinberger Foundation Grants, 3357
Rhode Island Foundation Grants, 3362
Robert F. Stoico/FIRSTFED Charitable Foundation Grants, 3382
Robert & Clara Milton Senior Housing Grants, 3384
Rockwell Fund, Inc. Grants, 3398
Ruth Anderson Foundation Grants, 3425
S.H. Cowell Foundation Grants, 3447
S. Mark Taper Foundation Grants, 3449
Salem Foundation Charitable Trust Grants, 3456
Schramm Foundation Grants, 3495
Seagate Tech Corp Capacity to Care Grants, 3502
Sid W. Richardson Foundation Grants, 3553
Southbury Community Trust Fund, 3625
Spencer County Community Foundation Grants, 3632
Textron Corporate Contributions Grants, 3712
Thompson Charitable Foundation Grants, 3727
TJX Foundation Grants, 3735
Union Bank, N.A. Corporate Sponsorships and Donations, 3787
Union Bank, N.A. Foundation Grants, 3788
Union Benevolent Association Grants, 3789
Vancouver Foundation Grants and Community Initiatives Program, 3849
Warrick County Community Foundation Grants, 3904
Wayne County Foundation Grants, 3909
Wilson-Wood Foundation Grants, 3960

Human Connectome
NIH Human Connectome Project Grants, 2881

Human Development
California Community Foundation Human Development Grants, 999
Grammy Foundation Grants, 1863
NIH Research on Sleep and Sleep Disorders, 2898
Vanderburgh Community Foundation Grants, 3852

Human Factors in Engineering
NLM Research Project Grants, 2951

Human Genome
AACR Team Science Award, 126
ASGH Award for Excellence in Human Genetics Education, 689
ASGH C.W. Cotterman Award, 690
ASGH Charles J. Epstein Trainee Awards for Excellence in Human Genetics Research, 691
ASGH McKusick Leadership Award, 692
ASGH William Allan Award, 693
ASGH William Curt Stern Award, 694
Gruber Foundation Gentics Prize, 1908
NHGRI Mentored Research Scientist Development Award, 2686
NIH Human Connectome Project Grants, 2881

Human Learning and Memory
Coca-Cola Foundation Grants, 1203
Fndn for Appalachian Ohio Bachtel Scholarships, 1692
Helen Bader Foundation Grants, 1992
NSF Perception, Action & Cognition Grants, 3019
SfN Swartz Prize for Theoretical and Computational Neuroscience, 3537

Human Population Genetics
ASGH C.W. Cotterman Award, 690
ASGH Charles J. Epstein Trainee Awards for Excellence in Human Genetics Research, 691
Gruber Foundation Rosalind Franklin Young Investigator Award, 1910

Human Reproduction/Fertility
Bernard F. and Alva B. Gimbel Foundation Reproductive Rights Grants, 883
Donald and Sylvia Robinson Family Foundation Grants, 1486
Educational Foundation of America Grants, 1553
Gardner Foundation Grants, 1772
General Service Foundation Human Rights and Economic Justice Grants, 1784
Huber Foundation Grants, 2071
IIE Hewlett Fnd/IIE Dissertation Fellowship, 2132
Lalor Foundation Anna Lalor Burdick Grants, 2335
MMAAP Foundation Fellowship Award in Reproductive Medicine, 2580
MMAAP Foundation Research Project Award in Reproductive Medicine, 2584
Packard Foundation Local Grants, 3105
Packard Foundation Population and Reproductive Health Grants, 3106
Playboy Foundation Grants, 3237
Robert Sterling Clark Foundation Reproductive Rights and Health Grants, 3388
Susan Vaughan Foundation Grants, 3695
USAID Family Planning and Reproductive Health Methods Grants, 3811
USAID Social Science Research in Population and Reproductive Health Grants, 3822
W.C. Griffith Foundation Grants, 3881
Wallace Alexander Gerbode Foundation Grants, 3896
WestWind Foundation Reproductive Health and Rights Grants, 3924

Human Resources
Bill and Melinda Gates Foundation Emergency Response Grants, 892
Bill and Melinda Gates Foundation Water, Sanitation and Hygiene Grants, 894
Mann T. Lowry Foundation Grants, 2419
Mayo Clinic Administrative Fellowship, 2460
Mayo Clinic Admin Fellowship - Eau Claire, 2461
Middlesex Savings Charitable Foundation Capacity Building Grants, 2551

Human Services
3M Company Fndn Community Giving Grants, 6
3M Fndn Health & Human Services Grants, 7
1675 Foundation Grants, 11
Abbott Fund Global AIDS Care Grants, 202
Abell Fndn Health & Human Services Grants, 209
ACF Native American Social and Economic Development Strategies Grants, 231
Ackerman Foundation Grants, 236
Adams and Reese Corporate Giving Grants, 299
Adelaide Breed Bayrd Foundation Grants, 306
Adler-Clark Electric Community Commitment Foundation Grants, 307
Albert W. Rice Charitable Foundation Grants, 411
Alfred and Tillie Shemanski Testamentary Trust Grants, 428
Alfred Bersted Foundation Grants, 429
Alfred E. Chase Charitable Foundation Grants, 430
Allan C. and Lelia J. Garden Foundation Grants, 438
Alpha Natural Resources Corporate Giving, 450
Alvin and Fanny Blaustein Thalheimer Foundation Baltimore Communal Grants, 459
Amelia Sillman Rockwell and Carlos Perry Rockwell Charities Fund Grants, 501
American Electric Power Foundation Grants, 530
Anna Fitch Ardenghi Trust Grants, 592
Anne J. Caudal Foundation Grants, 595
Arkansas Community Foundation Grants, 660
Arthur F. & Arnold M. Frankel Fndn Grants, 672
Assisi Fndn of Memphis Capital Project Grants, 786
Assisi Foundation of Memphis General Grants, 787
Autodesk Community Relations Grants, 809
Avista Foundation Economic and Cultural Vitality Grants, 812
Batchelor Foundation Grants, 837
Belvedere Community Foundation Grants, 874
Blackford County Community Foundation Grants, 906

758 / Human Services

Blanche and Julian Robertson Family Foundation Grants, 911
Boeing Company Contributions Grants, 930
BP Foundation Grants, 943
Bradley C. Higgins Foundation Grants, 945
Brown Advisory Charitable Foundation Grants, 969
Business Bank of Nevada Community Grants, 981
Caesars Foundation Grants, 995
Carl R. Hendrickson Family Foundation Grants, 1028
Cass County Community Foundation Grants, 1041
CFFVR Project Grants, 1101
Charles H. Hall Foundation, 1120
Charles H. Pearson Foundation Grants, 1121
Charles M. & Mary D. Grant Fndn Grants, 1124
Charles Nelson Robinson Fund Grants, 1125
Chazen Foundation Grants, 1129
Chemtura Corporation Contributions Grants, 1130
Chilkat Valley Community Foundation Grants, 1148
Christine & Katharina Pauly Trust Grants, 1150
Clara Blackford Smith and W. Aubrey Smith Charitable Foundation Grants, 1165
Clayton Fund Grants, 1180
Clinton County Community Foundation Grants, 1184
CNCS Senior Companion Program Grants, 1191
CNO Financial Group Community Grants, 1201
Colonel Stanley R. McNeil Foundation Grants, 1211
Community Foundation for Greater Atlanta Clayton County Fund Grants, 1240
Community Foundation for Greater Atlanta Morgan County Fund Grants, 1244
Community Foundation for Greater Atlanta Newton County Fund Grants, 1245
Community Foundation for Southeast Michigan Grants, 1258
Community Foundation of Boone County Grants, 1266
Community Foundation of Grant County Grants, 1271
Community Fndn of Jackson County Grants, 1287
Community Foundation of Louisville Morris and Esther Lee Fund Grants, 1298
Community Foundation of Muncie and Delaware County Grants, 1300
Community Foundation of Muncie and Delaware County Maxon Grants, 1301
Community Foundation of Riverside & San Bernardino County Impact Grants, 1303
Community Fndn of Riverside & San Bernardino County Irene S. Rockwell Grants, 1304
Community Foundation of St. Joseph County Special Project Challenge Grants, 1310
Community Foundation Partnerships - Lawrence County Grants, 1317
Community Foundation Serving Riverside and San Bernardino Counties Impact Grants, 1319
Connelly Foundation Grants, 1326
ConocoPhillips Foundation Grants, 1327
Covenant Educational Foundation Grants, 1348
Crown Point Community Foundation Grants, 1361
CSRA Community Foundation Grants, 1366
CSX Corporate Contributions Grants, 1368
D.V. and Ida J. McEachern Trust Grants, 1406
Dana Brown Charitable Trust Grants, 1419
David M. and Marjorie D. Rosenberg Foundation Grants, 1432
David N. Lane Trust Grants for Aged and Indigent Women, 1433
Dayton Power and Light Foundation Grants, 1437
Deborah Munroe Noonan Memorial Grants, 1443
Decatur County Community Foundation Large Project Grants, 1444
DeKalb County Community Foundation Grants, 1447
Denver Foundation Community Grants, 1461
DHHS Emerging Leaders Program Internships, 1473
Dominion Foundation Grants, 1485
Dr. & Mrs. Paul Pierce Memorial Fndn Grants, 1505
Drs. Bruce and Lee Foundation Grants, 1513
DTE Energy Foundation Health and Human Services Grants, 1516
Earl and Maxine Claussen Trust Grants, 1541
Edward and Helen Bartlett Foundation Grants, 1554

Edward N. and Della L. Thome Memorial Foundation Grants, 1557
Elaine Feld Stern Charitable Trust Grants, 1565
Elizabeth Morse Genius Charitable Trust Grants, 1569
Ensworth Charitable Foundation Grants, 1595
Eugene G. and Margaret M. Blackford Memorial Fund Grants, 1610
Eugene Straus Charitable Trust, 1612
Federal Express Corporate Contributions, 1657
Ferree Foundation Grants, 1658
Fifth Third Foundation Grants, 1661
Finance Factors Foundation Grants, 1666
Flextronics Foundation Disaster Relief Grants, 1674
Florian O. Bartlett Trust Grants, 1677
Foster Foundation Grants, 1689
Four County Community Fndn General Grants, 1710
Four County Community Foundation Healthy Senior/Healthy Youth Fund Grants, 1711
France-Merrick Foundation Health and Human Services Grants, 1715
Frank B. Hazard General Charity Fund Grants, 1721
Franklin H. Wells and Ruth L. Wells Foundation Grants, 1725
Frank Loomis Palmer Fund Grants, 1726
Frank Reed and Margaret Jane Peters Memorial Fund I Grants, 1727
Frank Reed and Margaret Jane Peters Memorial Fund II Grants, 1728
Frank W. & Carl S. Adams Memorial Grants, 1731
Fred & Gretel Biel Charitable Trust Grants, 1734
Frederick McDonald Trust Grants, 1740
Frederick W. Marzahl Memorial Fund Grants, 1741
G.A. Ackermann Memorial Fund Grants, 1767
Gardner Foundation Grants, 1772
Gardner W. & Joan G. Heidrick, Jr. Fndn Grants, 1773
Garland D. Rhoads Foundation, 1774
Genuardi Family Foundation Grants, 1787
George A. and Grace L. Long Foundation Grants, 1788
George F. Baker Trust Grants, 1793
George H.C. Ensworth Memorial Fund Grants, 1797
George Kress Foundation Grants, 1799
George W. Wells Foundation Grants, 1804
Gibson County Community Foundation Women's Fund Grants, 1817
Gil and Dody Weaver Foundation Grants, 1819
GNOF Bayou Communities Grants, 1835
GNOF Freeman Challenge Grants, 1837
GNOF IMPACT Grants for Health and Human Services, 1839
GNOF Maison Hospitaliere Grants, 1843
GNOF Norco Community Grants, 1845
Grace Bersted Foundation Grants, 1859
Graham Foundation Grants, 1862
Grand Rapids Community Foundation Lowell Area Fund Grants, 1872
Great-West Life Grants, 1879
Green Foundation Human Services Grants, 1894
Green River Area Community Fndn Grants, 1895
Grover Hermann Foundation Grants, 1906
Grundy Foundation Grants, 1912
Guy I. Bromley Trust Grants, 1919
Hackett Foundation Grants, 1926
Hampton Roads Community Foundation Health and Human Service Grants, 1941
Hancock County Community Foundation - Field of Interest Grants, 1943
Harold and Arlene Schnitzer CARE Foundation Grants, 1950
Harold R. Bechtel Testamentary Charitable Trust Grants, 1954
Harry B. and Jane H. Brock Foundation Grants, 1960
Harry Edison Foundation, 1962
Harvey Randall Wickes Foundation Grants, 1972
HCA Foundation Grants, 1980
Helen Irwin Littauer Educational Trust Grants, 1994
Helen Pumphrey Denit Charitable Trust Grants, 1996
Hendricks County Community Fndn Grants, 2000
Henrietta Lange Burk Fund Grants, 2001
Herbert A. & Adrian W. Woods Fndn Grants, 2007
Hilda and Preston Davis Foundation Grants, 2028

SUBJECT INDEX

Hill Crest Foundation Grants, 2030
Horace Moses Charitable Foundation Grants, 2047
Hormel Foundation Grants, 2051
Human Source Foundation Grants, 2078
Huntington County Community Foundation Make a Difference Grants, 2083
Hutton Foundation Grants, 2087
Ida S. Barter Trust Grants, 2116
IDPH Emergency Medical Services Assistance Fund Grants, 2118
IIE Chevron International REACH Scholarships, 2128
IIE New Leaders Group Award for Mutual Understanding, 2141
Ike and Roz Friedman Foundation Grants, 2145
Intergrys Corporation Grants, 2160
Irving S. Gilmore Foundation Grants, 2163
J.H. Robbins Foundation Grants, 2171
J.N. and Macie Edens Foundation Grants, 2174
Jack H. & William M. Light Trust Grants, 2179
Jackson County Community Foundation Unrestricted Grants, 2180
Jacob G. Schmidlapp Trust Grants, 2183
James Ford Bell Foundation Grants, 2187
James J. & Angelia M. Harris Fndn Grants, 2192
James J. & Joan A. Gardner Family Fndn Grants, 2193
Jane's Trust Grants, 2206
Jasper Foundation Grants, 2212
Jayne and Leonard Abess Foundation Grants, 2214
Jessica Stevens Community Foundation Grants, 2224
Johnson Controls Foundation Health and Social Services Grants, 2255
John W. and Anna H. Hanes Foundation Grants, 2261
John W. Speas and Effie E. Speas Memorial Trust Grants, 2264
Josephine Goodyear Foundation Grants, 2272
Josephine Schell Russell Charitable Trust Grants, 2274
Julius N. Frankel Foundation Grants, 2282
Katharine Matthies Foundation Grants, 2291
Katrine Menzing Deakins Trust Grants, 2295
Kent D. Steadley and Mary L. Steadley Memorial Trust Grants, 2301
Kettering Family Foundation Grants, 2304
Kimball International-Habig Foundation Health and Human Services Grants, 2308
Kovler Family Foundation Grants, 2323
Kuntz Foundation Grants, 2326
Lake County Community Fund Grants, 2333
Lena Benas Memorial Fund Grants, 2353
Lewis H. Humphreys Charitable Trust Grants, 2361
Liberty Bank Foundation Grants, 2362
Libra Foundation Grants, 2363
Lil and Julie Rosenberg Foundation Grants, 2364
Lincoln Financial Foundation Grants, 2367
Linford and Mildred White Charitable Grants, 2368
Lisa and Douglas Goldman Fund Grants, 2369
Lisa Higgins-Hussman Foundation Grants, 2370
Louetta M. Cowden Foundation Grants, 2377
Louis and Elizabeth Nave Flarsheim Charitable Foundation Grants, 2379
Lucy Downing Nisbet Charitable Fund Grants, 2387
Lydia deForest Charitable Trust Grants, 2393
Lynn and Rovena Alexander Family Foundation Grants, 2399
M.D. Anderson Foundation Grants, 2403
Mabel A. Horne Trust Grants, 2406
Mabel F. Hoffman Charitable Trust Grants, 2407
Madison County Community Foundation - City of Anderson Quality of Life Grant, 2412
Mann T. Lowry Foundation Grants, 2419
Marathon Petroleum Corporation Grants, 2422
Marie H. Bechtel Charitable Remainder Uni-Trust Grants, 2432
Marion Gardner Jackson Charitable Trust Grants, 2434
Marion I. and Henry J. Knott Foundation Discretionary Grants, 2435
Marion I. and Henry J. Knott Foundation Standard Grants, 2436
Marjorie Moore Charitable Foundation Grants, 2438
Marshall County Community Foundation Grants, 2439
Mary Wilmer Covey Charitable Trust Grants, 2448

SUBJECT INDEX

Humanities /759

McCallum Family Foundation Grants, 2489
McCune Foundation Human Services Grants, 2493
McGraw-Hill Companies Community Grants, 2494
Memorial Foundation Grants, 2516
Mercedes-Benz USA Corporate Contributions, 2517
Merrick Foundation Grants, 2522
Meta and George Rosenberg Foundation Grants, 2525
Metzger-Price Fund Grants, 2530
Meyer Memorial Trust Grassroots Grants, 2536
Meyer Memorial Trust Responsive Grants, 2537
Middlesex Savings Charitable Foundation Basic Human Needs Grants, 2550
Miller Foundation Grants, 2556
MMS and Alliance Charitable Foundation Grants for Community Action and Care for the Medically Uninsured, 2587
Nell J. Redfield Foundation Grants, 2673
Norfolk Southern Foundation Grants, 2982
Northrop Grumman Corporation Grants, 2997
Notsew Orm Sands Foundation Grants, 3002
Oppenstein Brothers Foundation Grants, 3078
Orange County Community Foundation Grants, 3079
Oscar Rennebohm Foundation Grants, 3085
Pacific Life Foundation Grants, 3103
Palmer Foundation Grants, 3112
Perpetual Trust for Charitable Giving Grants, 3141
Philip L. Graham Fund Health and Human Services Grants, 3190
Piedmont Natural Gas Corporate and Charitable Contributions, 3220
PMI Foundation Grants, 3239
PNC Foundation Community Services Grants, 3240
Pohlad Family Foundation, 3241
Pokagon Fund Grants, 3242
PPCF Community Grants, 3256
PPG Industries Foundation Grants, 3257
Priddy Foundation Program Grants, 3266
Pulido Walker Foundation, 3282
Putnam County Community Foundation Grants, 3283
R.S. Gernon Trust Grants, 3293
RCF General Community Grants, 3312
Reinberger Foundation Grants, 3357
Robert R. Meyer Foundation Grants, 3387
Robert W. Woodruff Foundation Grants, 3389
Rohm and Haas Company Grants, 3403
Rollins-Luetkemeyer Foundation Grants, 3404
Sandy Hill Foundation Grants, 3477
San Juan Island Community Foundation Grants, 3480
Santa Fe Community Foundation Seasonal Grants-Fall Cycle, 3482
Sarkeys Foundation Grants, 3485
Sartain Lanier Family Foundation Grants, 3486
Schlessman Family Foundation Grants, 3493
Seward Community Foundation Grants, 3514
Seward Community Foundation Mini-Grants, 3515
Shell Oil Company Foundation Community Development Grants, 3544
Simpson Lumber Charitable Contributions, 3566
Sioux Falls Area Community Foundation Community Fund Grants (Unrestricted), 3568
Sioux Falls Area Community Foundation Field-of-Interest and Donor-Advised Grants, 3569
Sioux Falls Area Community Foundation Spot Grants (Unrestricted), 3570
Solo Cup Foundation Grants, 3620
Sony Corporation of America Grants, 3623
Sophia Romero Trust Grants, 3624
Stan and Sandy Checketts Foundation, 3644
Starke County Community Foundation Grants, 3646
Sterling and Shelli Gardner Foundation Grants, 3653
Stewart Huston Charitable Trust Grants, 3655
Sunoco Foundation Grants, 3679
Swindells Charitable Foundation, 3697
Taubman Endowment for the Arts Fndn Grants, 3702
Texas Instruments Corporation Health and Human Services Grants, 3709
Texas Instruments Foundation Community Services Grants, 3710
Textron Corporate Contributions Grants, 3712
Thelma Braun & Bocklett Family Fndn Grants, 3713

Thelma Doelger Charitable Trust Grants, 3714
Thomas J. Long Foundation Community Grants, 3722
Thomas Jefferson Rosenberg Foundation Grants, 3723
Thorman Boyle Foundation Grants, 3730
Tipton County Foundation Grants, 3734
Toyota Motor Engineering & Manufacturing North America Grants, 3742
Toyota Motor Manufacturing of Alabama Grants, 3743
Toyota Motor Manufacturing of Indiana Grants, 3744
Toyota Motor Manuf of Mississippi Grants, 3746
Toyota Motor Manufacturing of Texas Grants, 3747
Toyota Motor Manuf of West Virginia Grants, 3748
Union Bank, N.A. Corporate Sponsorships and Donations, 3787
Union Pacific Foundation Health and Human Services Grants, 3794
United Technologies Corporation Grants, 3803
Unity Foundation Of LaPorte County Grants, 3804
US Airways Community Foundation Grants, 3826
Vanderburgh Community Foundation Grants, 3852
Victor E. Speas Foundation Grants, 3868
Vigneron Memorial Fund Grants, 3873
Virginia W. Kettering Foundation Grants, 3878
W.H. & Mary Ellen Cobb Trust Grants, 3884
W.P. and Bulah Luse Foundation Grants, 3892
Walker Area Community Foundation Grants, 3895
Walter L. Gross III Family Foundation Grants, 3901
Warren County Community Foundation Grants, 3902
Warren County Community Fndn Mini-Grants, 3903
Washington County Community Fndn Grants, 3906
Weaver Popcorn Foundation Grants, 3910
Wege Foundation Grants, 3912
Wells County Foundation Grants, 3920
Western New York Foundation Grants, 3922
White County Community Foundation - Women Giving Together Grants, 3933
White County Community Foundation Grants, 3934
WHO Foundation Volunteer Service Grants, 3940
Willary Foundation Grants, 3944
William Ray & Ruth E. Collins Fndn Grants, 3956
Willis C. Helm Charitable Trust Grants, 3959
Windham Foundation Grants, 3961
Winston-Salem Foundation Competitive Grants, 3965
Yawkey Foundation Grants, 3980
ZYTL Foundation Grants, 3985

Human Subjects Policy
Alternatives Research and Development Foundation Alternatives in Education Grants, 455
Alternatives Research and Development Foundation Grants, 456
NIH Research On Ethical Issues In Human Subjects Research, 2897

Humanitarianism
Bill and Melinda Gates Foundation Emergency Response Grants, 892
Community Foundation of Riverside & San Bernardino County Impact Grants, 1303
USAID Higher Education Partnerships for Innovation and Impact (HEPII) Grants, 3815

Humanities
AAUW American Dissertation Fellowships, 195
AAUW American Postdoctoral Research Leave Fellowships, 196
AFG Industries Grants, 340
Ahmanson Foundation Grants, 355
Alexander von Humboldt Foundation Georg Forster Fellowships for Experienced Researchers, 423
amfAR Fellowships, 567
amfAR Global Initiatives Grants, 568
amfAR Mathilde Krim Fellowships in Basic Biomedical Research, 569
amfAR Public Policy Grants, 570
amfAR Research Grants, 571
Arkansas Community Foundation Grants, 660
Atherton Family Foundation Grants, 794
Athwin Foundation Grants, 795
Aurora Foundation Grants, 802

Avista Foundation Economic and Cultural Vitality Grants, 812
Avon Products Foundation Grants, 814
Banfi Vintners Foundation Grants, 824
Barra Foundation Community Fund Grants, 833
Baton Rouge Area Foundation Grants, 838
Benton County Foundation Grants, 880
Berrien Community Foundation Grants, 884
Blue Grass Community Fndn Harrison Grants, 919
Blue Grass Community Foundation Hudson-Ellis Fund Grants, 920
Boston Globe Foundation Grants, 937
Brookdale Foundation Leadership in Aging Fellowships, 965
Bush Fndn Health & Human Services Grants, 979
Caleb C. and Julia W. Dula Educational and Charitable Foundation Grants, 997
Callaway Foundation Grants, 1002
Canada-U.S. Fulbright New Century Scholars Program Grants, 1008
Canada-U.S. Fulbright Senior Specialists Grants, 1009
Centerville-Washington Foundation Grants, 1078
Chula Vista Charitable Foundation Grants, 1154
Clark County Community Foundation Grants, 1175
CNO Financial Group Community Grants, 1201
Collins Foundation Grants, 1210
Columbus Foundation Competitive Grants, 1217
Community Foundation for Greater Atlanta Clayton County Fund Grants, 1240
Community Foundation for Greater Atlanta Morgan County Fund Grants, 1244
Community Foundation for Greater Atlanta Newton County Fund Grants, 1245
Community Fndn of Central Illinois Grants, 1267
Community Fndn of East Central Illinois Grants, 1268
Community Foundation of Greater Birmingham Grants, 1272
Community Foundation of Greater Flint Grants, 1273
Community Foundation of Greater Fort Wayne - Lilly Endowment Scholarships, 1275
Community Foundation of Louisville Anna Marble Memorial Fund for Princeton Grants, 1290
Community Fndn of Randolph County Grants, 1302
Community Foundation of Riverside & San Bernardino County Impact Grants, 1303
Community Fndn of Riverside & San Bernardino County Irene S. Rockwell Grants, 1304
Community Foundation of St. Joseph County Lilly Endowment Community Scholarship, 1308
Connecticut Community Foundation Grants, 1324
Cooke Foundation Grants, 1344
Cresap Family Foundation Grants, 1359
DAAD Research Stays for University Academics and Scientists, 1408
Delaware Community Foundation Grants, 1449
Deutsche Banc Alex Brown and Sons Charitable Foundation Grants, 1471
Donna K. Yundt Memorial Scholarship, 1488
Dr. John Maniotes Scholarship, 1506
Dr. R.T. White Scholarship, 1510
Dubois County Community Foundation Grants, 1517
Earl and Maxine Claussen Trust Grants, 1541
Elkhart County Foundation Lilly Endowment Community Scholarships, 1574
El Paso Community Foundation Grants, 1579
El Pomar Foundation Anna Keesling Ackerman Fund Grants, 1581
F.M. Kirby Foundation Grants, 1624
Fairfield County Community Foundation Grants, 1625
Flinn Foundation Scholarships, 1675
Foellinger Foundation Grants, 1684
Ford Motor Company Fund Grants Program, 1687
Forrest C. Lattner Foundation Grants, 1688
Fred and Louise Latshaw Scholarship, 1735
Fulbright Alumni Initiatives Awards, 1749
Fulbright Distinguished Chairs Awards, 1750
Fulbright New Century Scholars Grants, 1753
Fulbright Specialists Program Grants, 1754
Fulbright Scholars in Europe and Eurasia, 1755

/ Humanities

Fulbright Traditional Scholar Program in Sub-Saharan Africa, 1756
Fulbright Traditional Scholar Program in the East Asia/Pacific Region, 1757
Fulbright Traditional Scholar Program in the Near East and North Africa Region, 1758
Fulbright Traditional Scholar Program in the South and Central Asia Region, 1759
Fulbright Traditional Scholar Program in the Western Hemisphere, 1760
Fulton County Community Foundation 4Community Higher Education Scholarship, 1762
GNOF Freeman Challenge Grants, 1837
GNOF Norco Community Grants, 1845
Graham Foundation Grants, 1862
Grand Rapids Community Foundation Ionia County Youth Fund Grants, 1871
Grand Rapids Community Foundation Sparta Youth Fund Grants, 1876
Greenwall Foundation Bioethics Grants, 1897
Hallmark Corporate Foundation Grants, 1936
Hammond Common Council Scholarships, 1938
Harold and Arlene Schnitzer CARE Foundation Grants, 1950
Honda of America Manufacturing Fndn Grants, 2045
IBEW Local Union #697 Memorial Scholarships, 2096
Idaho Power Company Corporate Contributions, 2115
IIE 911 Armed Forces Scholarships, 2122
IIE AmCham Charitable Foundation U.S. Studies Scholarship, 2125
IIE Chevron International REACH Scholarships, 2128
IIE Klein Family Scholarship, 2135
IIE Leonora Lindsley Memorial Fellowships, 2136
IIE Lingnan Foundation W.T. Chan Fellowship, 2137
IIE Lotus Scholarships, 2138
IIE Mattel Global Scholarship, 2139
IIE New Leaders Group Award for Mutual Understanding, 2141
IIE Rockefeller Foundation Bellagio Center Residencies, 2142
IIE Western Union Family Scholarships, 2143
J.B. Reynolds Foundation Grants, 2168
JELD-WEN Foundation Grants, 2218
Johnson County Community Foundation Grants, 2256
John V. and George Primich Family Scholarship, 2259
John W. Anderson Foundation Grants, 2262
Joseph Drown Foundation Grants, 2268
Joshua Benjamin Cohen Memorial Scholarship, 2276
Katharine Matthies Foundation Grants, 2291
Kettering Fund Grants, 2305
Kinsman Foundation Grants, 2309
Knox County Community Foundation Grants, 2313
LaGrange County Lilly Endowment Community Scholarship, 2331
Lake County Athletic Officials Association Scholarships, 2332
Lake County Lilly Endowment Community Scholarships, 2334
Land O'Lakes Foundation Mid-Atlantic Grants, 2341
LEGENDS Scholarship, 2351
Lester E. Yeager Charitable Trust B Grants, 2359
Libra Foundation Grants, 2363
Lowell Berry Foundation Grants, 2382
Marion I. and Henry J. Knott Foundation Discretionary Grants, 2435
Marion I. and Henry J. Knott Foundation Standard Grants, 2436
Maxon Charitable Foundation Grants, 2458
Miami County Community Foundation - Operation Round Up Grants, 2543
Mimi and Peter Haas Fund Grants, 2557
Moline Foundation Community Grants, 2590
Morris & Gwendolyn Cafritz Fndn Grants, 2594
Nancy J. Pinnick Memorial Scholarship, 2607
Nelda C. and H.J. Lutcher Stark Fndn Grants, 2672
Noble County Community Foundation - Kathleen June Earley Memorial Scholarship, 2973
Noble County Community Foundation Grants, 2975
Norwin S. & Elizabeth N. Bean Fndn Grants, 3001
Oceanside Charitable Foundation Grants, 3063

OSF Law and Health Initiative Grants, 3092
Owen County Community Foundation Grants, 3100
Paul Ogle Foundation Grants, 3124
Perry County Community Foundation Grants, 3142
Pike County Community Foundation Grants, 3222
Pittsburgh Foundation Community Fund Grants, 3230
Porter County Foundation Lilly Endowment Community Scholarships, 3245
Pulaski County Community Foundation Grants, 3279
Pulaski County Community Fndn Scholarships, 3281
Rajiv Gandhi Foundation Grants, 3302
Ripley County Community Foundation Grants, 3376
Ripley County Community Foundation Small Project Grants, 3377
Robert R. Meyer Foundation Grants, 3387
Rush County Community Foundation Grants, 3424
Ruth and Vernon Taylor Foundation Grants, 3426
Schrage Family Foundation Scholarships, 3494
Schramm Foundation Grants, 3495
Sengupta Family Scholarship, 3512
South Madison Community Foundation Grants, 3626
Spencer County Community Foundation Grants, 3632
SSHRC-NSERC College and Community Innovation Enhancement Grants, 3636
SSHRC Banting Postdoctoral Fellowships, 3637
Steele-Reese Foundation Grants, 3650
Steuben County Community Foundation Grants, 3654
Taylor S. Abernathy and Patti Harding Abernathy Charitable Trust Grants, 3704
Vermont Community Foundation Grants, 3865
W. James Spicer Scholarship, 3885
Walker Area Community Foundation Grants, 3895
Warrick County Community Foundation Grants, 3904
Webster Cornwell Memorial Scholarship, 3911
Western New York Foundation Grants, 3922
Whitley County Community Foundation - Lilly Endowment Scholarship, 3935
Whitley County Community Fndn Scholarships, 3937
William Ray & Ruth E. Collins Fndn Grants, 3956

Humanities Education
Chatlos Foundation Grants Program, 1128
Guy I. Bromley Trust Grants, 1919
Harry Frank Guggenheim Fnd Research Grants, 1964
Honda of America Manufacturing Fndn Grants, 2045
John W. Speas and Effie E. Speas Memorial Trust Grants, 2264
Lewis H. Humphreys Charitable Trust Grants, 2361
Louetta M. Cowden Foundation Grants, 2377
Louis and Elizabeth Nave Flarsheim Charitable Foundation Grants, 2379
Natalie W. Furniss Charitable Trust Grants, 2626
SSHRC Banting Postdoctoral Fellowships, 3637

Hunger
ACE Charitable Foundation Grants, 229
Albertson's Charitable Giving Grants, 410
American Electric Power Foundation Grants, 530
American Express Community Service Grants, 531
Bill and Melinda Gates Foundation Emergency Response Grants, 892
Bill and Melinda Gates Foundation Policy and Advocacy Grants, 893
BJ's Charitable Foundation Grants, 904
Boston Jewish Community Women's Fund Grants, 938
Campbell Soup Foundation Grants, 1007
ConAgra Foods Fndn Community Impact Grants, 1321
ConAgra Foods Nourish our Community Grants, 1322
Coors Brewing Corporate Contributions Grants, 1346
Denver Broncos Charities Fund Grants, 1460
Dominion Foundation Grants, 1485
Farmers Insurance Corporate Giving Grants, 1634
Fourjay Foundation Grants, 1712
Giant Food Charitable Grants, 1816
Green Foundation Human Services Grants, 1894
John Deere Foundation Grants, 2234
John Edward Fowler Memorial Fndn Grants, 2235
Johnson Controls Foundation Health and Social Services Grants, 2255
Kroger Company Donations, 2324

SUBJECT INDEX

Land O'Lakes Foundation Mid-Atlantic Grants, 2341
Meyer Foundation Healthy Communities Grants, 2533
NYCT Girls and Young Women Grants, 3053
Peacock Foundation Grants, 3133
PepsiCo Foundation Grants, 3137
Quantum Foundation Grants, 3290
R.C. Baker Foundation Grants, 3292
Robert R. McCormick Trib Community Grants, 3385
Rockefeller Foundation Grants, 3396
S. Mark Taper Foundation Grants, 3449
Santa Barbara Foundation Strategy Grants, 3481
Shaw's Supermarkets Donations, 3542
Southbury Community Trust Fund, 3625
United Methodist Committee on Relief Global Health Grants, 3799
Walmart Foundation Northwest Arkansas Giving, 3899
WHO Foundation Volunteer Service Grants, 3940

Huntington's Disease
Hereditary Disease Foundation John J. Wasmuth Postdoctoral Fellowships, 2009
Hereditary Disease Fnd Lieberman Award, 2010
Hereditary Disease Foundation Research Grants, 2011
Huntington's Disease Society of America Research Fellowships, 2079
Huntington's Disease Society of America Research Grants, 2080

Hurricanes
ADA Foundation Disaster Assistance Grants, 290

Hyperparathyroidism
NIH Receptors and Signaling in Bone in Health and Disease, 2891

Hypertension
NHLBI Research Grants on the Relationship Between Hypertension and Inflammation, 2725
Premera Blue Cross Grants, 3258

Illegal Aliens
San Diego Foundation for Change Grants, 3470

Immigrants
Alfred and Tillie Shemanski Testamentary Trust Grants, 428
Boston Foundation Grants, 936
Cralle Foundation Grants, 1356
Dallas Women's Foundation Grants, 1417
Deborah Munroe Noonan Memorial Grants, 1443
Grotto Foundation Project Grants, 1905
IIE Western Union Family Scholarships, 2143
Marie C. and Joseph C. Wilson Foundation Rochester Small Grants, 2431
Michael Reese Health Trust Responsive Grants, 2545
Minneapolis Foundation Community Grants, 2558
Norman Foundation Grants, 2983
NYC Managed Care Consumer Assistance Workshop Re-Design Grants, 3047
San Diego Foundation for Change Grants, 3470
Sophia Romero Trust Grants, 3624
Strowd Roses Grants, 3664
Theodore Edson Parker Foundation Grants, 3715

Immune System
ASM Abbott Laboratories Award in Clinical and Diagnostic Immunology, 717
Lymphatic Education and Research Network Postdoctoral Fellowships, 2395
NHLBI Developmental Origins of Altered Lung Physiology and Immune Function Grants, 2698
NIH Bone and the Hematopoietic and Immune Systems Research Grants, 2868

Immune System Disorders
Fairlawn Foundation Grants, 1626
HRSA Ryan White HIV AIDS Drug Assistance Grants, 2070
Lymphatic Education and Research Network Postdoctoral Fellowships, 2395

SUBJECT INDEX

NIH Bone and the Hematopoietic and Immune Systems Research Grants, 2868
Scleroderma Foundation Established Investigator Grants, 3497

Immunization Programs
Bernard and Audre Rapoport Foundation Health Grants, 882
CDC Epidemic Intell Service Training Grants, 1055
CDC Epidemiology Elective Rotation, 1056
CDC Public Health Associates, 1065
CDC Public Health Associates Hosts, 1066
Grifols Community Outreach Grants, 1904
Henrietta Lange Burk Fund Grants, 2001
NAPNAP Foundation Shourd Parks Immunization Project Small Grants, 2617
NAPNAP Foundation Wyeth Pediatric Immunization Grant, 2618
Premera Blue Cross Grants, 3258
Western Union Foundation Grants, 3923

Immunogenetics
National Psoriasis Foundation Research Grants, 2636

Immunology
AAAAI Allied Health Prof Recognition Award, 19
AAAAI ARTrust Mini Grants for Allied Health, 20
AAAAI ARTrust Grants for Allied Health Travel, 21
AAAAI ARTrust Grants for Clinical Research, 22
AAAAI ARTrust Grants in Faculty Development, 23
AAAAI Distinguished Scientist Award, 26
AAAAI Distinguished Service Award, 27
AAAAI Fellows-in-Training Abstract Award, 28
AAAAI Fellows-in-Training Grants, 29
AAAAI Fellows-in-Training Travel Grants, 30
AAAAI Mentorship Award, 31
AAAAI Outstanding Vol Clinical Faculty Award, 32
AAAAI RSLAAIS Leadership Award, 33
AAAAI Special Recognition Award, 34
AAO-HNSF Health Services Research Grants, 165
AAO-HNSF Maureen Hannley Research Training Awards, 166
AAO-HNSF Percy Memorial Research Award, 167
AAO-HNSF Rande H. Lazar Health Services Research Grant, 168
AAO-HNSF Resident Research Awards, 169
AAOA Foundation/AAO-HNSF Combined Research Grants, 172
AFAR Research Grants, 337
AIHS Alberta/Pfizer Translat Research Grants, 374
ASM Eli Lilly and Company Research Award, 726
ASM TREK Diagnostic ABMM/ABMLI Professional Recognition Award, 754
Baxter International Corporate Giving Grants, 841
Bristol-Myers Squibb Clinical Research Grants, 955
Dana Foundation Science and Health Grants, 1420
FDHN Fellowship to Faculty Transition Awards, 1644
Hawaii Community Foundation Health Education and Research Grants, 1976
MBL Frederik B. and Betsy G. Bang Summer Fellowships, 2473
National Blood Foundation Research Grants, 2628
NCI Technologies and Software to Support Integrative Cancer Biology Research (SBIR) Grants, 2656
NIH Nonhuman Primate Immune Tolerance Cooperative Study Group Grants, 2890
NIH Ruth L. Kirschstein National Research Service Awards for Individual Postdoc Fellowships, 2902
Searle Scholars Program Grants, 3503
Streilein Foundation for Ocular Immunology Visiting Professorships, 3660
US CRDF Leishmaniasis: Collaborative Research Opportunities in N Africa & Middle East, 3828

Immunopathology
AAAAI Distinguished Clinician Award, 24
National Psoriasis Foundation Research Grants, 2636

Independent Living Programs
Arizona Public Service Corporate Giving Program Grants, 658

BBVA Compass Foundation Charitable Grants, 848
Charles H. Farnsworth Trust Grants, 1119
Clark and Ruby Baker Foundation Grants, 1174
CNCS Senior Companion Program Grants, 1191
Coleman Foundation Developmental Disabilities Grants, 1207
Daniels Fund Grants-Aging, 1424
Dept of Ed Rehabilitation Training Grants, 1463
Florence Hunt Maxwell Foundation Grants, 1676
HAF Senior Opportunities Grants, 1932
John Edward Fowler Memorial Fndn Grants, 2235
Lydia deForest Charitable Trust Grants, 2393
May and Stanley Smith Charitable Trust Grants, 2459
McLean Contributionship Grants, 2497
Mericos Foundation Grants, 2518
MetroWest Health Foundation Grants--Healthy Aging, 2527
Peter and Elizabeth C. Tower Foundation Annual Intellectual Disabilities Grants, 3144
Piper Trust Older Adults Grants, 3229
Priddy Foundation Program Grants, 3266
Reinberger Foundation Grants, 3357
Robert & Clara Milton Senior Housing Grants, 3384
S. Mark Taper Foundation Grants, 3449
Saginaw Community Foundation Senior Citizen Enrichment Fund, 3450
Sara Elizabeth O'Brien Trust Grants, 3483
U.S. Department of Education Rehabilitation Research Training Centers (RRTCs), 3770
Union Bank, N.A. Foundation Grants, 3788
W.P. and Bulah Luse Foundation Grants, 3892

India
Elizabeth Glaser Int'l Leadership Awards, 1566
Fulbright Traditional Scholar Program in the Near East and North Africa Region, 1758
Rajiv Gandhi Foundation Grants, 3302
Robert R. Meyer Foundation Grants, 3387
Sir Dorabji Tata Trust Grants for NGOs or Voluntary Organizations, 3572
Sir Dorabji Tata Trust Individual Medical Grants, 3573

Indiana
Community Fndn of Bloomington & Monroe County - Precision Health Network Cycle Grants, 1264
Community Foundation of Grant County Grants, 1271
Community Foundation of Greater Fort Wayne - Lilly Endowment Scholarships, 1275
Dearborn Community Foundation City of Lawrenceburg Community Grants, 1441
Dearborn Community Foundation County Progress Grants, 1442
Decatur County Community Foundation Large Project Grants, 1444
Fayette County Foundation Grants, 1636
Fulton County Community Foundation Grants, 1763
Greene County Foundation Grants, 1892
Henry County Community Foundation Grants, 2005
Johnson County Community Foundation Grants, 2256
Marshall County Community Foundation Grants, 2439
Noble County Community Foundation - Kathleen June Earley Memorial Scholarship, 2973
Steuben County Community Foundation Grants, 3654
Whitley County Community Foundation - Lilly Endowment Scholarship, 3935

Indigenous Cultures
RCPSC Tom Dignan Indigenous Health Award, 3321
Ringing Rocks Foundation Discretionary Grants, 3375

Indonesia
IIE Freeman Foundation Indonesia Internships, 2131
Tifa Foundation Grants, 3733

Industrial Engineering
AOCS Industrial Oil Products Division Student Award, 616
Canada Graduate Scholarships (CGS) and NSERC Postgraduate Scholarships (PGS), 1010
Mayo Clinic Business Consulting Fellowship, 2462

Infectious Diseases/Agents /761

NSF-NIST Interaction in Chemistry, Materials Research, Molecular Biosciences, Bioengineering, and Chemical Engineering, 3004
U.S. Department of Education Erma Byrd Scholarships, 3768

Industrial Hygiene
U.S. Department of Education Erma Byrd Scholarships, 3768

Industry
ACF Native American Social and Economic Development Strategies Grants, 231
Collective Brands Foundation Grants, 1208
NSF Grant Opportunities for Academic Liaison with Industry (GOALI), 3016
RSC Thomas W. Eadie Medal, 3423

Infants
AAP Nutrition Award, 187
Bernard and Audre Rapoport Foundation Health Grants, 882
CIGNA Foundation Grants, 1160
Duke University Ped Anesthesiology Fellowship, 1527
Foundation for Seacoast Health Grants, 1698
George W. Wells Foundation Grants, 1804
Gerber Foundation Grants, 1807
Health Fndn of Greater Indianapolis Grants, 1984
Johnson & Johnson Community Health Grants, 2252
Kimball International-Habig Foundation Health and Human Services Grants, 2308
March of Dimes Program Grants, 2426
NIH Research on Sleep and Sleep Disorders, 2898
USAID Comprehensive District-Based Support for Better HIV/TB Patient Outcomes Grants, 3807

Infectious Diseases/Agents
Alliance Healthcare Foundation Grants, 444
AMHPS Dr. James A. Ferguson Emerging Infectious Diseases Fellowships, 573
APHL Emerging Infectious Diseases Fellowships, 636
ASM-PAHO Infectious Diseases Epidemiology and Surveillance Fellowships, 714
ASM/CDC Fellowships in Infectious Disease and Public Health Microbiology, 716
ASM ICAAC Young Investigator Awards, 732
ASM Merck Irving S. Sigal Memorial Awards, 739
ASM Siemens Healthcare Diagnostics Young Investigator Award, 751
ASPH/CDC Allan Rosenfield Global Health Fellowships, 780
ASPH/CDC Public Health Fellowships, 781
Bristol-Myers Squibb Clinical Outcomes and Research Grants, 955
CDC Cooperative Agreement for Continuing Enhanced National Surveillance for Prion Diseases in the United States, 1049
CDC David J. Sencer Museum Adult Group Tour, 1052
CDC David J. Sencer Museum Teacher Professional Development Workshops, 1053
CDC Epidemic Intell Service Training Grants, 1055
CDC Epidemiology Elective Rotation, 1056
CDC Evaluation of the Use of Rapid Testing For Influenza in Outpatient Medical Settings, 1057
CDC Experience Epidemiology Fellowships, 1059
CDC Foundation Atlanta International Health Fellowships, 1060
CDC Preventive Med Residency & Fellowship, 1064
CDC Public Health Associates, 1065
CDC Public Health Associates Hosts, 1066
CDC Public Health Prev Service Fellowships, 1069
CDC Public Health Prevention Service Fellowship Sponsorships, 1070
CSTE CDC/CSTE Applied Epidemiology Fellowships, 1367
Fndn for Appalachian Ohio Bachtel Scholarships, 1692
Grifols Community Outreach Grants, 1904
J.N. and Macie Edens Foundation Grants, 2174
March of Dimes Program Grants, 2426
National Blood Foundation Research Grants, 2628

Infectious Diseases/Agents

NIAID Independent Scientist Award, 2765
Pfizer Anti Infective Research EU Grants, 3164
Pfizer ASPIRE EU Antifungal Research Awards, 3165
Pfizer ASPIRE EU Emerging Mechanisms of Resistance Antibacterial Research Awards, 3167
Pfizer ASPIRE EU MRSA Nosocomial Pneumonia & MRSA Complicated Skin & Soft Tissue Infections Antibacterial Research Awards, 3168
Pfizer ASPIRE North America Broad Spectrum Antibiotics for the Treatment of Gram-Negative or Polymicrobial Infections Research Awards, 3169
Pfizer ASPIRE North America Narrow Spectrum Antibiotics for the Treatment of MRSA Research Awards, 3171
Pfizer Healthcare Charitable Contributions, 3180
Pfizer Medical Education Track One Grants, 3182
Pfizer Medical Education Track Two Grants, 3183
Robert W. Woodruff Foundation Grants, 3389

Inflammation

AIHS Alberta/Pfizer Translat Research Grants, 374
Broad Foundation IBD Research Grants, 964
NHLBI Research Grants on the Relationship Between Hypertension and Inflammation, 2725
Pfizer Inflammation Competitive Research Awards (UK), 3181
Pfizer Medical Education Track One Grants, 3182

Informatics

CDC Public Health Informatics Fellowships, 1068
MLA T. Mark Hodges Int'l Service Award, 2575
NHLBI Mentored Quantitative Research Career Development Awards, 2711
NLM Limited Competition for Continuation of Biomedical Informatics/Bioinformatics Resource Grants, 2947
PhRMA Fndn Info Sabbatical Fellowships, 3209

Information Dissemination

Carla J. Funk Governmental Relations Award, 1020
CDC Cooperative Agreement for Partnership to Enhance Public Health Informatics, 1050
Foundation for Seacoast Health Grants, 1698
Healthcare Fndn for Orange County Grants, 1981
MLA Donald A.B. Lindberg Fellowship, 2562
MLA T. Mark Hodges Int'l Service Award, 2575
MLA Thomas Reuters / Frank Bradway Rogers Information Advancement Award, 2576
NCI Centers of Excellence in Cancer Communications Research, 2649
NLM Informatics Conference Grants, 2945
NLM Research Project Grants, 2951
Pfizer Medical Education Track Two Grants, 3183
SAMHSA Conference Grants, 3462
Sigma Theta Tau International / Council for the Advancement of Nursing Science Grants, 3557

Information Science Education

Hospital Libraries Section / MLA Professional Development Grants, 2052
Majors MLA Chapter Project of the Year, 2418
MLA Continuing Education Grants, 2559
MLA David A. Kronick Traveling Fellowship, 2561
MLA Donald A.B. Lindberg Fellowship, 2562
MLA Lucretia W. McClure Excellence in Education Award, 2570
MLA Section Project of the Year Award, 2574
MLA T. Mark Hodges Int'l Service Award, 2575
U.S. Department of Education Erma Byrd Scholarships, 3768

Information Science/Systems

Canada Graduate Scholarships (CGS) and NSERC Postgraduate Scholarships (PGS), 1010
CDC Public Health Informatics Fellowships, 1068
DaimlerChrysler Corporation Fund Grants, 1412
EBSCO / MLA Annual Meeting Grants, 1545
Hospital Libraries Section / MLA Professional Development Grants, 2052
Majors MLA Chapter Project of the Year, 2418
Mayo Clinic Administrative Fellowship, 2460
McLean Contributionship Grants, 2497
Medical Informatics Section/MLA Career Development Grant, 2508
MLA Continuing Education Grants, 2559
MLA Cunningham Memorial International Fellowship, 2560
MLA David A. Kronick Traveling Fellowship, 2561
MLA Donald A.B. Lindberg Fellowship, 2562
MLA Estelle Brodman Academic Medical Librarian of the Year Award, 2563
MLA Graduate Scholarship, 2564
MLA Grad Scholarship for Minority Students, 2565
MLA Ida and George Eliot Prize, 2566
MLA Lois Ann Colaianni Award for Excellence and Achievement in Hospital Librarianship, 2568
MLA Louise Darling Medal for Distinguished Achievement in Collection Development in the Health Sciences, 2569
MLA Research, Development, and Demonstration Project Grant, 2572
MLA Section Project of the Year Award, 2574
MLA T. Mark Hodges Int'l Service Award, 2575
MLA Thomas Reuters / Frank Bradway Rogers Information Advancement Award, 2576
NLM Internet Connection for Medical Institutions Grants, 2946
NLM Research Project Grants, 2951
Thomson Reuters / MLA Doctoral Fellowships, 3728
U.S. Department of Education Erma Byrd Scholarships, 3768
Virginia L. and William K. Beatty MLA Volunteer Service Award, 3877

Information Technology

EBSCO / MLA Annual Meeting Grants, 1545
Foundation for the Mid South Health and Wellness Grants, 1700
Hospital Libraries Section / MLA Professional Development Grants, 2052
IDPH Hosptial Capital Investment Grants, 2119
Majors MLA Chapter Project of the Year, 2418
Medical Informatics Section/MLA Career Development Grant, 2508
MLA Continuing Education Grants, 2559
MLA Donald A.B. Lindberg Fellowship, 2562
MLA Estelle Brodman Academic Medical Librarian of the Year Award, 2563
MLA Janet Doe Lectureship Award, 2567
MLA Lois Ann Colaianni Award for Excellence and Achievement in Hospital Librarianship, 2568
MLA Louise Darling Medal for Distinguished Achievement in Collection Development in the Health Sciences, 2569
MLA Research, Development, and Demonstration Project Grant, 2572
MLA Section Project of the Year Award, 2574
MLA T. Mark Hodges Int'l Service Award, 2575
NLM Express Research Grants in Biomedical Informatics, 2943
Peter and Elizabeth C. Tower Foundation Phase II Technology Initiative Grants, 3149
Peter and Elizabeth C. Tower Foundation Phase I Technology Initiative Grants, 3150
USDD Clinical Functional Assessment Investigator-Initiated Grants, 3840
USDD HIV/AIDS Prevention Program Information Systems Development Grants, 3842
Virginia L. and William K. Beatty MLA Volunteer Service Award, 3877

Information Theory

Hospital Libraries Section / MLA Professional Development Grants, 2052
MLA Continuing Education Grants, 2559
MLA David A. Kronick Traveling Fellowship, 2561
MLA Donald A.B. Lindberg Fellowship, 2562
NCI Centers of Excellence in Cancer Communications Research, 2649

Infrastructure

AES Research Infrastructure Awards, 321
Jack H. & William M. Light Trust Grants, 2179
San Juan Island Community Foundation Grants, 3480

Injury

ASPH/CDC Public Health Fellowships, 781
CDC Epidemic Intell Service Training Grants, 1055
CDC Epidemiology Elective Rotation, 1056
CDC Experience Epidemiology Fellowships, 1059
CDC Public Health Associates, 1065
CDC Public Health Associates Hosts, 1066
Community Foundation of Louisville Lee Look Fund for Spinal Injury Grants, 1297
CSTE CDC/CSTE Applied Epidemiology Fellowships, 1367
DOL Occupational Safety and Health--Susan Harwood Training Grants, 1484
Dorothea Haus Ross Foundation Grants, 1498
George E. Hatcher, Jr. and Ann Williams Hatcher Foundation Grants, 1792
Quantum Foundation Grants, 3290
Thomas C. Burke Foundation Grants, 3720
USDD Workplace Violence in Military Grants, 3847

Injury, Head

AHNS/AAO-HNSF Surgeon Scientist Combined Award, 357
AHNS/AAO-HNSF Young Investigator Combined Award, 358
AHNS Alando J. Ballantyne Resident Research Pilot Grant, 359
AHNS Pilot Grant, 360
NFL Charities Medical Grants, 2685
USDD Workplace Violence in Military Grants, 3847

Injury, Spinal Cord

Christopher & Dana Reeve Foundation Quality of Life Grants, 1152
Community Foundation of Louisville Lee Look Fund for Spinal Injury Grants, 1297
Gruber Foundation Neuroscience Prize, 1909
NFL Charities Medical Grants, 2685
PVA Education Foundation Grants, 3284
PVA Research Foundation Grants, 3285

Inner Cities

Adolph Coors Foundation Grants, 311
Arie and Ida Crown Memorial Grants, 654
Marie C. and Joseph C. Wilson Foundation Rochester Small Grants, 2431
Mercedes-Benz USA Corporate Contributions, 2517
Nina Mason Pulliam Charitable Trust Grants, 2921
Saint Luke's Foundation Grants, 3454

Instruction/Curriculum Development

AAAAI Outstanding Vol Clinical Faculty Award, 32
ASHG Public Health Genetics Fellowship, 712
MLA Lucretia W. McClure Excellence in Education Award, 2570

Instructional Materials and Practices

Blue Cross Blue Shield of Minnesota Fndn Healthy Equity: Public Libraries for Health Grants, 918
SfN Science Educator Award, 3535
Tri-State Community Twenty-first Century Endowment Fund Grants, 3756

Instrumentation, Medical

Covidien Partnership for Neighborhood Wellness, 1353
IDPH Hosptial Capital Investment Grants, 2119
Maddie's Fund Medical Equipment Grants, 2411

Instrumentation, Scientific

NSF Instrument Development for Bio Research, 3017
NSF Major Research Instrumentation Program (MRI) Grants, 3018

Integrated Science

Bravewell Leadership Award, 946

SUBJECT INDEX

Intellectual Freedom
AAAS Award for Scie Freedom & Responsibility, 36
Playboy Foundation Grants, 3237

Intercultural Studies
BMW of North America Charitable Contributions, 926

Interior Design
Phi Upsilon Omicron Frances Morton Holbrook Alumni Award, 3193
Phi Upsilon Omicron Geraldine Clewell Senior Awards, 3194
Phi Upsilon Lillian P. Schoephoerster Award, 3196
Phi Upsilon Orinne Johnson Writing Award, 3198
Phi Upsilon Omicron Undergraduate Karen P. Goebel Conclave Award, 3200

International Affairs
Bill and Melinda Gates Foundation Policy and Advocacy Grants, 893

International Agriculture
Bill and Melinda Gates Foundation Policy and Advocacy Grants, 893
PepsiCo Foundation Grants, 3137

International Economics
Bill and Melinda Gates Foundation Policy and Advocacy Grants, 893
BP Foundation Grants, 943
IIE Hewlett Fnd/IIE Dissertation Fellowship, 2132
IIE Nancy Petry Scholarship, 2140

International Education/Training
AAAAI Fellows-in-Training Grants, 29
Abundance Foundation International Grants, 218
Acumen East Africa Fellowship, 284
Acumen Global Fellowships, 285
ADA Foundation Thomas J. Zwemer Award, 297
AES Research Infrastructure Awards, 321
Alexander von Humboldt Foundation Research Fellowships for Postdoctoral Researchers, 425
ASM-PAHO Infectious Diseases Epidemiology and Surveillance Fellowships, 714
ASM-UNESCO Leadership Grant for International Educators, 715
ASM Int'l Fellowship for Latin America and the Caribbean, 734
ASM Int'l Fellowships for Asia and Africa, 735
ASM Int'l Professorship for Latin America, 736
ASM Int'l Professorships for Asia and Africa, 737
ASM Millis-Colwell Postgraduate Travel Grant, 741
Australian Academy of Science Grants, 807
Bill and Melinda Gates Foundation Policy and Advocacy Grants, 893
Coca-Cola Foundation Grants, 1203
DAAD Research Stays for University Academics and Scientists, 1408
Elizabeth Glaser Int'l Leadership Awards, 1566
H.B. Fuller Foundation Grants, 1922
HHMI-NIH Cloister Research Scholars, 2015
HHMI Physician-Scientist Early Career Award, 2023
IBRO Asia APRC Lecturer Exchange Grants, 2102
IBRO Latin America Regional Funding for Neuroscience Schools, 2105
IBRO Latin America Regional Funding for Short Research Stays, 2108
IIE Chevron International REACH Scholarships, 2128
IIE Whitaker Int'l Fellowships & Scholarships, 2144
Medtronic Foundation Strengthening Health Systems Grants, 2515
MMS and Alliance Charitable Foundation International Health Studies Grants, 2588
PAHO-ASM Int'l Professorship Latin America, 3107
PC Fred H. Bixby Fellowships, 3126
RCPSC International Residency Educator of the Year Award, 3325
RCPSC Int'l Resident Leadership Award, 3326
RCPSC McLaughlin-Gallie Visiting Professor, 3333

Susan G. Komen Breast Cancer Foundation Challenge Grants: Career Catalyst Research, 3691
U.S. Department of Education United States-Russia Program: Improving Research and Educational Activities in Higher Education, 3773
Women of the ELCA Opportunity Scholarships for Lutheran Laywomen, 3970

International Exchange Programs
ACSM Oded Bar-Or Int'l Scholar Awards, 272
Albuquerque Community Foundation Grants, 412
American College of Surgeons International Guest Scholarships, 525
ASM Millis-Colwell Postgraduate Travel Grant, 741
Coca-Cola Foundation Grants, 1203
IBRO-PERC InEurope Short Stay Grants, 2097
IBRO Asia Reg APRC Exchange Fellowships, 2101
SfN Japan Neuroscience Society Meeting Travel Awards, 3525
U.S. Department of Education United States-Russia Program: Improving Research and Educational Activities in Higher Education, 3773

International Justice
Abundance Foundation International Grants, 218
Bill and Melinda Gates Foundation Policy and Advocacy Grants, 893

International Organizations
Bill and Melinda Gates Foundation Policy and Advocacy Grants, 893
Harry Kramer Memorial Fund Grants, 1965
Medtronic Foundation CommunityLink Health Grants, 2510
Raskob Foundation for Catholic Activities Grants, 3306

International Planning/Policy
Bill and Melinda Gates Foundation Policy and Advocacy Grants, 893
John Merck Fund Grants, 2245
Lynde & Harry Bradley Foundation Fellowships, 2396
Lynde and Harry Bradley Foundation Grants, 2397
Lynde and Harry Bradley Foundation Prizes: Bradley Prizes, 2398

International Programs
Abundance Foundation International Grants, 218
Acumen Global Fellowships, 285
AFAR Paul Beeson Career Development Awards in Aging Research for the Island of Ireland, 336
Alcoa Foundation Grants, 414
Alexander von Humboldt Foundation Georg Forster Fellowships for Experienced Researchers, 423
Alexander von Humboldt Foundation Georg Forster Fellowships for Postdoctoral Researchers, 424
American Jewish World Service Grants, 533
Archer Daniels Midland Foundation Grants, 651
ASCO Long-Term International Fellowships, 681
ASM GlaxoSmithKline International Member of the Year Award, 731
ASPH/CDC Allan Rosenfield Global Health Fellowships, 780
Atran Foundation Grants, 797
Banfi Vintners Foundation Grants, 824
Baxter International Foundation Grants, 843
Beatrice Laing Trust Grants, 864
Bechtel Group Foundation Building Positive Community Relationships Grants, 866
Becton Dickinson and Company Grants, 869
Besser Foundation Grants, 887
Bill and Melinda Gates Foundation Emergency Response Grants, 892
Bill and Melinda Gates Foundation Policy and Advocacy Grants, 893
Bill and Melinda Gates Foundation Water, Sanitation and Hygiene Grants, 894
Boston Jewish Community Women's Fund Grants, 938
Bristol-Myers Squibb Foundation Health Disparities Grants, 958
Broad Foundation IBD Research Grants, 964

Burden Trust Grants, 976
Caddock Foundation Grants, 994
Cargill Citizenship Fund Corporate Giving, 1018
Carylon Foundation Grants, 1040
CDC-Hubert Global Health Fellowship, 1047
Charles Delmar Foundation Grants, 1115
Chase Paymentech Corporate Giving Grants, 1127
Chatlos Foundation Grants Program, 1128
CIGNA Foundation Grants, 1160
CNA Foundation Grants, 1189
Commonwealth Fund Australian-American Health Policy Fellowships, 1228
ConocoPhillips Foundation Grants, 1327
Conquer Cancer Foundation of ASCO International Development and Education Awards, 1333
Cystic Fibrosis Canada Visiting Allied Health Professional Awards, 1394
Cystic Fibrosis Canada Visiting Clinician Awards, 1395
Cystic Fibrosis Trust Research Grants, 1404
Danellie Foundation Grants, 1421
Duke University Clinical Cardiac Electrophysiology Fellowships, 1522
Duke Univ Interventional Cardio Fellowships, 1524
Elizabeth Glaser Scientist Award, 1567
H.J. Heinz Company Foundation Grants, 1923
Harold Simmons Foundation Grants, 1955
Helen Bader Foundation Grants, 1992
Hershey Company Grants, 2013
James H. Cummings Foundation Grants, 2190
John Deere Foundation Grants, 2234
Johnson & Johnson/SAH Arts & Healing Grants, 2254
Landon Foundation-AACR Innovator Award for International Collaboration, 2343
Lawrence Foundation Grants, 2347
Ludwick Family Foundation Grants, 2389
Lustgarten Foundation for Pancreatic Cancer Research Grants, 2392
Mead Johnson Nutritionals Charitable Giving, 2502
Mead Johnson Nutritionals Med Educ Grants, 2504
MGFA Post-Doctoral Research Fellowships, 2539
MGFA Student Fellowships, 2540
Nuffield Foundation Africa Program Grants, 3026
Nuffield Foundation Law and Society Grants, 3028
Nuffield Foundation Small Grants, 3031
OSF Mental Health Initiative Grants, 3093
Patrick and Anna M. Cudahy Fund Grants, 3120
Paul Balint Charitable Trust Grants, 3122
PDF International Research Grants, 3128
Pfizer Healthcare Charitable Contributions, 3180
Potts Memorial Foundation Grants, 3254
Procter and Gamble Fund Grants, 3274
RCPSC McLaughlin-Gallie Visiting Professor, 3333
Rockefeller Archive Center Research Grants, 3393
Rockefeller Foundation Grants, 3396
Rockwell Collins Charitable Corporation Grants, 3397
Rosalinde and Arthur Gilbert Foundation/AFAR New Investigator Awards in Alzheimer's Disease, 3406
Savoy Foundation Post-Doctoral and Clinical Research Fellowships, 3488
Savoy Foundation Research Grants, 3489
Savoy Foundation Studentships, 3490
SfN Chapter Grants, 3520
SfN Peter and Patricia Gruber International Research Award, 3533
Sigrid Juselius Foundation Grants, 3561
SNM/ Covidien Seed Grant in Molecular Imaging/ Nuclear Medicine Research, 3582
SNM Postdoc Molecular Imaging Scholar Grants, 3587
Trull Foundation Grants, 3761
United Technologies Corporation Grants, 3803
W.K. Kellogg Foundation Healthy Kids Grants, 3886

International Relations
AAAS Science and Technology Policy Fellowships: Global Health and Development, 41
American College of Surgeons International Guest Scholarships, 525
Archer Daniels Midland Foundation Grants, 651
Arie and Ida Crown Memorial Grants, 654
Coca-Cola Foundation Grants, 1203

764 / International Relations

Donald and Sylvia Robinson Family Foundation Grants, 1486
Harry Frank Guggenheim Foundation Dissertation Fellowships, 1963
Harry Kramer Memorial Fund Grants, 1965
Rajiv Gandhi Foundation Grants, 3302
Stella and Charles Guttman Foundation Grants, 3651
U.S. Department of Education United States-Russia Program: Improving Research and Educational Activities in Higher Education, 3773
United States Institute of Peace - Jennings Randolph Senior Fellowships, 3802

International Students
Acumen East Africa Fellowship, 284
Grass Foundation Marine Biological Laboratory Fellowships, 1878
IBRO Latin America Regional Funding for Neuroscience Schools, 2105
IBRO Regional Grants for Int'l Fellowships to U.S. Laboratory Summer Neuroscience Courses, 2110
IIE Adell and Hancock Scholarships, 2123
IIE Chevron International REACH Scholarships, 2128
IIE Eurobank EFG Scholarships, 2130
IIE Freeman Foundation Indonesia Internships, 2131
IIE Klein Family Scholarship, 2135
MBL Gruss Lipper Family Foundation Summer Fellowships, 2475
MBL Scholarships and Awards, 2486
MBL William Townsend Porter Summer Fellowships for Minority Investigators, 2488
SfN Japan Neuroscience Society Meeting Travel Awards, 3525

International Studies
AAP International Travel Grants, 184
Acumen East Africa Fellowship, 284
CDC-Hubert Global Health Fellowship, 1047
Coca-Cola Foundation Grants, 1203
DAAD Research Stays for University Academics and Scientists, 1408
IIE Adell and Hancock Scholarships, 2123
SRC Medical Research Grants, 3635

International Trade and Finance
Archer Daniels Midland Foundation Grants, 651

International and Comparative Law
Bill and Melinda Gates Foundation Policy and Advocacy Grants, 893

Internet
EDS Foundation Grants, 1552
MLA Rittenhouse Award, 2573
NLM Internet Connection for Medical Institutions Grants, 2946
RSC Thomas W. Eadie Medal, 3423
Verizon Foundation Maryland Grants, 3859
Verizon Foundation New York Grants, 3860
Verizon Foundation Northeast Region Grants, 3861
Verizon Foundation Pennsylvania Grants, 3862
Verizon Foundation Vermont Grants, 3863
Verizon Foundation Virginia Grants, 3864

Internship Programs
AFAR Medical Student Training in Aging Research Program, 335
AIHS/Mitacs Health Pilot Partnership Interns, 373
AMA-MSS Government Relations Internships, 476
ASPH/CDC Public Health Internships, 782
CDC Collegiate Leaders in Environmental Health Internships, 1048
CDC Epidemic Intell Service Training Grants, 1055
CDC Summer Graduate Environmental Health Internships, 1072
CDC Summer Program In Environmental Health Internships, 1073
DHHS Emerging Leaders Program Internships, 1473
George A. and Grace L. Long Foundation Grants, 1788
Independence Blue Cross Nursing Internships, 2152

Morehouse PHSI Project Imhotep Internships, 2592
NHLBI Summer Internship Program in Biomedical Research, 2735
Strowd Roses Grants, 3664
Susan Mott Webb Charitable Trust Grants, 3694
William Blair and Company Foundation Grants, 3948

Intervention Programs
Bodman Foundation Grants, 929
Cambridge Community Foundation Grants, 1004
CDC Public Health Prev Service Fellowships, 1069
CDC Public Health Prevention Service Fellowship Sponsorships, 1070
CDC Steven M. Teutsch Prevention Effectiveness Fellowships, 1071
CSTE CDC/CSTE Applied Epidemiology Fellowships, 1367
Dept of Ed Special Education--Personnel Development to Improve Services and Results for Children with Disabilities, 1465
DOJ Gang-Free Schools and Communities Intervention Grants, 1481
Hasbro Children's Fund Grants, 1973
NIH Biobehavioral Research for Effective Sleep, 2865
Peter and Elizabeth C. Tower Foundation Annual Mental Health Grants, 3145
Priddy Foundation Program Grants, 3266
SAMHSA Conference Grants, 3462
Union Bank, N.A. Foundation Grants, 3788
Visiting Nurse Foundation Grants, 3879
William T. Grant Foundation Research Grants, 3958

Invention and Innovation
AIHS Alberta/Pfizer Translat Research Grants, 374
AIHS Collaborative Research and Innovation Grants - Collaborative Project, 377
Caplow Applied Science Carcinogen Prize, 1015
Conquer Cancer Foundation of ASCO International Innovation Grant, 1334
GNOF New Orleans Works Grants, 1844
NHLBI Bioengineering Approaches to Energy Balance and Obesity Grants for SBIR, 2690
SfN Nemko Prize in Molecular Neuroscience, 3529
USAID Higher Education Partnerships for Innovation and Impact (HEPII) Grants, 3815

Investments and Securities
Bill and Melinda Gates Foundation Policy and Advocacy Grants, 893

Isotope/Radiation Technology
DSO Radiation Biodosimetry (RaBiD) Grants, 1515

Israel
Abramson Family Foundation Grants, 217
Atran Foundation Grants, 797
Boston Jewish Community Women's Fund Grants, 938
Chazen Foundation Grants, 1129
Donald and Sylvia Robinson Family Foundation Grants, 1486
Fishman Family Foundation Grants, 1672
Harry Kramer Memorial Fund Grants, 1965
Helen Bader Foundation Grants, 1992
Stella and Charles Guttman Foundation Grants, 3651

Japan
Fulbright International Education Administrators (IEA) Seminar Program Grants, 1752

Japanese Americans
California Endowment Innovative Ideas Challenge Grants, 1000

Japanese Art
Fulbright International Education Administrators (IEA) Seminar Program Grants, 1752

Jewish Culture
Harry Edison Foundation, 1962
Meyer and Stephanie Eglin Foundation Grants, 2531

Jewish Services
100 Mile Man Foundation Grants, 10
A/H Foundation Grants, 18
Aaron & Cecile Goldman Family Fndn Grants, 191
Aaron & Freda Glickman Foundation Grants, 192
Aaron Foundation Grants, 193
Abramson Family Foundation Grants, 217
Adolph Coors Foundation Grants, 311
Albuquerque Community Foundation Grants, 412
Alexis Gregory Foundation Grants, 426
Alfred and Tillie Shemanski Testamentary Trust Grants, 428
Alvin and Fanny Blaustein Thalheimer Foundation Baltimore Communal Grants, 459
Ansell, Zaro, Grimm & Aaron Foundation Grants, 603
Archer Daniels Midland Foundation Grants, 651
Arie and Ida Crown Memorial Grants, 654
Arthur and Rochelle Belfer Foundation Grants, 670
Atran Foundation Grants, 797
Audrey & Sydney Irmas Foundation Grants, 799
Barr Fund Grants, 835
Bender Foundation Grants, 876
Bildner Family Foundation Grants, 891
Blanche and Irving Laurie Foundation Grants, 910
Blumenthal Foundation Grants, 925
Boettcher Foundation Grants, 931
Boston Jewish Community Women's Fund Grants, 938
Caddock Foundation Grants, 994
Carl C. Icahn Foundation Grants, 1023
Charles H. Revson Foundation Grants, 1122
Chazen Foundation Grants, 1129
CRH Foundation Grants, 1360
David Geffen Foundation Grants, 1431
Donald and Sylvia Robinson Foundation Grants, 1486
Fishman Family Foundation Grants, 1672
Giant Eagle Foundation Grants, 1815
Greenspun Family Foundation Grants, 1896
H. Schaffer Foundation Grants, 1925
Harold and Arlene Schnitzer CARE Foundation Grants, 1950
Harry Edison Foundation, 1962
Harry Kramer Memorial Fund Grants, 1965
Healthcare Foundation of New Jersey Grants, 1982
Howard and Bush Foundation Grants, 2054
Irvin Stern Foundation Grants, 2164
Jewish Fund Grants, 2227
John J. Leidy Foundation Grants, 2241
Joseph Alexander Foundation Grants, 2265
Kovler Family Foundation Grants, 2323
Lisa and Douglas Goldman Fund Grants, 2369
Manuel D. & Rhoda Mayerson Fndn Grants, 2420
Meyer and Stephanie Eglin Foundation Grants, 2531
Michael Reese Health Trust Responsive Grants, 2545
Mt. Sinai Health Care Foundation Health of the Jewish Community Grants, 2599
New York University Steinhardt School of Education Fellowships, 2684
Oppenstein Brothers Foundation Grants, 3078
Paul Balint Charitable Trust Grants, 3122
Pollock Foundation Grants, 3244
Tension Envelope Foundation Grants, 3707

Jewish Studies
Aaron & Cecile Goldman Family Fndn Grants, 191
Aaron & Freda Glickman Foundation Grants, 192
Alexis Gregory Foundation Grants, 426
Ansell, Zaro, Grimm & Aaron Foundation Grants, 603
Arthur and Rochelle Belfer Foundation Grants, 670
Atran Foundation Grants, 797
Barr Fund Grants, 835
Blanche and Irving Laurie Foundation Grants, 910
Blumenthal Foundation Grants, 925
Fishman Family Foundation Grants, 1672
Harry Edison Foundation, 1962
Helen Bader Foundation Grants, 1992
Howard and Bush Foundation Grants, 2054
Irvin Stern Foundation Grants, 2164
M.B. and Edna Zale Foundation Grants, 2401
Nathan Cummings Foundation Grants, 2627
Oppenstein Brothers Foundation Grants, 3078

SUBJECT INDEX

Job Training Programs
3M Company Fndn Community Giving Grants, 6
ACF Native American Social and Economic Development Strategies Grants, 231
Achelis Foundation Grants, 234
Adaptec Foundation Grants, 304
Alfred E. Chase Charitable Foundation Grants, 430
AMD Corporate Contributions Grants, 500
Anschutz Family Foundation Grants, 602
Bayer Foundation Grants, 846
Bodman Foundation Grants, 929
Boeing Company Contributions Grants, 930
BP Foundation Grants, 943
CFFVR Jewelers Mutual Charitable Giving, 1099
CFFVR Schmidt Family G4 Grants, 1103
Citizens Bank Mid-Atlantic Charitable Foundation Grants, 1162
Coleman Foundation Developmental Disabilities Grants, 1207
Coors Brewing Corporate Contributions Grants, 1346
DaimlerChrysler Corporation Fund Grants, 1412
Dallas Women's Foundation Grants, 1417
DeKalb County Community Foundation - Garrett Hospital Aid Foundation Grants, 1446
DHHS Emerging Leaders Program Internships, 1473
eBay Foundation Community Grants, 1543
Fairfield County Community Foundation Grants, 1625
Ford Family Foundation Grants - Access to Health and Dental Services, 1686
Frank B. Hazard General Charity Fund Grants, 1721
General Mills Foundation Grants, 1782
Global Fund for Children Grants, 1832
GNOF New Orleans Works Grants, 1844
Grand Rapids Area Community Foundation Wyoming Youth Fund Grants, 1868
Grifols Community Outreach Grants, 1904
Hampton Roads Community Foundation Health and Human Service Grants, 1941
Harold Brooks Foundation Grants, 1952
Heineman Foundation for Research, Education, Charitable and Scientific Purposes, 1990
Helen Bader Foundation Grants, 1992
Highmark Corporate Giving Grants, 2025
Houston Endowment Grants, 2053
Janus Foundation Grants, 2211
John Edward Fowler Memorial Fndn Grants, 2235
John Merck Fund Grants, 2245
Johnson & Johnson Corporate Contributions, 2253
Katharine Matthies Foundation Grants, 2291
Liberty Bank Foundation Grants, 2362
Lincoln Financial Foundation Grants, 2367
M.B. and Edna Zale Foundation Grants, 2401
May and Stanley Smith Charitable Trust Grants, 2459
MGM Resorts Foundation Community Grants, 2541
Minneapolis Foundation Community Grants, 2558
OneFamily Foundation Grants, 3075
Pinellas County Grants, 3223
Piper Jaffray Fndn Communities Giving Grants, 3227
PMI Foundation Grants, 3239
Polk Bros. Foundation Grants, 3243
Portland General Electric Foundation Grants, 3250
Priddy Foundation Program Grants, 3266
Procter and Gamble Fund Grants, 3274
Richard King Mellon Foundation Grants, 3369
Rockefeller Foundation Grants, 3396
Rockwell Fund, Inc. Grants, 3398
S.H. Cowell Foundation Grants, 3447
Sierra Health Foundation Responsive Grants, 3555
Sioux Falls Area Community Foundation Community Fund Grants (Unrestricted), 3568
Sioux Falls Area Community Foundation Spot Grants (Unrestricted), 3570
Textron Corporate Contributions Grants, 3712
TJX Foundation Grants, 3735
Topfer Family Foundation Grants, 3739
U.S. Department of Education 21st Century Community Learning Centers, 3767
Union Bank, N.A. Foundation Grants, 3788
USDA Organic Agriculture Research Grants, 3830
Verizon Foundation Maryland Grants, 3859
Verizon Foundation New York Grants, 3860
Verizon Foundation Northeast Region Grants, 3861
Verizon Foundation Pennsylvania Grants, 3862
Verizon Foundation Vermont Grants, 3863
Verizon Foundation Virginia Grants, 3864
Waitt Family Foundation Grants, 3894

Journalism
AcademyHealth PHSR Research Article of the Year Awards, 226
AMA Virtual Mentor Theme Issue Editor Grants, 496
American Academy of Nursing Media Awards, 509
APSA Congressional Health and Aging Policy Fellowships, 644
Canada-U.S. Fulbright New Century Scholars Program Grants, 1008
Canada-U.S. Fulbright Senior Specialists Grants, 1009
Fulbright Alumni Initiatives Awards, 1749
Fulbright Distinguished Chairs Awards, 1750
Fulbright New Century Scholars Grants, 1753
Fulbright Specialists Program Grants, 1754
Fulbright Scholars in Europe and Eurasia, 1755
Fulbright Traditional Scholar Program in Sub-Saharan Africa, 1756
Fulbright Traditional Scholar Program in the East Asia/Pacific Region, 1757
Fulbright Traditional Scholar Program in the Near East and North Africa Region, 1758
Fulbright Traditional Scholar Program in the South and Central Asia Region, 1759
Fulbright Traditional Scholar Program in the Western Hemisphere, 1760
Harry Frank Guggenheim Foundation Dissertation Fellowships, 1963
Helen Irwin Littauer Educational Trust Grants, 1994
James M. Cox Foundation of Georgia Grants, 2196
John Ben Snow Memorial Trust Grants, 2231
NIHCM Foundation Health Care Print Journalism Awards, 2870
NIHCM Foundation Health Care Television and Radio Journalism Awards, 2872
NMF National Medical Association Awards for Medical Journalism, 2962
Playboy Foundation Grants, 3237
Prudential Foundation Education Grants, 3276
SfN Science Journalism Student Awards, 3536
White County Community Foundation - Annie Horton Scholarship, 3929

Journalism Education
Helen Irwin Littauer Educational Trust Grants, 1994

Judaism
David M. and Marjorie D. Rosenberg Foundation Grants, 1432
Ike and Roz Friedman Foundation Grants, 2145
Lil and Julie Rosenberg Foundation Grants, 2364
Taubman Endowment for the Arts Fndn Grants, 3702
Thomas Jefferson Rosenberg Foundation Grants, 3723

Junior High School Education
CDC David J. Sencer Museum Teacher Professional Development Workshops, 1053
CFFVR Shawano Area Community Foundation Grants, 1104
Colonel Stanley R. McNeil Foundation Grants, 1211
Dr. & Mrs. Paul Pierce Memorial Fndn Grants, 1505
Eisner Foundation Grants, 1564
Frank B. Hazard General Charity Fund Grants, 1721
Frank Loomis Palmer Fund Grants, 1726
Fred & Gretel Biel Charitable Trust Grants, 1734
Helen Bader Foundation Grants, 1992
John Clarke Trust Grants, 2232
Katrine Menzing Deakins Trust Grants, 2295
Linford and Mildred White Charitable Grants, 2368
Marjorie Moore Charitable Foundation Grants, 2438
R.S. Gernon Trust Grants, 3293
W.H. & Mary Ellen Cobb Trust Grants, 3884
William J. Brace Charitable Trust, 3955

Junior and Community Colleges
American Electric Power Foundation Grants, 530
Community Foundation of Greater Fort Wayne Scholarships, 1276
Doree Taylor Charitable Foundation, 1490
FirstEnergy Foundation Community Grants, 1668
GNOF New Orleans Works Grants, 1844
Guy I. Bromley Trust Grants, 1919
HAF Riley Frazel Memorial Scholarship, 1931
Harvey Randall Wickes Foundation Grants, 1972
IIE Western Union Family Scholarships, 2143
J.M. Long Foundation Grants, 2173
James G.K. McClure Educational and Development Fund Grants, 2188
Louetta M. Cowden Foundation Grants, 2377
Lynn and Rovena Alexander Family Foundation Grants, 2399
Marion Gardner Jackson Charitable Trust Grants, 2434
McGraw-Hill Companies Community Grants, 2494
Norfolk Southern Foundation Grants, 2982
PMI Foundation Grants, 3239
Pride Foundation Fellowships, 3267
Pride Foundation Scholarships, 3269
Sidgmore Family Foundation Grants, 3551
Victor E. Speas Foundation Grants, 3868
White County Community Foundation - Landis Memorial Scholarship, 3930
White County Community Foundation - Tri-County Educational Scholarships, 3932
Whitley County Community Fndn Scholarships, 3937
Windham Foundation Grants, 3961

Justice
Abell Foundation Criminal Justice and Addictions Grants, 208
Nuffield Foundation Africa Program Grants, 3026
Tifa Foundation Grants, 3733

Juvenile Correctional Facilities
Abell Foundation Criminal Justice and Addictions Grants, 208

Juvenile Delinquency
Abell Foundation Criminal Justice and Addictions Grants, 208
Barker Welfare Foundation Grants, 831
Beazley Foundation Grants, 865
Boston Foundation Grants, 936
Cambridge Community Foundation Grants, 1004
Connecticut Community Foundation Grants, 1324
DOJ Gang-Free Schools and Communities Intervention Grants, 1481
G.N. Wilcox Trust Grants, 1768
Mary Owen Borden Foundation Grants, 2446
McKesson Foundation Grants, 2496
Oppenstein Brothers Foundation Grants, 3078

Juvenile Law
NYCT Girls and Young Women Grants, 3053

Kazakhstan
Fulbright Traditional Scholar Program in the South and Central Asia Region, 1759

Kenya
USAID Accelerating Progress Against Tuberculosis in Kenya Grants, 3805

Kidney Diseases and Disorders
ASCO/UICC International Cancer Technology Transfer (ICRETT) Fellowships, 677
ASCO Young Investigator Award, 683
Austin S. Nelson Foundation Grants, 805
Baxter International Corporate Giving Grants, 841
CFFVR Robert and Patricia Endries Family Foundation Grants, 1102
Collins C. Diboll Private Foundation Grants, 1209
Conquer Cancer Foundation of ASCO Career Development Award, 1328

766 / Kidney Diseases and Disorders

NIA Renal Function and Chronic Kidney Disease in Aging Grants, 2781
NIDDK Advances in Polycystic Kidney Disease, 2797
NIDDK Ancillary Studies of Kidney Disease Accessing Information from Clinical Trials, Epidemiological Studies, and Databases Grants, 2799
NIDDK Calcium Oxalate Stone Diseases Grants, 2803
NIDDK Devel of Disease Biomarkers Grants, 2809
NIDDK Education Program Grants, 2813
NIDDK Enhancing Zebrafish Research with Research Tools and Techniques, 2815
NIDDK Health Disparities in NIDDK Diseases, 2820
NIDDK Mentored Research Scientist Development Award, 2825
NIDDK Multi-Center Clinical Study Cooperative Agreements, 2826
NIDDK Multi-Center Clinical Study Implementation Planning Grants, 2827
NIDDK Non-Invasive Methods for Diagnosis and Progression of Diabetes, Kidney, Urological, Hematological and Digestive Diseases, 2832
NIDDK Pilot and Feasibility Clinical Research Grants in Kidney or Urologic Diseases, 2835
NIDDK Proteomics: Diabetes, Obesity, And Endocrine, Digestive, Kidney, Urologic, And Hematologic Diseases, 2838
NIDDK Research Grants for Studies of Hepatitis C in the Setting of Renal Disease, 2839
NIDDK Secondary Analyses in Obesity, Diabetes and Digestive and Kidney Diseases Grants, 2842
NIDDK Small Grants for K08/K23 Recipients, 2846
NIDDK Small Grants for Underrepresented Minority Scientists in Diabetes and Digestive and Kidney Diseases, 2847
NIDDK Training in Clinical Investigation in Kidney and Urology Grants, 2849
NIH Nonhuman Primate Immune Tolerance Cooperative Study Group Grants, 2890
NIH Sarcoidosis: Research into the Cause of Multi-Organ Disease and Clinical Strategies for Therapy Grants, 2907
NKF Clinical Scientist Awards, 2935
NKF Franklin McDonald/Fresenius Medical Care Clinical Research Grants, 2936
NKF Professional Councils Research Grants, 2937
NKF Research Fellowships, 2938
NKF Young Investigator Grants, 2939
NYAM Edward N. Gibbs Memorial Lecture and Award in Nephrology, 3041
Pfizer ASPIRE North America Broad Spectrum Antibiotics for the Treatment of Gram-Negative or Polymicrobial Infections Research Awards, 3169
Pfizer ASPIRE Worldwide Endocrine Young Investigator Grants, 3173
Pfizer Healthcare Charitable Contributions, 3180
Pfizer Medical Education Track Two Grants, 3183
PKD Foundation Lillian Kaplan International Prize for the Advancement in the Understanding of Polycystic Kidney Disease, 3232
PKD Foundation Research Grants, 3233

Kinesiology
Canada Graduate Scholarships (CGS) and NSERC Postgraduate Scholarships (PGS), 1010
Foundation for Balance and Harmony Grants, 1695

Knowledge Acceleration
NLM Exceptional, Unconventional Research Grants Enabling Knowledge Acceleration, 2941

Korea
Fulbright International Education Administrators (IEA) Seminar Program Grants, 1752

Kyrgyzstan
Fulbright Traditional Scholar Program in the South and Central Asia Region, 1759
OSF Public Health Grants in Kyrgyzstan, 3094

Labor Law
Blue Cross Blue Shield of Minnesota Foundation - Healthy Equity: Health Impact Assessment Demonstration Project Grants, 916
Global Fund for Children Grants, 1832

Land Management
Bella Vista Foundation Grants, 873
Champlin Foundations Grants, 1110
Community Foundation for Southeast Michigan Grants, 1258
Fremont Area Community Fndn General Grants, 1746
Richard King Mellon Foundation Grants, 3369
Samuel S. Johnson Foundation Grants, 3468
USAID Land Use Change and Disease Emergence Grants, 3818
Washington Gas Charitable Giving, 3908
Weyerhaeuser Company Foundation Grants, 3925

Land Use Planning/Policy
ACF Native American Social and Economic Development Strategies Grants, 231
Blue Cross Blue Shield of Minnesota Foundation - Healthy Equity: Health Impact Assessment Demonstration Project Grants, 916
Blue Cross Blue Shield of Minnesota Foundation - Healthy Equity: Health Impact Assessment Program Grants, 917
DuPont Pioneer Community Giving Grants, 1532
Emma B. Howe Memorial Foundation Grants, 1593
Greater Kanawha Valley Foundation Grants, 1882
NIEHS Mentored Clinical Scientist Research Career Development Awards, 2857
USAID Land Use Change and Disease Emergence Grants, 3818
Washington Gas Charitable Giving, 3908

Language
George Foundation Grants, 1795

Language Acquisition and Development
ASHFoundation Student Research Grants in Early Childhood Language Development, 711
SfN Science Educator Award, 3535

Laryngology
ALVRE Casselberry Award, 460
ALVRE Grant, 461
ALVRE Seymour R. Cohen Award, 462

Latin America
ASM-UNESCO Leadership Grant for International Educators, 715
Charles Delmar Foundation Grants, 1115
IBRO Latin America Regional Funding for Short Research Stays, 2108

Law
AMA-MSS Government Relations Advocacy Fellowship, 475
AMA-MSS Government Relations Internships, 476
AMA-RFS Legislative Awareness Internships, 479
Bush Fndn Health & Human Services Grants, 979
Canada-U.S. Fulbright New Century Scholars Program Grants, 1008
Canada-U.S. Fulbright Senior Specialists Grants, 1009
Community Foundation of St. Joseph County Lilly Endowment Community Scholarship, 1308
E.L. Wiegand Foundation Grants, 1538
Elkhart County Foundation Lilly Endowment Community Scholarships, 1574
Fulbright Alumni Initiatives Awards, 1749
Fulbright Distinguished Chairs Awards, 1750
Fulbright New Century Scholars Awards, 1753
Fulbright Specialists Program Grants, 1754
Fulbright Scholars in Europe and Eurasia, 1755
Fulbright Traditional Scholar Program in Sub-Saharan Africa, 1756
Fulbright Traditional Scholar Program in the East Asia/Pacific Region, 1757

SUBJECT INDEX

Fulbright Traditional Scholar Program in the Near East and North Africa Region, 1758
Fulbright Traditional Scholar Program in the South and Central Asia Region, 1759
Fulbright Traditional Scholar Program in the Western Hemisphere, 1760
Harry Frank Guggenheim Foundation Dissertation Fellowships, 1963
IIE AmCham Charitable Foundation U.S. Studies Scholarship, 2125
IIE Iraq Scholars and Leaders Scholarships, 2133
IIE Klein Family Scholarship, 2135
IIE Lingnan Foundation W.T. Chan Fellowship, 2137
John Ben Snow Memorial Trust Grants, 2231
Porter County Foundation Lilly Endowment Community Scholarships, 3245
Prudential Foundation Education Grants, 3276
Rajiv Gandhi Foundation Grants, 3302
Seattle Fndn Doyne M. Green Scholarships, 3505
Special Olympics Health Profess Student Grants, 3629
Tifa Foundation Grants, 3733

Law Enforcement
Athwin Foundation Grants, 795
Baptist Community Ministries Grants, 828
Burlington Northern Santa Fe Foundation Grants, 978
Cemala Foundation Grants, 1077
FIU Global Civic Engagement Mini Grants, 1673
Hudson Webber Foundation Grants, 2072
Kenneth T. & Eileen L. Norris Fndn Grants, 2300
Robert R. Meyer Foundation Grants, 3387
Sonora Area Foundation Competitive Grants, 3622

Law and Social Change
USAID Higher Education Partnerships for Innovation and Impact (HEPII) Grants, 3815

Law and Society
Triangle Community Foundation Donor-Advised Grants, 3757

Leadership
AAAAI Distinguished Scientist Award, 26
AAAAI RSLAAIS Leadership Award, 33
AABB Hemphill-Jordan Leadership Award, 44
AACR-Minorities in Cancer Research Jane Cooke Wright Lectureship Awards, 94
AACR Foti Award for Leadership & Extraordinary Achievements in Cancer Research, 118
AAOHN Foundation New Investigator Researcher Grants, 174
AAOHN Foundation Professional Development Scholarships, 175
AAPD Paul G. Hearne Leadership Award, 183
Acumen East Africa Fellowship, 284
AMA-MSS Government Relations Advocacy Fellowship, 475
AMA-MSS Government Relations Internships, 476
American College of Surgeons Resident and Associate Society (RAS-ACS) Leadership Scholarships, 527
ASHA Minority Student Leadership Awards, 696
ASM bioMerieux Sonnenwirth Award for Leadership in Clinical Microbiology, 719
ASM Gen-Probe Joseph Public Health Award, 729
ASM GlaxoSmithKline International Member of the Year Award, 731
ASM TREK Diagnostic ABMM/ABMLI Professional Recognition Award, 754
Austin College Leadership Award, 804
Bank of America Fndn Volunteer Grants, 826
Bodman Foundation Grants, 929
Boeing Company Contributions Grants, 930
Bullitt Foundation Grants, 973
Carroll County Community Foundation Grants, 1039
CDC Preventive Med Residency & Fellowship, 1064
Central Carolina Community Foundation Community Impact Grants, 1079
Charles Lafitte Foundation Grants, 1123
Community Foundation for the National Capital Region Community Leadership Grants, 1260

SUBJECT INDEX

Community Foundation of the Verdugos Educational
  Endowment Fund Grants, 1314
Cowles Charitable Trust Grants, 1354
Dallas Mavericks Foundation Grants, 1416
Dorothy Rider Pool Health Care Grants, 1500
Educational Foundation of America Grants, 1553
Essex County Community Foundation Women's Fund
  Grants, 1605
FCD New American Children Grants, 1637
Frank Reed and Margaret Jane Peters Memorial Fund
  II Grants, 1728
George Family Foundation Grants, 1794
Guy I. Bromley Trust Grants, 1919
Harmony Project Grants, 1948
Heineman Foundation for Research, Education,
  Charitable and Scientific Purposes, 1990
Hutton Foundation Grants, 2087
IIE African Center of Excellence for Women's
  Leadership Grants, 2124
IIE Eurobank EFG Scholarships, 2130
John Edward Fowler Memorial Fndn Grants, 2235
John M. Lloyd Foundation Grants, 2244
Johnson & Johnson Corporate Contributions, 2253
Louetta M. Cowden Foundation Grants, 2377
Majors MLA Chapter Project of the Year, 2418
Meyer Fndn Management Assistance Grants, 2534
Miller Foundation Grants, 2556
MLA Estelle Brodman Academic Medical Librarian of
  the Year Award, 2563
MLA Lois Ann Colaianni Award for Excellence and
  Achievement in Hospital Librarianship, 2568
MLA Lucretia W. McClure Excellence in Education
  Award, 2570
MLA Section Project of the Year Award, 2574
NCI Cancer Education and Career Development
  Program, 2648
NMF Franklin C. McLean Award, 2955
NMF Hugh J. Andersen Memorial Scholarship, 2958
NMF Metropolitan Life Foundation Awards Program
  for Academic Excellence in Medicine, 2961
NMF National Medical Association Awards for
  Medical Journalism, 2962
NMF National Medical Association Emerging Scholar
  Awards, 2963
NMF Ralph W. Ellison Prize, 2966
NMF William and Charlotte Cadbury Award, 2967
Nordson Corporation Foundation Grants, 2981
Northwest Minnesota Foundation Women's Fund
  Grants, 3000
NSF Presidential Early Career Awards for Scientists
  and Engineers (PECASE) Grants, 3021
Paso del Norte Health Foundation Grants, 3119
PepsiCo Foundation Grants, 3137
Peter and Elizabeth C. Tower Foundation
  Organizational Scholarships, 3148
PeyBack Foundation Grants, 3159
Philadelphia Foundation Organizational Effectiveness
  Grants, 3189
Portland General Electric Foundation Grants, 3250
Prudential Foundation Education Grants, 3276
RCPSC James H. Graham Award of Merit, 3327
RCPSC Program Director of the Year Award, 3340
Rhode Island Foundation Grants, 3362
Richard Davoud Donchian Foundation Grants, 3367
Robert R. Meyer Foundation Grants, 3387
Rockefeller Brothers Peace & Security Grants, 3394
RWJF Community Health Leaders Awards, 3432
Saint Louis Rams Fndn Community Donations, 3453
Santa Barbara Foundation Strategy Grants - Core
  Support, 3481
SfN Louise Hanson Marshall Specific Recognition
  Award, 3527
Tifa Foundation Grants, 3733
Virginia L. and William K. Beatty MLA Volunteer
  Service Award, 3877
W.K. Kellogg Foundation Healthy Kids Grants, 3886
White County Community Foundation - Annie Horton
  Scholarship, 3929
Willary Foundation Grants, 3944

Learning Disabilities
AACAP Pilot Research Award for Learning
  Disabilities, Supported by the Elaine Schlosser
  Lewis Fund, 67
AACAP Sidney Berman Award for the School-Based
  Study and Intervention for Learning Disorders and
  Mental Illness, 73
ACL Partnerships in Employment Systems Change
  Grants, 246
ACL Training and Technical Assistance Center for
  State Intellectual and Developmental Disabilities
  Delivery Systems Grants, 248
Adams Rotary Memorial Fund A Grants, 302
CVS All Kids Can Grants, 1377
Elsie H. Wilcox Foundation Grants, 1583
FRAXA Research Foundation Program Grants, 1733
Grace and Franklin Bernsen Foundation Grants, 1858
Graham and Carolyn Holloway Family Foundation
  Grants, 1861
Peter and Elizabeth C. Tower Foundation Annual
  Intellectual Disabilities Grants, 3144
Peter and Elizabeth C. Tower Foundation Annual
  Mental Health Grants, 3145
Peter and Elizabeth C. Tower Foundation Learning
  Disability Grants, 3146
Peter and Elizabeth C. Tower Foundation Phase II
  Technology Initiative Grants, 3149
Peter and Elizabeth C. Tower Foundation Phase I
  Technology Initiative Grants, 3150
Pinkerton Foundation Grants, 3224
Roy & Christine Sturgis Charitable Grants, 3415
Thomas Sill Foundation Grants, 3724
UCT Scholarship Program, 3777

Learning Disabled, Education for
DeKalb County Community Foundation - Garrett
  Hospital Aid Foundation Grants, 1446
Educational Foundation of America Grants, 1553
Emily Hall Tremaine Foundation Learning Disabilities
  Grants, 1592
Peter and Elizabeth C. Tower Foundation Annual
  Intellectual Disabilities Grants, 3144

Lectureships
ANA F.E. Bennett Memorial Lectureship, 581
ANA Raymond D. Adams Lectureship, 583
ANA Soriano Lectureship, 584
IBRO-PERC Support for Site Lectures, 2099
IBRO Asia APRC Lecturer Exchange Grants, 2102
MLA Janet Doe Lectureship Award, 2567
RCPSC Harry S. Morton Lectureship in Surgery, 3323

Legal Education
CICF Indianapolis Fndn Community Grants, 1156

Legal Reform
Bernard F. and Alva B. Gimbel Foundation
  Reproductive Rights Grants, 883
DPA Promoting Policy Change Advocacy Grants, 1504
OSF Law and Health Initiative Grants, 3092

Legal Services
BancorpSouth Foundation Grants, 823
Bingham McHale LLP Pro Bono Services, 897
Bodenwein Public Benevolent Foundation Grants, 928
Charles H. Pearson Foundation Grants, 1121
David Geffen Foundation Grants, 1431
Fred & Gretel Biel Charitable Trust Grants, 1734
Georgiana Goddard Eaton Memorial Grants, 1805
Howard and Bush Foundation Grants, 2054
Marie C. and Joseph C. Wilson Foundation Rochester
  Small Grants, 2431
McCune Foundation Human Services Grants, 2493
Morris & Gwendolyn Cafritz Fndn Grants, 2594
OSF Law and Health Initiative Grants, 3092
OSF Public Health Grants in Kyrgyzstan, 3094
Otto Bremer Foundation Grants, 3099
Paul Rapoport Foundation Grants, 3125
Perry County Community Foundation Grants, 3142
Rhode Island Foundation Grants, 3362

Sonora Area Foundation Competitive Grants, 3622
Southbury Community Trust Fund, 3625

Legal Systems
Carrier Corporation Contributions Grants, 1038

Leukemia
ACGT Investigators Grants, 232
AMA Foundation Seed Grants for Research, 494
Angel Kiss Foundation Grants, 589
Austin S. Nelson Foundation Grants, 805
Children's Leukemia Research Association Research
  Grants, 1138
HAF Barry F. Phelps Fund Grants, 1927
NYCT Biomedical Research Grants, 3049
Pfizer Healthcare Charitable Contributions, 3180
Pfizer Medical Education Track Two Grants, 3183
Ryan Gibson Foundation Grants, 3445

Liberal Arts Education
Brown Advisory Charitable Foundation Grants, 969
W.M. Keck Fndn Undergrad Ed Grants, 3891

Libraries
Ahmanson Foundation Grants, 355
Air Products and Chemicals Grants, 399
Alcatel-Lucent Technologies Foundation Grants, 413
Alexis Gregory Foundation Grants, 426
Auburn Foundation Grants, 798
Bertha Russ Lytel Foundation Grants, 885
Blue Cross Blue Shield of Minnesota Fndn Healthy
  Equity: Public Libraries for Health Grants, 918
Bodenwein Public Benevolent Foundation Grants, 928
Boettcher Foundation Grants, 931
Booth-Bricker Fund Grants, 933
Brinson Foundation Grants, 953
Burlington Northern Santa Fe Foundation Grants, 978
Caleb C. and Julia W. Dula Educational and Charitable
  Foundation Grants, 997
Callaway Foundation Grants, 1002
Carnahan-Jackson Foundation Grants, 1032
Champlin Foundations Grants, 1110
Clark and Ruby Baker Foundation Grants, 1174
Community Foundation of Eastern Connecticut
  General Southeast Grants, 1269
Community Fndn of Randolph County Grants, 1302
Community Foundation of the Verdugos Educational
  Endowment Fund Grants, 1314
Constantin Foundation Grants, 1341
Cooper Industries Foundation Grants, 1345
Faye McBeath Foundation Grants, 1635
Fred C. & Katherine B. Andersen Grants, 1737
George Kress Foundation Grants, 1799
Gladys Brooks Foundation Grants, 1826
Greater Sitka Legacy Fund Grants, 1885
Green Bay Packers Foundation Grants, 1890
H. Schaffer Foundation Grants, 1925
Hall-Perrine Foundation Grants, 1935
Harry Bramhall Gilbert Charitable Trust Grants, 1961
Harvey Randall Wickes Foundation Grants, 1972
High Meadow Foundation Grants, 2027
Huie-Dellmon Trust Grants, 2075
Idaho Community Foundation Eastern Region
  Competitive Grants, 2114
J.L. Bedsole Foundation Grants, 2172
James G.K. McClure Educational and Development
  Fund Grants, 2188
James L. and Mary Jane Bowman Charitable Trust
  Grants, 2194
Janus Foundation Grants, 2211
John Ben Snow Memorial Trust Grants, 2231
Kenai Peninsula Foundation Grants, 2298
Ketchikan Community Foundation Grants, 2303
Leon and Thea Koerner Foundation Grants, 2355
Lucy Downing Nisbet Charitable Fund Grants, 2387
Maine Community Foundation Charity Grants, 2415
McLean Contributionship Grants, 2497
NHSCA Arts in Health Care Project Grants, 2741
NLM Internet Connection for Medical Institutions
  Grants, 2946

768 / Libraries   SUBJECT INDEX

Parkersburg Area Community Foundation Action Grants, 3116
Phelps County Community Foundation Grants, 3188
Pollock Foundation Grants, 3244
Portland General Electric Foundation Grants, 3250
Posey County Community Foundation Grants, 3252
Procter and Gamble Fund Grants, 3274
Rayonier Foundation Grants, 3311
Reinberger Foundation Grants, 3357
Rice Foundation Grants, 3363
Rochester Area Community Foundation Grants, 3391
Schlessman Family Foundation Grants, 3493
Shell Deer Park Grants, 3543
Sioux Falls Area Community Foundation Community Fund Grants (Unrestricted), 3568
Sioux Falls Area Community Foundation Spot Grants (Unrestricted), 3570
Susan Vaughan Foundation Grants, 3695
Thomas J. Long Foundation Community Grants, 3722
Thomas Sill Foundation Grants, 3724
W. C. Griffith Foundation Grants, 3882
W.C. Griffith Foundation Grants, 3881
William H. Hannon Foundation Grants, 3953

Libraries, Academic
Humana Foundation Grants, 2077
John Deere Foundation Grants, 2234
Kelvin and Eleanor Smith Foundation Grants, 2297
MLA Estelle Brodman Academic Medical Librarian of the Year Award, 2563
Richard King Mellon Foundation Grants, 3369
Roy J. Carver Charitable Trust Medical and Science Research Grants, 3416

Libraries, Medical
ASM USFCC/J. Roger Porter Award, 756
Carla J. Funk Governmental Relations Award, 1020
EBSCO / MLA Annual Meeting Grants, 1545
Hospital Libraries Section / MLA Professional Development Grants, 2052
Majors MLA Chapter Project of the Year, 2418
Medical Informatics Section/MLA Career Development Grant, 2508
MLA Continuing Education Grants, 2559
MLA Cunningham Memorial International Fellowship, 2560
MLA David A. Kronick Traveling Fellowship, 2561
MLA Donald A.B. Lindberg Fellowship, 2562
MLA Estelle Brodman Academic Medical Librarian of the Year Award, 2563
MLA Graduate Scholarship, 2564
MLA Grad Scholarship for Minority Students, 2565
MLA Ida and George Eliot Prize, 2566
MLA Janet Doe Lectureship Award, 2567
MLA Lois Ann Colaianni Award for Excellence and Achievement in Hospital Librarianship, 2568
MLA Louise Darling Medal for Distinguished Achievement in Collection Development in the Health Sciences, 2569
MLA Lucretia W. McClure Excellence in Education Award, 2570
MLA Murray Gottlieb Prize Essay Award, 2571
MLA Research, Development, and Demonstration Project Grant, 2572
MLA Rittenhouse Award, 2573
MLA Section Project of the Year Award, 2574
MLA T. Mark Hodges Int'l Service Award, 2575
MLA Thomas Reuters / Frank Bradway Rogers Information Advancement Award, 2576
NLM Exploratory/Developmental Grants, 2942
NLM Grants for Scholarly Works in Biomedicine and Health, 2944
Thomson Reuters / MLA Doctoral Fellowships, 3728
Virginia L. and William K. Beatty MLA Volunteer Service Award, 3877

Libraries, Public
Adelaide Breed Bayrd Foundation Grants, 306
Adler-Clark Electric Community Commitment Foundation Grants, 307

Blue Cross Blue Shield of Minnesota Fndn Healthy Equity: Public Libraries for Health Grants, 918
Champlin Foundations Grants, 1110
James L. and Mary Jane Bowman Charitable Trust Grants, 2194
John G. Martin Foundation Grants, 2237
Kent D. Steadley and Mary L. Steadley Memorial Trust Grants, 2301
Milagro Foundation Grants, 2552
Piedmont Natural Gas Corporate and Charitable Contributions, 3220

Libraries, Research
MBL Fred Karush Library Readership, 2474

Library Administration
MLA Continuing Education Grants, 2559
MLA David A. Kronick Traveling Fellowship, 2561
MLA Donald A.B. Lindberg Fellowship, 2562
MLA Janet Doe Lectureship Award, 2567
Petersburg Community Foundation Grants, 3156
Virginia L. and William K. Beatty MLA Volunteer Service Award, 3877

Library Automation
MLA Ida and George Eliot Prize, 2566

Library History
ALA Donald G. Davis Article Award, 401
MLA Janet Doe Lectureship Award, 2567

Library Science
Alexis Gregory Foundation Grants, 426
Canada-U.S. Fulbright New Century Scholars Program Grants, 1008
Canada-U.S. Fulbright Senior Specialists Grants, 1009
Fulbright Alumni Initiatives Awards, 1749
Fulbright Distinguished Chairs Awards, 1750
Fulbright New Century Scholars Grants, 1753
Fulbright Specialists Program Grants, 1754
Fulbright Scholars in Europe and Eurasia, 1755
Fulbright Traditional Scholar Program in Sub-Saharan Africa, 1756
Fulbright Traditional Scholar Program in the East Asia/Pacific Region, 1757
Fulbright Traditional Scholar Program in the Near East and North Africa Region, 1758
Fulbright Traditional Scholar Program in the South and Central Asia Region, 1759
Fulbright Traditional Scholar Program in the Western Hemisphere, 1760
Medical Informatics Section/MLA Career Development Grant, 2508
MLA Donald A.B. Lindberg Fellowship, 2562
MLA Graduate Scholarship, 2564
MLA Grad Scholarship for Minority Students, 2565
MLA Research, Development, and Demonstration Project Grant, 2572
Pollock Foundation Grants, 3244
Thomas J. Long Foundation Community Grants, 3722
W. C. Griffith Foundation Grants, 3882
W.C. Griffith Foundation Grants, 3881

Library Science Education
Gates Millennium Scholars Program, 1776

Life Sciences
AIHS ForeFront MBA Studentship Award, 381
Arnold and Mabel Beckman Foundation Young Investigators Grants, 666
CCFF Chairman's Dist Life Sciences Award, 1044
CCFF Life Sciences Student Awards, 1045
HHMI Grants and Fellowships Programs, 2020
Homer Foundation Grants, 2044
Kavli Foundation Research Grants, 2296
NHLBI Pathway to Independence Awards, 2718
NSF Biomedical Engineering and Engineering Healthcare Grants, 3009
RSC Thomas W. Eadie Medal, 3423
Seattle Fndn Hawkins Memorial Scholarship, 3506

W.M. Keck Foundation Science and Engineering Research Grants, 3889
Wellcome Trust Biomedical Science Grants, 3918

Life Skills Training
Albert W. Rice Charitable Foundation Grants, 411
Alfred E. Chase Charitable Foundation Grants, 430
Cargill Citizenship Fund Corporate Giving, 1018
Coleman Foundation Developmental Disabilities Grants, 1207
Coors Brewing Corporate Contributions Grants, 1346
Cruise Industry Charitable Foundation Grants, 1363
Frank Reed and Margaret Jane Peters Memorial Fund II Grants, 1728
George W. Wells Foundation Grants, 1804
Global Fund for Children Grants, 1832
May and Stanley Smith Charitable Trust Grants, 2459
Pinkerton Foundation Grants, 3224
Robert R. Meyer Foundation Grants, 3387
Union Bank, N.A. Foundation Grants, 3788

Linguistics/Philology
Australasian Institute of Judicial Administration Seed Funding Grants, 806
Canada-U.S. Fulbright New Century Scholars Program Grants, 1008
Canada-U.S. Fulbright Senior Specialists Grants, 1009
Fulbright Alumni Initiatives Awards, 1749
Fulbright Distinguished Chairs Awards, 1750
Fulbright New Century Scholars Grants, 1753
Fulbright Specialists Program Grants, 1754
Fulbright Scholars in Europe and Eurasia, 1755
Fulbright Traditional Scholar Program in Sub-Saharan Africa, 1756
Fulbright Traditional Scholar Program in the East Asia/Pacific Region, 1757
Fulbright Traditional Scholar Program in the Near East and North Africa Region, 1758
Fulbright Traditional Scholar Program in the South and Central Asia Region, 1759
Fulbright Traditional Scholar Program in the Western Hemisphere, 1760
George Foundation Grants, 1795
NLM Research Project Grants, 2951

Lipids
AOCS Alton E. Bailey Award, 611
AOCS Analytical Division Student Award, 612
AOCS Health & Nutrition Div Student Award, 614
AOCS Health and Nutrition Poster Competition, 615
AOCS Lipid Oxidation and Quality Division Poster Competition, 617
AOCS Holman Lifetime Achievement Award, 620
NIH Structural Bio of Membrane Proteins Grant, 2911
Pfizer Medical Education Track Two Grants, 3183
Robert B McMillen Foundation Grants, 3381

Literacy
AAFP Foundation Health Literacy State Grants, 132
Ahmanson Foundation Grants, 355
Alcatel-Lucent Technologies Foundation Grants, 413
Allen P. & Josephine B. Green Fndn Grants, 443
Allyn Foundation Grants, 448
AMD Corporate Contributions Grants, 500
Amgen Foundation Grants, 572
Anschutz Family Foundation Grants, 602
Arizona Cardinals Grants, 655
Arizona Diamondbacks Charities Grants, 657
Arkell Hall Foundation Grants, 661
Arlington Community Foundation Grants, 662
Ashland Corporate Contributions Grants, 713
Assisi Fndn of Memphis Capital Project Grants, 786
Assisi Foundation of Memphis General Grants, 787
Atlanta Foundation Grants, 796
Auburn Foundation Grants, 798
Bacon Family Foundation Grants, 817
Ball Brothers Foundation General Grants, 820
Barker Welfare Foundation Grants, 831
Battle Creek Community Foundation Grants, 839
Bayer Foundation Grants, 846

BBVA Compass Foundation Charitable Grants, 848
Benton Community Foundation Grants, 878
Berrien Community Foundation Grants, 884
Blue Cross Blue Shield of Minnesota Fndn Healthy Equity: Public Libraries for Health Grants, 918
Blue Mountain Community Foundation Grants, 921
Blue River Community Foundation Grants, 922
Bodenwein Public Benevolent Foundation Grants, 928
Bodman Foundation Grants, 929
Boettcher Foundation Grants, 931
Booth-Bricker Fund Grants, 933
Boston Foundation Grants, 936
Boston Globe Foundation Grants, 937
Brinson Foundation Grants, 953
Brown County Community Foundation Grants, 970
Cabot Corporation Foundation Grants, 993
Carrie Estelle Doheny Foundation Grants, 1037
Cemala Foundation Grants, 1077
CFFVR Jewelers Mutual Charitable Giving, 1099
CFFVR Schmidt Family G4 Grants, 1103
CFFVR Wisconsin King's Daus & Sons Grants, 1106
Charles Lafitte Foundation Grants, 1123
CIGNA Foundation Grants, 1160
CNO Financial Group Community Grants, 1201
Comerica Charitable Foundation Grants, 1224
Community Foundation Alliance City of Evansville Endowment Fund Grants, 1238
Community Foundation of Bartholomew County Heritage Fund Grants, 1262
Community Foundation of Bartholomew County James A. Henderson Award for Fundraising, 1263
Community Fndn of Central Illinois Grants, 1267
Community Foundation of Greater Birmingham Grants, 1272
Community Foundation of Greater Fort Wayne - Community Endowment and Clarke Endowment Grants, 1274
Community Foundation of Greenville Hollingsworth Funds Program/Project Grants, 1285
Community Foundation of the Verdugos Grants, 1315
Community Foundation Partnerships - Lawrence County Grants, 1317
Cooper Industries Foundation Grants, 1345
Coors Brewing Corporate Contributions Grants, 1346
Cornerstone Foundation of Northeastern Wisconsin Grants, 1347
Covenant Educational Foundation Grants, 1348
Cowles Charitable Trust Grants, 1354
Crail-Johnson Foundation Grants, 1355
Cruise Industry Charitable Foundation Grants, 1363
CVS All Kids Can Grants, 1377
Dayton Power and Light Foundation Grants, 1437
Deaconess Community Foundation Grants, 1439
Deborah Munroe Noonan Memorial Grants, 1443
Decatur County Community Foundation Large Project Grants, 1444
Edyth Bush Charitable Foundation Grants, 1562
El Paso Community Foundation Grants, 1579
Entergy Corporation Micro Grants, 1596
Essex County Community Foundation Merrimack Valley General Fund Grants, 1603
Evjue Foundation Grants, 1617
Farmers Insurance Corporate Giving Grants, 1634
Faye McBeath Foundation Grants, 1635
Field Foundation of Illinois Grants, 1660
Fourjay Foundation Grants, 1712
Frances L. and Edwin L. Cummings Grants, 1717
Fremont Area Community Foundation Amazing X Grants, 1744
G.N. Wilcox Trust Grants, 1768
GenCorp Foundation Grants, 1779
General Mills Foundation Grants, 1782
GlaxoSmithKline Corporate Grants, 1829
Grace and Franklin Bernsen Foundation Grants, 1858
Grand Rapids Community Foundation Ionia County Youth Fund Grants, 1871
Green Bay Packers Foundation Grants, 1890
Guido A. & Elizabeth H. Binda Fndn Grants, 1913
H.B. Fuller Foundation Grants, 1922
Hall-Perrine Foundation Grants, 1935

Hallmark Corporate Foundation Grants, 1936
Harold Simmons Foundation Grants, 1955
Harvest Foundation Grants, 1971
Hasbro Children's Fund Grants, 1973
Helen Steiner Rice Foundation Grants, 1998
Herbert A. & Adrian W. Woods Fndn Grants, 2007
Highmark Corporate Giving Grants, 2025
Hilda and Preston Davis Foundation Grants, 2028
Houston Endowment Grants, 2053
Howard and Bush Foundation Grants, 2054
Hugh J. Andersen Foundation Grants, 2074
Humana Foundation Grants, 2077
IIE Western Union Family Scholarships, 2143
Irvin Stern Foundation Grants, 2164
James A. and Faith Knight Foundation Grants, 2186
James S. Copley Foundation Grants, 2200
Jean and Louis Dreyfus Foundation Grants, 2215
Jim Moran Foundation Grants, 2228
John Edward Fowler Memorial Fndn Grants, 2235
John H. and Wilhelmina D. Harland Charitable Foundation Children and Youth Grants, 2238
John I. Smith Charities Grants, 2240
John P. McGovern Foundation Grants, 2247
Joseph H. & Florence A. Roblee Fndn Grants, 2269
Kosciusko County Community Foundation REMC Operation Round Up Grants, 2320
Leo Goodwin Foundation Grants, 2354
Leo Niessen Jr., Charitable Trust Grants, 2358
Lincoln Financial Foundation Grants, 2367
Lisa and Douglas Goldman Fund Grants, 2369
Lubrizol Foundation Grants, 2384
Mardag Foundation Grants, 2428
Marie C. and Joseph C. Wilson Foundation Rochester Small Grants, 2431
Mattel Children's Foundation Grants, 2454
Mattel International Grants Program, 2455
Milagro Foundation Grants, 2552
NLM Understanding and Promoting Health Literacy Research Grants, 2953
Norcliffe Foundation Grants, 2979
Nordson Corporation Foundation Grants, 2981
Parkersburg Area Community Action Grants, 3116
PepsiCo Foundation Grants, 3137
Percy B. Ferebee Endowment Grants, 3138
Peyton Anderson Foundation Grants, 3160
Pinkerton Foundation Grants, 3224
PMI Foundation Grants, 3239
Portland General Electric Foundation Grants, 3250
Rajiv Gandhi Foundation Grants, 3302
Reinberger Foundation Grants, 3357
Reynolds & Reynolds Associate Fndn Grants, 3360
RGk Foundation Grants, 3361
Richard Davoud Donchian Foundation Grants, 3367
Robert R. Meyer Foundation Grants, 3387
Robert W. Woodruff Foundation Grants, 3389
Saint Louis Rams Fndn Community Donations, 3453
Samuel S. Johnson Foundation Grants, 3468
San Antonio Area Foundation Grants, 3469
Self Foundation Grants, 3510
Sony Corporation of America Grants, 3623
Southbury Community Trust Fund, 3625
Stackpole-Hall Foundation Grants, 3643
Stocker Foundation Grants, 3656
Strake Foundation Grants, 3658
Textron Corporate Contributions Grants, 3712
Thomas Sill Foundation Grants, 3724
U.S. Department of Education 21st Century Community Learning Centers, 3767
Union Bank, N.A. Foundation Grants, 3788
Verizon Foundation Maryland Grants, 3859
Verizon Foundation New York Grants, 3860
Verizon Foundation Northeast Region Grants, 3861
Verizon Foundation Pennsylvania Grants, 3862
Verizon Foundation Vermont Grants, 3863
Verizon Foundation Virginia Grants, 3864
Wayne County Foundation Grants, 3909
Western Union Foundation Grants, 3923
WHO Foundation Volunteer Service Grants, 3940
William G. & Helen C. Hoffman Fndn Grants, 3950
Wilson-Wood Foundation Grants, 3960

Literary Arts
Lil and Julie Rosenberg Foundation Grants, 2364
Marion Gardner Jackson Charitable Trust Grants, 2434
Society for the Arts in Healthcare Grants, 3602
Strowd Roses Grants, 3664

Literature
Albert and Margaret Alkek Foundation Grants, 405
Arthur Ashley Williams Foundation Grants, 671
Canada-U.S. Fulbright New Century Scholars Program Grants, 1008
Canada-U.S. Fulbright Senior Specialists Grants, 1009
Cleveland-Cliffs Foundation Grants, 1181
Fulbright Alumni Initiatives Awards, 1749
Fulbright Distinguished Chairs Awards, 1750
Fulbright New Century Scholars Grants, 1753
Fulbright Specialists Program Grants, 1754
Fulbright Scholars in Europe and Eurasia, 1755
Fulbright Traditional Scholar Program in Sub-Saharan Africa, 1756
Fulbright Traditional Scholar Program in the East Asia/Pacific Region, 1757
Fulbright Traditional Scholar Program in the Near East and North Africa Region, 1758
Fulbright Traditional Scholar Program in the South and Central Asia Region, 1759
Fulbright Traditional Scholar Program in the Western Hemisphere, 1760
George Foundation Grants, 1795
Harry Frank Guggenheim Foundation Dissertation Fellowships, 1963
Heineman Foundation for Research, Education, Charitable and Scientific Purposes, 1990
Hoblitzelle Foundation Grants, 2038
Josiah W. and Bessie H. Kline Foundation Grants, 2278
McKesson Foundation Grants, 2496
Mercedes-Benz USA Corporate Contributions, 2517
Milagro Foundation Grants, 2552
SunTrust Bank Trusteed Foundations Florence C. and Harry L. English Memorial Fund Grants, 3680
SunTrust Bank Trusteed Foundations Greene-Sawtell Grants, 3681
SunTrust Bank Trusteed Foundations Harriet McDaniel Marshall Tust Grants, 3682
SunTrust Bank Trusteed Foundations Nell Warren Elkin and William Simpson Elkin Grants, 3683
SunTrust Bank Trusteed Foundations Thomas Guy Woolford Charitable Trust Grants, 3684
SunTrust Bank Trusteed Foundations Walter H. and Marjory M. Rich Memorial Fund Grants, 3685

Liver Diseases and Disorders
FDHN Fellowship to Faculty Transition Awards, 1644
FDHN Graduate Student Awards, 1646
FDHN Moti L. & Kamla Rustgi International Travel Awards, 1649
FDHN Research Scholar Awards, 1652
NIDDK Developmental Biology and Regeneration of the Liver Grants, 2807
NIDDK Devel of Disease Biomarkers Grants, 2809
NIDDK New Tech for Liver Disease SBIR, 2830
NIDDK New Tech for Liver Disease STTR, 2831
NLM New Tech for Liver Disease Grants, 2949

Local History
Beirne Carter Foundation Grants, 870
Parke County Community Foundation Grants, 3115
Sarkeys Foundation Grants, 3485
Turner Foundation Grants, 3763

Long-Term Care
Blanche and Irving Laurie Foundation Grants, 910
Brookdale Fnd National Group Respite Grants, 966
Del Mar Healthcare Fund Grants, 1453
George E. Hatcher, Jr. and Ann Williams Hatcher Foundation Grants, 1792
HRSA Nurse Education, Practice, Quality and Retention (NEPQR) Grants, 2068
NINR Acute and Chronic Care During Mechanical Ventilation Research Grants, 2926

770 / Long-Term Care

Retirement Research Foundation General Program Grants, 3359
Sensient Technologies Foundation Grants, 3513

Loss-Prevention Programs
Priddy Foundation Program Grants, 3266
RCF General Community Grants, 3312

Lung Disease
Austin S. Nelson Foundation Grants, 805
Medtronic Foundation Strengthening Health Systems Grants, 2515
NHLBI Ancillary Studies in Clinical Trials, 2688
NHLBI Career Transition Awards, 2694
NHLBI Microbiome of Lung & Respiratory in HIV Infected Individuals & Uninfected Controls, 2712
NIH Sarcoidosis: Research into the Cause of Multi-Organ Disease and Clinical Strategies for Therapy Grants, 2907
Pfizer Medical Education Track Two Grants, 3183

Lupus
Partnership for Cures Two Years To Cures Grants, 3118

Lymphatic System
Lymphatic Education and Research Network Additional Support Grants for NIH-funded F32 Postdoctoral Fellows, 2394
Lymphatic Education and Research Network Postdoctoral Fellowships, 2395
NHLBI Lymphatics in Health and Disease in the Digestive, Cardio & Pulmonary Systems, 2704
NHLBI Lymphatics in Health and Disease in the Digestive, Urinary, Cardiovascular and Pulmonary Systems, 2705

Lymphoma
ACGT Investigators Grants, 232
Pfizer Healthcare Charitable Contributions, 3180
Pfizer Medical Education Track Two Grants, 3183

Malaria
ASPH/CDC Allan Rosenfield Global Health Fellowships, 780
ExxonMobil Foundation Malaria Grants, 1621
United Methodist Committee on Relief Global Health Grants, 3799

Mammalogy
Lalor Foundation Postdoctoral Fellowships, 2336

Managed Care
ACL Business Acumen for Disability Organizations Grants, 238
Bush Foundation Medical Fellowships, 980
CMS Research and Demonstration Grants, 1188
Community Foundation for the National Capital Region Community Leadership Grants, 1260
Fallon OrNda Community Health Fund Grants, 1627
Mayo Clinic Administrative Fellowship, 2460

Management
Benton Community Foundation Grants, 878
Blue River Community Foundation Grants, 922
Brown County Community Foundation Grants, 970
Community Foundation for Greater Atlanta Managing For Excellence Award, 1242
Community Foundation of Bartholomew County Heritage Fund Grants, 1262
Community Foundation of Bartholomew County James A. Henderson Award for Fundraising, 1263
Community Foundation of Greater Fort Wayne - Community Endowment and Clarke Endowment Grants, 1274
Community Foundation of Greater Fort Wayne - Lilly Endowment Scholarships, 1275
Community Foundation of St. Joseph County Lilly Endowment Community Scholarship, 1308
Donna K. Yundt Memorial Scholarship, 1488
Dr. John Maniotes Scholarship, 1506

Dr. R.T. White Scholarship, 1510
Elkhart County Foundation Lilly Endowment Community Scholarships, 1574
Fred and Louise Latshaw Scholarship, 1735
Fulton County Community Foundation 4Community Higher Education Scholarship, 1762
GNOF Organizational Effectiveness Grants and Workshops, 1846
IBEW Local Union #697 Memorial Scholarships, 2096
IIE New Leaders Group Award for Mutual Understanding, 2141
John V. and George Primich Family Scholarship, 2259
Joshua Benjamin Cohen Memorial Scholarship, 2276
LaGrange County Lilly Endowment Community Scholarship, 2331
Lake County Athletic Officials Association Scholarships, 2332
Lake County Lilly Endowment Community Scholarships, 2334
LEGENDS Scholarship, 2351
Nancy J. Pinnick Memorial Scholarship, 2607
Noble County Community Foundation - Kathleen June Earley Memorial Scholarship, 2973
Porter County Foundation Lilly Endowment Community Scholarships, 3245
Richard and Helen DeVos Foundation Grants, 3364
Richard King Mellon Foundation Grants, 3369
Sengupta Family Scholarship, 3512
VDH Rescue Squad Assistance Fund Grants, 3856
W. James Spicer Scholarship, 3885
Webster Cornwell Memorial Scholarship, 3911
Whitley County Community Foundation - Lilly Endowment Scholarship, 3935
Whitley County Community Fndn Scholarships, 3937

Management Information Systems
USDD HIV/AIDS Prevention Program Information Systems Development Grants, 3842

Management Planning/Policy
John W. Speas and Effie E. Speas Memorial Trust Grants, 2264
Lewis H. Humphreys Charitable Trust Grants, 2361
Louis and Elizabeth Nave Flarsheim Charitable Foundation Grants, 2379

Management Sciences
AAFCS International Graduate Fellowships, 129
AAFCS National Graduate Fellowships, 130
AAFCS National Undergraduate Scholarships, 131
DaimlerChrysler Corporation Fund Grants, 1412
General Motors Foundation Grants, 1783
Johnson & Johnson Corporate Contributions, 2253

Manufacturing
CCFF Chairman's Dist Life Sciences Award, 1044
Indiana 21st Century Research and Technology Fund Awards, 2154

Manufacturing Processes
NSF-NIST Interaction in Chemistry, Materials Research, Molecular Biosciences, Bioengineering, and Chemical Engineering, 3004
NSF Chemistry Research Experiences for Undergraduates (REU), 3012

Manuscripts/Books/Music Scores
AAAS/Subaru SB&F Prize for Excellence in Science Books, 35
ASHA Advancing Academic-Research Award, 695
H. Schaffer Foundation Grants, 1925

Marine Engineering
IIE KAUST Graduate Fellowships, 2134

Marine Resources
Coastal Community Foundation of South Carolina Grants, 1202
Vancouver Foundation Grants and Community Initiatives Program, 3849

SUBJECT INDEX

Marine Sciences
AAAS Award for Scie Freedom & Responsibility, 36
Frederick Gardner Cottrell Foundation Grants, 1739
IIE KAUST Graduate Fellowships, 2134
Margaret T. Morris Foundation Grants, 2429
Maurice J. Masserini Charitable Trust Grants, 2456
MBL Frederik B. and Betsy G. Bang Summer Fellowships, 2473

Marketing Research
AOCS Industrial Oil Products Division Student Award, 616
Charles H. Farnsworth Trust Grants, 1119
John W. Speas and Effie E. Speas Memorial Trust Grants, 2264
Lewis H. Humphreys Charitable Trust Grants, 2361
Louis and Elizabeth Nave Flarsheim Charitable Foundation Grants, 2379

Marketing/Public Relations
HAF Technical Assistance Program Grants, 1933
John W. Speas and Effie E. Speas Memorial Trust Grants, 2264
Lewis H. Humphreys Charitable Trust Grants, 2361
Louis and Elizabeth Nave Flarsheim Charitable Foundation Grants, 2379
Western New York Foundation Grants, 3922

Mass Communication
Tifa Foundation Grants, 3733

Mass Media
American Academy of Nursing Media Awards, 509
Gill Foundation - Gay and Lesbian Fund Grants, 1821
OSF Health Media Initiative Grants, 3089
SfN Science Journalism Student Awards, 3536
Tifa Foundation Grants, 3733

Mass Spectrometry
APHL Emerging Infectious Diseases Fellowships, 636
Smithsonian Museum Conservation Institute Research Post-Doctorate/Post-Graduate Fellowships, 3581

Massage Therapy
Massage Therapy Foundation Community Service Grants, 2449
Massage Therapy Foundation Practitioner Case Report Contest, 2450
Massage Therapy Foundation Research Grants, 2451

Materials Acquisition (Books, Tapes, etc.)
HAF Phyllis Nilsen Leal Memorial Fund Gifts, 1930
MBL Fred Karush Library Readership, 2474
Porter County Health and Wellness Grant, 3246
Vancouver Foundation Grants and Community Initiatives Program, 3849

Materials Engineering
IIE KAUST Graduate Fellowships, 2134

Materials Sciences
Canada Graduate Scholarships (CGS) and NSERC Postgraduate Scholarships (PGS), 1010
HHMI-NIBIB Interfaces Initiative Grants, 2014
IIE KAUST Graduate Fellowships, 2134
NSF-NIST Interaction in Chemistry, Materials Research, Molecular Biosciences, Bioengineering, and Chemical Engineering, 3004
NSF Major Research Instrumentation Program (MRI) Grants, 3018

Materials Testing
AOCS Industrial Oil Products Division Student Award, 616

Materials, Metals/Alloys
NSF Small Business Innovation Research (SBIR) Grants, 3022

SUBJECT INDEX

Materials, Processing and Finishing
NSF-NIST Interaction in Chemistry, Materials Research, Molecular Biosciences, Bioengineering, and Chemical Engineering, 3004

Mathematics
Abdus Salam ICTP Federation Arrangement Scheme Grants, 205
Alfred P. Sloan Foundation Research Fellowships, 432
Alfred P. Sloan Foundation Science of Learning STEM Grants, 433
Baxter International Corporate Giving Grants, 841
BP Foundation Grants, 943
BWF Career Awards at the Scientific Interface, 983
Canada-U.S. Fulbright New Century Scholars Program Grants, 1008
Canada-U.S. Fulbright Senior Specialists Grants, 1009
Canada Graduate Scholarships (CGS) and NSERC Postgraduate Scholarships (PGS), 1010
Community Foundation of Greater Flint Grants, 1273
Fluor Foundation Grants, 1682
Fulbright Alumni Initiatives Awards, 1749
Fulbright Distinguished Chairs Awards, 1750
Fulbright New Century Scholars Awards, 1753
Fulbright Specialists Program Grants, 1754
Fulbright Scholars in Europe and Eurasia, 1755
Fulbright Traditional Scholar Program in Sub-Saharan Africa, 1756
Fulbright Traditional Scholar Program in the East Asia/Pacific Region, 1757
Fulbright Traditional Scholar Program in the Near East and North Africa Region, 1758
Fulbright Traditional Scholar Program in the South and Central Asia Region, 1759
Fulbright Traditional Scholar Program in the Western Hemisphere, 1760
Gladys Brooks Foundation Grants, 1826
Grand Rapids Community Foundation Grants, 1869
Gruber Foundation Cosmology Prize, 1907
HHMI-NIBIB Interfaces Initiative Grants, 2014
IIE 911 Armed Forces Scholarships, 2122
IIE AmCham Charitable Foundation U.S. Studies Scholarship, 2125
IIE Brazil Science Without Borders Undergraduate Scholarships, 2126
IIE Central Europe Summer Research Institute Summer Research Fellowship, 2127
IIE Chevron International REACH Scholarships, 2128
IIE David L. Boren Fellowships, 2129
MBL Eugene and Millicent Bell Fellowships, 2470
NHGRI Mentored Research Scientist Development Award, 2686
NHLBI Mentored Quantitative Research Career Development Awards, 2711
NIMH Curriculum Development Award in Neuroinformatics Research and Analysis, 2917
North Carolina GlaxoSmithKline Fndn Grants, 2991
NSF Alan T. Waterman Award, 3006
Qualcomm Grants, 3287
SfN Swartz Prize for Theoretical and Computational Neuroscience, 3537
Simone and Cino del Duca Grand Prix Awards, 3564
Texas Instruments Foundation STEM Education Grants, 3711
Toyota Motor Manufacturing of Kentucky Grants, 3745

Mathematics Education
3M Company Fndn Community Giving Grants, 6
Affymetrix Corporate Contributions Grants, 339
Alcoa Foundation Grants, 414
Alfred P. Sloan Foundation International Science Engagement Grants, 431
AMD Corporate Contributions Grants, 500
American Electric Power Foundation Grants, 530
ARCO Foundation Education Grants, 652
Baptist Community Ministries Grants, 828
Emily Davie and Joseph S. Kornfeld Foundation Grants, 1591
Ford Motor Company Fund Grants Program, 1687
Gates Millennium Scholars Program, 1776

Goodrich Corporation Foundation Grants, 1854
Grifols Community Outreach Grants, 1904
H.B. Fuller Foundation Grants, 1922
Houston Endowment Grants, 2053
James L. and Mary Jane Bowman Charitable Trust Grants, 2194
Lockheed Martin Corp Foundation Grants, 2374
Qualcomm Grants, 3287
Rathmann Family Foundation Grants, 3309
RGk Foundation Grants, 3361
Rockwell Collins Charitable Corporation Grants, 3397
Rohm and Haas Company Grants, 3403
TE Foundation Grants, 3705
Tellabs Foundation Grants, 3706
United Technologies Corporation Grants, 3803
Washington Gas Charitable Giving, 3908

Mathematics and Statistics
AAUW American Dissertation Fellowships, 195

Mathematics, Applied
Alfred P. Sloan Foundation Research Fellowships, 432
Canada Graduate Scholarships (CGS) and NSERC Postgraduate Scholarships (PGS), 1010
IIE KAUST Graduate Fellowships, 2134

Mechanical Engineering
IIE KAUST Graduate Fellowships, 2134
Lubrizol Foundation Grants, 2384

Mechanics
BMW of North America Charitable Contributions, 926

Media
New York University Steinhardt School of Education Fellowships, 2684
OSF Health Media Initiative Grants, 3089
Reinberger Foundation Grants, 3357
Willary Foundation Grants, 3944

Media Arts
Dept of Ed Special Education-National Activities-Technology and Media Services for Individuals with Disabilities, 1468
Grand Rapids Area Community Fndn Grants, 1865
Grand Rapids Area Community Foundation Nashwauk Area Endowment Fund Grants, 1866
Leon and Thea Koerner Foundation Grants, 2355
Pittsburgh Foundation Community Fund Grants, 3230
Reinberger Foundation Grants, 3357

Medical Devices Engineering
MMAAP Foundation Fellowship Award in Translational Medicine, 2581
MMAAP Foundation Research Project Award in Translational Medicine, 2585
W.M. Keck Foundation Med Research Grants, 3888

Medical Education
1st Source Foundation Grants, 2
AACAP Child and Adolescent Psychiatry (CAP) Teaching Scholarships, 51
AACAP Jeanne Spurlock Research Fellowship in Substance Abuse and Addiction for Minority Medical Students, 60
AACAP Life Members Mentorship Grants for Medical Students, 63
AACAP Summer Medical Student Fellowships, 75
AAFP Foundation Health Literacy State Grants, 132
AAFP Foundation Pfizer Teacher Devel Awards, 134
AAFP Foundation Wyeth Immunization Awards, 135
AAFP Research Stimulation Grants, 136
AAMC Abraham Flexner Award, 153
AAMC Alpha Omega Alpha Robert J. Glaser Distinguished Teacher Awards, 154
AAMC Caring for Community Grants, 156
AAMC Herbert W. Nickens Award, 158
AAMC Herbert W. Nickens Faculty Fellowship, 159
AAMC Herbert W. Nickens Medical Student Scholarships, 160

AAMC Humanism in Medicine Award, 161
AAO-HNS Medical Student Research Prize, 170
AAP Resident Research Grants, 190
Abbot and Dorothy H. Stevens Foundation Grants, 197
ACS Cancer Control Career Development Awards for Primary Care Physicians, 250
ACS Doc Degree Scholarships in Cancer Nursing, 252
ACS Doctoral Grants in Oncology Social Work, 254
ACS Scholarships in Cancer Nursing Practice, 255
ACS Physician Training Awards in Prevent Med, 275
Adams Rotary Memorial Fund B Scholarships, 303
Advocate HealthCare Post Graduate Administrative Fellowship, 314
Aetna Foundation National Medical Fellowship in Healthcare Leadership, 330
AIHS ForeFront MBA Studentship Award, 381
AIHS ForeFront MBT Studentship Awards, 382
Allyn Foundation Grants, 448
AlohaCare Believes in Me Scholarship, 449
AMA Fndn Arthur N. Wilson Scholarship, 482
AMA Foundation Joan F. Giambalvo Memorial Scholarships, 488
AMA Foundation Physicians of Tomorrow Scholarships, 492
AMDA Foundation Quality Improvement Award, 499
American College of Surgeons Co-Sponsored K08/K23 NIH Supplement Awards, 520
American College of Surgeons Health Policy Scholarships, 523
American College of Surgeons Health Policy Scholarships for General Surgeons, 524
American College of Surgeons International Guest Scholarships, 525
American College of Surgeons Resident Research Scholarships, 528
American Psychiatric Foundation Helping Hands Grants, 554
ANA Distinguished Neurology Teacher Award, 580
APA Young Investigator Grant Program, 634
Arkell Hall Foundation Grants, 661
Australian Academy of Science Grants, 807
AVDF Health Care Grants, 811
Benton County Foundation - Fitzgerald Family Scholarships, 879
Biogen Foundation Healthcare Professional Education Grants, 900
Biogen Foundation Patient Educational Grants, 901
Blackford County Community Foundation Hartford City Kiwanis Scholarship in Memory of Mike McDougall, 907
Blackford County Community Foundation Noble Memorial Scholarship, 908
Booth-Bricker Fund Grants, 933
Bright Family Foundation Grants, 951
Bush Foundation Medical Fellowships, 980
BWF Ad Hoc Grants, 982
BWF Career Awards for Medical Scientists, 984
California Wellness Foundation Work and Health Program Grants, 1001
Canadian Patient Safety Institute (CPSI) Patient Safety Studentships, 1012
Canadian Patient Safety Institute (CPSI) Research Grants, 1013
Catherine Holmes Wilkins Foundation Grants, 1042
CDC Epidemic Intell Service Training Grants, 1055
CDC Epidemiology Elective Rotation, 1056
Clarian Health Critical/Progressive Care Interns, 1168
Clarian Health Multi-Specialty Internships, 1169
Clarian Health OR Internships, 1170
Clarian Health Student Nurse Scholarships, 1172
Clowes ACS/AAST/NIGMS Jointly Sponsored Mentored Clinical Scientist Dev Award, 1185
Community Fndn for the Capital Region Grants, 1259
Conquer Cancer Foundation of ASCO International Development and Education Awards, 1333
Covenant Foundation of Waterloo Auxiliary Scholarships, 1349
Covidien Clinical Education Grants, 1351
Cowles Charitable Trust Grants, 1354
Daphne Seybolt Culpeper Fndn Grants, 1425

DAR Alice W. Rooke Scholarship, 1426
DAR Dr. Francis Anthony Beneventi Medical Scholarship, 1427
DAR Irene and Daisy MacGregor Memorial Scholarship, 1428
Denton A. Cooley Foundation Grants, 1459
Donald W. Reynolds Foundation Aging and Quality of Life Grants, 1487
Doris Duke Charitable Foundation Clinical Research Fellowships for Medical Students, 1493
Dr. John T. Macdonald Foundation Scholarships, 1508
Dr. Scholl Foundation Grants Program, 1511
Drs. Bruce and Lee Foundation Grants, 1513
Duke University Adult Cardiothoracic Anesthesia and Critical Care Medicine Fellowships, 1520
Duke University Ambulatory and Regional Anesthesia Fellowships, 1521
Duke University Clinical Cardiac Electrophysiology Fellowships, 1522
Duke University Hyperbaric Center Fellowships, 1523
Duke Univ Interventional Cardio Fellowships, 1524
Duke Univ Obstetric Anesthesia Fellowships, 1525
Duke University Pain Management Fellowships, 1526
Duke University Ped Anesthesiology Fellowship, 1527
Effie and Wofford Cain Foundation Grants, 1563
FMC Foundation Grants, 1683
Foundation for Appalachian Ohio Zelma Gray Medical School Scholarship, 1694
Goldman Philanthropic Partnerships Program, 1853
Harry A. & Margaret D. Towsley Fndn Grants, 1959
Hawaii Community Foundation Health Education and Research Grants, 1976
Healthcare Foundation of New Jersey Grants, 1982
Health Fndn of Greater Indianapolis Grants, 1984
Herman Goldman Foundation Grants, 2012
HHMI-NIH Cloister Research Scholars, 2015
HHMI Biomedical Research Grants for International Scientists: Infectious Diseases & Parasitology, 2017
HHMI Biomedical Research Grants for International Scientists in Canada and Latin America, 2018
HHMI Biomedical Research Grants for International Scientists in the Baltics, Central and Eastern Europe, Russia, and Ukraine, 2019
HHMI Med into Grad Initiative Grants, 2022
HHMI Physician-Scientist Early Career Award, 2023
HHMI Research Training Fellowships, 2024
J.W. Kieckhefer Foundation Grants, 2176
James H. Cummings Foundation Grants, 2190
James M. Collins Foundation Grants, 2195
John I. Smith Charities Grants, 2240
John S. Dunn Research Foundation Grants, 2250
Joseph Collins Foundation Scholarships, 2267
Josiah Macy Jr. Foundation Grants, 2277
Julius N. Frankel Foundation Grants, 2282
Lester Ray Fleming Scholarships, 2360
Lubbock Area Foundation Grants, 2383
Lynn and Rovena Alexander Family Foundation Grants, 2399
Mead Johnson Nutritionals Med Educ Grants, 2504
Mead Johnson Nutritionals Scholarships, 2505
MLA Donald A.B. Lindberg Fellowship, 2562
MMS and Alliance Charitable Foundation Grants for Community Action and Care for the Medically Uninsured, 2587
MMS and Alliance Charitable Foundation International Health Studies Grants, 2588
Mt. Sinai Health Care Foundation Academic Medicine and Bioscience Grants, 2598
Mt. Sinai Health Care Foundation Health of the Jewish Community Grants, 2599
NBME Stemmler Med Ed Research Grants, 2638
NCI Cancer Education and Career Development Program, 2648
Nesbitt Medical Student Foundation Scholarship, 2674
NHLBI Biomedical Research Training Program for Individuals from Underrepresented Groups, 2692
NHLBI Short-Term Research Education Program to Increase Diversity in Health-Related Research, 2733
NIH Mentored Clinical Scientist Research Career Development Award, 2887

NIH Ruth L. Kirschstein National Research Service Award Institutional Research Training Grants, 2900
NIH Ruth L. Kirschstein National Research Service Awards for Individual Predoctoral Fellows, 2903
NIH Ruth L. Kirschstein Nat Research Service Award Short-Term Institutional Training Grants, 2906
NLM Internet Connection for Medical Institutions Grants, 2946
NMF Aura E. Severinghaus Award, 2954
NMF Franklin C. McLean Award, 2955
NMF General Electric Med Fellowships, 2956
NMF Henry G. Halladay Awards, 2957
NMF Hugh J. Andersen Memorial Scholarship, 2958
NMF Irving Graef Memorial Scholarship, 2959
NMF Mary Ball Carrera Scholarship, 2960
NMF Metropolitan Life Foundation Awards Program for Academic Excellence in Medicine, 2961
NMF National Medical Association Awards for Medical Journalism, 2962
NMF National Medical Association Emerging Scholar Awards, 2963
NMF National Medical Association Patti LaBelle Award, 2964
NMF Need-Based Scholarships, 2965
NMF Ralph W. Ellison Prize, 2966
NMF William and Charlotte Cadbury Award, 2967
Nuffield Foundation Africa Program Grants, 3026
Nuffield Foundation Law and Society Grants, 3028
Nuffield Foundation Small Grants, 3031
NYAM David E. Rogers Fellowships, 3040
Ohio Learning Network Grants, 3070
Partnership for Cures Charles E. Culpeper Scholarships in Medical Science, 3117
Paso del Norte Health Foundation Grants, 3119
Pfizer/AAFP Foundation Visiting Professorship Program in Family Medicine, 3162
Pfizer Medical Education Track One Grants, 3182
Pfizer Medical Education Track Two Grants, 3183
PhRMA Foundation Informatics Pre Doctoral Fellowships, 3207
Piper Trust Healthcare & Med Research Grants, 3228
Potts Memorial Foundation Grants, 3254
Presbyterian Health Foundation Bridge, Seed and Equipment Grants, 3259
Rathmann Family Foundation Grants, 3309
RCPSC/AMS CanMEDs Research and Development Grants, 3316
RCPSC/AMS Donald Richards Wilson Award, 3317
RCPSC Continuing Proffesional Dev Grants, 3319
RCPSC Tom Dignan Indigenous Health Award, 3321
RCPSC Duncan Graham Award, 3322
RCPSC Harry S. Morton Lectureship in Surgery, 3323
RCPSC International Residency Educator of the Year Award, 3325
RCPSC Int'l Resident Leadership Award, 3326
RCPSC Janes Visiting Professorship in Surgery, 3328
RCPSC KJR Wightman Visiting Professorship in Medicine, 3331
RCPSC Kristin Sivertz Res Leadership Award, 3332
RCPSC McLaughlin-Gallie Visiting Professor, 3333
RCPSC Medical Education Research Grants, 3334
RCPSC Mentor of the Year Award, 3335
RCPSC Peter Warren Traveling Scholarship, 3336
RCPSC Prix D'excellence (Specialist of the Year) Award, 3337
RCPSC Program Administrator Award for Innovation and Excellence, 3339
RCPSC Program Director of the Year Award, 3340
RCPSC Robert Maudsley Fellowship for Studies in Medical Education, 3341
RCPSC Royal College Medal Award in Medicine, 3342
RCPSC Royal College Medal Award in Surgery, 3343
Regenstrief General Internal Medicine - General Pediatrics Research Fellowship Program, 3351
Regenstrief Geriatrics Fellowship, 3352
Regenstrief Master Of Science In Clinical Research/CITE Program, 3353
Rice Foundation Grants, 3363
Ruth Eleanor Bamberger and John Ernest Bamberger Memorial Foundation Grants, 3427

RWJF Harold Amos Medical Faculty Development Scholars, 3435
RWJF Nurse Faculty Scholars, 3441
Sain-Orr and Royak-DeForest Steadman Foundation Grants, 3452
Saint Luke's Health Initiatives Grants, 3455
Samueli Foundation Health Grants, 3465
San Antonio Area Foundation Grants, 3469
Schering-Plough Foundation Health Grants, 3492
Seattle Fndn Doyne M. Green Scholarships, 3505
SNM Student Fellowship Awards, 3588
Society for Imaging Informatics in Medicine (SIIM) Emeritus Mentor Grants, 3597
Society for Imaging Informatics in Medicine (SIIM) Micro Grants, 3598
Society for Imaging Informatics in Medicine (SIIM) Research Grants, 3599
Society for Imaging Informatics in Med Small Grant for Product Support Development, 3600
Society for Imaging Informatics in Medicine (SIIM) Small Training Grants, 3601
Special Olympics Health Profess Student Grants, 3629
Summer Medical and Dental Education Program: Interprofessional Pilot Grants, 3673
Vancouver Fndn Public Health Bursary Package, 3850
VDH Emergency Med Services Training Grants, 3855
VDH Rescue Squad Assistance Fund Grants, 3856
Victor E. Speas Foundation Grants, 3868
Yampa Valley Community Foundation Cody St. John Scholarships, 3977
Yampa Valley Community Foundation Volunteer Firemen Scholarships, 3979

Medical Ethics
AAMC Herbert W. Nickens Award, 158
AMDA Foundation Medical Director of the Year Award, 497
Greenwall Foundation Bioethics Grants, 1897
HHMI Grants and Fellowships Programs, 2020
NINR Acute and Chronic Care During Mechanical Ventilation Research Grants, 2926
RCPSC KJR Wightman Award for Scholarship in Ethics, 3330
Regenstrief Geriatrics Fellowship, 3352
RSC Abbyann D. Lynch Medal in Bioethics, 3419

Medical Informatics
AHIMA Dissertation Assistance Grants, 352
AHIMA Faculty Development Stipends, 353
AHIMA Grant-In-Aid Research Grants, 354
Carla J. Funk Governmental Relations Award, 1020
CDC Public Health Informatics Fellowships, 1068
Covidien Partnership for Neighborhood Wellness Grants, 1353
EBSCO / MLA Annual Meeting Grants, 1545
Fannie E. Rippel Foundation Grants, 1630
Foundation for Seacoast Health Grants, 1698
Healthcare Fndn for Orange County Grants, 1981
Majors MLA Chapter Project of the Year, 2418
Medical Informatics Section/MLA Career Development Grant, 2508
MLA Continuing Education Grants, 2559
MLA Cunningham Memorial International Fellowship, 2560
MLA David A. Kronick Traveling Fellowship, 2561
MLA Estelle Brodman Academic Medical Librarian of the Year Award, 2563
MLA Graduate Scholarship, 2564
MLA Grad Scholarship for Minority Students, 2565
MLA Ida and George Eliot Prize, 2566
MLA Lois Ann Colaianni Award for Excellence and Achievement in Hospital Librarianship, 2568
MLA Louise Darling Medal for Distinguished Achievement in Collection Development in the Health Sciences, 2569
MLA Lucretia W. McClure Excellence in Education Award, 2570
MLA Research, Development, and Demonstration Project Grant, 2572
MLA Rittenhouse Award, 2573

SUBJECT INDEX

MLA Section Project of the Year Award, 2574
MLA Thomas Reuters / Frank Bradway Rogers Information Advancement Award, 2576
NEI Ruth L. Kirschstein National Research Service Award Short-Term Institutional Research Training Grants, 2668
NHGRI Mentored Research Scientist Development Award, 2686
NIMH Curriculum Development Award in Neuroinformatics Research and Analysis, 2917
NIMH Short Courses in Neuroinformatics, 2920
NLM Exploratory/Developmental Grants, 2942
NLM Grants for Scholarly Works in Biomedicine and Health, 2944
NLM Internet Connection for Medical Institutions Grants, 2946
NLM Research Project Grants, 2951
NSF Postdoc Research Fellowships in Biology, 3020
PhRMA Foundation Informatics Post Doctoral Fellowships, 3206
PhRMA Foundation Informatics Pre Doctoral Fellowships, 3207
PhRMA Foundation Informatics Research Starter Grants, 3208
PhRMA Fndn Info Sabbatical Fellowships, 3209
PhRMA Foundation Pharmacology/Toxicology Post Doctoral Fellowships, 3215
PhRMA Foundation Pharmacology/Toxicology Sabbatical Fellowships, 3218
Regenstrief General Internal Medicine - General Pediatrics Research Fellowship Program, 3351
Regenstrief Geriatrics Fellowship, 3352
Regenstrief Master Of Science In Clinical Research/ CITE Program, 3353
Society for Imaging Informatics in Medicine (SIIM) Emeritus Mentor Grants, 3597
Society for Imaging Informatics in Medicine (SIIM) Micro Grants, 3598
Society for Imaging Informatics in Medicine (SIIM) Research Grants, 3599
Society for Imaging Informatics in Med Small Grant for Product Support Development, 3600
Society for Imaging Informatics in Medicine (SIIM) Small Training Grants, 3601
Thomson Reuters / MLA Doctoral Fellowships, 3728
USDD Clinical Functional Assessment Investigator-Initiated Grants, 3840
Virginia L. and William K. Beatty MLA Volunteer Service Award, 3877

Medical Physics
A Social Corporation Grants, 758
HRAMF Smith Family Awards for Excellence in Biomedical Research, 2063
NIH Support of Competitive Research (SCORE) Pilot Project Awards, 2913
NIH Support of Competitive Research (SCORE) Research Advancement Awards, 2914
Piedmont Natural Gas Foundation Health and Human Services Grants, 3221
W.M. Keck Foundation Med Research Grants, 3888

Medical Programs
AACAP Child and Adolescent Psychiatry (CAP) Teaching Scholarships, 51
AAP Community Access To Child Health (CATCH) Implementation Grants, 178
AAP Legislative Conference Scholarships, 185
ACL Empowering Seniors to Prevent Health Care Fraud Grants, 241
ACL Medicare Improvements for Patients & Providers Funding for Beneficiary Outreach & Assistance for Title VI Native Americans, 245
Addison H. Gibson Foundation Medical Grants, 305
Advocate HealthCare Post Graduate Administrative Fellowship, 314
Alaska Airlines Corporate Giving Medical Emergency and Research Grants, 402
American Medical Association Awards, 536
ASHG Public Health Genetics Fellowship, 712

Baptist-Trinity Lutheran Legacy Fndn Grants, 827
Baton Rouge Area Foundation Grants, 838
Bernard and Audre Rapoport Foundation Health Grants, 882
Blowitz-Ridgeway Foundation Early Childhood Development Research Award, 912
Bristol-Myers Squibb Foundation Product Donations Grants, 960
BWF Ad Hoc Grants, 982
BWF Career Awards for Medical Scientists, 984
Carl R. Hendrickson Family Foundation Grants, 1028
Carroll County Community Foundation Grants, 1039
CDC Evaluation of the Use of Rapid Testing For Influenza in Outpatient Medical Settings, 1057
CDC Preventive Med Residency & Fellowship, 1064
Clark and Ruby Baker Foundation Grants, 1174
Community Fndn AIDS Endowment Awards, 1237
Community Foundation of Riverside and San Bernardino County James Bernard and Mildred Jordan Tucker Grants, 1305
ConocoPhillips Foundation Grants, 1327
Conquer Cancer Foundation of ASCO Improving Cancer Care Grants, 1331
Covidien Clinical Education Grants, 1351
Cultural Society of Filipino Americans Grants, 1373
Deborah Munroe Noonan Memorial Grants, 1443
DeKalb County Community Foundation - Garrett Hospital Aid Foundation Grants, 1446
Denton A. Cooley Foundation Grants, 1459
Different Needz Foundation Grants, 1478
Edna G. Kynett Memorial Foundation Grants, 1551
Florence Hunt Maxwell Foundation Grants, 1676
Ford Family Foundation Grants - Access to Health and Dental Services, 1686
Frank B. Hazard General Charity Fund Grants, 1721
Frederick McDonald Trust Grants, 1740
Frederick W. Marzahl Memorial Fund Grants, 1741
George A. and Grace L. Long Foundation Grants, 1788
George S. & Dolores Dore Eccles Fndn Grants, 1801
Goldman Philanthropic Partnerships Program, 1853
Green River Area Community Fndn Grants, 1895
Hamilton Company Syringe Product Grant, 1937
Hampton Roads Community Foundation Health and Human Service Grants, 1941
Helen Irwin Littauer Educational Trust Grants, 1994
Henrietta Lange Burk Fund Grants, 2001
Henry L. Guenther Foundation Grants, 2006
Horace Moses Charitable Foundation Grants, 2047
HRK Foundation Health Grants, 2067
Huntington Beach Police Officers Fndn Grants, 2081
IDPH Emergency Medical Services Assistance Fund Grants, 2118
IDPH Local Health Department Public Health Emergency Response Grants, 2120
Independence Blue Cross Charitable Medical Care Grants, 2150
Independence Blue Cross Nursing Internships, 2152
James M. Collins Foundation Grants, 2195
John Edward Fowler Memorial Fndn Grants, 2235
Kimball International-Habig Foundation Health and Human Services Grants, 2308
Kosciuszko Foundation Dr. Marie E. Zakrzewski Medical Scholarship, 2321
Kuntz Foundation Grants, 2326
Leonard L. & Bertha U. Abess Fndn Grants, 2357
Lucile Packard Foundation for Children's Health Grants, 2386
Lucy Downing Nisbet Charitable Fund Grants, 2387
Lynn and Rovena Alexander Family Foundation Grants, 2399
M.D. Anderson Foundation Grants, 2403
Mabel A. Horne Trust Grants, 2406
Marcia and Otto Koehler Foundation Grants, 2427
Marjorie Moore Charitable Foundation Grants, 2438
Mead Johnson Nutritionals Med Educ Grants, 2504
Miami County Community Foundation - Operation Round Up Grants, 2543
Milken Family Foundation Grants, 2555
MMS and Alliance Charitable Foundation International Health Studies Grants, 2588

Medical Research /773

Nelda C. and H.J. Lutcher Stark Fndn Grants, 2672
NIH Support of Competitive Research (SCORE) Pilot Project Awards, 2913
NIH Support of Competitive Research (SCORE) Research Advancement Awards, 2914
OSF Access to Essential Medicines Initiative, 3086
OSF Public Health Grants in Kyrgyzstan, 3094
Pittsburgh Foundation Medical Research Grants, 3231
Posey County Community Foundation Grants, 3252
Rayonier Foundation Grants, 3311
RCPSC/AMS CanMEDs Research and Development Grants, 3316
RCPSC KJR Wightman Visiting Professorship in Medicine, 3331
RCPSC Professional Development Grants, 3338
RCPSC Royal College Medal Award in Medicine, 3342
Regenstrief Geriatrics Fellowship, 3352
Robert and Joan Dircks Foundation Grants, 3379
Robert B McMillen Foundation Grants, 3381
RWJF Harold Amos Medical Faculty Development Scholars, 3435
RWJF Nurse Faculty Scholars, 3441
Sain-Orr and Royak-DeForest Steadman Foundation Grants, 3452
Samueli Foundation Health Grants, 3465
Sarasota Memorial Healthcare Fndn Grants, 3484
Seabury Foundation Grants, 3501
Sick Kids Fndn Community Conference Grants, 3549
SOCFOC Catholic Ministries Grants, 3596
SRC Medical Research Grants, 3635
St. Joseph Community Health Foundation Burn Care and Prevention Grants, 3639
St. Joseph Community Health Foundation Improving Healthcare Access Grants, 3640
Straits Area Community Foundation Health Care Grants, 3657
STTI/ATI Educational Assessment Nursing Research Grants, 3665
Summer Medical and Dental Education Program: Interprofessional Pilot Grants, 3673
Tri-State Community Twenty-first Century Endowment Fund Grants, 3756
Unihealth Foundation Grants for Non-Profit Oranizations, 3783
UnitedHealthcare Children's Foundation Grants, 3795
USDD Care for the Critically Injured Burn Patient Grants, 3839
USDD Medical Practice Breadth of Medical Practice and Disease Frequency Exposure Grants, 3843
USDD Medical Practice Initiative Procedural Skill Decay and Maintenance Grants, 3844
USDD U.S. Army Medical Research and Materiel Command Broad Agency Announcement for Extramural Medical Research Grants, 3846
Vancouver Foundation Grants and Community Initiatives Program, 3849
Volkswagen of America Corporate Contributions, 3880
Walter L. Gross III Family Foundation Grants, 3901
WHO Foundation Volunteer Service Grants, 3940
Wilhelmina W. Jackson Trust Scholarships, 3941

Medical Research
AAAAI ARTrust Mini Grants for Allied Health, 20
AAAAI Distinguished Layperson Award, 25
AACAP Life Members Mentorship Grants for Medical Students, 63
Abracadabra Foundation Grants, 216
Acid Maltase Deficiency Association Helen Walker Research Grant, 235
ADA Foundation George C. Paffenbarger Student Research Awards, 291
AES-Grass Young Investigator Travel Awards, 317
AES Epilepsy Research Recognition Awards, 318
AES Research and Training Workshop Awards, 320
AES Robert S. Morison Fellowship, 323
AES William G. Lennox Award, 326
Aetna Foundation National Medical Fellowship in Healthcare Leadership, 330
Alaska Airlines Corporate Giving Medical Emergency and Research Grants, 402

774 / Medical Research

AMDA Foundation Quality Improvement Award, 499
AMHPS Dr. James A. Ferguson Emerging Infectious Diseases Fellowships, 573
ANA/Grass Foundation Award in Neuroscience, 578
ANA Derek Denny-Brown Neurological Scholar Award, 579
ANA Wolfe Neuropathy Research Prize, 585
APHL Emerging Infectious Diseases Fellowships, 636
ASPH/CDC Public Health Fellowships, 781
Australian Academy of Science Grants, 807
Biogen Foundation Scientific Research Grants, 902
Blowitz-Ridgeway Foundation Early Childhood Development Research Award, 912
Children's Leukemia Research Association Research Grants, 1138
Children's Tumor Fndn Clin Research Awards, 1140
Children's Tumor Foundation Drug Discovery Initiative Awards, 1141
Children's Tumor Foundation Schwannomatosis Awards, 1142
Children's Tumor Foundation Young Investigator Awards, 1143
CJ Foundation for SIDS Research Grants, 1164
Community Foundation of Louisville Bobbye M. Robinson Fund Grants, 1291
Community Foundation of Louisville Dr. W. Barnett Owen Memorial Fund for the Children of Louisville and Jefferson County Grants, 1294
Community Foundation of Louisville Lee Look Fund for Spinal Injury Grants, 1297
CSTE CDC/CSTE Applied Epidemiology Fellowships, 1367
Elizabeth Glaser Int'l Leadership Awards, 1566
Elizabeth Morse Genius Charitable Trust Grants, 1569
FAMRI Clinical Innovator Awards, 1628
FAMRI Young Clinical Scientist Awards, 1629
Foundation for Pharmaceutical Sciences Herb and Nina Demuth Grant, 1697
Foundation of Orthopedic Trauma Grants, 1707
Frank Reed and Margaret Jane Peters Memorial Fund II Grants, 1728
Giant Food Charitable Grants, 1816
Gilbert Memorial Fund Grants, 1820
Grass Foundation Marine Biological Laboratory Fellowships, 1878
Grover Hermann Foundation Grants, 1906
Hampton Roads Community Foundation Mental Health Research Grants, 1942
Hearst Foundations Health Grants, 1987
HRAMF Charles A. King Trust Postdoctoral Research Fellowships, 2056
HRAMF Jeffress Trust Awards in Interdisciplinary Research, 2060
HRAMF Ralph and Marian Falk Medical Research Trust Catalyst Awards, 2061
HRAMF Ralph and Marian Falk Medical Research Trust Transformational Awards, 2062
HRAMF Smith Family Awards for Excellence in Biomedical Research, 2063
HRAMF Thome Foundation Awards in Alzheimer's Disease Drug Discovery Research, 2066
HRK Foundation Health Grants, 2067
IBRO Latin America Regional Funding for Short Research Stays, 2108
IBRO Return Home Fellowships, 2112
Jeffrey Thomas Stroke Shield Foundation Research Grants, 2216
Kovler Family Foundation Grants, 2323
Kroger Foundation Women's Health Grants, 2325
Leo Goodwin Foundation Grants, 2354
Lisa Higgins-Hussman Foundation Grants, 2370
Lymphatic Education and Research Network Additional Support Grants for NIH-funded F32 Postdoctoral Fellows, 2394
Lymphatic Education and Research Network Postdoctoral Fellowships, 2395
Mary Kay Foundation Cancer Research Grants, 2444
MBL Fred Karush Library Readership, 2474
MBL Plum Fndn John E. Dowling Fellowships, 2484
Mericos Foundation Grants, 2518

Merrick Foundation Grants, 2521
Meyer and Stephanie Eglin Foundation Grants, 2531
Milken Family Foundation Grants, 2555
MMAAP Foundation Geriatric Project Awards, 2582
MMAAP Fndn Hematology Project Awards, 2583
MMAAP Foundation Research Project Award in Reproductive Medicine, 2584
Morehouse PHSI Project Imhotep Internships, 2592
NEAVS Fellowsips in Women's Health and Sex Differences, 2659
NFL Charities Medical Grants, 2685
NHLBI Ruth L. Kirschstein National Research Service Awards for Individual Predoctoral Fellowships to Promote Diversity in Health-Related Research, 2730
NHLBI Ruth L. Kirschstein National Research Service Awards for Individual Predoctoral MD/PhD Fellows and Other Dual Degree Fellows, 2731
NHLBI Ruth L. Kirschstein National Research Service Awards for Individual Senior Fellows, 2732
NIH Ruth L. Kirschstein National Research Service Award Institutional Research Training Grants, 2900
NIH Ruth L. Kirschstein National Research Service Awards for Individual Predoctoral Fellows, 2903
NIH Ruth L. Kirschstein Nat Research Service Award Short-Term Institutional Training Grants, 2906
Notsew Orm Sands Foundation Grants, 3002
Olympus Corporation of Americas Fellowships, 3074
Orthopaedic Trauma Association Kathy Cramer Young Clinician Memorial Fellowships, 3083
Orthopaedic Trauma Assoc Research Grants, 3084
Osteosynthesis and Trauma Care Foundation European Visiting Fellowships, 3097
Perpetual Trust for Charitable Giving Grants, 3141
Pfizer Advancing Research in Transplantation Science (ARTS) Research Awards, 3163
Pfizer ASPIRE EU Dupuytren's Contracture Research Awards, 3166
Pfizer ASPIRE North America Hemophilia Research Awards, 3170
Pfizer ASPIRE North America Narrow Spectrum Antibiotics for the Treatment of MRSA Research Awards, 3171
Pfizer Australia Cancer Research Grants, 3174
Pfizer Australia Neuroscience Research Grants, 3175
Pfizer Australia Paediatric Endocrine Care Research Grants, 3176
Pfizer Inflammation Competitive Research Awards (UK), 3181
Pfizer Research Initiative Rheumatology Grants (Germany), 3185
RCPSC Harry S. Morton Traveling Fellowship in Surgery, 3324
RCPSC Medical Education Research Grants, 3334
RCPSC Peter Warren Traveling Scholarship, 3336
RCPSC Robert Maudsley Fellowship for Studies in Medical Education, 3341
Reinberger Foundation Grants, 3357
Ruth Estrin Goldberg Memorial for Cancer Research Grants, 3428
Sara Elizabeth O'Brien Trust Grants, 3483
Schramm Foundation Grants, 3495
Scleroderma Foundation Established Investigator Grants, 3497
SfN Jacob P. Waletzky Award, 3523
SfN Janett Rosenberg Trubatch Career Development Award, 3524
SfN Julius Axelrod Prize, 3526
SfN Peter & Pat Gruber Int'l Research Award, 3533
SfN Young Investigator Award, 3541
Sick Kids Foundation New Investigator Research Grants, 3550
Sioux Falls Area Community Foundation Community Fund Grants (Unrestricted), 3568
Sioux Falls Area Community Foundation Spot Grants (Unrestricted), 3570
USDD Broad Agency Announcement Grants, 3837
USDD Broad Agency Announcement HIV/AIDS Prevention Grants, 3838
USDD Care for the Critically Injured Burn Patient Grants, 3839

USDD Clinical Functional Assessment Investigator-Initiated Grants, 3840
USDD HIV/AIDS Prevention: Military Specific HIV/AIDS Prevention, Care, and Treatment for Non-PEPFAR Funded Countries Grants, 3841
USDD HIV/AIDS Prevention Program Information Systems Development Grants, 3842
USDD Medical Practice Breadth of Medical Practice and Disease Frequency Exposure Grants, 3843
USDD Medical Practice Initiative Procedural Skill Decay and Maintenance Grants, 3844
USDD President's Emergency Plan for AIDS Relief Military Specific HIV Seroprevalence and Behavioral Epidemiology Risk Suvey Grants, 3845
USDD U.S. Army Medical Research and Materiel Command Broad Agency Announcement for Extramural Medical Research Grants, 3846
Victor E. Speas Foundation Grants, 3868
W.M. Keck Foundation Med Research Grants, 3888
Wellcome Trust New Investigator Awards, 3919

Medical Sciences
ASHFoundation New Century Scholars Program Doctoral Scholarships, 706
ASHFoundation New Century Scholars Research Grant, 707
ASHFoundation Grant for New Investigators, 708
Conquer Cancer Foundation of ASCO Improving Cancer Care Grants, 1331
DAR Irene and Daisy MacGregor Memorial Scholarship, 1428
Harmony Project Grants, 1948
HRAMF Jeffress Trust Awards in Interdisciplinary Research, 2060
IIE AmCham Charitable Foundation U.S. Studies Scholarship, 2125
IIE Chevron International REACH Scholarships, 2128
Pfizer Special Events Grants, 3186
Scleroderma Foundation Established Investigator Grants, 3497
USDD Clinical Functional Assessment Investigator-Initiated Grants, 3840
W.M. Keck Foundation Med Research Grants, 3888
W.M. Keck Foundation Science and Engineering Research Grants, 3889

Medical Technology
APHL Emerging Infectious Diseases Fellowships, 636
A Social Corporation Grants, 758
Bernard and Audre Rapoport Foundation Health Grants, 882
Foundation for the Mid South Health and Wellness Grants, 1700
Highmark Physician eHealth Grants, 2026
Independence Blue Cross Charitable Medical Care Grants, 2150
Maddie's Fund Medical Equipment Grants, 2411
Mead Johnson Nutritionals Med Educ Grants, 2504
NCCAM Translational Tools for Clinical Studies of CAM Interventions Grants, 2641
NCI Academic-Industrial Partnerships for Devel and Validation of In Vivo Imaging Systems and Methods for Cancer Investigations, 2644
NCI Application and Use of Transformative Emerging Technologies in Cancer Research, 2645
NIH Support of Competitive Research (SCORE) Pilot Project Awards, 2913
NIH Support of Competitive Research (SCORE) Research Advancement Awards, 2914
NLM Manufacturing Processes of Medical, Dental, and Biological Technologies (SBIR) Grants, 2948
Piedmont Natural Gas Foundation Health and Human Services Grants, 3221
RCPSC/AMS CanMEDs Research and Development Grants, 3316
Regenstrief General Internal Medicine - General Pediatrics Research Fellowship Program, 3351
Regenstrief Master Of Science In Clinical Research/CITE Program, 3353
Samueli Foundation Health Grants, 3465

SUBJECT INDEX

STTI Joan K. Stout RN Research Grants, 3668
Topeka Community Foundation Kansas Blood Services Scholarships, 3738
USDD Clinical Functional Assessment Investigator-Initiated Grants, 3840
W.M. Keck Foundation Med Research Grants, 3888

Medical/Diagnostics Imaging
A Social Corporation Grants, 758
Grass Foundation Marine Biological Laboratory Advanced Imaging Fellowships, 1877
Hilda and Preston Davis Foundation Postdoctoral Fellowships in Eating Disorders Research, 2029
IBCAT Screening Mammography Grants, 2095
Maddie's Fund Medical Equipment Grants, 2411
NCCAM Translational Tools for Clinical Studies of CAM Interventions Grants, 2641
NCI Cancer Education and Career Development Program, 2648
NHLBI Career Transition Awards, 2694
Pfizer ASPIRE EU Antifungal Research Awards, 3165
Pfizer Inflammation Competitive Research Awards (UK), 3181
Piper Trust Healthcare & Med Research Grants, 3228
Regenstrief General Internal Medicine - General Pediatrics Research Fellowship Program, 3351
Samueli Foundation Health Grants, 3465
SNM Molecular Imaging Research Grant For Junior Medical Faculty, 3583
SNM Postdoc Molecular Imaging Scholar Grants, 3587

Medicinal Chemistry
American Chemical Society Alfred Burger Award in Medicinal Chemistry, 510
Arnold and Mabel Beckman Foundation Scholars Grants, 665
E.B. Hershberg Award for Important Discoveries in Medicinally Active Substances, 1535
HRAMF Thome Foundation Awards in Alzheimer's Disease Drug Discovery Research, 2066

Medicine
Adams Rotary Memorial Fund B Scholarships, 303
Ahmanson Foundation Grants, 355
American Chemical Society Award in Separations Science and Technology, 514
Blowitz-Ridgeway Foundation Early Childhood Development Research Award, 912
CDC Public Health Informatics Fellowships, 1068
HHMI Biomedical Research Grants for International Scientists: Infectious Diseases & Parasitology, 2017
HHMI Biomedical Research Grants for International Scientists in Canada and Latin America, 2018
HHMI Biomedical Research Grants for International Scientists in the Baltics, Central and Eastern Europe, Russia, and Ukraine, 2019
HHMI Med into Grad Initiative Grants, 2022
JELD-WEN Foundation Grants, 2218
Kuntz Foundation Grants, 2326
OSF Access to Essential Medicines Initiative, 3086
Perkin Fund Grants, 3139
Pfizer Healthcare Charitable Contributions, 3180
Pfizer Medical Education Track One Grants, 3182
Piper Trust Healthcare & Med Research Grants, 3228
Searle Scholars Program Grants, 3503
Seattle Fndn Doyne M. Green Scholarships, 3505
Seattle Fndn Hawkins Memorial Scholarship, 3506
SfN Science Journalism Student Awards, 3536
UniHealth Foundation Workforce Development Grants, 3786
Union Labor Health Fndn Angel Fund Grants, 3791
Wolfe Associates Grants, 3969

Medicine, History of
AIHP Sonnedecker Visiting Scholar Grants, 371
AIHP Thesis Support Grants, 372
Lillian Sholtis Brunner Summer Fellowships for Historical Research in Nursing, 2366
MLA Murray Gottlieb Prize Essay Award, 2571

NLM Grants for Scholarly Works in Biomedicine and Health, 2944
NYAM Student Essay Grants, 3046
RSC Jason A. Hannah Medal, 3420

Medicine, Internal
AAHPM Hospice & Palliative Med Fellowships, 152
ACS Cancer Control Career Development Awards for Primary Care Physicians, 250
American Philosophical Society Daland Fellowships in Clinical Investigation, 537
ARS Foundation Grants, 667
Bonfils-Stanton Foundation Grants, 932
CFF First- and Second-Year Clinical Fellowships, 1085
CFF Leroy Matthews Physician-Scientist Awards, 1087
Deutsche Banc Alex Brown and Sons Charitable Foundation Grants, 1471
NCCAM Translational Tools for Clinical Studies of CAM Interventions Grants, 2641
NIA Diversity in Medication Use and Outcomes in Aging Populations Grants, 2756
OSF Access to Essential Medicines Initiative, 3086
RCPSC Royal College Medal Award in Medicine, 3342
Regenstrief General Internal Medicine - General Pediatrics Research Fellowship Program, 3351
Regenstrief Geriatrics Fellowship, 3352

Medicine, Palliative
Conquer Cancer Foundation of ASCO International Development and Education Award in Palliative Care, 1332
Daniels Fund Grants-Aging, 1424
OSF Access to Essential Medicines Initiative, 3086
OSF International Palliative Care Grants, 3091
RCPSC Balfour M. Mount Visiting Professorship in Palliative Medicine, 3318

Membranes
NIH Structural Bio of Membrane Proteins Grant, 2911
NSF Cell Systems Cluster Grants, 3011

Men
Charles Delmar Foundation Grants, 1115
Katherine Baxter Memorial Foundation Grants, 2292

Menopause
AFAR Research Grants, 337
Pfizer Healthcare Charitable Contributions, 3180
Pfizer Medical Education Track Two Grants, 3183

Mental Disorders
AACAP Rieger Service Award for Excellence, 71
AACAP Sidney Berman Award for the School-Based Study and Intervention for Learning Disorders and Mental Illness, 73
Able To Serve Grants, 212
Adolph Coors Foundation Grants, 311
Alexander and Margaret Stewart Trust Grants, 419
American Psychiatric Foundation Alexander Gralnick, MD, Award for Research in Schizophrenia, 550
American Psychiatric Foundation Awards for Advancing Minority Mental Health, 551
American Psychiatric Foundation Grantss, 552
American Psychiatric Foundation Disaster Recovery Fund for Psychiatrists, 553
American Psychiatric Foundation James H. Scully Jr., M.D., Educational Fund Grants, 555
American Psychiatric Foundation Jeanne Spurlock Congressional Fellowship, 556
American Psychiatric Foundation Minority Fellowships, 557
American Psychiatric Foundation Typical or Troubled School Mental Health Education Grants, 558
Annunziata Sanguinetti Foundation Grants, 600
Auburn Foundation Grants, 798
Ben B. Cheney Foundation Grants, 875
Catherine Holmes Wilkins Foundation Grants, 1042
CNCS Senior Companion Program Grants, 1191
Collins Foundation Grants, 1210
Cralle Foundation Grants, 1356

Mental Health /775

Dammann Fund Grants, 1418
Danellie Foundation Grants, 1421
Elizabeth Morse Genius Charitable Trust Grants, 1569
Elkhart County Community Foundation Fund for Elkhart County, 1572
Erie Community Foundation Grants, 1601
Eva L. & Joseph M. Bruening Fndn Grants, 1613
Fairlawn Foundation Grants, 1626
Health Foundation of Greater Cincinnati Grants, 1983
Hogg Foundation for Mental Health Grants, 2039
Howe Foundation of North Carolina Grants, 2055
Jacob and Hilda Blaustein Foundation Health and Mental Health Grants, 2181
James H. Cummings Foundation Grants, 2190
John W. Alden Trust Grants, 2260
Leonard and Helen R. Stulman Charitable Foundation Grants, 2356
NARSAD Distinguished Investigator Grants, 2620
NARSAD Independent Investigator Grants, 2622
NARSAD Ruane Prize for Child and Adolescent Psychiatric Research, 2624
NARSAD Young Investigator Grants, 2625
NIEHS Small Business Innovation Research (SBIR) Program Grants, 2860
NIMH Early Identification and Treatment of Mental Disorders in Children and Adolescents Grants, 2918
Nina Mason Pulliam Charitable Trust Grants, 2921
NYCT Mental Health and Retardation Grants, 3056
Oppenstein Brothers Foundation Grants, 3078
Peter and Elizabeth C. Tower Foundation Annual Mental Health Grants, 3145
Peter and Elizabeth C. Tower Foundation Mental Health Reference and Resource Materials Mini-Grants, 3147
Peter and Elizabeth C. Tower Foundation Phase II Technology Initiative Grants, 3149
Peter and Elizabeth C. Tower Foundation Phase I Technology Initiative Grants, 3150
Peter and Elizabeth C. Tower Foundation Social and Emotional Preschool Curriculum Grants, 3151
Piedmont Natural Gas Foundation Health and Human Services Grants, 3221
Robert and Joan Dircks Foundation Grants, 3379
Rosalynn Carter Institute John and Betty Pope Fellowships, 3408
SAMHSA Campus Suicide Prevention Grants, 3460
SAMHSA Conference Grants, 3462
Simple Advise Education Center Grants, 3565
Staunton Farm Foundation Grants, 3649
Strowd Roses Grants, 3664
Union County Community Foundation Grants, 3790
Vancouver Sun Children's Fund Grants, 3851
Weyerhaeuser Family Foundation Health Grants, 3926

Mental Health
AACAP Sidney Berman Award for the School-Based Study and Intervention for Learning Disorders and Mental Illness, 73
Abbott Fund Science Education Grants, 203
Able Trust Vocational Rehabilitation Grants for Individuals, 213
Adams Foundation Grants, 301
Adolph Coors Foundation Grants, 311
Allen P. & Josephine B. Green Fndn Grants, 443
Alliance Healthcare Foundation Grants, 444
American Psychiatric Foundation Alexander Gralnick, MD, Award for Research in Schizophrenia, 550
American Psychiatric Foundation Awards for Advancing Minority Mental Health, 551
American Psychiatric Foundation Grantss, 552
American Psychiatric Foundation Disaster Recovery Fund for Psychiatrists, 553
American Psychiatric Foundation Helping Hands Grants, 554
American Psychiatric Foundation James H. Scully Jr., M.D., Educational Fund Grants, 555
American Psychiatric Foundation Jeanne Spurlock Congressional Fellowship, 556
American Psychiatric Foundation Minority Fellowships, 557

776 / Mental Health   SUBJECT INDEX

American Psychiatric Foundation Typical or Troubled School Mental Health Education Grants, 558
American Sociological Association Minority Fellowships, 563
AON Foundation Grants, 623
Arizona Diamondbacks Charities Grants, 657
ASA Metlife Foundation MindAlert Awards, 676
Austin S. Nelson Foundation Grants, 805
Barr Fund Grants, 835
Baxter International Foundation Grants, 843
Ben B. Cheney Foundation Grants, 875
Blowitz-Ridgeway Foundation Early Childhood Development Research Award, 912
Bodenwein Public Benevolent Foundation Grants, 928
Boyd Gaming Corporation Contributions Program, 941
Brookdale Fnd National Group Respite Grants, 966
Brooklyn Community Foundation Caring Neighbors Grants, 968
Bush Fndn Health & Human Services Grants, 979
Caesars Foundation Grants, 995
California Endowment Innovative Ideas Challenge Grants, 1000
Callaway Golf Company Foundation Grants, 1003
Campbell Hoffman Foundation Grants, 1006
CFFVR Jewelers Mutual Charitable Giving, 1099
CFFVR Robert and Patricia Endries Family Foundation Grants, 1102
CFFVR Schmidt Family G4 Grants, 1103
Charles H. Pearson Foundation Grants, 1121
Children Affected by AIDS Foundation Domestic Grants, 1145
Community Fndn of Louisville Health Grants, 1295
Comprehensive Health Education Fndn Grants, 1320
Connecticut Community Foundation Grants, 1324
Connecticut Health Foundation Health Initiative Grants, 1325
Cornerstone Foundation of Northeastern Wisconsin Grants, 1347
D. W. McMillan Foundation Grants, 1407
Dammann Fund Grants, 1418
DHHS Adolescent Family Life Demonstration Projects, 1472
Dolan Children's Foundation Grants, 1482
Duchossois Family Foundation Grants, 1518
eBay Foundation Community Grants, 1543
Edwards Memorial Trust Grants, 1558
Edwin W. and Catherine M. Davis Fndn Grants, 1561
Eisner Foundation Grants, 1564
Elizabeth Morse Genius Charitable Trust Grants, 1569
El Paso Community Foundation Grants, 1579
Evjue Foundation Grants, 1617
Fairlawn Foundation Grants, 1626
FAR Fund Grants, 1631
Faye McBeath Foundation Grants, 1635
Florida BRAIVE Fund of Dade Community Foundation, 1678
Foundation for a Healthy Kentucky Grants, 1691
Foundation for Seacoast Health Grants, 1698
Fourjay Foundation Grants, 1712
Frank Stanley Beveridge Foundation Grants, 1730
Fremont Area Community Foundation Elderly Needs Grants, 1745
Gardner Foundation Grants, 1772
George H.C. Ensworth Memorial Fund Grants, 1797
Gheens Foundation Grants, 1814
GNOF IMPACT Grants for Health and Human Services, 1839
Greater Tacoma Community Foundation Ryan Alan Hade Endowment Fund, 1887
Gulf Coast Foundation of Community Operating Grants, 1916
Gulf Coast Foundation of Community Program Grants, 1917
Hampton Roads Community Foundation Mental Health Research Grants, 1942
Harold Brooks Foundation Grants, 1952
Hasbro Children's Fund Grants, 1973
Hawaii Community Foundation Health Education and Research Grants, 1976
HCA Foundation Grants, 1980

Hilton Head Island Foundation Grants, 2036
Hogg Foundation for Mental Health Grants, 2039
Irvin Stern Foundation Grants, 2164
Ittleson Foundation Mental Health Grants, 2167
Jacob and Hilda Blaustein Foundation Health and Mental Health Grants, 2181
James R. Thorpe Foundation Grants, 2199
Leonard and Helen R. Stulman Charitable Foundation Grants, 2356
Lydia deForest Charitable Trust Grants, 2393
M.J. Murdock Charitable Trust General Grants, 2405
Margaret T. Morris Foundation Grants, 2429
Mattel Children's Foundation Grants, 2454
Mattel International Grants Program, 2455
McCarthy Family Foundation Grants, 2490
Medtronic Foundation Patient Link Grants, 2514
MetroWest Health Foundation Grants to Schools to Conduct Mental Health Capacity Assessments, 2529
Morris & Gwendolyn Cafritz Fndn Grants, 2594
Morton K. and Jane Blaustein Foundation Health Grants, 2595
NARSAD Distinguished Investigator Grants, 2620
NARSAD Independent Investigator Grants, 2622
NARSAD Young Investigator Grants, 2625
Nick Traina Foundation Grants, 2792
NIH Research on Sleep and Sleep Disorders, 2898
North Dakota Community Foundation Grants, 2994
NYCT Girls and Young Women Grants, 3053
NYCT Mental Health and Retardation Grants, 3056
Ordean Foundation Grants, 3080
OSF Mental Health Initiative Grants, 3093
Paso del Norte Health Foundation Grants, 3119
Perry County Community Foundation Grants, 3142
Peter and Elizabeth C. Tower Foundation Annual Mental Health Grants, 3145
Peter and Elizabeth C. Tower Foundation Mental Health Reference and Resource Materials Mini-Grants, 3147
Peter and Elizabeth C. Tower Foundation Phase II Technology Initiative Grants, 3149
Peter and Elizabeth C. Tower Foundation Phase I Technology Initiative Grants, 3150
Peter and Elizabeth C. Tower Foundation Social and Emotional Preschool Curriculum Grants, 3151
Peter F. McManus Charitable Trust Grants, 3153
Philip L. Graham Fund Health and Human Services Grants, 3190
Piedmont Natural Gas Foundation Health and Human Services Grants, 3221
Pike County Community Foundation Grants, 3222
Posey Community Fndn Women's Fund Grants, 3251
Posey County Community Foundation Grants, 3252
Quantum Corporation Snap Server Grants, 3289
Rainbow Endowment Grants, 3300
Rayonier Foundation Grants, 3311
Retirement Research Foundation General Program Grants, 3359
Rosalynn Carter Institute Mattie J. T. Stepanek Caregiving Scholarships, 3409
SAMHSA Campus Suicide Prevention Grants, 3460
SAMHSA Child Mental Health Initiative (CMHI) Grants, 3461
SAMHSA Conference Grants, 3462
SAMHSA Drug Free Communities Support Program Grants, 3463
SAMHSA Strategic Prevention Framework State Incentive Grants, 3464
Seabury Foundation Grants, 3501
Sensient Technologies Foundation Grants, 3513
Sidgmore Family Foundation Grants, 3551
Sierra Health Foundation Responsive Grants, 3555
Simone and Cino del Duca Grand Prix Awards, 3564
Sioux Falls Area Community Foundation Community Fund Grants (Unrestricted), 3568
Sioux Falls Area Community Foundation Spot Grants (Unrestricted), 3570
Sisters of St. Joseph Healthcare Fndn Grants, 3576
Sophia Romero Trust Grants, 3624
Southbury Community Trust Fund, 3625
Spencer County Community Foundation Grants, 3632

St. Joseph Community Health Foundation Improving Healthcare Access Grants, 3640
Stackpole-Hall Foundation Grants, 3643
Staunton Farm Foundation Grants, 3649
Stella and Charles Guttman Foundation Grants, 3651
Thomas Thompson Trust Grants, 3725
Thomas W. Bradley Foundation Grants, 3726
USDD Workplace Violence in Military Grants, 3847
Warrick County Community Foundation Grants, 3904
Weyerhaeuser Family Foundation Health Grants, 3926
William T. Grant Foundation Research Grants, 3958

Mental Retardation
Able To Serve Grants, 212
Ben B. Cheney Foundation Grants, 875
Benton Community Foundation Grants, 878
Blue River Community Foundation Grants, 922
Brown County Community Foundation Grants, 970
Community Foundation of Bartholomew County Heritage Fund Grants, 1262
Community Foundation of Bartholomew County James A. Henderson Award for Fundraising, 1263
Community Foundation of Greater Fort Wayne - Community Endowment and Clarke Endowment Grants, 1274
Dept of Ed Rehabilitation Training Grants, 1463
Fragile X Syndrome Postdoc Research Fellows, 1714
George Foundation Grants, 1795
John W. Alden Trust Grants, 2260
Joseph P. Kennedy Jr. Foundation Grants, 2275
NYCT Mental Health and Retardation Grants, 3056
Peter and Elizabeth C. Tower Foundation Annual Mental Health Grants, 3145
Pinkerton Foundation Grants, 3224
Roeher Institute Research Grants, 3400
Saint Louis Rams Fndn Community Donations, 3453
Simple Advise Education Center Grants, 3565
Thomas Sill Foundation Grants, 3724
UCT Scholarship Program, 3777

Mentoring Programs
AAAAI Mentorship Award, 31
AACN Mentorship Grant, 85
AAMC Humanism in Medicine Award, 161
Aetna Foundation Minority Scholars Grants, 329
American Psychiatric Association Minority Medical Student Summer Mentoring Program, 546
Andrew Family Foundation Grants, 588
ASHA Advancing Academic-Research Award, 695
ASHA Research Mentoring-Pair Travel Award, 698
ASHA Students Preparing for Academic & Research Careers (SPARC) Award, 700
ASM D.C. White Research and Mentoring Award, 723
ASM Roche Diagnostics Alice C. Evans Award, 748
Boeing Company Contributions Grants, 930
CDC Preventive Med Residency & Fellowship, 1064
CDC Public Health Prevention Service Fellowship Sponsorships, 1070
Coca-Cola Foundation Grants, 1203
Conquer Cancer Foundation of ASCO Comparative Effectiveness Research Professorship in Breast Cancer, 1329
Conquer Cancer Foundation of ASCO International Development and Education Awards, 1333
Conquer Cancer Foundation of ASCO Medical Student Rotation Grants, 1335
Conquer Cancer Foundation of ASCO Translational Research Professorships, 1337
Constellation Energy Corporate Grants, 1342
Cooper Industries Foundation Grants, 1345
Cruise Industry Charitable Foundation Grants, 1363
CSTE CDC/CSTE Applied Epidemiology Fellowships, 1367
Cystic Fibrosis Lifestyle Foundation Mentored Recreation Grants, 1399
eBay Foundation Community Grants, 1543
Elizabeth Glaser Int'l Leadership Awards, 1566
Essex County Community Foundation Women's Fund Grants, 1605
Farmers Insurance Corporate Giving Grants, 1634

SUBJECT INDEX

Greater Milwaukee Foundation Grants, 1883
Highmark Corporate Giving Grants, 2025
John Edward Fowler Memorial Fndn Grants, 2235
Kansas Health Foundation Recognition Grants, 2289
Kosciusko County Community Foundation REMC Operation Round Up Grants, 2320
Liberty Bank Foundation Grants, 2362
Mericos Foundation Grants, 2518
NCI Mentored Career Development Award to Promote Diversity, 2653
NEI Mentored Clinical Scientist Dev Award, 2666
NHLBI Mentored Career Dev Award to Promote Faculty Diversity in Biomed Research, 2708
NHLBI Mentored Quantitative Research Career Development Awards, 2711
NHLBI Midcareer Investigator Award in Patient-Oriented Research, 2713
NHLBI Pathway to Independence Awards, 2718
NIA Mentored Clinical Scientist Research Career Development Awards, 2771
NIDDK Mentored Research Scientist Development Award, 2825
NIH Mentored Clinical Scientist Devel Award, 2886
NSF Undergraduate Research and Mentoring in the Biological Sciences (URM), 3025
OneFamily Foundation Grants, 3075
PC Fred H. Bixby Fellowships, 3126
Peyton Anderson Foundation Grants, 3160
Piedmont Natural Gas Foundation Health and Human Services Grants, 3221
Priddy Foundation Program Grants, 3266
Prudential Foundation Education Grants, 3276
Robert R. Meyer Foundation Grants, 3387
Saint Louis Rams Fndn Community Donations, 3453
Samuel S. Johnson Foundation Grants, 3468
Schlessman Family Foundation Grants, 3493
SfN Bernice Grafstein Award for Outstanding Accomplishments in Mentoring, 3518
SfN Julius Axelrod Prize, 3526
SfN Louise Hanson Marshall Specific Recognition Award, 3527
SfN Science Educator Award, 3535
SfN Science Journalism Student Awards, 3536
Sidgmore Family Foundation Grants, 3551
Society for Imaging Informatics in Medicine (SIIM) Emeritus Mentor Grants, 3597
Vancouver Foundation Grants and Community Initiatives Program, 3849
Weingart Foundation Grants, 3913

Metabolic Diseases
Astellas Foundation Research Grants, 790
Bristol-Myers Squibb Clinical Outcomes and Research Grants, 955
CFF Postdoctoral Research Fellowships, 1089
CFF Research Grants, 1090
MDA Neuromuscular Disease Research Grants, 2501
NIDDK Collaborative Interdisciplinary Team Science in Diabetes, Endocrinology and Metabolic Diseases Grants, 2806
NIDDK Devel of Disease Biomarkers Grants, 2809
NIDDK Diabetes, Endocrinology, and Metabolic Diseases NRSAs--Individual, 2810
NIDDK Feasibility Clinical Research Grants in Diabetes, Endocrine and Metabolic Diseases, 2834
NIDDK Seeding Collaborative Interdisciplinary Team Science in Diabetes, Endocrinology and Metabolic Diseases Grants, 2843
NIDDK Translational Research for the Prevention and Control of Diabetes and Obesity, 2850
Zellweger Baby Support Network Grants, 3984

Metabolism
AFAR Research Grants, 337
Arnold and Mabel Beckman Foundation Young Investigators Grants, 666
Astellas Foundation Research Grants, 790
CFF Postdoctoral Research Fellowships, 1089
CFF Research Grants, 1090

NCI Application of Metabolomics for Translational and Biological Research Grants, 2646
NHLBI Career Transition Awards, 2694
NHLBI Research on the Role of Cardiomyocyte Mitochondria in Heart Disease: An Integrated Approach, 2726
NIDDK Collaborative Interdisciplinary Team Science in Diabetes, Endocrinology and Metabolic Diseases Grants, 2806
NIDDK Diet Comp & Energy Balance Grants, 2812
NIDDK Small Grants for Underrepresented Minority Scientists in Diabetes and Digestive and Kidney Diseases, 2847
NIDDK Type 2 Diabetes in the Pediatric Population Research Grants, 2852
NSF Biomolecular Systems Cluster Grants, 3010

Metallurgy
Canada Graduate Scholarships (CGS) and NSERC Postgraduate Scholarships (PGS), 1010

Meteorology
Canada Graduate Scholarships (CGS) and NSERC Postgraduate Scholarships (PGS), 1010

Methodist Church
Barra Foundation Project Grants, 834
Bradley-Turner Foundation Grants, 944
Danellie Foundation Grants, 1421
Effie and Wofford Cain Foundation Grants, 1563
McCallum Family Foundation Grants, 2489
United Methodist Committee on Relief Global AIDS Fund Grants, 3798
United Methodist Committee on Relief Global Health Grants, 3799
United Methodist Child Mental Health Grants, 3800

Mexico
Fulbright Traditional Scholar Program in the Western Hemisphere, 1760

Microbiology
Abbott-ASM Lifetime Achievement Award, 198
APHL Emerging Infectious Diseases Fellowships, 636
ASM-PAHO Infectious Diseases Epidemiology and Surveillance Fellowships, 714
ASM-UNESCO Leadership Grant for International Educators, 715
ASM/CDC Fellowships in Infectious Disease and Public Health Microbiology, 716
ASM BD Award for Research in Clin Microbio, 718
ASM bioMerieux Sonnenwirth Award for Leadership in Clinical Microbiology, 719
ASM Congressional Science Fellowships, 720
ASM Corporate Activities Program Student Travel Grants, 721
ASMCUE/GM Travel Assistance Grants, 722
ASM D.C. White Research and Mentoring Award, 723
ASM Early-Career Faculty Travel Award, 724
ASM Eli Lilly and Company Research Award, 726
ASM Faculty Ehancement Travel Award, 727
ASM Founders Distinguished Service Award, 728
ASM Gen-Probe Joseph Public Health Award, 729
ASM General Meeting Minority Travel Grants, 730
ASM GlaxoSmithKline International Member of the Year Award, 731
ASM ICAAC Young Investigator Awards, 732
ASM Intel Awards, 733
ASM Int'l Fellowship for Latin America and the Caribbean, 734
ASM Int'l Fellowships for Asia and Africa, 735
ASM Int'l Professorship for Latin America, 736
ASM Int'l Professorships for Asia and Africa, 737
ASM Merck Irving S. Sigal Memorial Awards, 739
ASM Microbiology Undergraduate Research Fellowship, 740
ASM Millis-Colwell Postgraduate Travel Grant, 741
ASM Procter & Gamble Award in Applied and Environmental Microbiology, 742
ASM Promega Biotechnology Research Award, 743

Middle School Education /777

ASM Public Communications Award, 744
ASM Raymond W. Sarber Awards, 745
ASM Richard and Mary Finkelstein Travel Grant, 746
ASM Robert D. Watkins Graduate Research Fellowship, 747
ASM Roche Diagnostics Alice C. Evans Award, 748
ASM sanofi-aventis ICAAC Award, 749
ASM Scherago-Rubin Award, 750
ASM Siemens Healthcare Diagnostics Young Investigator Award, 751
ASM Student & Postdoc Fellow Travel Grants, 752
ASM Student Travel Grants, 753
ASM TREK Diagnostic ABMM/ABMLI Professional Recognition Award, 754
ASM Undergraduate Research Fellowship, 755
ASM USFCC/J. Roger Porter Award, 756
ASM William A Hinton Research Training Award, 757
HRAMF Jeffress Trust Awards in Interdisciplinary Research, 2060
MBL Albert & Ellen Grass Fellowships, 2463
MBL Associates Summer Fellowships, 2465
MBL Burr & Susie Steinbach Fellowship, 2467
MBL E.E. Just Summer Fellowship for Minority Scientists, 2468
MBL Erik B. Fries Summer Fellowships, 2469
MBL Eugene and Millicent Bell Fellowships, 2470
MBL Evelyn and Melvin Spiegal Fellowship, 2471
MBL Frank R. Lillie Summer Fellowship, 2472
MBL Fred Karush Library Readership, 2474
MBL Gruss Lipper Family Foundation Summer Fellowship, 2475
MBL H. Keffer Hartline and Edward F. MacNichol, Jr. Fellowships, 2476
MBL Herbert W. Rand Summer Fellowship, 2477
MBL James E. and Faith Miller Memorial Summer Fellowship, 2478
MBL John M. Arnold Award, 2479
MBL Lucy B. Lemann Summer Fellowships, 2481
MBL M.G.F. Fuortes Summer Fellowships, 2482
MBL Nikon Summer Fellowship, 2483
MBL Scholarships and Awards, 2486
MBL Stephen W. Kuffler Summer Fellowships, 2487
MBL William Townsend Porter Summer Fellowships for Minority Investigators, 2488
NIH Ruth L. Kirschstein National Research Service Awards for Individual Postdoc Fellowships, 2902
NSF Animal Developmental Mechanisms Grants, 3007
NSF Cell Systems Cluster Grants, 3011
NSF Genes and Genome Systems Cluster Grants, 3015
PAHO-ASM Int'l Professorship Latin America, 3107
Pfizer ASPIRE EU Antifungal Research Awards, 3165
Pfizer ASPIRE EU Emerging Mechanisms of Resistance Antibacterial Research Awards, 3167

Microbiome
NHLBI Microbiome of Lung & Respiratory in HIV Infected Individuals & Uninfected Controls, 2712
Pfizer ASPIRE EU Emerging Mechanisms of Resistance Antibacterial Research Awards, 3167

Microenterprises
Union Bank, N.A. Foundation Grants, 3788

Microscopy
Arnold and Mabel Beckman Foundation Young Investigators Grants, 666

Middle School
W.H. & Mary Ellen Cobb Trust Grants, 3884

Middle School Education
American Electric Power Foundation Grants, 530
Carnegie Corporation of New York Grants, 1033
CDC David J. Sencer Museum Teacher Professional Development Workshops, 1053
Colonel Stanley R. McNeil Foundation Grants, 1211
DaimlerChrysler Corporation Fund Grants, 1412
Dr. & Mrs. Paul Pierce Memorial Fndn Grants, 1505
Frank B. Hazard General Charity Fund Grants, 1721
Frank Loomis Palmer Fund Grants, 1726

778 / Middle School Education

Fred & Gretel Biel Charitable Trust Grants, 1734
Helen Bader Foundation Grants, 1992
Hutton Foundation Grants, 2087
J.C. Penney Company Grants, 2169
John Clarke Trust Grants, 2232
Katrine Menzing Deakins Trust Grants, 2295
Linford and Mildred White Charitable Grants, 2368
Marjorie Moore Charitable Foundation Grants, 2438
Miller Foundation Grants, 2556
R.S. Gernon Trust Grants, 3293
Raskob Foundation for Catholic Activities Grants, 3306
Reinberger Foundation Grants, 3357
Thelma Braun & Bocklett Family Fndn Grants, 3713
William J. Brace Charitable Trust, 3955
Wood Family Charitable Trust Grants, 3972

Midwifery
Virginia Department of Health Nurse Practitioner/Nurse Midwife Scholarships, 3876

Migrant Labor
Charles Delmar Foundation Grants, 1115
Union Bank, N.A. Foundation Grants, 3788

Migrants
California Endowment Innovative Ideas Challenge Grants, 1000
Dept of Ed College Assistance Migrant Grants, 1462
Farmers Insurance Corporate Giving Grants, 1634
IIE Western Union Family Scholarships, 2143
Union Bank, N.A. Foundation Grants, 3788

Migratory Animals and Birds
Emma Barnsley Foundation Grants, 1594
Lucy Downing Nisbet Charitable Fund Grants, 2387

Military History
Harry Frank Guggenheim Foundation Dissertation Fellowships, 1963

Military Personnel
Anne J. Caudal Foundation Grants, 595
CONSOL Military and Armed Services Grants, 1340
Northrop Grumman Corporation Grants, 2997
USDD Care for the Critically Injured Burn Patient Grants, 3839
USDD Workplace Violence in Military Grants, 3847

Military Sciences
Fisher House Fnd Newman's Own Awards, 1671
Northrop Grumman Corporation Grants, 2997
USDD Medical Practice Initiative Procedural Skill Decay and Maintenance Grants, 3844
USDD U.S. Army Medical Research and Materiel Command Broad Agency Announcement for Extramural Medical Research Grants, 3846

Military Training
Florida BRAIVE Fund of Dade Community Foundation, 1678
Gulf Coast Foundation of Community Operating Grants, 1916
Gulf Coast Foundation of Community Program Grants, 1917
Northrop Grumman Corporation Grants, 2997

Mineralogy
U.S. Department of Education Erma Byrd Scholarships, 3768

Mining
U.S. Department of Education Erma Byrd Scholarships, 3768

Ministry
Chapman Charitable Foundation Grants, 1112
Joseph and Luella Abell Trust Scholarships, 2266
Welborn Baptist Fndn Health Ministries Grants, 3915

Minorities
AACR-Minorities in Cancer Research Jane Cooke Wright Lectureship Awards, 94
AACR Minority-Serving Institution Faculty Scholar in Cancer Research Awards, 119
AACR Min Scholar in Cancer Research Award, 120
AAMC Herbert W. Nickens Faculty Fellowship, 159
AAMC Herbert W. Nickens Medical Student Scholarships, 160
Abbot and Dorothy H. Stevens Foundation Grants, 197
ACL Field Initiated Projects Program: Minority-Serving Institution Development Grants, 242
ACL Field Initiated Projects Program: Minority-Serving Institution Research Grants, 243
Adolph Coors Foundation Grants, 311
Aetna Foundation Minority Scholars Grants, 329
Alberto Culver Corporate Contributions Grants, 408
Alfred P. Sloan Foundation Research Fellowships, 432
ALSAM Foundation Grants, 454
American Psychiatric Foundation Awards for Advancing Minority Mental Health, 551
American Psychiatric Foundation Jeanne Spurlock Congressional Fellowship, 556
American Psychiatric Foundation Minority Fellowships, 557
American Sociological Association Minority Fellowships, 563
Anheuser-Busch Foundation Grants, 591
Archer Daniels Midland Foundation Grants, 651
Arlington Community Foundation Grants, 662
ASM General Meeting Minority Travel Grants, 730
ASM Microbiology Undergraduate Research Fellowship, 740
ASM William A Hinton Research Training Award, 757
Avon Products Foundation Grants, 814
Benton Community Foundation Grants, 878
Blue River Community Foundation Grants, 922
BMW of North America Charitable Contributions, 926
Bodenwein Public Benevolent Foundation Grants, 928
Boston Foundation Grants, 936
Boston Globe Foundation Grants, 937
Boyd Gaming Corporation Contributions Program, 941
Breast Cancer Fund Grants, 947
Brown County Community Foundation Grants, 970
Carnegie Corporation of New York Grants, 1033
Charles Delmar Foundation Grants, 1115
Christopher & Dana Reeve Foundation Quality of Life Grants, 1152
Community Foundation of Bartholomew County Heritage Fund Grants, 1262
Community Foundation of Bartholomew County James A. Henderson Award for Fundraising, 1263
Community Foundation of Greater Fort Wayne - Community Endowment and Clarke Endowment Grants, 1274
ConocoPhillips Foundation Grants, 1327
Conquer Cancer Foundation of ASCO Medical Student Rotation Grants, 1335
Conquer Cancer Foundation of ASCO Resident Travel Award for Underrepresented Populations, 1336
Cralle Foundation Grants, 1356
Daphne Seybolt Culpeper Fndn Grants, 1425
Dr. Stanley Pearle Scholarships, 1512
Edward W. and Stella C. Van Houten Memorial Fund Grants, 1559
Edwin S. Webster Foundation Grants, 1560
Eugene M. Lang Foundation Grants, 1611
Farmers Insurance Corporate Giving Grants, 1634
FDHN Student Research Fellowships, 1654
Ford Motor Company Fund Grants Program, 1687
Frederick McDonald Trust Grants, 1740
General Service Foundation Human Rights and Economic Justice Grants, 1784
George A. and Grace L. Long Foundation Grants, 1788
George W. Brackenridge Foundation Grants, 1802
George W. Wells Foundation Grants, 1804
Grotto Foundation Project Grants, 1905
H.J. Heinz Company Foundation Grants, 1923
Harold Brooks Foundation Grants, 1952
Health Fndn of Greater Indianapolis Grants, 1984

Helen Steiner Rice Foundation Grants, 1998
Jacob G. Schmidlapp Trust Grants, 2183
James Ford Bell Foundation Grants, 2187
Janirve Foundation Grants, 2209
John Clarke Trust Grants, 2232
John P. Murphy Foundation Grants, 2248
Joseph H. & Florence A. Roblee Fndn Grants, 2269
Katharine Matthies Foundation Grants, 2291
MBL E.E. Just Summer Fellowship for Minority Scientists, 2468
MBL William Townsend Porter Summer Fellowships for Minority Investigators, 2488
MLA Grad Scholarship for Minority Students, 2565
Morehouse PHSI Project Imhotep Internships, 2592
NHLBI Biomedical Research Training Program for Individuals from Underrepresented Groups, 2692
NHLBI Lymphatics in Health and Disease in the Digestive, Cardio & Pulmonary Systems, 2704
NHLBI Research Supplements to Promote Diversity in Health-Related Research, 2727
NHLBI Ruth L. Kirschstein National Research Service Awards for Individual Predoctoral Fellowships to Promote Diversity in Health-Related Research, 2730
NIDDK Small Grants for Underrepresented Minority Scientists in Diabetes and Digestive and Kidney Diseases, 2847
NIEHS Small Business Innovation Research (SBIR) Program Grants, 2860
NIH Enhancing Adherence to Diabetes Self-Management Behaviors, 2875
NIH Health Promotion Among Racial and Ethnic Minority Males, 2880
NINR Diabetes Self-Management in Minority Populations Grants, 2930
NMF Franklin C. McLean Award, 2955
NMF General Electric Med Fellowships, 2956
NSF Postdoc Research Fellowships in Biology, 3020
NSF Undergraduate Research and Mentoring in the Biological Sciences (URM), 3025
Oppenstein Brothers Foundation Grants, 3078
Pfizer ASPIRE North America Rheumatology Research Awards, 3172
Playboy Foundation Grants, 3237
Portland General Electric Foundation Grants, 3250
Powell Foundation Grants, 3255
Prudential Foundation Education Grants, 3276
Rayonier Foundation Grants, 3311
Rhode Island Foundation Grants, 3362
Richard & Susan Smith Family Fndn Grants, 3365
Robert R. Meyer Foundation Grants, 3387
RWJF Disparities Research for Change Grants, 3433
RWJF Harold Amos Medical Faculty Development Research Grants, 3434
RWJF Harold Amos Medical Faculty Development Scholars, 3435
Sensient Technologies Foundation Grants, 3513
Shell Oil Company Foundation Community Development Grants, 3544
SNM PDEF Mickey Williams Minority Student Scholarships, 3584
Sony Corporation of America Grants, 3623
Sophia Romero Trust Grants, 3624
Textron Corporate Contributions Grants, 3712
Thelma Braun & Bocklett Family Fndn Grants, 3713
Union Bank, N.A. Foundation Grants, 3788
Volkswagen of America Corporate Contributions, 3880
W. C. Griffith Foundation Grants, 3882
W.C. Griffith Foundation Grants, 3881
Wilson-Wood Foundation Grants, 3960
Winston-Salem Fndn Stokes County Grants, 3967
Z. Smith Reynolds Foundation Small Grants, 3982

Minorities, Ethnic
ACL Field Initiated Projects Program: Minority-Serving Institution Research Grants, 243
Aetna Foundation Minority Scholars Grants, 329
ANA Junior Academic Neurologist Scholarships, 582
BMW of North America Charitable Contributions, 926
Robert R. Meyer Foundation Grants, 3387
SfN Neuroscience Fellowships, 3530

SUBJECT INDEX

Minority Education
3M Company Fndn Community Giving Grants, 6
AACAP Jeanne Spurlock Minority Med Student Clinical Fellow in Child & Adolescent Psych, 59
AAMC Herbert W. Nickens Faculty Fellowship, 159
AAMC Herbert W. Nickens Medical Student Scholarships, 160
AAUW American Postdoctoral Research Leave Fellowships, 196
ACL Field Initiated Projects Program: Minority-Serving Institution Development Grants, 242
ACL Field Initiated Projects Program: Minority-Serving Institution Research Grants, 243
Aetna Foundation Minority Scholars Grants, 329
Air Products and Chemicals Grants, 399
Alcoa Foundation Grants, 414
American Psychiatric Association Minority Fellowships Program, 543
American Psychiatric Foundation Jeanne Spurlock Congressional Fellowship, 556
American Psychiatric Foundation Minority Fellowships, 557
Anheuser-Busch Foundation Grants, 591
ARCO Foundation Education Grants, 652
ASHFoundation Graduate Student Scholarships, 703
ASHFoundation Graduate Student Scholarships for Minority Students, 704
ASM Robert D. Watkins Graduate Research Fellowship, 747
AUPHA Corris Boyd Scholarships, 800
AUPHA Foster G. McGaw Scholarships, 801
BBVA Compass Foundation Charitable Grants, 848
BMW of North America Charitable Contributions, 926
CIGNA Foundation Grants, 1160
CMS Historically Black Colleges and Universities (HBCU) Health Services Research Grants, 1187
Coca-Cola Foundation Grants, 1203
Cruise Industry Charitable Foundation Grants, 1363
DaimlerChrysler Corporation Fund Grants, 1412
Edward W. and Stella C. Van Houten Memorial Fund Grants, 1559
FMC Foundation Grants, 1683
Gates Millennium Scholars Program, 1776
Graco Foundation Grants, 1860
Grifols Community Outreach Grants, 1904
Health Management Scholarships and Grants for Minorities, 1986
Indiana Minority Teacher/Special Services Scholarships, 2156
Johnson & Johnson Corporate Contributions, 2253
Josiah Macy Jr. Foundation Grants, 2277
MLA Grad Scholarship for Minority Students, 2565
NHLBI Mentored Career Award for Faculty at Institutions that Promote Diversity (K01), 2706
NHLBI Mentored Career Award For Faculty At Minority Institutions, 2707
NHLBI Short-Term Research Education Program to Increase Diversity in Health-Related Research, 2733
NHLBI Training Program Grants for Institutions that Promote Diversity, 2738
NIA Minority Dissert Aging Research Grants, 2773
NIH Health Promotion Among Racial and Ethnic Minority Males, 2880
NMF Aura E. Severinghaus Award, 2954
NMF General Electric Med Fellowships, 2956
NMF Henry G. Halladay Awards, 2957
NMF Hugh J. Andersen Memorial Scholarship, 2958
NMF Irving Graef Memorial Scholarship, 2959
NMF Mary Ball Carrera Scholarship, 2960
NMF Metropolitan Life Foundation Awards Program for Academic Excellence in Medicine, 2961
NMF National Medical Association Awards for Medical Journalism, 2962
NMF National Medical Association Emerging Scholar Awards, 2963
NMF National Medical Association Patti LaBelle Award, 2964
NMF Need-Based Scholarships, 2965
NMF Ralph W. Ellison Prize, 2966
NMF William and Charlotte Cadbury Award, 2967

NSF Undergraduate Research and Mentoring in the Biological Sciences (URM), 3025
RGk Foundation Grants, 3361
SfN Neuroscience Fellowships, 3530
Union Labor Health Fndn Community Grants, 3792
William J. Brace Charitable Trust, 3955

Minority Employment
ACL Field Initiated Projects Program: Minority-Serving Institution Development Grants, 242
ACL Field Initiated Projects Program: Minority-Serving Institution Research Grants, 243
Albert W. Rice Charitable Foundation Grants, 411
Charles H. Pearson Foundation Grants, 1121
George W. Wells Foundation Grants, 1804
GNOF New Orleans Works Grants, 1844
M.D. Anderson Foundation Grants, 2403
Textron Corporate Contributions Grants, 3712
USDA Organic Agriculture Research Grants, 3830

Minority Health
ACL Field Initiated Projects Program: Minority-Serving Institution Development Grants, 242
ACL Field Initiated Projects Program: Minority-Serving Institution Research Grants, 243
Aetna Foundation Integrated Health Care Grants, 328
Aetna Foundation Racial and Ethnic Health Care Equity Grants, 332
AHRQ Independent Scientist Award, 365
Aid for Starving Children Emergency Assistance Fund Grants, 367
American Psychiatric Foundation Awards for Advancing Minority Mental Health, 551
American Psychiatric Foundation Disaster Recovery Fund for Psychiatrists, 553
Anheuser-Busch Foundation Grants, 591
Avon Foundation Breast Care Fund Grants, 813
BCBSNC Foundation Grants, 860
California Endowment Innovative Ideas Challenge Grants, 1000
CDC Fnd Tobacco Network Lab Fellowship, 1062
Cigna Civic Affairs Sponsorships, 1159
Commonwealth Fund/Harvard University Fellowship in Minority Health Policy, 1226
Commonwealth Fund Harkness Fellowships in Health Care Policy and Practice, 1229
Community Foundation for the National Capital Region Community Leadership Grants, 1260
Community Fndn of Louisville Health Grants, 1295
Connecticut Health Foundation Health Initiative Grants, 1325
Foundations of East Chicago Health Grants, 1709
Frank W. & Carl S. Adams Memorial Grants, 1731
GNOF IMPACT Grants for Health and Human Services, 1839
Ms. Foundation for Women Health Grants, 2597
NMF General Electric Med Fellowships, 2956
Philip L. Graham Fund Health and Human Services Grants, 3190
Portland General Electric Foundation Grants, 3250
RWJF Disparities Research for Change Grants, 3433
Seattle Foundation Health and Wellness Grants, 3507
SfN Science Educator Award, 3535
Union Labor Health Fndn Community Grants, 3792

Minority Schools
Alcoa Foundation Grants, 414
Allyn Foundation Grants, 448
Coca-Cola Foundation Grants, 1203
NHLBI Mentored Career Award for Faculty at Institutions that Promote Diversity (K01), 2706
NHLBI Training Program Grants for Institutions that Promote Diversity, 2738
NIEHS Mentored Research Scientist Development Award, 2858
NINR Mentored Research Scientist Development Award, 2932
Procter and Gamble Fund Grants, 3274

Minority/Woman-Owned Business
NSF Small Business Innovation Research (SBIR) Grants, 3022

Molecular Probes
AAAS Eppendorf Science Prize for Neurobiology, 38
NIH Solicitation of Assays for High Throughput Screening (HTS) in the Molecular Libraries Probe Production Centers Network (MLPCN), 2910
NINDS Optimization of Small Molecule Probes for the Nervous System (SBIR) Grants, 2922
NINDS Optimization of Small Molecule Probes for the Nervous System (STTR) Grants, 2923
NINDS Optimization of Small Molecule Probes for the Nervous System Grants, 2924

Morphology
NSF Systematic Biology and Biodiversity Inventories Grants, 3024

Multiculturalism
American-Scandinavian Foundation Visiting Lectureship Grants, 502

Multiple Myeloma
ASCO/UICC International Cancer Technology Transfer (ICRETT) Fellowships, 677
ASCO Young Investigator Award, 683
Conquer Cancer Foundation of ASCO Career Development Award, 1328
Robert Leet & Clara Guthrie Patterson Grants, 3383

Multiple Sclerosis
Austin S. Nelson Foundation Grants, 805
Multiple Sclerosis Foundation Brighter Tomorrow Grants, 2602
NMSS Scholarships, 2968
Pfizer Healthcare Charitable Contributions, 3180
Pfizer Medical Education Track Two Grants, 3183
Pittsburgh Foundation Medical Research Grants, 3231
Seattle Foundation Medical Funds Grants, 3508

Muscular Dystrophy
Biogen Foundation Fellowships, 898
Doris and Victor Day Foundation Grants, 1491
MDA Development Grant, 2500
MDA Neuromuscular Disease Research Grants, 2501

Musculoskeletal Disorders
CDI Interdisciplinary Research Initiatives Grants, 1074
CDI Postdoctoral Fellowships, 1075
LAM Fnd Established Investigator Awards, 2337
LAM Foundation Pilot Project Grants, 2338
LAM Foundation Postdoctoral Fellowships, 2339
NIAMS Pilot and Feasibility Clinical Research Grants in Arthritis, Musculoskeletal & Skin Diseases, 2774
Pfizer ASPIRE EU Dupuytren's Contracture Research Awards, 3166

Musculoskeletal System
ACSM NASA Space Physiology Research Grants, 271
CDI Interdisciplinary Research Initiatives Grants, 1074
CDI Postdoctoral Fellowships, 1075

Museum Education
Colonel Stanley R. McNeil Foundation Grants, 1211
Shield-Ayres Foundation Grants, 3546
The Ray Charles Foundation Grants, 3716

Museums
1st Source Foundation Grants, 2
3M Company Fndn Community Giving Grants, 6
Abbot and Dorothy H. Stevens Foundation Grants, 197
Agnes Gund Foundation Grants, 345
Air Products and Chemicals Grants, 399
Albuquerque Community Foundation Grants, 412
Alexis Gregory Foundation Grants, 426
Andrew Family Foundation Grants, 588
Aratani Foundation Grants, 649
Arthur and Rochelle Belfer Foundation Grants, 670

780 / Museums

Arthur Ashley Williams Foundation Grants, 671
Atherton Family Foundation Grants, 794
Auburn Foundation Grants, 798
Ball Brothers Foundation General Grants, 820
Barra Foundation Project Grants, 834
Batchelor Foundation Grants, 837
Ben B. Cheney Foundation Grants, 875
Bender Foundation Grants, 876
Berrien Community Foundation Grants, 884
Bertha Russ Lytel Foundation Grants, 885
Booth-Bricker Fund Grants, 933
Bruce and Adele Greenfield Foundation Grants, 972
Burlington Industries Foundation Grants, 977
Burlington Northern Santa Fe Foundation Grants, 978
Byron W. & Alice L. Lockwood Fnd Grants, 991
Caleb C. and Julia W. Dula Educational and Charitable Foundation Grants, 997
Carl Gellert and Celia Berta Gellert Foundation Grants, 1024
Carylon Foundation Grants, 1040
Cessna Foundation Grants Program, 1081
Chamberlain Foundation Grants, 1108
Chazen Foundation Grants, 1129
Christensen Fund Regional Grants, 1149
Cleveland-Cliffs Foundation Grants, 1181
Cockrell Foundation Grants, 1204
Collins C. Diboll Private Foundation Grants, 1209
Colonel Stanley R. McNeil Foundation Grants, 1211
Community Foundation Alliance City of Evansville Endowment Fund Grants, 1238
Community Foundation of Eastern Connecticut General Southeast Grants, 1269
Connecticut Community Foundation Grants, 1324
Constantin Foundation Grants, 1341
Cooper Industries Foundation Grants, 1345
Cowles Charitable Trust Grants, 1354
Cralle Foundation Grants, 1356
Crystelle Waggoner Charitable Trust Grants, 1364
David Geffen Foundation Grants, 1431
Daywood Foundation Grants, 1438
DeRoy Testamentary Foundation Grants, 1469
Doris and Victor Day Foundation Grants, 1491
Dr. Scholl Foundation Grants Program, 1511
E. Rhodes & Leona B. Carpenter Grants, 1539
Elizabeth McGraw Foundation Grants, 1568
FirstEnergy Foundation Community Grants, 1668
Gardner Foundation Grants, 1772
George and Ruth Bradford Foundation Grants, 1789
George Foundation Grants, 1795
George W. Codrington Charitable Fndn Grants, 1803
Greater Green Bay Community Fndn Grants, 1881
Greater Worcester Community Foundation Discretionary Grants, 1888
Greenspun Family Foundation Grants, 1896
H. Leslie Hoffman and Elaine S. Hoffman Foundation Grants, 1924
Harold and Arlene Schnitzer CARE Foundation Grants, 1950
Harold R. Bechtel Charitable Remainder Uni-Trust Grants, 1953
High Meadow Foundation Grants, 2027
Hillman Foundation Grants, 2032
Hudson Webber Foundation Grants, 2072
Huffy Foundation Grants, 2073
Hugh J. Andersen Foundation Grants, 2074
Huie-Dellmon Trust Grants, 2075
Idaho Power Company Corporate Contributions, 2115
Ike and Roz Friedman Foundation Grants, 2145
Jessie Ball Dupont Fund Grants, 2226
Joseph Alexander Foundation Grants, 2265
Kenneth T. & Eileen L. Norris Fndn Grants, 2300
Lubbock Area Foundation Grants, 2383
Mabel Y. Hughes Charitable Trust Grants, 2409
Marcia and Otto Koehler Foundation Grants, 2427
Marie H. Bechtel Charitable Remainder Uni-Trust Grants, 2432
Maurice J. Masserini Charitable Trust Grants, 2456
Maxon Charitable Foundation Grants, 2458
McConnell Foundation Grants, 2492
McLean Contributionship Grants, 2497

Mericos Foundation Grants, 2518
Morris & Gwendolyn Cafritz Fndn Grants, 2594
Nicholas H. Noyes Jr. Memorial Fndn Grants, 2791
Oppenstein Brothers Foundation Grants, 3078
Parkersburg Area Community Foundation Action Grants, 3116
Perkins Charitable Foundation Grants, 3140
Piedmont Natural Gas Corporate and Charitable Contributions, 3220
PMI Foundation Grants, 3239
Price Family Charitable Fund Grants, 3262
R.C. Baker Foundation Grants, 3292
Reinberger Foundation Grants, 3357
Richard & Susan Smith Family Fndn Grants, 3365
Robert W. Woodruff Foundation Grants, 3389
Samuel S. Johnson Foundation Grants, 3468
Schlessman Family Foundation Grants, 3493
Schurz Communications Foundation Grants, 3496
Shell Deer Park Grants, 3543
Shield-Ayres Foundation Grants, 3546
Sid W. Richardson Foundation Grants, 3553
Sioux Falls Area Community Foundation Community Fund Grants (Unrestricted), 3568
Sioux Falls Area Community Foundation Spot Grants (Unrestricted), 3570
Strake Foundation Grants, 3658
Stranahan Foundation Grants, 3659
Sunderland Foundation Grants, 3675
SunTrust Bank Trusteed Foundations Florence C. and Harry L. English Memorial Fund Grants, 3680
SunTrust Bank Trusteed Foundations Greene-Sawtell Grants, 3681
SunTrust Bank Trusteed Foundations Harriet McDaniel Marshall Tust Grants, 3682
SunTrust Bank Trusteed Foundations Nell Warren Elkin and William Simpson Elkin Grants, 3683
SunTrust Bank Trusteed Foundations Walter H. and Marjory M. Rich Memorial Fund Grants, 3685
Susan Vaughan Foundation Grants, 3695
Taubman Endowment for the Arts Fndn Grants, 3702
Taylor S. Abernathy and Patti Harding Abernathy Charitable Trust Grants, 3704
Textron Corporate Contributions Grants, 3712
Thelma Doelger Charitable Trust Grants, 3714
The Ray Charles Foundation Grants, 3716
Thomas Sill Foundation Grants, 3724
Tri-State Community Twenty-first Century Endowment Fund Grants, 3756
Union Bank, N.A. Foundation Grants, 3788
Vermont Community Foundation Grants, 3865
W.C. Griffith Foundation Grants, 3881
W. C. Griffith Foundation Grants, 3882
Wayne County Foundation Grants, 3909
Wege Foundation Grants, 3912
William B. Dietrich Foundation Grants, 3946
William B. Stokely Jr. Foundation Grants, 3947
William G. Gilmore Foundation Grants, 3951
William J. & Tina Rosenberg Fndn Grants, 3954

Music
Adaptec Foundation Grants, 304
Agnes Gund Foundation Grants, 345
Alexis Gregory Foundation Grants, 426
Auburn Foundation Grants, 798
Barberton Community Foundation Grants, 829
Berks County Community Foundation Grants, 881
Blanche and Irving Laurie Foundation Grants, 910
Canada-U.S. Fulbright New Century Scholars Program Grants, 1008
Canada-U.S. Fulbright Senior Specialists Grants, 1009
Chazen Foundation Grants, 1129
Clarence E. Heller Charitable Foundation Grants, 1166
Community Fndn for the Capital Region Grants, 1259
Community Foundation of Bloomington and Monroe County Grants, 1265
Community Foundation of Mount Vernon and Knox County Grants, 1299
Cudd Foundation Grants, 1370
DAAD Research Stays for University Academics and Scientists, 1408

Edwin W. and Catherine M. Davis Fndn Grants, 1561
Ensworth Charitable Foundation Grants, 1595
FirstEnergy Foundation Community Grants, 1668
Fulbright Alumni Initiatives Awards, 1749
Fulbright Distinguished Chairs Awards, 1750
Fulbright New Century Scholars Grants, 1753
Fulbright Specialists Program Grants, 1754
Fulbright Scholars in Europe and Eurasia, 1755
Fulbright Traditional Scholar Program in Sub-Saharan Africa, 1756
Fulbright Traditional Scholar Program in the East Asia/Pacific Region, 1757
Fulbright Traditional Scholar Program in the Near East and North Africa Region, 1758
Fulbright Traditional Scholar Program in the South and Central Asia Region, 1759
Fulbright Traditional Scholar Program in the Western Hemisphere, 1760
Gibson Foundation Grants, 1818
Grammy Foundation Grants, 1863
Grand Rapids Area Community Fndn Grants, 1865
Grand Rapids Area Community Foundation Nashwauk Area Endowment Fund Grants, 1866
Greater Milwaukee Foundation Grants, 1883
Guitar Center Music Foundation Grants, 1914
Harold R. Bechtel Charitable Remainder Uni-Trust Grants, 1953
High Meadow Foundation Grants, 2027
I.A. O'Shaughnessy Foundation Grants, 2088
Idaho Community Foundation Eastern Region Competitive Grants, 2114
John Ben Snow Memorial Trust Grants, 2231
Katharine Matthies Foundation Grants, 2291
Leon and Thea Koerner Foundation Grants, 2355
Long Island Community Foundation Grants, 2375
M. Bastian Family Foundation Grants, 2402
Margaret T. Morris Foundation Grants, 2429
Maurice J. Masserini Charitable Trust Grants, 2456
New York University Steinhardt School of Education Fellowships, 2684
Olive Higgins Prouty Foundation Grants, 3072
Procter and Gamble Fund Grants, 3274
Richard D. Bass Foundation Grants, 3366
Thelma Braun & Bocklett Family Fndn Grants, 3713
The Ray Charles Foundation Grants, 3716
W.C. Griffith Foundation Grants, 3881
White County Community Foundation - Annie Horton Scholarship, 3929

Music Appreciation
Tri-State Community Twenty-first Century Endowment Fund Grants, 3756

Music Composition
IIE Rockefeller Foundation Bellagio Center Residencies, 2142
Mockingbird Foundation Grants, 2589

Music Conducting
Leon and Thea Koerner Foundation Grants, 2355

Music Education
Cemala Foundation Grants, 1077
Clarence E. Heller Charitable Foundation Grants, 1166
Deborah Munroe Noonan Memorial Grants, 1443
George A. and Grace L. Long Foundation Grants, 1788
Gibson Foundation Grants, 1818
Grand Rapids Area Community Fndn Grants, 1865
Grand Rapids Area Community Foundation Nashwauk Area Endowment Fund Grants, 1866
Guitar Center Music Foundation Grants, 1914
Heineman Foundation for Research, Education, Charitable and Scientific Purposes, 1990
Milagro Foundation Grants, 2552
Mockingbird Foundation Grants, 2589
Portland General Electric Foundation Grants, 3250
Southbury Community Trust Fund, 3625
Wood Family Charitable Trust Grants, 3972

SUBJECT INDEX

Music Therapy
Community Fndn for the Capital Region Grants, 1259
Deborah Munroe Noonan Memorial Grants, 1443
Guitar Center Music Foundation Grants, 1914

Music Video Industry
Daniel & Nanna Stern Family Fndn Grants, 1422

Music, Instrumental
Mockingbird Foundation Grants, 2589

Music, Vocal
Bonfils-Stanton Foundation Grants, 932
Mockingbird Foundation Grants, 2589
Peyton Anderson Foundation Grants, 3160

Musical Instruments
Gibson Foundation Grants, 1818
Guitar Center Music Foundation Grants, 1914

Musicology/Music Theory
Australasian Institute of Judicial Administration Seed Funding Grants, 806

Myasthenia Gravis
MDA Neuromuscular Disease Research Grants, 2501
MGFA Post-Doctoral Research Fellowships, 2539
MGFA Student Fellowships, 2540

Mycology
ASM/CDC Fellowships in Infectious Disease and Public Health Microbiology, 716

Myeloma
Robert Leet & Clara Guthrie Patterson Grants, 3383

Nanoscience
HRAMF Thome Foundation Awards in Alzheimer's Disease Drug Discovery Research, 2066
Royal Norwegian Embassy Kavli Prizes, 3414

National Disease Organizations
Alcatel-Lucent Technologies Foundation Grants, 413
Archer Daniels Midland Foundation Grants, 651
ArvinMeritor Foundation Health Grants, 674
CDC-Hubert Global Health Fellowship, 1047
CDC Epidemic Intell Service Training Grants, 1055
CDC Epidemiology Elective Rotation, 1056
Lockheed Martin Corp Foundation Grants, 2374

National Planning/Policy
AMA-MSS Government Relations Internships, 476
Carla J. Funk Governmental Relations Award, 1020

National Security
AAAS Science and Technology Policy Fellowships: Global Health and Development, 41
Carnegie Corporation of New York Grants, 1033
Florida BRAIVE Fund of Dade Community Foundation, 1678
Gulf Coast Foundation of Community Operating Grants, 1916
Gulf Coast Foundation of Community Program Grants, 1917
IIE Rockefeller Foundation Bellagio Center Residencies, 2142
United States Institute of Peace - Jennings Randolph Senior Fellowships, 3802

Native American Education
Burlington Northern Santa Fe Foundation Grants, 978
Dorothea Haus Ross Foundation Grants, 1498
General Motors Foundation Grants, 1783
John R. Oishei Foundation Grants, 2249
MLA Grad Scholarship for Minority Students, 2565
NHLBI Training Program Grants for Institutions that Promote Diversity, 2738
NMF Mary Ball Carrera Scholarship, 2960
NMF Need-Based Scholarships, 2965

U.S. Department of Education Rehabilitation Training - Rehabilitation Continuing Education - Institute on Rehabilitation Issues, 3771

Native Americans
ADA Found Minority Dental Scholarships, 292
Administration on Aging Senior Medicare Patrol Project Grants, 309
AMA Foundation Minority Scholars Awards, 491
American Sociological Association Minority Fellowships, 563
amfAR Fellowships, 567
amfAR Global Initiatives Grants, 568
amfAR Mathilde Krim Fellowships in Basic Biomedical Research, 569
amfAR Public Policy Grants, 570
amfAR Research Grants, 571
ASM Robert D. Watkins Graduate Research Fellowship, 747
CDC Increasing Breast and Cervical Cancer Screening Services for Urban American Indian/Alaska Native Women, 1063
Charles Delmar Foundation Grants, 1115
FDHN Student Research Fellowships, 1654
GNOF Bayou Communities Grants, 1835
Grotto Foundation Project Grants, 1905
Lotus 88 Fnd for Women & Children Grants, 2376
NMF Franklin C. McLean Award, 2955
Percy B. Ferebee Endowment Grants, 3138
RWJF Harold Amos Medical Faculty Development Research Grants, 3434
Sioux Falls Area Community Foundation Field-of-Interest and Donor-Advised Grants, 3569
SNM PDEF Mickey Williams Minority Student Scholarships, 3584
U.S. Department of Education Vocational Rehabilitation Services Projects for American Indians with Disabilities Grants, 3774

Natural History
NIDDK Type 2 Diabetes in the Pediatric Population Research Grants, 2852

Natural Products
Dean Foods Community Involvement Grants, 1440

Natural Resources
Abbot and Dorothy H. Stevens Foundation Grants, 197
Ahn Family Foundation Grants, 356
BBVA Compass Foundation Charitable Grants, 848
Beirne Carter Foundation Grants, 870
BP Foundation Grants, 943
Cessna Foundation Grants Program, 1081
Charles Delmar Foundation Grants, 1115
Clayton Fund Grants, 1180
Community Fndn for Greater Buffalo Grants, 1247
Community Foundation of Greater Birmingham Grants, 1272
Constellation Energy Corporate Grants, 1342
Dorrance Family Foundation Grants, 1501
Elliot Foundation Inc Grants, 1575
Fremont Area Community Fndn General Grants, 1746
George and Ruth Bradford Foundation Grants, 1789
H.A. & Mary K. Chapman Trust Grants, 1921
High Meadow Community Grants, 2027
Hoblitzelle Foundation Grants, 2038
Horace A. Kimball and S. Ella Kimball Foundation Grants, 2046
James Ford Bell Foundation Grants, 2187
Janirve Foundation Grants, 2209
John W. and Anna H. Hanes Foundation Grants, 2261
Knox County Community Foundation Grants, 2313
Kosciusko County Community Fndn Grants, 2319
Mary K. Chapman Foundation Grants, 2443
Maxon Charitable Foundation Grants, 2458
Owen County Community Foundation Grants, 3100
Perkins Charitable Foundation Grants, 3140
Pinellas County Grants, 3223
Posey County Community Foundation Grants, 3252
Pulaski County Community Foundation Grants, 3279

Rhode Island Foundation Grants, 3362
Ripley County Community Foundation Grants, 3376
Ripley County Community Foundation Small Project Grants, 3377
Samuel S. Johnson Foundation Grants, 3468
Seabury Foundation Grants, 3501
Sir Dorabji Tata Trust Grants for NGOs or Voluntary Organizations, 3572
Spencer County Community Foundation Grants, 3632
Texas Commission on the Arts Arts Respond Project Grants, 3708
Triangle Community Foundation Donor-Advised Grants, 3757
Vancouver Foundation Grants and Community Initiatives Program, 3849

Natural Sciences
AAUW American Dissertation Fellowships, 195
AAUW American Postdoctoral Research Leave Fellowships, 196
Charles A. Frueauff Foundation Grants, 1114
Cockrell Foundation Grants, 1204
DAAD Research Stays for University Academics and Scientists, 1408
Harry Frank Guggenheim Fnd Research Grants, 1964
HRAMF Jeffress Trust Awards in Interdisciplinary Research, 2060
IIE Rockefeller Foundation Bellagio Center Residencies, 2142
Reinberger Foundation Grants, 3357
SSHRC-NSERC College and Community Innovation Enhancement Grants, 3636

Nature Centers
Boston Foundation Grants, 936
Shield-Ayres Foundation Grants, 3546
Union Bank, N.A. Foundation Grants, 3788

Naval Sciences
DOL Occupational Safety and Health--Susan Harwood Training Grants, 1484

Neighborhood Revitalization
Blackford County Community Foundation Grants, 906
Blue Cross Blue Shield of Minnesota Foundation - Healthy Children: Growing Up Healthy Grants, 915
El Pomar Foundation Grants, 1582
George A. and Grace L. Long Foundation Grants, 1788
Hardin County Community Foundation Grants, 1946
Intergrys Corporation Grants, 2160
J.N. and Macie Edens Foundation Grants, 2174
Leon and Thea Koerner Foundation Grants, 2355
Mabel F. Hoffman Charitable Trust Grants, 2407
MGM Resorts Foundation Community Grants, 2541
Miller Foundation Grants, 2556
Textron Corporate Contributions Grants, 3712

Neighborhoods
Allstate Corporate Giving Grants, 446
Allstate Corp Hometown Commitment Grants, 447
Amerigroup Foundation Grants, 565
Bayer Foundation Grants, 846
Beckley Area Foundation Grants, 867
Brooklyn Community Foundation Caring Neighbors Grants, 968
Carnegie Corporation of New York Grants, 1033
CIT Corporate Giving Grants, 1161
Citizens Bank Mid-Atlantic Charitable Foundation Grants, 1162
Columbus Foundation Competitive Grants, 1217
Community Foundation for Greater New Haven Valley Neighborhood Grants, 1251
Community Fndn for Monterey County Grants, 1253
Community Foundation for the National Capital Region Community Leadership Grants, 1260
Cooper Industries Foundation Grants, 1345
Crail-Johnson Foundation Grants, 1355
CSRA Community Foundation Grants, 1366
DOJ Gang-Free Schools and Communities Intervention Grants, 1481

782 / Neighborhoods SUBJECT INDEX

Edward N. and Della L. Thome Memorial Foundation Grants, 1557
Erie Community Foundation Grants, 1601
Farmers Insurance Corporate Giving Grants, 1634
Four County Community Fndn General Grants, 1710
Frances L. and Edwin L. Cummings Memorial Fund Grants, 1717
George A. and Grace L. Long Foundation Grants, 1788
George Foundation Grants, 1795
George Gund Foundation Grants, 1796
Giant Food Charitable Grants, 1816
Graco Foundation Grants, 1860
Greater Milwaukee Foundation Grants, 1883
Greater Worcester Community Foundation Discretionary Grants, 1888
Hasbro Children's Fund Grants, 1973
Helen Bader Foundation Grants, 1992
Herbert A. & Adrian W. Woods Fndn Grants, 2007
Independence Community Foundation Community Quality of Life Grant, 2153
Intergrys Corporation Grants, 2160
Liberty Bank Foundation Grants, 2362
Lynde & Harry Bradley Foundation Fellowships, 2396
Lynde and Harry Bradley Foundation Grants, 2397
Lynde and Harry Bradley Foundation Prizes: Bradley Prizes, 2398
MGM Resorts Foundation Community Grants, 2541
Nordson Corporation Foundation Grants, 2981
Ralphs Food 4 Less Foundation Grants, 3305
RWJF New Jersey Health Initiatives Grants, 3440
San Diego Foundation for Change Grants, 3470
Skillman Fndn Good Neighborhoods Grants, 3577

Nepal
Fulbright Traditional Scholar Program in the South and Central Asia Region, 1759

Nephrology
NIDDK Training in Clinical Investigation in Kidney and Urology Grants, 2849
NKF Research Fellowships, 2938
NKF Young Investigator Grants, 2939
NYAM Edward N. Gibbs Memorial Lecture and Award in Nephrology, 3041
PKD Foundation Lillian Kaplan International Prize for the Advancement in the Understanding of Polycystic Kidney Disease, 3232

Nervous System
A-T Children's Project Grants, 13
A-T Children's Project Post Doctoral Fellowships, 14
Grass Foundation Marine Biological Laboratory Advanced Imaging Fellowships, 1877
Gruber Foundation Neuroscience Prize, 1909
NINDS Optimization of Small Molecule Probes for the Nervous System (SBIR) Grants, 2922
NINDS Optimization of Small Molecule Probes for the Nervous System (STTR) Grants, 2923
NINDS Optimization of Small Molecule Probes for the Nervous System Grants, 2924
PVA Education Foundation Grants, 3284
PVA Research Foundation Grants, 3285
SfN Jacob P. Waletzky Award, 3523
SfN Julius Axelrod Prize, 3526
SfN Next Generation Award, 3531
SfN Ralph W. Gerard Prize in Neuroscience, 3534
SfN Science Educator Award, 3535
SfN Swartz Prize for Theoretical and Computational Neuroscience, 3537
Simone and Cino del Duca Grand Prix Awards, 3564

Networking (Computers)
NLM Exploratory/Developmental Grants, 2942

Neuroanatomy
Grass Foundation Marine Biological Laboratory Advanced Imaging Fellowships, 1877
Grass Foundation Marine Biological Laboratory Fellowships, 1878
NIDDK Neuroimaging in Obesity Research, 2829

Neurobiology
AAAS Eppendorf Science Prize for Neurobiology, 38
AFAR Research Grants, 337
AHAF Alzheimer's Disease Research Grants, 349
Alfred P. Sloan Foundation International Science Engagement Grants, 431
ASHA Research Mentoring-Pair Travel Award, 698
Grass Foundation Marine Biological Laboratory Advanced Imaging Fellowships, 1877
Grass Foundation Marine Biological Laboratory Fellowships, 1878
Hereditary Disease Foundation John J. Wasmuth Postdoctoral Fellowships, 2009
Hereditary Disease Fnd Lieberman Award, 2010
Hereditary Disease Foundation Research Grants, 2011
Hilda and Preston Davis Foundation Postdoctoral Fellowships in Eating Disorders Research, 2029
John Merck Scholars Awards, 2246
Klarman Family Foundation Grants in Eating Disorders Research Grants, 2310
Lumosity Human Cognition Grant, 2390
MBL Albert & Ellen Grass Fellowships, 2463
MBL Associates Summer Fellowships, 2465
MBL Burr & Susie Steinbach Fellowship, 2467
MBL E.E. Just Summer Fellowship for Minority Scientists, 2468
MBL Erik B. Fries Summer Fellowships, 2469
MBL Evelyn and Melvin Spiegal Fellowship, 2471
MBL Frank R. Lillie Summer Fellowship, 2472
MBL Fred Karush Library Readership, 2474
MBL Gruss Lipper Family Foundation Summer Fellowship, 2475
MBL H. Keffer Hartline and Edward F. MacNichol, Jr. Fellowships, 2476
MBL Herbert W. Rand Summer Fellowship, 2477
MBL James E. and Faith Miller Memorial Summer Fellowship, 2478
MBL John M. Arnold Award, 2479
MBL Lucy B. Lemann Summer Fellowship, 2481
MBL M.G.F. Fuortes Summer Fellowships, 2482
MBL Nikon Summer Fellowship, 2483
MBL Plum Fndn John E. Dowling Fellowships, 2484
MBL Stephen W. Kuffler Summer Fellowship, 2487
MBL William Townsend Porter Summer Fellowships for Minority Investigators, 2488
NARSAD Goldman-Rakic Prize for Cognitive Neuroscience, 2621
National Center for Responsible Gaming Grants, 2629
National Center for Responsible Gaming Postdoctoral Fellowships, 2630
National Center for Resp Gaming Seed Grants, 2631
NHLBI Exploratory Studies in the Neurobiology of Pain in Sickle Cell Disease, 2699
NIA Role of Nuclear Receptors in Tissue and Organismal Aging Grants, 2783
NIDDK Neurobiology of Diabetic Complications Grants, 2828
NIH Research on Sleep and Sleep Disorders, 2898

Neurochemistry
Grass Foundation Marine Biological Laboratory Advanced Imaging Fellowships, 1877
SfN Julius Axelrod Prize, 3526

Neuroendocrinology
Grass Foundation Marine Biological Laboratory Advanced Imaging Fellowships, 1877
HRAMF Charles H. Hood Foundation Child Health Research Awards, 2057

Neurological Disorders
AMA Foundation Seed Grants for Research, 494
ANA Wolfe Neuropathy Research Prize, 585
Children's Tumor Foundation Drug Discovery Initiative Awards, 1141
Children's Tumor Foundation Schwannomatosis Awards, 1142
Dana Foundation Science and Health Grants, 1420
Fairlawn Foundation Grants, 1626
John Merck Scholars Awards, 2246

Judith and Jean Pape Adams Charitable Foundation ALS Grants, 2279
NARSAD Goldman-Rakic Prize for Cognitive Neuroscience, 2621
NIEHS Small Business Innovation Research (SBIR) Program Grants, 2860
NIH Research on Sleep and Sleep Disorders, 2898
NINDS Support of Scientific Meetings as Cooperative Agreements, 2925
PDF Summer Fellowships, 3131
Pittsburgh Foundation Medical Research Grants, 3231
SfN Julius Axelrod Prize, 3526

Neurology
American Philosophical Society Daland Fellowships in Clinical Investigation, 537
ANA Distinguished Neurology Teacher Award, 580
ANA Junior Academic Neurologist Scholarships, 582
ANS/AAO-HNSF Herbert Silverstein Otology and Neurotology Research Award, 601
ANS Neurotology Fellows Award, 604
ANS Nicholas Torok Vestibular Award, 605
ANS Trainee Award, 606
Biogen Foundation Healthcare Professional Education Grants, 900
Biogen Foundation Scientific Research Grants, 902
Grass Foundation Marine Biological Laboratory Advanced Imaging Fellowships, 1877
Grass Foundation Marine Biological Laboratory Fellowships, 1878
Joseph Collins Foundation Scholarships, 2267
NIA Role of Nuclear Receptors in Tissue and Organismal Aging Grants, 2783
NIDDK Neurobiology of Diabetic Complications Grants, 2828
NINDS Support of Scientific Meetings as Cooperative Agreements, 2925
NYAM Charles A. Elsberg Fellowship, 3039
PDF Postdoctoral Fellowships for Neurologists, 3130
Pfizer Healthcare Charitable Contributions, 3180

Neuromuscular Disorders
ANA Wolfe Neuropathy Research Prize, 585
Judith and Jean Pape Adams Charitable Foundation ALS Grants, 2279
MDA Development Grant, 2500
MDA Neuromuscular Disease Research Grants, 2501
MGFA Post-Doctoral Research Fellowships, 2539
MGFA Student Fellowships, 2540
NAF Fellowships, 2603
NAF Kyle Bryant Translational Research Award, 2604
NAF Research Grants, 2605
NAF Young Investigator Awards, 2606

Neuropharmacology
Grass Foundation Marine Biological Laboratory Advanced Imaging Fellowships, 1877
PVA Education Foundation Grants, 3284
SfN Julius Axelrod Prize, 3526

Neurophysiology
Grass Foundation Marine Biological Laboratory Advanced Imaging Fellowships, 1877
Lumosity Human Cognition Grant, 2390
NARSAD Goldman-Rakic Prize for Cognitive Neuroscience, 2621
SfN Ralph W. Gerard Prize in Neuroscience, 3534

Neuroscience
AES-Grass Young Investigator Travel Awards, 317
AIHS Alberta/Pfizer Translat Research Grants, 374
Alfred P. Sloan Foundation Research Fellowships, 432
ANA/Grass Foundation Award in Neuroscience, 578
ANA Derek Denny-Brown Neurological Scholar Award, 579
ANA Distinguished Neurology Teacher Award, 580
ANA F.E. Bennett Memorial Lectureship, 581
ANA Raymond D. Adams Lectureship, 583
ANA Soriano Lectureship, 584
ANA Wolfe Neuropathy Research Prize, 585

SUBJECT INDEX

Bristol-Myers Squibb Clinical Outcomes and Research Grants, 955
Dana Foundation Science and Health Grants, 1420
Dorothy Rider Pool Health Care Grants, 1500
Fndn for Appalachian Ohio Bachtel Scholarships, 1692
Grass Foundation Marine Biological Laboratory Advanced Imaging Fellowships, 1877
Grass Foundation Marine Biological Laboratory Fellowships, 1878
Gruber Foundation Neuroscience Prize, 1909
Harry Frank Guggenheim Foundation Dissertation Fellowships, 1963
IBRO-PERC InEurope Short Stay Grants, 2097
IBRO-PERC Support for European Workshops, Symposia and Meetings, 2098
IBRO-PERC Support for Site Lectures, 2099
IBRO/SfN International Travel Grants, 2100
IBRO Asia Reg APRC Exchange Fellowships, 2101
IBRO Asia APRC Lecturer Exchange Grants, 2102
IBRO Asia Regional APRC Travel Grants, 2103
IBRO International Travel Grants, 2104
IBRO Latin America Regional Funding for Neuroscience Schools, 2105
IBRO Latin America Regional Funding for PROLAB Collaborations, 2106
IBRO Latin America Regional Funding for Short Courses, Workshops, and Symposia, 2107
IBRO Latin America Regional Funding for Short Research Stays, 2108
IBRO Latin America Regional Travel Grants, 2109
IBRO Regional Grants for Int'l Fellowships to U.S. Laboratory Summer Neuroscience Courses, 2110
IBRO Research Fellowships, 2111
IBRO Return Home Fellowships, 2112
Lumosity Human Cognition Grant, 2390
MBL Ann E. Kammer Summer Fellowship, 2464
MBL James E. and Faith Miller Memorial Summer Fellowship, 2478
MBL Scholarships and Awards, 2486
NARSAD Goldman-Rakic Prize for Cognitive Neuroscience, 2621
National Center for Responsible Gaming Grants, 2629
National Center for Responsible Gaming Postdoctoral Fellowships, 2630
National Center for Resp Gaming Seed Grants, 2631
NIA Vulnerable Dendrites and Synapses in Aging and Alzheimer's Disease Grants, 2788
NIH Research on Sleep and Sleep Disorders, 2898
NIMH Curriculum Development Award in Neuroinformatics Research and Analysis, 2917
NIMH Jointly Sponsored Ruth L. Kirschstein National Research Service Award Institutional Predoctoral Training Program in the Neurosciences, 2919
NIMH Short Courses in Neuroinformatics, 2920
Pfizer Australia Neuroscience Research Grants, 3175
Rosalinde and Arthur Gilbert Foundation/AFAR New Investigator Awards in Alzheimer's Disease, 3406
Royal Norwegian Embassy Kavli Prizes, 3414
Searle Scholars Program Grants, 3503
SfN Award for Education in Neuroscience, 3517
SfN Bernice Grafstein Award for Outstanding Accomplishments in Mentoring, 3518
SfN Chapter-of-the-Year Award, 3519
SfN Chapter Grants, 3520
SfN Donald B. Lindsley Prize in Behavioral Neuroscience, 3521
SfN Federation of European Neuroscience Societies Forum Travel Awards, 3522
SfN Jacob P. Waletzky Award, 3523
SfN Janett Rosenberg Trubatch Career Development Award, 3524
SfN Japan Neuroscience Society Meeting Travel Awards, 3525
SfN Mika Salpeter Lifetime Achievement Award, 3528
SfN Nemko Prize in Cellular or Molecular Neuroscience, 3529
SfN Next Generation Award, 3531
SfN Patricia Goldman-Rakic Hall of Honor, 3532
SfN Peter and Patricia Gruber International Research Award, 3533

SfN Ralph W. Gerard Prize in Neuroscience, 3534
SfN Science Educator Award, 3535
SfN Swartz Prize for Theoretical and Computational Neuroscience, 3537
SfN Trainee Professional Development Awards, 3538
SfN Travel Awards for the International Brain Research Organization World Congress, 3539
SfN Undergrad Brain Awareness Travel Award, 3540
SfN Young Investigator Award, 3541
W.M. Keck Foundation Science and Engineering Research Grants, 3889

Neurosurgery
AANS Neurosurgery Research and Education Foundation/Spine Section Young Clinician Investigator Award, 162
AANS Neurosurgery Research & Educ Foundation Young Clinician Investigator Award, 163
American College of Surgeons and The Triological Society Clinical Scientist Development Awards, 518
American College of Surgeons Resident and Associate Society (RAS-ACS) Leadership Scholarships, 527
Community Foundation of Louisville Diller B. and Katherine P. Groff Fund for Pediatric Surgery Grants, 1293

Neurotology
ANS Neurotology Fellows Award, 604
ANS Nicholas Torok Vestibular Award, 605
ANS Trainee Award, 606

Neurotransmitters
Lumosity Human Cognition Grant, 2390
SfN Julius Axelrod Prize, 3526

New York
Northern Chautauqua Community Foundation Community Grants, 2995

New Zealand
American College of Surgeons Australia and New Zealand Chapter Travelling Fellowships, 519

Nonfiction
MLA Murray Gottlieb Prize Essay Award, 2571
MLA Rittenhouse Award, 2573
NLM Grants for Scholarly Works in Biomedicine and Health, 2944
Quantum Corporation Snap Server Grants, 3289

Nonprofit Organizations
ADA Foundation Samuel Harris Children's Dental Health Grants, 294
Altman Foundation Health Care Grants, 457
Antone & Edene Vidinha Charitable Trust Grants, 609
Battle Creek Community Foundation Grants, 839
Bayer Foundation Grants, 846
Blue Cross Blue Shield of Minnesota Foundation - Health Equity: Building Health Equity Together Grants, 914
Blue Cross Blue Shield of Minnesota Foundation - Healthy Children: Growing Up Healthy Grants, 915
Blue Cross Blue Shield of Minnesota Foundation - Healthy Equity: Health Impact Assessment Demonstration Project Grants, 916
Blumenthal Foundation Grants, 925
Brookdale Fnd National Group Respite Grants, 966
Callaway Golf Company Foundation Grants, 1003
Carl M. Freeman Foundation Grants, 1027
CFFVR Capital Credit Union Charitable Giving Grants, 1095
CFFVR Clintonville Area Foundation Grants, 1097
CFFVR Frank C. Shattuck Community Grants, 1098
CFFVR Shawano Area Community Foundation Grants, 1104
CFFVR Waupaca Area Community Foundation Grants, 1105
CFFVR Wisconsin King's Daus & Sons Grants, 1106
CFFVR Women's Fund for the Fox Valley Region Grants, 1107

Clarence T.C. Ching Foundation Grants, 1167
CNCS Social Innovation Grants, 1192
Columbus Foundation Central Benefits Health Care Foundation Grants, 1216
Community Foundation for Greater New Haven $5,000 and Under Grants, 1248
Community Foundation for Greater New Haven Responsive New Grants, 1249
Community Foundation for Greater New Haven Sponsorship Grants, 1250
Community Foundation of Abilene Celebration of Life Grants, 1261
Community Fndn of Bloomington & Monroe County - Precision Health Network Cycle Grants, 1264
Community Foundation of Jackson County Seymour Noon Lions Club Grant, 1288
Community Foundation of Muncie and Delaware County Maxon Grants, 1301
Community Foundation of St. Joseph County Special Project Challenge Grants, 1310
Community Foundation Partnerships - Lawrence County Grants, 1317
Cone Health Foundation Grants, 1323
Constellation Energy Corporate Grants, 1342
Covidien Medical Product Donations, 1352
David M. and Marjorie D. Rosenberg Foundation Grants, 1432
Deborah Munroe Noonan Memorial Grants, 1443
Decatur County Community Foundation Small Project Grants, 1445
Denver Foundation Community Grants, 1461
Dubois County Community Foundation Grants, 1517
Dyson Foundation Mid-Hudson Valley General Operating Support Grants, 1533
Dyson Foundation Mid-Hudson Valley Project Support Grants, 1534
Elizabeth Glaser Int'l Leadership Awards, 1566
Eugene Straus Charitable Trust, 1612
Franklin H. Wells and Ruth L. Wells Foundation Grants, 1725
George Foundation Grants, 1795
George Gund Foundation Grants, 1796
Gibson County Community Foundation Women's Fund Grants, 1817
GNOF Exxon-Mobil Grants, 1836
GNOF Freeman Challenge Grants, 1837
GNOF New Orleans Works Grants, 1844
GNOF Norco Community Grants, 1845
GNOF Organizational Effectiveness Grants and Workshops, 1846
GNOF Stand Up For Our Children Grants, 1847
Grand Rapids Area Community Foundation Wyoming Grants, 1867
Grand Rapids Community Foundation Southeast Ottawa Grants, 1873
Grand Rapids Community Fndn Sparta Grants, 1875
Great-West Life Grants, 1879
Greater Green Bay Community Fndn Grants, 1881
Grundy Foundation Grants, 1912
Guy I. Bromley Trust Grants, 1919
Harden Foundation Grants, 1945
Harold and Rebecca H. Gross Foundation Grants, 1951
Harvey Randall Wickes Foundation Grants, 1972
Helen Bader Foundation Grants, 1992
Hilton Head Island Foundation Grants, 2036
Human Source Foundation Grants, 2078
Jacobs Family Village Neighborhoods Grants, 2184
Janson Foundation Grants, 2210
Jeffris Wood Foundation Grants, 2217
Johnson Foundation Wingspread Conference Support Program, 2257
John W. Speas and Effie E. Speas Memorial Trust Grants, 2264
Kent D. Steadley and Mary L. Steadley Memorial Trust Grants, 2301
Lewis H. Humphreys Charitable Trust Grants, 2361
Libra Foundation Grants, 2363
Louetta M. Cowden Foundation Grants, 2377
Louis and Elizabeth Nave Flarsheim Charitable Foundation Grants, 2379

784 / Nonprofit Organizations

M.J. Murdock Charitable Trust General Grants, 2405
Manuel D. & Rhoda Mayerson Fndn Grants, 2420
Mary S. and David C. Corbin Foundation Grants, 2447
McCallum Family Foundation Grants, 2489
McConnell Foundation Grants, 2492
McLean Foundation Grants, 2498
Medtronic Foundation CommunityLink Health Grants, 2510
Meyer Foundation Benevon Grants, 2532
Meyer Fndn Management Assistance Grants, 2534
MGN Family Foundation Grants, 2542
Middlesex Savings Charitable Foundation Capacity Building Grants, 2551
Miller Foundation Grants, 2556
Montgomery County Community Fndn Grants, 2591
Morris & Gwendolyn Cafritz Fndn Grants, 2594
Ms. Fndn for Women Ending Violence Grants, 2596
Nina Mason Pulliam Charitable Trust Grants, 2921
Norfolk Southern Foundation Grants, 2982
Northern Chautauqua Community Foundation Community Grants, 2995
NWHF Health Advocacy Small Grants, 3033
NWHF Kaiser Permanente Community Grants, 3034
OSF Tackling Addiction Grants in Baltimore, 3096
Otto Bremer Foundation Grants, 3099
Percy B. Ferebee Endowment Grants, 3138
Peter and Elizabeth C. Tower Foundation Organizational Scholarships, 3148
Peter and Elizabeth C. Tower Foundation Phase II Technology Initiative Grants, 3149
Peter and Elizabeth C. Tower Foundation Phase I Technology Initiative Grants, 3150
Peter F. McManus Charitable Trust Grants, 3153
Pfizer Special Events Grants, 3186
Phoenix Suns Charities Grants, 3201
Piedmont Health Foundation Grants, 3219
Pohlad Family Foundation, 3241
Price Family Charitable Fund Grants, 3262
Priddy Foundation Program Grants, 3266
Rhode Island Foundation Grants, 3362
Richard King Mellon Foundation Grants, 3369
Robert R. McCormick Tribune Veterans Initiative Grants, 3386
Rockefeller Archive Center Research Grants, 3393
Rose Hills Foundation Grants, 3413
RRF Organizational Capacity Building Grants, 3418
San Francisco Fndn Community Health Grants, 3478
Santa Barbara Foundation Strategy Grants - Core Support, 3481
Shaw's Supermarkets Donations, 3542
Skoll Fndn Awards for Social Entrepreneurship, 3579
Stranahan Foundation Grants, 3659
Texas Instruments Community Services Grants, 3710
Tipton County Foundation Grants, 3734
Tri-State Community Twenty-first Century Endowment Fund Grants, 3756
Verizon Foundation Virginia Grants, 3864
Wallace Alexander Gerbode Foundation Grants, 3896
Wege Foundation Grants, 3912
William A. Badger Foundation Grants, 3945
Z. Smith Reynolds Foundation Small Grants, 3982

Nuclear Medicine
SNM/ Covidien Seed Grant in Molecular Imaging/ Nuclear Medicine Research, 3582
SNM PDEF Mickey Williams Minority Student Scholarships, 3584
SNM Pilot Research Grants, 3585
SNM Pilot Research Grants in Nuclear Medicine/ Molecular Imaging, 3586
SNM Student Fellowship Awards, 3588
SNM Tetalman Young Investigator Awards, 3589
SNMTS Clinical Advancement Scholarships, 3590
SNMTS Outstanding Educator Awards, 3591
SNMTS Outstanding Technologist Awards, 3592
SNMTS Paul Cole Scholarships, 3593

Nuclear Receptors
NIA Role of Nuclear Receptors in Tissue and Organismal Aging Grants, 2783

Nuclear Safety
IIE Rockefeller Foundation Bellagio Center Residencies, 2142

Nuclear Sciences
Canada Graduate Scholarships (CGS) and NSERC Postgraduate Scholarships (PGS), 1010

Nuclear Weapons
NIEHS Hazmat Training at Doe Nuclear Weapons Complex Grants, 2855

Nuclear/Radioactive Waste Disposal
NIEHS Hazmat Training at Doe Nuclear Weapons Complex Grants, 2855

Nucleic Acids
NSF Biomolecular Systems Cluster Grants, 3010

Nurse Practitioners
AACN-Sigma Theta Tau Critical Care Grant, 79
AACN Clinical Inquiry Fund Grants, 80
AACN End of Life/Palliative Care Small Projects Grants, 83
NAPNAP Fndn Elaine Gelman Scholarship, 2609
NAPNAP Foundation McNeil PNP Scholarships, 2613
NAPNAP Foundation McNeil Rural and Underserved Scholarships, 2614
NAPNAP Foundation Nursing Research Grants, 2615
NAPNAP Foundation Reckitt Benckiser Student Scholarship, 2616

Nursing
AACN-Edwards Lifesciences Nurse-Driven Clinical Practice Outcomes Grants, 77
AACN-Philips Medical Systems Clinical Outcomes Grants, 78
AACN Clinical Practice Grants, 81
AACN Critical Care Grants, 82
AACN Evidence-Based Clinical Practice Grants, 84
AACN Mentorship Grant, 85
AACN Physio-Control Small Projects Grants, 86
AAHN Grant for Historical Research, 146
AAHN Pre-Doctoral Research Grant, 147
Abington Foundation Grants, 211
ACS Doctoral Scholarships in Cancer Nursing, 253
Alice Fisher Society Fellowships, 436
AMDA Foundation Quality Improvement and Health Outcome Awards, 498
American Academy of Nursing Building Academic Geriatric Nursing Capacity Scholarships, 505
American Academy of Nursing Claire M. Fagin Fellowships, 506
American Academy of Nursing Mayday Grants, 507
American Academy of Nursing MBA Scholarships, 508
American Academy of Nursing Media Awards, 509
ASPEN Rhoads Research Foundation Abbott Nutrition Research Grants, 761
ASPEN Rhoads Research Foundation Baxter Parenteral Nutrition Research Grant, 762
ASPEN Rhoads Research Foundation C. Richard Fleming Grant, 763
ASPEN Rhoads Research Foundation Maurice Shils Grant, 764
ASPEN Rhoads Research Foundation Norman Yoshimura Grant, 765
AVDF Health Care Grants, 811
Brookdale Foundation Leadership in Aging Fellowships, 965
CDC Public Health Informatics Fellowships, 1068
Christy-Houston Foundation Grants, 1153
Clarian Health Scholarships for LPNs, 1171
DAR Nursing/Physical Therapy Scholarships, 1430
Foundation for Appalachian Ohio Susan K. Ipacs Nursing Legacy Scholarships, 1693
Foundation for Health Enhancement Grants, 1696
Fulton County Community Foundation Paul and Dorothy Arven Memorial Scholarship, 1765
Hampton Roads Community Foundation Faith Community Nursing Grants, 1940

Independence Blue Cross Nurse Scholars, 2151
Independence Blue Cross Nursing Internships, 2152
Indiana Nursing Scholarships, 2157
Jenkins Foundation: Improving the Health of Greater Richmond Grants, 2219
Joseph and Luella Abell Trust Scholarships, 2266
June Pangburn Memorial Scholarship, 2283
L. A. Hollinger Respiratory Therapy and Nursing Scholarships, 2327
Lillian Sholtis Brunner Summer Fellowships for Historical Research in Nursing, 2366
March of Dimes Agnes Higgins Award, 2423
March of Dimes Graduate Nursing Scholarships, 2424
NAPNAP Foundation Innovative Health Care Small Grant, 2611
NAPNAP Foundation McNeil Grant-in-Aid, 2612
NAPNAP Foundation Nursing Research Grants, 2615
NCI Stages of Breast Development: Normal to Metastatic Disease Grants, 2655
NHLBI Mentored Patient-Oriented Research Career Development Awards, 2710
NINR Acute and Chronic Care During Mechanical Ventilation Research Grants, 2926
NINR Mentored Research Scientist Development Award, 2932
NINR Ruth L. Kirschstein National Research Service Award for Predoc Fellows in Nursing, 2934
Noble County Community Foundation - Lolita J. Hornett Memorial Nursing Scholarship, 2974
NWHF Partners Investing in Nursing's Future, 3036
Oregon Community Foundation Better Nursing Home Care Grants, 3081
Pfizer Medical Education Track One Grants, 3182
PhRMA Foundation Health Outcomes Post Doctoral Fellowships, 3202
Pollock Foundation Grants, 3244
Porter County Health Occupations Scholarship, 3247
Prospect Burma Scholarships, 3275
Rehabilitation Nursing Fndn Research Grants, 3354
Roget Begnoche Scholarship, 3402
RWJF Interdisciplinary Nursing Quality Research Initiative Grants, 3438
RWJF New Careers in Nursing Scholarships, 3439
Sid W. Richardson Foundation Grants, 3553
Sigma Theta Tau International /American Nurses Foundation Grant, 3556
Sigma Theta Tau International / Council for the Advancement of Nursing Science Grants, 3557
Sigma Theta Tau International / Oncology Nursing Society Grant, 3558
Sigma Theta Tau International Doris Bloch Research Award, 3559
Sigma Theta Tau Small Grants, 3560
Special Olympics Health Profess Student Grants, 3629
STTI Emergency Nurses Association Foundation Grant, 3666
STTI Environment of Elder Care Nursing Research Grants, 3667
STTI Joan K. Stout RN Research Grants, 3668
STTI National League for Nursing Grants, 3669
STTI Rosemary Berkel Crisp Research Awards, 3670
STTI Virginia Henderson Research Grants, 3671
Trinity Lutheran School of Nursing Alumnae Scholarships, 3759
VDH Commonwealth of Virginia Nurse Educator Scholarships, 3854
Virginia Department of Health Mary Marshall Nursing Scholarships for Licensed Practical Nurses, 3874
Virginia Department of Health Mary Marshall Nursing Scholarships for Registered Nurses, 3875
Virginia Department of Health Nurse Practitioner/ Nurse Midwife Scholarships, 3876
Visiting Nurse Foundation Grants, 3879

Nursing Education
AAOHN Found Experienced Researcher Grants, 173
AAOHN Foundation New Investigator Researcher Grants, 174
AAOHN Foundation Professional Development Scholarships, 175

SUBJECT INDEX

Abell-Hanger Foundation Grants, 207
Adams Rotary Memorial Fund B Scholarships, 303
Ahmanson Foundation Grants, 355
American Academy of Nursing Building Academic Geriatric Nursing Capacity Scholarships, 505
American Academy of Nursing Claire M. Fagin Fellowships, 506
American Academy of Nursing Mayday Grants, 507
American Academy of Nursing MBA Scholarships, 508
Benton County Foundation - Fitzgerald Family Scholarships, 879
Bertha Russ Lytel Foundation Grants, 885
Burlington Industries Foundation Grants, 977
Clarian Health OR Internships, 1170
Clarian Health Scholarships for LPNs, 1171
Clarian Health Student Nurse Scholarships, 1172
Community Foundation of Greater Lafayette - Robert and Dorothy Hughes Scholarships, 1279
Daphne Seybolt Culpeper Fndn Grants, 1425
DAR Irene and Daisy MacGregor Memorial Scholarship, 1428
DAR Madeline Cogswell Nursing Scholarship, 1429
DAR Nursing/Physical Therapy Scholarships, 1430
Dr. John T. Macdonald Foundation Scholarships, 1508
Dr. Scholl Foundation Grants Program, 1511
E. Rhodes & Leona B. Carpenter Grants, 1539
Edward W. and Stella C. Van Houten Memorial Fund Grants, 1559
Effie and Wofford Cain Foundation Grants, 1563
Foundation for Appalachian Ohio Susan K. Ipacs Nursing Legacy Scholarships, 1693
Fulton County Community Foundation Paul and Dorothy Arven Memorial Scholarship, 1765
Health Fndn of Greater Indianapolis Grants, 1984
Highmark BCBS Challenge for Healthier Schools in Western Pennsylvania, 3696
HRSA Nurse Education, Practice, Quality and Retention (NEPQR) Grants, 2068
Independence Blue Cross Nurse Scholars, 2151
Independence Blue Cross Nursing Internships, 2152
Indiana Nursing Scholarships, 2157
Joseph and Luella Abell Trust Scholarships, 2266
Josiah Macy Jr. Foundation Grants, 2277
L. A. Hollinger Respiratory Therapy and Nursing Scholarships, 2327
Lubbock Area Foundation Grants, 2383
Lucy Downing Nisbet Charitable Fund Grants, 2387
March of Dimes Agnes Higgins Award, 2423
March of Dimes Graduate Nursing Scholarships, 2424
NAPNAP Fndn Elaine Gelman Scholarship, 2609
NAPNAP Fndn Grad Research Grant, 2610
NAPNAP Foundation McNeil PNP Scholarships, 2613
NAPNAP Foundation McNeil Rural and Underserved Scholarships, 2614
NAPNAP Foundation Nursing Research Grants, 2615
NAPNAP Foundation Reckitt Benckiser Student Scholarship, 2616
New York University Steinhardt School of Education Fellowships, 2684
NHLBI Short-Term Research Education Program to Increase Diversity in Health-Related Research, 2733
NIA Mentored Clinical Scientist Research Career Development Awards, 2771
NINR Ruth L. Kirschstein National Research Service Award for Predoc Fellows in Nursing, 2934
Oregon Community Foundation Better Nursing Home Care Grants, 3081
Porter County Health Occupations Scholarship, 3247
Rehabilitation Nursing Fndn Research Grants, 3354
Rehabilitation Nursing Foundation Scholarships, 3355
Ruth Eleanor Bamberger and John Ernest Bamberger Memorial Foundation Grants, 3427
RWJF Interdisciplinary Nursing Quality Research Initiative Grants, 3438
RWJF New Careers in Nursing Scholarships, 3439
Saint Luke's Health Initiatives Grants, 3455
Samuel S. Johnson Foundation Grants, 3468
San Antonio Area Foundation Grants, 3469
Stroke Association Allied Health Professional Research Bursaries, 3661
STTI/ATI Educational Assessment Nursing Research Grants, 3665
STTI Joan K. Stout RN Research Grants, 3668
STTI National League for Nursing Grants, 3669
Trinity Lutheran School of Nursing Alumnae Scholarships, 3759
VDH Commonwealth of Virginia Nurse Educator Scholarships, 3854
Virginia Department of Health Mary Marshall Nursing Scholarships for Licensed Practical Nurses, 3874
Virginia Department of Health Mary Marshall Nursing Scholarships for Registered Nurses, 3875
Virginia Department of Health Nurse Practitioner/Nurse Midwife Scholarships, 3876

Nursing Homes

AMDA Foundation Quality Improvement and Health Outcome Awards, 498
Arkell Hall Foundation Grants, 661
Burden Trust Grants, 976
Community Fndn for the Capital Region Grants, 1259
David N. Lane Trust Grants for Aged and Indigent Women, 1433
Jacob and Valeria Langeloth Foundation Grants, 2182
Lydia deForest Charitable Trust Grants, 2393
McLean Contributionship Grants, 2497
Mervin Bovaird Foundation Grants, 2523
MetroWest Health Foundation Grants--Healthy Aging, 2527
Oregon Community Foundation Better Nursing Home Care Grants, 3081
Reinberger Foundation Grants, 3357
Robert & Clara Milton Senior Housing Grants, 3384
Union Bank, N.A. Foundation Grants, 3788

Nutrition Education

Abbott Fund Access to Health Care Grants, 199
Abbott Fund Science Education Grants, 203
ACF Native American Social and Economic Development Strategies Grants, 231
ACSM-GSSI Young Scholar Travel Award, 257
Aid for Starving Children International Grants, 368
Allen Foundation Educational Nutrition Grants, 442
ASPEN Dudrick Research Scholar Award, 759
ASPEN Harry M. Vars Award, 760
ASPEN Rhoads Research Foundation Abbott Nutrition Research Grants, 761
ASPEN Rhoads Research Foundation Baxter Parenteral Nutrition Research Grant, 762
ASPEN Rhoads Research Foundation C. Richard Fleming Grant, 763
ASPEN Rhoads Research Foundation Maurice Shils Grant, 764
ASPEN Rhoads Research Foundation Norman Yoshimura Grant, 765
ASPEN Scientific Abstracts Awards for Papers or Posters, 766
ASPEN Scientific Abstracts Promising Investigator Awards, 767
BCBSM Building Healthy Communities Engaging Elementary Schools and Community Partners Grants, 849
Caesars Foundation Grants, 995
Campbell Soup Foundation Grants, 1007
CDC Public Health Prev Service Fellowships, 1069
CDC Public Health Prevention Service Fellowship Sponsorships, 1070
Coca-Cola Foundation Grants, 1203
Colonel Stanley R. McNeil Foundation Grants, 1211
Fargo-Moorhead Area Foundation Grants, 1632
Foundation for Seacoast Health Grants, 1698
General Mills Champs for Healthy Kids Grants, 1781
Great-West Life Grants, 1879
James R. Thorpe Foundation Grants, 2199
Medtronic Foundation CommunityLink Health Grants, 2510
Mt. Sinai Health Care Foundation Health of the Jewish Community Grants, 2599
Mt. Sinai Health Care Foundation Health of the Urban Community Grants, 2600
Nestle Foundation Large Research Grants, 2675
Nestle Foundation Pilot Grants, 2676
Nestle Foundation Re-entry Grants, 2677
Nestle Foundation Small Research Grants, 2678
Nestle Foundation Training Grant, 2679
Paso del Norte Health Foundation Grants, 3119
PepsiCo Foundation Grants, 3137
Perry County Community Foundation Grants, 3142
Phi Upsilon Omicron Florence Fallgatter Distinguished Service Award, 3192
Posey Community Fndn Women's Fund Grants, 3251
Posey County Community Foundation Grants, 3252
Robert R. Meyer Foundation Grants, 3387
Shaw's Supermarkets Donations, 3542
Sierra Health Foundation Responsive Grants, 3555
U.S. Department of Education 21st Century Community Learning Centers, 3767
USDA Organic Agriculture Research Grants, 3830
Vancouver Foundation Grants and Community Initiatives Program, 3849
Wilson-Wood Foundation Grants, 3960

Nutrition/Dietetics

AAFCS International Graduate Fellowships, 129
AAFCS National Graduate Fellowships, 130
AAFCS National Undergraduate Scholarships, 131
AAP Nutrition Award, 187
Abbott Fund Science Education Grants, 203
Administration on Aging Senior Medicare Patrol Project Grants, 309
AFAR Research Grants, 337
Albertson's Charitable Giving Grants, 410
Allen Foundation Educational Nutrition Grants, 442
ASPEN Dudrick Research Scholar Award, 759
ASPEN Harry M. Vars Award, 760
ASPEN Rhoads Research Foundation Abbott Nutrition Research Grants, 761
ASPEN Rhoads Research Foundation Baxter Parenteral Nutrition Research Grant, 762
ASPEN Rhoads Research Foundation C. Richard Fleming Grant, 763
ASPEN Rhoads Research Foundation Maurice Shils Grant, 764
ASPEN Rhoads Research Foundation Norman Yoshimura Grant, 765
ASPEN Scientific Abstracts Awards for Papers or Posters, 766
ASPEN Scientific Abstracts Promising Investigator Awards, 767
Assurant Foundation Grants, 788
Beazley Foundation Grants, 865
Boeing Company Contributions Grants, 930
Boston Foundation Grants, 936
Bristol-Myers Squibb/Meade Johnson Award for Dist Achievement in Nutrition Research, 954
Bristol-Myers Squibb Foundation Health Disparities Grants, 958
Caesars Foundation Grants, 995
Charles H. Farnsworth Trust Grants, 1119
Christy-Houston Foundation Grants, 1153
ConAgra Foods Fndn Community Impact Grants, 1321
Connecticut Health Foundation Health Initiative Grants, 1325
Denver Foundation Community Grants, 1461
DHHS Adolescent Family Life Demonstration Projects, 1472
Dole Food Company Charitable Contributions, 1483
FDHN Bridging Grants, 1638
FDHN Isenberg Int'l Research Scholar Award, 1647
FDHN Student Research Fellowships, 1654
Foundation for a Healthy Kentucky Grants, 1691
Fndn for Appalachian Ohio Bachtel Scholarships, 1692
General Mills Champs for Healthy Kids Grants, 1781
General Mills Foundation Grants, 1782
Gerber Foundation Grants, 1807
Grand Rapids Community Foundation Ionia County Youth Fund Grants, 1871
Grand Rapids Community Foundation Sparta Youth Fund Grants, 1876
H.J. Heinz Company Foundation Grants, 1923

786 / Nutrition/Dietetics

Healthcare Fndn for Orange County Grants, 1981
Health Fndn of Greater Indianapolis Grants, 1984
March of Dimes Agnes Higgins Award, 2423
March of Dimes Program Grants, 2426
Meyer Foundation Healthy Communities Grants, 2533
NCI Cancer Education and Career Development Program, 2648
NCI Technologies and Software to Support Integrative Cancer Biology Research (SBIR) Grants, 2656
NHLBI Nutrition and Diet in the Causation, Prevention, and Management of Heart Failure Research Grants, 2717
NIA Malnutrition in Older Persons Research, 2767
NIDDK Diet Comp & Energy Balance Grants, 2812
NIDDK Pilot and Feasibility Clinical Research Studies in Digestive Diseases and Nutrition, 2836
NIDDK Research Grants on Improving Health Care for Obese Patients, 2840
NIDDK Small Grants for Underrepresented Minority Scientists in Diabetes and Digestive and Kidney Diseases, 2847
NWHF Physical Activity and Nutrition Grants, 3037
Paso del Norte Health Foundation Grants, 3119
Sensient Technologies Foundation Grants, 3513
Sheltering Arms Fund Grants, 3545
Susan G. Komen Breast Cancer Foundation Challege Grants: Investigator Initiated Research, 3690
United Methodist Child Mental Health Grants, 3800
Western Union Foundation Grants, 3923

Nutritional Diseases and Disorders
Abbott Fund Science Education Grants, 203
ASPEN Dudrick Research Scholar Award, 759
ASPEN Harry M. Vars Award, 760
ASPEN Rhoads Research Foundation Abbott Nutrition Research Grants, 761
ASPEN Rhoads Research Foundation Baxter Parenteral Nutrition Research Grant, 762
ASPEN Rhoads Research Foundation C. Richard Fleming Grant, 763
ASPEN Rhoads Research Foundation Maurice Shils Grant, 764
ASPEN Rhoads Research Foundation Norman Yoshimura Grant, 765
ASPEN Scientific Abstracts Awards for Papers or Posters, 766
ASPEN Scientific Abstracts Promising Investigator Awards, 767
CDC David J. Sencer Museum Adult Group Tour, 1052
FDHN Bridging Grants, 1638
FDHN Fellow Abstract Prizes, 1643
FDHN Graduate Student Awards, 1646
FDHN Isenberg Int'l Research Scholar Award, 1647
Nestle Foundation Large Research Grants, 2675
Nestle Foundation Pilot Grants, 2676
Nestle Foundation Re-entry Grants, 2677
Nestle Foundation Small Research Grants, 2678
Nestle Foundation Training Grant, 2679
NIDDK Exploratory/Developmental Clinical Research Grants in Obesity, 2818
NIDDK Pilot and Feasibility Clinical Research Studies in Digestive Diseases and Nutrition, 2836

Obesity
Aetna Foundation Obesity Grants, 331
Bright Promises Foundation Grants, 952
Campbell Soup Foundation Grants, 1007
CDC David J. Sencer Museum Adult Group Tour, 1052
CDC Public Health Prev Service Fellowships, 1069
CDC Public Health Prevention Service Fellowship Sponsorships, 1070
Colonel Stanley R. McNeil Foundation Grants, 1211
Foundation for the Mid South Health and Wellness Grants, 1700
Gerber Foundation Grants, 1807
GNOF IMPACT Kahn-Oppenheim Grants, 1842
Grand Rapids Community Foundation Ionia County Youth Fund Grants, 1871
Healthcare Fndn for Orange County Grants, 1981

Highmark BCBS Challenge for Healthier Schools in Western Pennsylvania, 3696
Horizon Foundation for New Jersey Grants, 2048
HRAMF Charles H. Hood Foundation Child Health Research Awards, 2057
Institute for Agriculture and Trade Policy Food and Society Fellowships, 2159
Johnson & Johnson Community Health Grants, 2252
Mt. Sinai Health Care Foundation Health of the Jewish Community Grants, 2599
NHLBI Bioengineering and Obesity Grants, 2689
NHLBI Bioengineering Approaches to Energy Balance and Obesity Grants for SBIR, 2690
NHLBI Bioengineering Approaches to Energy Balance and Obesity Grants for STTR, 2691
NHLBI Targeted Approaches to Weight Control for Young Adults Grants, 2736
NHLBI Translating Basic Behavioral and Social Science Discoveries into Interventions to Reduce Obesity: Centers for Behavioral Intervention Development Grants, 2739
NIDDK Adverse Metabolic Side Effects of Second Generation Psychotropic Medications Leading to Obesity and Increased Diabetes Risk Grants, 2798
NIDDK Devel of Disease Biomarkers Grants, 2809
NIDDK Education Program Grants, 2813
NIDDK Exploratory/Developmental Clinical Research Grants in Obesity, 2818
NIDDK Identifying & Reducing Diabetes & Obesity Related Disparities in Healthcare Systems, 2821
NIDDK Neuroimaging in Obesity Research, 2829
NIDDK Planning Grants For Translational Research For The Prevention And Control Of Diabetes And Obesity, 2837
NIDDK Proteomics: Diabetes, Obesity, And Endocrine, Digestive, Kidney, Urologic, And Hematologic Diseases, 2838
NIDDK Research Grants on Improving Health Care for Obese Patients, 2840
NIDDK Role of Gastrointestinal Surgical Procedures in Amelioration of Obesity-Related Insulin Resistance & Diabetes Weight Loss, 2841
NIH School Interventions to Prevent Obesity, 2908
Northwestern Mutual Foundation Grants, 2999
NWHF Physical Activity and Nutrition Grants, 3037
Obesity Society Grants, 3060
Pajaro Valley Community Health Trust Diabetes and Contributing Factors Grants, 3109
PepsiCo Foundation Grants, 3137
Pfizer ASPIRE North America Broad Spectrum Antibiotics for the Treatment of Gram-Negative or Polymicrobial Infections Research Awards, 3169
Pfizer ASPIRE North America Narrow Spectrum Antibiotics for the Treatment of MRSA Research Awards, 3171
Piedmont Health Foundation Grants, 3219
RCF The Women's Fund Grants, 3315
RWJF Childhood Obesity Grants, 3431
RWJF Healthy Eating Research Grants, 3437

Obstetrics-Gynecology
ACS Cancer Control Career Development Awards for Primary Care Physicians, 250
Blanche and Irving Laurie Foundation Grants, 910
Duke Univ Obstetric Anesthesia Fellowships, 1525
Fairlawn Foundation Grants, 1626
HRAMF Charles H. Hood Foundation Child Health Research Awards, 2057
John S. Dunn Research Foundation Grants, 2250

Occupational Health and Safety
AAOHN Found Experienced Researcher Grants, 173
AAOHN Foundation New Investigator Researcher Grants, 174
AAOHN Foundation Professional Development Scholarships, 175
ASPH/CDC Public Health Fellowships, 781
California Wellness Foundation Work and Health Program Grants, 1001
CDC Epidemic Intell Service Training Grants, 1055

CDC Epidemiology Elective Rotation, 1056
CSTE CDC/CSTE Applied Epidemiology Fellowships, 1367
DOL Occupational Safety and Health--Susan Harwood Training Grants, 1484
DuPont Pioneer Community Giving Grants, 1532
IIE Rockefeller Foundation Bellagio Center Residencies, 2142
Morehouse PHSI Project Imhotep Internships, 2592
Susan G. Komen Breast Cancer Foundation Challege Grants: Investigator Initiated Research, 3690
U.S. Department of Education Erma Byrd Scholarships, 3768

Oceanography
Canada Graduate Scholarships (CGS) and NSERC Postgraduate Scholarships (PGS), 1010

Odontology
SRC Medical Research Grants, 3635

Oils and Fats
AOCS Health & Nutrition Div Student Award, 614
AOCS Health and Nutrition Poster Competition, 615
AOCS Industrial Oil Products Division Student Award, 616
AOCS Lipid Oxidation and Quality Division Poster Competition, 617
AOCS Processing Division Student Award, 618
AOCS Holman Lifetime Achievement Award, 620
AOCS Supelco/Nicholas Pelick-AOCS Research Award, 621
AOCS Surfactants and Detergents Division Student Award, 622

Oncology
AACR Clinical & Translat Research Fellowships, 110
Abbott Fund Science Education Grants, 203
ACS Doctoral Grants in Oncology Social Work, 254
ACS Master's Training Grants in Clinical Oncology Social Work, 258
ACS Research Scholar Grants in Basic, Preclinical, Clinical and Epidemiology Research, 280
AHNS/AAO-HNSF Surgeon Scientist Combined Award, 357
AHNS/AAO-HNSF Young Investigator Combined Award, 358
AHNS Alando J. Ballantyne Resident Research Pilot Grant, 359
AHNS Pilot Grant, 360
AMA Foundation Seed Grants for Research, 494
Ann and Robert H. Lurie Family Fndn Grants, 593
ASCO/UICC International Cancer Technology Transfer (ICRETT) Fellowships, 677
ASCO Advanced Clinical Research Award in Colorectal Cancer, 678
ASCO Advanced Clinical Research Awards in Breast Cancer, 679
ASCO Advanced Clinical Research Awards in Sarcoma, 680
ASCO Long-Term International Fellowships, 681
ASCO Merit Awards, 682
ASCO Young Investigator Award, 683
Bristol-Myers Squibb Clinical Outcomes and Research Grants, 955
Bristol-Myers Squibb Foundation Fellowships, 956
Conquer Cancer Foundation of ASCO Career Development Award, 1328
Conquer Cancer Foundation of ASCO Comparative Effectiveness Research Professorship in Breast Cancer, 1329
Conquer Cancer Foundation of ASCO Drug Development Research Professorship, 1330
Conquer Cancer Foundation of ASCO Improving Cancer Care Grants, 1331
Conquer Cancer Foundation of ASCO International Development and Education Award in Palliative Care, 1332
Conquer Cancer Foundation of ASCO International Development and Education Awards, 1333

SUBJECT INDEX

Conquer Cancer Foundation of ASCO International Innovation Grant, 1334
Conquer Cancer Foundation of ASCO Medical Student Rotation Grants, 1335
Conquer Cancer Foundation of ASCO Resident Travel Award for Underrepresented Populations, 1336
Conquer Cancer Foundation of ASCO Translational Research Professorships, 1337
Genentech Corp Charitable Contributions, 1780
NCI Technologies and Software to Support Integrative Cancer Biology Research (SBIR) Grants, 2656
Pediatric Cancer Research Foundation Grants, 3135
Pfizer Australia Cancer Research Grants, 3174
Pfizer Healthcare Charitable Contributions, 3180
Pfizer Medical Education Track One Grants, 3182
Pfizer Medical Education Track Two Grants, 3183
Sigma Theta Tau International / Oncology Nursing Society Grant, 3558
STTI Rosemary Berkel Crisp Research Awards, 3670
Thomas C. Burke Foundation Grants, 3720
UICC American Cancer Society International Fellowships for Beginning Investigators, 3778
UICC International Cancer Technology Transfer (ICRETT) Fellowships, 3779
UICC Raisa Gorbachev Memorial International Cancer Fellowships, 3780

Opera/Musical Theater
Adaptec Foundation Grants, 304
Bonfils-Stanton Foundation Grants, 932
Boston Foundation Grants, 936
Elizabeth McGraw Foundation Grants, 1568
Henrietta Lange Burk Fund Grants, 2001
Mabel Y. Hughes Charitable Trust Grants, 2409
Seaver Institute Grants, 3509

Operating Support
Abby's Legendary Pizza Foundation Grants, 204
Alfred E. Chase Charitable Foundation Grants, 430
Alliant Energy Corporation Contributions, 445
Anne J. Caudal Foundation Grants, 595
Blumenthal Foundation Grants, 925
Boyle Foundation Grants, 942
Brown Advisory Charitable Foundation Grants, 969
Carl R. Hendrickson Family Foundation Grants, 1028
Charles H. Farnsworth Trust Grants, 1119
CJ Foundation for SIDS Program Services Grants, 1163
Claude Bennett Family Foundation Grants, 1176
Community Foundation for Greater Atlanta Clayton County Fund Grants, 1240
Community Foundation for Greater Atlanta Common Good Funds Grants, 1241
Community Foundation for Greater Atlanta Metropolitan Atlanta An Extra Wish Grants, 1243
Community Foundation for Greater Atlanta Morgan County Fund Grants, 1244
Community Foundation for Greater Atlanta Newton County Fund Grants, 1245
E. Clayton and Edith P. Gengras, Jr. Foundation, 1536
Edward and Romell Ackley Foundation Grants, 1555
Expect Miracles Foundation Grants, 1619
Frank B. Hazard General Charity Fund Grants, 1721
Fred & Gretel Biel Charitable Trust Grants, 1734
Frederick McDonald Trust Grants, 1740
George and Sarah Buchanan Foundation Grants, 1790
George E. Hatcher, Jr. and Ann Williams Hatcher Foundation Grants, 1792
GNOF IMPACT Grants for Arts and Culture, 1838
GNOF IMPACT Grants for Health and Human Services, 1839
GNOF Maison Hospitaliere Grants, 1843
Green River Area Community Fndn Grants, 1895
Hardin County Community Foundation Grants, 1946
Harry S. Black and Allon Fuller Fund Grants, 1966
Helen Pumphrey Denit Charitable Trust Grants, 1996
Jack H. & William M. Light Trust Grants, 2179
Jacob and Hilda Blaustein Foundation Health and Mental Health Grants, 2181
Jayne and Leonard Abess Foundation Grants, 2214
John D. & Katherine A. Johnston Fndn Grants, 2233

John W. Boynton Fund Grants, 2263
Joseph Henry Edmondson Foundation Grants, 2270
Katherine Baxter Memorial Foundation Grants, 2292
Kinsman Foundation Grants, 2309
Mabel A. Horne Trust Grants, 2406
Marion I. and Henry J. Knott Foundation Discretionary Grants, 2435
Marion I. and Henry J. Knott Foundation Standard Grants, 2436
May and Stanley Smith Charitable Trust Grants, 2459
Michael Reese Health Trust Core Grants, 2544
Nell J. Redfield Foundation Grants, 2673
Olive Higgins Prouty Foundation Grants, 3072
Priddy Foundation Operating Grants, 3264
Robert R. Meyer Foundation Grants, 3387
Shield-Ayres Foundation Grants, 3546
ZYTL Foundation Grants, 3985

Operations Research
CDC Steven M. Teutsch Prevention Effectiveness Fellowships, 1071
Mayo Clinic Business Consulting Fellowship, 2462

Ophthalmology
AHAF National Glaucoma Research Grants, 351
Alcon Foundation Grants Program, 416
Alice C. A. Sibley Fund Grants, 435
CNIB Baker Applied Research Fund Grants, 1193
CNIB Baker Fellowships, 1194
CNIB Baker New Researcher Fund Grants, 1195
CNIB Barbara Tuck MacPhee Award, 1196
CNIB Canada Glaucoma Clinical Research Council Grants, 1197
CNIB Chanchlani Global Vision Research Award, 1198
CNIB E. (Ben) & Mary Hochhausen Access Technology Research Grants, 1199
CNIB Ross Purse Doctoral Fellowships, 1200
Dr. Stanley Pearle Scholarships, 1512
Frederick Gardner Cottrell Foundation Grants, 1739
HRAMF Thome Foundation Awards in Age-Related Macular Degeneration Research, 2065
Joseph Alexander Foundation Grants, 2265
NEI Clinical Vision Research Devel Award, 2664
NEI Vision Research Core Grants, 2671
OneSight Research Foundation Block Grants, 3076
Richmond Eye and Ear Fund Grants, 3372

Optometry
Alice C. A. Sibley Fund Grants, 435
Canadian Optometric Education Trust Grants, 1011
Dr. Stanley Pearle Scholarships, 1512
Eye-Bank for Sight Restoration and Fight for Sight Summer Student Research Fellowship, 1622
Fight for Sight-Streilein Foundation for Ocular Immunology Research Award, 1662
Fight for Sight Grants-in-Aid, 1663
Fight for Sight Post-Doctoral Awards, 1664
Fight for Sight Summer Student Fellowships, 1665
OneSight Research Foundation Block Grants, 3076
Special Olympics Health Profess Student Grants, 3629
Streilein Foundation for Ocular Immunology Visiting Professorships, 3660

Oral Diseases
DHHS Oral Health Promotion Research Across the Lifespan, 1474
Pajaro Valley Community Health Trust Oral Health: Prevention & Access Grants, 3110
USAID Ebola Response, Recovery and Resilience in West Africa Grants, 3809

Oral Health and Hygiene
ADA Foundation Samuel Harris Children's Dental Health Grants, 294
Aetna Foundation Regional Health Grants, 333
Connecticut Health Foundation Health Initiative Grants, 1325
DHHS Oral Health Promotion Research Across the Lifespan, 1474
Illinois Children's Healthcare Foundation Grants, 2146

Organizational Development /787

Mary Black Fndn Community Health Grants, 2441
Pajaro Valley Community Health Trust Oral Health: Prevention & Access Grants, 3110
United Methodist Child Mental Health Grants, 3800

Oral History
Australasian Institute of Judicial Administration Seed Funding Grants, 806
Turner Foundation Grants, 3763

Orchestras
Ahmanson Foundation Grants, 355
Alcatel-Lucent Technologies Foundation Grants, 413
Barr Fund Grants, 835
Clarence E. Heller Charitable Foundation Grants, 1166
Claude Worthington Benedum Fndn Grants, 1178
Community Foundation Alliance City of Evansville Endowment Fund Grants, 1238
Constantin Foundation Grants, 1341
El Paso Community Foundation Grants, 1579
Fremont Area Community Fndn General Grants, 1746
Gardner Foundation Grants, 1770
George W. Codrington Charitable Fndn Grants, 1803
Griffin Foundation Grants, 1903
Henrietta Lange Burk Fund Grants, 2001
Howard and Bush Foundation Grants, 2054
Katrine Menzing Deakins Trust Grants, 2295
Kenneth T. & Eileen L. Norris Fndn Grants, 2300
Lubbock Area Foundation Grants, 2383
Robert W. Woodruff Foundation Grants, 3389
San Antonio Area Foundation Grants, 3469
Sid W. Richardson Foundation Grants, 3553
Sioux Falls Area Community Foundation Field-of-Interest and Donor-Advised Grants, 3569
Taubman Endowment for the Arts Fndn Grants, 3702
Wayne County Foundation Grants, 3909

Organ Transplants
Cystic Fibrosis Canada Transplant Center Incentive Grants, 1392
Foundation of CVPH Melissa Lahtinen-Penfield Organ Donor Fund Grants, 1703
MBL Eugene and Millicent Bell Fellowships, 2470
Pfizer Advancing Research in Transplantation Science (ARTS) Research Awards, 3163
Pfizer ASPIRE North America Broad Spectrum Antibiotics for the Treatment of Gram-Negative or Polymicrobial Infections Research Awards, 3169
Pfizer Healthcare Charitable Contributions, 3180
Robert B McMillen Foundation Grants, 3381
Sid W. Richardson Foundation Grants, 3553
Young Ambassador Scholarship In Memory of Christopher Nordquist, 3981

Organizational Development
California Endowment Innovative Ideas Challenge Grants, 1000
CNCS Social Innovation Grants, 1192
Deborah Munroe Noonan Memorial Grants, 1443
GNOF Organizational Effectiveness Grants and Workshops, 1846
Lisa and Douglas Goldman Fund Grants, 2369
Mayo Clinic Administrative Fellowship, 2460
Meyer Foundation Benevon Grants, 2532
Meyer Fndn Management Assistance Grants, 2534
Middlesex Savings Charitable Foundation Capacity Building Grants, 2551
MLA T. Mark Hodges Int'l Service Award, 2575
NWHF Health Advocacy Small Grants, 3033
NWHF Kaiser Permanente Community Grants, 3034
Otto Bremer Foundation Grants, 3099
Peter and Elizabeth C. Tower Foundation Organizational Scholarships, 3148
Priddy Foundation Organizational Development Grants, 3265
SfN Louise Hanson Marshall Specific Recognition Award, 3527
UniHealth Foundation Healthcare Systems Enhancements Grants, 3784
Union Bank, N.A. Foundation Grants, 3788

Organizational Theory and Behavior
Jane Beattie Memorial Scholarship, 2207

Orthopedics
El Paso Community Foundation Grants, 1579
Fndn for Appalachian Ohio Bachtel Scholarships, 1692
Foundation of Orthopedic Trauma Grants, 1707
John S. Dunn Research Foundation Grants, 2250
Orthopaedic Trauma Association Kathy Cramer Young Clinician Memorial Fellowships, 3083
Orthopaedic Trauma Assoc Research Grants, 3084
Union Labor Health Fndn Angel Fund Grants, 3791

Osteopathic Medicine
New Jersey Osteopathic Education Foundation Scholarships, 2680
NMF Need-Based Scholarships, 2965
OSF Access to Essential Medicines Initiative, 3086
Scleroderma Foundation Established Investigator Grants, 3497

Osteoporosis
Fairlawn Foundation Grants, 1626
NIH Receptors and Signaling in Bone in Health and Disease, 2891
Premera Blue Cross Grants, 3258

Otolaryngic
AAO-HNSF Health Services Research Grants, 165
AAO-HNSF Maureen Hannley Research Training Awards, 166
AAO-HNSF Percy Memorial Research Award, 167
AAO-HNSF Rande H. Lazar Health Services Research Grant, 168
AAO-HNSF Resident Research Awards, 169
AAOA Foundation/AAO-HNSF Combined Research Grants, 172

Otolaryngology
AAO-HNSF CORE Research Grants, 164
AAO-HNS Medical Student Research Prize, 170
AAO-HNS Resident Research Prizes, 171
ASPO Daiichi Innovative Technology Grant, 783
ASPO Fellowships, 784
ASPO Research Grants, 785
Richmond Eye and Ear Fund Grants, 3372
Triological Society Research Career Development Awards, 3760
Xoran Technologies Resident Research Grant, 3976

Otology
AHRF Eugene L. Derlacki Research Grants, 361
AHRF Georgia Birtman Grant, 362
AHRF Wiley H. Harrison Research Award, 364
ANS/AAO-HNSF Herbert Silverstein Otology and Neurotology Research Award, 601
ANS Neurotology Fellows Award, 604
ANS Nicholas Torok Vestibular Award, 605
ANS Trainee Award, 606
Oticon Focus on People Awards, 3098
Triological Society Research Career Development Awards, 3760

Outpatient Care
CDC Evaluation of the Use of Rapid Testing For Influenza in Outpatient Medical Settings, 1057
Erie Community Foundation Grants, 1601
George A. and Grace L. Long Foundation Grants, 1788
HAF David (Davey) H. Somerville Medical Travel Fund Grants, 1928
USDD Medical Practice Breadth of Medical Practice and Disease Frequency Exposure Grants, 3843

Ovarian Cancer
ASCO/UICC International Cancer Technology Transfer (ICRETT) Fellowships, 677
ASCO Young Investigator Award, 683
Conquer Cancer Foundation of ASCO Career Development Award, 1328
Conquer Cancer Foundation of ASCO International Development and Education Awards, 1333
GNOF IMPACT Harold W. Newman, Jr. Charitable Trust Grants, 1841
Mary Kay Foundation Cancer Research Grants, 2444
Penny Severns Breast, Cervical and Ovarian Cancer Research Grants, 3136
Seattle Foundation Medical Funds Grants, 3508

Pacific Islanders
American Sociological Association Minority Fellowships, 563
ASM Robert D. Watkins Graduate Research Fellowship, 747
FDHN Student Research Fellowships, 1654
MLA Grad Scholarship for Minority Students, 2565
NHLBI Training Program Grants for Institutions that Promote Diversity, 2738
SNM PDEF Mickey Williams Minority Student Scholarships, 3584

Pacific Islands
Fulbright Traditional Scholar Program in the East Asia/Pacific Region, 1757

Pain
AAHPM Hospice & Palliative Med Fellowships, 152
Duke University Pain Management Fellowships, 1526
Duke University Ped Anesthesiology Fellowship, 1527
IAFF Burn Foundation Research Grants, 2089
NIA Mechanisms, Measurement, and Management of Pain in Aging: from Molecular to Clinical, 2768
NINR Mechanisms, Models, Measurement, & Management in Pain Research Grants, 2931
NINR Quality of Life for Individuals at the End of Life Grants, 2933
Pfizer Australia Neuroscience Research Grants, 3175
Pfizer Healthcare Charitable Contributions, 3180
Pfizer Medical Education Track One Grants, 3182

Pakistan
USAID Child, Newborn, and Maternal Health Project Grants, 3806
USAID Health System Strengthening Grants, 3814

Paleontology
NSF Systematic Biology and Biodiversity Inventories Grants, 3024

Palliative Care
AACN End of Life/Palliative Care Small Projects Grants, 83
Aetna Foundation Integrated Health Care Grants, 328
Aetna Foundation Regional Health Grants, 333
Austin S. Nelson Foundation Grants, 805
Conquer Cancer Foundation of ASCO International Development and Education Award in Palliative Care, 1332
Daniels Fund Grants-Aging, 1424
OSF International Palliative Care Grants, 3091
Piper Trust Healthcare & Med Research Grants, 3228
Sara Elizabeth O'Brien Trust Grants, 3483

Pancreas
AACR-FNAB Fellows Grant for Translational Pancreatic Cancer Research, 89
AACR-Pancreatic Cancer Action Network Career Development Award for Research, 97
AACR-Pancreatic Cancer Action Network Innovative Grants, 98
FDHN Designated Research Award in Research Related to Pancreatitis, 1642
FDHN Graduate Student Awards, 1646
NIAAA Mechanisms of Alcohol and Drug-Induced Pancreatitis Grants, 2743
NIDDK Pancreatic Development and Regeneration Grants: Cellular Therapies for Diabetes, 2833
NIDDK Small Business Innov Research to Develop New Therapeutics & Monitoring Tech for Type 1 Diabetes Towards an Artificial Pancreas, 2845

Pancreatic Cancer
ASCO/UICC International Cancer Technology Transfer (ICRETT) Fellowships, 677
ASCO Young Investigator Award, 683
Conquer Cancer Foundation of ASCO Career Development Award, 1328
Conquer Cancer Foundation of ASCO International Development and Education Awards, 1333
GNOF IMPACT Harold W. Newman, Jr. Charitable Trust Grants, 1841
Mary Kay Foundation Cancer Research Grants, 2444
Pfizer Healthcare Charitable Contributions, 3180
Pfizer Medical Education Track Two Grants, 3183
Seattle Foundation Medical Funds Grants, 3508

Parasitology
ASM/CDC Fellowships in Infectious Disease and Public Health Microbiology, 716
BWF Ad Hoc Grants, 982
BWF Investigators in the Pathogenesis of Infectious Disease Awards, 988
US CRDF Leishmaniasis: Collaborative Research Opportunities in N Africa & Middle East, 3828

Parent Education
3M Company Fndn Community Giving Grants, 6
A Fund for Women Grants, 342
Allen Foundation Educational Nutrition Grants, 442
Bender Foundation Grants, 876
Bodman Foundation Grants, 929
Children's Trust Fund of Oregon Fndn Grants, 1139
CJ Foundation for SIDS Program Services Grants, 1163
Community Foundation for Greater Atlanta Clayton County Fund Grants, 1240
Community Foundation for Greater Atlanta Morgan County Fund Grants, 1244
Community Foundation for Greater Atlanta Newton County Fund Grants, 1245
Comprehensive Health Education Fndn Grants, 1320
Connecticut Community Foundation Grants, 1324
DHHS Adolescent Family Life Demonstration Projects, 1472
FCD New American Children Grants, 1637
George W. Wells Foundation Grants, 1804
Gerber Foundation Grants, 1807
Grand Rapids Community Foundation Ionia County Youth Fund Grants, 1871
Mary Black Foundation Early Childhood Development Grants, 2442
Medtronic Foundation Community Link Human Services Grants, 2511
MGM Resorts Foundation Community Grants, 2541
OneFamily Foundation Grants, 3075
Philadelphia Foundation Organizational Effectiveness Grants, 3189
Piper Jaffray Fndn Communities Giving Grants, 3227
San Diego Foundation for Change Grants, 3470
Seattle Foundation Benjamin N. Phillips Memorial Fund Grants, 3504
Southbury Community Trust Fund, 3625
U.S. Department of Education 21st Century Community Learning Centers, 3767
U.S. Dept of Education Special Ed--National Activities--Parent Information Centers, 3772
Union Bank, N.A. Foundation Grants, 3788
Wood-Claeyssens Foundation Grants, 3971

Parent Involvement
Bella Vista Foundation Grants, 873
Bodman Foundation Grants, 929
GenCorp Foundation Grants, 1779
GNOF Stand Up For Our Children Grants, 1847
Harmony Project Grants, 1948
Healthcare Fndn for Orange County Grants, 1981
Louis H. Aborn Foundation Grants, 2380
Mary Black Foundation Early Childhood Development Grants, 2442
Prince Charitable Trusts Chicago Grants, 3270
Pulaski County Community Foundation Grants, 3279
Tommy Hilfiger Corporate Foundation Grants, 3737

SUBJECT INDEX

Vancouver Foundation Grants and Community Initiatives Program, 3849

Parenteral Nutrition
ASPEN Dudrick Research Scholar Award, 759
ASPEN Harry M. Vars Award, 760
ASPEN Rhoads Research Foundation Abbott Nutrition Research Grants, 761
ASPEN Rhoads Research Foundation Baxter Parenteral Nutrition Research Grant, 762
ASPEN Rhoads Research Foundation C. Richard Fleming Grant, 763
ASPEN Rhoads Research Foundation Maurice Shils Grant, 764
ASPEN Rhoads Research Foundation Norman Yoshimura Grant, 765
ASPEN Scientific Abstracts Awards for Papers or Posters, 766
ASPEN Scientific Abstracts Promising Investigator Awards, 767

Parkinson's Disease
Fairlawn Foundation Grants, 1626
James J. & Joan A. Gardner Family Fndn Grants, 2193
National Parkinson Foundation Clinical Research Fund Grants, 2635
Partnership for Cures Two Years To Cures Grants, 3118
PDF-AANF Clinician-Scientist Dev Awards, 3127
PDF International Research Grants, 3128
PDF Postdoctoral Fellowships for Basic Scientists, 3129
PDF Postdoctoral Fellowships for Neurologists, 3130
PDF Summer Fellowships, 3131
PSG Mentored Clinical Research Awards, 3277
SfN Julius Axelrod Prize, 3526

Parks
Arthur M. Blank Family Foundation Inspiring Spaces Grants, 673
Bacon Family Foundation Grants, 817
Blackford County Community Foundation Grants, 906
BMW of North America Charitable Contributions, 926
Clara Blackford Smith and W. Aubrey Smith Charitable Foundation Grants, 1165
Community Foundation of Bloomington and Monroe County Grants, 1265
Community Foundation of St. Joseph County Special Project Challenge Grants, 1310
Community Foundation Partnerships - Martin County Grants, 1318
ConocoPhillips Foundation Grants, 1327
Cumberland Community Foundation Grants, 1374
Del E. Webb Foundation Grants, 1450
Foundation for the Mid South Community Development Grants, 1699
Knox County Community Foundation Grants, 2313
Merrick Foundation Grants, 2522
Perry County Community Foundation Grants, 3142
Pike County Community Foundation Grants, 3222
Prince Charitable Trusts Chicago Grants, 3270
Pulaski County Community Foundation Grants, 3279
Sioux Falls Area Community Foundation Community Fund Grants (Unrestricted), 3568
Sioux Falls Area Community Foundation Spot Grants (Unrestricted), 3570
Union Bank, N.A. Foundation Grants, 3788

Pathogenesis
AHNS/AAO-HNSF Surgeon Scientist Combined Award, 357
AHNS/AAO-HNSF Young Investigator Combined Award, 358
AHNS Alando J. Ballantyne Resident Research Pilot Grant, 359
AHNS Pilot Grant, 360
ALVRE Casselberry Award, 460
ALVRE Grant, 461
ALVRE Seymour R. Cohen Award, 462
Alzheimer's Association Conference Grants, 463
Alzheimer's Association Development of New Cognitive and Functional Instruments Grants, 464

Alzheimer's Association Everyday Technologies for Alzheimer Care Grants, 465
Alzheimer's Association Mentored New Investigator Research Grants to Promote Diversity, 467
Alzheimer's Association Neuronal Hyper Excitability and Seizures in Alzheimer's Disease Grants, 468
Alzheimer's Association New Investigator Research Grants to Promote Diversity, 470
Alzheimer's Association U.S.-U.K. Young Investigator Exchange Fellowships, 471
Alzheimer's Association Zenith Fellows Awards, 472
APHL Emerging Infectious Diseases Fellowships, 636
ASM Maurice Hilleman/Merck Award, 738
BWF Ad Hoc Grants, 982
BWF Investigators in the Pathogenesis of Infectious Disease Awards, 988
DSO Controlling Pathogen Workshop Grants, 1514
Lymphatic Education and Research Network Postdoctoral Fellowships, 2395
NIA Alzheimer's Disease Core Centers Grants, 2748
Penny Severns Breast, Cervical and Ovarian Cancer Research Grants, 3136
Pfizer ASPIRE EU Emerging Mechanisms of Resistance Antibacterial Research Awards, 3167
Tourette Syndrome Association Post-Doctoral Fellowships, 3740
Tourette Syndrome Association Research Grants, 3741

Pathology
AHAF Alzheimer's Disease Research Grants, 349
Joseph Drown Foundation Grants, 2268
NYAM Edward N. Gibbs Memorial Lecture and Award in Nephrology, 3041
PhRMA Foundation Pharmacology/Toxicology Post Doctoral Fellowships, 3215
PhRMA Foundation Pharmacology/Toxicology Sabbatical Fellowships, 3218
Simone and Cino del Duca Grand Prix Awards, 3564
UICC Raisa Gorbachev Memorial International Cancer Fellowships, 3780

Pathophysiology
AHNS/AAO-HNSF Surgeon Scientist Combined Award, 357
AHNS/AAO-HNSF Young Investigator Combined Award, 358
AHNS Alando J. Ballantyne Resident Research Pilot Grant, 359
AHNS Pilot Grant, 360
ALVRE Casselberry Award, 460
ALVRE Grant, 461
ALVRE Seymour R. Cohen Award, 462
CFF Clinical Research Grants, 1084
NARSAD Colvin Prize for Outstanding Achievement in Mood Disorders Research, 2619
NARSAD Ruane Prize for Child and Adolescent Psychiatric Research, 2624
NHLBI Lymphatics in Health and Disease in the Digestive, Urinary, Cardiovascular and Pulmonary Systems, 2705
NIDDK Type 2 Diabetes in the Pediatric Population Research Grants, 2852
Tourette Syndrome Association Post-Doctoral Fellowships, 3740
Tourette Syndrome Association Research Grants, 3741

Patient Care and Education
AAAAI Distinguished Clinician Award, 24
AAAAI Distinguished Layperson Award, 25
AACN Physio-Control Small Projects Grants, 86
AAFP Foundation Joint Grants, 133
AES JK Penry Excellence in Epilepsy Care Award, 319
AMA Foundation Health Literacy Grants, 485
AMSSM Foundation Research Grants, 577
APA Young Investigator Grant Program, 634
AVDF Health Care Grants, 811
Becton Dickinson and Company Grants, 869
Biogen Foundation Patient Educational Grants, 901
BWF Institutional Program Unifying Population and Laboratory Based Sciences Grants, 987

Canadian Patient Safety Institute (CPSI) Patient Safety Studentships, 1012
Conquer Cancer Foundation of ASCO Improving Cancer Care Grants, 1331
Cystic Fibrosis Canada Clinic Incentive Grants, 1383
Everyone Breathe Asthma Education Grants, 1616
Fairlawn Foundation Grants, 1626
Four J Foundation Grants, 1713
Greenwall Foundation Bioethics Grants, 1897
Healthcare Foundation of New Jersey Grants, 1982
Jacob and Valeria Langeloth Foundation Grants, 2182
Mary Wilmer Covey Charitable Trust Grants, 2448
Mayo Clinic Admin Fellowship - Eau Claire, 2461
McCarthy Family Foundation Grants, 2490
Medtronic Foundation Patient Link Grants, 2514
MMS and Alliance Charitable Foundation Grants for Community Action and Care for the Medically Uninsured, 2587
Mt. Sinai Health Care Foundation Health of the Jewish Community Grants, 2599
NHLBI Mentored Patient-Oriented Research Career Development Awards, 2710
NIH Enhancing Adherence to Diabetes Self-Management Behaviors, 2875
NIH Mentored Patient-Oriented Research Career Development Award, 2888
NINR Acute and Chronic Care During Mechanical Ventilation Research Grants, 2926
NINR Quality of Life for Individuals at the End of Life Grants, 2933
Penny Severns Breast, Cervical and Ovarian Cancer Research Grants, 3136
Pfizer Healthcare Charitable Contributions, 3180
Pfizer Special Events Grants, 3186
PhRMA Foundation Health Outcomes Research Starter Grants, 3204
Presbyterian Patient Assistance Program, 3260
Rathmann Family Foundation Grants, 3309
RCPSC John G. Wade Visiting Professorship in Patient Safety & Simulation-Based Med Education, 3329
STTI Environment of Elder Care Nursing Research Grants, 3667
UniHealth Foundation Healthcare Systems Enhancements Grants, 3784
USDD Care for the Critically Injured Burn Patient Grants, 3839

Patient/Physician Relationship
AES JK Penry Excellence in Epilepsy Care Award, 319
AMA Foundation Health Literacy Grants, 485
AVDF Health Care Grants, 811
Conquer Cancer Foundation of ASCO Improving Cancer Care Grants, 1331
Jacob and Valeria Langeloth Foundation Grants, 2182
NHLBI Midcareer Investigator Award in Patient-Oriented Research, 2713
USDD Medical Practice Breadth of Medical Practice and Disease Frequency Exposure Grants, 3843
USDD Medical Practice Initiative Procedural Skill Decay and Maintenance Grants, 3844

Peace/Disarmament
Donald and Sylvia Robinson Family Grants, 1486
Elizabeth Morse Genius Charitable Trust Grants, 1569
Global Fund for Women Grants, 1833
S.H. Cowell Foundation Grants, 3447
United States Institute of Peace - Jennings Randolph Senior Fellowships, 3802

Pediatric Cancer
CDI Interdisciplinary Research Initiatives Grants, 1074
CDI Postdoctoral Fellowships, 1075
Conquer Cancer Foundation of ASCO International Development and Education Awards, 1333
GNOF IMPACT Harold W. Newman, Jr. Charitable Trust Grants, 1841
OSF International Palliative Care Grants, 3091
Sara Elizabeth O'Brien Trust Grants, 3483
Seattle Foundation Medical Funds Grants, 3508
Victor E. Speas Foundation Grants, 3868

Pediatrics

1976 Foundation Grants, 12
AACAP Rob Cancro Academic Leadership Award, 72
AACR Career Development Awards for Pediatric Cancer Research, 107
AAP Anne E. Dyson Child Advocacy Awards, 176
AAP Community Access To Child Health (CATCH) Advocacy Training Grants, 177
AAP Community Access To Child Health (CATCH) Implementation Grants, 178
AAP Community Access to Child Health (CATCH) Planning Grants, 179
AAP Community Access To Child Health (CATCH) Residency Training Grants, 180
AAP Community Access To Child Health (CATCH) Resident Grants, 181
AAP Legislative Conference Scholarships, 185
AAP Leonard P. Rome Community Access to Child Health (CATCH) Visiting Professorships, 186
AAP Nutrition Award, 187
AAP Resident Initiative Fund Grants, 189
AAP Resident Research Grants, 190
AcademyHealth Nemours Child Health Services Research Awards, 224
ACS Cancer Control Career Development Awards for Primary Care Physicians, 250
Alexander and Margaret Stewart Trust Grants, 419
American Philosophical Society Daland Fellowships in Clinical Investigation, 537
APA Young Investigator Grant Program, 634
ASPO Daiichi Innovative Technology Grant, 783
ASPO Fellowships, 784
ASPO Research Grants, 785
Austin S. Nelson Foundation Grants, 805
Blowitz-Ridgeway Foundation Grants, 913
CDI Interdisciplinary Research Initiatives Grants, 1074
CDI Postdoctoral Fellowships, 1075
CFF First- and Second-Year Clinical Fellowships, 1085
CFF Leroy Matthews Physician-Scientist Awards, 1087
Children's Brain Tumor Fnd Research Grants, 1136
Children's Cardiomyopathy Foundation Research Grants, 1137
Columbus Foundation Central Benefits Health Care Foundation Grants, 1216
Community Foundation of Louisville Diller B. and Katherine P. Groff Fund for Pediatric Surgery Grants, 1293
DHHS Adolescent Family Life Demonstration Projects, 1472
E.W. "Al" Thrasher Awards, 1540
Elizabeth Glaser Int'l Leadership Awards, 1566
Elizabeth Glaser Scientist Award, 1567
Fuji Film Grants, 1748
Gerber Foundation Grants, 1807
Health Fndn of Greater Indianapolis Grants, 1984
HRAMF Charles H. Hood Foundation Child Health Research Awards, 2057
Lowe Foundation Grants, 2381
Lucile Packard Foundation for Children's Health Grants, 2386
M. Bastian Family Foundation Grants, 2402
Mary S. and David C. Corbin Foundation Grants, 2447
MedImmune Charitable Grants, 2509
NAPNAP Fndn Elaine Gelman Scholarship, 2609
NAPNAP Fndn Grad Research Grant, 2610
NAPNAP Foundation Innovative Health Care Small Grant, 2611
NAPNAP Foundation McNeil Grant-in-Aid, 2612
NAPNAP Foundation McNeil PNP Scholarships, 2613
NAPNAP Foundation McNeil Rural and Underserved Scholarships, 2614
NAPNAP Foundation Nursing Research Grants, 2615
NAPNAP Foundation Reckitt Benckiser Student Scholarship, 2616
NHLBI Pediatric Cardiac Genomics Consortium Grants, 2719
NIDDK Planning Grants For Translational Research For The Prevention And Control Of Diabetes And Obesity, 2837
NIDDK Type 2 Diabetes in the Pediatric Population Research Grants, 2852
NYAM Brock Lecture, Award and Visiting Professorship in Pediatrics, 3038
OneSight Research Foundation Block Grants, 3076
Peabody Foundation Grants, 3132
Pediatric Brain Tumor Fndn Research Grants, 3134
Pediatric Cancer Research Foundation Grants, 3135
Pfizer Australia Paediatric Endocrine Care Research Grants, 3176
Piper Trust Healthcare & Med Research Grants, 3228
Ralph M. Parsons Foundation Grants, 3304
Regenstrief General Internal Medicine - General Pediatrics Research Fellowship Program, 3351
Saint Luke's Health Initiatives Grants, 3455
STTI Rosemary Berkel Crisp Research Awards, 3670
Thrasher Research Fund Early Career Awards, 3731
Victor E. Speas Foundation Grants, 3868

Penology/Correctional Institutions and Procedures

Baptist Community Ministries Grants, 828
Caddock Foundation Grants, 994
Cambridge Community Foundation Grants, 1004

Peptides

Ralph F. Hirschmann Award in Peptide Chem, 3303

Performance Art

Crystelle Waggoner Charitable Trust Grants, 1364
George S. & Dolores Dore Eccles Fndn Grants, 1801
Katharine Matthies Foundation Grants, 2291
Southbury Community Trust Fund, 3625

Performing Arts

Actors Addiction & Recovery Services Grants, 282
Actors Fund HIV/AIDS Initiative Grants, 283
Adams Foundation Grants, 301
Adaptec Foundation Grants, 304
Agnes Gund Foundation Grants, 345
Alcatel-Lucent Technologies Foundation Grants, 413
Alexis Gregory Foundation Grants, 426
Amgen Foundation Grants, 572
AMI Semiconductors Corporate Grants, 576
Arizona Public Service Corporate Giving Program Grants, 658
Arlington Community Foundation Grants, 662
Atherton Family Foundation Grants, 794
Athwin Foundation Grants, 795
Atlanta Foundation Grants, 796
Autzen Foundation Grants, 810
Avon Products Foundation Grants, 814
BancorpSouth Foundation Grants, 823
Bildner Family Foundation Grants, 891
Blanche and Irving Laurie Foundation Grants, 910
Blue Mountain Community Foundation Grants, 921
Bodenwein Public Benevolent Foundation Grants, 928
Boeing Company Contributions Grants, 930
Booth-Bricker Fund Grants, 933
Bruce and Adele Greenfield Foundation Grants, 972
Burlington Northern Santa Fe Foundation Grants, 978
Carnahan-Jackson Foundation Grants, 1032
Cemala Foundation Grants, 1077
Charles Delmar Foundation Grants, 1115
Colonel Stanley R. McNeil Foundation Grants, 1211
Connecticut Community Foundation Grants, 1324
ConocoPhillips Foundation Grants, 1327
Cowles Charitable Trust Grants, 1354
Cudd Foundation Grants, 1370
Cullen Foundation Grants, 1372
D.F. Halton Foundation Grants, 1405
Daniel & Nanna Stern Family Fndn Grants, 1422
Davis Family Foundation Grants, 1435
Deborah Munroe Noonan Memorial Grants, 1443
DeMatteis Family Foundation Grants, 1457
DeRoy Testamentary Foundation Grants, 1469
Donald and Sylvia Robinson Family Foundation Grants, 1486
E. Rhodes & Leona B. Carpenter Grants, 1539
Elizabeth McGraw Foundation Grants, 1568
Elizabeth Morse Genius Charitable Trust Grants, 1569

Elsie H. Wilcox Foundation Grants, 1583
Eugene B. Casey Foundation Grants, 1609
Eugene M. Lang Foundation Grants, 1611
FirstEnergy Foundation Community Grants, 1668
Frank Loomis Palmer Fund Grants, 1726
G.N. Wilcox Trust Grants, 1768
Gardner Foundation Grants, 1772
General Mills Foundation Grants, 1782
George A. and Grace L. Long Foundation Grants, 1788
George Gund Foundation Grants, 1796
George W. Brackenridge Foundation Grants, 1802
George W. Codrington Charitable Fndn Grants, 1803
Georgia Power Foundation Grants, 1806
Giant Eagle Foundation Grants, 1815
GNOF IMPACT Grants for Arts and Culture, 1838
Grand Rapids Area Community Fndn Grants, 1865
Grand Rapids Area Community Foundation Nashwauk Area Endowment Fund Grants, 1866
Griffin Foundation Grants, 1903
Guy I. Bromley Trust Grants, 1919
Harold R. Bechtel Charitable Remainder Uni-Trust Grants, 1953
Helen Gertrude Sparks Charitable Trust Grants, 1993
Helen S. Boylan Foundation Grants, 1997
Henrietta Lange Burk Fund Grants, 2001
Herman Goldman Foundation Grants, 2012
High Meadow Foundation Grants, 2027
Horace Moses Charitable Foundation Grants, 2047
Howard and Bush Foundation Grants, 2054
James M. Cox Foundation of Georgia Grants, 2196
James S. Copley Foundation Grants, 2200
Janus Foundation Grants, 2211
Jerome Robbins Foundation Grants, 2223
John I. Smith Charities Grants, 2240
John W. Speas and Effie E. Speas Memorial Trust Grants, 2264
Julius N. Frankel Foundation Grants, 2282
Katharine Matthies Foundation Grants, 2291
Kelvin and Eleanor Smith Foundation Grants, 2297
Leo Goodwin Foundation Grants, 2354
Lewis H. Humphreys Charitable Trust Grants, 2361
Linford and Mildred White Charitable Grants, 2368
Louetta M. Cowden Foundation Grants, 2377
Louis and Elizabeth Nave Flarsheim Charitable Foundation Grants, 2379
Lubrizol Foundation Grants, 2384
Lucile Horton Howe and Mitchell B. Howe Foundation Grants, 2385
Mabel Y. Hughes Charitable Trust Grants, 2409
Margaret T. Morris Foundation Grants, 2429
Marie H. Bechtel Charitable Remainder Uni-Trust Grants, 2432
Marjorie Moore Charitable Foundation Grants, 2438
McKesson Foundation Grants, 2496
McLean Contributionship Grants, 2497
Meyer and Stephanie Eglin Foundation Grants, 2531
Morris & Gwendolyn Cafritz Fndn Grants, 2594
Nicholas H. Noyes Jr. Memorial Fndn Grants, 2791
Northwestern Mutual Foundation Grants, 2999
Oppenstein Brothers Foundation Grants, 3078
Piedmont Natural Gas Corporate and Charitable Contributions, 3220
Powell Foundation Grants, 3255
Principal Financial Group Foundation Grants, 3273
Rainbow Fund Grants, 3301
Rayonier Foundation Grants, 3311
Reinberger Foundation Grants, 3357
Rhode Island Foundation Grants, 3362
Richard D. Bass Foundation Grants, 3366
Sain-Orr and Royak-DeForest Steadman Foundation Grants, 3452
San Juan Island Community Foundation Grants, 3480
Schurz Communications Foundation Grants, 3496
Sensient Technologies Foundation Grants, 3513
Shell Deer Park Grants, 3543
Sid W. Richardson Foundation Grants, 3553
Society for the Arts in Healthcare Grants, 3602
SunTrust Bank Trusteed Foundations Florence C. and Harry L. English Memorial Fund Grants, 3680

SUBJECT INDEX

SunTrust Bank Trusteed Foundations Greene-Sawtell Grants, 3681
SunTrust Bank Trusteed Foundations Harriet McDaniel Marshall Tust Grants, 3682
SunTrust Bank Trusteed Foundations Nell Warren Elkin and William Simpson Elkin Grants, 3683
SunTrust Bank Trusteed Foundations Thomas Guy Woolford Charitable Trust Grants, 3684
SunTrust Bank Trusteed Foundations Walter H. and Marjory M. Rich Memorial Fund Grants, 3685
Textron Corporate Contributions Grants, 3712
Thomas Sill Foundation Grants, 3724
Union Bank, N.A. Foundation Grants, 3788
W. C. Griffith Foundation Grants, 3882
W.C. Griffith Foundation Grants, 3881
Wege Foundation Grants, 3912
William G. Gilmore Foundation Grants, 3951
Winifred & Harry B. Allen Foundation Grants, 3962
Wood Family Charitable Trust Grants, 3972

Perinatal Disorders
HRAMF Charles H. Hood Foundation Child Health Research Awards, 2057

Periodontics
DHHS Oral Health Promotion Research Across the Lifespan, 1474

Personnel Training and Development
Charlotte County (FL) Community Foundation Grants, 1126
Children Affected by AIDS Foundation Camp Network Grants, 1144
Claude Worthington Benedum Fndn Grants, 1178
Dept of Ed Special Education--Personnel Development to Improve Services and Results for Children with Disabilities, 1465
ERC Starting Grants, 1599
Piper Trust Healthcare & Med Research Grants, 3228
Priddy Foundation Organizational Development Grants, 3265
UniHealth Workforce Development Grants, 3786
Union Bank, N.A. Foundation Grants, 3788

Pest Control
Conservation, Food, and Health Foundation Grants for Developing Countries, 1338

Pesticides
Clarence E. Heller Charitable Foundation Grants, 1166

Petroleum Science
NSF-NIST Interaction in Chemistry, Materials Research, Molecular Biosciences, Bioengineering, and Chemical Engineering, 3004

Pharmaceuticals
American Chemical Society GCI Pharmaceutical Roundtable Research Grants, 516
Campbell Hoffman Foundation Grants, 1006
CCFF Chairman's Dist Life Sciences Award, 1044
Conquer Cancer Foundation of ASCO Drug Development Research Professorship, 1330
Foundation for Pharmaceutical Sciences Herb and Nina Demuth Grant, 1697
Hawaii Community Foundation Health Education and Research Grants, 1976
NSF Chemistry Research Experiences for Undergraduates (REU), 3012
Pfizer Advancing Research in Transplantation Science (ARTS) Research Awards, 3163
Pfizer Anti Infective Research EU Grants, 3164
Pfizer ASPIRE EU Antifungal Research Awards, 3165
Pfizer ASPIRE EU Dupuytren's Contracture Research Awards, 3166
Pfizer ASPIRE EU Emerging Mechanisms of Resistance Antibacterial Research Awards, 3167
Pfizer ASPIRE EU MRSA Nosocomial Pneumonia & MRSA Complicated Skin & Soft Tissue Infections Antibacterial Research Awards, 3168

Pfizer ASPIRE North America Broad Spectrum Antibiotics for the Treatment of Gram-Negative or Polymicrobial Infections Research Awards, 3169
Pfizer ASPIRE North America Hemophilia Research Awards, 3170
Pfizer ASPIRE North America Narrow Spectrum Antibiotics for the Treatment of MRSA Research Awards, 3171
Pfizer ASPIRE North America Rheumatology Research Awards, 3172
Pfizer ASPIRE Worldwide Endocrine Young Investigator Grants, 3173
Pfizer Australia Cancer Research Grants, 3174
Pfizer Australia Neuroscience Research Grants, 3175
Pfizer Australia Paediatric Endocrine Care Research Grants, 3176
PFizer Compound Transfer Agreements, 3177
Pfizer Global Investigator Research Grants, 3178
Pfizer Global Research Awards for Nicotine Independence, 3179
Pfizer Healthcare Charitable Contributions, 3180
Pfizer Inflammation Competitive Research Awards (UK), 3181
Pfizer Medical Education Track One Grants, 3182
Pfizer Research Initiative Dermatology Grants (Germany), 3184
Pfizer Research Initiative Rheumatology Grants (Germany), 3185
PhRMA Foundation Health Outcomes Post Doctoral Fellowships, 3202
PhRMA Foundation Health Outcomes Research Starter Grants, 3204
PhRMA Foundation Informatics Post Doctoral Fellowships, 3206
PhRMA Foundation Pharmaceutics Postdoctoral Fellowships, 3211
PhRMA Foundation Pharmaceutics Pre Doctoral Fellowships, 3212
PhRMA Foundation Pharmaceutics Research Starter Grants, 3213
PhRMA Foundation Pharmaceutics Sabbatical Fellowships, 3214
Society of Cosmetic Chemists Allan B. Black Award Sponsored by Presperse, 3604
Society of Cosmetic Chemists Award Sponsored by McIntyre Group, Ltd., 3605
Society of Cosmetic Chemists Award Sponsored by The HallStar Company, 3606
Society of Cosmetic Chemists Chapter Best Speaker Award, 3607
Society of Cosmetic Chem Chapter Merit Award, 3608
Society of Cosmetic Chemists Frontier of Science Award, 3609
Society of Cosmetic Chem Hans Schaeffer Award, 3610
Society of Cosmetic Chem Jos Ciaudelli Award, 3611
Society of Cosmetic Chemists Keynote Award Lecture Sponsored by Ruger Chemical Corporation, 3612
Society of Cosmetic Chemists Literature Award, 3613
Society of Cosmetic Chemists Maison G. de Navarre Medal, 3614
Society of Cosmetic Chemists Merit Award, 3615
Society of Cosmetic Chemists Robert A. Kramer Lifetime Service Award, 3616
Society of Cosmetic Chem Shaw Mudge Award, 3617
Society of Cosmetic Chem Stud Poster Awards, 3618
Society of Cosmetic Chem You Scientist Awards, 3619
SRC Medical Research Grants, 3635

Pharmacogenomics
NIGMS Enabling Resources for Pharmacogenomics Grants, 2862
Pfizer Advancing Research in Transplantation Science (ARTS) Research Awards, 3163

Pharmacology
AHAF Alzheimer's Disease Research Grants, 349
AHAF National Glaucoma Research Grants, 351
AOCS George Schroepfer Medal, 613
ASM sanofi-aventis ICAAC Award, 749

ASPEN Rhoads Research Foundation Abbott Nutrition Research Grants, 761
ASPEN Rhoads Research Foundation Baxter Parenteral Nutrition Research Grant, 762
ASPEN Rhoads Research Foundation C. Richard Fleming Grant, 763
ASPEN Rhoads Research Foundation Maurice Shils Grant, 764
ASPEN Rhoads Research Foundation Norman Yoshimura Grant, 765
ASPET-Astellas Awards in Translational Pharmacology, 768
ASPET Benedict R. Lucchesi Award in Cardiac Pharmacology, 769
ASPET Brodie Award in Drug Metabolism, 770
ASPET Division for Drug Metabolism Early Career Achievement Award, 771
ASPET Epilepsy Research Award for Outstanding Contributions to the Pharmacology of Antiepileptic Drugs, 772
ASPET Goodman and Gilman Award in Drug Receptor Pharmacology, 773
ASPET John J. Abel Award in Pharmacology, 774
ASPET Julius Axelrod Award in Pharmacology, 775
ASPET P. B. Dews Lifetime Achievement Award for Research in Behavioral Pharmacology, 776
ASPET Paul M. Vanhoutte Award, 777
ASPET Torald Sollmann Award in Pharmacology, 778
ASPET Travel Awards, 779
Astellas Foundation Research Grants, 790
Fragile X Syndrome Postdoc Research Fellows, 1714
NHLBI Ruth L. Kirschstein National Research Service Awards for Individual Senior Fellows, 2732
NIDA Mentored Clinical Scientist Research Career Development Awards, 2793
NIGMS Enabling Resources for Pharmacogenomics Grants, 2862
NIH Ruth L. Kirschstein National Research Service Awards for Individual Postdoc Fellowships, 2902
Pfizer ASPIRE EU Antifungal Research Awards, 3165
Pfizer ASPIRE EU Emerging Mechanisms of Resistance Antibacterial Research Awards, 3167
Pfizer ASPIRE EU MRSA Nosocomial Pneumonia & MRSA Complicated Skin & Soft Tissue Infections Antibacterial Research Awards, 3168
Pfizer ASPIRE North America Hemophilia Research Awards, 3170
Pfizer ASPIRE North America Rheumatology Research Awards, 3172
Pfizer Inflammation Competitive Research Awards (UK), 3181
Pharmacia-ASPET Award for Experimental Therapeutics, 3187
PhRMA Foundation Health Outcomes Post Doctoral Fellowships, 3202
PhRMA Foundation Health Outcomes Pre-Doctoral Fellowships, 3203
PhRMA Foundation Health Outcomes Research Starter Grants, 3204
PhRMA Foundation Informatics Post Doctoral Fellowships, 3206
PhRMA Foundation Informatics Pre Doctoral Fellowships, 3207
PhRMA Foundation Informatics Research Starter Grants, 3208
PhRMA Fndn Info Sabbatical Fellowships, 3209
PhRMA Foundation Paul Calabresi Medical Student Research Fellowship, 3210
PhRMA Foundation Pharmaceutics Postdoctoral Fellowships, 3211
PhRMA Foundation Pharmaceutics Research Starter Grants, 3213
PhRMA Foundation Pharmaceutics Sabbatical Fellowships, 3214
PhRMA Foundation Pharmacology/Toxicology Post Doctoral Fellowships, 3215
PhRMA Foundation Pharmacology/Toxicology Pre Doctoral Fellowships, 3216
PhRMA Foundation Pharmacology/Toxicology Research Starter Grants, 3217

792 / Pharmacology

PhRMA Foundation Pharmacology/Toxicology Sabbatical Fellowships, 3218
PVA Research Foundation Grants, 3285
Searle Scholars Program Grants, 3503
Sigrid Juselius Foundation Grants, 3561
Simone and Cino del Duca Grand Prix Awards, 3564
Victor E. Speas Foundation Grants, 3868

Pharmacy
AIHP Sonnedecker Visiting Scholar Grants, 371
AIHP Thesis Support Grants, 372
Fischelis Grants for Research in the History of American Pharmacy, 1669
IIE New Leaders Group Award for Mutual Understanding, 2141
J.M. Long Foundation Grants, 2173
NHLBI Pathway to Independence Awards, 2718
PhRMA Foundation Informatics Pre Doctoral Fellowships, 3207
Schering-Plough Foundation Health Grants, 3492
Seattle Fndn Hawkins Memorial Scholarship, 3506

Pharmacy Education
North Carolina GlaxoSmithKline Fndn Grants, 2991
Pfizer Healthcare Charitable Contributions, 3180
Pfizer Medical Education Track One Grants, 3182
Schering-Plough Foundation Health Grants, 3492
Seattle Fndn Hawkins Memorial Scholarship, 3506

Philanthropy
Annunziata Sanguinetti Foundation Grants, 600
Aragona Family Foundation Grants, 648
Ashland Corporate Contributions Grants, 713
Beazley Foundation Grants, 865
Berrien Community Foundation Grants, 884
Carl Gellert and Celia Berta Gellert Grants, 1024
CICF Indianapolis Fndn Community Grants, 1156
Cleveland-Cliffs Foundation Grants, 1181
Columbus Foundation Competitive Grants, 1217
Community Fndn for Monterey County Grants, 1253
Community Foundation of Eastern Connecticut General Southeast Grants, 1269
Community Foundation of Greater Flint Grants, 1273
DaimlerChrysler Corporation Fund Grants, 1412
Estee Lauder Grants, 1606
Fargo-Moorhead Area Foundation Grants, 1632
George Foundation Grants, 1795
George W. Codrington Charitable Fndn Grants, 1803
Giant Eagle Foundation Grants, 1815
Global Fund for Women Grants, 1833
Hillsdale County Community General Adult Foundation Grants, 2034
Hoblitzelle Foundation Grants, 2038
Huffy Foundation Grants, 2073
Josiah W. and Bessie H. Kline Foundation Grants, 2278
Kansas Health Fndn Major Initiatives Grants, 2288
Leo Goodwin Foundation Grants, 2354
Libra Foundation Grants, 2363
McCombs Foundation Grants, 2491
Noble County Community Foundation - Delta Theta Tau Sorority IOTA IOTA Chapter Riecke Scholarship, 2971
OneFamily Foundation Grants, 3075
Parkersburg Area Community Foundation Action Grants, 3116
Pittsburgh Foundation Community Fund Grants, 3230
Rochester Area Community Foundation Grants, 3391
Rockefeller Archive Center Research Grants, 3393
Ruth Anderson Foundation Grants, 3425
Sioux Falls Area Community Foundation Spot Grants (Unrestricted), 3570
Stewart Huston Charitable Trust Grants, 3655
Wallace Alexander Gerbode Foundation Grants, 3896

Philosophy
Gruber Foundation Cosmology Prize, 1907
Harry Frank Guggenheim Foundation Dissertation Fellowships, 1963
James Hervey Johnson Charitable Educational Trust Grants, 2191

Photography
Meta and George Rosenberg Foundation Grants, 2525
Wood Family Charitable Trust Grants, 3972

Photojournalism
Wood Family Charitable Trust Grants, 3972

Physical Activity
ACSM Carl V. Gisolfi Memorial Fund Grant, 259
ACSM Coca-Cola Company Doctoral Student Grant on Behavior Research, 261
ACSM Dr. Raymond A. Weiss Research Endowment Grant, 262
ACSM International Student Awards, 269
ACSM Oded Bar-Or Int'l Scholar Awards, 272
ACSM Paffenbarger-Blair Fund for Epidemiological Research on Physical Activity Grants, 273
Collective Brands Foundation Grants, 1208

Physical Disability
ACL Business Acumen for Disability Organizations Grants, 238
ACL Partnerships in Employment Systems Change Grants, 246
ACL Training and Technical Assistance Center for State Intellectual and Developmental Disabilities Delivery Systems Grants, 248
Adams Rotary Memorial Fund A Grants, 302
Allan C. and Lelia J. Garden Foundation Grants, 438
Christopher & Dana Reeve Foundation Quality of Life Grants, 1152
Different Needz Foundation Grants, 1478
Elkhart County Community Foundation Fund for Elkhart County, 1572
George E. Hatcher, Jr. and Ann Williams Hatcher Foundation Grants, 1792
Grace Bersted Foundation Grants, 1859
Harold and Rebecca H. Gross Foundation Grants, 1951
HRAMF Deborah Munroe Noonan Memorial Research Grants, 2058
John D. & Katherine A. Johnston Fndn Grants, 2233
Margaret Wiegand Trust Grants, 2430
Reader's Digest Partners for Sight Fndn Grants, 3345
Sara Elizabeth O'Brien Trust Grants, 3483
Thomas C. Burke Foundation Grants, 3720
Thomas W. Bradley Foundation Grants, 3726

Physical Education
AAHPERD Abernathy Presidential Scholarships, 151
BCBSM Building Healthy Communities Engaging Elementary Schools and Community Partners Grants, 849
Homer Foundation Grants, 2044

Physical Growth/Retardation
Able To Serve Grants, 212
Columbus Foundation Traditional Grants, 1223
Vancouver Sun Children's Fund Grants, 3851
WHAS Crusade for Children Grants, 3927

Physical Medicine and Rehabilitation
AAPD Henry B. Betts Award, 182
ACSM Carl V. Gisolfi Memorial Fund Grant, 259
ACSM Coca-Cola Company Doctoral Student Grant on Behavior Research, 261
ACSM Dr. Raymond A. Weiss Research Endowment Grant, 262
ACSM Foundation Clinical Sports Medicine Endowment Grants, 265
ACSM Foundation Doctoral Research Grants, 266
ACSM Foundation Research Endowment Grants, 267
ACSM International Student Awards, 269
ACSM Oded Bar-Or Int'l Scholar Awards, 272
ACSM Paffenbarger-Blair Fund for Epidemiological Research on Physical Activity Grants, 273
Annunziata Sanguinetti Foundation Grants, 600
Disable American Veterans Charitable Grants, 1479
GEICO Public Service Awards, 1778
IATA Research Grants, 2093
IATA Scholarships, 2094

Jacob and Valeria Langeloth Foundation Grants, 2182
NHLBI Mentored Patient-Oriented Research Career Development Awards, 2710
Special Olympics Health Profess Student Grants, 3629
WHAS Crusade for Children Grants, 3927

Physical Sciences
AAAS Science and Technology Policy Fellowships: Global Health and Development, 41
Abdus Salam ICTP Federation Arrangement Scheme Grants, 205
APHL Emerging Infectious Diseases Fellowships, 636
BWF Career Awards at the Scientific Interface, 983
Canada-U.S. Fulbright New Century Scholars Program Grants, 1008
Canada-U.S. Fulbright Senior Specialists Grants, 1009
CDC Collegiate Leaders in Environmental Health Internships, 1048
Fulbright Alumni Initiatives Awards, 1749
Fulbright Distinguished Chairs Awards, 1750
Fulbright New Century Scholars Grants, 1753
Fulbright Specialists Program Grants, 1754
Fulbright Scholars in Europe and Eurasia, 1755
Fulbright Traditional Scholar Program in Sub-Saharan Africa, 1756
Fulbright Traditional Scholar Program in the East Asia/Pacific Region, 1757
Fulbright Traditional Scholar Program in the Near East and North Africa Region, 1758
Fulbright Traditional Scholar Program in the South and Central Asia Region, 1759
Fulbright Traditional Scholar Program in the Western Hemisphere, 1760
HHMI-NIBIB Interfaces Initiative Grants, 2014
IIE Klein Family Scholarship, 2135
Marion I. and Henry J. Knott Foundation Standard Grants, 2436
North Carolina GlaxoSmithKline Fndn Grants, 2991
NSF Alan T. Waterman Award, 3006
Society of Cosmetic Chemists Allan B. Black Award Sponsored by Presperse, 3604
Society of Cosmetic Chemists Award Sponsored by McIntyre Group, Ltd., 3605
Society of Cosmetic Chemists Award Sponsored by The HallStar Company, 3606
Society of Cosmetic Chemists Chapter Best Speaker Award, 3607
Society of Cosmetic Chem Chapter Merit Award, 3608
Society of Cosmetic Chemists Frontier of Science Award, 3609
Society of Cosmetic Chem Hans Schaeffer Award, 3610
Society of Cosmetic Chem Jos Ciaudelli Award, 3611
Society of Cosmetic Chemists Keynote Award Lecture Sponsored by Ruger Chemical Corporation, 3612
Society of Cosmetic Chemists Literature Award, 3613
Society of Cosmetic Chemists Maison G. de Navarre Medal, 3614
Society of Cosmetic Chemists Merit Award, 3615
Society of Cosmetic Chemists Robert A. Kramer Lifetime Service Award, 3616
Society of Cosmetic Chem Shaw Mudge Award, 3617
Society of Cosmetic Chem Stud Poster Awards, 3618
Society of Cosmetic Chem You Scientist Awards, 3619
W.M. Keck Foundation Science and Engineering Research Grants, 3889
Wellcome Trust Biomedical Science Grants, 3918

Physics
AACR Team Science Award, 126
Abdus Salam ICTP Federation Arrangement Scheme Grants, 205
Alfred P. Sloan Foundation International Science Engagement Grants, 431
Alfred P. Sloan Foundation Research Fellowships, 432
BWF Career Awards at the Scientific Interface, 983
Canada-U.S. Fulbright New Century Scholars Program Grants, 1008
Canada-U.S. Fulbright Senior Specialists Grants, 1009
Canada Graduate Scholarships (CGS) and NSERC Postgraduate Scholarships (PGS), 1010

Fulbright Alumni Initiatives Awards, 1749
Fulbright Distinguished Chairs Awards, 1750
Fulbright New Century Scholars Grants, 1753
Fulbright Specialists Program Grants, 1754
Fulbright Scholars in Europe and Eurasia, 1755
Fulbright Traditional Scholar Program in Sub-Saharan Africa, 1756
Fulbright Traditional Scholar Program in the East Asia/Pacific Region, 1757
Fulbright Traditional Scholar Program in the Near East and North Africa Region, 1758
Fulbright Traditional Scholar Program in the South and Central Asia Region, 1759
Fulbright Traditional Scholar Program in the Western Hemisphere, 1760
HHMI-NIBIB Interfaces Initiative Grants, 2014
HRAMF Smith Family Awards for Excellence in Biomedical Research, 2063
Marion I. and Henry J. Knott Foundation Standard Grants, 2436
MBL Eugene and Millicent Bell Fellowships, 2470
NHGRI Mentored Research Scientist Development Award, 2686
NHLBI Mentored Quantitative Research Career Development Awards, 2711
NIMH Curriculum Development Award in Neuroinformatics Research and Analysis, 2917
NSF Biological Physics (BP), 3008
Weingart Foundation Grants, 3913

Physics, Computational
SfN Swartz Prize for Theoretical and Computational Neuroscience, 3537

Physics, Theoretical
SfN Swartz Prize for Theoretical and Computational Neuroscience, 3537

Physiology
ACSM International Student Awards, 269
ACSM NASA Space Physiology Research Grants, 271
ACSM Oded Bar-Or Int'l Scholar Awards, 272
AHAF Alzheimer's Disease Research Grants, 349
AHAF National Glaucoma Research Grants, 351
AOCS George Schroepfer Medal, 613
AOCS Health & Nutrition Div Student Award, 614
AOCS Health and Nutrition Poster Competition, 615
AOCS Holman Lifetime Achievement Award, 620
Lalor Foundation Postdoctoral Fellowships, 2336
MBL Albert & Ellen Grass Fellowships, 2463
MBL Associates Summer Fellowships, 2465
MBL Burr & Susie Steinbach Fellowship, 2467
MBL E.E. Just Summer Fellowship for Minority Scientists, 2468
MBL Erik B. Fries Summer Fellowships, 2469
MBL Evelyn and Melvin Spiegal Fellowship, 2471
MBL Frank R. Lillie Summer Fellowship, 2472
MBL Fred Karush Library Readership, 2474
MBL Gruss Lipper Family Foundation Summer Fellowship, 2475
MBL H. Keffer Hartline and Edward F. MacNichol, Jr. Fellowships, 2476
MBL Herbert W. Rand Summer Fellowship, 2477
MBL James E. and Faith Miller Memorial Summer Fellowship, 2478
MBL John M. Arnold Award, 2479
MBL Lucy B. Lemann Summer Fellowship, 2481
MBL M.G.F. Fuortes Summer Fellowships, 2482
MBL Nikon Summer Fellowship, 2483
MBL Stephen W. Kuffler Summer Fellowships, 2487
MBL William Townsend Porter Summer Fellowships for Minority Investigators, 2488
NIH Biobehavioral Research for Effective Sleep, 2865

Physiology, Human
ACSM NASA Space Physiology Research Grants, 271
NHLBI Lymphatics in Health and Disease in the Digestive, Urinary, Cardiovascular and Pulmonary Systems, 2705

Piano
Thelma Braun & Bocklett Family Fndn Grants, 3713

Planning/Policy Studies
CDC Steven M. Teutsch Prevention Effectiveness Fellowships, 1071
IIE Rockefeller Foundation Bellagio Center Residencies, 2142
Meyer Foundation Healthy Communities Grants, 2533
RWJF Policies for Action: Policy and Law Research to Build a Culture of Health Grants, 3443

Plant Genetics
Pioneer Hi-Bred Society Fellowships, 3226

Plant Sciences
Janirve Foundation Grants, 2209
NSF Animal Developmental Mechanisms Grants, 3007
NSF Cell Systems Cluster Grants, 3011
Pioneer Hi-Bred Society Fellowships, 3226

Plastic Surgery
AAFPRS Ben Shuster Memorial Award, 137
AAFPRS Bernstein Grant, 138
AAFPRS Fellowships, 139
AAFPRS Investigator Development Grant, 140
AAFPRS John Orlando Roe Award, 142
AAFPRS Leslie Bernstein Res Research Grants, 143
AAFPRS Residency Travel Awards, 144
AAFPRS Sir Harold Delf Gillies Award, 145
Plastic Surgery Educational Foundation/AAO-HNSF Combined Grant, 3234
Plastic Surgery Educational Foundation Basic Research Grants, 3235
Plastic Surgery Educational Foundation Research Fellowship Grants, 3236
ROSE Fund Grants, 3412

Plasticity
SfN Swartz Prize for Theoretical and Computational Neuroscience, 3537

Playgrounds
American Academy of Dermatology Shade Structure Grants, 504
Clara Blackford Smith and W. Aubrey Smith Charitable Foundation Grants, 1165
Hasbro Children's Fund Grants, 1973
William Blair and Company Foundation Grants, 3948

Podiatry
Special Olympics Health Profess Student Grants, 3629

Poetry
Axe-Houghton Foundation Grants, 815
Olive Higgins Prouty Foundation Grants, 3072

Poland
Kosciuszko Fndn Grants for Polish Citizens, 2322

Polish Americans
Kosciuszko Foundation Dr. Marie E. Zakrzewski Medical Scholarship, 2321

Polish Language/Literature
Kosciuszko Fndn Grants for Polish Citizens, 2322

Political Behavior
Carnegie Corporation of New York Grants, 1033
Harry Frank Guggenheim Fnd Research Grants, 1964

Political Science
Canada-U.S. Fulbright New Century Scholars Program Grants, 1008
Canada-U.S. Fulbright Senior Specialists Grants, 1009
Fulbright Alumni Initiatives Awards, 1749
Fulbright Distinguished Chairs Awards, 1750
Fulbright New Century Scholars Grants, 1753
Fulbright Specialists Program Grants, 1754
Fulbright Scholars in Europe and Eurasia, 1755
Fulbright Traditional Scholar Program in Sub-Saharan Africa, 1756
Fulbright Traditional Scholar Program in the East Asia/Pacific Region, 1757
Fulbright Traditional Scholar Program in the Near East and North Africa Region, 1758
Fulbright Traditional Scholar Program in the South and Central Asia Region, 1759
Fulbright Traditional Scholar Program in the Western Hemisphere, 1760
Harry Frank Guggenheim Foundation Dissertation Fellowships, 1963

Political Science Education
Coca-Cola Foundation Grants, 1203

Pollution
Gamble Foundation Grants, 1769

Pollution Control
Sioux Falls Area Community Foundation Spot Grants (Unrestricted), 3570

Pollution, Air
Bullitt Foundation Grants, 973
Carrier Corporation Contributions Grants, 1038
San Diego Foundation for Change Grants, 3470
Washington Gas Charitable Giving, 3908

Pollution, Land
San Diego Foundation for Change Grants, 3470

Pollution, Noise
San Diego Foundation for Change Grants, 3470

Pollution, Water
Beldon Fund Grants, 871
Heineman Foundation for Research, Education, Charitable and Scientific Purposes, 1990
PepsiCo Foundation Grants, 3137
Rohm and Haas Company Grants, 3403
San Diego Foundation for Change Grants, 3470
Washington Gas Charitable Giving, 3908

Polymer Science
Arnold and Mabel Beckman Foundation Young Investigators Grants, 666

Population Control
Clayton Baker Trust Grants, 1179
General Service Foundation Human Rights and Economic Justice Grants, 1784
Huber Foundation Grants, 2071
IIE Hewlett Fnd/IIE Dissertation Fellowship, 2132
Lumpkin Family Fnd Healthy People Grants, 2391
Packard Foundation Local Grants, 3105
Packard Foundation Population and Reproductive Health Grants, 3106
PC Fred H. Bixby Fellowships, 3126
S.H. Cowell Foundation Grants, 3447
USAID Policy, Advocacy, and Communication Enhanced for Population and Reproductive Health Grants, 3820

Population Studies
IIE Hewlett Fnd/IIE Dissertation Fellowship, 2132
NCI Academic-Industrial Partnerships for Devel and Validation of In Vivo Imaging Systems and Methods for Cancer Investigations, 2644
NCI Application and Use of Transformative Emerging Technologies in Cancer Research, 2645
NCI Cancer Education and Career Development Program, 2648
USAID Policy, Advocacy, and Communication Enhanced for Population and Reproductive Health Grants, 3820

Poverty and the Poor
ACE Charitable Foundation Grants, 229
Achelis Foundation Grants, 234

794 / Poverty and the Poor

Alice Tweed Tuohy Foundation Grants Program, 437
Allen Foundation Educational Nutrition Grants, 442
Alliance Healthcare Foundation Grants, 444
American Express Community Service Grants, 531
Ann Arbor Area Community Foundation Grants, 594
Arizona Diamondbacks Charities Grants, 657
ASPH/CDC Allan Rosenfield Global Health Fellowships, 780
Austin S. Nelson Foundation Grants, 805
Avon Foundation Breast Care Fund Grants, 813
Barberton Community Foundation Grants, 829
Beatrice Laing Trust Grants, 864
Bernard and Audre Rapoport Foundation Health Grants, 882
Bill and Melinda Gates Foundation Emergency Response Grants, 892
Bill and Melinda Gates Foundation Policy and Advocacy Grants, 893
Bill and Melinda Gates Foundation Water, Sanitation and Hygiene Grants, 894
Birmingham Foundation Grants, 903
Bodman Foundation Grants, 929
Breast Cancer Fund Grants, 947
Burden Trust Grants, 976
Carnegie Corporation of New York Grants, 1033
Carrie Estelle Doheny Foundation Grants, 1037
CCHD Community Development Grants, 1046
CFFVR Basic Needs Giving Partnership Grants, 1094
CFFVR Jewelers Mutual Charitable Giving, 1099
CharityWorks Grants, 1113
CICF Indianapolis Fndn Community Grants, 1156
CNCS AmeriCorps VISTA Project Grants, 1190
CNCS Senior Companion Program Grants, 1191
Community Foundation of Greater Flint Grants, 1273
Community Foundation of Greenville Hollingsworth Funds Program/Project Grants, 1285
Connelly Foundation Grants, 1326
CRH Foundation Grants, 1360
Danellie Foundation Grants, 1421
Eden Hall Foundation Grants, 1549
Educational Foundation of America Grants, 1553
Edyth Bush Charitable Foundation Grants, 1562
Elizabeth Glaser Int'l Leadership Awards, 1566
Emma B. Howe Memorial Foundation Grants, 1593
Farmers Insurance Corporate Giving Grants, 1634
Four J Foundation Grants, 1713
General Service Foundation Human Rights and Economic Justice Grants, 1784
George Gund Foundation Grants, 1796
George P. Davenport Trust Fund Grants, 1800
Gertrude E. Skelly Charitable Foundation Grants, 1810
Ginn Foundation Grants, 1822
GNOF IMPACT Grants for Health and Human Services, 1839
GNOF Maison Hospitaliere Grants, 1843
GNOF Stand Up For Our Children Grants, 1847
Greater Milwaukee Foundation Grants, 1883
Greater Worcester Community Foundation Discretionary Grants, 1888
H.B. Fuller Foundation Grants, 1922
Hasbro Children's Fund Grants, 1973
Healthcare Fndn for Orange County Grants, 1981
Health Foundation of Greater Cincinnati Grants, 1983
Hilda and Preston Davis Foundation Grants, 2028
Hillcrest Foundation Grants, 2031
IBCAT Screening Mammography Grants, 2095
IIE David L. Boren Fellowships, 2129
IIE Hewlett Fnd/IIE Dissertation Fellowship, 2132
Independence Blue Cross Charitable Medical Care Grants, 2150
Irvin Stern Foundation Grants, 2164
J.L. Bedsole Foundation Grants, 2172
J.N. and Macie Edens Foundation Grants, 2174
Jacob G. Schmidlapp Trust Grants, 2183
Jenkins Foundation: Improving the Health of Greater Richmond Grants, 2219
Jessie Ball Dupont Fund Grants, 2226
John Edward Fowler Memorial Fndn Grants, 2235
Josephine Goodyear Foundation Grants, 2272
Kahuku Community Fund, 2286

Kroger Company Donations, 2324
Lynde & Harry Bradley Foundation Fellowships, 2396
Lynde and Harry Bradley Foundation Grants, 2397
Lynde and Harry Bradley Foundation Prizes: Bradley Prizes, 2398
MAP International Medical Fellowships, 2421
Marie C. and Joseph C. Wilson Foundation Rochester Small Grants, 2431
May and Stanley Smith Charitable Trust Grants, 2459
Medtronic Foundation Patient Link Grants, 2514
Meyer Foundation Healthy Communities Grants, 2533
Michael Reese Health Trust Core Grants, 2544
Michael Reese Health Trust Responsive Grants, 2545
Mid-Iowa Health Foundation Community Response Grants, 2549
Mockingbird Foundation Grants, 2589
Nathan Cummings Foundation Grants, 2627
Nationwide Insurance Foundation Grants, 2637
OneFamily Foundation Grants, 3075
Packard Foundation Children, Families, and Communities Grants, 3104
Paso del Norte Health Foundation Grants, 3119
PC Fred H. Bixby Fellowships, 3126
PepsiCo Foundation Grants, 3137
Pittsburgh Foundation Community Fund Grants, 3230
Playboy Foundation Grants, 3237
S.H. Cowell Foundation Grants, 3447
Santa Barbara Foundation Strategy Grants - Core Support, 3481
Shield-Ayres Foundation Grants, 3546
Sidgmore Family Foundation Grants, 3551
Sioux Falls Area Community Foundation Spot Grants (Unrestricted), 3570
Sir Dorabji Tata Trust Grants for NGOs or Voluntary Organizations, 3572
Stewart Huston Charitable Trust Grants, 3655
Thomas Sill Foundation Grants, 3724
TJX Foundation Grants, 3735
United Methodist Committee on Relief Global Health Grants, 3799
W.K. Kellogg Foundation Healthy Kids Grants, 3886
W.K. Kellogg Foundation Secure Families Grants, 3887

Pregnancy
Allen Foundation Educational Nutrition Grants, 442
Alliance Healthcare Foundation Grants, 444
General Service Foundation Human Rights and Economic Justice Grants, 1784
NHLBI Prematurity and Respiratory Outcomes Program (PROP) Grants, 2721
NIH Research on Sleep and Sleep Disorders, 2898
Portland Fndn Women's Giving Circle Grant, 3249
USAID Comprehensive District-Based Support for Better HIV/TB Patient Outcomes Grants, 3807
USAID Family Health Plus Project Grants, 3810
USAID Family Planning and Reproductive Health Methods Grants, 3811
USAID Global Health Development Innovation Accelerator Broad Agency Grants, 3813
USAID HIV Prevention with Key Populations - Mali Grants, 3816
W.P. and Bulah Luse Foundation Grants, 3892

Premature Births
BWF Preterm Birth Initiative Grants, 990
NHLBI Prematurity and Respiratory Outcomes Program (PROP) Grants, 2721

Prenatal Factors
Ford Family Foundation Grants - Access to Health and Dental Services, 1686
TJX Foundation Grants, 3735

Presbyterian Church
Barra Foundation Project Grants, 834
Booth-Bricker Fund Grants, 933
Bradley-Turner Foundation Grants, 944
Effie and Wofford Cain Foundation Grants, 1563

Preschool Education
Alabama Power Foundation Grants, 400
American Electric Power Foundation Grants, 530
Arizona Diamondbacks Charities Grants, 657
Benton Community Foundation Grants, 878
Blackford County Community Foundation Grants, 906
Blue River Community Foundation Grants, 922
Bodenwein Public Benevolent Foundation Grants, 928
Booth-Bricker Fund Grants, 933
Boston Foundation Grants, 936
Brown County Community Foundation Grants, 970
Carnegie Corporation of New York Grants, 1033
CFFVR Basic Needs Giving Partnership Grants, 1094
Clinton County Community Foundation Grants, 1184
CNO Financial Group Community Grants, 1201
Community Foundation Alliance City of Evansville Endowment Fund Grants, 1238
Community Foundation of Bartholomew County Heritage Fund Grants, 1262
Community Foundation of Bartholomew County James A. Henderson Award for Fundraising, 1263
Community Foundation of Greater Fort Wayne - Community Endowment and Clarke Endowment Grants, 1274
Community Foundation of Greenville Community Enrichment Grants, 1284
Community Foundation of Greenville Hollingsworth Funds Program/Project Grants, 1285
Cowles Charitable Trust Grants, 1354
Effie and Wofford Cain Foundation Grants, 1563
Eugene M. Lang Foundation Grants, 1611
FCD New American Children Grants, 1637
Fisher Foundation Grants, 1670
Four County Community Fndn General Grants, 1710
Frances and John L. Loeb Family Fund Grants, 1716
George Foundation Grants, 1795
Helen Steiner Rice Foundation Grants, 1998
Hutchinson Community Foundation Grants, 2085
J.C. Penney Company Grants, 2169
John Edward Fowler Memorial Fndn Grants, 2235
Johnson & Johnson Corporate Contributions, 2253
Laura B. Vogler Foundation Grants, 2345
Marie C. and Joseph C. Wilson Foundation Rochester Small Grants, 2431
Miller Foundation Grants, 2556
Mimi and Peter Haas Fund Grants, 2557
Oppenstein Brothers Foundation Grants, 3078
Packard Foundation Children, Families, and Communities Grants, 3104
Peter and Elizabeth C. Tower Foundation Social and Emotional Preschool Curriculum Grants, 3151
Phoenix Suns Charities Grants, 3201
Portland General Electric Foundation Grants, 3250
Quantum Foundation Grants, 3290
San Antonio Area Foundation Grants, 3469
Sid W. Richardson Foundation Grants, 3553
Union Bank, N.A. Foundation Grants, 3788
Wayne County Foundation Grants, 3909
Windham Foundation Grants, 3961

Preventive Medicine
ADA Foundation Thomas J. Zwemer Award, 297
Alexander and Margaret Stewart Trust Grants, 419
ALVRE Casselberry Award, 460
ALVRE Grant, 461
ALVRE Seymour R. Cohen Award, 462
Alzheimer's Association Conference Grants, 463
Alzheimer's Association Development of New Cognitive and Functional Instruments Grants, 464
Alzheimer's Association Everyday Technologies for Alzheimer Care Grants, 465
Alzheimer's Association Mentored New Investigator Research Grants to Promote Diversity, 467
Alzheimer's Association Neuronal Hyper Excitability and Seizures in Alzheimer's Disease Grants, 468
Alzheimer's Association New Investigator Research Grants to Promote Diversity, 470
Alzheimer's Association U.S.-U.K. Young Investigator Exchange Fellowships, 471
Alzheimer's Association Zenith Fellows Awards, 472

SUBJECT INDEX

APHL Emerging Infectious Diseases Fellowships, 636
Banfi Vintners Foundation Grants, 824
Baptist Community Ministries Grants, 828
Battle Creek Community Foundation Grants, 839
BCBSM Foundation Primary and Clinical Prevention Grants, 856
California Endowment Innovative Ideas Challenge Grants, 1000
California Wellness Foundation Work and Health Program Grants, 1001
CDC Preventive Med Residency & Fellowship, 1064
CDC Public Health Conference Support Grant, 1067
CNIB Baker Applied Research Fund Grants, 1193
CNIB Baker Fellowships, 1194
CNIB Baker New Researcher Fund Grants, 1195
CNIB Barbara Tuck MacPhee Award, 1196
CNIB Canada Glaucoma Clinical Research Council Grants, 1197
CNIB Chanchlani Global Vision Research Award, 1198
CNIB E. (Ben) & Mary Hochhausen Access Technology Research Grants, 1199
CNIB Ross Purse Doctoral Fellowships, 1200
Community Fndn AIDS Endowment Awards, 1237
Daniel Mendelsohn New Investigator Award, 1423
DHHS Adolescent Family Life Demonstration Projects, 1472
Dr. John T. Macdonald Foundation Grants, 1507
Eden Hall Foundation Grants, 1549
Edwards Memorial Trust Grants, 1558
Emma B. Howe Memorial Foundation Grants, 1593
Express Scripts Foundation Grants, 1620
Fannie E. Rippel Foundation Grants, 1630
Foundation for Health Enhancement Grants, 1696
Foundation for Seacoast Health Grants, 1698
Frances and John L. Loeb Family Fund Grants, 1716
Gates Award for Global Health, 1775
General Mills Foundation Grants, 1782
Healthcare Fndn for Orange County Grants, 1981
Health Foundation of Southern Florida Responsive Grants, 1985
Independence Community Foundation Community Quality of Life Grant, 2153
Johnson & Johnson Corporate Contributions, 2253
Kansas Health Foundation Recognition Grants, 2289
Leahi Fund, 2350
Mary Black Foundation Active Living Grants, 2440
Medtronic Foundation Patient Link Grants, 2514
Michael Reese Health Trust Responsive Grants, 2545
NIA Health Behaviors and Aging Grants, 2761
NIAMS Pilot and Feasibility Clinical Research Grants in Arthritis, Musculoskeletal & Skin Diseases, 2774
NIDA Mentored Clinical Scientist Research Career Development Awards, 2793
NIDDK Type 2 Diabetes in the Pediatric Population Research Grants, 2852
Paso del Norte Health Foundation Grants, 3119
Pfizer Healthcare Charitable Contributions, 3180
Premera Blue Cross Grants, 3258
Robert W. Woodruff Foundation Grants, 3389
SAMHSA Conference Grants, 3462
Self Foundation Grants, 3510
Sid W. Richardson Foundation Grants, 3553
UniHealth Foundation Community Health Improvement Grants, 3782
Unihealth Foundation Grants for Non-Profit Oranizations, 3783
USAID HIV Prevention with Key Populations - Mali Grants, 3816
USCM HIV/AIDS Prevention Grants, 3827
Visiting Nurse Foundation Grants, 3879
Wilhelm Sander-Stiftung Foundation Grants, 3942

Primary Care Services
ACS Cancer Control Career Development Awards for Primary Care Physicians, 250
CVS Community Grants, 1379
DHHS Adolescent Family Life Demonstration Projects, 1472
George Foundation Grants, 1795
Health Foundation of Greater Cincinnati Grants, 1983
Health Foundation of Southern Florida Responsive Grants, 1985
Josiah Macy Jr. Foundation Grants, 2277
Kansas Health Foundation Recognition Grants, 2289
RWJF New Jersey Health Initiatives Grants, 3440
Union Pacific Foundation Health and Human Services Grants, 3794
W.K. Kellogg Foundation Healthy Kids Grants, 3886

Primatology
NIH Nonhuman Primate Immune Tolerance Cooperative Study Group Grants, 2890

Print Media
APSA Congressional Health and Aging Policy Fellowships, 644
SfN Science Educator Award, 3535
SfN Science Journalism Student Awards, 3536

Prion Disease
CDC Cooperative Agreement for Continuing Enhanced National Surveillance for Prion Diseases in the United States, 1049

Prisoners
Achelis Foundation Grants, 234
Bodman Foundation Grants, 929
Jacob and Valeria Langeloth Foundation Grants, 2182
Mockingbird Foundation Grants, 2589

Private and Parochial Education
Bailey Foundation Grants, 818
Boettcher Foundation Grants, 931
Coca-Cola Foundation Grants, 1203
Community Foundation of Greater Fort Wayne Scholarships, 1276
ConocoPhillips Foundation Grants, 1327
Dr. Scholl Foundation Grants Program, 1511
Eugene B. Casey Foundation Grants, 1609
Frank G. and Freida K. Brotz Family Foundation Grants, 1723
Frank Reed and Margaret Jane Peters Memorial Fund II Grants, 1728
Huisking Foundation Grants, 2076
Kenneth T. & Eileen L. Norris Fndn Grants, 2300
Lubbock Area Foundation Grants, 2383
Lubrizol Foundation Grants, 2384
Lynde & Harry Bradley Foundation Fellowships, 2396
Lynde and Harry Bradley Foundation Grants, 2397
Lynde and Harry Bradley Foundation Prizes: Bradley Prizes, 2398
Marion I. & Henry J. Knott Discretionary Grants, 2435
Marion I. and Henry J. Knott Standard Grants, 2436
Mervin Bovaird Foundation Grants, 2523
MGN Family Foundation Grants, 2542
Piper Jaffray Fndn Communities Giving Grants, 3227
Richard King Mellon Foundation Grants, 3369
Roy & Christine Sturgis Charitable Grants, 3415
S.H. Cowell Foundation Grants, 3447
Strake Foundation Grants, 3658
Sunderland Foundation Grants, 3675

Problem Solving
NSF Grant Opportunities for Academic Liaison with Industry (GOALI), 3016
NSF Perception, Action & Cognition Grants, 3019

Process Simulation and Control
NSF-NIST Interaction in Chemistry, Materials Research, Molecular Biosciences, Bioengineering, and Chemical Engineering, 3004

Production/Operations Management
NSF Grant Opportunities for Academic Liaison with Industry (GOALI), 3016

Professional Associations
AAAAI Mentorship Award, 31
AOCS A. Richard Baldwin Award, 610
AOCS Analytical Division Student Award, 612
AOCS George Schroepfer Medal, 613
AOCS Health & Nutrition Div Student Award, 614
AOCS Health and Nutrition Poster Competition, 615
AOCS Industrial Oil Products Division Student Award, 616
AOCS Lipid Oxidation and Quality Division Poster Competition, 617
AOCS Processing Division Student Award, 618
AOCS Protein and Co-Products Division Student Poster Competition, 619
AOCS Holman Lifetime Achievement Award, 620
AOCS Surfactants and Detergents Division Student Award, 622
ASHA Minority Student Leadership Awards, 696
ASME H.R. Lissner Award, 725
Cemala Foundation Grants, 1077
Hospital Libraries Section / MLA Professional Development Grants, 2052
MLA Section Project of the Year Award, 2574
Pfizer Medical Education Track Two Grants, 3183
Pfizer Special Events Grants, 3186
Phi Upsilon Omicron Alumni Research Grant, 3191
Phi Upsilon Omicron Florence Fallgatter Distinguished Service Award, 3192
Phi Upsilon Omicron Frances Morton Holbrook Alumni Award, 3193
Phi Upsilon Omicron Geraldine Clewell Senior Awards, 3194
Phi Upsilon Omicron Janice Cory Bullock Collegiate Award, 3195
Phi Upsilon Lillian P. Schoephoerster Award, 3196
Phi Upsilon Omicron Margaret Jerome Sampson Scholarships, 3197
Phi Upsilon Omicron Orinne Johnson Writing Award, 3198
Phi Upsilon Omicron Sarah Thorniley Phillips Leader Awards, 3199
Phi Upsilon Omicron Undergraduate Karen P. Goebel Conclave Award, 3200
SfN Chapter-of-the-Year Award, 3519
SfN Japan Neuroscience Society Meeting Travel Awards, 3525
SfN Next Generation Award, 3531
SfN Travel Awards for the International Brain Research Organization World Congress, 3539

Professional Development
AAAAI Mentorship Award, 31
AES Epilepsy Research Recognition Awards, 318
AES JK Penry Excellence in Epilepsy Care Award, 319
AES Service Award, 324
AES William G. Lennox Award, 326
AMHPS Dr. James A. Ferguson Emerging Infectious Diseases Fellowships, 573
ANA/Grass Foundation Award in Neuroscience, 578
ANA Derek Denny-Brown Neurological Scholar Award, 579
ANA Distinguished Neurology Teacher Award, 580
ANA F.E. Bennett Memorial Lectureship, 581
ANA Raymond D. Adams Lectureship, 583
ANA Soriano Lectureship, 584
ANA Wolfe Neuropathy Research Prize, 585
AOCS A. Richard Baldwin Award, 610
AOCS Analytical Division Student Award, 612
AOCS George Schroepfer Medal, 613
AOCS Health & Nutrition Div Student Award, 614
AOCS Health and Nutrition Poster Competition, 615
AOCS Industrial Oil Products Division Student Award, 616
AOCS Lipid Oxidation and Quality Division Poster Competition, 617
AOCS Processing Division Student Award, 618
AOCS Protein and Co-Products Division Student Poster Competition, 619
AOCS Holman Lifetime Achievement Award, 620
AOCS Surfactants and Detergents Division Student Award, 622
ASPH/CDC Public Health Fellowships, 781
ASPH/CDC Public Health Internships, 782
BBVA Compass Foundation Charitable Grants, 848

796 / Professional Development

CDC David J. Sencer Museum Teacher Professional Development Workshops, 1053
CDC Public Health Associates, 1065
Community Fndn for the Capital Region Grants, 1259
CSTE CDC/CSTE Applied Epidemiology Fellowships, 1367
DHHS Emerging Leaders Program Internships, 1473
EBSCO / MLA Annual Meeting Grants, 1545
ERC Starting Grants, 1599
FAMRI Young Clinical Scientist Awards, 1629
Grass Foundation Marine Biological Laboratory Advanced Imaging Fellowships, 1877
Hearst Foundations Health Grants, 1987
Hospital Libraries Section / MLA Professional Development Grants, 2052
IBRO Regional Grants for Int'l Fellowships to U.S. Laboratory Summer Neuroscience Courses, 2110
Majors MLA Chapter Project of the Year, 2418
MBL Baxter Postdoctoral Summer Fellowship, 2466
MBL E.E. Just Summer Fellowship for Minority Scientists, 2468
MBL Eugene and Millicent Bell Fellowships, 2470
MBL Frederik B. and Betsy G. Bang Summer Fellowships, 2473
MBL H. Keffer Hartline and Edward F. MacNichol, Jr. Fellowships, 2476
MBL Plum Fndn John E. Dowling Fellowships, 2484
MBL Robert Day Allen Summer Fellowship, 2485
MBL William Townsend Porter Summer Fellowships for Minority Investigators, 2488
Medical Informatics Section/MLA Career Development Grant, 2508
MLA Continuing Education Grants, 2559
MLA Cunningham Memorial International Fellowship, 2560
MLA Section Project of the Year Award, 2574
MLA T. Mark Hodges Int'l Service Award, 2575
Pfizer Medical Education Track Two Grants, 3183
Phi Upsilon Omicron Florence Fallgatter Distinguished Service Award, 3192
Phi Upsilon Omicron Frances Morton Holbrook Alumni Award, 3193
Priddy Foundation Organizational Development Grants, 3265
RCPSC Professional Development Grants, 3338
SfN Award for Education in Neuroscience, 3517
SfN Bernice Grafstein Award for Outstanding Accomplishments in Mentoring, 3518
SfN Chapter-of-the-Year Award, 3519
SfN Janett Rosenberg Trubatch Career Development Award, 3524
SfN Japan Neuroscience Society Meeting Travel Awards, 3525
SfN Julius Axelrod Prize, 3526
SfN Louise Hanson Marshall Specific Recognition Award, 3527
SfN Mika Salpeter Lifetime Achievement Award, 3528
SfN Neuroscience Fellowships, 3530
SfN Next Generation Award, 3531
SfN Ralph W. Gerard Prize in Neuroscience, 3534
SfN Science Educator Award, 3535
SfN Swartz Prize for Theoretical and Computational Neuroscience, 3537
SfN Travel Awards for the International Brain Research Organization World Congress, 3539
SfN Young Investigator Award, 3541

Professorship

Cabot Corporation Foundation Grants, 993
RCPSC John G. Wade Visiting Professorship in Patient Safety & Simulation-Based Med Education, 3329
Roy & Christine Sturgis Charitable Grants, 3415
Streilein Foundation for Ocular Immunology Visiting Professorships, 3660

Program Evaluation

Alliant Energy Corporation Contributions, 445
CDC Preventive Med Residency & Fellowship, 1064
Dept of Ed Special Education--Studies and Evaluations, 1466
Priddy Foundation Program Grants, 3266
South Madison Community Foundation Grants, 3626

Prostate Gland

NIDDK Cell-Specific Delineation of Prostate and Genitourinary Development, 2804
NIH Biology, Development, and Progression of Malignant Prostate Disease Research Grants, 2866

Prosthetic Devices

U.S. Department of Education Rehabilitation Engineering Research Centers Grants, 3769

Protein Pathways

AOCS Protein and Co-Products Division Student Poster Competition, 619
NCRR Networks and Pathways Collaborative Research Projects Grants, 2657

Proteins and Macromolecules

AOCS Protein and Co-Products Division Student Poster Competition, 619
NCRR Networks and Pathways Collaborative Research Projects Grants, 2657
NHLBI Career Transition Awards, 2694
NHLBI Protein Interactions Governing Membrane Transport in Pulmonary Health Grants, 2723
NIH Structural Bio of Membrane Proteins Grant, 2911
NSF Biomolecular Systems Cluster Grants, 3010

Protestant Church

Aragona Family Foundation Grants, 648
Camp-Younts Foundation Grants, 1005
Charles Delmar Foundation Grants, 1115
Clark and Ruby Baker Foundation Grants, 1174
Collins C. Diboll Private Foundation Grants, 1209
Dora Roberts Foundation Grants, 1489
E. Rhodes & Leona B. Carpenter Grants, 1539
G.A. Ackermann Memorial Fund Grants, 1767
G.N. Wilcox Trust Grants, 1768
Huie-Dellmon Trust Grants, 2075
Notsew Orm Sands Foundation Grants, 3002
Stewart Huston Charitable Trust Grants, 3655

Psychiatry

AACAP Beatrix A. Hamburg Award for the Best New Research Poster by a Child and Adolescent Psychiatry Resident, 50
AACAP Child and Adolescent Psychiatry (CAP) Teaching Scholarships, 51
AACAP Educational Outreach Program for Child and Adolescent Psychiatry Residents, 52
AACAP Educational Outreach Program for General Psychiatry Residents, 53
AACAP Elaine Schlosser Lewis Award for Research on Attention-Deficit Disorder, 55
AACAP George Tarjan Award for Contributions in Developmental Disabilities, 56
AACAP Irving Philips Award for Prevention, 57
AACAP Jeanne Spurlock Lecture and Award on Diversity and Culture, 58
AACAP Klingenstein Third Generation Foundation Award for Research in Depression or Suicide, 62
AACAP Norbert and Charlotte Rieger Award for Scientific Achievement, 65
AACAP Pilot Research Award for Learning Disabilities, Supported by the Elaine Schlosser Lewis Fund, 67
AACAP Rieger Psychody Psychotherapy Award, 70
AACAP Rieger Service Award for Excellence, 71
AACAP Rob Cancro Academic Leadership Award, 72
AACAP Sidney Berman Award for the School-Based Study and Intervention for Learning Disorders and Mental Illness, 73
AACAP Simon Wile Leader in Consultation Award, 74
AACAP Systems of Care Special Scholarships, 76
American Philosophical Society Daland Fellowships in Clinical Investigation, 537
American Psychiatric Association/AstraZeneca Young Minds in Psychiatry International Awards, 538

SUBJECT INDEX

American Psychiatric Association/Bristol-Myers Squibb Fellowships in Public Psychiatry, 539
American Psychiatric Association/Janssen Resident Psychiatric Research Scholars, 540
American Psychiatric Association/Merck Co. Early Academic Career Research Award, 541
American Psychiatric Association Award for Research in Psychiatry, 542
American Psychiatric Association Minority Fellowships Program, 543
American Psychiatric Association Minority Medical Student Fellowship in HIV Psychiatry, 544
American Psychiatric Association Minority Medical Student Summer Externship in Addiction Psychiatry, 545
American Psychiatric Association Minority Medical Student Summer Mentoring Program, 546
American Psychiatric Association Program for Minority Research Training in Psychiatry, 547
American Psychiatric Association Research Colloquium for Junior Investigators, 548
American Psychiatric Association Travel Scholarships for Minority Medical Students, 549
American Psychiatric Foundation Alexander Gralnick, MD, Award for Research in Schizophrenia, 550
American Psychiatric Foundation Awards for Advancing Minority Mental Health, 551
American Psychiatric Foundation Grantss, 552
American Psychiatric Foundation Disaster Recovery Fund for Psychiatrists, 553
American Psychiatric Foundation James H. Scully Jr., M.D., Educational Fund Grants, 555
American Psychiatric Foundation Jeanne Spurlock Congressional Fellowship, 556
American Psychiatric Foundation Minority Fellowships, 557
American Psychiatric Foundation Typical or Troubled School Mental Health Education Grants, 558
APSAA Fellowships, 638
APSAA Foundation Grants, 639
APSAA Mini-Career Grants, 640
APSAA Research Grants, 641
APSAA Small Beginning Scholar Pre-Investigation Grants, 642
APSAA Small Beginning Scholar Visiting and Consulting Grants, 643
Baltimore Washington Center Adult Psychoanalysis Fellowships, 821
Baltimore Washington Center Child Psychotherapy Fellowships, 822
Blowitz-Ridgeway Foundation Grants, 913
Boston Psychoanalytic Society and Institute Fellowship in Child Psychoanalytic Psychotherapy, 939
Boston Psychoanalytic Society and Institute Fellowship in Psychoanalytic Psychotherapy, 940
Chicago Institute for Psychoanalysis Fellowships, 1135
Joseph Collins Foundation Scholarships, 2267
Michigan Psychoanalytic Institute and Society SATA Grants, 2546
Michigan Psychoanalytic Institute and Society Scholarships, 2547
SOBP A.E. Bennett Research Award, 3594
SOBP Ziskind-Somerfeld Research Awards, 3595

Psychodynamics

AACAP Rieger Psychody Psychotherapy Award, 70
APSAA Fellowships, 638

Psychology

ACSM-GSSI Young Scholar Travel Award, 257
APA Congressional Fellowships, 624
APA Culture of Service in the Psychological Sciences Award, 625
APA Dissertation Research Awards, 626
APA Distinguished Scientific Award for Early Career Contribution to Psychology, 627
APA Distinguished Scientific Award for the Applications of Psychology, 628
APA Distinguished Scientific Contribution Award, 629

SUBJECT INDEX

APA Distinguished Service to Psychological Science Award, 630
APA Meritorious Research Service Award, 631
APA Scientific Conferences Grants, 632
APA Travel Awards, 633
APF/COGDOP Graduate Research Scholarships in Psychology, 635
APSAA Fellowships, 638
APSAA Foundation Grants, 639
APSAA Mini-Career Grants, 640
APSAA Research Grants, 641
APSAA Small Beginning Scholar Pre-Investigation Grants, 642
APSAA Small Beginning Scholar Visiting and Consulting Grants, 643
Australasian Institute of Judicial Administration Seed Funding Grants, 806
Baltimore Washington Center Adult Psychoanalysis Fellowships, 821
Baltimore Washington Center Child Psychotherapy Fellowships, 822
Blowitz-Ridgeway Foundation Grants, 913
Boston Psychoanalytic Society and Institute Fellowship in Child Psychoanalytic Psychotherapy, 939
Boston Psychoanalytic Society and Institute Fellowship in Psychoanalytic Psychotherapy, 940
Canada-U.S. Fulbright New Century Scholars Program Grants, 1008
Canada-U.S. Fulbright Senior Specialists Grants, 1009
Canada Graduate Scholarships (CGS) and NSERC Postgraduate Scholarships (PGS), 1010
Chicago Institute for Psychoanalysis Fellowships, 1135
Edward Bangs Kelley and Elza Kelley Foundation Grants, 1556
Fulbright Alumni Initiatives Awards, 1749
Fulbright Distinguished Chairs Awards, 1750
Fulbright New Century Scholars Grants, 1753
Fulbright Specialists Program Grants, 1754
Fulbright Scholars in Europe and Eurasia, 1755
Fulbright Traditional Scholar Program in Sub-Saharan Africa, 1756
Fulbright Traditional Scholar Program in the East Asia/Pacific Region, 1757
Fulbright Traditional Scholar Program in the Near East and North Africa Region, 1758
Fulbright Traditional Scholar Program in the South and Central Asia Region, 1759
Fulbright Traditional Scholar Program in the Western Hemisphere, 1760
Harry Frank Guggenheim Foundation Dissertation Fellowships, 1963
International Positive Psychology Association Student Scholarships, 2161
James McKeen Cattell Fund Fellowships, 2197
Michigan Psychoanalytic Institute and Society SATA Grants, 2546
Michigan Psychoanalytic Institute and Society Scholarships, 2547
National Center for Responsible Gaming Grants, 2629
National Center for Responsible Gaming Postdoctoral Fellowships, 2630
National Center for Resp Gaming Seed Grants, 2631
New York University Steinhardt School of Education Fellowships, 2684
NIGMS Advancing Basic Behavioral and Social Research on Resilience: an Integrative Science Approach, 2861
NIH Biobehavioral Research for Effective Sleep, 2865
NSF Social Psychology Research Grants, 3023
UICC American Cancer Society International Fellowships for Beginning Investigators, 3778
USDD Workplace Violence in Military Grants, 3847

Psychology Education
APF/COGDOP Graduate Research Scholarships in Psychology, 635
Blowitz-Ridgeway Foundation Early Childhood Development Research Award, 912

Psychology of Aging
Edward N. and Della L. Thome Memorial Foundation Grants, 1557
NIA Health Behaviors and Aging Grants, 2761

Psychotherapy
AACAP Rieger Psychody Psychotherapy Award, 70
APSAA Foundation Grants, 639
APSAA Mini-Career Grants, 640
APSAA Research Grants, 641
APSAA Small Beginning Scholar Pre-Investigation Grants, 642
APSAA Small Beginning Scholar Visiting and Consulting Grants, 643
Baltimore Washington Center Adult Psychoanalysis Fellowships, 821
Baltimore Washington Center Child Psychotherapy Fellowships, 822
Boston Psychoanalytic Society and Institute Fellowship in Child Psychoanalytic Psychotherapy, 939
Boston Psychoanalytic Society and Institute Fellowship in Psychoanalytic Psychotherapy, 940
Chicago Institute for Psychoanalysis Fellowships, 1135
Michigan Psychoanalytic Institute and Society SATA Grants, 2546
Michigan Psychoanalytic Institute and Society Scholarships, 2547
NIA Mechanisms Underlying the Links between Psychosocial Stress, Aging, the Brain and the Body Grants, 2769
WHO Foundation Volunteer Service Grants, 3940

Public Administration
APSA Congressional Health and Aging Policy Fellowships, 644
APSA Robert Wood Johnson Foundation Health Policy Fellowships, 645
Arlington Community Foundation Grants, 662
Canada-U.S. Fulbright New Century Scholars Program Grants, 1008
Canada-U.S. Fulbright Senior Specialists Grants, 1009
Carpenter Foundation Grants, 1035
Commonwealth Fund/Harvard University Fellowship in Minority Health Policy, 1226
Community Foundation for the National Capital Region Community Leadership Grants, 1260
Effie and Wofford Cain Foundation Grants, 1563
Fulbright Alumni Initiatives Awards, 1749
Fulbright Distinguished Chairs Awards, 1750
Fulbright New Century Scholars Grants, 1753
Fulbright Specialists Program Grants, 1754
Fulbright Scholars in Europe and Eurasia, 1755
Fulbright Traditional Scholar Program in Sub-Saharan Africa, 1756
Fulbright Traditional Scholar Program in the East Asia/Pacific Region, 1757
Fulbright Traditional Scholar Program in the Near East and North Africa Region, 1758
Fulbright Traditional Scholar Program in the South and Central Asia Region, 1759
Fulbright Traditional Scholar Program in the Western Hemisphere, 1760
Huffy Foundation Grants, 2073
M.D. Anderson Foundation Grants, 2403
Peter Kiewit Foundation General Grants, 3154
Peter Kiewit Foundation Small Grants, 3155
Prospect Burma Scholarships, 3275
S. Mark Taper Foundation Grants, 3449
SfN Louise Hanson Marshall Specific Recognition Award, 3527
Sonora Area Foundation Competitive Grants, 3622
Special Olympics Health Profess Student Grants, 3629

Public Affairs
1st Source Foundation Grants, 2
AFG Industries Grants, 340
Allegheny Technologies Charitable Trust, 440
Amica Companies Foundation Grants, 574
Axe-Houghton Foundation Grants, 815
Banfi Vintners Foundation Grants, 824

Battle Creek Community Foundation Grants, 839
Berrien Community Foundation Grants, 884
Blue Grass Community Fndn Harrison Grants, 919
Blue Grass Community Foundation Hudson-Ellis Fund Grants, 920
Burlington Northern Santa Fe Foundation Grants, 978
Caesars Foundation Grants, 995
Carpenter Foundation Grants, 1035
Clark County Community Foundation Grants, 1175
Commonwealth Edison Grants, 1225
CSRA Community Foundation Grants, 1366
Cullen Foundation Grants, 1372
Cyrus Eaton Foundation Grants, 1380
Daisy Marquis Jones Foundation Grants, 1414
Earl and Maxine Claussen Trust Grants, 1541
Emerson Electric Company Contributions Grants, 1590
F.M. Kirby Foundation Grants, 1624
Field Foundation of Illinois Grants, 1660
FirstEnergy Foundation Community Grants, 1668
Fluor Foundation Grants, 1682
FMC Foundation Grants, 1683
Howard and Bush Foundation Grants, 2054
Idaho Community Foundation Eastern Region Competitive Grants, 2114
John Merck Fund Grants, 2245
Katharine Matthies Foundation Grants, 2291
Land O'Lakes Foundation Mid-Atlantic Grants, 2341
Libra Foundation Grants, 2363
Lockheed Martin Corp Foundation Grants, 2374
Lynde & Harry Bradley Foundation Fellowships, 2396
Lynde and Harry Bradley Foundation Grants, 2397
Lynde and Harry Bradley Foundation Prizes: Bradley Prizes, 2398
Marathon Petroleum Corporation Grants, 2422
Marjorie Moore Charitable Foundation Grants, 2438
Meyer Memorial Trust Grassroots Grants, 2536
Meyer Memorial Trust Responsive Grants, 2537
Mimi and Peter Haas Fund Grants, 2557
Noble County Community Foundation Grants, 2975
Pike County Community Foundation Grants, 3222
Putnam County Community Foundation Grants, 3283
Rajiv Gandhi Foundation Grants, 3302
Robert R. Meyer Foundation Grants, 3387
Ruth and Vernon Taylor Foundation Grants, 3426
Schramm Foundation Grants, 3495
Seattle Fndn Doyne M. Green Scholarships, 3505
Seaver Institute Grants, 3509
Shell Deer Park Grants, 3543
Shell Oil Company Foundation Community Development Grants, 3544
Vermont Community Foundation Grants, 3865
Vernon K. Krieble Foundation Grants, 3866
Warrick County Community Foundation Grants, 3904
Weyerhaeuser Company Foundation Grants, 3925
William B. Stokely Jr. Foundation Grants, 3947
William Blair and Company Foundation Grants, 3948
Winston-Salem Foundation Competitive Grants, 3965
Z. Smith Reynolds Foundation Small Grants, 3982

Public Broadcasting
Air Products and Chemicals Grants, 399
ASM Public Communications Award, 744
ConocoPhillips Foundation Grants, 1327
Cooper Industries Foundation Grants, 1345
Doree Taylor Charitable Foundation, 1490
Foellinger Foundation Grants, 1684
G.N. Wilcox Trust Grants, 1768
PMI Foundation Grants, 3239
Rasmuson Foundation Tier One Grants, 3307
Rasmuson Foundation Tier Two Grants, 3308
Textron Corporate Contributions Grants, 3712

Public Education
AAAAI Distinguished Layperson Award, 25
ConocoPhillips Foundation Grants, 1327
DPA Promoting Policy Change Advocacy Grants, 1504
Express Scripts Foundation Grants, 1620
HomeBanc Foundation Grants, 2042
John P. Murphy Foundation Grants, 2248

798 / Public Education

Komen Greater NYC Clinical Research Enrollment Grants, 2315
Komen Greater NYC Community Breast Health Grants, 2316
Minneapolis Foundation Community Grants, 2558
Rasmuson Foundation Tier Two Grants, 3308
Stewart Huston Charitable Trust Grants, 3655
William J. & Tina Rosenberg Fndn Grants, 3954

Public Finance
OSF Global Health Financing Initiative Grants, 3088

Public Health
AAAS Award for Scie Freedom & Responsibility, 36
AcademyHealth PHSR Research Article of the Year Awards, 226
ACSM Paffenbarger-Blair Fund for Epidemiological Research on Physical Activity Grants, 273
Aetna Foundation Racial and Ethnic Health Care Equity Grants, 332
AMA-MSS Government Relations Advocacy Fellowship, 475
AMA-MSS Government Relations Internships, 476
AMA Foundation Fund for Better Health Grants, 484
AMA Foundation Pride in the Profession Grants, 493
American Psychiatric Foundation Awards for Advancing Minority Mental Health, 551
American Psychiatric Foundation Disaster Recovery Fund for Psychiatrists, 553
AMHPS Dr. James A. Ferguson Emerging Infectious Diseases Fellowships, 573
APHL Emerging Infectious Diseases Fellowships, 636
Appalachian Regional Commission Health Grants, 637
ASHG Public Health Genetics Fellowship, 712
ASM/CDC Fellowships in Infectious Disease and Public Health Microbiology, 716
ASM Congressional Science Fellowships, 720
ASM Gen-Probe Joseph Public Health Award, 729
ASM Public Communications Award, 744
ASPH/CDC Allan Rosenfield Global Health Fellowships, 780
ASPH/CDC Public Health Fellowships, 781
ASPH/CDC Public Health Internships, 782
AUPHA Corris Boyd Scholarships, 800
AUPHA Foster G. McGaw Scholarships, 801
Baxter International Foundation Grants, 843
BCBSM Frank J. McDevitt, DO, Excellence in Research Awards, 859
Blue Cross Blue Shield of Minnesota Foundation - Healthy Equity: Health Impact Assessment Demonstration Project Grants, 916
Brinson Foundation Grants, 953
Bupa Foundation Medical Research Grants, 974
California Wellness Foundation Work and Health Program Grants, 1001
Canada-U.S. Fulbright New Century Scholars Program Grants, 1008
Canada-U.S. Fulbright Senior Specialists Grants, 1009
Carla J. Funk Governmental Relations Award, 1020
CDC-Hubert Global Health Fellowship, 1047
CDC Collegiate Leaders in Environmental Health Internships, 1048
CDC Cooperative Agreement for the Development, Operation, and Evaluation of an Entertainment Education Program, 1051
CDC David J. Sencer Museum Teacher Professional Development Workshops, 1053
CDC Epidemic Intell Service Training Grants, 1055
CDC Epidemiology Elective Rotation, 1056
CDC Experience Epidemiology Fellowships, 1059
CDC Fnd Tobacco Network Lab Fellowship, 1062
CDC Preventive Med Residency & Fellowship, 1064
CDC Public Health Associates, 1065
CDC Public Health Associates Hosts, 1066
CDC Public Health Conference Support Grant, 1067
CDC Public Health Informatics Fellowships, 1068
CDC Public Health Prev Service Fellowships, 1069
CDC Public Health Prevention Service Fellowship Sponsorships, 1070

CDC Steven M. Teutsch Prevention Effectiveness Fellowships, 1071
CDC Summer Graduate Environmental Health Internships, 1072
CDC Summer Program In Environmental Health Internships, 1073
Cigna Civic Affairs Sponsorships, 1159
Colonel Stanley R. McNeil Foundation Grants, 1211
Commonwealth Fund/Harvard University Fellowship in Minority Health Policy, 1226
Commonwealth Fund Small Grants, 1234
Community Fndn of Bloomington & Monroe County - Precision Health Network Cycle Grants, 1264
Community Fndn of Louisville Health Grants, 1295
Conservation, Food, and Health Foundation Grants for Developing Countries, 1338
Covidien Medical Product Donations, 1352
Cruise Industry Charitable Foundation Grants, 1363
CSTE CDC/CSTE Applied Epidemiology Fellowships, 1367
Deaconess Community Foundation Grants, 1439
DeRoy Testamentary Foundation Grants, 1469
DHHS Emerging Leaders Program Internships, 1473
E. Rhodes & Leona B. Carpenter Grants, 1539
Edward and Helen Bartlett Foundation Grants, 1554
Foundation for Seacoast Health Grants, 1698
Frances and John L. Loeb Family Fund Grants, 1716
Frank W. & Carl S. Adams Memorial Grants, 1731
Fred L. Emerson Foundation Grants, 1743
Fulbright Alumni Initiatives Awards, 1749
Fulbright Distinguished Chairs Awards, 1750
Fulbright New Century Scholars Grants, 1753
Fulbright Specialists Program Grants, 1754
Fulbright Scholars in Europe and Eurasia, 1755
Fulbright Traditional Scholar Program in Sub-Saharan Africa, 1756
Fulbright Traditional Scholar Program in the East Asia/Pacific Region, 1757
Fulbright Traditional Scholar Program in the Near East and North Africa Region, 1758
Fulbright Traditional Scholar Program in the South and Central Asia Region, 1759
Fulbright Traditional Scholar Program in the Western Hemisphere, 1760
Gates Award for Global Health, 1775
Gates Millennium Scholars Program, 1776
George W. Wells Foundation Grants, 1804
GFWC of Massachusetts Catherine E. Philbin Scholarship, 1811
GNOF IMPACT Grants for Health and Human Services, 1839
GNOF IMPACT Kahn-Oppenheim Grants, 1842
Greater Worcester Community Foundation Jeppson Memorial Fund for Brookfield Grants, 1889
Harry Frank Guggenheim Foundation Dissertation Fellowships, 1963
Henry and Ruth Blaustein Rosenberg Foundation Health Grants, 2004
HRAMF Charles A. King Trust Postdoctoral Research Fellowships, 2056
IDPH Emergency Medical Services Assistance Fund Grants, 2118
IDPH Local Health Department Public Health Emergency Response Grants, 2120
IIE Freeman Foundation Indonesia Internships, 2131
IIE Iraq Scholars and Leaders Scholarships, 2133
Jane's Trust Grants, 2206
Johnson & Johnson Community Health Grants, 2252
John W. Speas and Effie E. Speas Memorial Trust Grants, 2264
Kansas Health Foundation Recognition Grants, 2289
Knox County Community Foundation Grants, 2313
M-A-C AIDS Fund Grants, 2400
Mary Black Fndn Community Health Grants, 2441
Mayo Clinic Administrative Fellowship, 2460
Mayo Clinic Admin Fellowship - Eau Claire, 2461
McCarthy Family Foundation Grants, 2490
Michael Reese Health Trust Responsive Grants, 2545
MLA Donald A.B. Lindberg Fellowship, 2562
Morehouse PHSI Project Imhotep Internships, 2592

SUBJECT INDEX

Morris & Gwendolyn Cafritz Fndn Grants, 2594
NCHS National Center for Health Statistics Postdoctoral Research Appointments, 2642
Nestle Foundation Training Grant, 2679
NIA Mentored Clinical Scientist Research Career Development Awards, 2771
Norman Foundation Grants, 2983
Northwestern Mutual Foundation Grants, 2999
NYAM Student Essay Grants, 3046
OSF Accountability and Monitoring in Health Initiative Grants, 3087
OSF Health Media Initiative Grants, 3089
OSF Law and Health Initiative Grants, 3092
PC Fred H. Bixby Fellowships, 3126
Pfizer Healthcare Charitable Contributions, 3180
Pfizer Medical Education Track One Grants, 3182
PhRMA Foundation Health Outcomes Post Doctoral Fellowships, 3202
Piedmont Natural Gas Foundation Health and Human Services Grants, 3221
Pollock Foundation Grants, 3244
Premera Blue Cross Grants, 3258
Prospect Burma Scholarships, 3275
Quality Health Foundation Grants, 3288
Rose Community Foundation Aging Grants, 3410
RWJF Community Health Leaders Awards, 3432
RWJF Pioneer Portfolio Grants, 3442
Saint Luke's Health Initiatives Grants, 3455
Samueli Institute Scientific Research Grants, 3466
San Francisco Fndn Community Health Grants, 3478
Seattle Foundation Health and Wellness Grants, 3507
Sigma Theta Tau International Doris Bloch Research Award, 3559
Siragusa Foundation Health Services & Medical Research Grants, 3571
Sophia Romero Trust Grants, 3624
Special Olympics Health Profess Student Grants, 3629
Susan G. Komen Breast Cancer Foundation Challenge Grants: Career Catalyst Research, 3691
Texas Commission on the Arts Arts Respond Project Grants, 3708
Theodore Edson Parker Foundation Grants, 3715
United Methodist Committee on Relief Global Health Grants, 3799
USAID Devel Innovation Accelerator Grants, 3808
USAID Family Health Plus Project Grants, 3810
Vancouver Fndn Public Health Bursary Package, 3850

Public Planning/Policy
AAAS Award for Scie Freedom & Responsibility, 36
Able Trust Vocational Rehabilitation Grants for Individuals, 213
Achelis Foundation Grants, 234
Adolph Coors Foundation Grants, 311
AMA-MSS Government Relations Advocacy Fellowship, 475
American-Scandinavian Foundation Visiting Lectureship Grants, 502
Amerigroup Foundation Grants, 565
amfAR Fellowships, 567
amfAR Global Initiatives Grants, 568
amfAR Mathilde Krim Fellowships in Basic Biomedical Research, 569
amfAR Public Policy Grants, 570
amfAR Research Grants, 571
APA Congressional Fellowships, 624
Assurant Health Foundation Grants, 789
AT&T Foundation Civic and Community Service Program Grants, 793
Austin College Leadership Award, 804
Blue Cross Blue Shield of Minnesota Foundation - Healthy Equity: Health Impact Assessment Demonstration Project Grants, 916
Blue Cross Blue Shield of Minnesota Foundation - Healthy Equity: Health Impact Assessment Program Grants, 917
Bodman Foundation Grants, 929
Carla J. Funk Governmental Relations Award, 1020
Carnegie Corporation of New York Grants, 1033

SUBJECT INDEX

CDC Steven M. Teutsch Prevention Effectiveness Fellowships, 1071
Charles H. Revson Foundation Grants, 1122
CIGNA Foundation Grants, 1160
Claude Pepper Foundation Grants, 1177
Commonwealth Fund Patient-Centered Coordinated Care Program Grants, 1231
Community Foundation of Eastern Connecticut General Southeast Grants, 1269
Community Foundation of Mount Vernon and Knox County Grants, 1299
Community Foundation of Riverside & San Bernardino County Impact Grants, 1303
Conservation, Food, and Health Foundation Grants for Developing Countries, 1338
DPA Promoting Policy Change Advocacy Grants, 1504
E.L. Wiegand Foundation Grants, 1538
Eberly Foundation Grants, 1544
Emma B. Howe Memorial Foundation Grants, 1593
F.M. Kirby Foundation Grants, 1624
FCD New American Children Grants, 1637
Ferree Foundation Grants, 1658
Frances and John L. Loeb Family Fund Grants, 1716
General Motors Foundation Grants, 1783
George Gund Foundation Grants, 1796
Gill Foundation - Gay and Lesbian Fund Grants, 1821
Greenspun Family Foundation Grants, 1896
Grover Hermann Foundation Grants, 1906
Harry Frank Guggenheim Foundation Dissertation Fellowships, 1963
Hilton Hotels Corporate Giving Program Grants, 2037
IIE Iraq Scholars and Leaders Scholarships, 2133
Illinois Tool Works Foundation Grants, 2147
J.W. Kieckhefer Foundation Grants, 2176
JM Foundation Grants, 2229
Joseph P. Kennedy Jr. Foundation Grants, 2275
Lester Ray Fleming Scholarships, 2360
Lynde & Harry Bradley Foundation Fellowships, 2396
Lynde and Harry Bradley Foundation Grants, 2397
Lynde and Harry Bradley Foundation Prizes: Bradley Prizes, 2398
M.D. Anderson Foundation Grants, 2403
Meyer Foundation Healthy Communities Grants, 2533
Mimi and Peter Haas Fund Grants, 2557
Minneapolis Foundation Community Grants, 2558
Nordson Corporation Foundation Grants, 2981
Norman Foundation Grants, 2983
Oleonda Jameson Trust Grants, 3071
Pittsburgh Foundation Community Fund Grants, 3230
Playboy Foundation Grants, 3237
PMI Foundation Grants, 3239
Procter and Gamble Fund Grants, 3274
Richard and Helen DeVos Foundation Grants, 3364
RRF General Program Grants, 3417
RWJF Policies for Action: Policy and Law Research to Build a Culture of Health Grants, 3443
Seaver Institute Grants, 3509
Sensient Technologies Foundation Grants, 3513
Vernon K. Krieble Foundation Grants, 3866
Weyerhaeuser Company Foundation Grants, 3925
Windham Foundation Grants, 3961
Z. Smith Reynolds Foundation Small Grants, 3982

Public Policy Systems Analysis
CDC Steven M. Teutsch Prevention Effectiveness Fellowships, 1071
Grover Hermann Foundation Grants, 1906
IIE Iraq Scholars and Leaders Scholarships, 2133

Public Relations
AAAS Early Career Award for Public Engagement with Science, 37
Carrier Corporation Contributions Grants, 1038

Public Safety
ACF Native American Social and Economic Development Strategies Grants, 231
Bikes Belong Foundation Paul David Clark Bicycling Safety Grants, 890
Caesars Foundation Grants, 995
California Endowment Innovative Ideas Challenge Grants, 1000
Greater Sitka Legacy Fund Grants, 1885
Harrison County Community Foundation Signature Grants, 1958
IDPH Local Health Department Public Health Emergency Response Grants, 2120
Kenai Peninsula Foundation Grants, 2298
Ketchikan Community Foundation Grants, 2303
Kodiak Community Foundation Grants, 2314
Lil and Julie Rosenberg Foundation Grants, 2364
Meyer Foundation Healthy Communities Grants, 2533
OneStar Foundation AmeriCorps Grants, 3077
Petersburg Community Foundation Grants, 3156
Toyota Motor Manufacturing of Indiana Grants, 3744
Toyota Motor Manufacturing of Kentucky Grants, 3745

Public Safety Law
AMA Foundation Fund for Better Health Grants, 484
Baptist Community Ministries Grants, 828
Perry County Community Foundation Grants, 3142
Pike County Community Foundation Grants, 3222
Posey County Community Foundation Grants, 3252
Texas Commission on the Arts Arts Respond Project Grants, 3708

Publication
AAAS Early Career Award for Public Engagement with Science, 37
ASHA Advancing Academic-Research Award, 695
ASHA Student Research Travel Award, 699
George A Ohl Jr. Foundation Grants, 1791
Lynde & Harry Bradley Foundation Fellowships, 2396
Lynde and Harry Bradley Foundation Grants, 2397
Lynde and Harry Bradley Foundation Prizes: Bradley Prizes, 2398
MLA Lois Ann Colaianni Award for Excellence and Achievement in Hospital Librarianship, 2568
MLA Lucretia W. McClure Excellence in Education Award, 2570
NLM Exploratory/Developmental Grants, 2942
NLM Grants for Scholarly Works in Biomedicine and Health, 2944
Pfizer Global Research Awards for Nicotine Independence, 3179
Potts Memorial Foundation Grants, 3254
RSC Jason A. Hannah Medal, 3420
South Madison Community Foundation Grants, 3626

Publication Education
Bullitt Foundation Grants, 973

Publishing Industry
Victor Grifols i Lucas Foundation Prize for Journalistic Work on Bioethics, 3870

Publishing, Electronic
NLM Grants for Scholarly Works in Biomedicine and Health, 2944

Puerto Rico
American Sociological Association Minority Fellowships, 563
Chest Fnd Grant in Venous Thromboembolism, 1134
NMF Franklin C. McLean Award, 2955
NMF Need-Based Scholarships, 2965
RWJF Harold Amos Medical Faculty Development Research Grants, 3434

Pulmonary Diseases
AMA Foundation Seed Grants for Research, 494
CDI Interdisciplinary Research Initiatives Grants, 1074
CDI Postdoctoral Fellowships, 1075
CFF First- and Second-Year Clinical Fellowships, 1085
CFF Postdoctoral Research Fellowships, 1089
CFF Research Grants, 1090
CFF 3rd through 5th Year Clinical Fellowships, 1092
Duke University Adult Cardiothoracic Anesthesia and Critical Care Medicine Fellowships, 1520
Everyone Breathe Asthma Education Grants, 1616
Genentech Corp Charitable Contributions, 1780
Hawaii Community Foundation Health Education and Research Grants, 1976
HRAMF Harold S. Geneen Charitable Trust Awards for Coronary Heart Disease Research, 2059
Jacob and Valeria Langeloth Foundation Grants, 2182
LAM Fnd Established Investigator Awards, 2337
LAM Foundation Pilot Project Grants, 2338
LAM Foundation Postdoctoral Fellowships, 2339
LAM Foundation Research Grants, 2340
Leahi Fund, 2350
M. Bastian Family Foundation Grants, 2402
NHLBI Ancillary Studies in Clinical Trials, 2688
NHLBI Circadian-Coupled Cellular Function in Heart, Lung, and Blood Tissue Grants, 2696
NHLBI Developmental Origins of Altered Lung Physiology and Immune Function Grants, 2698
NHLBI Lymphatics in Health and Disease in the Digestive, Cardio & Pulmonary Systems, 2704
NHLBI Lymphatics in Health and Disease in the Digestive, Urinary, Cardiovascular and Pulmonary Systems, 2705
NHLBI Mentored Career Award for Faculty at Institutions that Promote Diversity (K01), 2706
NHLBI Mentored Career Award For Faculty At Minority Institutions, 2707
NHLBI Mentored Career Dev Award to Promote Faculty Diversity in Biomed Research, 2708
NHLBI Nutrition and Diet in the Causation, Prevention, and Management of Heart Failure Research Grants, 2717
NHLBI Phase II Clinical Trials of Novel Therapies for Lung Diseases, 2720
NHLBI Protein Interactions Governing Membrane Transport in Pulmonary Health Grants, 2723
NHLBI Ruth L. Kirschstein National Research Service Awards for Individual Postdoctoral Fellows, 2729
NHLBI Ruth L. Kirschstein National Research Service Awards for Individual Predoctoral MD/PhD Fellows and Other Dual Degree Fellows, 2731
NHLBI Short-Term Research Education Program to Increase Diversity in Health-Related Research, 2733
NHLBI Training Program Grants for Institutions that Promote Diversity, 2738
NHLBI Translational Grants in Lung Diseases, 2740
NIH Research on Sleep and Sleep Disorders, 2898
Pfizer ASPIRE North America Broad Spectrum Antibiotics for the Treatment of Gram-Negative or Polymicrobial Infections Research Awards, 3169
Pfizer ASPIRE North America Narrow Spectrum Antibiotics for the Treatment of MRSA Research Awards, 3171
Pfizer Healthcare Charitable Contributions, 3180
Pfizer Medical Education Track Two Grants, 3183

Quality of Life
Allen P. & Josephine B. Green Fndn Grants, 443
Bernard and Audre Rapoport Foundation Health Grants, 882
Bikes Belong Foundation Paul David Clark Bicycling Safety Grants, 890
Business Bank of Nevada Community Grants, 981
CharityWorks Grants, 1113
Christopher & Dana Reeve Foundation Quality of Life Grants, 1152
Coleman Foundation Developmental Disabilities Grants, 1207
Conquer Cancer Foundation of ASCO International Development and Education Award in Palliative Care, 1332
CSL Behring Local Empowerment for Advocacy Development (LEAD) Grants, 1365
Dennis & Phyllis Washington Fndn Grants, 1458
DIFFA/Chicago Grants, 1477
Donald W. Reynolds Foundation Aging and Quality of Life Grants, 1487
Ezra M. Cutting Trust Grants, 1623
Fifth Third Foundation Grants, 1661
Foster Foundation Grants, 1689
GNOF IMPACT Grants for Arts and Culture, 1838

800 / Quality of Life

HAF Senior Opportunities Grants, 1932
Hall-Perrine Foundation Grants, 1935
Harry S. Black and Allon Fuller Fund Grants, 1966
Hilton Head Island Foundation Grants, 2036
Horizon Foundation for New Jersey Grants, 2048
Jim Moran Foundation Grants, 2228
Kessler Fnd Signature Employment Grants, 2302
KeyBank Foundation Grants, 2307
Mary Wilmer Covey Charitable Trust Grants, 2448
May and Stanley Smith Charitable Trust Grants, 2459
Middlesex Savings Charitable Foundation Basic Human Needs Grants, 2550
Schlessman Family Foundation Grants, 3493
Simpson Lumber Charitable Contributions, 3566
Singing for Change Foundation Grants, 3567
Theodore Edson Parker Foundation Grants, 3715
Turner Foundation Grants, 3763
Walmart Foundation Northwest Arkansas Giving, 3899

Quality/Product Control
AMDA Foundation Quality Improvement Award, 499
NCHS National Center for Health Statistics Postdoctoral Research Appointments, 2642
U.S. Department of Education Erma Byrd Scholarships, 3768

Racism
Allstate Corporate Giving Grants, 446
Allstate Corp Hometown Commitment Grants, 447
Cowles Charitable Trust Grants, 1354
Georgia Power Foundation Grants, 1806
Joseph H. & Florence A. Roblee Fndn Grants, 2269
NEI Ruth L. Kirschstein National Research Service Award Short-Term Institutional Training, 2668
Pittsburgh Foundation Community Fund Grants, 3230
RWJF Disparities Research for Change Grants, 3433
San Diego Foundation for Change Grants, 3470
Sioux Falls Area Community Foundation Spot Grants (Unrestricted), 3570

Radiation
DSO Radiation Biodosimetry (RaBiD) Grants, 1515

Radiation Effects
American Academy of Dermatology Shade Structure Grants, 504
DSO Radiation Biodosimetry (RaBiD) Grants, 1515
Susan G. Komen Breast Cancer Foundation Challege Grants: Investigator Initiated Research, 3690

Radiation Instrumentation
DSO Radiation Biodosimetry (RaBiD) Grants, 1515

Radio
AIHS Media Summer Fellowship, 392
Albert Pick Jr. Fund Grants, 409
American Academy of Nursing Media Awards, 509
Doree Taylor Charitable Foundation, 1490
G.N. Wilcox Trust Grants, 1768
Gill Foundation - Gay and Lesbian Fund Grants, 1821
TE Foundation Grants, 3705

Radiology
Society for Imaging Informatics in Medicine (SIIM) Emeritus Mentor Grants, 3597
Society for Imaging Informatics in Medicine (SIIM) Micro Grants, 3598
Society for Imaging Informatics in Medicine (SIIM) Research Grants, 3599
Society for Imaging Informatics in Med Small Grant for Product Support Development, 3600
Society for Imaging Informatics in Medicine (SIIM) Small Training Grants, 3601

Rape/Sexual Assault
Meyer Foundation Healthy Communities Grants, 2533

Reading
Barker Welfare Foundation Grants, 831
Bodenwein Public Benevolent Foundation Grants, 928
Boettcher Foundation Grants, 931
Boston Foundation Grants, 936
Howard and Bush Foundation Grants, 2054
NSF Perception, Action & Cognition Grants, 3019
Robert R. Meyer Foundation Grants, 3387
Stocker Foundation Grants, 3656

Reading Education
Arlington Community Foundation Grants, 662
Baptist Community Ministries Grants, 828
DaimlerChrysler Corporation Fund Grants, 1412
Emily Davie and Joseph S. Kornfeld Foundation Grants, 1591
Stocker Foundation Grants, 3656

Reconstructive Surgery
AAFPRS Ben Shuster Memorial Award, 137
AAFPRS Bernstein Grant, 138
AAFPRS Fellowships, 139
AAFPRS Investigator Development Grant, 140
AAFPRS John Orlando Roe Award, 142
AAFPRS Leslie Bernstein Res Research Grants, 143
AAFPRS Residency Travel Awards, 144
AAFPRS Sir Harold Delf Gillies Award, 145
Dorr Institute for Arthritis Research & Educ, 1502
Plastic Surgery Educational Foundation/AAO-HNSF Combined Grant, 3234
Plastic Surgery Educational Foundation Basic Research Grants, 3235
Plastic Surgery Educational Foundation Research Fellowship Grants, 3236
Ronald McDonald House Charities Grants, 3405
ROSE Fund Grants, 3412

Recreation and Leisure
Adler-Clark Electric Community Commitment Foundation Grants, 307
Agnes M. Lindsay Trust Grants, 346
Alice Tweed Tuohy Foundation Grants Program, 437
Amador Community Foundation Grants, 481
American Academy of Dermatology Camp Discovery Scholarships, 503
Andre Agassi Charitable Foundation Grants, 587
Aratani Foundation Grants, 649
Argyros Foundation Grants Program, 653
Arthur M. Blank Family Foundation Inspiring Spaces Grants, 673
Atlanta Foundation Grants, 796
Bacon Family Foundation Grants, 817
Beazley Foundation Grants, 865
Beckley Area Foundation Grants, 867
Belvedere Community Foundation Grants, 874
Ben B. Cheney Foundation Grants, 875
Bikes Belong Foundation Paul David Clark Bicycling Safety Grants, 890
Blanche and Julian Robertson Family Foundation Grants, 911
Boston Foundation Grants, 936
Campbell Soup Foundation Grants, 1007
Carl C. Icahn Foundation Grants, 1023
Carls Foundation Grants, 1030
Carrie Estelle Doheny Foundation Grants, 1037
Carrier Corporation Contributions Grants, 1038
CFFVR Shawano Area Community Foundation Grants, 1104
Children Affected by AIDS Foundation Domestic Grants, 1145
Clara Blackford Smith and W. Aubrey Smith Charitable Foundation Grants, 1165
Community Foundation for Northeast Michigan Mini-Grants, 1255
Community Foundation of Boone County Grants, 1266
Community Foundation of Mount Vernon and Knox County Grants, 1299
Community Fndn of So Alabama Grants, 1306
Community Foundation of St. Joseph County Special Project Challenge Grants, 1310
Community Fndn of Wabash County Grants, 1316
Community Foundation Partnerships - Lawrence County Grants, 1317

Community Foundation Partnerships - Martin County Grants, 1318
Cystic Fibrosis Lifestyle Foundation Individual Recreation Grants, 1397
Cystic Fibrosis Lifestyle Foundation Loretta Morris Memorial Fund Grants, 1398
Cystic Fibrosis Lifestyle Foundation Mentored Recreation Grants, 1399
Cystic Fibrosis Lifestyle Foundation Peer Support Grants, 1400
Decatur County Community Foundation Large Project Grants, 1444
Decatur County Community Foundation Small Project Grants, 1445
Dubois County Community Foundation Grants, 1517
Emma B. Howe Memorial Foundation Grants, 1593
Fargo-Moorhead Area Foundation Grants, 1632
Florence Hunt Maxwell Foundation Grants, 1676
Foellinger Foundation Grants, 1684
Frank Stanley Beveridge Foundation Grants, 1730
Frederick W. Marzahl Memorial Fund Grants, 1741
Fremont Area Community Fndn General Grants, 1746
Fulton County Community Foundation Grants, 1763
Gardner W. & Joan G. Heidrick, Jr. Fndn Grants, 1773
George A. and Grace L. Long Foundation Grants, 1788
George and Ruth Bradford Foundation Grants, 1789
George A Ohl Jr. Foundation Grants, 1791
George Foundation Grants, 1795
George Kress Foundation Grants, 1799
Gil and Dody Weaver Foundation Grants, 1819
Golden Heart Community Foundation Grants, 1850
Grand Haven Area Community Fndn Grants, 1864
Grand Rapids Area Community Foundation Wyoming Youth Fund Grants, 1868
Grand Rapids Community Foundation Grants, 1869
Grand Rapids Community Foundation Lowell Area Fund Grants, 1872
Greater Kanawha Valley Foundation Grants, 1882
Greater Sitka Legacy Fund Grants, 1885
Greater Worcester Community Foundation Jeppson Memorial Fund for Brookfield Grants, 1889
Greenspun Family Foundation Grants, 1896
Hardin County Community Foundation Grants, 1946
Harrison County Community Foundation Grants, 1957
Harrison County Community Foundation Signature Grants, 1958
Harvey Randall Wickes Foundation Grants, 1972
Hasbro Children's Fund Grants, 1973
Heckscher Foundation for Children Grants, 1988
Helen S. Boylan Foundation Grants, 1997
Herbert H. & Grace A. Dow Fndn Grants, 2008
Holland/Zeeland Community Fndn Grants, 2041
Idaho Community Foundation Eastern Region Competitive Grants, 2114
Impact 100 Grants, 2148
James S. Copley Foundation Grants, 2200
Josephine S. Gumbiner Foundation Grants, 2273
Kahuku Community Fund, 2286
Kenai Peninsula Foundation Grants, 2298
Knox County Community Foundation Grants, 2313
Kosciusko County Community Fndn Grants, 2319
Lisa and Douglas Goldman Fund Grants, 2369
Lubrizol Foundation Grants, 2384
M-A-C AIDS Fund Grants, 2400
Marie H. Bechtel Charitable Remainder Uni-Trust Grants, 2432
Marjorie Moore Charitable Foundation Grants, 2438
Marshall County Community Foundation Grants, 2439
Maxon Charitable Foundation Grants, 2458
McCombs Foundation Grants, 2491
McConnell Foundation Grants, 2492
McKesson Foundation Grants, 2496
McMillen Foundation Grants, 2499
Miami County Community Foundation - Operation Round Up Grants, 2543
Miller Foundation Grants, 2556
North Dakota Community Foundation Grants, 2994
Owen County Community Foundation Grants, 3100
Paul Rapoport Foundation Grants, 3125
Perry County Community Foundation Grants, 3142

SUBJECT INDEX

Petersburg Community Foundation Grants, 3156
Phoenix Suns Charities Grants, 3201
Pike County Community Foundation Grants, 3222
Pohlad Family Foundation, 3241
Pokagon Fund Grants, 3242
Posey County Community Foundation Grants, 3252
Pride Foundation Grants, 3268
Principal Financial Group Foundation Grants, 3273
Putnam County Community Foundation Grants, 3283
Ralphs Food 4 Less Foundation Grants, 3305
Rayonier Foundation Grants, 3311
Reinberger Foundation Grants, 3357
Rochester Area Foundation Grants, 3392
Saint Louis Rams Fndn Community Donations, 3453
Scott County Community Foundation Grants, 3500
Seabury Foundation Grants, 3501
Seneca Foods Foundation Grants, 3511
Sidgmore Family Foundation Grants, 3551
Sioux Falls Area Community Foundation Community Fund Grants (Unrestricted), 3568
Sioux Falls Area Community Foundation Field-of-Interest and Donor-Advised Grants, 3569
Sioux Falls Area Community Foundation Spot Grants (Unrestricted), 3570
Southbury Community Trust Fund, 3625
Spencer County Community Foundation Grants, 3632
Springs Close Foundation Grants, 3633
Starke County Community Foundation Grants, 3646
Steuben County Community Foundation Grants, 3654
Strowd Roses Grants, 3664
Thomas Austin Finch, Sr. Foundation Grants, 3718
Turner Foundation Grants, 3763
U.S. Department of Education 21st Century Community Learning Centers, 3767
Vanderburgh Community Foundation Grants, 3852
Walker Area Community Foundation Grants, 3895
Warren County Community Foundation Grants, 3902
Warren County Community Fndn Mini-Grants, 3903
Warrick County Community Foundation Grants, 3904
Washington County Community Fndn Grants, 3906
Washington County Community Youth Grants, 3907
Wells County Foundation Grants, 3920
White County Community Foundation Grants, 3934
Whitley County Community Foundation Grants, 3936
Winston-Salem Foundation Competitive Grants, 3965
Winston-Salem Fndn Stokes County Grants, 3967
Yawkey Foundation Grants, 3980

Recreation and Leisure Studies
Huntington County Community Foundation Make a Difference Grants, 2083

Recycling
Giant Food Charitable Grants, 1816
Union Bank, N.A. Foundation Grants, 3788

Reference Materials
ASM USFCC/J. Roger Porter Award, 756
NLM Grants for Scholarly Works in Biomedicine and Health, 2944

Refugees
Blumenthal Foundation Grants, 925
Dallas Women's Foundation Grants, 1417
Deborah Munroe Noonan Memorial Grants, 1443
Jacob and Valeria Langeloth Foundation Grants, 2182
Michael Reese Health Trust Responsive Grants, 2545
Minneapolis Foundation Community Grants, 2558
Nathan Cummings Foundation Grants, 2627
Robert W. Woodruff Foundation Grants, 3389
San Diego Foundation for Change Grants, 3470
Theodore Edson Parker Foundation Grants, 3715

Regional Economics
Walmart Foundation State Giving Grants, 3900

Regional Planning/Policy
Blue Cross Blue Shield of Minnesota Foundation - Healthy Equity: Health Impact Assessment Demonstration Project Grants, 916

Blue Cross Blue Shield of Minnesota Foundation - Healthy Equity: Health Impact Assessment Program Grants, 917
Canada-U.S. Fulbright New Century Scholars Program Grants, 1008
Canada-U.S. Fulbright Senior Specialists Grants, 1009
Carla J. Funk Governmental Relations Award, 1020
Carpenter Foundation Grants, 1035
Community Foundation for the National Capital Region Community Leadership Grants, 1260
Dorothy Rider Pool Health Care Grants, 1500
Fulbright Alumni Initiatives Awards, 1749
Fulbright Distinguished Chairs Awards, 1750
Fulbright New Century Scholars Grants, 1753
Fulbright Specialists Program Grants, 1754
Fulbright Scholars in Europe and Eurasia, 1755
Fulbright Traditional Scholar Program in Sub-Saharan Africa, 1756
Fulbright Traditional Scholar Program in the East Asia/Pacific Region, 1757
Fulbright Traditional Scholar Program in the Near East and North Africa Region, 1758
Fulbright Traditional Scholar Program in the South and Central Asia Region, 1759
Fulbright Traditional Scholar Program in the Western Hemisphere, 1760
Lockheed Martin Corp Foundation Grants, 2374
Sarkeys Foundation Grants, 3485

Regional/Urban Design
United Technologies Corporation Grants, 3803

Rehabilitation/Therapy
AFB Rudolph Dillman Memorial Scholarship, 338
Alberto Culver Corporate Contributions Grants, 408
Cailloux Foundation Grants, 996
Carl W. and Carrie Mae Joslyn Trust Grants, 1031
Dammann Fund Grants, 1418
Dolan Children's Foundation Grants, 1482
Dr. John T. Macdonald Foundation Grants, 1507
Edward W. and Stella C. Van Houten Memorial Fund Grants, 1559
Fremont Area Community Foundation Amazing X Grants, 1744
Henry County Community Foundation Grants, 2005
IAFF Burn Foundation Research Grants, 2089
John W. Alden Trust Grants, 2260
Lisa Higgins-Hussman Foundation Grants, 2370
Lubrizol Foundation Grants, 2384
Massage Therapy Research Fund (MTRF) Grants, 2452
NIA Health Behaviors and Aging Grants, 2761
NIDRR Field-Initiated Projects, 2853
NYCT Girls and Young Women Grants, 3053
Ordean Foundation Grants, 3080
Paul Rapoport Foundation Grants, 3125
Peabody Foundation Grants, 3132
Phoenix Suns Charities Grants, 3201
Potts Memorial Foundation Grants, 3254
Robert B McMillen Foundation Grants, 3381
Robert R. Meyer Foundation Grants, 3387
Shell Deer Park Grants, 3543
U.S. Department of Education Rehabilitation Engineering Research Centers Grants, 3769
WHO Foundation Volunteer Service Grants, 3940

Rehabilitation/Therapy, Emotional/Social
Dept of Ed Rehabilitation Training Grants, 1463
Different Needz Foundation Grants, 1478
Fourjay Foundation Grants, 1712
Lisa Higgins-Hussman Foundation Grants, 2370
Military Ex-Prisoners of War Foundation Grants, 2554
Perry County Community Foundation Grants, 3142
Posey County Community Foundation Grants, 3252
Robert B McMillen Foundation Grants, 3381
Robert R. McCormick Tribune Veterans Initiative Grants, 3386
Seabury Foundation Grants, 3501
WHO Foundation Volunteer Service Grants, 3940

Rehabilitation/Therapy, Occupational/Vocational
Able Trust Vocational Rehabilitation Grants for Individuals, 213
Achelis Foundation Grants, 234
Bodman Foundation Grants, 929
Dept of Ed Rehabilitation Training Grants, 1463
Different Needz Foundation Grants, 1478
Robert R. Meyer Foundation Grants, 3387
U.S. Department of Education Rehabilitation Research Training Centers (RRTCs), 3770
U.S. Department of Education Vocational Rehabilitation Services Projects for American Indians with Disabilities Grants, 3774
WHO Foundation Volunteer Service Grants, 3940

Rehabilitation/Therapy, Physical
AAPD Henry B. Betts Award, 182
DAR Nursing/Physical Therapy Scholarships, 1430
Different Needz Foundation Grants, 1478
GFWC of Massachusetts Memorial Education Scholarship, 1813
PVA Education Foundation Grants, 3284
PVA Research Foundation Grants, 3285
Robert B McMillen Foundation Grants, 3381
Robert R. Meyer Foundation Grants, 3387
Seabury Foundation Grants, 3501
WHO Foundation Volunteer Service Grants, 3940

Religion
Aaron Foundation Grants, 193
Abel Foundation Grants, 206
Abell-Hanger Foundation Grants, 207
Able To Serve Grants, 212
Abramson Family Foundation Grants, 217
Adams County Community Foundation of Pennsylvania Grants, 300
A Friends' Foundation Trust Grants, 341
Ahmanson Foundation Grants, 355
Ahn Family Foundation Grants, 356
Albert and Margaret Alkek Foundation Grants, 405
Alcoa Foundation Grants, 414
Alfred and Tillie Shemanski Testamentary Trust Grants, 428
ALSAM Foundation Grants, 454
Amelia Sillman Rockwell and Carlos Perry Rockwell Charities Fund Grants, 501
American Foodservice Charitable Trust Grants, 532
Amerigroup Foundation Grants, 565
Anschutz Family Foundation Grants, 602
Anthony R. Abraham Foundation Grants, 608
Antone & Edene Vidinha Charitable Trust Grants, 609
Aragona Family Foundation Grants, 648
Aratani Foundation Grants, 649
Archer Daniels Midland Foundation Grants, 651
Argyros Foundation Grants Program, 653
Arkansas Community Foundation Grants, 660
Arkell Hall Foundation Grants, 661
Arthur Ashley Williams Foundation Grants, 671
Athwin Foundation Grants, 795
Audrey & Sydney Irmas Foundation Grants, 799
Babcock Charitable Trust Grants, 816
Bacon Family Foundation Grants, 817
Bailey Foundation Grants, 818
Banfi Vintners Foundation Grants, 824
Barra Foundation Project Grants, 834
Baton Rouge Area Foundation Grants, 838
Beatrice Laing Trust Grants, 864
Beazley Foundation Grants, 865
Belk Foundation Grants, 872
Bender Foundation Grants, 876
Besser Foundation Grants, 887
BibleLands Grants, 889
Bill Hannon Foundation Grants, 895
Blumenthal Foundation Grants, 925
Bodenwein Public Benevolent Foundation Grants, 928
Bodman Foundation Grants, 929
Borkee-Hagley Foundation Grants, 934
Bradley-Turner Foundation Grants, 944
Bright Family Foundation Grants, 951
Bruce and Adele Greenfield Foundation Grants, 972

802 / Religion

Burden Trust Grants, 976
Byron W. & Alice L. Lockwood Fnd Grants, 991
Caddock Foundation Grants, 994
Caleb C. and Julia W. Dula Educational and Charitable Foundation Grants, 997
Callaway Foundation Grants, 1002
Carl Gellert and Celia Berta Gellert Foundation Grants, 1024
Carl R. Hendrickson Family Foundation Grants, 1028
Carrie Estelle Doheny Foundation Grants, 1037
Carylon Foundation Grants, 1040
CFFVR Robert and Patricia Endries Family Foundation Grants, 1102
Chamberlain Foundation Grants, 1108
Chapman Charitable Foundation Grants, 1112
Chatlos Foundation Grants Program, 1128
Chiles Foundation Grants, 1147
Citizens Bank Mid-Atlantic Charitable Foundation Grants, 1162
Clark-Winchcole Foundation Grants, 1173
Clark and Ruby Baker Foundation Grants, 1174
Claude Bennett Family Foundation Grants, 1176
Cleveland-Cliffs Foundation Grants, 1181
Coastal Community Foundation of South Carolina Grants, 1202
Collins Foundation Grants, 1210
Community Foundation of Greenville Community Enrichment Grants, 1284
Community Foundation of Greenville Hollingsworth Funds Program/Project Grants, 1285
Community Foundation of Mount Vernon and Knox County Grants, 1299
Connelly Foundation Grants, 1326
Crane Foundation Grants, 1357
CRH Foundation Grants, 1360
Dade Community Foundation Grants, 1410
DaimlerChrysler Corporation Fund Grants, 1412
Dale and Edna Walsh Foundation Grants, 1415
Danellie Foundation Grants, 1421
Daywood Foundation Grants, 1438
Deaconess Community Foundation Grants, 1439
Della B. Gardner Fund Grants, 1451
Donald and Sylvia Robinson Family Foundation Grants, 1486
Dora Roberts Foundation Grants, 1489
Dorothy Hooper Beattie Foundation Grants, 1499
Dr. Scholl Foundation Grants Program, 1511
Earl and Maxine Claussen Trust Grants, 1541
Edwin W. and Catherine M. Davis Fndn Grants, 1561
Effie and Wofford Cain Foundation Grants, 1563
Elliot Foundation Inc Grants, 1575
Elsie H. Wilcox Foundation Grants, 1583
Emily Davie and Joseph S. Kornfeld Foundation Grants, 1591
Ensworth Charitable Foundation Grants, 1595
Erie Chapman Foundation Grants, 1600
Eugene B. Casey Foundation Grants, 1609
Evjue Foundation Grants, 1617
F.M. Kirby Foundation Grants, 1624
Frank G. and Freida K. Brotz Family Foundation Grants, 1723
Franklin County Community Foundation Grants, 1724
Frank S. Flowers Foundation Grants, 1729
Frank Stanley Beveridge Foundation Grants, 1730
Fraser-Parker Foundation Grants, 1732
Fred C. & Katherine B. Andersen Grants, 1737
Fritz B. Burns Foundation Grants, 1747
Fuller E. Callaway Foundation Grants, 1761
G.N. Wilcox Trust Grants, 1768
Garland D. Rhoads Foundation, 1774
George and Sarah Buchanan Foundation Grants, 1790
George Family Foundation Grants, 1794
George H.C. Ensworth Memorial Fund Grants, 1797
George Kress Foundation Grants, 1799
George P. Davenport Trust Fund Grants, 1800
Gheens Foundation Grants, 1814
Giant Eagle Foundation Grants, 1815
Grace and Franklin Bernsen Foundation Grants, 1858
Great-West Life Grants, 1879
Greater Saint Louis Community Fndn Grants, 1884

Greater Worcester Community Foundation Discretionary Grants, 1888
Gregory Family Foundation Grants (Florida), 1899
Griffin Family Foundation Grants, 1902
Grover Hermann Foundation Grants, 1906
Guido A. & Elizabeth H. Binda Fndn Grants, 1913
H. Leslie Hoffman and Elaine S. Hoffman Foundation Grants, 1924
Hackett Foundation Grants, 1926
Hagedorn Fund Grants, 1934
Harold Simmons Foundation Grants, 1955
Harry Frank Guggenheim Foundation Dissertation Fellowships, 1963
Harry Frank Guggenheim Fnd Research Grants, 1964
Harry W. Bass, Jr. Foundation Grants, 1968
Helen Bader Foundation Grants, 1992
Hillsdale Fund Grants, 2035
I.A. O'Shaughnessy Foundation Grants, 2088
Ida Alice Ryan Charitable Trust Grants, 2113
Irvin Stern Foundation Grants, 2164
J.C. Penney Company Grants, 2169
J.E. and L.E. Mabee Foundation Grants, 2170
James Hervey Johnson Charitable Educational Trust Grants, 2191
James J. & Angelia M. Harris Fndn Grants, 2192
James J. & Joan A. Gardner Family Fndn Grants, 2193
James M. Collins Foundation Grants, 2195
Janus Foundation Grants, 2211
Jayne and Leonard Abess Foundation Grants, 2214
Jessie Ball Dupont Fund Grants, 2226
Joe W. and Dorothy Dorsett Brown Fndn Grants, 2230
John G. Duncan Charitable Trust Grants, 2236
John H. and Wilhelmina D. Harland Charitable Foundation Children and Youth Grants, 2238
John J. Leidy Foundation Grants, 2241
John P. Murphy Foundation Grants, 2248
Joseph H. & Florence A. Roblee Fndn Grants, 2269
Katharine Matthies Foundation Grants, 2291
Kathryne Beynon Foundation Grants, 2294
Kluge Center David B. Larson Fellowship in Health and Spirituality, 2311
Knight Family Charitable and Educational Foundation Grants, 2312
Leo Niessen Jr., Charitable Trust Grants, 2358
Libra Foundation Grants, 2363
Lotus 88 Fnd for Women & Children Grants, 2376
Lowell Berry Foundation Grants, 2382
Lubbock Area Foundation Grants, 2383
Lucile Horton Howe and Mitchell B. Howe Foundation Grants, 2385
Lumpkin Family Fnd Healthy People Grants, 2391
Lydia deForest Charitable Trust Grants, 2393
M. Bastian Family Foundation Grants, 2402
Marion I. and Henry J. Knott Foundation Discretionary Grants, 2435
Marion I. and Henry J. Knott Foundation Standard Grants, 2436
Maxon Charitable Foundation Grants, 2458
McCallum Family Foundation Grants, 2489
McCombs Foundation Grants, 2491
Meriden Foundation Grants, 2519
Merkel Foundation Grants, 2520
Mervin Bovaird Foundation Grants, 2523
MGN Family Foundation Grants, 2542
Nathan Cummings Foundation Grants, 2627
Nell J. Redfield Foundation Grants, 2673
Norcliffe Foundation Grants, 2979
North Carolina Community Foundation Grants, 2990
Patrick and Anna M. Cudahy Fund Grants, 3120
Perkins Charitable Foundation Grants, 3140
Pittsburgh Foundation Community Fund Grants, 3230
Polk Bros. Foundation Grants, 3243
R.C. Baker Foundation Grants, 3292
Rainbow Fund Grants, 3301
Religion and Health: Effects, Mechanisms, and Interpretation RFP, 3358
Richard and Helen DeVos Foundation Grants, 3364
Richard D. Bass Foundation Grants, 3366
Ricks Family Charitable Trust Grants, 3373
Robins Foundation Grants, 3390

Roy J. Carver Charitable Trust Medical and Science Research Grants, 3416
Rush County Community Foundation Grants, 3424
S.H. Cowell Foundation Grants, 3447
Salem Foundation Grants, 3457
Siebert Lutheran Foundation Grants, 3554
Sioux Falls Area Community Foundation Community Fund Grants (Unrestricted), 3568
Sioux Falls Area Community Foundation Spot Grants (Unrestricted), 3570
Sisters of St. Joseph Healthcare Fndn Grants, 3576
SOCFOC Catholic Ministries Grants, 3596
Solo Cup Foundation Grants, 3620
Stewart Huston Charitable Trust Grants, 3655
Sunderland Foundation Grants, 3675
Support Our Aging Religious (SOAR) Grants, 3686
Susan Mott Webb Charitable Trust Grants, 3694
T. Spencer Shore Foundation Grants, 3701
Thomas and Dorothy Leavey Foundation Grants, 3717
Thomas Austin Finch, Sr. Foundation Grants, 3718
Thompson Charitable Foundation Grants, 3727
Todd Brock Family Foundation Grants, 3736
Trull Foundation Grants, 3761
Vancouver Foundation Grants and Community Initiatives Program, 3849
W. C. Griffith Foundation Grants, 3882
W.C. Griffith Foundation Grants, 3881
Wabash Valley Community Foundation Grants, 3893
Walter L. Gross III Family Foundation Grants, 3901
Weingart Foundation Grants, 3913
Welborn Baptist Fndn Health Ministries Grants, 3915
William B. Stokely Jr. Foundation Grants, 3947
William G. & Helen C. Hoffman Fndn Grants, 3950
William H. Hannon Foundation Grants, 3953
William Ray & Ruth E. Collins Fndn Grants, 3956
Woodward Fund Grants, 3973

Religious Studies
Able To Serve Grants, 212
Bailey Foundation Grants, 818
Blumenthal Foundation Grants, 925
Boettcher Foundation Grants, 931
Booth-Bricker Fund Grants, 933
Chatlos Foundation Grants Program, 1128
Cockrell Foundation Grants, 1204
E. Rhodes & Leona B. Carpenter Grants, 1539
FirstEnergy Foundation Community Grants, 1668
Frank G. and Freida K. Brotz Family Foundation Grants, 1723
George W. Brackenridge Foundation Grants, 1802
J.N. and Macie Edens Foundation Grants, 2174
Jayne and Leonard Abess Foundation Grants, 2214
John I. Smith Charities Grants, 2240
Nathan Cummings Foundation Grants, 2627
Rainbow Fund Grants, 3301
Richard and Helen DeVos Foundation Grants, 3364
Siebert Lutheran Foundation Grants, 3554
T. Spencer Shore Foundation Grants, 3701

Religious Welfare Programs
Able To Serve Grants, 212
Adams County Community Foundation of Pennsylvania Grants, 300
American Foodservice Charitable Trust Grants, 532
Babcock Charitable Trust Grants, 816
Blue Cross Blue Shield of Minnesota Foundation - Healthy Children: Growing Up Healthy Grants, 915
Bob & Delores Hope Foundation Grants, 927
Byron W. & Alice L. Lockwood Fnd Grants, 991
Carl R. Hendrickson Family Foundation Grants, 1028
Carrie E. and Lena V. Glenn Foundation Grants, 1036
Charles H. Hall Foundation, 1120
Clark and Ruby Baker Foundation Grants, 1174
Claude Bennett Family Foundation Grants, 1176
Community Foundation of Abilene Celebration of Life Grants, 1261
Crane Foundation Grants, 1357
Della B. Gardner Fund Grants, 1451
Earl and Maxine Claussen Trust Grants, 1541
Ensworth Charitable Foundation Grants, 1595

SUBJECT INDEX

Frederick W. Marzahl Memorial Fund Grants, 1741
George A. and Grace L. Long Foundation Grants, 1788
George E. Hatcher, Jr. and Ann Williams Hatcher Foundation Grants, 1792
Greenspun Family Foundation Grants, 1896
Gregory Family Foundation Grants (Florida), 1899
Grover Hermann Foundation Grants, 1906
Harry Kramer Memorial Fund Grants, 1965
Harvey Randall Wickes Foundation Grants, 1972
Helen Irwin Littauer Educational Trust Grants, 1994
Henrietta Lange Burk Fund Grants, 2001
Herbert A. & Adrian W. Woods Fndn Grants, 2007
Horace Moses Charitable Foundation Grants, 2047
Huisking Foundation Grants, 2076
Ida Alice Ryan Charitable Trust Grants, 2113
J.N. and Macie Edens Foundation Grants, 2174
James J. & Joan A. Gardner Family Fndn Grants, 2193
Jayne and Leonard Abess Foundation Grants, 2214
Meriden Community Grants, 2519
Parke County Community Foundation Grants, 3115
Priddy Foundation Program Grants, 3266
Richard and Helen DeVos Foundation Grants, 3364
Robbins Charitable Foundation Grants, 3378
Robert R. McCormick Trib Community Grants, 3385
Sierra Health Foundation Responsive Grants, 3555
Sioux Falls Area Community Foundation Field-of-Interest and Donor-Advised Grants, 3569
Solo Cup Foundation Grants, 3620
St. Joseph Community Health Foundation Improving Healthcare Access Grants, 3640
Swindells Charitable Foundation, 3697
T. Spencer Shore Foundation Grants, 3701
Todd Brock Family Foundation Grants, 3736
Triangle Community Foundation Donor-Advised Grants, 3757
Vancouver Foundation Grants and Community Initiatives Program, 3849
W. C. Griffith Foundation Grants, 3882
William J. & Tina Rosenberg Fndn Grants, 3954
William Ray & Ruth E. Collins Fndn Grants, 3956
Women of the ELCA Opportunity Scholarships for Lutheran Laywomen, 3970
ZYTL Foundation Grants, 3985

Remedial Education
Albert W. Rice Charitable Foundation Grants, 411
Alfred E. Chase Charitable Foundation Grants, 430
Amgen Foundation Grants, 572
Charles H. Pearson Foundation Grants, 1121
Community Fndn for the Capital Region Grants, 1259
Frank Reed and Margaret Jane Peters Memorial Fund II Grants, 1728
George W. Wells Foundation Grants, 1804
Harold Brooks Foundation Grants, 1952
Lewis H. Humphreys Charitable Trust Grants, 2361
Mary Wilmer Covey Charitable Trust Grants, 2448

Renewable Energy Sources
Carolyn Foundation Grants, 1034
Union Bank, N.A. Foundation Grants, 3788

Renovation
Charles H. Farnsworth Trust Grants, 1119
Eugene Straus Charitable Trust, 1612
GNOF Albert N. & Hattie M. McClure Grants, 1834
GNOF Exxon-Mobil Grants, 1836
GNOF Norco Community Grants, 1845
Hillcrest Foundation Grants, 2031
IDPH Hosptial Capital Investment Grants, 2119
Janson Foundation Grants, 2210
Katharine Matthies Foundation Grants, 2291
Priddy Foundation Capital Grants, 3263
Vigneron Memorial Fund Grants, 3873

Reproduction
Bernard F. and Alva B. Gimbel Foundation Reproductive Rights Grants, 883
Gardner Foundation Grants, 1772
IIE Hewlett Fnd/IIE Dissertation Fellowship, 2132
Lalor Foundation Anna Lalor Burdick Grants, 2335
MMAAP Foundation Fellowship Award in Reproductive Medicine, 2580
MMAAP Foundation Research Project Award in Reproductive Medicine, 2584
Packard Foundation Local Grants, 3105
USAID Family Planning and Reproductive Health Methods Grants, 3811
USAID Social Science Research in Population and Reproductive Health Grants, 3822
WestWind Foundation Reproductive Health and Rights Grants, 3924

Reproductive Disorders
USAID Family Planning and Reproductive Health Methods Grants, 3811

Reproductive Rights
Baxter International Corporate Giving Grants, 841
Bernard F. and Alva B. Gimbel Foundation Reproductive Rights Grants, 883
General Service Reproductive Justice Grants, 1785
Global Fund for Women Grants, 1833
Ms. Foundation for Women Health Grants, 2597
USAID Family Planning and Reproductive Health Methods Grants, 3811
WestWind Foundation Reproductive Health and Rights Grants, 3924

Research Participation
A-T Medical Research Foundation Grants, 15
AACR-Minorities in Cancer Research Jane Cooke Wright Lectureship Awards, 94
AACR-National Brain Tumor Fnd Fellows Grant, 95
AACR-Women in Cancer Research Charlotte Friend Memorial Lectureship Awards, 101
AACR Award for Outstanding Achievement in Cancer Research, 103
AACR Basic Cancer Research Fellowships, 105
AACR Clinical & Translat Research Fellowships, 110
AACR Minority-Serving Institution Faculty Scholar in Cancer Research Awards, 119
AACR Richard & Hinda Rosenthal Mem Awards, 123
ACS Doctoral Grants in Oncology Social Work, 254
ACS Physician Training Awards in Prevent Med, 275
AFAR Medical Student Training in Aging Research Program, 335
AFAR Paul Beeson Career Development Awards in Aging Research for the Island of Ireland, 336
AHRF Regular Research Grants, 363
AIHS Full-Time Health Research Studentships, 384
Alexander Eastman Foundation Grants, 420
AMA-MSS Research Poster Award, 477
AMA-WPC Joan F. Giambalvo Memorial Scholarships, 480
ASHFoundation Clinical Research Grants, 701
ASM-PAHO Infectious Diseases Epidemiology and Surveillance Fellowships, 714
ASM/CDC Fellowships in Infectious Disease and Public Health Microbiology, 716
ASM Abbott Laboratories Award in Clinical and Diagnostic Immunology, 717
ASM BD Award for Research in Clin Microbio, 718
ASM D.C. White Research and Mentoring Award, 723
ASM Eli Lilly and Company Research Award, 726
ASM ICAAC Young Investigator Awards, 732
ASM Merck Irving S. Sigal Memorial Awards, 739
ASM Microbiology Undergraduate Research Fellowship, 740
ASM Procter & Gamble Award in Applied and Environmental Microbiology, 742
ASM Promega Biotechnology Research Award, 743
ASM Raymond W. Sarber Awards, 745
ASM Robert D. Watkins Graduate Research Fellowship, 747
ASM sanofi-aventis ICAAC Award, 749
ASM Siemens Healthcare Diagnostics Young Investigator Award, 751
ASM Undergraduate Research Fellowship, 755
ASM William A Hinton Research Training Award, 757
BCBSM Foundation Excellence in Research Awards for Students, 853
BCBSM Foundation Primary and Clinical Prevention Grants, 856
Carrier Corporation Contributions Grants, 1038
CCFF Chairman's Dist Life Sciences Award, 1044
Cruise Industry Charitable Foundation Grants, 1363
Cystic Fibrosis Canada Clinical Fellowships, 1381
Cystic Fibrosis Canada Clinical Project Grants, 1382
Cystic Fibrosis Canada Clinic Incentive Grants, 1383
Cystic Fibrosis Canada Scholarships, 1386
DHHS Oral Health Promotion Research Across the Lifespan, 1474
Duke University Adult Cardiothoracic Anesthesia and Critical Care Medicine Fellowships, 1520
Duke University Ambulatory and Regional Anesthesia Fellowships, 1521
Duke University Clinical Cardiac Electrophysiology Fellowships, 1522
Duke University Hyperbaric Center Fellowships, 1523
Duke Univ Interventional Cardio Fellowships, 1524
Duke Univ Obstetric Anesthesia Fellowships, 1525
Duke University Pain Management Fellowships, 1526
Duke University Postdoctoral Research Fellowships in Aging, 1528
Edward W. and Stella C. Van Houten Memorial Fund Grants, 1559
Ellison Medical Foundation/AFAR Julie Martin Mid-Career Award in Aging Research, 1576
FAR Fund Grants, 1631
Foundation for Pharmaceutical Sciences Herb and Nina Demuth Grant, 1697
George A Ohl Jr. Foundation Grants, 1791
George H. Hitchings New Investigator Award in Health Research, 1798
Gertrude B. Elion Mentored Medical Student Research Awards, 1809
Gill Foundation - Gay and Lesbian Fund Grants, 1821
Goldman Philanthropic Partnerships Program, 1853
Harry W. Bass, Jr. Foundation Grants, 1968
Hawaii Community Foundation Health Education and Research Grants, 1976
Herbert H. & Grace A. Dow Fndn Grants, 2008
Hilton Hotels Corporate Giving Program Grants, 2037
Hoglund Foundation Grants, 2040
HomeBanc Foundation Grants, 2042
Ireland Family Foundation Grants, 2162
John G. Duncan Charitable Trust Grants, 2236
John W. Anderson Foundation Grants, 2262
Mabel H. Flory Charitable Trust Grants, 2408
McCarthy Family Foundation Grants, 2490
MGFA Post-Doctoral Research Fellowships, 2539
MGFA Student Fellowships, 2540
NEI Mentored Clinical Scientist Dev Award, 2666
NEI Research Grant For Secondary Data Analysis Grants, 2667
NEI Ruth L. Kirschstein National Research Service Award Short-Term Institutional Research Training Grants, 2668
Nelda C. and H.J. Lutcher Stark Fndn Grants, 2672
NHLBI Independent Scientist Award, 2701
NHLBI Research Supplements to Promote Diversity in Health-Related Research, 2727
NIA Independent Scientist Award, 2766
NIH Exploratory/Devel Research Grant, 2877
NSF Accelerating Innovation Research, 3005
NYAM Edward N. Gibbs Memorial Lecture and Award in Nephrology, 3041
NYCT Blood Disease Research Grants, 3051
Pancreatic Cancer Action Network Fellowship, 3113
Pancreatic Cancer Action Network-AACR Pathway to Leadership Grants, 3114
PDF International Research Grants, 3128
PDF Postdoctoral Fellowships for Neurologists, 3130
Peacock Foundation Grants, 3133
Percy B. Ferebee Endowment Grants, 3138
Pinkerton Foundation Grants, 3224
Pittsburgh Foundation Community Fund Grants, 3230
Pittsburgh Foundation Medical Research Grants, 3231
Pott Foundation Grants, 3253

Potts Memorial Foundation Grants, 3254
R.C. Baker Foundation Grants, 3292
Regenstrief General Internal Medicine - General Pediatrics Research Fellowship Program, 3351
Regenstrief Master Of Science In Clinical Research/CITE Program, 3353
Retirement Research Foundation General Program Grants, 3359
Robert B McMillen Foundation Grants, 3381
Rockefeller Archive Center Research Grants, 3393
Rosalinde and Arthur Gilbert Foundation/AFAR New Investigator Awards in Alzheimer's Disease, 3406
Savoy Foundation Post-Doctoral and Clinical Research Fellowships, 3488
Savoy Foundation Research Grants, 3489
Sigma Theta Tau International / Council for the Advancement of Nursing Science Grants, 3557
Sigma Theta Tau Small Grants, 3560
Sigrid Juselius Foundation Grants, 3561
SNM/ Covidien Seed Grant in Molecular Imaging/Nuclear Medicine Research, 3582
SNM Molecular Imaging Research Grant For Junior Medical Faculty, 3583
SNM Pilot Research Grants, 3585
SNM Pilot Research Grants in Nuclear Medicine/Molecular Imaging, 3586
SNM Postdoc Molecular Imaging Scholar Grants, 3587
SNM Student Fellowship Awards, 3588
SNM Tetalman Young Investigator Awards, 3589
STTI/ATI Educational Assessment Nursing Research Grants, 3665
STTI Emergency Nurses Association Foundation Grant, 3666
STTI Environment of Elder Care Nursing Research Grants, 3667
STTI Joan K. Stout RN Research Grants, 3668
STTI National League for Nursing Grants, 3669
STTI Rosemary Berkel Crisp Research Awards, 3670
STTI Virginia Henderson Research Grants, 3671
Vancouver Foundation Grants and Community Initiatives Program, 3849
W. C. Griffith Foundation Grants, 3882

Research Resources (Health/Safety/Medical)
Cystic Fibrosis Canada Clinic Incentive Grants, 1383
Cystic Fibrosis Canada Senior Scientist Research Training Awards, 1387
IHI Quality Improvement Fellowships, 2121
John W. Anderson Foundation Grants, 2262
NEI Research Grant For Secondary Data Analysis Grants, 2667
NSERC Brockhouse Canada Prize for Interdisciplinary Research in Science and Engineering Grant, 3003
Pittsburgh Foundation Medical Research Grants, 3231
Regenstrief General Internal Medicine - General Pediatrics Research Fellowship Program, 3351
Regenstrief Geriatrics Fellowship, 3352
USDA Organic Agriculture Research Grants, 3830
USDD Clinical Functional Assessment Investigator-Initiated Grants, 3840

Respiratory Diseases
CDC Epidemic Intell Service Training Grants, 1055
CDC Epidemiology Elective Rotation, 1056
CDC Public Health Prev Service Fellowships, 1069
CDC Public Health Prevention Service Fellowship Sponsorships, 1070
DOL Occupational Safety and Health--Susan Harwood Training Grants, 1484
Fairlawn Foundation Grants, 1626
Medtronic Foundation Strengthening Health Systems Grants, 2515
NHLBI Ancillary Studies in Clinical Trials, 2688
NHLBI Developmental Origins of Altered Lung Physiology and Immune Function Grants, 2698
NHLBI Microbiome of Lung & Respiratory in HIV Infected Individuals & Uninfected Controls, 2712
NHLBI Phase II Clinical Trials of Novel Therapies for Lung Diseases, 2720
NHLBI Prematurity and Respiratory Grants, 2721

NHLBI Right Heart Function in Health and Chronic Lung Diseases Grants, 2728
NHLBI Ruth L. Kirschstein National Research Service Awards for Individual Predoctoral MD/PhD Fellows and Other Dual Degree Fellows, 2731
NHLBI Translational Grants in Lung Diseases, 2740
NIEHS Small Business Innovation Research (SBIR) Program Grants, 2860
Pfizer Healthcare Charitable Contributions, 3180
Pfizer Medical Education Track Two Grants, 3183

Respiratory System
L. A. Hollinger Respiratory Therapy and Nursing Scholarships, 2327
NHLBI Microbiome of Lung & Respiratory in HIV Infected Individuals & Uninfected Controls, 2712
NHLBI Prematurity and Respiratory Outcomes Program (PROP) Grants, 2721
Trinity Lutheran School of Nursing Alumnae Scholarships, 3759

Restoration and Preservation
Albuquerque Community Foundation Grants, 412
Auburn Foundation Grants, 798
Barberton Community Foundation Grants, 829
Berrien Community Foundation Grants, 884
Blue Mountain Community Foundation Grants, 921
Boeing Company Contributions Grants, 930
Booth-Bricker Fund Grants, 933
Callaway Foundation Grants, 1002
Carls Foundation Grants, 1030
Champlin Foundations Grants, 1110
Charles Edison Fund Grants, 1116
Charlotte County (FL) Community Foundation Grants, 1126
CLIF Bar Family Foundation Grants, 1183
Community Foundation Alliance City of Evansville Endowment Fund Grants, 1238
Community Fndn for Monterey County Grants, 1253
Crown Point Community Foundation Grants, 1361
Cudd Foundation Grants, 1370
D.F. Halton Foundation Grants, 1405
Fayette County Foundation Grants, 1636
FirstEnergy Foundation Community Grants, 1668
Foundation for the Mid South Community Development Grants, 1699
Fred C. & Katherine B. Andersen Grants, 1737
George Foundation Grants, 1795
GNOF Bayou Communities Grants, 1835
GNOF Exxon-Mobil Grants, 1836
Grammy Foundation Grants, 1863
Greater Green Bay Community Fndn Grants, 1881
Greater Milwaukee Foundation Grants, 1883
High Meadow Foundation Grants, 2027
Hoblitzelle Foundation Grants, 2038
Holland/Zeeland Community Fndn Grants, 2041
Illinois Tool Works Foundation Grants, 2147
Jane's Trust Grants, 2206
Jasper Foundation Grants, 2212
Jennings County Community Foundation Grants, 2220
Jessie Ball Dupont Fund Grants, 2226
M.E. Raker Foundation Grants, 2404
Mary Black Foundation Active Living Grants, 2440
McLean Contributionship Grants, 2497
Norcliffe Foundation Grants, 2979
North Carolina Community Foundation Grants, 2990
Norwin S. & Elizabeth N. Bean Fndn Grants, 3001
Ogden Codman Trust Grants, 3066
Parkersburg Area Community Foundation Action Grants, 3116
Perkins Charitable Foundation Grants, 3140
R.T. Vanderbilt Trust Grants, 3294
Ripley County Community Foundation Small Project Grants, 3377
Roy & Christine Sturgis Charitable Grants, 3415
S.H. Cowell Foundation Grants, 3447
Southwest Florida Community Foundation Competitive Grants, 3627
Stewart Huston Charitable Trust Grants, 3655

SunTrust Bank Trusteed Foundations Florence C. and Harry L. English Memorial Fund Grants, 3680
SunTrust Bank Trusteed Foundations Greene-Sawtell Grants, 3681
SunTrust Bank Trusteed Foundations Harriet McDaniel Marshall Tust Grants, 3682
SunTrust Bank Trusteed Foundations Nell Warren Elkin and William Simpson Elkin Grants, 3683
SunTrust Bank Trusteed Foundations Thomas Guy Woolford Charitable Trust Grants, 3684
SunTrust Bank Trusteed Foundations Walter H. and Marjory M. Rich Memorial Fund Grants, 3685
Union Bank, N.A. Foundation Grants, 3788
Wayne County Foundation Grants, 3909

Restoration and Preservation, Art Works/Artifacts
E. Rhodes & Leona B. Carpenter Grants, 1539
Robert R. Meyer Foundation Grants, 3387

Restoration and Preservation, Structural/Architectural
Benton Community Foundation Grants, 878
Blue River Community Foundation Grants, 922
Brown County Community Foundation Grants, 970
Caleb C. and Julia W. Dula Educational and Charitable Foundation Grants, 997
Community Foundation of Bartholomew County Heritage Fund Grants, 1262
Community Foundation of Bartholomew County James A. Henderson Award for Fundraising, 1263
Community Foundation of Greater Fort Wayne - Community Endowment and Clarke Endowment Grants, 1274
Delaware Community Foundation Grants, 1449
GNOF Exxon-Mobil Grants, 1836
Grand Rapids Community Foundation Grants, 1869
Kosciusko County Community Foundation REMC Operation Round Up Grants, 2320
Rhode Island Foundation Grants, 3362
Ripley County Community Foundation Grants, 3376
Ripley County Community Foundation Small Project Grants, 3377
Robert R. Meyer Foundation Grants, 3387
Textron Corporate Contributions Grants, 3712
Trull Foundation Grants, 3761

Retirement
David N. Lane Trust Grants for Aged and Indigent Women, 1433
Robert & Clara Milton Senior Housing Grants, 3384
RRF General Program Grants, 3417

Rheumatic Diseases
AMA Foundation Seed Grants for Research, 494
NIAMS Pilot and Feasibility Clinical Research Grants in Arthritis, Musculoskeletal & Skin Diseases, 2774
Nuffield Foundation Oliver Bird Rheumatism Program Grants, 3029
Pfizer ASPIRE North America Rheumatology Research Awards, 3172
Pfizer Inflammation Competitive Research Awards (UK), 3181
Pfizer Research Initiative Rheumatology Grants (Germany), 3185

Rheumatology
American College of Rheumatology Fellows-in-Training Travel Scholarships, 517
Nuffield Foundation Oliver Bird Rheumatism Program Grants, 3029
Pfizer ASPIRE North America Rheumatology Research Awards, 3172
Pfizer Inflammation Comp Research Awards, 3181
Pfizer Research Initiative Rheumatology Grants (Germany), 3185

Rhinology
AAO-HNSF Health Services Research Grants, 165
AAO-HNSF Maureen Hannley Research Training Awards, 166

SUBJECT INDEX

AAO-HNSF Percy Memorial Research Award, 167
AAO-HNSF Rande H. Lazar Health Services Research Grant, 168
AAO-HNSF Resident Research Awards, 169
AAOA Foundation/AAO-HNSF Combined Research Grants, 172
ALVRE Casselberry Award, 460
ALVRE Grant, 461
ALVRE Seymour R. Cohen Award, 462
ARS New Investigator Award, 668
ARS Resident Research Grants, 669

Risk Factors/Analysis
American Chemical Society Award for Creative Advances in Environmental Science and Technology, 513
Glaucoma Research Pilot Project Grants, 1828

Roma
OSF Roma Health Project Grants, 3095

Roman Catholic Church
Dolan Children's Foundation Grants, 1482
G.A. Ackermann Memorial Fund Grants, 1767
Huisking Foundation Grants, 2076
Merkel Foundation Grants, 2520

Runaway Youth
Christine & Katharina Pauly Trust Grants, 1150
ConocoPhillips Foundation Grants, 1327
Denver Foundation Community Grants, 1461

Rural Areas
Agnes M. Lindsay Trust Grants, 346
Agway Foundation Grants, 348
Dean Foods Community Involvement Grants, 1440
Greenwall Foundation Bioethics Grants, 1897
J.L. Bedsole Foundation Grants, 2172
James G.K. McClure Educational and Development Fund Grants, 2188
Land O'Lakes Foundation Mid-Atlantic Grants, 2341
Peter Kiewit Foundation General Grants, 3154
Peter Kiewit Foundation Small Grants, 3155
Rochester Area Community Foundation Grants, 3391
Samuel S. Johnson Foundation Grants, 3468
Steele-Reese Foundation Grants, 3650
USDA Delta Health Care Services Grants, 3829
W.K. Kellogg Foundation Healthy Kids Grants, 3886

Rural Development
Peter Kiewit Foundation Small Grants, 3155
Priddy Foundation Organizational Development Grants, 3265
USDA Delta Health Care Services Grants, 3829

Rural Education
Coastal Community Foundation of South Carolina Grants, 1202
U.S. Department of Education 21st Century Community Learning Centers, 3767

Rural Health Care
Agway Foundation Grants, 348
Appalachian Regional Commission Health Grants, 637
Claude Worthington Benedum Fndn Grants, 1178
DuPont Pioneer Community Giving Grants, 1532
Fannie E. Rippel Foundation Grants, 1630
Foundation for the Mid South Health and Wellness Grants, 1700
IBCAT Screening Mammography Grants, 2095
Jenkins Foundation: Improving the Health of Greater Richmond Grants, 2219
Kansas Health Foundation Recognition Grants, 2289
Marathon Petroleum Corporation Grants, 2422
Mary Black Fndn Community Health Grants, 2441
Moline Foundation Community Grants, 2590
Rajiv Gandhi Foundation Grants, 3302
USDA Delta Health Care Services Grants, 3829

Rural Planning/Policy
Blue Cross Blue Shield of Minnesota Foundation - Healthy Equity: Health Impact Assessment Program Grants, 917

Russia
Carnegie Corporation of New York Grants, 1033
U.S. Department of Education United States-Russia Program: Improving Research and Educational Activities in Higher Education, 3773

Safety
AAAS Award for Scie Freedom & Responsibility, 36
ACF Native American Social and Economic Development Strategies Grants, 231
Air Products and Chemicals Grants, 399
Alcatel-Lucent Technologies Foundation Grants, 413
Allstate Corporate Giving Grants, 446
Allstate Corp Hometown Commitment Grants, 447
Alpine Winter Foundation Grants, 453
AMD Corporate Contributions Grants, 500
American Academy of Dermatology Shade Structure Grants, 504
American Electric Power Foundation Grants, 530
American Trauma Society, Pennsylvania Division Mini-Grants, 564
Beazley Foundation Grants, 865
Bikes Belong Foundation Paul David Clark Bicycling Safety Grants, 890
BJ's Charitable Foundation Grants, 904
Blue Cross Blue Shield of Minnesota Foundation - Health Equity: Building Health Equity Together Grants, 914
BMW of North America Charitable Contributions, 926
Canadian Patient Safety Institute (CPSI) Patient Safety Studentships, 1012
Canadian Patient Safety Institute (CPSI) Research Grants, 1013
Caring Foundation Grants, 1019
Cooper Industries Foundation Grants, 1345
DaimlerChrysler Corporation Fund Grants, 1412
Decatur County Community Foundation Large Project Grants, 1444
DuPont Pioneer Community Giving Grants, 1532
EPA Children's Health Protection Grants, 1598
Essex County Community Foundation Women's Fund Grants, 1605
Farmers Insurance Corporate Giving Grants, 1634
Faye McBeath Foundation Grants, 1635
Federal Express Corporate Contributions, 1657
Ford Motor Company Fund Grants Program, 1687
Frank Stanley Beveridge Foundation Grants, 1730
GEICO Public Service Awards, 1778
General Mills Foundation Grants, 1782
Golden Heart Community Foundation Grants, 1850
Goodyear Tire Grants, 1856
Greater Sitka Legacy Fund Grants, 1885
Greater Worcester Community Foundation Jeppson Memorial Fund for Brookfield Grants, 1889
Harrison County Community Foundation Grants, 1957
Harrison County Community Foundation Signature Grants, 1958
Harvest Foundation Grants, 1971
John Deere Foundation Grants, 2234
Kenai Peninsula Foundation Grants, 2298
Ketchikan Community Foundation Grants, 2303
Kodiak Community Foundation Grants, 2314
Noble County Community Foundation Grants, 2975
Petersburg Community Foundation Grants, 3156
PPG Industries Foundation Grants, 3257
RCPSC John G. Wade Visiting Professorship in Patient Safety & Simulation-Based Med Education, 3329
Robert R. Meyer Foundation Grants, 3387
Rohm and Haas Company Grants, 3403
Salt River Health & Human Services Grants, 3459
Toyota Motor Engineering & Manufacturing North America Grants, 3742
Toyota Motor Manufacturing of Alabama Grants, 3743
Toyota Motor Manufacturing of Indiana Grants, 3744
Toyota Motor Manufacturing of Kentucky Grants, 3745

Toyota Motor N America of New York Grants, 3749
Toyota Motor Sales, USA Grants, 3750

Safety Engineering
BMW of North America Charitable Contributions, 926

Sanitary Engineering
Bill and Melinda Gates Foundation Water, Sanitation and Hygiene Grants, 894

Sarcoidosis
NIH Sarcoidosis: Research into the Cause of Multi-Organ Disease and Clinical Strategies for Therapy Grants, 2907

Sarcoma
ASCO/UICC International Cancer Technology Transfer (ICRETT) Fellowships, 677
ASCO Young Investigator Award, 683
Conquer Cancer Foundation of ASCO Career Development Award, 1328

Schizophrenia
American Psychiatric Association/AstraZeneca Young Minds in Psychiatry International Awards, 538
American Psychiatric Association/Merck Co. Early Academic Career Research Award, 541
American Psychiatric Foundation Alexander Gralnick, MD, Award for Research in Schizophrenia, 550
NARSAD Distinguished Investigator Grants, 2620
NARSAD Independent Investigator Grants, 2622
NARSAD Lieber Prize for Schizo Research, 2623
NARSAD Young Investigator Grants, 2625
NIMH Early Identification and Treatment of Mental Disorders in Children and Adolescents Grants, 2918
Peter and Elizabeth C. Tower Foundation Annual Mental Health Grants, 3145
Peter and Elizabeth C. Tower Foundation Mental Health Reference and Resource Materials Mini-Grants, 3147

Scholarship Programs, General
AAHPERD-AAHE Barb A. Cooley Scholarships, 148
AAHPERD-AAHE Del Oberteuffer Scholarships, 149
AAHPERD-AAHE Undergraduate Scholarships, 150
AAHPERD Abernathy Presidential Scholarships, 151
AAP Legislative Conference Scholarships, 185
Abell-Hanger Foundation Grants, 207
Abney Foundation Grants, 214
AcademyHealth Alice Hersh Student Scholarships, 221
AcademyHealth PHSR Interest Group Student Scholarships, 225
Achelis Foundation Grants, 234
ADA Foundation Dental Student Scholarships, 288
ADA Found Minority Dental Scholarships, 292
ADA Foundation Thomas J. Zwemer Award, 297
Adams Rotary Memorial Fund B Scholarships, 303
Agnes M. Lindsay Trust Grants, 346
Albuquerque Community Foundation Grants, 412
Alliant Energy Corporation Contributions, 445
AlohaCare Believes in Me Scholarship, 449
Alpha Omega Foundation Grants, 451
AMA-WPC Joan F. Giambalvo Memorial Scholarships, 480
AMA Fndn Arthur N. Wilson Scholarship, 482
AMA Foundation Joan F. Giambalvo Memorial Scholarships, 488
AMA Foundation Minority Scholars Awards, 491
AMA Foundation Physicians of Tomorrow Scholarships, 492
American College of Surgeons Health Policy Scholarships, 523
American College of Surgeons Health Policy Scholarships for General Surgeons, 524
American Foodservice Charitable Trust Grants, 532
Andrew Family Foundation Grants, 588
Antone & Edene Vidinha Charitable Trust Grants, 609
Aratani Foundation Grants, 649
Archer Daniels Midland Foundation Grants, 651
Arizona Community Foundation Scholarships, 656

806 / Scholarship Programs, General

Arkansas Community Foundation Grants, 660
Arlington Community Foundation Grants, 662
ASM BD Award for Research in Clin Microbio, 718
ASM bioMerieux Sonnenwirth Award for Leadership in Clinical Microbiology, 719
ASM D.C. White Research and Mentoring Award, 723
ASM Eli Lilly and Company Research Award, 726
ASM Gen-Probe Joseph Public Health Award, 729
ASM ICAAC Young Investigator Awards, 732
ASM Maurice Hilleman/Merck Award, 738
ASM Merck Irving S. Sigal Memorial Awards, 739
ASM Procter & Gamble Award in Applied and Environmental Microbiology, 742
ASM Promega Biotechnology Research Award, 743
ASM Public Communications Award, 744
ASM Raymond W. Sarber Awards, 745
ASM sanofi-aventis ICAAC Award, 749
ASM Scherago-Rubin Award, 750
ASM Siemens Healthcare Diagnostics Young Investigator Award, 751
ASM USFCC/J. Roger Porter Award, 756
ASM William A Hinton Research Training Award, 757
AUPHA Corris Boyd Scholarships, 800
AUPHA Foster G. McGaw Scholarships, 801
Bailey Foundation Grants, 818
Barberton Community Foundation Grants, 829
Batts Foundation Grants, 840
Ben B. Cheney Foundation Grants, 875
Benton County Foundation - Fitzgerald Family Scholarships, 879
Bindley Family Foundation Grants, 896
Blackford County Community Foundation Hartford City Kiwanis Scholarship in Memory of Mike McDougall, 907
Blackford County Community Foundation Noble Memorial Scholarship, 908
Blue Mountain Community Foundation Grants, 921
Bodman Foundation Grants, 929
Boettcher Foundation Grants, 931
Bright Family Foundation Grants, 951
Brook J. Lenfest Foundation Grants, 967
Burlington Industries Foundation Grants, 977
Burlington Northern Santa Fe Foundation Grants, 978
Campbell Soup Foundation Grants, 1007
Carnahan-Jackson Foundation Grants, 1032
Carpenter Foundation Grants, 1035
Carrier Corporation Contributions Grants, 1038
Cetana Educational Foundation Scholarships, 1082
CFFVR Shawano Area Community Foundation Grants, 1104
CFFVR Women's Fund for the Fox Valley Region Grants, 1107
Champlin Foundations Grants, 1110
Chiles Foundation Grants, 1147
Christy-Houston Foundation Grants, 1153
Clarence T.C. Ching Foundation Grants, 1167
Clarian Health Critical/Progressive Care Interns, 1168
Clarian Health Multi-Specialty Internships, 1169
Clarian Health OR Internships, 1170
Clarian Health Scholarships for LPNs, 1171
Clarian Health Student Nurse Scholarships, 1172
Clark-Winchcole Foundation Grants, 1173
Cleveland-Cliffs Foundation Grants, 1181
Clinton County Community Foundation Grants, 1184
Coca-Cola Foundation Grants, 1203
Collins Foundation Grants, 1210
Comerica Charitable Foundation Grants, 1224
Community Fndn for Greater Buffalo Grants, 1247
Community Fndn for Muskegon County Grants, 1254
Community Foundation of Bloomington and Monroe County Grants, 1265
Community Foundation of Greater Fort Wayne - Lilly Endowment Scholarships, 1275
Community Foundation of Greater Fort Wayne Scholarships, 1276
Community Foundation of Greater New Britain Grants, 1281
Community Foundation of Mount Vernon and Knox County Grants, 1299
Community Fndn of Randolph County Grants, 1302

Community Fndn of South Puget Sound Grants, 1307
Community Foundation of St. Joseph County Lilly Endowment Community Scholarship, 1308
Community Foundation of St. Joseph County Scholarships, 1309
Community Foundation of the Verdugos Grants, 1315
Constellation Energy Corporate Grants, 1342
Cooper Industries Foundation Grants, 1345
Covenant Foundation of Waterloo Auxiliary Scholarships, 1349
Crown Point Community Fndn Scholarships, 1362
CSRA Community Foundation Grants, 1366
Cystic Fibrosis Canada Scholarships, 1386
Cystic Fibrosis Scholarships, 1403
DAAD Research Stays for University Academics and Scientists, 1408
Danellie Foundation Grants, 1421
Daphne Seybolt Culpeper Fndn Grants, 1425
DAR Alice W. Rooke Scholarship, 1426
DAR Dr. Francis Anthony Beneventi Medical Scholarship, 1427
DAR Irene and Daisy MacGregor Memorial Scholarship, 1428
DAR Madeline Cogswell Nursing Scholarship, 1429
DAR Nursing/Physical Therapy Scholarships, 1430
Dell Scholars Program Scholarships, 1452
Donna K. Yundt Memorial Scholarship, 1488
Dr. John T. Macdonald Foundation Scholarships, 1508
Dr. Stanley Pearle Scholarship, 1512
Dubois County Community Foundation Grants, 1517
Edward W. and Stella C. Van Houten Memorial Fund Grants, 1559
El Paso Community Foundation Grants, 1579
Elsie H. Wilcox Foundation Grants, 1583
Fairfield County Community Foundation Grants, 1625
Fargo-Moorhead Area Fndn Woman's Grants, 1633
FMC Foundation Grants, 1683
Fndn for Appalachian Ohio Bachtel Scholarships, 1692
Foundation for Appalachian Ohio Susan K. Ipacs Nursing Legacy Scholarships, 1693
Foundation for Appalachian Ohio Zelma Gray Medical School Scholarship, 1694
Fourjay Foundation Grants, 1712
Frances and John L. Loeb Family Fund Grants, 1716
Franklin County Community Foundation Grants, 1724
Frank S. Flowers Foundation Grants, 1729
Fritz B. Burns Foundation Grants, 1747
Fulton County Community Foundation 4Community Higher Education Scholarship, 1762
Fulton County Community Foundation Lilly Endowment Community Scholarships, 1764
Fulton County Community Foundation Paul and Dorothy Arven Memorial Scholarship, 1765
Gebbie Foundation Grants, 1777
General Mills Foundation Grants, 1782
George and Ruth Bradford Foundation Grants, 1789
George Family Foundation Grants, 1794
George Foundation Grants, 1795
George W. Brackenridge Foundation Grants, 1802
Gertrude E. Skelly Charitable Foundation Grants, 1810
GFWC of Massachusetts Catherine E. Philbin Scholarship, 1811
Gil and Dody Weaver Foundation Grants, 1819
Gladys Brooks Foundation Grants, 1826
Greater Saint Louis Community Fndn Grants, 1884
Green Bay Packers Foundation Grants, 1890
Griffin Foundation Grants, 1903
Grover Hermann Foundation Grants, 1906
Gulf Coast Community Foundation Grants, 1915
HAF Riley Frazel Memorial Scholarship, 1931
Harold Simmons Foundation Grants, 1955
Harry Bramhall Gilbert Charitable Trust Grants, 1961
Harry Kramer Memorial Fund Grants, 1965
Henry County Community Foundation Grants, 2005
High Meadow Foundation Grants, 2027
Hilton Hotels Corporate Giving Program Grants, 2037
Hoglund Foundation Grants, 2040
Holland/Zeeland Community Fndn Grants, 2041
HomeBanc Foundation Grants, 2042
Homer Foundation Grants, 2044

Hormel Foods Charitable Trust Grants, 2050
Huie-Dellmon Trust Grants, 2075
Huisking Foundation Grants, 2076
Humana Foundation Grants, 2077
IATA Scholarships, 2094
IBEW Local Union #697 Memorial Scholarships, 2096
Idaho Power Company Corporate Contributions, 2115
IIE Adell and Hancock Scholarships, 2123
IIE Chevron International REACH Scholarships, 2128
IIE Whitaker Int'l Fellowships & Scholarships, 2144
Indiana Minority Teacher/Special Services Scholarships, 2156
Indiana Nursing Scholarships, 2157
Isabel Allende Foundation Esperanza Grants, 2165
J.E. and L.E. Mabee Foundation Grants, 2170
J.L. Bedsole Foundation Grants, 2172
J. Willard Marriott, Jr. Foundation Grants, 2178
Jacob G. Schmidlapp Trust Grants, 2183
James & Abigail Campbell Family Fndn Grants, 2185
James G.K. McClure Educational and Development Fund Grants, 2188
James R. Thorpe Foundation Grants, 2199
Janus Foundation Grants, 2211
JELD-WEN Foundation Grants, 2218
John J. Leidy Foundation Grants, 2241
John Jewett and Helen Chandler Garland Foundation Grants, 2242
John R. Oishei Foundation Grants, 2249
John W. Anderson Foundation Grants, 2262
Joseph Alexander Foundation Grants, 2265
Joseph and Luella Abell Trust Scholarships, 2266
Joseph Collins Foundation Scholarships, 2267
Josephine G. Russell Trust Grants, 2271
Kathryne Beynon Foundation Grants, 2294
Kettering Fund Grants, 2305
Kosciuszko Foundation Dr. Marie E. Zakrzewski Medical Scholarship, 2321
L. A. Hollinger Respiratory Therapy and Nursing Scholarships, 2327
LaGrange County Community Foundation Scholarships, 2330
LaGrange County Lilly Endowment Community Scholarship, 2331
Leo Goodwin Foundation Grants, 2354
Lester Ray Fleming Scholarships, 2360
Lillian S. Wells Foundation Grants, 2365
Lowell Berry Foundation Grants, 2382
Lubbock Area Foundation Grants, 2383
Lubrizol Foundation Grants, 2384
Lumpkin Family Fnd Healthy People Grants, 2391
Lynde & Harry Bradley Foundation Fellowships, 2396
Lynde and Harry Bradley Foundation Grants, 2397
Lynde and Harry Bradley Foundation Prizes: Bradley Prizes, 2398
M. Bastian Family Foundation Grants, 2402
March of Dimes Graduate Nursing Scholarships, 2424
Maxon Charitable Foundation Grants, 2458
MBL Scholarships and Awards, 2486
McInerny Foundation Grants, 2495
Mead Johnson Nutritionals Scholarships, 2505
Mead Witter Foundation Grants, 2507
Medtronic Foundation Fellowships, 2512
MGN Family Foundation Grants, 2542
Military Ex-Prisoners of War Foundation Grants, 2554
MLA Graduate Scholarship, 2564
MLA Grad Scholarship for Minority Students, 2565
NAPNAP Fndn Elaine Gelman Scholarship, 2609
NAPNAP Foundation McNeil PNP Scholarships, 2613
NAPNAP Foundation McNeil Rural and Underserved Scholarships, 2614
Nelda C. and H.J. Lutcher Stark Fndn Grants, 2672
Nesbitt Medical Student Foundation Scholarship, 2674
Newton County Community Foundation Lilly Scholarships, 2682
Newton County Community Fndn Scholarships, 2683
NMF Hugh J. Andersen Memorial Scholarship, 2958
NMF Irving Graef Memorial Scholarship, 2959
NMF Mary Ball Carrera Scholarship, 2960
NMSS Scholarships, 2968

SUBJECT INDEX

Noble County Community Foundation - Arthur A. and Hazel S. Auer Scholarship, 2969
Noble County Community Foundation - Art Hutsell Scholarship, 2970
Noble County Community Foundation - Delta Theta Tau Sorority IOTA IOTA Chapter Riecke Scholarship, 2971
Noble County Community Foundation - Democrat Central Committee Scholarships, 2972
Noble County Community Foundation - Kathleen June Earley Memorial Scholarship, 2973
Noble County Community Foundation - Lolita J. Hornett Memorial Nursing Scholarship, 2974
Noble County Community Foundation Lilly Endowment Scholarship, 2976
Noble County Community Fndn Scholarships, 2977
North Carolina Community Foundation Grants, 2990
Oleonda Jameson Trust Grants, 3071
Ordean Foundation Grants, 3080
Pajaro Valley Community Health Trust Promoting Entry & Advancement in the Health Professions Grants, 3111
Parkersburg Area Community Foundation Action Grants, 3116
Patrick and Anna M. Cudahy Fund Grants, 3120
Percy B. Ferebee Endowment Grants, 3138
Phi Upsilon Omicron Margaret Jerome Sampson Scholarships, 3197
Porter County Health Occupations Scholarship, 3247
Pott Foundation Grants, 3253
Potts Memorial Foundation Grants, 3254
Price Family Charitable Fund Grants, 3262
Pride Foundation Scholarships, 3269
Prospect Burma Scholarships, 3275
Puerto Rico Community Foundation Grants, 3278
Pulaski County Community Foundation Lilly Endowment Community Scholarships, 3280
Pulaski County Community Fndn Scholarships, 3281
Rathmann Family Foundation Grants, 3309
Regenstrief Master Of Science In Clinical Research/ CITE Program, 3353
Rehabilitation Nursing Foundation Scholarships, 3355
Rhode Island Foundation Grants, 3362
Robert B. Adams Foundation Scholarships, 3380
Roy J. Carver Charitable Trust Medical and Science Research Grants, 3416
Ruth Eleanor Bamberger and John Ernest Bamberger Memorial Foundation Grants, 3427
RWJF New Careers in Nursing Scholarships, 3439
S. Livingston Mather Charitable Trust Grants, 3448
Samuel S. Johnson Foundation Grants, 3468
Seabury Foundation Grants, 3501
Seattle Fndn Doyne M. Green Scholarships, 3505
Seattle Fndn Hawkins Memorial Scholarship, 3506
Seneca Foods Foundation Grants, 3511
Simmons Foundation Grants, 3563
SNM PDEF Mickey Williams Minority Student Scholarships, 3584
SNMTS Clinical Advancement Scholarships, 3590
SNMTS Paul Cole Scholarships, 3593
Square D Foundation Grants, 3634
Starke County Community Foundation Lilly Endowment Community Scholarships, 3647
Starke County Community Fndn Scholarships, 3648
Strowd Roses Grants, 3664
Susan G. Komen Breast Cancer Foundation College Scholarships, 3692
Thomas and Dorothy Leavey Foundation Grants, 3717
Thomas Austin Finch, Sr. Foundation Grants, 3718
Thompson Charitable Foundation Grants, 3727
Trinity Lutheran School of Nursing Alumnae Scholarships, 3759
Trull Foundation Grants, 3761
Tylenol Future Care Scholarships, 3764
U.S. Department of Education Erma Byrd Scholarships, 3768
UCT Scholarship Program, 3777
VDH Commonwealth of Virginia Nurse Educator Scholarships, 3854
Virginia Department of Health Mary Marshall Nursing Scholarships for Licensed Practical Nurses, 3874
Virginia Department of Health Mary Marshall Nursing Scholarships for Registered Nurses, 3875
Virginia Department of Health Nurse Practitioner/ Nurse Midwife Scholarships, 3876
White County Community Foundation - Adam Krintz Memorial Scholarship, 3928
White County Community Foundation - Annie Horton Scholarship, 3929
White County Community Foundation - Landis Memorial Scholarship, 3930
White County Community Foundation - Lilly Endowment Scholarships, 3931
White County Community Foundation - Tri-County Educational Scholarships, 3932
Wilhelmina W. Jackson Trust Scholarships, 3941
William B. Stokely Jr. Foundation Grants, 3947
William G. & Helen C. Hoffman Fndn Grants, 3950
William G. Gilmore Foundation Grants, 3951
Williams Companies Foundation Homegrown Giving Grants, 3957
Women of the ELCA Opportunity Scholarships for Lutheran Laywomen, 3970
Wood-Claeyssens Foundation Grants, 3971
Yampa Valley Community Foundation Cody St. John Scholarships, 3977
Yampa Valley Community Foundation Volunteer Firemen Scholarships, 3979
Young Ambassador Scholarship In Memory of Christopher Nordquist, 3981
Z. Smith Reynolds Foundation Small Grants, 3982

School Dental Programs
ADA Foundation Bud Tarrson Dental School Student Community Leadership Awards, 287
ADA Foundation Dentsply International Research Fellowships, 289
ADA Foundation Samuel Harris Children's Dental Health Grants, 294
Community Foundation of Louisville Delta Dental of Kentucky Fund Grants, 1292
Healthcare Foundation of New Jersey Grants, 1982
Pollock Foundation Grants, 3244

School Food Programs
ConAgra Foods Fndn Community Impact Grants, 1321
ConAgra Foods Nourish our Community Grants, 1322

School Health Programs
Appalachian Regional Commission Health Grants, 637
Cigna Civic Affairs Sponsorships, 1159
Colonel Stanley R. McNeil Foundation Grants, 1211
Cone Health Foundation Grants, 1323
DHHS Oral Health Promotion Research Across the Lifespan, 1474
Dr. John T. Macdonald Foundation Grants, 1507
Fairlawn Foundation Grants, 1626
Foundations of East Chicago Health Grants, 1709
Gibson County Community Foundation Women's Fund Grants, 1817
Healthcare Foundation of New Jersey Grants, 1982
Health Foundation of Greater Cincinnati Grants, 1983
McKesson Foundation Grants, 2496
MetroWest Health Foundation Grants to Schools to Conduct Mental Health Capacity Assessments, 2529
Mt. Sinai Health Care Foundation Health of the Urban Community Grants, 2600
Philip L. Graham Fund Health and Human Services Grants, 3190
Piedmont Health Foundation Grants, 3219
Seattle Foundation Health and Wellness Grants, 3507
Union Labor Health Fndn Community Grants, 3792
United Methodist Health Ministry Fund Grants, 3801
Visiting Nurse Foundation Grants, 3879

School-to-Work Transition
Crail-Johnson Foundation Grants, 1355
Denver Foundation Community Grants, 1461
Hawaiian Electric Industries Charitable Foundation Grants, 1975
KeyBank Foundation Grants, 2307
Liberty Bank Foundation Grants, 2362
Peter and Elizabeth C. Tower Foundation Annual Intellectual Disabilities Grants, 3144
Rohm and Haas Company Grants, 3403
S.H. Cowell Foundation Grants, 3447
Textron Corporate Contributions Grants, 3712
Vectren Foundation Grants, 3857
Verizon Foundation Maryland Grants, 3859
Verizon Foundation New York Grants, 3860
Verizon Foundation Northeast Region Grants, 3861
Verizon Foundation Pennsylvania Grants, 3862
Verizon Foundation Vermont Grants, 3863
Verizon Foundation Virginia Grants, 3864

Science
AAAS/Subaru SB&F Prize for Excellence in Science Books, 35
AAAS Award for Scie Freedom & Responsibility, 36
AAAS Early Career Award for Public Engagement with Science, 37
AAAS Science Prize for Online Resources in Educ, 42
AHRQ Independent Scientist Award, 365
Albert and Margaret Alkek Foundation Grants, 405
Alcoa Foundation Grants, 414
Alfred P. Sloan Foundation International Science Engagement Grants, 431
Alfred P. Sloan Foundation Research Fellowships, 432
Alfred W. Bressler Prize in Vision Science, 434
Alton Ochsner Award, 458
American Schlafhorst Foundation Grants, 559
Angels Baseball Foundation Grants, 590
ARCO Foundation Education Grants, 652
Arizona Cardinals Grants, 655
Arthur Ashley Williams Foundation Grants, 671
ArvinMeritor Grants, 675
ASM Intel Awards, 733
ASU Graduate College Science Foundation Arizona Bisgrove Postdoctoral Scholars, 792
Australian Academy of Science Grants, 807
Banfi Vintners Foundation Grants, 824
Baxter International Corporate Giving Grants, 841
Blumenthal Foundation Grants, 925
Bonfils-Stanton Foundation Grants, 932
BP Foundation Grants, 943
Brinson Foundation Grants, 953
Bristol-Myers Squibb Foundation Science Education Grants, 961
BWF Ad Hoc Grants, 982
BWF Career Awards for Medical Scientists, 984
Cabot Corporation Foundation Grants, 993
Canada-U.S. Fulbright New Century Scholars Program Grants, 1008
Canada-U.S. Fulbright Senior Specialists Grants, 1009
Canada Graduate Scholarships (CGS) and NSERC Postgraduate Scholarships (PGS), 1010
Carl Gellert and Celia Berta Gellert Foundation Grants, 1024
Carylon Foundation Grants, 1040
Chamberlain Foundation Grants, 1108
Cleveland-Cliffs Foundation Grants, 1181
Community Foundation of Greater Flint Grants, 1273
Community Foundation of Greater Fort Wayne - Lilly Endowment Scholarships, 1275
Community Foundation of St. Joseph County Lilly Endowment Community Scholarship, 1308
Crane Foundation Grants, 1357
Cyrus Eaton Foundation Grants, 1380
DAAD Research Stays for University Academics and Scientists, 1408
DaimlerChrysler Corporation Fund Grants, 1412
Dana Foundation Science and Health Grants, 1420
Deutsche Banc Alex Brown and Sons Charitable Foundation Grants, 1471
Dolan Children's Foundation Grants, 1482
Donna K. Yundt Memorial Scholarship, 1488
E.L. Wiegand Foundation Grants, 1538

808 / Science

Elkhart County Foundation Lilly Endowment Community Scholarships, 1574
EMBO Installation Grants, 1586
Flinn Foundation Scholarships, 1675
Fluor Foundation Grants, 1682
Frank Stanley Beveridge Foundation Grants, 1730
Fred and Louise Latshaw Scholarship, 1735
Frederick Gardner Cottrell Foundation Grants, 1739
Fulbright Alumni Initiatives Awards, 1749
Fulbright Distinguished Chairs Awards, 1750
Fulbright German Studies Seminar Grants, 1751
Fulbright New Century Scholars Grants, 1753
Fulbright Specialists Program Grants, 1754
Fulbright Scholars in Europe and Eurasia, 1755
Fulbright Traditional Scholar Program in Sub-Saharan Africa, 1756
Fulbright Traditional Scholar Program in the East Asia/Pacific Region, 1757
Fulbright Traditional Scholar Program in the Near East and North Africa Region, 1758
Fulbright Traditional Scholar Program in the South and Central Asia Region, 1759
Fulbright Traditional Scholar Program in the Western Hemisphere, 1760
Fulton County Community Foundation 4Community Higher Education Scholarship, 1762
Genentech Corp Charitable Contributions, 1780
George A Ohl Jr. Foundation Grants, 1791
George Foundation Grants, 1795
Gladys Brooks Foundation Grants, 1826
Grace and Franklin Bernsen Foundation Grants, 1858
Grand Rapids Community Foundation Grants, 1869
Gruber Foundation Weizmann Institute Awards, 1911
H.B. Fuller Foundation Grants, 1922
Hammond Common Council Scholarships, 1938
Harry W. Bass, Jr. Foundation Grants, 1968
Heineman Foundation for Research, Education, Charitable and Scientific Purposes, 1990
Herbert H. & Grace A. Dow Fndn Grants, 2008
Hoblitzelle Foundation Grants, 2038
IBEW Local Union #697 Memorial Scholarships, 2096
Idaho Community Foundation Eastern Region Competitive Grants, 2114
IIE 911 Armed Forces Scholarships, 2122
IIE AmCham Charitable Foundation U.S. Studies Scholarship, 2125
IIE Brazil Science Without Borders Undergraduate Scholarships, 2126
IIE Chevron International REACH Scholarships, 2128
IIE Klein Family Scholarship, 2135
IIE Leonora Lindsley Memorial Fellowships, 2136
IIE Lingnan Foundation W.T. Chan Fellowship, 2137
IIE Lotus Scholarships, 2138
IIE Mattel Global Scholarship, 2139
IIE New Leaders Group Award for Mutual Understanding, 2141
IIE Rockefeller Foundation Bellagio Center Residencies, 2142
IIE Western Union Family Scholarships, 2143
John V. and George Primich Family Scholarship, 2259
Joshua Benjamin Cohen Memorial Scholarship, 2276
Josiah W. and Bessie H. Kline Foundation Grants, 2278
Kavli Foundation Research Grants, 2296
Kenneth T. & Eileen L. Norris Fndn Grants, 2300
LaGrange County Lilly Endowment Community Scholarship, 2331
Lake County Athletic Officials Association Scholarships, 2332
Lake County Lilly Endowment Community Scholarships, 2334
Land O'Lakes Foundation Mid-Atlantic Grants, 2341
LEGENDS Scholarship, 2351
Lil and Julie Rosenberg Foundation Grants, 2364
M.J. Murdock Charitable Trust General Grants, 2405
Marion Gardner Jackson Charitable Trust Grants, 2434
McCarthy Family Foundation Grants, 2490
Medtronic Foundation Fellowships, 2512
Mercedes-Benz USA Corporate Contributions, 2517
Nancy J. Pinnick Memorial Scholarship, 2607
NIH Independent Scientist Award, 2882

NIH Mentored Clinical Scientist Devel Award, 2886
NIH Mentored Research Scientist Dev Awards, 2889
NIH Support of Competitive Research (SCORE) Pilot Project Awards, 2913
NIH Support of Competitive Research (SCORE) Research Advancement Awards, 2914
Noble County Community Foundation - Kathleen June Earley Memorial Scholarship, 2973
North Carolina GlaxoSmithKline Fndn Grants, 2991
NSERC Brockhouse Canada Prize for Interdisciplinary Research in Science and Engineering Grant, 3003
NSF Presidential Early Career Awards for Scientists and Engineers (PECASE) Grants, 3021
NSF Small Business Innovation Research (SBIR) Grants, 3022
Nuffield Foundation Africa Program Grants, 3026
Nuffield Foundation Law and Society Grants, 3028
Nuffield Foundation Small Grants, 3031
Phoenix Suns Charities Grants, 3201
Porter County Foundation Lilly Endowment Community Scholarships, 3245
Prospect Burma Scholarships, 3275
Puerto Rico Community Foundation Grants, 3278
Pulaski County Community Fndn Scholarships, 3281
Qualcomm Grants, 3287
R.C. Baker Foundation Grants, 3292
Rajiv Gandhi Foundation Grants, 3302
Rayonier Foundation Grants, 3311
Robert Leet & Clara Guthrie Patterson Grants, 3383
Robins Foundation Grants, 3390
Rockefeller Foundation Grants, 3396
Roy J. Carver Charitable Trust Medical and Science Research Grants, 3416
RSC McNeil Medal for the Public Awareness of Science, 3422
RSC Thomas W. Eadie Medal, 3423
Saint Louis Rams Fndn Community Donations, 3453
San Diego Women's Foundation Grants, 3474
Schrage Family Foundation Scholarships, 3494
Schramm Foundation Grants, 3495
Seaver Institute Grants, 3509
Sengupta Family Scholarship, 3512
SfN Science Journalism Student Awards, 3536
Sidney Stern Memorial Trust Grants, 3552
Sid W. Richardson Foundation Grants, 3553
Stella and Charles Guttman Foundation Grants, 3651
Tri-State Community Twenty-first Century Endowment Fund Grants, 3756
W. James Spicer Scholarship, 3885
W.M. Keck Foundation Science and Engineering Research Grants, 3889
W.M. Keck Fndn Undergrad Ed Grants, 3891
Webster Cornwell Memorial Scholarship, 3911
Wellcome Trust Biomedical Science Grants, 3918
White County Community Foundation - Annie Horton Scholarship, 3929
Whitley County Community Foundation - Lilly Endowment Scholarship, 3935
Whitley County Community Fndn Scholarships, 3937

Science Education

3M Company Fndn Community Giving Grants, 6
Abbott Fund Global AIDS Care Grants, 202
Abell-Hanger Foundation Grants, 207
Affymetrix Corporate Contributions Grants, 339
Alcoa Foundation Grants, 414
Alfred P. Sloan Foundation International Science Engagement Grants, 431
Alfred P. Sloan Foundation Science of Learning STEM Grants, 433
AMD Corporate Contributions Grants, 500
American Electric Power Foundation Grants, 530
Amgen Foundation Grants, 572
ANA Distinguished Neurology Teacher Award, 580
ArvinMeritor Grants, 675
Baptist Community Ministries Grants, 828
Bayer Foundation Grants, 846
Beckman Coulter Foundation Grants, 868
Bodman Foundation Grants, 929

SUBJECT INDEX

Bristol-Myers Squibb Foundation Science Education Grants, 961
Cabot Corporation Foundation Grants, 993
CDC David J. Sencer Museum Adult Group Tour, 1052
CDC David J. Sencer Museum Teacher Professional Development Workshops, 1053
CDC Disease Detective Camp, 1054
Charles Edison Fund Grants, 1116
DaimlerChrysler Corporation Fund Grants, 1412
Emily Davie and Joseph S. Kornfeld Foundation Grants, 1591
Ford Motor Company Fund Grants Program, 1687
Gates Millennium Scholars Program, 1776
Genentech Corp Charitable Contributions, 1780
Goodrich Corporation Foundation Grants, 1854
Grifols Community Outreach Grants, 1904
Heineman Foundation for Research, Education, Charitable and Scientific Purposes, 1990
HHMI-NIBIB Interfaces Initiative Grants, 2014
HHMI Biomedical Research Grants for International Scientists: Infectious Diseases & Parasitology, 2017
HHMI Biomedical Research Grants for International Scientists in Canada and Latin America, 2018
HHMI Biomedical Research Grants for International Scientists in the Baltics, Central and Eastern Europe, Russia, and Ukraine, 2019
HHMI Grants and Fellowships Programs, 2020
HHMI Med into Grad Initiative Grants, 2022
Houston Endowment Grants, 2053
James L. and Mary Jane Bowman Charitable Trust Grants, 2194
KeyBank Foundation Grants, 2307
Lockheed Martin Corp Foundation Grants, 2374
McCarthy Family Foundation Grants, 2490
MedImmune Charitable Grants, 2509
Medtronic Foundation Patient Link Grants, 2514
Mercedes-Benz USA Corporate Contributions, 2517
Pittsburgh Foundation Community Fund Grants, 3230
Qualcomm Grants, 3287
Rathmann Family Foundation Grants, 3309
RGk Foundation Grants, 3361
Rockwell Collins Charitable Corporation Grants, 3397
Rohm and Haas Company Grants, 3403
Roy & Christine Sturgis Charitable Grants, 3415
RSC McNeil Medal for the Public Awareness of Science, 3422
Schering-Plough Foundation Health Grants, 3492
Seagate Tech Corp Capacity to Care Grants, 3502
SfN Science Educator Award, 3535
TE Foundation Grants, 3705
Tellabs Foundation Grants, 3706
United Technologies Corporation Grants, 3803
W.M. Keck Fndn Undergrad Ed Grants, 3891
Washington Gas Charitable Giving, 3908

Science Planning/Policy

Alfred P. Sloan Foundation Research Fellowships, 432
Carla J. Funk Governmental Relations Award, 1020

Science and Technology

AAUW American Dissertation Fellowships, 195
Alfred P. Sloan Foundation Science of Learning STEM Grants, 433
AOCS Surfactants and Detergents Division Student Award, 622
ASM Congressional Science Fellowships, 720
ASM Intel Awards, 733
ASM Int'l Fellowship for Latin America and the Caribbean, 734
ASM Int'l Professorship for Latin America, 736
ASM Int'l Professorships for Asia and Africa, 737
ASM Public Communications Award, 744
Autodesk Community Relations Grants, 809
Beckman Coulter Foundation Grants, 868
BP Foundation Grants, 943
IIE David L. Boren Fellowships, 2129
KeyBank Foundation Grants, 2307
NSF Biological Physics (BP), 3008
NSF Emerging Frontiers in Research and Innovation (EFRI) Grants, 3014

SUBJECT INDEX

Secondary Education /809

PAHO-ASM Int'l Professorship Latin America, 3107
Rohm and Haas Company Grants, 3403
STTI National League for Nursing Grants, 3669
Texas Instruments Foundation STEM Education Grants, 3711
USAID Higher Education Partnerships for Innovation and Impact (HEPII) Grants, 3815
USDA Organic Agriculture Research Grants, 3830
W.M. Keck Foundation Science and Engineering Research Grants, 3889

Science and Technology Centers
Marion Gardner Jackson Charitable Trust Grants, 2434

Scleroderma
Scleroderma Foundation Established Investigator Grants, 3497
Scleroderma Foundation New Investigator Grants, 3498

Secondary Education
3M Company Fndn Community Giving Grants, 6
A.O. Smith Foundation Community Grants, 17
Abby's Legendary Pizza Foundation Grants, 204
Abernethy Family Foundation Grants, 210
ACL Partnerships in Employment Systems Change Grants, 246
Adolph Coors Foundation Grants, 311
Advance Auto Parts Corporate Giving Grants, 312
Affymetrix Corporate Contributions Grants, 339
A Fund for Women Grants, 342
Ahmanson Foundation Grants, 355
AIHS Heritage Youth Researcher Summer Science Program, 387
Alabama Power Foundation Grants, 400
Alcatel-Lucent Technologies Foundation Grants, 413
Alcoa Foundation Grants, 414
Alpha Natural Resources Corporate Giving, 450
AMD Corporate Contributions Grants, 500
American Electric Power Foundation Grants, 530
American Schlafhorst Foundation Grants, 559
Amerigroup Foundation Grants, 565
Amica Companies Foundation Grants, 574
Amica Insurance Company Community Grants, 575
Andrew Family Foundation Grants, 588
Ann Jackson Family Foundation Grants, 598
Archer Daniels Midland Foundation Grants, 651
ARCO Foundation Education Grants, 652
Atherton Family Foundation Grants, 794
Auburn Foundation Grants, 798
Ball Brothers Foundation General Grants, 820
Baxter International Corporate Giving Grants, 841
Belvedere Community Foundation Grants, 874
Benton County Foundation Grants, 880
Berrien Community Foundation Grants, 884
Blackford County Community Foundation Grants, 906
BMW of North America Charitable Contributions, 926
Boeing Company Contributions Grants, 930
Boettcher Foundation Grants, 931
Booth-Bricker Fund Grants, 933
Boston Foundation Grants, 936
BP Foundation Grants, 943
Bristol-Myers Squibb Foundation Health Education Grants, 959
Bristol-Myers Squibb Foundation Science Education Grants, 961
Brown Advisory Charitable Foundation Grants, 969
Burlington Industries Foundation Grants, 977
Callaway Foundation Grants, 1002
Camp-Younts Foundation Grants, 1005
Carnegie Corporation of New York Grants, 1033
Carrie E. and Lena V. Glenn Foundation Grants, 1036
Carrie Estelle Doheny Foundation Grants, 1037
Carroll County Community Foundation Grants, 1039
CCFF Life Sciences Student Awards, 1045
CDC David J. Sencer Museum Teacher Professional Development Workshops, 1053
Chapman Charitable Foundation Grants, 1112
Charles H. Pearson Foundation Grants, 1121
Charles Nelson Robinson Fund Grants, 1125
CIGNA Foundation Grants, 1160

Clara Blackford Smith and W. Aubrey Smith Charitable Foundation Grants, 1165
Clarence E. Heller Charitable Foundation Grants, 1166
Clayton Fund Grants, 1180
Clinton County Community Foundation Grants, 1184
CMS Historically Black Colleges and Universities (HBCU) Health Services Research Grants, 1187
Coca-Cola Foundation Grants, 1203
Coeta and Donald Barker Foundation Grants, 1205
Colonel Stanley R. McNeil Foundation Grants, 1211
Community Fndn of Central Illinois Grants, 1267
Community Foundation of Greater Fort Wayne Scholarships, 1276
Connelly Foundation Grants, 1326
Constantin Foundation Grants, 1341
Cowles Charitable Trust Grants, 1354
Crail-Johnson Foundation Grants, 1355
Crane Foundation Grants, 1357
Cresap Family Foundation Grants, 1359
Cullen Foundation Grants, 1372
CVS All Kids Can Grants, 1377
DaimlerChrysler Corporation Fund Grants, 1412
Dana Brown Charitable Trust Grants, 1419
David M. and Marjorie D. Rosenberg Foundation Grants, 1432
Dean Foods Community Involvement Grants, 1440
Decatur County Community Foundation Large Project Grants, 1444
Dennis & Phyllis Washington Fndn Grants, 1458
Dept of Ed College Assistance Migrant Grants, 1462
Dickson Foundation Grants, 1476
Dr. & Mrs. Paul Pierce Memorial Fndn Grants, 1505
Dr. Scholl Foundation Grants Program, 1511
Dunspaugh-Dalton Foundation Grants, 1530
DuPont Pioneer Community Giving Grants, 1532
E.J. Grassmann Trust Grants, 1537
Earl and Maxine Claussen Trust Grants, 1541
Effie and Wofford Cain Foundation Grants, 1563
Eisner Foundation Grants, 1564
Elizabeth McGraw Foundation Grants, 1568
Elizabeth Nash Fnd Summer Research Awards, 1571
El Pomar Foundation Grants, 1582
Eva L. & Joseph M. Bruening Fndn Grants, 1613
Ferree Foundation Grants, 1658
Field Foundation of Illinois Grants, 1660
Firelight Foundation Grants, 1667
FirstEnergy Foundation Community Grants, 1668
Florida Division of Cultural Affairs Arts In Education Arts Partnership Grants, 1679
Florida High School/High Tech Project Grants, 1680
Fondren Foundation Grants, 1685
Four County Community Fndn General Grants, 1710
Four County Community Foundation Healthy Senior/ Healthy Youth Fund Grants, 1711
Four J Foundation Grants, 1713
Frances and John L. Loeb Family Fund Grants, 1716
Frances L. and Edwin L. Cummings Memorial Fund Grants, 1717
Francis L. Abreu Charitable Trust Grants, 1719
Francis T. & Louise T. Nichols Fndn Grants, 1720
Frank B. Hazard General Charity Fund Grants, 1721
Frank Loomis Palmer Fund Grants, 1726
Frank Reed and Margaret Jane Peters Memorial Fund II Grants, 1728
Fred & Gretel Biel Charitable Trust Grants, 1734
Fred C. & Katherine B. Andersen Grants, 1737
Gardner Foundation Grants, 1771
Gardner Foundation Grants, 1772
Gardner Foundation Grants, 1770
GenCorp Foundation Grants, 1779
General Mills Foundation Grants, 1782
George A Ohl Jr. Foundation Grants, 1791
George F. Baker Trust Grants, 1793
George Foundation Grants, 1795
GNOF Exxon-Mobil Grants, 1836
GNOF New Orleans Works Grants, 1844
GNOF Norco Community Grants, 1845
Goodrich Corporation Foundation Grants, 1854
Grace and Franklin Bernsen Foundation Grants, 1858
Greater Milwaukee Foundation Grants, 1883

Green River Area Community Fndn Grants, 1895
Greenspun Family Foundation Grants, 1896
Greenwall Foundation Bioethics Grants, 1897
Grundy Foundation Grants, 1912
Harold Alfond Foundation Grants, 1949
Harrison County Community Foundation Grants, 1957
Harrison County Community Foundation Signature Grants, 1958
Helen Bader Foundation Grants, 1992
Helen Gertrude Sparks Charitable Trust Grants, 1993
HHMI Grants and Fellowships Programs, 2020
Horace A. Kimball and S. Ella Kimball Foundation Grants, 2046
Houston Endowment Grants, 2053
Howe Foundation of North Carolina Grants, 2055
Huie-Dellmon Trust Grants, 2075
Huisking Foundation Grants, 2076
Hutchinson Community Foundation Grants, 2085
Hutton Foundation Grants, 2087
I.A. O'Shaughnessy Foundation Grants, 2088
Idaho Power Company Corporate Contributions, 2115
IIE Klein Family Scholarship, 2135
J.C. Penney Company Grants, 2169
J.L. Bedsole Foundation Grants, 2172
J. Walton Bissell Foundation Grants, 2177
Janirve Foundation Grants, 2209
Janus Foundation Grants, 2211
Jessica Stevens Community Foundation Grants, 2224
John Clarke Trust Grants, 2232
John H. and Wilhelmina D. Harland Charitable Foundation Children and Youth Grants, 2238
John H. Wellons Foundation Grants, 2239
John P. Murphy Foundation Grants, 2248
Joseph Henry Edmondson Foundation Grants, 2270
K and F Baxter Family Foundation Grants, 2287
Katharine Matthies Foundation Grants, 2291
Katrine Menzing Deakins Trust Grants, 2295
Kenneth T. & Eileen L. Norris Fndn Grants, 2300
Kent D. Steadley and Mary L. Steadley Memorial Trust Grants, 2301
Linford and Mildred White Charitable Grants, 2368
Lockheed Martin Corp Foundation Grants, 2374
Lubbock Area Foundation Grants, 2383
Lubrizol Foundation Grants, 2384
Lumpkin Family Fnd Healthy People Grants, 2391
Lynn and Rovena Alexander Family Foundation Grants, 2399
Marion I. and Henry J. Knott Foundation Discretionary Grants, 2435
Marion I. and Henry J. Knott Foundation Standard Grants, 2436
Marjorie Moore Charitable Foundation Grants, 2438
McCarthy Family Foundation Grants, 2490
McConnell Foundation Grants, 2492
McGraw-Hill Companies Community Grants, 2494
Mead Johnson Nutritionals Evansville-Area Organizations Grants, 2503
Mead Witter Foundation Grants, 2507
Medtronic Foundation Patient Link Grants, 2514
Mericos Foundation Grants, 2518
Meta and George Rosenberg Foundation Grants, 2525
Meyer Memorial Trust Grassroots Grants, 2536
Meyer Memorial Trust Responsive Grants, 2537
MGM Resorts Foundation Community Grants, 2541
Military Ex-Prisoners of War Foundation Grants, 2554
Newton County Community Fndn Scholarships, 2683
NHLBI Research Supplements to Promote Diversity in Health-Related Research, 2727
Nicholas H. Noyes Jr. Memorial Fndn Grants, 2791
Nina Mason Pulliam Charitable Trust Grants, 2921
Noble County Community Foundation Lilly Endowment Scholarship, 2976
Norcliffe Foundation Grants, 2979
Nordson Corporation Foundation Grants, 2981
Olive Higgins Prouty Foundation Grants, 3072
Oppenstein Brothers Foundation Grants, 3078
Parkersburg Area Community Foundation Action Grants, 3116
Perpetual Trust for Charitable Giving Grants, 3141
Pew Charitable Trusts Children & Youth Grants, 3158

810 / Secondary Education

Phoenix Suns Charities Grants, 3201
Piper Jaffray Fndn Communities Giving Grants, 3227
PMI Foundation Grants, 3239
PPCF Community Grants, 3256
Price Family Charitable Fund Grants, 3262
Procter and Gamble Fund Grants, 3274
Quantum Foundation Grants, 3290
R.S. Gernon Trust Grants, 3293
Raskob Foundation for Catholic Activities Grants, 3306
Reinberger Foundation Grants, 3357
Richard & Susan Smith Family Fndn Grants, 3365
Richland County Bank Grants, 3371
Robert R. McCormick Trib Community Grants, 3385
Robert R. Meyer Foundation Grants, 3387
Robert W. Woodruff Foundation Grants, 3389
Rollins-Luetkemeyer Foundation Grants, 3404
Roy & Christine Sturgis Charitable Grants, 3415
Ruth Eleanor Bamberger and John Ernest Bamberger Memorial Foundation Grants, 3427
Saigh Foundation Grants, 3451
Sain-Orr and Royak-DeForest Steadman Foundation Grants, 3452
Salem Foundation Grants, 3457
San Juan Island Community Foundation Grants, 3480
Sartain Lanier Family Foundation Grants, 3486
Schering-Plough Foundation Health Grants, 3492
Scott County Community Foundation Grants, 3500
Seabury Foundation Grants, 3501
Seagate Tech Corp Capacity to Care Grants, 3502
Seneca Foods Foundation Grants, 3511
Seward Community Foundation Grants, 3514
Seward Community Foundation Mini-Grants, 3515
SfN Science Educator Award, 3535
Shopko Fndn Community Charitable Grants, 3547
Sid W. Richardson Foundation Grants, 3553
Sioux Falls Area Community Foundation Community Fund Grants (Unrestricted), 3568
Sioux Falls Area Community Foundation Spot Grants (Unrestricted), 3570
Sonoco Foundation Grants, 3621
Sony Corporation of America Grants, 3623
Southwest Gas Corporation Foundation Grants, 3628
Stackpole-Hall Foundation Grants, 3643
Strake Foundation Grants, 3658
Strowd Roses Grants, 3664
Sunoco Foundation Grants, 3679
Susan Vaughan Foundation Grants, 3695
Taubman Endowment for the Arts Fndn Grants, 3702
TE Foundation Grants, 3705
Thelma Braun & Bocklett Family Fndn Grants, 3713
The Ray Charles Foundation Grants, 3716
Thomas and Dorothy Leavey Foundation Grants, 3717
Trull Foundation Grants, 3761
U.S. Department of Education 21st Century Community Learning Centers, 3767
Union Bank, N.A. Corporate Sponsorships and Donations, 3787
Union Bank, N.A. Foundation Grants, 3788
Union County Community Foundation Grants, 3790
Victor E. Speas Foundation Grants, 3868
Victor Grifols i Lucas Foundation Ethics and Science Awards for Educational Institutions, 3869
Victor Grifols i Lucas Foundation Secondary School Prizes, 3872
W. C. Griffith Foundation Grants, 3882
W.C. Griffith Foundation Grants, 3881
W.H. & Mary Ellen Cobb Trust Grants, 3884
Walker Area Community Foundation Grants, 3895
Washington Gas Charitable Giving, 3908
Weaver Popcorn Foundation Grants, 3910
Wege Foundation Grants, 3912
Weingart Foundation Grants, 3913
Welborn Baptist Fndn School Health Grants, 3917
William D. Laurie, Jr. Charitable Fndn Grants, 3949
William J. Brace Charitable Trust, 3955
Windham Foundation Grants, 3961
Wood Family Charitable Trust Grants, 3972
Z. Smith Reynolds Foundation Small Grants, 3982

Security
IIE Rockefeller Foundation Bellagio Center Residencies, 2142
Otto Bremer Foundation Grants, 3099
Rockefeller Brothers Peace & Security Grants, 3394
Skoll Fndn Awards for Social Entrepreneurship, 3579

Seminars
Acumen East Africa Fellowship, 284
Alaska Airlines Foundation Grants, 403
Blue Cross Blue Shield of Minnesota Fndn Healthy Equity: Public Libraries for Health Grants, 918
GNOF Organizational Effectiveness Grants and Workshops, 1846
Greater Worcester Community Foundation Jeppson Memorial Fund for Brookfield Grants, 1889
Meyer Foundation Benevon Grants, 2532
Oppenstein Brothers Foundation Grants, 3078
Potts Memorial Foundation Grants, 3254
St. Joseph Community Health Foundation Schneider Fellowships, 3642

Senile Dementia
AHAF Alzheimer's Disease Research Grants, 349
Alzheimer's Association Conference Grants, 463
Alzheimer's Association Development of New Cognitive and Functional Instruments Grants, 464
Alzheimer's Association Everyday Technologies for Alzheimer Care Grants, 465
Alzheimer's Association Investigator-Initiated Research Grants, 466
Alzheimer's Association Mentored New Investigator Research Grants to Promote Diversity, 467
Alzheimer's Association Neuronal Hyper Excitability and Seizures in Alzheimer's Disease Grants, 468
Alzheimer's Association New Investigator Research Grants, 469
Alzheimer's Association New Investigator Research Grants to Promote Diversity, 470
Alzheimer's Association U.S.-U.K. Young Investigator Exchange Fellowships, 471
Alzheimer's Association Zenith Fellows Awards, 472
Helen Bader Foundation Grants, 1992
Henrietta Lange Burk Fund Grants, 2001
Pfizer Healthcare Charitable Contributions, 3180

Senior Citizen Programs and Services
ACL Empowering Seniors to Prevent Health Care Fraud Grants, 241
Adelaide Breed Bayrd Foundation Grants, 306
Amelia Sillman Rockwell and Carlos Perry Rockwell Charities Fund Grants, 501
Arkell Hall Foundation Grants, 661
Bertha Russ Lytel Foundation Grants, 885
California Endowment Innovative Ideas Challenge Grants, 1000
Carlsbad Charitable Foundation Grants, 1029
Christine & Katharina Pauly Trust Grants, 1150
Clark and Ruby Baker Foundation Grants, 1174
CNCS Senior Companion Program Grants, 1191
Community Fndn for the Capital Region Grants, 1259
Community Foundation of Boone County Grants, 1266
Community Foundation of the Verdugos Grants, 1315
ConocoPhillips Foundation Grants, 1327
Daniels Fund Grants-Aging, 1424
David N. Lane Trust Grants for Aged and Indigent Women, 1433
Del Mar Healthcare Fund Grants, 1453
Edward N. and Della L. Thome Memorial Foundation Grants, 1557
Eisner Foundation Grants, 1564
Florence Hunt Maxwell Foundation Grants, 1676
Frank B. Hazard General Charity Fund Grants, 1721
Frank Reed and Margaret Jane Peters Memorial Fund I Grants, 1727
Frederick McDonald Trust Grants, 1740
Fremont Area Community Foundation Elderly Needs Grants, 1745
Fulton County Community Foundation Women's Giving Circle Grants, 1766
GNOF IMPACT Grants for Health and Human Services, 1839
HAF Senior Opportunities Grants, 1932
Harden Foundation Grants, 1945
Hearst Foundations Health Grants, 1987
Helen Bader Foundation Grants, 1992
Helen K. & Arthur E. Johnson Fndn Grants, 1995
Henrietta Lange Burk Fund Grants, 2001
Illinois Tool Works Foundation Grants, 2147
Jim Moran Foundation Grants, 2228
Kansas Health Foundation Recognition Grants, 2289
M.D. Anderson Foundation Grants, 2403
Marjorie Moore Charitable Foundation Grants, 2438
McLean Foundation Grants, 2498
Mericos Foundation Grants, 2518
Morris & Gwendolyn Cafritz Fndn Grants, 2594
Nell J. Redfield Foundation Grants, 2673
Oregon Community Fndn Community Grants, 3082
PMI Foundation Grants, 3239
Portland General Electric Foundation Grants, 3250
Priddy Foundation Program Grants, 3266
Robert & Clara Milton Senior Housing Grants, 3384
Saginaw Community Foundation Senior Citizen Enrichment Fund, 3450
Scott County Community Foundation Grants, 3500
Sierra Health Foundation Responsive Grants, 3555
Sophia Romero Trust Grants, 3624
Union Bank, N.A. Foundation Grants, 3788
Vigneron Memorial Fund Grants, 3873
White County Community Foundation Grants, 3934

Sensing Devices and Transducers
NSF Biomedical Engineering and Engineering Healthcare Grants, 3009

Sensory System
SfN Swartz Prize for Theoretical and Computational Neuroscience, 3537

Serbia
IIE Eurobank EFG Scholarships, 2130

Service Delivery Programs
Able To Serve Grants, 212
Administration on Aging Senior Medicare Patrol Project Grants, 309
AMD Corporate Contributions Grants, 500
APA Meritorious Research Service Award, 631
Avista Foundation Economic and Cultural Vitality Grants, 812
Carl R. Hendrickson Family Foundation Grants, 1028
CDC Public Health Associates Hosts, 1066
Chase Paymentech Corporate Giving Grants, 1127
Clark and Ruby Baker Foundation Grants, 1174
CMS Research and Demonstration Grants, 1188
CNCS Senior Companion Program Grants, 1191
Community Foundation of Louisville Anna Marble Memorial Fund for Princeton Grants, 1290
Consumers Energy Foundation, 1343
Fallon OrNda Community Health Fund Grants, 1627
Ferree Foundation Grants, 1658
Florence Hunt Maxwell Foundation Grants, 1676
Gertrude E. Skelly Charitable Foundation Grants, 1810
Goodrich Corporation Foundation Grants, 1854
Harold and Rebecca H. Gross Foundation Grants, 1951
Hasbro Children's Fund Grants, 1973
Health Foundation of Southern Florida Responsive Grants, 1985
Henry L. Guenther Foundation Grants, 2006
Hogg Foundation for Mental Health Grants, 2039
Howard and Bush Foundation Grants, 2054
Ittleson Foundation AIDS Grants, 2166
Jenkins Foundation: Improving the Health of Greater Richmond Grants, 2219
Jerome and Mildred Paddock Foundation Grants, 2222
JM Foundation Grants, 2229
Mericos Foundation Grants, 2518
Middlesex Savings Charitable Foundation Basic Human Needs Grants, 2550

SUBJECT INDEX

Middlesex Savings Charitable Foundation Capacity Building Grants, 2551
Qualcomm Grants, 3287
Richard & Susan Smith Family Fndn Grants, 3365
Robert and Joan Dircks Foundation Grants, 3379
Robert R. McCormick Tribune Veterans Initiative Grants, 3386
Seagate Tech Corp Capacity to Care Grants, 3502
Singing for Change Foundation Grants, 3567
Sunoco Foundation Grants, 3679
Symantec Community Relations and Corporate Philanthropy Grants, 3699
Toys R Us Children's Fund Grants, 3751
Union Pacific Foundation Health and Human Services Grants, 3794
Women of the ELCA Opportunity Scholarships for Lutheran Laywomen, 3970

Service Learning
AAAAI Distinguished Service Award, 27
ADA Foundation Thomas J. Zwemer Award, 297

Sex Education
DHHS Adolescent Family Life Demonstration Projects, 1472
General Service Reproductive Justice Grants, 1785
Ittleson Foundation AIDS Grants, 2166
Ms. Fndn for Women Ending Violence Grants, 2596
Ms. Foundation for Women Health Grants, 2597
Packard Foundation Population and Reproductive Health Grants, 3106

Sexism
Ms. Fndn for Women Ending Violence Grants, 2596
Sioux Falls Area Community Foundation Spot Grants (Unrestricted), 3570

Sexual Abuse
Austin S. Nelson Foundation Grants, 805
Baxter International Foundation Grants, 843
CFFVR Schmidt Family G4 Grants, 1103
Educational Foundation of America Grants, 1553
Global Fund for Children Grants, 1832
Global Fund for Women Grants, 1833
Grand Rapids Area Community Foundation Wyoming Youth Fund Grants, 1868
Greater Tacoma Community Foundation Ryan Alan Hade Endowment Fund, 1887
OneFamily Foundation Grants, 3075
WHO Foundation Volunteer Service Grants, 3940

Sexual Behavior
amfAR Fellowships, 567
amfAR Global Initiatives Grants, 568
amfAR Mathilde Krim Fellowships in Basic Biomedical Research, 569
amfAR Public Policy Grants, 570
amfAR Research Grants, 571
Community Foundation of Eastern Connecticut Northeast Women and Girls Grants, 1270
DHHS Adolescent Family Life Demonstration Projects, 1472
Playboy Foundation Grants, 3237
Pride Foundation Grants, 3268

Sexuality
Ms. Foundation for Women Health Grants, 2597

Sexually Transmitted Diseases
APHL Emerging Infectious Diseases Fellowships, 636
ASM/CDC Fellowships in Infectious Disease and Public Health Microbiology, 716
ASPH/CDC Allan Rosenfield Global Health Fellowships, 780
CDC-Hubert Global Health Fellowship, 1047
CDC Epidemic Intell Service Training Grants, 1055
CDC Epidemiology Elective Rotation, 1056
CDC Foundation Atlanta International Health Fellowships, 1060
CDC Public Health Associates, 1065

CDC Public Health Associates Hosts, 1066
Community Foundation of Eastern Connecticut Northeast Women and Girls Grants, 1270
Cone Health Foundation Grants, 1323
DHHS Adolescent Family Life Demonstration Projects, 1472
Ittleson Foundation AIDS Grants, 2166
Kate B. Reynolds Trust Health Care Grants, 2290
Pfizer Healthcare Charitable Contributions, 3180
Sexually Transmitted Diseases Postdoctoral Fellowships, 3516
USAID Comprehensive District-Based Support for Better HIV/TB Patient Outcomes Grants, 3807

Shade
American Academy of Dermatology Shade Structure Grants, 504

Shelters
Alberto Culver Corporate Contributions Grants, 408
Alpha Natural Resources Corporate Giving, 450
Amelia Sillman Rockwell and Carlos Perry Rockwell Charities Fund Grants, 501
BancorpSouth Foundation Grants, 823
Blowitz-Ridgeway Foundation Grants, 913
Brooklyn Community Foundation Caring Neighbors Grants, 968
Campbell Soup Foundation Grants, 1007
Cemala Foundation Grants, 1077
CFFVR Capital Credit Union Charitable Giving Grants, 1095
CFFVR Jewelers Mutual Charitable Giving, 1099
CFFVR Schmidt Family G4 Grants, 1103
Charles H. Hall Foundation, 1120
Charles Nelson Robinson Fund Grants, 1125
Clara Blackford Smith and W. Aubrey Smith Charitable Foundation Grants, 1165
Columbus Foundation Traditional Grants, 1223
Constantin Foundation Grants, 1341
Dallas Women's Foundation Grants, 1417
David N. Lane Trust Grants for Aged and Indigent Women, 1433
Dennis & Phyllis Washington Fndn Grants, 1458
Doree Taylor Charitable Foundation, 1490
Doris and Victor Day Foundation Grants, 1491
Edina Realty Foundation Grants, 1550
Frank B. Hazard General Charity Fund Grants, 1721
Frank Loomis Palmer Fund Grants, 1726
Frank Reed and Margaret Jane Peters Memorial Fund II Grants, 1728
George and Ruth Bradford Foundation Grants, 1789
George E. Hatcher, Jr. and Ann Williams Hatcher Foundation Grants, 1792
GNOF Maison Hospitaliere Grants, 1843
Harold Brooks Foundation Grants, 1952
Helen Irwin Littauer Educational Trust Grants, 1994
HomeBanc Foundation Grants, 2042
Jacob G. Schmidlapp Trust Grants, 2183
Jeffris Wood Foundation Grants, 2217
John W. Speas and Effie E. Speas Grants, 2264
Laura B. Vogler Foundation Grants, 2345
Lincoln Financial Foundation Grants, 2367
Lydia deForest Charitable Trust Grants, 2393
Mardag Foundation Grants, 2428
Maurice J. Masserini Charitable Trust Grants, 2456
Mericos Foundation Grants, 2518
Mervin Bovaird Foundation Grants, 2523
MGN Family Foundation Grants, 2542
Middlesex Savings Charitable Foundation Basic Human Needs Grants, 2550
Oleonda Jameson Trust Grants, 3071
OneFamily Foundation Grants, 3075
Otto Bremer Foundation Grants, 3099
Packard Foundation Local Grants, 3105
Perpetual Trust for Charitable Giving Grants, 3141
Pinkerton Foundation Grants, 3224
Porter County Women's Grant, 3248
Posey County Community Foundation Grants, 3252
R.S. Gernon Trust Grants, 3293
S. Mark Taper Foundation Grants, 3449

Samuel S. Johnson Foundation Grants, 3468
Santa Barbara Foundation Strategy Grants - Core Support, 3481
Spencer County Community Foundation Grants, 3632
Swindells Charitable Foundation, 3697
Textron Corporate Contributions Grants, 3712
Thomas Sill Foundation Grants, 3724
Thompson Charitable Foundation Grants, 3727
TJX Foundation Grants, 3735
Vancouver Foundation Grants and Community Initiatives Program, 3849
Warrick County Community Foundation Grants, 3904

Sickle Cell Disease
NHLBI Exploratory Studies in the Neurobiology of Pain in Sickle Cell Disease, 2699

Single-Parent Families
Arkell Hall Foundation Grants, 661
Austin-Bailey Health and Wellness Fndn Grants, 803
CFFVR Robert and Patricia Endries Family Foundation Grants, 1102
Community Foundation for Greater Atlanta Clayton County Fund Grants, 1240
Community Foundation for Greater Atlanta Morgan County Fund Grants, 1244
Community Foundation for Greater Atlanta Newton County Fund Grants, 1245
Crail-Johnson Foundation Grants, 1355
Cralle Foundation Grants, 1356
Eugene M. Lang Foundation Grants, 1611
Mary Black Foundation Early Childhood Development Grants, 2442
Medtronic Foundation Community Link Human Services Grants, 2511
Seattle Foundation Benjamin N. Phillips Memorial Fund Grants, 3504
Warrick County Community Foundation Women's Fund, 3905

Skin Diseases
American Academy of Dermatology Camp Discovery Scholarships, 503
MMAAP Foundation Dermatology Fellowships, 2577
MMAAP Fndn Dermatology Project Awards, 2578
National Psoriasis Foundation Research Grants, 2636
NIAMS Pilot and Feasibility Clinical Research Grants in Arthritis, Musculoskeletal & Skin Diseases, 2774
Pfizer Inflammation Comp Research Awards, 3181
Pfizer Research Initiative Dermatology Grants (Germany), 3184
Skin Cancer Foundation Research Grants, 3578

Sleep
NIH Biobehavioral Research for Effective Sleep, 2865

Sleep Disorders
American Psychiatric Association/Merck Co. Early Academic Career Research Award, 541
CJ Foundation for SIDS Program Services Grants, 1163
CJ Foundation for SIDS Research Grants, 1164
NHLBI Ancillary Studies in Clinical Trials, 2688
NHLBI Mentored Career Award For Faculty At Minority Institutions, 2707
NHLBI Ruth L. Kirschstein National Research Service Awards for Individual Postdoctoral Fellows, 2729
NHLBI Short-Term Research Education Program to Increase Diversity in Health-Related Research, 2733
NIH Biobehavioral Research for Effective Sleep, 2865
NIH Research on Sleep and Sleep Disorders, 2898

Small Businesses
Frederick McDonald Trust Grants, 1740
Indiana 21st Century Research and Technology Fund Awards, 2154
KeyBank Foundation Grants, 2307
Maine Community Foundation Charity Grants, 2415
NCI Academic-Industrial Partnerships for Devel and Validation of In Vivo Imaging Systems and Methods for Cancer Investigations, 2644

812 / Small Businesses

NHLBI Bioengineering Approaches to Energy Balance and Obesity Grants for SBIR, 2690
NHLBI Bioengineering Approaches to Energy Balance and Obesity Grants for STTR, 2691
NIEHS Small Business Innovation Research (SBIR) Program Grants, 2860
NIH Recovery Act Limited Competition: Small Business Catalyst Awards for Accelerating Innovative Research Grants, 2895
NSF Small Business Innovation Research (SBIR) Grants, 3022
SSHRC-NSERC College and Community Innovation Enhancement Grants, 3636
Union Bank, N.A. Foundation Grants, 3788

Smoking Behavior
Alton Ochsner Award, 458
American Legacy Fnd Small Innovative Grants, 535
Bupa Foundation Multi-Country Grant, 975
CDC Fnd Tobacco Network Lab Fellowship, 1062
Christine Gregoire Youth/Young Adult Award for Use of Tobacco Industry Documents, 1151
CTCRI Idea Grants, 1369
Foundation for a Healthy Kentucky Grants, 1691
GNOF IMPACT Kahn-Oppenheim Grants, 1842
Grand Rapids Area Community Foundation Wyoming Youth Fund Grants, 1868
Health Fndn of Greater Indianapolis Grants, 1984
Hillsdale County Community Foundation Healthy Senior/Healthy Youth Fund Grants, 2033
Pfizer Global Research Awards for Nicotine Independence, 3179
Pfizer Healthcare Charitable Contributions, 3180
Pfizer Medical Education Track One Grants, 3182
Premera Blue Cross Grants, 3258
Sybil G. Jacobs Award for Outstanding Use of Tobacco Industry Documents, 3698
TRDRP California Research Awards, 3753
TRDRP New Investigator Awards, 3755

Social Change
Acumen East Africa Fellowship, 284
Changemakers Innovation Awards, 1111
Echoing Green Fellowships, 1546
Educational Foundation of America Grants, 1553
Global Fund for Women Grants, 1833
Ittleson Foundation Mental Health Grants, 2167
Jay and Rose Phillips Family Foundation Grants, 2213
John M. Lloyd Foundation Grants, 2244
Minneapolis Foundation Community Grants, 2558
Norman Foundation Grants, 2983
Playboy Foundation Grants, 3237
San Diego Foundation for Change Grants, 3470
Singing for Change Foundation Grants, 3567
Skoll Fndn Awards for Social Entrepreneurship, 3579
USAID Higher Education Partnerships for Innovation and Impact (HEPII) Grants, 3815

Social History
Radcliffe Institute Carol K. Pforzheimer Student Fellowships, 3298
Turner Foundation Grants, 3763

Social Justice
Meyer Foundation Healthy Communities Grants, 2533
Skoll Fndn Awards for Social Entrepreneurship, 3579

Social Media
OSF Health Media Initiative Grants, 3089

Social Psychology
National Center for Responsible Gaming Grants, 2629
National Center for Responsible Gaming Postdoctoral Fellowships, 2630
National Center for Resp Gaming Seed Grants, 2631
NIGMS Advancing Basic Behavioral and Social Research on Resilience: an Integrative Science Approach, 2861
NSF Social Psychology Research Grants, 3023

Social Science Education
Blowitz-Ridgeway Foundation Early Childhood Development Research Award, 912
Radcliffe Institute Residential Fellowships, 3299
SSHRC Banting Postdoctoral Fellowships, 3637

Social Sciences
AAAS Science and Technology Policy Fellowships: Global Health and Development, 41
AAUW American Dissertation Fellowships, 195
AAUW American Postdoctoral Research Leave Fellowships, 196
Alzheimer's Association Investigator-Initiated Research Grants, 466
amfAR Fellowships, 567
amfAR Global Initiatives Grants, 568
amfAR Mathilde Krim Fellowships in Basic Biomedical Research, 569
amfAR Public Policy Grants, 570
amfAR Research Grants, 571
APA Congressional Fellowships, 624
Banfi Vintners Foundation Grants, 824
Blowitz-Ridgeway Foundation Early Childhood Development Research Award, 912
Blowitz-Ridgeway Foundation Grants, 913
Brookdale Foundation Leadership in Aging Fellowships, 965
Canada-U.S. Fulbright New Century Scholars Program Grants, 1008
Canada-U.S. Fulbright Senior Specialists Grants, 1009
CDC Collegiate Leaders in Environmental Health Internships, 1048
Community Foundation of St. Joseph County Lilly Endowment Community Scholarship, 1308
Community Fndn of Wabash County Grants, 1316
DAAD Research Stays for University Academics and Scientists, 1408
Duke University Postdoctoral Research Fellowships in Aging, 1528
Fulbright Alumni Initiatives Awards, 1749
Fulbright Distinguished Chairs Awards, 1750
Fulbright New Century Scholars Grants, 1753
Fulbright Specialists Program Grants, 1754
Fulbright Scholars in Europe and Eurasia, 1755
Fulbright Traditional Scholar Program in Sub-Saharan Africa, 1756
Fulbright Traditional Scholar Program in the East Asia/Pacific Region, 1757
Fulbright Traditional Scholar Program in the Near East and North Africa Region, 1758
Fulbright Traditional Scholar Program in the South and Central Asia Region, 1759
Fulbright Traditional Scholar Program in the Western Hemisphere, 1760
Fulton County Community Foundation 4Community Higher Education Scholarship, 1762
Harry Frank Guggenheim Fnd Research Grants, 1964
Health Foundation of Greater Cincinnati Grants, 1983
IIE 911 Armed Forces Scholarships, 2122
IIE AmCham Charitable Foundation U.S. Studies Scholarship, 2125
IIE Lingnan Foundation W.T. Chan Fellowship, 2137
IIE New Leaders Group Award for Mutual Understanding, 2141
IIE Rockefeller Foundation Bellagio Center Residencies, 2142
IIE Western Union Family Scholarships, 2143
John Ben Snow Memorial Trust Grants, 2231
John Lord Knight Foundation Grants, 2243
National Center for Responsible Gaming Grants, 2629
National Center for Responsible Gaming Postdoctoral Fellowships, 2630
National Center for Resp Gaming Seed Grants, 2631
NCI Stages of Breast Development: Normal to Metastatic Disease Grants, 2655
NHLBI Translating Basic Behavioral and Social Science Discoveries into Interventions to Reduce Obesity: Centers for Behavioral Intervention Development Grants, 2739

NIA Behavioral and Social Research Grants on Disasters and Health, 2754
NIA Network Infrastructure Support for Emerging Behavioral & Social Research in Aging, 2777
NLM Grants for Scholarly Works in Biomedicine and Health, 2944
NSF Alan T. Waterman Award, 3006
Nuffield Foundation Africa Program Grants, 3026
Nuffield Foundation Law and Society Grants, 3028
Nuffield Foundation Small Grants, 3031
OBSSR Behavioral and Social Science Research on Understanding and Reducing Health Disparities Grants, 3061
Oleonda Jameson Trust Grants, 3071
PC Fred H. Bixby Fellowships, 3126
Radcliffe Institute Residential Fellowships, 3299
Savoy Foundation Studentships, 3490
SfN Neuroscience Fellowships, 3530
SSHRC-NSERC College and Community Innovation Enhancement Grants, 3636
SSHRC Banting Postdoctoral Fellowships, 3637
Thomas Sill Foundation Grants, 3724
W. James Spicer Scholarship, 3885

Social Services
1st Source Foundation Grants, 2
100 Mile Man Foundation Grants, 10
A. Gary Anderson Family Foundation Grants, 16
A.O. Smith Foundation Community Grants, 17
A/H Foundation Grants, 18
Aaron Foundation Grants, 194
Aaron Foundation Grants, 193
Abbot and Dorothy H. Stevens Foundation Grants, 197
Abbott Fund Community Grants, 201
Abel Foundation Grants, 206
Abell Fndn Health & Human Services Grants, 209
Able To Serve Grants, 212
Abney Foundation Grants, 214
Abramson Family Foundation Grants, 217
Adams and Reese Corporate Giving Grants, 299
Administaff Community Affairs Grants, 308
Advanced Micro Devices Comm Affairs Grants, 313
AEC Trust Grants, 315
AFG Industries Grants, 340
A Friends' Foundation Trust Grants, 341
A Good Neighbor Foundation Grants, 347
Ahmanson Foundation Grants, 355
Air Products and Chemicals Grants, 399
Alabama Power Foundation Grants, 400
Alberto Culver Corporate Contributions Grants, 408
Albert Pick Jr. Fund Grants, 409
Albertson's Charitable Giving Grants, 410
Albuquerque Community Foundation Grants, 412
Alcatel-Lucent Technologies Foundation Grants, 413
Alcoa Foundation Grants, 414
Alexander & Baldwin Fnd Mainland Grants, 417
Alexander and Baldwin Foundation Hawaiian and Pacific Island Grants, 418
Alexander Eastman Foundation Grants, 420
Allegheny Technologies Charitable Trust, 440
Allyn Foundation Grants, 448
ALSAM Foundation Grants, 454
Amador Community Foundation Grants, 481
American Foodservice Charitable Trust Grants, 532
Amerigroup Foundation Grants, 565
Amgen Foundation Grants, 572
AMI Semiconductors Corporate Grants, 576
Andrew Family Foundation Grants, 588
Anheuser-Busch Foundation Grants, 591
Anna Fitch Ardenghi Trust Grants, 592
Ann Arbor Area Community Foundation Grants, 594
Ann Peppers Foundation Grants, 599
Annunziata Sanguinetti Foundation Grants, 600
Anthony R. Abraham Foundation Grants, 608
Aragona Family Foundation Grants, 648
Archer Daniels Midland Foundation Grants, 651
Argyros Foundation Grants Program, 653
Arizona Cardinals Grants, 655
Arizona Public Service Corporate Giving Program Grants, 658

SUBJECT INDEX

Social Services /813

Arkansas Community Foundation Grants, 660
ARS Foundation Grants, 667
Arthur Ashley Williams Foundation Grants, 671
Arthur F. & Arnold M. Frankel Fndn Grants, 672
AT&T Foundation Civic and Community Service Program Grants, 793
Athwin Foundation Grants, 795
Atlanta Foundation Grants, 796
Auburn Foundation Grants, 798
Aurora Foundation Grants, 802
Autzen Foundation Grants, 810
Avista Foundation Economic and Cultural Vitality Grants, 812
Bailey Foundation Grants, 818
Balfe Family Foundation Grants, 819
Ball Brothers Foundation General Grants, 820
BancorpSouth Foundation Grants, 823
Barberton Community Foundation Grants, 829
Barker Foundation Grants, 830
Barker Welfare Foundation Grants, 831
Barra Foundation Community Fund Grants, 833
Barra Foundation Project Grants, 834
Barr Fund Grants, 835
Batchelor Foundation Grants, 837
Baton Rouge Area Foundation Grants, 838
Battle Creek Community Foundation Grants, 839
Batts Foundation Grants, 840
Baxter International Foundation Grants, 843
Bayer Foundation Grants, 846
Beazley Foundation Grants, 865
Bechtel Group Foundation Building Positive Community Relationships Grants, 866
Becton Dickinson and Company Grants, 869
Ben B. Cheney Foundation Grants, 875
Bender Foundation Grants, 876
Benton Community Foundation Grants, 878
Berks County Community Foundation Grants, 881
Besser Foundation Grants, 887
Blanche and Irving Laurie Foundation Grants, 910
Blue Grass Community Fndn Harrison Grants, 919
Blue Grass Community Foundation Hudson-Ellis Fund Grants, 920
Blue Mountain Community Foundation Grants, 921
Blue River Community Foundation Grants, 922
Blum-Kovler Foundation Grants, 924
Blumenthal Foundation Grants, 925
Bob & Delores Hope Foundation Grants, 927
Bodenwein Public Benevolent Foundation Grants, 928
Bodman Foundation Grants, 929
Boeing Company Contributions Grants, 930
Boettcher Foundation Grants, 931
Booth-Bricker Fund Grants, 933
Borkee-Hagley Foundation Grants, 934
Bosque Foundation Grants, 935
Boston Foundation Grants, 936
Bridgestone/Firestone Trust Fund Grants, 949
Bright Family Foundation Grants, 951
Brookdale Fnd National Group Respite Grants, 966
Brooklyn Community Foundation Caring Neighbors Grants, 968
Brown County Community Foundation Grants, 970
Burlington Industries Foundation Grants, 977
Burlington Northern Santa Fe Foundation Grants, 978
Cambridge Community Foundation Grants, 1004
Camp-Younts Foundation Grants, 1005
Campbell Soup Foundation Grants, 1007
Carl & Eloise Pohlad Family Fndn Grants, 1021
Carl B. and Florence E. King Foundation Grants, 1022
Carl M. Freeman Foundation FACES Grants, 1026
Carl R. Hendrickson Family Foundation Grants, 1028
Carolyn Foundation Grants, 1034
Carpenter Foundation Grants, 1035
Carrie E. and Lena V. Glenn Foundation Grants, 1036
Carrie Estelle Doheny Foundation Grants, 1037
Catherine Kennedy Home Foundation Grants, 1043
Cemala Foundation Grants, 1077
Centerville-Washington Foundation Grants, 1078
Central Carolina Community Foundation Community Impact Grants, 1079
Cessna Foundation Grants Program, 1081

CFFVR Frank C. Shattuck Community Grants, 1098
CFFVR Project Grants, 1101
CFFVR Robert and Patricia Endries Family Foundation Grants, 1102
CFFVR Shawano Area Community Foundation Grants, 1104
CFFVR Waupaca Area Community Foundation Grants, 1105
Chamberlain Foundation Grants, 1108
Champlin Foundations Grants, 1110
Charles A. Frueauff Foundation Grants, 1114
Charles Delmar Foundation Grants, 1115
Charles F. Bacon Trust Grants, 1117
Charlotte County (FL) Community Foundation Grants, 1126
Chase Paymentech Corporate Giving Grants, 1127
Chatlos Foundation Grants Program, 1128
CICF Indianapolis Fndn Community Grants, 1156
CICF James Proctor Grant for Aged, 1157
CIGNA Foundation Grants, 1160
Clark-Winchcole Foundation Grants, 1173
Clark County Community Foundation Grants, 1175
Claude Worthington Benedum Fndn Grants, 1178
Cleveland-Cliffs Foundation Grants, 1181
CNA Foundation Grants, 1189
Coastal Community Foundation of South Carolina Grants, 1202
Cockrell Foundation Grants, 1204
Coeta and Donald Barker Foundation Grants, 1205
Collins Foundation Grants, 1210
Columbus Foundation Competitive Grants, 1217
Columbus Foundation J. Floyd Dixon Memorial Fund Grants, 1219
Columbus Foundation Mary Eleanor Morris Fund Grants, 1220
Columbus Foundation Paul G. Duke Grants, 1221
Columbus Foundation Traditional Grants, 1223
Comerica Charitable Foundation Grants, 1224
Commonwealth Edison Grants, 1225
Communities Foundation of Texas Grants, 1236
Community Foundation Alliance City of Evansville Endowment Fund Grants, 1238
Community Fndn for Monterey County Grants, 1253
Community Fndn for Muskegon County Grants, 1254
Community Foundation for Northeast Michigan Mini-Grants, 1255
Community Fndn for San Benito County Grants, 1257
Community Foundation of Bartholomew County Heritage Fund Grants, 1262
Community Foundation of Bartholomew County James A. Henderson Award for Fundraising, 1263
Community Foundation of Bloomington and Monroe County Grants, 1265
Community Fndn of Central Illinois Grants, 1267
Community Fndn of East Central Illinois Grants, 1268
Community Foundation of Eastern Connecticut General Southeast Grants, 1269
Community Foundation of Greater Birmingham Grants, 1272
Community Foundation of Greater Flint Grants, 1273
Community Foundation of Greater Fort Wayne - Community Endowment and Clarke Endowment Grants, 1274
Community Foundation of Greater Greensboro Community Grants, 1277
Community Foundation of Greater Greensboro Women to Women Fund Grants, 1278
Community Foundation of Greater New Britain Grants, 1281
Community Fndn of Greater Tampa Grants, 1282
Community Foundation of Greenville-Greenville Women Giving Grants, 1283
Community Foundation of Greenville Community Enrichment Grants, 1284
Community Foundation of Greenville Hollingsworth Funds Program/Project Grants, 1285
Community Fndn of Howard County Grants, 1286
Community Foundation of Louisville Anna Marble Memorial Fund for Princeton Grants, 1290

Community Foundation of Mount Vernon and Knox County Grants, 1299
Community Fndn of Riverside & San Bernardino County Irene S. Rockwell Grants, 1304
Community Fndn of So Alabama Grants, 1306
Community Foundation of Tampa Bay Grants, 1312
Community Foundation of the Verdugos Grants, 1315
Community Fndn of Wabash County Grants, 1316
ConAgra Foods Fndn Community Impact Grants, 1321
ConAgra Foods Nourish our Community Grants, 1322
Connelly Foundation Grants, 1326
ConocoPhillips Foundation Grants, 1327
Constantin Foundation Grants, 1341
Constellation Energy Corporate Grants, 1342
Consumers Energy Foundation, 1343
Cooke Foundation Grants, 1344
Cooper Industries Foundation Grants, 1345
Cornerstone Foundation of Northeastern Wisconsin Grants, 1347
Cowles Charitable Trust Grants, 1354
Crail-Johnson Foundation Grants, 1355
Cralle Foundation Grants, 1356
CRH Foundation Grants, 1360
Crystelle Waggoner Charitable Trust Grants, 1364
CSX Corporate Contributions Grants, 1368
Cuesta Foundation Grants, 1371
Cultural Society of Filipino Americans Grants, 1373
Cumberland Community Foundation Grants, 1374
CUNA Mutual Group Fndn Community Grants, 1375
Curtis Foundation Grants, 1376
Cyrus Eaton Foundation Grants, 1380
D.F. Halton Foundation Grants, 1405
D. W. McMillan Foundation Grants, 1407
Dade Community Foundation Grants, 1410
DaimlerChrysler Corporation Fund Grants, 1412
Danellie Foundation Grants, 1421
Daphne Seybolt Culpeper Fndn Grants, 1425
Dayton Power and Light Foundation Grants, 1437
Daywood Foundation Grants, 1438
Dearborn Community Foundation County Progress Grants, 1442
Delaware Community Foundation-Youth Philanthropy Board for Kent County, 1448
Delaware Community Foundation Grants, 1449
DeMatteis Family Foundation Grants, 1457
Denver Foundation Community Grants, 1461
DeRoy Testamentary Foundation Grants, 1469
DHHS Adolescent Family Life Demonstration Projects, 1472
Donald and Sylvia Robinson Family Foundation Grants, 1486
Dr. Scholl Foundation Grants Program, 1511
Duchossois Family Foundation Grants, 1518
Dunspaugh-Dalton Foundation Grants, 1530
E.J. Grassmann Trust Grants, 1537
Earl and Maxine Claussen Trust Grants, 1541
Eden Hall Foundation Grants, 1549
Edina Realty Foundation Grants, 1550
Edwards Memorial Trust Grants, 1558
Edwin W. and Catherine M. Davis Fndn Grants, 1561
Eisner Foundation Grants, 1564
Elizabeth Morse Genius Charitable Trust Grants, 1569
Elkhart County Community Foundation Grants, 1573
El Paso Community Foundation Grants, 1579
Elsie H. Wilcox Foundation Grants, 1583
Emerson Charitable Trust Grants, 1589
Ensworth Charitable Foundation Grants, 1595
Erie Community Foundation Grants, 1601
Estee Lauder Grants, 1606
Ethel Sergeant Clark Smith Foundation Grants, 1608
Eugene M. Lang Foundation Grants, 1611
Evan and Susan Bayh Foundation Grants, 1614
Evjue Foundation Grants, 1617
F.M. Kirby Foundation Grants, 1624
Fargo-Moorhead Area Foundation Grants, 1632
Fargo-Moorhead Area Fndn Woman's Grants, 1633
Farmers Insurance Corporate Giving Grants, 1634
Field Foundation of Illinois Grants, 1660
FirstEnergy Foundation Community Grants, 1668
Fisher Foundation Grants, 1670

814 / Social Services

Flextronics Foundation Disaster Relief Grants, 1674
Fluor Foundation Grants, 1682
FMC Foundation Grants, 1683
Foellinger Foundation Grants, 1684
Fondren Foundation Grants, 1685
Ford Motor Company Fund Grants Program, 1687
Forrest C. Lattner Foundation Grants, 1688
Fourjay Foundation Grants, 1712
Frances L. and Edwin L. Cummings Memorial Fund Grants, 1717
Frank B. Hazard General Charity Fund Grants, 1721
Frank S. Flowers Foundation Grants, 1729
Frank Stanley Beveridge Foundation Grants, 1730
Frank W. & Carl S. Adams Memorial Grants, 1731
Fred C. & Katherine B. Andersen Grants, 1737
Fred L. Emerson Foundation Grants, 1743
Fremont Area Community Foundation Elderly Needs Grants, 1745
Fremont Area Community Fndn General Grants, 1746
Fritz B. Burns Foundation Grants, 1747
Fuji Film Grants, 1748
Fuller E. Callaway Foundation Grants, 1761
G.A. Ackermann Memorial Fund Grants, 1767
G.N. Wilcox Trust Grants, 1768
Gebbie Foundation Grants, 1777
GenCorp Foundation Grants, 1779
George and Ruth Bradford Foundation Grants, 1789
George A Ohl Jr. Foundation Grants, 1791
George Gund Foundation Grants, 1796
George W. Brackenridge Foundation Grants, 1802
Georgiana Goddard Eaton Memorial Grants, 1805
Gill Foundation - Gay and Lesbian Fund Grants, 1821
Giving Sum Annual Grant, 1825
GNOF IMPACT Grants for Health and Human Services, 1839
GNOF Maison Hospitaliere Grants, 1843
GNOF Norco Community Grants, 1845
Grace and Franklin Bernsen Foundation Grants, 1858
Graco Foundation Grants, 1860
Grand Haven Area Community Fndn Grants, 1864
Grand Rapids Area Community Foundation Wyoming Grants, 1867
Grand Rapids Community Foundation Grants, 1869
Grand Rapids Community Foundation Ionia County Grants, 1870
Grand Rapids Community Foundation Lowell Area Fund Grants, 1872
Grand Rapids Community Foundation Southeast Ottawa Grants, 1873
Grand Rapids Community Fndn Sparta Grants, 1875
Greater Cincinnati Foundation Priority and Small Projects/Capacity-Building Grants, 1880
Greater Green Bay Community Fndn Grants, 1881
Greater Kanawha Valley Foundation Grants, 1882
Greater Milwaukee Foundation Grants, 1883
Greater Saint Louis Community Fndn Grants, 1884
Greater Tacoma Community Foundation Grants, 1886
Greater Worcester Community Foundation Discretionary Grants, 1888
Green Bay Packers Foundation Grants, 1890
Green Foundation Human Services Grants, 1894
Grotto Foundation Project Grants, 1905
Guido A. & Elizabeth H. Binda Fndn Grants, 1913
Gulf Coast Community Foundation Grants, 1915
H & R Foundation Grants, 1920
H.A. & Mary K. Chapman Trust Grants, 1921
H.B. Fuller Foundation Grants, 1922
H.J. Heinz Company Foundation Grants, 1923
H. Leslie Hoffman and Elaine S. Hoffman Foundation Grants, 1924
H. Schaffer Foundation Grants, 1925
Hagedorn Fund Grants, 1934
Hall-Perrine Foundation Grants, 1935
Hallmark Corporate Foundation Grants, 1936
Hampton Roads Community Foundation Health and Human Service Grants, 1941
Hannaford Charitable Foundation Grants, 1944
Harold and Arlene Schnitzer CARE Foundation Grants, 1950

Harold R. Bechtel Charitable Remainder Uni-Trust Grants, 1953
Harold Simmons Foundation Grants, 1955
Harris and Eliza Kempner Fund Grants, 1956
Harrison County Community Foundation Grants, 1957
Harrison County Community Foundation Signature Grants, 1958
Harry A. & Margaret D. Towsley Fndn Grants, 1959
Harry Kramer Memorial Fund Grants, 1965
Harry Sudakoff Foundation Grants, 1967
Hartford Courant Foundation Grants, 1969
Hartford Foundation Regular Grants, 1970
Hawn Foundation Grants, 1979
Health Fndn of Greater Indianapolis Grants, 1984
Heckscher Foundation for Children Grants, 1988
Hedco Foundation Grants, 1989
Helen K. & Arthur E. Johnson Fndn Grants, 1995
Helen S. Boylan Foundation Grants, 1997
Henrietta Tower Wurts Memorial Fndn Grants, 2002
Henry L. Guenther Foundation Grants, 2006
Hershey Company Grants, 2013
Highmark Corporate Giving Grants, 2025
High Meadow Foundation Grants, 2027
Hillman Foundation Grants, 2032
Hoblitzelle Foundation Grants, 2038
Hoglund Foundation Grants, 2040
Holland/Zeeland Community Fndn Grants, 2041
Homer Foundation Grants, 2044
Horace A. Kimball and S. Ella Kimball Foundation Grants, 2046
Horizons Community Issues Grants, 2049
Houston Endowment Grants, 2053
Howard and Bush Foundation Grants, 2054
Howe Foundation of North Carolina Grants, 2055
HRSA Ryan White HIV AIDS Drug Assistance Grants, 2070
Huffy Foundation Grants, 2073
Huie-Dellmon Trust Grants, 2075
Humana Foundation Grants, 2077
Huntington National Bank Community Grants, 2084
Hutchinson Community Foundation Grants, 2085
Hutton Foundation Grants, 2087
Illinois Tool Works Foundation Grants, 2147
Intergrys Corporation Grants, 2160
Ireland Family Foundation Grants, 2162
J.B. Reynolds Foundation Grants, 2168
J.C. Penney Company Grants, 2169
J.E. and L.E. Mabee Foundation Grants, 2170
J.L. Bedsole Foundation Grants, 2172
J.M. Long Foundation Grants, 2173
J.W. Kieckhefer Foundation Grants, 2176
Jacob G. Schmidlapp Trust Grants, 2183
James Graham Brown Foundation Grants, 2189
James H. Cummings Foundation Grants, 2190
James J. & Angelia M. Harris Fndn Grants, 2192
James M. Collins Foundation Grants, 2195
James R. Dougherty Jr. Foundation Grants, 2198
James S. Copley Foundation Grants, 2200
Janirve Foundation Grants, 2209
Janus Foundation Grants, 2211
Jean and Louis Dreyfus Foundation Grants, 2215
Jerome and Mildred Paddock Foundation Grants, 2222
Jessie Ball Dupont Fund Grants, 2226
John Deere Foundation Grants, 2234
John Edward Fowler Memorial Fndn Grants, 2235
John G. Duncan Charitable Trust Grants, 2236
John G. Martin Foundation Grants, 2237
John H. and Wilhelmina D. Harland Charitable Foundation Children and Youth Grants, 2238
John I. Smith Charities Grants, 2240
John J. Leidy Foundation Grants, 2241
John Jewett and Helen Chandler Garland Foundation Grants, 2242
John P. McGovern Foundation Grants, 2247
John P. Murphy Foundation Grants, 2248
John R. Oishei Foundation Grants, 2249
Johns Manville Fund Grants, 2251
John W. Boynton Fund Grants, 2263
Joseph Alexander Foundation Grants, 2265
Joseph Drown Foundation Grants, 2268

SUBJECT INDEX

Josephine G. Russell Trust Grants, 2271
Josiah W. and Bessie H. Kline Foundation Grants, 2278
Kahuku Community Fund, 2286
Katharine Matthies Foundation Grants, 2291
Kathryne Beynon Foundation Grants, 2294
Kimball International-Habig Foundation Health and Human Services Grants, 2308
Knight Family Charitable and Educational Foundation Grants, 2312
L. W. Pierce Family Foundation Grants, 2328
Land O'Lakes Foundation Mid-Atlantic Grants, 2341
Lands' End Corporate Giving Program, 2344
Laura B. Vogler Foundation Grants, 2345
Laura Moore Cunningham Foundation Grants, 2346
Lawrence S. Huntington Fund Grants, 2348
Leon and Thea Koerner Foundation Grants, 2355
Lillian S. Wells Foundation Grants, 2365
Lincoln Financial Foundation Grants, 2367
Lisa and Douglas Goldman Fund Grants, 2369
Lockheed Martin Corp Foundation Grants, 2374
Lotus 88 Fnd for Women & Children Grants, 2376
Louie M. & Betty M. Phillips Fndn Grants, 2378
Lowell Berry Foundation Grants, 2382
Lubbock Area Foundation Grants, 2383
Lubrizol Foundation Grants, 2384
Lucile Horton Howe and Mitchell B. Howe Foundation Grants, 2385
Lucy Gooding Charitable Fndn Grants, 2388
Lynde & Harry Bradley Foundation Fellowships, 2396
Lynde and Harry Bradley Foundation Grants, 2397
Lynde and Harry Bradley Foundation Prizes: Bradley Prizes, 2398
M.B. and Edna Zale Foundation Grants, 2401
M. Bastian Family Foundation Grants, 2402
M.E. Raker Foundation Grants, 2404
Mabel Y. Hughes Charitable Trust Grants, 2409
Maine Community Foundation Hospice Grants, 2416
Mann T. Lowry Foundation Grants, 2419
Marcia and Otto Koehler Foundation Grants, 2427
Margaret T. Morris Foundation Grants, 2429
Mary K. Chapman Foundation Grants, 2443
Mary L. Peyton Foundation Grants, 2445
Mary S. and David C. Corbin Foundation Grants, 2447
Matilda R. Wilson Fund Grants, 2453
McCallum Family Foundation Grants, 2489
McCarthy Family Foundation Grants, 2490
McConnell Foundation Grants, 2492
McCune Foundation Human Services Grants, 2493
McInerny Foundation Grants, 2495
McKesson Foundation Grants, 2496
McLean Foundation Grants, 2498
Mead Johnson Nutritionals Evansville-Area Organizations Grants, 2503
Mead Witter Foundation Grants, 2507
Meriden Foundation Grants, 2519
Merrick Foundation Grants, 2522
Mervin Bovaird Foundation Grants, 2523
Meyer Foundation Healthy Communities Grants, 2533
Meyer Memorial Trust Special Grants, 2538
MGN Family Foundation Grants, 2542
Mimi and Peter Haas Fund Grants, 2557
Moline Foundation Community Grants, 2590
Nelda C. and H.J. Lutcher Stark Fndn Grants, 2672
Nicholas H. Noyes Jr. Memorial Fndn Grants, 2791
Norcliffe Foundation Grants, 2979
Nord Family Foundation Grants, 2980
Nordson Corporation Foundation Grants, 2981
North Carolina Community Foundation Grants, 2990
Northern New York Community Fndn Grants, 2996
Northwestern Mutual Foundation Grants, 2999
Norwin S. & Elizabeth N. Bean Fndn Grants, 3001
Nuffield Foundation Africa Program Grants, 3026
Nuffield Foundation Law and Society Grants, 3028
Nuffield Foundation Open Door Grants, 3030
Nuffield Foundation Small Grants, 3031
NYCT Girls and Young Women Grants, 3053
Ogden Codman Trust Grants, 3066
Oleonda Jameson Trust Grants, 3071
Oppenstein Brothers Foundation Grants, 3078
Ordean Foundation Grants, 3080

SUBJECT INDEX

Owen County Community Foundation Grants, 3100
Pacific Life Foundation Grants, 3103
Parkersburg Area Community Foundation Action Grants, 3116
Patrick and Anna M. Cudahy Fund Grants, 3120
Paul Balint Charitable Trust Grants, 3122
Peacock Foundation Grants, 3133
Percy B. Ferebee Endowment Grants, 3138
Perkin Fund Grants, 3139
Perpetual Trust for Charitable Giving Grants, 3141
Peter Kiewit Foundation General Grants, 3154
Peter Kiewit Foundation Small Grants, 3155
Pew Charitable Trusts Children & Youth Grants, 3158
Peyton Anderson Foundation Grants, 3160
Phelps County Community Foundation Grants, 3188
Philadelphia Foundation Organizational Effectiveness Grants, 3189
Phi Upsilon Omicron Florence Fallgatter Distinguished Service Award, 3192
Phoenix Suns Charities Grants, 3201
Pinellas County Grants, 3223
Pittsburgh Foundation Community Fund Grants, 3230
Plough Foundation Grants, 3238
PNC Foundation Community Services Grants, 3240
Polk Bros. Foundation Grants, 3243
Pollock Foundation Grants, 3244
Pott Foundation Grants, 3253
Powell Foundation Grants, 3255
Pride Foundation Grants, 3268
Principal Financial Group Foundation Grants, 3273
Procter and Gamble Fund Grants, 3274
Puerto Rico Community Foundation Grants, 3278
Quaker Chemical Foundation Grants, 3286
Questar Corporate Contributions Grants, 3291
R.C. Baker Foundation Grants, 3292
Rajiv Gandhi Foundation Grants, 3302
Raskob Foundation for Catholic Activities Grants, 3306
Rasmuson Foundation Tier One Grants, 3307
Rasmuson Foundation Tier Two Grants, 3308
Rayonier Foundation Grants, 3311
RCF General Community Grants, 3312
RCF Individual Assistance Grants, 3313
Regence Fndn Access to Health Care Grants, 3346
Regence Foundation Health Care Community Awareness and Engagement Grants, 3347
Regence Fndn Health Care Connections Grants, 3348
Regence Fndn Improving End-of-Life Grants, 3349
Regence Fndn Tools & Technology Grants, 3350
Reinberger Foundation Grants, 3357
Rhode Island Foundation Grants, 3362
Richard and Helen DeVos Foundation Grants, 3364
Richard E. Griffin Family Foundation Grants, 3368
Richard King Mellon Foundation Grants, 3369
Ripley County Community Foundation Grants, 3376
Ripley County Community Foundation Small Project Grants, 3377
Robert and Joan Dircks Foundation Grants, 3379
Robert R. McCormick Tribune Veterans Initiative Grants, 3386
Robert R. Meyer Foundation Grants, 3387
Robert W. Woodruff Foundation Grants, 3389
Robins Foundation Grants, 3390
Rochester Area Community Foundation Grants, 3391
Rochester Area Foundation Grants, 3392
Rockwell International Corporate Trust Grants Program, 3399
Roeher Institute Research Grants, 3400
Rogers Family Foundation Grants, 3401
Ronald McDonald House Charities Grants, 3405
Roy & Christine Sturgis Charitable Grants, 3415
RRF General Program Grants, 3417
Ruth Anderson Foundation Grants, 3425
Ruth and Vernon Taylor Foundation Grants, 3426
S.H. Cowell Foundation Grants, 3447
S. Livingston Mather Charitable Trust Grants, 3448
Samuel S. Johnson Foundation Grants, 3468
San Antonio Area Foundation Grants, 3469
San Diego Foundation Health & Human Services Grants, 3471
Sandy Hill Foundation Grants, 3477
Santa Fe Community Foundation Seasonal Grants-Fall Cycle, 3482
Sarkeys Foundation Grants, 3485
Schlessman Family Foundation Grants, 3493
Seagate Tech Corp Capacity to Care Grants, 3502
Seattle Fndn Doyne M. Green Scholarships, 3505
Self Foundation Grants, 3510
Sensient Technologies Foundation Grants, 3513
Shell Deer Park Grants, 3543
Sidney Stern Memorial Trust Grants, 3552
Sid W. Richardson Foundation Grants, 3553
Siebert Lutheran Foundation Grants, 3554
Sioux Falls Area Community Foundation Community Fund Grants (Unrestricted), 3568
Sioux Falls Area Community Foundation Spot Grants (Unrestricted), 3570
Sisters of Mercy of North Carolina Fndn Grants, 3575
Sonora Area Foundation Competitive Grants, 3622
Southwest Florida Community Foundation Competitive Grants, 3627
Southwest Gas Corporation Foundation Grants, 3628
Stackpole-Hall Foundation Grants, 3643
Steele-Reese Foundation Grants, 3650
Stewart Huston Charitable Trust Grants, 3655
Sunoco Foundation Grants, 3679
SunTrust Bank Trusteed Foundations Florence C. and Harry L. English Memorial Fund Grants, 3680
SunTrust Bank Trusteed Foundations Greene-Sawtell Grants, 3681
SunTrust Bank Trusteed Foundations Harriet McDaniel Marshall Tust Grants, 3682
SunTrust Bank Trusteed Foundations Nell Warren Elkin and William Simpson Elkin Grants, 3683
SunTrust Bank Trusteed Foundations Thomas Guy Woolford Charitable Trust Grants, 3684
SunTrust Bank Trusteed Foundations Walter H. and Marjory M. Rich Memorial Fund Grants, 3685
Susan Mott Webb Charitable Trust Grants, 3694
Tauck Family Foundation Grants, 3703
Taylor S. Abernathy and Patti Harding Abernathy Charitable Trust Grants, 3704
Tellabs Foundation Grants, 3706
Textron Corporate Contributions Grants, 3712
Theodore Edson Parker Foundation Grants, 3715
Thomas and Dorothy Leavey Foundation Grants, 3717
Thomas C. Ackerman Foundation Grants, 3719
Thomas J. Atkins Memorial Trust Fund Grants, 3721
Thomas J. Long Foundation Community Grants, 3722
Thomas Thompson Trust Grants, 3725
Thompson Charitable Foundation Grants, 3727
TJX Foundation Grants, 3735
Tommy Hilfiger Corporate Foundation Grants, 3737
Topfer Family Foundation Grants, 3739
Triangle Community Foundation Donor-Advised Grants, 3757
Trull Foundation Grants, 3761
Tull Charitable Foundation Grants, 3762
U.S. Cellular Corporation Grants, 3766
U.S. Department of Education 21st Century Community Learning Centers, 3767
Union Bank, N.A. Foundation Grants, 3788
Union Benevolent Association Grants, 3789
Union Pacific Foundation Health and Human Services Grants, 3794
United Technologies Corporation Grants, 3803
US Airways Community Contributions, 3825
USG Foundation Grants, 3848
Vancouver Foundation Grants and Community Initiatives Program, 3849
Vermont Community Foundation Grants, 3865
W.H. & Mary Ellen Cobb Trust Grants, 3884
Wabash Valley Community Foundation Grants, 3893
Walter L. Gross III Family Foundation Grants, 3901
Washington County Community Fndn Grants, 3906
Washington County Community Foundation Youth Grants, 3907
Wayne County Foundation Grants, 3909
Weingart Foundation Grants, 3913
Weyerhaeuser Company Foundation Grants, 3925
WHO Foundation General Grants, 3939
Willard & Pat Walker Charitable Fndn Grants, 3943
William G. & Helen C. Hoffman Fndn Grants, 3950
William G. Gilmore Foundation Grants, 3951
William J. & Tina Rosenberg Fndn Grants, 3954
Willis C. Helm Charitable Trust Grants, 3959
Wilson-Wood Foundation Grants, 3960
Winston-Salem Foundation Competitive Grants, 3965
Wood-Claeyssens Foundation Grants, 3971
Z. Smith Reynolds Foundation Small Grants, 3982

Social Services Delivery
Aaron Foundation Grants, 193
Abbott Fund Community Grants, 201
Abel Foundation Grants, 206
Able To Serve Grants, 212
ACF Native American Social and Economic Development Strategies Grants, 231
Agnes B. Hunt Trust Grants, 344
Albert and Margaret Alkek Foundation Grants, 405
Alexander & Baldwin Fnd Mainland Grants, 417
Alexander and Baldwin Foundation Hawaiian and Pacific Island Grants, 418
Allegan County Community Foundation Grants, 439
American Schlafhorst Foundation Grants, 559
Amica Insurance Company Community Grants, 575
Andersen Corporate Foundation, 586
Anheuser-Busch Foundation Grants, 591
Ann Peppers Foundation Grants, 599
AptarGroup Foundation Grants, 647
Arlington Community Foundation Grants, 662
Arthur F. & Arnold M. Frankel Fndn Grants, 672
ArvinMeritor Grants, 675
Aurora Foundation Grants, 802
Autzen Foundation Grants, 810
Bacon Family Foundation Grants, 817
Balfe Family Foundation Grants, 819
Barra Foundation Community Fund Grants, 833
Bayer Foundation Grants, 846
Bill Hannon Foundation Grants, 895
Blackford County Community Foundation Grants, 906
Blanche and Julian Robertson Family Foundation Grants, 911
Blue Grass Community Fndn Harrison Grants, 919
Boston Jewish Community Women's Fund Grants, 938
Byron W. & Alice L. Lockwood Fnd Grants, 991
Catherine Holmes Wilkins Foundation Grants, 1042
Charles H. Hall Foundation, 1120
Charles Lafitte Foundation Grants, 1123
Charlotte County (FL) Community Foundation Grants, 1126
Chase Paymentech Corporate Giving Grants, 1127
CICF Legacy Fund Grants, 1158
CIT Corporate Giving Grants, 1161
Clinton County Community Foundation Grants, 1184
Columbus Foundation Mary Eleanor Morris Fund Grants, 1220
Community Foundation for Northeast Michigan Mini-Grants, 1255
Community Foundation of Louisville Anna Marble Memorial Fund for Princeton Grants, 1290
Community Fndn of Riverside & San Bernardino County Irene S. Rockwell Grants, 1304
Community Fndn of South Puget Sound Grants, 1307
Community Foundation of St. Joseph County Special Project Challenge Grants, 1310
Crystelle Waggoner Charitable Trust Grants, 1364
CSRA Community Foundation Grants, 1366
Cudd Foundation Grants, 1370
Cultural Society of Filipino Americans Grants, 1373
Dairy Queen Corporate Contributions Grants, 1413
Deaconess Community Foundation Grants, 1439
DeMatteis Family Foundation Grants, 1457
Dennis & Phyllis Washington Fndn Grants, 1458
DeRoy Testamentary Foundation Grants, 1469
DHL Charitable Shipment Support, 1475
Donald and Sylvia Robinson Family Foundation Grants, 1486
Dora Roberts Foundation Grants, 1489
Dr. Leon Bromberg Charitable Trust Grants, 1509
Drs. Bruce and Lee Foundation Grants, 1513

816 / Social Services Delivery SUBJECT INDEX

DuPage Community Foundation Grants, 1531
Earl and Maxine Claussen Trust Grants, 1541
Eastman Chemical Company Foundation Grants, 1542
EDS Foundation Grants, 1552
Eva L. & Joseph M. Bruening Fndn Grants, 1613
Evan and Susan Bayh Foundation Grants, 1614
Ewing Halsell Foundation Grants, 1618
Foster Foundation Grants, 1689
Francis L. Abreu Charitable Trust Grants, 1719
Frank B. Hazard General Charity Fund Grants, 1721
Fraser-Parker Foundation Grants, 1732
G.A. Ackermann Memorial Fund Grants, 1767
George Foundation Grants, 1795
GNOF IMPACT Grants for Health and Human Services, 1839
GNOF Maison Hospitaliere Grants, 1843
Goodyear Tire Grants, 1856
Greater Cincinnati Foundation Priority and Small Projects/Capacity-Building Grants, 1880
Green Foundation Human Services Grants, 1894
Grifols Community Outreach Grants, 1904
Hall-Perrine Foundation Grants, 1935
Hampton Roads Community Foundation Health and Human Service Grants, 1941
Harold Alfond Foundation Grants, 1949
Harrison County Community Foundation Grants, 1957
Harrison County Community Foundation Signature Grants, 1958
Harry W. Bass, Jr. Foundation Grants, 1968
Harvest Foundation Grants, 1971
Hasbro Children's Fund Grants, 1973
Hawn Foundation Grants, 1979
Henry L. Guenther Foundation Grants, 2006
Hillsdale Fund Grants, 2035
Homer Foundation Grants, 2044
Howard and Bush Foundation Grants, 2054
Hugh J. Andersen Foundation Grants, 2074
Ida Alice Ryan Charitable Trust Grants, 2113
Ittleson Foundation Mental Health Grants, 2167
James M. Collins Foundation Grants, 2195
Jay and Rose Phillips Family Foundation Grants, 2213
Jennings County Community Foundation Women's Giving Circle Grant, 2221
Jerome and Mildred Paddock Foundation Grants, 2222
Joe W. and Dorothy Dorsett Brown Fndn Grants, 2230
Kettering Family Foundation Grants, 2304
KeyBank Foundation Grants, 2307
Laura Moore Cunningham Foundation Grants, 2346
Louie M. & Betty M. Phillips Fndn Grants, 2378
Ludwick Family Foundation Grants, 2389
M-A-C AIDS Fund Grants, 2400
Mann T. Lowry Foundation Grants, 2419
Marie C. and Joseph C. Wilson Foundation Rochester Small Grants, 2431
Marion I. and Henry J. Knott Foundation Discretionary Grants, 2435
Marion I. and Henry J. Knott Foundation Standard Grants, 2436
Mary K. Chapman Foundation Grants, 2443
Max and Victoria Dreyfus Foundation Grants, 2457
McCune Foundation Human Services Grants, 2493
MeadWestvaco Foundation Sustainable Communities Grants, 2506
Memorial Foundation Grants, 2516
Meriden Foundation Grants, 2519
Moline Foundation Community Grants, 2590
Oscar Rennebohm Foundation Grants, 3085
Parke County Community Foundation Grants, 3115
Peter Kiewit Foundation Small Grants, 3155
Pew Charitable Trusts Children & Youth Grants, 3158
PNC Foundation Community Services Grants, 3240
Polk Bros. Foundation Grants, 3243
Pollock Foundation Grants, 3244
Portland Fndn Women's Giving Circle Grant, 3249
Powell Foundation Grants, 3255
Priddy Foundation Program Grants, 3266
Prince Charitable Trusts Chicago Grants, 3270
Rasmuson Foundation Tier One Grants, 3307
Rasmuson Foundation Tier Two Grants, 3308
Rayonier Foundation Grants, 3311

Regence Fndn Access to Health Care Grants, 3346
Regence Foundation Health Care Community Awareness and Engagement Grants, 3347
Regence Fndn Health Care Connections Grants, 3348
Regence Fndn Improving End-of-Life Grants, 3349
Regence Fndn Tools & Technology Grants, 3350
Richard & Susan Smith Family Fndn Grants, 3365
Ripley County Community Foundation Small Project Grants, 3377
Robert R. Meyer Foundation Grants, 3387
Rollins-Luetkemeyer Foundation Grants, 3404
Ronald McDonald House Charities Grants, 3405
San Diego Foundation Health & Human Services Grants, 3471
Sandy Hill Foundation Grants, 3477
San Francisco Fndn Community Health Grants, 3478
San Juan Island Community Foundation Grants, 3480
Santa Fe Community Foundation Seasonal Grants-Fall Cycle, 3482
Sarkeys Foundation Grants, 3485
Sartain Lanier Family Foundation Grants, 3486
Schramm Foundation Grants, 3495
Seagate Tech Corp Capacity to Care Grants, 3502
Shell Oil Company Foundation Community Development Grants, 3544
Sierra Health Foundation Responsive Grants, 3555
Sisters of Mercy of North Carolina Fndn Grants, 3575
Sisters of St. Joseph Healthcare Fndn Grants, 3576
Sonoco Foundation Grants, 3621
Sony Corporation of America Grants, 3623
Southwest Gas Corporation Foundation Grants, 3628
Square D Foundation Grants, 3634
Stella and Charles Guttman Foundation Grants, 3651
Summit Foundation Grants, 3674
Support Our Aging Religious (SOAR) Grants, 3686
Thomas J. Atkins Memorial Trust Fund Grants, 3721
Thomas J. Long Foundation Community Grants, 3722
Tipton County Foundation Grants, 3734
Trull Foundation Grants, 3761
U.S. Cellular Corporation Grants, 3766
United Methodist Child Mental Health Grants, 3800
US Airways Community Contributions, 3825
Virginia W. Kettering Foundation Grants, 3878
Volkswagen of America Corporate Contributions, 3880
W.C. Griffith Foundation Grants, 3881
Walter L. Gross III Family Foundation Grants, 3901
Western Union Foundation Grants, 3923
White County Community Foundation - Women Giving Together Grants, 3933
Whitney Foundation Grants, 3938
WHO Foundation General Grants, 3939
William H. Hannon Foundation Grants, 3953
Woodward Fund Grants, 3973

Social Work
ACS Doctoral Grants in Oncology Social Work, 254
APSAA Fellowships, 638
Canada-U.S. Fulbright New Century Scholars Program Grants, 1008
Canada-U.S. Fulbright Senior Specialists Grants, 1009
Elizabeth Morse Genius Charitable Trust Grants, 1569
Fulbright Alumni Initiatives Awards, 1749
Fulbright Distinguished Chairs Awards, 1750
Fulbright New Century Scholars Grants, 1753
Fulbright Specialists Program Grants, 1754
Fulbright Scholars in Europe and Eurasia, 1755
Fulbright Traditional Scholar Program in Sub-Saharan Africa, 1756
Fulbright Traditional Scholar Program in the East Asia/Pacific Region, 1757
Fulbright Traditional Scholar Program in the Near East and North Africa Region, 1758
Fulbright Traditional Scholar Program in the South and Central Asia Region, 1759
Fulbright Traditional Scholar Program in the Western Hemisphere, 1760

Social Work Education
ACS Master's Training Grants in Clinical Oncology Social Work, 258

Sociology
AAO-HNSF Rande H. Lazar Health Services Research Grant, 168
American Sociological Association Minority Fellowships, 563
Canada-U.S. Fulbright New Century Scholars Program Grants, 1008
Canada-U.S. Fulbright Senior Specialists Grants, 1009
Edward Bangs Kelley and Elza Kelley Foundation Grants, 1556
Fulbright Alumni Initiatives Awards, 1749
Fulbright Distinguished Chairs Awards, 1750
Fulbright New Century Scholars Grants, 1753
Fulbright Specialists Program Grants, 1754
Fulbright Scholars in Europe and Eurasia, 1755
Fulbright Traditional Scholar Program in Sub-Saharan Africa, 1756
Fulbright Traditional Scholar Program in the East Asia/Pacific Region, 1757
Fulbright Traditional Scholar Program in the Near East and North Africa Region, 1758
Fulbright Traditional Scholar Program in the South and Central Asia Region, 1759
Fulbright Traditional Scholar Program in the Western Hemisphere, 1760
Harry Frank Guggenheim Foundation Dissertation Fellowships, 1963
NIH Biobehavioral Research for Effective Sleep, 2865
NIH Enhancing Adherence to Diabetes Self-Management Behaviors, 2875

Soil Sciences, Soil Genesis
Land O'Lakes Foundation Mid-Atlantic Grants, 2341

Solid Waste Disposal
Estee Lauder Grants, 1606
Fremont Area Community Fndn General Grants, 1746

South Africa
USAID Comprehensive District-Based Support for Better HIV/TB Patient Outcomes Grants, 3807
USAID Integration of Care and Support within the Health System to Support Better Patient Outcomes Grants, 3817

South America
Fulbright Traditional Scholar Program in the Western Hemisphere, 1760

South Asia
USAID Support for International Family Planning Organizations Grants, 3823

Space
ACSM NASA Space Physiology Research Grants, 271

Space Sciences
Canada Graduate Scholarships (CGS) and NSERC Postgraduate Scholarships (PGS), 1010
GenCorp Foundation Grants, 1779
NIH Biomedical Research on the International Space Station, 2867

Spain
Good Samaritan Inc Grants, 1855

Special Education
Air Products and Chemicals Grants, 399
Annunziata Sanguinetti Foundation Grants, 600
Becton Dickinson and Company Grants, 869
Blackford County Community Foundation Grants, 906
Carroll County Community Foundation Grants, 1039
Clinton County Community Foundation Grants, 1184
CVS Community Grants, 1379
Dept of Ed Special Education--Personnel Development to Improve Services and Results for Children with Disabilities, 1465
Dept of Ed Special Education--Studies and Evaluations, 1466

SUBJECT INDEX

Dept of Ed Special Education--Technical Assistance and Dissemination to Improve Services and Results for Children with Disabilities, 1467
Dept of Ed Special Education-National Activities-Technology and Media Services for Individuals with Disabilities, 1468
Hasbro Children's Fund Grants, 1973
Henry County Community Foundation Grants, 2005
Herbert A. & Adrian W. Woods Fndn Grants, 2007
Indiana Minority Teacher/Special Services Scholarships, 2156
Marjorie Moore Charitable Foundation Grants, 2438
Parke County Community Foundation Grants, 3115
Peter and Elizabeth C. Tower Foundation Annual Intellectual Disabilities Grants, 3144
Roy & Christine Sturgis Charitable Grants, 3415
Scott County Community Foundation Grants, 3500
Stewart Huston Charitable Trust Grants, 3655
Thomas Sill Foundation Grants, 3724
U.S. Department of Education Rehabilitation Training - Rehabilitation Continuing Education - Institute on Rehabilitation Issues, 3771
U.S. Dept of Education Special Ed--National Activities--Parent Information Centers, 3772
UCT Scholarship Program, 3777

Special Populations
Alfred E. Chase Charitable Foundation Grants, 430
AMHPS Dr. James A. Ferguson Emerging Infectious Diseases Fellowships, 573
Bill and Melinda Gates Foundation Policy and Advocacy Grants, 893
Carl R. Hendrickson Family Foundation Grants, 1028
Charles H. Hall Foundation, 1120
Florence Hunt Maxwell Foundation Grants, 1676
Guy I. Bromley Trust Grants, 1919
Helen Irwin Littauer Educational Trust Grants, 1994
Henrietta Lange Burk Fund Grants, 2001
Herbert A. & Adrian W. Woods Fndn Grants, 2007
IIE AmCham Charitable Foundation U.S. Studies Scholarship, 2125
IIE Hewlett Fnd/IIE Dissertation Fellowship, 2132
John Clarke Trust Grants, 2232
Louetta M. Cowden Foundation Grants, 2377
Morehouse PHSI Project Imhotep Internships, 2592
Pride Foundation Scholarships, 3269
Roget Begnoche Scholarship, 3402
Roy & Christine Sturgis Charitable Grants, 3415
Vigneron Memorial Fund Grants, 3873

Specialized Museums
SunTrust Bank Trusteed Foundations Thomas Guy Woolford Charitable Trust Grants, 3684

Speech
Axe-Houghton Foundation Grants, 815
Claude Pepper Foundation Grants, 1177
Reinberger Foundation Grants, 3357

Speech Pathology
ASHA Multicultural Activities Projects Grants, 697
ASHA Student Research Travel Award, 699
ASHFoundation Graduate Student International Scholarship, 702
ASHFoundation Graduate Student Scholarships, 703
ASHFoundation Graduate Student Scholarships for Minority Students, 704
ASHFoundation Graduate Student with a Disability Scholarship, 705
ASHFoundation New Century Scholars Program Doctoral Scholarships, 706
ASHFoundation Grant for New Investigators, 708
ASHFoundation Research Grant in Speech, 709
Axe-Houghton Foundation Grants, 815
Edward W. and Stella C. Van Houten Memorial Fund Grants, 1559
Grass Foundation Marine Biological Laboratory Advanced Imaging Fellowships, 1877
Mabel H. Flory Charitable Trust Grants, 2408

Speech/Communication Education
Axe-Houghton Foundation Grants, 815
GFWC of Massachusetts Communication Disorder/Speech Therapy Scholarship, 1812
Mabel H. Flory Charitable Trust Grants, 2408

Sports
Abby's Legendary Pizza Foundation Grants, 204
ACSM Chas M. Tipton Student Research Award, 260
ACSM Dr. Raymond A. Weiss Research Endowment Grant, 262
ACSM International Student Award, 268
ACSM Steven M. Horvath Travel Award, 274
AMSSM Foundation Research Grants, 577
Angels Baseball Foundation Grants, 590
Barberton Community Foundation Grants, 829
Belvedere Community Foundation Grants, 874
Bert W. Martin Foundation Grants, 886
Callaway Golf Company Foundation Grants, 1003
Carl & Eloise Pohlad Family Fndn Grants, 1021
Carl C. Icahn Foundation Grants, 1023
CFFVR Robert and Patricia Endries Family Foundation Grants, 1102
Children Affected by AIDS Foundation Domestic Grants, 1145
Chiles Foundation Grants, 1147
Cleveland Browns Foundation Grants, 1182
Collins C. Diboll Private Foundation Grants, 1209
Denver Broncos Charities Fund Grants, 1460
Elizabeth Morse Genius Charitable Trust Grants, 1569
El Pomar Foundation Anna Keesling Ackerman Fund Grants, 1581
El Pomar Foundation Grants, 1582
El Pomar Foundation Grants Jr Jr Fndn General Grants, 1710
Four County Community Fndn General Grants, 1710
Gardner W. & Joan G. Heidrick, Jr. Fndn Grants, 1773
General Mills Champs for Healthy Kids Grants, 1781
George Kress Foundation Grants, 1799
Grand Rapids Area Community Foundation Wyoming Youth Fund Grants, 1868
Greater Worcester Community Foundation Jeppson Memorial Fund for Brookfield Grants, 1889
Gregory L. Gibson Charitable Fndn Grants, 1900
Helen Bader Foundation Grants, 1992
Henrietta Tower Wurts Memorial Fndn Grants, 2002
Homer Foundation Grants, 2044
IATA Research Grants, 2093
IATA Scholarships, 2094
Kevin P. and Sydney B. Knight Family Foundation Grants, 2306
Knox County Community Foundation Grants, 2313
Lisa and Douglas Goldman Fund Grants, 2369
Marion I. and Henry J. Knott Foundation Standard Grants, 2436
McCombs Foundation Grants, 2491
NOCSAE Research Grants, 2978
Perry County Community Foundation Grants, 3142
Posey County Community Foundation Grants, 3252
Richard and Helen DeVos Foundation Grants, 3364
Robert R. Meyer Foundation Grants, 3387
Saint Louis Rams Fndn Community Donations, 3453
Seneca Foods Foundation Grants, 3511
Sidgmore Family Foundation Grants, 3551
Summit Foundation Grants, 3674
U.S. Lacrosse EAD Grants, 3775

Sports Equipment
1st Touch Foundation Grants, 3
Abby's Legendary Pizza Foundation Grants, 204
El Pomar Foundation Grants, 1582
NOCSAE Research Grants, 2978
Sioux Falls Area Community Foundation Field-of-Interest and Donor-Advised Grants, 3569

Sports Medicine
ACSM-GSSI Young Scholar Travel Award, 257
ACSM Chas M. Tipton Student Research Award, 260
ACSM Dr. Raymond A. Weiss Research Endowment Grant, 262
ACSM Foundation Clinical Sports Medicine Endowment Grants, 265

ACSM Foundation Doctoral Research Grants, 266
ACSM Foundation Research Endowment Grants, 267
ACSM International Student Award, 268
ACSM Michael L. Pollock Student Scholarship, 270
ACSM Oded Bar-Or Int'l Scholar Awards, 272
ACSM Paffenbarger-Blair Fund for Epidemiological Research on Physical Activity Grants, 273
ACSM Steven M. Horvath Travel Award, 274
AMSSM Foundation Research Grants, 577
Angels Baseball Foundation Grants, 590
IATA Research Grants, 2093
IATA Scholarships, 2094
NOCSAE Research Grants, 2978
U.S. Lacrosse EAD Grants, 3775

Sports, Amateur
Abby's Legendary Pizza Foundation Grants, 204
Belvedere Community Foundation Grants, 874
Lil and Julie Rosenberg Foundation Grants, 2364

Sri Lanka
Fulbright Traditional Scholar Program in the South and Central Asia Region, 1759

Statistics
Canada Graduate Scholarships (CGS) and NSERC Postgraduate Scholarships (PGS), 1010
CDC Public Health Informatics Fellowships, 1068
CSTE CDC/CSTE Applied Epidemiology Fellowships, 1367
NCHS National Center for Health Statistics Postdoctoral Research Appointments, 2642
NEI Clinical Vision Research Devel Award, 2664
NHGRI Mentored Research Scientist Development Award, 2686
NHLBI Mentored Quantitative Research Career Development Awards, 2711

Stem Cell Therapy
Pediatric Cancer Research Foundation Grants, 3135

Stem Cell Transplantation
NCRR Novel Approaches to Enhance Animal Stem Cell Research, 2658
NIDDK Intestinal Stem Cell Consortium Grants, 2824
Pediatric Cancer Research Foundation Grants, 3135
Pfizer Advancing Research in Transplantation Science (ARTS) Research Awards, 3163

Strategic Planning
AES Research Infrastructure Awards, 321
Blue Cross Blue Shield of Minnesota Foundation - Healthy Equity: Health Impact Assessment Program Grants, 917
GNOF Organizational Effectiveness Grants and Workshops, 1846
Guy I. Bromley Trust Grants, 1919
HAF Technical Assistance Program Grants, 1933
John W. Speas and Effie E. Speas Memorial Trust Grants, 2264
Lewis H. Humphreys Charitable Trust Grants, 2361
Louetta M. Cowden Foundation Grants, 2377
Louis and Elizabeth Nave Flarsheim Charitable Foundation Grants, 2379
Meyer Fndn Management Assistance Grants, 2534
Middlesex Savings Charitable Foundation Capacity Building Grants, 2551
Peter and Elizabeth C. Tower Foundation Phase II Technology Initiative Grants, 3149
Peter and Elizabeth C. Tower Foundation Phase I Technology Initiative Grants, 3150

Stress
Grand Rapids Area Community Foundation Wyoming Youth Fund Grants, 1868
NIA Aging, Oxidative & Cell Death Grants, 2745
NIA Mechanisms Underlying the Links between Psychosocial Stress, Aging, the Brain and the Body Grants, 2769

NIGMS Advancing Basic Behavioral and Social
    Research on Resilience: an Integrative Science
    Approach, 2861
Peter and Elizabeth C. Tower Foundation Annual
    Mental Health Grants, 3145
Premera Blue Cross Grants, 3258

Stroke
CNCS Senior Companion Program Grants, 1191
Goddess Scholars Grants, 1848
Henrietta Lange Burk Fund Grants, 2001
Jeffrey Thomas Stroke Shield Foundation Research
    Grants, 2216
NIH Research on Sleep and Sleep Disorders, 2898
NINDS Support of Scientific Meetings as Cooperative
    Agreements, 2925
NINR Clinical Interventions for Managing the
    Symptoms of Stroke Research Grants, 2929
Premera Blue Cross Grants, 3258
Stroke Association Allied Health Professional Research
    Bursaries, 3661
Stroke Association Clinical Fellowships, 3662
Stroke Association Research Project Grants, 3663

Student Support (incl. Dissertation Support)
Cystic Fibrosis Canada Studentships, 1390
SfN Donald B. Lindsley Prize in Behavioral
    Neuroscience, 3521

Substance Abuse
Alpha Natural Resources Corporate Giving, 450
Austin S. Nelson Foundation Grants, 805
Children's Trust Fund of Oregon Fndn Grants, 1139
Cone Health Foundation Grants, 1323
Cralle Foundation Grants, 1356
CSTE CDC/CSTE Applied Epidemiology
    Fellowships, 1367
DPA Promoting Policy Change Advocacy Grants, 1504
Gamble Foundation Grants, 1769
George H.C. Ensworth Memorial Fund Grants, 1797
George W. Wells Foundation Grants, 1804
Hoblitzelle Foundation Grants, 2038
Lydia deForest Charitable Trust Grants, 2393
May and Stanley Smith Charitable Trust Grants, 2459
Memorial Foundation Grants, 2516
MetroWest Health Foundation Grants to Reduce
    the Incidence of High Risk Behaviors Among
    Adolescents, 2528
Paso del Norte Health Foundation Grants, 3119
Peter and Elizabeth C. Tower Foundation Mental
    Health Reference and Resource Materials Mini-
    Grants, 3147
Peter and Elizabeth C. Tower Foundation Phase II
    Technology Initiative Grants, 3149
Peter and Elizabeth C. Tower Foundation Phase I
    Technology Initiative Grants, 3150
Peter and Elizabeth C. Tower Foundation Substance
    Abuse Grants, 3152
Peter F. McManus Charitable Trust Grants, 3153
Piedmont Natural Gas Foundation Health and Human
    Services Grants, 3221
Reinberger Foundation Grants, 3357
SfN Jacob P. Waletzky Award, 3523
Sierra Health Foundation Responsive Grants, 3555
Sioux Falls Area Community Foundation Community
    Fund Grants (Unrestricted), 3568
Sioux Falls Area Community Foundation Spot Grants
    (Unrestricted), 3570
T.L.L. Temple Foundation Grants, 3700
Union Bank, N.A. Foundation Grants, 3788

Sudden Infant Death Syndrome
CJ Foundation for SIDS Program Services Grants, 1163
CJ Foundation for SIDS Research Grants, 1164

Suicide
AACAP Klingenstein Third Generation Foundation
    Award for Research in Depression or Suicide, 62
NARSAD Distinguished Investigator Grants, 2620
NARSAD Independent Investigator Grants, 2622

NARSAD Young Investigator Grants, 2625
SAMHSA Campus Suicide Prevention Grants, 3460

Suicide Prevention
AACAP Klingenstein Third Generation Foundation
    Award for Research in Depression or Suicide, 62
ACF Native American Social and Economic
    Development Strategies Grants, 231
Florida BRAIVE Fund of Dade Community
    Foundation, 1678
Gulf Coast Foundation of Community Operating
    Grants, 1916
Gulf Coast Foundation of Community Program
    Grants, 1917
Nick Traina Foundation Grants, 2792
RCF General Community Grants, 3312

Summer Camp
CDC Disease Detective Camp, 1054
Florence Hunt Maxwell Foundation Grants, 1676
Frederick W. Marzahl Memorial Fund Grants, 1741
Gil and Dody Weaver Foundation Grants, 1819
RCF Summertime Kids Grants, 3314
Zane's Foundation Grants, 3983

Supportive Housing Programs
Adelaide Breed Bayrd Foundation Grants, 306
Bill and Melinda Gates Foundation Emergency
    Response Grants, 892
Blue Cross Blue Shield of Minnesota Foundation -
    Healthy Children: Growing Up Healthy Grants, 915
Charles H. Farnsworth Trust Grants, 1119
David N. Lane Trust Grants for Aged and Indigent
    Women, 1433
Dominion Foundation Grants, 1485
Edward N. and Della L. Thome Memorial Foundation
    Grants, 1557
GNOF Maison Hospitaliere Grants, 1843
Home Building Industry Disaster Relief Fund, 2043
Lena Benas Memorial Fund Grants, 2353
Lydia deForest Charitable Trust Grants, 2393
May and Stanley Smith Charitable Trust Grants, 2459
Mericos Foundation Grants, 2518
Meyer Memorial Trust Responsive Grants, 2537
MGM Resorts Foundation Community Grants, 2541
Piedmont Natural Gas Foundation Health and Human
    Services Grants, 3221
Priddy Foundation Program Grants, 3266
Reinberger Foundation Grants, 3357
Robert R. McCormick Trib Community Grants, 3385
Rockwell Fund, Inc. Grants, 3398
Union Bank, N.A. Corporate Sponsorships and
    Donations, 3787
Union Bank, N.A. Foundation Grants, 3788

Surgery
AAFPRS Bernstein Grant, 138
AAFPRS Fellowships, 139
AAFPRS Investigator Development Grant, 140
AAFPRS Ira J. Tresley Research Award, 141
AAFPRS John Orlando Roe Award, 142
AAFPRS Leslie Bernstein Res Research Grants, 143
AAFPRS Residency Travel Awards, 144
AAFPRS Sir Harold Delf Gillies Award, 145
AANS Neurosurgery Research and Education
    Foundation/Spine Section Young Clinician
    Investigator Award, 162
AANS Neurosurgery Research & Educ Foundation
    Young Clinician Investigator Award, 163
AAO-HNSF CORE Research Grants, 164
AAO-HNSF Health Services Research Grants, 165
AAO-HNSF Maureen Hannley Research Training
    Awards, 166
AAO-HNSF Percy Memorial Research Award, 167
AAO-HNSF Rande H. Lazar Health Services
    Research Grant, 168
AAO-HNSF Resident Research Awards, 169
AAO-HNS Medical Student Research Prize, 170
AAO-HNS Resident Research Prizes, 171

AAOA Foundation/AAO-HNSF Combined Research
    Grants, 172
AHAF Alzheimer's Disease Research Grants, 349
AHNS/AAO-HNSF Surgeon Scientist Combined
    Award, 357
AHNS/AAO-HNSF Young Investigator Combined
    Award, 358
AHNS Alando J. Ballantyne Resident Research Pilot
    Grant, 359
AHNS Pilot Grant, 360
American College of Surgeons and The Triological
    Society Clinical Scientist Development Awards, 518
American College of Surgeons Co-Sponsored K08/K23
    NIH Supplement Awards, 520
American College of Surgeons Faculty Career
    Development Award for Neurological Surgeons, 521
American College of Surgeons Faculty Research
    Fellowships, 522
American College of Surgeons Health Policy
    Scholarships, 523
American College of Surgeons Health Policy
    Scholarships for General Surgeons, 524
American College of Surgeons International Guest
    Scholarships, 525
American College of Surgeons Nizar N. Oweida, MD,
    FACS, Scholarships, 526
American College of Surgeons Resident and Associate
    Society (RAS-ACS) Leadership Scholarships, 527
American College of Surgeons Resident Research
    Scholarships, 528
American College of Surgeons Wound Care
    Management Award, 529
American Philosophical Society Daland Fellowships in
    Clinical Investigation, 537
Clowes ACS/AAST/NIGMS Jointly Sponsored
    Mentored Clinical Scientist Dev Award, 1185
Community Foundation of Louisville Diller B. and
    Katherine P. Groff Fund for Pediatric Surgery
    Grants, 1293
Denton A. Cooley Foundation Grants, 1459
Dorr Institute for Arthritis Research & Educ, 1502
E.L. Wiegand Foundation Grants, 1538
Foundation for Health Enhancement Grants, 1696
HRAMF Charles H. Hood Foundation Child Health
    Research Awards, 2057
NYAM Charles A. Elsberg Fellowship, 3039
Pfizer ASPIRE EU Dupuytren's Contracture Research
    Awards, 3166
Pfizer Healthcare Charitable Contributions, 3180
RCPSC Detweiler Traveling Fellowships, 3320
RCPSC Harry S. Morton Lectureship in Surgery, 3323
RCPSC Harry S. Morton Traveling Fellowship in
    Surgery, 3324
RCPSC Janes Visiting Professorship in Surgery, 3328
RCPSC McLaughlin-Gallie Visiting Professor, 3333
RCPSC Royal College Medal Award in Surgery, 3343
Richmond Eye and Ear Fund Grants, 3372
Triological Society Research Career Development
    Awards, 3760
Xoran Technologies Resident Research Grant, 3976

Surveillance Systems
APHL Emerging Infectious Diseases Fellowships, 636
CSTE CDC/CSTE Applied Epidemiology
    Fellowships, 1367

Sustainable Development
American Jewish World Service Grants, 533
Boeing Company Contributions Grants, 930
Charlotte County (FL) Community Foundation
    Grants, 1126
Clarence E. Heller Charitable Foundation Grants, 1166
CLIF Bar Family Foundation Grants, 1183
Conservation, Food, and Health Foundation Grants for
    Developing Countries, 1338
IIE David L. Boren Fellowships, 2129
McGraw-Hill Companies Community Grants, 2494
Michael Reese Health Trust Core Grants, 2544
PepsiCo Foundation Grants, 3137

SUBJECT INDEX

Priddy Foundation Organizational Development Grants, 3265
Rockefeller Brothers Peace & Security Grants, 3394
Shaw's Supermarkets Donations, 3542
Skoll Fndn Awards for Social Entrepreneurship, 3579
United Technologies Corporation Grants, 3803

Symposiums
FDHN Non-Career Research Awards, 1650
IBRO-PERC Support for European Workshops, Symposia and Meetings, 2098
IBRO Latin America Regional Funding for Short Courses, Workshops, and Symposia, 2107
IBRO Latin America Regional Travel Grants, 2109
MLA Lucretia W. McClure Excellence in Education Award, 2570
SfN Science Educator Award, 3535

Systems Theory
RSC Thomas W. Eadie Medal, 3423

Tajikistan
Fulbright Traditional Scholar Program in the South and Central Asia Region, 1759

Taxonomy, Animal
NSF Systematic Biology and Biodiversity Inventories Grants, 3024

Taxonomy, Plant
NSF Systematic Biology and Biodiversity Inventories Grants, 3024

Teacher Certification
DaimlerChrysler Corporation Fund Grants, 1412

Teacher Education
Abell-Hanger Foundation Grants, 207
Allen Foundation Educational Nutrition Grants, 442
AMD Corporate Contributions Grants, 500
ASHA Advancing Academic-Research Award, 695
Baxter International Corporate Giving Grants, 841
Benton County Foundation Grants, 880
Carnegie Corporation of New York Grants, 1033
Clarence E. Heller Charitable Foundation Grants, 1166
Claude Worthington Benedum Fndn Grants, 1178
Florida Division of Cultural Affairs Arts In Education Arts Partnership Grants, 1679
Fritz B. Burns Foundation Grants, 1747
GenCorp Foundation Grants, 1779
Guy I. Bromley Trust Grants, 1919
Houston Endowment Grants, 2053
Indiana Minority Teacher/Special Services Scholarships, 2156
Johnson & Johnson Corporate Contributions, 2253
John W. Speas and Effie E. Speas Memorial Trust Grants, 2264
Joseph H. & Florence A. Roblee Fndn Grants, 2269
Lewis H. Humphreys Charitable Trust Grants, 2361
Louetta M. Cowden Foundation Grants, 2377
Louis and Elizabeth Nave Flarsheim Grants, 2379
Ohio Learning Network Grants, 3070
Procter and Gamble Fund Grants, 3274
Seagate Tech Corp Capacity to Care Grants, 3502
UCT Scholarship Program, 3777
Union County Community Foundation Grants, 3790

Teacher Education, Inservice
Baxter International Corporate Giving Grants, 841
BBVA Compass Foundation Charitable Grants, 848
Benton County Foundation Grants, 880
Coca-Cola Foundation Grants, 1203
eBay Foundation Community Grants, 1543
Guy I. Bromley Trust Grants, 1919
John W. Speas and Effie E. Speas Memorial Trust Grants, 2264
Lewis H. Humphreys Charitable Trust Grants, 2361
Louetta M. Cowden Foundation Grants, 2377
Louis and Elizabeth Nave Flarsheim Charitable Foundation Grants, 2379

Teacher Training
AAAAI Outstanding Vol Clinical Faculty Award, 32
Baxter International Corporate Giving Grants, 841
BBVA Compass Foundation Charitable Grants, 848
Grifols Community Outreach Grants, 1904
Guy I. Bromley Trust Grants, 1919
John W. Speas and Effie E. Speas Memorial Trust Grants, 2264
Louetta M. Cowden Foundation Grants, 2377
Louis and Elizabeth Nave Flarsheim Charitable Foundation Grants, 2379
Sidgmore Family Foundation Grants, 3551
Texas Instruments Foundation STEM Education Grants, 3711
W.H. & Mary Ellen Cobb Trust Grants, 3884

Technological Change
James S. McDonnell Foundation Complex Systems Collaborative Activity Awards, 2202
Majors MLA Chapter Project of the Year, 2418
USAID Higher Education Partnerships for Innovation and Impact (HEPII) Grants, 3815

Technology
AAAS Award for Scie Freedom & Responsibility, 36
Alfred P. Sloan Foundation International Science Engagement Grants, 431
Alfred P. Sloan Foundation Science of Learning STEM Grants, 433
American Chemical Society Award for Creative Advances in Environmental Science and Technology, 513
American Chemical Society Award in Separations Science and Technology, 514
ArvinMeritor Grants, 675
AT&T Foundation Civic and Community Service Program Grants, 793
Barberton Community Foundation Grants, 829
BP Foundation Grants, 943
Cabot Corporation Foundation Grants, 993
Cleveland-Cliffs Foundation Grants, 1181
Community Foundation of Greater Fort Wayne - Lilly Endowment Scholarships, 1275
Crail-Johnson Foundation Grants, 1355
Dolan Children's Foundation Grants, 1482
EDS Foundation Grants, 1552
Elkhart County Foundation Lilly Endowment Community Scholarships, 1574
Ford Motor Company Fund Grants Program, 1687
Fred and Louise Latshaw Scholarship, 1735
Fulton County Community Foundation 4Community Higher Education Scholarship, 1762
GNOF New Orleans Works Grants, 1844
Gruber Foundation Weizmann Institute Awards, 1911
H.B. Fuller Foundation Grants, 1922
IDPH Hosptial Capital Investment Grants, 2119
IIE 911 Armed Forces Scholarships, 2122
IIE Brazil Science Without Borders Undergraduate Scholarships, 2126
IIE Chevron International REACH Scholarships, 2128
Indiana 21st Century Research and Technology Fund Awards, 2154
Jessie Ball Dupont Fund Grants, 2226
John R. Oishei Foundation Grants, 2249
LaGrange County Lilly Endowment Community Scholarship, 2331
Lake County Lilly Endowment Community Scholarships, 2334
LEGENDS Scholarship, 2351
Marion I. and Henry J. Knott Foundation Discretionary Grants, 2435
Marion I. and Henry J. Knott Foundation Standard Grants, 2436
Mericos Foundation Grants, 2518
NHGRI Mentored Research Scientist Development Award, 2686
NIEHS SBIR E-Learning for HAZMAT and Emergency Response Grants, 2859
NIH Enhancing Adherence to Diabetes Self-Management Behaviors, 2875

Technology Education /819

NIH Recovery Act Limited Competition: Academic Research Enhancement Awards, 2892
NIH Recovery Act Limited Competition: Biomedical Research, Development, and Growth to Spur the Acceleration of New Technologies Grants, 2893
NIH Recovery Act Limited Competition: Small Business Catalyst Awards for Accelerating Innovative Research Grants, 2895
NIH Recovery Act Limited Competition: Supporting New Faculty Recruitment to Enhance Research Resources through Biomedical Research Core Centers Grants, 2896
NLM Informatics Conference Grants, 2945
Noble County Community Foundation - Kathleen June Earley Memorial Scholarship, 2973
North Carolina Biotechnology Center Technology Enhancement Grants, 2989
NSF Small Business Innovation Research (SBIR) Grants, 3022
Ohio Learning Network Grants, 3070
Pittsburgh Foundation Community Fund Grants, 3230
Porter County Foundation Lilly Endowment Community Scholarships, 3245
Puerto Rico Community Foundation Grants, 3278
Pulaski County Community Fndn Scholarships, 3281
Rajiv Gandhi Foundation Grants, 3302
Rayonier Foundation Grants, 3311
Rockefeller Foundation Grants, 3396
Rockwell International Corporate Trust Grants Program, 3399
San Diego Women's Foundation Grants, 3474
Schrage Family Foundation Scholarships, 3494
Schramm Foundation Grants, 3495
Seaver Institute Grants, 3509
Tellabs Foundation Grants, 3706
Texas Instruments Foundation STEM Education Grants, 3711
Tommy Hilfiger Corporate Foundation Grants, 3737
USAID Higher Education Partnerships for Innovation and Impact (HEPII) Grants, 3815
Verizon Foundation Maryland Grants, 3859
Verizon Foundation New York Grants, 3860
Verizon Foundation Northeast Region Grants, 3861
Verizon Foundation Pennsylvania Grants, 3862
Verizon Foundation Vermont Grants, 3863
Verizon Foundation Virginia Grants, 3864
W. James Spicer Scholarship, 3885
Waitt Family Foundation Grants, 3894
Webster Cornwell Memorial Scholarship, 3911
Western New York Foundation Grants, 3922
Whitley County Community Foundation - Lilly Endowment Scholarship, 3935
Whitley County Community Fndn Scholarships, 3937

Technology Education
AIHS ForeFront MBA Studentship Award, 381
American Electric Power Foundation Grants, 530
ArvinMeritor Grants, 675
Boeing Company Contributions Grants, 930
Charles Edison Fund Grants, 1116
Charles Lafitte Foundation Grants, 1123
Marie C. and Joseph C. Wilson Foundation Rochester Small Grants, 2431
Microsoft Research Cell Phone as a Platform for Healthcare Grants, 2548
Ohio Learning Network Grants, 3070
Procter and Gamble Fund Grants, 3274
Qualcomm Grants, 3287
RGk Foundation Grants, 3361
Roy & Christine Sturgis Charitable Grants, 3415
Sony Corporation of America Grants, 3623
United Technologies Corporation Grants, 3803
Verizon Foundation Maryland Grants, 3859
Verizon Foundation New York Grants, 3860
Verizon Foundation Northeast Region Grants, 3861
Verizon Foundation Pennsylvania Grants, 3862
Verizon Foundation Vermont Grants, 3863
Verizon Foundation Virginia Grants, 3864
Washington Gas Charitable Giving, 3908

820 / Technology Planning/Policy

Technology Planning/Policy
ALFJ Astraea U.S. and Int'l Movement Fund, 427
Carla J. Funk Governmental Relations Award, 1020
Gill Foundation - Gay and Lesbian Fund Grants, 1821
Peter and Elizabeth C. Tower Foundation Phase II Technology Initiative Grants, 3149
Peter and Elizabeth C. Tower Foundation Phase I Technology Initiative Grants, 3150

Technology Transfer
Boeing Company Contributions Grants, 930
Indiana 21st Century Research and Technology Fund Awards, 2154
NCI Academic-Industrial Partnerships for Devel and Validation of In Vivo Imaging Systems and Methods for Cancer Investigations, 2644
NHLBI Bioengineering Approaches to Energy Balance and Obesity Grants for SBIR, 2690
NHLBI Bioengineering Approaches to Energy Balance and Obesity Grants for STTR, 2691
SSHRC-NSERC College and Community Innovation Enhancement Grants, 3636
U.S. Department of Education Rehabilitation Engineering Research Centers Grants, 3769

Technology, Hardware and Software
Alice Tweed Tuohy Foundation Grants Program, 437
Central Carolina Community Foundation Community Impact Grants, 1079
Chatlos Foundation Grants Program, 1128
DaimlerChrysler Corporation Fund Grants, 1412
E.L. Wiegand Foundation Grants, 1538
Grand Haven Area Community Fndn Grants, 1864
Peter and Elizabeth C. Tower Foundation Phase II Technology Initiative Grants, 3149
Peter and Elizabeth C. Tower Foundation Phase I Technology Initiative Grants, 3150
Puerto Rico Community Foundation Grants, 3278
Weingart Foundation Grants, 3913
Western New York Foundation Grants, 3922

Teen Pregnancy
California Wellness Foundation Work and Health Program Grants, 1001
Community Foundation of Eastern Connecticut Northeast Women and Girls Grants, 1270
Cone Health Foundation Grants, 1323
DHHS Adolescent Family Life Demonstration Projects, 1472
Grand Rapids Area Community Foundation Wyoming Youth Fund Grants, 1868
Grand Rapids Community Foundation Ionia County Youth Fund Grants, 1871
Grand Rapids Community Foundation Southeast Ottawa Youth Fund Grants, 1874
Grand Rapids Community Foundation Sparta Youth Fund Grants, 1876
Health Fndn of Greater Indianapolis Grants, 1984
Joseph Drown Foundation Grants, 2268
Mary Black Foundation Early Childhood Development Grants, 2442
MetroWest Health Foundation Grants to Reduce the Incidence of High Risk Behaviors Among Adolescents, 2528
OneFamily Foundation Grants, 3075
Philadelphia Foundation Organizational Effectiveness Grants, 3189
Prince Charitable Trusts Chicago Grants, 3270
Puerto Rico Community Foundation Grants, 3278
Thompson Charitable Foundation Grants, 3727
USAID Family Planning and Reproductive Health Methods Grants, 3811

Telecommunications
John W. Speas and Effie E. Speas Memorial Trust Grants, 2264

Television
1st Source Foundation Grants, 2
American Academy of Nursing Media Awards, 509
CDC Cooperative Agreement for the Development, Operation, and Evaluation of an Entertainment Education Program, 1051
Daniel & Nanna Stern Family Fndn Grants, 1422

Television, Cable
Cable Positive's Tony Cox Community Grants, 992

Television, Children's
Farmers Insurance Corporate Giving Grants, 1634

Television, Public
Air Products and Chemicals Grants, 399
ConocoPhillips Foundation Grants, 1327
Doree Taylor Charitable Foundation, 1490
Farmers Insurance Corporate Giving Grants, 1634
George W. Codrington Charitable Fndn Grants, 1803
Hawaiian Electric Industries Charitable Foundation Grants, 1975
John Ben Snow Memorial Trust Grants, 2231
PMI Foundation Grants, 3239
TE Foundation Grants, 3705

Terminally Ill
Columbus Foundation Estrich Fund Grants, 1218
Cralle Foundation Grants, 1356
Florence Hunt Maxwell Foundation Grants, 1676
Graham and Carolyn Holloway Family Foundation Grants, 1861
Herbert A. & Adrian W. Woods Fndn Grants, 2007
May and Stanley Smith Charitable Trust Grants, 2459

Terrorism
Harry Kramer Memorial Fund Grants, 1965

Theater
Adams Foundation Grants, 301
Atherton Family Foundation Grants, 794
Bildner Family Foundation Grants, 891
Colonel Stanley R. McNeil Foundation Grants, 1211
Frank Loomis Palmer Fund Grants, 1726
Gardner Foundation Grants, 1770
Horace Moses Charitable Foundation Grants, 2047
Idaho Power Company Corporate Contributions, 2115
Jerome Robbins Foundation Grants, 2223
Leon and Thea Koerner Foundation Grants, 2355
Robert R. Meyer Foundation Grants, 3387
Sioux Falls Area Community Foundation Field-of-Interest and Donor-Advised Grants, 3569
Sioux Falls Area Community Foundation Spot Grants (Unrestricted), 3570
Whitney Foundation Grants, 3938

Theology
Boettcher Foundation Grants, 931
Booth-Bricker Fund Grants, 933
E. Rhodes & Leona B. Carpenter Grants, 1539
FirstEnergy Foundation Community Grants, 1668
John I. Smith Charities Grants, 2240
Rainbow Fund Grants, 3301

Therapy Evaluation
Baltimore Washington Center Adult Psychoanalysis Fellowships, 821
Baltimore Washington Center Child Psychotherapy Fellowships, 822
Boston Psychoanalytic Society and Institute Fellowship in Child Psychoanalytic Psychotherapy, 939
Boston Psychoanalytic Society and Institute Fellowship in Psychoanalytic Psychotherapy, 940
Chicago Institute for Psychoanalysis Fellowships, 1135
Michigan Psychoanalytic Institute and Society SATA Grants, 2546
Michigan Psychoanalytic Institute and Society Scholarships, 2547
NHLBI Airway Smooth Muscle Function and Targeted Therapeutics in Human Asthma Grants, 2687
NICHD Innovative Therapies and Clinical Studies for Screenable Disorders Grants, 2789

SUBJECT INDEX

Therapy/Rehabilitation
ASHFoundation New Century Scholars Program Doctoral Scholarships, 706
ASHFoundation New Century Scholars Research Grant, 707
ASHFoundation Grant for New Investigators, 708
ASHFoundation Student Research Grants in Audiology, 710
Different Needz Foundation Grants, 1478

Third World Nations
Carl R. Hendrickson Family Foundation Grants, 1028
Conservation, Food, and Health Foundation Grants for Developing Countries, 1338
Dorothea Haus Ross Foundation Grants, 1498
PepsiCo Foundation Grants, 3137

Thoracic Medicine
Foundation of the American Thoracic Society Research Grants, 1708
Thoracic Foundation Grants, 3729

Tissue Culture
NIA Role of Nuclear Receptors in Tissue and Organismal Aging Grants, 2783
NSF Biomedical Engineering and Engineering Healthcare Grants, 3009

Tobacco
American Legacy Foundation National Calls for Proposals Grants, 534
American Legacy Fnd Small Innovative Grants, 535
CDC Fnd Tobacco Network Lab Fellowship, 1062
Christine Gregoire Youth/Young Adult Award for Use of Tobacco Industry Documents, 1151
Community Foundation for Northeast Michigan Tobacco Settlement Grants, 1256
CTCRI Idea Grants, 1369
FAMRI Clinical Innovator Awards, 1628
FAMRI Young Clinical Scientist Awards, 1629
Health Fndn of Greater Indianapolis Grants, 1984
Hillsdale County Community Foundation Healthy Senior/Healthy Youth Fund Grants, 2033
Paso del Norte Health Foundation Grants, 3119
Pfizer Global Research Awards for Nicotine Independence, 3179
Sybil G. Jacobs Award for Outstanding Use of Tobacco Industry Documents, 3698
TRDRP California Research Awards, 3753
TRDRP Exploratory/Devel Research Awards, 3754
TRDRP New Investigator Awards, 3755

Tolerance
Elizabeth Morse Genius Charitable Trust Grants, 1569

Tolerance, Ethnic
Elizabeth Morse Genius Charitable Trust Grants, 1569
Harry Frank Guggenheim Foundation Dissertation Fellowships, 1963
Skoll Fndn Awards for Social Entrepreneurship, 3579

Tolerance, Religious
Alfred and Tillie Shemanski Testamentary Trust Grants, 1408
Elizabeth Morse Genius Charitable Trust Grants, 1569
Skoll Fndn Awards for Social Entrepreneurship, 3579

Tourette Syndrome
Tourette Syndrome Association Post-Doctoral Fellowships, 3740
Tourette Syndrome Association Research Grants, 3741

Tourism
Miller Foundation Grants, 2556
Principal Financial Group Foundation Grants, 3273

Toxic Substances
Beldon Fund Grants, 871
Bullitt Foundation Grants, 973
CDC Epidemic Intell Service Training Grants, 1055

SUBJECT INDEX

Travel /821

CDC Epidemiology Elective Rotation, 1056
Clarence E. Heller Charitable Foundation Grants, 1166

Toxicology
ASM sanofi-aventis ICAAC Award, 749
Canada Graduate Scholarships (CGS) and NSERC Postgraduate Scholarships (PGS), 1010
NIEHS Small Business Innovation Research (SBIR) Program Grants, 2860
PhRMA Foundation Health Outcomes Post Doctoral Fellowships, 3202
PhRMA Foundation Health Outcomes Research Starter Grants, 3204
PhRMA Foundation Informatics Post Doctoral Fellowships, 3206
PhRMA Foundation Pharmacology/Toxicology Post Doctoral Fellowships, 3215
PhRMA Foundation Pharmacology/Toxicology Pre Doctoral Fellowships, 3216
PhRMA Foundation Pharmacology/Toxicology Research Starter Grants, 3217
PhRMA Foundation Pharmacology/Toxicology Sabbatical Fellowships, 3218

Training and Development
Able Trust Vocational Rehabilitation Grants for Individuals, 213
ACF Native American Social and Economic Development Strategies Grants, 231
ACL University Centers for Excellence in Developmental Network Diversity and Inclusion Training Action Planning Grants, 249
AES Research and Training Workshop Awards, 320
AES Susan S. Spencer Clinical Research Epilepsy Training Fellowship, 325
Alberto Culver Corporate Contributions Grants, 408
Allen Foundation Educational Nutrition Grants, 442
APHL Emerging Infectious Diseases Fellowships, 636
Boeing Company Contributions Grants, 930
Burden Trust Grants, 976
CDC Public Health Associates, 1065
CDC Public Health Associates Hosts, 1066
Clarian Health Critical/Progressive Care Interns, 1168
Clarian Health Multi-Specialty Internships, 1169
Clarian Health OR Internships, 1170
Community Foundation of the Verdugos Educational Endowment Fund Grants, 1314
Conservation, Food, and Health Foundation Grants for Developing Countries, 1338
Covidien Clinical Education Grants, 1351
Cruise Industry Charitable Foundation Grants, 1363
Cystic Fibrosis Canada Special Travel Grants For Fellows and Students, 1389
DHHS Emerging Leaders Program Internships, 1473
DOL Occupational Safety and Health--Susan Harwood Training Grants, 1484
Elizabeth Glaser Int'l Leadership Awards, 1566
ERC Starting Grants, 1599
Fremont Area Community Foundation Amazing X Grants, 1744
GNOF Organizational Effectiveness Grants and Workshops, 1846
Greater Milwaukee Foundation Grants, 1883
Harmony Project Grants, 1948
IBRO Return Home Fellowships, 2112
Jacobs Family Village Neighborhoods Grants, 2184
MAP International Medical Fellowships, 2421
Mary Wilmer Covey Charitable Trust Grants, 2448
Medtronic Foundation HeartRescue Grants, 2513
MLA T. Mark Hodges Int'l Service Award, 2575
MMS and Alliance Charitable Foundation International Health Studies Grants, 2588
NEI Ruth L. Kirschstein National Research Service Award Short-Term Institutional Research Training Grants, 2668
NIDCD Mentored Research Scientist Development Award, 2796
NIH Ruth L. Kirschstein National Research Service Award Institutional Research Training Grants, 2900

NIH Ruth L. Kirschstein Nat Research Service Award Short-Term Institutional Training Grants, 2906
OneFamily Foundation Grants, 3075
Peter and Elizabeth C. Tower Foundation Social and Emotional Preschool Curriculum Grants, 3151
Priddy Foundation Organizational Development Grants, 3265
Prudential Foundation Education Grants, 3276
Raskob Foundation for Catholic Activities Grants, 3306
Regenstrief Geriatrics Fellowship, 3352
RRF General Program Grants, 3417
Saint Luke's Health Initiatives Grants, 3455
Scott County Community Foundation Grants, 3500
SfN Award for Education in Neuroscience, 3517
SfN Science Educator Award, 3535
Sioux Falls Area Community Foundation Community Fund Grants (Unrestricted), 3568
Sioux Falls Area Community Foundation Spot Grants (Unrestricted), 3570
South Madison Community Foundation Grants, 3626
Symantec Community Relations and Corporate Philanthropy Grants, 3699
U.S. Department of Education Rehabilitation Research Training Centers (RRTCs), 3770
UniHealth Foundation Workforce Development Grants, 3786
Union Bank, N.A. Foundation Grants, 3788
United Hospital Fund of New York Health Care Improvement Grants, 3797
USDA Organic Agriculture Research Grants, 3830
VDH Emergency Med Services Training Grants, 3855
Welborn Baptist Fndn Health Ministries Grants, 3915

Transexuals
Ms. Fndn for Women Ending Violence Grants, 2596
Ms. Foundation for Women Health Grants, 2597

Transfusion Medicine
AABB Dale A. Smith Memorial Award, 43
AABB Hemphill-Jordan Leadership Award, 44
AABB John Elliott Memorial Award, 45
AABB Karl Landsteiner Memorial Award and Lectureship, 46
AABB Sally Frank Award & Lectureship, 47
AABB Tibor Greenwalt Memorial Award and Lectureship, 48
National Blood Foundation Research Grants, 2628
NHLBI Immunomodulatory, Inflammatory, & Vasoregulatory Properties of Transfused Red Blood Cell Units as a Function of Preparation & Storage Grants, 2700
NHLBI Mentored Career Award For Faculty At Minority Institutions, 2707

Transitional Students
Guy I. Bromley Trust Grants, 1919
Lewis H. Humphreys Charitable Trust Grants, 2361
Louetta M. Cowden Foundation Grants, 2377

Translation
NCI Application of Metabolomics for Translational and Biological Research Grants, 2646

Transplantation Immunology
National Blood Foundation Research Grants, 2628
NIH Nonhuman Primate Immune Tolerance Cooperative Study Group Grants, 2890
Pfizer Advancing Research in Transplantation Science (ARTS) Research Awards, 3163

Transportation
Able To Serve Grants, 212
Barberton Community Foundation Grants, 829
Bullitt Foundation Grants, 973
Community Foundation of Greater Birmingham Grants, 1272
Elizabeth Morse Genius Charitable Trust Grants, 1569
Fallon OrNda Community Health Fund Grants, 1627
Farmers Insurance Corporate Giving Grants, 1634
Foellinger Foundation Grants, 1684

Fremont Area Community Foundation Amazing X Grants, 1744
Giving in Action Society Children & Youth with Special Needs Grants, 1823
Giving in Action Society Family Indep Grants, 1824
Helen Steiner Rice Foundation Grants, 1998
Hillcrest Foundation Grants, 2031
Rochester Area Community Foundation Grants, 3391
US Airways Community Contributions, 3825
Vigneron Memorial Fund Grants, 3873

Transportation Engineering
Elizabeth Morse Genius Charitable Trust Grants, 1569

Transportation Planning/Policy
Bikes Belong Foundation Paul David Clark Bicycling Safety Grants, 890
Blue Cross Blue Shield of Minnesota Foundation - Healthy Equity: Health Impact Assessment Demonstration Project Grants, 916
Blue Cross Blue Shield of Minnesota Foundation - Healthy Equity: Health Impact Assessment Program Grants, 917
Elizabeth Morse Genius Charitable Trust Grants, 1569

Trauma
American Trauma Society, Pennsylvania Division Mini-Grants, 564
Colorado Resource for Emergency Education and Trauma Grants, 1212
Foundation of Orthopedic Trauma Grants, 1707
Orthopaedic Trauma Assoc Research Grants, 3084
Osteosynthesis and Trauma Care Foundation European Visiting Fellowships, 3097
Peter and Elizabeth C. Tower Foundation Annual Mental Health Grants, 3145
Quantum Foundation Grants, 3290
Robert R. McCormick Tribune Veterans Initiative Grants, 3386
Trauma Center of America Finance Fellowship, 3752
USDD Workplace Violence in Military Grants, 3847

Travel
AAAAI ARTrust Grants for Allied Health Travel, 21
AAAAI Fellows-in-Training Travel Grants, 30
AACAP Educational Outreach Program for Child and Adolescent Psychiatry Residents, 52
AACAP Educational Outreach Program for General Psychiatry Residents, 53
AACAP Systems of Care Special Scholarships, 76
AACR-GlaxoSmithKline Clinical Cancer Research Scholar Awards, 93
AACR-NCI International Investigator Opportunity Grants, 96
AACR-WICR Scholar Awards, 100
AACR Brigid G. Leventhal Scholar in Cancer Research Awards, 106
AACR Min Scholar in Cancer Research Award, 120
AACR Scholar-in-Training Awards, 124
AACR Scholar in Training Awards Special Conferences, 125
AACR Thomas J. Bardos Science Ed Awards, 127
AAP International Travel Grants, 184
AAP Program Delegate Awards, 188
AcademyHealth Alice Hersh Student Scholarships, 221
AcademyHealth Awards, 222
AES-Grass Young Investigator Travel Awards, 317
AIHP Sonnedecker Visiting Scholar Grants, 371
ALFJ Astraea U.S. and Int'l Movement Fund, 427
AMA-MSS Research Poster Award, 477
AMA-WPC Joan F. Giambalvo Memorial Scholarships, 480
AMA Foundation Jordan Fieldman, MD, Awards, 489
AMA Foundation Leadership Awards, 490
AMA Foundation Pride in the Profession Grants, 493
AMA Virtual Mentor Theme Issue Editor Grants, 496
American College of Rheumatology Fellows-in-Training Travel Scholarships, 517
American College of Surgeons Australia and New Zealand Chapter Travelling Fellowships, 519

822 / Travel

American College of Surgeons Nizar N. Oweida, MD, FACS, Scholarships, 526
American Psychiatric Association Travel Scholarships for Minority Medical Students, 549
ASCO Merit Awards, 682
ASM Corporate Activities Program Student Travel Grants, 721
ASMCUE/GM Travel Assistance Grants, 722
ASM Early-Career Faculty Travel Award, 724
ASM Faculty Ehancement Travel Awards, 727
ASM Founders Distinguished Service Award, 728
ASM General Meeting Minority Travel Grants, 730
ASM GlaxoSmithKline International Member of the Year Award, 731
ASM Richard and Mary Finkelstein Travel Grant, 746
ASM Roche Diagnostics Alice C. Evans Award, 748
ASM Student & Postdoc Fellow Travel Grants, 752
ASM Student Travel Grants, 753
ASM TREK Diagnostic ABMM/ABMLI Professional Recognition Award, 754
ASM Undergraduate Research Fellowship, 755
ASPET Travel Awards, 779
BWF Collaborative Research Travel Grants, 985
Cystic Fibrosis Canada Special Travel Grants For Fellows and Students, 1389
Cystic Fibrosis Canada Transplant Center Incentive Grants, 1392
Cystic Fibrosis Canada Visiting Clinician Awards, 1395
FDHN Moti L. & Kamla Rustgi International Travel Awards, 1649
FDHN Non-Career Research Awards, 1650
Foundation of CVPH Travel Fund Grants, 1706
HAF David (Davey) H. Somerville Medical Travel Fund Grants, 1928
IBRO-PERC InEurope Short Stay Grants, 2097
IBRO Asia Reg APRC Exchange Fellowships, 2101
IBRO Asia Regional APRC Travel Grants, 2103
IBRO Latin America Regional Travel Grants, 2109
Jane Beattie Memorial Scholarship, 2207
MAP International Medical Fellowships, 2421
MBL Gruss Lipper Family Foundation Summer Fellowship, 2475
MBL Herbert W. Rand Summer Fellowship, 2477
Mead Johnson Nutritionals Scholarships, 2505
MLA David A. Kronick Traveling Fellowship, 2561
MMS and Alliance Charitable Foundation International Health Studies Grants, 2588
NAPNAP Foundation McNeil Grant-in-Aid, 2612
NAPNAP Foundation Reckitt Benckiser Student Scholarship, 2616
Northwest Airlines KidCares Medical Travel Assistance, 2998
Osteosynthesis and Trauma Care Foundation European Visiting Fellowships, 3097
Pinellas County Grants, 3223
Robert R. Meyer Foundation Grants, 3387
Sigrid Juselius Foundation Grants, 3561
SRC Medical Research Grants, 3635
St. Joseph Community Health Foundation Schneider Fellowships, 3642
UICC International Cancer Technology Transfer (ICRETT) Fellowships, 3779
UICC Raisa Gorbachev Memorial International Cancer Fellowships, 3780
Union Labor Health Fndn Angel Fund Grants, 3791

Tropical Medicine
Conservation, Food, and Health Foundation Grants for Developing Countries, 1338

Tuberculosis
Alliance Healthcare Foundation Grants, 444
ASM/CDC Fellowships in Infectious Disease and Public Health Microbiology, 716
ASPH/CDC Allan Rosenfield Global Health Fellowships, 780
CDC Epidemic Intell Service Training Grants, 1055
CDC Epidemiology Elective Rotation, 1056
CDC Public Health Associates, 1065
CDC Public Health Associates Hosts, 1066

Gates Award for Global Health, 1775
NYCT Biomedical Research Grants, 3049
Potts Memorial Foundation Grants, 3254
USAID Accelerating Progress Against Tuberculosis in Kenya Grants, 3805
USAID Comprehensive District-Based Support for Better HIV/TB Patient Outcomes Grants, 3807
USAID Systems Strengthening for Better HIV/TB Patient Outcomes Grants, 3824

Tumor Immunology
Children's Brain Tumor Fnd Research Grants, 1136
Children's Tumor Fndn Clin Research Awards, 1140
Children's Tumor Foundation Drug Discovery Initiative Awards, 1141
Children's Tumor Foundation Schwannomatosis Awards, 1142
Children's Tumor Foundation Young Investigator Awards, 1143

Undergraduate Education
AAHPERD Abernathy Presidential Scholarships, 151
AFB Rudolph Dillman Memorial Scholarship, 338
AIHS Media Summer Fellowship, 392
BMW of North America Charitable Contributions, 926
DAR Alice W. Rooke Scholarship, 1426
DAR Dr. Francis Anthony Beneventi Medical Scholarship, 1427
Elizabeth Nash Fnd Summer Research Awards, 1571
Gates Millennium Scholars Program, 1776
GFWC of Massachusetts Catherine E. Philbin Scholarship, 1811
HHMI Grants and Fellowships Programs, 2020
HRAMF Jeffress Trust Awards in Interdisciplinary Research, 2060
IIE AmCham Charitable Foundation U.S. Studies Scholarship, 2125
MBL William Townsend Porter Summer Fellowships for Minority Investigators, 2488
NBME Stemmler Med Ed Research Grants, 2638
North Carolina GlaxoSmithKline Fndn Grants, 2991
NSF Undergraduate Research and Mentoring in the Biological Sciences (URM), 3025
Phi Upsilon Omicron Geraldine Clewell Senior Awards, 3194
Phi Upsilon Omicron Janice Cory Bullock Collegiate Award, 3195
Phi Upsilon Omicron Margaret Jerome Sampson Scholarships, 3197
Phi Upsilon Orinne Johnson Writing Award, 3198
Phi Upsilon Omicron Sarah Thorniley Phillips Leader Awards, 3199
Phi Upsilon Omicron Undergraduate Karen P. Goebel Conclave Award, 3200
Pride Foundation Scholarships, 3269
SfN Neuroscience Fellowships, 3530
SfN Science Journalism Student Awards, 3536
SfN Undergrad Brain Awareness Travel Award, 3540
Thomas Jefferson Rosenberg Foundation Grants, 3723
U.S. Department of Education Erma Byrd Scholarships, 3768

United States History
ALA Donald G. Davis Article Award, 401
CDC David J. Sencer Museum Adult Group Tour, 1052

University/Industry Cooperative Activities
American Chemical Society GCI Pharmaceutical Roundtable Research Grants, 516
Medtronic Foundation Fellowships, 2512
NSF Grant Opportunities for Academic Liaison with Industry (GOALI), 3016
Shell Deer Park Grants, 3543
SSHRC-NSERC College and Community Innovation Enhancement Grants, 3636

Urban Affairs
Audrey & Sydney Irmas Foundation Grants, 799
Charles H. Revson Foundation Grants, 1122
Columbus Foundation Competitive Grants, 1217

Community Fndn for Muskegon County Grants, 1254
Community Foundation for the National Capital Region Community Leadership Grants, 1260
Community Fndn of East Central Illinois Grants, 1268
Community Foundation of St. Joseph County Special Project Challenge Grants, 1310
Howard and Bush Foundation Grants, 2054
Nordson Corporation Foundation Grants, 2981
Pittsburgh Foundation Community Fund Grants, 3230

Urban Areas
Achelis Foundation Grants, 234
Bodman Foundation Grants, 929
Commonwealth Fund Harkness Fellowships in Health Care Policy and Practice, 1229
CUNA Mutual Group Fndn Community Grants, 1375
E.L. Wiegand Foundation Grants, 1538
Elmer L. & Eleanor J. Andersen Fndn Grants, 1578
Hasbro Children's Fund Grants, 1973
Hudson Webber Foundation Grants, 2072
Irvin Stern Foundation Grants, 2164
Lynde & Harry Bradley Foundation Fellowships, 2396
Lynde and Harry Bradley Foundation Grants, 2397
Lynde and Harry Bradley Foundation Prizes: Bradley Prizes, 2398
Rajiv Gandhi Foundation Grants, 3302
Richard King Mellon Foundation Grants, 3369

Urban Education
Carnegie Corporation of New York Grants, 1033
Frances and John L. Loeb Family Fund Grants, 1716
U.S. Department of Education 21st Century Community Learning Centers, 3767

Urban Planning/Policy
Air Products and Chemicals Grants, 399
Blue Cross Blue Shield of Minnesota Foundation - Healthy Equity: Health Impact Assessment Program Grants, 917
Canada-U.S. Fulbright New Century Scholars Program Grants, 1008
Canada-U.S. Fulbright Senior Specialists Grants, 1009
Community Fndn for Muskegon County Grants, 1254
FMC Foundation Grants, 1683
Fulbright Alumni Initiatives Awards, 1749
Fulbright Distinguished Chairs Awards, 1750
Fulbright New Century Scholars Grants, 1753
Fulbright Specialists Program Grants, 1754
Fulbright Scholars in Europe and Eurasia, 1755
Fulbright Traditional Scholar Program in Sub-Saharan Africa, 1756
Fulbright Traditional Scholar Program in the East Asia/Pacific Region, 1757
Fulbright Traditional Scholar Program in the Near East and North Africa Region, 1758
Fulbright Traditional Scholar Program in the South and Central Asia Region, 1759
Fulbright Traditional Scholar Program in the Western Hemisphere, 1760
Meyer Foundation Healthy Communities Grants, 2533
Price Family Charitable Fund Grants, 3262
Sensient Technologies Foundation Grants, 3513
Western New York Foundation Grants, 3922

Urinary Tract
NHLBI Lymphatics in Health and Disease in the Digestive, Urinary, Cardiovascular and Pulmonary Systems, 2705
NIDDK Basic Research Grants in the Bladder and Lower Urinary Tract, 2802

Urogenital System
NIDDK Cell-Specific Delineation of Prostate and Genitourinary Development, 2804

Urologic Diseases
NHLBI Lymphatics in Health and Disease in the Digestive, Urinary, Cardiovascular and Pulmonary Systems, 2705
NIDDK Devel of Disease Biomarkers Grants, 2809

SUBJECT INDEX

NIDDK Education Program Grants, 2813
Pfizer Healthcare Charitable Contributions, 3180

Urology
NIDDK Basic and Clinical Studies of Congenital Urinary Tract Obstruction Grants, 2801
NIDDK Non-Invasive Methods for Diagnosis and Progression of Diabetes, Kidney, Urological, Hematological and Digestive Diseases, 2832
NIDDK Pilot and Feasibility Clinical Research Grants in Kidney or Urologic Diseases, 2835
NIDDK Proteomics: Diabetes, Obesity, And Endocrine, Digestive, Kidney, Urologic, And Hematologic Diseases, 2838
NIDDK Small Grants for Underrepresented Minority Scientists in Diabetes and Digestive and Kidney Diseases, 2847
NIDDK Training in Clinical Investigation in Kidney and Urology Grants, 2849
NKF Research Fellowships, 2938
NKF Young Investigator Grants, 2939
NYAM Edwin Beer Research Fellowship, 3042
NYAM Ferdinand C. Valentine Fellowship, 3043
Pfizer Healthcare Charitable Contributions, 3180
Pfizer Medical Education Track Two Grants, 3183

Utilities
Alexander Foundation Cancer, Catastrophic Illness and Injury Grants, 421
Barberton Community Foundation Grants, 829
Children Affected by AIDS Foundation Family Assistance Emergency Fund Grants, 1146
Delmarva Power and Light Contributions, 1454
MGM Resorts Foundation Community Grants, 2541
Presbyterian Patient Assistance Program, 3260

Uzbekistan
Fulbright Traditional Scholar Program in the South and Central Asia Region, 1759

Vaccines
Alternatives Research and Development Foundation Alternatives in Education Grants, 455
Alternatives Research and Development Foundation Grants, 456
ASM Maurice Hilleman/Merck Award, 738
Gates Award for Global Health, 1775
Grifols Community Outreach Grants, 1904
Pfizer Medical Education Track One Grants, 3182
Thrasher Research Fund Grants, 3732

Values/Moral Education
Cargill Citizenship Fund Corporate Giving, 1018

Venereal Diseases
ASPH/CDC Allan Rosenfield Global Health Fellowships, 780
CDC Epidemic Intell Service Training Grants, 1055
CDC Epidemiology Elective Rotation, 1056
CDC Foundation Atlanta International Health Fellowships, 1060
DHHS Adolescent Family Life Demo Projects, 1472

Venture Capital
Union Bank, N.A. Foundation Grants, 3788

Veterans
Anne J. Caudal Foundation Grants, 595
Arthur F. & Arnold M. Frankel Fndn Grants, 672
Balfe Family Endowment Grants, 819
Charles Delmar Foundation Grants, 1115
CNCS AmeriCorps VISTA Project Grants, 1190
CONSOL Military and Armed Services Grants, 1340
Disable American Veterans Charitable Grants, 1479
MGN Family Foundation Grants, 2542
Military Ex-Prisoners of War Foundation Grants, 2554
Northrop Grumman Corporation Grants, 2997
OceanFirst Foundation Major Grants, 3062
Robert R. McCormick Tribune Veterans Initiative Grants, 3386

Veterinary Medicine
ACVIM Foundation Clinical Investigation Grants, 286
APHL Emerging Infectious Diseases Fellowships, 636
Bodman Foundation Grants, 929
CDC-Hubert Global Health Fellowship, 1047
CDC Epidemic Intell Service Training Grants, 1055
CDC Epidemiology Elective Rotation, 1056
CDC Public Health Informatics Fellowships, 1068
Community Foundation of St. Joseph County Lilly Endowment Community Scholarship, 1308
Elkhart County Foundation Lilly Endowment Community Scholarships, 1574
Fulton County Community Foundation 4Community Higher Education Scholarship, 1762
IIE AmCham Charitable Foundation U.S. Studies Scholarship, 2125
IIE Chevron International REACH Scholarships, 2128
IIE Klein Family Scholarship, 2135
IIE Lingnan Foundation W.T. Chan Fellowship, 2137
IIE New Leaders Group Award for Mutual Understanding, 2141
Natalie W. Furniss Charitable Trust Grants, 2626
NHLBI Pathway to Independence Awards, 2718
NHLBI Ruth L. Kirschstein National Research Service Awards for Individual Senior Fellows, 2732
Porter County Foundation Lilly Endowment Community Scholarships, 3245
Posey County Community Foundation Grants, 3252
Pulaski County Community Fndn Scholarships, 3281
Scleroderma Established Investigator Grants, 3497
Sengupta Family Scholarship, 3512
Whitley County Community Foundation - Lilly Endowment Scholarship, 3935
Whitley County Community Fndn Scholarships, 3937
Winn Feline Foundation Grants, 3964

Veterinary Medicine Education
CDC-Hubert Global Health Fellowship, 1047
CDC Epidemic Intell Service Training Grants, 1055
CDC Epidemiology Elective Rotation, 1056
Maddie's Fund Medical Equipment Grants, 2411
MLA Donald A.B. Lindberg Fellowship, 2562
NHLBI Short-Term Research Education Program to Increase Diversity in Health-Related Research, 2733
NIH Ruth L. Kirschstein National Research Service Award Institutional Research Training Grants, 2900
NIH Ruth L. Kirschstein Nat Research Service Award Short-Term Institutional Training Grants, 2906
Pet Care Trust Sue Busch Memorial Award, 3143
Winn Feline Foundation Grants, 3964

Video Production
Daniel & Nanna Stern Family Fndn Grants, 1422

Violence
49ers Foundation Grants, 9
Ahmanson Foundation Grants, 355
Alliance Healthcare Foundation Grants, 444
Allstate Corporate Giving Grants, 446
Allstate Corp Hometown Commitment Grants, 447
California Endowment Innovative Ideas Challenge Grants, 1000
California Wellness Foundation Work and Health Program Grants, 1001
Carlisle Foundation Grants, 1025
Community Foundation for the National Capital Region Community Leadership Grants, 1260
Global Fund for Women Grants, 1833
Greater Tacoma Community Foundation Ryan Alan Hade Endowment Fund, 1887
Harry Frank Guggenheim Fnd Research Grants, 1964
Hasbro Children's Fund Grants, 1973
Isabel Allende Foundation Esperanza Grants, 2165
Joseph Drown Foundation Grants, 2268
Joseph H. & Florence A. Roblee Fndn Grants, 2269
Quantum Foundation Grants, 3290
San Diego Foundation for Change Grants, 3470
Sisters of St. Joseph Healthcare Fndn Grants, 3576
Sonoco Foundation Grants, 3621
Waitt Family Foundation Grants, 3894

Violence Prevention
Albert W. Rice Charitable Foundation Grants, 411
Alfred E. Chase Charitable Foundation Grants, 430
AMA Foundation Fund for Better Health Grants, 484
Charles H. Pearson Foundation Grants, 1121
Frank Reed and Margaret Jane Peters Memorial Fund II Grants, 1728
George W. Wells Foundation Grants, 1804
Linford and Mildred White Charitable Grants, 2368
Meyer Foundation Healthy Communities Grants, 2533
Ms. Fndn for Women Ending Violence Grants, 2596
RCF General Community Grants, 3312
Robert R. McCormick Trib Community Grants, 3385
Union Bank, N.A. Foundation Grants, 3788

Violence in Schools
49ers Foundation Grants, 9
DOJ Gang-Free Schools and Communities Intervention Grants, 1481

Violent Crime
Allstate Corporate Giving Grants, 446
Allstate Corp Hometown Commitment Grants, 447
Plough Foundation Grants, 3238

Viral Studies
CDC Evaluation of the Use of Rapid Testing For Influenza in Outpatient Medical Settings, 1057

Virology
APHL Emerging Infectious Diseases Fellowships, 636
ASM/CDC Fellowships in Infectious Disease and Public Health Microbiology, 716
Bristol-Myers Squibb Clinical Outcomes and Research Grants, 955
Bristol-Myers Squibb Foundation Fellowships, 956
Bristol-Myers Squibb Foundation Global HIV/AIDS Initiative Grants, 957
NSF Genes and Genome Systems Cluster Grants, 3015

Vision
1st Touch Foundation Grants, 3
Alcon Foundation Grants Program, 416
Alfred W. Bressler Prize in Vision Science, 434
Arnold and Mabel Beckman Foundation Beckman-Argyros Award in Vision Research, 664
Canadian Optometric Education Trust Grants, 1011
CNIB Baker Fellowships, 1194
CNIB Barbara Tuck MacPhee Award, 1196
CNIB Canada Glaucoma Clinical Research Council Grants, 1197
CNIB Chanchlani Global Vision Research Award, 1198
CNIB E. (Ben) & Mary Hochhausen Access Technology Research Grants, 1199
CNIB Ross Purse Doctoral Fellowships, 1200
Dr. Stanley Pearle Scholarships, 1512
Eye-Bank for Sight Restoration and Fight for Sight Summer Student Research Fellowship, 1622
Fight for Sight-Streilein Foundation for Ocular Immunology Research Award, 1662
Fight for Sight Grants-in-Aid, 1663
Fight for Sight Post-Doctoral Awards, 1664
Fight for Sight Summer Student Fellowships, 1665
Foundation Fighting Blindness Marjorie Carr Adams Women's Career Development Awards, 1690
HAF JoAllen K. Twiddy-Wood Memorial Fund Grants, 1929
Hawaii Community Foundation Health Education and Research Grants, 1976
HRAMF Thome Foundation Awards in Age-Related Macular Degeneration Research, 2065
Meyer Fndn Management Assistance Grants, 2534
NEI Mentored Clinical Scientist Dev Award, 2666
NEI Scholars Program, 2669
NEI Translational Research Program On Therapy For Visual Disorders, 2670
NEI Vision Research Core Grants, 2671
Nina Mason Pulliam Charitable Trust Grants, 2921
NYCT Blindness and Visual Disabilities Grants, 3050
OneSight Research Foundation Block Grants, 3076

824 / Vision

Streilein Foundation for Ocular Immunology Visiting Professorships, 3660
Young Ambassador Scholarship In Memory of Christopher Nordquist, 3981

Visual Arts
Amgen Foundation Grants, 572
Athwin Foundation Grants, 795
Booth-Bricker Fund Grants, 933
Christensen Fund Regional Grants, 1149
ConocoPhillips Foundation Grants, 1327
DAAD Research Stays for University Academics and Scientists, 1408
General Mills Foundation Grants, 1782
George Gund Foundation Grants, 1796
George S. & Dolores Dore Eccles Fndn Grants, 1801
Georgia Power Foundation Grants, 1806
GNOF IMPACT Grants for Arts and Culture, 1838
Grand Rapids Area Community Fndn Grants, 1865
Grand Rapids Area Community Foundation Nashwauk Area Endowment Fund Grants, 1866
Guy I. Bromley Trust Grants, 1919
Helen Gertrude Sparks Charitable Trust Grants, 1993
James M. Cox Foundation of Georgia Grants, 2196
John W. Speas and Effie E. Speas Memorial Trust Grants, 2264
Kelvin and Eleanor Smith Foundation Grants, 2297
Leon and Thea Koerner Foundation Grants, 2355
Lewis H. Humphreys Charitable Trust Grants, 2361
Louetta M. Cowden Foundation Grants, 2377
Louis and Elizabeth Nave Flarsheim Charitable Foundation Grants, 2379
Piedmont Natural Gas Corporate and Charitable Contributions, 3220
Powell Foundation Grants, 3255
Procter and Gamble Fund Grants, 3274
Reinberger Foundation Grants, 3357
Shell Deer Park Grants, 3543
Sid W. Richardson Foundation Grants, 3553
Society for the Arts in Healthcare Grants, 3602
Textron Corporate Contributions Grants, 3712
Thomas Sill Foundation Grants, 3724
Vigneron Memorial Fund Grants, 3873
Winifred & Harry B. Allen Foundation Grants, 3962
Wood-Claeyssens Foundation Grants, 3971
Wood Family Charitable Trust Grants, 3972

Visual Impairments
AFB Rudolph Dillman Memorial Scholarship, 338
Agnes M. Lindsay Trust Grants, 346
AHAF National Glaucoma Research Grants, 351
Alfred W. Bressler Prize in Vision Science, 434
Annunziata Sanguinetti Foundation Grants, 600
Arnold and Mabel Beckman Foundation Beckman-Argyros Award in Vision Research, 664
Atran Foundation Grants, 797
Carl W. and Carrie Mae Joslyn Trust Grants, 1031
Charles Delmar Foundation Grants, 1115
CNIB Baker Applied Research Fund Grants, 1193
CNIB Baker Fellowships, 1194
CNIB Baker New Researcher Fund Grants, 1195
CNIB Barbara Tuck MacPhee Award, 1196
CNIB Canada Glaucoma Clinical Research Council Grants, 1197
CNIB Chanchlani Global Vision Research Award, 1198
CNIB E. (Ben) & Mary Hochhausen Access Technology Research Grants, 1199
CNIB Ross Purse Doctoral Fellowships, 1200
Cornerstone Foundation of Northeastern Wisconsin Grants, 1347
Cralle Foundation Grants, 1356
Donald and Sylvia Robinson Family Foundation Grants, 1486
Dr. Stanley Pearle Scholarships, 1512
El Paso Community Foundation Grants, 1579
Eugene G. and Margaret M. Blackford Memorial Fund Grants, 1610
Florence Hunt Maxwell Foundation Grants, 1676
Foundation Fighting Blindness Marjorie Carr Adams Women's Career Development Awards, 1690

George Gund Foundation Grants, 1796
Hagedorn Fund Grants, 1934
J. Walton Bissell Foundation Grants, 2177
John W. Alden Trust Grants, 2260
Josiah W. and Bessie H. Kline Foundation Grants, 2278
Lydia deForest Charitable Trust Grants, 2393
Mabel H. Flory Charitable Trust Grants, 2408
Marjorie C. Adams Charitable Trust Grants, 2437
Mericos Foundation Grants, 2518
NEI Clinical Vision Research Devel Award, 2664
NEI Scholars Program, 2669
NEI Translational Research Program On Therapy For Visual Disorders, 2670
Nina Mason Pulliam Charitable Trust Grants, 2921
NYCT Blindness and Visual Disabilities Grants, 3050
OneSight Research Foundation Block Grants, 3076
Pfizer Healthcare Charitable Contributions, 3180
Philadelphia Foundation Organizational Effectiveness Grants, 3189
Reader's Digest Partners for Sight Fndn Grants, 3345
S. Mark Taper Foundation Grants, 3449
Sara Elizabeth O'Brien Trust Grants, 3483
Shell Deer Park Grants, 3543
The Ray Charles Foundation Grants, 3716
Thomas Sill Foundation Grants, 3724
U.S. Department of Education Rehabilitation Training - Rehabilitation Continuing Education - Institute on Rehabilitation Issues, 3771
Union County Community Foundation Grants, 3790
W.P. and Bulah Luse Foundation Grants, 3892
William G. & Helen C. Hoffman Fndn Grants, 3950

Vocational Counseling
Achelis Foundation Grants, 234
Bodman Foundation Grants, 929
DHHS Adolescent Family Life Demonstration Projects, 1472
GNOF New Orleans Works Grants, 1844
Graco Foundation Grants, 1860

Vocational Education
BMW of North America Charitable Contributions, 926
Charles Nelson Robinson Fund Grants, 1125
GNOF New Orleans Works Grants, 1844
Heckscher Foundation for Children Grants, 1988
IIE Mattel Global Scholarship, 2139
Richard King Mellon Foundation Grants, 3369

Vocational Services
GNOF New Orleans Works Grants, 1844

Vocational Training
Alfred E. Chase Charitable Foundation Grants, 430
CDC Public Health Associates, 1065
Charles Nelson Robinson Fund Grants, 1125
Gamble Foundation Grants, 1769
George W. Wells Foundation Grants, 1804
GNOF New Orleans Works Grants, 1844
IIE 911 Armed Forces Scholarships, 2122
KeyBank Foundation Grants, 2307
MGM Resorts Foundation Community Grants, 2541
PMI Foundation Grants, 3239
Pride Foundation Fellowships, 3267
Pride Foundation Scholarships, 3269
Sioux Falls Area Community Foundation Community Fund Grants (Unrestricted), 3568
Sioux Falls Area Community Foundation Spot Grants (Unrestricted), 3570
Union Bank, N.A. Foundation Grants, 3788
USDA Organic Agriculture Research Grants, 3830

Vocational/Technical Education
Besser Foundation Grants, 887
Burlington Industries Foundation Grants, 977
Charles Nelson Robinson Fund Grants, 1125
Constantin Foundation Grants, 1341
Cooper Industries Foundation Grants, 1345
Crail-Johnson Foundation Grants, 1355
D.F. Halton Foundation Grants, 1405
DaimlerChrysler Corporation Fund Grants, 1412

Dr. John Maniotes Scholarship, 1506
FirstEnergy Foundation Community Grants, 1668
Frances L. and Edwin L. Cummings Memorial Fund Grants, 1717
Fred and Louise Latshaw Scholarship, 1735
Fulton County Community Foundation 4Community Higher Education Scholarship, 1762
GNOF New Orleans Works Grants, 1844
Graco Foundation Grants, 1860
HAF Riley Frazel Memorial Scholarship, 1931
Hammond Common Council Scholarships, 1938
IBEW Local Union #697 Memorial Scholarships, 2096
IIE Mattel Global Scholarship, 2139
IIE Western Union Family Scholarships, 2143
Irvin Stern Foundation Grants, 2164
Janus Foundation Grants, 2211
LEGENDS Scholarship, 2351
Mary L. Peyton Foundation Grants, 2445
Nancy J. Pinnick Memorial Scholarship, 2607
Norcliffe Foundation Grants, 2979
Oppenstein Brothers Foundation Grants, 3078
Principal Financial Group Foundation Grants, 3273
Pulaski County Community Fndn Scholarships, 3281
Samuel S. Johnson Foundation Grants, 3468
Seabury Foundation Grants, 3501
Stackpole-Hall Foundation Grants, 3643
U.S. Department of Education Rehabilitation Research Training Centers (RRTCs), 3770
United Technologies Corporation Grants, 3803
W. James Spicer Scholarship, 3885
Webster Cornwell Memorial Scholarship, 3911
Whitley County Community Fndn Scholarships, 3937

Volunteers
AAAAI Distinguished Layperson Award, 25
ACE Charitable Foundation Grants, 229
ACL Empowering Seniors to Prevent Health Care Fraud Grants, 241
Agway Foundation Grants, 348
Alliant Energy Corporation Contributions, 445
AMA Foundation Jack B. McConnell, MD Awards for Excellence in Volunteerism, 487
Ann Arbor Area Community Foundation Grants, 594
Anschutz Family Foundation Grants, 602
Aragona Family Foundation Grants, 648
ASM Founders Distinguished Service Award, 728
Bank of America Fndn Volunteer Grants, 826
Benton Community Foundation Grants, 878
Blue River Community Foundation Grants, 922
Bodman Foundation Grants, 929
Brown County Community Foundation Grants, 970
Cambridge Community Foundation Grants, 1004
Campbell Soup Foundation Grants, 1007
CFFVR Waupaca Area Community Foundation Grants, 1105
CharityWorks Grants, 1113
Chemtura Corporation Contributions Grants, 1130
CNCS Senior Companion Program Grants, 1191
CNCS Social Innovation Grants, 1192
Comerica Charitable Foundation Grants, 1224
Commonwealth Edison Grants, 1225
Community Foundation of Bartholomew County Heritage Fund Grants, 1262
Community Foundation of Bartholomew County James A. Henderson Award for Fundraising, 1263
Community Foundation of Greater Fort Wayne - Community Endowment and Clarke Endowment Grants, 1274
Community Foundation of Greenville Community Enrichment Grants, 1284
Constellation Energy Corporate Grants, 1342
Cooper Industries Foundation Grants, 1345
Coors Brewing Corporate Contributions Grants, 1346
eBay Foundation Community Grants, 1543
Express Scripts Foundation Grants, 1620
Fargo-Moorhead Area Foundation Grants, 1632
Fidelity Foundation Grants, 1659
G.N. Wilcox Trust Grants, 1768
Giant Eagle Foundation Grants, 1815
Great-West Life Grants, 1879

SUBJECT INDEX

Harry B. and Jane H. Brock Foundation Grants, 1960
Helen Steiner Rice Foundation Grants, 1998
Hilton Hotels Corporate Giving Program Grants, 2037
Hormel Foods Charitable Trust Grants, 2050
J.C. Penney Company Grants, 2169
Janus Foundation Grants, 2211
Jessie Ball Dupont Fund Grants, 2226
John Ben Snow Memorial Trust Grants, 2231
Johns Manville Fund Grants, 2251
John W. Speas and Effie E. Speas Memorial Trust Grants, 2264
Lewis H. Humphreys Charitable Trust Grants, 2361
Lillian S. Wells Foundation Grants, 2365
Lockheed Martin Corp Foundation Grants, 2374
Louis and Elizabeth Nave Flarsheim Charitable Foundation Grants, 2379
McCombs Foundation Grants, 2491
MMS and Alliance Charitable Foundation Grants for Community Action and Care for the Medically Uninsured, 2587
Nordson Corporation Foundation Grants, 2981
North Carolina GlaxoSmithKline Fndn Grants, 2991
Oppenstein Brothers Foundation Grants, 3078
PepsiCo Foundation Grants, 3137
Phi Upsilon Omicron Florence Fallgatter Distinguished Service Award, 3192
Princeton Area Community Foundation Greater Mercer Grants, 3272
Procter and Gamble Fund Grants, 3274
Prudential Foundation Education Grants, 3276
Radcliffe Institute Carol K. Pforzheimer Student Fellowships, 3298
Rayonier Foundation Grants, 3311
Rhode Island Foundation Grants, 3362
Rohm and Haas Company Grants, 3403
Sensient Technologies Foundation Grants, 3513
Sisters of Mercy of North Carolina Fndn Grants, 3575
Sony Corporation of America Grants, 3623
Stackpole-Hall Foundation Grants, 3643
Symantec Community Relations and Corporate Philanthropy Grants, 3699
Tri-State Community Twenty-first Century Endowment Fund Grants, 3756
Triangle Community Foundation Donor-Advised Grants, 3757
Union Benevolent Association Grants, 3789
United Technologies Corporation Grants, 3803
W.K. Kellogg Foundation Healthy Kids Grants, 3886

Volunteers (Education)
ADA Foundation Bud Tarrson Dental School Student Community Leadership Awards, 287
ADA Foundation Thomas J. Zwemer Award, 297
Alliant Energy Corporation Contributions, 445
Daniels Fund Grants-Aging, 1424
Harry B. and Jane H. Brock Foundation Grants, 1960
Lewis H. Humphreys Charitable Trust Grants, 2361
Louis and Elizabeth Nave Flarsheim Charitable Foundation Grants, 2379
Radcliffe Institute Carol K. Pforzheimer Student Fellowships, 3298
Samuel S. Johnson Foundation Grants, 3468

Voter Educational Programs
Clayton Baker Trust Grants, 1179
Farmers Insurance Corporate Giving Grants, 1634

Voter Registration Programs
Clayton Baker Trust Grants, 1179

Waste Management
Bill and Melinda Gates Foundation Water, Sanitation and Hygiene Grants, 894

Waste Management/Fossil Energy
Rohm and Haas Company Grants, 3403

Wastewater Treatment
Bill and Melinda Gates Foundation Water, Sanitation and Hygiene Grants, 894

Water Resources
1675 Foundation Grants, 11
Beldon Fund Grants, 871
Bill and Melinda Gates Foundation Water, Sanitation and Hygiene Grants, 894
Bullitt Foundation Grants, 973
Carolyn Foundation Grants, 1034
Community Fndn for Muskegon County Grants, 1254
DuPont Pioneer Community Giving Grants, 1532
Fremont Area Community Fndn General Grants, 1746
Gebbie Foundation Grants, 1777
Greater Milwaukee Foundation Grants, 1883
PepsiCo Foundation Grants, 3137
Thompson Charitable Foundation Grants, 3727
Union Bank, N.A. Foundation Grants, 3788
W.K. Kellogg Foundation Healthy Kids Grants, 3886

Water Resources, Environmental Impacts
Bella Vista Foundation Grants, 873
Bill and Melinda Gates Foundation Water, Sanitation and Hygiene Grants, 894
GNOF Bayou Communities Grants, 1835
Land O'Lakes Foundation Mid-Atlantic Grants, 2341
PepsiCo Foundation Grants, 3137
Sioux Falls Area Community Foundation Community Fund Grants (Unrestricted), 3568
Sioux Falls Area Community Foundation Spot Grants (Unrestricted), 3570
White County Community Foundation Grants, 3934

Water Resources, Management/Planning
Bill and Melinda Gates Foundation Water, Sanitation and Hygiene Grants, 894
Blue Cross Blue Shield of Minnesota Foundation - Healthy Equity: Health Impact Assessment Demonstration Project Grants, 916
Blue Cross Blue Shield of Minnesota Foundation - Healthy Equity: Health Impact Assessment Program Grants, 917
PepsiCo Foundation Grants, 3137

Water Supply
Aid for Starving Children International Grants, 368
Bill and Melinda Gates Foundation Water, Sanitation and Hygiene Grants, 894
PepsiCo Foundation Grants, 3137

Water Treatment
Bill and Melinda Gates Foundation Water, Sanitation and Hygiene Grants, 894

Waterways and Harbors
BMW of North America Charitable Contributions, 926

Weapons
AAAS Science and Technology Policy Fellowships: Global Health and Development, 41

Welding
DOL Occupational Safety and Health--Susan Harwood Training Grants, 1484
IIE Mattel Global Scholarship, 2139

Welfare Reform
Legler Benbough Foundation Grants, 2352
Macquarie Bank Foundation Grants, 2410
Whitley County Community Foundation Grants, 3936
William J. & Tina Rosenberg Fndn Grants, 3954

Welfare-to-Work Programs
Benton County Foundation Grants, 880
Carl R. Hendrickson Family Foundation Grants, 1028
Cessna Foundation Grants Program, 1081
D. W. McMillan Foundation Grants, 1407
Elizabeth Morse Genius Charitable Trust Grants, 1569
Frank B. Hazard General Charity Fund Grants, 1721
Helen Bader Foundation Grants, 1992
Joseph Henry Edmondson Foundation Grants, 2270
Priddy Foundation Program Grants, 3266
Prudential Foundation Education Grants, 3276

Robert R. McCormick Trib Community Grants, 3385
Robert R. McCormick Tribune Veterans Initiative Grants, 3386
Textron Corporate Contributions Grants, 3712
Union Bank, N.A. Foundation Grants, 3788
Verizon Foundation Maryland Grants, 3859
Verizon Foundation New York Grants, 3860
Verizon Foundation Northeast Region Grants, 3861
Verizon Foundation Pennsylvania Grants, 3862
Verizon Foundation Vermont Grants, 3863
Verizon Foundation Virginia Grants, 3864
Whitley County Community Foundation Grants, 3936

Wellness
AIHS Alberta/Pfizer Translat Research Grants, 374
Bupa Foundation Medical Research Grants, 974
Bupa Foundation Multi-Country Grant, 975
CDC David J. Sencer Museum Adult Group Tour, 1052
CDC Preventive Med Residency & Fellowship, 1064
CFFVR Jewelers Mutual Charitable Giving, 1099
CICF City of Noblesville Community Grant, 1155
CICF James Proctor Grant for Aged, 1157
Cigna Civic Affairs Sponsorships, 1159
Cresap Family Foundation Grants, 1359
Daniels Fund Grants-Aging, 1424
Delta Air Lines Foundation Health and Wellness Grants, 1455
Foundation for the Mid South Health and Wellness Grants, 1700
Giant Food Charitable Grants, 1816
Hillsdale County Community General Adult Foundation Grants, 2034
Manuel D. & Rhoda Mayerson Fndn Grants, 2420
Mary Wilmer Covey Charitable Trust Grants, 2448
Mt. Sinai Health Care Foundation Health of the Urban Community Grants, 2600
NIGMS Advancing Basic Behavioral and Social Research on Resilience: an Integrative Science Approach, 2861
Obesity Society Grants, 3060
Oregon Community Fndn Community Grants, 3082
Robert R. McCormick Trib Community Grants, 3385
Samueli Institute Scientific Research Grants, 3466
Seattle Foundation Health and Wellness Grants, 3507
Walmart Foundation Community Giving Grants, 3897
Walmart Foundation National Giving Grants, 3898
Washington Gas Charitable Giving, 3908

Wetlands
Bella Vista Foundation Grants, 873

Wilderness
Boyle Foundation Grants, 942

Wildlife
Aragona Family Foundation Grants, 648
Arthur Ashley Williams Foundation Grants, 671
Banfi Vintners Foundation Grants, 824
Batchelor Foundation Grants, 837
Boyle Foundation Grants, 942
Bullitt Foundation Grants, 973
Cambridge Community Foundation Grants, 1004
Carl C. Icahn Foundation Grants, 1023
Collins Foundation Grants, 1210
Community Fndn of Greater Tampa Grants, 1282
Community Foundation of the Verdugos Grants, 1315
Cruise Industry Charitable Foundation Grants, 1363
Donald and Sylvia Robinson Family Foundation Grants, 1486
Elkhart County Community Foundation Fund for Elkhart County, 1572
El Paso Community Foundation Grants, 1579
George and Ruth Bradford Foundation Grants, 1789
Greygates Foundation Grants, 1901
H.A. & Mary K. Chapman Trust Grants, 1921
J.M. Long Foundation Grants, 2173
James Ford Bell Foundation Grants, 2187
James M. Cox Foundation of Georgia Grants, 2196
James S. Copley Foundation Grants, 2200
Kinsman Foundation Grants, 2309

826 / Wildlife

Knox County Community Foundation Grants, 2313
Lawrence Foundation Grants, 2347
Lucy Downing Nisbet Charitable Fund Grants, 2387
M. Bastian Family Foundation Grants, 2402
Mary K. Chapman Foundation Grants, 2443
Mericos Foundation Grants, 2518
Natalie W. Furniss Charitable Trust Grants, 2626
Nina Mason Pulliam Charitable Trust Grants, 2921
North Dakota Community Foundation Grants, 2994
Perkins Charitable Foundation Grants, 3140
Perry County Community Foundation Grants, 3142
Pike County Community Foundation Grants, 3222
Posey County Community Foundation Grants, 3252
Richard King Mellon Foundation Grants, 3369
Sioux Falls Area Community Foundation Community Fund Grants (Unrestricted), 3568
Sioux Falls Area Community Foundation Spot Grants (Unrestricted), 3570
Spencer County Community Foundation Grants, 3632
Susan Mott Webb Charitable Trust Grants, 3694
Thorman Boyle Foundation Grants, 3730
Trull Foundation Grants, 3761
Union Bank, N.A. Foundation Grants, 3788
Warrick County Community Foundation Grants, 3904
William D. Laurie, Jr. Charitable Fndn Grants, 3949
Winifred & Harry B. Allen Foundation Grants, 3962
Yawkey Foundation Grants, 3980

Women
AACR-WICR Scholar Awards, 100
ACE Charitable Foundation Grants, 229
AEC Trust Grants, 315
A Fund for Women Grants, 342
Alberto Culver Corporate Contributions Grants, 408
Alfred P. Sloan Foundation Research Fellowships, 432
Allstate Corporate Giving Grants, 446
Allstate Corp Hometown Commitment Grants, 447
AMA-WPC Joan F. Giambalvo Memorial Scholarships, 480
AMA Foundation Joan F. Giambalvo Memorial Scholarships, 488
American Jewish World Service Grants, 533
American Legacy Foundation National Calls for Proposals Grants, 534
Archer Daniels Midland Foundation Grants, 651
Arkell Hall Foundation Grants, 661
Arthur and Rochelle Belfer Foundation Grants, 670
ASM Roche Diagnostics Alice C. Evans Award, 748
Atran Foundation Grants, 797
Avon Products Foundation Grants, 814
Benton Community Fndn Cookie Jar Grant, 877
Bernard F. and Alva B. Gimbel Foundation Reproductive Rights Grants, 883
BMW of North America Charitable Contributions, 926
Bodenwein Public Benevolent Foundation Grants, 928
Boston Foundation Grants, 936
Boston Globe Foundation Grants, 937
Boston Jewish Community Women's Fund Grants, 938
Cambridge Community Foundation Grants, 1004
Carnegie Corporation of New York Grants, 1033
Carrie E. and Lena V. Glenn Foundation Grants, 1036
Catherine Holmes Wilkins Foundation Grants, 1042
Cemala Foundation Grants, 1077
CFFVR Schmidt Family G4 Grants, 1103
CFFVR Women's Fund for the Fox Valley Region Grants, 1107
Charles Delmar Foundation Grants, 1115
Community Foundation for Greater New Haven Women & Girls Grants, 1252
Community Foundation of Eastern Connecticut Northeast Women and Girls Grants, 1270
Community Foundation of Greater Greensboro Women to Women Fund Grants, 1278
ConocoPhillips Foundation Grants, 1327
Cowles Charitable Trust Grants, 1354
Cralle Foundation Grants, 1356
Daisy Marquis Jones Foundation Grants, 1414
Dallas Women's Foundation Grants, 1417
Daphne Seybolt Culpeper Fndn Grants, 1425

David N. Lane Trust Grants for Aged and Indigent Women, 1433
Dennis & Phyllis Washington Fndn Grants, 1458
Eckerd Corporation Foundation Grants, 1547
El Paso Community Foundation Grants, 1579
Essex County Community Foundation Women's Fund Grants, 1605
Fargo-Moorhead Area Fndn Woman's Grants, 1633
Fourjay Foundation Grants, 1712
Frederick McDonald Trust Grants, 1740
Fremont Area Community Fndn General Grants, 1746
General Service Foundation Human Rights and Economic Justice Grants, 1784
George Family Foundation Grants, 1794
Georgia Power Foundation Grants, 1806
Global Fund for Women Grants, 1833
GNOF Maison Hospitaliere Grants, 1843
Goddess Scholars Grants, 1848
Greater Worcester Community Foundation Discretionary Grants, 1888
H.J. Heinz Company Foundation Grants, 1923
Harry B. and Jane H. Brock Foundation Grants, 1960
Health Fndn of Greater Indianapolis Grants, 1984
Heineman Foundation for Research, Education, Charitable and Scientific Purposes, 1990
Helena Rubinstein Foundation Grants, 1991
Helen Steiner Rice Foundation Grants, 1998
Henrietta Tower Wurts Memorial Fndn Grants, 2002
Huffy Foundation Grants, 2073
IIE African Center of Excellence for Women's Leadership Grants, 2124
Isabel Allende Foundation Esperanza Grants, 2165
James A. and Faith Knight Foundation Grants, 2186
James Ford Bell Foundation Grants, 2187
James R. Dougherty Jr. Foundation Grants, 2198
Jean and Louis Dreyfus Foundation Grants, 2215
Joseph H. & Florence A. Roblee Fndn Grants, 2269
Josephine Goodyear Foundation Grants, 2272
Josephine S. Gumbiner Foundation Grants, 2273
Katharine Matthies Foundation Grants, 2291
Katherine Baxter Memorial Foundation Grants, 2292
Kimball International-Habig Foundation Health and Human Services Grants, 2308
Kosciusko County Community Fndn Grants, 2319
Lalor Foundation Anna Lalor Burdick Grants, 2335
Laura B. Vogler Foundation Grants, 2345
Long Island Community Foundation Grants, 2375
Lotus 88 Fnd for Women & Children Grants, 2376
Mardag Foundation Grants, 2428
MBL Ann E. Kammer Summer Fellowship, 2464
Mercedes-Benz USA Corporate Contributions, 2517
Nell J. Redfield Foundation Grants, 2673
Nina Mason Pulliam Charitable Trust Grants, 2921
NMF Mary Ball Carrera Scholarship, 2960
North Carolina Community Foundation Grants, 2990
NYCT Girls and Young Women Grants, 3053
OneFamily Foundation Grants, 3075
Peacock Foundation Grants, 3133
Perpetual Trust for Charitable Giving Grants, 3141
Playboy Foundation Grants, 3237
Posey Community Fndn Women's Fund Grants, 3251
Pott Foundation Grants, 3253
R.S. Gernon Trust Grants, 3293
Rajiv Gandhi Foundation Grants, 3302
Rayonier Foundation Grants, 3311
RCF The Women's Fund Grants, 3315
Robert R. Meyer Foundation Grants, 3387
Rochester Area Community Foundation Grants, 3391
Rochester Area Foundation Grants, 3392
ROSE Fund Grants, 3412
S. Mark Taper Foundation Grants, 3449
San Diego Foundation for Change Grants, 3470
Schramm Foundation Grants, 3495
SfN Bernice Grafstein Award for Outstanding Accomplishments in Mentoring, 3518
SfN Louise Hanson Marshall Specific Recognition Award, 3527
SfN Mika Salpeter Lifetime Achievement Award, 3528
SfN Patricia Goldman-Rakic Hall of Honor, 3532
Simmons Foundation Grants, 3563

SUBJECT INDEX

Sisters of Mercy of North Carolina Fndn Grants, 3575
Sophia Romero Trust Grants, 3624
Southwest Gas Corporation Foundation Grants, 3628
Stark Community Fndn Women's Grants, 3645
Strowd Roses Grants, 3664
Textron Corporate Contributions Grants, 3712
Theodore Edson Parker Foundation Grants, 3715
Thomas Sill Foundation Grants, 3724
USAID Comprehensive District-Based Support for Better HIV/TB Patient Outcomes Grants, 3807
USAID Family Planning and Reproductive Health Methods Grants, 3811
Victor E. Speas Foundation Grants, 3868
Weingart Foundation Grants, 3913
WHO Foundation General Grants, 3939
Wilson-Wood Foundation Grants, 3960
Z. Smith Reynolds Foundation Small Grants, 3982

Women's Education
AAUW American Dissertation Fellowships, 195
AAUW American Postdoctoral Research Leave Fellowships, 196
A Fund for Women Grants, 342
AMA Foundation Joan F. Giambalvo Memorial Scholarships, 488
Avon Products Foundation Grants, 814
Benton Community Fndn Cookie Jar Grant, 877
Blackford County Community Foundation - WOW Grants, 905
Boston Jewish Community Women's Fund Grants, 938
CFFVR Schmidt Family G4 Grants, 1103
Community Foundation for Greater New Haven Women & Girls Grants, 1252
Community Foundation of Eastern Connecticut Northeast Women and Girls Grants, 1270
Essex County Community Foundation Women's Fund Grants, 1605
GFWC of Massachusetts Memorial Education Scholarship, 1813
Gibson County Community Foundation Women's Fund Grants, 1817
Heineman Foundation for Research, Education, Charitable and Scientific Purposes, 1990
IIE African Center of Excellence for Women's Leadership Grants, 2124
James A. and Faith Knight Foundation Grants, 2186
PepsiCo Foundation Grants, 3137
Porter County Women's Grant, 3248
Posey Community Fndn Women's Fund Grants, 3251
Radcliffe Institute Residential Fellowships, 3299
RCF The Women's Fund Grants, 3315
RGk Foundation Grants, 3361
SfN Louise Hanson Marshall Specific Recognition Award, 3527
SfN Patricia Goldman-Rakic Hall of Honor, 3532
SfN Science Educator Award, 3535
Stark Community Fndn Women's Grants, 3645
Warrick County Community Foundation Women's Fund, 3905
WestWind Foundation Reproductive Health and Rights Grants, 3924
WHO Foundation General Grants, 3939
Women of the ELCA Opportunity Scholarships for Lutheran Laywomen, 3970

Women's Employment
Albert W. Rice Charitable Foundation Grants, 411
Anschutz Family Foundation Grants, 602
Avon Products Foundation Grants, 814
Benton Community Fndn Cookie Jar Grant, 877
CFFVR Schmidt Family G4 Grants, 1103
Charles H. Pearson Foundation Grants, 1121
Community Foundation for Greater New Haven Women & Girls Grants, 1252
Community Foundation of Eastern Connecticut Northeast Women and Girls Grants, 1270
Community Foundation of Greater Flint Grants, 1273
Essex County Community Foundation Women's Fund Grants, 1605
George W. Wells Foundation Grants, 1804

SUBJECT INDEX

Global Fund for Women Grants, 1833
M.D. Anderson Foundation Grants, 2403
Posey Community Fndn Women's Fund Grants, 3251
Radcliffe Institute Residential Fellowships, 3299
RCF The Women's Fund Grants, 3315
Stark Community Fndn Women's Grants, 3645
Textron Corporate Contributions Grants, 3712
USDA Organic Agriculture Research Grants, 3830

Women's Health
A Fund for Women Grants, 342
Alliance Healthcare Foundation Grants, 444
American Legacy Foundation National Calls for Proposals Grants, 534
Appalachian Regional Commission Health Grants, 637
ASPH/CDC Allan Rosenfield Global Health Fellowships, 780
Avon Foundation Breast Care Fund Grants, 813
Avon Products Foundation Grants, 814
Benton Community Fndn Cookie Jar Grant, 877
Bernard and Audre Rapoport Foundation Health Grants, 882
Bernard F. and Alva B. Gimbel Foundation Reproductive Rights Grants, 883
Blackford County Community Foundation - WOW Grants, 905
Blanche and Irving Laurie Foundation Grants, 910
Boston Jewish Community Women's Fund Grants, 938
Breast Cancer Fund Grants, 947
Bristol-Myers Squibb Foundation Women's Health Grants, 962
California Endowment Innovative Ideas Challenge Grants, 1000
California Wellness Foundation Work and Health Program Grants, 1001
CDC Fnd Tobacco Network Lab Fellowship, 1062
CDC Increasing Breast and Cervical Cancer Screening Services for Urban American Indian/Alaska Native Women, 1063
CFFVR Schmidt Family G4 Grants, 1103
Charles F. Bacon Trust Grants, 1117
Collective Brands Foundation Grants, 1208
Community Foundation for Greater New Haven Women & Girls Grants, 1252
Community Fndn of Bloomington & Monroe County - Precision Health Network Cycle Grants, 1264
Community Foundation of Eastern Connecticut Northeast Women and Girls Grants, 1270
Community Foundation of Greater Greensboro Women to Women Fund Grants, 1278
Community Fndn of Louisville Health Grants, 1295
David N. Lane Trust Grants for Aged and Indigent Women, 1433
Duke Univ Obstetric Anesthesia Fellowships, 1525
Essex County Community Foundation Women's Fund Grants, 1605
Fannie E. Rippel Foundation Grants, 1630
Foundation for Seacoast Health Grants, 1698
Foundations of East Chicago Health Grants, 1709
General Service Reproductive Justice Grants, 1785
Gibson County Community Foundation Women's Fund Grants, 1817
Global Fund for Women Grants, 1833
GNOF Maison Hospitaliere Grants, 1843
Goddess Scholars Grants, 1848
Harry B. and Jane H. Brock Foundation Grants, 1960
IBCAT Screening Mammography Grants, 2095
IIE African Center of Excellence for Women's Leadership Grants, 2124
James A. and Faith Knight Foundation Grants, 2186
Jennings County Community Foundation Women's Giving Circle Grant, 2221
Johnson & Johnson Community Health Grants, 2252
Kate B. Reynolds Trust Health Care Grants, 2290
Kosciusko County Community Foundation REMC Operation Round Up Grants, 2320
Kroger Foundation Women's Health Grants, 2325
Lalor Foundation Anna Lalor Burdick Grants, 2335
Lillian S. Wells Foundation Grants, 2365
Lowe Foundation Grants, 2381

Ms. Foundation for Women Health Grants, 2597
NCI Stages of Breast Development: Normal to Metastatic Disease Grants, 2655
NIH Building Interdisciplinary Research Careers in Women's Health Grants, 2869
Packard Foundation Population and Reproductive Health Grants, 3106
Pfizer Healthcare Charitable Contributions, 3180
Pfizer Medical Education Track One Grants, 3182
Pfizer Medical Education Track Two Grants, 3183
Philip L. Graham Fund Health and Human Services Grants, 3190
Porter County Health and Wellness Grant, 3246
Porter County Women's Grant, 3248
Posey Community Fndn Women's Fund Grants, 3251
Pride Foundation Grants, 3268
Radcliffe Institute Carol K. Pforzheimer Student Fellowships, 3298
Radcliffe Institute Residential Fellowships, 3299
Ralph M. Parsons Foundation Grants, 3304
RCF The Women's Fund Grants, 3315
Robert R. Meyer Foundation Grants, 3387
Seattle Foundation Health and Wellness Grants, 3507
SfN Science Educator Award, 3535
Stark Community Fndn Women's Grants, 3645
Stella and Charles Guttman Foundation Grants, 3651
STTI Rosemary Berkel Crisp Research Awards, 3670
Susan G. Komen Breast Cancer Foundation Challege Grants: Investigator Initiated Research, 3690
Susan G. Komen Breast Cancer Foundation Challenge Grants: Career Catalyst Research, 3691
Susan G. Komen Breast Cancer Foundation College Scholarships, 3692
Susan G. Komen Breast Cancer Foundation Investigator Initiated Research Grants, 3693
United Hospital Fund of New York Health Care Improvement Grants, 3797
USAID Comprehensive District-Based Support for Better HIV/TB Patient Outcomes Grants, 3807
USAID Family Health Plus Project Grants, 3810
USAID Family Planning and Reproductive Health Methods Grants, 3811
USAID Microbicides Research, Development, and Introduction Grants, 3819
W.P. and Bulah Luse Foundation Grants, 3892
Warrick County Community Foundation Women's Fund, 3905
WHO Foundation General Grants, 3939

Women's Rights
Baxter International Corporate Giving Grants, 841
Benton Community Fndn Cookie Jar Grant, 877
Bernard F. and Alva B. Gimbel Foundation Reproductive Rights Grants, 883
Blackford County Community Foundation - WOW Grants, 905
Elkhart County Community Foundation Fund for Elkhart County, 1572
General Service Reproductive Justice Grants, 1785
Global Fund for Women Grants, 1833
IIE African Center of Excellence for Women's Leadership Grants, 2124
Kimball International-Habig Foundation Health and Human Services Grants, 2308
Porter County Women's Grant, 3248
Radcliffe Institute Residential Fellowships, 3299
RCF The Women's Fund Grants, 3315
Warrick County Community Foundation Women's Fund, 3905
WestWind Foundation Reproductive Health and Rights Grants, 3924

Women's Studies
AMA-WPC Joan F. Giambalvo Memorial Scholarships, 480
Radcliffe Institute Residential Fellowships, 3299
Stark Community Fndn Women's Grants, 3645

Work Motivation
Elizabeth Morse Genius Charitable Trust Grants, 1569

Workforce Development
Alfred E. Chase Charitable Foundation Grants, 430
Boeing Company Contributions Grants, 930
CDC Public Health Associates, 1065
CDC Public Health Associates Hosts, 1066
Community Foundation for Southeast Michigan Grants, 1258
Elizabeth Morse Genius Charitable Trust Grants, 1569
Foundation for the Mid South Community Development Grants, 1699
Frederick W. Marzahl Memorial Fund Grants, 1741
GNOF New Orleans Works Grants, 1844
Harold Brooks Foundation Grants, 1952
Harry S. Black and Allon Fuller Fund Grants, 1966
Helen Bader Foundation Grants, 1992
KeyBank Foundation Grants, 2307
Lincoln Financial Foundation Grants, 2367
NWHF Partners Investing in Nursing's Future, 3036
Piper Trust Older Adults Grants, 3229
PMI Foundation Grants, 3239
Priddy Fndn Organizational Development Grants, 3265
RCF General Community Grants, 3312
Reinberger Foundation Grants, 3357
Robert R. McCormick Trib Community Grants, 3385
Robert R. McCormick Tribune Veterans Initiative Grants, 3386
UniHealth Fndn Workforce Development Grants, 3786
Union Bank, N.A. Foundation Grants, 3788
Walmart Foundation Community Giving Grants, 3897
Walmart Foundation National Giving Grants, 3898

Workshops
AES Research and Training Workshop Awards, 320
Blue Cross Blue Shield of Minnesota Fndn Healthy Equity: Public Libraries for Health Grants, 918
CDC David J. Sencer Museum Teacher Professional Development Workshops, 1053
CJ Foundation for SIDS Program Services Grants, 1163
GNOF Organizational Effectiveness Grants and Workshops, 1846
IBRO-PERC Support for European Workshops, Symposia and Meetings, 2098
IBRO Latin America Regional Funding for Short Courses, Workshops, and Symposia, 2107
IBRO Latin America Regional Travel Grants, 2109
Meyer Foundation Benevon Grants, 2532
Orthopaedic Trauma Association Kathy Cramer Young Clinician Memorial Fellowships, 3083
Pfizer/AAFP Foundation Visiting Professorship Program in Family Medicine, 3162
SfN Science Educator Award, 3535

Wound Healing
IAFF Burn Foundation Research Grants, 2089
Plastic Surgery Educational Foundation Basic Research Grants, 3235

Writers in Residence
IIE Rockefeller Foundation Bellagio Center Residencies, 2142

Writing
AAAS/Subaru SB&F Prize for Excellence in Science Books, 35
AIHS Media Summer Fellowship, 392
BCBSM Fndn Proposal Development Awards, 857
IIE Rockefeller Foundation Bellagio Center Residencies, 2142
Jayne and Leonard Abess Foundation Grants, 2214
MLA Murray Gottlieb Prize Essay Award, 2571
Olive Higgins Prouty Foundation Grants, 3072
Phi Upsilon Orinne Johnson Writing Award, 3198
SfN Science Journalism Student Awards, 3536
Victor Grifols i Lucas Foundation Prize for Journalistic Work on Bioethics, 3870

Writing/Composition Education
Baptist Community Ministries Grants, 828
Emily Davie and Joseph S. Kornfeld Foundation Grants, 1591

828 / Writing/Composition Education

Houston Endowment Grants, 2053
IIE Rockefeller Foundation Bellagio Center Residencies, 2142

Youth Programs
3M Company Fndn Community Giving Grants, 6
Aaron Foundation Grants, 193
Abbot and Dorothy H. Stevens Foundation Grants, 197
Abell-Hanger Foundation Grants, 207
ACF Native American Social and Economic Development Strategies Grants, 231
Adams and Reese Corporate Giving Grants, 299
Adaptec Foundation Grants, 304
Adelaide Breed Bayrd Foundation Grants, 306
Adolph Coors Foundation Grants, 311
Agnes M. Lindsay Trust Grants, 346
Agway Foundation Grants, 348
Ahmanson Foundation Grants, 355
Alberto Culver Corporate Contributions Grants, 408
Albert Pick Jr. Fund Grants, 409
Albertson's Charitable Giving Grants, 410
Albert W. Rice Charitable Foundation Grants, 411
Alcatel-Lucent Technologies Foundation Grants, 413
Alice Tweed Tuohy Foundation Grants Program, 437
Allegan County Community Foundation Grants, 439
Allen Foundation Educational Nutrition Grants, 442
Alliance Healthcare Foundation Grants, 444
Allyn Foundation Grants, 448
Alpha Natural Resources Corporate Giving, 450
Amelia Sillman Rockwell and Carlos Perry Rockwell Charities Fund Grants, 501
American Foodservice Charitable Trust Grants, 532
Andersen Corporate Foundation, 586
Andre Agassi Charitable Foundation Grants, 587
Andrew Family Foundation Grants, 588
Ann and Robert H. Lurie Family Fndn Grants, 593
Ann Arbor Area Community Foundation Grants, 594
Annenberg Foundation Grants, 596
Anschutz Family Foundation Grants, 602
Anthony R. Abraham Foundation Grants, 608
Antone & Edene Vidinha Charitable Trust Grants, 609
Aragona Family Foundation Grants, 648
Archer Daniels Midland Foundation Grants, 651
Arizona Diamondbacks Charities Grants, 657
Arkansas Community Foundation Arkansas Black Hall of Fame Grants, 659
AT&T Foundation Civic and Community Service Program Grants, 793
Athwin Foundation Grants, 795
Atlanta Foundation Grants, 796
Autzen Foundation Grants, 810
Bacon Family Foundation Grants, 817
Bailey Foundation Grants, 818
BancorpSouth Foundation Grants, 823
Barker Welfare Foundation Grants, 831
Batchelor Foundation Grants, 837
Beazley Foundation Grants, 865
Beirne Carter Foundation Grants, 870
Belk Foundation Grants, 872
Ben B. Cheney Foundation Grants, 875
Bender Foundation Grants, 876
Benton Community Foundation Grants, 878
Benton County Foundation Grants, 880
Berks County Community Foundation Grants, 881
Berrien Community Foundation Grants, 884
Birmingham Foundation Grants, 903
Blanche and Julian Robertson Family Foundation Grants, 911
Blue River Community Foundation Grants, 922
Blumenthal Foundation Grants, 925
Bodenwein Public Benevolent Foundation Grants, 928
Bodman Foundation Grants, 929
Boston Foundation Grants, 936
Boston Globe Foundation Grants, 937
Bright Family Foundation Grants, 951
Brown County Community Foundation Grants, 970
Browning-Kimball Foundation Grants, 971
Bullitt Foundation Grants, 973
Burlington Industries Foundation Grants, 977
Bush Fndn Health & Human Services Grants, 979
Caddock Foundation Grants, 994
Callaway Golf Company Foundation Grants, 1003
Camp-Younts Foundation Grants, 1005
Cardinal Health Foundation Grants, 1017
Cargill Citizenship Fund Corporate Giving, 1018
Carl & Eloise Pohlad Family Fndn Grants, 1021
Carl B. and Florence E. King Foundation Grants, 1022
Carl R. Hendrickson Family Foundation Grants, 1028
Carnahan-Jackson Foundation Grants, 1032
Carnegie Corporation of New York Grants, 1033
Carrie E. and Lena V. Glenn Foundation Grants, 1036
Carrie Estelle Doheny Foundation Grants, 1037
Carrier Corporation Contributions Grants, 1038
Carroll County Community Foundation Grants, 1039
Cemala Foundation Grants, 1077
Cessna Foundation Grants Program, 1081
CFFVR Jewelers Mutual Charitable Giving, 1099
CFFVR Myra M. and Robert L. Vandehey Foundation Grants, 1100
CFFVR Robert and Patricia Endries Family Foundation Grants, 1102
CFFVR Schmidt Family G4 Grants, 1103
CFFVR Shawano Area Community Foundation Grants, 1104
Champlin Foundations Grants, 1110
Christine & Katharina Pauly Trust Grants, 1150
Christine Gregoire Youth/Young Adult Award for Use of Tobacco Industry Documents, 1151
Clark-Winchcole Foundation Grants, 1173
Clark and Ruby Baker Foundation Grants, 1174
CLIF Bar Family Foundation Grants, 1183
Coca-Cola Foundation Grants, 1203
Cockrell Foundation Grants, 1204
Collins C. Diboll Private Foundation Grants, 1209
Collins Foundation Grants, 1210
Colonel Stanley R. McNeil Foundation Grants, 1211
Communities Foundation of Texas Grants, 1236
Community Foundation for Greater New Haven Women & Girls Grants, 1252
Community Foundation for Northeast Michigan Tobacco Settlement Grants, 1256
Community Fndn for San Benito County Grants, 1257
Community Foundation of Bartholomew County Heritage Fund Grants, 1262
Community Foundation of Bartholomew County James A. Henderson Award for Fundraising, 1263
Community Foundation of Boone County Grants, 1266
Community Fndn of East Central Illinois Grants, 1268
Community Foundation of Eastern Connecticut General Southeast Grants, 1269
Community Foundation of Greater Fort Wayne - Community Endowment and Clarke Endowment Grants, 1274
Community Fndn of Greater Tampa Grants, 1282
Community Foundation of Jackson County Seymour Noon Lions Club Grant, 1288
Community Foundation of Louisville Dr. W. Barnett Owen Memorial Fund for the Children of Louisville and Jefferson County Grants, 1294
Community Foundation of Mount Vernon and Knox County Grants, 1299
Community Fndn of Randolph County Grants, 1302
Community Foundation of Riverside & San Bernardino County Impact Grants, 1303
Community Foundation of St. Joseph County Special Project Challenge Grants, 1310
Community Fndn of Switzerland County Grants, 1311
Community Foundation of the Verdugos Educational Endowment Fund Grants, 1314
Community Foundation of the Verdugos Grants, 1315
Community Foundation Serving Riverside and San Bernardino Counties Impact Grants, 1319
ConAgra Foods Nourish our Community Grants, 1322
Connecticut Community Foundation Grants, 1324
ConocoPhillips Foundation Grants, 1327
Constantin Foundation Grants, 1341
Constellation Energy Corporate Grants, 1342
Cooke Foundation Grants, 1344
Cooper Industries Foundation Grants, 1345
Cornerstone Foundation of Northeastern Wisconsin Grants, 1347
Cowles Charitable Trust Grants, 1354
Cralle Foundation Grants, 1356
Cresap Family Foundation Grants, 1359
Cruise Industry Charitable Foundation Grants, 1363
Cudd Foundation Grants, 1370
CUNA Mutual Group Fndn Community Grants, 1375
Cyrus Eaton Foundation Grants, 1380
D.F. Halton Foundation Grants, 1405
D. W. McMillan Foundation Grants, 1407
Dade Community Foundation Grants, 1410
Dairy Queen Corporate Contributions Grants, 1413
Dallas Women's Foundation Grants, 1417
Dana Brown Charitable Trust Grants, 1419
Daphne Seybolt Culpeper Fndn Grants, 1425
Dayton Power and Light Company Foundation Signature Grants, 1436
Daywood Foundation Grants, 1438
Dearborn Community Foundation City of Lawrenceburg Community Grants, 1441
Deborah Munroe Noonan Memorial Grants, 1443
Decatur County Community Foundation Large Project Grants, 1444
Decatur County Community Foundation Small Project Grants, 1445
DeKalb County Community Foundation - Garrett Hospital Aid Foundation Grants, 1446
DeKalb County Community Foundation Grants, 1447
Delaware Community Foundation-Youth Philanthropy Board for Kent County, 1448
Delmarva Power and Light Contributions, 1454
Dennis & Phyllis Washington Fndn Grants, 1458
Denver Broncos Charities Fund Grants, 1460
DeRoy Testamentary Foundation Grants, 1469
Dickson Foundation Grants, 1476
Do Something Awards, 1503
Dubois County Community Foundation Grants, 1517
Dunspaugh-Dalton Foundation Grants, 1530
Eberly Foundation Grants, 1544
Edward and Romell Ackley Foundation Grants, 1555
Edwin S. Webster Foundation Grants, 1560
Edwin W. and Catherine M. Davis Fndn Grants, 1561
Eisner Foundation Grants, 1564
El Paso Community Foundation Grants, 1579
Elsie Lee Garthwaite Memorial Fndn Grants, 1584
Emerson Charitable Trust Grants, 1589
Ensworth Charitable Foundation Grants, 1595
Essex County Community Foundation Discretionary Fund Grants, 1602
Essex County Community Foundation Webster Family Fund Grants, 1604
Evjue Foundation Grants, 1617
Ewing Halsell Foundation Grants, 1618
F.M. Kirby Foundation Grants, 1624
Fargo-Moorhead Area Foundation Grants, 1632
Fargo-Moorhead Area Fndn Woman's Grants, 1633
Farmers Insurance Corporate Giving Grants, 1634
Fayette County Foundation Grants, 1636
Ferree Foundation Grants, 1658
Field Foundation of Illinois Grants, 1660
FirstEnergy Foundation Community Grants, 1668
Florence Hunt Maxwell Foundation Grants, 1676
Foellinger Foundation Grants, 1684
Fondren Foundation Grants, 1685
Ford Motor Company Fund Grants Program, 1687
Four County Community Fndn General Grants, 1710
Four County Community Foundation Healthy Senior/Healthy Youth Fund Grants, 1711
Francis T. & Louise T. Nichols Fndn Grants, 1720
Frank B. Hazard General Charity Fund Grants, 1721
Frank G. and Freida K. Brotz Family Foundation Grants, 1723
Frank Reed and Margaret Jane Peters Memorial Fund I Grants, 1727
Frank Stanley Beveridge Foundation Grants, 1730
Frederick W. Marzahl Memorial Fund Grants, 1741
Fremont Area Community Foundation Amazing X Grants, 1744
Fremont Area Community Fndn General Grants, 1746

SUBJECT INDEX

Fuller E. Callaway Foundation Grants, 1761
G.N. Wilcox Trust Grants, 1768
Gamble Foundation Grants, 1769
General Mills Champs for Healthy Kids Grants, 1781
George A. and Grace L. Long Foundation Grants, 1788
George and Ruth Bradford Foundation Grants, 1789
George Foundation Grants, 1795
George W. Brackenridge Foundation Grants, 1802
George W. Wells Foundation Grants, 1804
Gertrude and William C. Wardlaw Fund Grants, 1808
Gil and Dody Weaver Foundation Grants, 1819
GNOF Norco Community Grants, 1845
Godfrey Foundation Grants, 1849
Golden Heart Community Foundation Grants, 1850
Grace and Franklin Bernsen Foundation Grants, 1858
Graco Foundation Grants, 1860
Grand Haven Area Community Fndn Grants, 1864
Grand Rapids Area Community Foundation Wyoming Youth Fund Grants, 1868
Grand Rapids Community Foundation Grants, 1869
Grand Rapids Community Foundation Ionia County Youth Fund Grants, 1871
Grand Rapids Community Foundation Southeast Ottawa Youth Fund Grants, 1874
Grand Rapids Community Foundation Sparta Youth Fund Grants, 1876
Greater Milwaukee Foundation Grants, 1883
Greater Tacoma Community Foundation Grants, 1886
Greater Tacoma Community Foundation Ryan Alan Hade Endowment Fund, 1887
Greater Worcester Community Foundation Discretionary Grants, 1888
Green Bay Packers Foundation Grants, 1890
Green River Area Community Fndn Grants, 1895
Grover Hermann Foundation Grants, 1906
H.B. Fuller Foundation Grants, 1922
H.J. Heinz Company Foundation Grants, 1923
H. Leslie Hoffman and Elaine S. Hoffman Foundation Grants, 1924
Hagedorn Fund Grants, 1934
Harold and Arlene Schnitzer CARE Foundation Grants, 1950
Harold Simmons Foundation Grants, 1955
Harris and Eliza Kempner Fund Grants, 1956
Harry Bramhall Gilbert Charitable Trust Grants, 1961
Hartford Courant Foundation Grants, 1969
Hasbro Corporation Gift of Play Hospital and Pediatric Health Giving, 1974
Hawaii Community Foundation Reverend Takie Okumura Family Grants, 1977
HCA Foundation Grants, 1980
Healthcare Fndn for Orange County Grants, 1981
Helena Rubinstein Foundation Grants, 1991
Helen Irwin Littauer Educational Trust Grants, 1994
Helen K. & Arthur E. Johnson Fndn Grants, 1995
Helen S. Boylan Foundation Grants, 1997
Hendricks County Community Fndn Grants, 2000
Henrietta Tower Wurts Memorial Fndn Grants, 2002
Hill Crest Foundation Grants, 2030
Hillman Foundation Grants, 2032
Hilton Hotels Corporate Giving Program Grants, 2037
Hoglund Foundation Grants, 2040
Holland/Zeeland Community Fndn Grants, 2041
Horace A. Kimball and S. Ella Kimball Foundation Grants, 2046
Horace Moses Charitable Foundation Grants, 2047
Howard and Bush Foundation Grants, 2054
Huffy Foundation Grants, 2073
Humana Foundation Grants, 2077
Human Source Foundation Grants, 2078
Hut Foundation Grants, 2086
Hutton Foundation Grants, 2087
Illinois Tool Works Foundation Grants, 2147
J.M. Long Foundation Grants, 2173
J.W. Kieckhefer Foundation Grants, 2176
J. Walton Bissell Foundation Grants, 2177
Jack H. & William M. Light Trust Grants, 2179
Jacobs Family Village Neighborhoods Grants, 2184
James & Abigail Campbell Family Fndn Grants, 2185
James Graham Brown Foundation Grants, 2189

James J. & Angelia M. Harris Fndn Grants, 2192
James R. Thorpe Foundation Grants, 2199
James S. Copley Foundation Grants, 2200
Janirve Foundation Grants, 2209
Janus Foundation Grants, 2211
Jean and Louis Dreyfus Foundation Grants, 2215
Jeffris Wood Foundation Grants, 2217
JELD-WEN Foundation Grants, 2218
John Edward Fowler Memorial Fndn Grants, 2235
John G. Martin Foundation Grants, 2237
John H. and Wilhelmina D. Harland Charitable Foundation Children and Youth Grants, 2238
John Jewett and Helen Chandler Garland Foundation Grants, 2242
John W. and Anna H. Hanes Foundation Grants, 2261
John W. Anderson Foundation Grants, 2262
Joseph H. & Florence A. Roblee Fndn Grants, 2269
Josephine G. Russell Trust Grants, 2271
Josephine Goodyear Foundation Grants, 2272
Josephine Schell Russell Charitable Trust Grants, 2274
Josiah W. and Bessie H. Kline Foundation Grants, 2278
Judith Clark-Morrill Foundation Grants, 2281
Julius N. Frankel Foundation Grants, 2282
Katharine Matthies Foundation Grants, 2291
Katherine John Murphy Foundation Grants, 2293
Kathryne Beynon Foundation Grants, 2294
Kenneth T. & Eileen L. Norris Fndn Grants, 2300
Kimball International-Habig Foundation Health and Human Services Grants, 2308
Knox County Community Foundation Grants, 2313
Kosciusko County Community Fndn Grants, 2319
Kosciusko County Community Foundation REMC Operation Round Up Grants, 2320
L. W. Pierce Family Foundation Grants, 2328
Land O'Lakes Foundation Mid-Atlantic Grants, 2341
Leo Goodwin Foundation Grants, 2354
Liberty Bank Foundation Grants, 2362
Libra Foundation Grants, 2363
Lillian S. Wells Foundation Grants, 2365
Linford and Mildred White Charitable Grants, 2368
Lubrizol Foundation Grants, 2384
Lucile Horton Howe and Mitchell B. Howe Foundation Grants, 2385
Lucy Gooding Charitable Fndn Grants, 2388
Lumpkin Family Fnd Healthy People Grants, 2391
M.E. Raker Foundation Grants, 2404
M.J. Murdock Charitable Trust General Grants, 2405
Mabel A. Horne Trust Grants, 2406
Mabel F. Hoffman Charitable Trust Grants, 2407
Mabel Y. Hughes Charitable Trust Grants, 2409
Mardag Foundation Grants, 2428
Marie C. and Joseph C. Wilson Foundation Rochester Small Grants, 2431
Marie H. Bechtel Charitable Remainder Uni-Trust Grants, 2432
Mary Black Foundation Early Childhood Development Grants, 2442
Mary Owen Borden Foundation Grants, 2446
Mary S. and David C. Corbin Foundation Grants, 2447
Matilda R. Wilson Fund Grants, 2453
Max and Victoria Dreyfus Foundation Grants, 2457
Maxon Charitable Foundation Grants, 2458
McCombs Foundation Grants, 2491
McInerny Foundation Grants, 2495
McKesson Foundation Grants, 2496
McLean Contributionship Grants, 2497
McMillen Foundation Grants, 2499
Mead Witter Foundation Grants, 2507
Mericos Foundation Grants, 2518
Meriden Foundation Grants, 2519
Merrick Foundation Grants, 2521
Mervin Bovaird Foundation Grants, 2523
Meyer Memorial Trust Special Grants, 2538
MGN Family Foundation Grants, 2542
Mid-Iowa Health Foundation Community Response Grants, 2549
Middlesex Savings Charitable Foundation Capacity Building Grants, 2551
Minneapolis Foundation Community Grants, 2558
Nicholas H. Noyes Jr. Memorial Fndn Grants, 2791

Youth Programs /829

Noble County Community Foundation Grants, 2975
Norcliffe Foundation Grants, 2979
North Carolina Community Foundation Grants, 2990
North Dakota Community Foundation Grants, 2994
Northwestern Mutual Foundation Grants, 2999
NYCT Girls and Young Women Grants, 3053
Oleonda Jameson Trust Grants, 3071
OneFamily Foundation Grants, 3075
Oppenstein Brothers Foundation Grants, 3078
Ordean Foundation Grants, 3080
Parke County Community Foundation Grants, 3115
Patrick and Anna M. Cudahy Fund Grants, 3120
Paul Rapoport Foundation Grants, 3125
Peacock Foundation Grants, 3133
Perkins Charitable Foundation Grants, 3140
Perpetual Trust for Charitable Giving Grants, 3141
Perry County Community Foundation Grants, 3142
Peter and Elizabeth C. Tower Foundation Annual Intellectual Disabilities Grants, 3144
Peter and Elizabeth C. Tower Foundation Annual Mental Health Grants, 3145
Peter and Elizabeth C. Tower Foundation Learning Disability Grants, 3146
Peter and Elizabeth C. Tower Foundation Mental Health Reference and Resource Materials Mini-Grants, 3147
Peter and Elizabeth C. Tower Foundation Phase II Technology Initiative Grants, 3149
Peter and Elizabeth C. Tower Foundation Phase I Technology Initiative Grants, 3150
Peter and Elizabeth C. Tower Foundation Social and Emotional Preschool Curriculum Grants, 3151
Peter and Elizabeth C. Tower Foundation Substance Abuse Grants, 3152
Peter Kiewit Foundation General Grants, 3154
Peter Kiewit Foundation Small Grants, 3155
PeyBack Foundation Grants, 3159
Piedmont Natural Gas Corporate and Charitable Contributions, 3220
Piedmont Natural Gas Foundation Health and Human Services Grants, 3221
Pike County Community Foundation Grants, 3222
Pinkerton Foundation Grants, 3224
Pinnacle Foundation Grants, 3225
Piper Jaffray Fndn Communities Giving Grants, 3227
Plough Foundation Grants, 3238
PMI Foundation Grants, 3239
Pohlad Family Foundation, 3241
Polk Bros. Foundation Grants, 3243
Pollock Foundation Grants, 3244
Porter County Health and Wellness Grant, 3246
Portland Fndn Women's Giving Circle Grant, 3249
Posey Community Fndn Women's Fund Grants, 3251
Posey County Community Foundation Grants, 3252
Pott Foundation Grants, 3253
Price Chopper's Golub Foundation Grants, 3261
Priddy Foundation Program Grants, 3266
Pride Foundation Grants, 3268
Prince Charitable Trusts DC Grants, 3271
Procter and Gamble Fund Grants, 3274
Prudential Foundation Education Grants, 3276
R.C. Baker Foundation Grants, 3292
Ralph M. Parsons Foundation Grants, 3304
Ralphs Food 4 Less Foundation Grants, 3305
Raskob Foundation for Catholic Activities Grants, 3306
Rhode Island Foundation Grants, 3362
Rice Foundation Grants, 3363
Richard King Mellon Foundation Grants, 3369
Robert B McMillen Foundation Grants, 3381
Robert R. McCormick Trib Community Grants, 3385
Rochester Area Foundation Grants, 3392
Rockwell Collins Charitable Corporation Grants, 3397
Rockwell International Corporate Trust Grants Program, 3399
Rohm and Haas Company Grants, 3403
Ronald McDonald House Charities Grants, 3405
Rose Hills Foundation Grants, 3413
Roy & Christine Sturgis Charitable Grants, 3415
Roy J. Carver Charitable Trust Medical and Science Research Grants, 3416

830 / Youth Programs

Rush County Community Foundation Grants, 3424
Ruth Eleanor Bamberger and John Ernest Bamberger Memorial Foundation Grants, 3427
S. Livingston Mather Charitable Trust Grants, 3448
Saint Louis Rams Fndn Community Donations, 3453
Salmon Foundation Grants, 3458
SAMHSA Strategic Prevention Framework State Incentive Grants, 3464
Samuel S. Johnson Foundation Grants, 3468
San Diego Foundation for Change Grants, 3470
Sands Foundation Grants, 3476
Schlessman Family Foundation Grants, 3493
Schurz Communications Foundation Grants, 3496
Scott County Community Foundation Grants, 3500
Seabury Foundation Grants, 3501
Seattle Foundation Benjamin N. Phillips Memorial Fund Grants, 3504
Self Foundation Grants, 3510
Seneca Foods Foundation Grants, 3511
Sid W. Richardson Foundation Grants, 3553
Sierra Health Foundation Responsive Grants, 3555
Sonoco Foundation Grants, 3621
Southbury Community Trust Fund, 3625
Southwest Gas Corporation Foundation Grants, 3628
Spencer County Community Foundation Grants, 3632
Stackpole-Hall Foundation Grants, 3643
Sterling and Shelli Gardner Foundation Grants, 3653
Strowd Roses Grants, 3664
Sunderland Foundation Grants, 3675
SunTrust Bank Trusteed Foundations Florence C. and Harry L. English Memorial Fund Grants, 3680
SunTrust Bank Trusteed Foundations Greene-Sawtell Grants, 3681
SunTrust Bank Trusteed Foundations Harriet McDaniel Marshall Tust Grants, 3682
SunTrust Bank Trusteed Foundations Nell Warren Elkin and William Simpson Elkin Grants, 3683
SunTrust Bank Trusteed Foundations Thomas Guy Woolford Charitable Trust Grants, 3684
SunTrust Bank Trusteed Foundations Walter H. and Marjory M. Rich Memorial Fund Grants, 3685
Susan Mott Webb Charitable Trust Grants, 3694
Tauck Family Foundation Grants, 3703
Tension Envelope Foundation Grants, 3707
Thomas and Dorothy Leavey Foundation Grants, 3717
Thomas Austin Finch, Sr. Foundation Grants, 3718
Thomas C. Ackerman Foundation Grants, 3719
Thomas J. Long Foundation Community Grants, 3722
Thompson Charitable Foundation Grants, 3727
Thorman Boyle Foundation Grants, 3730
Toyota Motor Manuf of Mississippi Grants, 3746
Trull Foundation Grants, 3761
Tull Charitable Foundation Grants, 3762
Union Bank, N.A. Foundation Grants, 3788
Union Benevolent Association Grants, 3789
Union Labor Health Foundation Dental Angel Fund Grants, 3793
United Healthcare Comm Grants in Michigan, 3796
Vancouver Foundation Grants and Community Initiatives Program, 3849
Vanderburgh Community Foundation Grants, 3852
Victor E. Speas Foundation Grants, 3868
W.K. Kellogg Foundation Healthy Kids Grants, 3886
Walker Area Community Foundation Grants, 3895
Walmart Foundation National Giving Grants, 3898
Warrick County Community Foundation Grants, 3904
Washington County Community Fndn Grants, 3906
Weaver Popcorn Foundation Grants, 3910
Wege Foundation Grants, 3912
WestWind Foundation Reproductive Health and Rights Grants, 3924
Weyerhaeuser Company Foundation Grants, 3925
William B. Stokely Jr. Foundation Grants, 3947
William Blair and Company Foundation Grants, 3948
William G. Gilmore Foundation Grants, 3951
William T. Grant Foundation Research Grants, 3958
Wilson-Wood Foundation Grants, 3960
Wood-Claeyssens Foundation Grants, 3971
Yawkey Foundation Grants, 3980

Youth Services
ACF Native American Social and Economic Development Strategies Grants, 231
ACL Partnerships in Employment Systems Change Grants, 246
Adams and Reese Corporate Giving Grants, 299
Alfred and Tillie Shemanski Testamentary Trust Grants, 428
Allan C. and Lelia J. Garden Foundation Grants, 438
Allen Foundation Educational Nutrition Grants, 442
Amelia Sillman Rockwell and Carlos Perry Rockwell Charities Fund Grants, 501
Baxter International Corporate Giving Grants, 841
Ben B. Cheney Foundation Grants, 875
Browning-Kimball Foundation Grants, 971
Caesars Foundation Grants, 995
CFFVR Jewelers Mutual Charitable Giving, 1099
Charles H. Pearson Foundation Grants, 1121
Christine & Katharina Pauly Trust Grants, 1150
Community Fndn for San Benito County Grants, 1257
Community Foundation of Louisville Diller B. and Katherine P. Groff Fund for Pediatric Surgery Grants, 1293
Community Foundation of Louisville Dr. W. Barnett Owen Memorial Fund for the Children of Louisville and Jefferson County Grants, 1294
Community Fndn of Riverside & San Bernardino County Irene S. Rockwell Grants, 1304
Crane Fund Grants, 1358
CUNA Mutual Group Fndn Community Grants, 1375
Deborah Munroe Noonan Memorial Grants, 1443
Decatur County Community Foundation Large Project Grants, 1444
Decatur County Community Foundation Small Project Grants, 1445
Delmarva Power and Light Contributions, 1454
Edward and Romell Ackley Foundation Grants, 1555
Eugene Straus Charitable Trust, 1612
Ezra M. Cutting Trust Grants, 1623
Four County Community Fndn General Grants, 1710
Four County Community Foundation Healthy Senior/Healthy Youth Fund Grants, 1711
Frederick McDonald Trust Grants, 1740
George H.C. Ensworth Memorial Fund Grants, 1797
George W. Wells Foundation Grants, 1804
Gil and Dody Weaver Foundation Grants, 1819
GNOF Norco Community Grants, 1845
Golden Heart Community Foundation Grants, 1850
Green River Area Community Fndn Grants, 1895
Greenspun Family Foundation Grants, 1896
Grover Hermann Foundation Grants, 1906
Jack H. & William M. Light Trust Grants, 2179
John W. Speas and Effie E. Speas Memorial Trust Grants, 2264
Josephine Schell Russell Charitable Trust Grants, 2274
Judith Clark-Morrill Foundation Grants, 2281
Julius N. Frankel Foundation Grants, 2282
Kimball International-Habig Foundation Health and Human Services Grants, 2308
Kovler Family Foundation Grants, 2323
Lake County Community Fund Grants, 2333
M.D. Anderson Foundation Grants, 2403
Mary Black Foundation Early Childhood Development Grants, 2442
McCune Foundation Human Services Grants, 2493
Merrick Foundation Grants, 2521
Middlesex Savings Charitable Foundation Capacity Building Grants, 2551
Oppenstein Brothers Foundation Grants, 3078
Peter and Elizabeth C. Tower Foundation Annual Intellectual Disabilities Grants, 3144
Peter and Elizabeth C. Tower Foundation Annual Mental Health Grants, 3145
Peter and Elizabeth C. Tower Foundation Learning Disability Grants, 3146
Peter and Elizabeth C. Tower Foundation Mental Health Reference and Resource Materials Mini-Grants, 3147
Peter and Elizabeth C. Tower Foundation Phase II Technology Initiative Grants, 3149

Peter and Elizabeth C. Tower Foundation Phase I Technology Initiative Grants, 3150
Peter and Elizabeth C. Tower Foundation Social and Emotional Preschool Curriculum Grants, 3151
Peter and Elizabeth C. Tower Foundation Substance Abuse Grants, 3152
Piedmont Natural Gas Foundation Health and Human Services Grants, 3221
PMI Foundation Grants, 3239
Porter County Health and Wellness Grant, 3246
Portland Fndn Women's Giving Circle Grant, 3249
Price Chopper's Golub Foundation Grants, 3261
Reinberger Foundation Grants, 3357
Richard Davoud Donchian Foundation Grants, 3367
RWJF Childhood Obesity Grants, 3431
Sandy Hill Foundation Grants, 3477
Seattle Foundation Benjamin N. Phillips Memorial Fund Grants, 3504
Sterling and Shelli Gardner Foundation Grants, 3653
Thorman Boyle Foundation Grants, 3730
Toyota Motor Manuf of Mississippi Grants, 3746
Toyota Motor Manufacturing of Texas Grants, 3747
Union Labor Health Foundation Dental Angel Fund Grants, 3793
Union Pacific Foundation Health and Human Services Grants, 3794
United Healthcare Comm Grants in Michigan, 3796
Vanderburgh Community Foundation Grants, 3852
Walker Area Community Foundation Grants, 3895
Walmart Foundation National Giving Grants, 3898
Washington County Community Fndn Grants, 3906
Weaver Popcorn Foundation Grants, 3910
Wolfe Associates Grants, 3969
Wood-Claeyssens Foundation Grants, 3971

Youth Violence
Alliance Healthcare Foundation Grants, 444
Gamble Foundation Grants, 1769
Kansas Health Foundation Recognition Grants, 2289
Ms. Fndn for Women Ending Violence Grants, 2596
Ordean Foundation Grants, 3080
Piedmont Natural Gas Foundation Health and Human Services Grants, 3221
William T. Grant Foundation Research Grants, 3958

Zoology
DeRoy Testamentary Foundation Grants, 1469

Zoos
Blum-Kovler Foundation Grants, 924
Cessna Foundation Grants Program, 1081
Chamberlain Foundation Grants, 1108
DeRoy Testamentary Foundation Grants, 1469
FirstEnergy Foundation Community Grants, 1668
George F. Baker Trust Grants, 1793
Katherine John Murphy Foundation Grants, 2293
Oscar Rennebohm Foundation Grants, 3085
Perry County Community Foundation Grants, 3142
Pike County Community Foundation Grants, 3222
PMI Foundation Grants, 3239
Posey County Community Foundation Grants, 3252
Procter and Gamble Fund Grants, 3274
Reinberger Foundation Grants, 3357
Saint Louis Rams Fndn Community Donations, 3453
Schlessman Family Foundation Grants, 3493
Shell Deer Park Grants, 3543
Sioux Falls Area Community Foundation Field-of-Interest and Donor-Advised Grants, 3569
Spencer County Community Foundation Grants, 3632
Thelma Doelger Charitable Trust Grants, 3714
Warrick County Community Foundation Grants, 3904

Program Type Index

NOTE: Numbers refer to entry numbers

Adult Basic Education
ACE Charitable Foundation Grants, 229
Achelis Foundation Grants, 234
Adolph Coors Foundation Grants, 311
Ahmanson Foundation Grants, 355
Albert W. Rice Charitable Foundation Grants, 411
Allyn Foundation Grants, 448
Arkell Hall Foundation Grants, 661
Arlington Community Foundation Grants, 662
Atlanta Foundation Grants, 796
Auburn Foundation Grants, 798
Autauga Area Community Foundation Grants, 808
Ball Brothers Foundation General Grants, 820
Battle Creek Community Foundation Grants, 839
Beckman Coulter Foundation Grants, 868
Benton Community Foundation Grants, 878
Benton County Foundation Grants, 880
Blue Cross Blue Shield of Minnesota Fndn Healthy Equity: Public Libraries for Health Grants, 918
Blue Mountain Community Foundation Grants, 921
Blue River Community Foundation Grants, 922
Bodenwein Public Benevolent Foundation Grants, 928
Bodman Foundation Grants, 929
Boettcher Foundation Grants, 931
Booth-Bricker Fund Grants, 933
Boston Foundation Grants, 936
Boston Globe Foundation Grants, 937
Bridgestone/Firestone Trust Fund Grants, 949
Brook J. Lenfest Foundation Grants, 967
Brown County Community Foundation Grants, 970
Carl R. Hendrickson Family Foundation Grants, 1028
Carrie Estelle Doheny Foundation Grants, 1037
Cass County Community Foundation Grants, 1041
CCHD Community Development Grants, 1046
Cemala Foundation Grants, 1077
Charles Nelson Robinson Fund Grants, 1125
Charlotte County (FL) Community Foundation Grants, 1126
Chilkat Valley Community Foundation Grants, 1148
CIGNA Foundation Grants, 1160
Clinton County Community Foundation Grants, 1184
CNCS AmeriCorps VISTA Project Grants, 1190
Columbus Foundation Competitive Grants, 1217
Community Fndn for the Capital Region Grants, 1259
Community Foundation of Bartholomew County Heritage Fund Grants, 1262
Community Foundation of Boone County Grants, 1266
Community Fndn of Central Illinois Grants, 1267
Community Foundation of Eastern Connecticut General Southeast Grants, 1269
Community Foundation of Greater Fort Wayne - Community Endowment and Clarke Endowment Grants, 1274
Community Foundation of Greater New Britain Grants, 1281
Constantin Foundation Grants, 1341
Coors Brewing Corporate Contributions Grants, 1346
Cornerstone Foundation of Northeastern Wisconsin Grants, 1347
Cowles Charitable Trust Grants, 1354
Daisy Marquis Jones Foundation Grants, 1414
Dayton Power and Light Foundation Grants, 1437
Dean Foods Community Involvement Grants, 1440
Dr. Scholl Foundation Grants Program, 1511
Edward W. and Stella C. Van Houten Memorial Fund Grants, 1559
Elizabeth Morse Genius Charitable Trust Grants, 1569
El Pomar Foundation Anna Keesling Ackerman Fund Grants, 1581
El Pomar Foundation Grants, 1582
Evjue Foundation Grants, 1617
Field Foundation of Illinois Grants, 1660
Florida Division of Cultural Affairs Arts In Education Arts Partnership Grants, 1679
Four County Community Fndn General Grants, 1710
Frances L. and Edwin L. Cummings Memorial Fund Grants, 1717

Frank Reed and Margaret Jane Peters Memorial Fund II Grants, 1728
G.N. Wilcox Trust Grants, 1768
George Foundation Grants, 1795
George W. Wells Foundation Grants, 1804
Goodrich Corporation Foundation Grants, 1854
Greater Sitka Legacy Fund Grants, 1885
Greygates Foundation Grants, 1901
Guido A. & Elizabeth H. Binda Fndn Grants, 1913
H.B. Fuller Foundation Grants, 1922
Hallmark Corporate Foundation Grants, 1936
Harden Foundation Grants, 1945
Harold Brooks Foundation Grants, 1952
Harold R. Bechtel Testamentary Charitable Trust Grants, 1954
Harold Simmons Foundation Grants, 1955
Helen Steiner Rice Foundation Grants, 1998
Hoblitzelle Foundation Grants, 2038
Howard and Bush Foundation Grants, 2054
Idaho Power Company Corporate Contributions, 2115
James A. and Faith Knight Foundation Grants, 2186
James Ford Bell Foundation Grants, 2187
James S. Copley Foundation Grants, 2200
Jessica Stevens Community Foundation Grants, 2224
John H. and Wilhelmina D. Harland Charitable Foundation Children and Youth Grants, 2238
John I. Smith Charities Grants, 2240
Joseph H. & Florence A. Roblee Fndn Grants, 2269
Joseph Henry Edmondson Foundation Grants, 2270
Ketchikan Community Foundation Grants, 2303
Kodiak Community Foundation Grants, 2314
Kuntz Foundation Grants, 2326
Leo Niessen Jr., Charitable Trust Grants, 2358
Liberty Bank Foundation Grants, 2362
Lubrizol Foundation Grants, 2384
Lydia deForest Charitable Trust Grants, 2393
Mardag Foundation Grants, 2428
Mary Wilmer Covey Charitable Trust Grants, 2448
May and Stanley Smith Charitable Trust Grants, 2459
McCarthy Family Foundation Grants, 2490
Nina Mason Pulliam Charitable Trust Grants, 2921
Norcliffe Foundation Grants, 2979
Oppenstein Brothers Foundation Grants, 3078
Oregon Community Fndn Community Grants, 3082
Parkersburg Area Community Foundation Action Grants, 3116
PepsiCo Foundation Grants, 3137
Peyton Anderson Foundation Grants, 3160
Piper Jaffray Fndn Communities Giving Grants, 3227
PMI Foundation Grants, 3239
Portland General Electric Foundation Grants, 3250
Princeton Area Community Foundation Greater Mercer Grants, 3272
Principal Financial Group Foundation Grants, 3273
Putnam County Community Foundation Grants, 3283
RCF Individual Assistance Grants, 3313
Reinberger Foundation Grants, 3357
Robert R. McCormick Trib Community Grants, 3385
Robert R. McCormick Tribune Veterans Initiative Grants, 3386
Robert R. Meyer Foundation Grants, 3387
Rose Hills Foundation Grants, 3413
San Antonio Area Foundation Grants, 3469
Seward Community Foundation Grants, 3514
Seward Community Foundation Mini-Grants, 3515
SfN Science Educator Award, 3535
Shield-Ayres Foundation Grants, 3546
Sisters of Charity Foundation of Cleveland Reducing Health and Educational Disparities in the Central Neighborhood Grants, 3574
Sony Corporation of America Grants, 3623
Texas Commission on the Arts Arts Respond Project Grants, 3708
Thomas Sill Foundation Grants, 3724
TJX Foundation Grants, 3735
Tri-State Community Twenty-first Century Endowment Fund Grants, 3756

U.S. Department of Education Rehabilitation Training - Rehabilitation Continuing Education - Institute on Rehabilitation Issues, 3771
USDA Organic Agriculture Research Grants, 3830
Vanderburgh Community Fndn Women's Fund, 3853
Vectren Foundation Grants, 3857
Verizon Foundation New York Grants, 3860
Verizon Foundation Virginia Grants, 3864
Walker Area Community Foundation Grants, 3895
Warrick County Community Foundation Women's Fund, 3905
Wayne County Foundation Grants, 3909
WHO Foundation General Grants, 3939
WHO Foundation Volunteer Service Grants, 3940
Wilson-Wood Foundation Grants, 3960

Adult/Family Literacy Training
AAFP Foundation Health Literacy State Grants, 132
Able Trust Vocational Rehabilitation Grants for Individuals, 213
ACE Charitable Foundation Grants, 229
Achelis Foundation Grants, 234
Adolph Coors Foundation Grants, 311
Ahmanson Foundation Grants, 355
Albert W. Rice Charitable Foundation Grants, 411
Alcatel-Lucent Technologies Foundation Grants, 413
Allyn Foundation Grants, 448
Alpha Natural Resources Corporate Giving, 450
Altman Foundation Health Care Grants, 457
Amador Community Foundation Grants, 481
Amgen Foundation Grants, 572
Andersen Corporate Foundation, 586
Annenberg Foundation Grants, 596
Anschutz Family Foundation Grants, 602
Arizona Cardinals Grants, 655
Arkell Hall Foundation Grants, 661
Arlington Community Foundation Grants, 662
ASHA Multicultural Activities Projects Grants, 697
Assisi Foundation of Memphis General Grants, 787
Atlanta Foundation Grants, 796
Auburn Foundation Grants, 798
Autauga Area Community Foundation Grants, 808
Bacon Family Foundation Grants, 817
Ball Brothers Foundation General Grants, 820
Barra Foundation Community Fund Grants, 833
Battle Creek Community Foundation Grants, 839
Bayer Foundation Grants, 846
BBVA Compass Foundation Charitable Grants, 848
Beazley Foundation Grants, 865
Beckman Coulter Foundation Grants, 868
Benton Community Foundation Grants, 878
Blue Cross Blue Shield of Minnesota Fndn Healthy Equity: Public Libraries for Health Grants, 918
Blue Mountain Community Foundation Grants, 921
Blue River Community Foundation Grants, 922
Bodenwein Public Benevolent Foundation Grants, 928
Bodman Foundation Grants, 929
Boettcher Foundation Grants, 931
Booth-Bricker Fund Grants, 933
Boston Foundation Grants, 936
Boston Globe Foundation Grants, 937
Brown County Community Foundation Grants, 970
Cabot Corporation Foundation Grants, 993
Caesars Foundation Grants, 995
Carl B. and Florence E. King Foundation Grants, 1022
Carl R. Hendrickson Family Foundation Grants, 1028
Carrie Estelle Doheny Foundation Grants, 1037
Cass County Community Foundation Grants, 1041
Cemala Foundation Grants, 1077
Central Okanagan Foundation Grants, 1080
CFFVR Basic Needs Giving Partnership Grants, 1094
CFFVR Schmidt Family G4 Grants, 1103
CFFVR Wisconsin King's Daus & Sons Grants, 1106
Charles Delmar Foundation Grants, 1115
Charles Nelson Robinson Fund Grants, 1125
Charlotte County (FL) Community Foundation Grants, 1126
Chilkat Valley Community Foundation Grants, 1148

CIGNA Foundation Grants, 1160
Cleveland-Cliffs Foundation Grants, 1181
Clinton County Community Foundation Grants, 1184
CNA Foundation Grants, 1189
CNCS AmeriCorps VISTA Project Grants, 1190
CNO Financial Group Community Grants, 1201
Coca-Cola Foundation Grants, 1203
Columbus Foundation Competitive Grants, 1217
Comerica Charitable Foundation Grants, 1224
Community Foundation for Northeast Michigan Mini-Grants, 1255
Community Fndn for the Capital Region Grants, 1259
Community Foundation for the National Capital Region Community Leadership Grants, 1260
Community Foundation of Bartholomew County Heritage Fund Grants, 1262
Community Foundation of Boone County Grants, 1266
Community Fndn of Central Illinois Grants, 1267
Community Foundation of Eastern Connecticut General Southeast Grants, 1269
Community Foundation of Greater Birmingham Grants, 1272
Community Foundation of Greater Fort Wayne - Community Endowment and Clarke Endowment Grants, 1274
Community Foundation of Greenville Hollingsworth Funds Program/Project Grants, 1285
Community Foundation of Muncie and Delaware County Maxon Grants, 1301
Community Foundation of the Verdugos Educational Endowment Fund Grants, 1314
Community Foundation of the Verdugos Grants, 1315
Cooper Industries Foundation Grants, 1345
Coors Brewing Corporate Contributions Grants, 1346
Cornerstone Foundation of Northeastern Wisconsin Grants, 1347
Cowles Charitable Trust Grants, 1354
Crail-Johnson Foundation Grants, 1355
Cruise Industry Charitable Foundation Grants, 1363
Cumberland Community Foundation Grants, 1374
Daisy Marquis Jones Foundation Grants, 1414
Dallas Women's Foundation Grants, 1417
Danellie Foundation Grants, 1421
Dayton Power and Light Foundation Grants, 1437
Deaconess Community Foundation Grants, 1439
Denver Foundation Community Grants, 1461
Dr. Scholl Foundation Grants Program, 1511
Edward W. and Stella C. Van Houten Memorial Fund Grants, 1559
Elizabeth Morse Genius Charitable Trust Grants, 1569
El Paso Community Foundation Grants, 1579
Entergy Corporation Micro Grants, 1596
Essex County Community Foundation Merrimack Valley General Fund Grants, 1603
Evjue Foundation Grants, 1617
Farmers Insurance Corporate Giving Grants, 1634
Faye McBeath Foundation Grants, 1635
Field Foundation of Illinois Grants, 1660
Fisher Foundation Grants, 1670
Four County Community Fndn General Grants, 1710
Fourjay Foundation Grants, 1712
Frances L. and Edwin L. Cummings Memorial Fund Grants, 1717
Frank Reed and Margaret Jane Peters Memorial Fund II Grants, 1728
Fred & Gretel Biel Charitable Trust Grants, 1734
Fremont Area Community Foundation Amazing X Grants, 1744
G.N. Wilcox Trust Grants, 1768
GenCorp Foundation Grants, 1779
Genuardi Family Foundation Grants, 1787
George Foundation Grants, 1795
George W. Wells Foundation Grants, 1804
Gibson Foundation Grants, 1818
Grand Rapids Community Foundation Ionia County Youth Fund Grants, 1871
Greater Sitka Legacy Fund Grants, 1885
Green Bay Packers Foundation Grants, 1890
Guido A. & Elizabeth H. Binda Fndn Grants, 1913
H.B. Fuller Foundation Grants, 1922

Hallmark Corporate Foundation Grants, 1936
Hancock County Community Foundation - Field of Interest Grants, 1943
Harold Brooks Foundation Grants, 1952
Harold Simmons Foundation Grants, 1955
Harry S. Black and Allon Fuller Fund Grants, 1966
Harvest Foundation Grants, 1971
Hasbro Children's Fund Grants, 1973
Heckscher Foundation for Children Grants, 1988
Helen Bader Foundation Grants, 1992
Helen Steiner Rice Foundation Grants, 1998
Highmark Corporate Giving Grants, 2025
Hilda and Preston Davis Foundation Grants, 2028
Hoblitzelle Foundation Grants, 2038
Hoglund Foundation Grants, 2040
Howard and Bush Foundation Grants, 2054
Humana Foundation Grants, 2077
Irvin Stern Foundation Grants, 2164
Jacobs Family Village Neighborhoods Grants, 2184
James & Abigail Campbell Family Fndn Grants, 2185
James A. and Faith Knight Foundation Grants, 2186
James S. Copley Foundation Grants, 2200
Jean and Louis Dreyfus Foundation Grants, 2215
Jessica Stevens Community Foundation Grants, 2224
Jessie Ball Dupont Fund Grants, 2226
Jim Moran Foundation Grants, 2228
John Edward Fowler Memorial Fndn Grants, 2235
John H. and Wilhelmina D. Harland Charitable Foundation Children and Youth Grants, 2238
John I. Smith Charities Grants, 2240
John Merck Fund Grants, 2245
John P. McGovern Foundation Grants, 2247
Joseph H. & Florence A. Roblee Fndn Grants, 2269
Joseph Henry Edmondson Foundation Grants, 2270
Kenai Peninsula Foundation Grants, 2298
Ketchikan Community Foundation Grants, 2303
Knight Family Charitable and Educational Foundation Grants, 2312
Kodiak Community Foundation Grants, 2314
Kuntz Foundation Grants, 2326
Leo Goodwin Foundation Grants, 2354
Leo Niessen Jr., Charitable Trust Grants, 2358
Lubrizol Foundation Grants, 2384
Lydia deForest Charitable Trust Grants, 2393
Macquarie Bank Foundation Grants, 2410
Mardag Foundation Grants, 2428
Marie C. and Joseph C. Wilson Foundation Rochester Small Grants, 2431
Mary Wilmer Covey Charitable Trust Grants, 2448
May and Stanley Smith Charitable Trust Grants, 2459
McGraw-Hill Companies Community Grants, 2494
Montgomery County Community Fndn Grants, 2591
Morris & Gwendolyn Cafritz Fndn Grants, 2594
Nationwide Insurance Foundation Grants, 2637
NLM Understanding and Promoting Health Literacy Research Grants, 2953
Norcliffe Foundation Grants, 2979
Nordson Corporation Foundation Grants, 2981
Northern Chautauqua Community Foundation Community Grants, 2995
Oregon Community Fndn Community Grants, 3082
PacifiCare Foundation Grants, 3102
Parkersburg Area Community Foundation Action Grants, 3116
PepsiCo Foundation Grants, 3137
Percy B. Ferebee Endowment Grants, 3138
Petersburg Community Foundation Grants, 3156
Pew Charitable Trusts Children & Youth Grants, 3158
Peyton Anderson Foundation Grants, 3160
Pinkerton Foundation Grants, 3224
PMI Foundation Grants, 3239
Price Chopper's Golub Foundation Grants, 3261
Principal Financial Group Foundation Grants, 3273
RCF Individual Assistance Grants, 3313
Reinberger Foundation Grants, 3357
Reynolds & Reynolds Associate Fndn Grants, 3360
RGk Foundation Grants, 3361
Richard Davoud Donchian Foundation Grants, 3367
Robert R. McCormick Trib Community Grants, 3385
Robert R. Meyer Foundation Grants, 3387

Rochester Area Community Foundation Grants, 3391
Rose Hills Foundation Grants, 3413
Saint Louis Rams Fndn Community Donations, 3453
Samuel S. Johnson Foundation Grants, 3468
San Antonio Area Foundation Grants, 3469
San Diego Women's Foundation Grants, 3474
Seattle Foundation Benjamin N. Phillips Memorial Fund Grants, 3504
Seaver Institute Grants, 3509
Seward Community Foundation Grants, 3514
Seward Community Foundation Mini-Grants, 3515
Shield-Ayres Foundation Grants, 3546
Sidney Stern Memorial Trust Grants, 3552
Sierra Health Foundation Responsive Grants, 3555
Sisters of Charity Foundation of Cleveland Reducing Health and Educational Disparities in the Central Neighborhood Grants, 3574
Sony Corporation of America Grants, 3623
Southbury Community Trust Fund, 3625
Stackpole-Hall Foundation Grants, 3643
Stark Community Fndn Women's Grants, 3645
Sterling-Turner Charitable Foundation Grants, 3652
Stocker Foundation Grants, 3656
Strowd Roses Grants, 3664
Textron Corporate Contributions Grants, 3712
Thomas Sill Foundation Grants, 3724
TJX Foundation Grants, 3735
Todd Brock Family Foundation Grants, 3736
Tri-State Community Twenty-first Century Endowment Fund Grants, 3756
Union Bank, N.A. Foundation Grants, 3788
US Airways Community Contributions, 3825
USDA Organic Agriculture Research Grants, 3830
Vanderburgh Community Fndn Women's Fund, 3853
Vectren Foundation Grants, 3857
Verizon Foundation Maryland Grants, 3859
Verizon Foundation New York Grants, 3860
Verizon Foundation Northeast Region Grants, 3861
Verizon Foundation Pennsylvania Grants, 3862
Verizon Foundation Vermont Grants, 3863
Verizon Foundation Virginia Grants, 3864
Walker Area Community Foundation Grants, 3895
Warrick County Community Foundation Women's Fund, 3905
Wayne County Foundation Grants, 3909
Western New York Foundation Grants, 3922
Western Union Foundation Grants, 3923
WHO Foundation General Grants, 3939
WHO Foundation Volunteer Service Grants, 3940
Wilson-Wood Foundation Grants, 3960
Winston-Salem Fndn Stokes County Grants, 3967

Awards/Prizes

AAAAI Allied Health Prof Recognition Award, 19
AAAAI Distinguished Layperson Award, 25
AAAAI Distinguished Scientist Award, 26
AAAAI Distinguished Service Award, 27
AAAAI Fellows-in-Training Abstract Award, 28
AAAAI Mentorship Award, 31
AAAAI Outstanding Vol Clinical Faculty Award, 32
AAAAI RSLAAIS Leadership Award, 33
AAAAI Special Recognition Award, 34
AAAS/Subaru SB&F Prize for Excellence in Science Books, 35
AAAS Award for Scie Freedom & Responsibility, 36
AAAS Early Career Award for Public Engagement with Science, 37
AAAS Eppendorf Science Prize for Neurobiology, 38
AAAS GE & Science Prize for Yng Life Scientists, 39
AAAS Martin and Rose Wachtel Cancer Research Award, 40
AAAS Science Prize for Online Resources in Educ, 42
AABB Dale A. Smith Memorial Award, 43
AABB Hemphill-Jordan Leadership Award, 44
AABB John Elliott Memorial Award, 45
AABB Karl Landsteiner Memorial Award and Lectureship, 46
AABB Sally Frank Award & Lectureship, 47
AABB Tibor Greenwalt Memorial Award and Lectureship, 48

PROGRAM TYPE INDEX

Awards/Prizes / 833

AACAP Beatrix A. Hamburg Award for the Best New Research Poster by a Child and Adolescent Psychiatry Resident, 50
AACAP Elaine Schlosser Lewis Award for Research on Attention-Deficit Disorder, 55
AACAP George Tarjan Award for Contributions in Developmental Disabilities, 56
AACAP Irving Philips Award for Prevention, 57
AACAP Jeanne Spurlock Lecture and Award on Diversity and Culture, 58
AACAP Klingenstein Third Generation Foundation Award for Research in Depression or Suicide, 62
AACAP Norbert and Charlotte Rieger Award for Scientific Achievement, 65
AACAP Pilot Research Awards, Supported by Eli Lilly and Company, 68
AACAP Rieger Psychody Psychotherapy Award, 70
AACAP Rieger Service Award for Excellence, 71
AACAP Rob Cancro Academic Leadership Award, 72
AACAP Sidney Berman Award for the School-Based Study and Intervention for Learning Disorders and Mental Illness, 73
AACAP Simon Wile Leader in Consultation Award, 74
AACR-American Cancer Society Award for Research Excellence in Epidemiology and Prevention, 87
AACR-Minorities in Cancer Research Jane Cooke Wright Lectureship Awards, 94
AACR-Prevent Cancer Foundation Award for Excellence in Cancer Prevention Research, 99
AACR-Women in Cancer Research Charlotte Friend Memorial Lectureship Awards, 101
AACR Award for Lifetime Achievement in Cancer Research, 102
AACR Award for Outstanding Achievement in Cancer Research, 103
AACR Award for Outstanding Achievement in Chemistry in Cancer Research, 104
AACR G.H.A. Clowes Memorial Award, 112
AACR Joseph H. Burchenal Memorial Award, 115
AACR Foti Award for Leadership & Extraordinary Achievements in Cancer Research, 118
AACR Outstanding Investigator Award for Breast Cancer Research, 121
AACR Princess Takamatsu Memorial Lectureship, 122
AACR Richard & Hinda Rosenthal Mem Awards, 123
AACR Scholar-in-Training Awards, 124
AACR Scholar-in-Training Awards - Special Conferences, 125
AACR Team Science Award, 126
AAFP Foundation Wyeth Immunization Awards, 135
AAFPRS Ben Shuster Memorial Award, 137
AAFPRS Bernstein Grant, 138
AAFPRS Ira J. Tresley Research Award, 141
AAFPRS John Orlando Roe Award, 142
AAFPRS Leslie Bernstein Res Research Grants, 143
AAFPRS Residency Travel Awards, 144
AAFPRS Sir Harold Delf Gillies Award, 145
AAMC Abraham Flexner Award, 153
AAMC Alpha Omega Alpha Robert J. Glaser Distinguished Teacher Awards, 154
AAMC Award for Distinguished Research, 155
AAMC David E. Rogers Award, 157
AAMC Herbert W. Nickens Award, 158
AAMC Humanism in Medicine Award, 161
AAO-HNS Medical Student Research Prize, 170
AAO-HNS Resident Research Prizes, 171
AAP Anne E. Dyson Child Advocacy Awards, 176
AAPD Henry B. Betts Award, 182
AAPD Paul G. Hearne Leadership Award, 183
AAP Nutrition Award, 187
Abbott-ASM Lifetime Achievement Award, 198
AcademyHealth Alice S. Hersh New Investigator Awards, 220
AcademyHealth Awards, 222
AcademyHealth HSR Impact Awards, 223
AcademyHealth Nemours Child Health Services Research Awards, 224
AcademyHealth PHSR Research Article of the Year Awards, 226
ACS Doctoral Grants in Oncology Social Work, 254

ACSM-GSSI Young Scholar Travel Award, 257
ACSM Chas M. Tipton Student Research Award, 260
ACSM International Student Award, 268
ACSM Steven M. Horvath Travel Award, 274
ACS Physician Training Awards in Prevent Med, 275
ADA Foundation Bud Tarrson Dental School Student Community Leadership Awards, 287
ADA Foundation George C. Paffenbarger Student Research Awards, 291
AES Epilepsy Research Recognition Awards, 318
AES JK Penry Excellence in Epilepsy Care Award, 319
AES Research and Training Workshop Awards, 320
AES Research Infrastructure Awards, 321
AES Research Initiative Awards, 322
AES Service Award, 324
AES William G. Lennox Award, 326
AFAR Medical Student Training in Aging Research Program, 335
AIHS ForeFront MBA Studentship Award, 381
AIHS ForeFront MBT Studentship Awards, 382
AIHS Research Prize, 396
ALA Donald G. Davis Article Award, 401
Albany Medical Center Prize in Medicine and Biomedical Research, 404
Alfred and Tillie Shemanski Testamentary Trust Grants, 428
Alfred W. Bressler Prize in Vision Science, 434
Alton Ochsner Award, 458
ALVRE Casselberry Award, 460
ALVRE Seymour R. Cohen Award, 462
AMA-MSS Chapter of the Year (COTY) Award, 474
AMA-MSS Research Poster Award, 477
AMA Foundation Dr. Nathan Davis International Awards in Medicine, 483
AMA Foundation Jordan Fieldman, MD, Awards, 489
AMA Foundation Leadership Awards, 490
AMA Foundation Pride in the Profession Grants, 493
AMDA Foundation Medical Director of the Year Award, 497
AMDA Foundation Quality Improvement and Health Outcome Awards, 498
AMDA Foundation Quality Improvement Award, 499
American Academy of Nursing Media Awards, 509
American Chemical Society Alfred Burger Award in Medicinal Chemistry, 510
American Chemical Society ANYL Arthur F. Findeis Award for Achievements by a Young Analytical Scientist, 511
American Chemical Society ANYL Award for Distinguished Service in the Advancement of Analytical Chemistry, 512
American Chemical Society Award for Creative Advances in Environmental Science and Technology, 513
American Chemical Society Award in Separations Science and Technology, 514
American Chem Society Claude Hudson Awards, 515
American Medical Association Awards, 536
American Psychiatric Association/AstraZeneca Young Minds in Psychiatry International Awards, 538
American Psychiatric Association Award for Research in Psychiatry, 542
American Psychiatric Foundation Alexander Gralnick, MD, Award for Research in Schizophrenia, 550
American Psychiatric Foundation Awards for Advancing Minority Mental Health, 551
American Society on Aging Mental Health and Aging Awards, 561
American Society on Aging NOMA Award for Excellence in Multicultural Aging, 562
ANA/Grass Foundation Award in Neuroscience, 578
ANA Derek Denny-Brown Neurological Scholar Award, 579
ANA Distinguished Neurology Teacher Award, 580
ANA Wolfe Neuropathy Research Prize, 585
Anne J. Caudal Foundation Grants, 595
ANS Neurotology Fellows Award, 604
ANS Nicholas Torok Vestibular Award, 605
ANS Trainee Award, 606
AOCS A. Richard Baldwin Award, 610

AOCS Alton E. Bailey Award, 611
AOCS Analytical Division Student Award, 612
AOCS George Schroepfer Medal, 613
AOCS Health & Nutrition Div Student Award, 614
AOCS Health and Nutrition Poster Competition, 615
AOCS Industrial Oil Products Division Student Award, 616
AOCS Lipid Oxidation and Quality Division Poster Competition, 617
AOCS Processing Division Student Award, 618
AOCS Protein and Co-Products Division Student Poster Competition, 619
AOCS Holman Lifetime Achievement Award, 620
AOCS Supelco/Nicholas Pelick-AOCS Research Award, 621
AOCS Surfactants and Detergents Division Student Award, 622
APA Culture of Service in the Psychological Sciences Award, 625
APA Distinguished Scientific Award for Early Career Contribution to Psychology, 627
APA Distinguished Scientific Award for the Applications of Psychology, 628
APA Distinguished Scientific Contribution Award, 629
APA Distinguished Service to Psychological Science Award, 630
APA Meritorious Research Service Award, 631
Arnold and Mabel Beckman Foundation Beckman-Argyros Award in Vision Research, 664
ASA Metlife Foundation MindAlert Awards, 676
ASGE / ConMed Award for Outstanding Manuscript by a Fellow/Resident, 684
ASGE Don Wilson Award, 685
ASGH Award for Excellence in Human Genetics Education, 689
ASGH C.W. Cotterman Award, 690
ASGH Charles J. Epstein Trainee Awards for Excellence in Human Genetics Research, 691
ASGH McKusick Leadership Award, 692
ASGH William Allan Award, 693
ASGH William Curt Stern Award, 694
ASHA Advancing Academic-Research Award, 695
ASHA Minority Student Leadership Awards, 696
ASHA Research Mentoring-Pair Travel Awards, 698
ASHA Student Research Travel Award, 699
ASHA Students Preparing for Academic & Research Careers (SPARC) Award, 700
ASM Abbott Laboratories Award in Clinical and Diagnostic Immunology, 717
ASM BD Award for Research in Clin Microbio, 718
ASM bioMerieux Sonnenwirth Award for Leadership in Clinical Microbiology, 719
ASM D.C. White Research and Mentoring Award, 723
ASME H.R. Lissner Award, 725
ASM Eli Lilly and Company Research Award, 726
ASM Founders Distinguished Service Award, 728
ASM Gen-Probe Joseph Public Health Award, 729
ASM GlaxoSmithKline International Member of the Year Award, 731
ASM ICAAC Young Investigator Awards, 732
ASM Intel Awards, 733
ASM Maurice Hilleman/Merck Award, 738
ASM Merck Irving S. Sigal Memorial Awards, 739
ASM Procter & Gamble Award in Applied and Environmental Microbiology, 742
ASM Promega Biotechnology Research Award, 743
ASM Public Communications Award, 744
ASM Raymond W. Sarber Awards, 745
ASM Richard and Mary Finkelstein Travel Grant, 746
ASM Roche Diagnostics Alice C. Evans Award, 748
ASM sanofi-aventis ICAAC Award, 749
ASM Scherago-Rubin Award, 750
ASM Siemens Healthcare Diagnostics Young Investigator Award, 751
ASM TREK Diagnostic ABMM/ABMLI Professional Recognition Award, 754
ASM USFCC/J. Roger Porter Award, 756
ASM William A Hinton Research Training Award, 757
ASPEN Dudrick Research Scholar Award, 759
ASPEN Harry M. Vars Award, 760

834 / Awards/Prizes

ASPEN Scientific Abstracts Awards for Papers or Posters, 766
ASPET Benedict R. Lucchesi Award in Cardiac Pharmacology, 769
ASPET Brodie Award in Drug Metabolism, 770
ASPET Division for Drug Metabolism Early Career Achievement Award, 771
ASPET Epilepsy Research Award for Outstanding Contributions to the Pharmacology of Antiepileptic Drugs, 772
ASPET Goodman and Gilman Award in Drug Receptor Pharmacology, 773
ASPET John J. Abel Award in Pharmacology, 774
ASPET Julius Axelrod Award in Pharmacology, 775
ASPET P. B. Dews Lifetime Achievement Award for Research in Behavioral Pharmacology, 776
ASPET Paul M. Vanhoutte Award, 777
ASPET Torald Sollmann Award in Pharmacology, 778
Austin College Leadership Award, 804
Baxter International Foundation Foster G. McGaw Prize, 842
Baxter International Foundation William B. Graham Prize for Health Services Research, 844
BCBSM Claude Pepper Award, 850
BCBSM Foundation Excellence in Research Awards for Students, 853
BCBSM Fndn Proposal Development Awards, 857
BCBSM Frank J. McDevitt, DO, Excellence in Research Awards, 859
Bravewell Leadership Award, 946
Bristol-Myers Squibb/Meade Johnson Award for Dist Achievement in Nutrition Research, 954
Caplow Applied Science Carcinogen Prize, 1015
Caplow Applied Science Children's Prize, 1016
Carla J. Funk Governmental Relations Award, 1020
CCFF Chairman's Dist Life Sciences Award, 1044
CCFF Life Sciences Student Awards, 1045
Changemakers Innovation Awards, 1111
Christine Gregoire Youth/Young Adult Award for Use of Tobacco Industry Documents, 1151
Claude Pepper Foundation Grants, 1177
CNIB Chanchlani Global Vision Research Award, 1198
Community Fndn AIDS Endowment Awards, 1237
Community Foundation for Greater Atlanta Managing For Excellence Award, 1242
Community Foundation of Bartholomew County James A. Henderson Award for Fundraising, 1263
Conquer Cancer Foundation of ASCO International Innovation Grant, 1334
Cystic Fibrosis Canada Visiting Allied Health Professional Awards, 1394
Daniel Mendelsohn New Investigator Award, 1423
Delta Air Lines Foundation Prize for Global Understanding, 1456
Do Something Awards, 1503
E.B. Hershberg Award for Important Discoveries in Medicinally Active Substances, 1535
Elizabeth Glaser Int'l Leadership Awards, 1566
El Pomar Foundation Anna Keesling Ackerman Fund Grants, 1581
Everyone Breathe Asthma Education Grants, 1616
FDHN Designated Outcomes Award in Geriatric Gastroenterology, 1640
FDHN Designated Research Award in Research Related to Pancreatitis, 1642
FDHN Fellow Abstract Prizes, 1643
FDHN Funderburg Research Scholar Award in Gastric Biology Related to Cancer, 1645
FDHN Graduate Student Awards, 1646
FDHN Isenberg Int'l Research Scholar Award, 1647
FDHN Moti L. & Kamla Rustgi International Travel Awards, 1649
FDHN Research Scholar Awards, 1652
FDHN Student Abstract Prizes, 1653
Frederic Stanley Kipping Award in Silicon Chem, 1742
Gates Award for Global Health, 1775
GEICO Public Service Awards, 1778
GlaxoSmithKline Corporate Grants, 1829
GlaxoSmithKline Foundation IMPACT Awards, 1830
Google Grants Beta, 1857

Gruber Foundation Cosmology Prize, 1907
Gruber Foundation Genetics Prize, 1908
Gruber Foundation Neuroscience Prize, 1909
Gruber Foundation Rosalind Franklin Young Investigator Award, 1910
Gruber Foundation Weizmann Institute Awards, 1911
Hereditary Disease Fnd Lieberman Award, 2010
HRAMF Smith Family Awards for Excellence in Biomedical Research, 2063
Landon Foundation-AACR Innovator Award for Cancer Prevention Research, 2342
Landon Foundation-AACR Innovator Award for International Collaboration, 2343
Layne Beachley Aim for the Stars Fnd Grants, 2349
Lynde and Harry Bradley Foundation Prizes: Bradley Prizes, 2398
Majors MLA Chapter Project of the Year, 2418
March of Dimes Agnes Higgins Award, 2423
March of Dimes Newborn Screening Awards, 2425
Massage Therapy Foundation Practitioner Case Report Contest, 2450
MLA Estelle Brodman Academic Medical Librarian of the Year Award, 2563
MLA Ida and George Eliot Prize, 2566
MLA Janet Doe Lectureship Award, 2567
MLA Lois Ann Colaianni Award for Excellence and Achievement in Hospital Librarianship, 2568
MLA Louise Darling Medal for Distinguished Achievement in Collection Development in the Health Sciences, 2569
MLA Lucretia W. McClure Excellence in Education Award, 2570
MLA Murray Gottlieb Prize Essay Award, 2571
MLA Rittenhouse Award, 2573
MLA Section Project of the Year Award, 2574
MLA T. Mark Hodges Int'l Service Award, 2575
MLA Thomas Reuters / Frank Bradway Rogers Information Advancement Award, 2576
NARSAD Colvin Prize for Outstanding Achievement in Mood Disorders Research, 2619
NARSAD Goldman-Rakic Prize for Cognitive Neuroscience, 2621
NARSAD Lieber Prize for Schizo Research, 2623
NARSAD Ruane Prize for Child and Adolescent Psychiatric Research, 2624
NHLBI Career Transition Awards, 2694
NHLBI National Research Service Award Programs in Cardiovascular Epidemiology and Biostatistics, 2714
NIA Mentored Clinical Scientist Research Career Development Awards, 2771
NIHCM Foundation Health Care Print Journalism Awards, 2870
NIHCM Fndn Health Care Research Awards, 2871
NIHCM Foundation Health Care Television and Radio Journalism Awards, 2872
NKF Clinical Scientist Awards, 2935
NMF Aura E. Severinghaus Award, 2954
NMF Franklin C. McLean Award, 2955
NMF Henry G. Halladay Awards, 2957
NMF National Medical Association Awards for Medical Journalism, 2962
NMF National Medical Association Emerging Scholar Awards, 2963
NMF National Medical Association Patti LaBelle Award, 2964
NMF Ralph W. Ellison Prize, 2966
NMF William and Charlotte Cadbury Award, 2967
NSERC Brockhouse Canada Prize for Interdisciplinary Research in Science and Engineering Grant, 3003
NSF Alan T. Waterman Award, 3006
NSF Systematic Biology and Biodiversity Inventories Grants, 3024
NYAM Brock Lecture, Award and Visiting Professorship in Pediatrics, 3038
NYAM Edward N. Gibbs Memorial Lecture and Award in Nephrology, 3041
NYAM Lewis Rudin Glaucoma Grants, 3044
NYAM Student Essay Grants, 3046
Oticon Focus on People Awards, 3098
Paul Marks Prizes for Cancer Research, 3123

PROGRAM TYPE INDEX

Pet Care Trust Sue Busch Memorial Award, 3143
Pezcoller Foundation-AACR International Award for Cancer Research, 3161
Pfizer Special Events Grants, 3186
Pharmacia-ASPET Award for Experimental Therapeutics, 3187
Phi Upsilon Omicron Florence Fallgatter Distinguished Service Award, 3192
Phi Upsilon Omicron Frances Morton Holbrook Alumni Award, 3193
Phi Upsilon Omicron Geraldine Clewell Senior Awards, 3194
Phi Upsilon Omicron Janice Cory Bullock Collegiate Award, 3195
Phi Upsilon Omicron Lillian P. Schoephoerster Award, 3196
Phi Upsilon Orinne Johnson Writing Award, 3198
Phi Upsilon Omicron Sarah Thorniley Phillips Leader Awards, 3199
Phi Upsilon Omicron Undergraduate Karen P. Goebel Conclave Award, 3200
Portland General Electric Foundation Grants, 3250
Prince Charitable Trusts Chicago Grants, 3270
RACGP Nat Asthma Council Research Award, 3296
Rajiv Gandhi Foundation Grants, 3302
Ralph F. Hirschmann Award in Peptide Chem, 3303
RCPSC/AMS Donald Richards Wilson Award, 3317
RCPSC Duncan Graham Award, 3322
RCPSC International Residency Educator of the Year Award, 3325
RCPSC Int'l Resident Leadership Award, 3326
RCPSC James H. Graham Award of Merit, 3327
RCPSC KJR Wightman Award for Scholarship in Ethics, 3330
RCPSC Kristin Sivertz Res Leadership Award, 3332
RCPSC Mentor of the Year Award, 3335
RCPSC Program Administrator Award for Innovation and Excellence, 3339
RCPSC Program Director of the Year Award, 3340
RCPSC Royal College Medal Award in Medicine, 3342
RCPSC Royal College Medal Award in Surgery, 3343
RCPSC Teasdale-Corti Humanitarian Award, 3344
Rosalynn Carter Institute Georgia Caregiver of the Year Awards, 3407
Royal Norwegian Embassy Kavli Prizes, 3414
RSC Abbyann D. Lynch Medal in Bioethics, 3419
RSC Jason A. Hannah Medal, 3420
RSC McLaughlin Medal, 3421
RSC McNeil Medal for the Public Awareness of Science, 3422
RSC Thomas W. Eadie Medal, 3423
RWJF Community Health Leaders Awards, 3432
SfN Award for Education in Neuroscience, 3517
SfN Bernice Grafstein Award for Outstanding Accomplishments in Mentoring, 3518
SfN Chapter-of-the-Year Award, 3519
SfN Donald B. Lindsley Prize in Behavioral Neuroscience, 3521
SfN Jacob P. Waletzky Award, 3523
SfN Japan Neuroscience Society Meeting Travel Awards, 3525
SfN Julius Axelrod Prize, 3526
SfN Louise Hanson Marshall Specific Recognition Award, 3527
SfN Mika Salpeter Lifetime Achievement Award, 3528
SfN Nemko Prize in Cellular or Molecular Neuroscience, 3529
SfN Next Generation Award, 3531
SfN Patricia Goldman-Rakic Hall of Honor, 3532
SfN Peter and Patricia Gruber International Research Award, 3533
SfN Ralph W. Gerard Prize in Neuroscience, 3534
SfN Science Educator Award, 3535
SfN Science Journalism Student Awards, 3536
SfN Swartz Prize for Theoretical and Computational Neuroscience, 3537
SfN Trainee Professional Development Awards, 3538
SfN Young Investigator Award, 3541
Simone and Cino del Duca Grand Prix Awards, 3564
Skoll Fndn Awards for Social Entrepreneurship, 3579
SNM Tetalman Young Investigator Awards, 3589

PROGRAM TYPE INDEX

SNMTS Outstanding Educator Awards, 3591
SNMTS Outstanding Technologist Awards, 3592
SOBP A.E. Bennett Research Award, 3594
SOBP Ziskind-Somerfeld Research Awards, 3595
Society of Cosmetic Chemists Allan B. Black Award Sponsored by Presperse, 3604
Society of Cosmetic Chemists Award Sponsored by McIntyre Group, Ltd., 3605
Society of Cosmetic Chemists Award Sponsored by The HallStar Company, 3606
Society of Cosmetic Chemists Chapter Best Speaker Award, 3607
Society of Cosmetic Chem Chapter Merit Award, 3608
Society of Cosmetic Chemists Frontier of Science Award, 3609
Society of Cosmetic Chem Hans Schaeffer Award, 3610
Society of Cosmetic Chem Jos Ciaudelli Award, 3611
Society of Cosmetic Chemists Keynote Award Lecture Sponsored by Ruger Chemical Corporation, 3612
Society of Cosmetic Chemists Literature Award, 3613
Society of Cosmetic Chemists Maison G. de Navarre Medal, 3614
Society of Cosmetic Chemists Merit Award, 3615
Society of Cosmetic Chemists Robert A. Kramer Lifetime Service Award, 3616
Society of Cosmetic Chem Shaw Mudge Award, 3617
Society of Cosmetic Chem Stud Poster Awards, 3618
Society of Cosmetic Chem You Scientist Awards, 3619
Sybil G. Jacobs Award for Outstanding Use of Tobacco Industry Documents, 3698
TRDRP California Research Awards, 3753
TRDRP Exploratory/Devel Research Awards, 3754
TRDRP New Investigator Awards, 3755
Victor Grifols i Lucas Foundation Ethics and Science Awars for Educational Institutions, 3869
Victor Grifols i Lucas Foundation Prize for Journalistic Work on Bioethics, 3870
Victor Grifols i Lucas Foundation Secondary School Prizes, 3872
Virginia L. and William K. Beatty MLA Volunteer Service Award, 3877
Washington County Community Foundation Youth Grants, 3907
World of Children Health Award, 3974
World of Children Humanitarian Award, 3975

Basic Research
1 in 9: Long Island Breast Cancer Action Coalition Grants, 1
1st Touch Foundation Grants, 3
25th Anniversary Foundation Grants, 8
100 Mile Man Foundation Grants, 10
1976 Foundation Grants, 12
A-T Children's Project Grants, 13
A-T Children's Project Post Doctoral Fellowships, 14
A-T Medical Research Foundation Grants, 15
A.O. Smith Foundation Community Grants, 17
AAAAI ARTrust Mini Grants for Allied Health, 20
AAAAI ARTrust Grants for Clinical Research, 22
AAAAI ARTrust Grants in Faculty Development, 23
AAAAI Distinguished Clinician Award, 24
AAAAI Distinguished Layperson Award, 25
AAAAI Distinguished Scientist Award, 26
AAAAI Distinguished Service Award, 27
AAAAI Fellows-in-Training Grants, 29
AAAS Eppendorf Science Prize for Neurobiology, 38
AAAS GE & Science Prize for Yng Life Scientists, 39
AAAS Martin and Rose Wachtel Cancer Research Award, 40
AACAP-NIDA K12 Career Development Awards, 49
AACAP Beatrix A. Hamburg Award for the Best New Research Poster by a Child and Adolescent Psychiatry Resident, 50
AACAP Educational Outreach Program for General Psychiatry Residents, 53
AACAP Educational Outreach Program for Residents in Alcohol Research, 54
AACAP Elaine Schlosser Lewis Award for Research on Attention-Deficit Disorder, 55

AACAP George Tarjan Award for Contributions in Developmental Disabilities, 56
AACAP Irving Philips Award for Prevention, 57
AACAP Jeanne Spurlock Lecture and Award on Diversity and Culture, 58
AACAP Jeanne Spurlock Minority Med Student Clinical Fellow in Child & Adolescent Psych, 59
AACAP Jeanne Spurlock Research Fellowship in Substance Abuse and Addiction for Minority Medical Students, 60
AACAP Junior Investigator Awards, 61
AACAP Klingenstein Third Generation Foundation Award for Research in Depression or Suicide, 62
AACAP Life Members Mentorship Grants for Medical Students, 63
AACAP Mary Crosby Congressional Fellowships, 64
AACAP Pilot Research Award for Attention-Deficit Disorder, 66
AACAP Pilot Research Award for Learning Disabilities, Supported by the Elaine Schlosser Lewis Fund, 67
AACAP Pilot Research Awards, Supported by Eli Lilly and Company, 68
AACAP Quest for the Test Bipolar Disorder Pilot Research Award, 69
AACAP Rieger Psychody Psychotherapy Award, 70
AACAP Rieger Service Award for Excellence, 71
AACAP Rob Cancro Academic Leadership Award, 72
AACAP Sidney Berman Award for the School-Based Study and Intervention for Learning Disorders and Mental Ilness, 73
AACAP Summer Medical Student Fellowships, 75
AACN-Edwards Lifesciences Nurse-Driven Clinical Practice Outcomes Grants, 77
AACN-Philips Medical Systems Clinical Outcomes Grants, 78
AACN-Sigma Theta Tau Critical Care Grant, 79
AACN Clinical Inquiry Fund Grants, 80
AACN Clinical Practice Grants, 81
AACN Critical Care Grants, 82
AACN End of Life/Palliative Care Small Projects Grants, 83
AACN Evidence-Based Clinical Practice Grants, 84
AACN Physio-Control Small Projects Grants, 86
AACR-American Cancer Society Award for Research Excellence in Epidemiology and Prevention, 87
AACR-Colorectal Cancer Coalition Fellows Grant, 88
AACR-FNAB Fellows Grant for Translational Pancreatic Cancer Research, 89
AACR-FNAB Foundation Career Development Award for Translational Cancer Research, 90
AACR-Genentech BioOncology Career Development Award for Cancer Research on the HER Family Pathway, 91
AACR-Genentech BioOncology Fellowship for Cancer Research in Angiogenesis, 92
AACR-Minorities in Cancer Research Jane Cooke Wright Lectureship Awards, 94
AACR-National Brain Tumor Fnd Fellows Grant, 95
AACR-NCI International Investigator Opportunity Grants, 96
AACR-Pancreatic Cancer Action Network Career Development Award for Research, 97
AACR-Pancreatic Cancer Action Network Innovative Grants, 98
AACR-Prevent Cancer Foundation Award for Excellence in Cancer Prevention Research, 99
AACR-Women in Cancer Research Charlotte Friend Memorial Lectureship Awards, 101
AACR Award for Lifetime Achievement in Cancer Research, 102
AACR Award for Outstanding Achievement in Cancer Research, 103
AACR Award for Outstanding Achievement in Chemistry in Cancer Research, 104
AACR Basic Cancer Research Fellowships, 105
AACR Brigid G. Leventhal Scholar in Cancer Research Awards, 106
AACR Career Development Awards for Pediatric Cancer Research, 107

AACR Centennial Postdoctoral Fellowships in Cancer Research, 108
AACR Centennial Pre-doctoral Fellowships in Cancer Research, 109
AACR Fellows Grants, 111
AACR Shepard Bladder Cancer Research Grants, 114
AACR Joseph H. Burchenal Memorial Award, 115
AACR Judah Folkman Career Development Award for Anti-Angiogenesis Research, 116
AACR Kirk A. Landon and Dorothy P. Landon Foundation Prizes, 117
AACR Foti Award for Leadership & Extraordinary Achievements in Cancer Research, 118
AACR Minority-Serving Institution Faculty Scholar in Cancer Research Awards, 119
AACR Min Scholar in Cancer Research Award, 120
AACR Outstanding Investigator Award for Breast Cancer Research, 121
AACR Richard & Hinda Rosenthal Mem Awards, 123
AACR Team Science Award, 126
AACR Thomas J. Bardos Science Ed Awards, 127
AAFA Investigator Research Grants, 128
AAFP Foundation Joint Grants, 133
AAFP Foundation Wyeth Immunization Awards, 135
AAFP Research Stimulation Grants, 136
AAFPRS Bernstein Grant, 138
AAFPRS Fellowships, 139
AAFPRS Investigator Development Grant, 140
AAFPRS Leslie Bernstein Res Research Grants, 143
AAHN Grant for Historical Research, 146
AAHN Pre-Doctoral Research Grant, 147
AAMC Award for Distinguished Research, 155
AANS Neurosurgery Research and Education Foundation/Spine Section Young Clinician Investigator Award, 162
AAO-HNSF CORE Research Grants, 164
AAO-HNSF Health Services Research Grants, 165
AAO-HNSF Maureen Hannley Research Training Awards, 166
AAO-HNSF Percy Memorial Research Award, 167
AAO-HNSF Rande H. Lazar Health Services Research Grant, 168
AAO-HNSF Resident Research Awards, 169
AAO-HNS Medical Student Research Prize, 170
AAO-HNS Resident Research Prizes, 171
AAOA Foundation/AAO-HNSF Combined Research Grants, 172
AAOHN Found Experienced Researcher Grants, 173
AAOHN Foundation New Investigator Researcher Grants, 174
Abbott Fund Access to Health Care Grants, 199
Abbott Fund Science Education Grants, 203
Abdus Salam ICTP Federation Arrangement Scheme Grants, 205
Abell-Hanger Foundation Grants, 207
Abington Foundation Grants, 211
Abracadabra Foundation Grants, 216
ACAAI Foundation Research Grants, 219
AcademyHealth HSR Impact Awards, 223
ACGT Investigators Grants, 232
ACGT Young Investigator Grants, 233
Achelis Foundation Grants, 234
Acid Maltase Deficiency Association Helen Walker Research Grant, 235
ACL Field Initiated Projects Program: Minority-Serving Institution Research Grants, 243
ACS Clinical Research Professor Grants, 251
ACS Doctoral Grants in Oncology Social Work, 254
ACS Institutional Research Grants, 256
ACSM Carl V. Gisolfi Memorial Fund Grant, 259
ACSM Chas M. Tipton Student Research Award, 260
ACSM Coca-Cola Company Doctoral Student Grant on Behavior Research, 261
ACSM Dr. Raymond A. Weiss Research Endowment Grant, 262
ACS Mentored Research Scholar Grant in Applied and Clinical Research, 264
ACSM Foundation Clinical Sports Medicine Endowment Grants, 265
ACSM Foundation Doctoral Research Grants, 266

836 / Basic Research PROGRAM TYPE INDEX

ACSM Foundation Research Endowment Grants, 267
ACSM International Student Awards, 269
ACSM Michael L. Pollock Student Scholarship, 270
ACSM NASA Space Physiology Research Grants, 271
ACSM Oded Bar-Or Int'l Scholar Awards, 272
ACSM Paffenbarger-Blair Fund for Epidemiological Research on Physical Activity Grants, 273
ACS Physician Training Awards in Prevent Med, 275
ACS Pilot and Exploratory Projects in Palliative Care of Cancer Patients and Their Families Grants, 276
ACS Research Professor Grants, 278
ACS Research Scholar Grants for Health Services and Health Policy Research, 279
ACS Research Scholar Grants in Basic, Preclinical, Clinical and Epidemiology Research, 280
ACS Research Scholar Grants in Psychosocial and Behavioral and Cancer Control Research, 281
ACVIM Foundation Clinical Investigation Grants, 286
Adams County Community Foundation of Pennsylvania Grants, 300
Adaptec Foundation Grants, 304
Administration on Aging Senior Medicare Patrol Project Grants, 309
AEC Trust Grants, 315
AES Research Infrastructure Awards, 321
AES Research Initiative Awards, 322
Aetna Foundation Integrated Health Care Grants, 328
Aetna Foundation Minority Scholars Grants, 329
AFAR CART Fund Grants, 334
AFAR Paul Beeson Career Development Awards in Aging Research for the Island of Ireland, 336
AFAR Research Grants, 337
Affymetrix Corporate Contributions Grants, 339
AHAF Alzheimer's Disease Research Grants, 349
AHAF Macular Degeneration Research Grants, 350
AHAF National Glaucoma Research Grants, 351
AHIMA Dissertation Assistance Grants, 352
AHIMA Faculty Development Stipends, 353
AHIMA Grant-In-Aid Research Grants, 354
AHNS/AAO-HNSF Surgeon Scientist Combined Award, 357
AHNS/AAO-HNSF Young Investigator Combined Award, 358
AHNS Alando J. Ballantyne Resident Research Pilot Grant, 359
AHNS Pilot Grant, 360
AHRF Eugene L. Derlacki Research Grants, 361
AHRF Georgia Birtman Grant, 362
AHRF Regular Research Grants, 363
AHRF Wiley H. Harrison Research Award, 364
AHRQ Independent Scientist Award, 365
AHRQ Individual Awards for Postdoc Fellows Ruth L. Kirschstein National Research Service Awards, 366
AIHP Sonnedecker Visiting Scholar Grants, 371
AIHS/Mitacs Health Pilot Partnership Interns, 373
AIHS Alberta/Pfizer Translat Research Grants, 374
AIHS Clinical Fellowships, 375
AIHS Collaborative Research and Innovation Grants - Collaborative Program, 376
AIHS Collaborative Research and Innovation Grants - Collaborative Project, 377
AIHS Collaborative Research and Innovation Grants - Collaborative Team, 378
AIHS Fast-Track Fellowships, 379
AIHS ForeFront Internships, 380
AIHS ForeFront MBA Studentship Award, 381
AIHS ForeFront MBT Studentship Awards, 382
AIHS Full-Time Health Research Studentships, 384
AIHS Full-Time Studentships, 386
AIHS Interdisciplinary Team Grants, 388
AIHS Knowledge Exchange - Visiting Scientists, 391
AIHS Part-Time Fellowships, 393
AIHS Proposals for Special Initiatives, 395
AIHS Research Prize, 396
AIHS Summer Studentships, 397
AIHS Sustainability Grants, 398
Alaska Airlines Corporate Giving Medical Emergency and Research Grants, 402
Albany Medical Center Prize in Medicine and Biomedical Research, 404

Albert and Margaret Alkek Foundation Grants, 405
Albert and Mary Lasker Foundation Awards, 406
Albert and Mary Lasker Foundation Clinical Research Scholars, 407
Alcoa Foundation Grants, 414
Alcohol Misuse and Alcoholism Research Grants, 415
Alexander and Margaret Stewart Trust Grants, 419
Alexander von Humboldt Foundation Georg Forster Fellowships for Experienced Researchers, 423
Alexander von Humboldt Foundation Georg Forster Fellowships for Postdoctoral Researchers, 424
Alexander von Humboldt Foundation Research Fellowships for Postdoctoral Researchers, 425
Alfred P. Sloan Foundation International Science Engagement Grants, 431
Alfred P. Sloan Foundation Research Fellowships, 432
Alfred P. Sloan Foundation Science of Learning STEM Grants, 433
Alfred W. Bressler Prize in Vision Science, 434
Allyn Foundation Grants, 448
Alpha Omega Foundation Grants, 451
Alpha Research Foundation Grants, 452
ALSAM Foundation Grants, 454
Alternatives Research and Development Foundation Alternatives in Education Grants, 455
Alternatives Research and Development Foundation Grants, 456
Alton Ochsner Award, 458
ALVRE Grant, 461
Alzheimer's Association Conference Grants, 463
Alzheimer's Association Development of New Cognitive and Functional Instruments Grants, 464
Alzheimer's Association Everyday Technologies for Alzheimer Care Grants, 465
Alzheimer's Association Investigator-Initiated Research Grants, 466
Alzheimer's Association Mentored New Investigator Research Grants to Promote Diversity, 467
Alzheimer's Association Neuronal Hyper Excitability and Seizures in Alzheimer's Disease Grants, 468
Alzheimer's Association New Investigator Research Grants, 469
Alzheimer's Association New Investigator Research Grants to Promote Diversity, 470
Alzheimer's Association U.S.-U.K. Young Investigator Exchange Fellowships, 471
Alzheimer's Association Zenith Fellows Awards, 472
AMA-WPC Joan F. Giambalvo Memorial Scholarships, 480
AMA Foundation Joan F. Giambalvo Memorial Scholarships, 488
AMA Foundation Seed Grants for Research, 494
American Academy of Nursing Building Academic Geriatric Nursing Capacity Scholarships, 505
American Academy of Nursing Claire M. Fagin Fellowships, 506
American Academy of Nursing Mayday Grants, 507
American Chemical Society Alfred Burger Award in Medicinal Chemistry, 510
American Chemical Society ANYL Arthur F. Findeis Award for Achievements by a Young Analytical Scientist, 511
American Chemical Society ANYL Award for Distinguished Service in the Advancement of Analytical Chemistry, 512
American Chemical Society Award for Creative Advances in Environmental Science and Technology, 513
American Chemical Society Award in Separations Science and Technology, 514
American Chemical Society GCI Pharmaceutical Roundtable Research Grants, 516
American College of Rheumatology Fellows-in-Training Travel Scholarships, 517
American College of Surgeons and The Triological Society Clinical Scientist Development Awards, 518
American College of Surgeons Australia and New Zealand Chapter Travelling Fellowships, 519
American College of Surgeons Co-Sponsored K08/K23 NIH Supplement Awards, 520

American College of Surgeons Faculty Research Fellowships, 522
American College of Surgeons Nizar N. Oweida, MD, FACS, Scholarships, 526
American College of Surgeons Wound Care Management Award, 529
American Legacy Foundation National Calls for Proposals Grants, 534
American Legacy Fnd Small Innovative Grants, 535
American Psychiatric Association/AstraZeneca Young Minds in Psychiatry International Awards, 538
American Psychiatric Association/Janssen Resident Psychiatric Research Scholars, 540
American Psychiatric Association/Merck Co. Early Academic Career Research Award, 541
American Psychiatric Association Award for Research in Psychiatry, 542
American Psychiatric Association Minority Medical Student Fellowship in HIV Psychiatry, 544
American Psychiatric Association Minority Medical Student Summer Externship in Addiction Psychiatry, 545
American Psychiatric Association Minority Medical Student Summer Mentoring Program, 546
American Psychiatric Association Program for Minority Research Training in Psychiatry, 547
American Psychiatric Association Research Colloquium for Junior Investigators, 548
American Psychiatric Foundation Grantss, 552
American Psychiatric Foundation James H. Scully Jr., M.D., Educational Fund Grants, 555
American Psychiatric Foundation Jeanne Spurlock Congressional Fellowship, 556
American Psychiatric Foundation Minority Fellowships, 557
American Schlafhorst Foundation Grants, 559
American Society on Aging Graduate Student Research Award, 560
American Sociological Association Minority Fellowships, 563
amfAR Fellowships, 567
amfAR Mathilde Krim Fellowships in Basic Biomedical Research, 569
amfAR Research Grants, 571
Amgen Foundation Grants, 572
AMI Semiconductors Corporate Grants, 576
AMSSM Foundation Research Grants, 577
ANA/Grass Foundation Award in Neuroscience, 578
ANA Junior Academic Neurologist Scholarships, 582
Andrew Family Foundation Grants, 588
Angels Baseball Foundation Grants, 590
Ann and Robert H. Lurie Family Fndn Grants, 593
Ann Arbor Area Community Foundation Grants, 594
Ann Peppers Foundation Grants, 599
ANS/AAO-HNSF Herbert Silverstein Otology and Neurotology Research Award, 601
ANS Neurotology Fellows Award, 604
Anthony R. Abraham Foundation Grants, 608
AOCS Health & Nutrition Div Student Award, 614
APA Culture of Service in the Psychological Sciences Award, 625
APA Distinguished Scientific Award for Early Career Contribution to Psychology, 627
APA Distinguished Scientific Award for the Applications of Psychology, 628
APA Distinguished Scientific Contribution Award, 629
APA Distinguished Service to Psychological Science Award, 630
APA Young Investigator Grant Program, 634
APHL Emerging Infectious Diseases Fellowships, 636
APSAA Fellowships, 638
APSAA Mini-Career Grants, 640
APSAA Research Grants, 641
APSAA Small Beginning Scholar Pre-Investigation Grants, 642
APSAA Small Beginning Scholar Visiting and Consulting Grants, 643
APSA Congressional Health and Aging Policy Fellowships, 644
Arcadia Foundation Grants, 650

PROGRAM TYPE INDEX

Arie and Ida Crown Memorial Grants, 654
Arizona Cardinals Grants, 655
Arizona Public Service Corporate Giving Program Grants, 658
Armstrong McDonald Foundation Health Grants, 663
Arnold and Mabel Beckman Foundation Beckman-Argyros Award in Vision Research, 664
Arnold and Mabel Beckman Foundation Scholars Grants, 665
Arnold and Mabel Beckman Foundation Young Investigators Grants, 666
ARS New Investigator Award, 668
ARS Resident Research Grants, 669
Arthur Ashley Williams Foundation Grants, 671
ArvinMeritor Foundation Health Grants, 674
ASCO/UICC International Cancer Technology Transfer (ICRETT) Fellowships, 677
ASCO Advanced Clinical Research Award in Colorectal Cancer, 678
ASCO Advanced Clinical Research Awards in Breast Cancer, 679
ASCO Advanced Clinical Research Awards in Sarcoma, 680
ASCO Merit Awards, 682
ASCO Young Investigator Award, 683
ASGE / ConMed Award for Outstanding Manuscript by a Fellow/Resident, 684
ASGE Don Wilson Award, 685
ASGE Endoscopic Research Awards, 686
ASGE Endoscopic Research Career Development Awards, 687
ASGE Given Capsule Endoscopy Research Award, 688
ASHA Advancing Academic-Research Award, 695
ASHA Research Mentoring-Pair Travel Award, 698
ASHA Student Research Travel Award, 699
ASHA Students Preparing for Academic & Research Careers (SPARC) Award, 700
ASHFoundation Clinical Research Grants, 701
ASHFoundation New Century Scholars Program Doctoral Scholarships, 706
ASHFoundation New Century Scholars Research Grant, 707
ASHFoundation Grant for New Investigators, 708
ASHFoundation Research Grant in Speech, 709
ASHFoundation Student Research Grants in Audiology, 710
ASHFoundation Student Research Grants in Early Childhood Language Development, 711
ASM-PAHO Infectious Diseases Epidemiology and Surveillance Fellowships, 714
ASM-UNESCO Leadership Grant for International Educators, 715
ASM/CDC Fellowships in Infectious Disease and Public Health Microbiology, 716
ASM Abbott Laboratories Award in Clinical and Diagnostic Immunology, 717
ASM Int'l Fellowship for Latin America and the Caribbean, 734
ASM Microbiology Undergraduate Research Fellowship, 740
ASM Millis-Colwell Postgraduate Travel Grant, 741
ASM Robert D. Watkins Graduate Research Fellowship, 747
ASM Undergraduate Research Fellowship, 755
A Social Corporation Grants, 758
ASPEN Rhoads Research Foundation Abbott Nutrition Research Grants, 761
ASPEN Rhoads Research Foundation Baxter Parenteral Nutrition Research Grant, 762
ASPEN Rhoads Research Foundation C. Richard Fleming Grant, 763
ASPEN Rhoads Research Foundation Maurice Shils Grant, 764
ASPEN Rhoads Research Foundation Norman Yoshimura Grant, 765
ASPET-Astellas Awards in Translational Pharmacology, 768
ASPET Brodie Award in Drug Metabolism, 770
ASPET Division for Drug Metabolism Early Career Achievement Award, 771
ASPET Epilepsy Research Award for Outstanding Contributions to the Pharmacology of Antiepileptic Drugs, 772
ASPET P. B. Dews Lifetime Achievement Award for Research in Behavioral Pharmacology, 776
ASPET Paul M. Vanhoutte Award, 777
ASPET Torald Sollmann Award in Pharmacology, 778
ASPO Daiichi Innovative Technology Grant, 783
ASPO Fellowships, 784
ASPO Research Grants, 785
Assisi Foundation of Memphis General Grants, 787
Assurant Health Foundation Grants, 789
Astellas Foundation Research Grants, 790
Atran Foundation Grants, 797
Australasian Institute of Judicial Administration Seed Funding Grants, 806
Australian Academy of Science Grants, 807
Avon Foundation Breast Care Fund Grants, 813
Ball Brothers Foundation General Grants, 820
Baltimore Washington Center Adult Psychoanalysis Fellowships, 821
Baltimore Washington Center Child Psychotherapy Fellowships, 822
Barra Foundation Project Grants, 834
Barth Syndrome Foundation Research Grants, 836
Batchelor Foundation Grants, 837
Batts Foundation Grants, 840
Baxter International Foundation Grants, 843
Bayer Foundation Grants, 846
BCBSM Foundation Investigator Initiated Research Grants, 854
BCBSM Foundation Physician Investigator Research Awards, 855
BCBSM Foundation Primary and Clinical Prevention Grants, 856
BCBSM Foundation Student Award Program, 858
BCBSM Frank J. McDevitt, DO, Excellence in Research Awards, 859
BCBSNC Foundation Grants, 860
BCRF-AACR Grants for Translational Breast Cancer Research, 861
BCRF Research Grants, 862
Beckman Coulter Foundation Grants, 868
Becton Dickinson and Company Grants, 869
Benton Community Foundation Grants, 878
Berks County Community Foundation Grants, 881
Bill and Melinda Gates Foundation Policy and Advocacy Grants, 893
Bill and Melinda Gates Foundation Water, Sanitation and Hygiene Grants, 894
Biogen Foundation Fellowships, 898
Biogen Foundation Scientific Research Grants, 902
Blowitz-Ridgeway Foundation Early Childhood Development Research Award, 912
Blowitz-Ridgeway Foundation Grants, 913
Blue Shield of California Grants, 923
Blum-Kovler Foundation Grants, 924
BMW of North America Charitable Contributions, 926
Bodenwein Public Benevolent Foundation Grants, 928
Bodman Foundation Grants, 929
Booth-Bricker Fund Grants, 933
Bosque Foundation Grants, 935
Boston Psychoanalytic Society and Institute Fellowship in Child Psychoanalytic Psychotherapy, 939
Boston Psychoanalytic Society and Institute Fellowship in Psychoanalytic Psychotherapy, 940
Boyle Foundation Grants, 942
Breast Cancer Fund Grants, 947
Bridgestone/Firestone Trust Fund Grants, 949
Bristol-Myers Squibb Clinical Outcomes and Research Grants, 955
Bristol-Myers Squibb Foundation Fellowships, 956
Bristol-Myers Squibb Foundation Global HIV/AIDS Initiative Grants, 957
Bristol-Myers Squibb Foundation Health Disparities Grants, 958
Bristol-Myers Squibb Foundation Women's Health Grants, 962
Broad Foundation IBD Research Grants, 964
Brookdale Foundation Leadership in Aging Fellowships, 965
Brown Advisory Charitable Foundation Grants, 969
Bupa Foundation Medical Research Grants, 974
Bupa Foundation Multi-Country Grant, 975
Burden Trust Grants, 976
Burlington Industries Foundation Grants, 977
BWF Ad Hoc Grants, 982
BWF Career Awards at the Scientific Interface, 983
BWF Career Awards for Medical Scientists, 984
BWF Collaborative Research Travel Grants, 985
BWF Innovation in Regulatory Science Grants, 986
BWF Institutional Program Unifying Population and Laboratory Based Sciences Grants, 987
BWF Investigators in the Pathogenesis of Infectious Disease Awards, 988
BWF Preterm Birth Initiative Grants, 990
Byron W. & Alice L. Lockwood Fnd Grants, 991
Cabot Corporation Foundation Grants, 993
Caesars Foundation Grants, 995
Callaway Golf Company Foundation Grants, 1003
Campbell Soup Foundation Grants, 1007
Canada Graduate Scholarships (CGS) and NSERC Postgraduate Scholarships (PGS), 1010
Canadian Optometric Education Trust Grants, 1011
Canadian Patient Safety Institute (CPSI) Research Grants, 1013
Cargill Citizenship Fund Corporate Giving, 1018
Carl & Eloise Pohlad Family Fndn Grants, 1021
Carl Gellert and Celia Berta Gellert Foundation Grants, 1024
Carl M. Freeman Foundation Grants, 1027
Carl W. and Carrie Mae Joslyn Trust Grants, 1031
Carnegie Corporation of New York Grants, 1033
Carrier Corporation Contributions Grants, 1038
Catherine Holmes Wilkins Foundation Grants, 1042
CCFF Chairman's Dist Life Sciences Award, 1044
CDC Cooperative Agreement for Continuing Enhanced National Surveillance for Prion Diseases in the United States, 1049
CDC Cooperative Agreement for the Development, Operation, and Evaluation of an Entertainment Education Program, 1051
CDC Evaluation of the Use of Rapid Testing For Influenza in Outpatient Medical Settings, 1057
CDC Evidence-Based Laboratory Medicine: Quality/Performance Measure Evaluation, 1058
CDC Preventive Med Residency & Fellowship, 1064
CDC Public Health Conference Support Grant, 1067
CDC Steven M. Teutsch Prevention Effectiveness Fellowships, 1071
CDI Interdisciplinary Research Initiatives Grants, 1074
CDI Postdoctoral Fellowships, 1075
Cedar Tree Foundation David H. Smith Conservation Research Fellowship, 1076
CFF-NIH Funding Grants, 1083
CFF Clinical Research Grants, 1084
CFF Shwachman Clinical Investigator Award, 1086
CFF Pilot and Feasibility Awards, 1088
CFF Research Grants, 1090
CFF Student Traineeships, 1091
Champ-A Champion Fur Kids Grants, 1109
Chapman Charitable Foundation Grants, 1112
Charles Edison Fund Grants, 1116
Charles H. Farnsworth Trust Grants, 1119
Charles H. Revson Foundation Grants, 1122
Charles Lafitte Foundation Grants, 1123
Chest Foundation/LUNGevity Foundation Clinical Research in Lung Cancer Grants, 1131
Chest Fndn Eli Lilly & Company Distinguished Scholar in Critical Care Med Award, 1132
Chest Foundation Geriatric Development Research Awards, 1133
Chest Fnd Grant in Venous Thromboembolism, 1134
Chicago Institute for Psychoanalysis Fellowships, 1135
Children's Brain Tumor Fnd Research Grants, 1136
Children's Cardiomyopathy Foundation Research Grants, 1137
Children's Leukemia Research Association Research Grants, 1138

838 / Basic Research

Children's Tumor Fndn Clin Research Awards, 1140
Children's Tumor Foundation Drug Discovery Initiative Awards, 1141
Children's Tumor Foundation Schwannomatosis Awards, 1142
Children's Tumor Foundation Young Investigator Awards, 1143
Chiles Foundation Grants, 1147
Christine Gregoire Youth/Young Adult Award for Use of Tobacco Industry Documents, 1151
CICF Indianapolis Fndn Community Grants, 1156
CJ Foundation for SIDS Research Grants, 1164
Clarence E. Heller Charitable Foundation Grants, 1166
Claude Pepper Foundation Grants, 1177
Cleveland-Cliffs Foundation Grants, 1181
Clowes ACS/AAST/NIGMS Jointly Sponsored Mentored Clinical Scientist Dev Award, 1185
CMS Hispanic Health Services Research Grants, 1186
CMS Historically Black Colleges and Universities (HBCU) Health Services Research Grants, 1187
CMS Research and Demonstration Grants, 1188
CNIB Baker Applied Research Fund Grants, 1193
CNIB Baker New Researcher Fund Grants, 1195
CNIB Barbara Tuck MacPhee Award, 1196
CNIB Canada Glaucoma Clinical Research Council Grants, 1197
CNIB E. (Ben) & Mary Hochhausen Access Technology Research Grants, 1199
CNIB Ross Purse Doctoral Fellowships, 1200
Coeta and Donald Barker Foundation Grants, 1205
Coleman Foundation Cancer Care Grants, 1206
Colonel Stanley R. McNeil Foundation Grants, 1211
Columbus Foundation Ann Ellis Fund Grants, 1215
Commonwealth Fund Affordable Health Insurance Grants, 1227
Commonwealth Fund Australian-American Health Policy Fellowships, 1228
Commonwealth Fund Harkness Fellowships in Health Care Policy and Practice, 1229
Commonwealth Fund Health Care Quality Improvement and Efficiency Grants, 1230
Commonwealth Payment System Reform Grants, 1232
Commonwealth Fund Quality of Care for Frail Elders Grants, 1233
Commonwealth Fund Small Grants, 1234
Commonwealth Fund State High Performance Health Systems Grants, 1235
Communities Foundation of Texas Grants, 1236
Community Fndn AIDS Endowment Awards, 1237
Community Foundation for Greater Atlanta AIDS Fund Grants, 1239
Community Fndn for Greater Buffalo Grants, 1247
Community Fndn for Muskegon County Grants, 1254
Community Foundation of Boone County Grants, 1266
Community Foundation of Greater Fort Wayne - Community Endowment and Clarke Endowment Grants, 1274
Community Foundation of Louisville Bobbye M. Robinson Fund Grants, 1291
Community Foundation of Louisville Diller B. and Katherine P. Groff Fund for Pediatric Surgery Grants, 1293
Community Foundation of Louisville Dr. W. Barnett Owen Memorial Fund for the Children of Louisville and Jefferson County Grants, 1294
Community Fndn of Louisville Health Grants, 1295
Community Foundation of Louisville Irving B. Klempner Fund Grants, 1296
Community Foundation of Louisville Lee Look Fund for Spinal Injury Grants, 1297
Community Foundation of Mount Vernon and Knox County Grants, 1299
Community Foundation of Muncie and Delaware County Grants, 1300
Community Fndn of So Alabama Grants, 1306
Community Fndn of Wabash County Grants, 1316
Conquer Cancer Foundation of ASCO Career Development Award, 1328

Conquer Cancer Foundation of ASCO Comparative Effectiveness Research Professorship in Breast Cancer, 1329
Conquer Cancer Foundation of ASCO Drug Development Research Professorship, 1330
Conquer Cancer Foundation of ASCO Improving Cancer Care Grants, 1331
Conquer Cancer Foundation of ASCO International Development and Education Award in Palliative Care, 1332
Conquer Cancer Foundation of ASCO International Innovation Grant, 1334
Conquer Cancer Foundation of ASCO Medical Student Rotation Grants, 1335
Conquer Cancer Foundation of ASCO Translational Research Professorships, 1337
Conservation, Food, and Health Foundation Grants for Developing Countries, 1338
Coors Brewing Corporate Contributions Grants, 1346
Cowles Charitable Trust Grants, 1354
Crane Foundation Grants, 1357
CSTE CDC/CSTE Applied Epidemiology Fellowships, 1367
CTCRI Idea Grants, 1369
Cudd Foundation Grants, 1370
Cystic Fibrosis Canada Clinical Fellowships, 1381
Cystic Fibrosis Canada Clinical Project Grants, 1382
Cystic Fibrosis Canada Clinic Incentive Grants, 1383
Cystic Fibrosis Canada Fellowships, 1384
Cystic Fibrosis Canada Research Grants, 1385
Cystic Fibrosis Canada Scholarships, 1386
Cystic Fibrosis Canada Senior Scientist Research Training Awards, 1387
Cystic Fibrosis Canada Small Conference Grants, 1388
Cystic Fibrosis Canada Studentships, 1390
Cystic Fibrosis Canada Summer Studentships, 1391
Cystic Fibrosis Canada Travel Supplement Grants, 1393
Cystic Fibrosis Canada Visiting Allied Health Professional Awards, 1394
Cystic Fibrosis Canada Visiting Clinician Awards, 1395
Cystic Fibrosis Canada Visiting Scientist Awards, 1396
Cystic Fibrosis Research Eliz Nash Fellowships, 1401
Cystic Fibrosis Research New Horizons Campaign Grants, 1402
Cystic Fibrosis Trust Research Grants, 1404
D.F. Halton Foundation Grants, 1405
DAAD Research Stays for University Academics and Scientists, 1408
Daimler and Benz Foundation Fellowships, 1411
Dana Foundation Science and Health Grants, 1420
Daniel Mendelsohn New Investigator Award, 1423
Davis Family Foundation Grants, 1435
Daywood Foundation Grants, 1438
Del E. Webb Foundation Grants, 1450
Delta Air Lines Foundation Health and Wellness Grants, 1455
DeMatteis Family Foundation Grants, 1457
Denton A. Cooley Foundation Grants, 1459
DHHS Oral Health Promotion Research Across the Lifespan, 1474
DOD HBCU/MI Partnership Training Award, 1480
Donald and Sylvia Robinson Family Foundation Grants, 1486
Donald W. Reynolds Foundation Aging and Quality of Life Grants, 1487
Doris Duke Charitable Foundation Clinical Interfaces Award Program, 1492
Doris Duke Charitable Foundation Clinical Research Fellowships for Medical Students, 1493
Doris Duke Charitable Foundation Clinical Scientist Development Award, 1494
Doris Duke Charitable Foundation Clinical Scientist Development Award (CSDA) Bridge Grants, 1495
Doris Duke Charitable Foundation Distinguished Clinical Scientist Award Program, 1496
Doris Duke Charitable Foundation Operations Research on AIDS Care and Treatment in Africa (ORACTA) Grants, 1497
Dorothea Haus Ross Foundation Grants, 1498
Dorothy Rider Pool Health Care Grants, 1500

Dorr Institute for Arthritis Research & Educ, 1502
Drs. Bruce and Lee Foundation Grants, 1513
DSO Controlling Pathogen Workshop Grants, 1514
DSO Radiation Biodosimetry (RaBiD) Grants, 1515
Duchossois Family Foundation Grants, 1518
Duke Endowment Health Care Grants, 1519
Duke University Ambulatory and Regional Anesthesia Fellowships, 1521
Duke University Clinical Cardiac Electrophysiology Fellowships, 1522
Duke University Hyperbaric Center Fellowships, 1523
Duke Univ Interventional Cardio Fellowships, 1524
Duke Univ Obstetric Anesthesia Fellowships, 1525
Duke University Pain Management Fellowships, 1526
Duke University Postdoctoral Research Fellowships in Aging, 1528
E.W. "Al" Thrasher Awards, 1540
Eberly Foundation Grants, 1544
Educational Foundation of America Grants, 1553
Edward Bangs Kelley and Elza Kelley Foundation Grants, 1556
Edward W. and Stella C. Van Houten Memorial Fund Grants, 1559
Edwin S. Webster Foundation Grants, 1560
Edwin W. and Catherine M. Davis Fndn Grants, 1561
Elizabeth Glaser Int'l Leadership Awards, 1566
Elizabeth Glaser Scientist Award, 1567
Elizabeth McGraw Foundation Grants, 1568
Elizabeth Morse Genius Charitable Trust Grants, 1569
Elizabeth Nash Fnd Summer Research Awards, 1571
Ellison Medical Foundation/AFAR Julie Martin Mid-Career Award in Aging Research, 1576
Ellison Medical Foundation/AFAR Postdoctoral Fellows in Aging Research Program, 1577
Elmer L. & Eleanor J. Andersen Fndn Grants, 1578
Elton John AIDS Foundation Grants, 1585
EMBO Installation Grants, 1586
Emerson Charitable Trust Grants, 1589
Emma Barnsley Foundation Grants, 1594
ERC Starting Grants, 1599
Ewing Halsell Foundation Grants, 1618
ExxonMobil Foundation Malaria Grants, 1621
Eye-Bank for Sight Restoration and Fight for Sight Summer Student Research Fellowship, 1622
F.M. Kirby Foundation Grants, 1624
Fairlawn Foundation Grants, 1626
FAMRI Clinical Innovator Awards, 1628
FAMRI Young Clinical Scientist Awards, 1629
Fannie E. Rippel Foundation Grants, 1630
FAR Fund Grants, 1631
Farmers Insurance Corporate Giving Grants, 1634
FCD New American Children Grants, 1637
FDHN Bridging Grants, 1638
FDHN Centocor International Research Fellowship in Gastrointestinal Inflammation & Immunology, 1639
FDHN Designated Outcomes Award in Geriatric Gastroenterology, 1640
FDHN Designated Research Award in Geriatric Gastroenterology, 1641
FDHN Designated Research Award in Research Related to Pancreatitis, 1642
FDHN Fellow Abstract Prizes, 1643
FDHN Funderburg Research Scholar Award in Gastric Biology Related to Cancer, 1645
FDHN Graduate Student Awards, 1646
FDHN Isenberg Int'l Research Scholar Award, 1647
FDHN June & Donald O. Castell MD, Esophageal Clinical Research Award, 1648
FDHN Moti L. & Kamla Rustgi International Travel Awards, 1649
FDHN Non-Career Research Awards, 1650
FDHN Non-Career Research Grants, 1651
FDHN Research Scholar Awards, 1652
FDHN TAP Endowed Designated Research Award in Acid-Related Diseases, 1655
FDHN Translational Research Awards, 1656
Fidelity Foundation Grants, 1659
Fight for Sight-Streilein Foundation for Ocular Immunology Research Award, 1662
Fight for Sight Grants-in-Aid, 1663

PROGRAM TYPE INDEX

Basic Research / 839

Fight for Sight Post-Doctoral Awards, 1664
Fight for Sight Summer Student Fellowships, 1665
FirstEnergy Foundation Community Grants, 1668
Fischelis Grants for Research in the History of American Pharmacy, 1669
Fishman Family Foundation Grants, 1672
Forrest C. Lattner Foundation Grants, 1688
Foundation Fighting Blindness Marjorie Carr Adams Women's Career Development Awards, 1690
Foundation for Pharmaceutical Sciences Herb and Nina Demuth Grant, 1697
Foundation for Seacoast Health Grants, 1698
Foundation of Orthopedic Trauma Grants, 1707
Foundation of the American Thoracic Society Research Grants, 1708
Fragile X Syndrome Postdoc Research Fellows, 1714
Francis L. Abreu Charitable Trust Grants, 1719
Frank G. and Freida K. Brotz Family Foundation Grants, 1723
Frank Reed and Margaret Jane Peters Memorial Fund II Grants, 1728
Frank Stanley Beveridge Foundation Grants, 1730
FRAXA Research Foundation Program Grants, 1733
Frederick Gardner Cottrell Foundation Grants, 1739
Fritz B. Burns Foundation Grants, 1747
Fulbright Alumni Initiatives Awards, 1749
Fulbright New Century Scholars Grants, 1753
Garland D. Rhoads Foundation, 1774
Gebbie Foundation Grants, 1777
General Motors Foundation Grants, 1783
George Foundation Grants, 1795
George Gund Foundation Grants, 1796
George H. Hitchings New Investigator Award in Health Research, 1798
George S. & Dolores Dore Eccles Fndn Grants, 1801
George W. Brackenridge Foundation Grants, 1802
George W. Codrington Charitable Fndn Grants, 1803
Georgia Power Foundation Grants, 1806
Gerber Foundation Grants, 1807
Gertrude B. Elion Mentored Medical Student Research Awards, 1809
Gil and Dody Weaver Foundation Grants, 1819
Gilbert Memorial Fund Grants, 1820
Gill Foundation - Gay and Lesbian Fund Grants, 1821
Glaucoma Foundation Grants, 1827
Glaucoma Research Pilot Project Grants, 1828
Glenn/AFAR Gerontology Awards, 1831
Goddess Scholars Grants, 1848
Goldhirsh Fnd Brain Tumor Research Grants, 1852
Goldman Philanthropic Partnerships Program, 1853
Grammy Foundation Grants, 1863
Grass Foundation Marine Biological Laboratory Fellowships, 1878
Green Diamond Charitable Contributions, 1891
Greenwall Foundation Bioethics Grants, 1897
Grifols Community Outreach Grants, 1904
Gruber Foundation Cosmology Prize, 1907
Gruber Foundation Gentics Prize, 1908
Gruber Foundation Neuroscience Prize, 1909
Gruber Foundation Rosalind Franklin Young Investigator Award, 1910
Gruber Foundation Weizmann Institute Awards, 1911
H.A. & Mary K. Chapman Trust Grants, 1921
H. Leslie Hoffman and Elaine S. Hoffman Foundation Grants, 1924
Hampton Roads Community Foundation Mental Health Research Grants, 1942
Harmony Project Grants, 1948
Harold R. Bechtel Charitable Remainder Uni-Trust Grants, 1953
Harold R. Bechtel Testamentary Charitable Trust Grants, 1954
Harold Simmons Foundation Grants, 1955
Harry A. & Margaret D. Towsley Fndn Grants, 1959
Harry Edison Foundation, 1962
Harry Frank Guggenheim Foundation Dissertation Fellowships, 1963
Harry Frank Guggenheim Fnd Research Grants, 1964
Hawaii Community Foundation Health Education and Research Grants, 1976

Hawn Foundation Grants, 1979
Healthcare Foundation of New Jersey Grants, 1982
Health Foundation of Southern Florida Responsive Grants, 1985
Hearst Foundations Health Grants, 1987
Heineman Foundation for Research, Education, Charitable and Scientific Purposes, 1990
Helen Bader Foundation Grants, 1992
Hendrick Foundation for Children Grants, 1999
Henrietta Lange Burk Fund Grants, 2001
Henry L. Guenther Foundation Grants, 2006
Herbert A. & Adrian W. Woods Fndn Grants, 2007
Herbert H. & Grace A. Dow Fndn Grants, 2008
Hereditary Disease Fnd Lieberman Award, 2010
Hereditary Disease Foundation Research Grants, 2011
Herman Goldman Foundation Grants, 2012
HHMI-NIBIB Interfaces Initiative Grants, 2014
HHMI-NIH Cloister Research Scholars, 2015
HHMI/EMBO Start-up Grants for C Europe, 2016
HHMI Biomedical Research Grants for International Scientists: Infectious Diseases & Parasitology, 2017
HHMI Biomedical Research Grants for International Scientists in Canada and Latin America, 2018
HHMI Biomedical Research Grants for International Scientists in the Baltics, Central and Eastern Europe, Russia, and Ukraine, 2019
HHMI Grants and Fellowships Programs, 2020
HHMI International Research Scholars Program, 2021
HHMI Med into Grad Initiative Grants, 2022
HHMI Physician-Scientist Early Career Award, 2023
Hilda and Preston Davis Foundation Grants, 2028
Hilda and Preston Davis Foundation Postdoctoral Fellowships in Eating Disorders Research, 2029
Hill Crest Foundation Grants, 2030
Hilton Hotels Corporate Giving Program Grants, 2037
Hoglund Foundation Grants, 2040
HomeBanc Foundation Grants, 2042
Hormel Foundation Grants, 2051
Hospital Libraries Section / MLA Professional Development Grants, 2052
HRAMF Charles A. King Trust Postdoctoral Research Fellowships, 2056
HRAMF Charles H. Hood Foundation Child Health Research Awards, 2057
HRAMF Deborah Munroe Noonan Memorial Research Grants, 2058
HRAMF Harold S. Geneen Charitable Trust Awards for Coronary Heart Disease Research, 2059
HRAMF Jeffress Trust Awards in Interdisciplinary Research, 2060
HRAMF Ralph and Marian Falk Medical Research Trust Catalyst Awards, 2061
HRAMF Ralph and Marian Falk Medical Research Trust Transformational Awards, 2062
HRAMF Smith Family Awards for Excellence in Biomedical Research, 2063
HRAMF Taub Fnd Grants for MDS Research, 2064
HRAMF Thome Foundation Awards in Age-Related Macular Degeneration Research, 2065
HRAMF Thome Foundation Awards in Alzheimer's Disease Drug Discovery Research, 2066
HRSA Nurse Education, Practice, Quality and Retention (NEPQR) Grants, 2068
Huffy Foundation Grants, 2073
Huie-Dellmon Trust Grants, 2075
Huisking Foundation Grants, 2076
Huntington's Disease Society of America Research Grants, 2080
Huntington County Community Foundation Make a Difference Grants, 2083
Hutchinson Community Foundation Grants, 2085
IAFF Burn Foundation Research Grants, 2089
IARC Expertise Transfer Fellowship, 2090
IARC Postdoctoral Fellowships for Training in Cancer Research, 2091
IARC Visiting Award for Senior Scientists, 2092
IATA Research Grants, 2093
IBRO Asia Regional APRC Travel Grants, 2103
IBRO Latin America Regional Funding for Short Research Stays, 2108

IIE Whitaker Int'l Fellowships & Scholarships, 2144
Impact 100 Grants, 2148
Indiana 21st Century Research and Technology Fund Awards, 2154
Infinity Foundation Grants, 2158
Intergrys Corporation Grants, 2160
Ireland Family Foundation Grants, 2162
Ittleson Foundation AIDS Grants, 2166
Ittleson Foundation Mental Health Grants, 2167
J.B. Reynolds Foundation Grants, 2168
J.W. Kieckhefer Foundation Grants, 2176
Jack H. & William M. Light Trust Grants, 2179
Jacob and Valeria Langeloth Foundation Grants, 2182
James H. Cummings Foundation Grants, 2190
James L. and Mary Jane Bowman Charitable Trust Grants, 2194
James M. Collins Foundation Grants, 2195
James R. Dougherty Jr. Foundation Grants, 2198
James S. Copley Foundation Grants, 2200
James S. McDonnell Foundation Brain Cancer Research Collaborative Activity Awards, 2201
James S. McDonnell Foundation Complex Systems Collaborative Activity Awards, 2202
James S. McDonnell Fnd Research Grants, 2203
James S. McDonnell Foundation Scholar Awards, 2204
James S. McDonnell Foundation Understanding Human Cognition Awards, 2205
Jane Beattie Memorial Scholarship, 2207
Jane Bradley Pettit Foundation Health Grants, 2208
Jeffrey Thomas Stroke Shield Foundation Research Grants, 2216
Jerome Robbins Foundation Grants, 2223
Jessie Ball Dupont Fund Grants, 2226
JM Foundation Grants, 2229
Joe W. and Dorothy Dorsett Brown Fndn Grants, 2230
John Deere Foundation Grants, 2234
John Merck Scholars Awards, 2246
John P. McGovern Foundation Grants, 2247
John S. Dunn Research Foundation Grants, 2250
Johnson & Johnson Corporate Contributions, 2253
John W. Alden Trust Grants, 2260
Joseph Alexander Foundation Grants, 2265
Josephine Goodyear Foundation Grants, 2272
Josiah W. and Bessie H. Kline Foundation Grants, 2278
Judith and Jean Pape Adams Charitable Foundation ALS Grants, 2279
Kathryne Beynon Foundation Grants, 2294
Kavli Foundation Research Grants, 2296
Kenneth T. & Eileen L. Norris Fndn Grants, 2300
Kettering Fund Grants, 2305
Kevin P. and Sydney B. Knight Family Foundation Grants, 2306
Klarman Family Foundation Grants in Eating Disorders Research Grants, 2310
Komen Greater NYC Clinical Research Enrollment Grants, 2315
Komen Greater NYC Community Breast Health Grants, 2316
Kosciuszko Fndn Grants for Polish Citizens, 2322
Kovler Family Foundation Grants, 2323
Lalor Foundation Postdoctoral Fellowships, 2336
LAM Fnd Established Investigator Awards, 2337
LAM Foundation Pilot Project Grants, 2338
LAM Foundation Postoctoral Fellowships, 2339
LAM Foundation Research Grants, 2340
Land O'Lakes Foundation Mid-Atlantic Grants, 2341
Lawrence S. Huntington Fund Grants, 2348
Leahi Fund, 2350
Leo Goodwin Foundation Grants, 2354
Libra Foundation Grants, 2363
Lillian S. Wells Foundation Grants, 2365
Lisa Higgins-Hussman Foundation Grants, 2370
Littlefield-AACR Grants in Metastatic Colon Cancer Research, 2371
Little Life Foundation Grants, 2372
Lloyd A. Fry Foundation Health Grants, 2373
Lotus 88 Fnd for Women & Children Grants, 2376
Lubbock Area Foundation Grants, 2383
Lubrizol Foundation Grants, 2384

840 / Basic Research

Lucile Horton Howe and Mitchell B. Howe Foundation Grants, 2385
Lumosity Human Cognition Grant, 2390
Lumpkin Family Fnd Healthy People Grants, 2391
Lustgarten Foundation for Pancreatic Cancer Research Grants, 2392
Lymphatic Education and Research Network Additional Support Grants for NIH-funded F32 Postdoctoral Fellows, 2394
Lymphatic Education and Research Network Postdoctoral Fellowships, 2395
Lynde & Harry Bradley Foundation Fellowships, 2396
Lynde and Harry Bradley Foundation Grants, 2397
Lynn and Rovena Alexander Family Foundation Grants, 2399
M. Bastian Family Foundation Grants, 2402
M.J. Murdock Charitable Trust General Grants, 2405
Mabel H. Flory Charitable Trust Grants, 2408
Mabel Y. Hughes Charitable Trust Grants, 2409
Marcia and Otto Koehler Foundation Grants, 2427
Margaret T. Morris Foundation Grants, 2429
Marjorie C. Adams Charitable Trust Grants, 2437
Mary K. Chapman Foundation Grants, 2443
Mary Kay Foundation Cancer Research Grants, 2444
Mary S. and David C. Corbin Foundation Grants, 2447
Massage Therapy Foundation Practitioner Case Report Contest, 2450
Massage Therapy Foundation Research Grants, 2451
Massage Therapy Research Fund (MTRF) Grants, 2452
Matilda R. Wilson Fund Grants, 2453
Mattel Children's Foundation Grants, 2454
Mattel International Grants Program, 2455
Max and Victoria Dreyfus Foundation Grants, 2457
MBL Fred Karush Library Readership, 2474
McCarthy Family Foundation Grants, 2490
McCombs Foundation Grants, 2491
McGraw-Hill Companies Community Grants, 2494
McLean Contributionship Grants, 2497
MDA Development Grant, 2500
MDA Neuromuscular Disease Research Grants, 2501
Mead Johnson Nutritionals Med Educ Grants, 2504
Mead Johnson Nutritionals Scholarships, 2505
Mead Witter Foundation Grants, 2507
MedImmune Charitable Grants, 2509
Mericos Foundation Grants, 2518
Merkel Foundation Grants, 2520
Merrick Foundation Grants, 2521
Mervin Bovaird Foundation Grants, 2523
Mesothelioma Applied Research Fndn Grants, 2524
Meta and George Rosenberg Foundation Grants, 2525
Meyer and Stephanie Eglin Foundation Grants, 2531
Meyer Memorial Trust Responsive Grants, 2537
Meyer Memorial Trust Special Grants, 2538
MGFA Post-Doctoral Research Fellowships, 2539
MGFA Student Fellowships, 2540
MGM Resorts Foundation Community Grants, 2541
MGN Family Foundation Grants, 2542
Microsoft Research Cell Phone as a Platform for Healthcare Grants, 2548
Milken Family Foundation Grants, 2555
Mimi and Peter Haas Fund Grants, 2557
MLA Cunningham Memorial International Fellowship, 2560
MLA David A. Kronick Traveling Fellowship, 2561
MLA Donald A.B. Lindberg Fellowship, 2562
MLA Research, Development, and Demonstration Project Grant, 2572
MMAAP Foundation Dermatology Fellowships, 2577
MMAAP Foundation Fellowship Award in Hematology, 2579
MMAAP Foundation Fellowship Award in Reproductive Medicine, 2580
MMAAP Foundation Fellowship Award in Translational Medicine, 2581
MMAAP Foundation Geriatric Project Awards, 2582
MMAAP Fndn Hematology Project Awards, 2583
MMAAP Foundation Research Project Award in Reproductive Medicine, 2584
MMAAP Foundation Research Project Award in Translational Medicine, 2585

MMAAP Foundation Senior Health Fellowship Award in Geriatrics Medicine and Aging Research, 2586
MMS and Alliance Charitable Foundation International Health Studies Grants, 2588
Montgomery County Community Fndn Grants, 2591
Morgan Adams Foundation Grants, 2593
NAF Kyle Bryant Translational Research Award, 2604
NAF Research Grants, 2605
NAF Young Investigator Awards, 2606
NAPNAP Fndn Grad Research Grant, 2610
NAPNAP Foundation Nursing Research Grants, 2615
NARSAD Distinguished Investigator Grants, 2620
NARSAD Independent Investigator Grants, 2622
NARSAD Lieber Prize for Schizo Research, 2623
NARSAD Ruane Prize for Child and Adolescent Psychiatric Research, 2624
NARSAD Young Investigator Grants, 2625
National Blood Foundation Research Grants, 2628
National Center for Responsible Gaming Grants, 2629
National Center for Resp Gaming Seed Grants, 2631
National Headache Foundation Research Grants, 2632
National Headache Foundation Seymour Diamond Clinical Fellowship in Headache Education, 2633
National MPS Society Grants, 2634
National Parkinson Foundation Clinical Research Fund Grants, 2635
National Psoriasis Foundation Research Grants, 2636
NBME Stemmler Med Ed Research Grants, 2638
NCCAM Exploratory Developmental Grants for Complementary and Alternative Medicine (CAM) Studies of Humans, 2639
NCCAM Translational Tools for Clinical Studies of CAM Interventions Grants, 2641
NCHS National Center for Health Statistics Postdoctoral Research Appointments, 2642
NCI/NCCAM Quick-Trials for Novel Cancer Therapies, 2643
NCI Academic-Industrial Partnerships for Devel and Validation of In Vivo Imaging Systems and Methods for Cancer Investigations, 2644
NCI Application and Use of Transformative Emerging Technologies in Cancer Research, 2645
NCI Application of Metabolomics for Translational and Biological Research Grants, 2646
NCI Basic Cancer Research in Cancer Health Disparities Grants, 2647
NCI Centers of Excellence in Cancer Communications Research, 2649
NCI Diet, Epigenetic Events, and Cancer Prevention Grants, 2650
NCI Exploratory Grants for Behavioral Research in Cancer Control, 2651
NCI Improving Diet and Physical Activity Assessment Grants, 2652
NCI Mentored Career Development Award to Promote Diversity, 2653
NCI Ruth L. Kirschstein National Research Service Award Institutional Training Grants, 2654
NCI Stages of Breast Development: Normal to Metastatic Disease Grants, 2655
NCRR Networks and Pathways Collaborative Research Projects Grants, 2657
NCRR Novel Approaches to Enhance Animal Stem Cell Research, 2658
NEAVS Fellowsips in Women's Health and Sex Differences, 2659
NEDA/AED Charron Family Research Grant, 2660
NEDA/AED Joan Wismer Research Grant, 2661
NEDA/AED Tampa Bay Eating Disorders Task Force Award, 2662
NEI Clinical Study Planning Grant, 2663
NEI Clinical Vision Research Devel Award, 2664
NEI Research Grant For Secondary Data Analysis Grants, 2667
NEI Ruth L. Kirschstein National Research Service Award Short-Term Institutional Research Training Grants, 2668
NEI Translational Research Program On Therapy For Visual Disorders, 2670
NEI Vision Research Core Grants, 2671

PROGRAM TYPE INDEX

Nestle Foundation Large Research Grants, 2675
Nestle Foundation Pilot Grants, 2676
Nestle Foundation Re-entry Grants, 2677
Nestle Foundation Small Research Grants, 2678
Nestle Foundation Training Grant, 2679
NFL Charities Medical Grants, 2685
NHGRI Mentored Research Scientist Development Award, 2686
NHLBI Airway Smooth Muscle Function and Targeted Therapeutics in Human Asthma Grants, 2687
NHLBI Ancillary Studies in Clinical Trials, 2688
NHLBI Bioengineering and Obesity Grants, 2689
NHLBI Bioengineering Approaches to Energy Balance and Obesity Grants for SBIR, 2690
NHLBI Bioengineering Approaches to Energy Balance and Obesity Grants for STTR, 2691
NHLBI Biomedical Research Training Program for Individuals from Underrepresented Groups, 2692
NHLBI Cardiac Devel Consortium Grants, 2693
NHLBI Career Transition Awards, 2694
NHLBI Characterizing the Blood Stem Cell Niche Grants, 2695
NHLBI Circadian-Coupled Cellular Function in Heart, Lung, and Blood Tissue Grants, 2696
NHLBI Clinical Centers for the NHLBI Asthma Network (AsthmaNet) Grants, 2697
NHLBI Developmental Origins of Altered Lung Physiology and Immune Function Grants, 2698
NHLBI Exploratory Studies in the Neurobiology of Pain in Sickle Cell Disease, 2699
NHLBI Immunomodulatory, Inflammatory, & Vasoregulatory Properties of Transfused Red Blood Cell Units as a Function of Preparation & Storage Grants, 2700
NHLBI Independent Scientist Award, 2701
NHLBI Intramural Research Training Awards, 2702
NHLBI Investigator Initiated Multi-Site Clinical Trials, 2703
NHLBI Lymphatics in Health and Disease in the Digestive, Cardio & Pulmonary Systems, 2704
NHLBI Lymphatics in Health and Disease in the Digestive, Urinary, Cardiovascular and Pulmonary Systems, 2705
NHLBI Mentored Career Award for Faculty at Institutions that Promote Diversity (K01), 2706
NHLBI Mentored Career Award For Faculty At Minority Institutions, 2707
NHLBI Mentored Career Dev Award to Promote Faculty Diversity in Biomed Research, 2708
NHLBI Mentored Clinical Scientist Research Career Development Awards, 2709
NHLBI Mentored Patient-Oriented Research Career Development Awards, 2710
NHLBI Mentored Quantitative Research Career Development Awards, 2711
NHLBI Microbiome of Lung & Respiratory in HIV Infected Individuals & Uninfected Controls, 2712
NHLBI Midcareer Investigator Award in Patient-Oriented Research, 2713
NHLBI National Research Service Award Programs in Cardiovascular Epidemiology and Biostatistics, 2714
NHLBI New Approaches to Arrhythmia Detection and Treatment Grants for SBIR, 2715
NHLBI New Approaches to Arrhythmia Detection and Treatment Grants for STTR, 2716
NHLBI Nutrition and Diet in the Causation, Prevention, and Management of Heart Failure Research Grants, 2717
NHLBI Pathway to Independence Awards, 2718
NHLBI Pediatric Cardiac Genomics Consortium Grants, 2719
NHLBI Phase II Clinical Trials of Novel Therapies for Lung Diseases, 2720
NHLBI Prematurity and Respiratory Outcomes Program (PROP) Grants, 2721
NHLBI Progenitor Cell Biology Consortium Administrative Coordinating Center Grants, 2722
NHLBI Protein Interactions Governing Membrane Transport in Pulmonary Health Grants, 2723

PROGRAM TYPE INDEX

Basic Research / 841

NHLBI Research Demonstration and Dissemination Grants, 2724
NHLBI Research Grants on the Relationship Between Hypertension and Inflammation, 2725
NHLBI Research on the Role of Cardiomyocyte Mitochondria in Heart Disease: An Integrated Approach, 2726
NHLBI Research Supplements to Promote Diversity in Health-Related Research, 2727
NHLBI Right Heart Function in Health and Chronic Lung Diseases Grants, 2728
NHLBI Ruth L. Kirschstein National Research Service Awards for Individual Postdoctoral Fellows, 2729
NHLBI Ruth L. Kirschstein National Research Service Awards for Individual Predoctoral Fellowships to Promote Diversity in Health-Related Research, 2730
NHLBI Ruth L. Kirschstein National Research Service Awards for Individual Senior Fellows, 2732
NHLBI Short-Term Research Education Program to Increase Diversity in Health-Related Research, 2733
NHLBI Summer Inst for Training in Biostatistics, 2734
NHLBI Summer Internship Program in Biomedical Research, 2735
NHLBI Targeted Approaches to Weight Control for Young Adults Grants, 2736
NHLBI Targeting Calcium Regulatory Molecules for Arrhythmia Prevention Grants, 2737
NHLBI Translating Basic Behavioral and Social Science Discoveries into Interventions to Reduce Obesity: Centers for Behavioral Intervention Development Grants, 2739
NHLBI Translational Grants in Lung Diseases, 2740
NIAAA Independent Scientist Award, 2742
NIAAA Mechanisms of Alcohol and Drug-Induced Pancreatitis Grants, 2743
NIAAA Mentored Clinical Scientist Research Career Development Awards, 2744
NIA Aging, Oxidative & Cell Death Grants, 2745
NIA Aging Research Dissertation Awards to Increase Diversity, 2746
NIA AIDS and Aging: Behavioral Sciences Prevention Research Grants, 2747
NIA Alzheimer's Disease Core Centers Grants, 2748
NIA Alzheimer's Disease Drug Development Program Grants, 2749
NIA Alzheimer's Disease Pilot Clinical Trials, 2750
NIA Alzheimer's Disease Centers Grants, 2751
NIA Archiving and Development of Socialbehavioral Datasets in Aging Related Studies Grants, 2752
NIA Awards to Support Research on the Biology of Aging in Invertebrates Grants, 2753
NIA Behavioral and Social Research Grants on Disasters and Health, 2754
NIA Bioenergetics, Fatigability, and Activity Limitations in Aging, 2755
NIA Diversity in Medication Use and Outcomes in Aging Populations Grants, 2756
NIA Effects of Gene-Social Environment Interplay on Health and Behavior in Later Life Grants, 2757
NIA Factors Affecting Cognitive Function in Adults with Down Syndrome Grants, 2758
NIA Grants for Alzheimer's Disease Drugs, 2759
NIA Harmonization of Longitudinal Cross-National Surveys of Aging Grants, 2760
NIA Health Behaviors and Aging Grants, 2761
NIA Healthy Aging through Behavioral Economic Analyses of Situations Grants, 2762
NIA Higher-Order Cognitive Functioning and Aging Grants, 2763
NIA Human Biospecimen Resources for Aging Research Grants, 2764
NIAID Independent Scientist Award, 2765
NIA Independent Scientist Award, 2766
NIA Malnutrition in Older Persons Research, 2767
NIA Mechanisms, Measurement, and Management of Pain in Aging: from Molecular to Clinical, 2768
NIA Mechanisms Underlying the Links between Psychosocial Stress, Aging, the Brain and the Body Grants, 2769

NIA Medical Management of Older Patients with HIV/AIDS Grants, 2770
NIA Mentored Clinical Scientist Research Career Development Awards, 2771
NIA Minority Dissert Aging Research Grants, 2773
NIAMS Pilot and Feasibility Clinical Research Grants in Arthritis, Musculoskeletal & Skin Diseases, 2774
NIA Multicenter Study on Exceptional Survival in Families: The Long Life Family Study (LLFS) Grants, 2775
NIA Network Infrastructure Support for Emerging Behavioral & Social Research in Aging, 2777
NIA Postdoctoral Research on Aging in Canada, 2779
NIA Promoting Careers in Aging and Health Disparities Research Grants, 2780
NIA Renal Function and Chronic Kidney Disease in Aging Grants, 2781
NIA Role of Apolipoprotein E, Lipoprotein Receptors and CNS Lipid Homeostasis in Brain Aging and Alzheimer's Disease, 2782
NIA Role of Nuclear Receptors in Tissue and Organismal Aging Grants, 2783
NIA Thyroid in Aging Grants, 2785
NIA Transdisciplinary Research on Fatigue and Fatigability in Aging Grants, 2786
NIA Translational Research at the Aging/Cancer Interface (TRACI) Grants, 2787
NIA Vulnerable Dendrites and Synapses in Aging and Alzheimer's Disease Grants, 2788
NICHD Innovative Therapies and Clinical Studies for Screenable Disorders Grants, 2789
NICHD Ruth L. Kirschstein National Research Service Award (NRSA) Institutional Predoctoral Training Program in Systems Biology of Developmental Biology & Birth Defects, 2790
NIDA Pilot and Feasibility Studies in Preparation for Drug Abuse Prevention Trials, 2794
NIDCD Mentored Clinical Scientist Research Career Development Award, 2795
NIDCD Mentored Research Scientist Development Award, 2796
NIDDK Advances in Polycystic Kidney Disease, 2797
NIDDK Adverse Metabolic Side Effects of Second Generation Psychotropic Medications Leading to Obesity and Increased Diabetes Risk Grants, 2798
NIDDK Ancillary Studies of Kidney Disease Accessing Information from Clinical Trials, Epidemiological Studies, and Databases Grants, 2799
NIDDK Ancillary Studies to Major Ongoing NIDDK and NHLBI Clinical Research Studies, 2800
NIDDK Basic and Clinical Studies of Congenital Urinary Tract Obstruction Grants, 2801
NIDDK Basic Research Grants in the Bladder and Lower Urinary Tract, 2802
NIDDK Calcium Oxalate Stone Diseases Grants, 2803
NIDDK Cell-Specific Delineation of Prostate and Genitourinary Development, 2804
NIDDK Co-Activators and Co-Repressors in Gene Expression Grants, 2805
NIDDK Collaborative Interdisciplinary Team Science in Diabetes, Endocrinology and Metabolic Diseases Grants, 2806
NIDDK Developmental Biology and Regeneration of the Liver Grants, 2807
NIDDK Development of Assays for High-Throughput Drug Screening Grants, 2808
NIDDK Devel of Disease Biomarkers Grants, 2809
NIDDK Diabetes, Endocrinology, and Metabolic Diseases NRSAs--Individual, 2810
NIDDK Diabetes Research Centers, 2811
NIDDK Diet Comp & Energy Balance Grants, 2812
NIDDK Education Program Grants, 2813
NIDDK Endoscopic Clinical Research Grants In Pancreatic And Biliary Diseases, 2814
NIDDK Enhancing Zebrafish Research with Research Tools and Techniques, 2815
NIDDK Erythroid Lineage Molecular Grants, 2816
NIDDK Erythropoiesis: Components and Mechanisms Grants, 2817

NIDDK Exploratory/Developmental Clinical Research Grants in Obesity, 2818
NIDDK Grants for Basic Research in Glomerular Diseases, 2819
NIDDK Health Disparities in NIDDK Diseases, 2820
NIDDK Identifying & Reducing Diabetes & Obesity Related Disparities in Healthcare Systems, 2821
NIDDK Insulin Signaling and Receptor Cross Talk Grants, 2822
NIDDK Intestinal Failure, Short Gut Syndrome and Small Bowel Transplantation Grants, 2823
NIDDK Intestinal Stem Cell Consortium Grants, 2824
NIDDK Mentored Research Scientist Development Award, 2825
NIDDK Multi-Center Clinical Study Cooperative Agreements, 2826
NIDDK Multi-Center Clinical Study Implementation Planning Grants, 2827
NIDDK Neurobiology of Diabetic Complications Grants, 2828
NIDDK Neuroimaging in Obesity Research, 2829
NIDDK New Tech for Liver Disease SBIR, 2830
NIDDK New Tech for Liver Disease STTR, 2831
NIDDK Non-Invasive Methods for Diagnosis and Progression of Diabetes, Kidney, Urological, Hematological and Digestive Diseases, 2832
NIDDK Pancreatic Development and Regeneration Grants: Cellular Therapies for Diabetes, 2833
NIDDK Feasibility Clinical Research Grants in Diabetes, Endocrine and Metabolic Diseases, 2834
NIDDK Pilot and Feasibility Clinical Research Grants in Kidney or Urologic Diseases, 2835
NIDDK Pilot and Feasibility Clinical Research Studies in Digestive Diseases and Nutrition, 2836
NIDDK Planning Grants For Translational Research For The Prevention And Control Of Diabetes And Obesity, 2837
NIDDK Proteomics: Diabetes, Obesity, And Endocrine, Digestive, Kidney, Urologic, And Hematologic Diseases, 2838
NIDDK Research Grants for Studies of Hepatitis C in the Setting of Renal Disease, 2839
NIDDK Research Grants on Improving Health Care for Obese Patients, 2840
NIDDK Role of Gastrointestinal Surgical Procedures in Amelioration of Obesity-Related Insulin Resistance & Diabetes Weight Loss, 2841
NIDDK Secondary Analyses in Obesity, Diabetes and Digestive and Kidney Diseases Grants, 2842
NIDDK Seeding Collaborative Interdisciplinary Team Science in Diabetes, Endocrinology and Metabolic Diseases Grants, 2843
NIDDK Silvio O. Conte Digestive Diseases Research Core Centers Grants, 2844
NIDDK Small Business Innov Research to Develop New Therapeutics & Monitoring Tech for Type 1 Diabetes Towards an Artificial Pancreas, 2845
NIDDK Small Grants for K08/K23 Recipients, 2846
NIDDK Small Grants for Underrepresented Minority Scientists in Diabetes and Digestive and Kidney Diseases, 2847
NIDDK Traditional Conference Grants, 2848
NIDDK Training in Clinical Investigation in Kidney and Urology Grants, 2849
NIDDK Translational Research for the Prevention and Control of Diabetes and Obesity, 2850
NIDDK Transmission of Human Immunodeficiency Virus (HIV) In Semen, 2851
NIDDK Type 2 Diabetes in the Pediatric Population Research Grants, 2852
NIDRR Field-Initiated Projects, 2853
NIEHS Hazardous Materials Worker Health and Safety Training Grants, 2854
NIEHS Hazmat Training at Doe Nuclear Weapons Complex Grants, 2855
NIEHS Independent Scientist Award, 2856
NIEHS SBIR E-Learning for HAZMAT and Emergency Response Grants, 2859
NIEHS Small Business Innovation Research (SBIR) Program Grants, 2860

NIGMS Advancing Basic Behavioral and Social Research on Resilience: an Integrative Science Approach, 2861
NIGMS Enabling Resources for Pharmacogenomics Grants, 2862
NIGMS Ruth L. Kirschstein National Research Service Awards for Individual Predoctoral Fellows in PharmD/PhD Grants, 2863
NIH Academic Research Enhancement Awards, 2864
NIH Biobehavioral Research for Effective Sleep, 2865
NIH Biology, Development, and Progression of Malignant Prostate Disease Research Grants, 2866
NIH Biomedical Research on the International Space Station, 2867
NIH Bone and the Hematopoietic and Immune Systems Research Grants, 2868
NIH Building Interdisciplinary Research Careers in Women's Health Grants, 2869
NIH Dietary Supplement Research Centers: Botanicals Grants, 2873
NIH Earth-Based Research Relevant to the Space Environment, 2874
NIH Enhancing Adherence to Diabetes Self-Management Behaviors, 2875
NIH Exceptional, Unconventional Research Enabling Knowledge Acceleration (EUREKA) Grants, 2876
NIH Exploratory/Devel Research Grant, 2877
NIH Exploratory Innovations in Biomedical Computational Science & Technology Grants, 2878
NIH Fine Mapping and Function of Genes for Type 1 Diabetes Grants, 2879
NIH Health Promotion Among Racial and Ethnic Minority Males, 2880
NIH Human Connectome Project Grants, 2881
NIH Independent Scientist Award, 2882
NIH Innovations in Biomedical Computational Science and Technology Grants, 2883
NIH Innovations in Biomedical Computational Science and Technology Initiative Grants for SBIR, 2884
NIH Innovations in Biomedical Computational Science and Technology Initiative Grants for STTR, 2885
NIH Mentored Clinical Scientist Research Career Development Award, 2887
NIH Mentored Patient-Oriented Research Career Development Award, 2888
NIH Mentored Research Scientist Dev Awards, 2889
NIH Nonhuman Primate Immune Tolerance Cooperative Study Group Grants, 2890
NIH Receptors and Signaling in Bone in Health and Disease, 2891
NIH Recovery Act Limited Competition: Academic Research Enhancement Awards, 2892
NIH Recovery Act Limited Competition: Biomedical Research, Development, and Growth to Spur the Acceleration of New Technologies Grants, 2893
NIH Recovery Act Limited Competition: Building Sustainable Community-Linked Infrastructure to Enable Health Science Research, 2894
NIH Recovery Act Limited Competition: Small Business Catalyst Awards for Accelerating Innovative Research Grants, 2895
NIH Recovery Act Limited Competition: Supporting New Faculty Recruitment to Enhance Research Resources through Biomedical Research Core Centers Grants, 2896
NIH Research On Ethical Issues In Human Subjects Research, 2897
NIH Research on Sleep and Sleep Disorders, 2898
NIH Research Project Grants, 2899
NIH Ruth L. Kirschstein National Research Service Award Institutional Research Training Grants, 2900
NIH Ruth L. Kirschstein National Research Service Awards (NRSA) for Individual Senior Fellows, 2901
NIH Ruth L. Kirschstein National Research Service Awards for Individual Predoctoral Fellows, 2903
NIH Ruth L. Kirschstein National Research Service Awards for Individual Predoctoral Fellowships to Promote Diversity in Research, 2904

NIH Ruth L. Kirschstein National Research Service Awards for Individual Predoctoral MD/PhD and Other Dual Doctoral Degree Fellows, 2905
NIH Ruth L. Kirschstein Nat Research Service Award Short-Term Institutional Training Grants, 2906
NIH Sarcoidosis: Research into the Cause of Multi-Organ Disease and Clinical Strategies for Therapy Grants, 2907
NIH School Interventions to Prevent Obesity, 2908
NIH Self-Management Strategies Across Chronic Diseases Grants, 2909
NIH Solicitation of Assays for High Throughput Screening (HTS) in the Molecular Libraries Probe Production Centers Network (MLPCN), 2910
NIH Structural Bio of Membrane Proteins Grant, 2911
NIH Summer Internship Program in Biomedical Research, 2912
NIH Support of Competitive Research (SCORE) Pilot Project Awards, 2913
NIH Support of Competitive Research (SCORE) Research Advancement Awards, 2914
NIMH AIDS and Aging: Behavioral Sciences Prevention Research Grants, 2915
NIMH Basic and Translational Research in Emotion Grants, 2916
NIMH Curriculum Development Award in Neuroinformatics Research and Analysis, 2917
NIMH Early Identification and Treatment of Mental Disorders in Children and Adolescents Grants, 2918
NIMH Jointly Sponsored Ruth L. Kirschstein National Research Service Award Institutional Predoctoral Training Program in the Neurosciences, 2919
NIMH Short Courses in Neuroinformatics, 2920
NINDS Optimization of Small Molecule Probes for the Nervous System (SBIR) Grants, 2922
NINDS Optimization of Small Molecule Probes for the Nervous System (STTR) Grants, 2923
NINDS Optimization of Small Molecule Probes for the Nervous System Grants, 2924
NINDS Support of Scientific Meetings as Cooperative Agreements, 2925
NINR Acute and Chronic Care During Mechanical Ventilation Research Grants, 2926
NINR AIDS and Aging: Behavioral Sciences Prevention Research Grants, 2927
NINR Chronic Illness Self-Management in Children and Adolescents Grants, 2928
NINR Clinical Interventions for Managing the Symptoms of Stroke Research Grants, 2929
NINR Diabetes Self-Management in Minority Populations Grants, 2930
NINR Mechanisms, Models, Measurement, & Management in Pain Research Grants, 2931
NINR Mentored Research Scientist Development Award, 2932
NINR Quality of Life for Individuals at the End of Life Grants, 2933
NINR Ruth L. Kirschstein National Research Service Award for Predoc Fellows in Nursing, 2934
NKF Clinical Scientist Awards, 2935
NKF Franklin McDonald/Fresenius Medical Care Clinical Research Grants, 2936
NKF Professional Councils Research Grants, 2937
NKF Research Fellowships, 2938
NKF Young Investigator Grants, 2939
NLM Academic Research Enhancement Awards, 2940
NLM Exceptional, Unconventional Research Grants Enabling Knowledge Acceleration, 2941
NLM Exploratory/Developmental Grants, 2942
NLM Express Research Grants in Biomedical Informatics, 2943
NLM Grants for Scholarly Works in Biomedicine and Health, 2944
NLM Limited Competition for Continuation of Biomedical Informatics/Bioinformatics Resource Grants, 2947
NLM Manufacturing Processes of Medical, Dental, and Biological Technologies (SBIR) Grants, 2948
NLM New Tech for Liver Disease Grants, 2949

NLM Predictive Multiscale Models of the Physiome in Health and Disease Grants, 2950
NLM Research Project Grants, 2951
NLM Research Supplements to Promote Reentry in Health-Related Research, 2952
NMF General Electric Med Fellowships, 2956
NMF National Medical Association Awards for Medical Journalism, 2962
NMF Ralph W. Ellison Prize, 2966
Noble County Community Foundation Grants, 2975
NOCSAE Research Grants, 2978
Norcliffe Foundation Grants, 2979
Norfolk Southern Foundation Grants, 2982
Northwestern Mutual Foundation Grants, 2999
Notsew Orm Sands Foundation Grants, 3002
NSF-NIST Interaction in Chemistry, Materials Research, Molecular Biosciences, Bioengineering, and Chemical Engineering, 3004
NSF Accelerating Innovation Research, 3005
NSF Alan T. Waterman Award, 3006
NSF Animal Developmental Mechanisms Grants, 3007
NSF Biological Physics (BP), 3008
NSF Biomedical Engineering and Engineering Healthcare Grants, 3009
NSF Biomolecular Systems Cluster Grants, 3010
NSF Cell Systems Cluster Grants, 3011
NSF Chemistry Research Experiences for Undergraduates (REU), 3012
NSF Doctoral Dissertation Improvement Grants in the Directorate for Biological Sciences (DDIG), 3013
NSF Emerging Frontiers in Research and Innovation (EFRI) Grants, 3014
NSF Genes and Genome Systems Cluster Grants, 3015
NSF Perception, Action & Cognition Grants, 3019
NSF Postdoc Research Fellowships in Biology, 3020
NSF Presidential Early Career Awards for Scientists and Engineers (PECASE) Grants, 3021
NSF Small Business Innovation Research (SBIR) Grants, 3022
NSF Social Psychology Research Grants, 3023
NSF Systematic Biology and Biodiversity Inventories Grants, 3024
NSF Undergraduate Research and Mentoring in the Biological Sciences (URM), 3025
Nuffield Foundation Children & Families Grants, 3027
Nuffield Foundation Law and Society Grants, 3028
Nuffield Foundation Oliver Bird Rheumatism Program Grants, 3029
Nuffield Foundation Open Door Grants, 3030
Nuffield Foundation Small Grants, 3031
NWHF Community-Based Participatory Research Grants, 3032
NWHF Mark O. Hatfield Research Fellowship, 3035
NYAM Charles A. Elsberg Fellowship, 3039
NYAM David E. Rogers Fellowships, 3040
NYAM Edward N. Gibbs Memorial Lecture and Award in Nephrology, 3041
NYAM Edwin Beer Research Fellowship, 3042
NYAM Ferdinand C. Valentine Fellowship, 3043
NYAM Mary and David Hoar Fellowship, 3045
NYCT Biomedical Research Grants, 3049
NYCT Blood Disease Research Grants, 3051
Obesity Society Grants, 3060
OBSSR Behavioral and Social Science Research on Understanding and Reducing Health Disparities Grants, 3061
Ohio Learning Network Grants, 3070
Olympus Corporation of Americas Corporate Giving Grants, 3073
Olympus Corporation of Americas Fellowships, 3074
OneFamily Foundation Grants, 3075
OneSight Research Foundation Block Grants, 3076
Oppenstein Brothers Foundation Grants, 3078
Orthopaedic Trauma Association Kathy Cramer Young Clinician Memorial Fellowships, 3083
Orthopaedic Trauma Assoc Research Grants, 3084
Oscar Rennebohm Foundation Grants, 3085
Osteosynthesis and Trauma Care Foundation European Visiting Fellowships, 3097
Pancreatic Cancer Action Network Fellowship, 3113

PROGRAM TYPE INDEX

Basic Research / 843

Pancreatic Cancer Action Network-AACR Pathway to Leadership Grants, 3114
Partnership for Cures Charles E. Culpeper Scholarships in Medical Science, 3117
Partnership for Cures Two Years To Cures Grants, 3118
Patron Saints Foundation Grants, 3121
Paul Rapoport Foundation Grants, 3125
PDF International Research Grants, 3128
Peabody Foundation Grants, 3132
Peacock Foundation Grants, 3133
Pediatric Brain Tumor Fndn Research Grants, 3134
Pediatric Cancer Research Foundation Grants, 3135
Penny Severns Breast, Cervical and Ovarian Cancer Research Grants, 3136
Percy B. Ferebee Endowment Grants, 3138
Perkin Fund Grants, 3139
Perpetual Trust for Charitable Giving Grants, 3141
Peter F. McManus Charitable Trust Grants, 3153
Peter Kiewit Foundation General Grants, 3154
Peter Kiewit Foundation Small Grants, 3155
Pew Charitable Trusts Biomed Research Grants, 3157
Peyton Anderson Foundation Grants, 3160
Pezcoller Foundation-AACR International Award for Cancer Research, 3161
Pfizer Advancing Research in Transplantation Science (ARTS) Research Awards, 3163
Pfizer Anti Infective Research EU Grants, 3164
Pfizer ASPIRE EU Antifungal Research Awards, 3165
Pfizer ASPIRE EU Dupuytren's Contracture Research Awards, 3166
Pfizer ASPIRE EU Emerging Mechanisms of Resistance Antibacterial Research Awards, 3167
Pfizer ASPIRE EU MRSA Nosocomial Pneumonia & MRSA Complicated Skin & Soft Tissue Infections Antibacterial Research Awards, 3168
Pfizer ASPIRE North America Broad Spectrum Antibiotics for the Treatment of Gram-Negative or Polymicrobial Infections Research Awards, 3169
Pfizer ASPIRE North America Hemophilia Research Awards, 3170
Pfizer ASPIRE North America Narrow Spectrum Antibiotics for the Treatment of MRSA Research Awards, 3171
Pfizer ASPIRE North America Rheumatology Research Awards, 3172
Pfizer ASPIRE Worldwide Endocrine Young Investigator Grants, 3173
Pfizer Australia Cancer Research Grants, 3174
Pfizer Australia Neuroscience Research Grants, 3175
Pfizer Australia Paediatric Endocrine Care Research Grants, 3176
PFizer Compound Transfer Agreements, 3177
Pfizer Global Investigator Research Grants, 3178
Pfizer Global Research Awards for Nicotine Independence, 3179
Pfizer Inflammation Competitive Research Awards (UK), 3181
Pfizer Medical Education Track Two Grants, 3183
Pfizer Research Initiative Dermatology Grants (Germany), 3184
Pfizer Research Initiative Rheumatology Grants (Germany), 3185
Phelps County Community Foundation Grants, 3188
Philip L. Graham Fund Health and Human Services Grants, 3190
Phi Upsilon Omicron Alumni Research Grant, 3191
Phoenix Suns Charities Grants, 3201
PhRMA Foundation Health Outcomes Post Doctoral Fellowships, 3202
PhRMA Foundation Health Outcomes Pre-Doctoral Fellowships, 3203
PhRMA Foundation Health Outcomes Research Starter Grants, 3204
PhRMA Foundation Health Outcomes Sabbatical Fellowships, 3205
PhRMA Foundation Informatics Research Starter Grants, 3208
PhRMA Fndn Info Sabbatical Fellowships, 3209
PhRMA Foundation Paul Calabresi Medical Student Research Fellowship, 3210

PhRMA Foundation Pharmaceutics Postdoctoral Fellowships, 3211
PhRMA Foundation Pharmaceutics Pre Doctoral Fellowships, 3212
PhRMA Foundation Pharmaceutics Research Starter Grants, 3213
PhRMA Foundation Pharmaceutics Sabbatical Fellowships, 3214
PhRMA Foundation Pharmacology/Toxicology Post Doctoral Fellowships, 3215
PhRMA Foundation Pharmacology/Toxicology Pre Doctoral Fellowships, 3216
PhRMA Foundation Pharmacology/Toxicology Research Starter Grants, 3217
PhRMA Foundation Pharmacology/Toxicology Sabbatical Fellowships, 3218
Pinkerton Foundation Grants, 3224
Pioneer Hi-Bred Society Fellowships, 3226
Piper Jaffray Fndn Communities Giving Grants, 3227
Piper Trust Healthcare & Med Research Grants, 3228
Pittsburgh Foundation Medical Research Grants, 3231
PKD Foundation Lillian Kaplan International Prize for the Advancement in the Understanding of Polycystic Kidney Disease, 3232
PKD Foundation Research Grants, 3233
Plastic Surgery Educational Foundation/AAO-HNSF Combined Grant, 3234
Plastic Surgery Educational Foundation Basic Research Grants, 3235
Plastic Surgery Educational Foundation Research Fellowship Grants, 3236
Potts Memorial Foundation Grants, 3254
PPG Industries Foundation Grants, 3257
Presbyterian Health Foundation Bridge, Seed and Equipment Grants, 3259
Price Family Charitable Fund Grants, 3262
Prudential Foundation Education Grants, 3276
PVA Research Foundation Grants, 3285
R.C. Baker Foundation Grants, 3292
RACGP Cardiovascular Research Grants in General Practice, 3295
RACGP Vicki Kotsirilos Integrative Med Grants, 3297
Rajiv Gandhi Foundation Grants, 3302
Ralph M. Parsons Foundation Grants, 3304
Rayonier Foundation Grants, 3311
RCPSC/AMS CanMEDs Research and Development Grants, 3316
RCPSC Detweiler Traveling Fellowships, 3320
RCPSC Tom Dignan Indigenous Health Award, 3321
RCPSC Harry S. Morton Traveling Fellowship in Surgery, 3324
RCPSC Medical Education Research Grants, 3334
RCPSC Peter Warren Traveling Scholarship, 3336
RCPSC Prix D'excellence (Specialist of the Year) Award, 3337
RCPSC Robert Maudsley Fellowship for Studies in Medical Education, 3341
Regenstrief General Internal Medicine - General Pediatrics Research Fellowship Program, 3351
Regenstrief Geriatrics Fellowship, 3352
Regenstrief Master Of Science In Clinical Research/ CITE Program, 3353
Rehabilitation Nursing Fndn Research Grants, 3354
Reinberger Foundation Grants, 3357
Religion and Health: Effects, Mechanisms, and Interpretation RFP, 3358
RGk Foundation Grants, 3361
Richard & Susan Smith Family Fndn Grants, 3365
Richard Davoud Donchian Foundation Grants, 3367
Robert B McMillen Foundation Grants, 3381
Robert Leet & Clara Guthrie Patterson Grants, 3383
Robert & Clara Milton Senior Housing Grants, 3384
Robert Sterling Clark Foundation Reproductive Rights and Health Grants, 3388
Rockefeller Archive Center Research Grants, 3393
Rockefeller Foundation Grants, 3396
Rockwell International Corporate Trust Grants Program, 3399
Roeher Institute Research Grants, 3400
Rohm and Haas Company Grants, 3403

Rollins-Luetkemeyer Foundation Grants, 3404
Ronald McDonald House Charities Grants, 3405
Rosalinde and Arthur Gilbert Foundation/AFAR New Investigator Awards in Alzheimer's Disease, 3406
Roy & Christine Sturgis Charitable Grants, 3415
Roy J. Carver Charitable Trust Medical and Science Research Grants, 3416
RRF General Program Grants, 3417
RSC Abbyann D. Lynch Medal in Bioethics, 3419
RSC Jason A. Hannah Medal, 3420
RSC McLaughlin Medal, 3421
Ruth Estrin Goldberg Memorial for Cancer Research Grants, 3428
RWJF Changes in Health Care Financing and Organization Grants, 3430
RWJF Childhood Obesity Grants, 3431
RWJF Disparities Research for Change Grants, 3433
RWJF Harold Amos Medical Faculty Development Scholars, 3435
RWJF Health and Society Scholars, 3436
RWJF Healthy Eating Research Grants, 3437
RWJF Interdisciplinary Nursing Quality Research Initiative Grants, 3438
RWJF Nurse Faculty Scholars, 3441
RWJF Pioneer Portfolio Grants, 3442
RWJF Policies for Action: Policy and Law Research to Build a Culture of Health Grants, 3443
Ryan Gibson Foundation Grants, 3445
S. Mark Taper Foundation Grants, 3449
Saint Luke's Foundation Grants, 3454
Samueli Foundation Health Grants, 3465
Samueli Institute Scientific Research Grants, 3466
San Antonio Area Foundation Grants, 3469
Sandler Program for Asthma Research Grants, 3475
Sara Elizabeth O'Brien Trust Grants, 3483
Sarasota Memorial Healthcare Fndn Grants, 3484
Sarkeys Foundation Grants, 3485
Savoy Foundation Post-Doctoral and Clinical Research Fellowships, 3488
Savoy Foundation Research Grants, 3489
Savoy Foundation Studentships, 3490
Schramm Foundation Grants, 3495
Scleroderma Foundation Established Investigator Grants, 3497
Scleroderma Foundation New Investigator Grants, 3498
Seattle Foundation Medical Funds Grants, 3508
Seaver Institute Grants, 3509
Self Foundation Grants, 3510
Sensient Technologies Foundation Grants, 3513
SfN Chapter Grants, 3520
Sick Kids Foundation New Investigator Research Grants, 3550
Sidgmore Family Foundation Grants, 3551
Sid W. Richardson Foundation Grants, 3553
Sigma Theta Tau International /American Nurses Foundation Grant, 3556
Sigma Theta Tau International / Council for the Advancement of Nursing Science Grants, 3557
Sigma Theta Tau International / Oncology Nursing Society Grant, 3558
Sigma Theta Tau International Doris Bloch Research Award, 3559
Sigma Theta Tau Small Grants, 3560
Sigrid Juselius Foundation Grants, 3561
Simeon J. Fortin Charitable Foundation Grants, 3562
Simone and Cino del Duca Grand Prix Awards, 3564
Sioux Falls Area Community Foundation Community Fund Grants (Unrestricted), 3568
Sioux Falls Area Community Foundation Spot Grants (Unrestricted), 3570
Siragusa Foundation Health Services & Medical Research Grants, 3571
Skin Cancer Foundation Research Grants, 3578
Smithsonian Biodiversity Genomics and Bioinformatics Postdoctoral Fellowships, 3580
Smithsonian Museum Conservation Institute Research Post-Doctorate/Post-Graduate Fellowships, 3581
SNM/ Covidien Seed Grant in Molecular Imaging/ Nuclear Medicine Research, 3582

844 / Basic Research

SNM Molecular Imaging Research Grant For Junior Medical Faculty, 3583
SNM Pilot Research Grants, 3585
SNM Pilot Research Grants in Nuclear Medicine/Molecular Imaging, 3586
SNM Postdoc Molecular Imaging Scholar Grants, 3587
SNM Tetalman Young Investigator Awards, 3589
SOBP A.E. Bennett Research Award, 3594
Society for Imaging Informatics in Medicine (SIIM) Emeritus Mentor Grants, 3597
Society for Imaging Informatics in Medicine (SIIM) Micro Grants, 3598
Society for Imaging Informatics in Medicine (SIIM) Research Grants, 3599
Society for Imaging Informatics in Med Small Grant for Product Support Development, 3600
Society for Imaging Informatics in Medicine (SIIM) Small Training Grants, 3601
Society for the Arts in Healthcare Environmental Arts Research Grant, 3603
Sony Corporation of America Grants, 3623
Southwest Gas Corporation Foundation Grants, 3628
SRC Medical Research Grants, 3635
SSHRC-NSERC College and Community Innovation Enhancement Grants, 3636
SSHRC Banting Postdoctoral Fellowships, 3637
SSRC Collaborative Research Grants on Environment and Health in China, 3638
Steuben County Community Foundation Grants, 3654
Strake Foundation Grants, 3658
Streilein Foundation for Ocular Immunology Visiting Professorships, 3660
Stroke Association Allied Health Professional Research Bursaries, 3661
Stroke Association Clinical Fellowships, 3662
Stroke Association Research Project Grants, 3663
STTI/ATI Educational Assessment Nursing Research Grants, 3665
STTI Emergency Nurses Association Foundation Grant, 3666
STTI Environment of Elder Care Nursing Research Grants, 3667
STTI Joan K. Stout RN Research Grants, 3668
STTI National League for Nursing Grants, 3669
STTI Rosemary Berkel Crisp Research Awards, 3670
STTI Virginia Henderson Research Grants, 3671
Susan G. Komen Breast Cancer Foundation Brinker Awards for Scientific Distinction, 3687
Susan G. Komen Breast Cancer Foundation Career Catalyst Research Grants, 3688
Susan G. Komen Breast Cancer Foundation Challege Grants: Breast Cancer and the Environment, 3689
Susan G. Komen Breast Cancer Foundation Challege Grants: Investigator Initiated Research, 3690
Susan G. Komen Breast Cancer Foundation Challenge Grants: Career Catalyst Research, 3691
Susan G. Komen Breast Cancer Foundation Investigator Initiated Research Grants, 3693
Sybil G. Jacobs Award for Outstanding Use of Tobacco Industry Documents, 3698
Tellabs Foundation Grants, 3706
Theodore Edson Parker Foundation Grants, 3715
The Ray Charles Foundation Grants, 3716
Thomas and Dorothy Leavey Foundation Grants, 3717
Thomas Austin Finch, Sr. Foundation Grants, 3718
Thomas Thompson Trust Grants, 3725
Thomas W. Bradley Foundation Grants, 3726
Thompson Charitable Foundation Grants, 3727
Thomson Reuters / MLA Doctoral Fellowships, 3728
Thoracic Foundation Grants, 3729
Thrasher Research Fund Early Career Awards, 3731
Thrasher Research Fund Grants, 3732
Tourette Syndrome Association Post-Doctoral Fellowships, 3740
Tourette Syndrome Association Research Grants, 3741
TRDRP California Research Awards, 3753
TRDRP Exploratory/Devel Research Awards, 3754
TRDRP New Investigator Awards, 3755
Tri-State Community Twenty-first Century Endowment Fund Grants, 3756

Triological Society Research Career Development Awards, 3760
U.S. Department of Education Rehabilitation Engineering Research Centers Grants, 3769
U.S. Department of Education Rehabilitation Research Training Centers (RRTCs), 3770
U.S. Department of Education United States-Russia Program: Improving Research and Educational Activities in Higher Education, 3773
UCLA / RAND Corporation Post-Doctoral Fellowships, 3776
UICC American Cancer Society International Fellowships for Beginning Investigators, 3778
UICC International Cancer Technology Transfer (ICRETT) Fellowships, 3779
UICC Yamagiwa-Yoshida Memorial International Cancer Study Grants, 3781
Union Pacific Foundation Health and Human Services Grants, 3794
United States Institute of Peace - Jennings Randolph Senior Fellowships, 3802
USAID Comprehensive District-Based Support for Better HIV/TB Patient Outcomes Grants, 3807
USAID Family Planning and Reproductive Health Methods Grants, 3811
USAID Fighting Ebola Grants, 3812
USAID Microbicides Research, Development, and Introduction Grants, 3819
USAID Policy, Advocacy, and Communication Enhanced for Population and Reproductive Health Grants, 3820
USAID Research and Innovation for Health Supply Chain Systems & Commodity Security Grants, 3821
USAID Social Science Research in Population and Reproductive Health Grants, 3822
US CRDF Leishmaniasis: Collaborative Research Opportunities in N Africa & Middle East, 3828
USDD Breast Cancer Clinical Translational Research Grants, 3831
USDD Breast Cancer Era of Hope Grants, 3832
USDD Breast Cancer Idea Grants, 3833
USDD Breast Cancer Impact Grants, 3834
USDD Breast Cancer Innovation Grants, 3835
USDD Breast Cancer Transformative Grant, 3836
USDD Broad Agency Announcement HIV/AIDS Prevention Grants, 3838
USDD Care for the Critically Injured Burn Patient Grants, 3839
USDD Clinical Functional Assessment Investigator-Initiated Grants, 3840
USDD HIV/AIDS Prevention: Military Specific HIV/AIDS Prevention, Care, and Treatment for Non-PEPFAR Funded Countries Grants, 3841
USDD HIV/AIDS Prevention Program Information Systems Development Grants, 3842
USDD Medical Practice Breadth of Medical Practice and Disease Frequency Exposure Grants, 3843
USDD Medical Practice Initiative Procedural Skill Decay and Maintenance Grants, 3844
USDD President's Emergency Plan for AIDS Relief Military Specific HIV Seroprevalence and Behavioral Epidemiology Risk Suvey Grants, 3845
USDD U.S. Army Medical Research and Materiel Command Broad Agency Announcement for Extramural Medical Research Grants, 3846
USDD Workplace Violence in Military Grants, 3847
Vancouver Foundation Grants and Community Initiatives Program, 3849
Victor E. Speas Foundation Grants, 3868
Victor Grifols I Lucas Fndn Research Grants, 3871
Victor Grifols i Lucas Foundation Secondary School Prizes, 3872
Virginia W. Kettering Foundation Grants, 3878
Visiting Nurse Foundation Grants, 3879
Volkswagen of America Corporate Contributions, 3880
W.M. Keck Foundation Med Research Grants, 3888
W.M. Keck Foundation Science and Engineering Research Grants, 3889
W.M. Keck Fndn Undergrad Ed Grants, 3891
Wellcome Trust Biomedical Science Grants, 3918

PROGRAM TYPE INDEX

Wellcome Trust New Investigator Awards, 3919
Weyerhaeuser Company Foundation Grants, 3925
Wilhelm Sander-Stiftung Foundation Grants, 3942
Willard & Pat Walker Charitable Fndn Grants, 3943
Willary Foundation Grants, 3944
William Blair and Company Foundation Grants, 3948
William G. & Helen C. Hoffman Fndn Grants, 3950
William H. Adams Foundation for ALS Grants, 3952
William H. Hannon Foundation Grants, 3953
William T. Grant Foundation Research Grants, 3958
Winn Feline Foundation/AVMF Excellence In Feline Research Award, 3963
Winn Feline Foundation Grants, 3964
Xoran Technologies Resident Research Grant, 3976

Building Construction and/or Renovation
A.O. Smith Foundation Community Grants, 17
Abbot and Dorothy H. Stevens Foundation Grants, 197
Abbott Fund Global AIDS Care Grants, 202
Abel Foundation Grants, 206
Achelis Foundation Grants, 234
Adam Richter Charitable Trust Grants, 298
Adams County Community Foundation of Pennsylvania Grants, 300
Adelaide Breed Bayrd Foundation Grants, 306
Adolph Coors Foundation Grants, 311
Agnes M. Lindsay Trust Grants, 346
Ahmanson Foundation Grants, 355
AIHS Full-Time M.D./Ph.D. Studentships, 385
AIHS Part-Time Studentships, 394
Air Products and Chemicals Grants, 399
Alabama Power Foundation Grants, 400
Albertson's Charitable Giving Grants, 410
Alcoa Foundation Grants, 414
Alexander & Baldwin Fnd Mainland Grants, 417
Alexander and Baldwin Foundation Hawaiian and Pacific Island Grants, 418
Alice Tweed Tuohy Foundation Grants Program, 437
Allegan County Community Foundation Grants, 439
Allegheny Technologies Charitable Trust, 440
Allen P. & Josephine B. Green Fndn Grants, 443
Allyn Foundation Grants, 448
Alvin and Fanny Blaustein Thalheimer Foundation Baltimore Communal Grants, 459
American Schlafhorst Foundation Grants, 559
Amerigroup Foundation Grants, 565
Andre Agassi Charitable Foundation Grants, 587
Anheuser-Busch Foundation Grants, 591
Annenberg Foundation Grants, 596
Ann Jackson Family Foundation Grants, 598
Ann Peppers Foundation Grants, 599
Anthony R. Abraham Foundation Grants, 608
Antone & Edene Vidinha Charitable Trust Grants, 609
Aratani Foundation Grants, 649
Arcadia Foundation Grants, 650
Arie and Ida Crown Memorial Grants, 654
Arkell Hall Foundation Grants, 661
Assisi Foundation of Memphis General Grants, 787
Atherton Family Foundation Grants, 794
Atlanta Foundation Grants, 796
Auburn Foundation Grants, 798
Autzen Foundation Grants, 810
Babcock Charitable Trust Grants, 816
Bacon Family Foundation Grants, 817
Bailey Foundation Grants, 818
Ball Brothers Foundation General Grants, 820
Bank of America Fndn Volunteer Grants, 826
Barker Foundation Grants, 830
Barker Welfare Foundation Grants, 831
Barra Foundation Project Grants, 834
Battle Creek Community Foundation Grants, 839
Beazley Foundation Grants, 865
Beckley Area Foundation Grants, 867
Becton Dickinson and Company Grants, 869
Belk Foundation Grants, 872
Ben B. Cheney Foundation Grants, 875
Bender Foundation Grants, 876
Berrien Community Foundation Grants, 884
Bertha Russ Lytel Foundation Grants, 885
Besser Foundation Grants, 887

PROGRAM TYPE INDEX   Building Construction and/or Renovation   / 845

Bikes Belong Foundation Paul David Clark Bicycling Safety Grants, 890
Bill and Melinda Gates Foundation Emergency Response Grants, 892
Bill and Melinda Gates Foundation Water, Sanitation and Hygiene Grants, 894
Bill Hannon Foundation Grants, 895
Blanche and Irving Laurie Foundation Grants, 910
Blanche and Julian Robertson Family Foundation Grants, 911
Blowitz-Ridgeway Foundation Early Childhood Development Research Award, 912
Blue Cross Blue Shield of Minnesota Fndn Healthy Equity: Public Libraries for Health Grants, 918
Blue Grass Community Fndn Harrison Grants, 919
Blue Grass Community Foundation Hudson-Ellis Fund Grants, 920
Blue Mountain Community Foundation Grants, 921
Blue River Community Foundation Grants, 922
Blumenthal Foundation Grants, 925
Bob & Delores Hope Foundation Grants, 927
Bodenwein Public Benevolent Foundation Grants, 928
Bodman Foundation Grants, 929
Boeing Company Contributions Grants, 930
Boettcher Foundation Grants, 931
Booth-Bricker Fund Grants, 933
Borkee-Hagley Foundation Grants, 934
Bosque Foundation Grants, 935
Boston Globe Foundation Grants, 937
Bright Family Foundation Grants, 951
Brinson Foundation Grants, 953
Burlington Northern Santa Fe Foundation Grants, 978
Byron W. & Alice L. Lockwood Fnd Grants, 991
Cabot Corporation Foundation Grants, 993
Cailloux Foundation Grants, 996
Callaway Foundation Grants, 1002
Cambridge Community Foundation Grants, 1004
Campbell Soup Foundation Grants, 1007
Capital Region Community Foundation Grants, 1014
Cargill Citizenship Fund Corporate Giving, 1018
Carl & Eloise Pohlad Family Fndn Grants, 1021
Carl C. Icahn Foundation Grants, 1023
Carl Gellert and Celia Berta Gellert Foundation Grants, 1024
Carl M. Freeman Foundation FACES Grants, 1026
Carl M. Freeman Foundation Grants, 1027
Carls Foundation Grants, 1030
Carnahan-Jackson Foundation Grants, 1032
Carolyn Foundation Grants, 1034
Carrie E. and Lena V. Glenn Foundation Grants, 1036
Carrie Estelle Doheny Foundation Grants, 1037
Cemala Foundation Grants, 1077
Cessna Foundation Grants Program, 1081
Champlin Foundations Grants, 1110
Charles A. Frueauff Foundation Grants, 1114
Charles Delmar Foundation Grants, 1115
Charles H. Farnsworth Trust Grants, 1119
Charles M. & Mary D. Grant Fndn Grants, 1124
Chatlos Foundation Grants Program, 1128
Chazen Foundation Grants, 1129
Chemtura Corporation Contributions Grants, 1130
Chiles Foundation Grants, 1147
Christensen Fund Regional Grants, 1149
Christopher & Dana Reeve Foundation Quality of Life Grants, 1152
Christy-Houston Foundation Grants, 1153
Citizens Bank Mid-Atlantic Charitable Foundation Grants, 1162
Clara Blackford Smith and W. Aubrey Smith Charitable Foundation Grants, 1165
Clarence T.C. Ching Foundation Grants, 1167
Clark-Winchcole Foundation Grants, 1173
Clark and Ruby Baker Foundation Grants, 1174
Clark County Community Foundation Grants, 1175
Claude Worthington Benedum Fndn Grants, 1178
Clayton Baker Trust Grants, 1179
Clayton Fund Grants, 1180
Cleveland-Cliffs Foundation Grants, 1181
Coastal Community Foundation of South Carolina Grants, 1202

Cockrell Foundation Grants, 1204
Coeta and Donald Barker Foundation Grants, 1205
Collins C. Diboll Private Foundation Grants, 1209
Collins Foundation Grants, 1210
Colonel Stanley R. McNeil Foundation Grants, 1211
Columbus Foundation Competitive Grants, 1217
Communities Foundation of Texas Grants, 1236
Community Foundation Alliance City of Evansville Endowment Fund Grants, 1238
Community Fndn for Greater Buffalo Grants, 1247
Community Fndn for Monterey County Grants, 1253
Community Fndn for Muskegon County Grants, 1254
Community Foundation of Bartholomew County Heritage Fund Grants, 1262
Community Foundation of Bloomington and Monroe County Grants, 1265
Community Fndn of East Central Illinois Grants, 1268
Community Foundation of Eastern Connecticut General Southeast Grants, 1269
Community Foundation of Greater Birmingham Grants, 1272
Community Foundation of Greater Flint Grants, 1273
Community Foundation of Greater New Britain Grants, 1281
Community Foundation of Mount Vernon and Knox County Grants, 1299
Community Foundation of the Verdugos Educational Endowment Fund Grants, 1314
Connecticut Community Foundation Grants, 1324
ConocoPhillips Foundation Grants, 1327
Constantin Foundation Grants, 1341
Constellation Energy Corporate Grants, 1342
Consumers Energy Foundation, 1343
Cooke Foundation Grants, 1344
Cooper Industries Foundation Grants, 1345
Cornerstone Foundation of Northeastern Wisconsin Grants, 1347
Cowles Charitable Trust Grants, 1354
Crail-Johnson Foundation Grants, 1355
Cudd Foundation Grants, 1370
Cullen Foundation Grants, 1372
Cumberland Community Foundation Grants, 1374
CVS Community Grants, 1379
D.V. and Ida J. McEachern Trust Grants, 1406
DaimlerChrysler Corporation Fund Grants, 1412
Daisy Marquis Jones Foundation Grants, 1414
Danellie Foundation Grants, 1421
Daphne Seybolt Culpeper Fndn Grants, 1425
Dayton Power and Light Company Foundation Signature Grants, 1436
Daywood Foundation Grants, 1438
Delaware Community Foundation Grants, 1449
Del E. Webb Foundation Grants, 1450
DeMatteis Family Foundation Grants, 1457
Dennis & Phyllis Washington Fndn Grants, 1458
Denver Foundation Community Grants, 1461
DeRoy Testamentary Foundation Grants, 1469
Deutsche Banc Alex Brown and Sons Charitable Foundation Grants, 1471
Dolan Children's Foundation Grants, 1482
Donald and Sylvia Robinson Family Foundation Grants, 1486
Doris and Victor Day Foundation Grants, 1491
Dorothea Haus Ross Foundation Grants, 1498
Dr. John T. Macdonald Foundation Grants, 1507
Dr. Leon Bromberg Charitable Trust Grants, 1509
Drs. Bruce and Lee Foundation Grants, 1513
DuPont Pioneer Community Giving Grants, 1532
E.J. Grassmann Trust Grants, 1537
E.L. Wiegand Foundation Grants, 1538
Eastman Chemical Company Foundation Grants, 1542
Eberly Foundation Grants, 1544
Edwards Memorial Trust Grants, 1558
Edward W. and Stella C. Van Houten Memorial Fund Grants, 1559
Edwin S. Webster Foundation Grants, 1560
Edyth Bush Charitable Foundation Grants, 1562
Effie and Wofford Cain Foundation Grants, 1563
Eisner Foundation Grants, 1564
Elmer L. & Eleanor J. Andersen Fndn Grants, 1578

El Paso Corporate Foundation Grants, 1580
El Pomar Foundation Anna Keesling Ackerman Fund Grants, 1581
El Pomar Foundation Grants, 1582
Elsie H. Wilcox Foundation Grants, 1583
Erie Community Foundation Grants, 1601
Ethel S. Abbott Charitable Foundation Grants, 1607
Eugene B. Casey Foundation Grants, 1609
Eugene Straus Charitable Trust, 1612
Eva L. & Joseph M. Bruening Fndn Grants, 1613
Ewing Halsell Foundation Grants, 1618
Fairfield County Community Foundation Grants, 1625
Fallon OrNda Community Health Fund Grants, 1627
Fargo-Moorhead Area Foundation Grants, 1632
Faye McBeath Foundation Grants, 1635
Fayette County Foundation Grants, 1636
Fidelity Foundation Grants, 1659
Field Foundation of Illinois Grants, 1660
Fifth Third Foundation Grants, 1661
Fluor Foundation Grants, 1682
FMC Foundation Grants, 1683
Foellinger Foundation Grants, 1684
Ford Family Foundation Grants - Access to Health and Dental Services, 1686
Ford Motor Company Fund Grants Program, 1687
Foster Foundation Grants, 1689
Four County Community Fndn General Grants, 1710
Four County Community Foundation Healthy Senior/Healthy Youth Fund Grants, 1711
France-Merrick Foundation Health and Human Services Grants, 1715
Frank E. and Seba B. Payne Foundation Grants, 1722
Frank Loomis Palmer Fund Grants, 1726
Fred L. Emerson Foundation Grants, 1743
Fremont Area Community Foundation Amazing X Grants, 1744
Fremont Area Community Fndn General Grants, 1746
Fritz B. Burns Foundation Grants, 1747
Fuller E. Callaway Foundation Grants, 1761
G.N. Wilcox Trust Grants, 1768
Gebbie Foundation Grants, 1777
GenCorp Foundation Grants, 1779
General Motors Foundation Grants, 1783
Genesis Foundation Grants, 1786
George and Ruth Bradford Foundation Grants, 1789
George Foundation Grants, 1795
George Gund Foundation Grants, 1796
George Kress Foundation Grants, 1799
George S. & Dolores Dore Eccles Fndn Grants, 1801
George W. Codrington Charitable Fndn Grants, 1803
Gheens Foundation Grants, 1814
Gil and Dody Weaver Foundation Grants, 1819
Giving in Action Society Children & Youth with Special Needs Grants, 1823
Giving in Action Society Family Indep Grants, 1824
Gladys Brooks Foundation Grants, 1826
GNOF Albert N. & Hattie M. McClure Grants, 1834
GNOF Bayou Communities Grants, 1835
GNOF Exxon-Mobil Grants, 1836
GNOF Norco Community Grants, 1845
Goodrich Corporation Foundation Grants, 1854
Grace and Franklin Bernsen Foundation Grants, 1858
Graco Foundation Grants, 1860
Grand Rapids Area Community Foundation Wyoming Grants, 1868
Grand Rapids Community Foundation Ionia County Grants, 1870
Grand Rapids Community Foundation Lowell Area Fund Grants, 1872
Grand Rapids Community Foundation Southeast Ottawa Grants, 1873
Grand Rapids Community Fndn Sparta Grants, 1875
Greater Saint Louis Community Fndn Grants, 1884
Greater Tacoma Community Foundation Grants, 1886
Greater Worcester Community Foundation Jeppson Memorial Fund for Brookfield Grants, 1889
Green Diamond Charitable Contributions, 1891
Greenspun Family Foundation Grants, 1896
Grover Hermann Foundation Grants, 1906
Guido A. & Elizabeth H. Binda Fndn Grants, 1913

Building Construction and/or Renovation

Guy's and St. Thomas' Charity Grants, 1918
H & R Foundation Grants, 1920
H.A. & Mary K. Chapman Trust Grants, 1921
H.J. Heinz Company Foundation Grants, 1923
Harold Alfond Foundation Grants, 1949
Harold and Arlene Schnitzer CARE Foundation Grants, 1950
Harold Simmons Foundation Grants, 1955
Harris and Eliza Kempner Fund Grants, 1956
Harry A. & Margaret D. Towsley Fndn Grants, 1959
Harry Bramhall Gilbert Charitable Trust Grants, 1961
Harry Edison Foundation, 1962
Harry Kramer Memorial Fund Grants, 1965
Harry W. Bass, Jr. Foundation Grants, 1968
Hartford Foundation Regular Grants, 1970
Harvest Foundation Grants, 1971
Harvey Randall Wickes Foundation Grants, 1972
Health Fndn of Greater Indianapolis Grants, 1984
Hedco Foundation Grants, 1989
Helen Bader Foundation Grants, 1992
Helen S. Boylan Foundation Grants, 1997
Henry L. Guenther Foundation Grants, 2006
High Meadow Foundation Grants, 2027
Hillcrest Foundation Grants, 2031
Hill Crest Foundation Grants, 2030
Hillman Foundation Grants, 2032
Hilton Hotels Corporate Giving Program Grants, 2037
Hoblitzelle Foundation Grants, 2038
Hoglund Foundation Grants, 2040
Home Building Industry Disaster Relief Fund, 2043
Horace A. Kimball and S. Ella Kimball Foundation Grants, 2046
Hormel Foundation Grants, 2051
Howard and Bush Foundation Grants, 2054
Hudson Webber Foundation Grants, 2072
Huie-Dellmon Trust Grants, 2075
Huisking Foundation Grants, 2076
I.A. O'Shaughnessy Foundation Grants, 2088
Idaho Power Company Corporate Contributions, 2115
IDPH Hosptial Capital Investment Grants, 2119
Illinois Tool Works Foundation Grants, 2147
Inasmuch Foundation Grants, 2149
Independence Community Foundation Community Quality of Life Grant, 2153
J.B. Reynolds Foundation Grants, 2168
J.E. and L.E. Mabee Foundation Grants, 2170
J.N. and Macie Edens Foundation Grants, 2174
J.W. Kieckhefer Foundation Grants, 2176
Jacobs Family Village Neighborhoods Grants, 2184
James & Abigail Campbell Family Fndn Grants, 2185
James A. and Faith Knight Foundation Grants, 2186
James Graham Brown Foundation Grants, 2189
James H. Cummings Foundation Grants, 2190
James J. & Angelia M. Harris Fndn Grants, 2192
James M. Cox Foundation of Georgia Grants, 2196
James R. Dougherty Jr. Foundation Grants, 2198
Jane's Trust Grants, 2206
Janirve Foundation Grants, 2209
Janson Foundation Grants, 2210
Jay and Rose Phillips Family Foundation Grants, 2213
JELD-WEN Foundation Grants, 2218
Jenkins Foundation: Improving the Health of Greater Richmond Grants, 2219
Jennings County Community Foundation Grants, 2220
Jessica Stevens Community Foundation Grants, 2224
Jessie Ball Dupont Fund Grants, 2226
John Ben Snow Memorial Trust Grants, 2231
John D. & Katherine A. Johnston Fndn Grants, 2233
John Deere Foundation Grants, 2234
John Edward Fowler Memorial Fndn Grants, 2235
John G. Duncan Charitable Trust Grants, 2236
John G. Martin Foundation Grants, 2237
John H. and Wilhelmina D. Harland Charitable Foundation Children and Youth Grants, 2238
John J. Leidy Foundation Grants, 2241
John Jewett and Helen Chandler Garland Foundation Grants, 2242
Johnson & Johnson/SAH Arts & Healing Grants, 2254
John Stauffer Charitable Trust Grants, 2258
John W. and Anna H. Hanes Foundation Grants, 2261
Joseph Alexander Foundation Grants, 2265
Joseph H. & Florence A. Roblee Fndn Grants, 2269
Joseph Henry Edmondson Foundation Grants, 2270
Josephine Goodyear Foundation Grants, 2272
Josephine Schell Russell Charitable Trust Grants, 2274
Josiah W. and Bessie H. Kline Foundation Grants, 2278
Katharine Matthies Foundation Grants, 2291
Katherine John Murphy Foundation Grants, 2293
Kelvin and Eleanor Smith Foundation Grants, 2297
Kenneth T. & Eileen L. Norris Fndn Grants, 2300
KeyBank Foundation Grants, 2307
Kinsman Foundation Grants, 2309
L. W. Pierce Family Foundation Grants, 2328
Laura B. Vogler Foundation Grants, 2345
Lewis H. Humphreys Charitable Trust Grants, 2361
Lillian S. Wells Foundation Grants, 2365
Louie M. & Betty M. Phillips Fndn Grants, 2378
Lowe Foundation Grants, 2381
Lubbock Area Foundation Grants, 2383
Lubrizol Foundation Grants, 2384
Lucy Gooding Charitable Fndn Grants, 2388
Ludwick Family Foundation Grants, 2389
M.D. Anderson Foundation Grants, 2403
M.E. Raker Foundation Grants, 2404
M.J. Murdock Charitable Trust General Grants, 2405
Mabel Y. Hughes Charitable Trust Grants, 2409
Macquarie Bank Foundation Grants, 2410
Marcia and Otto Koehler Foundation Grants, 2427
Margaret T. Morris Foundation Grants, 2429
Marion I. and Henry J. Knott Foundation Discretionary Grants, 2435
Marion I. and Henry J. Knott Foundation Standard Grants, 2436
Marjorie Moore Charitable Foundation Grants, 2438
Mary Black Foundation Active Living Grants, 2440
Mary K. Chapman Foundation Grants, 2443
Mary S. and David C. Corbin Foundation Grants, 2447
Matilda R. Wilson Fund Grants, 2453
Maurice J. Masserini Charitable Trust Grants, 2456
McInerny Foundation Grants, 2495
McLean Contributionship Grants, 2497
McMillen Foundation Grants, 2499
Mericos Foundation Grants, 2518
Mervin Bovaird Foundation Grants, 2523
MetroWest Health Foundation Capital Grants for Health-Related Facilities, 2526
Meyer Memorial Trust Grassroots Grants, 2536
Meyer Memorial Trust Responsive Grants, 2537
Meyer Memorial Trust Special Grants, 2538
Minneapolis Foundation Community Grants, 2558
Mt. Sinai Health Care Foundation Academic Medicine and Bioscience Grants, 2598
Mt. Sinai Health Care Foundation Health of the Jewish Community Grants, 2599
Mt. Sinai Health Care Foundation Health of the Urban Community Grants, 2600
NCI Technologies and Software to Support Integrative Cancer Biology Research (SBIR) Grants, 2656
Nell J. Redfield Foundation Grants, 2673
Nina Mason Pulliam Charitable Trust Grants, 2921
Norcliffe Foundation Grants, 2979
Nordson Corporation Foundation Grants, 2981
North Carolina Biotechnology Centers of Innovation Grants, 2988
North Central Health Services Grants, 2992
Northern New York Community Fndn Grants, 2996
Northwestern Mutual Foundation Grants, 2999
Norwin S. & Elizabeth N. Bean Fndn Grants, 3001
Oleonda Jameson Trust Grants, 3071
Oppenstein Brothers Foundation Grants, 3078
Oregon Community Fndn Community Grants, 3082
Oscar Rennebohm Foundation Grants, 3085
Otto Bremer Foundation Grants, 3099
PACCAR Foundation Grants, 3101
Parkersburg Area Community Foundation Action Grants, 3116
Patrick and Anna M. Cudahy Fund Grants, 3120
Patron Saints Foundation Grants, 3121
Paul Ogle Foundation Grants, 3124
Perkin Fund Grants, 3139
Peter and Elizabeth C. Tower Foundation Phase II Technology Initiative Grants, 3149
Piper Trust Healthcare & Med Research Grants, 3228
Pittsburgh Foundation Community Fund Grants, 3230
Plough Foundation Grants, 3238
Pohlad Family Foundation, 3241
Porter County Health and Wellness Grant, 3246
Potts Memorial Foundation Grants, 3254
PPG Industries Foundation Grants, 3257
Price Chopper's Golub Foundation Grants, 3261
Price Family Charitable Fund Grants, 3262
Priddy Foundation Capital Grants, 3263
Prince Charitable Trusts Chicago Grants, 3270
Prince Charitable Trusts DC Grants, 3271
Principal Financial Group Foundation Grants, 3273
Procter and Gamble Fund Grants, 3274
Puerto Rico Community Foundation Grants, 3278
Quantum Corporation Snap Server Grants, 3289
Quantum Foundation Grants, 3290
R.C. Baker Foundation Grants, 3292
R.T. Vanderbilt Trust Grants, 3294
Radcliffe Institute Residential Fellowships, 3299
Ralph M. Parsons Foundation Grants, 3304
Rathmann Family Foundation Grants, 3309
Rayonier Foundation Grants, 3311
RCF General Community Grants, 3312
Reinberger Foundation Grants, 3357
Rhode Island Foundation Grants, 3362
Richard and Helen DeVos Foundation Grants, 3364
Richard D. Bass Foundation Grants, 3366
Richard King Mellon Foundation Grants, 3369
Riley Foundation Grants, 3374
Robert R. Meyer Foundation Grants, 3387
Robert W. Woodruff Foundation Grants, 3389
Robins Foundation Grants, 3390
Rochester Area Community Foundation Grants, 3391
Rockwell International Corporate Trust Grants Program, 3399
Ronald McDonald House Charities Grants, 3405
Rose Community Foundation Health Grants, 3411
Rose Hills Foundation Grants, 3413
Roy & Christine Sturgis Charitable Grants, 3415
Roy J. Carver Charitable Trust Medical and Science Research Grants, 3416
Rush Community Foundation Grants, 3424
Ruth Anderson Foundation Grants, 3425
Ruth H. & Warren A. Ellsworth Fndn Grants, 3429
S.D. Bechtel, Jr. Fndn/Stephen Bechtel Preventive Healthcare and Selected Research Grants, 3446
S.H. Cowell Foundation Grants, 3447
S. Livingston Mather Charitable Trust Grants, 3448
S. Mark Taper Foundation Grants, 3449
Salem Foundation Grants, 3457
Samuel S. Johnson Foundation Grants, 3468
San Antonio Area Foundation Grants, 3469
San Juan Island Community Foundation Grants, 3480
Sarkeys Foundation Grants, 3485
Sartain Lanier Family Foundation Grants, 3486
Schering-Plough Foundation Community Initiatives Grants, 3491
Schramm Foundation Grants, 3495
Scott B. & Annie P. Appleby Charitable Grants, 3499
Seabury Foundation Grants, 3501
Seneca Foods Foundation Grants, 3511
Sensient Technologies Foundation Grants, 3513
Seward Community Foundation Grants, 3514
Seward Community Foundation Mini-Grants, 3515
Shell Deer Park Grants, 3543
Shield-Ayres Foundation Grants, 3546
Sidney Stern Memorial Trust Grants, 3552
Sid W. Richardson Foundation Grants, 3553
Siebert Lutheran Foundation Grants, 3554
Simmons Foundation Grants, 3563
Sisters of Mercy of North Carolina Fndn Grants, 3575
Solo Cup Foundation Grants, 3620
Sonora Area Foundation Competitive Grants, 3622
Sony Corporation of America Grants, 3623
Southbury Community Trust Fund, 3625
Southwest Gas Corporation Foundation Grants, 3628
Spencer County Community Foundation Grants, 3632

PROGRAM TYPE INDEX

Stackpole-Hall Foundation Grants, 3643
Stewart Huston Charitable Trust Grants, 3655
Strake Foundation Grants, 3658
Stranahan Foundation Grants, 3659
Strowd Roses Grants, 3664
Subaru of Indiana Automotive Foundation Grants, 3672
Summit Foundation Grants, 3674
Sunderland Foundation Grants, 3675
Sunflower Foundation Walking Trails Grants, 3678
Sunoco Foundation Grants, 3679
SunTrust Bank Trusteed Foundations Florence C. and Harry L. English Memorial Fund Grants, 3680
SunTrust Bank Trusteed Foundations Greene-Sawtell Grants, 3681
SunTrust Bank Trusteed Foundations Harriet McDaniel Marshall Tust Grants, 3682
SunTrust Bank Trusteed Foundations Nell Warren Elkin and William Simpson Elkin Grants, 3683
SunTrust Bank Trusteed Foundations Thomas Guy Woolford Charitable Trust Grants, 3684
SunTrust Bank Trusteed Foundations Walter H. and Marjory M. Rich Memorial Fund Grants, 3685
Support Our Aging Religious (SOAR) Grants, 3686
Susan Mott Webb Charitable Trust Grants, 3694
Susan Vaughan Foundation Grants, 3695
T.L.L. Temple Foundation Grants, 3700
TE Foundation Grants, 3705
Tellabs Foundation Grants, 3706
Textron Corporate Contributions Grants, 3712
The Ray Charles Foundation Grants, 3716
Thomas Austin Finch, Sr. Foundation Grants, 3718
Thompson Charitable Foundation Grants, 3727
Triangle Community Foundation Donor-Advised Grants, 3757
Tull Charitable Foundation Grants, 3762
Turner Foundation Grants, 3763
Union Bank, N.A. Foundation Grants, 3788
Union Pacific Foundation Health and Human Services Grants, 3794
United Technologies Corporation Grants, 3803
US Airways Community Foundation Grants, 3826
Victor E. Speas Foundation Grants, 3868
Vigneron Memorial Fund Grants, 3873
W. C. Griffith Foundation Grants, 3882
W.C. Griffith Foundation Grants, 3881
Walker Area Community Foundation Grants, 3895
Walter L. Gross III Family Foundation Grants, 3901
Warrick County Community Foundation Grants, 3904
Weingart Foundation Grants, 3913
Weyerhaeuser Company Foundation Grants, 3925
Willard & Pat Walker Charitable Fndn Grants, 3943
William Blair and Company Foundation Grants, 3948
William G. Gilmore Foundation Grants, 3951
William H. Hannon Foundation Grants, 3953
Wilson-Wood Foundation Grants, 3960
Windham Foundation Grants, 3961
Winston-Salem Foundation Competitive Grants, 3965
Woodward Fund Grants, 3973
Z. Smith Reynolds Foundation Small Grants, 3982

Capital Campaigns
A. Gary Anderson Family Foundation Grants, 16
Aaron Foundation Grants, 193
Abbot and Dorothy H. Stevens Foundation Grants, 197
Abbott Fund Global AIDS Care Grants, 202
Abel Foundation Grants, 206
Abell-Hanger Foundation Grants, 207
ACGT Investigators Grants, 232
Achelis Foundation Grants, 234
Ackerman Foundation Grants, 236
Adams County Community Foundation of Pennsylvania Grants, 300
Adelaide Breed Bayrd Foundation Grants, 306
Adolph Coors Foundation Grants, 311
AEC Trust Grants, 315
Agnes M. Lindsay Trust Grants, 346
Ahmanson Foundation Grants, 355
Air Products and Chemicals Grants, 399
Alabama Power Foundation Grants, 400
Alberto Culver Corporate Contributions Grants, 408

Albert W. Rice Charitable Foundation Grants, 411
Alcoa Foundation Grants, 414
Alexander & Baldwin Fnd Mainland Grants, 417
Alexander and Baldwin Foundation Hawaiian and Pacific Island Grants, 418
Alexander Eastman Foundation Grants, 420
Alice Tweed Tuohy Foundation Grants Program, 437
Allen P. & Josephine B. Green Fndn Grants, 443
Allyn Foundation Grants, 448
Alpha Natural Resources Corporate Giving, 450
Alvin and Fanny Blaustein Thalheimer Foundation Baltimore Communal Grants, 459
American Electric Power Foundation Grants, 530
American Schlafhorst Foundation Grants, 559
Amgen Foundation Grants, 572
Andrew Family Foundation Grants, 588
Angels Baseball Foundation Grants, 590
Anheuser-Busch Foundation Grants, 591
Ann Jackson Family Foundation Grants, 598
Ann Peppers Foundation Grants, 599
Aratani Foundation Grants, 649
Arcadia Foundation Grants, 650
Arie and Ida Crown Memorial Grants, 654
Arizona Public Service Corporate Giving Program Grants, 658
Arkell Hall Foundation Grants, 661
Arthur Ashley Williams Foundation Grants, 671
Assisi Fndn of Memphis Capital Project Grants, 786
Assurant Health Foundation Grants, 789
Athwin Foundation Grants, 795
Atran Foundation Grants, 797
Auburn Foundation Grants, 798
Avon Products Foundation Grants, 814
Bacon Family Foundation Grants, 817
Bailey Foundation Grants, 818
Ball Brothers Foundation General Grants, 820
Barker Foundation Grants, 830
Barker Welfare Foundation Grants, 831
Barra Foundation Project Grants, 834
Batchelor Foundation Grants, 837
Baton Rouge Area Foundation Grants, 838
Batts Foundation Grants, 840
Beazley Foundation Grants, 865
Ben B. Cheney Foundation Grants, 875
Bender Foundation Grants, 876
Benton Community Foundation Grants, 878
Berks County Community Foundation Grants, 881
Berrien Community Foundation Grants, 884
Besser Foundation Grants, 887
Bill and Melinda Gates Foundation Emergency Response Grants, 892
Bill and Melinda Gates Foundation Water, Sanitation and Hygiene Grants, 894
Bill Hannon Foundation Grants, 895
Blanche and Irving Laurie Foundation Grants, 910
Blanche and Julian Robertson Family Foundation Grants, 911
Blowitz-Ridgeway Foundation Early Childhood Development Research Award, 912
Blowitz-Ridgeway Foundation Grants, 913
Blue Mountain Community Foundation Grants, 921
Blue River Community Foundation Grants, 922
Blumenthal Foundation Grants, 925
BMW of North America Charitable Contributions, 926
Bodenwein Public Benevolent Foundation Grants, 928
Bodman Foundation Grants, 929
Boeing Company Contributions Grants, 930
Boettcher Foundation Grants, 931
Bonfils-Stanton Foundation Grants, 932
Booth-Bricker Fund Grants, 933
Borkee-Hagley Foundation Grants, 934
Bosque Foundation Grants, 935
Bradley-Turner Foundation Grants, 944
Brooklyn Community Foundation Caring Neighbors Grants, 968
Brown County Community Foundation Grants, 970
Burlington Industries Foundation Grants, 977
Burlington Northern Santa Fe Foundation Grants, 978
Byron W. & Alice L. Lockwood Fnd Grants, 991
Cabot Corporation Foundation Grants, 993

Capital Campaigns / 847

Cailloux Foundation Grants, 996
Callaway Foundation Grants, 1002
Cargill Citizenship Fund Corporate Giving, 1018
Carl & Eloise Pohlad Family Fndn Grants, 1021
Carl B. and Florence E. King Foundation Grants, 1022
Carl C. Icahn Foundation Grants, 1023
Carl Gellert and Celia Berta Gellert Foundation Grants, 1024
Carlisle Foundation Grants, 1025
Carl M. Freeman Foundation Grants, 1027
Carls Foundation Grants, 1030
Carl W. and Carrie Mae Joslyn Trust Grants, 1031
Carnahan-Jackson Foundation Grants, 1032
Carpenter Foundation Grants, 1035
Carrie Estelle Doheny Foundation Grants, 1037
Carrier Corporation Contributions Grants, 1038
Cemala Foundation Grants, 1077
Cessna Foundation Grants Program, 1081
CFFVR Basic Needs Giving Partnership Grants, 1094
CFFVR Clintonville Area Foundation Grants, 1097
CFFVR Frank C. Shattuck Community Grants, 1098
CFFVR Robert and Patricia Endries Family Foundation Grants, 1102
CFFVR Schmidt Family G4 Grants, 1103
Champlin Foundations Grants, 1110
Chapman Charitable Foundation Grants, 1112
Charles A. Frueauff Foundation Grants, 1114
Charles Delmar Foundation Grants, 1115
Charles H. Revson Foundation Grants, 1122
Charles M. & Mary D. Grant Fndn Grants, 1124
Chatlos Foundation Grants Program, 1128
Chazen Foundation Grants, 1129
CICF Indianapolis Fndn Community Grants, 1156
Clark and Ruby Baker Foundation Grants, 1174
Claude Worthington Benedum Fndn Grants, 1178
Clayton Fund Grants, 1180
Cleveland-Cliffs Foundation Grants, 1181
Coastal Community Foundation of South Carolina Grants, 1202
Cockrell Foundation Grants, 1204
Collins C. Diboll Private Foundation Grants, 1209
Colonel Stanley R. McNeil Foundation Grants, 1211
Columbus Foundation Competitive Grants, 1217
Commonwealth Edison Grants, 1225
Communities Foundation of Texas Grants, 1236
Community Fndn for Greater Buffalo Grants, 1247
Community Fndn for Monterey County Grants, 1253
Community Fndn for Muskegon County Grants, 1254
Community Fndn for the Capital Region Grants, 1259
Community Foundation of Bartholomew County Heritage Fund Grants, 1262
Community Foundation of Boone County Grants, 1266
Community Fndn of Central Illinois Grants, 1267
Community Foundation of Greater Birmingham Grants, 1272
Community Foundation of Greater Flint Grants, 1273
Community Foundation of Greater Fort Wayne - Community Endowment and Clarke Endowment Grants, 1274
Community Fndn of Greater Lafayette Grants, 1280
Community Foundation of Greater New Britain Grants, 1281
Community Foundation of Greenville Hollingsworth Funds Program/Project Grants, 1285
Community Foundation of Mount Vernon and Knox County Grants, 1299
Community Fndn of So Alabama Grants, 1306
Community Foundation of St. Joseph County Special Project Challenge Grants, 1310
Community Foundation of the Verdugos Educational Endowment Fund Grants, 1314
Connecticut Community Foundation Grants, 1324
Connelly Foundation Grants, 1326
Constantin Foundation Grants, 1341
Constellation Energy Corporate Grants, 1342
Consumers Energy Foundation, 1343
Cooke Foundation Grants, 1344
Cooper Industries Foundation Grants, 1345
Cornerstone Foundation of Northeastern Wisconsin Grants, 1347

848 / Capital Campaigns   PROGRAM TYPE INDEX

Covidien Partnership for Neighborhood Wellness Grants, 1353
Cowles Charitable Trust Grants, 1354
Crail-Johnson Foundation Grants, 1355
Cralle Foundation Grants, 1356
CSRA Community Foundation Grants, 1366
Cudd Foundation Grants, 1370
Cullen Foundation Grants, 1372
Cumberland Community Foundation Grants, 1374
D.F. Halton Foundation Grants, 1405
Dale and Edna Walsh Foundation Grants, 1415
Danellie Foundation Grants, 1421
Daniel & Nanna Stern Family Fndn Grants, 1422
Daniels Fund Grants-Aging, 1424
Daphne Seybolt Culpeper Fndn Grants, 1425
David Geffen Foundation Grants, 1431
Daywood Foundation Grants, 1438
Deaconess Community Foundation Grants, 1439
Delaware Community Foundation Grants, 1449
DeMatteis Family Foundation Grants, 1457
Denton A. Cooley Foundation Grants, 1459
Deutsche Banc Alex Brown and Sons Charitable Foundation Grants, 1471
Dolan Children's Foundation Grants, 1482
Donald and Sylvia Robinson Family Foundation Grants, 1486
Dorrance Family Foundation Grants, 1501
Dr. Leon Bromberg Charitable Trust Grants, 1509
Drs. Bruce and Lee Foundation Grants, 1513
Duchossois Family Foundation Grants, 1518
Dunspaugh-Dalton Foundation Grants, 1530
DuPage Community Foundation Grants, 1531
Dyson Foundation Mid-Hudson Valley Project Support Grants, 1534
E. Clayton and Edith P. Gengras, Jr. Foundation, 1536
E.J. Grassmann Trust Grants, 1537
E. Rhodes & Leona B. Carpenter Grants, 1539
Eastman Chemical Company Foundation Grants, 1542
Eden Hall Foundation Grants, 1549
Edwards Memorial Trust Grants, 1558
Edward W. and Stella C. Van Houten Memorial Fund Grants, 1559
Edwin S. Webster Foundation Grants, 1560
Edwin W. and Catherine M. Davis Fndn Grants, 1561
Edyth Bush Charitable Foundation Grants, 1562
Effie and Wofford Cain Foundation Grants, 1563
Eisner Foundation Grants, 1564
Elmer L. & Eleanor J. Andersen Fndn Grants, 1578
Erie Community Foundation Grants, 1601
Essex County Community Foundation Webster Family Fund Grants, 1604
Ethel S. Abbott Charitable Foundation Grants, 1607
Ethel Sergeant Clark Smith Foundation Grants, 1608
Eugene Straus Charitable Trust, 1612
Eva L. & Joseph M. Bruening Fndn Grants, 1613
Evjue Foundation Grants, 1617
Ewing Halsell Foundation Grants, 1618
Fallon OrNda Community Health Fund Grants, 1627
Fargo-Moorhead Area Foundation Grants, 1632
Faye McBeath Foundation Grants, 1635
Ferree Foundation Grants, 1658
Fidelity Foundation Grants, 1659
Field Foundation of Illinois Grants, 1660
FirstEnergy Foundation Community Grants, 1668
Fluor Foundation Grants, 1682
FMC Foundation Grants, 1683
Foellinger Foundation Grants, 1684
Foundation for the Mid South Health and Wellness Grants, 1700
Four County Community Fndn General Grants, 1710
Four County Community Foundation Healthy Senior/Healthy Youth Fund Grants, 1711
Fourjay Foundation Grants, 1712
France-Merrick Foundation Health and Human Services Grants, 1715
Francis L. Abreu Charitable Trust Grants, 1719
Franklin County Community Foundation Grants, 1724
Frank Stanley Beveridge Foundation Grants, 1730
Frank W. & Carl S. Adams Memorial Grants, 1731
Fred Baldwin Memorial Foundation Grants, 1736

Fred C. & Katherine B. Andersen Grants, 1737
Freddie Mac Foundation Grants, 1738
Frederick McDonald Trust Grants, 1740
Fred L. Emerson Foundation Grants, 1743
Fritz B. Burns Foundation Grants, 1747
Fuller E. Callaway Foundation Grants, 1761
G.N. Wilcox Trust Grants, 1768
Gardner Foundation Grants, 1770
Gardner Foundation Grants, 1772
Gebbie Foundation Grants, 1777
GenCorp Foundation Grants, 1779
General Mills Foundation Grants, 1782
Genesis Foundation Grants, 1786
Genuardi Family Foundation Grants, 1787
George Foundation Grants, 1795
George Gund Foundation Grants, 1796
George Kress Foundation Grants, 1799
George S. & Dolores Dore Eccles Fndn Grants, 1801
George W. Brackenridge Foundation Grants, 1802
George W. Codrington Charitable Fndn Grants, 1803
George W. Wells Foundation Grants, 1804
Georgia Power Foundation Grants, 1806
Gheens Foundation Grants, 1814
Gil and Dody Weaver Foundation Grants, 1819
Gill Foundation - Gay and Lesbian Fund Grants, 1821
Giving in Action Society Children & Youth with Special Needs Grants, 1823
Giving in Action Society Family Indep Grants, 1824
Gladys Brooks Foundation Grants, 1826
GNOF Exxon-Mobil Grants, 1836
GNOF Norco Community Grants, 1845
Goodrich Corporation Foundation Grants, 1854
Grace and Franklin Bernsen Foundation Grants, 1858
Graco Foundation Grants, 1860
Graham Foundation Grants, 1862
Grand Haven Area Community Fndn Grants, 1864
Grand Rapids Area Community Foundation Wyoming Grants, 1867
Grand Rapids Community Foundation Grants, 1869
Grand Rapids Community Foundation Ionia County Grants, 1870
Grand Rapids Community Foundation Southeast Ottawa Grants, 1873
Grand Rapids Community Foundation Sparta Grants, 1875
Greater Tacoma Community Foundation Grants, 1886
Greater Worcester Community Foundation Jeppson Memorial Fund for Brookfield Grants, 1889
Grover Hermann Foundation Grants, 1906
Grundy Foundation Grants, 1912
Guido A. & Elizabeth H. Binda Fndn Grants, 1913
Guy's and St. Thomas' Charity Grants, 1918
H & R Foundation Grants, 1920
H.A. & Mary K. Chapman Trust Grants, 1921
H.J. Heinz Company Foundation Grants, 1923
H. Schaffer Foundation Grants, 1925
Hagedorn Fund Grants, 1934
Hallmark Corporate Foundation Grants, 1936
Hannaford Charitable Foundation Grants, 1944
Harold and Arlene Schnitzer CARE Foundation Grants, 1950
Harold Simmons Foundation Grants, 1955
Harris and Eliza Kempner Fund Grants, 1956
Harry A. & Margaret D. Towsley Fndn Grants, 1959
Harry B. and Jane H. Brock Foundation Grants, 1960
Harry Bramhall Gilbert Charitable Trust Grants, 1961
Harry Edison Foundation, 1962
Harry Kramer Memorial Fund Grants, 1965
Harry W. Bass, Jr. Foundation Grants, 1968
Hartford Courant Foundation Grants, 1969
Hartford Foundation Regular Grants, 1970
HCA Foundation Grants, 1980
Hearst Foundations Health Grants, 1987
Heckscher Foundation for Children Grants, 1988
Hedco Foundation Grants, 1989
Helen Bader Foundation Grants, 1992
Helen K. & Arthur E. Johnson Fndn Grants, 1995
Helen S. Boylan Foundation Grants, 1997
Henry and Ruth Blaustein Rosenberg Foundation Health Grants, 2004
Herbert A. & Adrian W. Woods Fndn Grants, 2007

Hershey Company Grants, 2013
High Meadow Foundation Grants, 2027
Hill Crest Foundation Grants, 2030
Hillcrest Foundation Grants, 2031
Hillman Foundation Grants, 2032
Hoglund Foundation Grants, 2040
Holland/Zeeland Community Fndn Grants, 2041
Horace A. Kimball and S. Ella Kimball Foundation Grants, 2046
Hormel Foundation Grants, 2051
Houston Endowment Grants, 2053
HRK Foundation Health Grants, 2067
Hudson Webber Foundation Grants, 2072
Huffy Foundation Grants, 2073
Hugh J. Andersen Foundation Grants, 2074
Huie-Dellmon Trust Grants, 2075
Huisking Foundation Grants, 2076
Huntington County Community Foundation Make a Difference Grants, 2083
Ida Alice Ryan Charitable Trust Grants, 2113
Idaho Power Company Corporate Contributions, 2115
Illinois Tool Works Foundation Grants, 2147
Impact 100 Grants, 2148
Inasmuch Foundation Grants, 2149
Independence Community Foundation Community Quality of Life Grant, 2153
Intergrys Corporation Grants, 2160
Irving S. Gilmore Foundation Grants, 2163
J.B. Reynolds Foundation Grants, 2168
Jack H. & William M. Light Trust Grants, 2179
Jackson County Community Foundation Unrestricted Grants, 2180
Jacob G. Schmidlapp Trust Grants, 2183
James A. and Faith Knight Foundation Grants, 2186
James Graham Brown Foundation Grants, 2189
James H. Cummings Foundation Grants, 2190
James J. & Angelia M. Harris Fndn Grants, 2192
James M. Cox Foundation of Georgia Grants, 2196
James R. Dougherty Jr. Foundation Grants, 2198
James R. Thorpe Foundation Grants, 2199
James S. Copley Foundation Grants, 2200
Jane's Trust Grants, 2206
Janirve Foundation Grants, 2209
Janson Foundation Grants, 2210
Jay and Rose Phillips Family Foundation Grants, 2213
JELD-WEN Foundation Grants, 2218
Jenkins Foundation: Improving the Health of Greater Richmond Grants, 2219
Jessie Ball Dupont Fund Grants, 2226
John G. Martin Foundation Grants, 2237
John H. and Wilhelmina D. Harland Charitable Foundation Children and Youth Grants, 2238
John I. Smith Charities Grants, 2240
John Jewett and Helen Chandler Garland Foundation Grants, 2242
John R. Oishei Foundation Grants, 2249
John Stauffer Charitable Trust Grants, 2258
John W. and Anna H. Hanes Foundation Grants, 2261
John W. Anderson Foundation Grants, 2262
Joseph Alexander Foundation Grants, 2265
Joseph H. & Florence A. Roblee Fndn Grants, 2269
Joseph Henry Edmondson Foundation Grants, 2270
Josephine Goodyear Foundation Grants, 2272
Josephine Schell Russell Charitable Trust Grants, 2274
Josiah W. and Bessie H. Kline Foundation Grants, 2278
Judith and Jean Pape Adams Charitable Foundation Tulsa Area Grants, 2280
Kansas Health Foundation Recognition Grants, 2289
Katharine Matthies Foundation Grants, 2291
Katherine John Murphy Foundation Grants, 2293
Kathryne Beynon Foundation Grants, 2294
Kelvin and Eleanor Smith Foundation Grants, 2297
Kenneth T. & Eileen L. Norris Fndn Grants, 2300
Kettering Fund Grants, 2305
Kinsman Foundation Grants, 2309
Kroger Foundation Women's Health Grants, 2325
L. W. Pierce Family Foundation Grants, 2328
Land O'Lakes Foundation Mid-Atlantic Grants, 2341
Leo Goodwin Foundation Grants, 2354
Lewis H. Humphreys Charitable Trust Grants, 2361

PROGRAM TYPE INDEX					Centers: Research/Demonstration/Service / 849

Lillian S. Wells Foundation Grants, 2365
Louie M. & Betty M. Phillips Fndn Grants, 2378
Louis and Elizabeth Nave Flarsheim Charitable Foundation Grants, 2379
Lowe Foundation Grants, 2381
Lubbock Area Foundation Grants, 2383
Lubrizol Foundation Grants, 2384
Lucy Gooding Charitable Fndn Grants, 2388
Lumpkin Family Fnd Healthy People Grants, 2391
Lynde and Harry Bradley Foundation Grants, 2397
M.E. Raker Foundation Grants, 2404
Mabel Y. Hughes Charitable Trust Grants, 2409
Macquarie Bank Foundation Grants, 2410
Mardag Foundation Grants, 2428
Margaret T. Morris Foundation Grants, 2429
Marion Gardner Jackson Charitable Trust Grants, 2434
Marion I. and Henry J. Knott Foundation Discretionary Grants, 2435
Marion I. and Henry J. Knott Foundation Standard Grants, 2436
Marjorie Moore Charitable Foundation Grants, 2438
Mary K. Chapman Foundation Grants, 2443
Mary Owen Borden Foundation Grants, 2446
Matilda R. Wilson Fund Grants, 2453
McCune Foundation Human Services Grants, 2493
McInerny Foundation Grants, 2495
McLean Contributionship Grants, 2497
McMillen Foundation Grants, 2499
Mead Witter Foundation Grants, 2507
MetroWest Health Foundation Capital Grants for Health-Related Facilities, 2526
Meyer Foundation Healthy Communities Grants, 2533
Mimi and Peter Haas Fund Grants, 2557
Minneapolis Foundation Community Grants, 2558
Nationwide Insurance Foundation Grants, 2637
Nell J. Redfield Foundation Grants, 2673
Nicholas H. Noyes Jr. Memorial Fndn Grants, 2791
Nina Mason Pulliam Charitable Trust Grants, 2921
Noble County Community Foundation Grants, 2975
Norcliffe Foundation Grants, 2979
Nordson Corporation Foundation Grants, 2981
Norfolk Southern Foundation Grants, 2982
North Carolina Community Foundation Grants, 2990
North Central Health Services Grants, 2992
North Central Health Services Grants, 2993
Northern Chautauqua Community Foundation Community Grants, 2995
Northern New York Community Fndn Grants, 2996
Northwestern Mutual Foundation Grants, 2999
Norwin S. & Elizabeth N. Bean Fndn Grants, 3001
Notsew Orm Sands Foundation Grants, 3002
Oak Foundation Child Abuse Grants, 3059
Oleonda Jameson Trust Grants, 3071
Olive Higgins Prouty Foundation Grants, 3072
OneFamily Foundation Grants, 3075
Oppenstein Brothers Foundation Grants, 3078
Oregon Community Fndn Community Grants, 3082
OSF International Harm Reduction Development Program Grants, 3090
Otto Bremer Foundation Grants, 3099
Owen County Community Foundation Grants, 3100
PACCAR Foundation Grants, 3101
Pacific Life Foundation Grants, 3103
Parkersburg Area Community Foundation Action Grants, 3116
Patrick and Anna M. Cudahy Fund Grants, 3120
Perkin Fund Grants, 3139
Peter and Elizabeth C. Tower Foundation Phase II Technology Initiative Grants, 3149
Peter Kiewit Foundation General Grants, 3154
Peter Kiewit Foundation Small Grants, 3155
Pfizer Special Events Grants, 3186
Phoenix Suns Charities Grants, 3201
Pinellas County Grants, 3223
Piper Jaffray Fndn Communities Giving Grants, 3227
Piper Trust Healthcare & Med Research Grants, 3228
Plough Foundation Grants, 3238
Pohlad Family Foundation, 3241
Powell Foundation Grants, 3255
PPG Industries Foundation Grants, 3257

Presbyterian Health Foundation Bridge, Seed and Equipment Grants, 3259
Price Chopper's Golub Foundation Grants, 3261
Priddy Foundation Capital Grants, 3263
Principal Financial Group Foundation Grants, 3273
Prudential Foundation Education Grants, 3276
R.C. Baker Foundation Grants, 3292
Raskob Foundation for Catholic Activities Grants, 3306
Rasmuson Foundation Tier One Grants, 3307
Rasmuson Foundation Tier Two Grants, 3308
Rathmann Family Foundation Grants, 3309
Rayonier Foundation Grants, 3311
RCF General Community Grants, 3312
Reinberger Foundation Grants, 3357
Rhode Island Foundation Grants, 3362
Richard and Helen DeVos Foundation Grants, 3364
Richard & Susan Smith Family Fndn Grants, 3365
Richard D. Bass Foundation Grants, 3366
Robert F. Stoico/FIRSTFED Charitable Foundation Grants, 3382
Robert W. Woodruff Foundation Grants, 3389
Robins Foundation Grants, 3390
Rockwell International Corporate Trust Grants Program, 3399
Roy & Christine Sturgis Charitable Grants, 3415
Ruth and Vernon Taylor Foundation Grants, 3426
Ruth Eleanor Bamberger and John Ernest Bamberger Memorial Foundation Grants, 3427
S. Livingston Mather Charitable Trust Grants, 3448
S. Mark Taper Foundation Grants, 3449
Salem Foundation Grants, 3457
Salmon Foundation Grants, 3458
San Juan Island Community Foundation Grants, 3480
Sarkeys Foundation Grants, 3485
Sartain Lanier Family Foundation Grants, 3486
Sasco Foundation Grants, 3487
Schering-Plough Foundation Community Initiatives Grants, 3491
Schering-Plough Foundation Health Grants, 3492
Schlessman Family Foundation Grants, 3493
Schurz Communications Foundation Grants, 3496
Scott B. & Annie P. Appleby Charitable Grants, 3499
Self Foundation Grants, 3510
Sensient Technologies Foundation Grants, 3513
Shell Deer Park Grants, 3543
Shield-Ayres Foundation Grants, 3546
Sid W. Richardson Foundation Grants, 3553
Simpson Lumber Charitable Contributions, 3566
Sisters of Charity Foundation of Cleveland Reducing Health and Educational Disparities in the Central Neighborhood Grants, 3574
Sisters of Mercy of North Carolina Fndn Grants, 3575
Solo Cup Foundation Grants, 3620
Sonora Area Foundation Competitive Grants, 3622
Sony Corporation of America Grants, 3623
Southwest Gas Corporation Foundation Grants, 3628
Square D Foundation Grants, 3634
St. Joseph Community Health Foundation Burn Care and Prevention Grants, 3639
St. Joseph Community Health Foundation Pfeiffer Fund Grants, 3641
Stackpole-Hall Foundation Grants, 3643
Steele-Reese Foundation Grants, 3650
Stewart Huston Charitable Trust Grants, 3655
Strake Foundation Grants, 3658
Stranahan Foundation Grants, 3659
Strowd Roses Grants, 3664
Subaru of Indiana Automotive Foundation Grants, 3672
Summit Foundation Grants, 3674
Sunderland Foundation Grants, 3675
SunTrust Bank Trusteed Foundations Florence C. and Harry L. English Memorial Fund Grants, 3680
SunTrust Bank Trusteed Foundations Greene-Sawtell Grants, 3681
SunTrust Bank Trusteed Foundations Harriet McDaniel Marshall Tust Grants, 3682
SunTrust Bank Trusteed Foundations Nell Warren Elkin and William Simpson Elkin Grants, 3683
SunTrust Bank Trusteed Foundations Thomas Guy Woolford Charitable Trust Grants, 3684

SunTrust Bank Trusteed Foundations Walter H. and Marjory M. Rich Memorial Fund Grants, 3685
Support Our Aging Religious (SOAR) Grants, 3686
Susan Mott Webb Charitable Trust Grants, 3694
Susan Vaughan Foundation Grants, 3695
T.L.L. Temple Foundation Grants, 3700
TE Foundation Grants, 3705
Tension Envelope Foundation Grants, 3707
Texas Instruments Foundation Community Services Grants, 3710
Textron Corporate Contributions Grants, 3712
Thomas Sill Foundation Grants, 3724
Thompson Charitable Foundation Grants, 3727
Tull Charitable Foundation Grants, 3762
Union Bank, N.A. Foundation Grants, 3788
Union Pacific Foundation Health and Human Services Grants, 3794
US Airways Community Foundation Grants, 3826
Vanderburgh Community Fndn Women's Fund, 3853
Visiting Nurse Foundation Grants, 3879
W. C. Griffith Foundation Grants, 3882
W.C. Griffith Foundation Grants, 3881
W.K. Kellogg Foundation Healthy Kids Grants, 3886
Walker Area Community Foundation Grants, 3895
Weingart Foundation Grants, 3913
Willard & Pat Walker Charitable Fndn Grants, 3943
William A. Badger Foundation Grants, 3945
William G. & Helen C. Hoffman Fndn Grants, 3950
William G. Gilmore Foundation Grants, 3951
Wilson-Wood Foundation Grants, 3960
Winston-Salem Foundation Competitive Grants, 3965
Winston-Salem Fndn Elkin/Tri-County Grants, 3966
Woodward Fund Grants, 3973

Centers: Research/Demonstration/Service
AAAS Science and Technology Policy Fellowships: Global Health and Development, 41
ACSM Carl V. Gisolfi Memorial Fund Grant, 259
ACSM Coca-Cola Company Doctoral Student Grant on Behavior Research, 261
ACSM Dr. Raymond A. Weiss Research Endowment Grant, 262
ACSM Foundation Clinical Sports Medicine Endowment Grants, 265
ACSM Foundation Research Endowment Grants, 267
ACSM International Student Awards, 269
ACSM NASA Space Physiology Research Grants, 271
ACSM Oded Bar-Or Int'l Scholar Awards, 272
ACSM Paffenbarger-Blair Fund for Epidemiological Research on Physical Activity Grants, 273
Administration on Aging Senior Medicare Patrol Project Grants, 309
AMA Foundation Fund for Better Health Grants, 484
APHL Emerging Infectious Diseases Fellowships, 636
AVDF Health Care Grants, 811
Bearemy's Kennel Pals Grants, 863
Bildner Family Foundation Grants, 891
Blue Cross Blue Shield of Minnesota Fndn Healthy Equity: Public Libraries for Health Grants, 918
Cambridge Community Foundation Grants, 1004
CCHD Community Development Grants, 1046
CDC Cooperative Agreement for Continuing Enhanced National Surveillance for Prion Diseases in the United States, 1049
CDC Cooperative Agreement for Partnership to Enhance Public Health Informatics, 1050
CDC Evaluation of the Use of Rapid Testing For Influenza in Outpatient Medical Settings, 1057
CDC Evidence-Based Laboratory Medicine: Quality/Performance Measure Evaluation, 1058
CDI Interdisciplinary Research Initiatives Grants, 1074
CDI Postdoctoral Fellowships, 1075
Champ-A Champion Fur Kids Grants, 1109
CMS Hispanic Health Services Research Grants, 1186
CMS Historically Black Colleges and Universities (HBCU) Health Services Research Grants, 1187
CMS Research and Demonstration Grants, 1188
Collins C. Diboll Private Foundation Grants, 1209
Colonel Stanley R. McNeil Foundation Grants, 1211
ConocoPhillips Foundation Grants, 1327

Conquer Cancer Foundation of ASCO Drug Development Research Professorship, 1330
Conquer Cancer Foundation of ASCO Improving Cancer Care Grants, 1331
Cystic Fibrosis Canada Transplant Center Incentive Grants, 1392
Dallas Women's Foundation Grants, 1417
Deborah Munroe Noonan Memorial Grants, 1443
Dr. John T. Macdonald Foundation Grants, 1507
Duke Endowment Health Care Grants, 1519
EMBO Installation Grants, 1586
Evjue Foundation Grants, 1617
Expect Miracles Foundation Grants, 1619
Foundation for the Mid South Community Development Grants, 1699
Frank Loomis Palmer Fund Grants, 1726
Frederick McDonald Trust Grants, 1740
G.A. Ackermann Memorial Fund Grants, 1767
George W. Wells Foundation Grants, 1804
GNOF Maison Hospitaliere Grants, 1843
Health Fndn of Greater Indianapolis Grants, 1984
Health Foundation of Southern Florida Responsive Grants, 1985
Hearst Foundations Health Grants, 1987
Herbert A. & Adrian W. Woods Fndn Grants, 2007
Hilda and Preston Davis Foundation Postdoctoral Fellowships in Eating Disorders Research, 2029
HRAMF Deborah Munroe Noonan Memorial Research Grants, 2058
HRAMF Jeffress Trust Awards in Interdisciplinary Research, 2060
HRAMF Taub Fnd Grants for MDS Research, 2064
HRAMF Thome Foundation Awards in Age-Related Macular Degeneration Research, 2065
HRAMF Thome Foundation Awards in Alzheimer's Disease Drug Discovery Research, 2066
Hudson Webber Foundation Grants, 2072
IAFF Burn Foundation Research Grants, 2089
IBCAT Screening Mammography Grants, 2095
IDPH Emergency Medical Services Assistance Fund Grants, 2118
IDPH Hosptial Capital Investment Grants, 2119
IDPH Local Health Department Public Health Emergency Response Grants, 2120
IIE African Center of Excellence for Women's Leadership Grants, 2124
IIE Rockefeller Foundation Bellagio Center Residencies, 2142
John Merck Fund Grants, 2245
John W. and Anna H. Hanes Foundation Grants, 2261
Josephine Goodyear Foundation Grants, 2272
Joseph P. Kennedy Jr. Foundation Grants, 2275
Katharine Matthies Foundation Grants, 2291
Kessler Fnd Signature Employment Grants, 2302
Klarman Family Foundation Grants in Eating Disorders Research Grants, 2310
Leo Goodwin Foundation Grants, 2354
Lymphatic Education and Research Network Additional Support Grants for NIH-funded F32 Postdoctoral Fellows, 2394
Lymphatic Education and Research Network Postdoctoral Fellowships, 2395
Marion Gardner Jackson Charitable Trust Grants, 2434
Marjorie Moore Charitable Foundation Grants, 2438
McGraw-Hill Companies Community Grants, 2494
MGN Family Foundation Grants, 2542
Michael Reese Health Trust Core Grants, 2544
Michael Reese Health Trust Responsive Grants, 2545
Mid-Iowa Health Foundation Community Response Grants, 2549
Miles of Hope Breast Cancer Foundation Grants, 2553
MMS and Alliance Charitable Foundation Grants for Community Action and Care for the Medically Uninsured, 2587
Mt. Sinai Health Care Foundation Academic Medicine and Bioscience Grants, 2598
Mt. Sinai Health Care Foundation Health of the Jewish Community Grants, 2599
Mt. Sinai Health Care Foundation Health of the Urban Community Grants, 2600

Mt. Sinai Health Care Foundation Health Policy Grants, 2601
NCI Centers of Excellence in Cancer Communications Research, 2649
NEI Mentored Clinical Scientist Dev Award, 2666
NFL Charities Medical Grants, 2685
NHLBI Ancillary Studies in Clinical Trials, 2688
NHLBI Career Transition Awards, 2694
NHLBI Investigator Initiated Multi-Site Clinical Trials, 2703
NHLBI Lymphatics in Health and Disease in the Digestive, Urinary, Cardiovascular and Pulmonary Systems, 2705
NHLBI Mentored Career Award for Faculty at Institutions that Promote Diversity (K01), 2706
NHLBI Mentored Career Dev Award to Promote Faculty Diversity in Biomed Research, 2708
NHLBI Ruth L. Kirschstein National Research Service Awards for Individual Predoctoral MD/PhD Fellows and Other Dual Degree Fellows, 2731
NHLBI Training Program Grants for Institutions that Promote Diversity, 2738
NIA Alzheimer's Disease Core Centers Grants, 2748
NIA Nathan Shock Centers of Excellence in Basic Biology of Aging Grants, 2776
NIDDK Diabetes Research Centers, 2811
NIDDK Multi-Center Clinical Study Cooperative Agreements, 2826
NIDDK Multi-Center Clinical Study Implementation Planning Grants, 2827
NIDDK Pilot and Feasibility Clinical Research Grants in Kidney or Urologic Diseases, 2835
NIDDK Pilot and Feasibility Clinical Research Studies in Digestive Diseases and Nutrition, 2836
NIDDK Seeding Collaborative Interdisciplinary Team Science in Diabetes, Endocrinology and Metabolic Diseases Grants, 2843
NIH Biology, Development, and Progression of Malignant Prostate Disease Research Grants, 2866
NIH Nonhuman Primate Immune Tolerance Cooperative Study Group Grants, 2890
North Carolina Biotechnology Centers of Innovation Grants, 2988
NSERC Brockhouse Canada Prize for Interdisciplinary Research in Science and Engineering Grant, 3003
NSF Emerging Frontiers in Research and Innovation (EFRI) Grants, 3014
NSF Major Research Instrumentation Program (MRI) Grants, 3018
OSF Tackling Addiction Grants in Baltimore, 3096
Piper Trust Healthcare & Med Research Grants, 3228
Piper Trust Older Adults Grants, 3229
Pollock Foundation Grants, 3244
Quality Health Foundation Grants, 3288
Richard Davoud Donchian Foundation Grants, 3367
Rockefeller Brothers Peace & Security Grants, 3394
Rockwell International Corporate Trust Grants Program, 3399
Salem Foundation Charitable Trust Grants, 3456
Samueli Foundation Health Grants, 3465
San Diego Women's Foundation Grants, 3474
Schering-Plough Foundation Health Grants, 3492
Shell Deer Park Grants, 3543
Sidgmore Family Foundation Grants, 3551
Sisters of Mercy of North Carolina Fndn Grants, 3575
Sisters of St. Joseph Healthcare Fndn Grants, 3576
Southwest Florida Community Foundation Competitive Grants, 3627
Stewart Huston Charitable Trust Grants, 3655
Stocker Foundation Grants, 3656
T.L.L. Temple Foundation Grants, 3700
Tommy Hilfiger Corporate Foundation Grants, 3737
Trauma Center of America Finance Fellowship, 3752
U.S. Department of Education 21st Century Community Learning Centers, 3767
U.S. Department of Education Rehabilitation Engineering Research Centers Grants, 3769
U.S. Department of Education Rehabilitation Research Training Centers (RRTCs), 3770

U.S. Dept of Education Special Ed--National Activities--Parent Information Centers, 3772
UCLA / RAND Corporation Post-Doctoral Fellowships, 3776
United States Institute of Peace - Jennings Randolph Senior Fellowships, 3802
Vigneron Memorial Fund Grants, 3873
W.M. Keck Foundation Science and Engineering Research Grants, 3889
W.P. and Bulah Luse Foundation Grants, 3892
Walter L. Gross III Family Foundation Grants, 3901
Weingart Foundation Grants, 3913

Citizenship Instruction
Adolph Coors Foundation Grants, 311
AT&T Foundation Civic and Community Service Program Grants, 793
Axe-Houghton Foundation Grants, 815
Blue Cross Blue Shield of Minnesota Fndn Healthy Equity: Public Libraries for Health Grants, 918
Brinson Foundation Grants, 953
CCHD Community Development Grants, 1046
Harvest Foundation Grants, 1971
Helen Irwin Littauer Educational Trust Grants, 1994
IIE Hewlett Fnd/IIE Dissertation Fellowship, 2132
John Edward Fowler Memorial Fndn Grants, 2235
Lynde and Harry Bradley Foundation Grants, 2397
Mardag Foundation Grants, 2428
Miles of Hope Breast Cancer Foundation Grants, 2553
TJX Foundation Grants, 3735
Vanderburgh Community Fndn Women's Fund, 3853
Warrick County Community Foundation Women's Fund, 3905

Community Development
1st Source Foundation Grants, 2
1st Touch Foundation Grants, 3
2 Depot Square Ipswich Charitable Fndn Grants, 4
3M Company Fndn Community Giving Grants, 6
3M Fndn Health & Human Services Grants, 7
100 Mile Man Foundation Grants, 10
A.O. Smith Foundation Community Grants, 17
A/H Foundation Grants, 18
AAP Community Access To Child Health (CATCH) Implementation Grants, 178
AAP Resident Initiative Fund Grants, 189
Aaron & Cecile Goldman Family Fndn Grants, 191
Aaron & Freda Glickman Foundation Grants, 192
Aaron Foundation Grants, 194
Aaron Foundation Grants, 193
Abbot and Dorothy H. Stevens Foundation Grants, 197
Abbott Fund Access to Health Care Grants, 199
Abbott Fund Community Grants, 201
Abbott Fund Global AIDS Care Grants, 202
Abbott Fund Science Education Grants, 203
Abby's Legendary Pizza Foundation Grants, 204
Abel Foundation Grants, 206
Abell-Hanger Foundation Grants, 207
Abell Foundation Criminal Justice and Addictions Grants, 208
Abell Fndn Health & Human Services Grants, 209
Abington Foundation Grants, 211
Able To Serve Grants, 212
Able Trust Vocational Rehabilitation Grants for Individuals, 213
Abney Foundation Grants, 214
Abundance Foundation International Grants, 218
ACE Charitable Foundation Grants, 229
ACF Foundation Grants, 230
ACF Native American Social and Economic Development Strategies Grants, 231
Achelis Foundation Grants, 234
Ackerman Foundation Grants, 236
ACL Alzheimer's Disease Initiative Grants, 237
ACL Business Acumen for Disability Organizations Grants, 238
ACL Centers for Independent Living Competition Grants, 239
ACL Diversity Community of Practice Grants, 240

PROGRAM TYPE INDEX

Community Development

ACL Empowering Seniors to Prevent Health Care Fraud Grants, 241
ACL Field Initiated Projects Program: Minority-Serving Institution Development Grants, 242
ACL Field Initiated Projects Program: Minority-Serving Institution Research Grants, 243
ACL Learning Collaboratives for Advanced Business Acumen Skills Grants, 244
ACL Partnerships in Employment Systems Change Grants, 246
ACL Self-Advocacy Resource Center Grants, 247
ACL University Centers for Excellence in Developmental Network Diversity and Inclusion Training Action Planning Grants, 249
Adam Richter Charitable Trust Grants, 298
Adams and Reese Corporate Giving Grants, 299
Adams Foundation Grants, 301
Adelaide Breed Bayrd Foundation Grants, 306
Adler-Clark Electric Community Commitment Foundation Grants, 307
Administaff Community Affairs Grants, 308
Adobe Community Investment Grants, 310
Adolph Coors Foundation Grants, 311
Advance Auto Parts Corporate Giving Grants, 312
Advanced Micro Devices Comm Affairs Grants, 313
AEC Trust Grants, 315
AEGON Transamerica Foundation Health and Welfare Grants, 316
Aetna Foundation Health Grants in Connecticut, 327
Aetna Foundation Integrated Health Care Grants, 328
Aetna Foundation Obesity Grants, 331
Aetna Foundation Racial and Ethnic Health Care Equity Grants, 332
AFG Industries Grants, 340
A Friends' Foundation Trust Grants, 341
A Fund for Women Grants, 342
Agnes M. Lindsay Trust Grants, 346
A Good Neighbor Foundation Grants, 347
Agway Foundation Grants, 348
Ahmanson Foundation Grants, 355
Ahn Family Foundation Grants, 356
Aid for Starving Children Emergency Assistance Fund Grants, 367
Aid for Starving Children International Grants, 368
AIG Disaster Relief Fund Grants, 370
Air Products and Chemicals Grants, 399
Alabama Power Foundation Grants, 400
Alaska Airlines Corporate Giving Medical Emergency and Research Grants, 402
Alaska Airlines Foundation Grants, 403
Albert and Margaret Alkek Foundation Grants, 405
Alberto Culver Corporate Contributions Grants, 408
Albert Pick Jr. Fund Grants, 409
Albertson's Charitable Giving Grants, 410
Albert W. Rice Charitable Foundation Grants, 411
Albuquerque Community Foundation Grants, 412
Alcatel-Lucent Technologies Foundation Grants, 413
Alcoa Foundation Grants, 414
Alcon Foundation Grants Program, 416
Alexander & Baldwin Fnd Mainland Grants, 417
Alexander and Baldwin Foundation Hawaiian and Pacific Island Grants, 418
Alexander and Margaret Stewart Trust Grants, 419
Alexander Eastman Foundation Grants, 420
Alexander Foundation Cancer, Catastrophic Illness and Injury Grants, 421
Alexis Gregory Foundation Grants, 426
Alfred and Tillie Shemanski Testamentary Trust Grants, 428
Alfred Bersted Foundation Grants, 429
Alice C. A. Sibley Fund Grants, 435
Alice Tweed Tuohy Foundation Grants Program, 437
Allan C. and Lelia J. Garden Foundation Grants, 438
Allegan County Community Foundation Grants, 439
Allegheny Technologies Charitable Trust, 440
Allen P. & Josephine B. Green Fndn Grants, 443
Alliant Energy Corporation Contributions, 445
Allstate Corporate Giving Grants, 446
Allstate Corp Hometown Commitment Grants, 447
Allyn Foundation Grants, 448

Alpha Natural Resources Corporate Giving, 450
ALSAM Foundation Grants, 454
Altman Foundation Health Care Grants, 457
Alvin and Fanny Blaustein Thalheimer Foundation Baltimore Communal Grants, 459
AMA-MSS Chapter Involvement Grants, 473
Amador Community Foundation Grants, 481
AMA Foundation Fund for Better Health Grants, 484
AMA Foundation Healthy Communities/Healthy America Grants, 486
AMA Foundation Jack B. McConnell, MD Awards for Excellence in Volunteerism, 487
AMD Corporate Contributions Grants, 500
Amelia Sillman Rockwell and Carlos Perry Rockwell Charities Fund Grants, 501
American Academy of Dermatology Shade Structure Grants, 504
American Express Community Service Grants, 531
American Foodservice Charitable Trust Grants, 532
American Jewish World Service Grants, 533
American Schlafhorst Foundation Grants, 559
Amerigroup Foundation Grants, 565
amfAR Global Initiatives Grants, 568
amfAR Public Policy Grants, 570
Amgen Foundation Grants, 572
Amica Companies Foundation Grants, 574
AMI Semiconductors Corporate Grants, 576
Andersen Corporate Foundation, 586
Andre Agassi Charitable Foundation Grants, 587
Andrew Family Foundation Grants, 588
Angels Baseball Foundation Grants, 590
Anheuser-Busch Foundation Grants, 591
Anna Fitch Ardenghi Trust Grants, 592
Ann and Robert H. Lurie Family Fndn Grants, 593
Ann Arbor Area Community Foundation Grants, 594
Anne J. Caudal Foundation Grants, 595
Annie Sinclair Knudsen Memorial Fund/Kaua'i Community Grants Program, 597
Annunziata Sanguinetti Foundation Grants, 600
Ansell, Zaro, Grimm & Aaron Foundation Grants, 603
Antone & Edene Vidinha Charitable Trust Grants, 609
Appalachian Regional Commission Health Grants, 637
APSAA Foundation Grants, 639
APS Foundation Grants, 646
Aragona Family Foundation Grants, 648
Aratani Foundation Grants, 649
Arcadia Foundation Grants, 650
Archer Daniels Midland Foundation Grants, 651
Argyros Foundation Grants Program, 653
Arizona Cardinals Grants, 655
Arizona Diamondbacks Charities Grants, 657
Arizona Public Service Corporate Giving Program Grants, 658
Arkansas Community Foundation Arkansas Black Hall of Fame Grants, 659
Arkansas Community Foundation Grants, 660
Arkell Hall Foundation Grants, 661
Arlington Community Foundation Grants, 662
Armstrong McDonald Foundation Health Grants, 663
Arthur Ashley Williams Foundation Grants, 671
Arthur F. & Arnold M. Frankel Fndn Grants, 672
Arthur M. Blank Family Foundation Inspiring Spaces Grants, 673
ArvinMeritor Foundation Health Grants, 674
ArvinMeritor Grants, 675
Ashland Corporate Contributions Grants, 713
Assisi Foundation of Memphis General Grants, 787
Assurant Foundation Grants, 788
Assurant Health Foundation Grants, 789
AT&T Foundation Civic and Community Service Program Grants, 793
Atherton Family Foundation Grants, 794
Athwin Foundation Grants, 795
Atlanta Foundation Grants, 796
Auburn Foundation Grants, 798
Audrey & Sydney Irmas Foundation Grants, 799
Aurora Foundation Grants, 802
Austin S. Nelson Foundation Grants, 805
Autauga Area Community Foundation Grants, 808
Autodesk Community Relations Grants, 809

Avista Foundation Economic and Cultural Vitality Grants, 812
Babcock Charitable Trust Grants, 816
Bacon Family Foundation Grants, 817
Bailey Foundation Grants, 818
Balfe Family Foundation Grants, 819
Ball Brothers Foundation General Grants, 820
BancorpSouth Foundation Grants, 823
Banfi Vintners Foundation Grants, 824
Bank of America Fndn Matching Gifts, 825
Bank of America Fndn Volunteer Grants, 826
Baptist-Trinity Lutheran Legacy Fndn Grants, 827
Baptist Community Ministries Grants, 828
Barberton Community Foundation Grants, 829
Barker Foundation Grants, 830
Barker Welfare Foundation Grants, 831
Barnes Group Foundation Grants, 832
Barra Foundation Community Fund Grants, 833
Barra Foundation Project Grants, 834
Barr Fund Grants, 835
Batchelor Foundation Grants, 837
Baton Rouge Area Foundation Grants, 838
Battle Creek Community Foundation Grants, 839
Batts Foundation Grants, 840
Baxter International Corporate Giving Grants, 841
Baxter International Foundation Grants, 843
Bayer Foundation Grants, 846
BBVA Compass Foundation Charitable Grants, 848
BCBSM Building Healthy Communities Engaging Elementary Schools and Community Partners Grants, 849
BCBSM Corporate Contributions Grants, 851
BCBSM Foundation Community Health Matching Grants, 852
BCBSM Foundation Investigator Initiated Research Grants, 854
BCBSM Fndn Proposal Development Awards, 857
BCBSNC Foundation Grants, 860
Bearemy's Kennel Pals Grants, 863
Beazley Foundation Grants, 865
Bechtel Group Foundation Building Positive Community Relationships Grants, 866
Beckley Area Foundation Grants, 867
Beckman Coulter Foundation Grants, 868
Beirne Carter Foundation Grants, 870
Beldon Fund Grants, 871
Belvedere Community Foundation Grants, 874
Ben B. Cheney Foundation Grants, 875
Bender Foundation Grants, 876
Benton Community Fndn Cookie Jar Grant, 877
Benton Community Foundation Grants, 878
Benton County Foundation Grants, 880
Berks County Community Foundation Grants, 881
Bernard and Audre Rapoport Foundation Health Grants, 882
Bernard F. and Alva B. Gimbel Foundation Reproductive Rights Grants, 883
Berrien Community Foundation Grants, 884
Besser Foundation Grants, 887
BHHS Legacy Foundation Grants, 888
Bikes Belong Foundation Paul David Clark Bicycling Safety Grants, 890
Bill and Melinda Gates Foundation Emergency Response Grants, 892
Bill and Melinda Gates Foundation Policy and Advocacy Grants, 893
Bill and Melinda Gates Foundation Water, Sanitation and Hygiene Grants, 894
Bingham McHale LLP Pro Bono Services, 897
Biogen Foundation General Donations, 899
Biogen Foundation Patient Educational Grants, 901
Birmingham Foundation Grants, 903
BJ's Charitable Foundation Grants, 904
Blackford County Community Foundation Grants, 906
Black River Falls Area Foundation Grants, 909
Blanche and Irving Laurie Foundation Grants, 910
Blanche and Julian Robertson Family Foundation Grants, 911

Community Development

Blue Cross Blue Shield of Minnesota Foundation - Health Equity: Building Health Equity Together Grants, 914
Blue Cross Blue Shield of Minnesota Foundation - Healthy Equity: Health Impact Assessment Program Grants, 917
Blue Cross Blue Shield of Minnesota Fndn Healthy Equity: Public Libraries for Health Grants, 918
Blue Grass Community Fndn Harrison Grants, 919
Blue Grass Community Foundation Hudson-Ellis Fund Grants, 920
Blue Mountain Community Foundation Grants, 921
Blue River Community Foundation Grants, 922
Blum-Kovler Foundation Grants, 924
Blumenthal Foundation Grants, 925
Bodenwein Public Benevolent Foundation Grants, 928
Bodman Foundation Grants, 929
Boeing Company Contributions Grants, 930
Boettcher Foundation Grants, 931
Bonfils-Stanton Foundation Grants, 932
Boston Foundation Grants, 936
Boston Globe Foundation Grants, 937
Boyd Gaming Corporation Contributions Program, 941
Boyle Foundation Grants, 942
BP Foundation Grants, 943
Bradley-Turner Foundation Grants, 944
Bradley C. Higgins Foundation Grants, 945
Brian G. Dyson Foundation Grants, 948
Bridgestone/Firestone Trust Fund Grants, 949
Bright Family Foundation Grants, 951
Bright Promises Foundation Grants, 952
Bristol-Myers Squibb Foundation Health Disparities Grants, 958
Bristol-Myers Squibb Foundation Product Donations Grants, 960
Brookdale Fnd National Group Respite Grants, 966
Brook J. Lenfest Foundation Grants, 967
Brooklyn Community Foundation Caring Neighbors Grants, 968
Brown Advisory Charitable Foundation Grants, 969
Brown County Community Foundation Grants, 970
Browning-Kimball Foundation Grants, 971
Bruce and Adele Greenfield Foundation Grants, 972
Burlington Industries Foundation Grants, 977
Burlington Northern Santa Fe Foundation Grants, 978
Bush Fndn Health & Human Services Grants, 979
Bush Foundation Medical Fellowships, 980
Business Bank of Nevada Community Grants, 981
Cable Positive's Tony Cox Community Grants, 992
Cabot Corporation Foundation Grants, 993
Caddock Foundation Grants, 994
Caesars Foundation Grants, 995
Caleb C. and Julia W. Dula Educational and Charitable Foundation Grants, 997
California Community Foundation Health Care Grants, 998
California Community Foundation Human Development Grants, 999
California Endowment Innovative Ideas Challenge Grants, 1000
Callaway Foundation Grants, 1002
Callaway Golf Company Foundation Grants, 1003
Cambridge Community Foundation Grants, 1004
Camp-Younts Foundation Grants, 1005
Campbell Hoffman Foundation Grants, 1006
Campbell Soup Foundation Grants, 1007
Capital Region Community Foundation Grants, 1014
Cargill Citizenship Fund Corporate Giving, 1018
Caring Foundation Grants, 1019
Carl & Eloise Pohlad Family Fndn Grants, 1021
Carl B. and Florence E. King Foundation Grants, 1022
Carl C. Icahn Foundation Grants, 1023
Carl Gellert and Celia Berta Gellert Foundation Grants, 1024
Carlisle Foundation Grants, 1025
Carl M. Freeman Foundation FACES Grants, 1026
Carl M. Freeman Foundation Grants, 1027
Carl R. Hendrickson Family Foundation Grants, 1028
Carlsbad Charitable Foundation Grants, 1029
Carl W. and Carrie Mae Joslyn Trust Grants, 1031

Carnahan-Jackson Foundation Grants, 1032
Carnegie Corporation of New York Grants, 1033
Carolyn Foundation Grants, 1034
Carpenter Foundation Grants, 1035
Carrie E. and Lena V. Glenn Foundation Grants, 1036
Carrie Estelle Doheny Foundation Grants, 1037
Carrier Corporation Contributions Grants, 1038
Carroll County Community Foundation Grants, 1039
Cass County Community Foundation Grants, 1041
Catherine Kennedy Home Foundation Grants, 1043
CCHD Community Development Grants, 1046
CDC Foundation Emergency Response Grants, 1061
Cemala Foundation Grants, 1077
Centerville-Washington Foundation Grants, 1078
Central Carolina Community Foundation Community Impact Grants, 1079
Central Okanagan Foundation Grants, 1080
Cessna Foundation Grants Program, 1081
CFFVR Alcoholism and Drug Abuse Grants, 1093
CFFVR Basic Needs Giving Partnership Grants, 1094
CFFVR Capital Credit Union Charitable Giving Grants, 1095
CFFVR Chilton Area Community Fndn Grants, 1096
CFFVR Clintonville Area Foundation Grants, 1097
CFFVR Frank C. Shattuck Community Grants, 1098
CFFVR Jewelers Mutual Charitable Giving, 1099
CFFVR Myra M. and Robert L. Vandehey Foundation Grants, 1100
CFFVR Project Grants, 1101
CFFVR Robert and Patricia Endries Family Foundation Grants, 1102
CFFVR Schmidt Family G4 Grants, 1103
CFFVR Shawano Area Community Foundation Grants, 1104
CFFVR Waupaca Area Community Foundation Grants, 1105
CFFVR Wisconsin King's Daus & Sons Grants, 1106
CFFVR Women's Fund for the Fox Valley Region Grants, 1107
Chamberlain Foundation Grants, 1108
Champ-A Champion Fur Kids Grants, 1109
Champlin Foundations Grants, 1110
Chapman Charitable Foundation Grants, 1112
CharityWorks Grants, 1113
Charles A. Frueauff Foundation Grants, 1114
Charles Delmar Foundation Grants, 1115
Charles F. Bacon Trust Grants, 1117
Charles H. Dater Foundation Grants, 1118
Charles H. Farnsworth Trust Grants, 1119
Charles H. Hall Foundation, 1120
Charles H. Pearson Foundation Grants, 1121
Charles H. Revson Foundation Grants, 1122
Charles Lafitte Foundation Grants, 1123
Charles M. & Mary D. Grant Fndn Grants, 1124
Charles Nelson Robinson Fund Grants, 1125
Charlotte County (FL) Community Foundation Grants, 1126
Chatlos Foundation Grants Program, 1128
Chemtura Corporation Contributions Grants, 1130
Children's Trust Fund of Oregon Fndn Grants, 1139
Children Affected by AIDS Foundation Camp Network Grants, 1144
Children Affected by AIDS Foundation Domestic Grants, 1145
Children Affected by AIDS Foundation Family Assistance Emergency Fund Grants, 1146
Chiles Foundation Grants, 1147
Chilkat Valley Community Foundation Grants, 1148
Christine & Katharina Pauly Trust Grants, 1150
Christopher & Dana Reeve Foundation Quality of Life Grants, 1152
Christy-Houston Foundation Grants, 1153
Chula Vista Charitable Foundation Grants, 1154
CICF City of Noblesville Community Grant, 1155
CICF Indianapolis Fndn Community Grants, 1156
CICF James Proctor Grant for Aged, 1157
CICF Legacy Fund Grants, 1158
Cigna Civic Affairs Sponsorships, 1159
CIT Corporate Giving Grants, 1161

Citizens Bank Mid-Atlantic Charitable Foundation Grants, 1162
Clara Blackford Smith and W. Aubrey Smith Charitable Foundation Grants, 1165
Clarence E. Heller Charitable Foundation Grants, 1166
Clarence T.C. Ching Foundation Grants, 1167
Clark-Winchcole Foundation Grants, 1173
Clark and Ruby Baker Foundation Grants, 1174
Clark County Community Foundation Grants, 1175
Claude Bennett Family Foundation Grants, 1176
Claude Worthington Benedum Fndn Grants, 1178
Clayton Baker Trust Grants, 1179
Clayton Fund Grants, 1180
Cleveland-Cliffs Foundation Grants, 1181
CLIF Bar Family Foundation Grants, 1183
Clinton County Community Foundation Grants, 1184
CNA Foundation Grants, 1189
CNCS AmeriCorps VISTA Project Grants, 1190
CNCS Senior Companion Program Grants, 1191
CNCS Social Innovation Grants, 1192
CNO Financial Group Community Grants, 1201
Coastal Community Foundation of South Carolina Grants, 1202
Coca-Cola Foundation Grants, 1203
Cockrell Foundation Grants, 1204
Coeta and Donald Barker Foundation Grants, 1205
Coleman Foundation Cancer Care Grants, 1206
Coleman Foundation Developmental Disabilities Grants, 1207
Collective Brands Foundation Grants, 1208
Collins C. Diboll Private Foundation Grants, 1209
Collins Foundation Grants, 1210
Colonel Stanley R. McNeil Foundation Grants, 1211
Colorado Resource for Emergency Education and Trauma Grants, 1212
Colorado Trust Grants, 1213
Columbus Foundation Competitive Grants, 1217
Columbus Foundation Paul G. Duke Grants, 1221
Columbus Foundation Robert E. and Genevieve B. Schaefer Fund Grants, 1222
Columbus Foundation Traditional Grants, 1223
Comerica Charitable Foundation Grants, 1224
Commonwealth Edison Grants, 1225
Commonwealth Fund Patient-Centered Coordinated Care Program Grants, 1231
Communities Foundation of Texas Grants, 1236
Community Fndn AIDS Endowment Awards, 1237
Community Foundation Alliance City of Evansville Endowment Fund Grants, 1238
Community Foundation for Greater Atlanta AIDS Fund Grants, 1239
Community Foundation for Greater Atlanta Clayton County Fund Grants, 1240
Community Foundation for Greater Atlanta Common Good Funds Grants, 1241
Community Foundation for Greater Atlanta Managing For Excellence Award, 1242
Community Foundation for Greater Atlanta Metropolitan Atlanta An Extra Wish Grants, 1243
Community Foundation for Greater Atlanta Morgan County Fund Grants, 1244
Community Foundation for Greater Atlanta Newton County Fund Grants, 1245
Community Foundation for Greater Atlanta Strategic Restructuring Fund Grants, 1246
Community Fndn for Greater Buffalo Grants, 1247
Community Foundation for Greater New Haven $5,000 and Under Grants, 1248
Community Foundation for Greater New Haven Responsive New Grants, 1249
Community Foundation for Greater New Haven Sponsorship Grants, 1250
Community Foundation for Greater New Haven Valley Neighborhood Grants, 1251
Community Foundation for Greater New Haven Women & Girls Grants, 1252
Community Fndn for Monterey County Grants, 1253
Community Fndn for Muskegon County Grants, 1254
Community Foundation for Northeast Michigan Mini-Grants, 1255

PROGRAM TYPE INDEX   Community Development   / 853

Community Foundation for Northeast Michigan Tobacco Settlement Grants, 1256
Community Fndn for San Benito County Grants, 1257
Community Foundation for Southeast Michigan Grants, 1258
Community Fndn for the Capital Region Grants, 1259
Community Foundation for the National Capital Region Community Leadership Grants, 1260
Community Foundation of Bartholomew County Heritage Fund Grants, 1262
Community Fndn of Bloomington & Monroe County - Precision Health Network Cycle Grants, 1264
Community Foundation of Bloomington and Monroe County Grants, 1265
Community Foundation of Boone County Grants, 1266
Community Fndn of Central Illinois Grants, 1267
Community Fndn of East Central Illinois Grants, 1268
Community Foundation of Eastern Connecticut General Southeast Grants, 1269
Community Foundation of Eastern Connecticut Northeast Women and Girls Grants, 1270
Community Foundation of Grant County Grants, 1271
Community Foundation of Greater Birmingham Grants, 1272
Community Foundation of Greater Flint Grants, 1273
Community Foundation of Greater Fort Wayne - Community Endowment and Clarke Endowment Grants, 1274
Community Foundation of Greater Greensboro Community Grants, 1277
Community Foundation of Greater Greensboro Women to Women Fund Grants, 1278
Community Fndn of Greater Lafayette Grants, 1280
Community Foundation of Greater New Britain Grants, 1281
Community Fndn of Greater Tampa Grants, 1282
Community Foundation of Greenville-Greenville Women Giving Grants, 1283
Community Foundation of Greenville Community Enrichment Grants, 1284
Community Foundation of Greenville Hollingsworth Funds Program/Project Grants, 1285
Community Fndn of Howard County Grants, 1286
Community Fndn of Jackson County Grants, 1287
Community Foundation of Louisville AIDS Project Fund Grants, 1289
Community Foundation of Louisville Anna Marble Memorial Fund for Princeton Grants, 1290
Community Foundation of Louisville Bobbye M. Robinson Fund Grants, 1291
Community Foundation of Louisville Delta Dental of Kentucky Fund Grants, 1292
Community Foundation of Louisville Diller B. and Katherine P. Groff Fund for Pediatric Surgery Grants, 1293
Community Foundation of Louisville Dr. W. Barnett Owen Memorial Fund for the Children of Louisville and Jefferson County Grants, 1294
Community Fndn of Louisville Health Grants, 1295
Community Foundation of Louisville Morris and Esther Lee Fund Grants, 1298
Community Foundation of Mount Vernon and Knox County Grants, 1299
Community Foundation of Muncie and Delaware County Grants, 1300
Community Foundation of Muncie and Delaware County Maxon Grants, 1301
Community Fndn of Randolph County Grants, 1302
Community Foundation of Riverside & San Bernardino County Impact Grants, 1303
Community Fndn of Riverside & San Bernardino County Irene S. Rockwell Grants, 1304
Community Fndn of So Alabama Grants, 1306
Community Fndn of South Puget Sound Grants, 1307
Community Foundation of St. Joseph County Special Project Challenge Grants, 1310
Community Fndn of Switzerland County Grants, 1311
Community Foundation of Tampa Bay Grants, 1312
Community Foundation of the Eastern Shore Community Needs Grants, 1313

Community Foundation of the Verdugos Educational Endowment Fund Grants, 1314
Community Foundation of the Verdugos Grants, 1315
Community Foundation Partnerships - Lawrence County Grants, 1317
Community Foundation Partnerships - Martin County Grants, 1318
Community Foundation Serving Riverside and San Bernardino Counties Impact Grants, 1319
ConAgra Foods Fndn Community Impact Grants, 1321
ConAgra Foods Nourish our Community Grants, 1322
Cone Health Foundation Grants, 1323
Connecticut Community Foundation Grants, 1324
Connelly Foundation Grants, 1326
ConocoPhillips Foundation Grants, 1327
Conservation, Food, and Health Foundation Grants for Developing Countries, 1338
CONSOL Energy Community Dev Grants, 1339
CONSOL Military and Armed Services Grants, 1340
Constantin Foundation Grants, 1341
Constellation Energy Corporate Grants, 1342
Consumers Energy Foundation, 1343
Cooke Foundation Grants, 1344
Cooper Industries Foundation Grants, 1345
Coors Brewing Corporate Contributions Grants, 1346
Cornerstone Foundation of Northeastern Wisconsin Grants, 1347
Covenant Educational Foundation Grants, 1348
Covidien Medical Product Donations, 1352
Covidien Partnership for Neighborhood Wellness Grants, 1353
Crail-Johnson Foundation Grants, 1355
Cralle Foundation Grants, 1356
Crane Foundation Grants, 1357
Crane Fund Grants, 1358
Cresap Family Foundation Grants, 1359
Crown Point Community Foundation Grants, 1361
Cruise Industry Charitable Foundation Grants, 1363
Crystelle Waggoner Charitable Trust Grants, 1364
CSL Behring Local Empowerment for Advocacy Development (LEAD) Grants, 1365
CSRA Community Foundation Grants, 1366
CSX Corporate Contributions Grants, 1368
Cuesta Foundation Grants, 1371
Cullen Foundation Grants, 1372
Cultural Society of Filipino Americans Grants, 1373
Cumberland Community Foundation Grants, 1374
CUNA Mutual Group Fndn Community Grants, 1375
Curtis Foundation Grants, 1376
CVS All Kids Can Grants, 1377
CVS Caremark Charitable Trust Grants, 1378
CVS Community Grants, 1379
Cyrus Eaton Foundation Grants, 1380
D.F. Halton Foundation Grants, 1405
D. W. McMillan Foundation Grants, 1407
Dade Community Foundation Community AIDS Partnership Grants, 1409
Dade Community Foundation Grants, 1410
DaimlerChrysler Corporation Fund Grants, 1412
Daisy Marquis Jones Foundation Grants, 1414
Dale and Edna Walsh Foundation Grants, 1415
Dallas Mavericks Foundation Grants, 1416
Dallas Women's Foundation Grants, 1417
Dammann Fund Grants, 1418
Dana Brown Charitable Trust Grants, 1419
Danellie Foundation Grants, 1421
Daniel & Nanna Stern Family Fndn Grants, 1422
Daniels Fund Grants-Aging, 1424
Daphne Seybolt Culpeper Fndn Grants, 1425
David Geffen Foundation Grants, 1431
David M. and Marjorie D. Rosenberg Foundation Grants, 1432
David N. Lane Trust Grants for Aged and Indigent Women, 1433
Daviess County Community Foundation Health Grants, 1434
Dayton Power and Light Company Foundation Signature Grants, 1436
Dayton Power and Light Foundation Grants, 1437
Daywood Foundation Grants, 1438

Deaconess Community Foundation Grants, 1439
Dean Foods Community Involvement Grants, 1440
Dearborn Community Foundation City of Lawrenceburg Community Grants, 1441
Dearborn Community Foundation County Progress Grants, 1442
Deborah Munroe Noonan Memorial Grants, 1443
Decatur County Community Foundation Large Project Grants, 1444
Decatur County Community Foundation Small Project Grants, 1445
DeKalb County Community Foundation - Garrett Hospital Aid Foundation Grants, 1446
DeKalb County Community Foundation Grants, 1447
Delaware Community Foundation-Youth Philanthropy Board for Kent County, 1448
Delaware Community Foundation Grants, 1449
Della B. Gardner Fund Grants, 1451
Del Mar Healthcare Fund Grants, 1453
Delmarva Power and Light Contributions, 1454
DeMatteis Family Foundation Grants, 1457
Dennis & Phyllis Washington Fndn Grants, 1458
Denver Broncos Charities Fund Grants, 1460
Denver Foundation Community Grants, 1461
Dept of Ed Safe and Drug-Free Schools and Communities State Grants, 1464
DeRoy Testamentary Foundation Grants, 1469
Detlef Schrempf Foundation Grants, 1470
DIFFA/Chicago Grants, 1477
DOJ Gang-Free Schools and Communities Intervention Grants, 1481
Dole Food Company Charitable Contributions, 1483
Dominion Foundation Grants, 1485
Dora Roberts Foundation Grants, 1489
Doree Taylor Charitable Foundation, 1490
Doris and Victor Day Foundation Grants, 1491
Dorothea Haus Ross Foundation Grants, 1498
Dorothy Rider Pool Health Care Grants, 1500
Dorrance Family Foundation Grants, 1501
Do Something Awards, 1503
DPA Promoting Policy Change Advocacy Grants, 1504
Dr. & Mrs. Paul Pierce Memorial Fndn Grants, 1505
Dr. John T. Macdonald Foundation Grants, 1507
Dr. Scholl Foundation Grants Program, 1511
Drs. Bruce and Lee Foundation Grants, 1513
DTE Energy Foundation Health and Human Services Grants, 1516
Dubois County Community Foundation Grants, 1517
Duneland Health Council Incorporated Grants, 1529
Dunspaugh-Dalton Foundation Grants, 1530
DuPage Community Foundation Grants, 1531
DuPont Pioneer Community Giving Grants, 1532
Dyson Foundation Mid-Hudson Valley Project Support Grants, 1534
E. Clayton and Edith P. Gengras, Jr. Foundation Grants, 1536
E.J. Grassmann Trust Grants, 1537
E.L. Wiegand Foundation Grants, 1538
E. Rhodes & Leona B. Carpenter Grants, 1539
Earl and Maxine Claussen Trust Grants, 1541
Eastman Chemical Company Foundation Grants, 1542
eBay Foundation Community Grants, 1543
Eberly Foundation Grants, 1544
Eddie C. & Sylvia Brown Family Fndn Grants, 1548
Eden Hall Foundation Grants, 1549
Edina Realty Foundation Grants, 1550
Edward and Helen Bartlett Foundation Grants, 1554
Edward and Romell Ackley Foundation Grants, 1555
Edward Bangs Kelley and Elza Kelley Foundation Grants, 1556
Edward N. and Della L. Thome Memorial Foundation Grants, 1557
Edwards Memorial Trust Grants, 1558
Edward W. and Stella C. Van Houten Memorial Fund Grants, 1559
Edwin S. Webster Foundation Grants, 1560
Edyth Bush Charitable Foundation Grants, 1562
Effie and Wofford Cain Foundation Grants, 1563
Eisner Foundation Grants, 1564
Elaine Feld Stern Charitable Trust Grants, 1565
Elizabeth Morse Genius Charitable Trust Grants, 1569

854 / Community Development

Elkhart County Community Foundation Fund for Elkhart County, 1572
Elkhart County Community Foundation Grants, 1573
Elliot Foundation Inc Grants, 1575
Elmer L. & Eleanor J. Andersen Fndn Grants, 1578
El Paso Community Foundation Grants, 1579
El Paso Corporate Foundation Grants, 1580
El Pomar Foundation Anna Keesling Ackerman Fund Grants, 1581
El Pomar Foundation Grants, 1582
Elsie H. Wilcox Foundation Grants, 1583
Elsie Lee Garthwaite Memorial Fndn Grants, 1584
Emily Davie and Joseph S. Kornfeld Foundation Grants, 1591
Emily Hall Tremaine Foundation Learning Disabilities Grants, 1592
Emma B. Howe Memorial Foundation Grants, 1593
Emma Barnsley Foundation Grants, 1594
Ensworth Charitable Foundation Grants, 1595
Entergy Corporation Micro Grants, 1596
Entergy Corporation Open Grants for Healthy Families, 1597
EPA Children's Health Protection Grants, 1598
Erie Chapman Foundation Grants, 1600
Erie Community Foundation Grants, 1601
Essex County Community Foundation Discretionary Fund Grants, 1602
Essex County Community Foundation Merrimack Valley General Fund Grants, 1603
Essex County Community Foundation Webster Family Fund Grants, 1604
Essex County Community Foundation Women's Fund Grants, 1605
Ethel S. Abbott Charitable Foundation Grants, 1607
Ethel Sergeant Clark Smith Foundation Grants, 1608
Eugene M. Lang Foundation Grants, 1611
Eugene Straus Charitable Trust, 1612
Eva L. & Joseph M. Bruening Fndn Grants, 1613
Evan and Susan Bayh Foundation Grants, 1614
Evanston Community Foundation Grants, 1615
Evjue Foundation Grants, 1617
Expect Miracles Foundation Grants, 1619
Ezra M. Cutting Trust Grants, 1623
F.M. Kirby Foundation Grants, 1624
Fairfield County Community Foundation Grants, 1625
Fallon OrNda Community Health Fund Grants, 1627
FAR Fund Grants, 1631
Fargo-Moorhead Area Foundation Grants, 1632
Fargo-Moorhead Area Fndn Woman's Grants, 1633
Farmers Insurance Corporate Giving Grants, 1634
Faye McBeath Foundation Grants, 1635
Fayette County Foundation Grants, 1636
Federal Express Corporate Contributions, 1657
Ferree Foundation Grants, 1658
Fidelity Foundation Grants, 1659
Field Foundation of Illinois Grants, 1660
Fifth Third Foundation Grants, 1661
Finance Factors Foundation Grants, 1666
FirstEnergy Foundation Community Grants, 1668
Fisher Foundation Grants, 1670
Fisher House Fnd Newman's Own Awards, 1671
FIU Global Civic Engagement Mini Grants, 1673
Flextronics Foundation Disaster Relief Grants, 1674
Florence Hunt Maxwell Foundation Grants, 1676
Florian O. Bartlett Trust Grants, 1677
Florida BRAIVE Fund of Dade Community Foundation, 1678
Floyd A. & Kathleen C. Cailloux Fnd Grants, 1681
Fluor Foundation Grants, 1682
FMC Foundation Grants, 1683
Foellinger Foundation Grants, 1684
Fondren Foundation Grants, 1685
Ford Family Foundation Grants - Access to Health and Dental Services, 1686
Ford Motor Company Fund Grants Program, 1687
Forrest C. Lattner Foundation Grants, 1688
Foster Foundation Grants, 1689
Foundation for Health Enhancement Grants, 1696
Foundation for Seacoast Health Grants, 1698
Foundation for the Mid South Community Development Grants, 1699
Foundation for the Mid South Health and Wellness Grants, 1700
Foundation of CVPH Rabin Fund Grants, 1704
Foundation of CVPH Roger Senecal Endowment Fund Grants, 1705
Foundations of East Chicago Health Grants, 1709
Four County Community Fndn General Grants, 1710
Four County Community Foundation Healthy Senior/Healthy Youth Fund Grants, 1711
Fourjay Foundation Grants, 1712
Four J Foundation Grants, 1713
Frances and John L. Loeb Family Fund Grants, 1716
Frances L. and Edwin L. Cummings Memorial Fund Grants, 1717
Frances W. Emerson Foundation Grants, 1718
Francis L. Abreu Charitable Trust Grants, 1719
Francis T. & Louise T. Nichols Fndn Grants, 1720
Frank B. Hazard General Charity Fund Grants, 1721
Frank G. and Freida K. Brotz Family Foundation Grants, 1723
Franklin County Community Foundation Grants, 1724
Franklin H. Wells and Ruth L. Wells Foundation Grants, 1725
Frank Reed and Margaret Jane Peters Memorial Fund I Grants, 1727
Frank Reed and Margaret Jane Peters Memorial Fund II Grants, 1728
Frank Stanley Beveridge Foundation Grants, 1730
Frank W. & Carl S. Adams Memorial Grants, 1731
Fred & Gretel Biel Charitable Trust Grants, 1734
Fred Baldwin Memorial Foundation Grants, 1736
Freddie Mac Foundation Grants, 1738
Frederick McDonald Trust Grants, 1740
Frederick W. Marzahl Memorial Fund Grants, 1741
Fred L. Emerson Foundation Grants, 1743
Fremont Area Community Foundation Amazing X Grants, 1744
Fremont Area Community Foundation Elderly Needs Grants, 1745
Fremont Area Community Fndn General Grants, 1746
Fritz B. Burns Foundation Grants, 1747
Fuller E. Callaway Foundation Grants, 1761
Fulton County Community Foundation Grants, 1763
Fulton County Community Foundation Women's Giving Circle Grants, 1766
G.A. Ackermann Memorial Fund Grants, 1767
Gamble Foundation Grants, 1769
Gardner Foundation Grants, 1771
Gardner Foundation Grants, 1770
Gardner Foundation Grants, 1772
Garland D. Rhoads Foundation, 1774
Gebbie Foundation Grants, 1777
GenCorp Foundation Grants, 1779
Genentech Corp Charitable Contributions, 1780
General Mills Foundation Grants, 1782
General Motors Foundation Grants, 1783
General Service Foundation Human Rights and Economic Justice Grants, 1784
General Service Reproductive Justice Grants, 1785
Genuardi Family Foundation Grants, 1787
George A. and Grace L. Long Foundation Grants, 1788
George and Ruth Bradford Foundation Grants, 1789
George and Sarah Buchanan Foundation Grants, 1790
George A Ohl Jr. Foundation Grants, 1791
George F. Baker Trust Grants, 1793
George Family Foundation Grants, 1794
George Foundation Grants, 1795
George Gund Foundation Grants, 1796
George H.C. Ensworth Memorial Fund Grants, 1797
George P. Davenport Trust Fund Grants, 1800
George S. & Dolores Dore Eccles Fndn Grants, 1801
George W. Brackenridge Foundation Grants, 1802
George W. Codrington Charitable Fndn Grants, 1803
George W. Wells Foundation Grants, 1804
Georgiana Goddard Eaton Memorial Grants, 1805
Georgia Power Foundation Grants, 1806
Gertrude and William C. Wardlaw Fund Grants, 1808
Gertrude E. Skelly Charitable Foundation Grants, 1810

PROGRAM TYPE INDEX

Gheens Foundation Grants, 1814
Giant Eagle Foundation Grants, 1815
Giant Food Charitable Grants, 1816
Gibson County Community Foundation Women's Fund Grants, 1817
Gibson Foundation Grants, 1818
Gil and Dody Weaver Foundation Grants, 1819
Gill Foundation - Gay and Lesbian Fund Grants, 1821
Ginn Foundation Grants, 1822
Giving Sum Annual Grant, 1825
Gladys Brooks Foundation Grants, 1826
GlaxoSmithKline Corporate Grants, 1829
Global Fund for Children Grants, 1832
Global Fund for Women Grants, 1833
GNOF Albert N. & Hattie M. McClure Grants, 1834
GNOF Bayou Communities Grants, 1835
GNOF Exxon-Mobil Grants, 1836
GNOF IMPACT Grants for Arts and Culture, 1838
GNOF IMPACT Grants for Health and Human Services, 1839
GNOF IMPACT Gulf States Eye Surgery Fund, 1840
GNOF IMPACT Harold W. Newman, Jr. Charitable Trust Grants, 1841
GNOF IMPACT Kahn-Oppenheim Grants, 1842
GNOF Maison Hospitaliere Grants, 1843
GNOF New Orleans Works Grants, 1844
GNOF Norco Community Grants, 1845
GNOF Organizational Effectiveness Grants and Workshops, 1846
GNOF Stand Up For Our Children Grants, 1847
Godfrey Foundation Grants, 1849
Golden Heart Community Foundation Grants, 1850
Golden State Warriors Foundation Grants, 1851
Goodrich Corporation Foundation Grants, 1854
Goodyear Tire Grants, 1856
Grace and Franklin Bernsen Foundation Grants, 1858
Grace Bersted Foundation Grants, 1859
Graco Foundation Grants, 1860
Graham and Carolyn Holloway Family Foundation Grants, 1861
Graham Foundation Grants, 1862
Grand Haven Area Community Fndn Grants, 1864
Grand Rapids Area Community Fndn Grants, 1865
Grand Rapids Area Community Foundation Nashwauk Area Endowment Fund Grants, 1866
Grand Rapids Area Community Foundation Wyoming Grants, 1867
Grand Rapids Area Community Foundation Wyoming Youth Fund Grants, 1868
Grand Rapids Community Foundation Grants, 1869
Grand Rapids Community Foundation Ionia County Grants, 1870
Grand Rapids Community Foundation Ionia County Youth Fund Grants, 1871
Grand Rapids Community Foundation Lowell Area Fund Grants, 1872
Grand Rapids Community Foundation Southeast Ottawa Grants, 1873
Grand Rapids Community Foundation Southeast Ottawa Youth Fund Grants, 1874
Grand Rapids Community Fndn Sparta Grants, 1875
Grand Rapids Community Foundation Sparta Youth Fund Grants, 1876
Great-West Life Grants, 1879
Greater Cincinnati Foundation Priority and Small Projects/Capacity-Building Grants, 1880
Greater Green Bay Community Fndn Grants, 1881
Greater Kanawha Valley Foundation Grants, 1882
Greater Milwaukee Foundation Grants, 1883
Greater Saint Louis Community Fndn Grants, 1884
Greater Sitka Legacy Fund Grants, 1885
Greater Tacoma Community Foundation Grants, 1886
Greater Worcester Community Foundation Discretionary Grants, 1888
Greater Worcester Community Foundation Jeppson Memorial Fund for Brookfield Grants, 1889
Green Bay Packers Foundation Grants, 1890
Green Diamond Charitable Contributions, 1891
Greene County Foundation Grants, 1892
Green Foundation Human Services Grants, 1894

PROGRAM TYPE INDEX   Community Development   / 855

Green River Area Community Fndn Grants, 1895
Greenwall Foundation Bioethics Grants, 1897
Gregory B. Davis Foundation Grants, 1898
Greygates Foundation Grants, 1901
Grotto Foundation Project Grants, 1905
Grundy Foundation Grants, 1912
Guido A. & Elizabeth H. Binda Fndn Grants, 1913
Gulf Coast Community Foundation Grants, 1915
Gulf Coast Foundation of Community Operating Grants, 1916
Gulf Coast Foundation of Community Program Grants, 1917
H & R Foundation Grants, 1920
H.A. & Mary K. Chapman Trust Grants, 1921
H.B. Fuller Foundation Grants, 1922
H.J. Heinz Company Foundation Grants, 1923
H. Leslie Hoffman and Elaine S. Hoffman Foundation Grants, 1924
H. Schaffer Foundation Grants, 1925
Hackett Foundation Grants, 1926
HAF Technical Assistance Program Grants, 1933
Hagedorn Fund Grants, 1934
Hall-Perrine Foundation Grants, 1935
Hallmark Corporate Foundation Grants, 1936
Hampton Roads Community Foundation Faith Community Nursing Grants, 1940
Hampton Roads Community Foundation Health and Human Service Grants, 1941
Hancock County Community Foundation - Field of Interest Grants, 1943
Hannaford Charitable Foundation Grants, 1944
Harden Foundation Grants, 1945
Hardin County Community Foundation Grants, 1946
Harley Davidson Foundation Grants, 1947
Harmony Project Grants, 1948
Harold Alfond Foundation Grants, 1949
Harold and Arlene Schnitzer CARE Foundation Grants, 1950
Harold and Rebecca H. Gross Foundation Grants, 1951
Harold Brooks Foundation Grants, 1952
Harold R. Bechtel Charitable Remainder Uni-Trust Grants, 1953
Harold R. Bechtel Testamentary Charitable Trust Grants, 1954
Harold Simmons Foundation Grants, 1955
Harris and Eliza Kempner Fund Grants, 1956
Harrison County Community Foundation Grants, 1957
Harrison County Community Foundation Signature Grants, 1958
Harry B. and Jane H. Brock Foundation Grants, 1960
Harry Bramhall Gilbert Charitable Trust Grants, 1961
Harry Edison Foundation, 1962
Harry Kramer Memorial Fund Grants, 1965
Harry S. Black and Allon Fuller Fund Grants, 1966
Harry Sudakoff Foundation Grants, 1967
Harry W. Bass, Jr. Foundation Grants, 1968
Hartford Courant Foundation Grants, 1969
Hartford Foundation Regular Grants, 1970
Harvest Foundation Grants, 1971
Harvey Randall Wickes Foundation Grants, 1972
Hasbro Children's Fund Grants, 1973
Hawaiian Electric Industries Charitable Foundation Grants, 1975
Hawaii Community Foundation Reverend Takie Okumura Family Grants, 1977
Hawaii Community Foundation West Hawaii Fund Grants, 1978
HCA Foundation Grants, 1980
Healthcare Fndn for Orange County Grants, 1981
Healthcare Foundation of New Jersey Grants, 1982
Health Foundation of Greater Cincinnati Grants, 1983
Health Fndn of Greater Indianapolis Grants, 1984
Health Foundation of Southern Florida Responsive Grants, 1985
Heckscher Foundation for Children Grants, 1988
Hedco Foundation Grants, 1989
Helena Rubinstein Foundation Grants, 1991
Helen Bader Foundation Grants, 1992
Helen Gertrude Sparks Charitable Trust Grants, 1993
Helen Irwin Littauer Educational Trust Grants, 1994

Helen K. & Arthur E. Johnson Fndn Grants, 1995
Helen Pumphrey Denit Charitable Trust Grants, 1996
Helen S. Boylan Foundation Grants, 1997
Helen Steiner Rice Foundation Grants, 1998
Hendrick Foundation for Children Grants, 1999
Hendricks County Community Fndn Grants, 2000
Henrietta Tower Wurts Memorial Fndn Grants, 2002
Henry A. and Mary J. MacDonald Foundation, 2003
Henry and Ruth Blaustein Rosenberg Foundation Health Grants, 2004
Henry County Community Foundation Grants, 2005
Henry L. Guenther Foundation Grants, 2006
Herbert A. & Adrian W. Woods Fndn Grants, 2007
Herbert H. & Grace A. Dow Fndn Grants, 2008
Herman Goldman Foundation Grants, 2012
Hershey Company Grants, 2013
Highmark BCBS Challenge for Healthier Schools in Western Pennsylvania, 3696
Highmark Corporate Giving Grants, 2025
High Meadow Foundation Grants, 2027
Hilda and Preston Davis Foundation Grants, 2028
Hillcrest Foundation Grants, 2031
Hill Crest Foundation Grants, 2030
Hillman Foundation Grants, 2032
Hillsdale County Community Foundation Healthy Senior/Healthy Youth Fund Grants, 2033
Hillsdale County Community General Adult Foundation Grants, 2034
Hilton Head Island Foundation Grants, 2036
Hilton Hotels Corporate Giving Program Grants, 2037
Hoblitzelle Foundation Grants, 2038
Hogg Foundation for Mental Health Grants, 2039
Hoglund Foundation Grants, 2040
Holland/Zeeland Community Fndn Grants, 2041
HomeBanc Foundation Grants, 2042
Home Building Industry Disaster Relief Fund, 2043
Homer Foundation Grants, 2044
Honda of America Manufacturing Fndn Grants, 2045
Horace A. Kimball and S. Ella Kimball Foundation Grants, 2046
Horace Moses Charitable Foundation Grants, 2047
Horizon Foundation for New Jersey Grants, 2048
Horizons Community Issues Grants, 2049
Houston Endowment Grants, 2053
Howard and Bush Foundation Grants, 2054
Howe Foundation of North Carolina Grants, 2055
HRK Foundation Health Grants, 2067
HRSA Ryan White HIV AIDS Drug Assistance Grants, 2070
Huber Foundation Grants, 2071
Hudson Webber Foundation Grants, 2072
Huffy Foundation Grants, 2073
Hugh J. Andersen Foundation Grants, 2074
Huie-Dellmon Trust Grants, 2075
Humana Foundation Grants, 2077
Huntington Beach Police Officers Fndn Grants, 2081
Huntington Clinical Foundation Grants, 2082
Huntington County Community Foundation Make a Difference Grants, 2083
Hutchinson Community Foundation Grants, 2085
Hut Foundation Grants, 2086
Hutton Foundation Grants, 2087
I.A. O'Shaughnessy Foundation Grants, 2088
IBCAT Screening Mammography Grants, 2095
Ida Alice Ryan Charitable Trust Grants, 2113
Idaho Community Foundation Eastern Region Competitive Grants, 2114
Idaho Power Company Corporate Contributions, 2115
IDPH Carolyn Adams Ticket for the Cure Community Grants, 2117
IDPH Emergency Medical Services Assistance Fund Grants, 2118
IDPH Hosptial Capital Investment Grants, 2119
IDPH Local Health Department Public Health Emergency Response Grants, 2120
IIE David L. Boren Fellowships, 2129
IIE Freeman Foundation Indonesia Internships, 2131
Ike and Roz Friedman Foundation Grants, 2145
Illinois Tool Works Foundation Grants, 2147
Impact 100 Grants, 2148

Inasmuch Foundation Grants, 2149
Independence Blue Cross Charitable Medical Care Grants, 2150
Independence Community Foundation Community Quality of Life Grant, 2153
Indiana 21st Century Research and Technology Fund Awards, 2154
Indiana AIDS Fund Grants, 2155
Ireland Family Foundation Grants, 2162
Irving S. Gilmore Foundation Grants, 2163
Irvin Stern Foundation Grants, 2164
Isabel Allende Foundation Esperanza Grants, 2165
Ittleson Foundation AIDS Grants, 2166
Ittleson Foundation Mental Health Grants, 2167
J.B. Reynolds Foundation Grants, 2168
J.C. Penney Company Grants, 2169
J.E. and L.E. Mabee Foundation Grants, 2170
J.H. Robbins Foundation Grants, 2171
J.L. Bedsole Foundation Grants, 2172
J.M. Long Foundation Grants, 2173
J.N. and Macie Edens Foundation Grants, 2174
J. Spencer Barnes Memorial Foundation Grants, 2175
J.W. Kieckhefer Foundation Grants, 2176
J. Walton Bissell Foundation Grants, 2177
J. Willard Marriott, Jr. Foundation Grants, 2178
Jack H. & William M. Light Trust Grants, 2179
Jackson County Community Foundation Unrestricted Grants, 2180
Jacob and Hilda Blaustein Foundation Health and Mental Health Grants, 2181
Jacob G. Schmidlapp Trust Grants, 2183
Jacobs Family Village Neighborhoods Grants, 2184
James & Abigail Campbell Family Fndn Grants, 2185
James A. and Faith Knight Foundation Grants, 2186
James Ford Bell Foundation Grants, 2187
James G.K. McClure Educational and Development Fund Grants, 2188
James Graham Brown Foundation Grants, 2189
James J. & Angelia M. Harris Fndn Grants, 2192
James J. & Joan A. Gardner Family Fndn Grants, 2193
James M. Collins Foundation Grants, 2195
James M. Cox Foundation of Georgia Grants, 2196
James R. Dougherty Jr. Foundation Grants, 2198
James R. Thorpe Foundation Grants, 2199
James S. Copley Foundation Grants, 2200
Jane's Trust Grants, 2206
Jane Bradley Pettit Foundation Health Grants, 2208
Janirve Foundation Grants, 2209
Janson Foundation Grants, 2210
Janus Foundation Grants, 2211
Jasper Foundation Grants, 2212
Jean and Louis Dreyfus Foundation Grants, 2215
Jeffris Wood Foundation Grants, 2217
JELD-WEN Foundation Grants, 2218
Jenkins Foundation: Improving the Health of Greater Richmond Grants, 2219
Jennings County Community Foundation Grants, 2220
Jennings County Community Foundation Women's Giving Circle Grant, 2221
Jerome and Mildred Paddock Foundation Grants, 2222
Jerome Robbins Foundation Grants, 2223
Jessica Stevens Community Foundation Grants, 2224
Jessie B. Cox Charitable Trust Grants, 2225
Jessie Ball Dupont Fund Grants, 2226
Jewish Fund Grants, 2227
Jim Moran Foundation Grants, 2228
JM Foundation Grants, 2229
Joe W. and Dorothy Dorsett Brown Fndn Grants, 2230
John Ben Snow Memorial Trust Grants, 2231
John Deere Foundation Grants, 2234
John Edward Fowler Memorial Fndn Grants, 2235
John G. Duncan Charitable Trust Grants, 2236
John H. Wellons Foundation Grants, 2239
John I. Smith Charities Grants, 2240
John J. Leidy Foundation Grants, 2241
John Jewett and Helen Chandler Garland Foundation Grants, 2242
John Lord Knight Foundation Grants, 2243
John Merck Fund Grants, 2245
John P. McGovern Foundation Grants, 2247

John R. Oishei Foundation Grants, 2249
Johns Manville Fund Grants, 2251
Johnson County Community Foundation Grants, 2256
John Stauffer Charitable Trust Grants, 2258
John W. Alden Trust Grants, 2260
John W. and Anna H. Hanes Foundation Grants, 2261
John W. Anderson Foundation Grants, 2262
John W. Boynton Fund Grants, 2263
John W. Speas and Effie E. Speas Memorial Trust Grants, 2264
Joseph Drown Foundation Grants, 2268
Joseph H. & Florence A. Roblee Fndn Grants, 2269
Joseph Henry Edmondson Foundation Grants, 2270
Josephine G. Russell Trust Grants, 2271
Josephine Goodyear Foundation Grants, 2272
Josephine Schell Russell Charitable Trust Grants, 2274
Joseph P. Kennedy Jr. Foundation Grants, 2275
Josiah W. and Bessie H. Kline Foundation Grants, 2278
Judith and Jean Pape Adams Charitable Foundation Tulsa Area Grants, 2280
Judith Clark-Morrill Foundation Grants, 2281
Julius N. Frankel Foundation Grants, 2282
K21 Health Foundation Cancer Care Grants, 2284
K21 Health Foundation Grants, 2285
Kahuku Community Fund, 2286
Kansas Health Fndn Major Initiatives Grants, 2288
Kansas Health Foundation Recognition Grants, 2289
Kate B. Reynolds Trust Health Care Grants, 2290
Katharine Matthies Foundation Grants, 2291
Katherine Baxter Memorial Foundation Grants, 2292
Kathryne Beynon Foundation Grants, 2294
Katrine Menzing Deakins Trust Grants, 2295
Kelvin and Eleanor Smith Foundation Grants, 2297
Kenai Peninsula Foundation Grants, 2298
Kendrick Foundation Grants, 2299
Kent D. Steadley and Mary L. Steadley Memorial Trust Grants, 2301
Kessler Fnd Signature Employment Grants, 2302
Ketchikan Community Foundation Grants, 2303
Kettering Fund Grants, 2305
Kevin P. and Sydney B. Knight Family Foundation Grants, 2306
KeyBank Foundation Grants, 2307
Kimball International-Habig Foundation Health and Human Services Grants, 2308
Knox County Community Foundation Grants, 2313
Kodiak Community Foundation Grants, 2314
Komen Greater NYC Community Breast Health Grants, 2316
Komen Greater NYC Small Grants, 2317
Kosair Charities Grants, 2318
Kosciusko County Community Fndn Grants, 2319
Kosciusko County Community Foundation REMC Operation Round Up Grants, 2320
Kovler Family Foundation Grants, 2323
Kroger Company Donations, 2324
Kuntz Foundation Grants, 2326
L. W. Pierce Family Foundation Grants, 2328
LaGrange County Community Foundation Grants, 2329
Lake County Community Fund Grants, 2333
Land O'Lakes Foundation Mid-Atlantic Grants, 2341
Lands' End Corporate Giving Program, 2344
Laura B. Vogler Foundation Grants, 2345
Laura Moore Cunningham Foundation Grants, 2346
Lawrence Foundation Grants, 2347
Legler Benbough Foundation Grants, 2352
Lena Benas Memorial Fund Grants, 2353
Leo Goodwin Foundation Grants, 2354
Leon and Thea Koerner Foundation Grants, 2355
Leonard and Helen R. Stulman Charitable Foundation Grants, 2356
Leo Niessen Jr., Charitable Trust Grants, 2358
Lester E. Yeager Charitable Trust B Grants, 2359
Lewis H. Humphreys Charitable Trust Grants, 2361
Liberty Bank Foundation Grants, 2362
Libra Foundation Grants, 2363
Lil and Julie Rosenberg Foundation Grants, 2364
Lillian S. Wells Foundation Grants, 2365
Lincoln Financial Foundation Grants, 2367
Lisa and Douglas Goldman Fund Grants, 2369

Lloyd A. Fry Foundation Health Grants, 2373
Lockheed Martin Corp Foundation Grants, 2374
Long Island Community Foundation Grants, 2375
Lotus 88 Fnd for Women & Children Grants, 2376
Louetta M. Cowden Foundation Grants, 2377
Louie M. & Betty M. Phillips Fndn Grants, 2378
Louis and Elizabeth Nave Flarsheim Charitable Foundation Grants, 2379
Louis H. Aborn Foundation Grants, 2380
Lowe Foundation Grants, 2381
Lowell Berry Foundation Grants, 2382
Lubbock Area Foundation Grants, 2383
Lubrizol Foundation Grants, 2384
Lucile Horton Howe and Mitchell B. Howe Foundation Grants, 2385
Lucile Packard Foundation for Children's Health Grants, 2386
Lucy Downing Nisbet Charitable Fund Grants, 2387
Lucy Gooding Charitable Fndn Grants, 2388
Lumpkin Family Fnd Healthy People Grants, 2391
Lydia deForest Charitable Trust Grants, 2393
Lynde and Harry Bradley Foundation Grants, 2397
Lynn and Rovena Alexander Family Foundation Grants, 2399
M.B. and Edna Zale Foundation Grants, 2401
M. Bastian Family Foundation Grants, 2402
M.E. Raker Foundation Grants, 2404
M.J. Murdock Charitable Trust General Grants, 2405
Mabel A. Horne Trust Grants, 2406
Mabel F. Hoffman Charitable Trust Grants, 2407
Macquarie Bank Foundation Grants, 2410
Madison County Community Foundation - City of Anderson Quality of Life Grant, 2412
Madison County Community Foundation General Grants, 2413
Maine Community Foundation Charity Grants, 2415
Maine Community Foundation Penobscot Valley Health Association Grants, 2417
Mann T. Lowry Foundation Grants, 2419
Manuel D. & Rhoda Mayerson Fndn Grants, 2420
Marathon Petroleum Corporation Grants, 2422
March of Dimes Program Grants, 2426
Marcia and Otto Koehler Foundation Grants, 2427
Mardag Foundation Grants, 2428
Margaret T. Morris Foundation Grants, 2429
Marie C. and Joseph C. Wilson Foundation Rochester Small Grants, 2431
Marie H. Bechtel Charitable Remainder Uni-Trust Grants, 2432
Marin Community Foundation Improving Community Health Grants, 2433
Marion Gardner Jackson Charitable Trust Grants, 2434
Marion I. and Henry J. Knott Foundation Discretionary Grants, 2435
Marion I. and Henry J. Knott Foundation Standard Grants, 2436
Marjorie Moore Charitable Foundation Grants, 2438
Marshall County Community Foundation Grants, 2439
Mary Black Foundation Active Living Grants, 2440
Mary Black Fndn Community Health Grants, 2441
Mary Black Foundation Early Childhood Development Grants, 2442
Mary K. Chapman Foundation Grants, 2443
Mary L. Peyton Foundation Grants, 2445
Mary Owen Borden Foundation Grants, 2446
Mary S. and David C. Corbin Foundation Grants, 2447
Massage Therapy Foundation Community Service Grants, 2449
Maurice J. Masserini Charitable Trust Grants, 2456
Maxon Charitable Foundation Grants, 2458
May and Stanley Smith Charitable Trust Grants, 2459
McCarthy Family Foundation Grants, 2490
McConnell Foundation Grants, 2492
McCune Foundation Human Services Grants, 2493
McGraw-Hill Companies Community Grants, 2494
McInerny Foundation Grants, 2495
McKesson Foundation Grants, 2496
McLean Contributionship Grants, 2497
McLean Foundation Grants, 2498
Mead Johnson Nutritionals Charitable Giving, 2502

Mead Johnson Nutritionals Evansville-Area Organizations Grants, 2503
Mead Johnson Nutritionals Med Educ Grants, 2504
Mead Witter Foundation Grants, 2507
MedImmune Charitable Grants, 2509
Medtronic Foundation CommunityLink Health Grants, 2510
Mericos Foundation Grants, 2518
Meriden Foundation Grants, 2519
Merrick Foundation Grants, 2521
Merrick Foundation Grants, 2522
Mervin Bovaird Foundation Grants, 2523
Meta and George Rosenberg Foundation Grants, 2525
MetroWest Health Foundation Capital Grants for Health-Related Facilities, 2526
MetroWest Health Foundation Grants--Healthy Aging, 2527
MetroWest Health Foundation Grants to Reduce the Incidence of High Risk Behaviors Among Adolescents, 2528
Metzger-Price Fund Grants, 2530
Meyer and Stephanie Eglin Foundation Grants, 2531
Meyer Foundation Benevon Grants, 2532
Meyer Foundation Healthy Communities Grants, 2533
Meyer Fndn Management Assistance Grants, 2534
Meyer Memorial Trust Grassroots Grants, 2536
Meyer Memorial Trust Responsive Grants, 2537
Meyer Memorial Trust Special Grants, 2538
MGM Resorts Foundation Community Grants, 2541
MGN Family Foundation Grants, 2542
Miami County Community Foundation - Operation Round Up Grants, 2543
Michael Reese Health Trust Core Grants, 2544
Michael Reese Health Trust Responsive Grants, 2545
Mid-Iowa Health Foundation Community Response Grants, 2549
Middlesex Savings Charitable Foundation Basic Human Needs Grants, 2550
Middlesex Savings Charitable Foundation Capacity Building Grants, 2551
Military Ex-Prisoners of War Foundation Grants, 2554
Milken Family Foundation Grants, 2555
Miller Foundation Grants, 2556
Mimi and Peter Haas Fund Grants, 2557
Minneapolis Foundation Community Grants, 2558
MMS and Alliance Charitable Foundation Grants for Community Action and Care for the Medically Uninsured, 2587
MMS and Alliance Charitable Foundation International Health Studies Grants, 2588
Moline Foundation Community Grants, 2590
Montgomery County Community Fndn Grants, 2591
Morris & Gwendolyn Cafritz Fndn Grants, 2594
Morton K. and Jane Blaustein Foundation Health Grants, 2595
Ms. Fndn for Women Ending Violence Grants, 2596
Ms. Foundation for Women Health Grants, 2597
Mt. Sinai Health Care Foundation Health of the Jewish Community Grants, 2599
Mt. Sinai Health Care Foundation Health of the Urban Community Grants, 2600
Mt. Sinai Health Care Foundation Health Policy Grants, 2601
Natalie W. Furniss Charitable Trust Grants, 2626
Nationwide Insurance Foundation Grants, 2637
NEI Innovative Patient Outreach Programs And Ocular Screening Technologies To Improve Detection Of Diabetic Retinopathy Grants, 2665
Nelda C. and H.J. Lutcher Stark Fndn Grants, 2672
Nell J. Redfield Foundation Grants, 2673
Newton County Community Foundation Grants, 2681
NHLBI Bioengineering and Obesity Grants, 2689
NHLBI Lymphatics in Health & Disease in Digestive, Urinary, Cardio & Pulmonary Systems, 2705
Nicholas H. Noyes Jr. Memorial Fndn Grants, 2791
Nick Traina Foundation Grants, 2792
NIH Exploratory Innovations in Biomedical Computational Science & Technology Grants, 2878
NIH Innovations in Biomedical Computational Science and Technology Grants, 2883

PROGRAM TYPE INDEX — Community Development / 857

NIH Recovery Act Limited Competition: Building Sustainable Community-Linked Infrastructure to Enable Health Science Research, 2894
Nina Mason Pulliam Charitable Trust Grants, 2921
NLM Understanding and Promoting Health Literacy Research Grants, 2953
Noble County Community Foundation Grants, 2975
Norcliffe Foundation Grants, 2979
Nord Family Foundation Grants, 2980
Nordson Corporation Foundation Grants, 2981
Norfolk Southern Foundation Grants, 2982
Norman Foundation Grants, 2983
North Carolina Biotechnology Center Event Sponsorship Grants, 2984
North Carolina Biotech Center Meeting Grants, 2986
North Carolina Biotechnology Center Regional Development Grants, 2987
North Carolina Community Foundation Grants, 2990
North Carolina GlaxoSmithKline Fndn Grants, 2991
North Central Health Services Grants, 2992
North Central Health Services Grants, 2993
North Dakota Community Foundation Grants, 2994
Northern Chautauqua Community Foundation Community Grants, 2995
Northern New York Community Fndn Grants, 2996
Northwest Airlines KidCares Medical Travel Assistance, 2998
Northwestern Mutual Foundation Grants, 2999
Northwest Minnesota Foundation Women's Fund Grants, 3000
Norwin S. & Elizabeth N. Bean Fndn Grants, 3001
NSERC Brockhouse Canada Prize for Interdisciplinary Research in Science and Engineering Grant, 3003
Nuffield Foundation Africa Program Grants, 3026
Nuffield Foundation Children & Families Grants, 3027
Nuffield Foundation Small Grants, 3031
NWHF Community-Based Participatory Research Grants, 3032
NWHF Health Advocacy Small Grants, 3033
NWHF Kaiser Permanente Community Grants, 3034
NWHF Partners Investing in Nursing's Future, 3036
NWHF Physical Activity and Nutrition Grants, 3037
NYCT AIDS/HIV Grants, 3048
NYCT Children/Youth with Disabilities Grants, 3052
NYCT Girls and Young Women Grants, 3053
NYCT Grants for the Elderly, 3054
NYCT Health Care Services, Systems, and Policies Grants, 3055
NYCT Mental Health and Retardation Grants, 3056
NYCT Technical Assistance Grants, 3058
Oak Foundation Child Abuse Grants, 3059
OceanFirst Foundation Major Grants, 3062
Oceanside Charitable Foundation Grants, 3063
Office Depot Corporation Community Relations Grants, 3065
Ogden Codman Trust Grants, 3066
Ohio County Community Foundation Board of Directors Grants, 3067
Ohio County Community Foundation Grants, 3068
Oleonda Jameson Trust Grants, 3071
Olive Higgins Prouty Foundation Grants, 3072
Olympus Corporation of Americas Corporate Giving Grants, 3073
OneFamily Foundation Grants, 3075
OneStar Foundation AmeriCorps Grants, 3077
Oppenstein Brothers Foundation Grants, 3078
Orange County Community Foundation Grants, 3079
Ordean Foundation Grants, 3080
Oregon Community Foundation Better Nursing Home Care Grants, 3081
Oregon Community Fndn Community Grants, 3082
OSF Mental Health Initiative Grants, 3093
OSF Public Health Grants in Kyrgyzstan, 3094
OSF Tackling Addiction Grants in Baltimore, 3096
Otto Bremer Foundation Grants, 3099
Owen County Community Foundation Grants, 3100
PacifiCare Foundation Grants, 3102
Pacific Life Foundation Grants, 3103
Packard Foundation Children, Families, and Communities Grants, 3104

Packard Foundation Local Grants, 3105
Packard Foundation Population and Reproductive Health Grants, 3106
Pajaro Valley Community Health Health Trust Insurance/Coverage & Education on Using the System Grants, 3108
Pajaro Valley Community Health Trust Diabetes and Contributing Factors Grants, 3109
Parke County Community Foundation Grants, 3115
Parkersburg Area Community Foundation Action Grants, 3116
Paso del Norte Health Foundation Grants, 3119
Patrick and Anna M. Cudahy Fund Grants, 3120
Patron Saints Foundation Grants, 3121
Paul Ogle Foundation Grants, 3124
Paul Rapoport Foundation Grants, 3125
Peacock Foundation Grants, 3133
PepsiCo Foundation Grants, 3137
Percy B. Ferebee Endowment Grants, 3138
Perkins Charitable Foundation Grants, 3140
Perpetual Trust for Charitable Giving Grants, 3141
Perry County Community Foundation Grants, 3142
Pet Care Trust Sue Busch Memorial Award, 3143
Peter and Elizabeth C. Tower Foundation Annual Intellectual Disabilities Grants, 3144
Peter and Elizabeth C. Tower Foundation Annual Mental Health Grants, 3145
Peter and Elizabeth C. Tower Foundation Learning Disability Grants, 3146
Peter and Elizabeth C. Tower Foundation Mental Health Reference and Resource Materials Mini-Grants, 3147
Peter and Elizabeth C. Tower Foundation Organizational Scholarships, 3148
Peter and Elizabeth C. Tower Foundation Phase II Technology Initiative Grants, 3149
Peter and Elizabeth C. Tower Foundation Phase I Technology Initiative Grants, 3150
Peter and Elizabeth C. Tower Foundation Social and Emotional Preschool Curriculum Grants, 3151
Peter and Elizabeth C. Tower Foundation Substance Abuse Grants, 3152
Peter Kiewit Foundation General Grants, 3154
Peter Kiewit Foundation Small Grants, 3155
Petersburg Community Foundation Grants, 3156
Pew Charitable Trusts Children & Youth Grants, 3158
Peyton Anderson Foundation Grants, 3160
Pfizer Healthcare Charitable Contributions, 3180
Pfizer Special Events Grants, 3186
Phelps County Community Foundation Grants, 3188
Philadelphia Foundation Organizational Effectiveness Grants, 3189
Philip L. Graham Fund Health and Human Services Grants, 3190
Phoenix Suns Charities Grants, 3201
Piedmont Health Foundation Grants, 3219
Piedmont Natural Gas Corporate and Charitable Contributions, 3220
Piedmont Natural Gas Foundation Health and Human Services Grants, 3221
Pike County Community Foundation Grants, 3222
Pinellas County Grants, 3223
Pinkerton Foundation Grants, 3224
Pinnacle Foundation Grants, 3225
Piper Jaffray Fndn Communities Giving Grants, 3227
Piper Trust Healthcare & Med Research Grants, 3228
Piper Trust Older Adults Grants, 3229
Pittsburgh Foundation Community Fund Grants, 3230
Playboy Foundation Grants, 3237
Plough Foundation Grants, 3238
PMI Foundation Grants, 3239
PNC Foundation Community Services Grants, 3240
Pohlad Family Foundation, 3241
Pokagon Fund Grants, 3242
Polk Bros. Foundation Grants, 3243
Pollock Foundation Grants, 3244
Porter County Health and Wellness Grant, 3246
Porter County Women's Grant, 3248
Portland Fndn Women's Giving Circle Grant, 3249
Portland General Electric Foundation Grants, 3250

Posey Community Fndn Women's Fund Grants, 3251
Posey County Community Foundation Grants, 3252
Pott Foundation Grants, 3253
Powell Foundation Grants, 3255
PPCF Community Grants, 3256
PPG Industries Foundation Grants, 3257
Premera Blue Cross Grants, 3258
Presbyterian Health Foundation Bridge, Seed and Equipment Grants, 3259
Presbyterian Patient Assistance Program, 3260
Price Chopper's Golub Foundation Grants, 3261
Price Family Charitable Fund Grants, 3262
Priddy Foundation Organizational Development Grants, 3265
Pride Foundation Grants, 3268
Prince Charitable Trusts Chicago Grants, 3270
Prince Charitable Trusts DC Grants, 3271
Princeton Area Community Foundation Greater Mercer Grants, 3272
Principal Financial Group Foundation Grants, 3273
Procter and Gamble Fund Grants, 3274
Puerto Rico Community Foundation Grants, 3278
Pulaski County Community Foundation Grants, 3279
Pulido Walker Foundation, 3282
Putnam County Community Foundation Grants, 3283
Quaker Chemical Foundation Grants, 3286
Quality Health Foundation Grants, 3288
Quantum Corporation Snap Server Grants, 3289
Quantum Foundation Grants, 3290
Questar Corporate Contributions Grants, 3291
R.C. Baker Foundation Grants, 3292
R.S. Gernon Trust Grants, 3293
Rainbow Endowment Grants, 3300
Rainbow Fund Grants, 3301
Ralphs Food 4 Less Foundation Grants, 3305
Raskob Foundation for Catholic Activities Grants, 3306
Rasmuson Foundation Tier One Grants, 3307
Rasmuson Foundation Tier Two Grants, 3308
Rathmann Family Foundation Grants, 3309
Raymond John Wean Foundation Grants, 3310
Rayonier Foundation Grants, 3311
RCF General Community Grants, 3312
RCF Individual Assistance Grants, 3313
RCF Summertime Kids Grants, 3314
RCF The Women's Fund Grants, 3315
Regence Fndn Access to Health Care Grants, 3346
Regence Foundation Health Care Community Awareness and Engagement Grants, 3347
Regence Fndn Health Care Connections Grants, 3348
Regence Fndn Improving End-of-Life Grants, 3349
Regence Fndn Tools & Technology Grants, 3350
Reinberger Foundation Grants, 3357
Retirement Research Foundation General Program Grants, 3359
Reynolds & Reynolds Associate Fndn Grants, 3360
Rhode Island Foundation Grants, 3362
Rice Foundation Grants, 3363
Richard and Helen DeVos Foundation Grants, 3364
Richard & Susan Smith Family Fndn Grants, 3365
Richard D. Bass Foundation Grants, 3366
Richard Davoud Donchian Foundation Grants, 3367
Richard M. Fairbanks Foundation Grants, 3370
Richland County Bank Grants, 3371
Richmond Eye and Ear Fund Grants, 3372
Riley Foundation Grants, 3374
Ripley County Community Foundation Grants, 3376
Robbins Charitable Foundation Grants, 3378
Robert and Joan Dircks Foundation Grants, 3379
Robert B McMillen Foundation Grants, 3381
Robert F. Stoico/FIRSTFED Charitable Foundation Grants, 3382
Robert & Clara Milton Senior Housing Grants, 3384
Robert R. McCormick Trib Community Grants, 3385
Robert R. McCormick Tribune Veterans Initiative Grants, 3386
Robert R. Meyer Foundation Grants, 3387
Robert Sterling Clark Foundation Reproductive Rights and Health Grants, 3388
Robert W. Woodruff Foundation Grants, 3389
Robins Foundation Grants, 3390

Community Development

Rochester Area Community Foundation Grants, 3391
Rochester Area Foundation Grants, 3392
Rockefeller Brothers Peace & Security Grants, 3394
Rockefeller Foundation Grants, 3396
Rockwell Fund, Inc. Grants, 3398
Rockwell International Corporate Trust Grants Program, 3399
Rohm and Haas Company Grants, 3403
Ronald McDonald House Charities Grants, 3405
Rose Community Foundation Aging Grants, 3410
Rose Community Foundation Health Grants, 3411
Rose Hills Foundation Grants, 3413
Roy & Christine Sturgis Charitable Grants, 3415
RRF General Program Grants, 3417
Rush County Community Foundation Grants, 3424
Ruth Anderson Foundation Grants, 3425
Ruth and Vernon Taylor Foundation Grants, 3426
Ruth Eleanor Bamberger and John Ernest Bamberger Memorial Foundation Grants, 3427
Ruth H. & Warren A. Ellsworth Fndn Grants, 3429
RWJF Changes in Health Care Financing and Organization Grants, 3430
RWJF Community Health Leaders Awards, 3432
RWJF Healthy Eating Research Grants, 3437
RWJF New Jersey Health Initiatives Grants, 3440
RWJF Pioneer Portfolio Grants, 3442
RWJF Vulnerable Populations Portfolio Grants, 3444
S.D. Bechtel, Jr. Fndn/Stephen Bechtel Preventive Healthcare and Selected Research Grants, 3446
S.H. Cowell Foundation Grants, 3447
S. Mark Taper Foundation Grants, 3449
Saginaw Community Foundation Senior Citizen Enrichment Fund, 3450
Saigh Foundation Grants, 3451
Sain-Orr and Royak-DeForest Steadman Foundation Grants, 3452
Saint Louis Rams Fndn Community Donations, 3453
Saint Luke's Foundation Grants, 3454
Saint Luke's Health Initiatives Grants, 3455
Salem Foundation Charitable Trust Grants, 3456
Salmon Foundation Grants, 3458
Salt River Health & Human Services Grants, 3459
SAMHSA Child Mental Health Initiative (CMHI) Grants, 3461
SAMHSA Drug Free Communities Support Program Grants, 3463
SAMHSA Strategic Prevention Framework State Incentive Grants, 3464
Samuel N. and Mary Castle Foundation Grants, 3467
Samuel S. Johnson Foundation Grants, 3468
San Antonio Area Foundation Grants, 3469
San Diego Foundation for Change Grants, 3470
San Diego Foundation Health & Human Services Grants, 3471
San Diego Foundation Paradise Valley Hospital Community Fund Grants, 3472
San Diego HIV Funding Collaborative Grants, 3473
San Diego Women's Foundation Grants, 3474
Sands Foundation Grants, 3476
Sandy Hill Foundation Grants, 3477
San Francisco Fndn Community Health Grants, 3478
San Juan Island Community Foundation Grants, 3480
Santa Barbara Foundation Strategy Grants - Core Support, 3481
Santa Fe Community Foundation Seasonal Grants-Fall Cycle, 3482
Sarasota Memorial Healthcare Fndn Grants, 3484
Sarkeys Foundation Grants, 3485
Sartain Lanier Family Foundation Grants, 3486
Sasco Foundation Grants, 3487
Schering-Plough Foundation Community Initiatives Grants, 3491
Schering-Plough Foundation Health Grants, 3492
Schlessman Family Foundation Grants, 3493
Schramm Foundation Grants, 3495
Schurz Communications Foundation Grants, 3496
Scott B. & Annie P. Appleby Charitable Grants, 3499
Scott County Community Foundation Grants, 3500
Seabury Foundation Grants, 3501
Seattle Foundation Benjamin N. Phillips Grants, 3504
Seattle Foundation Health and Wellness Grants, 3507
Seattle Foundation Medical Funds Grants, 3508
Seaver Institute Grants, 3509
Self Foundation Grants, 3510
Seneca Foods Foundation Grants, 3511
Sensient Technologies Foundation Grants, 3513
Seward Community Foundation Grants, 3514
Seward Community Foundation Mini-Grants, 3515
Shaw's Supermarkets Donations, 3542
Shell Deer Park Grants, 3543
Shell Oil Company Foundation Community Development Grants, 3544
Sheltering Arms Fund Grants, 3545
Shield-Ayres Foundation Grants, 3546
Shopko Fndn Community Charitable Grants, 3547
Shopko Foundation Green Bay Area Community Grants, 3548
Sidgmore Family Foundation Grants, 3551
Sidney Stern Memorial Trust Grants, 3552
Sid W. Richardson Foundation Grants, 3553
Siebert Lutheran Foundation Grants, 3554
Sierra Health Foundation Responsive Grants, 3555
Simmons Foundation Grants, 3563
Simple Advise Education Center Grants, 3565
Simpson Lumber Charitable Contributions, 3566
Singing for Change Foundation Grants, 3567
Sioux Falls Area Community Foundation Community Fund Grants (Unrestricted), 3568
Sioux Falls Area Community Foundation Field-of-Interest and Donor-Advised Grants, 3569
Sioux Falls Area Community Foundation Spot Grants (Unrestricted), 3570
Siragusa Foundation Health Services & Medical Research Grants, 3571
Sir Dorabji Tata Trust Grants for NGOs or Voluntary Organizations, 3572
Sisters of Charity Foundation of Cleveland Reducing Health and Educational Disparities in the Central Neighborhood Grants, 3574
Sisters of Mercy of North Carolina Fndn Grants, 3575
Sisters of St. Joseph Healthcare Fndn Grants, 3576
Skillman Fndn Good Neighborhoods Grants, 3577
Skoll Fndn Awards for Social Entrepreneurship, 3579
SOCFOC Catholic Ministries Grants, 3596
Society for the Arts in Healthcare Grants, 3602
Society for the Arts in Healthcare Environmental Arts Research Grant, 3603
Solo Cup Foundation Grants, 3620
Sonoco Foundation Grants, 3621
Sonora Area Foundation Competitive Grants, 3622
Sony Corporation of America Grants, 3623
Sophia Romero Trust Grants, 3624
Southbury Community Trust Fund, 3625
South Madison Community Foundation Grants, 3626
Southwest Florida Community Foundation Competitive Grants, 3627
Southwest Gas Corporation Foundation Grants, 3628
Spencer County Community Foundation Grants, 3632
Springs Close Foundation Grants, 3633
Square D Foundation Grants, 3634
St. Joseph Community Health Foundation Improving Healthcare Access Grants, 3640
Stackpole-Hall Foundation Grants, 3643
Stark Community Fndn Women's Grants, 3645
Starke County Community Foundation Grants, 3646
Steele-Reese Foundation Grants, 3650
Stella and Charles Guttman Foundation Grants, 3651
Sterling-Turner Charitable Foundation Grants, 3652
Sterling and Shelli Gardner Foundation Grants, 3653
Steuben County Community Foundation Grants, 3654
Stewart Huston Charitable Trust Grants, 3655
Stocker Foundation Grants, 3656
Straits Area Community Foundation Health Care Grants, 3657
Strake Foundation Grants, 3658
Stranahan Foundation Grants, 3659
Strowd Roses Grants, 3664
Subaru of Indiana Automotive Foundation Grants, 3672
Summit Foundation Grants, 3674
Sunderland Foundation Grants, 3675
Sunflower Foundation Capacity Building Grants, 3677
Sunflower Foundation Walking Trails Grants, 3678
Sunoco Foundation Grants, 3679
SunTrust Bank Trusteed Foundations Florence C. and Harry L. English Memorial Fund Grants, 3680
SunTrust Bank Trusteed Foundations Greene-Sawtell Grants, 3681
SunTrust Bank Trusteed Foundations Harriet McDaniel Marshall Tust Grants, 3682
SunTrust Bank Trusteed Foundations Nell Warren Elkin and William Simpson Elkin Grants, 3683
SunTrust Bank Trusteed Foundations Thomas Guy Woolford Charitable Trust Grants, 3684
SunTrust Bank Trusteed Foundations Walter H. and Marjory M. Rich Memorial Fund Grants, 3685
Susan Mott Webb Charitable Trust Grants, 3694
Susan Vaughan Foundation Grants, 3695
Swindells Charitable Foundation, 3697
T.L.L. Temple Foundation Grants, 3700
T. Spencer Shore Foundation Grants, 3701
Taubman Endowment for the Arts Fndn Grants, 3702
Tauck Family Foundation Grants, 3703
Taylor S. Abernathy and Patti Harding Abernathy Charitable Trust Grants, 3704
TE Foundation Grants, 3705
Tension Envelope Foundation Grants, 3707
Texas Commission on the Arts Arts Respond Project Grants, 3708
Texas Instruments Corporation Health and Human Services Grants, 3709
Texas Instruments Foundation Community Services Grants, 3710
Textron Corporate Contributions Grants, 3712
Thelma Braun & Bocklett Family Fndn Grants, 3713
Thelma Doelger Charitable Trust Grants, 3714
Theodore Edson Parker Foundation Grants, 3715
Thomas Austin Finch, Sr. Foundation Grants, 3718
Thomas C. Ackerman Foundation Grants, 3719
Thomas J. Atkins Memorial Trust Fund Grants, 3721
Thomas J. Long Foundation Community Grants, 3722
Thomas Jefferson Rosenberg Foundation Grants, 3723
Thomas Sill Foundation Grants, 3724
Thomas Thompson Trust Grants, 3725
Thomas W. Bradley Foundation Grants, 3726
Thompson Charitable Foundation Grants, 3727
Thorman Boyle Foundation Grants, 3730
Tifa Foundation Grants, 3733
Tipton County Foundation Grants, 3734
TJX Foundation Grants, 3735
Todd Brock Family Foundation Grants, 3736
Topeka Community Foundation Kansas Blood Services Scholarships, 3738
Toyota Motor Engineering & Manufacturing North America Grants, 3742
Toyota Motor Manufacturing of Alabama Grants, 3743
Toyota Motor Manufacturing of Indiana Grants, 3744
Toyota Motor Manufacturing of Kentucky Grants, 3745
Toyota Motor Manuf of Mississippi Grants, 3746
Toyota Motor Manufacturing of Texas Grants, 3747
Toyota Motor Manuf of West Virginia Grants, 3748
Toyota Motor N America of New York Grants, 3749
Toyota Motor Sales, USA Grants, 3750
Tri-State Community Twenty-first Century Endowment Fund Grants, 3756
Triangle Community Foundation Donor-Advised Grants, 3757
Triangle Community Foundation Shaver-Hitchings Scholarship, 3758
Trull Foundation Grants, 3761
Tull Charitable Foundation Grants, 3762
Turner Foundation Grants, 3763
U.S. Cellular Corporation Grants, 3766
U.S. Department of Education 21st Century Community Learning Centers, 3767
U.S. Department of Education Rehabilitation Training - Rehabilitation Continuing Education - Institute on Rehabilitation Issues, 3771
U.S. Dept of Education Special Ed--National Activities--Parent Information Centers, 3772

PROGRAM TYPE INDEX

U.S. Department of Education Vocational Rehabilitation Services Projects for American Indians with Disabilities Grants, 3774
Union Bank, N.A. Corporate Sponsorships and Donations, 3787
Union Bank, N.A. Foundation Grants, 3788
Union Benevolent Association Grants, 3789
Union County Community Foundation Grants, 3790
Union Labor Health Fndn Community Grants, 3792
Union Pacific Foundation Health and Human Services Grants, 3794
United Healthcare Comm Grants in Michigan, 3796
United Hospital Fund of New York Health Care Improvement Grants, 3797
United Methodist Committee on Relief Global Health Grants, 3799
United Methodist Health Ministry Fund Grants, 3801
United Technologies Corporation Grants, 3803
Unity Foundation Of LaPorte County Grants, 3804
USAID Accelerating Progress Against Tuberculosis in Kenya Grants, 3805
USAID Comprehensive District-Based Support for Better HIV/TB Patient Outcomes Grants, 3807
USAID Devel Innovation Accelerator Grants, 3808
USAID Ebola Response, Recovery and Resilience in West Africa Grants, 3809
USAID Family Health Plus Project Grants, 3810
USAID Global Health Development Innovation Accelerator Broad Agency Grants, 3813
USAID Higher Education Partnerships for Innovation and Impact (HEPII) Grants, 3815
USAID HIV Prevention with Key Populations - Mali Grants, 3816
USAID Integration of Care and Support within the Health System to Support Better Patient Outcomes Grants, 3817
USAID Land Use Change and Disease Emergence Grants, 3818
USAID Policy, Advocacy, and Communication Enhanced for Population and Reproductive Health Grants, 3820
USAID Research and Innovation for Health Supply Chain Systems & Commodity Security Grants, 3821
USAID Support for International Family Planning Organizations Grants, 3823
USAID Systems Strengthening for Better HIV/TB Patient Outcomes Grants, 3824
US Airways Community Contributions, 3825
US Airways Community Foundation Grants, 3826
USCM HIV/AIDS Prevention Grants, 3827
USDA Delta Health Care Services Grants, 3829
Vancouver Foundation Grants and Community Initiatives Program, 3849
Vancouver Sun Children's Fund Grants, 3851
Vanderburgh Community Foundation Grants, 3852
Vanderburgh Community Fndn Women's Fund, 3853
Vectren Foundation Grants, 3857
Verizon Foundation Health Care and Accessibility Grants, 3858
Verizon Foundation Maryland Grants, 3859
Verizon Foundation New York Grants, 3860
Verizon Foundation Northeast Region Grants, 3861
Verizon Foundation Pennsylvania Grants, 3862
Verizon Foundation Vermont Grants, 3863
Verizon Foundation Virginia Grants, 3864
Vermont Community Foundation Grants, 3865
Vernon K. Krieble Foundation Grants, 3866
Victor E. Speas Foundation Grants, 3868
Vigneron Memorial Fund Grants, 3873
Virginia W. Kettering Foundation Grants, 3878
Visiting Nurse Foundation Grants, 3879
Volkswagen of America Corporate Contributions, 3880
W.C. Griffith Foundation Grants, 3881
W.K. Kellogg Foundation Healthy Kids Grants, 3886
W.K. Kellogg Foundation Secure Families Grants, 3887
W.M. Keck Fndn So California Grants, 3890
W.P. and Bulah Luse Foundation Grants, 3892
Wabash Valley Community Foundation Grants, 3893
Waitt Family Foundation Grants, 3894
Walker Area Community Foundation Grants, 3895
Wallace Alexander Gerbode Foundation Grants, 3896
Walmart Foundation Community Giving Grants, 3897
Walmart Foundation National Giving Grants, 3898
Walmart Foundation Northwest Arkansas Giving, 3899
Walmart Foundation State Giving Grants, 3900
Walter L. Gross III Family Foundation Grants, 3901
Warren County Community Foundation Grants, 3902
Warren County Community Fndn Mini-Grants, 3903
Warrick County Community Foundation Grants, 3904
Warrick County Community Foundation Women's Fund, 3905
Washington County Community Fndn Grants, 3906
Washington County Community Foundation Youth Grants, 3907
Washington Gas Charitable Giving, 3908
Wayne County Foundation Grants, 3909
Weaver Popcorn Foundation Grants, 3910
Weingart Foundation Grants, 3913
Welborn Baptist Fndn General Op Grants, 3914
Welborn Baptist Foundation Improvements to Community Health Status Grants, 3916
Wells County Foundation Grants, 3920
Western Indiana Community Foundation Grants, 3921
Western New York Foundation Grants, 3922
Weyerhaeuser Company Foundation Grants, 3925
Weyerhaeuser Family Foundation Health Grants, 3926
WHAS Crusade for Children Grants, 3927
White County Community Foundation - Women Giving Together Grants, 3933
White County Community Foundation Grants, 3934
Whitley County Community Foundation Grants, 3936
Whitney Foundation Grants, 3938
WHO Foundation General Grants, 3939
WHO Foundation Volunteer Service Grants, 3940
Willard & Pat Walker Charitable Fndn Grants, 3943
Willary Foundation Grants, 3944
William A. Badger Foundation Grants, 3945
William B. Dietrich Foundation Grants, 3946
William B. Stokely Jr. Foundation Grants, 3947
William Blair and Company Foundation Grants, 3948
William D. Laurie, Jr. Charitable Fndn Grants, 3949
William G. & Helen C. Hoffman Fndn Grants, 3950
William G. Gilmore Foundation Grants, 3951
William J. & Tina Rosenberg Fndn Grants, 3954
William J. Brace Charitable Trust, 3955
William Ray & Ruth E. Collins Fndn Grants, 3956
Williams Companies Foundation Homegrown Giving Grants, 3957
Willis C. Helm Charitable Trust Grants, 3959
Wilson-Wood Foundation Grants, 3960
Windham Foundation Grants, 3961
Winston-Salem Foundation Competitive Grants, 3965
Winston-Salem Fndn Elkin/Tri-County Grants, 3966
Winston-Salem Fndn Stokes County Grants, 3967
Wolfe Associates Grants, 3969
Wood-Claeyssens Foundation Grants, 3971
Yawkey Foundation Grants, 3980
Z. Smith Reynolds Foundation Small Grants, 3982
ZYTL Foundation Grants, 3985

Consulting/Visiting Personnel

AAAS Science and Technology Policy Fellowships: Global Health and Development, 41
AAP Leonard P. Rome Community Access to Child Health (CATCH) Visiting Professorships, 186
Abbott Fund Access to Health Care Grants, 199
Abbott Fund Global AIDS Care Grants, 202
Able To Serve Grants, 212
ALFJ Astraea U.S. and Int'l Movement Fund, 427
American-Scandinavian Foundation Visiting Lectureship Grants, 502
American College of Surgeons International Guest Scholarships, 525
APSAA Small Beginning Scholar Visiting and Consulting Grants, 643
ASM Int'l Professorship for Latin America, 736
ASM Int'l Professorships for Asia and Africa, 737
Assurant Health Foundation Grants, 789
Avon Products Foundation Grants, 814
BCBSM Fndn Proposal Development Awards, 857

Consulting/Visiting Personnel / 859

Beckley Area Foundation Grants, 867
Bill Hannon Foundation Grants, 895
Bodenwein Public Benevolent Foundation Grants, 928
CJ Foundation for SIDS Program Services Grants, 1163
Clarence E. Heller Charitable Foundation Grants, 1166
Claude Pepper Foundation Grants, 1177
Claude Worthington Benedum Fndn Grants, 1178
Columbus Foundation Traditional Grants, 1223
Community Fndn for Greater Buffalo Grants, 1247
Community Fndn for Monterey County Grants, 1253
Community Fndn for Muskegon County Grants, 1254
Community Fndn of East Central Illinois Grants, 1268
Community Foundation of Greater Fort Wayne - Community Endowment and Clarke Endowment Grants, 1274
ConAgra Foods Nourish our Community Grants, 1322
Connecticut Community Foundation Grants, 1324
Conquer Cancer Foundation of ASCO International Development and Education Award in Palliative Care, 1332
Conquer Cancer Foundation of ASCO International Development and Education Awards, 1333
Cystic Fibrosis Canada Special Travel Grants For Fellows and Students, 1389
Cystic Fibrosis Canada Transplant Center Incentive Grants, 1392
Cystic Fibrosis Canada Visiting Allied Health Professional Awards, 1394
Cystic Fibrosis Canada Visiting Clinician Awards, 1395
D. W. McMillan Foundation Grants, 1407
Dade Community Foundation Community AIDS Partnership Grants, 1409
Delaware Community Foundation-Youth Philanthropy Board for Kent County, 1448
Drs. Bruce and Lee Foundation Grants, 1513
Eddie C. & Sylvia Brown Family Fndn Grants, 1548
Edina Realty Foundation Grants, 1550
Eisner Foundation Grants, 1564
Emily Davie and Joseph S. Kornfeld Foundation Grants, 1591
FAR Fund Grants, 1631
Florida BRAVE Fund of Dade Community Foundation, 1678
Foellinger Foundation Grants, 1684
Frances L. and Edwin L. Cummings Memorial Fund Grants, 1717
Fremont Area Community Foundation Amazing X Grants, 1744
Fremont Area Community Fndn General Grants, 1746
Grand Rapids Community Foundation Ionia County Youth Fund Grants, 1871
Gulf Coast Foundation of Community Operating Grants, 1916
Gulf Coast Foundation of Community Program Grants, 1917
Hawaii Community Foundation Health Education and Research Grants, 1976
Helen Steiner Rice Foundation Grants, 1998
Hogg Foundation for Mental Health Grants, 2039
HRSA Nurse Education, Practice, Quality and Retention (NEPQR) Grants, 2068
James & Abigail Campbell Family Fndn Grants, 2185
James R. Thorpe Foundation Grants, 2199
Jennings County Community Foundation Grants, 2220
Jessie Ball Dupont Fund Grants, 2226
John G. Martin Foundation Grants, 2237
Kinsman Foundation Grants, 2309
Leo Niessen Jr., Charitable Trust Grants, 2358
Meyer Fndn Management Assistance Grants, 2534
MLA Janet Doe Lectureship Award, 2567
Montgomery County Community Fndn Grants, 2591
NCI Technologies and Software to Support Integrative Cancer Biology Research (SBIR) Grants, 2656
Noble County Community Foundation Grants, 2975
North Carolina Biotechnology Center Grantsmanship Training Grants, 2985
North Carolina Community Foundation Grants, 2990
OneFamily Foundation Grants, 3075
Ordean Foundation Grants, 3080
PAHO-ASM Int'l Professorship Latin America, 3107

860 / Consulting/Visiting Personnel

Patrick and Anna M. Cudahy Fund Grants, 3120
Pfizer/AAFP Foundation Visiting Professorship Program in Family Medicine, 3162
PhRMA Foundation Health Outcomes Sabbatical Fellowships, 3205
Pike County Community Foundation Grants, 3222
Pinellas County Grants, 3223
Pittsburgh Foundation Community Fund Grants, 3230
Prudential Foundation Education Grants, 3276
Puerto Rico Community Foundation Grants, 3278
RCPSC Harry S. Morton Lectureship in Surgery, 3323
RCPSC Harry S. Morton Traveling Fellowship in Surgery, 3324
RCPSC Janes Visiting Professorship in Surgery, 3328
RCPSC John G. Wade Visiting Professorship in Patient Safety & Simulation-Based Med Education, 3329
RCPSC KJR Wightman Visiting Professorship in Medicine, 3331
RCPSC McLaughlin-Gallie Visiting Professor, 3333
RCPSC Professional Development Grants, 3338
Rhode Island Foundation Grants, 3362
Rochester Area Community Foundation Grants, 3391
Ronald McDonald House Charities Grants, 3405
Samuel S. Johnson Foundation Grants, 3468
Seabury Foundation Grants, 3501
Shell Oil Company Foundation Community Development Grants, 3544
Society for the Arts in Healthcare Grants, 3602
St. Joseph Community Health Foundation Schneider Fellowships, 3642
Streilein Foundation for Ocular Immunology Visiting Professorships, 3660
Texas Commission on the Arts Arts Respond Project Grants, 3708
US CRDF Leishmaniasis: Collaborative Research Opportunities in N Africa & Middle East, 3828

Cultural Outreach
3M Company Fndn Community Giving Grants, 6
A. Gary Anderson Family Foundation Grants, 16
A.O. Smith Foundation Community Grants, 17
Aaron Foundation Grants, 193
Abbot and Dorothy H. Stevens Foundation Grants, 197
Abbott Fund Community Grants, 201
Abby's Legendary Pizza Foundation Grants, 204
Abell-Hanger Foundation Grants, 207
Abington Foundation Grants, 211
Achelis Foundation Grants, 234
Ackerman Foundation Grants, 236
ACL Diversity Community of Practice Grants, 240
Adams Foundation Grants, 301
Adaptec Foundation Grants, 304
Adelaide Breed Bayrd Foundation Grants, 306
Adolph Coors Foundation Grants, 311
Advanced Micro Devices Comm Affairs Grants, 313
AEC Trust Grants, 315
AFG Industries Grants, 340
A Friends' Foundation Trust Grants, 341
Agnes Gund Foundation Grants, 345
Agway Foundation Grants, 348
Ahmanson Foundation Grants, 355
Ahn Family Foundation Grants, 356
Air Products and Chemicals Grants, 399
Albert and Margaret Alkek Foundation Grants, 405
Alberto Culver Corporate Contributions Grants, 408
Albert Pick Jr. Fund Grants, 409
Albertson's Charitable Giving Grants, 410
Albuquerque Community Foundation Grants, 412
Alcatel-Lucent Technologies Foundation Grants, 413
Alexander & Baldwin Fnd Mainland Grants, 417
Alexander and Baldwin Foundation Hawaiian and Pacific Island Grants, 418
ALFJ Astraea U.S. and Int'l Movement Fund, 427
Allen P. & Josephine B. Green Fndn Grants, 443
Allstate Corp Hometown Commitment Grants, 447
Alpha Natural Resources Corporate Giving, 450
Altman Foundation Health Care Grants, 457
Alvin and Fanny Blaustein Thalheimer Foundation Baltimore Communal Grants, 459
Amador Community Foundation Grants, 481

Amerigroup Foundation Grants, 565
Amgen Foundation Grants, 572
Amica Insurance Company Community Grants, 575
Anheuser-Busch Foundation Grants, 591
Annenberg Foundation Grants, 596
Annie Sinclair Knudsen Memorial Fund/Kaua'i Community Grants Program, 597
Ann Peppers Foundation Grants, 599
Anschutz Family Foundation Grants, 602
AON Foundation Grants, 623
Aratani Foundation Grants, 649
Arcadia Foundation Grants, 650
Archer Daniels Midland Foundation Grants, 651
Argyros Foundation Grants Program, 653
Arie and Ida Crown Memorial Grants, 654
Arizona Cardinals Grants, 655
Arizona Public Service Corporate Giving Program Grants, 658
Arkansas Community Foundation Grants, 660
ARS Foundation Grants, 667
Arthur Ashley Williams Foundation Grants, 671
Arthur F. & Arnold M. Frankel Fndn Grants, 672
ArvinMeritor Grants, 675
ASHA Multicultural Activities Projects Grants, 697
Ashland Corporate Contributions Grants, 713
Assisi Foundation of Memphis General Grants, 787
Assurant Health Foundation Grants, 789
AT&T Foundation Civic and Community Service Program Grants, 793
Atherton Family Foundation Grants, 794
Athwin Foundation Grants, 795
Atlanta Foundation Grants, 796
Auburn Foundation Grants, 798
Audrey & Sydney Irmas Foundation Grants, 799
Aurora Foundation Grants, 802
Autauga Area Community Foundation Grants, 808
Autzen Foundation Grants, 810
Avista Foundation Economic and Cultural Vitality Grants, 812
Avon Products Foundation Grants, 814
Axe-Houghton Foundation Grants, 815
Babcock Charitable Trust Grants, 816
Bacon Family Foundation Grants, 817
Bailey Foundation Grants, 818
Ball Brothers Foundation General Grants, 820
Banfi Vintners Foundation Grants, 824
Bank of America Fndn Matching Gifts, 825
Bank of America Fndn Volunteer Grants, 826
Barberton Community Foundation Grants, 829
Barker Welfare Foundation Grants, 831
Barnes Group Foundation Grants, 832
Barra Foundation Community Fund Grants, 833
Barra Foundation Project Grants, 834
Barr Fund Grants, 835
Batchelor Foundation Grants, 837
Baton Rouge Area Foundation Grants, 838
Battle Creek Community Foundation Grants, 839
Batts Foundation Grants, 840
Baxter International Corporate Giving Grants, 841
Baxter International Foundation Grants, 843
BCBSNC Foundation Grants, 860
Bechtel Group Foundation Building Positive Community Relationships Grants, 866
Beckley Area Foundation Grants, 867
Beckman Coulter Foundation Grants, 868
Belk Foundation Grants, 872
Benton County Foundation Grants, 880
Berks County Community Foundation Grants, 881
Berrien Community Foundation Grants, 884
Besser Foundation Grants, 887
Bildner Family Foundation Grants, 891
Bill and Melinda Gates Foundation Emergency Response Grants, 892
Bill and Melinda Gates Foundation Policy and Advocacy Grants, 893
Blackford County Community Foundation Grants, 906
Blanche and Irving Laurie Foundation Grants, 910
Blanche and Julian Robertson Family Foundation Grants, 911

PROGRAM TYPE INDEX

Blue Cross Blue Shield of Minnesota Fndn Healthy Equity: Public Libraries for Health Grants, 918
Blue Grass Community Fndn Harrison Grants, 919
Blue Grass Community Foundation Hudson-Ellis Fund Grants, 920
Blue Mountain Community Foundation Grants, 921
Blum-Kovler Foundation Grants, 924
Blumenthal Foundation Grants, 925
Bob & Delores Hope Foundation Grants, 927
Bodenwein Public Benevolent Foundation Grants, 928
Bodman Foundation Grants, 929
Boeing Company Contributions Grants, 930
Boettcher Foundation Grants, 931
Bonfils-Stanton Foundation Grants, 932
Booth-Bricker Fund Grants, 933
Boston Foundation Grants, 936
Boston Globe Foundation Grants, 937
Boston Jewish Community Women's Fund Grants, 938
Bradley-Turner Foundation Grants, 944
Bradley C. Higgins Foundation Grants, 945
Bridgestone/Firestone Trust Fund Grants, 949
Browning-Kimball Foundation Grants, 971
Burlington Industries Foundation Grants, 977
Burlington Northern Santa Fe Foundation Grants, 978
Business Bank of Nevada Community Grants, 981
Byron W. & Alice L. Lockwood Fnd Grants, 991
Caesars Foundation Grants, 995
Caleb C. and Julia W. Dula Educational and Charitable Foundation Grants, 997
Cambridge Community Foundation Grants, 1004
Campbell Hoffman Foundation Grants, 1006
Campbell Soup Foundation Grants, 1007
Canada-U.S. Fulbright New Century Scholars Program Grants, 1008
Canada-U.S. Fulbright Senior Specialists Grants, 1009
Carl & Eloise Pohlad Family Fndn Grants, 1021
Carl B. and Florence E. King Foundation Grants, 1022
Carl C. Icahn Foundation Grants, 1023
Carl Gellert and Celia Berta Gellert Foundation Grants, 1024
Carl M. Freeman Foundation FACES Grants, 1026
Carl M. Freeman Foundation Grants, 1027
Carnegie Corporation of New York Grants, 1033
Carolyn Foundation Grants, 1034
Carpenter Foundation Grants, 1035
Carrie Estelle Doheny Foundation Grants, 1037
Carroll County Community Foundation Grants, 1039
Cemala Foundation Grants, 1077
Centerville-Washington Foundation Grants, 1078
Central Carolina Community Foundation Community Impact Grants, 1079
Central Okanagan Foundation Grants, 1080
Cessna Foundation Grants Program, 1081
CFFVR Chilton Area Community Fndn Grants, 1096
CFFVR Clintonville Area Foundation Grants, 1097
CFFVR Jewelers Mutual Charitable Giving, 1099
CFFVR Project Grants, 1101
CFFVR Shawano Area Community Foundation Grants, 1104
Chamberlain Foundation Grants, 1108
Champlin Foundations Grants, 1110
Charles H. Dater Foundation Grants, 1118
Charles H. Hall Foundation, 1120
Charles Lafitte Foundation Grants, 1123
Charlotte County (FL) Community Foundation Grants, 1126
Chatlos Foundation Grants Program, 1128
Chazen Foundation Grants, 1129
Chiles Foundation Grants, 1147
Chilkat Valley Community Foundation Grants, 1148
Christy-Houston Foundation Grants, 1153
CICF City of Noblesville Community Grant, 1155
CICF Indianapolis Fndn Community Grants, 1156
CICF Legacy Fund Grants, 1158
CIGNA Foundation Grants, 1160
Citizens Bank Mid-Atlantic Charitable Foundation Grants, 1162
Clarence E. Heller Charitable Foundation Grants, 1166
Clark-Winchcole Foundation Grants, 1173
Cleveland Browns Foundation Grants, 1182

PROGRAM TYPE INDEX

Cultural Outreach / 861

CNA Foundation Grants, 1189
CNO Financial Group Community Grants, 1201
Coastal Community Foundation of South Carolina Grants, 1202
Coca-Cola Foundation Grants, 1203
Cockrell Foundation Grants, 1204
Collins Foundation Grants, 1210
Colonel Stanley R. McNeil Foundation Grants, 1211
Columbus Foundation Competitive Grants, 1217
Comerica Charitable Foundation Grants, 1224
Commonwealth Edison Grants, 1225
Community Foundation Alliance City of Evansville Endowment Fund Grants, 1238
Community Foundation for Greater Atlanta Clayton County Fund Grants, 1240
Community Foundation for Greater Atlanta Morgan County Fund Grants, 1244
Community Foundation for Greater Atlanta Newton County Fund Grants, 1245
Community Fndn for Monterey County Grants, 1253
Community Fndn for Muskegon County Grants, 1254
Community Foundation for Northeast Michigan Mini-Grants, 1255
Community Fndn for San Benito County Grants, 1257
Community Foundation for Southeast Michigan Grants, 1258
Community Fndn for the Capital Region Grants, 1259
Community Foundation for the National Capital Region Community Leadership Grants, 1260
Community Foundation of Bartholomew County Heritage Fund Grants, 1262
Community Foundation of Bloomington and Monroe County Grants, 1265
Community Fndn of Central Illinois Grants, 1267
Community Fndn of East Central Illinois Grants, 1268
Community Foundation of Eastern Connecticut General Southeast Grants, 1269
Community Foundation of Grant County Grants, 1271
Community Foundation of Greater Birmingham Grants, 1272
Community Foundation of Greater Flint Grants, 1273
Community Foundation of Greater Fort Wayne - Community Endowment and Clarke Endowment Grants, 1274
Community Foundation of Greater Greensboro Community Grants, 1277
Community Fndn of Greater Tampa Grants, 1282
Community Foundation of Greenville-Greenville Women Giving Grants, 1283
Community Foundation of Greenville Hollingsworth Funds Program/Project Grants, 1285
Community Fndn of Howard County Grants, 1286
Community Fndn of Jackson County Grants, 1287
Community Foundation of Mount Vernon and Knox County Grants, 1299
Community Foundation of Muncie and Delaware County Grants, 1300
Community Foundation of Muncie and Delaware County Maxon Grants, 1301
Community Fndn of Randolph County Grants, 1302
Community Foundation of Riverside & San Bernardino County Impact Grants, 1303
Community Fndn of Riverside & San Bernardino County Irene S. Rockwell Grants, 1304
Community Fndn of So Alabama Grants, 1306
Community Fndn of South Puget Sound Grants, 1307
Community Foundation of the Eastern Shore Community Needs Grants, 1313
Community Foundation of the Verdugos Educational Endowment Fund Grants, 1314
Community Foundation of the Verdugos Grants, 1315
Community Foundation Partnerships - Martin County Grants, 1318
Community Foundation Serving Riverside and San Bernardino Counties Impact Grants, 1319
ConAgra Foods Fndn Community Impact Grants, 1321
Connecticut Community Foundation Grants, 1324
Connelly Foundation Grants, 1326
ConocoPhillips Foundation Grants, 1327
Constantin Foundation Grants, 1341

Constellation Energy Corporate Grants, 1342
Consumers Energy Foundation, 1343
Cooke Foundation Grants, 1344
Cooper Industries Foundation Grants, 1345
Cornerstone Foundation of Northeastern Wisconsin Grants, 1347
Covenant Educational Foundation Grants, 1348
Cowles Charitable Trust Grants, 1354
Crane Foundation Grants, 1357
Crane Fund Grants, 1358
Cresap Family Foundation Grants, 1359
Crystelle Waggoner Charitable Trust Grants, 1364
CSRA Community Foundation Grants, 1366
CSX Corporate Contributions Grants, 1368
Cudd Foundation Grants, 1370
Cullen Foundation Grants, 1372
Cultural Society of Filipino Americans Grants, 1373
Cumberland Community Foundation Grants, 1374
CUNA Mutual Group Fndn Community Grants, 1375
Cyrus Eaton Foundation Grants, 1380
D.F. Halton Foundation Grants, 1405
Dade Community Foundation Grants, 1410
Dairy Queen Corporate Contributions Grants, 1413
Dale and Edna Walsh Foundation Grants, 1415
Dana Brown Charitable Trust Grants, 1419
Daniel & Nanna Stern Family Fndn Grants, 1422
Daniels Fund Grants-Aging, 1424
Davis Family Foundation Grants, 1435
Dayton Power and Light Foundation Grants, 1437
Daywood Foundation Grants, 1438
Deborah Munroe Noonan Memorial Grants, 1443
Delaware Community Foundation-Youth Philanthropy Board for Kent County, 1448
Delaware Community Foundation Grants, 1449
DeMatteis Family Foundation Grants, 1457
Denver Broncos Charities Fund Grants, 1460
Denver Foundation Community Grants, 1461
DeRoy Testamentary Foundation Grants, 1469
Dole Food Company Charitable Contributions, 1483
Donald and Sylvia Robinson Family Foundation Grants, 1486
Dorothy Rider Pool Health Care Grants, 1500
Dorrance Family Foundation Grants, 1501
Dr. Leon Bromberg Charitable Trust Grants, 1509
Dr. Scholl Foundation Grants Program, 1511
Drs. Bruce and Lee Foundation Grants, 1513
Dunspaugh-Dalton Foundation Grants, 1530
DuPage Community Foundation Grants, 1531
E. Clayton and Edith P. Gengras, Jr. Foundation, 1536
E.L. Wiegand Foundation Grants, 1538
E. Rhodes & Leona B. Carpenter Grants, 1539
Earl and Maxine Claussen Trust Grants, 1541
Eastman Chemical Company Foundation Grants, 1542
Eberly Foundation Grants, 1544
Eddie C. & Sylvia Brown Family Fndn Grants, 1548
Eden Hall Foundation Grants, 1549
Edina Realty Foundation Grants, 1550
Educational Foundation of America Grants, 1553
Edward Bangs Kelley and Elza Kelley Foundation Grants, 1556
Edwin S. Webster Foundation Grants, 1560
Edwin W. and Catherine M. Davis Fndn Grants, 1561
Edyth Bush Charitable Foundation Grants, 1562
Eisner Foundation Grants, 1564
Elizabeth Morse Genius Charitable Trust Grants, 1569
Elkhart County Community Foundation Grants, 1573
Elmer L. & Eleanor J. Andersen Fndn Grants, 1578
El Paso Community Foundation Grants, 1579
El Pomar Foundation Anna Keesling Ackerman Fund Grants, 1581
El Pomar Foundation Grants, 1582
Elsie H. Wilcox Foundation Grants, 1583
Emerson Charitable Trust Grants, 1589
Emily Davie and Joseph S. Kornfeld Foundation Grants, 1591
Emma B. Howe Memorial Foundation Grants, 1593
Ensworth Charitable Foundation Grants, 1595
Entergy Corporation Micro Grants, 1596
Erie Community Foundation Grants, 1601

Essex County Community Foundation Discretionary Fund Grants, 1602
Essex County Community Foundation Merrimack Valley General Fund Grants, 1603
Essex County Community Foundation Webster Family Fund Grants, 1604
Estee Lauder Grants, 1606
Ethel Sergeant Clark Smith Foundation Grants, 1608
Eugene M. Lang Foundation Grants, 1611
Evanston Community Foundation Grants, 1615
Evjue Foundation Grants, 1617
F.M. Kirby Foundation Grants, 1624
Fallon OrNda Community Health Fund Grants, 1627
FAR Fund Grants, 1631
Fargo-Moorhead Area Foundation Grants, 1632
Farmers Insurance Corporate Giving Grants, 1634
Faye McBeath Foundation Grants, 1635
Federal Express Corporate Contributions, 1657
Ferree Foundation Grants, 1658
Field Foundation of Illinois Grants, 1660
Fishman Family Foundation Grants, 1672
Florida BRAIVE Fund of Dade Community Foundation, 1678
Florida Division of Cultural Affairs Arts In Education Arts Partnership Grants, 1679
Floyd A. & Kathleen C. Cailloux Fnd Grants, 1681
Fluor Foundation Grants, 1682
Foellinger Foundation Grants, 1684
Fondren Foundation Grants, 1685
Ford Motor Company Fund Grants Program, 1687
Forrest C. Lattner Foundation Grants, 1688
Foster Foundation Grants, 1689
Foundation for the Mid South Community Development Grants, 1699
Four County Community Fndn General Grants, 1710
Four J Foundation Grants, 1713
Francis L. Abreu Charitable Trust Grants, 1719
Francis T. & Louise T. Nichols Fndn Grants, 1720
Frank Stanley Beveridge Foundation Grants, 1730
Fred Baldwin Memorial Foundation Grants, 1736
Fred L. Emerson Foundation Grants, 1743
Fremont Area Community Foundation Amazing X Grants, 1744
Fremont Area Community Foundation Elderly Needs Grants, 1745
Fremont Area Community Fndn General Grants, 1746
Fuji Film Grants, 1748
Fulbright Alumni Initiatives Awards, 1749
Fulbright Distinguished Chairs Awards, 1750
Fulbright German Studies Seminar Grants, 1751
Fulbright International Education Administrators (IEA) Seminar Program Grants, 1752
Fulbright New Century Scholars Grants, 1753
Fulbright Specialists Program Grants, 1754
Fulbright Scholars in Europe and Eurasia, 1755
Fulbright Traditional Scholar Program in Sub-Saharan Africa, 1756
Fulbright Traditional Scholar Program in the East Asia/Pacific Region, 1757
Fulbright Traditional Scholar Program in the Near East and North Africa Region, 1758
Fulbright Traditional Scholar Program in the South and Central Asia Region, 1759
Fulbright Traditional Scholar Program in the Western Hemisphere, 1760
G.N. Wilcox Trust Grants, 1768
Gardner Foundation Grants, 1770
Gardner Foundation Grants, 1771
Gebbie Foundation Grants, 1777
GenCorp Foundation Grants, 1779
General Mills Foundation Grants, 1782
General Motors Foundation Grants, 1783
Genesis Foundation Grants, 1786
Genuardi Family Foundation Grants, 1787
George A. and Grace L. Long Foundation Grants, 1788
George and Ruth Bradford Foundation Grants, 1789
George A Ohl Jr. Foundation Grants, 1791
George Family Foundation Grants, 1794
George Foundation Grants, 1795
George Gund Foundation Grants, 1796

Cultural Outreach

George H.C. Ensworth Memorial Fund Grants, 1797
George Kress Foundation Grants, 1799
George S. & Dolores Dore Eccles Fndn Grants, 1801
Georgia Power Foundation Grants, 1806
Gertrude and William C. Wardlaw Fund Grants, 1808
Giant Eagle Foundation Grants, 1815
Giant Food Charitable Grants, 1816
Gibson Foundation Grants, 1818
Gil and Dody Weaver Foundation Grants, 1819
Gill Foundation - Gay and Lesbian Fund Grants, 1821
GlaxoSmithKline Corporate Grants, 1829
GNOF Exxon-Mobil Grants, 1836
GNOF IMPACT Grants for Arts and Culture, 1838
GNOF Norco Community Grants, 1845
Golden Heart Community Foundation Grants, 1850
Golden State Warriors Foundation Grants, 1851
Goodrich Corporation Foundation Grants, 1854
Goodyear Tire Grants, 1856
Grace and Franklin Bernsen Foundation Grants, 1858
Graham Foundation Grants, 1862
Grammy Foundation Grants, 1863
Grand Rapids Area Community Fndn Grants, 1865
Grand Rapids Area Community Foundation Nashwauk Area Endowment Fund Grants, 1866
Grand Rapids Area Community Foundation Wyoming Youth Fund Grants, 1868
Grand Rapids Community Foundation Ionia County Grants, 1870
Grand Rapids Community Foundation Ionia County Youth Fund Grants, 1871
Grand Rapids Community Foundation Lowell Area Fund Grants, 1872
Greater Cincinnati Foundation Priority and Small Projects/Capacity-Building Grants, 1880
Greater Green Bay Community Fndn Grants, 1881
Greater Kanawha Valley Foundation Grants, 1882
Greater Milwaukee Foundation Grants, 1883
Greater Saint Louis Community Fndn Grants, 1884
Greater Sitka Legacy Fund Grants, 1885
Greater Tacoma Community Foundation Grants, 1886
Greater Worcester Community Foundation Discretionary Grants, 1888
Greater Worcester Community Foundation Jeppson Memorial Fund for Brookfield Grants, 1889
Green Bay Packers Foundation Grants, 1890
Green Diamond Charitable Contributions, 1891
Green River Area Community Fndn Grants, 1895
Grundy Foundation Grants, 1912
Guido A. & Elizabeth H. Binda Fndn Grants, 1913
Gulf Coast Community Foundation Grants, 1915
Guy I. Bromley Trust Grants, 1919
H & R Foundation Grants, 1920
H.A. & Mary K. Chapman Trust Grants, 1921
H.B. Fuller Foundation Grants, 1922
H.J. Heinz Company Foundation Grants, 1923
H. Leslie Hoffman and Elaine S. Hoffman Foundation Grants, 1924
Hallmark Corporate Foundation Grants, 1936
Hardin County Community Foundation Grants, 1946
Harley Davidson Foundation Grants, 1947
Harold Alfond Foundation Grants, 1949
Harold and Arlene Schnitzer CARE Foundation Grants, 1950
Harold R. Bechtel Charitable Remainder Uni-Trust Grants, 1953
Harold Simmons Foundation Grants, 1955
Harrison County Community Foundation Grants, 1957
Harrison County Community Foundation Signature Grants, 1958
Harry Bramhall Gilbert Charitable Trust Grants, 1961
Harry Frank Guggenheim Fnd Research Grants, 1964
Harry W. Bass, Jr. Foundation Grants, 1968
Hartford Foundation Regular Grants, 1970
Hawaiian Electric Industries Charitable Foundation Grants, 1975
Hawaii Community Foundation Reverend Takie Okumura Family Grants, 1977
Hawaii Community Foundation West Hawaii Fund Grants, 1978
Hawn Foundation Grants, 1979
HCA Foundation Grants, 1980
Heckscher Foundation for Children Grants, 1988
Heineman Foundation for Research, Education, Charitable and Scientific Purposes, 1990
Helena Rubinstein Foundation Grants, 1991
Helen Bader Foundation Grants, 1992
Helen Gertrude Sparks Charitable Trust Grants, 1993
Helen Irwin Littauer Educational Trust Grants, 1994
Helen K. & Arthur E. Johnson Fndn Grants, 1995
Helen Pumphrey Denit Charitable Trust Grants, 1996
Henrietta Tower Wurts Memorial Fndn Grants, 2002
Herbert A. & Adrian W. Woods Fndn Grants, 2007
Herbert H. & Grace A. Dow Fndn Grants, 2008
Herman Goldman Foundation Grants, 2012
Hershey Company Grants, 2013
Highmark Corporate Giving Grants, 2025
Hill Crest Foundation Grants, 2030
Hillman Foundation Grants, 2032
Hillsdale County Community General Adult Foundation Grants, 2034
Hillsdale Fund Grants, 2035
Hilton Head Island Foundation Grants, 2036
Hoblitzelle Foundation Grants, 2038
Hoglund Foundation Grants, 2040
Holland/Zeeland Community Fndn Grants, 2041
Homer Foundation Grants, 2044
Honda of America Manufacturing Fndn Grants, 2045
Horace A. Kimball and S. Ella Kimball Foundation Grants, 2046
Horace Moses Charitable Foundation Grants, 2047
Horizon Foundation for New Jersey Grants, 2048
Horizons Community Issues Grants, 2049
Houston Endowment Grants, 2053
Howard and Bush Foundation Grants, 2054
Hudson Webber Foundation Grants, 2072
Huffy Foundation Grants, 2073
Hugh J. Andersen Foundation Grants, 2074
Huie-Dellmon Trust Grants, 2075
Humana Foundation Grants, 2077
Huntington National Bank Community Grants, 2084
Hut Foundation Grants, 2086
Hutton Foundation Grants, 2087
I.A. O'Shaughnessy Foundation Grants, 2088
IBEW Local Union #697 Memorial Scholarships, 2096
Idaho Community Foundation Eastern Region Competitive Grants, 2114
Idaho Power Company Corporate Contributions, 2115
IIE African Center of Excellence for Women's Leadership Grants, 2124
IIE David L. Boren Fellowships, 2129
IIE Freeman Foundation Indonesia Internships, 2131
Ike and Roz Friedman Foundation Grants, 2145
Impact 100 Grants, 2148
Inasmuch Foundation Grants, 2149
Infinity Foundation Grants, 2158
Intergrys Corporation Grants, 2160
Irving S. Gilmore Foundation Grants, 2163
Irvin Stern Foundation Grants, 2164
J.B. Reynolds Foundation Grants, 2168
J.C. Penney Company Grants, 2169
J.L. Bedsole Foundation Grants, 2172
J.M. Long Foundation Grants, 2173
J. Walton Bissell Foundation Grants, 2177
J. Willard Marriott, Jr. Foundation Grants, 2178
Jackson County Community Foundation Unrestricted Grants, 2180
Jacob G. Schmidlapp Trust Grants, 2183
Jacobs Family Village Neighborhoods Grants, 2184
James & Abigail Campbell Family Fndn Grants, 2185
James Ford Bell Foundation Grants, 2187
James G.K. McClure Educational and Development Fund Grants, 2188
James Graham Brown Foundation Grants, 2189
James J. & Angelia M. Harris Fndn Grants, 2192
James L. and Mary Jane Bowman Charitable Trust Grants, 2194
James M. Collins Foundation Grants, 2195
James M. Cox Foundation of Georgia Grants, 2196
James R. Thorpe Foundation Grants, 2199
James S. Copley Foundation Grants, 2200
Jane's Trust Grants, 2206
Janus Foundation Grants, 2211
Jay and Rose Phillips Family Foundation Grants, 2213
Jayne and Leonard Abess Foundation Grants, 2214
Jean and Louis Dreyfus Foundation Grants, 2215
JELD-WEN Foundation Grants, 2218
Jennings County Community Foundation Women's Giving Circle Grant, 2221
Jerome Robbins Foundation Grants, 2223
Jessica Stevens Community Foundation Grants, 2224
Jessie B. Cox Charitable Trust Grants, 2225
Jessie Ball Dupont Fund Grants, 2226
Jim Moran Foundation Grants, 2228
John Ben Snow Memorial Trust Grants, 2231
John Deere Foundation Grants, 2234
John G. Duncan Charitable Trust Grants, 2236
John I. Smith Charities Grants, 2240
John J. Leidy Foundation Grants, 2241
John P. Murphy Foundation Grants, 2248
John R. Oishei Foundation Grants, 2249
Johns Manville Fund Grants, 2251
Johnson & Johnson/SAH Arts & Healing Grants, 2254
John W. and Anna H. Hanes Foundation Grants, 2261
John W. Speas and Effie E. Speas Memorial Trust Grants, 2264
Joseph Drown Foundation Grants, 2268
Joseph H. & Florence A. Roblee Fndn Grants, 2269
Joseph Henry Edmondson Foundation Grants, 2270
Judith and Jean Pape Adams Charitable Foundation Tulsa Area Grants, 2280
Julius N. Frankel Foundation Grants, 2282
Kahuku Community Fund, 2286
Katharine Matthies Foundation Grants, 2291
Kelvin and Eleanor Smith Foundation Grants, 2297
Kenai Peninsula Foundation Grants, 2298
Kenneth T. & Eileen L. Norris Fndn Grants, 2300
Kent D. Steadley and Mary L. Steadley Memorial Trust Grants, 2301
Ketchikan Community Foundation Grants, 2303
Kettering Family Foundation Grants, 2304
KeyBank Grants, 2307
Kinsman Foundation Grants, 2309
Kodiak Community Foundation Grants, 2314
Kovler Family Foundation Grants, 2323
Kuntz Foundation Grants, 2326
L. W. Pierce Family Foundation Grants, 2328
Land O'Lakes Foundation Mid-Atlantic Grants, 2341
Legler Benbough Foundation Grants, 2352
Leo Goodwin Foundation Grants, 2354
Leon and Thea Koerner Foundation Grants, 2355
Leo Niessen Jr., Charitable Trust Grants, 2358
Lewis H. Humphreys Charitable Trust Grants, 2361
Liberty Bank Foundation Grants, 2362
Libra Foundation Grants, 2363
Lillian S. Wells Foundation Grants, 2365
Lincoln Financial Foundation Grants, 2367
Lisa and Douglas Goldman Fund Grants, 2369
Lotus 88 Fnd for Women & Children Grants, 2376
Louie M. & Betty M. Phillips Fndn Grants, 2378
Louis and Elizabeth Nave Flarsheim Charitable Foundation Grants, 2379
Lubrizol Foundation Grants, 2384
Lucy Downing Nisbet Charitable Fund Grants, 2387
Ludwick Family Foundation Grants, 2389
Lynde and Harry Bradley Foundation Grants, 2397
M. Bastian Family Foundation Grants, 2402
M.J. Murdock Charitable Trust General Grants, 2405
Mabel F. Hoffman Charitable Trust Grants, 2407
Mabel Y. Hughes Charitable Trust Grants, 2409
Macquarie Bank Foundation Grants, 2410
Madison County Community Foundation General Grants, 2413
Maine Community Foundation Charity Grants, 2415
Manuel D. & Rhoda Mayerson Fndn Grants, 2420
Marcia and Otto Koehler Foundation Grants, 2427
Margaret T. Morris Foundation Grants, 2429
Marion Gardner Jackson Charitable Trust Grants, 2434
Marion I. and Henry J. Knott Foundation Discretionary Grants, 2435

PROGRAM TYPE INDEX

Cultural Outreach / 863

Marion I. and Henry J. Knott Foundation Standard Grants, 2436
Mary K. Chapman Foundation Grants, 2443
Mary Owen Borden Foundation Grants, 2446
Matilda R. Wilson Fund Grants, 2453
Mattel Children's Foundation Grants, 2454
Maurice J. Masserini Charitable Trust Grants, 2456
Max and Victoria Dreyfus Foundation Grants, 2457
Maxon Charitable Foundation Grants, 2458
McGraw-Hill Companies Community Grants, 2494
McInerny Foundation Grants, 2495
McKesson Foundation Grants, 2496
McLean Contributionship Grants, 2497
Mead Johnson Nutritionals Evansville-Area Organizations Grants, 2503
MeadWestvaco Foundation Sustainable Communities Grants, 2506
Mead Witter Foundation Grants, 2507
Medtronic Foundation CommunityLink Health Grants, 2510
Mercedes-Benz USA Corporate Contributions, 2517
Mericos Foundation Grants, 2518
Meriden Foundation Grants, 2519
Merrick Foundation Grants, 2522
Meyer and Stephanie Eglin Foundation Grants, 2531
Meyer Memorial Trust Grassroots Grants, 2536
Meyer Memorial Trust Responsive Grants, 2537
Meyer Memorial Trust Special Grants, 2538
MGN Family Foundation Grants, 2542
Miami County Community Foundation - Operation Round Up Grants, 2543
Milagro Foundation Grants, 2552
Mimi and Peter Haas Fund Grants, 2557
Mockingbird Foundation Grants, 2589
Morris & Gwendolyn Cafritz Fndn Grants, 2594
Nathan Cummings Foundation Grants, 2627
Nell J. Redfield Foundation Grants, 2673
Nicholas H. Noyes Jr. Memorial Fndn Grants, 2791
Noble County Community Foundation Grants, 2975
Norcliffe Foundation Grants, 2979
Nord Family Foundation Grants, 2980
Nordson Corporation Foundation Grants, 2981
Norfolk Southern Foundation Grants, 2982
Norman Foundation Grants, 2983
North Carolina Community Foundation Grants, 2990
North Dakota Community Foundation Grants, 2994
Northern Chautauqua Community Foundation Community Grants, 2995
Northwestern Mutual Foundation Grants, 2999
Norwin S. & Elizabeth N. Bean Fndn Grants, 3001
Nuffield Foundation Africa Program Grants, 3026
OceanFirst Foundation Major Grants, 3062
Ogden Codman Trust Grants, 3066
Oleonda Jameson Trust Grants, 3071
Olive Higgins Prouty Foundation Grants, 3072
Oppenstein Brothers Foundation Grants, 3078
Orange County Community Foundation Grants, 3079
Ordean Foundation Grants, 3080
Oregon Community Fndn Community Grants, 3082
Otto Bremer Foundation Grants, 3099
PACCAR Foundation Grants, 3101
Pacific Life Foundation Grants, 3103
Packard Foundation Local Grants, 3105
Palmer Foundation Grants, 3112
Parkersburg Area Community Foundation Action Grants, 3116
Patrick and Anna M. Cudahy Fund Grants, 3120
Percy B. Ferebee Endowment Grants, 3138
Perkins Charitable Foundation Grants, 3140
Peter Kiewit Foundation General Grants, 3154
Peter Kiewit Foundation Small Grants, 3155
Petersburg Community Foundation Grants, 3156
Peyton Anderson Foundation Grants, 3160
Phelps County Community Foundation Grants, 3188
Philadelphia Foundation Organizational Effectiveness Grants, 3189
Phoenix Suns Charities Grants, 3201
Piedmont Natural Gas Corporate and Charitable Contributions, 3220
Pittsburgh Foundation Community Fund Grants, 3230

PMI Foundation Grants, 3239
Pokagon Fund Grants, 3242
Polk Bros. Foundation Grants, 3243
Pollock Foundation Grants, 3244
Portland General Electric Foundation Grants, 3250
Posey Community Fndn Women's Fund Grants, 3251
Powell Foundation Grants, 3255
PPCF Community Grants, 3256
PPG Industries Foundation Grants, 3257
Price Chopper's Golub Foundation Grants, 3261
Priddy Foundation Program Grants, 3266
Pride Foundation Grants, 3268
Prince Charitable Trusts Chicago Grants, 3270
Prince Charitable Trusts DC Grants, 3271
Principal Financial Group Foundation Grants, 3273
Procter and Gamble Fund Grants, 3274
Puerto Rico Community Foundation Grants, 3278
Pulido Walker Foundation, 3282
Putnam County Community Foundation Grants, 3283
Quaker Chemical Foundation Grants, 3286
Qualcomm Grants, 3287
R.C. Baker Foundation Grants, 3292
Rajiv Gandhi Foundation Grants, 3302
Ralph M. Parsons Foundation Grants, 3304
Ralphs Food 4 Less Foundation Grants, 3305
Rasmuson Foundation Tier One Grants, 3307
Rasmuson Foundation Tier Two Grants, 3308
Rathmann Family Foundation Grants, 3309
RCF General Community Grants, 3312
Regence Fndn Access to Health Care Grants, 3346
Reinberger Foundation Grants, 3357
Rhode Island Foundation Grants, 3362
Rice Foundation Grants, 3363
Richard & Susan Smith Family Fndn Grants, 3365
Richard King Mellon Foundation Grants, 3369
Richland County Bank Grants, 3371
Riley Foundation Grants, 3374
Ripley County Community Foundation Grants, 3376
Ripley County Community Foundation Small Project Grants, 3377
Robbins Charitable Foundation Grants, 3378
Robert B McMillen Foundation Grants, 3381
Robert F. Stoico/FIRSTFED Charitable Foundation Grants, 3382
Robert R. Meyer Foundation Grants, 3387
Robert W. Woodruff Foundation Grants, 3389
Rochester Area Community Foundation Grants, 3391
Rochester Area Foundation Grants, 3392
Rockefeller Brothers Peace & Security Grants, 3394
Rockefeller Foundation Grants, 3396
Rockwell International Corporate Trust Grants Program, 3399
Rohm and Haas Company Grants, 3403
Ronald McDonald House Charities Grants, 3405
Rose Hills Foundation Grants, 3413
Roy & Christine Sturgis Charitable Grants, 3415
Ruth and Vernon Taylor Foundation Grants, 3426
S. Mark Taper Foundation Grants, 3449
Saigh Foundation Grants, 3451
Sain-Orr and Royak-DeForest Steadman Foundation Grants, 3452
Salem Foundation Charitable Trust Grants, 3456
Samuel N. and Mary Castle Foundation Grants, 3467
Samuel S. Johnson Foundation Grants, 3468
San Antonio Area Foundation Grants, 3469
San Diego Foundation for Change Grants, 3470
San Diego Women's Foundation Grants, 3474
San Juan Island Community Foundation Grants, 3480
Santa Fe Community Foundation Seasonal Grants-Fall Cycle, 3482
Sarkeys Foundation Grants, 3485
Sartain Lanier Family Foundation Grants, 3486
Sasco Foundation Grants, 3487
Schering-Plough Foundation Community Initiatives Grants, 3491
Schlessman Family Foundation Grants, 3493
Schramm Foundation Grants, 3495
Scott B. & Annie P. Appleby Charitable Grants, 3499
Scott County Community Foundation Grants, 3500
Seagate Tech Corp Capacity to Care Grants, 3502

Sensient Technologies Foundation Grants, 3513
Seward Community Foundation Grants, 3514
Seward Community Foundation Mini-Grants, 3515
Shell Deer Park Grants, 3543
Shield-Ayres Foundation Grants, 3546
Shopko Fndn Community Charitable Grants, 3547
Shopko Foundation Green Bay Area Community Grants, 3548
Siebert Lutheran Foundation Grants, 3554
Simmons Foundation Grants, 3563
Sioux Falls Area Community Foundation Community Fund Grants (Unrestricted), 3568
Sioux Falls Area Community Foundation Field-of-Interest and Donor-Advised Grants, 3569
Sioux Falls Area Community Foundation Spot Grants (Unrestricted), 3570
Skoll Fndn Awards for Social Entrepreneurship, 3579
SOCFOC Catholic Ministries Grants, 3596
Sonoco Foundation Grants, 3621
Sonora Area Foundation Competitive Grants, 3622
Sony Corporation of America Grants, 3623
Southbury Community Trust Fund, 3625
Southwest Florida Community Foundation Competitive Grants, 3627
Spencer County Community Foundation Grants, 3632
Square D Foundation Grants, 3634
Stackpole-Hall Foundation Grants, 3643
Stella and Charles Guttman Foundation Grants, 3651
Sterling-Turner Charitable Foundation Grants, 3652
Sterling and Shelli Gardner Foundation Grants, 3653
Stewart Huston Charitable Trust Grants, 3655
Strake Foundation Grants, 3658
Stranahan Foundation Grants, 3659
Subaru of Indiana Automotive Foundation Grants, 3672
Summit Foundation Grants, 3674
SunTrust Bank Trusteed Foundations Florence C. and Harry L. English Memorial Fund Grants, 3680
SunTrust Bank Trusteed Foundations Greene-Sawtell Grants, 3681
SunTrust Bank Trusteed Foundations Harriet McDaniel Marshall Tust Grants, 3682
SunTrust Bank Trusteed Foundations Nell Warren Elkin and William Simpson Elkin Grants, 3683
SunTrust Bank Trusteed Foundations Thomas Guy Woolford Charitable Trust Grants, 3684
SunTrust Bank Trusteed Foundations Walter H. and Marjory M. Rich Memorial Fund Grants, 3685
Susan Mott Webb Charitable Trust Grants, 3694
T.L.L. Temple Foundation Grants, 3700
Taubman Endowment for the Arts Fndn Grants, 3702
Tauck Family Foundation Grants, 3703
TE Foundation Grants, 3705
Tension Envelope Foundation Grants, 3707
Texas Commission on the Arts Arts Respond Project Grants, 3708
Textron Corporate Contributions Grants, 3712
Thelma Braun & Bocklett Family Fndn Grants, 3713
The Ray Charles Foundation Grants, 3716
Thomas C. Ackerman Foundation Grants, 3719
Thomas J. Long Foundation Community Grants, 3722
Thomas Sill Foundation Grants, 3724
Thompson Charitable Foundation Grants, 3727
Tommy Hilfiger Corporate Foundation Grants, 3737
Toyota Motor Manuf of Mississippi Grants, 3746
Toyota Motor Manufacturing of Texas Grants, 3747
Toyota Motor Manuf of West Virginia Grants, 3748
Toyota Motor Sales, USA Grants, 3750
Triangle Community Foundation Donor-Advised Grants, 3757
Trull Foundation Grants, 3761
Tull Charitable Foundation Grants, 3762
U.S. Cellular Corporation Grants, 3766
U.S. Department of Education United States-Russia Program: Improving Research and Educational Activities in Higher Education, 3773
Union Bank, N.A. Corporate Sponsorships and Donations, 3787
Union Bank, N.A. Foundation Grants, 3788
United Technologies Corporation Grants, 3803

864 / Cultural Outreach

Vancouver Foundation Grants and Community Initiatives Program, 3849
Vanderburgh Community Fndn Women's Fund, 3853
Vigneron Memorial Fund Grants, 3873
Virginia W. Kettering Foundation Grants, 3878
W.C. Griffith Foundation Grants, 3881
W.M. Keck Fndn So California Grants, 3890
Wabash Valley Community Foundation Grants, 3893
Waitt Family Foundation Grants, 3894
Walker Area Community Foundation Grants, 3895
Wallace Alexander Gerbode Foundation Grants, 3896
Walter L. Gross III Family Foundation Grants, 3901
Warrick County Community Foundation Women's Fund, 3905
Washington County Community Fndn Grants, 3906
Washington County Community Foundation Youth Grants, 3907
Wayne County Foundation Grants, 3909
Weingart Foundation Grants, 3913
Western Indiana Community Foundation Grants, 3921
Western New York Foundation Grants, 3922
Weyerhaeuser Company Foundation Grants, 3925
White County Community Foundation - Women Giving Together Grants, 3933
White County Community Foundation Grants, 3934
Whitney Foundation Grants, 3938
Willard & Pat Walker Charitable Fndn Grants, 3943
Willary Foundation Grants, 3944
William B. Dietrich Foundation Grants, 3946
William B. Stokely Jr. Foundation Grants, 3947
William Blair and Company Foundation Grants, 3948
William G. Gilmore Foundation Grants, 3951
William J. & Tina Rosenberg Fndn Grants, 3954
William Ray & Ruth E. Collins Fndn Grants, 3956
Williams Companies Foundation Homegrown Giving Grants, 3957
Windham Foundation Grants, 3961
Winston-Salem Foundation Competitive Grants, 3965
Winston-Salem Fndn Stokes County Grants, 3967
Wolfe Associates Grants, 3969
Wood-Claeyssens Foundation Grants, 3971
Z. Smith Reynolds Foundation Small Grants, 3982

Curriculum Development/Teacher Training
3M Company Fndn Community Giving Grants, 6
AAAAI RSLAAIS Leadership Award, 33
Abbott Fund Science Education Grants, 203
Abington Foundation Grants, 211
Advance Auto Parts Corporate Giving Grants, 312
Advanced Micro Devices Comm Affairs Grants, 313
Affymetrix Corporate Contributions Grants, 339
A Fund for Women Grants, 342
Albertson's Charitable Giving Grants, 410
Alpha Natural Resources Corporate Giving, 450
AMD Corporate Contributions Grants, 500
American Electric Power Foundation Grants, 530
American Foodservice Charitable Trust Grants, 532
American Psychiatric Foundation Typical or Troubled School Mental Health Education Grants, 558
American Trauma Society, Pennsylvania Division Mini-Grants, 564
Amerigroup Foundation Grants, 565
Amgen Foundation Grants, 572
Angels Baseball Foundation Grants, 590
Annenberg Foundation Grants, 596
Aratani Foundation Grants, 649
ASHA Advancing Academic-Research Award, 695
Autauga Area Community Foundation Grants, 808
Babcock Charitable Trust Grants, 816
Ball Brothers Foundation General Grants, 820
Barra Foundation Community Fund Grants, 833
Baxter International Corporate Giving Grants, 841
Bayer Foundation Grants, 846
Beckman Coulter Foundation Grants, 868
Benton Community Foundation Grants, 878
Blanche and Julian Robertson Family Foundation Grants, 911
Blue River Community Foundation Grants, 922
Boeing Company Contributions Grants, 930
Bonfils-Stanton Foundation Outrants, 932

BP Foundation Grants, 943
Bristol-Myers Squibb Foundation Health Education Grants, 959
Bristol-Myers Squibb Foundation Science Education Grants, 961
Brown County Community Foundation Grants, 970
Cabot Corporation Foundation Grants, 993
Cambridge Community Foundation Grants, 1004
Carl M. Freeman Foundation Grants, 1027
Carnahan-Jackson Foundation Grants, 1032
Carnegie Corporation of New York Grants, 1033
Carolyn Foundation Grants, 1034
Carrie Estelle Doheny Foundation Grants, 1037
CFFVR Alcoholism and Drug Abuse Grants, 1093
CFFVR Clintonville Area Foundation Grants, 1097
Chapman Charitable Foundation Grants, 1112
Chilkat Valley Community Foundation Grants, 1148
Christensen Fund Regional Grants, 1149
Clarence E. Heller Charitable Foundation Grants, 1166
Clark County Community Foundation Grants, 1175
Clinton County Community Foundation Grants, 1184
Coca-Cola Foundation Grants, 1203
Community Foundation Alliance City of Evansville Endowment Fund Grants, 1238
Community Fndn for the Capital Region Grants, 1259
Community Foundation of Bartholomew County Heritage Fund Grants, 1262
Community Foundation of Boone County Grants, 1266
Community Foundation of Greater Birmingham Grants, 1272
Community Fndn of Riverside & San Bernardino County Irene S. Rockwell Grants, 1304
Community Foundation of the Verdugos Educational Endowment Fund Grants, 1314
Community Fndn of Wabash County Grants, 1316
Comprehensive Health Education Fndn Grants, 1320
ConocoPhillips Foundation Grants, 1327
Crail-Johnson Foundation Grants, 1355
Cudd Foundation Grants, 1370
Dayton Power and Light Foundation Grants, 1437
Dean Foods Community Involvement Grants, 1440
Dubois County Community Foundation Grants, 1517
Earl and Maxine Claussen Trust Grants, 1541
eBay Foundation Community Grants, 1543
Eddie C. & Sylvia Brown Family Fndn Grants, 1548
Edward W. and Stella C. Van Houten Memorial Fund Grants, 1559
Effie and Wofford Cain Foundation Grants, 1563
Emily Davie and Joseph S. Kornfeld Foundation Grants, 1591
Faye McBeath Foundation Grants, 1635
Field Foundation of Illinois Grants, 1660
Ford Motor Company Fund Grants Program, 1687
Foundation for the Mid South Community Development Grants, 1699
Four County Community Fndn General Grants, 1710
Frank Stanley Beveridge Foundation Grants, 1730
Fremont Area Community Fndn General Grants, 1746
Garland D. Rhoads Foundation, 1774
GenCorp Foundation Grants, 1779
Genuardi Family Foundation Grants, 1787
Gheens Foundation Grants, 1814
Gibson Foundation Grants, 1818
GNOF Exxon-Mobil Grants, 1836
GNOF IMPACT Kahn-Oppenheim Grants, 1842
Golden Heart Community Foundation Grants, 1850
Goodrich Corporation Foundation Grants, 1854
Graham Foundation Grants, 1862
Grand Rapids Area Community Fndn Grants, 1865
Grand Rapids Area Community Foundation Nashwauk Area Endowment Fund Grants, 1866
Greater Sitka Legacy Fund Grants, 1885
Greater Tacoma Community Foundation Ryan Alan Hade Endowment Fund, 1887
Greater Worcester Community Foundation Jeppson Memorial Fund for Brookfield Grants, 1889
Green River Area Community Fndn Grants, 1895
Grifols Community Outreach Grants, 1904
Guido A. & Elizabeth H. Binda Fndn Grants, 1913
Hardin County Community Foundation Grants, 1946

PROGRAM TYPE INDEX

Harmony Project Grants, 1948
Harrison County Community Foundation Signature Grants, 1958
Heineman Foundation for Research, Education, Charitable and Scientific Purposes, 1990
Hillsdale County Community General Adult Foundation Grants, 2034
Hilton Hotels Corporate Giving Program Grants, 2037
Hoglund Foundation Grants, 2040
Honda of America Manufacturing Fndn Grants, 2045
Hormel Foundation Grants, 2051
Huntington Beach Police Officers Fndn Grants, 2081
Huntington County Community Foundation Make a Difference Grants, 2083
J.C. Penney Company Grants, 2169
Jack H. & William M. Light Trust Grants, 2179
James & Abigail Campbell Family Fndn Grants, 2185
James R. Dougherty Jr. Foundation Grants, 2198
Janus Foundation Grants, 2211
Jennings County Community Foundation Grants, 2220
Jessica Stevens Community Foundation Grants, 2224
Jessie Ball Dupont Fund Grants, 2226
Joe W. and Dorothy Dorsett Brown Fndn Grants, 2230
John Jewett and Helen Chandler Garland Foundation Grants, 2242
John P. McGovern Foundation Grants, 2247
John W. Speas and Effie E. Speas Memorial Trust Grants, 2264
Joseph Alexander Foundation Grants, 2265
Joseph H. & Florence A. Roblee Fndn Grants, 2269
Joseph Henry Edmondson Foundation Grants, 2270
Josephine Goodyear Foundation Grants, 2272
Josiah W. and Bessie H. Kline Foundation Grants, 2278
K and F Baxter Family Foundation Grants, 2287
Katherine Baxter Memorial Foundation Grants, 2292
Kenai Peninsula Foundation Grants, 2298
Ketchikan Community Foundation Grants, 2303
Kinsman Foundation Grants, 2309
Knight Family Charitable and Educational Foundation Grants, 2312
Kodiak Community Foundation Grants, 2314
Kuntz Foundation Grants, 2326
Lincoln Financial Foundation Grants, 2367
Lotus 88 Fnd for Women & Children Grants, 2376
Louis and Elizabeth Nave Flarsheim Charitable Foundation Grants, 2379
Louis H. Aborn Foundation Grants, 2380
Lubbock Area Foundation Grants, 2383
Mattel Children's Foundation Grants, 2454
Mead Johnson Nutritionals Evansville-Area Organizations Grants, 2503
Mead Johnson Nutritionals Med Educ Grants, 2504
Meyer Memorial Trust Responsive Grants, 2537
Meyer Memorial Trust Special Grants, 2538
Mid-Iowa Health Foundation Community Response Grants, 2549
Mimi and Peter Haas Fund Grants, 2557
Mockingbird Foundation Grants, 2589
Montgomery County Community Fndn Grants, 2591
Mt. Sinai Health Care Foundation Academic Medicine and Bioscience Grants, 2598
NIH School Interventions to Prevent Obesity, 2908
Northern Chautauqua Community Foundation Community Grants, 2995
Ohio County Community Foundation Board of Directors Grants, 3067
Oppenstein Brothers Foundation Grants, 3078
Oregon Community Fndn Community Grants, 3082
PacifiCare Foundation Grants, 3102
Petersburg Community Foundation Grants, 3156
Pew Charitable Trusts Children & Youth Grants, 3158
Phelps County Community Foundation Grants, 3188
Pokagon Fund Grants, 3242
Polk Bros. Foundation Grants, 3243
Procter and Gamble Fund Grants, 3274
Rathmann Family Foundation Grants, 3309
Richard & Susan Smith Family Fndn Grants, 3365
Richard Davoud Donchian Foundation Grants, 3367
Richland County Bank Grants, 3371

Ripley County Community Foundation Small Project Grants, 3377
Robert R. Meyer Foundation Grants, 3387
San Diego Women's Foundation Grants, 3474
Schlessman Family Foundation Grants, 3493
Seabury Foundation Grants, 3501
Seattle Foundation Health and Wellness Grants, 3507
Seward Community Foundation Grants, 3514
Seward Community Foundation Mini-Grants, 3515
SfN Science Educator Award, 3535
Shell Oil Company Foundation Community Development Grants, 3544
Shield-Ayres Foundation Grants, 3546
Sidgmore Family Foundation Grants, 3551
Sonora Area Foundation Competitive Grants, 3622
Sony Corporation of America Grants, 3623
South Madison Community Foundation Grants, 3626
Strowd Roses Grants, 3664
STTI/ATI Educational Assessment Nursing Research Grants, 3665
Summit Foundation Grants, 3674
Tellabs Foundation Grants, 3706
Texas Instruments Foundation STEM Education Grants, 3711
Thomas Sill Foundation Grants, 3724
Toyota Motor Manufacturing of Alabama Grants, 3743
Toyota Motor Manufacturing of Indiana Grants, 3744
Toyota Motor Manuf of Mississippi Grants, 3746
Toyota Motor Manufacturing of Texas Grants, 3747
Toyota Motor Manuf of West Virginia Grants, 3748
Toyota Motor Sales, USA Grants, 3750
UCT Scholarship Program, 3777
Union Bank, N.A. Foundation Grants, 3788
Union County Community Foundation Grants, 3790
Vancouver Foundation Grants and Community Initiatives Program, 3849
Vectren Foundation Grants, 3857
Verizon Foundation Maryland Grants, 3859
Verizon Foundation Vermont Grants, 3863
Verizon Foundation Virginia Grants, 3864
W.H. & Mary Ellen Cobb Trust Grants, 3884
Weaver Popcorn Foundation Grants, 3910
Weingart Foundation Grants, 3913
Western Indiana Community Foundation Grants, 3921
Weyerhaeuser Company Foundation Grants, 3925
William Ray & Ruth E. Collins Fndn Grants, 3956
Winston-Salem Fndn Elkin/Tri-County Grants, 3966
Winston-Salem Fndn Stokes County Grants, 3967

Demonstration Grants
ACF Native American Social and Economic Development Strategies Grants, 231
Aetna Foundation Racial and Ethnic Health Care Equity Grants, 332
Alliant Energy Corporation Contributions, 445
American Legacy Fnd Small Innovative Grants, 535
amfAR Public Policy Grants, 570
Bank of America Fndn Matching Gifts, 825
Barra Foundation Project Grants, 834
BCBSM Foundation Physician Investigator Research Awards, 855
Berks County Community Foundation Grants, 881
Blue Cross Blue Shield of Minnesota Foundation - Healthy Equity: Health Impact Assessment Demonstration Project Grants, 916
Bullitt Foundation Grants, 973
Caesars Foundation Grants, 995
Campbell Hoffman Foundation Grants, 1006
Carlisle Foundation Grants, 1025
CDC Evidence-Based Laboratory Medicine: Quality/Performance Measure Evaluation, 1058
Children's Cardiomyopathy Foundation Research Grants, 1137
Cigna Civic Affairs Sponsorships, 1159
Clayton Fund Grants, 1180
CNCS Social Innovation Grants, 1192
Columbus Foundation Competitive Grants, 1217
Commonwealth Fund Patient-Centered Coordinated Care Program Grants, 1231

Commonwealth Fund Quality of Care for Frail Elders Grants, 1233
Community Foundation for Greater Atlanta Metropolitan Atlanta An Extra Wish Grants, 1243
Community Foundation for Northeast Michigan Mini-Grants, 1255
Community Foundation for Northeast Michigan Tobacco Settlement Grants, 1256
Community Foundation of Greater Fort Wayne - Community Endowment and Clarke Endowment Grants, 1274
Community Foundation of Louisville Anna Marble Memorial Fund for Princeton Grants, 1290
Community Fndn of So Alabama Grants, 1306
CONSOL Military and Armed Services Grants, 1340
Delta Air Lines Foundation Health and Wellness Grants, 1455
Denver Foundation Community Grants, 1461
DHHS Adolescent Family Life Demonstration Projects, 1472
DHHS Oral Health Promotion Research Across the Lifespan, 1474
Dyson Foundation Mid-Hudson Valley Project Support Grants, 1534
Educational Foundation of America Grants, 1553
Fight for Sight Grants-in-Aid, 1663
G.A. Ackermann Memorial Fund Grants, 1767
Harold R. Bechtel Charitable Remainder Uni-Trust Grants, 1953
Health Foundation of Greater Cincinnati Grants, 1983
HRSA Resource and Technical Assistance Center for HIV Prevention and Care for Black Men who Have Sex with Men Cooperative Agreement, 2069
Ittleson Foundation AIDS Grants, 2166
Ittleson Foundation Mental Health Grants, 2167
James S. McDonnell Foundation Complex Systems Collaborative Activity Awards, 2202
James S. McDonnell Foundation Scholar Awards, 2204
MLA Research, Development, and Demonstration Project Grant, 2572
Mt. Sinai Health Care Foundation Health of the Jewish Community Grants, 2599
Mt. Sinai Health Care Foundation Health Policy Grants, 2601
National Psoriasis Foundation Research Grants, 2636
NIGMS Advancing Basic Behavioral and Social Research on Resilience: an Integrative Science Approach, 2861
NLM Exploratory/Developmental Grants, 2942
Northern Chautauqua Community Foundation Community Grants, 2995
OceanFirst Foundation Major Grants, 3062
Oregon Community Fndn Community Grants, 3082
Philip L. Graham Fund Health and Human Services Grants, 3190
PhRMA Foundation Health Outcomes Research Starter Grants, 3204
Pinkerton Foundation Grants, 3224
Pittsburgh Foundation Community Fund Grants, 3230
Plough Foundation Grants, 3238
Principal Financial Group Foundation Grants, 3273
RRF General Program Grants, 3417
RWJF New Jersey Health Initiatives Grants, 3440
Saint Luke's Health Initiatives Grants, 3455
SAMHSA Campus Suicide Prevention Grants, 3460
Seattle Foundation Benjamin N. Phillips Memorial Fund Grants, 3504
Self Foundation Grants, 3510
Skin Cancer Foundation Research Grants, 3578
Sonora Area Foundation Competitive Grants, 3622
Taylor S. Abernathy and Patti Harding Abernathy Charitable Trust Grants, 3704
Union Bank, N.A. Foundation Grants, 3788
Unity Foundation Of LaPorte County Grants, 3804
USAID Child, Newborn, and Maternal Health Project Grants, 3806
USAID Family Health Plus Project Grants, 3810
USAID Health System Strengthening Grants, 3814
Weyerhaeuser Family Foundation Health Grants, 3926

Development (Institutional/Departmental)
ACF Native American Social and Economic Development Strategies Grants, 231
ACL Field Initiated Projects Program: Minority-Serving Institution Development Grants, 242
Aetna Foundation Racial and Ethnic Health Care Equity Grants, 332
Albuquerque Community Foundation Grants, 412
Alfred and Tillie Shemanski Testamentary Trust Grants, 428
Alvin and Fanny Blaustein Thalheimer Foundation Baltimore Communal Grants, 459
AMA Foundation Seed Grants for Research, 494
Amerigroup Foundation Grants, 565
amfAR Global Initiatives Grants, 568
Anheuser-Busch Foundation Grants, 591
Ann and Robert H. Lurie Family Fndn Grants, 593
Ann Arbor Area Community Foundation Grants, 594
AptarGroup Foundation Grants, 647
ASHA Multicultural Activities Projects Grants, 697
Assisi Foundation of Memphis General Grants, 787
Athwin Foundation Grants, 795
AVDF Health Care Grants, 811
Bank of America Fndn Volunteer Grants, 826
Baxter International Foundation Grants, 843
BCBSM Fndn Proposal Development Awards, 857
BCBSNC Foundation Grants, 860
Bill and Melinda Gates Foundation Policy and Advocacy Grants, 893
Blowitz-Ridgeway Foundation Early Childhood Development Research Award, 912
Blue Cross Blue Shield of Minnesota Foundation - Healthy Children: Growing Up Healthy Grants, 915
Blumenthal Foundation Grants, 925
Bonfils-Stanton Foundation Grants, 932
Boyd Gaming Corporation Contributions Program, 941
Brown Advisory Charitable Foundation Grants, 969
Burlington Industries Foundation Grants, 977
Bush Fndn Health & Human Services Grants, 979
California Endowment Innovative Ideas Challenge Grants, 1000
California Wellness Foundation Work and Health Program Grants, 1001
Cambridge Community Foundation Grants, 1004
Caring Foundation Grants, 1019
Carl C. Icahn Foundation Grants, 1023
Carl M. Freeman Foundation Grants, 1027
Carnegie Corporation of New York Grants, 1033
Carolyn Foundation Grants, 1034
CFF Pilot and Feasibility Awards, 1088
CFFVR Clintonville Area Foundation Grants, 1097
Charlotte County (FL) Community Foundation Grants, 1126
Children Affected by AIDS Foundation Camp Network Grants, 1144
CLIF Bar Family Foundation Grants, 1183
CNCS Social Innovation Grants, 1192
Community Foundation for Greater Atlanta Strategic Restructuring Fund Grants, 1246
Community Foundation for Greater New Haven Responsive New Grants, 1249
Community Foundation of St. Joseph County Special Project Challenge Grants, 1310
Conquer Cancer Foundation of ASCO Drug Development Research Professorship, 1330
Conquer Cancer Foundation of ASCO Improving Cancer Care Grants, 1331
Covenant Matters Foundation Grants, 1350
CSL Behring Local Empowerment for Advocacy Development (LEAD) Grants, 1365
Cystic Fibrosis Canada Clinic Incentive Grants, 1383
Dallas Women's Foundation Grants, 1417
Daniels Fund Grants-Aging, 1424
Daphne Seybolt Culpeper Fndn Grants, 1425
Deborah Munroe Noonan Memorial Grants, 1443
DOD HBCU/MI Partnership Training Award, 1480
DOJ Gang-Free Schools and Communities Intervention Grants, 1481

866 / Development (Institutional/Departmental) PROGRAM TYPE INDEX

Doris Duke Charitable Foundation Operations Research on AIDS Care and Treatment in Africa (ORACTA) Grants, 1497
Dubois County Community Foundation Grants, 1517
Duke Endowment Health Care Grants, 1519
Edward Bangs Kelley and Elza Kelley Foundation Grants, 1556
Evjue Foundation Grants, 1617
Fannie E. Rippel Foundation Grants, 1630
FAR Fund Grants, 1631
Farmers Insurance Corporate Giving Grants, 1634
Fidelity Foundation Grants, 1659
Fifth Third Foundation Grants, 1661
Finance Factors Foundation Grants, 1666
Foundation for the Mid South Community Development Grants, 1699
France-Merrick Foundation Health and Human Services Grants, 1715
Ginn Foundation Grants, 1822
GNOF Bayou Communities Grants, 1835
GNOF New Orleans Works Grants, 1844
GNOF Organizational Effectiveness Grants and Workshops, 1846
GNOF Stand Up For Our Children Grants, 1847
Grand Rapids Community Foundation Grants, 1869
Grand Rapids Community Foundation Lowell Area Fund Grants, 1872
Greater Cincinnati Foundation Priority and Small Projects/Capacity-Building Grants, 1880
Greater Saint Louis Community Fndn Grants, 1884
Greygates Foundation Grants, 1901
Grover Hermann Foundation Grants, 1906
Guy's and St. Thomas' Charity Grants, 1918
HAF Technical Assistance Program Grants, 1933
Harry A. & Margaret D. Towsley Fndn Grants, 1959
Harry Kramer Memorial Fund Grants, 1965
Harvest Foundation Grants, 1971
Health Foundation of Southern Florida Responsive Grants, 1985
Hilton Head Island Foundation Grants, 2036
HRAMF Deborah Munroe Noonan Memorial Research Grants, 2058
Hudson Webber Foundation Grants, 2072
IDPH Hosptial Capital Investment Grants, 2119
Indiana 21st Century Research and Technology Fund Awards, 2154
Intergrys Corporation Grants, 2160
Jacob and Valeria Langeloth Foundation Grants, 2182
James Ford Bell Foundation Grants, 2187
Janus Foundation Grants, 2211
Jay and Rose Phillips Family Foundation Grants, 2213
Jessie Ball Dupont Fund Grants, 2226
John J. Leidy Foundation Grants, 2241
John Merck Fund Grants, 2245
Johnson & Johnson/SAH Arts & Healing Grants, 2254
John W. and Anna H. Hanes Foundation Grants, 2261
John W. Speas and Effie E. Speas Memorial Trust Grants, 2264
Kinsman Foundation Grants, 2309
Long Island Community Foundation Grants, 2375
Lotus 88 Fnd for Women & Children Grants, 2376
Louis and Elizabeth Nave Flarsheim Charitable Foundation Grants, 2379
Lucy Gooding Charitable Fndn Grants, 2388
Lynn and Rovena Alexander Family Foundation Grants, 2399
Marion I. and Henry J. Knott Foundation Discretionary Grants, 2435
Marion I. and Henry J. Knott Foundation Standard Grants, 2436
Maurice J. Masserini Charitable Trust Grants, 2456
May and Stanley Smith Charitable Trust Grants, 2459
Mead Johnson Nutritionals Charitable Giving, 2502
Meyer Foundation Benevon Grants, 2532
Meyer Foundation Healthy Communities Grants, 2533
Meyer Fndn Management Assistance Grants, 2534
Michael Reese Health Trust Core Grants, 2544
Mid-Iowa Health Foundation Community Response Grants, 2549

Middlesex Savings Charitable Foundation Capacity Building Grants, 2551
Miles of Hope Breast Cancer Foundation Grants, 2553
MLA David A. Kronick Traveling Fellowship, 2561
MMS and Alliance Charitable Foundation Grants for Community Action and Care for the Medically Uninsured, 2587
MMS and Alliance Charitable Foundation International Health Studies Grants, 2588
Mt. Sinai Health Care Foundation Academic Medicine and Bioscience Grants, 2598
Mt. Sinai Health Care Foundation Health of the Jewish Community Grants, 2599
Mt. Sinai Health Care Foundation Health of the Urban Community Grants, 2600
Mt. Sinai Health Care Foundation Health Policy Grants, 2601
NCI Cancer Education and Career Development Program, 2648
NIH Biology, Development, and Progression of Malignant Prostate Disease Research Grants, 2866
NIH Ruth L. Kirschstein Nat Research Service Award Short-Term Institutional Training Grants, 2906
Noble County Community Foundation Grants, 2975
Norman Foundation Grants, 2983
North Carolina Biotechnology Center Regional Development Grants, 2987
North Carolina Biotechnology Centers of Innovation Grants, 2988
North Carolina Biotechnology Center Technology Enhancement Grants, 2989
NSF Chemistry Research Experiences for Undergraduates (REU), 3012
NSF Undergraduate Research and Mentoring in the Biological Sciences (URM), 3025
NWHF Partners Investing in Nursing's Future, 3036
Oregon Community Foundation Better Nursing Home Care Grants, 3081
Otto Bremer Foundation Grants, 3099
Pajaro Valley Community Health Health Trust Insurance/Coverage & Education on Using the System Grants, 3108
Pajaro Valley Community Health Trust Diabetes and Contributing Factors Grants, 3109
Pajaro Valley Community Health Trust Oral Health: Prevention & Access Grants, 3110
Perkin Fund Grants, 3139
Peter and Elizabeth C. Tower Foundation Annual Intellectual Disabilities Grants, 3144
Peter and Elizabeth C. Tower Foundation Annual Mental Health Grants, 3145
Peter and Elizabeth C. Tower Foundation Mental Health Reference and Resource Materials Mini-Grants, 3147
Peter and Elizabeth C. Tower Foundation Organizational Scholarships, 3148
Peter and Elizabeth C. Tower Foundation Phase II Technology Initiative Grants, 3149
Peter and Elizabeth C. Tower Foundation Phase I Technology Initiative Grants, 3150
Pfizer/AAFP Foundation Visiting Professorship Program in Family Medicine, 3162
Philadelphia Foundation Organizational Effectiveness Grants, 3189
Phi Upsilon Omicron Alumni Research Grant, 3191
Piper Trust Healthcare & Med Research Grants, 3228
Piper Trust Older Adults Grants, 3229
Pittsburgh Foundation Community Fund Grants, 3230
Price Chopper's Golub Foundation Grants, 3261
Priddy Foundation Organizational Development Grants, 3265
Prince Charitable Trusts Chicago Grants, 3270
Prince Charitable Trusts DC Grants, 3271
Quality Health Foundation Grants, 3288
Rhode Island Foundation Grants, 3362
Richard Davoud Donchian Foundation Grants, 3367
Rockwell International Corporate Trust Grants Program, 3399
Rohm and Haas Company Grants, 3403
RRF Organizational Capacity Building Grants, 3418

RWJF Vulnerable Populations Portfolio Grants, 3444
Saint Luke's Health Initiatives Grants, 3455
Salem Foundation Grants, 3457
Santa Barbara Foundation Strategy Grants - Core Support, 3481
Sartain Lanier Family Foundation Grants, 3486
Schering-Plough Foundation Health Grants, 3492
Seattle Foundation Health and Wellness Grants, 3507
Sensient Technologies Foundation Grants, 3513
Sierra Health Foundation Responsive Grants, 3555
Sioux Falls Area Community Foundation Community Fund Grants (Unrestricted), 3568
Sioux Falls Area Community Foundation Spot Grants (Unrestricted), 3570
Sisters of Mercy of North Carolina Fndn Grants, 3575
Sisters of St. Joseph Healthcare Fndn Grants, 3576
Solo Cup Foundation Grants, 3620
Sonora Area Foundation Competitive Grants, 3622
Stewart Huston Charitable Trust Grants, 3655
Streilein Foundation for Ocular Immunology Visiting Professorships, 3660
Strowd Roses Grants, 3664
Sunflower Foundation Bridge Grants, 3676
Sunflower Foundation Capacity Building Grants, 3677
Susan Mott Webb Charitable Trust Grants, 3694
Taubman Endowment for the Arts Fndn Grants, 3702
TJX Foundation Grants, 3735
Trull Foundation Grants, 3761
U.S. Department of Education 21st Century Community Learning Centers, 3767
U.S. Department of Education Rehabilitation Research Training Centers (RRTCs), 3770
U.S. Department of Education Rehabilitation Training - Rehabilitation Continuing Education - Institute on Rehabilitation Issues, 3771
U.S. Department of Education Vocational Rehabilitation Services Projects for American Indians with Disabilities Grants, 3774
Union Bank, N.A. Foundation Grants, 3788
United Hospital Fund of New York Health Care Improvement Grants, 3797
Vancouver Foundation Grants and Community Initiatives Program, 3849
Visiting Nurse Foundation Grants, 3879
Waitt Family Foundation Grants, 3894
Z. Smith Reynolds Foundation Small Grants, 3982

Dissertation/Thesis Research Support
AAHN Pre-Doctoral Research Grant, 147
AHIMA Dissertation Assistance Grants, 352
AIHP Thesis Support Grants, 372
APA Dissertation Research Awards, 626
Daimler and Benz Foundation Fellowships, 1411
Dale and Edna Walsh Foundation Grants, 1415
FDHN Graduate Student Awards, 1646
Harry Frank Guggenheim Foundation Dissertation Fellowships, 1963
Nestle Foundation Training Grant, 2679
NHLBI Ruth L. Kirschstein National Research Service Awards for Individual Predoctoral Fellowships to Promote Diversity in Health-Related Research, 2730
NIA Aging Research Dissertation Awards to Increase Diversity, 2746
NIA Minority Dissert Aging Research Grants, 2773
NSF Doctoral Dissertation Improvement Grants in the Directorate for Biological Sciences (DDIG), 3013
Pfizer Australia Neuroscience Research Grants, 3175
PhRMA Foundation Health Outcomes Post Doctoral Fellowships, 3202
PhRMA Foundation Health Outcomes Pre-Doctoral Fellowships, 3203
PhRMA Foundation Informatics Pre Doctoral Fellowships, 3207
PhRMA Foundation Pharmaceutics Pre Doctoral Fellowships, 3212
SfN Donald B. Lindsley Prize in Behavioral Neuroscience, 3521
Sigma Theta Tau International /American Nurses Foundation Grant, 3556

PROGRAM TYPE INDEX   Educational Programs   / 867

Sigma Theta Tau International Doris Bloch Research Award, 3559
Sigma Theta Tau Small Grants, 3560
Thomson Reuters / MLA Doctoral Fellowships, 3728

Educational Programs
1st Source Foundation Grants, 2
2 Depot Square Ipswich Charitable Fndn Grants, 4
3M Company Fndn Community Giving Grants, 6
3M Fndn Health & Human Services Grants, 7
49ers Foundation Grants, 9
1675 Foundation Grants, 11
A. Gary Anderson Family Foundation Grants, 16
A.O. Smith Foundation Community Grants, 17
AAAAI RSLAAIS Leadership Award, 33
AAAS Science Prize for Online Resources in Educ, 42
AACAP Child and Adolescent Psychiatry (CAP) Teaching Scholarships, 51
AAFA Investigator Research Grants, 128
AAFP Foundation Health Literacy State Grants, 132
AAFP Foundation Pfizer Teacher Devel Awards, 134
AAFP Foundation Wyeth Immunization Awards, 135
AAHN Pre-Doctoral Research Grant, 147
AAHPERD-AAHE Barb A. Cooley Scholarships, 148
AAHPERD-AAHE Del Oberteuffer Scholarships, 149
AAHPERD-AAHE Undergraduate Scholarships, 150
AAHPERD Abernathy Presidential Scholarships, 151
AAHPM Hospice & Palliative Med Fellowships, 152
AAMC Alpha Omega Alpha Robert J. Glaser Distinguished Teacher Awards, 154
AAMC Caring for Community Grants, 156
AAMC Herbert W. Nickens Faculty Fellowship, 159
AAMC Herbert W. Nickens Medical Student Scholarships, 160
AAP Resident Initiative Fund Grants, 189
Aaron & Cecile Goldman Family Fndn Grants, 191
Aaron Foundation Grants, 193
Abbot and Dorothy H. Stevens Foundation Grants, 197
Abbott Fund Access to Health Care Grants, 199
Abbott Fund Community Grants, 201
Abbott Fund Global AIDS Care Grants, 202
Abbott Fund Science Education Grants, 203
Abby's Legendary Pizza Foundation Grants, 204
Abel Foundation Grants, 206
Abell-Hanger Foundation Grants, 207
Abernethy Family Foundation Grants, 210
Abington Foundation Grants, 211
Able To Serve Grants, 212
Abney Foundation Grants, 214
Abramson Family Foundation Grants, 217
Abundance Foundation International Grants, 218
ACE Charitable Foundation Grants, 229
ACF Foundation Grants, 230
ACF Native American Social and Economic Development Strategies Grants, 231
Achelis Foundation Grants, 234
Ackerman Foundation Grants, 236
ACL Business Acumen for Disability Organizations Grants, 238
ACL University Centers for Excellence in Developmental Network Diversity and Inclusion Training Action Planning Grants, 249
ACS Doc Degree Scholarships in Cancer Nursing, 252
ACS Scholarships in Cancer Nursing Practice, 255
ACSM International Student Award, 268
ACVIM Foundation Clinical Investigation Grants, 286
ADA Foundation Dentsply International Research Fellowships, 289
Adam Richter Charitable Trust Grants, 298
Adams and Reese Corporate Giving Grants, 299
Adams County Community Foundation of Pennsylvania Grants, 300
Adams Foundation Grants, 301
Adaptec Foundation Grants, 304
Adelaide Breed Bayrd Foundation Grants, 306
Adler-Clark Electric Community Commitment Foundation Grants, 307
Administaff Community Affairs Grants, 308
Administration on Aging Senior Medicare Patrol Project Grants, 309

Adobe Community Investment Grants, 310
Adolph Coors Foundation Grants, 311
Advance Auto Parts Corporate Giving Grants, 312
Advanced Micro Devices Comm Affairs Grants, 313
AEC Trust Grants, 315
AEGON Transamerica Foundation Health and Welfare Grants, 316
AES Research and Training Workshop Awards, 320
AES William G. Lennox Award, 326
Aetna Foundation Minority Scholars Grants, 329
Aetna Foundation National Medical Fellowship in Healthcare Leadership, 330
Affymetrix Corporate Contributions Grants, 339
AFG Industries Grants, 340
A Friends' Foundation Trust Grants, 341
A Fund for Women Grants, 342
Agnes B. Hunt Trust Grants, 344
Agnes Gund Foundation Grants, 345
Agnes M. Lindsay Trust Grants, 346
A Good Neighbor Foundation Grants, 347
Agway Foundation Grants, 348
AHIMA Grant-In-Aid Research Grants, 354
Ahmanson Foundation Grants, 355
Ahn Family Foundation Grants, 356
AIDS Vaccine Advocacy Coalition (AVAC) Fund Grants, 369
AIHS Heritage Youth Researcher Summer Science Program, 387
Air Products and Chemicals Grants, 399
Alaska Airlines Foundation Grants, 403
Albert and Margaret Alkek Foundation Grants, 405
Alberto Culver Corporate Contributions Grants, 408
Albert Pick Jr. Fund Grants, 409
Albertson's Charitable Giving Grants, 410
Albert W. Rice Charitable Foundation Grants, 411
Albuquerque Community Foundation Grants, 412
Alcatel-Lucent Technologies Foundation Grants, 413
Alcoa Foundation Grants, 414
Alcohol Misuse and Alcoholism Research Grants, 415
Alcon Foundation Grants Program, 416
Alexander & Baldwin Fnd Mainland Grants, 417
Alexander and Baldwin Foundation Hawaiian and Pacific Island Grants, 418
Alexander and Margaret Stewart Trust Grants, 419
Alexander Eastman Foundation Grants, 420
Alexis Gregory Foundation Grants, 426
Alfred and Tillie Shemanski Testamentary Trust Grants, 428
Alfred Bersted Foundation Grants, 429
Alfred E. Chase Charitable Foundation Grants, 430
Allan C. and Lelia J. Garden Foundation Grants, 438
Allegan County Community Foundation Grants, 439
Allegheny Technologies Charitable Trust, 440
Allegis Group Foundation Grants, 441
Allen Foundation Educational Nutrition Grants, 442
Allen P. & Josephine B. Green Fndn Grants, 443
Alliance Healthcare Foundation Grants, 444
Allstate Corporate Giving Grants, 446
Allstate Corp Hometown Commitment Grants, 447
Allyn Foundation Grants, 448
Alpha Natural Resources Corporate Giving, 450
Alpha Omega Foundation Grants, 451
Altman Foundation Health Care Grants, 457
Alvin and Fanny Blaustein Thalheimer Foundation Baltimore Communal Grants, 459
AMA-MSS Chapter Involvement Grants, 473
AMA-MSS Government Relations Advocacy Fellowship, 475
AMA-RFS and AMA Foundation Medical Student & Resident/Fellow Elective-Medicine & Media, 478
AMA-RFS Legislative Awareness Internships, 479
Amador Community Foundation Grants, 481
AMA Fndn Arthur N. Wilson Scholarship, 482
AMA Foundation Fund for Better Health Grants, 484
AMA Foundation Joan F. Giambalvo Memorial Scholarships, 488
AMA Foundation Minority Scholars Awards, 491
AMA Foundation Physicians of Tomorrow Scholarships, 492
AMDA Foundation Quality Improvement Award, 499

AMD Corporate Contributions Grants, 500
Amelia Sillman Rockwell and Carlos Perry Rockwell Charities Fund Grants, 501
American-Scandinavian Foundation Visiting Lectureship Grants, 502
American Academy of Nursing Building Academic Geriatric Nursing Capacity Scholarships, 505
American Academy of Nursing Claire M. Fagin Fellowships, 506
American Academy of Nursing MBA Scholarships, 508
American Chemical Society GCI Pharmaceutical Roundtable Research Grants, 516
American College of Surgeons and The Triological Society Clinical Scientist Development Awards, 518
American College of Surgeons Resident and Associate Society (RAS-ACS) Leadership Scholarships, 527
American Electric Power Foundation Grants, 530
American Foodservice Charitable Trust Grants, 532
American Jewish World Service Grants, 533
American Legacy Foundation National Calls for Proposals Grants, 534
American Psychiatric Association/Merck Co. Early Academic Career Research Award, 541
American Psychiatric Foundation Grantss, 552
American Psychiatric Foundation Helping Hands Grants, 554
American Psychiatric Foundation James H. Scully Jr., M.D., Educational Fund Grants, 555
American Psychiatric Foundation Jeanne Spurlock Congressional Fellowship, 556
American Psychiatric Foundation Typical or Troubled School Mental Health Education Grants, 558
American Society on Aging Graduate Student Research Award, 560
American Trauma Society, Pennsylvania Division Mini-Grants, 564
Amerigroup Foundation Grants, 565
Amerisure Insurance Community Service Grants, 566
AMHPS Dr. James A. Ferguson Emerging Infectious Diseases Fellowships, 573
Amica Companies Foundation Grants, 574
Amica Insurance Company Community Grants, 575
AMI Semiconductors Corporate Grants, 576
ANA Distinguished Neurology Teacher Award, 580
Andersen Corporate Foundation, 586
Andre Agassi Charitable Foundation Grants, 587
Andrew Family Foundation Grants, 588
Angels Baseball Foundation Grants, 590
Anheuser-Busch Foundation Grants, 591
Anna Fitch Ardenghi Trust Grants, 592
Ann and Robert H. Lurie Family Fndn Grants, 593
Ann Arbor Area Community Foundation Grants, 594
Anne J. Caudal Foundation Grants, 595
Annenberg Foundation Grants, 596
Annie Sinclair Knudsen Memorial Fund/Kaua'i Community Grants Program, 597
Annunziata Sanguinetti Foundation Grants, 600
Anschutz Family Foundation Grants, 602
Anthem Blue Cross and Blue Shield Grants, 607
Anthony R. Abraham Foundation Grants, 608
Antone & Edene Vidinha Charitable Trust Grants, 609
AON Foundation Grants, 623
APF/COGDOP Graduate Research Scholarships in Psychology, 635
APHL Emerging Infectious Diseases Fellowships, 636
Appalachian Regional Commission Health Grants, 637
APSAA Foundation Grants, 639
APSAA Mini-Career Grants, 640
APSAA Small Beginning Scholar Pre-Investigation Grants, 642
APSAA Small Beginning Scholar Visiting and Consulting Grants, 643
APS Foundation Grants, 646
Aragona Family Foundation Grants, 648
Aratani Foundation Grants, 649
Arcadia Foundation Grants, 650
Archer Daniels Midland Foundation Grants, 651
ARCO Foundation Education Grants, 652
Arie and Ida Crown Memorial Grants, 654
Arizona Cardinals Grants, 655

868 / Educational Programs PROGRAM TYPE INDEX

Arizona Diamondbacks Charities Grants, 657
Arizona Public Service Corporate Giving Program Grants, 658
Arkansas Community Foundation Arkansas Black Hall of Fame Grants, 659
Arkansas Community Foundation Grants, 660
Arkell Hall Foundation Grants, 661
Arlington Community Foundation Grants, 662
ARS Foundation Grants, 667
Arthur and Rochelle Belfer Foundation Grants, 670
Arthur Ashley Williams Foundation Grants, 671
ArvinMeritor Grants, 675
ASCO Long-Term International Fellowships, 681
ASHA Advancing Academic-Research Award, 695
ASHA Minority Student Leadership Awards, 696
ASHA Multicultural Activities Projects Grants, 697
ASHA Research Mentoring-Pair Travel Award, 698
ASHA Student Research Travel Award, 699
ASHA Students Preparing for Academic & Research Careers (SPARC) Award, 700
ASHFoundation New Century Scholars Program Doctoral Scholarships, 706
Ashland Corporate Contributions Grants, 713
ASM Int'l Fellowships for Asia and Africa, 735
ASM Int'l Professorship for Latin America, 736
ASM Int'l Professorships for Asia and Africa, 737
ASM Robert D. Watkins Graduate Research Fellowship, 747
ASPH/CDC Allan Rosenfield Global Health Fellowships, 780
ASPH/CDC Public Health Fellowships, 781
ASPH/CDC Public Health Internships, 782
Assisi Foundation of Memphis General Grants, 787
Assurant Foundation Grants, 788
Assurant Health Foundation Grants, 789
AT&T Foundation Civic and Community Service Program Grants, 793
Atherton Family Foundation Grants, 794
Athwin Foundation Grants, 795
Atlanta Foundation Grants, 796
Atran Foundation Grants, 797
Auburn Foundation Grants, 798
Audrey & Sydney Irmas Foundation Grants, 799
Aurora Foundation Grants, 802
Austin-Bailey Health and Wellness Fndn Grants, 803
Austin S. Nelson Foundation Grants, 805
Australian Academy of Science Grants, 807
Autauga Area Community Foundation Grants, 808
Autodesk Community Relations Grants, 809
Autzen Foundation Grants, 810
Avista Foundation Economic and Cultural Vitality Grants, 812
Axe-Houghton Foundation Grants, 815
Babcock Charitable Trust Grants, 816
Bacon Family Foundation Grants, 817
Bailey Foundation Grants, 818
Balfe Family Foundation Grants, 819
Ball Brothers Foundation General Grants, 820
BancorpSouth Foundation Grants, 823
Banfi Vintners Foundation Grants, 824
Bank of America Fndn Matching Gifts, 825
Bank of America Fndn Volunteer Grants, 826
Baptist Community Ministries Grants, 828
Barberton Community Foundation Grants, 829
Barker Foundation Grants, 830
Barker Welfare Foundation Grants, 831
Barnes Group Foundation Grants, 832
Barra Foundation Community Fund Grants, 833
Barra Foundation Project Grants, 834
Barr Fund Grants, 835
Batchelor Foundation Grants, 837
Baton Rouge Area Foundation Grants, 838
Battle Creek Community Foundation Grants, 839
Batts Foundation Grants, 840
Baxter International Corporate Giving Grants, 841
Baxter International Foundation Grants, 843
Bayer Clinical Scholarship Award, 845
Bayer Foundation Grants, 846
Bayer Hemophilia Caregivers Education Award, 847
BBVA Compass Foundation Charitable Grants, 848

BCBSM Building Healthy Communities Engaging Elementary Schools and Community Partners Grants, 849
BCBSM Foundation Student Award Program, 858
BCBSNC Foundation Grants, 860
Bearemy's Kennel Pals Grants, 863
Beazley Foundation Grants, 865
Bechtel Group Foundation Building Positive Community Relationships Grants, 866
Beckley Area Foundation Grants, 867
Beckman Coulter Foundation Grants, 868
Beirne Carter Foundation Grants, 870
Belk Foundation Grants, 872
Belvedere Community Foundation Grants, 874
Ben B. Cheney Foundation Grants, 875
Bender Foundation Grants, 876
Benton Community Fndn Cookie Jar Grant, 877
Benton Community Foundation Grants, 878
Benton County Foundation - Fitzgerald Family Scholarships, 879
Benton County Foundation Grants, 880
Berks County Community Foundation Grants, 881
Bernard F. and Alva B. Gimbel Foundation Reproductive Rights Grants, 883
Berrien Community Foundation Grants, 884
Bert W. Martin Foundation Grants, 886
Besser Foundation Grants, 887
Bikes Belong Foundation Paul David Clark Bicycling Safety Grants, 890
Bill and Melinda Gates Foundation Policy and Advocacy Grants, 893
Bill and Melinda Gates Foundation Water, Sanitation and Hygiene Grants, 894
Bill Hannon Foundation Grants, 895
Bindley Family Foundation Grants, 896
Biogen Foundation Healthcare Professional Education Grants, 900
Biogen Foundation Patient Educational Grants, 901
BJ's Charitable Foundation Grants, 904
Blackford County Community Foundation - WOW Grants, 905
Blackford County Community Foundation Grants, 906
Black River Falls Area Foundation Grants, 909
Blanche and Irving Laurie Foundation Grants, 910
Blanche and Julian Robertson Family Foundation Grants, 911
Blowitz-Ridgeway Foundation Early Childhood Development Research Award, 912
Blue Cross Blue Shield of Minnesota Foundation - Health Equity: Building Health Equity Together Grants, 914
Blue Cross Blue Shield of Minnesota Foundation - Healthy Equity: Health Impact Assessment Demonstration Project Grants, 916
Blue Cross Blue Shield of Minnesota Foundation - Healthy Equity: Health Impact Assessment Program Grants, 917
Blue Cross Blue Shield of Minnesota Fndn Healthy Equity: Public Libraries for Health Grants, 918
Blue Mountain Community Foundation Grants, 921
Blue River Community Foundation Grants, 922
Blue Shield of California Grants, 923
Blum-Kovler Foundation Grants, 924
Blumenthal Foundation Grants, 925
BMW of North America Charitable Contributions, 926
Bob & Delores Hope Foundation Grants, 927
Bodenwein Public Benevolent Foundation Grants, 928
Bodman Foundation Grants, 929
Boeing Company Contributions Grants, 930
Boettcher Foundation Grants, 931
Booth-Bricker Fund Grants, 933
Bosque Foundation Grants, 935
Boston Foundation Grants, 936
Boston Globe Foundation Grants, 937
Boston Jewish Community Women's Fund Grants, 938
Boyd Gaming Corporation Contributions Program, 941
Boyle Foundation Grants, 942
BP Foundation Grants, 943
Bradley-Turner Foundation Grants, 944
Bradley C. Higgins Foundation Grants, 945

Brian G. Dyson Foundation Grants, 948
Bridgestone/Firestone Trust Fund Grants, 949
Brighter Tomorrow Foundation Grants, 950
Bright Family Foundation Grants, 951
Bright Promises Foundation Grants, 952
Brinson Foundation Grants, 953
Bristol-Myers Squibb Foundation Global HIV/AIDS Initiative Grants, 957
Bristol-Myers Squibb Foundation Health Disparities Grants, 958
Bristol-Myers Squibb Foundation Health Education Grants, 959
Bristol-Myers Squibb Foundation Science Education Grants, 961
Bristol-Myers Squibb Foundation Women's Health Grants, 962
Brook J. Lenfest Foundation Grants, 967
Brown Advisory Charitable Foundation Grants, 969
Brown County Community Foundation Grants, 970
Bruce and Adele Greenfield Foundation Grants, 972
Bullitt Foundation Grants, 973
Burlington Industries Foundation Grants, 977
Burlington Northern Santa Fe Foundation Grants, 978
Bush Fndn Health & Human Services Grants, 979
Business Bank of Nevada Community Grants, 981
BWF Ad Hoc Grants, 982
BWF Career Awards for Medical Scientists, 984
BWF Postdoctoral Enrichment Grants, 989
Byron W. & Alice L. Lockwood Fnd Grants, 991
Cable Positive's Tony Cox Community Grants, 992
Caddock Foundation Grants, 994
Caesars Foundation Grants, 995
California Endowment Innovative Ideas Challenge Grants, 1000
California Wellness Foundation Work and Health Program Grants, 1001
Callaway Foundation Grants, 1002
Callaway Golf Company Foundation Grants, 1003
Cambridge Community Foundation Grants, 1004
Camp-Younts Foundation Grants, 1005
Campbell Soup Foundation Grants, 1007
Canada-U.S. Fulbright New Century Scholars Program Grants, 1008
Canada-U.S. Fulbright Senior Specialists Grants, 1009
Canada Graduate Scholarships (CGS) and NSERC Postgraduate Scholarships (PGS), 1010
Canadian Optometric Education Trust Grants, 1011
Canadian Patient Safety Institute (CPSI) Patient Safety Studentships, 1012
Capital Region Community Foundation Grants, 1014
Cardinal Health Foundation Grants, 1017
Cargill Citizenship Fund Corporate Giving, 1018
Caring Foundation Grants, 1019
Carla J. Funk Governmental Relations Award, 1020
Carl & Eloise Pohlad Family Fndn Grants, 1021
Carl B. and Florence E. King Foundation Grants, 1022
Carl C. Icahn Foundation Grants, 1023
Carl M. Freeman Foundation FACES Grants, 1026
Carl M. Freeman Foundation Grants, 1027
Carl R. Hendrickson Family Foundation Grants, 1028
Carls Foundation Grants, 1030
Carnegie Corporation of New York Grants, 1033
Carolyn Foundation Grants, 1034
Carpenter Foundation Grants, 1035
Carrie E. and Lena V. Glenn Foundation Grants, 1036
Carrie Estelle Doheny Foundation Grants, 1037
Carrier Corporation Contributions Grants, 1038
Carroll County Community Foundation Grants, 1039
Carylon Foundation Grants, 1040
Cass County Community Foundation Grants, 1041
Catherine Holmes Wilkins Foundation Grants, 1042
CCFF Life Sciences Student Awards, 1045
CCHD Community Development Grants, 1046
CDC-Hubert Global Health Fellowship, 1047
CDC Collegiate Leaders in Environmental Health Internships, 1048
CDC Cooperative Agreement for Partnership to Enhance Public Health Informatics, 1050

PROGRAM TYPE INDEX   Educational Programs   / 869

CDC Cooperative Agreement for the Development, Operation, and Evaluation of an Entertainment Education Program, 1051
CDC David J. Sencer Museum Adult Group Tour, 1052
CDC David J. Sencer Museum Teacher Professional Development Workshops, 1053
CDC Disease Detective Camp, 1054
CDC Epidemiology Elective Rotation, 1056
CDC Experience Epidemiology Fellowships, 1059
CDC Public Health Associates, 1065
CDC Public Health Associates Hosts, 1066
CDC Public Health Informatics Fellowships, 1068
CDC Public Health Prev Service Fellowships, 1069
CDC Steven M. Teutsch Prevention Effectiveness Fellowships, 1071
CDC Summer Graduate Environmental Health Internships, 1072
CDC Summer Program In Environmental Health Internships, 1073
Cedar Tree Foundation David H. Smith Conservation Research Fellowship, 1076
Cemala Foundation Grants, 1077
Centerville-Washington Foundation Grants, 1078
Central Carolina Community Foundation Community Impact Grants, 1079
Central Okanagan Foundation Grants, 1080
Cessna Foundation Grants Program, 1081
Cetana Educational Foundation Scholarships, 1082
CFFVR Alcoholism and Drug Abuse Grants, 1093
CFFVR Basic Needs Giving Partnership Grants, 1094
CFFVR Chilton Area Community Fndn Grants, 1096
CFFVR Clintonville Area Foundation Grants, 1097
CFFVR Frank C. Shattuck Community Grants, 1098
CFFVR Jewelers Mutual Charitable Giving, 1099
CFFVR Myra M. and Robert L. Vandehey Foundation Grants, 1100
CFFVR Project Grants, 1101
CFFVR Robert and Patricia Endries Family Foundation Grants, 1102
CFFVR Schmidt Family G4 Grants, 1103
CFFVR Shawano Area Community Foundation Grants, 1104
CFFVR Waupaca Area Community Foundation Grants, 1105
CFFVR Wisconsin King's Daus & Sons Grants, 1106
CFFVR Women's Fund for the Fox Valley Region Grants, 1107
Chamberlain Foundation Grants, 1108
Champ-A Champion Fur Kids Grants, 1109
Champlin Foundations Grants, 1110
Chapman Charitable Foundation Grants, 1112
CharityWorks Grants, 1113
Charles A. Frueauff Foundation Grants, 1114
Charles Delmar Foundation Grants, 1115
Charles Edison Fund Grants, 1116
Charles F. Bacon Trust Grants, 1117
Charles H. Dater Foundation Grants, 1118
Charles H. Hall Foundation, 1120
Charles H. Pearson Foundation Grants, 1121
Charles H. Revson Foundation Grants, 1122
Charles Lafitte Foundation Grants, 1123
Charles M. & Mary D. Grant Fndn Grants, 1124
Charles Nelson Robinson Fund Grants, 1125
Charlotte County (FL) Community Foundation Grants, 1126
Chase Paymentech Corporate Giving Grants, 1127
Chatlos Foundation Grants Program, 1128
Chazen Foundation Grants, 1129
Chiles Foundation Grants, 1147
Chilkat Valley Community Foundation Grants, 1148
Christensen Fund Regional Grants, 1149
Christine & Katharina Pauly Trust Grants, 1150
Christopher & Dana Reeve Foundation Quality of Life Grants, 1152
Christy-Houston Foundation Grants, 1153
Chula Vista Charitable Foundation Grants, 1154
CICF Indianapolis Fndn Community Grants, 1156
CICF Legacy Fund Grants, 1158
Cigna Civic Affairs Sponsorships, 1159
CIGNA Foundation Grants, 1160

CIT Corporate Giving Grants, 1161
Citizens Bank Mid-Atlantic Charitable Foundation Grants, 1162
CJ Foundation for SIDS Program Services Grants, 1163
Clara Blackford Smith and W. Aubrey Smith Charitable Foundation Grants, 1165
Clarence E. Heller Charitable Foundation Grants, 1166
Clarence T.C. Ching Foundation Grants, 1167
Clarian Health Critical/Progressive Care Interns, 1168
Clarian Health Multi-Specialty Internships, 1169
Clarian Health OR Internships, 1170
Clark-Winchcole Foundation Grants, 1173
Clark County Community Foundation Grants, 1175
Claude Bennett Family Foundation Grants, 1176
Claude Pepper Foundation Grants, 1177
Claude Worthington Benedum Fndn Grants, 1178
Clayton Baker Trust Grants, 1179
Clayton Fund Grants, 1180
Cleveland-Cliffs Foundation Grants, 1181
Cleveland Browns Foundation Grants, 1182
Clinton County Community Foundation Grants, 1184
CNA Foundation Grants, 1189
CNO Financial Group Community Grants, 1201
Coastal Community Foundation of South Carolina Grants, 1202
Coca-Cola Foundation Grants, 1203
Cockrell Foundation Grants, 1204
Coeta and Donald Barker Foundation Grants, 1205
Collins C. Diboll Private Foundation Grants, 1209
Collins Foundation Grants, 1210
Colonel Stanley R. McNeil Foundation Grants, 1211
Colorado Resource for Emergency Education and Trauma Grants, 1212
Columbus Foundation Competitive Grants, 1217
Columbus Foundation J. Floyd Dixon Memorial Fund Grants, 1219
Columbus Foundation Mary Eleanor Morris Fund Grants, 1220
Columbus Foundation Paul G. Duke Grants, 1221
Columbus Foundation Traditional Grants, 1223
Comerica Charitable Foundation Grants, 1224
Commonwealth Edison Grants, 1225
Commonwealth Fund/Harvard University Fellowship in Minority Health Policy, 1226
Community Fndn AIDS Endowment Awards, 1237
Community Foundation Alliance City of Evansville Endowment Fund Grants, 1238
Community Foundation for Greater Atlanta AIDS Fund Grants, 1239
Community Foundation for Greater Atlanta Clayton County Fund Grants, 1240
Community Foundation for Greater Atlanta Metropolitan Atlanta An Extra Wish Grants, 1243
Community Foundation for Greater Atlanta Morgan County Fund Grants, 1244
Community Foundation for Greater Atlanta Newton County Fund Grants, 1245
Community Fndn for Greater Buffalo Grants, 1247
Community Foundation for Greater New Haven Women & Girls Grants, 1252
Community Fndn for Monterey County Grants, 1253
Community Fndn for Muskegon County Grants, 1254
Community Foundation for Northeast Michigan Mini-Grants, 1255
Community Foundation for Northeast Michigan Tobacco Settlement Grants, 1256
Community Fndn for San Benito County Grants, 1257
Community Fndn for the Capital Region Grants, 1259
Community Foundation for the National Capital Region Community Leadership Grants, 1260
Community Foundation of Abilene Celebration of Life Grants, 1261
Community Foundation of Bartholomew County Heritage Fund Grants, 1262
Community Foundation of Bloomington and Monroe County Grants, 1265
Community Foundation of Boone County Grants, 1266
Community Fndn of Central Illinois Grants, 1267
Community Fndn of East Central Illinois Grants, 1268

Community Foundation of Eastern Connecticut General Southeast Grants, 1269
Community Foundation of Eastern Connecticut Northeast Women and Girls Grants, 1270
Community Foundation of Grant County Grants, 1271
Community Foundation of Greater Birmingham Grants, 1272
Community Foundation of Greater Flint Grants, 1273
Community Foundation of Greater Fort Wayne - Community Endowment and Clarke Endowment Grants, 1274
Community Foundation of Greater New Britain Grants, 1281
Community Fndn of Greater Tampa Grants, 1282
Community Foundation of Greenville-Greenville Women Giving Grants, 1283
Community Foundation of Greenville Hollingsworth Funds Program/Project Grants, 1285
Community Fndn of Howard County Grants, 1286
Community Fndn of Jackson County Grants, 1287
Community Foundation of Louisville AIDS Project Fund Grants, 1289
Community Foundation of Louisville Anna Marble Memorial Fund for Princeton Grants, 1290
Community Foundation of Louisville Delta Dental of Kentucky Fund Grants, 1292
Community Foundation of Louisville Dr. W. Barnett Owen Memorial Fund for the Children of Louisville and Jefferson County Grants, 1294
Community Fndn of Louisville Health Grants, 1295
Community Foundation of Louisville Morris and Esther Lee Fund Grants, 1298
Community Foundation of Mount Vernon and Knox County Grants, 1299
Community Foundation of Muncie and Delaware County Grants, 1300
Community Foundation of Muncie and Delaware County Maxon Grants, 1301
Community Fndn of Randolph County Grants, 1302
Community Foundation of Riverside & San Bernardino County Impact Grants, 1303
Community Fndn of Riverside & San Bernardino County Irene S. Rockwell Grants, 1304
Community Fndn of So Alabama Grants, 1306
Community Fndn of South Puget Sound Grants, 1307
Community Foundation of St. Joseph County Special Project Challenge Grants, 1310
Community Fndn of Switzerland County Grants, 1311
Community Foundation of Tampa Bay Grants, 1312
Community Foundation of the Verdugos Educational Endowment Fund Grants, 1314
Community Foundation of the Verdugos Grants, 1315
Community Foundation Partnerships - Lawrence County Grants, 1317
Community Foundation Partnerships - Martin County Grants, 1318
Community Foundation Serving Riverside and San Bernardino Counties Impact Grants, 1319
Comprehensive Health Education Fndn Grants, 1320
ConAgra Foods Fndn Community Impact Grants, 1321
ConAgra Foods Nourish our Community Grants, 1322
Cone Health Foundation Grants, 1323
Connecticut Community Foundation Grants, 1324
Connelly Foundation Grants, 1326
ConocoPhillips Foundation Grants, 1327
Conquer Cancer Foundation of ASCO Career Development Award, 1328
Conquer Cancer Foundation of ASCO Comparative Effectiveness Research Professorship in Breast Cancer, 1329
Conquer Cancer Foundation of ASCO Medical Student Rotation Grants, 1335
Conservation, Food, and Health Foundation Grants for Developing Countries, 1338
Constantin Foundation Grants, 1341
Constellation Energy Corporate Grants, 1342
Consumers Energy Foundation, 1343
Cooper Industries Foundation Grants, 1345
Coors Brewing Corporate Contributions Grants, 1346

Cornerstone Foundation of Northeastern Wisconsin Grants, 1347
Covidien Clinical Education Grants, 1351
Covidien Partnership for Neighborhood Wellness Grants, 1353
Cowles Charitable Trust Grants, 1354
Crail-Johnson Foundation Grants, 1355
Cralle Foundation Grants, 1356
Crane Foundation Grants, 1357
Crane Fund Grants, 1358
Cresap Family Foundation Grants, 1359
Crown Point Community Foundation Grants, 1361
Cruise Industry Charitable Foundation Grants, 1363
Crystelle Waggoner Charitable Trust Grants, 1364
CSRA Community Foundation Grants, 1366
CSX Corporate Contributions Grants, 1368
Cudd Foundation Grants, 1370
Cullen Foundation Grants, 1372
Cultural Society of Filipino Americans Grants, 1373
Cumberland Community Foundation Grants, 1374
CUNA Mutual Group Fndn Community Grants, 1375
Curtis Foundation Grants, 1376
CVS All Kids Can Grants, 1377
CVS Caremark Charitable Trust Grants, 1378
CVS Community Grants, 1379
Cyrus Eaton Foundation Grants, 1380
Cystic Fibrosis Canada Clinical Fellowships, 1381
Cystic Fibrosis Canada Clinic Incentive Grants, 1383
D.F. Halton Foundation Grants, 1405
D.V. and Ida J. McEachern Trust Grants, 1406
DAAD Research Stays for University Academics and Scientists, 1408
Dade Community Foundation Community AIDS Partnership Grants, 1409
Dade Community Foundation Grants, 1410
Dairy Queen Corporate Contributions Grants, 1413
Daisy Marquis Jones Foundation Grants, 1414
Dale and Edna Walsh Foundation Grants, 1415
Dallas Mavericks Foundation Grants, 1416
Dallas Women's Foundation Grants, 1417
Dammann Fund Grants, 1418
Dana Foundation Science and Health Grants, 1420
Daniel & Nanna Stern Family Fndn Grants, 1422
Daniel Mendelsohn New Investigator Award, 1423
Daphne Seybolt Culpeper Fndn Grants, 1425
David M. and Marjorie D. Rosenberg Foundation Grants, 1432
Davis Family Foundation Grants, 1435
Dayton Power and Light Company Foundation Signature Grants, 1436
Dayton Power and Light Foundation Grants, 1437
Daywood Foundation Grants, 1438
Deaconess Community Foundation Grants, 1439
Dean Foods Community Involvement Grants, 1440
Dearborn Community Foundation County Progress Grants, 1442
Deborah Munroe Noonan Memorial Grants, 1443
DeKalb County Community Foundation Grants, 1447
Delaware Community Foundation-Youth Philanthropy Board for Kent County, 1448
Delaware Community Foundation Grants, 1449
Del E. Webb Foundation Grants, 1450
Delmarva Power and Light Contributions, 1454
DeMatteis Family Foundation Grants, 1457
Dennis & Phyllis Washington Fndn Grants, 1458
Denver Broncos Charities Fund Grants, 1460
Dept of Ed College Assistance Migrant Grants, 1462
Dept of Ed Safe and Drug-Free Schools and Communities State Grants, 1464
Dept of Ed Special Education--Personnel Development to Improve Services and Results for Children with Disabilities, 1465
Dept of Ed Special Education--Studies and Evaluations, 1466
Dept of Ed Special Education--Technical Assistance and Dissemination to Improve Services and Results for Children with Disabilities, 1467
Dept of Ed Special Education-National Activities-Technology and Media Services for Individuals with Disabilities, 1468
DeRoy Testamentary Foundation Grants, 1469
Dickson Foundation Grants, 1476
DOD HBCU/MI Partnership Training Award, 1480
Dole Food Company Charitable Contributions, 1483
Donald W. Reynolds Foundation Aging and Quality of Life Grants, 1487
Dora Roberts Foundation Grants, 1489
Doree Taylor Charitable Foundation, 1490
Doris and Victor Day Foundation Grants, 1491
Dorothea Haus Ross Foundation Grants, 1498
Dorothy Hooper Beattie Foundation Grants, 1499
Dorothy Rider Pool Health Care Grants, 1500
Dorrance Family Foundation Grants, 1501
Do Something Awards, 1503
DPA Promoting Policy Change Advocacy Grants, 1504
Dr. & Mrs. Paul Pierce Memorial Fndn Grants, 1505
Dr. Leon Bromberg Charitable Trust Grants, 1509
Drs. Bruce and Lee Foundation Grants, 1513
DTE Energy Foundation Health and Human Services Grants, 1516
Dubois County Community Foundation Grants, 1517
Duke University Adult Cardiothoracic Anesthesia and Critical Care Medicine Fellowships, 1520
Duke University Ambulatory and Regional Anesthesia Fellowships, 1521
Duke University Clinical Cardiac Electrophysiology Fellowships, 1522
Duke University Hyperbaric Center Fellowships, 1523
Duke Univ Interventional Cardio Fellowships, 1524
Duke Univ Obstetric Anesthesia Fellowships, 1525
Duke University Pain Management Fellowships, 1526
Duke University Ped Anesthesiology Fellowship, 1527
Duneland Health Council Incorporated Grants, 1529
Dunspaugh-Dalton Foundation Grants, 1530
DuPage Community Foundation Grants, 1531
DuPont Pioneer Community Giving Grants, 1532
Dyson Foundation Mid-Hudson Valley Project Support Grants, 1534
E. Clayton and Edith P. Gengras, Jr. Foundation, 1536
E.J. Grassmann Trust Grants, 1537
E.L. Wiegand Foundation Grants, 1538
E. Rhodes & Leona B. Carpenter Grants, 1539
Earl and Maxine Claussen Trust Grants, 1541
Eastman Chemical Company Foundation Grants, 1542
eBay Foundation Community Grants, 1543
Eberly Foundation Grants, 1544
Eckerd Corporation Foundation Grants, 1547
Eddie C. & Sylvia Brown Family Fndn Grants, 1548
Eden Hall Foundation Grants, 1549
Edina Realty Foundation Grants, 1550
Edna G. Kynett Memorial Foundation Grants, 1551
EDS Foundation Grants, 1552
Educational Foundation of America Grants, 1553
Edward and Helen Bartlett Foundation Grants, 1554
Edward Bangs Kelley and Elza Kelley Foundation Grants, 1556
Edward W. and Stella C. Van Houten Memorial Fund Grants, 1559
Edwin S. Webster Foundation Grants, 1560
Edwin W. and Catherine M. Davis Fndn Grants, 1561
Edyth Bush Charitable Foundation Grants, 1562
Effie and Wofford Cain Foundation Grants, 1563
Eisner Foundation Grants, 1564
Elaine Feld Stern Charitable Trust Grants, 1565
Elizabeth McGraw Foundation Grants, 1568
Elizabeth Morse Genius Charitable Trust Grants, 1569
Elizabeth Nash Foundation Scholarships, 1570
Elizabeth Nash Fnd Summer Research Awards, 1571
Elkhart County Community Foundation Grants, 1573
Elliot Foundation Inc Grants, 1575
Elmer L. & Eleanor J. Andersen Fndn Grants, 1578
El Paso Community Foundation Grants, 1579
El Paso Corporate Foundation Grants, 1580
El Pomar Foundation Anna Keesling Ackerman Fund Grants, 1581
El Pomar Foundation Grants, 1582
Elsie H. Wilcox Foundation Grants, 1583
Elsie Lee Garthwaite Memorial Fndn Grants, 1584
Emerson Charitable Trust Grants, 1589
Emily Davie and Joseph S. Kornfeld Foundation Grants, 1591
Emily Hall Tremaine Foundation Learning Disabilities Grants, 1592
Emma B. Howe Memorial Foundation Grants, 1593
Ensworth Charitable Foundation Grants, 1595
Entergy Corporation Micro Grants, 1596
Entergy Corporation Open Grants for Healthy Families, 1597
EPA Children's Health Protection Grants, 1598
Erie Community Foundation Grants, 1601
Essex County Community Foundation Discretionary Fund Grants, 1602
Essex County Community Foundation Merrimack Valley General Fund Grants, 1603
Essex County Community Foundation Women's Fund Grants, 1605
Estee Lauder Grants, 1606
Ethel S. Abbott Charitable Foundation Grants, 1607
Ethel Sergeant Clark Smith Foundation Grants, 1608
Eugene B. Casey Foundation Grants, 1609
Eugene G. and Margaret M. Blackford Memorial Fund Grants, 1610
Eugene M. Lang Foundation Grants, 1611
Eva L. & Joseph M. Bruening Fndn Grants, 1613
Evan and Susan Bayh Foundation Grants, 1614
Evanston Community Foundation Grants, 1615
Everyone Breathe Asthma Education Grants, 1616
Evjue Foundation Grants, 1617
Expect Miracles Foundation Grants, 1619
Express Scripts Foundation Grants, 1620
Ezra M. Cutting Trust Grants, 1623
F.M. Kirby Foundation Grants, 1624
Fairfield County Community Foundation Grants, 1625
Fairlawn Foundation Grants, 1626
FAR Fund Grants, 1631
Fargo-Moorhead Area Foundation Grants, 1632
Farmers Insurance Corporate Giving Grants, 1634
Faye McBeath Foundation Grants, 1635
Fayette County Foundation Grants, 1636
FCD New American Children Grants, 1637
Federal Express Corporate Contributions, 1657
Ferree Foundation Grants, 1658
Fidelity Foundation Grants, 1659
Field Foundation of Illinois Grants, 1660
Fifth Third Foundation Grants, 1661
Finance Factors Foundation Grants, 1666
Firelight Foundation Grants, 1667
FirstEnergy Foundation Community Grants, 1668
Fisher Foundation Grants, 1670
Fishman Family Foundation Grants, 1672
FIU Global Civic Engagement Mini Grants, 1673
Flinn Foundation Scholarships, 1675
Florian O. Bartlett Trust Grants, 1677
Florida BRAIVE Fund of Dade Community Foundation, 1678
Florida Division of Cultural Affairs Arts In Education Arts Partnership Grants, 1679
Florida High School/High Tech Project Grants, 1680
Floyd A. & Kathleen C. Cailloux Fnd Grants, 1681
Fluor Foundation Grants, 1682
FMC Foundation Grants, 1683
Foellinger Foundation Grants, 1684
Fondren Foundation Grants, 1685
Ford Motor Company Fund Grants Program, 1687
Forrest C. Lattner Foundation Grants, 1688
Foster Foundation Grants, 1689
Foundation for a Healthy Kentucky Grants, 1691
Foundation for Balance and Harmony Grants, 1695
Foundation for Health Enhancement Grants, 1696
Foundation for Seacoast Health Grants, 1698
Foundation for the Mid South Community Development Grants, 1699
Foundation for the Mid South Health and Wellness Grants, 1700
Foundations of East Chicago Health Grants, 1709
Four County Community Fndn General Grants, 1710
Four County Community Foundation Healthy Senior/Healthy Youth Fund Grants, 1711
Fourjay Foundation Grants, 1712

PROGRAM TYPE INDEX — Educational Programs / 871

Four J Foundation Grants, 1713
Frances and John L. Loeb Family Fund Grants, 1716
Frances L. and Edwin L. Cummings Memorial Fund Grants, 1717
Frances W. Emerson Foundation Grants, 1718
Francis L. Abreu Charitable Trust Grants, 1719
Francis T. & Louise T. Nichols Fndn Grants, 1720
Frank B. Hazard General Charity Fund Grants, 1721
Frank E. and Seba B. Payne Foundation Grants, 1722
Frank G. and Freida K. Brotz Family Foundation Grants, 1723
Franklin County Community Foundation Grants, 1724
Franklin H. Wells and Ruth L. Wells Foundation Grants, 1725
Frank Loomis Palmer Fund Grants, 1726
Frank Reed and Margaret Jane Peters Memorial Fund I Grants, 1727
Frank Reed and Margaret Jane Peters Memorial Fund II Grants, 1728
Frank Stanley Beveridge Foundation Grants, 1730
Frank W. & Carl S. Adams Memorial Grants, 1731
Fraser-Parker Foundation Grants, 1732
Fred & Gretel Biel Charitable Trust Grants, 1734
Fred Baldwin Memorial Foundation Grants, 1736
Fred C. & Katherine B. Andersen Grants, 1737
Freddie Mac Foundation Grants, 1738
Frederick Gardner Cottrell Foundation Grants, 1739
Frederick McDonald Trust Grants, 1740
Frederick W. Marzahl Memorial Fund Grants, 1741
Fred L. Emerson Foundation Grants, 1743
Fremont Area Community Foundation Amazing X Grants, 1744
Fremont Area Community Foundation Elderly Needs Grants, 1745
Fremont Area Community Fndn General Grants, 1746
Fritz B. Burns Foundation Grants, 1747
Fuji Film Grants, 1748
Fulbright Alumni Initiatives Awards, 1749
Fulbright Distinguished Chairs Awards, 1750
Fulbright German Studies Seminar Grants, 1751
Fulbright International Education Administrators (IEA) Seminar Program Grants, 1752
Fulbright Specialists Program Grants, 1754
Fulbright Scholars in Europe and Eurasia, 1755
Fulbright Traditional Scholar Program in Sub-Saharan Africa, 1756
Fulbright Traditional Scholar Program in the East Asia/Pacific Region, 1757
Fulbright Traditional Scholar Program in the Near East and North Africa Region, 1758
Fulbright Traditional Scholar Program in the South and Central Asia Region, 1759
Fulbright Traditional Scholar Program in the Western Hemisphere, 1760
Fulton County Community Foundation Grants, 1763
G.A. Ackermann Memorial Fund Grants, 1767
G.N. Wilcox Trust Grants, 1768
Gamble Foundation Grants, 1769
Gardner Foundation Grants, 1771
Gardner Foundation Grants, 1770
Gardner W. & Joan G. Heidrick, Jr. Fndn Grants, 1773
Garland D. Rhoads Foundation, 1774
Gebbie Foundation Grants, 1777
GEICO Public Service Awards, 1778
GenCorp Foundation Grants, 1779
Genentech Corp Charitable Contributions, 1780
General Mills Champs for Healthy Kids Grants, 1781
General Mills Foundation Grants, 1782
General Motors Foundation Grants, 1783
Genesis Foundation Grants, 1786
Genuardi Family Foundation Grants, 1787
George A. and Grace L. Long Foundation Grants, 1788
George and Ruth Bradford Foundation Grants, 1789
George A Ohl Jr. Foundation Grants, 1791
George F. Baker Trust Grants, 1793
George Family Foundation Grants, 1794
George Foundation Grants, 1795
George Gund Foundation Grants, 1796
George H.C. Ensworth Memorial Fund Grants, 1797

George H. Hitchings New Investigator Award in Health Research, 1798
George Kress Foundation Grants, 1799
George P. Davenport Trust Fund Grants, 1800
George S. & Dolores Dore Eccles Fndn Grants, 1801
George W. Brackenridge Foundation Grants, 1802
George W. Codrington Charitable Fndn Grants, 1803
George W. Wells Foundation Grants, 1804
Georgiana Goddard Eaton Memorial Grants, 1805
Georgia Power Foundation Grants, 1806
Gertrude and William C. Wardlaw Fund Grants, 1808
Gertrude E. Skelly Charitable Foundation Grants, 1810
Gheens Foundation Grants, 1814
Giant Eagle Foundation Grants, 1815
Gibson County Community Foundation Women's Fund Grants, 1817
Gibson Foundation Grants, 1818
Gil and Dody Weaver Foundation Grants, 1819
Gill Foundation - Gay and Lesbian Fund Grants, 1821
Ginn Foundation Grants, 1822
Giving Sum Annual Grant, 1825
Gladys Brooks Foundation Grants, 1826
GlaxoSmithKline Corporate Grants, 1829
Global Fund for Children Grants, 1832
Global Fund for Women Grants, 1833
GNOF Exxon-Mobil Grants, 1836
GNOF IMPACT Grants for Arts and Culture, 1838
GNOF IMPACT Grants for Health and Human Services, 1839
GNOF IMPACT Kahn-Oppenheim Grants, 1842
GNOF New Orleans Works Grants, 1844
GNOF Norco Community Grants, 1845
GNOF Organizational Effectiveness Grants and Workshops, 1846
GNOF Stand Up For Our Children Grants, 1847
Godfrey Foundation Grants, 1849
Golden Heart Community Foundation Grants, 1850
Golden State Warriors Foundation Grants, 1851
Goodrich Corporation Foundation Grants, 1854
Good Samaritan Inc Grants, 1855
Goodyear Tire Grants, 1856
Grace and Franklin Bernsen Foundation Grants, 1858
Grace Bersted Foundation Grants, 1859
Graco Foundation Grants, 1860
Grand Haven Area Community Fndn Grants, 1864
Grand Rapids Area Community Fndn Grants, 1865
Grand Rapids Area Community Foundation Nashwauk Area Endowment Fund Grants, 1866
Grand Rapids Area Community Foundation Wyoming Grants, 1867
Grand Rapids Area Community Foundation Wyoming Youth Fund Grants, 1868
Grand Rapids Community Foundation Grants, 1869
Grand Rapids Community Foundation Ionia County Grants, 1870
Grand Rapids Community Foundation Ionia County Youth Fund Grants, 1871
Grand Rapids Community Foundation Lowell Area Fund Grants, 1872
Grand Rapids Community Foundation Southeast Ottawa Grants, 1873
Grand Rapids Community Foundation Southeast Ottawa Youth Fund Grants, 1874
Grand Rapids Community Fndn Sparta Grants, 1875
Grand Rapids Community Foundation Sparta Youth Fund Grants, 1876
Grass Foundation Marine Biological Laboratory Advanced Imaging Fellowships, 1877
Great-West Life Grants, 1879
Greater Cincinnati Foundation Priority and Small Projects/Capacity-Building Grants, 1880
Greater Green Bay Community Fndn Grants, 1881
Greater Kanawha Valley Foundation Grants, 1882
Greater Milwaukee Foundation Grants, 1883
Greater Saint Louis Community Fndn Grants, 1884
Greater Sitka Legacy Fund Grants, 1885
Greater Tacoma Community Foundation Grants, 1886
Greater Tacoma Community Foundation Ryan Alan Hade Endowment Fund, 1887

Greater Worcester Community Foundation Discretionary Grants, 1888
Greater Worcester Community Foundation Jeppson Memorial Fund for Brookfield Grants, 1889
Green Bay Packers Foundation Grants, 1890
Green Diamond Charitable Contributions, 1891
Greene County Foundation Grants, 1892
Green River Area Community Fndn Grants, 1895
Greenspun Family Foundation Grants, 1896
Grifols Community Outreach Grants, 1904
Grotto Foundation Project Grants, 1905
Gruber Foundation Neuroscience Prize, 1909
Grundy Foundation Grants, 1912
Guido A. & Elizabeth H. Binda Fndn Grants, 1913
Guitar Center Music Foundation Grants, 1914
Gulf Coast Community Foundation Grants, 1915
Gulf Coast Foundation of Community Operating Grants, 1916
Gulf Coast Foundation of Community Program Grants, 1917
Guy I. Bromley Trust Grants, 1919
H & R Foundation Grants, 1920
H.A. & Mary K. Chapman Trust Grants, 1921
H.B. Fuller Foundation Grants, 1922
H.J. Heinz Company Foundation Grants, 1923
H. Leslie Hoffman and Elaine S. Hoffman Foundation Grants, 1924
Hackett Foundation Grants, 1926
Hagedorn Fund Grants, 1934
Hall-Perrine Foundation Grants, 1935
Hallmark Corporate Foundation Grants, 1936
Hancock County Community Foundation - Field of Interest Grants, 1943
Hannaford Charitable Foundation Grants, 1944
Hardin County Community Foundation Grants, 1946
Harley Davidson Foundation Grants, 1947
Harmony Project Grants, 1948
Harold Alfond Foundation Grants, 1949
Harold and Arlene Schnitzer CARE Foundation Grants, 1950
Harold Brooks Foundation Grants, 1952
Harold R. Bechtel Charitable Remainder Uni-Trust Grants, 1953
Harold R. Bechtel Testamentary Charitable Trust Grants, 1954
Harris and Eliza Kempner Fund Grants, 1956
Harrison County Community Foundation Grants, 1957
Harrison County Community Foundation Signature Grants, 1958
Harry B. and Jane H. Brock Foundation Grants, 1960
Harry Bramhall Gilbert Charitable Trust Grants, 1961
Harry Frank Guggenheim Fnd Research Grants, 1964
Harry S. Black and Allon Fuller Fund Grants, 1966
Harry Sudakoff Foundation Grants, 1967
Harry W. Bass, Jr. Foundation Grants, 1968
Hartford Courant Foundation Grants, 1969
Hartford Foundation Regular Grants, 1970
Harvest Foundation Grants, 1971
Harvey Randall Wickes Foundation Grants, 1972
Hasbro Children's Fund Grants, 1973
Hawaiian Electric Industries Charitable Foundation Grants, 1975
Hawaii Community Foundation Health Education and Research Grants, 1976
Hawaii Community Foundation Reverend Takie Okumura Family Grants, 1977
Hawaii Community Foundation West Hawaii Fund Grants, 1978
Hawn Foundation Grants, 1979
HCA Foundation Grants, 1980
Healthcare Foundation of New Jersey Grants, 1982
Health Foundation of Greater Cincinnati Grants, 1983
Health Foundation of Southern Florida Responsive Grants, 1985
Hearst Foundations Health Grants, 1987
Heckscher Foundation for Children Grants, 1988
Hedco Foundation Grants, 1989
Heineman Foundation for Research, Education, Charitable and Scientific Purposes, 1990
Helena Rubinstein Foundation Grants, 1991

872 / Educational Programs

Helen Bader Foundation Grants, 1992
Helen Gertrude Sparks Charitable Trust Grants, 1993
Helen Irwin Littauer Educational Trust Grants, 1994
Helen K. & Arthur E. Johnson Fndn Grants, 1995
Helen Pumphrey Denit Charitable Trust Grants, 1996
Helen S. Boylan Foundation Grants, 1997
Helen Steiner Rice Foundation Grants, 1998
Hendrick Foundation for Children Grants, 1999
Hendricks County Community Fndn Grants, 2000
Henrietta Tower Wurts Memorial Fndn Grants, 2002
Herbert A. & Adrian W. Woods Fndn Grants, 2007
Herbert H. & Grace A. Dow Fndn Grants, 2008
Hershey Company Grants, 2013
HHMI-NIBIB Interfaces Initiative Grants, 2014
HHMI Biomedical Research Grants for International Scientists in Canada and Latin America, 2018
HHMI Biomedical Research Grants for International Scientists in the Baltics, Central and Eastern Europe, Russia, and Ukraine, 2019
HHMI Grants and Fellowships Programs, 2020
HHMI Med into Grad Initiative Grants, 2022
Highmark BCBS Challenge for Healthier Schools in Western Pennsylvania, 3696
Highmark Corporate Giving Grants, 2025
High Meadow Foundation Grants, 2027
Hilda and Preston Davis Foundation Grants, 2028
Hill Crest Foundation Grants, 2030
Hillcrest Foundation Grants, 2031
Hillman Foundation Grants, 2032
Hillsdale County Community Foundation Healthy Senior/Healthy Youth Fund Grants, 2033
Hillsdale County Community General Adult Foundation Fund Grants, 2034
Hillsdale Fund Grants, 2035
Hilton Head Island Foundation Grants, 2036
Hilton Hotels Corporate Giving Program Grants, 2037
Hoblitzelle Foundation Grants, 2038
Hogg Foundation for Mental Health Grants, 2039
Hoglund Foundation Grants, 2040
Holland/Zeeland Community Fndn Grants, 2041
HomeBanc Foundation Grants, 2042
Home Building Industry Disaster Relief Fund, 2043
Homer Foundation Grants, 2044
Honda of America Manufacturing Fndn Grants, 2045
Horace A. Kimball and S. Ella Kimball Foundation Grants, 2046
Horace Moses Charitable Foundation Grants, 2047
Hospital Libraries Section / MLA Professional Development Grants, 2052
Houston Endowment Grants, 2053
Howard and Bush Foundation Grants, 2054
Howe Foundation of North Carolina Grants, 2055
HRAMF Harold S. Geneen Charitable Trust Awards for Coronary Heart Disease Research, 2059
HRSA Nurse Education, Practice, Quality and Retention (NEPQR) Grants, 2068
Huber Foundation Grants, 2071
Hudson Webber Foundation Grants, 2072
Hugh J. Andersen Foundation Grants, 2074
Huie-Dellmon Trust Grants, 2075
Huisking Foundation Grants, 2076
Humana Foundation Grants, 2077
Human Source Foundation Grants, 2078
Huntington Beach Police Officers Fndn Grants, 2081
Huntington Clinical Foundation Grants, 2082
Huntington County Community Foundation Make a Difference Grants, 2083
Huntington National Bank Community Grants, 2084
Hutchinson Community Foundation Grants, 2085
Hut Foundation Grants, 2086
Hutton Foundation Grants, 2087
I.A. O'Shaughnessy Foundation Grants, 2088
IBRO Latin America Regional Funding for Neuroscience Schools, 2105
IBRO Latin America Regional Funding for PROLAB Collaborations, 2106
Ida Alice Ryan Charitable Trust Grants, 2113
Idaho Community Foundation Eastern Region Competitive Grants, 2114
Idaho Power Company Corporate Contributions, 2115

Ida S. Barter Trust Grants, 2116
IIE 911 Armed Forces Scholarships, 2122
IIE African Center of Excellence for Women's Leadership Grants, 2124
IIE Chevron International REACH Scholarships, 2128
IIE David L. Boren Fellowships, 2129
IIE Freeman Foundation Indonesia Internships, 2131
IIE New Leaders Group Award for Mutual Understanding, 2141
Ike and Roz Friedman Foundation Grants, 2145
Illinois Tool Works Foundation Grants, 2147
Impact 100 Grants, 2148
Inasmuch Foundation Grants, 2149
Independence Blue Cross Nurse Scholars, 2151
Independence Blue Cross Nursing Internships, 2152
Independence Community Foundation Community Quality of Life Grant, 2153
Indiana AIDS Fund Grants, 2155
Infinity Foundation Grants, 2158
Ireland Family Foundation Grants, 2162
Irving S. Gilmore Foundation Grants, 2163
Irvin Stern Foundation Grants, 2164
Isabel Allende Foundation Esperanza Grants, 2165
Ittleson Foundation AIDS Grants, 2166
Ittleson Foundation Mental Health Grants, 2167
J.B. Reynolds Foundation Grants, 2168
J.C. Penney Company Grants, 2169
J.E. and L.E. Mabee Foundation Grants, 2170
J.L. Bedsole Foundation Grants, 2172
J.M. Long Foundation Grants, 2173
J.N. and Macie Edens Foundation Grants, 2174
J. Walton Bissell Foundation Grants, 2177
J. Willard Marriott, Jr. Foundation Grants, 2178
Jack H. & William M. Light Trust Grants, 2179
Jackson County Community Foundation Unrestricted Grants, 2180
Jacob and Valeria Langeloth Foundation Grants, 2182
Jacob G. Schmidlapp Trust Grants, 2183
Jacobs Family Village Neighborhoods Grants, 2184
James & Abigail Campbell Family Fndn Grants, 2185
James A. and Faith Knight Foundation Grants, 2186
James Ford Bell Foundation Grants, 2187
James Graham Brown Foundation Grants, 2189
James H. Cummings Foundation Grants, 2190
James J. & Angelia M. Harris Fndn Grants, 2192
James J. & Joan A. Gardner Family Fndn Grants, 2193
James L. and Mary Jane Bowman Charitable Trust Grants, 2194
James M. Collins Foundation Grants, 2195
James M. Cox Foundation of Georgia Grants, 2196
James R. Thorpe Foundation Grants, 2199
James S. Copley Foundation Grants, 2200
Jane's Trust Grants, 2206
Janirve Foundation Grants, 2209
Janus Foundation Grants, 2211
Jasper Foundation Grants, 2212
Jay and Rose Phillips Family Foundation Grants, 2213
Jayne and Leonard Abess Foundation Grants, 2214
Jean and Louis Dreyfus Foundation Grants, 2215
Jeffris Wood Foundation Grants, 2217
JELD-WEN Foundation Grants, 2218
Jenkins Foundation: Improving the Health of Greater Richmond Grants, 2219
Jennings County Community Foundation Grants, 2220
Jennings County Community Foundation Women's Giving Circle Grant, 2221
Jessica Stevens Community Foundation Grants, 2224
Jessie B. Cox Charitable Trust Grants, 2225
Jessie Ball Dupont Fund Grants, 2226
Jim Moran Foundation Grants, 2228
JM Foundation Grants, 2229
Joe W. and Dorothy Dorsett Brown Fndn Grants, 2230
John Clarke Trust Grants, 2232
John Deere Foundation Grants, 2234
John Edward Fowler Memorial Fndn Grants, 2235
John G. Duncan Charitable Trust Grants, 2236
John H. and Wilhelmina D. Harland Charitable Foundation Children and Youth Grants, 2238
John I. Smith Charities Grants, 2240
John J. Leidy Foundation Grants, 2241

PROGRAM TYPE INDEX

John Jewett and Helen Chandler Garland Foundation Grants, 2242
John Lord Knight Foundation Grants, 2243
John M. Lloyd Foundation Grants, 2244
John Merck Fund Grants, 2245
John P. McGovern Foundation Grants, 2247
John P. Murphy Foundation Grants, 2248
John R. Oishei Foundation Grants, 2249
Johns Manville Fund Grants, 2251
Johnson & Johnson Corporate Contributions, 2253
Johnson Controls Foundation Health and Social Services Grants, 2255
Johnson Foundation Wingspread Conference Support Program, 2257
John W. Alden Trust Grants, 2260
John W. and Anna H. Hanes Foundation Grants, 2261
John W. Anderson Foundation Grants, 2262
John W. Speas and Effie E. Speas Memorial Trust Grants, 2264
Joseph Alexander Foundation Grants, 2265
Joseph Collins Foundation Scholarships, 2267
Joseph Drown Foundation Grants, 2268
Joseph H. & Florence A. Roblee Fndn Grants, 2269
Joseph Henry Edmondson Foundation Grants, 2270
Josephine G. Russell Trust Grants, 2271
Josephine Goodyear Foundation Grants, 2272
Josephine S. Gumbiner Foundation Grants, 2273
Josiah Macy Jr. Foundation Grants, 2277
Josiah W. and Bessie H. Kline Foundation Grants, 2278
Judith and Jean Pape Adams Charitable Foundation Tulsa Area Grants, 2280
Julius N. Frankel Foundation Grants, 2282
Kahuku Community Fund, 2286
K and F Baxter Family Foundation Grants, 2287
Kansas Health Fndn Major Initiatives Grants, 2288
Katharine Matthies Foundation Grants, 2291
Katherine Baxter Memorial Foundation Grants, 2292
Kathryne Beynon Foundation Grants, 2294
Katrine Menzing Deakins Trust Grants, 2295
Kavli Foundation Research Grants, 2296
Kelvin and Eleanor Smith Foundation Grants, 2297
Kenai Peninsula Foundation Grants, 2298
Kendrick Foundation Grants, 2299
Kenneth T. & Eileen L. Norris Fndn Grants, 2300
Kent D. Steadley and Mary L. Steadley Memorial Trust Grants, 2301
Kessler Fnd Signature Employment Grants, 2302
Ketchikan Community Foundation Grants, 2303
Kettering Family Foundation Grants, 2304
Kevin P. and Sydney B. Knight Family Foundation Grants, 2306
Kluge Center David B. Larson Fellowship in Health and Spirituality, 2311
Knight Family Charitable and Educational Foundation Grants, 2312
Knox County Community Foundation Grants, 2313
Kodiak Community Foundation Grants, 2314
Komen Greater NYC Clinical Research Enrollment Grants, 2315
Komen Greater NYC Community Breast Health Grants, 2316
Kosciusko County Community Fndn Grants, 2319
Kovler Family Foundation Grants, 2323
Kroger Foundation Women's Health Grants, 2325
Kuntz Foundation Grants, 2326
L. W. Pierce Family Foundation Grants, 2328
LaGrange County Community Fndn Grants, 2329
Lalor Foundation Anna Lalor Burdick Grants, 2335
Lalor Foundation Postdoctoral Fellowships, 2336
Land O'Lakes Foundation Mid-Atlantic Grants, 2341
Lands' End Corporate Giving Program, 2344
Laura B. Vogler Foundation Grants, 2345
Laura Moore Cunningham Foundation Grants, 2346
Lawrence Foundation Grants, 2347
Leahi Fund, 2350
Leo Goodwin Foundation Grants, 2354
Leon and Thea Koerner Foundation Grants, 2355
Leonard and Helen R. Stulman Charitable Foundation Grants, 2356
Leonard L. & Bertha U. Abess Fndn Grants, 2357

PROGRAM TYPE INDEX   Educational Programs / 873

Leo Niessen Jr., Charitable Trust Grants, 2358
Lester E. Yeager Charitable Trust B Grants, 2359
Lester Ray Fleming Scholarships, 2360
Lewis H. Humphreys Charitable Trust Grants, 2361
Liberty Bank Foundation Grants, 2362
Libra Foundation Grants, 2363
Lincoln Financial Foundation Grants, 2367
Lisa and Douglas Goldman Fund Grants, 2369
Long Island Community Foundation Grants, 2375
Lotus 88 Fnd for Women & Children Grants, 2376
Louetta M. Cowden Foundation Grants, 2377
Louie M. & Betty M. Phillips Fndn Grants, 2378
Louis and Elizabeth Nave Flarsheim Charitable Foundation Grants, 2379
Louis H. Aborn Foundation Grants, 2380
Lowe Foundation Grants, 2381
Lubbock Area Foundation Grants, 2383
Lubrizol Foundation Grants, 2384
Lucile Horton Howe and Mitchell B. Howe Foundation Grants, 2385
Lucy Downing Nisbet Charitable Fund Grants, 2387
Lucy Gooding Charitable Fndn Grants, 2388
Lydia deForest Charitable Trust Grants, 2393
Lynde and Harry Bradley Foundation Grants, 2397
Lynn and Rovena Alexander Family Foundation Grants, 2399
M-A-C AIDS Fund Grants, 2400
M.B. and Edna Zale Foundation Grants, 2401
M. Bastian Family Foundation Grants, 2402
M.E. Raker Foundation Grants, 2404
M.J. Murdock Charitable Trust General Grants, 2405
Mabel A. Horne Trust Grants, 2406
Mabel F. Hoffman Charitable Trust Grants, 2407
Mabel Y. Hughes Charitable Trust Grants, 2409
Macquarie Bank Foundation Grants, 2410
Madison County Community Foundation General Grants, 2413
Majors MLA Chapter Project of the Year, 2418
Manuel D. & Rhoda Mayerson Fndn Grants, 2420
March of Dimes Agnes Higgins Award, 2423
March of Dimes Program Grants, 2426
Marcia and Otto Koehler Foundation Grants, 2427
Mardag Foundation Grants, 2428
Margaret T. Morris Foundation Grants, 2429
Marie C. and Joseph C. Wilson Foundation Rochester Small Grants, 2431
Marie H. Bechtel Charitable Remainder Uni-Trust Grants, 2432
Marin Community Foundation Improving Community Health Grants, 2433
Marion Gardner Jackson Charitable Trust Grants, 2434
Marion I. and Henry J. Knott Foundation Discretionary Grants, 2435
Marion I. and Henry J. Knott Foundation Standard Grants, 2436
Marjorie Moore Charitable Foundation Grants, 2438
Marshall County Community Foundation Grants, 2439
Mary Black Foundation Active Living Grants, 2440
Mary Black Fndn Community Health Grants, 2441
Mary Black Foundation Early Childhood Development Grants, 2442
Mary K. Chapman Foundation Grants, 2443
Mary L. Peyton Foundation Grants, 2445
Mary Owen Borden Foundation Grants, 2446
Mary S. and David C. Corbin Foundation Grants, 2447
Mary Wilmer Covey Charitable Trust Grants, 2448
Matilda R. Wilson Fund Grants, 2453
Mattel Children's Foundation Grants, 2454
Mattel International Grants Program, 2455
Maurice J. Masserini Charitable Trust Grants, 2456
Max and Victoria Dreyfus Foundation Grants, 2457
Maxon Charitable Foundation Grants, 2458
May and Stanley Smith Charitable Trust Grants, 2459
McCallum Family Foundation Grants, 2489
McCarthy Family Foundation Grants, 2490
McCombs Foundation Grants, 2491
McConnell Foundation Grants, 2492
McGraw-Hill Companies Community Grants, 2494
McInerny Foundation Grants, 2495
McKesson Foundation Grants, 2496

McLean Contributionship Grants, 2497
McLean Foundation Grants, 2498
McMillen Foundation Grants, 2499
Mead Johnson Nutritionals Evansville-Area Organizations Grants, 2503
Mead Johnson Nutritionals Med Educ Grants, 2504
Mead Witter Foundation Grants, 2507
Medical Informatics Section/MLA Career Development Grant, 2508
MedImmune Charitable Grants, 2509
Medtronic Foundation CommunityLink Health Grants, 2510
Medtronic Foundation HeartRescue Grants, 2513
Medtronic Foundation Patient Link Grants, 2514
Medtronic Foundation Strengthening Health Systems Grants, 2515
Mercedes-Benz USA Corporate Contributions, 2517
Mericos Foundation Grants, 2518
Meriden Foundation Grants, 2519
Merkel Foundation Grants, 2520
Merrick Foundation Grants, 2522
Merrick Foundation Grants, 2521
Meta and George Rosenberg Foundation Grants, 2525
MetroWest Health Foundation Grants to Reduce the Incidence of High Risk Behaviors Among Adolescents, 2528
MetroWest Health Foundation Grants to Schools to Conduct Mental Health Capacity Assessments, 2529
Metzger-Price Fund Grants, 2530
Meyer Foundation Healthy Communities Grants, 2533
Meyer Memorial Trust Grassroots Grants, 2536
Meyer Memorial Trust Responsive Grants, 2537
Meyer Memorial Trust Special Grants, 2538
MGFA Post-Doctoral Research Fellowships, 2539
MGM Resorts Foundation Community Grants, 2541
MGN Family Foundation Grants, 2542
Miami County Community Foundation - Operation Round Up Grants, 2543
Milagro Foundation Grants, 2552
Miles of Hope Breast Cancer Foundation Grants, 2553
Military Ex-Prisoners of War Foundation Grants, 2554
Milken Family Foundation Grants, 2555
Miller Foundation Grants, 2556
Mimi and Peter Haas Fund Grants, 2557
Minneapolis Foundation Community Grants, 2558
MLA Continuing Education Grants, 2559
MLA Cunningham Memorial International Fellowship, 2560
MLA Grad Scholarship for Minority Students, 2565
MLA Section Project of the Year Award, 2574
MMAAP Fndn Dermatology Project Awards, 2578
MMS and Alliance Charitable Foundation Grants for Community Action and Care for the Medically Uninsured, 2587
Mockingbird Foundation Grants, 2589
Moline Foundation Community Grants, 2590
Montgomery County Community Fndn Grants, 2591
Morehouse PHSI Project Imhotep Internships, 2592
Morris & Gwendolyn Cafritz Fndn Grants, 2594
Ms. Fndn for Women Ending Violence Grants, 2596
Ms. Foundation for Women Health Grants, 2597
Mt. Sinai Health Care Foundation Academic Medicine and Bioscience Grants, 2598
Mt. Sinai Health Care Foundation Health of the Urban Community Grants, 2600
NAPNAP Fndn Elaine Gelman Scholarship, 2609
NAPNAP Fndn Grad Research Grant, 2610
NAPNAP Foundation McNeil PNP Scholarships, 2613
NAPNAP Foundation McNeil Rural and Underserved Scholarships, 2614
NAPNAP Foundation Shourd Parks Immunization Project Small Grants, 2617
NAPNAP Foundation Wyeth Pediatric Immunization Grant, 2618
Natalie W. Furniss Charitable Trust Grants, 2626
Nathan Cummings Foundation Grants, 2627
Nelda C. and H.J. Lutcher Stark Fndn Grants, 2672
Nell J. Redfield Foundation Grants, 2673
NHLBI Investigator Initiated Multi-Site Clinical Trials, 2703

NHLBI Mentored Career Award for Faculty at Institutions that Promote Diversity (K01), 2706
NHLBI Research Supplements to Promote Diversity in Health-Related Research, 2727
NHLBI Summer Internship Program in Biomedical Research, 2735
NHSCA Arts in Health Care Project Grants, 2741
Nicholas H. Noyes Jr. Memorial Fndn Grants, 2791
NIEHS Hazardous Materials Worker Health and Safety Training Grants, 2854
NIEHS Hazmat Training at Doe Nuclear Weapons Complex Grants, 2855
NIEHS SBIR E-Learning for HAZMAT and Emergency Response Grants, 2859
NIH Academic Research Enhancement Awards, 2864
NIH School Interventions to Prevent Obesity, 2908
NIH Summer Internship Program in Biomedical Research, 2912
NIMH Jointly Sponsored Ruth L. Kirschstein National Research Service Award Institutional Predoctoral Training Program in the Neurosciences, 2919
NLM Understanding and Promoting Health Literacy Research Grants, 2953
NMF General Electric Med Fellowships, 2956
NMF National Medical Association Awards for Medical Journalism, 2962
NMF National Medical Association Emerging Scholar Awards, 2963
NMF National Medical Association Patti LaBelle Award, 2964
NMF William and Charlotte Cadbury Award, 2967
NMSS Scholarships, 2968
Noble County Community Foundation Grants, 2975
Norcliffe Foundation Grants, 2979
Nord Family Foundation Grants, 2980
Nordson Corporation Foundation Grants, 2981
Norfolk Southern Foundation Grants, 2982
Norman Foundation Grants, 2983
North Carolina Biotechnology Center Event Sponsorship Grants, 2984
North Carolina Community Foundation Grants, 2990
North Carolina GlaxoSmithKline Fndn Grants, 2991
North Central Health Services Grants, 2992
North Central Health Services Grants, 2993
North Dakota Community Foundation Grants, 2994
Northern Chautauqua Community Foundation Community Grants, 2995
Northern New York Community Fndn Grants, 2996
Northrop Grumman Corporation Grants, 2997
Northwestern Mutual Foundation Grants, 2999
Northwest Minnesota Foundation Women's Fund Grants, 3000
Norwin S. & Elizabeth N. Bean Fndn Grants, 3001
Notsew Orm Sands Foundation Grants, 3002
NSF Chemistry Research Experiences for Undergraduates (REU), 3012
NSF Instrument Development for Bio Research, 3017
NSF Postdoc Research Fellowships in Biology, 3020
NSF Presidential Early Career Awards for Scientists and Engineers (PECASE) Grants, 3021
NSF Undergraduate Research and Mentoring in the Biological Sciences (URM), 3025
Nuffield Foundation Africa Program Grants, 3026
NYCT AIDS/HIV Grants, 3048
NYCT Girls and Young Women Grants, 3053
NYCT Substance Abuse Grants, 3057
Oak Foundation Child Abuse Grants, 3059
Oceanside Charitable Foundation Grants, 3063
Office Depot Corporation Community Relations Grants, 3065
Ohio County Community Foundation Board of Directors Grants, 3067
Ohio County Community Foundation Grants, 3068
Ohio County Community Fndn Mini-Grants, 3069
Ohio Learning Network Grants, 3070
Olenda Jameson Trust Grants, 3071
Olive Higgins Prouty Foundation Grants, 3072
OneFamily Foundation Grants, 3075
OneSight Research Foundation Block Grants, 3076
Oppenstein Brothers Foundation Grants, 3078

Orange County Community Foundation Grants, 3079
Ordean Foundation Grants, 3080
Oregon Community Fndn Community Grants, 3082
Oscar Rennebohm Foundation Grants, 3085
OSF Health Media Initiative Grants, 3089
OSF International Palliative Care Grants, 3091
OSF Tackling Addiction Grants in Baltimore, 3096
Otto Bremer Foundation Grants, 3099
Owen County Community Foundation Grants, 3100
PACCAR Foundation Grants, 3101
PacifiCare Foundation Grants, 3102
Pacific Life Foundation Grants, 3103
Packard Foundation Children, Families, and Communities Grants, 3104
Packard Foundation Local Grants, 3105
Pajaro Valley Community Health Trust Oral Health: Prevention & Access Grants, 3110
Pajaro Valley Community Health Trust Promoting Entry & Advancement in the Health Professions Grants, 3111
Parkersburg Area Community Foundation Action Grants, 3116
Partnership for Cures Charles E. Culpeper Scholarships in Medical Science, 3117
Paso del Norte Health Foundation Grants, 3119
Patrick and Anna M. Cudahy Fund Grants, 3120
Paul Ogle Foundation Grants, 3124
Peacock Foundation Grants, 3133
PepsiCo Foundation Grants, 3137
Percy B. Ferebee Endowment Grants, 3138
Perkins Charitable Foundation Grants, 3140
Perpetual Trust for Charitable Giving Grants, 3141
Perry County Community Foundation Grants, 3142
Peter and Elizabeth C. Tower Foundation Annual Intellectual Disabilities Grants, 3144
Peter and Elizabeth C. Tower Foundation Annual Mental Health Grants, 3145
Peter and Elizabeth C. Tower Foundation Learning Disability Grants, 3146
Peter and Elizabeth C. Tower Foundation Mental Health Reference and Resource Materials Mini-Grants, 3147
Peter and Elizabeth C. Tower Foundation Organizational Scholarships, 3148
Peter and Elizabeth C. Tower Foundation Phase II Technology Initiative Grants, 3149
Peter and Elizabeth C. Tower Foundation Phase I Technology Initiative Grants, 3150
Peter and Elizabeth C. Tower Foundation Social and Emotional Preschool Curriculum Grants, 3151
Peter and Elizabeth C. Tower Foundation Substance Abuse Grants, 3152
Peter Kiewit Foundation General Grants, 3154
Peter Kiewit Foundation Small Grants, 3155
Petersburg Community Foundation Grants, 3156
Pew Charitable Trusts Children & Youth Grants, 3158
PeyBack Foundation Grants, 3159
Peyton Anderson Foundation Grants, 3160
Pfizer/AAFP Foundation Visiting Professorship Program in Family Medicine, 3162
Pfizer Healthcare Charitable Contributions, 3180
Pfizer Medical Education Track One Grants, 3182
Pfizer Medical Education Track Two Grants, 3183
Phelps County Community Foundation Grants, 3188
Philadelphia Foundation Organizational Effectiveness Grants, 3189
Philip L. Graham Fund Health and Human Services Grants, 3190
Phi Upsilon Omicron Geraldine Clewell Senior Awards, 3194
Phi Upsilon Omicron Janice Cory Bullock Collegiate Award, 3195
Phi Upsilon Lillian P. Schoephoerster Award, 3196
Phi Upsilon Orinne Johnson Writing Award, 3198
Phi Upsilon Omicron Sarah Thorniley Phillips Leader Awards, 3199
Phi Upsilon Omicron Undergraduate Karen P. Goebel Conclave Award, 3200
Phoenix Suns Charities Grants, 3201

PhRMA Foundation Informatics Pre Doctoral Fellowships, 3207
PhRMA Foundation Informatics Research Starter Grants, 3208
PhRMA Fndn Info Sabbatical Fellowships, 3209
PhRMA Foundation Paul Calabresi Medical Student Research Fellowship, 3210
PhRMA Foundation Pharmaceutics Postdoctoral Fellowships, 3211
PhRMA Foundation Pharmaceutics Pre Doctoral Fellowships, 3212
PhRMA Foundation Pharmaceutics Research Starter Grants, 3213
PhRMA Foundation Pharmaceutics Sabbatical Fellowships, 3214
PhRMA Foundation Pharmacology/Toxicology Sabbatical Fellowships, 3218
Piedmont Health Foundation Grants, 3219
Piedmont Natural Gas Corporate and Charitable Contributions, 3220
Pike County Community Foundation Grants, 3222
Pinellas County Grants, 3223
Pinnacle Foundation Grants, 3225
Piper Trust Older Adults Grants, 3229
Pittsburgh Foundation Community Fund Grants, 3230
Playboy Foundation Grants, 3237
Plough Foundation Grants, 3238
PMI Foundation Grants, 3239
PNC Foundation Community Services Grants, 3240
Pohlad Family Foundation, 3241
Pokagon Fund Grants, 3242
Polk Bros. Foundation Grants, 3243
Pollock Foundation Grants, 3244
Porter County Health and Wellness Grant, 3246
Porter County Women's Grant, 3248
Portland Fndn Women's Giving Circle Grant, 3249
Portland General Electric Foundation Grants, 3250
Posey Community Fndn Women's Fund Grants, 3251
Posey County Community Foundation Grants, 3252
Pott Foundation Grants, 3253
Potts Memorial Foundation Grants, 3254
Powell Foundation Grants, 3255
PPCF Community Grants, 3256
PPG Industries Foundation Grants, 3257
Premera Blue Cross Grants, 3258
Presbyterian Health Foundation Bridge, Seed and Equipment Grants, 3259
Price Chopper's Golub Foundation Grants, 3261
Price Family Charitable Fund Grants, 3262
Priddy Foundation Program Grants, 3266
Pride Foundation Grants, 3268
Prince Charitable Trusts Chicago Grants, 3270
Prince Charitable Trusts DC Grants, 3271
Princeton Area Community Foundation Greater Mercer Grants, 3272
Principal Financial Group Foundation Grants, 3273
Procter and Gamble Fund Grants, 3274
Prudential Foundation Education Grants, 3276
PSG Mentored Clinical Research Awards, 3277
Puerto Rico Community Foundation Grants, 3278
Pulaski County Community Foundation Grants, 3279
Pulido Walker Foundation, 3282
Putnam County Community Foundation Grants, 3283
PVA Education Foundation Grants, 3284
Quaker Chemical Foundation Grants, 3286
Qualcomm Grants, 3287
Quality Health Foundation Grants, 3288
Quantum Foundation Grants, 3290
Questar Corporate Contributions Grants, 3291
R.C. Baker Foundation Grants, 3292
R.S. Gernon Trust Grants, 3293
Rainbow Fund Grants, 3301
Rajiv Gandhi Foundation Grants, 3302
Ralphs Food 4 Less Foundation Grants, 3305
Raskob Foundation for Catholic Activities Grants, 3306
Rasmuson Foundation Tier One Grants, 3307
Rasmuson Foundation Tier Two Grants, 3308
Rathmann Family Foundation Grants, 3309
Raymond John Wean Foundation Grants, 3310
RCF General Community Grants, 3312

RCF Individual Assistance Grants, 3313
RCF Summertime Kids Grants, 3314
RCF The Women's Fund Grants, 3315
RCPSC Balfour M. Mount Visiting Professorship in Palliative Medicine, 3318
RCPSC Tom Dignan Indigenous Health Award, 3321
RCPSC Janes Visiting Professorship in Surgery, 3328
RCPSC KJR Wightman Visiting Professorship in Medicine, 3331
RCPSC McLaughlin-Gallie Visiting Professor, 3333
RCPSC Medical Education Research Grants, 3334
RCPSC Professional Development Grants, 3338
Regence Foundation Health Care Community Awareness and Engagement Grants, 3347
Regence Fndn Improving End-of-Life Grants, 3349
Rehabilitation Nursing Foundation Scholarships, 3355
Reinberger Foundation Grants, 3357
Retirement Research Foundation General Program Grants, 3359
Reynolds & Reynolds Associate Fndn Grants, 3360
RGk Foundation Grants, 3361
Rhode Island Foundation Grants, 3362
Rice Foundation Grants, 3363
Richard & Susan Smith Family Fndn Grants, 3365
Richard D. Bass Foundation Grants, 3366
Richard Davoud Donchian Foundation Grants, 3367
Richard King Mellon Foundation Grants, 3369
Richland County Bank Grants, 3371
Richmond Eye and Ear Fund Grants, 3372
Ricks Family Charitable Trust Grants, 3373
Riley Foundation Grants, 3374
Ripley County Community Foundation Grants, 3376
Ripley County Community Foundation Small Project Grants, 3377
Robert and Joan Dircks Foundation Grants, 3379
Robert B. Adams Foundation Scholarships, 3380
Robert & Clara Milton Senior Housing Grants, 3384
Robert R. McCormick Trib Community Grants, 3385
Robert R. McCormick Tribune Veterans Initiative Grants, 3386
Robert R. Meyer Foundation Grants, 3387
Robert W. Woodruff Foundation Grants, 3389
Robins Foundation Grants, 3390
Rochester Area Community Foundation Grants, 3391
Rochester Area Foundation Grants, 3392
Rockefeller Brothers Fund Pivotal Places Grants: South Africa, 3395
Rockwell Collins Charitable Corporation Grants, 3397
Rockwell Fund, Inc. Grants, 3398
Rohm and Haas Company Grants, 3403
Rollins-Luetkemeyer Foundation Grants, 3404
Ronald McDonald House Charities Grants, 3405
Rosalynn Carter Institute John and Betty Pope Fellowships, 3408
Rosalynn Carter Institute Mattie J. T. Stepanek Caregiving Scholarships, 3409
Rose Hills Foundation Grants, 3413
Roy & Christine Sturgis Charitable Grants, 3415
RRF General Program Grants, 3417
RSC McNeil Medal for the Public Awareness of Science, 3422
Rush County Community Foundation Grants, 3424
Ruth Anderson Foundation Grants, 3425
Ruth and Vernon Taylor Foundation Grants, 3426
Ruth Eleanor Bamberger and John Ernest Bamberger Memorial Foundation Grants, 3427
Ruth H. & Warren A. Ellsworth Fndn Grants, 3429
RWJF Health and Society Scholars, 3436
RWJF New Careers in Nursing Scholarships, 3439
S.H. Cowell Foundation Grants, 3447
S. Mark Taper Foundation Grants, 3449
Saigh Foundation Grants, 3451
Sain-Orr and Royak-DeForest Steadman Foundation Grants, 3452
Salem Foundation Charitable Trust Grants, 3456
Salmon Foundation Grants, 3458
Samueli Foundation Health Grants, 3465
Samuel N. and Mary Castle Foundation Grants, 3467
Samuel S. Johnson Foundation Grants, 3468
San Antonio Area Foundation Grants, 3469

PROGRAM TYPE INDEX / Educational Programs / 875

San Diego Foundation for Change Grants, 3470
San Diego HIV Funding Collaborative Grants, 3473
San Diego Women's Foundation Grants, 3474
Sandy Hill Foundation Grants, 3477
San Francisco Fndn Community Health Grants, 3478
San Juan Island Community Foundation Grants, 3480
Santa Fe Community Foundation Seasonal Grants-Fall Cycle, 3482
Sarkeys Foundation Grants, 3485
Sartain Lanier Family Foundation Grants, 3486
Sasco Foundation Grants, 3487
Schering-Plough Foundation Health Grants, 3492
Schlessman Family Foundation Grants, 3493
Schramm Foundation Grants, 3495
Scott B. & Annie P. Appleby Charitable Grants, 3499
Scott County Community Foundation Grants, 3500
Seabury Foundation Grants, 3501
Seagate Tech Corp Capacity to Care Grants, 3502
Seattle Foundation Benjamin N. Phillips Memorial Fund Grants, 3504
Seaver Institute Grants, 3509
Self Foundation Grants, 3510
Seneca Foods Foundation Grants, 3511
Sensient Technologies Foundation Grants, 3513
Seward Community Foundation Grants, 3514
Seward Community Foundation Mini-Grants, 3515
SfN Award for Education in Neuroscience, 3517
SfN Science Educator Award, 3535
Shaw's Supermarkets Donations, 3542
Shell Deer Park Grants, 3543
Shell Oil Company Foundation Community Development Grants, 3544
Sheltering Arms Fund Grants, 3545
Shield-Ayres Foundation Grants, 3546
Shopko Fndn Community Charitable Grants, 3547
Shopko Foundation Green Bay Area Community Grants, 3548
Sidgmore Family Foundation Grants, 3551
Sidney Stern Memorial Trust Grants, 3552
Sid W. Richardson Foundation Grants, 3553
Siebert Lutheran Foundation Grants, 3554
Sierra Health Foundation Responsive Grants, 3555
Simmons Foundation Grants, 3563
Simpson Lumber Charitable Contributions, 3566
Singing for Change Foundation Grants, 3567
Sioux Falls Area Community Foundation Community Fund Grants (Unrestricted), 3568
Sioux Falls Area Community Foundation Field-of-Interest and Donor-Advised Grants, 3569
Sioux Falls Area Community Foundation Spot Grants (Unrestricted), 3570
Sir Dorabji Tata Trust Grants for NGOs or Voluntary Organizations, 3572
Sisters of Charity Foundation of Cleveland Reducing Health and Educational Disparities in the Central Neighborhood Grants, 3574
Sisters of Mercy of North Carolina Fndn Grants, 3575
Sisters of St. Joseph Healthcare Fndn Grants, 3576
Skoll Fndn Awards for Social Entrepreneurship, 3579
SNM PDEF Mickey Williams Minority Student Scholarships, 3584
SNMTS Paul Cole Scholarships, 3593
Society for Imaging Informatics in Medicine (SIIM) Emeritus Mentor Grants, 3597
Society for Imaging Informatics in Medicine (SIIM) Micro Grants, 3598
Society for Imaging Informatics in Medicine (SIIM) Research Grants, 3599
Society for Imaging Informatics in Medicine (SIIM) Small Training Grants, 3601
Sonoco Foundation Grants, 3621
Sonora Area Foundation Competitive Grants, 3622
Sony Corporation of America Grants, 3623
Southbury Community Trust Fund, 3625
South Madison Community Foundation Grants, 3626
Southwest Florida Community Foundation Competitive Grants, 3627
Special Olympics Health Profess Student Grants, 3629
Special People in Need Grants, 3630
Spencer County Community Foundation Grants, 3632

Springs Close Foundation Grants, 3633
Square D Foundation Grants, 3634
St. Joseph Community Health Foundation Burn Care and Prevention Grants, 3639
St. Joseph Community Health Foundation Schneider Fellowships, 3642
Stackpole-Hall Foundation Grants, 3643
Stark Community Fndn Women's Grants, 3645
Staunton Farm Foundation Grants, 3649
Steele-Reese Foundation Grants, 3650
Sterling-Turner Charitable Foundation Grants, 3652
Sterling and Shelli Gardner Foundation Grants, 3653
Steuben County Community Foundation Grants, 3654
Stewart Huston Charitable Trust Grants, 3655
Stocker Foundation Grants, 3656
Strake Foundation Grants, 3658
Stranahan Foundation Grants, 3659
Streilein Foundation for Ocular Immunology Visiting Professorships, 3660
Strowd Roses Grants, 3664
Subaru of Indiana Automotive Foundation Grants, 3672
Summer Medical and Dental Education Program: Interprofessional Pilot Grants, 3673
Summit Foundation Grants, 3674
Sunderland Foundation Grants, 3675
Sunoco Foundation Grants, 3679
SunTrust Bank Trusteed Foundations Florence C. and Harry L. English Memorial Fund Grants, 3680
SunTrust Bank Trusteed Foundations Greene-Sawtell Grants, 3681
SunTrust Bank Trusteed Foundations Harriet McDaniel Marshall Tust Grants, 3682
SunTrust Bank Trusteed Foundations Nell Warren Elkin and William Simpson Elkin Grants, 3683
SunTrust Bank Trusteed Foundations Thomas Guy Woolford Charitable Trust Grants, 3684
SunTrust Bank Trusteed Foundations Walter H. and Marjory M. Rich Memorial Fund Grants, 3685
Support Our Aging Religious (SOAR) Grants, 3686
Susan G. Komen Breast Cancer Foundation College Scholarships, 3692
Susan Mott Webb Charitable Trust Grants, 3694
Susan Vaughan Foundation Grants, 3695
Symantec Community Relations and Corporate Philanthropy Grants, 3699
T.L.L. Temple Foundation Grants, 3700
T. Spencer Shore Foundation Grants, 3701
Tauck Family Foundation Grants, 3703
Taylor S. Abernathy and Patti Harding Abernathy Charitable Trust Grants, 3704
TE Foundation Grants, 3705
Tellabs Foundation Grants, 3706
Tension Envelope Foundation Grants, 3707
Texas Commission on the Arts Arts Respond Project Grants, 3708
Texas Instruments Foundation STEM Education Grants, 3711
Textron Corporate Contributions Grants, 3712
Thelma Braun & Bocklett Family Fndn Grants, 3713
Thelma Doelger Charitable Trust Grants, 3714
Theodore Edson Parker Foundation Grants, 3715
The Ray Charles Foundation Grants, 3716
Thomas and Dorothy Leavey Foundation Grants, 3717
Thomas Austin Finch, Sr. Foundation Grants, 3718
Thomas C. Ackerman Foundation Grants, 3719
Thomas J. Atkins Memorial Trust Fund Grants, 3721
Thomas J. Long Foundation Community Grants, 3722
Thomas Jefferson Rosenberg Foundation Grants, 3723
Thomas Sill Foundation Grants, 3724
Thomas Thompson Trust Grants, 3725
Thompson Charitable Foundation Grants, 3727
Tifa Foundation Grants, 3733
Tipton County Foundation Grants, 3734
TJX Foundation Grants, 3735
Todd Brock Family Foundation Grants, 3736
Tommy Hilfiger Corporate Foundation Grants, 3737
Topeka Community Foundation Kansas Blood Services Scholarships, 3738
Toyota Motor Engineering & Manufacturing North America Grants, 3742

Toyota Motor Manufacturing of Alabama Grants, 3743
Toyota Motor Manufacturing of Indiana Grants, 3744
Toyota Motor Manufacturing of Kentucky Grants, 3745
Toyota Motor Manuf of Mississippi Grants, 3746
Toyota Motor Manufacturing of Texas Grants, 3747
Toyota Motor Manuf of West Virginia Grants, 3748
Toyota Motor N America of New York Grants, 3749
Toyota Motor Sales, USA Grants, 3750
Toys R Us Children's Fund Grants, 3751
Trauma Center of America Finance Fellowship, 3752
Tri-State Community Twenty-first Century Endowment Fund Grants, 3756
Triangle Community Foundation Donor-Advised Grants, 3757
Triangle Community Foundation Shaver-Hitchings Scholarship, 3758
Trull Foundation Grants, 3761
Tull Charitable Foundation Grants, 3762
Turner Foundation Grants, 3763
Tylenol Future Care Scholarships, 3764
U.S. Cellular Corporation Grants, 3766
U.S. Department of Education 21st Century Community Learning Centers, 3767
U.S. Department of Education Erma Byrd Scholarships, 3768
U.S. Department of Education Rehabilitation Engineering Research Centers Grants, 3769
U.S. Department of Education Rehabilitation Research Training Centers (RRTCs), 3770
U.S. Department of Education Rehabilitation Training - Rehabilitation Continuing Education - Institute on Rehabilitation Issues, 3771
U.S. Dept of Education Special Ed--National Activities--Parent Information Centers, 3772
U.S. Department of Education United States-Russia Program: Improving Research and Educational Activities in Higher Education, 3773
U.S. Department of Education Vocational Rehabilitation Services Projects for American Indians with Disabilities Grants, 3774
UCLA / RAND Corporation Post-Doctoral Fellowships, 3776
UniHealth Foundation Workforce Development Grants, 3786
Union Bank, N.A. Corporate Sponsorships and Donations, 3787
Union Bank, N.A. Foundation Grants, 3788
Union Benevolent Association Grants, 3789
Union County Community Foundation Grants, 3790
United Methodist Committee on Relief Global AIDS Fund Grants, 3798
United Methodist Health Ministry Fund Grants, 3801
United Technologies Corporation Grants, 3803
Unity Foundation Of LaPorte County Grants, 3804
USAID Child, Newborn, and Maternal Health Project Grants, 3806
USAID Family Health Plus Project Grants, 3810
USAID Health System Strengthening Grants, 3814
USAID HIV Prevention with Key Populations - Mali Grants, 3816
USAID Microbicides Research, Development, and Introduction Grants, 3819
USAID Policy, Advocacy, and Communication Enhanced for Population and Reproductive Health Grants, 3820
US Airways Community Contributions, 3825
US Airways Community Foundation Grants, 3826
USDA Delta Health Care Services Grants, 3829
USG Foundation Grants, 3848
Vancouver Foundation Grants and Community Initiatives Program, 3849
Vancouver Fndn Public Health Bursary Package, 3850
Vanderburgh Community Foundation Grants, 3852
Vanderburgh Community Fndn Women's Fund, 3853
VDH Commonwealth of Virginia Nurse Educator Scholarships, 3854
VDH Emergency Med Services Training Grants, 3855
Vectren Foundation Grants, 3857
Verizon Foundation Maryland Grants, 3859
Verizon Foundation New York Grants, 3860

876 / Educational Programs

Verizon Foundation Northeast Region Grants, 3861
Verizon Foundation Pennsylvania Grants, 3862
Verizon Foundation Vermont Grants, 3863
Verizon Foundation Virginia Grants, 3864
Vermont Community Foundation Grants, 3865
Victor E. Speas Foundation Grants, 3868
Victor Grifols i Lucas Foundation Ethics and Science Awars for Educational Institutions, 3869
Victor Grifols i Lucas Foundation Secondary School Prizes, 3872
Vigneron Memorial Fund Grants, 3873
Virginia Department of Health Mary Marshall Nursing Scholarships for Licensed Practical Nurses, 3874
Virginia Department of Health Mary Marshall Nursing Scholarships for Registered Nurses, 3875
Virginia Department of Health Nurse Practitioner/ Nurse Midwife Scholarships, 3876
Virginia W. Kettering Foundation Grants, 3878
Visiting Nurse Foundation Grants, 3879
Volkswagen of America Corporate Contributions, 3880
W.C. Griffith Foundation Grants, 3881
W.H. & Mary Ellen Cobb Trust Grants, 3884
W.K. Kellogg Foundation Healthy Kids Grants, 3886
W.K. Kellogg Foundation Secure Families Grants, 3887
W.M. Keck Fndn So California Grants, 3890
W.M. Keck Fndn Undergrad Ed Grants, 3891
W.P. and Bulah Luse Foundation Grants, 3892
Wabash Valley Community Foundation Grants, 3893
Walker Area Community Foundation Grants, 3895
Walmart Foundation Community Giving Grants, 3897
Walmart Foundation National Giving Grants, 3898
Walmart Foundation Northwest Arkansas Giving, 3899
Walmart Foundation State Giving Grants, 3900
Walter L. Gross III Family Foundation Grants, 3901
Warrick County Community Foundation Grants, 3904
Warrick County Community Foundation Women's Fund, 3905
Washington County Community Fndn Grants, 3906
Washington County Community Foundation Youth Grants, 3907
Washington Gas Charitable Giving, 3908
Wayne County Foundation Grants, 3909
Weaver Popcorn Foundation Grants, 3910
Weingart Foundation Grants, 3913
Welborn Baptist Fndn General Op Grants, 3914
Welborn Baptist Fndn School Health Grants, 3917
Wells County Foundation Grants, 3920
Western Indiana Community Foundation Grants, 3921
Western New York Foundation Grants, 3922
Western Union Foundation Grants, 3923
WestWind Foundation Reproductive Health and Rights Grants, 3924
Weyerhaeuser Company Foundation Grants, 3925
White County Community Foundation - Women Giving Together Grants, 3933
White County Community Foundation Grants, 3934
Whitney Foundation Grants, 3938
WHO Foundation General Grants, 3939
WHO Foundation Volunteer Service Grants, 3940
Willary Foundation Grants, 3944
William A. Badger Foundation Grants, 3945
William B. Stokely Jr. Foundation Grants, 3947
William Blair and Company Foundation Grants, 3948
William D. Laurie, Jr. Charitable Fndn Grants, 3949
William G. & Helen C. Hoffman Fndn Grants, 3950
William G. Gilmore Foundation Grants, 3951
William H. Hannon Foundation Grants, 3953
William J. & Tina Rosenberg Fndn Grants, 3954
William J. Brace Charitable Trust, 3955
William Ray & Ruth E. Collins Fndn Grants, 3956
William T. Grant Foundation Research Grants, 3958
Willis C. Helm Charitable Trust, 3959
Wilson-Wood Foundation Grants, 3960
Windham Foundation Grants, 3961
Winifred & Harry B. Allen Foundation Grants, 3962
Winston-Salem Foundation Competitive Grants, 3965
Winston-Salem Fndn Elkin/Tri-County Grants, 3966
Winston-Salem Fndn Stokes County Grants, 3967
Wolfe Associates Grants, 3969

Women of the ELCA Opportunity Scholarships for Lutheran Laywomen, 3970
Wood-Claeyssens Foundation Grants, 3971
Wood Family Charitable Trust Grants, 3972
Woodward Fund Grants, 3973
Yampa Valley Community Foundation Cody St. John Scholarships, 3977
Yampa Valley Community Foundation Grants, 3978
Yampa Valley Community Foundation Volunteer Firemen Scholarships, 3979
Yawkey Foundation Grants, 3980
Z. Smith Reynolds Foundation Small Grants, 3982

Emergency Programs
Able Trust Vocational Rehabilitation Grants for Individuals, 213
Abundance Foundation International Grants, 218
ADA Foundation Relief Grants, 293
Adaptec Foundation Grants, 304
Adelaide Breed Bayrd Foundation Grants, 306
Adler-Clark Electric Community Commitment Foundation Grants, 307
Advance Auto Parts Corporate Giving Grants, 312
AEGON Transamerica Foundation Health and Welfare Grants, 316
Aid for Starving Children Emergency Assistance Fund Grants, 367
Aid for Starving Children International Grants, 368
AIG Disaster Relief Fund Grants, 370
Alcatel-Lucent Technologies Foundation Grants, 413
Allan C. and Lelia J. Garden Foundation Grants, 438
American Express Community Service Grants, 531
American Foodservice Charitable Trust Grants, 532
American Psychiatric Foundation Disaster Recovery Fund for Psychiatrists, 553
Anheuser-Busch Foundation Grants, 591
Auburn Foundation Grants, 798
Austin S. Nelson Foundation Grants, 805
Avon Products Foundation Grants, 814
Bank of America Fndn Matching Gifts, 825
Bank of America Fndn Volunteer Grants, 826
Baxter International Corporate Giving Grants, 841
Beckman Coulter Foundation Grants, 868
Becton Dickinson and Company Grants, 869
Bill and Melinda Gates Foundation Emergency Response Grants, 892
Biogen Foundation General Donations, 899
Blanche and Julian Robertson Family Foundation Grants, 911
Blowitz-Ridgeway Foundation Grants, 913
Blum-Kovler Foundation Grants, 924
Blumenthal Foundation Grants, 925
Boeing Company Contributions Grants, 930
Borkee-Hagley Foundation Grants, 934
BP Foundation Grants, 943
Brighter Tomorrow Foundation Grants, 950
Brooklyn Community Foundation Caring Neighbors Grants, 968
Byron W. & Alice L. Lockwood Fnd Grants, 991
Callaway Golf Company Foundation Grants, 1003
Campbell Soup Foundation Grants, 1007
Cargill Citizenship Fund Corporate Giving, 1018
Carl R. Hendrickson Family Foundation Grants, 1028
Carpenter Foundation Grants, 1035
CDC Foundation Emergency Response Grants, 1061
CDC Public Health Prevention Service Fellowship Sponsorships, 1070
Cemala Foundation Grants, 1077
Central Okanagan Foundation Grants, 1080
Charles H. Hall Foundation, 1120
Charles Nelson Robinson Fund Grants, 1125
Children Affected by AIDS Foundation Family Assistance Emergency Fund Grants, 1146
Clara Blackford Smith and W. Aubrey Smith Charitable Foundation Grants, 1165
CNCS AmeriCorps VISTA Project Grants, 1190
Coastal Community Foundation of South Carolina Grants, 1202
Coca-Cola Foundation Grants, 1203
Cockrell Foundation Grants, 1204

PROGRAM TYPE INDEX

Colorado Resource for Emergency Education and Trauma Grants, 1212
Communities Foundation of Texas Grants, 1236
Community Fndn for Monterey County Grants, 1253
Community Fndn for Muskegon County Grants, 1254
Community Foundation of Eastern Connecticut General Southeast Grants, 1269
Community Foundation of Greater Birmingham Grants, 1272
Community Foundation of Greater Fort Wayne - Community Endowment and Clarke Endowment Grants, 1274
Community Foundation of Louisville Anna Marble Memorial Fund for Princeton Grants, 1290
Community Foundation of Louisville Morris and Esther Lee Fund Grants, 1298
Community Foundation of Mount Vernon and Knox County Grants, 1299
Connelly Foundation Grants, 1326
Cornerstone Foundation of Northeastern Wisconsin Grants, 1347
Covidien Medical Product Donations, 1352
Cruise Industry Charitable Foundation Grants, 1363
Cudd Foundation Grants, 1370
Cullen Foundation Grants, 1372
Daywood Foundation Grants, 1438
DIFFA/Chicago Grants, 1477
Dominion Foundation Grants, 1485
Donald and Sylvia Robinson Family Foundation Grants, 1486
Doree Taylor Charitable Foundation, 1490
Doris and Victor Day Foundation Grants, 1491
Doris Duke Charitable Foundation Operations Research on AIDS Care and Treatment in Africa (ORACTA) Grants, 1497
Dr. & Mrs. Paul Pierce Memorial Fndn Grants, 1505
Drs. Bruce and Lee Foundation Grants, 1513
Edyth Bush Charitable Foundation Grants, 1562
Erie Chapman Foundation Grants, 1600
Erie Community Foundation Grants, 1601
FAR Fund Grants, 1631
Flextronics Foundation Disaster Relief Grants, 1674
Frank B. Hazard General Charity Fund Grants, 1721
Franklin H. Wells and Ruth L. Wells Foundation Grants, 1725
Frank Stanley Beveridge Foundation Grants, 1730
Fred L. Emerson Foundation Grants, 1743
Fremont Area Community Fndn General Grants, 1746
Gardner Foundation Grants, 1772
George H.C. Ensworth Memorial Fund Grants, 1797
Gill Foundation - Gay and Lesbian Fund Grants, 1821
GNOF Albert N. & Hattie M. McClure Grants, 1834
GNOF Bayou Communities Grants, 1835
GNOF IMPACT Grants for Health and Human Services, 1839
GNOF IMPACT Harold W. Newman, Jr. Charitable Trust Grants, 1841
GNOF Maison Hospitaliere Grants, 1843
Grace Bersted Foundation Grants, 1859
Green Foundation Human Services Grants, 1894
Green River Area Community Fndn Grants, 1895
Gulf Coast Community Foundation Grants, 1915
Gulf Coast Foundation of Community Operating Grants, 1916
Gulf Coast Foundation of Community Program Grants, 1917
H.J. Heinz Company Foundation Grants, 1923
Hardin County Community Foundation Grants, 1946
Harold and Arlene Schnitzer CARE Foundation Grants, 1950
Harold Brooks Foundation Grants, 1952
Harrison County Community Foundation Signature Grants, 1958
Henrietta Tower Wurts Memorial Fndn Grants, 2002
Herbert A. & Adrian W. Woods Fndn Grants, 2007
High Meadow Foundation Grants, 2027
Hillcrest Foundation Grants, 2031
Hoglund Foundation Grants, 2040
Home Building Industry Disaster Relief Fund, 2043
Homer Foundation Grants, 2044

PROGRAM TYPE INDEX

Horace Moses Charitable Foundation Grants, 2047
Humana Foundation Grants, 2077
IDPH Local Health Department Public Health Emergency Response Grants, 2120
Ike and Roz Friedman Foundation Grants, 2145
Indiana AIDS Fund Grants, 2155
J.H. Robbins Foundation Grants, 2171
J.N. and Macie Edens Foundation Grants, 2174
James & Abigail Campbell Family Fndn Grants, 2185
James Graham Brown Foundation Grants, 2189
James R. Thorpe Foundation Grants, 2199
Jane's Trust Grants, 2206
John Deere Foundation Grants, 2234
John G. Duncan Charitable Trust Grants, 2236
John I. Smith Charities Grants, 2240
John Jewett and Helen Chandler Garland Foundation Grants, 2242
John P. McGovern Foundation Grants, 2247
Johns Manville Fund Grants, 2251
John W. and Anna H. Hanes Foundation Grants, 2261
John W. Boynton Fund Grants, 2263
John W. Speas and Effie E. Speas Memorial Trust Grants, 2264
Josephine G. Russell Trust Grants, 2271
Josiah W. and Bessie H. Kline Foundation Grants, 2278
Judith Clark-Morrill Foundation Grants, 2281
Kimball International-Habig Foundation Health and Human Services Grants, 2308
Kuntz Foundation Grants, 2326
Lawrence Foundation Grants, 2347
Lena Benas Memorial Fund Grants, 2353
Louis and Elizabeth Nave Flarsheim Charitable Foundation Grants, 2379
Lucy Downing Nisbet Charitable Fund Grants, 2387
Lydia deForest Charitable Trust Grants, 2393
Marion I. and Henry J. Knott Foundation Discretionary Grants, 2435
McCune Foundation Human Services Grants, 2493
Merkel Foundation Grants, 2520
Meyer Foundation Healthy Communities Grants, 2533
Meyer Memorial Trust Emergency Grants, 2535
Middlesex Savings Charitable Foundation Basic Human Needs Grants, 2550
Morris & Gwendolyn Cafritz Fndn Grants, 2594
Nationwide Insurance Foundation Grants, 2637
Nordson Corporation Foundation Grants, 2981
Northwest Airlines KidCares Medical Travel Assistance, 2998
Northwestern Mutual Foundation Grants, 2999
Oleonda Jameson Trust Grants, 3071
Oppenstein Brothers Foundation Grants, 3078
Oregon Community Fndn Community Grants, 3082
OSF Access to Essential Medicines Initiative, 3086
OSF Accountability and Monitoring in Health Initiative Grants, 3087
Packard Foundation Local Grants, 3105
Piedmont Natural Gas Foundation Health and Human Services Grants, 3221
Pinkerton Foundation Grants, 3224
PMI Foundation Grants, 3239
Pohlad Family Foundation, 3241
Prince Charitable Trusts DC Grants, 3271
Puerto Rico Community Foundation Grants, 3278
R.S. Gernon Trust Grants, 3293
RCF Individual Assistance Grants, 3313
Reinberger Foundation Grants, 3357
Rhode Island Foundation Grants, 3362
Robert R. Meyer Foundation Grants, 3387
S. Livingston Mather Charitable Trust Grants, 3448
Salt River Health & Human Services Grants, 3459
Samuel S. Johnson Foundation Grants, 3468
San Francisco Foundation Disability Rights Advocate Fund Emergency Grants, 3479
Seattle Foundation Health and Wellness Grants, 3507
Seneca Foods Foundation Grants, 3511
Shield-Ayres Foundation Grants, 3546
Sioux Falls Area Community Foundation Community Fund Grants (Unrestricted), 3568
Sioux Falls Area Community Foundation Field-of-Interest and Donor-Advised Grants, 3569

Sioux Falls Area Community Foundation Spot Grants (Unrestricted), 3570
Sony Corporation of America Grants, 3623
Sunoco Foundation Grants, 3679
Swindells Charitable Foundation, 3697
Texas Commission on the Arts Arts Respond Project Grants, 3708
Textron Corporate Contributions Grants, 3712
TJX Foundation Grants, 3735
Todd Brock Family Foundation Grants, 3736
Union Bank, N.A. Corporate Sponsorships and Donations, 3787
Union Bank, N.A. Foundation Grants, 3788
Union Labor Health Foundation Dental Angel Fund Grants, 3793
Union Pacific Foundation Health and Human Services Grants, 3794
UnitedHealthcare Children's Foundation Grants, 3795
USAID Fighting Ebola Grants, 3812
USAID Higher Education Partnerships for Innovation and Impact (HEPII) Grants, 3815
VHA Health Foundation Grants, 3867
Victor E. Speas Foundation Grants, 3868
W.H. & Mary Ellen Cobb Trust Grants, 3884
Walker Area Community Foundation Grants, 3895
Washington Gas Charitable Giving, 3908
Welborn Baptist Foundation Improvements to Community Health Status Grants, 3916
Western Union Foundation Grants, 3923
William B. Stokely Jr. Foundation Grants, 3947
Winston-Salem Fndn Victim Assistance Grants, 3968
Zellweger Baby Support Network Grants, 3984

Endowments

Achelis Foundation Grants, 234
Adams Foundation Grants, 301
Alfred and Tillie Shemanski Testamentary Trust Grants, 428
Blanche and Irving Laurie Foundation Grants, 910
Blum-Kovler Foundation Grants, 924
Blumenthal Foundation Grants, 925
BMW of North America Charitable Contributions, 926
Bodman Foundation Grants, 929
Caesars Foundation Grants, 995
Clayton Fund Grants, 1180
Cralle Foundation Grants, 1356
Gardner Foundation Grants, 1770
GNOF Freeman Challenge Grants, 1837
Graham Foundation Grants, 1862
Grover Hermann Foundation Grants, 1906
Hearst Foundations Health Grants, 1987
J.W. Kieckhefer Foundation Grants, 2176
Jack H. & William M. Light Trust Grants, 2179
Kovler Family Foundation Grants, 2323
Margaret T. Morris Foundation Grants, 2429
McLean Contributionship Grants, 2497
Norcliffe Foundation Grants, 2979
Paso del Norte Health Foundation Grants, 3119
Perkin Fund Grants, 3139
PMI Foundation Grants, 3239
RCF The Women's Fund Grants, 3315
Reinberger Foundation Grants, 3357
Robert R. Meyer Foundation Grants, 3387
Roy & Christine Sturgis Charitable Grants, 3415
Saigh Foundation Grants, 3451
Shield-Ayres Foundation Grants, 3546
Sony Corporation of America Grants, 3623
Textron Corporate Contributions Grants, 3712
Union Bank, N.A. Foundation Grants, 3788
W.P. and Bulah Luse Foundation Grants, 3892
William Blair and Company Foundation Grants, 3948

Environmental Programs

3M Company Fndn Community Giving Grants, 6
1675 Foundation Grants, 11
Abbot and Dorothy H. Stevens Foundation Grants, 197
Abbott Fund Community Grants, 201
Abel Foundation Grants, 206
Abell-Hanger Foundation Grants, 207
ACE Charitable Foundation Grants, 229

Environmental Programs / 877

ACF Native American Social and Economic Development Strategies Grants, 231
Achelis Foundation Grants, 234
ACS Doctoral Grants in Oncology Social Work, 254
Administaff Community Affairs Grants, 308
Adobe Community Investment Grants, 310
Ahn Family Foundation Grants, 356
Aid for Starving Children International Grants, 368
Air Products and Chemicals Grants, 399
Alaska Airlines Foundation Grants, 403
Alberto Culver Corporate Contributions Grants, 408
Albert Pick Jr. Fund Grants, 409
Albuquerque Community Foundation Grants, 412
Alcoa Foundation Grants, 414
Alexander & Baldwin Fnd Mainland Grants, 417
Alexander and Baldwin Foundation Hawaiian and Pacific Island Grants, 418
Allegan County Community Foundation Grants, 439
Allen Foundation Educational Nutrition Grants, 442
Allen P. & Josephine B. Green Fndn Grants, 443
Alliant Energy Corporation Contributions, 445
Allstate Corporate Giving Grants, 446
Allstate Corp Hometown Commitment Grants, 447
Alpha Natural Resources Corporate Giving, 450
Alpine Winter Foundation Grants, 453
ALSAM Foundation Grants, 454
Amador Community Foundation Grants, 481
American Chemical Society GCI Pharmaceutical Roundtable Research Grants, 516
American College of Surgeons International Guest Scholarships, 525
American Electric Power Foundation Grants, 530
American Psychiatric Association Minority Medical Student Summer Externship in Addiction Psychiatry, 545
Amgen Foundation Grants, 572
Andrew Family Foundation Grants, 588
Angels Baseball Foundation Grants, 590
Anheuser-Busch Foundation Grants, 591
Ann Arbor Area Community Foundation Grants, 594
Annie Sinclair Knudsen Memorial Fund/Kaua'i Community Grants Program, 597
APS Foundation Grants, 646
Arcadia Foundation Grants, 650
Archer Daniels Midland Foundation Grants, 651
Arizona Public Service Corporate Giving Program Grants, 658
Arkansas Community Foundation Grants, 660
Arthur Ashley Williams Foundation Grants, 671
Ashland Corporate Contributions Grants, 713
AT&T Foundation Civic and Community Service Program Grants, 793
Athwin Foundation Grants, 795
Autauga Area Community Foundation Grants, 808
Autodesk Community Relations Grants, 809
Autzen Foundation Grants, 810
Avista Foundation Economic and Cultural Vitality Grants, 812
Babcock Charitable Trust Grants, 816
Bacon Family Foundation Grants, 817
Banfi Vintners Foundation Grants, 824
Bank of America Fndn Matching Gifts, 825
Bank of America Fndn Volunteer Grants, 826
Barker Welfare Foundation Grants, 831
Barra Foundation Community Fund Grants, 833
Batchelor Foundation Grants, 837
Baton Rouge Area Foundation Grants, 838
Baxter International Corporate Giving Grants, 841
Beazley Foundation Grants, 865
Beckman Coulter Foundation Grants, 868
Beirne Carter Foundation Grants, 870
Beldon Fund Grants, 871
Belk Foundation Grants, 872
Bella Vista Foundation Grants, 873
Belvedere Community Foundation Grants, 874
Bender Foundation Grants, 876
Benton Community Foundation Grants, 878
Benton County Foundation Grants, 880
Berks County Community Foundation Grants, 881

Bill and Melinda Gates Foundation Emergency Response Grants, 892
Bill and Melinda Gates Foundation Policy and Advocacy Grants, 893
Blanche and Julian Robertson Family Foundation Grants, 911
Blue Grass Community Fndn Harrison Grants, 919
Blue Mountain Community Foundation Grants, 921
Blue River Community Foundation Grants, 922
Blum-Kovler Foundation Grants, 924
Blumenthal Foundation Grants, 925
Bodenwein Public Benevolent Foundation Grants, 928
Bodman Foundation Grants, 929
Boeing Company Contributions Grants, 930
Boston Foundation Grants, 936
Boston Globe Foundation Grants, 937
Boyd Gaming Corporation Contributions Program, 941
BP Foundation Grants, 943
Brian G. Dyson Foundation Grants, 948
Bridgestone/Firestone Trust Fund Grants, 949
Brown County Community Foundation Grants, 970
Browning-Kimball Foundation Grants, 971
Bruce and Adele Greenfield Foundation Grants, 972
Bullitt Foundation Grants, 973
Burlington Northern Santa Fe Foundation Grants, 978
Cabot Corporation Foundation Grants, 993
California Wellness Foundation Work and Health Program Grants, 1001
Callaway Foundation Grants, 1002
Cambridge Community Foundation Grants, 1004
Capital Region Community Foundation Grants, 1014
Cargill Citizenship Fund Corporate Giving, 1018
Carl & Eloise Pohlad Family Fndn Grants, 1021
Carl C. Icahn Foundation Grants, 1023
Carl M. Freeman Foundation FACES Grants, 1026
Carl M. Freeman Foundation Grants, 1027
Carls Foundation Grants, 1030
Carolyn Foundation Grants, 1034
Carpenter Foundation Grants, 1035
Carrie E. and Lena V. Glenn Foundation Grants, 1036
Carrier Corporation Contributions Grants, 1038
Cass County Community Foundation Grants, 1041
Centerville-Washington Foundation Grants, 1078
Central Okanagan Foundation Grants, 1080
Cessna Foundation Grants Program, 1081
CFFVR Project Grants, 1101
CFFVR Shawano Area Community Foundation Grants, 1104
Champlin Foundations Grants, 1110
Chapman Charitable Foundation Grants, 1112
Charles Delmar Foundation Grants, 1115
Charles H. Hall Foundation, 1120
Charles M. & Mary D. Grant Fndn Grants, 1124
Charlotte County (FL) Community Foundation Grants, 1126
Chilkat Valley Community Foundation Grants, 1148
Christensen Fund Regional Grants, 1149
Chula Vista Charitable Foundation Grants, 1154
CICF Indianapolis Fndn Community Grants, 1156
Citizens Bank Mid-Atlantic Charitable Foundation Grants, 1162
Clarence E. Heller Charitable Foundation Grants, 1166
Clark County Community Foundation Grants, 1175
Claude Worthington Benedum Fndn Grants, 1178
Clayton Baker Trust Grants, 1179
Cleveland-Cliffs Foundation Grants, 1181
CLIF Bar Family Foundation Grants, 1183
Clinton County Community Foundation Grants, 1184
CNCS AmeriCorps VISTA Project Grants, 1190
Coastal Community Foundation of South Carolina Grants, 1202
Coca-Cola Foundation Grants, 1203
Coeta and Donald Barker Foundation Grants, 1205
Collective Brands Foundation Grants, 1208
Collins Foundation Grants, 1210
Columbus Foundation Competitive Grants, 1217
Columbus Foundation Mary Eleanor Morris Fund Grants, 1220
Community Foundation Alliance City of Evansville Endowment Fund Grants, 1238

Community Foundation for Greater Atlanta Metropolitan Atlanta An Extra Wish Grants, 1243
Community Fndn for Greater Buffalo Grants, 1247
Community Fndn for Monterey County Grants, 1253
Community Fndn for Muskegon County Grants, 1254
Community Foundation for Northeast Michigan Mini-Grants, 1255
Community Fndn for San Benito County Grants, 1257
Community Fndn for the Capital Region Grants, 1259
Community Foundation for the National Capital Region Community Leadership Grants, 1260
Community Foundation of Bartholomew County Heritage Fund Grants, 1262
Community Foundation of Boone County Grants, 1266
Community Fndn of East Central Illinois Grants, 1268
Community Foundation of Eastern Connecticut General Southeast Grants, 1269
Community Foundation of Greater Birmingham Grants, 1272
Community Foundation of Greater Flint Grants, 1273
Community Foundation of Greater Greensboro Community Grants, 1277
Community Foundation of Greater Greensboro Women to Women Fund Grants, 1278
Community Fndn of Greater Tampa Grants, 1282
Community Foundation of Greenville-Greenville Women Giving Grants, 1283
Community Foundation of Mount Vernon and Knox County Grants, 1299
Community Foundation of Muncie and Delaware County Grants, 1300
Community Foundation of Muncie and Delaware County Maxon Grants, 1301
Community Fndn of Randolph County Grants, 1302
Community Fndn of So Alabama Grants, 1306
Community Fndn of South Puget Sound Grants, 1307
Community Fndn of Switzerland County Grants, 1311
Community Foundation of Tampa Bay Grants, 1312
Community Foundation of the Eastern Shore Community Needs Grants, 1313
Community Foundation of the Verdugos Grants, 1315
Connecticut Community Foundation Grants, 1324
ConocoPhillips Foundation Grants, 1327
Conservation, Food, and Health Foundation Grants for Developing Countries, 1338
Constellation Energy Corporate Grants, 1342
Cooke Foundation Grants, 1344
Cooper Industries Foundation Grants, 1345
Coors Brewing Corporate Contributions Grants, 1346
Cowles Charitable Trust Grants, 1354
Cresap Family Foundation Grants, 1359
Cruise Industry Charitable Foundation Grants, 1363
Crystelle Waggoner Charitable Trust Grants, 1364
CSRA Community Foundation Grants, 1366
CSX Corporate Contributions Grants, 1368
Cudd Foundation Grants, 1370
Cumberland Community Foundation Grants, 1374
Cyrus Eaton Foundation Grants, 1380
Dade Community Foundation Grants, 1410
DaimlerChrysler Corporation Fund Grants, 1412
Dayton Power and Light Company Foundation Signature Grants, 1436
Dean Foods Community Involvement Grants, 1440
Dearborn Community Foundation County Progress Grants, 1442
DeKalb County Community Foundation Grants, 1447
Delaware Community Foundation Grants, 1449
Delmarva Power and Light Contributions, 1454
Dole Food Company Charitable Contributions, 1483
Doree Taylor Charitable Foundation, 1490
Dorothy Hooper Beattie Foundation Grants, 1499
Do Something Awards, 1503
Dr. Scholl Foundation Grants Program, 1511
Drs. Bruce and Lee Foundation Grants, 1513
Dubois County Community Foundation Grants, 1517
DuPont Pioneer Community Giving Grants, 1532
Dyson Foundation Mid-Hudson Valley Project Support Grants, 1534
E. Clayton and Edith P. Gengras, Jr. Foundation, 1536
E.J. Grassmann Trust Grants, 1537

Earl and Maxine Claussen Trust Grants, 1541
eBay Foundation Community Grants, 1543
Educational Foundation of America Grants, 1553
Edward and Helen Bartlett Foundation Grants, 1554
Edward Bangs Kelley and Elza Kelley Foundation Grants, 1556
Edwin W. and Catherine M. Davis Fndn Grants, 1561
Elliot Foundation Inc Grants, 1575
Elmer L. & Eleanor J. Andersen Fndn Grants, 1578
El Paso Community Foundation Grants, 1579
El Pomar Foundation Anna Keesling Ackerman Fund Grants, 1581
El Pomar Foundation Grants, 1582
EPA Children's Health Protection Grants, 1598
Erie Community Foundation Grants, 1601
Essex County Community Foundation Discretionary Fund Grants, 1602
Estee Lauder Grants, 1606
Eugene B. Casey Foundation Grants, 1609
Evjue Foundation Grants, 1617
Ewing Halsell Foundation Grants, 1618
Fairfield County Community Foundation Grants, 1625
Fargo-Moorhead Area Foundation Grants, 1632
Ferree Foundation Grants, 1658
FirstEnergy Foundation Community Grants, 1668
FIU Global Civic Engagement Mini Grants, 1673
Forrest C. Lattner Foundation Grants, 1688
Foundation for the Mid South Community Development Grants, 1699
Four County Community Fndn General Grants, 1710
Frances and John L. Loeb Family Fund Grants, 1716
Francis T. & Louise T. Nichols Fndn Grants, 1720
Franklin County Community Foundation Grants, 1724
Frank Stanley Beveridge Foundation Grants, 1730
Fraser-Parker Foundation Grants, 1732
Fred Baldwin Memorial Foundation Grants, 1736
Frederick W. Marzahl Memorial Fund Grants, 1741
Fremont Area Community Fndn General Grants, 1746
Fulbright German Studies Seminar Grants, 1751
Fulton County Community Foundation Grants, 1763
G.N. Wilcox Trust Grants, 1768
Gamble Foundation Grants, 1769
Gardner Foundation Grants, 1771
Gebbie Foundation Grants, 1777
General Motors Foundation Grants, 1783
Genuardi Family Foundation Grants, 1787
George and Ruth Bradford Foundation Grants, 1789
George A Ohl Jr. Foundation Grants, 1791
George Gund Foundation Grants, 1796
George H.C. Ensworth Memorial Fund Grants, 1797
George S. & Dolores Dore Eccles Fndn Grants, 1801
Georgia Power Foundation Grants, 1806
Gerber Foundation Grants, 1807
Giving Sum Annual Grant, 1825
Global Fund for Women Grants, 1833
GNOF Bayou Communities Grants, 1835
GNOF Exxon-Mobil Grants, 1836
GNOF Norco Community Grants, 1845
Goodrich Corporation Foundation Grants, 1854
Good Samaritan Inc Grants, 1855
Grace and Franklin Bernsen Foundation Grants, 1858
Grand Haven Area Community Fndn Grants, 1864
Grand Rapids Area Community Foundation Nashwauk Area Endowment Fund Grants, 1866
Grand Rapids Area Community Foundation Wyoming Grants, 1867
Grand Rapids Community Foundation Grants, 1869
Grand Rapids Community Foundation Ionia County Grants, 1870
Grand Rapids Community Foundation Lowell Area Fund Grants, 1872
Grand Rapids Community Foundation Southeast Ottawa Grants, 1873
Grand Rapids Community Fndn Sparta Grants, 1875
Greater Cincinnati Foundation Priority and Small Projects/Capacity-Building Grants, 1880
Greater Green Bay Community Fndn Grants, 1881
Greater Milwaukee Foundation Grants, 1883
Greater Saint Louis Community Fndn Grants, 1884
Greater Sitka Legacy Fund Grants, 1885

PROGRAM TYPE INDEX   Environmental Programs / 879

Greater Tacoma Community Foundation Grants, 1886
Greater Worcester Community Foundation Discretionary Grants, 1888
Green Diamond Charitable Contributions, 1891
Green River Area Community Fndn Grants, 1895
Greygates Foundation Grants, 1901
Grundy Foundation Grants, 1912
H.A. & Mary K. Chapman Trust Grants, 1921
Hannaford Charitable Foundation Grants, 1944
Harden Foundation Grants, 1945
Hardin County Community Foundation Grants, 1946
Harrison County Community Foundation Grants, 1957
Harrison County Community Foundation Signature Grants, 1958
Harry A. & Margaret D. Towsley Fndn Grants, 1959
Harry Frank Guggenheim Fnd Research Grants, 1964
Hawaiian Electric Industries Charitable Foundation Grants, 1975
Hawaii Community Foundation West Hawaii Fund Grants, 1978
Heineman Foundation for Research, Education, Charitable and Scientific Purposes, 1990
Helen S. Boylan Foundation Grants, 1997
Hendricks County Community Fndn Grants, 2000
Herbert A. & Adrian W. Woods Fndn Grants, 2007
Hershey Company Grants, 2013
High Meadow Foundation Grants, 2027
Hillman Foundation Grants, 2032
Hillsdale County Community General Adult Foundation Grants, 2034
Hilton Head Island Foundation Grants, 2036
Hoblitzelle Foundation Grants, 2038
Holland/Zeeland Community Fndn Grants, 2041
Homer Foundation Grants, 2044
Honda of America Manufacturing Fndn Grants, 2045
Horace A. Kimball and S. Ella Kimball Foundation Grants, 2046
Houston Endowment Grants, 2053
Huntington County Community Foundation Make a Difference Grants, 2083
Idaho Power Company Corporate Contributions, 2115
IIE David L. Boren Fellowships, 2129
Impact 100 Grants, 2148
Inasmuch Foundation Grants, 2149
J.C. Penney Company Grants, 2169
J.W. Kieckhefer Foundation Grants, 2176
Jackson County Community Foundation Unrestricted Grants, 2180
Jacobs Family Village Neighborhoods Grants, 2184
James & Abigail Campbell Family Fndn Grants, 2185
James Ford Bell Foundation Grants, 2187
James M. Cox Foundation of Georgia Grants, 2196
James S. Copley Foundation Grants, 2200
Jane's Trust Grants, 2206
Janirve Foundation Grants, 2209
Jasper Foundation Grants, 2212
Jeffris Wood Foundation Grants, 2217
Jennings County Community Foundation Grants, 2220
Jennings County Community Foundation Women's Giving Circle Grant, 2221
Jessica Stevens Community Foundation Grants, 2224
Jessie B. Cox Charitable Trust Grants, 2225
John Deere Foundation Grants, 2234
John W. and Anna H. Hanes Foundation Grants, 2261
Joseph Henry Edmondson Foundation Grants, 2270
Kavli Foundation Research Grants, 2296
Kelvin and Eleanor Smith Foundation Grants, 2297
Kenai Peninsula Foundation Grants, 2298
Ketchikan Community Foundation Grants, 2303
Kettering Family Foundation Grants, 2304
Kettering Fund Grants, 2305
Knox County Community Foundation Grants, 2313
Kodiak Community Foundation Grants, 2314
Kosciusko County Community Fndn Grants, 2319
LaGrange County Community Fndn Grants, 2329
Land O'Lakes Foundation Mid-Atlantic Grants, 2341
Lands' End Corporate Giving Program, 2344
Lawrence Foundation Grants, 2347
Lawrence S. Huntington Fund Grants, 2348
Lester E. Yeager Charitable Trust B Grants, 2359

Libra Foundation Grants, 2363
Lisa and Douglas Goldman Fund Grants, 2369
Long Island Community Foundation Grants, 2375
Lotus 88 Fnd for Women & Children Grants, 2376
Lubrizol Foundation Grants, 2384
Lucy Downing Nisbet Charitable Fund Grants, 2387
Lynde and Harry Bradley Foundation Grants, 2397
M.E. Raker Foundation Grants, 2404
Macquarie Bank Foundation Grants, 2410
Margaret T. Morris Foundation Grants, 2429
Marie C. and Joseph C. Wilson Foundation Rochester Small Grants, 2431
Marie H. Bechtel Charitable Remainder Uni-Trust Grants, 2432
Marjorie Moore Charitable Foundation Grants, 2438
Marshall County Community Foundation Grants, 2439
Mary K. Chapman Foundation Grants, 2443
Mary Owen Borden Foundation Grants, 2446
Maxon Charitable Foundation Grants, 2458
McConnell Foundation Grants, 2492
McInerny Foundation Grants, 2495
McKesson Foundation Grants, 2496
McLean Contributionship Grants, 2497
Mead Johnson Nutritionals Evansville-Area Organizations Grants, 2503
MeadWestvaco Foundation Sustainable Communities Grants, 2506
Mead Witter Foundation Grants, 2507
Merrick Foundation Grants, 2522
Meyer Memorial Trust Grassroots Grants, 2536
Meyer Memorial Trust Responsive Grants, 2537
Meyer Memorial Trust Special Grants, 2538
Miami County Community Foundation - Operation Round Up Grants, 2543
Mimi and Peter Haas Fund Grants, 2557
Montgomery County Community Fndn Grants, 2591
Morris & Gwendolyn Cafritz Fndn Grants, 2594
Natalie W. Furniss Charitable Trust Grants, 2626
Nathan Cummings Foundation Grants, 2627
Nelda C. and H.J. Lutcher Stark Fndn Grants, 2672
NIEHS Hazardous Materials Worker Health and Safety Training Grants, 2854
NIEHS Hazmat Training at Doe Nuclear Weapons Complex Grants, 2855
NIEHS SBIR E-Learning for HAZMAT and Emergency Response Grants, 2859
Nina Mason Pulliam Charitable Trust Grants, 2921
Norcliffe Foundation Grants, 2979
Norfolk Southern Foundation Grants, 2982
Norman Foundation Grants, 2983
North Carolina Community Foundation Grants, 2990
North Dakota Community Foundation Grants, 2994
Northern Chautauqua Community Foundation Community Grants, 2995
Northrop Grumman Corporation Grants, 2997
Norwin S. & Elizabeth N. Bean Fndn Grants, 3001
Ogden Codman Trust Grants, 3066
Ohio County Community Foundation Board of Directors Grants, 3067
Ohio County Community Foundation Grants, 3068
Oppenstein Brothers Foundation Grants, 3078
Oregon Community Fndn Community Grants, 3082
Oscar Rennebohm Foundation Grants, 3085
Otto Bremer Foundation Grants, 3099
Owen County Community Foundation Grants, 3100
Pacific Life Foundation Grants, 3103
Packard Foundation Local Grants, 3105
Patrick and Anna M. Cudahy Fund Grants, 3120
Peacock Foundation Grants, 3133
PepsiCo Foundation Grants, 3137
Perkins Charitable Foundation Grants, 3140
Perry County Community Foundation Grants, 3142
Petersburg Community Foundation Grants, 3156
Piedmont Natural Gas Corporate and Charitable Contributions, 3220
Pike County Community Foundation Grants, 3222
Pinellas County Grants, 3223
Pittsburgh Foundation Community Fund Grants, 3230
Pokagon Fund Grants, 3242
Posey County Community Foundation Grants, 3252

Powell Foundation Grants, 3255
PPCF Community Grants, 3256
Premera Blue Cross Grants, 3258
Prince Charitable Trusts Chicago Grants, 3270
Prince Charitable Trusts DC Grants, 3271
Princeton Area Community Foundation Greater Mercer Grants, 3272
Principal Financial Group Foundation Grants, 3273
Procter and Gamble Fund Grants, 3274
Pulaski County Community Foundation Grants, 3279
Putnam County Community Foundation Grants, 3283
Rajiv Gandhi Foundation Grants, 3302
Rathmann Family Foundation Grants, 3309
RCF General Community Grants, 3312
Rhode Island Foundation Grants, 3362
Rice Foundation Grants, 3363
Richard King Mellon Foundation Grants, 3369
Richland County Bank Grants, 3371
Riley Foundation Grants, 3374
Ripley County Community Foundation Grants, 3376
Ripley County Community Foundation Small Project Grants, 3377
Robbins Charitable Foundation Grants, 3378
Robert R. Meyer Foundation Grants, 3387
Robert W. Woodruff Foundation Grants, 3389
Rochester Area Community Foundation Grants, 3391
Rohm and Haas Company Grants, 3403
Ronald McDonald House Charities Grants, 3405
Ruth Anderson Foundation Grants, 3425
Ruth and Vernon Taylor Foundation Grants, 3426
Ruth Eleanor Bamberger and John Ernest Bamberger Memorial Foundation Grants, 3427
S. Livingston Mather Charitable Trust Grants, 3448
S. Mark Taper Foundation Grants, 3449
Samuel S. Johnson Foundation Grants, 3468
San Diego Foundation for Change Grants, 3470
San Juan Island Community Foundation Grants, 3480
Sartain Lanier Family Foundation Grants, 3486
Sasco Foundation Grants, 3487
Schering-Plough Foundation Community Initiatives Grants, 3491
Scott County Community Foundation Grants, 3500
Seabury Foundation Grants, 3501
Seagate Tech Corp Capacity to Care Grants, 3502
Seward Community Foundation Grants, 3514
Seward Community Foundation Mini-Grants, 3515
Shaw's Supermarkets Donations, 3542
Shell Deer Park Grants, 3543
Shell Oil Company Foundation Community Development Grants, 3544
Singing for Change Foundation Grants, 3567
Sioux Falls Area Community Foundation Community Fund Grants (Unrestricted), 3568
Sioux Falls Area Community Foundation Field-of-Interest and Donor-Advised Grants, 3569
Sioux Falls Area Community Foundation Spot Grants (Unrestricted), 3570
Sir Dorabji Tata Trust Grants for NGOs or Voluntary Organizations, 3572
Skoll Fndn Awards for Social Entrepreneurship, 3579
Sonora Area Foundation Competitive Grants, 3622
Sony Corporation of America Grants, 3623
Southbury Community Trust Fund, 3625
Southwest Florida Community Foundation Competitive Grants, 3627
Spencer County Community Foundation Grants, 3632
Springs Close Foundation Grants, 3633
SSRC Collaborative Research Grants on Environment and Health in China, 3638
Stackpole-Hall Foundation Grants, 3643
Steuben County Community Foundation Grants, 3654
Strowd Roses Grants, 3664
Summit Foundation Grants, 3674
Sunderland Foundation Grants, 3675
Sunflower Foundation Walking Trails Grants, 3678
Sunoco Foundation Grants, 3679
Susan Mott Webb Charitable Trust Grants, 3694
Susan Vaughan Foundation Grants, 3695
T.L.L. Temple Foundation Grants, 3700
Tellabs Foundation Grants, 3706

880 / Environmental Programs

Texas Commission on the Arts Arts Respond Project Grants, 3708
Theodore Edson Parker Foundation Grants, 3715
Thomas J. Long Foundation Community Grants, 3722
Thomas Sill Foundation Grants, 3724
Thompson Charitable Foundation Grants, 3727
Thorman Boyle Foundation Grants, 3730
Toyota Motor Engineering & Manufacturing North America Grants, 3742
Toyota Motor Manufacturing of Alabama Grants, 3743
Toyota Motor Manufacturing of Indiana Grants, 3744
Toyota Motor Manufacturing of Kentucky Grants, 3745
Toyota Motor Manuf of Mississippi Grants, 3746
Toyota Motor Manufacturing of Texas Grants, 3747
Toyota Motor Manuf of West Virginia Grants, 3748
Toyota Motor N America of New York Grants, 3749
Toyota Motor Sales, USA Grants, 3750
Triangle Community Foundation Donor-Advised Grants, 3757
Trull Foundation Grants, 3761
Turner Foundation Grants, 3763
U.S. Cellular Corporation Grants, 3766
U.S. Department of Education United States-Russia Program: Improving Research and Educational Activities in Higher Education, 3773
Union Bank, N.A. Corporate Sponsorships and Donations, 3787
Union Bank, N.A. Foundation Grants, 3788
Union Benevolent Association Grants, 3789
United Technologies Corporation Grants, 3803
Unity Foundation Of LaPorte County Grants, 3804
USAID Ebola Response, Recovery and Resilience in West Africa Grants, 3809
USAID Higher Education Partnerships for Innovation and Impact (HEPII) Grants, 3815
USAID Land Use Change and Disease Emergence Grants, 3818
USDA Organic Agriculture Research Grants, 3830
Vancouver Foundation Grants and Community Initiatives Program, 3849
Vanderburgh Community Foundation Grants, 3852
Vanderburgh Community Fndn Women's Fund, 3853
Vermont Community Foundation Grants, 3865
Virginia W. Kettering Foundation Grants, 3878
W.C. Griffith Foundation Grants, 3881
Walker Area Community Foundation Grants, 3895
Wallace Alexander Gerbode Foundation Grants, 3896
Walmart Foundation Community Giving Grants, 3897
Walmart Foundation National Giving Grants, 3898
Walmart Foundation State Giving Grants, 3900
Walter L. Gross III Family Foundation Grants, 3901
Warrick County Community Foundation Grants, 3904
Washington County Community Fndn Grants, 3906
Washington County Community Foundation Youth Grants, 3907
Washington Gas Charitable Giving, 3908
Wells County Foundation Grants, 3920
Western Indiana Community Foundation Grants, 3921
Weyerhaeuser Company Foundation Grants, 3925
White County Community Foundation Grants, 3934
William B. Stokely Jr. Foundation Grants, 3947
William G. & Helen C. Hoffman Fndn Grants, 3950
William Ray & Ruth E. Collins Fndn Grants, 3956
Windham Foundation Grants, 3961
Winifred & Harry B. Allen Foundation Grants, 3962
Winston-Salem Fndn Stokes County Grants, 3967
Yampa Valley Community Foundation Grants, 3978
Z. Smith Reynolds Foundation Small Grants, 3982

Exchange Programs
AES Research Infrastructure Awards, 321
IBRO Asia Reg APRC Exchange Fellowships, 2101
SSRC Collaborative Research Grants on Environment and Health in China, 3638

Exhibitions, Collections, Performances, Video/Film Production
3M Company Fndn Community Giving Grants, 6
Aaron Foundation Grants, 193
Abell-Hanger Foundation Grants, 207

Adaptec Foundation Grants, 304
Agnes Gund Foundation Grants, 345
Alaska Airlines Foundation Grants, 403
Alcatel-Lucent Technologies Foundation Grants, 413
AMA-RFS and AMA Foundation Medical Student & Resident/Fellow Elective-Medicine & Media, 478
Anna Fitch Ardenghi Trust Grants, 592
Anne J. Caudal Foundation Grants, 595
Arthur Ashley Williams Foundation Grants, 671
Autauga Area Community Foundation Grants, 808
Bacon Family Foundation Grants, 817
Bank of America Fndn Matching Gifts, 825
Beckman Coulter Foundation Grants, 868
Benton Community Foundation Grants, 878
Blanche and Julian Robertson Family Foundation Grants, 911
Blue Mountain Community Foundation Grants, 921
Blue River Community Foundation Grants, 922
Boeing Company Contributions Grants, 930
Bonfils-Stanton Foundation Grants, 932
Cable Positive's Tony Cox Community Grants, 992
Caesars Foundation Grants, 995
Carl M. Freeman Foundation FACES Grants, 1026
Carl M. Freeman Foundation Grants, 1027
CDC David J. Sencer Museum Adult Group Tour, 1052
CFFVR Project Grants, 1101
Chilkat Valley Community Foundation Grants, 1148
Coca-Cola Foundation Grants, 1203
Collins Foundation Grants, 1210
Community Fndn for Muskegon County Grants, 1254
Community Foundation for Northeast Michigan Mini-Grants, 1255
Community Fndn for the Capital Region Grants, 1259
Community Foundation of Eastern Connecticut General Southeast Grants, 1269
Community Foundation of Riverside & San Bernardino County Impact Grants, 1303
Community Fndn of Riverside & San Bernardino County Irene S. Rockwell Grants, 1304
Community Foundation Partnerships - Martin County Grants, 1318
Cowles Charitable Trust Grants, 1354
Cralle Foundation Grants, 1356
Cresap Family Foundation Grants, 1359
Crystelle Waggoner Charitable Trust Grants, 1364
CSRA Community Foundation Grants, 1366
D.F. Halton Foundation Grants, 1405
Daniel & Nanna Stern Family Fndn Grants, 1422
Donald and Sylvia Robinson Family Foundation Grants, 1486
Dorrance Family Foundation Grants, 1501
Dr. Leon Bromberg Charitable Trust Grants, 1509
Drs. Bruce and Lee Foundation Grants, 1513
Earl and Maxine Claussen Trust Grants, 1541
Eddie C. & Sylvia Brown Family Fndn Grants, 1548
Edyth Bush Charitable Foundation Grants, 1562
Elizabeth McGraw Foundation Grants, 1568
Elmer L. & Eleanor J. Andersen Fndn Grants, 1578
Ensworth Charitable Foundation Grants, 1595
Fifth Third Foundation Grants, 1661
Florida Division of Cultural Affairs Arts In Education Arts Partnership Grants, 1679
Foster Foundation Grants, 1689
Four County Community Fndn General Grants, 1710
Frank Loomis Palmer Fund Grants, 1726
George A Ohl Jr. Foundation Grants, 1791
George S. & Dolores Dore Eccles Fndn Grants, 1801
Gil and Dody Weaver Foundation Grants, 1819
Global Fund for Children Grants, 1832
GNOF Exxon-Mobil Grants, 1836
GNOF IMPACT Grants for Arts and Culture, 1838
Golden Heart Community Foundation Grants, 1850
Goodrich Corporation Foundation Grants, 1854
Grand Rapids Area Community Fndn Grants, 1865
Greater Sitka Legacy Fund Grants, 1885
Greater Worcester Community Foundation Jeppson Memorial Fund for Brookfield Grants, 1889
Harold R. Bechtel Charitable Remainder Uni-Trust Grants, 1953
Health Fndn of Greater Indianapolis Grants, 1984

PROGRAM TYPE INDEX

Helen Gertrude Sparks Charitable Trust Grants, 1993
Helen Pumphrey Denit Charitable Trust Grants, 1996
Honda of America Manufacturing Fndn Grants, 2045
Horace Moses Charitable Foundation Grants, 2047
Horizons Community Issues Grants, 2049
Janus Foundation Grants, 2211
Jessica Stevens Community Foundation Grants, 2224
Johnson & Johnson/SAH Arts & Healing Grants, 2254
John W. Speas and Effie E. Speas Memorial Trust Grants, 2264
Joseph Henry Edmondson Foundation Grants, 2270
Judith and Jean Pape Adams Charitable Foundation Tulsa Area Grants, 2280
Kenai Peninsula Foundation Grants, 2298
Ketchikan Community Foundation Grants, 2303
KeyBank Foundation Grants, 2307
Kodiak Community Foundation Grants, 2314
Kuntz Foundation Grants, 2326
Leo Goodwin Foundation Grants, 2354
Libra Foundation Grants, 2363
Louis and Elizabeth Nave Flarsheim Charitable Foundation Grants, 2379
McCombs Foundation Grants, 2491
Meriden Foundation Grants, 2519
Meyer and Stephanie Eglin Foundation Grants, 2531
MLA Louise Darling Medal for Distinguished Achievement in Collection Development in the Health Sciences, 2569
Norcliffe Foundation Grants, 2979
OceanFirst Foundation Major Grants, 3062
Oppenstein Brothers Foundation Grants, 3078
Oregon Community Fndn Community Grants, 3082
Petersburg Community Foundation Grants, 3156
Piedmont Natural Gas Corporate and Charitable Contributions, 3220
Pollock Foundation Grants, 3244
PPCF Community Grants, 3256
Price Chopper's Golub Foundation Grants, 3261
Price Family Charitable Fund Grants, 3262
Rasmuson Foundation Tier Two Grants, 3308
Reinberger Foundation Grants, 3357
Richard D. Bass Foundation Grants, 3366
Robbins Charitable Foundation Grants, 3378
Robert F. Stoico/FIRSTFED Charitable Foundation Grants, 3382
Robert R. Meyer Foundation Grants, 3387
Rockwell Collins Charitable Corporation Grants, 3397
Rohm and Haas Company Grants, 3403
San Diego Foundation for Change Grants, 3470
Santa Fe Community Foundation Seasonal Grants-Fall Cycle, 3482
Schurz Communications Foundation Grants, 3496
Scott B. & Annie P. Appleby Charitable Grants, 3499
Seward Community Foundation Grants, 3514
Seward Community Foundation Mini-Grants, 3515
Shield-Ayres Foundation Grants, 3546
Sioux Falls Area Community Foundation Community Fund Grants (Unrestricted), 3568
Sioux Falls Area Community Foundation Spot Grants (Unrestricted), 3570
Sony Corporation of America Grants, 3623
Southbury Community Trust Fund, 3625
Stewart Huston Charitable Trust Grants, 3655
SunTrust Bank Trusteed Foundations Florence C. and Harry L. English Memorial Fund Grants, 3680
SunTrust Bank Trusteed Foundations Greene-Sawtell Grants, 3681
SunTrust Bank Trusteed Foundations Harriet McDaniel Marshall Tust Grants, 3682
SunTrust Bank Trusteed Foundations Nell Warren Elkin and William Simpson Elkin Grants, 3683
SunTrust Bank Trusteed Foundations Thomas Guy Woolford Charitable Trust Grants, 3684
SunTrust Bank Trusteed Foundations Walter H. and Marjory M. Rich Memorial Fund Grants, 3685
Taylor S. Abernathy and Patti Harding Abernathy Charitable Trust Grants, 3704
Thomas J. Long Foundation Community Grants, 3722
Union Bank, N.A. Foundation Grants, 3788
United Technologies Corporation Grants, 3803

PROGRAM TYPE INDEX

Faculty/Professional Development / 881

W.C. Griffith Foundation Grants, 3881
W.M. Keck Fndn So California Grants, 3890
Wabash Valley Community Foundation Grants, 3893
Walker Area Community Foundation Grants, 3895
Walter L. Gross III Family Foundation Grants, 3901
William Ray & Ruth E. Collins Fndn Grants, 3956
Winston-Salem Foundation Competitive Grants, 3965

Faculty/Professional Development
AAAAI Distinguished Clinician Award, 24
AAAAI Fellows-in-Training Abstract Award, 28
AACN Physio-Control Small Projects Grants, 86
AACR Career Development Awards for Pediatric Cancer Research, 107
AACR Gertrude Elion Cancer Research Award, 113
AACR Judah Folkman Career Development Award for Anti-Angiogenesis Research, 116
AAFP Foundation Pfizer Teacher Devel Awards, 134
AANS Neurosurgery Research & Educ Foundation Young Clinician Investigator Award, 163
Able Trust Vocational Rehabilitation Grants for Individuals, 213
ACS Cancer Control Career Development Awards for Primary Care Physicians, 250
ACS MEN2 Thyroid Cancer Professorship Grants, 263
AFAR Paul Beeson Career Development Awards in Aging Research for the Island of Ireland, 336
AHIMA Faculty Development Stipends, 353
AIHP Sonnedecker Visiting Scholar Grants, 371
AIHS Collaborative Research and Innovation Grants - Collaborative Project, 377
AIHS Interdisciplinary Team Grants, 388
AIHS Knowledge Exchange - Visiting Scientists, 391
AIHS Research Prize, 396
Alexander von Humboldt Foundation Research Fellowships for Postdoctoral Researchers, 425
Alfred P. Sloan Foundation Research Fellowships, 432
AMA Foundation Leadership Awards, 490
AMDA Foundation Quality Improvement Award, 499
American College of Surgeons Faculty Career Development Award for Neurological Surgeons, 521
American Psychiatric Association/AstraZeneca Young Minds in Psychiatry International Awards, 538
Amerigroup Foundation Grants, 565
AMI Semiconductors Corporate Grants, 576
ANA Distinguished Neurology Teacher Award, 580
ANA F.E. Bennett Memorial Lectureship, 581
ANA Raymond D. Adams Lectureship, 583
ANA Soriano Lectureship, 584
Anheuser-Busch Foundation Grants, 591
Annenberg Foundation Grants, 596
Arnold and Mabel Beckman Foundation Young Investigators Grants, 666
ASHA Research Mentoring-Pair Travel Award, 698
ASHA Students Preparing for Academic & Research Careers (SPARC) Award, 700
Assisi Foundation of Memphis General Grants, 787
AVDF Health Care Grants, 811
Ball Brothers Foundation General Grants, 820
Baxter International Foundation Grants, 843
Boeing Company Contributions Grants, 930
Boyd Gaming Corporation Contributions Program, 941
Brookdale Foundation Leadership in Aging Fellowships, 965
Bush Foundation Medical Fellowships, 980
BWF Career Awards at the Scientific Interface, 983
BWF Institutional Program Unifying Population and Laboratory Based Sciences Grants, 987
Byron W. & Alice L. Lockwood Fnd Grants, 991
Cabot Corporation Foundation Grants, 993
Canada-U.S. Fulbright New Century Scholars Program Grants, 1008
Canada-U.S. Fulbright Senior Specialists Grants, 1009
Canadian Optometric Education Trust Grants, 1011
Carl B. and Florence E. King Foundation Grants, 1022
Carl M. Freeman Foundation Grants, 1027
Carpenter Foundation Grants, 1035
Carrie Estelle Doheny Foundation Grants, 1037
Carrier Corporation Contributions Grants, 1038

CDC David J. Sencer Museum Teacher Professional Development Workshops, 1053
CFF Shwachman Clinical Investigator Award, 1086
CFF Leroy Matthews Physician-Scientist Awards, 1087
CFF Research Grants, 1090
CFFVR Basic Needs Giving Partnership Grants, 1094
Charlotte County (FL) Community Foundation Grants, 1126
Chazen Foundation Grants, 1129
Christensen Fund Regional Grants, 1149
Clarence E. Heller Charitable Foundation Grants, 1166
Claude Worthington Benedum Fndn Grants, 1178
Cleveland-Cliffs Foundation Grants, 1181
Clowes ACS/AAST/NIGMS Jointly Sponsored Mentored Clinical Scientist Dev Award, 1185
Cockrell Foundation Grants, 1204
Columbus Foundation Competitive Grants, 1217
Community Foundation for Greater New Haven Valley Neighborhood Grants, 1251
Community Fndn for Muskegon County Grants, 1254
Community Foundation of Greater Birmingham Grants, 1272
Community Foundation of Greater New Britain Grants, 1281
Community Foundation of the Verdugos Educational Endowment Fund Grants, 1314
ConAgra Foods Nourish our Community Grants, 1322
Conquer Cancer Foundation of ASCO Career Development Award, 1328
Conquer Cancer Foundation of ASCO Comparative Effectiveness Research Professorship in Breast Cancer, 1329
Conquer Cancer Foundation of ASCO Drug Development Research Professorship, 1330
Conquer Cancer Foundation of ASCO International Development and Education Award in Palliative Care, 1332
Conquer Cancer Foundation of ASCO International Development and Education Awards, 1333
Conquer Cancer Foundation of ASCO Medical Student Rotation Grants, 1335
Conquer Cancer Foundation of ASCO Resident Travel Award for Underrepresented Populations, 1336
Conquer Cancer Foundation of ASCO Translational Research Professorships, 1337
Constellation Energy Corporate Grants, 1342
Cowles Charitable Trust Grants, 1354
Cystic Fibrosis Canada Senior Scientist Research Training Awards, 1387
Denton A. Cooley Foundation Grants, 1459
Dept of Ed Special Education--Personnel Development to Improve Services and Results for Children with Disabilities, 1465
Dept of Ed Special Education--Technical Assistance and Dissemination to Improve Services and Results for Children with Disabilities, 1467
Dr. Leon Bromberg Charitable Trust Grants, 1509
Drs. Bruce and Lee Foundation Grants, 1513
Dubois County Community Foundation Grants, 1517
Dunspaugh-Dalton Foundation Grants, 1530
Eberly Foundation Grants, 1544
Edward W. and Stella C. Van Houten Memorial Fund Grants, 1559
Edwin S. Webster Foundation Grants, 1560
Effie and Wofford Cain Foundation Grants, 1563
El Paso Community Foundation Grants, 1579
Emily Davie and Joseph S. Kornfeld Foundation Grants, 1591
Eugene M. Lang Foundation Grants, 1611
Evjue Foundation Grants, 1617
FAMRI Young Clinical Scientist Awards, 1629
FAR Fund Grants, 1631
Fargo-Moorhead Area Fndn Woman's Grants, 1633
Faye McBeath Foundation Grants, 1635
FDHN Fellowship to Faculty Transition Awards, 1644
FDHN Research Scholar Awards, 1652
France-Merrick Foundation Health and Human Services Grants, 1715
Frances L. and Edwin L. Cummings Memorial Fund Grants, 1717

Fremont Area Community Foundation Amazing X Grants, 1744
Fremont Area Community Fndn General Grants, 1746
Fritz B. Burns Foundation Grants, 1747
Fulbright Distinguished Chairs Awards, 1750
Fulbright International Education Administrators (IEA) Seminar Program Grants, 1752
Fulbright New Century Scholars Grants, 1753
Fulbright Specialists Program Grants, 1754
Fulbright Scholars in Europe and Eurasia, 1755
Fulbright Traditional Scholar Program in Sub-Saharan Africa, 1756
Fulbright Traditional Scholar Program in the East Asia/Pacific Region, 1757
Fulbright Traditional Scholar Program in the Near East and North Africa Region, 1758
Fulbright Traditional Scholar Program in the South and Central Asia Region, 1759
Fulbright Traditional Scholar Program in the Western Hemisphere, 1760
George A Ohl Jr. Foundation Grants, 1791
George Foundation Grants, 1795
George Kress Foundation Grants, 1799
George S. & Dolores Dore Eccles Fndn Grants, 1801
Good Samaritan Inc Grants, 1855
Grifols Community Outreach Grants, 1904
HAF Technical Assistance Program Grants, 1933
Harry A. & Margaret D. Towsley Fndn Grants, 1959
Harry Frank Guggenheim Fnd Research Grants, 1964
Health Fndn of Greater Indianapolis Grants, 1984
Hearst Foundations Health Grants, 1987
HHMI/EMBO Start-up Grants for C Europe, 2016
Hill Crest Foundation Grants, 2030
Hillman Foundation Grants, 2032
Hospital Libraries Section / MLA Professional Development Grants, 2052
Independence Blue Cross Charitable Medical Care Grants, 2150
J.M. Long Foundation Grants, 2173
James McKeen Cattell Fund Fellowships, 2197
James R. Dougherty Jr. Foundation Grants, 2198
Jessie Ball Dupont Fund Grants, 2226
John P. McGovern Foundation Grants, 2247
John S. Dunn Research Foundation Grants, 2250
Johnson & Johnson Corporate Contributions, 2253
Joseph Alexander Foundation Grants, 2265
Josiah Macy Jr. Foundation Grants, 2277
Kenneth T. & Eileen L. Norris Fndn Grants, 2300
Kosciusko County Community Fndn Grants, 2319
Landon Foundation-AACR Innovator Award for International Collaboration, 2343
Lowe Foundation Grants, 2381
Lynde and Harry Bradley Foundation Grants, 2397
M.B. and Edna Zale Foundation Grants, 2401
Mead Johnson Nutritionals Med Educ Grants, 2504
Mead Witter Foundation Grants, 2507
Medical Informatics Section/MLA Career Development Grant, 2508
MLA Continuing Education Grants, 2559
MLA Cunningham Memorial International Fellowship, 2560
MMS and Alliance Charitable Foundation International Health Studies Grants, 2588
NCI Cancer Education and Career Development Program, 2648
NCI Mentored Career Development Award to Promote Diversity, 2653
NHGRI Mentored Research Scientist Development Award, 2686
NHLBI Mentored Career Award for Faculty at Institutions that Promote Diversity (K01), 2706
NHLBI Mentored Career Award For Faculty At Minority Institutions, 2707
NHLBI Mentored Career Dev Award to Promote Faculty Diversity in Biomed Research, 2708
NHLBI Mentored Patient-Oriented Research Career Development Awards, 2710
NHLBI Mentored Quantitative Research Career Development Awards, 2711
NHLBI Pathway to Independence Awards, 2718

882 / Faculty/Professional Development

NHLBI Ruth L. Kirschstein National Research Service Awards for Individual Predoctoral Fellowships to Promote Diversity in Health-Related Research, 2730
NHLBI Ruth L. Kirschstein National Research Service Awards for Individual Senior Fellows, 2732
NHLBI Short-Term Research Education Program to Increase Diversity in Health-Related Research, 2733
NHLBI Training Program Grants for Institutions that Promote Diversity, 2738
NIAAA Independent Scientist Award, 2742
NIA Independent Scientist Award, 2766
NIA Mentored Research Scientist Devel Award, 2772
NIA Pilot Research Grant Program, 2778
NIDA Mentored Clinical Scientist Research Career Development Awards, 2793
NIDCD Mentored Clinical Scientist Research Career Development Award, 2795
NIEHS Independent Scientist Award, 2856
NIEHS Mentored Clinical Scientist Research Career Development Awards, 2857
NIEHS Mentored Research Scientist Development Award, 2858
NIH Mentored Clinical Scientist Devel Award, 2886
NIH Mentored Clinical Scientist Research Career Development Award, 2887
NINR Mentored Research Scientist Development Award, 2932
NKF Research Fellowships, 2938
NKF Young Investigator Grants, 2939
Noble County Community Foundation Grants, 2975
North Carolina Biotechnology Center Grantsmanship Training Grants, 2985
North Carolina Biotechnology Center Regional Development Grants, 2987
North Carolina GlaxoSmithKline Fndn Grants, 2991
Northern Chautauqua Community Foundation Community Grants, 2995
NSF Postdoc Research Fellowships in Biology, 3020
NSF Presidential Early Career Awards for Scientists and Engineers (PECASE) Grants, 3021
NSF Undergraduate Research and Mentoring in the Biological Sciences (URM), 3025
Pancreatic Cancer Action Network-AACR Pathway to Leadership Grants, 3114
PDF-AANF Clinician-Scientist Dev Awards, 3127
Pfizer ASPIRE North America Hemophilia Research Awards, 3170
Pfizer Australia Cancer Research Grants, 3174
Pfizer Australia Neuroscience Research Grants, 3175
Pfizer Medical Education Track Two Grants, 3183
Philadelphia Foundation Organizational Effectiveness Grants, 3189
Phi Upsilon Omicron Florence Fallgatter Distinguished Service Award, 3192
Phi Upsilon Omicron Frances Morton Holbrook Alumni Award, 3193
PhRMA Foundation Health Outcomes Post Doctoral Fellowships, 3202
PhRMA Foundation Informatics Post Doctoral Fellowships, 3206
PhRMA Foundation Informatics Research Starter Grants, 3208
PhRMA Fndn Info Sabbatical Fellowships, 3209
PhRMA Foundation Pharmaceutics Postdoctoral Fellowships, 3211
PhRMA Foundation Pharmaceutics Research Starter Grants, 3213
PhRMA Foundation Pharmaceutics Sabbatical Fellowships, 3214
PhRMA Foundation Pharmacology/Toxicology Post Doctoral Fellowships, 3215
PhRMA Foundation Pharmacology/Toxicology Pre Doctoral Fellowships, 3216
PhRMA Foundation Pharmacology/Toxicology Research Starter Grants, 3217
PhRMA Foundation Pharmacology/Toxicology Sabbatical Fellowships, 3218
Presbyterian Health Foundation Bridge, Seed and Equipment Grants, 3259
Prince Charitable Trusts DC Grants, 3271

PSG Mentored Clinical Research Awards, 3277
Puerto Rico Community Foundation Grants, 3278
Pulaski County Community Foundation Grants, 3279
RCPSC Continuing Proffesional Dev Grants, 3319
RCPSC Professional Development Grants, 3338
Rhode Island Foundation Grants, 3362
Robert & Clara Milton Senior Housing Grants, 3384
Rohm and Haas Company Grants, 3403
Ronald McDonald House Charities Grants, 3405
Roy J. Carver Charitable Trust Medical and Science Research Grants, 3416
RRF General Program Grants, 3417
RRF Organizational Capacity Building Grants, 3418
RWJF Harold Amos Medical Faculty Development Scholars, 3435
RWJF Nurse Faculty Scholars, 3441
San Antonio Area Foundation Grants, 3469
Sarkeys Foundation Grants, 3485
Schering-Plough Foundation Health Grants, 3492
Searle Scholars Program Grants, 3503
SfN Award for Education in Neuroscience, 3517
SfN Janett Rosenberg Trubatch Career Development Award, 3524
SfN Science Educator Award, 3535
SNM Molecular Imaging Research Grant For Junior Medical Faculty, 3583
SNM Postdoc Molecular Imaging Scholar Grants, 3587
Sonora Area Foundation Competitive Grants, 3622
Square D Foundation Grants, 3634
St. Joseph Community Health Foundation Burn Care and Prevention Grants, 3639
St. Joseph Community Health Foundation Improving Healthcare Access Grants, 3640
Steele-Reese Foundation Grants, 3650
Topeka Community Foundation Kansas Blood Services Scholarships, 3738
UCT Scholarship Program, 3777
UICC Yamagiwa-Yoshida Memorial International Cancer Study Grants, 3781
Union County Community Foundation Grants, 3790
Vancouver Foundation Grants and Community Initiatives Program, 3849
VDH Rescue Squad Assistance Fund Grants, 3856
Weingart Foundation Grants, 3913
World of Children Humanitarian Award, 3975

Fellowships
3M Company Fndn Community Giving Grants, 6
A-T Children's Project Post Doctoral Fellowships, 14
AAAAI Fellows-in-Training Grants, 29
AAAS Science and Technology Policy Fellowships: Global Health and Development, 41
AACAP-NIDA K12 Career Development Awards, 49
AACAP Educational Outreach Program for Residents in Alcohol Research, 54
AACAP Jeanne Spurlock Minority Med Student Clinical Fellow in Child & Adolescent Psych, 59
AACAP Jeanne Spurlock Research Fellowship in Substance Abuse and Addiction for Minority Medical Students, 60
AACAP Life Members Mentorship Grants for Medical Students, 63
AACAP Mary Crosby Congressional Fellowships, 64
AACAP Pilot Research Awards, Supported by Eli Lilly and Company, 68
AACAP Summer Medical Student Fellowships, 75
AACR-Colorectal Cancer Coalition Fellows Grant, 88
AACR-FNAB Fellows Grant for Translational Pancreatic Cancer Research, 89
AACR-Genentech BioOncology Fellowship for Cancer Research in Angiogenesis, 92
AACR-National Brain Tumor Fnd Fellows Grant, 95
AACR Basic Cancer Research Fellowships, 105
AACR Centennial Postdoctoral Fellowships in Cancer Research, 108
AACR Centennial Pre-doctoral Fellowships in Cancer Research, 109
AACR Clinical & Translat Research Fellowships, 110
AACR Fellows Grants, 111
AAFA Investigator Research Grants, 128

PROGRAM TYPE INDEX

AAFCS International Graduate Fellowships, 129
AAFCS National Graduate Fellowships, 130
AAFPRS Fellowships, 139
AAHPM Hospice & Palliative Med Fellowships, 152
AAMC Herbert W. Nickens Faculty Fellowship, 159
AAUW American Dissertation Fellowships, 195
AAUW American Postdoctoral Research Leave Fellowships, 196
Abbott Fund Science Education Grants, 203
Abell-Hanger Foundation Grants, 207
Abney Foundation Grants, 214
ACAAI Foundation Research Grants, 219
ACS Postdoctoral Fellowships, 277
Acumen East Africa Fellowship, 284
Acumen Global Fellowships, 285
ADA Foundation Dentsply International Research Fellowships, 289
ADA Foundation Scientist in Training Fellowship, 295
ADA Foundation Summer Scholars Fellowships, 296
Advocate HealthCare Post Graduate Administrative Fellowship, 314
AES-Grass Young Investigator Travel Awards, 317
AES Robert S. Morison Fellowship, 323
AES Susan S. Spencer Clinical Research Epilepsy Training Fellowship, 325
Aetna Foundation Minority Scholars Grants, 329
Aetna Foundation National Medical Fellowship in Healthcare Leadership, 330
AFAR Medical Student Training in Aging Research Program, 335
AGHE Graduate Scholarships and Fellowships, 343
AHRQ Individual Awards for Postdoc Fellows Ruth L. Kirschstein National Research Service Awards, 366
AIHS Clinical Fellowships, 375
AIHS Fast-Track Fellowships, 379
AIHS Full-Time Fellowships, 383
AIHS Media Summer Fellowship, 392
AIHS Part-Time Fellowships, 393
Albert and Mary Lasker Foundation Clinical Research Scholars, 407
Alexander von Humboldt Foundation Georg Forster Fellowships for Experienced Researchers, 423
Alexander von Humboldt Foundation Georg Forster Fellowships for Postdoctoral Researchers, 424
Alexander von Humboldt Foundation Research Fellowships for Postdoctoral Researchers, 425
Alfred P. Sloan Foundation Research Fellowships, 432
Alzheimer's Association U.S.-U.K. Young Investigator Exchange Fellowships, 471
Alzheimer's Association Zenith Fellows Awards, 472
AMA-MSS Government Relations Advocacy Fellowship, 475
AMA Foundation Joan F. Giambalvo Memorial Scholarships, 488
American Academy of Nursing Claire M. Fagin Fellowships, 506
American College of Surgeons Faculty Research Fellowships, 522
American College of Surgeons Nizar N. Oweida, MD, FACS, Scholarships, 526
American College of Surgeons Wound Care Management Award, 529
American Philosophical Society Daland Fellowships in Clinical Investigation, 537
American Psychiatric Association/Bristol-Myers Squibb Fellowships in Public Psychiatry, 539
American Psychiatric Association Minority Fellowships Program, 543
American Psychiatric Association Minority Medical Student Fellowship in HIV Psychiatry, 544
American Psychiatric Foundation Jeanne Spurlock Congressional Fellowship, 556
American Psychiatric Foundation Minority Fellowships, 557
American Sociological Association Minority Fellowships, 563
amfAR Fellowships, 567
amfAR Mathilde Krim Fellowships in Basic Biomedical Research, 569
amfAR Research Grants, 571

PROGRAM TYPE INDEX

Fellowships

AMHPS Dr. James A. Ferguson Emerging Infectious Diseases Fellowships, 573
APA Congressional Fellowships, 624
APHL Emerging Infectious Diseases Fellowships, 636
APSAA Fellowships, 638
APSA Congressional Health and Aging Policy Fellowships, 644
APSA Robert Wood Johnson Foundation Health Policy Fellowships, 645
Aratani Foundation Grants, 649
Arie and Ida Crown Memorial Grants, 654
Arthur and Rochelle Belfer Foundation Grants, 670
ArvinMeritor Foundation Health Grants, 674
ASCO/UICC International Cancer Technology Transfer (ICRETT) Fellowships, 677
ASCO Long-Term International Fellowships, 681
ASHG Public Health Genetics Fellowship, 712
ASM-PAHO Infectious Diseases Epidemiology and Surveillance Fellowships, 714
ASM/CDC Fellowships in Infectious Disease and Public Health Microbiology, 716
ASM Congressional Science Fellowships, 720
ASM Int'l Fellowship for Latin America and the Caribbean, 734
ASM Int'l Fellowships for Asia and Africa, 735
ASM Microbiology Undergraduate Research Fellowship, 740
ASM Robert D. Watkins Graduate Research Fellowship, 747
ASM Undergraduate Research Fellowship, 755
ASPH/CDC Allan Rosenfield Global Health Fellowships, 780
ASPH/CDC Public Health Fellowships, 781
ASPO Fellowships, 784
ASU Grad College Reach for the Stars Fellowship, 791
ASU Graduate College Science Foundation Arizona Bisgrove Postdoctoral Scholars, 792
Australian Academy of Science Grants, 807
Ball Brothers Foundation General Grants, 820
Baltimore Washington Center Adult Psychoanalysis Fellowships, 821
Baltimore Washington Center Child Psychotherapy Fellowships, 822
Biogen Foundation Fellowships, 898
Boeing Company Contributions Grants, 930
Boston Psychoanalytic Society and Institute Fellowship in Child Psychoanalytic Psychotherapy, 939
Boston Psychoanalytic Society and Institute Fellowship in Psychoanalytic Psychotherapy, 940
Bristol-Myers Squibb Foundation Fellowships, 956
Brookdale Foundation Leadership in Aging Fellowships, 965
BWF Institutional Program Unifying Population and Laboratory Based Sciences Grants, 987
Cabot Corporation Foundation Grants, 993
Caddock Foundation Grants, 994
Caesars Foundation Grants, 995
Carrier Corporation Contributions Grants, 1038
Carroll County Community Foundation Grants, 1039
CDC-Hubert Global Health Fellowship, 1047
CDC Experience Epidemiology Fellowships, 1059
CDC Foundation Atlanta International Health Fellowships, 1060
CDC Fnd Tobacco Network Lab Fellowship, 1062
CDC Preventive Med Residency & Fellowship, 1064
CDC Public Health Informatics Fellowships, 1068
CDC Public Health Prev Service Fellowships, 1069
CDC Public Health Prevention Service Fellowship Sponsorships, 1070
CDC Steven M. Teutsch Prevention Effectiveness Fellowships, 1071
Cedar Tree Foundation David H. Smith Conservation Research Fellowship, 1076
Cemala Foundation Grants, 1077
CFF First- and Second-Year Clinical Fellowships, 1085
CFF Postdoctoral Research Fellowships, 1089
CFF 3rd through 5th Year Clinical Fellowships, 1092
Charles H. Revson Foundation Grants, 1122
Chicago Institute for Psychoanalysis Fellowships, 1135
Christensen Fund Regional Grants, 1149

CIGNA Foundation Grants, 1160
CNIB Baker Applied Research Fund Grants, 1193
CNIB Baker Fellowships, 1194
CNIB Baker New Researcher Fund Grants, 1195
CNIB E. (Ben) & Mary Hochhausen Access Technology Research Grants, 1199
CNIB Ross Purse Doctoral Fellowships, 1200
Coca-Cola Foundation Grants, 1203
Commonwealth Fund/Harvard University Fellowship in Minority Health Policy, 1226
Commonwealth Fund Australian-American Health Policy Fellowships, 1228
Commonwealth Fund Harkness Fellowships in Health Care Policy and Practice, 1229
CSTE CDC/CSTE Applied Epidemiology Fellowships, 1367
Cystic Fibrosis Canada Clinical Fellowships, 1381
Cystic Fibrosis Canada Fellowships, 1384
Cystic Fibrosis Canada Senior Scientist Research Training Awards, 1387
Daimler and Benz Foundation Fellowships, 1411
Delta Air Lines Foundation Health and Wellness Grants, 1455
Doris Duke Charitable Foundation Clinical Research Fellowships for Medical Students, 1493
Duke University Adult Cardiothoracic Anesthesia and Critical Care Medicine Fellowships, 1520
Duke University Ambulatory and Regional Anesthesia Fellowships, 1521
Duke University Clinical Cardiac Electrophysiology Fellowships, 1522
Duke University Hyperbaric Center Fellowships, 1523
Duke Univ Interventional Cardio Fellowships, 1524
Duke Univ Obstetric Anesthesia Fellowships, 1525
Duke University Pain Management Fellowships, 1526
Duke University Ped Anesthesiology Fellowship, 1527
Duke University Postdoctoral Research Fellowships in Aging, 1528
Echoing Green Fellowships, 1546
Edward W. and Stella C. Van Houten Memorial Fund Grants, 1559
Edwin S. Webster Foundation Grants, 1560
Edwin W. and Catherine M. Davis Fndn Grants, 1561
Effie and Wofford Cain Foundation Grants, 1563
Eugene M. Lang Foundation Grants, 1611
Eye-Bank for Sight Restoration and Fight for Sight Summer Student Research Fellowship, 1622
FAMRI Young Clinical Scientist Awards, 1629
FDHN Centocor International Research Fellowship in Gastrointestinal Inflammation & Immunology, 1639
FDHN Student Research Fellowships, 1654
Fight for Sight-Streilein Foundation for Ocular Immunology Research Award, 1662
Fight for Sight Post-Doctoral Awards, 1664
Fight for Sight Summer Student Fellowships, 1665
Foundation of the American Thoracic Society Research Grants, 1708
Fragile X Syndrome Postdoc Research Fellows, 1714
Fritz B. Burns Foundation Grants, 1747
Fulbright Traditional Scholar Program in the East Asia/Pacific Region, 1757
Fulbright Traditional Scholar Program in the Near East and North Africa Region, 1758
Fulbright Traditional Scholar Program in the Western Hemisphere, 1760
Gertrude E. Skelly Charitable Foundation Grants, 1810
Grass Foundation Marine Biological Laboratory Advanced Imaging Fellowships, 1877
Grass Foundation Marine Biological Laboratory Fellowships, 1878
Gruber Foundation Rosalind Franklin Young Investigator Award, 1910
H.A. & Mary K. Chapman Trust Grants, 1921
H. Schaffer Foundation Grants, 1925
Harry Frank Guggenheim Foundation Dissertation Fellowships, 1963
Hereditary Disease Foundation John J. Wasmuth Postdoctoral Fellowships, 2009
Hereditary Disease Foundation Research Grants, 2011
HHMI Grants and Fellowships Programs, 2020

HHMI Research Training Fellowships, 2024
HRAMF Charles A. King Trust Postdoctoral Research Fellowships, 2056
Huntington's Disease Society of America Research Fellowships, 2079
IARC Expertise Transfer Fellowship, 2090
IARC Postdoctoral Fellowships for Training in Cancer Research, 2091
IBRO Asia Reg APRC Exchange Fellowships, 2101
IBRO Regional Grants for Int'l Fellowships to U.S. Laboratory Summer Neuroscience Courses, 2110
IBRO Research Fellowships, 2111
IBRO Return Home Fellowships, 2112
IHI Quality Improvement Fellowships, 2121
IIE Central Europe Summer Research Institute Summer Research Fellowship, 2127
IIE David L. Boren Fellowships, 2129
IIE Hewlett Fnd/IIE Dissertation Fellowship, 2132
IIE KAUST Graduate Fellowships, 2134
IIE Leonora Lindsley Memorial Fellowships, 2136
IIE Lingnan Foundation W.T. Chan Fellowship, 2137
IIE Whitaker Int'l Fellowships & Scholarships, 2144
Institute for Agriculture and Trade Policy Food and Society Fellowships, 2159
James McKeen Cattell Fund Fellowships, 2197
John Deere Foundation Grants, 2234
John Merck Scholars Awards, 2246
Johnson & Johnson Corporate Contributions, 2253
John Stauffer Charitable Trust Grants, 2258
Kansas Health Foundation Recognition Grants, 2289
Kluge Center David B. Larson Fellowship in Health and Spirituality, 2311
Lalor Foundation Postdoctoral Fellowships, 2336
LAM Foundation Postdoctoral Fellowships, 2339
Lillian Sholtis Brunner Summer Fellowships for Historical Research in Nursing, 2366
Lubrizol Foundation Grants, 2384
Lymphatic Education and Research Network Postdoctoral Fellowships, 2395
Lynde & Harry Bradley Foundation Fellowships, 2396
Lynde and Harry Bradley Foundation Grants, 2397
MAP International Medical Fellowships, 2421
Mary K. Chapman Foundation Grants, 2443
Mayo Clinic Administrative Fellowship, 2460
Mayo Clinic Admin Fellowship - Eau Claire, 2461
Mayo Clinic Business Consulting Fellowship, 2462
MBL Albert & Ellen Grass Fellowships, 2463
MBL Ann E. Kammer Summer Fellowship, 2464
MBL Associates Summer Fellowships, 2465
MBL Baxter Postdoctoral Summer Fellowship, 2466
MBL Burr & Susie Steinbach Fellowship, 2467
MBL E.E. Just Summer Fellowship for Minority Scientists, 2468
MBL Erik B. Fries Summer Fellowships, 2469
MBL Eugene and Millicent Bell Fellowships, 2470
MBL Evelyn and Melvin Spiegal Fellowship, 2471
MBL Frank R. Lillie Summer Fellowship, 2472
MBL Frederik B. and Betsy G. Bang Summer Fellowships, 2473
MBL Fred Karush Library Readership, 2474
MBL Gruss Lipper Family Foundation Summer Fellowship, 2475
MBL H. Keffer Hartline and Edward F. MacNichol, Jr. Fellowships, 2476
MBL Herbert W. Rand Summer Fellowship, 2477
MBL James E. and Faith Miller Memorial Summer Fellowship, 2478
MBL John M. Arnold Award, 2479
MBL Laura and Arthur Colwin Fellowships, 2480
MBL Lucy B. Lemann Summer Fellowships, 2481
MBL M.G.F. Fuortes Summer Fellowships, 2482
MBL Nikon Summer Fellowship, 2483
MBL Plum Fndn John E. Dowling Fellowships, 2484
MBL Robert Day Allen Summer Fellowship, 2485
MBL Stephen W. Kuffler Summer Fellowships, 2487
MBL William Townsend Porter Summer Fellowships for Minority Investigators, 2488
Mead Johnson Nutritionals Scholarships, 2505
MedImmune Charitable Grants, 2509
Medtronic Foundation Fellowships, 2512

884 / Fellowships

Mericos Foundation Grants, 2518
MGFA Post-Doctoral Research Fellowships, 2539
MGFA Student Fellowships, 2540
Michigan Psychoanalytic Institute and Society Scholarships, 2547
MLA Cunningham Memorial International Fellowship, 2560
MLA David A. Kronick Traveling Fellowship, 2561
MLA Donald A.B. Lindberg Fellowship, 2562
MMAAP Foundation Dermatology Fellowships, 2577
MMAAP Foundation Fellowship Award in Hematology, 2579
MMAAP Foundation Fellowship Award in Reproductive Medicine, 2580
MMAAP Foundation Fellowship Award in Translational Medicine, 2581
MMAAP Foundation Senior Health Fellowship Award in Geriatrics Medicine and Aging Research, 2586
Morris & Gwendolyn Cafritz Fndn Grants, 2594
NAF Fellowships, 2603
National Center for Responsible Gaming Postdoctoral Fellowships, 2630
National Headache Foundation Seymour Diamond Clinical Fellowship in Headache Education, 2633
National MPS Society Grants, 2634
NCCAM Ruth L. Kirschstein National Research Service Awards for Postdoctoral Training in Complementary and Alternative Medicine, 2640
NCHS National Center for Health Statistics Postdoctoral Research Appointments, 2642
NEAVS Fellowsips in Women's Health and Sex Differences, 2659
New York University Steinhardt School of Education Fellowships, 2684
NHLBI Ruth L. Kirschstein National Research Service Awards for Individual Predoctoral Fellowships to Promote Diversity in Health-Related Research, 2730
NHLBI Ruth L. Kirschstein National Research Service Awards for Individual Predoctoral MD/PhD Fellows and Other Dual Degree Fellows, 2731
NHLBI Ruth L. Kirschstein National Research Service Awards for Individual Senior Fellows, 2732
NIGMS Ruth L. Kirschstein National Research Service Awards for Individual Predoctoral Fellows in PharmD/PhD Grants, 2863
NIH Ruth L. Kirschstein National Research Service Awards (NRSA) for Individual Senior Fellows, 2901
NIH Ruth L. Kirschstein National Research Service Awards for Individual Postdoc Fellowships, 2902
NIH Ruth L. Kirschstein National Research Service Awards for Individual Predoctoral Fellows, 2903
NIH Ruth L. Kirschstein National Research Service Awards for Individual Predoctoral Fellowships to Promote Diversity in Research, 2904
NIH Ruth L. Kirschstein National Research Service Awards for Individual Predoctoral MD/PhD and Other Dual Doctoral Degree Fellows, 2905
NINR Ruth L. Kirschstein National Research Service Award for Predoc Fellows in Nursing, 2934
NKF Research Fellowships, 2938
NMF General Electric Med Fellowships, 2956
Norcliffe Foundation Grants, 2979
NSF Grant Opportunities for Academic Liaison with Industry (GOALI), 3016
NSF Postdoc Research Fellowships in Biology, 3020
Nuffield Foundation Law and Society Grants, 3028
NWHF Mark O. Hatfield Research Fellowship, 3035
NYAM Charles A. Elsberg Fellowship, 3039
NYAM David E. Rogers Fellowships, 3040
NYAM Edwin Beer Research Fellowship, 3042
NYAM Ferdinand C. Valentine Fellowship, 3043
NYAM Mary and David Hoar Fellowship, 3045
Olympus Corporation of Americas Fellowships, 3074
Orthopaedic Trauma Association Kathy Cramer Young Clinician Memorial Fellowships, 3083
Pancreatic Cancer Action Network Fellowship, 3113
Pancreatic Cancer Action Network-AACR Pathway to Leadership Grants, 3114
Patrick and Anna M. Cudahy Fund Grants, 3120
PC Fred H. Bixby Fellowships, 3126

PROGRAM TYPE INDEX

PDF-AANF Clinician-Scientist Dev Awards, 3127
PDF Postdoctoral Fellowships for Basic Scientists, 3129
PDF Postdoctoral Fellowships for Neurologists, 3130
PDF Summer Fellowships, 3131
Pediatric Brain Tumor Fndn Research Grants, 3134
Penny Severns Breast, Cervical and Ovarian Cancer Research Grants, 3136
Pew Charitable Trusts Biomed Research Grants, 3157
PhRMA Foundation Health Outcomes Post Doctoral Fellowships, 3202
PhRMA Foundation Health Outcomes Pre-Doctoral Fellowships, 3203
PhRMA Foundation Health Outcomes Sabbatical Fellowships, 3205
PhRMA Foundation Informatics Post Doctoral Fellowships, 3206
PhRMA Foundation Informatics Pre Doctoral Fellowships, 3207
PhRMA Fndn Info Sabbatical Fellowships, 3209
PhRMA Foundation Paul Calabresi Medical Student Research Fellowship, 3210
PhRMA Foundation Pharmaceutics Postdoctoral Fellowships, 3211
PhRMA Foundation Pharmaceutics Pre Doctoral Fellowships, 3212
PhRMA Foundation Pharmaceutics Sabbatical Fellowships, 3214
PhRMA Foundation Pharmacology/Toxicology Post Doctoral Fellowships, 3215
PhRMA Foundation Pharmacology/Toxicology Pre Doctoral Fellowships, 3216
PhRMA Foundation Pharmacology/Toxicology Sabbatical Fellowships, 3218
Pioneer Hi-Bred Society Fellowships, 3226
Plastic Surgery Educational Foundation Research Fellowship Grants, 3236
Potts Memorial Foundation Grants, 3254
Price Family Charitable Fund Grants, 3262
Pride Foundation Fellowships, 3267
PSG Mentored Clinical Research Awards, 3277
PVA Research Foundation Grants, 3285
R.C. Baker Foundation Grants, 3292
Radcliffe Institute Carol K. Pforzheimer Student Fellowships, 3298
Radcliffe Institute Residential Fellowships, 3299
Rajiv Gandhi Foundation Grants, 3302
Ralph M. Parsons Foundation Grants, 3304
Rathmann Family Foundation Grants, 3309
RCPSC Robert Maudsley Fellowship for Studies in Medical Education, 3341
Regenstrief General Internal Medicine - General Pediatrics Research Fellowship Program, 3351
Regenstrief Geriatrics Fellowship, 3352
Rhode Island Foundation Grants, 3362
Rosalynn Carter Institute John and Betty Pope Fellowships, 3408
Roy J. Carver Charitable Trust Medical and Science Research Grants, 3416
RWJF Harold Amos Medical Faculty Development Research Grants, 3434
RWJF Harold Amos Medical Faculty Development Scholars, 3435
RWJF Health and Society Scholars, 3436
RWJF Nurse Faculty Scholars, 3441
Sara Elizabeth O'Brien Trust Grants, 3483
Savoy Foundation Post-Doctoral and Clinical Research Fellowships, 3488
Sexually Transmitted Diseases Postdoctoral Fellowships, 3516
SfN Neuroscience Fellowships, 3530
Shell Oil Company Foundation Community Development Grants, 3544
Siragusa Foundation Health Services & Medical Research Grants, 3571
Smithsonian Biodiversity Genomics and Bioinformatics Postdoctoral Fellowships, 3580
Smithsonian Museum Conservation Institute Research Post-Doctorate/Post-Graduate Fellowships, 3581
SNM Student Fellowship Awards, 3588
SRC Medical Research Grants, 3635

St. Joseph Community Health Foundation Schneider Fellowships, 3642
Stroke Association Clinical Fellowships, 3662
Thomson Reuters / MLA Doctoral Fellowships, 3728
Tourette Syndrome Association Post-Doctoral Fellowships, 3740
Trauma Center of America Finance Fellowship, 3752
UCLA / RAND Corporation Post-Doctoral Fellowships, 3776
UICC American Cancer Society International Fellowships for Beginning Investigators, 3778
UICC International Cancer Technology Transfer (ICRETT) Fellowships, 3779
United States Institute of Peace - Jennings Randolph Senior Fellowships, 3802
W.M. Keck Foundation Med Research Grants, 3888
Weyerhaeuser Company Foundation Grants, 3925

General Operating Support

1st Source Foundation Grants, 2
2COBS Private Charitable Foundation Grants, 5
A/H Foundation Grants, 18
Abbot and Dorothy H. Stevens Foundation Grants, 197
Abbott Fund Access to Health Care Grants, 199
Abbott Fund Community Grants, 201
Abbott Fund Global AIDS Care Grants, 202
Abbott Fund Science Education Grants, 203
Abby's Legendary Pizza Foundation Grants, 204
Abel Foundation Grants, 206
Abell-Hanger Foundation Grants, 207
Abell Foundation Criminal Justice and Addictions Grants, 208
Abell Fndn Health & Human Services Grants, 209
Abernethy Family Foundation Grants, 210
Able To Serve Grants, 212
Aboudane Family Foundation Grants, 215
Abundance Foundation International Grants, 218
Achelis Foundation Grants, 234
Ackerman Foundation Grants, 236
Adam Richter Charitable Trust Grants, 298
Adams and Reese Corporate Giving Grants, 299
Adams County Community Foundation of Pennsylvania Grants, 300
Adams Foundation Grants, 301
Adolph Coors Foundation Grants, 311
Advance Auto Parts Corporate Giving Grants, 312
AEC Trust Grants, 315
AEGON Transamerica Foundation Health and Welfare Grants, 316
AFG Industries Grants, 340
Agnes Gund Foundation Grants, 345
AIHS Alberta/Pfizer Translat Research Grants, 374
Air Products and Chemicals Grants, 399
Alabama Power Foundation Grants, 400
Alberto Culver Corporate Contributions Grants, 408
Albert Pick Jr. Fund Grants, 409
Albertson's Charitable Giving Grants, 410
Albert W. Rice Charitable Foundation Grants, 411
Albuquerque Community Foundation Grants, 412
Alcatel-Lucent Technologies Foundation Grants, 413
Alcoa Foundation Grants, 414
Alcon Foundation Grants Program, 416
Alexander & Baldwin Fnd Mainland Grants, 417
Alexander and Baldwin Foundation Hawaiian and Pacific Island Grants, 418
ALFJ Astraea U.S. and Int'l Movement Fund, 427
Alfred and Tillie Shemanski Testamentary Trust Grants, 428
Alfred Bersted Foundation Grants, 429
Alfred E. Chase Charitable Foundation Grants, 430
Allegheny Technologies Charitable Trust, 440
Alliant Energy Corporation Contributions, 445
Alpha Natural Resources Corporate Giving, 450
Alpine Winter Foundation Grants, 453
Alvin and Fanny Blaustein Thalheimer Foundation Baltimore Communal Grants, 459
Amador Community Foundation Grants, 481
AMA Foundation Healthy Communities/Healthy America Grants, 486

PROGRAM TYPE INDEX

General Operating Support / 885

Amelia Sillman Rockwell and Carlos Perry Rockwell Charities Fund Grants, 501
American Express Community Service Grants, 531
American Foodservice Charitable Trust Grants, 532
American Psychiatric Foundation Disaster Recovery Fund for Psychiatrists, 553
American Schlafhorst Foundation Grants, 559
Amerigroup Foundation Grants, 565
Amgen Foundation Grants, 572
AMI Semiconductors Corporate Grants, 576
Andersen Corporate Foundation, 586
Andre Agassi Charitable Foundation Grants, 587
Andrew Family Foundation Grants, 588
Anheuser-Busch Foundation Grants, 591
Anne J. Caudal Foundation Grants, 595
Ann Jackson Family Foundation Grants, 598
Ann Peppers Foundation Grants, 599
Anschutz Family Foundation Grants, 602
Anthony R. Abraham Foundation Grants, 608
Antone & Edene Vidinha Charitable Trust Grants, 609
Aragona Family Foundation Grants, 648
Aratani Foundation Grants, 649
Arcadia Foundation Grants, 650
ARCO Foundation Education Grants, 652
Arie and Ida Crown Memorial Grants, 654
Arizona Diamondbacks Charities Grants, 657
Arizona Public Service Corporate Giving Program Grants, 658
Arkell Hall Foundation Grants, 661
Arlington Community Foundation Grants, 662
ARS Foundation Grants, 667
Arthur and Rochelle Belfer Foundation Grants, 670
Assurant Foundation Grants, 788
Atherton Family Foundation Grants, 794
Athwin Foundation Grants, 795
Atlanta Foundation Grants, 796
Atran Foundation Grants, 797
Auburn Foundation Grants, 798
Audrey & Sydney Irmas Foundation Grants, 799
Austin S. Nelson Foundation Grants, 805
Autauga Area Community Foundation Grants, 808
Avon Products Foundation Grants, 814
Babcock Charitable Trust Grants, 816
Bacon Family Foundation Grants, 817
Bailey Foundation Grants, 818
Balfe Family Foundation Grants, 819
Ball Brothers Foundation General Grants, 820
BancorpSouth Foundation Grants, 823
Banfi Vintners Foundation Grants, 824
Bank of America Fndn Matching Gifts, 825
Bank of America Fndn Volunteer Grants, 826
Baptist-Trinity Lutheran Legacy Fndn Grants, 827
Barker Foundation Grants, 830
Barker Welfare Foundation Grants, 831
Barr Fund Grants, 835
Batchelor Foundation Grants, 837
Batts Foundation Grants, 840
Baxter International Corporate Giving Grants, 841
Baxter International Foundation Grants, 843
BCBSM Foundation Primary and Clinical Prevention Grants, 856
Bearemy's Kennel Pals Grants, 863
Beazley Foundation Grants, 865
Bechtel Group Foundation Building Positive Community Relationships Grants, 866
Beldon Fund Grants, 871
Belk Foundation Grants, 872
Belvedere Community Foundation Grants, 874
Bender Foundation Grants, 876
Berks County Community Foundation Grants, 881
Bernard F. and Alva B. Gimbel Foundation Reproductive Rights Grants, 883
Berrien Community Foundation Grants, 884
Bertha Russ Lytel Foundation Grants, 885
Besser Foundation Grants, 887
Bildner Family Foundation Grants, 891
Bill and Melinda Gates Foundation Emergency Response Grants, 892
Bill and Melinda Gates Foundation Water, Sanitation and Hygiene Grants, 894

Bindley Family Foundation Grants, 896
Blanche and Irving Laurie Foundation Grants, 910
Blanche and Julian Robertson Family Foundation Grants, 911
Blowitz-Ridgeway Foundation Grants, 913
Blue Cross Blue Shield of Minnesota Fndn Healthy Equity: Public Libraries for Health Grants, 918
Blue Mountain Community Foundation Grants, 921
Blum-Kovler Foundation Grants, 924
Blumenthal Foundation Grants, 925
BMW of North America Charitable Contributions, 926
Bodman Foundation Grants, 929
Boeing Company Contributions Grants, 930
Boettcher Foundation Grants, 931
Bonfils-Stanton Foundation Grants, 932
Borkee-Hagley Foundation Grants, 934
Boston Foundation Grants, 936
Boston Globe Foundation Grants, 937
Boyd Gaming Corporation Contributions Program, 941
Boyle Foundation Grants, 942
Bradley-Turner Foundation Grants, 944
Bradley C. Higgins Foundation Grants, 945
Bright Family Foundation Grants, 951
Bright Promises Foundation Grants, 952
Bristol-Myers Squibb Foundation Health Disparities Grants, 958
Brookdale Fnd National Group Respite Grants, 966
Brooklyn Community Foundation Caring Neighbors Grants, 968
Brown Advisory Charitable Foundation Grants, 969
Bruce and Adele Greenfield Foundation Grants, 972
Bullitt Foundation Grants, 973
Burlington Northern Santa Fe Foundation Grants, 978
Bush Fndn Health & Human Services Grants, 979
Byron W. & Alice L. Lockwood Fnd Grants, 991
Cabot Corporation Foundation Grants, 993
Caddock Foundation Grants, 994
Caesars Foundation Grants, 995
Cailloux Foundation Grants, 996
Caleb C. and Julia W. Dula Educational and Charitable Foundation Grants, 997
California Community Foundation Health Care Grants, 998
California Wellness Foundation Work and Health Program Grants, 1001
Callaway Foundation Grants, 1002
Callaway Golf Company Foundation Grants, 1003
Cambridge Community Foundation Grants, 1004
Camp-Younts Foundation Grants, 1005
Campbell Hoffman Foundation Grants, 1006
Campbell Soup Foundation Grants, 1007
Cargill Citizenship Fund Corporate Giving, 1018
Caring Foundation Grants, 1019
Carl & Eloise Pohlad Family Fndn Grants, 1021
Carl C. Icahn Foundation Grants, 1023
Carl Gellert and Celia Berta Gellert Foundation Grants, 1024
Carl M. Freeman Foundation FACES Grants, 1026
Carl M. Freeman Foundation Grants, 1027
Carl R. Hendrickson Family Foundation Grants, 1028
Carl W. and Carrie Mae Joslyn Trust Grants, 1031
Carnahan-Jackson Foundation Grants, 1032
Carnegie Corporation of New York Grants, 1033
Carolyn Foundation Grants, 1034
Carpenter Foundation Grants, 1035
Carrie E. and Lena V. Glenn Foundation Grants, 1036
Carrie Estelle Doheny Foundation Grants, 1037
Carrier Corporation Contributions Grants, 1038
Carylon Foundation Grants, 1040
CDC Cooperative Agreement for Continuing Enhanced National Surveillance for Prion Diseases in the United States, 1049
CDC Increasing Breast and Cervical Cancer Screening Services for Urban American Indian/Alaska Native Women, 1063
Cemala Foundation Grants, 1077
CFFVR Alcoholism and Drug Abuse Grants, 1093
CFFVR Basic Needs Giving Partnership Grants, 1094
CFFVR Capital Credit Union Charitable Giving Grants, 1095

CFFVR Clintonville Area Foundation Grants, 1097
CFFVR Frank C. Shattuck Community Grants, 1098
CFFVR Schmidt Family G4 Grants, 1103
CFFVR Shawano Area Community Foundation Grants, 1104
CFFVR Wisconsin King's Daus & Sons Grants, 1106
CFFVR Women's Fund for the Fox Valley Region Grants, 1107
Chamberlain Foundation Grants, 1108
Champ-A Champion Fur Kids Grants, 1109
Chapman Charitable Foundation Grants, 1112
Charles A. Frueauff Foundation Grants, 1114
Charles Delmar Foundation Grants, 1115
Charles H. Farnsworth Trust Grants, 1119
Charles H. Pearson Foundation Grants, 1121
Charles M. & Mary D. Grant Fndn Grants, 1124
Charles Nelson Robinson Fund Grants, 1125
Charlotte County (FL) Community Foundation Grants, 1126
Chatlos Foundation Grants Program, 1128
Chazen Foundation Grants, 1129
Chemtura Corporation Contributions Grants, 1130
Chiles Foundation Grants, 1147
Chilkat Valley Community Foundation Grants, 1148
Christensen Fund Regional Grants, 1149
Christine & Katharina Pauly Trust Grants, 1150
CICF Indianapolis Fndn Community Grants, 1156
CIGNA Foundation Grants, 1160
CJ Foundation for SIDS Program Services Grants, 1163
Clara Blackford Smith and W. Aubrey Smith Charitable Foundation Grants, 1165
Clarence E. Heller Charitable Foundation Grants, 1166
Clarence T.C. Ching Foundation Grants, 1167
Clark-Winchcole Foundation Grants, 1173
Clark and Ruby Baker Foundation Grants, 1174
Claude Bennett Family Foundation Grants, 1176
Claude Worthington Benedum Fndn Grants, 1178
Clayton Baker Trust Grants, 1179
Cleveland-Cliffs Foundation Grants, 1181
CNA Foundation Grants, 1189
CNCS Senior Companion Program Grants, 1191
Coastal Community Foundation of South Carolina Grants, 1202
Coca-Cola Foundation Grants, 1203
Coeta and Donald Barker Foundation Grants, 1205
Coleman Foundation Cancer Care Grants, 1206
Coleman Foundation Developmental Disabilities Grants, 1207
Collins C. Diboll Private Foundation Grants, 1209
Collins Foundation Grants, 1210
Colonel Stanley R. McNeil Foundation Grants, 1211
Colorado Trust Grants, 1213
Columbus Foundation Central Benefits Health Care Foundation Grants, 1216
Columbus Foundation Estrich Fund Grants, 1218
Columbus Foundation Mary Eleanor Morris Fund Grants, 1220
Columbus Foundation Traditional Grants, 1223
Community Foundation for Greater Atlanta Clayton County Fund Grants, 1240
Community Foundation for Greater Atlanta Common Good Funds Grants, 1241
Community Foundation for Greater Atlanta Morgan County Fund Grants, 1244
Community Foundation for Greater Atlanta Newton County Fund Grants, 1245
Community Fndn for Greater Buffalo Grants, 1247
Community Foundation for Greater New Haven $5,000 and Under Grants, 1248
Community Foundation for Greater New Haven Responsive New Grants, 1249
Community Foundation for Greater New Haven Women & Girls Grants, 1252
Community Fndn for Monterey County Grants, 1253
Community Fndn for the Capital Region Grants, 1259
Community Foundation for the National Capital Region Community Leadership Grants, 1260
Community Foundation of Bartholomew County Heritage Fund Grants, 1262
Community Fndn of Central Illinois Grants, 1267

Community Foundation of Grant County Grants, 1271
Community Foundation of Greater Birmingham Grants, 1272
Community Foundation of Greater Flint Grants, 1273
Community Foundation of Greater Greensboro Community Grants, 1277
Community Foundation of Greenville Hollingsworth Funds Program/Project Grants, 1285
Community Foundation of Muncie and Delaware County Grants, 1300
Community Fndn of So Alabama Grants, 1306
Community Fndn of South Puget Sound Grants, 1307
Community Foundation of St. Joseph County Special Project Challenge Grants, 1310
Community Fndn of Switzerland County Grants, 1311
Community Foundation of the Verdugos Educational Endowment Fund Grants, 1314
Community Foundation of the Verdugos Grants, 1315
Community Fndn of Wabash County Grants, 1316
Comprehensive Health Education Fndn Grants, 1320
ConAgra Foods Fndn Community Impact Grants, 1321
ConAgra Foods Nourish our Community Grants, 1322
Connelly Foundation Grants, 1326
ConocoPhillips Foundation Grants, 1327
Constellation Energy Corporate Grants, 1342
Consumers Energy Foundation, 1343
Cooke Foundation Grants, 1344
Cooper Industries Foundation Grants, 1345
Cornerstone Foundation of Northeastern Wisconsin Grants, 1347
Cowles Charitable Trust Grants, 1354
Crail-Johnson Foundation Grants, 1355
Cralle Foundation Grants, 1356
Crane Foundation Grants, 1357
Crane Fund Grants, 1358
Cresap Family Foundation Grants, 1359
CRH Foundation Grants, 1360
Cuesta Foundation Grants, 1371
Cullen Foundation Grants, 1372
Cystic Fibrosis Canada Senior Scientist Research Training Awards, 1387
Cystic Fibrosis Canada Transplant Center Incentive Grants, 1392
D.F. Halton Foundation Grants, 1405
D. W. McMillan Foundation Grants, 1407
Dade Community Foundation Community AIDS Partnership Grants, 1409
Dade Community Foundation Grants, 1410
DaimlerChrysler Corporation Fund Grants, 1412
Dale and Edna Walsh Foundation Grants, 1415
Dallas Women's Foundation Grants, 1417
Danellie Foundation Grants, 1421
Daniel & Nanna Stern Family Fndn Grants, 1422
Daniels Fund Grants-Aging, 1424
Daphne Seybolt Culpeper Fndn Grants, 1425
David Geffen Foundation Grants, 1431
David M. and Marjorie D. Rosenberg Foundation Grants, 1432
Dayton Power and Light Foundation Grants, 1437
Daywood Foundation Grants, 1438
Deaconess Community Foundation Grants, 1439
Deborah Munroe Noonan Memorial Grants, 1443
DeKalb County Community Foundation Grants, 1447
Della B. Gardner Fund Grants, 1451
Delmarva Power and Light Contributions, 1454
Denton A. Cooley Foundation Grants, 1459
DeRoy Testamentary Foundation Grants, 1469
Deutsche Banc Alex Brown and Sons Charitable Foundation Grants, 1471
DHHS Oral Health Promotion Research Across the Lifespan, 1474
DHL Charitable Shipment Support, 1475
Dolan Children's Foundation Grants, 1482
Donald and Sylvia Robinson Family Foundation Grants, 1486
Dora Roberts Foundation Grants, 1489
Doree Taylor Charitable Foundation, 1490
Doris and Victor Day Foundation Grants, 1491
Dorothy Hooper Beattie Foundation Grants, 1499
Dorrance Family Foundation Grants, 1501

DPA Promoting Policy Change Advocacy Grants, 1504
DTE Energy Foundation Health and Human Services Grants, 1516
Dubois County Community Foundation Grants, 1517
Duchossois Family Foundation Grants, 1518
Duke Endowment Health Care Grants, 1519
Dunspaugh-Dalton Foundation Grants, 1530
DuPont Pioneer Community Giving Grants, 1532
Dyson Foundation Mid-Hudson Valley General Operating Support Grants, 1533
eBay Foundation Community Grants, 1543
Edina Realty Foundation Grants, 1550
Edward and Romell Ackley Foundation Grants, 1555
Edward N. and Della L. Thome Memorial Foundation Grants, 1557
Edwards Memorial Trust Grants, 1558
Edward W. and Stella C. Van Houten Memorial Fund Grants, 1559
Edwin S. Webster Foundation Grants, 1560
Edwin W. and Catherine M. Davis Fndn Grants, 1561
Effie and Wofford Cain Foundation Grants, 1563
Eisner Foundation Grants, 1564
Elaine Feld Stern Charitable Trust Grants, 1565
Elizabeth Morse Genius Charitable Trust Grants, 1569
Elliot Foundation Inc Grants, 1575
Elmer L. & Eleanor J. Andersen Fndn Grants, 1578
El Paso Community Foundation Grants, 1579
El Pomar Foundation Anna Keesling Ackerman Fund Grants, 1581
El Pomar Foundation Grants, 1582
Elsie H. Wilcox Foundation Grants, 1583
Elsie Lee Garthwaite Memorial Fndn Grants, 1584
EMBO Installation Grants, 1586
Emerson Charitable Trust Grants, 1589
Emerson Electric Company Contributions Grants, 1590
Emily Hall Tremaine Foundation Learning Disabilities Grants, 1592
Emma B. Howe Memorial Foundation Grants, 1593
Essex County Community Foundation Discretionary Fund Grants, 1602
Ethel S. Abbott Charitable Foundation Grants, 1607
Ethel Sergeant Clark Smith Foundation Grants, 1608
Eugene B. Casey Foundation Grants, 1609
Eugene M. Lang Foundation Grants, 1611
Expect Miracles Foundation Grants, 1619
Fairfield County Community Foundation Grants, 1625
Fallon OrNda Community Health Fund Grants, 1627
Fannie E. Rippel Foundation Grants, 1630
FAR Fund Grants, 1631
Faye McBeath Foundation Grants, 1635
Ferree Foundation Grants, 1658
Field Foundation of Illinois Grants, 1660
FirstEnergy Foundation Community Grants, 1668
Fisher Foundation Grants, 1670
Florian O. Bartlett Trust Grants, 1677
Florida BRAIVE Fund of Dade Community Foundation, 1678
Fluor Foundation Grants, 1682
FMC Foundation Grants, 1683
Foellinger Foundation Grants, 1684
Fondren Foundation Grants, 1685
Ford Family Foundation Grants - Access to Health and Dental Services, 1686
Ford Motor Company Fund Grants Program, 1687
Foster Foundation Grants, 1689
Four County Community Fndn General Grants, 1710
Four County Community Foundation Healthy Senior/ Healthy Youth Fund Grants, 1711
Fourjay Foundation Grants, 1712
Frances W. Emerson Foundation Grants, 1718
Francis L. Abreu Charitable Trust Grants, 1719
Francis T. & Louise T. Nichols Fndn Grants, 1720
Frank B. Hazard General Charity Fund Grants, 1721
Frank E. and Seba B. Payne Foundation Grants, 1722
Frank G. and Freida K. Brotz Family Foundation Grants, 1723
Franklin County Community Foundation Grants, 1724
Frank Reed and Margaret Jane Peters Memorial Fund I Grants, 1727

Frank Reed and Margaret Jane Peters Memorial Fund II Grants, 1728
Frank Stanley Beveridge Foundation Grants, 1730
Frank W. & Carl S. Adams Memorial Grants, 1731
Fraser-Parker Foundation Grants, 1732
Fred & Gretel Biel Charitable Trust Grants, 1734
Fred C. & Katherine B. Andersen Grants, 1737
Freddie Mac Foundation Grants, 1738
Frederick McDonald Trust Grants, 1740
Fremont Area Community Foundation Amazing X Grants, 1744
Fremont Area Community Foundation Elderly Needs Grants, 1745
Fremont Area Community Fndn General Grants, 1746
Fuller E. Callaway Foundation Grants, 1761
G.N. Wilcox Trust Grants, 1768
Gamble Foundation Grants, 1769
Gardner Foundation Grants, 1771
Gardner Foundation Grants, 1772
Gardner W. & Joan G. Heidrick, Jr. Fndn Grants, 1773
Gebbie Foundation Grants, 1777
GenCorp Foundation Grants, 1779
Genentech Corp Charitable Contributions, 1780
General Mills Foundation Grants, 1782
General Motors Foundation Grants, 1783
Genesis Foundation Grants, 1786
Genuardi Family Foundation Grants, 1787
George and Ruth Bradford Foundation Grants, 1789
George and Sarah Buchanan Foundation Grants, 1790
George A Ohl Jr. Foundation Grants, 1791
George E. Hatcher, Jr. and Ann Williams Hatcher Foundation Grants, 1792
George F. Baker Trust Grants, 1793
George Foundation Grants, 1795
George Gund Foundation Grants, 1796
George Kress Foundation Grants, 1799
George P. Davenport Trust Fund Grants, 1800
George S. & Dolores Dore Eccles Fndn Grants, 1801
George W. Codrington Charitable Fndn Grants, 1803
George W. Wells Foundation Grants, 1804
Georgiana Goddard Eaton Memorial Grants, 1805
Gertrude and William C. Wardlaw Fund Grants, 1808
Gheens Foundation Grants, 1814
Gibson Foundation Grants, 1818
Gil and Dody Weaver Foundation Grants, 1819
Ginn Foundation Grants, 1822
Global Fund for Children Grants, 1832
Global Fund for Women Grants, 1833
GNOF Bayou Communities Grants, 1835
GNOF IMPACT Grants for Arts and Culture, 1838
GNOF IMPACT Grants for Health and Human Services, 1839
GNOF IMPACT Harold W. Newman, Jr. Charitable Trust Grants, 1841
GNOF Maison Hospitaliere Grants, 1843
Godfrey Foundation Grants, 1849
Golden Heart Community Foundation Grants, 1850
Goodrich Corporation Foundation Grants, 1854
Good Samaritan Inc Grants, 1855
Google Grants Beta, 1857
Grace and Franklin Bernsen Foundation Grants, 1858
Grace Bersted Foundation Grants, 1859
Graham and Carolyn Holloway Family Foundation Grants, 1861
Graham Foundation Grants, 1862
Grand Rapids Area Community Fndn Grants, 1865
Grand Rapids Area Community Foundation Nashwauk Area Endowment Fund Grants, 1866
Greater Saint Louis Community Fndn Grants, 1884
Greater Tacoma Community Foundation Grants, 1886
Greater Worcester Community Foundation Jeppson Memorial Fund for Brookfield Grants, 1889
Greene County Foundation Grants, 1892
Greenfield Foundation of Maine Grants, 1893
Green River Area Community Fndn Grants, 1895
Greenspun Family Foundation Grants, 1896
Gregory B. Davis Foundation Grants, 1898
Gregory L. Gibson Charitable Fndn Grants, 1900
Greygates Foundation Grants, 1901
Griffin Family Foundation Grants, 1902

PROGRAM TYPE INDEX   General Operating Support / 887

Griffin Foundation Grants, 1903
Grotto Foundation Project Grants, 1905
Grover Hermann Foundation Grants, 1906
Grundy Foundation Grants, 1912
Gulf Coast Foundation of Community Operating Grants, 1916
Guy's and St. Thomas' Charity Grants, 1918
Guy I. Bromley Trust Grants, 1919
H & R Foundation Grants, 1920
H.A. & Mary K. Chapman Trust Grants, 1921
H.J. Heinz Company Foundation Grants, 1923
H. Leslie Hoffman and Elaine S. Hoffman Foundation Grants, 1924
H. Schaffer Foundation Grants, 1925
Hagedorn Fund Grants, 1934
Hallmark Corporate Foundation Grants, 1936
Harden Foundation Grants, 1945
Hardin County Community Foundation Grants, 1946
Harold and Arlene Schnitzer CARE Foundation Grants, 1950
Harold Simmons Foundation Grants, 1955
Harris and Eliza Kempner Fund Grants, 1956
Harry B. and Jane H. Brock Foundation Grants, 1960
Harry Bramhall Gilbert Charitable Trust Grants, 1961
Harry Edison Foundation, 1962
Harry Frank Guggenheim Fnd Research Grants, 1964
Harry S. Black and Allon Fuller Fund Grants, 1966
Harry W. Bass, Jr. Foundation Grants, 1968
Hartford Foundation Regular Grants, 1970
Harvest Foundation Grants, 1971
Hawaii Community Foundation Health Education and Research Grants, 1976
HCA Foundation Grants, 1980
Health Fndn of Greater Indianapolis Grants, 1984
Health Foundation of Southern Florida Responsive Grants, 1985
Heckscher Foundation for Children Grants, 1988
Helena Rubinstein Foundation Grants, 1991
Helen Bader Foundation Grants, 1992
Helen K. & Arthur E. Johnson Fndn Grants, 1995
Helen Pumphrey Denit Charitable Trust Grants, 1996
Helen S. Boylan Foundation Grants, 1997
Helen Steiner Rice Foundation Grants, 1998
Henrietta Tower Wurts Memorial Fndn Grants, 2002
Henry A. and Mary J. MacDonald Foundation, 2003
Henry and Ruth Blaustein Rosenberg Foundation Health Grants, 2004
Herbert A. & Adrian W. Woods Fndn Grants, 2007
Herbert H. & Grace A. Dow Fndn Grants, 2008
Herman Goldman Foundation Grants, 2012
High Meadow Foundation Grants, 2027
Hillman Foundation Grants, 2032
Hillsdale County Community Foundation Healthy Senior/Healthy Youth Fund Grants, 2033
Hilton Hotels Corporate Giving Program Grants, 2037
Hoglund Foundation Grants, 2040
Horace A. Kimball and S. Ella Kimball Foundation Grants, 2046
Horace Moses Charitable Foundation Grants, 2047
Horizon Foundation for New Jersey Grants, 2048
Horizons Community Issues Grants, 2049
Hormel Foods Charitable Trust Grants, 2050
Hormel Foundation Grants, 2051
Howe Foundation of North Carolina Grants, 2055
HRK Foundation Health Grants, 2067
Huber Foundation Grants, 2071
Hudson Webber Foundation Grants, 2072
Huffy Foundation Grants, 2073
Hugh J. Andersen Foundation Grants, 2074
Huie-Dellmon Trust Grants, 2075
Huisking Foundation Grants, 2076
Human Source Foundation Grants, 2078
Huntington Clinical Foundation Grants, 2082
I.A. O'Shaughnessy Foundation Grants, 2088
Idaho Power Company Corporate Contributions, 2115
Ida S. Barter Trust Grants, 2116
Ike and Roz Friedman Foundation Grants, 2145
Illinois Tool Works Foundation Grants, 2147
Independence Blue Cross Charitable Medical Care Grants, 2150

Independence Community Foundation Community Quality of Life Grant, 2153
Indiana 21st Century Research and Technology Fund Awards, 2154
Intergrys Corporation Grants, 2160
Irving S. Gilmore Foundation Grants, 2163
Irvin Stern Foundation Grants, 2164
Ittleson Foundation Mental Health Grants, 2167
J.B. Reynolds Foundation Grants, 2168
J. Spencer Barnes Memorial Foundation Grants, 2175
J.W. Kieckhefer Foundation Grants, 2176
J. Walton Bissell Foundation Grants, 2177
Jack H. & William M. Light Trust Grants, 2179
Jacob and Hilda Blaustein Foundation Health and Mental Health Grants, 2181
Jacobs Family Village Neighborhoods Grants, 2184
James & Abigail Campbell Family Fndn Grants, 2185
James A. and Faith Knight Foundation Grants, 2186
James Ford Bell Foundation Grants, 2187
James J. & Joan A. Gardner Family Fndn Grants, 2193
James L. and Mary Jane Bowman Charitable Trust Grants, 2194
James R. Dougherty Jr. Foundation Grants, 2198
James R. Thorpe Foundation Grants, 2199
Jane's Trust Grants, 2206
Jayne and Leonard Abess Foundation Grants, 2214
Jean and Louis Dreyfus Foundation Grants, 2215
Jeffris Wood Foundation Grants, 2217
JELD-WEN Foundation Grants, 2218
Jenkins Foundation: Improving the Health of Greater Richmond Grants, 2219
Jessie Ball Dupont Fund Grants, 2226
Jewish Fund Grants, 2227
Jim Moran Foundation Grants, 2228
Joe W. and Dorothy Dorsett Brown Fndn Grants, 2230
John D. & Katherine A. Johnston Fndn Grants, 2233
John Deere Foundation Grants, 2234
John Edward Fowler Memorial Fndn Grants, 2235
John H. and Wilhelmina D. Harland Charitable Foundation Children and Youth Grants, 2238
John I. Smith Charities Grants, 2240
John J. Leidy Foundation Grants, 2241
John Jewett and Helen Chandler Garland Foundation Grants, 2242
John P. McGovern Foundation Grants, 2247
Johns Manville Fund Grants, 2251
Johnson & Johnson Corporate Contributions, 2253
Johnson Controls Foundation Health and Social Services Grants, 2255
John W. Anderson Foundation Grants, 2262
John W. Boynton Fund Grants, 2263
John W. Speas and Effie E. Speas Memorial Trust Grants, 2264
Joseph Alexander Foundation Grants, 2265
Joseph Drown Foundation Grants, 2268
Joseph H. & Florence A. Roblee Fndn Grants, 2269
Joseph Henry Edmondson Foundation Grants, 2270
Josephine G. Russell Trust Grants, 2271
Josephine S. Gumbiner Foundation Grants, 2273
Judith and Jean Pape Adams Charitable Foundation Tulsa Area Grants, 2280
Julius N. Frankel Foundation Grants, 2282
K21 Health Foundation Cancer Care Grants, 2284
K and F Baxter Family Foundation Grants, 2287
Kansas Health Foundation Recognition Grants, 2289
Kate B. Reynolds Trust Health Care Grants, 2290
Katharine Matthies Foundation Grants, 2291
Katherine Baxter Memorial Foundation Grants, 2292
Kathryne Beynon Foundation Grants, 2294
Kelvin and Eleanor Smith Foundation Grants, 2297
Kenneth T. & Eileen L. Norris Fndn Grants, 2300
Kimball International-Habig Foundation Health and Human Services Grants, 2308
Kinsman Foundation Grants, 2309
Kovler Family Foundation Grants, 2323
Kroger Foundation Women's Health Grants, 2325
L. W. Pierce Family Foundation Grants, 2328
Land O'Lakes Foundation Mid-Atlantic Grants, 2341
Lena Benas Memorial Fund Grants, 2353
Leonard L. & Bertha U. Abess Fndn Grants, 2357

Leo Niessen Jr., Charitable Trust Grants, 2358
Lester E. Yeager Charitable Trust B Grants, 2359
Lewis H. Humphreys Charitable Trust Grants, 2361
Lil and Julie Rosenberg Foundation Grants, 2364
Lillian S. Wells Foundation Grants, 2365
Lisa Higgins-Hussman Foundation Grants, 2370
Lotus 88 Fnd for Women & Children Grants, 2376
Louie M. & Betty M. Phillips Fndn Grants, 2378
Louis and Elizabeth Nave Flarsheim Charitable Foundation Grants, 2379
Lowe Foundation Grants, 2381
Lowell Berry Foundation Grants, 2382
Lubbock Area Foundation Grants, 2383
Lubrizol Foundation Grants, 2384
Lucile Horton Howe and Mitchell B. Howe Foundation Grants, 2385
Lucy Downing Nisbet Charitable Fund Grants, 2387
Lumpkin Family Fnd Healthy People Grants, 2391
Lydia deForest Charitable Trust Grants, 2393
Lynde and Harry Bradley Foundation Grants, 2397
Lynn and Rovena Alexander Family Foundation Grants, 2399
M.B. and Edna Zale Foundation Grants, 2401
M. Bastian Family Foundation Grants, 2402
M.E. Raker Foundation Grants, 2404
Mabel Y. Hughes Charitable Trust Grants, 2409
Macquarie Bank Foundation Grants, 2410
Madison County Community Foundation General Grants, 2413
Mann T. Lowry Foundation Grants, 2419
Manuel D. & Rhoda Mayerson Fndn Grants, 2420
Marathon Petroleum Corporation Grants, 2422
Marcia and Otto Koehler Foundation Grants, 2427
Mardag Foundation Grants, 2428
Margaret T. Morris Foundation Grants, 2429
Marie C. and Joseph C. Wilson Foundation Rochester Small Grants, 2431
Marion Gardner Jackson Charitable Trust Grants, 2434
Marion I. and Henry J. Knott Foundation Discretionary Grants, 2435
Marion I. and Henry J. Knott Foundation Standard Grants, 2436
Marjorie Moore Charitable Foundation Grants, 2438
Mary Black Foundation Active Living Grants, 2440
Mary K. Chapman Foundation Grants, 2443
Mary Owen Borden Foundation Grants, 2446
Matilda R. Wilson Fund Grants, 2453
Mattel International Grants Program, 2455
Maurice J. Masserini Charitable Trust Grants, 2456
Maxon Charitable Foundation Grants, 2458
May and Stanley Smith Charitable Trust Grants, 2459
McCarthy Family Foundation Grants, 2490
McCune Foundation Human Services Grants, 2493
McInerny Foundation Grants, 2495
McKesson Foundation Grants, 2496
Mead Johnson Nutritionals Charitable Giving, 2502
MeadWestvaco Foundation Sustainable Communities Grants, 2506
Mead Witter Foundation Grants, 2507
Medtronic Foundation HeartRescue Grants, 2513
Memorial Foundation Grants, 2516
Mericos Foundation Grants, 2518
Meriden Foundation Grants, 2519
Merkel Foundation Grants, 2520
Merrick Foundation Grants, 2522
Merrick Foundation Grants, 2521
Mervin Bovaird Foundation Grants, 2523
Meta and George Rosenberg Foundation Grants, 2525
Metzger-Price Fund Grants, 2530
Meyer Foundation Healthy Communities Grants, 2533
Meyer Memorial Trust Grassroots Grants, 2536
Meyer Memorial Trust Responsive Grants, 2537
Meyer Memorial Trust Special Grants, 2538
MGN Family Foundation Grants, 2542
Michael Reese Health Trust Core Grants, 2544
Michael Reese Health Trust Responsive Grants, 2545
Miles of Hope Breast Cancer Foundation Grants, 2553
Mimi and Peter Haas Fund Grants, 2557
Minneapolis Foundation Community Grants, 2558
Montgomery County Community Fndn Grants, 2591

Morris & Gwendolyn Cafritz Fndn Grants, 2594
Mt. Sinai Health Care Foundation Academic Medicine and Bioscience Grants, 2598
Mt. Sinai Health Care Foundation Health of the Urban Community Grants, 2600
Natalie W. Furniss Charitable Trust Grants, 2626
Nationwide Insurance Foundation Grants, 2637
NCI Technologies and Software to Support Integrative Cancer Biology Research (SBIR) Grants, 2656
Nell J. Redfield Foundation Grants, 2673
Nicholas H. Noyes Jr. Memorial Fndn Grants, 2791
NIH Biology, Development, and Progression of Malignant Prostate Disease Research Grants, 2866
Nina Mason Pulliam Charitable Trust Grants, 2921
Noble County Community Foundation Grants, 2975
Norcliffe Foundation Grants, 2979
Nordson Corporation Foundation Grants, 2981
Norfolk Southern Foundation Grants, 2982
Norman Foundation Grants, 2983
North Carolina GlaxoSmithKline Fndn Grants, 2991
Northern Chautauqua Community Foundation Community Grants, 2995
Northwestern Mutual Foundation Grants, 2999
Norwin S. & Elizabeth N. Bean Fndn Grants, 3001
NSF Presidential Early Career Awards for Scientists and Engineers (PECASE) Grants, 3021
NYCT AIDS/HIV Grants, 3048
NYCT Blood Disease Research Grants, 3051
NYCT Girls and Young Women Grants, 3053
NYCT Health Care Services, Systems, and Policies Grants, 3055
OceanFirst Foundation Major Grants, 3062
Olive Higgins Prouty Foundation Grants, 3072
OneFamily Foundation Grants, 3075
Oppenstein Brothers Foundation Grants, 3078
Ordean Foundation Grants, 3080
Otto Bremer Foundation Grants, 3099
Owen County Community Foundation Grants, 3100
PacifiCare Foundation Grants, 3102
Pacific Life Foundation Grants, 3103
Packard Foundation Local Grants, 3105
Pajaro Valley Community Health Health Trust Insurance/Coverage & Education on Using the System Grants, 3108
Pajaro Valley Community Health Trust Diabetes and Contributing Factors Grants, 3109
Pajaro Valley Community Health Trust Oral Health: Prevention & Access Grants, 3110
Palmer Foundation Grants, 3112
Patrick and Anna M. Cudahy Fund Grants, 3120
Paul Rapoport Foundation Grants, 3125
Percy B. Ferebee Endowment Grants, 3138
Perkin Fund Grants, 3139
Perpetual Trust for Charitable Giving Grants, 3141
Peter Kiewit Foundation General Grants, 3154
Peter Kiewit Foundation Small Grants, 3155
Phelps County Community Foundation Grants, 3188
Philadelphia Foundation Organizational Effectiveness Grants, 3189
Philip L. Graham Fund Health and Human Services Grants, 3190
Phoenix Suns Charities Grants, 3201
Piedmont Health Foundation Grants, 3219
Piedmont Natural Gas Foundation Health and Human Services Grants, 3221
Pinellas County Grants, 3223
Pinkerton Foundation Grants, 3224
Piper Jaffray Fndn Communities Giving Grants, 3227
Playboy Foundation Grants, 3237
PMI Foundation Grants, 3239
Pollock Foundation Grants, 3244
Porter County Health and Wellness Grant, 3246
Powell Foundation Grants, 3255
PPG Industries Foundation Grants, 3257
Price Chopper's Golub Foundation Grants, 3261
Price Family Charitable Fund Grants, 3262
Priddy Foundation Operating Grants, 3264
Prince Charitable Trusts Chicago Grants, 3270
Prince Charitable Trusts DC Grants, 3271
Principal Financial Group Foundation Grants, 3273
Procter and Gamble Fund Grants, 3274
Prudential Foundation Education Grants, 3276
Puerto Rico Community Foundation Grants, 3278
Pulido Walker Foundation, 3282
Putnam County Community Foundation Grants, 3283
R.C. Baker Foundation Grants, 3292
Ralph M. Parsons Foundation Grants, 3304
Ralphs Food 4 Less Foundation Grants, 3305
Raskob Foundation for Catholic Activities Grants, 3306
Rathmann Family Foundation Grants, 3309
Rayonier Foundation Grants, 3311
RCF General Community Grants, 3312
Reinberger Foundation Grants, 3357
Rhode Island Foundation Grants, 3362
Rice Foundation Grants, 3363
Richard and Helen DeVos Foundation Grants, 3364
Richard & Susan Smith Family Fndn Grants, 3365
Richard D. Bass Foundation Grants, 3366
Richard E. Griffin Family Foundation Grants, 3368
Richard King Mellon Foundation Grants, 3369
Richland County Bank Grants, 3371
Ricks Family Charitable Trust Grants, 3373
Riley Foundation Grants, 3374
Ringing Rocks Foundation Discretionary Grants, 3375
Ripley County Community Foundation Grants, 3376
Robbins Charitable Foundation Grants, 3378
Robert B McMillen Foundation Grants, 3381
Robert R. McCormick Tribune Veterans Initiative Grants, 3386
Robert R. Meyer Foundation Grants, 3387
Robins Foundation Grants, 3390
Rochester Area Community Foundation Grants, 3391
Rockefeller Brothers Fund Pivotal Places Grants: South Africa, 3395
Rockwell International Corporate Trust Grants Program, 3399
Rogers Family Foundation Grants, 3401
Rollins-Luetkemeyer Foundation Grants, 3404
Rose Hills Foundation Grants, 3413
Roy & Christine Sturgis Charitable Grants, 3415
RRF General Program Grants, 3417
RRF Organizational Capacity Building Grants, 3418
Ruth Anderson Foundation Grants, 3425
Ruth and Vernon Taylor Foundation Grants, 3426
Ruth Eleanor Bamberger and John Ernest Bamberger Memorial Foundation Grants, 3427
Ruth H. & Warren A. Ellsworth Fndn Grants, 3429
S.H. Cowell Foundation Grants, 3447
S. Livingston Mather Charitable Trust Grants, 3448
S. Mark Taper Foundation Grants, 3449
Sain-Orr and Royak-DeForest Steadman Foundation Grants, 3452
Saint Louis Rams Fndn Community Donations, 3453
Salem Foundation Charitable Trust Grants, 3456
Salem Foundation Grants, 3457
Salmon Foundation Grants, 3458
Samuel S. Johnson Foundation Grants, 3468
San Antonio Area Foundation Grants, 3469
San Diego HIV Funding Collaborative Grants, 3473
San Diego Women's Foundation Grants, 3474
San Francisco Fndn Community Health Grants, 3478
Santa Barbara Foundation Strategy Grants - Core Support, 3481
Santa Fe Community Foundation Seasonal Grants-Fall Cycle, 3482
Sartain Lanier Family Foundation Grants, 3486
Sasco Foundation Grants, 3487
Schering-Plough Foundation Community Initiatives Grants, 3491
Schering-Plough Foundation Health Grants, 3492
Schlessman Family Foundation Grants, 3493
Schramm Foundation Grants, 3495
Schurz Communications Foundation Grants, 3496
Scott B. & Annie P. Appleby Charitable Grants, 3499
Seattle Foundation Benjamin N. Phillips Memorial Fund Grants, 3504
Seattle Foundation Health and Wellness Grants, 3507
Seattle Foundation Medical Funds Grants, 3508
Seneca Foods Foundation Grants, 3511
Sensient Technologies Foundation Grants, 3513
Sheltering Arms Fund Grants, 3545
Shield-Ayres Foundation Grants, 3546
Sidgmore Family Foundation Grants, 3551
Sidney Stern Memorial Trust Grants, 3552
Sid W. Richardson Foundation Grants, 3553
Siebert Lutheran Foundation Grants, 3554
Simmons Foundation Grants, 3563
Sioux Falls Area Community Foundation Community Fund Grants (Unrestricted), 3568
Sioux Falls Area Community Foundation Spot Grants (Unrestricted), 3570
Sisters of Charity Foundation of Cleveland Reducing Health and Educational Disparities in the Central Neighborhood Grants, 3574
Sisters of Mercy of North Carolina Fndn Grants, 3575
Sisters of St. Joseph Healthcare Fndn Grants, 3576
Skillman Fndn Good Neighborhoods Grants, 3577
Skoll Fndn Awards for Social Entrepreneurship, 3579
SOCFOC Catholic Ministries Grants, 3596
Solo Cup Foundation Grants, 3620
Sonoco Foundation Grants, 3621
Sony Corporation of America Grants, 3623
Sophia Romero Trust Grants, 3624
Southwest Gas Corporation Foundation Grants, 3628
Square D Foundation Grants, 3634
St. Joseph Community Health Foundation Improving Healthcare Access Grants, 3640
Stan and Sandy Checketts Foundation, 3644
Staunton Farm Foundation Grants, 3649
Steele-Reese Foundation Grants, 3650
Stella and Charles Guttman Foundation Grants, 3651
Sterling-Turner Charitable Foundation Grants, 3652
Sterling and Shelli Gardner Foundation Grants, 3653
Stewart Huston Charitable Trust Grants, 3655
Strake Foundation Grants, 3658
Stranahan Foundation Grants, 3659
Strowd Roses Grants, 3664
Susan Mott Webb Charitable Trust Grants, 3694
Susan Vaughan Foundation Grants, 3695
Swindells Charitable Foundation, 3697
Taubman Endowment for the Arts Fndn Grants, 3702
TE Foundation Grants, 3705
Texas Commission on the Arts Arts Respond Project Grants, 3708
Textron Corporate Contributions Grants, 3712
Thelma Doelger Charitable Trust Grants, 3714
The Ray Charles Foundation Grants, 3716
Thomas and Dorothy Leavey Foundation Grants, 3717
Thomas Austin Finch, Sr. Foundation Grants, 3718
Thomas C. Ackerman Foundation Grants, 3719
Thomas Jefferson Rosenberg Foundation Grants, 3723
Thomas Sill Foundation Grants, 3724
Thomas Thompson Trust Grants, 3725
Thomas W. Bradley Foundation Grants, 3726
Thompson Charitable Foundation Grants, 3727
Thorman Boyle Foundation Grants, 3730
Todd Brock Family Foundation Grants, 3736
Toyota Motor N America of New York Grants, 3749
Triangle Community Foundation Donor-Advised Grants, 3757
UniHealth Foundation Community Health Improvement Grants, 3782
UniHealth Foundation Healthcare Systems Enhancements Grants, 3784
Union Bank, N.A. Foundation Grants, 3788
Union Pacific Foundation Health and Human Services Grants, 3794
United Technologies Corporation Grants, 3803
VDH Rescue Squad Assistance Fund Grants, 3856
Victor E. Speas Foundation Grants, 3868
Vigneron Memorial Fund Grants, 3873
Visiting Nurse Foundation Grants, 3879
W. C. Griffith Foundation Grants, 3882
W. Clarke Swanson, Jr. Foundation Grants, 3883
Walter L. Gross III Family Foundation Grants, 3901
Weaver Popcorn Foundation Grants, 3910
WestWind Foundation Reproductive Health and Rights Grants, 3924
Weyerhaeuser Company Foundation Grants, 3925
WHAS Crusade for Children Grants, 3927

PROGRAM TYPE INDEX

Willard & Pat Walker Charitable Fndn Grants, 3943
William A. Badger Foundation Grants, 3945
William Blair and Company Foundation Grants, 3948
William D. Laurie, Jr. Charitable Fndn Grants, 3949
William G. & Helen C. Hoffman Fndn Grants, 3950
William G. Gilmore Foundation Grants, 3951
William J. & Tina Rosenberg Fndn Grants, 3954
William J. Brace Charitable Trust, 3955
William Ray & Ruth E. Collins Fndn Grants, 3956
Wilson-Wood Foundation Grants, 3960
Winifred & Harry B. Allen Foundation Grants, 3962
Wolfe Associates Grants, 3969
Yampa Valley Community Foundation Grants, 3978
Z. Smith Reynolds Foundation Small Grants, 3982
ZYTL Foundation Grants, 3985

Graduate Assistantships
AAAS GE & Science Prize for Yng Life Scientists, 39
Abbott Fund Science Education Grants, 203
ASHA Minority Student Leadership Awards, 696
Evjue Foundation Grants, 1617
Health Management Scholarships and Grants for Minorities, 1986
NAPNAP Fndn Grad Research Grant, 2610
Vancouver Foundation Grants and Community Initiatives Program, 3849

Grants to Individuals
Actors Addiction & Recovery Services Grants, 282
Actors Fund HIV/AIDS Initiative Grants, 283
ADA Foundation Relief Grants, 293
Adams Rotary Memorial Fund A Grants, 302
Aid for Starving Children Emergency Assistance Fund Grants, 367
AIG Disaster Relief Fund Grants, 370
Alaska Airlines Corporate Giving Medical Emergency and Research Grants, 402
Bingham McHale LLP Pro Bono Services, 897
Caplow Applied Science Carcinogen Prize, 1015
Carla J. Funk Governmental Relations Award, 1020
CDC David J. Sencer Museum Teacher Professional Development Workshops, 1053
CDC Disease Detective Camp, 1054
Cystic Fibrosis Lifestyle Foundation Individual Recreation Grants, 1397
Cystic Fibrosis Lifestyle Foundation Loretta Morris Memorial Fund Grants, 1398
Cystic Fibrosis Lifestyle Foundation Mentored Recreation Grants, 1399
Cystic Fibrosis Lifestyle Foundation Peer Support Grants, 1400
Delmarva Power and Light Contributions, 1454
Different Needz Foundation Grants, 1478
EBSCO / MLA Annual Meeting Grants, 1545
Everyone Breathe Asthma Education Grants, 1616
HAF Barry F. Phelps Fund Grants, 1927
HAF David (Davey) H. Somerville Medical Travel Fund Grants, 1928
HAF Phyllis Nilsen Leal Memorial Fund Gifts, 1930
Maggie Welby Foundation Grants, 2414
Margaret Wiegand Trust Grants, 2430
Mary L. Peyton Foundation Grants, 2445
Medical Informatics Section/MLA Career Development Grant, 2508
MLA Continuing Education Grants, 2559
MLA Cunningham Memorial International Fellowship, 2560
MLA David A. Kronick Traveling Fellowship, 2561
MLA Donald A.B. Lindberg Fellowship, 2562
MLA Estelle Brodman Academic Medical Librarian of the Year Award, 2563
MLA Graduate Scholarship, 2564
MLA Grad Scholarship for Minority Students, 2565
MLA Ida and George Eliot Prize, 2566
MLA Janet Doe Lectureship Award, 2567
MLA Lois Ann Colaianni Award for Excellence and Achievement in Hospital Librarianship, 2568
MLA Louise Darling Medal for Distinguished Achievement in Collection Development in the Health Sciences, 2569

MLA Lucretia W. McClure Excellence in Education Award, 2570
MLA Murray Gottlieb Prize Essay Award, 2571
MLA Research, Development, and Demonstration Project Grant, 2572
MLA Rittenhouse Award, 2573
MLA T. Mark Hodges Int'l Service Award, 2575
MLA Thomas Reuters / Frank Bradway Rogers Information Advancement Award, 2576
MMS and Alliance Charitable Foundation International Health Studies Grants, 2588
Phi Upsilon Omicron Florence Fallgatter Distinguished Service Award, 3192
Phi Upsilon Omicron Frances Morton Holbrook Alumni Award, 3193
Phi Upsilon Omicron Geraldine Clewell Senior Awards, 3194
Phi Upsilon Lillian P. Schoephoerster Award, 3196
Phi Upsilon Orinne Johnson Writing Award, 3198
Phi Upsilon Omicron Sarah Thorniley Phillips Leader Awards, 3199
Phi Upsilon Omicron Undergraduate Karen P. Goebel Conclave Award, 3200
Stan and Sandy Checketts Foundation, 3644
Thomson Reuters / MLA Doctoral Fellowships, 3728
UCT Scholarship Program, 3777
Union Labor Health Fndn Angel Fund Grants, 3791
Union Labor Health Foundation Dental Angel Fund Grants, 3793
UnitedHealthcare Children's Foundation Grants, 3795
USDD Breast Cancer Era of Hope Grants, 3832
USDD Breast Cancer Impact Grants, 3834
Virginia L. and William K. Beatty MLA Volunteer Service Award, 3877
Willary Foundation Grants, 3944
Zane's Foundation Grants, 3983
Zellweger Baby Support Network Grants, 3984

International Exchange Programs
ACSM Oded Bar-Or Int'l Scholar Awards, 272
American College of Surgeons International Guest Scholarships, 525
Aratani Foundation Grants, 649
ASM Int'l Fellowship for Latin America and the Caribbean, 734
ASM Int'l Fellowships for Asia and Africa, 735
ASM Millis-Colwell Postgraduate Travel Grant, 741
Canada-U.S. Fulbright Senior Specialists Grants, 1009
Coca-Cola Foundation Grants, 1203
Fulbright Alumni Initiatives Awards, 1749
Fulbright Distinguished Chairs Awards, 1750
Fulbright German Studies Seminar Grants, 1751
Fulbright Specialists Program Grants, 1754
Fulbright Scholars in Europe and Eurasia, 1755
Fulbright Traditional Scholar Program in Sub-Saharan Africa, 1756
Fulbright Traditional Scholar Program in the East Asia/Pacific Region, 1757
Fulbright Traditional Scholar Program in the Near East and North Africa Region, 1758
Fulbright Traditional Scholar Program in the South and Central Asia Region, 1759
Fulbright Traditional Scholar Program in the Western Hemisphere, 1760
IBRO-PERC InEurope Short Stay Grants, 2097
IBRO Asia APRC Lecturer Exchange Grants, 2102
NHLBI Investigator Initiated Multi-Site Clinical Trials, 2703
Nuffield Foundation Open Door Grants, 3030
SfN Japan Neuroscience Society Meeting Travel Awards, 3525
Shell Deer Park Grants, 3543
Summit Foundation Grants, 3674
U.S. Department of Education United States-Russia Program: Improving Research and Educational Activities in Higher Education, 3773
UICC American Cancer Society International Fellowships for Beginning Investigators, 3778

International Grants
A-T Medical Research Foundation Grants, 15
AAAAI RSLAAIS Leadership Award, 33
AAAS GE & Science Prize for Yng Life Scientists, 39
AAAS Science and Technology Policy Fellowships: Global Health and Development, 41
AACR Gertrude Elion Cancer Research Award, 113
AACR Kirk A. Landon and Dorothy P. Landon Foundation Prizes, 117
AAFCS International Graduate Fellowships, 129
Abbott-ASM Lifetime Achievement Award, 198
Abbott Fund Global AIDS Care Grants, 202
Abdus Salam ICTP Federation Arrangement Scheme Grants, 205
Abundance Foundation International Grants, 218
ACGT Young Investigator Grants, 233
ACSM Carl V. Gisolfi Memorial Fund Grant, 259
ACSM Coca-Cola Company Doctoral Student Grant on Behavior Research, 261
ACSM Dr. Raymond A. Weiss Research Endowment Grant, 262
ACSM Foundation Clinical Sports Medicine Endowment Grants, 265
ACSM Foundation Research Endowment Grants, 267
ACSM International Student Awards, 269
ACSM Oded Bar-Or Int'l Scholar Awards, 272
AFAR Paul Beeson Career Development Awards in Aging Research for the Island of Ireland, 336
Aid for Starving Children International Grants, 368
AIG Disaster Relief Fund Grants, 370
AIHS Proposals for Special Initiatives, 395
Air Products and Chemicals Grants, 399
Albany Medical Center Prize in Medicine and Biomedical Research, 404
Albert and Mary Lasker Foundation Awards, 406
Alcoa Foundation Grants, 414
Alcohol Misuse and Alcoholism Research Grants, 415
Alexander von Humboldt Foundation Georg Forster Fellowships for Experienced Researchers, 423
Alexander von Humboldt Foundation Georg Forster Fellowships for Postdoctoral Researchers, 424
Alexander von Humboldt Foundation Research Fellowships for Postdoctoral Researchers, 425
Alfred P. Sloan Foundation International Science Engagement Grants, 431
Alternatives Research and Development Foundation Alternatives in Education Grants, 455
Alternatives Research and Development Foundation Grants, 456
Alzheimer's Association Investigator-Initiated Research Grants, 466
Alzheimer's Association New Investigator Research Grants, 469
Alzheimer's Association U.S.-U.K. Young Investigator Exchange Fellowships, 471
AMD Corporate Contributions Grants, 500
American-Scandinavian Foundation Visiting Lectureship Grants, 502
American Jewish World Service Grants, 533
American Philosophical Society Daland Fellowships in Clinical Investigation, 537
American Psychiatric Association/AstraZeneca Young Minds in Psychiatry International Awards, 538
amfAR Global Initiatives Grants, 568
amfAR Research Grants, 571
Amgen Foundation Grants, 572
AON Foundation Grants, 623
Archer Daniels Midland Foundation Grants, 651
Arnold and Mabel Beckman Foundation Beckman-Argyros Award in Vision Research, 664
Arnold and Mabel Beckman Foundation Young Investigators Grants, 666
ASGE Don Wilson Award, 685
ASGE Endoscopic Research Awards, 686
ASGE Given Capsule Endoscopy Research Award, 688
ASHFoundation Graduate Student International Scholarship, 702
ASM-PAHO Infectious Diseases Epidemiology and Surveillance Fellowships, 714

International Grants

ASM-UNESCO Leadership Grant for International Educators, 715
ASM Int'l Fellowships for Asia and Africa, 735
ASM Millis-Colwell Postgraduate Travel Grant, 741
ASPET Paul M. Vanhoutte Award, 777
ASPH/CDC Allan Rosenfield Global Health Fellowships, 780
Atran Foundation Grants, 797
Australasian Institute of Judicial Administration Seed Funding Grants, 806
Australian Academy of Science Grants, 807
Avon Foundation Breast Care Fund Grants, 813
Avon Products Foundation Grants, 814
Banfi Vintners Foundation Grants, 824
Bank of America Fndn Matching Gifts, 825
Barth Syndrome Foundation Research Grants, 836
Baxter International Corporate Giving Grants, 841
Baxter International Foundation Grants, 843
Bayer Hemophilia Caregivers Education Award, 847
Beatrice Laing Trust Grants, 864
Becton Dickinson and Company Grants, 869
Besser Foundation Grants, 887
BibleLands Grants, 889
Bill and Melinda Gates Foundation Emergency Response Grants, 892
Bill and Melinda Gates Foundation Policy and Advocacy Grants, 893
Bill and Melinda Gates Foundation Water, Sanitation and Hygiene Grants, 894
Boston Jewish Community Women's Fund Grants, 938
BP Foundation Grants, 943
Bristol-Myers Squibb Foundation Global HIV/AIDS Initiative Grants, 957
Bristol-Myers Squibb Foundation Health Disparities Grants, 958
Bristol-Myers Squibb Foundation Product Donations Grants, 960
Burden Trust Grants, 976
BWF Career Awards at the Scientific Interface, 983
BWF Institutional Program Unifying Population and Laboratory Based Sciences Grants, 987
Cabot Corporation Foundation Grants, 993
Caddock Foundation Grants, 994
Caesars Foundation Grants, 995
Canada-U.S. Fulbright New Century Scholars Program Grants, 1008
Canada-U.S. Fulbright Senior Specialists Grants, 1009
Canada Graduate Scholarships (CGS) and NSERC Postgraduate Scholarships (PGS), 1010
Canadian Optometric Education Trust Grants, 1011
Cargill Citizenship Fund Corporate Giving, 1018
Carnegie Corporation of New York Grants, 1033
Carylon Foundation Grants, 1040
CDC Foundation Atlanta International Health Fellowships, 1060
Central Okanagan Foundation Grants, 1080
Cetana Educational Foundation Scholarships, 1082
Changemakers Innovation Awards, 1111
Charles Delmar Foundation Grants, 1115
Charles H. Revson Foundation Grants, 1122
Chase Paymentech Corporate Giving Grants, 1127
Chatlos Foundation Grants Program, 1128
Chazen Foundation Grants, 1129
Children's Cardiomyopathy Foundation Research Grants, 1137
Chiles Foundation Grants, 1147
Christensen Fund Regional Grants, 1149
Christopher & Dana Reeve Foundation Quality of Life Grants, 1152
CIGNA Foundation Grants, 1160
CNIB Baker Applied Research Fund Grants, 1193
CNIB Baker Fellowships, 1194
CNIB Baker New Researcher Fund Grants, 1195
CNIB Barbara Tuck MacPhee Award, 1196
CNIB Canada Glaucoma Clinical Research Council Grants, 1197
CNIB E. (Ben) & Mary Hochhausen Access Technology Research Grants, 1199
CNIB Ross Purse Doctoral Fellowships, 1200
Coca-Cola Foundation Grants, 1203

Commonwealth Fund Harkness Fellowships in Health Care Policy and Practice, 1229
Commonwealth Fund Small Grants, 1234
ConocoPhillips Foundation Grants, 1327
Conquer Cancer Foundation of ASCO Drug Development Research Professorship, 1330
Conquer Cancer Foundation of ASCO International Development and Education Award in Palliative Care, 1332
Conquer Cancer Foundation of ASCO International Development and Education Awards, 1333
Conservation, Food, and Health Foundation Grants for Developing Countries, 1338
Cooper Industries Foundation Grants, 1345
CTCRI Idea Grants, 1369
Cultural Society of Filipino Americans Grants, 1373
Cystic Fibrosis Canada Clinical Project Grants, 1382
Cystic Fibrosis Canada Fellowships, 1384
Cystic Fibrosis Canada Research Grants, 1385
Cystic Fibrosis Canada Scholarships, 1386
Cystic Fibrosis Canada Studentships, 1390
Cystic Fibrosis Canada Visiting Allied Health Professional Awards, 1394
Cystic Fibrosis Canada Visiting Clinician Awards, 1395
Cystic Fibrosis Trust Research Grants, 1404
Daimler and Benz Foundation Fellowships, 1411
Dale and Edna Walsh Foundation Grants, 1415
Danellie Foundation Grants, 1421
Doris Duke Charitable Foundation Operations Research on AIDS Care and Treatment in Africa (ORACTA) Grants, 1497
Duke University Adult Cardiothoracic Anesthesia and Critical Care Medicine Fellowships, 1520
Duke University Clinical Cardiac Electrophysiology Fellowships, 1522
Duke Univ Interventional Cardio Fellowships, 1524
eBay Foundation Community Grants, 1543
EBSCO / MLA Annual Meeting Grants, 1545
Echoing Green Fellowships, 1546
Elizabeth Glaser Int'l Leadership Awards, 1566
Elizabeth Glaser Scientist Award, 1567
Elton John AIDS Foundation Grants, 1585
EMBO Installation Grants, 1586
EMBO Long-Term Fellowships, 1587
EMBO Short-Term Fellowships, 1588
ExxonMobil Foundation Malaria Grants, 1621
FAMRI Clinical Innovator Awards, 1628
FDHN Moti L. & Kamla Rustgi International Travel Awards, 1649
Federal Express Corporate Contributions, 1657
Firelight Foundation Grants, 1667
Fishman Family Foundation Grants, 1672
Fluor Foundation Grants, 1682
FMC Foundation Grants, 1683
Fulbright Alumni Initiatives Awards, 1749
Fulbright Distinguished Chairs Awards, 1750
Fulbright German Studies Seminar Grants, 1751
Fulbright International Education Administrators (IEA) Seminar Program Grants, 1752
Fulbright New Century Scholars Grants, 1753
Fulbright Specialists Program Grants, 1754
Fulbright Scholars in Europe and Eurasia, 1755
Fulbright Traditional Scholar Program in Sub-Saharan Africa, 1756
Fulbright Traditional Scholar Program in the East Asia/Pacific Region, 1757
Fulbright Traditional Scholar Program in the Near East and North Africa Region, 1758
Fulbright Traditional Scholar Program in the South and Central Asia Region, 1759
Fulbright Traditional Scholar Program in the Western Hemisphere, 1760
Genesis Foundation Grants, 1786
Gerber Foundation Grants, 1807
Glaucoma Foundation Grants, 1827
Global Fund for Children Grants, 1832
Global Fund for Women Grants, 1833
Greater Saint Louis Community Fndn Grants, 1884
Greygates Foundation Grants, 1901
Gruber Foundation Cosmology Prize, 1907

H.B. Fuller Foundation Grants, 1922
H.J. Heinz Company Foundation Grants, 1923
Harold Simmons Foundation Grants, 1955
Harry Frank Guggenheim Fnd Research Grants, 1964
Helen Bader Foundation Grants, 1992
Hershey Company Grants, 2013
HHMI/EMBO Start-up Grants for C Europe, 2016
HHMI Grants and Fellowships Programs, 2020
HHMI International Research Scholars Program, 2021
Hospital Libraries Section / MLA Professional Development Grants, 2052
IARC Postdoctoral Fellowships for Training in Cancer Research, 2091
IBRO-PERC InEurope Short Stay Grants, 2097
IIE 911 Armed Forces Scholarships, 2122
IIE African Center of Excellence for Women's Leadership Grants, 2124
IIE David L. Boren Fellowships, 2129
Isabel Allende Foundation Esperanza Grants, 2165
James H. Cummings Foundation Grants, 2190
James S. McDonnell Fnd Research Grants, 2203
John Deere Foundation Grants, 2234
Johns Manville Fund Grants, 2251
Johnson & Johnson Community Health Grants, 2252
Johnson & Johnson Corporate Contributions, 2253
Johnson & Johnson/SAH Arts & Healing Grants, 2254
Kosciuszko Fndn Grants for Polish Citizens, 2322
LAM Fnd Established Investigator Awards, 2337
LAM Foundation Research Grants, 2340
Landon Foundation-AACR Innovator Award for International Collaboration, 2343
Ludwick Family Foundation Grants, 2389
Lustgarten Foundation for Pancreatic Cancer Research Grants, 2392
M-A-C AIDS Fund Grants, 2400
Majors MLA Chapter Project of the Year, 2418
MAP International Medical Fellowships, 2421
Mattel International Grants Program, 2455
May and Stanley Smith Charitable Trust Grants, 2459
Mead Johnson Nutritionals Charitable Giving, 2502
Mead Johnson Nutritionals Med Educ Grants, 2504
Medical Informatics Section/MLA Career Development Grant, 2508
Mesothelioma Applied Research Fndn Grants, 2524
MGFA Post-Doctoral Research Fellowships, 2539
Milagro Foundation Grants, 2552
MLA Cunningham Memorial International Fellowship, 2560
MLA Estelle Brodman Academic Medical Librarian of the Year Award, 2563
MLA Ida and George Eliot Prize, 2566
MLA Janet Doe Lectureship Award, 2567
MLA Lois Ann Colaianni Award for Excellence and Achievement in Hospital Librarianship, 2568
MLA Louise Darling Medal for Distinguished Achievement in Collection Development in the Health Sciences, 2569
MLA Lucretia W. McClure Excellence in Education Award, 2570
MLA Murray Gottlieb Prize Essay Award, 2571
MLA Section Project of the Year Award, 2574
MLA T. Mark Hodges Int'l Service Award, 2575
MLA Thomas Reuters / Frank Bradway Rogers Information Advancement Award, 2576
MMAAP Fndn Dermatology Project Awards, 2578
MMAAP Foundation Fellowship Award in Hematology, 2579
MMAAP Foundation Fellowship Award in Reproductive Medicine, 2580
MMAAP Foundation Fellowship Award in Translational Medicine, 2581
MMAAP Fndn Hematology Project Awards, 2583
MMAAP Foundation Research Project Award in Reproductive Medicine, 2584
MMAAP Foundation Research Project Award in Translational Medicine, 2585
MMS and Alliance Charitable Foundation International Health Studies Grants, 2588
National Center for Responsible Gaming Grants, 2629

PROGRAM TYPE INDEX

Job Training/Adult Vocational Programs / 891

National Parkinson Foundation Clinical Research Fund Grants, 2635
Nestle Foundation Training Grant, 2679
NIA Malnutrition in Older Persons Research, 2767
NIDDK Co-Activators and Co-Repressors in Gene Expression Grants, 2805
NIDDK Neurobiology of Diabetic Complications Grants, 2828
NIH Biology, Development, and Progression of Malignant Prostate Disease Research Grants, 2866
NIH Bone and the Hematopoietic and Immune Systems Research Grants, 2868
NIH Structural Bio of Membrane Proteins Grant, 2911
NIMH Early Identification and Treatment of Mental Disorders in Children and Adolescents Grants, 2918
NINR Acute and Chronic Care During Mechanical Ventilation Research Grants, 2926
NINR Clinical Interventions for Managing the Symptoms of Stroke Research Grants, 2929
NINR Quality of Life for Individuals at the End of Life Grants, 2933
Norman Foundation Grants, 2983
Notsew Orm Sands Foundation Grants, 3002
NSERC Brockhouse Canada Prize for Interdisciplinary Research in Science and Engineering Grant, 3003
Nuffield Foundation Children & Families Grants, 3027
Nuffield Foundation Law and Society Grants, 3028
Nuffield Foundation Small Grants, 3031
Oak Foundation Child Abuse Grants, 3059
OECD Sexual Health and Rights Project Grants, 3064
Olympus Corporation of Americas Corporate Giving Grants, 3073
OSF Access to Essential Medicines Initiative, 3086
OSF Accountability and Monitoring in Health Initiative Grants, 3087
OSF Global Health Financing Initiative Grants, 3088
OSF Health Media Initiative Grants, 3089
OSF International Harm Reduction Development Program Grants, 3090
OSF International Palliative Care Grants, 3091
OSF Law and Health Initiative Grants, 3092
OSF Mental Health Initiative Grants, 3093
OSF Public Health Grants in Kyrgyzstan, 3094
OSF Roma Health Project Grants, 3095
Packard Foundation Population and Reproductive Health Grants, 3106
PAHO-ASM Int'l Professorship Latin America, 3107
Palmer Foundation Grants, 3112
Patrick and Anna M. Cudahy Fund Grants, 3120
Paul Balint Charitable Trust Grants, 3122
PDF International Research Grants, 3128
PepsiCo Foundation Grants, 3137
Pfizer ASPIRE Worldwide Endocrine Young Investigator Grants, 3173
Pfizer Healthcare Charitable Contributions, 3180
Pfizer Medical Education Track One Grants, 3182
Pfizer Medical Education Track Two Grants, 3183
Pfizer Special Events Grants, 3186
PKD Foundation Lillian Kaplan International Prize for the Advancement in the Understanding of Polycystic Kidney Disease, 3232
Potts Memorial Foundation Grants, 3254
Prospect Burma Scholarships, 3275
Radcliffe Institute Carol K. Pforzheimer Student Fellowships, 3298
Radcliffe Institute Residential Fellowships, 3299
Rajiv Gandhi Foundation Grants, 3302
Raskob Foundation for Catholic Activities Grants, 3306
RCPSC Detweiler Traveling Fellowships, 3320
RCPSC Harry S. Morton Traveling Fellowship in Surgery, 3324
RCPSC McLaughlin-Gallie Visiting Professor, 3333
RCPSC Robert Maudsley Fellowship for Studies in Medical Education, 3341
Ringing Rocks Foundation Discretionary Grants, 3375
Rockefeller Archive Center Research Grants, 3393
Rockefeller Brothers Peace & Security Grants, 3394
Rockefeller Brothers Fund Pivotal Places Grants: South Africa, 3395
Rockefeller Foundation Grants, 3396

Rockwell Collins Charitable Corporation Grants, 3397
Rohm and Haas Company Grants, 3403
Ronald McDonald House Charities Grants, 3405
Rosalinde and Arthur Gilbert Foundation/AFAR New Investigator Awards in Alzheimer's Disease, 3406
RSC Abbyann D. Lynch Medal in Bioethics, 3419
RSC Jason A. Hannah Medal, 3420
RSC McLaughlin Medal, 3421
RSC McNeil Medal for the Public Awareness of Science, 3422
RSC Thomas W. Eadie Medal, 3423
Scleroderma Foundation Established Investigator Grants, 3497
Seagate Tech Corp Capacity to Care Grants, 3502
SfN Chapter-of-the-Year Award, 3519
SfN Chapter Grants, 3520
Sigma Theta Tau International /American Nurses Foundation Grant, 3556
Sigma Theta Tau International / Council for the Advancement of Nursing Science Grants, 3557
Sigma Theta Tau International / Oncology Nursing Society Grant, 3558
Sigma Theta Tau International Doris Bloch Research Award, 3559
Sigma Theta Tau Small Grants, 3560
Sigrid Juselius Foundation Grants, 3561
Simone and Cino del Duca Grand Prix Awards, 3564
Sir Dorabji Tata Trust Grants for NGOs or Voluntary Organizations, 3572
Skoll Fndn Awards for Social Entrepreneurship, 3579
SNM/ Covidien Seed Grant in Molecular Imaging/ Nuclear Medicine Research, 3582
SNM Postdoc Molecular Imaging Scholar Grants, 3587
SNM Student Fellowship Awards, 3588
Special Olympics Health Profess Student Grants, 3629
SRC Medical Research Grants, 3635
SSHRC-NSERC College and Community Innovation Enhancement Grants, 3636
SSHRC Banting Postdoctoral Fellowships, 3637
Stella and Charles Guttman Foundation Grants, 3651
Susan G. Komen Breast Cancer Foundation Brinker Awards for Scientific Distinction, 3687
Susan G. Komen Breast Cancer Foundation Career Catalyst Research Grants, 3688
Susan G. Komen Breast Cancer Foundation Challenge Grants: Career Catalyst Research, 3691
TE Foundation Grants, 3705
Thrasher Research Fund Grants, 3732
Tifa Foundation Grants, 3733
Trull Foundation Grants, 3761
U.S. Department of Education United States-Russia Program: Improving Research and Educational Activities in Higher Education, 3773
UICC American Cancer Society International Fellowships for Beginning Investigators, 3778
UICC International Cancer Technology Transfer (ICRETT) Fellowships, 3779
UICC Yamagiwa-Yoshida Memorial International Cancer Study Grants, 3781
United Methodist Committee on Relief Global Health Grants, 3799
United States Institute of Peace - Jennings Randolph Senior Fellowships, 3802
United Technologies Corporation Grants, 3803
USAID Accelerating Progress Against Tuberculosis in Kenya Grants, 3805
USAID Child, Newborn, and Maternal Health Project Grants, 3806
USAID Comprehensive District-Based Support for Better HIV/TB Patient Outcomes Grants, 3807
USAID Devel Innovation Accelerator Grants, 3808
USAID Family Health Plus Project Grants, 3810
USAID Fighting Ebola Grants, 3812
USAID Global Health Development Innovation Accelerator Broad Agency Grants, 3813
USAID Health System Strengthening Grants, 3814
USAID Higher Education Partnerships for Innovation and Impact (HEPII) Grants, 3815
USAID HIV Prevention with Key Populations - Mali Grants, 3816

USAID Integration of Care and Support within the Health System to Support Better Patient Outcomes Grants, 3817
USAID Land Use Change and Disease Emergence Grants, 3818
USAID Microbicides Research, Development, and Introduction Grants, 3819
USAID Research and Innovation for Health Supply Chain Systems & Commodity Security Grants, 3821
USAID Support for International Family Planning Organizations Grants, 3823
USDD Broad Agency Announcement Grants, 3837
USDD Broad Agency Announcement HIV/AIDS Prevention Grants, 3838
USDD HIV/AIDS Prevention: Military Specific HIV/ AIDS Prevention, Care, and Treatment for Non-PEPFAR Funded Countries Grants, 3841
Virginia L. and William K. Beatty MLA Volunteer Service Award, 3877
W.K. Kellogg Foundation Healthy Kids Grants, 3886
WestWind Foundation Reproductive Health and Rights Grants, 3924
Wilhelm Sander-Stiftung Foundation Grants, 3942
World of Children Humanitarian Award, 3975

Job Training/Adult Vocational Programs
3M Company Fndn Community Giving Grants, 6
Able Trust Vocational Rehabilitation Grants for Individuals, 213
ACE Charitable Foundation Grants, 229
ACF Native American Social and Economic Development Strategies Grants, 231
Achelis Foundation Grants, 234
Adaptec Foundation Grants, 304
Adolph Coors Foundation Grants, 311
A Fund for Women Grants, 342
Alaska Airlines Foundation Grants, 403
Alcoa Foundation Grants, 414
Alliant Energy Corporation Contributions, 445
Allstate Corporate Giving Grants, 446
Alpha Natural Resources Corporate Giving, 450
AMD Corporate Contributions Grants, 500
Anschutz Family Foundation Grants, 602
ARCO Foundation Education Grants, 652
ASHA Students Preparing for Academic & Research Careers (SPARC) Award, 700
Ashland Corporate Contributions Grants, 713
Assisi Foundation of Memphis General Grants, 787
Assurant Health Foundation Grants, 789
AT&T Foundation Civic and Community Service Program Grants, 793
Autauga Area Community Foundation Grants, 808
BBVA Compass Foundation Charitable Grants, 848
Beckman Coulter Foundation Grants, 868
Benton Community Foundation Grants, 878
Berks County Community Foundation Grants, 881
Blue River Community Foundation Grants, 922
Bodman Foundation Grants, 929
Boeing Company Contributions Grants, 930
Boston Foundation Grants, 936
Boston Jewish Community Women's Fund Grants, 938
BP Foundation Grants, 943
Bridgestone/Firestone Trust Fund Grants, 949
Brooklyn Community Foundation Caring Neighbors Grants, 968
Brown County Community Foundation Grants, 970
Burlington Industries Foundation Grants, 977
Bush Fndn Health & Human Services Grants, 979
Cargill Citizenship Fund Corporate Giving, 1018
Carl M. Freeman Foundation Grants, 1027
Carrie Estelle Doheny Foundation Grants, 1037
CCHD Community Development Grants, 1046
CFFVR Jewelers Mutual Charitable Giving, 1099
CFFVR Schmidt Family G4 Grants, 1103
Charlotte County (FL) Community Foundation Grants, 1126
Citizens Bank Mid-Atlantic Charitable Foundation Grants, 1162
Cleveland Browns Foundation Grants, 1182
CNA Foundation Grants, 1189

892 / Job Training/Adult Vocational Programs PROGRAM TYPE INDEX

Community Fndn for the Capital Region Grants, 1259
Community Foundation of Bartholomew County Heritage Fund Grants, 1262
Community Foundation of Eastern Connecticut Northeast Women and Girls Grants, 1270
Community Foundation of Louisville Anna Marble Memorial Fund for Princeton Grants, 1290
Community Foundation of Muncie and Delaware County Maxon Grants, 1301
Community Fndn of Riverside & San Bernardino County Irene S. Rockwell Grants, 1304
Constellation Energy Corporate Grants, 1342
Coors Brewing Corporate Contributions Grants, 1346
Cruise Industry Charitable Foundation Grants, 1363
DaimlerChrysler Corporation Fund Grants, 1412
Dallas Women's Foundation Grants, 1417
Denver Foundation Community Grants, 1461
eBay Foundation Community Grants, 1543
Essex County Community Foundation Women's Fund Grants, 1605
Evjue Foundation Grants, 1617
Faye McBeath Foundation Grants, 1635
Field Foundation of Illinois Grants, 1660
Fisher Foundation Grants, 1670
Ford Motor Company Fund Grants Program, 1687
Four County Community Fndn General Grants, 1710
Frances L. and Edwin L. Cummings Memorial Fund Grants, 1717
George Family Foundation Grants, 1794
George Gund Foundation Grants, 1796
GNOF New Orleans Works Grants, 1844
Grand Rapids Area Community Foundation Wyoming Youth Fund Grants, 1868
Greater Milwaukee Foundation Grants, 1883
Greater Sitka Legacy Fund Grants, 1885
Green River Area Community Fndn Grants, 1895
Grifols Community Outreach Grants, 1904
Hallmark Corporate Foundation Grants, 1936
Hampton Roads Community Foundation Health and Human Service Grants, 1941
Harry S. Black and Allon Fuller Fund Grants, 1966
Harvest Foundation Grants, 1971
Heckscher Foundation for Children Grants, 1988
Helen Bader Foundation Grants, 1992
Highmark Corporate Giving Grants, 2025
Hudson Webber Foundation Grants, 2072
Hutchinson Community Foundation Grants, 2085
Irvin Stern Foundation Grants, 2164
Isabel Allende Foundation Esperanza Grants, 2165
Jacobs Family Village Neighborhoods Grants, 2184
Jessica Stevens Community Foundation Grants, 2224
JM Foundation Grants, 2229
John Edward Fowler Memorial Fndn Grants, 2235
John Merck Fund Grants, 2245
Johnson & Johnson Corporate Contributions, 2253
Joseph Henry Edmondson Foundation Grants, 2270
Katharine Matthies Foundation Grants, 2291
Kessler Fnd Signature Employment Grants, 2302
Ketchikan Community Foundation Grants, 2303
Kodiak Community Foundation Grants, 2314
Kuntz Foundation Grants, 2326
Liberty Bank Foundation Grants, 2362
M.B. and Edna Zale Foundation Grants, 2401
Mary L. Peyton Foundation Grants, 2445
May and Stanley Smith Charitable Trust Grants, 2459
Merrick Foundation Grants, 2522
Miller Foundation Grants, 2556
Montgomery County Community Fndn Grants, 2591
Morris & Gwendolyn Cafritz Fndn Grants, 2594
Norcliffe Foundation Grants, 2979
Norman Foundation Grants, 2983
OneFamily Foundation Grants, 3075
Oppenstein Brothers Foundation Grants, 3078
Oregon Community Foundation Better Nursing Home Care Grants, 3081
Oregon Community Fndn Community Grants, 3082
PacifiCare Foundation Grants, 3102
Patrick and Anna M. Cudahy Fund Grants, 3120
Perry County Community Foundation Grants, 3142
Piedmont Natural Gas Corporate and Charitable Contributions, 3220
Piper Jaffray Fndn Communities Giving Grants, 3227
Polk Bros. Foundation Grants, 3243
Portland General Electric Foundation Grants, 3250
Price Chopper's Golub Foundation Grants, 3261
Principal Financial Group Foundation Grants, 3273
Pulaski County Community Foundation Grants, 3279
Reinberger Foundation Grants, 3357
Richard King Mellon Foundation Grants, 3369
Robert & Clara Milton Senior Housing Grants, 3384
Robert R. McCormick Tribune Veterans Initiative Grants, 3386
Rockefeller Foundation Grants, 3396
Rockwell Fund, Inc. Grants, 3398
S. Mark Taper Foundation Grants, 3449
Salmon Foundation Grants, 3458
Samuel S. Johnson Foundation Grants, 3468
Seneca Foods Foundation Grants, 3511
Seward Community Foundation Grants, 3514
Seward Community Foundation Mini-Grants, 3515
Shield-Ayres Foundation Grants, 3546
Sid W. Richardson Foundation Grants, 3553
Sierra Health Foundation Responsive Grants, 3555
Sioux Falls Area Community Foundation Community Fund Grants (Unrestricted), 3568
Sioux Falls Area Community Foundation Spot Grants (Unrestricted), 3570
Stackpole-Hall Foundation Grants, 3643
Strake Foundation Grants, 3658
Textron Corporate Contributions Grants, 3712
Thomas Sill Foundation Grants, 3724
Topfer Family Foundation Grants, 3739
Toyota Motor Manuf of Mississippi Grants, 3746
Toyota Motor Manuf of West Virginia Grants, 3748
Tri-State Community Twenty-first Century Endowment Fund Grants, 3756
U.S. Department of Education Rehabilitation Research Training Centers (RRTCs), 3770
U.S. Department of Education Vocational Rehabilitation Services Projects for American Indians with Disabilities Grants, 3774
Union Bank, N.A. Foundation Grants, 3788
Vanderburgh Community Fndn Women's Fund, 3853
Vectren Foundation Grants, 3857
Verizon Foundation Maryland Grants, 3859
Verizon Foundation New York Grants, 3860
Verizon Foundation Pennsylvania Grants, 3862
Waitt Family Foundation Grants, 3894
Walker Area Community Foundation Grants, 3895

Land Acquisition
G.A. Ackermann Memorial Fund Grants, 1767
Harrison County Community Foundation Signature Grants, 1958
Janson Foundation Grants, 2210
Norcliffe Foundation Grants, 2979
North Central Health Services Grants, 2992
Oregon Community Fndn Community Grants, 3082
Priddy Foundation Capital Grants, 3263
Robert R. Meyer Foundation Grants, 3387
Shield-Ayres Foundation Grants, 3546
The Ray Charles Foundation Grants, 3716

Matching/Challenge Funds
3M Company Fndn Community Giving Grants, 6
A.O. Smith Foundation Community Grants, 17
Abbot and Dorothy H. Stevens Foundation Grants, 197
Abbott Fund Global AIDS Care Grants, 202
Abell-Hanger Foundation Grants, 207
ACE Charitable Foundation Grants, 229
ACF Native American Social and Economic Development Strategies Grants, 231
Adaptec Foundation Grants, 304
AEC Trust Grants, 315
Agnes M. Lindsay Trust Grants, 346
Ahmanson Foundation Grants, 355
Air Products and Chemicals Grants, 399
Alabama Power Foundation Grants, 400
Alberto Culver Corporate Contributions Grants, 408
Albertson's Charitable Giving Grants, 410
Alcoa Foundation Grants, 414
Alice Tweed Tuohy Foundation Grants Program, 437
Allegan County Community Foundation Grants, 439
Allegheny Technologies Charitable Trust, 440
Allen P. & Josephine B. Green Fndn Grants, 443
Allstate Corporate Giving Grants, 446
Allstate Corp Hometown Commitment Grants, 447
Allyn Foundation Grants, 448
Alpha Natural Resources Corporate Giving, 450
Amica Insurance Company Community Grants, 575
AMI Semiconductors Corporate Grants, 576
Anheuser-Busch Foundation Grants, 591
Ann Arbor Area Community Foundation Grants, 594
Ann Peppers Foundation Grants, 599
Appalachian Regional Commission Health Grants, 637
ARCO Foundation Education Grants, 652
Arie and Ida Crown Memorial Grants, 654
Arizona Public Service Corporate Giving Program Grants, 658
Arkell Hall Foundation Grants, 661
Arthur Ashley Williams Foundation Grants, 671
Ashland Corporate Contributions Grants, 713
Assisi Fndn of Memphis Capital Project Grants, 786
Assurant Health Foundation Grants, 789
Atran Foundation Grants, 797
Autzen Foundation Grants, 810
Avon Products Foundation Grants, 814
Bailey Foundation Grants, 818
Ball Brothers Foundation General Grants, 820
Bank of America Fndn Matching Gifts, 825
Barker Welfare Foundation Grants, 831
Barra Foundation Project Grants, 834
Batchelor Foundation Grants, 837
Baton Rouge Area Foundation Grants, 838
Battle Creek Community Foundation Grants, 839
Batts Foundation Grants, 840
Baxter International Corporate Giving Grants, 841
Baxter International Foundation Grants, 843
BCBSM Foundation Community Health Matching Grants, 852
Beckley Area Foundation Grants, 867
Becton Dickinson and Company Grants, 869
Bender Foundation Grants, 876
Benton Community Foundation Grants, 878
Bertha Russ Lytel Foundation Grants, 885
Besser Foundation Grants, 887
Blanche and Julian Robertson Family Foundation Grants, 911
Blue Cross Blue Shield of Minnesota Foundation - Healthy Equity: Health Impact Assessment Program Grants, 917
Blue Grass Community Fndn Harrison Grants, 919
Blue Grass Community Foundation Hudson-Ellis Fund Grants, 920
Blue River Community Foundation Grants, 922
Blumenthal Foundation Grants, 925
Bodenwein Public Benevolent Foundation Grants, 928
Boeing Company Contributions Grants, 930
Boettcher Foundation Grants, 931
Boston Foundation Grants, 936
Bridgestone/Firestone Trust Fund Grants, 949
Bristol-Myers Squibb Foundation Health Disparities Grants, 958
Bullitt Foundation Grants, 973
Burlington Industries Foundation Grants, 977
Burlington Northern Santa Fe Foundation Grants, 978
Cabot Corporation Foundation Grants, 993
Cailloux Foundation Grants, 996
Callaway Foundation Grants, 1002
Callaway Golf Company Foundation Grants, 1003
Campbell Soup Foundation Grants, 1007
Cardinal Health Foundation Grants, 1017
Cargill Citizenship Fund Corporate Giving, 1018
Carl & Eloise Pohlad Family Fndn Grants, 1021
Carl C. Icahn Foundation Grants, 1023
Carlisle Foundation Grants, 1025
Carl M. Freeman Foundation Grants, 1027
Carnahan-Jackson Foundation Grants, 1032
Carpenter Foundation Grants, 1035

PROGRAM TYPE INDEX   Matching/Challenge Funds   / 893

Carrie E. and Lena V. Glenn Foundation Grants, 1036
Carrie Estelle Doheny Foundation Grants, 1037
Cass County Community Foundation Grants, 1041
CCHD Community Development Grants, 1046
CDC Public Health Conference Support Grant, 1067
Cemala Foundation Grants, 1077
Central Carolina Community Foundation Community Impact Grants, 1079
Cessna Foundation Grants Program, 1081
CFFVR Clintonville Area Foundation Grants, 1097
CFFVR Robert and Patricia Endries Family Foundation Grants, 1102
CFFVR Women's Fund for the Fox Valley Region Grants, 1107
Charles A. Frueauff Foundation Grants, 1114
Charles H. Revson Foundation Grants, 1122
Chatlos Foundation Grants Program, 1128
Chemtura Corporation Contributions Grants, 1130
Chilkat Valley Community Foundation Grants, 1148
Christensen Fund Regional Grants, 1149
Christy-Houston Foundation Grants, 1153
CICF City of Noblesville Community Grant, 1155
CIGNA Foundation Grants, 1160
Claude Worthington Benedum Fndn Grants, 1178
Cleveland-Cliffs Foundation Grants, 1181
CNA Foundation Grants, 1189
CNCS Senior Companion Program Grants, 1191
CNCS Social Innovation Grants, 1192
Coca-Cola Foundation Grants, 1203
Cockrell Foundation Grants, 1204
Collins Foundation Grants, 1210
Colonel Stanley R. McNeil Foundation Grants, 1211
Columbus Foundation Competitive Grants, 1217
Columbus Foundation Paul G. Duke Grants, 1221
Commonwealth Fund Small Grants, 1234
Communities Foundation of Texas Grants, 1236
Community Fndn for Greater Buffalo Grants, 1247
Community Fndn for Monterey County Grants, 1253
Community Fndn for Muskegon County Grants, 1254
Community Fndn for the Capital Region Grants, 1259
Community Foundation of Bartholomew County Heritage Fund Grants, 1262
Community Foundation of Boone County Grants, 1266
Community Foundation of Greater Birmingham Grants, 1272
Community Foundation of Greater Fort Wayne - Community Endowment and Clarke Endowment Grants, 1274
Community Fndn of Greater Lafayette Grants, 1280
Community Fndn of Greater Tampa Grants, 1282
Community Foundation of Greenville-Greenville Women Giving Grants, 1283
Community Foundation of Greenville Community Enrichment Grants, 1284
Community Foundation of Mount Vernon and Knox County Grants, 1299
Community Foundation of Muncie and Delaware County Grants, 1300
Community Fndn of So Alabama Grants, 1306
Community Fndn of South Puget Sound Grants, 1307
Community Foundation of St. Joseph County Special Project Challenge Grants, 1310
Community Foundation of the Verdugos Educational Endowment Fund Grants, 1314
Community Fndn of Wabash County Grants, 1316
Connecticut Community Foundation Grants, 1324
Connelly Foundation Grants, 1326
ConocoPhillips Foundation Grants, 1327
Constantin Foundation Grants, 1341
Constellation Energy Corporate Grants, 1342
Cooke Foundation Grants, 1344
Cooper Industries Foundation Grants, 1345
Cornerstone Foundation of Northeastern Wisconsin Grants, 1347
Cowles Charitable Trust Grants, 1354
Cralle Foundation Grants, 1356
CSRA Community Foundation Grants, 1366
Cumberland Community Foundation Grants, 1374
DaimlerChrysler Corporation Fund Grants, 1412
Dale and Edna Walsh Foundation Grants, 1415

Dana Foundation Science and Health Grants, 1420
Daniels Fund Grants-Aging, 1424
Daphne Seybolt Culpeper Fndn Grants, 1425
Daywood Foundation Grants, 1438
Decatur County Community Foundation Large Project Grants, 1444
Delmarva Power and Light Contributions, 1454
DHHS Adolescent Family Life Demonstration Projects, 1472
Dolan Children's Foundation Grants, 1482
DOL Occupational Safety and Health--Susan Harwood Training Grants, 1484
Dorothea Haus Ross Foundation Grants, 1498
Dr. John T. Macdonald Foundation Grants, 1507
Drs. Bruce and Lee Foundation Grants, 1513
Duchossois Family Foundation Grants, 1518
Dunspaugh-Dalton Foundation Grants, 1530
DuPage Community Foundation Grants, 1531
Dyson Foundation Mid-Hudson Valley Project Support Grants, 1534
Eberly Foundation Grants, 1544
Educational Foundation of America Grants, 1553
Edward Bangs Kelley and Elza Kelley Foundation Grants, 1556
Edward W. and Stella C. Van Houten Memorial Fund Grants, 1559
Edwin S. Webster Foundation Grants, 1560
Edyth Bush Charitable Foundation Grants, 1562
Effie and Wofford Cain Foundation Grants, 1563
Eisner Foundation Grants, 1564
Elizabeth Morse Genius Charitable Trust Grants, 1569
Elkhart County Community Foundation Grants, 1573
El Paso Community Foundation Grants, 1579
Emerson Charitable Trust Grants, 1589
Emily Davie and Joseph S. Kornfeld Foundation Grants, 1591
Erie Community Foundation Grants, 1601
Ewing Halsell Foundation Grants, 1618
Fairfield County Community Foundation Grants, 1625
Fannie E. Rippel Foundation Grants, 1630
Fargo-Moorhead Area Fndn Woman's Grants, 1633
Farmers Insurance Corporate Giving Grants, 1634
Fayette County Foundation Grants, 1636
Federal Express Corporate Contributions, 1657
Fidelity Foundation Grants, 1659
Field Foundation of Illinois Grants, 1660
FirstEnergy Foundation Community Grants, 1668
Fluor Foundation Grants, 1682
FMC Foundation Grants, 1683
Foellinger Foundation Grants, 1684
Foster Foundation Grants, 1689
Foundation for Seacoast Health Grants, 1698
France-Merrick Foundation Health and Human Services Grants, 1715
Frances L. and Edwin L. Cummings Memorial Fund Grants, 1717
Francis L. Abreu Charitable Trust Grants, 1719
Franklin County Community Foundation Grants, 1724
Fred L. Emerson Foundation Grants, 1743
Fremont Area Community Fndn General Grants, 1746
Fuller E. Callaway Foundation Grants, 1761
G.N. Wilcox Trust Grants, 1768
Gebbie Foundation Grants, 1777
GenCorp Foundation Grants, 1779
Genentech Corp Charitable Contributions, 1780
General Mills Foundation Grants, 1782
George F. Baker Trust Grants, 1793
George Foundation Grants, 1795
George Gund Foundation Grants, 1796
George P. Davenport Trust Fund Grants, 1800
George S. & Dolores Dore Eccles Fndn Grants, 1801
Georgia Power Foundation Grants, 1806
Gertrude E. Skelly Charitable Foundation Grants, 1810
GNOF Freeman Challenge Grants, 1837
Golden Heart Community Foundation Grants, 1850
Goldman Philanthropic Partnerships Program, 1853
Goodrich Corporation Foundation Grants, 1854
Grace and Franklin Bernsen Foundation Grants, 1858
Graco Foundation Grants, 1860
Grand Haven Area Community Fndn Grants, 1864

Grand Rapids Area Community Fndn Grants, 1865
Grand Rapids Area Community Foundation Nashwauk Area Endowment Fund Grants, 1866
Grand Rapids Area Community Foundation Wyoming Grants, 1867
Grand Rapids Community Foundation Grants, 1869
Grand Rapids Community Foundation Ionia County Grants, 1870
Grand Rapids Community Foundation Southeast Ottawa Grants, 1873
Grand Rapids Community Fndn Sparta Grants, 1875
Great-West Life Grants, 1879
Greater Cincinnati Foundation Priority and Small Projects/Capacity-Building Grants, 1880
Greater Worcester Community Foundation Discretionary Grants, 1888
Green River Area Community Fndn Grants, 1895
Gulf Coast Community Foundation Grants, 1915
H & R Foundation Grants, 1920
H.A. & Mary K. Chapman Trust Grants, 1921
Hardin County Community Foundation Grants, 1946
Harold and Arlene Schnitzer CARE Foundation Grants, 1950
Harris and Eliza Kempner Fund Grants, 1956
Harry A. & Margaret D. Towsley Fndn Grants, 1959
Hartford Courant Foundation Grants, 1969
Hartford Foundation Regular Grants, 1970
Harvey Randall Wickes Foundation Grants, 1972
HCA Foundation Grants, 1980
Health Fndn of Greater Indianapolis Grants, 1984
Hedco Foundation Grants, 1989
Herbert A. & Adrian W. Woods Fndn Grants, 2007
Herbert H. & Grace A. Dow Fndn Grants, 2008
High Meadow Foundation Grants, 2027
Hill Crest Foundation Grants, 2030
Hillcrest Foundation Grants, 2031
Holland/Zeeland Community Fndn Grants, 2041
Honda of America Manufacturing Fndn Grants, 2045
Horace A. Kimball and S. Ella Kimball Foundation Grants, 2046
Hormel Foods Charitable Trust Grants, 2050
Howard and Bush Foundation Grants, 2054
HRK Foundation Health Grants, 2067
Huffy Foundation Grants, 2073
Hugh J. Andersen Foundation Grants, 2074
Huie-Dellmon Trust Grants, 2075
Humana Foundation Grants, 2077
Hutton Foundation Grants, 2087
Idaho Power Company Corporate Contributions, 2115
Illinois Tool Works Foundation Grants, 2147
Indiana 21st Century Research and Technology Fund Awards, 2154
Intergrys Corporation Grants, 2160
J.W. Kieckhefer Foundation Grants, 2176
James H. Cummings Foundation Grants, 2190
James J. & Angelia M. Harris Fndn Grants, 2192
James McKeen Cattell Fund Fellowships, 2197
James R. Dougherty Jr. Foundation Grants, 2198
James R. Thorpe Foundation Grants, 2199
James S. Copley Foundation Grants, 2200
Jane's Trust Grants, 2206
Janus Foundation Grants, 2211
Jean and Louis Dreyfus Foundation Grants, 2215
JELD-WEN Foundation Grants, 2218
Jennings County Community Foundation Grants, 2220
Jessica Stevens Community Foundation Grants, 2224
Jessie Ball Dupont Fund Grants, 2226
JM Foundation Grants, 2229
John Ben Snow Memorial Trust Grants, 2231
John Deere Foundation Grants, 2234
John G. Martin Foundation Grants, 2237
John Jewett and Helen Chandler Garland Foundation Grants, 2242
John P. McGovern Foundation Grants, 2247
John R. Oishei Foundation Grants, 2249
John Stauffer Charitable Trust Grants, 2258
John W. Alden Trust Grants, 2260
John W. and Anna H. Hanes Foundation Grants, 2261
Joseph Drown Foundation Grants, 2268
Josephine Goodyear Foundation Grants, 2272

894 / Matching/Challenge Funds

Josiah W. and Bessie H. Kline Foundation Grants, 2278
Kenneth T. & Eileen L. Norris Fndn Grants, 2300
Land O'Lakes Foundation Mid-Atlantic Grants, 2341
Lillian S. Wells Foundation Grants, 2365
Lotus 88 Fnd for Women & Children Grants, 2376
Lubbock Area Foundation Grants, 2383
Lubrizol Foundation Grants, 2384
Lucy Downing Nisbet Charitable Fund Grants, 2387
Lumpkin Family Fnd Healthy People Grants, 2391
Lymphatic Education and Research Network Additional Support Grants for NIH-funded F32 Postdoctoral Fellows, 2394
Lynde and Harry Bradley Foundation Grants, 2397
M.D. Anderson Foundation Grants, 2403
M.E. Raker Foundation Grants, 2404
Manuel D. & Rhoda Mayerson Fndn Grants, 2420
Mardag Foundation Grants, 2428
Margaret T. Morris Foundation Grants, 2429
Mary K. Chapman Foundation Grants, 2443
Mary Owen Borden Foundation Grants, 2446
Mary S. and David C. Corbin Foundation Grants, 2447
Matilda R. Wilson Fund Grants, 2453
Maurice J. Masserini Charitable Trust Grants, 2456
McCarthy Family Foundation Grants, 2490
McConnell Foundation Grants, 2492
McGraw-Hill Companies Community Grants, 2494
McKesson Foundation Grants, 2496
McMillen Foundation Grants, 2499
Mead Witter Foundation Grants, 2507
Medtronic Foundation Patient Link Grants, 2514
Mercedes-Benz USA Corporate Contributions, 2517
Mericos Foundation Grants, 2518
Mervin Bovaird Foundation Grants, 2523
MetroWest Health Foundation Capital Grants for Health-Related Facilities, 2526
Meyer Fndn Management Assistance Grants, 2534
Meyer Memorial Trust Grassroots Grants, 2536
Meyer Memorial Trust Responsive Grants, 2537
Meyer Memorial Trust Special Grants, 2538
Mimi and Peter Haas Fund Grants, 2557
Morris & Gwendolyn Cafritz Fndn Grants, 2594
Nelda C. and H.J. Lutcher Stark Fndn Grants, 2672
NHLBI Midcareer Investigator Award in Patient-Oriented Research, 2713
NHSCA Arts in Health Care Project Grants, 2741
Nina Mason Pulliam Charitable Trust Grants, 2921
Noble County Community Foundation Grants, 2975
Norcliffe Foundation Grants, 2979
Nordson Corporation Foundation Grants, 2981
Norfolk Southern Foundation Grants, 2982
North Central Health Services Grants, 2992
Northern Chautauqua Community Foundation Community Grants, 2995
Northwestern Mutual Foundation Grants, 2999
NSF Grant Opportunities for Academic Liaison with Industry (GOALI), 3016
NWHF Partners Investing in Nursing's Future, 3036
Oppenstein Brothers Foundation Grants, 3078
Ordean Foundation Grants, 3080
Oregon Community Fndn Community Grants, 3082
Otto Bremer Foundation Grants, 3099
Palmer Foundation Grants, 3112
Parkersburg Area Community Foundation Action Grants, 3116
Patron Saints Foundation Grants, 3121
Paul Rapoport Foundation Grants, 3125
Perkin Fund Grants, 3139
Peter Kiewit Foundation General Grants, 3154
Peter Kiewit Foundation Small Grants, 3155
PeyBack Foundation Grants, 3159
Phelps County Community Foundation Grants, 3188
Philadelphia Foundation Organizational Effectiveness Grants, 3189
Phoenix Suns Charities Grants, 3201
PhRMA Foundation Health Outcomes Sabbatical Fellowships, 3205
PhRMA Fndn Info Sabbatical Fellowships, 3209
PhRMA Foundation Pharmaceutics Sabbatical Fellowships, 3214

PhRMA Foundation Pharmacology/Toxicology Sabbatical Fellowships, 3218
Piedmont Natural Gas Corporate and Charitable Contributions, 3220
Pinkerton Foundation Grants, 3224
Piper Jaffray Fndn Communities Giving Grants, 3227
Playboy Foundation Grants, 3237
Plough Foundation Grants, 3238
PMI Foundation Grants, 3239
Price Chopper's Golub Foundation Grants, 3261
Principal Financial Group Foundation Grants, 3273
Prudential Foundation Education Grants, 3276
Puerto Rico Community Foundation Grants, 3278
Pulaski County Community Foundation Grants, 3279
Putnam County Community Foundation Grants, 3283
Quaker Chemical Foundation Grants, 3286
Quantum Foundation Grants, 3290
R.C. Baker Foundation Grants, 3292
Ralph M. Parsons Foundation Grants, 3304
Raskob Foundation for Catholic Activities Grants, 3306
Rathmann Family Foundation Grants, 3309
Rayonier Foundation Grants, 3311
RCF General Community Grants, 3312
RCF Individual Assistance Grants, 3313
Reinberger Foundation Grants, 3357
Rhode Island Foundation Grants, 3362
Richard and Helen DeVos Foundation Grants, 3364
Richard King Mellon Foundation Grants, 3369
Ripley County Community Foundation Grants, 3376
Robert W. Woodruff Foundation Grants, 3389
Rochester Area Foundation Grants, 3392
Rockwell International Corporate Trust Grants Program, 3399
Rohm and Haas Company Grants, 3403
Rollins-Luetkemeyer Foundation Grants, 3404
Ronald McDonald House Charities Grants, 3405
Rose Community Foundation Aging Grants, 3410
Rose Hills Foundation Grants, 3413
Roy & Christine Sturgis Charitable Grants, 3415
Roy J. Carver Charitable Trust Medical and Science Research Grants, 3416
RRF General Program Grants, 3417
Rush County Community Foundation Grants, 3424
Ruth Anderson Foundation Grants, 3425
S. Mark Taper Foundation Grants, 3449
Samuel S. Johnson Foundation Grants, 3468
San Diego Women's Foundation Grants, 3474
San Juan Island Community Foundation Grants, 3480
Sara Elizabeth O'Brien Trust Grants, 3483
Sarkeys Foundation Grants, 3485
Schering-Plough Foundation Health Grants, 3492
Schlessman Family Foundation Grants, 3493
Schramm Foundation Grants, 3495
Scott County Community Foundation Grants, 3500
Self Foundation Grants, 3510
Seward Community Foundation Grants, 3514
Seward Community Foundation Mini-Grants, 3515
Shell Deer Park Grants, 3543
Shell Oil Company Foundation Community Development Grants, 3544
Sid W. Richardson Foundation Grants, 3553
Sisters of St. Joseph Healthcare Fndn Grants, 3576
Sonoco Foundation Grants, 3621
Sony Corporation of America Grants, 3623
South Madison Community Foundation Grants, 3626
Southwest Florida Community Foundation Competitive Grants, 3627
Square D Foundation Grants, 3634
St. Joseph Community Health Foundation Improving Healthcare Access Grants, 3640
Stackpole-Hall Foundation Grants, 3643
Steele-Reese Foundation Grants, 3650
Stella and Charles Guttman Foundation Grants, 3651
Steuben County Community Foundation Grants, 3654
Stewart Huston Charitable Trust Grants, 3655
Strake Foundation Grants, 3658
Strowd Roses Grants, 3664
Summit Foundation Grants, 3674
Sunderland Foundation Grants, 3675
Susan Vaughan Foundation Grants, 3695

PROGRAM TYPE INDEX

TE Foundation Grants, 3705
Texas Commission on the Arts Arts Respond Project Grants, 3708
Textron Corporate Contributions Grants, 3712
Theodore Edson Parker Foundation Grants, 3715
Thomas Austin Finch, Sr. Foundation Grants, 3718
Thomas C. Ackerman Foundation Grants, 3719
Toyota Motor Manuf of West Virginia Grants, 3748
Toyota Motor N America of New York Grants, 3749
Trull Foundation Grants, 3761
Union Bank, N.A. Foundation Grants, 3788
Union Pacific Foundation Health and Human Services Grants, 3794
United Hospital Fund of New York Health Care Improvement Grants, 3797
United Technologies Corporation Grants, 3803
USAID Support for International Family Planning Organizations Grants, 3823
VDH Rescue Squad Assistance Fund Grants, 3856
Verizon Foundation Virginia Grants, 3864
Victor E. Speas Foundation Grants, 3868
Wabash Valley Community Foundation Grants, 3893
Walker Area Community Foundation Grants, 3895
Washington County Community Fndn Grants, 3906
Washington County Community Foundation Youth Grants, 3907
Weingart Foundation Grants, 3913
Weyerhaeuser Company Foundation Grants, 3925
Willard & Pat Walker Charitable Fndn Grants, 3943
Williams Companies Foundation Homegrown Giving Grants, 3957
Winston-Salem Foundation Competitive Grants, 3965
Z. Smith Reynolds Foundation Small Grants, 3982

Materials/Equipment Acquisition (Computers, Books, Videos, etc.)
1st Touch Foundation Grants, 3
AACN-Edwards Lifesciences Nurse-Driven Clinical Practice Outcomes Grants, 77
AACN Clinical Inquiry Fund Grants, 80
AACN End of Life/Palliative Care Small Projects Grants, 83
AACN Physio-Control Small Projects Grants, 86
Abbot and Dorothy H. Stevens Foundation Grants, 197
Abdus Salam ICTP Federation Arrangement Scheme Grants, 205
Abell-Hanger Foundation Grants, 207
Abington Foundation Grants, 211
Able To Serve Grants, 212
Able Trust Vocational Rehabilitation Grants for Individuals, 213
ACGT Investigators Grants, 232
Achelis Foundation Grants, 234
ACSM Foundation Doctoral Research Grants, 266
Adams County Community Foundation of Pennsylvania Grants, 300
Adelaide Breed Bayrd Foundation Grants, 306
Administaff Community Affairs Grants, 308
Advance Auto Parts Corporate Giving Grants, 312
AEC Trust Grants, 315
Agnes M. Lindsay Trust Grants, 346
Ahmanson Foundation Grants, 355
AIDS Vaccine Advocacy Coalition (AVAC) Fund Grants, 369
AIHS Alberta/Pfizer Translat Research Grants, 374
Air Products and Chemicals Grants, 399
Albert W. Rice Charitable Foundation Grants, 411
Alexander and Margaret Stewart Trust Grants, 419
ALFJ Astraea U.S. and Int'l Movement Fund, 427
Alfred P. Sloan Foundation International Science Engagement Grants, 431
Alice Tweed Tuohy Foundation Grants Program, 437
Allegan County Community Foundation Grants, 439
Allegheny Technologies Charitable Trust, 440
Allen P. & Josephine B. Green Fndn Grants, 443
Allyn Foundation Grants, 448
AMA Foundation Worldscopes Program, 495
Amelia Sillman Rockwell and Carlos Perry Rockwell Charities Fund Grants, 501

PROGRAM TYPE INDEX — Materials/Equipment Acquisition (Computers, Books, Videos, etc.)

American Academy of Dermatology Shade Structure Grants, 504
American Schlafhorst Foundation Grants, 559
Amgen Foundation Grants, 572
Anheuser-Busch Foundation Grants, 591
Ann Arbor Area Community Foundation Grants, 594
Annunziata Sanguinetti Foundation Grants, 600
Antone & Edene Vidinha Charitable Trust Grants, 609
Arcadia Foundation Grants, 650
ARCO Foundation Education Grants, 652
Arie and Ida Crown Memorial Grants, 654
Arkell Hall Foundation Grants, 661
Armstrong McDonald Foundation Health Grants, 663
ASHFoundation Research Grant in Speech, 709
Assisi Fndn of Memphis Capital Project Grants, 786
Assisi Foundation of Memphis General Grants, 787
AT&T Foundation Civic and Community Service Program Grants, 793
Atlanta Foundation Grants, 796
Auburn Foundation Grants, 798
Aurora Foundation Grants, 802
Autauga Area Community Foundation Grants, 808
Avista Foundation Economic and Cultural Vitality Grants, 812
Bailey Foundation Grants, 818
Barker Welfare Foundation Grants, 831
Barnes Group Foundation Grants, 832
Battle Creek Community Foundation Grants, 839
BCBSM Foundation Primary and Clinical Prevention Grants, 856
BCBSM Fndn Proposal Development Awards, 857
BCBSNC Foundation Grants, 860
Bearemy's Kennel Pals Grants, 863
Beazley Foundation Grants, 865
Beckley Area Foundation Grants, 867
Beckman Coulter Foundation Grants, 868
Ben B. Cheney Foundation Grants, 875
Benton Community Foundation Grants, 878
Biogen Foundation General Donations, 899
Blanche and Irving Laurie Foundation Grants, 910
Blanche and Julian Robertson Family Foundation Grants, 911
Blowitz-Ridgeway Foundation Early Childhood Development Research Award, 912
Blue Cross Blue Shield of Minnesota Fndn Healthy Equity: Public Libraries for Health Grants, 918
Blue Grass Community Foundation Hudson-Ellis Fund Grants, 920
Blue River Community Foundation Grants, 922
Blumenthal Foundation Grants, 925
Bodenwein Public Benevolent Foundation Grants, 928
Bodman Foundation Grants, 929
Boeing Company Contributions Grants, 930
Boettcher Foundation Grants, 931
Booth-Bricker Fund Grants, 933
Borkee-Hagley Foundation Grants, 934
Bosque Foundation Grants, 935
Boston Globe Foundation Grants, 937
Brighter Tomorrow Foundation Grants, 950
Bristol-Myers Squibb Foundation Product Donations Grants, 960
Brown County Community Foundation Grants, 970
Byron W. & Alice L. Lockwood Fnd Grants, 991
Cabot Corporation Foundation Grants, 993
Caesars Foundation Grants, 995
Cailloux Foundation Grants, 996
California Endowment Innovative Ideas Challenge Grants, 1000
Callaway Foundation Grants, 1002
Callaway Golf Company Foundation Grants, 1003
Cambridge Community Foundation Grants, 1004
Campbell Hoffman Foundation Grants, 1006
Campbell Soup Foundation Grants, 1007
Capital Region Community Foundation Grants, 1014
Cargill Citizenship Fund Corporate Giving, 1018
Carl Gellert and Celia Berta Gellert Foundation Grants, 1024
Carl M. Freeman Foundation FACES Grants, 1026
Carl M. Freeman Foundation Grants, 1027
Carls Foundation Grants, 1030

Carl W. and Carrie Mae Joslyn Trust Grants, 1031
Carnahan-Jackson Foundation Grants, 1032
Carpenter Foundation Grants, 1035
Carrie E. and Lena V. Glenn Foundation Grants, 1036
Carrie Estelle Doheny Foundation Grants, 1037
Cemala Foundation Grants, 1077
Centerville-Washington Foundation Grants, 1078
Central Carolina Community Foundation Community Impact Grants, 1079
Central Okanagan Foundation Grants, 1080
Cessna Foundation Grants Program, 1081
CFFVR Alcoholism and Drug Abuse Grants, 1093
CFFVR Basic Needs Giving Partnership Grants, 1094
CFFVR Capital Credit Union Charitable Giving Grants, 1095
CFFVR Clintonville Area Foundation Grants, 1097
CFFVR Shawano Area Community Foundation Grants, 1104
Chamberlain Foundation Grants, 1108
Champ-A Champion Fur Kids Grants, 1109
Champlin Foundations Grants, 1110
Charles A. Frueauff Foundation Grants, 1114
Charles H. Dater Foundation Grants, 1118
Charles H. Farnsworth Trust Grants, 1119
Charles H. Pearson Foundation Grants, 1121
Chatlos Foundation Grants Program, 1128
Children Affected by AIDS Foundation Family Assistance Emergency Fund Grants, 1146
Chiles Foundation Grants, 1147
Chilkat Valley Community Foundation Grants, 1148
Christopher & Dana Reeve Foundation Quality of Life Grants, 1152
Christy-Houston Foundation Grants, 1153
CICF Indianapolis Fndn Community Grants, 1156
Clara Blackford Smith and W. Aubrey Smith Charitable Foundation Grants, 1165
Clarence T.C. Ching Foundation Grants, 1167
Clark County Community Foundation Grants, 1175
Claude Worthington Benedum Fndn Grants, 1178
CNA Foundation Grants, 1189
CNCS Senior Companion Program Grants, 1191
Coeta and Donald Barker Foundation Grants, 1205
Collins Foundation Grants, 1210
Comerica Charitable Foundation Grants, 1224
Communities Foundation of Texas Grants, 1236
Community Foundation for Greater Atlanta Metropolitan Atlanta An Extra Wish Grants, 1243
Community Fndn for Greater Buffalo Grants, 1247
Community Foundation for Greater New Haven Women & Girls Grants, 1252
Community Fndn for Monterey County Grants, 1253
Community Fndn for Muskegon County Grants, 1254
Community Foundation for Northeast Michigan Mini-Grants, 1255
Community Foundation of Bartholomew County Heritage Fund Grants, 1262
Community Foundation of Bloomington and Monroe County Grants, 1265
Community Foundation of Boone County Grants, 1266
Community Fndn of Central Illinois Grants, 1267
Community Fndn of East Central Illinois Grants, 1268
Community Foundation of Eastern Connecticut General Southeast Grants, 1269
Community Foundation of Greater Birmingham Grants, 1272
Community Foundation of Greater Fort Wayne - Community Endowment and Clarke Endowment Grants, 1274
Community Fndn of Greater Tampa Grants, 1282
Community Foundation of Greenville-Greenville Women Giving Grants, 1283
Community Foundation of Mount Vernon and Knox County Grants, 1299
Community Foundation of Muncie and Delaware County Grants, 1300
Community Foundation of Riverside & San Bernardino County Impact Grants, 1303
Community Fndn of Riverside & San Bernardino County Irene S. Rockwell Grants, 1304

Community Foundation of Riverside and San Bernardino County James Bernard and Mildred Jordan Tucker Grants, 1305
Community Fndn of South Puget Sound Grants, 1307
Community Foundation of the Verdugos Educational Endowment Fund Grants, 1314
Community Foundation of the Verdugos Grants, 1315
Community Fndn of Wabash County Grants, 1316
ConAgra Foods Fndn Community Impact Grants, 1321
ConAgra Foods Nourish our Community Grants, 1322
Connecticut Community Foundation Grants, 1324
Connelly Foundation Grants, 1326
ConocoPhillips Foundation Grants, 1327
Constantin Foundation Grants, 1341
Constellation Energy Corporate Grants, 1342
Consumers Energy Foundation, 1343
Coors Brewing Corporate Contributions Grants, 1346
Cornerstone Foundation of Northeastern Wisconsin Grants, 1347
Cowles Charitable Trust Grants, 1354
Crail-Johnson Foundation Grants, 1355
Cralle Foundation Grants, 1356
Cresap Family Foundation Grants, 1359
CSRA Community Foundation Grants, 1366
Cudd Foundation Grants, 1370
Cullen Foundation Grants, 1372
Cystic Fibrosis Canada Research Grants, 1385
Cystic Fibrosis Canada Senior Scientist Research Training Awards, 1387
D.V. and Ida J. McEachern Trust Grants, 1406
Dade Community Foundation Community AIDS Partnership Grants, 1409
Daimler and Benz Foundation Fellowships, 1411
DaimlerChrysler Corporation Fund Grants, 1412
Daisy Marquis Jones Foundation Grants, 1414
Dallas Women's Foundation Grants, 1417
Daphne Seybolt Culpeper Fndn Grants, 1425
Daywood Foundation Grants, 1438
Deaconess Community Foundation Grants, 1439
Deborah Munroe Noonan Memorial Grants, 1443
Delaware Community Foundation Grants, 1449
DeMatteis Family Foundation Grants, 1457
Denver Foundation Community Grants, 1461
Deutsche Banc Alex Brown and Sons Charitable Foundation Grants, 1471
Different Needz Foundation Grants, 1478
DOJ Gang-Free Schools and Communities Intervention Grants, 1481
Dolan Children's Foundation Grants, 1482
Doree Taylor Charitable Foundation, 1490
Doris and Victor Day Foundation Grants, 1491
Dorothea Haus Ross Foundation Grants, 1498
Dr. Leon Bromberg Charitable Trust Grants, 1509
Drs. Bruce and Lee Foundation Grants, 1513
Dubois County Community Foundation Grants, 1517
DuPont Pioneer Community Giving Grants, 1532
E.J. Grassmann Trust Grants, 1537
E.L. Wiegand Foundation Grants, 1538
E. Rhodes & Leona B. Carpenter Grants, 1539
Eastman Chemical Company Foundation Grants, 1542
EDS Foundation Grants, 1552
Edward N. and Della L. Thome Memorial Foundation Grants, 1557
Edwards Memorial Trust Grants, 1558
Edward W. and Stella C. Van Houten Memorial Fund Grants, 1559
Edwin S. Webster Foundation Grants, 1560
Edyth Bush Charitable Foundation Grants, 1562
Effie and Wofford Cain Foundation Grants, 1563
Eisner Foundation Grants, 1564
Ellison Medical Foundation/AFAR Postdoctoral Fellows in Aging Research Program, 1577
El Paso Community Foundation Grants, 1579
El Pomar Foundation Anna Keesling Ackerman Fund Grants, 1581
El Pomar Foundation Grants, 1582
Elsie H. Wilcox Foundation Grants, 1583
Erie Community Foundation Grants, 1601
Eugene B. Casey Foundation Grants, 1609
Eva L. & Joseph M. Bruening Fndn Grants, 1613

Materials/Equipment Acquisition (Computers, Books, Videos, etc.)

Everyone Breathe Asthma Education Grants, 1616
Evjue Foundation Grants, 1617
Ewing Halsell Foundation Grants, 1618
F.M. Kirby Foundation Grants, 1624
Fairfield County Community Foundation Grants, 1625
Fannie E. Rippel Foundation Grants, 1630
Faye McBeath Foundation Grants, 1635
Field Foundation of Illinois Grants, 1660
Fight for Sight Grants-in-Aid, 1663
FirstEnergy Foundation Community Grants, 1668
Flextronics Foundation Disaster Relief Grants, 1674
FMC Foundation Grants, 1683
Foellinger Foundation Grants, 1684
Ford Family Foundation Grants - Access to Health and Dental Services, 1686
Ford Motor Company Fund Grants Program, 1687
Foundation for Seacoast Health Grants, 1698
Foundation for the Mid South Health and Wellness Grants, 1700
Four County Community Fndn General Grants, 1710
Four County Community Foundation Healthy Senior/Healthy Youth Fund Grants, 1711
France-Merrick Foundation Health and Human Services Grants, 1715
Francis L. Abreu Charitable Trust Grants, 1719
Frank E. and Seba B. Payne Foundation Grants, 1722
Frank G. and Freida K. Brotz Family Foundation Grants, 1723
Frank Reed and Margaret Jane Peters Memorial Fund I Grants, 1727
Frank Reed and Margaret Jane Peters Memorial Fund II Grants, 1728
Frank Stanley Beveridge Foundation Grants, 1730
Fred C. & Katherine B. Andersen Grants, 1737
Frederick McDonald Trust Grants, 1740
Fremont Area Community Foundation Amazing X Grants, 1744
Fremont Area Community Foundation Elderly Needs Grants, 1745
Fremont Area Community Fndn General Grants, 1746
Fritz B. Burns Foundation Grants, 1747
Fuller E. Callaway Foundation Grants, 1761
G.N. Wilcox Trust Grants, 1768
Gamble Foundation Grants, 1769
Gardner Foundation Grants, 1770
Gebbie Foundation Grants, 1777
General Motors Foundation Grants, 1783
Genesis Foundation Grants, 1786
George A Ohl Jr. Foundation Grants, 1791
George Foundation Grants, 1795
George Kress Foundation Grants, 1799
George W. Codrington Charitable Fndn Grants, 1803
George W. Wells Foundation Grants, 1804
Gil and Dody Weaver Foundation Grants, 1819
Giving in Action Society Children & Youth with Special Needs Grants, 1823
Giving in Action Society Family Indep Grants, 1824
Gladys Brooks Foundation Grants, 1826
GNOF Albert N. & Hattie M. McClure Grants, 1834
Grace and Franklin Bernsen Foundation Grants, 1858
Graco Foundation Grants, 1860
Graham Foundation Grants, 1862
Grand Rapids Area Community Foundation Wyoming Grants, 1867
Grand Rapids Area Community Foundation Wyoming Youth Fund Grants, 1868
Grand Rapids Community Foundation Grants, 1869
Grand Rapids Community Foundation Ionia County Grants, 1870
Grand Rapids Community Foundation Lowell Area Fund Grants, 1872
Grand Rapids Community Foundation Southeast Ottawa Grants, 1873
Grand Rapids Community Foundation Southeast Ottawa Youth Fund Grants, 1874
Grand Rapids Community Fndn Sparta Grants, 1875
Grand Rapids Community Foundation Sparta Youth Fund Grants, 1876
Greater Saint Louis Community Fndn Grants, 1884
Greater Sitka Legacy Fund Grants, 1885
Greater Tacoma Community Foundation Grants, 1886
Greater Worcester Community Foundation Jeppson Memorial Fund for Brookfield Grants, 1889
Green Bay Packers Foundation Grants, 1890
Green River Area Community Fndn Grants, 1895
Greygates Foundation Grants, 1901
Grover Hermann Foundation Grants, 1906
Grundy Foundation Grants, 1912
Guitar Center Music Foundation Grants, 1914
H & R Foundation Grants, 1920
H.A. & Mary K. Chapman Trust Grants, 1921
Hackett Foundation Grants, 1926
HAF Phyllis Nilsen Leal Memorial Fund Gifts, 1930
Hamilton Company Syringe Product Grant, 1937
Hampton Roads Community Foundation Faith Community Nursing Grants, 1940
Harold and Arlene Schnitzer CARE Foundation Grants, 1950
Harris and Eliza Kempner Fund Grants, 1956
Harrison County Community Foundation Grants, 1957
Harrison County Community Foundation Signature Grants, 1958
Harry W. Bass, Jr. Foundation Grants, 1968
Hartford Foundation Regular Grants, 1970
Harvest Foundation Grants, 1971
Hawaii Community Foundation Health Education and Research Grants, 1976
HCA Foundation Grants, 1980
Health Fndn of Greater Indianapolis Grants, 1984
Hearst Foundations Health Grants, 1987
Heckscher Foundation for Children Grants, 1988
Helen Bader Foundation Grants, 1992
Helen Irwin Littauer Educational Trust Grants, 1994
Helen Pumphrey Denit Charitable Trust Grants, 1996
Helen S. Boylan Foundation Grants, 1997
Henrietta Tower Wurts Memorial Fndn Grants, 2002
Henry L. Guenther Foundation Grants, 2006
Highmark Physician eHealth Grants, 2026
High Meadow Foundation Grants, 2027
Hillcrest Foundation Grants, 2031
Hill Crest Foundation Grants, 2030
Hillman Foundation Grants, 2032
Hilton Hotels Corporate Giving Program Grants, 2037
Homer Foundation Grants, 2044
Horace Moses Charitable Foundation Grants, 2047
Hormel Foundation Grants, 2051
Howard and Bush Foundation Grants, 2054
HRAMF Jeffress Trust Awards in Interdisciplinary Research, 2060
HRK Foundation Health Grants, 2067
Hudson Webber Foundation Grants, 2072
Huie-Dellmon Trust Grants, 2075
Huntington Clinical Foundation Grants, 2082
I.A. O'Shaughnessy Foundation Grants, 2088
Idaho Power Company Corporate Contributions, 2115
IDPH Hosptial Capital Investment Grants, 2119
IDPH Local Health Department Public Health Emergency Response Grants, 2120
Independence Blue Cross Charitable Medical Care Grants, 2150
Irvin Stern Foundation Grants, 2164
J.B. Reynolds Foundation Grants, 2168
J.M. Long Foundation Grants, 2173
J.W. Kieckhefer Foundation Grants, 2176
Jack H. & William M. Light Trust Grants, 2179
Jacob G. Schmidlapp Trust Grants, 2183
James & Abigail Campbell Family Fndn Grants, 2185
James Graham Brown Foundation Grants, 2189
James Hervey Johnson Charitable Educational Trust Grants, 2191
James L. and Mary Jane Bowman Charitable Trust Grants, 2194
James R. Thorpe Foundation Grants, 2199
James S. Copley Foundation Grants, 2200
Janirve Foundation Grants, 2209
Janson Foundation Grants, 2210
Jay and Rose Phillips Family Foundation Grants, 2213
JELD-WEN Foundation Grants, 2218
Jenkins Foundation: Improving the Health of Greater Richmond Grants, 2219
Jennings County Community Foundation Grants, 2220
Jessie Ball Dupont Fund Grants, 2226
John Ben Snow Memorial Trust Grants, 2231
John D. & Katherine A. Johnston Fndn Grants, 2233
John G. Duncan Charitable Trust Grants, 2236
John H. and Wilhelmina D. Harland Charitable Foundation Children and Youth Grants, 2238
John J. Leidy Foundation Grants, 2241
John Jewett and Helen Chandler Garland Foundation Grants, 2242
John P. Murphy Foundation Grants, 2248
John S. Dunn Research Foundation Grants, 2250
John W. and Anna H. Hanes Foundation Grants, 2261
John W. Speas and Effie E. Speas Memorial Trust Grants, 2264
Joseph Alexander Foundation Grants, 2265
Joseph H. & Florence A. Roblee Fndn Grants, 2269
Joseph Henry Edmondson Foundation Grants, 2270
Josephine Goodyear Foundation Grants, 2272
Josephine Schell Russell Charitable Trust Grants, 2274
Josiah W. and Bessie H. Kline Foundation Grants, 2278
Katharine Matthies Foundation Grants, 2291
Katherine John Murphy Foundation Grants, 2293
Kelvin and Eleanor Smith Foundation Grants, 2297
Kenai Peninsula Foundation Grants, 2298
Kenneth T. & Eileen L. Norris Fndn Grants, 2300
Ketchikan Community Foundation Grants, 2303
Kinsman Foundation Grants, 2309
Kodiak Community Foundation Grants, 2314
Kuntz Foundation Grants, 2326
Land O'Lakes Foundation Mid-Atlantic Grants, 2341
Lewis H. Humphreys Charitable Trust Grants, 2361
Linford and Mildred White Charitable Grants, 2368
Lotus 88 Fnd for Women & Children Grants, 2376
Louie M. & Betty M. Phillips Fndn Grants, 2378
Louis and Elizabeth Nave Flarsheim Charitable Foundation Grants, 2379
Lubbock Area Foundation Grants, 2383
Lubrizol Foundation Grants, 2384
Lucy Downing Nisbet Charitable Fund Grants, 2387
Lucy Gooding Charitable Fndn Grants, 2388
Ludwick Family Foundation Grants, 2389
Lumpkin Family Fnd Healthy People Grants, 2391
Lydia deForest Charitable Trust Grants, 2393
Lymphatic Education and Research Network Additional Support Grants for NIH-funded F32 Postdoctoral Fellows, 2394
Lynde and Harry Bradley Foundation Grants, 2397
M.D. Anderson Foundation Grants, 2403
M.J. Murdock Charitable Trust General Grants, 2405
Mabel A. Horne Trust Grants, 2406
Mabel Y. Hughes Charitable Trust Grants, 2409
Macquarie Bank Foundation Grants, 2410
Maddie's Fund Medical Equipment Grants, 2411
Marcia and Otto Koehler Foundation Grants, 2427
Mardag Foundation Grants, 2428
Marie C. and Joseph C. Wilson Foundation Rochester Small Grants, 2431
Marion Gardner Jackson Charitable Trust Grants, 2434
Marion I. and Henry J. Knott Foundation Discretionary Grants, 2435
Marion I. and Henry J. Knott Foundation Standard Grants, 2436
Mary Black Foundation Active Living Grants, 2440
Mary K. Chapman Foundation Grants, 2443
Mary Owen Borden Foundation Grants, 2446
Mary S. and David C. Corbin Foundation Grants, 2447
Matilda R. Wilson Fund Grants, 2453
Maurice J. Masserini Charitable Trust Grants, 2456
Maxon Charitable Foundation Grants, 2458
May and Stanley Smith Charitable Trust Grants, 2459
MBL Fred Karush Library Readership, 2474
McConnell Foundation Grants, 2492
McInerny Foundation Grants, 2495
McKesson Foundation Grants, 2496
McLean Contributionship Grants, 2497
Mead Witter Foundation Grants, 2507
Memorial Foundation Grants, 2516
Mericos Foundation Grants, 2518
Meriden Foundation Grants, 2519

PROGRAM TYPE INDEX   Preservation/Restoration   / 897

Merrick Foundation Grants, 2522
MetroWest Health Foundation Capital Grants for Health-Related Facilities, 2526
Meyer Memorial Trust Grassroots Grants, 2536
Meyer Memorial Trust Responsive Grants, 2537
Meyer Memorial Trust Special Grants, 2538
MGN Family Foundation Grants, 2542
Milagro Foundation Grants, 2552
Mimi and Peter Haas Fund Grants, 2557
Minneapolis Foundation Community Grants, 2558
Mockingbird Foundation Grants, 2589
Montgomery County Community Fndn Grants, 2591
Morris & Gwendolyn Cafritz Fndn Grants, 2594
Mt. Sinai Health Care Foundation Academic Medicine and Bioscience Grants, 2598
Multiple Sclerosis Foundation Brighter Tomorrow Grants, 2602
Natalie W. Furniss Charitable Trust Grants, 2626
NCI Technologies and Software to Support Integrative Cancer Biology Research (SBIR), 2656
Nell J. Redfield Foundation Grants, 2673
NIGMS Advancing Basic Behavioral and Social Research on Resilience: an Integrative Science Approach, 2861
NLM Internet Connection for Medical Institutions Grants, 2946
Noble County Community Foundation Grants, 2975
Norcliffe Foundation Grants, 2979
Nordson Corporation Foundation Grants, 2981
Norfolk Southern Foundation Grants, 2982
North Central Health Services Grants, 2993
Northern Chautauqua Community Foundation Community Grants, 2995
Norwin S. & Elizabeth N. Bean Fndn Grants, 3001
NSF-NIST Interaction in Chemistry, Materials Research, Molecular Biosciences, Bioengineering, and Chemical Engineering, 3004
NSF Instrument Development for Bio Research, 3017
NYCT Blood Disease Research Grants, 3051
NYCT Technical Assistance Grants, 3058
OceanFirst Foundation Major Grants, 3062
Ohio County Community Foundation Board of Directors Grants, 3067
Oleonda Jameson Trust Grants, 3071
OneFamily Foundation Grants, 3075
OneSight Research Foundation Block Grants, 3076
Oppenstein Brothers Foundation Grants, 3078
Oregon Community Fndn Community Grants, 3082
Oscar Rennebohm Foundation Grants, 3085
Parkersburg Area Community Foundation Action Grants, 3116
Patrick and Anna M. Cudahy Fund Grants, 3120
Patron Saints Foundation Grants, 3121
Paul Ogle Foundation Grants, 3124
Perpetual Trust for Charitable Giving Grants, 3141
Peter and Elizabeth C. Tower Foundation Phase II Technology Initiative Grants, 3149
Peter and Elizabeth C. Tower Foundation Social and Emotional Preschool Curriculum Grants, 3151
Peter Kiewit Foundation General Grants, 3154
Peter Kiewit Foundation Small Grants, 3155
Petersburg Community Foundation Grants, 3156
Pew Charitable Trusts Children & Youth Grants, 3158
Pfizer Special Events Grants, 3186
Phelps County Community Foundation Grants, 3188
Pinellas County Grants, 3223
Piper Trust Healthcare & Med Research Grants, 3228
Piper Trust Older Adults Grants, 3229
Pittsburgh Foundation Community Fund Grants, 3230
Pokagon Fund Grants, 3242
Portland General Electric Foundation Grants, 3250
Posey Community Fndn Women's Fund Grants, 3251
Potts Memorial Foundation Grants, 3254
PPCF Community Grants, 3256
PPG Industries Foundation Grants, 3257
Presbyterian Health Foundation Bridge, Seed and Equipment Grants, 3259
Price Family Charitable Fund Grants, 3262
Priddy Foundation Capital Grants, 3263
Prince Charitable Trusts Chicago Grants, 3270

Prince Charitable Trusts DC Grants, 3271
Principal Financial Group Foundation Grants, 3273
Prudential Foundation Education Grants, 3276
Puerto Rico Community Foundation Grants, 3278
R.C. Baker Foundation Grants, 3292
Ralph M. Parsons Foundation Grants, 3304
Raskob Foundation for Catholic Activities Grants, 3306
Rasmuson Foundation Tier One Grants, 3307
Rasmuson Foundation Tier Two Grants, 3308
Rathmann Family Foundation Grants, 3309
Rayonier Foundation Grants, 3311
RCF General Community Grants, 3312
Reinberger Foundation Grants, 3357
Rhode Island Foundation Grants, 3362
Richard & Susan Smith Family Fndn Grants, 3365
Ripley County Community Foundation Grants, 3376
Ripley County Community Foundation Small Project Grants, 3377
Robert R. Meyer Foundation Grants, 3387
Robert W. Woodruff Foundation Grants, 3389
Robins Foundation Grants, 3390
Rochester Area Community Foundation Grants, 3391
Rochester Area Foundation Grants, 3392
Ronald McDonald House Charities Grants, 3405
Roy & Christine Sturgis Charitable Grants, 3415
Rush County Community Foundation Grants, 3424
Ruth Eleanor Bamberger and John Ernest Bamberger Memorial Foundation Grants, 3427
Ruth H. & Warren A. Ellsworth Fndn Grants, 3429
Saint Luke's Health Initiatives Grants, 3455
Samuel S. Johnson Foundation Grants, 3468
San Antonio Area Foundation Grants, 3469
San Juan Island Community Foundation Grants, 3480
Santa Barbara Foundation Strategy Grants - Core Support, 3481
Santa Fe Community Foundation Seasonal Grants-Fall Cycle, 3482
Sarkeys Foundation Grants, 3485
Schering-Plough Foundation Health Grants, 3492
Schlessman Family Foundation Grants, 3493
Schramm Foundation Grants, 3495
Seabury Foundation Grants, 3501
Seattle Foundation Medical Funds Grants, 3508
Sensient Technologies Foundation Grants, 3513
Shell Deer Park Grants, 3543
Sheltering Arms Fund Grants, 3545
Shield-Ayres Foundation Grants, 3546
Shopko Foundation Green Bay Area Community Grants, 3548
Sidney Stern Memorial Trust Grants, 3552
Sid W. Richardson Foundation Grants, 3553
Siebert Lutheran Foundation Grants, 3554
Sierra Health Foundation Responsive Grants, 3555
Simmons Foundation Grants, 3563
Sioux Falls Area Community Foundation Community Fund Grants (Unrestricted), 3568
Sioux Falls Area Community Foundation Spot Grants (Unrestricted), 3570
Sisters of Mercy of North Carolina Fndn Grants, 3575
Sisters of St. Joseph Healthcare Fndn Grants, 3576
SOCFOC Catholic Ministries Grants, 3596
Solo Cup Foundation Grants, 3620
Sonora Area Foundation Competitive Grants, 3622
Sony Corporation of America Grants, 3623
Square D Foundation Grants, 3634
St. Joseph Community Health Foundation Burn Care and Prevention Grants, 3639
St. Joseph Community Health Foundation Improving Healthcare Access Grants, 3640
St. Joseph Community Health Foundation Pfeiffer Fund Grants, 3641
Stackpole-Hall Foundation Grants, 3643
Steele-Reese Foundation Grants, 3650
Sterling-Turner Charitable Foundation Grants, 3652
Stewart Huston Charitable Trust Grants, 3655
Stocker Foundation Grants, 3656
Strake Foundation Grants, 3658
Strowd Roses Grants, 3664
Subaru of Indiana Automotive Foundation Grants, 3672
Summit Foundation Grants, 3674

Sunderland Foundation Grants, 3675
Sunoco Foundation Grants, 3679
Support Our Aging Religious (SOAR) Grants, 3686
Susan Mott Webb Charitable Trust Grants, 3694
T.L.L. Temple Foundation Grants, 3700
Taubman Endowment for the Arts Fndn Grants, 3702
TE Foundation Grants, 3705
Textron Corporate Contributions Grants, 3712
Theodore Edson Parker Foundation Grants, 3715
The Ray Charles Foundation Grants, 3716
Thomas Austin Finch, Sr. Foundation Grants, 3718
Thomas J. Long Foundation Community Grants, 3722
Thomas Sill Foundation Grants, 3724
Thomas Thompson Trust Grants, 3725
TJX Foundation Grants, 3735
Toyota Motor Engineering & Manufacturing North America Grants, 3742
Toyota Motor Manufacturing of Alabama Grants, 3743
Toyota Motor Manuf of West Virginia Grants, 3748
Toyota Motor Sales, USA Grants, 3750
Tri-State Community Twenty-first Century Endowment Fund Grants, 3756
Trull Foundation Grants, 3761
U.S. Department of Education Rehabilitation Engineering Research Centers Grants, 3769
U.S. Lacrosse EAD Grants, 3775
Union Bank, N.A. Foundation Grants, 3788
United Methodist Committee on Relief Global Health Grants, 3799
United Technologies Corporation Grants, 3803
US Airways Community Foundation Grants, 3826
Vancouver Foundation Grants and Community Initiatives Program, 3849
VDH Rescue Squad Assistance Fund Grants, 3856
Vectren Foundation Grants, 3857
Verizon Foundation Health Care and Accessibility Grants, 3858
Victor E. Speas Foundation Grants, 3868
Vigneron Memorial Fund Grants, 3873
Visiting Nurse Foundation Grants, 3879
W.M. Keck Foundation Med Research Grants, 3888
W.M. Keck Foundation Science and Engineering Research Grants, 3889
W.M. Keck Fndn So California Grants, 3890
Waitt Family Foundation Grants, 3894
Walker Area Community Foundation Grants, 3895
Walter L. Gross III Family Foundation Grants, 3901
Wayne County Foundation Grants, 3909
Wege Foundation Grants, 3912
Weingart Foundation Grants, 3913
Western Indiana Community Foundation Grants, 3921
Western New York Foundation Grants, 3922
Weyerhaeuser Company Foundation Grants, 3925
Willard & Pat Walker Charitable Fndn Grants, 3943
William G. & Helen C. Hoffman Fndn Grants, 3950
William G. Gilmore Foundation Grants, 3951
Wilson-Wood Foundation Grants, 3960
Wood Family Charitable Trust Grants, 3972
Woodward Fund Grants, 3973
Z. Smith Reynolds Foundation Small Grants, 3982

Preservation/Restoration
Abell-Hanger Foundation Grants, 207
Albuquerque Community Foundation Grants, 412
Alpha Natural Resources Corporate Giving, 450
Anne J. Caudal Foundation Grants, 595
Annenberg Foundation Grants, 596
Anschutz Family Foundation Grants, 602
APA Distinguished Service to Psychological Science Award, 630
Ashland Corporate Contributions Grants, 713
Avista Foundation Economic and Cultural Vitality Grants, 812
Bank of America Fndn Matching Gifts, 825
Barker Welfare Foundation Grants, 831
Batchelor Foundation Grants, 837
Beckman Coulter Foundation Grants, 868
Blue Mountain Community Foundation Grants, 921
Blue River Community Foundation Grants, 922
Booth-Bricker Fund Grants, 933

898 / Preservation/Restoration

Callaway Foundation Grants, 1002
Carl M. Freeman Foundation Grants, 1027
Cessna Foundation Grants Program, 1081
Champlin Foundations Grants, 1110
Charles Edison Fund Grants, 1116
Charlotte County (FL) Community Foundation Grants, 1126
Chula Vista Charitable Foundation Grants, 1154
Clarence E. Heller Charitable Foundation Grants, 1166
CLIF Bar Family Foundation Grants, 1183
Community Foundation Alliance City of Evansville Endowment Fund Grants, 1238
Community Fndn for Greater Buffalo Grants, 1247
Community Fndn for Monterey County Grants, 1253
Community Fndn of Greater Tampa Grants, 1282
Community Foundation Partnerships - Martin County Grants, 1318
ConocoPhillips Foundation Grants, 1327
Consumers Energy Foundation, 1343
Cudd Foundation Grants, 1370
D.F. Halton Foundation Grants, 1405
Daniel & Nanna Stern Family Fndn Grants, 1422
E. Rhodes & Leona B. Carpenter Grants, 1539
Earl and Maxine Claussen Trust Grants, 1541
Edwin S. Webster Foundation Grants, 1560
El Pomar Foundation Anna Keesling Ackerman Fund Grants, 1581
El Pomar Foundation Grants, 1582
Fargo-Moorhead Area Foundation Grants, 1632
Ferree Foundation Grants, 1658
Forrest C. Lattner Foundation Grants, 1688
Four County Community Fndn General Grants, 1710
George Kress Foundation Grants, 1799
GNOF Exxon-Mobil Grants, 1836
Grammy Foundation Grants, 1863
Greater Green Bay Community Fndn Grants, 1881
Greater Milwaukee Foundation Grants, 1883
Greater Tacoma Community Foundation Grants, 1886
Green Diamond Charitable Contributions, 1891
Green River Area Community Fndn Grants, 1895
Harold R. Bechtel Charitable Remainder Uni-Trust Grants, 1953
Harrison County Community Foundation Grants, 1957
Harrison County Community Foundation Signature Grants, 1958
High Meadow Foundation Grants, 2027
Hillcrest Foundation Grants, 2031
Hoblitzelle Foundation Grants, 2038
Holland/Zeeland Community Fndn Grants, 2041
Huntington County Community Foundation Make a Difference Grants, 2083
J.N. and Macie Edens Foundation Grants, 2174
J.W. Kieckhefer Foundation Grants, 2176
Jennings County Community Foundation Grants, 2220
Jessica Stevens Community Foundation Grants, 2224
John Ben Snow Memorial Trust Grants, 2231
John J. Leidy Foundation Grants, 2241
John W. and Anna H. Hanes Foundation Grants, 2261
Joseph Henry Edmondson Foundation Grants, 2270
Kinsman Foundation Grants, 2309
Knox County Community Foundation Grants, 2313
Lawrence S. Huntington Fund Grants, 2348
Lotus 88 Fnd for Women & Children Grants, 2376
M.E. Raker Foundation Grants, 2404
Marcia and Otto Koehler Foundation Grants, 2427
Mary Black Foundation Active Living Grants, 2440
McCombs Foundation Grants, 2491
McLean Contributionship Grants, 2497
Merrick Foundation Grants, 2522
Nina Mason Pulliam Charitable Trust Grants, 2921
Norcliffe Foundation Grants, 2979
Norfolk Southern Foundation Grants, 2982
North Carolina Community Foundation Grants, 2990
OceanFirst Foundation Major Grants, 3062
Ogden Codman Trust Grants, 3066
Oregon Community Fndn Community Grants, 3082
Parkersburg Area Community Foundation Action Grants, 3116
Perkins Charitable Foundation Grants, 3140
Pokagon Fund Grants, 3242

Posey County Community Foundation Grants, 3252
Prince Charitable Trusts Chicago Grants, 3270
Prince Charitable Trusts DC Grants, 3271
Putnam County Community Foundation Grants, 3283
Rasmuson Foundation Tier Two Grants, 3308
Rhode Island Foundation Grants, 3362
Ripley County Community Foundation Grants, 3376
Ripley County Community Foundation Small Project Grants, 3377
Robert R. Meyer Foundation Grants, 3387
Rollins-Luetkemeyer Foundation Grants, 3404
Salem Foundation Charitable Trust Grants, 3456
Schering-Plough Foundation Community Initiatives Grants, 3491
Seward Community Foundation Grants, 3514
Seward Community Foundation Mini-Grants, 3515
Shield-Ayres Foundation Grants, 3546
Sioux Falls Area Community Foundation Community Fund Grants (Unrestricted), 3568
Sioux Falls Area Community Foundation Spot Grants (Unrestricted), 3570
Skillman Fndn Good Neighborhoods Grants, 3577
Southwest Florida Community Foundation Competitive Grants, 3627
Spencer County Community Foundation Grants, 3632
Stewart Huston Charitable Trust Grants, 3655
Strake Foundation Grants, 3658
Taylor S. Abernathy and Patti Harding Abernathy Charitable Trust Grants, 3704
Textron Corporate Contributions Grants, 3712
Thomas Austin Finch, Sr. Foundation Grants, 3718
Toyota Motor Manuf of Mississippi Grants, 3746
Tri-State Community Twenty-first Century Endowment Fund Grants, 3756
Trull Foundation Grants, 3761
Turner Foundation Grants, 3763
Union Bank, N.A. Foundation Grants, 3788
Vanderburgh Community Fndn Women's Fund, 3853
Vermont Community Foundation Grants, 3865
Walker Area Community Foundation Grants, 3895
Washington County Community Fndn Grants, 3906
Washington County Community Foundation Youth Grants, 3907
Western Indiana Community Foundation Grants, 3921
Winston-Salem Fndn Stokes County Grants, 3967

Professorships
AIHS Knowledge Exchange Visiting Professors, 390
Conquer Cancer Foundation of ASCO Drug Development Research Professorship, 1330
Roy & Christine Sturgis Charitable Grants, 3415
Streilein Foundation for Ocular Immunology Visiting Professorships, 3660

Publishing/Editing/Translating
AACN-Edwards Lifesciences Nurse-Driven Clinical Practice Outcomes Grants, 77
AACN Clinical Inquiry Fund Grants, 80
AACN End of Life/Palliative Care Small Projects Grants, 83
AACN Physio-Control Small Projects Grants, 86
AEC Trust Grants, 315
Ahn Family Foundation Grants, 356
Albuquerque Community Foundation Grants, 412
AMA Virtual Mentor Theme Issue Editor Grants, 496
Ann Arbor Area Community Foundation Grants, 594
ASHA Advancing Academic-Research Award, 695
ASHA Student Research Travel Award, 699
Atran Foundation Grants, 797
Ball Brothers Foundation General Grants, 820
Battle Creek Community Foundation Grants, 839
Bodenwein Public Benevolent Foundation Grants, 928
Carl Gellert and Celia Berta Gellert Foundation Grants, 1024
Chatlos Foundation Grants Program, 1128
Christine Gregoire Youth/Young Adult Award for Use of Tobacco Industry Documents, 1151
CJ Foundation for SIDS Program Services Grants, 1163
Clarence E. Heller Charitable Foundation Grants, 1166

Coastal Community Foundation of South Carolina Grants, 1202
Columbus Foundation Competitive Grants, 1217
Community Fndn for Muskegon County Grants, 1254
Community Fndn of East Central Illinois Grants, 1268
Community Foundation of Eastern Connecticut General Southeast Grants, 1269
Community Foundation of Greater Birmingham Grants, 1272
Community Foundation of the Verdugos Educational Endowment Fund Grants, 1314
Connecticut Community Foundation Grants, 1324
Conservation, Food, and Health Foundation Grants for Developing Countries, 1338
Dyson Foundation Mid-Hudson Valley Project Support Grants, 1534
Elmer L. & Eleanor J. Andersen Fndn Grants, 1578
Ewing Halsell Foundation Grants, 1618
Frank Stanley Beveridge Foundation Grants, 1730
General Motors Foundation Grants, 1783
George A Ohl Jr. Foundation Grants, 1791
George Gund Foundation Grants, 1796
HHMI International Research Scholars Program, 2021
Hill Crest Foundation Grants, 2030
Huber Foundation Grants, 2071
J.B. Reynolds Foundation Grants, 2168
J.W. Kieckhefer Foundation Grants, 2176
Jessie Ball Dupont Fund Grants, 2226
JM Foundation Grants, 2229
John P. McGovern Foundation Grants, 2247
Josephine G. Russell Trust Grants, 2271
Kinsman Foundation Grants, 2309
Lubbock Area Foundation Grants, 2383
Lynde and Harry Bradley Foundation Grants, 2397
McLean Contributionship Grants, 2497
MLA Murray Gottlieb Prize Essay Award, 2571
National Headache Foundation Seymour Diamond Clinical Fellowship in Headache Education, 2633
NIHCM Foundation Health Care Print Journalism Awards, 2870
NLM Grants for Scholarly Works in Biomedicine and Health, 2944
Norcliffe Foundation Grants, 2979
Ohio County Community Foundation Board of Directors Grants, 3067
Paul Rapoport Foundation Grants, 3125
Phelps County Community Foundation Grants, 3188
Playboy Foundation Grants, 3237
Potts Memorial Foundation Grants, 3254
Puerto Rico Community Foundation Grants, 3278
Reinberger Foundation Grants, 3357
Rochester Area Community Foundation Grants, 3391
RSC Jason A. Hannah Medal, 3420
SfN Science Educator Award, 3535
Shell Oil Company Foundation Community Development Grants, 3544
Sid W. Richardson Foundation Grants, 3553
Sigma Theta Tau International / Council for the Advancement of Nursing Science Grants, 3557
South Madison Community Foundation Grants, 3626
Susan Mott Webb Charitable Trust Grants, 3694
Textron Corporate Contributions Grants, 3712
Thoracic Foundation Grants, 3729
United Hospital Fund of New York Health Care Improvement Grants, 3797
United States Institute of Peace - Jennings Randolph Senior Fellowships, 3802
Victor Grifols i Lucas Foundation Prize for Journalistic Work on Bioethics, 3870
Wayne County Foundation Grants, 3909
Weyerhaeuser Company Foundation Grants, 3925

Religious Programs
100 Mile Man Foundation Grants, 10
A/H Foundation Grants, 18
Aaron & Cecile Goldman Family Fndn Grants, 191
Aaron & Freda Glickman Foundation Grants, 192
Aaron Foundation Grants, 193
Abel Foundation Grants, 206
Abell-Hanger Foundation Grants, 207

PROGRAM TYPE INDEX

Religious Programs / 899

Able To Serve Grants, 212
Aboudane Family Foundation Grants, 215
Abramson Family Foundation Grants, 217
Achelis Foundation Grants, 234
Adaptec Foundation Grants, 304
Aetna Foundation Regional Health Grants, 333
A Friends' Foundation Trust Grants, 341
Ahmanson Foundation Grants, 355
Alabama Power Foundation Grants, 400
Albert and Margaret Alkek Foundation Grants, 405
Albertson's Charitable Giving Grants, 410
Albuquerque Community Foundation Grants, 412
Alcoa Foundation Grants, 414
Alfred and Tillie Shemanski Testamentary Trust Grants, 428
Allyn Foundation Grants, 448
ALSAM Foundation Grants, 454
Alvin and Fanny Blaustein Thalheimer Foundation Baltimore Communal Grants, 459
American Foodservice Charitable Trust Grants, 532
American Jewish World Service Grants, 533
Amerigroup Foundation Grants, 565
Andre Agassi Charitable Foundation Grants, 587
Anheuser-Busch Foundation Grants, 591
Ann Arbor Area Community Foundation Grants, 594
Anschutz Family Foundation Grants, 602
Ansell, Zaro, Grimm & Aaron Foundation Grants, 603
Anthony R. Abraham Foundation Grants, 608
Antone & Edene Vidinha Charitable Trust Grants, 609
Aragona Family Foundation Grants, 648
Aratani Foundation Grants, 649
Arcadia Foundation Grants, 650
Archer Daniels Midland Foundation Grants, 651
Argyros Foundation Grants Program, 653
Arie and Ida Crown Memorial Grants, 654
Arkansas Community Foundation Grants, 660
Arkell Hall Foundation Grants, 661
Arlington Community Foundation Grants, 662
Arthur and Rochelle Belfer Foundation Grants, 670
Arthur Ashley Williams Foundation Grants, 671
ArvinMeritor Grants, 675
Ashland Corporate Contributions Grants, 713
Assisi Foundation of Memphis General Grants, 787
Athwin Foundation Grants, 795
Atran Foundation Grants, 797
Audrey & Sydney Irmas Foundation Grants, 799
Babcock Charitable Trust Grants, 816
Bacon Family Foundation Grants, 817
Bailey Foundation Grants, 818
Banfi Vintners Foundation Grants, 824
Baptist Community Ministries Grants, 828
Barra Foundation Project Grants, 834
Barr Fund Grants, 835
Baton Rouge Area Foundation Grants, 838
Beatrice Laing Trust Grants, 864
Beazley Foundation Grants, 865
Beldon Fund Grants, 871
Belk Foundation Grants, 872
Bender Foundation Grants, 876
Benton County Foundation Grants, 880
Besser Foundation Grants, 887
BibleLands Grants, 889
Bildner Family Foundation Grants, 891
Bill Hannon Foundation Grants, 895
Blanche and Irving Laurie Foundation Grants, 910
Blowitz-Ridgeway Foundation Early Childhood Development Research Award, 912
Blowitz-Ridgeway Foundation Grants, 913
Blue Shield of California Grants, 923
Blum-Kovler Foundation Grants, 924
Blumenthal Foundation Grants, 925
Bob & Delores Hope Foundation Grants, 927
Bodenwein Public Benevolent Foundation Grants, 928
Bodman Foundation Grants, 929
Boettcher Foundation Grants, 931
Booth-Bricker Fund Grants, 933
Borkee-Hagley Foundation Grants, 934
Boston Foundation Grants, 936
Boston Jewish Community Women's Fund Grants, 938
Bradley-Turner Foundation Grants, 944

Bright Family Foundation Grants, 951
Bright Promises Foundation Grants, 952
Bristol-Myers Squibb Foundation Health Disparities Grants, 958
Burden Trust Grants, 976
Byron W. & Alice L. Lockwood Fnd Grants, 991
Cabot Corporation Foundation Grants, 993
Caddock Foundation Grants, 994
Caleb C. and Julia W. Dula Educational and Charitable Foundation Grants, 997
Callaway Foundation Grants, 1002
Callaway Golf Company Foundation Grants, 1003
Cambridge Community Foundation Grants, 1004
Camp-Younts Foundation Grants, 1005
Campbell Hoffman Foundation Grants, 1006
Cargill Citizenship Fund Corporate Giving, 1018
Carl C. Icahn Foundation Grants, 1023
Carl Gellert and Celia Berta Gellert Foundation Grants, 1024
Carl R. Hendrickson Family Foundation Grants, 1028
Carnahan-Jackson Foundation Grants, 1032
Carolyn Foundation Grants, 1034
Carrie E. and Lena V. Glenn Foundation Grants, 1036
Carrie Estelle Doheny Foundation Grants, 1037
Carylon Foundation Grants, 1040
Catherine Kennedy Home Foundation Grants, 1043
CCHD Community Development Grants, 1046
Central Okanagan Foundation Grants, 1080
CFFVR Alcoholism and Drug Abuse Grants, 1093
CFFVR Basic Needs Giving Partnership Grants, 1094
CFFVR Capital Credit Union Charitable Giving Grants, 1095
CFFVR Myra M. and Robert L. Vandehey Foundation Grants, 1100
CFFVR Robert and Patricia Endries Family Foundation Grants, 1102
Chamberlain Foundation Grants, 1108
Champlin Foundations Grants, 1110
Chapman Charitable Foundation Grants, 1112
Charles Delmar Foundation Grants, 1115
Charles H. Hall Foundation, 1120
Charles H. Revson Foundation Grants, 1122
Charles Nelson Robinson Fund Grants, 1125
Chatlos Foundation Grants Program, 1128
Chazen Foundation Grants, 1129
Chiles Foundation Grants, 1147
CIT Corporate Giving Grants, 1161
Citizens Bank Mid-Atlantic Charitable Foundation Grants, 1162
Clark-Winchcole Foundation Grants, 1173
Clark and Ruby Baker Foundation Grants, 1174
Claude Worthington Benedum Fndn Grants, 1178
Cleveland-Cliffs Foundation Grants, 1181
Cleveland Browns Foundation Grants, 1182
CNA Foundation Grants, 1189
CNCS Senior Companion Program Grants, 1191
Coastal Community Foundation of South Carolina Grants, 1202
Cockrell Foundation Grants, 1204
Coeta and Donald Barker Foundation Grants, 1205
Collins C. Diboll Private Foundation Grants, 1209
Collins Foundation Grants, 1210
Columbus Foundation Competitive Grants, 1217
Commonwealth Fund Small Grants, 1234
Communities Foundation of Texas Grants, 1236
Community Fndn for Greater Buffalo Grants, 1247
Community Fndn for Monterey County Grants, 1253
Community Fndn for Muskegon County Grants, 1254
Community Foundation of Abilene Celebration of Life Grants, 1261
Community Fndn of Central Illinois Grants, 1267
Community Fndn of East Central Illinois Grants, 1268
Community Foundation of Greater Flint Grants, 1273
Community Foundation of Greater Fort Wayne - Community Endowment and Clarke Endowment Grants, 1274
Community Foundation of Greenville Hollingsworth Funds Program/Project Grants, 1285
Community Foundation of Mount Vernon and Knox County Grants, 1299

Connecticut Health Foundation Health Initiative Grants, 1325
Connelly Foundation Grants, 1326
ConocoPhillips Foundation Grants, 1327
Covidien Partnership for Neighborhood Wellness Grants, 1353
Crane Foundation Grants, 1357
CRH Foundation Grants, 1360
CSRA Community Foundation Grants, 1366
CUNA Mutual Group Fndn Community Grants, 1375
Dade Community Foundation Grants, 1410
DaimlerChrysler Corporation Fund Grants, 1412
Danellie Foundation Grants, 1421
David Geffen Foundation Grants, 1431
David M. and Marjorie D. Rosenberg Foundation Grants, 1432
Daywood Foundation Grants, 1438
Delaware Community Foundation Grants, 1449
Dennis & Phyllis Washington Fndn Grants, 1458
DHL Charitable Shipment Support, 1475
Dolan Children's Foundation Grants, 1482
Donald and Sylvia Robinson Family Foundation Grants, 1486
Dora Roberts Foundation Grants, 1489
Dorothy Hooper Beattie Foundation Grants, 1499
Dorrance Family Foundation Grants, 1501
Dr. Scholl Foundation Grants Program, 1511
E. Clayton and Edith P. Gengras, Jr. Foundation, 1536
E. Rhodes & Leona B. Carpenter Grants, 1539
Earl and Maxine Claussen Trust Grants, 1541
eBay Foundation Community Grants, 1543
Eden Hall Foundation Grants, 1549
Edwin W. and Catherine M. Davis Fndn Grants, 1561
Edyth Bush Charitable Foundation Grants, 1562
Effie and Wofford Cain Foundation Grants, 1563
Elliot Foundation Inc Grants, 1575
El Pomar Foundation Anna Keesling Ackerman Fund Grants, 1581
El Pomar Foundation Grants, 1582
Elsie H. Wilcox Foundation Grants, 1583
Emily Davie and Joseph S. Kornfeld Foundation Grants, 1591
Ensworth Charitable Foundation Grants, 1595
Entergy Corporation Micro Grants, 1596
Erie Chapman Foundation Grants, 1600
Essex County Community Foundation Discretionary Fund Grants, 1602
Eugene B. Casey Foundation Grants, 1609
Evjue Foundation Grants, 1617
F.M. Kirby Foundation Grants, 1624
FirstEnergy Foundation Community Grants, 1668
Fishman Family Foundation Grants, 1672
Floyd A. & Kathleen C. Cailloux Fnd Grants, 1681
Foundation for a Healthy Kentucky Grants, 1691
Frank B. Hazard General Charity Fund Grants, 1721
Frank G. and Freida K. Brotz Family Foundation Grants, 1723
Franklin County Community Foundation Grants, 1724
Frank Reed and Margaret Jane Peters Memorial Fund II Grants, 1728
Frank Stanley Beveridge Foundation Grants, 1730
Fred L. Emerson Foundation Grants, 1743
Fritz B. Burns Foundation Grants, 1747
Fuller E. Callaway Foundation Grants, 1761
G.A. Ackermann Memorial Fund Grants, 1767
G.N. Wilcox Trust Grants, 1768
Garland D. Rhoads Foundation, 1774
George and Sarah Buchanan Foundation Grants, 1790
George E. Hatcher, Jr. and Ann Williams Hatcher Foundation Grants, 1792
George Family Foundation Grants, 1794
George H.C. Ensworth Memorial Fund Grants, 1797
George Kress Foundation Grants, 1799
George P. Davenport Trust Fund Grants, 1800
George W. Brackenridge Foundation Grants, 1802
Gertrude E. Skelly Charitable Foundation Grants, 1810
Gheens Foundation Grants, 1814
Giant Eagle Foundation Grants, 1815
Giant Food Charitable Grants, 1816
Ginn Foundation Grants, 1822

900 / Religious Programs   PROGRAM TYPE INDEX

Golden State Warriors Foundation Grants, 1851
Grace and Franklin Bernsen Foundation Grants, 1858
Great-West Life Grants, 1879
Greater Cincinnati Foundation Priority and Small Projects/Capacity-Building Grants, 1880
Greater Saint Louis Community Fndn Grants, 1884
Greater Tacoma Community Foundation Grants, 1886
Greater Worcester Community Foundation Discretionary Grants, 1888
Green Bay Packers Foundation Grants, 1890
Greenspun Family Foundation Grants, 1896
Gregory Family Foundation Grants (Florida), 1899
Guido A. & Elizabeth H. Binda Fndn Grants, 1913
Guy's and St. Thomas' Charity Grants, 1918
H. Leslie Hoffman and Elaine S. Hoffman Foundation Grants, 1924
H. Schaffer Foundation Grants, 1925
Hackett Foundation Grants, 1926
Hagedorn Fund Grants, 1934
Harold and Arlene Schnitzer CARE Foundation Grants, 1950
Harold Simmons Foundation Grants, 1955
Harrison County Community Foundation Grants, 1957
Harrison County Community Foundation Signature Grants, 1958
Harry Frank Guggenheim Fnd Research Grants, 1964
Harry Kramer Memorial Fund Grants, 1965
Harry W. Bass, Jr. Foundation Grants, 1968
Harvey Randall Wickes Foundation Grants, 1972
Hawaii Community Foundation Reverend Takie Okumura Family Grants, 1977
HCA Foundation Grants, 1980
Healthcare Fndn for Orange County Grants, 1981
Healthcare Foundation of New Jersey Grants, 1982
Helen Bader Foundation Grants, 1992
Helen Steiner Rice Foundation Grants, 1998
Henrietta Lange Burk Fund Grants, 2001
Henry L. Guenther Foundation Grants, 2006
Herbert A. & Adrian W. Woods Fndn Grants, 2007
Herbert H. & Grace A. Dow Fndn Grants, 2008
Highmark Corporate Giving Grants, 2025
Hilda and Preston Davis Foundation Grants, 2028
Hillsdale Fund Grants, 2035
Howe Foundation of North Carolina Grants, 2055
Huffy Foundation Grants, 2073
Huie-Dellmon Trust Grants, 2075
Huisking Foundation Grants, 2076
I.A. O'Shaughnessy Foundation Grants, 2088
Ida Alice Ryan Charitable Trust Grants, 2113
Ike and Roz Friedman Foundation Grants, 2145
Infinity Foundation Grants, 2158
Intergrys Corporation Grants, 2160
Irvin Stern Foundation Grants, 2164
J.C. Penney Company Grants, 2169
J.E. and L.E. Mabee Foundation Grants, 2170
J.N. and Macie Edens Foundation Grants, 2174
James Hervey Johnson Charitable Educational Trust Grants, 2191
James J. & Angelia M. Harris Fndn Grants, 2192
James J. & Joan A. Gardner Family Fndn Grants, 2193
James L. and Mary Jane Bowman Charitable Trust Grants, 2194
James M. Collins Foundation Grants, 2195
Janus Foundation Grants, 2211
Jayne and Leonard Abess Foundation Grants, 2214
Jeffris Wood Foundation Grants, 2217
Jessie Ball Dupont Fund Grants, 2226
Jewish Fund Grants, 2227
Joe W. and Dorothy Dorsett Brown Fndn Grants, 2230
John Clarke Trust Grants, 2232
John Edward Fowler Memorial Fndn Grants, 2235
John G. Duncan Charitable Trust Grants, 2236
John H. and Wilhelmina D. Harland Charitable Foundation Children and Youth Grants, 2238
John I. Smith Charities Grants, 2240
John P. Murphy Foundation Grants, 2248
Johns Manville Fund Grants, 2251
Joseph Alexander Foundation Grants, 2265
Joseph and Luella Abell Trust Scholarships, 2266
Joseph H. & Florence A. Roblee Fndn Grants, 2269

Judith Clark-Morrill Foundation Grants, 2281
Katharine Matthies Foundation Grants, 2291
Kevin P. and Sydney B. Knight Family Foundation Grants, 2306
Kimball International-Habig Foundation Health and Human Services Grants, 2308
Kluge Center David B. Larson Fellowship in Health and Spirituality, 2311
Knight Family Charitable and Educational Foundation Grants, 2312
Kovler Family Foundation Grants, 2323
Kroger Foundation Women's Health Grants, 2325
Kuntz Foundation Grants, 2326
Lands' End Corporate Giving Program, 2344
Leonard L. & Bertha U. Abess Fndn Grants, 2357
Leo Niessen Jr., Charitable Trust Grants, 2358
Liberty Bank Foundation Grants, 2362
Libra Foundation Grants, 2363
Lil and Julie Rosenberg Foundation Grants, 2364
Lotus 88 Fnd for Women & Children Grants, 2376
Lowell Berry Foundation Grants, 2382
Lubbock Area Foundation Grants, 2383
Lucile Horton Howe and Mitchell B. Howe Foundation Grants, 2385
Lumpkin Family Fnd Healthy People Grants, 2391
Lydia deForest Charitable Trust Grants, 2393
Lynde and Harry Bradley Foundation Grants, 2397
Lynn and Rovena Alexander Family Foundation Grants, 2399
M. Bastian Family Foundation Grants, 2402
M.J. Murdock Charitable Trust General Grants, 2405
Manuel D. & Rhoda Mayerson Fndn Grants, 2420
Mardag Foundation Grants, 2428
Marion Gardner Jackson Charitable Trust Grants, 2434
Marion I. and Henry J. Knott Foundation Discretionary Grants, 2435
Marion I. and Henry J. Knott Foundation Standard Grants, 2436
Mary Black Foundation Active Living Grants, 2440
Maurice J. Masserini Charitable Trust Grants, 2456
Maxon Charitable Foundation Grants, 2458
McCallum Family Foundation Grants, 2489
McCune Foundation Human Services Grants, 2493
McKesson Foundation Grants, 2496
Mead Johnson Nutritionals Evansville-Area Organizations Grants, 2503
Mead Witter Foundation Grants, 2507
Medtronic Foundation CommunityLink Health Grants, 2510
Medtronic Foundation HeartRescue Grants, 2513
Memorial Foundation Grants, 2516
Meriden Foundation Grants, 2519
Merkel Foundation Grants, 2520
Mervin Bovaird Foundation Grants, 2523
Meyer and Stephanie Eglin Foundation Grants, 2531
MGN Family Foundation Grants, 2542
Michael Reese Health Trust Core Grants, 2544
Michael Reese Health Trust Responsive Grants, 2545
Morris & Gwendolyn Cafritz Fndn Grants, 2594
Mt. Sinai Health Care Foundation Health of the Jewish Community Grants, 2599
Mt. Sinai Health Care Foundation Health of the Urban Community Grants, 2600
Nathan Cummings Foundation Grants, 2627
Nell J. Redfield Foundation Grants, 2673
New York University Steinhardt School of Education Fellowships, 2684
Nina Mason Pulliam Charitable Trust Grants, 2921
Norcliffe Foundation Grants, 2979
North Carolina Community Foundation Grants, 2990
Oppenstein Brothers Foundation Grants, 3078
Otto Bremer Foundation Grants, 3099
PacifiCare Foundation Grants, 3102
Patrick and Anna M. Cudahy Fund Grants, 3120
Paul Balint Charitable Trust Grants, 3122
Perkins Charitable Foundation Grants, 3140
Perpetual Trust for Charitable Giving Grants, 3141
Phelps County Community Foundation Grants, 3188
Philadelphia Foundation Organizational Effectiveness Grants, 3189

Pinkerton Foundation Grants, 3224
Pittsburgh Foundation Community Fund Grants, 3230
Polk Bros. Foundation Grants, 3243
Pollock Foundation Grants, 3244
Powell Foundation Grants, 3255
Premera Blue Cross Grants, 3258
Priddy Foundation Organizational Development Grants, 3265
Priddy Foundation Program Grants, 3266
Procter and Gamble Fund Grants, 3274
Quality Health Foundation Grants, 3288
R.C. Baker Foundation Grants, 3292
Rainbow Endowment Grants, 3300
Raskob Foundation for Catholic Activities Grants, 3306
Rathmann Family Foundation Grants, 3309
Religion and Health: Effects, Mechanisms, and Interpretation RFP, 3358
RGk Foundation Grants, 3361
Richard and Helen DeVos Foundation Grants, 3364
Richard D. Bass Foundation Grants, 3366
Richard King Mellon Foundation Grants, 3369
Ricks Family Charitable Trust Grants, 3373
Robbins Charitable Foundation Grants, 3378
Robert R. McCormick Trib Community Grants, 3385
Robert W. Woodruff Foundation Grants, 3389
Robins Foundation Grants, 3390
Rockwell Collins Charitable Corporation Grants, 3397
Rockwell International Corporate Trust Grants Program, 3399
Rollins-Luetkemeyer Foundation Grants, 3404
Ronald McDonald House Charities Grants, 3405
Rose Community Foundation Aging Grants, 3410
Roy J. Carver Charitable Trust Medical and Science Research Grants, 3416
RRF Organizational Capacity Building Grants, 3418
S.H. Cowell Foundation Grants, 3447
Saint Luke's Foundation Grants, 3454
San Antonio Area Foundation Grants, 3469
San Diego Foundation for Change Grants, 3470
Sandy Hill Foundation Grants, 3477
Schlessman Family Foundation Grants, 3493
Scott B. & Annie P. Appleby Charitable Grants, 3499
Seagate Tech Corp Capacity to Care Grants, 3502
Shell Deer Park Grants, 3543
Sidney Stern Memorial Trust Grants, 3552
Siebert Lutheran Foundation Grants, 3554
Sierra Health Foundation Responsive Grants, 3555
Sioux Falls Area Community Foundation Community Fund Grants (Unrestricted), 3568
Sioux Falls Area Community Foundation Field-of-Interest and Donor-Advised Grants, 3569
Sioux Falls Area Community Foundation Spot Grants (Unrestricted), 3570
Sisters of Charity Foundation of Cleveland Reducing Health and Educational Disparities in the Central Neighborhood Grants, 3574
Sisters of St. Joseph Healthcare Fndn Grants, 3576
Skoll Fndn Awards for Social Entrepreneurship, 3579
SOCFOC Catholic Ministries Grants, 3596
Solo Cup Foundation Grants, 3620
Square D Foundation Grants, 3634
St. Joseph Community Health Foundation Improving Healthcare Access Grants, 3640
Stan and Sandy Checketts Foundation, 3644
Stella and Charles Guttman Foundation Grants, 3651
Stewart Huston Charitable Trust Grants, 3655
Strake Foundation Grants, 3658
Strowd Roses Grants, 3664
Sunderland Foundation Grants, 3675
SunTrust Bank Trusteed Foundations Florence C. and Harry L. English Memorial Fund Grants, 3680
SunTrust Bank Trusteed Foundations Greene-Sawtell Grants, 3681
SunTrust Bank Trusteed Foundations Harriet McDaniel Marshall Tust Grants, 3682
SunTrust Bank Trusteed Foundations Nell Warren Elkin and William Simpson Elkin Grants, 3683
SunTrust Bank Trusteed Foundations Thomas Guy Woolford Charitable Trust Grants, 3684

PROGRAM TYPE INDEX   Scholarships   / 901

SunTrust Bank Trusteed Foundations Walter H. and
　　Marjory M. Rich Memorial Fund Grants, 3685
Support Our Aging Religious (SOAR) Grants, 3686
Susan Mott Webb Charitable Trust Grants, 3694
Symantec Community Relations and Corporate
　　Philanthropy Grants, 3699
T.L.L. Temple Foundation Grants, 3700
Tension Envelope Foundation Grants, 3707
Thomas and Dorothy Leavey Foundation Grants, 3717
Thomas Austin Finch, Sr. Foundation Grants, 3718
Thomas C. Ackerman Foundation Grants, 3719
Thomas Jefferson Rosenberg Foundation Grants, 3723
Thompson Charitable Foundation Grants, 3727
Todd Brock Family Foundation Grants, 3736
Triangle Community Foundation Donor-Advised
　　Grants, 3757
Trull Foundation Grants, 3761
Tyler Aaron Bookman Memorial Foundation Trust
　　Grants, 3765
U.S. Cellular Corporation Grants, 3766
United Methodist Committee on Relief Global AIDS
　　Fund Grants, 3798
United Methodist Committee on Relief Global Health
　　Grants, 3799
United Methodist Child Mental Health Grants, 3800
United Methodist Health Ministry Fund Grants, 3801
Vancouver Foundation Grants and Community
　　Initiatives Program, 3849
Vanderburgh Community Fndn Women's Fund, 3853
W.C. Griffith Foundation Grants, 3881
W.K. Kellogg Foundation Secure Families Grants, 3887
W.P. and Bulah Luse Foundation Grants, 3892
Walter L. Gross III Family Foundation Grants, 3901
Warrick County Community Foundation Women's
　　Fund, 3905
Washington Gas Charitable Giving, 3908
Welborn Baptist Fndn Health Ministries Grants, 3915
Weyerhaeuser Company Foundation Grants, 3925
William G. & Helen C. Hoffman Fndn Grants, 3950
William H. Hannon Foundation Grants, 3953
William Ray & Ruth E. Collins Fndn Grants, 3956
Women of the ELCA Opportunity Scholarships for
　　Lutheran Laywomen, 3970
Woodward Fund Grants, 3973

Scholarships
A.O. Smith Foundation Community Grants, 17
AACAP Child and Adolescent Psychiatry (CAP)
　　Teaching Scholarships, 51
AAFCS National Undergraduate Scholarships, 131
AAHPERD-AAHE Barb A. Cooley Scholarships, 148
AAHPERD-AAHE Del Oberteuffer Scholarships, 149
AAHPERD-AAHE Undergraduate Scholarships, 150
AAHPERD Abernathy Presidential Scholarships, 151
AAMC Herbert W. Nickens Medical Student
　　Scholarships, 160
AAOHN Foundation Professional Development
　　Scholarships, 175
AAP Resident Research Grants, 190
Abbot and Dorothy H. Stevens Foundation Grants, 197
Abbott Fund CFCareForward Scholarships, 200
Abbott Fund Science Education Grants, 203
Abell-Hanger Foundation Grants, 207
Abramson Family Foundation Grants, 217
AcademyHealth Alice Hersh Student Scholarships, 221
AcademyHealth PHSR Interest Group Student
　　Scholarships, 225
Achelis Foundation Grants, 234
ACS Doc Degree Scholarships in Cancer Nursing, 252
ACS Doctoral Scholarships in Cancer Nursing, 253
ACS Scholarships in Cancer Nursing Practice, 255
ADA Foundation Dental Student Scholarships, 288
ADA Found Minority Dental Scholarships, 292
ADA Foundation Thomas J. Zwemer Award, 297
Adams County Community Foundation of
　　Pennsylvania Grants, 300
Adams Rotary Memorial Fund B Scholarships, 303
Adaptec Foundation Grants, 304
AFB Rudolph Dillman Memorial Scholarship, 338
AGHE Graduate Scholarships and Fellowships, 343

Agnes M. Lindsay Trust Grants, 346
Ahmanson Foundation Grants, 355
AIHS Full-Time M.D./Ph.D. Studentships, 385
AIHS Part-Time Studentships, 394
Alabama Power Foundation Grants, 400
Albert and Margaret Alkek Foundation Grants, 405
Albuquerque Community Foundation Grants, 412
Alcatel-Lucent Technologies Foundation Grants, 413
Alcoa Foundation Grants, 414
Alexis Gregory Foundation Grants, 426
Alfred and Tillie Shemanski Testamentary Trust
　　Grants, 428
Alice Fisher Society Fellowships, 436
Alice Tweed Tuohy Foundation Grants Program, 437
Alliant Energy Corporation Contributions, 445
Allyn Foundation Grants, 448
AlohaCare Believes in Me Scholarship, 449
Alpha Natural Resources Corporate Giving, 450
Alpha Omega Foundation Grants, 451
Alpine Winter Foundation Grants, 453
ALSAM Foundation Grants, 454
AMA-WPC Joan F. Giambalvo Memorial
　　Scholarships, 480
AMA Fndn Arthur N. Wilson Scholarship, 482
AMA Foundation Joan F. Giambalvo Memorial
　　Scholarships, 488
AMA Foundation Minority Scholars Awards, 491
AMA Foundation Physicians of Tomorrow
　　Scholarships, 492
American Academy of Dermatology Camp Discovery
　　Scholarships, 503
American Academy of Nursing Building Academic
　　Geriatric Nursing Capacity Scholarships, 505
American Academy of Nursing Mayday Grants, 507
American Academy of Nursing MBA Scholarships, 508
American College of Surgeons and The Triological
　　Society Clinical Scientist Development Awards, 518
American College of Surgeons Health Policy
　　Scholarships, 523
American College of Surgeons Health Policy
　　Scholarships for General Surgeons, 524
American College of Surgeons Resident Research
　　Scholarships, 528
American Foodservice Charitable Trust Grants, 532
American Psychiatric Association Travel Scholarships
　　for Minority Medical Students, 549
American Schlafhorst Foundation Grants, 559
AMI Semiconductors Corporate Grants, 576
Andre Agassi Charitable Foundation Grants, 587
Andrew Family Foundation Grants, 588
Angels Baseball Foundation Grants, 590
Anheuser-Busch Foundation Grants, 591
Ann Arbor Area Community Foundation Grants, 594
Ann Peppers Foundation Grants, 599
Antone & Edene Vidinha Charitable Trust Grants, 609
APF/COGDOP Graduate Research Scholarships in
　　Psychology, 635
Aratani Foundation Grants, 649
Arcadia Foundation Grants, 650
Archer Daniels Midland Foundation Grants, 651
ARCO Foundation Education Grants, 652
Argyros Foundation Grants Program, 653
Arie and Ida Crown Memorial Grants, 654
Arizona Community Foundation Scholarships, 656
Arkansas Community Foundation Grants, 660
Arkell Hall Foundation Grants, 661
Arlington Community Foundation Grants, 662
Arnold and Mabel Beckman Foundation Scholars
　　Grants, 665
ASHA Minority Student Leadership Awards, 696
ASHA Students Preparing for Academic & Research
　　Careers (SPARC) Award, 700
ASHFoundation Graduate Student International
　　Scholarship, 702
ASHFoundation Graduate Student Scholarships, 703
ASHFoundation Graduate Student Scholarships for
　　Minority Students, 704
ASHFoundation Graduate Student with a Disability
　　Scholarship, 705

ASHFoundation New Century Scholars Program
　　Doctoral Scholarships, 706
ASHFoundation Student Research Grants in Early
　　Childhood Language Development, 711
Assurant Foundation Grants, 788
Assurant Health Foundation Grants, 789
Atran Foundation Grants, 797
AUPHA Corris Boyd Scholarships, 800
AUPHA Foster G. McGaw Scholarships, 801
Australian Academy of Science Grants, 807
Avon Products Foundation Grants, 814
Bailey Foundation Grants, 818
Barberton Community Foundation Grants, 829
Battle Creek Community Foundation Grants, 839
Batts Foundation Grants, 840
Bayer Clinical Scholarship Award, 845
Beazley Foundation Grants, 865
Beckley Area Foundation Grants, 867
Becton Dickinson and Company Grants, 869
Ben B. Cheney Foundation Grants, 875
Bender Foundation Grants, 876
Benton County Foundation - Fitzgerald Family
　　Scholarships, 879
Berks County Community Foundation Grants, 881
Bertha Russ Lytel Foundation Grants, 885
Besser Foundation Grants, 887
Bill Hannon Foundation Grants, 895
Bindley Family Foundation Grants, 896
Blackford County Community Foundation Hartford
　　City Kiwanis Scholarship in Memory of Mike
　　McDougall, 907
Blackford County Community Foundation Noble
　　Memorial Scholarship, 908
Blanche and Irving Laurie Foundation Grants, 910
Blowitz-Ridgeway Foundation Early Childhood
　　Development Research Award, 912
Blue Grass Community Foundation Hudson-Ellis Fund
　　Grants, 920
Blue Mountain Community Foundation Grants, 921
Bodenwein Public Benevolent Foundation Grants, 928
Bodman Foundation Grants, 929
Boeing Company Contributions Grants, 930
Bright Family Foundation Grants, 951
Brook J. Lenfest Foundation Grants, 967
Burlington Northern Santa Fe Foundation Grants, 978
Cabot Corporation Foundation Grants, 993
California Wellness Foundation Work and Health
　　Program Grants, 1001
Cambridge Community Foundation Grants, 1004
Campbell Soup Foundation Grants, 1007
Canada Graduate Scholarships (CGS) and NSERC
　　Postgraduate Scholarships (PGS), 1010
Cargill Citizenship Fund Corporate Giving, 1018
Carl & Eloise Pohlad Family Fndn Grants, 1021
Carl Gellert and Celia Berta Gellert Foundation
　　Grants, 1024
Carnahan-Jackson Foundation Grants, 1032
Carolyn Foundation Grants, 1034
Carpenter Foundation Grants, 1035
Carrie Estelle Doheny Foundation Grants, 1037
Carrier Corporation Contributions Grants, 1038
Carroll County Community Foundation Grants, 1039
Cemala Foundation Grants, 1077
Central Carolina Community Foundation Community
　　Impact Grants, 1079
Central Okanagan Foundation Grants, 1080
Cessna Foundation Grants Program, 1081
Cetana Educational Foundation Scholarships, 1082
CFFVR Shawano Area Community Foundation
　　Grants, 1104
CFFVR Women's Fund for the Fox Valley Region
　　Grants, 1107
Champlin Foundations Grants, 1110
Charles A. Frueauff Foundation Grants, 1114
Charles Delmar Foundation Grants, 1115
Chatlos Foundation Grants Program, 1128
Chazen Foundation Grants, 1129
Chiles Foundation Grants, 1147
Christensen Fund Regional Grants, 1149
Christy-Houston Foundation Grants, 1153

Scholarships

CIGNA Foundation Grants, 1160
Clarence E. Heller Charitable Foundation Grants, 1166
Clarence T.C. Ching Foundation Grants, 1167
Clarian Health Scholarships for LPNs, 1171
Clarian Health Student Nurse Scholarships, 1172
Clark-Winchcole Foundation Grants, 1173
Claude Pepper Foundation Grants, 1177
Cleveland-Cliffs Foundation Grants, 1181
Coastal Community Foundation of South Carolina Grants, 1202
Coeta and Donald Barker Foundation Grants, 1205
Collins Foundation Grants, 1210
Columbus Foundation Competitive Grants, 1217
Comerica Charitable Foundation Grants, 1224
Community Fndn for Greater Buffalo Grants, 1247
Community Fndn for Muskegon County Grants, 1254
Community Foundation of Bloomington and Monroe County Grants, 1265
Community Fndn of East Central Illinois Grants, 1268
Community Foundation of Greater Flint Grants, 1273
Community Foundation of Greater Fort Wayne - Community Endowment and Clarke Endowment Grants, 1274
Community Foundation of Greater Fort Wayne - Lilly Endowment Scholarships, 1275
Community Foundation of Greater Fort Wayne Scholarships, 1276
Community Foundation of Greater Lafayette - Robert and Dorothy Hughes Scholarships, 1279
Community Foundation of Greater New Britain Grants, 1281
Community Foundation of Mount Vernon and Knox County Grants, 1299
Community Fndn of Randolph County Grants, 1302
Community Fndn of So Alabama Grants, 1306
Community Fndn of South Puget Sound Grants, 1307
Community Foundation of St. Joseph County Lilly Endowment Community Scholarship, 1308
Community Foundation of St. Joseph County Scholarships, 1309
Community Foundation of the Verdugos Grants, 1315
Community Fndn of Wabash County Grants, 1316
Connecticut Community Foundation Grants, 1324
Connecticut Health Foundation Health Initiative Grants, 1325
Constellation Energy Corporate Grants, 1342
Cooper Industries Foundation Grants, 1345
Covenant Foundation of Waterloo Auxiliary Scholarships, 1349
Cralle Foundation Grants, 1356
Crown Point Community Fndn Scholarships, 1362
Cruise Industry Charitable Foundation Grants, 1363
CSRA Community Foundation Grants, 1366
Cudd Foundation Grants, 1370
Cultural Society of Filipino Americans Grants, 1373
Cystic Fibrosis Canada Scholarships, 1386
Cystic Fibrosis Scholarships, 1403
D.F. Halton Foundation Grants, 1405
DAAD Research Stays for University Academics and Scientists, 1408
DaimlerChrysler Corporation Fund Grants, 1412
Danellie Foundation Grants, 1421
Daphne Seybolt Culpeper Fndn Grants, 1425
DAR Alice W. Rooke Scholarship, 1426
DAR Dr. Francis Anthony Beneventi Medical Scholarship, 1427
DAR Irene and Daisy MacGregor Memorial Scholarship, 1428
DAR Madeline Cogswell Nursing Scholarship, 1429
DAR Nursing/Physical Therapy Scholarships, 1430
Dell Scholars Program Scholarships, 1452
Dennis & Phyllis Washington Fndn Grants, 1458
DeRoy Testamentary Foundation Grants, 1469
Deutsche Banc Alex Brown and Sons Charitable Foundation Grants, 1471
Donna K. Yundt Memorial Scholarship, 1488
Doris and Victor Day Foundation Grants, 1491
Do Something Awards, 1503
Dr. John Maniotes Scholarship, 1506
Dr. John T. Macdonald Foundation Scholarships, 1508

Dr. Leon Bromberg Charitable Trust Grants, 1509
Dr. R.T. White Scholarship, 1510
Dr. Stanley Pearle Scholarships, 1512
DuPage Community Foundation Grants, 1531
eBay Foundation Community Grants, 1543
EBSCO / MLA Annual Meeting Grants, 1545
Eden Hall Foundation Grants, 1549
Edward Bangs Kelley and Elza Kelley Foundation Grants, 1556
Edward W. and Stella C. Van Houten Memorial Fund Grants, 1559
Edwin S. Webster Foundation Grants, 1560
Edwin W. and Catherine M. Davis Fndn Grants, 1561
Effie and Wofford Cain Foundation Grants, 1563
Elizabeth Nash Foundation Scholarships, 1570
Elkhart County Foundation Lilly Endowment Community Scholarships, 1574
El Paso Community Foundation Grants, 1579
Elsie H. Wilcox Foundation Grants, 1583
Emerson Charitable Trust Grants, 1589
Emerson Electric Company Contributions Grants, 1590
Erie Community Foundation Grants, 1601
Eugene B. Casey Foundation Grants, 1609
Eugene M. Lang Foundation Grants, 1611
Evan and Susan Bayh Foundation Grants, 1614
Everyone Breathe Asthma Education Grants, 1616
Evjue Foundation Grants, 1617
Fairfield County Community Foundation Grants, 1625
Fargo-Moorhead Area Fndn Woman's Grants, 1633
FDHN Isenberg Int'l Research Scholar Award, 1647
Fishman Family Foundation Grants, 1672
Flinn Foundation Scholarships, 1675
Fluor Foundation Grants, 1682
FMC Foundation Grants, 1683
Fndn for Appalachian Ohio Bachtel Scholarships, 1692
Foundation for Appalachian Ohio Susan K. Ipacs Nursing Legacy Scholarships, 1693
Foundation for Appalachian Ohio Zelma Gray Medical School Scholarship, 1694
Foundation for Seacoast Health Grants, 1698
Four J Foundation Grants, 1713
Frances and John L. Loeb Family Fund Grants, 1716
Fred and Louise Latshaw Scholarship, 1735
Fred L. Emerson Foundation Grants, 1743
Fremont Area Community Fndn General Grants, 1746
Fritz B. Burns Foundation Grants, 1747
Fuller E. Callaway Foundation Grants, 1761
Fulton County Community Foundation 4Community Higher Education Scholarship, 1762
Fulton County Community Foundation Lilly Endowment Community Scholarships, 1764
Fulton County Community Foundation Paul and Dorothy Arven Memorial Scholarship, 1765
G.N. Wilcox Trust Grants, 1768
Gates Millennium Scholars Program, 1776
Genuardi Family Foundation Grants, 1787
George and Ruth Bradford Foundation Grants, 1789
George Family Foundation Grants, 1794
George Foundation Grants, 1795
George S. & Dolores Dore Eccles Fndn Grants, 1801
George W. Brackenridge Foundation Grants, 1802
Gertrude B. Elion Mentored Medical Student Research Awards, 1809
Gertrude E. Skelly Charitable Foundation Grants, 1810
GFWC of Massachusetts Catherine E. Philbin Scholarship, 1811
GFWC of Massachusetts Communication Disorder/Speech Therapy Scholarship, 1812
GFWC of Massachusetts Memorial Education Scholarship, 1813
Giant Eagle Foundation Grants, 1815
Gladys Brooks Foundation Grants, 1826
Global Fund for Children Grants, 1832
Graco Foundation Grants, 1860
Grand Haven Area Community Fndn Grants, 1864
Greater Saint Louis Community Fndn Grants, 1884
Greater Worcester Community Foundation Discretionary Grants, 1888
Green Bay Packers Foundation Grants, 1890
Griffin Foundation Grants, 1903

Grover Hermann Foundation Grants, 1906
Guido A. & Elizabeth H. Binda Fndn Grants, 1913
Gulf Coast Community Foundation Grants, 1915
H & R Foundation Grants, 1920
H. Schaffer Foundation Grants, 1925
HAF Riley Frazel Memorial Scholarship, 1931
Hammond Common Council Scholarships, 1938
Harold and Arlene Schnitzer CARE Foundation Grants, 1950
Harrison County Community Foundation Grants, 1957
Harry Bramhall Gilbert Charitable Trust Grants, 1961
Harry Kramer Memorial Fund Grants, 1965
Health Management Scholarships and Grants for Minorities, 1986
Helena Rubinstein Foundation Grants, 1991
Helen Bader Foundation Grants, 1992
Helen Irwin Littauer Educational Trust Grants, 1994
High Meadow Foundation Grants, 2027
Hillsdale County Community General Adult Foundation Grants, 2034
Hilton Hotels Corporate Giving Program Grants, 2037
HomeBanc Foundation Grants, 2042
Homer Foundation Grants, 2044
Hormel Foods Charitable Trust Grants, 2050
Hormel Foundation Grants, 2051
Houston Endowment Grants, 2053
Huie-Dellmon Trust Grants, 2075
Huisking Foundation Grants, 2076
Humana Foundation Grants, 2077
IATA Scholarships, 2094
IBEW Local Union #697 Memorial Scholarships, 2096
IIE 911 Armed Forces Scholarships, 2122
IIE Adell and Hancock Scholarships, 2123
IIE AmCham Charitable Foundation U.S. Studies Scholarship, 2125
IIE Brazil Science Without Borders Undergraduate Scholarships, 2126
IIE Chevron International REACH Scholarships, 2128
IIE Eurobank EFG Scholarships, 2130
IIE Iraq Scholars and Leaders Scholarships, 2133
IIE Klein Family Scholarship, 2135
IIE Lotus Scholarships, 2138
IIE Mattel Global Scholarship, 2139
IIE Nancy Petry Scholarship, 2140
IIE Western Union Family Scholarships, 2143
IIE Whitaker Int'l Fellowships & Scholarships, 2144
Illinois Tool Works Foundation Grants, 2147
Independence Blue Cross Nurse Scholars, 2151
Indiana Minority Teacher/Special Services Scholarships, 2156
Indiana Nursing Scholarships, 2157
Isabel Allende Foundation Esperanza Grants, 2165
J.C. Penney Company Grants, 2169
J.E. and L.E. Mabee Foundation Grants, 2170
J.L. Bedsole Foundation Grants, 2172
J. Willard Marriott, Jr. Foundation Grants, 2178
James & Abigail Campbell Family Fndn Grants, 2185
James G.K. McClure Educational and Development Fund Grants, 2188
James J. & Angelia M. Harris Fndn Grants, 2192
James R. Dougherty Jr. Foundation Grants, 2198
James R. Thorpe Foundation Grants, 2199
James S. Copley Foundation Grants, 2200
Jane Beattie Memorial Scholarship, 2207
Janus Foundation Grants, 2211
JELD-WEN Foundation Grants, 2218
Joe W. and Dorothy Dorsett Brown Fndn Grants, 2230
John Ben Snow Memorial Trust Grants, 2231
John Deere Foundation Grants, 2234
John H. and Wilhelmina D. Harland Charitable Foundation Children and Youth Grants, 2238
John I. Smith Charities Grants, 2240
John J. Leidy Foundation Grants, 2241
John Jewett and Helen Chandler Garland Foundation Grants, 2242
John P. McGovern Foundation Grants, 2247
Johns Manville Fund Grants, 2251
John Stauffer Charitable Trust Grants, 2258
John V. and George Primich Family Scholarship, 2259
John W. Anderson Foundation Grants, 2262

PROGRAM TYPE INDEX

Scholarships / 903

Joseph Alexander Foundation Grants, 2265
Joseph and Luella Abell Trust Scholarships, 2266
Joseph Collins Foundation Scholarships, 2267
Joseph Drown Foundation Grants, 2268
Josephine G. Russell Trust Grants, 2271
Joshua Benjamin Cohen Memorial Scholarship, 2276
Josiah W. and Bessie H. Kline Foundation Grants, 2278
June Pangburn Memorial Scholarship, 2283
Kathryne Beynon Foundation Grants, 2294
Kenneth T. & Eileen L. Norris Fndn Grants, 2300
Kettering Fund Grants, 2305
Kosciuszko Foundation Dr. Marie E. Zakrzewski Medical Scholarship, 2321
L. A. Hollinger Respiratory Therapy and Nursing Scholarships, 2327
LaGrange County Community Foundation Scholarships, 2330
LaGrange County Lilly Endowment Community Scholarship, 2331
Lake County Athletic Officials Association Scholarships, 2332
Lake County Lilly Endowment Community Scholarships, 2334
LEGENDS Scholarship, 2351
Leo Goodwin Foundation Grants, 2354
Lester Ray Fleming Scholarships, 2360
Lillian S. Wells Foundation Grants, 2365
Lowell Berry Foundation Grants, 2382
Lubbock Area Foundation Grants, 2383
Lubrizol Foundation Grants, 2384
Lynde and Harry Bradley Foundation Grants, 2397
M. Bastian Family Foundation Grants, 2402
Marathon Petroleum Corporation Grants, 2422
March of Dimes Graduate Nursing Scholarships, 2424
Margaret T. Morris Foundation Grants, 2429
Mary L. Peyton Foundation Grants, 2445
Matilda R. Wilson Fund Grants, 2453
Maxon Charitable Foundation Grants, 2458
MBL Scholarships and Awards, 2486
McGraw-Hill Companies Community Grants, 2494
McInerny Foundation Grants, 2495
McKesson Foundation Grants, 2496
Mead Johnson Nutritionals Scholarships, 2505
Mead Witter Foundation Grants, 2507
Meriden Foundation Grants, 2519
Mervin Bovaird Foundation Grants, 2523
MGN Family Foundation Grants, 2542
Michigan Psychoanalytic Institute and Society Scholarships, 2547
Military Ex-Prisoners of War Foundation Grants, 2554
MLA Continuing Education Grants, 2559
MLA Graduate Scholarship, 2564
MLA Grad Scholarship for Minority Students, 2565
Morris & Gwendolyn Cafritz Fndn Grants, 2594
Nancy J. Pinnick Memorial Scholarship, 2607
NAPNAP Fndn Elaine Gelman Scholarship, 2609
NAPNAP Foundation McNeil PNP Scholarships, 2613
NAPNAP Foundation McNeil Rural and Underserved Scholarships, 2614
NAPNAP Foundation Nursing Research Grants, 2615
Nell J. Redfield Foundation Grants, 2673
Nesbitt Medical Student Foundation Scholarship, 2674
New Jersey Osteopathic Education Foundation Scholarships, 2680
Newton County Community Foundation Lilly Scholarships, 2682
Newton County Community Fndn Scholarships, 2683
Nicholas H. Noyes Jr. Memorial Fndn Grants, 2791
NMF Aura E. Severinghaus Award, 2954
NMF Henry G. Halladay Awards, 2957
NMF Hugh J. Andersen Memorial Scholarship, 2958
NMF Irving Graef Memorial Scholarship, 2959
NMF Mary Ball Carrera Scholarship, 2960
NMF Metropolitan Life Foundation Awards Program for Academic Excellence in Medicine, 2961
NMF National Medical Association Emerging Scholar Awards, 2963
NMF National Medical Association Patti LaBelle Award, 2964
NMF Need-Based Scholarships, 2965

NMSS Scholarships, 2968
Noble County Community Foundation - Arthur A. and Hazel S. Auer Scholarship, 2969
Noble County Community Foundation - Art Hutsell Scholarship, 2970
Noble County Community Foundation - Delta Theta Tau Sorority IOTA IOTA Chapter Riecke Scholarship, 2971
Noble County Community Foundation - Democrat Central Committee Scholarships, 2972
Noble County Community Foundation - Kathleen June Earley Memorial Scholarship, 2973
Noble County Community Foundation - Lolita J. Hornett Memorial Nursing Scholarship, 2974
Noble County Community Foundation Lilly Endowment Scholarship, 2976
Noble County Community Fndn Scholarships, 2977
Norcliffe Foundation Grants, 2979
Nordson Corporation Foundation Grants, 2981
Norfolk Southern Foundation Grants, 2982
North Carolina Community Foundation Grants, 2990
North Carolina GlaxoSmithKline Fndn Grants, 2991
Northwestern Mutual Foundation Grants, 2999
Oleonda Jameson Trust Grants, 3071
Ordean Foundation Grants, 3080
Pajaro Valley Community Health Trust Promoting Entry & Advancement in the Health Professions Grants, 3111
Partnership for Cures Charles E. Culpeper Scholarships in Medical Science, 3117
Patrick and Anna M. Cudahy Fund Grants, 3120
Percy B. Ferebee Endowment Grants, 3138
Peter and Elizabeth C. Tower Foundation Organizational Scholarships, 3148
Peter Kiewit Foundation General Grants, 3154
Peter Kiewit Foundation Small Grants, 3155
Peyton Anderson Foundation Grants, 3160
Phelps County Community Foundation Grants, 3188
Phi Upsilon Omicron Margaret Jerome Sampson Scholarships, 3197
Piper Jaffray Fndn Communities Giving Grants, 3227
Porter County Foundation Lilly Endowment Community Scholarships, 3245
Porter County Health Occupations Scholarship, 3247
Portland General Electric Foundation Grants, 3250
Pott Foundation Grants, 3253
Potts Memorial Foundation Grants, 3254
PPG Industries Foundation Grants, 3257
Price Chopper's Golub Foundation Grants, 3261
Price Family Charitable Fund Grants, 3262
Pride Foundation Scholarships, 3269
Principal Financial Group Foundation Grants, 3273
Prudential Foundation Education Grants, 3276
Pulaski County Community Foundation Lilly Endowment Community Scholarships, 3280
Pulaski County Community Fndn Scholarships, 3281
Putnam County Community Foundation Grants, 3283
Quaker Chemical Foundation Grants, 3286
R.C. Baker Foundation Grants, 3292
Rajiv Gandhi Foundation Grants, 3302
Ralph M. Parsons Foundation Grants, 3304
Rathmann Family Foundation Grants, 3309
Rayonier Foundation Grants, 3311
Regenstrief Master Of Science In Clinical Research/ CITE Program, 3353
Rehabilitation Nursing Foundation Scholarships, 3355
Reinberger Foundation Grants, 3357
Rhode Island Foundation Grants, 3362
Richland County Bank Grants, 3371
Robert B. Adams Foundation Scholarships, 3380
Robert R. Meyer Foundation Grants, 3387
Rockwell International Corporate Trust Grants Program, 3399
Roget Begnoche Scholarship, 3402
Ronald McDonald House Charities Grants, 3405
Rosalynn Carter Institute Mattie J. T. Stepanek Caregiving Scholarships, 3409
Roy J. Carver Charitable Trust Medical and Science Research Grants, 3416

Ruth Eleanor Bamberger and John Ernest Bamberger Memorial Foundation Grants, 3427
RWJF New Careers in Nursing Scholarships, 3439
S. Livingston Mather Charitable Trust Grants, 3448
S. Mark Taper Foundation Grants, 3449
Samuel S. Johnson Foundation Grants, 3468
San Antonio Area Foundation Grants, 3469
San Juan Island Community Foundation Grants, 3480
Sartain Lanier Family Foundation Grants, 3486
Schering-Plough Foundation Health Grants, 3492
Schrage Family Foundation Scholarships, 3494
Schramm Foundation Grants, 3495
Scott B. & Annie P. Appleby Charitable Grants, 3499
Seabury Foundation Grants, 3501
Seattle Fndn Doyne M. Green Scholarships, 3505
Seattle Fndn Hawkins Memorial Scholarship, 3506
Seneca Foods Foundation Grants, 3511
Sengupta Family Scholarship, 3512
Shell Oil Company Foundation Community Development Grants, 3544
Simmons Foundation Grants, 3563
SNM PDEF Mickey Williams Minority Student Scholarships, 3584
SNMTS Clinical Advancement Scholarships, 3590
SNMTS Paul Cole Scholarships, 3593
Solo Cup Foundation Grants, 3620
Sony Corporation of America Grants, 3623
Special People in Need Scholarships, 3631
Square D Foundation Grants, 3634
Stackpole-Hall Foundation Grants, 3643
Starke County Community Foundation Lilly Endowment Community Scholarships, 3647
Starke County Community Fndn Scholarships, 3648
Steele-Reese Foundation Grants, 3650
Strake Foundation Grants, 3658
Stranahan Foundation Grants, 3659
Strowd Roses Grants, 3664
Summit Foundation Grants, 3674
Susan G. Komen Breast Cancer Foundation College Scholarships, 3692
Thomas and Dorothy Leavey Foundation Grants, 3717
Thomas Austin Finch, Sr. Foundation Grants, 3718
Thomas C. Ackerman Foundation Grants, 3719
Thompson Charitable Foundation Grants, 3727
Topeka Community Foundation Kansas Blood Services Scholarships, 3738
Trinity Lutheran School of Nursing Alumnae Scholarships, 3759
Trull Foundation Grants, 3761
Tull Charitable Foundation Grants, 3762
Tylenol Future Care Scholarships, 3764
U.S. Department of Education Erma Byrd Scholarships, 3768
UCT Scholarship Program, 3777
Union Bank, N.A. Foundation Grants, 3788
VDH Commonwealth of Virginia Nurse Educator Scholarships, 3854
Virginia Department of Health Mary Marshall Nursing Scholarships for Licensed Practical Nurses, 3874
Virginia Department of Health Mary Marshall Nursing Scholarships for Registered Nurses, 3875
Virginia Department of Health Nurse Practitioner/ Nurse Midwife Scholarships, 3876
W. James Spicer Scholarship, 3885
W.P. and Bulah Luse Foundation Grants, 3892
Wayne County Foundation Grants, 3909
Webster Cornwell Memorial Scholarship, 3911
White County Community Foundation - Adam Krintz Memorial Scholarship, 3928
White County Community Foundation - Annie Horton Scholarship, 3929
White County Community Foundation - Landis Memorial Scholarship, 3930
White County Community Foundation - Lilly Endowment Scholarships, 3931
White County Community Foundation - Tri-County Educational Scholarships, 3932
Whitley County Community Foundation - Lilly Endowment Scholarship, 3935
Whitley County Community Fndn Scholarships, 3937

904 / Scholarships

Wilhelmina W. Jackson Trust Scholarships, 3941
William Blair and Company Foundation Grants, 3948
William G. & Helen C. Hoffman Fndn Grants, 3950
William G. Gilmore Foundation Grants, 3951
Williams Companies Foundation Homegrown Giving Grants, 3957
Winston-Salem Foundation Competitive Grants, 3965
Wolfe Associates Grants, 3969
Women of the ELCA Opportunity Scholarships for Lutheran Laywomen, 3970
Yampa Valley Community Foundation Cody St. John Scholarships, 3977
Yampa Valley Community Foundation Grants, 3978
Yampa Valley Community Foundation Volunteer Firemen Scholarships, 3979
Young Ambassador Scholarship In Memory of Christopher Nordquist, 3981
Z. Smith Reynolds Foundation Small Grants, 3982

Seed Grants
3M Fndn Health & Human Services Grants, 7
100 Mile Man Foundation Grants, 10
A-T Medical Research Foundation Grants, 15
AAAAI RSLAAIS Leadership Award, 33
AAFPRS Bernstein Grant, 138
Abbot and Dorothy H. Stevens Foundation Grants, 197
Abbott Fund Community Grants, 201
Abby's Legendary Pizza Foundation Grants, 204
Abell-Hanger Foundation Grants, 207
Achelis Foundation Grants, 234
ACS Institutional Research Grants, 256
Adams County Community Foundation of Pennsylvania Grants, 300
Adams Foundation Grants, 301
Adelaide Breed Bayrd Foundation Grants, 306
Adolph Coors Foundation Grants, 311
AFAR CART Fund Grants, 334
A Friends' Foundation Trust Grants, 341
Agnes M. Lindsay Trust Grants, 346
Ahmanson Foundation Grants, 355
Air Products and Chemicals Grants, 399
Alabama Power Foundation Grants, 400
Alberto Culver Corporate Contributions Grants, 408
Albuquerque Community Foundation Grants, 412
Alcoa Foundation Grants, 414
Alcon Foundation Grants Program, 416
Alexander and Margaret Stewart Trust Grants, 419
Alfred and Tillie Shemanski Testamentary Trust Grants, 428
Allen P. & Josephine B. Green Fndn Grants, 443
Alliance Healthcare Foundation Grants, 444
Allstate Corporate Giving Grants, 446
Allstate Corp Hometown Commitment Grants, 447
Allyn Foundation Grants, 448
Alpha Natural Resources Corporate Giving, 450
Alpha Omega Foundation Grants, 451
AMA-WPC Joan F. Giambalvo Memorial Scholarships, 480
Amador Community Foundation Grants, 481
AMA Foundation Fund for Better Health Grants, 484
AMA Foundation Seed Grants for Research, 494
American Legacy Foundation National Calls for Proposals Grants, 534
American Psychiatric Foundation Disaster Recovery Fund for Psychiatrists, 553
American Schlafhorst Foundation Grants, 559
amfAR Research Grants, 571
Angels Baseball Foundation Grants, 590
Anna Fitch Ardenghi Trust Grants, 592
Ann Arbor Area Community Foundation Grants, 594
Annenberg Foundation Grants, 596
Anschutz Family Foundation Grants, 602
Aratani Foundation Grants, 649
ARCO Foundation Education Grants, 652
Arkansas Community Foundation Grants, 660
Arkell Hall Foundation Grants, 661
Arlington Community Foundation Grants, 662
Arthur Ashley Williams Foundation Grants, 671
Arthur F. & Arnold M. Frankel Fndn Grants, 672

Arthur M. Blank Family Foundation Inspiring Spaces Grants, 673
AT&T Foundation Civic and Community Service Program Grants, 793
Atlanta Foundation Grants, 796
Atran Foundation Grants, 797
Austin S. Nelson Foundation Grants, 805
Autauga Area Community Foundation Grants, 808
Autzen Foundation Grants, 810
Axe-Houghton Foundation Grants, 815
Babcock Charitable Trust Grants, 816
Bacon Family Foundation Grants, 817
Ball Brothers Foundation General Grants, 820
Bank of America Fndn Matching Gifts, 825
Bank of America Fndn Volunteer Grants, 826
Barker Welfare Foundation Grants, 831
Barra Foundation Community Fund Grants, 833
Barra Foundation Project Grants, 834
Barth Syndrome Foundation Research Grants, 836
Baton Rouge Area Foundation Grants, 838
Battle Creek Community Foundation Grants, 839
Baxter International Corporate Giving Grants, 841
Baxter International Foundation Grants, 843
BCBSM Foundation Physician Investigator Research Awards, 855
Beckley Area Foundation Grants, 867
Beckman Coulter Foundation Grants, 868
Becton Dickinson and Company Grants, 869
Beldon Fund Grants, 871
Bender Foundation Grants, 876
Berks County Community Foundation Grants, 881
Bertha Russ Lytel Foundation Grants, 885
Biogen Foundation Healthcare Professional Education Grants, 900
Blue Grass Community Fndn Harrison Grants, 919
Blue Grass Community Foundation Hudson-Ellis Fund Grants, 920
Blue Mountain Community Foundation Grants, 921
Blue River Community Foundation Grants, 922
Blumenthal Foundation Grants, 925
Bodenwein Public Benevolent Foundation Grants, 928
Bodman Foundation Grants, 929
Boeing Company Contributions Grants, 930
Boettcher Foundation Grants, 931
Boston Foundation Grants, 936
Boston Globe Foundation Grants, 937
BP Foundation Grants, 943
Brookdale Fnd National Group Respite Grants, 966
Brown Advisory Charitable Foundation Grants, 969
Browning-Kimball Foundation Grants, 971
Bullitt Foundation Grants, 973
Cabot Corporation Foundation Grants, 993
Caesars Foundation Grants, 995
California Community Foundation Human Development Grants, 999
Cambridge Community Foundation Grants, 1004
Campbell Soup Foundation Grants, 1007
Capital Region Community Foundation Grants, 1014
Carlisle Foundation Grants, 1025
Carl M. Freeman Foundation Grants, 1027
Carls Foundation Grants, 1030
Carnahan-Jackson Foundation Grants, 1032
Carpenter Foundation Grants, 1035
Carrie E. and Lena V. Glenn Foundation Grants, 1036
Carrie Estelle Doheny Foundation Grants, 1037
CCHD Community Development Grants, 1046
Cemala Foundation Grants, 1077
Central Carolina Community Foundation Community Impact Grants, 1079
Central Okanagan Foundation Grants, 1080
CFFVR Alcoholism and Drug Abuse Grants, 1093
CFFVR Clintonville Area Foundation Grants, 1097
CFFVR Schmidt Family G4 Grants, 1103
CFFVR Women's Fund for the Fox Valley Region Grants, 1107
Chapman Charitable Foundation Grants, 1112
Charles Delmar Foundation Grants, 1115
Charles Edison Fund Grants, 1116
Charles H. Farnsworth Trust Grants, 1119

Charlotte County (FL) Community Foundation Grants, 1126
Children's Cardiomyopathy Foundation Research Grants, 1137
Children's Leukemia Research Association Research Grants, 1138
Chilkat Valley Community Foundation Grants, 1148
Christensen Fund Regional Grants, 1149
Cigna Civic Affairs Sponsorships, 1159
Clarence E. Heller Charitable Foundation Grants, 1166
Clark County Community Foundation Grants, 1175
Claude Worthington Benedum Fndn Grants, 1178
Clayton Baker Trust Grants, 1179
Coastal Community Foundation of South Carolina Grants, 1202
Collins C. Diboll Private Foundation Grants, 1209
Colonel Stanley R. McNeil Foundation Grants, 1211
Columbus Foundation Competitive Grants, 1217
Columbus Foundation Traditional Grants, 1223
Communities Foundation of Texas Grants, 1236
Community Foundation Alliance City of Evansville Endowment Fund Grants, 1238
Community Foundation for Greater Atlanta Clayton County Fund Grants, 1240
Community Foundation for Greater Atlanta Metropolitan Atlanta An Extra Wish Grants, 1243
Community Foundation for Greater Atlanta Morgan County Fund Grants, 1244
Community Foundation for Greater Atlanta Newton County Fund Grants, 1245
Community Foundation for Greater Atlanta Strategic Restructuring Fund Grants, 1246
Community Fndn for Greater Buffalo Grants, 1247
Community Fndn for Monterey County Grants, 1253
Community Fndn for Muskegon County Grants, 1254
Community Foundation for Northeast Michigan Mini-Grants, 1255
Community Fndn for the Capital Region Grants, 1259
Community Foundation of Bartholomew County Heritage Fund Grants, 1262
Community Foundation of Bloomington and Monroe County Grants, 1265
Community Foundation of Boone County Grants, 1266
Community Fndn of Central Illinois Grants, 1267
Community Foundation of Eastern Connecticut General Southeast Grants, 1269
Community Foundation of Greater Birmingham Grants, 1272
Community Foundation of Greater Flint Grants, 1273
Community Foundation of Greater Fort Wayne - Community Endowment and Clarke Endowment Grants, 1274
Community Foundation of Greater Greensboro Community Grants, 1277
Community Foundation of Greater Greensboro Women to Women Fund Grants, 1278
Community Fndn of Howard County Grants, 1286
Community Foundation of Louisville Anna Marble Memorial Fund for Princeton Grants, 1290
Community Foundation of Mount Vernon and Knox County Grants, 1299
Community Foundation of Muncie and Delaware County Grants, 1300
Community Fndn of Randolph County Grants, 1302
Community Fndn of So Alabama Grants, 1306
Community Foundation of Tampa Bay Grants, 1312
Community Foundation of the Eastern Shore Community Needs Grants, 1313
Community Fndn of Wabash County Grants, 1316
Connecticut Community Foundation Grants, 1324
Conservation, Food, and Health Foundation Grants for Developing Countries, 1338
CONSOL Energy Community Dev Grants, 1339
Cooke Foundation Grants, 1344
Cooper Industries Foundation Grants, 1345
Cowles Charitable Trust Grants, 1354
Cralle Foundation Grants, 1356
Crane Fund Grants, 1358
CSRA Community Foundation Grants, 1366
Cystic Fibrosis Canada Clinical Project Grants, 1382

PROGRAM TYPE INDEX

Seed Grants / 905

Dayton Power and Light Company Foundation Signature Grants, 1436
Daywood Foundation Grants, 1438
Decatur County Community Foundation Large Project Grants, 1444
Doris and Victor Day Foundation Grants, 1491
Dorothea Haus Ross Foundation Grants, 1498
Dr. John T. Macdonald Foundation Grants, 1507
Drs. Bruce and Lee Foundation Grants, 1513
DuPont Pioneer Community Giving Grants, 1532
Dyson Foundation Mid-Hudson Valley Project Support Grants, 1534
E. Clayton and Edith P. Gengras, Jr. Foundation, 1536
Echoing Green Fellowships, 1546
Eddie C. & Sylvia Brown Family Fndn Grants, 1548
Educational Foundation of America Grants, 1553
Edward Bangs Kelley and Elza Kelley Foundation Grants, 1556
Edyth Bush Charitable Foundation Grants, 1562
Effie and Wofford Cain Foundation Grants, 1563
Eisner Foundation Grants, 1564
Elizabeth Morse Genius Charitable Trust Grants, 1569
Elmer L. & Eleanor J. Andersen Fndn Grants, 1578
El Paso Community Foundation Grants, 1579
El Pomar Foundation Anna Keesling Ackerman Fund Grants, 1581
El Pomar Foundation Grants, 1582
Elton John AIDS Foundation Grants, 1585
Emily Davie and Joseph S. Kornfeld Foundation Grants, 1591
Emma B. Howe Memorial Foundation Grants, 1593
Erie Community Foundation Grants, 1601
Essex County Community Foundation Discretionary Fund Grants, 1602
Eugene M. Lang Foundation Grants, 1611
Eva L. & Joseph M. Bruening Fndn Grants, 1613
Evan and Susan Bayh Foundation Grants, 1614
Evanston Community Foundation Grants, 1615
Evjue Foundation Grants, 1617
Ewing Halsell Foundation Grants, 1618
F.M. Kirby Foundation Grants, 1624
Fargo-Moorhead Area Foundation Grants, 1632
Federal Express Corporate Contributions, 1657
Field Foundation of Illinois Grants, 1660
Florida Division of Cultural Affairs Arts In Education Arts Partnership Grants, 1679
Foellinger Foundation Grants, 1684
Fondren Foundation Grants, 1685
Foundation for Seacoast Health Grants, 1698
Four County Community Fndn General Grants, 1710
Four County Community Foundation Healthy Senior/Healthy Youth Fund Grants, 1711
Four J Foundation Grants, 1713
Frances L. and Edwin L. Cummings Memorial Fund Grants, 1717
Francis L. Abreu Charitable Trust Grants, 1719
Franklin H. Wells and Ruth L. Wells Foundation Grants, 1725
Frank Stanley Beveridge Foundation Grants, 1730
Fremont Area Community Fndn General Grants, 1746
Fulton County Community Foundation Grants, 1763
G.N. Wilcox Trust Grants, 1768
Gardner Foundation Grants, 1770
Gebbie Foundation Grants, 1777
General Motors Foundation Grants, 1783
George Foundation Grants, 1795
George Gund Foundation Grants, 1796
George H. Hitchings New Investigator Award in Health Research, 1798
George P. Davenport Trust Fund Grants, 1800
George W. Wells Foundation Grants, 1804
Georgiana Goddard Eaton Memorial Grants, 1805
Gil and Dody Weaver Foundation Grants, 1819
GlaxoSmithKline Corporate Grants, 1829
GNOF Exxon-Mobil Grants, 1836
GNOF Norco Community Grants, 1845
Golden Heart Community Foundation Grants, 1850
Good Samaritan Inc Grants, 1855
Graham Foundation Grants, 1862
Grand Haven Area Community Fndn Grants, 1864

Grand Rapids Area Community Foundation Wyoming Grants, 1867
Grand Rapids Community Foundation Grants, 1869
Grand Rapids Community Foundation Ionia County Grants, 1870
Grand Rapids Community Foundation Southeast Ottawa Grants, 1873
Grand Rapids Community Fndn Sparta Grants, 1875
Greater Saint Louis Community Fndn Grants, 1884
Greater Sitka Legacy Fund Grants, 1885
Greater Tacoma Community Foundation Grants, 1886
Greater Worcester Community Foundation Discretionary Grants, 1888
Green Foundation Human Services Grants, 1894
Green River Area Community Fndn Grants, 1895
Grover Hermann Foundation Grants, 1906
Guido A. & Elizabeth H. Binda Fndn Grants, 1913
Gulf Coast Foundation of Community Program Grants, 1917
H.J. Heinz Company Foundation Grants, 1923
H. Schaffer Foundation Grants, 1925
Hardin County Community Foundation Grants, 1946
Harold R. Bechtel Testamentary Charitable Trust Grants, 1954
Harold Simmons Foundation Grants, 1955
Harris and Eliza Kempner Fund Grants, 1956
Harry A. & Margaret D. Towsley Fndn Grants, 1959
Harry Edison Foundation, 1962
Harry Kramer Memorial Fund Grants, 1965
Hartford Courant Foundation Grants, 1969
Hartford Foundation Regular Grants, 1970
Harvey Randall Wickes Foundation Grants, 1972
Hasbro Children's Fund Grants, 1973
Health Fndn of Greater Indianapolis Grants, 1984
Heineman Foundation for Research, Education, Charitable and Scientific Purposes, 1990
Helen S. Boylan Foundation Grants, 1997
Helen Steiner Rice Foundation Grants, 1998
Henrietta Tower Wurts Memorial Fndn Grants, 2002
Henry and Ruth Blaustein Rosenberg Foundation Health Grants, 2004
Herbert H. & Grace A. Dow Fndn Grants, 2008
Hereditary Disease Foundation Research Grants, 2011
Herman Goldman Foundation Grants, 2012
Hill Crest Foundation Grants, 2030
Hilton Head Island Foundation Grants, 2036
Holland/Zeeland Community Fndn Grants, 2041
Horace A. Kimball and S. Ella Kimball Foundation Grants, 2046
Howard and Bush Foundation Grants, 2054
HRAMF Charles H. Hood Foundation Child Health Research Awards, 2057
HRK Foundation Health Grants, 2067
Huber Foundation Grants, 2071
Huffy Foundation Grants, 2073
Huntington's Disease Society of America Research Grants, 2080
Idaho Community Foundation Eastern Region Competitive Grants, 2114
Ike and Roz Friedman Foundation Grants, 2145
Illinois Tool Works Foundation Grants, 2147
Impact 100 Grants, 2148
Indiana 21st Century Research and Technology Fund Awards, 2154
Ittleson Foundation AIDS Grants, 2166
Ittleson Foundation Mental Health Grants, 2167
J.N. and Macie Edens Foundation Grants, 2174
J. Spencer Barnes Memorial Foundation Grants, 2175
J. Walton Bissell Foundation Grants, 2177
Jack H. & William M. Light Trust Grants, 2179
Jackson County Community Foundation Unrestricted Grants, 2180
Jacob G. Schmidlapp Trust Grants, 2183
Jacobs Family Village Neighborhoods Grants, 2184
James Ford Bell Foundation Grants, 2187
James H. Cummings Foundation Grants, 2190
James J. & Angelia M. Harris Fndn Grants, 2192
James L. and Mary Jane Bowman Charitable Trust Grants, 2194
James R. Dougherty Jr. Foundation Grants, 2198

JELD-WEN Foundation Grants, 2218
Jessica Stevens Community Foundation Grants, 2224
Jessie Ball Dupont Fund Grants, 2226
Jewish Fund Grants, 2227
JM Foundation Grants, 2229
John Ben Snow Memorial Trust Grants, 2231
John Deere Foundation Grants, 2234
John G. Duncan Charitable Trust Grants, 2236
John Merck Fund Grants, 2245
John P. Murphy Foundation Grants, 2248
Johns Manville Fund Grants, 2251
Johnson County Community Foundation Grants, 2256
John W. Alden Trust Grants, 2260
John W. and Anna H. Hanes Foundation Grants, 2261
Joseph Drown Foundation Grants, 2268
Josephine Goodyear Foundation Grants, 2272
Josephine Schell Russell Charitable Trust Grants, 2274
Joseph P. Kennedy Jr. Foundation Grants, 2275
Judith Clark-Morrill Foundation Grants, 2281
Kate B. Reynolds Trust Health Care Grants, 2290
Katharine Matthies Foundation Grants, 2291
Katherine John Murphy Foundation Grants, 2293
Kenai Peninsula Foundation Grants, 2298
Ketchikan Community Foundation Grants, 2303
Kettering Fund Grants, 2305
Kinsman Foundation Grants, 2309
Kodiak Community Foundation Grants, 2314
Komen Greater NYC Small Grants, 2317
Kosciusko County Community Fndn Grants, 2319
Kroger Foundation Women's Health Grants, 2325
Kuntz Foundation Grants, 2326
LAM Foundation Research Grants, 2340
Land O'Lakes Foundation Mid-Atlantic Grants, 2341
Laura B. Vogler Foundation Grants, 2345
Layne Beachley Aim for the Stars Fnd Grants, 2349
Lubbock Area Foundation Grants, 2383
M.B. and Edna Zale Foundation Grants, 2401
M.D. Anderson Foundation Grants, 2403
Mabel Y. Hughes Charitable Trust Grants, 2409
Macquarie Bank Foundation Grants, 2410
Manuel D. & Rhoda Mayerson Fndn Grants, 2420
Marathon Petroleum Corporation Grants, 2422
Mardag Foundation Grants, 2428
Marie C. and Joseph C. Wilson Foundation Rochester Small Grants, 2431
Mary Owen Borden Foundation Grants, 2446
McCarthy Family Foundation Grants, 2490
McConnell Foundation Grants, 2492
McCune Foundation Human Services Grants, 2493
McInerny Foundation Grants, 2495
McKesson Foundation Grants, 2496
McLean Contributionship Grants, 2497
MeadWestvaco Foundation Sustainable Communities Grants, 2506
Mead Witter Foundation Grants, 2507
Memorial Foundation Grants, 2516
Merrick Foundation Grants, 2522
Meyer Memorial Trust Grassroots Grants, 2536
Meyer Memorial Trust Responsive Grants, 2537
Meyer Memorial Trust Special Grants, 2538
Morris & Gwendolyn Cafritz Fndn Grants, 2594
Morton K. and Jane Blaustein Foundation Health Grants, 2595
NAF Research Grants, 2605
Nancy R. Gelman Foundation Breast Cancer Seed Grants, 2608
National Center for Resp Gaming Seed Grants, 2631
National Psoriasis Foundation Research Grants, 2636
NIGMS Advancing Basic Behavioral and Social Research on Resilience: an Integrative Science Approach, 2861
Nina Mason Pulliam Charitable Trust Grants, 2921
Noble County Community Foundation Grants, 2975
Norcliffe Foundation Grants, 2979
Nordson Corporation Foundation Grants, 2981
Norfolk Southern Foundation Grants, 2982
North Carolina GlaxoSmithKline Fndn Grants, 2991
Northern Chautauqua Community Foundation Community Grants, 2995
Northern New York Community Fndn Grants, 2996

906 / Seed Grants

Northwest Minnesota Foundation Women's Fund Grants, 3000
Notsew Orm Sands Foundation Grants, 3002
NWHF Health Advocacy Small Grants, 3033
Oak Foundation Child Abuse Grants, 3059
Oppenstein Brothers Foundation Grants, 3078
Oregon Community Fndn Community Grants, 3082
PacifiCare Foundation Grants, 3102
Patrick and Anna M. Cudahy Fund Grants, 3120
Patron Saints Foundation Grants, 3121
Paul Rapoport Foundation Grants, 3125
Pediatric Brain Tumor Fndn Research Grants, 3134
Peter Kiewit Foundation General Grants, 3154
Peter Kiewit Foundation Small Grants, 3155
Petersburg Community Foundation Grants, 3156
Peyton Anderson Foundation Grants, 3160
Phelps County Community Foundation Grants, 3188
Philadelphia Foundation Organizational Effectiveness Grants, 3189
Philip L. Graham Fund Health and Human Services Grants, 3190
PhRMA Foundation Health Outcomes Research Starter Grants, 3204
PhRMA Foundation Informatics Research Starter Grants, 3208
PhRMA Foundation Pharmaceutics Research Starter Grants, 3213
PhRMA Foundation Pharmacology/Toxicology Research Starter Grants, 3217
Piedmont Health Foundation Grants, 3219
Pinkerton Foundation Grants, 3224
Plastic Surgery Educational Foundation Basic Research Grants, 3235
Playboy Foundation Grants, 3237
Plough Foundation Grants, 3238
Pokagon Fund Grants, 3242
Potts Memorial Foundation Grants, 3254
Price Chopper's Golub Foundation Grants, 3261
Priddy Foundation Organizational Development Grants, 3265
Principal Financial Group Foundation Grants, 3273
Prudential Foundation Education Grants, 3276
Quantum Foundation Grants, 3290
Ralph M. Parsons Foundation Grants, 3304
Raskob Foundation for Catholic Activities Grants, 3306
Rathmann Family Foundation Grants, 3309
Rayonier Foundation Grants, 3311
RCF General Community Grants, 3312
Rhode Island Foundation Grants, 3362
Richard and Helen DeVos Foundation Grants, 3364
Richard & Susan Smith Family Fndn Grants, 3365
Richard King Mellon Foundation Grants, 3369
Richland County Bank Grants, 3371
Ripley County Community Foundation Grants, 3376
Ripley County Community Foundation Small Project Grants, 3377
Robert R. Meyer Foundation Grants, 3387
Rochester Area Community Foundation Grants, 3391
Rohm and Haas Company Grants, 3403
Ronald McDonald House Charities Grants, 3405
Roy & Christine Sturgis Charitable Grants, 3415
Roy J. Carver Charitable Trust Medical and Science Research Grants, 3416
Ruth Anderson Foundation Grants, 3425
Ruth H. & Warren A. Ellsworth Fndn Grants, 3429
RWJF Policies for Action: Policy and Law Research to Build a Culture of Health Grants, 3443
S. Livingston Mather Charitable Trust Grants, 3448
S. Mark Taper Foundation Grants, 3449
Saigh Foundation Grants, 3451
Salem Foundation Charitable Trust Grants, 3456
Samuel S. Johnson Foundation Grants, 3468
San Antonio Area Foundation Grants, 3469
San Diego Women's Foundation Grants, 3474
San Francisco Fndn Community Health Grants, 3478
Sartain Lanier Family Foundation Grants, 3486
Schering-Plough Foundation Health Grants, 3492
Scott County Community Foundation Grants, 3500
Seabury Foundation Grants, 3501
Seattle Foundation Health and Wellness Grants, 3507

Seaver Institute Grants, 3509
Sensient Technologies Foundation Grants, 3513
Seward Community Foundation Grants, 3514
Seward Community Foundation Mini-Grants, 3515
Shield-Ayres Foundation Grants, 3546
Sidney Stern Memorial Trust Grants, 3552
Sid W. Richardson Foundation Grants, 3553
Siebert Lutheran Foundation Grants, 3554
Sioux Falls Area Community Foundation Community Fund Grants (Unrestricted), 3568
Sioux Falls Area Community Foundation Spot Grants (Unrestricted), 3570
Sisters of Mercy of North Carolina Fndn Grants, 3575
Sisters of St. Joseph Healthcare Fndn Grants, 3576
SNM/ Covidien Seed Grant in Molecular Imaging/ Nuclear Medicine Research, 3582
Sonora Area Foundation Competitive Grants, 3622
Sony Corporation of America Grants, 3623
St. Joseph Community Health Foundation Improving Healthcare Access Grants, 3640
Stackpole-Hall Foundation Grants, 3643
Stewart Huston Charitable Trust Grants, 3655
Stranahan Foundation Grants, 3659
Strowd Roses Grants, 3664
Summit Foundation Grants, 3674
Sunoco Foundation Grants, 3679
Susan G. Komen Breast Cancer Foundation Challenge Grants: Career Catalyst Research, 3691
Taylor S. Abernathy and Patti Harding Abernathy Charitable Trust Grants, 3704
Textron Corporate Contributions Grants, 3712
Thomas Austin Finch, Sr. Foundation Grants, 3718
Thomas C. Ackerman Foundation Grants, 3719
Todd Brock Family Foundation Grants, 3736
Toyota Motor Engineering & Manufacturing North America Grants, 3742
Toyota Motor Manufacturing of Alabama Grants, 3743
Toyota Motor Manufacturing of Indiana Grants, 3744
Toyota Motor Manufacturing of Kentucky Grants, 3745
Toyota Motor Manuf of Mississippi Grants, 3746
Toyota Motor Manufacturing of Texas Grants, 3747
Toyota Motor Manuf of West Virginia Grants, 3748
Toyota Motor N America of New York Grants, 3749
Tri-State Community Twenty-first Century Endowment Fund Grants, 3756
U.S. Lacrosse EAD Grants, 3775
Union County Community Foundation Grants, 3790
United Healthcare Comm Grants in Michigan, 3796
Unity Foundation Of LaPorte County Grants, 3804
USAID Social Science Research in Population and Reproductive Health Grants, 3822
Verizon Foundation Vermont Grants, 3863
Victor E. Speas Foundation Grants, 3868
Virginia W. Kettering Foundation Grants, 3878
W.K. Kellogg Foundation Secure Families Grants, 3887
W.M. Keck Fndn So California Grants, 3890
Wabash Valley Community Foundation Grants, 3893
Walker Area Community Foundation Grants, 3895
Wayne County Foundation Grants, 3909
Weingart Foundation Grants, 3913
Weyerhaeuser Company Foundation Grants, 3925
William J. & Tina Rosenberg Fndn Grants, 3954
William Ray & Ruth E. Collins Fndn Grants, 3956
Winston-Salem Foundation Competitive Grants, 3965
Z. Smith Reynolds Foundation Small Grants, 3982

Service Delivery Programs
1 in 9: Long Island Breast Cancer Action Coalition Grants, 1
1st Source Foundation Grants, 2
2 Depot Square Ipswich Charitable Fndn Grants, 4
2COBS Private Charitable Foundation Grants, 5
3M Company Fndn Community Giving Grants, 6
3M Fndn Health & Human Services Grants, 7
49ers Foundation Grants, 9
1976 Foundation Grants, 12
A. Gary Anderson Family Foundation Grants, 16
A.O. Smith Foundation Community Grants, 17
AAHPERD Abernathy Presidential Scholarships, 151
AAHPM Hospice & Palliative Med Fellowships, 152

PROGRAM TYPE INDEX

AAMC Caring for Community Grants, 156
AAP Community Access To Child Health (CATCH) Advocacy Training Grants, 177
AAP Community Access To Child Health (CATCH) Implementation Grants, 178
AAP Community Access to Child Health (CATCH) Planning Grants, 179
AAP Community Access To Child Health (CATCH) Residency Training Grants, 180
AAP Community Access To Child Health (CATCH) Resident Grants, 181
AAP Leonard P. Rome Community Access to Child Health (CATCH) Visiting Professorships, 186
AAP Resident Initiative Fund Grants, 189
Aaron & Freda Glickman Foundation Grants, 192
Aaron Foundation Grants, 193
Aaron Foundation Grants, 194
Abbot and Dorothy H. Stevens Foundation Grants, 197
Abbott Fund Access to Health Care Grants, 199
Abbott Fund Community Grants, 201
Abbott Fund Global AIDS Care Grants, 202
Abbott Fund Science Education Grants, 203
Abel Foundation Grants, 206
Abell-Hanger Foundation Grants, 207
Abell Foundation Criminal Justice and Addictions Grants, 208
Abell Fndn Health & Human Services Grants, 209
Abernethy Family Foundation Grants, 210
Abington Foundation Grants, 211
Able To Serve Grants, 212
Able Trust Vocational Rehabilitation Grants for Individuals, 213
Aboudane Family Foundation Grants, 215
Abramson Family Foundation Grants, 217
ACE Charitable Foundation Grants, 229
ACF Foundation Grants, 230
ACF Native American Social and Economic Development Strategies Grants, 231
Achelis Foundation Grants, 234
ACL Alzheimer's Disease Initiative Grants, 237
ACL Business Acumen for Disability Organizations Grants, 238
ACL Centers for Independent Living Competition Grants, 239
ACL Empowering Seniors to Prevent Health Care Fraud Grants, 241
ACL Field Initiated Projects Program: Minority-Serving Institution Development Grants, 242
ACL Field Initiated Projects Program: Minority-Serving Institution Research Grants, 243
ACL Learning Collaboratives for Advanced Business Acumen Skills Grants, 244
ACL Medicare Improvements for Patients & Providers Funding for Beneficiary Outreach & Assistance for Title VI Native Americans, 245
ACL Partnerships in Employment Systems Change Grants, 246
ACL Self-Advocacy Resource Center Grants, 247
Actors Addiction & Recovery Services Grants, 282
Actors Fund HIV/AIDS Initiative Grants, 283
ADA Foundation Disaster Assistance Grants, 290
ADA Foundation Samuel Harris Children's Dental Health Grants, 294
Adam Richter Charitable Trust Grants, 298
Adams County Community Foundation of Pennsylvania Grants, 300
Adams Rotary Memorial Fund A Grants, 302
Adaptec Foundation Grants, 304
Addison H. Gibson Foundation Medical Grants, 305
Adelaide Breed Bayrd Foundation Grants, 306
Adler-Clark Electric Community Commitment Foundation Grants, 307
Administaff Community Affairs Grants, 308
Administration on Aging Senior Medicare Patrol Project Grants, 309
Adobe Community Investment Grants, 310
Advance Auto Parts Corporate Giving Grants, 312
Advanced Micro Devices Comm Affairs Grants, 313
Advocate HealthCare Post Graduate Administrative Fellowship, 314

PROGRAM TYPE INDEX

Service Delivery Programs / 907

AEC Trust Grants, 315
AEGON Transamerica Foundation Health and Welfare Grants, 316
Aetna Foundation Health Grants in Connecticut, 327
Aetna Foundation Obesity Grants, 331
Aetna Foundation Racial and Ethnic Health Care Equity Grants, 332
Aetna Foundation Regional Health Grants, 333
AFG Industries Grants, 340
A Friends' Foundation Trust Grants, 341
A Fund for Women Grants, 342
Agnes B. Hunt Trust Grants, 344
Agnes M. Lindsay Trust Grants, 346
A Good Neighbor Foundation Grants, 347
Agway Foundation Grants, 348
Ahmanson Foundation Grants, 355
Ahn Family Foundation Grants, 356
AIDS Vaccine Advocacy Coalition (AVAC) Fund Grants, 369
Air Products and Chemicals Grants, 399
Alaska Airlines Corporate Giving Medical Emergency and Research Grants, 402
Alaska Airlines Foundation Grants, 403
Albert and Margaret Alkek Foundation Grants, 405
Alberto Culver Corporate Contributions Grants, 408
Albert Pick Jr. Fund Grants, 409
Albertson's Charitable Giving Grants, 410
Albert W. Rice Charitable Foundation Grants, 411
Albuquerque Community Foundation Grants, 412
Alcatel-Lucent Technologies Foundation Grants, 413
Alcoa Foundation Grants, 414
Alexander & Baldwin Fnd Mainland Grants, 417
Alexander and Baldwin Foundation Hawaiian and Pacific Island Grants, 418
Alexander and Margaret Stewart Trust Grants, 419
Alexander Eastman Foundation Grants, 420
Alexander Foundation Cancer, Catastrophic Illness and Injury Grants, 421
Alexander Fndn Insurance Continuation Grants, 422
Alfred and Tillie Shemanski Testamentary Trust Grants, 428
Alfred E. Chase Charitable Foundation Grants, 430
Alice C. A. Sibley Fund Grants, 435
Allan C. and Lelia J. Garden Foundation Grants, 438
Allegan County Community Foundation Grants, 439
Allegheny Technologies Charitable Trust, 440
Allegis Group Foundation Grants, 441
Allen P. & Josephine B. Green Fndn Grants, 443
Alliance Healthcare Foundation Grants, 444
Alliant Energy Corporation Contributions, 445
Allstate Corporate Giving Grants, 446
Allstate Corp Hometown Commitment Grants, 447
Allyn Foundation Grants, 448
Alpha Natural Resources Corporate Giving, 450
ALSAM Foundation Grants, 454
Alvin and Fanny Blaustein Thalheimer Foundation Baltimore Communal Grants, 459
Amador Community Foundation Grants, 481
AMA Foundation Fund for Better Health Grants, 484
AMA Foundation Health Literacy Grants, 485
AMA Foundation Healthy Communities/Healthy America Grants, 486
AMA Foundation Seed Grants for Research, 494
AMD Corporate Contributions Grants, 500
Amelia Sillman Rockwell and Carlos Perry Rockwell Charities Fund Grants, 501
American Academy of Dermatology Camp Discovery Scholarships, 503
American Electric Power Foundation Grants, 530
American Express Community Service Grants, 531
American Jewish World Service Grants, 533
American Legacy Foundation National Calls for Proposals Grants, 534
American Psychiatric Foundation Grantss, 552
American Psychiatric Foundation Disaster Recovery Fund for Psychiatrists, 553
American Psychiatric Foundation Helping Hands Grants, 554
Amerigroup Foundation Grants, 565
Amerisure Insurance Community Service Grants, 566

amfAR Global Initiatives Grants, 568
Amgen Foundation Grants, 572
Amica Companies Foundation Grants, 574
Amica Insurance Company Community Grants, 575
AMI Semiconductors Corporate Grants, 576
Andre Agassi Charitable Foundation Grants, 587
Angel Kiss Foundation Grants, 589
Angels Baseball Foundation Grants, 590
Anheuser-Busch Foundation Grants, 591
Anna Fitch Ardenghi Trust Grants, 592
Ann and Robert H. Lurie Family Fndn Grants, 593
Ann Arbor Area Community Foundation Grants, 594
Anne J. Caudal Foundation Grants, 595
Annie Sinclair Knudsen Memorial Fund/Kaua'i Community Grants Program, 597
Ann Peppers Foundation Grants, 599
Annunziata Sanguinetti Foundation Grants, 600
Ansell, Zaro, Grimm & Aaron Foundation Grants, 603
Anthem Blue Cross and Blue Shield Grants, 607
Anthony R. Abraham Foundation Grants, 608
AON Foundation Grants, 623
APA Culture of Service in the Psychological Sciences Award, 625
APA Meritorious Research Service Award, 631
Appalachian Regional Commission Health Grants, 637
APSAA Foundation Grants, 639
APS Foundation Grants, 646
Aratani Foundation Grants, 649
Arcadia Foundation Grants, 650
Archer Daniels Midland Foundation Grants, 651
ARCO Foundation Education Grants, 652
Argyros Foundation Grants Program, 653
Arie and Ida Crown Memorial Grants, 654
Arizona Cardinals Grants, 655
Arizona Diamondbacks Charities Grants, 657
Arkansas Community Foundation Arkansas Black Hall of Fame Grants, 659
Arkansas Community Foundation Grants, 660
Arkell Hall Foundation Grants, 661
Arlington Community Foundation Grants, 662
ARS Foundation Grants, 667
Arthur and Rochelle Belfer Foundation Grants, 670
Arthur Ashley Williams Foundation Grants, 671
Arthur F. & Arnold M. Frankel Fndn Grants, 672
ArvinMeritor Foundation Health Grants, 674
ArvinMeritor Grants, 675
ASHA Advancing Academic-Research Award, 695
ASHA Multicultural Activities Projects Grants, 697
Ashland Corporate Contributions Grants, 713
Assisi Foundation of Memphis General Grants, 787
Assurant Foundation Grants, 788
AT&T Foundation Civic and Community Service Program Grants, 793
Atherton Family Foundation Grants, 794
Athwin Foundation Grants, 795
Atlanta Foundation Grants, 796
Atran Foundation Grants, 797
Auburn Foundation Grants, 798
Audrey & Sydney Irmas Foundation Grants, 799
Austin-Bailey Health and Wellness Fndn Grants, 803
Austin S. Nelson Foundation Grants, 805
Autauga Area Community Foundation Grants, 808
Autodesk Community Relations Grants, 809
Autzen Foundation Grants, 810
AVDF Health Care Grants, 811
Avista Foundation Economic and Cultural Vitality Grants, 812
Avon Foundation Breast Care Fund Grants, 813
Avon Products Foundation Grants, 814
Babcock Charitable Trust Grants, 816
Bailey Foundation Grants, 818
Balfe Family Foundation Grants, 819
Ball Brothers Foundation General Grants, 820
Banfi Vintners Foundation Grants, 824
Bank of America Fndn Matching Gifts, 825
Baptist-Trinity Lutheran Legacy Fndn Grants, 827
Baptist Community Ministries Grants, 828
Barberton Community Foundation Grants, 829
Barker Foundation Grants, 830
Barker Welfare Foundation Grants, 831

Barra Foundation Community Fund Grants, 833
Barra Foundation Project Grants, 834
Barr Fund Grants, 835
Batchelor Foundation Grants, 837
Baton Rouge Area Foundation Grants, 838
Battle Creek Community Foundation Grants, 839
Baxter International Corporate Giving Grants, 841
Baxter International Foundation Grants, 843
BBVA Compass Foundation Charitable Grants, 848
BCBSM Claude Pepper Award, 850
BCBSM Foundation Community Health Matching Grants, 852
BCBSM Fndn Proposal Development Awards, 857
BCBSNC Foundation Grants, 860
Bearemy's Kennel Pals Grants, 863
Beazley Foundation Grants, 865
Beckley Area Foundation Grants, 867
Beckman Coulter Foundation Grants, 868
Becton Dickinson and Company Grants, 869
Belk Foundation Grants, 872
Belvedere Community Foundation Grants, 874
Ben B. Cheney Foundation Grants, 875
Bender Foundation Grants, 876
Benton Community Foundation Grants, 878
Benton County Foundation Grants, 880
Berks County Community Foundation Grants, 881
Besser Foundation Grants, 887
BHHS Legacy Foundation Grants, 888
BibleLands Grants, 889
Bildner Family Foundation Grants, 891
Bill Hannon Foundation Grants, 895
Birmingham Foundation Grants, 903
BJ's Charitable Foundation Grants, 904
Black River Falls Area Foundation Grants, 909
Blanche and Irving Laurie Foundation Grants, 910
Blowitz-Ridgeway Foundation Grants, 913
Blue Cross Blue Shield of Minnesota Foundation - Health Equity: Building Health Equity Together Grants, 914
Blue Cross Blue Shield of Minnesota Foundation - Healthy Children: Growing Up Healthy Grants, 915
Blue Grass Community Fndn Harrison Grants, 919
Blue Grass Community Foundation Hudson-Ellis Fund Grants, 920
Blue Mountain Community Foundation Grants, 921
Blue River Community Foundation Grants, 922
Blue Shield of California Grants, 923
Blum-Kovler Foundation Grants, 924
Blumenthal Foundation Grants, 925
Bodenwein Public Benevolent Foundation Grants, 928
Bodman Foundation Grants, 929
Boettcher Foundation Grants, 931
Booth-Bricker Fund Grants, 933
Boston Foundation Grants, 936
Boston Globe Foundation Grants, 937
Boston Jewish Community Women's Fund Grants, 938
BP Foundation Grants, 943
Bradley-Turner Foundation Grants, 944
Bradley C. Higgins Foundation Grants, 945
Brian G. Dyson Foundation Grants, 948
Bridgestone/Firestone Trust Fund Grants, 949
Brighter Tomorrow Foundation Grants, 950
Bright Family Foundation Grants, 951
Bright Promises Foundation Grants, 952
Bristol-Myers Squibb Foundation Global HIV/AIDS Initiative Grants, 957
Bristol-Myers Squibb Foundation Health Disparities Grants, 958
Bristol-Myers Squibb Foundation Women's Health Grants, 962
Bristol-Myers Squibb Patient Assistance Grants, 963
Brookdale Foundation Leadership in Aging Fellowships, 965
Brown County Community Foundation Grants, 970
Browning-Kimball Foundation Grants, 971
Burden Trust Grants, 976
Burlington Northern Santa Fe Foundation Grants, 978
Bush Fndn Health & Human Services Grants, 979
Business Bank of Nevada Community Grants, 981
Byron W. & Alice L. Lockwood Fnd Grants, 991

Service Delivery Programs

Cable Positive's Tony Cox Community Grants, 992
Caddock Foundation Grants, 994
Caesars Foundation Grants, 995
Caleb C. and Julia W. Dula Educational and Charitable Foundation Grants, 997
California Community Foundation Health Care Grants, 998
California Community Foundation Human Development Grants, 999
California Endowment Innovative Ideas Challenge Grants, 1000
California Wellness Foundation Work and Health Program Grants, 1001
Callaway Golf Company Foundation Grants, 1003
Cambridge Community Foundation Grants, 1004
Camp-Younts Foundation Grants, 1005
Campbell Hoffman Foundation Grants, 1006
Canadian Optometric Education Trust Grants, 1011
Cardinal Health Foundation Grants, 1017
Cargill Citizenship Fund Corporate Giving, 1018
Carl & Eloise Pohlad Family Fndn Grants, 1021
Carl C. Icahn Foundation Grants, 1023
Carlisle Foundation Grants, 1025
Carl M. Freeman Foundation FACES Grants, 1026
Carl M. Freeman Foundation Grants, 1027
Carl R. Hendrickson Family Foundation Grants, 1028
Carlsbad Charitable Foundation Grants, 1029
Carl W. and Carrie Mae Joslyn Trust Grants, 1031
Carnahan-Jackson Foundation Grants, 1032
Carolyn Foundation Grants, 1034
Carpenter Foundation Grants, 1035
Carrie Estelle Doheny Foundation Grants, 1037
Cass County Community Foundation Grants, 1041
Catherine Holmes Wilkins Foundation Grants, 1042
Catherine Kennedy Home Foundation Grants, 1043
CCHD Community Development Grants, 1046
CDC Evaluation of the Use of Rapid Testing For Influenza in Outpatient Medical Settings, 1057
CDC Increasing Breast and Cervical Cancer Screening Services for Urban American Indian/Alaska Native Women, 1063
Cemala Foundation Grants, 1077
Centerville-Washington Foundation Grants, 1078
Central Carolina Community Foundation Community Impact Grants, 1079
Cessna Foundation Grants Program, 1081
CFFVR Basic Needs Giving Partnership Grants, 1094
CFFVR Project Grants, 1101
CFFVR Shawano Area Community Foundation Grants, 1104
CFFVR Waupaca Area Community Foundation Grants, 1105
Champ-A Champion Fur Kids Grants, 1109
Champlin Foundations Grants, 1110
Chapman Charitable Foundation Grants, 1112
CharityWorks Grants, 1113
Charles A. Frueauff Foundation Grants, 1114
Charles F. Bacon Trust Grants, 1117
Charles H. Dater Foundation Grants, 1118
Charles H. Farnsworth Trust Grants, 1119
Charles H. Hall Foundation, 1120
Charles H. Pearson Foundation Grants, 1121
Charles Lafitte Foundation Grants, 1123
Charles M. & Mary D. Grant Fndn Grants, 1124
Charles Nelson Robinson Fund Grants, 1125
Charlotte County (FL) Community Foundation Grants, 1126
Chase Paymentech Corporate Giving Grants, 1127
Chatlos Foundation Grants Program, 1128
Chazen Foundation Grants, 1129
Children's Trust Fund of Oregon Fndn Grants, 1139
Children Affected by AIDS Foundation Camp Network Grants, 1144
Children Affected by AIDS Foundation Domestic Grants, 1145
Children Affected by AIDS Foundation Family Assistance Emergency Fund Grants, 1146
Chilkat Valley Community Foundation Grants, 1148
Christine & Katharina Pauly Trust Grants, 1150

Christopher & Dana Reeve Foundation Quality of Life Grants, 1152
Christy-Houston Foundation Grants, 1153
Chula Vista Charitable Foundation Grants, 1154
CICF City of Noblesville Community Grant, 1155
CICF Indianapolis Fndn Community Grants, 1156
CICF Legacy Fund Grants, 1158
Cigna Civic Affairs Sponsorships, 1159
CIGNA Foundation Grants, 1160
CIT Corporate Giving Grants, 1161
Citizens Bank Mid-Atlantic Charitable Foundation Grants, 1162
CJ Foundation for SIDS Program Services Grants, 1163
Clara Blackford Smith and W. Aubrey Smith Charitable Foundation Grants, 1165
Clarence E. Heller Charitable Foundation Grants, 1166
Clark-Winchcole Foundation Grants, 1173
Clark and Ruby Baker Foundation Grants, 1174
Clark County Community Foundation Grants, 1175
Claude Bennett Family Foundation Grants, 1176
Claude Worthington Benedum Fndn Grants, 1178
Clayton Baker Trust Grants, 1179
Clayton Fund Grants, 1180
Cleveland-Cliffs Foundation Grants, 1181
Cleveland Browns Foundation Grants, 1182
CLIF Bar Family Foundation Grants, 1183
Clinton County Community Foundation Grants, 1184
CMS Research and Demonstration Grants, 1188
CNA Foundation Grants, 1189
CNCS Senior Companion Program Grants, 1191
CNCS Social Innovation Grants, 1192
CNO Financial Group Community Grants, 1201
Coastal Community Foundation of South Carolina Grants, 1202
Cockrell Foundation Grants, 1204
Coleman Foundation Developmental Disabilities Grants, 1207
Collins Foundation Grants, 1210
Colonel Stanley R. McNeil Foundation Grants, 1211
Columbus Foundation Allen Eiry Fund Grants, 1214
Columbus Foundation Competitive Grants, 1217
Columbus Foundation Estrich Fund Grants, 1218
Columbus Foundation J. Floyd Dixon Memorial Fund Grants, 1219
Columbus Foundation Mary Eleanor Morris Fund Grants, 1220
Columbus Foundation Paul G. Duke Grants, 1221
Columbus Foundation Robert E. and Genevieve B. Schaefer Fund Grants, 1222
Columbus Foundation Traditional Grants, 1223
Comerica Charitable Foundation Grants, 1224
Commonwealth Edison Grants, 1225
Commonwealth Fund Small Grants, 1234
Community Fndn AIDS Endowment Awards, 1237
Community Foundation Alliance City of Evansville Endowment Fund Grants, 1238
Community Foundation for Greater Atlanta Clayton County Fund Grants, 1240
Community Foundation for Greater Atlanta Morgan County Fund Grants, 1244
Community Foundation for Greater Atlanta Newton County Fund Grants, 1245
Community Fndn for Monterey County Grants, 1253
Community Fndn for Muskegon County Grants, 1254
Community Foundation for Northeast Michigan Mini-Grants, 1255
Community Fndn for the Capital Region Grants, 1259
Community Foundation of Abilene Celebration of Life Grants, 1261
Community Foundation of Bartholomew County Heritage Fund Grants, 1262
Community Foundation of Bloomington and Monroe County Grants, 1265
Community Foundation of Boone County Grants, 1266
Community Fndn of Central Illinois Grants, 1267
Community Fndn of East Central Illinois Grants, 1268
Community Foundation of Eastern Connecticut General Southeast Grants, 1269
Community Foundation of Eastern Connecticut Northeast Women and Girls Grants, 1270

Community Foundation of Greater Birmingham Grants, 1272
Community Foundation of Greater Flint Grants, 1273
Community Foundation of Greater Fort Wayne - Community Endowment and Clarke Endowment Grants, 1274
Community Foundation of Greater Greensboro Community Grants, 1277
Community Fndn of Greater Tampa Grants, 1282
Community Foundation of Greenville Hollingsworth Funds Program/Project Grants, 1285
Community Foundation of Jackson County Seymour Noon Lions Club Grant, 1288
Community Foundation of Louisville AIDS Project Fund Grants, 1289
Community Foundation of Louisville Anna Marble Memorial Fund for Princeton Grants, 1290
Community Foundation of Louisville Diller B. and Katherine P. Groff Fund for Pediatric Surgery Grants, 1293
Community Fndn of Louisville Health Grants, 1295
Community Foundation of Louisville Lee Look Fund for Spinal Injury Grants, 1297
Community Foundation of Mount Vernon and Knox County Grants, 1299
Community Foundation of Muncie and Delaware County Grants, 1300
Community Foundation of Muncie and Delaware County Maxon Grants, 1301
Community Foundation of Riverside & San Bernardino County Impact Grants, 1303
Community Fndn of Riverside & San Bernardino County Irene S. Rockwell Grants, 1304
Community Foundation of Riverside and San Bernardino County James Bernard and Mildred Jordan Tucker Grants, 1305
Community Fndn of South Puget Sound Grants, 1307
Community Foundation of St. Joseph County Special Project Challenge Grants, 1310
Community Foundation of the Verdugos Grants, 1315
Community Foundation Partnerships - Lawrence County Grants, 1317
Community Foundation Partnerships - Martin County Grants, 1318
Community Foundation Serving Riverside and San Bernardino Counties Impact Grants, 1319
Comprehensive Health Education Fndn Grants, 1320
Cone Health Foundation Grants, 1323
Connecticut Community Foundation Grants, 1324
Connecticut Health Foundation Health Initiative Grants, 1325
Connelly Foundation Grants, 1326
ConocoPhillips Foundation Grants, 1327
Conquer Cancer Foundation of ASCO Improving Cancer Care Grants, 1331
Conservation, Food, and Health Foundation Grants for Developing Countries, 1338
CONSOL Energy Community Dev Grants, 1339
CONSOL Military and Armed Services Grants, 1340
Constantin Foundation Grants, 1341
Consumers Energy Foundation, 1343
Cooke Foundation Grants, 1344
Cooper Industries Foundation Grants, 1345
Coors Brewing Corporate Contributions Grants, 1346
Cornerstone Foundation of Northeastern Wisconsin Grants, 1347
Covenant Educational Foundation Grants, 1348
Covenant Matters Foundation Grants, 1350
Covidien Medical Product Donations, 1352
Covidien Partnership for Neighborhood Wellness Grants, 1353
Cowles Charitable Trust Grants, 1354
Crail-Johnson Foundation Grants, 1355
Cralle Foundation Grants, 1356
Crane Fund Grants, 1358
Cresap Family Foundation Grants, 1359
CRH Foundation Grants, 1360
Cruise Industry Charitable Foundation Grants, 1363
Crystelle Waggoner Charitable Trust Grants, 1364

PROGRAM TYPE INDEX

Service Delivery Programs / 909

CSL Behring Local Empowerment for Advocacy Development (LEAD) Grants, 1365
CSRA Community Foundation Grants, 1366
CSX Corporate Contributions Grants, 1368
Cudd Foundation Grants, 1370
Cuesta Foundation Grants, 1371
Cullen Foundation Grants, 1372
Cultural Society of Filipino Americans Grants, 1373
CUNA Mutual Group Fndn Community Grants, 1375
CVS All Kids Can Grants, 1377
CVS Caremark Charitable Trust Grants, 1378
CVS Community Grants, 1379
Cyrus Eaton Foundation Grants, 1380
D.F. Halton Foundation Grants, 1405
D.V. and Ida J. McEachern Trust Grants, 1406
DaimlerChrysler Corporation Fund Grants, 1412
Dairy Queen Corporate Contributions Grants, 1413
Dallas Mavericks Foundation Grants, 1416
Dammann Fund Grants, 1418
Dana Brown Charitable Trust Grants, 1419
Daniels Fund Grants-Aging, 1424
Daphne Seybolt Culpeper Fndn Grants, 1425
David Geffen Foundation Grants, 1431
David M. and Marjorie D. Rosenberg Foundation Grants, 1432
David N. Lane Trust Grants for Aged and Indigent Women, 1433
Daviess County Community Foundation Health Grants, 1434
Dayton Power and Light Company Foundation Signature Grants, 1436
Dayton Power and Light Foundation Grants, 1437
Daywood Foundation Grants, 1438
Deaconess Community Foundation Grants, 1439
Dean Foods Community Involvement Grants, 1440
Delaware Community Foundation Grants, 1449
Della B. Gardner Fund Grants, 1451
Del Mar Healthcare Fund Grants, 1453
Delmarva Power and Light Contributions, 1454
DeMatteis Family Foundation Grants, 1457
Dennis & Phyllis Washington Fndn Grants, 1458
Denver Broncos Charities Fund Grants, 1460
Denver Foundation Community Grants, 1461
DeRoy Testamentary Foundation Grants, 1469
Detlef Schrempf Foundation Grants, 1470
DHHS Adolescent Family Life Demonstration Projects, 1472
DHHS Oral Health Promotion Research Across the Lifespan, 1474
Dickson Foundation Grants, 1476
DIFFA/Chicago Grants, 1477
Disable American Veterans Charitable Grants, 1479
DOJ Gang-Free Schools and Communities Intervention Grants, 1481
Dolan Children's Foundation Grants, 1482
Dole Food Company Charitable Contributions, 1483
Dominion Foundation Grants, 1485
Donald and Sylvia Robinson Family Foundation Grants, 1486
Dora Roberts Foundation Grants, 1489
Doree Taylor Charitable Foundation, 1490
Doris and Victor Day Foundation Grants, 1491
Doris Duke Charitable Foundation Operations Research on AIDS Care and Treatment in Africa (ORACTA) Grants, 1497
Dorothy Hooper Beattie Foundation Grants, 1499
Dorothy Rider Pool Health Care Grants, 1500
Dorrance Family Foundation Grants, 1501
Do Something Awards, 1503
Dr. & Mrs. Paul Pierce Memorial Fndn Grants, 1505
Dr. John T. Macdonald Foundation Grants, 1507
Dr. Scholl Foundation-Grants Program, 1511
Drs. Bruce and Lee Foundation Grants, 1513
DTE Energy Foundation Health and Human Services Grants, 1516
Duke Endowment Health Care Grants, 1519
Dunspaugh-Dalton Foundation Grants, 1530
DuPage Community Foundation Grants, 1531
DuPont Pioneer Community Giving Grants, 1532
E. Clayton and Edith P. Gengras, Jr. Foundation, 1536

E.J. Grassmann Trust Grants, 1537
E. Rhodes & Leona B. Carpenter Grants, 1539
Earl and Maxine Claussen Trust Grants, 1541
Eastman Chemical Company Foundation Grants, 1542
Eckerd Corporation Foundation Grants, 1547
Eddie C. & Sylvia Brown Family Fndn Grants, 1548
Eden Hall Foundation Grants, 1549
Edina Realty Foundation Grants, 1550
EDS Foundation Grants, 1552
Edward and Helen Bartlett Foundation Grants, 1554
Edward and Romell Ackley Foundation Grants, 1555
Edward Bangs Kelley and Elza Kelley Foundation Grants, 1556
Edward N. and Della L. Thome Memorial Foundation Grants, 1557
Edwards Memorial Trust Grants, 1558
Edwin W. and Catherine M. Davis Fndn Grants, 1561
Edyth Bush Charitable Foundation Grants, 1562
Effie and Wofford Cain Foundation Grants, 1563
Eisner Foundation Grants, 1564
Elaine Feld Stern Charitable Trust Grants, 1565
Elizabeth Morse Genius Charitable Trust Grants, 1569
Elkhart County Community Foundation Grants, 1573
Elmer L. & Eleanor J. Andersen Fndn Grants, 1578
El Paso Community Foundation Grants, 1579
El Pomar Foundation Anna Keesling Ackerman Fund Grants, 1581
El Pomar Foundation Grants, 1582
Elsie H. Wilcox Foundation Grants, 1583
Emerson Charitable Trust Grants, 1589
Emily Davie and Joseph S. Kornfeld Foundation Grants, 1591
Emma B. Howe Memorial Foundation Grants, 1593
Emma Barnsley Foundation Grants, 1594
Entergy Corporation Micro Grants, 1596
Entergy Corporation Open Grants for Healthy Families, 1597
EPA Children's Health Protection Grants, 1598
Erie Chapman Foundation Grants, 1600
Erie Community Foundation Grants, 1601
Essex County Community Foundation Discretionary Fund Grants, 1602
Essex County Community Foundation Merrimack Valley General Fund Grants, 1603
Essex County Community Foundation Women's Fund Grants, 1605
Estee Lauder Grants, 1606
Ethel S. Abbott Charitable Foundation Grants, 1607
Ethel Sergeant Clark Smith Foundation Grants, 1608
Eugene G. and Margaret M. Blackford Memorial Fund Grants, 1610
Eugene M. Lang Foundation Grants, 1611
Eva L. & Joseph M. Bruening Fndn Grants, 1613
Evan and Susan Bayh Foundation Grants, 1614
Evanston Community Foundation Grants, 1615
Expect Miracles Foundation Grants, 1619
Express Scripts Foundation Grants, 1620
ExxonMobil Foundation Malaria Grants, 1621
F.M. Kirby Foundation Grants, 1624
Fairfield County Community Foundation Grants, 1625
Fairlawn Foundation Grants, 1626
Fallon OrNda Community Health Fund Grants, 1627
Fannie E. Rippel Foundation Grants, 1630
Fargo-Moorhead Area Foundation Grants, 1632
Farmers Insurance Corporate Giving Grants, 1634
Faye McBeath Foundation Grants, 1635
FCD New American Children Grants, 1637
Federal Express Corporate Contributions, 1657
Ferree Foundation Grants, 1658
Fidelity Foundation Grants, 1659
Field Foundation of Illinois Grants, 1660
Fifth Third Foundation Grants, 1661
Firelight Foundation Grants, 1667
FirstEnergy Foundation Community Grants, 1668
Fisher Foundation Grants, 1670
Flextronics Foundation Disaster Relief Grants, 1674
Florence Hunt Maxwell Foundation Grants, 1676
Floyd A. & Kathleen C. Cailloux Fnd Grants, 1681
Foellinger Foundation Grants, 1684
Fondren Foundation Grants, 1685

Ford Family Foundation Grants - Access to Health and Dental Services, 1686
Ford Motor Company Fund Grants Program, 1687
Forrest C. Lattner Foundation Grants, 1688
Foundation for a Healthy Kentucky Grants, 1691
Foundation for Health Enhancement Grants, 1696
Foundation for Seacoast Health Grants, 1698
Foundation for the Mid South Community Development Grants, 1699
Foundation for the Mid South Health and Wellness Grants, 1700
Foundation of CVPH April LaValley Grants, 1701
Foundation of CVPH Chelsea's Rainbow Grants, 1702
Foundation of CVPH Melissa Lahtinen-Penfield Organ Donor Fund Grants, 1703
Foundation of CVPH Rabin Fund Grants, 1704
Foundation of CVPH Roger Senecal Endowment Fund Grants, 1705
Foundations of East Chicago Health Grants, 1709
Four County Community Fndn General Grants, 1710
Four County Community Foundation Healthy Senior/Healthy Youth Fund Grants, 1711
Fourjay Foundation Grants, 1712
Four J Foundation Grants, 1713
France-Merrick Foundation Health and Human Services Grants, 1715
Frances and John L. Loeb Family Fund Grants, 1716
Frances L. and Edwin L. Cummings Memorial Fund Grants, 1717
Francis L. Abreu Charitable Trust Grants, 1719
Francis T. & Louise T. Nichols Fndn Grants, 1720
Frank B. Hazard General Charity Fund Grants, 1721
Frank E. and Seba B. Payne Foundation Grants, 1722
Frank G. and Freida K. Brotz Family Foundation Grants, 1723
Franklin H. Wells and Ruth L. Wells Foundation Grants, 1725
Frank Reed and Margaret Jane Peters Memorial Fund I Grants, 1727
Frank Reed and Margaret Jane Peters Memorial Fund II Grants, 1728
Frank Stanley Beveridge Foundation Grants, 1730
Frank W. & Carl S. Adams Memorial Grants, 1731
Fraser-Parker Foundation Grants, 1732
Fred & Gretel Biel Charitable Trust Grants, 1734
Fred Baldwin Memorial Foundation Grants, 1736
Freddie Mac Foundation Grants, 1738
Frederick McDonald Trust Grants, 1740
Fred L. Emerson Foundation Grants, 1743
Fremont Area Community Foundation Amazing X Grants, 1744
Fremont Area Community Fndn General Grants, 1746
Fuji Film Grants, 1748
Fuller E. Callaway Foundation Grants, 1761
G.A. Ackermann Memorial Fund Grants, 1767
G.N. Wilcox Trust Grants, 1768
Gardner Foundation Grants, 1771
Gardner W. & Joan G. Heidrick, Jr. Fndn Grants, 1773
Garland D. Rhoads Foundation, 1774
Gebbie Foundation Grants, 1777
GEICO Public Service Awards, 1778
GenCorp Foundation Grants, 1779
Genentech Corp Charitable Contributions, 1780
General Mills Champs for Healthy Kids Grants, 1781
General Mills Foundation Grants, 1782
Genuardi Family Foundation Grants, 1787
George A. and Grace L. Long Foundation Grants, 1788
George and Ruth Bradford Foundation Grants, 1789
George E. Hatcher, Jr. and Ann Williams Hatcher Foundation Grants, 1792
George F. Baker Trust Grants, 1793
George Family Foundation Grants, 1794
George Foundation Grants, 1795
George Gund Foundation Grants, 1796
George H.C. Ensworth Memorial Fund Grants, 1797
George Kress Foundation Grants, 1799
George P. Davenport Trust Fund Grants, 1800
George W. Codrington Charitable Fndn Grants, 1803
George W. Wells Foundation Grants, 1804
Georgiana Goddard Eaton Memorial Grants, 1805

Georgia Power Foundation Grants, 1806
Gerber Foundation Grants, 1807
Gheens Foundation Grants, 1814
Giant Food Charitable Grants, 1816
Gibson County Community Foundation Women's Fund Grants, 1817
Gibson Foundation Grants, 1818
Gil and Dody Weaver Foundation Grants, 1819
Gill Foundation - Gay and Lesbian Fund Grants, 1821
Ginn Foundation Grants, 1822
Giving Sum Annual Grant, 1825
GlaxoSmithKline Corporate Grants, 1829
Global Fund for Children Grants, 1832
GNOF Albert N. & Hattie M. McClure Grants, 1834
GNOF Bayou Communities Grants, 1835
GNOF Exxon-Mobil Grants, 1836
GNOF IMPACT Grants for Health and Human Services, 1839
GNOF IMPACT Gulf States Eye Surgery Fund, 1840
GNOF IMPACT Harold W. Newman, Jr. Charitable Trust Grants, 1841
GNOF Maison Hospitaliere Grants, 1843
GNOF Norco Community Grants, 1845
GNOF Organizational Effectiveness Grants and Workshops, 1846
GNOF Stand Up For Our Children Grants, 1847
Godfrey Foundation Grants, 1849
Golden Heart Community Foundation Grants, 1850
Golden State Warriors Foundation Grants, 1851
Goodrich Corporation Foundation Grants, 1854
Goodyear Tire Grants, 1856
Grace and Franklin Bernsen Foundation Grants, 1858
Grace Bersted Foundation Grants, 1859
Graco Foundation Grants, 1860
Graham and Carolyn Holloway Family Foundation Grants, 1861
Graham Foundation Grants, 1862
Grand Rapids Area Community Fndn Grants, 1865
Grand Rapids Area Community Foundation Nashwauk Area Endowment Fund Grants, 1866
Grand Rapids Community Foundation Grants, 1869
Grand Rapids Community Foundation Lowell Area Fund Grants, 1872
Great-West Life Grants, 1879
Greater Cincinnati Foundation Priority and Small Projects/Capacity-Building Grants, 1880
Greater Green Bay Community Fndn Grants, 1881
Greater Kanawha Valley Foundation Grants, 1882
Greater Milwaukee Foundation Grants, 1883
Greater Saint Louis Community Fndn Grants, 1884
Greater Sitka Legacy Fund Grants, 1885
Greater Tacoma Community Foundation Grants, 1886
Greater Tacoma Community Foundation Ryan Alan Hade Endowment Fund, 1887
Greater Worcester Community Foundation Discretionary Grants, 1888
Green Bay Packers Foundation Grants, 1890
Green Diamond Charitable Contributions, 1891
Green Bader Human Services Grants, 1894
Green River Area Community Fndn Grants, 1895
Greenspun Family Foundation Grants, 1896
Greenwall Foundation Bioethics Grants, 1897
Gregory Family Foundation Grants (Florida), 1899
Griffin Foundation Grants, 1903
Grifols Community Outreach Grants, 1904
Grotto Foundation Project Grants, 1905
Grundy Foundation Grants, 1912
Guido A. & Elizabeth H. Binda Fndn Grants, 1913
Guitar Center Music Foundation Grants, 1914
Gulf Coast Community Foundation Grants, 1915
Gulf Coast Foundation of Community Program Grants, 1917
Guy's and St. Thomas' Charity Grants, 1918
Guy I. Bromley Trust Grants, 1919
H & R Foundation Grants, 1920
H.A. & Mary K. Chapman Trust Grants, 1921
H.B. Fuller Foundation Grants, 1922
H.J. Heinz Company Foundation Grants, 1923
H. Leslie Hoffman and Elaine S. Hoffman Foundation Grants, 1924

H. Schaffer Foundation Grants, 1925
Hackett Foundation Grants, 1926
HAF David (Davey) H. Somerville Medical Travel Fund Grants, 1928
HAF JoAllen K. Twiddy-Wood Memorial Fund Grants, 1929
HAF Phyllis Nilsen Leal Memorial Fund Gifts, 1930
HAF Senior Opportunities Grants, 1932
Hagedorn Fund Grants, 1934
Hall-Perrine Foundation Grants, 1935
Hallmark Corporate Foundation Grants, 1936
Hampton Roads Community Foundation Developmental Disabilities Grants, 1939
Hampton Roads Community Foundation Health and Human Service Grants, 1941
Hancock County Community Foundation - Field of Interest Grants, 1943
Hannaford Charitable Foundation Grants, 1944
Hardin County Community Foundation Grants, 1946
Harley Davidson Foundation Grants, 1947
Harold Alfond Foundation Grants, 1949
Harold and Arlene Schnitzer CARE Foundation Grants, 1950
Harold and Rebecca H. Gross Foundation Grants, 1951
Harold Brooks Foundation Grants, 1952
Harold R. Bechtel Charitable Remainder Uni-Trust Grants, 1953
Harold R. Bechtel Testamentary Charitable Trust Grants, 1954
Harold Simmons Foundation Grants, 1955
Harrison County Community Foundation Grants, 1957
Harrison County Community Foundation Signature Grants, 1958
Harry Sudakoff Foundation Grants, 1967
Harry W. Bass, Jr. Foundation Grants, 1968
Hartford Foundation Regular Grants, 1970
Harvest Foundation Grants, 1971
Harvey Randall Wickes Foundation Grants, 1972
Hasbro Children's Fund Grants, 1973
Hasbro Corporation Gift of Play Hospital and Pediatric Health Giving, 1974
Hawaiian Electric Industries Charitable Foundation Grants, 1975
Hawaii Community Foundation Reverend Takie Okumura Family Grants, 1977
Hawaii Community Foundation West Hawaii Fund Grants, 1978
Hawn Foundation Grants, 1979
HCA Foundation Grants, 1980
Healthcare Fndn for Orange County Grants, 1981
Healthcare Foundation of New Jersey Grants, 1982
Health Foundation of Greater Cincinnati Grants, 1983
Health Fndn of Greater Indianapolis Grants, 1984
Health Foundation of Southern Florida Responsive Grants, 1985
Hearst Foundations Health Grants, 1987
Heckscher Foundation for Children Grants, 1988
Helena Rubinstein Foundation Grants, 1991
Helen Bader Foundation Grants, 1992
Helen Gertrude Sparks Charitable Trust Grants, 1993
Helen Irwin Littauer Educational Trust Grants, 1994
Helen K. & Arthur E. Johnson Fndn Grants, 1995
Helen Pumphrey Denit Charitable Trust Grants, 1996
Helen Steiner Rice Foundation Grants, 1998
Hendrick Foundation for Children Grants, 1999
Henrietta Lange Burk Fund Grants, 2001
Henry and Ruth Blaustein Rosenberg Foundation Health Grants, 2004
Henry L. Guenther Foundation Grants, 2006
Herbert A. & Adrian W. Woods Fndn Grants, 2007
Herbert H. & Grace A. Dow Fndn Grants, 2008
Herman Goldman Foundation Grants, 2012
Hershey Company Grants, 2013
Highmark Corporate Giving Grants, 2025
Hilda and Preston Davis Foundation Grants, 2028
Hilda and Preston Davis Foundation Postdoctoral Fellowships in Eating Disorders Research, 2029
Hillcrest Foundation Grants, 2031
Hill Crest Foundation Grants, 2030
Hillman Foundation Grants, 2032

Hillsdale County Community Foundation Healthy Senior/Healthy Youth Fund Grants, 2033
Hillsdale County Community General Adult Foundation Grants, 2034
Hillsdale Fund Grants, 2035
Hilton Head Island Foundation Grants, 2036
Hoblitzelle Foundation Grants, 2038
Hogg Foundation for Mental Health Grants, 2039
Hoglund Foundation Grants, 2040
Holland/Zeeland Community Fndn Grants, 2041
HomeBanc Foundation Grants, 2042
Home Building Industry Disaster Relief Fund, 2043
Homer Foundation Grants, 2044
Honda of America Manufacturing Fndn Grants, 2045
Horace Moses Charitable Foundation Grants, 2047
Horizon Foundation for New Jersey Grants, 2048
Horizons Community Issues Grants, 2049
Houston Endowment Grants, 2053
Howard and Bush Foundation Grants, 2054
HRAMF Deborah Munroe Noonan Memorial Research Grants, 2058
HRAMF Taub Fnd Grants for MDS Research, 2064
HRSA Ryan White HIV AIDS Drug Assistance Grants, 2070
Huber Foundation Grants, 2071
Hudson Webber Foundation Grants, 2072
Huffy Foundation Grants, 2073
Hugh J. Andersen Foundation Grants, 2074
Huie-Dellmon Trust Grants, 2075
Humana Foundation Grants, 2077
Human Source Foundation Grants, 2078
Huntington Beach Police Officers Fndn Grants, 2081
Huntington County Community Foundation Make a Difference Grants, 2083
Huntington National Bank Community Grants, 2084
Hutchinson Community Foundation Grants, 2085
Hut Foundation Grants, 2086
Hutton Foundation Grants, 2087
I.A. O'Shaughnessy Foundation Grants, 2088
IATA Research Grants, 2093
IBCAT Screening Mammography Grants, 2095
Ida Alice Ryan Charitable Trust Grants, 2113
Idaho Community Foundation Eastern Region Competitive Grants, 2114
Idaho Power Company Corporate Contributions, 2115
IDPH Emergency Medical Services Assistance Fund Grants, 2118
IDPH Hosptial Capital Investment Grants, 2119
IDPH Local Health Department Public Health Emergency Response Grants, 2120
IIE African Center of Excellence for Women's Leadership Grants, 2124
Ike and Roz Friedman Foundation Grants, 2145
Illinois Children's Healthcare Foundation Grants, 2146
Illinois Tool Works Foundation Grants, 2147
Inasmuch Foundation Grants, 2149
Independence Blue Cross Charitable Medical Care Grants, 2150
Independence Community Foundation Community Quality of Life Grant, 2153
Indiana AIDS Fund Grants, 2155
Intergrys Corporation Grants, 2160
Irving S. Gilmore Foundation Grants, 2163
Irvin Stern Foundation Grants, 2164
Isabel Allende Foundation Esperanza Grants, 2165
Ittleson Foundation AIDS Grants, 2166
Ittleson Foundation Mental Health Grants, 2167
J.B. Reynolds Foundation Grants, 2168
J.C. Penney Company Grants, 2169
J.E. and L.E. Mabee Foundation Grants, 2170
J.H. Robbins Foundation Grants, 2171
J.L. Bedsole Foundation Grants, 2172
J.M. Long Foundation Grants, 2173
J.N. and Macie Edens Foundation Grants, 2174
J.W. Kieckhefer Foundation Grants, 2176
J. Walton Bissell Foundation Grants, 2177
Jack H. & William M. Light Trust Grants, 2179
Jackson County Community Foundation Unrestricted Grants, 2180

PROGRAM TYPE INDEX

Service Delivery Programs / 911

Jacob and Hilda Blaustein Foundation Health and Mental Health Grants, 2181
Jacob and Valeria Langeloth Foundation Grants, 2182
Jacob G. Schmidlapp Trust Grants, 2183
Jacobs Family Village Neighborhoods Grants, 2184
James & Abigail Campbell Family Fndn Grants, 2185
James A. and Faith Knight Foundation Grants, 2186
James Ford Bell Foundation Grants, 2187
James J. & Joan A. Gardner Family Fndn Grants, 2193
James L. and Mary Jane Bowman Charitable Trust Grants, 2194
James M. Collins Foundation Grants, 2195
James M. Cox Foundation of Georgia Grants, 2196
James R. Thorpe Foundation Grants, 2199
James S. Copley Foundation Grants, 2200
Jane's Trust Grants, 2206
Jay and Rose Phillips Family Foundation Grants, 2213
Jayne and Leonard Abess Foundation Grants, 2214
Jean and Louis Dreyfus Foundation Grants, 2215
Jeffris Wood Foundation Grants, 2217
JELD-WEN Foundation Grants, 2218
Jenkins Foundation: Improving the Health of Greater Richmond Grants, 2219
Jennings County Community Foundation Women's Giving Circle Grant, 2221
Jerome and Mildred Paddock Foundation Grants, 2222
Jerome Robbins Foundation Grants, 2223
Jessica Stevens Community Foundation Grants, 2224
Jessie Ball Dupont Fund Grants, 2226
Jewish Fund Grants, 2227
Jim Moran Foundation Grants, 2228
JM Foundation Grants, 2229
Joe W. and Dorothy Dorsett Brown Fndn Grants, 2230
John Clarke Trust Grants, 2232
John D. & Katherine A. Johnston Fndn Grants, 2233
John Deere Foundation Grants, 2234
John Edward Fowler Memorial Fndn Grants, 2235
John J. Leidy Foundation Grants, 2241
John Lord Knight Foundation Grants, 2243
John Merck Fund Grants, 2245
John P. Murphy Foundation Grants, 2248
John R. Oishei Foundation Grants, 2249
Johns Manville Fund Grants, 2251
Johnson & Johnson Community Health Grants, 2252
Johnson & Johnson Corporate Contributions, 2253
Johnson & Johnson/SAH Arts & Healing Grants, 2254
Johnson Controls Foundation Health and Social Services Grants, 2255
John W. and Anna H. Hanes Foundation Grants, 2261
John W. Speas and Effie E. Speas Memorial Trust Grants, 2264
Joseph H. & Florence A. Roblee Fndn Grants, 2269
Joseph Henry Edmondson Foundation Grants, 2270
Josephine Goodyear Foundation Grants, 2272
Josephine S. Gumbiner Foundation Grants, 2273
Josephine Schell Russell Charitable Trust Grants, 2274
Joseph P. Kennedy Jr. Foundation Grants, 2275
Josiah W. and Bessie H. Kline Foundation Grants, 2278
Judith and Jean Pape Adams Charitable Foundation Tulsa Area Grants, 2280
Judith Clark-Morrill Foundation Grants, 2281
Julius N. Frankel Foundation Grants, 2282
Kahuku Community Fund, 2286
Kansas Health Foundation Recognition Grants, 2289
Kate B. Reynolds Trust Health Care Grants, 2290
Katharine Matthies Foundation Grants, 2291
Kathryne Beynon Foundation Grants, 2294
Katrine Menzing Deakins Trust Grants, 2295
Kelvin and Eleanor Smith Foundation Grants, 2297
Kenai Peninsula Foundation Grants, 2298
Kendrick Foundation Grants, 2299
Kenneth T. & Eileen L. Norris Fndn Grants, 2300
Kent D. Steadley and Mary L. Steadley Memorial Trust Grants, 2301
Kessler Fnd Signature Employment Grants, 2302
Ketchikan Community Foundation Grants, 2303
Kettering Family Foundation Grants, 2304
Kettering Fund Grants, 2305
KeyBank Foundation Grants, 2307

Kimball International-Habig Foundation Health and Human Services Grants, 2308
Klarman Family Foundation Grants in Eating Disorders Research Grants, 2310
Knight Family Charitable and Educational Foundation Grants, 2312
Kodiak Community Foundation Grants, 2314
Kovler Family Foundation Grants, 2323
Kroger Foundation Women's Health Grants, 2325
Kuntz Foundation Grants, 2326
L. W. Pierce Family Foundation Grants, 2328
LaGrange County Community Fndn Grants, 2329
Land O'Lakes Foundation Mid-Atlantic Grants, 2341
Lands' End Corporate Giving Program, 2344
Laura B. Vogler Foundation Grants, 2345
Leo Goodwin Foundation Grants, 2354
Leon and Thea Koerner Foundation Grants, 2355
Leonard and Helen R. Stulman Charitable Foundation Grants, 2356
Leo Niessen Jr., Charitable Trust Grants, 2358
Lester Ray Fleming Scholarships, 2360
Liberty Bank Foundation Grants, 2362
Libra Foundation Grants, 2363
Lil and Julie Rosenberg Foundation Grants, 2364
Lincoln Financial Foundation Grants, 2367
Little Life Foundation Grants, 2372
Lloyd A. Fry Foundation Health Grants, 2373
Louetta M. Cowden Foundation Grants, 2377
Louie M. & Betty M. Phillips Fndn Grants, 2378
Louis and Elizabeth Nave Flarsheim Charitable Foundation Grants, 2379
Louis H. Aborn Foundation Grants, 2380
Lowe Foundation Grants, 2381
Lowell Berry Foundation Grants, 2382
Lubbock Area Foundation Grants, 2383
Lucile Horton Howe and Mitchell B. Howe Foundation Grants, 2385
Ludwick Family Foundation Grants, 2389
Lydia deForest Charitable Trust Grants, 2393
Lymphatic Education and Research Network Additional Support Grants for NIH-funded F32 Postdoctoral Fellows, 2394
Lymphatic Education and Research Network Postdoctoral Fellowships, 2395
Lynde and Harry Bradley Foundation Grants, 2397
Lynn and Rovena Alexander Family Foundation Grants, 2399
M-A-C AIDS Fund Grants, 2400
M.B. and Edna Zale Foundation Grants, 2401
M. Bastian Family Foundation Grants, 2402
M.E. Raker Foundation Grants, 2404
Mabel A. Horne Trust Grants, 2406
Mabel Y. Hughes Charitable Trust Grants, 2409
Macquarie Bank Foundation Grants, 2410
Maggie Welby Foundation Grants, 2414
Maine Community Foundation Hospice Grants, 2416
Manuel D. & Rhoda Mayerson Fndn Grants, 2420
MAP International Medical Fellowships, 2421
Marathon Petroleum Corporation Grants, 2422
March of Dimes Program Grants, 2426
Marcia and Otto Koehler Foundation Grants, 2427
Margaret T. Morris Foundation Grants, 2429
Marie C. and Joseph C. Wilson Foundation Rochester Small Grants, 2431
Marie H. Bechtel Charitable Remainder Uni-Trust Grants, 2432
Marin Community Foundation Improving Community Health Grants, 2433
Marion Gardner Jackson Charitable Trust Grants, 2434
Marion I. and Henry J. Knott Foundation Discretionary Grants, 2435
Marion I. and Henry J. Knott Foundation Standard Grants, 2436
Marshall County Community Foundation Grants, 2439
Mary Black Foundation Active Living Grants, 2440
Mary Black Fndn Community Health Grants, 2441
Mary Black Foundation Early Childhood Development Grants, 2442
Mary K. Chapman Foundation Grants, 2443
Mary Owen Borden Foundation Grants, 2446

Mary S. and David C. Corbin Foundation Grants, 2447
Mary Wilmer Covey Charitable Trust Grants, 2448
Massage Therapy Foundation Community Service Grants, 2449
Matilda R. Wilson Fund Grants, 2453
Mattel Children's Foundation Grants, 2454
Mattel International Grants Program, 2455
Maurice J. Masserini Charitable Trust Grants, 2456
Max and Victoria Dreyfus Foundation Grants, 2457
May and Stanley Smith Charitable Trust Grants, 2459
McCallum Family Foundation Grants, 2489
McCarthy Family Foundation Grants, 2490
McCombs Foundation Grants, 2491
McConnell Foundation Grants, 2492
McCune Foundation Human Services Grants, 2493
McGraw-Hill Companies Community Grants, 2494
McInerny Foundation Grants, 2495
McLean Contributionship Grants, 2497
McLean Foundation Grants, 2498
Mead Johnson Nutritionals Charitable Giving, 2502
Mead Johnson Nutritionals Evansville-Area Organizations Grants, 2503
Mead Johnson Nutritionals Med Educ Grants, 2504
MeadWestvaco Foundation Sustainable Communities Grants, 2506
Mead Witter Foundation Grants, 2507
MedImmune Charitable Grants, 2509
Medtronic Foundation CommunityLink Health Grants, 2510
Medtronic Foundation Community Link Human Services Grants, 2511
Medtronic Foundation Patient Link Grants, 2514
Medtronic Foundation Strengthening Health Systems Grants, 2515
Memorial Foundation Grants, 2516
Mercedes-Benz USA Corporate Contributions, 2517
Mericos Foundation Grants, 2518
Meriden Foundation Grants, 2519
Merrick Action Grants, 2521
Merrick Foundation Grants, 2522
Mervin Bovaird Foundation Grants, 2523
Meta and George Rosenberg Foundation Grants, 2525
MetroWest Health Foundation Capital Grants for Health-Related Facilities, 2526
MetroWest Health Foundation Grants--Healthy Aging, 2527
MetroWest Health Foundation Grants to Reduce the Incidence of High Risk Behaviors Among Adolescents, 2528
Metzger-Price Fund Grants, 2530
Meyer Foundation Healthy Communities Grants, 2533
Meyer Memorial Trust Grassroots Grants, 2536
Meyer Memorial Trust Responsive Grants, 2537
Meyer Memorial Trust Special Grants, 2538
MGM Resorts Foundation Community Grants, 2541
MGN Family Foundation Grants, 2542
Michael Reese Health Trust Core Grants, 2544
Michael Reese Health Trust Responsive Grants, 2545
Mid-Iowa Health Foundation Community Response Grants, 2549
Middlesex Savings Charitable Foundation Basic Human Needs Grants, 2550
Middlesex Savings Charitable Foundation Capacity Building Grants, 2551
Miles of Hope Breast Cancer Foundation Grants, 2553
Miller Foundation Grants, 2556
Mimi and Peter Haas Fund Grants, 2557
Minneapolis Foundation Community Grants, 2558
MMS and Alliance Charitable Foundation Grants for Community Action and Care for the Medically Uninsured, 2587
MMS and Alliance Charitable Foundation International Health Studies Grants, 2588
Moline Foundation Community Grants, 2590
Montgomery County Community Fndn Grants, 2591
Morris & Gwendolyn Cafritz Fndn Grants, 2594
Morton K. and Jane Blaustein Foundation Health Grants, 2595
Ms. Fndn for Women Ending Violence Grants, 2596
Ms. Foundation for Women Health Grants, 2597

Mt. Sinai Health Care Foundation Health of the Jewish Community Grants, 2599
Mt. Sinai Health Care Foundation Health of the Urban Community Grants, 2600
Mt. Sinai Health Care Foundation Health Policy Grants, 2601
NAPNAP Foundation Innovative Health Care Small Grant, 2611
NAPNAP Foundation McNeil Grant-in-Aid, 2612
NAPNAP Foundation Nursing Research Grants, 2615
NAPNAP Foundation Shourd Parks Immunization Project Small Grants, 2617
NAPNAP Foundation Wyeth Pediatric Immunization Grant, 2618
Nathan Cummings Foundation Grants, 2627
Nationwide Insurance Foundation Grants, 2637
Nelda C. and H.J. Lutcher Stark Fndn Grants, 2672
Nell J. Redfield Foundation Grants, 2673
NHLBI Bioengineering and Obesity Grants, 2689
NHLBI Investigator Initiated Multi-Site Clinical Trials, 2703
NHLBI Lymphatics in Health and Disease in the Digestive, Urinary, Cardiovascular and Pulmonary Systems, 2705
NHLBI Midcareer Investigator Award in Patient-Oriented Research, 2713
NHLBI Research Supplements to Promote Diversity in Health-Related Research, 2727
NHLBI Short-Term Research Education Program to Increase Diversity in Health-Related Research, 2733
NHSCA Arts in Health Care Project Grants, 2741
Nicholas H. Noyes Jr. Memorial Fndn Grants, 2791
Nina Mason Pulliam Charitable Trust Grants, 2921
NMF General Electric Med Fellowships, 2956
Norcliffe Foundation Grants, 2979
Nord Family Foundation Grants, 2980
Nordson Corporation Foundation Grants, 2981
North Carolina Community Foundation Grants, 2990
North Carolina GlaxoSmithKline Fndn Grants, 2991
North Central Health Services Grants, 2992
Northern Chautauqua Community Foundation Community Grants, 2995
Northern New York Community Fndn Grants, 2996
Northwestern Mutual Foundation Grants, 2999
Northwest Minnesota Foundation Women's Fund Grants, 3000
Norwin S. & Elizabeth N. Bean Fndn Grants, 3001
NSF Chemistry Research Experiences for Undergraduates (REU), 3012
NSF Emerging Frontiers in Research and Innovation (EFRI) Grants, 3014
NSF Instrument Development for Bio Research, 3017
NWHF Physical Activity and Nutrition Grants, 3037
NYC Managed Care Consumer Assistance Workshop Re-Design Grants, 3047
NYCT AIDS/HIV Grants, 3048
NYCT Blindness and Visual Disabilities Grants, 3050
NYCT Children/Youth with Disabilities Grants, 3052
NYCT Girls and Young Women Grants, 3053
NYCT Grants for the Elderly, 3054
NYCT Health Care Services, Systems, and Policies Grants, 3055
NYCT Mental Health and Retardation Grants, 3056
NYCT Substance Abuse Grants, 3057
NYCT Technical Assistance Grants, 3058
Oak Foundation Child Abuse Grants, 3059
OceanFirst Foundation Major Grants, 3062
OECD Sexual Health and Rights Project Grants, 3064
Office Depot Corporation Community Relations Grants, 3065
Ogden Codman Trust Grants, 3066
Ohio County Community Foundation Board of Directors Grants, 3067
Ohio County Community Foundation Grants, 3068
Oleonda Jameson Trust Grants, 3071
Olive Higgins Prouty Foundation Grants, 3072
OneFamily Foundation Grants, 3075
OneSight Research Foundation Block Grants, 3076
OneStar Foundation AmeriCorps Grants, 3077
Oppenstein Brothers Foundation Grants, 3078
Orange County Community Foundation Grants, 3079
Ordean Foundation Grants, 3080
Oregon Community Foundation Better Nursing Home Care Grants, 3081
Oregon Community Fndn Community Grants, 3082
Oscar Rennebohm Foundation Grants, 3085
OSF Access to Essential Medicines Initiative, 3086
OSF Accountability and Monitoring in Health Initiative Grants, 3087
OSF Global Health Financing Initiative Grants, 3088
OSF Health Media Initiative Grants, 3089
OSF International Harm Reduction Development Program Grants, 3090
OSF International Palliative Care Grants, 3091
OSF Law and Health Initiative Grants, 3092
OSF Mental Health Initiative Grants, 3093
OSF Public Health Grants in Kyrgyzstan, 3094
OSF Roma Health Project Grants, 3095
OSF Tackling Addiction Grants in Baltimore, 3096
Otto Bremer Foundation Grants, 3099
PACCAR Foundation Grants, 3101
PacifiCare Foundation Grants, 3102
Pacific Life Foundation Grants, 3103
Packard Foundation Local Grants, 3105
Pajaro Valley Community Health Trust Insurance/Coverage & Education on Using the System Grants, 3108
Pajaro Valley Community Health Trust Diabetes and Contributing Factors Grants, 3109
Pajaro Valley Community Health Trust Oral Health: Prevention & Access Grants, 3110
Palmer Foundation Grants, 3112
Parkersburg Area Community Foundation Action Grants, 3116
Paso del Norte Health Foundation Grants, 3119
Patrick and Anna M. Cudahy Fund Grants, 3120
Patron Saints Foundation Grants, 3121
Paul Balint Charitable Trust Grants, 3122
Paul Rapoport Foundation Grants, 3125
Peabody Foundation Grants, 3132
Percy B. Ferebee Endowment Grants, 3138
Perpetual Trust for Charitable Giving Grants, 3141
Peter and Elizabeth C. Tower Foundation Annual Intellectual Disabilities Grants, 3144
Peter and Elizabeth C. Tower Foundation Annual Mental Health Grants, 3145
Peter and Elizabeth C. Tower Foundation Learning Disability Grants, 3146
Peter and Elizabeth C. Tower Foundation Mental Health Reference and Resource Materials Mini-Grants, 3147
Peter and Elizabeth C. Tower Foundation Phase II Technology Initiative Grants, 3149
Peter and Elizabeth C. Tower Foundation Phase I Technology Initiative Grants, 3150
Peter and Elizabeth C. Tower Foundation Social and Emotional Preschool Curriculum Grants, 3151
Peter and Elizabeth C. Tower Foundation Substance Abuse Grants, 3152
Peter Kiewit Foundation General Grants, 3154
Peter Kiewit Foundation Small Grants, 3155
Petersburg Community Foundation Grants, 3156
Pew Charitable Trusts Children & Youth Grants, 3158
PeyBack Foundation Grants, 3159
Peyton Anderson Foundation Grants, 3160
Pfizer/AAFP Foundation Visiting Professorship Program in Family Medicine, 3162
Pfizer Healthcare Charitable Contributions, 3180
Pfizer Medical Education Track One Grants, 3182
Phelps County Community Foundation Grants, 3188
Philadelphia Foundation Organizational Effectiveness Grants, 3189
Philip L. Graham Fund Health and Human Services Grants, 3190
Phoenix Suns Charities Grants, 3201
Piedmont Health Foundation Grants, 3219
Piedmont Natural Gas Corporate and Charitable Contributions, 3220
Piedmont Natural Gas Foundation Health and Human Services Grants, 3221
Pinellas County Grants, 3223
Pinkerton Foundation Grants, 3224
Pinnacle Foundation Grants, 3225
Piper Jaffray Fndn Communities Giving Grants, 3227
Piper Trust Healthcare & Med Research Grants, 3228
Piper Trust Older Adults Grants, 3229
Pittsburgh Foundation Community Fund Grants, 3230
Plough Foundation Grants, 3238
PMI Foundation Grants, 3239
PNC Foundation Community Services Grants, 3240
Pokagon Fund Grants, 3242
Polk Bros. Foundation Grants, 3243
Pollock Foundation Grants, 3244
Portland General Electric Foundation Grants, 3250
Posey County Community Foundation Grants, 3252
PPCF Community Grants, 3256
PPG Industries Foundation Grants, 3257
Premera Blue Cross Grants, 3258
Price Chopper's Golub Foundation Grants, 3261
Priddy Foundation Program Grants, 3266
Prince Charitable Trusts Chicago Grants, 3270
Prince Charitable Trusts DC Grants, 3271
Princeton Area Community Foundation Greater Mercer Grants, 3272
Principal Financial Group Foundation Grants, 3273
Procter and Gamble Fund Grants, 3274
Puerto Rico Community Foundation Grants, 3278
Putnam County Community Foundation Grants, 3283
Quaker Chemical Foundation Grants, 3286
Qualcomm Grants, 3287
Quality Health Foundation Grants, 3288
Quantum Corporation Snap Server Grants, 3289
Quantum Foundation Grants, 3290
Questar Corporate Contributions Grants, 3291
R.C. Baker Foundation Grants, 3292
R.S. Gernon Trust Grants, 3293
Rainbow Fund Grants, 3301
Ralph M. Parsons Foundation Grants, 3304
Ralphs Food 4 Less Foundation Grants, 3305
Raskob Foundation for Catholic Activities Grants, 3306
Rathmann Family Foundation Grants, 3309
Raymond John Wean Foundation Grants, 3310
RCF General Community Grants, 3312
RCF Individual Assistance Grants, 3313
Reader's Digest Partners for Sight Fndn Grants, 3345
Regence Fndn Access to Health Care Grants, 3346
Regence Foundation Health Care Community Awareness and Engagement Grants, 3347
Regence Fndn Health Care Connections Grants, 3348
Regence Fndn Improving End-of-Life Grants, 3349
Regence Fndn Tools & Technology Grants, 3350
Rehab Therapy Foundation Grants, 3356
Reinberger Foundation Grants, 3357
Reynolds & Reynolds Associate Fndn Grants, 3360
RGk Foundation Grants, 3361
Rhode Island Foundation Grants, 3362
Rice Foundation Grants, 3363
Richard and Helen DeVos Foundation Grants, 3364
Richard & Susan Smith Family Fndn Grants, 3365
Richard Davoud Donchian Foundation Grants, 3367
Richard King Mellon Foundation Grants, 3369
Richland County Bank Grants, 3371
Richmond Eye and Ear Fund Grants, 3372
Ricks Family Charitable Trust Grants, 3373
Riley Foundation Grants, 3374
Ringing Rocks Foundation Discretionary Grants, 3375
Robbins Charitable Foundation Grants, 3378
Robert and Joan Dircks Foundation Grants, 3379
Robert & Clara Milton Senior Housing Grants, 3384
Robert R. McCormick Trib Community Grants, 3385
Robert R. McCormick Tribune Veterans Initiative Grants, 3386
Robert R. Meyer Foundation Grants, 3387
Robert W. Woodruff Foundation Grants, 3389
Robins Foundation Grants, 3390
Rochester Area Community Foundation Grants, 3391
Rochester Area Foundation Grants, 3392
Rockefeller Brothers Fund Pivotal Places Grants: South Africa, 3395
Rockefeller Foundation Grants, 3396

PROGRAM TYPE INDEX

Service Delivery Programs

Rockwell Collins Charitable Corporation Grants, 3397
Rockwell International Corporate Trust Grants Program, 3399
Rohm and Haas Company Grants, 3403
Rollins-Luetkemeyer Foundation Grants, 3404
Ronald McDonald House Charities Grants, 3405
Rosalynn Carter Institute Georgia Caregiver of the Year Awards, 3407
Rosalynn Carter Institute John and Betty Pope Fellowships, 3408
Rose Community Foundation Aging Grants, 3410
Rose Community Foundation Health Grants, 3411
ROSE Fund Grants, 3412
Rose Hills Foundation Grants, 3413
Roy & Christine Sturgis Charitable Grants, 3415
RRF General Program Grants, 3417
RRF Organizational Capacity Building Grants, 3418
RSC McNeil Medal for the Public Awareness of Science, 3422
Ruth Anderson Foundation Grants, 3425
Ruth and Vernon Taylor Foundation Grants, 3426
Ruth Eleanor Bamberger and John Ernest Bamberger Memorial Foundation Grants, 3427
Ruth H. & Warren A. Ellsworth Fndn Grants, 3429
RWJF Community Health Leaders Awards, 3432
RWJF Disparities Research for Change Grants, 3433
RWJF Harold Amos Medical Faculty Development Scholars, 3435
RWJF New Jersey Health Initiatives Grants, 3440
RWJF Pioneer Portfolio Grants, 3442
RWJF Vulnerable Populations Portfolio Grants, 3444
S. Mark Taper Foundation Grants, 3449
Saigh Foundation Grants, 3451
Sain-Orr and Royak-DeForest Steadman Foundation Grants, 3452
Saint Louis Rams Fndn Community Donations, 3453
Saint Luke's Foundation Grants, 3454
Saint Luke's Health Initiatives Grants, 3455
Salem Foundation Charitable Trust Grants, 3456
Salem Foundation Grants, 3457
Salmon Foundation Grants, 3458
SAMHSA Campus Suicide Prevention Grants, 3460
SAMHSA Child Mental Health Initiative (CMHI) Grants, 3461
Samueli Foundation Health Grants, 3465
Samuel S. Johnson Foundation Grants, 3468
San Antonio Area Foundation Grants, 3469
San Diego Foundation for Change Grants, 3470
San Diego Foundation Health & Human Services Grants, 3471
San Diego Foundation Paradise Valley Hospital Community Fund Grants, 3472
San Diego HIV Funding Collaborative Grants, 3473
San Diego Women's Foundation Grants, 3474
Sands Foundation Grants, 3476
San Francisco Fndn Community Health Grants, 3478
San Francisco Foundation Disability Rights Advocate Fund Emergency Grants, 3479
San Juan Island Community Foundation Grants, 3480
Santa Barbara Foundation Strategy Grants - Core Support, 3481
Santa Fe Community Foundation Seasonal Grants-Fall Cycle, 3482
Sara Elizabeth O'Brien Trust Grants, 3483
Sarkeys Foundation Grants, 3485
Sartain Lanier Family Foundation Grants, 3486
Schering-Plough Foundation Community Initiatives Grants, 3491
Schering-Plough Foundation Health Grants, 3492
Schlessman Family Foundation Grants, 3493
Schramm Foundation Grants, 3495
Scott B. & Annie P. Appleby Charitable Grants, 3499
Seagate Tech Corp Capacity to Care Grants, 3502
Seattle Foundation Benjamin N. Phillips Memorial Fund Grants, 3504
Seaver Institute Grants, 3509
Self Foundation Grants, 3510
Sensient Technologies Foundation Grants, 3513
Seward Community Foundation Grants, 3514
Seward Community Foundation Mini-Grants, 3515

Shaw's Supermarkets Donations, 3542
Shell Oil Company Foundation Community Development Grants, 3544
Sheltering Arms Fund Grants, 3545
Shield-Ayres Foundation Grants, 3546
Shopko Fndn Community Charitable Grants, 3547
Shopko Foundation Green Bay Area Community Grants, 3548
Sidgmore Family Foundation Grants, 3551
Sidney Stern Memorial Trust Grants, 3552
Sid W. Richardson Foundation Grants, 3553
Siebert Lutheran Foundation Grants, 3554
Sierra Health Foundation Responsive Grants, 3555
Simmons Foundation Grants, 3563
Simpson Lumber Charitable Contributions, 3566
Singing for Change Foundation Grants, 3567
Sioux Falls Area Community Foundation Community Fund Grants (Unrestricted), 3568
Sioux Falls Area Community Foundation Field-of-Interest and Donor-Advised Grants, 3569
Sioux Falls Area Community Foundation Spot Grants (Unrestricted), 3570
Sir Dorabji Tata Trust Grants for NGOs or Voluntary Organizations, 3572
Sir Dorabji Tata Trust Individual Medical Grants, 3573
Sisters of Mercy of North Carolina Fndn Grants, 3575
Sisters of St. Joseph Healthcare Fndn Grants, 3576
Solo Cup Foundation Grants, 3620
Sonoco Foundation Grants, 3621
Sonora Area Foundation Competitive Grants, 3622
Sony Corporation of America Grants, 3623
Sophia Romero Trust Grants, 3624
Southbury Community Trust Fund, 3625
Southwest Florida Community Foundation Competitive Grants, 3627
Special People in Need Grants, 3630
Square D Foundation Grants, 3634
Stackpole-Hall Foundation Grants, 3643
Stan and Sandy Checketts Foundation, 3644
Staunton Farm Foundation Grants, 3649
Steele-Reese Foundation Grants, 3650
Stella and Charles Guttman Foundation Grants, 3651
Sterling-Turner Charitable Foundation Grants, 3652
Sterling and Shelli Gardner Foundation Grants, 3653
Stewart Huston Charitable Trust Grants, 3655
Stocker Foundation Grants, 3656
Straits Area Community Foundation Health Care Grants, 3657
Strake Foundation Grants, 3658
Stranahan Foundation Grants, 3659
Subaru of Indiana Automotive Foundation Grants, 3672
Summit Foundation Grants, 3674
Sunflower Foundation Bridge Grants, 3676
Sunflower Foundation Capacity Building Grants, 3677
Sunflower Foundation Walking Trails Grants, 3678
Sunoco Foundation Grants, 3679
SunTrust Bank Trusteed Foundations Florence C. and Harry L. English Memorial Fund Grants, 3680
SunTrust Bank Trusteed Foundations Greene-Sawtell Grants, 3681
SunTrust Bank Trusteed Foundations Harriet McDaniel Marshall Tust Grants, 3682
SunTrust Bank Trusteed Foundations Nell Warren Elkin and William Simpson Elkin Grants, 3683
SunTrust Bank Trusteed Foundations Thomas Guy Woolford Charitable Trust Grants, 3684
SunTrust Bank Trusteed Foundations Walter H. and Marjory M. Rich Memorial Fund Grants, 3685
Support Our Aging Religious (SOAR) Grants, 3686
Swindells Charitable Foundation, 3697
Symantec Community Relations and Corporate Philanthropy Grants, 3699
T.L.L. Temple Foundation Grants, 3700
T. Spencer Shore Foundation Grants, 3701
Taubman Endowment for the Arts Fndn Grants, 3702
Tauck Family Foundation Grants, 3703
Taylor S. Abernathy and Patti Harding Abernathy Charitable Trust Grants, 3704
TE Foundation Grants, 3705
Tellabs Foundation Grants, 3706

Tension Envelope Foundation Grants, 3707
Textron Corporate Contributions Grants, 3712
Thelma Doelger Charitable Trust Grants, 3714
Theodore Edson Parker Foundation Grants, 3715
Thomas C. Ackerman Foundation Grants, 3719
Thomas C. Burke Foundation Grants, 3720
Thomas J. Atkins Memorial Trust Fund Grants, 3721
Thomas J. Long Foundation Community Grants, 3722
Thomas Sill Foundation Grants, 3724
Thomas Thompson Trust Grants, 3725
Thompson Charitable Foundation Grants, 3727
Thoracic Foundation Grants, 3729
Tifa Foundation Grants, 3733
TJX Foundation Grants, 3735
Todd Brock Family Foundation Grants, 3736
Topfer Family Foundation Grants, 3739
Toyota Motor Engineering & Manufacturing North America Grants, 3742
Toyota Motor Manufacturing of Alabama Grants, 3743
Toyota Motor Manufacturing of Indiana Grants, 3744
Toyota Motor Manufacturing of Kentucky Grants, 3745
Toyota Motor Manuf of Mississippi Grants, 3746
Toyota Motor Manufacturing of Texas Grants, 3747
Toyota Motor Manuf of West Virginia Grants, 3748
Toyota Motor N America of New York Grants, 3749
Toyota Motor Sales, USA Grants, 3750
Toys R Us Children's Fund Grants, 3751
Tri-State Community Twenty-first Century Endowment Fund Grants, 3756
Triangle Community Foundation Donor-Advised Grants, 3757
Trull Foundation Grants, 3761
Tull Charitable Foundation Grants, 3762
U.S. Cellular Corporation Grants, 3766
U.S. Department of Education 21st Century Community Learning Centers, 3767
U.S. Department of Education Rehabilitation Engineering Research Centers Grants, 3769
U.S. Department of Education Rehabilitation Research Training Centers (RRTCs), 3770
U.S. Department of Education Rehabilitation Training - Rehabilitation Continuing Education - Institute on Rehabilitation Issues, 3771
U.S. Dept of Education Special Ed--National Activities--Parent Information Centers, 3772
U.S. Department of Education Vocational Rehabilitation Services Projects for American Indians with Disabilities Grants, 3774
UCLA / RAND Corporation Post-Doctoral Fellowships, 3776
UniHealth Foundation Community Health Improvement Grants, 3782
Unihealth Foundation Grants for Non-Profit Oranizations, 3783
UniHealth Foundation Healthcare Systems Enhancements Grants, 3784
UniHealth Foundation Innovation Fund Grants, 3785
UniHealth Foundation Workforce Development Grants, 3786
Union Bank, N.A. Corporate Sponsorships and Donations, 3787
Union Bank, N.A. Foundation Grants, 3788
Union Benevolent Association Grants, 3789
Union Labor Health Fndn Community Grants, 3792
Union Pacific Foundation Health and Human Services Grants, 3794
UnitedHealthcare Children's Foundation Grants, 3795
United Healthcare Comm Grants in Michigan, 3796
United Hospital Fund of New York Health Care Improvement Grants, 3797
United Methodist Committee on Relief Global AIDS Fund Grants, 3798
United Methodist Committee on Relief Global Health Grants, 3799
United Methodist Child Mental Health Grants, 3800
United States Institute of Peace - Jennings Randolph Senior Fellowships, 3802
United Technologies Corporation Grants, 3803
USAID Child, Newborn, and Maternal Health Project Grants, 3806

914 / Service Delivery Programs

USAID Fighting Ebola Grants, 3812
USAID Health System Strengthening Grants, 3814
USAID Integration of Care and Support within the Health System to Support Better Patient Outcomes Grants, 3817
USAID Microbicides Research, Development, and Introduction Grants, 3819
US Airways Community Contributions, 3825
US Airways Community Foundation Grants, 3826
USDA Delta Health Care Services Grants, 3829
Vancouver Foundation Grants and Community Initiatives Program, 3849
Vanderburgh Community Fndn Women's Fund, 3853
Vectren Foundation Grants, 3857
Verizon Foundation Health Care and Accessibility Grants, 3858
Verizon Foundation New York Grants, 3860
Verizon Foundation Northeast Region Grants, 3861
Verizon Foundation Virginia Grants, 3864
VHA Health Foundation Grants, 3867
Victor E. Speas Foundation Grants, 3868
Vigneron Memorial Fund Grants, 3873
Virginia W. Kettering Foundation Grants, 3878
Visiting Nurse Foundation Grants, 3879
Volkswagen of America Corporate Contributions, 3880
W.C. Griffith Foundation Grants, 3881
W.H. & Mary Ellen Cobb Trust Grants, 3884
W.K. Kellogg Foundation Healthy Kids Grants, 3886
W.M. Keck Fndn So California Grants, 3890
W.P. and Bulah Luse Foundation Grants, 3892
Waitt Family Foundation Grants, 3894
Walker Area Community Foundation Grants, 3895
Walmart Foundation Northwest Arkansas Giving, 3899
Walter L. Gross III Family Foundation Grants, 3901
Warrick County Community Women's Fund, 3905
Washington Gas Charitable Giving, 3908
Wayne County Foundation Grants, 3909
Weaver Popcorn Foundation Grants, 3910
Weingart Foundation Grants, 3913
Welborn Baptist Fndn General Op Grants, 3914
Welborn Baptist Foundation Improvements to Community Health Status Grants, 3916
Western Indiana Community Foundation Grants, 3921
Western Union Foundation Grants, 3923
WestWind Foundation Reproductive Health and Rights Grants, 3924
Weyerhaeuser Company Foundation Grants, 3925
Weyerhaeuser Family Foundation Health Grants, 3926
WHAS Crusade for Children Grants, 3927
WHO Foundation General Grants, 3939
WHO Foundation Volunteer Service Grants, 3940
Willard & Pat Walker Charitable Fndn Grants, 3943
Willary Foundation Grants, 3944
William B. Stokely Jr. Foundation Grants, 3947
William Blair and Company Foundation Grants, 3948
William J. & Tina Rosenberg Fndn Grants, 3954
William Ray & Ruth E. Collins Fndn Grants, 3956
Williams Companies Homegrown Giving Grants, 3957
Willis C. Helm Charitable Trust Grants, 3959
Wilson-Wood Foundation Grants, 3960
Windham Foundation Grants, 3961
Winston-Salem Foundation Competitive Grants, 3965
Winston-Salem Fndn Elkin/Tri-County Grants, 3966
Winston-Salem Fndn Stokes County Grants, 3967
Winston-Salem Fndn Victim Assistance Grants, 3968
Wolfe Associates Grants, 3969
Woodward Fund Grants, 3973
Z. Smith Reynolds Foundation Small Grants, 3982
Zane's Foundation Grants, 3983

Symposiums, Conferences, Workshops, Seminars
AACAP Systems of Care Special Scholarships, 76
AACR Scholar-in-Training Awards - Special Conferences, 125
AAP Community Access To Child Health (CATCH) Advocacy Training Grants, 177
AcademyHealth Alice Hersh Student Scholarships, 221
AcademyHealth Awards, 222
AcademyHealth PHSR Interest Group Student Scholarships, 225
ACSM International Student Awards, 269
AES Research and Training Workshop Awards, 320
AIHS Knowledge Exchange - Conference Grants, 389
Alaska Airlines Foundation Grants, 403
Alcoa Foundation Grants, 414
ALFJ Astraea U.S. and Int'l Movement Fund, 427
Alfred P. Sloan Foundation International Science Engagement Grants, 431
Allen P. & Josephine B. Green Fndn Grants, 443
Alzheimer's Association Conference Grants, 463
AMA-MSS Chapter Involvement Grants, 473
AMA Foundation Dr. Nathan Davis International Awards in Medicine, 483
AMA Foundation Jack B. McConnell, MD Awards for Excellence in Volunteerism, 487
AMA Foundation Jordan Fieldman, MD, Awards, 489
AMA Foundation Leadership Awards, 490
AMA Foundation Pride in the Profession Grants, 493
American College of Rheumatology Fellows-in-Training Travel Scholarships, 517
American College of Surgeons Nizar N. Oweida, MD, FACS, Scholarships, 526
American College of Surgeons Resident and Associate Society (RAS-ACS) Leadership Scholarships, 527
AMI Semiconductors Corporate Grants, 576
Ann Arbor Area Community Foundation Grants, 594
AOCS Industrial Oil Products Division Student Award, 616
APA Scientific Conferences Grants, 632
Aratani Foundation Grants, 649
Archer Daniels Midland Foundation Grants, 651
Arizona Public Service Corporate Giving Program Grants, 658
ASHA Minority Student Leadership Awards, 696
ASHA Research Mentoring-Pair Travel Award, 698
ASHA Student Research Travel Award, 699
ASHA Students Preparing for Academic & Research Careers (SPARC) Award, 700
ASM-UNESCO Leadership Grant for International Educators, 715
ASM Millis-Colwell Postgraduate Travel Grant, 741
Atran Foundation Grants, 797
Avon Foundation Breast Care Fund Grants, 813
Ball Brothers Foundation General Grants, 820
Battle Creek Community Foundation Grants, 839
Berks County Community Foundation Grants, 881
Blanche and Julian Robertson Family Foundation Grants, 911
Blue Cross Blue Shield of Minnesota Foundation - Healthy Equity: Health Impact Assessment Program Grants, 917
Blue Cross Blue Shield of Minnesota Fndn Healthy Equity: Public Libraries for Health Grants, 918
Blumenthal Foundation Grants, 925
Bodenwein Public Benevolent Foundation Grants, 928
Boeing Company Contributions Grants, 930
Bullitt Foundation Grants, 973
Caddock Foundation Grants, 994
Canadian Patient Safety Institute (CPSI) Patient Safety Studentships, 1012
Carl M. Freeman Foundation Grants, 1027
CDC Cooperative Agreement for the Development, Operation, and Evaluation of an Entertainment Education Program, 1051
CDC Public Health Conference Support Grant, 1067
CFFVR Alcoholism and Drug Abuse Grants, 1093
Charles Delmar Foundation Grants, 1115
Charles Lafitte Foundation Grants, 1123
CIGNA Foundation Grants, 1160
CJ Foundation for SIDS Program Services Grants, 1163
Claude Worthington Benedum Fndn Grants, 1178
Cleveland-Cliffs Foundation Grants, 1181
Community Fndn for Greater Buffalo Grants, 1247
Community Fndn for Muskegon County Grants, 1254
Community Foundation of Eastern Connecticut General Southeast Grants, 1269
Community Foundation of Muncie and Delaware County Grants, 1300
Connecticut Community Foundation Grants, 1324

PROGRAM TYPE INDEX

Conquer Cancer Foundation of ASCO International Development and Education Award in Palliative Care, 1332
Conquer Cancer Foundation of ASCO International Development and Education Awards, 1333
Conquer Cancer Foundation of ASCO Resident Travel Award for Underrepresented Populations, 1336
Covidien Partnership for Neighborhood Wellness Grants, 1353
Cystic Fibrosis Canada Special Travel Grants For Fellows and Students, 1389
Cystic Fibrosis Canada Visiting Clinician Awards, 1395
Daimler and Benz Foundation Fellowships, 1411
Dallas Women's Foundation Grants, 1417
Denver Foundation Community Grants, 1461
Dr. Leon Bromberg Charitable Trust Grants, 1509
DSO Controlling Pathogen Workshop Grants, 1514
Dubois County Community Foundation Grants, 1517
Dyson Foundation Mid-Hudson Valley Project Support Grants, 1534
EBSCO / MLA Annual Meeting Grants, 1545
Eugene M. Lang Foundation Grants, 1611
Evanston Community Foundation Grants, 1615
Expect Miracles Foundation Grants, 1619
Faye McBeath Foundation Grants, 1635
Fayette County Foundation Grants, 1636
FDHN Non-Career Research Awards, 1650
Finance Factors Foundation Grants, 1666
Foellinger Foundation Grants, 1684
Fremont Area Community Fndn General Grants, 1746
Fulbright German Studies Seminar Grants, 1751
Fulton County Community Foundation Grants, 1763
George Gund Foundation Grants, 1796
Gill Foundation - Gay and Lesbian Fund Grants, 1821
Global Fund for Women Grants, 1833
GNOF Organizational Effectiveness Grants and Workshops, 1846
Greater Saint Louis Community Fndn Grants, 1884
Greater Worcester Community Foundation Jeppson Memorial Fund for Brookfield Grants, 1889
Health Fndn of Greater Indianapolis Grants, 1984
Helen Bader Foundation Grants, 1992
IBRO-PERC Support for European Workshops, Symposia and Meetings, 2098
IBRO-PERC Support for Site Lectures, 2099
IBRO Asia APRC Lecturer Exchange Grants, 2102
IBRO Latin America Regional Funding for Short Courses, Workshops, and Symposia, 2107
International Positive Psychology Association Student Scholarships, 2161
J.W. Kieckhefer Foundation Grants, 2176
Jacobs Family Village Neighborhoods Grants, 2184
Jerome Robbins Foundation Grants, 2223
John P. McGovern Foundation Grants, 2247
Johnson Foundation Wingspread Conference Support Program, 2257
John W. Speas and Effie E. Speas Memorial Trust Grants, 2264
Joseph Alexander Foundation Grants, 2265
Kinsman Foundation Grants, 2309
Komen Greater NYC Small Grants, 2317
Landon Foundation-AACR Innovator Award for International Collaboration, 2343
Lillian S. Wells Foundation Grants, 2365
Louis and Elizabeth Nave Flarsheim Charitable Foundation Grants, 2379
Lowell Berry Foundation Grants, 2382
Lynde and Harry Bradley Foundation Grants, 2397
Manuel D. & Rhoda Mayerson Fndn Grants, 2420
Marie C. and Joseph C. Wilson Foundation Rochester Small Grants, 2431
Mary Black Foundation Active Living Grants, 2440
McLean Contributionship Grants, 2497
Meyer Foundation Benevon Grants, 2532
MGN Family Foundation Grants, 2542
Michigan Psychoanalytic Institute and Society SATA Grants, 2546
MLA Lucretia W. McClure Excellence in Education Award, 2570
NAPNAP Foundation McNeil Grant-in-Aid, 2612

PROGRAM TYPE INDEX

NAPNAP Foundation Reckitt Benckiser Student Scholarship, 2616
NAPNAP Foundation Wyeth Pediatric Immunization Grant, 2618
NIA Support of Scientific Meetings as Cooperative Agreements, 2784
NIDDK Traditional Conference Grants, 2848
NINDS Support of Scientific Meetings as Cooperative Agreements, 2925
NLM Informatics Conference Grants, 2945
Noble County Community Foundation Grants, 2975
Norcliffe Foundation Grants, 2979
North Carolina Biotechnology Center Event Sponsorship Grants, 2984
North Carolina Biotechnology Center Grantsmanship Training Grants, 2985
North Carolina Biotech Center Meeting Grants, 2986
North Carolina Community Foundation Grants, 2990
North Carolina GlaxoSmithKline Fndn Grants, 2991
NYAM Brock Lecture, Award and Visiting Professorship in Pediatrics, 3038
Oppenstein Brothers Foundation Grants, 3078
Oregon Community Fndn Community Grants, 3082
Orthopaedic Trauma Association Kathy Cramer Young Clinician Memorial Fellowships, 3083
Paul Rapoport Foundation Grants, 3125
Peter and Elizabeth C. Tower Foundation Organizational Scholarships, 3148
Pfizer/AAFP Foundation Visiting Professorship Program in Family Medicine, 3162
Pfizer Medical Education Track One Grants, 3182
Pfizer Medical Education Track Two Grants, 3183
Pinellas County Grants, 3223
Potts Memorial Foundation Grants, 3254
Principal Financial Group Foundation Grants, 3273
Prudential Foundation Education Grants, 3276
Puerto Rico Community Foundation Grants, 3278
PVA Education Foundation Grants, 3284
PVA Research Foundation Grants, 3285
Quantum Corporation Snap Server Grants, 3289
Raskob Foundation for Catholic Activities Grants, 3306
Rathmann Family Foundation Grants, 3309
RCPSC Balfour M. Mount Visiting Professorship in Palliative Medicine, 3318
RCPSC Harry S. Morton Lectureship in Surgery, 3323
RCPSC John G. Wade Visiting Professorship in Patient Safety & Simulation-Based Med Education, 3329
RCPSC Professional Development Grants, 3338
RGk Foundation Grants, 3361
Rhode Island Foundation Grants, 3362
Rochester Area Community Foundation Grants, 3391
Rochester Area Foundation Grants, 3392
Rockwell International Corporate Trust Grants Program, 3399
Rose Community Foundation Aging Grants, 3410
RRF Organizational Capacity Building Grants, 3418
SAMHSA Conference Grants, 3462
San Antonio Area Foundation Grants, 3469
SfN Science Educator Award, 3535
SfN Science Journalism Student Awards, 3536
Sick Kids Fndn Community Conference Grants, 3549
Sid W. Richardson Foundation Grants, 3553
Sigma Theta Tau International / Council for the Advancement of Nursing Science Grants, 3557
South Madison Community Foundation Grants, 3626
St. Joseph Community Health Foundation Schneider Fellowships, 3642
Strowd Roses Grants, 3664
Summit Foundation Grants, 3674
Taylor S. Abernathy and Patti Harding Abernathy Charitable Trust Grants, 3704
Textron Corporate Contributions Grants, 3712
Thoracic Foundation Grants, 3729
Topeka Community Foundation Kansas Blood Services Scholarships, 3738
W.M. Keck Foundation Science and Engineering Research Grants, 3889
Wayne County Foundation Grants, 3909
Weyerhaeuser Company Foundation Grants, 3925
Z. Smith Reynolds Foundation Small Grants, 3982

Technical Assistance
AAAS Science and Technology Policy Fellowships: Global Health and Development, 41
Abbot and Dorothy H. Stevens Foundation Grants, 197
Abbott Fund Access to Health Care Grants, 199
Abbott Fund Global AIDS Care Grants, 202
ACL Partnerships in Employment Systems Change Grants, 246
ACL Self-Advocacy Resource Center Grants, 247
ACL Training and Technical Assistance Center for State Intellectual and Developmental Disabilities Delivery Systems Grants, 248
AEC Trust Grants, 315
Albert Pick Jr. Fund Grants, 409
Albuquerque Community Foundation Grants, 412
ALFJ Astraea U.S. and Int'l Movement Fund, 427
Alliant Energy Corporation Contributions, 445
Allstate Corporate Giving Grants, 446
Alpha Natural Resources Corporate Giving, 450
Altman Foundation Health Care Grants, 457
AMA-RFS and AMA Foundation Medical Student & Resident/Fellow Elective-Medicine & Media, 478
Amador Community Foundation Grants, 481
Anschutz Family Foundation Grants, 602
ARCO Foundation Education Grants, 652
Arie and Ida Crown Memorial Grants, 654
Arlington Community Foundation Grants, 662
Assisi Foundation of Memphis General Grants, 787
Autauga Area Community Foundation Grants, 808
Avon Products Foundation Grants, 814
Ball Brothers Foundation General Grants, 820
Baxter International Corporate Giving Grants, 841
BCBSM Fndn Proposal Development Awards, 857
Beckman Coulter Foundation Grants, 868
Beldon Fund Grants, 871
Bernard F. and Alva B. Gimbel Foundation Reproductive Rights Grants, 883
Blanche and Julian Robertson Family Foundation Grants, 911
Blue Cross Blue Shield of Minnesota Foundation - Healthy Equity: Health Impact Assessment Program Grants, 917
Blue River Community Foundation Grants, 922
Boeing Company Contributions Grants, 930
Boston Foundation Grants, 936
Boston Globe Foundation Grants, 937
BP Foundation Grants, 943
Bullitt Foundation Grants, 973
Burlington Industries Foundation Grants, 977
Caesars Foundation Grants, 995
Cailloux Foundation Grants, 996
Cambridge Community Foundation Grants, 1004
Canadian Patient Safety Institute (CPSI) Patient Safety Studentships, 1012
Carl M. Freeman Foundation Grants, 1027
Carpenter Foundation Grants, 1035
Cass County Community Foundation Grants, 1041
CFFVR Basic Needs Giving Partnership Grants, 1094
Charles H. Farnsworth Trust Grants, 1119
Chilkat Valley Community Foundation Grants, 1148
Clarence E. Heller Charitable Foundation Grants, 1166
CNCS Senior Companion Program Grants, 1191
Coastal Community Foundation of South Carolina Grants, 1202
Columbus Foundation Competitive Grants, 1217
Columbus Foundation Traditional Grants, 1223
Community Foundation Alliance City of Evansville Endowment Fund Grants, 1238
Community Foundation for Greater Atlanta Metropolitan Atlanta An Extra Wish Grants, 1243
Community Foundation for Greater Atlanta Strategic Restructuring Fund Grants, 1246
Community Fndn for Greater Buffalo Grants, 1247
Community Foundation for Greater New Haven Responsive New Grants, 1249
Community Foundation for Greater New Haven Valley Neighborhood Grants, 1251
Community Foundation for Greater New Haven Women & Girls Grants, 1252
Community Fndn for Monterey County Grants, 1253

Community Fndn for the Capital Region Grants, 1259
Community Foundation for the National Capital Region Community Leadership Grants, 1260
Community Foundation of Boone County Grants, 1266
Community Foundation of Eastern Connecticut General Southeast Grants, 1269
Community Foundation of Greater Flint Grants, 1273
Community Foundation of Greater Fort Wayne - Community Endowment and Clarke Endowment Grants, 1274
Community Foundation of Greater Greensboro Community Grants, 1277
Community Foundation of Muncie and Delaware County Grants, 1300
Community Foundation of Riverside & San Bernardino County Impact Grants, 1303
Community Fndn of Wabash County Grants, 1316
Connecticut Community Foundation Grants, 1324
ConocoPhillips Foundation Grants, 1327
Conservation, Food, and Health Foundation Grants for Developing Countries, 1338
Consumers Energy Foundation, 1343
Cooke Foundation Grants, 1344
Covenant Matters Foundation Grants, 1350
Crail-Johnson Foundation Grants, 1355
Daisy Marquis Jones Foundation Grants, 1414
Dallas Women's Foundation Grants, 1417
Deborah Munroe Noonan Memorial Grants, 1443
Dept of Ed Special Education--Technical Assistance and Dissemination to Improve Services and Results for Children with Disabilities, 1467
Dorothea Haus Ross Foundation Grants, 1498
Dyson Foundation Mid-Hudson Valley Project Support Grants, 1534
Echoing Green Fellowships, 1546
Edina Realty Foundation Grants, 1550
Educational Foundation of America Grants, 1553
Edward Bangs Kelley and Elza Kelley Foundation Grants, 1556
Edyth Bush Charitable Foundation Grants, 1562
Elizabeth Morse Genius Charitable Trust Grants, 1569
Elkhart County Community Foundation Grants, 1573
Elmer L. & Eleanor J. Andersen Fndn Grants, 1578
El Paso Community Foundation Grants, 1579
Elsie Lee Garthwaite Memorial Fndn Grants, 1584
Ewing Halsell Foundation Grants, 1618
Expect Miracles Foundation Grants, 1619
Fairfield County Community Foundation Grants, 1625
Fargo-Moorhead Area Foundation Grants, 1632
Faye McBeath Foundation Grants, 1635
Fidelity Foundation Grants, 1659
Field Foundation of Illinois Grants, 1660
Florida BRAIVE Fund of Dade Community Foundation, 1678
Fluor Foundation Grants, 1682
Four County Community Fndn General Grants, 1710
Four County Community Foundation Healthy Senior/Healthy Youth Fund Grants, 1711
France-Merrick Foundation Health and Human Services Grants, 1715
Frances L. and Edwin L. Cummings Memorial Fund Grants, 1717
Frank Stanley Beveridge Foundation Grants, 1730
Frederick McDonald Trust Grants, 1740
Fremont Area Community Fndn General Grants, 1746
General Motors Foundation Grants, 1783
GNOF Organizational Effectiveness Grants and Workshops, 1846
Grand Rapids Area Community Foundation Wyoming Youth Fund Grants, 1868
Grand Rapids Community Foundation Ionia County Youth Fund Grants, 1871
Greater Tacoma Community Foundation Grants, 1886
Greater Worcester Community Foundation Discretionary Grants, 1888
Green River Area Community Fndn Grants, 1895
Gulf Coast Community Foundation Grants, 1915
Gulf Coast Foundation of Community Operating Grants, 1916

916 / Technical Assistance

Gulf Coast Foundation of Community Program Grants, 1917
H.J. Heinz Company Foundation Grants, 1923
HAF Technical Assistance Program Grants, 1933
Hardin County Community Foundation Grants, 1946
Health Fndn of Greater Indianapolis Grants, 1984
Hill Crest Foundation Grants, 2030
HRSA Resource and Technical Assistance Center for HIV Prevention and Care for Black Men who Have Sex with Men Cooperative Agreement, 2069
HRSA Ryan White HIV AIDS Drug Assistance Grants, 2070
Jacob G. Schmidlapp Trust Grants, 2183
Jacobs Family Village Neighborhoods Grants, 2184
James R. Dougherty Jr. Foundation Grants, 2198
Jessica Stevens Community Foundation Grants, 2224
JM Foundation Grants, 2229
Joseph H. & Florence A. Roblee Fndn Grants, 2269
Joseph Henry Edmondson Foundation Grants, 2270
Josephine Goodyear Foundation Grants, 2272
Kansas Health Foundation Recognition Grants, 2289
Kinsman Foundation Grants, 2309
Kuntz Foundation Grants, 2326
L. W. Pierce Family Foundation Grants, 2328
Lewis H. Humphreys Charitable Trust Grants, 2361
Linford and Mildred White Charitable Grants, 2368
Louetta M. Cowden Foundation Grants, 2377
Manuel D. & Rhoda Mayerson Fndn Grants, 2420
Merrick Foundation Grants, 2522
MetroWest Health Foundation Grants to Schools to Conduct Mental Health Capacity Assessments, 2529
MGN Family Foundation Grants, 2542
Michael Reese Health Trust Core Grants, 2544
Mimi and Peter Haas Fund Grants, 2557
Minneapolis Foundation Community Grants, 2558
Montgomery County Community Fndn Grants, 2591
Morton K. and Jane Blaustein Foundation Health Grants, 2595
Ms. Foundation for Women Health Grants, 2597
Mt. Sinai Health Care Foundation Health Policy Grants, 2601
NIDDK Small Business Innov Research to Develop New Therapeutics & Monitoring Tech for Type 1 Diabetes Towards an Artificial Pancreas, 2845
NIEHS SBIR E-Learning for HAZMAT and Emergency Response Grants, 2859
NIGMS Advancing Basic Behavioral and Social Research on Resilience: an Integrative Science Approach, 2861
Norcliffe Foundation Grants, 2979
Nordson Corporation Foundation Grants, 2981
North Carolina Biotechnology Center Technology Enhancement Grants, 2989
NWHF Health Advocacy Small Grants, 3033
NWHF Physical Activity and Nutrition Grants, 3037
NYCT Technical Assistance Grants, 3058
Ohio County Community Foundation Board of Directors Grants, 3067
OneFamily Foundation Grants, 3075
Oregon Community Foundation Better Nursing Home Care Grants, 3081
Oregon Community Fndn Community Grants, 3082
OSF Law and Health Initiative Grants, 3092
OSF Mental Health Initiative Grants, 3093
Paso del Norte Health Foundation Grants, 3119
Paul Rapoport Foundation Grants, 3125
Peter and Elizabeth C. Tower Foundation Phase II Technology Initiative Grants, 3149
Peter and Elizabeth C. Tower Foundation Phase I Technology Initiative Grants, 3150
Peter and Elizabeth C. Tower Foundation Social and Emotional Preschool Curriculum Grants, 3151
Pfizer/AAFP Foundation Visiting Professorship Program in Family Medicine, 3162
Philadelphia Foundation Organizational Effectiveness Grants, 3189
Pinkerton Foundation Grants, 3224
PNC Foundation Community Services Grants, 3240
Pokagon Fund Grants, 3242
PPCF Community Grants, 3256

Prudential Foundation Education Grants, 3276
Puerto Rico Community Foundation Grants, 3278
Putnam County Community Foundation Grants, 3283
Rainbow Fund Grants, 3301
RCPSC Balfour M. Mount Visiting Professorship in Palliative Medicine, 3318
RCPSC Janes Visiting Professorship in Surgery, 3328
Rhode Island Foundation Grants, 3362
Rochester Area Community Foundation Grants, 3391
RRF Organizational Capacity Building Grants, 3418
RWJF Policies for Action: Policy and Law Research to Build a Culture of Health Grants, 3443
San Diego Foundation for Change Grants, 3470
Seabury Foundation Grants, 3501
Seattle Foundation Health and Wellness Grants, 3507
Seattle Foundation Medical Funds Grants, 3508
Seward Community Foundation Grants, 3514
Seward Community Foundation Mini-Grants, 3515
Shield-Ayres Foundation Grants, 3546
Solo Cup Foundation Grants, 3620
South Madison Community Foundation Grants, 3626
St. Joseph Community Health Foundation Burn Care and Prevention Grants, 3639
St. Joseph Community Health Foundation Improving Healthcare Access Grants, 3640
Stewart Huston Charitable Trust Grants, 3655
Streilein Foundation for Ocular Immunology Visiting Professorships, 3660
Susan Mott Webb Charitable Trust Grants, 3694
Taylor S. Abernathy and Patti Harding Abernathy Charitable Trust Grants, 3704
Textron Corporate Contributions Grants, 3712
Theodore Edson Parker Foundation Grants, 3715
Toyota Motor Manuf of West Virginia Grants, 3748
Toyota Motor N America of New York Grants, 3749
Toyota Motor Sales, USA Grants, 3750
Tri-State Community Twenty-first Century Endowment Fund Grants, 3756
U.S. Department of Education Rehabilitation Training - Rehabilitation Continuing Education - Institute on Rehabilitation Issues, 3771
Union Bank, N.A. Foundation Grants, 3788
Union Pacific Foundation Health and Human Services Grants, 3794
Vancouver Foundation Grants and Community Initiatives Program, 3849
Verizon Foundation Vermont Grants, 3863
W.M. Keck Fndn So California Grants, 3890
Walker Area Community Foundation Grants, 3895
Welborn Baptist Fndn Health Ministries Grants, 3915
Western Indiana Community Foundation Grants, 3921
Weyerhaeuser Company Foundation Grants, 3925
William G. Gilmore Foundation Grants, 3951
William Ray & Ruth E. Collins Fndn Grants, 3956

Training Programs/Internships
AACAP Child and Adolescent Psychiatry (CAP) Teaching Scholarships, 51
AACN Mentorship Grant, 85
AAP Community Access To Child Health (CATCH) Advocacy Training Grants, 177
AAP Community Access To Child Health (CATCH) Residency Training Grants, 180
ACCP Anticoagulation Training Program Grants, 227
ACCP Heart Failure Training Program Grants, 228
ACF Native American Social and Economic Development Strategies Grants, 231
Achelis Foundation Grants, 234
ACL Training and Technical Assistance Center for State Intellectual and Developmental Disabilities Delivery Systems Grants, 248
ACL University Centers for Excellence in Developmental Network Diversity and Inclusion Training Action Planning Grants, 249
ACS Master's Training Grants in Clinical Oncology Social Work, 258
ACS Physician Training Awards in Prevent Med, 275
Adolph Coors Foundation Grants, 311
Aetna Foundation Minority Scholars Grants, 329

PROGRAM TYPE INDEX

AFAR Medical Student Training in Aging Research Program, 335
AHIMA Faculty Development Stipends, 353
AHNS/AAO-HNSF Surgeon Scientist Combined Award, 357
AHNS/AAO-HNSF Young Investigator Combined Award, 358
AIHS/Mitacs Health Pilot Partnership Interns, 373
AIHS ForeFront Internships, 380
Allen Foundation Educational Nutrition Grants, 442
AMA-MSS Government Relations Internships, 476
AMA-RFS Legislative Awareness Internships, 479
AMD Corporate Contributions Grants, 500
American-Scandinavian Foundation Visiting Lectureship Grants, 502
American Psychiatric Association/Janssen Resident Psychiatric Research Scholars, 540
American Psychiatric Association Minority Medical Student Summer Externship in Addiction Psychiatry, 545
American Psychiatric Association Minority Medical Student Summer Mentoring Program, 546
Anschutz Family Foundation Grants, 602
ASGE Endoscopic Research Awards, 686
ASHA Research Mentoring-Pair Travel Award, 698
ASHA Students Preparing for Academic & Research Careers (SPARC) Award, 700
ASPH/CDC Public Health Internships, 782
Baxter International Corporate Giving Grants, 841
Blowitz-Ridgeway Foundation Grants, 913
Bodman Foundation Grants, 929
Canada-U.S. Fulbright New Century Scholars Program Grants, 1008
Canadian Patient Safety Institute (CPSI) Patient Safety Studentships, 1012
Carrie Estelle Doheny Foundation Grants, 1037
Carrier Corporation Contributions Grants, 1038
CDC Collegiate Leaders in Environmental Health Internships, 1048
CDC Epidemic Intell Service Training Grants, 1055
CDC Public Health Associates, 1065
CDC Public Health Associates Hosts, 1066
CDC Summer Graduate Environmental Health Internships, 1072
CDC Summer Program In Environmental Health Internships, 1073
CFF Student Traineeships, 1091
Charles A. Frueauff Foundation Grants, 1114
Charles Delmar Foundation Grants, 1115
Charles H. Revson Foundation Grants, 1122
Clarian Health Critical/Progressive Care Interns, 1168
Clarian Health Multi-Specialty Internships, 1169
Clarian Health OR Internships, 1170
CNCS Social Innovation Grants, 1192
Collins Foundation Grants, 1210
Columbus Foundation Competitive Grants, 1217
Community Fndn for Greater Buffalo Grants, 1247
Conservation, Food, and Health Foundation Grants for Developing Countries, 1338
Cruise Industry Charitable Foundation Grants, 1363
Dept of Ed College Assistance Migrant Grants, 1462
Dept of Ed Rehabilitation Training Grants, 1463
DHHS Adolescent Family Life Demonstration Projects, 1472
DHHS Emerging Leaders Program Internships, 1473
DOD HBCU/MI Partnership Training Award, 1480
DOJ Gang-Free Schools and Communities Intervention Grants, 1481
DOL Occupational Safety and Health--Susan Harwood Training Grants, 1484
Donald W. Reynolds Foundation Aging and Quality of Life Grants, 1487
Dr. John T. Macdonald Foundation Grants, 1507
E.L. Wiegand Foundation Grants, 1538
Earl and Maxine Claussen Trust Grants, 1541
Educational Foundation of America Grants, 1553
Edwin S. Webster Foundation Grants, 1560
Effie and Wofford Cain Foundation Grants, 1563
Essex County Community Foundation Merrimack Valley General Fund Grants, 1603

PROGRAM TYPE INDEX

Essex County Community Foundation Women's Fund Grants, 1605
Eugene M. Lang Foundation Grants, 1611
Evjue Foundation Grants, 1617
Faye McBeath Foundation Grants, 1635
FCD New American Children Grants, 1637
Foundation for Seacoast Health Grants, 1698
Foundation for the Mid South Community Development Grants, 1699
Frederick W. Marzahl Memorial Fund Grants, 1741
Fred L. Emerson Foundation Grants, 1743
Fulbright Alumni Initiatives Awards, 1749
Fulbright International Education Administrators (IEA) Seminar Program Grants, 1752
Fulbright New Century Scholars Grants, 1753
Fulbright Traditional Scholar Program in the Near East and North Africa Region, 1758
Genesis Foundation Grants, 1786
George Family Foundation Grants, 1794
George Gund Foundation Grants, 1796
GNOF New Orleans Works Grants, 1844
Green Bay Packers Foundation Grants, 1890
Hearst Foundations Health Grants, 1987
Helen Bader Foundation Grants, 1992
Helen Irwin Littauer Educational Trust Grants, 1994
Hogg Foundation for Mental Health Grants, 2039
Holland/Zeeland Community Fndn Grants, 2041
IIE Freeman Foundation Indonesia Internships, 2131
Independence Blue Cross Nursing Internships, 2152
JM Foundation Grants, 2229
John Merck Fund Grants, 2245
Johnson & Johnson Corporate Contributions, 2253
Kessler Fnd Signature Employment Grants, 2302
Kinsman Foundation Grants, 2309
Lynde and Harry Bradley Foundation Grants, 2397
Macquarie Bank Foundation Grants, 2410
Max and Victoria Dreyfus Foundation Grants, 2457
May and Stanley Smith Charitable Trust Grants, 2459
McLean Contributionship Grants, 2497
Mead Johnson Nutritionals Scholarships, 2505
Montgomery County Community Fndn Grants, 2591
Morehouse PHSI Project Imhotep Internships, 2592
Morton K. and Jane Blaustein Foundation Health Grants, 2595
Mt. Sinai Health Care Foundation Academic Medicine and Bioscience Grants, 2598
NEI Ruth L. Kirschstein National Research Service Award Short-Term Institutional Research Training Grants, 2668
NEI Scholars Program, 2669
NHLBI Biomedical Research Training Program for Individuals from Underrepresented Groups, 2692
NHLBI Mentored Career Award for Faculty at Institutions that Promote Diversity (K01), 2706
NHLBI Research Supplements to Promote Diversity in Health-Related Research, 2727
NHLBI Short-Term Research Education Program to Increase Diversity in Health-Related Research, 2733
NHLBI Summer Internship Program in Biomedical Research, 2735
NHLBI Training Program Grants for Institutions that Promote Diversity, 2738
NIDDK Mentored Research Scientist Development Award, 2825
NIDDK Training in Clinical Investigation in Kidney and Urology Grants, 2849
NIH Ruth L. Kirschstein National Research Service Award Institutional Research Training Grants, 2900
NIH Ruth L. Kirschstein Nat Research Service Award Short-Term Institutional Training Grants, 2906
NIH Summer Internship Program in Biomedical Research, 2912
NLM Exploratory/Developmental Grants, 2942
Otto Bremer Foundation Grants, 3099
Peter Kiewit Foundation General Grants, 3154
Peter Kiewit Foundation Small Grants, 3155
Pfizer Medical Education Track One Grants, 3182
PhRMA Foundation Health Outcomes Pre-Doctoral Fellowships, 3203
Pinkerton Foundation Grants, 3224

PPG Industries Foundation Grants, 3257
Principal Financial Group Foundation Grants, 3273
Prudential Foundation Education Grants, 3276
Ralph M. Parsons Foundation Grants, 3304
Raskob Foundation for Catholic Activities Grants, 3306
Rathmann Family Foundation Grants, 3309
RCPSC Janes Visiting Professorship in Surgery, 3328
Regenstrief General Internal Medicine - General Pediatrics Research Fellowship Program, 3351
Regenstrief Geriatrics Fellowship, 3352
Regenstrief Master Of Science In Clinical Research/ CITE Program, 3353
Ruth Eleanor Bamberger and John Ernest Bamberger Memorial Foundation Grants, 3427
Saint Luke's Health Initiatives Grants, 3455
Salmon Foundation Grants, 3458
Savoy Foundation Studentships, 3490
Sidgmore Family Foundation Grants, 3551
Sierra Health Foundation Responsive Grants, 3555
Sioux Falls Area Community Foundation Community Fund Grants (Unrestricted), 3568
Sioux Falls Area Community Foundation Spot Grants (Unrestricted), 3570
Society for Imaging Informatics in Medicine (SIIM) Small Training Grants, 3601
Sony Corporation of America Grants, 3623
Strowd Roses Grants, 3664
Summer Medical and Dental Education Program: Interprofessional Pilot Grants, 3673
Susan Mott Webb Charitable Trust Grants, 3694
TE Foundation Grants, 3705
Textron Corporate Contributions Grants, 3712
UniHealth Foundation Healthcare Systems Enhancements Grants, 3784
Union Bank, N.A. Foundation Grants, 3788
United Hospital Fund of New York Health Care Improvement Grants, 3797
USDA Organic Agriculture Research Grants, 3830
Vectren Foundation Grants, 3857
Verizon Foundation New York Grants, 3860
Walker Area Community Foundation Grants, 3895
Weingart Foundation Grants, 3913
Welborn Baptist Fndn Health Ministries Grants, 3915

Travel Grants
AAAAI ARTrust Grants for Allied Health Travel, 21
AAAAI Fellows-in-Training Travel Grants, 30
AACAP Educational Outreach Program for Child and Adolescent Psychiatry Residents, 52
AACAP Educational Outreach Program for General Psychiatry Residents, 53
AACAP Educational Outreach Program for Residents in Alcohol Research, 54
AACAP Life Members Mentorship Grants for Medical Students, 63
AACAP Systems of Care Special Scholarships, 76
AACR-GlaxoSmithKline Clinical Cancer Research Scholar Awards, 93
AACR NCI International Investigator Opportunity Grants, 96
AACR-WICR Scholar Awards, 100
AACR Brigid G. Leventhal Scholar in Cancer Research Awards, 106
AACR Gertrude Elion Cancer Research Award, 113
AACR Minority-Serving Institution Faculty Scholar in Cancer Research Awards, 119
AACR Min Scholar in Cancer Research Award, 120
AACR Scholar-in-Training Awards, 124
AACR Scholar-in-Training Awards - Special Conferences, 125
AACR Thomas J. Bardos Science Ed Awards, 127
AAFPRS Residency Travel Awards, 144
AAP Anne E. Dyson Child Advocacy Awards, 176
AAP International Travel Grants, 184
AAP Program Delegate Awards, 188
Abbott-ASM Lifetime Achievement Award, 198
AcademyHealth Alice Hersh Student Scholarships, 221
AcademyHealth Awards, 222
AcademyHealth PHSR Interest Group Student Scholarships, 225

Travel Grants / 917

ACCP Anticoagulation Training Program Grants, 227
ACCP Heart Failure Training Program Grants, 228
ACSM-GSSI Young Scholar Travel Award, 257
ACSM International Student Awards, 269
AES-Grass Young Investigator Travel Awards, 317
AIHP Sonnedecker Visiting Scholar Grants, 371
ALFJ Astraea U.S. and Int'l Movement Fund, 427
AMA-MSS Research Poster Award, 477
AMA-WPC Joan F. Giambalvo Memorial Scholarships, 480
AMA Foundation Dr. Nathan Davis International Awards in Medicine, 483
AMA Foundation Jack B. McConnell, MD Awards for Excellence in Volunteerism, 487
AMA Foundation Jordan Fieldman, MD, Awards, 489
AMA Foundation Leadership Awards, 490
AMA Foundation Pride in the Profession Grants, 493
AMA Virtual Mentor Theme Issue Editor Grants, 496
American Academy of Dermatology Camp Discovery Scholarships, 503
American College of Rheumatology Fellows-in-Training Travel Scholarships, 517
American College of Surgeons Australia and New Zealand Chapter Travelling Fellowships, 519
American College of Surgeons Nizar N. Oweida, MD, FACS, Scholarships, 526
American Psychiatric Association Travel Scholarships for Minority Medical Students, 549
amfAR Research Grants, 571
ANA F.E. Bennett Memorial Lectureship, 581
ANA Junior Academic Neurologist Scholarships, 582
AOCS Industrial Oil Products Division Student Award, 616
AOCS Lipid Oxidation and Quality Division Poster Competition, 617
AOCS Surfactants and Detergents Division Student Award, 622
APA Travel Awards, 633
ASCO Merit Awards, 682
ASGE Don Wilson Award, 685
ASHA Minority Student Leadership Awards, 696
ASHA Research Mentoring-Pair Travel Award, 698
ASHA Student Research Travel Award, 699
ASHA Students Preparing for Academic & Research Careers (SPARC) Award, 700
ASHFoundation Research Grant in Speech, 709
ASM-UNESCO Leadership Grant for International Educators, 715
ASM Corporate Activities Program Student Travel Grants, 721
ASMCUE/GM Travel Assistance Grants, 722
ASM Early-Career Faculty Travel Award, 724
ASM Faculty Ehancement Travel Awards, 727
ASM Founders Distinguished Service Award, 728
ASM General Meeting Minority Travel Grants, 730
ASM GlaxoSmithKline International Member of the Year Award, 731
ASM Int'l Professorship for Latin America, 736
ASM Int'l Professorships for Asia and Africa, 737
ASM Millis-Colwell Postgraduate Travel Grant, 741
ASM Richard and Mary Finkelstein Travel Grant, 746
ASM Roche Diagnostics Alice C. Evans Award, 748
ASM Student & Postdoc Fellow Travel Grants, 752
ASM Student Travel Grants, 753
ASM TREK Diagnostic ABMM/ABMLI Professional Recognition Award, 754
ASPEN Scientific Abstracts Awards for Papers or Posters, 766
ASPEN Scientific Abstracts Promising Investigator Awards, 767
ASPET Travel Awards, 779
Australian Academy of Science Grants, 807
Bush Foundation Medical Fellowships, 980
BWF Collaborative Research Travel Grants, 985
Canada-U.S. Fulbright Senior Specialists Grants, 1009
Carl M. Freeman Foundation Grants, 1027
CDC-Hubert Global Health Fellowship, 1047
CDC Foundation Atlanta International Health Fellowships, 1060
CDC Public Health Conference Support Grant, 1067

918 / Travel Grants

CJ Foundation for SIDS Program Services Grants, 1163
Conquer Cancer Foundation of ASCO International Development and Education Award in Palliative Care, 1332
Conquer Cancer Foundation of ASCO International Development and Education Awards, 1333
Conquer Cancer Foundation of ASCO Resident Travel Award for Underrepresented Populations, 1336
Cystic Fibrosis Canada Senior Scientist Research Training Awards, 1387
Cystic Fibrosis Canada Transplant Center Incentive Grants, 1392
Cystic Fibrosis Canada Travel Supplement Grants, 1393
Cystic Fibrosis Canada Visiting Allied Health Professional Awards, 1394
Cystic Fibrosis Canada Visiting Clinician Awards, 1395
Cystic Fibrosis Canada Visiting Scientist Awards, 1396
Daimler and Benz Foundation Fellowships, 1411
Decatur County Community Foundation Large Project Grants, 1444
EBSCO / MLA Annual Meeting Grants, 1545
Elizabeth Glaser Int'l Leadership Awards, 1566
FDHN Moti L. & Kamla Rustgi International Travel Awards, 1649
FDHN Non-Career Research Awards, 1650
Foundation of CVPH Travel Fund Grants, 1706
Fulbright Alumni Initiatives Awards, 1749
Fulbright Distinguished Chairs Awards, 1750
Fulbright German Studies Seminar Grants, 1751
Fulbright International Education Administrators (IEA) Seminar Program Grants, 1752
Fulbright Specialists Program Grants, 1754
Fulbright Scholars in Europe and Eurasia, 1755
Fulbright Traditional Scholar Program in Sub-Saharan Africa, 1756
Fulbright Traditional Scholar Program in the East Asia/Pacific Region, 1757
Fulbright Traditional Scholar Program in the South and Central Asia Region, 1759
Fulbright Traditional Scholar Program in the Western Hemisphere, 1760
Fulton County Community Foundation Grants, 1763
Global Fund for Women Grants, 1833
GNOF Organizational Effectiveness Grants and Workshops, 1846
HAF David (Davey) H. Somerville Medical Travel Fund Grants, 1928
IBRO-PERC InEurope Short Stay Grants, 2097
IBRO/SfN International Travel Grants, 2100
IBRO Asia Regional APRC Travel Grants, 2103
IBRO International Travel Grants, 2104
IBRO Latin America Regional Travel Grants, 2109
IIE Whitaker Int'l Fellowships & Scholarships, 2144
International Positive Psychology Association Student Scholarships, 2161
Jane Beattie Memorial Scholarship, 2207
Komen Greater NYC Small Grants, 2317
MAP International Medical Fellowships, 2421
MBL M.G.F. Fuortes Summer Fellowships, 2482
Medical Informatics Section/MLA Career Development Grant, 2508
Meyer Foundation Benevon Grants, 2532
Michigan Psychoanalytic Institute and Society SATA Grants, 2546
MLA Cunningham Memorial International Fellowship, 2560
MLA David A. Kronick Traveling Fellowship, 2561
MLA Janet Doe Lectureship Award, 2567
MMS and Alliance Charitable Foundation International Health Studies Grants, 2588
NAPNAP Foundation McNeil Grant-in-Aid, 2612
NAPNAP Foundation Reckitt Benckiser Student Scholarship, 2616
NCI Technologies and Software to Support Integrative Cancer Biology Research (SBIR) Grants, 2656
NIDDK Traditional Conference Grants, 2848
Northwest Airlines KidCares Medical Travel Assistance, 2998

NSF-NIST Interaction in Chemistry, Materials Research, Molecular Biosciences, Bioengineering, and Chemical Engineering, 3004
NSF Grant Opportunities for Academic Liaison with Industry (GOALI), 3016
NYAM Edward N. Gibbs Memorial Lecture and Award in Nephrology, 3041
Oregon Community Fndn Community Grants, 3082
Osteosynthesis and Trauma Care Foundation European Visiting Fellowships, 3097
PAHO-ASM Int'l Professorship Latin America, 3107
Phi Upsilon Omicron Florence Fallgatter Distinguished Service Award, 3192
Phi Upsilon Omicron Frances Morton Holbrook Alumni Award, 3193
Phi Upsilon Omicron Undergraduate Karen P. Goebel Conclave Award, 3200
Pinellas County Grants, 3223
Pokagon Fund Grants, 3242
RCPSC Balfour M. Mount Visiting Professorship in Palliative Medicine, 3318
RCPSC Detweiler Traveling Fellowships, 3320
RCPSC Harry S. Morton Lectureship in Surgery, 3323
RCPSC Janes Visiting Professorship in Surgery, 3328
RCPSC John G. Wade Visiting Professorship in Patient Safety & Simulation-Based Med Education, 3329
RCPSC KJR Wightman Visiting Professorship in Medicine, 3331
RCPSC McLaughlin-Gallie Visiting Professor, 3333
Rockefeller Archive Center Research Grants, 3393
SfN Federation of European Neuroscience Societies Forum Travel Awards, 3522
SfN Japan Neuroscience Society Meeting Travel Awards, 3525
SfN Louise Hanson Marshall Specific Recognition Award, 3527
SfN Science Educator Award, 3535
SfN Science Journalism Student Awards, 3536
SfN Trainee Professional Development Awards, 3538
SfN Travel Awards for the International Brain Research Organization World Congress, 3539
SfN Undergrad Brain Awareness Travel Award, 3540
Sigrid Juselius Foundation Grants, 3561
SRC Medical Research Grants, 3635
St. Joseph Community Health Foundation Schneider Fellowships, 3642
Streilein Foundation for Ocular Immunology Visiting Professorships, 3660
Thomson Reuters / MLA Doctoral Fellowships, 3728
Topeka Community Foundation Kansas Blood Services Scholarships, 3738
UICC American Cancer Society International Fellowships for Beginning Investigators, 3778
UICC International Cancer Technology Transfer (ICRETT) Fellowships, 3779
UICC Yamagiwa-Yoshida Memorial International Cancer Study Grants, 3781
United States Institute of Peace - Jennings Randolph Senior Fellowships, 3802
US Airways Community Contributions, 3825
US CRDF Leishmaniasis: Collaborative Research Opportunities in N Africa & Middle East, 3828
Walker Area Community Foundation Grants, 3895

Vocational Education
Achelis Foundation Grants, 234
Adolph Coors Foundation Grants, 311
Alpha Natural Resources Corporate Giving, 450
AMD Corporate Contributions Grants, 500
ArvinMeritor Grants, 675
Bodman Foundation Grants, 929
BP Foundation Grants, 943
Bush Fndn Health & Human Services Grants, 979
Carl M. Freeman Foundation Grants, 1027
Carrier Corporation Contributions Grants, 1038
Charlotte County (FL) Community Foundation Grants, 1126
Chilkat Valley Community Foundation Grants, 1148
Claude Pepper Foundation Grants, 1177
Cleveland Browns Foundation Grants, 1182

CNA Foundation Grants, 1189
ConocoPhillips Foundation Grants, 1327
Consumers Energy Foundation, 1343
Crail-Johnson Foundation Grants, 1355
D.F. Halton Foundation Grants, 1405
Danellie Foundation Grants, 1421
Elizabeth Morse Genius Charitable Trust Grants, 1569
Four County Community Fndn General Grants, 1710
Frank Reed and Margaret Jane Peters Memorial Fund II Grants, 1728
Gamble Foundation Grants, 1769
George W. Wells Foundation Grants, 1804
GNOF New Orleans Works Grants, 1844
Goodrich Corporation Foundation Grants, 1854
Grand Rapids Area Community Foundation Wyoming Youth Fund Grants, 1868
Greater Sitka Legacy Fund Grants, 1885
Harvest Foundation Grants, 1971
IIE Chevron International REACH Scholarships, 2128
Illinois Tool Works Foundation Grants, 2147
Irvin Stern Foundation Grants, 2164
Janus Foundation Grants, 2211
Jessica Stevens Community Foundation Grants, 2224
John Merck Fund Grants, 2245
Joseph Henry Edmondson Foundation Grants, 2270
Kessler Fnd Signature Employment Grants, 2302
Ketchikan Community Foundation Grants, 2303
Kodiak Community Foundation Grants, 2314
Leo Goodwin Foundation Grants, 2354
May and Stanley Smith Charitable Trust Grants, 2459
Oppenstein Brothers Foundation Grants, 3078
Oregon Community Fndn Community Grants, 3082
Reinberger Foundation Grants, 3357
Robert R. McCormick Tribune Veterans Initiative Grants, 3386
Robert R. Meyer Foundation Grants, 3387
Samuel S. Johnson Foundation Grants, 3468
Seabury Foundation Grants, 3501
Seward Community Foundation Grants, 3514
Seward Community Foundation Mini-Grants, 3515
Shield-Ayres Foundation Grants, 3546
Textron Corporate Contributions Grants, 3712
Thomas Sill Foundation Grants, 3724
U.S. Department of Education Rehabilitation Research Training Centers (RRTCs), 3770
U.S. Department of Education Vocational Rehabilitation Services Projects for American Indians with Disabilities Grants, 3774
Union Bank, N.A. Foundation Grants, 3788
USDA Organic Agriculture Research Grants, 3830
Vancouver Foundation Grants and Community Initiatives Program, 3849
Vectren Foundation Grants, 3857
Verizon Foundation Pennsylvania Grants, 3862
Walker Area Community Foundation Grants, 3895
Women of the ELCA Opportunity Scholarships for Lutheran Laywomen, 3970

Geographic Index

Note: This index lists grants for which applicants must be residents of or located in a specific geographic area. Numbers refer to entry numbers.

United States

All States

1st Touch Foundation Grants, 3
2COBS Private Charitable Foundation Grants, 5
25th Anniversary Foundation Grants, 8
100 Mile Man Foundation Grants, 10
A-T Children's Project Grants, 13
A-T Children's Project Post Doctoral Fellowships, 14
A-T Medical Research Foundation Grants, 15
A.O. Smith Foundation Community Grants, 17
AAAAI Allied Health Prof Recognition Award, 19
AAAAI ARTrust Mini Grants for Allied Health, 20
AAAAI ARTrust Grants for Allied Health Travel, 21
AAAAI ARTrust Grants for Clinical Research, 22
AAAAI ARTrust Grants in Faculty Development, 23
AAAAI Distinguished Clinician Award, 24
AAAAI Distinguished Layperson Award, 25
AAAAI Distinguished Scientist Award, 26
AAAAI Distinguished Service Award, 27
AAAAI Fellows-in-Training Abstract Award, 28
AAAAI Fellows-in-Training Grants, 29
AAAAI Fellows-in-Training Travel Grants, 30
AAAAI Mentorship Award, 31
AAAAI Outstanding Vol Clinical Faculty Award, 32
AAAAI RSLAAIS Leadership Award, 33
AAAAI Special Recognition Award, 34
AAAS/Subaru SB&F Prize for Excellence in Science Books, 35
AAAS Award for Scie Freedom & Responsibility, 36
AAAS Early Career Award for Public Engagement with Science, 37
AAAS Eppendorf Science Prize for Neurobiology, 38
AAAS GE & Science Prize for Yng Life Scientists, 39
AAAS Martin and Rose Wachtel Cancer Research Award, 40
AAAS Science and Technology Policy Fellowships: Global Health and Development, 41
AAAS Science Prize for Online Resources in Educ, 42
AABB Dale A. Smith Memorial Award, 43
AABB Hemphill-Jordan Leadership Award, 44
AABB John Elliott Memorial Award, 45
AABB Karl Landsteiner Memorial Award and Lectureship, 46
AABB Sally Frank Award & Lectureship, 47
AABB Tibor Greenwalt Memorial Award and Lectureship, 48
AACAP-NIDA K12 Career Development Awards, 49
AACAP Beatrix A. Hamburg Award for the Best New Research Poster by a Child and Adolescent Psychiatry Resident, 50
AACAP Child and Adolescent Psychiatry (CAP) Teaching Scholarships, 51
AACAP Educational Outreach Program for Child and Adolescent Psychiatry Residents, 52
AACAP Educational Outreach Program for General Psychiatry Residents, 53
AACAP Educational Outreach Program for Residents in Alcohol Research, 54
AACAP Elaine Schlosser Lewis Award for Research on Attention-Deficit Disorder, 55
AACAP George Tarjan Award for Contributions in Developmental Disabilities, 56
AACAP Irving Philips Award for Prevention, 57
AACAP Jeanne Spurlock Lecture and Award on Diversity and Culture, 58
AACAP Jeanne Spurlock Minority Med Student Clinical Fellow in Child & Adolescent Psych, 59
AACAP Jeanne Spurlock Research Fellowship in Substance Abuse and Addiction for Minority Medical Students, 60
AACAP Junior Investigator Awards, 61
AACAP Klingenstein Third Generation Foundation Award for Research in Depression or Suicide, 62
AACAP Life Members Mentorship Grants for Medical Students, 63
AACAP Mary Crosby Congressional Fellowships, 64
AACAP Norbert and Charlotte Rieger Award for Scientific Achievement, 65
AACAP Pilot Research Award for Attention-Deficit Disorder, 66
AACAP Pilot Research Award for Learning Disabilities, Supported by the Elaine Schlosser Lewis Fund, 67
AACAP Pilot Research Awards, Supported by Eli Lilly and Company, 68
AACAP Quest for the Test Bipolar Disorder Pilot Research Award, 69
AACAP Rieger Psychody Psychotherapy Award, 70
AACAP Rieger Service Award for Excellence, 71
AACAP Rob Cancro Academic Leadership Award, 72
AACAP Sidney Berman Award for the School-Based Study and Intervention for Learning Disorders and Mental Illness, 73
AACAP Simon Wile Leader in Consultation Award, 74
AACAP Summer Medical Student Fellowships, 75
AACAP Systems of Care Special Scholarships, 76
AACN-Edwards Lifesciences Nurse-Driven Clinical Practice Outcomes Grants, 77
AACN-Philips Medical Systems Clinical Outcomes Grants, 78
AACN-Sigma Theta Tau Critical Care Grant, 79
AACN Clinical Inquiry Fund Grants, 80
AACN Clinical Practice Grants, 81
AACN Critical Care Grants, 82
AACN End of Life/Palliative Care Small Projects Grants, 83
AACN Evidence-Based Clinical Practice Grants, 84
AACN Mentorship Grant, 85
AACN Physio-Control Small Projects Grants, 86
AACR-American Cancer Society Award for Research Excellence in Epidemiology and Prevention, 87
AACR-Colorectal Cancer Coalition Fellows Grant, 88
AACR-FNAB Fellows Grant for Translational Pancreatic Cancer Research, 89
AACR-FNAB Foundation Career Development Award for Translational Cancer Research, 90
AACR-Genentech BioOncology Career Development Award for Cancer Research on the HER Family Pathway, 91
AACR-Genentech BioOncology Fellowship for Cancer Research in Angiogenesis, 92
AACR-GlaxoSmithKline Clinical Cancer Research Scholar Awards, 93
AACR-Minorities in Cancer Research Jane Cooke Wright Lectureship Awards, 94
AACR-National Brain Tumor Fnd Fellows Grant, 95
AACR-NCI International Investigator Opportunity Grants, 96
AACR-Pancreatic Cancer Action Network Career Development Award for Research, 97
AACR-Pancreatic Cancer Action Network Innovative Grants, 98
AACR Prevent Cancer Foundation Award for Excellence in Cancer Prevention Research, 99
AACR-WICR Scholar Awards, 100
AACR-Women in Cancer Research Charlotte Friend Memorial Lectureship Awards, 101
AACR Award for Lifetime Achievement in Cancer Research, 102
AACR Award for Outstanding Achievement in Cancer Research, 103
AACR Award for Outstanding Achievement in Chemistry in Cancer Research, 104
AACR Basic Cancer Research Fellowships, 105
AACR Brigid G. Leventhal Scholar in Cancer Research Awards, 106
AACR Career Development Awards for Pediatric Cancer Research, 107
AACR Centennial Postdoctoral Fellowships in Cancer Research, 108
AACR Centennial Pre-doctoral Fellowships in Cancer Research, 109
AACR Clinical & Translat Research Fellowships, 110
AACR Fellows Grants, 111
AACR G.H.A. Clowes Memorial Award, 112
AACR Gertrude Elion Cancer Research Award, 113
AACR Shepard Bladder Cancer Research Grants, 114
AACR Joseph H. Burchenal Memorial Award, 115
AACR Judah Folkman Career Development Award for Anti-Angiogenesis Research, 116
AACR Kirk A. Landon and Dorothy P. Landon Foundation Prizes, 117
AACR Foti Award for Leadership & Extraordinary Achievements in Cancer Research, 118
AACR Minority-Serving Institution Faculty Scholar in Cancer Research Awards, 119
AACR Min Scholar in Cancer Research Award, 120
AACR Outstanding Investigator Award for Breast Cancer Research, 121
AACR Princess Takamatsu Memorial Lectureship, 122
AACR Richard & Hinda Rosenthal Mem Awards, 123
AACR Scholar-in-Training Awards, 124
AACR Scholar-in-Training Awards - Special Conferences, 125
AACR Team Science Award, 126
AACR Thomas J. Bardos Science Ed Awards, 127
AAFA Investigator Research Grants, 128
AAFCS International Graduate Fellowships, 129
AAFCS National Graduate Fellowships, 130
AAFCS National Undergraduate Scholarships, 131
AAFP Foundation Health Literacy State Grants, 132
AAFP Foundation Joint Grants, 133
AAFP Foundation Pfizer Teacher Devel Awards, 134
AAFP Foundation Wyeth Immunization Awards, 135
AAFP Research Stimulation Grants, 136
AAFPRS Ben Shuster Memorial Award, 137
AAFPRS Bernstein Grant, 138
AAFPRS Fellowships, 139
AAFPRS Investigator Development Grant, 140
AAFPRS Ira J. Tresley Research Award, 141
AAFPRS John Orlando Roe Award, 142
AAFPRS Leslie Bernstein Res Research Grants, 143
AAFPRS Residency Travel Awards, 144
AAFPRS Sir Harold Delf Gillies Award, 145
AAHN Grant for Historical Research, 146
AAHN Pre-Doctoral Research Grant, 147
AAHPERD-AAHE Barb A. Cooley Scholarships, 148
AAHPERD-AAHE Del Oberteuffer Scholarships, 149
AAHPERD-AAHE Undergraduate Scholarships, 150
AAHPERD Abernathy Presidential Scholarships, 151
AAHPM Hospice & Palliative Med Fellowships, 152
AAMC Abraham Flexner Award, 153
AAMC Alpha Omega Alpha Robert J. Glaser Distinguished Teacher Awards, 154
AAMC Award for Distinguished Research, 155
AAMC Caring for Community Grants, 156
AAMC David E. Rogers Award, 157
AAMC Herbert W. Nickens Award, 158
AAMC Herbert W. Nickens Faculty Fellowship, 159
AAMC Herbert W. Nickens Medical Student Scholarships, 160
AAMC Humanism in Medicine Award, 161
AANS Neurosurgery Research and Education Foundation/Spine Section Young Clinician Investigator Award, 162
AANS Neurosurgery Research & Educ Foundation Young Clinician Investigator Award, 163
AAO-HNSF CORE Research Grants, 164
AAO-HNSF Health Services Research Grants, 165
AAO-HNSF Maureen Hannley Research Training Awards, 166
AAO-HNSF Percy Memorial Research Award, 167
AAO-HNSF Rande H. Lazar Health Services Research Grant, 168
AAO-HNSF Resident Research Awards, 169
AAO-HNS Medical Student Research Prize, 170
AAO-HNS Resident Research Prizes, 171
AAOA Foundation/AAO-HNSF Combined Research Grants, 172
AAOHN Found Experienced Researcher Grants, 173
AAOHN Foundation New Investigator Researcher Grants, 174

920 / All States GEOGRAPHIC INDEX

AAOHN Foundation Professional Development Scholarships, 175
AAP Anne E. Dyson Child Advocacy Awards, 176
AAP Community Access To Child Health (CATCH) Advocacy Training Grants, 177
AAP Community Access To Child Health (CATCH) Implementation Grants, 178
AAP Community Access to Child Health (CATCH) Planning Grants, 179
AAP Community Access To Child Health (CATCH) Residency Training Grants, 180
AAP Community Access To Child Health (CATCH) Resident Grants, 181
AAPD Henry B. Betts Award, 182
AAPD Paul G. Hearne Leadership Award, 183
AAP International Travel Grants, 184
AAP Legislative Conference Scholarships, 185
AAP Leonard P. Rome Community Access to Child Health (CATCH) Visiting Professorships, 186
AAP Nutrition Award, 187
AAP Program Delegate Awards, 188
AAP Resident Initiative Fund Grants, 189
AAP Resident Research Grants, 190
AAUW American Dissertation Fellowships, 195
AAUW American Postdoctoral Research Leave Fellowships, 196
Abbott-ASM Lifetime Achievement Award, 198
Abbott Fund CFCareForward Scholarships, 200
Aboudane Family Foundation Grants, 215
Abramson Family Foundation Grants, 217
Abundance Foundation International Grants, 218
ACAAI Foundation Research Grants, 219
AcademyHealth Alice S. Hersh New Investigator Awards, 220
AcademyHealth Alice Hersh Student Scholarships, 221
AcademyHealth Awards, 222
AcademyHealth HSR Impact Awards, 223
AcademyHealth Nemours Child Health Services Research Awards, 224
AcademyHealth PHSR Interest Group Student Scholarships, 225
AcademyHealth PHSR Research Article of the Year Awards, 226
ACCP Anticoagulation Training Program Grants, 227
ACCP Heart Failure Training Program Grants, 228
ACF Native American Social and Economic Development Strategies Grants, 231
ACGT Investigators Grants, 232
ACGT Young Investigator Grants, 233
Acid Maltase Deficiency Association Helen Walker Research Grant, 235
Ackerman Foundation Grants, 236
ACL Alzheimer's Disease Initiative Grants, 237
ACL Business Acumen for Disability Organizations Grants, 238
ACL Centers for Independent Living Competition Grants, 239
ACL Diversity Community of Practice Grants, 240
ACL Field Initiated Projects Program: Minority-Serving Institution Development Grants, 242
ACL Field Initiated Projects Program: Minority-Serving Institution Research Grants, 243
ACL Learning Collaboratives for Advanced Business Acumen Skills Grants, 244
ACL Medicare Improvements for Patients & Providers Funding for Beneficiary Outreach & Assistance for Title VI Native Americans, 245
ACL Partnerships in Employment Systems Change Grants, 246
ACL Self-Advocacy Resource Center Grants, 247
ACL Training and Technical Assistance Center for State Intellectual and Developmental Disabilities Delivery Systems Grants, 248
ACL University Centers for Excellence in Developmental Network Diversity and Inclusion Training Action Planning Grants, 249
ACS Cancer Control Career Development Awards for Primary Care Physicians, 250
ACS Clinical Research Professor Grants, 251
ACS Doc Degree Scholarships in Cancer Nursing, 252

ACS Doctoral Scholarships in Cancer Nursing, 253
ACS Doctoral Grants in Oncology Social Work, 254
ACS Scholarships in Cancer Nursing Practice, 255
ACS Institutional Research Grants, 256
ACSM-GSSI Young Scholar Travel Award, 257
ACS Master's Training Grants in Clinical Oncology Social Work, 258
ACSM Carl V. Gisolfi Memorial Fund Grant, 259
ACSM Chas M. Tipton Student Research Award, 260
ACSM Coca-Cola Company Doctoral Student Grant on Behavior Research, 261
ACSM Dr. Raymond A. Weiss Research Endowment Grant, 262
ACS MEN2 Thyroid Cancer Professorship Grants, 263
ACS Mentored Research Scholar Grant in Applied and Clinical Research, 264
ACSM Foundation Clinical Sports Medicine Endowment Grants, 265
ACSM Foundation Doctoral Research Grants, 266
ACSM Foundation Research Endowment Grants, 267
ACSM International Student Award, 268
ACSM Michael L. Pollock Student Scholarship, 270
ACSM NASA Space Physiology Research Grants, 271
ACSM Oded Bar-Or Int'l Scholar Awards, 272
ACSM Paffenbarger-Blair Fund for Epidemiological Research on Physical Activity Grants, 273
ACSM Steven M. Horvath Travel Award, 274
ACS Physician Training Awards in Prevent Med, 275
ACS Pilot and Exploratory Projects in Palliative Care of Cancer Patients and Their Families Grants, 276
ACS Postdoctoral Fellowships, 277
ACS Research Professor Grants, 278
ACS Research Scholar Grants for Health Services and Health Policy Research, 279
ACS Research Scholar Grants in Basic, Preclinical, Clinical and Epidemiology Research, 280
ACS Research Scholar Grants in Psychosocial and Behavioral and Cancer Control Research, 281
Actors Addiction & Recovery Services Grants, 282
Actors Fund HIV/AIDS Initiative Grants, 283
Acumen Global Fellowships, 285
ACVIM Foundation Clinical Investigation Grants, 286
ADA Foundation Bud Tarrson Dental School Student Community Leadership Awards, 287
ADA Foundation Dental Student Scholarships, 288
ADA Foundation Dentsply International Research Fellowships, 289
ADA Foundation Disaster Assistance Grants, 290
ADA Foundation George C. Paffenbarger Student Research Awards, 291
ADA Found Minority Dental Scholarships, 292
ADA Foundation Relief Grants, 293
ADA Foundation Samuel Harris Children's Dental Health Grants, 294
ADA Foundation Scientist in Training Fellowship, 295
ADA Foundation Summer Scholars Fellowships, 296
ADA Foundation Thomas J. Zwemer Award, 297
Adaptec Foundation Grants, 304
Administaff Community Affairs Grants, 308
Administration on Aging Senior Medicare Patrol Project Grants, 309
Adolph Coors Foundation Grants, 311
Advance Auto Parts Corporate Giving Grants, 312
Advocate HealthCare Post Graduate Administrative Fellowship, 314
AES-Grass Young Investigator Travel Awards, 317
AES Epilepsy Research Recognition Awards, 318
AES JK Penry Excellence in Epilepsy Care Award, 319
AES Research and Training Workshop Awards, 320
AES Research Infrastructure Awards, 321
AES Research Initiative Awards, 322
AES Robert S. Morison Fellowship, 323
AES Service Award, 324
AES Susan S. Spencer Clinical Research Epilepsy Training Fellowship, 325
AES William G. Lennox Award, 326
Aetna Foundation Integrated Health Care Grants, 328
Aetna Foundation Minority Scholars Grants, 329
Aetna Foundation National Medical Fellowship in Healthcare Leadership, 330

Aetna Foundation Obesity Grants, 331
Aetna Foundation Racial and Ethnic Health Care Equity Grants, 332
AFAR CART Fund Grants, 334
AFAR Medical Student Training in Aging Research Program, 335
AFAR Paul Beeson Career Development Awards in Aging Research for the Island of Ireland, 336
AFAR Research Grants, 337
AFB Rudolph Dillman Memorial Scholarship, 338
Affymetrix Corporate Contributions Grants, 339
AGHE Graduate Scholarships and Fellowships, 343
AHAF Alzheimer's Disease Research Grants, 349
AHAF Macular Degeneration Research Grants, 350
AHAF National Glaucoma Research Grants, 351
AHIMA Dissertation Assistance Grants, 352
AHIMA Faculty Development Stipends, 353
AHIMA Grant-In-Aid Research Grants, 354
AHNS/AAO-HNSF Surgeon Scientist Combined Award, 357
AHNS/AAO-HNSF Young Investigator Combined Award, 358
AHNS Alando J. Ballantyne Resident Research Pilot Grant, 359
AHNS Pilot Grant, 360
AHRF Eugene L. Derlacki Research Grants, 361
AHRF Georgia Birtman Grant, 362
AHRF Regular Research Grants, 363
AHRF Wiley H. Harrison Research Award, 364
AHRQ Independent Scientist Grants, 365
AHRQ Individual Awards for Postdoc Fellows Ruth L. Kirschstein National Research Service Awards, 366
Aid for Starving Children Emergency Assistance Fund Grants, 367
Aid for Starving Children International Grants, 368
AIDS Vaccine Advocacy Coalition (AVAC) Fund Grants, 369
AIG Disaster Relief Fund Grants, 370
AIHP Sonnedecker Visiting Scholar Grants, 371
AIHP Thesis Support Grants, 372
AIHS/Mitacs Health Pilot Partnership Interns, 373
AIHS Alberta/Pfizer Translat Research Grants, 374
AIHS Collaborative Research and Innovation Grants - Collaborative Program, 376
AIHS Collaborative Research and Innovation Grants - Collaborative Team, 378
AIHS ForeFront Internships, 380
AIHS ForeFront MBA Studentship Award, 381
AIHS ForeFront MBT Studentship Awards, 382
AIHS Heritage Youth Researcher Summer Science Program, 387
AIHS Interdisciplinary Team Grants, 388
AIHS Knowledge Exchange - Conference Grants, 389
AIHS Knowledge Exchange Visiting Professors, 390
AIHS Knowledge Exchange - Visiting Scientists, 391
AIHS Media Summer Fellowship, 392
AIHS Proposals for Special Initiatives, 395
AIHS Research Prize, 396
AIHS Summer Studentships, 397
AIHS Sustainability Grants, 398
Air Products and Chemicals Grants, 399
ALA Donald G. Davis Article Award, 401
Albany Medical Center Prize in Medicine and Biomedical Research, 404
Albert and Mary Lasker Foundation Awards, 406
Albert and Mary Lasker Foundation Clinical Research Scholars, 407
Alberto Culver Corporate Contributions Grants, 408
Alcatel-Lucent Technologies Foundation Grants, 413
Alcoa Foundation Grants, 414
Alcohol Misuse and Alcoholism Research Grants, 415
Alcon Foundation Grants Program, 416
Alexander & Baldwin Fnd Mainland Grants, 417
Alexander von Humboldt Foundation Georg Forster Fellowships for Postdoctoral Researchers, 424
Alexander von Humboldt Foundation Research Fellowships for Postdoctoral Researchers, 425
ALFJ Astraea U.S. and Int'l Movement Fund, 427
Alfred P. Sloan Foundation International Science Engagement Grants, 431

GEOGRAPHIC INDEX

Alfred P. Sloan Foundation Research Fellowships, 432
Alfred P. Sloan Foundation Science of Learning STEM Grants, 433
Alfred W. Bressler Prize in Vision Science, 434
Alice Fisher Society Fellowships, 436
Allegis Group Foundation Grants, 441
Allen Foundation Educational Nutrition Grants, 442
Allstate Corporate Giving Grants, 446
Alpha Omega Foundation Grants, 451
ALSAM Foundation Grants, 454
Alternatives Research and Development Foundation Alternatives in Education Grants, 455
Alternatives Research and Development Foundation Grants, 456
Alton Ochsner Award, 458
ALVRE Casselberry Award, 460
ALVRE Grant, 461
ALVRE Seymour R. Cohen Award, 462
Alzheimer's Association Conference Grants, 463
Alzheimer's Association Development of New Cognitive and Functional Instruments Grants, 464
Alzheimer's Association Everyday Technologies for Alzheimer Care Grants, 465
Alzheimer's Association Investigator-Initiated Research Grants, 466
Alzheimer's Association Mentored New Investigator Research Grants to Promote Diversity, 467
Alzheimer's Association Neuronal Hyper Excitability and Seizures in Alzheimer's Disease Grants, 468
Alzheimer's Association New Investigator Research Grants, 469
Alzheimer's Association New Investigator Research Grants to Promote Diversity, 470
Alzheimer's Association U.S.-U.K. Young Investigator Exchange Fellowships, 471
Alzheimer's Association Zenith Fellows Awards, 472
AMA-MSS Chapter Involvement Grants, 473
AMA-MSS Chapter of the Year (COTY) Award, 474
AMA-MSS Government Relations Advocacy Fellowship, 475
AMA-MSS Government Relations Internships, 476
AMA-MSS Research Poster Award, 477
AMA-RFS and AMA Foundation Medical Student & Resident/Fellow Elective-Medicine & Media, 478
AMA-RFS Legislative Awareness Internships, 479
AMA-WPC Joan F. Giambalvo Memorial Scholarships, 480
AMA Foundation Dr. Nathan Davis International Awards in Medicine, 483
AMA Foundation Fund for Better Health Grants, 484
AMA Foundation Health Literacy Grants, 485
AMA Foundation Healthy Communities/Healthy America Grants, 486
AMA Foundation Jack B. McConnell, MD Awards for Excellence in Volunteerism, 487
AMA Foundation Joan F. Giambalvo Memorial Scholarships, 488
AMA Foundation Jordan Fieldman, MD, Awards, 489
AMA Foundation Leadership Awards, 490
AMA Foundation Minority Scholars Awards, 491
AMA Foundation Physicians of Tomorrow Scholarships, 492
AMA Foundation Pride in the Profession Grants, 493
AMA Foundation Seed Grants for Research, 494
AMA Foundation Worldscopes Program, 495
AMA Virtual Mentor Theme Issue Editor Grants, 496
AMDA Foundation Medical Director of the Year Award, 497
AMDA Foundation Quality Improvement and Health Outcome Awards, 498
AMDA Foundation Quality Improvement Award, 499
AMD Corporate Contributions Grants, 500
American-Scandinavian Foundation Visiting Lectureship Grants, 502
American Academy of Dermatology Camp Discovery Scholarships, 503
American Academy of Dermatology Shade Structure Grants, 504
American Academy of Nursing Building Academic Geriatric Nursing Capacity Scholarships, 505

American Academy of Nursing Claire M. Fagin Fellowships, 506
American Academy of Nursing Mayday Grants, 507
American Academy of Nursing MBA Scholarships, 508
American Academy of Nursing Media Awards, 509
American Chemical Society Alfred Burger Award in Medicinal Chemistry, 510
American Chemical Society ANYL Arthur F. Findeis Award for Achievements by a Young Analytical Scientist, 511
American Chemical Society ANYL Award for Distinguished Service in the Advancement of Analytical Chemistry, 512
American Chemical Society Award for Creative Advances in Environmental Science and Technology, 513
American Chemical Society Award in Separations Science and Technology, 514
American Chem Society Claude Hudson Awards, 515
American Chemical Society GCI Pharmaceutical Roundtable Research Grants, 516
American College of Rheumatology Fellows-in-Training Travel Scholarships, 517
American College of Surgeons and The Triological Society Clinical Scientist Development Awards, 518
American College of Surgeons Australia and New Zealand Chapter Travelling Fellowships, 519
American College of Surgeons Co-Sponsored K08/K23 NIH Supplement Awards, 520
American College of Surgeons Faculty Career Development Award for Neurological Surgeons, 521
American College of Surgeons Faculty Research Fellowships, 522
American College of Surgeons Health Policy Scholarships, 523
American College of Surgeons Health Policy Scholarships for General Surgeons, 524
American College of Surgeons International Guest Scholarships, 525
American College of Surgeons Nizar N. Oweida, MD, FACS, Scholarships, 526
American College of Surgeons Resident and Associate Society (RAS-ACS) Leadership Scholarships, 527
American College of Surgeons Resident Research Scholarships, 528
American College of Surgeons Wound Care Management Award, 529
American Express Community Service Grants, 531
American Jewish World Service Grants, 533
American Legacy Foundation National Calls for Proposals Grants, 534
American Legacy Fnd Small Innovative Grants, 535
American Medical Association Awards, 536
American Philosophical Society Daland Fellowships in Clinical Investigation, 537
American Psychiatric Association/AstraZeneca Young Minds in Psychiatry International Awards, 538
American Psychiatric Association/Bristol-Myers Squibb Fellowships in Public Psychiatry, 539
American Psychiatric Association/Janssen Resident Psychiatric Research Scholars, 540
American Psychiatric Association/Merck Co. Early Academic Career Research Award, 541
American Psychiatric Association Award for Research in Psychiatry, 542
American Psychiatric Association Minority Fellowships Program, 543
American Psychiatric Association Minority Medical Student Fellowship in HIV Psychiatry, 544
American Psychiatric Association Minority Medical Student Summer Externship in Addiction Psychiatry, 545
American Psychiatric Association Minority Medical Student Summer Mentoring Program, 546
American Psychiatric Association Program for Minority Research Training in Psychiatry, 547
American Psychiatric Association Research Colloquium for Junior Investigators, 548
American Psychiatric Association Travel Scholarships for Minority Medical Students, 549

American Psychiatric Foundation Alexander Gralnick, MD, Award for Research in Schizophrenia, 550
American Psychiatric Foundation Awards for Advancing Minority Mental Health, 551
American Psychiatric Foundation Grantss, 552
American Psychiatric Foundation Disaster Recovery Fund for Psychiatrists, 553
American Psychiatric Foundation Helping Hands Grants, 554
American Psychiatric Foundation James H. Scully Jr., M.D., Educational Fund Grants, 555
American Psychiatric Foundation Jeanne Spurlock Congressional Fellowship, 556
American Psychiatric Foundation Minority Fellowships, 557
American Psychiatric Foundation Typical or Troubled School Mental Health Education Grants, 558
American Society on Aging Graduate Student Research Award, 560
American Society on Aging Mental Health and Aging Awards, 561
American Society on Aging NOMA Award for Excellence in Multicultural Aging, 562
American Sociological Association Minority Fellowships, 563
Amerisure Insurance Community Service Grants, 566
amfAR Fellowships, 567
amfAR Global Initiatives Grants, 568
amfAR Mathilde Krim Fellowships in Basic Biomedical Research, 569
amfAR Public Policy Grants, 570
amfAR Research Grants, 571
Amgen Foundation Grants, 572
AMHPS Dr. James A. Ferguson Emerging Infectious Diseases Fellowships, 573
AMI Semiconductors Corporate Grants, 576
AMSSM Foundation Research Grants, 577
ANA/Grass Foundation Award in Neuroscience, 578
ANA Derek Denny-Brown Neurological Scholar Award, 579
ANA Distinguished Neurology Teacher Award, 580
ANA F.E. Bennett Memorial Lectureship, 581
ANA Junior Academic Neurologist Scholarships, 582
ANA Raymond D. Adams Lectureship, 583
ANA Soriano Lectureship, 584
ANA Wolfe Neuropathy Research Prize, 585
Angel Kiss Foundation Grants, 589
Anne J. Caudal Foundation Grants, 595
Annenberg Foundation Grants, 596
ANS/AAO-HNSF Herbert Silverstein Otology and Neurotology Research Award, 601
ANS Neurotology Fellows Award, 604
ANS Nicholas Torok Vestibular Award, 605
ANS Trainee Award, 606
Anthony R. Abraham Foundation Grants, 608
AOCS A. Richard Baldwin Award, 610
AOCS Alton E. Bailey Award, 611
AOCS Analytical Division Student Award, 612
AOCS George Schroepfer Medal, 613
AOCS Health & Nutrition Div Student Award, 614
AOCS Health and Nutrition Poster Competition, 615
AOCS Industrial Oil Products Division Student Award, 616
AOCS Lipid Oxidation and Quality Division Poster Competition, 617
AOCS Processing Division Student Award, 618
AOCS Protein and Co-Products Division Student Poster Competition, 619
AOCS Holman Lifetime Achievement Award, 620
AOCS Supelco/Nicholas Pelick-AOCS Research Award, 621
AOCS Surfactants and Detergents Division Student Award, 622
AON Foundation Grants, 623
APA Congressional Fellowships, 624
APA Culture of Service in the Psychological Sciences Award, 625
APA Dissertation Research Awards, 626
APA Distinguished Scientific Award for Early Career Contribution to Psychology, 627

922 / All States GEOGRAPHIC INDEX

APA Distinguished Scientific Award for the Applications of Psychology, 628
APA Distinguished Scientific Contribution Award, 629
APA Distinguished Service to Psychological Science Award, 630
APA Meritorious Research Service Award, 631
APA Scientific Conferences Grants, 632
APA Travel Awards, 633
APA Young Investigator Grant Program, 634
APF/COGDOP Graduate Research Scholarships in Psychology, 635
APHL Emerging Infectious Diseases Fellowships, 636
APSAA Fellowships, 638
APSAA Foundation Grants, 639
APSAA Mini-Career Grants, 640
APSAA Research Grants, 641
APSAA Small Beginning Scholar Pre-Investigation Grants, 642
APSAA Small Beginning Scholar Visiting and Consulting Grants, 643
APSA Congressional Health and Aging Policy Fellowships, 644
APSA Robert Wood Johnson Foundation Health Policy Fellowships, 645
Armstrong McDonald Foundation Health Grants, 663
Arnold and Mabel Beckman Foundation Beckman-Argyros Award in Vision Research, 664
Arnold and Mabel Beckman Foundation Scholars Grants, 665
Arnold and Mabel Beckman Foundation Young Investigators Grants, 666
ARS New Investigator Award, 668
ARS Resident Research Grants, 669
Arthur Ashley Williams Foundation Grants, 671
ArvinMeritor Foundation Health Grants, 674
ArvinMeritor Grants, 675
ASA Metlife Foundation MindAlert Awards, 676
ASCO/UICC International Cancer Technology Transfer (ICRETT) Fellowships, 677
ASCO Advanced Clinical Research Award in Colorectal Cancer, 678
ASCO Advanced Clinical Research Awards in Breast Cancer, 679
ASCO Advanced Clinical Research Awards in Sarcoma, 680
ASCO Long-Term International Fellowships, 681
ASCO Merit Awards, 682
ASCO Young Investigator Award, 683
ASGE / ConMed Award for Outstanding Manuscript by a Fellow/Resident, 684
ASGE Don Wilson Award, 685
ASGE Endoscopic Research Awards, 686
ASGE Endoscopic Research Career Development Awards, 687
ASGE Given Capsule Endoscopy Research Award, 688
ASGH Award for Excellence in Human Genetics Education, 689
ASGH C.W. Cotterman Award, 690
ASGH Charles J. Epstein Trainee Awards for Excellence in Human Genetics Research, 691
ASGH McKusick Leadership Award, 692
ASGH William Allan Award, 693
ASGH William Curt Stern Award, 694
ASHA Advancing Academic-Research Award, 695
ASHA Minority Student Leadership Awards, 696
ASHA Multicultural Activities Projects Grants, 697
ASHA Research Mentoring-Pair Travel Award, 698
ASHA Student Research Travel Award, 699
ASHA Students Preparing for Academic & Research Careers (SPARC) Award, 700
ASHFoundation Clinical Research Grants, 701
ASHFoundation Graduate Student International Scholarship, 702
ASHFoundation Graduate Student Scholarships, 703
ASHFoundation Graduate Student Scholarships for Minority Students, 704
ASHFoundation Graduate Student with a Disability Scholarship, 705
ASHFoundation New Century Scholars Program Doctoral Scholarships, 706

ASHFoundation New Century Scholars Research Grant, 707
ASHFoundation Grant for New Investigators, 708
ASHFoundation Research Grant in Speech, 709
ASHFoundation Student Research Grants in Audiology, 710
ASHFoundation Student Research Grants in Early Childhood Language Development, 711
ASHG Public Health Genetics Fellowship, 712
ASM-PAHO Infectious Diseases Epidemiology and Surveillance Fellowships, 714
ASM-UNESCO Leadership Grant for International Educators, 715
ASM/CDC Fellowships in Infectious Disease and Public Health Microbiology, 716
ASM Abbott Laboratories Award in Clinical and Diagnostic Immunology, 717
ASM BD Award for Research in Clin Microbio, 718
ASM bioMerieux Sonnenwirth Award for Leadership in Clinical Microbiology, 719
ASM Congressional Science Fellowships, 720
ASM Corporate Activities Program Student Travel Grants, 721
ASMCUE/GM Travel Assistance Grants, 722
ASM D.C. White Research and Mentoring Award, 723
ASM Early-Career Faculty Travel Award, 724
ASME H.R. Lissner Award, 725
ASM Eli Lilly and Company Research Award, 726
ASM Faculty Ehancement Travel Awards, 727
ASM Founders Distinguished Service Award, 728
ASM Gen-Probe Joseph Public Health Award, 729
ASM General Meeting Minority Travel Grants, 730
ASM GlaxoSmithKline International Member of the Year Award, 731
ASM ICAAC Young Investigator Awards, 732
ASM Intel Awards, 733
ASM Int'l Fellowship for Latin America and the Caribbean, 734
ASM Int'l Fellowships for Asia and Africa, 735
ASM Int'l Professorship for Latin America, 736
ASM Int'l Professorships for Asia and Africa, 737
ASM Maurice Hilleman/Merck Award, 738
ASM Merck Irving S. Sigal Memorial Awards, 739
ASM Microbiology Undergraduate Research Fellowship, 740
ASM Millis-Colwell Postgraduate Travel Grant, 741
ASM Procter & Gamble Award in Applied and Environmental Microbiology, 742
ASM Promega Biotechnology Research Award, 743
ASM Public Communications Award, 744
ASM Raymond W. Sarber Awards, 745
ASM Richard and Mary Finkelstein Travel Grant, 746
ASM Robert D. Watkins Graduate Research Fellowship, 747
ASM Roche Diagnostics Alice C. Evans Award, 748
ASM sanofi-aventis ICAAC Award, 749
ASM Scherago-Rubin Award, 750
ASM Siemens Healthcare Diagnostics Young Investigator Award, 751
ASM Student & Postdoc Fellow Travel Grants, 752
ASM Student Travel Grants, 753
ASM TREK Diagnostic ABMM/ABMLI Professional Recognition Award, 754
ASM Undergraduate Research Fellowship, 755
ASM USFCC/J. Roger Porter Award, 756
ASM William A Hinton Research Training Award, 757
ASPEN Dudrick Research Scholar Award, 759
ASPEN Harry M. Vars Award, 760
ASPEN Rhoads Research Foundation Abbott Nutrition Research Grants, 761
ASPEN Rhoads Research Foundation Baxter Parenteral Nutrition Research Grant, 762
ASPEN Rhoads Research Foundation C. Richard Fleming Grant, 763
ASPEN Rhoads Research Foundation Maurice Shils Grant, 764
ASPEN Rhoads Research Foundation Norman Yoshimura Grant, 765
ASPEN Scientific Abstracts Awards for Papers or Posters, 766

ASPEN Scientific Abstracts Promising Investigator Awards, 767
ASPET-Astellas Awards in Translational Pharmacology, 768
ASPET Benedict R. Lucchesi Award in Cardiac Pharmacology, 769
ASPET Brodie Award in Drug Metabolism, 770
ASPET Division for Drug Metabolism Early Career Achievement Award, 771
ASPET Epilepsy Research Award for Outstanding Contributions to the Pharmacology of Antiepileptic Drugs, 772
ASPET Goodman and Gilman Award in Drug Receptor Pharmacology, 773
ASPET John J. Abel Award in Pharmacology, 774
ASPET Julius Axelrod Award in Pharmacology, 775
ASPET P. B. Dews Lifetime Achievement Award for Research in Behavioral Pharmacology, 776
ASPET Paul M. Vanhoutte Award, 777
ASPET Torald Sollmann Award in Pharmacology, 778
ASPET Travel Awards, 779
ASPH/CDC Allan Rosenfield Global Health Fellowships, 780
ASPH/CDC Public Health Fellowships, 781
ASPH/CDC Public Health Internships, 782
ASPO Daiichi Innovative Technology Grant, 783
ASPO Fellowships, 784
ASPO Research Grants, 785
Astellas Foundation Research Grants, 790
ASU Grad College Reach for the Stars Fellowship, 791
ASU Graduate College Science Foundation Arizona Bisgrove Postdoctoral Scholars, 792
AT&T Foundation Civic and Community Service Program Grants, 793
Atran Foundation Grants, 797
AUPHA Corris Boyd Scholarships, 800
AUPHA Foster G. McGaw Scholarships, 801
Austin College Leadership Award, 804
Australasian Institute of Judicial Administration Seed Funding Grants, 806
Australian Academy of Science Grants, 807
Autodesk Community Relations Grants, 809
AVDF Health Care Grants, 811
Avon Foundation Breast Care Fund Grants, 813
Avon Products Foundation Grants, 814
Balfe Family Foundation Grants, 819
Baltimore Washington Center Adult Psychoanalysis Fellowships, 821
Baltimore Washington Center Child Psychotherapy Fellowships, 822
Bank of America Fndn Matching Gifts, 825
Bank of America Fndn Volunteer Grants, 826
Barth Syndrome Foundation Research Grants, 836
Baxter International Corporate Giving Grants, 841
Baxter International Foundation Foster G. McGaw Prize, 842
Baxter International Foundation Grants, 843
Baxter International Foundation William B. Graham Prize for Health Services Research, 844
Bayer Clinical Scholarship Award, 845
Bayer Hemophilia Caregivers Education Award, 847
BCRF-AACR Grants for Translational Breast Cancer Research, 861
BCRF Research Grants, 862
Bearemy's Kennel Pals Grants, 863
Beatrice Laing Trust Grants, 864
Beckman Coulter Foundation Grants, 868
Becton Dickinson and Company Grants, 869
Bernard and Audre Rapoport Foundation Health Grants, 882
Bikes Belong Foundation Paul David Clark Bicycling Safety Grants, 890
Bill and Melinda Gates Foundation Water, Sanitation and Hygiene Grants, 894
Biogen Foundation Fellowships, 898
Biogen Foundation General Donations, 899
Biogen Foundation Healthcare Professional Education Grants, 900
Biogen Foundation Patient Educational Grants, 901
Biogen Foundation Scientific Research Grants, 902

GEOGRAPHIC INDEX

Blanche and Irving Laurie Foundation Grants, 910
Blowitz-Ridgeway Foundation Early Childhood Development Research Award, 912
Blue Cross Blue Shield of Minnesota Foundation - Healthy Equity: Health Impact Assessment Demonstration Project Grants, 916
Blue Cross Blue Shield of Minnesota Foundation - Healthy Equity: Health Impact Assessment Program Grants, 917
BMW of North America Charitable Contributions, 926
Boston Psychoanalytic Society and Institute Fellowship in Child Psychoanalytic Psychotherapy, 939
Boston Psychoanalytic Society and Institute Fellowship in Psychoanalytic Psychotherapy, 940
BP Foundation Grants, 943
Bravewell Leadership Award, 946
Breast Cancer Fund Grants, 947
Brighter Tomorrow Foundation Grants, 950
Brinson Foundation Grants, 953
Bristol-Myers Squibb/Meade Johnson Award for Dist Achievement in Nutrition Research, 954
Bristol-Myers Squibb Clinical Outcomes and Research Grants, 955
Bristol-Myers Squibb Foundation Fellowships, 956
Bristol-Myers Squibb Foundation Global HIV/AIDS Initiative Grants, 957
Bristol-Myers Squibb Foundation Health Education Grants, 959
Bristol-Myers Squibb Foundation Product Donations Grants, 960
Bristol-Myers Squibb Foundation Science Education Grants, 961
Bristol-Myers Squibb Foundation Women's Health Grants, 962
Bristol-Myers Squibb Patient Assistance Grants, 963
Broad Foundation IBD Research Grants, 964
Brookdale Foundation Leadership in Aging Fellowships, 965
Brookdale Fnd National Group Respite Grants, 966
Bupa Foundation Medical Research Grants, 974
Bupa Foundation Multi-Country Grant, 975
Burden Trust Grants, 976
Burlington Northern Santa Fe Foundation Grants, 978
BWF Ad Hoc Grants, 982
BWF Career Awards at the Scientific Interface, 983
BWF Career Awards for Medical Scientists, 984
BWF Collaborative Research Travel Grants, 985
BWF Innovation in Regulatory Science Grants, 986
BWF Institutional Program Unifying Population and Laboratory Based Sciences Grants, 987
BWF Investigators in the Pathogenesis of Infectious Disease Awards, 988
BWF Postdoctoral Enrichment Grants, 989
BWF Preterm Birth Initiative Grants, 990
Byron W. & Alice L. Lockwood Fnd Grants, 991
Caddock Foundation Grants, 994
Cailloux Foundation Grants, 996
Caleb C. and Julia W. Dula Educational and Charitable Foundation Grants, 997
Canada-U.S. Fulbright New Century Scholars Program Grants, 1008
Canada-U.S. Fulbright Senior Specialists Grants, 1009
Canada Graduate Scholarships (CGS) and NSERC Postgraduate Scholarships (PGS), 1010
Canadian Optometric Education Trust Grants, 1011
Canadian Patient Safety Institute (CPSI) Patient Safety Studentships, 1012
Canadian Patient Safety Institute (CPSI) Research Grants, 1013
Caplow Applied Science Carcinogen Prize, 1015
Caplow Applied Science Children's Prize, 1016
Cardinal Health Foundation Grants, 1017
Cargill Citizenship Fund Corporate Giving, 1018
Carla J. Funk Governmental Relations Award, 1020
Carnegie Corporation of New York Grants, 1033
Carrie Estelle Doheny Foundation Grants, 1037
Carylon Foundation Grants, 1040
CCFF Chairman's Dist Life Sciences Award, 1044
CCFF Life Sciences Student Awards, 1045
CCHD Community Development Grants, 1046

CDC-Hubert Global Health Fellowship, 1047
CDC Collegiate Leaders in Environmental Health Internships, 1048
CDC Cooperative Agreement for Continuing Enhanced National Surveillance for Prion Diseases in the United States, 1049
CDC Cooperative Agreement for Partnership to Enhance Public Health Informatics, 1050
CDC Cooperative Agreement for the Development, Operation, and Evaluation of an Entertainment Education Program, 1051
CDC David J. Sencer Museum Adult Group Tour, 1052
CDC David J. Sencer Museum Teacher Professional Development Workshops, 1053
CDC Disease Detective Camp, 1054
CDC Epidemic Intell Service Training Grants, 1055
CDC Epidemiology Elective Rotation, 1056
CDC Evaluation of the Use of Rapid Testing For Influenza in Outpatient Medical Settings, 1057
CDC Evidence-Based Laboratory Medicine: Quality/ Performance Measure Evaluation, 1058
CDC Experience Epidemiology Fellowships, 1059
CDC Foundation Atlanta International Health Fellowships, 1060
CDC Foundation Emergency Response Grants, 1061
CDC Fnd Tobacco Network Lab Fellowship, 1062
CDC Increasing Breast and Cervical Cancer Screening Services for Urban American Indian/Alaska Native Women, 1063
CDC Preventive Med Residency & Fellowship, 1064
CDC Public Health Associates, 1065
CDC Public Health Associates Hosts, 1066
CDC Public Health Conference Support Grant, 1067
CDC Public Health Informatics Fellowships, 1068
CDC Public Health Prev Service Fellowships, 1069
CDC Public Health Prevention Service Fellowship Sponsorships, 1070
CDC Steven M. Teutsch Prevention Effectiveness Fellowships, 1071
CDC Summer Graduate Environmental Health Internships, 1072
CDC Summer Program In Environmental Health Internships, 1073
Cedar Tree Foundation David H. Smith Conservation Research Fellowship, 1076
Central Okanagan Foundation Grants, 1080
Cessna Foundation Grants Program, 1081
Cetana Educational Foundation Scholarships, 1082
CFF-NIH Funding Grants, 1083
CFF Clinical Research Grants, 1084
CFF First- and Second-Year Clinical Fellowships, 1085
CFF Shwachman Clinical Investigator Award, 1086
CFF Leroy Matthews Physician-Scientist Awards, 1087
CFF Pilot and Feasibility Awards, 1088
CFF Postdoctoral Research Fellowships, 1089
CFF Research Grants, 1090
CFF Student Traineeships, 1091
CFF 3rd through 5th Year Clinical Fellowships, 1092
Champ A Champion Fur Kids Grants, 1109
Changemakers Innovation Awards, 1111
Charles H. Revson Foundation Grants, 1122
Charles Lafitte Foundation Grants, 1123
Chatlos Foundation Grants Program, 1128
Chazen Foundation Grants, 1129
Chest Foundation/LUNGevity Foundation Clinical Research in Lung Cancer Grants, 1131
Chest Fndn Eli Lilly & Company Distinguished Scholar in Critical Care Med Award, 1132
Chest Foundation Geriatric Development Research Awards, 1133
Chest Fnd Grant in Venous Thromboembolism, 1134
Chicago Institute for Psychoanalysis Fellowships, 1135
Children's Brain Tumor Fnd Research Grants, 1136
Children's Cardiomyopathy Foundation Research Grants, 1137
Children's Leukemia Research Association Research Grants, 1138
Children's Tumor Fndn Clin Research Awards, 1140
Children's Tumor Foundation Drug Discovery Initiative Awards, 1141

Children's Tumor Foundation Schwannomatosis Awards, 1142
Children's Tumor Foundation Young Investigator Awards, 1143
Children Affected by AIDS Foundation Camp Network Grants, 1144
Children Affected by AIDS Foundation Domestic Grants, 1145
Children Affected by AIDS Foundation Family Assistance Emergency Fund Grants, 1146
Christensen Fund Regional Grants, 1149
Christine Gregoire Youth/Young Adult Award for Use of Tobacco Industry Documents, 1151
Christopher & Dana Reeve Foundation Quality of Life Grants, 1152
Cigna Civic Affairs Sponsorships, 1159
CIGNA Foundation Grants, 1160
CIT Corporate Giving Grants, 1161
CJ Foundation for SIDS Program Services Grants, 1163
CJ Foundation for SIDS Research Grants, 1164
Claude Pepper Foundation Grants, 1177
CLIF Bar Family Foundation Grants, 1183
Clowes ACS/AAST/NIGMS Jointly Sponsored Mentored Clinical Scientist Dev Award, 1185
CMS Hispanic Health Services Research Grants, 1186
CMS Historically Black Colleges and Universities (HBCU) Health Services Research Grants, 1187
CMS Research and Demonstration Grants, 1188
CNA Foundation Grants, 1189
CNCS AmeriCorps VISTA Project Grants, 1190
CNCS Senior Companion Program Grants, 1191
CNCS Social Innovation Grants, 1192
CNIB Baker Fellowships, 1194
CNIB Barbara Tuck MacPhee Award, 1196
CNIB Canada Glaucoma Clinical Research Council Grants, 1197
CNIB Chanchlani Global Vision Research Award, 1198
CNIB E. (Ben) & Mary Hochhausen Access Technology Research Grants, 1199
CNIB Ross Purse Doctoral Fellowships, 1200
CNO Financial Group Community Grants, 1201
Coca-Cola Foundation Grants, 1203
Collective Brands Foundation Grants, 1208
Columbus Foundation Ann Ellis Fund Grants, 1215
Commonwealth Fund/Harvard University Fellowship in Minority Health Policy, 1226
Commonwealth Fund Affordable Health Insurance Grants, 1227
Commonwealth Fund Australian-American Health Policy Fellowships, 1228
Commonwealth Fund Harkness Fellowships in Health Care Policy and Practice, 1229
Commonwealth Fund Health Care Quality Improvement and Efficiency Grants, 1230
Commonwealth Fund Patient-Centered Coordinated Care Program Grants, 1231
Commonwealth Payment System Reform Grants, 1232
Commonwealth Fund Quality of Care for Frail Elders Grants, 1233
Commonwealth Fund Small Grants, 1234
Commonwealth Fund State High Performance Health Systems Grants, 1235
Conquer Cancer Foundation of ASCO Career Development Award, 1328
Conquer Cancer Foundation of ASCO Comparative Effectiveness Research Professorship in Breast Cancer, 1329
Conquer Cancer Foundation of ASCO Drug Development Research Professorship, 1330
Conquer Cancer Foundation of ASCO Improving Cancer Care Grants, 1331
Conquer Cancer Foundation of ASCO International Development and Education Award in Palliative Care, 1332
Conquer Cancer Foundation of ASCO International Development and Education Awards, 1333
Conquer Cancer Foundation of ASCO International Innovation Grant, 1334
Conquer Cancer Foundation of ASCO Medical Student Rotation Grants, 1335

Conquer Cancer Foundation of ASCO Resident Travel Award for Underrepresented Populations, 1336
Conquer Cancer Foundation of ASCO Translational Research Professorships, 1337
Conservation, Food, and Health Foundation Grants for Developing Countries, 1338
Constellation Energy Corporate Grants, 1342
Covenant Matters Foundation Grants, 1350
Covidien Clinical Education Grants, 1351
Covidien Medical Product Donations, 1352
Covidien Partnership for Neighborhood Wellness Grants, 1353
Crane Foundation Grants, 1357
Crane Fund Grants, 1358
CRH Foundation Grants, 1360
Cruise Industry Charitable Foundation Grants, 1363
CSL Behring Local Empowerment for Advocacy Development (LEAD) Grants, 1365
CSTE CDC/CSTE Applied Epidemiology Fellowships, 1367
CTCRI Idea Grants, 1369
Cultural Society of Filipino Americans Grants, 1373
CVS All Kids Can Grants, 1377
CVS Caremark Charitable Trust Grants, 1378
CVS Community Grants, 1379
Cystic Fibrosis Canada Visiting Allied Health Professional Awards, 1394
Cystic Fibrosis Canada Visiting Clinician Awards, 1395
Cystic Fibrosis Lifestyle Foundation Individual Recreation Grants, 1397
Cystic Fibrosis Lifestyle Foundation Loretta Morris Memorial Fund Grants, 1398
Cystic Fibrosis Lifestyle Foundation Mentored Recreation Grants, 1399
Cystic Fibrosis Lifestyle Foundation Peer Support Grants, 1400
Cystic Fibrosis Research New Horizons Campaign Grants, 1402
Cystic Fibrosis Scholarships, 1403
Cystic Fibrosis Trust Research Grants, 1404
DAAD Research Stays for University Academics and Scientists, 1408
DaimlerChrysler Corporation Fund Grants, 1412
Dale and Edna Walsh Foundation Grants, 1415
Dana Foundation Science and Health Grants, 1420
Daniel Mendelsohn New Investigator Award, 1423
DAR Alice W. Rooke Scholarship, 1426
DAR Dr. Francis Anthony Beneventi Medical Scholarship, 1427
DAR Irene and Daisy MacGregor Memorial Scholarship, 1428
DAR Madeline Cogswell Nursing Scholarship, 1429
DAR Nursing/Physical Therapy Scholarships, 1430
Dell Scholars Program Scholarships, 1452
Delta Air Lines Foundation Health and Wellness Grants, 1455
Delta Air Lines Foundation Prize for Global Understanding, 1456
Dept of Ed College Assistance Migrant Grants, 1462
Dept of Ed Rehabilitation Training Grants, 1463
Dept of Ed Safe and Drug-Free Schools and Communities State Grants, 1464
Dept of Ed Special Education--Personnel Development to Improve Services and Results for Children with Disabilities, 1465
Dept of Ed Special Education--Studies and Evaluations, 1466
Dept of Ed Special Education--Technical Assistance and Dissemination to Improve Services and Results for Children with Disabilities, 1467
Dept of Ed Special Education-National Activities-Technology and Media Services for Individuals with Disabilities, 1468
DHHS Adolescent Family Life Demonstration Projects, 1472
DHHS Emerging Leaders Program Internships, 1473
DHHS Oral Health Promotion Research Across the Lifespan, 1474
DHL Charitable Shipment Support, 1475
Different Needz Foundation Grants, 1478

Disable American Veterans Charitable Grants, 1479
DOD HBCU/MI Partnership Training Award, 1480
DOJ Gang-Free Schools and Communities Intervention Grants, 1481
Dole Food Company Charitable Contributions, 1483
DOL Occupational Safety and Health--Susan Harwood Training Grants, 1484
Donald W. Reynolds Foundation Aging and Quality of Life Grants, 1487
Doree Taylor Charitable Foundation, 1490
Doris Duke Charitable Foundation Clinical Interfaces Award Program, 1492
Doris Duke Charitable Foundation Clinical Research Fellowships for Medical Students, 1493
Doris Duke Charitable Foundation Clinical Scientist Development Award, 1494
Doris Duke Charitable Foundation Clinical Scientist Development Award (CSDA) Bridge Grants, 1495
Doris Duke Charitable Foundation Distinguished Clinical Scientist Award Program, 1496
Doris Duke Charitable Foundation Operations Research on AIDS Care and Treatment in Africa (ORACTA) Grants, 1497
Dorothea Haus Ross Foundation Grants, 1498
Dorr Institute for Arthritis Research & Educ, 1502
Do Something Awards, 1503
DPA Promoting Policy Change Advocacy Grants, 1504
Dr. Leon Bromberg Charitable Trust Grants, 1509
Dr. Scholl Foundation Grants Program, 1511
Dr. Stanley Pearle Scholarships, 1512
DSO Controlling Pathogen Workshop Grants, 1514
DSO Radiation Biodosimetry (RaBiD) Grants, 1515
Duke University Adult Cardiothoracic Anesthesia and Critical Care Medicine Fellowships, 1520
Duke University Ambulatory and Regional Anesthesia Fellowships, 1521
Duke University Clinical Cardiac Electrophysiology Fellowships, 1522
Duke University Hyperbaric Center Fellowships, 1523
Duke Univ Interventional Cardio Fellowships, 1524
Duke Univ Obstetric Anesthesia Fellowships, 1525
Duke University Pain Management Fellowships, 1526
Duke University Ped Anesthesiology Fellowship, 1527
Duke University Postdoctoral Research Fellowships in Aging, 1528
DuPont Pioneer Community Giving Grants, 1532
E.B. Hershberg Award for Important Discoveries in Medicinally Active Substances, 1535
E. Clayton and Edith P. Gengras, Jr. Foundation, 1536
E. Rhodes & Leona B. Carpenter Grants, 1539
E.W. "Al" Thrasher Awards, 1540
Earl and Maxine Claussen Trust Grants, 1541
EBSCO / MLA Annual Meeting Grants, 1545
Echoing Green Fellowships, 1546
Eckerd Corporation Foundation Grants, 1547
Educational Foundation of America Grants, 1553
Elizabeth Glaser Scientist Award, 1567
Elizabeth Nash Foundation Scholarships, 1570
Elizabeth Nash Fnd Summer Research Awards, 1571
Ellison Medical Foundation/AFAR Julie Martin Mid-Career Award in Aging Research, 1576
Ellison Medical Foundation/AFAR Postdoctoral Fellows in Aging Research Program, 1577
Elton John AIDS Foundation Grants, 1585
EMBO Installation Grants, 1586
EMBO Long-Term Fellowships, 1587
EMBO Short-Term Fellowships, 1588
Emerson Charitable Trust Grants, 1589
Emerson Electric Company Contributions Grants, 1590
Emily Davie and Joseph S. Kornfeld Foundation Grants, 1591
Emily Hall Tremaine Foundation Learning Disabilities Grants, 1592
EPA Children's Health Protection Grants, 1598
Estee Lauder Grants, 1606
Everyone Breathe Asthma Education Grants, 1616
Express Scripts Foundation Grants, 1620
ExxonMobil Foundation Malaria Grants, 1621
Eye-Bank for Sight Restoration and Fight for Sight Summer Student Research Fellowship, 1622

FAMRI Clinical Innovator Awards, 1628
FAMRI Young Clinical Scientist Awards, 1629
Farmers Insurance Corporate Giving Grants, 1634
FCD New American Children Grants, 1637
FDHN Bridging Grants, 1638
FDHN Centocor International Research Fellowship in Gastrointestinal Inflammation & Immunology, 1639
FDHN Designated Outcomes Award in Geriatric Gastroenterology, 1640
FDHN Designated Research Award in Geriatric Gastroenterology, 1641
FDHN Designated Research Award in Research Related to Pancreatitis, 1642
FDHN Fellow Abstract Prizes, 1643
FDHN Fellowship to Faculty Transition Awards, 1644
FDHN Funderburg Research Scholar Award in Gastric Biology Related to Cancer, 1645
FDHN Graduate Student Awards, 1646
FDHN Isenberg Int'l Research Scholar Award, 1647
FDHN June & Donald O. Castell MD, Esophageal Clinical Research Award, 1648
FDHN Moti L. & Kamla Rustgi International Travel Awards, 1649
FDHN Non-Career Research Awards, 1650
FDHN Non-Career Research Grants, 1651
FDHN Research Scholar Awards, 1652
FDHN Student Abstract Prizes, 1653
FDHN Student Research Fellowships, 1654
FDHN TAP Endowed Designated Research Award in Acid-Related Diseases, 1655
FDHN Translational Research Awards, 1656
Federal Express Corporate Contributions, 1657
Fidelity Foundation Grants, 1659
Fight for Sight-Streilein Foundation for Ocular Immunology Research Award, 1662
Fight for Sight Grants-in-Aid, 1663
Fight for Sight Post-Doctoral Awards, 1664
Fight for Sight Summer Student Fellowships, 1665
Fischelis Grants for Research in the History of American Pharmacy, 1669
Fisher House Fnd Newman's Own Awards, 1671
Fishman Family Foundation Grants, 1672
FIU Global Civic Engagement Mini Grants, 1673
Flextronics Foundation Disaster Relief Grants, 1674
Floyd A. & Kathleen C. Cailloux Fnd Grants, 1681
Ford Motor Company Fund Grants Program, 1687
Foundation Fighting Blindness Marjorie Carr Adams Women's Career Development Awards, 1690
Foundation for Balance and Harmony Grants, 1695
Foundation for Pharmaceutical Sciences Herb and Nina Demuth Grant, 1697
Foundation of Orthopedic Trauma Grants, 1707
Foundation of the American Thoracic Society Research Grants, 1708
Fragile X Syndrome Postdoc Research Fellows, 1714
Frances and John L. Loeb Family Fund Grants, 1716
Fraser-Parker Foundation Grants, 1732
FRAXA Research Foundation Program Grants, 1733
Fred C. & Katherine B. Andersen Grants, 1737
Frederick Gardner Cottrell Foundation Grants, 1739
Frederic Stanley Kipping Award in Silicon Chem, 1742
Fuji Film Grants, 1748
Fulbright Alumni Initiatives Awards, 1749
Fulbright Distinguished Chairs Awards, 1750
Fulbright German Studies Seminar Grants, 1751
Fulbright International Education Administrators (IEA) Seminar Program Grants, 1752
Fulbright New Century Scholars Grants, 1753
Fulbright Specialists Program Grants, 1754
Fulbright Scholars in Europe and Eurasia, 1755
Fulbright Traditional Scholar Program in Sub-Saharan Africa, 1756
Fulbright Traditional Scholar Program in the East Asia/Pacific Region, 1757
Fulbright Traditional Scholar Program in the Near East and North Africa Region, 1758
Fulbright Traditional Scholar Program in the South and Central Asia Region, 1759
Fulbright Traditional Scholar Program in the Western Hemisphere, 1760

GEOGRAPHIC INDEX

Gates Award for Global Health, 1775
Gates Millennium Scholars Program, 1776
GEICO Public Service Awards, 1778
General Mills Champs for Healthy Kids Grants, 1781
General Motors Foundation Grants, 1783
General Service Foundation Human Rights and Economic Justice Grants, 1784
General Service Reproductive Justice Grants, 1785
George Gund Foundation Grants, 1796
Gerber Foundation Grants, 1807
Gertrude E. Skelly Charitable Foundation Grants, 1810
Gibson Foundation Grants, 1818
Gilbert Memorial Fund Grants, 1820
Giving in Action Society Children & Youth with Special Needs Grants, 1823
Giving in Action Society Family Indep Grants, 1824
Glaucoma Foundation Grants, 1827
Glaucoma Research Pilot Project Grants, 1828
Glenn/AFAR Gerontology Awards, 1831
Global Fund for Children Grants, 1832
GNOF IMPACT Grants for Arts and Culture, 1838
GNOF IMPACT Grants for Health and Human Services, 1839
GNOF IMPACT Gulf States Eye Surgery Fund, 1840
GNOF IMPACT Harold W. Newman, Jr. Charitable Trust Grants, 1841
GNOF IMPACT Kahn-Oppenheim Grants, 1842
GNOF New Orleans Works Grants, 1844
Goddess Scholars Grants, 1848
Goldhirsh Fnd Brain Tumor Research Grants, 1852
Goldman Philanthropic Partnerships Program, 1853
Goodrich Corporation Foundation Grants, 1854
Good Samaritan Inc Grants, 1855
Goodyear Tire Grants, 1856
Google Grants Beta, 1857
Grammy Foundation Grants, 1863
Grand Rapids Area Community Foundation Wyoming Grants, 1867
Grass Foundation Marine Biological Laboratory Advanced Imaging Fellowships, 1877
Grass Foundation Marine Biological Laboratory Fellowships, 1878
Great-West Life Grants, 1879
Green Foundation Human Services Grants, 1894
Greenwall Foundation Bioethics Grants, 1897
Gregory B. Davis Foundation Grants, 1898
Greygates Foundation Grants, 1901
Gruber Foundation Cosmology Prize, 1907
Gruber Foundation Genetics Prize, 1908
Gruber Foundation Neuroscience Prize, 1909
Gruber Foundation Rosalind Franklin Young Investigator Award, 1910
Gruber Foundation Weizmann Institute Awards, 1911
Guy's and St. Thomas' Charity Grants, 1918
Guy I. Bromley Trust Grants, 1919
H.J. Heinz Company Foundation Grants, 1923
Hamilton Company Syringe Product Grant, 1937
Harold and Rebecca H. Gross Foundation Grants, 1951
Harry Frank Guggenheim Foundation Dissertation Fellowships, 1963
Harry Frank Guggenheim Fnd Research Grants, 1964
Harry Kramer Memorial Fund Grants, 1965
Health Management Scholarships and Grants for Minorities, 1986
Hearst Foundations Health Grants, 1987
Heineman Foundation for Research, Education, Charitable and Scientific Purposes, 1990
Helen Bader Foundation Grants, 1992
Helen Irwin Littauer Educational Trust Grants, 1994
Hendrick Foundation for Children Grants, 1999
Henry A. and Mary J. MacDonald Foundation, 2003
Hereditary Disease Foundation John J. Wasmuth Postdoctoral Fellowships, 2009
Hereditary Disease Fnd Lieberman Award, 2010
Hereditary Disease Foundation Research Grants, 2011
HHMI-NIBIB Interfaces Initiative Grants, 2014
HHMI-NIH Cloister Research Scholars, 2015
HHMI/EMBO Start-up Grants for C Europe, 2016
HHMI Biomedical Research Grants for International Scientists: Infectious Diseases & Parasitology, 2017
HHMI Biomedical Research Grants for International Scientists in Canada and Latin America, 2018
HHMI Biomedical Research Grants for International Scientists in the Baltics, Central and Eastern Europe, Russia, and Ukraine, 2019
HHMI Grants and Fellowships Programs, 2020
HHMI International Research Scholars Program, 2021
HHMI Med into Grad Initiative Grants, 2022
HHMI Physician-Scientist Early Career Award, 2023
HHMI Research Training Fellowships, 2024
Hilda and Preston Davis Foundation Postdoctoral Fellowships in Eating Disorders Research, 2029
Home Building Industry Disaster Relief Fund, 2043
Hospital Libraries Section / MLA Professional Development Grants, 2052
HRAMF Harold S. Geneen Charitable Trust Awards for Coronary Heart Disease Research, 2059
HRAMF Ralph and Marian Falk Medical Research Trust Catalyst Awards, 2061
HRAMF Ralph and Marian Falk Medical Research Trust Transformational Awards, 2062
HRAMF Taub Fnd Grants for MDS Research, 2064
HRAMF Thome Foundation Awards in Age-Related Macular Degeneration Research, 2065
HRAMF Thome Foundation Awards in Alzheimer's Disease Drug Discovery Research, 2066
HRSA Nurse Education, Practice, Quality and Retention (NEPQR) Grants, 2068
HRSA Resource and Technical Assistance Center for HIV Prevention and Care for Black Men who Have Sex with Men Cooperative Agreement, 2069
HRSA Ryan White HIV AIDS Drug Assistance Grants, 2070
Huber Foundation Grants, 2071
Huisking Foundation Grants, 2076
Huntington's Disease Society of America Research Fellowships, 2079
Huntington's Disease Society of America Research Grants, 2080
IAFF Burn Foundation Research Grants, 2089
IARC Expertise Transfer Fellowship, 2090
IARC Postdoctoral Fellowships for Training in Cancer Research, 2091
IARC Visiting Award for Senior Scientists, 2092
IHI Quality Improvement Fellowships, 2121
IIE 911 Armed Forces Scholarships, 2122
IIE Adell and Hancock Scholarships, 2123
IIE Central Europe Summer Research Institute Summer Research Fellowship, 2127
IIE David L. Boren Fellowships, 2129
IIE Freeman Foundation Indonesia Internships, 2131
IIE Hewlett Fnd/IIE Dissertation Fellowship, 2132
IIE KAUST Graduate Fellowships, 2134
IIE Leonora Lindsley Memorial Fellowships, 2136
IIE Mattel Global Scholarship, 2139
IIE New Leaders Group Award for Mutual Understanding, 2141
IIE Rockefeller Foundation Bellagio Center Residencies, 2142
IIE Whitaker Int'l Fellowships & Scholarships, 2144
Illinois Tool Works Foundation Grants, 2147
Infinity Foundation Grants, 2158
Institute for Agriculture and Trade Policy Food and Society Fellowships, 2159
International Positive Psychology Association Student Scholarships, 2161
Irvin Stern Foundation Grants, 2164
Ittleson Foundation AIDS Grants, 2166
Ittleson Foundation Mental Health Grants, 2167
J.C. Penney Company Grants, 2169
J.W. Kieckhefer Foundation Grants, 2176
James Hervey Johnson Charitable Educational Trust Grants, 2191
James J. & Joan A. Gardner Family Fndn Grants, 2193
James M. Cox Foundation of Georgia Grants, 2196
James McKeen Cattell Fund Fellowships, 2197
James S. McDonnell Foundation Brain Cancer Research Collaborative Activity Awards, 2201
James S. McDonnell Foundation Complex Systems Collaborative Activity Awards, 2202
James S. McDonnell Fnd Research Grants, 2203
James S. McDonnell Foundation Scholar Awards, 2204
James S. McDonnell Foundation Understanding Human Cognition Awards, 2205
Jane Beattie Memorial Scholarship, 2207
Janus Foundation Grants, 2211
Jayne and Leonard Abess Foundation Grants, 2214
Jeffrey Thomas Stroke Shield Foundation Research Grants, 2216
JELD-WEN Foundation Grants, 2218
Jerome and Mildred Paddock Foundation Grants, 2222
Jerome Robbins Foundation Grants, 2223
JM Foundation Grants, 2229
John Ben Snow Memorial Trust Grants, 2231
John M. Lloyd Foundation Grants, 2244
John Merck Fund Grants, 2245
John Merck Scholars Awards, 2246
Johnson & Johnson Corporate Contributions, 2253
Johnson & Johnson/SAH Arts & Healing Grants, 2254
Johnson Controls Foundation Health and Social Services Grants, 2255
Johnson Foundation Wingspread Conference Support Program, 2257
Joseph Collins Foundation Scholarships, 2267
Joseph P. Kennedy Jr. Foundation Grants, 2275
Josiah Macy Jr. Foundation Grants, 2277
Judith and Jean Pape Adams Charitable Foundation ALS Grants, 2279
K and F Baxter Family Foundation Grants, 2287
Kavli Foundation Research Grants, 2296
Kenai Peninsula Foundation Grants, 2298
Kessler Fnd Signature Employment Grants, 2302
Kettering Family Foundation Grants, 2304
Klarman Family Foundation Grants in Eating Disorders Research Grants, 2310
Kluge Center David B. Larson Fellowship in Health and Spirituality, 2311
Knight Family Charitable and Educational Foundation Grants, 2312
Kroger Company Donations, 2324
Kroger Foundation Women's Health Grants, 2325
L. A. Hollinger Respiratory Therapy and Nursing Scholarships, 2327
Lalor Foundation Anna Lalor Burdick Grants, 2335
Lalor Foundation Postdoctoral Fellowships, 2336
LAM Fnd Established Investigator Awards, 2337
LAM Foundation Pilot Project Grants, 2338
LAM Foundation Postdoctoral Fellowships, 2339
LAM Foundation Research Grants, 2340
Landon Foundation-AACR Innovator Award for Cancer Prevention Research, 2342
Landon Foundation-AACR Innovator Award for International Collaboration, 2343
Lawrence Foundation Grants, 2347
Layne Beachley Aim for the Stars Fnd Grants, 2349
Leonard and Helen R. Stulman Charitable Foundation Grants, 2356
Lillian Sholtis Brunner Summer Fellowships for Historical Research in Nursing, 2366
Lisa and Douglas Goldman Fund Grants, 2369
Lisa Higgins-Hussman Foundation Grants, 2370
Littlefield-AACR Grants in Metastatic Colon Cancer Research, 2371
Lotus 88 Fnd for Women & Children Grants, 2376
Ludwick Family Foundation Grants, 2389
Lumosity Human Cognition Grant, 2390
Lumpkin Family Fnd Healthy People Grants, 2391
Lustgarten Foundation for Pancreatic Cancer Research Grants, 2392
Lymphatic Education and Research Network Additional Support Grants for NIH-funded F32 Postdoctoral Fellows, 2394
Lymphatic Education and Research Network Postdoctoral Fellowships, 2395
Lynde & Harry Bradley Foundation Fellowships, 2396
Lynde and Harry Bradley Foundation Grants, 2397
Lynde and Harry Bradley Foundation Prizes: Bradley Prizes, 2398
M-A-C AIDS Fund Grants, 2400
M. Bastian Family Foundation Grants, 2402

Macquarie Bank Foundation Grants, 2410
Maddie's Fund Medical Equipment Grants, 2411
Maggie Welby Foundation Grants, 2414
Majors MLA Chapter Project of the Year, 2418
MAP International Medical Fellowships, 2421
March of Dimes Agnes Higgins Award, 2423
March of Dimes Graduate Nursing Scholarships, 2424
March of Dimes Newborn Screening Awards, 2425
March of Dimes Program Grants, 2426
Mary Kay Foundation Cancer Research Grants, 2444
Massage Therapy Foundation Community Service Grants, 2449
Massage Therapy Foundation Practitioner Case Report Contest, 2450
Massage Therapy Foundation Research Grants, 2451
Massage Therapy Research Fund (MTRF) Grants, 2452
Mattel Children's Foundation Grants, 2454
Mattel International Grants Program, 2455
Max and Victoria Dreyfus Foundation Grants, 2457
May and Stanley Smith Charitable Trust Grants, 2459
Mayo Clinic Administrative Fellowship, 2460
Mayo Clinic Admin Fellowship - Eau Claire, 2461
Mayo Clinic Business Consulting Fellowship, 2462
MBL Albert & Ellen Grass Fellowships, 2463
MBL Ann E. Kammer Summer Fellowship, 2464
MBL Associates Summer Fellowships, 2465
MBL Baxter Postdoctoral Summer Fellowship, 2466
MBL Burr & Susie Steinbach Fellowship, 2467
MBL E.E. Just Summer Fellowship for Minority Scientists, 2468
MBL Erik B. Fries Summer Fellowships, 2469
MBL Eugene and Millicent Bell Fellowships, 2470
MBL Evelyn and Melvin Spiegal Fellowship, 2471
MBL Frank R. Lillie Summer Fellowship, 2472
MBL Frederik B. and Betsy G. Bang Summer Fellowships, 2473
MBL Fred Karush Library Readership, 2474
MBL H. Keffer Hartline and Edward F. MacNichol, Jr. Fellowships, 2476
MBL Herbert W. Rand Summer Fellowship, 2477
MBL James E. and Faith Miller Memorial Summer Fellowship, 2478
MBL John M. Arnold Award, 2479
MBL Laura and Arthur Colwin Fellowships, 2480
MBL Lucy B. Lemann Summer Fellowship, 2481
MBL M.G.F. Fuortes Summer Fellowships, 2482
MBL Nikon Summer Fellowship, 2483
MBL Plum Fndn John E. Dowling Fellowships, 2484
MBL Robert Day Allen Summer Fellowship, 2485
MBL Scholarships and Awards, 2486
MBL Stephen W. Kuffler Summer Fellowships, 2487
MBL William Townsend Porter Summer Fellowships for Minority Investigators, 2488
McCune Foundation Human Services Grants, 2493
McGraw-Hill Companies Community Grants, 2494
McKesson Foundation Grants, 2496
MDA Development Grant, 2500
MDA Neuromuscular Disease Research Grants, 2501
Mead Johnson Nutritionals Charitable Giving, 2502
Mead Johnson Nutritionals Med Educ Grants, 2504
Mead Johnson Nutritionals Scholarships, 2505
Medical Informatics Section/MLA Career Development Grant, 2508
MedImmune Charitable Grants, 2509
Medtronic Foundation Fellowships, 2512
Medtronic Foundation HeartRescue Grants, 2513
Medtronic Foundation Patient Link Grants, 2514
Mesothelioma Applied Research Fndn Grants, 2524
Meyer Foundation Healthy Communities Grants, 2533
MGFA Post-Doctoral Research Fellowships, 2539
MGFA Student Fellowships, 2540
MGN Family Foundation Grants, 2542
Michigan Psychoanalytic Institute and Society SATA Grants, 2546
Michigan Psychoanalytic Institute and Society Scholarships, 2547
Microsoft Research Cell Phone as a Platform for Healthcare Grants, 2548
Milagro Foundation Grants, 2552
Military Ex-Prisoners of War Foundation Grants, 2554

Milken Family Foundation Grants, 2555
MLA Continuing Education Grants, 2559
MLA David A. Kronick Traveling Fellowship, 2561
MLA Donald A.B. Lindberg Fellowship, 2562
MLA Estelle Brodman Academic Medical Librarian of the Year Award, 2563
MLA Graduate Scholarship, 2564
MLA Grad Scholarship for Minority Students, 2565
MLA Ida and George Eliot Prize, 2566
MLA Janet Doe Lectureship Award, 2567
MLA Lois Ann Colaianni Award for Excellence and Achievement in Hospital Librarianship, 2568
MLA Louise Darling Medal for Distinguished Achievement in Collection Development in the Health Sciences, 2569
MLA Lucretia W. McClure Excellence in Education Award, 2570
MLA Murray Gottlieb Prize Essay Award, 2571
MLA Research, Development, and Demonstration Project Grant, 2572
MLA Rittenhouse Award, 2573
MLA Section Project of the Year Award, 2574
MLA Thomas Reuters / Frank Bradway Rogers Information Advancement Award, 2576
MMAAP Foundation Dermatology Fellowships, 2577
MMAAP Fndn Dermatology Project Awards, 2578
MMAAP Foundation Fellowship Award in Hematology, 2579
MMAAP Foundation Fellowship Award in Reproductive Medicine, 2580
MMAAP Foundation Fellowship Award in Translational Medicine, 2581
MMAAP Foundation Geriatric Project Awards, 2582
MMAAP Fndn Hematology Project Awards, 2583
MMAAP Foundation Research Project Award in Reproductive Medicine, 2584
MMAAP Foundation Research Project Award in Translational Medicine, 2585
MMAAP Foundation Senior Health Fellowship Award in Geriatrics Medicine and Aging Research, 2586
Mockingbird Foundation Grants, 2589
Morehouse PHSI Project Imhotep Internships, 2592
Morgan Adams Foundation Grants, 2593
Ms. Fndn for Women Ending Violence Grants, 2596
Ms. Foundation for Women Health Grants, 2597
Multiple Sclerosis Foundation Brighter Tomorrow Grants, 2602
NAF Fellowships, 2603
NAF Kyle Bryant Translational Research Award, 2604
NAF Research Grants, 2605
NAF Young Investigator Awards, 2606
Nancy R. Gelman Foundation Breast Cancer Seed Grants, 2608
NAPNAP Fndn Elaine Gelman Scholarship, 2609
NAPNAP Fndn Grad Research Grant, 2610
NAPNAP Foundation Innovative Health Care Small Grant, 2611
NAPNAP Foundation McNeil Grant-in-Aid, 2612
NAPNAP Foundation McNeil PNP Scholarships, 2613
NAPNAP Foundation McNeil Rural and Underserved Scholarships, 2614
NAPNAP Foundation Nursing Research Grants, 2615
NAPNAP Foundation Reckitt Benckiser Student Scholarship, 2616
NAPNAP Foundation Shourd Parks Immunization Project Small Grants, 2617
NAPNAP Foundation Wyeth Pediatric Immunization Grant, 2618
NARSAD Colvin Prize for Outstanding Achievement in Mood Disorders Research, 2619
NARSAD Distinguished Investigator Grants, 2620
NARSAD Goldman-Rakic Prize for Cognitive Neuroscience, 2621
NARSAD Independent Investigator Grants, 2622
NARSAD Lieber Prize for Schizo Research, 2623
NARSAD Ruane Prize for Child and Adolescent Psychiatric Research, 2624
NARSAD Young Investigator Grants, 2625
Nathan Cummings Foundation Grants, 2627
National Blood Foundation Research Grants, 2628

National Center for Responsible Gaming Grants, 2629
National Center for Responsible Gaming Postdoctoral Fellowships, 2630
National Center for Resp Gaming Seed Grants, 2631
National Headache Foundation Research Grants, 2632
National Headache Foundation Seymour Diamond Clinical Fellowship in Headache Education, 2633
National MPS Society Grants, 2634
National Parkinson Foundation Clinical Research Fund Grants, 2635
National Psoriasis Foundation Research Grants, 2636
NBME Stemmler Med Ed Research Grants, 2638
NCCAM Exploratory Developmental Grants for Complementary and Alternative Medicine (CAM) Studies of Humans, 2639
NCCAM Ruth L. Kirschstein National Research Service Awards for Postdoctoral Training in Complementary and Alternative Medicine, 2640
NCCAM Translational Tools for Clinical Studies of CAM Interventions Grants, 2641
NCHS National Center for Health Statistics Postdoctoral Research Appointments, 2642
NCI/NCCAM Quick-Trials for Novel Cancer Therapies, 2643
NCI Academic-Industrial Partnerships for Devel and Validation of In Vivo Imaging Systems and Methods for Cancer Investigations, 2644
NCI Application and Use of Transformative Emerging Technologies in Cancer Research, 2645
NCI Application of Metabolomics for Translational and Biological Research Grants, 2646
NCI Basic Cancer Research in Cancer Health Disparities Grants, 2647
NCI Cancer Education and Career Development Program, 2648
NCI Centers of Excellence in Cancer Communications Research, 2649
NCI Diet, Epigenetic Events, and Cancer Prevention Grants, 2650
NCI Exploratory Grants for Behavioral Research in Cancer Control, 2651
NCI Improving Diet and Physical Activity Assessment Grants, 2652
NCI Mentored Career Development Award to Promote Diversity, 2653
NCI Ruth L. Kirschstein National Research Service Award Institutional Training Grants, 2654
NCI Stages of Breast Development: Normal to Metastatic Disease Grants, 2655
NCI Technologies and Software to Support Integrative Cancer Biology Research (SBIR) Grants, 2656
NCRR Networks and Pathways Collaborative Research Projects Grants, 2657
NCRR Novel Approaches to Enhance Animal Stem Cell Research, 2658
NEAVS Fellowsips in Women's Health and Sex Differences, 2659
NEDA/AED Charron Family Research Grant, 2660
NEDA/AED Joan Wismer Research Grant, 2661
NEDA/AED Tampa Bay Eating Disorders Task Force Award, 2662
NEI Clinical Study Planning Grant, 2663
NEI Clinical Vision Research Devel Award, 2664
NEI Innovative Patient Outreach Programs And Ocular Screening Technologies To Improve Detection Of Diabetic Retinopathy Grants, 2665
NEI Mentored Clinical Scientist Dev Award, 2666
NEI Research Grant For Secondary Data Analysis Grants, 2667
NEI Ruth L. Kirschstein National Research Service Award Short-Term Institutional Research Training Grants, 2668
NEI Scholars Program, 2669
NEI Translational Research Program On Therapy For Visual Disorders, 2670
NEI Vision Research Core Grants, 2671
Nelda C. and H.J. Lutcher Stark Fndn Grants, 2672
New York University Steinhardt School of Education Fellowships, 2684
NFL Charities Medical Grants, 2685

NHGRI Mentored Research Scientist Development Award, 2686
NHLBI Airway Smooth Muscle Function and Targeted Therapeutics in Human Asthma Grants, 2687
NHLBI Ancillary Studies in Clinical Trials, 2688
NHLBI Bioengineering and Obesity Grants, 2689
NHLBI Bioengineering Approaches to Energy Balance and Obesity Grants for SBIR, 2690
NHLBI Bioengineering Approaches to Energy Balance and Obesity Grants for STTR, 2691
NHLBI Biomedical Research Training Program for Individuals from Underrepresented Groups, 2692
NHLBI Cardiac Devel Consortium Grants, 2693
NHLBI Career Transition Awards, 2694
NHLBI Characterizing the Blood Stem Cell Niche Grants, 2695
NHLBI Circadian-Coupled Cellular Function in Heart, Lung, and Blood Tissue Grants, 2696
NHLBI Clinical Centers for the NHLBI Asthma Network (AsthmaNet) Grants, 2697
NHLBI Developmental Origins of Altered Lung Physiology and Immune Function Grants, 2698
NHLBI Exploratory Studies in the Neurobiology of Pain in Sickle Cell Disease, 2699
NHLBI Immunomodulatory, Inflammatory, & Vasoregulatory Properties of Transfused Red Blood Cell Units as a Function of Preparation & Storage Grants, 2700
NHLBI Independent Scientist Award, 2701
NHLBI Intramural Research Training Awards, 2702
NHLBI Investigator Initiated Multi-Site Clinical Trials, 2703
NHLBI Lymphatics in Health and Disease in the Digestive, Cardio & Pulmonary Systems, 2704
NHLBI Lymphatics in Health and Disease in the Digestive, Urinary, Cardiovascular and Pulmonary Systems, 2705
NHLBI Mentored Career Award for Faculty at Institutions that Promote Diversity (K01), 2706
NHLBI Mentored Career Award For Faculty At Minority Institutions, 2707
NHLBI Mentored Career Dev Award to Promote Faculty Diversity in Biomed Research, 2708
NHLBI Mentored Clinical Scientist Research Career Development Awards, 2709
NHLBI Mentored Patient-Oriented Research Career Development Awards, 2710
NHLBI Mentored Quantitative Research Career Development Awards, 2711
NHLBI Microbiome of Lung & Respiratory in HIV Infected Individuals & Uninfected Controls, 2712
NHLBI Midcareer Investigator Award in Patient-Oriented Research, 2713
NHLBI National Research Service Award Programs in Cardiovascular Epidemiology and Biostatistics, 2714
NHLBI New Approaches to Arrhythmia Detection and Treatment Grants for SBIR, 2715
NHLBI New Approaches to Arrhythmia Detection and Treatment Grants for STTR, 2716
NHLBI Nutrition and Diet in the Causation, Prevention, and Management of Heart Failure Research Grants, 2717
NHLBI Pathway to Independence Awards, 2718
NHLBI Pediatric Cardiac Genomics Consortium Grants, 2719
NHLBI Phase II Clinical Trials of Novel Therapies for Lung Diseases, 2720
NHLBI Prematurity and Respiratory Outcomes Program (PROP) Grants, 2721
NHLBI Progenitor Cell Biology Consortium Administrative Coordinating Center Grants, 2722
NHLBI Protein Interactions Governing Membrane Transport in Pulmonary Health Grants, 2723
NHLBI Research Demonstration and Dissemination Grants, 2724
NHLBI Research Grants on the Relationship Between Hypertension and Inflammation, 2725
NHLBI Research on the Role of Cardiomyocyte Mitochondria in Heart Disease: An Integrated Approach, 2726
NHLBI Research Supplements to Promote Diversity in Health-Related Research, 2727
NHLBI Right Heart Function in Health and Chronic Lung Diseases Grants, 2728
NHLBI Ruth L. Kirschstein National Research Service Awards for Individual Postdoctoral Fellows, 2729
NHLBI Ruth L. Kirschstein National Research Service Awards for Individual Predoctoral Fellowships to Promote Diversity in Health-Related Research, 2730
NHLBI Ruth L. Kirschstein National Research Service Awards for Individual Predoctoral MD/PhD Fellows and Other Dual Degree Fellows, 2731
NHLBI Ruth L. Kirschstein National Research Service Awards for Individual Senior Fellows, 2732
NHLBI Short-Term Research Education Program to Increase Diversity in Health-Related Research, 2733
NHLBI Summer Inst for Training in Biostatistics, 2734
NHLBI Summer Internship Program in Biomedical Research, 2735
NHLBI Targeted Approaches to Weight Control for Young Adults Grants, 2736
NHLBI Targeting Calcium Regulatory Molecules for Arrhythmia Prevention Grants, 2737
NHLBI Training Program Grants for Institutions that Promote Diversity, 2738
NHLBI Translating Basic Behavioral and Social Science Discoveries into Interventions to Reduce Obesity: Centers for Behavioral Intervention Development Grants, 2739
NHLBI Translational Grants in Lung Diseases, 2740
NIAAA Independent Scientist Award, 2742
NIAAA Mechanisms of Alcohol and Drug-Induced Pancreatitis Grants, 2743
NIAAA Mentored Clinical Scientist Research Career Development Awards, 2744
NIA Aging, Oxidative & Cell Death Grants, 2745
NIA Aging Research Dissertation Awards to Increase Diversity, 2746
NIA AIDS and Aging: Behavioral Sciences Prevention Research Grants, 2747
NIA Alzheimer's Disease Core Centers Grants, 2748
NIA Alzheimer's Disease Drug Development Program Grants, 2749
NIA Alzheimer's Disease Pilot Clinical Trials, 2750
NIA Alzheimer's Disease Centers Grants, 2751
NIA Archiving and Development of Socialbehavioral Datasets in Aging Related Studies Grants, 2752
NIA Awards to Support Research on the Biology of Aging in Invertebrates Grants, 2753
NIA Behavioral and Social Research Grants on Disasters and Health, 2754
NIA Bioenergetics, Fatigability, and Activity Limitations in Aging, 2755
NIA Diversity in Medication Use and Outcomes in Aging Populations Grants, 2756
NIA Effects of Gene-Social Environment Interplay on Health and Behavior in Later Life Grants, 2757
NIA Factors Affecting Cognitive Function in Adults with Down Syndrome Grants, 2758
NIA Grants for Alzheimer's Disease Drugs, 2759
NIA Harmonization of Longitudinal Cross-National Surveys of Aging Grants, 2760
NIA Health Behaviors and Aging Grants, 2761
NIA Healthy Aging through Behavioral Economic Analyses of Situations Grants, 2762
NIA Higher-Order Cognitive Functioning and Aging Grants, 2763
NIA Human Biospecimen Resources for Aging Research Grants, 2764
NIAID Independent Scientist Award, 2765
NIA Independent Scientist Award, 2766
NIA Malnutrition in Older Persons Research, 2767
NIA Mechanisms, Measurement, and Management of Pain in Aging: from Molecular to Clinical, 2768
NIA Mechanisms Underlying the Links between Psychosocial Stress, Aging, the Brain and the Body Grants, 2769
NIA Medical Management of Older Patients with HIV/AIDS Grants, 2770
NIA Mentored Clinical Scientist Research Career Development Awards, 2771
NIA Mentored Research Scientist Devel Award, 2772
NIA Minority Dissert Aging Research Grants, 2773
NIAMS Pilot and Feasibility Clinical Research Grants in Arthritis, Musculoskeletal & Skin Diseases, 2774
NIA Multicenter Study on Exceptional Survival in Families: The Long Life Family Study (LLFS) Grants, 2775
NIA Nathan Shock Centers of Excellence in Basic Biology of Aging Grants, 2776
NIA Network Infrastructure Support for Emerging Behavioral & Social Research in Aging, 2777
NIA Pilot Research Grant Program, 2778
NIA Postdoctoral Research on Aging in Canada, 2779
NIA Promoting Careers in Aging and Health Disparities Research Grants, 2780
NIA Renal Function and Chronic Kidney Disease in Aging Grants, 2781
NIA Role of Apolipoprotein E, Lipoprotein Receptors and CNS Lipid Homeostasis in Brain Aging and Alzheimer's Disease, 2782
NIA Role of Nuclear Receptors in Tissue and Organismal Aging Grants, 2783
NIA Support of Scientific Meetings as Cooperative Agreements, 2784
NIA Thyroid in Aging Grants, 2785
NIA Transdisciplinary Research on Fatigue and Fatigability in Aging Grants, 2786
NIA Translational Research at the Aging/Cancer Interface (TRACI) Grants, 2787
NIA Vulnerable Dendrites and Synapses in Aging and Alzheimer's Disease Grants, 2788
NICHD Innovative Therapies and Clinical Studies for Screenable Disorders Grants, 2789
NICHD Ruth L. Kirschstein National Research Service Award (NRSA) Institutional Predoctoral Training Program in Systems Biology of Developmental Biology & Birth Defects, 2790
NIDA Mentored Clinical Scientist Research Career Development Awards, 2793
NIDA Pilot and Feasibility Studies in Preparation for Drug Abuse Prevention Trials, 2794
NIDCD Mentored Clinical Scientist Research Career Development Award, 2795
NIDCD Mentored Research Scientist Development Award, 2796
NIDDK Advances in Polycystic Kidney Disease, 2797
NIDDK Adverse Metabolic Side Effects of Second Generation Psychotropic Medications Leading to Obesity and Increased Diabetes Risk Grants, 2798
NIDDK Ancillary Studies of Kidney Disease Accessing Information from Clinical Trials, Epidemiological Studies, and Databases Grants, 2799
NIDDK Ancillary Studies to Major Ongoing NIDDK and NHLBI Clinical Research Studies, 2800
NIDDK Basic and Clinical Studies of Congenital Urinary Tract Obstruction Grants, 2801
NIDDK Basic Research Grants in the Bladder and Lower Urinary Tract, 2802
NIDDK Calcium Oxalate Stone Diseases Grants, 2803
NIDDK Cell-Specific Delineation of Prostate and Genitourinary Development, 2804
NIDDK Co-Activators and Co-Repressors in Gene Expression Grants, 2805
NIDDK Collaborative Interdisciplinary Team Science in Diabetes, Endocrinology and Metabolic Diseases Grants, 2806
NIDDK Developmental Biology and Regeneration of the Liver Grants, 2807
NIDDK Development of Assays for High-Throughput Drug Screening Grants, 2808
NIDDK Devel of Disease Biomarkers Grants, 2809
NIDDK Diabetes, Endocrinology, and Metabolic Diseases NRSAs--Individual, 2810
NIDDK Diabetes Research Centers, 2811
NIDDK Diet Comp & Energy Balance Grants, 2812
NIDDK Education Program Grants, 2813
NIDDK Endoscopic Clinical Research Grants In Pancreatic And Biliary Diseases, 2814

NIDDK Enhancing Zebrafish Research with Research Tools and Techniques, 2815
NIDDK Erythroid Lineage Molecular Grants, 2816
NIDDK Erythropoiesis: Components and Mechanisms Grants, 2817
NIDDK Exploratory/Developmental Clinical Research Grants in Obesity, 2818
NIDDK Grants for Basic Research in Glomerular Diseases, 2819
NIDDK Health Disparities in NIDDK Diseases, 2820
NIDDK Identifying & Reducing Diabetes & Obesity Related Disparities in Healthcare Systems, 2821
NIDDK Insulin Signaling and Receptor Cross Talk Grants, 2822
NIDDK Intestinal Failure, Short Gut Syndrome and Small Bowel Transplantation Grants, 2823
NIDDK Intestinal Stem Cell Consortium Grants, 2824
NIDDK Mentored Research Scientist Development Award, 2825
NIDDK Multi-Center Clinical Study Cooperative Agreements, 2826
NIDDK Multi-Center Clinical Study Implementation Planning Grants, 2827
NIDDK Neurobiology of Diabetic Complications Grants, 2828
NIDDK Neuroimaging in Obesity Research, 2829
NIDDK New Tech for Liver Disease SBIR, 2830
NIDDK New Tech for Liver Disease STTR, 2831
NIDDK Non-Invasive Methods for Diagnosis and Progression of Diabetes, Kidney, Urological, Hematological and Digestive Diseases, 2832
NIDDK Pancreatic Development and Regeneration Grants: Cellular Therapies for Diabetes, 2833
NIDDK Feasibility Clinical Research Grants in Diabetes, Endocrine and Metabolic Diseases, 2834
NIDDK Pilot and Feasibility Clinical Research Grants in Kidney or Urologic Diseases, 2835
NIDDK Pilot and Feasibility Clinical Research Studies in Digestive Diseases and Nutrition, 2836
NIDDK Planning Grants For Translational Research For The Prevention And Control Of Diabetes And Obesity, 2837
NIDDK Proteomics: Diabetes, Obesity, And Endocrine, Digestive, Kidney, Urologic, And Hematologic Diseases, 2838
NIDDK Research Grants for Studies of Hepatitis C in the Setting of Renal Disease, 2839
NIDDK Research Grants on Improving Health Care for Obese Patients, 2840
NIDDK Role of Gastrointestinal Surgical Procedures in Amelioration of Obesity-Related Insulin Resistance & Diabetes Weight Loss, 2841
NIDDK Secondary Analyses in Obesity, Diabetes and Digestive and Kidney Diseases Grants, 2842
NIDDK Seeding Collaborative Interdisciplinary Team Science in Diabetes, Endocrinology and Metabolic Diseases Grants, 2843
NIDDK Silvio O. Conte Digestive Diseases Research Core Centers Grants, 2844
NIDDK Small Business Innov Research to Develop New Therapeutics & Monitoring Tech for Type 1 Diabetes Towards an Artificial Pancreas, 2845
NIDDK Small Grants for K08/K23 Recipients, 2846
NIDDK Small Grants for Underrepresented Minority Scientists in Diabetes and Digestive and Kidney Diseases, 2847
NIDDK Traditional Conference Grants, 2848
NIDDK Training in Clinical Investigation in Kidney and Urology Grants, 2849
NIDDK Translational Research for the Prevention and Control of Diabetes and Obesity, 2850
NIDDK Transmission of Human Immunodeficiency Virus (HIV) In Semen, 2851
NIDDK Type 2 Diabetes in the Pediatric Population Research Grants, 2852
NIDRR Field-Initiated Projects, 2853
NIEHS Hazardous Materials Worker Health and Safety Training Grants, 2854
NIEHS Hazmat Training at Doe Nuclear Weapons Complex Grants, 2855
NIEHS Independent Scientist Award, 2856
NIEHS Mentored Clinical Scientist Research Career Development Awards, 2857
NIEHS Mentored Research Scientist Development Award, 2858
NIEHS SBIR E-Learning for HAZMAT and Emergency Response Grants, 2859
NIEHS Small Business Innovation Research (SBIR) Program Grants, 2860
NIGMS Advancing Basic Behavioral and Social Research on Resilience: an Integrative Science Approach, 2861
NIGMS Enabling Resources for Pharmacogenomics Grants, 2862
NIGMS Ruth L. Kirschstein National Research Service Awards for Individual Predoctoral Fellows in PharmD/PhD Grants, 2863
NIH Academic Research Enhancement Awards, 2864
NIH Biobehavioral Research for Effective Sleep, 2865
NIH Biology, Development, and Progression of Malignant Prostate Disease Research Grants, 2866
NIH Biomedical Research on the International Space Station, 2867
NIH Bone and the Hematopoietic and Immune Systems Research Grants, 2868
NIH Building Interdisciplinary Research Careers in Women's Health Grants, 2869
NIHCM Foundation Health Care Print Journalism Awards, 2870
NIHCM Fndn Health Care Research Awards, 2871
NIHCM Foundation Health Care Television and Radio Journalism Awards, 2872
NIH Dietary Supplement Research Centers: Botanicals Grants, 2873
NIH Earth-Based Research Relevant to the Space Environment, 2874
NIH Enhancing Adherence to Diabetes Self-Management Behaviors, 2875
NIH Exceptional, Unconventional Research Enabling Knowledge Acceleration (EUREKA) Grants, 2876
NIH Exploratory/Devel Research Grant, 2877
NIH Exploratory Innovations in Biomedical Computational Science & Technology Grants, 2878
NIH Fine Mapping and Function of Genes for Type 1 Diabetes Grants, 2879
NIH Health Promotion Among Racial and Ethnic Minority Males, 2880
NIH Human Connectome Project Grants, 2881
NIH Independent Scientist Award, 2882
NIH Innovations in Biomedical Computational Science and Technology Grants, 2883
NIH Innovations in Biomedical Computational Science and Technology Initiative Grants for SBIR, 2884
NIH Innovations in Biomedical Computational Science and Technology Initiative Grants for STTR, 2885
NIH Mentored Clinical Scientist Devel Award, 2886
NIH Mentored Clinical Scientist Research Career Development Award, 2887
NIH Mentored Patient-Oriented Research Career Development Award, 2888
NIH Mentored Research Scientist Dev Awards, 2889
NIH Nonhuman Primate Immune Tolerance Cooperative Study Group Grants, 2890
NIH Receptors and Signaling in Bone in Health and Disease, 2891
NIH Recovery Act Limited Competition: Academic Research Enhancement Awards, 2892
NIH Recovery Act Limited Competition: Biomedical Research, Development, and Growth to Spur the Acceleration of New Technologies Grants, 2893
NIH Recovery Act Limited Competition: Building Sustainable Community-Linked Infrastructure to Enable Health Science Research, 2894
NIH Recovery Act Limited Competition: Small Business Catalyst Awards for Accelerating Innovative Research Grants, 2895
NIH Recovery Act Limited Competition: Supporting New Faculty Recruitment to Enhance Research Resources through Biomedical Research Core Centers Grants, 2896
NIH Research On Ethical Issues In Human Subjects Research, 2897
NIH Research on Sleep and Sleep Disorders, 2898
NIH Research Project Grants, 2899
NIH Ruth L. Kirschstein National Research Service Award Institutional Research Training Grants, 2900
NIH Ruth L. Kirschstein National Research Service Awards (NRSA) for Individual Senior Fellows, 2901
NIH Ruth L. Kirschstein National Research Service Awards for Individual Postdoc Fellowships, 2902
NIH Ruth L. Kirschstein National Research Service Awards for Individual Predoctoral Fellows, 2903
NIH Ruth L. Kirschstein National Research Service Awards for Individual Predoctoral Fellowships to Promote Diversity in Research, 2904
NIH Ruth L. Kirschstein National Research Service Awards for Individual Predoctoral MD/PhD and Other Dual Doctoral Degree Fellows, 2905
NIH Ruth L. Kirschstein Nat Research Service Award Short-Term Institutional Training Grants, 2906
NIH Sarcoidosis: Research into the Cause of Multi-Organ Disease and Clinical Strategies for Therapy Grants, 2907
NIH School Interventions to Prevent Obesity, 2908
NIH Self-Management Strategies Across Chronic Diseases Grants, 2909
NIH Solicitation of Assays for High Throughput Screening (HTS) in the Molecular Libraries Probe Production Centers Network (MLPCN), 2910
NIH Structural Bio of Membrane Proteins Grant, 2911
NIH Summer Internship Program in Biomedical Research, 2912
NIH Support of Competitive Research (SCORE) Pilot Project Awards, 2913
NIH Support of Competitive Research (SCORE) Research Advancement Awards, 2914
NIMH AIDS and Aging: Behavioral Sciences Prevention Research Grants, 2915
NIMH Basic and Translational Research in Emotion Grants, 2916
NIMH Curriculum Development Award in Neuroinformatics Research and Analysis, 2917
NIMH Early Identification and Treatment of Mental Disorders in Children and Adolescents Grants, 2918
NIMH Jointly Sponsored Ruth L. Kirschstein National Research Service Award Institutional Predoctoral Training Program in the Neurosciences, 2919
NIMH Short Courses in Neuroinformatics, 2920
NINDS Optimization of Small Molecule Probes for the Nervous System (SBIR) Grants, 2922
NINDS Optimization of Small Molecule Probes for the Nervous System (STTR) Grants, 2923
NINDS Optimization of Small Molecule Probes for the Nervous System Grants, 2924
NINDS Support of Scientific Meetings as Cooperative Agreements, 2925
NINR Acute and Chronic Care During Mechanical Ventilation Research Grants, 2926
NINR AIDS and Aging: Behavioral Sciences Prevention Research Grants, 2927
NINR Chronic Illness Self-Management in Children and Adolescents Grants, 2928
NINR Clinical Interventions for Managing the Symptoms of Stroke Research Grants, 2929
NINR Diabetes Self-Management in Minority Populations Grants, 2930
NINR Mechanisms, Models, Measurement, & Management in Pain Research Grants, 2931
NINR Mentored Research Scientist Development Award, 2932
NINR Quality of Life for Individuals at the End of Life Grants, 2933
NINR Ruth L. Kirschstein National Research Service Award for Predoc Fellows in Nursing, 2934
NKF Clinical Scientist Awards, 2935
NKF Franklin McDonald/Fresenius Medical Care Clinical Research Grants, 2936
NKF Professional Councils Research Grants, 2937
NKF Research Fellowships, 2938
NKF Young Investigator Grants, 2939

GEOGRAPHIC INDEX

NLM Academic Research Enhancement Awards, 2940
NLM Exceptional, Unconventional Research Grants Enabling Knowledge Acceleration, 2941
NLM Exploratory/Developmental Grants, 2942
NLM Express Research Grants in Biomedical Informatics, 2943
NLM Grants for Scholarly Works in Biomedicine and Health, 2944
NLM Informatics Conference Grants, 2945
NLM Internet Connection for Medical Institutions Grants, 2946
NLM Limited Competition for Continuation of Biomedical Informatics/Bioinformatics Resource Grants, 2947
NLM Manufacturing Processes of Medical, Dental, and Biological Technologies (SBIR) Grants, 2948
NLM New Tech for Liver Disease Grants, 2949
NLM Predictive Multiscale Models of the Physiome in Health and Disease Grants, 2950
NLM Research Project Grants, 2951
NLM Research Supplements to Promote Reentry in Health-Related Research, 2952
NLM Understanding and Promoting Health Literacy Research Grants, 2953
NMF Franklin C. McLean Award, 2955
NMF General Electric Med Fellowships, 2956
NMF Henry G. Halladay Awards, 2957
NMF Irving Graef Memorial Scholarship, 2959
NMF Mary Ball Carrera Scholarship, 2960
NMF Metropolitan Life Foundation Awards Program for Academic Excellence in Medicine, 2961
NMF National Medical Association Awards for Medical Journalism, 2962
NMF National Medical Association Emerging Scholar Awards, 2963
NMF National Medical Association Patti LaBelle Award, 2964
NMF Need-Based Scholarships, 2965
NMF Ralph W. Ellison Prize, 2966
NMF William and Charlotte Cadbury Award, 2967
NMSS Scholarships, 2968
NOCSAE Research Grants, 2978
Norman Foundation Grants, 2983
Northrop Grumman Corporation Grants, 2997
Northwest Airlines KidCares Medical Travel Assistance, 2998
Notsew Orm Sands Foundation Grants, 3002
NSERC Brockhouse Canada Prize for Interdisciplinary Research in Science and Engineering Grant, 3003
NSF-NIST Interaction in Chemistry, Materials Research, Molecular Biosciences, Bioengineering, and Chemical Engineering, 3004
NSF Accelerating Innovation Research, 3005
NSF Alan T. Waterman Award, 3006
NSF Animal Developmental Mechanisms Grants, 3007
NSF Biological Physics (BP), 3008
NSF Biomedical Engineering and Engineering Healthcare Grants, 3009
NSF Biomolecular Systems Cluster Grants, 3010
NSF Cell Systems Cluster Grants, 3011
NSF Chemistry Research Experiences for Undergraduates (REU), 3012
NSF Doctoral Dissertation Improvement Grants in the Directorate for Biological Sciences (DDIG), 3013
NSF Emerging Frontiers in Research and Innovation (EFRI) Grants, 3014
NSF Genes and Genome Systems Cluster Grants, 3015
NSF Grant Opportunities for Academic Liaison with Industry (GOALI), 3016
NSF Instrument Development for Bio Research, 3017
NSF Major Research Instrumentation Program (MRI) Grants, 3018
NSF Perception, Action & Cognition Grants, 3019
NSF Postdoc Research Fellowships in Biology, 3020
NSF Presidential Early Career Awards for Scientists and Engineers (PECASE) Grants, 3021
NSF Small Business Innovation Research (SBIR) Grants, 3022
NSF Social Psychology Research Grants, 3023

NSF Systematic Biology and Biodiversity Inventories Grants, 3024
NSF Undergraduate Research and Mentoring in the Biological Sciences (URM), 3025
Nuffield Foundation Africa Program Grants, 3026
Nuffield Foundation Children & Families Grants, 3027
Nuffield Foundation Law and Society Grants, 3028
Nuffield Foundation Oliver Bird Rheumatism Program Grants, 3029
Nuffield Foundation Open Door Grants, 3030
Nuffield Foundation Small Grants, 3031
NWHF Mark O. Hatfield Research Fellowship, 3035
NWHF Partners Investing in Nursing's Future, 3036
NYAM Brock Lecture, Award and Visiting Professorship in Pediatrics, 3038
NYAM Charles A. Elsberg Fellowship, 3039
NYAM David E. Rogers Fellowships, 3040
NYAM Edward N. Gibbs Memorial Lecture and Award in Nephrology, 3041
NYAM Edwin Beer Research Fellowship, 3042
NYAM Ferdinand C. Valentine Fellowship, 3043
NYAM Lewis Rudin Glaucoma Grants, 3044
NYAM Mary and David Hoar Fellowship, 3045
NYAM Student Essay Grants, 3046
NYC Managed Care Consumer Assistance Workshop Re-Design Grants, 3047
Oak Foundation Child Abuse Grants, 3059
Obesity Society Grants, 3060
OBSSR Behavioral and Social Science Research on Understanding and Reducing Health Disparities Grants, 3061
OECD Sexual Health and Rights Project Grants, 3064
Office Depot Corporation Community Relations Grants, 3065
Ogden Codman Trust Grants, 3066
Olive Higgins Prouty Foundation Grants, 3072
Olympus Corporation of Americas Corporate Giving Grants, 3073
Olympus Corporation of Americas Fellowships, 3074
OneSight Research Foundation Block Grants, 3076
OneStar Foundation AmeriCorps Grants, 3077
Orthopaedic Trauma Association Kathy Cramer Young Clinician Memorial Fellowships, 3083
Orthopaedic Trauma Assoc Research Grants, 3084
OSF Access to Essential Medicines Initiative, 3086
OSF Accountability and Monitoring in Health Initiative Grants, 3087
OSF Global Health Financing Initiative Grants, 3088
OSF Health Media Initiative Grants, 3089
OSF International Harm Reduction Development Program Grants, 3090
OSF International Palliative Care Grants, 3091
OSF Law and Health Initiative Grants, 3092
OSF Public Health Grants in Kyrgyzstan, 3094
OSF Roma Health Project Grants, 3095
Osteosynthesis and Trauma Care Foundation European Visiting Fellowships, 3097
Oticon Focus on People Awards, 3098
Pacific Life Foundation Grants, 3103
Packard Foundation Children, Families, and Communities Grants, 3104
Packard Foundation Population and Reproductive Health Grants, 3106
PAHO-ASM Int'l Professorship Latin America, 3107
Palmer Foundation Grants, 3112
Pancreatic Cancer Action Network Fellowship, 3113
Pancreatic Cancer Action Network-AACR Pathway to Leadership Grants, 3114
Partnership for Cures Charles E. Culpeper Scholarships in Medical Science, 3117
Partnership for Cures Two Years To Cures Grants, 3118
Paul Balint Charitable Trust Grants, 3122
Paul Marks Prizes for Cancer Research, 3123
PDF-AANF Clinician-Scientist Dev Awards, 3127
PDF International Research Grants, 3128
PDF Postdoctoral Fellowships for Basic Scientists, 3129
PDF Postdoctoral Fellowships for Neurologists, 3130
PDF Summer Fellowships, 3131
Pediatric Brain Tumor Fndn Research Grants, 3134
Pediatric Cancer Research Foundation Grants, 3135

PepsiCo Foundation Grants, 3137
Perkins Charitable Foundation Grants, 3140
Pet Care Trust Sue Busch Memorial Award, 3143
Pew Charitable Trusts Biomed Research Grants, 3157
Pew Charitable Trusts Children & Youth Grants, 3158
Pezcoller Foundation-AACR International Award for Cancer Research, 3161
Pfizer/AAFP Foundation Visiting Professorship Program in Family Medicine, 3162
Pfizer Advancing Research in Transplantation Science (ARTS) Research Awards, 3163
Pfizer ASPIRE North America Broad Spectrum Antibiotics for the Treatment of Gram-Negative or Polymicrobial Infections Research Awards, 3169
Pfizer ASPIRE North America Hemophilia Research Awards, 3170
Pfizer ASPIRE North America Narrow Spectrum Antibiotics for the Treatment of MRSA Research Awards, 3171
Pfizer ASPIRE North America Rheumatology Research Awards, 3172
Pfizer ASPIRE Worldwide Endocrine Young Investigator Grants, 3173
PFizer Compound Transfer Agreements, 3177
Pfizer Global Investigator Research Grants, 3178
Pfizer Global Research Awards for Nicotine Independence, 3179
Pfizer Healthcare Charitable Contributions, 3180
Pfizer Medical Education Track One Grants, 3182
Pfizer Medical Education Track Two Grants, 3183
Pfizer Special Events Grants, 3186
Pharmacia-ASPET Award for Experimental Therapeutics, 3187
Phi Upsilon Omicron Alumni Research Grant, 3191
Phi Upsilon Omicron Florence Fallgatter Distinguished Service Award, 3192
Phi Upsilon Omicron Frances Morton Holbrook Alumni Award, 3193
Phi Upsilon Omicron Geraldine Clewell Senior Awards, 3194
Phi Upsilon Omicron Janice Cory Bullock Collegiate Award, 3195
Phi Upsilon Lillian P. Schoephoerster Award, 3196
Phi Upsilon Omicron Margaret Jerome Sampson Scholarships, 3197
Phi Upsilon Orinne Johnson Writing Award, 3198
Phi Upsilon Omicron Sarah Thorniley Phillips Leader Awards, 3199
Phi Upsilon Omicron Undergraduate Karen P. Goebel Conclave Award, 3200
PhRMA Foundation Health Outcomes Post Doctoral Fellowships, 3202
PhRMA Foundation Health Outcomes Pre-Doctoral Fellowships, 3203
PhRMA Foundation Health Outcomes Research Starter Grants, 3204
PhRMA Foundation Health Outcomes Sabbatical Fellowships, 3205
PhRMA Foundation Informatics Post Doctoral Fellowships, 3206
PhRMA Foundation Informatics Pre Doctoral Fellowships, 3207
PhRMA Foundation Informatics Research Starter Grants, 3208
PhRMA Fndn Info Sabbatical Fellowships, 3209
PhRMA Foundation Paul Calabresi Medical Student Research Fellowship, 3210
PhRMA Foundation Pharmaceutics Postdoctoral Fellowships, 3211
PhRMA Foundation Pharmaceutics Pre Doctoral Fellowships, 3212
PhRMA Foundation Pharmaceutics Research Starter Grants, 3213
PhRMA Foundation Pharmaceutics Sabbatical Fellowships, 3214
PhRMA Foundation Pharmacology/Toxicology Post Doctoral Fellowships, 3215
PhRMA Foundation Pharmacology/Toxicology Pre Doctoral Fellowships, 3216

PhRMA Foundation Pharmacology/Toxicology Research Starter Grants, 3217
PhRMA Foundation Pharmacology/Toxicology Sabbatical Fellowships, 3218
Pinkerton Foundation Grants, 3224
Pinnacle Foundation Grants, 3225
Pioneer Hi-Bred Society Fellowships, 3226
PKD Foundation Lillian Kaplan International Prize for the Advancement in the Understanding of Polycystic Kidney Disease, 3232
PKD Foundation Research Grants, 3233
Plastic Surgery Educational Foundation/AAO-HNSF Combined Grant, 3234
Plastic Surgery Educational Foundation Basic Research Grants, 3235
Plastic Surgery Educational Foundation Research Fellowship Grants, 3236
Playboy Foundation Grants, 3237
PMI Foundation Grants, 3239
Potts Memorial Foundation Grants, 3254
PPG Industries Foundation Grants, 3257
Presbyterian Patient Assistance Program, 3260
Pride Foundation Fellowships, 3267
Procter and Gamble Fund Grants, 3274
Prospect Burma Scholarships, 3275
PSG Mentored Clinical Research Awards, 3277
PVA Education Foundation Grants, 3284
PVA Research Foundation Grants, 3285
Quaker Chemical Foundation Grants, 3286
RACGP Cardiovascular Research Grants in General Practice, 3295
RACGP Nat Asthma Council Research Award, 3296
RACGP Vicki Kotsirilos Integrative Med Grants, 3297
Radcliffe Institute Carol K. Pforzheimer Student Fellowships, 3298
Radcliffe Institute Residential Fellowships, 3299
Rainbow Endowment Grants, 3300
Rajiv Gandhi Foundation Grants, 3302
Ralph F. Hirschmann Award in Peptide Chem, 3303
Raskob Foundation for Catholic Activities Grants, 3306
RCPSC/AMS CanMEDs Research and Development Grants, 3316
RCPSC McLaughlin-Gallie Visiting Professor, 3333
RCPSC Medical Education Research Grants, 3334
Reader's Digest Partners for Sight Fndn Grants, 3345
Regenstrief General Internal Medicine - General Pediatrics Research Fellowship Program, 3351
Regenstrief Geriatrics Fellowship, 3352
Regenstrief Master Of Science In Clinical Research/CITE Program, 3353
Rehabilitation Nursing Fndn Research Grants, 3354
Rehabilitation Nursing Foundation Scholarships, 3355
Religion and Health: Effects, Mechanisms, and Interpretation RFP, 3358
Retirement Research Foundation General Program Grants, 3359
RGk Foundation Grants, 3361
Richard D. Bass Foundation Grants, 3366
Richard E. Griffin Family Foundation Grants, 3368
Ringing Rocks Foundation Discretionary Grants, 3375
Robert Sterling Clark Foundation Reproductive Rights and Health Grants, 3388
Rockefeller Archive Center Research Grants, 3393
Rockefeller Brothers Peace & Security Grants, 3394
Rockefeller Brothers Fund Pivotal Places Grants: South Africa, 3395
Rockefeller Foundation Grants, 3396
Rockwell Collins Charitable Corporation Grants, 3397
Rockwell International Corporate Trust Grants Program, 3399
Roeher Institute Research Grants, 3400
Rohm and Haas Company Grants, 3403
Ronald McDonald House Charities Grants, 3405
Rosalinde and Arthur Gilbert Foundation/AFAR New Investigator Awards in Alzheimer's Disease, 3406
Rosalynn Carter Institute John and Betty Pope Fellowships, 3408
Rosalynn Carter Institute Mattie J. T. Stepanek Caregiving Scholarships, 3409
Royal Norwegian Embassy Kavli Prizes, 3414

RSC Abbyann D. Lynch Medal in Bioethics, 3419
RSC Jason A. Hannah Medal, 3420
RSC McLaughlin Medal, 3421
RSC McNeil Medal for the Public Awareness of Science, 3422
RSC Thomas W. Eadie Medal, 3423
RWJF Changes in Health Care Financing and Organization Grants, 3430
RWJF Childhood Obesity Grants, 3431
RWJF Community Health Leaders Awards, 3432
RWJF Disparities Research for Change Grants, 3433
RWJF Harold Amos Medical Faculty Development Research Grants, 3434
RWJF Harold Amos Medical Faculty Development Scholars, 3435
RWJF Health and Society Scholars, 3436
RWJF Healthy Eating Research Grants, 3437
RWJF Interdisciplinary Nursing Quality Research Initiative Grants, 3438
RWJF New Careers in Nursing Scholarships, 3439
RWJF Nurse Faculty Scholars, 3441
RWJF Pioneer Portfolio Grants, 3442
RWJF Policies for Action: Policy and Law Research to Build a Culture of Health Grants, 3443
RWJF Vulnerable Populations Portfolio Grants, 3444
Ryan Gibson Foundation Grants, 3445
SAMHSA Campus Suicide Prevention Grants, 3460
SAMHSA Child Mental Health Initiative (CMHI) Grants, 3461
SAMHSA Conference Grants, 3462
SAMHSA Drug Free Communities Support Program Grants, 3463
SAMHSA Strategic Prevention Framework State Incentive Grants, 3464
Samueli Foundation Health Grants, 3465
Samueli Institute Scientific Research Grants, 3466
San Diego Foundation for Change Grants, 3470
Sandler Program for Asthma Research Grants, 3475
Savoy Foundation Post-Doctoral and Clinical Research Fellowships, 3488
Savoy Foundation Research Grants, 3489
Savoy Foundation Studentships, 3490
Schering-Plough Foundation Community Initiatives Grants, 3491
Schering-Plough Foundation Health Grants, 3492
Scleroderma Foundation Established Investigator Grants, 3497
Scleroderma Foundation New Investigator Grants, 3498
Scott B. & Annie P. Appleby Charitable Grants, 3499
Searle Scholars Program Grants, 3503
Seaver Institute Grants, 3509
Seneca Foods Foundation Grants, 3511
Sexually Transmitted Diseases Postdoctoral Fellowships, 3516
SfN Award for Education in Neuroscience, 3517
SfN Bernice Grafstein Award for Outstanding Accomplishments in Mentoring, 3518
SfN Chapter-of-the-Year Award, 3519
SfN Chapter Grants, 3520
SfN Donald B. Lindsley Prize in Behavioral Neuroscience, 3521
SfN Federation of European Neuroscience Societies Forum Travel Awards, 3522
SfN Jacob P. Waletzky Award, 3523
SfN Janett Rosenberg Trubatch Career Development Award, 3524
SfN Japan Neuroscience Society Meeting Travel Awards, 3525
SfN Julius Axelrod Prize, 3526
SfN Louise Hanson Marshall Specific Recognition Award, 3527
SfN Mika Salpeter Lifetime Achievement Award, 3528
SfN Nemko Prize in Cellular or Molecular Neuroscience, 3529
SfN Neuroscience Fellowships, 3530
SfN Next Generation Award, 3531
SfN Patricia Goldman-Rakic Hall of Honor, 3532
SfN Peter and Patricia Gruber International Research Award, 3533
SfN Ralph W. Gerard Prize in Neuroscience, 3534

SfN Science Educator Award, 3535
SfN Science Journalism Student Awards, 3536
SfN Swartz Prize for Theoretical and Computational Neuroscience, 3537
SfN Trainee Professional Development Awards, 3538
SfN Travel Awards for the International Brain Research Organization World Congress, 3539
SfN Undergrad Brain Awareness Travel Award, 3540
SfN Young Investigator Award, 3541
Shell Oil Company Foundation Community Development Grants, 3544
Sidgmore Family Foundation Grants, 3551
Sigma Theta Tau International /American Nurses Foundation Grant, 3556
Sigma Theta Tau International / Council for the Advancement of Nursing Science Grants, 3557
Sigma Theta Tau International / Oncology Nursing Society Grant, 3558
Sigma Theta Tau International Doris Bloch Research Award, 3559
Sigma Theta Tau Small Grants, 3560
Sigrid Juselius Foundation Grants, 3561
Simone and Cino del Duca Grand Prix Awards, 3564
Singing for Change Foundation Grants, 3567
Sir Dorabji Tata Trust Grants for NGOs or Voluntary Organizations, 3572
Sir Dorabji Tata Trust Individual Medical Grants, 3573
Skin Cancer Foundation Research Grants, 3578
Skoll Fndn Awards for Social Entrepreneurship, 3579
Smithsonian Biodiversity Genomics and Bioinformatics Postdoctoral Fellowships, 3580
Smithsonian Museum Conservation Institute Research Post-Doctorate/Post-Graduate Fellowships, 3581
SNM/ Covidien Seed Grant in Molecular Imaging/Nuclear Medicine Research, 3582
SNM Molecular Imaging Research Grant For Junior Medical Faculty, 3583
SNM PDEF Mickey Williams Minority Student Scholarships, 3584
SNM Pilot Research Grants, 3585
SNM Pilot Research Grants in Nuclear Medicine/Molecular Imaging, 3586
SNM Postdoc Molecular Imaging Scholar Grants, 3587
SNM Student Fellowship Awards, 3588
SNM Tetalman Young Investigator Awards, 3589
SNMTS Clinical Advancement Scholarships, 3590
SNMTS Outstanding Educator Awards, 3591
SNMTS Outstanding Technologist Awards, 3592
SNMTS Paul Cole Scholarships, 3593
SOBP A.E. Bennett Research Award, 3594
SOBP Ziskind-Somerfeld Research Awards, 3595
Society for Imaging Informatics in Medicine (SIIM) Emeritus Mentor Grants, 3597
Society for Imaging Informatics in Medicine (SIIM) Micro Grants, 3598
Society for Imaging Informatics in Medicine (SIIM) Research Grants, 3599
Society for Imaging Informatics in Med Small Grant for Product Support Development, 3600
Society for Imaging Informatics in Medicine (SIIM) Small Training Grants, 3601
Society for the Arts in Healthcare Grants, 3602
Society for the Arts in Healthcare Environmental Arts Research Grant, 3603
Society of Cosmetic Chemists Allan B. Black Award Sponsored by Presperse, 3604
Society of Cosmetic Chemists Award Sponsored by McIntyre Group, Ltd., 3605
Society of Cosmetic Chemists Award Sponsored by The HallStar Company, 3606
Society of Cosmetic Chemists Chapter Best Speaker Award, 3607
Society of Cosmetic Chem Chapter Merit Award, 3608
Society of Cosmetic Chemists Frontier of Science Award, 3609
Society of Cosmetic Chem Hans Schaeffer Award, 3610
Society of Cosmetic Chem Jos Ciaudelli Award, 3611
Society of Cosmetic Chemists Keynote Award Lecture Sponsored by Ruger Chemical Corporation, 3612
Society of Cosmetic Chemists Literature Award, 3613

GEOGRAPHIC INDEX

Society of Cosmetic Chemists Maison G. de Navarre Medal, 3614
Society of Cosmetic Chemists Merit Award, 3615
Society of Cosmetic Chemists Robert A. Kramer Lifetime Service Award, 3616
Society of Cosmetic Chem Shaw Mudge Award, 3617
Society of Cosmetic Chem Stud Poster Awards, 3618
Society of Cosmetic Chem You Scientist Awards, 3619
Sony Corporation of America Grants, 3623
Special Olympics Health Profess Student Grants, 3629
Special People in Need Grants, 3630
Special People in Need Scholarships, 3631
SRC Medical Research Grants, 3635
SSHRC Banting Postdoctoral Fellowships, 3637
Strake Foundation Grants, 3658
Stranahan Foundation Grants, 3659
Streilein Foundation for Ocular Immunology Visiting Professorships, 3660
Stroke Association Allied Health Professional Research Bursaries, 3661
Stroke Association Clinical Fellowships, 3662
Stroke Association Research Project Grants, 3663
STTI/ATI Educational Assessment Nursing Research Grants, 3665
STTI Emergency Nurses Association Foundation Grant, 3666
STTI Environment of Elder Care Nursing Research Grants, 3667
STTI Joan K. Stout RN Research Grants, 3668
STTI National League for Nursing Grants, 3669
STTI Rosemary Berkel Crisp Research Awards, 3670
STTI Virginia Henderson Research Grants, 3671
Summer Medical and Dental Education Program: Interprofessional Pilot Grants, 3673
SunTrust Bank Trusteed Foundations Nell Warren Elkin and William Simpson Elkin Grants, 3683
Support Our Aging Religious (SOAR) Grants, 3686
Susan G. Komen Breast Cancer Foundation Brinker Awards for Scientific Distinction, 3687
Susan G. Komen Breast Cancer Foundation Career Catalyst Research Grants, 3688
Susan G. Komen Breast Cancer Foundation Challenge Grants: Breast Cancer and the Environment, 3689
Susan G. Komen Breast Cancer Foundation Challenge Grants: Investigator Initiated Research, 3690
Susan G. Komen Breast Cancer Foundation Challenge Grants: Career Catalyst Research, 3691
Susan G. Komen Breast Cancer Foundation College Scholarships, 3692
Susan G. Komen Breast Cancer Foundation Investigator Initiated Research Grants, 3693
Sybil G. Jacobs Award for Outstanding Use of Tobacco Industry Documents, 3698
Tellabs Foundation Grants, 3706
The Ray Charles Foundation Grants, 3716
Thomson Reuters / MLA Doctoral Fellowships, 3728
Thorman Boyle Foundation Grants, 3730
Thrasher Research Fund Early Career Awards, 3731
Thrasher Research Fund Grants, 3732
Tifa Foundation Grants, 3733
TJX Foundation Grants, 3735
Tommy Hilfiger Corporate Foundation Grants, 3737
Tourette Syndrome Association Post-Doctoral Fellowships, 3740
Tourette Syndrome Association Research Grants, 3741
Toyota Motor Manuf of Mississippi Grants, 3746
Toyota Motor N America of New York Grants, 3749
Toys R Us Children's Fund Grants, 3751
Trauma Center of America Finance Fellowship, 3752
TRDRP Exploratory/Devel Research Awards, 3754
Trinity Lutheran School of Nursing Alumnae Scholarships, 3759
Triological Society Research Career Development Awards, 3760
Trull Foundation Grants, 3761
Tylenol Future Care Scholarships, 3764
U.S. Department of Education 21st Century Community Learning Centers, 3767
U.S. Department of Education Erma Byrd Scholarships, 3768
U.S. Department of Education Rehabilitation Engineering Research Centers Grants, 3769
U.S. Department of Education Rehabilitation Research Training Centers (RRTCs), 3770
U.S. Dept of Education Special Ed--National Activities--Parent Information Centers, 3772
U.S. Department of Education United States-Russia Program: Improving Research and Educational Activities in Higher Education, 3773
U.S. Department of Education Vocational Rehabilitation Services Projects for American Indians with Disabilities Grants, 3774
U.S. Lacrosse EAD Grants, 3775
UCLA / RAND Corporation Post-Doctoral Fellowships, 3776
UCT Scholarship Program, 3777
UICC American Cancer Society International Fellowships for Beginning Investigators, 3778
UICC International Cancer Technology Transfer (ICRETT) Fellowships, 3779
UICC Raisa Gorbachev Memorial International Cancer Fellowships, 3780
UICC Yamagiwa-Yoshida Memorial International Cancer Study Grants, 3781
UnitedHealthcare Children's Foundation Grants, 3795
United Healthcare Comm Grants in Michigan, 3796
United Methodist Committee on Relief Global AIDS Fund Grants, 3798
United Methodist Committee on Relief Global Health Grants, 3799
United States Institute of Peace - Jennings Randolph Senior Fellowships, 3802
United Technologies Corporation Grants, 3803
USAID Child, Newborn, and Maternal Health Project Grants, 3806
USAID Devel Innovation Accelerator Grants, 3808
USAID Ebola Response, Recovery and Resilience in West Africa Grants, 3809
USAID Family Planning and Reproductive Health Methods Grants, 3811
USAID Fighting Ebola Grants, 3812
USAID Global Health Development Innovation Accelerator Broad Agency Grants, 3813
USAID Health System Strengthening Grants, 3814
USAID Higher Education Partnerships for Innovation and Impact (HEPII) Grants, 3815
USAID Land Use Change and Disease Emergence Grants, 3818
USAID Microbicides Research, Development, and Introduction Grants, 3819
USAID Policy, Advocacy, and Communication Enhanced for Population and Reproductive Health Grants, 3820
USAID Research and Innovation for Health Supply Chain Systems & Commodity Security Grants, 3821
USAID Social Science Research in Population and Reproductive Health Grants, 3822
USAID Support for International Family Planning Organizations Grants, 3823
USCM HIV/AIDS Prevention Grants, 3827
US CRDF Leishmaniasis: Collaborative Research Opportunities in N Africa & Middle East, 3828
USDA Organic Agriculture Research Grants, 3830
USDD Breast Cancer Clinical Translational Research Grants, 3831
USDD Breast Cancer Era of Hope Grants, 3832
USDD Breast Cancer Idea Grants, 3833
USDD Breast Cancer Impact Grants, 3834
USDD Breast Cancer Innovation Grants, 3835
USDD Breast Cancer Transformative Grant, 3836
USDD Broad Agency Announcement Grants, 3837
USDD Broad Agency Announcement HIV/AIDS Prevention Grants, 3838
USDD Care for the Critically Injured Burn Patient Grants, 3839
USDD Clinical Functional Assessment Investigator-Initiated Grants, 3840
USDD HIV/AIDS Prevention: Military Specific HIV/AIDS Prevention, Care, and Treatment for Non-PEPFAR Funded Countries Grants, 3841
USDD HIV/AIDS Prevention Program Information Systems Development Grants, 3842
USDD Medical Practice Breadth of Medical Practice and Disease Frequency Exposure Grants, 3843
USDD Medical Practice Initiative Procedural Skill Decay and Maintenance Grants, 3844
USDD President's Emergency Plan for AIDS Relief Military Specific HIV Seroprevalence and Behavioral Epidemiology Risk Suvey Grants, 3845
USDD U.S. Army Medical Research and Materiel Command Broad Agency Announcement for Extramural Medical Research Grants, 3846
USDD Workplace Violence in Military Grants, 3847
USG Foundation Grants, 3848
Vancouver Foundation Grants and Community Initiatives Program, 3849
Vancouver Fndn Public Health Bursary Package, 3850
Vancouver Sun Children's Fund Grants, 3851
Verizon Foundation Health Care and Accessibility Grants, 3858
Vernon K. Krieble Foundation Grants, 3866
VHA Health Foundation Grants, 3867
Victor Grifols i Lucas Foundation Prize for Journalistic Work on Bioethics, 3870
Victor Grifols I Lucas Fndn Research Grants, 3871
Virginia L. and William K. Beatty MLA Volunteer Service Award, 3877
Volkswagen of America Corporate Contributions, 3880
W.C. Griffith Foundation Grants, 3881
W. Clarke Swanson, Jr. Foundation Grants, 3883
W.K. Kellogg Foundation Healthy Kids Grants, 3886
W.K. Kellogg Foundation Secure Families Grants, 3887
W.M. Keck Foundation Med Research Grants, 3888
W.M. Keck Foundation Science and Engineering Research Grants, 3889
Walmart Foundation Community Giving Grants, 3897
Walmart Foundation National Giving Grants, 3898
Walmart Foundation State Giving Grants, 3900
Welborn Baptist Fndn General Op Grants, 3914
Wellcome Trust Biomedical Science Grants, 3918
WestWind Foundation Reproductive Health and Rights Grants, 3924
Weyerhaeuser Family Foundation Health Grants, 3926
WHO Foundation General Grants, 3939
WHO Foundation Volunteer Service Grants, 3940
William A. Badger Foundation Grants, 3945
William B. Dietrich Foundation Grants, 3946
William G. & Helen C. Hoffman Fndn Grants, 3950
William H. Adams Foundation for ALS Grants, 3952
William T. Grant Foundation Research Grants, 3958
Willis C. Helm Charitable Trust Grants, 3959
Winn Feline Foundation/AVMF Excellence In Feline Research Award, 3963
Winn Feline Foundation Grants, 3964
Women of the ELCA Opportunity Scholarships for Lutheran Laywomen, 3970
World of Children Health Award, 3974
World of Children Humanitarian Award, 3975
Xoran Technologies Resident Research Grant, 3976
Yampa Valley Community Foundation Cody St. John Scholarships, 3977
Zellweger Baby Support Network Grants, 3984

Alabama

3M Company Fndn Community Giving Grants, 6
3M Fndn Health & Human Services Grants, 7
Adams and Reese Corporate Giving Grants, 299
Alabama Power Foundation Grants, 400
Appalachian Regional Commission Health Grants, 637
Autauga Area Community Foundation Grants, 808
BBVA Compass Foundation Charitable Grants, 848
Belk Foundation Grants, 872
Boeing Company Contributions Grants, 930
Bridgestone/Firestone Trust Fund Grants, 949
Caring Foundation Grants, 1019
Carrier Corporation Contributions Grants, 1038
Charles A. Frueauff Foundation Grants, 1114
Charles M. & Mary D. Grant Fndn Grants, 1124
Claude Bennett Family Foundation Grants, 1176
Cleveland-Cliffs Foundation Grants, 1181

Alabama

Community Foundation of Greater Birmingham Grants, 1272
Community Fndn of So Alabama Grants, 1306
Cooper Industries Foundation Grants, 1345
CSX Corporate Contributions Grants, 1368
D. W. McMillan Foundation Grants, 1407
Dean Foods Community Involvement Grants, 1440
El Paso Corporate Foundation Grants, 1580
GenCorp Foundation Grants, 1779
Harley Davidson Foundation Grants, 1947
Harry B. and Jane H. Brock Foundation Grants, 1960
Hill Crest Foundation Grants, 2030
J.L. Bedsole Foundation Grants, 2172
Johnson & Johnson Community Health Grants, 2252
MeadWestvaco Foundation Sustainable Communities Grants, 2506
Mercedes-Benz USA Corporate Contributions, 2517
Norfolk Southern Foundation Grants, 2982
PNC Foundation Community Services Grants, 3240
Robert B. Adams Foundation Scholarships, 3380
Robert R. Meyer Foundation Grants, 3387
Salmon Foundation Grants, 3458
Sunoco Foundation Grants, 3679
Susan Mott Webb Charitable Trust Grants, 3694
Toyota Motor Manufacturing of Alabama Grants, 3743
USDA Delta Health Care Services Grants, 3829
Walker Area Community Foundation Grants, 3895
Weyerhaeuser Company Foundation Grants, 3925
Williams Companies Foundation Homegrown Giving Grants, 3957
Woodward Fund Grants, 3973

Alaska

3M Company Fndn Community Giving Grants, 6
3M Fndn Health & Human Services Grants, 7
Alaska Airlines Corporate Giving Medical Emergency and Research Grants, 402
Alaska Airlines Foundation Grants, 403
AMA Fndn Arthur N. Wilson Scholarship, 482
ARCO Foundation Education Grants, 652
Avista Foundation Economic and Cultural Vitality Grants, 812
Bullitt Foundation Grants, 973
Chilkat Valley Community Foundation Grants, 1148
ConocoPhillips Foundation Grants, 1327
Fluor Foundation Grants, 1682
Foster Foundation Grants, 1689
Golden Heart Community Foundation Grants, 1850
Greater Sitka Legacy Fund Grants, 1885
Homer Foundation Grants, 2044
Jessica Stevens Community Foundation Grants, 2224
Ketchikan Community Foundation Grants, 2303
KeyBank Foundation Grants, 2307
Kodiak Community Foundation Grants, 2314
M.J. Murdock Charitable Trust General Grants, 2405
Petersburg Community Foundation Grants, 3156
Pride Foundation Grants, 3268
Pride Foundation Scholarships, 3269
Rasmuson Foundation Tier One Grants, 3307
Rasmuson Foundation Tier Two Grants, 3308
Robert B McMillen Foundation Grants, 3381
Seward Community Foundation Grants, 3514
Seward Community Foundation Mini-Grants, 3515
W.M. Keck Fndn Undergrad Ed Grants, 3891

American Samoa

ACF Native American Social and Economic Development Strategies Grants, 231
Bank of America Fndn Matching Gifts, 825
Bank of America Fndn Volunteer Grants, 826
NHLBI Ancillary Studies in Clinical Trials, 2688
NHLBI Bioengineering and Obesity Grants, 2689
NHLBI Career Transition Awards, 2694
NHLBI Mentored Career Dev Award to Promote Faculty Diversity in Biomed Research, 2708
NHLBI Mentored Patient-Oriented Research Career Development Awards, 2710
NHLBI Mentored Quantitative Research Career Development Awards, 2711
NHLBI Pathway to Independence Awards, 2718
NHLBI Ruth L. Kirschstein National Research Service Awards for Individual Senior Fellows, 2732
NIGMS Advancing Basic Behavioral and Social Research on Resilience: an Integrative Science Approach, 2861
NIGMS Enabling Resources for Pharmacogenomics Grants, 2862
NIH Mentored Clinical Scientist Research Career Development Award, 2887
NIH Ruth L. Kirschstein National Research Service Award Institutional Research Training Grants, 2900
NIH Ruth L. Kirschstein National Research Service Awards for Individual Postdoc Fellowships, 2902
NIH Ruth L. Kirschstein National Research Service Awards for Individual Predoctoral Fellows, 2903
NIH Ruth L. Kirschstein Nat Research Service Award Short-Term Institutional Training Grants, 2906
NIH Support of Competitive Research (SCORE) Pilot Project Awards, 2913
NIH Support of Competitive Research (SCORE) Research Advancement Awards, 2914
Wellcome Trust New Investigator Awards, 3919

Arizona

Abbott Fund Access to Health Care Grants, 199
Abbott Fund Community Grants, 201
Abbott Fund Global AIDS Care Grants, 202
Abbott Fund Science Education Grants, 203
ACE Charitable Foundation Grants, 229
Aetna Foundation Regional Health Grants, 333
Albertson's Charitable Giving Grants, 410
Amerigroup Foundation Grants, 565
Amica Insurance Company Community Grants, 575
APS Foundation Grants, 646
ARCO Foundation Education Grants, 652
Arizona Cardinals Grants, 655
Arizona Community Foundation Scholarships, 656
Arizona Diamondbacks Charities Grants, 657
Arizona Public Service Corporate Giving Program Grants, 658
BBVA Compass Foundation Charitable Grants, 848
Bechtel Group Foundation Building Positive Community Relationships Grants, 866
Bert W. Martin Foundation Grants, 886
BHHS Legacy Foundation Grants, 888
Boeing Company Contributions Grants, 930
Caesars Foundation Grants, 995
Carrier Corporation Contributions Grants, 1038
Chase Paymentech Corporate Giving Grants, 1127
Comerica Charitable Foundation Grants, 1224
ConAgra Foods Fndn Community Impact Grants, 1321
ConAgra Foods Nourish our Community Grants, 1322
Del E. Webb Foundation Grants, 1450
Dorrance Family Foundation Grants, 1501
E.L. Wiegand Foundation Grants, 1538
Flinn Foundation Scholarships, 1675
FMC Foundation Grants, 1683
General Mills Foundation Grants, 1782
Humana Foundation Grants, 2077
Kevin P. and Sydney B. Knight Family Foundation Grants, 2306
Margaret T. Morris Foundation Grants, 2429
Medtronic Foundation CommunityLink Health Grants, 2510
Medtronic Foundation Community Link Human Services Grants, 2511
Nationwide Insurance Foundation Grants, 2637
Nina Mason Pulliam Charitable Trust Grants, 2921
PacifiCare Health Systems Grants, 3102
Phoenix Suns Charities Grants, 3201
Piper Jaffray Fndn Communities Giving Grants, 3227
Piper Trust Healthcare & Med Research Grants, 3228
Piper Trust Older Adults Grants, 3229
Prudential Foundation Education Grants, 3276
Saint Luke's Health Initiatives Grants, 3455
Salt River Health & Human Services Grants, 3459
Southwest Gas Corporation Foundation Grants, 3628
Stocker Foundation Grants, 3656
Union Pacific Foundation Health and Human Services Grants, 3794
US Airways Community Contributions, 3825
US Airways Community Foundation Grants, 3826
W.M. Keck Fndn Undergrad Ed Grants, 3891
Weyerhaeuser Company Foundation Grants, 3925

Arkansas

3M Company Fndn Community Giving Grants, 6
3M Fndn Health & Human Services Grants, 7
ACE Charitable Foundation Grants, 229
AEGON Transamerica Foundation Health and Welfare Grants, 316
Albertson's Charitable Giving Grants, 410
American Electric Power Foundation Grants, 530
Archer Daniels Midland Foundation Grants, 651
Arkansas Community Foundation Arkansas Black Hall of Fame Grants, 659
Arkansas Community Foundation Grants, 660
Assisi Fndn of Memphis Capital Project Grants, 786
Assisi Foundation of Memphis General Grants, 787
BancorpSouth Foundation Grants, 823
Belk Foundation Grants, 872
Bridgestone/Firestone Trust Fund Grants, 949
Carl B. and Florence E. King Foundation Grants, 1022
Charles A. Frueauff Foundation Grants, 1114
ConAgra Foods Fndn Community Impact Grants, 1321
ConAgra Foods Nourish our Community Grants, 1322
Entergy Corporation Micro Grants, 1596
Entergy Corporation Open Grants for Healthy Families, 1597
Foundation for the Mid South Community Development Grants, 1699
Foundation for the Mid South Health and Wellness Grants, 1700
GenCorp Foundation Grants, 1779
General Mills Foundation Grants, 1782
J.E. and L.E. Mabee Foundation Grants, 2170
Johnson & Johnson Community Health Grants, 2252
MeadWestvaco Foundation Sustainable Communities Grants, 2506
Piper Jaffray Fndn Communities Giving Grants, 3227
Roy & Christine Sturgis Charitable Grants, 3415
Sunderland Foundation Grants, 3675
T.L.L. Temple Foundation Grants, 3700
Union Pacific Foundation Health and Human Services Grants, 3794
USDA Delta Health Care Services Grants, 3829
Walmart Foundation Northwest Arkansas Giving, 3899
Weyerhaeuser Company Foundation Grants, 3925
Willard & Pat Walker Charitable Fndn Grants, 3943

California

3M Company Fndn Community Giving Grants, 6
3M Fndn Health & Human Services Grants, 7
49ers Foundation Grants, 9
A. Gary Anderson Family Foundation Grants, 16
A/H Foundation Grants, 18
Aaron Foundation Grants, 194
Abbott Fund Access to Health Care Grants, 199
Abbott Fund Community Grants, 201
Abbott Fund Global AIDS Care Grants, 202
Abbott Fund Science Education Grants, 203
ACE Charitable Foundation Grants, 229
Adam Richter Charitable Trust Grants, 298
Adobe Community Investment Grants, 310
Advanced Micro Devices Comm Affairs Grants, 313
AEGON Transamerica Foundation Health and Welfare Grants, 316
Aetna Foundation Regional Health Grants, 333
Ahmanson Foundation Grants, 355
Albertson's Charitable Giving Grants, 410
Alice Tweed Tuohy Foundation Grants Program, 437
Alliance Healthcare Foundation Grants, 444
Alpine Winter Foundation Grants, 453
Amador Community Foundation Grants, 481
Amerigroup Foundation Grants, 565
Amica Insurance Company Community Grants, 575
Angels Baseball Foundation Grants, 590
Anheuser-Busch Foundation Grants, 591
Ann Jackson Family Foundation Grants, 598
Ann Peppers Foundation Grants, 599

GEOGRAPHIC INDEX

California

Annunziata Sanguinetti Foundation Grants, 600
Aratani Foundation Grants, 649
Archer Daniels Midland Foundation Grants, 651
ARCO Foundation Education Grants, 652
Argyros Foundation Grants Program, 653
Arthur F. & Arnold M. Frankel Fndn Grants, 672
Audrey & Sydney Irmas Foundation Grants, 799
Avista Foundation Economic and Cultural Vitality Grants, 812
Bayer Foundation Grants, 846
BBVA Compass Foundation Charitable Grants, 848
Bechtel Group Foundation Building Positive Community Relationships Grants, 866
Bella Vista Foundation Grants, 873
Belvedere Community Foundation Grants, 874
Ben B. Cheney Foundation Grants, 875
Bertha Russ Lytel Foundation Grants, 885
Bill Hannon Foundation Grants, 895
Blue Shield of California Grants, 923
Bob & Delores Hope Foundation Grants, 927
Boeing Company Contributions Grants, 930
Bright Family Foundation Grants, 951
Caesars Foundation Grants, 995
California Community Foundation Health Care Grants, 998
California Community Foundation Human Development Grants, 999
California Endowment Innovative Ideas Challenge Grants, 1000
California Wellness Foundation Work and Health Program Grants, 1001
Callaway Golf Company Foundation Grants, 1003
Carl Gellert and Celia Berta Gellert Foundation Grants, 1024
Carlsbad Charitable Foundation Grants, 1029
Chapman Charitable Foundation Grants, 1112
Chiles Foundation Grants, 1147
Chula Vista Charitable Foundation Grants, 1154
Clarence E. Heller Charitable Foundation Grants, 1166
Coeta and Donald Barker Foundation Grants, 1205
Comerica Charitable Foundation Grants, 1224
Community Fndn for Monterey County Grants, 1253
Community Fndn for San Benito County Grants, 1257
Community Foundation of Riverside & San Bernardino County Impact Grants, 1303
Community Fndn of Riverside & San Bernardino County Irene S. Rockwell Grants, 1304
Community Foundation of Riverside and San Bernardino County James Bernard and Mildred Jordan Tucker Grants, 1305
Community Foundation of the Verdugos Educational Endowment Fund Grants, 1314
Community Foundation of the Verdugos Grants, 1315
Community Foundation Serving Riverside and San Bernardino Counties Impact Grants, 1319
ConAgra Foods Fndn Community Impact Grants, 1321
ConAgra Foods Nourish our Community Grants, 1322
ConocoPhillips Foundation Grants, 1327
Crail-Johnson Foundation Grants, 1355
Cudd Foundation Grants, 1370
Cystic Fibrosis Research Eliz Nash Fellowships, 1401
David Geffen Foundation Grants, 1431
David M. and Marjorie D. Rosenberg Foundation Grants, 1432
Dean Foods Community Involvement Grants, 1440
Del E. Webb Foundation Grants, 1450
Dorrance Family Foundation Grants, 1501
Dunspaugh-Dalton Foundation Grants, 1530
E.L. Wiegand Foundation Grants, 1538
eBay Foundation Community Grants, 1543
EDS Foundation Grants, 1552
Eisner Foundation Grants, 1564
Expect Miracles Foundation Grants, 1619
Fluor Foundation Grants, 1682
FMC Foundation Grants, 1683
Ford Family Foundation Grants - Access to Health and Dental Services, 1686
Fritz B. Burns Foundation Grants, 1747
Gamble Foundation Grants, 1769
GenCorp Foundation Grants, 1779

Genentech Corp Charitable Contributions, 1780
General Mills Foundation Grants, 1782
George and Ruth Bradford Foundation Grants, 1789
Gil and Dody Weaver Foundation Grants, 1819
Golden State Warriors Foundation Grants, 1851
Green Diamond Charitable Contributions, 1891
Grifols Community Outreach Grants, 1904
Grover Hermann Foundation Grants, 1906
H. Leslie Hoffman and Elaine S. Hoffman Foundation Grants, 1924
HAF Barry F. Phelps Fund Grants, 1927
HAF David (Davey) H. Somerville Medical Travel Fund Grants, 1928
HAF JoAllen K. Twiddy-Wood Memorial Fund Grants, 1929
HAF Phyllis Nilsen Leal Memorial Fund Gifts, 1930
HAF Riley Frazel Memorial Scholarship, 1931
HAF Senior Opportunities Grants, 1932
HAF Technical Assistance Program Grants, 1933
Harden Foundation Grants, 1945
Hasbro Children's Fund Grants, 1973
Hasbro Corporation Gift of Play Hospital and Pediatric Health Giving, 1974
Healthcare Fndn for Orange County Grants, 1981
Hedco Foundation Grants, 1989
Henry L. Guenther Foundation Grants, 2006
Hilda and Preston Davis Foundation Grants, 2028
Hilton Hotels Corporate Giving Program Grants, 2037
Horizons Community Issues Grants, 2049
Hormel Foods Charitable Trust Grants, 2050
Huffy Foundation Grants, 2073
Huntington Beach Police Officers Fndn Grants, 2081
Hut Foundation Grants, 2086
Hutton Foundation Grants, 2087
IIE Western Union Family Scholarships, 2143
Ireland Family Foundation Grants, 2162
Isabel Allende Foundation Esperanza Grants, 2165
J.H. Robbins Foundation Grants, 2171
J.M. Long Foundation Grants, 2173
Jacobs Family Village Neighborhoods Grants, 2184
James S. Copley Foundation Grants, 2200
John Jewett and Helen Chandler Garland Foundation Grants, 2242
Johnson & Johnson Community Health Grants, 2252
John Stauffer Charitable Trust Grants, 2258
Joseph Drown Foundation Grants, 2268
Josephine S. Gumbiner Foundation Grants, 2273
Katherine Baxter Memorial Foundation Grants, 2292
Kathryne Beynon Foundation Grants, 2294
Kenneth T. & Eileen L. Norris Fndn Grants, 2300
Kimball International-Habig Foundation Health and Human Services Grants, 2308
Legler Benbough Foundation Grants, 2352
Lockheed Martin Corp Foundation Grants, 2374
Lowell Berry Foundation Grants, 2382
Lucile Horton Howe and Mitchell B. Howe Foundation Grants, 2385
Lucile Packard Foundation for Children's Health Grants, 2386
Manuel D. & Rhoda Mayerson Fndn Grants, 2420
Marin Community Foundation Improving Community Health Grants, 2433
Maurice J. Masserini Charitable Trust Grants, 2456
McCarthy Family Foundation Grants, 2490
McConnell Foundation Grants, 2492
McLean Foundation Grants, 2498
MeadWestvaco Foundation Sustainable Communities Grants, 2506
Medtronic Foundation CommunityLink Health Grants, 2510
Medtronic Foundation Community Link Human Services Grants, 2511
Mercedes-Benz USA Corporate Contributions, 2517
Mericos Foundation Grants, 2518
Meta and George Rosenberg Foundation Grants, 2525
Mimi and Peter Haas Fund Grants, 2557
Nationwide Insurance Foundation Grants, 2637
Nick Traina Foundation Grants, 2792
Nordson Corporation Foundation Grants, 2981
Oceanside Charitable Foundation Grants, 3063

Orange County Community Foundation Grants, 3079
PacifiCare Foundation Grants, 3102
Packard Foundation Local Grants, 3105
Pajaro Valley Community Health Health Trust Insurance/Coverage & Education on Using the System Grants, 3108
Pajaro Valley Community Health Trust Diabetes and Contributing Factors Grants, 3109
Pajaro Valley Community Health Trust Oral Health: Prevention & Access Grants, 3110
Pajaro Valley Community Health Trust Promoting Entry & Advancement in the Health Professions Grants, 3111
Patron Saints Foundation Grants, 3121
Peter F. McManus Charitable Trust Grants, 3153
Peter Kiewit Foundation General Grants, 3154
Peter Kiewit Foundation Small Grants, 3155
Piper Jaffray Fndn Communities Giving Grants, 3227
Price Family Charitable Fund Grants, 3262
Prudential Foundation Education Grants, 3276
Pulido Walker Foundation, 3282
Qualcomm Grants, 3287
Quantum Corporation Snap Server Grants, 3289
Ralph M. Parsons Foundation Grants, 3304
Ralphs Food 4 Less Foundation Grants, 3305
Rathmann Family Foundation Grants, 3309
Robert R. McCormick Trib Community Grants, 3385
Rose Hills Foundation Grants, 3413
S.D. Bechtel, Jr. Fndn/Stephen Bechtel Preventive Healthcare and Selected Research Grants, 3446
S.H. Cowell Foundation Grants, 3447
S. Mark Taper Foundation Grants, 3449
Salmon Foundation Grants, 3458
San Diego Foundation Health & Human Services Grants, 3471
San Diego Foundation Paradise Valley Hospital Community Fund Grants, 3472
San Diego HIV Funding Collaborative Grants, 3473
San Diego Women's Foundation Grants, 3474
San Francisco Fndn Community Health Grants, 3478
San Francisco Foundation Disability Rights Advocate Fund Emergency Grants, 3479
Santa Barbara Foundation Strategy Grants - Core Support, 3481
Seagate Tech Corp Capacity to Care Grants, 3502
Shopko Fndn Community Charitable Grants, 3547
Sidney Stern Memorial Trust Grants, 3552
Sierra Health Foundation Responsive Grants, 3555
Sisters of St. Joseph Healthcare Fndn Grants, 3576
Sonora Area Foundation Competitive Grants, 3622
Southwest Gas Corporation Foundation Grants, 3628
Stocker Foundation Grants, 3656
Symantec Community Relations and Corporate Philanthropy Grants, 3699
TE Foundation Grants, 3705
Tension Envelope Foundation Grants, 3707
Thelma Doelger Charitable Trust Grants, 3714
Thomas and Dorothy Leavey Foundation Grants, 3717
Thomas C. Ackerman Foundation Grants, 3719
Thomas J. Long Foundation Community Grants, 3722
Thomas Jefferson Rosenberg Foundation Grants, 3723
Toyota Motor Sales, USA Grants, 3750
TRDRP California Research Awards, 3753
TRDRP New Investigator Awards, 3755
U.S. Cellular Corporation Grants, 3766
UniHealth Foundation Community Health Improvement Grants, 3782
Unihealth Foundation Grants for Non-Profit Oranizations, 3783
UniHealth Foundation Healthcare Systems Enhancements Grants, 3784
UniHealth Foundation Innovation Fund Grants, 3785
UniHealth Foundation Workforce Development Grants, 3786
Union Bank, N.A. Corporate Sponsorships and Donations, 3787
Union Bank, N.A. Foundation Grants, 3788
Union Labor Health Fndn Angel Fund Grants, 3791
Union Labor Health Fndn Community Grants, 3792

934 / California

Union Labor Health Foundation Dental Angel Fund Grants, 3793
Union Pacific Foundation Health and Human Services Grants, 3794
W.M. Keck Fndn So California Grants, 3890
W.M. Keck Fndn Undergrad Ed Grants, 3891
Waitt Family Foundation Grants, 3894
Wallace Alexander Gerbode Foundation Grants, 3896
Weingart Foundation Grants, 3913
Weyerhaeuser Company Foundation Grants, 3925
William G. Gilmore Foundation Grants, 3951
William H. Hannon Foundation Grants, 3953
Winifred & Harry B. Allen Foundation Grants, 3962
Wood-Claeyssens Foundation Grants, 3971

Colorado

ACE Charitable Foundation Grants, 229
AEC Trust Grants, 315
Albertson's Charitable Giving Grants, 410
Alexander Foundation Cancer, Catastrophic Illness and Injury Grants, 421
Alexander Fndn Insurance Continuation Grants, 422
Amerigroup Foundation Grants, 565
Amica Insurance Company Community Grants, 575
Anheuser-Busch Foundation Grants, 591
Anschutz Family Foundation Grants, 602
Anthem Blue Cross and Blue Shield Grants, 607
ARCO Foundation Education Grants, 652
Bacon Family Foundation Grants, 817
BBVA Compass Foundation Charitable Grants, 848
Boeing Company Contributions Grants, 930
Boettcher Foundation Grants, 931
Bonfils-Stanton Foundation Grants, 932
Bridgestone/Firestone Trust Fund Grants, 949
Carl W. and Carrie Mae Joslyn Trust Grants, 1031
Chamberlain Foundation Grants, 1108
Charles A. Frueauff Foundation Grants, 1114
Claude Bennett Family Foundation Grants, 1176
Colorado Resource for Emergency Education and Trauma Grants, 1212
Colorado Trust Grants, 1213
ConAgra Foods Fndn Community Impact Grants, 1321
ConAgra Foods Nourish our Community Grants, 1322
Coors Brewing Corporate Contributions Grants, 1346
D.F. Halton Foundation Grants, 1405
Daniels Fund Grants-Aging, 1424
Dean Foods Community Involvement Grants, 1440
Denver Broncos Charities Fund Grants, 1460
Denver Foundation Community Grants, 1461
EDS Foundation Grants, 1552
El Paso Corporate Foundation Grants, 1580
El Pomar Foundation Anna Keesling Ackerman Fund Grants, 1581
El Pomar Foundation Grants, 1582
Emma Barnsley Foundation Grants, 1594
Gil and Dody Weaver Foundation Grants, 1819
Gill Foundation - Gay and Lesbian Fund Grants, 1821
Griffin Foundation Grants, 1903
Helen K. & Arthur E. Johnson Fndn Grants, 1995
Humana Foundation Grants, 2077
IIE Nancy Petry Scholarship, 2140
IIE Western Union Family Scholarships, 2143
Inasmuch Foundation Grants, 2149
John G. Duncan Charitable Trust Grants, 2236
Johns Manville Fund Grants, 2251
Joseph Henry Edmondson Foundation Grants, 2270
KeyBank Foundation Grants, 2307
Lockheed Martin Corp Foundation Grants, 2374
Mabel Y. Hughes Charitable Trust Grants, 2409
Medtronic Foundation CommunityLink Health Grants, 2510
Medtronic Foundation Community Link Human Services Grants, 2511
Nationwide Insurance Foundation Grants, 2637
PacifiCare Foundation Grants, 3102
Packard Foundation Local Grants, 3105
PeyBack Foundation Grants, 3159
Piper Jaffray Fndn Communities Giving Grants, 3227
PPCF Community Grants, 3256
Qualcomm Grants, 3287

Questar Corporate Contributions Grants, 3291
Robert R. McCormick Trib Community Grants, 3385
Rose Community Foundation Aging Grants, 3410
Rose Community Foundation Health Grants, 3411
Ruth and Vernon Taylor Foundation Grants, 3426
Salmon Foundation Grants, 3458
Schlessman Family Foundation Grants, 3493
Schramm Foundation Grants, 3495
Seagate Tech Corp Capacity to Care Grants, 3502
Summit Foundation Grants, 3674
Union Pacific Foundation Health and Human Services Grants, 3794
W.M. Keck Fndn Undergrad Ed Grants, 3891
Western Union Foundation Grants, 3923
Weyerhaeuser Company Foundation Grants, 3925
William G. Gilmore Foundation Grants, 3951
William Ray & Ruth E. Collins Fndn Grants, 3956
Yampa Valley Community Foundation Grants, 3978
Yampa Valley Community Foundation Volunteer Firemen Scholarships, 3979
ZYTL Foundation Grants, 3985

Connecticut

3M Company Fndn Community Giving Grants, 6
3M Fndn Health & Human Services Grants, 7
Aaron Foundation Grants, 193
ACE Charitable Foundation Grants, 229
Aetna Foundation Health Grants in Connecticut, 327
Aetna Foundation Regional Health Grants, 333
Agway Foundation Grants, 348
Ahn Family Foundation Grants, 356
Amerigroup Foundation Grants, 565
Amica Insurance Company Community Grants, 575
Anna Fitch Ardenghi Trust Grants, 592
Anthem Blue Cross and Blue Shield Grants, 607
ARS Foundation Grants, 667
Barnes Group Foundation Grants, 832
BJ's Charitable Foundation Grants, 904
Bodenwein Public Benevolent Foundation Grants, 928
Bridgestone/Firestone Trust Fund Grants, 949
Bristol-Myers Squibb Foundation Health Disparities Grants, 958
Campbell Soup Foundation Grants, 1007
Carlisle Foundation Grants, 1025
Carolyn Foundation Grants, 1034
Carrier Corporation Contributions Grants, 1038
Charles A. Frueauff Foundation Grants, 1114
Charles Nelson Robinson Fund Grants, 1125
Chemtura Corporation Contributions Grants, 1130
Community Foundation for Greater New Haven $5,000 and Under Grants, 1248
Community Foundation for Greater New Haven Responsive New Grants, 1249
Community Foundation for Greater New Haven Sponsorship Grants, 1250
Community Foundation for Greater New Haven Valley Neighborhood Grants, 1251
Community Foundation for Greater New Haven Women & Girls Grants, 1252
Community Foundation of Eastern Connecticut General Southeast Grants, 1269
Community Foundation of Eastern Connecticut Northeast Women and Girls Grants, 1270
Community Foundation of Greater New Britain Grants, 1281
ConAgra Foods Nourish our Community Grants, 1322
Connecticut Community Foundation Grants, 1324
Connecticut Health Foundation Health Initiative Grants, 1325
CSX Corporate Contributions Grants, 1368
Dammann Fund Grants, 1418
Daphne Seybolt Culpeper Fndn Grants, 1425
David N. Lane Trust Grants for Aged and Indigent Women, 1433
Dean Foods Community Involvement Grants, 1440
Dominion Foundation Grants, 1485
Ensworth Charitable Foundation Grants, 1595
Eugene G. and Margaret M. Blackford Memorial Fund Grants, 1610
Expect Miracles Foundation Grants, 1619

Fairfield County Community Foundation Grants, 1625
Fannie E. Rippel Foundation Grants, 1630
Fisher Foundation Grants, 1670
Frank Loomis Palmer Fund Grants, 1726
Frederick W. Marzahl Memorial Fund Grants, 1741
George A. and Grace L. Long Foundation Grants, 1788
George F. Baker Trust Grants, 1793
George H.C. Ensworth Memorial Fund Grants, 1797
Gladys Brooks Foundation Grants, 1826
Hallmark Corporate Foundation Grants, 1936
Harold and Rebecca H. Gross Foundation Grants, 1951
Hartford Courant Foundation Grants, 1969
Hartford Foundation Regular Grants, 1970
Hilda and Preston Davis Foundation Grants, 2028
HRAMF Charles H. Hood Foundation Child Health Research Awards, 2057
J. Walton Bissell Foundation Grants, 2177
Jessie B. Cox Charitable Trust Grants, 2225
John G. Martin Foundation Grants, 2237
Katharine Matthies Foundation Grants, 2291
Kosciuszko Foundation Dr. Marie E. Zakrzewski Medical Scholarship, 2321
Lena Benas Memorial Fund Grants, 2353
Liberty Bank Foundation Grants, 2362
Lil and Julie Rosenberg Foundation Grants, 2364
Lincoln Financial Foundation Grants, 2367
Linford and Mildred White Charitable Grants, 2368
Louis H. Aborn Foundation Grants, 2380
Mabel F. Hoffman Charitable Trust Grants, 2407
Marjorie Moore Charitable Foundation Grants, 2438
Meriden Foundation Grants, 2519
Norfolk Southern Foundation Grants, 2982
Perkin Fund Grants, 3139
Peter F. McManus Charitable Trust Grants, 3153
Price Chopper's Golub Foundation Grants, 3261
Prudential Foundation Education Grants, 3276
R.S. Gernon Trust Grants, 3293
R.T. Vanderbilt Trust Grants, 3294
Richard Davoud Donchian Foundation Grants, 3367
Robert and Joan Dircks Foundation Grants, 3379
Robert Leet & Clara Guthrie Patterson Grants, 3383
ROSE Fund Grants, 3412
Salmon Foundation Grants, 3458
Sasco Foundation Grants, 3487
Shaw's Supermarkets Donations, 3542
Southbury Community Trust Fund, 3625
Stocker Foundation Grants, 3656
Sunoco Foundation Grants, 3679
Swindells Charitable Foundation, 3697
T. Spencer Shore Foundation Grants, 3701
Tauck Family Foundation Grants, 3703
Thomas J. Atkins Memorial Trust Fund Grants, 3721
Thomas Jefferson Rosenberg Foundation Grants, 3723
Yawkey Foundation Grants, 3980

Delaware

ACE Charitable Foundation Grants, 229
Albertson's Charitable Giving Grants, 410
BJ's Charitable Foundation Grants, 904
Borkee-Hagley Foundation Grants, 934
Brook J. Lenfest Foundation Grants, 967
Carl M. Freeman Foundation FACES Grants, 1026
Carl M. Freeman Foundation Grants, 1027
Charles A. Frueauff Foundation Grants, 1114
CSX Corporate Contributions Grants, 1368
Delaware Community Foundation-Youth Philanthropy Board for Kent County, 1448
Delaware Community Foundation Grants, 1449
Delmarva Power and Light Contributions, 1454
FMC Foundation Grants, 1683
Giant Food Charitable Grants, 1816
Gladys Brooks Foundation Grants, 1826
Highmark Corporate Giving Grants, 2025
Hilda and Preston Davis Foundation Grants, 2028
Norfolk Southern Foundation Grants, 2982
PNC Foundation Community Services Grants, 3240
Principal Financial Group Foundation Grants, 3273
Richard Davoud Donchian Foundation Grants, 3367
Sunoco Foundation Grants, 3679

GEOGRAPHIC INDEX

District of Columbia
Aaron & Cecile Goldman Family Fndn Grants, 191
ACE Charitable Foundation Grants, 229
Adams and Reese Corporate Giving Grants, 299
Aetna Foundation Regional Health Grants, 333
Alexander and Margaret Stewart Trust Grants, 419
Aratani Foundation Grants, 649
Bank of America Fndn Matching Gifts, 825
Bank of America Fndn Volunteer Grants, 826
Bender Foundation Grants, 876
Blum-Kovler Foundation Grants, 924
Boeing Company Contributions Grants, 930
CharityWorks Grants, 1113
Charles A. Frueauff Foundation Grants, 1114
Charles Delmar Foundation Grants, 1115
Clark-Winchcole Foundation Grants, 1173
Community Foundation for the National Capital Region Community Leadership Grants, 1260
ConAgra Foods Fndn Community Impact Grants, 1321
ConAgra Foods Nourish our Community Grants, 1322
David M. and Marjorie D. Rosenberg Grants, 1432
Deutsche Banc Alex Brown and Sons Charitable Foundation Grants, 1471
E.L. Wiegand Foundation Grants, 1538
Eugene B. Casey Foundation Grants, 1609
Expect Miracles Foundation Grants, 1619
Freddie Mac Foundation Grants, 1738
Genesis Foundation Grants, 1786
Giant Food Charitable Grants, 1816
Ginn Foundation Grants, 1822
Hilda and Preston Davis Foundation Grants, 2028
IIE Western Union Family Scholarships, 2143
J. Willard Marriott, Jr. Foundation Grants, 2178
John Edward Fowler Memorial Fndn Grants, 2235
Mabel H. Flory Charitable Trust Grants, 2408
MeadWestvaco Foundation Sustainable Communities Grants, 2506
Meyer Foundation Benevon Grants, 2532
Meyer Fndn Management Assistance Grants, 2534
Morris & Gwendolyn Cafritz Fndn Grants, 2594
Morton K. and Jane Blaustein Foundation Health Grants, 2595
NIGMS Enabling Resources for Pharmacogenomics Grants, 2862
NIH Mentored Clinical Scientist Research Career Development Award, 2887
NIH Ruth L. Kirschstein National Research Service Award Institutional Research Training Grants, 2900
NIH Ruth L. Kirschstein National Research Service Awards for Individual Postdoc Fellowships, 2902
NIH Ruth L. Kirschstein National Research Service Awards for Individual Predoctoral Fellows, 2903
NIH Support of Competitive Research (SCORE) Pilot Project Awards, 2913
NIH Support of Competitive Research (SCORE) Research Advancement Awards, 2914
Norfolk Southern Foundation Grants, 2982
Philip L. Graham Fund Health and Human Services Grants, 3190
PNC Foundation Community Services Grants, 3240
Prince Charitable Trusts DC Grants, 3271
Quality Health Foundation Grants, 3288
Richard Davoud Donchian Foundation Grants, 3367
Robert R. McCormick Trib Community Grants, 3385
Salmon Foundation Grants, 3458
Sunoco Foundation Grants, 3679
Thomas Jefferson Rosenberg Foundation Grants, 3723
Thoracic Foundation Grants, 3729
U.S. Department of Education Rehabilitation Training - Rehabilitation Continuing Education - Institute on Rehabilitation Issues, 3771
US Airways Community Contributions, 3825
Washington Gas Charitable Giving, 3908
Weyerhaeuser Company Foundation Grants, 3925
Winifred & Harry B. Allen Foundation Grants, 3962

Florida
Abernethy Family Foundation Grants, 210
Able Trust Vocational Rehabilitation Grants for Individuals, 213
ACE Charitable Foundation Grants, 229
Adams and Reese Corporate Giving Grants, 299
AEC Trust Grants, 315
AEGON Transamerica Foundation Health and Welfare Grants, 316
Aetna Foundation Regional Health Grants, 333
A Friends' Foundation Trust Grants, 341
Albertson's Charitable Giving Grants, 410
Amerigroup Foundation Grants, 565
Anheuser-Busch Foundation Grants, 591
Aratani Foundation Grants, 649
Babcock Charitable Trust Grants, 816
Batchelor Foundation Grants, 837
BBVA Compass Foundation Charitable Grants, 848
Beldon Fund Grants, 871
Belk Foundation Grants, 872
Bert W. Martin Foundation Grants, 886
BJ's Charitable Foundation Grants, 904
Boeing Company Contributions Grants, 930
Bridgestone/Firestone Trust Fund Grants, 949
Bruce and Adele Greenfield Foundation Grants, 972
Camp-Younts Foundation Grants, 1005
Campbell Soup Foundation Grants, 1007
Charles A. Frueauff Foundation Grants, 1114
Charles M. & Mary D. Grant Fndn Grants, 1124
Charlotte County (FL) Community Foundation Grants, 1126
Chase Paymentech Corporate Giving Grants, 1127
Comerica Charitable Foundation Grants, 1224
Community Fndn of Greater Tampa Grants, 1282
Community Foundation of Tampa Bay Grants, 1312
ConAgra Foods Fndn Community Impact Grants, 1321
ConAgra Foods Nourish our Community Grants, 1322
Cowles Charitable Trust Grants, 1354
CSX Corporate Contributions Grants, 1368
D. W. McMillan Foundation Grants, 1407
Dade Community Foundation Community AIDS Partnership Grants, 1409
Dade Community Foundation Grants, 1410
Daphne Seybolt Culpeper Fndn Grants, 1425
Dean Foods Community Involvement Grants, 1440
Dr. John T. Macdonald Foundation Grants, 1507
Dr. John T. Macdonald Foundation Scholarships, 1508
Dunspaugh-Dalton Foundation Grants, 1530
Edyth Bush Charitable Foundation Grants, 1562
Elizabeth McGraw Foundation Grants, 1568
Erie Chapman Foundation Grants, 1600
Fifth Third Foundation Grants, 1661
Florida BRAIVE Fund of Dade Community Foundation, 1678
Florida Division of Cultural Affairs Arts In Education Arts Partnership Grants, 1679
Florida High School/High Tech Project Grants, 1680
FMC Foundation Grants, 1683
Forrest C. Lattner Foundation Grants, 1688
Gardner Foundation Grants, 1771
Genesis Foundation Grants, 1786
George F. Baker Trust Grants, 1793
Gladys Brooks Foundation Grants, 1826
Gregory Family Foundation Grants (Florida), 1899
Gulf Coast Foundation of Community Operating Grants, 1916
Gulf Coast Foundation of Community Program Grants, 1917
Harold Alfond Foundation Grants, 1949
Harry Sudakoff Foundation Grants, 1967
Health Foundation of Southern Florida Responsive Grants, 1985
HomeBanc Foundation Grants, 2042
Humana Foundation Grants, 2077
IIE Western Union Family Scholarships, 2143
Jane's Trust Grants, 2206
Jim Moran Foundation Grants, 2228
John Lord Knight Foundation Grants, 2243
Johnson & Johnson Community Health Grants, 2252
Joseph H. & Florence A. Roblee Fndn Grants, 2269
Kimball International-Habig Foundation Health and Human Services Grants, 2308
L. W. Pierce Family Foundation Grants, 2328
Leo Goodwin Foundation Grants, 2354
Leonard L. & Bertha U. Abess Fndn Grants, 2357
Lillian S. Wells Foundation Grants, 2365
Lockheed Martin Corp Foundation Grants, 2374
Lucy Gooding Charitable Fndn Grants, 2388
M.B. and Edna Zale Foundation Grants, 2401
Manuel D. & Rhoda Mayerson Fndn Grants, 2420
McLean Contributionship Grants, 2497
MeadWestvaco Foundation Sustainable Communities Grants, 2506
Medtronic Foundation CommunityLink Health Grants, 2510
Medtronic Foundation Community Link Human Services Grants, 2511
Mercedes-Benz USA Corporate Contributions, 2517
Nationwide Insurance Foundation Grants, 2637
Norfolk Southern Foundation Grants, 2982
Peacock Foundation Grants, 3133
Pinellas County Grants, 3223
PNC Foundation Community Services Grants, 3240
Prudential Foundation Education Grants, 3276
Quantum Foundation Grants, 3290
Rainbow Fund Grants, 3301
Rayonier Foundation Grants, 3311
Richard and Helen DeVos Foundation Grants, 3364
Richard Davoud Donchian Foundation Grants, 3367
Robert R. McCormick Trib Community Grants, 3385
RRF General Program Grants, 3417
Ruth Anderson Foundation Grants, 3425
Sain-Orr and Royak-DeForest Steadman Foundation Grants, 3452
Salem Foundation Grants, 3457
Sarasota Memorial Healthcare Fndn Grants, 3484
Southwest Florida Community Foundation Competitive Grants, 3627
Sunoco Foundation Grants, 3679
Symantec Community Relations and Corporate Philanthropy Grants, 3699
Taubman Endowment for the Arts Fndn Grants, 3702
Thomas Jefferson Rosenberg Foundation Grants, 3723
Western Union Foundation Grants, 3923
William J. & Tina Rosenberg Fndn Grants, 3954
Wilson-Wood Foundation Grants, 3960
Woodward Fund Grants, 3973

Georgia
3M Company Fndn Community Giving Grants, 6
3M Fndn Health & Human Services Grants, 7
ACE Charitable Foundation Grants, 229
AEC Trust Grants, 315
AEGON Transamerica Foundation Health and Welfare Grants, 316
Aetna Foundation Regional Health Grants, 333
Agnes B. Hunt Trust Grants, 344
Albertson's Charitable Giving Grants, 410
Allan C. and Lelia J. Garden Foundation Grants, 438
American Foodservice Charitable Trust Grants, 532
Amerigroup Foundation Grants, 565
Amica Insurance Company Community Grants, 575
Anheuser-Busch Foundation Grants, 591
Appalachian Regional Commission Health Grants, 637
Archer Daniels Midland Foundation Grants, 651
Arthur M. Blank Family Foundation Inspiring Spaces Grants, 673
Atlanta Foundation Grants, 796
BBVA Compass Foundation Charitable Grants, 848
Belk Foundation Grants, 872
BJ's Charitable Foundation Grants, 904
Boeing Company Contributions Grants, 930
Bradley-Turner Foundation Grants, 944
Brian G. Dyson Foundation Grants, 948
Cabot Corporation Foundation Grants, 993
Callaway Foundation Grants, 1002
Camp-Younts Foundation Grants, 1005
Carrier Corporation Contributions Grants, 1038
Charles A. Frueauff Foundation Grants, 1114
Charles M. & Mary D. Grant Fndn Grants, 1124
Chemtura Corporation Contributions Grants, 1130
Clark and Ruby Baker Foundation Grants, 1174
Community Foundation for Greater Atlanta AIDS Fund Grants, 1239

936 / Georgia

Community Foundation for Greater Atlanta Clayton County Fund Grants, 1240
Community Foundation for Greater Atlanta Common Good Funds Grants, 1241
Community Foundation for Greater Atlanta Managing For Excellence Award, 1242
Community Foundation for Greater Atlanta Metropolitan Atlanta An Extra Wish Grants, 1243
Community Foundation for Greater Atlanta Morgan County Fund Grants, 1244
Community Foundation for Greater Atlanta Newton County Fund Grants, 1245
Community Foundation for Greater Atlanta Strategic Restructuring Fund Grants, 1246
ConAgra Foods Fndn Community Impact Grants, 1321
ConAgra Foods Nourish our Community Grants, 1322
Cooper Industries Foundation Grants, 1345
CSRA Community Foundation Grants, 1366
CSX Corporate Contributions Grants, 1368
Dean Foods Community Involvement Grants, 1440
E.J. Grassmann Trust Grants, 1537
EDS Foundation Grants, 1552
Florence Hunt Maxwell Foundation Grants, 1676
Forrest C. Lattner Foundation Grants, 1688
Francis L. Abreu Charitable Trust Grants, 1719
Fuller E. Callaway Foundation Grants, 1761
General Mills Foundation Grants, 1782
George E. Hatcher, Jr. and Ann Williams Hatcher Foundation Grants, 1792
Georgia Power Foundation Grants, 1806
Gertrude and William C. Wardlaw Fund Grants, 1808
H.B. Fuller Foundation Grants, 1922
Hallmark Corporate Foundation Grants, 1936
HomeBanc Foundation Grants, 2042
Hormel Foods Charitable Trust Grants, 2050
Humana Foundation Grants, 2077
Ida Alice Ryan Charitable Trust Grants, 2113
James J. & Angelia M. Harris Fndn Grants, 2192
John Deere Foundation Grants, 2234
John H. and Wilhelmina D. Harland Charitable Foundation Children and Youth Grants, 2238
Katherine John Murphy Foundation Grants, 2293
Lockheed Martin Corp Foundation Grants, 2374
Mary Wilmer Covey Charitable Trust Grants, 2448
MeadWestvaco Foundation Sustainable Communities Grants, 2506
Nationwide Insurance Foundation Grants, 2637
Nordson Corporation Foundation Grants, 2981
Norfolk Southern Foundation Grants, 2982
Peyton Anderson Foundation Grants, 3160
PNC Foundation Community Services Grants, 3240
Prudential Foundation Education Grants, 3276
Qualcomm Grants, 3287
Rainbow Fund Grants, 3301
Rayonier Foundation Grants, 3311
Robert W. Woodruff Foundation Grants, 3389
Rosalynn Carter Institute Georgia Caregiver of the Year Awards, 3407
Sartain Lanier Family Foundation Grants, 3486
Steele-Reese Foundation Grants, 3650
Stewart Huston Charitable Trust Grants, 3655
Sunoco Foundation Grants, 3679
SunTrust Bank Trusteed Foundations Florence C. and Harry L. English Memorial Fund Grants, 3680
SunTrust Bank Trusteed Foundations Greene-Sawtell Grants, 3681
SunTrust Bank Trusteed Foundations Harriet McDaniel Marshall Tust Grants, 3682
SunTrust Bank Trusteed Foundations Nell Warren Elkin and William Simpson Elkin Grants, 3683
SunTrust Bank Trusteed Foundations Thomas Guy Woolford Charitable Trust Grants, 3684
SunTrust Bank Trusteed Foundations Walter H. and Marjory M. Rich Memorial Fund Grants, 3685
Textron Corporate Contributions Grants, 3712
Thomas C. Burke Foundation Grants, 3720
Tull Charitable Foundation Grants, 3762
Weyerhaeuser Company Foundation Grants, 3925
Winifred & Harry B. Allen Foundation Grants, 3962
Woodward Fund Grants, 3973

Guam
ACF Native American Social and Economic Development Strategies Grants, 231
Bank of America Fndn Matching Gifts, 825
Bank of America Fndn Volunteer Grants, 826
NHLBI Ancillary Studies in Clinical Trials, 2688
NHLBI Bioengineering and Obesity Grants, 2689
NHLBI Career Transition Awards, 2694
NHLBI Lymphatics in Health and Disease in the Digestive, Cardio & Pulmonary Systems, 2704
NHLBI Lymphatics in Health and Disease in the Digestive, Urinary, Cardiovascular and Pulmonary Systems, 2705
NHLBI Mentored Career Dev Award to Promote Faculty Diversity in Biomed Research, 2708
NHLBI Mentored Patient-Oriented Research Career Development Awards, 2710
NHLBI Mentored Quantitative Research Career Development Awards, 2711
NHLBI Pathway to Independence Awards, 2718
NHLBI Research on the Role of Cardiomyocyte Mitochondria in Heart Disease: An Integrated Approach, 2726
NHLBI Ruth L. Kirschstein National Research Service Awards for Individual Postdoctoral Fellows, 2729
NHLBI Ruth L. Kirschstein National Research Service Awards for Individual Senior Fellows, 2732
NHLBI Short-Term Research Education Program to Increase Diversity in Health-Related Research, 2733
NHLBI Training Program Grants for Institutions that Promote Diversity, 2738
NIGMS Advancing Basic Behavioral and Social Research on Resilience: an Integrative Science Approach, 2861
NIH Mentored Clinical Scientist Research Career Development Award, 2887
NIH Ruth L. Kirschstein National Research Service Award Institutional Research Training Grants, 2900
NIH Ruth L. Kirschstein National Research Service Awards for Individual Postdoc Fellowships, 2902
NIH Ruth L. Kirschstein National Research Service Awards for Individual Predoctoral Fellows, 2903
NIH Ruth L. Kirschstein Nat Research Service Award Short-Term Institutional Training Grants, 2906
NIH Support of Competitive Research (SCORE) Pilot Project Awards, 2913
NIH Support of Competitive Research (SCORE) Research Advancement Awards, 2914

Hawaii
3M Company Fndn Community Giving Grants, 6
3M Fndn Health & Human Services Grants, 7
ACE Charitable Foundation Grants, 229
Alaska Airlines Foundation Grants, 403
Alexander and Baldwin Foundation Hawaiian and Pacific Island Grants, 418
AlohaCare Believes in Me Scholarship, 449
Anheuser-Busch Foundation Grants, 591
Annie Sinclair Knudsen Memorial Fund/Kaua'i Community Grants Program, 597
Antone & Edene Vidinha Charitable Trust Grants, 609
Atherton Family Foundation Grants, 794
Boeing Company Contributions Grants, 930
Boyd Gaming Corporation Contributions Program, 941
Clarence T.C. Ching Foundation Grants, 1167
Cooke Foundation Grants, 1344
Dean Foods Community Involvement Grants, 1440
Dorrance Family Foundation Grants, 1501
Elsie H. Wilcox Foundation Grants, 1583
Finance Factors Foundation Grants, 1666
Fred Baldwin Memorial Foundation Grants, 1736
G.N. Wilcox Trust Grants, 1768
Hawaiian Electric Industries Charitable Foundation Grants, 1975
Hawaii Community Foundation Health Education and Research Grants, 1976
Hawaii Community Foundation Reverend Takie Okumura Family Grants, 1977
Hawaii Community Foundation West Hawaii Fund Grants, 1978

GEOGRAPHIC INDEX

Hershey Company Grants, 2013
J.M. Long Foundation Grants, 2173
James & Abigail Campbell Family Fndn Grants, 2185
Kahuku Community Fund, 2286
Leahi Fund, 2350
McInerny Foundation Grants, 2495
Samuel N. and Mary Castle Foundation Grants, 3467
W.M. Keck Fndn Undergrad Ed Grants, 3891
Wallace Alexander Gerbode Foundation Grants, 3896

Idaho
Albertson's Charitable Giving Grants, 410
Avista Foundation Economic and Cultural Vitality Grants, 812
Bullitt Foundation Grants, 973
ConAgra Foods Fndn Community Impact Grants, 1321
ConAgra Foods Nourish our Community Grants, 1322
Dean Foods Community Involvement Grants, 1440
Detlef Schrempf Foundation Grants, 1470
E.L. Wiegand Foundation Grants, 1538
Foster Foundation Grants, 1689
Four J Foundation Grants, 1713
Idaho Community Foundation Eastern Region Competitive Grants, 2114
Idaho Power Company Corporate Contributions, 2115
KeyBank Foundation Grants, 2307
Kimball International-Habig Foundation Health and Human Services Grants, 2308
Laura Moore Cunningham Foundation Grants, 2346
M.J. Murdock Charitable Trust General Grants, 2405
NWHF Health Advocacy Small Grants, 3033
Piper Jaffray Fndn Communities Giving Grants, 3227
Pride Foundation Grants, 3268
Pride Foundation Scholarships, 3269
Regence Fndn Access to Health Care Grants, 3346
Regence Foundation Health Care Community Awareness and Engagement Grants, 3347
Regence Fndn Health Care Connections Grants, 3348
Regence Fndn Improving End-of-Life Grants, 3349
Regence Fndn Tools & Technology Grants, 3350
Shopko Fndn Community Charitable Grants, 3547
Steele-Reese Foundation Grants, 3650
Sunderland Foundation Grants, 3675
Union Pacific Foundation Health and Human Services Grants, 3794
W.M. Keck Fndn Undergrad Ed Grants, 3891
Weyerhaeuser Company Foundation Grants, 3925

Illinois
3M Company Fndn Community Giving Grants, 6
3M Fndn Health & Human Services Grants, 7
Abbott Fund Access to Health Care Grants, 199
Abbott Fund Community Grants, 201
Abbott Fund Global AIDS Care Grants, 202
Abbott Fund Science Education Grants, 203
ACE Charitable Foundation Grants, 229
Aetna Foundation Regional Health Grants, 333
Albert Pick Jr. Fund Grants, 409
Albertson's Charitable Giving Grants, 410
Alfred Bersted Foundation Grants, 429
Alliant Energy Corporation Contributions, 445
Allstate Corp Hometown Commitment Grants, 447
Alpha Natural Resources Corporate Giving, 450
Amica Insurance Company Community Grants, 575
Andrew Family Foundation Grants, 588
Ann and Robert H. Lurie Family Fndn Grants, 593
AptarGroup Foundation Grants, 647
Archer Daniels Midland Foundation Grants, 651
Arie and Ida Crown Memorial Grants, 654
Aurora Foundation Grants, 802
Barr Fund Grants, 835
Baxter International Foundation Grants, 843
Bindley Family Foundation Grants, 896
Blowitz-Ridgeway Foundation Grants, 913
Blum-Kovler Foundation Grants, 924
Boeing Company Contributions Grants, 930
Boyd Gaming Corporation Contributions Program, 941
Bridgestone/Firestone Trust Fund Grants, 949
Bright Promises Foundation Grants, 952
Cabot Corporation Foundation Grants, 993

GEOGRAPHIC INDEX　　　　　　　　　　　　　　　　　　　　　　　　　　　　　　　　　　　　　　Indiana　/ 937

Caesars Foundation Grants, 995
Campbell Soup Foundation Grants, 1007
Carl R. Hendrickson Family Foundation Grants, 1028
Carrier Corporation Contributions Grants, 1038
Charles A. Frueauff Foundation Grants, 1114
Chemtura Corporation Contributions Grants, 1130
Coleman Foundation Cancer Care Grants, 1206
Coleman Foundation Developmental Disabilities Grants, 1207
Colonel Stanley R. McNeil Foundation Grants, 1211
Commonwealth Edison Grants, 1225
Community Fndn of Central Illinois Grants, 1267
Community Fndn of East Central Illinois Grants, 1268
ConAgra Foods Fndn Community Impact Grants, 1321
ConAgra Foods Nourish our Community Grants, 1322
ConocoPhillips Foundation Grants, 1327
Cooper Industries Foundation Grants, 1345
CSX Corporate Contributions Grants, 1368
Dean Foods Community Involvement Grants, 1440
DIFFA/Chicago Grants, 1477
Dominion Foundation Grants, 1485
Doris and Victor Day Foundation Grants, 1491
Duchossois Family Foundation Grants, 1518
DuPage Community Foundation Grants, 1531
EDS Foundation Grants, 1552
Elizabeth Morse Genius Charitable Trust Grants, 1569
Evanston Community Foundation Grants, 1615
Expect Miracles Foundation Grants, 1619
Field Foundation of Illinois Grants, 1660
Fifth Third Foundation Grants, 1661
FMC Foundation Grants, 1683
Foundation for Health Enhancement Grants, 1696
Frank E. and Seba B. Payne Foundation Grants, 1722
G.A. Ackermann Memorial Fund Grants, 1767
Gardner W. & Joan G. Heidrick, Jr. Fndn Grants, 1773
General Mills Foundation Grants, 1782
Ginn Foundation Grants, 1822
Gladys Brooks Foundation Grants, 1826
Grace Bersted Foundation Grants, 1859
Greater Saint Louis Community Fndn Grants, 1884
Grover Hermann Foundation Grants, 1906
H.B. Fuller Foundation Grants, 1922
Hallmark Corporate Foundation Grants, 1936
Harry S. Black and Allon Fuller Fund Grants, 1966
Henrietta Lange Burk Fund Grants, 2001
Hershey Company Grants, 2013
Hormel Foods Charitable Trust Grants, 2050
Humana Foundation Grants, 2077
I.A. O'Shaughnessy Foundation Grants, 2088
IATA Research Grants, 2093
IATA Scholarships, 2094
IDPH Carolyn Adams Ticket for the Cure Community Grants, 2117
IDPH Emergency Medical Services Assistance Fund Grants, 2118
IDPH Hosptial Capital Investment Grants, 2119
IDPH Local Health Department Public Health Emergency Response Grants, 2120
IIE Western Union Family Scholarships, 2143
Illinois Children's Healthcare Foundation Grants, 2146
Intergrys Corporation Grants, 2160
James S. Copley Foundation Grants, 2200
John Deere Foundation Grants, 2234
Joseph H. & Florence A. Roblee Fndn Grants, 2269
Julius N. Frankel Foundation Grants, 2282
Kovler Family Foundation Grants, 2323
Lillian S. Wells Foundation Grants, 2365
Lincoln Financial Foundation Grants, 2367
Lloyd A. Fry Foundation Health Grants, 2373
Marathon Petroleum Corporation Grants, 2422
Marion Gardner Jackson Charitable Trust Grants, 2434
MeadWestvaco Foundation Sustainable Communities Grants, 2506
Mercedes-Benz USA Corporate Contributions, 2517
Michael Reese Health Trust Core Grants, 2544
Michael Reese Health Trust Responsive Grants, 2545
Moline Foundation Community Grants, 2590
Nesbitt Medical Student Foundation Scholarship, 2674
Norfolk Southern Foundation Grants, 2982
Patrick and Anna M. Cudahy Fund Grants, 3120

Penny Severns Breast, Cervical and Ovarian Cancer Research Grants, 3136
Piper Jaffray Fndn Communities Giving Grants, 3227
PNC Foundation Community Services Grants, 3240
Polk Bros. Foundation Grants, 3243
Prince Charitable Trusts Chicago Grants, 3270
Prudential Foundation Education Grants, 3276
Rice Foundation Grants, 3363
Robert R. McCormick Trib Community Grants, 3385
Robert R. McCormick Tribune Veterans Initiative Grants, 3386
Roy J. Carver Charitable Trust Medical and Science Research Grants, 3416
RRF General Program Grants, 3417
RRF Organizational Capacity Building Grants, 3418
Ruth and Vernon Taylor Foundation Grants, 3426
Seabury Foundation Grants, 3501
Shopko Fndn Community Charitable Grants, 3547
Siragusa Foundation Health Services & Medical Research Grants, 3571
Solo Cup Foundation Grants, 3620
Square D Foundation Grants, 3634
Textron Corporate Contributions Grants, 3712
Topfer Family Foundation Grants, 3739
Toyota Motor Manufacturing of Indiana Grants, 3744
U.S. Cellular Corporation Grants, 3766
USDA Delta Health Care Services Grants, 3829
Visiting Nurse Foundation Grants, 3879
Welborn Baptist Fndn Health Ministries Grants, 3915
Welborn Baptist Foundation Improvements to Community Health Status Grants, 3916
Welborn Baptist Fndn School Health Grants, 3917
Weyerhaeuser Company Foundation Grants, 3925
William Blair and Company Foundation Grants, 3948

Indiana
1st Source Foundation Grants, 2
3M Company Fndn Community Giving Grants, 6
3M Fndn Health & Human Services Grants, 7
ACE Charitable Foundation Grants, 229
Adams Rotary Memorial Fund A Grants, 302
Adams Rotary Memorial Fund B Scholarships, 303
Albertson's Charitable Giving Grants, 410
Allstate Corp Hometown Commitment Grants, 447
American Electric Power Foundation Grants, 530
Amerigroup Foundation Grants, 565
Anthem Blue Cross and Blue Shield Grants, 607
Archer Daniels Midland Foundation Grants, 651
Ball Brothers Foundation General Grants, 820
Barker Welfare Foundation Grants, 831
Benton Community Fndn Cookie Jar Grant, 877
Benton Community Foundation Grants, 878
Benton County Foundation - Fitzgerald Family Scholarships, 879
Bindley Family Foundation Grants, 896
Bingham McHale LLP Pro Bono Services, 897
Blackford County Community Foundation - WOW Grants, 905
Blackford County Community Foundation Grants, 906
Blackford County Community Foundation Hartford City Kiwanis Scholarship in Memory of Mike McDougall, 907
Blackford County Community Foundation Noble Memorial Scholarship, 908
Blue River Community Foundation Grants, 922
Boyd Gaming Corporation Contributions Program, 941
Bridgestone/Firestone Trust Fund Grants, 949
Bristol-Myers Squibb Foundation Health Disparities Grants, 958
Brown County Community Foundation Grants, 970
Caesars Foundation Grants, 995
Carrier Corporation Contributions Grants, 1038
Carroll County Community Foundation Grants, 1039
Cass County Community Foundation Grants, 1041
Charles A. Frueauff Foundation Grants, 1114
Charles H. Dater Foundation Grants, 1118
Chemtura Corporation Contributions Grants, 1130
CICF City of Noblesville Community Grant, 1155
CICF Indianapolis Fndn Community Grants, 1156
CICF James Proctor Grant for Aged, 1157

CICF Legacy Fund Grants, 1158
Clarian Health Critical/Progressive Care Interns, 1168
Clarian Health Multi-Specialty Internships, 1169
Clarian Health OR Internships, 1170
Clarian Health Scholarships for LPNs, 1171
Clarian Health Student Nurse Scholarships, 1172
Clinton County Community Foundation Grants, 1184
Coleman Foundation Cancer Care Grants, 1206
Coleman Foundation Developmental Disabilities Grants, 1207
Community Foundation Alliance City of Evansville Endowment Fund Grants, 1238
Community Foundation of Bartholomew County Heritage Fund Grants, 1262
Community Foundation of Bartholomew County James A. Henderson Award for Fundraising, 1263
Community Fndn of Bloomington & Monroe County - Precision Health Network Cycle Grants, 1264
Community Foundation of Bloomington and Monroe County Grants, 1265
Community Foundation of Boone County Grants, 1266
Community Foundation of Grant County Grants, 1271
Community Foundation of Greater Fort Wayne - Community Endowment and Clarke Endowment Grants, 1274
Community Foundation of Greater Fort Wayne - Lilly Endowment Scholarships, 1275
Community Foundation of Greater Fort Wayne Scholarships, 1276
Community Foundation of Greater Lafayette - Robert and Dorothy Hughes Scholarships, 1279
Community Fndn of Greater Lafayette Grants, 1280
Community Fndn of Howard County Grants, 1286
Community Fndn of Jackson County Grants, 1287
Community Foundation of Jackson County Seymour Noon Lions Club Grant, 1288
Community Foundation of Muncie and Delaware County Grants, 1300
Community Foundation of Muncie and Delaware County Maxon Grants, 1301
Community Fndn of Randolph County Grants, 1302
Community Foundation of St. Joseph County Lilly Endowment Community Scholarship, 1308
Community Foundation of St. Joseph County Scholarships, 1309
Community Foundation of St. Joseph County Special Project Challenge Grants, 1310
Community Fndn of Switzerland County Grants, 1311
Community Fndn of Wabash County Grants, 1316
Community Foundation Partnerships - Lawrence County Grants, 1317
Community Foundation Partnerships - Martin County Grants, 1318
ConAgra Foods Fndn Community Impact Grants, 1321
ConAgra Foods Nourish our Community Grants, 1322
Crown Point Community Foundation Grants, 1361
Crown Point Community Fndn Scholarships, 1362
CSX Corporate Contributions Grants, 1368
Daviess County Community Foundation Health Grants, 1434
Dean Foods Community Involvement Grants, 1440
Dearborn Community Foundation City of Lawrenceburg Community Grants, 1441
Dearborn Community Foundation County Progress Grants, 1442
Decatur County Community Foundation Large Project Grants, 1444
Decatur County Community Foundation Small Project Grants, 1445
DeKalb County Community Foundation - Garrett Hospital Aid Foundation Grants, 1446
DeKalb County Community Foundation Grants, 1447
Dominion Foundation Grants, 1485
Donna K. Yundt Memorial Scholarship, 1488
Dr. John Maniotes Scholarship, 1506
Dr. R.T. White Scholarship, 1510
Dubois County Community Foundation Grants, 1517
Duneland Health Council Incorporated Grants, 1529
Elkhart County Community Foundation Fund for Elkhart County, 1572

938 / Indiana

Elkhart County Community Foundation Grants, 1573
Elkhart County Foundation Lilly Endowment Community Scholarships, 1574
Elliot Foundation Inc Grants, 1575
Evan and Susan Bayh Foundation Grants, 1614
Fayette County Foundation Grants, 1636
Fifth Third Foundation Grants, 1661
Foellinger Foundation Grants, 1684
Foundations of East Chicago Health Grants, 1709
Franklin County Community Foundation Grants, 1724
Fred and Louise Latshaw Scholarship, 1735
Fulton County Community Foundation 4Community Higher Education Scholarship, 1762
Fulton County Community Foundation Grants, 1763
Fulton County Community Foundation Lilly Endowment Community Scholarships, 1764
Fulton County Community Foundation Paul and Dorothy Arven Memorial Scholarship, 1765
Fulton County Community Foundation Women's Giving Circle Grants, 1766
General Mills Foundation Grants, 1782
Gibson County Community Foundation Women's Fund Grants, 1817
Giving Sum Annual Grant, 1825
Gladys Brooks Foundation Grants, 1826
Greater Cincinnati Foundation Priority and Small Projects/Capacity-Building Grants, 1880
Greene County Foundation Grants, 1892
Gregory L. Gibson Charitable Fndn Grants, 1900
Hammond Common Council Scholarships, 1938
Hancock County Community Foundation - Field of Interest Grants, 1943
Harrison County Community Foundation Grants, 1957
Harrison County Community Foundation Signature Grants, 1958
Health Foundation of Greater Cincinnati Grants, 1983
Health Fndn of Greater Indianapolis Grants, 1984
Hendricks County Community Fndn Grants, 2000
Henry County Community Foundation Grants, 2005
Humana Foundation Grants, 2077
Huntington County Community Foundation Make a Difference Grants, 2083
Huntington National Bank Community Grants, 2084
IBCAT Screening Mammography Grants, 2095
IBEW Local Union #697 Memorial Scholarships, 2096
Impact 100 Grants, 2148
Indiana 21st Century Research and Technology Fund Awards, 2154
Indiana AIDS Fund Grants, 2155
Indiana Minority Teacher/Special Services Scholarships, 2156
Indiana Nursing Scholarships, 2157
Jacob G. Schmidlapp Trust Grants, 2183
Jasper Foundation Grants, 2212
Jennings County Community Foundation Grants, 2220
Jennings County Community Foundation Women's Giving Circle Grant, 2221
Johnson County Community Foundation Grants, 2256
John V. and George Primich Family Scholarship, 2259
John W. Anderson Foundation Grants, 2262
Joseph and Luella Abell Trust Scholarships, 2266
Joshua Benjamin Cohen Memorial Scholarship, 2276
Judith Clark-Morrill Foundation Grants, 2281
June Pangburn Memorial Scholarship, 2283
K21 Health Foundation Cancer Care Grants, 2284
K21 Health Foundation Grants, 2285
Kendrick Foundation Grants, 2299
KeyBank Foundation Grants, 2307
Kimball International-Habig Foundation Health and Human Services Grants, 2308
Knox County Community Foundation Grants, 2313
Kosair Charities Grants, 2318
Kosciusko County Community Fndn Grants, 2319
Kosciusko County Community Foundation REMC Operation Round Up Grants, 2320
LaGrange County Community Fndn Grants, 2329
LaGrange County Community Foundation Scholarships, 2330
LaGrange County Lilly Endowment Community Scholarship, 2331

Lake County Athletic Officials Association Scholarships, 2332
Lake County Community Fund Grants, 2333
Lake County Lilly Endowment Community Scholarships, 2334
LEGENDS Scholarship, 2351
Lester E. Yeager Charitable Trust B Grants, 2359
Lincoln Financial Foundation Grants, 2367
M.E. Raker Foundation Grants, 2404
Madison County Community Foundation - City of Anderson Quality of Life Grant, 2412
Madison County Community Foundation General Grants, 2413
Marathon Petroleum Corporation Grants, 2422
Marshall County Community Foundation Grants, 2439
Maxon Charitable Foundation Grants, 2458
McMillen Foundation Grants, 2499
Mead Johnson Nutritionals Evansville-Area Organizations Grants, 2503
Medtronic CommunityLink Health Grants, 2510
Medtronic Foundation Community Link Human Services Grants, 2511
Miami County Community Foundation - Operation Round Up Grants, 2543
Montgomery County Community Fndn Grants, 2591
Nancy J. Pinnick Memorial Scholarship, 2607
Newton County Community Foundation Grants, 2681
Newton County Community Foundation Lilly Scholarships, 2682
Newton County Community Fndn Scholarships, 2683
Nicholas H. Noyes Jr. Memorial Fndn Grants, 2791
Nina Mason Pulliam Charitable Trust Grants, 2921
Noble County Community Foundation - Arthur A. and Hazel S. Auer Scholarship, 2969
Noble County Community Foundation - Art Hutsell Scholarship, 2970
Noble County Community Foundation - Delta Theta Tau Sorority IOTA IOTA Chapter Riecke Scholarship, 2971
Noble County Community Foundation - Democrat Central Committee Scholarships, 2972
Noble County Community Foundation - Kathleen June Earley Memorial Scholarship, 2973
Noble County Community Foundation - Lolita J. Hornett Memorial Nursing Scholarship, 2974
Noble County Community Foundation Grants, 2975
Noble County Community Foundation Lilly Endowment Scholarship, 2976
Noble County Community Fndn Scholarships, 2977
Norfolk Southern Foundation Grants, 2982
North Central Health Services Grants, 2992
North Central Health Services Grants, 2993
Ohio County Community Foundation Board of Directors Grants, 3067
Ohio County Community Foundation Grants, 3068
Ohio County Community Fndn Mini-Grants, 3069
Owen County Community Foundation Grants, 3100
Parke County Community Foundation Grants, 3115
Paul Ogle Foundation Grants, 3124
Perry County Community Foundation Grants, 3142
PeyBack Foundation Grants, 3159
Pike County Community Foundation Grants, 3222
PNC Foundation Community Services Grants, 3240
Porter County Foundation Lilly Endowment Community Scholarships, 3245
Porter County Health and Wellness Grant, 3246
Porter County Health Occupations Scholarship, 3247
Porter County Women's Grant, 3248
Portland Fndn Women's Giving Circle Grant, 3249
Posey Community Fndn Women's Fund Grants, 3251
Posey County Community Foundation Grants, 3252
Pulaski County Community Foundation Grants, 3279
Pulaski County Community Foundation Lilly Endowment Community Scholarships, 3280
Pulaski County Community Fndn Scholarships, 3281
Putnam County Community Foundation Grants, 3283
Richard M. Fairbanks Foundation Grants, 3370
Ripley County Community Foundation Grants, 3376
Ripley County Community Foundation Small Project Grants, 3377

Robert & Clara Milton Senior Housing Grants, 3384
Roget Begnoche Scholarship, 3402
RRF General Program Grants, 3417
Rush County Community Foundation Grants, 3424
Schrage Family Foundation Scholarships, 3494
Schurz Communications Foundation Grants, 3496
Scott County Community Foundation Grants, 3500
Sengupta Family Scholarship, 3512
Sensient Technologies Foundation Grants, 3513
Shopko Fndn Community Charitable Grants, 3547
South Madison Community Foundation Grants, 3626
Spencer County Community Foundation Grants, 3632
Square D Foundation Grants, 3634
St. Joseph Community Health Foundation Burn Care and Prevention Grants, 3639
St. Joseph Community Health Foundation Improving Healthcare Access Grants, 3640
St. Joseph Community Health Foundation Pfeiffer Fund Grants, 3641
St. Joseph Community Health Foundation Schneider Fellowships, 3642
Starke County Community Foundation Grants, 3646
Starke County Community Foundation Lilly Endowment Community Scholarships, 3647
Starke County Community Fndn Scholarships, 3648
Steuben County Community Foundation Grants, 3654
Subaru of Indiana Automotive Foundation Grants, 3672
Sunoco Foundation Grants, 3679
Tipton County Foundation Grants, 3734
Toyota Motor Engineering & Manufacturing North America Grants, 3742
Toyota Motor Manufacturing of Indiana Grants, 3744
U.S. Cellular Corporation Grants, 3766
Unity Foundation Of LaPorte County Grants, 3804
Vanderburgh Community Foundation Grants, 3852
Vanderburgh Community Fndn Women's Fund, 3853
Vectren Foundation Grants, 3857
W. C. Griffith Foundation Grants, 3882
W. James Spicer Scholarship, 3885
Wabash Valley Community Foundation Grants, 3893
Warren County Community Foundation Grants, 3902
Warren County Community Fndn Mini-Grants, 3903
Warrick County Community Foundation Grants, 3904
Warrick County Community Foundation Women's Fund, 3905
Washington County Community Fndn Grants, 3906
Washington County Community Youth Grants, 3907
Wayne County Foundation Grants, 3909
Weaver Popcorn Foundation Grants, 3910
Webster Cornwell Memorial Scholarship, 3911
Welborn Baptist Fndn Health Ministries Grants, 3915
Welborn Baptist Foundation Improvements to Community Health Status Grants, 3916
Welborn Baptist Fndn School Health Grants, 3917
Wells County Foundation Grants, 3920
Western Indiana Community Foundation Grants, 3921
WHAS Crusade for Children Grants, 3927
White County Community Foundation - Adam Krintz Memorial Scholarship, 3928
White County Community Foundation - Annie Horton Scholarship, 3929
White County Community Foundation - Landis Memorial Scholarship, 3930
White County Community Foundation - Lilly Endowment Scholarships, 3931
White County Community Foundation - Tri-County Educational Scholarships, 3932
White County Community Foundation - Women Giving Together Grants, 3933
White County Community Foundation Grants, 3934
Whitley Community Foundation - Lilly Endowment Scholarship, 3935
Whitley County Community Foundation Grants, 3936
Whitley County Community Fndn Scholarships, 3937

Iowa
3M Company Fndn Community Giving Grants, 6
3M Fndn Health & Human Services Grants, 7
AEGON Transamerica Foundation Health and Welfare Grants, 316

GEOGRAPHIC INDEX

Albertson's Charitable Giving Grants, 410
Alliant Energy Corporation Contributions, 445
Andersen Corporate Foundation, 586
Archer Daniels Midland Foundation Grants, 651
Bridgestone/Firestone Trust Fund Grants, 949
Caesars Foundation Grants, 995
Coleman Foundation Cancer Care Grants, 1206
Coleman Foundation Developmental Disabilities Grants, 1207
ConAgra Foods Fndn Community Impact Grants, 1321
ConAgra Foods Nourish our Community Grants, 1322
Covenant Foundation of Waterloo Auxiliary Scholarships, 1349
CUNA Mutual Group Fndn Community Grants, 1375
Doris and Victor Day Foundation Grants, 1491
General Mills Foundation Grants, 1782
Hall-Perrine Foundation Grants, 1935
Harold R. Bechtel Charitable Remainder Uni-Trust Grants, 1953
Harold R. Bechtel Testamentary Charitable Trust Grants, 1954
Hormel Foods Charitable Trust Grants, 2050
John Deere Foundation Grants, 2234
Marie H. Bechtel Charitable Remainder Uni-Trust Grants, 2432
Meta and George Rosenberg Foundation Grants, 2525
Mid-Iowa Health Foundation Community Response Grants, 2549
Moline Foundation Community Grants, 2590
Nationwide Insurance Foundation Grants, 2637
Norfolk Southern Foundation Grants, 2982
Peter Kiewit Foundation General Grants, 3154
Peter Kiewit Foundation Small Grants, 3155
Piper Jaffray Fndn Communities Giving Grants, 3227
Principal Financial Group Foundation Grants, 3273
Prudential Foundation Education Grants, 3276
Roy J. Carver Charitable Trust Medical and Science Research Grants, 3416
RRF General Program Grants, 3417
Shopko Fndn Community Charitable Grants, 3547
Square D Foundation Grants, 3634
Sunderland Foundation Grants, 3675
Tension Envelope Foundation Grants, 3707
U.S. Cellular Corporation Grants, 3766
Union Pacific Foundation Health and Human Services Grants, 3794
Waitt Family Foundation Grants, 3894

Kansas
Abbott Fund Access to Health Care Grants, 199
Abbott Fund Community Grants, 201
Abbott Fund Global AIDS Care Grants, 202
Abbott Fund Science Education Grants, 203
ACE Charitable Foundation Grants, 229
Albertson's Charitable Giving Grants, 410
Amerigroup Foundation Grants, 565
Archer Daniels Midland Foundation Grants, 651
Bayer Foundation Grants, 846
Boeing Company Contributions Grants, 930
Charles A. Frueauff Foundation Grants, 1114
Forrest C. Lattner Foundation Grants, 1688
H & R Foundation Grants, 1920
Hallmark Corporate Foundation Grants, 1936
Hormel Foods Charitable Trust Grants, 2050
Humana Foundation Grants, 2077
Hutchinson Community Foundation Grants, 2085
I.A. O'Shaughnessy Foundation Grants, 2088
J.E. and L.E. Mabee Foundation Grants, 2170
John Deere Foundation Grants, 2234
Kansas Health Fndn Major Initiatives Grants, 2288
Kansas Health Foundation Recognition Grants, 2289
Lewis H. Humphreys Charitable Trust Grants, 2361
MeadWestvaco Foundation Sustainable Communities Grants, 2506
Piper Jaffray Fndn Communities Giving Grants, 3227
Shopko Fndn Community Charitable Grants, 3547
Sunderland Foundation Grants, 3675
Sunflower Foundation Bridge Grants, 3676
Sunflower Foundation Capacity Building Grants, 3677
Sunflower Foundation Walking Trails Grants, 3678

Taylor S. Abernathy and Patti Harding Abernathy Charitable Trust Grants, 3704
Tension Envelope Foundation Grants, 3707
Textron Corporate Contributions Grants, 3712
Topeka Community Foundation Kansas Blood Services Scholarships, 3738
Union Pacific Foundation Health and Human Services Grants, 3794
United Methodist Child Mental Health Grants, 3800
United Methodist Health Ministry Fund Grants, 3801
W.M. Keck Fndn Undergrad Ed Grants, 3891

Kentucky
3M Company Fndn Community Giving Grants, 6
3M Fndn Health & Human Services Grants, 7
AEGON Transamerica Foundation Health and Welfare Grants, 316
A Good Neighbor Foundation Grants, 347
Alpha Natural Resources Corporate Giving, 450
American Electric Power Foundation Grants, 530
Amerigroup Foundation Grants, 565
Anheuser-Busch Foundation Grants, 591
Anthem Blue Cross and Blue Shield Grants, 607
Appalachian Regional Commission Health Grants, 637
Archer Daniels Midland Foundation Grants, 651
Ashland Corporate Contributions Grants, 713
Belk Foundation Grants, 872
Blue Grass Community Fndn Harrison Grants, 919
Blue Grass Community Foundation Hudson-Ellis Fund Grants, 920
Bridgestone/Firestone Trust Fund Grants, 949
Charles A. Frueauff Foundation Grants, 1114
Charles H. Dater Foundation Grants, 1118
Charles M. & Mary D. Grant Fndn Grants, 1124
Clark County Community Foundation Grants, 1175
Community Foundation of Louisville AIDS Project Fund Grants, 1289
Community Foundation of Louisville Anna Marble Memorial Fund for Princeton Grants, 1290
Community Foundation of Louisville Bobbye M. Robinson Fund Grants, 1291
Community Foundation of Louisville Delta Dental of Kentucky Fund Grants, 1292
Community Foundation of Louisville Diller B. and Katherine P. Groff Fund for Pediatric Surgery Grants, 1293
Community Foundation of Louisville Dr. W. Barnett Owen Memorial Fund for the Children of Louisville and Jefferson County Grants, 1294
Community Fndn of Louisville Health Grants, 1295
Community Foundation of Louisville Irving B. Klempner Fund Grants, 1296
Community Foundation of Louisville Lee Look Fund for Spinal Injury Grants, 1297
Community Foundation of Louisville Morris and Esther Lee Fund Grants, 1298
Cralle Foundation Grants, 1356
CSX Corporate Contributions Grants, 1368
Dean Foods Community Involvement Grants, 1440
Fifth Third Foundation Grants, 1661
Foundation for a Healthy Kentucky Grants, 1691
Gheens Foundation Grants, 1814
Greater Cincinnati Foundation Priority and Small Projects/Capacity-Building Grants, 1880
Green River Area Community Fndn Grants, 1895
H.B. Fuller Foundation Grants, 1922
Harmony Project Grants, 1948
Health Foundation of Greater Cincinnati Grants, 1983
Humana Foundation Grants, 2077
Huntington National Bank Community Grants, 2084
Impact 100 Grants, 2148
Jacob G. Schmidlapp Trust Grants, 2183
James Graham Brown Foundation Grants, 2189
Josephine Schell Russell Charitable Trust Grants, 2274
KeyBank Foundation Grants, 2307
Kimball International-Habig Foundation Health and Human Services Grants, 2308
Kosair Charities Grants, 2318
Lester E. Yeager Charitable Trust B Grants, 2359

Louisiana / 939

Lynn and Rovena Alexander Family Foundation Grants, 2399
Marathon Petroleum Corporation Grants, 2422
Norfolk Southern Foundation Grants, 2982
Paul Ogle Foundation Grants, 3124
Piper Jaffray Fndn Communities Giving Grants, 3227
PNC Foundation Community Services Grants, 3240
Rainbow Fund Grants, 3301
RRF General Program Grants, 3417
Shopko Fndn Community Charitable Grants, 3547
Square D Foundation Grants, 3634
Steele-Reese Foundation Grants, 3650
Sunoco Foundation Grants, 3679
Thomas Jefferson Rosenberg Foundation Grants, 3723
Thompson Charitable Foundation Grants, 3727
Toyota Motor Engineering & Manufacturing North America Grants, 3742
Toyota Motor Manufacturing of Indiana Grants, 3744
Toyota Motor Manufacturing of Kentucky Grants, 3745
Tri-State Community Twenty-first Century Endowment Fund Grants, 3756
USDA Delta Health Care Services Grants, 3829
Walter L. Gross III Family Foundation Grants, 3901
Welborn Baptist Fndn Health Ministries Grants, 3915
Welborn Baptist Foundation Improvements to Community Health Status Grants, 3916
Welborn Baptist Fndn School Health Grants, 3917
Weyerhaeuser Company Foundation Grants, 3925
WHAS Crusade for Children Grants, 3927

Louisiana
ACE Charitable Foundation Grants, 229
Adams and Reese Corporate Giving Grants, 299
Albertson's Charitable Giving Grants, 410
American Electric Power Foundation Grants, 530
Amerigroup Foundation Grants, 565
Baptist Community Ministries Grants, 828
Baton Rouge Area Foundation Grants, 838
Belk Foundation Grants, 872
Booth-Bricker Fund Grants, 933
Boyd Gaming Corporation Contributions Program, 941
Bridgestone/Firestone Trust Fund Grants, 949
Cabot Corporation Foundation Grants, 993
Caesars Foundation Grants, 995
Charles A. Frueauff Foundation Grants, 1114
Collins C. Diboll Private Foundation Grants, 1209
ConAgra Foods Fndn Community Impact Grants, 1321
ConAgra Foods Nourish our Community Grants, 1322
ConocoPhillips Foundation Grants, 1327
CSX Corporate Contributions Grants, 1368
Cudd Foundation Grants, 1370
Dean Foods Community Involvement Grants, 1440
Entergy Corporation Micro Grants, 1596
Entergy Corporation Open Grants for Healthy Families, 1597
FAR Fund Grants, 1631
Fluor Foundation Grants, 1682
FMC Foundation Grants, 1683
Foundation for the Mid South Community Development Grants, 1699
Foundation for the Mid South Health and Wellness Grants, 1700
Gheens Foundation Grants, 1814
Gil and Dody Weaver Foundation Grants, 1819
Gladys Brooks Foundation Grants, 1826
GNOF Albert N. & Hattie M. McClure Grants, 1834
GNOF Bayou Communities Grants, 1835
GNOF Exxon-Mobil Grants, 1836
GNOF Freeman Challenge Grants, 1837
GNOF Maison Hospitaliere Grants, 1843
GNOF Norco Community Grants, 1845
GNOF Organizational Effectiveness Grants and Workshops, 1846
GNOF Stand Up For Our Children Grants, 1847
Huie-Dellmon Trust Grants, 2075
Humana Foundation Grants, 2077
Joe W. and Dorothy Dorsett Brown Fndn Grants, 2230
John Deere Foundation Grants, 2234
Lockheed Martin Corp Foundation Grants, 2374
Marathon Petroleum Corporation Grants, 2422

940 / Louisiana

MeadWestvaco Foundation Sustainable Communities Grants, 2506
Norfolk Southern Foundation Grants, 2982
PeyBack Foundation Grants, 3159
Prudential Foundation Education Grants, 3276
Textron Corporate Contributions Grants, 3712
Union Pacific Foundation Health and Human Services Grants, 3794
USDA Delta Health Care Services Grants, 3829
W.M. Keck Fndn Undergrad Ed Grants, 3891
Weyerhaeuser Company Foundation Grants, 3925
Williams Companies Foundation Homegrown Giving Grants, 3957

Maine
Aetna Foundation Regional Health Grants, 333
Agnes M. Lindsay Trust Grants, 346
Agway Foundation Grants, 348
Albertson's Charitable Giving Grants, 410
Amerigroup Foundation Grants, 565
Amica Insurance Company Community Grants, 575
Anthem Blue Cross and Blue Shield Grants, 607
Barnes Group Foundation Grants, 832
BJ's Charitable Foundation Grants, 904
Carlisle Foundation Grants, 1025
Charles A. Frueauff Foundation Grants, 1114
ConAgra Foods Nourish our Community Grants, 1322
Davis Family Foundation Grants, 1435
Dean Foods Community Involvement Grants, 1440
Doree Taylor Charitable Foundation, 1490
Expect Miracles Foundation Grants, 1619
Fannie E. Rippel Foundation Grants, 1630
FMC Foundation Grants, 1683
Foundation for Seacoast Health Grants, 1698
Frances W. Emerson Foundation Grants, 1718
Francis T. & Louise T. Nichols Fndn Grants, 1720
George P. Davenport Trust Fund Grants, 1800
Gladys Brooks Foundation Grants, 1826
Greenfield Foundation of Maine Grants, 1893
Hannaford Charitable Foundation Grants, 1944
Harold Alfond Foundation Grants, 1949
HRAMF Charles H. Hood Foundation Child Health Research Awards, 2057
Jane's Trust Grants, 2206
Jessie B. Cox Charitable Trust Grants, 2225
KeyBank Foundation Grants, 2307
Kosciuszko Foundation Dr. Marie E. Zakrzewski Medical Scholarship, 2321
Libra Foundation Grants, 2363
Maine Community Foundation Charity Grants, 2415
Maine Community Foundation Hospice Grants, 2416
Maine Community Foundation Penobscot Valley Health Association Grants, 2417
Norfolk Southern Foundation Grants, 2982
R.T. Vanderbilt Trust Grants, 3294
Richard Davoud Donchian Foundation Grants, 3367
Robert and Joan Dircks Foundation Grants, 3379
ROSE Fund Grants, 3412
Salem Foundation Grants, 3457
Sasco Foundation Grants, 3487
Shaw's Supermarkets Donations, 3542
Simmons Foundation Grants, 3563
Sunoco Foundation Grants, 3679
T. Spencer Shore Foundation Grants, 3701
Yawkey Foundation Grants, 3980

Marshall Islands
ACF Native American Social and Economic Development Strategies Grants, 231
Bank of America Fndn Matching Gifts, 825
Bank of America Fndn Volunteer Grants, 826
NHLBI Ancillary Studies in Clinical Trials, 2688
NHLBI Bioengineering and Obesity Grants, 2689
NHLBI Career Transition Awards, 2694
NHLBI Lymphatics in Health and Disease in the Digestive, Cardio & Pulmonary Systems, 2704
NHLBI Lymphatics in Health and Disease in the Digestive, Urinary, Cardiovascular and Pulmonary Systems, 2705

NHLBI Mentored Career Dev Award to Promote Faculty Diversity in Biomed Research, 2708
NHLBI Mentored Patient-Oriented Research Career Development Awards, 2710
NHLBI Mentored Quantitative Research Career Development Awards, 2711
NHLBI Pathway to Independence Awards, 2718
NHLBI Research on the Role of Cardiomyocyte Mitochondria in Heart Disease: An Integrated Approach, 2726
NHLBI Ruth L. Kirschstein National Research Service Awards for Individual Postdoctoral Fellows, 2729
NHLBI Ruth L. Kirschstein National Research Service Awards for Individual Senior Fellows, 2732
NHLBI Short-Term Research Education Program to Increase Diversity in Health-Related Research, 2733
NHLBI Training Program Grants for Institutions that Promote Diversity, 2738
NIGMS Advancing Basic Behavioral and Social Research on Resilience: an Integrative Science Approach, 2861
NIGMS Enabling Resources for Pharmacogenomics Grants, 2862
NIH Mentored Clinical Scientist Research Career Development Award, 2887
NIH Ruth L. Kirschstein National Research Service Award Institutional Research Training Grants, 2900
NIH Ruth L. Kirschstein National Research Service Awards for Individual Postdoc Fellowships, 2902
NIH Ruth L. Kirschstein National Research Service Awards for Individual Predoctoral Fellows, 2903
NIH Ruth L. Kirschstein Nat Research Service Award Short-Term Institutional Training Grants, 2906
NIH Support of Competitive Research (SCORE) Pilot Project Awards, 2913
NIH Support of Competitive Research (SCORE) Research Advancement Awards, 2914

Maryland
Abell Foundation Criminal Justice and Addictions Grants, 208
Abell Fndn Health & Human Services Grants, 209
ACE Charitable Foundation Grants, 229
AEGON Transamerica Foundation Health and Welfare Grants, 316
Aetna Foundation Regional Health Grants, 333
Albertson's Charitable Giving Grants, 410
Alpha Research Foundation Grants, 452
Alvin and Fanny Blaustein Thalheimer Foundation Baltimore Communal Grants, 459
Amerigroup Foundation Grants, 565
Amica Insurance Company Community Grants, 575
Appalachian Regional Commission Health Grants, 637
Babcock Charitable Trust Grants, 816
Bechtel Group Foundation Building Positive Community Relationships Grants, 866
Belk Foundation Grants, 872
Bender Foundation Grants, 876
BJ's Charitable Foundation Grants, 904
Boeing Company Contributions Grants, 930
Brown Advisory Charitable Foundation Grants, 969
Carl M. Freeman Foundation FACES Grants, 1026
Carl M. Freeman Foundation Grants, 1027
Charles A. Frueauff Foundation Grants, 1114
Charles Delmar Foundation Grants, 1115
Chase Paymentech Corporate Giving Grants, 1127
Clayton Baker Trust Grants, 1179
Clayton Fund Grants, 1180
Community Foundation for the National Capital Region Community Leadership Grants, 1260
Community Foundation of the Eastern Shore Community Needs Grants, 1313
CSX Corporate Contributions Grants, 1368
Danellie Foundation Grants, 1421
Dean Foods Community Involvement Grants, 1440
Delmarva Power and Light Contributions, 1454
Deutsche Banc Alex Brown and Sons Charitable Foundation Grants, 1471
Dominion Foundation Grants, 1485
Eddie C. & Sylvia Brown Family Fndn Grants, 1548

Edward N. and Della L. Thome Memorial Foundation Grants, 1557
Eugene B. Casey Foundation Grants, 1609
Expect Miracles Foundation Grants, 1619
FMC Foundation Grants, 1683
France-Merrick Foundation Health and Human Services Grants, 1715
Freddie Mac Foundation Grants, 1738
General Mills Foundation Grants, 1782
Giant Food Charitable Grants, 1816
Gladys Brooks Foundation Grants, 1826
Helen Pumphrey Denit Charitable Trust Grants, 1996
Henry and Ruth Blaustein Rosenberg Foundation Health Grants, 2004
Hilda and Preston Davis Foundation Grants, 2028
J. Willard Marriott, Jr. Foundation Grants, 2178
Jacob and Hilda Blaustein Foundation Health and Mental Health Grants, 2181
John Edward Fowler Memorial Fndn Grants, 2235
John J. Leidy Foundation Grants, 2241
Kosciuszko Foundation Dr. Marie E. Zakrzewski Medical Scholarship, 2321
Land O'Lakes Foundation Mid-Atlantic Grants, 2341
Lockheed Martin Corp Foundation Grants, 2374
Mabel H. Flory Charitable Trust Grants, 2408
Marion I. and Henry J. Knott Foundation Discretionary Grants, 2435
Marion I. and Henry J. Knott Foundation Standard Grants, 2436
Marjorie C. Adams Charitable Trust Grants, 2437
Mercedes-Benz USA Corporate Contributions, 2517
Meyer Foundation Benevon Grants, 2532
Meyer Fndn Management Assistance Grants, 2534
Morris & Gwendolyn Cafritz Fndn Grants, 2594
Morton K. and Jane Blaustein Foundation Health Grants, 2595
Nationwide Insurance Foundation Grants, 2637
Norfolk Southern Foundation Grants, 2982
OSF Tackling Addiction Grants in Baltimore, 3096
Peter F. McManus Charitable Trust Grants, 3153
Philip L. Graham Fund Health and Human Services Grants, 3190
PNC Foundation Community Services Grants, 3240
Quality Health Foundation Grants, 3288
Rathmann Family Foundation Grants, 3309
Richard Davoud Donchian Foundation Grants, 3367
Rollins-Luetkemeyer Foundation Grants, 3404
Salmon Foundation Grants, 3458
Sunoco Foundation Grants, 3679
Taubman Endowment for the Arts Fndn Grants, 3702
Textron Corporate Contributions Grants, 3712
Thomas W. Bradley Foundation Grants, 3726
U.S. Cellular Corporation Grants, 3766
Verizon Foundation Maryland Grants, 3859
Washington Gas Charitable Giving, 3908
Weyerhaeuser Company Foundation Grants, 3925

Massachusetts
2 Depot Square Ipswich Charitable Fndn Grants, 4
3M Company Fndn Community Giving Grants, 6
3M Fndn Health & Human Services Grants, 7
1675 Foundation Grants, 11
Aaron Foundation Grants, 193
Abbot and Dorothy H. Stevens Foundation Grants, 197
Abbott Fund Access to Health Care Grants, 199
Abbott Fund Community Grants, 201
Abbott Fund Global AIDS Care Grants, 202
Abbott Fund Science Education Grants, 203
ACE Charitable Foundation Grants, 229
Adelaide Breed Bayrd Foundation Grants, 306
AEC Trust Grants, 315
Agnes M. Lindsay Trust Grants, 346
Agway Foundation Grants, 348
Ahn Family Foundation Grants, 356
Albertson's Charitable Giving Grants, 410
Albert W. Rice Charitable Foundation Grants, 411
Alfred E. Chase Charitable Foundation Grants, 430
Alice C. A. Sibley Fund Grants, 435
Amelia Sillman Rockwell and Carlos Perry Rockwell Charities Fund Grants, 501

GEOGRAPHIC INDEX

Amerigroup Foundation Grants, 565
Amica Insurance Company Community Grants, 575
Anheuser-Busch Foundation Grants, 591
Archer Daniels Midland Foundation Grants, 651
Auburn Foundation Grants, 798
Babcock Charitable Trust Grants, 816
Banfi Vintners Foundation Grants, 824
Barnes Group Foundation Grants, 832
BJ's Charitable Foundation Grants, 904
Boston Foundation Grants, 936
Boston Globe Foundation Grants, 937
Boston Jewish Community Women's Fund Grants, 938
Boyle Foundation Grants, 942
Bradley C. Higgins Foundation Grants, 945
Bristol-Myers Squibb Foundation Health Disparities Grants, 958
Cabot Corporation Foundation Grants, 993
Cambridge Community Foundation Grants, 1004
Carlisle Foundation Grants, 1025
Charles A. Frueauff Foundation Grants, 1114
Charles F. Bacon Trust Grants, 1117
Charles H. Farnsworth Trust Grants, 1119
Charles H. Hall Foundation, 1120
Charles H. Pearson Foundation Grants, 1121
ConAgra Foods Fndn Community Impact Grants, 1321
ConAgra Foods Nourish our Community Grants, 1322
CSX Corporate Contributions Grants, 1368
Dean Foods Community Involvement Grants, 1440
Deborah Munroe Noonan Memorial Grants, 1443
Dominion Foundation Grants, 1485
Edward Bangs Kelley and Elza Kelley Foundation Grants, 1556
Edwin S. Webster Foundation Grants, 1560
Elizabeth McGraw Foundation Grants, 1568
Entergy Corporation Micro Grants, 1596
Entergy Corporation Open Grants for Healthy Families, 1597
Essex County Community Foundation Discretionary Fund Grants, 1602
Essex County Community Foundation Merrimack Valley General Fund Grants, 1603
Essex County Community Foundation Webster Family Fund Grants, 1604
Essex County Community Foundation Women's Fund Grants, 1605
Expect Miracles Foundation Grants, 1619
Ezra M. Cutting Trust Grants, 1623
Fairlawn Foundation Grants, 1626
Fallon OrNda Community Health Fund Grants, 1627
Fannie E. Rippel Foundation Grants, 1630
Florian O. Bartlett Trust Grants, 1677
Frances W. Emerson Foundation Grants, 1718
Frank Reed and Margaret Jane Peters Memorial Fund I Grants, 1727
Frank Reed and Margaret Jane Peters Memorial Fund II Grants, 1728
Frank Stanley Beveridge Foundation Grants, 1730
Frank W. & Carl S. Adams Memorial Grants, 1731
General Mills Foundation Grants, 1782
George F. Baker Trust Grants, 1793
George W. Wells Foundation Grants, 1804
Georgiana Goddard Eaton Memorial Grants, 1805
GFWC of Massachusetts Catherine E. Philbin Scholarship, 1811
GFWC of Massachusetts Communication Disorder/Speech Therapy Scholarship, 1812
GFWC of Massachusetts Memorial Education Scholarship, 1813
Gladys Brooks Foundation Grants, 1826
Greater Worcester Community Foundation Discretionary Grants, 1888
Greater Worcester Community Foundation Jeppson Memorial Fund for Brookfield Grants, 1889
Hannaford Charitable Foundation Grants, 1944
Harold Brooks Foundation Grants, 1952
Hasbro Children's Fund Grants, 1973
Hasbro Corporation Gift of Play Hospital and Pediatric Health Giving, 1974
High Meadow Foundation Grants, 2027
Hilda and Preston Davis Foundation Grants, 2028
Horace Moses Charitable Foundation Grants, 2047
HRAMF Charles A. King Trust Postdoctoral Research Fellowships, 2056
HRAMF Charles H. Hood Foundation Child Health Research Awards, 2057
HRAMF Deborah Munroe Noonan Memorial Research Grants, 2058
HRAMF Smith Family Awards for Excellence in Biomedical Research, 2063
Ida S. Barter Trust Grants, 2116
Jane's Trust Grants, 2206
Jessie B. Cox Charitable Trust Grants, 2225
Johnson & Johnson Community Health Grants, 2252
John W. Alden Trust Grants, 2260
John W. Boynton Fund Grants, 2263
Josephine G. Russell Trust Grants, 2271
Kosciuszko Foundation Dr. Marie E. Zakrzewski Medical Scholarship, 2321
Mabel A. Horne Trust Grants, 2406
McCallum Family Foundation Grants, 2489
Medtronic Foundation CommunityLink Health Grants, 2510
Medtronic Foundation Community Link Human Services Grants, 2511
MetroWest Health Foundation Capital Grants for Health-Related Facilities, 2526
MetroWest Health Foundation Grants--Healthy Aging, 2527
MetroWest Health Foundation Grants to Reduce the Incidence of High Risk Behaviors Among Adolescents, 2528
MetroWest Health Foundation Grants to Schools to Conduct Mental Health Capacity Assessments, 2529
Middlesex Savings Charitable Foundation Basic Human Needs Grants, 2550
Middlesex Savings Charitable Foundation Capacity Building Grants, 2551
MMS and Alliance Charitable Foundation Grants for Community Action and Care for the Medically Uninsured, 2587
MMS and Alliance Charitable Foundation International Health Studies Grants, 2588
Nordson Corporation Foundation Grants, 2981
Norfolk Southern Foundation Grants, 2982
Peabody Foundation Grants, 3132
Perkin Fund Grants, 3139
Perpetual Trust for Charitable Giving Grants, 3141
Peter and Elizabeth C. Tower Foundation Annual Intellectual Disabilities Grants, 3144
Peter and Elizabeth C. Tower Foundation Annual Mental Health Grants, 3145
Peter and Elizabeth C. Tower Foundation Learning Disability Grants, 3146
Peter and Elizabeth C. Tower Foundation Mental Health Reference and Resource Materials Mini-Grants, 3147
Peter and Elizabeth C. Tower Foundation Organizational Scholarships, 3148
Peter and Elizabeth C. Tower Foundation Phase II Technology Initiative Grants, 3149
Peter and Elizabeth C. Tower Foundation Phase I Technology Initiative Grants, 3150
Peter and Elizabeth C. Tower Foundation Social and Emotional Preschool Curriculum Grants, 3151
Peter and Elizabeth C. Tower Foundation Substance Abuse Grants, 3152
Peter F. McManus Charitable Trust Grants, 3153
Price Chopper's Golub Foundation Grants, 3261
Richard & Susan Smith Family Fndn Grants, 3365
Robbins Charitable Foundation Grants, 3378
Robert and Joan Dircks Foundation Grants, 3379
Robert F. Stoico/FIRSTFED Charitable Foundation Grants, 3382
Rogers Family Foundation Grants, 3401
ROSE Fund Grants, 3412
Ruth H. & Warren A. Ellsworth Fndn Grants, 3429
Sara Elizabeth O'Brien Trust Grants, 3483
Seagate Tech Corp Capacity to Care Grants, 3502
Shaw's Supermarkets Donations, 3542
Simeon J. Fortin Charitable Foundation Grants, 3562
Sophia Romero Trust Grants, 3624
Sunoco Foundation Grants, 3679
T. Spencer Shore Foundation Grants, 3701
TE Foundation Grants, 3705
Textron Corporate Contributions Grants, 3712
Theodore Edson Parker Foundation Grants, 3715
Thomas Jefferson Rosenberg Foundation Grants, 3723
Thoracic Foundation Grants, 3729
US Airways Community Contributions, 3825
Verizon Foundation Northeast Region Grants, 3861
Wilhelmina W. Jackson Trust Scholarships, 3941
Winifred & Harry B. Allen Foundation Grants, 3962
Yawkey Foundation Grants, 3980

Michigan

3M Company Fndn Community Giving Grants, 6
3M Fndn Health & Human Services Grants, 7
Abbott Fund Access to Health Care Grants, 199
Abbott Fund Community Grants, 201
Abbott Fund Global AIDS Care Grants, 202
Abbott Fund Science Education Grants, 203
ACE Charitable Foundation Grants, 229
Albertson's Charitable Giving Grants, 410
Allegan County Community Foundation Grants, 439
American Electric Power Foundation Grants, 530
Amica Insurance Company Community Grants, 575
Ann and Robert H. Lurie Family Fndn Grants, 593
Ann Arbor Area Community Foundation Grants, 594
Archer Daniels Midland Foundation Grants, 651
Battle Creek Community Foundation Grants, 839
Batts Foundation Grants, 840
BCBSM Building Healthy Communities Engaging Elementary Schools and Community Partners Grants, 849
BCBSM Claude Pepper Award, 850
BCBSM Corporate Contributions Grants, 851
BCBSM Foundation Community Health Matching Grants, 852
BCBSM Foundation Excellence in Research Awards for Students, 853
BCBSM Foundation Investigator Initiated Research Grants, 854
BCBSM Foundation Physician Investigator Research Awards, 855
BCBSM Foundation Primary and Clinical Prevention Grants, 856
BCBSM Fndn Proposal Development Awards, 857
BCBSM Foundation Student Award Program, 858
BCBSM Frank J. McDevitt, DO, Excellence in Research Awards, 859
Beldon Fund Grants, 871
Berrien Community Foundation Grants, 884
Besser Foundation Grants, 887
Bridgestone/Firestone Trust Fund Grants, 949
Campbell Soup Foundation Grants, 1007
Capital Region Community Foundation Grants, 1014
Carls Foundation Grants, 1030
Carrier Corporation Contributions Grants, 1038
Cleveland-Cliffs Foundation Grants, 1181
Coleman Foundation Cancer Care Grants, 1206
Coleman Foundation Developmental Disabilities Grants, 1207
Comerica Charitable Foundation Grants, 1224
Community Fndn for Muskegon County Grants, 1254
Community Foundation for Northeast Michigan Mini-Grants, 1255
Community Foundation for Northeast Michigan Tobacco Settlement Grants, 1256
Community Foundation for Southeast Michigan Grants, 1258
Community Foundation of Greater Flint Grants, 1273
ConAgra Foods Fndn Community Impact Grants, 1321
ConAgra Foods Nourish our Community Grants, 1322
Consumers Energy Foundation, 1343
CSX Corporate Contributions Grants, 1368
Dean Foods Community Involvement Grants, 1440
DeRoy Testamentary Foundation Grants, 1469
DTE Energy Foundation Health and Human Services Grants, 1516
EDS Foundation Grants, 1552

Michigan

Edward N. and Della L. Thome Memorial Foundation Grants, 1557
Entergy Corporation Micro Grants, 1596
Entergy Corporation Open Grants for Healthy Families, 1597
Fifth Third Foundation Grants, 1661
Four County Community Fndn General Grants, 1710
Four County Community Foundation Healthy Senior/Healthy Youth Fund Grants, 1711
Fremont Area Community Foundation Amazing X Grants, 1744
Fremont Area Community Foundation Elderly Needs Grants, 1745
Fremont Area Community Fndn General Grants, 1746
General Mills Foundation Grants, 1782
Grand Haven Area Community Fndn Grants, 1864
Grand Rapids Area Community Foundation Wyoming Youth Fund Grants, 1868
Grand Rapids Community Foundation Grants, 1869
Grand Rapids Community Foundation Ionia County Grants, 1870
Grand Rapids Community Foundation Ionia County Youth Fund Grants, 1871
Grand Rapids Community Foundation Lowell Area Fund Grants, 1872
Grand Rapids Community Foundation Southeast Ottawa Grants, 1873
Grand Rapids Community Foundation Southeast Ottawa Youth Fund Grants, 1874
Grand Rapids Community Fndn Sparta Grants, 1875
Grand Rapids Community Foundation Sparta Youth Fund Grants, 1876
Guido A. & Elizabeth H. Binda Fndn Grants, 1913
H.B. Fuller Foundation Grants, 1922
Harry A. & Margaret D. Towsley Fndn Grants, 1959
Harvey Randall Wickes Foundation Grants, 1972
Herbert H. & Grace A. Dow Fndn Grants, 2008
Hillsdale County Community Foundation Healthy Senior/Healthy Youth Fund Grants, 2033
Hillsdale County Community General Adult Foundation Grants, 2034
Holland/Zeeland Community Fndn Grants, 2041
Hudson Webber Foundation Grants, 2072
Humana Foundation Grants, 2077
Huntington National Bank Community Grants, 2084
Intergrys Corporation Grants, 2160
Irving S. Gilmore Foundation Grants, 2163
J. Spencer Barnes Memorial Foundation Grants, 2175
Jackson County Community Foundation Unrestricted Grants, 2180
Jacob G. Schmidlapp Trust Grants, 2183
James A. and Faith Knight Foundation Grants, 2186
Jewish Fund Grants, 2227
KeyBank Foundation Grants, 2307
Marathon Petroleum Corporation Grants, 2422
Matilda R. Wilson Fund Grants, 2453
MGM Resorts Foundation Community Grants, 2541
Miller Foundation Grants, 2556
Norfolk Southern Foundation Grants, 2982
PNC Foundation Community Services Grants, 3240
Pokagon Fund Grants, 3242
Richard and Helen DeVos Foundation Grants, 3364
Saginaw Community Foundation Senior Citizen Enrichment Fund, 3450
Shopko Fndn Community Charitable Grants, 3547
Skillman Fndn Good Neighborhoods Grants, 3577
Straits Area Community Foundation Health Care Grants, 3657
Sunoco Foundation Grants, 3679
Taubman Endowment for the Arts Fndn Grants, 3702
TE Foundation Grants, 3705
Wege Foundation Grants, 3912
Weyerhaeuser Company Foundation Grants, 3925

Minnesota

3M Company Fndn Community Giving Grants, 6
3M Fndn Health & Human Services Grants, 7
ACE Charitable Foundation Grants, 229
Albertson's Charitable Giving Grants, 410
Alliant Energy Corporation Contributions, 445
Amica Insurance Company Community Grants, 575
Andersen Corporate Foundation, 586
Archer Daniels Midland Foundation Grants, 651
Athwin Foundation Grants, 795
Beldon Fund Grants, 871
Blue Cross Blue Shield of Minnesota Foundation - Health Equity: Building Health Equity Together Grants, 914
Blue Cross Blue Shield of Minnesota Foundation - Healthy Children: Growing Up Healthy Grants, 915
Blue Cross Blue Shield of Minnesota Fndn Healthy Equity: Public Libraries for Health Grants, 918
Bridgestone/Firestone Trust Fund Grants, 949
Bush Fndn Health & Human Services Grants, 979
Bush Foundation Medical Fellowships, 980
Carl & Eloise Pohlad Family Fndn Grants, 1021
Carolyn Foundation Grants, 1034
Cleveland-Cliffs Foundation Grants, 1181
ConAgra Foods Fndn Community Impact Grants, 1321
ConAgra Foods Nourish our Community Grants, 1322
Dairy Queen Corporate Contributions Grants, 1413
Dean Foods Community Involvement Grants, 1440
Edina Realty Foundation Grants, 1550
Edwards Memorial Trust Grants, 1558
Elmer L. & Eleanor J. Andersen Fndn Grants, 1578
Emma B. Howe Memorial Foundation Grants, 1593
Fargo-Moorhead Area Foundation Grants, 1632
Fargo-Moorhead Area Fndn Woman's Grants, 1633
General Mills Foundation Grants, 1782
George Family Foundation Grants, 1794
Ginn Foundation Grants, 1822
Graco Foundation Grants, 1860
Grand Rapids Area Community Fndn Grants, 1865
Grand Rapids Area Community Foundation Nashwauk Area Endowment Fund Grants, 1866
Grotto Foundation Project Grants, 1905
H.B. Fuller Foundation Grants, 1922
Hormel Foods Charitable Trust Grants, 2050
Hormel Foundation Grants, 2051
HRK Foundation Health Grants, 2067
Hugh J. Andersen Foundation Grants, 2074
I.A. O'Shaughnessy Foundation Grants, 2088
Intergrys Corporation Grants, 2160
James Ford Bell Foundation Grants, 2187
James R. Thorpe Foundation Grants, 2199
Jay and Rose Phillips Family Foundation Grants, 2213
Lockheed Martin Corp Foundation Grants, 2374
Mardag Foundation Grants, 2428
MeadWestvaco Foundation Sustainable Communities Grants, 2506
Medtronic Foundation CommunityLink Health Grants, 2510
Medtronic Foundation Community Link Human Services Grants, 2511
Minneapolis Foundation Community Grants, 2558
NMF Hugh J. Andersen Memorial Scholarship, 2958
Northwest Minnesota Foundation Women's Fund Grants, 3000
Ordean Foundation Grants, 3080
Otto Bremer Foundation Grants, 3099
Piper Jaffray Fndn Communities Giving Grants, 3227
Pohlad Family Foundation, 3241
Prudential Foundation Education Grants, 3276
Rathmann Family Foundation Grants, 3309
Rochester Area Foundation Grants, 3392
Salem Foundation Grants, 3457
Seagate Tech Corp Capacity to Care Grants, 3502
Shopko Fndn Community Charitable Grants, 3547
Symantec Community Relations and Corporate Philanthropy Grants, 3699
Tension Envelope Foundation Grants, 3707
Union Pacific Foundation Health and Human Services Grants, 3794
Weyerhaeuser Company Foundation Grants, 3925
Whitney Foundation Grants, 3938

Mississippi

Adams and Reese Corporate Giving Grants, 299
Albertson's Charitable Giving Grants, 410
Appalachian Regional Commission Health Grants, 637
Archer Daniels Midland Foundation Grants, 651
Assisi Fndn of Memphis Capital Project Grants, 786
Assisi Foundation of Memphis General Grants, 787
BancorpSouth Foundation Grants, 823
Belk Foundation Grants, 872
Boyd Gaming Corporation Contributions Program, 941
Bridgestone/Firestone Trust Fund Grants, 949
Caesars Foundation Grants, 995
Charles A. Frueauff Foundation Grants, 1114
Charles M. & Mary D. Grant Fndn Grants, 1124
ConAgra Foods Fndn Community Impact Grants, 1321
ConAgra Foods Nourish our Community Grants, 1322
CSX Corporate Contributions Grants, 1368
Entergy Corporation Micro Grants, 1596
Entergy Corporation Open Grants for Healthy Families, 1597
Foundation for the Mid South Community Development Grants, 1699
Foundation for the Mid South Health and Wellness Grants, 1700
Gil and Dody Weaver Foundation Grants, 1819
Gulf Coast Community Foundation Grants, 1915
Joe W. and Dorothy Dorsett Brown Fndn Grants, 2230
Lockheed Martin Corp Foundation Grants, 2374
MGM Resorts Foundation Community Grants, 2541
Norfolk Southern Foundation Grants, 2982
Rainbow Fund Grants, 3301
Riley Foundation Grants, 3374
USDA Delta Health Care Services Grants, 3829
W.M. Keck Fndn Undergrad Ed Grants, 3891
Weyerhaeuser Company Foundation Grants, 3925

Missouri

3M Company Fndn Community Giving Grants, 6
3M Fndn Health & Human Services Grants, 7
ACF Foundation Grants, 230
Albertson's Charitable Giving Grants, 410
Allen P. & Josephine B. Green Fndn Grants, 443
Amerigroup Foundation Grants, 565
Anheuser-Busch Foundation Grants, 591
Anthem Blue Cross and Blue Shield Grants, 607
Archer Daniels Midland Foundation Grants, 651
Baptist-Trinity Lutheran Legacy Fndn Grants, 827
Boeing Company Contributions Grants, 930
Caesars Foundation Grants, 995
CDI Interdisciplinary Research Initiatives Grants, 1074
CDI Postdoctoral Fellowships, 1075
Charles A. Frueauff Foundation Grants, 1114
Christine & Katharina Pauly Trust Grants, 1150
ConAgra Foods Fndn Community Impact Grants, 1321
ConAgra Foods Nourish our Community Grants, 1322
Cooper Industries Foundation Grants, 1345
Dana Brown Charitable Trust Grants, 1419
Edward N. and Della L. Thome Memorial Foundation Grants, 1557
Elaine Feld Stern Charitable Trust Grants, 1565
FMC Foundation Grants, 1683
General Mills Foundation Grants, 1782
Greater Saint Louis Community Fndn Grants, 1884
H & R Foundation Grants, 1920
Hallmark Corporate Foundation Grants, 1936
Harley Davidson Foundation Grants, 1947
Harry Edison Foundation, 1962
Helen S. Boylan Foundation Grants, 1997
Herbert A. & Adrian W. Woods Fndn Grants, 2007
J.B. Reynolds Foundation Grants, 2168
J.E. and L.E. Mabee Foundation Grants, 2170
John Deere Foundation Grants, 2234
John W. Speas and Effie E. Speas Memorial Trust Grants, 2264
Joseph H. & Florence A. Roblee Fndn Grants, 2269
Kent D. Steadley and Mary L. Steadley Memorial Trust Grants, 2301
Louetta M. Cowden Foundation Grants, 2377
Louis and Elizabeth Nave Flarsheim Charitable Foundation Grants, 2379
Norfolk Southern Foundation Grants, 2982
Oppenstein Brothers Foundation Grants, 3078
Piper Jaffray Fndn Communities Giving Grants, 3227
PNC Foundation Community Services Grants, 3240

GEOGRAPHIC INDEX

Pott Foundation Grants, 3253
RRF General Program Grants, 3417
Saigh Foundation Grants, 3451
Saint Louis Rams Fndn Community Donations, 3453
Sensient Technologies Foundation Grants, 3513
Shopko Fndn Community Charitable Grants, 3547
Sunderland Foundation Grants, 3675
Taylor S. Abernathy and Patti Harding Abernathy Charitable Trust Grants, 3704
Tension Envelope Foundation Grants, 3707
U.S. Cellular Corporation Grants, 3766
Union Pacific Foundation Health and Human Services Grants, 3794
USDA Delta Health Care Services Grants, 3829
Victor E. Speas Foundation Grants, 3868
Weyerhaeuser Company Foundation Grants, 3925
William J. Brace Charitable Trust, 3955

Montana
Albertson's Charitable Giving Grants, 410
Archer Daniels Midland Foundation Grants, 651
Avista Foundation Economic and Cultural Vitality Grants, 812
Browning-Kimball Foundation Grants, 971
Bullitt Foundation Grants, 973
ConocoPhillips Foundation Grants, 1327
Dean Foods Community Involvement Grants, 1440
Dennis & Phyllis Washington Fndn Grants, 1458
Foster Foundation Grants, 1689
General Mills Foundation Grants, 1782
M.J. Murdock Charitable Trust General Grants, 2405
Piper Jaffray Fndn Communities Giving Grants, 3227
Pride Foundation Grants, 3268
Pride Foundation Scholarships, 3269
Ruth and Vernon Taylor Foundation Grants, 3426
Shopko Fndn Community Charitable Grants, 3547
Steele-Reese Foundation Grants, 3650
Sunderland Foundation Grants, 3675
Union Pacific Foundation Health and Human Services Grants, 3794
W.M. Keck Fndn Undergrad Ed Grants, 3891

Nebraska
3M Company Fndn Community Giving Grants, 6
3M Fndn Health & Human Services Grants, 7
Abel Foundation Grants, 206
Albertson's Charitable Giving Grants, 410
Archer Daniels Midland Foundation Grants, 651
Charles A. Frueauff Foundation Grants, 1114
Dean Foods Community Involvement Grants, 1440
Ethel S. Abbott Charitable Foundation Grants, 1607
Gardner Foundation Grants, 1770
Hormel Foods Charitable Trust Grants, 2050
Ike and Roz Friedman Foundation Grants, 2145
Lincoln Financial Foundation Grants, 2367
Merrick Foundation Grants, 2522
Nationwide Insurance Foundation Grants, 2637
Peter Kiewit Foundation General Grants, 3154
Peter Kiewit Foundation Small Grants, 3155
Phelps County Community Foundation Grants, 3188
Piper Jaffray Fndn Communities Giving Grants, 3227
Principal Financial Group Foundation Grants, 3273
Shopko Fndn Community Charitable Grants, 3547
Square D Foundation Grants, 3634
Sunderland Foundation Grants, 3675
U.S. Cellular Corporation Grants, 3766
Union Pacific Foundation Health and Human Services Grants, 3794
Waitt Family Foundation Grants, 3894
Western Union Foundation Grants, 3923

Nevada
ACE Charitable Foundation Grants, 229
Albertson's Charitable Giving Grants, 410
Amerigroup Foundation Grants, 565
Amica Insurance Company Community Grants, 575
Andre Agassi Charitable Foundation Grants, 587
Anthem Blue Cross and Blue Shield Grants, 607
ARCO Foundation Education Grants, 652
Boeing Company Contributions Grants, 930

Boyd Gaming Corporation Contributions Program, 941
Bridgestone/Firestone Trust Fund Grants, 949
Business Bank of Nevada Community Grants, 981
Caesars Foundation Grants, 995
Carrier Corporation Contributions Grants, 1038
Dean Foods Community Involvement Grants, 1440
Del E. Webb Foundation Grants, 1450
E.L. Wiegand Foundation Grants, 1538
Greenspun Family Foundation Grants, 1896
MeadWestvaco Foundation Sustainable Communities Grants, 2506
MGM Resorts Foundation Community Grants, 2541
Nell J. Redfield Foundation Grants, 2673
PacifiCare Foundation Grants, 3102
Piper Jaffray Fndn Communities Giving Grants, 3227
Sands Foundation Grants, 3476
Southwest Gas Corporation Foundation Grants, 3628
Union Pacific Foundation Health and Human Services Grants, 3794
US Airways Community Contributions, 3825
W.M. Keck Fndn Undergrad Ed Grants, 3891

New Hampshire
Agnes M. Lindsay Trust Grants, 346
Agway Foundation Grants, 348
Albertson's Charitable Giving Grants, 410
Alexander Eastman Foundation Grants, 420
Amerigroup Foundation Grants, 565
Amica Insurance Company Community Grants, 575
Anheuser-Busch Foundation Grants, 591
Anthem Blue Cross and Blue Shield Grants, 607
Barker Foundation Grants, 830
Barnes Group Foundation Grants, 832
BJ's Charitable Foundation Grants, 904
Carlisle Foundation Grants, 1025
Charles A. Frueauff Foundation Grants, 1114
Chase Paymentech Corporate Giving Grants, 1127
ConAgra Foods Nourish our Community Grants, 1322
Entergy Corporation Micro Grants, 1596
Entergy Corporation Open Grants for Healthy Families, 1597
Expect Miracles Foundation Grants, 1619
Fannie E. Rippel Foundation Grants, 1630
Foundation for Seacoast Health Grants, 1698
Gladys Brooks Foundation Grants, 1826
Hannaford Charitable Foundation Grants, 1944
Hilda and Preston Davis Foundation Grants, 2028
HRAMF Charles H. Hood Foundation Child Health Research Awards, 2057
J. Willard Marriott, Jr. Foundation Grants, 2178
Jane's Trust Grants, 2206
Jessie B. Cox Charitable Trust Grants, 2225
Kosciuszko Foundation Dr. Marie E. Zakrzewski Medical Scholarship, 2321
Lincoln Financial Foundation Grants, 2367
McLean Contributionship Grants, 2497
NHSCA Arts in Health Care Project Grants, 2741
Norfolk Southern Foundation Grants, 2982
Norwin S. & Elizabeth N. Bean Fndn Grants, 3001
Oleonda Jameson Trust Grants, 3071
Price Chopper's Golub Foundation Grants, 3261
Robert and Joan Dircks Foundation Grants, 3379
Rogers Family Foundation Grants, 3401
ROSE Fund Grants, 3412
Salmon Foundation Grants, 3458
Shaw's Supermarkets Donations, 3542
Sunoco Foundation Grants, 3679
T. Spencer Shore Foundation Grants, 3701
Weyerhaeuser Company Foundation Grants, 3925
Yawkey Foundation Grants, 3980

New Jersey
1 in 9: Long Island Breast Cancer Action Coalition Grants, 1
3M Fndn Health & Human Services Grants, 7
Abbott Fund Access to Health Care Grants, 199
Abbott Fund Community Grants, 201
Abbott Fund Global AIDS Care Grants, 202
Abbott Fund Science Education Grants, 203
ACE Charitable Foundation Grants, 229

Aetna Foundation Regional Health Grants, 333
Albertson's Charitable Giving Grants, 410
Amerigroup Foundation Grants, 565
Amica Insurance Company Community Grants, 575
Anheuser-Busch Foundation Grants, 591
Ansell, Zaro, Grimm & Aaron Foundation Grants, 603
Archer Daniels Midland Foundation Grants, 651
Bayer Foundation Grants, 846
Bildner Family Foundation Grants, 891
BJ's Charitable Foundation Grants, 904
Bodman Foundation Grants, 929
Boyd Gaming Corporation Contributions Program, 941
Bristol-Myers Squibb Foundation Health Disparities Grants, 958
Brook J. Lenfest Foundation Grants, 967
Caesars Foundation Grants, 995
Campbell Soup Foundation Grants, 1007
Carl C. Icahn Foundation Grants, 1023
Charles A. Frueauff Foundation Grants, 1114
Charles Edison Fund Grants, 1116
ConocoPhillips Foundation Grants, 1327
CSX Corporate Contributions Grants, 1368
Danellie Foundation Grants, 1421
Dean Foods Community Involvement Grants, 1440
E.J. Grassmann Trust Grants, 1537
Edward W. and Stella C. Van Houten Memorial Fund Grants, 1559
Expect Miracles Foundation Grants, 1619
F.M. Kirby Foundation Grants, 1624
FirstEnergy Foundation Community Grants, 1668
Fluor Foundation Grants, 1682
FMC Foundation Grants, 1683
Frances L. and Edwin L. Cummings Memorial Fund Grants, 1717
Frank S. Flowers Foundation Grants, 1729
General Mills Foundation Grants, 1782
George A Ohl Jr. Foundation Grants, 1791
Gladys Brooks Foundation Grants, 1826
Hackett Foundation Grants, 1926
Healthcare Foundation of New Jersey Grants, 1982
Hilda and Preston Davis Foundation Grants, 2028
Horizon Foundation for New Jersey Grants, 2048
Independence Community Foundation Community Quality of Life Grant, 2153
Johnson & Johnson Community Health Grants, 2252
Land O'Lakes Foundation Mid-Atlantic Grants, 2341
Little Life Foundation Grants, 2372
Lockheed Martin Corp Foundation Grants, 2374
Lydia deForest Charitable Trust Grants, 2393
Mary Owen Borden Foundation Grants, 2446
MeadWestvaco Foundation Sustainable Communities Grants, 2506
Mercedes-Benz USA Corporate Contributions, 2517
Natalie W. Furniss Charitable Trust Grants, 2626
New Jersey Osteopathic Education Foundation Scholarships, 2680
Norfolk Southern Foundation Grants, 2982
OceanFirst Foundation Major Grants, 3062
PNC Foundation Community Services Grants, 3240
Princeton Area Community Foundation Greater Mercer Grants, 3272
Prudential Foundation Education Grants, 3276
Qualcomm Grants, 3287
Richard Davoud Donchian Foundation Grants, 3367
Robert Leet & Clara Guthrie Patterson Grants, 3383
Ruth and Vernon Taylor Foundation Grants, 3426
Ruth Estrin Goldberg Memorial for Cancer Research Grants, 3428
RWJF New Jersey Health Initiatives Grants, 3440
Sunoco Foundation Grants, 3679
Tyler Aaron Bookman Memorial Foundation Trust Grants, 3765
Weyerhaeuser Company Foundation Grants, 3925
Williams Companies Foundation Homegrown Giving Grants, 3957

New Mexico
Albertson's Charitable Giving Grants, 410
Albuquerque Community Foundation Grants, 412
Amerigroup Foundation Grants, 565

944 / New Mexico

BBVA Compass Foundation Charitable Grants, 848
Boeing Company Contributions Grants, 930
Cabot Corporation Foundation Grants, 993
Charles A. Frueauff Foundation Grants, 1114
ConAgra Foods Fndn Community Impact Grants, 1321
ConAgra Foods Nourish our Community Grants, 1322
Cudd Foundation Grants, 1370
Daniels Fund Grants-Aging, 1424
Dean Foods Community Involvement Grants, 1440
GenCorp Foundation Grants, 1779
General Mills Foundation Grants, 1782
Gil and Dody Weaver Foundation Grants, 1819
J.E. and L.E. Mabee Foundation Grants, 2170
Lockheed Martin Corp Foundation Grants, 2374
Santa Fe Community Foundation Seasonal Grants-Fall Cycle, 3482
Stocker Foundation Grants, 3656
Union Pacific Foundation Health and Human Services Grants, 3794
W.M. Keck Fndn Undergrad Ed Grants, 3891
Weyerhaeuser Company Foundation Grants, 3925

New York
1 in 9: Long Island Breast Cancer Action Coalition Grants, 1
3M Company Fndn Community Giving Grants, 6
3M Fndn Health & Human Services Grants, 7
Aaron & Freda Glickman Foundation Grants, 192
Abbott Fund Access to Health Care Grants, 199
Abbott Fund Community Grants, 201
Abbott Fund Global AIDS Care Grants, 202
Abbott Fund Science Education Grants, 203
ACE Charitable Foundation Grants, 229
Achelis Foundation Grants, 234
Adams Foundation Grants, 301
AEGON Transamerica Foundation Health and Welfare Grants, 316
Aetna Foundation Regional Health Grants, 333
Agnes Gund Foundation Grants, 345
Ahn Family Foundation Grants, 356
Alexis Gregory Foundation Grants, 426
Allyn Foundation Grants, 448
Altman Foundation Health Care Grants, 457
Amerigroup Foundation Grants, 565
Amica Insurance Company Community Grants, 575
Anheuser-Busch Foundation Grants, 591
Appalachian Regional Commission Health Grants, 637
Aratani Foundation Grants, 649
Archer Daniels Midland Foundation Grants, 651
Arkell Hall Foundation Grants, 661
ARS Foundation Grants, 667
Arthur and Rochelle Belfer Foundation Grants, 670
Arthur F. & Arnold M. Frankel Fndn Grants, 672
Assurant Foundation Grants, 788
Axe-Houghton Foundation Grants, 815
Babcock Charitable Trust Grants, 816
Banfi Vintners Foundation Grants, 824
Barker Welfare Foundation Grants, 831
Bernard F. and Alva B. Gimbel Foundation Reproductive Rights Grants, 883
Bildner Family Foundation Grants, 891
BJ's Charitable Foundation Grants, 904
Bodman Foundation Grants, 929
Bristol-Myers Squibb Foundation Health Disparities Grants, 958
Brooklyn Community Foundation Caring Neighbors Grants, 968
Carl C. Icahn Foundation Grants, 1023
Carnahan-Jackson Foundation Grants, 1032
Carrier Corporation Contributions Grants, 1038
Charles A. Frueauff Foundation Grants, 1114
Charles Edison Fund Grants, 1116
Clayton Fund Grants, 1180
Community Fndn for Greater Buffalo Grants, 1247
Community Fndn for the Capital Region Grants, 1259
Cooper Industries Foundation Grants, 1345
Cowles Charitable Trust Grants, 1354
CSX Corporate Contributions Grants, 1368
Daisy Marquis Jones Foundation Grants, 1414
Dammann Fund Grants, 1418

Daniel & Nanna Stern Family Fndn Grants, 1422
David Geffen Foundation Grants, 1431
Dean Foods Community Involvement Grants, 1440
DeMatteis Family Foundation Grants, 1457
Dolan Children's Foundation Grants, 1482
Dyson Foundation Mid-Hudson Valley General Operating Support Grants, 1533
Dyson Foundation Mid-Hudson Valley Project Support Grants, 1534
E.L. Wiegand Foundation Grants, 1538
Elizabeth McGraw Foundation Grants, 1568
Emma Barnsley Foundation Grants, 1594
Entergy Corporation Micro Grants, 1596
Entergy Corporation Open Grants for Healthy Families, 1597
Eugene M. Lang Foundation Grants, 1611
Expect Miracles Foundation Grants, 1619
Fluor Foundation Grants, 1682
FMC Foundation Grants, 1683
Foundation of CVPH April LaValley Grants, 1701
Foundation of CVPH Chelsea's Rainbow Grants, 1702
Foundation of CVPH Melissa Lahtinen-Penfield Organ Donor Fund Grants, 1703
Foundation of CVPH Rabin Fund Grants, 1704
Foundation of CVPH Roger Senecal Endowment Fund Grants, 1705
Foundation of CVPH Travel Fund Grants, 1706
Frances L. and Edwin L. Cummings Memorial Fund Grants, 1717
Frederick McDonald Trust Grants, 1740
Fred L. Emerson Foundation Grants, 1743
G.A. Ackermann Memorial Fund Grants, 1767
Gebbie Foundation Grants, 1777
General Mills Foundation Grants, 1782
Genesis Foundation Grants, 1786
George F. Baker Trust Grants, 1793
Gladys Brooks Foundation Grants, 1826
Gregory Family Foundation Grants (Florida), 1899
Griffin Family Foundation Grants, 1902
H. Schaffer Foundation Grants, 1925
Hackett Foundation Grants, 1926
Hagedorn Fund Grants, 1934
Hannaford Charitable Foundation Grants, 1944
Harry S. Black and Allon Fuller Fund Grants, 1966
Heckscher Foundation for Children Grants, 1988
Helena Rubinstein Foundation Grants, 1991
Herman Goldman Foundation Grants, 2012
Hilda and Preston Davis Foundation Grants, 2028
Howard and Bush Foundation Grants, 2054
IIE Western Union Family Scholarships, 2143
Independence Community Foundation Community Quality of Life Grant, 2153
Jacob and Valeria Langeloth Foundation Grants, 2182
James H. Cummings Foundation Grants, 2190
Jean and Louis Dreyfus Foundation Grants, 2215
John R. Oishei Foundation Grants, 2249
Joseph Alexander Foundation Grants, 2265
Josephine Goodyear Foundation Grants, 2272
KeyBank Foundation Grants, 2307
Komen Greater NYC Clinical Research Enrollment Grants, 2315
Komen Greater NYC Community Breast Health Grants, 2316
Komen Greater NYC Small Grants, 2317
Land O'Lakes Foundation Mid-Atlantic Grants, 2341
Laura B. Vogler Foundation Grants, 2345
Lawrence S. Huntington Fund Grants, 2348
Lockheed Martin Corp Foundation Grants, 2374
Long Island Community Foundation Grants, 2375
Louis H. Aborn Foundation Grants, 2380
M.B. and Edna Zale Foundation Grants, 2401
Marie C. and Joseph C. Wilson Foundation Rochester Small Grants, 2431
Marjorie C. Adams Charitable Trust Grants, 2437
MeadWestvaco Foundation Sustainable Communities Grants, 2506
Metzger-Price Fund Grants, 2530
Miles of Hope Breast Cancer Foundation Grants, 2553
Morton K. and Jane Blaustein Foundation Health Grants, 2595

GEOGRAPHIC INDEX

Nationwide Insurance Foundation Grants, 2637
NMF Aura E. Severinghaus Award, 2954
Norfolk Southern Foundation Grants, 2982
Northern Chautauqua Community Foundation Community Grants, 2995
Northern New York Community Fndn Grants, 2996
NYCT AIDS/HIV Grants, 3048
NYCT Biomedical Research Grants, 3049
NYCT Blindness and Visual Disabilities Grants, 3050
NYCT Blood Disease Research Grants, 3051
NYCT Children/Youth with Disabilities Grants, 3052
NYCT Girls and Young Women Grants, 3053
NYCT Grants for the Elderly, 3054
NYCT Health Care Services, Systems, and Policies Grants, 3055
NYCT Mental Health and Retardation Grants, 3056
NYCT Substance Abuse Grants, 3057
NYCT Technical Assistance Grants, 3058
Paul Rapoport Foundation Grants, 3125
Perkin Fund Grants, 3139
Peter and Elizabeth C. Tower Foundation Annual Intellectual Disabilities Grants, 3144
Peter and Elizabeth C. Tower Foundation Annual Mental Health Grants, 3145
Peter and Elizabeth C. Tower Foundation Learning Disability Grants, 3146
Peter and Elizabeth C. Tower Foundation Mental Health Reference and Resource Materials Mini-Grants, 3147
Peter and Elizabeth C. Tower Foundation Organizational Scholarships, 3148
Peter and Elizabeth C. Tower Foundation Phase II Technology Initiative Grants, 3149
Peter and Elizabeth C. Tower Foundation Phase I Technology Initiative Grants, 3150
Peter and Elizabeth C. Tower Foundation Social and Emotional Preschool Curriculum Grants, 3151
Peter and Elizabeth C. Tower Foundation Substance Abuse Grants, 3152
Price Chopper's Golub Foundation Grants, 3261
Prudential Foundation Education Grants, 3276
R.T. Vanderbilt Trust Grants, 3294
Richard Davoud Donchian Foundation Grants, 3367
Robert Leet & Clara Guthrie Patterson Grants, 3383
Robert R. McCormick Trib Community Grants, 3385
Rochester Area Community Foundation Grants, 3391
Ruth and Vernon Taylor Foundation Grants, 3426
Ruth Estrin Goldberg Memorial for Cancer Research Grants, 3428
Sandy Hill Foundation Grants, 3477
Sasco Foundation Grants, 3487
Stella and Charles Guttman Foundation Grants, 3651
Sunoco Foundation Grants, 3679
SunTrust Bank Trusteed Foundations Nell Warren Elkin and William Simpson Elkin Grants, 3683
T. Spencer Shore Foundation Grants, 3701
Taubman Endowment for the Arts Fndn Grants, 3702
Textron Corporate Contributions Grants, 3712
Thomas Thompson Trust Grants, 3725
United Hospital Fund of New York Health Care Improvement Grants, 3797
US Airways Community Contributions, 3825
Verizon Foundation New York Grants, 3860
Western New York Foundation Grants, 3922
Western Union Foundation Grants, 3923
Winifred & Harry B. Allen Foundation Grants, 3962
Young Ambassador Scholarship In Memory of Christopher Nordquist, 3981

North Carolina
Abbott Fund Access to Health Care Grants, 199
Abbott Fund Community Grants, 201
Abbott Fund Global AIDS Care Grants, 202
Abbott Fund Science Education Grants, 203
Able To Serve Grants, 212
ACE Charitable Foundation Grants, 229
Aetna Foundation Regional Health Grants, 333
American Schlafhorst Foundation Grants, 559
Amica Insurance Company Community Grants, 575
Appalachian Regional Commission Health Grants, 637

GEOGRAPHIC INDEX

Archer Daniels Midland Foundation Grants, 651
Bayer Foundation Grants, 846
BCBSNC Foundation Grants, 860
Beldon Fund Grants, 871
Belk Foundation Grants, 872
BJ's Charitable Foundation Grants, 904
Blanche and Julian Robertson Family Foundation Grants, 911
Blumenthal Foundation Grants, 925
Bridgestone/Firestone Trust Fund Grants, 949
Burlington Industries Foundation Grants, 977
Caesars Foundation Grants, 995
Camp-Younts Foundation Grants, 1005
Campbell Soup Foundation Grants, 1007
Carrie E. and Lena V. Glenn Foundation Grants, 1036
Carrier Corporation Contributions Grants, 1038
Catherine Kennedy Home Foundation Grants, 1043
Cemala Foundation Grants, 1077
Charles A. Frueauff Foundation Grants, 1114
Charles M. & Mary D. Grant Fndn Grants, 1124
Community Foundation of Greater Greensboro Community Grants, 1277
Community Foundation of Greater Greensboro Women to Women Fund Grants, 1278
ConAgra Foods Fndn Community Impact Grants, 1321
ConAgra Foods Nourish our Community Grants, 1322
Cone Health Foundation Grants, 1323
Cooper Industries Foundation Grants, 1345
Covenant Educational Foundation Grants, 1348
CSX Corporate Contributions Grants, 1368
Cumberland Community Foundation Grants, 1374
D.F. Halton Foundation Grants, 1405
Dean Foods Community Involvement Grants, 1440
Dickson Foundation Grants, 1476
Dominion Foundation Grants, 1485
Duke Endowment Health Care Grants, 1519
Dunspaugh-Dalton Foundation Grants, 1530
F.M. Kirby Foundation Grants, 1624
Fluor Foundation Grants, 1682
FMC Foundation Grants, 1683
Gardner W. & Joan G. Heidrick, Jr. Fndn Grants, 1773
George H. Hitchings New Investigator Award in Health Research, 1798
Gertrude B. Elion Mentored Medical Student Research Awards, 1809
Graham and Carolyn Holloway Family Foundation Grants, 1861
Hillsdale Fund Grants, 2035
Howe Foundation of North Carolina Grants, 2055
Ireland Family Foundation Grants, 2162
James G.K. McClure Educational and Development Fund Grants, 2188
James H. Cummings Foundation Grants, 2190
James J. & Angelia M. Harris Fndn Grants, 2192
Janirve Foundation Grants, 2209
John Deere Foundation Grants, 2234
John H. Wellons Foundation Grants, 2239
John W. and Anna H. Hanes Foundation Grants, 2261
Kate B. Reynolds Trust Health Care Grants, 2290
Lester Ray Fleming Scholarships, 2360
Lincoln Financial Foundation Grants, 2367
MeadWestvaco Foundation Sustainable Communities Grants, 2506
Nationwide Insurance Foundation Grants, 2637
Norfolk Southern Foundation Grants, 2982
North Carolina Biotechnology Center Event Sponsorship Grants, 2984
North Carolina Biotechnology Center Grantsmanship Training Grants, 2985
North Carolina Biotech Center Meeting Grants, 2986
North Carolina Biotech Center Regional Development Grants, 2987
North Carolina Biotechnology Centers of Innovation Grants, 2988
North Carolina Biotechnology Center Technology Enhancement Grants, 2989
North Carolina Community Foundation Grants, 2990
North Carolina GlaxoSmithKline Fndn Grants, 2991
Percy B. Ferebee Endowment Grants, 3138

Piedmont Natural Gas Corporate and Charitable Contributions, 3220
Piedmont Natural Gas Foundation Health and Human Services Grants, 3221
PNC Foundation Community Services Grants, 3240
Qualcomm Grants, 3287
Rehab Therapy Foundation Grants, 3356
Richard Davoud Donchian Foundation Grants, 3367
Ricks Family Charitable Trust Grants, 3373
Simple Advise Education Center Grants, 3565
Sisters of Mercy of North Carolina Fndn Grants, 3575
Square D Foundation Grants, 3634
Steele-Reese Foundation Grants, 3650
Strowd Roses Grants, 3664
Sunoco Foundation Grants, 3679
TE Foundation Grants, 3705
Tension Envelope Foundation Grants, 3707
Textron Corporate Contributions Grants, 3712
Thomas Austin Finch, Sr. Foundation Grants, 3718
Triangle Community Foundation Donor-Advised Grants, 3757
Triangle Community Foundation Shaver-Hitchings Scholarship, 3758
Tyler Aaron Bookman Memorial Foundation Trust Grants, 3765
U.S. Cellular Corporation Grants, 3766
Union County Community Foundation Grants, 3790
US Airways Community Contributions, 3825
US Airways Community Foundation Grants, 3826
Weyerhaeuser Company Foundation Grants, 3925
Williams Companies Foundation Homegrown Giving Grants, 3957
Winston-Salem Foundation Competitive Grants, 3965
Winston-Salem Fndn Elkin/Tri-County Grants, 3966
Winston-Salem Fndn Stokes County Grants, 3967
Winston-Salem Fndn Victim Assistance Grants, 3968
Woodward Fund Grants, 3973
Z. Smith Reynolds Foundation Small Grants, 3982

North Dakota
Albertson's Charitable Giving Grants, 410
Archer Daniels Midland Foundation Grants, 651
Bush Fndn Health & Human Services Grants, 979
Bush Foundation Medical Fellowships, 980
Dean Foods Community Involvement Grants, 1440
Fargo-Moorhead Area Foundation Grants, 1632
Fargo-Moorhead Area Fndn Woman's Grants, 1633
North Dakota Community Foundation Grants, 2994
Otto Bremer Foundation Grants, 3099
Piper Jaffray Fndn Communities Giving Grants, 3227
Shopko Fndn Community Charitable Grants, 3547

Northern Mariana Islands
ACF Native American Social and Economic Development Strategies Grants, 231
Bank of America Fndn Matching Gifts, 825
Bank of America Fndn Volunteer Grants, 826
NHLBI Ancillary Studies in Clinical Trials, 2688
NHLBI Bioengineering and Obesity Grants, 2689
NHLBI Career Transition Awards, 2694
NHLBI Lymphatics in Health and Disease in the Digestive, Cardio & Pulmonary Systems, 2704
NHLBI Lymphatics in Health and Disease in the Digestive, Urinary, Cardiovascular and Pulmonary Systems, 2705
NHLBI Mentored Career Dev Award to Promote Faculty Diversity in Biomed Research, 2708
NHLBI Mentored Patient-Oriented Research Career Development Awards, 2710
NHLBI Mentored Quantitative Research Career Development Awards, 2711
NHLBI Pathway to Independence Awards, 2718
NHLBI Research on the Role of Cardiomyocyte Mitochondria in Heart Disease: An Integrated Approach, 2726
NHLBI Ruth L. Kirschstein National Research Service Awards for Individual Postdoctoral Fellowships, 2729
NHLBI Ruth L. Kirschstein National Research Service Awards for Individual Senior Fellows, 2732

NHLBI Short-Term Research Education Program to Increase Diversity in Health-Related Research, 2733
NHLBI Training Program Grants for Institutions that Promote Diversity, 2738
NIGMS Advancing Basic Behavioral and Social Research on Resilience: an Integrative Science Approach, 2861
NIGMS Enabling Resources for Pharmacogenomics Grants, 2862
NIH Mentored Clinical Scientist Research Career Development Award, 2887
NIH Ruth L. Kirschstein National Research Service Award Institutional Research Training Grants, 2900
NIH Ruth L. Kirschstein National Research Service Awards for Individual Postdoc Fellowships, 2902
NIH Ruth L. Kirschstein National Research Service Awards for Individual Predoctoral Fellows, 2903
NIH Ruth L. Kirschstein Nat Research Service Award Short-Term Institutional Training Grants, 2906
NIH Support of Competitive Research (SCORE) Pilot Project Awards, 2913
NIH Support of Competitive Research (SCORE) Research Advancement Awards, 2914

Ohio
3M Company Fndn Community Giving Grants, 6
3M Fndn Health & Human Services Grants, 7
Abbott Fund Access to Health Care Grants, 199
Abbott Fund Community Grants, 201
Abbott Fund Global AIDS Care Grants, 202
Abbott Fund Science Education Grants, 203
Abington Foundation Grants, 211
ACE Charitable Foundation Grants, 229
Aetna Foundation Regional Health Grants, 333
A Good Neighbor Foundation Grants, 347
American Electric Power Foundation Grants, 530
Amerigroup Foundation Grants, 565
Amica Insurance Company Community Grants, 575
Anheuser-Busch Foundation Grants, 591
Anthem Blue Cross and Blue Shield Grants, 607
Appalachian Regional Commission Health Grants, 637
Archer Daniels Midland Foundation Grants, 651
Ashland Corporate Contributions Grants, 713
Austin-Bailey Health and Wellness Fndn Grants, 803
Barberton Community Foundation Grants, 829
BJ's Charitable Foundation Grants, 904
Boeing Company Contributions Grants, 930
Bridgestone/Firestone Trust Fund Grants, 949
Campbell Soup Foundation Grants, 1007
Centerville-Washington Foundation Grants, 1078
Charles H. Dater Foundation Grants, 1118
Chase Paymentech Corporate Giving Grants, 1127
Cleveland-Cliffs Foundation Grants, 1181
Cleveland Browns Foundation Grants, 1182
Coleman Foundation Cancer Care Grants, 1206
Coleman Foundation Developmental Disabilities Grants, 1207
Columbus Foundation Allen Eiry Fund Grants, 1214
Columbus Foundation Central Benefits Health Care Foundation Grants, 1216
Columbus Foundation Competitive Grants, 1217
Columbus Foundation Estrich Fund Grants, 1218
Columbus Foundation J. Floyd Dixon Memorial Fund Grants, 1219
Columbus Foundation Mary Eleanor Morris Fund Grants, 1220
Columbus Foundation Paul G. Duke Grants, 1221
Columbus Foundation Robert E. and Genevieve B. Schaefer Fund Grants, 1222
Columbus Foundation Traditional Grants, 1223
Community Foundation of Mount Vernon and Knox County Grants, 1299
ConAgra Foods Fndn Community Impact Grants, 1321
ConAgra Foods Nourish our Community Grants, 1322
CONSOL Energy Community Dev Grants, 1339
CONSOL Military and Armed Services Grants, 1340
CSX Corporate Contributions Grants, 1368
Cyrus Eaton Foundation Grants, 1380
Dayton Power and Light Company Foundation Signature Grants, 1436

946 / Ohio   GEOGRAPHIC INDEX

Dayton Power and Light Foundation Grants, 1437
Deaconess Community Foundation Grants, 1439
Dean Foods Community Involvement Grants, 1440
Della B. Gardner Fund Grants, 1451
Del Mar Healthcare Fund Grants, 1453
Dominion Foundation Grants, 1485
Eva L. & Joseph M. Bruening Fndn Grants, 1613
Fifth Third Foundation Grants, 1661
FirstEnergy Foundation Community Grants, 1668
Fndn for Appalachian Ohio Bachtel Scholarships, 1692
Foundation for Appalachian Ohio Susan K. Ipacs Nursing Legacy Scholarships, 1693
Foundation for Appalachian Ohio Zelma Gray Medical School Scholarship, 1694
General Mills Foundation Grants, 1782
George W. Codrington Charitable Fndn Grants, 1803
Giant Eagle Foundation Grants, 1815
Ginn Foundation Grants, 1822
Gladys Brooks Foundation Grants, 1826
Graco Foundation Grants, 1860
Greater Cincinnati Foundation Priority and Small Projects/Capacity-Building Grants, 1880
Hardin County Community Foundation Grants, 1946
Harmony Project Grants, 1948
Health Foundation of Greater Cincinnati Grants, 1983
Helen Steiner Rice Foundation Grants, 1998
Honda of America Manufacturing Fndn Grants, 2045
Huffy Foundation Grants, 2073
Humana Foundation Grants, 2077
Huntington National Bank Community Grants, 2084
Impact 100 Grants, 2148
Jacob G. Schmidlapp Trust Grants, 2183
James S. Copley Foundation Grants, 2200
John Lord Knight Foundation Grants, 2243
John P. Murphy Foundation Grants, 2248
Johnson & Johnson Community Health Grants, 2252
Josephine Schell Russell Charitable Trust Grants, 2274
Kelvin and Eleanor Smith Foundation Grants, 2297
Kettering Fund Grants, 2305
KeyBank Foundation Grants, 2307
Kuntz Foundation Grants, 2326
Lockheed Martin Corp Foundation Grants, 2374
Lubrizol Foundation Grants, 2384
Manuel D. & Rhoda Mayerson Fndn Grants, 2420
Marathon Petroleum Corporation Grants, 2422
Mary S. and David C. Corbin Foundation Grants, 2447
MeadWestvaco Foundation Sustainable Communities Grants, 2506
Mt. Sinai Health Care Foundation Academic Medicine and Bioscience Grants, 2598
Mt. Sinai Health Care Foundation Health of the Jewish Community Grants, 2599
Mt. Sinai Health Care Foundation Health of the Urban Community Grants, 2600
Mt. Sinai Health Care Foundation Health Policy Grants, 2601
Nationwide Insurance Foundation Grants, 2637
Nord Family Foundation Grants, 2980
Nordson Corporation Foundation Grants, 2981
Norfolk Southern Foundation Grants, 2982
Ohio Learning Network Grants, 3070
Parkersburg Area Community Foundation Action Grants, 3116
Piper Jaffray Fndn Communities Giving Grants, 3227
PNC Foundation Community Services Grants, 3240
Raymond John Wean Foundation Grants, 3310
RCF General Community Grants, 3312
RCF Individual Assistance Grants, 3313
RCF Summertime Kids Grants, 3314
RCF The Women's Fund Grants, 3315
Reinberger Foundation Grants, 3357
Reynolds & Reynolds Associate Fndn Grants, 3360
S. Livingston Mather Charitable Trust Grants, 3448
Saint Luke's Foundation Grants, 3454
Shopko Fndn Community Charitable Grants, 3547
Sisters of Charity Foundation of Cleveland Reducing Health and Educational Disparities in the Central Neighborhood Grants, 3574
SOCFOC Catholic Ministries Grants, 3596
Stark Community Fndn Women's Grants, 3645

Stocker Foundation Grants, 3656
Sunoco Foundation Grants, 3679
T. Spencer Shore Foundation Grants, 3701
Toyota Motor Engineering & Manufacturing North America Grants, 3742
Tri-State Community Twenty-first Century Endowment Fund Grants, 3756
Turner Foundation Grants, 3763
Virginia W. Kettering Foundation Grants, 3878
Walter L. Gross III Family Foundation Grants, 3901
Weyerhaeuser Company Foundation Grants, 3925
Wolfe Associates Grants, 3969
Zane's Foundation Grants, 3983

Oklahoma
Albertson's Charitable Giving Grants, 410
American Electric Power Foundation Grants, 530
Anheuser-Busch Foundation Grants, 591
Archer Daniels Midland Foundation Grants, 651
Boeing Company Contributions Grants, 930
Cable Positive's Tony Cox Community Grants, 992
Charles A. Frueauff Foundation Grants, 1114
ConocoPhillips Foundation Grants, 1327
Cresap Family Foundation Grants, 1359
Cudd Foundation Grants, 1370
Cuesta Foundation Grants, 1371
Dean Foods Community Involvement Grants, 1440
Edward and Helen Bartlett Foundation Grants, 1554
General Mills Foundation Grants, 1782
Gil and Dody Weaver Foundation Grants, 1819
Grace and Franklin Bernsen Foundation Grants, 1858
Guitar Center Music Foundation Grants, 1914
H.A. & Mary K. Chapman Trust Grants, 1921
Inasmuch Foundation Grants, 2149
J.E. and L.E. Mabee Foundation Grants, 2170
Judith and Jean Pape Adams Charitable Foundation Tulsa Area Grants, 2280
Mary K. Chapman Foundation Grants, 2443
Merrick Foundation Grants, 2521
Mervin Bovaird Foundation Grants, 2523
PACCAR Foundation Grants, 3101
PacifiCare Foundation Grants, 3102
Presbyterian Health Foundation Bridge, Seed and Equipment Grants, 3259
Priddy Foundation Capital Grants, 3263
Priddy Foundation Operating Grants, 3264
Priddy Foundation Organizational Development Grants, 3265
Priddy Foundation Program Grants, 3266
Questar Corporate Contributions Grants, 3291
Sarkeys Foundation Grants, 3485
Seagate Tech Corp Capacity to Care Grants, 3502
U.S. Cellular Corporation Grants, 3766
Union Pacific Foundation Health and Human Services Grants, 3794
W.M. Keck Fndn Undergrad Ed Grants, 3891
Weyerhaeuser Company Foundation Grants, 3925

Oregon
Abby's Legendary Pizza Foundation Grants, 204
Abracadabra Foundation Grants, 216
ACE Charitable Foundation Grants, 229
Alaska Airlines Corporate Giving Medical Emergency and Research Grants, 402
Albertson's Charitable Giving Grants, 410
Amica Insurance Company Community Grants, 575
Aratani Foundation Grants, 649
Autzen Foundation Grants, 810
Avista Foundation Economic and Cultural Vitality Grants, 812
Bella Vista Foundation Grants, 873
Ben B. Cheney Foundation Grants, 875
Benton County Foundation Grants, 880
Blue Mountain Community Foundation Grants, 921
Boeing Company Contributions Grants, 930
Bullitt Foundation Grants, 973
Cable Positive's Tony Cox Community Grants, 992
Carpenter Foundation Grants, 1035
Children's Trust Fund of Oregon Fndn Grants, 1139
Chiles Foundation Grants, 1147

Coeta and Donald Barker Foundation Grants, 1205
Collins Foundation Grants, 1210
ConAgra Foods Fndn Community Impact Grants, 1321
ConAgra Foods Nourish our Community Grants, 1322
Detlef Schrempf Foundation Grants, 1470
E.L. Wiegand Foundation Grants, 1538
Edward and Romell Ackley Foundation Grants, 1555
Ford Family Foundation Grants - Access to Health and Dental Services, 1686
Foster Foundation Grants, 1689
Guitar Center Music Foundation Grants, 1914
Harold and Arlene Schnitzer CARE Foundation Grants, 1950
Idaho Power Company Corporate Contributions, 2115
KeyBank Foundation Grants, 2307
Kinsman Foundation Grants, 2309
M.B. and Edna Zale Foundation Grants, 2401
M.J. Murdock Charitable Trust General Grants, 2405
Meyer Memorial Trust Emergency Grants, 2535
Meyer Memorial Trust Grassroots Grants, 2536
Meyer Memorial Trust Responsive Grants, 2537
Meyer Memorial Trust Special Grants, 2538
NWHF Community-Based Participatory Research Grants, 3032
NWHF Health Advocacy Small Grants, 3033
NWHF Kaiser Permanente Community Grants, 3034
NWHF Physical Activity and Nutrition Grants, 3037
Oregon Community Foundation Better Nursing Home Care Grants, 3081
Oregon Community Fndn Community Grants, 3082
PacifiCare Foundation Grants, 3102
Piper Jaffray Fndn Communities Giving Grants, 3227
Portland General Electric Foundation Grants, 3250
Pride Foundation Grants, 3268
Pride Foundation Scholarships, 3269
Regence Fndn Access to Health Care Grants, 3346
Regence Foundation Health Care Community Awareness and Engagement Grants, 3347
Regence Fndn Health Care Connections Grants, 3348
Regence Fndn Improving End-of-Life Grants, 3349
Regence Fndn Tools & Technology Grants, 3350
Salem Foundation Charitable Trust Grants, 3456
Samuel S. Johnson Foundation Grants, 3468
Shopko Fndn Community Charitable Grants, 3547
Sunderland Foundation Grants, 3675
U.S. Cellular Corporation Grants, 3766
Union Bank, N.A. Corporate Sponsorships and Donations, 3787
Union Bank, N.A. Foundation Grants, 3788
Union Pacific Foundation Health and Human Services Grants, 3794
W.M. Keck Fndn Undergrad Ed Grants, 3891
Weyerhaeuser Company Foundation Grants, 3925
William G. Gilmore Foundation Grants, 3951
Wood Family Charitable Trust Grants, 3972

Pennsylvania
1 in 9: Long Island Breast Cancer Action Coalition Grants, 1
1675 Foundation Grants, 11
1976 Foundation Grants, 12
ACE Charitable Foundation Grants, 229
Adams County Community Foundation of Pennsylvania Grants, 300
Adams Foundation Grants, 301
Addison H. Gibson Foundation Medical Grants, 305
AEGON Transamerica Foundation Health and Welfare Grants, 316
Aetna Foundation Regional Health Grants, 333
Albertson's Charitable Giving Grants, 410
Allegheny Technologies Charitable Trust, 440
Alpha Natural Resources Corporate Giving, 450
American Foodservice Charitable Trust Grants, 532
American Trauma Society, Pennsylvania Division Mini-Grants, 564
Amica Insurance Company Community Grants, 575
Appalachian Regional Commission Health Grants, 637
Arcadia Foundation Grants, 650
Archer Daniels Midland Foundation Grants, 651
Babcock Charitable Trust Grants, 816

GEOGRAPHIC INDEX

Barra Foundation Community Fund Grants, 833
Barra Foundation Project Grants, 834
Bayer Foundation Grants, 846
Berks County Community Foundation Grants, 881
Birmingham Foundation Grants, 903
BJ's Charitable Foundation Grants, 904
Boeing Company Contributions Grants, 930
Bridgestone/Firestone Trust Fund Grants, 949
Brook J. Lenfest Foundation Grants, 967
Bruce and Adele Greenfield Foundation Grants, 972
Cable Positive's Tony Cox Community Grants, 992
Cabot Corporation Foundation Grants, 993
Caesars Foundation Grants, 995
Campbell Soup Foundation Grants, 1007
Charles A. Frueauff Foundation Grants, 1114
Chemtura Corporation Contributions Grants, 1130
Claude Worthington Benedum Fndn Grants, 1178
ConAgra Foods Fndn Community Impact Grants, 1321
ConAgra Foods Nourish our Community Grants, 1322
Connelly Foundation Grants, 1326
ConocoPhillips Foundation Grants, 1327
CONSOL Energy Community Dev Grants, 1339
CONSOL Military and Armed Services Grants, 1340
CSX Corporate Contributions Grants, 1368
David M. and Marjorie D. Rosenberg Foundation Grants, 1432
Dean Foods Community Involvement Grants, 1440
Dominion Foundation Grants, 1485
Donald and Sylvia Robinson Family Foundation Grants, 1486
Dorothy Rider Pool Health Care Grants, 1500
Eastman Chemical Company Foundation Grants, 1542
Eberly Foundation Grants, 1544
Eden Hall Foundation Grants, 1549
Edna G. Kynett Memorial Foundation Grants, 1551
Elsie Lee Garthwaite Memorial Fndn Grants, 1584
Erie Community Foundation Grants, 1601
Ethel Sergeant Clark Smith Foundation Grants, 1608
Eugene M. Lang Foundation Grants, 1611
Expect Miracles Foundation Grants, 1619
F.M. Kirby Foundation Grants, 1624
Ferree Foundation Grants, 1658
FirstEnergy Foundation Community Grants, 1668
Fluor Foundation Grants, 1682
FMC Foundation Grants, 1683
Fourjay Foundation Grants, 1712
Frank E. and Seba B. Payne Foundation Grants, 1722
Franklin H. Wells and Ruth L. Wells Foundation Grants, 1725
Frank S. Flowers Foundation Grants, 1729
General Mills Foundation Grants, 1782
Genuardi Family Foundation Grants, 1787
Giant Eagle Foundation Grants, 1815
Gladys Brooks Foundation Grants, 1826
GlaxoSmithKline Corporate Grants, 1829
GlaxoSmithKline Foundation IMPACT Awards, 1830
Graham Foundation Grants, 1862
Grifols Community Outreach Grants, 1904
Grundy Foundation Grants, 1912
Guitar Center Music Foundation Grants, 1914
Hackett Foundation Grants, 1926
Harley Davidson Foundation Grants, 1947
Henrietta Tower Wurts Memorial Fndn Grants, 2002
Hershey Company Grants, 2013
Highmark BCBS Challenge for Healthier Schools in Western Pennsylvania, 3696
Highmark Corporate Giving Grants, 2025
Highmark Physician eHealth Grants, 2026
Hilda and Preston Davis Foundation Grants, 2028
Hillman Foundation Grants, 2032
Huffy Foundation Grants, 2073
Independence Blue Cross Charitable Medical Care Grants, 2150
Independence Blue Cross Nurse Scholars, 2151
Independence Blue Cross Nursing Internships, 2152
Josiah W. and Bessie H. Kline Foundation Grants, 2278
L. W. Pierce Family Foundation Grants, 2328
Land O'Lakes Foundation Mid-Atlantic Grants, 2341
Leo Niessen Jr., Charitable Trust Grants, 2358
Lincoln Financial Foundation Grants, 2367
Lockheed Martin Corp Foundation Grants, 2374
McLean Contributionship Grants, 2497
Meyer and Stephanie Eglin Foundation Grants, 2531
Nationwide Insurance Foundation Grants, 2637
Norfolk Southern Foundation Grants, 2982
Philadelphia Foundation Organizational Effectiveness Grants, 3189
Pittsburgh Foundation Community Fund Grants, 3230
Pittsburgh Foundation Medical Research Grants, 3231
PNC Foundation Community Services Grants, 3240
PPG Industries Foundation Grants, 3257
Price Chopper's Golub Foundation Grants, 3261
Prudential Foundation Education Grants, 3276
Rathmann Family Foundation Grants, 3309
Raymond John Wean Foundation Grants, 3310
Richard King Mellon Foundation Grants, 3369
Ruth and Vernon Taylor Foundation Grants, 3426
Ruth Estrin Goldberg Memorial for Cancer Research Grants, 3428
Salmon Foundation Grants, 3458
Sands Foundation Grants, 3476
Stackpole-Hall Foundation Grants, 3643
Staunton Farm Foundation Grants, 3649
Stewart Huston Charitable Trust Grants, 3655
Sunoco Foundation Grants, 3679
SunTrust Bank Trusteed Foundations Nell Warren Elkin and William Simpson Elkin Grants, 3683
TE Foundation Grants, 3705
Textron Corporate Contributions Grants, 3712
Tyler Aaron Bookman Memorial Foundation Trust Grants, 3765
Union Benevolent Association Grants, 3789
US Airways Community Contributions, 3825
US Airways Community Foundation Grants, 3826
Verizon Foundation Pennsylvania Grants, 3862
Weyerhaeuser Company Foundation Grants, 3925
Willary Foundation Grants, 3944
Williams Companies Foundation Homegrown Giving Grants, 3957

Puerto Rico
Abbott Fund Access to Health Care Grants, 199
Abbott Fund Community Grants, 201
Abbott Fund Global AIDS Care Grants, 202
Abbott Fund Science Education Grants, 203
ACE Charitable Foundation Grants, 229
Archer Daniels Midland Foundation Grants, 651
Bank of America Fndn Matching Gifts, 825
Bank of America Fndn Volunteer Grants, 826
Elton John AIDS Foundation Grants, 1585
Guitar Center Music Foundation Grants, 1914
IBRO Latin America Regional Funding for Neuroscience Schools, 2105
IBRO Latin America Regional Funding for PROLAB Collaborations, 2106
IBRO Latin America Regional Funding for Short Courses, Workshops, and Symposia, 2107
IBRO Latin America Regional Funding for Short Research Stays, 2108
IBRO Latin America Regional Travel Grants, 2109
Medtronic Foundation CommunityLink Health Grants, 2510
NHLBI Ancillary Studies in Clinical Trials, 2688
NHLBI Bioengineering and Obesity Grants, 2689
NHLBI Career Transition Awards, 2694
NHLBI Lymphatics in Health and Disease in the Digestive, Cardio & Pulmonary Systems, 2704
NHLBI Lymphatics in Health and Disease in the Digestive, Urinary, Cardiovascular and Pulmonary Systems, 2705
NHLBI Mentored Career Dev Award to Promote Faculty Diversity in Biomed Research, 2708
NHLBI Mentored Patient-Oriented Research Career Development Awards, 2710
NHLBI Mentored Quantitative Research Career Development Awards, 2711
NHLBI Pathway to Independence Awards, 2718
NHLBI Research on the Role of Cardiomyocyte Mitochondria in Heart Disease: An Integrated Approach, 2726
NHLBI Ruth L. Kirschstein National Research Service Awards for Individual Postdoctoral Fellows, 2729
NHLBI Ruth L. Kirschstein National Research Service Awards for Individual Senior Fellows, 2732
NHLBI Short-Term Research Education Program to Increase Diversity in Health-Related Research, 2733
NHLBI Training Program Grants for Institutions that Promote Diversity, 2738
NIGMS Advancing Basic Behavioral and Social Research on Resilience: an Integrative Science Approach, 2861
NIGMS Enabling Resources for Pharmacogenomics Grants, 2862
NIH Ruth L. Kirschstein National Research Service Award Institutional Research Training Grants, 2900
NIH Ruth L. Kirschstein National Research Service Awards for Individual Postdoc Fellowships, 2902
NIH Ruth L. Kirschstein National Research Service Awards for Individual Predoctoral Fellows, 2903
NIH Ruth L. Kirschstein Nat Research Service Award Short-Term Institutional Training Grants, 2906
NIH Support of Competitive Research (SCORE) Pilot Project Awards, 2913
NIH Support of Competitive Research (SCORE) Research Advancement Awards, 2914
Puerto Rico Community Foundation Grants, 3278
SfN Award for Education in Neuroscience, 3517
SfN Bernice Grafstein Award for Outstanding Accomplishments in Mentoring, 3518
SfN Donald B. Lindsley Prize in Behavioral Neuroscience, 3521
SfN Jacob P. Waletzky Award, 3523
SfN Janett Rosenberg Trubatch Career Development Award, 3524
SfN Julius Axelrod Prize, 3526
SfN Louise Hanson Marshall Specific Recognition Award, 3527
SfN Mika Salpeter Lifetime Achievement Award, 3528
SfN Next Generation Award, 3531
SfN Peter and Patricia Gruber International Research Award, 3533
SfN Ralph W. Gerard Prize in Neuroscience, 3534
SfN Science Educator Award, 3535
SfN Swartz Prize for Theoretical and Computational Neuroscience, 3537
SfN Young Investigator Award, 3541
U.S. Department of Education Rehabilitation Training - Rehabilitation Continuing Education - Institute on Rehabilitation Issues, 3771
WHO Foundation General Grants, 3939
WHO Foundation Volunteer Service Grants, 3940

Rhode Island
Aaron Foundation Grants, 193
Agway Foundation Grants, 348
Amica Companies Foundation Grants, 574
Amica Insurance Company Community Grants, 575
Aratani Foundation Grants, 649
Barnes Group Foundation Grants, 832
BJ's Charitable Foundation Grants, 904
Cable Positive's Tony Cox Community Grants, 992
Carlisle Foundation Grants, 1025
Champlin Foundations Grants, 1110
Charles A. Frueauff Foundation Grants, 1114
Citizens Bank Mid-Atlantic Charitable Foundation Grants, 1162
ConAgra Foods Nourish our Community Grants, 1322
Dominion Foundation Grants, 1485
Dora Roberts Foundation Grants, 1489
Expect Miracles Foundation Grants, 1619
Fannie E. Rippel Foundation Grants, 1630
Forrest C. Lattner Foundation Grants, 1688
Frank B. Hazard General Charity Fund Grants, 1721
Gladys Brooks Foundation Grants, 1826
Guitar Center Music Foundation Grants, 1914
Hasbro Children's Fund Grants, 1973
Hasbro Corporation Gift of Play Hospital and Pediatric Health Giving, 1974
Hilda and Preston Davis Foundation Grants, 2028

948 / Rhode Island

Horace A. Kimball and S. Ella Kimball Foundation Grants, 2046
HRAMF Charles H. Hood Foundation Child Health Research Awards, 2057
Jessie B. Cox Charitable Trust Grants, 2225
John Clarke Trust Grants, 2232
John D. & Katherine A. Johnston Fndn Grants, 2233
Nordson Corporation Foundation Grants, 2981
Norfolk Southern Foundation Grants, 2982
Rhode Island Foundation Grants, 3362
Robert and Joan Dircks Foundation Grants, 3379
Robert F. Stoico/FIRSTFED Charitable Foundation Grants, 3382
ROSE Fund Grants, 3412
Shaw's Supermarkets Donations, 3542
Sunoco Foundation Grants, 3679
T. Spencer Shore Foundation Grants, 3701
Textron Corporate Contributions Grants, 3712
Verizon Foundation Northeast Region Grants, 3861
Vigneron Memorial Fund Grants, 3873
William D. Laurie, Jr. Charitable Fndn Grants, 3949
Yawkey Foundation Grants, 3980

South Carolina
3M Company Fndn Community Giving Grants, 6
3M Fndn Health & Human Services Grants, 7
Abney Foundation Grants, 214
ACE Charitable Foundation Grants, 229
Adams and Reese Corporate Giving Grants, 299
Amerigroup Foundation Grants, 565
Amica Insurance Company Community Grants, 575
Appalachian Regional Commission Health Grants, 637
Archer Daniels Midland Foundation Grants, 651
Bailey Foundation Grants, 818
Belk Foundation Grants, 872
BJ's Charitable Foundation Grants, 904
Boeing Company Contributions Grants, 930
Bridgestone/Firestone Trust Fund Grants, 949
Burlington Industries Foundation Grants, 977
Cable Positive's Tony Cox Community Grants, 992
Campbell Soup Foundation Grants, 1007
Carrier Corporation Contributions Grants, 1038
Central Carolina Community Foundation Community Impact Grants, 1079
Charles A. Frueauff Foundation Grants, 1114
Charles M. & Mary D. Grant Fndn Grants, 1124
Coastal Community Foundation of South Carolina Grants, 1202
Community Foundation of Greenville-Greenville Women Giving Grants, 1283
Community Foundation of Greenville Community Enrichment Grants, 1284
Community Foundation of Greenville Hollingsworth Funds Program/Project Grants, 1285
ConAgra Foods Fndn Community Impact Grants, 1321
ConAgra Foods Nourish our Community Grants, 1322
Cooper Industries Foundation Grants, 1345
CSX Corporate Contributions Grants, 1368
Dean Foods Community Involvement Grants, 1440
Dorothy Hooper Beattie Foundation Grants, 1499
Drs. Bruce and Lee Foundation Grants, 1513
Duke Endowment Health Care Grants, 1519
Eastman Chemical Company Foundation Grants, 1542
Fluor Foundation Grants, 1682
Gardner Foundation Grants, 1770
Guitar Center Music Foundation Grants, 1914
Hilton Head Island Foundation Grants, 2036
John I. Smith Charities Grants, 2240
Lockheed Martin Corp Foundation Grants, 2374
Mary Black Foundation Active Living Grants, 2440
Mary Black Fndn Community Health Grants, 2441
Mary Black Foundation Early Childhood Development Grants, 2442
MeadWestvaco Foundation Sustainable Communities Grants, 2506
Norfolk Southern Foundation Grants, 2982
Piedmont Health Foundation Grants, 3219
Piedmont Natural Gas Corporate and Charitable Contributions, 3220
Piedmont Natural Gas Foundation Health and Human Services Grants, 3221
PNC Foundation Community Services Grants, 3240
Pulido Walker Foundation, 3282
Richard Davoud Donchian Foundation Grants, 3367
Self Foundation Grants, 3510
Sisters of Mercy of North Carolina Fndn Grants, 3575
Sonoco Foundation Grants, 3621
Springs Close Foundation Grants, 3633
Sunoco Foundation Grants, 3679
TE Foundation Grants, 3705
Weyerhaeuser Company Foundation Grants, 3925
Williams Companies Foundation Homegrown Giving Grants, 3957
Woodward Fund Grants, 3973
Yawkey Foundation Grants, 3980

South Dakota
3M Company Fndn Community Giving Grants, 6
3M Fndn Health & Human Services Grants, 7
Albertson's Charitable Giving Grants, 410
Bush Fndn Health & Human Services Grants, 979
Bush Foundation Medical Fellowships, 980
Cable Positive's Tony Cox Community Grants, 992
Charles A. Frueauff Foundation Grants, 1114
Dean Foods Community Involvement Grants, 1440
Graco Foundation Grants, 1860
Guitar Center Music Foundation Grants, 1914
Piper Jaffray Fndn Communities Giving Grants, 3227
Shopko Fndn Community Charitable Grants, 3547
Sioux Falls Area Community Foundation Community Fund Grants (Unrestricted), 3568
Sioux Falls Area Community Foundation Field-of-Interest and Donor-Advised Grants, 3569
Sioux Falls Area Community Foundation Spot Grants (Unrestricted), 3570
Waitt Family Foundation Grants, 3894

Tennessee
Adams and Reese Corporate Giving Grants, 299
Aetna Foundation Regional Health Grants, 333
AFG Industries Grants, 340
Albertson's Charitable Giving Grants, 410
American Electric Power Foundation Grants, 530
Amerigroup Foundation Grants, 565
Amica Insurance Company Community Grants, 575
Appalachian Regional Commission Health Grants, 637
Archer Daniels Midland Foundation Grants, 651
Assisi Fndn of Memphis Capital Project Grants, 786
Assisi Foundation of Memphis General Grants, 787
BancorpSouth Foundation Grants, 823
Bechtel Group Foundation Building Positive Community Relationships Grants, 866
Belk Foundation Grants, 872
Bridgestone/Firestone Trust Fund Grants, 949
Cable Positive's Tony Cox Community Grants, 992
Carrier Corporation Contributions Grants, 1038
Charles A. Frueauff Foundation Grants, 1114
Charles M. & Mary D. Grant Fndn Grants, 1124
Christy-Houston Foundation Grants, 1153
ConAgra Foods Fndn Community Impact Grants, 1321
ConAgra Foods Nourish our Community Grants, 1322
Coors Brewing Corporate Contributions Grants, 1346
CSX Corporate Contributions Grants, 1368
Dean Foods Community Involvement Grants, 1440
Eastman Chemical Company Foundation Grants, 1542
Erie Chapman Foundation Grants, 1600
Fifth Third Foundation Grants, 1661
Fluor Foundation Grants, 1682
FMC Foundation Grants, 1683
GenCorp Foundation Grants, 1779
General Mills Foundation Grants, 1782
Gladys Brooks Foundation Grants, 1826
Graham and Carolyn Holloway Family Foundation Grants, 1861
Guitar Center Music Foundation Grants, 1914
HCA Foundation Grants, 1980
Hershey Company Grants, 2013
Hilton Hotels Corporate Giving Program Grants, 2037
Humana Foundation Grants, 2077
John Deere Foundation Grants, 2234
Johns Manville Fund Grants, 2251
Louie M. & Betty M. Phillips Fndn Grants, 2378
Medtronic Foundation CommunityLink Health Grants, 2510
Medtronic Foundation Community Link Human Services Grants, 2511
Memorial Foundation Grants, 2516
Nationwide Insurance Foundation Grants, 2637
Norfolk Southern Foundation Grants, 2982
PeyBack Foundation Grants, 3159
Piedmont Natural Gas Corporate and Charitable Contributions, 3220
Piedmont Natural Gas Foundation Health and Human Services Grants, 3221
Piper Jaffray Fndn Communities Giving Grants, 3227
Plough Foundation Grants, 3238
Salmon Foundation Grants, 3458
Square D Foundation Grants, 3634
Sunoco Foundation Grants, 3679
Tension Envelope Foundation Grants, 3707
Thompson Charitable Foundation Grants, 3727
U.S. Cellular Corporation Grants, 3766
Union Pacific Foundation Health and Human Services Grants, 3794
USDA Delta Health Care Services Grants, 3829
William B. Stokely Jr. Foundation Grants, 3947
Woodward Fund Grants, 3973

Texas
3M Company Fndn Community Giving Grants, 6
3M Fndn Health & Human Services Grants, 7
Abbott Fund Access to Health Care Grants, 199
Abbott Fund Community Grants, 201
Abbott Fund Global AIDS Care Grants, 202
Abbott Fund Science Education Grants, 203
Abell-Hanger Foundation Grants, 207
ACE Charitable Foundation Grants, 229
Adams and Reese Corporate Giving Grants, 299
Advanced Micro Devices Comm Affairs Grants, 313
AEGON Transamerica Foundation Health and Welfare Grants, 316
Aetna Foundation Regional Health Grants, 333
Albert and Margaret Alkek Foundation Grants, 405
Albertson's Charitable Giving Grants, 410
American Electric Power Foundation Grants, 530
American Foodservice Charitable Trust Grants, 532
Amerigroup Foundation Grants, 565
Amica Insurance Company Community Grants, 575
Anheuser-Busch Foundation Grants, 591
Aragona Family Foundation Grants, 648
Archer Daniels Midland Foundation Grants, 651
ARCO Foundation Education Grants, 652
Bayer Foundation Grants, 846
BBVA Compass Foundation Charitable Grants, 848
Bechtel Group Foundation Building Positive Community Relationships Grants, 866
Belk Foundation Grants, 872
Bob & Delores Hope Foundation Grants, 927
Boeing Company Contributions Grants, 930
Bosque Foundation Grants, 935
Bridgestone/Firestone Trust Fund Grants, 949
Cable Positive's Tony Cox Community Grants, 992
Cabot Corporation Foundation Grants, 993
Campbell Soup Foundation Grants, 1007
Carl B. and Florence E. King Foundation Grants, 1022
Carrier Corporation Contributions Grants, 1038
Charles A. Frueauff Foundation Grants, 1114
Chase Paymentech Corporate Giving Grants, 1127
Clara Blackford Smith and W. Aubrey Smith Charitable Foundation Grants, 1165
Clayton Fund Grants, 1180
Cockrell Foundation Grants, 1204
Comerica Charitable Foundation Grants, 1224
Communities Foundation of Texas Grants, 1236
Community Foundation of Abilene Celebration of Life Grants, 1261
ConAgra Foods Fndn Community Impact Grants, 1321
ConAgra Foods Nourish our Community Grants, 1322
ConocoPhillips Foundation Grants, 1327

GEOGRAPHIC INDEX

Constantin Foundation Grants, 1341
Cooper Industries Foundation Grants, 1345
Crystelle Waggoner Charitable Trust Grants, 1364
Cullen Foundation Grants, 1372
CUNA Mutual Group Fndn Community Grants, 1375
Curtis Foundation Grants, 1376
Dallas Mavericks Foundation Grants, 1416
Dallas Women's Foundation Grants, 1417
Dean Foods Community Involvement Grants, 1440
Denton A. Cooley Foundation Grants, 1459
Dominion Foundation Grants, 1485
Dr. & Mrs. Paul Pierce Memorial Fndn Grants, 1505
Eastman Chemical Company Foundation Grants, 1542
EDS Foundation Grants, 1552
Effie and Wofford Cain Foundation Grants, 1563
El Paso Community Foundation Grants, 1579
El Paso Corporate Foundation Grants, 1580
Entergy Corporation Micro Grants, 1596
Entergy Corporation Open Grants for Healthy Families, 1597
Eugene Straus Charitable Trust, 1612
Ewing Halsell Foundation Grants, 1618
Fluor Foundation Grants, 1682
FMC Foundation Grants, 1683
Fondren Foundation Grants, 1685
Forrest C. Lattner Foundation Grants, 1688
Gardner W. & Joan G. Heidrick, Jr. Fndn Grants, 1773
Garland D. Rhoads Foundation, 1774
George Foundation Grants, 1795
George W. Brackenridge Foundation Grants, 1802
Gil and Dody Weaver Foundation Grants, 1819
Graham and Carolyn Holloway Family Foundation Grants, 1861
Guitar Center Music Foundation Grants, 1914
Hallmark Corporate Foundation Grants, 1936
Harold Simmons Foundation Grants, 1955
Harris and Eliza Kempner Fund Grants, 1956
Harry W. Bass, Jr. Foundation Grants, 1968
Hawn Foundation Grants, 1979
Helen Gertrude Sparks Charitable Trust Grants, 1993
Helen S. Boylan Foundation Grants, 1997
Hillcrest Foundation Grants, 2031
Hoblitzelle Foundation Grants, 2038
Hogg Foundation for Mental Health Grants, 2039
Hoglund Foundation Grants, 2040
Houston Endowment Grants, 2053
Humana Foundation Grants, 2077
Human Source Foundation Grants, 2078
I.A. O'Shaughnessy Foundation Grants, 2088
J.E. and L.E. Mabee Foundation Grants, 2170
J.N. and Macie Edens Foundation Grants, 2174
Jack H. & William M. Light Trust Grants, 2179
James M. Collins Foundation Grants, 2195
James R. Dougherty Jr. Foundation Grants, 2198
John P. McGovern Foundation Grants, 2247
John S. Dunn Research Foundation Grants, 2250
Katrine Menzing Deakins Trust Grants, 2295
Lockheed Martin Corp Foundation Grants, 2374
Lowe Foundation Grants, 2381
Lubbock Area Foundation Grants, 2383
Lubrizol Foundation Grants, 2384
Lynn and Rovena Alexander Family Foundation Grants, 2399
M.B. and Edna Zale Foundation Grants, 2401
M.D. Anderson Foundation Grants, 2403
Marathon Petroleum Corporation Grants, 2422
Marcia and Otto Koehler Foundation Grants, 2427
Mary L. Peyton Foundation Grants, 2445
McCombs Foundation Grants, 2491
MeadWestvaco Foundation Sustainable Communities Grants, 2506
Medtronic Foundation CommunityLink Health Grants, 2510
Medtronic Foundation Community Link Human Services Grants, 2511
Mercedes-Benz USA Corporate Contributions, 2517
Nationwide Insurance Foundation Grants, 2637
Norfolk Southern Foundation Grants, 2982
PACCAR Foundation Grants, 3101
PacifiCare Foundation Grants, 3102

Paso del Norte Health Foundation Grants, 3119
Pollock Foundation Grants, 3244
Powell Foundation Grants, 3255
Priddy Foundation Capital Grants, 3263
Priddy Foundation Operating Grants, 3264
Priddy Foundation Organizational Development Grants, 3265
Priddy Foundation Program Grants, 3266
Prudential Foundation Education Grants, 3276
Qualcomm Grants, 3287
Rainbow Fund Grants, 3301
Rockwell Fund, Inc. Grants, 3398
Roy & Christine Sturgis Charitable Grants, 3415
Ruth and Vernon Taylor Foundation Grants, 3426
San Antonio Area Foundation Grants, 3469
Shell Deer Park Grants, 3543
Shield-Ayres Foundation Grants, 3546
Sid W. Richardson Foundation Grants, 3553
Square D Foundation Grants, 3634
Steele-Reese Foundation Grants, 3650
Sterling-Turner Charitable Foundation Grants, 3652
Susan Vaughan Foundation Grants, 3695
T.L.L. Temple Foundation Grants, 3700
TE Foundation Grants, 3705
Tension Envelope Foundation Grants, 3707
Texas Commission on the Arts Arts Respond Project Grants, 3708
Texas Instruments Corporation Health and Human Services Grants, 3709
Texas Instruments Foundation Community Services Grants, 3710
Texas Instruments Foundation STEM Education Grants, 3711
Textron Corporate Contributions Grants, 3712
Thelma Braun & Bocklett Family Fndn Grants, 3713
Thomas Jefferson Rosenberg Foundation Grants, 3723
Todd Brock Family Foundation Grants, 3736
Topfer Family Foundation Grants, 3739
Toyota Motor Manufacturing of Texas Grants, 3747
U.S. Cellular Corporation Grants, 3766
Union Pacific Foundation Health and Human Services Grants, 3794
W.H. & Mary Ellen Cobb Trust Grants, 3884
W.M. Keck Fndn Undergrad Ed Grants, 3891
W.P. and Bulah Luse Foundation Grants, 3892
Western Union Foundation Grants, 3923
Weyerhaeuser Company Foundation Grants, 3925
Williams Companies Foundation Homegrown Giving Grants, 3957

US Virgin Islands
Bank of America Fndn Matching Gifts, 825
Bank of America Fndn Volunteer Grants, 826
Elton John AIDS Foundation Grants, 1585
NHLBI Ancillary Studies in Clinical Trials, 2688
NHLBI Bioengineering and Obesity Grants, 2689
NHLBI Career Transition Awards, 2694
NHLBI Lymphatics in Health and Disease in the Digestive, Cardio & Pulmonary Systems, 2704
NHLBI Lymphatics in Health and Disease in the Digestive, Urinary, Cardiovascular and Pulmonary Systems, 2705
NHLBI Mentored Career Dev Award to Promote Faculty Diversity in Biomed Research, 2708
NHLBI Mentored Patient-Oriented Research Career Development Awards, 2710
NHLBI Mentored Quantitative Research Career Development Awards, 2711
NHLBI Pathway to Independence Awards, 2718
NHLBI Research on the Role of Cardiomyocyte Mitochondria in Heart Disease: An Integrated Approach, 2726
NHLBI Ruth L. Kirschstein National Research Service Awards for Individual Postdoctoral Fellows, 2729
NHLBI Ruth L. Kirschstein National Research Service Awards for Individual Senior Fellows, 2732
NHLBI Short-Term Research Education Program to Increase Diversity in Health-Related Research, 2733
NHLBI Training Program Grants for Institutions that Promote Diversity, 2738

NIGMS Advancing Basic Behavioral and Social Research on Resilience: an Integrative Science Approach, 2861
NIGMS Enabling Resources for Pharmacogenomics Grants, 2862
NIH Mentored Clinical Scientist Research Career Development Award, 2887
NIH Ruth L. Kirschstein National Research Service Award Institutional Research Training Grants, 2900
NIH Ruth L. Kirschstein National Research Service Awards for Individual Postdoc Fellowships, 2902
NIH Ruth L. Kirschstein National Research Service Awards for Individual Predoctoral Fellows, 2903
NIH Ruth L. Kirschstein Nat Research Service Award Short-Term Institutional Training Grants, 2906
NIH Support of Competitive Research (SCORE) Pilot Project Awards, 2913
NIH Support of Competitive Research (SCORE) Research Advancement Awards, 2914
U.S. Department of Education Rehabilitation Training - Rehabilitation Continuing Education - Institute on Rehabilitation Issues, 3771

Utah
3M Company Fndn Community Giving Grants, 6
3M Fndn Health & Human Services Grants, 7
Abbott Fund Access to Health Care Grants, 199
Abbott Fund Community Grants, 201
Abbott Fund Global AIDS Care Grants, 202
Abbott Fund Science Education Grants, 203
Albertson's Charitable Giving Grants, 410
Boeing Company Contributions Grants, 930
Bridgestone/Firestone Trust Fund Grants, 949
Cable Positive's Tony Cox Community Grants, 992
Campbell Soup Foundation Grants, 1007
Daniels Fund Grants-Aging, 1424
Dean Foods Community Involvement Grants, 1440
E.L. Wiegand Foundation Grants, 1538
eBay Foundation Community Grants, 1543
GenCorp Foundation Grants, 1779
George S. & Dolores Dore Eccles Fndn Grants, 1801
Guitar Center Music Foundation Grants, 1914
Humana Foundation Grants, 2077
KeyBank Foundation Grants, 2307
Piper Jaffray Fndn Communities Giving Grants, 3227
Questar Corporate Contributions Grants, 3291
Regence Fndn Access to Health Care Grants, 3346
Regence Foundation Health Care Community Awareness and Engagement Grants, 3347
Regence Fndn Health Care Connections Grants, 3348
Regence Fndn Improving End-of-Life Grants, 3349
Regence Fndn Tools & Technology Grants, 3350
Ruth Eleanor Bamberger and John Ernest Bamberger Memorial Foundation Grants, 3427
Shopko Fndn Community Charitable Grants, 3547
Stan and Sandy Checketts Foundation, 3644
Sterling and Shelli Gardner Foundation Grants, 3653
Sunderland Foundation Grants, 3675
Union Pacific Foundation Health and Human Services Grants, 3794
W.M. Keck Fndn Undergrad Ed Grants, 3891
Weyerhaeuser Company Foundation Grants, 3925

Vermont
Agnes M. Lindsay Trust Grants, 346
Agway Foundation Grants, 348
Albertson's Charitable Giving Grants, 410
Barnes Group Foundation Grants, 832
Cable Positive's Tony Cox Community Grants, 992
Carlisle Foundation Grants, 1025
Charles A. Frueauff Foundation Grants, 1114
ConAgra Foods Nourish our Community Grants, 1322
Entergy Corporation Micro Grants, 1596
Entergy Corporation Open Grants for Healthy Families, 1597
Expect Miracles Foundation Grants, 1619
Fannie E. Rippel Foundation Grants, 1630
Gladys Brooks Foundation Grants, 1826
Guitar Center Music Foundation Grants, 1914
Hannaford Charitable Foundation Grants, 1944

950 / Vermont   GEOGRAPHIC INDEX

Hilda and Preston Davis Foundation Grants, 2028
HRAMF Charles H. Hood Foundation Child Health Research Awards, 2057
Jane's Trust Grants, 2206
Jessie B. Cox Charitable Trust Grants, 2225
KeyBank Foundation Grants, 2307
Kosciuszko Foundation Dr. Marie E. Zakrzewski Medical Scholarship, 2321
Lucy Downing Nisbet Charitable Fund Grants, 2387
Norfolk Southern Foundation Grants, 2982
Price Chopper's Golub Foundation Grants, 3261
Richard Davoud Donchian Foundation Grants, 3367
Robert and Joan Dircks Foundation Grants, 3379
ROSE Fund Grants, 3412
Shaw's Supermarkets Donations, 3542
Sunoco Foundation Grants, 3679
T. Spencer Shore Foundation Grants, 3701
Thomas Thompson Trust Grants, 3725
Verizon Foundation Vermont Grants, 3863
Vermont Community Foundation Grants, 3865
Windham Foundation Grants, 3961
Yawkey Foundation Grants, 3980

Virginia
Abbott Fund Access to Health Care Grants, 199
Abbott Fund Community Grants, 201
Abbott Fund Global AIDS Care Grants, 202
Abbott Fund Science Education Grants, 203
ACE Charitable Foundation Grants, 229
Aetna Foundation Regional Health Grants, 333
Alpha Natural Resources Corporate Giving, 450
American Electric Power Foundation Grants, 530
Amerigroup Foundation Grants, 565
Amica Insurance Company Community Grants, 575
Andersen Corporate Foundation, 586
Anheuser-Busch Foundation Grants, 591
Anthem Blue Cross and Blue Shield Grants, 607
Appalachian Regional Commission Health Grants, 637
Arlington Community Foundation Grants, 662
Beazley Foundation Grants, 865
Bechtel Group Foundation Building Positive Community Relationships Grants, 866
Beirne Carter Foundation Grants, 870
Belk Foundation Grants, 872
BJ's Charitable Foundation Grants, 904
Burlington Industries Foundation Grants, 977
Cable Positive's Tony Cox Community Grants, 992
Camp-Younts Foundation Grants, 1005
Campbell Hoffman Foundation Grants, 1006
Charles A. Frueauff Foundation Grants, 1114
Charles Delmar Foundation Grants, 1115
Charles M. & Mary D. Grant Fndn Grants, 1124
Community Fndn AIDS Endowment Awards, 1237
Community Foundation for the National Capital Region Community Leadership Grants, 1260
CONSOL Energy Community Dev Grants, 1339
CONSOL Military and Armed Services Grants, 1340
Coors Brewing Corporate Contributions Grants, 1346
CSX Corporate Contributions Grants, 1368
Dammann Fund Grants, 1418
Dean Foods Community Involvement Grants, 1440
Dominion Foundation Grants, 1485
EDS Foundation Grants, 1552
Fluor Foundation Grants, 1682
Freddie Mac Foundation Grants, 1738
GenCorp Foundation Grants, 1779
George and Sarah Buchanan Foundation Grants, 1790
Giant Food Charitable Grants, 1816
Guitar Center Music Foundation Grants, 1914
Hampton Roads Community Foundation Developmental Disabilities Grants, 1939
Hampton Roads Community Foundation Faith Community Nursing Grants, 1940
Hampton Roads Community Foundation Health and Human Service Grants, 1941
Hampton Roads Community Foundation Mental Health Research Grants, 1942
Harry Bramhall Gilbert Charitable Trust Grants, 1961
Harvest Foundation Grants, 1971
Hershey Company Grants, 2013

Hilda and Preston Davis Foundation Grants, 2028
HRAMF Jeffress Trust Awards in Interdisciplinary Research, 2060
Hut Foundation Grants, 2086
James L. and Mary Jane Bowman Charitable Trust Grants, 2194
Jenkins Foundation: Improving the Health of Greater Richmond Grants, 2219
John Edward Fowler Memorial Fndn Grants, 2235
Land O'Lakes Foundation Mid-Atlantic Grants, 2341
Lockheed Martin Corp Foundation Grants, 2374
Mann T. Lowry Foundation Grants, 2419
Mary Wilmer Covey Charitable Trust Grants, 2448
MeadWestvaco Foundation Sustainable Communities Grants, 2506
Meyer Foundation Benevon Grants, 2532
Meyer Fndn Management Assistance Grants, 2534
Morris & Gwendolyn Cafritz Fndn Grants, 2594
Nationwide Insurance Foundation Grants, 2637
Norfolk Southern Foundation Grants, 2982
Philip L. Graham Fund Health and Human Services Grants, 3190
PNC Foundation Community Services Grants, 3240
Richard Davoud Donchian Foundation Grants, 3367
Richmond Eye and Ear Fund Grants, 3372
Robins Foundation Grants, 3390
Salmon Foundation Grants, 3458
Sheltering Arms Fund Grants, 3545
Sunoco Foundation Grants, 3679
Symantec Community Relations and Corporate Philanthropy Grants, 3699
TE Foundation Grants, 3705
Thompson Charitable Foundation Grants, 3727
U.S. Cellular Corporation Grants, 3766
VDH Commonwealth of Virginia Nurse Educator Scholarships, 3854
VDH Emergency Med Services Training Grants, 3855
VDH Rescue Squad Assistance Fund Grants, 3856
Verizon Foundation Virginia Grants, 3864
Virginia Department of Health Mary Marshall Nursing Scholarships for Licensed Practical Nurses, 3874
Virginia Department of Health Mary Marshall Nursing Scholarships for Registered Nurses, 3875
Virginia Department of Health Nurse Practitioner/Nurse Midwife Scholarships, 3876
Washington Gas Charitable Giving, 3908
Weyerhaeuser Company Foundation Grants, 3925
Williams Companies Foundation Homegrown Giving Grants, 3957

Washington
Abby's Legendary Pizza Foundation Grants, 204
ACE Charitable Foundation Grants, 229
Adobe Community Investment Grants, 310
Aetna Foundation Regional Health Grants, 333
Alaska Airlines Corporate Giving Medical Emergency and Research Grants, 402
Alaska Airlines Foundation Grants, 403
Albertson's Charitable Giving Grants, 410
Alfred and Tillie Shemanski Testamentary Trust Grants, 428
Amerigroup Foundation Grants, 565
Amica Insurance Company Community Grants, 575
Aratani Foundation Grants, 649
Archer Daniels Midland Foundation Grants, 651
ARCO Foundation Education Grants, 652
Autzen Foundation Grants, 810
Avista Foundation Economic and Cultural Vitality Grants, 812
Bechtel Group Foundation Building Positive Community Relationships Grants, 866
Ben B. Cheney Foundation Grants, 875
Blue Mountain Community Foundation Grants, 921
Boeing Company Contributions Grants, 930
Bullitt Foundation Grants, 973
Cable Positive's Tony Cox Community Grants, 992
Campbell Soup Foundation Grants, 1007
Catherine Holmes Wilkins Foundation Grants, 1042
Community Fndn of South Puget Sound Grants, 1307
Comprehensive Health Education Fndn Grants, 1320

ConAgra Foods Fndn Community Impact Grants, 1321
ConAgra Foods Nourish our Community Grants, 1322
ConocoPhillips Foundation Grants, 1327
D.V. and Ida J. McEachern Trust Grants, 1406
Detlef Schrempf Foundation Grants, 1470
E.L. Wiegand Foundation Grants, 1538
Edwin W. and Catherine M. Davis Fndn Grants, 1561
Fluor Foundation Grants, 1682
Foster Foundation Grants, 1689
Fred & Gretel Biel Charitable Trust Grants, 1734
GenCorp Foundation Grants, 1779
Greater Tacoma Community Foundation Grants, 1886
Greater Tacoma Community Foundation Ryan Alan Hade Endowment Fund, 1887
Green Diamond Charitable Contributions, 1891
Grifols Community Outreach Grants, 1904
Guitar Center Music Foundation Grants, 1914
H.B. Fuller Foundation Grants, 1922
Harold and Arlene Schnitzer CARE Foundation Grants, 1950
Hasbro Children's Fund Grants, 1973
Hasbro Corporation Gift of Play Hospital and Pediatric Health Giving, 1974
Janson Foundation Grants, 2210
Jeffris Wood Foundation Grants, 2217
KeyBank Foundation Grants, 2307
Kinsman Foundation Grants, 2309
M.J. Murdock Charitable Trust General Grants, 2405
Medtronic Foundation CommunityLink Health Grants, 2510
Medtronic Foundation Community Link Human Services Grants, 2511
Meyer Memorial Trust Emergency Grants, 2535
Meyer Memorial Trust Grassroots Grants, 2536
Meyer Memorial Trust Responsive Grants, 2537
Meyer Memorial Trust Special Grants, 2538
Norcliffe Foundation Grants, 2979
NWHF Community-Based Participatory Research Grants, 3032
NWHF Health Advocacy Small Grants, 3033
NWHF Kaiser Permanente Community Grants, 3034
NWHF Physical Activity and Nutrition Grants, 3037
OneFamily Foundation Grants, 3075
PACCAR Foundation Grants, 3101
PacifiCare Foundation Grants, 3102
Piper Jaffray Fndn Communities Giving Grants, 3227
Premera Blue Cross Grants, 3258
Pride Foundation Grants, 3268
Pride Foundation Scholarships, 3269
Principal Financial Group Foundation Grants, 3273
Rathmann Family Foundation Grants, 3309
Rayonier Foundation Grants, 3311
Regence Fndn Access to Health Care Grants, 3346
Regence Foundation Health Care Community Awareness and Engagement Grants, 3347
Regence Fndn Health Care Connections Grants, 3348
Regence Fndn Improving End-of-Life Grants, 3349
Regence Fndn Tools & Technology Grants, 3350
Robert B McMillen Foundation Grants, 3381
Samuel S. Johnson Foundation Grants, 3468
San Juan Island Community Foundation Grants, 3480
Seattle Foundation Benjamin N. Phillips Memorial Fund Grants, 3504
Seattle Fndn Doyne M. Green Scholarships, 3505
Seattle Fndn Hawkins Memorial Scholarship, 3506
Seattle Foundation Health and Wellness Grants, 3507
Seattle Foundation Medical Funds Grants, 3508
Shopko Fndn Community Charitable Grants, 3547
Simpson Lumber Charitable Contributions, 3566
Stocker Foundation Grants, 3656
Sunderland Foundation Grants, 3675
U.S. Cellular Corporation Grants, 3766
Union Bank, N.A. Corporate Sponsorships and Donations, 3787
Union Bank, N.A. Foundation Grants, 3788
Union Pacific Foundation Health and Human Services Grants, 3794
W.M. Keck Fndn Undergrad Ed Grants, 3891
Weyerhaeuser Company Foundation Grants, 3925

GEOGRAPHIC INDEX

All Countries / 951

West Virginia
Alpha Natural Resources Corporate Giving, 450
American Electric Power Foundation Grants, 530
Amerigroup Foundation Grants, 565
Appalachian Regional Commission Health Grants, 637
Ashland Corporate Contributions Grants, 713
Beckley Area Foundation Grants, 867
Belk Foundation Grants, 872
Cable Positive's Tony Cox Community Grants, 992
Cabot Corporation Foundation Grants, 993
Carl M. Freeman Foundation FACES Grants, 1026
Carl M. Freeman Foundation Grants, 1027
Charles A. Frueauff Foundation Grants, 1114
Charles Delmar Foundation Grants, 1115
Charles M. & Mary D. Grant Fndn Grants, 1124
Claude Worthington Benedum Fndn Grants, 1178
Cleveland-Cliffs Foundation Grants, 1181
CONSOL Energy Community Dev Grants, 1339
CONSOL Military and Armed Services Grants, 1340
CSX Corporate Contributions Grants, 1368
Daywood Foundation Grants, 1438
Dominion Foundation Grants, 1485
Fifth Third Foundation Grants, 1661
FMC Foundation Grants, 1683
Greater Kanawha Valley Foundation Grants, 1882
Guitar Center Music Foundation Grants, 1914
Highmark Corporate Giving Grants, 2025
Hilda and Preston Davis Foundation Grants, 2028
Huntington Clinical Foundation Grants, 2082
Huntington National Bank Community Grants, 2084
Marathon Petroleum Corporation Grants, 2422
MeadWestvaco Foundation Sustainable Communities Grants, 2506
Norfolk Southern Foundation Grants, 2982
Parkersburg Area Community Foundation Action Grants, 3116
Sunoco Foundation Grants, 3679
Toyota Motor Manuf of West Virginia Grants, 3748
Tri-State Community Twenty-first Century Endowment Fund Grants, 3756
U.S. Cellular Corporation Grants, 3766
Weyerhaeuser Company Foundation Grants, 3925

Wisconsin
3M Company Fndn Community Giving Grants, 6
3M Fndn Health & Human Services Grants, 7
ACL Empowering Seniors to Prevent Health Care Fraud Grants, 241
Adler-Clark Electric Community Commitment Foundation Grants, 307
A Fund for Women Grants, 342
Albertson's Charitable Giving Grants, 410
Alliant Energy Corporation Contributions, 445
Amerigroup Foundation Grants, 565
Amica Insurance Company Community Grants, 575
Andersen Corporate Foundation, 586
Anthem Blue Cross and Blue Shield Grants, 607
Archer Daniels Midland Foundation Grants, 651
A Social Corporation Grants, 758
Assurant Health Foundation Grants, 789
Athwin Foundation Grants, 795
Babcock Charitable Trust Grants, 816
Beldon Fund Grants, 871
Black River Falls Area Foundation Grants, 909
Bridgestone/Firestone Trust Fund Grants, 949
Cable Positive's Tony Cox Community Grants, 992
Campbell Soup Foundation Grants, 1007
CFFVR Alcoholism and Drug Abuse Grants, 1093
CFFVR Basic Needs Giving Partnership Grants, 1094
CFFVR Capital Credit Union Charitable Giving Grants, 1095
CFFVR Chilton Area Community Fndn Grants, 1096
CFFVR Clintonville Area Foundation Grants, 1097
CFFVR Frank C. Shattuck Community Grants, 1098
CFFVR Jewelers Mutual Charitable Giving, 1099
CFFVR Myra M. and Robert L. Vandehey Foundation Grants, 1100
CFFVR Project Grants, 1101
CFFVR Robert and Patricia Endries Family Foundation Grants, 1102
CFFVR Schmidt Family G4 Grants, 1103
CFFVR Shawano Area Community Foundation Grants, 1104
CFFVR Waupaca Area Community Foundation Grants, 1105
CFFVR Wisconsin King's Daus & Sons Grants, 1106
CFFVR Women's Fund for the Fox Valley Region Grants, 1107
Coleman Foundation Cancer Care Grants, 1206
Coleman Foundation Developmental Disabilities Grants, 1207
ConAgra Foods Fndn Community Impact Grants, 1321
ConAgra Foods Nourish our Community Grants, 1322
Cooper Industries Foundation Grants, 1345
Cornerstone Foundation of Northeastern Wisconsin Grants, 1347
CUNA Mutual Group Fndn Community Grants, 1375
Dean Foods Community Involvement Grants, 1440
Evjue Foundation Grants, 1617
Faye McBeath Foundation Grants, 1635
Frank G. and Freida K. Brotz Family Foundation Grants, 1723
Gardner Foundation Grants, 1772
General Mills Foundation Grants, 1782
George Kress Foundation Grants, 1799
Godfrey Foundation Grants, 1849
Greater Green Bay Community Fndn Grants, 1881
Greater Milwaukee Foundation Grants, 1883
Green Bay Packers Foundation Grants, 1890
Guitar Center Music Foundation Grants, 1914
Harley Davidson Foundation Grants, 1947
Hormel Foods Charitable Trust Grants, 2050
HRK Foundation Health Grants, 2067
Huffy Foundation Grants, 2073
Hugh J. Andersen Foundation Grants, 2074
Humana Foundation Grants, 2077
Intergrys Corporation Grants, 2160
Jane Bradley Pettit Foundation Health Grants, 2208
John Deere Foundation Grants, 2234
Lands' End Corporate Giving Program, 2344
Margaret Wiegand Trust Grants, 2430
Mead Witter Foundation Grants, 2507
Merkel Foundation Grants, 2520
Nationwide Insurance Foundation Grants, 2637
Norfolk Southern Foundation Grants, 2982
Northwestern Mutual Foundation Grants, 2999
Oscar Rennebohm Foundation Grants, 3085
Otto Bremer Foundation Grants, 3099
Patrick and Anna M. Cudahy Fund Grants, 3120
Piper Jaffray Fndn Communities Giving Grants, 3227
PNC Foundation Community Services Grants, 3240
Richland County Bank Grants, 3371
RRF General Program Grants, 3417
Sensient Technologies Foundation Grants, 3513
Shopko Fndn Community Charitable Grants, 3547
Shopko Fndn Green Bay Community Grants, 3548
Siebert Lutheran Foundation Grants, 3554
Square D Foundation Grants, 3634
U.S. Cellular Corporation Grants, 3766
Union Pacific Foundation Health and Human Services Grants, 3794
Weyerhaeuser Company Foundation Grants, 3925

Wyoming
Albertson's Charitable Giving Grants, 410
Cable Positive's Tony Cox Community Grants, 992
Daniels Fund Grants-Aging, 1424
FMC Foundation Grants, 1683
Griffin Foundation Grants, 1903
Guitar Center Music Foundation Grants, 1914
Peter Kiewit Foundation General Grants, 3154
Peter Kiewit Foundation Small Grants, 3155
Piper Jaffray Fndn Communities Giving Grants, 3227
Questar Corporate Contributions Grants, 3291
Ruth and Vernon Taylor Foundation Grants, 3426
Shopko Fndn Community Charitable Grants, 3547
Steele-Reese Foundation Grants, 3650
Union Pacific Foundation Health and Human Services Grants, 3794
W.M. Keck Fndn Undergrad Ed Grants, 3891

Foreign Countries

All Countries
AAAAI Fellows-in-Training Grants, 29
AAAS Award for Scie Freedom & Responsibility, 36
Abdus Salam ICTP Federation Arrangement Scheme Grants, 205
ACSM Carl V. Gisolfi Memorial Fund Grant, 259
ACSM Coca-Cola Company Doctoral Student Grant on Behavior Research, 261
ACSM Dr. Raymond A. Weiss Research Endowment Grant, 262
ACSM Foundation Clinical Sports Medicine Endowment Grants, 265
ACSM Foundation Doctoral Research Grants, 266
ACSM Foundation Research Endowment Grants, 267
ACSM International Student Awards, 269
ACSM Paffenbarger-Blair Fund for Epidemiological Research on Physical Activity Grants, 273
Acumen Global Fellowships, 285
AES Epilepsy Research Recognition Awards, 318
AES Robert S. Morison Fellowship, 323
AIG Disaster Relief Fund Grants, 370
AIHS Collaborative Research and Innovation Grants - Collaborative Team, 378
Alcatel-Lucent Technologies Foundation Grants, 413
Alexander von Humboldt Foundation Georg Forster Fellowships for Experienced Researchers, 423
Alexander von Humboldt Foundation Research Fellowships for Postdoctoral Researchers, 425
ALFJ Astraea U.S. and Int'l Movement Fund, 427
Alfred P. Sloan Foundation International Science Engagement Grants, 431
Alpha Omega Foundation Grants, 451
American Express Community Service Grants, 531
ANA F.E. Bennett Memorial Lectureship, 581
ANA Raymond D. Adams Lectureship, 583
ANA Soriano Lectureship, 584
AOCS A. Richard Baldwin Award, 610
AOCS Alton E. Bailey Award, 611
AOCS Analytical Division Student Award, 612
AOCS George Schroepfer Medal, 613
AOCS Health & Nutrition Div Student Award, 614
AOCS Health and Nutrition Poster Competition, 615
AOCS Industrial Oil Products Division Student Award, 616
AOCS Lipid Oxidation and Quality Division Poster Competition, 617
AOCS Processing Division Student Award, 618
AOCS Protein and Co-Products Division Student Poster Competition, 619
AOCS Holman Lifetime Achievement Award, 620
AOCS Supelco/Nicholas Pelick-AOCS Research Award, 621
AOCS Surfactants and Detergents Division Student Award, 622
ASHG Public Health Genetics Fellowship, 712
ASM/CDC Fellowships in Infectious Disease and Public Health Microbiology, 716
ASPH/CDC Allan Rosenfield Global Health Fellowships, 780
Australasian Institute of Judicial Administration Seed Funding Grants, 806
Baxter International Corporate Giving Grants, 841
Baxter International Foundation Foster G. McGaw Prize, 842
Baxter International Foundation Grants, 843
Baxter International Foundation William B. Graham Prize for Health Services Research, 844
Bill and Melinda Gates Foundation Emergency Response Grants, 892
Bill and Melinda Gates Foundation Policy and Advocacy Grants, 893
Bill and Melinda Gates Foundation Water, Sanitation and Hygiene Grants, 894
Biogen Foundation Fellowships, 898
Bupa Foundation Medical Research Grants, 974
Caplow Applied Science Carcinogen Prize, 1015
CDC Epidemic Intell Service Training Grants, 1055

952 / All Countries

CDC Foundation Atlanta International Health Fellowships, 1060
CDC Fnd Tobacco Network Lab Fellowship, 1062
CDC Public Health Informatics Fellowships, 1068
CDC Steven M. Teutsch Prevention Effectiveness Fellowships, 1071
Changemakers Innovation Awards, 1111
Children's Tumor Foundation Drug Discovery Initiative Awards, 1141
Christopher & Dana Reeve Foundation Quality of Life Grants, 1152
CNIB Barbara Tuck MacPhee Award, 1196
CNIB Chanchlani Global Vision Research Award, 1198
CNIB E. (Ben) & Mary Hochhausen Access Technology Research Grants, 1199
CNIB Ross Purse Doctoral Fellowships, 1200
Conquer Cancer Foundation of ASCO International Development and Education Award in Palliative Care, 1332
Conquer Cancer Foundation of ASCO International Development and Education Awards, 1333
Conquer Cancer Foundation of ASCO International Innovation Grant, 1334
Conservation, Food, and Health Foundation Grants for Developing Countries, 1338
Covidien Medical Product Donations, 1352
Cystic Fibrosis Canada Visiting Allied Health Professional Awards, 1394
Cystic Fibrosis Canada Visiting Clinician Awards, 1395
Cystic Fibrosis Trust Research Grants, 1404
Delta Air Lines Foundation Prize for Global Understanding, 1456
E.W. "Al" Thrasher Awards, 1540
EBSCO / MLA Annual Meeting Grants, 1545
Elizabeth Glaser Int'l Leadership Awards, 1566
Elizabeth Glaser Scientist Award, 1567
FAMRI Clinical Innovator Awards, 1628
FAMRI Young Clinical Scientist Awards, 1629
Gates Award for Global Health, 1775
Global Fund for Women Grants, 1833
Grass Foundation Marine Biological Laboratory Advanced Imaging Fellowships, 1877
Grass Foundation Marine Biological Laboratory Fellowships, 1878
Greygates Foundation Grants, 1901
Gruber Foundation Cosmology Prize, 1907
Gruber Foundation Neuroscience Prize, 1909
Gruber Foundation Rosalind Franklin Young Investigator Award, 1910
Gruber Foundation Weizmann Institute Awards, 1911
Harry Frank Guggenheim Foundation Dissertation Fellowships, 1963
Harry Frank Guggenheim Fnd Research Grants, 1964
Helen Bader Foundation Grants, 1992
IBRO/SfN International Travel Grants, 2100
IBRO International Travel Grants, 2104
IBRO Regional Grants for Int'l Fellowships to U.S. Laboratory Summer Neuroscience Courses, 2110
IBRO Research Fellowships, 2111
IBRO Return Home Fellowships, 2112
IIE Adell and Hancock Scholarships, 2123
IIE KAUST Graduate Fellowships, 2134
IIE New Leaders Group Award for Mutual Understanding, 2141
International Positive Psychology Association Student Scholarships, 2161
Lawrence Foundation Grants, 2347
Lymphatic Education and Research Network Postdoctoral Fellowships, 2395
Macquarie Bank Foundation Grants, 2410
Majors MLA Chapter Project of the Year, 2418
MBL Albert & Ellen Grass Fellowships, 2463
MBL Ann E. Kammer Summer Fellowship, 2464
MBL Associates Summer Fellowships, 2465
MBL Baxter Postdoctoral Summer Fellowship, 2466
MBL Burr & Susie Steinback Fellowship, 2467
MBL E.E. Just Summer Fellowship for Minority Scientists, 2468
MBL Erik B. Fries Summer Fellowships, 2469
MBL Eugene and Millicent Bell Fellowships, 2470

MBL Evelyn and Melvin Spiegal Fellowship, 2471
MBL Frank R. Lillie Summer Fellowship, 2472
MBL Frederik B. and Betsy G. Bang Summer Fellowships, 2473
MBL Fred Karush Library Readership, 2474
MBL H. Keffer Hartline and Edward F. MacNichol, Jr. Fellowships, 2476
MBL Herbert W. Rand Summer Fellowship, 2477
MBL James E. and Faith Miller Memorial Summer Fellowship, 2478
MBL John M. Arnold Award, 2479
MBL Laura and Arthur Colwin Fellowships, 2480
MBL Lucy B. Lemann Summer Fellowship, 2481
MBL M.G.F. Fuortes Summer Fellowships, 2482
MBL Nikon Summer Fellowship, 2483
MBL Plum Fndn John E. Dowling Fellowships, 2484
MBL Robert Day Allen Summer Fellowship, 2485
MBL Scholarships and Awards, 2486
MBL Stephen W. Kuffler Summer Fellowships, 2487
MBL William Townsend Porter Summer Fellowships for Minority Investigators, 2488
MeadWestvaco Foundation Sustainable Communities Grants, 2506
Medical Informatics Section/MLA Career Development Grant, 2508
MLA Cunningham Memorial International Fellowship, 2560
MLA Estelle Brodman Academic Medical Librarian of the Year Award, 2563
MLA Ida and George Eliot Prize, 2566
MLA Janet Doe Lectureship Award, 2567
MLA Lois Ann Colaianni Award for Excellence and Achievement in Hospital Librarianship, 2568
MLA Louise Darling Medal for Distinguished Achievement in Collection Development in the Health Sciences, 2569
MLA Lucretia W. McClure Excellence in Education Award, 2570
MLA Murray Gottlieb Prize Essay Award, 2571
MLA Section Project of the Year Award, 2574
MLA T. Mark Hodges Int'l Service Award, 2575
MLA Thomas Reuters / Frank Bradway Rogers Information Advancement Award, 2576
Morehouse PHSI Project Imhotep Internships, 2592
National Parkinson Foundation Clinical Research Fund Grants, 2635
NEAVS Fellowsips in Women's Health and Sex Differences, 2659
Nestle Foundation Large Research Grants, 2675
Nestle Foundation Pilot Grants, 2676
Nestle Foundation Re-entry Grants, 2677
Nestle Foundation Small Research Grants, 2678
Nestle Foundation Training Grant, 2679
NHLBI Ancillary Studies in Clinical Trials, 2688
NHLBI Investigator Initiated Multi-Site Clinical Trials, 2703
NHLBI Lymphatics in Health and Disease in the Digestive, Cardio & Pulmonary Systems, 2704
NHLBI Lymphatics in Health and Disease in the Digestive, Urinary, Cardiovascular and Pulmonary Systems, 2705
NHLBI Research on the Role of Cardiomyocyte Mitochondria in Heart Disease: An Integrated Approach, 2726
NHLBI Ruth L. Kirschstein National Research Service Awards for Individual Postdoctoral Fellows, 2729
NIGMS Advancing Basic Behavioral and Social Research on Resilience: an Integrative Science Approach, 2861
NLM Express Research Grants in Biomedical Informatics, 2943
NSF Social Psychology Research Grants, 3023
OECD Sexual Health and Rights Project Grants, 3064
OSF Access to Essential Medicines Initiative, 3086
OSF Accountability and Monitoring in Health Initiative Grants, 3087
OSF Global Health Financing Initiative Grants, 3088
OSF Health Media Initiative Grants, 3089
OSF International Harm Reduction Development Program Grants, 3090

GEOGRAPHIC INDEX

OSF International Palliative Care Grants, 3091
OSF Law and Health Initiative Grants, 3092
PC Fred H. Bixby Fellowships, 3126
PepsiCo Foundation Grants, 3137
Pfizer ASPIRE Worldwide Endocrine Young Investigator Grants, 3173
Pfizer Australia Paediatric Endocrine Care Research Grants, 3176
PFizer Compound Transfer Agreements, 3177
Pfizer Global Investigator Research Grants, 3178
Pfizer Global Research Awards for Nicotine Independence, 3179
Pfizer Healthcare Charitable Contributions, 3180
Pfizer Medical Education Track One Grants, 3182
Pfizer Medical Education Track Two Grants, 3183
Pfizer Special Events Grants, 3186
Radcliffe Institute Residential Fellowships, 3299
Raskob Foundation for Catholic Activities Grants, 3306
RCPSC McLaughlin-Gallie Visiting Professor, 3333
Royal Norwegian Embassy Kavli Prizes, 3414
Scleroderma Foundation Established Investigator Grants, 3497
SfN Chapter-of-the-Year Award, 3519
SfN Chapter Grants, 3520
SfN Patricia Goldman-Rakic Hall of Honor, 3532
SfN Trainee Professional Development Awards, 3538
Sigma Theta Tau International /American Nurses Foundation Grant, 3556
Sigma Theta Tau International / Council for the Advancement of Nursing Science Grants, 3557
Sigma Theta Tau International / Oncology Nursing Society Grant, 3558
Sigma Theta Tau International Doris Bloch Research Award, 3559
Sigma Theta Tau Small Grants, 3560
SOBP A.E. Bennett Research Award, 3594
SSHRC Banting Postdoctoral Fellowships, 3637
Susan G. Komen Breast Cancer Foundation Career Catalyst Research Grants, 3688
Susan G. Komen Breast Cancer Foundation Challenge Grants: Career Catalyst Research, 3691
Susan G. Komen Breast Cancer Foundation Investigator Initiated Research Grants, 3693
Thrasher Research Fund Early Career Awards, 3731
Thrasher Research Fund Grants, 3732
Tourette Syndrome Association Post-Doctoral Fellowships, 3740
Tourette Syndrome Association Research Grants, 3741
United Methodist Committee on Relief Global Health Grants, 3799
United States Institute of Peace - Jennings Randolph Senior Fellowships, 3802
United Technologies Corporation Grants, 3803
USAID Child, Newborn, and Maternal Health Project Grants, 3806
USAID Devel Innovation Accelerator Grants, 3808
USAID Fighting Ebola Grants, 3812
USAID Global Health Development Innovation Accelerator Broad Agency Grants, 3813
USAID Health System Strengthening Grants, 3814
USAID Land Use Change and Disease Emergence Grants, 3818
USAID Microbicides Research, Development, and Introduction Grants, 3819
USAID Social Science Research in Population and Reproductive Health Grants, 3822
USAID Support for International Family Planning Organizations Grants, 3823
Virginia L. and William K. Beatty MLA Volunteer Service Award, 3877
Wellcome Trust Biomedical Science Grants, 3918

Abkhazia
Fulbright Traditional Scholar Program in the East Asia/Pacific Region, 1757
H.B. Fuller Foundation Grants, 1922

Afghanistan
Fulbright Traditional Scholar Program in the East Asia/Pacific Region, 1757

GEOGRAPHIC INDEX

H.B. Fuller Foundation Grants, 1922
IBRO Asia Reg APRC Exchange Fellowships, 2101
IBRO Asia APRC Lecturer Exchange Grants, 2102
IBRO Asia Regional APRC Travel Grants, 2103
Wellcome Trust New Investigator Awards, 3919

Albania
Advanced Micro Devices Comm Affairs Grants, 313
Beatrice Laing Trust Grants, 864
Cargill Citizenship Fund Corporate Giving, 1018
Charles Delmar Foundation Grants, 1115
EMBO Installation Grants, 1586
ERC Starting Grants, 1599
Fluor Foundation Grants, 1682
Fulbright Scholars in Europe and Eurasia, 1755
HHMI Biomedical Research Grants for International Scientists in the Baltics, Central and Eastern Europe, Russia, and Ukraine, 2019
IBRO-PERC InEurope Short Stay Grants, 2097
OSF Mental Health Initiative Grants, 3093
OSF Roma Health Project Grants, 3095
Simone and Cino del Duca Grand Prix Awards, 3564
Sir Dorabji Tata Trust Grants for NGOs or Voluntary Organizations, 3572
Wellcome Trust New Investigator Awards, 3919

Algeria
Aid for Starving Children International Grants, 368
Beatrice Laing Trust Grants, 864
Cargill Citizenship Fund Corporate Giving, 1018
Fulbright Traditional Scholar Program in Sub-Saharan Africa, 1756
Fulbright Traditional Scholar Program in the Near East and North Africa Region, 1758
Nuffield Foundation Africa Program Grants, 3026
Rockefeller Brothers Fund Pivotal Places Grants: South Africa, 3395
Wellcome Trust New Investigator Awards, 3919

Andorra
Advanced Micro Devices Comm Affairs Grants, 313
Beatrice Laing Trust Grants, 864
Cargill Citizenship Fund Corporate Giving, 1018
Charles Delmar Foundation Grants, 1115
EMBO Installation Grants, 1586
ERC Starting Grants, 1599
Fluor Foundation Grants, 1682
Fulbright Scholars in Europe and Eurasia, 1755
HHMI Biomedical Research Grants for International Scientists in the Baltics, Central and Eastern Europe, Russia, and Ukraine, 2019
IBRO-PERC InEurope Short Stay Grants, 2097
Simone and Cino del Duca Grand Prix Awards, 3564
Sir Dorabji Tata Trust Grants for NGOs or Voluntary Organizations, 3572

Angola
Aid for Starving Children International Grants, 368
Beatrice Laing Trust Grants, 864
Cargill Citizenship Fund Corporate Giving, 1018
Fulbright Traditional Scholar Program in Sub-Saharan Africa, 1756
Fulbright Traditional Scholar Program in the Near East and North Africa Region, 1758
IIE Chevron International REACH Scholarships, 2128
IIE Hewlett Fnd/IIE Dissertation Fellowship, 2132
Nuffield Foundation Africa Program Grants, 3026
Rockefeller Brothers Fund Pivotal Places Grants: South Africa, 3395
Wellcome Trust New Investigator Awards, 3919

Anguilla
Elton John AIDS Foundation Grants, 1585

Antigua & Barbuda
ASM Int'l Fellowship for Latin America and the Caribbean, 734
Elton John AIDS Foundation Grants, 1585
Fulbright Traditional Scholar Program in the Western Hemisphere, 1760

IBRO Latin America Regional Funding for Neuroscience Schools, 2105
IBRO Latin America Regional Funding for PROLAB Collaborations, 2106
IBRO Latin America Regional Funding for Short Courses, Workshops, and Symposia, 2107
IBRO Latin America Regional Funding for Short Research Stays, 2108
IBRO Latin America Regional Travel Grants, 2109

Argentina
ASM-PAHO Infectious Diseases Epidemiology and Surveillance Fellowships, 714
ASM Int'l Fellowship for Latin America and the Caribbean, 734
Charles Delmar Foundation Grants, 1115
Elton John AIDS Foundation Grants, 1585
Fulbright Traditional Scholar Program in the Western Hemisphere, 1760
IBRO Latin America Regional Funding for Neuroscience Schools, 2105
IBRO Latin America Regional Funding for PROLAB Collaborations, 2106
IBRO Latin America Regional Funding for Short Courses, Workshops, and Symposia, 2107
IBRO Latin America Regional Funding for Short Research Stays, 2108
IBRO Latin America Regional Travel Grants, 2109
IIE Chevron International REACH Scholarships, 2128
Katherine John Murphy Foundation Grants, 2293
Olympus Corporation of Americas Corporate Giving Grants, 3073
SfN Award for Education in Neuroscience, 3517
SfN Bernice Grafstein Award for Outstanding Accomplishments in Mentoring, 3518
SfN Donald B. Lindsley Prize in Behavioral Neuroscience, 3521
SfN Jacob P. Waletzky Award, 3523
SfN Janett Rosenberg Trubatch Career Development Award, 3524
SfN Julius Axelrod Prize, 3526
SfN Louise Hanson Marshall Recognition Award, 3527
SfN Mika Salpeter Lifetime Achievement Award, 3528
SfN Next Generation Award, 3531
SfN International Research Award, 3533
SfN Ralph W. Gerard Prize in Neuroscience, 3534
SfN Science Educator Award, 3535
SfN Swartz Prize for Theoretical and Computational Neuroscience, 3537
SfN Young Investigator Award, 3541
Wellcome Trust New Investigator Awards, 3919

Armenia
Advanced Micro Devices Comm Affairs Grants, 313
Beatrice Laing Trust Grants, 864
Cargill Citizenship Fund Corporate Giving, 1018
Charles Delmar Foundation Grants, 1115
EMBO Installation Grants, 1586
ERC Starting Grants, 1599
Fluor Foundation Grants, 1682
Fulbright Scholars in Europe and Eurasia, 1755
Fulbright Traditional Scholar Program in the East Asia/Pacific Region, 1757
H.B. Fuller Foundation Grants, 1922
HHMI Biomedical Research Grants for International Scientists in the Baltics, Central and Eastern Europe, Russia, and Ukraine, 2019
IBRO-PERC InEurope Short Stay Grants, 2097
Simone and Cino del Duca Grand Prix Awards, 3564
Sir Dorabji Tata Trust Grants for NGOs or Voluntary Organizations, 3572
Wellcome Trust New Investigator Awards, 3919

Aruba
Elton John AIDS Foundation Grants, 1585

Australia
Australian Academy of Science Grants, 807
Bechtel Group Foundation Building Positive Community Relationships Grants, 866

Boeing Company Contributions Grants, 930
Bupa Foundation Multi-Country Grant, 975
ERC Starting Grants, 1599
Fluor Foundation Grants, 1682
Fulbright Traditional Scholar Program in the East Asia/Pacific Region, 1757
IIE Chevron International REACH Scholarships, 2128
Layne Beachley Aim for the Stars Fnd Grants, 2349
May and Stanley Smith Charitable Trust Grants, 2459
PACCAR Foundation Grants, 3101
Pediatric Brain Tumor Fndn Research Grants, 3134
Pfizer Australia Cancer Research Grants, 3174
Pfizer Australia Neuroscience Research Grants, 3175
RACGP Cardiovascular Research Grants in General Practice, 3295
RACGP Nat Asthma Council Research Award, 3296
RACGP Vicki Kotsirilos Integrative Med Grants, 3297
SfN Award for Education in Neuroscience, 3517
SfN Bernice Grafstein Award for Outstanding Accomplishments in Mentoring, 3518
SfN Donald B. Lindsley Prize in Behavioral Neuroscience, 3521
SfN Jacob P. Waletzky Award, 3523
SfN Janett Rosenberg Trubatch Career Development Award, 3524
SfN Julius Axelrod Prize, 3526
SfN Louise Hanson Marshall Specific Recognition Award, 3527
SfN Mika Salpeter Lifetime Achievement Award, 3528
SfN Next Generation Award, 3531
SfN Peter and Patricia Gruber International Research Award, 3533
SfN Ralph W. Gerard Prize in Neuroscience, 3534
SfN Science Educator Award, 3535
SfN Swartz Prize for Theoretical and Computational Neuroscience, 3537
SfN Young Investigator Award, 3541

Austria
Advanced Micro Devices Comm Affairs Grants, 313
Beatrice Laing Trust Grants, 864
BP Foundation Grants, 943
Cargill Citizenship Fund Corporate Giving, 1018
Charles Delmar Foundation Grants, 1115
EMBO Installation Grants, 1586
Fluor Foundation Grants, 1682
Fulbright Scholars in Europe and Eurasia, 1755
HHMI/EMBO Start-up Grants for C Europe, 2016
HHMI Biomedical Research Grants for International Scientists in the Baltics, Central and Eastern Europe, Russia, and Ukraine, 2019
IBRO-PERC InEurope Short Stay Grants, 2097
IBRO-PERC Support for European Workshops, Symposia and Meetings, 2098
IBRO-PERC Support for Site Lectures, 2099
Medtronic Foundation Strengthening Health Systems Grants, 2515
Pfizer ASPIRE EU Antifungal Research Awards, 3165
Pfizer ASPIRE EU Dupuytren's Contracture Research Awards, 3166
Pfizer ASPIRE EU Emerging Mechanisms of Resistance Antibacterial Research Awards, 3167
Pfizer ASPIRE EU MRSA Nosocomial Pneumonia & MRSA Complicated Skin & Soft Tissue Infections Antibacterial Research Awards, 3168
Simone and Cino del Duca Grand Prix Awards, 3564
Sir Dorabji Tata Trust Grants for NGOs or Voluntary Organizations, 3572

Azerbaijan
Advanced Micro Devices Comm Affairs Grants, 313
Beatrice Laing Trust Grants, 864
Cargill Citizenship Fund Corporate Giving, 1018
Charles Delmar Foundation Grants, 1115
EMBO Installation Grants, 1586
ERC Starting Grants, 1599
Fluor Foundation Grants, 1682
Fulbright Scholars in Europe and Eurasia, 1755
Fulbright Traditional Scholar Program in the East Asia/Pacific Region, 1757

954 / Azerbaijan

H.B. Fuller Foundation Grants, 1922
HHMI Biomedical Research Grants for International Scientists in the Baltics, Central and Eastern Europe, Russia, and Ukraine, 2019
IBRO-PERC InEurope Short Stay Grants, 2097
IIE Chevron International REACH Scholarships, 2128
Simone and Cino del Duca Grand Prix Awards, 3564
Sir Dorabji Tata Trust Grants for NGOs or Voluntary Organizations, 3572
Wellcome Trust New Investigator Awards, 3919

Bahamas
ASM Int'l Fellowship for Latin America and the Caribbean, 734
Elton John AIDS Foundation Grants, 1585
Fulbright Traditional Scholar Program in the Western Hemisphere, 1760
IBRO Latin America Regional Funding for Neuroscience Schools, 2105
IBRO Latin America Regional Funding for PROLAB Collaborations, 2106
IBRO Latin America Regional Funding for Short Courses, Workshops, and Symposia, 2107
IBRO Latin America Regional Funding for Short Research Stays, 2108
IBRO Latin America Regional Travel Grants, 2109
May and Stanley Smith Charitable Trust Grants, 2459

Bahrain
Fulbright Traditional Scholar Program in the East Asia/Pacific Region, 1757
H.B. Fuller Foundation Grants, 1922
IBRO Asia Reg APRC Exchange Fellowships, 2101
IBRO Asia APRC Lecturer Exchange Grants, 2102
IBRO Asia Regional APRC Travel Grants, 2103

Bangladesh
Fulbright Traditional Scholar Program in the East Asia/Pacific Region, 1757
H.B. Fuller Foundation Grants, 1922
IBRO Asia Reg APRC Exchange Fellowships, 2101
IBRO Asia APRC Lecturer Exchange Grants, 2102
IBRO Asia Regional APRC Travel Grants, 2103
IIE Chevron International REACH Scholarships, 2128
Wellcome Trust New Investigator Awards, 3919

Barbados
ASM Int'l Fellowship for Latin America and the Caribbean, 734
Elton John AIDS Foundation Grants, 1585
Fulbright Traditional Scholar Program in the Western Hemisphere, 1760
IBRO Latin America Regional Funding for Neuroscience Schools, 2105
IBRO Latin America Regional Funding for PROLAB Collaborations, 2106
IBRO Latin America Regional Funding for Short Courses, Workshops, and Symposia, 2107
IBRO Latin America Regional Funding for Short Research Stays, 2108
IBRO Latin America Regional Travel Grants, 2109

Belarus
Advanced Micro Devices Comm Affairs Grants, 313
Beatrice Laing Trust Grants, 864
Cargill Citizenship Fund Corporate Giving, 1018
Charles Delmar Foundation Grants, 1115
EMBO Installation Grants, 1586
ERC Starting Grants, 1599
Fluor Foundation Grants, 1682
Fulbright Scholars in Europe and Eurasia, 1755
HHMI Biomedical Research Grants for International Scientists in the Baltics, Central and Eastern Europe, Russia, and Ukraine, 2019
IBRO-PERC InEurope Short Stay Grants, 2097
John Deere Foundation Grants, 2234
Medtronic Foundation Strengthening Health Systems Grants, 2515
OSF Roma Health Project Grants, 3095
Simone and Cino del Duca Grand Prix Awards, 3564
Sir Dorabji Tata Trust Grants for NGOs or Voluntary Organizations, 3572
Wellcome Trust New Investigator Awards, 3919

Belgium
Advanced Micro Devices Comm Affairs Grants, 313
Beatrice Laing Trust Grants, 864
BP Foundation Grants, 943
Cabot Corporation Foundation Grants, 993
Cargill Citizenship Fund Corporate Giving, 1018
Charles Delmar Foundation Grants, 1115
EMBO Installation Grants, 1586
ERC Starting Grants, 1599
Fluor Foundation Grants, 1682
Fulbright Scholars in Europe and Eurasia, 1755
HHMI Biomedical Research Grants for International Scientists in the Baltics, Central and Eastern Europe, Russia, and Ukraine, 2019
IBRO-PERC InEurope Short Stay Grants, 2097
IBRO-PERC Support for European Workshops, Symposia and Meetings, 2098
IBRO-PERC Support for Site Lectures, 2099
Medtronic Foundation Patient Link Grants, 2514
Pfizer ASPIRE EU Antifungal Research Awards, 3165
Pfizer ASPIRE EU Dupuytren's Contracture Research Awards, 3166
Pfizer ASPIRE EU Emerging Mechanisms of Resistance Antibacterial Research Awards, 3167
Pfizer ASPIRE EU MRSA Nosocomial Pneumonia & MRSA Complicated Skin & Soft Tissue Infections Antibacterial Research Awards, 3168
Simone and Cino del Duca Grand Prix Awards, 3564
Sir Dorabji Tata Trust Grants for NGOs or Voluntary Organizations, 3572

Belize
ASM Int'l Fellowship for Latin America and the Caribbean, 734
Elton John AIDS Foundation Grants, 1585
Fulbright Traditional Scholar Program in the Western Hemisphere, 1760
IBRO Latin America Regional Funding for Neuroscience Schools, 2105
IBRO Latin America Regional Funding for PROLAB Collaborations, 2106
IBRO Latin America Regional Funding for Short Research Stays, 2108
IBRO Latin America Regional Travel Grants, 2109
Wellcome Trust New Investigator Awards, 3919

Benin
Aid for Starving Children International Grants, 368
Beatrice Laing Trust Grants, 864
Cargill Citizenship Fund Corporate Giving, 1018
Fulbright Traditional Scholar Program in Sub-Saharan Africa, 1756
Fulbright Traditional Scholar Program in the Near East and North Africa Region, 1758
IIE Hewlett Fnd/IIE Dissertation Fellowship, 2132
Nuffield Foundation Africa Program Grants, 3026
Rockefeller Brothers Fund Pivotal Places Grants: South Africa, 3395
Wellcome Trust New Investigator Awards, 3919

Bhutan
Fulbright Traditional Scholar Program in the East Asia/Pacific Region, 1757
H.B. Fuller Foundation Grants, 1922
IBRO Asia Reg APRC Exchange Fellowships, 2101
IBRO Asia APRC Lecturer Exchange Grants, 2102
IBRO Asia Regional APRC Travel Grants, 2103
Wellcome Trust New Investigator Awards, 3919

Bolivia
ASM-PAHO Infectious Diseases Epidemiology and Surveillance Fellowships, 714
ASM Int'l Fellowship for Latin America and the Caribbean, 734
Charles Delmar Foundation Grants, 1115
Elton John AIDS Foundation Grants, 1585
Fulbright Traditional Scholar Program in the Western Hemisphere, 1760
IBRO Latin America Regional Funding for Neuroscience Schools, 2105
IBRO Latin America Regional Funding for PROLAB Collaborations, 2106
IBRO Latin America Regional Funding for Short Courses, Workshops, and Symposia, 2107
IBRO Latin America Regional Funding for Short Research Stays, 2108
IBRO Latin America Regional Travel Grants, 2109
Katherine John Murphy Foundation Grants, 2293
Olympus Corporation of Americas Corporate Giving Grants, 3073
Wellcome Trust New Investigator Awards, 3919

Bosnia & Herzegovina
Advanced Micro Devices Comm Affairs Grants, 313
Beatrice Laing Trust Grants, 864
Cargill Citizenship Fund Corporate Giving, 1018
Charles Delmar Foundation Grants, 1115
EMBO Installation Grants, 1586
ERC Starting Grants, 1599
Fluor Foundation Grants, 1682
Fulbright Scholars in Europe and Eurasia, 1755
HHMI Biomedical Research Grants for International Scientists in the Baltics, Central and Eastern Europe, Russia, and Ukraine, 2019
IBRO-PERC InEurope Short Stay Grants, 2097
OSF Mental Health Initiative Grants, 3093
OSF Roma Health Project Grants, 3095
Simone and Cino del Duca Grand Prix Awards, 3564
Sir Dorabji Tata Trust Grants for NGOs or Voluntary Organizations, 3572
Wellcome Trust New Investigator Awards, 3919

Botswana
Aid for Starving Children International Grants, 368
Beatrice Laing Trust Grants, 864
Cargill Citizenship Fund Corporate Giving, 1018
Fulbright Traditional Scholar Program in Sub-Saharan Africa, 1756
Fulbright Traditional Scholar Program in the Near East and North Africa Region, 1758
IIE Hewlett Fnd/IIE Dissertation Fellowship, 2132
Nuffield Foundation Africa Program Grants, 3026
Rockefeller Brothers Fund Pivotal Places Grants: South Africa, 3395
Wellcome Trust New Investigator Awards, 3919

Brazil
Abbott Fund Community Grants, 201
Abbott Fund Science Education Grants, 203
ASM-PAHO Infectious Diseases Epidemiology and Surveillance Fellowships, 714
ASM Int'l Fellowship for Latin America and the Caribbean, 734
Bechtel Group Foundation Building Positive Community Relationships Grants, 866
Charles Delmar Foundation Grants, 1115
Elton John AIDS Foundation Grants, 1585
Fulbright Traditional Scholar Program in the Western Hemisphere, 1760
Hershey Company Grants, 2013
IBRO Latin America Regional Funding for Neuroscience Schools, 2105
IBRO Latin America Regional Funding for PROLAB Collaborations, 2106
IBRO Latin America Regional Funding for Short Courses, Workshops, and Symposia, 2107
IBRO Latin America Regional Funding for Short Research Stays, 2108
IIE Brazil Science Without Borders Undergraduate Scholarships, 2126
IIE Chevron International REACH Scholarships, 2128
John Deere Foundation Grants, 2234
Katherine John Murphy Foundation Grants, 2293
Medtronic Foundation Strengthening Health Systems Grants, 2515
Oak Foundation Child Abuse Grants, 3059

GEOGRAPHIC INDEX

Olympus Corporation of Americas Corporate Giving Grants, 3073
SfN Award for Education in Neuroscience, 3517
SfN Bernice Grafstein Award for Outstanding Accomplishments in Mentoring, 3518
SfN Donald B. Lindsley Prize in Behavioral Neuroscience, 3521
SfN Jacob P. Waletzky Award, 3523
SfN Janett Rosenberg Trubatch Career Development Award, 3524
SfN Julius Axelrod Prize, 3526
SfN Louise Hanson Marshall Specific Recognition Award, 3527
SfN Mika Salpeter Lifetime Achievement Award, 3528
SfN Next Generation Award, 3531
SfN Peter and Patricia Gruber International Research Award, 3533
SfN Ralph W. Gerard Prize in Neuroscience, 3534
SfN Science Educator Award, 3535
SfN Swartz Prize for Theoretical and Computational Neuroscience, 3537
SfN Young Investigator Award, 3541
Wellcome Trust New Investigator Awards, 3919

British Indian Ocean Territory
Fulbright Traditional Scholar Program in the East Asia/Pacific Region, 1757
H.B. Fuller Foundation Grants, 1922

British Virgin Islands
Elton John AIDS Foundation Grants, 1585

Brunei
Fulbright Traditional Scholar Program in the East Asia/Pacific Region, 1757
H.B. Fuller Foundation Grants, 1922
IBRO Asia Reg APRC Exchange Fellowships, 2101
IBRO Asia APRC Lecturer Exchange Grants, 2102
IBRO Asia Regional APRC Travel Grants, 2103

Bulgaria
Advanced Micro Devices Comm Affairs Grants, 313
Beatrice Laing Trust Grants, 864
Cargill Citizenship Fund Corporate Giving, 1018
Charles Delmar Foundation Grants, 1115
EMBO Installation Grants, 1586
ERC Starting Grants, 1599
Fluor Foundation Grants, 1682
Fulbright Scholars in Europe and Eurasia, 1755
HHMI Biomedical Research Grants for International Scientists in the Baltics, Central and Eastern Europe, Russia, and Ukraine, 2019
IBRO-PERC InEurope Short Stay Grants, 2097
Medtronic Foundation Strengthening Health Systems Grants, 2515
Oak Foundation Child Abuse Grants, 3059
OSF Mental Health Initiative Grants, 3093
Simone and Cino del Duca Grand Prix Awards, 3564
Sir Dorabji Tata Trust Grants for NGOs or Voluntary Organizations, 3572
Wellcome Trust New Investigator Awards, 3919

Burkina Faso
Aid for Starving Children International Grants, 368
Beatrice Laing Trust Grants, 864
Cargill Citizenship Fund Corporate Giving, 1018
Fulbright Traditional Scholar Program in Sub-Saharan Africa, 1756
Fulbright Traditional Scholar Program in the Near East and North Africa Region, 1758
IIE Hewlett Fnd/IIE Dissertation Fellowship, 2132
Nuffield Foundation Africa Program Grants, 3026
Rockefeller Brothers Fund Pivotal Places Grants: South Africa, 3395
Wellcome Trust New Investigator Awards, 3919

Burma (Myanmar)
Fulbright Traditional Scholar Program in the East Asia/Pacific Region, 1757
H.B. Fuller Foundation Grants, 1922

Burundi
Acumen East Africa Fellowship, 284
Aid for Starving Children International Grants, 368
Beatrice Laing Trust Grants, 864
Cargill Citizenship Fund Corporate Giving, 1018
Fulbright Traditional Scholar Program in Sub-Saharan Africa, 1756
Fulbright Traditional Scholar Program in the Near East and North Africa Region, 1758
Nuffield Foundation Africa Program Grants, 3026
Rockefeller Brothers Fund Pivotal Places Grants: South Africa, 3395
Wellcome Trust New Investigator Awards, 3919

Cambodia
Fulbright Traditional Scholar Program in the East Asia/Pacific Region, 1757
H.B. Fuller Foundation Grants, 1922
IBRO Asia Reg APRC Exchange Fellowships, 2101
IBRO Asia APRC Lecturer Exchange Grants, 2102
IBRO Asia Regional APRC Travel Grants, 2103
IIE Chevron International REACH Scholarships, 2128
Wellcome Trust New Investigator Awards, 3919

Cameroon
Aid for Starving Children International Grants, 368
Beatrice Laing Trust Grants, 864
Cargill Citizenship Fund Corporate Giving, 1018
Fulbright Traditional Scholar Program in Sub-Saharan Africa, 1756
Fulbright Traditional Scholar Program in the Near East and North Africa Region, 1758
IIE Hewlett Fnd/IIE Dissertation Fellowship, 2132
Nuffield Foundation Africa Program Grants, 3026
Rockefeller Brothers Fund Pivotal Places Grants: South Africa, 3395
Wellcome Trust New Investigator Awards, 3919

Canada
Abbott Fund Community Grants, 201
Abbott Fund Science Education Grants, 203
ACSM Oded Bar-Or Int'l Scholar Awards, 272
Adobe Community Investment Grants, 310
AIHS/Mitacs Health Pilot Partnership Interns, 373
AIHS Alberta/Pfizer Translat Research Grants, 374
AIHS Clinical Fellowships, 375
AIHS Collaborative Research and Innovation Grants - Collaborative Program, 376
AIHS Collaborative Research and Innovation Grants - Collaborative Project, 377
AIHS Fast-Track Fellowships, 379
AIHS ForeFront Internships, 380
AIHS ForeFront MBA Studentship Award, 381
AIHS ForeFront MBT Studentship Awards, 382
AIHS Full-Time Fellowships, 383
AIHS Full-Time Health Research Studentships, 384
AIHS Full-Time M.D./Ph.D. Studentships, 385
AIHS Full-Time Studentships, 386
AIHS Heritage Youth Researcher Summer Science Program, 387
AIHS Interdisciplinary Team Grants, 388
AIHS Knowledge Exchange - Conference Grants, 389
AIHS Knowledge Exchange Visiting Professors, 390
AIHS Knowledge Exchange - Visiting Scientists, 391
AIHS Media Summer Fellowship, 392
AIHS Part-Time Fellowships, 393
AIHS Part-Time Studentships, 394
AIHS Proposals for Special Initiatives, 395
AIHS Research Prize, 396
AIHS Summer Studentships, 397
AIHS Sustainability Grants, 398
Alaska Airlines Corporate Giving Medical Emergency and Research Grants, 402
Alfred P. Sloan Foundation Research Fellowships, 432
ANA Distinguished Neurology Teacher Award, 580
Andersen Corporate Foundation, 586
Austin S. Nelson Foundation Grants, 805
Bank of America Fndn Matching Gifts, 825
Bank of America Fndn Volunteer Grants, 826
Bearemy's Kennel Pals Grants, 863

Bechtel Group Foundation Building Positive Community Relationships Grants, 866
Boeing Company Contributions Grants, 930
BP Foundation Grants, 943
Bullitt Foundation Grants, 973
BWF Ad Hoc Grants, 982
BWF Career Awards at the Scientific Interface, 983
BWF Career Awards for Medical Scientists, 984
BWF Collaborative Research Travel Grants, 985
BWF Innovation in Regulatory Science Grants, 986
BWF Institutional Program Unifying Population and Laboratory Based Sciences Grants, 987
BWF Investigators in the Pathogenesis of Infectious Disease Awards, 988
BWF Postdoctoral Enrichment Grants, 989
BWF Preterm Birth Initiative Grants, 990
Cabot Corporation Foundation Grants, 993
Canada-U.S. Fulbright New Century Scholars Program Grants, 1008
Canada-U.S. Fulbright Senior Specialists Grants, 1009
Canada Graduate Scholarships (CGS) and NSERC Postgraduate Scholarships (PGS), 1010
Canadian Optometric Education Trust Grants, 1011
Canadian Patient Safety Institute (CPSI) Patient Safety Studentships, 1012
Canadian Patient Safety Institute (CPSI) Research Grants, 1013
Central Okanagan Foundation Grants, 1080
Champ-A Champion Fur Kids Grants, 1109
Chase Paymentech Corporate Giving Grants, 1127
Children's Cardiomyopathy Foundation Research Grants, 1137
CNIB Baker Applied Research Fund Grants, 1193
CNIB Baker Fellowships, 1194
CNIB Baker New Researcher Fund Grants, 1195
CNIB Ross Purse Doctoral Fellowships, 1200
CTCRI Idea Grants, 1369
Cystic Fibrosis Canada Clinical Fellowships, 1381
Cystic Fibrosis Canada Clinical Project Grants, 1382
Cystic Fibrosis Canada Clinic Incentive Grants, 1383
Cystic Fibrosis Canada Fellowships, 1384
Cystic Fibrosis Canada Research Grants, 1385
Cystic Fibrosis Canada Scholarships, 1386
Cystic Fibrosis Canada Senior Scientist Research Training Awards, 1387
Cystic Fibrosis Canada Small Conference Grants, 1388
Cystic Fibrosis Canada Special Travel Grants For Fellows and Students, 1389
Cystic Fibrosis Canada Studentships, 1390
Cystic Fibrosis Canada Summer Studentships, 1391
Cystic Fibrosis Canada Transplant Center Incentive Grants, 1392
Cystic Fibrosis Canada Travel Supplement Grants, 1393
Cystic Fibrosis Canada Visiting Scientist Awards, 1396
DPA Promoting Policy Change Advocacy Grants, 1504
Elton John AIDS Foundation Grants, 1585
Fluor Foundation Grants, 1682
FMC Foundation Grants, 1683
Fulbright Traditional Scholar Program in the Western Hemisphere, 1760
Giving in Action Society Children & Youth with Special Needs Grants, 1823
Giving in Action Society Family Indep Grants, 1824
Hospital Libraries Section / MLA Professional Development Grants, 2052
IAFF Burn Foundation Research Grants, 2089
IIE Chevron International REACH Scholarships, 2128
IIE Hewlett Fnd/IIE Dissertation Fellowship, 2132
James H. Cummings Foundation Grants, 2190
John Deere Foundation Grants, 2234
Johns Manville Fund Grants, 2251
Johnson & Johnson/SAH Arts & Healing Grants, 2254
Klarman Family Foundation Grants in Eating Disorders Research Grants, 2310
Leon and Thea Koerner Foundation Grants, 2355
Lockheed Martin Corp Foundation Grants, 2374
Massage Therapy Research Fund (MTRF) Grants, 2452
May and Stanley Smith Charitable Trust Grants, 2459
Medtronic Foundation CommunityLink Health Grants, 2510

956 / Canada

Medtronic Foundation Patient Link Grants, 2514
MLA Continuing Education Grants, 2559
MLA Donald A.B. Lindberg Fellowship, 2562
MLA Graduate Scholarship, 2564
MLA Grad Scholarship for Minority Students, 2565
MLA Research, Development, and Demonstration Project Grant, 2572
Nelda C. and H.J. Lutcher Stark Fndn Grants, 2672
NSERC Brockhouse Canada Prize for Interdisciplinary Research in Science and Engineering Grant, 3003
Oak Foundation Child Abuse Grants, 3059
Olympus Corporation of Americas Corporate Giving Grants, 3073
PACCAR Foundation Grants, 3101
Pediatric Brain Tumor Fndn Research Grants, 3134
PVA Education Foundation Grants, 3284
PVA Research Foundation Grants, 3285
RCPSC/AMS Donald Richards Wilson Award, 3317
RCPSC Balfour M. Mount Visiting Professorship in Palliative Medicine, 3318
RCPSC Continuing Proffesional Dev Grants, 3319
RCPSC Detweiler Traveling Fellowships, 3320
RCPSC Tom Dignan Indigenous Health Award, 3321
RCPSC Duncan Graham Award, 3322
RCPSC Harry S. Morton Lectureship in Surgery, 3323
RCPSC Harry S. Morton Traveling Fellowship in Surgery, 3324
RCPSC International Residency Educator of the Year Award, 3325
RCPSC Int'l Resident Leadership Award, 3326
RCPSC James H. Graham Award of Merit, 3327
RCPSC Janes Visiting Professorship in Surgery, 3328
RCPSC John G. Wade Visiting Professorship in Patient Safety & Simulation-Based Med Education, 3329
RCPSC KJR Wightman Award for Scholarship in Ethics, 3330
RCPSC KJR Wightman Visiting Professorship in Medicine, 3331
RCPSC Kristin Sivertz Res Leadership Award, 3332
RCPSC Medical Education Research Grants, 3334
RCPSC Mentor of the Year Award, 3335
RCPSC Peter Warren Traveling Scholarship, 3336
RCPSC Prix D'excellence (Specialist of the Year) Award, 3337
RCPSC Professional Development Grants, 3338
RCPSC Program Administrator Award for Innovation and Excellence, 3339
RCPSC Program Director of the Year Award, 3340
RCPSC Robert Maudsley Fellowship for Studies in Medical Education, 3341
RCPSC Royal College Medal Award in Medicine, 3342
RCPSC Royal College Medal Award in Surgery, 3343
RCPSC Teasdale-Corti Humanitarian Award, 3344
RSC Abbyann D. Lynch Medal in Bioethics, 3419
RSC Jason A. Hannah Medal, 3420
RSC McLaughlin Medal, 3421
RSC McNeil Medal for the Public Awareness of Science, 3422
RSC Thomas W. Eadie Medal, 3423
Savoy Foundation Post-Doctoral and Clinical Research Fellowships, 3488
Savoy Foundation Research Grants, 3489
Savoy Foundation Studentships, 3490
SfN Award for Education in Neuroscience, 3517
SfN Bernice Grafstein Award for Outstanding Accomplishments in Mentoring, 3518
SfN Donald B. Lindsley Prize in Behavioral Neuroscience, 3521
SfN Federation of European Neuroscience Societies Forum Travel Awards, 3522
SfN Jacob P. Waletzky Award, 3523
SfN Janett Rosenberg Trubatch Career Development Award, 3524
SfN Japan Neuroscience Society Meeting Travel Awards, 3525
SfN Julius Axelrod Prize, 3526
SfN Louise Hanson Marshall Specific Recognition Award, 3527
SfN Mika Salpeter Lifetime Achievement Award, 3528
SfN Next Generation Award, 3531
SfN Peter and Patricia Gruber International Research Award, 3533
SfN Ralph W. Gerard Prize in Neuroscience, 3534
SfN Science Educator Award, 3535
SfN Swartz Prize for Theoretical and Computational Neuroscience, 3537
SfN Travel Awards for the International Brain Research Organization World Congress, 3539
SfN Young Investigator Award, 3541
Sick Kids Fndn Community Conference Grants, 3549
Sick Kids Foundation New Investigator Research Grants, 3550
SSHRC-NSERC College and Community Innovation Enhancement Grants, 3636
SSHRC Banting Postdoctoral Fellowships, 3637
Thomas Sill Foundation Grants, 3724
Thomson Reuters / MLA Doctoral Fellowships, 3728
UCT Scholarship Program, 3777
Vancouver Foundation Grants and Community Initiatives Program, 3849
Vancouver Fndn Public Health Bursary Package, 3850
Vancouver Sun Children's Fund Grants, 3851

Cape Verde
Aid for Starving Children International Grants, 368
Beatrice Laing Trust Grants, 864
Cargill Citizenship Fund Corporate Giving, 1018
Fulbright Traditional Scholar Program in Sub-Saharan Africa, 1756
Fulbright Traditional Scholar Program in the Near East and North Africa Region, 1758
IIE Hewlett Fnd/IIE Dissertation Fellowship, 2132
Nuffield Foundation Africa Program Grants, 3026
Rockefeller Brothers Fund Pivotal Places Grants: South Africa, 3395
Wellcome Trust New Investigator Awards, 3919

Caribbean
Elton John AIDS Foundation Grants, 1585
Fluor Foundation Grants, 1682
Olympus Corporation of Americas Corporate Giving Grants, 3073

Cayman Islands
Elton John AIDS Foundation Grants, 1585

Central African Republic
Aid for Starving Children International Grants, 368
Beatrice Laing Trust Grants, 864
Cargill Citizenship Fund Corporate Giving, 1018
Fulbright Traditional Scholar Program in Sub-Saharan Africa, 1756
Fulbright Traditional Scholar Program in the Near East and North Africa Region, 1758
IIE Hewlett Fnd/IIE Dissertation Fellowship, 2132
Nuffield Foundation Africa Program Grants, 3026
Rockefeller Brothers Fund Pivotal Places Grants: South Africa, 3395
Wellcome Trust New Investigator Awards, 3919

Chad
Aid for Starving Children International Grants, 368
Beatrice Laing Trust Grants, 864
Cargill Citizenship Fund Corporate Giving, 1018
Fulbright Trad Scholar in Sub-Saharan Africa, 1756
Fulbright Traditional Scholar Program in the Near East and North Africa Region, 1758
IIE Chevron International REACH Scholarships, 2128
IIE Hewlett Fnd/IIE Dissertation Fellowship, 2132
Nuffield Foundation Africa Program Grants, 3026
Rockefeller Brothers Fund Pivotal Places Grants: South Africa, 3395
Wellcome Trust New Investigator Awards, 3919

Chile
ASM-PAHO Infectious Diseases Epidemiology and Surveillance Fellowships, 714
ASM Int'l Fellowship for Latin America and the Caribbean, 734
Charles Delmar Foundation Grants, 1115
Elton John AIDS Foundation Grants, 1585
Fulbright Traditional Scholar Program in the Western Hemisphere, 1760
IBRO Latin America Regional Funding for Neuroscience Schools, 2105
IBRO Latin America Regional Funding for PROLAB Collaborations, 2106
IBRO Latin America Regional Funding for Short Courses, Workshops, and Symposia, 2107
IBRO Latin America Regional Funding for Short Research Stays, 2108
IBRO Latin America Regional Travel Grants, 2109
Isabel Allende Foundation Esperanza Grants, 2165
Katherine John Murphy Foundation Grants, 2293
Olympus Corporation of Americas Corporate Giving Grants, 3073

China
Bechtel Group Foundation Building Positive Community Relationships Grants, 866
BP Foundation Grants, 943
Cabot Corporation Foundation Grants, 993
Fluor Foundation Grants, 1682
Fulbright Traditional Scholar Program in the East Asia/Pacific Region, 1757
H.B. Fuller Foundation Grants, 1922
Hershey Company Grants, 2013
IBRO Asia Reg APRC Exchange Fellowships, 2101
IBRO Asia APRC Lecturer Exchange Grants, 2102
IBRO Asia Regional APRC Travel Grants, 2103
IIE Chevron International REACH Scholarships, 2128
IIE Lingnan Foundation W.T. Chan Fellowship, 2137
Kimball International-Habig Foundation Health and Human Services Grants, 2308
Medtronic Foundation Strengthening Health Systems Grants, 2515
MMAAP Foundation Dermatology Fellowships, 2577
MMAAP Fndn Dermatology Project Awards, 2578
MMAAP Foundation Fellowship Award in Hematology, 2579
MMAAP Foundation Fellowship Award in Reproductive Medicine, 2580
MMAAP Foundation Fellowship Award in Translational Medicine, 2581
MMAAP Foundation Geriatric Project Awards, 2582
MMAAP Fndn Hematology Project Awards, 2583
MMAAP Foundation Research Project Award in Reproductive Medicine, 2584
MMAAP Foundation Research Project Award in Translational Medicine, 2585
MMAAP Foundation Senior Health Fellowship Award in Geriatrics Medicine and Aging Research, 2586
Seagate Tech Corp Capacity to Care Grants, 3502
SfN Award for Education in Neuroscience, 3517
SfN Bernice Grafstein Award for Outstanding Accomplishments in Mentoring, 3518
SfN Donald B. Lindsley Prize in Behavioral Neuroscience, 3521
SfN Janett Rosenberg Trubatch Career Development Award, 3524
SfN Julius Axelrod Prize, 3526
SfN Louise Hanson Marshall Specific Recognition Award, 3527
SfN Mika Salpeter Lifetime Achievement Award, 3528
SfN Next Generation Award, 3531
SfN Peter and Patricia Gruber International Research Award, 3533
SfN Ralph W. Gerard Prize in Neuroscience, 3534
SfN Science Educator Award, 3535
SfN Swartz Prize for Theoretical and Computational Neuroscience, 3537
SfN Young Investigator Award, 3541
SSRC Collaborative Research Grants on Environment and Health in China, 3638
Wellcome Trust New Investigator Awards, 3919

Christmas Island
Fulbright Traditional Scholar Program in the East Asia/Pacific Region, 1757
H.B. Fuller Foundation Grants, 1922

GEOGRAPHIC INDEX

Cocos
Fulbright Traditional Scholar Program in the East Asia/Pacific Region, 1757
H.B. Fuller Foundation Grants, 1922

Colombia
ASM-PAHO Infectious Diseases Epidemiology and Surveillance Fellowships, 714
ASM Int'l Fellowship for Latin America and the Caribbean, 734
Charles Delmar Foundation Grants, 1115
Elton John AIDS Foundation Grants, 1585
Fulbright Traditional Scholar Program in the Western Hemisphere, 1760
Genesis Foundation Grants, 1786
IBRO Latin America Regional Funding for Neuroscience Schools, 2105
IBRO Latin America Regional Funding for PROLAB Collaborations, 2106
IBRO Latin America Regional Funding for Short Courses, Workshops, and Symposia, 2107
IBRO Latin America Regional Funding for Short Research Stays, 2108
IBRO Latin America Regional Travel Grants, 2109
IIE Chevron International REACH Scholarships, 2128
Katherine John Murphy Foundation Grants, 2293
Olympus Corporation of Americas Corporate Giving Grants, 3073
Wellcome Trust New Investigator Awards, 3919

Comoros
Aid for Starving Children International Grants, 368
Beatrice Laing Trust Grants, 864
Cargill Citizenship Fund Corporate Giving, 1018
Fulbright Traditional Scholar Program in Sub-Saharan Africa, 1756
Fulbright Traditional Scholar Program in the Near East and North Africa Region, 1758
IIE Hewlett Fnd/IIE Dissertation Fellowship, 2132
Nuffield Foundation Africa Program Grants, 3026
Rockefeller Brothers Fund Pivotal Places Grants: South Africa, 3395
Wellcome Trust New Investigator Awards, 3919

Congo
Aid for Starving Children International Grants, 368
Beatrice Laing Trust Grants, 864
Cargill Citizenship Fund Corporate Giving, 1018
Fulbright Traditional Scholar Program in Sub-Saharan Africa, 1756
Fulbright Traditional Scholar Program in the Near East and North Africa Region, 1758
IIE Hewlett Fnd/IIE Dissertation Fellowship, 2132
Nuffield Foundation Africa Program Grants, 3026
Rockefeller Brothers Fund Pivotal Places Grants: South Africa, 3395
Wellcome Trust New Investigator Awards, 3919

Congo, Democratic Republic of
Aid for Starving Children International Grants, 368
Beatrice Laing Trust Grants, 864
Cargill Citizenship Fund Corporate Giving, 1018
Fulbright Traditional Scholar Program in Sub-Saharan Africa, 1756
Fulbright Traditional Scholar Program in the Near East and North Africa Region, 1758
IIE Hewlett Fnd/IIE Dissertation Fellowship, 2132
Nuffield Foundation Africa Program Grants, 3026
Rockefeller Brothers Fund Pivotal Places Grants: South Africa, 3395
Wellcome Trust New Investigator Awards, 3919

Costa Rica
ASM Int'l Fellowship for Latin America and the Caribbean, 734
Elton John AIDS Foundation Grants, 1585
Fulbright Traditional Scholar Program in the Western Hemisphere, 1760
IBRO Latin America Regional Funding for Neuroscience Schools, 2105

IBRO Latin America Regional Funding for PROLAB Collaborations, 2106
IBRO Latin America Regional Funding for Short Courses, Workshops, and Symposia, 2107
IBRO Latin America Regional Funding for Short Research Stays, 2108
IBRO Latin America Regional Travel Grants, 2109
Katherine John Murphy Foundation Grants, 2293
Olympus Corporation of Americas Corporate Giving Grants, 3073
Wellcome Trust New Investigator Awards, 3919

Cote d' Ivoire (Ivory Coast)
Aid for Starving Children International Grants, 368
Beatrice Laing Trust Grants, 864
Cargill Citizenship Fund Corporate Giving, 1018
Fulbright Traditional Scholar Program in Sub-Saharan Africa, 1756
Fulbright Traditional Scholar Program in the Near East and North Africa Region, 1758
IIE Hewlett Fnd/IIE Dissertation Fellowship, 2132
Nuffield Foundation Africa Program Grants, 3026
Rockefeller Brothers Fund Pivotal Places Grants: South Africa, 3395
Wellcome Trust New Investigator Awards, 3919

Croatia
Advanced Micro Devices Comm Affairs Grants, 313
Beatrice Laing Trust Grants, 864
Cargill Citizenship Fund Corporate Giving, 1018
Charles Delmar Foundation Grants, 1115
EMBO Installation Grants, 1586
ERC Starting Grants, 1599
Fluor Foundation Grants, 1682
Fulbright Scholars in Europe and Eurasia, 1755
HHMI Biomedical Research Grants for International Scientists in the Baltics, Central and Eastern Europe, Russia, and Ukraine, 2019
IBRO-PERC InEurope Short Stay Grants, 2097
Medtronic Foundation Strengthening Health Systems Grants, 2515
OSF Mental Health Initiative Grants, 3093
OSF Roma Health Project Grants, 3095
Simone and Cino del Duca Grand Prix Awards, 3564
Sir Dorabji Tata Trust Grants for NGOs or Voluntary Organizations, 3572

Cuba
ASM Int'l Fellowship for Latin America and the Caribbean, 734
Elton John AIDS Foundation Grants, 1585
Fulbright Traditional Scholar Program in the Western Hemisphere, 1760
IBRO Latin America Regional Funding for Neuroscience Schools, 2105
IBRO Latin America Regional Funding for PROLAB Collaborations, 2106
IBRO Latin America Regional Funding for Short Courses, Workshops, and Symposia, 2107
IBRO Latin America Regional Funding for Short Research Stays, 2108
IBRO Latin America Regional Travel Grants, 2109
Katherine John Murphy Foundation Grants, 2293
Wellcome Trust New Investigator Awards, 3919

Cyprus
Advanced Micro Devices Comm Affairs Grants, 313
Beatrice Laing Trust Grants, 864
Cargill Citizenship Fund Corporate Giving, 1018
Charles Delmar Foundation Grants, 1115
EMBO Installation Grants, 1586
ERC Starting Grants, 1599
Fluor Foundation Grants, 1682
Fulbright Scholars in Europe and Eurasia, 1755
Fulbright Traditional Scholar Program in the East Asia/Pacific Region, 1757
H.B. Fuller Foundation Grants, 1922
HHMI Biomedical Research Grants for International Scientists in the Baltics, Central and Eastern Europe, Russia, and Ukraine, 2019

IBRO-PERC InEurope Short Stay Grants, 2097
Simone and Cino del Duca Grand Prix Awards, 3564
Sir Dorabji Tata Trust Grants for NGOs or Voluntary Organizations, 3572

Czech Republic
Advanced Micro Devices Comm Affairs Grants, 313
Beatrice Laing Trust Grants, 864
Cargill Citizenship Fund Corporate Giving, 1018
Charles Delmar Foundation Grants, 1115
EMBO Installation Grants, 1586
ERC Starting Grants, 1599
Fluor Foundation Grants, 1682
Fulbright Scholars in Europe and Eurasia, 1755
HHMI/EMBO Start-up Grants for C Europe, 2016
HHMI Biomedical Research Grants for International Scientists in the Baltics, Central and Eastern Europe, Russia, and Ukraine, 2019
IBRO-PERC InEurope Short Stay Grants, 2097
Medtronic Foundation Strengthening Health Systems Grants, 2515
OSF Mental Health Initiative Grants, 3093
OSF Roma Health Project Grants, 3095
Simone and Cino del Duca Grand Prix Awards, 3564
Sir Dorabji Tata Trust Grants for NGOs or Voluntary Organizations, 3572

Denmark
Advanced Micro Devices Comm Affairs Grants, 313
Beatrice Laing Trust Grants, 864
BP Foundation Grants, 943
Cargill Citizenship Fund Corporate Giving, 1018
Charles Delmar Foundation Grants, 1115
EMBO Installation Grants, 1586
ERC Starting Grants, 1599
Fluor Foundation Grants, 1682
Fulbright Scholars in Europe and Eurasia, 1755
HHMI Biomedical Research Grants for International Scientists in the Baltics, Central and Eastern Europe, Russia, and Ukraine, 2019
IBRO-PERC InEurope Short Stay Grants, 2097
IBRO-PERC Support for European Workshops, Symposia and Meetings, 2098
IBRO-PERC Support for Site Lectures, 2099
Medtronic Foundation Patient Link Grants, 2514
Pfizer ASPIRE EU Antifungal Research Awards, 3165
Pfizer ASPIRE EU Dupuytren's Contracture Research Awards, 3166
Pfizer ASPIRE EU Emerging Mechanisms of Resistance Antibacterial Research Awards, 3167
Pfizer ASPIRE EU MRSA Nosocomial Pneumonia & MRSA Complicated Skin & Soft Tissue Infections Antibacterial Research Awards, 3168
SfN Award for Education in Neuroscience, 3517
SfN Bernice Grafstein Award for Outstanding Accomplishments in Mentoring, 3518
SfN Donald B. Lindsley Prize in Behavioral Neuroscience, 3521
SfN Jacob P. Waletzky Award, 3523
SfN Janett Rosenberg Trubatch Career Development Award, 3524
SfN Julius Axelrod Prize, 3526
SfN Louise Hanson Marshall Specific Recognition Award, 3527
SfN Mika Salpeter Lifetime Achievement Award, 3528
SfN Next Generation Award, 3531
SfN Peter and Patricia Gruber International Research Award, 3533
SfN Ralph W. Gerard Prize in Neuroscience, 3534
SfN Science Educator Award, 3535
SfN Swartz Prize for Theoretical and Computational Neuroscience, 3537
SfN Young Investigator Award, 3541
Simone and Cino del Duca Grand Prix Awards, 3564
Sir Dorabji Tata Trust Grants for NGOs or Voluntary Organizations, 3572

Djibouti
Acumen East Africa Fellowship, 284
Aid for Starving Children International Grants, 368

Djibouti

Beatrice Laing Trust Grants, 864
Cargill Citizenship Fund Corporate Giving, 1018
Fulbright Traditional Scholar Program in Sub-Saharan Africa, 1756
Fulbright Traditional Scholar Program in the Near East and North Africa Region, 1758
IIE Hewlett Fnd/IIE Dissertation Fellowship, 2132
Nuffield Foundation Africa Program Grants, 3026
Rockefeller Brothers Fund Pivotal Places Grants: South Africa, 3395
Wellcome Trust New Investigator Awards, 3919

Dominica

Elton John AIDS Foundation Grants, 1585
Fulbright Traditional Scholar Program in the Western Hemisphere, 1760
IBRO Latin America Regional Funding for Neuroscience Schools, 2105
IBRO Latin America Regional Funding for PROLAB Collaborations, 2106
IBRO Latin America Regional Funding for Short Courses, Workshops, and Symposia, 2107
IBRO Latin America Regional Funding for Short Research Stays, 2108
IBRO Latin America Regional Travel Grants, 2109
Wellcome Trust New Investigator Awards, 3919

Dominican Republic

Elton John AIDS Foundation Grants, 1585
Fulbright Traditional Scholar Program in the Western Hemisphere, 1760
IBRO Latin America Regional Funding for Neuroscience Schools, 2105
IBRO Latin America Regional Funding for PROLAB Collaborations, 2106
IBRO Latin America Regional Funding for Short Courses, Workshops, and Symposia, 2107
IBRO Latin America Regional Funding for Short Research Stays, 2108
IBRO Latin America Regional Travel Grants, 2109
Katherine John Murphy Foundation Grants, 2293
Wellcome Trust New Investigator Awards, 3919

East Timor

IBRO Asia Reg APRC Exchange Fellowships, 2101
IBRO Asia APRC Lecturer Exchange Grants, 2102
IBRO Asia Regional APRC Travel Grants, 2103

Ecuador

ASM-PAHO Infectious Diseases Epidemiology and Surveillance Fellowships, 714
ASM Int'l Fellowship for Latin America and the Caribbean, 734
Charles Delmar Foundation Grants, 1115
Elton John AIDS Foundation Grants, 1585
Fulbright Traditional Scholar Program in the Western Hemisphere, 1760
IBRO Latin America Regional Funding for Neuroscience Schools, 2105
IBRO Latin America Regional Funding for PROLAB Collaborations, 2106
IBRO Latin America Regional Funding for Short Courses, Workshops, and Symposia, 2107
IBRO Latin America Regional Funding for Short Research Stays, 2108
IBRO Latin America Regional Travel Grants, 2109
Katherine John Murphy Foundation Grants, 2293
Olympus Corporation of Americas Corporate Giving Grants, 3073
Wellcome Trust New Investigator Awards, 3919

Egypt

Aid for Starving Children International Grants, 368
Beatrice Laing Trust Grants, 864
BibleLands Grants, 889
Cargill Citizenship Fund Corporate Giving, 1018
Fulbright Traditional Scholar Program in Sub-Saharan Africa, 1756
Fulbright Traditional Scholar Program in the Near East and North Africa Region, 1758
IIE Lotus Scholarships, 2138
Nuffield Foundation Africa Program Grants, 3026
Rockefeller Brothers Fund Pivotal Places Grants: South Africa, 3395
Wellcome Trust New Investigator Awards, 3919

El Salvador

Elton John AIDS Foundation Grants, 1585
Fulbright Traditional Scholar Program in the Western Hemisphere, 1760
IBRO Latin America Regional Funding for Neuroscience Schools, 2105
IBRO Latin America Regional Funding for PROLAB Collaborations, 2106
IBRO Latin America Regional Funding for Short Courses, Workshops, and Symposia, 2107
IBRO Latin America Regional Funding for Short Research Stays, 2108
IBRO Latin America Regional Travel Grants, 2109
Katherine John Murphy Foundation Grants, 2293
Olympus Corporation of Americas Corporate Giving Grants, 3073
Wellcome Trust New Investigator Awards, 3919

Equatorial Guinea

Aid for Starving Children International Grants, 368
Beatrice Laing Trust Grants, 864
Cargill Citizenship Fund Corporate Giving, 1018
Fulbright Traditional Scholar Program in Sub-Saharan Africa, 1756
Fulbright Traditional Scholar Program in the Near East and North Africa Region, 1758
IIE Hewlett Fnd/IIE Dissertation Fellowship, 2132
Nuffield Foundation Africa Program Grants, 3026
Rockefeller Brothers Fund Pivotal Places Grants: South Africa, 3395

Eritrea

Acumen East Africa Fellowship, 284
Aid for Starving Children International Grants, 368
Beatrice Laing Trust Grants, 864
Cargill Citizenship Fund Corporate Giving, 1018
Fulbright Traditional Scholar Program in Sub-Saharan Africa, 1756
Fulbright Traditional Scholar Program in the Near East and North Africa Region, 1758
IIE Hewlett Fnd/IIE Dissertation Fellowship, 2132
Nuffield Foundation Africa Program Grants, 3026
Rockefeller Brothers Fund Pivotal Places Grants: South Africa, 3395
Wellcome Trust New Investigator Awards, 3919

Estonia

Advanced Micro Devices Comm Affairs Grants, 313
Beatrice Laing Trust Grants, 864
BP Foundation Grants, 943
Cargill Citizenship Fund Corporate Giving, 1018
Charles Delmar Foundation Grants, 1115
EMBO Installation Grants, 1586
ERC Starting Grants, 1599
Fluor Foundation Grants, 1682
Fulbright Scholars in Europe and Eurasia, 1755
HHMI Biomedical Research Grants for International Scientists in the Baltics, Central and Eastern Europe, Russia, and Ukraine, 2019
IBRO-PERC InEurope Short Stay Grants, 2097
John Deere Foundation Grants, 2234
OSF Mental Health Initiative Grants, 3093
Simone and Cino del Duca Grand Prix Awards, 3564
Sir Dorabji Tata Trust Grants for NGOs or Voluntary Organizations, 3572

Ethiopia

Acumen East Africa Fellowship, 284
Aid for Starving Children International Grants, 368
Beatrice Laing Trust Grants, 864
Cargill Citizenship Fund Corporate Giving, 1018
Firelight Foundation Grants, 1667
Fulbright Traditional Scholar Program in Sub-Saharan Africa, 1756
Fulbright Traditional Scholar Program in the Near East and North Africa Region, 1758
IIE African Center of Excellence for Women's Leadership Grants, 2124
IIE Hewlett Fnd/IIE Dissertation Fellowship, 2132
Nuffield Foundation Africa Program Grants, 3026
Oak Foundation Child Abuse Grants, 3059
Rockefeller Brothers Fund Pivotal Places Grants: South Africa, 3395
Wellcome Trust New Investigator Awards, 3919

Fiji

Wellcome Trust New Investigator Awards, 3919

Finland

Advanced Micro Devices Comm Affairs Grants, 313
Beatrice Laing Trust Grants, 864
BP Foundation Grants, 943
Cargill Citizenship Fund Corporate Giving, 1018
Charles Delmar Foundation Grants, 1115
EMBO Installation Grants, 1586
ERC Starting Grants, 1599
Fluor Foundation Grants, 1682
Fulbright Scholars in Europe and Eurasia, 1755
HHMI Biomedical Research Grants for International Scientists in the Baltics, Central and Eastern Europe, Russia, and Ukraine, 2019
IBRO-PERC InEurope Short Stay Grants, 2097
IBRO-PERC Support for European Workshops, Symposia and Meetings, 2098
IBRO-PERC Support for Site Lectures, 2099
Medtronic Foundation Patient Link Grants, 2514
Pfizer ASPIRE EU Antifungal Research Awards, 3165
Pfizer ASPIRE EU Dupuytren's Contracture Research Awards, 3166
Pfizer ASPIRE EU Emerging Mechanisms of Resistance Antibacterial Research Awards, 3167
Pfizer ASPIRE EU MRSA Nosocomial Pneumonia & MRSA Complicated Skin & Soft Tissue Infections Antibacterial Research Awards, 3168
Simone and Cino del Duca Grand Prix Awards, 3564
Sir Dorabji Tata Trust Grants for NGOs or Voluntary Organizations, 3572

France

Advanced Micro Devices Comm Affairs Grants, 313
Beatrice Laing Trust Grants, 864
BP Foundation Grants, 943
Cargill Citizenship Fund Corporate Giving, 1018
Charles Delmar Foundation Grants, 1115
EMBO Installation Grants, 1586
ERC Starting Grants, 1599
Fluor Foundation Grants, 1682
Fulbright Scholars in Europe and Eurasia, 1755
HHMI Biomedical Research Grants for International Scientists in the Baltics, Central and Eastern Europe, Russia, and Ukraine, 2019
IBRO-PERC InEurope Short Stay Grants, 2097
IBRO-PERC Support for European Workshops, Symposia and Meetings, 2098
IBRO-PERC Support for Site Lectures, 2099
IIE Leonora Lindsley Memorial Fellowships, 2136
Medtronic Foundation Patient Link Grants, 2514
Pfizer ASPIRE EU Antifungal Research Awards, 3165
Pfizer ASPIRE EU Dupuytren's Contracture Research Awards, 3166
Pfizer ASPIRE EU Emerging Mechanisms of Resistance Antibacterial Research Awards, 3167
Pfizer ASPIRE EU MRSA Nosocomial Pneumonia & MRSA Complicated Skin & Soft Tissue Infections Antibacterial Research Awards, 3168
Simone and Cino del Duca Grand Prix Awards, 3564
Sir Dorabji Tata Trust Grants for NGOs or Voluntary Organizations, 3572

Gabon

Aid for Starving Children International Grants, 368
Beatrice Laing Trust Grants, 864
Cargill Citizenship Fund Corporate Giving, 1018

GEOGRAPHIC INDEX

Fulbright Traditional Scholar Program in Sub-Saharan Africa, 1756
Fulbright Traditional Scholar Program in the Near East and North Africa Region, 1758
IIE Hewlett Fnd/IIE Dissertation Fellowship, 2132
Nuffield Foundation Africa Program Grants, 3026
Rockefeller Brothers Fund Pivotal Places Grants: South Africa, 3395
Wellcome Trust New Investigator Awards, 3919

Gambia
Aid for Starving Children International Grants, 368
Beatrice Laing Trust Grants, 864
Cargill Citizenship Fund Corporate Giving, 1018
Fulbright Traditional Scholar Program in Sub-Saharan Africa, 1756
Fulbright Traditional Scholar Program in the Near East and North Africa Region, 1758
IIE Hewlett Fnd/IIE Dissertation Fellowship, 2132
Nuffield Foundation Africa Program Grants, 3026
Rockefeller Brothers Fund Pivotal Places Grants: South Africa, 3395
Wellcome Trust New Investigator Awards, 3919

Georgia
Advanced Micro Devices Comm Affairs Grants, 313
Beatrice Laing Trust Grants, 864
Cargill Citizenship Fund Corporate Giving, 1018
Charles Delmar Foundation Grants, 1115
EMBO Installation Grants, 1586
Fluor Foundation Grants, 1682
Fulbright Scholars in Europe and Eurasia, 1755
Fulbright Traditional Scholar Program in the East Asia/Pacific Region, 1757
HHMI Biomedical Research Grants for International Scientists in the Baltics, Central and Eastern Europe, Russia, and Ukraine, 2019
Simone and Cino del Duca Grand Prix Awards, 3564
Sir Dorabji Tata Trust Grants for NGOs or Voluntary Organizations, 3572

Germany
Abbott Fund Community Grants, 201
Abbott Fund Science Education Grants, 203
Advanced Micro Devices Comm Affairs Grants, 313
Beatrice Laing Trust Grants, 864
BP Foundation Grants, 943
Cargill Citizenship Fund Corporate Giving, 1018
Charles Delmar Foundation Grants, 1115
Chiles Foundation Grants, 1147
Daimler and Benz Foundation Fellowships, 1411
EMBO Installation Grants, 1586
EMBO Long-Term Fellowships, 1587
EMBO Short-Term Fellowships, 1588
ERC Starting Grants, 1599
Fluor Foundation Grants, 1682
Fulbright German Studies Seminar Grants, 1751
Fulbright Scholars in Europe and Eurasia, 1755
HHMI/EMBO Start-up Grants for C Europe, 2016
HHMI Biomedical Research Grants for International Scientists in the Baltics, Central and Eastern Europe, Russia, and Ukraine, 2019
IBRO-PERC InEurope Short Stay Grants, 2097
IBRO-PERC Support for European Workshops, Symposia and Meetings, 2098
IBRO-PERC Support for Site Lectures, 2099
Medtronic Foundation Patient Link Grants, 2514
Medtronic Foundation Strengthening Health Systems Grants, 2515
OSF Mental Health Initiative Grants, 3093
Pfizer ASPIRE EU Antifungal Research Awards, 3165
Pfizer ASPIRE EU Dupuytren's Contracture Research Awards, 3166
Pfizer ASPIRE EU Emerging Mechanisms of Resistance Antibacterial Research Awards, 3167
Pfizer ASPIRE EU MRSA Nosocomial Pneumonia & MRSA Complicated Skin & Soft Tissue Infections Antibacterial Research Awards, 3168
Pfizer Research Initiative Dermatology Grants (Germany), 3184

Pfizer Research Initiative Rheumatology Grants (Germany), 3185
Simone and Cino del Duca Grand Prix Awards, 3564
Sir Dorabji Tata Trust Grants for NGOs or Voluntary Organizations, 3572
Textron Corporate Contributions Grants, 3712
Wilhelm Sander-Stiftung Foundation Grants, 3942

Ghana
Aid for Starving Children International Grants, 368
Beatrice Laing Trust Grants, 864
Cargill Citizenship Fund Corporate Giving, 1018
Fulbright Traditional Scholar Program in Sub-Saharan Africa, 1756
Fulbright Traditional Scholar Program in the Near East and North Africa Region, 1758
IIE Hewlett Fnd/IIE Dissertation Fellowship, 2132
Nuffield Foundation Africa Program Grants, 3026
Rockefeller Brothers Fund Pivotal Places Grants: South Africa, 3395
Wellcome Trust New Investigator Awards, 3919

Great Britain
Abbott Fund Community Grants, 201
Abbott Fund Science Education Grants, 203
BP Foundation Grants, 943
Medtronic Foundation Patient Link Grants, 2514
Textron Corporate Contributions Grants, 3712

Greece
Advanced Micro Devices Comm Affairs Grants, 313
Beatrice Laing Trust Grants, 864
BP Foundation Grants, 943
Cargill Citizenship Fund Corporate Giving, 1018
Charles Delmar Foundation Grants, 1115
EMBO Installation Grants, 1586
ERC Starting Grants, 1599
Fluor Foundation Grants, 1682
Fulbright Scholars in Europe and Eurasia, 1755
HHMI Biomedical Research Grants for International Scientists in the Baltics, Central and Eastern Europe, Russia, and Ukraine, 2019
IBRO-PERC InEurope Short Stay Grants, 2097
IBRO-PERC Support for European Workshops, Symposia and Meetings, 2098
IBRO-PERC Support for Site Lectures, 2099
Pfizer ASPIRE EU Antifungal Research Awards, 3165
Pfizer ASPIRE EU Dupuytren's Contracture Research Awards, 3166
Pfizer ASPIRE EU Emerging Mechanisms of Resistance Antibacterial Research Awards, 3167
Pfizer ASPIRE EU MRSA Nosocomial Pneumonia & MRSA Complicated Skin & Soft Tissue Infections Antibacterial Research Awards, 3168
Simone and Cino del Duca Grand Prix Awards, 3564
Sir Dorabji Tata Trust Grants for NGOs or Voluntary Organizations, 3572

Grenada
Elton John AIDS Foundation Grants, 1585
Fulbright Traditional Scholar Program in the Western Hemisphere, 1760
IBRO Latin America Regional Funding for Neuroscience Schools, 2105
IBRO Latin America Regional Funding for PROLAB Collaborations, 2106
IBRO Latin America Regional Funding for Short Courses, Workshops, and Symposia, 2107
IBRO Latin America Regional Funding for Short Research Stays, 2108
IBRO Latin America Regional Travel Grants, 2109
Wellcome Trust New Investigator Awards, 3919

Guadeloupe
Elton John AIDS Foundation Grants, 1585

Guatemala
Elton John AIDS Foundation Grants, 1585
Fulbright Traditional Scholar Program in the Western Hemisphere, 1760

IBRO Latin America Regional Funding for Neuroscience Schools, 2105
IBRO Latin America Regional Funding for PROLAB Collaborations, 2106
IBRO Latin America Regional Funding for Short Courses, Workshops, and Symposia, 2107
IBRO Latin America Regional Funding for Short Research Stays, 2108
IBRO Latin America Regional Travel Grants, 2109
Katherine John Murphy Foundation Grants, 2293
Olympus Corporation of Americas Corporate Giving Grants, 3073
Wellcome Trust New Investigator Awards, 3919

Guinea
Aid for Starving Children International Grants, 368
Beatrice Laing Trust Grants, 864
Cargill Citizenship Fund Corporate Giving, 1018
Fulbright Traditional Scholar Program in Sub-Saharan Africa, 1756
Fulbright Traditional Scholar Program in the Near East and North Africa Region, 1758
IIE Hewlett Fnd/IIE Dissertation Fellowship, 2132
Nuffield Foundation Africa Program Grants, 3026
Rockefeller Brothers Fund Pivotal Places Grants: South Africa, 3395
Wellcome Trust New Investigator Awards, 3919

Guinea-Bissau
Aid for Starving Children International Grants, 368
Beatrice Laing Trust Grants, 864
Cargill Citizenship Fund Corporate Giving, 1018
Fulbright Traditional Scholar Program in Sub-Saharan Africa, 1756
Fulbright Traditional Scholar Program in the Near East and North Africa Region, 1758
IIE Hewlett Fnd/IIE Dissertation Fellowship, 2132
Nuffield Foundation Africa Program Grants, 3026
Rockefeller Brothers Fund Pivotal Places Grants: South Africa, 3395
Wellcome Trust New Investigator Awards, 3919

Guyana
ASM-PAHO Infectious Diseases Epidemiology and Surveillance Fellowships, 714
ASM Int'l Fellowship for Latin America and the Caribbean, 734
Charles Delmar Foundation Grants, 1115
Elton John AIDS Foundation Grants, 1585
Fulbright Traditional Scholar Program in the Western Hemisphere, 1760
IBRO Latin America Regional Funding for Neuroscience Schools, 2105
IBRO Latin America Regional Funding for PROLAB Collaborations, 2106
IBRO Latin America Regional Funding for Short Courses, Workshops, and Symposia, 2107
IBRO Latin America Regional Funding for Short Research Stays, 2108
IBRO Latin America Regional Travel Grants, 2109
Wellcome Trust New Investigator Awards, 3919

Haiti
Abundance Foundation International Grants, 218
Elton John AIDS Foundation Grants, 1585
Fulbright Traditional Scholar Program in the Western Hemisphere, 1760
IBRO Latin America Regional Funding for Neuroscience Schools, 2105
IBRO Latin America Regional Funding for PROLAB Collaborations, 2106
IBRO Latin America Regional Funding for Short Courses, Workshops, and Symposia, 2107
IBRO Latin America Regional Funding for Short Research Stays, 2108
IBRO Latin America Regional Travel Grants, 2109
Katherine John Murphy Foundation Grants, 2293
Wellcome Trust New Investigator Awards, 3919

Honduras
Elton John AIDS Foundation Grants, 1585
Fulbright Traditional Scholar Program in the Western Hemisphere, 1760
IBRO Latin America Regional Funding for Neuroscience Schools, 2105
IBRO Latin America Regional Funding for PROLAB Collaborations, 2106
IBRO Latin America Regional Funding for Short Courses, Workshops, and Symposia, 2107
IBRO Latin America Regional Funding for Short Research Stays, 2108
IBRO Latin America Regional Travel Grants, 2109
Katherine John Murphy Foundation Grants, 2293
Olympus Corporation of Americas Corporate Giving Grants, 3073
Wellcome Trust New Investigator Awards, 3919

Hong Kong
BP Foundation Grants, 943
Fulbright Traditional Scholar Program in the East Asia/Pacific Region, 1757
H.B. Fuller Foundation Grants, 1922
IIE AmCham Charitable Foundation U.S. Studies Scholarship, 2125
IIE Lingnan Foundation W.T. Chan Fellowship, 2137
May and Stanley Smith Charitable Trust Grants, 2459

Hungary
Advanced Micro Devices Comm Affairs Grants, 313
Beatrice Laing Trust Grants, 864
Cargill Citizenship Fund Corporate Giving, 1018
Charles Delmar Foundation Grants, 1115
EMBO Installation Grants, 1586
ERC Starting Grants, 1599
Fluor Foundation Grants, 1682
Fulbright Scholars in Europe and Eurasia, 1755
HHMI/EMBO Start-up Grants for C Europe, 2016
HHMI Biomedical Research Grants for International Scientists in the Baltics, Central and Eastern Europe, Russia, and Ukraine, 2019
HHMI Biomedical Research Grants for International Scientists in the Baltics, Central and Eastern Europe, Russia, and Ukraine, 2019
IBRO-PERC InEurope Short Stay Grants, 2097
IIE Klein Family Scholarship, 2135
Medtronic Foundation Strengthening Health Systems Grants, 2515
OSF Mental Health Initiative Grants, 3093
OSF Roma Health Project Grants, 3095
Simone and Cino del Duca Grand Prix Awards, 3564
Sir Dorabji Tata Trust Grants for NGOs or Voluntary Organizations, 3572

Iceland
Advanced Micro Devices Comm Affairs Grants, 313
Beatrice Laing Trust Grants, 864
Cargill Citizenship Fund Corporate Giving, 1018
Charles Delmar Foundation Grants, 1115
EMBO Installation Grants, 1586
ERC Starting Grants, 1599
Fluor Foundation Grants, 1682
Fulbright Scholars in Europe and Eurasia, 1755
HHMI Biomedical Research Grants for International Scientists in the Baltics, Central and Eastern Europe, Russia, and Ukraine, 2019
IBRO-PERC InEurope Short Stay Grants, 2097
IBRO-PERC Support for European Workshops, Symposia and Meetings, 2098
IBRO-PERC Support for Site Lectures, 2099
Medtronic Foundation Patient Link Grants, 2514
Simone and Cino del Duca Grand Prix Awards, 3564
Sir Dorabji Tata Trust Grants for NGOs or Voluntary Organizations, 3572

India
Bupa Foundation Multi-Country Grant, 975
Fulbright Traditional Scholar Program in the East Asia/Pacific Region, 1757
H.B. Fuller Foundation Grants, 1922
IBRO Asia Reg APRC Exchange Fellowships, 2101
IBRO Asia APRC Lecturer Exchange Grants, 2102
IBRO Asia Regional APRC Travel Grants, 2103
Medtronic Foundation Strengthening Health Systems Grants, 2515
SfN Award for Education in Neuroscience, 3517
SfN Bernice Grafstein Award for Outstanding Accomplishments in Mentoring, 3518
SfN Donald B. Lindsley Prize in Behavioral Neuroscience, 3521
SfN Jacob P. Waletzky Award, 3523
SfN Janett Rosenberg Trubatch Career Development Award, 3524
SfN Julius Axelrod Prize, 3526
SfN Louise Hanson Marshall Specific Recognition Award, 3527
SfN Mika Salpeter Lifetime Achievement Award, 3528
SfN Next Generation Award, 3531
SfN Peter and Patricia Gruber International Research Award, 3533
SfN Ralph W. Gerard Prize in Neuroscience, 3534
SfN Science Educator Award, 3535
SfN Swartz Prize for Theoretical and Computational Neuroscience, 3537
SfN Young Investigator Award, 3541
Wellcome Trust New Investigator Awards, 3919

Indonesia
Fulbright Traditional Scholar Program in the East Asia/Pacific Region, 1757
H.B. Fuller Foundation Grants, 1922
IBRO Asia Reg APRC Exchange Fellowships, 2101
IBRO Asia APRC Lecturer Exchange Grants, 2102
IBRO Asia Regional APRC Travel Grants, 2103
IIE Chevron International REACH Scholarships, 2128
IIE Freeman Foundation Indonesia Internships, 2131
Wellcome Trust New Investigator Awards, 3919

Iran
Fulbright Traditional Scholar Program in the East Asia/Pacific Region, 1757
H.B. Fuller Foundation Grants, 1922
IBRO Asia Reg APRC Exchange Fellowships, 2101
IBRO Asia APRC Lecturer Exchange Grants, 2102
IBRO Asia Regional APRC Travel Grants, 2103
Wellcome Trust New Investigator Awards, 3919

Iraq
Fulbright Traditional Scholar Program in the East Asia/Pacific Region, 1757
H.B. Fuller Foundation Grants, 1922
IBRO Asia Reg APRC Exchange Fellowships, 2101
IBRO Asia APRC Lecturer Exchange Grants, 2102
IBRO Asia Regional APRC Travel Grants, 2103
IIE Iraq Scholars and Leaders Scholarships, 2133
Wellcome Trust New Investigator Awards, 3919

Ireland
Abbott Fund Community Grants, 201
Abbott Fund Science Education Grants, 203
Advanced Micro Devices Comm Affairs Grants, 313
Beatrice Laing Trust Grants, 864
BP Foundation Grants, 943
Cargill Citizenship Fund Corporate Giving, 1018
Charles Delmar Foundation Grants, 1115
EMBO Installation Grants, 1586
ERC Starting Grants, 1599
Fluor Foundation Grants, 1682
Fulbright Scholars in Europe and Eurasia, 1755
HHMI Biomedical Research Grants for International Scientists in the Baltics, Central and Eastern Europe, Russia, and Ukraine, 2019
IBRO-PERC InEurope Short Stay Grants, 2097
IBRO-PERC Support for European Workshops, Symposia and Meetings, 2098
IBRO-PERC Support for Site Lectures, 2099
Medtronic Foundation CommunityLink Health Grants, 2510
Medtronic Foundation Patient Link Grants, 2514
Pfizer ASPIRE EU Antifungal Research Awards, 3165
Pfizer ASPIRE EU Dupuytren's Contracture Research Awards, 3166
Pfizer ASPIRE EU Emerging Mechanisms of Resistance Antibacterial Research Awards, 3167
Pfizer ASPIRE EU MRSA Nosocomial Pneumonia & MRSA Complicated Skin & Soft Tissue Infections Antibacterial Research Awards, 3168
Seagate Tech Corp Capacity to Care Grants, 3502
Simone and Cino del Duca Grand Prix Awards, 3564
Sir Dorabji Tata Trust Grants for NGOs or Voluntary Organizations, 3572
Wellcome Trust New Investigator Awards, 3919

Israel
BibleLands Grants, 889
Boston Jewish Community Women's Fund Grants, 938
David Geffen Foundation Grants, 1431
Fulbright Traditional Scholar Program in the East Asia/Pacific Region, 1757
Fulbright Traditional Scholar Program in the Near East and North Africa Region, 1758
H.B. Fuller Foundation Grants, 1922
Harry Kramer Memorial Fund Grants, 1965
IBRO Asia Reg APRC Exchange Fellowships, 2101
IBRO Asia APRC Lecturer Exchange Grants, 2102
IBRO Asia Regional APRC Travel Grants, 2103
Klarman Family Foundation Grants in Eating Disorders Research Grants, 2310
Manuel D. & Rhoda Mayerson Fndn Grants, 2420
MBL Gruss Lipper Family Foundation Summer Fellowship, 2475
Rosalinde and Arthur Gilbert Foundation/AFAR New Investigator Awards in Alzheimer's Disease, 3406
SfN Award for Education in Neuroscience, 3517
SfN Bernice Grafstein Award for Outstanding Accomplishments in Mentoring, 3518
SfN Donald B. Lindsley Prize in Behavioral Neuroscience, 3521
SfN Jacob P. Waletzky Award, 3523
SfN Janett Rosenberg Trubatch Career Development Award, 3524
SfN Julius Axelrod Prize, 3526
SfN Louise Hanson Marshall Specific Recognition Award, 3527
SfN Mika Salpeter Lifetime Achievement Award, 3528
SfN Next Generation Award, 3531
SfN Peter and Patricia Gruber International Research Award, 3533
SfN Ralph W. Gerard Prize in Neuroscience, 3534
SfN Science Educator Award, 3535
SfN Swartz Prize for Theoretical and Computational Neuroscience, 3537
SfN Young Investigator Award, 3541

Italy
Abbott Fund Community Grants, 201
Abbott Fund Science Education Grants, 203
Advanced Micro Devices Comm Affairs Grants, 313
Beatrice Laing Trust Grants, 864
BP Foundation Grants, 943
Cargill Citizenship Fund Corporate Giving, 1018
Charles Delmar Foundation Grants, 1115
EMBO Installation Grants, 1586
ERC Starting Grants, 1599
Fluor Foundation Grants, 1682
Fulbright Scholars in Europe and Eurasia, 1755
HHMI Biomedical Research Grants for International Scientists in the Baltics, Central and Eastern Europe, Russia, and Ukraine, 2019
IBRO-PERC InEurope Short Stay Grants, 2097
IBRO-PERC Support for European Workshops, Symposia and Meetings, 2098
IBRO-PERC Support for Site Lectures, 2099
Medtronic Foundation Patient Link Grants, 2514
Pfizer ASPIRE EU Antifungal Research Awards, 3165
Pfizer ASPIRE EU Dupuytren's Contracture Research Awards, 3166
Pfizer ASPIRE EU Emerging Mechanisms of Resistance Antibacterial Research Awards, 3167

GEOGRAPHIC INDEX

Pfizer ASPIRE EU MRSA Nosocomial Pneumonia & MRSA Complicated Skin & Soft Tissue Infections Antibacterial Research Awards, 3168
Simone and Cino del Duca Grand Prix Awards, 3564
Sir Dorabji Tata Trust Grants for NGOs or Voluntary Organizations, 3572

Jamaica
Elton John AIDS Foundation Grants, 1585
Fulbright Traditional Scholar Program in the Western Hemisphere, 1760
IBRO Latin America Regional Funding for Neuroscience Schools, 2105
IBRO Latin America Regional Funding for PROLAB Collaborations, 2106
IBRO Latin America Regional Funding for Short Courses, Workshops, and Symposia, 2107
IBRO Latin America Regional Funding for Short Research Stays, 2108
IBRO Latin America Regional Travel Grants, 2109
Wellcome Trust New Investigator Awards, 3919

Japan
Abbott Fund Community Grants, 201
Abbott Fund Science Education Grants, 203
Bechtel Group Foundation Building Positive Community Relationships Grants, 866
BP Foundation Grants, 943
Fluor Foundation Grants, 1682
Fulbright Traditional Scholar Program in the East Asia/Pacific Region, 1757
H.B. Fuller Foundation Grants, 1922
IBRO Asia Reg APRC Exchange Fellowships, 2101
IBRO Asia APRC Lecturer Exchange Grants, 2102
IBRO Asia Regional APRC Travel Grants, 2103
Medtronic Fndn CommunityLink Health Grants, 2510
Medtronic Foundation Patient Link Grants, 2514
SfN Japan Neuroscience Society Meeting Travel Awards, 3525

Jordan
Fulbright Traditional Scholar Program in the East Asia/Pacific Region, 1757
H.B. Fuller Foundation Grants, 1922
IBRO Asia Reg APRC Exchange Fellowships, 2101
IBRO Asia APRC Lecturer Exchange Grants, 2102
IBRO Asia Regional APRC Travel Grants, 2103
Wellcome Trust New Investigator Awards, 3919

Kazakhstan
Fulbright Traditional Scholar Program in the East Asia/Pacific Region, 1757
Fulbright Traditional Scholar Program in the South and Central Asia Region, 1759
H.B. Fuller Foundation Grants, 1922
IBRO Asia Reg APRC Exchange Fellowships, 2101
IBRO Asia APRC Lecturer Exchange Grants, 2102
IBRO Asia Regional APRC Travel Grants, 2103
IIE Chevron International REACH Scholarships, 2128
Wellcome Trust New Investigator Awards, 3919

Kenya
Acumen East Africa Fellowship, 284
Aid for Starving Children International Grants, 368
Beatrice Laing Trust Grants, 864
Cargill Citizenship Fund Corporate Giving, 1018
Firelight Foundation Grants, 1667
Fulbright Traditional Scholar Program in Sub-Saharan Africa, 1756
Fulbright Traditional Scholar Program in the Near East and North Africa Region, 1758
IIE African Center of Excellence for Women's Leadership Grants, 2124
IIE Hewlett Fnd/IIE Dissertation Fellowship, 2132
Nuffield Foundation Africa Program Grants, 3026
Rockefeller Brothers Fund Pivotal Places Grants: South Africa, 3395
USAID Accelerating Progress Against Tuberculosis in Kenya Grants, 3805
Wellcome Trust New Investigator Awards, 3919

Kiribati
Wellcome Trust New Investigator Awards, 3919

Korea
Wellcome Trust New Investigator Awards, 3919

Kosovo
Advanced Micro Devices Comm Affairs Grants, 313
Beatrice Laing Trust Grants, 864
Cargill Citizenship Fund Corporate Giving, 1018
Charles Delmar Foundation Grants, 1115
EMBO Installation Grants, 1586
ERC Starting Grants, 1599
Fluor Foundation Grants, 1682
Fulbright Scholars in Europe and Eurasia, 1755
HHMI Biomedical Research Grants for International Scientists in the Baltics, Central and Eastern Europe, Russia, and Ukraine, 2019
OSF Mental Health Initiative Grants, 3093
OSF Roma Health Project Grants, 3095
Simone and Cino del Duca Grand Prix Awards, 3564
Sir Dorabji Tata Trust Grants for NGOs or Voluntary Organizations, 3572
Wellcome Trust New Investigator Awards, 3919

Kuwait
Fulbright Traditional Scholar Program in the East Asia/Pacific Region, 1757
H.B. Fuller Foundation Grants, 1922
IBRO Asia Reg APRC Exchange Fellowships, 2101
IBRO Asia APRC Lecturer Exchange Grants, 2102
IBRO Asia Regional APRC Travel Grants, 2103
IIE Chevron International REACH Scholarships, 2128

Kyrgyzstan
Fulbright Traditional Scholar Program in the East Asia/Pacific Region, 1757
H.B. Fuller Foundation Grants, 1922
IBRO Asia Reg APRC Exchange Fellowships, 2101
IBRO Asia APRC Lecturer Exchange Grants, 2102
IBRO Asia Regional APRC Travel Grants, 2103
OSF Public Health Grants in Kyrgyzstan, 3094
Wellcome Trust New Investigator Awards, 3919

Laos
Fulbright Traditional Scholar Program in the East Asia/Pacific Region, 1757
H.B. Fuller Foundation Grants, 1922
IBRO Asia Reg APRC Exchange Fellowships, 2101
IBRO Asia APRC Lecturer Exchange Grants, 2102
IBRO Asia Regional APRC Travel Grants, 2103
Wellcome Trust New Investigator Awards, 3919

Latvia
Advanced Micro Devices Comm Affairs Grants, 313
Beatrice Laing Trust Grants, 864
Cargill Citizenship Fund Corporate Giving, 1018
Charles Delmar Foundation Grants, 1115
EMBO Installation Grants, 1586
ERC Starting Grants, 1599
Fluor Foundation Grants, 1682
Fulbright Scholars in Europe and Eurasia, 1755
HHMI Biomedical Research Grants for International Scientists in the Baltics, Central and Eastern Europe, Russia, and Ukraine, 2019
IBRO-PERC InEurope Short Stay Grants, 2097
John Deere Foundation Grants, 2234
Oak Foundation Child Abuse Grants, 3059
OSF Mental Health Initiative Grants, 3093
Simone and Cino del Duca Grand Prix Awards, 3564
Sir Dorabji Tata Trust Grants for NGOs or Voluntary Organizations, 3572

Lebanon
BibleLands Grants, 889
Fulbright Traditional Scholar Program in the East Asia/Pacific Region, 1757
H.B. Fuller Foundation Grants, 1922
IBRO Asia Reg APRC Exchange Fellowships, 2101
IBRO Asia APRC Lecturer Exchange Grants, 2102
IBRO Asia Regional APRC Travel Grants, 2103
Wellcome Trust New Investigator Awards, 3919

Lesotho
Aid for Starving Children International Grants, 368
Beatrice Laing Trust Grants, 864
Cargill Citizenship Fund Corporate Giving, 1018
Firelight Foundation Grants, 1667
Fulbright Traditional Scholar Program in Sub-Saharan Africa, 1756
Fulbright Traditional Scholar Program in the Near East and North Africa Region, 1758
IIE Hewlett Fnd/IIE Dissertation Fellowship, 2132
Nuffield Foundation Africa Program Grants, 3026
Rockefeller Brothers Fund Pivotal Places Grants: South Africa, 3395
Wellcome Trust New Investigator Awards, 3919

Liberia
Aid for Starving Children International Grants, 368
Beatrice Laing Trust Grants, 864
Cargill Citizenship Fund Corporate Giving, 1018
Fulbright Traditional Scholar Program in Sub-Saharan Africa, 1756
Fulbright Traditional Scholar Program in the Near East and North Africa Region, 1758
IIE Hewlett Fnd/IIE Dissertation Fellowship, 2132
Nuffield Foundation Africa Program Grants, 3026
Rockefeller Brothers Fund Pivotal Places Grants: South Africa, 3395
Wellcome Trust New Investigator Awards, 3919

Libya
Aid for Starving Children International Grants, 368
Beatrice Laing Trust Grants, 864
Cargill Citizenship Fund Corporate Giving, 1018
Fulbright Traditional Scholar Program in Sub-Saharan Africa, 1756
Fulbright Traditional Scholar Program in the Near East and North Africa Region, 1758
Nuffield Foundation Africa Program Grants, 3026
Rockefeller Brothers Fund Pivotal Places Grants: South Africa, 3395
Wellcome Trust New Investigator Awards, 3919

Liechtenstein
Advanced Micro Devices Comm Affairs Grants, 313
Beatrice Laing Trust Grants, 864
Cargill Citizenship Fund Corporate Giving, 1018
Charles Delmar Foundation Grants, 1115
EMBO Installation Grants, 1586
ERC Starting Grants, 1599
Fluor Foundation Grants, 1682
Fulbright Scholars in Europe and Eurasia, 1755
HHMI/EMBO Start-up Grants for C Europe, 2016
HHMI Biomedical Research Grants for International Scientists in the Baltics, Central and Eastern Europe, Russia, and Ukraine, 2019
IBRO-PERC InEurope Short Stay Grants, 2097
Medtronic Foundation Strengthening Health Systems Grants, 2515
Simone and Cino del Duca Grand Prix Awards, 3564
Sir Dorabji Tata Trust Grants for NGOs or Voluntary Organizations, 3572

Lithuania
Advanced Micro Devices Comm Affairs Grants, 313
Beatrice Laing Trust Grants, 864
Cargill Citizenship Fund Corporate Giving, 1018
Charles Delmar Foundation Grants, 1115
EMBO Installation Grants, 1586
ERC Starting Grants, 1599
Fluor Foundation Grants, 1682
Fulbright Scholars in Europe and Eurasia, 1755
HHMI Biomedical Research Grants for International Scientists in the Baltics, Central and Eastern Europe, Russia, and Ukraine, 2019
IBRO-PERC InEurope Short Stay Grants, 2097
John Deere Foundation Grants, 2234
Medtronic Foundation Patient Link Grants, 2514

962 / Lithuania

OSF Mental Health Initiative Grants, 3093
Simone and Cino del Duca Grand Prix Awards, 3564
Sir Dorabji Tata Trust Grants for NGOs or Voluntary Organizations, 3572

Luxembourg
Advanced Micro Devices Comm Affairs Grants, 313
Beatrice Laing Trust Grants, 864
BP Foundation Grants, 943
Cargill Citizenship Fund Corporate Giving, 1018
Charles Delmar Foundation Grants, 1115
EMBO Installation Grants, 1586
ERC Starting Grants, 1599
Fluor Foundation Grants, 1682
Fulbright Scholars in Europe and Eurasia, 1755
HHMI Biomedical Research Grants for International Scientists in the Baltics, Central and Eastern Europe, Russia, and Ukraine, 2019
IBRO-PERC InEurope Short Stay Grants, 2097
IBRO-PERC Support for European Workshops, Symposia and Meetings, 2098
IBRO-PERC Support for Site Lectures, 2099
Medtronic Foundation Patient Link Grants, 2514
Simone and Cino del Duca Grand Prix Awards, 3564
Sir Dorabji Tata Trust Grants for NGOs or Voluntary Organizations, 3572

Macau
Fulbright Traditional Scholar Program in the East Asia/Pacific Region, 1757
H.B. Fuller Foundation Grants, 1922
Sands Foundation Grants, 3476

Macedonia
Advanced Micro Devices Comm Affairs Grants, 313
Beatrice Laing Trust Grants, 864
Cargill Citizenship Fund Corporate Giving, 1018
Charles Delmar Foundation Grants, 1115
EMBO Installation Grants, 1586
ERC Starting Grants, 1599
Fluor Foundation Grants, 1682
Fulbright Scholars in Europe and Eurasia, 1755
HHMI Biomedical Research Grants for International Scientists in the Baltics, Central and Eastern Europe, Russia, and Ukraine, 2019
IBRO-PERC InEurope Short Stay Grants, 2097
OSF Mental Health Initiative Grants, 3093
OSF Roma Health Project Grants, 3095
Simone and Cino del Duca Grand Prix Awards, 3564
Sir Dorabji Tata Trust Grants for NGOs or Voluntary Organizations, 3572
Wellcome Trust New Investigator Awards, 3919

Madagascar
Aid for Starving Children International Grants, 368
Beatrice Laing Trust Grants, 864
Cargill Citizenship Fund Corporate Giving, 1018
Fulbright Trad Scholar in Sub-Saharan Africa, 1756
Fulbright Traditional Scholar Program in the Near East and North Africa Region, 1758
IIE Hewlett Fnd/IIE Dissertation Fellowship, 2132
Nuffield Foundation Africa Program Grants, 3026
Rockefeller Brothers Fund Pivotal Places Grants: South Africa, 3395
Wellcome Trust New Investigator Awards, 3919

Malawi
Aid for Starving Children International Grants, 368
Beatrice Laing Trust Grants, 864
Cargill Citizenship Fund Corporate Giving, 1018
Firelight Foundation Grants, 1667
Fulbright Traditional Scholar Program in Sub-Saharan Africa, 1756
Fulbright Traditional Scholar Program in the Near East and North Africa Region, 1758
IIE Hewlett Fnd/IIE Dissertation Fellowship, 2132
Nuffield Foundation Africa Program Grants, 3026
Rockefeller Brothers Fund Pivotal Places Grants: South Africa, 3395
Wellcome Trust New Investigator Awards, 3919

Malaysia
Fulbright Traditional Scholar Program in the East Asia/Pacific Region, 1757
H.B. Fuller Foundation Grants, 1922
IBRO Asia Reg APRC Exchange Fellowships, 2101
IBRO Asia APRC Lecturer Exchange Grants, 2102
IBRO Asia Regional APRC Travel Grants, 2103
Seagate Tech Corp Capacity to Care Grants, 3502
SfN Award for Education in Neuroscience, 3517
SfN Bernice Grafstein Award for Outstanding Accomplishments in Mentoring, 3518
SfN Donald B. Lindsley Prize in Behavioral Neuroscience, 3521
SfN Jacob P. Waletzky Award, 3523
SfN Janett Rosenberg Trubatch Career Development Award, 3524
SfN Julius Axelrod Prize, 3526
SfN Louise Hanson Marshall Specific Recognition Award, 3527
SfN Mika Salpeter Lifetime Achievement Award, 3528
SfN Next Generation Award, 3531
SfN Peter and Patricia Gruber International Research Award, 3533
SfN Ralph W. Gerard Prize in Neuroscience, 3534
SfN Science Educator Award, 3535
SfN Swartz Prize for Theoretical and Computational Neuroscience, 3537
SfN Young Investigator Award, 3541
Wellcome Trust New Investigator Awards, 3919

Maldives
Fulbright Traditional Scholar Program in the East Asia/Pacific Region, 1757
H.B. Fuller Foundation Grants, 1922
IBRO Asia Reg APRC Exchange Fellowships, 2101
IBRO Asia APRC Lecturer Exchange Grants, 2102
IBRO Asia Regional APRC Travel Grants, 2103
Wellcome Trust New Investigator Awards, 3919

Mali
Aid for Starving Children International Grants, 368
Beatrice Laing Trust Grants, 864
Cargill Citizenship Fund Corporate Giving, 1018
Fulbright Traditional Scholar Program in Sub-Saharan Africa, 1756
Fulbright Traditional Scholar Program in the Near East and North Africa Region, 1758
IIE Hewlett Fnd/IIE Dissertation Fellowship, 2132
Nuffield Foundation Africa Program Grants, 3026
Rockefeller Brothers Fund Pivotal Places Grants: South Africa, 3395
USAID HIV Prevention with Key Populations - Mali Grants, 3816
Wellcome Trust New Investigator Awards, 3919

Malta
Advanced Micro Devices Comm Affairs Grants, 313
Beatrice Laing Trust Grants, 864
Cargill Citizenship Fund Corporate Giving, 1018
Charles Delmar Foundation Grants, 1115
EMBO Installation Grants, 1586
ERC Starting Grants, 1599
Fluor Foundation Grants, 1682
Fulbright Scholars in Europe and Eurasia, 1755
HHMI Biomedical Research Grants for International Scientists in the Baltics, Central and Eastern Europe, Russia, and Ukraine, 2019
IBRO-PERC InEurope Short Stay Grants, 2097
Simone and Cino del Duca Grand Prix Awards, 3564
Sir Dorabji Tata Trust Grants for NGOs or Voluntary Organizations, 3572

Martinique
Elton John AIDS Foundation Grants, 1585

Mauritania
Aid for Starving Children International Grants, 368
Beatrice Laing Trust Grants, 864
Cargill Citizenship Fund Corporate Giving, 1018
Fulbright Trad Scholar in Sub-Saharan Africa, 1756

GEOGRAPHIC INDEX

Fulbright Traditional Scholar Program in the Near East and North Africa Region, 1758
IIE Hewlett Fnd/IIE Dissertation Fellowship, 2132
Nuffield Foundation Africa Program Grants, 3026
Rockefeller Brothers Fund Pivotal Places Grants: South Africa, 3395
Wellcome Trust New Investigator Awards, 3919

Mauritius
Aid for Starving Children International Grants, 368
Beatrice Laing Trust Grants, 864
Cargill Citizenship Fund Corporate Giving, 1018
Fulbright Traditional Scholar Program in Sub-Saharan Africa, 1756
Fulbright Traditional Scholar Program in the Near East and North Africa Region, 1758
IIE Hewlett Fnd/IIE Dissertation Fellowship, 2132
Nuffield Foundation Africa Program Grants, 3026
Rockefeller Brothers Fund Pivotal Places Grants: South Africa, 3395
Wellcome Trust New Investigator Awards, 3919

Mexico
Alaska Airlines Corporate Giving Medical Emergency and Research Grants, 402
BP Foundation Grants, 943
Elton John AIDS Foundation Grants, 1585
Fluor Foundation Grants, 1682
Fulbright Traditional Scholar Program in the Western Hemisphere, 1760
General Service Foundation Human Rights and Economic Justice Grants, 1784
Hershey Company Grants, 2013
IBRO Latin America Regional Funding for Neuroscience Schools, 2105
IBRO Latin America Regional Funding for PROLAB Collaborations, 2106
IBRO Latin America Regional Funding for Short Courses, Workshops, and Symposia, 2107
IBRO Latin America Regional Funding for Short Research Stays, 2108
IBRO Latin America Regional Travel Grants, 2109
Katherine John Murphy Foundation Grants, 2293
Kimball International-Habig Foundation Health and Human Services Grants, 2308
Oak Foundation Child Abuse Grants, 3059
Olympus Corp of Americas Giving Grants, 3073
PACCAR Foundation Grants, 3101
Paso del Norte Health Foundation Grants, 3119
Seagate Tech Corp Capacity to Care Grants, 3502
SfN Award for Education in Neuroscience, 3517
SfN Bernice Grafstein Award for Outstanding Accomplishments in Mentoring, 3518
SfN Donald B. Lindsley Prize in Behavioral Neuroscience, 3521
SfN Federation of European Neuroscience Societies Forum Travel Awards, 3522
SfN Jacob P. Waletzky Award, 3523
SfN Janett Rosenberg Trubatch Career Development Award, 3524
SfN Japan Neuroscience Society Meeting Travel Awards, 3525
SfN Julius Axelrod Prize, 3526
SfN Louise Hanson Marshall Specific Recognition Award, 3527
SfN Mika Salpeter Lifetime Achievement Award, 3528
SfN Next Generation Award, 3531
SfN Peter and Patricia Gruber International Research Award, 3533
SfN Ralph W. Gerard Prize in Neuroscience, 3534
SfN Science Educator Award, 3535
SfN Swartz Prize for Theoretical and Computational Neuroscience, 3537
SfN Travel Awards for the International Brain Research Organization World Congress, 3539
SfN Young Investigator Award, 3541
Wellcome Trust New Investigator Awards, 3919

Micronesia
Wellcome Trust New Investigator Awards, 3919

GEOGRAPHIC INDEX

Moldova
Advanced Micro Devices Comm Affairs Grants, 313
Beatrice Laing Trust Grants, 864
Cargill Citizenship Fund Corporate Giving, 1018
Charles Delmar Foundation Grants, 1115
EMBO Installation Grants, 1586
ERC Starting Grants, 1599
Fluor Foundation Grants, 1682
Fulbright Scholars in Europe and Eurasia, 1755
HHMI Biomedical Research Grants for International Scientists in the Baltics, Central and Eastern Europe, Russia, and Ukraine, 2019
IBRO-PERC InEurope Short Stay Grants, 2097
John Deere Foundation Grants, 2234
Medtronic Foundation Strengthening Health Systems Grants, 2515
Oak Foundation Child Abuse Grants, 3059
OSF Roma Health Project Grants, 3095
Simone and Cino del Duca Grand Prix Awards, 3564
Sir Dorabji Tata Trust Grants for NGOs or Voluntary Organizations, 3572
Wellcome Trust New Investigator Awards, 3919

Monaco
Advanced Micro Devices Comm Affairs Grants, 313
Beatrice Laing Trust Grants, 864
Cargill Citizenship Fund Corporate Giving, 1018
Charles Delmar Foundation Grants, 1115
EMBO Installation Grants, 1586
ERC Starting Grants, 1599
Fluor Foundation Grants, 1682
Fulbright Scholars in Europe and Eurasia, 1755
HHMI Biomedical Research Grants for International Scientists in the Baltics, Central and Eastern Europe, Russia, and Ukraine, 2019
IBRO-PERC InEurope Short Stay Grants, 2097
Simone and Cino del Duca Grand Prix Awards, 3564
Sir Dorabji Tata Trust Grants for NGOs or Voluntary Organizations, 3572

Mongolia
Fulbright Traditional Scholar Program in the East Asia/Pacific Region, 1757
H.B. Fuller Foundation Grants, 1922
IBRO Asia Reg APRC Exchange Fellowships, 2101
IBRO Asia APRC Lecturer Exchange Grants, 2102
IBRO Asia Regional APRC Travel Grants, 2103
Wellcome Trust New Investigator Awards, 3919

Montenegro
Advanced Micro Devices Comm Affairs Grants, 313
Beatrice Laing Trust Grants, 864
Cargill Citizenship Fund Corporate Giving, 1018
Charles Delmar Foundation Grants, 1115
EMBO Installation Grants, 1586
ERC Starting Grants, 1599
Fluor Foundation Grants, 1682
Fulbright Scholars in Europe and Eurasia, 1755
HHMI Biomedical Research Grants for International Scientists in the Baltics, Central and Eastern Europe, Russia, and Ukraine, 2019
IBRO-PERC InEurope Short Stay Grants, 2097
OSF Mental Health Initiative Grants, 3093
OSF Roma Health Project Grants, 3095
Simone and Cino del Duca Grand Prix Awards, 3564
Sir Dorabji Tata Trust Grants for NGOs or Voluntary Organizations, 3572
Wellcome Trust New Investigator Awards, 3919

Montserrat
Elton John AIDS Foundation Grants, 1585

Morocco
Aid for Starving Children International Grants, 368
Beatrice Laing Trust Grants, 864
Cargill Citizenship Fund Corporate Giving, 1018
Fulbright Traditional Scholar Program in Sub-Saharan Africa, 1756
Fulbright Traditional Scholar Program in the Near East and North Africa Region, 1758

Nuffield Foundation Africa Program Grants, 3026
Rockefeller Brothers Fund Pivotal Places Grants: South Africa, 3395
Wellcome Trust New Investigator Awards, 3919

Mozambique
Aid for Starving Children International Grants, 368
Beatrice Laing Trust Grants, 864
Cargill Citizenship Fund Corporate Giving, 1018
Fulbright Traditional Scholar Program in Sub-Saharan Africa, 1756
Fulbright Traditional Scholar Program in the Near East and North Africa Region, 1758
IIE Hewlett Fnd/IIE Dissertation Fellowship, 2132
Nuffield Foundation Africa Program Grants, 3026
Rockefeller Brothers Fund Pivotal Places Grants: South Africa, 3395
Wellcome Trust New Investigator Awards, 3919

Myanmar (Burma)
IBRO Asia Reg APRC Exchange Fellowships, 2101
IBRO Asia APRC Lecturer Exchange Grants, 2102
IBRO Asia Regional APRC Travel Grants, 2103
Wellcome Trust New Investigator Awards, 3919

Nagorno-Karabakh
Fulbright Traditional Scholar Program in the East Asia/Pacific Region, 1757
H.B. Fuller Foundation Grants, 1922

Namibia
Aid for Starving Children International Grants, 368
Beatrice Laing Trust Grants, 864
Cargill Citizenship Fund Corporate Giving, 1018
Fulbright Traditional Scholar Program in Sub-Saharan Africa, 1756
Fulbright Traditional Scholar Program in the Near East and North Africa Region, 1758
IIE Hewlett Fnd/IIE Dissertation Fellowship, 2132
Nuffield Foundation Africa Program Grants, 3026
Rockefeller Brothers Fund Pivotal Places Grants: South Africa, 3395
Wellcome Trust New Investigator Awards, 3919

Nepal
Fulbright Traditional Scholar Program in the East Asia/Pacific Region, 1757
H.B. Fuller Foundation Grants, 1922
IBRO Asia Reg APRC Exchange Fellowships, 2101
IBRO Asia APRC Lecturer Exchange Grants, 2102
IBRO Asia Regional APRC Travel Grants, 2103
Wellcome Trust New Investigator Awards, 3919

Netherlands
Abbott Fund Community Grants, 201
Abbott Fund Science Education Grants, 203
BP Foundation Grants, 943
IBRO-PERC InEurope Short Stay Grants, 2097
IBRO-PERC Support for European Workshops, Symposia and Meetings, 2098
IBRO-PERC Support for Site Lectures, 2099
IIE Chevron International REACH Scholarships, 2128
Medtronic Foundation CommunityLink Health Grants, 2510
Medtronic Foundation Patient Link Grants, 2514
Oak Foundation Child Abuse Grants, 3059
PACCAR Foundation Grants, 3101
Pfizer ASPIRE EU Antifungal Research Awards, 3165
Pfizer ASPIRE EU Dupuytren's Contracture Research Awards, 3166
Pfizer ASPIRE EU Emerging Mechanisms of Resistance Antibacterial Research Awards, 3167
Pfizer ASPIRE EU MRSA Nosocomial Pneumonia & MRSA Complicated Skin & Soft Tissue Infections Antibacterial Research Awards, 3168

Netherlands Antilles
Olympus Corporation of Americas Corporate Giving Grants, 3073

New Zealand
Bupa Foundation Multi-Country Grant, 975
Fluor Foundation Grants, 1682
IIE Chevron International REACH Scholarships, 2128
SfN Award for Education in Neuroscience, 3517
SfN Bernice Grafstein Award for Outstanding Accomplishments in Mentoring, 3518
SfN Donald B. Lindsley Prize in Behavioral Neuroscience, 3521
SfN Jacob P. Waletzky Award, 3523
SfN Janett Rosenberg Trubatch Career Development Award, 3524
SfN Julius Axelrod Prize, 3526
SfN Louise Hanson Marshall Specific Recognition Award, 3527
SfN Mika Salpeter Lifetime Achievement Award, 3528
SfN Next Generation Award, 3531
SfN Peter and Patricia Gruber International Research Award, 3533
SfN Ralph W. Gerard Prize in Neuroscience, 3534
SfN Science Educator Award, 3535
SfN Swartz Prize for Theoretical and Computational Neuroscience, 3537
SfN Young Investigator Award, 3541

Nicaragua
Elton John AIDS Foundation Grants, 1585
Fulbright Traditional Scholar Program in the Western Hemisphere, 1760
IBRO Latin America Regional Funding for Neuroscience Schools, 2105
IBRO Latin America Regional Funding for PROLAB Collaborations, 2106
IBRO Latin America Regional Funding for Short Courses, Workshops, and Symposia, 2107
IBRO Latin America Regional Funding for Short Research Stays, 2108
IBRO Latin America Regional Travel Grants, 2109
Katherine John Murphy Foundation Grants, 2293
Olympus Corporation of Americas Corporate Giving Grants, 3073
Wellcome Trust New Investigator Awards, 3919

Niger
Aid for Starving Children International Grants, 368
Beatrice Laing Trust Grants, 864
Cargill Citizenship Fund Corporate Giving, 1018
Fulbright Traditional Scholar Program in Sub-Saharan Africa, 1756
Fulbright Traditional Scholar Program in the Near East and North Africa Region, 1758
IIE Hewlett Fnd/IIE Dissertation Fellowship, 2132
Nuffield Foundation Africa Program Grants, 3026
Rockefeller Brothers Fund Pivotal Places Grants: South Africa, 3395
Wellcome Trust New Investigator Awards, 3919

Nigeria
Aid for Starving Children International Grants, 368
Beatrice Laing Trust Grants, 864
Cargill Citizenship Fund Corporate Giving, 1018
Fulbright Traditional Scholar Program in Sub-Saharan Africa, 1756
Fulbright Traditional Scholar Program in the Near East and North Africa Region, 1758
IIE Chevron International REACH Scholarships, 2128
IIE Hewlett Fnd/IIE Dissertation Fellowship, 2132
Nuffield Foundation Africa Program Grants, 3026
Rockefeller Brothers Fund Pivotal Places Grants: South Africa, 3395
SfN Award for Education in Neuroscience, 3517
SfN Bernice Grafstein Award for Outstanding Accomplishments in Mentoring, 3518
SfN Donald B. Lindsley Prize in Behavioral Neuroscience, 3521
SfN Jacob P. Waletzky Award, 3523
SfN Janett Rosenberg Trubatch Career Development Award, 3524
SfN Julius Axelrod Prize, 3526

964 / Nigeria

SfN Louise Hanson Marshall Specific Recognition Award, 3527
SfN Mika Salpeter Lifetime Achievement Award, 3528
SfN Next Generation Award, 3531
SfN Peter and Patricia Gruber International Research Award, 3533
SfN Ralph W. Gerard Prize in Neuroscience, 3534
SfN Science Educator Award, 3535
SfN Swartz Prize for Theoretical and Computational Neuroscience, 3537
SfN Young Investigator Award, 3541
USAID Family Health Plus Project Grants, 3810
Wellcome Trust New Investigator Awards, 3919

North Korea
Fulbright Traditional Scholar Program in the East Asia/Pacific Region, 1757
H.B. Fuller Foundation Grants, 1922
IBRO Asia Reg APRC Exchange Fellowships, 2101
IBRO Asia APRC Lecturer Exchange Grants, 2102
IBRO Asia Regional APRC Travel Grants, 2103

Northern Cyprus
Fulbright Traditional Scholar Program in the East Asia/Pacific Region, 1757
H.B. Fuller Foundation Grants, 1922

Norway
Advanced Micro Devices Comm Affairs Grants, 313
Beatrice Laing Trust Grants, 864
BP Foundation Grants, 943
Cargill Citizenship Fund Corporate Giving, 1018
Charles Delmar Foundation Grants, 1115
EMBO Installation Grants, 1586
ERC Starting Grants, 1599
Fluor Foundation Grants, 1682
Fulbright Scholars in Europe and Eurasia, 1755
HHMI Biomedical Research Grants for International Scientists in the Baltics, Central and Eastern Europe, Russia, and Ukraine, 2019
IBRO-PERC InEurope Short Stay Grants, 2097
IBRO-PERC Support for European Workshops, Symposia and Meetings, 2098
IBRO-PERC Support for Site Lectures, 2099
Medtronic Foundation Patient Link Grants, 2514
Pfizer ASPIRE EU Antifungal Research Awards, 3165
Pfizer ASPIRE EU Dupuytren's Contracture Research Awards, 3166
Pfizer ASPIRE EU Emerging Mechanisms of Resistance Antibacterial Research Awards, 3167
Pfizer ASPIRE EU MRSA Nosocomial Pneumonia & MRSA Complicated Skin & Soft Tissue Infections Antibacterial Research Awards, 3168
SfN Award for Education in Neuroscience, 3517
SfN Bernice Grafstein Award for Outstanding Accomplishments in Mentoring, 3518
SfN Donald B. Lindsley Prize in Behavioral Neuroscience, 3521
SfN Jacob P. Waletzky Award, 3523
SfN Janett Rosenberg Trubatch Career Development Award, 3524
SfN Julius Axelrod Prize, 3526
SfN Louise Hanson Marshall Specific Recognition Award, 3527
SfN Mika Salpeter Lifetime Achievement Award, 3528
SfN Next Generation Award, 3531
SfN Peter and Patricia Gruber International Research Award, 3533
SfN Ralph W. Gerard Prize in Neuroscience, 3534
SfN Science Educator Award, 3535
SfN Swartz Prize for Theoretical and Computational Neuroscience, 3537
SfN Young Investigator Award, 3541
Simone and Cino del Duca Grand Prix Awards, 3564
Sir Dorabji Tata Trust Grants for NGOs or Voluntary Organizations, 3572

Oman
Fulbright Traditional Scholar Program in the East Asia/Pacific Region, 1757

H.B. Fuller Foundation Grants, 1922
IBRO Asia Reg APRC Exchange Fellowships, 2101
IBRO Asia APRC Lecturer Exchange Grants, 2102
IBRO Asia Regional APRC Travel Grants, 2103

Pakistan
Fulbright Traditional Scholar Program in the East Asia/Pacific Region, 1757
H.B. Fuller Foundation Grants, 1922
IBRO Asia Reg APRC Exchange Fellowships, 2101
IBRO Asia APRC Lecturer Exchange Grants, 2102
IBRO Asia Regional APRC Travel Grants, 2103
Wellcome Trust New Investigator Awards, 3919

Palau
Wellcome Trust New Investigator Awards, 3919

Palestinian Authority
BibleLands Grants, 889
Fulbright Traditional Scholar Program in the East Asia/Pacific Region, 1757
H.B. Fuller Foundation Grants, 1922

Palestinian Territory
BibleLands Grants, 889

Panama
Elton John AIDS Foundation Grants, 1585
IBRO Latin America Regional Funding for Neuroscience Schools, 2105
IBRO Latin America Regional Funding for PROLAB Collaborations, 2106
IBRO Latin America Regional Funding for Short Courses, Workshops, and Symposia, 2107
IBRO Latin America Regional Funding for Short Research Stays, 2108
IBRO Latin America Regional Travel Grants, 2109
Katherine John Murphy Foundation Grants, 2293
Olympus Corporation of Americas Corporate Giving Grants, 3073
Wellcome Trust New Investigator Awards, 3919

Papua New Guinea
Wellcome Trust New Investigator Awards, 3919

Paraguay
ASM-PAHO Infectious Diseases Epidemiology and Surveillance Fellowships, 714
ASM Int'l Fellowship for Latin America and the Caribbean, 734
Charles Delmar Foundation Grants, 1115
Elton John AIDS Foundation Grants, 1585
Fulbright Traditional Scholar Program in the Western Hemisphere, 1760
IBRO Latin America Regional Funding for Neuroscience Schools, 2105
IBRO Latin America Regional Funding for PROLAB Collaborations, 2106
IBRO Latin America Regional Funding for Short Courses, Workshops, and Symposia, 2107
IBRO Latin America Regional Funding for Short Research Stays, 2108
IBRO Latin America Regional Travel Grants, 2109
Katherine John Murphy Foundation Grants, 2293
Olympus Corporation of Americas Corporate Giving Grants, 3073
Wellcome Trust New Investigator Awards, 3919

Peru
ASM-PAHO Infectious Diseases Epidemiology and Surveillance Fellowships, 714
ASM Int'l Fellowship for Latin America and the Caribbean, 734
Bechtel Group Foundation Building Positive Community Relationships Grants, 866
Charles Delmar Foundation Grants, 1115
Elton John AIDS Foundation Grants, 1585
Fluor Foundation Grants, 1682
Fulbright Traditional Scholar Program in the Western Hemisphere, 1760

GEOGRAPHIC INDEX

IBRO Latin America Regional Funding for Neuroscience Schools, 2105
IBRO Latin America Regional Funding for PROLAB Collaborations, 2106
IBRO Latin America Regional Funding for Short Courses, Workshops, and Symposia, 2107
IBRO Latin America Regional Funding for Short Research Stays, 2108
IBRO Latin America Regional Travel Grants, 2109
Katherine John Murphy Foundation Grants, 2293
Olympus Corporation of Americas Corporate Giving Grants, 3073
Wellcome Trust New Investigator Awards, 3919

Philippines
Bechtel Group Foundation Building Positive Community Relationships Grants, 866
BP Foundation Grants, 943
Fluor Foundation Grants, 1682
Fulbright Traditional Scholar Program in the East Asia/Pacific Region, 1757
H.B. Fuller Foundation Grants, 1922
IBRO Asia Reg APRC Exchange Fellowships, 2101
IBRO Asia APRC Lecturer Exchange Grants, 2102
IBRO Asia Regional APRC Travel Grants, 2103
IIE Chevron International REACH Scholarships, 2128
Wellcome Trust New Investigator Awards, 3919

Poland
Advanced Micro Devices Comm Affairs Grants, 313
Beatrice Laing Trust Grants, 864
Bechtel Group Foundation Building Positive Community Relationships Grants, 866
BP Foundation Grants, 943
Cargill Citizenship Fund Corporate Giving, 1018
Charles Delmar Foundation Grants, 1115
EMBO Installation Grants, 1586
ERC Starting Grants, 1599
Fluor Foundation Grants, 1682
Fluor Foundation Grants, 1682
Fulbright Scholars in Europe and Eurasia, 1755
HHMI/EMBO Start-up Grants for C Europe, 2016
HHMI Biomedical Research Grants for International Scientists in the Baltics, Central and Eastern Europe, Russia, and Ukraine, 2019
IBRO-PERC InEurope Short Stay Grants, 2097
Kimball International-Habig Foundation Health and Human Services Grants, 2308
Kosciuszko Fndn Grants for Polish Citizens, 2322
Medtronic Foundation Patient Link Grants, 2514
Medtronic Foundation Strengthening Health Systems Grants, 2515
OSF Mental Health Initiative Grants, 3093
OSF Roma Health Project Grants, 3095
Simone and Cino del Duca Grand Prix Awards, 3564
Sir Dorabji Tata Trust Grants for NGOs or Voluntary Organizations, 3572

Portugal
Advanced Micro Devices Comm Affairs Grants, 313
Beatrice Laing Trust Grants, 864
BP Foundation Grants, 943
Cargill Citizenship Fund Corporate Giving, 1018
Charles Delmar Foundation Grants, 1115
EMBO Installation Grants, 1586
ERC Starting Grants, 1599
Fluor Foundation Grants, 1682
Fulbright Scholars in Europe and Eurasia, 1755
HHMI Biomedical Research Grants for International Scientists in the Baltics, Central and Eastern Europe, Russia, and Ukraine, 2019
IBRO-PERC InEurope Short Stay Grants, 2097
IBRO-PERC Support for European Workshops, Symposia and Meetings, 2098
IBRO-PERC Support for Site Lectures, 2099
Medtronic Foundation Patient Link Grants, 2514
Pfizer ASPIRE EU Antifungal Research Awards, 3165
Pfizer ASPIRE EU Dupuytren's Contracture Research Awards, 3166

GEOGRAPHIC INDEX

Pfizer ASPIRE EU Emerging Mechanisms of
　Resistance Antibacterial Research Awards, 3167
Pfizer ASPIRE EU MRSA Nosocomial Pneumonia &
　MRSA Complicated Skin & Soft Tissue Infections
　Antibacterial Research Awards, 3168
Simone and Cino del Duca Grand Prix Awards, 3564
Sir Dorabji Tata Trust Grants for NGOs or Voluntary
　Organizations, 3572

Qatar
Fulbright Traditional Scholar Program in the East
　Asia/Pacific Region, 1757
H.B. Fuller Foundation Grants, 1922
IBRO Asia Reg APRC Exchange Fellowships, 2101
IBRO Asia APRC Lecturer Exchange Grants, 2102
IBRO Asia Regional APRC Travel Grants, 2103

Reunion
IIE Hewlett Fnd/IIE Dissertation Fellowship, 2132

Romania
Advanced Micro Devices Comm Affairs Grants, 313
Beatrice Laing Trust Grants, 864
Cargill Citizenship Fund Corporate Giving, 1018
Charles Delmar Foundation Grants, 1115
EMBO Installation Grants, 1586
ERC Starting Grants, 1599
Fluor Foundation Grants, 1682
Fulbright Scholars in Europe and Eurasia, 1755
HHMI Biomedical Research Grants for International
　Scientists in the Baltics, Central and Eastern
　Europe, Russia, and Ukraine, 2019
IBRO-PERC InEurope Short Stay Grants, 2097
Medtronic Foundation Strengthening Health Systems
　Grants, 2515
OSF Mental Health Initiative Grants, 3093
OSF Roma Health Project Grants, 3095
Simone and Cino del Duca Grand Prix Awards, 3564
Sir Dorabji Tata Trust Grants for NGOs or Voluntary
　Organizations, 3572
Wellcome Trust New Investigator Awards, 3919

Russia
Advanced Micro Devices Comm Affairs Grants, 313
Beatrice Laing Trust Grants, 864
Bechtel Group Foundation Building Positive
　Community Relationships Grants, 866
BP Foundation Grants, 943
Cargill Citizenship Fund Corporate Giving, 1018
Charles Delmar Foundation Grants, 1115
EMBO Installation Grants, 1586
ERC Starting Grants, 1599
Fluor Foundation Grants, 1682
Fluor Foundation Grants, 1682
Fulbright Scholars in Europe and Eurasia, 1755
Fulbright Traditional Scholar Program in the East
　Asia/Pacific Region, 1757
H.B. Fuller Foundation Grants, 1922
HHMI Biomedical Research Grants for International
　Scientists in the Baltics, Central and Eastern
　Europe, Russia, and Ukraine, 2019
HHMI Biomedical Research Grants for International
　Scientists in the Baltics, Central and Eastern
　Europe, Russia, and Ukraine, 2019
IBRO-PERC InEurope Short Stay Grants, 2097
IBRO Asia Reg APRC Exchange Fellowships, 2101
IBRO Asia APRC Lecturer Exchange Grants, 2102
IBRO Asia Regional APRC Travel Grants, 2103
IIE Chevron International REACH Scholarships, 2128
Medtronic Foundation Strengthening Health Systems
　Grants, 2515
OSF Mental Health Initiative Grants, 3093
OSF Roma Health Project Grants, 3095
Simone and Cino del Duca Grand Prix Awards, 3564
Sir Dorabji Tata Trust Grants for NGOs or Voluntary
　Organizations, 3572
U.S. Department of Education United States-Russia
　Program: Improving Research and Educational
　Activities in Higher Education, 3773

Rwanda
Acumen East Africa Fellowship, 284
Aid for Starving Children International Grants, 368
Beatrice Laing Trust Grants, 864
Cargill Citizenship Fund Corporate Giving, 1018
Firelight Foundation Grants, 1667
Fulbright Traditional Scholar Program in Sub-Saharan
　Africa, 1756
Fulbright Traditional Scholar Program in the Near East
　and North Africa Region, 1758
IIE African Center of Excellence for Women's
　Leadership Grants, 2124
IIE Hewlett Fnd/IIE Dissertation Fellowship, 2132
Nuffield Foundation Africa Program Grants, 3026
Rockefeller Brothers Fund Pivotal Places Grants: South
　Africa, 3395
Wellcome Trust New Investigator Awards, 3919

Saint Kitts And Nevis
Elton John AIDS Foundation Grants, 1585
IBRO Latin America Regional Funding for
　Neuroscience Schools, 2105
IBRO Latin America Regional Funding for PROLAB
　Collaborations, 2106
IBRO Latin America Regional Funding for Short
　Courses, Workshops, and Symposia, 2107
IBRO Latin America Regional Funding for Short
　Research Stays, 2108
IBRO Latin America Regional Travel Grants, 2109

San Marino
Advanced Micro Devices Comm Affairs Grants, 313
Beatrice Laing Trust Grants, 864
Cargill Citizenship Fund Corporate Giving, 1018
Charles Delmar Foundation Grants, 1115
EMBO Installation Grants, 1586
ERC Starting Grants, 1599
Fluor Foundation Grants, 1682
Fulbright Scholars in Europe and Eurasia, 1755
HHMI Biomedical Research Grants for International
　Scientists in the Baltics, Central and Eastern
　Europe, Russia, and Ukraine, 2019
IBRO-PERC InEurope Short Stay Grants, 2097
Simone and Cino del Duca Grand Prix Awards, 3564
Sir Dorabji Tata Trust Grants for NGOs or Voluntary
　Organizations, 3572

Sao Tome & Principe
Aid for Starving Children International Grants, 368
Beatrice Laing Trust Grants, 864
Cargill Citizenship Fund Corporate Giving, 1018
Fulbright Traditional Scholar Program in Sub-Saharan
　Africa, 1756
Fulbright Traditional Scholar Program in the Near East
　and North Africa Region, 1758
IIE Hewlett Fnd/IIE Dissertation Fellowship, 2132
Nuffield Foundation Africa Program Grants, 3026
Rockefeller Brothers Fund Pivotal Places Grants: South
　Africa, 3395
Wellcome Trust New Investigator Awards, 3919

Saudi Arabia
Fulbright Traditional Scholar Program in the East
　Asia/Pacific Region, 1757
H.B. Fuller Foundation Grants, 1922
IBRO Asia Reg APRC Exchange Fellowships, 2101
IBRO Asia APRC Lecturer Exchange Grants, 2102
IBRO Asia Regional APRC Travel Grants, 2103
IIE Chevron International REACH Scholarships, 2128

Senegal
Aid for Starving Children International Grants, 368
Beatrice Laing Trust Grants, 864
Cargill Citizenship Fund Corporate Giving, 1018
Fulbright Traditional Scholar Program in Sub-Saharan
　Africa, 1756
Fulbright Traditional Scholar Program in the Near East
　and North Africa Region, 1758
IIE Hewlett Fnd/IIE Dissertation Fellowship, 2132
Nuffield Foundation Africa Program Grants, 3026

Rockefeller Brothers Fund Pivotal Places Grants: South
　Africa, 3395
Wellcome Trust New Investigator Awards, 3919

Serbia
Advanced Micro Devices Comm Affairs Grants, 313
Beatrice Laing Trust Grants, 864
Cargill Citizenship Fund Corporate Giving, 1018
Charles Delmar Foundation Grants, 1115
EMBO Installation Grants, 1586
ERC Starting Grants, 1599
Fluor Foundation Grants, 1682
Fulbright Scholars in Europe and Eurasia, 1755
HHMI Biomedical Research Grants for International
　Scientists in the Baltics, Central and Eastern
　Europe, Russia, and Ukraine, 2019
IBRO-PERC InEurope Short Stay Grants, 2097
IIE Eurobank EFG Scholarships, 2130
Medtronic Foundation Strengthening Health Systems
　Grants, 2515
OSF Mental Health Initiative Grants, 3093
OSF Roma Health Project Grants, 3095
Simone and Cino del Duca Grand Prix Awards, 3564
Sir Dorabji Tata Trust Grants for NGOs or Voluntary
　Organizations, 3572
Wellcome Trust New Investigator Awards, 3919

Seychelles
Aid for Starving Children International Grants, 368
Beatrice Laing Trust Grants, 864
Cargill Citizenship Fund Corporate Giving, 1018
Fulbright Traditional Scholar Program in Sub-Saharan
　Africa, 1756
Fulbright Traditional Scholar Program in the Near East
　and North Africa Region, 1758
IIE Hewlett Fnd/IIE Dissertation Fellowship, 2132
Nuffield Foundation Africa Program Grants, 3026
Rockefeller Brothers Fund Pivotal Places Grants: South
　Africa, 3395
Wellcome Trust New Investigator Awards, 3919

Sierra Leone
Aid for Starving Children International Grants, 368
Beatrice Laing Trust Grants, 864
Cargill Citizenship Fund Corporate Giving, 1018
Fulbright Traditional Scholar Program in Sub-Saharan
　Africa, 1756
Fulbright Traditional Scholar Program in the Near East
　and North Africa Region, 1758
IIE Hewlett Fnd/IIE Dissertation Fellowship, 2132
Nuffield Foundation Africa Program Grants, 3026
Rockefeller Brothers Fund Pivotal Places Grants: South
　Africa, 3395
Wellcome Trust New Investigator Awards, 3919

Singapore
BP Foundation Grants, 943
Fulbright Traditional Scholar Program in the East
　Asia/Pacific Region, 1757
H.B. Fuller Foundation Grants, 1922
IBRO Asia Reg APRC Exchange Fellowships, 2101
IBRO Asia APRC Lecturer Exchange Grants, 2102
IBRO Asia Regional APRC Travel Grants, 2103
IIE Chevron International REACH Scholarships, 2128
Sands Foundation Grants, 3476
Seagate Tech Corp Capacity to Care Grants, 3502
SfN Award for Education in Neuroscience, 3517
SfN Bernice Grafstein Award for Outstanding
　Accomplishments in Mentoring, 3518
SfN Donald B. Lindsley Prize in Behavioral
　Neuroscience, 3521
SfN Jacob P. Waletzky Award, 3523
SfN Janett Rosenberg Trubatch Career Development
　Award, 3524
SfN Julius Axelrod Prize, 3526
SfN Louise Hanson Marshall Specific Recognition
　Award, 3527
SfN Mika Salpeter Lifetime Achievement Award, 3528
SfN Next Generation Award, 3531

SfN Peter and Patricia Gruber International Research Award, 3533
SfN Ralph W. Gerard Prize in Neuroscience, 3534
SfN Science Educator Award, 3535
SfN Swartz Prize for Theoretical and Computational Neuroscience, 3537
SfN Young Investigator Award, 3541

Slovakia
Advanced Micro Devices Comm Affairs Grants, 313
Beatrice Laing Trust Grants, 864
Cargill Citizenship Fund Corporate Giving, 1018
Charles Delmar Foundation Grants, 1115
EMBO Installation Grants, 1586
ERC Starting Grants, 1599
Fluor Foundation Grants, 1682
Fulbright Scholars in Europe and Eurasia, 1755
HHMI/EMBO Start-up Grants for C Europe, 2016
HHMI Biomedical Research Grants for International Scientists in the Baltics, Central and Eastern Europe, Russia, and Ukraine, 2019
IBRO-PERC InEurope Short Stay Grants, 2097
Medtronic Foundation Strengthening Health Systems Grants, 2515
OSF Mental Health Initiative Grants, 3093
OSF Roma Health Project Grants, 3095
Simone and Cino del Duca Grand Prix Awards, 3564
Sir Dorabji Tata Trust Grants for NGOs or Voluntary Organizations, 3572

Slovenia
Advanced Micro Devices Comm Affairs Grants, 313
Beatrice Laing Trust Grants, 864
Cargill Citizenship Fund Corporate Giving, 1018
Charles Delmar Foundation Grants, 1115
EMBO Installation Grants, 1586
ERC Starting Grants, 1599
Fluor Foundation Grants, 1682
Fulbright Scholars in Europe and Eurasia, 1755
HHMI/EMBO Start-up Grants for C Europe, 2016
HHMI Biomedical Research Grants for International Scientists in the Baltics, Central and Eastern Europe, Russia, and Ukraine, 2019
IBRO-PERC InEurope Short Stay Grants, 2097
Medtronic Foundation Strengthening Health Systems Grants, 2515
OSF Mental Health Initiative Grants, 3093
Simone and Cino del Duca Grand Prix Awards, 3564
Sir Dorabji Tata Trust Grants for NGOs or Voluntary Organizations, 3572

Solomon Islands
Wellcome Trust New Investigator Awards, 3919

Somalia
Acumen East Africa Fellowship, 284
Aid for Starving Children International Grants, 368
Beatrice Laing Trust Grants, 864
Cargill Citizenship Fund Corporate Giving, 1018
Fulbright Traditional Scholar Program in Sub-Saharan Africa, 1756
Fulbright Traditional Scholar Program in the Near East and North Africa Region, 1758
IIE Hewlett Fnd/IIE Dissertation Fellowship, 2132
Nuffield Foundation Africa Program Grants, 3026
Rockefeller Brothers Fund Pivotal Places Grants: South Africa, 3395
Wellcome Trust New Investigator Awards, 3919

South Africa
Aid for Starving Children International Grants, 368
Beatrice Laing Trust Grants, 864
Cargill Citizenship Fund Corporate Giving, 1018
Firelight Foundation Grants, 1667
Fulbright Traditional Scholar Program in Sub-Saharan Africa, 1756
Fulbright Traditional Scholar Program in the Near East and North Africa Region, 1758
IIE Chevron International REACH Scholarships, 2128
IIE Hewlett Fnd/IIE Dissertation Fellowship, 2132
Medtronic Foundation Strengthening Health Systems Grants, 2515
Nuffield Foundation Africa Program Grants, 3026
Oak Foundation Child Abuse Grants, 3059
Rockefeller Brothers Fund Pivotal Places Grants: South Africa, 3395
USAID Comprehensive District-Based Support for Better HIV/TB Patient Outcomes Grants, 3807
USAID Integration of Care and Support within the Health System to Support Better Patient Outcomes Grants, 3817
USAID Systems Strengthening for Better HIV/TB Patient Outcomes Grants, 3824
Wellcome Trust New Investigator Awards, 3919

South Korea
Fulbright Traditional Scholar Program in the East Asia/Pacific Region, 1757
H.B. Fuller Foundation Grants, 1922
IBRO Asia Reg APRC Exchange Fellowships, 2101
IBRO Asia APRC Lecturer Exchange Grants, 2102
IBRO Asia Regional APRC Travel Grants, 2103
IIE Chevron International REACH Scholarships, 2128
SfN Award for Education in Neuroscience, 3517
SfN Bernice Grafstein Award for Outstanding Accomplishments in Mentoring, 3518
SfN Donald B. Lindsley Prize in Behavioral Neuroscience, 3521
SfN Jacob P. Waletzky Award, 3523
SfN Janett Rosenberg Trubatch Career Development Award, 3524
SfN Julius Axelrod Prize, 3526
SfN Louise Hanson Marshall Specific Recognition Award, 3527
SfN Mika Salpeter Lifetime Achievement Award, 3528
SfN Next Generation Award, 3531
SfN Peter and Patricia Gruber International Research Award, 3533
SfN Science Educator Award, 3535
SfN Swartz Prize for Theoretical and Computational Neuroscience, 3537
SfN Young Investigator Award, 3541

South Ossetia
Fulbright Traditional Scholar Program in the East Asia/Pacific Region, 1757
H.B. Fuller Foundation Grants, 1922

South Sudan
Wellcome Trust New Investigator Awards, 3919

Spain
Advanced Micro Devices Comm Affairs Grants, 313
Beatrice Laing Trust Grants, 864
BP Foundation Grants, 943
Bupa Foundation Multi-Country Grant, 975
Cargill Citizenship Fund Corporate Giving, 1018
Charles Delmar Foundation Grants, 1115
EMBO Installation Grants, 1586
ERC Starting Grants, 1599
Fluor Foundation Grants, 1682
Fulbright Scholars in Europe and Eurasia, 1755
HHMI Biomedical Research Grants for International Scientists in the Baltics, Central and Eastern Europe, Russia, and Ukraine, 2019
IBRO-PERC InEurope Short Stay Grants, 2097
IBRO-PERC Support for European Workshops, Symposia and Meetings, 2098
IBRO-PERC Support for Site Lectures, 2099
Medtronic Foundation Patient Link Grants, 2514
Pfizer ASPIRE EU Antifungal Research Awards, 3165
Pfizer ASPIRE EU Dupuytren's Contracture Research Awards, 3166
Pfizer ASPIRE EU Emerging Mechanisms of Resistance Antibacterial Research Awards, 3167
Pfizer ASPIRE EU MRSA Nosocomial Pneumonia & MRSA Complicated Skin & Soft Tissue Infections Antibacterial Research Awards, 3168
SfN Award for Education in Neuroscience, 3517
SfN Bernice Grafstein Award for Outstanding Accomplishments in Mentoring, 3518
SfN Donald B. Lindsley Prize in Behavioral Neuroscience, 3521
SfN Jacob P. Waletzky Award, 3523
SfN Janett Rosenberg Trubatch Career Development Award, 3524
SfN Julius Axelrod Prize, 3526
SfN Louise Hanson Marshall Specific Recognition Award, 3527
SfN Mika Salpeter Lifetime Achievement Award, 3528
SfN Next Generation Award, 3531
SfN Peter and Patricia Gruber International Research Award, 3533
SfN Ralph W. Gerard Prize in Neuroscience, 3534
SfN Science Educator Award, 3535
SfN Swartz Prize for Theoretical and Computational Neuroscience, 3537
SfN Young Investigator Award, 3541
Simone and Cino del Duca Grand Prix Awards, 3564
Sir Dorabji Tata Trust Grants for NGOs or Voluntary Organizations, 3572
Victor Grifols i Lucas Foundation Ethics and Science Awars for Educational Institutions, 3869
Victor Grifols i Lucas Foundation Prize for Journalistic Work on Bioethics, 3870
Victor Grifols I Lucas Fndn Research Grants, 3871
Victor Grifols i Lucas Foundation Secondary School Prizes, 3872

Sri Lanka
Fulbright Traditional Scholar Program in the East Asia/Pacific Region, 1757
H.B. Fuller Foundation Grants, 1922
IBRO Asia Reg APRC Exchange Fellowships, 2101
IBRO Asia APRC Lecturer Exchange Grants, 2102
IBRO Asia Regional APRC Travel Grants, 2103
Wellcome Trust New Investigator Awards, 3919

St. Lucia
Elton John AIDS Foundation Grants, 1585
IBRO Latin America Regional Funding for Neuroscience Schools, 2105
IBRO Latin America Regional Funding for PROLAB Collaborations, 2106
IBRO Latin America Regional Funding for Short Courses, Workshops, and Symposia, 2107
IBRO Latin America Regional Funding for Short Research Stays, 2108
IBRO Latin America Regional Travel Grants, 2109
Wellcome Trust New Investigator Awards, 3919

St. Vincent and the Grenadines
Elton John AIDS Foundation Grants, 1585
IBRO Latin America Regional Funding for Neuroscience Schools, 2105
IBRO Latin America Regional Funding for PROLAB Collaborations, 2106
IBRO Latin America Regional Funding for Short Courses, Workshops, and Symposia, 2107
IBRO Latin America Regional Funding for Short Research Stays, 2108
IBRO Latin America Regional Travel Grants, 2109
Wellcome Trust New Investigator Awards, 3919

Sudan
Acumen East Africa Fellowship, 284
Aid for Starving Children International Grants, 368
Beatrice Laing Trust Grants, 864
Cargill Citizenship Fund Corporate Giving, 1018
Fulbright Traditional Scholar Program in Sub-Saharan Africa, 1756
Fulbright Traditional Scholar Program in the Near East and North Africa Region, 1758
IIE Hewlett Fnd/IIE Dissertation Fellowship, 2132
Nuffield Foundation Africa Program Grants, 3026
Rockefeller Brothers Fund Pivotal Places Grants: South Africa, 3395

GEOGRAPHIC INDEX

Suriname
Elton John AIDS Foundation Grants, 1585
IBRO Latin America Regional Funding for Neuroscience Schools, 2105
IBRO Latin America Regional Funding for PROLAB Collaborations, 2106
IBRO Latin America Regional Funding for Short Courses, Workshops, and Symposia, 2107
IBRO Latin America Regional Funding for Short Research Stays, 2108
IBRO Latin America Regional Travel Grants, 2109
Wellcome Trust New Investigator Awards, 3919

Swaziland
Aid for Starving Children International Grants, 368
Beatrice Laing Trust Grants, 864
Cargill Citizenship Fund Corporate Giving, 1018
Fulbright Traditional Scholar Program in Sub-Saharan Africa, 1756
Fulbright Traditional Scholar Program in the Near East and North Africa Region, 1758
IIE Hewlett Fnd/IIE Dissertation Fellowship, 2132
Nuffield Foundation Africa Program Grants, 3026
Rockefeller Brothers Fund Pivotal Places Grants: South Africa, 3395
Wellcome Trust New Investigator Awards, 3919

Sweden
Advanced Micro Devices Comm Affairs Grants, 313
Beatrice Laing Trust Grants, 864
BP Foundation Grants, 943
Cargill Citizenship Fund Corporate Giving, 1018
Charles Delmar Foundation Grants, 1115
EMBO Installation Grants, 1586
ERC Starting Grants, 1599
Fluor Foundation Grants, 1682
Fulbright Scholars in Europe and Eurasia, 1755
HHMI Biomedical Research Grants for International Scientists in the Baltics, Central and Eastern Europe, Russia, and Ukraine, 2019
IBRO-PERC InEurope Short Stay Grants, 2097
IBRO-PERC Support for European Workshops, Symposia and Meetings, 2098
IBRO-PERC Support for Site Lectures, 2099
Medtronic Foundation Patient Link Grants, 2514
Pfizer ASPIRE EU Antifungal Research Awards, 3165
Pfizer ASPIRE EU Dupuytren's Contracture Research Awards, 3166
Pfizer ASPIRE EU Emerging Mechanisms of Resistance Antibacterial Research Awards, 3167
Pfizer ASPIRE EU MRSA Nosocomial Pneumonia & MRSA Complicated Skin & Soft Tissue Infections Antibacterial Research Awards, 3168
Simone and Cino del Duca Grand Prix Awards, 3564
Sir Dorabji Tata Trust Grants for NGOs or Voluntary Organizations, 3572
Wilhelm Sander-Stiftung Foundation Grants, 3942

Switzerland
Advanced Micro Devices Comm Affairs Grants, 313
Beatrice Laing Trust Grants, 864
BP Foundation Grants, 943
Cabot Corporation Foundation Grants, 993
Cargill Citizenship Fund Corporate Giving, 1018
Charles Delmar Foundation Grants, 1115
EMBO Installation Grants, 1586
ERC Starting Grants, 1599
Fluor Foundation Grants, 1682
Fulbright Scholars in Europe and Eurasia, 1755
HHMI/EMBO Start-up Grants for C Europe, 2016
HHMI Biomedical Research Grants for International Scientists in the Baltics, Central and Eastern Europe, Russia, and Ukraine, 2019
IBRO-PERC InEurope Short Stay Grants, 2097
IBRO-PERC Support for European Workshops, Symposia and Meetings, 2098
IBRO-PERC Support for Site Lectures, 2099
Medtronic Foundation CommunityLink Health Grants, 2510
Medtronic Foundation Patient Link Grants, 2514

Medtronic Foundation Strengthening Health Systems Grants, 2515
Oak Foundation Child Abuse Grants, 3059
Pfizer ASPIRE EU Antifungal Research Awards, 3165
Pfizer ASPIRE EU Dupuytren's Contracture Research Awards, 3166
Pfizer ASPIRE EU Emerging Mechanisms of Resistance Antibacterial Research Awards, 3167
Pfizer ASPIRE EU MRSA Nosocomial Pneumonia & MRSA Complicated Skin & Soft Tissue Infections Antibacterial Research Awards, 3168
SfN Award for Education in Neuroscience, 3517
SfN Bernice Grafstein Award for Outstanding Accomplishments in Mentoring, 3518
SfN Donald B. Lindsley Prize in Behavioral Neuroscience, 3521
SfN Jacob P. Waletzky Award, 3523
SfN Janett Rosenberg Trubatch Career Development Award, 3524
SfN Julius Axelrod Prize, 3526
SfN Louise Hanson Marshall Specific Recognition Award, 3527
SfN Mika Salpeter Lifetime Achievement Award, 3528
SfN Next Generation Award, 3531
SfN Peter and Patricia Gruber International Research Award, 3533
SfN Ralph W. Gerard Prize in Neuroscience, 3534
SfN Science Educator Award, 3535
SfN Swartz Prize for Theoretical and Computational Neuroscience, 3537
SfN Young Investigator Award, 3541
Simone and Cino del Duca Grand Prix Awards, 3564
Sir Dorabji Tata Trust Grants for NGOs or Voluntary Organizations, 3572

Syria
Fulbright Traditional Scholar Program in the East Asia/Pacific Region, 1757
H.B. Fuller Foundation Grants, 1922
IBRO Asia Reg APRC Exchange Fellowships, 2101
IBRO Asia APRC Lecturer Exchange Grants, 2102
IBRO Asia Regional APRC Travel Grants, 2103

Syrian Arab Republic
Wellcome Trust New Investigator Awards, 3919

Taiwan
Fulbright Traditional Scholar Program in the East Asia/Pacific Region, 1757
H.B. Fuller Foundation Grants, 1922
IBRO Asia Reg APRC Exchange Fellowships, 2101
IBRO Asia APRC Lecturer Exchange Grants, 2102
IBRO Asia Regional APRC Travel Grants, 2103

Tajikistan
Fulbright Traditional Scholar Program in the East Asia/Pacific Region, 1757
H.B. Fuller Foundation Grants, 1922
IBRO Asia Reg APRC Exchange Fellowships, 2101
IBRO Asia APRC Lecturer Exchange Grants, 2102
IBRO Asia Regional APRC Travel Grants, 2103
Wellcome Trust New Investigator Awards, 3919

Tanzania
Acumen East Africa Fellowship, 284
Firelight Foundation Grants, 1667
IIE Hewlett Fnd/IIE Dissertation Fellowship, 2132
Oak Foundation Child Abuse Grants, 3059
Wellcome Trust New Investigator Awards, 3919

Thailand
Fulbright Traditional Scholar Program in the East Asia/Pacific Region, 1757
H.B. Fuller Foundation Grants, 1922
IBRO Asia Reg APRC Exchange Fellowships, 2101
IBRO Asia APRC Lecturer Exchange Grants, 2102
IBRO Asia Regional APRC Travel Grants, 2103
IIE Chevron International REACH Scholarships, 2128
Seagate Tech Corp Capacity to Care Grants, 3502
Wellcome Trust New Investigator Awards, 3919

The Netherlands
Advanced Micro Devices Comm Affairs Grants, 313
Beatrice Laing Trust Grants, 864
BP Foundation Grants, 943
Cargill Citizenship Fund Corporate Giving, 1018
Charles Delmar Foundation Grants, 1115
EMBO Installation Grants, 1586
ERC Starting Grants, 1599
Fluor Foundation Grants, 1682
Fulbright Scholars in Europe and Eurasia, 1755
HHMI Biomedical Research Grants for International Scientists in the Baltics, Central and Eastern Europe, Russia, and Ukraine, 2019
Medtronic Foundation Patient Link Grants, 2514
Simone and Cino del Duca Grand Prix Awards, 3564
Sir Dorabji Tata Trust Grants for NGOs or Voluntary Organizations, 3572

Timor-Lester
Fulbright Traditional Scholar Program in the East Asia/Pacific Region, 1757
H.B. Fuller Foundation Grants, 1922
Wellcome Trust New Investigator Awards, 3919

Togo
IIE Hewlett Fnd/IIE Dissertation Fellowship, 2132
Wellcome Trust New Investigator Awards, 3919

Tonga
Wellcome Trust New Investigator Awards, 3919

Trinidad and Tobago
Elton John AIDS Foundation Grants, 1585
IBRO Latin America Regional Funding for Neuroscience Schools, 2105
IBRO Latin America Regional Funding for PROLAB Collaborations, 2106
IBRO Latin America Regional Funding for Short Courses, Workshops, and Symposia, 2107
IBRO Latin America Regional Funding for Short Research Stays, 2108
IBRO Latin America Regional Travel Grants, 2109
IIE Chevron International REACH Scholarships, 2128

Tunisia
Wellcome Trust New Investigator Awards, 3919

Turkey
Advanced Micro Devices Comm Affairs Grants, 313
Beatrice Laing Trust Grants, 864
Cargill Citizenship Fund Corporate Giving, 1018
Charles Delmar Foundation Grants, 1115
EMBO Installation Grants, 1586
ERC Starting Grants, 1599
Fluor Foundation Grants, 1682
Fulbright Scholars in Europe and Eurasia, 1755
Fulbright Traditional Scholar Program in the East Asia/Pacific Region, 1757
H.B. Fuller Foundation Grants, 1922
HHMI Biomedical Research Grants for International Scientists in the Baltics, Central and Eastern Europe, Russia, and Ukraine, 2019
IBRO-PERC InEurope Short Stay Grants, 2097
IBRO Asia Reg APRC Exchange Fellowships, 2101
IBRO Asia APRC Lecturer Exchange Grants, 2102
IBRO Asia Regional APRC Travel Grants, 2103
OSF Roma Health Project Grants, 3095
SfN Award for Education in Neuroscience, 3517
SfN Bernice Grafstein Award for Outstanding Accomplishments in Mentoring, 3518
SfN Donald B. Lindsley Prize in Behavioral Neuroscience, 3521
SfN Jacob P. Waletzky Award, 3523
SfN Janett Rosenberg Trubatch Career Development Award, 3524
SfN Julius Axelrod Prize, 3526
SfN Louise Hanson Marshall Specific Recognition Award, 3527
SfN Mika Salpeter Lifetime Achievement Award, 3528
SfN Next Generation Award, 3531

Turkey

SfN Peter and Patricia Gruber International Research Award, 3533
SfN Ralph W. Gerard Prize in Neuroscience, 3534
SfN Science Educator Award, 3535
SfN Swartz Prize for Theoretical and Computational Neuroscience, 3537
SfN Young Investigator Award, 3541
Simone and Cino del Duca Grand Prix Awards, 3564
Sir Dorabji Tata Trust Grants for NGOs or Voluntary Organizations, 3572
Wellcome Trust New Investigator Awards, 3919

Turkmenistan

Fulbright Traditional Scholar Program in the East Asia/Pacific Region, 1757
H.B. Fuller Foundation Grants, 1922
IBRO Asia Reg APRC Exchange Fellowships, 2101
IBRO Asia APRC Lecturer Exchange Grants, 2102
IBRO Asia Regional APRC Travel Grants, 2103
Wellcome Trust New Investigator Awards, 3919

Turks and Caicos Islands

Elton John AIDS Foundation Grants, 1585

Tuvalu

Wellcome Trust New Investigator Awards, 3919

Uganda

Acumen East Africa Fellowship, 284
Firelight Foundation Grants, 1667
IIE African Center of Excellence for Women's Leadership Grants, 2124
IIE Hewlett Fnd/IIE Dissertation Fellowship, 2132
Oak Foundation Child Abuse Grants, 3059
Wellcome Trust New Investigator Awards, 3919

Ukraine

Advanced Micro Devices Comm Affairs Grants, 313
Beatrice Laing Trust Grants, 864
Cargill Citizenship Fund Corporate Giving, 1018
Charles Delmar Foundation Grants, 1115
EMBO Installation Grants, 1586
ERC Starting Grants, 1599
Fluor Foundation Grants, 1682
Fulbright Scholars in Europe and Eurasia, 1755
HHMI Biomedical Research Grants for International Scientists in the Baltics, Central and Eastern Europe, Russia, and Ukraine, 2019
HHMI Biomedical Research Grants for International Scientists in the Baltics, Central and Eastern Europe, Russia, and Ukraine, 2019
IBRO-PERC InEurope Short Stay Grants, 2097
John Deere Foundation Grants, 2234
Medtronic Foundation Strengthening Health Systems Grants, 2515
OSF Roma Health Project Grants, 3095
SfN Award for Education in Neuroscience, 3517
SfN Bernice Grafstein Award for Outstanding Accomplishments in Mentoring, 3518
SfN Donald B. Lindsley Prize in Behavioral Neuroscience, 3521
SfN Jacob P. Waletzky Award, 3523
SfN Janett Rosenberg Trubatch Career Development Award, 3524
SfN Julius Axelrod Prize, 3526
SfN Louise Hanson Marshall Specific Recognition Award, 3527
SfN Mika Salpeter Lifetime Achievement Award, 3528
SfN Next Generation Award, 3531
SfN Peter and Patricia Gruber International Research Award, 3533
SfN Ralph W. Gerard Prize in Neuroscience, 3534
SfN Science Educator Award, 3535
SfN Swartz Prize for Theoretical and Computational Neuroscience, 3537
SfN Young Investigator Award, 3541
Simone and Cino del Duca Grand Prix Awards, 3564
Sir Dorabji Tata Trust Grants for NGOs or Voluntary Organizations, 3572
Wellcome Trust New Investigator Awards, 3919

United Arab Emirates

Fulbright Traditional Scholar Program in the East Asia/Pacific Region, 1757
H.B. Fuller Foundation Grants, 1922
IBRO Asia Reg APRC Exchange Fellowships, 2101
IBRO Asia APRC Lecturer Exchange Grants, 2102
IBRO Asia Regional APRC Travel Grants, 2103

United Kingdom

Advanced Micro Devices Comm Affairs Grants, 313
Alzheimer's Association U.S.-U.K. Young Investigator Exchange Fellowships, 471
Bank of America Fndn Matching Gifts, 825
Bank of America Fndn Volunteer Grants, 826
Beatrice Laing Trust Grants, 864
Bechtel Group Foundation Building Positive Community Relationships Grants, 866
BP Foundation Grants, 943
Bupa Foundation Multi-Country Grant, 975
Cabot Corporation Foundation Grants, 993
Cargill Citizenship Fund Corporate Giving, 1018
Charles Delmar Foundation Grants, 1115
Commonwealth Fund Harkness Fellowships in Health Care Policy and Practice, 1229
Cooper Industries Foundation Grants, 1345
EMBO Installation Grants, 1586
ERC Starting Grants, 1599
ExxonMobil Foundation Malaria Grants, 1621
Fluor Foundation Grants, 1682
Fulbright Scholars in Europe and Eurasia, 1755
Guy's and St. Thomas' Charity Grants, 1918
HHMI Biomedical Research Grants for International Scientists in the Baltics, Central and Eastern Europe, Russia, and Ukraine, 2019
IBRO-PERC InEurope Short Stay Grants, 2097
IBRO-PERC Support for European Workshops, Symposia and Meetings, 2098
IBRO-PERC Support for Site Lectures, 2099
IIE Chevron International REACH Scholarships, 2128
Lincoln Financial Foundation Grants, 2367
May and Stanley Smith Charitable Trust Grants, 2459
Medtronic Foundation Fellowships, 2512
Medtronic Foundation Patient Link Grants, 2514
Nuffield Foundation Africa Program Grants, 3026
Nuffield Foundation Children & Families Grants, 3027
Nuffield Foundation Law and Society Grants, 3028
Nuffield Foundation Oliver Bird Rheumatism Program Grants, 3029
Nuffield Foundation Open Door Grants, 3030
Nuffield Foundation Small Grants, 3031
Oak Foundation Child Abuse Grants, 3059
PACCAR Foundation Grants, 3101
Pfizer Anti Infective Research EU Grants, 3164
Pfizer ASPIRE EU Antifungal Research Awards, 3165
Pfizer ASPIRE EU Dupuytren's Contracture Research Awards, 3166
Pfizer ASPIRE EU Emerging Mechanisms of Resistance Antibacterial Research Awards, 3167
Pfizer ASPIRE EU MRSA Nosocomial Pneumonia & MRSA Complicated Skin & Soft Tissue Infections Antibacterial Research Awards, 3168
Pfizer Inflammation Competitive Research Awards (UK), 3181
SfN Award for Education in Neuroscience, 3517
SfN Bernice Grafstein Award for Outstanding Accomplishments in Mentoring, 3518
SfN Donald B. Lindsley Prize in Behavioral Neuroscience, 3521
SfN Jacob P. Waletzky Award, 3523
SfN Janett Rosenberg Trubatch Career Development Award, 3524
SfN Julius Axelrod Prize, 3526
SfN Louise Hanson Marshall Specific Recognition Award, 3527
SfN Mika Salpeter Lifetime Achievement Award, 3528
SfN Next Generation Award, 3531
SfN Peter and Patricia Gruber International Research Award, 3533
SfN Ralph W. Gerard Prize in Neuroscience, 3534
SfN Science Educator Award, 3535
SfN Swartz Prize for Theoretical and Computational Neuroscience, 3537
SfN Young Investigator Award, 3541
Simone and Cino del Duca Grand Prix Awards, 3564
Sir Dorabji Tata Trust Grants for NGOs or Voluntary Organizations, 3572
Stroke Association Allied Health Professional Research Bursaries, 3661
Stroke Association Clinical Fellowships, 3662
Stroke Association Research Project Grants, 3663
Wellcome Trust New Investigator Awards, 3919

Uruguay

Elton John AIDS Foundation Grants, 1585
IBRO Latin America Regional Funding for Neuroscience Schools, 2105
IBRO Latin America Regional Funding for PROLAB Collaborations, 2106
IBRO Latin America Regional Funding for Short Courses, Workshops, and Symposia, 2107
IBRO Latin America Regional Funding for Short Research Stays, 2108
IBRO Latin America Regional Travel Grants, 2109
Katherine John Murphy Foundation Grants, 2293
Olympus Corporation of Americas Corporate Giving Grants, 3073

Uzbekistan

Fulbright Traditional Scholar Program in the East Asia/Pacific Region, 1757
Fulbright Traditional Scholar Program in the Near East and North Africa Region, 1758
Fulbright Traditional Scholar Program in the South and Central Asia Region, 1759
H.B. Fuller Foundation Grants, 1922
IBRO Asia Reg APRC Exchange Fellowships, 2101
IBRO Asia APRC Lecturer Exchange Grants, 2102
IBRO Asia Regional APRC Travel Grants, 2103
Wellcome Trust New Investigator Awards, 3919

Vanuatu

Wellcome Trust New Investigator Awards, 3919

Vatican City

Advanced Micro Devices Comm Affairs Grants, 313
Beatrice Laing Trust Grants, 864
Cargill Citizenship Fund Corporate Giving, 1018
Charles Delmar Foundation Grants, 1115
EMBO Installation Grants, 1586
ERC Starting Grants, 1599
Fluor Foundation Grants, 1682
Fulbright Scholars in Europe and Eurasia, 1755
HHMI Biomedical Research Grants for International Scientists in the Baltics, Central and Eastern Europe, Russia, and Ukraine, 2019
IBRO-PERC InEurope Short Stay Grants, 2097
Medtronic Foundation Patient Link Grants, 2514
Simone and Cino del Duca Grand Prix Awards, 3564
Sir Dorabji Tata Trust Grants for NGOs or Voluntary Organizations, 3572

Venezuela

Elton John AIDS Foundation Grants, 1585
Fluor Foundation Grants, 1682
IBRO Latin America Regional Funding for Neuroscience Schools, 2105
IBRO Latin America Regional Funding for PROLAB Collaborations, 2106
IBRO Latin America Regional Funding for Short Courses, Workshops, and Symposia, 2107
IBRO Latin America Regional Funding for Short Research Stays, 2108
IBRO Latin America Regional Travel Grants, 2109
IIE Chevron International REACH Scholarships, 2128
Katherine John Murphy Foundation Grants, 2293
Olympus Corporation of Americas Corporate Giving Grants, 3073
Wellcome Trust New Investigator Awards, 3919

GEOGRAPHIC INDEX

Vietnam
Fulbright Traditional Scholar Program in the East Asia/Pacific Region, 1757
H.B. Fuller Foundation Grants, 1922
IBRO Asia Reg APRC Exchange Fellowships, 2101
IBRO Asia APRC Lecturer Exchange Grants, 2102
IBRO Asia Regional APRC Travel Grants, 2103
IIE Chevron International REACH Scholarships, 2128
Wellcome Trust New Investigator Awards, 3919

Western Sahara
IIE Hewlett Fnd/IIE Dissertation Fellowship, 2132

Yemen
Fulbright Traditional Scholar Program in the East Asia/Pacific Region, 1757
H.B. Fuller Foundation Grants, 1922
IBRO Asia Reg APRC Exchange Fellowships, 2101
IBRO Asia APRC Lecturer Exchange Grants, 2102
IBRO Asia Regional APRC Travel Grants, 2103
Wellcome Trust New Investigator Awards, 3919

Zambia
Firelight Foundation Grants, 1667
IIE Hewlett Fnd/IIE Dissertation Fellowship, 2132
Wellcome Trust New Investigator Awards, 3919

Zimbabwe
Firelight Foundation Grants, 1667
IIE Hewlett Fnd/IIE Dissertation Fellowship, 2132
Wellcome Trust New Investigator Awards, 3919

CPSIA information can be obtained
at www.ICGtesting.com
Printed in the USA
BVHW010609180519
548021BV00007B/4/P